AF655650

A COURSE IN ELECTRICAL POWER

[Generation and Economic Considerations; Transmission and Distribution of Electrical Power; Switchgear and Protection Including Power System Analysis; and Utilization of Electric Power and Electric Traction]

J.B. GUPTA

OTHER USEFUL BOOKS

J.B. Gupta	*A Course in Electrical Technology*
J.B. Gupta	*Basic Electrical Engineering [A Course in Electrical Technology Vol. 1]*
J.B. Gupta	*Electrical Machines [A Course in Electrical Technology Vol. 2]*
J.B. Gupta	*Electronic Devices and Circuits [A Course in Electrical Technology Vol. 3]*
J.B. Gupta	*Theory and Performance of Electrical Machines [DC Machines, AC Machines and Polyphase Circuits]*
J.B. Gupta	*Electrical Machines-I [DC Machines]*
J.B. Gupta	*Electrical Machines-II [AC Machines]*
J.B. Gupta	*An Integrated Course in Electronics Engineering*
J.B. Gupta	*Fundamentals of Electrical and Electronics Engineering*
J.B. Gupta	*A Course in Electrical Power [Generation & Economic Considerations; Transmission & Distribution; Switchgear & Protection; Utilization of Electrical Power & Electric Traction]*
J.B. Gupta	*Generation and Economic Considerations [Electrical Power Systems Part I]*
J.B. Gupta	*Transmission and Distribution [Electrical Power Systems Part II]*
J.B. Gupta	*Switchgear and Protection [Electrical Power Systems Part III]*
J.B. Gupta	*Electrical Power Systems [Vol. I of A Course in Electrical Power]*
J.B. Gupta	*Utilization of Electrical Power and Electric Traction [Vol. II of A Course in Electrical Power]*
J.B. Gupta	*A Course in Electrical and Electronic Measurements and Instrumentation*
J.B. Gupta	*Question Bank in Electrical Engineering*
J.B. Gupta	*An Integrated Course in Electrical Engineering [Objective Type Questions With Answers & Notes]*
J.B. Gupta	*Electrical Installation Estimating & Costing*
J.B. Gupta	*Basic Electrical and Electronics Engineering [First Semester R.T.U.]*
J.B. Gupta	*Electrical Engineering [G.B.T.U.; Mahamaya T.U. and U.T.U.]*
J.B. Gupta	*Electrical Science [G.G.S.I.P. Univ.]*
J.B. Gupta	*Basic Electrical and Electronics Engineering [Pb. Technical University]*
J.B. Gupta	*A Textbook of Basic Electrical and Electronics Engineering [RGTU, Bhopal]*
J.B. Gupta	*Electrical Technology [M.D. Univ. and DCRUSTM]*
J.B. Gupta	*Electrical Science [AMITY Univ.]*
J.B. Gupta	*Electronics Engineering [G.B.; Mahamaya and Uttarakhand Technical Universities]*
J.B. Gupta	*Electrical Measurements and Measuring Instruments [M.D.U., P.T.U.; G.B.; Mahamaya and Uttarakhand Technical Universities]*
J.B. Gupta	*Electronic Measurements and Instrumentation [4th Semester Pb. Technical Univ.]*
J.B. Gupta	*Electronic Instrumentation and Measurements [4th Semester G.B. and Mahamaya Technical Universities]*
J.B. Gupta	*Electronic Devices and Circuits [3rd Semester Pb. Technical Univ.]*
J.B. Gupta	*Electrical Machines*
J.B. Gupta	*Electrical and Electronics Engineering Materials [4th Semester G.B. and Mahamaya Technical Universities]*
J.B. Gupta	*Electromechanical Energy Conversion-I [4th Semester G.B. and Mahamaya Technical Universities]*
J.B. Gupta	*Analog Electronics [4th Semester Pb. Technical Univ.]*
J.B. Gupta	*Pulse and Digital Switching Circuits [5th Semester Pb. Technical Univ.]*
J.B. Gupta	*Linear Integrated Circuits [5th Semester Pb. Technical Univ.]*
J.B. Gupta	*Electromechanical Energy Conversion-II [5th Semester G.B. and Mahamaya Technical Universities]*
J.B. Gupta	*Electronics Measurements and Instrumentation [5th Semester M.D. Univ.]*
J.B. Gupta	*Electromechanical Energy Conversion [3rd Semester M.D.U.]*
J.B. Gupta	*Analog Electronics [4th Semester M.D. Univ.]*
J.B. Gupta	*Analog Electronic Circuits [5th Semester M.D. Univ.]*
J.B. Gupta	*Basic Electronics [M.D. Univ.]*
J.B. Gupta	*Fundamentals of Electronic Devices [3rd Semester G.B. and Mahamaya Technical Universities]*
J.B. Gupta	*A Text Book of Analog Electronics [4th Semester R.T.U.]*
J.B. Gupta	*Elements of Electrical Engineering [Gujarat Technical Univ.]*
J.B. Gupta	*Electronic Devices and Circuits [3rd Semester M.D. Univ.]*
J.B. Gupta	*Electrical Machines–I [3rd Semester M.D. Univ.]*
J.B. Gupta	*Elements of Analog Electronics [3rd Semester G.G.S.I.P. Univ. Delhi]*
J.B. Gupta	*Elements of Power System [5th Semester G.B. and Mahamaya Technical Universities]*
J.B. Gupta	*Elements of Electronic Devices and Circuits [3rd Semester R.T.U.]*
J.B. Gupta	*Power System Analysis [6th Semester G.B. and Mahamaya Technical Universities]*
J.B. Gupta	*Switchgear and Protection [7th Semester G.B. and Mahamaya Technical Universities]*
J.B. Gupta	*Electrical Machines-II [5th Semester M.D. University]*
J.B. Gupta	*Basic Electronics [First Semester Him. T.U.]*
J.B. Gupta	*Analog Electronics-I [3rd Semester G.G.S.I.P. Univ. Delhi]*
J.B. Gupta	*Analog Electronics-II [4th Semester G.G.S.I.P. Univ. Delhi]*
J.B. Gupta	*Elements of Electric Drives*
J.B. Gupta	*Utilization of Electrical Energy and Traction [8th Semester G.B.T.U.]*
J.B. Gupta	*Electrical Engineering Materials & Semiconductor Devices [3rd Semester DCRUSTM]*
J.B. Gupta	*Analog Devices and Circuits [3rd Semester Pb. Technical Univ.]*
J.B. Gupta	*Basic Electrical and Electronics Engineering [First Semester Him. T.U.]*
J.B. Gupta	*Pulse Waveshaping and Switching [4th Semester Pb. Technical Univ.]*

A COURSE IN ELECTRICAL POWER

[Generation and Economic Considerations; Transmission and Distribution of Electrical Power; Switchgear and Protection Including Power System Analysis; and Utilization of Electric Power and Electric Traction]

for B.E., B.Tech; B.Sc. (Engineering); Sec. B of A.M.I.E. (India) U.P.S.C. I.E.S.; GATE and other Various Competitive Examinations

By

J.B. Gupta

B.Sc. (Elec. Engg.); F.I.E. (India);
C. Eng. (I); M.I.S.T.E.

S.K. KATARIA & SONS®

Publisher of Engineering Books

4885/109, Prakash Mahal, Dr. Subhash Bhargav Lane,
Opp. Delhi Medical Association, Daryaganj, New Delhi-110002 (India)
Tel.: +91-11-43551243, Mobile: +91-9871775858
E-mail: katariabook@gmail.com, katariabooks@yahoo.com
Website: www.skkatariaandsons.com

(e-books, subscriptions and bulk buying also available on our website)

Published By:
S.K. Kataria & Sons®
4885/109, Prakash Mahal, Dr. Subhash Bhargav Lane,
Opp. Delhi Medical Association, Daryaganj, New Delhi-110002 (India)
Tel.: +91-11-43551243, Mobile: +91-9871775858
E-mail: katariabook@gmail.com, katariabooks@yahoo.com
Website: www.skkatariaandsons.com

Edited By : Rajeev Manglik
B.E. (Elec.) Pb. University
and
Rohit Manglik
B.E. (Elec.) University of Roorkee

© ALL RIGHTS STRICTLY RESERVED

STATUTORY WARNING

Information contained in this book has been obtained by author from sources believed to be reliable and are correct to the best of his knowledge. Every effort has been made to avoid errors and omissions and ensure accuracy. Any error or omission noted may be brought to the notice of the publisher which shall be taken care of in the forthcoming edition of this book. However, neither the publisher nor the author guarantee the accuracy or completeness of any information publisted herein, and neither the publisher nor author take any responsibility or liability for any inconvenience, expenses, losses or damage to any one resulting from the contents of the book.

The book is meant for educational and learning purposes and there is no attempt on the part of publisher/author to render engineering and other professional services. If such services are required, the assistance of an appropriate professional should be sought.

The author of the book has taken all possible care to ensure that the contents of the book do not violate any existing copyright or other intellectual property rights of any person in any manner whatsoever. In the event, the author has been unable to track any source and if any copyright has been inadvertently infringed, the facts may be brought to the notice of the publisher in writing for corrective action.

Send all correspondence to : M/s S.K. Kataria & Sons, New Delhi
Email: katariabooks@yahoo.com

First Edition—July 1968, Fifteenth Edition—January 2013; Reprint: 2024-2025

Laser Typesetting by :
AISA ENTERPRISES, Delhi

Printed at :
Al-Aziz Books Binding

PREFACE TO THE FIFTEENTH EDITION

The author feels great pleasure in presenting the Fifteenth edition of the book with marked improvements over the previous one. The book has been thoroughly revised, enlarged with additions of the chapters and topics, rewritten and so designed as to meet the needs of Degree; A.M.I.E. Sec. B and Practising Engineers. The subject matter has been presented in an easily comprehensible manner, logical sequence using simple language and to provide detailed exposition of fundamentals. A large number of old diagrams have been replaced by new ones and numerous new diagrams have been added to make the subject more illustrative and clear. A large number of workedout examples, theoretical exercises and numerical problems have been deleted and replaced from latest examination papers of Institution of Engineers (India), Various Indian Universities, Technical Universities, U.P.S.C., I.E.S., GATE and other equivalent examining bodies. An earnest effort has been made to provide comprehensive coverage of the course as prescribed by the Institution of Engineers (India), Indian and foreign universities, Technical universities and various agencies conducting competitive examinations.

The book has been divided in four parts viz. "Generation and Economic Considerations", "Transmission and Distribution of Electrical Power", "Switchgear and Protection" including "Power System Analysis" and Utilization of Electric Power and Electric Traction (Electric Drives and Industrial Applications and Electric Power Utilization)". It consists of 67 chapters covering syllabus prescribed by various Indian universities/Technical universities, Institution of Engineers (India) and various agencies conducting competitive examinations. The book contains adequate materials for the papers of Power Plant Engineering; Power Station Practice; Power Station and Substation Practice; Transmission and Distribution; Switchgear and Protection; Switchgear and Protective Devices; Generation and Economic Considerations; Generation, Protection, Switchgear and Economics: Power Systems, Power System Analysis; Power System Analysis and Stability; Electric Drives and Control; Electric Drives and Their Control; Electric Drives and Utilization; Electric Drives and Utilization of Electrical Energy; Utilization of Electric Power; Utilization of Electrical Energy; Utilization and Traction; Utilization of Electric Power and Electric Traction; Traction and Utilization of Electric Power, Electric Traction; Illumination Engineering etc. Short answer type questions with answers given at the end of almost each chapter help the students in preparating for viva voce/practical examinations and interviews.

Most of the agencies conducting competitive examinations nowadays follow the pattern of objective evaluation. Even Institution of Engineers (India) and other examining bodies (Universities and Technical Universities) are adopting the pattern of objective evaluation (partly) for their regular examinations. The book contains 3,744 objective questions (2,165 Multiple Choice Type, 122 Assertion/Reason Type, 54 Matching Type, 755 True/False Type and 648 Fill in the Blanks Types) designed to bring out the hidden aspects and also test knowledge, understanding and application aspect of learning. Most of the objective type questions included in the book are from examination papers of various Universities/Technical Universities; A.M.I.E. Section B; U.P.S.C. (I.E.S.); and GATE or are of equivalent standard. These questions will help the students to score better in their regular as well as competitive examinations.

It is earnestly hoped that with the extensive additions, revisions, modifications and improvements, the present edition will facilitate the students not only in preparing for their regular examinations but also in preparing themselves for competitive examinations and prove more useful than the earlier edition.

Every care has been taken to eliminate misprints, omissions and errors but it is too much to expect that no inaccuracy/misprint/error has crept in and the author would be grateful to the readers for bringing to his notice any such errors/misprints/omissions they may come across while going through the book.

The author shall feel satisfied if the book meets the needs of the students for whom it is meant. Suggestions and criticism for the improvement of the book shall be welcomed.

The author will like to thank all the Scientists, Engineers, Authors and Publishers whose works and text have been the source of enlightenment, inspiration and guidance in presenting this new edition.

Lastly, but not the least, the author is thankful to his publisher for bringing out the book in such a short time and for pricing it moderately despite heavy cost of paper and printing.

January, 2013 — J.B. Gupta

CONTENTS

VOLUME-I: Power Systems

Part-I : Generation and Economic Considerations

Part-II : Transmission and Distribution of Electrical Power

Part-III : Switchgear and Protection

VOLUME-II: Utilization of Electric Power and Electric Traction

VOLUME-I

POWER SYSTEMS

Part-I : Generation and Economic Considerations

Part-II : Transmission and Distribution of Electrical Power

Part-III : Switchgear and Protection

PART I

GENERATION AND ECONOMIC CONSIDERATIONS

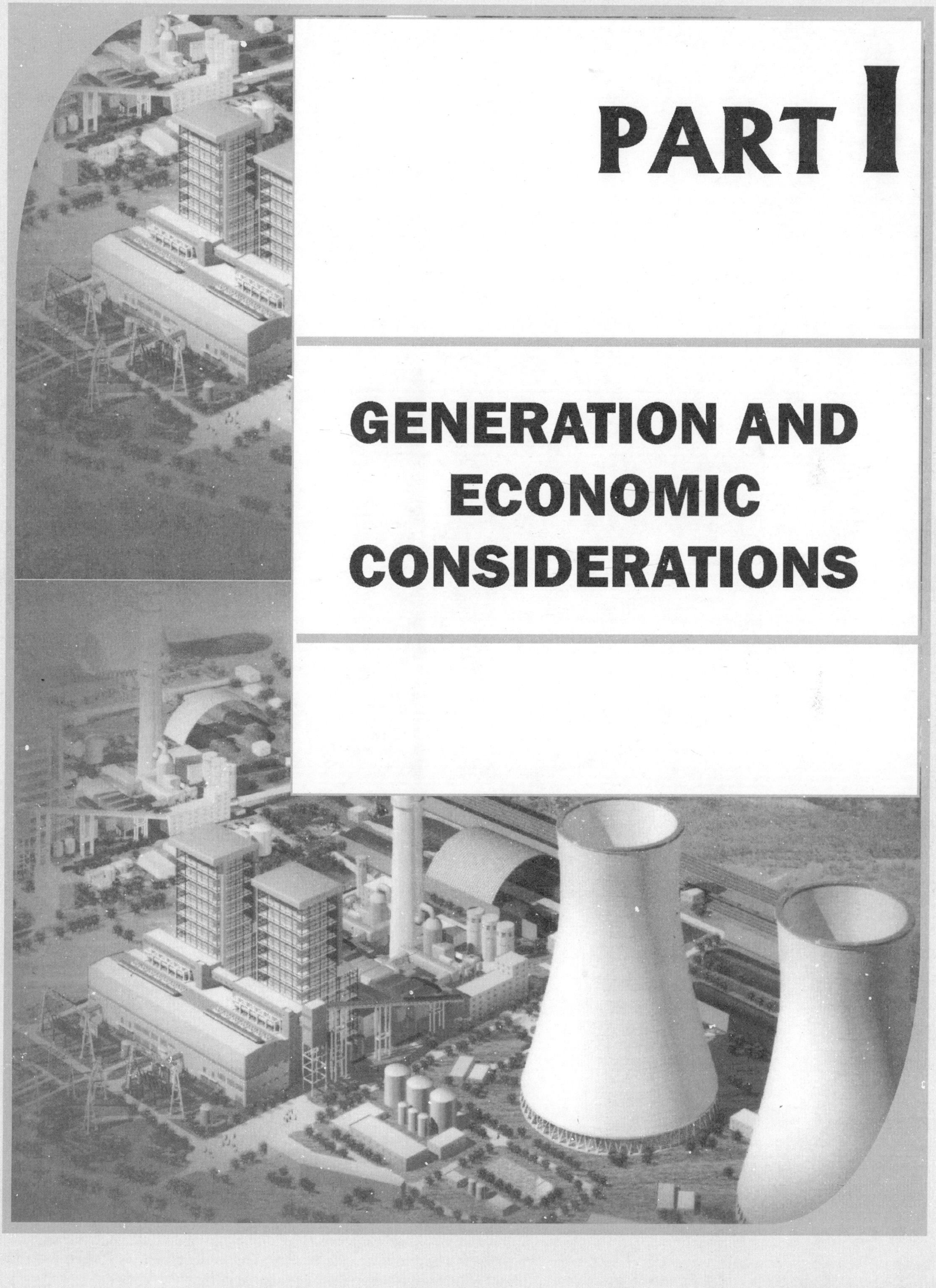

CHAPTER 1

Introductory

INSIDE THIS CHAPTER

1.1. IMPORTANCE OF ELECTRICAL ENERGY

Both the historical and the present-day civilization of mankind are closely interwoven with energy, and there is no reason to doubt but that in the future our existence will be more and more dependent upon the energy. Electrical energy occupies the top position in the energy hierarchy. It finds innumerable uses in home, industry, agriculture and even in transport. Besides its use for domestic, commercial and industrial purposes it is required for increasing defence and agricultural production. In agriculture, it is used for pumping water for irrigation and for improving the methods of production and numerous other operations. Electrical energy is a convenient form of energy because it can be generated centrally in bulk and transmitted economically over long distances and is almost pollution free at the consumer level. Further, it can be adopted conveniently in the domestic, industrial and agriculture fields. The process of modernisation, increase in productivity in industry and agriculture and improvement in the quality of life of the people depend so much upon the supply of electrical energy that the annual per capita consumption of electrical energy has emerged these days as an accepted yardstick to measure the prosperity of a nation. Some of the advanced and developed nations of North America and Europe have a very high annual per capita consumption of electrical energy, say from 8 to 13 thousand kWh, while in most of Africa, Asia and Latin America it is too low to be considered. India had an annual per capita consumption of electrical energy of 15.5 kWh in 1950, 105 kWh in 1975, 131 kWh in 1979, 154 kWh in 1984, 299 kWh in 1993, 349 kWh in 1997 and 631 kWh in 2006. The annual per capita consumption of electrical energy in some of the countries is : Canada 17,000 kWh, America 13,000 kWh, Japan 8,000 kWh, UK 6,200 kWh and Russia 5,600 kWh. The United States has only 6% of world population but accounts for over 30% of the electrical consumption of the world. The average per capita consumption of electricity in India is estimated to be 704 kWh during 2008-09. However, this is a fairly low when compared to that of some of the developed and emerging nations such as US ($\simeq$ 15,000 kWh) and China ($\simeq$ 1,800 kWh). The world average stands at 2,300 kWh.

1.2. SUPERIORITY OF ELECTRICAL ENERGY

Electrical energy is considered superior to all other forms (chemical, heat, light, sound, or mechanical) of energy due to the following reasons:

(*i*) *Cheapness.* It is much cheaper than that in other forms and therefore, it is economical to use energy in this form for domestic, commercial, industrial and agricultural purposes.

(*ii*) *Convenient and Efficient Transmission.* The electrical energy can be transmitted conveniently and efficiently from the generating stations, usually located quite away from the centres of usage, through conductors of suitable size.

(*iii*) *Easy Control.* Electrically operated machines have simple and convenient starting, control and operation. For example. an electric motor can be started or stopped by making the switch on or off and its speed can be conveniently controlled over a wider range with simple arrangements.

(*iv*) *Cleanliness.* Use of electricity (electric drive or electric heating) does not produce smoke, fumes, dust or poisonous gases and therefore, its use ensures cleanliness and pollution free conditions.

(*v*) *Greater Flexibility.* Electrical energy offers greater flexibility as it can be taken to any corner of the house, factory, street, hospital, farm, mine etc. through solid, stranded or flexible conductors.

(*vi*) *Versatile Form.* Electrical energy is a very convenient form of energy and it can be easily converted into other forms of energy—heat, light, mechanical, sound or chemical. For example, electric lamps, specially fluorescent tubes give rise to pleasant and cheaper light as compared to that produced by lamps of other types and can be located at any desired place where other lamps cannot be kept due to the danger of the fire or due to some other reasons.

Though at present about three-fourth of the total energy is still used in non-electrical form (transport, residential heating and industrial heating use energy mostly in non-electrical form) but because of numerous advantages, mentioned above, electricity will account for a greater and greater portion of total energy consumption in the coming years. It is expected that the electricity demand will continue to go up for many more years to come, even in developed countries.

1.3. GENERATION OF ELECTRICAL ENERGY

Electrical energy is generated by conversion of energy available in different forms from different natural sources such as kinetic energy of blowing winds, pressure head of water, chemical energy of fuels (either in solid, liquid or gaseous form) and nuclear energy of radioactive substances.

Conventional methods of power generation make use of prime movers (such as petrol engines, diesel engines, steam engines, steam turbines, gas turbines or hydraulic turbines) for driving electrical machines (generators or alternators) which convert mechanical energy into electrical energy. The electrical machines employed for generating dc are called the '*generators*' whereas those employed for generating ac are called the '*alternators*'. Popular methods of power generation by conventional methods are (*i*) thermal (*ii*) hydro and (*iii*) nuclear.

The alternate methods of generating electrical energy without the use of prime movers are called the *non-conventional methods of power generation*. The magneto-hydrodynamic (MHD) power generation is one of the example of a new unique method of power generation. The other non-conventional methods of power generation may be such as solar cells, fuel cells, thermoelectric generation, thermionic converters, solar power generation, wind power generation, geothermal energy generation, tidal power generation etc. All of these methods are still in experimental stages and of little commercial or industrial importance.

1.4. SOURCES OF ENERGY

The main sources of energy are the Sun, wind, terrestrial heat, ocean tides and waves, water, fuels and radioactive substances. The last three sources *i.e.*, water, fuels and radioactive substances are most dependable and are commonly used for generation of electrical energy.

1. *The Sun.* The Sun is the primary source of energy. The energy radiated by the Sun is in the form of electromagnetic waves, which include the heat, light and a lot of ultraviolet radiations. The radiated heat energy by the Sun can be utilised for the generation of electric power. The steam can be produced by focussing the heat over a boiler by use of some reflectors. This method is inconvenient as it requires a large area for generation of even small amount of electric power. This method cannot be adopted in cloudy sky or at night. This method is quite obsolete for generation of electric power because of uncertainty of weather. This method is used in U.A.R. for supplying irrigation and other pumping loads.

In our country conditions for utilisation of solar energy are favourable since for nearly six months of the year. Sunshine is uninterrupted during the day and useful energy estimated at noon is 12,000 kW/km^2 while in the other six months cloudy weather and rain provide conditions suitable for water power. Thus a co-ordination of solar energy and water power can be a useful and workable plan for most of the places in our country.

During the last few years more than 50 small power plants have been installed in U.P. and other states. The size of such plants is in the range of 2 to 10 kW. These plants are supplying power to the remote villages where power from grid is not available. Efforts are also being made to install large size power plants in some states.

2. *The Wind.* The winds produced by the Sun got sufficient energy which can be utilised in wind mills to drive small generators. Such generators are used for charging batteries for continuous use. The great advantage of this source of energy is that it is plentiful, inexhaustible, non-polluting and it does not require any operator. It also does not require any maintenance and repairs for long intervals.

The Indian Wind Energy already has the 5th largest installed capacity in the world and is set to grow at a rapid pace driven by investments from the private sector attracted by the generous wind subsidies from the state and central government in India. Most of the Installed Wind Capacity in India is located in the southern and industrial states of the country. The other states like UP, Bihar lacks sufficient Wind Energy Capacity despite having high power tariffs and substantial electricity deficits. It can be said that Wind Power in India is being developed only in the progressive states like Gujarat, Maharashtra and others. To continue to grow at the same rate, the industry needs to diversify geographically but that would need the backward state governments to pull up their socks.

Wind generators of capacity ranging between 225 kW and 2.1 MW are best suited for wind conditions prevailing in our country. There are two types of windmills namely gear type windmills suited for installation in regions of high speed winds and direct acting windmills suited for regions of low and moderate speed winds.

This method is unreliable since the production of electrical energy depends largely upon the pressure of the wind.

During past 30 years a few hundred small windmills have been installed in South India for pumping water. The total number of wind turbine generators in operation in India is more than 6,000 with a total installed capacity of about 14,157 as on March 2011. More wind turbine generator installations are in hand. The average capital cost per MW is Rs 5 to 7 crores and generation cost per unit is about Rs 4.50.

3. *Terrestrial Heat or Geothermal Energy.* The earth has a molten core. During volcanic action the material that comes

out from the bowls of earth to form volcanic explusions also produces steam vents and hot springs. This steam has been used for many years in Italy, Iceland and France for the generation of electric power and space heating. There are a number of hot springs in India, but the total exploitable energy potential seems to be very little. There is no operational plant, based on geothermal energy, in India.

4. *Ocean Tides and Waves.* There is a tremendous energy in ocean tides and waves but it is very difficult to harness this power for generation of electrical energy. Only at a few favourable points where the geography of an inlet or bay favours the construction of a large scale hydroelectrical plant, their use can be made for generation of electrical energy. A small dam with larger gates can be made across the mouth of the bay and low head hydraulic turbines are installed in it. At the time of high tide the gates are opened water is stored in the tidal basin and then the gates are closed. After the tide has receded, there is a working hydraulic head between the basin water and open ocean and the stored water is now allowed to flow back to the ocean through hydraulic turbines installed in the dam. In case use of reversible hydraulic turbines is made power generation is possible continuously, both during high tide and low tide.

The above source for generation of electrical energy has been vigorously investigated in some countries like France, Germany, UK, Canada and USA. Probably the best known and the most recent tidal scheme which utilises the tidal energy for generation of electrical energy has been constructed in France at the Rance Estuary where a power station of 240 MW capacity has been in operation since 1967.

In India potential for tidal energy exists 1,200 MW in Gulf of Kutch and 100 MW in Sunderban of West Bengal. Hydrological studies have indicated that power plants can be established on rivers Malta and Curzon Creek in this region. Potential of tidal power (7,000 MW) also exists in the Gulf of Cambay, near Bhavnagar, where tidal range is of the order of 11 to 12 metres.

5. *Water.* Water at a high pressure or flowing with a high velocity can be used to run turbines or waterwheels coupled to generators and, therefore for generation of electric power. This method of generation of electric power is becoming more and more popular as it is reliable and requires least maintenance and care. Though this method is costlier in initial cost but cheapest in maintenance and production cost. This source of generation of electric power will be discussed later on.

6. *Fuels.* The main source of energy is fuels. The fuels may be solid, liquid or gas such as coal, oil and coal gas.

The coal is available in huge quantities in almost all industrialised countries. The coal is fired in the boiler, which produces steam. The steam so produced is utilised in steam engines or turbines coupled to generators and therefore, for generation of electric power. The average calorific value of coal is 6,700 cal/kg. Though this method is most inefficient because of overall efficiency as low as 20 to 25% but due to availability of coal in huge quantity the cost per unit generated is not so high. The disadvantage of this source of energy for generation of electric power is that it requires a large space for storing the coal and involves heavy cost of transportation where the coal mines are away from the generating station.

Liquid fuel has a high calorific value say 8,300 k. calories/kg for light oil and 11,000 k. calories/kg for heavy oil but being costly, it is used for power production only where small power is required. The power stations employing liquid fuels are adopted either as standby power stations or as peak load power stations in interconnected grid systems.

Natural gas available in large quantity can be used to run gas engines (or gas turbines) or can be burnt to produce steam and steam turbines can there after run. Average calorific value of coal gas is 4.6 k. calories/litre and of producer gas varies from 0.11 k. calories/litre to 0.15 k. calories/litre. Gas engines are employed either near oilfields, or near gas works *i.e.,* where gas is the cheapest available fuel or where the process concerned does not require steam and a boiler plant would be uneconomical. The gas turbines have found great use in aircraft engines. In conjunction with hydroelectric stations gas turbo-sets are used to meet the power demand during peak load hours or as standby units.

7. *The Atomic Energy.* The tremendous amount of heat energy is liberated by fission of nuclear disintegration of uranium and other similar fissionable materials. It is estimated that 1 kg of nuclear fuel is equivalent to about 2,500 tonnes of coal. The heat energy so liberated in atomic reactors is extracted by pumping fluid or molten metal like liquid sodium or gas through the pile. The heated metal or gas is then allowed to exchange its heat to the heat exchanger by circulation. In the heat exchanger the gas is heated or steam is generated which is utilised to drive gas or steam turbines coupled to alternators thereby generating electrical energy.

The high capital cost (Rs. 15,000-60,000 per kW of installed capacity) of such generating stations, dearth of well trained personnel to man them and the difficulties associated with availability of suitable nuclear fuel and disposal of radioactive waste are the main hindrences in the development of nuclear power stations.

These power stations are best suited in areas far away from collieries (more than 800 km away from coal mines) and where alternative hydropower is not available as in Western UP, Northern and Western Rajasthan, Panjab and Haryana.

1.5. COMPARISON OF SOURCES OF ENERGY

The following factors are kept in view while comparing the various sources of power generation.

(*i*) *Initial Cost.* The initial cost of erecting a power plant for using water as the source, is very high as compared to the plant employing fuel as the sources of energy. This is due to large amount of excavation work and heavy cost of transportation of the plant and machinery in case of power plant using water as the source of energy. But the cost of erecting hydroelectric power plant is less than that for nuclear power plant. Thus the highest initial cost is involved in erecting nuclear power plant.

(*ii*) *Running Cost.* The running cost of hydroelectric power plant is less than that of steam or diesel power plant.

(*iii*) *Limit of Sources of Energy.* Of all the three sources for generating electric power viz. water, coal and nuclear, the nuclear source is inexhaustible. The availability of water as a source of power generation is not dependable because it depends upon the rainfall, which is at the whim of nature. Source of fuel *i.e.,* coal reserves all over the world is considered to be limited and coal mines are being exhausted and the time may come, possible after a few centuries, when all of these might finish. The raw materials for producing nuclear energy *i.e.,* fissionable materials in large deposits are available all over the world.

(*iv*) *Simplicity and Cleanliness.* The hydroelectric power plants are cleanest and simplest. There is no danger of any radiation hazards etc. and no nuisance of smoke, ashes etc.

(*v*) *Reliability.* Hydroelectric power plant is simplest, robust and more reliable. The nuclear power plant is also reliable.

(*vi*) *Field of Application.* If there is a perennial water source close by, hydroelectric power plant is the best. The steam (coal based) power plants are adopted only when coal supply is available in plenty, large amount of power is required to be generated and financial, climatic and geographical conditions are not favourable to hydroelectric power stations. According to the energy commission the generation from the coal based stations located within 800 km of the coal mines is cheaper than that from nuclear power plant.

The present contribution by various types of power plants are given below:

Type of power plants	Contribution
Hydroelectric power plants	22%
Steam power plants	65%
Nuclear power plants	2.7%
Renewable energy sources (RES)	10.3%

The relative contribution of hydro and steam power plants has remained almost the same during the past few years and is likely to continue. However, the steam power plant contribution may slightly increase due to large pit head plants being proposed. The nuclear power contribution is not likely to increase appreciably due to difficult position of nuclear fuel. The contribution of diesel and gas power plants is infinitesimal inspite of the fact that some gas turbine plants have been set up and some more are proposed in Gujarat and other states. This is because of escalating prices and poor availability of diesel and gas. Nevertheless a very large number of industries in small and medium sectors, have installed diesel power plants to serve as standby units. There are rare prospects of large scale utilisation of non-conventional sources of energy, such as Sun, wind, terrestrial heat, ocean tides and waves etc. for generation of electrical energy in the near future.

1.6. SOURCES OF ENERGY IN INDIA

Kerala, Karnataka and Jammu and Kashmir are very rich in water power resources and therefore, power development programmes should be based on this source.

In Assam and Maharashtra the major source of power is water and therefore, power development programmes should be based on it. However, in Assam the backing can be provided by gas turbine power plants using natural gas available locally and in Maharashtra by nuclear power plants.

In Punjab, Haryana and Himachal Pradesh sizeable potential of water is available, Expansion of base load requirements should be met by nuclear power plants located in this area in preference to conventional steam power plants, being uneconomical. However, this area can be backed from steam power plants already located in Delhi, Haryana and Panjab.

U.P. has sufficient water power resources and, therefore, this should be the main source for power generation. The water power can be supplemented by conventional steam power plants which should be preferably located in eastern part of the state.

Tamil Nadu is deficient in coal resources and it has already exhausted its limited water power resources. The future programmes should, therefore, be based largely on nuclear power plants.

Madhya Pradesh and Orissa have sufficient coal reserves and water power resources. Therefore the power development programmes should be based on a combination of these resources.

Gujarat and Rajasthan are deficient both in water power and coal resources, and therefore, the nuclear power should be the main source for power generation. However in Gujarat the small water potential and natural gas available can be utilised to meet with the peak loads.

Bihar has abundant coal reserves. The major power development programmes should, therefore be based on conventional steam power plants. Small water potential can be utilised to meet with peak loads.

West Bengal is rich in coal resources and deficient in water potential. The power development programmes should, therefore, be primarily based on conventional steam power plants.

The state/UT wise installed capacity of power utilities with generation mode break-up as on 31-03-2011 in million of watts (MW), as published by the Ministry of Power, is given below in tabular form.

Table 1.1

Rank	State/ Union Territory	Total Installed Capacity	Thermo	Nuclear	Hydro	RES (Renewable Energy Sources)
—	India	**173,626.40**	**112,824.48**	**4,780.00**	**37,567.40**	**18,454.52**
1	Maharashtra	22,645.29	15,813.98	690.14	3,331.84	2,809.33
2	Gujarat	15,722.65	12,390.86	559.32	772.00	2,000.47
3	Tamil Nadu	15,515.43	7,056.62	524.00	2,122.20	5,812.61
4	Andhra Pradesh	15,035.58	10,297.28	275.78	3,695.53	766.99
5	Karnataka	11,546.14	5,075.29	254.86	3,599.80	2,616.19
6	Uttar Pradesh	9,646.73	4,057.09	190.90	3,518.20	1,880.54
7	West Bengal	8,113.27	4,534.01	92.88	3,223.67	262.71

Rank	State/ Union Territory	Total Installed Capacity	Thermo	Nuclear	Hydro	RES (Renewable Energy Sources)
8	Madhya Pradesh	6,780.01	3,435.93	151.04	3,031.57	161.47
9	Punjab	6,426.15	3,774.03	469.00	1,456.82	726.30
10	Rajasthan	8,471.69	6,470.14	740.00	1,162.00	99.55
11	Haryana	8,530.29	7,054.03	76.16	1,331.40	68.70
12	Orissa	4,072.46	2,865.23	0.00	1,174.93	32.30
13	DVC	3,386.00	3,190.00	0.00	196.00	0.00
14	Delhi Territory	3,677.34	3,045.20	47.08	585.06	0.00
15	Chhattisgarh	3,607.05	3,312.90	0.00	120.00	174.15
16	Kerala	3,514.05	1,545.82	80.09	1,769.10	119.04
17	Uttarakhand	2,383.03	301.05	16.28	1,955.73	109.97
18	Jammu & Kashmir	2,158.95	509.62	68.00	1,469.50	111.83
19	Jharkhand	2,152.57	1,972.52	0.00	176.00	4.05
20	Bihar	1,969.99	1,846.59	0.00	73.00	50.40
21	Himachal Pradesh	1,896.47	156.43	14.08	1,540.84	185.12
22	Assam	980.30	522.19	0.00	431.00	27.11
23	Goa	357.23	327.18	0.00	0.00	30.05
24	Meghalaya	288.08	28.05	0.00	229.00	31.03
25	Pondicherry Territory	256.62	239.51	17.09	0.00	0.02
26	Tripura	243.36	165.35	0.00	62.00	16.01
27	Sikkim	193.09	76.98	0.00	75.00	41.11
28	Arunachal Pradesh	180.14	36.88	0.00	98.00	45.26
29	Manipur	158.86	71.41	0.00	81.00	5.45
30	Mizoram	119.33	67.86	0.00	34.00	17.47
31	Nagaland	102.67	21.00	0.00	53.00	28.67
32	NLC	100.17	100.17	0.00	0.00	0.00
33	Chandigarh Territory	93.46	41.58	4.84	47.04	0.00
34	Dadra and Nagar Haveli Territory	80.78	78.80	1.98	0.00	0.00
35	Daman and Diu Territory	71.10	69.12	1.98	0.00	0.00
36	Andaman and Nicobar Islands Territory	65.40	60.05	0.00	0.00	5.35
37	Lakshadweep Territory	10.73	9.97	0.00	0.00	0.76

Renewable Energy Sources (RES) include SHP, BG, BP, U&I and Wind Energy.

1.7. GROWTH OF POWER SYSTEMS IN INDIA

India is fairly rich in natural sources like coal and lignite and has immense water power resources of which only about 25% have been utilised so far. Some oil reserves have already been discovered and intense exploration is being undertaken in various states. Though India is deficient in uranium, but has rich deposits of thorium which can be utilised in future. Since independence, the country has made tremendous progress in the development of electric power and today India has the largest power system among developing countries and fourteenth largest power system in the world.

Electricity has been available in India for the past 100 years. The first station commissioned was a micro-hydel scheme near Darjeeling at the end of the 19th century. The first major effect in the public sector was the commissioning of the first stage of the Sivasmudram hydroelectric scheme, with an initial installed capacity of 4.5 MW in 1902. Thereafter, small power stations continued to be set up, largely around important urban centres, mainly through the efforts of private companies. By 1915 the installed capacity had increased to 105 MW. In 1939 the installed capacity in India was 1,200 MW.

At the time of independence the installed capacity was about 1,900 MW, the number of electrified villages was about 1,500 and the number of energised agricultural pumping sets was about 6,400. This was due to the fact that rural development had not been considered important in the pre-independence era. In the early stages of the growth of power plants, the major portion of the generation was through steam power stations, but due to economical reasons hydroelectric power development received special attention in areas like Kerala, Mysore, J & K, Punjab, UP, Assam, Maharastra, Tamil Nadu etc.

In the beginning of the First Five Year Plan, the installed generating capacity in the country was around 2,300 MW (Hydro-560 MW, Steam-1,004 MW, Oil-149 MW and non-utilities-587 MW) which raised to 3,420 MW at the end of the plan. The second Five Year Plan placed special emphasis on the development of public sector and heavy industries and correspondingly on the provision of necessary power generation and distribution facilities. At the end of the second Five Year Plan the installed generating capacity increased to 5,650 MW. At the end of the second Five Year Plan the number of villages electrified and the number of agricultural pumping sets energised in our country increased to 21,800 and 199,000 respectively. At the end of the Third Five Year Plan these figures regarding installed generating capacity, electrified villages and energised agricultural pumping sets increased to 10,170 MW, 45,100 and 513,000 respectively. The energy generated in 1965-66 was 36,825 million kWh.

During last 30 years, emphasis continued to be laid on achieving a rapid growth not only in the power generation facilities but also in strengthening and modernising the transmission and distribution systems. Along with this, rural electrification was given very high priority. During the last 30 years the installed generating capacity and energy generation has increased to 6-7 times whereas the number of electrified villages has increased to 5 times. The progress in installed generating capacity, energy generation, electrification of villages and annual per capita consumption of electricity during first 50 years after independence is given in Table 1.2.

Table 1.2. Growth of Electrical Energy in India

Year	1950-51	1955-56	1960-61	1965-66	1970-71	1973-74	1978-79	1987-88	1990-91	1993-94	1995-96	1997-98
Installed generating capacity in MW	2,300	3,420	5,650	10,170	14,710	16,663	26,680	54,195	66,090	76,750	83,290	89,090
Energy generation in million kWh	7,500	12,050	20,120	36,825	61,210	66,690	102,522	202,145	264,610	324,160	379,890	421,830
Total number of electrified villages in thousands	4	9.85	21.8	45.1	104	156.7	232.7	435.6	481	494	501.83	502.04
Annual per capita consumption of electricity in kWh	16	24	37	61	91	98	131	201	263	299	335	349

The total installed generating capacity by the end of the year 1998-99 became 94,055 MW while total generation became 448,000 million kWh.

Table 1.3. Installed Capacity in India, 1997-2005 and As on 31.10.2010 (in Thousands of MW)

	1997	2002	2003	2004	2005	As on 31.10.2010
Hydroelectric	21.65	26.76	26.76	29.50	30.94	37,3284
Nuclear	2.22	2.72	2.72	2.72	2.77	4.56
Geothermal/ Solar/Wind/ Biomass	1.27	1.51	1.74	1.87	3.81	16.78698*
Conventional Thermal	59.64	74.55	76.65	77.97	80.90	108.60298
Total Capacity	84.78	105.54	107.87	112.06	118.42	167.29

RES (MNRE) include small hydroelectric project (SHP), Biomass gasifier (BP), Biomass power (BP), Urban and industrial waste power (U&I).

India has the fifth largest generation capacity in the world with an insalled capacity of 152 GW as on 30.9.2009, which is about 4 per cent of the global power generation. The Indian Government has set ambitious goals in the 11th plan for power sector owing to which the power sector is poised for significant expansion in order to provide availability of over 1,000 units of per capita electricity by year 2012, it has been estimated that need-based capacity addition of more than 100 GW would be required. It is proposed to add 23,000 MW during 12th Five Year Plan (2012-17).

Operation performance of generating stations in the country during the year 2010-11 is given below in Tabular form.

Electric Energy Generation Target for the year 2010-11	830.8 BU
Actual Electric Energy Generation during the year	811.1 BU
Growth in generation during 2010-11	5.55%

The details of generation and growth rates are given in Table 1.4.

Table 1.4. Annual Electric Energy Generation Targets and Achievement

Category	Target 2010-11 (BU)	Actual 2010-11* (BU)	% of Target	Actual Last Year 2009-10 (BU)	Growth (%)
Thermal	690.9	664.9	96.24	640.5	3.81
Nuclear	22.0	26.3	119.48	18.6	41.04
Hydro	111.4	114.3	102.64	103.9	10.01
Bhutan Import	6.5	5.6	85.68	5.4	4.69
Total	830.8	811.1	97.63	768.4	5.55

*Generation excludes generation from plants up to 25 MW capacity.

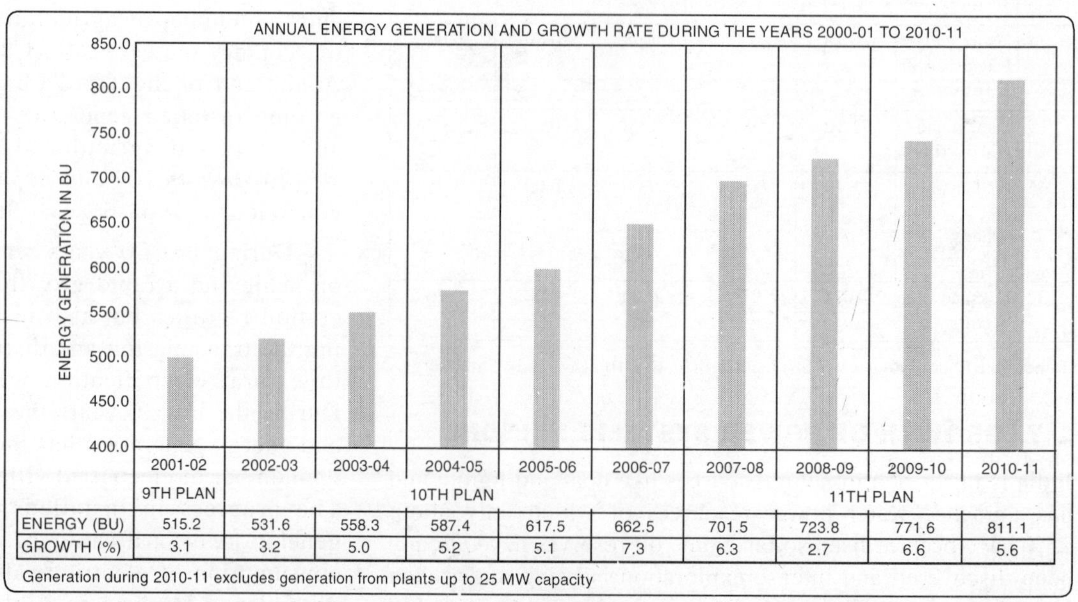

Fig. 1.1

The annual growth in the energy generation during the year has been 5.55% against the CAGR of 5.17% during the period 2001-02 to 2010-11. The year wise energy generation and annual growth rates during the period are presented in Fig. 1.1.

Significant progress has also been made in the expansion of transmission and distribution facilities. The total length of the transmission and distribution lines above 500 V has increased from about 250,000 km in 1965-66 to 860,000 km in 1973-74. Out of this, the increase in the main transmission lines of 66 kV and above was from 26,700 km to 83,225 km. The LT distribution lines up to 500 V registered an increase from about 290,000 km to 840,000 km in the same period. The transmission voltage adopted for bulk transmission of power has also been raised from 220 kV to 765 kV.

As a result of ambitious goals set by the Indian Government in the 11th plan for power sector, massive addition plans are being proposed in the sub-sectors of Generation, Transmission and Distribution.

The current installed transmission capacity is only 13% of the total installed generation capacity. With focus on increasing generation capacity over the next 8-10 years, the corresponding investments in the transmission sector is also expected to augment. The Ministry of power plans to establish an integrated National Power Grid in the country by 2012 with close to 200 GW generation capacity and 37.7 GW of inter-regional power transfer capacity, the current inter-regional power transfer capacity is 20.75 GW.

The total length of transmission lines has become, as on 31.03.2009, 222,746 circuit kilometres (220 kV–122,960 circuit kilometers, 400 kV–89,496 circuit kilometer, 765 kV–3,118 circuit kilometer and HVDC–7,172 circuit kilometer).

While some progress has been made at reducing transmission and distribution (T&D) losses, these still remain substantially higher than the global benchmarks, at approximately 33%. In order to address some of the issues in this segment, reforms have been undertaken through unbounding the State Electricity Boards into separate Generation, Transmission and Distribution units and privatization of power distribution has been initiated either through the outright privatization or the franchisee routes; results of these initiatives have been somewhat mixed. While there has been a slow and gradual improvement in metering, billing and collection efficiency, the current loss levels still pose a significant challenge for distribution companies going forward.

1.8. POWER CRISIS IN INDIA

The electricity requirements in India have grown tremendously and the demand has been running ahead of supply. Electricity generation and transmission progresses in the country are very inefficient in comparison with those of some advanced countries. Generating capacity in India is utilised on an average for 3,600 hours, while in Japan it is used for more than 5,100 hours in a year. The transmission losses in India are very high as compared to developed countries. The transmission loss in 1997-98 on a national basis were 23.68% consisting of both technical losses in transmission lines and transformers, and also non-technical losses caused by energy thefts and meters not being read correctly. It should be possible to achieve considerable saving by reducing this loss to 15% by using well known ways and means and by adopting sound commercial practices. The forced outage of power plant equipment has been rendering generating units idle for long periods. The plant outages have been due to poor quality of power plant equipment, improper operation and maintenance. The commissioning of many power projects has been delayed considerably due to various reasons such as non-availability of funds in time, shortage of building material, labour problems, shortage of trained personnel, delay in delivery of equipment, lack of coordination in construction activities, interstate disputes etc. Many a time the monsoons have been less than normal causing reduction in generation of hydro-energy. There had been also periodic closure of steam power stations due to shortage of coal, (either due to difficulties in coal mines or in transportation). These are the causes (*i.e.,* sharp increase in demand, poor utilisation of electrical equipment, high transmission losses including energy theft, the delay in commissioning of power projects, erratic monsoons, plant outages, shortage of coal, faulty planning), which are responsible for power crisis in the country.

1.8.1. Power Sector Reforms in India

To meet up with the large gap between demand and supply, various steps are required to be taken up. For this, instead of planning and building power plants in isolation, studies must be made to optimal generating capacity expansion. Attempt should be made to fix responsibility for delay. Modern management techniques have not been applied to remove the bottlenecks and streamline the system, this situation needs to get reformed and more modern techniques should be applied to the electricity production and transmission. Coal production would have to be stepped up.

Every attempt should be made to improve system load factors by flattening the load curve by giving proper tariff incentives and taking other administrative measures.

The government target of 20,000 MW of nuclear capacity by the year 2025 needs accelerated efforts to get achieved.

The renewable sources and combined power plants, with the help of modern technologies should get supported both by public and private sectors. The electric supply undertakings needs to analyse the system, spot the weaknesses and take corrective measures to improve reliability and conditions responsible for power crisis.

1.9. FUTURE TRENDS

The electricity requirement of India will continue to increase tremendously in the coming years. Hence certain concrete steps will have to be taken to increase the generating capacity so as to meet the increasing demand. Government of India has fixed a target of 20,000 MW of nuclear capacity by the year 2025, against present generating installed capacity of 4,560 MW of nuclear power plants in the country. Hydroelectric power will continue to remain cheaper than other types for the next two decades. As mentioned earlier the country has developed only about 45% of its estimated total hydroelectric potential of 84,000 MW. Long delayed hydroelectric projects are likely to come up.

During the early years, the emphasis has been on development of hydroelectric power but during last two decades there has been increasing tilt towards coal based steam power stations. Average coal production per man shift is only 0.5 tonne in India while it is more than 12 tonnes in America. Hence coal production would have to be stepped up in order to keep the superthermal power stations in operation.

Because of greater availability of natural gas, there has been a renewed interest in the development of gas turbine plants for supplying peak loads. The generating capacity of gas turbine plants in 1981-82 was 320 MW, which has increased to 24 times in last two decades and has become 7,660 MW in year 1997-98. More gas turbine power plants are likely to come up.

Many industrial undertakings (such as steel plants, fertilizer plants etc.) have installed captive power plants on their premises. Government has started permitting private industries to install captive plants. It is likely that more and more industries may set up captive plants to supply their own needs and possibly the adjoining areas.

Apart from the development of power plants based on conventional sources, such as coal, oil, water and nuclear, technological advances are in progress for generation of electrical energy by direct means, such as magneto-hydrodynamics (MHD), fuel cells etc. Besides this, concreted efforts are also to be made for development of solar energy and other non-conventional sources of energy such as geothermal energy, tidal power generation and wind power generation.

There is also a need to improve the utilization of generating equipment, take corrective measures to improve reliability of the electric power systems and take care of the environmental aspects of energy generation.

1.10. PRIVATE SECTOR IN ENERGY MANAGEMENT AND CAPTIVE POWER PLANTS

Huge finances are required in setting up of central power plants and for providing infrastructure for the transmission and distribution of generated electric power. Because of financial constraints, the central and state goverments, the various state electricity boards, and the various public sector organizations (like NTPC, NHPC, Nuclear Power Corporations) do not find them in a position to undertake the setting up of the large power plants in order to meet the country needs in this sector. Because of this, the private sector participation in the energy management of the country is gaining popularity. In many advanced countries, the energy sector is mainly in the hands of private sector. Such power through the national grid will be subjected to specified grid conditions. Alternatively, the private sector can lay its own transmission lines.

Captive power refers to generation from a unit set up by industry for its exclusive consumption. The estimates on captive power capacity in the country vary with the Central Electricity Authority (CEA) putting the figure at about 11,600 MW while industry experts feel that it is much more, about 20,000 MW.

Industrial sector is one of the largest consumers of electrical energy in India. However, a number of industries are now increasingly relying on their own generation (captive and cogeneration) rather than on grid supply, primarily for the following reasons :

1. Non-availability of adequate grid supply.
2. Poor quality and reliability of grid supply.
3. High tariff as a result of heavy cross-subsidisation.

Need For Captive Power Plants (C.P.P.)

1. Huge gap between the demand and supply from central power stations.
2. Frequent long and seasonal power cuts.
3. Frequent increase in the power tariffs by the utilities.

Advantages

1. The overall cost per unit may be low.
2. No energy theft.
3. No infrastructure (transmission and distribution of power).
4. No transmission losses.

Various Systems of Captive Power Generation

Captive power generation has the following options :

1. Power generation by an industry for
   (*a*) its use only
   (*b*) its use as well as supply of excess power to the neighbouring industry
   (*c*) its use and supply of excess power to the utility
   (*d*) its use and supply of excess power to the utility as well as the neighbouring industry.
2. Captive power plants set up by a co-operative society of companies for their own use.
3. Captive power plant set up by a private generating company for supply of power to the neighbouring industries.

Types of Captive Power Plants

The captive power plants may be steam, diesel, gas, cogeneration, hydro plants.

1. *Steam Power Plants:* This option is suitable for heavy industries such as steel, aluminium etc., which already have access to coal and the required infrastructure for purchasing and storing coal. Generally such captive power plants have generating capacity of about 100 MW or more. These plants are usually condensing type. Examples are: 247.5 MW coal based Visakhapatnam steel plant in Andhra Pradesh; 302 MW coal based Bokaro steel plant onwed by TISCO, in Bihar and 210 MW coal based captive power plant at Jamshedpur (Bihar).

   Small industries (power, textile, plastic, breweries etc.) has capacities 1 to 25 MW and the power plants are usually non-condensing type.

   Many sugar industries in A.P., Bihar, Gujarat, Haryana, Karnataka, MP, Maharashtra, Punjab, UP etc. use "bagasse (waste dry sugarcane left after the cane juice has been extracted)" as the fuel for their captive power plants.
2. *Diesel Power Plants.* Such power plants are by far the largest number of captive power plants. These are also used as peak load and standby plants. They have the capacities of 1 to 10 MW (sometimes up to 30 MW).

They are more popular in the states of AP, Haryana, Karnataka, MP, Tamil Nadu. Some typical diesel engine captive power plants are : IOC (Panipat in Haryana)—75 MW; Rajshree Cement (Gulbarga in Karnataka)—42 MW; TISCO (Jamshedpur in Bihar)—30 MW, India Cement (Cuddapah in AP)—30 MW; Rayalseema Alkalies (Kurnool in AP)—35 MW; Hindustan Zinc Limited, HZL (in Visakhapatnam in AP)—10 MW.

3. *Gas Turbine Power Plants.* Such captive power plants are popular with medium scale and large industries. Some industries, like, petrochemicals have easy access to gas and liquid fuels and, therefore, prefer captive plants using gas turbines. Also useful in combined power plants. Some typical gas-based captive power plants are : IOC (Jawaharnagar in Gujarat)—95 MW; Maruti Udyog (Gurgaon in Haryana)—60 MW; ONGC (Hazira in Gujarat)—59 MW; Reliance Industries (Narodan in Gujarat)—47.5 MW; ONGC (Gandhar in Gujarat)—31 MW; Essar Steels (Hazira in Gujarat)—30 MW; Nagarjuna Fertilizers (Kakinda in AP)—18 MW.
4. *Cogeneration Plants.* Cogeneration produces two or more types of useful energy *e.g.*, electrical energy and steam. Though the initial cost of a cogeneration plant is more than that of an electric power plant, but such plant has very high efficiency and leads to overall economy and conservation of fuel. A cogeneration plant may be steam-gas turbine plant with waste heat boiler, combined cycle cogeneration plant, biomass fuel plant, diesel engine plant etc. Cogeneration plants produce lesser pollution than other plants.

   Many state government and other organisations provide incentive for installation of cogeneration plants.

   Arvind Mills (Ahmedabad) has a combined cycle cogeneration captive power plant. It has two 27 MW gas turbines. TASL (Tamil Nadu) has a 28 MW cogeneration captive power plant. It can be operated on bagasse as well lignite.
5. *Mini Hydroelectric Plant* can be a very attractive option for industries located near a site suitable for such a plant. A 12 MW mini hydro plant in Kerala and a 10 MW mini hydro plant in Karnataka are typical hydro captive power plants.
6. *Captive Power Plants For Rural India.* It is well known that the economy of India is an agriculture-based. About 70 per cent of GDP comes from agricultural sector. So, this sector needs special attention for the economic growth of the country. For this, the energy requirements of this sector must be met on priority basis. However, a major portion of power generated is being consumed by the urban and industrial consumers, and there is a perennial shortage of power in the agriculture sector. This problem, however, can be overcome by installing small rural based captive power plants, where the energy source will be biomass, in the form of agri-residues. Such small captive power plants can meet the power needs of 25 to 40 nearby villages. This will solve the dual problem of disposal of agro-residues and shortage of power in rural area. Agriwaste, the cause of smoky clouds, can enlighten the gloomy rural area.

   The state-wise captive capacity in 1998 is given below in the tabular form.

Table 1.5. State-Wise Captive Capacity

State	Installed Capacity (MW)	Captive Capacity (MW)
Andhra Pradesh	8,204	1,220
Assam	1,078	—
Bihar	4,656	614
Delhi	1,436	—
Gujarat	8,376	1,505
Haryana	882	335
Himachal Pradesh	3,570	32
Jammu & Kashmir	1,536	3
Karnataka	3,462	1,045
Kerala	1,766	151
Madhya Pradesh	7,173	1,333
Maharashtra	11,072	570
Meghalaya	239	—
Orissa	3,243	1,544
Punjab	2,620	311
Rajasthan	2,176	528
Tamil Nadu	8,271	1,107
Uttar Pradesh	12,473	1,240
West Bengal	6,515	786
Total	89,167	12,322

Source: Power Line Research

Based on the fuel type used for captive power generation about 45% of power generated is from steam, 40% from diesel and 15% from Gas/Naphtha. Captive contribution by various industry type is as follows :

Table 1.6. Industry-Wise Contribution Towards Captive Power Generation

Industry	Capacity (MW)	% Share
Cement	1,223	9.9
Chemicals	2,076	16.8
Electronics	59	0.5
Engineering	2,479	20.1
Jute	207	1.7
Metals & Minerals	2,404	19.5
Miscellaneous	784	6.4
Paper	473	3.8
Services	80	0.6
Sugar	706	5.7
Textile	1,303	10.6
Unclassified	530	4.3
Total	12,322	100

Source: Power Line Research

Indian Power Sector at a glance as on 31.10.2010 is given below:

(Table 1.7)–Source Central Electricity Authority (CEA)

Sector	MW	%age of Total Installed Capacity
State sector	82,227.05	52.5
Central sector	51,867.63	34.0
Private sector	33,183.68	13.5
Total	167,278.36	

Exercises

1. The annual per capita income of electrical energy is often taken an index of the economic development, prosperity and standard of living of a nation. Discuss.
2. What are the various sources of energy for power generation and discuss their merits and demerits ?
[M.D. Univ. Electro Power Generation May 2011]
3. Discuss the different sources of energy available in nature.
[Pb. Technical Univ. Generation of Electric Power Dec. 2010]
4. Enumerate the various sources of energy for power generation and discuss their merits and demerits.
[U.P. Technical Univ. Power Station Practice 2009-10]
5. What are the sources of energy in nature and how are these sources utilised by the mankind to generate electrical power ?
[U.P. Technical Univ. Power Station Practice 2007-08]
6. What is the current generation of electricity from all possible sources in India ? What percentage of nuclear power energy is sharing the market.
[U.P. Technical Univ. Power Station Practice 2007]
7. Discuss electric energy demand and its growth in Indian power sector. List various electric energy sources.
[U.P. Technical Univ. Power Station Practice 2006-07]
8. What is the likely energy demand in India in the year 2010 ? Describe the development of power in India.
[G.B. Technical Univ. Power Station Practice 2010-11]
9. Discuss in brief the reasons of present power crisis in India. How the conditions can be improved ?
[Rajasthan Technical Univ. Generation of Electrical Power 2011]
10. Discuss how the power crisis can be solved by joint venture of public and private sectors.
[U.P. Technical Univ. Power Station Practice 2008-09]
11. Explain the power crisis situation and how it can be overcome ? What is the role of private sectors in managing the future energy demand ?
[G.B. Technical Univ. Power Station Practice 2011-12]
12. Discuss the organisation of power sector in India.
[Pb. Technical Univ. Generation of Electric Power May 2008]
13. Discuss the role of private sectors in energy management.
[U.P. Technical Univ. Power Station Practice 2007-08]
14. List various sources of electric power. Why some of them are called conventional and others are called non-conventional ? Discuss their importance.
[A.M.I.E. Sec. B. Power Systems Summer 2000]
15. What are the recent trends in electrical power generation ?
[M.D. Univ. Electro Power Generation May 2011]
16. Write short notes on the following :
   (*i*) Energy resources and their availability in India.
   [M.D. Univ. May 2007; December 2009]
   (*ii*) Non-conventional methods of power generation.
   [A.M.I.E. Sec. B Power System Winter 1993]
   (*iii*) Recent developments of power generation in India.
   [Pb. Univ. December 1993]
   (*iv*) Power crisis in India.

Short Answer Type Questions With Answers

Q. 1. What is the per capita consumption of electricity in India ?
Ans. 631 kWh in 2006.

Q. 2. Why electrical energy is more useful than other forms of energy ? **[U.P. Technical Univ. 2004-05]**
Ans. Electrical energy is considered more useful than all other forms of energy owing to (*i*) cleanliness, (*ii*) cheapness, (*iii*) easy control, (*iv*) convenient and efficient transmission, (*v*) greater flexibility and (*vi*) versatility.

Q. 3. What is the total installed generating capacity in India ?
Ans. 173,626.40 MW as on 31.3.2011.

Q. 4. What are the various sources of energy that are utilized for generation of electrical energy.
Ans. Various sources of energy for generation of electrical energy are fuel (solid, liquid or gaseous), water and atomic energy.

Q. 5. Give the present contribution by different types of power plants.
Ans. The present contribution by various types of power plants is as given below:

Type of power plants	Contribution
Hydroelectric power plants	22%
Steam power plants	65%
Nuclear power plants	2.7%
Renewable energy sources (RES)	10.3%

Q. 6. What is the percentage of power generated by hydro plants in India ? Which states have maximum hydro generation ? **[U.P. Technical Univ. 2005-06]**
Ans. Hydro power generation in India is about 22% of total power generation. Kerala (50%), Uttarakhand (82%), J&K (68%), Himachal Pradesh (81%), Meghalaya (79.5%), Arunachal Pradesh (54.5), Manipur (51%) and Nagaland (52%) have maximum hydro generation.

Q. 7. What do you mean by green power ? **[U.P. Technical Univ. 2004-05]**
Ans. Power derived from any renewable energy source (such as biomass, hydro, solar, wind, tide and geothermal) is termed as green power. This is due to negligible effect on GHG (green house gases) emissions.

Q. 8. What are the reasons of power crisis in India ? **[U.P. Technical Univ. 2009-10]**
Ans. The causes, which are responsible for power crises in India, are sharp increase in demand, poor utilisation of electrical equipment, high transmission losses, the delay in commissioning of power projects, erratic monsoons, shortage of coal, faulty planning and plant outages.

Q. 9. What do you mean by captive power ?
Ans. Captive power refers to generation from a unit set up by industry for its exclusive consumption.

Q. 10. Why are captive power plants (C.P.P.) required ?
Ans. Captive power plants are required as (*i*) there is a huge gap between the demand and supply from central power stations, (*ii*) there is frequent long and seasonal power cuts and (*iii*) there is frequent increase in the power tariff by utilities.

Q. 11. What is the percentage of captive power generation based on the fuel type used ?
Ans. Based on the fuel type used for captive power generation the contribution is as follows :
Steam—about 45%, Diesel—about 40% and Gas/Naphtha—about 15%.

Q. 12. Give the Indian power sector at a glance as on 31.10.2010.
Ans. Indian Power Sector at a glance as on 31.10.2010 is given below:

Sector	MW	%age of Total Installed Capacity
State sector	82,227.05	52.5
Central sector	51,867.63	34.0
Private sector	33,183.68	13.5
Total	167,278.36	

❑❑❑

CHAPTER 2

Hydroelectric Power Plants

INSIDE THIS CHAPTER

2.1. INTRODUCTION

Hydroelectric power is the power obtained from the energy of falling water whereas hydroelectric power plant is the power plant utilising the potential energy of water at a high level for the generation of electrical energy.

The chief source of energy is the Sun. Solar energy in the form of heat is absorbed by the oceans, lakes and rivers resulting in the formation of clouds eventual precipitation over the land. Thus the work done on water during the hydrologic cycle, by the energies of the Sun, wind and the cooling forces places it upland, impregnating it with the huge potential energy, which may be extracted from it, during its descent back to the oceans. Most of this energy, if not harnessed, dissipates itself into heat before reaching the ocean. If harnessed, on the other hand, this energy is converted into kinetic energy and this kinetic energy is converted into the mechanical energy by allowing the water to flow through the hydraulic turbine runner. This mechanical energy is utilized to run an electric generator which is coupled to the turbine shaft. The electrical power, P developed in this way is given as

$$P = wQH\eta \times 9.81 \times 10^{-3} \text{ kW} \quad ...(2.1)$$

where w is the specific weight of water in kg/m^3, Q is the rate of flow of water in m^3/s, H is the height of fall or head in metres and η is the overall efficiency of operation.

In a hydroelectric power station, water head is created by constructing a dam across a river or lake. The pressure head of water or kinetic energy of water is utilized to drive the water turbines coupled to alternators and, therefore, generation of electrical power.

There are two reasons for the extensive development of water power. One is the tremendous increase in demand of electric power for industrial, agricultural, commercial and domestic purposes and the other is high cost of fuels (coal and oils) and the limited resources. A water power site is usually developed to supply electric power to a newly and a specially established industry or town or to provide additional power to an already existing or proposed interconnected electric power system. Hydroelectric power stations are usually located in high hilly areas, where the dam can be built easily and large reservoirs can be obtained. Generally such projects are multipurpose *i.e.,* in addition to generation of electric power they are used for irrigation, flood control and navigation etc. Irrigation can be remarkedly improved by routing the canals in the downstream of hydro plants, draught and flood both can be effectively controlled utilising the dam water and discharge water of hydro plants.

According to the Central Electricity Authority (CEA), the viable hydro potential in India is 84,000 MW at 60 per cent load factor which is equivalent to 148,700 MW installed capacity. In addition to it, around 10,000 MW from about 2,000 small hydro schemes can also be economically exploited. Most of the high head power sites naturally lie in hilly areas such as the sub-Himalayan region of North, the Western Ghats, the Nilgiries etc. The medium and low head site developments, combining power and irrigation together, usually occupy the steep plateaus and the plains. The list of presently operating hydropower plants is given in Table 2.1.

However, before a water power site is considered for development, the following factors are required to be thoroughly analysed.

1. Capital cost of the plant.
2. Capital cost of erecting and maintaining the transmission lines and the annual energy loss in transformation and transmission of electric power. This is because the hydroelectric power plants are usually located in hilly areas, far away from the load centre.
3. Energy generation cost compared with those in case of steam, oil or gas plants which can be conveniently set up near the load centre.

Table 2.1. List of Presently Operating Hydropower Plants

Station	State	Operator	Generator units	Capacity (MW)
Tehri Dam	Uttarakhand	THDC India Ltd.	4 × 250, 4 × 100, 4 × 250	2,400
Koyna	Maharashtra	MahaGenco	4 × 65, 4 × 75, 2 × 20, 4 × 80, 4 × 250	1,920
Srisailam Dam	Andhra Pradesh	APGenco	6 × 150, 7 × 110	1,670
Nathpa Jhakri	Himachal Pradesh	SJVNL	6 × 250	1,500
Sharavathi	Karnataka	KPCL	10 × 103.5, 2 × 27.5, 4 × 60	1,469
Sardar Sarovar	Gujarat	SSNNL	6 × 200, 5 × 140	1,450
Bhakra Dam	Punjab	BBMB	5 × 108, 5 × 157	1,325
Kalinadi	Karnataka	KPCL	2 × 50, 1 × 135, 5 × 150, 3 × 50, 3 × 40	1,240
Indira Sagar	Madhya Pradesh	NHPC	8 × 125	1,000
Dehar (Pandoh)	Himachal Pradesh	BBMB	6 × 165	990
Nagarjunasagar	Andhra Pradesh	APGenco	1 × 110, 7 × 100.8, 5 × 30	965
Idukki	Kerala	KSEB	6 × 130	780
Salal	Jammu & Kashmir	NHPC	6 × 115	690
Chamera-I	Himachal Pradesh	NHPC	3 × 180	540
Omkareshwar	Madhya Pradesh	NHPC	8 × 65	520
Teesta-V	Sikkim	NHPC	3 × 170	510
Uri Hydroelectric Dam	Jammu & Kashmir	NHPC	4 × 120	480
Bansagar Dam	Madhya Pradesh			425
Pong	Himachal Pradesh	BBMB	6 × 66	396
Dulhasti	Jammu & Kashmir	NHPC	3 × 130	390
Chamera-II	Himachal Pradesh	NHPC	3 × 100	300
Baspa-II	Himachal Pradesh	JHPL	3 × 100	300
Dhauliganga-I	Uttarakhand	NHPC	4 × 70	280
Ghatghar Pumped Storage Scheme	Maharashtra	MahaGenco	125 × 2	250
Baira Suil	Himachal Pradesh	NHPC	3 × 60	180
Mulshi Dam	Maharashtra		6 × 25	150

Station	State	Operator	Generator units	Capacity (MW)
Tanakpur	Uttarakhand	NHPC	3 × 40	120
Loktak	Manipur	NHPC	3 × 35	105
Bargi Dam	Madhya Pradesh			105
Madikheda Dam	Madhya Pradesh			60
Rangeet	Sikkim	NHPC	3 × 20	60
Lainganamakki Dam	Karnataka			55
Jayakwadi Dam	Maharashtra			12

2.1.1. Hydrological or Rain Evaporation Cycle

It is a process by which moisture from the surface of water bodies covering the earth's surface transferred to the land and back to the water bodies again. This cycle is shown in Fig. 2.1. The input to this cycle is the solar energy. Due to this water from the water bodies evaporates and on cooling these water vapours form clouds. Further cooling makes the clouds to fall down in the form of rain, snow, hail or sleet etc. known as ***precipitation.*** Precipitation includes all water that falls from the atmosphere to the earth's surface in any form.

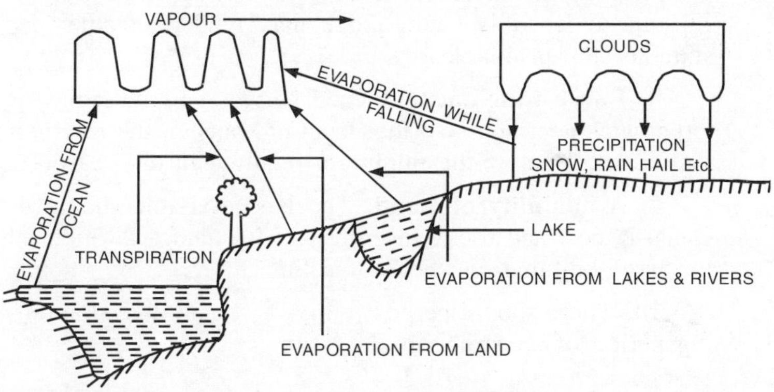

Fig. 2.1. Hydrological Cycle

Major portion (about 2/3rd) of this precipitation, which reaches the land surface is returned to the atmosphere by evaporation from water surfaces, soil and vegetation and through transpiration by plants. The remaining precipitation returns ultimately to the sea or ocean through surface or underground channels. Thus cycle is completed. The amount of rainfall which runs off the earth's land surface to form streams or rivers is useful for generation of electric power. The precipitation that falls on hills and mountains in the form of snow, melts during warmer weather as run-off and converges to form streams can also be used for power generation.

2.2. MERITS AND DEMERITS OF HYDROELECTRIC POWER PLANTS

Hydroelectric power plants offer many distinct advantages over other power plants.

These advantages can be summarised as under :

(*i*) No fuel is required by such plants as water is the source of energy. Hence operating costs are low and there are no problems of handling and storage of fuel and disposal of ash.

(*ii*) The plant is highly reliable and it is cheapest in operation and maintenance.

(*iii*) The plant can be run up and synchronised in a few minutes.

(*iv*) The load can be varied quickly and the rapidly changing load demands can be met without any difficulty.

(*v*) Very accurate governing is possible with water turbines so such power plants have constant speed and hence constant frequency.

(*vi*) There are no standby losses in such plants.

(*vii*) Such plants are robust and have got longer life (around 50 years).

(*viii*) The efficiency of such plants does not fall with the age.

(*ix*) It is very neat and clean plant because no smoke or ash is produced.

(*x*) Highly skilled engineers are required only at the time of construction but later on only a few experienced persons will be required.

(*xi*) Such plants, in addition to generation of electric power, also serve other purposes such as irrigation, flood control and navigation.

(*xii*) Hydroelectric plants are usually located in remote areas where land is available at cheaper rates.

However, the hydroelectric power plants have the following demerits also:

(*i*) It requires large area.

(*ii*) Its construction cost is enormously high and takes a long time for erection (owing to involvement of huge civil engineering works).

(*iii*) Long transmission lines are required as the plants are located in hilly areas which are quite away from the load centre.

(*iv*) The output of such plants is never constant owing to vagaries of monsoons and their dependence on the rate of water flow in a river. Long dry season may affect the power supply.

(*v*) The firm capacity of hydroelectric plants is low and so backup by steam plants is essential.

(*vi*) Hydroelectric power plant reservoir submerges huge areas, uproots large population and creates social and other problems.

2.3. SELECTION OF SITE FOR HYDROELECTRIC POWER PLANTS

While selecting a suitable site for a hydroelectric power plant, if a good system of natural storage—lakes at high altitudes and with large catchment areas can be located, the plant will be comparatively economical. Anyhow the essential characteristics of a good site are: large catchment area, high average rainfall, steep gradients and a favourable place for constructing the storage or reservoir. For this purpose, the geological, geographical and meteorological conditions of a site need careful investigation.

The following factors should be considered in the selection of site for hydroelectric power plants.

1. Availability of Water. Since in such power stations potential energy of waterfall or kinetic energy of flowing stream is utilized for generation of electric power, therefore such stations should be built where there is adequate water available at good head or huge quantity of water is flowing across a given point.

The plant capacity and the estimated output as well as the need for storage is governed by the average flow. The primary or dependable power which is available at all times depends upon the minimum flow. These conditions may also fix the capacity of the standby plant. The maximum of flood flow governs the size of the dam to be built with adequate spillway.

For estimation of energy available from a given stream or river, the discharge flowing and its variation with time over a number of years must be known. The estimates of the average quantity of water available is prepared on the basis of actual measurements of stream flow over as long a period as possible. Previous records of rainfall must be studied and minimum and maximum quantity of water available during the year should be estimated. After allowing for losses due to evaporation and seepage the net quantity of water available for power generation can be estimated.

2. Water Storage. Since storage of water in a suitable reservoir at a height or building of dam across the river is essential in order to have continuous and perennial supply during the dry season, therefore, convenient accommodation for the erection of a dam or reservoir must be available. The storage capacity can be determined from the hydrograph or from mass curve or by using analytical methods. For selection of site for the dam a careful study of the geology and topography of the catchment area is required to see whether the natural foundations can be used to the best advantage.

3. Water Head. The available water head depends upon the topography of the area. Availability of head of water has considerable effect on the cost and economy of power generation. An increase in effective head reduces the quantity of water to be stored and handled by penstocks, screens and turbines and, therefore, capital cost of the plant is reduced. In order to determine the most effective and economical head it is necessary to consider all possible factors which affect it.

4. Distance From Load Centre. As already mentioned, hydroelectric power plant is usually located far away from the load centre. Hence for economical transmission of electric power, the routes and the distances need active considerations.

5. Accessibility of The Site. Adequate transportation facilities must be available or there should be possibility of providing the same so that the necessary equipment and machinery could be easily transported.

6. Water Pollution. Polluted water may cause excessive corrosion and damage to the metallic structures. Hence availability of good quality water is essential.

7. Sedimentation. Gradual deposition of silt may reduce the capacity of the storage reservoir and may also cause damage to the turbine blades. Silting from forest areas is negligible but the regions subject to violent storms and not protected by vegetation contribute lot of silt to the run-off. In some cases, this factor alone may render an otherwise suitable site unsuitable.

8. Large Catchment Area. The reservoir must have a large catchment area* so that level of water in the reservoir may not fall below the minimum required in dry season.

9. Availability of Land. The land available should be cheap in cost and rocky in order to withstand the weight of the large building and heavy machinery.

10. There should be possibility of stream diversion during period of construction.

2.4. HYDROLOGY

The hydrography or hydrology deals with the occurrence and distribution of water over and under the earth surface. In other words it is the study of hydrological cycle, in which the water rises by evaporation from sea, is carried over the land, where it falls as precipitation and then rises again by evaporation or transpiration or flows back to ocean.

Precipitation is one of the three major phases (evaporation, precipitation and stream flow) of the hydrological cycle. Rainfall (used in general sense) or *precipitation* may be defined as the total condensation of moisture that reach the earth in any form. It includes all forms of rains, ice, hail, sleet dew and frost. Dew and frost result from cooling of moist air by direct contact with cold surfaces. The

* The catchment area is the area, bounded by watersheds, which drains into a stream or river across which the dam has been built at a suitable place.

cause of precipitation other than dew and frost is the adiabatic expansion of air *i.e.,* without any addition or subtraction of heat.

Evaporation represents practically all of that portion of the rainfall that does not reach the point of ultimate use as stream flow. So, evaporation includes all the rainfall that is returned to the atmosphere from the land and water surfaces. Thus total evaporation includes: (*i*) evaporation from land and water surfaces, (*ii*) evaporation by transpiration which is vaporisation of water from the breathing pores of vegetable matter and (*iii*) atmospheric evaporation (evaporation while precipitation is falling).

Run-off can be defined as that part of the precipitation which is available as stream flow.

i.e., Run-off = Total precipitation – total evaporation

Part of the precipitation is absorbed by the soil and seeps or percolates into the ground and will ultimately reach the catchment area through the underground channels. Thus,

Total run-off = Direct run-off over the land surface
+ run-off through seepage.

The factors affecting run-off are rainfall pattern, geology of the area, size and shape of catchment area, the topography and nature of soil in the catchment area, amount of vegetation and the weather conditions in the catchment area.

The unit of run-off are m^3/s or day-second metre.

One day-second metre

= Discharge collected in the catchment area at the rate of 1 m^3/s for one day

$= 1 \times 24 \times 60 \times 60 = 86{,}400\ m^3/day$

The run-off can also be expressed in cm of water on the drainage area feeding the river site for a stated period or km^2-cm of water per unit of time.

Hydrograph is a graphical representation between discharge or flow with time. It is plotted with flow as ordinates and time intervals as abscissas. The ordinates may be plotted in terms of the gauge height, the number of cubic metres per second per square kilometre, the power in kW that can be developed theoretically per metre of fall, or the energy in kWh recorded at the switchboard. The abscissae may be taken in terms of hours, days, weeks or months. It also indicates the power available from the stream at different times of the day or year. Hydrograph is similar to the load curve used in the study of electrical power.

A hydrograph provides the following informations:

1. The discharge at any time during the period under consideration.

2. The maximum and minimum run-off during the period.

3. The average run-off during the period.

4. Total volume of discharge up to any time, given by the area under the curve up to that time.

Flow duration curve is a very convenient form of hydrograph for determining the available power at site. It gives the relation between flows and lengths of time during which they are available. The flows are plotted as the ordinates and lengths of time as abscissas. Flow duration curve is a rearrangement of all the stream flow elements of a hydrograph in a descending order and is similar to load duration curve. Each point on a flow duration curve shows the percentage time during the period when the flow was equal to or greater than the given flow. Yearly flow duration curve may be obtained from mean monthly flows or daily flows. The area under a flow duration curve represents the total quantity of run-off during the period.

A flow duration curve may be used for determination of minimum and maximum conditions of flow. Maximum capacity of a plant may be decided on the basis of minimum flow conditions. However, a flow duration curve can be improved upon by storing water on the upstream side and using it during off peak periods.

Mass curve is a curve which indicates the total volume of run-off in m^3 up to a certain time. The abscissae may be a day, month or year. This curve is obtained from records of average monthly flows. The slope of this curve at any point gives the rate of flow at that time. If rainfall is uniform throughout the year, the mass curve would be a straight line having uniform slope. These curves are used in determination of the capacity of storage reservoir in hydro-projects. The capacity of the reservoir is given by the maximum ordinate between the mass curve and the demand line. The ordinate of the mass curve may be taken in terms of day-second metre.

By storage the flow of water can be modified according to the needs of power plant increasing the generating capacity of the power plant for the same mass flow.

Storage. In most of the streams, the flow in excess during one season and deficient during another and for its regulation artificial storage is needed. The purpose of providing storage is to make more water available during deficient flow times and thus increase the firm capacity of the power plant and also increase the energy generated. Availability of storage may render an otherwise uneconomical hydro-project economical, as increase in firm capacity means dispensing with alternative steam capacity.

The amount of storage can be determined with the help of hydrograph. If a horizontal line giving the average flow demand is drawn, the portion of the curve above this line will indicate the amount of water that should be stored in the reservoir for use during lean period. But, hydrograph varies from year to year and, therefore, determination of storage capacity with the help of hydrograph may lead to considerable error in some cases. So the storage capacity should be determined with the help of the mass curve.

Pondage. In case power plant is away from the storage reservoir, a small pond is required near the power plant in order to meet the hourly changes in power demand. The capacity of the pond should be sufficient to cope with hourly changes for 24 hours. In case the load conditions are very different during the week, the capacity of pond is determined from a study of load and stream flow for one week.

The pond capacity can be determined in a manner similar to that for determination of reservoir capacity. However, the demand in this case is represented by chronological load curve.

Necessity of pondage arise because of plant operation at a load factor of less than 100%. In an interconnected system, operation of one or two plants at 100% load factor may be possible.

Example 2.1. A hydroelectric power plant operates under an effective head of 50 m and a discharge of 94 m^3/s. Determine the power developed.

Solution : Discharge, Q = 94 m^3/s

Head, H = 50 m

Density of water, w = 1,000 kg/m^3

Power developed, P = wQH × 9.81 × 10^{-3} kW

= 1,000 × 94 × 50 × 9.81 × 10^{-3}

= 46,107 kW or 46.107 MW **Ans.**

Example 2.2. The mean monthly discharge at a particular site is given below:

Month	Discharge in m^3/s	Month	Discharge in m^3/s
January	200	July	1,600
February	400	August	1,200
March	600	September	2,000
April	2,400	October	1,200
May	1,200	November	800
June	1,800	December	400

Draw the hydrograph, flow duration curve and mass curve.

Determine the average inflow and the power that can be developed at an effective head of 90 m. Determine the capacity of the storage reservoir based on above one year data neglecting the losses due to seepage, evaporation etc. Assume overall generation efficiency to be 80%.

Solution : The average inflow is obtained by adding the 12 monthly discharges and dividing by 12.

i.e., Average inflow,

Q = 1/12 (200 + 400 + 600 + 2,400 + 1,200 + 1,800 + 1,600 + 1,200 + 2,000 + 1,200 + 800 + 400)

= 1,150 m^3/s **Ans.**

The power, that can be developed,

P = wQH × η × 9.81 × 10^{-3}

= 1,000 × 1,150 × 90 × 0.8 × 9.81 × 10^{-3} kW

= 812,268 kW or 812.268 MW **Ans.**

The hydrograph is plotted in Fig. 2.2. From this, the flow duration curve can be drawn by finding the lengths of time, during which certain flows are available *e.g.,* flow of 200 m^3/s is available for all the 12 months, flow of 400 m^3/s for 11 months and so on. The information is tabulated as below and the flow duration curve is plotted in Fig. 2.3.

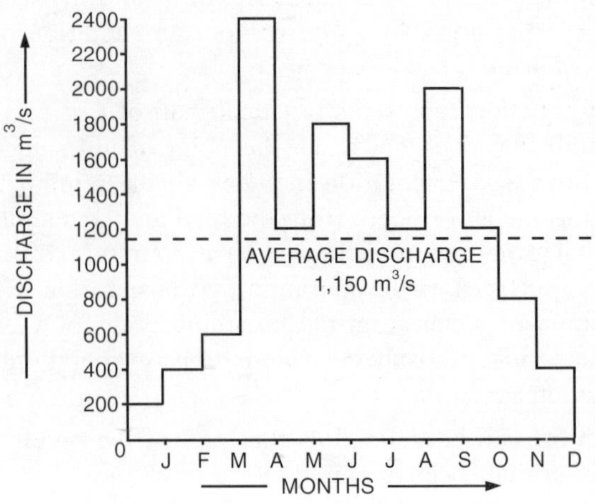

Fig. 2.2. *Hydrograph*

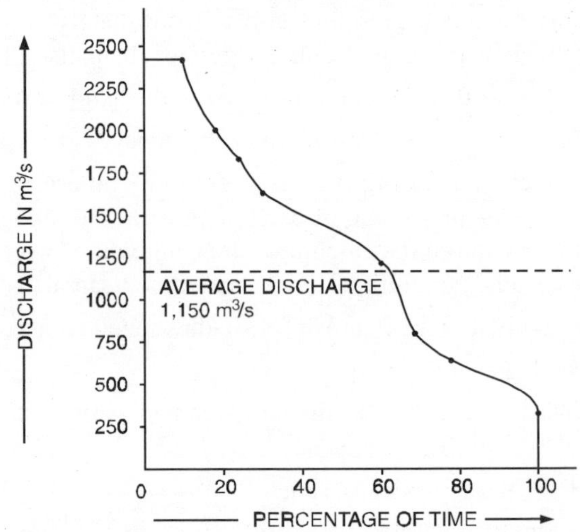

Fig. 2.3. *Flow Duration Curve*

Discharge in m^3/s	Duration in months	Percentage Time
200	12	100
400	11	91.66
600	9	75
800	8	66.66
1,200	7	58.33
1,600	4	33.33
1,800	3	25.00
2,000	2	16.67
2,400	1	8.33

During January, February and March, the discharge being less 950, 750 and 550 second-metre-month respectively have to be drawn from the reservoir (out of previous year storage). The excess discharge during April, May, June, July, August, September and October would completely fill the reservoir so that the deficiency during November and December of the year and January, February and March of next year can be made up by drawing water from reservoir.

The storage capacity of the reservoir is evidently the excess flow during April, May, June, July, August, September and October *i.e.,*

Total of 3,350 second-metre-months or 8.6832 × 10^9 m^3 **Ans.**

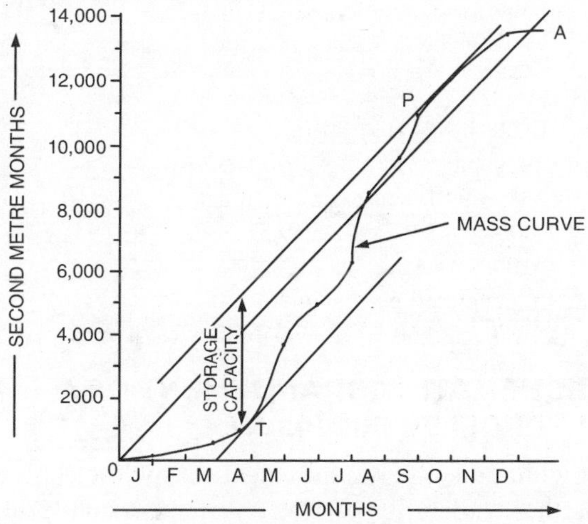

Fig. 2.4. ***Mass Curve***

To obtain mass curve, the total second-metre months at the end of various months of the year are determined as below:

Month	Discharge in m^3/s	Accumulative Run-off in second-metre-months
January	200	200
February	400	600
March	600	1,200
April	2,400	3,600
May	1,200	4,800
June	1,800	6,600
July	1,600	8,200
August	1,200	9,400
September	2,000	11,400
October	1,200	12,600
November	800	13,400
December	400	13,800

The summation of the volume of water that can be stored is plotted against months as shown in Fig. 2.4 giving mass curve.

For determining the storage capacity of the reservoir a line OA is drawn representing the average demand of 1,150 m^3/s and also draw lines parallel to this demand line and tangent to the mass curve from the peak P and the trough T. The vertical distance between the two tangents gives the total storage capacity of the reservoir which is equal to 3,350 second-metre-months.

Example 2.3. A hydroelectric power station is supplied from a reservoir of capacity 3×10^7 m^3 at a head of 150 m. Determine the total energy available in kWh if the overall efficiency of the plant is 70%.

Solution : Water head, H = 150 m

Volume of water, $v = 3 \times 10^7\ m^3$

Density of water, $w = 1{,}000\ kg/m^3$

Overall efficiency of the plant, $\eta = 70\% = 0.7$

Electrical energy available $= wvH\eta \times 9.81$ watt-secs

$= 1{,}000 \times 3 \times 10^7 \times 150 \times 9.81 \times 0.7$

$= 3{,}090.15 \times 10^{10}$ watt-secs or joules

$= 3{,}090.15 \times 10^{10}/(3{,}600 \times 1{,}000)$ kWh

$= 8.58375 \times 10^6$ kWh **Ans.**

Example 2.4. A hydroelectric power station is supplied from a reservoir having an area of 50 km² and a head of 50 m. If the overall efficiency of the plant be 60%, find the rate at which the water level will fall when the station is generating 30,000 kW.

Solution : Catchment area, $A = 50\ km^2 = 50 \times 10^6\ m^2$

Water head, H = 50 m

Density of water, $w = 1{,}000\ kg/m^3$

Average power generated, P = 30,000 kW

Also, average power generated, $P = wvH \times 9.81 \times 10^{-3}\ \eta$ kW

∴ Volume of water required per second,

$v = P/(w \times H \times 9.81 \times 10^{-3} \times \eta)$

$= 30{,}000/(1{,}000 \times 50 \times 9.81 \times 10^{-3} \times 0.6) = 101.9\ m^3/s$

Rate of fall of water level

$= v/A = 101.9/(50 \times 10^6)$ m/s

$= \dfrac{101.9}{50 \times 10^6} \times 3{,}600 \times 1{,}000 = 7.339$ mm/hour **Ans.**

Example 2.5. An hydroelectric station has to operate with a mean head of 50 m. It makes use of water collected over a catchment area of 200 km² over which the average annual rainfall is 420 cm with a 30% loss due to evaporation. Assuming the turbine efficiency as 85% and the alternator efficiency as 80% calculate the average power that can be generated.

Solution : Volume of water available for utilisation per annum,

v = Catchment area × average annual rainfall [1 – loss due to evaporation]

$= 200 \times 10^6 \times \dfrac{420}{100}(1 - 0.3) = 588 \times 10^6\ m^3$/annum

Water head, H = 50 m

Power plant efficiency,

η = Alternator efficiency × turbine efficiency = 0.8 × 0.85 = 0.68

Net energy available per annum

$= \dfrac{w \times v \times H \times 9.81 \times \eta}{60 \times 60 \times 1{,}000}$ kWh

$= \dfrac{1{,}000 \times 588 \times 10^6 \times 50 \times 9.81 \times 0.68}{60 \times 60 \times 1{,}000} = 54.4782 \times 10^6$ kWh

Available continuous power or average power,

$P = \dfrac{\text{Net energy available in kWh}}{24 \times 365} = \dfrac{54.4782 \times 10^6}{8{,}760}$

= 6,219 kW or 6.22 MW **Ans.**

Example 2.6. An hydroelectric station is to be designed to operate at a mean head of 205 m and supplied from a reservoir having a catchment area of 1,000 km² with average annual rainfall of 125 cm of which 80% is available for power generation. The expected load factor at the plant is 75%. Allowing a head loss of 5 m and assuming efficiency of turbine and generator to be respectively 90% and 95%, calculate suitable MW rating of the station. Comment also on the type of turbine to be installed.

Solution : Catchment area $= 1{,}000\ km^2 = 1{,}000 \times 10^6\ m^2$

Average annual rainfall = 125 cm = 1.25 m

Yield factor = 80% or 0.8

Volume of water available for utilisation per annum

v_{ANNUM} = Catchment area × average annual rainfall × yield factor

$= 1{,}000 \times 10^6 \times 1.25 \times 0.8 = 1{,}000 \times 10^6\ m^3$/annum

Volume of water available for utilisation per second

$$v_{SECOND} = \frac{v_{ANNUM}}{8,760 \times 60 \times 60} = \frac{1,000 \times 10^6}{8,760 \times 60 \times 60} = 31.71 \text{ m}^3/\text{s}$$

Available head, H = Gross head – head loss = 205 – 5 = 200 m

Power plant efficiency, $\eta = \eta_{TURBINE} \times \eta_{GENERATOR}$
$= 0.90 \times 0.95 = 0.855$

Average power, $P = wv_{SECOND}\, H \times 9.81 \times \eta \times 10^{-6}$ MW
$= 1,000 \times 31.71 \times 200 \times 9.81 \times 0.855 \times 10^{-6}$
$= 53.2$ MW

This is the average output of the generator units. With a load factor of 75%, overall capacity of the station for which it should be designed

$$= \frac{P}{\text{Load factor}} = \frac{53.2}{0.75} = 70.9 \text{ MW } \textbf{Ans.}$$

The type of turbine used for a head of 200 m recommended is Francis turbine.

Example 2.7. A hydroelectric station is to be designed for a catchment area of 400 km^2, rainfall for which is 125 cm/annum. The head available is 30 m. If 75% of the total rainfall is available, rest is lost to evaporation. Penstock efficiency 95%. Turbine efficiency 85% and generator efficiency is 90% and load factor 40%.

Determine the capacity of the station and suggest suitable machines.

Solution : Volume of water available for utilization per second,

$$v = \frac{\text{Catchment area in km}^2 \times 10^6 \times \text{average annual rainfall in } m \times \text{yield factor}}{365 \times 24 \times 60 \times 60}$$

$$= \frac{400 \times 10^6 \times 1.25 \times 0.75}{8,760 \times 3,600} = 11.89 \text{ m}^3/\text{s}$$

Available head, H = 30 m

Overall efficiency of power plant,

η = Penstock efficiency × turbine efficiency × generator efficiency
$= 0.95 \times 0.85 \times 0.9 = 0.72675$

Average power,

$P = wvH \times 9.81 \times \eta \times 10^{-6}$ MW
$= 1,000 \times 11.89 \times 30 \times 9.81 \times 0.72675 \times 10^{-6} = 2.543$ MW

$$\text{Capacity of power station} = \frac{P}{\text{Load factor}} = \frac{2.543}{0.4} = 6.36 \text{ MW } \textbf{Ans.}$$

Example 2.8. A hydroelectric station has a mean head of 30 m and receives inflows from a catchment area of 300 km^2 having an average rainfall of 200 cm per annum. The rainfall occurs during the monsoon period of 90 days in a year. The storage capacity of the station is limited to half the total inflows. If 70 per cent rainfall reaches the reservoir, determine the minimum capacity of the station. Assume that no head is lost in pipes etc. and the efficiencies of the turbine and generator are 85% and 95% respectively. Determine also the total energy produced and the plant use factor.

Solution : Total inflows per annum

= Catchment area × average annual rainfall

$$= 300 \times 10^6 \times \frac{200}{100} = 600 \times 10^6 \text{ m}^3/\text{annum}$$

Storage capacity of the station reservoir

$$= \frac{1}{2} \times 600 \times 10^6 = 300 \times 10^6 \text{ m}^3$$

Total volume of water available (300×10^6 m^3) is to be used during lean period of (365 – 90) *i.e.,* 275 days.

Water available for generation per second,

$$v = \frac{300 \times 10^6 \times 0.7}{275 \times 24 \times 60 \times 60} = 8.838 \text{ m}^3/\text{s}$$

Available head, H = 30 m

Capacity of plant,

$P = wvH \times 9.81 \times \eta_{TURBINE} \times \eta_{GENERATOR} \times 10^{-3}$ kW
$= 1,000 \times 8.838 \times 30 \times 9.81 \times 0.85 \times 0.95 \times 10^{-3}$
$= 2,100$ kW **Ans.**

Total energy generated per annum

$= 2,100 \times 8,760 = 18.396$ million kWh **Ans.**

2.5. SCHEMATIC ARRANGEMENT OF A HYDROELECTRIC PLANT

We know that the chief requirement for hydroelectric power plant is the availability of water in huge quantity at sufficient head and this requirement can be met by constructing a dam across a river or lake. The schematic arrangement of a typical hydroelectric plant is shown in Fig. 2.5(*a*). An artificial storage reservoir is formed by constructing a dam

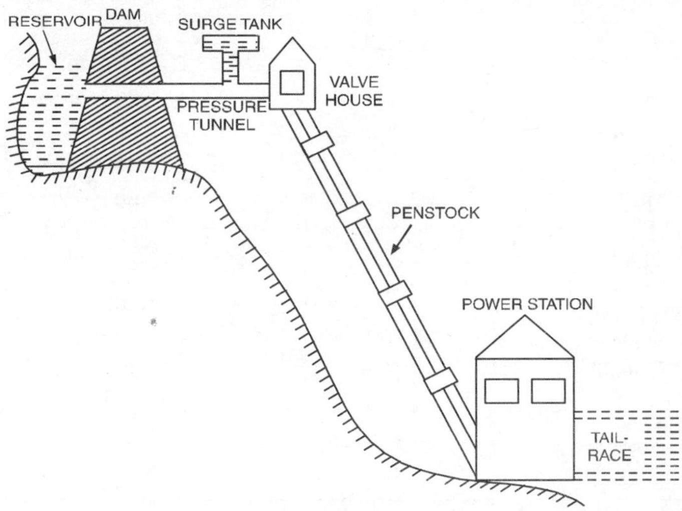

Fig. 2.5(*a*)

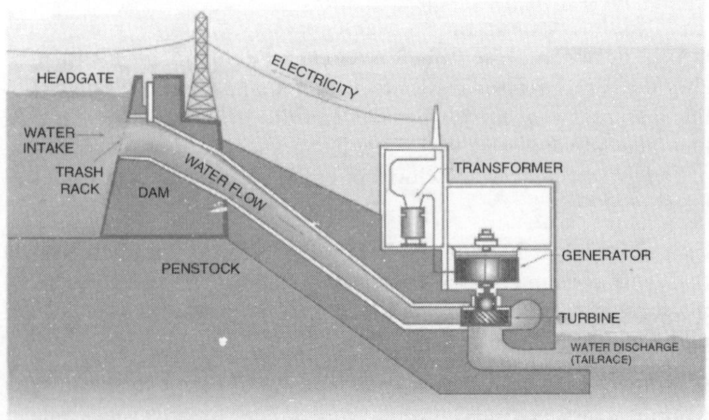

Fig. 2.5(*b*) *Schematic Arrangement of a Hydroelectric Power Plant*

across a river (or lake) and a pressure tunnel is taken off from the reservoir to the valve house at the start of the penstock. The valve house contains main sluice valves for controlling water flow to the power station and automatic isolating valves for cutting off water supply in case the penstock bursts. A surge tank is also provided just before the valve house for better regulation of water pressure in the system. From the reservoir the water is carried to valve house through pressure tunnel and from valve house to the water turbine through pipes of large diameter made of steel or reinforced concrete, called the *penstock*. The water turbine converts hydraulic energy into mechanical energy and the alternator coupled to the water turbine converts mechanical energy into electrical energy. Water after doing useful work is discharged to the tailrace.

2.6. ELEMENTS OF HYDROELECTRIC POWER PLANT

A hydroelectric plant consists of a reservoir for storage of water, a diversion dam, an intake structure for controlling and regulating the flow of water, a conduit system to carry the water from the intake to the waterwheel, the turbines coupled with generators, the draft tube for conveying water from waterwheel to the tailrace, the tailrace and a power house *i.e.,* the building to contain the turbines, generators, the accessories and other miscellaneous items. The size, location, and type of each of these essential elements depends upon the topography and geological conditions and the amount of water to be used. The height to which the dam may be built is usually limited by the extent of flowage damage. Pondage may have great value, particularly for peak load power plants, warranting the purchase of extensive flowage rights. The spillway section of the dam must be long enough to pass safely the maximum amount of water to be expected. Likewise the abutments and other short structures must be built to withstand successfully the greatest freshet conceivable on the river.

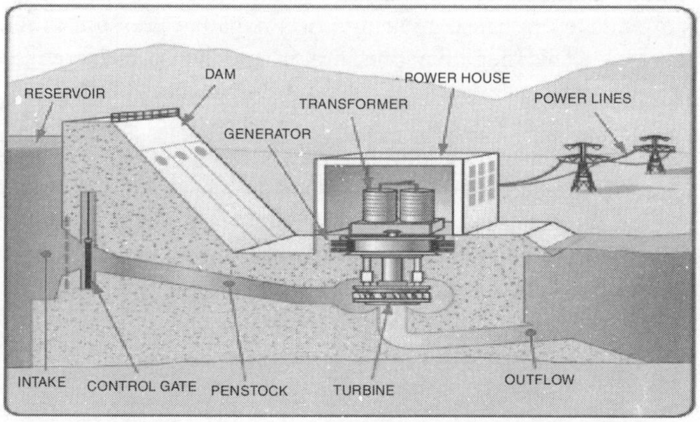

Fig. 2.6. ***Elements of Hydroelectric Power Plant***

1. *Storage Reservoir.* It is the basic requirement of a hydroelectric plant. Its purpose is to store water during excess flow periods (*i.e.,* rainy season) and supply the same during lean flow periods (*i.e.,* dry season) and thus it helps in supplying water to the turbines according to the load on the power plant. A reservoir can be either natural or artificial. A natural reservoir is a lake in high mountains and an artificial reservoir is made by constructing a dam across the river. Low head plants require very large storage reservoir. The capacity of reservoir depends on the difference between run-offs during high and lean flows.

2. *Dam.* The function of dam is not only to raise the water surface of the stream to create an artificial head but also to provide the pondage, storage or the facility of diversion into conduits. A dam is the most expensive and important part of a hydro-project. Dams are built of concrete or stone masonary, earth or rock fill. The type and arrangement depends upon the topography of the site. A masonary dam may be built in a narrow canyon. An earth dam may be best suited for a wide valley. The choice of dam also depends upon the foundation conditions, local materials and transportation available, occurrence of earth quakes and other hazards.

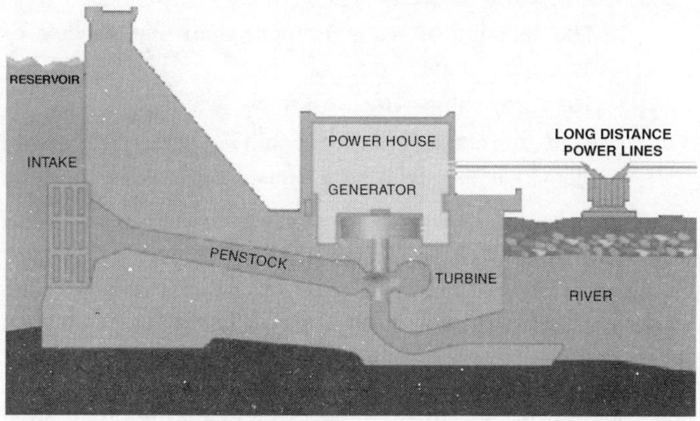

Fig. 2.7. ***Hydroelectric Dam***

Mansory dams are of three major classes viz solid gravity, buttress and the arch dams.

Solid gravity dam shown in Fig. 2.8 (*a*), is made of concrete and is suitable for most sites. The height of the dam, which cannot be very high, depends on the strength of subsoil strata. *Arch dam*, shown in Fig. 2.8 (*b*) is a curved dam and transmits a major portion of its water pressure horizontally to the abutments by arch action. An arch dam is preferred where a narrow canyon width is available. This dam has the inherent stability against sliding. The *buttress* or *deck dam* has an inclined upstream face, so that water pressure creates a large downward force which provides stability against over turning or sliding. Such a dam is more suitable for weak foundations and earth quake prone sites.

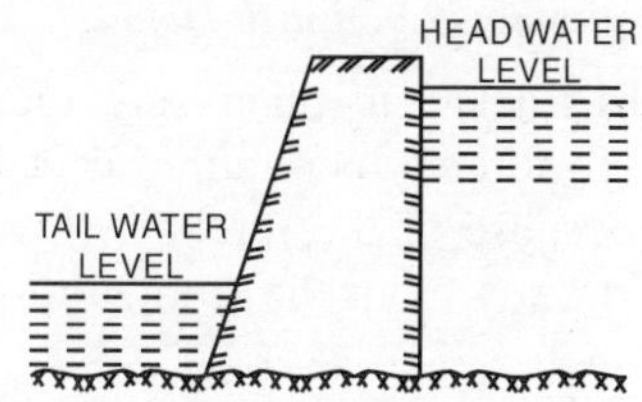

(a) Solid Gravity Dam

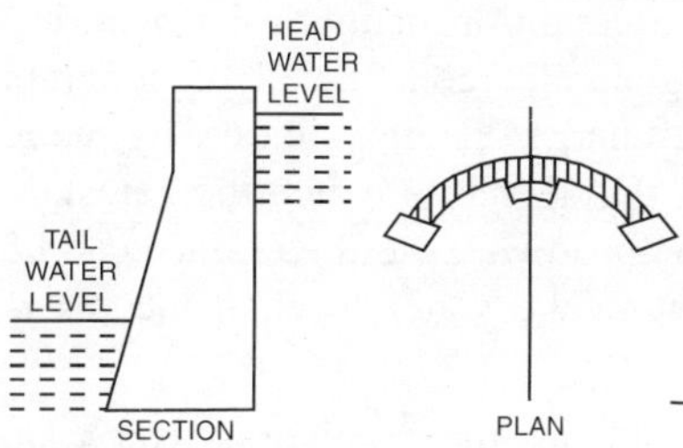

(b) Arch Dam

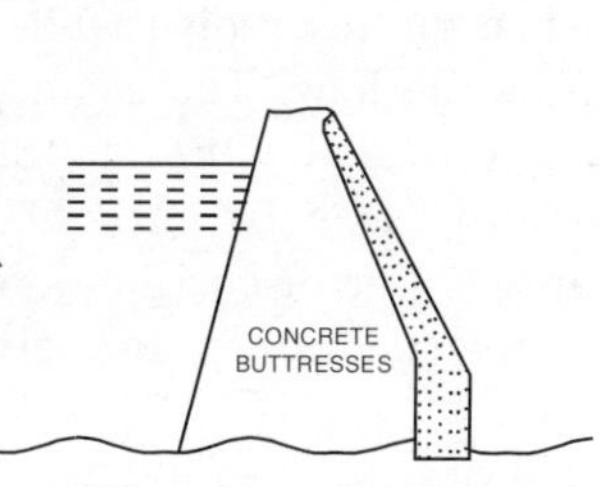

(c) Buttress Dam

Fig. 2.8. ***Masonary Dams***

An *earth dam* has a very wide base as compared to its height. Such dams are quite suitable for a pervious foundation because the wide base provides a long seepage path. The earth dams have got the following advantages.

1. They are cheaper than masonary dams.
2. They fit best in natural surroundings.
3. Such a dam provides the most permanent type of structure if protected against erosion.

However, earth dams have the following disadvantages.

1. The seepage of water is more than that in case of masonary dams.
2. They are subject to erosion by water.
3. They are not suitable for a spillway, therefore, a supplementary spillway of adequate capacity is required.

3. *Forebay.* The forebay serves as a regulating reservoir storing water temporarily during light load period and providing the same for initial increase on account of increasing load during which water in the canal is being accelerated. In short a forebay may be considered as an enlarged body of water just above the intake to store water temporarily to meet the hourly load fluctuations. This may either be a pond behind the diversion dam or an enlarged section of a canal spread out to accommodate the required widths of intake.

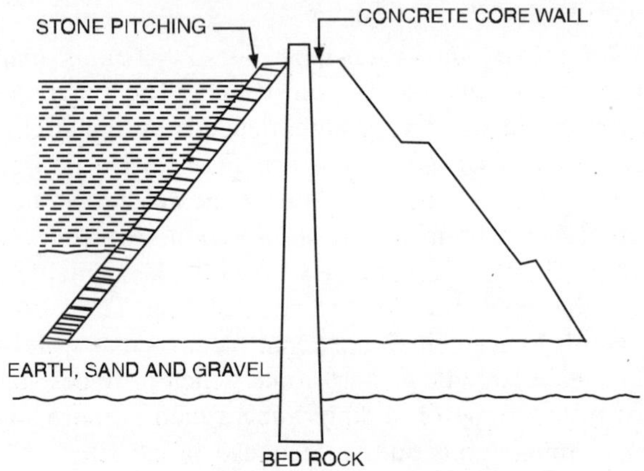

Fig. 2.9. ***Earth Dam***

Where the hydroelectric plants are located just at the base of the dam, no forebay is required because the reservoir itself serves the purpose of the forebay. However, where the plants are situated away from the storage reservoir a forebay is provided.

4. *Spillway.* This is constructed to act as a safety valve. It discharges the overflow water to the down stream side when the reservoir is full, a condition mainly arising during flood periods. These are generally constructed of concrete and provided with water discharge opening shut off by metal control gates. By changing the degree to which the gates are opened, the discharge of the head water to the tailrace can be regulated in order to maintain the water level in the reservoir.

5. *Intake.* The intake includes the headworks which are the structures at the intake of conduits, tunnels or flumes. These structures include booms, screens or trash racks, sluices to divert and prevent entry of debris and ice into the turbines. Booms prevent the ice and floating logs from going into the intake by diverting them to a bypass chute. Screens or trash racks are fitted directly at the intake to prevent the debris from going into the take. Debris cleaning devices should also be fitted on the trash racks.

Intake structures can be classified into *high pressure intakes* used in case of large storage reservoirs and *low pressure intakes* used in case of small ponds provided for storing small amount of water for daily or weekly load variations.

6. *Surge Tank.* A reduction in load on the generator causes the governor to close the turbine gates and thus create an increased pressure in the penstock. This may result in *water hammer phenomenon* and may need pipe of extraordinary strength to withstand it otherwise the penstock may burst. To avoid this positive water hammer pressure, some means are required to be provided for taking the rejected flow. This may be accomplished by providing a small storage reservoir or tank (open at the top) for receiving the rejected flow and thus relieving the conduit pipe of excessive water hammer pressure. This storage reservoir, called the *surge tank* is usually located as close to the power station as possible, preferably on ground to reduce the height of the tower. A decrease in load demand causes a rise in water level in the surge tank. This produces a retarding head and reduces the velocity of water in the penstock. The reduction in flow velocity to the desired level, makes the water in the tank to fall and rise until damped out by friction.

Increase in load on the plant causes the governor to open the turbine gates in order to allow more water to flow through the penstock to supply the increased load and there is a tendency to cause a vacuum or a negative pressure in the penstock. This negative pressure in the penstock provides the necessary accelerating force and is objectionable for very long conduits due to difficult turbine regulation. Again under such conditions, the additional water flows out of the surge tank. As a result the water level in the surge tank falls, an accelerating head is created and flow of water in the penstock is increased. Thus surge tank helps in stabilising the velocity and pressure in the penstock and reduces water hammer and negative pressure or vacuum.

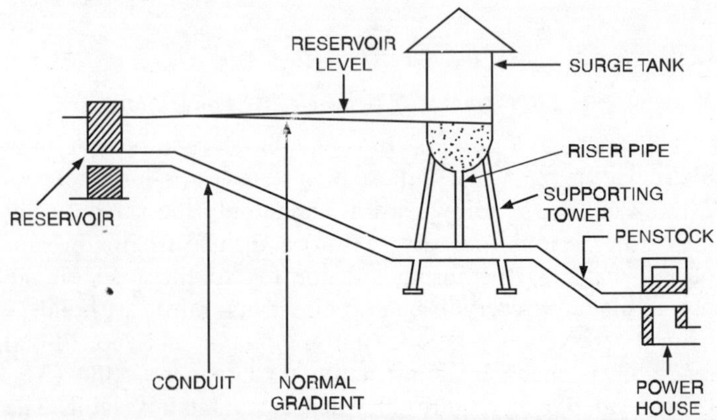

Fig. 2.10. ***Surge Tank***

Though by providing a relief valve at the turbine inlet rejected flow can be dealt in a better manner, but it cannot provide excess water required by the turbine when the load demand increases. Open conduits leading water to the turbine require no protection but when closed conduits are employed, protection becomes necessary to limit the abnormal pressure in the conduit. For this reason, close conduits are always provided with a surge tank. A forebay also serves the function of a surge tank.

The ideal location of a surge tank is at the turbine inlet but in the case of medium and high head power plants, the height of the surge tank will become excessive. Because of this reason, the surge tanks are usually provided at the junction of the pressure tunnel and the penstock.

Several designs of surge tanks have been adopted in hydroelectric power plants, the important considerations, being the amount of water to be stored, the magnitude of pressure to be relieved of and the availability of space at the construction site.

Surge tanks may be ***simple surge tank*** (Fig. 2.10), ***restricted orifice surge tank or differential surge tank.*** Simple surge tank is very sluggish in action and needs the largest volume. So this is the most expensive and is seldom used, except in special cases. The restricted orifice surge tank is more efficient and economical than the first one, but its main drawback is that the desirable sudden creation of accelerating and retarding heads in the conduits also results in correspondingly sudden fluctuations of head on the turbine, which the governors may have difficulty to accommodate. Differential surge tank is the compromise between the simple surge tank and the restricted orifice surge tank.

7. *Penstock.* It is a closed conduit which connects the forebay or surge tank to the scroll case of the turbine. In case of medium head power plants each unit is usually provided with its own penstock. In case of high head plants, a single penstock is frequently used, and branch connections are provided at the lower end to supply two or more units. Penstocks are built of steel or reinforced concrete. Steel penstocks are almost always welded on the longitudinal seam. The circumferential seam may be welded also. In long penstocks great care must be taken to protect the conduit against water hammer. The thickness must be adequate to withstand both the normal hydrostatic pressure and also the sudden surges both above and below normal caused by fluctuations in load and by emergency conditions.

8. *Valves and Gates.* In low head plants gates at the entrance to the turbine casing are usually all that is needed to shut off the flow and provide for unwatering the turbine for inspection and repairs. Individual hoist-operated gates are provided in cases where frequent shutdowns may be called for and where the time available for inspection is limited. Other plants employ stop gates or stop logs which are placed in sections by means of travelling crane. For installations employing medium or longer length penstocks or employing a common penstock for more than one unit, it is necessary to install valves at or near the entrance to the turbine casing. These are usually of the butterfly or pivot type for low and medium heads.

9. *Trash Racks.* These are built up from long, flat bars set vertically or nearly so and spaced in accordance with the minimum width of water passage through the turbine. The clear space between the bars varies from 25 mm or 40 mm to 150 or 200 mm on very large installations. These are to prevent the ingress of floating and other material to the turbine. In some cases where large diameter turbines are employed, the racks are omitted, but provision is usually made for skimmer walls or booms to prevent ice and other material from entering the unit.

10. *Tailrace.* The water after having done its useful work in the turbine is discharged to the tailrace which may lead it to the same stream or to another one. The design and size of tailrace should be such that water has a free exit and the jet of water, after it leaves the turbine, has unimpeded passage.

11. *Draft Tubes.* An airtight pipe of suitable diameter attached to the runner outlet and conducting water down from the wheel and discharging it under the surface of the water in the tailrace is known as *draft tube.*

If there is no draft tube and the water discharges freely from the turbine exit, then the turbine operates under a head equal to the height of the headrace water level above the runner exit. By installing draft tube, the operating head is increased by an amount equal to the height of the runner outlet above the tailrace. This creates a negative pressure head at the runner exit. This makes it possible to install the turbine above the tailrace without loss of head.

By installing the draft tube and increasing its section from runner exit to the tailrace, some of the kinetic energy possessed by the water leaving the runner outlet is converted into pressure energy and the water leaves at the tailrace at a much reduced velocity. This results again in the kinetic head which increases the negative pressure at the runner exit. This in turn increases the operating head on the turbine increasing its output and efficiency.

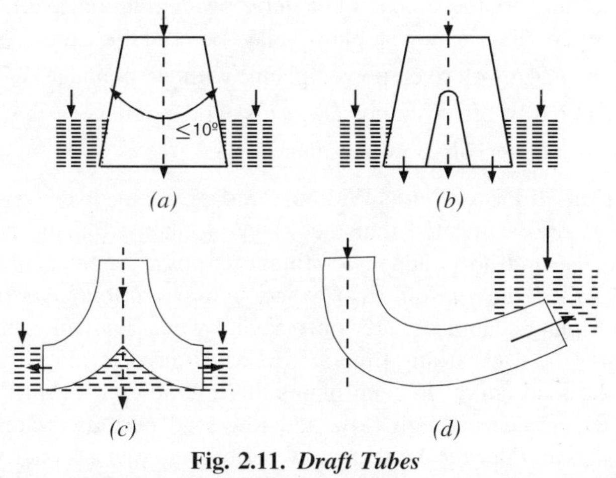

Fig. 2.11. ***Draft Tubes***

The height and type of tube used depends upon two factors. The pressure at the turbine exit or inlet of the draft tube should not be less than one-third of the atmospheric pressure. This is essential to avoid *cavitation**. Also to maintain continuity of flow without vapourisation, the pressure at any point in the tube should not fall below the vapour pressure of water. Further, to avoid separation of flow, the included angle should not exceed 10°.

* When the pressure in any portion of turbine drops below vapour pressure vapour bubbles or cavities are formed. This phenomenon is called the cavitation.

Various types of draft tubes are shown in Fig. 2.11. ***The straight conical type draft tube***, shown in Fig. 2.11 (*a*), has an efficiency of about 90% and is employed for low specific speed, vertical shaft Francis turbine. ***Vertical bell shaped draft tube*** is shown in Fig. 2.11 (*b*). Where there is a little head room available, the ***bent draft tubes***, shown in Figs. 2.11 (*c*) and 2.11 (*d*) are used. In Fig. 2.11 (*d*), the horizontal portion of the tube is gradually bent upwards to lead the water gradually to the tailrace and to prevent entry of air from the outlet end. The exit end of the tube must always be immersed in water.

12. *Prime Movers or Water Turbines.* In hydroelectric power plants water turbines are used as prime movers and their function is to convert the kinetic energy of water into mechanical energy which is further utilised to drive the alternators generating electrical energy. We will discuss the various types of water turbines used in hydroelectric power plants in detail in Art 2.8.

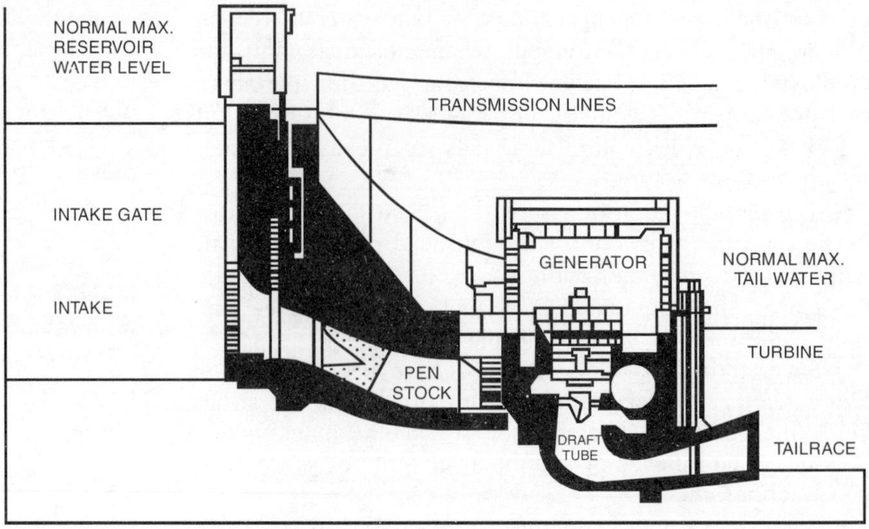

Fig. 2.12. ***Hydroelectric Power Station***

2.7. CLASSIFICATION OF HYDROELECTRIC POWER PLANTS

The hydroelectric power plants may be classified according to

(*i*) the extent of water flow regulation available

(*ii*) the availability of water head and

(*iii*) the type of load they supply.

2.7.1. Classification According To the Extent of Water Flow Regulation Available

According to the extent of water flow regulation available the hydroelectric power plants may be classified into:

(*a*) Run-off river power plants without pondage

(*b*) Run-off river power plants with pondage

(*c*) Reservoir power plants.

1. Run-off Power Plants Without Pondage. Some hydro power plants are so located that the water is taken from the river directly, and no pondage or storage is possible. Such plants are called the *run-off river power plants without pondage.* Such plants can use water only as and when available; these cannot be used at any time at will or fit any desired portion of the load curve. In such plants there is no control on flow of water. During high flow and low load periods, water is wasted and during the lean flow periods the plant capacity is very low. As such these plants have a very little firm capacity. At such places, the water is mainly used for irrigation or navigation and power generation is only incidental. Such plants can be built at a considerably low cost but the head available and the amount of power generated are usually very low. During floods, the tail water level may become excessive rendering the plant inoperative. The main objective of such plants is to use whatever flow is available for generation of energy and thus save coal that otherwise be necessary for the steam plants. During the high flow periods such plants can be employed to supply a substantial portion of base load.

2. Run-off River Power Plants With Pondage. The usefulness of run-off river power plants is increased by pondage. Pondage refers to storage at the plant which makes it possible to cope, hour to hour, with fluctuations of load throughout a week or some longer period depending on the size of pondage. With enough pondage, the firm capacity of the power plant is increased. Such type of power plants can be used on parts of the load curve as required, within certain limitations and is more useful than a plant without pondage. Such power plants are comparatively more reliable and its generating capacity is less dependent on available rate of flow of water. Such power plants can serve as base load or peak load power plants depending on the flow of stream. During high flow periods these plants may be used as base load and during lean flow periods these plants may be used to supply peak loads only. When providing pondage tailrace condition should be such that floods do not raise the tailrace water level, thus reducing the head on the plant and impairing its effectiveness. Such plants offer maximum conservation of coal when operated in conjunction with steam power plants.

3. Reservoir Power Plants. When water is stored in a big reservoir behind a dam, it is possible to control the flow of water and use it most effectively. Storage increases the firm capacity of the plant and it can be used efficiently throughout the year. Such a plant can be used as a base load or as a peak load plant as per requirement. It can also be used on any portion of the load curve in a grid system. Most of the hydroelectric power plants everywhere in the world are of this type.

2.7.2. Classification According to Availability of Water Head

According to availability of water head the hydroelectric power plants may be classified into:

(*a*) low head

(*b*) medium head and

(*c*) high head power plants.

Though there is no definite line of demarcation for low, medium and high heads but the head below 30 metres is considered low head, the head above 30 metres and below 300 metres is considered as medium head and above 300 metres is considered as high head.

(*a*) **Low Head Hydroelectric Power Plants.** Such a plant is shown in Fig. 2.13. A typical low head installation on a river consists essentially of a dam across the stream to back up the river and create a fall, the water flowing through the turbines and rejoining the river below the dam. A dam or barrage constructed across the river creates the necessary head. The power plant is located near the dam and therefore, no surge tank is required. Either one half of the barrage has regulating gates for discharge of surplus water while the plant is in front of second half or the plant is constructed by the side of the river.

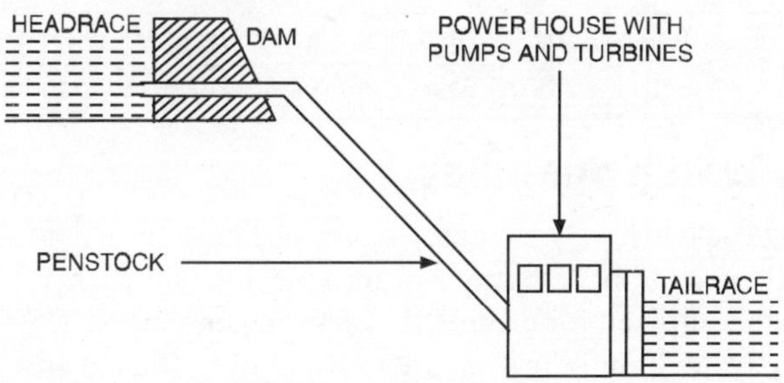

Fig. 2.13. *Arrangement of Low Head Hydroelectric Power Plant*

In low head power plants Francis, propeller or Kaplan turbines are employed. Since for given output, large quantity of water is required, head being low, therefore pipes of large diameter and short length are required in low head plants. Structure of such plants is extensive and expensive. Generators employed in such plants are of low speed and large diameter.

(*b*) **Medium Head Hydroelectric Power Plants.** In these power plants, the river water is usually tapped off to a forebay on one bank of the river as in case of a low head plant. From the forebay the water is led to the turbines through penstocks. The forebay provided at the beginning of penstock serves as a water reservoir for such power plants. In these plants water is usually carried in open channel from main reservoir to the forebay and then to the turbines through the penstock. The forebay itself serves as the surge tank in this case. In these plants horizontal shaft Francis, propeller or Kaplan turbines are used. The arrangement is shown in Fig. 2.14.

(*c*) **High Head Hydroelectric Power Plants.** If high head is available, a site may be chosen, where a stream descending a steep lateral valley can be dammed and a reservoir for storage of water is formed. A pressure tunnel is constructed between reservoir to valve house at the start of penstock to carry water from reservoir to valve house. Surge tank (a tank open from the top) is built just before the valve house so that the severity of water hammer effect on penstock can be reduced in case of sudden closing of fixed gates of the water turbine. Surge tank also serves as a ready reservoir from which the turbine can draw water temporarily when there is sudden increase in demand. The valve house consists of main sluice valves and automatic isolating valves, which operate on bursting of penstock and cut off further supply of water to penstock. Penstocks are pipes and carry the water from the valve house to the turbines.

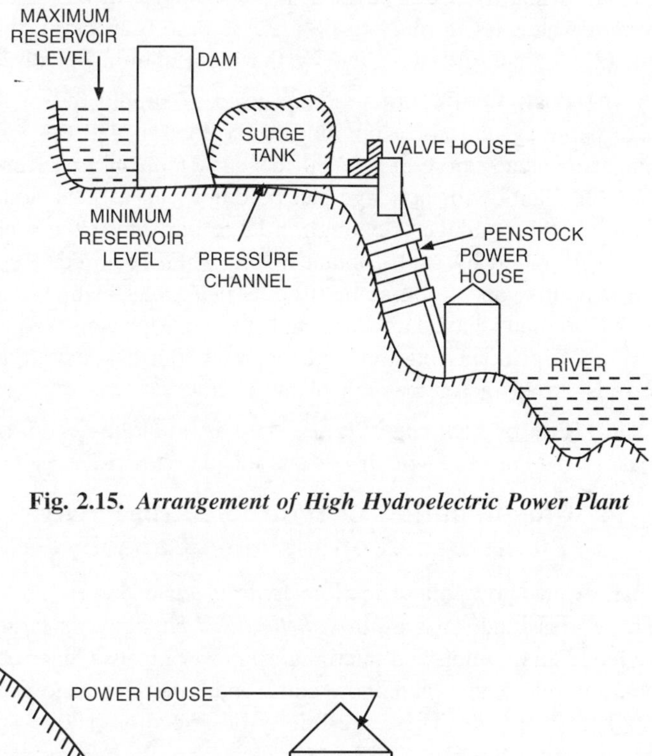

Fig. 2.15. *Arrangement of High Hydroelectric Power Plant*

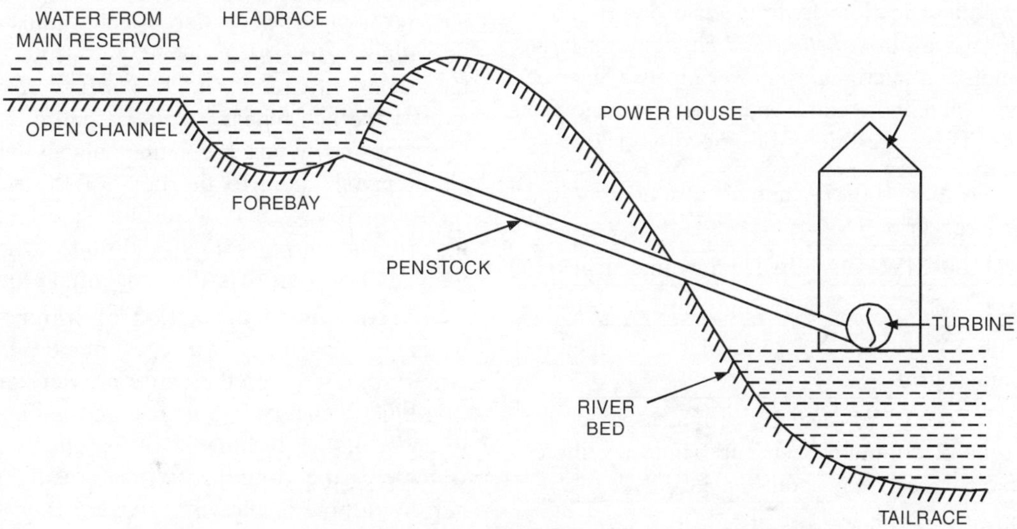

Fig. 2.14. *Arrangement of Medium Head Hydroelectric Power Plant*

For heads above 500 m Pelton wheels are used while for lower heads Francis turbines are employed. The generators used are of high speed and small diameter. Penstocks are of large length and comparatively smaller cross section.

2.7.3. Classification According to Type of Load Supplied

According to the load supplied hydroelectric power stations may be classified into:

(*a*) Base load

(*b*) Peak load and

(*c*) Pumped storage plants for the peak load.

(*a*) Base Load Plants. The plants, which can take up load on the base portion of the load curve of the power system, are called the *base load power plants.* Such plants are usually of large capacity. Since such plants are kept running practically on block load (*i.e.,* the load that is practically constant), load factor of such plants is therefore high. Run-off river plants without pondage and reservoir plants are used as base load plants. Plants having large storage can best be used as base load plants and particularly in rainy seasons, when the water level of the reservoir will be raised by rain water. For a plant to be used as base load plant, the unit cost of energy generated by the plant should be low.

(*b*) Peak Load Plants. Plants used to supply the peak load of the system corresponding to the load at the top portion of the load curve are called the *peak load plants*. Run-off river plants with pondage can be employed as peak load plants. If the pondage is enough, a large portion of the load can be supplied by such a plant if and when required. Reservoir plants can of course be used as peak load plants also. Peak load plants have large seasonal storage. They store water during off-peak periods and are operated during peak load periods. Load factor of such plants is low.

(*c*) Pumped Storage Plants. This is a unique design of peak load plant and will be discussed in Art. 2.13.

2.7.4. Classification of Hydroelectric Power Plants Based on Installed Capacity

Apart from above classification, hydroelectric power plants can be classified, on the *basis of installed capacity,* as large, medium, small, mini, and micro hydro power plants. Generally the mini, micro, and pico hydro come under the subcategory of small hydro plants. These are briefly described as below:

Large hydro	Exceeding 100 MW and usually feeding into a large grid.
Medium hydro	15-100 MW—usually feeding into a grid
Small hydro	1-15 MW—usually feeding into a grid
Mini hydro	100 kW-1 MW—either isolated or feeding into a grid
Micro hydro	5 kW-100 kW—usually provides power for a small community or rural industry in remote areas away from the grid.
Pico hydro	From few hundred watts up to 5 kW.

Apart from the above said classification, there is also a class of very large hydro power plants coming up with capacity ranging from more than 5,000 MW up to 10,000 MW due to the large scale investment and better technology available. However, as far as small hydro is concerned the upper and lower limits varies from country to country while defining the small hydro. There is a general tendency all over the world to define small hydro by power output. Different countries are following different norms keeping the upper limit ranging from 5 to 50 MW. In India small hydro schemes are further classified by the Central Electricity Authority (CEA) as follows :

Table 2.2. Classification of Small Hydro Power Plants

S. No.	Type	Plant Capacity	Unit Rating
1.	Micro	Up to 100 kW	Up to 100 kW
2.	Mini	101 kW to 2,000 kW	101 kW to 1,000 kW
3.	Small	2,001 kW to 15,000 kW	1,001 kW to 5,000 kW

2.8. WATER TURBINES

In hydroelectric power plants water turbines are used to convert the energy of falling water into mechanical energy, which is further utilised for driving the electric generators. The water turbines are simple in construction, highly efficient in operation (about 90% on full load), easily controllable and pick up the load in a very short time. They are built in various sizes up to 1,000,000 hp with speeds varying from 100 rpm to 1,000 rpm depending on their size.

2.8.1. Classification of Water Turbines

There are many ways of classifying water or hydraulic turbines, for example, according to the type of flow of water, according to action of water on moving blades, according to head and quantity of water available, after the name of originator and according to the specific speed of the machine.

According to the type of flow of water, the water turbines used as prime movers in hydroelectric power stations are of four types namely **(*i*)** ***axial flow turbines*** having flow of water along the shaft axis **(*ii*)** ***inward radial flow turbines*** having flow of water along the radius **(*iii*)** ***tangential or peripheral flow turbines*** having flow of water along the tangential directions and **(*iv*)** ***mixed flow*** (radial inlet and axial outlet) turbines. **Kaplan turbine** is an axial flow turbine and has adjustable runner blades which can be rotated about pivots fixed to the boss of the runner. If the runner blades of the axial flow turbines are fixed, these are called the *propeller turbines*. Francis turbine is the mixed flow turbine while Pelton wheel is the tangential flow turbine.

According to the action of water on moving blades, water turbines are of two types, namely *impulse* and *reaction* type *turbines*. When the entire pressure of water is converted into kinetic energy in a nozzle and the jet thus formed drives the wheel, the turbine is of *impulse type*, whereas if the water pressure combined with its velocity work on the runner the turbine is known as the *reaction type turbine*. Pelton wheel is an impulse turbine, in which the water flowing over the turbine rotor blades remains constant. In reaction turbines, the rotor of the turbine operates while submerged in water, the turbine casing being full of water. As the water flows through the rotor blades its pressure changes. Francis turbines, Kaplan turbines and propeller turbines are reaction turbines.

According to the head and quantity of water available, the water turbines are of two types viz *high head and low flow and low to medium head and high to medium discharge turbines.*

According to the name of originator, water turbines are of three types, namely *Pelton wheel, Francis turbine and Kaplan turbine*. Pelton wheel is an impulse turbine and is suited to high head and low flow plants. Francis turbine is a reaction turbine and is suited to medium head and medium flow plants. Kaplan turbine is a special type of propeller turbine having adjustable blades and is suited to low head and high flow plants.

Specific Speed. The specific speed of the turbine is defined as the speed of a geometrically similar turbine that would develop one metric horse power under a head of one metre. All geometrically similar turbines (irrespective of their sizes) have the same specific speed when operating under the same conditions of head and flow.

The specific speed for a turbine is given as

$$N_s = \frac{N\sqrt{P_t}}{H^{1.25}} \quad ...(2.2)$$

whereas N_s is the specific rotational speed in metric units, N is the actual rotational speed of turbine in rpm, P_t is output in metric *hp* and H is the effective head in metres.

A classification of turbines according to the range of head and specific speed is given below in the tabular form.

Table 2.3

Type of Turbine	Specific Speed in Metric Units	Head and Other Conditions
Kaplan or Movable Vane Propeller	300 – 1,000	Head less than 30 m and for variable load operation
Fixed Vane Propeller	300 – 1,000	Head less than 70 m and for fairly constant load operation
Francis	60 – 300	Head 30 m – 200 m
Pelton	10 – 50	Head above 200 m

The hydraulic turbines may have either horizontal or vertical shafts. A horizontal configuration is more accessible and one alternator can be driven by two turbines, one turbine on each side. Moreover the horizontal turbines operate at higher speeds, therefore a smaller and lighter alternator is required. A vertical machine needs a thrust bearing capable of carrying a heavy load and of operating at runaway speed. Small turbines (mostly under 10 MW) may have horizontal shafts, and even fairly large bulb-type turbines up to 100 MW or so may be horizontal. Very large Francis and Kaplan machines usually have vertical shafts because this makes best use of the available head, and makes installation of a generator more economical. Pelton wheels may be either vertical or horizontal shaft because the size of the machine is so much less than the available head. Some impulse turbines use multiple water jets per runner to increase specific speed and balance shaft thrust.

2.8.2. Pelton Wheel

A simple view of the Pelton wheel is shown in Fig. 2.16 (*a*).

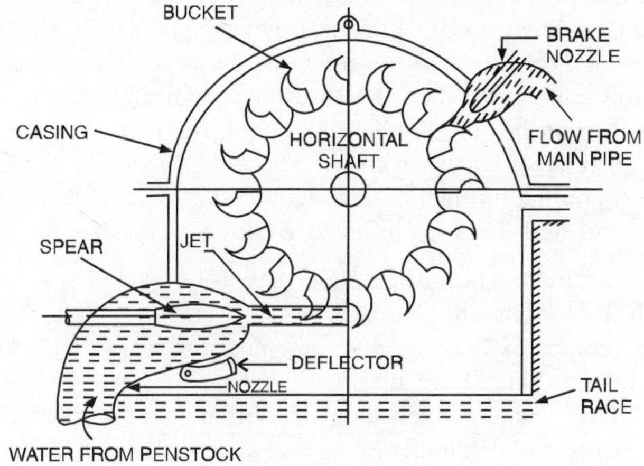

Fig. 2.16 (*a*)

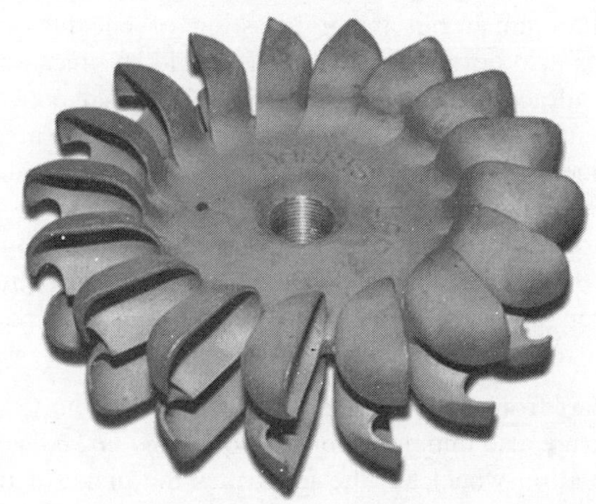

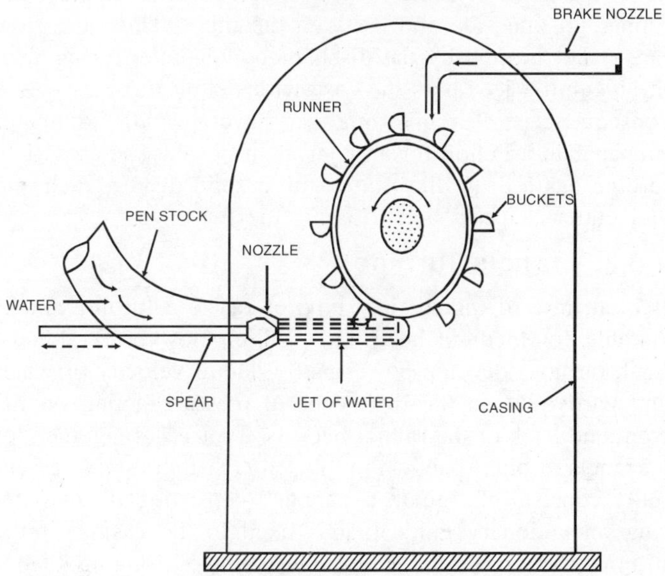

Fig. 2.16(*b*). *Pelton Turbine*

As already stated, it is an impulse turbine and is suitable for high head and low flow plants. The potential energy of water in the penstock is converted into kinetic energy in a water jet issuing from a nozzle. The inside pressure is atmospheric one. It consists of a rotor equipped with elliptical shaped buckets along the periphery of the turbine. The water jet impinges on the buckets and causes the motion of the rotor. After doing useful work, the water discharges into the tailrace. The quantity of water discharged by the nozzle is controlled by controlling the nozzle opening by means of a needle placed in the nozzle tip. The movement of the needle is controlled by the governor. When the load on the turbine reduces the governor pushes the needle into the nozzle, thereby reducing the quantity of water striking the buckets. In case of increase of load on the turbine, reverse action takes place. For control of speed deflectors are used in addition to the needle.

The Pelton turbines usually have one jet but machines with 2 or 4 jets are also being used. Majority of the Pelton turbines are of horizontal shaft type. Horizontal shaft type impluse turbines have usually two nozzles since there will be interference between them if their number is increased. For using more number of nozzles, vertical shaft arrangement will be required. The rotor of the runner is made of cast steel, bronze or stainless steel. The buckets are bolted on to the runner but integral casting of buckets with the runner is also possible. The runner should be placed as close to the tailrace level as possible since the head between the nozzle and the tailrace is ineffective but at the same time, the runner should not submerge in the water. Impulse turbines have long penstocks due to high heads. If such turbines are used in low head power plants, the Pelton wheel runner of large and unwieldy diameter will be required for the given output. As such the turbine of this type is not suitable for water heads below 200 metres.

Apart from Pelton wheel, the Turgo, and the Crossflow turbines are also impulse turbines. The *Turgo turbine* is similar to the Pelton wheel but the jet strikes the plane of the runner at an angle (typically 20°) so that the water enters the runner on one side and exits on the other. Thus, the flow rate is not limited by the discharged fluid interferring with the incoming jet (as is the case with Pelton turbines.) As a consequence, a Turgo turbine can have a smaller diameter runner than a Pelton for an equivalent power. The crossflow turbine has a drum-like rotor with a solid disk at each end and gutter-shaped "slots" joining the two discs.

2.8.3. Francis Turbine

It is an inward mixed flow type of reaction turbine and is suitable for medium head and medium flow power plants. Such turbines develop power partly due to velocity of water and partly due to the difference in pressure acting on the front and back of the runner buckets. Typical arrangement of a Francis turbine is shown in Fig. 2.17. Such a turbine essentially consists of "guide apparatus" consisting of an outer ring of stationary guide blades fixed to the casing of the turbine and an "inner ring" consisting of rotating blades forming a wheel or a runner.

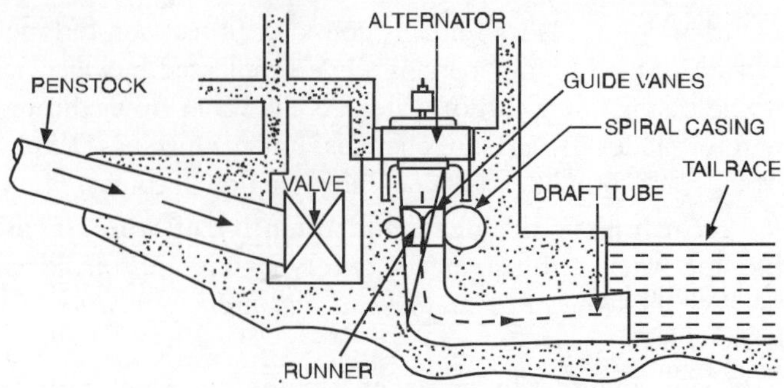

Fig. 2.17*(a)*

Fig. 2.17*(b)* *Francis Turbine*

In these turbines water glides over the blades with a small and fairly constant velocity and exerts a pressure, varying from maximum at the top to a small value at the bottom. The water flows radially inwards and changes to a downward direction while passing through the runner. As the water passes over the rotating blades of the runner, both pressure and velocity of water are reduced causing a reaction force driving the turbine. After doing work, water is discharged to the tailrace through a closed tube of increasing cross section, called the draft tube. It increases the effective head by an amount equal to difference of runner outlet level and tailrace level. Moreover the water leaving the runner still possesses kinetic energy and this kinetic energy is reduced as the water flows through the draft tube to the tailrace and results in setting up of negative pressure head at the runner outlet. Both of these factors increase the net effective head and, thereby the output and the efficiency of the turbine.

The guide blades of the turbine are each pivoted about an axis in parallel with the turbine axis so that the quantity of water entering the turbine may be regulated by turning them simultaneously in one direction or the other. Like Pelton wheel their motion is automatically controlled by governor. As in all reaction turbines, cavitation is a serious problem in a Francis turbine.

The full-load efficiency of this type of turbine is about 92 per cent but with the part load the efficiency decreases to

65%, at half-full load. Francis type turbines can be constructed in vertical or horizontal forms. The horizontal constructions are more accessible and have higher speed, but for large machines generally vertical construction is preferred to effect economy in space. The alternator is mounted above the turbine and thus is free from flooding.

As compared to Pelton wheel a Francis turbine offers the advantage of high efficiency at full load and at 75% of full load. Since this turbine can be designed for a higher speed than Pelton wheel, the dimensions and weights are smaller together with the reduced generator cost.

2.8.4. Kaplan Turbine

It is also a reaction type turbine and has gate and governing mechanism similar to that of a Francis turbine. The difference between Kaplan turbine and Francis turbine is that in the former runner the water strikes the turbine blades axially whereas the latter receives water radially. Water flows radially inwards through regulating gates all rounds the sides, changing direction in the runner to axial flow and causing a reaction force which drives the turbine. Practically all Kaplan turbines are of single runner vertical shaft type. A Kaplan turbine is shown in Fig. 2.18.

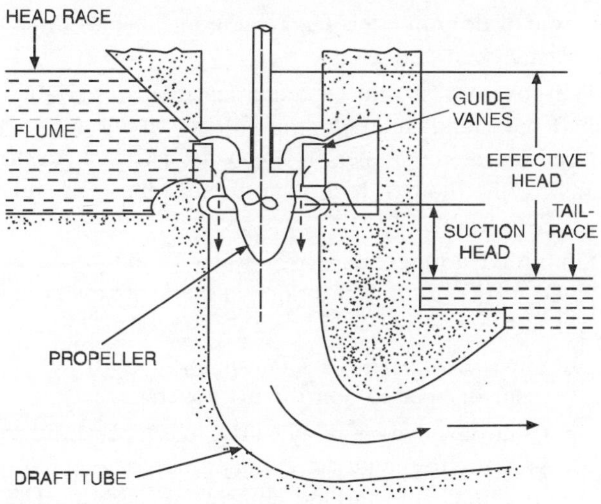

Fig. 2.18*(a)*

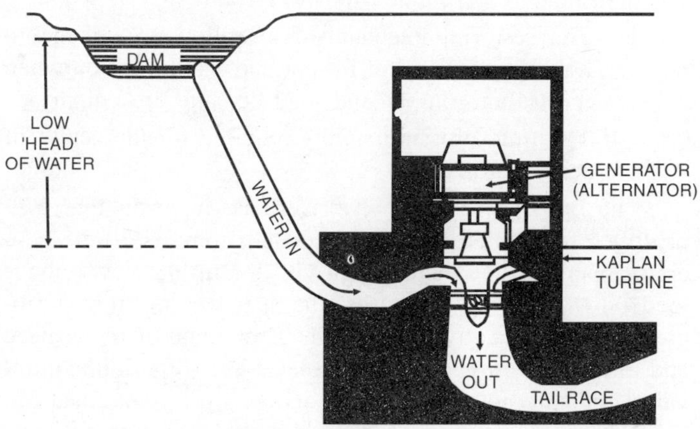

Fig. 2.18*(b)*. *Kaplan Turbine*

Kaplan turbine, because of its highest specific speed (2 to 3 times that of a Francis turbine), is suitable for low head and large flow plants. In this type of turbine the drawback of considerable loss at low loads due to rotary motion of water in Francis turbines is overcome and uniform efficiency at all loads is maintained. Thus part load efficiency of Kaplan turbine is much higher. Kaplan turbine gives high speed (in the range of 400-1,500 rpm) than ordinary Francis turbine resulting in lower cost of runner and alternator and power house structure. The Kaplan turbine, because of adjustable pitch of blades, is capable of operating in a wide range of heads.

All parts except runner of a Kaplan turbine (such as spiral casing, guide mechanism and draft tube) are similar to those of a Francis turbines. The runner blades in Kaplan turbines are less in number (3 to 6) as compared to 16-24 for a Francis turbine. The reduction in the number of blades in Kaplan turbine result in less friction between the blades and water and hence more efficiency. The characteristic feature of Kaplan turbine is that the gate opening and blade angle are adjusted simultaneously by governing mechanism. Its efficiency is about 90% at all loads. The cavitation problem in Kaplan turbine is more serious than in Francis turbines.

The special feature of Kaplan turbine is that its runner is capable of reverse operation as a pump and is, therefore, ideally suited to pumped storage plants.

2.8.5. Propeller Turbine

It is an axial flow reaction type turbine and has got no provision for changing the runner blade angle while turbine is in motion. In such a turbine the blades are cast integrally with the hub. Though its construction is simple and easy but its efficiency falls sharply at reduced loads and, therefore, such a turbine is kept fully loaded for efficient operation. Its efficiency is about 92% at full load and drops to 65% at half full-load.

2.8.6. Characteristics of Water Turbines

1. Head. Reaction turbines of various types can be used for operating heads up to 500 m and Pelton turbines are used for operating heads above 500 m.

2. Specific Speed. Already explained in Art. 2.8.1.

3. Turbine Setting. A Pelton wheel is always set at a higher level than the highest tailrace level (usually 2 m above) while a Francis turbine runner is placed at a level very near or below the lowest tailrace level.

4. Runaway speed. This is the maximum speed at which a turbine wheel would run under the worst* operating conditions with all gates open so as to allow all possible water inflow under maximum head. The generator coupled to the turbine must be capable of withstanding the full runaway speed of turbine under permissible head.

* The worst operating conditions means the generator load being suddenly thrown off and governor failure to act.

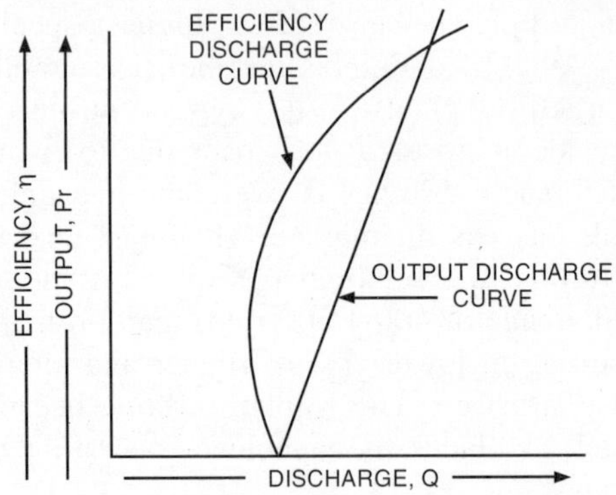

Fig. 2.19. *Efficiency-Discharge Curve*

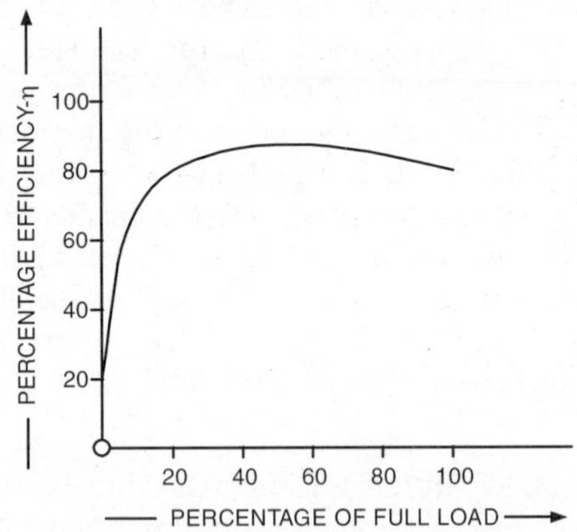

Fig. 2.20. *Efficiency Curve of Impulse Turbine*

5. Constant Speed Curves. In hydroelectric power plants, the turbines operate at constant speed and, therefore, variables are operating head H and discharge Q. As the discharge and head vary so as to keep the speed constant, the turbine output P_t is measured by brake arrangement. The turbine efficiency η is then calculated for various values of Q and H. Now the output discharge (P_t – Q), efficiency-discharge (η – Q) curves, as shown in Fig. 2.19 and efficiency-percentage full load curves are drawn as shown in Figs. 2.20

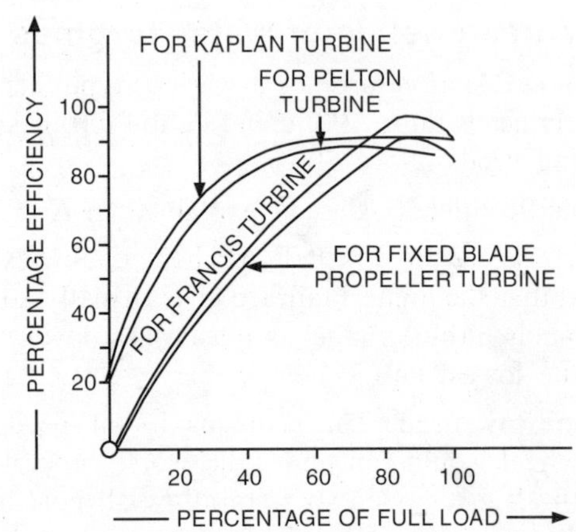

Fig. 2.21. *Efficiency Curve of Reaction Turbines*

and 2.21. From the curves drawn it can be concluded that the Kaplan and Pelton turbines perform well at part loads but Francis and Propeller turbines do not.

2.9. GOVERNING OF WATER TURBINES

In order to have electrical output of constant frequency it is necessary to maintain speed of the alternator driven by the turbine constant. This is achieved by controlling the flow of water entering the turbine by the automatic adjustment of guide vanes in case of reaction turbines and of the nozzle needle in the case of impulse turbines. Such an operation of speed regulation is called the governing, and it is attained automatically by means of a governor. In case of impulse turbine the governor also operates the auxiliary relief valves or jet deflectors.

For the regulation of water below the penstock connection, at the time of decrease in load on the impulse turbines, the governor reduces the water flow from the power nozzle and the surplus water is diverted with the help of auxiliary relief nozzles. In the case of multi-nozzle turbines, a deflector plate deflects some water from the runner buckets by swinging into the water jet from each nozzle. With the movement of deflector plate out of the path of water jets, the needles slowly reduces the flow of water so as to keep the output of the turbine constant at the level of new load.

In the case of Francis turbine, there are pressure regulators for discharging the water from the casing to the tailrace at the time of drop in load. The regulators close as fast as the guide vanes open and vice versa.

The governor should be quite sensitive to variations in the shaft speed and should be rapid in action but not so rapid as to cause water hammer in the penstock. The governing systems for the modern hydraulic turbines have a regulating time of 3–5 seconds.

Simplified arrangement of a water turbine governor is illustrated in Fig. 2.22. The principal elements of the governor are:

1. The speed-responsive element—usually flyball mechanism or speed (centrifugal) governor.
2. Control valve or relay valve to supply fluid under pressure to the power cylinder (servomotor) in order to actuate the turbine control mechanism. The use of control valve and servomotor is to amplify the small force created by the flyballs.
3. The restoring mechanism or follow-up linkage to hold the servomotor in required fixed position when the trubine output and load demand are equalised.
4. The fluid pressure supply required for the action of servomotor.

The flyballs may be belt driven, as shown in Fig. 2.22 or driven by a small electric motor fed from a separate generator operated in synchronism with the turbine. When the load on the turbine decreases, the speed of the turbine increases, consequently, the flyballs also rotate at high speed and move outwards. The floating lever gets lifted up, control valve is displaced upwards from its central or dead beat position, the upper port is uncovered and the oil flows from a pressure tank through the port into the right hand end of the servomotor cylinder. The piston moves to the left and closes the nozzle with the help of a spear in the case of Pelton wheels and adjusts the guide vanes in the case of reaction turbines. In case of increase of load on the generator, the speed of the turbine will decrease and the reverse action would take place. The restoring or follow up linkage resets the relay, pilot or control valve after the servomotor piston has adjusted the water control mechanism.

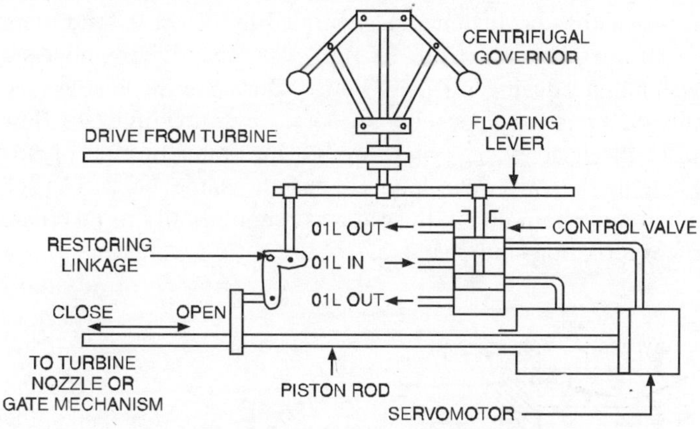

Fig. 2.22. *Governing of Water Turbines*

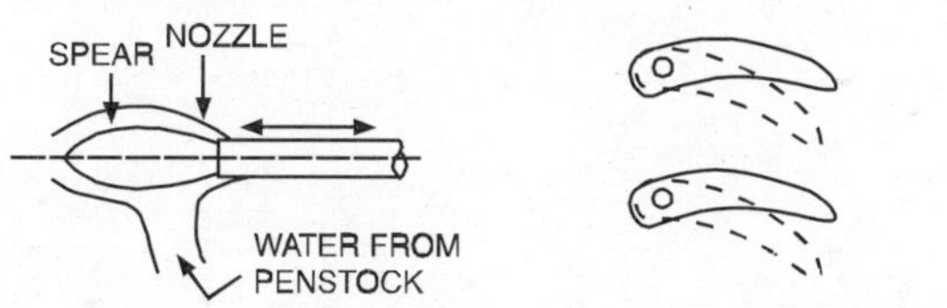

(a) Impulse Turbine *(b) Reaction Turbine*

Fig. 2.23

In case of Pelton wheel a combined spear and deflector regulation is employed in order to avoid water hammer in the penstock. In case of decrease of load on turbine, the deflector, which is usually a plate connected to the servomotor by means of levers, is brought in between the nozzle and buckets, thereby, diverting water away from the runner and directing into the tailrace. In the mean time, the spear has been adjusted to the new position of equilibrium and the deflector plate is moved out of the path of water nozzle.

2.10. SELECTION OF WATER TURBINE

The selection of a water turbine depends upon various factors such as working head, available discharge, speed, output and nature of load. The effective head under which the turbine is to operate gives the first guide to the selection of the type of turbine. For very high head *i.e.,* 500 metres and above, Pelton turbine is usually employed. Reaction turbines are not suitable for such high heads as due to high velocity of water, there will be rapid wear and friction losses. For medium heads *i.e.,* above 30 metres and below 500 metres, Francis turbine is usually adopted but impulse turbine may also be adopted even at the expense of efficiency. For low heads *i.e.,* below 70 metres, propeller type of turbine is used. In case of considerable variations in head and load Kaplan turbine gives improved efficiency as compared to the fixed blade propeller turbine. Even for the same head, two different types of turbines may be employed. For example, for a head of 200 m either of the Pelton or Francis turbine can be used. Similarly for a head of 30 m, either of the Kaplan or Francis turbine can be employed. It is a general practice to select a runner of higher specific speed. A runner of high specific speed will generate more power for the same head resulting in small size of the turbo-alternator and the power house. For larger output but at low head, a runner of high specific speed is used to keep the size of the turbo-alternator and power house smaller and economical. For high head and even with medium output plants, even a low specific speed runner will attain high rotational speed and a high specific speed runner need not be selected.

2.11. UNDERGROUND HYDRO POWER PLANT

An underground hydro power plant is one in which whole of the generating equipment is placed in an underground chamber. An underground layout of hydro power plant may be favoured due to technical and economical considerations.

The underground layout of hydro power plant have the advantages of better rocking bearing properties, greater security in war time, lower maintenance costs, shorter conduit lengths and therefore smaller surge tanks, minimum problem of land acquisition, easier design of tailrace tunnel in comparison to design of headrace pressure tunnel, no interference of surface topography in planning, design and location of the principal works, lower initial cost in some cases, no risk of avalanches and land slips, that may cause difficulties in site selection in steep and rugged countryside, no risk of forest fire hazard, protection to the staff as well as to plant against severe winter climate, no effect of bad weather on the progress of construction work, and maintenance of temperature control inside the plant with adequate control of ventilation irrespective of outside weather conditions.

However, underground hydro power plants have some shortcomings also such as increased cost of construction of power house and accessories, excessive cost of lighting, special ventilation and airconditioning, additional cost of underground location of transformers and switchgears, costly tailrace tunnels and additional surge chambers.

There are two types of layouts for underground power plants viz head development and tail development. In first case, the plant is situated near the intake and has a long tailrace tunnel. So the advantages of lower cost of intake structure and possibility of quicker control of turbines are available in the former case. In the second case, the plant is located at the end of a long pressure tunnel and has a short tailrace. The decision is influenced by the topography and character of the rock.

In case of high head power plants, consideration of access dictates the location of power house and it should be located at the down-stream end of the aqueduct system. On the other hand, if the head is low or medium and the rock condition is sound, the power house is situated close under the reservoir and the tail water is led to the stream by long unlined tailrace tunnel. The water from the head pond is led to the turbines by a vertical or nearly a vertical shaft.

860 MW Koyna underground hydro power plant in Maharastra and 105 MW Salal semi-underground hydro power plant are important hydro power plants in India.

2.12. SMALL HYDROELECTRIC POWER PLANTS

Mini hydro plants of capacity around 101-2,000 kW and micro hydro plants of capacity up to 100 kW fall under this category. Such plants operate under heads of a few metres. Such plants are becoming more and more popular due to rising fuel cost of thermal plants on one side and long construction period and heavy cost of civil works for large hydroelectric power plants on the other side. Small hydro power plants are simple in operation, reliable, need minimal maintenance and more effective than large plants in saving fuel.

In mini hydro and micro hydro schemes, the civil works are simple, and can be built with the local labour in a very short period. High equipment cost was the only hindrance in the development of small hydro power plants. But now due to newly developed ranges of hydroelectric units and their availability in market in standardised packaged units easily adoptable to different site conditions this problem has also been solved.

Small hydro schemes are very useful for small isolated groups of consumers and are likely to be technically and economically feasible at any site where adequate flow and head are available.

Global installed capacity of small hydro is about 47,000 MW against the estimated potential of 200,000 MW. India has a history of more than 100 years in small hydro. An estimated potential of 15,000 MW of small hydro exists in India. However, nearly 5,000 MW have been actually identified through more than 2,000 sites in 13 states of India and most of the sites are located in Himalayan region. Himalayan region is endowned with one of the world's largest water resources mostly rain and snow fed rivers, rivelets and streams with perennial flows.

A turbine unit, used in small hydro plants, consists of a runner connected to a shaft that converts the potential energy in falling water into mechanical or shaft power. A turbine and generator unit is shown in Fig. 2.24.

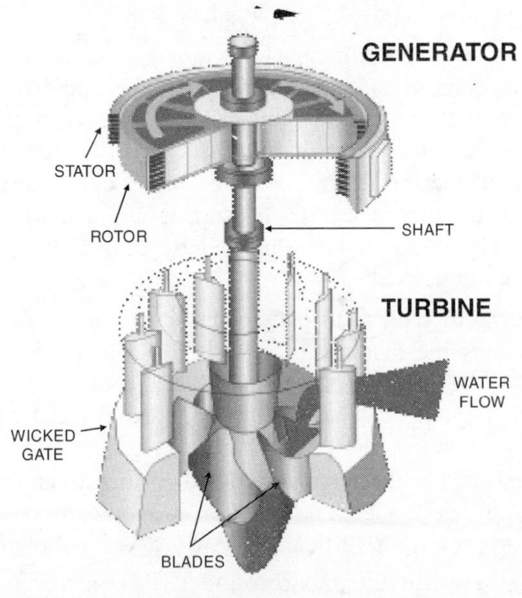

Fig. 2.24. ***Turbine and Generator Unit in Small Hydro Power Plants***

2.13. PUMPED STORAGE POWER PLANTS

This is a unique design of peak load plant in which the plant pumps back all or a portion of its water supply during low load period. The usual construction is a tail water pond and a head water pond connected through a penstock. The generating pumping plant is at the lower end. The plant utilises some of the surplus energy generated by the base load plant to pump the water from the tail water pond into the head water pond during off-peak hours. During peak load period this water is used to generate power by allowing it to flow from the head water pond through the water turbine of this plant to the tail water pond. Thus the same water is used again and again and extra water is required only to take care of evaporation and seepage.

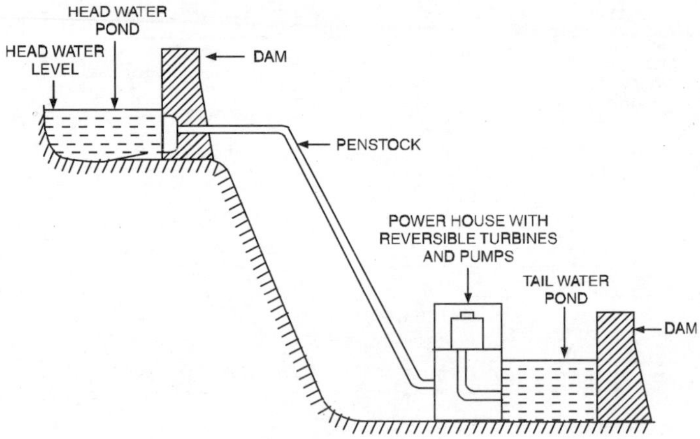

Fig. 2.25(*a*)

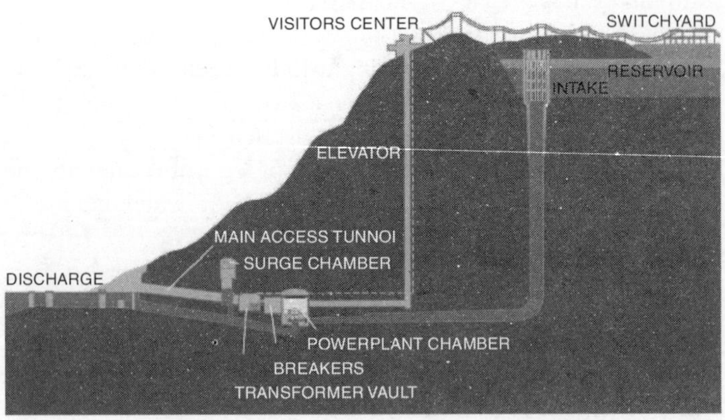

Fig. 2.25(*b*) ***Pumped Storage Power Plant***

The capacity of the pond should be such that the plant can supply peak load for 4 to 10 hours. A general arrangement of a pumped storage power plant is shown in Fig. 2.25. The plants can be used in conjunction with hydro, steam and IC engine plants. This plant is also called a hydraulic accumulator system. Such plants can have either vertical shaft arrangement or horizontal shaft arrangement. In the older plants, there used to be separate motor driven pumps and turbine driven generators. The improvement was the pump and turbine on the same shaft with electrical element acting as either generator or motor. The recent development in this field is to use a Francis turbine which is just the reverse of centrifugal pump. During peak hours the turbine drives the generator and the plant generates electrical energy while during off-peak hours the generator operates as a motor and drives the turbine which now works as a pump raising the water from the tail water pond to the head water pond. The power for driving the generator as a motor is taken from the system. This arrangement reduces the capital cost of the plant and improves the operating efficiency and thus results in economical operation.

The efficiency of such plants is around 60-70 per cent. Some water may evaporate from the head water pond resulting in the reduction in the stored energy or there might be run off through the soil. Also there will be some energy loss in the generating and pumping equipment and in power transmission. However, the energy loss to the system (the difference of energy input to the pumped storage plants and their output) is quite negligible in comparison to the substantial savings in fuel when these plants are operated in a mixed system for supplying peak loads.

The most obvious economies are in the replacement of standby thermal plants. When an old steam power plant is employed for peak load supply as an alternative to a pumped storage plant, it is a costly affair and, moreover, it needs large starting time and shutdown time. Hence these are not at all convenient for supplying peak loads. These steam power plants cannot be operated in the reversible mode as is possible with pumped storage plants.

Such plants can be operated only in interconnected systems where other generating plants (such as steam, nuclear, hydro power plants) are available. The steam and nuclear plants operate most economically when operated continuously at or near full load. The major problem with these plants is the peak load. Covering of short peak load demands by steam power plant is economically undesirable and is not adoptable due to slow pick-up. Pumped storage plants for peak load supply in interconnected systems are most suitable where the quantity of water available for power generation is insufficient but natural site for construction of high dam exists.

Pumped storage power plants have some very important advantages. Some of these are given below:

1. Peak loads can be supplied at a lower cost than that when supplied by steam and nuclear power plants.
2. The steam and nuclear power plants can be operated at almost unity load factor which ensures their most efficient and economic operation.
3. Because of their ability to take up loads in a very short time (pumped storage plants need a starting time of only 2-3 seconds and can be loaded fully in about 15 seconds), the spinning reserve requirement of the system is reduced.
4. In the event of an extra demand coming up suddenly on the system, such plants can be immediately switched on to meet this extra demand.
5. They can be employed for load frequency control.

It has been estimated that a pumped storage plant can be set up at half the cost of a nuclear power plant of the same size. The pumped storage schemes, because of their inherent benefits, are likely to play an important role.

The pumped storage plant essentially consists of a head water pond and a tail water pond. The site selection for such plant should, therefore, take into consideration of natural pond facilities in hilly area.

56 number of pumped storage projects have been identified with probable installed capacity of 94,000 MW. The first scheme was commissioned about 30 years back. At present five schemes (*Paithon* on Godavari river with installed capacity of 12 MW in Maharastra, *Kadamparai* on Kadamparai river with installed capacity of 400 MW in Tamilnadu, *Nagarjunsagar* on Krishna river with installed capacity of 700 MW in AP, *Panchet hill unit* on Damodar river with installed capacity of 40 MW and *Kadana stage I* on Mahi river with installed capacity of 120 MW in Gujarat) are in operation. Out of these only first two schemes (Paithon and Kadamparai) are being operated as pumped storage plants and the rest three are being operated only in generating mode due to non-construction of tail-water dams.

A number of other pumped schemes (such as 900 MW Srisailam in AP, 120 MW Kadana stage II, 1200 MW Sardar Sarover, both in Gujarat, 1000 MW Tehri Stage I in UP, 90 MW Bhivpuri, 150 MW Bhira, 250 MW Ghatgash, and 12 MW Ujjani, all in Maharashtra) are under planning.

2.14. HYDRO POTENTIAL IN INDIA

India is blessed with immense amount of hydroelectric potential and ranks 5th in terms of exploitable hydro potential on global scenario. According to the accessment made by CEA, India is endowed with economically exploitable hydro power potential to the tune of 148,700 MW of installed capacity. The basin wise assessed potential is as below:

Table 2.4. Basin Wise Assessed Hydro Potential

Basin/Rivers	Probable Installed Capacity (MW)
Indus Basin	33,832
Ganga Basin	20,711
Central Indian River System	4,152
Western Flowing Rivers of Southern India	9,430
Eastern Flowing Rivers of Southern India	14,511
Brahmaputra Basin	66,065
Total	**1,48,701**

In addition, 56 number of pumped storage projects have also been identified with probable installed capacity of 94,000 MW. In addition to this, hydro potential from small, mini and micro schemes has been estimated as 6,782 MW from 1512 sites. Thus, in totality India is endowned with hydro potential of about 250,000 MW. However, exploitation of hydro potential has not been up to the desired level due to various constraints confronting the sector.

In 1998, Government of India announced "Policy on Hydro Power Development" under which impetus is given to development of hydropower in the country. This was a welcome step towards effective utilization of our water resources in the direction of hydropower development. During October 2001, Central Electricity Authority (CEA) came out with a ranking study which prioritized and ranked the future executable projects. As per the study, 399 hydro schemes with an aggregate installed capacity of 106,910 MW were ranked in A, B and C categories depending upon their inter-se attractiveness. During May 2003, Govt. of India launched 50,000 MW hydro initiative in which preparation of Pre Feasibility Reports of 162 projects totalling to 50,000 MW was taken up by CEA through various agencies. The PFRs for all these projects have already been prepared and projects with low tariff (first year tariff less than Rs 2.50/kWh) have been identified for preparation of DPR.

2.15. AUTOMATIC AND REMOTE CONTROL OF HYDRO PLANTS

Control, automation and monitoring system in a hydro power plant is associated with start and stop sequence for units and optimum running control of power (true and reactive), voltage and frequency. In addition, many protective devices (such as overload relays, reverse power relays, negative phase sequence relays, generator short-circuit protection relays, generator winding temperature relays, bearing temperature relays etc.), are also required. For efficient and fast operation of protective devices, their automatic operation is essential so as to detect the abnormal conditions which call for their operation and shut down the unit as and when required.

Control functions of control in a hydro power plant may be categorized into the following:

Turbine Control. This is the speed/load control of turbine in which the governor adjusts the flow of water through the turbine to balance the input power with the load (output power). With an isolated system, the governor controls the frequency. In interconnected system, the governor may be used to regulate the unit load and may contribute to the system frequency control. In case of micro hydro plants in the range of micro hydel (100 kW unit size), load control is also used, where excess load is diverted to the dummy load to maintain constant speed.

Generator Control. This is the excitation control of synchronous generator. The excitation is an integral part of a synchronous generator which is used to regulate the operation of the generator.

Plant Control. Plant control deals with the operation of the plant. It includes sequential operations like start up of the machine, excitation control, synchronisation, loading of unit under specified operating conditions, normal and emergency shutdown, etc. The mode of control may be manual, semi-automatic or automatic and may be controlled locally or from remote location. Plant control usually includes monitoring and display of the plant conditions.

Control of different units in a power station is done from a centrally situated control room, from where all the operations including starting, synchronising, loading and shutdown of units are carried out.

Modern hydro plants are always operated as a part of an integrated power system consisting of many power plants. The whole system is remote controlled from an area control centre. The control engineer, also called the power controller, (in the case of manual load dispatching) or an on line computer (in the case of automatic load dispatching) determines the generation needs of various plants and transmits the information to the engineers-incharge of the plants for taking necessary steps. The communication between the control centre and power plants is through carrier current communication.

The control of the hydro plants is very much less complicated. As already stated, the various methods of starting and stopping of machines can be grouped under three headings:

1. Fully manual control, 2. Semi-automatic and 3. Fully automatic.

The fully manual control is still common. In this system, every operation from starting to stoping is performed by hand. This may be achieved either mechanically or by push buttons. In semi-automatic control system, only a single impulse is required for starting the machine and then the machine is brought to the condition where it is ready to synchronise on the system. Similarly, the machine can be stopped by single impulse. In fully automatic control, sequence of starting the machine, synchronizing and loading to some predetermined extent are all done automatically. Periodic cleaning, inspection and usual routine maintenance are the only needs to be met by the plant staff. The machines can be started in least possible time. There will also be savings in the salaries and wages of the staff. The extra cost of equipment is fully met by other savings.

A fully automatic plant may be controlled by means of either of three methods: time switch, float switch and a load sensitive device. In time switch technique, the clock timings preset and the plant is started and stopped accordingly. In the float switch technique, the output is dependent upon the water level in the reservoir. In load sensitive device, the device is operated by the load or power demand of the area and the output of the generators, varies automatically. If a hydro plant is controlled from a distance either from a control centre or a large hydro plant it is said to be remote controlled. The plant is controlled with the help of signals sent from the remote control centre through the automatic telephone equipment. Small hydro plants which are not justified to be developed for manual control, can usually be justified by the application of remote automatic control.

2.16. HYDROELECTRIC GENERATOR

The generators employed in hydroelectric power plants are three phase alternating current synchronous generators, called the alternators. Alternator consists essentially of two parts namely armature and field magnet system.

The armature of an alternator is an iron ring, formed of laminations of special magnetic iron or steel alloy (silicon steel) having slots on its inner periphery to accommodate armature conductors and is known as stator. The whole structure is held in a frame which may be of cast iron or welded steel plates. For minimising eddy current losses due to rotation of field structure in between the stator, the stator core is laminated. The laminations (usually of thickness 0.5 mm or less) are stamped out in complete rings (for small machines or in segments (for larger machines) and insulated from each other with paper or varnish. The stampings also have openings which make axial and radial ventilating ducts to provide efficient cooling. The open slots are more commonly used because the coils can be form wound and insulated prior to being placed in the slots giving least expenditure and more satisfactory winding method. Such slots also facilitates in removal and replacement of defective coils.

The field structure is the largest and heaviest component of alternator (in large machines, it may be 15 m in diameter and 1,000 ton in weight), and is called the rotor. The rotor houses the dc excitation winding and the exciting current is supplied to the rotor through two slip rings and brushes. The field windings are connected in series to form the excitation winding which is supplied with dc at 110/220/300 V. Earlier, the excitation was usually supplied by small dc generators (shunt or compound wound type) driven through pilot exciters from the turbo-generator shaft. The recent generators make use of static excitation systems.

The alternators employed in hydroelectric power plants have the following features.

The machines may have horizontal or vertical configuration. The alternators employed in conjunction with impulse turbines are usually of horizontal configuration while those employed with Francis and Kaplan turbines are of vertical configuration. Low speeds (about 50—500 rpm in case of vertical machines and 100—1,000 rpm for the horizontal machines). Protection against runaway speeds are to be provided. The machines are of large diameter and small lengths. The machines are usually of salient pole type and the number of poles, they consist of, varies from 6—120. These machines should be capable of supplying heavy line charging currents since hydroelectric power plants are usually located at considerable distance from load end. A higher value of short-circuit ratio (around unity) is therefore required.

The specifications of the generator for a hydro power plant includes MVA rating, number of phases (always 3), frequency (50 Hz in India), connections of stator windings (nearly always star), voltage rating (3.3/6.6/11 kV), power factor, current rating, speed, method of cooling, temperature rise, type of excitation, excitation voltage, different reactances, short-circuit ratio, position of shaft, type of voltage regulator and its response time, efficiency, field current at full load and rated power factor, regulation at full load, mechanical details like runaway speed, direction of rotation, overall size, flywheel effect, number and location of bearings etc. and details of accessories.

The machines are usually air cooled. The air is diverted through the machine by different methods (radial, axial and circumferential). Air is taken in at one or both ends, forced through ducts between the windings and sections of core and discharged again at the periphery. An intake may be taken from points over the tailrace at a safe distance above the highest water level in it. The amount of air requirement is about 3 m^3/minute/kW of loss in the generator, and air velocity should be restricted to 475 m/minute. The larger machines are completely enclosed and the air is recirculated in the housing after being cooled by means of water-cooled heat exchangers. Closed-circuit cooling has the advantage of contamination free air permitting the machine to run for much longer periods without cleaning. The moisture exclusion is also advantageous. The restricted amount of air available gives a natural protection against extensive damage by fire, in the event of electrical breakdown.

2.17. CHOICE OF SIZE AND NUMBER OF GENERATING UNITS

The load on the power station is never constant but varies at different timings of the day owing to varying demands and the generating plant should have the capacity to meet the maximum demand. In case one unit is taken of such a size as to meet the maximum demand of the power station, the plant will be operating on full load only for a short duration and will be running light or even practically on no load for rest of the day. The generating unit would, therefore, not be running at all times, under conditions best suited for its operation giving maximum efficiency. In case of an isolated station, in order to maintain reliability and continuity of power supply at all times, another one unit of equal capacity will be required. Thus the capital cost would be for two units, each of capacity corresponding to the maximum load on the power station. Alternatively number of small sized generating units can be chosen so as to fit the load curve as closely as possible *i.e.,* the generating units be taken of such sizes and in such a number that they work on suitable portions of the load curve in such a way that each unit will operate on full load or at the load giving maximum efficiency. In this case the reserve required would only be one unit of the largest size chosen and this unit would be much smaller than the maximum load capacity required in the former case. This alternative will require a large number of generating units and, therefore, the area required for the building will also increase and so the cost of the building. It will also require more frequent starting, stopping and parallel operation of the equipment. The maintenance cost will also increase.

Hence a compromise is to be made in the selection of the size and the number of generating units in the generating station in order to avoid both the extremes mentioned above. The aim should be to have a small number of units and to fit them as well as possible on the load curve. Neither we should go for a single generating unit of larger capacity nor for a large number of generating units of smaller sizes.

2.18. PLAN LAYOUT OF HYDRO POWER PLANT

The general layout of the hydro power plant is determined by its type. For plants employing vertical turbines, the most convenient and economical layout will be with turbines installed in a line parallel to the length of the turbine house, as illustrated in Fig. 2.26. The spacing between the machines will depend upon the size of scroll case, width of flume, or by the overall diameters of the alternators. In case of turbines with horizontal shaft arrangement, the most suitable layout will be placement of turbines at right angles to the length of turbine house. The horizontal machines can also be placed parallel to the longitudinal axis of the turbine house. A repair bay should be provided at one end of the turbine house near the workshop with ample space for dismantling and re-erection of machines.

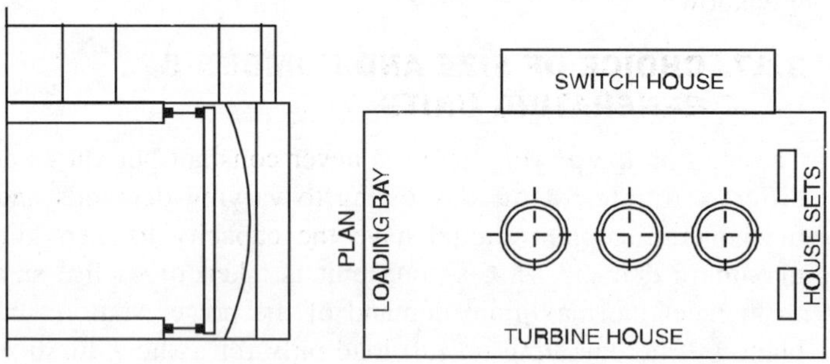

Fig. 2.26. *Typical Plan Layout of Hydro Power Plant with Vertical Reaction Turbines*

2.19. HYDRO PLANT AUXILIARIES

The auxiliaries essentially required for hydroelectric plant are governor, cranes, lubricating oil pumps, air compressors, high pressure oil pumps for generator rotor jacking system, fans, cooling water pumps, drainage and dewatering pumps, gate hoists, valves, battery charging units, CO_2 cylinders etc. These auxiliaries are generally electrically driven. Water may be used to cool the bearings of the turbines and generators and the transformers and is circulated through water pumps. Air compressors maintain a supply of air under pressure for operation of generator brakes and other uses in the power station. Fans are required for ventilation of the turbine and switchgear room or for cooling of transformers. Oil pumps handle transformer oil through the cleaning and cooling system. Cranes are required to lift heavy parts or place them in position during repairs. Water pumps are required for unwatering of turbine pits during repairs or inspection. Storage batteries are required to supply low voltage dc power for switchgear control. These batteries are constantly charged through a battery charging equipment using a rectifier or motor-generator set. Carbon dioxide cylinders and other fire extinguishing equipment are required in case the fire breaks out. The supply for the above auxiliaries is usually obtained from the station transformer which is installed solely for this purpose.

2.20. ENVIRONMENTAL IMPACTS OF HYDRO POWER PLANTS

Though the hydro power plants are considered clean and harmless but they have the following environmental impacts.

Hydro power plants need huge amount of water storage and therefore, large dams are to be constructed. This leads to displacement of the inhabitants of the area. The displacment of a large number of inhabitants may create a great variety of social and economic problems. The dam's lakes change markedly the local ecological conditions, vary the pressure applied to the land and the groundwater level, which adversely affects plant and animal life in the nearby region. The construction of dams for hydroelectric power plants slows down the flow of rivers and thus causes the pollution of water, growth of deleterious blue-green algae, encourages the reproduction of epidemic-carrying bacteria, checks the flood water with the result that water meadows cease to exist and sometimes salinization of soil occurs. Large area acquisition means destruction of forest cover which is harmful for environments. In addition to environmental degradation and disturbing ecological balance, the hydro power plants disturb the demographic balance also. Large number of workers required for construction are brought into the area and they disturb the very nature of local population.

Exercises

1. Write the advantages, disadvantages and applications of hydroelectric power plant.
[M.D. Univ. Power Plant Engineering May 2007]
2. Explain the criterions for hydro power plant site selection.
[Pb. Technical Univ. Generation of Electric Power May 2008]
3. What are the main considerations for selection of site for a hydroelectric power station.
[U.P. Technical Univ. Power Station Practice 2007-08]
4. Discuss the various factors which affect the location of site of a hydro power station.
[U.P. Technical Univ. Power Station Practice 2009-10]
5. What are the factors which determine the location and site of a hydro plant ?
[G.B. Technical Univ. Power Station Practice 2011-12]
6. What do you understand by hydrology ? How does the study of hydrology help in selecting the site for a hydroelectric power plant ? **[M.D. Univ. Power Plant Engineering May 2009]**
7. What is hydrograph ? What informations does it provide ?
[Rajasthan Technical Univ. Generation of Electrical Power 2010]
8. Discuss the utility of hydrograph and flow duration curve for the power plants.
[Pb. Univ. Power Station Practice May 1995]
9. Define run-off. How is it measured. List the factors which affect run-off. **[M.D. Univ. Power Plant Engineering May 2007]**
10. What is a mass curve and what information does it provide ? What is its use ?
[Pb. Univ. Electrical Power-II (Generation) Dec. 1995]
11. Explain the load duration curve and mass curve with their complete information. **[Rajasthan Univ. 2007]**
12. Draw and explain single line diagram of hydro plant.
[Pb. Technical Univ. Generation of Electric Power May 2010]

13. Give a layout of hydroelectric power plant. Also explain the principle.

[G.B. Technical Univ. Power Station Practice 2010-11]

14. Give a general layout of a hydroelectric power plant. Explain the functions of different components in storage reservoir plants.

[U.P. Technical Univ. Power Station Practice 2009-10]

15. With the help of a simple diagram explain the essential features of hydro power plant.

[Power Station Practice Pb. Univ. Dec. 1994; U.P. Technical Univ. 2007-08]

16. Explain with neat sketch the working of a hydroelectric power plant station and explain the function of each component in it.

[M.D. Univ. Electric Power Generation May 2011]

17. Water is available in a reservoir at 150 m head. The distance of the reservoir is approximately 2 km from the site. Give all the components of a hydro power plant to be installed at this site. Discuss the utility of each component.

[M.D. Univ. Power Plant Engineering May 2009]

18. What is the function of surge tank ?

[Rajasthan Technical Univ. Generation of Electrical Power 2010]

19. Explain the functions of the following for hydro power plants: (*i*) Dam (*ii*) Spillways (*iii*) Surge tank (*iv*) Draft tube. **[U.P. Technical Univ. Power Station Practice 2006-07]**

20. Briefly discuss the function of surge tank in a hydroelectric power station.

[U.P. Technical Univ. Power Station Practice 2008-09]

21. Briefly discuss the function of penstock in a hydroelectric power station.

[U.P. Technical Univ. Power Station Practice 2008-09]

22. Explain the phenomenon 'water hammer' in hydroelectric power station. State the procedure to overcome this problem.

[A.M.I.E. Sec. B. Power Plant Enginering Summer 1996]

23. How a surge tank helps in reducing water hammer effect? Describe various types of surge tanks.

[Pb. Univ. Power Station Practice January 1993]

24. What is the function of a draft tube. Describe the various types of draft tubes.

[Pb. Univ. Power Station Practice January 1993]

25. Classify hydroelectric plants. What factors are taken in location and site selection of this plant ?

[U.P. Technical Univ. Power Station Practice 2006-07]

26. Classify the hydroelectric power plants based on: (*i*) Plant capacity (*ii*) Head (*iii*) Hydraulic features.

[U.P. Technical Univ. Power Station Practice 2004-05]

27. Classify the hydroelectric plant on the basis of available water head and discuss.

[U.P. Technical Univ. Power Station Practice 2007]

28. How are hydroelectric power plants classified ?

[Rajasthan Technical Univ. Generation of Electrical Power 2008; 2011]

29. Sketch a typical layout for medium head hydro power plant using Francis turbines. Explain the salient features.

[Pb. Univ. Power Station Practice December 1995]

30. Describe with a neat diagram the run-off river plant with pondage. **[Pb. Univ. Electrical Power-II (Generation) May 1995]**

31. Explain run-off river power plant. How its performance is increased by introducing a pondage in the plant ?

[M.D. Univ. Power Plant Engineering May 2008]

32. What are the different types of hydro turbines ? What are their applications ? Discuss their main constructional features.

[Rajasthan Technical Univ. Generation of Electrical Power 2010]

33. What are the different types of hydro turbines ? Discuss their main constructional details.

[G.B. Technical Univ. Power Station Practice 2010-11]

34. Describe different turbines and their use in hydroelectric plants. **[U.P. Technical Univ. Power Station Practice 2006-07]**

35. Briefly describe impulse turbine used in hydroelectric station. **[U.P. Technical Univ. Power Station Practice 2008-09]**

36. Briefly describe Kaplan turbine used in hydroelectric power station. **[U.P. Technical Univ. Power Station Practice 2007-08]**

37. Bring out the comparison between impulse, reaction, Kaplan and Francis turbines.

[G.B. Technical Univ. Power Station Practice 2011-12]

38. Differentiate between the operation of a Pelton turbine and a Francis turbine.

[U.P. Technical Univ. Power Station Practice 2009-10]

39. What are the characteristics of a water turbine ? Explain the term "governing of turbines".

[U.P. Technical Univ. Power Station Practice 2003-04]

40. Draw the operating characteristic (constant speed) curves of a water turbine and explain how these are obtained ?

[Pb. Univ. Power Station Practice June 1992]

41. Discuss the following in details.

(*i*) Operating characteristics of hydraulic turbines.

(*ii*) Comparison of impluse and reaction turbines.

[Pb. Univ. Power Station Practice December 1995]

42. Explain the governing mechanism of Pelton turbine.

[Pb. Univ. Power Station Practice May 1995]

43. What factors should be considered for selection of hydraulic turbines for hydro plants?

[Pb. Univ. Power Station Practice May 1995]

44. Clearly bring out the difference between above ground and underground hydroelectric power plants.

45. What is a small hydro plant. Give its merits.

46. With the help of a schematic diagram, explain the working of a "pumped storage plant". Discuss the advantages and disadvantages of the plant.

[U.P.S.C. I.E.S. Electrical Engineering-II, 1997]

47. Draw a pumped storage hydroelectric power plant. Discuss the functions of each component of the plant.

[M.D. Univ. Power Plant Engineering May 2007]

48. Explain with a neat sketch a pumped storage plant.

[Rajasthan Technical Univ. Generation of Electrical Power 2008, 2010]

49. What are pumped storage plants ? Describe with neat sketches the principle of operation of such a plant. Also, discuss the role of the plant in a large interconnected power system. **[U.P.S.C. I.E.S. Electrical Engineering-II, 2008]**

50. Sketch and describe layout of hydro power plants. Also mention three hydro power plants installed in the country with installed capacity.

[M.D. Univ. Power Plant Engineering Dec. 2009]

51. Discuss features of pumped storage plant.

[M.D. Univ. Power Plant Engineering Dec. 2009]

52. Discuss in brief about hydroelectric power potential in India.

[U.P. Technical Univ. Power Station Practice 2007-08]

53. Discuss the special features of alternators used in hydro plants. **[Pb. Univ. Electrical Power-II (Generation) Jan. 1993; G.B. Technical Univ. Power Station Practice 2010-11]**

54. How the number of units are decided in a hydroelectric plant ? **[Agra Univ. Generation & Switchgear Practice 1981]**

55. Explain the environmental impacts of hydro power plants. **[Pb. Technical Univ. Generation of Electric Power May 2008]**

56. Write short notes on the following.

(*i*) Hydrological cycle. **[Pb. Univ. June 1991]**
(*ii*) Site selection for hydro power station.
(*iii*) Hydrograph and flow duration curve. **[Pb. Univ. April 1993]**
(*iv*) Hydraulic dams. **[Pb. Univ. May 1994]**
(*v*) Storage and pondage. **[Pb. Univ. May 1994]**
(*vi*) Pumped storage hydel plant.
(*vii*) Selection of water turbine.
(*viii*) Hydro potential in India. **[U.P. Technical Univ. 2006-07]**
(*ix*) Automatic and remcte control of hydro plants.
(*x*) Hydro plant auxiliaries.

Short Answer Type Questions With Answers

Q. 1. What is power generating station ? **[Pb. Technical Univ. Dec. 2010]**

Ans. Power generation station generates electrical energy by cor.version of energy available in different forms from natural sources such as kinetic energy of blowing winds, pressure head of water, chemical energy of fuels (either in solid, liquid or gaseous form) and nuclear energy of radioactive substances.

Q. 2. What is a hydrological cycle ?

Ans. Hydrological cycle is a process by which moisture from the surface of water bodies covering the earth's surface transferred to the land and back to the water bodies again. This cycle is shown in Fig. 2.1. The input to this cycle is the solar energy. Due to this water from the water bodies evaporates and on cooling these water vapours form clouds. Further cooling makes the clouds to fall down in the form of rain, snow, hail or sleet etc. known as ***precipitation.*** Precipitation includes all water that falls from the atmosphere to the earth's surface in any form.

Q. 3. What is the meaning of basic rule curve ? **[Pb. Technical Univ. Dec. 2009]**

Ans. Basic rule curve represents accumulation of water in reverse order of time of deficiency between firm load and energy available from stream flow in critically dry periods.

Q. 4. On what factors does the power output of a hydro power plant depend ?

Ans. The power developed by a hydro plant depends upon the discharge and head and is directly proportional to their (discharge and head) product.

Q. 5. What is the estimated hydro potential in India ?

Ans. About 84,000 MW.

Q. 6. Explain the essential factors which influence the choice of site for a hydroelectric plant. **[Pb. Technical Univ. Dec. 2010]**

or

What are the factors of selection of site for hydro power plant ? **[Pb. Technical Univ. Dec. 2009]**

Ans. The factors governing the site selection for a hydroelectric plant are : (*i*) Availability of water (*ii*) Water storage (*iii*) Water head (*iv*) Distance from load centre (*v*) Accessibility of the site (*vi*) Water pollution, (*vii*) Sedimentation (*viii*) Large catchment area (*ix*) Availability of land (*x*) Possibility of stream diversion.

Q. 7. What is meant by catchment area ?

Ans. The catchment area is the area, bounded by watersheds, which drains into a stream or river across which the dam has been built at a suitable place.

Q. 8. What is hydrograph ?

Ans. A hydrograph is a graphical representation between discharge or flow with time.

Q. 9. What is a mass curve ?

Ans. Mass curve is a curve which indicates the total volume of run-off in m^3 up to a certain time.

Q. 10. Between which quantities is mass curve plotted ? **[Pb. Technical Univ. Dec. 2008; May 2009]**

Ans. The abscissae for the mass curve may be a day, month or year and ordinate may be plotted in terms of second-metre-day or second-meter-month which means the flow collected at the rate of 1 cubic metre per second for one day or one month respectively. It indicates the total volume of run-off in m^3 up to a certain time.

Q. 11. What is the use of mass curve ?

Ans. Mass curve is used in determination of the capacity of the storage reservoir in hydro projects.

Q. 12. What for pondage is provided ?

Ans. In case power plant is away from the storage reservoir, a small pond is provided near the power plant in order to meet the hourly changes in power demand.

Q. 13. What for dam is provided ?

Ans. The functions of a dam in hydro power plants are to raise the water surface of the stream to create an artificial head and also to provide the pondage, storage or the facility of diversion into conduits.

Q. 14. What civil engineering works are necessary for a hydro-electric project ?

Ans. The civil engineering works for an hydroelectric project include reservoir dam, forebay, spillway, intake structures, surge tank, penstock, tailrace etc.

Q. 15. What is forebay and what for it is provided ?

Ans. The forebay serves as a regulating reservoir storing water temporarily during light load period and providing the same for initial increase on account of increasing load during which water in the canal is being accelerated. In short a forebay may be considered as an enlarged body of water just above the intake to store water temporarily to meet the hourly load fluctuations. This may either be a pond behind the diversion dam or an enlarged section of a canal spread out to accommodate the required widths of intake.

Q. 16. What is meant by the phenomenon "water hammer" ?

Ans. A reduction in load on the generator causes the governor to close the turbine gates and thus create an increased pressure in the penstock. This may result in *water hammer phenomenon* and may need pipe of extraordinary strength to withstand it otherwise the penstock may burst.

Q. 17. What are the functions of surge tank in hydro plants ? **[Pb. Technical Univ. May 2010]**

Ans. Surge tank is provided to absorb sudden changes in water requirements and reduce water hammer and negative pressure in penstock.

Q. 18. How a surge tank helps in reducing water hammer effect?

Ans. For reducing water hammer effect, some means are required to be provided for taking the rejected flow. This may be accomplished by providing a small storage reservoir or tank (open at the top) for receiving the rejected flow and thus relieving the conduit pipe of excessive water hammer pressure. This storage reservoir, called the surge tank is usually located as close to the power plant as possible, preferably on ground to reduce the height of the tower.

Q. 19. What is draft tube and why is its installation necessary ?

Ans. An airtight pipe of suitable diameter attached to the runner outlet and conducting water down from the wheel and discharging it under the surface of the water in the tailrace is known as *draft tube*.

If there is no draft tube and the water discharges freely from the turbine exit, then the turbine operates under a head equal to the height of the headrace water level above the runner exit. By installing draft tube, the operating head is increased by an amount equal to the height of the runner outlet above the tailrace. This creates a negative pressure head at the runner exit. This makes it possible to install the turbine above the tailrace without loss of head.

Q. 20. How are hydroelectric power plants classified ?

Ans. The hydroelectric power plants may be classified according to

(*i*) the extent of water flow regulation available

(*ii*) the availability of water head and

(*iii*) the type of load they supply.

Apart from above classification, hydroelectric power plants are also classified on the basis of installed capacity such as large, medium, small, mini and micro hydro power plants.

Q. 21. Explain about run-off river plants without pondage ?

[Pb. Technical Univ. December 2009]

Ans. Refer to Art. 2.7.1(1).

Q. 22. What are the different types of turbines ?

Ans. Turbines employed in hydroelectric power plants are Pelton wheel, Francis turbine and Kaplan turbine.

Q. 23. How are hydro turbines classified ?

Ans. There are many ways of classifying water or hydraulic turbines, for example, according to the type of flow of water, according to action of water on moving blades, according to head and quantity of water available, after the name of originator and according to the specific speed of the machine.

Q. 24. What is specific speed of a turbine ?

Ans. The specific speed n_s of a turbine characterizes the turbine's shape in a way that is not related to its size. This allows a new turbine design to be scaled from an existing design of known performance. The specific speed is also the main criteria for matching a specific hydro site with the correct turbine type. The specific speed is the speed with which the turbine turns for a particular discharge Q, with unit head and thereby is able to produce unit power.

Q. 25. What are the advantages of reaction turbine over the Pelton wheel with respect to efficiency, size, cost and maintenance ?

[U.P. Technical Univ. 2007-08]

Ans. As compared to Pelton wheel a Francis turbine offers the advantage of high efficiency at full load and at 75% of full load. Since this turbine can be designed for a higher speed than Pelton wheel, the dimensions and weights are smaller together with the reduced generator cost.

Q. 26. For what deflectors are provided in Pelton wheels ?

Ans. Deflectors are provided in Pelton wheels to control the speed.

Q. 27. How are reaction turbines distinguished from the impulse type ?

Ans. The reaction type turbines are distinguished from the impulse type by having a runner that always functions within a completely water-filled casing.

Q. 28. Compare Kaplan and Francis turbines.

Ans.

1. Kaplan turbine is more efficient, more compact in cross section and has higher rotational speed to that of an identical Francis turbine.
2. Kaplan turbine has less frictional loss, higher (2 to 3 times more), specific speed, and lesser number of runner blades.
3. In Kaplan turbine runner the water strikes the turbine blades axially while in Francis turbine runner receives the water radially.
4. Francis turbines are used in medium head plants while Kaplan turbines are used in low head plants.

Q. 29. Which turbine is ideally suited to pumped storage plants ?

Ans. Kaplan turbine is ideally suited to pumped storage plants because its runner is capable of reverse operation as a pump.

Q. 30. Why is governing of turbines essential ?

Ans. In order to have electrical output of constant frequency it is necessary to maintain speed of the alternator driven by the turbine constant. This is achieved by controlling the flow of water entering the turbine by the automatic adjustment of guide vanes in case of reaction turbines and of the nozzle needle in the case of impulse turbines. Such an operation of speed regulation is called the governing, and it is attained automatically by means of a governor. In case of impulse turbine the governor also operates the auxiliary relief valves or jet deflectors.

Q. 31. On what factors does the selection of a water turbine depend ?

Ans. The selection of a water turbine depends upon various factors such as working head, available discharge, speed, output and nature of load.

Q. 32. Why are small hydroelectric power plants becoming more and more popular ?

Ans. Small hydroelectric power plants are becoming more and more popular because they offer several advantages such as simple operation, more reliable, minimal maintenance, quick construction, simple civil engineering and small financial needs.

Q. 33. What is hydro potential in India ?

Ans. India is blessed with immense amount of hydroelectric potential and ranks 5th in terms of exploitable hydro potential on global scenario. According to the accessment made by CEA, India is endowed with economically exploitable hydro power potential to the tune of 148,700 MW of installed capacity.

In addition, 56 number of pumped storage projects have also been identified with probable installed capacity of 94,000 MW. In addition to this, hydro potential from small, mini and micro schemes has been estimated as 6,782 MW from 1512 sites. Thus, in totality India is endowned with hydro potential of about 250,000 MW.

Q. 34. What is the estimated potential of small hydro power schemes in India ?

Ans. About 15,000 MW.

Q. 35. Can an old steam power plant be an alternative to a pumped storage plant for peak load operations ?

Ans. No, because it is a costly proposition and moreover it (steam power plant) needs large starting time and shutdown time.

Problems

1. A hydro station has to operate with a mean head of 30 metres and is supplied from a reservoir lake at the rate of 6.93 m^3 per second. Calculate the power generated in kW. Assume density of water 1,000 kg/m^3, load factor of the station 80%, mechanical efficiency of water turbine 90%. Neglect head loss in pipes, penstocks etc. [**Ans.** 2.0 MW]

2. It has been estimated that a minimum run-off of approximately 94 m^3/s will be available at hydroelectric project with a head of 39 metres. Assuming the overall efficiency to be 65%, estimate the total rating of the generators and suggest the number and types of turbines to be installed. **[Allahabad Univ. Power System-II, 1975]**
[**Ans.** 24,000 kW; 4 units of 6,000 kW each; Propeller type turbine]

3. A proposed hydroelectric station has a available head of 25 metres, a catchment area of 5×10^7 m^2, the rainfall for which is 120 cm per annum. If 75% of the total rainfall can be collected, calculate the power that could be generated. Assume overall efficiency as 65%.
[**Ans.** 2×10^6 kWh]

4. The annual power duration curve for an hydroelectric station shows that 438×10^4 kWh of energy is available during the year. The station is to be used as peak load plant for an annual load factor of 20%. What is the capacity of the station ? If the plant capacity factor is 15%, what is the reserve capacity of the plant ?
[Agra Univ. Generation & Switchgear Practice 1978]
[**Ans.** 3,333 kW, 833 kW]

5. A hydroelectric station operates under a mean head of 50 m. The reservoir employed has a catchment area of 500 km^2. The average rainfall in this area is 150 cm per annum. Determine the capacity of the station for which it should be designed. Assume that 25 per cent of the rainfall is lost due to evaporation etc., 7.5 per cent of head is lost in penstock, turbine efficiency as 85%, alternator efficiency as 95% and load factor as 60%. **[A.M.I.E. Sec. B. Power System Winter 1993]**
[**Ans.** 10.89 MW]

6. The mean monthly discharge for 12 months at a particular site of a river is given below:

Month	Discharge (in millions of m^3 per month)
January	1,000
February	800
March	600
April	500
May	200
June	1,500
July	2,500
August	3,000
September	2,400
October	2,000
November	1,500
December	1,500

Draw hydrograph for the given discharge and find the average monthly flow. Calculate the power available at mean flow of water if the available head is 90 m and overall efficiency of generation is 80%.
[Pb. Univ. Power Station Practice December 1993]
[**Ans.** $1{,}458.33 \times 10^6$ m^3/month; 397.4 MW]

7. A power development site has the average estimated flows, as shown in table below. Find the reservoir capacity required for producing 18,000 hp continuously throughout the year at an overall efficiency of 80% with head of 100 m.

Month	Flow (in $m^3 \times 10^6$)
January	66
February	34
March	74
April	51
May	17
June	14
July	31
August	83
September	89
October	43
November	54
December	86

[Pb. Univ. Power Station Practice April 1993]
[**Ans.** 69.22×10^6 m^3]

[**Hint.** Average flow required is 43.74×10^6 m^3. The deficient flows during the months of May, June and July would cause $[43.74 \times 10^6 \times 3 - (17 + 14 + 31)]$ *i.e.,* 69.22×10^6 m^3 to be drawn from the reservoir to supply the constant demand of 43.74×10^6 m^3 per month]

8. The run-off data at a particular site for 12 weeks is tabulated below. Draw the mass curve and find the size of reservoir and possible rate of flow which would be available after the reservoir has been built.

Week	Weekly flow in m^3/s
1	6,000
2	5,000
3	4,000
4	2,000
5	1,000
6	500
7	1,000
8	4,000
9	8,000
10	5,000
11	2,000
12	1,000

[M.D. Univ. Power Plant Engineering May 2008]
[**Ans.** 25,000 day-second-meters; 1,800 m^3/sec]

9. At a particular site the mean monthly discharge is as follows :

Month	Discharge m^3/s
January	100
February	225
March	300
April	600
May	750
June	800
July	1,000
August	1,200
September	900
October	600
November	400
December	200

[M.D. Univ. Power Plant Engineering Dec. 2008]

10. A medium capacity storage type hydroelectric power plant covers 1,200 km^2 area. The annual rainfall in catchment area is 160 cm. The head available at the power site is 360 m. Assuming 25% of the rainfall is lost in evaporation and percolation, find the average power developed by the power plant and maximum demand. Take overall efficiency of the plant as 75% and load factor 0.5.
[A.M.I.E. Sec B. Power Plant Engineering Winter 1995]
[**Ans.** 120.95 MW; 241.9 MW]

❑❑❑

CHAPTER 3

Steam Power Plants

INSIDE THIS CHAPTER

3.1. INTRODUCTION

The fact that thermal energy is the major source of power generation itself shows the importance of thermal power generation in India—more than 50 per cent of electric power is produced by steam plants in India. This position is likely to continue due to large pit head plants being set up. Larger sizes of units due to overall increase in demand for power and because of necessity of keeping down the cost of power generation with increasing fuel prices are the developing trends in large steam power plants.

As on July 31, 2010, and as per the Central Electricity Authority the total installed capacity of Coal and Lignite based power plants in India is 87,093.38 MW. List of some of presently operating steam power plants of capacities 1,000 MW and above) is given below.

List of Presently Operating Steam Power Plants of Capacity 1000 MW and Above

Name	Operator	Location	State	Sector	Region	Units	Capacity (MW)
Mundra Thermal Power Station	Adani Power	Mundra (Kutch)	Gujarat	Private	Western	4 × 330, 5 × 660	4,620
Vindhyachal Super Thermal Power Station	NTPC	Vindhya Nagar (Sidhi)	Madhya Pradesh	Central	Western	6 × 210, 4 × 500	3,260
Talcher Super Thermal Power Station	NTPC	Kaniha (Angul)	Orissa	Central	Eastern	6 × 500	3,000
Amravati Thermal Power Plant	INDIA-BULLS	Nandgaonpeth (Amravati)	Maharasthra	State	Western	10 × 270	2,700
NTPC Ramagundam	NTPC	Jyothi Nagar (Karimnagar)	Andhra Pradesh	Central	Southern	3 × 200, 4 × 500	2,600
Jharsuguda TPP	Vedanta	Jharsuguda	Orissa	Private IPP	Eastern	4 × 600	2,400
Kahalgaon Super Thermal Power Station	NTPC	Kahalgaon (Bhagalpur)	Bihar	Central	Eastern	4 × 210, 3 × 500	2,340
Chandrapur Super Thermal Power Station	MAHA-GENCO	Urjanagar (Chandrapur)	Maharasthra	State	Western	4 × 210, 3 × 500	2,340
Mejia Thermal Power Station	DVC	Durlavpur (Bankura)	West Bengal	Central	Eastern	4 × 210, 2 × 250, 2 × 500	2,340
Korba Super Thermal Power Plant	NTPC	Jamani Palli (Korba)	Chattisgarh	Central	Western	3 × 200, 3 × 500	2,100

Name	Operator	Location	State	Sector	Region	Units	Capacity (MW)
Singrauli Super Thermal Power Station	NTPC	Shaktinagar (Sonebhadra)	Uttar Pradesh	Central	Northern	5 × 200, 2 × 500	2,000.00
Rihand Thermal Power Station	NTPC	Rihand Nagar (Sonebhadra)	Uttar Pradesh	Central	Northern	4 × 500	2,000.00
NTPC Dadri	NTPC	Vidyut Nagar (Gautam Budh Nagar)	Uttar Pradesh	Central	Northern	4 × 210, 2 × 490	1,820.00
Dr Narla Tatarao TPS	APGENCO	Ibrahimpatnam (Krishna)	Andhra Pradesh	State	Southern	6 × 210, 1 × 500	1,760
Raichur Thermal Power Station	KPCL	Raichur	Karnataka	State	Southern	7 × 210, 1 × 250	1,720
Anpara Thermal Power Station	UPRVUNL	Anpara (Sonebhadra)	Uttar Pradesh	State	Northern	3 × 210, 2 × 500	1,630.00
Farakka Super Thermal Power Station	NTPC	Nagarun (Murshidabad)	West Bengal	Central	Eastern	3 × 200, 2 × 500	1,600
Suratgarh Super Thermal Power Plant	RVUNL	Suratgarh (Sri Ganganagar)	Rajasthan	State	Northern	6 × 250	1,500.00
Wanakbori Thermal Power Station	GSECL	Wanakbori (Kheda)	Gujarat	State	Western	7 × 210	1,470
Neyveli Thermal Power Station - II	NLC	Neyveli (Cuddalore)	Tamilnadu	Central	Southern	7 × 210	1,470
Trombay Thermal Power Station	Tata	Trombay (Mumbai)	Maharasthra	Private	Western	1 × 150, 2 × 500, 1 × 250	1,400
Sanjay Gandhi Thermal Power Station	MPPGCL	Birsinghpur (Umaria)	Madhya Pradesh	State	Western	4 × 210, 1 × 500	1,340
Obra Thermal Power Station	UPRVUNL	Obra (Sonebhadra)	Uttar Pradesh	State	Northern	1 × 40, 3 × 94, 5 × 200	1,322.00
Kolaghat Thermal Power Station	WBPDCL	Mecheda (East Midnapore)	West Bengal	State	Eastern	6 × 210	1,260
Guru Gobind Singh Super Thermal Power Plant	PSPCL	Ghanauli (Rupnagar)	Punjab	State	Northern	6 × 210	1,260.00
Chandrapura Thermal Power Station	DVC	Chandrapura (Bokaro)	Jharkhand	Central	Eastern	3 × 130, 3 × 120, 2 × 250	1,250
Kota Super Thermal Power Plant	RVUNL	Kota	Rajasthan	State	Northern	2 × 110, 3 × 210, 2 × 195	1,240.00
Vedanta Aluminim CPP	Vedanta	Jharsuguda	Orissa	Private CPP	Eastern	9 × 135	1,215
Parli Thermal Power Station	MAHA-GENCO	Parli-Vaijnath (Beed)	Maharasthra	State	Western	3 × 210, 2 × 250	1,130
Bakreshwar Thermal Power Station	WBPDCL	Suri (Birbhum)	West Bengal	State	Eastern	5 × 210	1,050
Feroj Gandhi Unchahar Thermal Power Plant	NTPC	Unchahar (Raebareli)	Uttar Pradesh	Central	Northern	5 × 210	1,050.00
Tuticorin Thermal Power Station	TNEB	Tuticorin (Tuticorin)	Tamilnadu	State	Southern	5 × 210	1,050
Neyveli Thermal Power Station - 1	NLC	Neyveli (Cuddalore)	Tamilnadu	Central	Southern	6 × 50, 3 × 100, 2 × 210	1,020
Satpura Thermal Power Station	MPPGCL	Sarni (Betul)	Madhya Pradesh	State	Western	5 × 37.5, 1 × 200, 3 × 210	1,017.5
Simhadri Super Thermal Power Plant	NTPC	Simhadri (Visakhapatnam)	Andhra Pradesh	Central	Southern	2 × 500	1,000
Jindal Megha Power Plant	Jindal	Tamnar (Raigarh)	Chattisgarh	Private	Western	4 × 250	1,000
Sipat Thermal Power Plant	NTPC	Sipat (Bilaspur)	Chattisgarh	Central	Western	2 × 500	1,000

In steam power plants, the heat of combustion of fossil fuels (coal, oil or gas) is utilised by the boilers to raise steam at high pressure and temperature. The steam so produced is used in driving the steam turbines or sometimes steam engines coupled to generators and thus in generating electrical energy, as illustrated in Figs. 3.1 (*a*) and 3.1 (*b*). Steam turbines or steam engines used in steam power plants not only act as prime movers but also as drives for auxiliary equipment, such as pumps, stokers fans etc.

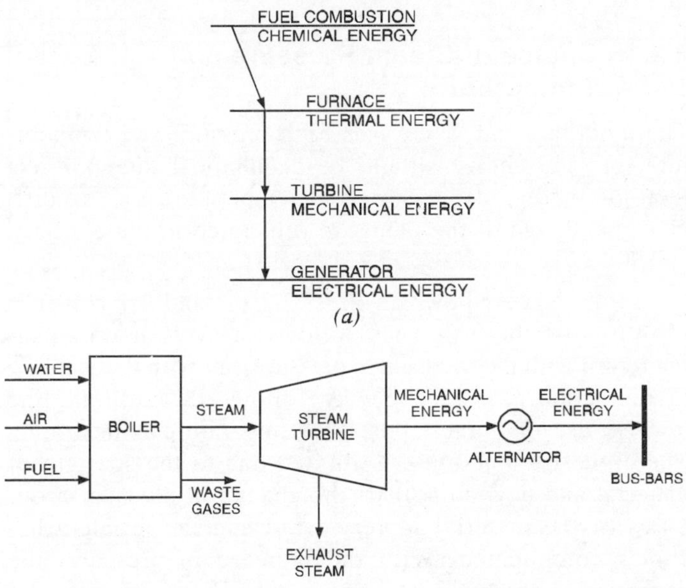

Fig. 3.1

Steam power plant may be installed either to generate electrical energy only or generate electrical energy along with the generation of steam for industrial purposes such as in paper mills, textile mills, sugar mills and refineries, chemical works, breweries, plastic manufacture, food manufacture etc. The steam for process purposes is extracted from a certain section of turbine and the remaining steam is allowed to expand in the turbine. Alternatively the exhaust steam may be used for process purposes.

Steam power plants may be either condensing or non-condensing type. In condensing type power plants the exhaust steam is discharged into a condenser, which creates a suction at very low pressure and allows the expansion of steam in the turbine to a very low pressure and thus increases the efficiency. In non-condensing type power plants, the steam exhausted from the turbine is discharged at atmospheric pressure or at a pressure greater than atmospheric. The principal advantages of condensing operation are the increased amount of energy extracted per kg of steam and the greater amount of power developed by a given size of turbine or engine. Moreover, in non-condensing type power plants, a continuous supply of fresh feed water is required whereas in condensing type power plants steam condensed into the water in the condenser, can be recirculated to the boilers with the help of pumps. This point becomes very important at places where there is a shortage of pure water.

According to use such plants can be classified into (*i*) industrial power plants or captive power plants and (*ii*) central power plants.

1. *Industrial Power Plants.* The industries requiring steam for process purposes may use steam turbines for generation of electrical energy also for its own use. The steam for process purposes may be tapped from extraction bleeding of the turbines at a pressure of 1.4-7 kg/cm^2. Such plants are usually of non-condensing type. Industrial turbo-generator plants are usually of small capacity (say up to 10 MW).

Industrial turbo-generator plants, also known as *captive power plants*, otherwise also becoming more and more popular nowadays. The reasons for it are poor reliability of electric supply from the grid, frequent upward revision of power tariff charged by electric utilities and long power cuts. In the recent past many state governments have revised their policies as regards captive power generation and are encouraging setting up of captive power plants.

2. *Central Power Plants.* Steam power plants are mainly used as central plants to generate electrical energy for supply to various consumers (industrial, agricultural, commercial, domestic etc.). Such plants are usually of condensing type. The size of the generating sets may be from 10-1,000 MW or even higher. When such power plants are used in a system having both hydro and steam power plants they can be used to supply load economically and to ensure reliability of power supply at all times.

Steam power plant for generation of electrical power is preferred, where large amount of power is required to be generated and financial, climatic, and geographical conditions do not permit the installation of hydroelectric power plants and coal is available in plenty.

3.2. EFFICIENCY OF STEAM POWER PLANTS

The thermal efficiency of steam power plants, defined as the ratio of the heat equivalent of mechanical energy transmitted to the turbine shaft and the heat of combustion is quite low (about 30%). Overall efficiency of the power plant, defined as the ratio of heat equivalent of electrical output to the heat of combustion, is about 29 per cent. The overall efficiency is determined by multiplying the thermal efficiency of power plant by the efficiency of generation (or electrical efficiency). In case of most modern supercritical pressure steam plants employing many heat saving devices, the plant overall efficiency may reach the value of 50 per cent.

Losses occurring in steam power plants may be summarized as follows:

(*a*) Boiler house losses

(*i*) To dry flue gases	5%
(*ii*) To moisture in gases	5%
(*iii*) To ash and unburnt carbon	1.0%
(*iv*) To radiation and leakage	2.5%
(*v*) Unknown losses	2.5%
Total	16.0%

(*b*) Turbine losses

Heat rejected to condenser	54%
Alternator losses	1%
Thus output is about	29%

From the above mentioned figures of various losses occurring in steam power plants it is obvious that more than 50 per cent of total heat of combustion is lost as heat rejected to the condenser. This loss of heat energy is unavoidable as heat energy cannot be converted into mechanical energy without a drop in temperature, and the steam in condenser is at the lowest temperature. The thermal efficiency of the plant mainly depends upon three factors (*i*) pressure and (*ii*) temperature of the steam entering the turbine and the pressure in the condenser. The thermal efficiency increases with the increase in temperature and pressure of the steam entering the turbine. For this reason high pressures and temperatures are used. The thermal efficiency is effectively increased by decreasing the pressure in the condenser. Pressure in the condenser is kept very low usually 0.04 kg/cm^2.

The thermal efficiency also increases by reheating the steam between turbine stages, but is somewhat inconvenient. Bleeding of steam also affects the thermal efficiency.

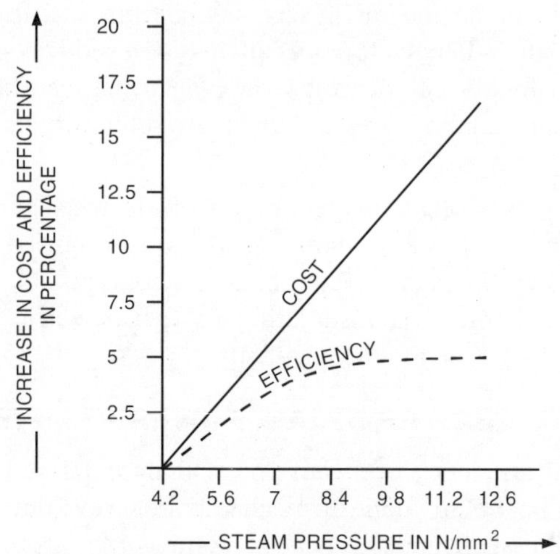

(a) Effect of Steam Pressure on Cost and Efficiency

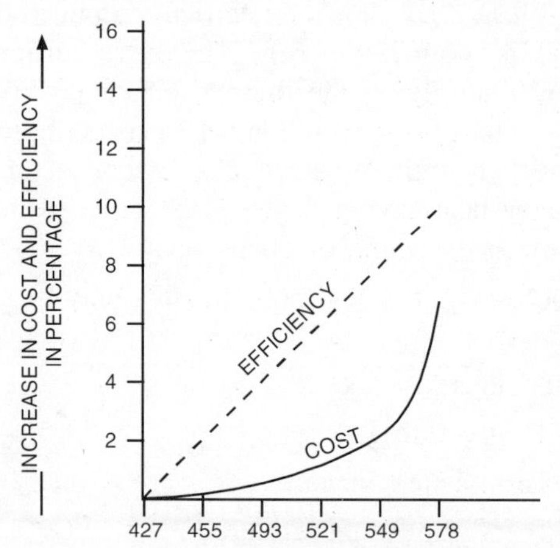

(b) Effect of Steam Temperature on Cost and Efficiency

Fig. 3.2

The overall thermal efficiency of 1st steam plant (50 MW units at Bokaro) commissioned in year 1952-53 in India was 28% while for the largest and most efficient thermal power plant having unit size of 160 MW at Trombay commissioned in 1965 it was 37% with heat rate 2,330 kcal/kWh of electrical energy produced. Each of the three major elements of a thermal power plant—the boiler, the turbine and the alternator have undergone intensive development and as a result the efficiency of electric generation has gone up to nearly 42% in 1980's.

3.2.1. Choice of Steam Pressure and Temperature

The modern trend is towards high pressure and temperature, but the choice should be economical one. The effects of increased pressure and temperature on the efficiency and cost of the plant are illustrated in Figs. 3.2 (*a*) and 3.2 (*b*).

From curves shown in Figs. 3.2 (*a*) and 3.2 (*b*), it is obvious that the efficiency follows the law of diminishing return with the increase in pressure, but with the increase in temperature, the efficiency follows the straight line law. So use of highest possible temperature is desirable. The highest temperature is limited due to the strength of material and beyond 480°C, the change in physical properties of the material is very rapid and the problem becomes complicated. With the increase in pressure, the degree of superheat is to be reduced so as to keep within limits the total temperature. The present practice is to use steam pressures of about 6.5 N/mm^2 for entirely new plant.

3.2.2. Supercritical Technology

At a temperature of about 600ºC and pressure of 30 N/mm^2, water enters a supercritical phase and has properties between those of liquid and gas. Water in supercritical stage can dissolve a number of organic compounds and gases and on addition of hydrogen peroxide and liquid oxygen combustion process starts. The steam power plants operating on this principle are called *supercritical plants*.

The advantages of such plants are that low grade fossil fuels (*e.g.,* lignite) can be used, NO_2 emissions are completely eliminated and SO_2 emissions are reduced and complete burning of coal occurs. So the plant has no need of desulphurisation and denitrification equipment and soot collector. With this system the cost of processing flue gas emissions (electrostatic precipitator etc.) is eliminated and cooling water requirements are also reduced, so the system becomes more economical and efficient. Supercritical power plants, these days have an overall efficiency of just over 40%. With the use of temperatures around 700ºC (known as ultra supercritical condition), the overall efficiency of the system may be improved to around 50%.

3.3. MERITS AND DEMERITS OF STEAM POWER PLANTS

The steam power plants have the following merits and demerits over other plants:

Merits. (*i*) Fuel used is cheaper.

(*ii*) Less space is required in comparison with that for hydroelectric plants.

(*iii*) Cheaper in initial cost in comparison with other types of power plants of same capacity.

(*iv*) Cheaper in production cost in comparison with that of diesel power plants.

(*v*) Such plants can be installed at any place irrespective of the existence of fuels, while hydroelectric plants can be developed only at the source of water power.

(*vi*) Such plants can be located near the load centres, while the hydroelectric plants have essentially to be installed at source of water power which is usually isolated from urban areas, transmission costs are, therefore, reduced.

(*vii*) Able to respond to rapidly changing loads without difficulty.

(*viii*) Steam engines and turbines can work under 25% of overload continuously.

(*ix*) A portion of steam raised can be used as process steam in various industries (paper mills, textile mills, sugar mills, refineries, chemical works etc.)

Demerits. (*i*) High maintenance and operating costs.

(*ii*) Pollution of atmosphere due to fumes and residues from pulverised fuels.

(*iii*) Requirement of water in huge quantity.

(*iv*) Handling of coal and disposal of ash is quite difficult.

(*v*) Troubles from smoke and heat from the plant.

(*vi*) With the increase in the operating temperature and pressure, the plant cost increases.

(*vii*) Requires long time for erection and put into action.

(*viii*) Efficiency falls rapidly below 75% of full load.

(*ix*) Costlier in operating cost in comparison with that of hydro and nuclear power plants.

3.4. SELECTION OF SITE FOR STEAM POWER PLANTS

The factors to be considered while selecting a site for a steam power plant for economical and efficient generation are:

(*i*) *Nearness To The Load Centre.* The power plant should be as near as possible to the centre of the load so that the transmission cost and losses are minimum. This factor is most important when dc supply system is adopted. However, in case of ac supply system when transformation of energy from lower voltage to higher voltage and vice versa is possible, power plant can be erected at places other than that of centre of load provided other conditions are favourable.

(*ii*) *Supply of Water.* Large quantity of water is required in a steam power plant. It is required (*i*) to raise the steam in boilers (*ii*) for cooling purposes such as in condensers (*iii*) as a carrying medium such as disposal of ash and (*iv*) for drinking purposes.

In steam power plants approximately 1.26×10^6 kcals of heat per MW per hour has to be disposed off in the condenser. In case of direct circulation from the source of water 120 m^3 of water is required per MW per hour for this purpose. In case cooling towers are used about 2.4 m^3 of water per MW per hour will be required in addition as make-up water. With pond cooling slightly large quantity of water is required to meet evaporation and drift losses etc. but cooling effect is better. The efficiency of direct cooled plant is about 0.5% higher than that of the plant cooled by cooling towers. This means a saving of about Rs 7.5 lacs per year in fuel cost for a 2,000 MW station. Hence it is clear that the plant should be located near a river, sea or a lake.

(*iii*) *Availability of Coal.* Huge amounts of coal is required for raising the steam (20,000 tonnes per day for a 2,000 MW station). Since the Government policy is to use only low grade coal with 30 to 40% ash content for power generation purposes, the steam power plant should be located near the coal mines to avoid the transport of coal and ash.

(*iv*) *Land Requirement.* The land is required not only for setting up the plant but for other purposes also such as staff colony, coal storage, ash disposal etc. For a 2,000 MW plant, the land requirement may be of the order of 200-250 acres. As the cost of the land adds up to the final cost of the plant, it should be available at a reasonable price. Land should be also available for future extension.

(*v*) *Type of Land.* Land should be of good bearing capacity since it is to withstand not only the dead load of the plant but also the forces transmitted to the foundations due to the operation of the plant and this total load may amount to about 7 kg per sq. cm. Moreover the land should be reasonably level and not low lying.

(*vi*) *Transportation Facilities.* The facilities must be available for transportation of heavy equipment and fuels *e.g.,* near railway station.

(*vii*) *Labour Supplies.* Skilled and unskilled labourers should be available at reasonable rates near the site of the plant.

(*viii*) *Ash Disposal.* Ash is the main waste product of the steam power plant and with low grade coal, it may be 3.5 tonnes per day, hence some suitable means for disposal of ash should be thought of. It may be purchased by building contractors, or it can be used for brick making near the plant site. If the site is near the coal mine, it can be dumped into the disused mines. In case of site located near a river, sea or lake, ash can be dumped into it. If the waste land is available near the site, then an area of 40 hectares excavated to a depth of 6.5 metres will be required per year for a 2,000 MW plant.

(*ix*) *Distance From Populated Area.* Since most of the modern generating stations employ pulverised fuel-residues and fumes from them are quite unhealthy, therefore the site for the power station should be away from the populated area.

All the factors given above but first one make us to select the site for the thermal power station away from the towns. Nowadays more importance is given to the facilities for generation than those of distribution and site for steam power station near by river side, where ample water is available, no pollution of atmosphere occurs and fuel can be transported easily, is an ideal choice.

Studies have revealed that it is more economical to locate the steam power plant at or near pit head rather than at load centre. For loads exceeding 200 MW at 220 kV and 375 MW at 400 kV, transmission of electrical energy is more economical in comparison to transport of coal for distances of the order of 100 to 700 km and even more.

3.5. WORKING OF STEAM POWER PLANT

Steam power plant basically operates on the Rankine cycle. Coal is burnt in a boiler, which converts water into steam. The steam is expanded in a turbine, which produces mechanical power driving the alternator coupled to the turbine. The steam after expansion in prime mover (turbine) is usually condensed in a condenser to be fed into the boiler again. In practice, however, a large number of modifications and improvements have been made so as to affect economy and improve the thermal efficiency of the plant.

The working of a modern coal-fired steam power plant can be studied conveniently with the help of various cycles. Schematic arrangement of a modern coal-fired steam power plant is shown in Fig. 3.3. The entire arrangement for the sake of simplicity may be divided into four main circuits namely (*i*) Fuel and ash circuit (*ii*) Air and fuel gas circuit (*iii*) Feed water and steam circuit and (*iv*) Cooling water circuit.

1. Fuel and Ash Circuit. Coal is delivered from the supply points to the storage site by road, rail or water. By road coal is transported in trucks and for small stations such inland transport may be enough. Where power plants are situated close to a waterway, such as a canal, river or sea transportation by boat or ship may be quite effective. However, in most cases transportation of coal by rail, road is the most common. In the case of small power plants the quantity of coal being small, manual unloading from rail car may be used but for large power stations the unloading from the railway siding is done with the help of wagon tipplers and then the coal is conveyed to the coal handling plant. From the coal handling plant, the coal having been good enough to be burnt into boilers is taken into the boiler bunkers by means of bucket conveyers. The coal is thus stored in the bunkers from where it falls into the hoppers by gravity and finally the requisite quantity of coal either goes on falling directly on the grate, or where the coal spreaders are provided, coal is spread in the grate up to the rear end. When use of spreaders is made, most of the coal burns in air and remaining falls at the rear end of the grate. Any unburnt coal particles in the

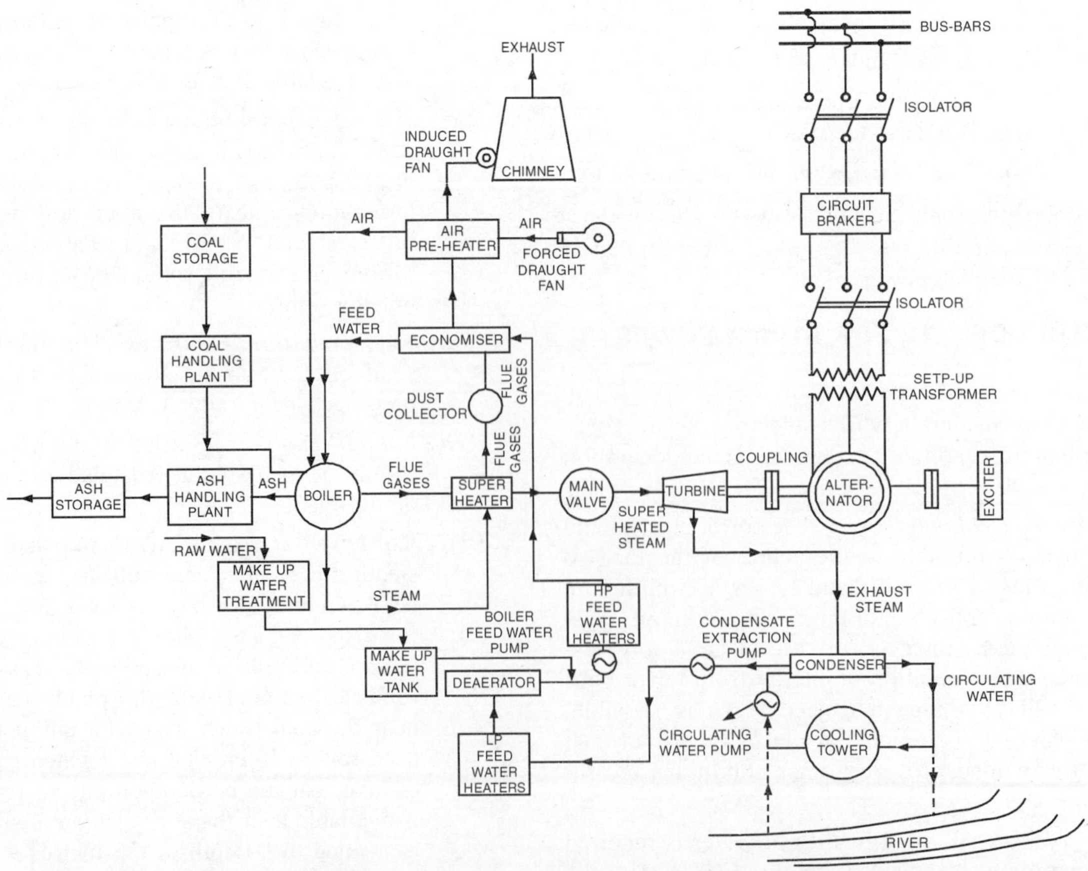

Fig. 3.3. *Schematic Arrangement of a Modern Coal Fired Power Plant*

middle of the grate are collected in a pipe and are again refired by cinder-refiring fan. The grate in such types of boilers, where use of spreaders is made move from rear end to front end, and without spreaders, the movement of the grate is from front end to rear end. Combustion is controlled by controlling the grate speed, quantity of coal entering the grate, the damper openings. The ash resulting after complete combustion of fuel collects at the back of the boiler and is removed to the ash storage by means of scrap conveyors.

2. Air and Fuel Gas Circuit. Air is drawn from the atmosphere by a forced draught fan or induced draught fan through the air preheater, in which it is heated by the heat of flue gases passing to chimney and then admitted to the furnaces. The flue gases after passing around boiler tubes and superheater tubes are drawn by the induced draught fan through dust collector (or precipitator), economiser and air preheater and finally exhausted to the atmosphere through chimney.

3. Feed Water and Steam Circuit. The steam coming out of the turbine is condensed and the condensate is extracted from the condenser by the condensate extraction pump and is forced to the low pressure feed water heaters (usually three in numbers) where its temperature is raised by the heat from bled steam (steam extracted from the lowest pressure extraction point of the turbine). The feed water is now pumped through deaerator to high pressure feed water heaters (usually 2 or 3 in numbers) where it gets heated by the heat from bled steam extracted at suitable point of the steam turbine. The function of *deaerator* is to reduce dissolved oxygen content in the condensate (*i.e.,* in the feed water). The feed water is then pumped into boiler through economiser in which it is further heated by the heat of flue gases passing through it on the way to chimney. A small part (about 1 per cent) of steam and water in passing through the different components of the system is lost. Therefore, water is added in the feed water system as make-up water, as shown in Fig. 3.3. In boiler water is converted into high pressure steam, which is wet. Wet steam is passed through superheater, where it is dried and further superheated, and then supplied to the steam turbine through the main valve. After giving out its heat energy to the turbine it is exhausted to the condenser where its latent heat is extracted and steam is converted into feed water. At one or more stages a quantity of steam is bled or withdrawn for heating of feed water. Making-up water for boiler is taken through the evaporator, where it is heated by low pressure steam extracted at suitable point of turbine.

4. Cooling Water Circuit. Cooling water is supplied from a natural source of supply such as river, canal, sea or lake or cooling towers through screens to remove the matter, that might choke the condenser tubes. It is circulated through the condenser for condensing the steam and finally discharged to the suitable position near the source of supply. During the passage its temperature rises and in the case of cooling towers the heat must be extracted before the water is again pumped to the condenser. The circulation of cooling water to the condenser help in maintaining a low pressure in the condenser.

3.6. CONSTITUENTS OF STEAM POWER PLANT AND LAYOUT

Coal-fired power plants are shown in Fig. 3.4 (*a*) and (*b*).

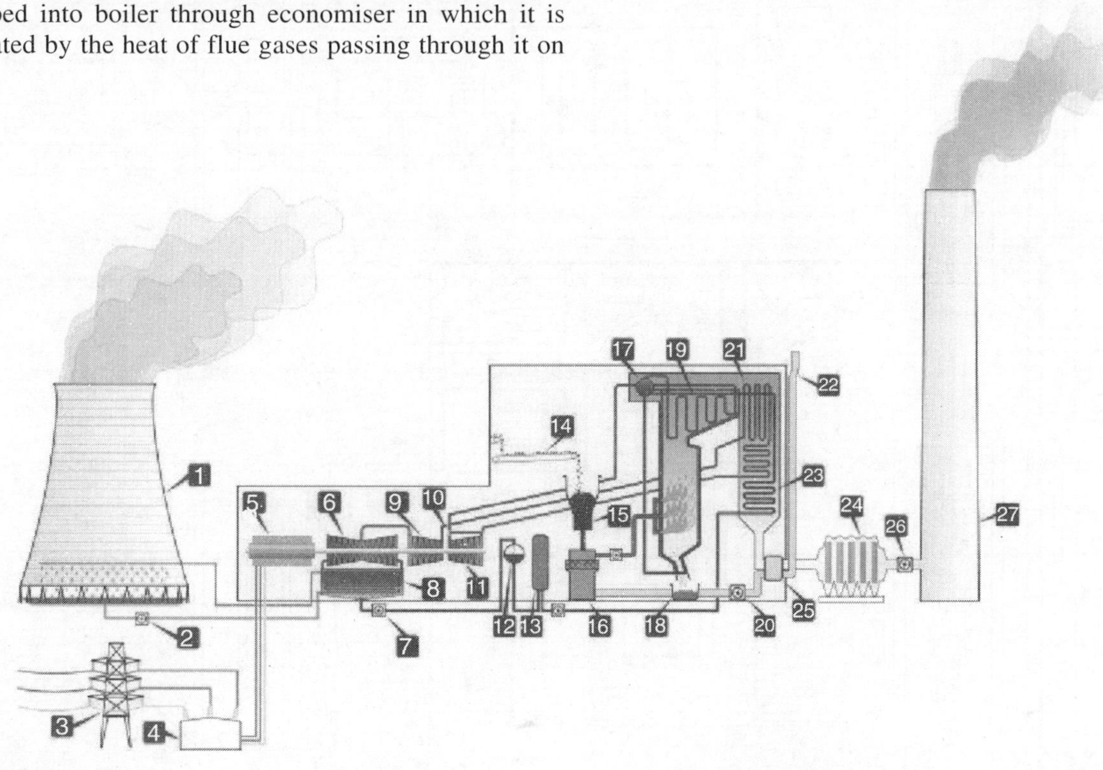

1. Cooling tower
2. Cooling water pump
3. Transmission line (3-phase)
4. Unit transformer (3-phase)
5. Electric generator (3-phase)
6. Low pressure turbine
7. Boiler feed pump
8. Condensor
9. Intermediate pressure turbine
10. Steam governor valve
11. High pressure turbine
12. Deaerator
13. Feed heater
14. Coal conveyor
15. Coal hopper
16. Pulverised fuel mill
17. Boiler drum
18. Ash hopper
19. Superheater
20. Forced draught fan
21. Reheater
22. Air intake
23. Economiser
24. Air preheater
25. Precipitator
26. Induced draught fan
27. Chimney Stack

Fig. 3.4(*a*). *Coal-Fired Power Plant*

① BOILER 485 TONS/hr 190 atm. GAUGE, ② 150 MW TURBO ALTERNATORS, ③ MAIN CONDENSER,
④ GLAND STEAM CONDENSER, ⑤ LP HEATER, ⑥ DEAERATER, ⑦ FEED WATER TANK 150 m^3
⑧ BOILER FEED PUMPS 310 TONS/HOUR EACH, ⑨ HP HEATER, ⑩ FLASH TANK FOR START UP STEAM,
⑪ SEPARATING TANK, ⑫ COLD CONDENSATE STORAGE TANK, ⑬ CONDENSATE FLASH TANK,
⑭ HOT CONDENSATE STORAGE TANK, ⑮ CHEMICAL WATER TREATMENT PLANT, ⑯ MAKE UP WATER DEAERATER,
⑰ MAKE UP WATER TANK, ⑱ AUXILIARY BOILER, 25 TONS/HOUR 10 ATMOS GAUGE,
⑲ AUXILIARY BOILER FEED WATER DEAERATER TANK, ⑳ AUXILIARY BOILER FEED WATER TANK,
㉑ STEAM MANI FOLD, ㉒ AIR PREHEATER, ㉓ ELECTROSTATIC PRECIPITATOR, ㉔ INDUCED DRAUGHT FANS,
㉕ THERMAL CONTROL ROOM, ㉖ ELECTRICAL CONTROL ROOM, ㉗ L.V. SWITCHGEAR, ㉘ MAIN TRANSFORMER 220 MVA 11/220 KV, ㉙ TRANSFORMERS FOR STATION AUXILIARIES, ㉚ STARTING TRANSFORMERS

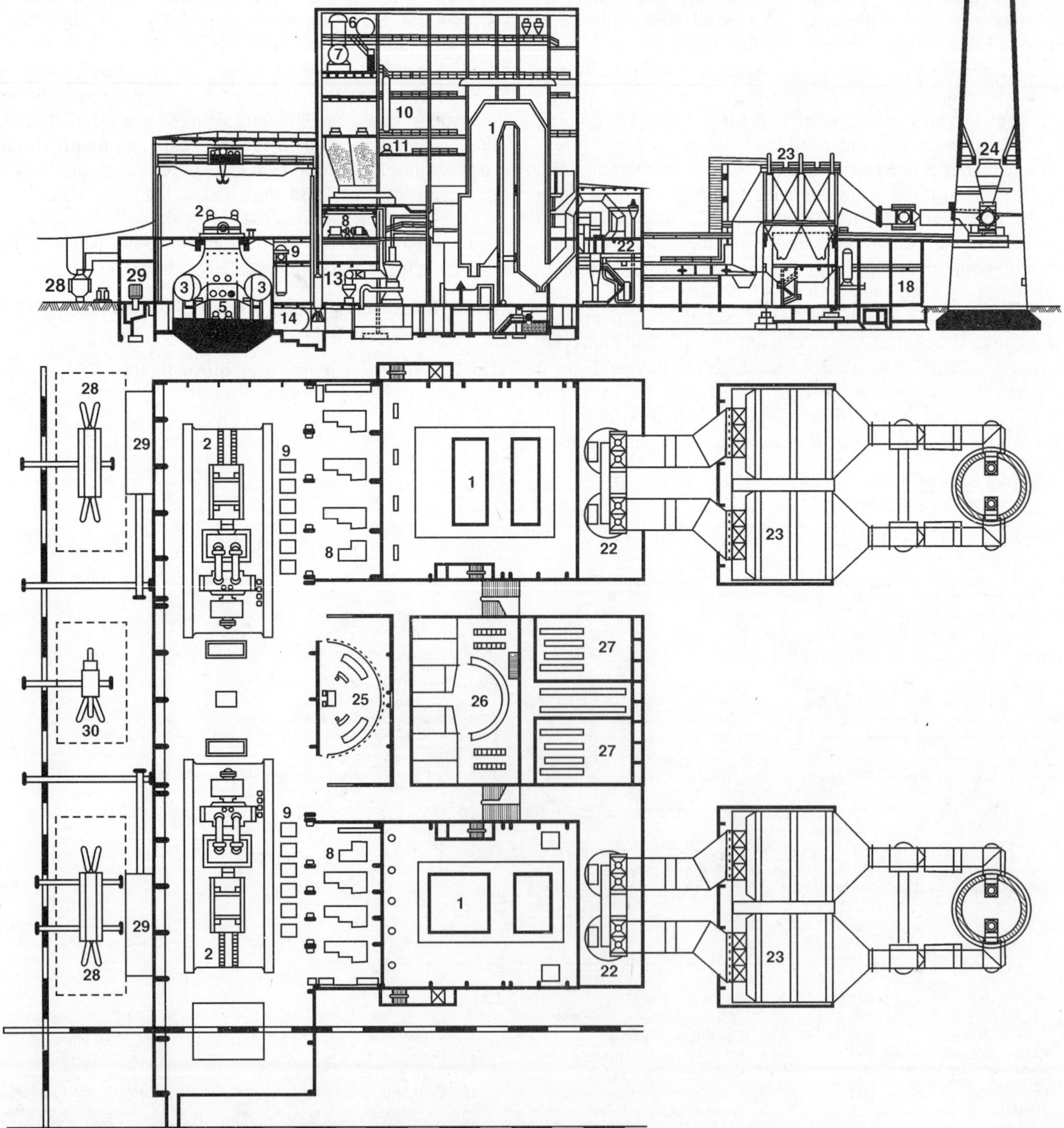

Fig. 3.4(*b*). ***300 MW Thermal (Coal-Fired) Power Station***

The important parts and auxiliaries of steam power plants are discussed in brief as follows:

1. Steam Generating Equipment.

(a) Boilers. Boilers or steam generators convert water into steam and form one of the major equipments in a steam power plant. Boilers used in steam power plants are of two types namely fire tube boilers and water tube boilers. In fire tube boilers the tubes containing hot gases of combustion inside are surrounded with water while in water tube boilers the water is inside the tubes and hot gases outside the tubes. Fire tube boilers are compact in size, have low initial cost and have the ability to raise rapidly large quantities of steam per unit area of fire grate but have the following drawbacks.

As water and steam, both are in the same shell, higher pressure of steam are not possible, the maximum pressure which can be had is about 17.5 kg/cm^2 and with a capacity of 15,000 kg of steam per hour. For higher pressures or higher rates of evaporation, the shell and fire tube boilers become extremely heavy and unwieldy. In the event of a sudden and major tube failure, steam explosions may be caused in the furnace due to rush of high pressure water into the hot combustion chamber which may generate large quantities of steam in the furnace.

Fire tube boilers use is, therefore, limited to low cost, small size and low pressure (to about 10 kg/cm^2) plants.

For central steam power plants of large capacity water tube boilers are used. Water tube boilers essentially consist of drums and tubes. The tubes are always external to drum. In comparison to fire tube boilers the drum in such boilers do not contain any tubular heating surface, so they can be built in smaller diameters and consequently they will withstand high pressure. The water tube boilers have got following advantages over the fire tube boilers.

1. High evaporation capacity due to availability of large heating surface.
2. Better heat transfer to the mass of water and better efficiency of plant owing to rapid and uniform circulation of water in tubes.
3. High working pressure due to small size of drum.
4. Quick raising of steam owing to large ratio of heating surface to water volume.
5. Safety in operation.
6. Less space occupied.
7. Better overall control.
8. Easy removal of scale from inside the tubes.

The design of boiler depends upon several service factors such as weight, height, portability, safety, bulk, character of operating labour, life, efficiency and cost. The boilers may be straight or bent tube of diameter ranging from 25 mm to 100 mm; longitudinal or cross drum, horizontal, vertical or inclined tube, forced or natural circulation, single or multi-drum, sectional or box header, cross or parallel baffles, marine or stationary. Header boilers, using straight tubes have been superseded by curved-tube drum type. Natural circulation (by density difference) of water is employed with pressure up to 175 kg/cm^2. Pumps giving controlled circulation, are often preferred on design delivering more than 340,000 kg of steam per hour at pressures above 100 kg/cm^2.

Research and development during last 60 years has led to large steam generating capacity boilers having improved design and construction. Today boilers are more efficient, compact and lighter in weight in comparison to old time boilers and can burn low grade fuels. Supercritical pressure boilers with pressures as high as 350 atm temperature 600ºC and capacity 1,000,000 kg/hour are available nowadays.

The selection of the size and type of boiler depends upon (*i*) the output required in terms of amount of steam per hour, operating temperature and pressure (*ii*) availability of fuel and water and (*iii*) the probable load factor. The other factors which influence the choice of a boiler are availability, initial cost, maintenance costs, labour costs, fuel costs and the space requirement.

Water tube boilers are used where large amounts of steam are to be produced at high temperature and pressure and weight and space considerations are important. To meet a required demand, the choice between two boilers will be based on economic considerations *i.e.,* total annual cost (fixed cost plus running cost). The noteworthy point is that the total cost of the fuel used by the boiler in its lifetime may be 3 to 4 times the initial investment.

(b) Boiler Furnaces. A boiler furnace is a chamber in which fuel is burnt to liberate the heat energy. In addition, it provides support and enclosures for the combustion equipment *i.e.,* burners. The boiler furnace walls are made of refractory materials such as fire clay, silica, kaolin etc. Such materials have the property of resisting change of shape, weight or physical properties at high temperatures. The construction of boiler furnace varies from plain refractory walls to completely water cooled walls, depending upon characteristics of fuel used and ash produced, firing methods, nature of load demand, combustion space required, excess air used, operating temperature, initial and operating costs.

The *plain refractory walls* are suitable for small plants where the furnace temperature may not be high. The arrangement may consist of a single section of homogeneous refractory or an insulation may be placed in between the refractory and the casing.

For larger plants, where the furnace temperature is quite high, refractory walls are made hollow and air is circulated through hollow space to keep the temperature of the furnace walls low. This type of arrangement is no more preferred.

Partially water cooled walls are similar to the plain refractory type but with a portion of the surface covered by water tubes. A proper balance can be made between the water cooled section and the refractory section to give best results. This type is used both for stoker fired and pulverised fuel fired boilers.

The recent development is to use *water walls.* Water walls are built of tubes of diameter ranging from 25 mm to 100 mm variously spaced with or without fins or studs and bare or with different thickness of moldable refractory on the inner face. Heat transfer rates run from 0.5×10^6 to 1.4×10^6 kilocalories per hour per cubic metre of surface. To meet these requirements of heat transmission, circulation on the water side must be adequate, obtained by convection or by pumps. This type is suitable for pulverised fuel fired boilers and high steaming rates can be maintained.

(c) Superheaters and Reheaters. A superheater is a device which removes the last traces of moisture from the saturated steam leaving the boiler tubes and also increases its temperature above the saturation temperature. For this purpose, the heat of combustion gases from the furnace is utilised.

Superheaters consist of groups of tubes made of steel (carbon steel for steam temperature up to 950°F, carbon-molybdenum steel for steam temperature of 1,050°F and stainless steel for steam temperature of 1,200°F) with an outside diameter ranging from 25 mm to 64 mm. Tube bundle location and arrangement, with counter current, and/or parallel flow is dictated by type of firing, required steam temperature, and steam-temperature characteristic. The superheater tubes are heated by the heat of combustion gases during their passage from the furnace to the chimney.

Superheaters are classified as radiant, convection or the combination of both.

Radiant superheater is located in the furnace between the furnace water walls and absorbs heat from the burning fuel through radiation. It has two main disadvantages firstly, owing to high furnace temperature, it may get overheated and, therefore, requires a careful design. Secondly it gives drooping characteristics *i.e.,* the temperature of superheat falls with the increase in steam output, because with the increase in steam output, the furnace temperature rises at a much less rapid rate than the steam output and the radiant heat transfer being a function of furnace temperature increases slowly with steam flow or the steam temperature falls.

Convection superheater is located well back in the boiler tube bank, receives its heat entirely from fuel gases through convection. It gives rising characteristics *i.e.,* the temperature of superheat increases with the increase in steam output because with the increase in steam output both gas flow over the superheater tubes and steam flow with in the tubes increase which causes increase in the rate of heat transfer and mean temperature difference. Convection superheaters are more commonly used.

Desired control of characteristic is obtained by

(*i*) proportioning and locating surfaces in series,

(*ii*) using internal dampers on boiler gas side,

(*iii*) attemperating by water or

(*iv*) supplementary burners. Heat transfer rates of 30 to 35 kilocalories per hour per square metre per degree temperature difference are representative.

The steam is superheated to the highest economical temperature not only to increase the efficiency but also to have following advantages.

(*i*) Reduction in requirement of steam quantity for a given output of energy owing to its high internal energy reduces the turbine size.

(*ii*) Superheated steam being dry, turbine blades remain dry so the mechanical resistance to the flow of steam over them is small resulting in high efficiency.

(*iii*) No corrosion and pitting at the turbine blades occur owing to dryness of steam.

The function of the reheaters is to resuperheat the partly expanded steam from the turbine. This is done so that the steam remains dry as far as possible through the last stage of the turbine. Modern plants have reheaters as well as superheaters in the same gas passage of the boiler. They can also be of combination type using both radiant and convective heating.

(d) Economiser and Air Preheaters. When the combustion gases leave the boiler after giving most of their heat to water tubes, superheater tubes and reheater tubes, they still possess lot of heat which if not recovered by means of some devices, would go waste (Fig. 3.5). Economiser and air preheater are such devices which recover the heat from the flue gases on their way to chimney and raise the temperature of feed water and air supplied for combustion respectively.

Economiser raises boiler efficiency (by 10-12%), causes saving in fuel consumption (5-15%) and reduces temperature stresses in boiler joints because of higher temperature of feed water, but involves extra cost of installation, maintenance and regular cleaning and additional requirement of space. High cost of fuel consumption, high load factor and high pressure and temperature conditions, all justify the use of economiser. For pressure of 70 kg/cm^2 or more the economiser becomes a necessity. Economiser tubes are made of steel either smooth or covered with fins to increase the heat transfer surface area. The tubes can be arranged in parallel continuous loops welded to and running between a pair of water headers, or in return bend design with horizontal tubes connected at their ends by welded or gasketed return bends outside the gas path. Such an economiser is shown in Fig. 3.6. The feed water flows through the tubes and the flue gases outside the tubes across them. The heat transfer from flue gases to feed water is by convection. The feed water should be sufficiently pure not to cause forming of scales and cause internal corrosion, and under boiler pressure. The temperature of feed water entering the economiser should be high enough so that moisture from the flue gases does not condense on the economiser tubes, which may absorb SO_2 and CO_2 from the flue gases and form acid to corrode the tubes. The temperature of the feed water entering the economiser is usually kept above 84°C. In a modern economiser, the temperature of feed water is raised from about 247°C to 276°C.

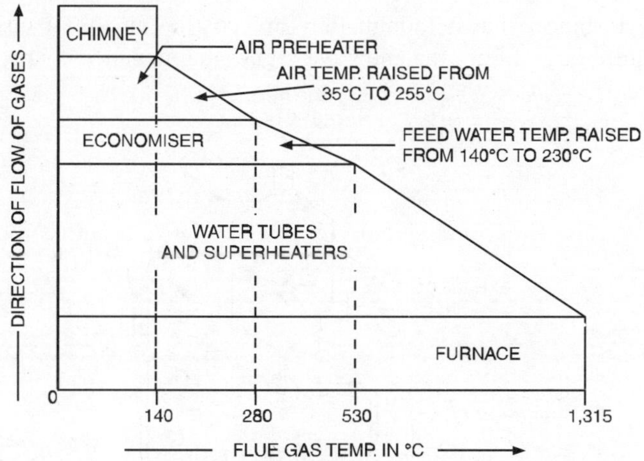

Fig. 3.5

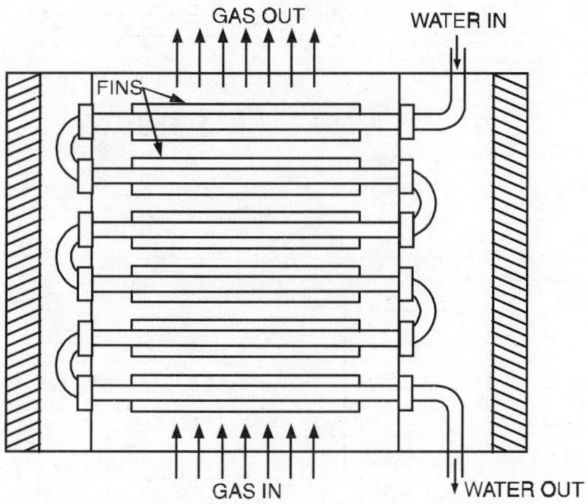

Fig. 3.6. ***Straight Tube Returns Bend Economiser***

The economiser chamber should be leak proof against air infiltration, otherwise boiler draught may be adversely affected.

Air preheaters are employed to recover the heat from the flue gases leaving the economiser and heat the incoming air required for combustion. This raises the temperature of the furnace gases, improves combustion rates and efficiency, and lowers the stack temperature, thus improving the overall efficiency of the boiler. It has been found that a drop of 20-22°C in the fuel gas temperature increases the boiler efficiency by about 1%. The use of air preheater is more economical with pulverised fuel boilers because the temperature of flue gases going out is sufficiently large and high air temperature (250-350°C) is always desirable for better combustion. The use of water cooled furnaces and feed water heaters has also resulted in the increasing use of air preheaters.

An air preheater should have high thermal efficiency, reliability of operation, less maintenance charges, should occupy small space, should be reasonable in initial cost and should be accessible. In order to avoid corrosion of the air preheaters, the flue gases should not be cooled below the dew point. This can be achieved by passing some air around the heaters or by recirculating air from the heater outlet to the forced draught fan inlet.

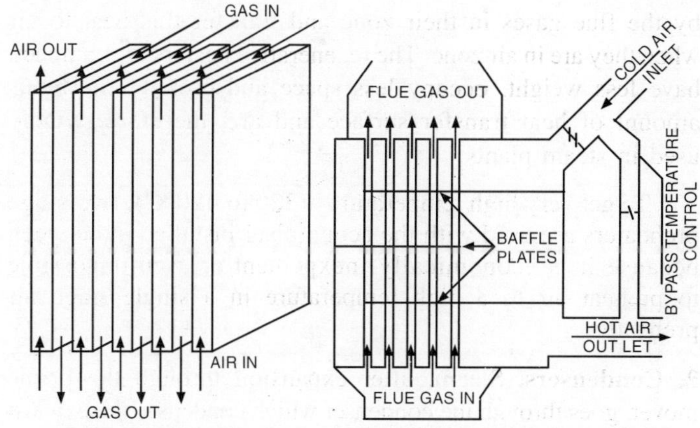

(*a*) *Plate Type Air Preheater* (*b*) *Tubular Type Air Preheater*

Fig. 3.7. ***Recuperative Type Air Preheaters***

The air preheaters are basically of two types viz. recuperative (plate type or tubular type) and regenerative types. Recuperative type air preheaters are continuous in action while the regenerative type are discontinuous in action and operates on cycle. In recuperative type of heaters, the two fluids are separated by heat transfer surface, one fluid flowing constantly on one side and the other fluid on the other side of the surface. In the recuperative type of heaters, the rate of heat transfer is low, space occupied is large and cleaning of surface is difficult. The plate type recuperative heater consists of rectangular flat plates spaced from 12.5 mm to 25 mm apart, leaving alternate air and gas passages [Fig. 3.7 (*a*)]. In tubular type recuperative heaters, shown in Fig. 3.7 (*b*), there is a counter flow, the flue gases pass through the tubes and air flows across the tubes or vice versa, but the first method is preferred owing to convenience in cleaning of tubes. Horizontal baffle plates are used so that maximum heat transfer takes place. The diameter of steel tubes is about 40-75 mm.

A regenerative heater is shown in Fig. 3.8. It consists of a rotor made up of corrugated elements. The rotor is placed in a drum which has been divided into two compartments, air

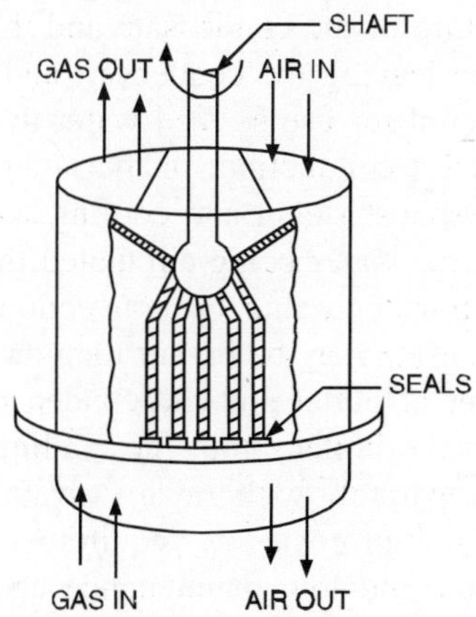

Fig. 3.8. ***Regenerative Preheater***

and gas compartments. To avoid leakage from one compartment to the other seals are provided. The rotor rotates at a very slow speed of 3-4 rpm. As the rotor rotates, it alternately passes through flue gases and air zones. The rotor elements are heated by the flue gases in their zone and transfer this heat to air when they are in air zone. The regenerative heaters are compact, have less weight, occupy less space and provide enormous amount of heat transfer surface and are, therefore, widely used in steam plants.

To get very high temperatures (320 to 420°C), two stage preheaters are used with the economiser installed in between because it is economically inexpedient or even impossible to preheat air to a high temperature in a single stage air preheater.

2. Condensers. Steam, after expansion through the prime mover, goes through the condenser which condenses the exhaust steam and also removes air and other non-condensable gases from steam while passing through them. The recovery of exhaust steam in the condenser reduces the make-up feed water that must be added to the system, from 100% when exhausted to atmosphere, to about 1-5% and thereby reduces considerably the capacity of water treatment plant. The exhaust pressure may be lowered from the standard atmospheric pressure to about 25 mm of Hg absolute and thereby permitting expansion of steam, in the prime mover, to a very low pressure and increasing plant efficiency. Maintenance of high vacuum in the condenser is essential for efficient operation. Any leakage of air into the condenser destroys the vacuum and causes (*i*) an increase in the condenser pressure which limits the useful heat drop in the prime mover and (*ii*) a lowering of the partial pressure of the steam and of the saturation temperature along with it. This means that the latent heat increases and therefore, more cooling water is required. Also, the under cooling of the condensate is likely to be more severe. This will result in lower efficiency. As it is not possible to eliminate air leakage completely, a vacuum pump is necessary to remove the air leaking into the condenser.

Condensers are of two types namely jet or *contact condensers* and *surface condensers*. The essential difference between a jet condenser and a surface condenser is that in the former, the exhaust steam mixes with the cooling water and the temperature of the condensate and the cooling water is the same when leaving the condenser; and the condensate cannot be recovered for use as feed water to the boiler; heat transfer is by direct conduction; in the latter *i.e.*, in surface condenser, the exhaust steam and cooling water do not mix with each other, the water being circulated through a nest of tubes, the heat transfer being by convection. The temperature of the condensate may be higher than the temperature of the cooling water at outlet and the condensate is recovered as feed water to the boiler. Both the cooling water and the condensate are separately withdrawn. Advantages of jet condensers are low initial cost, low requirements of floor area and cooling water and low maintenance charges. However, its disadvantages are that the condensate is wasted and high power is required for pumping water. The use of jet condensers is, therefore, limited to small industrial applications (1,000 kW) where high vacuum is not required (50 mm to 125 mm Hg abs). Modern power plants mostly use surface condensers from which condensate can be used as feed water requiring less pumping power and high vacuum is created at the turbine exhaust though they are costlier in initial cost, require large floor area and have higher maintenance charges.

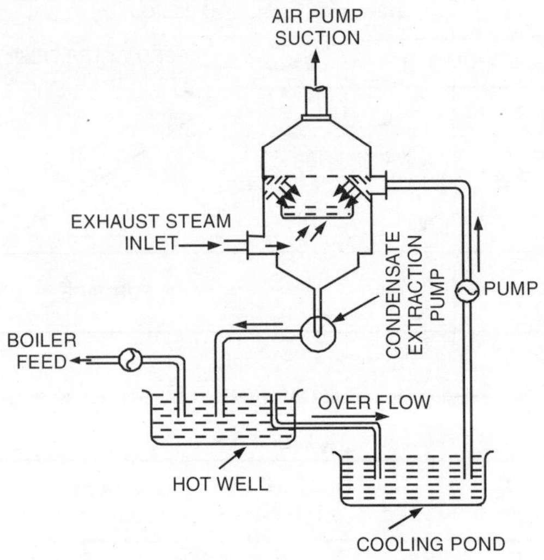

Fig. 3.9(*a*) *Low Level Jet Condenser*

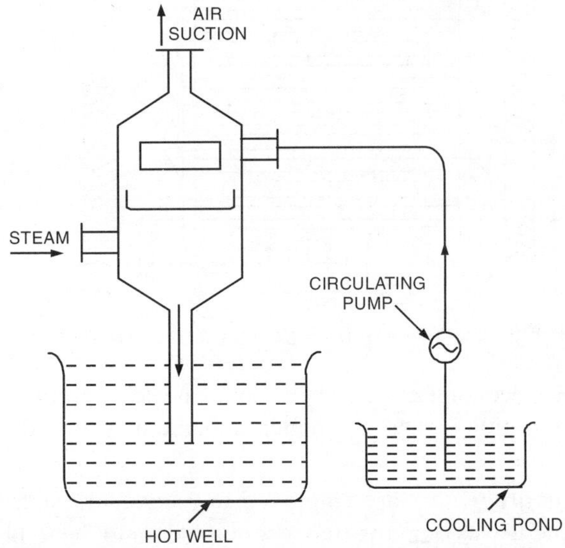

Fig. 3.9(*b*) *High Level or Barometric Jet Condenser*

The jet condensers may be further classified as (*i*) *parallel flow type jet condensers* and (*ii*) *counter flow type jet condensers*. In parallel flow type jet condensers the steam and cooling water enter at the top of the condenser and flow downwards in parallel. The coldest water is thus in contact with the hottest steam and, therefore, efficiency is low. In counter flow type jet condensers, the steam flows upwards through the condenser, meeting the cooling water flowing downwards from the top. The air is removed at the top and the condensate and water separately, at the bottom. In this type, since the hottest steam is in contact with the hottest cooling water, it is thermodynamically the most efficient, because heat transfer approximates towards reversibility. Also, the cooling of air is most effective and it will reduce the capacity of the air suction pump. The counter flow type jet condensers are of two designs namely, low level and high level counter flow type jet condensers. In low level jet condenser the supply of cold cooling water is drawn into the condenser shell due to the vacuum head created in the shell by the air pump. The water is sprayed downward in the shell into the up flowing steam. The condensed steam and cooling water flowing downward is discharged into the hot well. The high level jet condenser is also called the barometric jet condenser [Fig. 3.9(*b*)]. In case the bottom of the condenser is not less than, say 10.5 metres above the level of water in collecting tank (hot well), condensate extraction pump is not required and the condenser is self discharging. But, a pump is required to inject the cooling water into the condenser shell, from the cooling pond. In ejector condenser (Fig. 3.10) the cooling water enters the condenser at the top under a head of 4.5-6 metres and flows downward through a number of coaxial guide cones in a tube. As the water rushes across the gap between the central parts of nozzles (cones), it drags in the exhaust steam and air. The steam gets condensed in contact with cooling water and the air is carried forward with the water. The condenser thus acts as an air pump as well as a condenser.

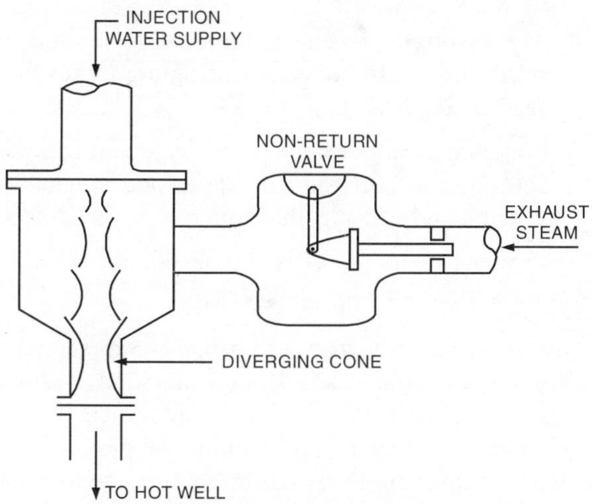

Fig. 3.10. ***Ejector Condenser***

The usual construction of the surface condenser is that there is a cast iron or steel shell fitted with a tube plate at each end. A large number of tubes extend between these end plates to form the cooling surface. Surface condensers can be classified depending upon whether the water flows through the tubes or steam flows through the tubes. The usual flow pattern is that the cooling water flows through the tubes and the exhausted steam is circulated around the tubes as the outside of the tubes is not contaminated by the clean steam. The steam enters the condenser through an opening in the top of the shell. The steam after being condensed leaves the condenser through a hole at the bottom of the shell (Fig. 3.11). The condensers may be single pass or two pass. In single pass condenser, the cooling water flows in one direction only through all the tubes and in the two-pass type (Fig. 3.11) the cooling water flows in one direction through part of the tubes and returns through the remaining of the tubes. Surface condensers are also classified as parallel flow, counter flow or cross flow depending upon the direction of flow of the condensate relative to the tubes. They can be further classified as down flow type, central flow type and inverted flow type. In the down flow type (Fig. 3.11) the steam enters at the top of the condenser and flows downwards over the tubes

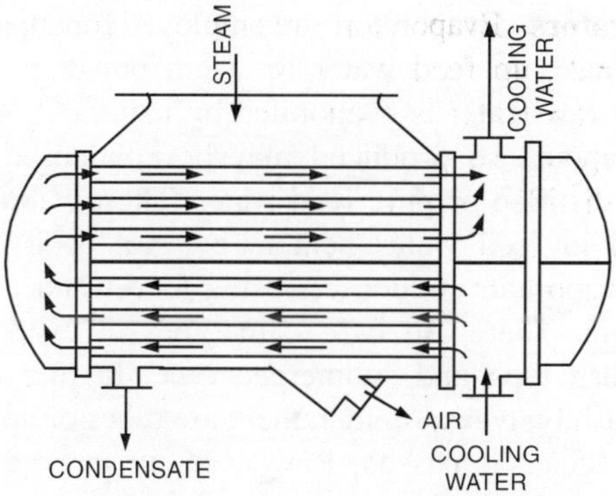

Fig. 3.11. ***Down Flow Surface Condenser***

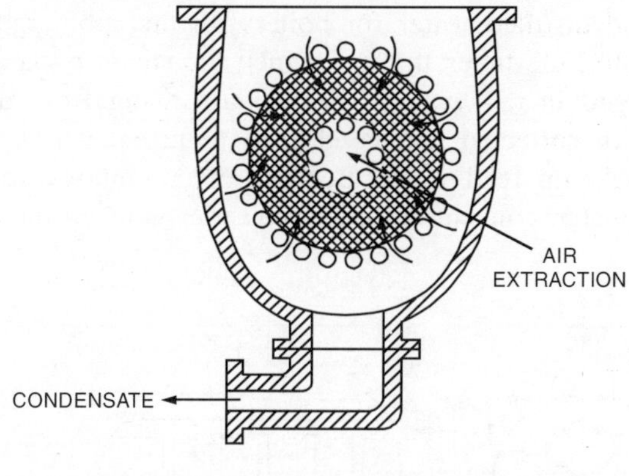

Fig. 3.12. ***Central Flow Type Surface Condenser***

through which cooling water flows as the extraction pump is at the bottom. The cooling water flows in one direction through the lower half of the tube nest and return in the reverse direction through the upper half of the tube nest. The air associated with the steam is also extracted from the bottom of the condenser where the temperature is lowest, so that the work of the air pump is reduced. In order to keep the velocity of steam across the tubes, approximately uniform, the cross section of the shell of the condenser is gradually reduced in width towards the bottom. Also the tubes are generally placed close together in the lower part. In central flow type (Fig. 3.12), the suction pipe of the air pump is located at the centre of the tube nest. The condensate then leaves at the bottom where the condensate extraction pump is placed. In this type, the steam comes into close contact with the whole periphery of the tubes. In the inverted type, the air suction pump is at the top. The steam flows upwards and then condensate returns to the bottom of the condenser by flowing near the outer surface. The condensate pump is at the bottom of the condenser.

The condenser is attached to the turbine by

(*i*) direct bolting to turbine exhaust flange (up to 20 MW)

(*ii*) direct bolting to turbine exhaust flange with spring supports under the condenser.

(*iii*) solid support of condenser with expansion joint between the turbine and the condenser.

3. Evaporators. Evaporators are employed for supplying pure water as make-up feed water in steam power plants. In an evaporator raw water is evaporated by using extracted steam and the vapours so produced may be condensed to give a supply of distilled or pure feed water. These vapours can be condensed in feed water heaters by the feed water or in separate evaporator condensers using feed water as the cooling medium. There are two main types of evaporators viz film or flash type and submerged type. In the former one (film or flash type evaporator), there are tubes or coils through which the steam is passed. Raw water is sprayed by means of nozzles on the surface of these tubes and some of the raw water will be converted into vapours. These vapours are collected from the evaporator and are condensed to give a pure and distilled water for boilers. In the submerged type evaporator, the tubes through which the steam is passed are submerged in raw water. The vapours rising from the raw water are collected and condensed to provide a supply of pure make-up feed water. Because of continuous operation of raw water, concentration of impurities goes on increasing,

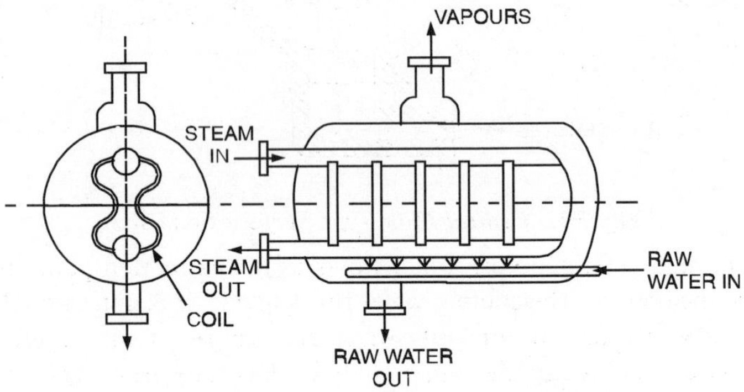

Fig. 3.13. *Flash Type Evaporator*

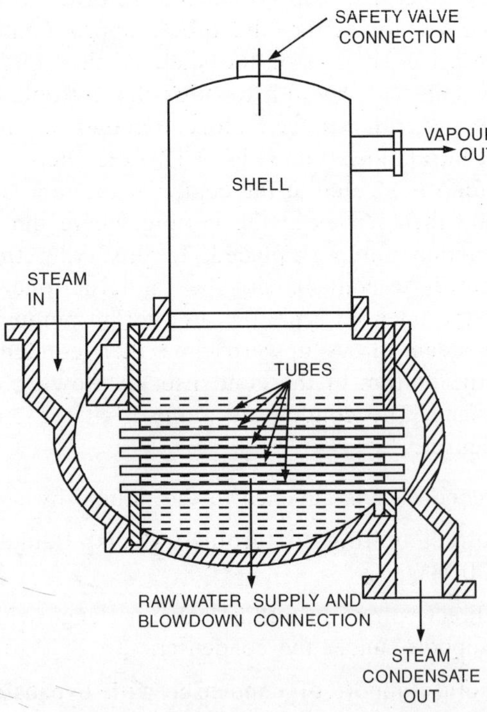

Fig. 3.14. *Submerged Type Evaporator*

so periodic blowing down of raw water is essential. Scales formed on the surface of the tubes will retard the heat transfer rate and so its removal is also necessary. This is removed by draining the raw water from the shell and then spraying the tubes with cold water while the tubes are kept hot by flow of steam through them. The scale is cracked off and is washed away by the spray.

4. Feed Water Heaters. These heaters are used to heat the feed water by means of bled steam before it is supplied to the boiler. Necessity of heating feed water before feeding it back to the boiler arises due to the following reasons:

1. Overall plant efficiency is improved.
2. Thermal stresses due to cold water entering the drum of boiler are avoided.
3. There is an increase in the quantity of steam produced by the boiler.
4. The dissolved oxygen and carbon dioxide which would otherwise cause boiler corrosion are removed in the feed water heaters.
5. Some other impurities carried by steam and condensate, due to corrosion in boiler and condenser, are precipitated outside the boiler.

Feed water heaters are of two types viz. contact or open heaters and surface or closed heaters.

Open or contact heaters are usually constructed to remove non-condensable gases from water and steam along with raising the feed water temperature. Such heaters are also called the *deaerator*. The amount of gas dissolved in water depends upon its temperature. This decreases sharply with the increasing temperatures and falls to almost zero at the boiling point. Such feed water heaters are used in small power plants.

Closed or surface heater consists of closed shell in which there are tubes or coils through which either steam or water is circulated. Usually, the water is circulated through the tubes and the steam and water may flow either in the same direction or in opposite directions. Such heaters may be either vertical or horizontal type. In such heaters, the feed water can never be heated to the temperature of steam. For maintaining a high overall heat transfer for the heater, the water velocity should be high but pumping costs limit the velocity to about 1-2.5 m/s.

5. Prime Movers. The prime mover converts the stored energy in steam into rotational mechanical energy. Steam prime movers are either reciprocating engines or turbines. Nowadays steam engines have become obsolete due to their reciprocating motion and the steam turbines are usually employed as prime movers. Steam turbines give high speed (standard speeds are 3,000 rpm and 1,500 rpm), maximum size (1,000 MW), minimum floor space, bulk and weight, maximum efficiencies in large sizes, suitability for highest steam pressures and steam temperatures. All large units (above 1 MW) are steam turbines.

Steam turbines will be discussed in Art. 3.7.

6. Spray Ponds and Cooling Towers. In the modern fossil-fuel steam power plants, about 10-15% of the heat input is rejected to the atmosphere through boiler chimneys, while 48-52% of the heat input is rejected to cooling water system through the steam condensers. The quantity of cooling water needed for condensing the steam in the condenser is, therefore, quite large. Roughly each kg of steam needs 100 kg of cooling water for the condenser. A typical 2,000 MW power plant requires about 300 × 10^6 litres of cooling water per hour. Such large requirement of cooling water can be conveniently met if the plant is located near a natural or artificial source of water such as river, sea, lake or canal. In such a case, water is pumped from the source of water to an open channel and taken to the circulating water pumps through strainers. The circulating water is then pumped through the condenser tubes. For such arrangements, a settling tank and syphon system is used before the circulating water pump at the inlet. The circulating water takes up the heat of the exhaust steam and itself becomes hot. This hot water coming out from the condensers is discharged at a suitable location down the source of water. In case natural or artificial source of water is not available as a source of cooling water for condensers, the warm water coming out of the condenser has to be cooled and reused. In this case water is first obtained from a tubewell or some other source and stored in a tank. From the tank it is pumped into the condenser, usually by a certifugal pump. The water there absorbs the latent heat from the exhaust steam and gets warm, the warm water is then cooled by means of *spray ponds or cooling towers*. Small plants use spray ponds and medium and large plants use cooling towers.

(a) Spray Pond. Spray pond consists of a water tank in which hot water is distributed by pipes throughout and is sprayed in air through nozzles at a suitable pressure. The water is cooled by both convection and evaporation. The sprayed water comes in contact with the atmospheric air and is cooled. Mostly the water is cooled by evaporation as the

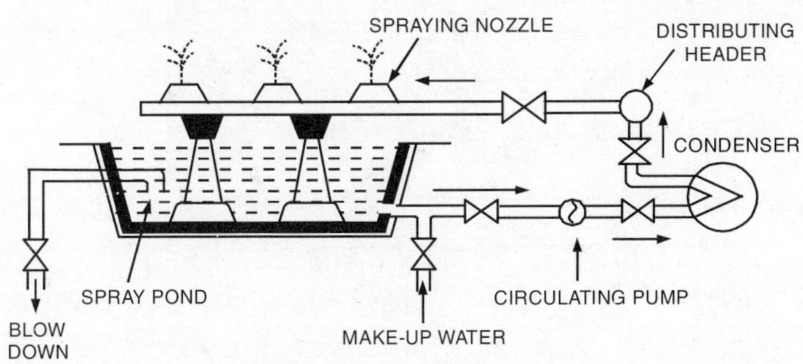

Fig. 3.15. *Spray Pond*

heat for evaporation is withdrawn from the water itself with the result that it is cooled. A small amount of cooling is due to conduction and radiation also. Since the cooling is mainly affected by evaporation it is, therefore, necessary to expose as large surface as possible. In the design of spray ponds the most important part is nozzle and it is necessary that it should produce spray, not sheet of water. The spacing between nozzles should be around 2.5-3.0 metres and about 1.2-2.5 metres above the surface of water. Further the nozzles should be adjustable for load and climatic conditions. The drawbacks of cooling ponds are that considerable quantity of water may be carried away in suspension in air when its velocity is high and loss due to evaporation and the large ground area required especially in cases of large sized power plants.

(b) Cooling Tower. A cooling tower is a wooden or metallic rectangular structure inside of which is packed with baffling devices. The hot water is led to the tower top and falls down through the tower and is broken into small particles while passing over the baffling devices. Air enters the tower from the bottom and flows upward. The air vaporises a small percentage of water, thereby cooling the remaining water. The air gets heated and leaves the tower at the top. The cooled water falls down into a tank below the tower from where it is pumped to the condenser and cycle is repeated (Fig. 3.16). The splitting of water into small droplets, the draught provided by the tower and the large evaporating surface help to cool water very quickly-practically during the time while it is descending. Although eliminators are provided at the top of the tower to prevent escape of water particles with air but even then there is a loss of water to the extent of around 5 per cent and this loss has to be made up by water drawn from well or any other source. Air can be circulated in cooling towers through natural draught, mechanical draught or mechanical natural draught. In natural draught, air movement is induced by a large chimney and the difference of densities between the air inside and outside the chimney. In mechanical draught, air is moved by fans, either of the induced type that pull the air through the tower, or the forced type that push the air through.

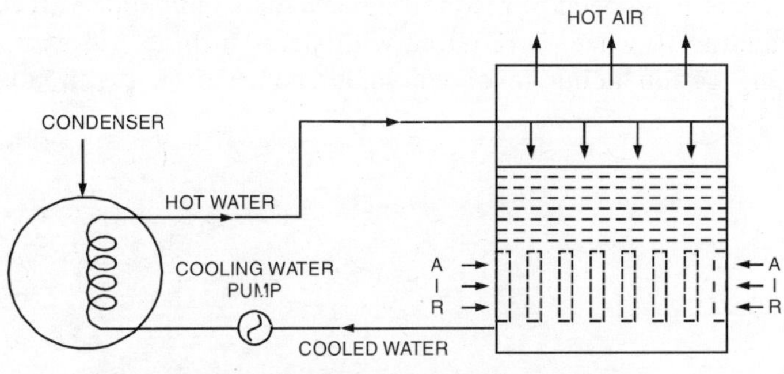

Fig. 3.16. *Arrangement of Cooling Tower*

Combination of mechanical and natural draught incorporate both a chimney and fans. The natural draught towers are being gradually replaced by mechanical draught type because the latter requires less pumping head, less space and less windage loss. Induced draught tower is considered to be better than the forced draught tower because the latter type requires more power and involves costlier maintenance of fans. The induced draught tower occupies less space as the fan drives are placed at the top of the tower. Moreover, since air is drawn from all the sides of the tower, the cooling effect is distributed across the entire cross section of the tower. Also, since the fans handle warm air, they are non-freezing. Again, as the air leaves the tower at a high speed, this type is non-circulating.

7. Control Room. The control room, in case of remote control, houses all the necessary measuring instruments for each panel of alternator and feeder, synchronising gear, protective gear, automatic voltage regulator, communication arrangement etc. A separate battery room and a motor-generator set or a rectifier is also installed for supplying to make and trip circuit of switchgear. In case of outdoor switchgear normally compressed air is used for operation.

8. Switchyard. The switchyard houses transformers, circuit breakers and switches for connecting and disconnecting the transformers and circuit breakers. It also has lightning arrestors for the protection of the power station against lightning strokes.

The supply to the bus-bars from alternators is taken through transformers and circuit breakers of suitable ratings.

The power station over and above this requires workshop, store, labour canteen, library and also residential accommodation for the operating staff.

The layout of a central coal-fired steam power plant is shown in Fig. 3.17.

3.7. STEAM TURBINES

The operation of a steam turbine is based upon the principle that the steam issuing from a small opening attains a high velocity. This velocity attained during the expansion of the steam depends on the initial and final heat contents of the steam. The difference of the two (the initial and final heat contents) represents the heat energy converted into kinetic energy.

The basic construction of a steam turbine is simple. There is no need of piston rod mechanism and slide valves and no flywheel is required with steam turbine. No wearing action being involved, maintenance of a steam turbine is comparatively much simpler. Problems of vibrations is also much less since high operating speeds result in lower weight of rotating parts for the same power. Governors are used to maintain speed constant when there is a change in load on the system. The governors may be centrifugal or hydraulic type.

3.7.1. Classification of Steam Turbines

There are different ways of classifying steam turbines and according to the action of steam on moving blades, the steam turbines, are of two types namely impulse and reaction types. These two types differ in working.

In an *impulse turbine*, the steam expands completely in the stationary nozzles, there being no pressure drop over the moving blades or runner. In doing so, the steam attains a high velocity and impinges against the blades fixed on the rotor periphery. This results in the impulsive force on the moving blades which sets the rotor rotating. The rotor may be a built-up rotor or integral rotor. A built-up rotor consists of a forged steel shaft on which separate forged steel discs are shrunk and keyed. In an integral rotor the wheels and shaft are formed from one solid forging. The built-up rotor is cheaper and easier to manufacture. The high pressure and intermediate pressure rotors are always of integral type. The blades have symmetrical profile. Impulse turbine has high speed and provides an ample clearance in between rotor or runner and stationary blades or casing (stationary blades are mounted in the casing). It also gives optimum utilisation of steam with a simple clear design. Owing to its simple and sturdy construction of all steam carrying parts, it has a long life.

In a *reaction turbine*, the steam does not expand in nozzles but expands as flows over the rotor blades, the blade will, therefore, act also as nozzles. The expansion of steam as it flows over the blades is adiabatic, any friction losses between the steam and blades are converted into heat, which,

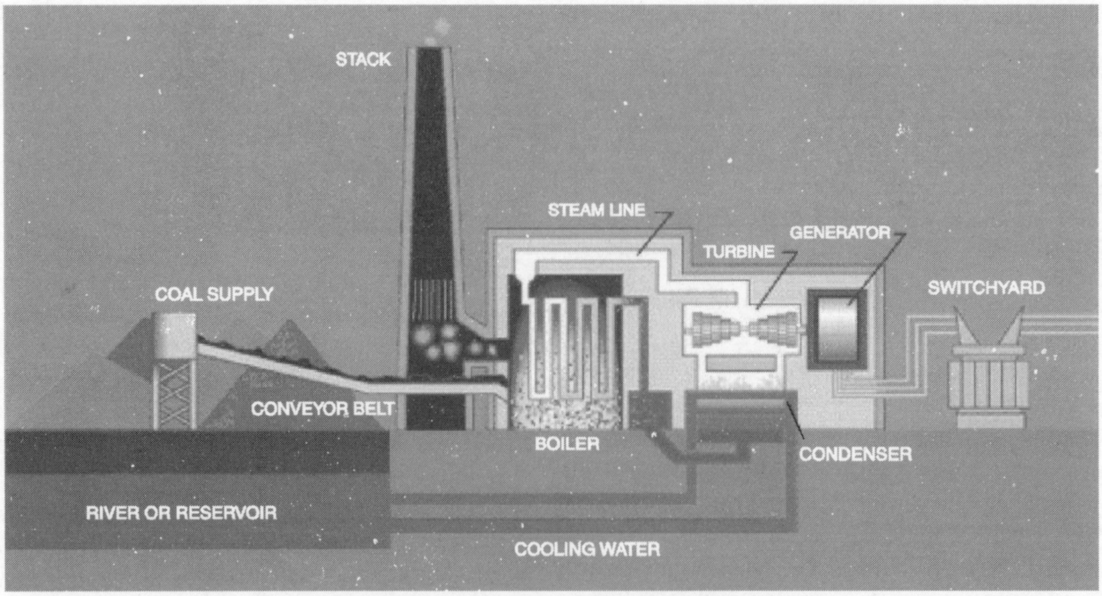

Fig. 3.17. *Layout of Coil-Fired Steam Power Plant*

in turn will reheat the steam. The effect of this is to dry or superheat the steam as it flows over the blades. Reaction turbines are characterised by a relatively low rpm.

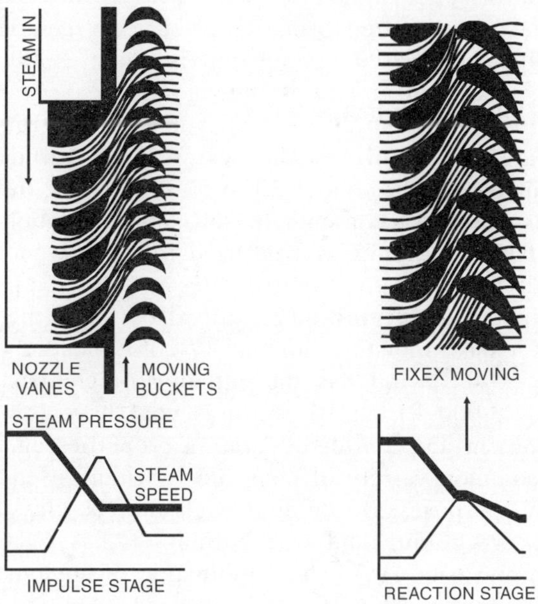

Fig. 3.18

Commercial turbines use series combination of impulse and reaction types because steam can be used more efficiently by using impulse and reaction blading on the same shaft. The steam is expanded through the turbine from a high pressure at the throttle valve to a back pressure corresponding to a vacuum of 71 to 73.5 cm Hg or an absolute pressure of 5 to 2.5 cm Hg. The standard speeds are 3,000 rpm and 1,500 rpm for coupling to 50 Hz alternators. Although all modern turbines are basically impulse-reaction turbines (partial expansion taking place in a nozzle) but are designated as reaction turbines.

According to the *type of flow of steam*, the steam turbines used are of two types namely axial flow and radial flow type. In axial flow type turbines the steam flows over the blades in a direction parallel to the axis of the wheel. In radial flow turbines the blades are arranged radially so that the steam enters at the blade tip nearest the axis of the wheel and flows towards the circumference. Majority of the steam turbines are of 'axial flow' type.

Steam turbines can also be classified in other ways, as given below:

Steam turbines are classified as condensing or non-condensing depending upon whether the back pressure is below or above atmospheric pressure.

Where there is no use of exhaust steam, the turbine is built as a pure *condensing turbine* in which steam is reduced in pressure down to a vacuum pressure which is in accordance with the cooling water temperature. The steam inlet in this case is regulated, to suit the output required for the machine, by means of a governor in such a way that the oil pressure equipment opens and closes as required, a number of nozzle valves which admit steam to individual groups of nozzles of first turbine stage *i.e.,* regulating stage.

Central-Station Turbines. These turbines are employed for driving alternators at synchronous speed (usually 3,000 rpm) and have capacities ranging from 16 to 1,500 MW.

Reheating Turbines. In these turbines, steam is returned after partial expansion to the boiler for resuperheating and then allowed to expand to back pressure.

Superposed or Topping Turbines. These are high pressure non-condensing turbines installed in existing low pressure steam plants. They exhaust into the existing low pressure turbines, thus, increasing plant capacity and overall thermal efficiency.

Back Pressure Type Steam Turbines. Where process heating steam is required at low pressures back pressure steam turbines are used, which utilizes the pressure difference (*i.e.,* heat gradient) available between live steam and back pressure of the process steam. The output of such a turbine is dictated by the amount of process steam required and the boiler steam condition which are determined by economical considerations. Back pressure turbines are controlled by a *pressure governor* which operates in the same way as the *speed governor* of a condensing turbine, being affected by the volume of back pressure and opening or closing the group of valves regulating the steam supply to the turbine.

Bleeder or Extraction Turbines. In such turbines part of the steam leaves the turbine casing before the exhaust, for feed water heating.

Extraction-Induction Turbines. Such turbines have provisions for both removing and introducing steam into the turbine at intermediate points.

As regards the size of the units to be installed, it depends upon several factors such as capital cost, spares required, load factor and the peak load of the system.

3.7.2. Compounding of Steam Turbines

In a single stage turbine, the steam is expanded in a single ring of nozzles from the boiler pressure down to the exhaust pressure, resulting in a extremely high velocity (of the order of 30,000 rpm). With such a turbine, the blade tip stresses and the disc friction losses will be very high, resulting in poor efficiencies. In thermal power plants where the generators usually run at 3,000 rpm, single stage turbines are undesirable. **Compounding** is, therefore, necessary for obtaining reasonable blade tip speeds in turbines. In compounding, a number of rotors in series, keyed on the same shaft, are employed and the steam pressure or the jet velocity is absorbed in steps as it flows over the moving blades.

There are two types of compounding—velocity compounding and pressure compounding. In **velocity compounding**, the steam is expanded from boiler pressure to condenser pressure in one set of stationary blades or nozzles. However, the total flow energy is absorbed in a number of rows (not in one) of moving blades (2,3 or even 4), with a row of fixed guide blades between every pair of them. The turbine is impulse one and is called the 'Curtis stage', because the pressure remains constant as the steam flows over the blades. **Pressure compounding** is equivalent to a number of simple impulse stages in series. The pressure drop occurring in each stage is only a portion of total pressure drop.

Combined velocity-pressure compounding makes use of both types of compounding discussed above. The overall pressure drop of steam is divided into stages and the velocity attained in each stage is also compounded.

3.7.3. Steam Nozzles

The function of a steam nozzle is to convert the heat energy of steam into kinetic energy and the chief use of a steam nozzle is to develop a high velocity jet of steam for driving a steam turbine, This is achieved by allowing the steam to expand from a region of high pressure at the inlet to a region of low pressure at the outlet. With the expansion of steam through the nozzles, its velocity and specific volume both increase. Its dryness fraction will also vary due to condensation of steam. Since the weight of steam per second flowing across any section of the nozzle is constant, the cross section of the nozzle will vary in accordance with velocity, specific volume and, the dryness fraction of steam. Depending upon its shape, the nozzles are of two types viz convergent-divergent nozzle and convergent nozzle. The nozzle is designed for discharge of maximum weight of steam for a given pressure drop.

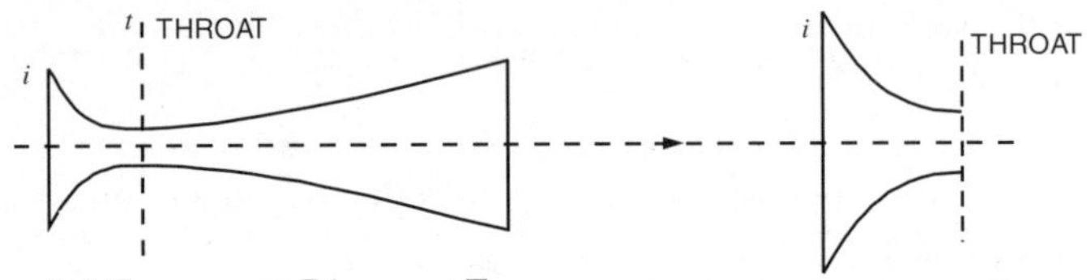

(a) Convergent-Divergent Type *(b) Convergent Type*

Fig. 3.19. ***Types of Nozzles***

In the *convergent-divergent nozzles*, the x-section first decreases until it reaches the minimum at a section called the nozzle throat and then it increases as shown in Fig. 3.19 (*a*), the largest x-section being at the exit end. In the *convergent type nozzles*, the nozzle exit is the throat itself, as shown in Fig. 3.19 (*b*).

3.8. FUELS

Fuels may be classified as solid, liquid and gaseous and as natural or prepared. The various fuels commonly used for combustion in the thermal power plants are coal, oil and gas. Gaseous fuel is rarely economical except when the power plant is located near natural gas field or gas manufacturing industries. Oil is used only where it is plentiful and cheap. Coal is the most commonly used fuel in thermal power plants.

3.8.1. Classification of Coal

Coal occurs naturally in seams and is the result of decomposition of vegetation accumulated in the earth, millions of years ago having got transformed by the action of intense heat and pressure. The conversion of vegetation into coal requires ages of time. Coal contains moisture, carbon, hydrogen, sulphur, nitrogen, oxygen and ash.

The coals can be classified in increasing order of heat or calorific value, namely peat, lignite, sub-bituminous, bituminous, semi-bituminous, semi-anthracite, anthracite an super anthracite coals. **Peat** is the first stage in the process of transformation of buried vegetation into coal. Its enthalpy of combustion is about 3,000 kJ/kg. Owing to a high moisture content (60 to 90%) and low carbon content (5 to 20%), peat is not suitable for use in power plants. **Lignite** is the next stage in the development of coal. It has about 30 to 50% moisture and 20 to 40% carbon and has a enthalpy of combustion of 13,800-17,600 kJ/kg. Lignite can be used for power generation but only locally owing to difficulties encountered in transportation. **Sub-bituminous coal,** also called the **black lignite,** has moisture content 17-20% and the volatile matter 35-45% and has the enthalpy of combustion of 18,800-23,000 kJ/kg. **Bituminous coal** have low moisture content and non-disintegrating properties and is the most common variety of coal and is widely used for almost all purposes. Bituminous coals may be classified as caking or cooking and free burning. It has 60 to 80% carbon and 6 to 10% ash and enthalpy of combustion of 23,000-34,000 kJ/kg. It is widely used in power plants. **Semi-bituminous coals** are the highest grade of bituminous coals having high fixed carbon content and the highest heating values. These have volatile matter between 14 and 22% and enthalpy of combustion between 27,000 and 35,000 kJ/kg. These coals are one of the best for power generation. **Semi-anthracite** coals are harder than the bituminous coals, found in small quantities and are, therefore, costlier for power generation. These coals have enthalpy of combustion between 33,500 and 34,750 kJ/kg. Anthracite coals have the highest carbon content but do not have the highest calorific value. Because of its high cost and scarcity, it is used mainly for domestic purposes. **Super-anthracite coals** are very hard with a shiny black surface. It is non-coking and is very difficult to ignite and so has little importance in the power generation.

3.8.2. Selection of Coal for Power Plants

The selection of coal for a power plant depends on a number of factors such as calorific value, weatherability, sulphur content, ash content, particle size, grindability index and caking characteristics.

Calorific value represents the amount of energy in a given mass and coal with a higher calorific value is preferred. **Weatherability** is a measure of the ability of coal to withstand exposure to environment without excessive crumbling. From the point of view of storage, the coal with higher weatherability is, obviously preferable. **Sulphur** is one of the combustible elements in the coal and produces energy but its primary combustion product, sulphur dioxide, is a health hazard. The sulphur content of the coal should, therefore, be low (preferably below 1%). **Ash** is an impurity, produces no heat and is to be removed from the furnace and disposed off. However, a minimum of 4 to 6 per cent ash is required in the coal burnt on hand fire grates and some stokers for protection of grates against overheating. The **size requirements** for different equipments vary widely. The **grindability index** is inversely proportional to the power required to grind the coal to a certain fineness. Higher the grindability index, better the coal is.

3.8.3. Indian Coals

India has large deposits of coals (around 170 billion tonnes) in Bengal, Bihar, MP, Orissa etc. In general, Indian coals have high ash content (as high as 40%). The high ash content of coal reduces the boiler thermal efficiency because heat loss through unburnt carbon, excessive clinker formation and heat in ashes is considerably more. It also adds to the difficulty in the disposal of hot ash. Thus the use of coal with high ash content increases the plant size and transportation cost and reduces the thermal efficiency. In addition it makes the control difficult because of irregular combustion. High ash content coals can be used more economically in pulverized form because in that form the thermal efficiency may be as high as 90% and controls could be made simpler just by adjustment of position of burners in pulverized fuel boilers. All new thermal power plants use pulverized coal.

Use of coal for power generation causes immense environmental pollution because sulphur dioxide having detrimental effect on our respiratory system is produced due to sulphur content in the coal. Carbon dioxide, carbon monoxide and particulates emissions from coal based thermal power plants are other important environmental pollutants.

3.8.4. Imported Coals

Imported coal has higher calorific value and lower ash content (less than 10%). From detailed cost calculations, it is concluded that if the calorific value, ash content, transportation costs and all other costs are considered, the cost of Indian coal and imported coal per unit of energy generated is almost the same for power plants situated near the coastal areas. As such imported coal is likely to play a vital role in power generation in the near future. BSES's Dahanu power plant in Maharashtra meets about 30% of its coal consumption through imports. Many proposed thermal power plants of IPPs (independent power producers) are likely to be based on imported coal. However, the imported coal has higher sulphur content (up to 1.2%) as compared to Indian coals (less than 1%).

3.8.5. Liquid Fuels

Oil can be used in a boiler furnace for generating steam. Practically all the commercially used liquid fuels are furnished by petroleum and its by-products. Fuel oils used in practice may be distillates or residuals, and either cracked or straight run. Fuel oil has become a very important fuel in power generation since the handling costs are less, load fluctuations can be met easily and few operations are needed and there is no disposal of ash. In the marine field it has almost replaced the coal due to its cleanliness, small pipes etc. However, the main drawback of liquid fuel is that the heat produced is costly in comparison to coal or gas. Moreover, in our country natural resources of oil are in short supply and use of oils is, therefore, limited for generation of electrical power. Further, as the fuel oil has a much higher percentage of hydrogen as compared to coal and, therefore, there will be much higher moisture loss in the flue gases resulting in lower overall efficiencies as compared to coal burning. Superheaters will be subjected to hard and adhesive oil slag which is often difficult to remove, but the other heating surfaces are almost trouble free.

3.8.6. Gaseous Fuels

The gaseous fuel may either be natural or a manufactured one. The later being too costly, usually the natural gas is used for power generation. Gaseous fuels have all the advantages of oil fuels except for ease of storage. The major limitation for using natural gas as fuel for power generation is location of power plant near the natural gas fields otherwise the transportation cost will be high.

Cost of transportation of fuel is an important consideration in selecting the fuel for a thermal power plant.

3.9. FUEL HANDLING

As already mentioned in foregoing section coal is the most commonly used fuel in thermal power plants. Therefore, we will restrict our discussion only to coal under this article. As enormous amount of coal is burnt in a thermal power plant (for example, a 200 MW power plant may require about 2,000 tonnes of coal per day) and the cost of coal may be as high as 60-70 per cent of the total operation cost of the plant, therefore, problems of coal handling in a thermal power plant require special consideration.

Coal can be handled manually or mechanically. Mechanical handling is usually adopted as it is reliable, expeditious and economical. Owing to large quantity of coal required to be handled everyday, mechanical handling has become absolutely necessary.

The main requirements of a coal handling plant are reliability, soundness and simplicity requiring a minimum of operatives and minimum of maintenance. Besides, the plant should be able to deliver the required quantity of coal at destination during peak hours.

3.9.1. Transportation or Delivery of Coal

The steam power plant should be located near the coal fields as far as possible in order to reduce the transportation charges. In case the plant is located near a coal pit, transportation cost by rail, road, sea or river is avoided. If a coal mine is very near, sometimes aerial ropeways are employed to carry coal to the site.

There are three ways of transporting coal from coal mines to the site of power plant *i.e.,* by sea or river, by road and by rail. If the power plant is situated on the bank of a river or near the seashore, it is often economical to transport coal in boats or barges, unload mechanically by cranes or grab buckets and place in the storage yard or directly to the conveyor system to be carried to the power plant. Transportation by road is possible for small and medium size plants only.

The chief advantage of this system is the possibility of carrying coal directly into the power house up to the point of consumption. Moreover due to less traffic restrictions it is considered better system in comparison to rail transport. Transportation of coal by rail, particularly for stations located interior, is still the most important means of transportation in common use.

A railway siding line is taken into the power station and the coal is delivered either in the storage yard or close to the point of utility. It can be unloaded by means of cranes, grab buckets or wagon tipplers. The capacity of the railway siding should be equal to three days requirement of the plant.

3.9.2. Methods of Coal Handling

Irrespective of the method of transportation of coal adopted, the coal has to be carried to the boiler stokers or the coal preparation plant in the case of pulverised fuel firing. The various stages in coal handling are (*i*) Unloading (*ii*) Transfer to dead storage or preparation plant (*iii*) Dead storage (*iv*) Reclaimation (*v*) Live storage (*vi*) Inplant handling and (*vii*) Coal weighing. The various stages are shown in Fig. 3.20.

The coal is unloaded from the point of delivery by means of (*i*) coal shakers or coal accelerators (*ii*) rotary car dumpers or wagon tipplers and (*iii*) grab buckets. The choice of equipment will depend upon the method of transportation adopted.

The main equipment employed for taking the coal from the unloading site to the dead storage are (*i*) belt conveyors (*ii*) screw conveyors (*iii*) bucket elevators (*iv*) skip hoist (*v*) grab bucket conveyors and (*vi*) flight conveyors.

Storage of coal is essential to insure against complete shutdown of a power plant which may arise from failure of normal coal supply (owing to eventualities like strikes in coal mines, failure of transport system, shortage of coal etc.). Storage also permits some choice of the date of purchase, allowing the purchasers to take advantage of seasonal market conditions. However, there are a number of factors, such as risk of spontaneous combustion, possible deterioration of coal during storage, interest on capital blocked, cost of insurance etc., which make storage of coal undesirable. The dead storage can be open space in one corner of the power station. Amount of coal to be stored depends on the available space for storage, transportation facilities and nearness of the power plant to coal mines. The usual practice is to store coal for one or two month requirements depending upon the distance of the power plant from the coal mines.

Reclaimation is the process of taking coal from dead storage for preparation or further feeding to hoppers or live storage. This may be accomplished by scrapers, belt conveyors, bull dozers. In case of vertical movement it can be done by skip hoists or buckets elevators.

Live Storage consists of about one day requirement of coal of power plant and is usually a covered storage in the power station near the boiler furnace. It can be provided with bunkers and coal bins.

Input handling refers to handling of coal between the live storage and firing equipment. In case of simple stoker firing only chutes may be required to feed the coal from storage bunkers to the firing units. However, gates and valves are included in the system for controlling the flow of coal according to the load on the power plant. In case of pulverised fuel firing system equipments such as chutes, pulverising mills, feeders, weighing equipment and many others for inplant handling would be required.

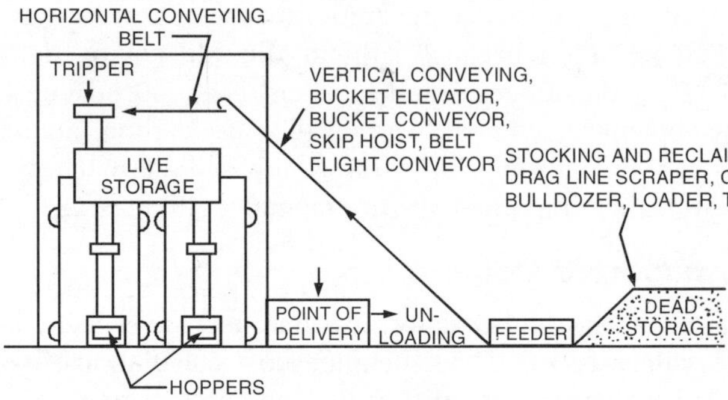

Fig. 3.21. ***Layout of Coal Handling Plant***

Coal weighing enables one to have an idea of total quantity of coal delivered at the site and also whether or not proper quantity has been burnt as per load on the plant. It can be accomplished by (*i*) weighing bridge, (*ii*) belt scale and (*iii*) automatic recording system. Figure 3.21 illustrates a typical layout of the coal handling plant of modern steam power plant. As mentioned in the foregoing paragraphs, coal is delivered to the power plants in

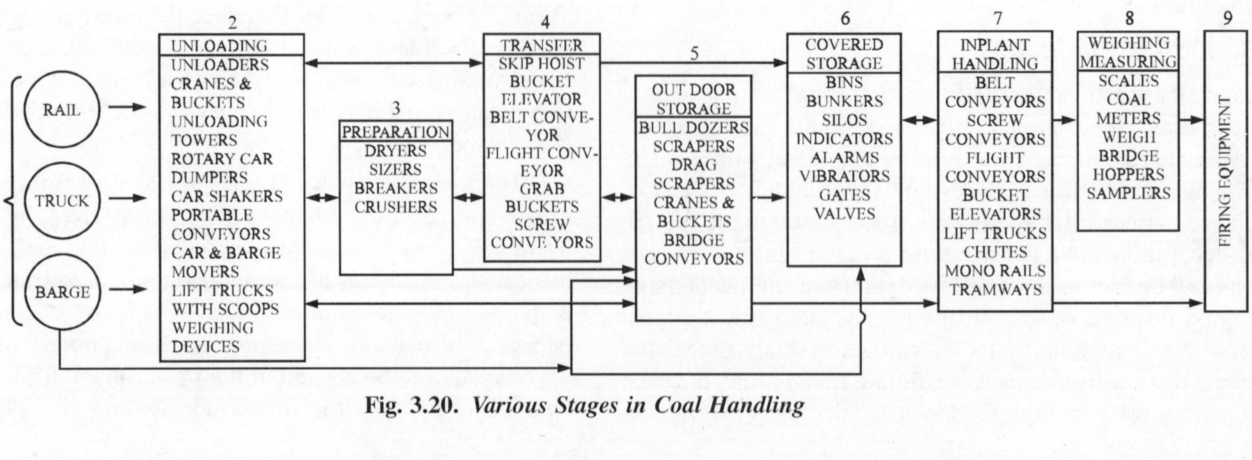

Fig. 3.20. ***Various Stages in Coal Handling***

trucks or railway wagons. It is weighed on balance and then unloaded into underground hoppers or bunkers. The wagon can be unloaded either manually or by using rotary wagon tipplers. In the later case, the wagon is moved into a rotor and is secured in it. Then the rotor is turned by electric drive and coal is unloaded from the open wagon. This considerably reduces the unloading time. From the bunkers, the coal is lifted by conveyor to the transfer tower from where it can be delivered either to the coal storage or by the conveyor to a crusher. After passing through magnetic separators and screens, coal is crushed in crushers into pieces 25-30 mm in size for stoker firing and 10-20 mm when pulverized fuel is fired in boiler furnaces. The crushed coal is then milled to a fine powder in the later case and then the fuel is carried through automatic weigher to a transfer tower where fuel is lifted to the level of the bunker gallery and distributed between boiler hoppers by a conveyor.

3.10. FUEL COMBUSTION AND COMBUSTION EQUIPMENT

Fuel is burnt in a confined space called the *furnace.* An efficient combustion of fuel is essential for economical operation of a power plant. The necessary requirements for efficient combustion of fuels are as follows.

(*i*) The proper quantity of primary or secondary air required for complete combustion.

(*ii*) Adequate stoker or grate area required for burning particular quantity of fuel.

(*iii*) Minimum operating and maintenance cost.

(*iv*) Attainment of proper designed temperature.

(*v*) There should be no formation of caking during burning of fuel.

(*vi*) The system should be easy to handle and dependable.

Although there are various designs and methods of firing but mainly there are two ways of firing namely *solid fuel firing and pulverised fuel firing.*

In case of solid fuel firing two general methods, namely hand firing and stoker or mechanical firing, are available.

In hand firing the coal is put into the furnace frequently by shovels. The primary and secondary air needed for combustion is regulated by dampers. Generally the grate of such boilers consists of bars over which the coal is put. This method of firing is simple and does not require any capital investment. This method can however be used only in small installations as uniformity of combustion is difficult to control in this type of firing. Further, adjustments for the supply of air are to be made every time coal is fed to the furnace.

3.10.1. Stoker Firing

Stokers give mechanical feeding of coal. Mechanical stokers receive fuel by gravity, carry it to the furnace for combustion and after combustion discharge the ash at the appropriate point. The advantages of stoker or mechanical firing over hand firing are given below.

(*i*) Uniform feeding of fuel into furnace.

(*ii*) Easy control of firing resulting in better regulation and efficient combustion.

(*iii*) Saving in labour cost.

(*iv*) Possibility of use of poor grades of coal due to better control.

(*v*) Greater combustion capacity.

(*vi*) Fluctuation of load demands can be met because of control of combustion.

(*vii*) Practically immune from explosions.

(*viii*) Very reliable, and maintenance charges are reasonably low.

However, the stoker or mechanical firing have the following drawbacks.

(*i*) Complicated construction.

(*ii*) Trouble due to slagging and clinkering of combustion chamber walls.

(*iii*) Loss of fuel in the form of riddlings through the grates.

(*iv*) Banking and standby losses are always present.

(*v*) Sudden variations in the steam demand cannot be met to the same degree.

(*vi*) With very large units the initial cost may be rather high than with pulverized fuel.

Mechanical stokers are of two types namely ***underfeed stokers*** *and* ***overfeed stokers.*** The two differ in the manner of feeding of coal below or above the level at which primary air is admitted in the furnace. The distinction between underfeed and overfeed stokers is made clear in Figs. 3.22 (*a*) and 3.22 (*b*). In the case of underfeed stokers, fuel is supplied into the furnace below the point of air admission, as shown in Fig. 3.22 (*a*), while in the case of overfeed stokers, the coal is supplied into the grate above the point of admission of air.

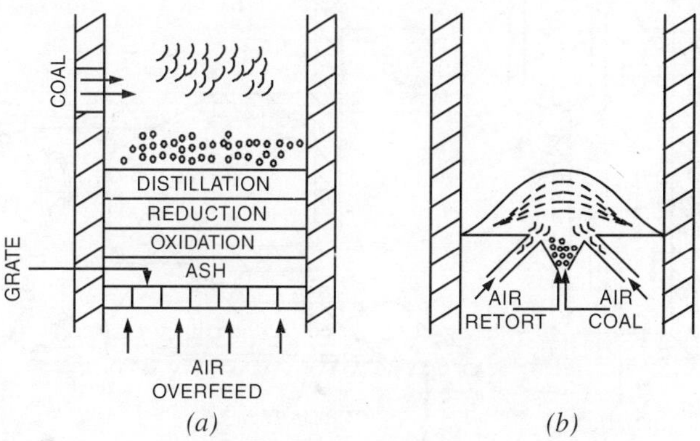

Fig. 3.22. ***Difference Between Underfeed and Overfeed Stoker Firing***

Underfeed Stokers. As already mentioned above, the coal is fed into the furnace below the point of admission of air. Coal from the hopper is pushed into the retort by means of reciprocating plunger. When the coal gets heated, all the volatiles in it are distilled and when coal reaches the zone of active combustion, it is in the form of coke and ash. The ash discharge plates are at the back of the furnace and by the time coal is pushed down on these plates, all the combustion

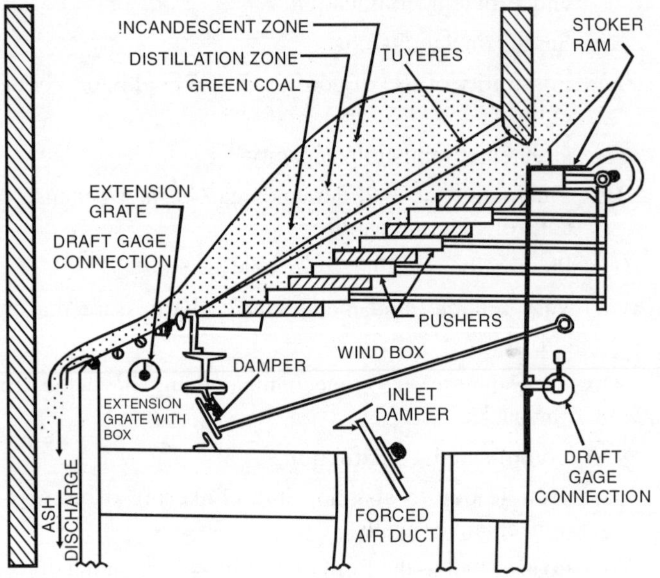

Fig. 3.23. ***Underfeed Stoker***

has been completed. Air is admitted into the furnace through holes in the retort sides. The coal is continuously agitated by the plunger and also by three pusher plates along the retort bottom. As such the fuel bed remains porous and free from clinkers. Underfeed stokers are stable for non-clinkering, high-volatile coals having caking properties and low ash content. Underfeed stoker is shown in Fig. 3.23.

Overfeed Stokers can be further classified as chain grate stokers, travelling grate stokers an spreader stokers.

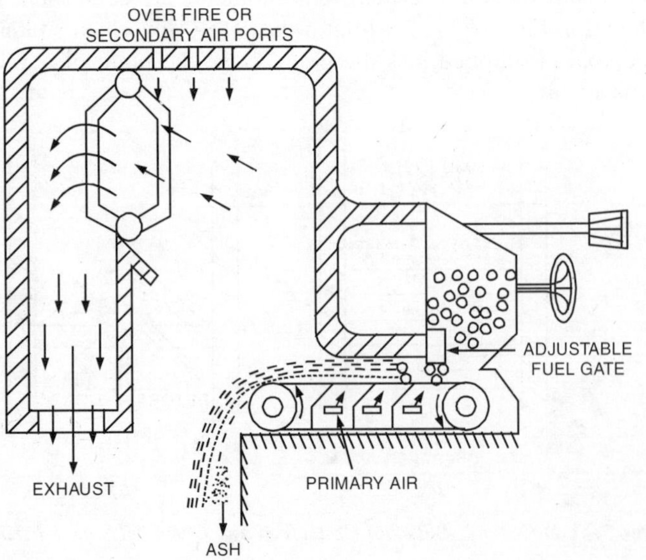

Fig. 3.24. ***Chain Grate Stoker***

Chain grate and *travelling grate stokers* are similar in appearance but differ in the grate construction. Chain grate consist of flexible endless chain as in crawlers of heavy machines and travelling grates consist of grate bars carried by steel chains. A chain grate stoker is shown in Fig. 3.24. The speed of the stoker is 127 mm to 508 mm per minute. An index plate with pointer indicates the thickness of coal bed at all times. This can be regulated either by adjusting the opening of the fuel gate or by the speed control of the stoker driving motor. The air is admitted from the underside of the grate which is divided into several compartments each connected to an air duct. The grate is to be protected against overheating and it is achieved by using coal having sufficient ash content. Since there is practically no agitation of the fuel bed, non-caking coals are best suited for chain grate stokers. Such stokers can burn about 150 kg of coal/m^2/ hour with natural draught and 200-300 kg of coal/m^2/ hour with forced draught.

In the ***spreader stoker*** the coal is not fed into the furnace by means of grate. The function of the grate is only to support a bed of ash and move it out of the furnace. From the coal hopper, coal is fed into the path of a rotor by means of a conveyor, and is thrown into the furnace by the rotor and is burnt in suspension. The air for combustion is supplied through the holes in the grate. Over fire air or secondary air to create turbulence and supply oxygen for the thorough combustion of coal is supplied through nozzles located directly above the ignition arch. Unburnt coal and ash are deposited on the grate which can be moved periodically to remove ash out of the furnace. Spreader stokers can burn any type of coal from lignite to semi-anthracite, whether they are free burning or coking. Their disadvantage is that fly ash is much more. Such stokers can be employed for boiler capacities from 70,000 kg per hour of steam to about 140,000 kg per hour.

Pulverised Fuel Firing. The use of pulverised fuel firing has revolutionized the design of boilers as regards their capacity, pressure of steam and its temperature. *Pulverization* is a means of exposing a large surface area to the action of oxygen and consequently helping the combustion.

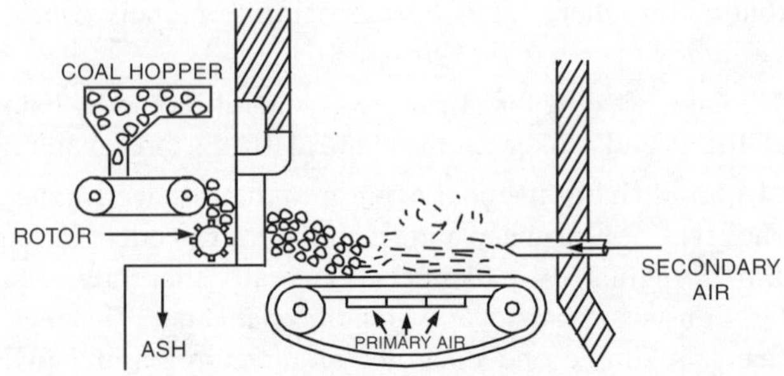

Fig. 3.25. ***Spreader Stoker***

Solid fuels can be used in a powdered form and burn like oil and gases. The coal is first dried usually by the gases, ground to a fine powder in a pulverised mill and then projected into the combustion chambers by means of a current of hot air. A further volume of preheated air to make up the necessary amount for combustion is blown in separately and the resulting turbulence in the high pressure combustion chamber helps in thorough combustion of the fuel. The air used to dry the coal and convey the powdered fuel to the furnace is called the primary air and the air blown in separately to complete the combustion is called the secondary air. The amount of primary air may range from as little as 10% to almost the full combustion air requirements depending on the type of the pulveriser and the load.

The pulverised fuel firing has the following advantages :

(*i*) In pulverised fuel firing better combustion is achieved due to use of hot air at temperature ranging from about 260ºC to 370ºC.

(*ii*) Surface exposure is increased resulting in rapid combustion without the use of large amounts of excess air.

(*iii*) Low grade of coal can be used since it is used in powdered form. Even fine wet coal can be used provided the conveying equipment can carry it to the pulverizing mill.

(*iv*) The rate of feed of the fuel can be regulated properly resulting in fuel economy.

(*v*) The surface area is increased in almost the ratio 400:1, therefore, high rates of combustion are possible. Moreover, a smaller quantity of air is required than when the fuel is burnt in lump form.

(*vi*) The smaller quantity of excess air and thorough mixing of air and fuel produce a high furnace temperature with little smoke.

(*vii*) Increased rate of evaporation and higher boiler efficiency because of almost complete combustion of fuel.

(*viii*) The system is practically free from slagging and clinkering troubles.

(*ix*) Ash removing troubles are reduced.

(*x*) No standby losses due to banked fires.

(*xi*) Firing of boiler becomes easy. The boiler can be started from cold conditions very rapidly.

(*xii*) Fluctuations of loads can be easily met.

(*xiii*) Automatic combustion control can be employed.

(*xiv*) Preheated air can be used successfully.

(*xv*) Larger capacity to meet peak loads.

(*xvi*) The pulverizing equipment is outside the furnace, therefore, it can be repaired without cooling down the unit.

However, the pulverised fuel firing have the following disadvantages:

(*i*) Coal preparation plant is required which makes the installation expensive in initial cost.

(*ii*) Lot of fly ash in the exhaust makes the removal of fine dust uneconomical.

(*iii*) There is a risk of explosion as coal is burnt like a gas. So skilled personnel are required.

(*iv*) Special equipment is required for starting this system.

(*v*) Higher combustion temperatures cause higher thermal losses in the flue gases.

(*vi*) The operating cost of pulverisation plant from energy utilisation point of view is also quite high.

(*vii*) The high furnace temperatures, unburnt fuel and the fluxing effect of ash etc. deteriorates the refractory material.

(*viii*) The plant requires electrostatic precipitator to reduce stack emissions to acceptable level.

The advantages of using pulverized fuel outweigh the disadvantages and all modern plants use pulverized fuel.

There are two systems of preparation of coal into a fine powder namely *unit* system and the *bin* or storage or *central system.* In unit system a separate pulverising unit is provided for each furnace and the fuel is fired directly into the furnace without being stored whereas in central system the fuel is pulverised in a central plant and stored in bins or bunkers. From the bunkers, the pulverised fuel is distributed to various burners through separate feeders in accordance with the load demands. Each system consists of crushers, magnetic separators, driers, pulverising mills, storage bins, conveyors and feeders.

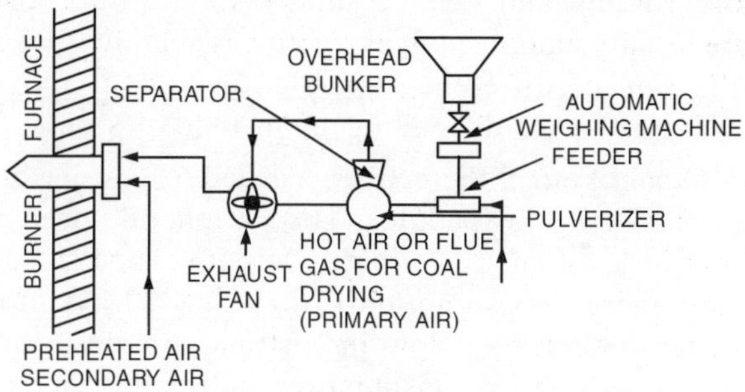

Fig. 3.26. ***Unit System***

In unit system (Fig. 3.26) the raw coal falls by gravity from the overhead bunker into a feeder where it is dried with the help of hot air. From the feeder, the coal passes on to the pulverising mill where it is crushed into a fine powder. The fine powder is extracted from the mill by means of an exhaust fan via a separator and blown into the furnace along with the hot air. In the separator, the big particles of coal are separated from the fine dust and these again fall down into the mill. It is simple, cheaper in installation and operation and easy in regulation. The drawbacks of the system are requirement of a reserve plant for each unit making the system more expensive. Moreover, the pulverising unit has to work under fluctuating load demands and the flow of fuel through the burners is not uniform.

In central system (Fig. 3.27) the crushed coal passes from the pulveriser to a separator where oversize coal particles are separated from the fine powder. The coal laden air then flows into a cyclon where the air is separated from the coal powder and returned to the mill circuit by means of an exhaust fan. From the cyclon the pulverised coal is conveyed to bunker by means of a screw conveyor. From the bunker the fuel is fed to the various burners through separate feeders. This plant is more reliable. The quantities of fuel and air can be regulated accurately. Moreover the pulveriser can work under constant speed irrespective of load demand. Also, the exhaust fan handles comparatively clean air. The drawbacks are requirement of large capital and large building space and higher operation and maintenance charges. Due to this, the central system has been superseded by the unit system.

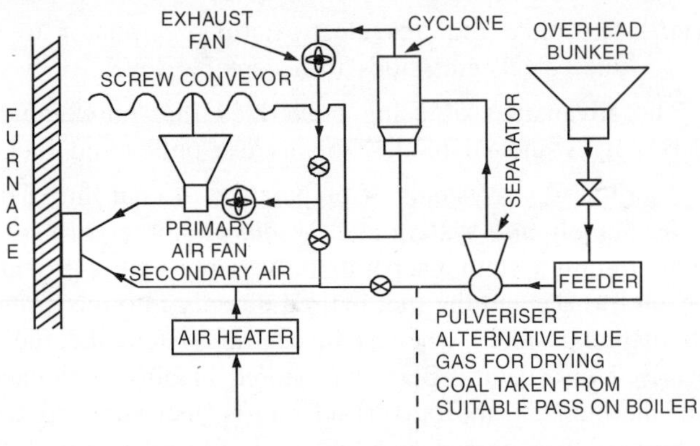

Fig. 3.27. ***Central System***

Pulverisers. There are three principles, which are employed in pulverising of coal, namely (*i*) impact (*ii*) attrition or grinding and (*iii*) crushing. All the three principles are usually employed in every pulverising mill, degrees of utilisation of different principles in various pulverising mills may vary depending upon their design.

In *impact system* the coal is crushed to powder due to hammering action of paddles. The peripheral speed is of the order of 5,400 metres/minute. Output is about 5 tonnes/hour and energy consumption is about 18-20 kWh/tonne. The system is compact, low in cost and simple in operation. However energy consumption and maintenance cost are high and it is difficult to maintain the finess of coal after pulverisation.

In *attrition* or *grinding of coal,* coal is grounded by the movement of a revolving roller against a race. The peripheral speed is about 180-210 metres/minute. The output is about 10 tonnes/hour and energy consumption is roughly 10-15 kWh per tonne. The plant occupies less space and is quiet in action. However, the initial cost is higher.

In *crushing system*, steel balls are allowed to move in a steel cylinder lined with renewable cast iron liner plates on the inside. The drum is supported at each end on bearings. Owing to continuous tumbling action, the steel balls strike against coal pieces reducing them to powder by combined impact, attrition and crushing action. Peripheral speed is about 120-135 metres/minute. Output is 20 tonnes/hour and energy consumption is about 20-25 kWh/tonne. The system is simple in operation, cheaper in initial cost, built costly in operation cost. There is considerable quantity of coal in the mill which acts as a reservoir.

All the above three principles are illustrated in Fig. 3.28.

The pulverising mills may also be classified according to the manner of air supply. In one type, the exhaust fan handles both air and powdered coal and is on the outlet side of the mill. In other type, the fan is on the inlet side of the mill and provides the necessary air current. In this system, the fan handles only clean air.

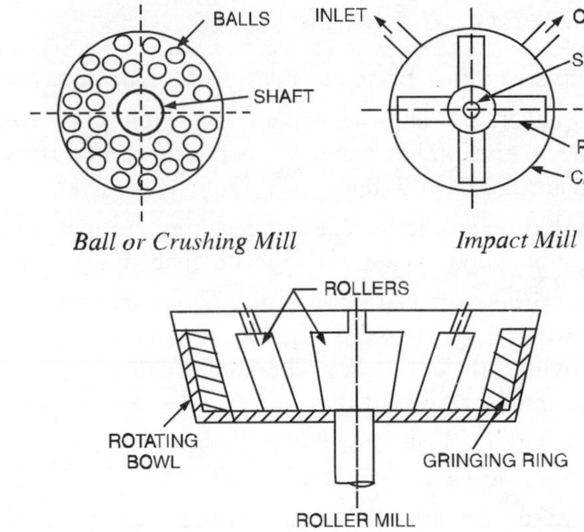

Fig. 3.28. ***Pulverisers***

3.11. COAL BURNERS

The coal burners are employed to fire the pulverised coal along with primary air into the furnace. The secondary air is admitted separately below the burner, around the burner or elsewhere in the furnace. The main requirement for a good coal burner is capability of producing uniform and stable flame with almost complete combustion of the fuel. Coal burners can be classified according to their design and by their arrangement in the furnace. In one type of firing, say in *opposed firing*, the burners are placed on the opposite walls of the furnace and they fire directly against each other causing intimate mixing of the fuel and air. In second type of firing, say *cross firing*, the burners fire in the vertical direction and in horizontal direction and the fuel streams intersect with each other. And in third type of firing, called the *tangential firing*, the burners are placed in the corners of the furnace and they send horizontal streams of air and fuel tangent to an imaginary circle in the centre of the furnace (Fig. 3.29). Tangential firing results in intense turbulence and thorough mixing of the fuel and air. All the fuel and air nozzles can be titled 24° above and below the horizontal.

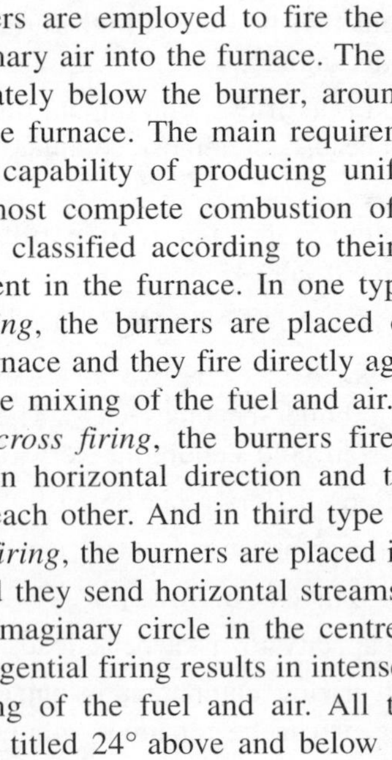

Fig. 3.29. ***Tangential Firing***

For firing crushed coal (not pulverised fuel) *cyclon burners* are employed. In such burners the crushed coal (max 6 mm size) from the feeder, and the primary air enter with a vortex motion at the centre of cyclon The secondary air is admitted separately in the vortex motion. The fuel is quickly burnt and ash in the form of molten slag drains down the inner wall of the cyclon. Hot flue gases with 10-20% of the ash in the coal in the form of fly ash enter the furnace. Due to centrifugal action, most of this fly ash is thrown against the walls of the furnace and is drained away along with the molten film of slag. The flue gases leaving the furnace are quite clean to flow through the rest of heat exchanger passages. Thus better heat transfer and good combustion is obtained. Also frequent cleaning requirement of furnace and fly ash trouble are very much reduced.

3.12. FLUIDIZED BED COMBUSTION

Fluidization is a new method of mixing fuel and air for obtaining combustion. A fluidized bed may be defined as the bed of solid particles behaving as a fluid. It operates on the following principle.

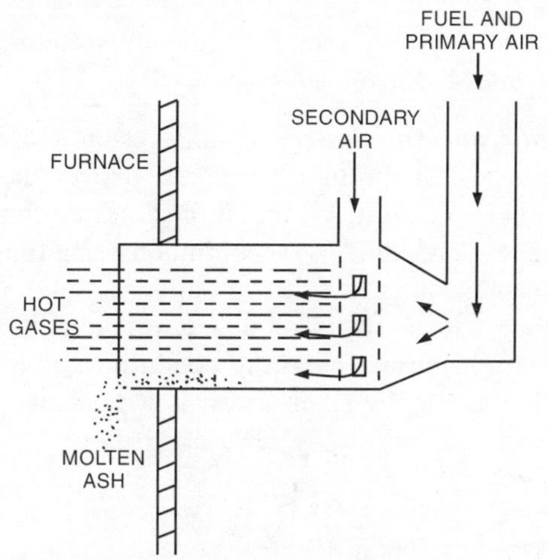

Fig. 3.30. *Cyclone Furnace*

When an evenly distributed air is passed upward through a finely divided bed of solid particles at low velocity, the particles remain undisturbed but if the velocity is steadily increased, a stage is reached when the individual particles are suspended in the air stream. If the air velocity is further increased, the bed becomes highly turbulent and rapid mixing of particles occur which appear like formation of bubbles in a boiling liquid and a bed is said to be fluidized.

The velocity of air causing fluidization depends on a given particle size, range and density. In a fluidized bed combustion, rapid mixing ensures uniformity of temperature. The main advantage of such a combustion system is that municipal waste, sewage plant sludge, biomass, agricultural waste (such as rice straw) and other high moisture fuels can be used for generating heat.

A fluidized furnace has an enclosed space with a base having openings to admit air. Crushed coal, ash and either crushed dolomite or limestone are mixed in the bed furnace and high velocity combustion air is then passed through the bed, entering from the furnace bottom. With the steady increase in the velocity of air, a stage will be reached when the pressure drop across the bed becomes equal to the weight per unit cross section of the bed and this critical velocity is called the *minimum fluidizing* velocity. With the further increase in velocity of air, the bed will expand and allow passage of additional air, in the form of bubbles. When the air

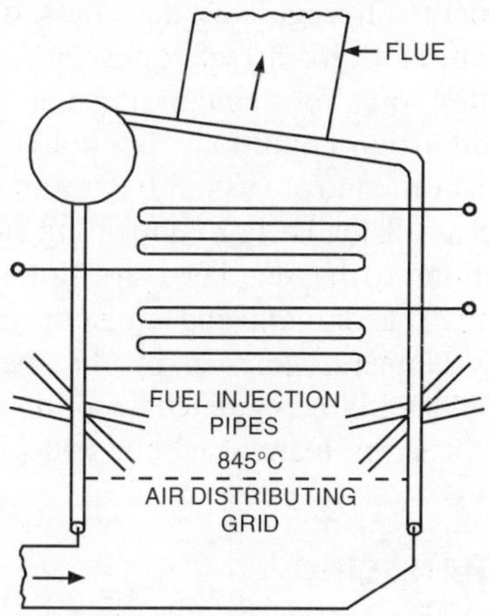

Fig. 3.31. *Fluidized Bed Combustion System*

velocity becomes 3-5 times the critical velocity, the bed resembles a violently boiling liquid. The evaporator tubes of boiler are directly immersed in the fluidized bed and the tubes, being in direct contact with the burning coal particles, produce very high heat transfer rates. Because of this, the unit size is reduced and also very low combustion temperatures are produced. Vertical type fluidized bed combustion (FBC) boilers are available up to 6 tonnes per hour, whereas the horizontal type FBC boilers are available in a capacity of 8 to 60 tonnes per hour.

There are two basic systems of fluidized bed combustion namely *atmospheric* FBC and *pressurized* FBC. The pressurized FBC system has the advantages of high burning rates, reduced volume, increased coal loading in comparison to atmospheric FBC, but has the drawback of difficult controls and short plant life.

The advantages of fluidized bed combustion are high thermal efficiency, easy ash removal system, short erection and commissioning period, fully automatic and safe operation, multi-fuel operation, reduced maintenance, uniformity of temperature, efficient operation at temperatures down to 150°C (well below the ash fusion temperature), reduced unit size and capital costs due to high heat transfer rates, reduced fouling and corrosion of tubes due to low combustion temperatures, reduced coal crushing cost etc.

The system can respond rapidly to changes in load demand, due to quick establishment of thermal equilibrium between air and fuel particles in the bed.

The operation of fluidized bed furnace at lower temperature helps in reducing air pollution. The low temperature operation also reduces the formation of nitrogen oxides. By adding either dolomite (a calcium-magnesium carbonate) or lime stone (calcium carbonate) to the furnace the discharge of sulphur oxides to the atmosphere can also be reduced if desired.

The major drawback of this system is that the fan power is considerably increased since the air is to be supplied at a high pressure for supporting the bed.

3.13. AUTOMATIC COMBUSTION CONTROL

Automatic combustion control is now in general use for large power plants. It regulates automatically changes in demand of steam and also affects quick and stable changes in other variables so as to maintain constant steam pressure and proper combustion conditions. It is obvious that an automatic combustion control system has to maintain constant steam pressure at all loads. For controlling rate of combustion the conditions to be regulated are draught, air supply and fuel supply. The draught and air supply may be regulated either by dampers, vanes or by the speed of the fans or both. The fuel supply is regulated according to the method of firing used. It saves manual labour and increases operation efficiencies.

3.14. ASH HANDLING

Coal contains a considerable amount of ash. The percentage of ash in the coal varies from about 5% in good quality coals to about 40% in poor quality coals. Generally poor quality coal is used in steam power plants and, therefore, a steam power plant produces hundreds of tonnes of ash daily (a modern 2,000 MW steam power plant produces about 5,000 tonnes of ash daily). For removal of ash from the boilers and its disposal to the suitable site is quite difficult and quite elaborate equipment is required. Ash handling comprises the following operations:

(*i*) Removal of ash from the furnace ash hoppers.

(*ii*) Transfer of this ash to a fill or storage and

(*iii*) disposal of stored ash.

The ash can be disposed off in the following ways.

(*i*) Waste land sites may be reserved for the disposal of ash.

(*ii*) Building contractors may utilise it to fill the low lying areas.

(*iii*) Disused quarries within reasonable distance of the power plant may be employed for dumping the ash into the evacuated land.

(*iv*) Deep ponds may be made and the ash can be dumped into these ponds to fill them completely. When such ponds are completely filled, they may be covered with soil and seeded with grass.

(*v*) When seaborne coal is used, barges may take the ash to sea for disposal into a water grave.

The **difficulties experienced** in handling and disposal of ash are as follows :

(*i*) The ash coming out of the furnace is very hot.

(*ii*) The ash is dusty and so irritating and annoying in handling.

(*iii*) The ash is abrasive and wears out the containers.

(*iv*) It produces poisonous gases and corrosive acids when mixed with water.

(*v*) It forms clinkers by fusing together in lumps.

The **chief requirements** of a good ash handling plant are :

(*i*) Capability to operate with minimum personal attention and handle large clinkers as well as dust and soot.

(*ii*) Minimum operation and maintenance charges.

(*iii*) Adequate capacity to deal with the ultimate plant capacity.

(*iv*) Speedy disposal of ash.

(*v*) Ability of handling both dry and wet ash.

(*vi*) Operation with little noise and minimum dust menace.

Theoretically, whole of the ash from the furnace should get deposited in the ash hoppers, but actually 5 to 40% of it leaves the furnace with the outgoing gases. The small power stations use some conveyor arrangement to carry ash to dump sites directly or for carrying and loading it to trucks and wagons which transport it to the disposal site. Large power stations use more elaborate arrangements.

The modern ash handling systems usually used in large steam power plants are :

1. Belt conveyor system 2. Pneumatic system 3. Hydraulic system and 4. Steam jet system.

1. *Belt Conveyor System*. In this system (Fig. 3.32) the ash is made to fall through a water seal over the belt conveyor in order to cool it down and then carried to a dumping site over the belt. This is a continuous handling and the power consumption is low. It can deliver around 3 tonnes of ash per hour with a speed of 0.3 m/minute. Life of belt is about 5 years in normal working conditions. This system of ash handling is employed in small power plants.

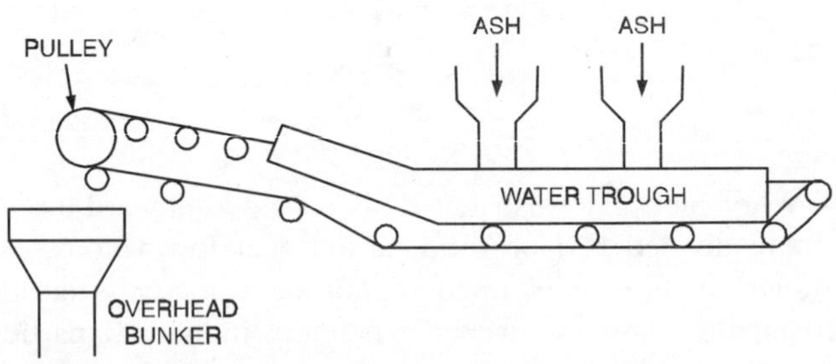

Fig. 3.32. *Belt Conveyor System*

2. *Pneumatic System.* In this system air is employed as the medium for driving the ash through a pipe over long distances. The ash is passed on to the crushers from the boilers and then into the conveying pipe. Air is sucked through the delivery end which makes the ash to flow into the separators where the ash is collected in ash hoppers. The dusty air is filtered and exhausted to atmosphere through the exhaust fan. The system can handle 5–30 tonnes of ash per hour. The advantages of this system are that it can carry ash through long distances and dust menace is reduced. Disadvantages of this system are that large amount of wear and tear results in the conveying pipe, high labour and maintenance charges and noisy operation. This system is usually employed for disposal of fly ash. This system is shown in Fig. 3.33.

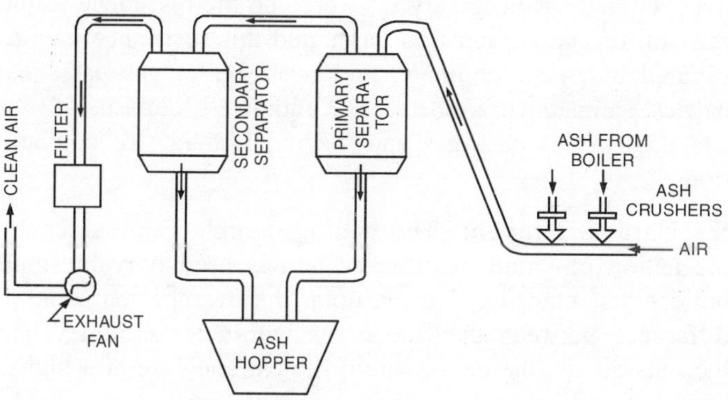

Fig. 3.33. ***Pneumatic Conveying System***

3. ***Hydraulic System.*** In this system a stream of water carries ash along with it in a closed channel and disposes it off to the proper site. This system can be used for large capacity power plants where the ash is to be disposed off over long distances. This is a healthy, clean, dustless and completely enclosed system. This system can also handle molten ash breaking it into small sizes.

Hydraulic systems are of two types namely (*i*) high pressure system and (*ii*) low pressure system. High pressure system operates intermittently whereas low pressure system is continuous one. In a high pressure system the hoppers below the boilers are fitted with water nozzles both at the top and on sides. The top nozzles quench the ash while the side ones provide the driving force for the ash. The ash and water then flow along a trough to a receiving hopper or sump where the ash is separated from the water. The water is used again while the ash is sent out through carriages. High pressure system is more expensive to install and operate.

In a low pressure system a trough or drain is provided below the boilers and water is made to flow through the trough. The ash directly falls into the trough and is carried by the water to the sumps and tanks. In the sump, the water and ash are made to pass through a screen in order to separate them out. The water is pumped back to the trough and used again while the ash is removed through carrier units and sent out to the dumping site.

4. ***Steam Jet System.*** This system employs jets of high pressure blowing in the direction of ash travel through a conveying pipe in which the ash from the boiler ash hoppers is fed. Chilled iron or nickel alloy are used in lining the pipe as ash is abrasive and travels at a high speed. This system is useful where installation of other ash handling plants and conveyors is not possible because of lack of space or where the path of travel of ash is not straight. This system is employed in small and medium size plants. Steam consumption is about 110 kg per tonne of material conveyed.

3.15. DUST COLLECTION

The exhaust gases leaving the boiler contain particles of solid matter in suspension-smoke, dust, soot, fly ash or carbon as material called '*cinder*'. The quantity of these solid particles largely depends upon the method of fuel firing. Flue dust is greatest with pulverised fuel and spreader stoker firing systems and is much less with underfeed stoker systems. In case of pulverized fuel firing, 60 to 80 per cent of the total ash produced in the furnace, escapes through the chimney as flue dust. Removal of dust from the exhaust gases is very important. The environment gets polluted and suspended matter can fall on the neighbour residential premises and create legal liabilities for the authorities. Cleaning of gas poses difficult problems because of large percentage of silica in the ash and finess of typical fly ash.

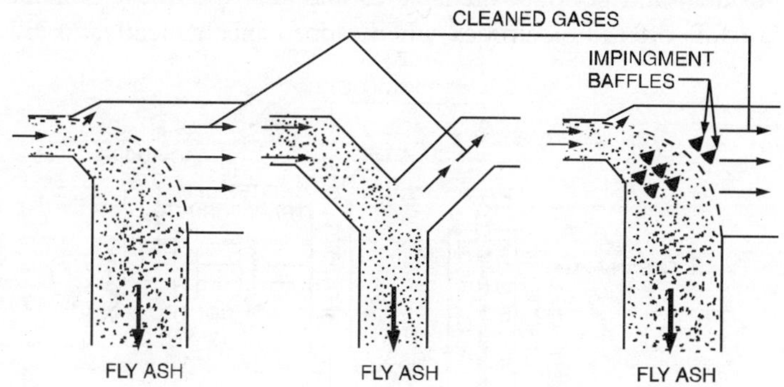

(*a*) (*b*) (*c*)

Fig. 3.34. ***Dry Type Mechanical Dust Collectors***

Gas cleaning devices make use of certain physical/electrical properties of the particular matter of the gas stream. Basically gas cleaning devices called the *dust collectors* may be classified into mechanical and electrical ones (electrostatic precipitators). Mechanical dust collectors have efficiency increasing with load while the efficiency of electrostatic precipitators falls with the increase in load. Combination of the two, giving constant efficiency characteristic, is often employed. Heavier dust particles are removed by mechanical *dust collectors* and finer particles are eliminated by electrostatic precipitator.

Mechanical dust collectors can be further classified as wet and dry dust collectors. In wet type units, dust is washed away from the flue gases by spraying water on it. This system is usually not used because it need large amounts of water. The principles employed in case of dry type mechanical dust collectors are shown in Fig. 3.34.

In Fig. 3.34 (*a*) the dust area is increased causing reduction in gas velocity and thus settle out heavier dust particles. In Fig. 3.34 (*b*) there is a sudden change in the direction of flow of the gas resulting in the settling out of the heavier particles which cannot flow along the gas. In Fig. 3.34 (*c*) the dust particles strike the baffles placed in the path of the exhaust gases and the dust particles will settle out.

3.15.1. Electrostatic Precipitator

It essentially consists of two sets of electrodes which are completely insulated from each other and a high voltage electrostatic field is maintained across them. One set, called the emitting or discharge electrode, is in the form of thin wires and the other set is called the collecting electrode. The emitting or discharge electrodes are placed in the centre of a pipe in case of tubular type precipitator (or midway between two plates in case of plate type precipitator) and are connected to negative polarity of HVDC source (25 to 100 kV) while the collecting electrodes are connected to the positive polarity of the source and are earthed. High electrostatic field thus set up between the two sets of electrodes creates corona discharge and ionises the gas molecules as the flue gas flows through the tube or in between the plates. The dust particles in the gas acquire negative charge and are attracted to the electrodes connected to the positive polarity (collecting electrodes) and get deposited there. The deposited dust is made to fall off the electrodes when rapped mechanically.

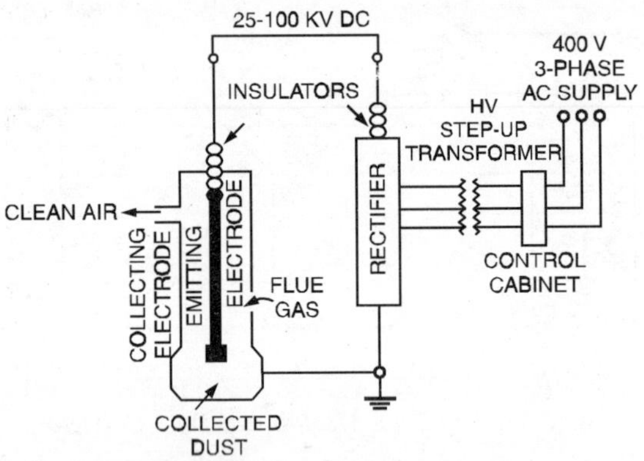

Fig. 3.35. *Electrostatic Precipitator*

The electrostatic precipitator may be (*i*) plate or tubular type (*ii*) horizontal flow or vertical flow type (*iii*) dry or wet type. In case of wet or irrigated precipitator, the deposited dust is removed by a water film flowing down on the inner side of the collecting electrodes. Usually, a dry type electrostatic precipitator is employed.

The electrostatic precipitator has the advantages of high efficiency (up to 99.5%), low pressure drop, easy removal of collected particles and capacity of handling large volume of flue gases.

The limitations of this system are high capital and operating costs, and requirement of more space. The maximum collecting efficiency is maintained only for one value of gas velocity for which the system is designed.

3.16. DRAUGHT SYSTEMS

In a boiler the combustion of the fuel requires supply of sufficient quantity of air and removal of exhaust gases and this is achieved by *draught system*.

The circulation of air is caused by a difference in pressure, known as *draught*. Thus the draught is the difference in pressure between the two points *i.e.*, atmosphere and inside the boiler. It is measured in mm of water. A differential in draught is required to cause flow of air or gas through the boiler setting. This required differential is proportional to the square of the rate of flow.

The draught may be produced either naturally or artificially (mechanical). Natural draught is produced thermally and the mechanical draught is produced by means of a fan or a steam jet. Steam jet is mostly used in steam locomotives. Draught is necessary to overcome the resistance to the flow of air and combustion gases and this resistance is contributed by ducts, chimney, fuel beds, boiler gas passages, baffles, dampers, air preheaters, economisers, deposits of soot and dust in gas passages and the required rate of combustion.

Natural Draught. The natural draught is provided by the action of chimney or stack and is used only in small boilers. Its intensity depends upon the average temperature difference between the flue gases within the chimney and the outside air (the gases within the chimney are at a higher temperature than that of the surrounding air) and also on the height of the chimney above the level of the furnace grate. Its intensity is also affected by weather conditions and boiler operating conditions. Chimney, in addition of providing natural draught, helps in reducing air pollution too, as it delivers the products of combustion and fly ash to a high altitude.

Mechanical Draught. Artificial or mechanical draught is provided when the natural draught caused by a chimney is not sufficient or where a certain draught is required to be maintained irrespective of weather conditions or boiler operating conditions. In case of large steam boilers where economisers and air preheaters are employed, the exit temperature of the flue gases is sufficiently lowered and also the volume of air required is tremendously high. In such cases the height of the chimney to cause the required draught may be excessive in height and cost. Under such conditions, mechanical draught is essential. The boilers with mechanical draught do not require a chimney of such a height as is necessary with natural draught.

In a mechanical draught system, the movement of air is due to the action of a fan. A mechanical draught may consist of induced draught or forced draught or both. The fans, in all these cases, have high efficiency aerofoil blades inclined backward to the direction of rotation. In an *induced draught system*, the blower is installed near the base of the chimney and the burnt gases are sucked out of the boiler, reducing the pressure inside the boiler to less than atmospheric one. This induces fresh air to enter the furnace. In case of a forced draught system the blower is installed near the base of the boiler. In this system of draught the air pressure throughout the system is above the atmospheric pressure and air is forced to flow through the system (furnace, economiser, air preheater and chimney).

In most of the cases, a forced or induced draught alone is not sufficient and to achieve advantages of both the systems a *balanced draught system* (combination of induced draught and forced draught systems) is used. Figure 3.36 illustrates the flow diagram of flue gases with a balanced draught system. In this system, the pressure of air below the boiler grate is above atmospheric and that above the grate is below atmospheric. The induced draught blower helps in sucking the products of combustion out of the system whereas the forced draught helps in overcoming the resistance inside the boiler, thus enabling supply of sufficient air to the fuel bed for proper and complete combustion.

Mechanical draught, though needs additional investment and maintenance, is essential for proper operation of modern large size boilers.

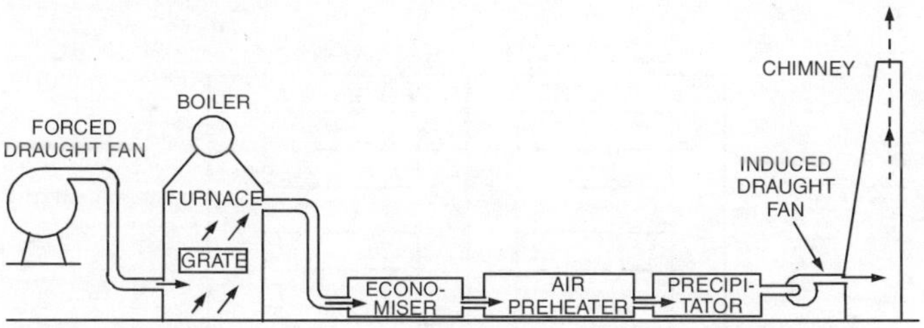

Fig. 3.36. *Flue Gas Flow Diagram with Balanced Draught*

3.17. FEED WATER

The steam coming out of turbine is condensed and the condensate is fed back to the boiler as feed water, as shown in Fig. 3.3. Some water may be lost due to blow-down, leakage etc. and to make up these losses additional water, called the *make-up water*, is required to be fed to the boiler. The make-up water in a modern thermal plant is about 1–4%. In a large size plant this quantity may be around a few hundred tonnes per hour.

The source of boiler feed water is generally a river or lake which may contain suspended and dissolved impurities, dissolved gases etc. The suspended impurities include mud, silt, clay and silica either in suspension or in colloidal form. The dissolved impurities include carbonates, bicarbonates, sulphates and chlorides of calcium, magnesium and sodium, iron oxide, silica. Dissolved gases include oxygen and carbon dioxide. These impurities in feed water may lead to scale formation, corrosion, carry over and embrittlement in boiler and other apparatus. Scale formation reduces the heat transmission through the heating surfaces and causes the overheating of the boiler tubes and the shell plates. Corrosion produces pits, grooves and cracks or a general wastage of material. Embrittlement leads to accelerated cracks in the edges of rivetted boiler plates, rivets and the metal around opening which has been highly stressed during its fabrication and operation. *Foaming and priming* of the boiler water forms the carry over. It causes filling up of steam passages with a mass of frothy bubbles.

It is necessary to heat and purify the water before feeding to the boiler.

The heating of feed water (*i*) improves the overall efficiency of the plant (*ii*) removes dissolved oxygen and carbon dioxide (*iii*) causes precipitation of other impurities carried by steam and condensate outside the boiler and (*iv*) avoids thermal stresses owing to entry of cold water into the boiler. Two types of heaters are used, namely open or contact type heaters and closed or surface type heaters for heating of feed water. Open or contact type heaters are used in small plants. Such heaters receive steam from back pressure turbine or engine driving auxiliaries. In large modern plants heaters bleed steam from the main turbine and use it for heating of feed water. The feed water is heated, put under pressure and then further heated so that its temperature approaches and pressure exceeds that of water in the boiler.

The water is treated for removal of suspended and soluble solids and removal of gases. The various methods used for water treatment are :

(*i*) Mechanical (sedimentation and filtration) (*ii*) thermal (distillation and deaerative heating) and (*iii*) chemical (lime treatment, soda treatment, lime soda treatment, zeolite treatment and demineralisation).

Sedimentation involves the water to standstill or flow with a very low velocity and causes the solid matter to settle down which can be removed periodically or continually. Clear water drains from the surface of the settling chamber. In *filtration* the dirty water is allowed to flow downward through the beds of material like graded sand or anthrafilt. The suspended matter adheres to the filter material and the clear water drains from the bottom. In *distillation* the raw water is evaporated with the help of steam, vapours are collected and condensed to give supply of pure feed water. In *dearerative* heating the water is broken up into fine droplets and heated to produce vapour within the deaerator and in the process air, CO_2 and other gases are separated from the vapour as it condenses. Steam is used for heating in deaerators.

Lime treatment is suitable for the treatment of carbonate hardness, carbon dioxide in the water, either in a free state or in bicarbonate combination. In the process lime is taken up in hydrated form and relatively insoluble precipitate of calcium carbonate and magnesium hydroxide are formed. The process is best carried on in large tanks from which the softened water may be taken out and the sludge removed from the treating plant. *Soda treatment* process is suitable for water containing non-carbonate hardness and soda ash or sodium carbonate are usually employed in the process. The decomposition of the sulphate occurs together with the precipitation of insoluble calcium carbonates and sodium sulphate. *Lime soda treatment* is used when both carbonate and non-carbonate hardness exist. The lime takes up the carbon dioxide not affected by the soda used in the treatment of non-carbonate hardness. *Zeolite treatment* is one of the *ion-exchanging* process wherein cations of two or more positive charges are exchanged for sodium or hydrogen and zeolite is either a natural or an artificial material that has the property of being able to exchange its base. Natural zeolites have a sodium base that can be exchanged for calcium or magnesium in raw water and the water will be softened. The action of sodium zeolite, Na_2Z on water are :

$$CaSO_4 + Na_2Z = Na_2SO_4 + CaZ$$
$$Mg(HCO_3)_2 + Na_2Z = 2NaHCO_3 + MgZ$$
$$Ca(HCO_3)_2 + Na_2Z = 2NaHCO_3 + CaZ$$

Thus as the raw water flows into a zeolite bed and emerges as soft water, there is loss of sodium and gain of calcium or magnesium by zeolite. Therefore the bed will become exhausted and fail to soften the water. However, zeolite has the property of regeneration of its strength through back washing with a sodium solution. The advantage of this process is that hardness is disposed off as a solution instead of solid requiring filtration. In *demineralisation* water may be freed of its mineral content by evaporation or by a series of cation and anion exchanges to produce essentially distilled water. It is usually the most economical method to produce make-up water for high pressure boilers.

3.18. STEAM POWER PLANT CONTROLS

Until few years back, even in case of large power plants the various controls used to be accomplished manually on the basis of instrument readings. However, now the various controls involved in the power plant operation have been completely automated resulting in (*i*) increased labour productivity (*ii*) improvement in the safety of operation and reliable functioning of the various instruments and equipment.

A number of controls, such at the boiler, turbine and generator unit are provided in a steam power plant so as to maintain the best conditions at all loads. Automatic combustion control has already been discussed in Art. 3.13.

Turbine governing is affected by throttling the steam at the main valve (thus reducing the steam pressure and, hence mass flow also) or by reducing only the steam mass flow by cutting off one or more nozzles through which the seam enters the blades. The first method of governing, known as *throttle governing* or *qualitative governing*, is used in case of small turbines and the second method of governing, known as *nozzle governing* or *cut-off governing* or *quantitative governing*, is used for large turbines. Maintenance of proper vacuum in the condenser, enough circulating water, a number of pumps, oil pressure for control of circuits, steam bleeding if any and the heater and feedwater control are other requirements for the turbine.

In case of an isolated generating unit, increase in load causes reduction in the speed of the unit and hence reduction in frequency. However, in case of generator connected to infinite bus-bars the load shared by the unit can be adjusted by adjusting the turbine speed. In this case frequency remains constant.

In general, centralized control is employed for modern steam power plants, the boiler and turbine control being at one place in the turbine room and the generator and feeder controls in the control room, in some cases all controls are centralized in one room, called the control room.

3.19. LAYOUT OF STEAM PLANT MAIN BUILDING

The layout of equipment in the boiler house and the machine room in the main building should be such that the following needs are met with:

1. Convenient erection, maintenance and supervision of the equipment and auxiliaries.
2. Efficient performance and minimum construction cost.
3. Good illumination and ventilation of the workplaces.
4. Dependable and trouble-free operation of the plant.
5. Provision for future expansion.

In modern steam power plants, a closed equipment layout is used. The boiler room and the machine room (where the turbo-generators are installed) adjoin each other and are arranged in parallel. In the boiler room, the boilers are usually installed in one row. As regards the installation of turbo-generators in the machine room, there are two possible arrangements as shown in Figs. 3.37 and 3.38.

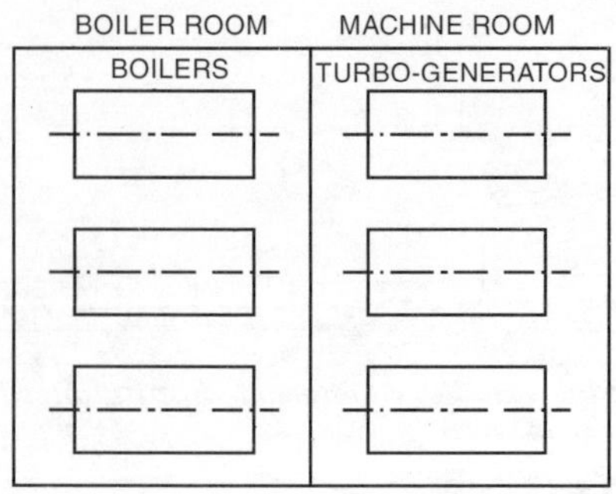

Fig. 3.37

In a perpendicular arrangement (Fig. 3.37), the turbo-generators are installed perpendicular to the centre line of the machine room. With such an arrangement, the span of the machine room is increased but the room length is reduced. Also, the throttle valve of each turbine can be fitted near the partition wall. This reduces the steam pipe length. This arrangement is common practice when unit operation (one boiler for each turbine) is envisaged. The boilers can be installed on the turbine centre line.

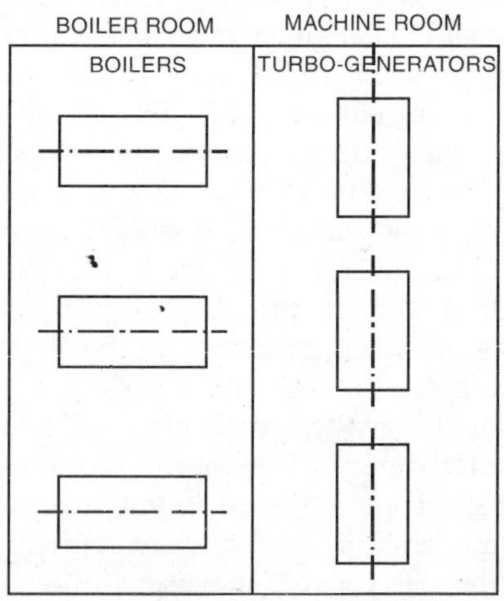

Fig. 3.38

The alternative arrangement is parallel arrangement (Fig. 3.38) where turbo-generators are placed parallel to the partition wall. Here, the turbine sets are placed much close together. The width of the machine room is less as compared to perpendicular arrangement. Also, more turbo-generators can be installed without increasing the width of the machine room.

The large capacity fuel bunkers holding enough fuel for two or three shifts are arranged either in a separate building or between the boiler and machine rooms. In the former case, the distance between the boiler units and turbines gets reduced significantly, but the crossing of the gas flues with pulverised-fuel ducts becomes rather involved. The ash room usually accommodates the coal pulverising mills and the other associated equipment.

The auxiliary equipment of boiler room and the machine room is sometimes installed in a special multistorey building constructed between them. The draught equipment and the ash collectors are usually located on the level of zero datum.

3.20. STEAM POWER PLANT AUXILIARIES

The equipments which help in the proper functioning of the plant are called *plant auxiliaries.* The various plant auxiliaries can be grouped under the subheadings of boiler auxiliaries, coal and ash auxiliaries, turbo-alternator auxiliaries and miscellaneous ones.

The **boiler auxiliaries** depend upon the type of fuel firing employed. In the case of *stoker firing*, stoker drives are essential along the forced and induced draught fans, boiler feed pumps, secondary air fans, air preheaters, soot blowers etc. In the case of *pulverized fuel firing*, the various auxiliaries are primary air fans, forced and induced draught fans, feed pumps, pulverized fuel conveyors and feeders, pulverizing mills, exhausters, air heaters and soot blowers etc. For *oil firing* the auxiliaries are fuel oil pumps and associated auxiliaries.

Coal and ash auxiliaries include wagon tipplers, elevators, skip hoists, conveyors, cranes, pumps, exhausters etc.

Turbo-alternator auxiliaries include circulating water pumps, condensate extraction pumps, governor control, evaporator, sludge and distillate pumps, ventilating fans, oil pumps, oil purifier, exciter, exciter field rheostat, turning gear etc.

Miscellaneous auxiliaries include air compressors, water and fire service pumps, workshop machinery and equipment.

All the above plant auxiliaries can be divided into two categories namely essential or continuous auxiliaries and non-essential or non-continuous auxiliaries. The essential auxiliaries are those which are associated with the running of a unit and whose loss would cause an immediate reduction in the output of the unit. The non-essential auxiliaries are those which may be put out of operation for some time.

The essential auxiliaries include forced and induced draught fans, feed water pumps, circulating water pumps, condensate pumps, auxiliary oil pump, ventilating fans for alternators, lighting and switchgear tripping circuits. In addition to these essential auxiliaries, the other essential auxiliaries required depend upon the type of boiler firing employed such as stoker drive in case of stoker firing, primary air fans and pulverized fuel feeders in case of central system of pulverized fuel firing, mills and feeders in case of unit system of pulverized fuel firing while fuel oil pumps and associated auxiliaries with oil fuel firing.

Non-essential or service auxiliaries include coal and ash handling plants, service pumps, oil pumps, exciter field rheostats, hoists and overhead cranes, valves, governor control, evaporator auxiliaries etc.

Auxiliaries in a steam power plant account for 4 to 8 per cent of station load.

3.20.1. Auxiliary Supply Features

There should be two main sources of supply (*i*) from the grid via the station transformers and (*ii*) from the main generators through the unit transformers. The essential auxiliaries should be supplied from the unit transformers while non-essential auxiliaries from the station transformers. Each generating unit should have a separate supply and failure of one unit should not affect the other. The unit board and station board should be interconnected for starting purposes and to provide a supply in case of failure of unit supply. A standby supply should also be provided so that a unit can be brought back to service in the event of failure of one of the components in any of the normal supply circuits.

For a 200 MW unit, the rating of unit transformer is about 12.5 MVA and the rating of station transformer is about 20 MVA. Such a station transformer can supply starting load of one unit (about 8 MVA), essential auxiliary load of one unit (about 8 MVA) and non-essential auxiliary load of about 4 MVA.

3.21. TURBO-ALTERNATORS

In central power stations, the steam turbine and alternator are directly coupled to avoid transmission losses. Turbo-alternators are high speed machines (3,000 or 1,500 rpm)* for 50 Hz systems. These machines have horizontal configurations and smooth cylindrical (or non-salient pole) type field structure wound usually for 2 or 4 poles. To reduce the peripheral speed (maximum peripheral speed should not exceed 175 m/s) the diameter of the rotor is kept small and axial length is increased. The ratio of diameter to axial length ranges from 1/3 to 1/2.

Due to high peripheral speed, the rotating part of the turbo-alternator is subjected to high mechanical stresses. As a result the rotor of large turbo-alternator is normally built from solid steel forging. Chromium-nickel steel or special chrome-nickel-molybdenum steel is used for rotors of turbo-alternators. The forging has radial slots in which the field copper, usually in the form of strips, is placed. The coils are held in place by steel or bronze wedges and the coil ends are fastened by metal rings. Normally two-thirds of the rotor is slotted for the field winding and one-third is left without slots so as to form the pole faces. Rectangular slots with tapered teeth are milled out in the rotor so that rectangular conductors can be used for the field windings. Each slot is provided with a ventilation hole at the bottom. To reduce harmful tooth ripples, either stator slots or poles are skewed.

The non-salient field structure used in turbo-alternators has the following special features :

(*i*) They are of smaller diameter (maximum 1 m in 2-pole machine) and of very long axial length.

(*ii*) Robust construction and noiseless operation.

(*iii*) Less windage (air-resistance) loss.

* 3,000 rpm for central power stations and 1,500 rpm for industrial power stations.

(*iv*) Better in dynamic balancing.

(*v*) High operating speed (3,000 or 1,500 rpm).

(*vi*) Nearly sinusoidal flux distribution around the periphery, and therefore, gives a better emf waveform than obtainable with salient pole field structure.

(*vii*) There is no need of providing damper windings (except in special cases to assist in synchronising) because the solid field poles themselves act as efficient dampers.

500 MW units generally use hollow stator conductors. The short-circuit ratio is 0.4 to 0.6.

Turbo-alternators are usually rated at 11 kV with 3-phase star-connection of stator windings. Standard frequency of generation is 50 Hz. Generally turbo-alternators are rated at 0.8 power factor lagging.

Main exciters are dc compound generators of 125 or 250 V rating capable of supplying required excitation for main generators on full load and with an overload of about 20% at rated power factor. Minimum of 10% overload capacity under normal operating conditions is provided for in the case of turbo-generators. Static exciters are also now in much use.

The small machines (of rating up to 50 MW) are air cooled. However, hydrogen cooling is invariably used for both stator and rotor of medium and large size turbo-generators. There are several advantages of using hydrogen as coolant in place of air in closed circuit ventilation system such as improved efficiency, increased output rating, increased life, elimination of fire hazard, less noise, smaller size of coolers etc.

The machines are built in size from 10 MW to over 1,500 MW. 500 MW units have been installed at Singrauli, Anpara, Chandarpur, Ramagundem, Trombay, Farakka and Korba.

Example 3.1. In a power plant, the efficiencies of the electric generator, turbine (mechanical), boiler, cycle and the overall plant are 0.97, 0.95, 0.92, 0.42 and 0.33 respectively. What percentage of the total electricity generated is consumed in running the auxiliaries of the plant ?

[*A.M.I.E. Sec. B. Power Plant Engineering Summer 2000*]

Solution : Let x be the percentage of the total electricity generated consumed in operating the auxiliaries of the plant.

Now since overall plant efficiency

= Generator efficiency × turbine efficiency × boiler efficiency

$$\times \text{ cycle efficiency} \times \left(1-\frac{x}{100}\right)$$

$$\therefore\ 0.33 = 0.97 \times 0.95 \times 0.92 \times 0.42 \times \left(1-\frac{x}{100}\right)$$

$$\text{or} \quad x = \left(1-\frac{0.33}{0.97\times0.95\times0.92\times0.42}\right) \times 100 = 7.32\% \ \textbf{Ans.}$$

Example 3.2. A thermal station has an overall efficiency of 21% and 0.75 kg of coal is burnt per kWh of generated energy. Determine the calorific value of coal.

Solution : Heat produced per 0.75 kg of coal

$$= \frac{\text{Output in heat units}}{\text{Overall efficiency}} = \frac{1\times 860}{0.21} = 4{,}095 \text{ kilocalories}$$

$\because$ 1 kWh = 860 kilocalories

$$\text{Hence, calorific value of coal} = \frac{4{,}095}{0.75} = 5{,}460 \text{ kilocalories } \textbf{Ans.}$$

Example 3.3. A certain power station consumes 0.7 kg of coal to generate one unit of electrical energy. If the calorific value of the fuel is 7,421 kilocalories per kg, find the thermal efficiency of the station.

Solution : Heat input for each kWh output from the power station

= 0.7 × 7,421 = 5,194.7 kilocalories

$$\text{Equivalent electrical energy for this input} = \frac{5{,}194.7}{860} = 6.04 \text{ kWh}$$

Since plant overall efficiency

= Thermal efficiency × turbine efficiency × alternator efficiency

$$\text{So} \quad \frac{1}{6.04} = \eta_{\text{Thermal}} \times \eta_{\text{Turbine}} \times \eta_{\text{alternator}}$$

Assuming turbine efficiency to be 0.95 and alternator efficiency to be 0.96 we have thermal efficiency,

$$\eta_{\text{Thermal}} = \frac{1}{6.04\times0.95\times0.96} = 0.1815 \text{ or } 18.15\% \ \textbf{Ans.}$$

Example 3.4. A thermal power plant spends Rs 25 lakhs in one year as coal consumption. The coal has heating value of 5,000 kcal/kg and costs Rs 500/ton. If the thermal efficiency is 35% and electrical efficiency is 90%, find the average load on the power plant. [*G.B. Technical Univ. Power Station Practice 2011-12*]

Solution : Coal consumption per annum

$$= \frac{\text{Coal bill per annum}}{\text{Cost of coal per ton}}$$

$$= \frac{25\times10^5}{500} = 5{,}000 \text{ ton or } 5 \times 10^6 \text{ kg}$$

Heat of combustion

= Coal consumption per annum × calorific value of coal

= 5 × 10^6 × 5,000 = 2.5 × 10^{10} kilocalories

Heat output

= Heat of combustion × thermal efficiency × electrical efficiency

$$= 2.5 \times 10^{10} \times \frac{35}{100}\times\frac{90}{100} = 0.7875 \times 10^{10} \text{ kilocalories}$$

Electrical energy generated per annum

$$= \frac{0.7875\times10^{10}}{860} = 9.157 \times 10^6 \text{ kWh}$$

$\because$ Heat equivalent of 1 kWh = 860 kilocalories

Average load on the power station

$$= \frac{\text{Electrical energy generated / annum in kWh}}{365\times24}$$

$$= \frac{9.157\times10^6}{365\times24} = 1{,}045 \text{ kW } \textbf{Ans.}$$

Example 3.5. A steam power station of 100 MW capacity uses coal of calorific value of 6,400 kcal/kg. The thermal efficiency of the station is 30% and electrical generation efficiency is 92%. Determine the coal required per hour when the plant is working at full load. [*A.M.I.E. Sec. B. Power Systems Winter 1996; Power Plant Engineering Winter 1995*]

Solution : Overall efficiency of power station

= Thermal efficiency × electrical efficiency = 0.30 × 0.92 = 0.276

Units generated per hour at full rated output = 100 × 1,000 kWh

$$\text{Heat produced per hour} = \frac{100 \times 1{,}000 \times 860}{0.276} = 311.6 \times 10^6 \text{ kcals.}$$

Coal consumption per hour

= Heat produced per hour/calorific value of coal

$$= \frac{311.6 \times 10^6}{6{,}400} = 48{,}687.00 \text{ kg} = 48.687 \text{ tonnes}$$ **Ans.**

Exercises

1. Discuss the factors which affect the thermal efficiency of a thermal power station.

[Power Station Prractice; Pb Univ. May 1994; U.P. Technical Univ. 2007-08]

2. Discuss why overall efficiency of thermal power plant is very low ?

[Rajasthan Technical Univ. Generation of Electrical Power 2010]

3. Write different methods which are used for increasing the efficiency of a thermal power station. Explain any one in detail. **[A.P.S. Univ. Rewa Electrical Power Generation, Transmission & Distribution 1977]**

4. Discuss the merits and demerits of steam power plants.

5. Discuss the factors to be taken into account while selecting the site for a thermal power station.

[G.B. Technical Univ. Power Station Practice 2010-11]

6. What are the main considerations for selection of site for a thermal power station ?

[U.P. Technical Univ. Power Station Practice 2008-09]

7. Draw the schematic diagram of a modern steam power station and explain its operation with its important components.

[Rajasthan Technical Univ. Generation of Electrical Power 2011[

8. Draw the schematic layout of coal-fired power station. Explain main parts and their working.

[G.B. Technical Univ. Power Station Practice 2011-12]

9. Justify that thermal power stations are normally located away from the load centers. Draw the feed water and steam flow diagram of a large thermal power station.

[U.P. Technical Univ. Power Station Practice 2004-05]

10. Draw a general layout of modern thermal power plant and explain the working of different circuits (fuel, air, steam, water, cooling, etc.)

[M.D. Univ. Power Plant Engineering May 2008]

11. Explain with a neat sketch the working of a thermal electric power plant station and explain the functions of major components in it. **[U.P.S.C. I.E.S. Electrical Engineering-II, 1995]**

12. Draw a labelled schematic diagram of a typical thermal power station. How do you compare thermal plants with hydroelectric plants from the economic point of view.

[U.P.S.C. I.E.S. Electrical Engineering-II, 2002]

13. Draw and explain single line diagram of thermal and hydroplants.

[Pb. Technical Univ. Generation of Electric Power May 2010]

14. What are the factors to be considered for selection of site for a thermal power plant ? Draw a typical layout and explain the working of a plant.

[U.P. Technical Univ. Power Station Practice 2003-04]

15. Explain the working of stream power station with diagram. What are the important constituents of a steam power station.

[Pb. Technical Univ. Generation of Electric Power Dec. 2010]

16. Describe the use of different parts of a thermal power plant with the help of neat sketch.

[U.P. Technical Univ. Power Station Practice 2006-07]

17. Explain briefly, the functions of (*i*) Reheaters (*ii*) Condensers, (*iii*) Cooling towers.

[U.P. Technical Univ. Power Station Practice 2004-05]

18. Briefly discuss the function of economizer in a thermal power plant.

[U.P. Technical Univ. Power Station Practice 2007-08]

19. Briefly discuss the function of air preheater in a thermal power plant.

[U.P. Technical Univ. Power Station Practice 2007-08]

20. Give the classification of steam turbines used in power plants.

21. Explain the principle of impulse turbine and how it differs from reaction turbine?

[Pb. Univ Power Station Practice April 1993]

22. What are the different methods of compounding of steam turbines. **[Pb. Univ Power Station Practice December 1994]**

23. Explain the various types of nozzles used in steam turbines.

[Pb. Univ Power Station Practice April 1993]

24. Give the classification of coal. On what factors does the selection of coal for a power plant depend ? What type of coal is used in steam power plants? How imported coal is superior to Indian coal ?

25. Draw a neat line diagram of inplant coal handling system and explain the working with neat diagram of equipments used at different stages.

[M.D. Univ. Power Plant Engineering May 2007]

26. Draw a neat line diagram of inplant coal handling system and explain screw conveyors used to transfer the coal.

[M.D. Univ. Power Plant Engineering December 2009]

27. Suggest a coal handling system for a large size thermal plant, located at about 25 km from coal mines.

[Pb. Univ Power Station Practice December 1995]

28. Explain the difference between overfeed and underfeed systems of burning solid coal. Which is preferred for high volatile coal and why ?

[A.M.I.E. Sec. B. Power Systems Winter 1999]

29. What are the advantages of mechanical handling of coal ?

[Pb. Univ Power Station Practice December 1993]

30. Explain the working of underfeed stokers with the help of a diagram.

[Pb. Univ Power Station Practice December 1995]

31. Why pulverized fuel is preferred ? What are the types of pulverized fuel burners ?

[G.B. Technical Univ. Power Station Practice 2011-12]

32. What is pulverized fuel ? Compare the pulverized firing and stoker firing. Discuss the characteristics of bin and unit systems of pulverized fuel firing.

[Pb. Univ Power Station Practice May 1994]

33. Describe briefly with flow sheet the pulverized coal preparation systems and the method of firing of pulverized fuel. Discuss briefly the advantages and disadvantages of pulverized coal firing.

[M.D. Univ. Power Plant Engineering May 2009]

34. Describe different types of burners for pulverized coal firing. How is degree of superheat controlled by tilting burner ?

[A.M.I.E. Sec B. Power Plant Engineering Winter 1994]

35. What do you understand by fluidized bed combustion ? Explain working principle with a neat line sketch. What are the major disadvantages of this system over conventional furnace/boiler system ?

[A.M.I.E. Sec B. Power Plant Engineering Winter 1993]

36. Briefly discuss the function of pulverizing plant in a thermal power station.

[U.P. Technical Univ. Power Station Practice 2008-09]

37. Write short notes on ash handling plant and steam turbines.

[Rajasthan Technical Univ. Generation of Electrical Power 2010]

38. Draw a typical layout of a thermal power plant and describe the function of a coal and ash handling plant.

[Pb. Univ. Electric Power-II (Generation) May 1994]

39. Explain about ash handling waste.

[Pb. Technical Univ. Generation of Electric Power Dec. 2009]

40. Draw a line diagram of pneumatic ash handling system used for steam power plant. Explain its working and difficulties encountered in design.

[M.D. Univ. Power Plant Engineering Dec. 2009]

41. List difficulties encountered in the ash handling plant layout. Mention the places of disposal of ash.

[M.D. Univ. Power Plant Engineering May 2009]

42. What are the principal requirements of an ash handling plant ?

[Pb. Univ. Power Station Practice December 1993]

43. How will you classify modern ash handling systems ? Describe each one of them briefly with the help of neat sketches. **[Pb. Univ. Power Station Practice December 1994]**

44. What are the methods of arresting ash from the flue gases ? Explain any one method.

[A.M.I.E. Sec. B. Power Plant Engineering Summer 1998]

45. Explain with the help of a line sketch the working of an electrostatic precipitator and give its outstanding features over other collectors.

[A.M.I.E. Sec. B. Power Plant Engineering Winter 1998]

46. Explain the functions of FD fan and ID fan. What is balanced draught ?

[A.M.I.E. Sec. B. Power Plant Engineering Winter 2000]

47. Discuss the following in case of steam power plant:

(*a*) Preparation of coal (*b*) Belt conveyor (*c*) Electrostatic precipitator (*d*) Comparison of unit and control systems of pulverized fuel firing.

[M.D. Univ. Power Plant Engineering Dec. 2008]

48. Briefly discuss the functions of economizers, superheaters and boilers in a thermal power plant.

What are the draught equipments used in steam power plant ?

[U.P. Technical Univ. Power Station Practice 2003-04]

49. Explain the process of treatment of feed water. Why is it necessary ? Describe the zeolite process for softening of water.

50. List the typical essential auxiliaries for a unit in a steam station.

[Pb. Univ. Electric Power-II (Generation) May 1995]

51. Discuss the characteristics of any one type of turbo-alternators. **[A.M.I.E. Sec B. Power Plant Engineering Winter 1996]**

52. Write short notes on the following:

(*i*) Site selection of thermal power plants.

[M.D. Univ. December 2008]

(*ii*) Superheater in a boiler. **[A.M.I.E. Sec. B. Winter 1995]**

(*iii*) Economiser in a boiler. **[A.M.I.E. Sec. B. Winter 1995]**

(*iv*) Recuperative and regenerative air preheaters.

[A.M.I.E. Sec. B. Summer 2000]

(*v*) Feed water heaters. **[Pb. Univ. May 1995]**

(*vi*) Preparation of fuel for steam power plant.

[M.D. Univ. December 2009]

(*vii*) Coal handling plant. **[A.M.I.E. Sec. B. Winter 1998]**

(*viii*) Ash handling. **[Pb. Univ. December 1994]**

(*ix*) Ash disposal and dust collection. **[Pb. Univ. May 1994]**

(*x*) Dust collectors used in steam power plants.

[M.D. Univ. May 2009]

(*xi*) Electrostatic precipitator.

[A.M.I.E. Sec. B. Summer 2000; M.D. Univ. May 2007]

(*xii*) Feedwater treatment. **[A.M.I.E. Sec. B. Summer 2000]**

(*xiii*) Supply to power station auxiliaries.

[A.M.I.E. Sec. B. Winter 1993]

(*xiv*) Turbo-alternators.

Short Answer Type Questions With Answers

Q. 1. Why overall efficiency of a thermal power plant is very low ? **[R.T.U. 2010]**

Ans. In a steam power station, more than 50% of total heat of combustion is lost as heat rejected to the condenser and this loss is unavoidable as heat energy cannot be converted into mechanical energy without a drop in temperature, and the steam in condenser is at the lowest temperature. This is the reason that the thermal efficiency of a power plant is quite low.

Q. 2. How is thermal efficiency of a steam power plant improved ?

Ans. The thermal efficiency of the plant mainly depends upon three factors (*i*) pressure and (*ii*) temperature of the steam entering the turbine and the pressure in the condenser. The thermal efficiency increases with the increase in temperature and pressure of the steam entering the turbine. For this reason high pressures and temperatures are used. The thermal efficiency is effectively increased by decreasing the pressure in the condenser. Pressure in the condenser is kept very low usually 0.04 kg/cm^2.

The thermal efficiency also increases by reheating the steam between turbine stages, but is somewhat inconvenient. Bleeding of steam also affects the thermal efficiency.

Q. 3. What factors determine the choice of site of a thermal power plant ?

Ans. The factors governing choice of site for a thermal power plant for economical and efficient generation are (*i*) Availability of supply of water (*ii*) Availability of coal (*iii*) Availability of land at a reasonable cost (*iv*) Type of land (*v*) Transportation facilities (*vi*) Facilities for disposal of ash (*vii*) Distance from populated area and (*viii*) Nearness to load centre.

Q. 4. Why thermal power stations are always situated by the side of a river or lake ?

Ans. The thermal power stations are always situated by the side of a river or lake so as to meet large quantity water requirement. Water is required in a steam power station (*i*) to raise the steam in boilers. (*ii*) for cooling purposes such as in condensers (*iii*) as a carrying medium such as in disposal of ash and (*iv*) for drinking purposes.

Q. 5. Why is the use of fire tube boilers restricted to low cost, small size and low pressure plants ?

Ans. Fire tube boilers are compact in size, have low initial cost and have the ability to raise rapidly large quantities of steam per unit area of fire grate but have the following drawbacks.

As water and steam, both are in the same shell, higher pressure of steam are not possible, the maximum pressure which can be had is about 17.5 kg/cm^2 and with a capacity of 15,000 kg of steam per hour. For higher pressures or higher rates of evaporation, the shell and fire tube boilers become extremely heavy and unwieldy. In the event of a sudden and major tube failure, steam explosions may be caused in the furnace due to rush of high pressure water into the hot combustion chamber which may generate large quantities of steam in the furnance.

Fire tube boilers use is, therefore, limited to low cost, small size and low pressure (to about 10 kg/cm^2) plants.

Q. 6. What factors determine the size and type of boiler ?

Ans. The selection of the size and type of boiler depends upon (*i*) the output required in terms of amount of steam per hour, operating temperature and pressure (*ii*) availability of fuel and water and (*iii*) the probable load factor. The other factors which influence the choice of a boiler are availability, initial cost, maintenance costs, labour costs, fuel costs and the space requirement.

Water tube boilers are used where large amounts of steam are to be produced at high temperature and pressure and weight and space considerations are important. To meet a required demand, the choice between two boilers will be based on economic considerations *i.e.,* total annual cost (fixed cost plus running cost). The noteworthy point is that the total cost of the fuel used by the boiler in its lifetime may be 3 to 4 times the initial investment.

Q. 7. What is the function of the deaerator in a steam power plant ?

Ans. The function of deaerator is to reduce dissolved oxygen content in the condensate (feedwater) by heating the feed water on the way from low pressure feed water heaters to the high pressure feed water heaters by the heat from bled steam.

Q. 8. Discuss the necessity of superheated steam in thermal power plant. **[R.T.U. 2008; 2010]**

Ans. Superheated steam not only improves the efficiency but also have the following advantages:

(*i*) Reduction in requirement of steam quantity for a given output of energy owing to its high internal energy reduces the turbine size.

(*ii*) Superheated steam being dry, turbine blades remain dry so the mechanical resistance to the flow of steam over them is small resulting in high efficiency.

(*iii*) No corrosion and pitting at the turbine blades occur owing to dryness of steam.

Q. 9. Why are superheaters and reheaters used in steam power plants ?

Ans. Superheaters are used in steam power plants to remove the last traces of moisture from the saturated steam leaving the boiler tubes and also to increase its temperature above the saturation temperature from the heat of combustion gases from the furnace.

Reheaters are used in steam power plants to resuperheat the partly expanded steam from the turbine so that the steam remains dry as far as possible through the last stage of the turbine.

Q. 10. Why economiser and air preheaters are used in steam power plants ?

Ans. Economiser and air preheaters are such devices which recover the heat from the flue gases on their way to the chimney and raise the temperature of feed water and air supplied for combustion respectively and thus improve the efficiency of steam plants.

Q. 11. How does air preheating save fuel ?

[A.M.I.E. Sec. B. Summer 2000]

Ans. In preheating of the incoming air the heat recovered from the flue gases leaving the economiser is utilized, which otherwise would have gone waste. This raises the temperature of the furnace gases, improves combustion rates and efficiency, and lowers the stack temperature, thus improving the overall efficiency of the boiler.

Q. 12. Why the heating of water is necessary before it is supplied to the boiler ?

Ans. Heating of water, before being supplied to the boiler is necessary because of the following factors:

(*i*) Overall plant efficiency is improved.

(*ii*) Thermal stresses due to cold water entering the drum of boiler are avoided.

(*iii*) There is an increase in the quantity of steam produced by the boiler.

(*iv*) The dissolved oxygen and carbon dioxide which would otherwise cause boiler corrosion are removed in the feed water heaters.

(*v*) Some other impurities carried by steam and condensate, due to corrosion in boiler and condenser, are precipitated outside the boiler.

Q. 13. What is the function of the condenser in a steam power plant ? **[A.M.I.E. Sec. B. Summer 1998]**

Ans. The condenser does the job of condensing the steam exhausted from turbine and thus it helps in maintaining low pressure (below atmospheric) at the exhaust, thereby permitting expansion of steam in the turbine to a very low pressure. This improves the plant efficiency.

Q. 14. Why is it necessary to maintain high vacuum in a condenser ?

Ans. Maintenance of high vacuum in the condenser is essential for efficient operation. Any leakage of air into the condenser destroys the vacuum and causes (*i*) an increase in the condenser pressure which limits the useful heat drop in the prime mover and (*ii*) a lowering of the partial pressure of the steam and of the saturation temperature along with it. This means that the latent heat increases and therefore, more cooling water is required. Also, the under cooling of the condensate is likely to be more severe. This will result in lower efficiency.

Q. 15. How does the jet condenser differ from a surface condenser ?

Ans. The essential difference between a jet condenser and a surface condenser is that in the former, the exhaust steam mixes with the cooling water and the temperature of the condensate and the cooling water is the same when leaving the condenser; and the condensate cannot be recovered for use as feed water to the boiler; heat transfer is by direct conduction; in the latter *i.e.,* in surface condenser, the exhaust steam and cooling water do not mix with each other, the water being circulated through a nest of tubes, the heat transfer being by convection. The temperature of the condensate may be higher than the temperature of the cooling water at outlet and the condensate is recovered as feed water to the boiler. Both the cooling water and the condensate are separately withdrawn. Advantages of jet condensers are low initial cost, low requirements of floor area and cooling water and low maintenance charges. However, its disadvantages are that the condensate is wasted and high power is required for pumping water.

Q. 16. What for evaporators are used in steam power plants ?

Ans. Evaportors are used for supplying pure water as make-up feed water in steam power plants.

Q. 17. Why are cooling towers used in thermal power stations ?

Ans. Cooling towers are provided to cool the hot water coming out from the condenser.

Q. 18. Draw the layout diagram of a thermal power plant.

[Pb. Technical Univ. May 2008]

OR

Draw a general layout of a modern thermal power plant.

[Gautam Budh Technical Univ. 2010-11]

Ans. Refer to Fig. 3.17.

Q. 19. What is the function of steam nozzle ?

Ans. The function of a steam nozzle is to convert the heat energy of steam into kinetic energy and the chief use of a steam nozzle is to develop a high velocity jet of steam for driving a steam turbine.

Q. 20. On what factors does the selection of coal for a steam power plant depend ?

Ans. The selection of coal for a steam power plant depends on calorific value, weatherability, sulphur content, ash content, particle size, grindability index, caking characteristics etc.

Q. 21. Why are semi-bituminous coals one of the best coals for power generation ?

Ans. Semi-bituminous coals are the highest grade of bituminous coals having high fixed carbon content and the highest calorific values. These have volatile matter between 14 and 22 per cent and enthalpy of combustion between 27,000 and 35,000 kJ/kg. So semi-bituminous coals are one of the best coals for power generation.

Q. 22. Give the necessary requirements for efficient combustion of fuel.

Ans. The necessary requirements for efficient combustion of fuels are as follows.

(*i*) The proper quantity of primary or secondary air required for complete combustion.

(*ii*) Adequate stoker or grate area required for burning particular quantity of fuel.

(*iii*) Minimum operating and maintenance cost.

(*iv*) Attainment of proper designed temperature.

(*v*) There should be no formation of caking during burning of fuel.

(*vi*) The system should be easy to handle and dependable.

Q. 23. How does overfeed and underfeed stokers differ ?

Ans. Overfeed stoker and underfeed stoker differ in the manner of feeding coal below or above the level at which primary air is admitted in the furnace. In the case of underfeed stokers, fuel is supplied into the furnace below the point of air admission while in the case of overfeed stokers, the coal is supplied into the grate above the point of admission of air.

Q. 24. What is tangential firing ? **[A.M.I.E. Sec. B. Summer 2000]**

Ans. In tangential firing, the burners are placed in the corners of the furnace and they send horizontal streams of air and fuel tangent to an imaginary circle in the centre of the furnace.

Q. 25. What are the functions of primary air and secondary air in a pulverized coal fired boiler ?

[A.M.I.E. Sec B. Summer 2000]

Ans. The function of primary air in a pulverized coal fired boiler is to dry the coal and convey the powdered fuel to the furnace while that of secondary air is to complete the combustion.

Q. 26. Why is 'excess oxygen' required in steam power generation ?

Ans. The oxygen of heated air in combustion chamber reacts with the carbon of coal and form carbon dioxide during the process of combustion. Carbon dioxide so formed is taken away as draught through chimney. As coal is the basic fuel and for combustion of coal oxygen is necessary hence excess oxygen is required in steam power generation.

Q. 27. What do you understand by fluidized bed combustion ?

Ans. Fluidization is a new method of mixing fuel and air for obtaining combustion. A fluidized bed may be defined as the bed of solid particles behaving as a fluid.

Q. 28. What for draught system is employed in steam power plants ?

Ans. Draught system is employed to admit an amount of air into the furnace for the combustion of fuel and to cause the flow of the products of combustion from a chimney.

Q. 29. How does the induced draught differ from forced draught ?

Ans. In an *induced draught system*, the blower is installed near the base of the chimney and the burnt gases are sucked out of the boiler, reducing the pressure inside the boiler to less than atmospheric one. This induces fresh air to enter the furnace. In case of a forced draught system the blower is installed near the base of the boiler. In this system of draught the air pressure throughout the system is above the atmospheric pressure and air is forced to flow through the system (furnace, economiser, air preheater and chimney).

Q. 30. What is balanced draught ?

Ans. A balanced draught is a combination of induced draught and forced draught. In balanced draught system, the pressure of air below the boiler grate is above atmospheric and that above the grate is below atmospheric. The induced draught blower helps in sucking the products of combustion out of the system where as the forced draught helps in overcoming the resistance inside the boiler, thus enabling supply of sufficient air to the fuel bed for proper and complete combustion.

Q. 31. In a coal-fired power station mention three distinct uses of water. **[A.M.I.E. Sec. B. Summer 1993]**

Ans. Three distinct uses of water in a coal-fired thermal power station are (*i*) for generating steam in the boiler (*ii*) for condensing steam in the condenser and (*iii*) for washing the coal so as to reduce the ash content.

Q. 32. Why is treatment of feed water essential ?

Ans. Treatment of feed water is essential so as to remove the suspended and soluble solids and gases. These impurities in feed water may cause scale formation, corrosion, carry over and embrittlement in boiler and other apparatus.

Q. 33. What is zeolite process and how is it advantageous to other processes of treatment of water ?

Ans. Zeolite treatment is one of the ion exchanging process wherein cations of two or more positive charges are exchanged for sodium or hydrogen and zeolite is either a natural or artificial material that has the property of being able to exchange its base. Natural zeolites have a sodium base that can be exchanged for calcium or magnesium in raw water and the water will be softened. The advantage of this process is that the hardness is disposed off as a solution instead of solid requiring filteration.

Q. 34. Why is it necessary to employ non-salient pole field structure in a turbo-alternator ?

Ans. Turbo-alternators are driven by steam turbines which have high efficiency at large speeds, and therefore, synchronous machines driven by steams (*i.e.*, turbo-alternators) are high speed machines. Non-salient pole field structure is used as the salient pole construction is impractical owing to large mechanical forces.

Q. 35. What is the largest size of thermal unit being used in India ?

[Pb. Technical Univ. December 2007]

Ans. The largest size of thermal unit being used in India is 500 MW.

Q. 36. Why thermal plants are not suitable for supplying fluctuating loads ? **[A.M.I.E. Sec. B. Winter 1996]**

Ans. Thermal power plants are not suitable for supplying fluctuating loads because any change in load demand requires corresponding change in the output energy. In the thermal plants input energy is produced by burning of coal and hence there is always a time lag between the change in energy at the output and input, which is not desirable and hence such power stations are used only as base load plants, for supplying constant power.

Q. 37. What is the running speed of a turbo-alternator in a central thermal power station ?

Ans. 3,000 rpm.

Problems

1. A thermal station has an efficiency of 15% and 10 kg of coal is burnt for every kWh generated. Determine the calorific value of coal. [**Ans.** 5,773 k.cals/kg of coal]
2. A Thermal power station has four generators of 50 MW each. Find out the following:
(*i*) rating of each turbine (*ii*) energy produced per annum (*iii*) cost of fuel per annum.
Given that, Maximum demand : 175 MW; Average load factor: 65%; Generator efficiency : 96%; Overall thermal efficiency : 20%; Calorific value of fuel : 6,400 k.cals/kg; Cost of fuel Rs 100 per metric ton. 1 kWh equivalent to 860 k.cals.
[**Ans.** (*i*) 52 MW (*ii*) 9.96×10^8 kWh (*iii*) Rs 66.95 lacs]

❑❑❑

CHAPTER 4

Nuclear Power Plants

INSIDE THIS CHAPTER

4.1. INTRODUCTION

The need of nuclear power plant lies in the fact that the hunger for electricity is virtually unending and after each decade the world demand for electricity is doubled owing to booming increase in the population and industrial growth. Moreover, the reserves of fossil fuels *i.e.,* coal, oil and gas are fast depleting. Again it would be better to utilise these minerals, particularly oil, for chemical industry as raw materials. Thus there is tendency to seek alternative source of energy and the nuclear power is the only alternative source which can meet the future energy demands of the world. One of its main attractions is the huge amount of energy that can be released from a small quantity of active material. The energy obtainable by completely burning of 1 kg of uranium would give energy equivalent 3,000 tonnes of high grade coal *i.e.,* uranium has 3 million times the energy of coal. The possible energy reserves in the form of uranium and thorium is many times greater than that of fossil fuels.

The findings of the committee appointed by the Govt. of India in 1984 for examining the comparative costs for nuclear and thermal power plants commissioned in 1983 are that the nuclear power would be cheaper by 8 to 10 paise per kWh than from thermal plants, at a distance of 800 km from pit heads. This difference goes on increasing as the distance from pit heads increases. The trend in favour of nuclear power becomes stronger as time passes.

The areas most suitable for nuclear power plants are Western UP, Northern and Western Rajasthan, Panjab and Haryana.

Though the capital cost of nuclear power plant, because of heavy costs of land having strong substrata,* foundation, nuclear reactors** and nuclear fuel*** is pretty high as compared to that of a conventional thermal power plant, but the total operating costs per kWh of a nuclear power plants are less than those of a coal-fired thermal plant. Nuclear power is the cheapest non-hydroelectric power in India.

The development of nuclear power station is due to attention of scientists towards the peaceful applications of atomic energy. The world's first atomic power plant was commissioned in U.S.S.R. on June 27, 1954 and after that a number of atomic power plants have been commissioned in many countries like U.S.A., Canada, Great Britain, Japan and France.

It will be noteworthy here that steam turbines employed in nuclear power plants will be of design quite different from those used in coal-fired steam power plants. The reasons are operating pressure and temperature, in case of nuclear power plants, being quite different than those in case of conventional steam power plants (turbines for the modern coal-fired steam power plants are designed for operation with superheated steam at temperatures about 565° and pressure about 15 MPa).

4.1.1. Nuclear Fuel Reserves

India domestic uranium reserves are small and the country is dependent on uranium imports to supply its nuclear power industry. Since early 1990s, Russia has been a major supplier of nuclear fuel to our country. Because of dwindling domestic uranium reserves, gernation of electrical energy from nuclear power in India declined by 12.83% from 2006 to 2008. Following a waiver from the Nuclear Suppliers Group in September 2008 which allowed it to commence international nuclear trade, India has signed bilateral deals on civilian

* Nuclear reactors are very heavy (about 1×10^5 tonnes) and so impose bearing pressures of about 50 tonnes/m^2. Hence strong substrata site is required.

** Very few countries possess the technology for manufacturing of nuclear reactors and therefore, the capital cost of nuclear reactors is very high.

*** Nuclear fuel may remain in a reactor for more than 5 years and, therefore, the cost of fuel injected initially is considered as a capital cost and may be a few crore rupees.

nuclear energy technology cooperation with several other countries, including France, the United States, the United Kingdom, Canada and South Korea. India has also uranium supply agreements with Russia, Mongolia, Kazakstan, Argentina and Nimbia. An Indian private company won a uranium exploration contract in Niger.

Large deposits of natural uranium, which promises to be one of the top 20 of the world's reserve have been found in the Tummalapalle belt in southern part of the Kadapa basin Andhra Pradesh in March 2011. The Atomic Minerals Directorate for Exploration and Research (AMD) of India which explores uranium in the country, has so far discovered 4,400 tonnes of natural uranium (U308) in just 15 km of the 160 km long belt.

4.1.2. Economics of Nuclear Power Plants

The economic analysis of nuclear power plants is in many respects similar to that for coal-fired thermal power plants and hydro power plants. However, the relative costs of different components are somewhat different in this case. Broadly speaking, the points for considerations are capital investment, plant life, fixed, operation and maintenance costs, plant operating condition such as plant load factor.

The capital investment of a nuclear power plant includes the cost of land, cost of nuclear reactor, heat exchanger, steam turbines, turbo-alternators etc., and the cost of design and planning.

The site for nuclear power plant should have substrata strong enough to support the heavy reactors which may weigh as high as 100,000 tonnes and impose bearing pressures around 50 tonnes/m^2. The cost of foundation is also quite high. The capital cost of nuclear reactors is very high because very few countries have the technology of manufacturing nuclear reactors. The cost of cooling towers is also high because of very large requirements of cooling water in case of steam power plants. Nuclear fuel may remain in a reactor for more than 5 years. So, the cost of fuel injected initially is taken as a capital cost and may be a few crore rupees.

The nuclear power station has a much higher capital investment than a thermal power station. Some years back, this difference was much higher than it is today because it had to cover the initial development of reactors of the Calder Hall Type.

Fixed Costs. The life of a reactor plant may be taken as between 15 and 20 years. For the other parts of the plant equipment, the life may be taken as 30 years. The fixed cost would be interest, depreciation, taxes and insurance charges. The depreciation would be determined on the life of the equipment in the plant as well as building etc. The insurance charges will be higher than for conventional power plants.

Operation and Maintenance Costs. The fuel, operation and maintenance costs will consist of the cost of the nuclear material, the cost of the processing material, the cost of fabricating the fuel elements, the cost of disposal of waste, and the salaries and wages of operating and maintenance staff. The total operating cost per kWh of a nuclear power plant are less than that of a coal-fired thermal power plant.

The overall efficiency of a nuclear power plant is around 30 to 40 per cent. The efficiency is higher at high load factors. Therefore, a nuclear power plant is always operated as a base load plant. Though the initial investment and capital cost of a nuclear power plant is higher than that of a conventional thermal power plant, but the cost of transport and handling of coal for conventional thermal power plant is much higher than the cost of nuclear fuel. With further developments, the cost of nuclear power plants likely to come down and they will soon be competitive.

4.1.3. Nuclear Power Growth in India

India now envisages to increase the contribution of nuclear power to overall electricity generation capacity from 2.8% to 9% within 25 years. By 2017, India's installed nuclear power generation capacity will increase to 10,080 MW. As of 2009, India stands 9th in the world in terms of number of operational nuclear power reactors. Indigenous atomic reactors include TAPS-3, and -4, both of which are 540 MW reactors. India's US$717 million fast breeder reactor project is expected to be operational by 2012-13.

The Indian nuclear power industry is expected to undergo a significant expansion in the coming years thanks in part to the passing of the U.S.-India Civil Nuclear Agreement. This agreement will allow India to carry out trade of nuclear fuel and technologies with other countries and significantly enhance its power generation capacity. When the agreement goes through, India is expected to generate an additional 25,000 MW of nuclear power by 2020, bringing total estimated nuclear power generation to 45,000 MW.

India has already been using imported enriched uranium for light-water reactors that are currently under IAEA safeguards, but it has developed other aspects of the nuclear fuel cycle to support its reactors. Development of select technologies has been strongly affected by limited imports. Use of heavy water reactors has been particularly attractive for the nation because it allows Uranium to be burnt with little to no enrichment capabilities. India has also done a great amount of work in the development of a thorium centered fuel cycle. While Uranium deposits in the nation are limited (see next paragraph) there are much greater reserves of thorium and it could provide hundreds of times the energy with the same mass of fuel. The fact that thorium can theoretically be utilized in heavy water reactors has tied the development of the two. A prototype reactor that would burn Uranium-Plutonium fuel while irradiating a thorium blanket is under construction at the Madras/Kalpakkam Atomic Power Station.

Uranium used for the weapons program has been separate from the power program, using uranium from indigenous reserves. This domestic reserve of 80,000 to 112,000 tons of uranium (approx 1% of global uranium reserves) is large enough to supply all of India's commercial and military reactors as well as supply all the needs of India's nuclear weapons arsenal. Currently, India's nuclear power reactors consume, at most, 478 tonnes of uranium per year. Even if India were quadruple its nuclear power output (and reactor base) to 20 GW by 2020, nuclear power generation would only consume

2000 tonnes of uranium per annum. Based on India's known commercially viable reserves of 80,000 to 112,000 tons of uranium, this represents a 40–50 years uranium supply for India's nuclear power reactors (note with reprocessing and breeder reactor technology, this supply could be stretched out many times over). Furthermore, the uranium requirements of India's Nuclear Arsenal are only a fifteenth (1/15) of that required for power generation (approx. 32 tonnes), meaning that India's domestic fissile material supply is more than enough to meet all needs for its strategic nuclear arsenal. Therefore, India has sufficient uranium resources to meet its strategic and power requirements for the foreseeable future.

Nuclear power is the fourth-largest source of electricity in India after thermal, hydroelectric and renewable sources of electricity. As of 2010, India has 20 nuclear power reactors in operation in 6 nuclear plants, generating 4,780 MW (2.9% of total installed base) while nine other reactors are under construction and are expected to generate an additional 6,700 MW.

Power Plants in Operation

Power Plant	Operator	State	Type	Units	Total Capacity (MW)
Kaiga	NPCIL	Karnataka	PHWR	220 × 4	880
Kakrapar	NPCIL	Gujarat	PHWR	220 × 2	440
Kalpakkam	NPCIL	Tamil Nadu	PHWR	220 × 2	440
Narora	NPCIL	Uttar Pradesh	PHWR	220 × 2	440
Rawatbhata	NPCIL	Rajasthan	PHWR	100 × 1 200 × 1 220 × 4	1,180
Tarapur	NPCIL	Maharashtra	BWR (PHWR)	160 × 2 540 × 2	1,400
			Total	**20**	**4,780**

Power Plants Under Construction

Power Plant	Operator	State	Type	Units	Total Capacity (MW)
Kudankulam	NPCIL	Tamil Nadu	VVER-1000	1,000 × 2	2,000
Kalpakkam	BHAVINI	Tamil Nadu	PFBR	500 × 1	500
Kakrapar	NPCIL	Gujarat	PHWR	700 × 2	1,400
Rawatbhata	NPCIL	Rajasthan	PHWR	700 × 2	1,400
Banswara	NPCIL	Rajasthan	PHWR	700 × 2	1,400
			Total	**9**	**6,700**

4.1.4. Future of Nuclear Power

The future of nuclear power is very bright as the reserves of fossil fuels are fast depleting and hydro power has also a fixed limit up to which it can be exploited. The three items—namely cost of power generation, availability of nuclear fuel and safety of operation of the power plants, need consideration in this respect. Though at present the generation cost is more but with the research in the nuclear technology, the generation cost is bound to come down to a value comparable with that for conventional power plants. As regards the availability of nuclear fuel we may face shortage of cheap uranium in future but this problem has got the solution and ultimately, we will have to switch over to breeder reactor in which fissionable fuel is produced at the same time they are consumed. About the safety of the nuclear power plant, it is noteworthy here that the nuclear reactor contains only a small amount of fissionable material whereas an atom bomb contains 90% fissionable fuel in its core. The nuclear power plant is designed so as not to explode like a nuclear bomb. So nuclear power plants can be taken as safe in operation also.

4.2. MERITS AND DEMERITS OF NUCLEAR POWER PLANTS

The chief merits of nuclear power plants over other conventional power plants are given below:

(*i*) The amount of fuel required is quite small, therefore, there is no problem of transportation, storage etc.

(*ii*) The demand for coal, oil and gas is reduced which are tending to rise in cost as the stocks are becoming depleted.

(*iii*) These plants need less area as compared to any other plant of the same size. A 2,000 MW nuclear power plant needs about 80 acres whereas the coal-fired steam power plant of the same capacity needs about 250 acres of land.

(*iv*) In addition to producing large amounts of power, the nuclear power plant can produce valuable fissile material, which is extracted when the fuel has to be renewed.

(*v*) These plants, because of the negligible cost of transportation of fuel, can be located near the load centres, therefore, primary distribution cost is reduced.

(*vi*) These plants are most economical in large capacity (100 MVA and more).

(*vii*) The output control is extremely flexible *i.e.*, the output can be instantaneously adjusted from zero to an upper limit. The limit is set by the capacity of the heat removal system to prevent overheating of the pile.

(*viii*) There are large deposits of nuclear fuel available all over the world. Therefore such plants can ensure continued supply of electrical energy for thousands of years.

(*ix*) A coal-fired steam power plant needs thousands of tonnes of coal per day and usually looks like a mad house of trains, trucks, coal and ashes, whereas a nuclear power plant will be very neat and clean and is hospital quiet.

(*x*) The operating cost is quite low and once the installation is completed, the loading of the power plant will have no effect on the generation cost. Therefore, a nuclear power plant is always operated as a base load plant. The nuclear power plants are usually not operated at a load factor less than 0.8.

However the nuclear power plants have the following drawbacks:

(*i*) The initial capital cost is very high as compared to other types of power plants.

(*ii*) The erection and commissioning of the plant requires greater technical know-how.

(*iii*) The fission by-products are generally radioactive and may cause a dangerous amount of radioactive pollution.

(*iv*) Nuclear power plants are not well suited for varying loads since the reactor does not respond to the fluctuations of load efficiently.

(*v*) The fuel used is expensive and is difficult to recover.

(*vi*) Maintenance charges are high owing to lack of standardisation. Salary bill of the maintenance staff is also high as specially trained personnels are required to handle the plant.

(*vii*) The disposal of the products, which are radioactive, is a big problem. They have either to be disposed off in a deep trench or in a sea away from seashore.

(*viii*) The cooling water requirements of a nuclear power plant are very heavy (more than twice the water required for the same size coal-fired steam power plant). Hence cooling towers required for nuclear power plants are larger and costlier than those for conventional steam power plants.

4.3. SELECTION OF SITE FOR NUCLEAR POWER PLANTS

The factors to be considered while selecting a site for nuclear power plant for economical and efficient generation are:

(*i*) *Availability of Water Supply.* As sufficient water (more than twice the water required for the coal plant of the same size) is required for cooling etc., therefore, the site selected for the plant should be near a river, or lake or by sea side.

(*ii*) *Distance From Populated Area.* There should be a reasonable distance between the nuclear power plant and the nearest populated area from the point of view of safety as there is a danger of presence of radioactivity in the atmosphere near the plant. However, as precautionary measure, a dome is used in the plant which does not allow the radioactivity to spread by wind or underground waterways.

It is highly undesirable to choose a site adjacent to chemical industries, oil refineries, PWD Works, hospitals and schools.

(*iii*) *Transportation Facilities.* As nuclear power plant needs very little fuel, hence it does not require direct rail facilities for fuel transport. However, transportation facilities are required during the construction stage.

(*iv*) *Nearness to Load Centre.* Such plants should be located as near to the load centre as possible, consistent with safety considerations in order to reduce the transmission costs.

(*v*) *Availability of Space for Disposal of Waste.* The site selected for such power plants should have adequate space and arrangement for the disposal of radioactive waste.

(*vi*) *Accessibility.* There should be reasonable accessibility for plant personnel, hauling in equipment, and dispatching and receiving heavily shielded radioactive materials.

(*vii*) *Type of Land.* The substrata must be strong enough to support the heavy reactors which may weigh as high as 100,000 tonnes and impose bearing pressures around 50 tonnes/m^2. Areas remote from coal fields and hydro sites are preferable so as to improve the reliability of supply over the area.

Generally hydel power plants are located far away from load centre. In conventional thermal power plants, enormous fuel shipment is required. Location of a nuclear power plant is practically independent of geographical factors, the only requirement being that there should be good supply of water. The ideal choice for a nuclear power plant would be near a sea, river or lake and away from thickly populated area.

4.4. NUCLEAR PHYSICS

Matter has been defined any thing that occupies space and has weight such as copper, iron, water, air, smoke, aluminium, silver etc. Matter is made up of extremely small particles known as molecules. A *molecule* is the smallest particle into which a given substance may be divided without changing its identity. Molecules can further be subdivided chemically into still minute particles known as atoms. An *atom* may be regarded as the smallest particle of an element that can exist.

4.4.1. Structure of Atom

John Dalton (1805) considered that *all matters were composed of small particles called atoms.* He visualised the atom as a hard solid individual particle incapable of subdivision. At the end of the nineteenth century there accumulated enough experimental evidence to show that the atom is made of still small particles. These subatomic particles are called the *fundamental particles.* The number of subatomic particles now known is very large. For us, the three most important are the *protons, neutrons and electrons.* How these fundamental particles go to make the internal structure of the atom is a fascinating story.

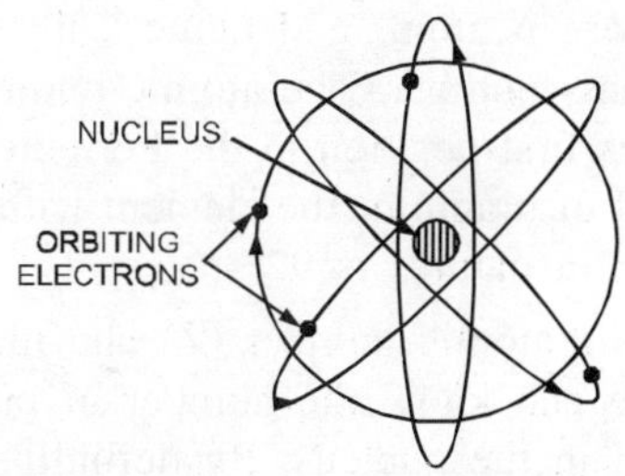

Fig. 4.1. ***Electrons Orbiting Around Nucleus***

According to modern electron theory, the atoms of all elements are composed of two parts, namely the *central nucleus* and the surrounding or orbital electrons, as illustrated in Fig. 4.1.

Thus the atom may be compared to a planet with orbiting satellites. Just as satellites are held in orbit by an attractive force of gravity due to the mass of the planet, so each electron is held in orbit by an *electrostatic force* of attraction between it and the nucleus.

The electrons, each has negative electric charge of 1.602×10^{-19} C and some particles within the nucleus have a positive charge of same magnitude. Since opposite charges attract each other, a force of attraction exists between the oppositely charged electrons and nucleus. Compared to the mass of nucleus, electrons are relatively tiny particles of almost negligible mass (each electron weighing 9.107×10^{-31} kg). Thus an electron may be defined, as a *subatomic particle bearing an electric charge of* 1.602×10^{-19} C *and having mass of* 9.107×10^{-31} kg *i.e.,* $\frac{1}{1,835}$ *th of the mass of an hydrogen atom.*

The nucleus of an atom is largely a cluster of two types of particles, called the *protons* and *neutrons.* Protons, each has a positive electrical charge, equal in magnitude (but opposite in polarity) to the negative charge of an electron. A neutron has no charge at all. Protons and neutrons each have mass about 1,800 times the mass of an electron. For a given atom, the number of protons in the nucleus normally is equal to the number of orbiting electrons. Thus the atom is electrically neutral. If an atom loses an electron, it loses some negative charge and becomes a positive ion. Similarly if an atom gains an additional electron, it becomes a negative ion.

The actual mass of proton is 1.672×10^{-27} kg. Since proton was the lightest positive particle found in atomic beams in the discharge tube, it was thought to be a unit present in all other atoms. Protons were also obtained in a variety of nuclear reactions indicating further that all atoms contain protons. Thus *a proton is defined as a subatomic particle which has a mass of 1 amu (atomic mass unit) and charge + 1 elementary charge unit (one unit positive charge).* 1 amu = 1.66×10^{-27} kg.

Neutron has almost the same mass (1.674×10^{-27} kg) as that of a proton and has no charge.

4.4.2. Atomic Number and Mass Number

The number of protons (or electrons) in an atom is referred to as the *atomic number (Z)* of the atom. The *atomic mass number (A)* is equal to the total number of protons and neutrons in the nucleus of the atom. The *atomic weight* is the actual weight of an atom, and numerically slightly differs from atomic mass number. The atomic number of hydrogen (which occupies first position in the Periodic Table) is 1, of helium is 2 and of uranium (the element with highest atomic weight existing in nature) is 92.

Knowing the atomic number (Z) and mass number (A) of an atom, we can know the number of protons and neutrons contained in the nucleus. By definition

Atomic number, Z = Number of protons

Mass number, A = Number of protons + number of neutrons

So the number of neutrons, $N = A - Z$ ...(4.1)

For example the number of electrons, protons and neutrons in an atom of Uranium having atomic number 92 and atomic weight 238.029 can be determined as follows:

Number of electrons = Number of protons

= Atomic number = 92

Number of neutrons, $N = A - Z = 238 - 92 = 146$

Thus uranium has 92 electrons, 92 protons and 146 neutrons.

4.4.3. Isotopes

The chemical nature of an element depends on its number of electrons which is equal to the atomic number of that element. As a result atoms with nuclei containing the same number of protons *i.e.,* same atomic number but with different number of neutrons (*i.e.,* with different mass numbers) are chemically identical but they exhibit marked differences of nuclear stability. The atoms of an element having the same number of protons and different number of neutrons are called *isotopes.* Alternatively, isotopes may be defined as *the atoms of an element which have the same atomic number but different atomic masses or mass numbers.*

There are three isotopes of hydrogen: protium (${}_1^1H$), *deuterium* (${}_1^2H$ or D) and tritium (${}_1^3H$ or T) having no neutron, one neutron and two neutrons respectively. Protium is by far the most abundant in natural hydrogen, deuterium about 0.015% and tritium only one out of 10,000,000 hydrogen atoms.

Oxygen has three isotopes ${}_8^{16}O$, ${}_8^{17}O$ and ${}_8^{18}O$. These are found with the relative abundances of 99.759, 0.037 and 0.204 respectively.

There are three isotopes of uranium: ${}_{92}^{238}U$, ${}_{92}^{235}U$, and ${}_{92}^{234}U$. Natural uranium consists almost entirely (99.23%) of ${}^{238}U$ with about 0.72% of ${}^{235}U$ and 0.006% of ${}^{234}U$. These isotopes are particularly important in atomic energy. The atomic number of the three isotopes of uranium is 92 and their mass numbers are 238, 235 and 234.

4.4.4. Radioactivity

A number of elements such as uranium, radium and thorium are unstable. Their atomic nucleus breaks of its own accord to form a smaller atomic nucleus of another element. The protons and neutrons in unstable nucleus regroup to give the new nucleus. This causes the release of excess particles and energy from the original nucleus, which we call *radiation.* The elements whose atomic nucleus emits radiation are said to be *radioactive.* The spontaneous breaking down of the unstable atoms is termed *radioactive disintegration or radioactive decay.*

The disintegration or decay of unstable atoms accompanied by emission of radiation is called radioactivity.

The active radiations are of three types namely α (alpha) particles, β (beta) particles and γ (gamma) rays.

Alpha particle has a mass of 4 with 2 protons and 2 neutrons *i.e.,* nuclei of helium and may be represented as ${}_2^4\alpha$ or ${}_2^4He$. α particles are ejected from radioactive nuclei with very high velocity (3×10^7 m/s), about one-tenth that of light. They have positive charge and when emitted from the nucleus of atom can travel only a short distance in air. If an α-particle emits from the nucleus of uranium atom, then thorium will be produced as shown below :

$$^{238}_{92}\text{U} \rightarrow {}^{4}_{2}\text{He} + {}^{234}_{90}\text{Th} \quad ...(4.2)$$

Beta particles result due to the disintegration of a neutron into proton. β-particles are identical with electrons and have very small mass $\left(\frac{1}{1,827}\text{ amu}\right)$ and charge of – 1. A β-particle is symbolized as $_{-1}^{0}\beta$ or $_{-1}^{0}e$. They travel about 10 times faster than α-particles *i.e.,* their velocity is about the same as of light. β-particles are more penetrating than α-particles. Overexposure to β-radiations may cause skin burns and repeated overexposure may lead to malignant growths. β-particles can be shielded with a thin sheet of metal or piece of wood.

Gamma rays are a form of *electromagnetic radiation* of shorter wavelength than *x*-rays. They could be thought of as high-energy photons released by the nucleus during α- or β-emissions. They have no mass or charge and may be symbolized as $_{0}^{0}\gamma$. Like all forms of electromagnetic radiation, γ rays travel with the velocity of light. Because of their high velocity and non-material nature, γ rays are most penetrating. They cannot be stopped even by a 5 cm thick sheet of lead or several metres thick layer of concrete. Exposure to γ rays may cause blood diseases and undesirable genetic effects.

The above three radiations are called *nuclear radiations.*

In addition to the above three radiations there is one more, called the *neutron emission.* The emission of a neutron can take place only when the nucleus of a radioactive isotope remains in a highly excited state. When a neutron is emitted, the element remains the same but a different isotope results.

4.4.5. Rate of Radioactive Decay

The decay of a radioactive isotope takes place by disintegration of the atomic nucleus. It is not influenced by any external conditions. Therefore the rate of decay is characteristic of an isotope and depends only on the number of atoms present. If N be the number of undecayed atoms of an isotope present in a sample of the isotope, at time *t*,

$$-\frac{dN}{dt} \propto N$$

or

$$-\frac{dN}{dt} = \lambda N \quad ...(4.3)$$

where $-\frac{dN}{dt}$ means the rate of decrease in the number of radioactive atoms in the sample, and λ is the proportionality factor. λ is known as the *decay* or *disintegration constant.* Substituting $dt = 1$ in Eq. (4.3), we have

$$-\frac{dN}{N} = \lambda \quad ...(4.4)$$

Thus decay constant may be defined as the proportion of atoms of an isotope decaying per second.

The standard unit of radioactivity or rate of disintegration is *Curie (c).* A curie is a quantity of radioactive material decaying at the same rate as 1 g of radium (3.7×10^{10} dps). Rutherford is a more recent unit. 1 Rutherford = 10^6 dps. SI unit is beequerel where 1 Bq = 1 dps.

4.4.6. Half Life

The *half life* or *half life period* of a radioactive isotope is the time required for one half of the isotope to decay or, it may be defined as the time for the radioactivity of an isotope to be reduced to half of its original value. *Half life* period is the characteristic of a radioactive element. For example, the half life of radium is 1,620 years. This means that 1 g of radium will be reduced to 0.5 g in 1,620 years and so on. Some other radioactive elements may have half life of a fraction of a second and for others it may be million of years. The unit of half life period is time^{-1}.

Determination of Half Life. Integrating Eq. (4.3), we have

$$N = N_0 \cdot e^{-\lambda t} \quad ...(4.5)$$

or $\lambda t = \log_e\left(\frac{N_0}{N}\right)$

where N_0 is the number of atoms at time $t = 0$

At half lifetime $(t_{1/2})$, $N = \frac{N_0}{2}$

$$\therefore \quad \lambda t_{1/2} = 2.303 \log_{10}\frac{N_0}{\frac{1}{2}N_0} = 2.303 \log_{10} 2$$

$$\text{or} \quad t_{1/2} = \frac{0.693}{\lambda} \quad ...(4.6)$$

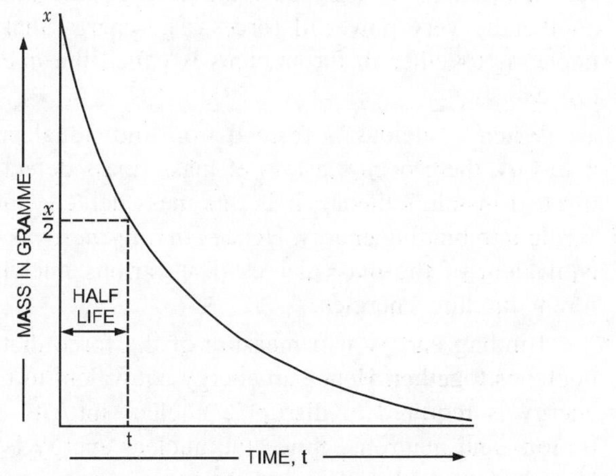

Fig. 4.2. ***Decay Curve of a Radioactive Isotope of Mass x***

4.4.7. Average Life

In a radioactive substance, some atoms decay earlier and others survive longer. The statistical average of the lives of all atoms present at any time is called the *average* or *mean life* T_{av}. It has been shown to be reciprocal of decay constant λ.

$$i.e., \quad T_{av} = \frac{1}{\lambda} = \frac{\text{Half life}}{0.693} \quad ...(4.7)$$

4.4.8. Mass Defect

We know that atomic nucleus consists of protons and neutrons; collectively known as *nucleons.* It is found that the measured mass of a nucleus is always less than the sum of masses of the individual protons and neutrons which make it up. Let us take the example of helium $^4_2$H. It consists of two protons and two neutrons. Its mass may be calculated as

$$\text{Mass of protons} = 2 \times 1.00815 = 2.01630$$
$$\text{Mass of neutrons} = 2 \times 1.00899 = \underline{2.01798}$$
$$4.03428$$

However, the experimental mass of the helium nucleus is only 4.00388. This is less by 0.03040 amu than that the calculated above. This is called the *mass defect* of helium nucleus.

Thus the difference between the experimental and calculated masses of the nucleus is called the *mass defect* or *mass deficit.*

The mass defect of a given nucleus can be calculated as follows:

$$\text{Mass defect} = Zm_p + (A - Z)\, m_n - \text{nuclear mass} \quad ...(4.8)$$

where m_p and m_n are the masses of a proton and a neutron respectively, Z is the atomic number or number of protons in the nucleus and A is the atomic mass number or total number of protons and neutrons in the nucleus.

4.4.9. Binding Energy

Atomic nucleus is made of protons and neutrons closely packed in a small volume. Although there exist intensive repulsive forces between the component protons, the nucleus is not split apart. This is so because the nucleons are bound to one another by very powerful forces. The energy that binds the nucleons together in the nucleus is called the *nuclear binding energy.*

When a nucleus is formed from individual protons and neutrons, there occurs a loss of mass (mass defect). According to Einstein's theory, it is this mass defect which is converted into binding energy. Hence *binding energy* is the energy equivalent of the mass defect. The various nuclei have different binding energies.

Binding energy is a measure of the force that holds the nucleons together. Hence an energy equivalent to the binding energy is required to disrupt a nucleus into its constituent protons and neutrons. Since the nuclear energy is of an extremely high order, it is not easy to fission a nucleus.

The binding energy of a nucleus can be calculated from its mass defect by using Einstein's equation, $\Delta E = \Delta m \times c^2$. For example, the binding energy in $^{11}_{5}B$ nucleus with mass defect 0.08181 amu comes out to be:

$$\Delta E = (0.8181 \text{ g/mole}) \frac{(3 \times 10^{10} \text{ cm/s})^2}{6.02 \times 10^{23}}$$
$$= 1.223 \times 10^{-4} \text{ ergs/nucleus}$$

$\because$ $\Delta m = 0.08181$ g/mole, $c = 3 \times 10^{10}$ cm/s and number of nuclei in one mole is 6.02×10^{23} (Avagadro's Law)

Binding Energy Per Nucleon. It can be calculated by dividing the total binding energy by the sum of the number of protons and neutrons present in the nucleus *i.e.,*

Binding energy per nucleon

$$= \frac{\text{Binding energy of nucleus}}{\text{Number of protons} + \text{number of neutrons}}$$

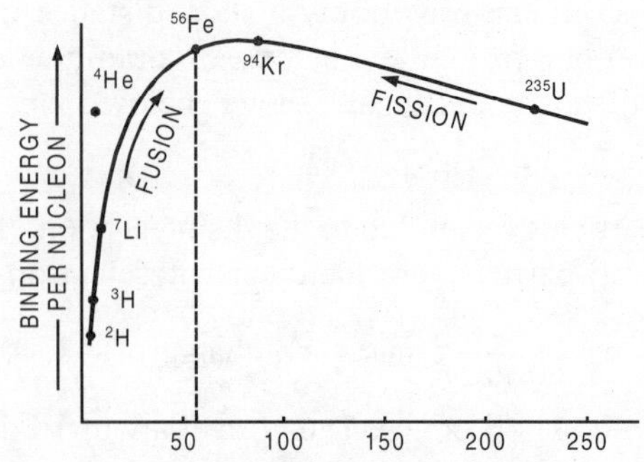

Fig. 4.3. ***A Curve of Binding Energy Per Nucleon Versus Mass Number***

By plotting the binding energy per nucleon against the mass number, we get the graph as illustrated in Fig. 4.3. This shows the relatively stability of the various nuclei. *The greater the binding energy per nucleon the more stable is the nucleus.* Thus the nuclei of about 60 atomic mass having maximum energy per nucleon are most stable *e.g.,* ^{56}Fe. The nuclei that are heavier or lighter than this have lower binding energies per nucleon and are less stable. Thus ^{235}U undergoes fission into lighter and more stable isotopes as ^{139}Ba and ^{94}Kr with the release of energy. Similarly two or more lighter nuclei (^2H, ^3H) with lower binding energy per nucleon combine or fuse together into a heavier and more stable nucleus. This is also accompanied by release of energy. This gives two fundamental ways of obtaining nuclear energy.

(*i*) Fission of heavy elements into lighter elements.

(*ii*) Fusion of light elements into heavier elements.

The nuclear fission process is used in nuclear power plants for generation of electrical energy. The nuclear fusion process has still not been exploited commercially.

4.4.10. Mass Energy Equivalence

Since 1 amu is exactly equal to $\frac{1}{12}$th of the mass of C^{12} atom, therefore

$$1 \text{ amu} = \frac{1}{12} \times \text{mass of } C^{12} \text{ atom}$$
$$= \frac{\frac{1}{12} \times \text{gram atomic mass of } C^{12}}{\text{Avagadro's number}}$$
$$= \frac{1}{12} \times \frac{12}{6.02 \times 10^{23}} \text{ gm}$$
$$= 1.66 \times 10^{-24} \text{ gm or } 1.66 \times 10^{-27} \text{ kg}$$

Also, $E = mc^2 = 1.66 \times 10^{-27} \times (3 \times 10^8)^2$
$= 1.494 \times 10^{-10}$ J

$\because$ c = Velocity of light $= 3 \times 10^8$ m/s

or $$E = \frac{1.494 \times 10^{-10}}{1.602 \times 10^{-19}} = 931.5 \times 10^6 \text{ eV} = 931.5 \text{ MeV}$$

$\because$ $1 \text{ eV} = 1.602 \times 10^{-19}$ J

Thus 1 amu = 931.5 MeV

4.5. NUCLEAR REACTION

Here, we shall consider the interaction between a projectile which may be a neutron, a proton or an α-particle and a target nucleus which is usually a heavy element. In nuclear reactions, the atomic numbers and mass numbers must balance on both sides of the equation. However, the masses of isotopes on both sides of the equation do not balance *i.e.,* there is, either an increase or decrease in mass between reactants and products. The mass can be converted into energy by Einstein's equation. Thus a *nuclear reaction is associated* with a release or absorption of energy. If there is a decrease in total mass after the reaction, then there will be release of energy and vice versa. In nuclear reaction, the nuclei of elements starting reaction do not show up as products. The products may consist of either isotopes of reactants or completely different.

The interaction between the projectile and target nucleus may result in elastic scattering, inelastic scattering and capture or absorption. If a neutron collides with a nucleus, and the nucleus remains unchanged in its isotopic composition and the neutron undergoes a change in its direction of motion (with or without change in energy) then the process is called *scattering.* If the neutron disappears from the system, the process is called *capture* or *absorption.*

In *elastic scattering,* the classical laws of conservation of mass, energy and momentum hold. The total kinetic energy of the struck nucleus and the projectile remains unchanged. There is neither release nor absorption of energy, there being only redistribution of kinetic energy of the colliding particles according to the laws of hard-ball collisions. The compound nucleus may emit a similar particle to one captured by the nucleus. Thus the elastic scattering of neutron with carbon is written as $^{12}C\ (n.n)\ ^{12}C$.

In inelastic scattering, the energy of the starting particle is quite large and the target nucleus is raised to an excited state. The compound nucleus attains its ground state by emitting gamma radiation. Thus the total kinetic energy of the colliding particles reduces after the collision, the difference being the γ radiation. There is neither absorption nor release of energy and the incident particle and the ejected particle are identical. This process occurs, *e.g.,* when a fast neutron hits a ^{238}U nucleus, the reaction being written as $^{238}U\ (n, n\gamma)\ ^{238}U$.

4.6. NUCLEAR FISSION PROCESS

This is the process in which heavy nucleus is split when it is bombarded by certain particles. Such a reaction is possible only with heavy nuclei such as $^{233}_{92}U$, $^{235}_{92}U$ and $^{239}_{94}Pu$. Although the fission can be caused bombarding with high energy particles such as protons, neutrons and *x*-rays, as well as with neutrons, the latter are most suited for fission. The main advantage of neutrons is that they are neutral (having no charge) and therefore they can make their way through the shells of electrons and then through the nucleus at low energy. This is the practical method used in all modern fission reactors.

In 1939, Hahn and Stassmann discovered that a heavy atomic nucleus as of uranium-235 upon bombardment by a neutron splits apart into two or more nuclei. U-235 first absorbs a neutron to form an unstable 'compound nucleus'. The excited 'compound nucleus' then divides into two daughter nuclei* with the release of neutrons and large amount of energy.

The splitting of a heavy nucleus into two or more smaller nuclei is termed nuclear fission. The process of fission is always accompanied by the ejection of two or more neutrons and liberation of vast energy.

A given large nucleus can fission in many ways forming a variety of products. Thus the fission of U-235 occurs in about 35 ways. Two of these are given below in the form of equations.

$$^{235}_{92}U + ^{1}_{0}n \rightarrow ^{139}_{56}Ba + ^{94}_{36}Kr + 3\,^{1}_{0}n + \text{energy}$$

$$^{235}_{92}U + ^{1}_{0}n \rightarrow ^{106}_{42}Mo + ^{128}_{50}Sn + 2\,^{1}_{0}n + \text{energy}$$

In these fission reactions, the mass of the products is less than that of the reactant. A loss of about 0.2 amu per uranium atom occurs. This mass is converted into a fantastic quantity of energy (0.2 × 931.5 *i.e.,* 187 MeV). Generally it is assumed that 1 fission of $^{235}_{92}U$ causes a release of 200 MeV of energy.

$$200 \text{ MeV} = 200 \times 1.6 \times 10^{-13}$$
$$= 3.2 \times 10^{-11} \text{ joules (or watt-seconds)}$$

Thus 1 watt (or 1 J/s) requires $\dfrac{1}{3.2 \times 10^{-11}}$ or 3.1×10^{10} fissions per second.

If all the atoms of 1 kg of pure $^{235}_{92}U$ (containing 25.64×10^{23} atoms) were fissioned, the energy released would be equivalent to that contained in 3×10^6 kg of coal with a calorific value of 6,000 K cal/kg. Natural uranium contains only 0.7% ^{235}U. If fission efficiency is 50% (*i.e.,* only half of the total atoms take part in fission) fission of 1 kg of natural uranium would give energy equivalent to $\dfrac{0.7}{100} \times 3 \times 10^6 \times 0.5$ or 10,500 kg of coal.

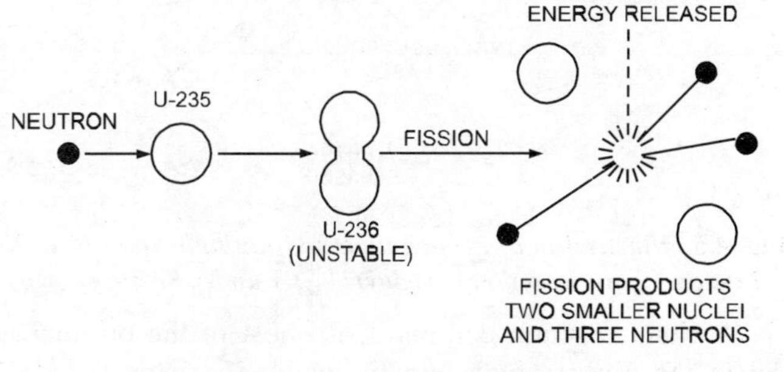

Fig. 4.4. ***Illustration of Fission Process***

* The nucleus of a radioactive atom is unstable and it undergoes decay or disintegration by spontaneous emission of an α or β particle. This results in the change of proton-neutron composition of the nucleus to form a more stable nucleus. The original nucleus is called the *parent nucleus* and the product is called the *daughter nucleus.*

Characteristics of nuclear fission may be summed up as below.

(*i*) Upon capturing a neutron, a heavy nucleus cleaves into two or more nuclei.

(*ii*) Two or more neutrons are produced by fission of each nucleus.

(*iii*) Vast quantities of energy (3 million times of that produced by equivalent amount of coal) are produced as a result of conversion of small mass into energy.

(*iv*) All the fission products are radioactive, giving off β and γ radiations.

4.7. NUCLEAR CHAIN REACTION

We know that U-235 nucleus when hit by a neutron undergoes the reaction,

$$^{235}_{92}\text{U} + {}^{1}_{0}n \rightarrow {}^{139}_{56}\text{Ba} + {}^{94}_{36}\text{Kr} + 3{}^{1}_{0}n$$

Each of the three neutrons produced in the reaction strikes another U-235 nucleus, thus causing nine subsequent reactions. These nine reactions, in turn, further give rise to twenty seven reactions. This process of propagation of the reaction by multiplication in threes at each fission, as shown in Fig. 4.5, is referred to as a *chain reaction.* Heavy unstable isotopes, exhibit a chain reaction by release of two or three neutrons at each fission. It may be defined as a *fission reaction where the neutrons from a previous step continue to propagate and repeat the reaction.*

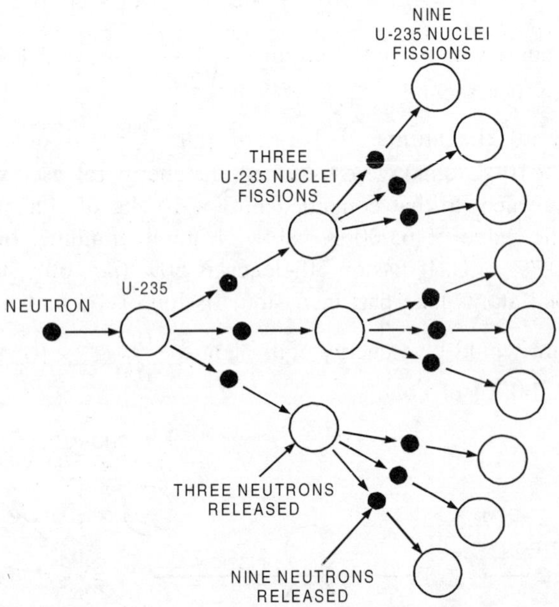

Fig. 4.5. ***Illustration of Propagation and Multiplication of U-235 Fission Chain Reaction (Products $^{139}_{56}$Ba and $^{94}_{36}$Kr not shown)***

A chain reaction continues till most of the original nuclei in the given sample are fissioned. However, it may be noted that all the neutrons released in the reaction are not used up in propagating the chain reaction. Some of these are lost to the surroundings. Thus for a chain reaction to occur, the sample of the fissionable material should be large enough to capture the neutron internally. If the sample is too small, most neutrons will escape from its surface, thereby breaking the chain. The minimum mass of fissionable material required to sustain a chain reaction is called the *critical mass.* The critical mass varies for each reaction. For U-235 fission reaction it is about 10 kg.

As already mentioned, even a single fission reaction produces a large amount of energy. A chain reaction that consists of innumerable fission reactions will, therefore, generate many times greater energy.

Materials fissionable by thermal or low-speed neutrons are U-233, U-235 and Pu-239.

4.8. NUCLEAR ENERGY

A heavy isotope as uranium-235 (or plutonium-239) can undergo nuclear chain reaction yielding vast amounts of energy. The energy released by the fission of nuclei is called *nuclear fission energy* or *nuclear energy.* Sometimes, it is incorrectly referred to as *atomic energy.*

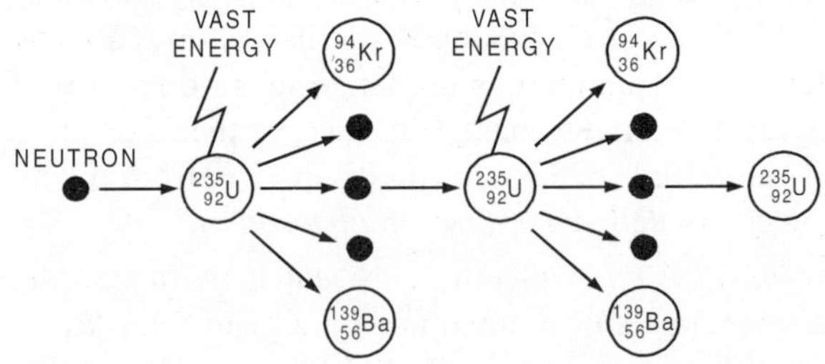

Fig. 4.6. ***Chain Reaction of Uranium-235 Producing Energy***

The fission of U-235 or Pu-239 occurs instantaneously, producing incomprehensible quantities of energy in the form of heat and radiation. If the reaction is uncontrolled, it is accompanied by explosive violence and can be used in atomic bombs. However, when controlled in a reactor, the fission of U-235 is harnessed to generate electricity.

4.9. NUCLEAR FUELS

The fuels mainly used are natural uranium (U-235), enriched uranium, plutonium (secondary fuel) and U-233 (secondary fuel available from breeder reactor). Natural uranium is the parent material.

The materials U-235, U-233 and Pu-239 are called *fissionable materials.* The only fissionable nuclear fuel occurring in nature is uranium, of which 99.3% is $^{238}_{92}$U and 0.7% is $^{235}_{92}$U and $^{234}_{92}$U is only a trace. Out of these isotopes only $^{235}_{92}$U will fission in a chain reaction, The other two fissionable materials $\left(^{239}_{94}\text{Pu and } ^{233}_{92}\text{U}\right)$ can be produced artificially from $^{238}_{92}$U and $^{232}_{90}$Th which occur in nature and are called *fertile materials.* Out of the three fissionable materials, U-235 has some advantage over the other two due to its higher fission percentage.

Fissionable materials $^{239}_{94}$Pu and $^{233}_{92}$U are formed in the nuclear reactors during fission process from $^{238}_{92}$U and

$^{232}_{90}$Th respectively due to absorption of neutrons without fission. The processes are given below:

$$^{238}_{92}\text{U} + ^{1}_{0}n \rightarrow ^{239}_{92}\text{U} + \gamma$$

$$^{239}_{92}\text{U} \rightarrow ^{239}_{93}\text{Np} + ^{0}_{-1}e$$

$$^{239}_{93}\text{Np} \rightarrow ^{239}_{94}\text{Pu} + ^{0}_{-1}e$$

The above process is called *conversion.* Absorption of a neutron by U-238 produces U-239, which is unstable with half life period of 23 minutes and decays into neptunium with emission of an electron. $^{239}N_p$ which has a half life period of 2.3 days is likewise transformed into Pu-239 which is a long lived fissionable isotope of plutonium.

Fissionable U-233 is produced in the following manner :

$$^{232}_{90}\text{Th} + ^{1}_{0}n \rightarrow ^{233}_{90}\text{Th} + \gamma$$

$$^{233}_{90}\text{Th} \xrightarrow{23.3\text{ min}} ^{233}_{91}\text{Pa} + ^{0}_{-1}e$$

$$^{233}_{91}\text{Pa} \xrightarrow{27.4\text{ day}} ^{233}_{92}\text{U} + ^{0}_{-1}e$$

The above process is called *breeding*.

4.10. CONSTITUENTS OF NUCLEAR POWER PLANT AND LAYOUT

The concepts of nuclear power generation are much similar to that of conventional steam power generation. The difference lies only in the steam generation part *i.e.,* coal or oil burning furnace and the boiler are replaced by nuclear reactor and heat exchanger. The schematic arrangement of a nuclear power plant is shown in Fig. 4.7(*a*) and (*b*).

Thus a nuclear power plant consists of a nuclear reactor (for heat generation), heat exchanger (for converting water into steam by using the heat generated in nuclear reactor), steam turbine, alternator, condenser etc. As in a conventional steam power plant, water for raising steam forms a closed feed system. However, the reactor and the cooling circuit have to be heavily shielded to eliminate radiation hazards.

The tremendous amount of heat energy produced in breaking of atoms of uranium or other similar metals of large atomic weight into metals of lower atomic weight by fission process in an atomic reactor is extracted by pumping fluid or molten metal like liquid sodium or gas through the pile. The heated metal or gas is then allowed to exchange its heat to the heat exchanger by circulation. In heat exchanger the gas is heated or steam is generated which are utilised to drive gas turbine or steam turbine coupled to an alternator thereby, generating electrical energy.

While deciding the layout of a nuclear power plant due consideration should be given to safety, operating convenience and capital economy. One of the important operational areas in a reactor building is the charge hall which is used for the refuelling operation. This is located directly over the reactor core. As far as the main parts of a nuclear power plant are concerned the reactor, turbine and generator—the layout is quite simple. A main control room is provided in a central location and consists of all the necessary equipment for controlling normal and emergency operation of the reactors as well as controls of boilers and turbo-alternators. All other ancillary rooms such as charge room, maintenance room, store, switchyard, railway siding, machine shop, instrument shop, office etc. are suitably located for convenient operation.

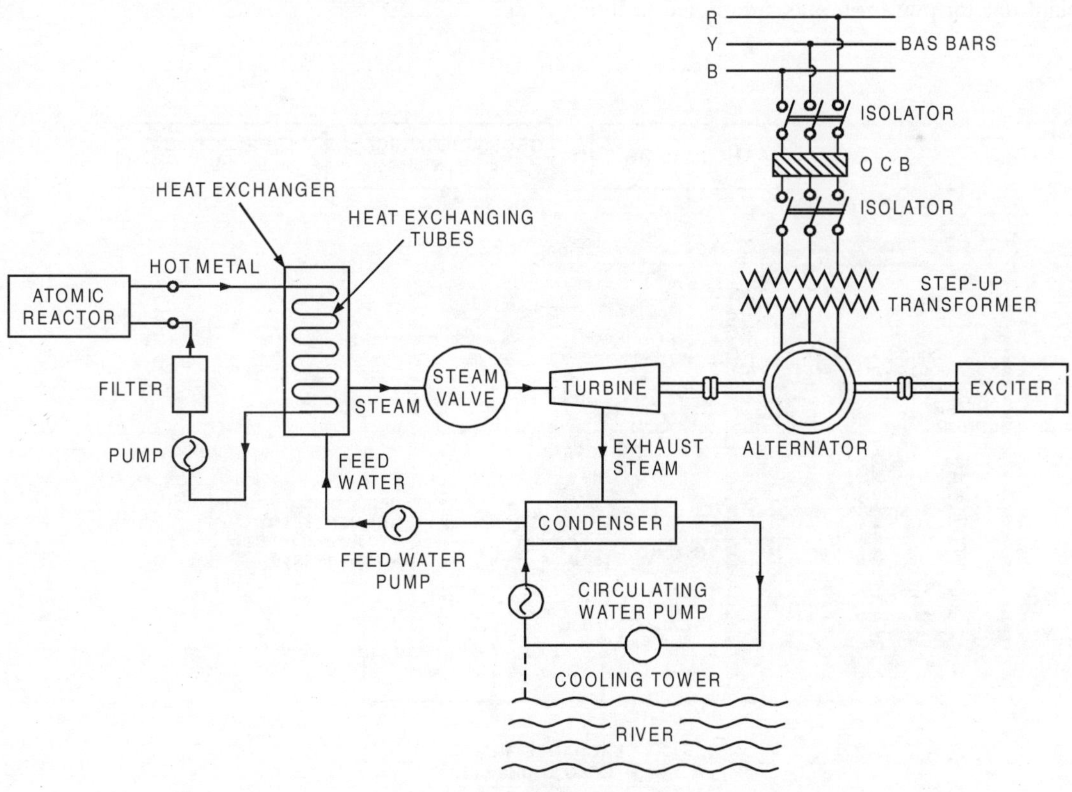

Fig. 4.7(*a*).

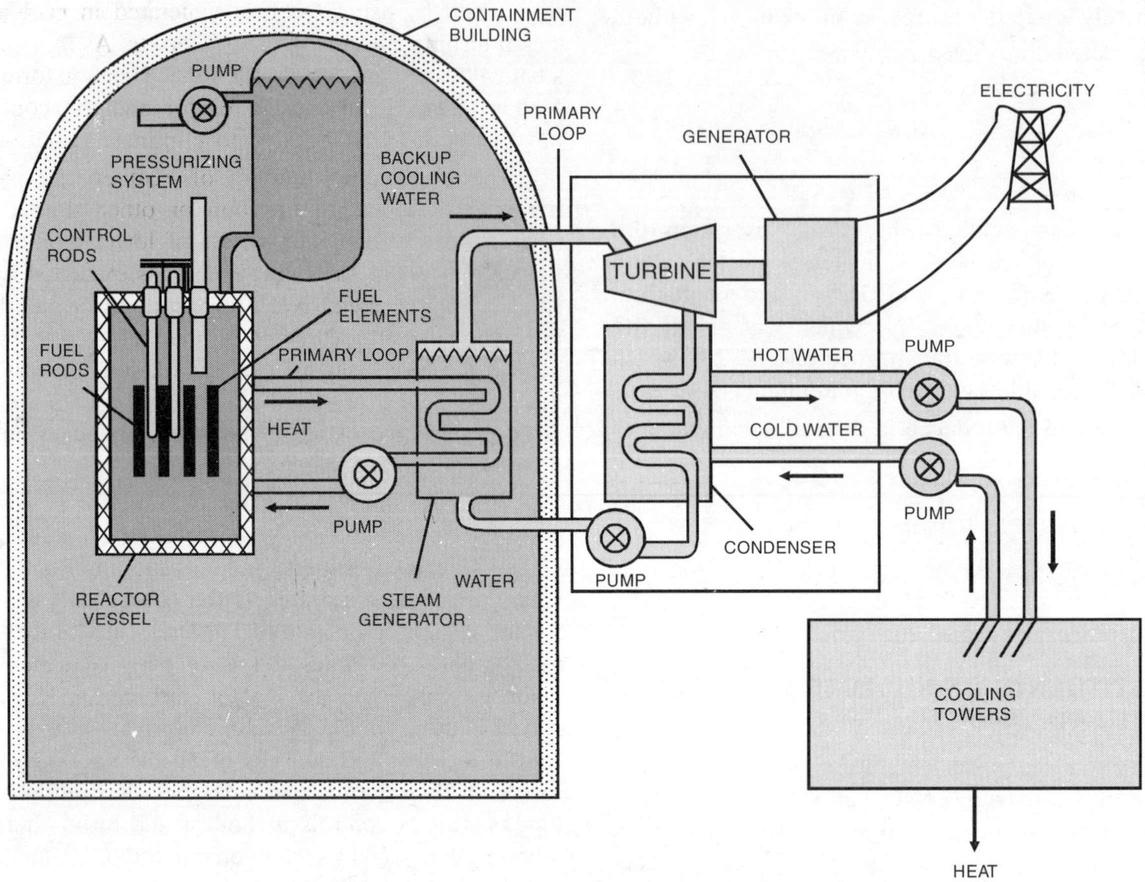

Fig. 4.7(*b*) ***Schematic Arrangement of Nuclear Power Plant***

4.11. POWER RATING OF A NUCLEAR POWER PLANT

Nuclear power plant reactor power outputs are quoted in three ways:

Thermal MW, which depends on the design of the actual nuclear reactor itself, and relates to the quantity and quality of the steam it produces.

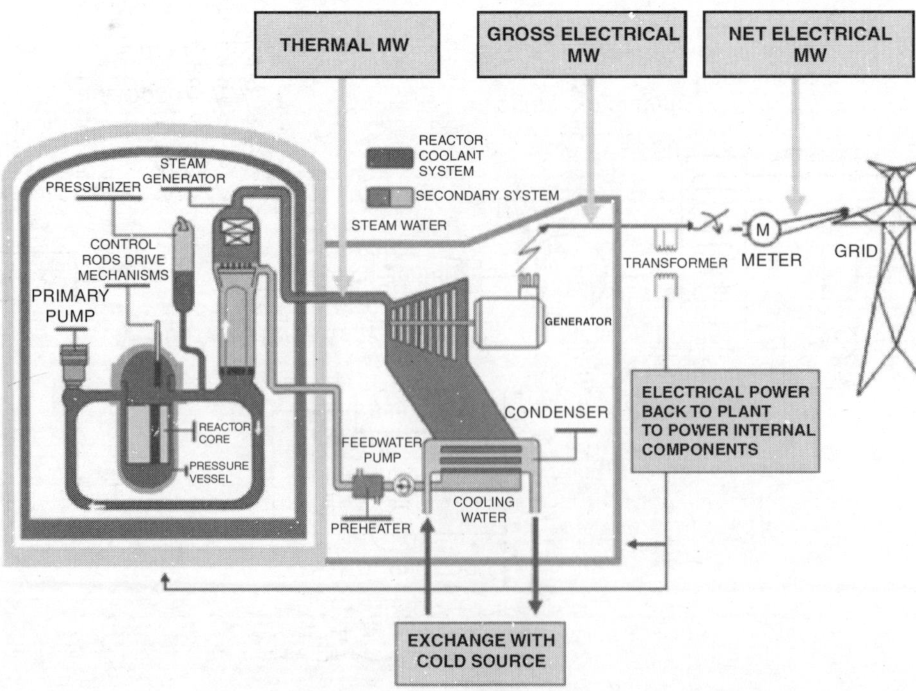

Fig. 4.8

Gross electrical MW indicates the power produced by the attached steam turbine and generator, and also takes into account the ambient temperature for the condenser circuit (cooler means more electric power, warmer means less). Rated gross power assumes certain conditions with both.

Net electrical MW, which is the power available to be sent out from the plant to the grid, after deducting the electrical power needed to run the reactor (cooling and feed water pumps, etc.) and the rest of the plant.*

The relationship between these is expressed in two ways:

- Percentage thermal efficiency is the ratio of gross electrical MW to thermal MW. This relates to the difference in temperature between the steam from the reactor and the cooling water. It is often 33-37%.
- Percentage net efficiency is the ratio of net electrical MW achieved to thermal MW. This is a little lower, and allows for plant usage.

In WNA papers and figures and WNN items, generally net MW is used for operating plants, and gross MW for those under construction or planned/proposed.

4.12. NUCLEAR REACTOR—MAIN PARTS AND THEIR FUNCTIONS

Reactor is that part of nuclear power plant where nuclear fuel is subjected to nuclear fission and the energy released in the process is utilised to heat the coolant which may in turn generate steam or be used in a gas turbine.

The main function of the reactor is to control the emission and absorption of neutrons. The neutrons emitted as a result of chain reaction are accounted for in the following ways:

1. Non-fission absorption by control rod materials, moderator material, fission fragments and impurities.
2. Non-fission absorption by U-238 present in natural uranium to form fissionable Pu-239.
3. Fission absorption by U-235, Pu-239 and U-233.
4. Escape from the fissionable material.

A nuclear reactor consists of the following basic components [Fig. 4.9(*a*)].

1. *Reactor Core.* It contains a number of fuel rods made of fissile material. They may be diluted with non-fissionable material for better control of the reaction or to reduce the damage from fission product poisoning. As the uranium gets oxidised rapidly, the fuel rods should be clad with aluminium, stainless steel or zirconium. The size of core, just sufficient to maintain a chain reaction is the "critical" size. It can be brought down by using enriched uranium as fuel.

It is desirable to use core as cubical or cylindrical in shape rather than spherical, as it facilitates the refuelling operation and simplifies the process of circulation of coolant through the core. With this configuration, the core has a series of parallel fuel elements in the form of thin plates or small rods, with coolant flowing axially and additional moderator or reflector material surrounding the assembly.

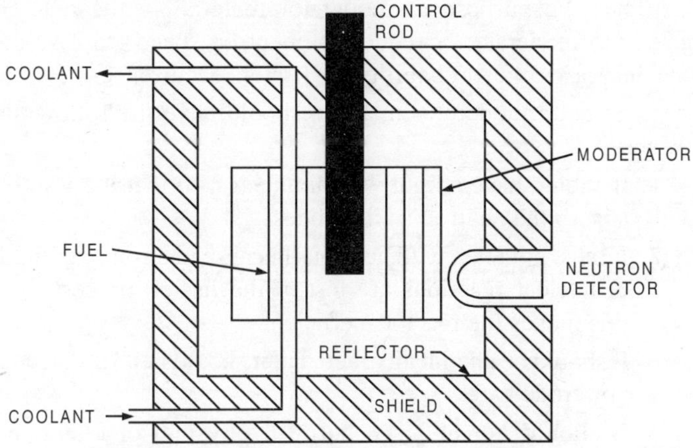

Fig. 4.9(*a*) *Basic Components of a Nuclear Reactor*

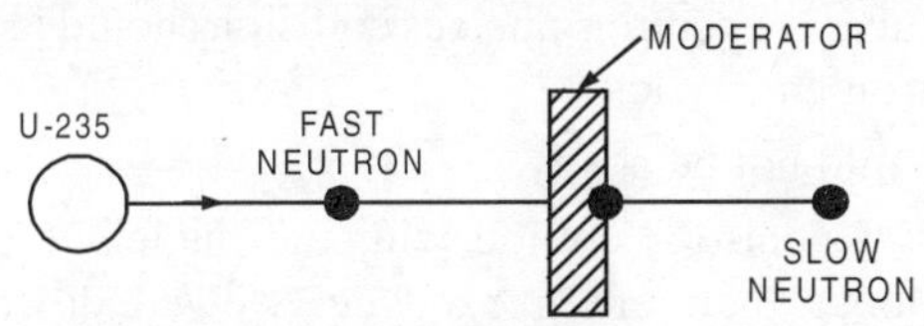

Fig. 4.9(*b*) *Moderator Slows Down a Fast Neutron*

For using the reactor to convert the fertile material into fissionable material, the material to be converted should be put around the core so that the neutrons, which otherwise would escape the core, would be utilized for conversion. This arrangement also simplifies the process of separation of the converted material during fuelling reprocessing.

2. *Moderator.* Neutrons produced by the fission process are ejected from the nucleus at a very high velocity of about 1.5×10^7 m/s and therefore, have a very large kinetic energy and are termed as *fast neutrons.* The elements which can undergo a fission reaction with fast neutrons are U-233, U-235 and Pu-239.

Natural uranium contains only 0.7% U-235. Fast neutrons are slowed down by elastic scattering process and chain reaction can still occur. But during this process, there is a possibility of their getting absorbed by U-238 and the chain reaction may not be maintained. If the proportion of U-235 in the metal is increased to more than 10%, the above absorption effect can be overcome and a chain reaction is possible. This occurs in fast reactors but the enriching process is expensive.

For more effective use in nuclear reactor, it is desirable to slow down the fast neutrons to speeds corresponding to the speed of molecules in a gas at NTP (*i.e.,* to a speed of about 2.2×10^3 m/s). Such neutrons are known as *slow* or *thermal neutrons.* The absorption properties of U-238 are very much reduced with thermal neutrons. Thus, if natural uranium is bombarded by thermal or slow neutrons, the chain

* This (as also actual gross electrical MW) varies slightly from summer to winter, so normally the lower summer figure, or an average figure, is used. If the summer figure is quoted plants may show a capacity factor greater than 100% in cooler times. Some design options, such as powering the main large feed water pumps with electric motors (as in EPR) rather than steam turbines (taking steam before it gets to the main turbine-generator), explains some gross to net differences between different reactor types. The EPR has a relatively large drop from gross to net MW for this reason.

reaction can be maintained. This is accomplished with the help of 'moderator' which is mixed with the fissile material in a suitable manner.

Thus the purpose of moderator material in the reactor core is to moderate, or reduce the neutron speeds to a value that increases the probability of fission occurrence.

The good moderator material should have the following properties:

1. It should have a light weight nucleus, so that it does not absorb the neutron as it collides.
2. It must not react with neutrons because neutrons captured in nuclear reactions are lost to the fission process and this makes the reactor inefficient.
3. It should be chemically inert and it should neither corrode nor erode.
4. It should not undergo harmful physical or chemical changes when bombarded by neutrons.
5. The average neutron-nucleus collision should lead to large neutron energy loss.
6. It should not be costly.

The fast neutrons collide with the nuclei of moderator material, loose their energy and get slowed down. As per simple laws of mechanics, if a neutron collides with a nucleus of equal mass it will loose all its energy and it will come to standstill, whereas if it collides with a heavier nucleus it will loose very little energy and there will be not much change in the magnitude of velocity but its direction will change. Thus it is evident that only elements at the top of periodic table or compounds with small molecular weights are suitable as moderator materials. Such elements are: Hydrogen, Deuterium, Helium, Lithium, Berylium, Boron, Carbon, Nitrogen, Oxygen. Other properties of good moderator are:

(*i*) High scattering cross section.

(*ii*) Low neutron absorption cross section.

Out of the elements having small atomic mass, gases (oxygen, nitrogen, hydrogen and deuterium) are unsuitable owing to their low density and the consequent small number of collisions. Helium and berylium are costly. Boron and lithium have high absorption cross section. At present, the common moderator materials are graphite, ordinary water and heavy water.

Graphite is simple to fabricate and handle and does not pose any containment problem. However, if continued bombing is maintained, this may create some stress problem. Ordinary water is cheap but it has high neutron absorption and can be used only with enriched uranium. This can be used as a coolant at moderate temperature and pressure. Heavy water is costlier per unit weight, as compared to graphite or ordinary water, as a result containment is a serious problem for heavy water than for ordinary water. For the same power output, the size of the reactor using heavy water is more compact as compared to one using ordinary water. It can be used with ordinary uranium. It is used in many reactors inspite of its heavy cost.

The moderator and the fuel can either be intimately mixed or the fuel may be scattered through out the moderator in discrete lumps. These two arrangements are called *homogenous* and *hetrogenous* arrangements respectively.

3. *Control Rods.* Control rods are meant for controlling the rate of fission of U-235. These are made of boron-10, cadmium or hafnium, that absorb some of the slowed neutrons.

In a reactor, nuclear chain reaction has to be initiated when started from cold, and the chain reaction is to be maintained at a steady value during the operation of reactor. Also the reactor must be able to shut down automatically under emergency conditions. All this requires a control of reactor so as to prevent the melting of fuel roads, disintegration of coolant and destruction of reactor as the amount of energy released is enormous.

Chain reaction is controlled either by removing the fuel rods or by inserting neutron absorbing materials. The materials used for control rods must have very high absorption capacity for neutrons.

The control rods are inserted into the reactor core from the top of the reactor vessel. These rods regulate the fissioning in the reactor by absorbing the excess neutrons. These rods can be moved in and out of the holes in the reactor core assembly. If the fissioning rate of the chain reaction is to be increased, the control rods are moved out slightly so that they absorb less number of neutrons and vice versa.

4. *Coolant.* It is a medium through which the heat generated in the reactor is transferred to the heat exchanger for further utilisation in power generation. Sometimes when water is used as a coolant it takes up heat and gets converted into steam in the reactor which is directly used for driving steam turbines. Coolant flows through and around the reactor core. It performs the additional function of keeping the interior of reactor at the desired temperature. Sometimes, the same medium is used as the coolant as well as the moderator though separate materials are used more commonly.

A good coolant should not absorb neutrons, should be non-oxidising, non-toxic and non-corrosive and have high chemical and radiation stability and good heat transfer capability.

Air, helium, hydrogen and CO_2 amongst the gases, light and heavy water amongst the liquids, and the molten sodium and lithium amongst the metals are the materials used as coolants.

Ordinary water is used both as coolant and moderator in boiling water reactors and pressurised water is used both as coolant and moderator in pressurised water reactors. Water has good thermal capacity per unit volume and is, therefore, a good heat transport medium. In reactors using water as coolant, the power consumption of circulating pump is low. Heavy water (D_2O) is even more efficient than light water.

Liquid metals (*e.g.*, sodium and potassium) are used as coolant in fast reactors which have large heat release from a small core. They have high heat transfer capability and low vapour pressure. Reactors employing liquid metal coolants can operate at high temperature. They are employed in liquid metal fuelled reactors.

Carbon dioxide is colourless and odourless and has low neutron absorption cross section. When dry, it does not react with mild steel of the pressure vessel and the supports of the core. However, it does react with graphite and therefore, special steps are to be taken in the design of the reactor so as to inhibit the reaction between CO_2 and graphite. It is used in Magnox and advanced gas cooled reactors.

5. *Reflector.* A neutron reflector is placed around the core and used to avoid the leakage of neutrons from the core. This completely surrounds the reactor core within the thermal shielding arrangement and bounces back most of the neutrons that escape from the fuel core. This conserves the nuclear fuel, as the low speed neutrons thus returned are useful in continuing the chain reaction. The reflector gets heated due to collision of neutrons with its atom, therefore, its cooling is essential. The reflector should have good neutron scattering properties and preferably a small tendency to absorb neutrons. It is often a moderating material and sometime the same material is used both for moderator and reflector.

6. *Thermal Shielding.* The shielding is usually constructed from iron and help in giving protection from the deadly α- and β-particle radiations and γ rays as well as neutrons given off by the process of fission with in the reactor. In this manner it gets heated and prevents the reactor wall from getting heated. Coolant flows over the shielding to take away the heat.

7. *Reactor Vessel.* The reactor core, reflector and thermal shielding are all enclosed in the main body of the reactor and is called the *reactor vessel* or *tank.* It is a strong walled container and provides the entrance and exit for the coolant and also the passages for its flow through and around the reactor core. There are holes at the top to allow the control rods to pass through them. The reactor core (fuel and moderator assembly) is usually placed at the bottom of the vessel. The reactor vessel has to withstand high pressures (up to 21 MPa).

8. *Biological Shield.* The whole of the reactor is enclosed in a biological shield to prevent the escape or leak away of the fast neutrons, slow neutrons, β-particles and γ rays as these radiations are very harmful for living organisms. Lead iron or dense concrete shields are used for this purpose.

4.13. REACTOR CONTROL

All power plant reactors are provided with the means to regulate the fission process so that energy is generated according to the load requirements and in an emergency the reactor can be quickly shut down. Fission control is affected by regulating the neutron population or flux as per power requirement by providing for absorption of excess neutrons through such substances which have high neutron absorption coefficient. These are called the *poisons.* Cadmium and carbon are two such substances which are inserted with the help of adjustable control rod. The position of the control rods is automatically regulated by electrochemical and electronic sensing objects which measure the neutron flux density in the reactor and actuate the control rods to regulate power generation.

As mentioned in Art. 4.7, all the neutrons released in the fission reaction are not used up in propagating the chain reaction but some of these are lost to the surroundings. For maintaining chain reaction, it is therefore, essential that the number of neutrons after the fission should be slightly more than the number before it to allow for the escape or leak of neutrons from the reactor core. The ratio is known as *multiplication* factor. The multiplication factor k for any reactor in defined as

$$k = \frac{\text{Number of neutrons produced in one generation}}{\text{Number of neutrons produced in the preceding generation}}$$

Unity value of k indicates that the chain reaction will continue at a steady rate (critical). If k is less than unity, the chain reaction will stop and the system is called *subcritical.* While for k exceeding unity, the chain reaction will build up and the system is called *super critical.*

The desirable requirement of power reactors is that the system should be critical (*i.e.,* k = 1). The critical size of a thermal reactor is one that produces neutrons just enough to balance those lost and absorbed and at the same time maintains the chain reaction.

For reactor control, the value of k is to be controlled. At the time of starting of reactor, value of k is kept above unity so as to build up the chain reaction. This increases the power level. Once the required power level has been attained, k is reduced to unity and is kept at this value as long as the output rate is to be maintained. For decreasing the output (power level) k is reduced to slightly less than unity till the required power level is attained and at this point k is brought back to unity and maintained. Similarly for shutting down the reactor k is reduced below unity; the chain reaction will stop.

4.14. CLASSIFICATION OF NUCLEAR REACTORS

Nuclear reactors may be classified in several ways *i.e.,* on basis of their applications, type of fission, fuel used, state of fuel, fuel cycle, arrangement of fissile and fertile material, arrangement of fuel and moderator, moderator material, cooling system employed, coolant used etc.

I. According to the applications the reactors are classified as
   (*a*) *Research and Development Reactors.* These reactors are used for testing new reactor designs and research.
   (*b*) *Production.* These reactors are used for converting fertile materials into fissile materials.
   (*c*) *Power.* These reactors are used for generation of electrical energy.

II. According to the type of fission the reactors are classified as fast reactors, slow reactors and intermediate reactors.

In fast reactors the fission is caused by fast neutrons whereas in thermal reactors it is caused by slow or thermal neutrons. Initially all the neutrons are fast when emitted in a reactor and in thermal reactor their speed is reduced with the help of moderator. For a natural uranium, graphite moderated reactor, the ratio of moderator to fuel volume is between 50 and 80; for heavy water moderated reactor this ratio lies between 20 and 40 while for enriched uranium light water moderated reactor this ratio is between 1.5 and 2.5. Thus the reactor core of a thermal reactor is very much larger than that of a fast reactor which has no moderator. Obviously heat generated per unit volume of the reactor core in a thermal reactor is very much less than that in a fast reactor with the result that the cooling problems are much simpler in a thermal reactor.

In intermediate reactors, most of fission events are caused by neutrons in the process of slowing down. Such systems are very difficult to design because of the resonance cross-section structure in this energy range. Slightly enriched fuels are used in these reactors.

The main advantages and disadvantages of fast reactors are given below.

Advantages. 1. Fast reactors can convert more fertile material to fissile material with the result that the net fuel consumption for such reactors is much less. As a matter of fact more fissile material could be produced in a fast reactor than would be consumed by it.

2. These reactors are small and compact and so easier to shield.
3. Since absorption cross sections are small, any structural material can be used for reactor core.

Disadvantages. 1. Heat transfer and cooling problems in the core are complicated due to high power density (thermal power to core volume ratio in kW/m^3)

2. High fuel loading requirement.
3. The core of a fast reactor requires high enrichment (above 10% of fissile material).
4. Radiation damage (embrittlement and swelling) of the structural materials in the core due to fast or high energy neutrons.
5. The fast neutrons having much shorter "neutron lifetimes" than thermal neutrons cause some control problems under certain conditions.

The advantages and disadvantages of thermal reactors are enumerated as below:

Advantages. 1. Ease of control because of relatively low power densities and longer neutron lifetimes.

2. Greater inherent safety.
3. Low fuel loading.

Disadvantages. 1. Very much restricted choice of fuel when uranium is used as fuel.

2. Higher size and weight of reactor per unit power due to low power density.
3. More fissile material consumption.
4. Requirements of small absorption cross section structural materials.

III. According to the type of fuel used the reactors may be classified as:

(*a*) natural uranium (*b*) enriched uranium (*c*) plutonium.

IV. According to the state of fuel the reactors may be classified as:

(*a*) solid (*b*) liquid.

V. According to the fuel cycle the reactors may be classified as:

(*a*) *Burner (Thermal) Reactor*. Such reactors are designed for generating heat only without any recovery of converted fertile material.

(*b*) *Converter Reactor*. Such reactors convert fertile material into fissile material different from the one initially fed into the reactor core. We know that natural uranium as the nuclear fuel U-238 is converted into plutonium.

(*c*) *Breeder Reactor*. Such reactors convert fertile material into a fissile material, which is similar to one initially supplied to the reactor core. A breeder reactor is also one in which the fertile material is converted into a fissile material at a rate higher than at which the fissile material is consumed.

VI. As per arrangement of fissile and fertile material the reactors may be classified as:

(*a*) One region (fissile and fertile material mixed).

(*b*) Two region (fissile and fertile material separate).

VII. According to the arrangement of fuel and moderator the reactors may be classified as:

(*a*) homogeneous and (*b*) heterogeneous.

In homogeneous reactors, the nuclear fuel and the moderator represent a uniform mixture in the fluid form, including gases, liquids and slurries whereas in heterogeneous reactors, separate fuel slugs or rods are inserted in the moderator in some sort of regular arrangement forming a so called lattice. In a homogeneous reactor, the mixture of nuclear fuel and moderator is circulated from the reactor to an external heat exchanger, then to a pump and back to the reactor. The major drawback of homogeneous reactor is that the fuel solution also contains the highly radioactive fission products. Any leakage or component failure in the primary-reactor coolant system are extremely difficult to repair because of the presence of these fission products. Otherwise, a homogeneous reactor has some significant advantages such as excellent in core heat transfer because of generation of the fission energy in the fuel-coolant solution itself. Also, the reactor fuel can be added, removed and reprocessed during reactor operation without shutting it down. Most of the present reactors are heterogeneous type.

VIII. On the basis of moderator material used the reactors my be classified as:

(*a*) heavy water (*b*) graphite (*c*) ordinary water (*d*) beryllium and (*e*) organic reactors.

The most commonly used moderator materials are graphite, ordinary or natural water and heavy water. Graphite has got higher atomic weight than water and, therefore, the reactors employing graphite as moderator will be very bulky. Natural water gives a small and compact reactor, but the reactor would have to be pressurized and use enriched fuel. With heavy water, ordinary natural fuel can be used but it is very expensive.

IX. On the basis of coolant used the reactors may be classified as:

(*a*) gas (*b*) water (*c*) heavy water and (*d*) liquid metal reactors.

X. On the basis of cooling system employed the reactors may be classified as:

(*a*) direct and (*b*) indirect reactors.

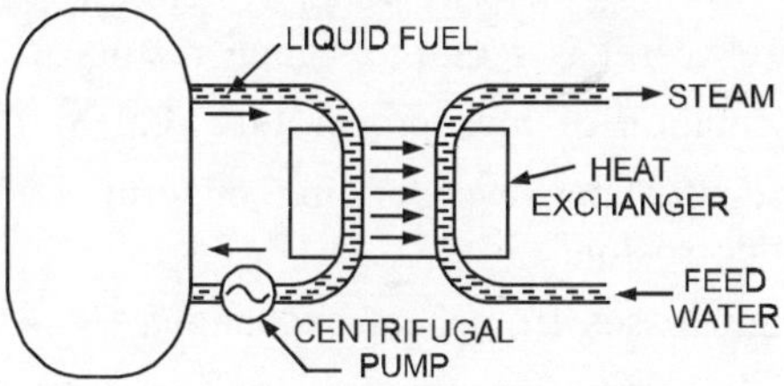

(*a*) *Direct Cooling System*

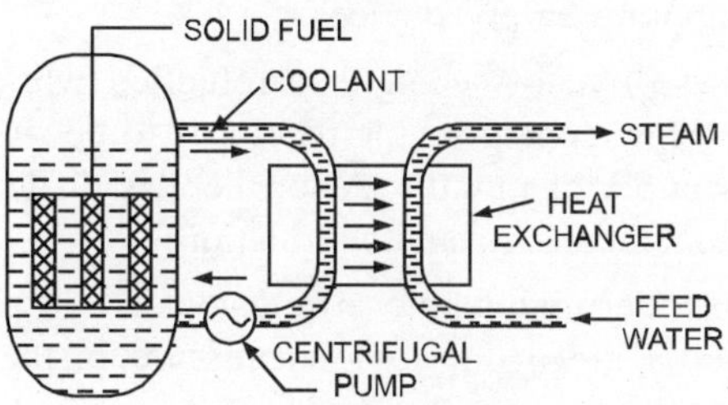

(*b*) *Indirect Cooling System*

Fig. 4.10. ***Systems of Cooling***

In *direct system of cooling, the* fuel is in the liquid form and it acts as a coolant. It is circulated through the reactor core and a heat exchanger in which its heat is transferred to the circulating water to produce steam. In this system reactor may be either of the aqueous homogeneous type or liquid metal fuelled type and the heat exchanger should be located within the biological shield because the circulating liquid fuel is highly radioactive.

In indirect system of cooling, the coolant may be a gas, water (light or heavy), a liquid metal or an organic coolant. In this system of cooling, coolant is pumped through pipes, as illustrated in Fig. 4.10 (*b*).

4.15. POWER REACTORS IN COMMON USE

Power reactors commonly employed in nuclear power plants are briefly described as below:

1. Boiling Water Reactor (BWR). This is the simplest type of water reactor. It has a steel pressure vessel surrounded by a concrete shield. Fuel used is enriched uranium oxide. Ordinary water is used both as moderator and coolant. The steam is generated in the reactor itself. Feed water enters the reactor vessel at the bottom and takes the heat produced due to fission of fuel and gets converted into steam. This steam leaves the reactor at the top and after passing through turbine and condenser returns to the reactor. Uranium fuel elements are arranged in a particular lattice form inside the pressure vessel containing water. A BWR assembly comprises 90-100 fuel rods and there are up to 750 assemblies in a core holding up to 140 tonnes of uranium. The secondary control system involves restricting water flow through the core so that more steam in top part reduces moderation. Most of the radioactivity in the water is very short lived (mostly N-16, with a 7 second half life), so the turbine hall can be entered soon after the reactor is shut down.

Advantages of this reactor include a small size pressure vessel, high steam pressure, simple construction and elimination of heat exchanger circuit resulting in reduction in cost and gain in thermal efficiency. Overall efficiency is about 33%.

Disadvantages. 1. In view of the direct cycle there is a danger of radioactive contamination of steam, therefore, more elaborate safety measures are to be provided for piping and turbine. These add to cost.

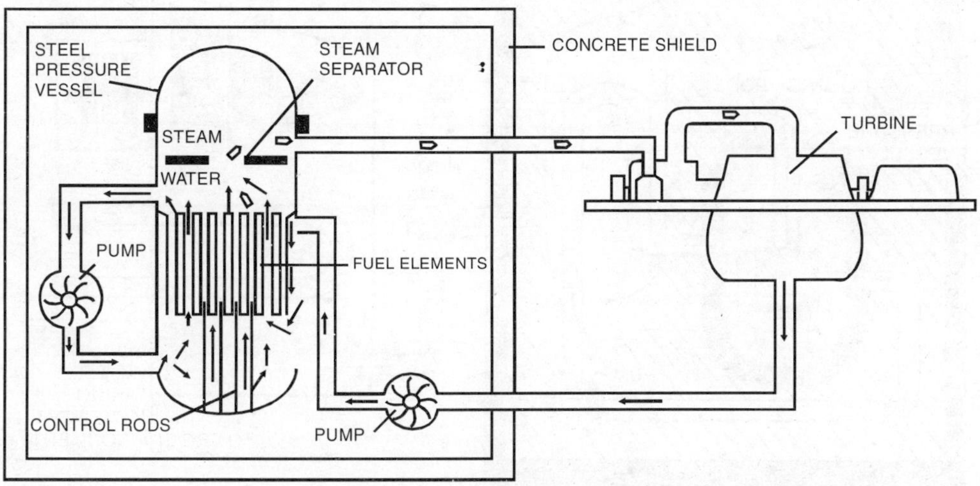

Fig. 4.11. ***Boiling Water Reactor (BWR)***

2. Because of the danger of small amounts of fissile materials passing through along with the coolant, more biological protection is required.
3. Wastage of steam results in lowering of thermal efficiency on part load operation.
4. It cannot meet a sudden increase in load.

The 420 MW power station, at Tarapur (India), consists of two enriched uranium reactors of the boiling water type. These reactors were built with the help of G.E.C. of the United States and became operational on 1.4.1969.

2. Pressurized Water Reactor (PWR). It is a thermal reactor, using enriched uranium oxide, clad in zircalloy as fuel. A PWR has fuel assemblies of 200-300 rods each, arranged vertically in the core, and a large reactor would have about 150-250 fuel assemblies with 80-100 tonnes of uranium. The pressure vessel is of steel. Water under pressure is used both as coolant and moderator. The pressure vessel and the heat exchanger are surrounded by a concrete shield. In this reactor, bulk boiling water is prevented as the water is pressurized to about 150 atmosphere.

The hot water from the reactor flows to a heat exchanger (or steam generator) where its heat is transferred to the feed water to generate steam. The secondary cooling operates at a low pressure. The primary coolant then flows from the heat exchanger to the primary circulating pump which pumps it back to the reactor. The steam is condensed in the condenser and the condensate returns to heat exchanger forming a closed circuit. The primary circuit of a pressurized water reactor (PWR) contains a 'pressurizer'. This is simply a pressure vessel with an electric heating coil at the bottom and a water spray at the top. The top of the vessel is filled with steam at primary circuit pressure. When the primary circuit pressure decreases, the heating coil gets energized and boils the water to form steam resulting in increase in steam content in the vessel. This results in the increase in pressure of the primary circuit. In case the steam pressure of the primary circuit becomes too high cold water is sprayed into the steam in the pressurizer. The steam is condensed and therefore primary circuit pressure is reduced. The steam generated is of rather poor quality, temperature around 250ºC and pressure 42 kg/cm^2.

Its advantages are (*i*) compactness (*ii*) possibility of breeding plutonium (*iii*) isolation of radioactive materials from the main steam system (*iv*) cheap light water can be used as coolant-cum-moderator and (*v*) high power density, (*vi*) The reactor responds to supply more power when the load increases. The positive power demand coefficient makes this almost automatic.

However, it suffers from the following drawbacks:

(*i*) Use of high pressure water system. So a strong pressure vessel is required which results in high cost.

(*ii*) Formation of low temperature (250º C) steam.

(*iii*) Use of expensive cladding material for prevention of corrosion.

(*iv*) High losses from heat exchanger.

(*v*) High power consumption by auxiliaries.

(*vi*) In comparison to other types, requirement of more elaborate safety devices.

(*vii*) These reactors cannot be refuelled while operating and for recharging the reactor is to be shut down for a couple of months. Also, there is difficulty in fuel element design and fabrication.

(*viii*) The thermodynamic efficiency of this plant is low (about 20%) due to low pressure in the secondary circuit.

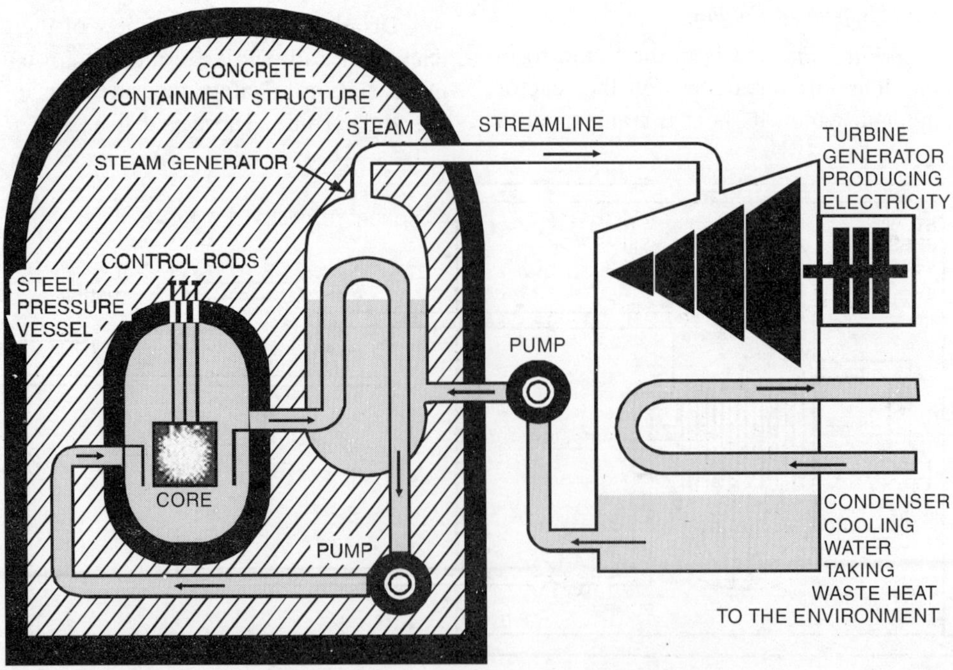

Fig. 4.12. ***Pressurized Water Reactor (PWR)***

The reactors installed at Rajasthan Atomic Power Station, Madras Atomic Power Station and Narora Atomic Power Project are of pressurized water reactor type. However they use heavy water as coolant and moderator. Such reactors are known as *pressurized heavy water reactors* (PHWR).

3. Gas Cooled Reactor. This type of reactor employs a gas (CO_2 or helium) in place of water as the coolant and graphite as the moderator. A heat exchanger is necessarily required. Gas is circulated through the reactor core and the heat exchanger by means of a blower or a gas compressor. Even though gas is inferior to water from the point of view of heat transfer properties but it offers numerous advantages which are not available with water. A large quantity of gas is required, due to poor heat transfer qualities, for circulation resulting in increased power consumption for auxiliaries. Thus, advantage of high thermal efficiency is to a large extent lost and overall plant efficiency is low. Graphite, as moderator is less effective than water and would require a large volume core in such reactors, the heat removal by gas cooling will be better. The gas is circulated at a pressure of 14-28 kg/cm^2. The tubes in the heat exchanger through which water is circulated should have fins on their surface so as to improve the rate of heat transfer.

Its advantages are:

(*i*) Less severe corrosion problems.

(*ii*) Possibility of use of natural uranium as fuel.

(*iii*) Greater safety in comparison with water cooled reactors.

(*iv*) Contamination problems are moderate.

(*v*) Low pressure coolant and relatively high reactor temperature.

Its drawbacks are (*i*) relatively large size of reactor because of use of natural fuel and graphite moderator (*ii*) extremely low power density (*iii*) low steam pressure and temperature and (*iv*) large energy consumption by gas blowers because of poor heat transfer characteristics of gases.

4. Heavy Water Cooled and Moderated (CANDU TYPE) Reactor. This reactor was first developed by Canada and is, therefore, known as CANDU type reactor. The word CANDU stands for *Canadian Deuterium Uranium.* These reactors make use of heavy water, composed of the heavy hydrogen isotope, $_1H^2$, as moderator to have maximum neutron economy and as coolant also. Such reactors are meant for those countries which do not have uranium enrichment facilities. Enrichment of uranium is costly affair and such reactors use natural uranium as fuel.

The primary and secondary circuits are similar to pressurized water reactor (PWR)—the coolant heavy water is circulated in the primary circuit and the steam is produced in the secondary circuit transferring the heat in the heat exchanger. Heavy hydrogen exists in nature in the ratio 1 : 6700 as compared to ordinary hydrogen and therefore, heavy water is very difficult and expensive to separate from ordinary water. However, it is simpler to accomplish in comparison to enrichment of uranium. Hence in some designs, heavy water is used as moderator and light water is used in the secondary circuit. Control rods are not required in such reactors as the reactor control is achieved by varying the moderator level in the reactor. For rapid shut down purposes the moderator can be dumped through a very large area into a tank provided below the reactor (not shown in the figure).

The most important advantage of such a reactor is that the heavy water has a very low absorption cross section and it can be used as a moderator in natural uranium thermal reactors and, therefore, the fuel need not be enriched. Other advantages are simpler reactor control because of absence of control rods, high multiplication factor, low fuel consumption and much more effectiveness in slowing down neutrons because of moderator being at low temperature. It is worthmentioning here that a major part of the equipment for this reactor can be manufactured in the shop and period required for site construction is also comparatively smaller.

Its main drawbacks are the heavy cost of heavy water, problems of leakage and very high standard design etc.

5. Liquid Metal Cooled Reactors. This type of reactor has been developed to avoid difficulties faced in pressurization of water as in PWR and at the same time retaining the advantage of having high temperature. Metals in liquid state have good thermal conductivity and high temperature can be had at moderate pressure. However, handling of sodium introduces

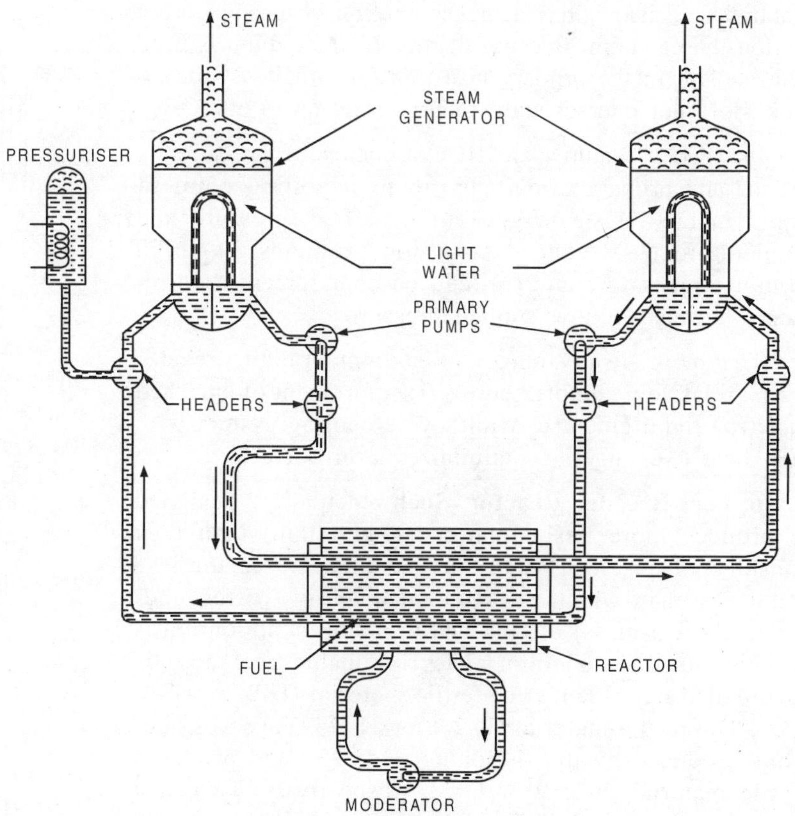

Fig. 4.13. ***CANDU Type Reactor***

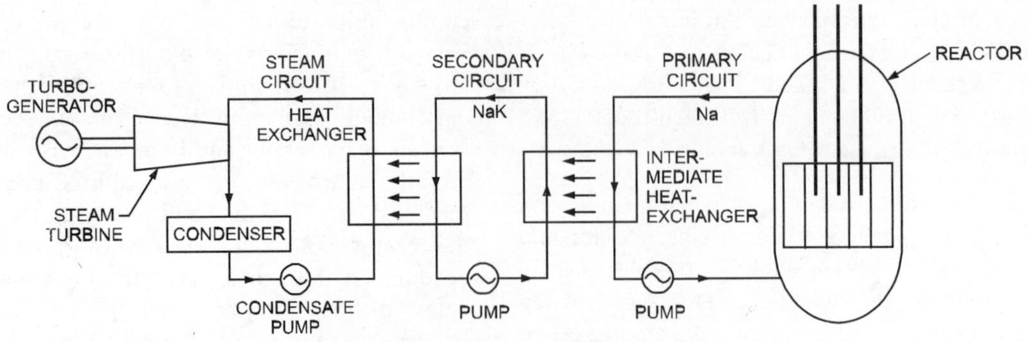

Fig. 4.14. ***Sodium Graphite Reactor***

difficulties because of its activity in reactor core. It is therefore necessary to employ two heat transfer circuits so that radioactive sodium does not come in contact with the steam circuit. A sodium graphite reactor is shown in Fig. 4.14.

Sodium is circulated through the reactor core and an intermediate heat exchanger where the heat from sodium (Na) is transferred to the NaK (an alloy of sodium and potassium) liquid metal which gives up heat in the heat exchanger to generate steam. Because of violent reaction of sodium with air and water the whole system should be leak tight. Charging and draining from one of the two loops should be done in an inert atmosphere to avoid the contact of Na or NaK with air. Sodium graphite reactor (SGR) uses slightly enriched uranium alloy or uranium carbide clad with stainless steel as fuel, graphite as moderator and liquid sodium as coolant.

NaK alloy has a lower melting point and therefore allows a higher heat absorption. But potassium may react with graphite and also it has a higher neutron absorption cross section than sodium. Because of this, liquid sodium is used as a coolant in the primary heat transfer circuit. Again sodium does not interact with stainless steel up to 600ºC.

Its special features are (*i*) elimination of pressure on reactor and primary circuit due to high boiling point of liquid metal (*ii*) steam generation at high pressure and temperature (*iii*) reduced corrosion problems (*iv*) high reactor temperature and (*v*) reduced containment requirements because of low coolant pressure.

The main disadvantages of sodium graphite reactor are (*i*) relatively complex core (*ii*) requirement of enriched fuel (*iii*) requirement of triple cycle cooling system with dual heat exchangers to minimize hazards etc.

6. Fast Breeder Reactor. Such reactors are designed to produce more fissile material (Plutonium) than they consume (Thorium Th-232). A fast breeder reactor is a small vessel in which the required quantity (corresponding to critical mass) of enriched uranium or plutonium is kept without a moderator. The fissionable fuel core is surrounded by a blanket of fertile material (U-238 or Th-232). The fertile material (U-238 or Th-232) absorbs neutrons produced by the fissioning of U-235 and produces fissile material Pu-239 or U-233 respectively. Two heat exchangers are used. The reactor core is cooled by liquid metal (sodium or potassium). In the second heat exchanger the coolant is again liquid sodium/potassium which transfers heat to feed water to generate steam.

This prevents the possibility of a sodium-water reaction with the radioactive sodium. The arrangement of liquid metal fast breeder reactor (LMFBR) is given in Fig. 4.15.

In fast breeder reactors neutron shielding is provided by using boron, light water, oil or graphite. Gamma-ray shielding is accomplished by lead, concrete with added magnetite or barium etc.

The core of a fast reactor needs high enrichment (above 10% of fissile material). To reduce the fuel cost effect, it is imperative to employ high ratings. The core consists of 30% fuel, 50% coolant and 20% canning and structural material by volume.

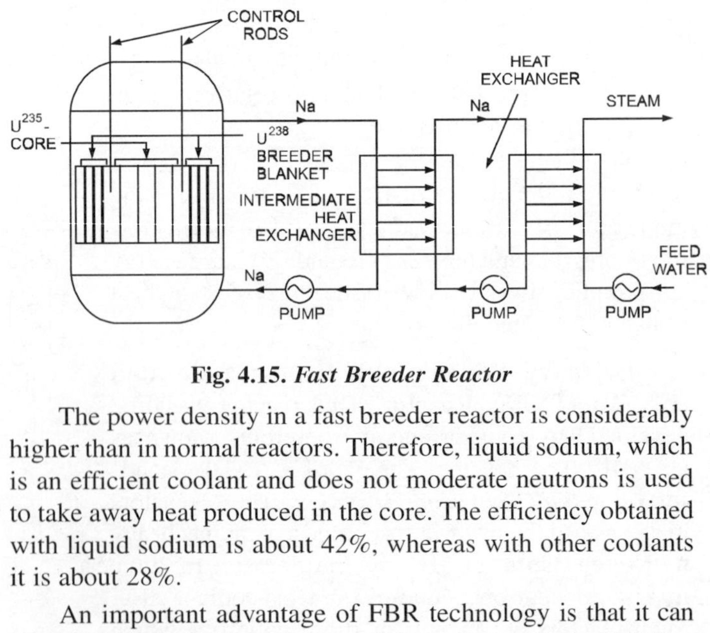

Fig. 4.15. ***Fast Breeder Reactor***

The power density in a fast breeder reactor is considerably higher than in normal reactors. Therefore, liquid sodium, which is an efficient coolant and does not moderate neutrons is used to take away heat produced in the core. The efficiency obtained with liquid sodium is about 42%, whereas with other coolants it is about 28%.

An important advantage of FBR technology is that it can use Thorium (as fertile material) which gets converted to U-233, a fissile isotope. This holds great promise for India as we have one of the world's largest deposit of Thorium—about 450,000 tons in form of sand dunes in Kerala and along Gopalpur Chattarpur coast of Orissa.

Typical power densities (MW/m^3) in fission reactor cores are : Gas cooled 0.53; High temperature gas cooled 7.75; Heavy water 18.0; Boiling water 29.0; Pressurized water 54.75 and Fast breeder reactor 760.0.

4.15.1. Comparison of Thermal and Fast Breeder Reactors

Thermal reactors have the following advantages and disadvantages as compared to fast breeder reactors.

Advantages. 1. Heat developed per unit volume of core or per unit area of fuel surface is less.

2. Ease of control.
3. Greater inherent safety.

Disadvantages. 1. Severely limited choice of fuel from the point of view of neutron economy when fuel used is uranium.

2. Much larger size and weight of reactor per unit power.
3. More fissile material consumption than could be automatically replaced.

Fast breeder reactors can converet more fertile material to fissile material and therefore, net fuel consumption is much less. As a matter of fact more fissile material could be produced than would be consumed by it (fast breeder reactor).

4.16. SAFETY

The reactor cannot explode like an atom bomb because of the following reasons:

1. In the atom bomb, the nuclear material is almost pure and highly fissionable. The interior mechanism of the bomb is designed so as to bring the material together instantly and compress into a dense mass. But in the power reactor, the nuclear fuel is always in the form of a chemical compound or alloy. Internal design of the power reactor is also such that it would be impossible for the fissionable material in the reactor core to be concentrated and held in place such that it could explode like a bomb.
2. Another inherent safeguard lies in the nature of fuel itself. Absorption of neutrons by U-238 is much larger at high temperature. So, in case of a occurrence of a sudden increase in reactivity, with a consequent increase in neutron flux and heat generation, the U-238 in the urinium alloy core immediately steals neutrons otherwise available for fissionable U-235 in the core and the fission rate falls rapidly. The self-sustaining reaction is thus no longer possible and the reactor automatically shut down, without any assistance from control rods or other devices. This effect is referred to as "Doppler effect or coefficient."

Along with these natural safeguards, there are various sensitive and sophisticated systems that monitor and control the operation of the reactor.

Reactor Operating Faults Affecting Reactor Safety. These faults may be

1. Uncontrolled reactivity release.
2. Loss of coolant or coolant flow.
3. Local overheating leading to a fire in the core.

The "fail to safety" principle *i.e.*, loss of power or malfunctioning of equipment results in its operating to a trip condition. The safety protection is provided in triplicate and is arranged to trip the reactor on a two out of three operating coincidences. The following safety protections are provided.

(*i*) *Fuel Element Temperature Trip.* An acceptable trip setting is 40°C above the nominal maximum core temperature.

(*ii*) *Outlet Duct Gas Temperature Trips.* Thermocouples are used as the detecting elements.

(*iii*) *Cooling Flow Failure Detection.* Loss of coolant flow will result in because of failure of one or more circulators or closure of an inlet or outlet duct valve, the most probable cause of the former being loss of electric supply. The current in each circulator motor is monitored by means of a CT (current transformer) operating a relay. The loss of two out of eight blowers results in slight temperature rise, and therefore, failure of any two blowers is arranged to trip the reactor.

(*iv*) *Rate of Change of Pressure Trips.*

(*v*) *Excess Flux Trip.* Trip level is set at 20% above the operating level.

In case all the above mentioned accident prevention systems should all fail, there are a series of mechanical barriers between the reactor core and the outside world:

1. Fuel element cladding.
2. Reactor vessel.
3. Reactor shielding.
4. Containment shell.

4.17. POLLUTION FROM NUCLEAR POWER PLANTS

In nuclear power stations, the combustion of fuel is low. Harmful effluents from such power stations in the atmosphere are insignificant. However, isotopes formed in nuclear power reactors have a high toxicity and their effect on living organisms may be accumulative. That is why, the *problems of disposal, transport and storage of liquid radioactive wastes* are extremely significant. Such power plants will produce practically no harmful effect on the biosphere provided the radioactive waste storage problem is safely solved.

A nuclear reactor produces α-rays, β-rays, γ-rays and neutrons which can disturb the normal working of living organisms, and thus calls for special safety measures. α-rays are heavy particles carrying positive charge and can cause internal hazard if ingested. β-rays have greater penetrating power, as compared to α-rays, due to their smaller size. Overexposure to β-rays can cause skin burns and repeated overexposure may result in malignant growth. γ-rays are electromagnetic radiations of very short wavelength, have high energy and are very penetrating. They are capable of causing considerable damage, especially to organic materials. Overexposure to γ-rays causes the blood diseases, undesirable genetic effects, anaemia etc. Larger exposure may cause death within hours of exposure. However, the effects of slow exposure may become apparent after several years. Neutrons are produced in fission with a very wide range of energies up to about 10 MeV. Though they have no charge but are highly penetrating. Their effects are similar to those of γ-rays.

The biological effects of nuclear radiations depend upon (*i*) amount of dose absorbed (*ii*) time duration of exposure (*iii*) sensitivity and recovery of recipient organism and (*iv*) distribution of active material within body. A long time exposure to even a small dose may not cause any immediate effect but leads to delayed effects such as shortening of life, leukomia, genetic effects etc.

Nuclear power stations are surrounded by a sanitary protective zone to minimize the risk of irradiation of the population within such a zone. It is prohibited to build residential buildings, children houses and auxiliary buildings not related to the concerned power plant. The level of radiation of this territory is checked periodically.

In nuclear power plants, there are three main sources of radioactive contaminating of air. The first source is the fission of nuclei of solid or gaseous nuclear fuels. Gaseous fission fragments which are more likely to enter the air include inert gases, such as Xenon, Crypton etc., and radioactive iodine. The second source is due to the effect of neutron fluxes on the heat carrier in the primary cooling system and on the ambient air. Among the components of air, an inert gas, argon-40 is the most prone to activation. It may form a radioactive isotope argon-41 with the half life period of 1-82 hours. Induced activity can appear in the dust present in the air. The third source is damage of shells of fuel elements or the presence of activated inert gases and aerosols in heat carrier leakages.

The processes in nuclear power plants include various types of technological blow-off of gases. At some periods (say, during overload operation) blown-off gases may have an elevated activity and need special decontaminating measured. Other potential sources of radioactivity are various auxiliary structures and elements (cooling ponds, reactor blow-off system, tanks for collecting radioactive leakages etc.) They may evolve radioactive inert gases.

Within any of the groups, elements may be ejected in any proportion, but their sum should not exceed the given value. Simultaneously daily emissions of all groups of isotopes is also allowed, provided that the given values for each of the groups are not exceeded.

4.17.1. Disposal of Nuclear Waste and Effluent

Solid radioactive wastes arise from used filters, sludge from the cooling ponds, pieces of discarded fuel element cans, splitters etc. These, along with discarded items of plant such as control rods have to be stored on site in shielded concrete vaults.

There are many ways for disposing of the solid fission products. The storing in shielded storage vaults consists in fixing the solid waste in borosilicate glass and then storage of this glass in leak tight capsules. These capsules or vaults can then be stored in deep salt mines or in deep wells drilled in the stable ocean floor. Deep salt mines are suggested because the presence of the salt pockets indicates that there has been no ground water in the vicinity for thousands of years. Sometimes, suitable containers are filled with radioactive waste and sunk to the bottom of seas and oceans. However, this method does not completely prevent the radioactivity from leading into the water. Another way of disposal is the separation and transmutation of the long-lived isotopes to short-lived or stable products following neutron absorption in a breeder or fusion reactor. The possibility of firing these long-lived products into the sun or into a long-term stable orbit is also being considered.

Radioactive liquid effluents arise from the laundry, personal decontamination, etc. together with activity accumulating from the corrosion of the irradiated fuel elements in the storage ponds. Before discharging to sea where enormous dilution takes place, the effluent is passed through ion exchange resins which absorb a large proportion of activity. The final levels of any particular isotope in the sea will be well below the maximum drinking water level.

It is safe enough to store radioactive waste underground in liquid form in suitable tanks or in reduction to clinker. Clinkering serves a two fold purpose of improving the protection and reducing the volume of waste. A promising method is known as "solidifying" the liquid radioactive waste through heat up and evaporation. The current technology enables 1,000 litres of highly radioactive liquid waste to be processed into less than 0.01 m^3 of solid waste. The solid waste is put into sealed metal containers suitable for storage in deep salt mines.

Gaseous effluents are filtered before discharging into atmosphere. Moreover, the filtered gas is discharged at high levels so that it is dispersed properly. The probability of fire in the reactor fuel channel is extremely low. However, if fire breaks out, large volumes of gaseous fission products may be released. So it is necessary to have a clean up plant through which these products can be passed for removal of radioactive iodine which is the major hazard.

It is essential to monitor the loss of CO_2 from the reactor to ensure that this loss does not exceed about 1 ton per day. It is also necessary to check the concentration of CO_2 in the atmosphere near the reactor. Proper precautions against toxic and radiological hazards are necessary.

4.17.2. Shielding

Adequate shielding is necessary to guard personnel and delicate instruments. The various materials used for shielding are lead, concrete, steel and cadmium. **Lead** is a common shielding material and is invariably employed due to its low cost. **Concrete** is another shielding material having efficiency lesser than that of lead. **Steel** is not an efficient shielding material but has good structural properties and is sometimes employed as an attenuating shield. **Cadmium** is capable of absorbing slow neutrons by a nuclear reaction. The effectiveness of a shielding material depends mostly on the density of material (lead-11,300 kg/m^3; concrete-2,400 kg/m^3; steel-7,800 kg/m^3; cadmium-8,650 kg/m^3).

No single material is effective in shielding radiations of different kinds. A material containing hydrogen *e.g.,* water or polythene is used to slow down fast neutrons, boron or steel is employed for absorption of thermal neutrons. A heavy material like lead is required to act as a thermal shield and to absorb gamma rays.

In nuclear power reactors a thermal shield of thickness of several cms of steel surrounded by about 3 m thick concrete is used. Water, in concrete, slows down fast neutrons while iron, barium or steel turnings are mixed in concrete to attenuate gamma rays and absorb thermal neutrons.

Exercises

1. Discuss the advantages and disadvantages of a nuclear plant as compared to other conventional power plants.
[U.P.S.C. I.E.S. Elec. Engineering II, 1996; G.B. Technical Univ. 2010-11; U.P. Technical Univ. 2007-08]
2. What are the advantages of nuclear generation ? What points are considered in the site selection for nuclear plants ?
[Rajasthan Technical Univ. Generation of Electrical Power 2011]
3. What considerations have to be kept in mind while selecting site of nuclear power plants ?
[U.P. Technical Univ. Power Station Practice 2008-09]
4. What are the factors to be considered for site selection of (*a*) Nuclear power plant (*b*) Hydroelectric power plant ?
[W.B. Univ. of Technology Power System II, 2010]
5. What do you mean by mass defect and binding energy ? What is radioactive decay ? What is half life?
[A.M.I.E. Sec. B. Power Plant Engineering, Summer 2000]
6. Explain the following terms :
(*i*) half life (*ii*) mass defect (*iii*) binding energy.
[Pb. Univ. Power Station Practice, May 1995]
7. What is a fissile material and a fertile material ? Give examples.
[Rajasthan Technical Univ. Generation of Electrical Power 2008, 2009; U.P. Technical Univ. Power Station Practice 2009-10]
8. With examples explain the difference between a fissile material and a fertile material.
[U.P. Technical Univ. Power Station Practice 2007-08]
9. Explain with a neat sketch the working of a nuclear power station. Discuss the merits and problems associated with nuclear power stations.
[U.P.S.C. I.E.S. Elec. Engineering II, 1994]
10. Draw a block diagram of a nuclear power plant showing basic components. Discuss in brief the merits and problems associated with nuclear power stations.
[U.P. Technical Univ. Power Station Practice 2009-10]
11. Explain general layout and operation of a nuclear power plant. Why nuclear power stations are becoming very popular ?
[U.P. Technical Univ. Power Station Practice 2006-07]
12. Give general layout of nuclear power plant. Identify the best operating mode of nuclear power plant along with justification.
[U.P. Technical Univ. Power Station Practice 2005-06]
13. Draw a neat line diagram of a nuclear power plant showing basic components.
Discuss the advantages of nuclear power plant compared with thermal power plants.
Discuss in brief the type of coolant used. What properties are required for a good coolant ?
[Pb. Univ. Power Station Practice, Dec. 1993]
14. What materials are used for fuel, coolant and moderator in nuclear power plants ? Give a schematic view of a nuclear plant. Describe merits and problems of nuclear power plants as compared to conventional thermal plants.
[Rajasthan Technical Univ. Generation of Electrical Power 2010]
15. Explain the following terms as applicable to nuclear power plants:
(*i*) binding energy (*ii*) critical size (*iii*) multiplication factor.
[A.M.I.E. Sec. B. Power Systems, Summer 1996]
16. Explain with a neat diagram various parts of a nuclear reactor, mentioning clearly the each part.
[G.B. Technical Univ. Power Station Practice 2010-11]
17. What different moderators are used in practice ? What different properties make them suitable moderator ?
[M.D. Univ. Power Plant Engineering May 2009]
18. Why is a moderator necessary in a nuclear reactor ? What materials are suitable as moderator materials in a nuclear reactor ? **[U.P. Technical Univ. Power Station Practice 2007-08]**
19. What is a moderator ? Name common moderators and discuss their advantages and limitations.
[M.D. Univ. Power Plant Engineering May 2007]
20. What are the functions of moderator and control rods in a nuclear reactor ? Explain.
[G.B. Technical Univ. Power Station Practice 2010-11]
21. Explain the function of moderator. How is a moderator selected ?
[U.P. Technical Univ. Power Station Practice 2009-10]
22. Give the schematic diagram of nuclear reactor and mention the cooling mechanism.
[U.P. Technical Univ. Power Station Practice 2007]
23. Explain the role of moderators and reflectors in getting maximum power output of nuclear power plant.
[U.P. Technical Univ. Power Station Practice 2007]
24. Discuss the boiling water reactor, mentioning its advantages and disadvantages.
Explain the terms (*i*) moderator (*ii*) heavy water.
[Pb. Univ. Power Station Practice Dec. 1995]
25. What is a nuclear reactor ? How are nuclear reactors classified ? **[M.D. Univ. Power Plant Engineering May 2007]**
26. Explain the working of nuclear power plant having following reactors:
(*a*) PWR (*b*) Gas cooled reactor.
[M.D. Univ. Power Plant Engineering May 2008]
27. Write advantages and disadvantages of pressurized water reactor. **[U.P. Technical Univ. Power Station Practice 2004-05]**
28. Draw a neat diagram of organic liquid cooled and moderated reactor power plant and list out its advantages and disadvantages over other reactors.
[M.D. Univ. Power Plant Engineering May 2009]
29. Explain the working, with neat diagram of a reactor in which moderator is not used.
[M.D. Univ. Power Plant Engineering May 2007]
30. Differentiate sodium graphite and fast breeder reactors.
[M.D. Univ. Power Plant Engineering Dec. 2008]
31. Discuss the following in reference to nuclear power plant: (*a*) CANDU reactor (*b*) Waste disposal (*c*) Advantages and limitations. **[M.D. Univ. Power Plant Engineering Dec. 2009]**

32. Describe the different types of fuels used in a nuclear power plant and discuss the problem of nuclear waste disposal.
[Pb. Univ. Power Station Practice, June 1992]

33. Discuss briefly the radioactive pollution of environment by nuclear power plants.
[Pb. Univ. Electrical Power-II (Generation) Dec. 1995]

34. Describe the working of moderator material. How disposal of nuclear waste in done ? Explain role of shielding.
[U.P. Technical Univ. Power Station Practice 2006-07]

35. What methods are used for disposal of nuclear waste material ?
[Rajasthan Technical Univ. Generation of Electrical Power 2010]

36. Write short notes on the following :
(*i*) India's nuclear power programme.
[A.M.I.E. Sec. B. Summer 2000]
(*ii*) Merits and demerits of nuclear power plants.
(*iii*) Site selection for nuclear power plants.
[A.M.I.E. Sec. B. Summer 1985]
(*iv*) Nuclear reactor.
[Pb. Univ. Power Station Practice, December 1994]
(*v*) Classification of nuclear reactors.
(*vi*) Radioactive pollution.
[Pb. Univ. Electrical Power-II (Generation) May 1994]
(*vii*) Nuclear waste disposal.
[Pb. Univ. Electrical Power-II (Generation) January 1993]

Short Answer Type Questions With Answers

Q. 1. Why are nuclear power stations becoming popular ?
[Rajasthan Univ. 2004, 2005]

Ans. The need of nuclear power plant lies in the fact that the hunger for electricity is virtually unending and after each decade the world demand for electricity is doubled owing to booming increase in the population and industrial growth. Moreover, the reserves of fossil fuels *i.e.,* coal, oil and gas are fast depleting. Again it would be better to utilise these minerals, particularly oil, for chemical industry as raw materials. Thus there is tendency to seek alternative source of energy and the nuclear power is the only alternative source which can meet the future energy demands of the world. One of its main attractions is the huge amount of energy that can be released from a small quantity of active material. The energy obtainable by completely burning of 1 kg of uranium would give energy equivalent 3,000 tonnes of high grade coal *i.e.,* uranium has 3 million times the energy of coal. The possible energy reserves in the form of uranium and thorium is many times greater than that of fossil fuels.

The factors which are in favour of nuclear energy are: it is practically independent of geographical factors, it has no combustion products and it is clean source of power which does not contribute to air pollution. Further it does not require fuel transportation mechanism and large storage facilities.

Q. 2. Why nuclear power plants are used as base load plants ?
[A.M.I.E. Sec. B Summer 1998]

Ans. The operating cost of nuclear power plants is quite low and once the installation is completed, the loading of the power plant will have no effect on the generation cost. Therefore, a nuclear power plant is always operated as a base load plant. The nuclear power plants are usually not operated at a load factor less than 0.8.

Q. 3. Name three important factors in choosing a site for atomic power plant. **[A.M.I.E. Sec. B. Power Systems, Summer 1993]**

Ans. Three important factors in choosing a site for atomic power plants are :
(*i*) availability of water supply (*ii*) distance from populated area and (*iii*) transportation facilities.

Q. 4. Why the site near a sea or river or lake and away from thickly populated area is considered an ideal site for nuclear power plants ?

Ans. Keeping in view the huge requirement of cooling water and facility for disposal of nuclear waste and effluents and safety from danger of presence of radioactivity in the atmosphere near the plant the site near a sea or river or lake and away from thickly populated area is considered an ideal site for nuclear power plants.

Q. 5. Name three different types of fuels used in nuclear reactors.
[A.M.I.E. Sec. B. Power Systems, Summer 1994]

Ans. Three different types of fuels used in nuclear reactor are :
(*i*) natural uranium U-235 (*ii*) plutonium (*iii*) enriched uranium.

Q. 6. How much power is generated by a fissioning of 1 gm of U^{235} per day ?
[Pb. Univ. Power Station Practice, December 1994]

Ans. $\dfrac{2.57}{2.68}$ MW

Q. 7. What is meant by radioactivity ?

Ans. The disintegration or decay of unstable atoms accompanied by emission of radiation is called the radioactivity.

Q. 8. What is decay constant ?

Ans. The decay or disintegration constant may be defined as the proportion of atoms of an isotope decaying per second.

Q. 9. What is half life ?

Ans. The half life or half life period of a radioactive isotope is the time required for one half of the isotope to decay.

Q. 10. What is meant by mass defect ?

Ans. The difference between the experimental mass and calculated mass of the nucleus is called the mass defect or mass deficit.

Q. 11. What is binding energy ?

Ans. The energy that binds the nucleus together in the nucleus is called the nuclear binding energy. Binding energy is the energy equivalent of the mass defect.

Q. 12. What is mass energy equivalence ?

Ans. 1 amu = 931.5 MeV; amu stands for atomic mass unit.

Q. 13. What is nuclear fission ?

Ans. The splitting of a heavy nucleus into two or more smaller nuclei is termed nuclear fission.

Q. 14. What is critical mass ?

Ans. The minimum mass of fissionable material required to sustain a chain reaction is called the critical mass.

Q. 15. Distinguish between fusion and fission.
[Rajasthan Univ. 2001]

Ans. When two or more lighter nuclei (H_2 or H_3) with lower binding energy per nucleon are combined or fused together into a heavier and more stable nucleus, there is a release of energy. This process of releasing energy is known as fusion.

The splitting of a heavy nucleus into two or more smaller nuclei is termed nuclear fission.

Q. 16. Distinguish between fissile and fertile materials.

[Rajasthan Univ. 2001]

Ans. The materials which are known to undergo neutron fission are U-235, U-233 and Pu-239. These are fissile materials. There are materials ($^{238}_{92}U$ and $^{232}_{90}Th$) which occur in nature, are called fertile materials. Fissionable materials $^{238}_{92}Pu$ and $^{233}_{92}U$ are formed in the nuclear reactors during fission process from $^{238}_{92}U$ and $^{232}_{90}Th$ respectively due to absorption of neutrons without fission.

Q. 17. How does a nuclear power plant differ from a conventional thermal power plant ?

Ans. The concepts of nuclear power generation are much similar to that of conventional steam power generation. The difference lies only in the steam generation part *i.e.,* coal or oil burning furnace and the boiler are replaced by nuclear reactor and heat exchanger.

Q. 18. Name the main components of nuclear power plants.

[Pb. Technical Univ. May 2008]

Ans. A nuclear power plant consists of a nuclear reactor (for heat generation), heat exchanger (for converting water into steam by using heat generated in the nuclear reactor), steam turbine, alternator, condenser etc.

Q. 19. What is nuclear reactor and what are its basic parts ?

Ans. Reactor is that part of nuclear power plant where nuclear fuel is subjected to nuclear fission and the energy released in the process is utilised to heat the coolant which may in turn generate steam or be used in a gas turbine.

A nuclear reactor consists of reactor core, moderator, control rods, coolant, reflector, thermal shielding, reactor vessel and biological shield.

Q. 20. In a nuclear reactor, multiplication factor is kept almost equal to one. Why ? **[A.M.I.E. Sec. B Summer 1998]**

Ans. As mentioned in Art. 4.7, all the neutrons released in the fission reaction are not used up in propagating the chain reaction but some of these are lost to the surroundings. For maintaining chain reaction, it is therefore, essential that the number of neutrons after the fission should be slightly more than the number before it to allow for the escape or leak of neutrons from the reactor core. The ratio is known as *multiplication* factor. The multiplication factor k for any reactor in defined as

$$k = \frac{\text{Number of neutrons produced in one generation}}{\text{Number of neutrons produced in the preceding generation}}$$

Unity value of k indicates that the chain reaction will continue at a steady rate (critical). If k is less than unity, the chain reaction will stop and the system is called *subcritical.* While for k exceeding unity, the chain reaction will build up and the system is called *super critical.*

Thus, for keeping chain reaction continuous multiplication factor in a nuclear reactor is kept almost equal to one.

Q. 21. How are reactor classified ?

Ans. Nuclear reactors may be classified in several ways *i.e.,* on basis of their applications, type of fission, fuel used, state of fuel, fuel cycle, arrangement of fissile and fertile material, arrangement of fuel and moderator, moderator material, cooling system employed, coolant used etc.

Q. 22. How are reactors classified according to type of coolant used ? **[A.M.I.E. Sec. B Summer 1998]**

Ans. On the basis of coolant used the reactors may be classified as (*i*) gas (*ii*) water (*iii*) heavy water and (*iv*) liquid metal reactors.

Q. 23. What is the purpose of a moderator in a nuclear power reactor ? **[A.M.I.E. Sec. B. Power Systems, Summer 1994]**

Ans. The purpose of a moderator in a nuclear power reactor is to slowdown the neutrons.

Q. 24. Name three moderators commonly used.

[A.M.I.E. Sec. B. Power Systems, Summer 1994]

Ans. Three commonly used moderators are: (*i*) graphite (*ii*) ordinary or natural water and (*iii*) heavy water.

Q. 25. What is the purpose of a control rod in a nuclear power reactor ? **[A.M.I.E. Sec. B. Summer 1994]**

Ans. Control rods are meant for controlling the rate of fission of U-235.

Q. 26. Name any two materials used for control rods.

[A.M.I.E. Sec. B. Summer 1994]

Ans. The materials used for control rods must have very high absorption capacity for neutrons. The commonly used materials for control rod are cadmium, boron or hafnium.

Q. 27. Distinguish between fast and thermal reactors.

[Rajasthan Univ. 2001]

Ans. In fast reactors, the fission is affected by fast neutrons without use of moderator but in thermal reactors, the fast neutrons are slowed with the use of moderators.

Q. 28. What is CANDU type reactor ?

[A.M.I.E. Sec. B, Summer 2000]

Ans. CANDU reactor was first developed by Canada and is, therefore, known as CANDU type reactor. The word CANDU stands for *Canadian Deuterium Uranium.* These reactors make use of heavy water, composed of the heavy hydrogen isotope, $_1H^2$, as moderator to have maximum neutron economy and as coolant also. Such reactors are meant for those countries which do not have uranium enrichment facilities. Enrichment of uranium is costly affair and such reactors use natural uranium as fuel.

Q. 29. Enumerate the main disadvantages of sodium graphite reactor.

Ans. The main disadvantages of sodium graphite reactor are (*i*) relatively complex core (*ii*) requirement of enriched fuel (*iii*) requirement of triple cycle cooling system with dual heat exchangers to minimize hazards etc.

Q. 30. How is nuclear waste disposed of ?

[Pb. Technical Univ. Dec. 2007]

Ans. The nuclear waste is disposed of as follows:

1. Solid waste is stored in suitable containers, which are sunk to the bottom of seas and oceans.
2. Liquid waste is diluted and disposed in river and sea after treatment.
3. Gaseous waste is discharged to atmosphere at high level after filteration and treatment.

❑❑❑

CHAPTER 5

Diesel Electric Power Plants

INSIDE THIS CHAPTER

5.1. INTRODUCTION

Diesel electric power plant is a power plant in which a diesel engine is used as the prime mover for the generation of electrical energy. The diesel engine obtains energy from a liquid fuel usually called the *diesel oil* and converts it into mechanical energy. An alternator or a dc generator mechanically coupled to the diesel engine converts the mechanical energy into electrical energy. It is noteworthy that the main difference between a diesel engine and a steam turbine is that in the first case the chemical energy of combustion is released inside the cylinder whereas in the second case the energy developed during combustion of fuel is first transformed into steam which in turn develops mechanical power in the turbine.

As on July 31, 2010, and as per the Central Electricity Authority the total installed capacity of Diesel based power plants in India is 1,199.75 MW. Normally the diesel based power plants are either operated from remote locations or operated to cater peak load demands. Here is some list of presently operating plants.

Power Station	Operator	Location	State	Region	Installed Capacity (MW)	Under Construction (MW)
Bemina Diesel Power Station	J & K Govt		Jammu & Kashmir	Northern	1 × 5	5.00
Kamah Diesel Power Station	J & K Govt		Jammu & Kashmir	Northern	1 × 0.06	0.06
Leh Diesel Power Station	J & K Govt		Jammu & Kashmir	Northern	1 × 2.18	2.18
Upper Sindh Diesel Power Station	J & K Govt		Jammu & Kashmir	Northern	1 × 1.7	1.70
GMR VASAVI Diesel Power Plant	TNEB		Tamil Nadu	Southern	4 × 50	200
Yelahanka Diesel Power Station	KPCL	Yelahanka	Karnataka	Southern	6 × 21.32	127.92
Brahmapuram Diesel Power Station	KSEB	Brahmapuram	Kerala	Southern	5 × 21.32	106.60
Kozhikode Diesel Power Station	KSEB	Kozhikode	Kerala	Southern	8 × 16.00	128.00
Gangtok Diesel Power Station	Sikkim Govt	Gangtok	Sikkim	Eastern		4.00
Ranipool Diesel Power Station	Sikkim Govt	Ranipool	Sikkim	Southern		1.00
Suryachakra Diesel Power Station	SPCL (Private)	A & N	Andaman & Nicobar	Islands		20
Total				**11**	**27**	**596.46**

5.2. MERITS AND DEMERITS

Diesel electric power plants have many advantages over other types of power plants, as given below:

(*i*) The design and installation of such power plants is very simple.

(*ii*) Such plants can be located at any place.

(*iii*) Such plants can be quickly procured, installed and commissioned.

(*iv*) The layout, design and construction of foundations and buildings for such power plants are simple and cheap.

(*v*) Such plants require less space for fuel storage and are free from ash handling problems.

(*vi*) Such plants can respond to varying loads without any difficulty.

(*vii*) Such plants occupy less space because of minimum auxiliaries.

(*viii*) No standby losses.

(*ix*) The quantity of water required for cooling is limited.

(*x*) Such plants can be started and put on load quickly.

(*xi*) Such plants maintain high operation efficiency irrespective of load.

(*xii*) The overall capital cost including installation per unit of installed capacity is lesser than that for steam power plants.

(*xiii*) The efficiency of such power plants falls off very little with the usage in comparison with steam power plants.

(*xiv*) Thermal efficiency of such power plants is higher (about 40%) than that of a steam power plant.

(*xv*) Variety of fuels, such as residual fuel oil, low sulphur heavy stock etc. can be used.

(*xvi*) Less fire hazard.

(*xvii*) Such plants being simpler in operation require less operating and supervising staff as compared to that for steam power plants.

However, the diesel electric power plants have some drawbacks as given below:

(*i*) Operating cost, due to high cost of diesel oil as fuel, is very high.

(*ii*) Maintenance and lubrication cost is also high as compared in the case of other plants.

(*iii*) Diesel plants cannot supply overloads continuously whereas steam plants can work under 25% overload continuously.

(*iv*) Diesel units capacity is limited. These cannot be constructed in large size.

(*v*) Noise from the exhaust is a serious problem.

(*vi*) Their useful life is very short (say, about 10 years).

5.3. APPLICATIONS

The size limitations not withstanding, the escalating oil costs have made the operation of diesel electric power plants very uneconomical. In view of this, such plants play a very minor role in electric power generation. Their principal fields of use are given below:

1. Central Power Stations: Diesel power plants are installed where supply of coal and water is not available in sufficient quantity or where power is to be generated in small quantity. Power stations of this type in common use are of capacities up to 10 MW. The limit will be decided by the economics of the cost of the plant and the local conditions such as fuel costs, availability of water supply etc.

2. Standby Power Stations: Diesel power plants may be used as standby plants where continuity of power supply is essential such as in hospitals, telephone exchanges, radio stations, cinemas etc.

3. Peak Load Plants: Diesel power plant can be employed to supply the peak load on the power system while the base load is supplied by a nuclear or hydro power plant. The base load factor will thus be improved and cost of electrical energy per unit will be reduced. Diesel power plant is particularly suitable for supplying a load of small load factor. Its characteristics of quick starting and absence of standby losses make such plants very suitable for supplying peak loads.

4. Emergency Plants: A small diesel power plant may be installed in a large power station to supply essential auxiliaries in case of failure of main supply. Arrangements can be made to start the diesel plant automatically.

5. Private Power Plant For Small Industries: Diesel electric power plants can be employed as private industrial plants for supply of electric power specially if the capacity requirements are within the limits, set by the size of the diesel units available. The diesel power plants have certain advantages over the steam power plants. The diesel power plant needs much less water than a steam plant of equal capacity, its other advantages are:

1. A diesel power plant can be started and stopped quickly, as and when required.
2. It does not require any warming period, and need to be kept running for a long time before picking up load. There are thus no standby losses, as in steam power plants.
3. Diesel power plant does not require a large amount of cooling.
4. The diesel engines have a lower installation cost per kW as compared with the same size of the steam unit.
5. The diesel power plant needs less operating and supervising staff than a steam power plant.
6. There are fewer auxiliaries as compared to a steam power plant.
7. Buildings and foundations are relatively simple.
8. The diesel engines maintain their efficiency even at fractional loads whereas the efficiency of a steam power plant decreases rapidly below at 75% load.

6. Power Plants For Sparsely Populated Areas: A diesel plant may be employed to supply a sparsely populated area where only one transmission line is economically justifiable, to maintain continuity of supply in case the line should fail. Continuity of supply can be maintained by starting the diesel plant as soon as the transmission line fails or the voltage drops below the permissible limit. In either event the diesel power plant can be made to start automatically.

7. Nursery Power Plants: In India, a large number of power projects are being developed under various five year plans, and they are interconnected wherever possible to form a grid system. This assists to make the power available to large area and to a number of places. Unless there is a enough load, it is, however, not possible to extend the grid immediately to every village and corner of the country. During the process of development, a small diesel unit may be temporarily installed to supply the load of a village or town including the farm load. When the load develops and the grid supply becomes available, the concerned town or village can be connected to the grid and the diesel engine set can be moved to another area which requires power on a small scale and where the grid supply is not available. The diesel power plants used in this way are known as 'nursery power plants'.

8. Mobile Plants: Mobile diesel electric power plants mounted on trailers are employed for use in emergency and for temporary purposes.

5.4. SELECTION OF SITE FOR DIESEL ELECTRIC POWER PLANTS

The following factors should be considered while selecting a site for diesel electric power plant:

1. *Distance from the load centre*. The site should be as near to the load centre as possible in order to avoid transmission costs and losses.
2. *Availability of land*. The land should be available at cheap rate to keep the capital cost of the plant to the reasonable one.
3. *Availability of fuel*. The fuel should be easily available and at reasonable rate.
4. *Availability of transportation facilities*. The transportation facilities should be available.
5. *Availability of water*. Water should be available in sufficient quantity for cooling purposes.
6. *Distance from populated area*. The site should be away from thickly populated area because of noise and nuisance caused from exhaust.
7. *Type of land*. The land should be of high bearing capacity to withstand the load of the plant and also vibrations transmitted to the foundations from compressors and diesel engines.

5.5. ELEMENTS OF DIESEL ELECTRIC POWER PLANTS

A diesel electric power plant comprises of the engine proper and the auxiliary equipment, as detailed below. A schematic arrangement for a diesel electric power plant with a diesel engine and all auxiliary equipment is illustrated in Fig. 5.1.

1. Diesel Engine: This is the main component of the power plant which develops power. The diesel engines employed for diesel electric power plants may be four or two stroke engines. In a 4-stroke engine the complete cycle of

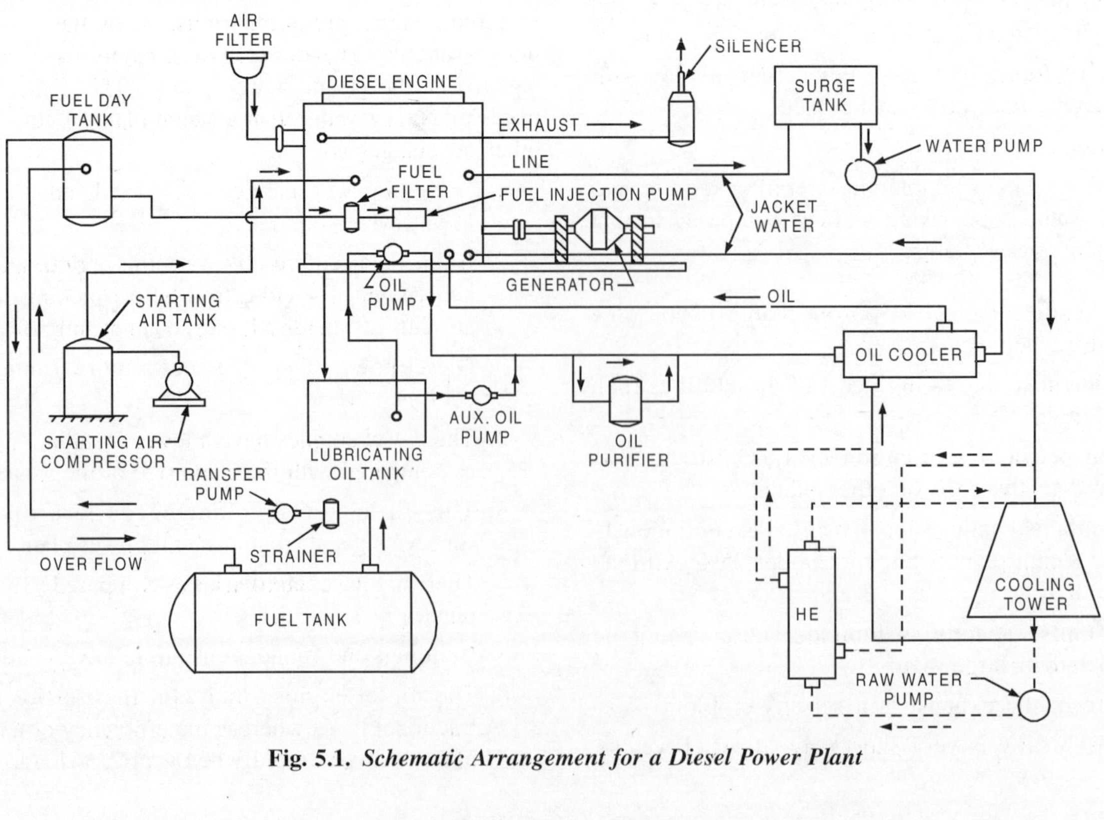

Fig. 5.1. *Schematic Arrangement for a Diesel Power Plant*

operations is performed in four strokes namely suction, compression, expansion (working or power) and exhaust strokes and two revolutions of the engine. In the case of a two-stroke diesel engine all the four operations are completed within two strokes of the piston or during one revolution of the crank shaft—the power and exhaust operations are completed during the travel of the piston in backward direction (*i.e.,* toward crank end of the cylinder) while intake and compression operations are completed during the forward travel of the piston. A two-stroke engine is rugged and compact in construction, simple in mechanical design, easily reversible, cheaper in cost and develops more power for the same speed and piston displacement because of elimination of two idle strokes. Flywheel required is also smaller one because of development of more uniform torque. However, a four-stroke engine has lower specific fuel consumption and more effective lubrication, more flexibility, less noisy exhaust, simple and better cooling, better scavenging and higher efficiency than a two-stroke engine.

The engine may be horizontal or vertical type. Horizontal engines are used for comparatively smaller output while vertical engines with multi-cylinder construction are employed for larger outputs. The arrangement of cylinders is important as it affects the foundations, building space and maintenance problems. Vertical in-line arrangement of cylinders is the most commonly used. To make the engine more compact, the cylinders may be arranged in V-shape (used for medium and large aircrafts). Depending upon the requirement of power the number of cylinders is decided. Six to eight cylinder engines are commonly used in power stations.

The ignition may be through compression ignition or spark ignition. In the compression-ignition engine, the combustion starts by spontaneous ignition of fuel and air because of high temperature developed by the compression of the air while in the spark-ignition engine, air-fuel mixture is compressed in the cylinder and at the end of compression, the mixture is ignited with the help of a spark. The diesel engines are compression-ignition type.

The diesel engines are available in low (lesser than 350 rpm), medium (lying between 350 and 1,200 rpm) and high speed (exceeding 1,200 rpm) ranges. It has been found that a higher rotational speed means a greater degree of compactness resulting in reduction of costs of both engines and generators and simultaneously in building requirements. Speeds commonly used are in the range from 500 to 1,000 rpm.

The rating of an engine which should be specified correctly is the net output in kW developed continuously at the crankshaft coupling by the engine in good operating condition, at a height of less than 450 m, at a temperature of 32°C and a barometric pressure of 717.5 mm Hg. Diesel engines are available in sizes from 75 kW to 3,750 kW.

Ambient conditions differing from the above have a marked effect on the output power. The power output is reduced by the high temperature firstly due to reason of diminished cooling effect of air and secondly by the expansion of air due to higher temperatures and its consequent thinning resulting in less fuel energy being converted into mechanical energy during each working cycle in a cylinder of a diesel engine.

At high altitudes the rarefied air produces an even greater reduction of output in the manner just described and in addition, its cooling power is reduced further by rarefication over and above the reduction produced by higher temperature.

The selection of engine size depends on the plant capacity. As already mentioned in Arts. 5.2 and 5.3, the overload capacity of the plant is very low; its operation at loads less than full load is uneconomical and its use is limited to emergency and special services, the determination of plant capacity is very important.

The size and number of engines depends upon plant capacity, purpose for which plant is being set up and the load characteristics. For small scale industries only one or two engines are sufficient. The total space available in the installation is also an important consideration in determination of number and size of units.

Diesel engines, large capacity industrial engines, have efficiencies in the range of 35-42%.

2. Engine Air Intake including Air Filters, Ducts and Supercharger (Integral) With The Engine. Air intake system is provided to supply necessary air to engine for fuel combustion. Air requirements of large diesel electric power plants are considerable (about 4-8 m^3 per kWh). The air filters are provided to remove dust and other suspended impurities from the air to be supplied to the engine. The supercharger is usually employed to increase the pressure of intake air above atmospheric one in order to develop an increased power output. The air is drawn from outside the engine room and delivered to the inlet manifold through ducts or pipes and filters. In cold season, the intake air is heated by the heat from exhaust gases, before injection. The filters are to be cleaned periodically. Superchargers are usually driven by the engines. Sometimes, a silencer is required in between the engine and the intake since the noise may be transmitted back to the outside air via the air intake system. There should be minimum pressure loss in the air intake system, otherwise specific fuel consumption will increase and the engine capacity will be reduced.

3. Engine Fuel System including Fuel Storage Tanks, Fuel Transfer Pumps, Strainers, Heaters and Connecting Pipe Work. Fuel transfer pumps are required to transfer fuel from delivery point to storage tanks and from storage tanks to daily consumption tanks and then to engine. Strainers are provided to remove suspended impurities and thus ensure clean fuel supply to the engine. Heaters are required to heat the oil specially during winter season.

The fuel oil is supplied at the plant site by rail or road. The fuel oil is stored in bulk storage tanks which are usually situated outdoor for safety. The fuel oil is transferred from these tanks to the small storage tanks called day *tanks* daily or at suitable intervals by means of transfer pumps through filters. From the day tank the fuel oil is passed through the strainers and fired into the engine through fuel filter and fuel injection pump. The capacity of the daily consumption tank should be atleast the 8 hour requirement of the plant. Nowadays all diesel power plants employ mechanical injection system in which fuel at a pressure of about 100 kg/cm^2 is delivered mechanically, through the atomizer nozzle, into the compre-ssed air.

4. Engine Exhaust System including Silencers and Connecting Ducts. This system is provided to discharge the engine exhaust to the atmosphere outside the building. The exhaust manifold connects the engine cylinder exhaust to the exhaust pipe provided with a muffler to reduce pressure in exhaust line and eliminate most of the noise which may result if the waste gases were discharged directly into the atmosphere. As the temperature of the exhaust gases is sufficiently high, therefore, the heat of these gases is utilised in heating oil or air supplied to the engine. In case of diesel electric generating units installed in steam power plants exhaust system also includes water heaters and steam boilers in order to utilise the heat of waste gases.

5. Engine Cooling System including Cooling Pumps, Cooling Towers or Spray Ponds, Water Treatment or Filteration Plant and Connecting Pipe. It is known that the heat generated by the burning of fuel in the engine cylinder is partially converted into useful work. The remainder is wasted as heat in the outgoing exhaust gases and in heating the engine, and if not removed may disintegrate the lubricating oil film on the cylinder walls and damage the cylinder liners, heads, walls, piston and rings. Small engines may be air-cooled but large engines mostly make use of forced water cooling. In the forced water cooling cold water is sent through the cylinder jacket with the help of a pump. The hot water is cooled in spray pond and is recirculated. The water used for cooling engine clyinder is softened by water treatment or water filteration plant in order to avoid formation of scales etc. in it. Hot well is provided in the surge tank to make a provision for expansion of jacket water. It is also necessary to keep the exit temperature of the cooling water around 70°C. In case it is too low, the lubricating oil film will not spread properly and will result in wear of cylinder and piston and if it is too high, the lubricating oil will burn. This requires a control of the flow of cooling water according to the load on the engine.

6. Engine Lubricating Oil System including Lubricating Oil Pumps, Oil Tanks, Filters, Coolers, Purifiers and Connecting Pipe Work. This system is of great importance and is provided for lubricating the moving parts, removing the heat from the cylinders and the bearings, helping the piston rings to seal the gases in the cylinders and for carrying away the solid dirt particles from the rubbing parts. The life of the engine and its efficiency depend largely on its lubrication system. The parts of the engine requiring lubrication are piston and cylinders, gears, crankshaft and connecting rod, bearings etc. Piston and cylinder require special lubricating oil. The forced feed lubrication is mostly employed. In this system of lubrication the lubrication oil is drawn from the sump by means of a pump and is passed through a strainer and then through a filter in order to remove impurities. Usually the lubricating oil is hot when drawn from a sump, if it is not, it should be heated, first before passing through the filters, in order to decrease its viscosity and make the filteration easy. The oil is then cooled through a heat exchanger and then sent into the engine. The lubricating oil costs of a diesel power plant are about 30% of the fuel costs.

7. Engine Starting including Battery, Starter, Compressed Air Supply etc. This system is provided to rotate the engine initially, while starting, until the firing starts and the unit runs under its own power. Small sets are usually started manually by handles but for sets of large capacity, say above 75 kW, the compressed air system is mostly employed for starting diesel engines. Battery driven motors can also be used for starting the diesel engine sets. In compressed air system of starting, the compressed air, from a tank, at high pressure (say 20 atmospheric pressure) is admitted to a few of the engine cylinders making them to act as reciprocating air motors to turn the engine shaft. The fuel is admitted to the remaining cylinders which causes the engine to start under its own power. A motor driven compressor is employed for supply of air to the compressed air tank.

8. Diesel Engine Generator (Alternator). The alternators used in diesel electric power plants are of rotating field, salient pole construction, speed ranging from 214 to 1,000 rpm (poles 28 to 6) and capacities ranging from 25 to 5,000 kVA at 0.8 pf lagging. Their output voltages are 440 V in case of small machines and as high as 2,200 V in case of large machines. Voltage regulation is about 30%.

They are directly coupled to the diesel engines. They are provided with automatic voltage regulators to allow close voltage regulation and satisfactory parallel operation. The excitation is usually provided at 115 or 230 V from a dc exciter of rating about 2-4% of the alternator rating, usually coupled to the engine shaft either directly or through a belt.

In addition to above, the diesel power plants are provided with different instruments such as *thermometers* or *pyrometers* (for measuring temperatures of lubricating oil, exhaust, jacket water, fuel oil, injection air etc.), *pressure gauges* (for measuring pressures of interstage and final stage of starting air system, lubricating oil system before and after cooling, cooling water, jacket water and fuel oil before and after straining), *ammeters* (for measuring excitation and output currents), *voltmeters* (for measuring excitation and output voltage), *wattmeters* (for measuring power outputs), *synchronizing equipment, visible* and *audible alarms* for giving indications of high jacket water temperature, high lubricating oil temperature, low lubricating oil pressure, low jacket water pressure, incorrect starting, blower motor overcurrent etc.

5.6. SUPERCHARGING

Supercharging, also known as *boosting,* can be defined as the admittance of an air charge into the cylinder larger than that obtainable as a result of the regular suction stroke. It is accomplished by external means as compared to regular suction, called the *natural aspiration.* The main object of supercharging is to increase the power output of the diesel engine without increasing its rotational speed or the cylinder dimensions. This is achieved by increasing the charge of air causing burning of more fuel and resulting in increase in mean effective pressure. The charge or weight of air for a given volume of air depends solely on pressure (varies directly as pressure) and temperature (varies inversely as temperature)*. Thus an air charge can be increased either by increasing pressure P or reducing the temperature T for a given volume. It can be shown that a large temperature drop is required to have a reasonable increase in weight—a drop of at least 38° would be required to give a 20% weight increase. Lowering of compression temperature will also make the starting very difficult. On the other hand, quite moderate pressure increase will give the desired result. An increase of about 0.003 MPa gauge results in an increase of weight by 27%. It is this method which is utilised in supercharging for increasing the weight of the air.

Supercharging offers advantages of an *increased power output* (30-50%) at the given speed, *fuel economy* due to better combustion, *higher mechanical efficiency, better scavenging action* (in the case of two-stroke engines), *increased volumetric efficiency* and *reduced possibility of knocking* in diesel engines.

Supercharging is done by means of blowers either of positive displacement type (such as piston cylinder type, root blowers or vane blowers), or of the centrifugal type or of the turbo-type. The blowers are usually driven from the engine crank shaft by means of spur, helical or herring-bone gears, silent chains or V-belts at a speed of 2-3 times the engine speed. In the case of a turbo-charger, the super-charger is coupled to a gas turbine in which the exhaust of the engine is expanded.

The positive displacement blowers are employed for low and medium speed engines with speeds not exceeding 4,000 rpm. Centrifugal blowers are employed both on low and high speed engines. Centrifugal blowers driven by exhaust gas turbines are small and light and are employed for stationary, locomotive, marine and aircraft engines.

There is an economical limit to the supercharging, which is about 40% at present. With the exceeding of limit of supercharging, the power absorbed by the engine driven blowers increases rapidly without providing appreciable gain. The other limiting factors are ability of engine to withstand the increased pressure and temperature stresses. However, these difficulties can be overcome by providing better cooling system, improving methods of lubrication and making metallurgical improvements.

The use of supercharging on the diesel engines has progressed rapidly as increase in compression pressure does not cause diesel knock in the engine. Actually it results in smoother running of the engine.

5.7. MAINTENANCE OF RECORDS (LOG SHEETS)

A correct record of instrument readings (ammeter in each phase, voltmeter, wattmeter) and conditions of operation (oil pressure, water pressure, inlet and outlet temperatures of water and oil etc.) is maintained at regular intervals, say every half an hour. Such records form *log sheets.* A separate log sheet is used for each unit. The data so recorded enables the personnel to check satisfactorily working of the plant. Such records also assist in evaluation of performance of the plant.

Maintenance includes cleaning of fuel oil from dirt and impurities by means of filters. Occasionally all the fuel is drained and the fuel tanks are cleaned thoroughly. The temperature and flow of coolant, lubricating oil and exhaust gases are also checked at regular intervals and recorded.

5.8. DIESEL ELECTRIC PLANT LAYOUT

For planning the layout of the diesel electric power plant the following points require consideration:

(*i*) Approximate dimensions of the equipment.

(*ii*) Location of equipment and minimum clearances between its parts.

(*iii*) Provision of natural lighting and proper ventilation.

(*iv*) Possibility of future additions to the building with a minimum changes.

(*v*) The spacing required between two units, the distance between their centre lines, the distance between the centre line of the end unit and the wall, the distance between the head end of the engine and the wall, and the generator end and the wall.

(*vi*) The location of switchboard, station auxiliary transformers, battery room, fuel-oil tank, compressed-air cylinder bottle for starting of engine, compressors, lubricating oil system and the cooling arrangements for cylinder jackets and suction and exhaust arrangements for the engine.

(*vii*) Provision for a small workshop, stores, office etc.

(*viii*) Provision for installation of at least one spare unit in future.

(*ix*) Provision for bringing the equipment into the building.

(*x*) Provision of space for erection work, and for handling the equipment for maintenance.

---

* $w = \frac{PV}{RT} = \text{Constant} \times P \times \frac{1}{T}$.

The general layout of a medium size diesel electric power plant is illustrated in Fig. 5.2. The units are usually placed parallel to each other in order to keep the length of electrical connections for generators to control board and air ducts and exhaust pipes minimum. The generating units (diesel engine-generator sets) are placed on large concrete slabs preferably reinforced. The foundation should be firm and subsoil solid. The air intake and filters and the exhaust mufflers should not be located in the engine room.

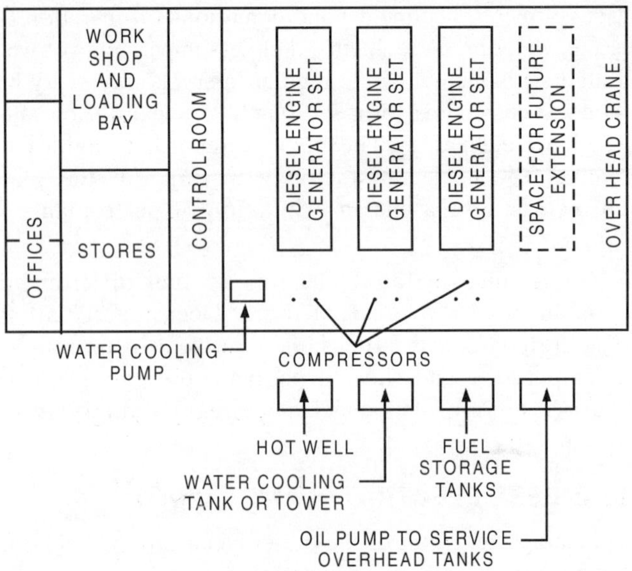

Fig. 5.2. *Layout of a Medium Size Diesel Electric Power Plant*

5.9. PERFORMANCE AND THERMAL EFFICIENCY OF DIESEL ELECTRIC POWER PLANTS

Only 34 per cent of the energy supplied to the diesel electric power plants in the form of fuel is converted into useful work, remaining is lost in cooling water (about 30%), through exhaust gases (about 25%) and in radiation, friction etc. (11%). Thus the full-load thermal efficiency of a diesel power plant comes out to be about 34%. At part load, the thermal efficiency decreases appreciably and specific fuel consumption increases, as shown in Fig. 5.3. So part-load operation of diesel power plants is not economical. Specific fuel consumption at full load is roughly 0.23 kg per kWh.

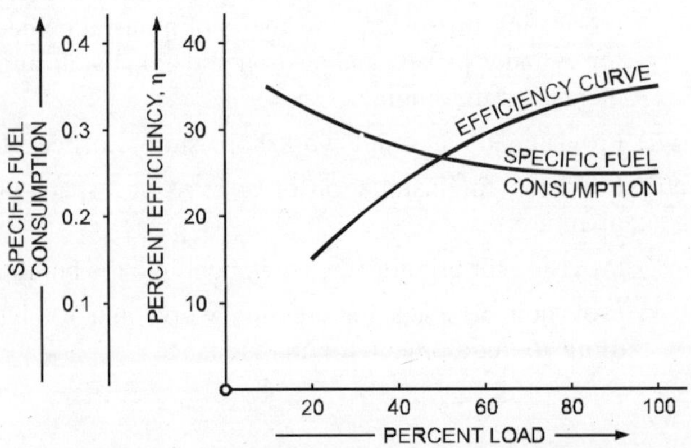

Fig. 5.3. *Performance Curves of Diesel Power Plant*

Exercises

1. Discuss the advantages and disadvantages of a diesel power plant.

[U.P. Technical Univ. Power Station Practice 2008-09]

2. What are the advantages of diesel plants over the thermal plants ? Why diesel plants are not used for high capacity ?

[U.P. Technical Univ. Power Station Practice 2009-10]

3. Identify the drawbacks of diesel power plants. How can these plants be effectively used as base/peak load plants ? Examine the best suited mode.

[U.P. Technical Univ. Power Station Practice 2005-06]

4. What are the various factors to be considered while selecting site for diesel power plant. Discuss the advantages and disadvantages of a diesel power plant.

[Power Station Practice Pb. Univ. June 1992; U.P. Technical 2007-08]

5. Why are diesel engines used as standby plants in steam power plants ?

[Pb. Univ. Power Station Practice, Dec. 1995]

6. Explain the functions of different components in diesel plant. Give its role and applications.

[U.P. Technical Univ. Power Station Practice 2006-07]

7. Draw the flow diagram of diesel power plant in brief.

[U.P. Technical Univ. Power Station Practice 2007-08]

8. Draw a layout of diesel power plant, showing the various systems, including cooling, lubrication, starting, intake and exhaust systems.

[Pb. Univ. Power Station Practice, May 1995]

9. Make a neat layout of a diesel power plant, showing fuel circuit, water circuit and lubricating circuit on the layout.

[Pb. Univ. Power Station Practice, January 1993]

10. Draw a layout of diesel power plant, showing the various systems.

[G.B. Technical Univ. Power Station Practice 2010-11]

11. Draw the plan layout of a diesel power plant.

[U.P. Technical Univ. Power Station Practice 2008-09]

12. Write short notes on the following :

(*i*) Merits and demerits of diesel electric power plants.

(*ii*) Performance and applications of diesel power plants.

[A.M.I.E. Sec. B. Winter 1984]

(*iii*) Applications of diesel electric power plants.

(*iv*) Essential elements of diesel power plants.

(*v*) Layout of a typical diesel power plant.

(*vi*) Supercharging and its advantages.

Short Answer Type Questions With Answers

Q. 1. What is the basic difference between a diesel engine and steam turbine ?

Ans. The basic difference between a diesel engine and a steam turbine is that in the first case the chemical energy of combustion is released inside the cylinder whereas in the second case the energy developed during combustion of fuel is first transformed into steam which in turn develops mechanical power in the turbine.

Q. 2. Inspite of various advantages of diesel power plants over other power plants their use is limited. Why ?

Ans. The size limitations and the escalating fuel cost have made the operation of diesel power plants very uneconomical and thereby their use limited. The other drawbacks of diesel power plants are high maintenance and lubrication costs, limited useful life, noise from exhaust and incapability of supplying overloads.

Q. 3. Why cannot diesel power stations be employed to generate bulk power ? **[U.P. Technical Univ. 2007-08]**

Ans. Because of limited capacity, escalating fuel cost, high maintenance and lubrication costs, limited useful life, noise from exhaust and incapability of supplying overloads, diesel power stations cannot be employed to generate bulk power.

Q. 4. Why is supercharger used in the air intake system of a diesel engine ?

Ans. The supercharger is usually employed to increase the pressure of intake air above atmospheric one in order to develop an increase power output.

Q. 5. Why are strainers provided in the fuel supply system of a diesel engine ?

Ans. Strainers are provided to remove suspended impurities and thus ensure clean fuel supply to the engine.

Q. 6. Why is exhaust system provided in the diesel power plant ?

Ans. The exhaust system is provided to discharge the engine exhaust to the atmosphere outside the building.

Q. 7. Why is cooling essential for a diesel engine ?

Ans. It is known that the heat generated by the burning of fuel in the engine cylinder is partially converted into useful work. The remainder is wasted as heat in the outgoing exhaust gases and in heating the engine, and if not removed may disintegrate the lubricating oil film on the cylinder walls and damage the cylinder liners, heads, walls, piston and rings. That is why cooling is essential for a diesel engine.

Q. 8. Why it is necessary to maintain the exit temperature of cooling water at about 70°C ?

Ans. It is necessary to maintain the exit temperature of cooling water at about 70°C because in case it is too low, the lubricating oil film will not spread properly and will result in wear of cylinder and piston and if it is too high, the lubricating oil will burn.

Q. 9. Why is lubrication of a diesel engine necessary ?

Ans. The engine lubricating oil system is of great importance and is provided for lubricating the moving parts, removing the heat from the cylinders and the bearings, helping the piston rings to seal the gases in the cylinders and for carrying away the solid dirt particles from the rubbing parts. The life of the engine and its efficiency depend largely on its lubrication system. The parts of the engine requiring lubrication are piston and cylinders, gears, crankshaft and connecting rod, bearings etc. Piston and cylinder require special lubricating oil.

Q. 10. Why are the generating units usually placed parallel to each other ?

Ans. The generating units are usually placed parallel to each other in order to keep the length of electrical connections for generator to control board and air ducts and exhaust pipes minimum.

❑❑❑

CHAPTER 6

Gas Turbine Power Plants

INSIDE THIS CHAPTER

6.1. INTRODUCTION

Gas turbine electric power plant is a power plant in which a gas turbine is used as the prime mover for the generation of electrical energy.

In steam power plants, the products of combustion do not form the working medium. These are utilised to produce an intermediate fluid, called the steam, which is expanded in the turbine. In case the intermediate step of converting water to steam by means of gases is eliminated, the arrangement would be far simple and less wasteful. This principle is employed in gas turbine electric power plants where the gas is directly expanded in the turbine. The gas turbine is different from the diesel engine in this respect that the combustion region is external to the prime mover.

In a gas turbine, the working medium is either a mixture of combustion products and air or heated air at a certain pressure and a higher initial temperature. With respect to design, a gas turbine is quite similar to a steam turbine. In the turbine blading, working gas expands and the heat energy is converted first into the kinetic energy and then to the work of the turbine shaft rotation.

The principle of a gas turbine power plant is that a turbo-compressor compresses working medium to a high pressure and then it is burnt in the combustion chamber where the combustion takes place at a constant pressure and temperature of the working medium is raised. This high pressure and high temperature working medium is then expanded in a gas turbine coupled to the generator (or alternator). The overall efficiency of a gas turbine as a prime mover is limited due to the fact that a large portion of the power developed by the turbine is used in driving the compressor and also by the temperature safely attainable. Thus two factors that need improvement are the design of the compressor and design of turbine with suitable materials to withstand high temperatures. A lot of research work is in progress in this direction.

As a result of advances in the design of gas turbines, the advanced gas turbines with special features (high firing temperatures, advanced cooling systems, advanced materials to withstand higher temperatures and more efficient compressors with transonic blades) have been developed. These gas turbines have high efficiency (more than 60%) less emission of NO_2, carbon monoxide and unburnt hydrocarbon, operate at higher temperature of about 1,400°C.

Aeroderivative gas turbines, based on aircraft engine technology, are now widely employed in the field of generation of electrical power. The propulsion engines are modified as per needs of power generation. They do not need large lubricating oil reservoirs, coolers and pumps. The main advantages of such turbines are compact size, modular design and quick installation (6 to 9 months), fast starting (about 3 minutes), easier and quicker maintenance, lesser space requirement, high operating efficiency (fully-load as well as part load) and high availability (about 95%). However, such turbines are costlier (about 10–15% more than heavy duty gas turbines) and need clean fuel involving higher fuel treatment costs.

Natural gas fired (including LNG fired) power plants account for almost 20% of the world's electrical power generation. These power plants use gas turbines or gas turbine based combined cycles. Gas turbines in the simple cycle mode (only gas turbines running), have an efficiency of 32 to 38%. The most important parameter that governs the efficiency is the maximum gas temperature possible. The latest gas turbines with technological advances in materials and aerodynamics has efficiencies up to 38%.

6.2. MERITS AND DEMERITS

The gas turbine power plant has got some inherent advantages such as simplicity of design and installation, high reliability, simple lubrication system, clean exhaust requiring no stack, compactness, low initial cost, no standby losses, requiring small building space and light foundations and requiring little cooling water. The delivery and installation time for such power plants is much less than that for steam power plants. The gas turbine can be started quickly and can be put to share full load within a few minutes and efficiency can be improved considerably by using heat economy devices. Gas turbines may be operated by remote control and need little or no personnel while operating and while shut-down. Though the fuel costs in such power plants are usually higher than in other power plants but maintenance costs are lower than for diesel power plants. Such plants have higher availability rate (about 90–95% as compared to about 80% for conventional steam power plants).

The gas turbine has got some limitations also, which probably will be overcome or minimised with more experience and research, such as inability of using coal or heavy residual petroleums as fuel, low net output (a greater part of power developed by the turbine being used in driving the compressor), low overall efficiency (the exhaust gases from the turbine contain sufficient heat), noisy operation, high specific fuel consumption and limited unit capacity. The gas turbine, like steam turbine, is not readily reversible. In summer, some pre-cooling of the air to the compressor is desirable and possibly essential since at temperatures higher than 35°C the density of air may be so low that oxygen starvation may occur. The temperature of combustion chamber is quite high 1,650°C and therefore its life is comparatively reduced. The unit has got starting problems as the compressor has to be started before starting of turbine and the former requires power from some external source. However, once the unit starts, the external power is not required as the turbine itself supplies the necessary motive power to the compressor.

6.3. APPLICATIONS

1. Majority of such power plants are usually used for driving generators and supplying peak loads in other types of power plants (steam and hydro electric power plants), because they can be started and loaded quickly. Other reasons for it are low initial costs and higher fuel costs.
2. Such a plant is used as a starting plant for driving auxiliaries in other power plants.
3. A large number of such plants are used as standby power plants.
4. Gas turbine power plants are also used as base load power plants where fuel oil or natural gas are cheap and easily available, water supply is scarce and load factor is very low, say 15–18%. Base load gas turbine power plants are likely to play significant role in future.
5. Such plants can be used in parallel with tidal power plants.
6. These plants are used to operate as combination plants with conventional steam boilers.

Because of recent advances in gas turbines and their inherent advantages, the gas turbine power plants are finding wide spread use, even as base load plants. As on June 30, 2011, and as per Central Electricity Authority the total installed capacity of gas based power plant in India is 17,706.35 MW. This accounts for 10% of total installed capacity. GAIL is the main source of fuel for most of these plants. Here is some list of presently oeprating plants.

Power Station	Operator	Location	District	State	Sector	Region	Unit Wise Capacity (MW)	Installed Capacity (MW)
IPGCL Gas Turbing Power Station	IPGCL		New Delhi	NCT Delhi	State	Northern	9 × 30	270.00
Pragati Gas Power Station	PPCL		New Delhi	NCT Delhi	State	Northern	2 × 104.6, 1 × 121.2	330.40
Pampore Gas Turbine Station I	J & K Govt	Pampore	Pulwama	Jammu & Kashmir	State	Nothern	3 × 25	75.00
Pampore Gas Turbine Station II	J & K Govt	Pampore	Pulwama	Jammu & Kashmir	State	Northern	4 × 25	100.00
Ramgarh Gas Thermal Power Station	RVUNL		Ramgarh	Rajasthan	State	Northern	1 × 3, 1 × 35.5, 1 × 37.5, 1 × 37.8	113.80
Dholpur Combined Cycle Power Station	RVUNL	Purani Chaoni	Dholpur	Rajasthan	State	Northern	3 × 110	330.00
Anta Thermal Power Station	NTPC	Anta	Baran	Rajasthan	Central	Northern	3 × 88, 1 × 149	413.00
Auraiya Thermal Power Station	NTPC	Dibiyapur	Auraiya	Uttar Pradesh	Central	Northern	4 × 110, 2 × 106	652.00

Power Station	Operator	Location	District	State	Sector	Region	Unit Wise Capacity (MW)	Installed Capacity (MW)
Faridabad Thermal Power Plant	NTPC	Mujedi	Faridabad	Haryana	Central	Northern	2 × 143, 1 × 144	430.00
National Capital TPP	NTPC	Vidyutnagar	Gautam Budh Nagar	Uttar Pradesh	Central	Northern	4 × 131, 2 × 146.5	817.00
Dhuvaran Gas Based CCPP-I	GSECL	Khambhat	Anand	Gujarat	State	Western	1 × 67.85, 1 × 38.77	106.62
Dhuvaran Gas Based CCPP-II	GSECL	Khambhat	Anand	Gujarat	State	Western	1 × 72.51, 1 × 39.34	112.45
Utran Gas Based CCPP	GSECL	Utran	Surat	Gujarat	State	Western	3 × 30, 1 × 45, 1 × 228	363.00
Vadodra Gas Based CCPP-I	GIPCL	Vadodara	Vadodara	Gujarat	State	Western	3 × 32, 1 × 49	145.00
Vadodra Gas Based CCPP-II	GIPCL	Vadodara	Vadodara	Gujarat	State	Western	1 × 111, 1 × 54	165.00
Uran Gas Turbine Power Station	Mahagenco	Bokadvira	Raigarh	Maharasthra	State	Western	4 × 108, 2 × 120	672.00
Kawas TPS	NTPC	Adityanagar	Surat	Gujarat	Central	Western	4 × 106, 2 × 110.5	645.00
Jhanor Gandhar TPS	NTPC	Urjanagar	Bharuch	Gujarat	Central	Western	3 × 131, 1 × 255	648.00
Goa Gas Power Station	RSPCL	Zuarinagar	Goa	Goa	Private	Western	1 × 32, 1 × 16	48.00
Vatva Combined Cycle Power Plant	Torrent	Vatva	Ahamada-bad	Gujarat	Private	Western	2 × 32.5, 1 × 35	100.00
SUGEN Combined Cycle Power Plant	Torrent	Akhakhol	Surat	Gujarat	Private	Western	3 × 382.5	1147.50
Essar Combined Cycle Power Plant	Essar	Hazira	Surat	Gujarat	Private	Western	3 × 110, 1 × 185	515.00
GSEG Combined Cycle Power Plant	GSEG	Hazira	Surat	Gujarat	Private	Western	3 × 52	156.00
GPEC Combined Cycle Power Plant	GPEC	Paguthan	Bharuch	Gujarat	Private	Western	3 × 135, 1 × 250	655.00
Trombay Gas Power Station	Tata	Trombay	Mumbai	Maharasthra	Private	Western	1 × 120, 1 × 60	180.00
Basin Bridge Gas Turbine Power Station	TNEB	Chennai	Chennai	Tamil Nadu	State	Southern	4 × 30	120.00
Thirumakottai Gas Turbine Power Station	TNEB	Thirumakottai	Thiruvarur	Tamil Nadu	State	Southern	1 × 70, 1 × 38.88	108.88
Kuttalam Gas Turbine Power Station	TNEB	Maruthur	Naga-pattinam	Tamil Nadu	State	Southern	1 × 64, 1 × 37	101.00

Power Station	Operator	Location	District	State	Sector	Region	Unit Wise Capacity (MW)	Installed Capacity (MW)
Valathur Gas Turbine Power Station-I	TNEB	Valathur	Ramanatha-puram	Tamil Nadu	State	Southern	1 × 61, 1 × 34	95.00
Valathur Gas Turbine Power Station-II	TNEB	Valathur	Ramanatha-puram	Tamil Nadu	State	Southern	1 × 59.8, 1 × 32.3	92.10
Karaikal Gas Turbine Power Station	PPCL	Karikal	Pondicherry	Pondicherry	State	Southern	1 × 22.9, 1 × 9.6	32.50
Rajiv Gandhi CCPP	NTPC	Kayamkulam	Alappuzha	Kerala	Central	Southern	2 × 115.20, 1 × 129.18	359.58
Jegurupadu Combined Cycle Power Plant-I	GVK	Jegurupadu	E Godavari	Andhra Pradesh	Private	Southern	2 × 46, 1 × 49, 1 × 75	216.00
APGPCL Plant-I	APGPCL	Vijjeswaram	W Godavari	Andhra Pradesh	Joint	Southern	2 × 33, 1 × 34	100.00
Kusalava Power Plant, GMK Formly	Triveni	Gollapallem	E Godavari	Andhra Pradesh	Private	Southern	2 × 1.1, 1 × 1	3.2
Spectrum Combined Cycle Power Plant	Spectrum		E Godavari	Andhra Pradesh	Private	Southern	2 × 46, 1 × 47, 1 × 70	209.00
Gautami Combined Cycle Power Plant	Maytas	Peddapuram	E Godavari	Andhra Pradesh	Private	Southern	2 × 145, 1 × 174	464.00
Konaseema Combined Cycle Power Plant	KGPL	Ravulapalem	E Godavari	Andhra Pradesh	Private	Southern	2 × 140, 1 × 165	445.00
Lanco Kondapalli Power Plant-1	Lanco	Kondapalli	Krishna	Andhra Pradesh	Private	Southern	2 × 119.57, 1 × 128.99	368.13
Lanco Kondapalli Power Plant-II	Lanco	Kondapalli	Krishna	Andhra Pradesh	Private	Southern	1 × 233, 1 × 133	366.00
Vemagiri Combined Cycle Power Plant	GMR	Vemagiri	E Godavari	Andhra Pradesh	Private	Southern	1 × 137, 1 × 233	370.00
Samarlakota Combined Cycle Power Plant	Reliance	Samarlakota	E Godavari	Andhra Pradesh	Private	Southern	1 × 140, 1 × 80	220.00
Aban Combined Cycle Power Plant	Lanco	Karuppur	Tanjore	Tamil Nadu	Private	Southern	1 × 74.41, 1 × 38.80	113.21
Kochi Combined Cycle Power Station	BSES	Kochi		Kerala	Private	Southern	8 × 40.50, 1 × 35.5	157.00
Lakwa Thermal Power Station	APGCL	Maibella	Sivasagar	Assam	State	N Eastern	4 × 15, 3 × 20	120.00
Namrup Thermal Power Station	APGCL		Dibrugarh	Assam	State	N Eastern	1 × 20, 2 × 21, 1 × 11, 1 × 24, 1 × 14	111.00
Calcutta Electric Supply Corporation								2,000
Total						**45**	**160**	**13,711.27**

6.4. SELECTION OF SITE FOR GAS TURBINE POWER PLANTS

The following factors should be considered while selecting a site for gas turbine power plant:

1. *Distance From Load Centre.* The site should be as near to the load centre as possible so that the transmission costs and losses are minimised.
2. *Availability of Land.* The land should be available at cheap rate in order to keep the capital cost of the plant low.
3. *Availability of Fuel.* The fuel should be easily available and at reasonable rate.
4. *Availability of Transportation Facilities.* The transportation facilities should be available.
5. *Distance From Populated Area.* The site should be away from thickly populated area because of noisy operation.
6. *Type of Land.* The land should be of high bearing capacity to withstand the load of the plant and also the vibrations transmitted to the foundations from compressors and turbines.

6.5. FUELS FOR GAS TURBINES

The most important requirement of the fuel used in a gas turbine is that its products of combustion should contain a minimum of matter which might get deposited on the turbine blades or the heat exchanger.

A variety of fuels—solid, liquid and gaseous are available for use in gas turbines. Though coal and peat can be employed as gas turbine fuels but these are not used mainly because of coal handling and ash handling problems. The petroleum fuels such as kerosene, gas oil, diesel oil, residual oil, are quite suitable for use in gas turbines but their use is limited because of their higher costs. Natural gas which is mainly methane, has a very high calorific value and is generally used for auxiliary power generation in oilfields.

Nowadays the gas turbine power plants use mainly natural gas and liquid petroleum fuels. Natural gas contains 80% methane and small fraction of other gases.

6.6. ELEMENTS OF SIMPLE GAS TURBINE POWER PLANTS

A simple gas turbine power plant essentially consists of the reaction type non-condensing turbine, a compressor mounted on the same shaft and coupled to the turbine, the combustion chamber, an alternator coupled to the turbine itself and auxiliaries such as starting motor, auxiliary lubrication pumps, oil system, fuel system and duct system etc. A schematic arrangement for simple gas turbine power plant is illustrated in Fig. 6.1.

The *compressor* used in the gas turbine power plant is usually of rotary type. The air is sucked in by the compressor from the atmosphere through the filter, which removes the dust from the air. The rotary blades of the compressor push the air between the stationary blades to raise its pressure to 4–5 atmospheres. Thus air is available at high pressure at the output of the compressor.

The *combustion chamber* consists of a vessel into which the air at high pressure from the compressor is fed. Fuel is injected into the combustion chamber and burnt in the stream of air supplied from the compressor. A part of the air is delivered ahead of the burning fuel in order to cool the very hot combustion products (at temperature of around 1,650°C) and bring them to a temperature around 800°C when these can be allowed to enter the turbine without causing damage to the first few rows of blades.

The products of combustion comprising of mixture of gases at high temperature and pressure are passed through the gas turbine. These gases in passing over the turbine blades expand and thus result in motion of rotor and finally discharged to the atmosphere at a temperature of about 540°C.

Part of the power developed by the gases in passing through the turbine is used to run the compressor and auxiliaries and the rest is used to generate electrical energy driving the generator.

The alternator is coupled to the gas turbine, converts mechanical energy of the turbine into electrical energy and feeds it to the bus-bars through transformers, circuit breakers and isolators etc.

An electric motor used for starting the compressor (compressor is required to be started before starting of turbine) is mounted on the same shaft as that of the turbine. Once the unit starts, a part of mechanical power output of the turbine drives the compressor and there is no need of motor now. The motor is energised by the batteries.

The gas turbine has to drive the compressor as well as the alternator. In some cases, these two functions of gas turbines are separated and two different turbines are employed—a high pressure turbine drives the compressor and a

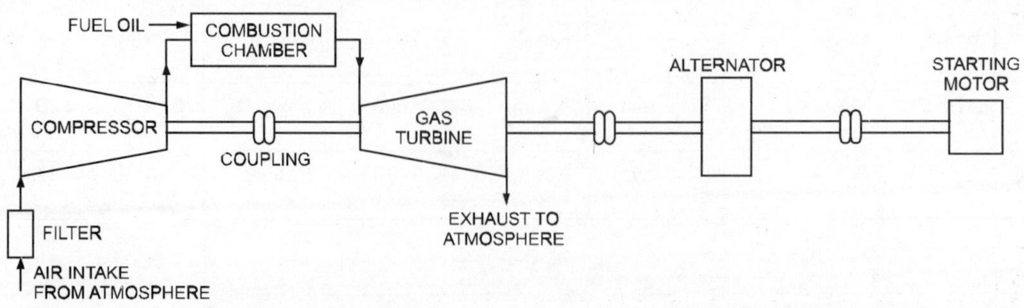

Fig. 6.1. ***Schematic Arrangement of a Simple Gas Turbine Power Plant***

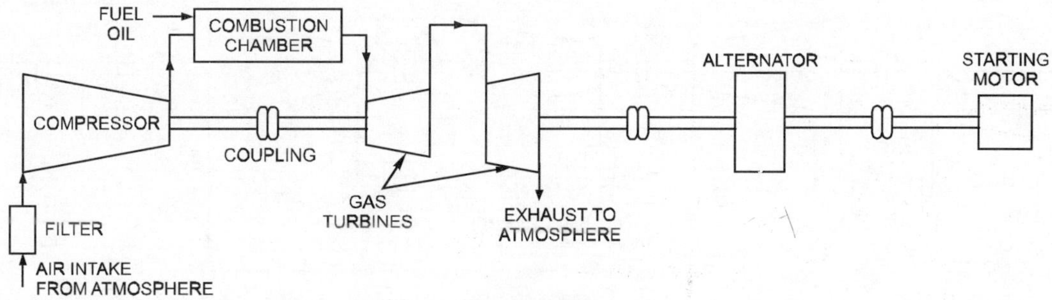

Fig. 6.2

low pressure turbine drives the alternator (Fig. 6.2). This arrangement has the advantage that the speed of the power turbine (the turbine driving the alternator) can be kept constant at synchronous speed while the speed of the turbine driving the compressor can be varied according to the output required.

The overall efficiency of such power plants *called the open cycle plants is very low* as a greater part (around 65%) of the power developed by the turbine is used in driving the compressor. The fact that the hot gases from the combustion chamber are cooled to a safe temperature before admitting it to the turbine, also reduces the efficiency. The overall efficiency of the plant, owing to above reasons seldom exceeds 20%.

Merits and Demerits of Open Cycle Gas Turbine Power Plants. Open cycle gas turbine power plants are simple in design, small in size (lower volume space per kW output), light in weight (lower specific weight), do not need any warm up time, possibility of great variety in the system arrangement and self contained*. The combustion chamber can be designed to burn any of the hydrocarbon fuels including solid fuels.

However, the open cycle gas turbine power plants have drawbacks of higher air rate in comparison to other prime movers, sensitivity to changes in component efficiencies (due to dust or dirt deposited on the compressor blades, by carbon or other foreign deposits from combustion in the combustion chamber, turbine, and regenerator) and erosion and corrosion of turbine blades.

6.7. METHODS OF IMPROVING THERMAL EFFICIENCY OF A SIMPLE GAS TURBINE POWER PLANT

The efficiency of a simple gas turbine power plant can be improved by employing regenerator, intercooler and reheater.

Regenerator is usually of shell and tube construction. The exhaust gases are made to flow inside the nest of tubes while air flows outside the tubes in the shell in the counter flow and heated up by the heat given out by the exhaust gases. Thus the regenerator utilises the heat of exhaust gases to heat the compressed air before it is sent to the combustion chamber, reduces the fuel consumption of the plant and improves the cycle thermal efficiency. It is noteworthy that addition of regenerator in the circuit makes no change in the duties/work of the compressor and turbine but the quantity of fuel supplied is substantially reduced as the temperature of the air entering the combustion chamber is raised. The other noteworthy point is that the gain in efficiency is greater at lower pressure ratios.

The heat transfer from the regenerator can be improved either by increasing the surface area or by increasing the flow turbulence. Increasing of surface area of regenerator involves higher initial cost, while the increasing of flow turbulence involves increased pressure drop. Thus the design of a regenerator is a compromise between the gain in heat recovery on the one hand and higher initial cost and operating cost on the other.

However, for short time operation such as peak loads, the cost of regenerator may not justify its use in gas cycle.

As already mentioned, a greater part (about two-third) of power developed by the turbine is used in driving the compressor. This requirement of power, however, can be reduced if the compression of air could be done in two or more stages and an **intercooler** is introduced between the two. This is because of reduction in volume of air due to cooling in the intercooler.

Intercooling means the removal of heat from compressed air between the stages of compression. This necessitates the use of compressors with two stages viz low pressure and high pressure compressors. The intercooler is heat exchanger which cools the partly compressed air in order to reduce volume and increase density. The intercooling results in improvement of thermal efficiency, air rate and work ratio. By use of intercooling the size of turbine and compressor for the same output is reduced.

The number of stages to be employed for compression, is generally decided on the basis of the saving in work done and additional investment required for the equipment (compressors and intercoolers). Maximum advantage of intercooling is had when pressure ratio for each stage is the same and high, compressor efficiency is low and regeneration is employed.

In **reheating** the combustion gases are not expanded in one turbine only but in two turbines. The exhaust of the high pressure turbine is reheated in a reheater and then expanded in a low pressure turbine. Reheating improves the output

* In the absence of intercooler, no cooling water is required. In case the turbine can be started by a powder charge or some other mechanical device, no electric power or battery is required for starting motor.

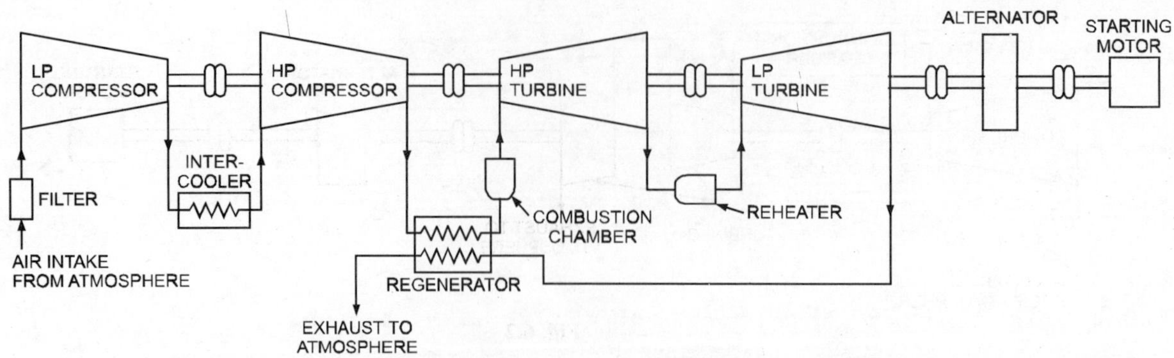

Fig. 6.3. *Open Cycle Gas Turbine Power Plant With Regenerator, Intercooler and Reheater*

from the turbines due to multiple heating in the same way as intercooling improves the performance of the compressor. However, the cost of additional fuel may be heavy unless a heat exchanger is also employed.

This process necessitates a compounding of the turbine. It is specially advantageous for improving the efficiency of plants having high pressure ratio and moderate cycle temperature.

Open cycle gas turbine power plant with a regenerator, an intercooler and a reheater is illustrated in Fig. 6.3. Both compressors (high pressure compressor and low pressure compressor) and both the turbines (HP turbine and LP turbine) and mounted on a common shaft. The alternator is coupled to the main shaft through gears so that the speed is reduced to the synchronous speed of the alternator. The efficiency of the plant can be improved to about 30% by using regenerator, intercooler and reheater.

6.8. CLOSED CYCLE GAS TURBINE POWER PLANTS

As mentioned in Art. 6.6, in open cycle gas turbine power plant, the fuel is mixed with air in the combustion chamber and the combustion gases are expanded in the gas turbine, which causes erosion and corrosion of turbine blades and, therefore, it becomes essential to use fuel of superior quality in the combustion chamber. This drawback is overcome in case closed cycle power plant is adopted, in which fuel is not mixed with the working medium (air or any other gas such as helium, argon, hydrogen and neon). The other drawback of the open cycle plant is that the turbine exhaust is discharged into the atmosphere resulting in rejection of heat of exhaust gases to the atmosphere. In closed cycle plant this drawback is overcome by rejecting heat of exhaust gases in a heat exchanger or recooler.

In a closed cycle plant the medium is heated externally and is continuously circulated through the compressors, heat exchangers, intercoolers, reheaters and gas recoolers. The arrangement is illustrated in Fig. 6.4. The pressure ratios, the temperatures and air velocities in the system remain practically constant.

The advantages of a closed cycle power plant are given below.

1. The risk of corrosion and abrasion of the interiors of turbines is eliminated as the turbines are kept free from the combustion products.
2. Since the working medium is heated externally and the fuel is not mixed with it, any fuel (inferior type of fuel or solid fuel) can be used.
3. The working medium is at relatively high internal pressure. So, specific volume is less and the dimensions of the compressors and turbines can be reduced and the maximum unit capacity can be increased.

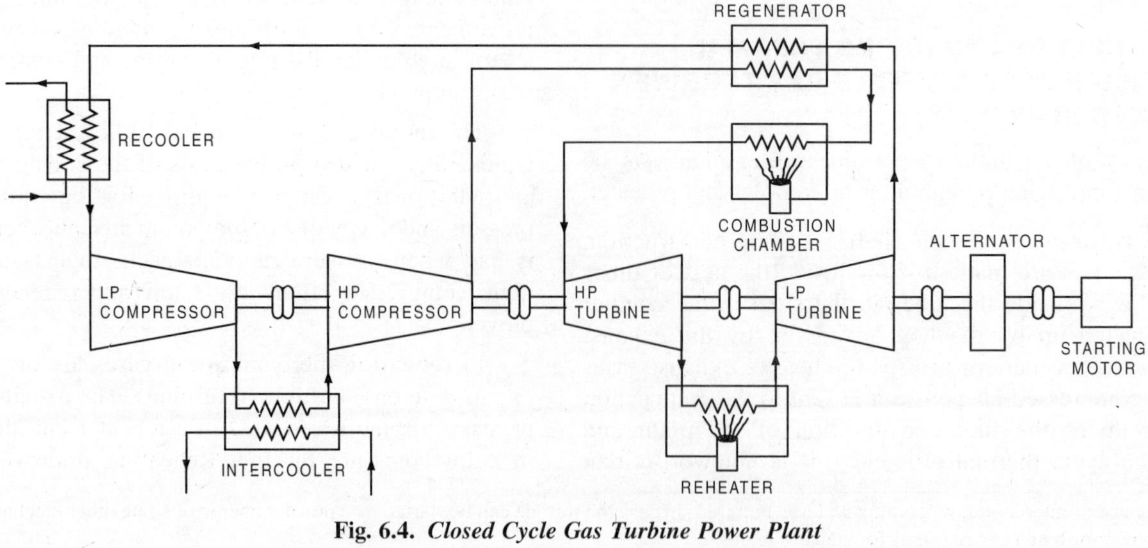

Fig. 6.4. *Closed Cycle Gas Turbine Power Plant*

4. A working medium with physical properties superior to those of air (such as helium, argon, hydrogen, neon) may be used. The heat conductivity of hydrogen is about 6.8 times that of air, so with the use of hydrogen as working medium, the size of heat exchangers is reduced.
5. There is an improvement in the rate of heat transmission.
6. There is a reduction in fluid friction loss due to higher Reynold's number.
7. These plants can be operated with highest efficiencies in comparison to open cycle plants at an equal initial temperature of working fluid.
8. The power output at constant speed can be varied by adding or subtracting the working fluid and thus altering the weight of the charge. This gives improved part-load efficiency as compared to open cycle gas turbine.

However, all the advantages of closed cycle operation are offset by the additional complexity of the plant which include externally fired heater, gas recooler and various heat exchangers.

If the working medium is other than air, it is necessary that the system be gas tight. This adds to the cost and increases the engineering problems.

Closed cycle gas turbine plant would be preferable over open cycle one only when inferior type of fuel or solid fuel is to be used and ample cooling water at site is available. Closed cycle plants have not as yet been employed for generation of electrical power because of requirement of large size heat exchangers.

Some gas plants have been designed with a combination of open and closed cycles. Such plants are known as semi-closed cycle plants.

6.9. OPERATION AND CONTROL OF GAS TURBINE POWER PLANT

Gas turbines are not self starting and do not become self sustaining until they attain around 50% speed and therefore, need means to crank the unit to firing speed and to assist acceleration after firing. Temperature control is also required to be provided to limit the maximum temperature in order to prevent damage to the turbine. Turbines that normally operate on bunker C oil must be started on diesel oil and transferred to bunker C at or above idle speed.

The starting and stopping of gas turbine power plant needs a certain procedure, which has to be properly controlled in correct sequence. This can be accomplished automatically. Operation of master switch initiates the starting or stopping sequence, which then proceeds automatically through the various steps required.

In starting of gas turbines the following steps are required:

1. Cranking of unit to around 10% speed to establish air flow through the combustion chamber.
2. Admitting correct amount of fuel, applying ignition and booster air to initiate combustion.
3. Accelerating of unit after firing assisted by the starting motor.
4. Disengagement of starting motor on attainment of self sustained speed by the gas turbine.
5. Transfer of turbine from diesel oil to bunker C.

Stopping of gas turbine requires the following steps:

1. Transfer of turbine from bunker C to diesel oil and operation on diesel oil for sufficient time in order to remove the heavy bunker C from the fuel pump, fuel lines, and fuel nozzles.
2. Stoppage of the supply of fuel to the combustion chambers.
3. Rotating of unit after shutdown to allow proper cooling of various parts.

In case of an open cycle gas turbine power plant the regulation is obtained by controlling the flow of fuel to the combustion chamber. The governor is located on the output shaft and regulates a valve in the fuel supply line by its action. In a closed cycle gas turbine power plant, the load is controlled by controlling the density of air in the closed circuit. This is obtained by passing air into or out of the circuit from reservoirs under pressure. The flow of fuel is also regulated simultaneously with the air density.

6.10. COMBINED STEAM AND GAS TURBINE POWER PLANTS

As already mentioned, the gas turbine power plants are mainly used for supplying peak loads in other types of power plants *e.g.,* in steam and hydroelectric power plants. The heat content of gas turbine exhaust is quite substantial. The temperature of gas turbine exhaust is about 540ºC and the oxygen content in this exhaust is roughly 16%. Essentially a combined gas turbine-steam turbine cycle aims at improving the overall plant efficiency by using the heat of exhaust gases from the gas turbine as a heat source for a steam plant cycle. There are three arrangements of combined cycles usually employed as given below.

1. Use of Exhaust Gases of Gas Turbine Power Plant For Heating of Feed Water. The heat energy of exhaust gases from the gas turbine can be utilized in heat exchanger for raising the temperature of the feed water coming from the steam turbine condenser, before it is delivered to the steam boiler, as illustrated in Fig. 6.5. In this arrangement, bleeding of steam from the steam turbine required for heating of feed water is avoided and the full steam supply to the steam turbine is available for expansion and producing mechanical power.

2. Use of Exhaust Gases From Gas Turbine As Combustion Air in Steam Boiler. In this arrangement the gas turbine exhaust is used as preheated air for the boiler of the steam plant, as illustrated in Fig. 6.6. The gas turbine exhaust has about 16% oxygen which is enough to support combustion in the boiler. The boiler can be supplied with supplementary fuel and air and would be larger than the conventional boiler. Plant heat rate can be raised by about 5% by using combined cycles.

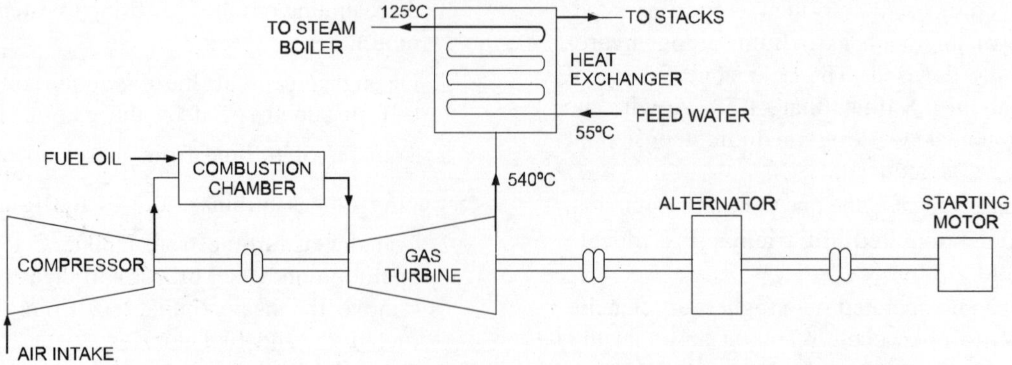

Fig. 6.5. *Use of Exhaust Gases From Gas Turbine for Heating of Feed Water*

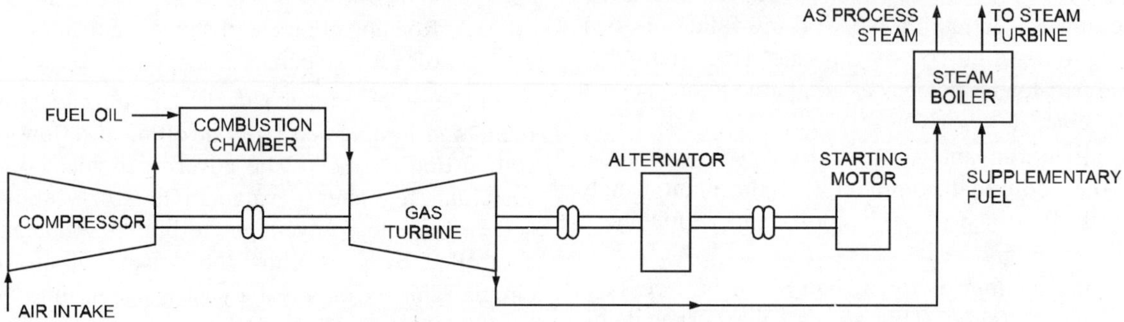

Fig. 6.6. *Use of Exhaust Gases From Gas Turbine as Combustion Air in Steam Boiler*

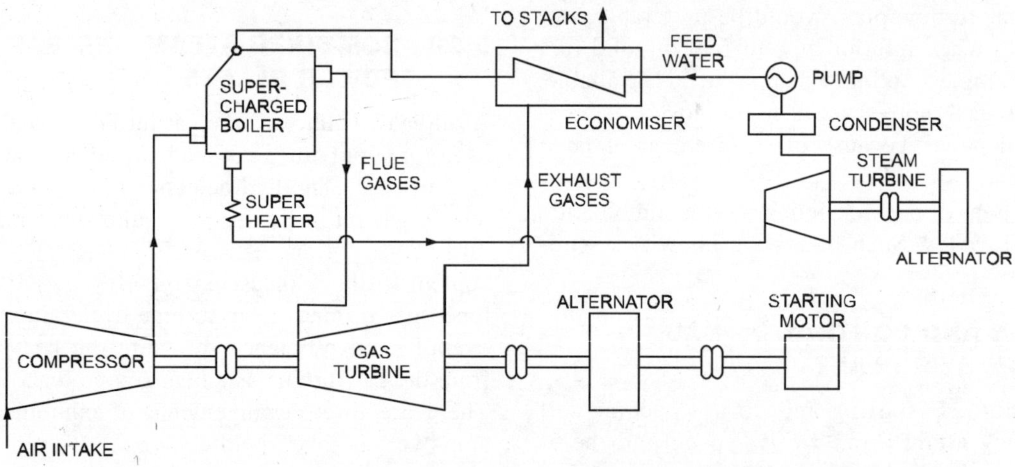

Fig. 6.7. *Use of Gases From a Supercharged Boiler For Expansion in Gas Turbine*

3. Use of Gases From a Supercharged Boiler For Expansion in Gas Turbine. In this arrangement the high pressure air from the compressor is supplied to the steam boiler and the exhaust gases from the steam boiler (at about 700°C) are expanded in the gas turbine and then through an economiser to heat the feed water before leaving out through stack, as illustrated in Fig. 6.7. The heat transfer rates in the steam boiler are increased due to the high density of air and therefore the weight of the boiler gets reduced by about 50%. Heat rate also gets improved by 7–8%. The capacity of the power plant is also increased and there is only a slight increase in the requirements of cooling water.

The advantages of combined cycle operation are enumerated as below:

1. Saving in exhaust heat of the gas turbine and, therefore, increase in its heat rate.
2. Reduction of stack emissions.
3. Reduction in space requirement in comparison with conventional generating units of a given capacity.
4. Reduction in requirements of condensing water by 60% as compared to a fossil fuelled plant of given capacity.
5. Reduction in starting time. The gas turbine can be brought to full load within 20 minutes from initial start giving around 60% of total plant output and subsequently the steam-turbo alternator can be synchronised and brought to full load giving 100% of plant output in less than an hour following an overnight shutdown.

The efficiency of a steam cycle can be increased by about 2%, with a combined cycle arrangement. Although there will be some loss of efficiency of the gas turbine owing to the back pressure created by the boiler furnace, there is still a useful overall improvement in cycle efficiency.

Capital costs per kW of these combined cycle plants are about 40% lower than those of conventional steam power plants.

6.11. PLANT LAYOUT

The plant layout of a gas turbine power plant has a significant effect on the performance of the plant and great care has therefore to be taken in the design and layout of the air as well as gas circuits, otherwise there will be unnecessary wastage of power in interconnecting ducts with a large number of sharp bends.

In a gas turbine power plant, turbine house accommodating the major portion of the plant and the auxiliaries is the main building.

The fuel oil storage tanks of capacity sufficient to operate the plant for about 100 hours are usually arranged outside but adjoining the turbine house. In some installations even heat exchangers are placed outside the main building. Intercoolers, combustion chambers, heat exchangers, waste heat boilers and interconnecting duct work occupies the major portion of the building. The rail siding is also provided for easy transport of fuel oil.

6.12. COMPARISON OF GAS TURBINE POWER PLANT WITH DIESEL ENGINE AND STEAM POWER PLANTS

6.12.1. Comparison With Diesel Engine Power Plants

The gas turbine power plants have numerous inherent advantages over diesel power plants such as smaller site area, lower cost of buildings, low initial cost, better heat rate, less water requirement, easier maintenance, reduced attendance charges, improved reliability, less vibrations, no cyclic variations etc. There are no problems of fuel injection and cylinder lubrication as with diesel engines. Possibilities of use of lower grades of fuel oil or low grade waste gases with gas turbines are also there.

The only drawback of gas turbine power plant in comparison to diesel power plants is higher fuel consumption (0.2 – 0.25 kg per kWh output) which is insignificant against their inherent advantages, mentioned above.

6.12.2. Comparison With Steam Power Plants

Advantages of gas turbine power plants over steam power plants are lower capital cost, smaller site area, simple and cheap foundations and buildings, smaller installation period, fewer auxiliaries, no standby losses, lesser water requirement, short starting time and quick load sharing, simple control and possibility of partial or full automation, lesser operational personnel (about one-third), possibility of location near the load centre (even in cities and towns) etc. Storage of fuel is

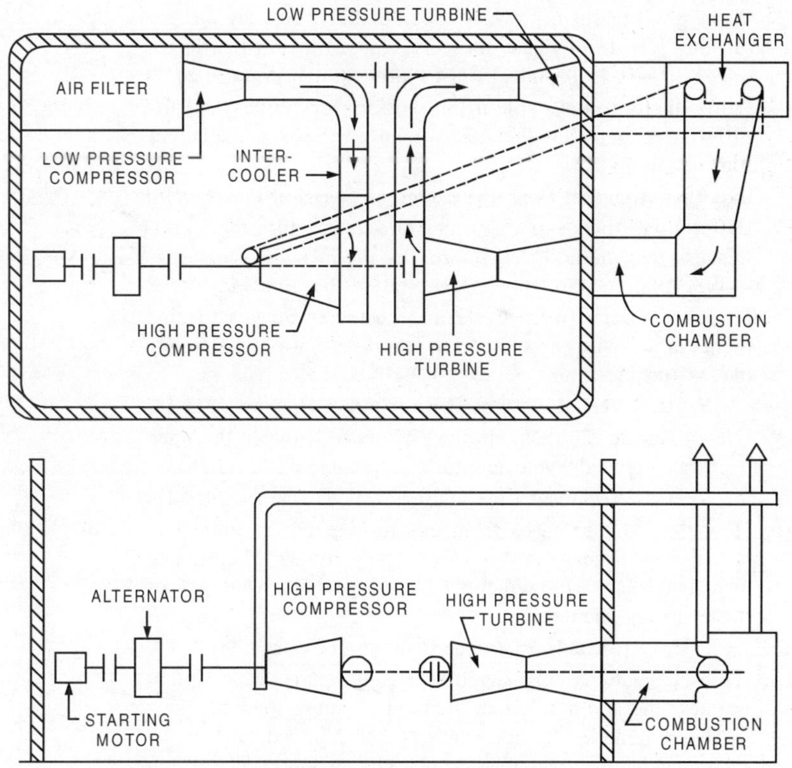

Fig. 6.8. *Plan Layout of a Gas Turbine Power Plant*

much smaller and its handling is easy and as such problems of coal and ash handling are completely eliminated in case of open cycle gas turbine power plants using gas or liquid fuel. Moreover, the heat rate of a gas turbine is usually higher than that of a steam turbine. The efficiency of a gas turbine plant is almost the same as that of a steam power plant using liquid fuel.

The drawbacks of gas turbine power plants over steam power plants are:

1. The operating costs of a gas turbine power plants are much higher than those of a steam power plant.
2. Gas turbines can be employed only in small sizes (about 50 MW or so).
3. Like steam turbines, the gas turbine is not readily reversible.
4. In case of a gas turbine power plant, the compressor has to be started before starting of turbine and the former requires power from some external source.

Exercises

1. Discuss the advantages and disadvantages of a gas turbine plant. **[U.P. Technical Univ. Power Station Practice 2006-07]**
2. Discuss the operation principle of gas turbine plant. **[U.P. Technical Univ. Power Station Practice 2008-09]**
3. With the help of a suitable diagram explain the working of a gas turbine plant. What are its merits and demerits ? **[U.P. Technical Univ. Power Station Practice 2009-10]**

4. Discuss in brief the methods of improving thermal efficiency of gas turbine power plants.
[U.P. Technical Univ. Power Station Practice 2009-10]
5. What are the factors which decide the efficiency of gas turbine ? How to make these turbines, the most efficient one ? **[U.P. Technical Univ. Power Station Practice 2007]**
6. Draw diagram and explain the working principles and the differences between the open cycle and closed cycle gas turbine plants.
[Rajasthan Technical Univ. Generation of Electrical Power 2010]
7. Differentiate between open and closed gas turbine cycles. Discuss two methods for improving the efficiency of a gas turbine plant. **[Pb. Univ. Power Station Practice, May 1995]**
8. Discuss in detail open cycle and closed cycle gas turbine plants along with the method to improve thermal efficiency of gas turbine plant.
[G.B. Technical Univ. Power Station Practice 2011-12]
9. Draw diagrams and explain the difference between the open cycle and closed cycle gas turbine plants.
[G.B. Technical Univ. Power Station Practice 2010-11]
10. How the exhaust gases from gas turbine power plant can be used in a steam power plant, show by line diagram at least two such arrangements ? Also discuss advantages of these arrangements.
[M.D. Univ. Power Plant Engineering May 2009]
11. A power plant comprising of gas and steam turbine plant is operating for a category of load. A huge load block is suddenly added. Mention the way the load added will be shared and more over which unit will respond quickly and why ? **[U.P. Technical Univ. Power Station Practice 2007]**
12. Explain the working of gas turbine plants. What are the different types of gas turbine plants ? Discuss the layout of any one of them.
[U.P. Technical Univ. Power Station Practice 2006-07]
13. Compare the diesel power plants and gas turbine plants of the same capacity.
[Pb. Univ. Power Station Practice, October 1991]
14. What are the comparative benefits associated with gas power plants to that of thermal power plants ?
What do you understand by open cycle and closed cycle gas turbine power plants ?
[U.P. Technical Univ. Power Station Practice 2005-06]
15. Explain the working of a gas turbine plant. What are its merits and demerits compared to the steam power plants ?
[U.P. Technical Univ. Power Station Practice 2010-11]
16. Write short notes on the following :
(*i*) Merits and demerits of gas turbine power plants.
(*ii*) Applications of gas turbine power plants.
(*iii*) Gas turbine fuels.
(*iv*) Open cycle gas turbine power plants.
(*v*) Methods of improving thermal efficiency of gas turbine power plants.
(*vi*) Combination gas turbine power plants.

Short Answer Type Questions With Answers

Q. 1. What for a starting motor is required in a gas turbine power plant ?

Ans. An electric motor is required for starting the compressor, which is required to be started before starting of turbine.

Q. 2. Why are majority of gas turbine power plants employed as peak load plants ?

Ans. Gas turbine power plants are usually employed as peak load plants because these plants can be started and loaded quickly and their initial costs are small but fuel costs are high.

Q. 3. Why coal and peat are not used as fuels in gas turbine power plants ?

Ans. Coal and peat are not used as fuels in gas turbine power plants because of coal handling and ash handling problems.

Q. 4. Why is it advantageous to use high pressure turbine for driving compressor and low pressure turbine for driving alternator in gas turbine power plants ?

Ans. High pressure turbine is used for driving compressor and a low pressure turbine is used in driving the alternator (Fig. 6.2) because this arrangement has the advantage that the speed of the power turbine (the turbine driving the alternator) can be kept constant at synchronous speed while the speed of the turbine driving the compressor can be varied according to the output required.

Q. 5. Why the efficiency of open cycle gas turbine power plant is quite low ?

Ans. The overall efficiency of the *open cycle plants is very low* as a greater part (around 65%) of the power developed by the turbine is used in driving the compressor. The fact that the hot gases from the combustion chamber are cooled to a safe temperature before admitting it to the turbine, also reduces the efficiency. The overall efficiency of the plant, owing to above reasons seldom exceeds 20%.

Q. 6. How can the thermal efficiency of a simple gas turbine power plant be improved ?

Ans. The efficiency of a simple gas turbine power plant can be improved by employing regenerator, intercooler and reheater.

The regenerator utilizes the heat of exhaust gases to heat the compressed air before it is sent to the combustion chamber, reduces the fuel consumption of the plant and improves the thermal efficiency.

The intercooler is heat exchanger which cools the partly compressed air in order to reduce volume and increase density. The intercooling results in improvement of thermal efficiency.

Reheating improves the output from the turbines due to multiple heating in the same way as intercooling improves the performance of the compressor.

Q. 7. How does the use of regenerator, intercooler and reheater in gas turbine power plants help in improvement of thermal efficiency ?

Ans. Refer to Answer of Q. 6.

Q. 8. In what respects gas turbine power plants are inferior to steam power plants ?

Ans. The drawbacks of gas turbine power plants over steam power plants are:

1. The operating costs of a gas turbine power plants are much higher than those of a steam power plant.
2. Gas turbines can be employed only in small sizes (about 50 MW or so).
3. Like steam turbines, the gas turbine is not readily reversible.
4. In case of a gas turbine power plant, the compressor has to be started before starting of turbine and the former requires power from some external source.

❑❑❑

CHAPTER 7

Non-Conventional Methods of Power Generation

INSIDE THIS CHAPTER

7.1. INTRODUCTION

The electricity requirements of the world including India are increasing at alarming rate and the power demand has been running ahead of supply. It is also now widely recognised that the fossil fuels (*i.e.*, coal, petroleum and natural gas) and other conventional resources, presently being used for generation of electrical energy, may not be either sufficient or suitable to keep pace with ever increasing demand of the electrical energy of the world. Also generation of electrical power by coal based steam power plants or nuclear power plants causes pollution, which is likely to be more acute in future due to large generating capacity on one side and greater awareness of the people in this respect. The recent severe energy crisis has forced the world to develop new and alternative methods of power generation, which could not be adopted so far due to various reasons. The best alternative of conventional energy sources are the *non-conventional energy sources*, also called the *renewable sources of energy,* which are continuously replenished by natural process. Solar energy, wind energy, bioenergy (biofuels grown sustainably), hydropower, etc. are some of the renewable energy sources.

A renewable energy system converts the energy of the sunlight, wind, falling water, sea waves, geothermal heat or biomass into a form, which can be used such as heat or electricity. Most of the renewable energy comes either directly or indirectly from sun and wind which can never be exhausted and therefore they are called renewable. Renewable energy sources are essentially flow of energy, whereas the fossil and nuclear fuels are essentially stock of energy. Various forms of renewable energy are: (1) solar energy (2) wind energy (3) hydroenergy (4) bioenergy (5) geothermal energy (6) wave and tidal energy.

In 2008, approximately 19% of global final energy consumption came from renewable energy sources with 13% coming from traditional biomass, which is mainly used for heat, and 3.2% from hydroelectricity. New renewable (small hydro, modern biomass, wind, solar, geothermal and biofuels) accounted for another 2.7% and are growing very fast. The share of renewable energy sources in electrical energy generation is about 18%, with 15% of global electricity coming from hydroelectricity and 3% from new renewable energy sources. Wind power is growing at the rate of 30%, annually, with a worldwide installed capacity of 158,000 MW in 2009, and is widely used in Europe, Asia and the United States. At the end of 2009, cumulative global photovoltaic (PV) installation surpassed 21,000 MW and PV power stations are popular in Germany and Spain. Solar thermal power stations are operating in Spain and USA, and the largest of these is the 354 MW SEGS power plant in the Mojave Desert. The world's largest geothermal power installation is The Geysers in California, with a rated capacity of 750 MW.

The major sources of energy in the world are oil, coal, natural gas, hydroenergy, nuclear energy, renewable combustible wastes (animal products, biomass and industrial wastes) and other energy sources. The contribution of different energy sources to the supply of energy in world are:

Oil	: 35.1%
Coal	: 23.5%
Natural gas	: 20.7%
Renewable combustible wastes	: 11.1%
Nuclear	: 6.8%
Hydro	: 2.3%
Other sources	: 0.5%

Oil is the most significant and abundant source of energy in the world but it is also the most highly consumed. However, crude oil price is very volatile and supply is driven by price. *Coal* is second abundant source of energy in the world and is highly used in electrical power generation. Coal ranks very low in terms of consumption, however, its consumption has increased in certain regions. A concern for many developing countries has been the poor quality of coal. *Natural gas* has been the energy source with highest rates of growth in recent years. The high end-use efficiency of natural gas has made it a popular choice for electrical power generation projects. Renewable energy sources are gaining popularity.

Present Energy Scenario in India: India is said to be one of the seven largest consumer of energy, but the growing gap between the demand and generation is a cause of concern. India's share in global oil reserves is approximately 0.5 per cent, whereas its share in global consumption is approximately 3 per cent. India is still dependent to the extent of 30 to 35 per cent on non-commercial fuel sources like cowdung, firewood, agricultural waste etc. The growing energy requirements of the emerging economics, specifically India, risks enhanced environmental damage from conventional carbon based sources of energy. The pressure on petrol is mounting and we have to concentrate on conservation of petroleum. Towards conservation of petroleum consumption Government has to improve power generation, focus on alternative sources of energy such as solar, wind and biofuels, set up standards for all vehicles and a mass awareness for conservation.

Power and energy sector is in a positive mood and is leaving no missed opportunity to make half of it, while the sun shines. India has set up a target of 20,000 MW of installed capacity by 2022 for harnessing solar technology. It is leaving no stone unturned to become a solar hub in the world.

Renewable energy technologies like solar, biomass, hydro etc. are deployed both in rural and urban areas to curb the growing gap between the demand and supply of power, which is due to increase in the per capita consumption and significantly, the much hyped climate change concerns. At 10,464 MW, India presently ranks fifth in the world in wind power generation.

The future of solar photovoltaic development in India seems to be very bright. India's solar emission envisages the promotion of solar energy to harness and distribute environment-friendly power available with high scalability, for sustainable economic growth by empowering natural energy security.

Indian clean development mechanism projects broadly cover a range of sectors viz., power generation from renewable energy, particularly wind and hydropower, biomass applications, waste heat and energy recycling. Accelerated growth is expected in renewable energy sector, particularly solar energy sector, wind energy sector, biofuels sector etc., with favourable conditions in terms of potential, technical support facilities, policy framework and regulatory environment, robust manufacturing base, and investors confidence in the country.

7.2. MAGNETO HYDRODYNAMIC (MHD) POWER GENERATION

The magneto hydrodynamic (MHD) power generation is one of the examples of a new unique method of power generation and provides a way of generating electrical energy directly from a fast moving stream of ionized gases without the need for any moving mechanical parts—no turbines and no rotary generators.

The basic principle of MHD generation is the same as that of a conventional electrical generator *i.e.,* the motion of a conductor through a magnetic field induces an emf in it—called the *Faraday's law* of electromagnetic induction. In convnetional steam power plants, the heat released by combustion of fuel is transformed into internal energy of steam. The steam turbine, then, converts steam energy into mechanical energy used in driving a generator. Thus, the mechanical energy is converted into electrical energy. The repeated conversion of various forms of energy involves losses and, therefore, the overall efficiency of thermal power plants is inherently very low. In MHD technology, electrical energy is directly generated from the hot gases produced by the combustion of fuel without mechanical moving parts.

In an MHD generator, electrically conducting gas at a very high temperature is passed at very high velocity through a strong magnetic field at right angle to the direction of its flow, thereby generating electrical energy. The electrical energy is then collected from stationary electrodes placed on the opposite sides of channel (Fig. 7.1). The current so obtained is direct current which can be converted into an ac by an inverter.

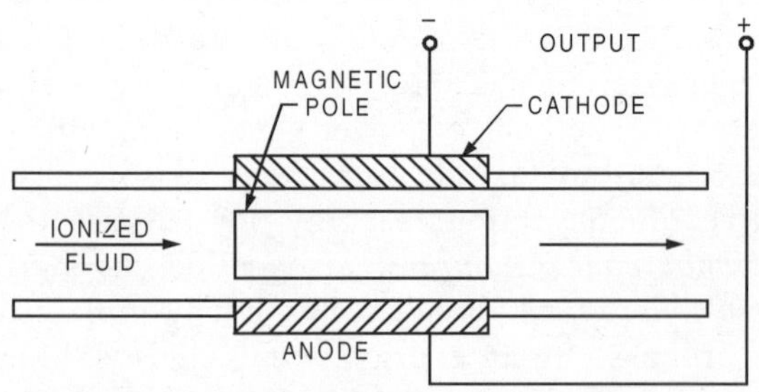

Fig. 7.1. ***Principle of MHD Power Generation***

Ionized gas can be produced by heating it to a high temperature. On heating of a gas, the outer electrons escape out from its atoms or molecules. The particles acquire an electric charge and the gases passes into the state of plasma. However, to achieve thermal ionization of products of combustion of fossil fuels or inert gases, extremely high temperatures are necessary. Air becomes highly ionized at temperature of 5,000° to 6,000°C. To have a reasonable value of electrical conductivity of gases at temperatures around 2,000 to 3,000 K by reasonable ionization, the gases are seeded with additives of easily ionizing materials (alkali metals) such as cesium or potassium.

MHD system may be an *open cycle system* or a *closed cycle system*. In an open cycle system, the working fluid after doing useful work (generating electrical energy) is discharged to the atmosphere through a stack while in a closed cycle system the working fluid is recycled to the heat source and thus used again and again. The operation of MHD generators directly on combustion products is an open cycle system using air as working fluid. In closed cycle systems gases used on the working fluid are helium or argon.

The use of a nuclear reactor employing solid fuel elements to supply heat energy for an MHD process needs that the working fluid should have the following properties.

(*i*) It should be capable of providing heat transfer under reactor working conditions.

(*ii*) It should not require excessive compressor work.

(*iii*) It should not be rendered active within the reactor.

Almost all of the above three requirements are fulfilled by helium.

1. Open Cycle MHD System. Figure 7.2 shows an elementary open cycle MHD system in which a high pressure, high temperature combustion gas is forced through a strong magnetic field.

Coal is processed and burnt in the *combustor* at a high temperature of about 2,600°C and pressure of about 12 atmosphere with preheated air to form the *plasma*. Then a seeding material, such as potassium carbonate, is injected to the plasma in order to increase the electrical conductivity. The resulting mixture having an electrical conductivity of about 10 siemens/m, is expanded through a nozzle, so as to have high velocity, and then passed through the strong magnetic field (5-7 T) of the MHD generator. During the expansion of gas at high temperature, the positive and negative ions move to the electrodes and so constitute an electric current. This current is dc and an inverter is employed for its conversion into ac. The gas leaving the MHD generator is still very hot. The heat from the exhaust gases of the MHD generator is utilised in preheating the air supplied to the combustor. The seed material is recovered from the gas for successive use and harmful emissions (such as nitrogen and sulphur) are removed from the gas, for pollution control, and the gas is finally discharged to the atmosphere through a stack.

The open cycle MHD system explained above is not suitable for commercial use. For making this process efficient, it is necessary to combine the MHD unit with steam turbine-alternator unit. In this system the heat from the exhaust gases of the MHD generator is used to raise steam which generates additional energy in a steam turbine-alternator units. A part of this steam is also used in a steam turbine driving a compressor for compressing air for the MHD cycle. Such a cycle is called the *hybrid, binary or topping* MHD steam plant cycle.

The electrodes are usually made of graphite and the duct of boron nitride.

Any type of fossil fuel (coal, oil, natural gas) can be used in MHD generator but a direct coal fired MHD generator has the following advantages.

(*i*) Slag from coal combustion coats the generator electrodes and protects from electrical and mechanical corrosion that otherwise occurs in a clean fuel generator. However, the electrodes may be short circuited by molten ash.

(*ii*) Coal contains less hydrogen and, therefore, the sink for electrons in the flow created by the presence of OH ions is reduced.

Char, having almost no hydrogen, is better than coal even. It is easier to handle and feed in comparison to coal. It results in a 25% increase in the performance of the generator.

Power generated by the MHD system is given as

$$\rho = \sigma B^2 v^2 K (1 - K) \text{ W/m}^3 \quad ...(7.1)$$

where σ is the specific electrical conductivity of gas in siemen/metre, B is magnetic field strength in Tesla (Wb/m^2), v is the velocity of gas in m/s and K is the ratio of external load voltage to open-circuit voltage.

2. Closed Cycle MHD System. As the name suggests, the working fluid, in a closed cycle, is circulated in a closed loop. The closed cycle MHD system may be either a *plasma converter* or a *liquid metal converter*. The plasma converter uses an ionized gas (helium or argon seeded with cesium) and the liquid metal converter uses the vapour of the metal or the metal in a liquid form (the metal may be an alkali or some other metal).

Liquid metal system has the basic advantage of high electrical conductivity. So high temperatures are not required in this system to achieve the high electrical conductivity and the systems are normally designed to operate at temperatures below 1,400 K (a much lower temperature in comparison to plasma converters). This temperature is low enough that the energy can be supplied by a nuclear reactor or fossil fueled system.

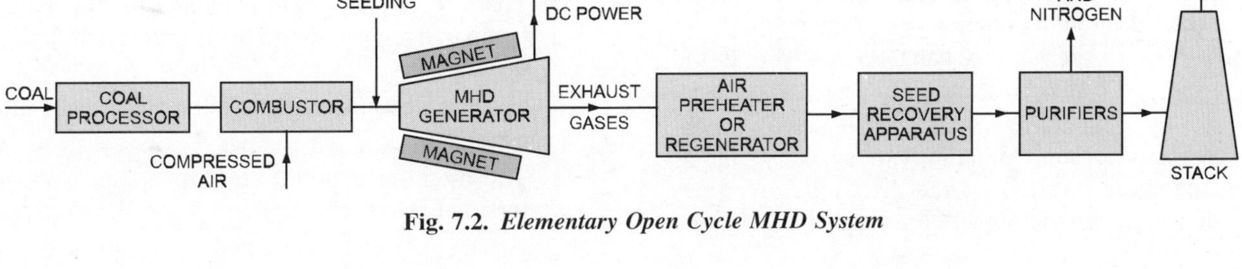

Fig. 7.2. ***Elementary Open Cycle MHD System***

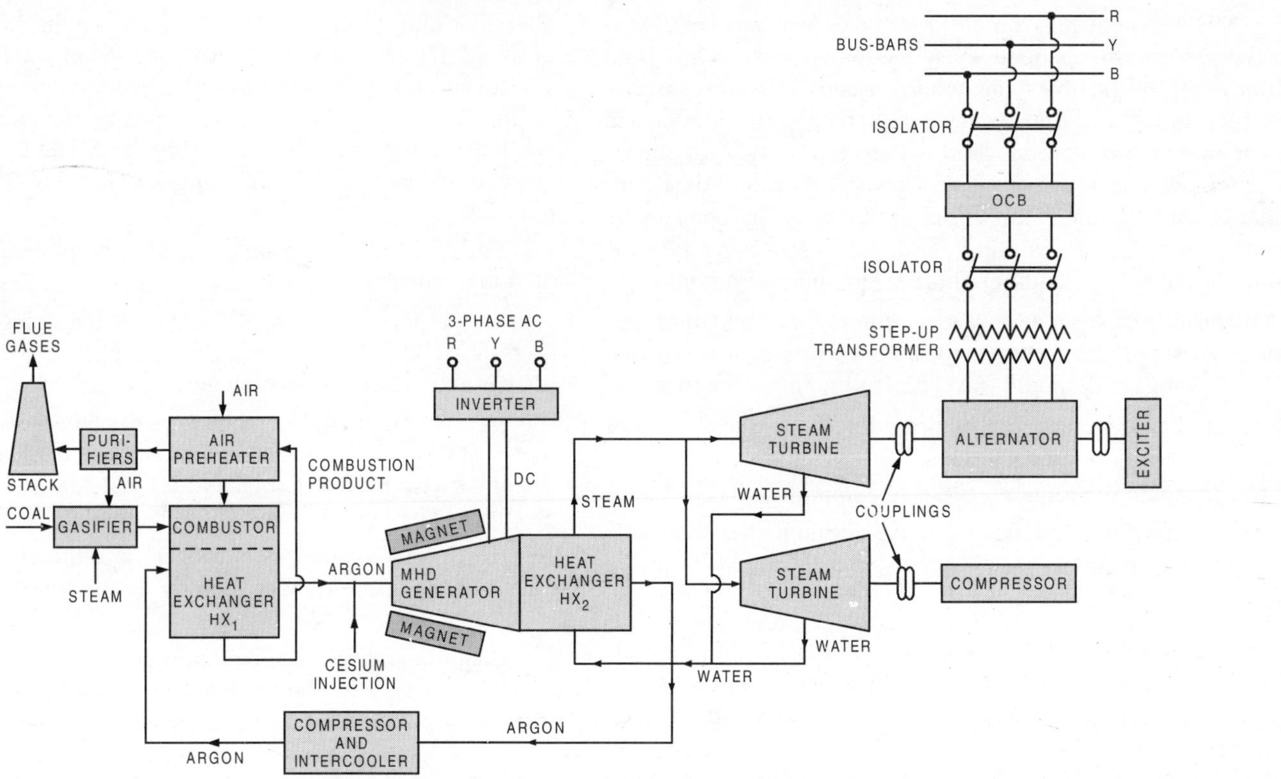

Fig. 7.3. *Closed Cycle Plasma MHD System*

Closed cycle plasma MHD system is shown in Fig. 7.3. The complete system can be divided into three distinct but interlocking loops.

Coal is gasified and the gas so produced has heat value of about 5.35 MJ/kg and a temperature of about 520°C. This gas is burnt in a combustor to generate heat. This heat is transferred to the working fluid (argon) of the MHD cycle in a heat exchanger (HX_1). The combustion products are passed through the air preheater (for recovery of a part of heat of combustion products) and purifiers and then discharged to the atmosphere. This is loop I and may be called an external heating loop.

The hot argon gas is seeded with cesium and passed through the MHD generator, that produces direct current (*dc*). This dc output is converted into ac by means of inverter and then supplied to the grid. This is the loop II and may be called an MHD loop.

For further recovery of heat from the working fluid and its use in generating of steam, the working fluid is slowed down, in a diffuser, to a low subsonic speed and then this fluid is passed through the heat exchanger HX_2. In heat exchanger HX_2, the fluid imparts it heat to water and so generates steam. This steam is used partly for driving a steam turbine operating the compressor and partly expanded in a steam turbine driving a three-phase alternator. The working fluid is then passed through compressor and intercooler and then returned back to the heat exchanger HX_1. This is the loop III and may be called a steam loop.

The closed cycle liquid metal MHD system is shown in Fig. 7.4. The superheated metallic vapour is expanded through a supersonic nozzle into the drift tube or mixer. Atomized subcooled liquid droplets are accelerated by the vapour. The vapours also condense on the liquid droplets so that the fluid entering the MHD generator is essentially a liquid. The resulting velocity of the fluid is more than 150 metres per second.

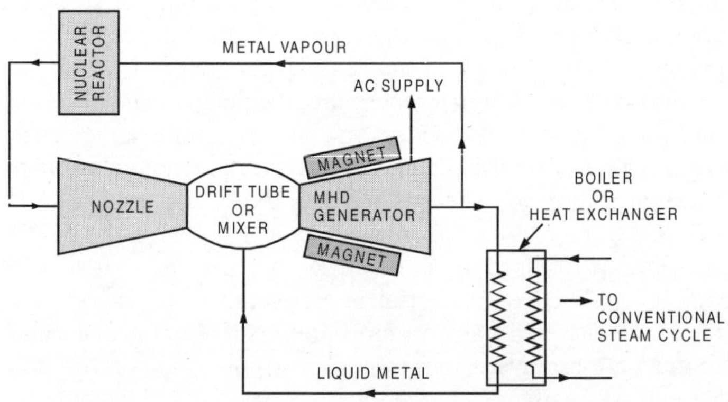

Fig. 7.4. *Closed Cycle Liquid Metal MHD System*

The advantages of liquid metal system are as follows :

1. This system can use nuclear energy as high temperature is not the requirement of this system as in case of the plasma converter.
2. It can easily provide ac power supply directly, while it is almost impossible to do so in a plasma system.
3. The size of the system including that of magnets is comparatively smaller. This is because of high power density.

However, the liquid metal system has the following limitations :

1. The metallic vapours are poor electrical conductors.

2. High velocities cannot be obtained by expansion in this system while it is much easier to achieve a high fluid velocity employing a gas and a nozzle. This is because the liquids are practically incompressible.
3. The overall conversion efficiencies obtainable with liquid metal system are quite below to that of plasma system.

A closed cycle system can provide more useful power conversion at lower temperatures of about 1,600°C but this system has not taken practical shape so far. The difficulties with such a system are the design of heat exchanger (the heat exchanger operates up to the highest temperature of the gas), requirement of absolute purity of the working fluid, the problems posed by electrical stability of the flow in the generator (the gas is subject to electric fields approaching breakdown conditions) etc.

Advantages of MHD Power Generation. MHD power generation offers several advantages over other conventional methods of power generation. Some of these are given below.

(*i*) Since high temperatures are involved, operation efficiency is high. MHD system is normally designed to be a topping power system to a conventional steam power plant. At present, the conversion efficiency of an MHD system is around 50% which can be increased to 60% with the improvements in experience and technology.

(*ii*) No moving part, so more reliable.

(*iii*) Conceptually such generators are much simpler.

(*iv*) As there is no limitation to the size of the duct, so high capacity generators are possible.

(*v*) The walls can be cooled below temperature of working gas.

(*vi*) Direct conversion of heat into electrical energy results in elimination of the gas turbine (compared with a gas turbine power plant) and both the boiler and the turbine (compared with a conventional steam power plant) and thus in reduction of energy losses.

(*vii*) Ability of reaching the full power level instantly.

(*viii*) The more efficient heat utilisation reduces the amount of heat discharged to environments and thus the cooling water requirements are reduced.

(*ix*) The MHD process is industrially attractive because of the reduced cooling water requirements and atmospheric pollution.

(*x*) MHD power generation process is applicable to all kinds of heat sources such as oil, coal, gas, nuclear, solar and thermonuclear fusion.

(*xi*) MHD power generation offers the flexibility of operation in different modes such as base load, peak load or semi-peak load.

(*xii*) The capital costs of the MHD plants are estimated to be competitive with those of coal fired steam power plants.

(*xiii*) The overall costs of the MHD power generation are also estimated to be lower (roughly 20%) than those of conventional power plants. This is because of higher efficiency of MHD power generation.

(*xiv*) The reduced fuel consumption, that is obtained because of higher efficiency or better fuel utilisation, offer additional economic and social benefits and also lead to conservation of energy sources.

Inspite of numerous inherent advantages the MHD system has not been accepted commercially because of the following limitations:

1. The efficiencies attained so far have been relatively low.
2. The power output of MHD generator is proportional to the square of the magnetic field density. The electromagnets need very large power for creating strong magnetic fields. The MHD technology is waiting for development of superconducting materials which will need very little power even at ambient temperatures.
3. The combustor, MHD duct, electrodes, and air preheaters are exposed to very corrosive combustion gases at very high temperatures. So, the life of these equipments has been reduced.
4. The ash (or slag) residue from the burning coal is carried over with the combustion gases and tends to cause erosion of exposed surfaces. However, deposition of the slag on such surfaces may also provide some protection.
5. There is a serious problem of separation of seed material from the fly ash and reconversion of potassium sulphate to potassium carbonate.
6. Special fuel gas and preheating of air are required to provide adequate working fluid temperatures.
7. There are serious problems associated with the fabrication of MHD duct, high temperature and high pressure heat exchangers and reactors.

Development of MHD programmes has been undertaken by different countries during the last two decades. In India also considerable studies have been carried out in this field by a team of scientists under the National Council of Science and Technology (NCST). The Department of Science and Technology of Govt. of India has sponsored research and development programmes on coal based MHD power generation. Bhabha Atomic Research Centre in collaboration with Bharat Heavy Electricals Ltd and Institute of High Temperature (USSR) is also executing Research and Development programmes in this field.

The specifications of a Japanese pilot plant (Tokyo) are given below:

Thermal input	: 24 MW
Electrical power output	: 1 MW
Mass flow of gas	: 2.8 kg/s
Flow velocity	: 900 m/s
Gas temperature	: 2,900 K
Duct length	: 1.2 m
Inlet	: 8 × 10 cm

Outlet	: 8 × 25 cm
Electrodes	: 30 pairs
Inlet pressure	: 4 bar
Magnetic flux density	: 3.5 T (Wb/m^2)
Materials for electrodes	: Graphite, water cooled copper

Russia has been the pioneer of this new technology. The world's first U-02 MHD unit was designed by the Russian scientists and power engineers in 1964.

Under intergovernmental agreement, Indian scientists from BARC and power engineers from BHEL worked on a pilot plant of MHD of 5 MW capacity with close interaction with the scientists of the High Temperature Science Institute of Moscow. The pilot plant was set up at BHEL, Trichy. BARC has developed special magnet MHD duct and power tap-off system. BHEL was responsible for the development and installation of fuel gas generator, oxygen plant, water treatment plant, high temperature regeneration air heaters and valves, the main combustion chamber, nozzles as well as seed injection and seed recovery system.

The main specifications of the pilot plant are as below:

Thermal input: I stage	: 5 MW
II stage	: 15 MW
Gas temperature	: 2,600°C
Flux density	: 3 T (Wb/m^2)
Seeding agent	: Potassium carbonate
Electrodes	: Water cooled copper

Besides the use of MHD system for commercial electrical power generation, it has got other special uses. A major effort was made in USA to use MHD as the conversion system in a nuclear electrical system for space crafts. MHD conversion has also been considered for ship propulsion, airborne applications, hypersonic wind tunnel experiments and for many other defence applications.

Voltage and Power Output of MHD Generator. Lorenz law describing the effects of a charged particle moving in a constant magnetic field can be stated as

$$\mathbf{F} = \mathrm{Q}v\mathbf{B} \qquad ...(7.2)$$

where F is the force acting on the charged particle, Q is charge of particle, v is velocity of particle and B is magnetic field.

The force on a charged particle moving in an electric field as well as magnetic field will be given as

$$\mathbf{F} = \mathrm{Q}(\mathbf{E} + v\mathbf{B}) \qquad ...(7.3)$$

where F, E, v and B are vectors.

The velocity in above equation is vector sum of gas velocity v and particle drift velocity u, so force F may be given as

$$\mathbf{F} = \mathrm{Q}(\mathbf{E} + v\mathbf{B} + \boldsymbol{u} \times \mathbf{B}) \qquad ...(7.4)$$

$$= \mathrm{Q}(\mathbf{E}' + \boldsymbol{u} \times \mathbf{B}) \qquad ...(7.5) \quad \text{where } \mathbf{E}' = \mathbf{E} + v\mathbf{B}$$

Consider Fig. 7.5, the motion of gas is in x direction, magnetic field B is in y direction and force on the particle in z direction. A load resistance R_L is connected across the electrodes (plates P_1 and P_2). When a current I flows through the load resistance R_L, then electric field intensity between the electrodes is given as

$$\mathbf{E}_z = -\frac{V}{d} \qquad ...(7.6)$$ where d is the distance between the plates.

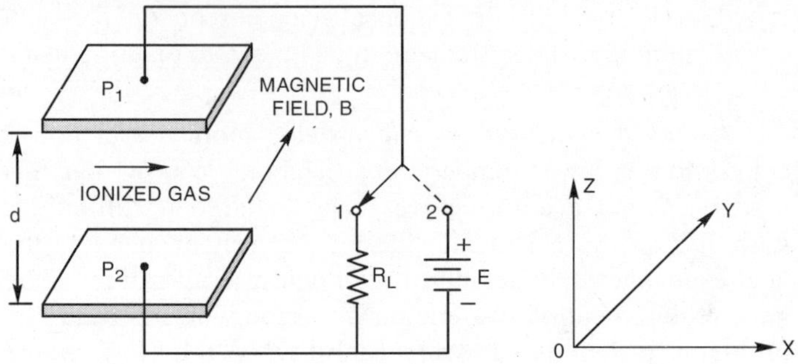

Fig. 7.5

Total electric field, $\mathbf{E}'_z = \mathbf{E}_z + \mathrm{B}v$

$$= -\frac{V}{d} + \mathrm{B}v = \frac{1}{d}(\mathrm{B}vd - \mathrm{V}) \qquad ...(7.7)$$

The electromagnetic field E_z and B acting on the moving gas, develops the same force on the ions as electromagnetic field E'_z and B develops on a gas with zero average velocity. Obviously the term Bvd provides the internal emf or open-circuit voltage E_0 of the MHD power generator *i.e.*,

$$E_0 = \mathrm{B}vd \qquad ...(7.8)$$

If R_G is the internal resistance of the generator, then maximum power output will be obtained when $R_G = R_L$.

Then, power output, $P_{out} = I^2R_L$

$$= \left(\frac{E_0}{R_G + R_L}\right)^2 R_L \quad \because \quad I = \frac{E_0}{R_G + R_L}$$

Substituting $R_L = R_G$ in above equation, we have

$$P_{out\,(max)} = \left(\frac{E_0}{R_G + R_G}\right)^2 R_G = \frac{E_0^2}{4R_G} \qquad ...(7.9)$$

The internal resistance of the generator, R_G is equal to $\frac{d}{\sigma A}$ where σ is the conductivity of the gas and A is the area of plates, then from Eqs. (7.8) and (7.9), we have

$$P_{out\,(max)} = \frac{B^2v^2d^2}{4d/\sigma A} = \frac{B^2v^2dA\sigma}{4} \qquad ...(7.10)$$

Maximum power per unit volume

$$= \frac{B^2v^2\sigma}{4} \qquad ...(7.11)$$

Substituting $E_0 = \mathrm{B}vd$ in Eq. (7.9), we have

$$P_{out\,(max)} = \frac{B^2v^2d^2}{4R_G} \qquad ...(7.12)$$

MHD system is reversible process. If load resistance R_L is replaced by emf E greater than E_0 by shifting the switch from position 1 to position 2, the directions of flow of current and force experienced on the ions will be reversed and the system will accelerate the gas particles because energy would be supplied to gas. Thus, the ejected gas would be at a higher velocity than the inlet gas. The reaction force experienced on the magnet would tend to push the MHD engine in the negative

x direction and thus the electrical energy would be converted into mechanical energy.

Example 7.1. Determine the open-circuit voltage and maximum power output of an MHD engine with following parameters:

Plate area = 0.2 m^2; Flux density = 2T (Wb/m^2)

Distance between plates = 0.4 m,

Average gas velocity = 1,200 m/s

Gas conductivity = 10 S/m.

Solution : Generator internal resistance,

$$R_G = \frac{d}{\sigma A} = \frac{0.4}{10 \times 0.2} = 0.2\ \Omega$$

Open-circuit voltage,

$$E_0 = Bvd = 2 \times 1{,}200 \times 0.4 = 960\ \text{V} \quad \textbf{Ans.}$$

Maximum power output,

$$P_{out\,(max)} = \frac{E_0^2}{4R_G} = \frac{(960)^2}{4 \times 0.2} = 1{,}152{,}000\ \text{W or } 1{,}152\ \text{kW} \quad \textbf{Ans.}$$

Parameters Controlling Power Output. The output of an MHD engine is given by an expression

$$P_{out} = \left(\frac{E_0}{R_G + R_L}\right)^2 R_L = \frac{E_0^2}{R_G} \times \frac{K}{(1+K)^2}$$

when K is equal to ratio of R_L and R_G *i.e.,* $K = \dfrac{R_L}{R_G}$

Substituting $E_0 = Bvd$ from Eq. (7.8) and $R_G = \dfrac{d}{\sigma A}$ in above equation, we have

$$P_{out} = \frac{(Bvd)^2}{\frac{d}{\sigma A}} \times \frac{K}{(1+K)^2} = \frac{B^2 v^2 d\sigma A \times K}{(1+K)^2} \qquad ...(7.13)$$

From above expression for power output, we see that for a given plate area and spacing and the ratio of $\dfrac{R_L}{R_G}$ the output power is governed by three factors viz. flux density B, gas velocity v and gaseous conductivity σ.

The power output varies as square of flux density *i.e.,* $P \propto B^2$. In conventional magnets flux density up to 3T (Wb/m^2) can be had. However, with superconducting materials, flux densities around 10T (Wb/m^2) can be had but the cost will become enormous.

The power output also varies as square of gas velocity *i.e.,* $P \propto v^2$. High velocity of gas can be achieved by converting thermal energy of gas emerging from furnace into directed kinetic energy. The conversion is accomplished through the conventional nozzle expansion method. Gas velocities of the order of 1,000 m/s can possibly be achieved.

The third factor which governs the power output of MHD generator is gaseous conductivity as power output is directly proportional to gaseous conductivity *i.e.,* $P \propto \sigma$. MHD generator will provide reasonable power output if σ lies between 10 and 100 siemens (or mho) per metre. This parameter is the most troublesome to achieve and has impeded the MHD generation progress. Introduction of seeding material, such as cesium and potassium, into the gas is only the solution. The seeding material ionizes more readily than the gas and increases the conductivity. This is because of lower ionization potential of seeding material than gas atoms. Cesium having ionization potential of 3.89 eV is the best seeding material but very costly. The next best seeding material is potassium with ionization potential of 4.34 eV.

7.3. THERMOELECTRIC POWER GENERATION

This process is based on the Seebeck effect which states that loop of dissimilar metals will develop an emf if the two junctions are kept at different temperatures. This principle is already being used in thermocouples for measurement of temperature. However the recent advances in material technology have made possible experimental thermoelectric generators of a few kW size. The simplest thermoelectric power generator consists of a thermocouple, comprising a P-type and N-type thermoelement connected electrically in series and thermally in parallel. Heat is pumped into one side of the couple and rejected from the opposite side. An electrical current is produced, proportional to the temperature gradient between the hot and cold junctions.

The net useful power output is given as

$$P_{out} = I^2R \text{ watts} \qquad ...(7.14)$$

where I is the current flowing through the circuit in amperes and R is the external load resistance connected across output terminals of the thermoelectric generator in ohms.

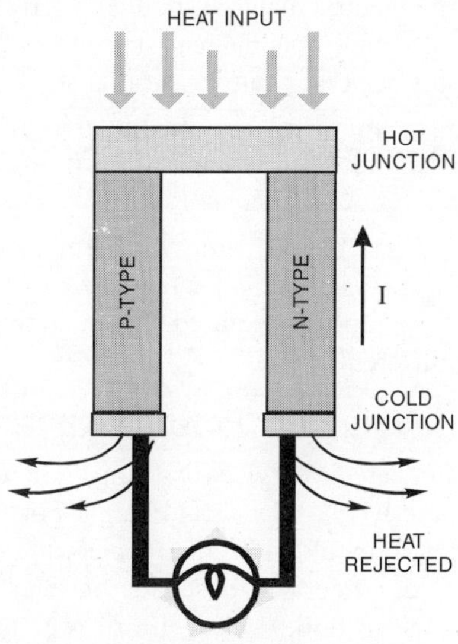

Fig. 7.6. ***Electrical Power Output of Thermoelectric Power Generator***

The current in the circuit is given by

$$I = \frac{\alpha \Delta T}{(R_{in} + R)} \text{ amperes} \qquad ...(7.15)$$

where α is Seebeck coefficient in V/k, ΔT is the temperature difference between hot and cold junctions in degrees absolute, K and R_{in} is the internal resistance of thermoelectric generator in ohms.

The magnitude of potential difference depends on the pair of conductor materials and on the temperature difference of the junctions.

For a loop made of copper and constantan wire, the value of Seebeck coefficient is 0.04 mV/k. For a temperature difference of 600 K between the junction, a voltage of 24 mV will be produced.

7.3.1. Thermoelectric Materials

The efficiency of thermoelectric generator depends upon suitable properties of the elements. It was later in 1909 and 1911 that Altenkirch showed that good thermoelectric materials should have large Seebeck coefficients, high electrical conductivity and low thermal conductivity. A high electrical conductivity is necessary to minimise Joule heating, whilst a low thermal conductivity helps to retain heat at junctions and maintain a large temperature gradients. These three properties were later embodied in the so-called figure-of-merit, Z. Since Z varies with temperature, a useful dimensionless figure-of-merits can be defined as ZT. The thermal conductivity of thermoelectric materials can be reduced by introducing suitable impurities:

The other requirements of thermoelectric materials are:

1. The mobility of current carriers (electrons or holes) should be as high as possible. The electrical conductivity can be raised by introducing suitable impurities.
2. One element should be purely P-type and the other N-type. The semiconductor material should have low ionization energy and narrow forbidden band.
3. The thermo elements should have variable impurity content so that electron concentration should depend upon the operating temperature.
4. The thermoelectric material should be corrosion resistant, should have high mechanical strength and elasticity so that it does not crack due to thermal stresses.
5. The bridge material should have high thermal and electrical conductivities and stability against thermal stresses.
6. The use of variable properties of thermoelectric elements becomes very significant when thermoelectric pile or cascaded operation is required. The operating temperature may be different.

7.3.2. Thermoelectric Power Generator

Thermogenerators are the devices that convert heat (temperature differences) directly into electrical energy. For the most part, this term is synonymous with "thermoelectric generator" and rarely used. Older Seebeck-based devices used bimetallic junctions and were bulky while more recent devices use semiconductor P-N junctions that can have thicknesses in the millimeter range. These are called solid state devices and unlike dynamos have no moving parts other than sometimes a fan. Fuel such as natural gas, propane or kerosene can thus be used for generating dc which can be converted into ac by using inverter.

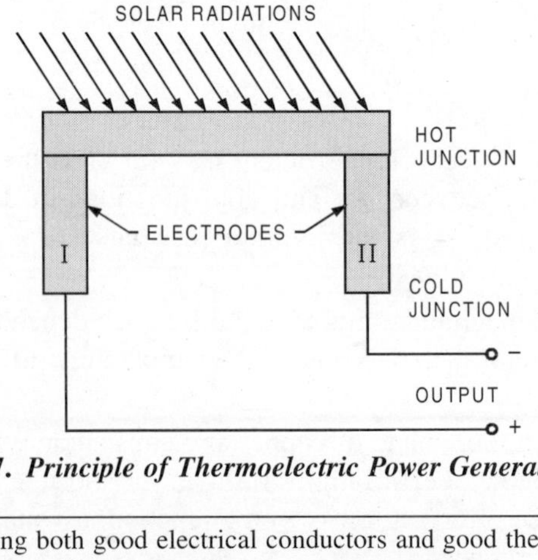

Fig. 7.7. ***Principle of Thermoelectric Power Generator***

The principle of thermoelectric generation is shown in Fig. 7.7. The electrodes I and II are bridged at hot junction and are connected to the output terminals at the cold junction. The hot junction is kept at high temperature by concentrating sun rays on it. The cold junction is kept cold by water cooling. The electric current is set up due to difference in temperatures of hot and cold junctions and completes its path through the external load circuit. Though it is based on direct conversion of heat into electric power but its efficiency is quite low (1 to 3%). Like any heat engine its efficiency depends upon the temperatures of the hot and cold junctions. A two stage device with hot and cold junction temperatures of 1,500 K and 300 K has been developed to give an overall efficiency of 13%. A lot of research is being done to find new materials capable of working at high temperatures. Special attention is being paid to semiconductor materials.

The voltage and electrical power output can be increased by increasing the temperature difference between the hot and cold ends. In order to achieve higher potential difference many generators have to be connected in parallel. For increasing the useful power output, parallel and series connections are used as shown in Fig. 7.8. The direct current generated can be changed into alternating current by using inverter. The alternating voltage is then increased to the desired value with a transformer.

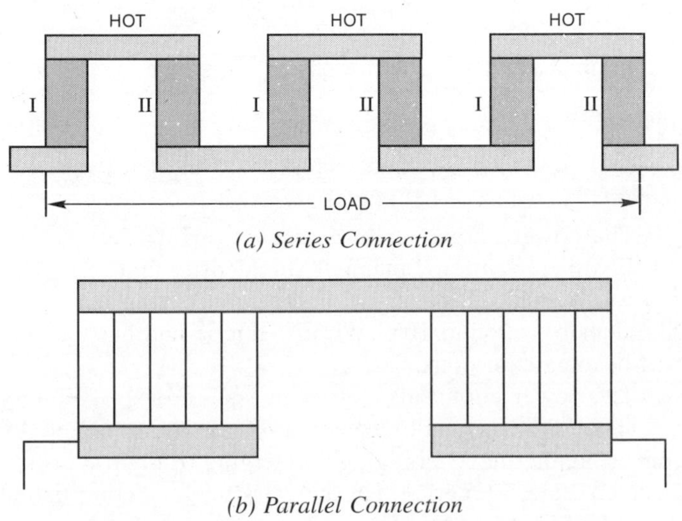

(a) Series Connection

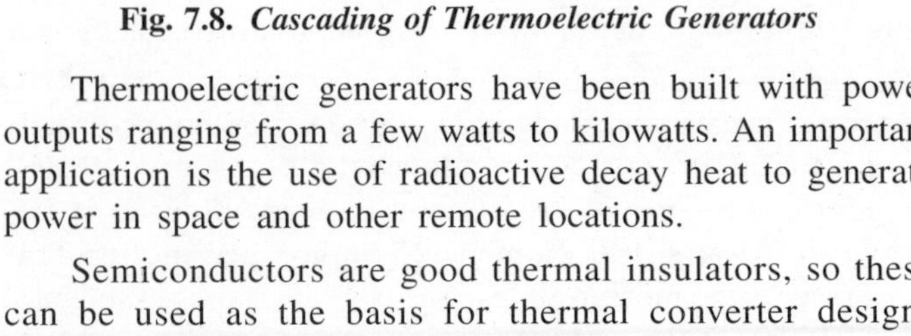

(b) Parallel Connection

Fig. 7.8. ***Cascading of Thermoelectric Generators***

Thermoelectric generators have been built with power outputs ranging from a few watts to kilowatts. An important application is the use of radioactive decay heat to generate power in space and other remote locations.

Semiconductors are good thermal insulators, so these can be used as the basis for thermal converter designs capable of withstanding large temperature gradients and hence generating larger thermo emfs*. At present, semi-

* Metals being both good electrical conductors and good thermal conductors raises the rate of heat transfer from hot to cold junction, equalizes the temperature, and, consequently reduces the thermo emf.

conductor materials like lead telluside, zinc antimonide, bismuth telluride, germanium silicide and the various selenides produce power with an efficiency of about 10%. Ternary compounds (such as silver-antimony telluride, lead-tin telluride) and quaternary compounds composed of bismuth, tellurium, selenium and antimony, called as *neelium*, are promising ones and higher efficiencies are expected to be achieved in the future, Extensive studies are currently being carried out with the aim to develop semiconductors capable of operating at high temperatures. Basically, thermal converters can employ liquid semiconductors. However, the main drawback of using liquid semiconductors is that the rate of heat transfer from the hot to cold junction is increased by convection.

Energy used for heating the hot junction of the thermal converter may be energy content of chemical fuels or heat generated in nuclear reactors besides solar radiant energy.

The main advantages of thermal converters are absence of moving parts, non-involvement of high pressures and use of any heat source. They are readily serviceable and quiet in operation. Thermal converters are preferred over other power sources from the point of view of reliability, compactness, portability and ease of use. These are widely employed as power sources on spacecrafts, submarines, missiles, beacons, and other equipment of different kinds.

The main drawbacks of thermal converters are high cost, low efficiency and low outputs.

7.3.3. Applications

The thermoelectric generators are recognized as very convenient direct energy conversion systems. These are quite suitable for remote and space applications and promising systems for waste heat recovery.

1. ***Nuclear Reactor With Thermoelectric Fuel Elements.*** A thermoelectric generator, as incorporated in the fuel elements of a nuclear reactor, is shown in Fig. 7.9. This will help in obtaining large outputs.

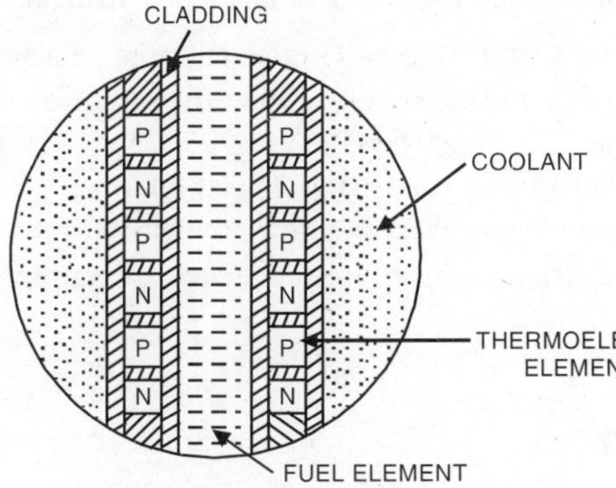

Fig. 7.9. *Thermoelectric Fuel Elements*

2. ***Combined Thermoelectric and Steam Power Plant.*** Thermoelectric generator can be employed as topping plant to a steam power plant. The overall efficiency of the combined plant will increase due to higher source temperature. The scheme is shown in Fig. 7.10.

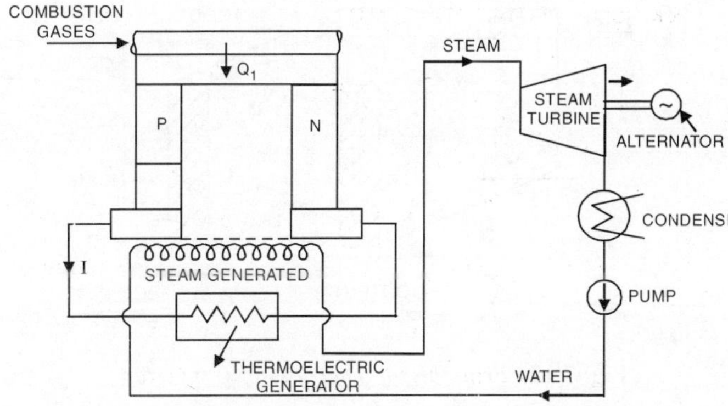

Fig. 7.10. *Combined Thermoelectric and Steam Power Plant*

3. ***Thermoelectric Waste Heat Stack.*** The waste heat from gas turbines, diesel engines and stack gases can be used to generate electricity by a thermoelectric generator. The metal stack consists of a series of rings of two alternate metals connected at the inner and outer annular edges alternately. These rings are thermally and electrically insulated. A schematic diagram is shown in Fig. 7.11.

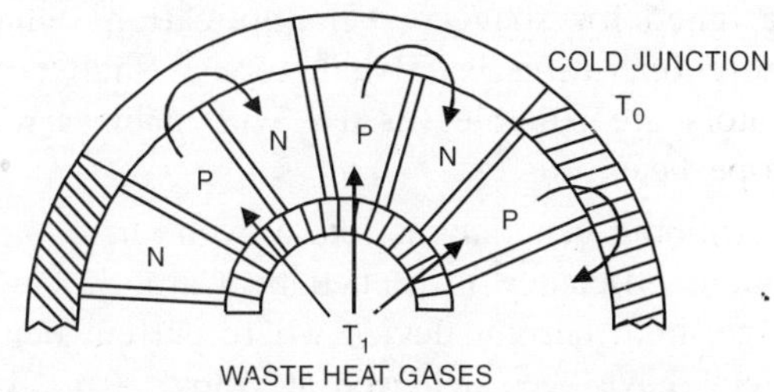

Fig. 7.11. *Thermoelectric Waste Heat Stack*

4. ***Decay Heat of Radioactive Isotopes.*** The decay heat of radioisotopes has been used for the operation of small (0.1 kW) thermoelectric generators. Based on heat of decay of Strontium 85, remote generators for signalling have been used.

5. ***Solar Energy.*** A combination of thermoelectric generator and solar collector can be used to generate electrical energy from solar energy.

7.4. THERMIONIC CONVERTER

A thermionic generator (converter) converts heat energy directly to electrical energy by utilizing thermionic emission effect. All metals and some oxides have free electrons which are released on heating. In a thermionic converter, electrons act as the working fluid in place of a vapour or gas. In this device electrons are emitted from the surface of heated metal. The energy required to extract an electron from the metal is known as *work function* and expressed in electron volts (eV). The work function depends upon the nature of metal and its surface condition.

The principle of thermionic converters is illustrated in Fig. 7.12. In a thermionic converter two electrodes are placed in a container containing an ionised gas or cesium vapour to reduce the space charge. The cathode is heated by concentrating the rays on it. On heating of cathode the electrons are emitted from it and travel to anode. The cathode and anode are connected externally through the load circuit. The electrons

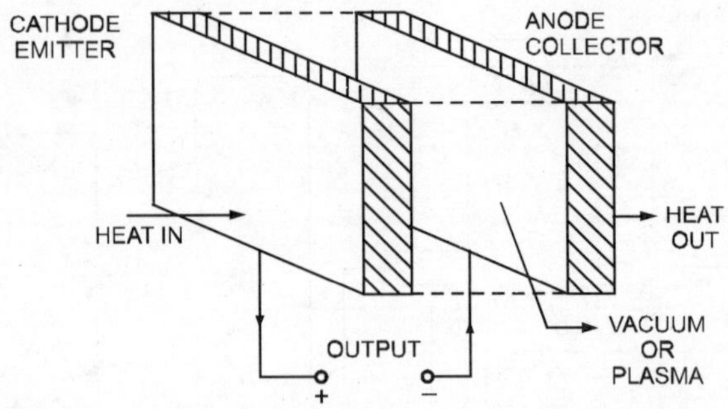

Fig. 7.12. ***Principle of Thermionic Converter***

return back to cathode through the external circuit and the current flows through the external circuit as shown in the figure. The heat energy is converted directly into electrical energy through a process similar to that in a steam plant where water is evaporated in a boiler and the steam is condensed after doing the useful work in an engine. With the discovery of materials that provide adequate electron emission rates without getting melted, the performance of the system has been largely improved. This device is being designed for space power applications where high temperature operation is advantageous. Such converters or generators are suitable for use with nuclear reactor or radioisotope heat sources.

A thermionic generator is like a cyclic heat engine and its maximum efficiency is limited by Carnot's law. It is a low-voltage, high current device where current densities of 20-50 A/cm^2 have been achieved at voltage from 1 to 2 V. Thermal efficiencies of 10-20% have been realized. Higher values are possible in future.

Thermionic generators can be used for large power generation. The fuel elements of a nuclear reactor may be very suitable high temperature heat source for thermionic generator. This can be surrounded by cooled anode and the in between space can be filled with ionized cesium vapour.

The energy of high temperature gases can be partly converted into electricity if the riser tubes of the boiler are provided with cathode and anode of a thermionic generator with the interspace filled with ionized cesium vapour.

In principle any heat source, fossil or nuclear fuel, a radioactive material or solar energy can be used in a thermionic generator.

Many applications have been suggested for remote locations on the earth and in space as the device is robust and reliable in unattended operation.

7.4.1. Thermionic Emitter (Cathode)

The metal used for emission of electrons is known as an emitter or cathode. The cathode is heated in an evacuated space to emit electrons. If the cathode is heated in open air it would burn because of presence of oxygen in air. Therefore, cathode should have the following properties:

1. The work function of the cathode should be low, so that small amount of energy is required for the emission of electrons.
2. Cathode should have high melting point, so that it may withstand high temperature without melting.
3. Cathode should have high mechanical strength.

7.4.2. Limitations

Thermionic converter is a direct energy conversion device with no moving parts. It is very robust, and does not require any supervision for operation. It is very suitable for remote and space applications. It suffers from the following limitations:

1. The operating temperature of cathode is very high, so costly materials like tungsten, rhenium are required for cathode.
2. Special shields of ceramic are required for protection of cathode from corrosive combustion gases.
3. The collector may also have to be made of molybdenum coated with cesium.
4. Ionised cesium vapour has to be filled in the interspace to reduce the space charge barrier to promote electron emission from the cathode.

7.4.3. Applications

1. Thermionic Converter in a Nuclear Reactor. The fuel element containing the fissile material carries the cathode surrounded by the anode. The interspace is filled with ionized cesium gas. The anode is cooled by the coolant from outside. Some of the energy released by nuclear fission is directly converted into electrical energy by thermionic conversion. The remaining heat is used in conventional bottoming steam plant. The overall efficiency of the plant, therefore, increases.

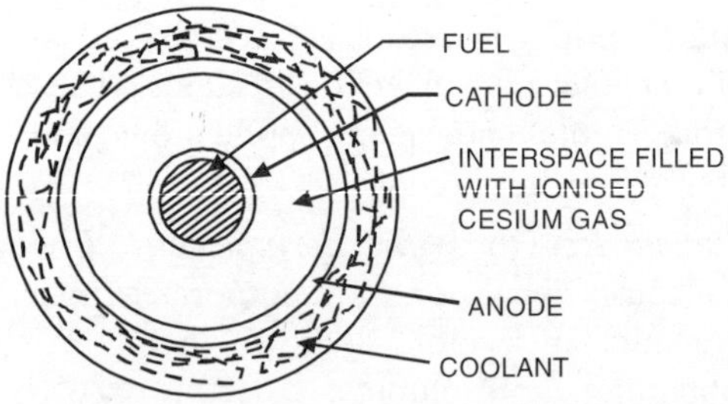

Fig. 7.13. ***Thermionic Converter in a Nuclear Reactor***

2. Thermionic Converter in The Riser Tube of a Boiler. The riser tubes of a boiler receive heat by radiation from combustion gases. The riser tube is provided by a cathode and anode of a thermionic converter. The interspace is filled with ionized cesium vapour. The use of the hot combustion gases to produce extra power before the steam cycle improves the overall plant efficiency.

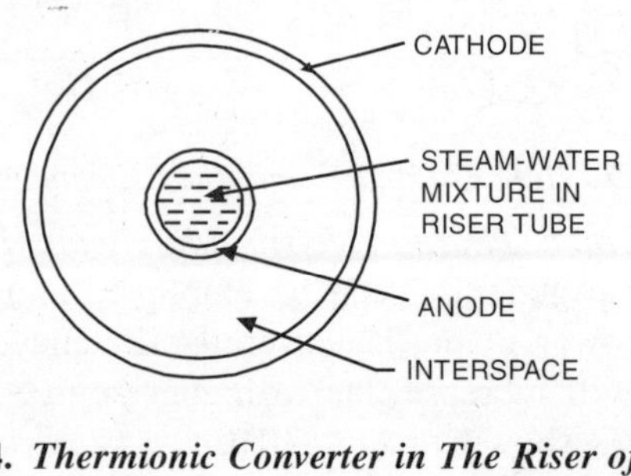

Fig. 7.14. ***Thermionic Converter in The Riser of a Boiler***

3. MHD Thermionic-Steam Power Plant. The waste heat from the MHD generator at about 1,900°C is used to heat the cathode of thermionic converter. The heat from the anode is used in the boiler of a steam power plant. The overall efficiency of the plant will, therefore increase.

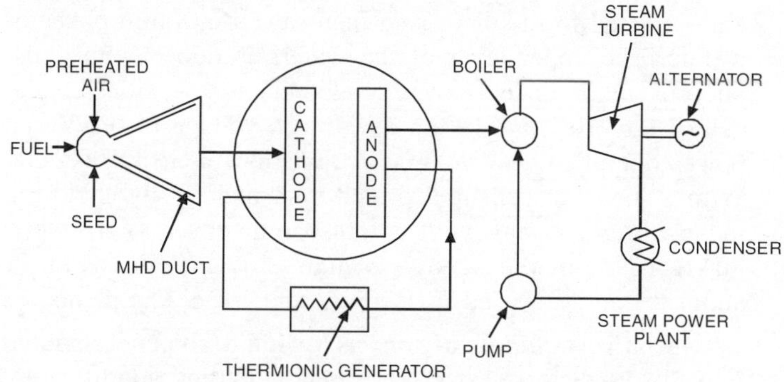

Fig. 7.15. *MHD Thermionic-Steam Power Plant*

7.5. PHOTOVOLTAIC OR SOLAR CELLS

It is possible to convert solar energy directly into electrical energy by means of silicon wafer *photovoltaic cells*, also called the *solar cells,* without any intermediate thermodynamic cycle. The solar cells operate on the principle of *photovoltaic* effect, which is a process of generating an emf as a result of the absorption of ionizing radiation. Thus a solar cell is a transducer, which converts the sun's radiant energy directly into electrical energy and is basically a semiconductor diode capable of developing a voltage of 0.5-1 volt and a current density of 20-40 mA/cm^2 depending on the materials used and the conditions of sunlight. The efficiency of the solar cells is as low as 15%, but that does not matter as the solar energy is basically free of cost. The main problem faced is that cost (Rs 1,400 to Rs 7,000 per watt) of the solar cells and their maintenance. With the likelihood of a breakthrough in the large scale production of solar cells at low cost, this technology may compete with conventional methods of generation of electrical power, particularly as conventional sources of energy become scarce.

The photovoltaic effect can be observed in nature in a variety of materials but the materials having the best performance in sunlight are the semiconductors. In a piece of pure semiconductor like silicon, there is no free charge carrier at ordinary temperatures, but if this piece of silicon is doped with phosphorous or arsenic there will be one extra electron per atom of the impurity leading to N-type (negative type) semiconductor. Similarly, if another piece of pure silicon is doped with boron (having one electron less than silicon) there will be deficiency of electrons (or excess of holes) leading to P-type (positive type) semiconductor. If these two pieces of silicon containing N-type and P-type impurities are connected by some means, a junction, at which the nature of the current carrier changes, is created. In fact, a potential energy gap (E_g) is created at the junction.

When a photon of energy '*hv*' is allowed to fall on the P-region, it is absorbed by an electron in the valence bond. If '*hv*' *exceeds* energy gap E_g, the electron will migrate to the N-region. Similarly if '*hv*' is less than E_g in the N-region, the photon will be absorbed by a hole which will migrate to P-region. This charge separation creates an electric field opposite to the electric field created by the diffusion of free electrons of the N-region and in case the field created by charge separation predominates the electric field created by the diffusion of free electrons from N-region to P-region and holes from P-region to N-region current will start flowing in the circuit, as shown in Fig. 7.16.

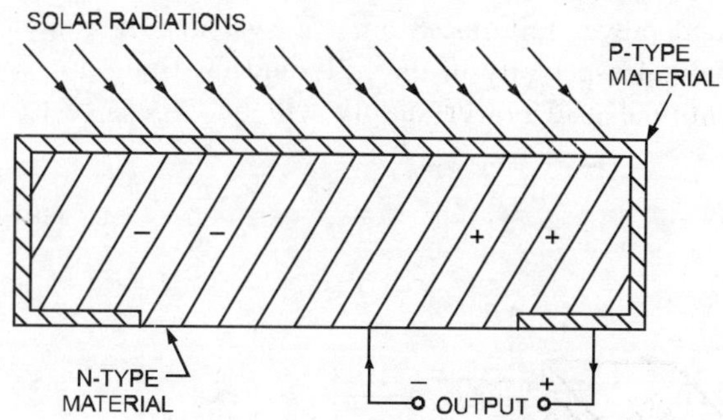

Fig. 7.16. *Photovoltaic or Solar Cell*

Photovoltaic cells generate a voltage proportional to electromagnetic radiation intensity and are called as such because of their voltage generating capability.

Single crystal silicon is most highly developed material for photovoltaic conversion. The physical properties of single crystal silicon are well known and the raw material is abundant. Single crystal silicon cells have been used for many years as power sources for spacecrafts in sizes from a few watts to over 20 kW per satellite. However they are still very costly and many attempts have been initiated in Japan, France, West Germany, USA etc. toward reduction of cost.

The cadmium sulphide/cuperous sulphide solar cell is the only other commercially available solar cell. It is not a simple P-N junction. It provides a charge separation field by the junction of two dissimilar materials that have different band structures. The advantage of this cell is that it can be made very thin (about 20 μm) using chemical processing techniques and does not require single crystal material. Poor efficiency is their main drawback.

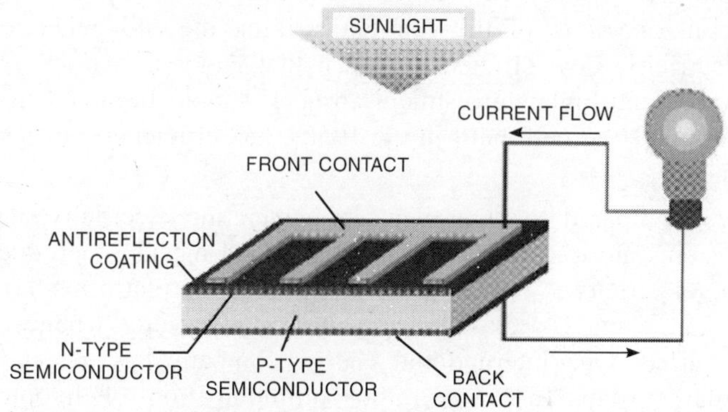

Fig. 7.17. *Photovoltaic Module*

Typically, one cell produces about 1.5 watts of power. Individual cells are connected together to form a *solar panel* or *module,* capable of developing 3 to 110 W power. Pannels be connected together in series and parallel to make a solar array which can produce any amount of wattage as space will permit. Modules are usually designed to supply electricity at 12 V. Photovoltaic (PV) modules are rated by their peak watt output at solar noon on a clear day.

Some applications for PV systems are lighting for commercial buildings, outdoor (street) lighting, rural and village lighting, etc. Solar electric power systems can offer independence from the utility grid and offer protection during extended power failures. Solar PV systems are found to be economical especially in the hilly and far flung areas where conventional grid power supply will be expensive to reach.

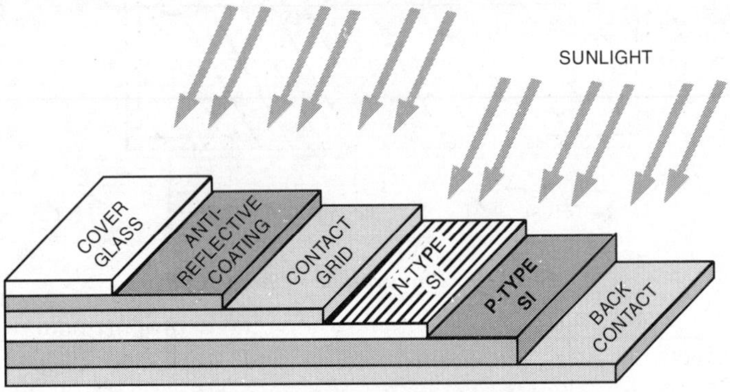

Fig. 7.18. ***Photovoltaic Structure***

PV tracking system is an alternative to the fixed, stationary PV panels. PV tracking systems are mounted and provided with tracking mechanisms to follow the sun as it moves through the sky. These tracking system run entirely on their own power and can increase output by 40%. *Backup systems* are necessary since PV systems only produce electricity during sunshine. The two most common methods of backing up solar electric systems are by connecting the system to the utility grid or storing excess electricity in batteries for use at night or on cloudy days.

The performance of a solar cell is expressed in terms of its efficiency in conversion of sunlight into electricity. Only sunlight of certain energy will work efficiently to produce electricity and much of it is reflected or absorbed by the material that make up the cell. Because of this, a typical commercial solar cell has an efficiency of 15% *i.e.*, only about one-sixth of the sunlight striking the cell generates electricity. Low efficiencies mean that larger arrays are required and higher investment costs. It should be noted that the first solar cell, built in the 1950s, had efficiencies of less than 4%.

Compared to other ways of generating electricity, PV systems are expensive, but they are good means of producing electricity in remote areas. Some offshore platforms have begun using PV systems to generate electricity whenever required. One international energy company has designed solar systems to power radio communications, helicopter landing pad lights and navigation warning lights on offshore platforms. Ocean buoys and other monitoring equipment also have PV cells as a power source.

Different Types of Solar Cells. According to type of crystal, the solar cells are of three types:

1. Monocrystalline silicon solar cells (band gap 1.12 eV)
2. Polycrystalline silicon solar cells (band gap 1.12 eV)
3. Thin film or amorphous silicon solar cells (band gap 1.75 eV)

In *monocrystalline silicon cells,* silicon is doped with boron to produce P-type semiconductor. Monocrystalline rods are extracted from silicon and then sawed into thin plates or wafers. The upper layer of the wafers is doped with phosphorous to produce N-type semiconductor. A silicon solar cell of size 10 cm × 10 cm develops a voltage of 0.5 V and power output of 1 W at a solar radiation intensity of 1,000 W/m^2. The solar cells are formed into modules by enclosing in an airtight casing with a transparent cover of synthetic glass. These modules possess high efficiency between 15 and 18% and are used in medium and large size plants.

In *polycrystalline silicon cells,* liquid silicon is poured into blocks that are sawed into plates. During solidification of the material, crystal structures of varying sizes are formed. The size of crystals mainly depends upon the cooling condition. If the molten silicon is cooled very slowly, the crystals of larger size are obtained. The silicon solar cells made from polycrystalline silicon are low cost but low efficiency. Maximum efficiency is 17-18%.

If a silicon film is deposited on glass or another subsrate material, this is so-called amorphous or thin layer cell. The layer thickness is less than 1 μm, so production costs are lower due to the low material costs. However, the efficiency of amorphous cells is much lower than that of other cells. Because of this, they are primarily used in low power equipment such as watches, pocket calculators etc. Maximum efficiency is 13%. Thin film solar cells are also manufactured from gallium arsenide (GaAs), Cadmium telliride (CdTe) and copper-indium-selenide (CuInSe).

Two systems namely roof top array system and satellite system for generation of electric power from solar cells are being considered.

1. ***Roof Top Array System.*** This system is an earth-based solar cell system mounted on roofs. The main difficulty is the problem of energy storage since this system will work when there is sunshine. The various possible alternatives for energy storage are:

(*a*) Electrochemical storage, but batteries of adequate capacity that can withstand frequent charging and discharging for several years are yet to be developed.

(*b*) Hydro storage in which water is pumped uphill when power is abundant and allowed to flow through hydro generators at a time of peak demand. This possibility is suitable only to a few regions of the country.

(*c*) Mechanical storage in high speed flywheels.

(*d*) To store energy in the form of hydrogen, which could be re-converted into electricity in fuel cells. Hydrogen is obtained by the electrolysis of water by the output of the solar cells.

2. ***Satellite System.*** A more long range system that would also avoid the need for major storage of power is a space *power station* in synchronous orbit around the earth. According to the proposal put forward by Dr. Glaser of Massachusetts (USA), a solar collector of 8 × 8 km will be fixed on the surface of the satellite. Electricity so produced will be used to produce a microwave beam. This microwave power will be transmitted to antennae on earth, and converted back into electric power. Although this system could provide large amount of power, but the problems regarding the endurance of the components, the control of large structures in space, and the safety of the microwave radiation, are still to be solved. Such a system is shown in Fig. 7.19.

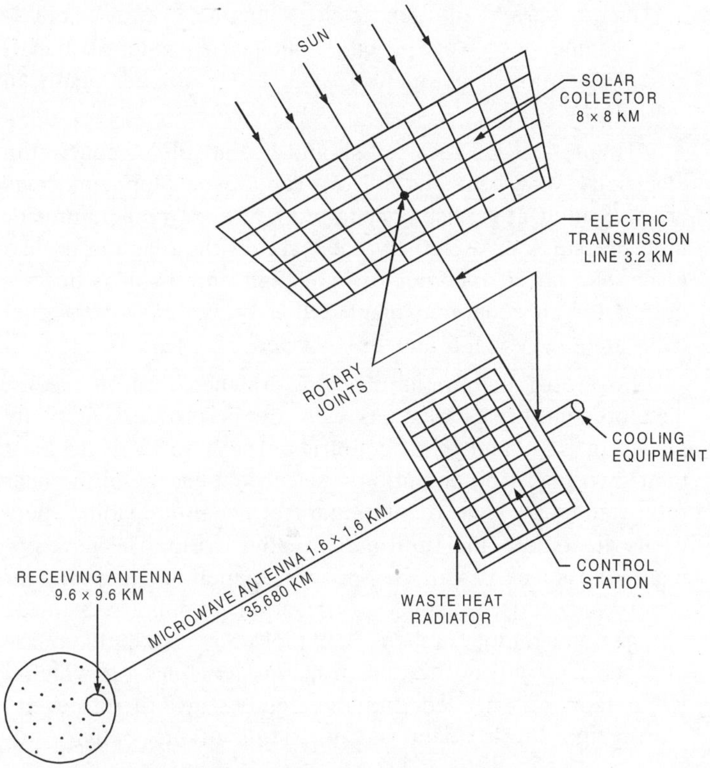

Fig. 7.19. ***Satellite Solar Power Station***

Advantages of Photovoltaic Solar Systems

1. Absence of moving parts.
2. Direct conversion of light to electricity at room temperature.
3. Can function unattended for long time.
4. Modular design: voltage and power outputs can be manipulated by integration.
5. Low maintenance cost.
6. No environmental pollution.
7. Very long life.
8. Highly reliable.
9. Solar energy is free and no fuel is required.
10. Can be started easily as no starting time is involved.
11. Solar cells can be made from microwatts to megawatts. These can be used to feed the utility grid with power conditioning circuitry.
12. Easy to fabricate.
13. These have high power-to-weight ratio, therefore very useful for space application.
14. System is noiseless and cheap.
15. Modularity in operation.
16. These can be employed with or without sun tracking.
17. Decentralized power generation at the point of power consumption can save power transmission and distribution costs.

Limitations of PV Solar Systems

1. Manufacture of silicon crystals is labour and energy intensive.
2. The insolation is unreliable and therefore, storage batteries are required.
3. Solar power plants need very large land areas.
4. The principal limitation is high cost, which is being reduced through various technological innovations.
5. Electrical generation cost is very high.
6. Low efficiency.
7. The initial cost of the plant is very high and still needs a long gastation period.
8. The energy spent in the manufacture of solar cells is very high. The plant operation period during which photovoltaic plant recovers the span energy varies from 4 to 7 years.

Applications of Solar Photovoltaic Systems. The solar photovoltaic systems are classified according to field of application, type of system, rated capacity, etc.

1. Autonomous System:

1. Amorphous silicon solar cells of very small capacity are employed in watches, pocket calculators, etc.
2. Small capacity solar systems from 50 W to 50 kW capacity are used for remote houses and villages for lighting for domestic use, street lighting, telecommunications, community development, water pumping, etc.
3. Roof-mounted solar system of 1 kW to 5 kW can be employed for residential houses.
4. In developing countries, solar systems can be used for lighting, recreation centres, radios, TV sets, small refrigerators etc.
5. In developing countries, solar system can find applications for drinking water supply, irrigation, vaccine refrigeration, milk chilling, rural power supply.
6. Solar generators can be employed for boats, lighting towers, radio buoys, traffic signals, parking lights etc.
7. Developing countries have high solar energy potential and solar generators are employed for water pumping, domestic power supply, hospitals, schools, farmhouses etc.
8. Other applications like catholic protection of oil pipelines, weather monitoring, railway signalling, battery charging etc.

2. Solar Water Pumps. The major applications of PV systems lies in water pumping for drinking water, irrigation in rural areas, cattle stock watering. The solar generators required for such purpose should have power capacity from 10 W to

10 kW. The water discharge of pump ranges between 1 to 40 m^3/hour with delivery head from 2.5 m to 120 m. These generators usually operate without storage batteries.

The total cost of the system including solar PV array, on/off switch, motor-pump set, accessories including suction and delivery valves and piping is about Rs 40,000, which is highly subsidised and installed at a cost of Rs 5,000 only for a small or marginal farmer.

3. Central Power Generation. A solar PV power plant of 16 MW capacity has been developed to supply electrical energy to 2,300 households. Solar power plants with a total capacity of 100 MW are planned for future installation.

4. Space Satellite Power Stations. The space satellite power station (SSPS) has been conceived as the future solar power supply system. Large surface area of PV panels will be mounted on a space satellite which will be synchronically moving with the earth orbit so that it will appear stationary from every point on earth. The energy produced will be converted into microwave energy and transmitted to the earth. It will be collected with antennas of a few square kilometer area and then converted into commercial frequency electric power.

Satellite mounted solar power plants can be designed for power outputs from 3 to 20 GW. The plant will require solar battery of 20 km^2 total area. The transmitting antenna of 1 km diameter and receiving antenna of 7 to 10 km diameter. The overall efficiency is expected to be 77%.

7.6. FUEL CELLS

A fuel cell is a device in which the chemical energy is converted directly into electrical energy and heat without combustion. The chemical energy is the free energy of the reactants employed. The basic feature of the fuel cell is that the fuel and its oxidant are combined in the form of ions rather than neutral molecules. Fuel cells are different from conventional batteries in that they consume reactant from an external source, which must be replenished—a thermodynamically open system. By contrast batteries store electrical energy chemically and hence represent a thermodynamically closed system. The first practical fuel cell was demonstrated in 1959 by Francis T. Bacin and J.C. Frost of Cambridge University.

In conventional steam power plants the chemical energy of the fuel is converted into heat energy by burning and the heat energy is, then, converted into electrical energy. The efficiency of this conversion process is limited by the limitations of Carnot cycle. In fuel cells the chemical energy of the reactants is converted into electrical energy as an isothermal process. Thus heat is not involved in the conversion process and a high conversion efficiency is possible. Another reason for the interest in fuel cells is that their efficiency and cost per kW of power are independent of size (or rating) of the fuel cell. This advantage makes the prospects of fuel cells very attractive as portable power plants for space crafts, locomotives etc. The other advantages of fuel cells are:

(*i*) The unit is lighter and smaller and requires little maintenance because of absence of mechanical parts.

(*ii*) They cause little pollution and little noise.

(*iii*) No overhead line is required.

(*iv*) Fuel can be used more effectively than in a central power plant.

(*v*) A fuel cell gives a few times more electrical energy per unit weight as compared to a turbo-generator or storage battery.

(*vi*) They can become remarkable home units.

(*vii*) A variety of fuels such as methane, ethane, ethylene, acetylene, propane, butane, benzene, methanol, ammonia, hydrazine, LPG, biogas or coal gas can be used.

Theoretically a fuel cell should be capable of generating electricity very efficiently. However the development costs are very high. It is necessary to work at high temperatures or high pressures or use costly catalysts for the reaction to take place at a speed in order to give high current densities required for an economic plant. Other drawbacks of the fuel cells are low voltage and low service life.

Inspite of the limitations and drawbacks of fuel cells, development works of fuel cells are in progress at many places in USA and other countries. They are likely to have their own place in generation of electrical energy in the near future and will have revolutionary effect in spreading electricity in remote and rural areas of the world. The development of fuel cells will be specially beneficial to India for supply of electrical energy to irrigation pumping sets in the villages and remote areas as the fuel cells as supply source of electricity will not require transmission lines for which a lot of money is needed. Indian scientists are, therefore, exploring the development and utilisation of fuel cells.

Fuel cells can be manufactured as small or as large as required for a particular power application. Presently, there are micro fuel cells that are of the size of a pencil eraser and generates few mW of power while there are others large enough to provide large amount of power. The power output of fuel cells is fully scalable by varying the cross-sectional area of each cell to provide desired current and by stacking multiples cells in series to provide the required voltage.

A fuel cell normally contains two electrodes separated by an electrolytic solution. A fuel reactant, usually hydrogen or carbon monoxide is fed into one porous electrode and oxygen or air is fed into the other porous electrode. The electrodes should be capable of passing through both fuel and electrolyte and also to conduct electrons to the terminal. The electrodes must contain a chemical catalyst that breaks the fuel compound into atoms so that they are more reactive. The most commonly used catalysts are platinum and sintered nickel. The electrodes should neither have pores of too large size to cause bubbles of fuel gas nor of too small size to cause insufficient contact between the reactant and the electrolyte. The electrolyte solution must be highly permeable to either a H^+ or OH^- ion which is produced as an intermediate product at one of the electrodes. The same ion is transferred through the electrolyte to the other electrode where it combines with the other reactant.

The working of a fuel cell using hydrogen and oxygen is explained below.

When two permeable nickel electrodes are immersed in a well conducting electrolyte (say a solution of H_2SO_4 or KOH), negative electrode is fed with hydrogen, bubbled around it through the solution and positive electrode is fed with oxygen and the electrodes are connected together through an external circuit, then for every molecule of hydrogen consumed, two electrons pass from negative to positive electrode, where they react with absorbed oxygen. The operation of fuel cell can be summarized as follows.

Reaction at –ve electrode is $2H_2 \rightarrow 4H^+ + 4e^-$

Every hydrogen molecule brought to the electrode surface is dissociated into two atoms by virtue of the catalytic properties of the surface. These enter the solution as hydrogen ions leaving behind two electrons which pass through the external circuit to the positive electrode. Reaction at positive electrode is $O_2 + 4H^+ + 4e^- \rightarrow 2H_2O$.

The oxygen supplied to the positive electrode reacts with hydrogen ions from the electrolyte and the electrons to give water. Thus water is the waste product of the cell. For the above case the electrolyte is acidic and the intermediate ion is H^+.

In case of alkaline electrolyte (a typical 40% KOH solution), the intermediate ion will be OH^- and the chemical reaction in the cell will be as follows.

At cathode $2H_2 + 4OH^- \rightarrow 4H_2O + 4e^-$

At anode $2H_2O + O_2 + 4e^- \rightarrow 4OH^-$

So, when the electrolyte is acidic, water is formed at the anode and when the electrolyte is alkaline, it is formed at cathode.

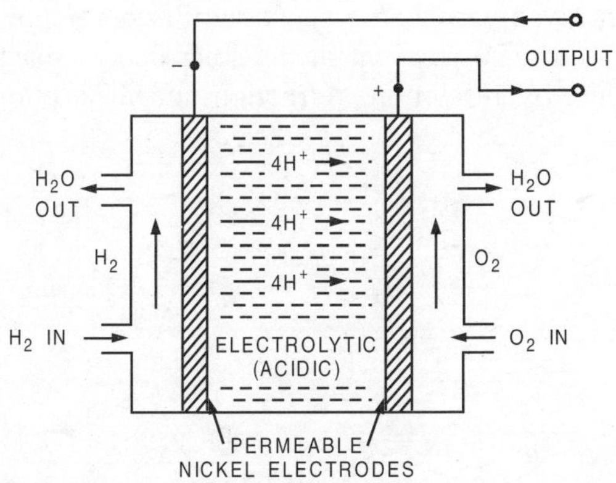

Fig. 7.20. *Fuel Cell*

If the electrodes are on open circuit, negative charges accumulate at hydrogen electrode. These negative charges attract potassium ions, K^+ of the electrolyte producing a double layer. Similarly the loss of electrons from the oxygen electrode results in a layer of positive charges which in turn attracts hydroxyl ions, OH^- from the electrolyte and form a double layer. These electrical double layers build up at the electrodes until the potentials are such that they inhibit any further reaction between the electrolyte and the fuel gases. An open-circuit voltage of 1.23 V at one atmospheric pressure at 25°C is developed. If the circuit is closed through an external load resistance, the electrons flow from the hydrogen electrode, through the external circuit to the oxygen electrode and take part in the reactions as mentioned above. The movement of electrons constitutes a current flowing through the external load circuit. The electrons movement is from the hydrogen electrode to oxygen electrode. Thus hydrogen electrode serves as cathode and the oxygen electrode as anode.

Fuel cells can be adopted to a variety of fuels by changing the catalyst, but 'hydrox' fuel cells using hydrogen and oxygen as fuel are the most efficient and most highly developed cells. Hydrogen can be obtained from natural gas, by catalytic cracking of ammonia or hydrocarbons or as a by-product from some processes. Oxygen may be obtained from the air or from decomposition of peroxides. A fuel cell power plant generally also contains a reformer and an inverter. The reformer uses chemical processes to convert the fuel to form that can be utilised by the cell while an inverter is used for converting output direct current into alternating current.

A single 'Hydrox' fuel cell can produce an emf of 1.23 volts at one atmospheric pressure and 25°C, as already mentioned. However, it is possible to create useful potentials of 100 to 1,000 volts and power level of 1 kW to 100 MW by connecting a number of cells in series-parallel combination. The current depends upon the physical size of the cell. The output of the fuel cell varies directly with pressure, so to increase the cell output, the gas pressure is raised. The optimum size of the cell at present is about 0.027 cubic metre per kW.

Hydrox cells are of two types namely low temperature cell and high pressure cell.

Fuel cells are particularly suited for low voltage and high current applications. Apollo Astronauts going to the Moon used fuel cells to convert hydrogen and oxygen to electricity. The power cells were located in the Apollo service module and provided the primary power source to operate life, support communication, guidance and other electrical system.

One type of fuel cell considered suitable for fuel cell power generation system is *phosphoric acid fuel cell* that operate at a temperature of about 190°C. The fuel used in this cell is high calorific value gas (with methane as the principal constituent), oxidizer is air, electrolyte is phosphoric acid and electrodes are made of carbon catalyzed by platinum. The fuel cell voltage is 0.7 V, current density 200 mA/cm^2 and expected life 10,000 hours.

Choice of Fuel For Fuel Cells. The choice of fuel for a fuel cell is governed by cost, availability, volume, transportability, etc.

Amongst the fuels used in a fuel cell, hydrogen is the most important. This is because hydrogen and oxygen are capable of releasing more energy per unit weight than most other oxidizer combinations. Such fuel cells are widely used in spacecraft power supplies.

Hydrocarbons (such as methane, ethane, acetylene, benzene etc.) are less reactive than hydrogen, much more difficult to oxidize and their by-products are usually undesirable.

Compromise fuels (such as methanol, ammonia, hydrazine etc.) have reactivity in between that of hydrogen and the hydrocarbons. They are easy to use. Hydrazine is highly reactive at normal temperatures and does not require any catalyst. Hydrazine is used in fuel cells employed in military systems and submarines though it is very costly and poisonous. Ammonia is much cheaper than hydrazine, is readily available and easier to handle but its reactivity is low. Fuel cells using ammonia are considered quite suitable for specialized remote, low power applications.

Different fuels with voltage at 25°C are given below in tabular form.

Table 7.1. Fuels For Fuel Cells

S. No.	Fuel	Voltage at 25°C
1.	Ammonia NH_3	1.17 V
2.	Hydrogen H_2	1.23 V
3.	Hydrazine N_2H_4	1.61 V
4.	Methanol	1.21 V
5.	Carbon monoxide	1.29 V
6.	Propane	1.09 V

Various types of fuel cells with their operating temperature are given below in tabular form.

Table 7.2. Type of Fuel Cells and Their Operating Temperatures

S. No.	Type of Fuel Cell	Operating Temperature
1.	Alkali	50–250°C
2.	Solid Polymer	50–100°C
3.	Phosphoric	~ 220°C
4.	Molten Carbonate	~ 600°C
5.	Solid Oxide	500–1,000°C

Advantages, Limitations and Applications of Fuel Cells

Fuel cell have the following *advantages*:

1. Fuel cells have high efficiency (exceeding 50%) in full-load as well as part-load operation. These are potentially ideal sources of power generation. Efficiencies of the order of 40% have already been achieved. An overall efficiency of more than 80% can be achieved in cogeneration plants if heat generated in the fuel cell can also be utilized in addition to electrical energy produced.
2. No pollution emissions. When fuel cells are used for power generation and transport sectors, NO_x will be reduced by 50% to 90% and CO_2 by 50% in comparison to present conventional technologies.
3. Water is bi-product of reaction. This is a useful product in space and remote applications.
4. The noiseless operation of fuel cells make them quite suitable for military and other strategic applications. This is because of absence of moving parts.

The main *limitations* hindering the growth of fuel cells are:

1. High capital cost of fuel cells is the main limitation against commercialization. However, this limitation can be overcome by development of new materials as the materials used presently are very costly. Search for new applications and markets can help in reducing the costs by taking the advantage of scale of production.
2. Short life span of fuel cells due to heavy corrosion of electrodes. This can be overcome again by new material technologies.

A 50 kW fuel cell stack has been developed by Plug Power LLC, Latham, New York. It generates electrical energy from different fuels like gasoline, ethanol, methanol and natural gas. A fuel processor has been developed to convert raw fuel into hydrogen, which is supplied to a fuel cell for generation of electrical engineering. The carbon dioxide removal system has also been developed. The technology can be used for powering vehicles and in homes and apartments to provide heat and electricity.

A fuel cell of 1.5 kW capacity has been developed in Australia to meet the power requirements of a house.

7.7. SOLAR POWER GENERATION

As already mentioned in chapter 1, the sun is the primary source of energy. Solar energy appears to be the most promising among the non-conventional sources of energy. The sun's great energy release is the result of an ellaborate chemical process in the sun's core—a process of thermonuclear fusion like the reaction in hydrogen bomb. Sun radiates energy of about 3.5×10^{23} kW into space and only 2×10^{14} kW reaches the earth. Converting even a part of the solar energy at a very very low efficiency can produce in far more energy than could conceivably be harnessed or utillised for power generation. Even if 90 per cent of the solar energy reaching the earth is lost by reflection, refraction and absorption in the

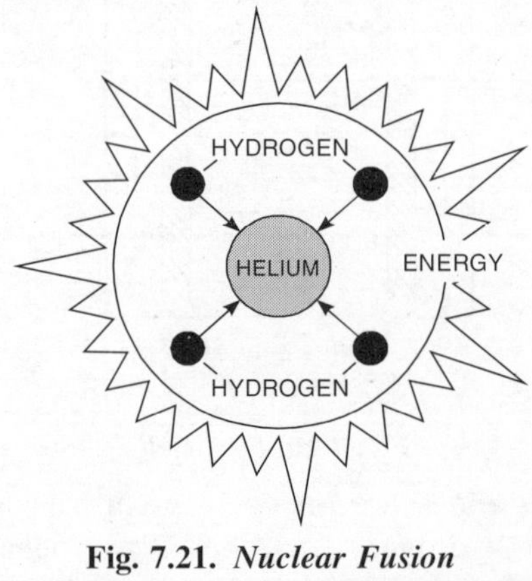

Fig. 7.21. *Nuclear Fusion*

outer layers of atmosphere, the quantity available at the surface is 2×10^{13} kW, which is equivalent to the burning of some 17 million tonne of coal. This stupendous solar energy, which is non-exhaustible and completely pollution free, could drive the civilisation for ever, if it could be properly and economically harnessed. However, it has got some serious drawbacks—energy density per unit area is very low, it is available for only a part of the day, and cloudy and hazy atmospheric conditions largely reduce the energy received. Therefore, in harnessing solar energy for generation of electrical power large areas to collect sufficient amount of energy and the means to store it will be required.

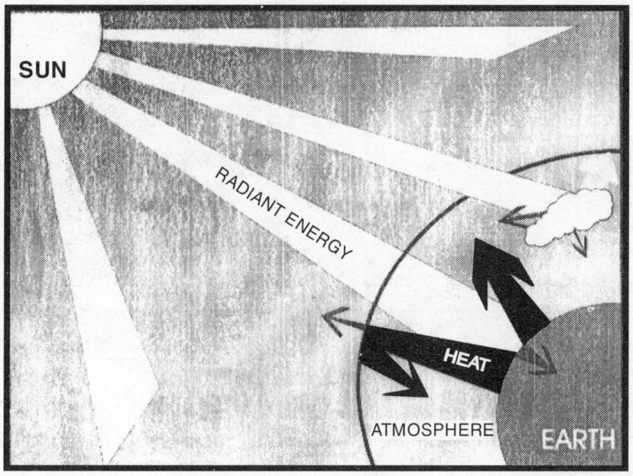

Fig. 7.22. *Solar Energy*

Solar energy reaching the earth in tropical zones is about 1 kW/m^2 giving approximately 5 to 10 kWh/m^2 per day. In countries within 3,200 km of equator, use of such energy can be economically significant. The use of solar energy is, therefore of special importance for India. Most of the areas in India receive high intensity fairly uniform sunshine for 6 to 8 hours a day and roughly 10 months a year. Solar energy is already being used in many advanced countries, both for domestic and commercial purposes such as water heating, water distillation, drying etc. The various methods, which can be adopted for converting solar energy into electrical energy are illustrated in Fig. 7.23.

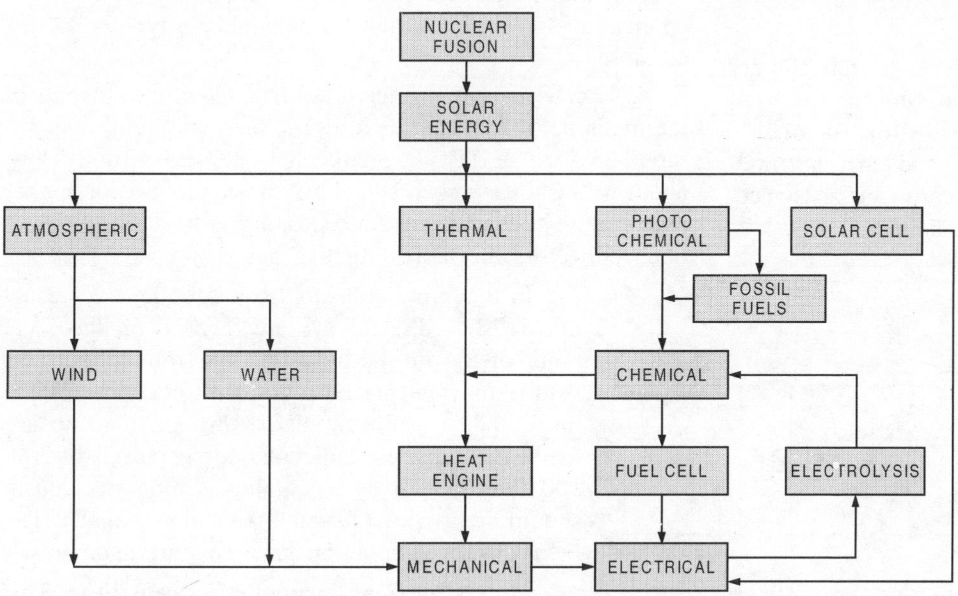

Fig. 7.23. *Methods of Converting Solar Energy into Electrical Energy*

The generation of electrical energy from solar energy is of special importance for India. The reasons are:

(*i*) There is a need for conservation of fossil fuel resources, such as gas, oil coal etc.

(*ii*) There is a need for reducing atmospheric and thermal pollution which are having serious detrimental effects on our environments.

(*iii*) There is a need for supply of electrical energy in remote areas.

Solar energy has got the additional advantages of availability in abundance and at no cost and no problem of transportation or transmission.

Solar Collectors. Since solar energy is widely spread, there is a need to make it available in concentrated form for which solar collectors are required. A solar collector absorbs the incident solar radiation and converts it into useful heat energy which is used for heating a collector fluid such as water, oil or air. The surface of a solar collector is designed for high absorption and low emission. Solar collectors are of two types namely flat plate type (or non-concentrating type) and focussing (concentrating) collectors.

Flat plate collectors are used where temperatures below 100°C are required. These collectors could be of liquid heating or air heating types. The liquid heating collectors are often used for heating water whereas air heating collectors are used for drying of agricultural products, heating of green houses etc. *Concentrating collectors* use optical systems in the form of reflectors or refractor to concentrate the energy of solar direct radiation on the absorbing surface. The reflectors may be flat mirrors or in the shape of a parabolic trough or paraboloidal dish.

1. Flat Plate Collectors. The important parts of a typical flat plate collector are shown in Fig. 7.24. The absorber plate with several parallel tubes is fabricated from copper tube and sheet by soft soldering. The plate is blackened in order to absorb maximum amount of sunlight. The plate is enclosed in a box to insulate it on the sides and bottom so as to prevent losses and thereby attain high temperatures. Also one or more transparent glass or plastic sheets are placed on the top of the blackened sheet so as to avoid heat losses by re-radiation. The complete structure is placed at a proper inclination to the sun so as to absorb solar radiations. The heat absorbed by plate is removed by circulating water or fluid or air in contact with it or through the tubes. The flat plate collector gives a temperature of only about 60ºC above ambient, *i.e.*, less than 100ºC with an efficiency of 30-50%. So flat plate collector is mainly used for water heating, space heating, space cooling and drying. However, by evaporating a low boiling organic fluid, a vapour turbine can be used to produce electric power (Fig. 7.25).

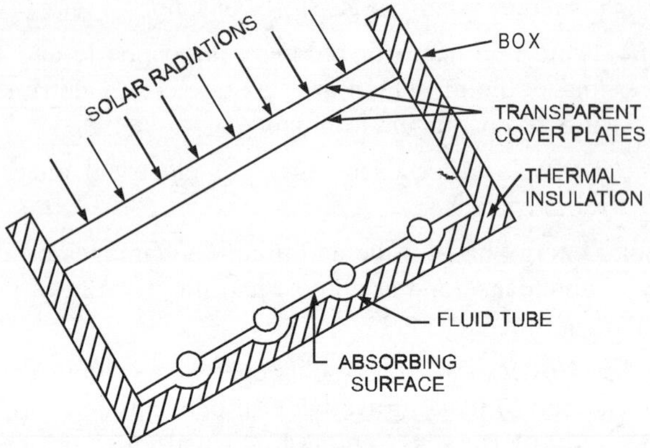

Fig. 7.24

The collection efficiency of flat plate collector varies from 40 to 60% for a temperature rise of about 15ºC but drops to 30% or even less for a 50ºC temperature rise. However, the absorbing efficiency of a flat plate collector can be increased by applying a selecting coating to the collecting surface, instead of flat black paint. The efficiency can also be improved by applying anti-reflective coatings on the transparent covers. The flat plate collector absorbs both diffuse and direct radiations. However, a tracking or a tilting arrangement is usually provided to track the sun from morning to sunset, as shown in Fig. 7.25. Here the heat absorbed is utilised to evaporate some working medium, butane, in a heat exchanger. The vapours are then expanded in a vapour turbine to get the output.

Different types of flat plate collectors are shown in Fig. 7.26. Pipe and fin type collectors are expensive but have good corrosion resistance. In this type liquid flows only in pipe and hence have low wetted area and liquid capacity. In sandwich type both the wetted area and capacity of water are high. They are cheap and light. These are used for low temperature requirement, such as swimming pool.

The properties of the materials used for collectors can be classified by thermophysical properties (such as thermal conductivity, heat capacity etc.); physical properties (like density, tensile strength, melting point etc.) and environmental properties (such as moisture penetration, corrosion resistance and degradation due to pollutants in atmosphere.)

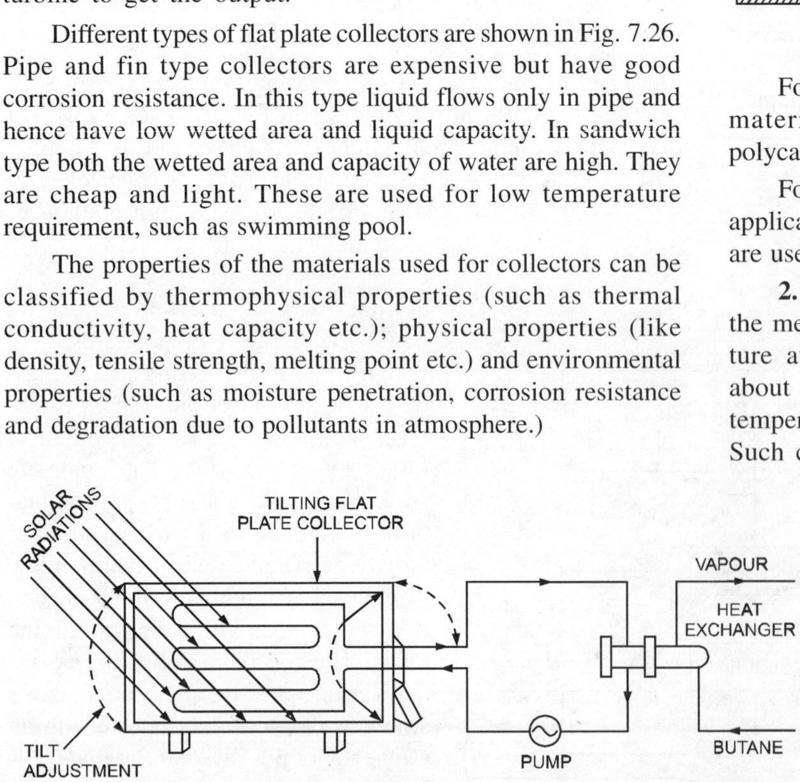

Fig. 7.25

The material for absorber plate should have high thermal conductivity, adequate tensile strength and good corrosion resistance. The most common material used for absorber plate is copper because of high conductivity and resistance to corrosion. Other materials used for absorber plate are aluminium, iron, brass, silver, tin and zinc.

The material for thermal insulation should have low thermal conductivity, should be stable at high temperature. Some commonly used materials are crown white wool, glass wool, calcium silicate, cellular form etc.

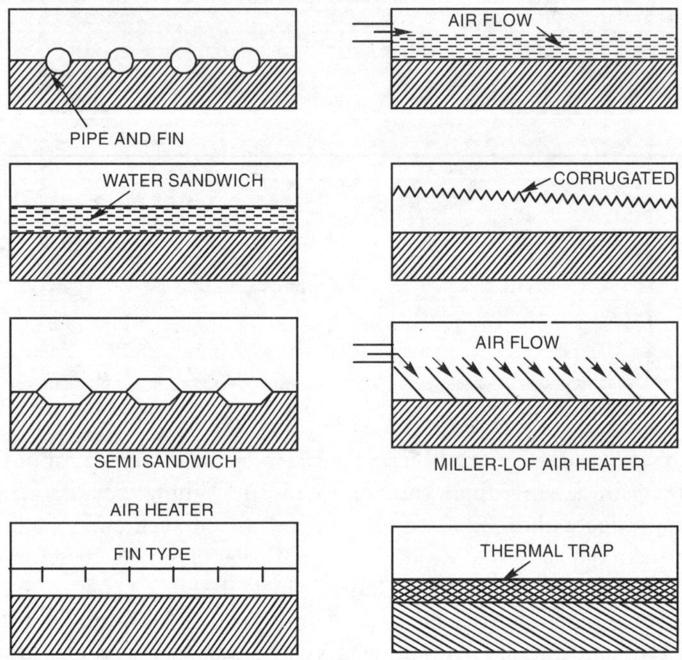

Fig. 7.26. ***Different Types of Flat Plate Collectors***

For cover plate tempered glass is most commonly used material. Transparent plastic materials such as acrylic polycarbonate plastic, polyvinyl fluoride are used for cover plate.

For higher temperature requirement, such as in domestic applications or for industrial use pipe and fin type collectors are used.

2. Focussing or Concentrating Collectors. Inspite of the methods of reducing heat losses, the maximum temperature at which a flat plate collector operates is quite low, about 100ºC in summer and 40ºC in winter. So for higher temperature focussing or concentrating collectors are used. Such collectors are more effective but very costly.

In focussing collectors, a parabolic or a fresnel mirror is used. Sun rays are focused on the focal point of the mirror by reflection from its surface. A tube may be placed along the focal line of the mirror and a fluid circulated through it to absorb the heat. With these collectors, temperatures of 200-300ºC or above may be obtained. Some mechanism should be provided to track the sun seasonally. The focussing collectors can have two arrangements.

(*a*) Cylindrical Parabolic Concentrator. This is a medium range temperature concentrator. It can give temperature in the range of 100-200ºC. This type of arrangement is generally employed for vapour engines and turbines, process heating, cooking etc.

(*b*) Paraboloids, Mirror Arrays. This is high temperature arrangement (gives temperature above 200°C). It can be used for steam engines and turbines, sterling engines and thermoelectric generator.

Mirrors and Fresnel lenses are imaging concentrators which are capable of focussing only beam radiation. In order to continuously collect beam radiation these concentrators have to be rotated by a tracking mechanism about one axis (linear concentrators) or two axes (point-focus concentrators) to follow the motion of the sun. High operating temperatures such as 1,000°C or even higher can be achieved when point focus concentrators with high concentration ratio are used.

Various types of concentrator solar collectors are shown in Fig. 7.27.

Figure 7.27(*a*) shows a paraboloidal reflector. Its polar axis is aligned E.W. A collecting tube receives a useful fraction of energy falling on the reflector.

Figure 7.27(*b*) shows a parabolic trough concentrator. This system has a parabolic mirror and receiver at its focal point. The concentration ratio is very high, and therefore, can be used where high temperatures are required.

Fresnel lens concentrator is shown in Fig. 7.27(*c*). It is a linear Fresnel lens solar concentrator. It consists of linear grooves on one surface of the refracting material. This concentrator may be installed with either the grooves facing the sun or the grooves facing downwards. Both glass and plastic can be used as refracting materials for Fresnel lenses.

In Fig. 7.27(*d*), the central stationary receiver, receives the solar radiations on a central receiver, which is stationary, concentration ratio is high, about 3,000. A combination of heliostat and paraboloida reflector may also be used. The heliostat can be adjusted to account for the change in sun's position.

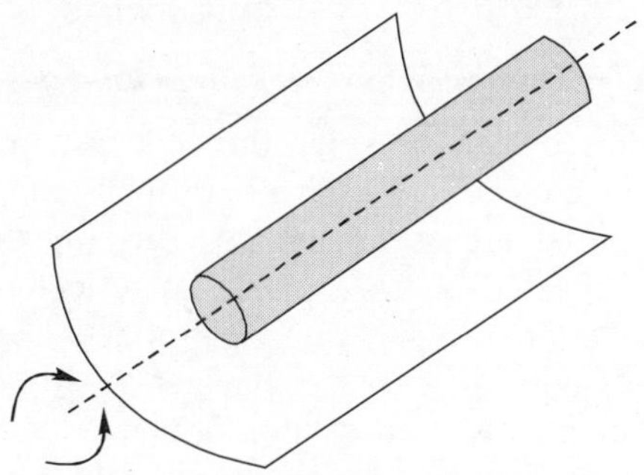

(a) Paraboloidal Reflector

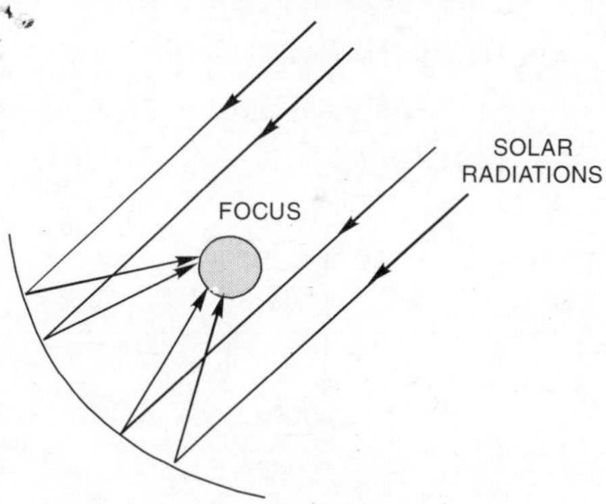

(b) Parabolic Trough Concentrator

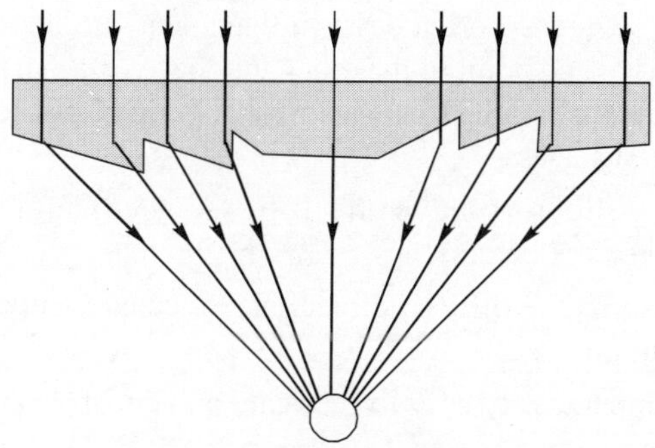

(c) Fresnel Lens Concentrator

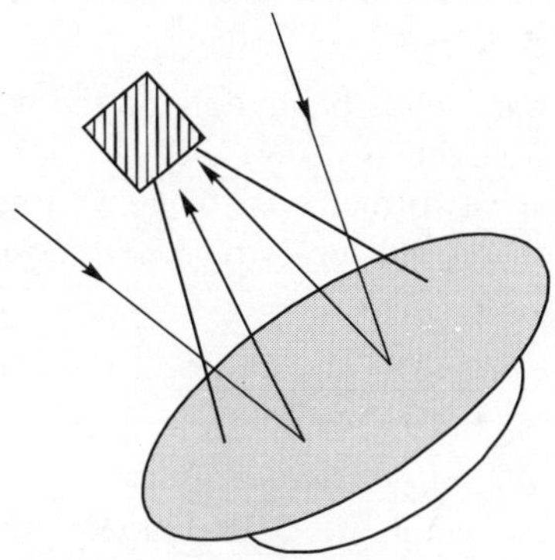

(d) Parabolic Dish Concentrator

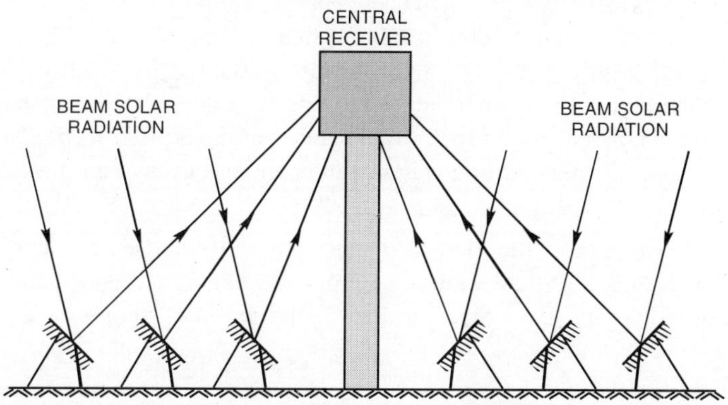

(e) Distributed Heliostat Point Focussing Reflector or Central Receiver With Heliostat Field

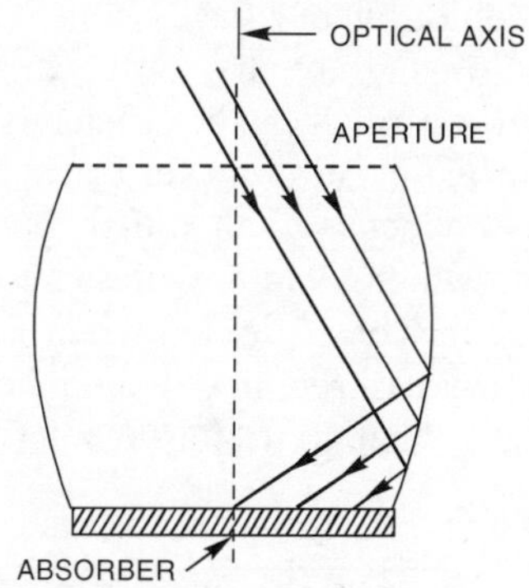

(f) Compound Parabolic Concentrator

Fig. 7.27. *Concentrator Solar Collector*

Heliostat have been found to be very suitable as collectors for solar power plants. The absorbed energy can be extracted from the receiver and delivered at a temperature and pressure suitable for driving turbines for power generation.

Figure 7.27(*f*) shows the compound parabolic concentrator. It was designed by Winston and Baranov. It consists of two parabolic segments, oriented such that focus of one is located at the bottom end point of the other and vice versa. The receiver is a flat surface parallel to the aperture joining of two foci of the reflecting surfaces.

Flat collectors have the following advantages over concentrator collectors:

1. Absorbs the diffuse, direct and reflected components of radition.
2. Comparatively easy to fabricate and cheaper; and
3. Since these are usually fixed in tilt and orientation, tracking is not required—this makes them maintenance free, except for surface cleaning.

Expensive research is being done in the various methods of converting solar energy into electrical energy. Broadly speaking there are two methods for converting solar energy into electrical energy namely *direct conversion method and conventional boiler method.*

The direct conversion methods have already been discussed in Arts. 7.3, 7.4 & 7.5.

Conventional Boiler Method. Basic elements of a solar power plant are shown in Fig. 7.28. Large parabolic collectors are employed for collecting solar energy, which is used to heat a fluid (usually water). This heat energy is finally transferred to feed water which is converted into steam. This steam is utilised to run a prime mover (steam turbine) coupled to an electric generator, which generates electric power. Steam is condensed in the condenser and feed water returns to the boiler for re-use. The heat of the cooling water of the condenser may be utilized for some other purposes.

Regarding the size of generating units some engineers prefer use of small units located at the place of use whereas others propose large centralised solar thermal plants.

Small generating units are best suited for India. According to the report of the National Council of Science and Technology (NCST) solar energy panel, 50 kW electrical generation plants are required to meet the energy needs of a village of 500 people. Such villages constitute almost 80% of the settlements in the country. Solar energy is the only possible energy source which can be efficiently used for inland areas remote from the existing power grids. For Indian conditions, flat plate collectors with mirror boosters and non-selective absorbers which yield low pressure steam at 150°C, are best. An efficiency of 15 to 18% can be aimed in such systems. Collector areas required would be 20 m^2/kW of peak generation and about 40 m^2 for the rated capacity. Small vapour turbines to generate electrical energy have been developed in Israel by Harry Tabor. According to him, monochloro-benzene is the best working fluid for use in small turbines operating at 150°C and 18,000 rpm. Other vapour engines using SO_2, NH_3 etc. have also been built. A miniature solar power plant in Senegal is already in operation.

Under the large centralised units for power generation there are two types of systems used namely *solar furnace system* and *solar farm system.* In solar furnace system the sun rays reflected from many different heliostats are concentrated on a single heat exchanger whereas in solar farm system a large number of linear reflectors focus sunrays on long pipes which collect heat.

There are two designs which can be used in solar furnace system and these are shown in Figs. 7.29 and 7.30. Figure 7.29 illustrates the tower concept for power generation. A boiler is mounted on the top of the tower located near the centre of the field of large mirrors called the *heliostats.* Sun rays after getting reflected from the heliostats are focused on the boiler. Thus boiler generates high temperature steam which is used to drive a steam turbine coupled to an electric generator. A 50 kW power plant based on this design has been built and operated in Italy. However, the world's largest power plant of 10 MW has been commissioned in 1982 in California (U.S.A.).

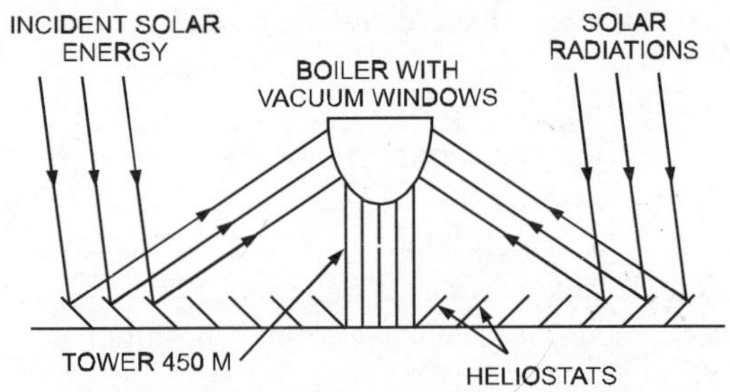

Fig. 7.29. ***Solar Furnace (Tower Concept)***

In another design of solar furnace system, shown in Fig. 7.30, arrays of heliostat mirrors to focus sun rays into a cavity type boiler are used. Sun rays striking the heliostat mirrors are reflected on to a parabolic reflector. From the parabolic reflector, the sun rays get reflected and concentrated in the cavity of a heat exchanger. A 1,000 kW solar furnace power plant, based on this principle, is being operated at Odeillo (France).

In a **solar farm system** the sun's heat is to be trapped in extensive arrays of steel pipes spread out in panels.

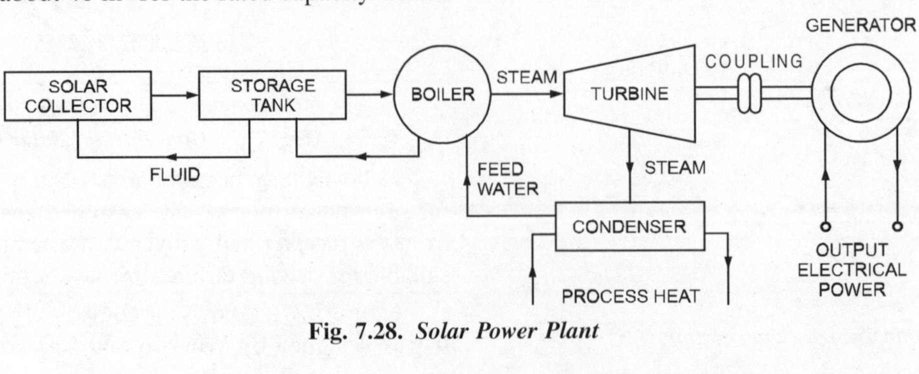

Fig. 7.28. ***Solar Power Plant***

For focussing the sun rays, cylindrical concave mirrors with parabolic cross section are used. The tubes are placed along the focal line of the mirrors. Heliostats are pivoted and are slowly rotated at a rate of 15° per hour to match the daily course of sun. During all these movements of mirrors, the tubes remain fixed. The tubes are coated with selective coatings (mixture of silicon and silver) and are enclosed in an evacuated glass chamber so that only a small proportion of the absorbed energy is emitted and conductive and convective heat losses are reduced. Nitrogen gas is pumped through the tubes at about 4 m/s to transfer the heat from the collectors to a central storage unit. The heat storage medium is a eutectic mixture of salts, mostly sodium nitrate, the heat would be used to raise steam. An overall efficiency of about 6% has been obtained at present. According to Dr. Menial of University of Arizona, 20 km^2 of cloud free land using this technique could produce a million kW of clean, pollution free power.

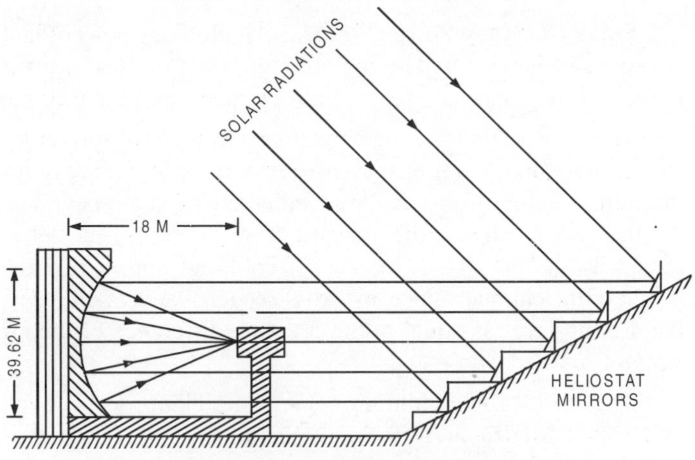

Fig. 7.30. ***Solar Furnace***

Fig. 7.31. ***Solar Trough Systems***

India's first solar thermal power plant based upon solar farm system was commissioned in May 1989 at Gawal Pahari, distt Gurgaon (Haryana). The total system consists of a solar collector field, an oil storage tank, a steam generator, air cooled steam condenser and a demineralised water plant. The capacity of the plant is 50 kW.

Solar Trough System. Trough systems predominate among today's commercial solar power plants. Trough systems convert the heat from the sun into electricity. Because of their parabolic shape, troughs can focus the sun at 30 to 60 times its normal intensity on a receiver pipe located along the focal line of the trough. Synthetic oil captures this heat as the oil circulates through the pipe, reaching temperatures as high as 390°C (735°F). The hot oil is pumped to a generating station and routed through a heat exchanger to produce steam. Finally, electricity is produced in a conventional steam turbine.

Solar Dish/Engine Systems. These systems, with net solar-to-electric conversion efficiencies reaching 30% can operate as stand-alone units in remote locations or can be linked together in groups to provide utility-scale power.

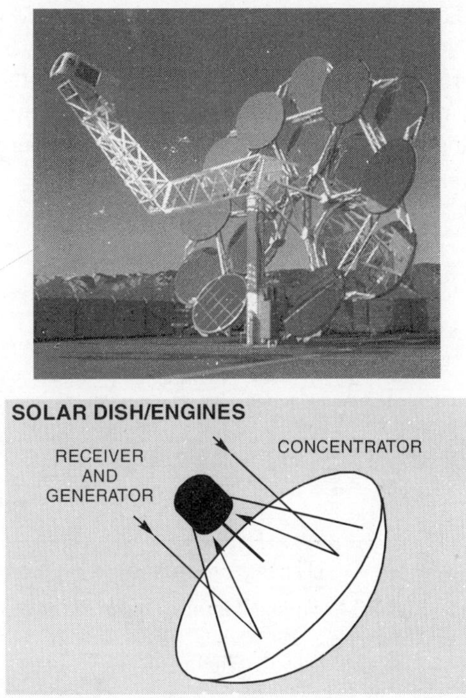

Fig. 7.32. ***The Sun's Energy is Concentrated on a Receiver and Generator Located at The Focal Points of The Parabolically Shaped Dish***

Solar dish/engine systems convert the heat energy of the sun into electricity at a very high efficiency. Using a mirror array formed into the shape of a dish, the solar dish focuses the sun's rays onto a receiver. The receiver transmits the energy to an engine, typically a kinematic sterling engine (although Brayton-cycle engines are also being considered), that generates electrical energy.

Because of high concentration ratios achievable with parabolic dishes and the small size of the receiver, solar dishes are efficient for collecting solar energy at very high temperatures. Tests of prototype systems and components at

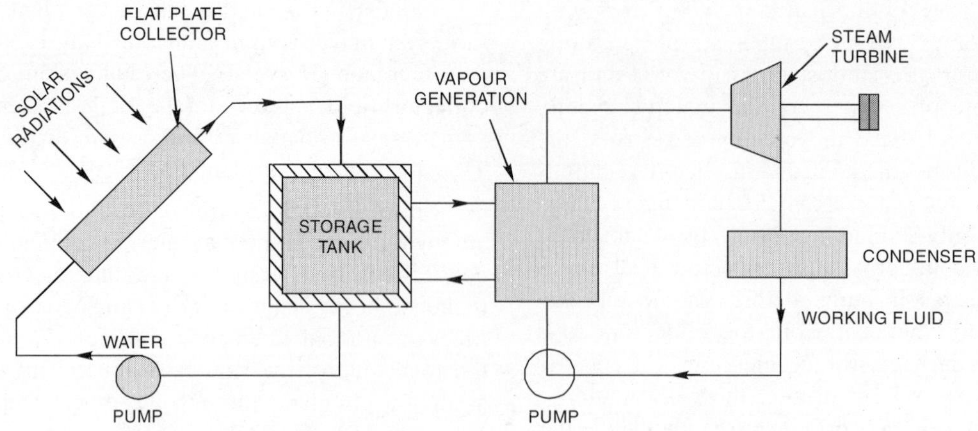

Fig. 7.33. *Low Temperature Power Generation Using Flat Plate Collectors*

locations throughout the United States have demonstrated net solar-to-electric conversion efficiencies as high as 30%. This is significantly higher than any other solar technology.

Application of Flat Plate Collectors in Power Generation. Low temperature power generation using flat plate collectors is shown in Fig. 7.33. The solar radiations are received by flat plate collectors. The heat energy is collected by water. The hot water is stored in insulated storage tank. From here, it flows through vapour generator through which the working fluid of Rankine cycle is also passed. The working fluid has a low boiling point. Vapour at 90°C leaves the vapour generator.

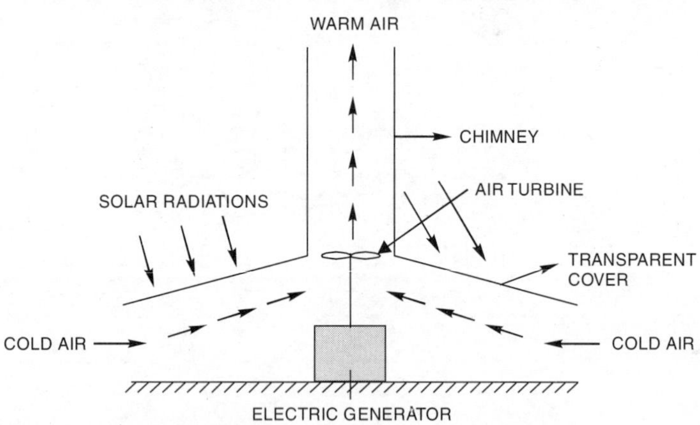

Fig. 7.34. *Solar Chimney Power Plant*

The vapour then executes a regular Rankine cycle by flowing through prime mover condenser and pump.

Solar Chimney Power Plant. Solar chimney power plant is shown in Fig. 7.34. The main components of such a plant are air heater of large area, a high chimney and air turbine.

The air stream is heated by solar radiation absorbed by the ground under a transparent cover of large size, flows through the air turbine rotor installed in the chimney. The driving force causing the upward flow of the air is determined by the density difference of cold and hot air and the height of the chimney. For 1 m^2 cross-sectional area of chimney, the driving force is equal to the pressure difference of the air stream.

The world's only chimney solar power plant was installed in 1981 at Manzanares in South Spain.

Solar Ponds. Solar ponds are also called *solar salt ponds.* Natural ponds convert solar radiation into heat, but the heat is quickly lost through convection in the pond and evaporation from its surface. A solar pond is designed to reduce convective and evaporative heat losses by reversing the temperature gradient with the help of nonuniform vertical concentration of salts. A greater salt concentration at the bottom than at top causes bottom water to have greater density and remains at the bottom and is also hotter. The

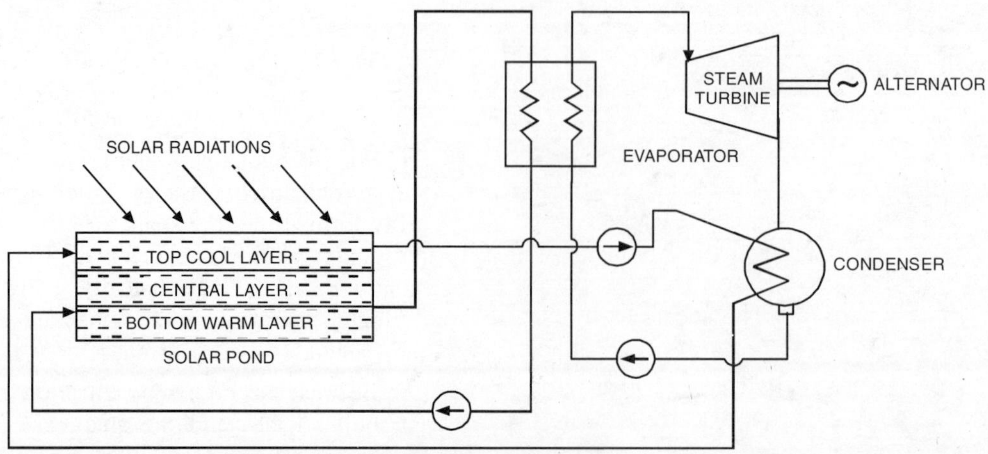

Fig. 7.35. *Solar Pond Power Plant*

solar energy is absorbed in deep layers and is usefully trapped.

The solar ponds are useful in two ways:

1. The conversion of solar energy to useful work as a result of the temperature difference between bottom and top layers of the pond.
2. The use of pond as a thermal storage medium.

A solar pond is a combination of solar collector and storage medium.

A schematic of artificial solar pond and conversion system is shown in Fig. 7.35. The pond is divided into three layers:

1. *Top Layer:* Low density convective layer.
2. *Central Layer*: Non-convective layer and contains the required salinity and density gradient with salinity greatest at the bottom.
3. *Bottom Layer*: High density-salinity convective storage layer.

The bottom warm water is used as the heat source and top cool water as the heat sink. The hot bottom water is pumped through an evaporator and back to the bottom. An organic fluid is evaporated and drives a turbo-generator. The turbine exhaust is condensed and the condensate is fed back to evaporator. The cooling water for condenser is taken from the top cool layer of the pond.

A leak-proof pond is constructed 330 kg of salt per square metre of pond area is added. The bottom layer can achieve a temperature of 93°C. The salinity gradient is maintained in the pond by injecting salt to the bottom layer and flushing the top layer with fresh water periodically.

The efficiency is very low but these plants have large potential for tapping solar energy.

The main applications of solar ponds are: (*i*) Power generation (*ii*) Space heating and cooling, (*iii*) Crop drying, (*iv*) Desalination, (*v*) Process heat.

The main limitations are: (*i*) Sunny climate, (*ii*) Requirement of large land area, (*iii*) Availability of water, (*iv*) Availability of salt (brackish water).

A solar pond power plant of 5 MW capacity is in operation in Dead sea, Israel.

Solar Lighting. Daylighting systems collect and distribute sunlight to provide interior illumination. This passive technology directly offsets energy use by replacing artificial lighting and indirectly offsets non-solar energy use by reducing the requirement for air-conditioning. Although difficult to quantify, the use of natural lighting also offers physiological and psychological benefits compared to artificial lighting. Daylighting design implies careful selection of window types, sizes and orientation; exterior shading devices may be considered as well. On implementation of daylighting features properly, light-related energy requirements can be reduced by 25%.

Hybrid solar lighting (HSL) is an active solar method of providing interior illumination. HSL systems collect sunlight using focus mirrors that track the Sun and use optical fibers for its transmission to inside the building for supplementing conventional lighting.

Solar Home Lighting System. In this system, solar cells are employed for supply of power. The electrical energy is stored in batteries and used for the lighting purpose when needed. Such systems are useful in non-electrified rural areas and as reliable emergency lighting system for important domestic, commercial and industrial applications. Solar photovoltaic systems have found important use in the dairy industry for lighting milk collection/chilling centers mostly located in rural areas.

Fig. 7.36. ***Solar Day Lighting***

The *solar home lighting system* is a fixed installation for domestic application. It comprises of solar PV module (solar cells), charge controller, battery and lighting system (lamps and fans). Figure 7.37 depicts the schematic of the solar home lighting system (HLS). The solar module is installed in the open on roof/terrace-exposed to sunlight and the charge controller and battery are kept inside a protected place in the house. The solar module needs periodic dusting for effective performance. The systems are designed to provide a daily working time of 3-4 hours with a fully charged battery. The system provides for buffer storage for 1-2 non-sunny/cloudy days.

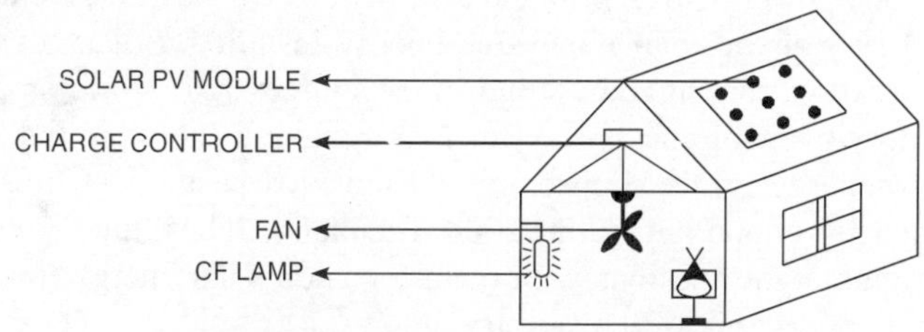

Fig. 7.37. ***Solar Home Lighting System***

Solar-Powered Pump. A solar powered pump is a pump running on the solar energy. It can be more environmentally friendly and economical in its operation. It consists of three parts: water pump, engine and energy source being powered by the sun.

The solar water pumping system is a stand-alone system operating on power generated by use of PV system. The energy produced by solar cells is used in operation of dc surface centrifugal mono-block pump set for lifting water from bore/ open well or water reservoir for minor irrigation and drinking water purpose.

Figure 7.38 depicts the schematic of the system. The system needs a shadow-free area for installation of the solar panel. The system is provided with 1,800 W solar PV panel (24 nos × 75 W power) and 2 hp centrifugal dc mono-block/ac submersible with inverter. The average water delivery of such a pump will be about 1.38-1.40 lakh litres per day, for a suction head of 6 m and dynamic head of 10 m. The size of suction and delivery lines is about 62.5 mm. Solar water pump sets have the advantages of zero fuel cost, nil conventional grid electricity requirement, long operating life, highly reliable and durable performance, easy operation and maintenance, eco-friendship and saving of conventional diesel fuel.

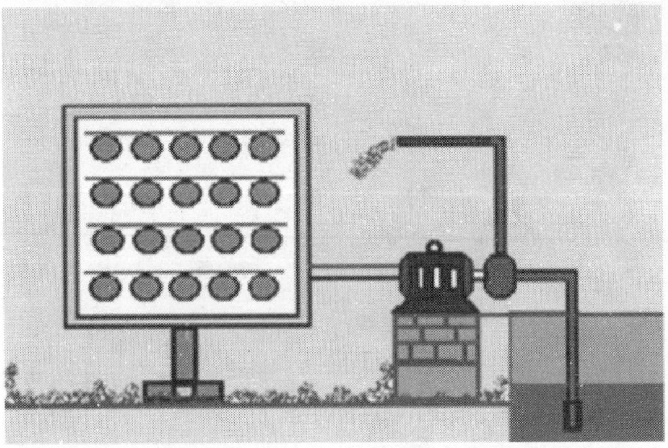

Fig. 7.38. ***Solar Water Pump***

These systems are employed for irrigation and drinking water in India. The majority of the pumps fitted with 200-3,000 watt motors are powered with 1,800 W power PV array that can deliver about 140,000 litres of water/day from a total head of 10 m.

Solar Power in India. As on 2010, India power sector has a total installed capacity of around 167,000 MW of which 54% is coal-based, 25% hydro, 8% is renewable and the balance is the nuclear and gas-based. Power shortages are estimated about 11% of total energy and 15% of peak capacity requirements and likely to increase in the coming years. Today the contribution of solar power with an installed capacity of 9.84 MW is a fraction (<0.1%) of the total renewable energy installed. Solar power generation has lagged behind other sources like wind, small hydropower, biomass etc. But now realizing the potential of solar energy, Prime Minister of India unveiled a National Climate Change Action Plan in June 2008. The plan will be implemented through eight missions with main focus on solar energy in the total energy mix of the country.

Development of alternate energy has been part of India's strategy for expanding energy supply and meeting decentralized requirements of the rural sector. The strategy is administrated through India's Ministry of New Renewable Energy (MNRE), Energy development agencies in various states and the Indian Renewable Energy Development Agency Limited (IREDA).

Involvement of various players in the energy sector, such as local industries. The private construction and operation contactors, Central Electricity Authority (CEA), MNRE and others has helped in increasing the capacity and capability of local technical expertise and further sustain the development of power in India in the longer term.

Solar industry in India gained momentum with the contribution of private organizations. Many Indian companies have planned major investments in this sector. The government has approved the projects of Chandradeep solar (for an R&D unit), Neotech Solutions, Photon Energy Systems, Surana Ventures and Ram Terra Solar Pvt Ltd., Tata Power and BP solar joint venture had been the leading Solar Photovoltaic (PV) manufacturers for the last many years. Moser Baer India Ltd. has entered the solar sector in a big way with both crystalline silicon cell technology and thin-film technology. Khandelwal Solar Power Ltd. (KSPL), Reliance Industries, Titan Energy Systems, Nano Tech Silicon India and XL Telecom & Energy also proposed to invest in the solar industry.

Present Status. As a result of efforts made during the last quarter century, a number of devices like solar water heaters, solar cookers, solar lanterns, solar street lights, solar water pumps have been developed and have become commercially viable. India has started wide solar photovoltaic program for about 2 decades and has installed an aggregate of 1.3 million systems. However, now the focus of the 11th Five Year Plan is on the grid connected power generation. India's integrated rural energy program using solar energy had served 300 districts and about 2,300 villages.

Research and Development Programmes. The Research and Development (R&D) programmes in the Solar PV technology have been aimed at development of materials employed in fabrication of solar cells and modules, different types of solar cell device structures, module designs, components, subsystems and systems, with a view to reduce the cost and improve the overall efficiency at different stages. The ministry has been sponsoring research and development projects on different aspects of the PV technology in academic and research institutions, national laboratories, IITs and industry, for development of new materials, processes, systems and production and testing equipment for solar cells and modules and electronics used in the PV systems. Number of R&D projects are going on solar PV program in India.

Challenges and Constraints

1. High Capital Cost. The hunt for better, cheaper cells is due in India. Despite the fact that the price of solar PV technology has been coming down over the years, it still remains economically unviable for power generation purposes. The average cost of solar PV modules was about Rs 2 lakhs/kW. Hoever, the estimated unit cost of generation of electrical energy from solar PV and solar thermal route is in the range of Rs 12-20 per kWh and Rs 10-15 per kWh respectively in India. With present level of technology, solar electrical energy produced through the PV conversion route is 4-5 times costlier than the electrical energy obtained from conventional fossil fuels.

2. Manufacturing Process. Solar PV cell manufacturing technology is a intensive process that needs high expertise and know-how. Besides, the technology landscape in the solar industry PV space is changing quite rapidly with innovations and research and development. It is challenging for new entrants to replicate the success of companies having a long standing in the solar PV market.

Raw Material and Waste Products. Some of the raw materials, such as cadmium, used for producing solar PV cells are hazardous and other raw materials like plastics used for the packaging of the cells are non-biodegradable, thereby having impacts on environments. Although some of the wastage produced during the manufacturing process is recyclable (silicon), not all other materials are recyclable and disposal of the same is a challenging process.

Environment Costs. Another concern area is installation of solar cells on the land area. The huge area of land required for utility–scale solar power plants–around one square kilometre for every 20-60 MW generated–poses an additional problem in India. Instead, solar energy in particular requires unique, massive applications in the agricultural sector, where farmers require electrical energy exclusive in the daytime. This could be the primary demand driver for solar energy in India.

In the very near future, breakthroughs in nano technologies promise significant increase in solar cell efficiencies from current 15% values to over 50% levels. These would in turn reduce the cost of solar energy production. However, capital costs have substantially declined over the past two decades, with solar PV costs declining by a factor of two. PV is projected to continue its current rapid cost reductions for the next decades to compete with fossil fuel. However, the realization of cost reductions is naturally closely linked to market development, government policies and support for research and development.

7.8. WIND POWER GENERATION

Winds are essentially caused by the solar heating of the atmosphere. They carry enormous quantity of energy. Before the development of electric power on large scale, wind power has served many countries as source of power in early days and were called as *windmills*. Wind power has been used for centuries to sail vessels, pump water, and grind wheat and corn. At the turn of this century, around 30,000 house windmills capable of generating a total of 100 MW and 3,000 industrial windmills generating another 100 MW were in operation in Denmark. Every year the winds provide 26,000 trillion kWh of which one-third is recoverable.

Wind as a source of power is very attractive because it is plentiful, inexhaustible, renewable, and non-polluting. There is no depletion of scarce resources. In large portion of the world, wind blows for 320 days in a year and this gives them an advantage over sunlight in direct conversion programmes. Operating cost of a windmill is negligible. Further, it does not impose extra heat burden on the environment.

Unfortunately this source of power is non-steady and unreliable. There are wide variations in the speed and direction of winds. Also it does not possess the basic requirements of any energy source *i.e.*, readily availability and long lasting supply. It makes necessary to store wind energy in some other forms during periods of high winds for use later on during calm–such as in the forms of chemical energy in rechargeable batteries, potential energy in high level reservoirs, mechanical energy in flywheels etc.

Another drawback of wind power is the unwieldy size, high structural area and quite large finance requirement. Also the varying wind speed creates problem in the case of windmills employed for electric power generation (ac output). Due to this the speed of the windmill will vary resulting in fluctuating voltage and frequency. However, methods have been devised to generate constant frequency power with varying wind speeds and consequently varying speeds of windmill propellers. In one of the mechanism, the pitch of the wind blades is controlled. Any change in wind speed is sensed by a governor and the blade pitch is automatically adjusted to regulate the output.

The kinetic energy per unit volume is given as $0.5\ \rho v^2$ where ρ is the density of air particles in kg/m^3 and v is the velocity of wind in m/s. For a surface area A, the kinetic energy comes out to be $0.5\ \rho v^2 A$ per unit distance in the direction of wind. The wind energy is convected along at velocity v. This means that kinetic energy available per second *i.e.*, power available is $0.5\ \rho v^3$. All this power cannot be extracted because for this the wind velocity would have to be reduced to zero which means that the wind mill would accumulate static air around it which would prevent the windmill operation.

It is clear that the power output of a windmill varies as the cube of the wind velocity, so the windmill produces maximum power at high wind velocity. The modern windmills are usually 50 m above the ground. However, the real power available is between 0.5 and 5 km above the ground level.

The ideal maximum efficiency using Froud momentum theory is equal to 59% but an overall efficiency of 30% could be had due to aerodynamic and other mechanical losses. This gives a power of about 0.3 kW/m^2 for a wind velocity of 10 m/s.

Worldwide, five nations—Germany, USA, Denmark, Spain and India—account for 80% of the world's installed wind energy capacity. Total worldwide wind power installed capacity is 79,300 MW. Kinetic energy available in the wind is converted into electrical energy by using rotor, gearbox and generator. The wind flows over the airfoil-shaped blades causing lift, like the effect on airplane wings, causing them to turn. The blades are connected to a drive shaft that power a pump or turns an electric generator, producing electricity. Larger blades capture more wind. As the diameter of the circle formed by the blades doubles, the power increases four times. Wind turbines are being installed on offshore oil and gas platforms in many areas to develop power for operating the equipment.

For wind machines to be economical, there must be winds that blow consistently above 16-24 km per hour. Many offshore areas have ideal wind conditions for wind machines. Denmark and UK have installed large offshore wind parks to take benefit of the consistent winds. Several offshore parks are planned for the United States in the near future.

Wind Characteristics. 1. Wind speed increases roughly as the 1/7th power of height. Typical tower heights are about 20-30 metres.

2. Energy pattern factor—the ratio of actual energy in varying wind to energy computed from the cube of mean wind speed, is always greater than unity *i.e.*, energy estimates based on mean (hourly) speed are pessimistic.

Utilization Aspects. There are three broad categories of utilization of wind energy.

1. *Isolated continuous duty systems* which require suitable energy storage and reconversion systems.
2. *Fuel-supplement systems* in conjunction with power grid or isolated conjunction conventional generating units.
3. *Small rural systems* which can use energy when wind is available.

Category 2 is the most predominant in use as it saves fuel and is fast growing particularly in energy deficient grids. Category 3 has applications in developing countries with large rural areas.

7.8.1. Wind Turbine

A wind turbine is a rotary device that extracts energy from the wind and converts into rotary mechanical energy. Blowing wind spins the blades on a wind turbine. When the wind blows, a pocket of low pressure air forms on the down-wind side of the blade. The low pressure air pockets then pulls the blade toward it, causing the rotor to turn. This is called lift. The force of the lift is actually much stronger than the wind's force against the front side of the blade, which is called drag. The combination of lift and drag causes the rotor to spin like a propeller. The rotor of the turbine are attached to a hub that is mounted on a rotating shaft. The shaft goes through a gear transmission box where the rotating speed is increased. The transmission is attached to a high speed shaft which turns a generator that produces electric power. The electric power may be dc or ac depending on the end user's need. Direct current can be generated by driving brushless dc generator attached to the windmill's rotor.

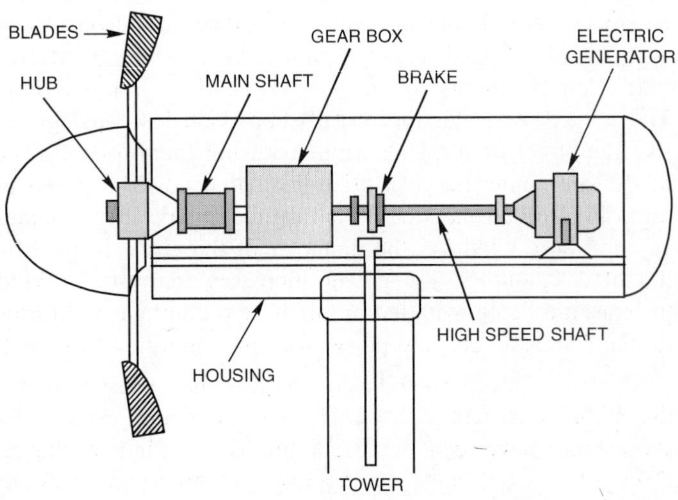

Fig. 7.39. ***Wind Turbine***

Simply stated, a wind turbine is the opposite of a fan. Instead of using electrical energy to produce wind, like a fan, wind turbines use wind to generate electrical energy.

Wind turbines can be used as stand-alone applications, or they can be connected to a utility power grid or even combined with photovoltaic system. For utility-scale sources of wind energy, a large number of wind turbines are usually built closer together to form a wind plant. Stand-alone wind turbines are used for water pumping or communications.

The performance of a wind power plant is characterized by the annual energy generation related to 1 m^2 of area swept by the rotor of wind turbine. Depending upon the average wind speed, the annual energy generation lies between 250 to 600 kWh per m^2 rotor area.

7.8.2. Difference Between Wind Turbines and Windmills

As already mentioned, a wind turbine is a rotary device that extracts energy from the wind and converts into rotary mechanical energy. If the mechanical energy is used directly by machinery, such as for pumping water or grinding stones, the machine is called a *windmill*. In case the mechanical energy is instead converted into electricity, the machine is known as wind generator, *wind turbine generator (WTG), wind power unit (WPU), wind energy converter (WEC) or aerogenerator.*

The following points further make the difference between wind turbines and windmills.

1. *Windmills* have been in use since the 9th century and use the kinetic energy of wind to do mechanical work, such as, pumping water, grain crushing.
2. *Wind turbines* use generators to convert the kinetic energy of the wind into electrical energy without producing any greenhouse gases.
3. *Windmills* are much smaller than wind turbines and usually have many blades. The blades catch more wind causing windmill to be able to do more physical work. The propeller blades are connected to an axle with gears. The gears are connected to a vertical shaft that runs down the length of the tower and is connected to other mechanical equipment. Windmills do work such as pump water or grind grain.

7.8.3. Type of Wind Turbines or Windmills

Basically, the wind turbines are of two types namely *horizontal axis wind turbines,* such as traditional farm windmills used for pumping water and the *vertical axis wind turbines,* such as the eggbeater-style Darrieus model, named after its French inventer.

Most large modern wind turbines are horizontal-axis turbines.

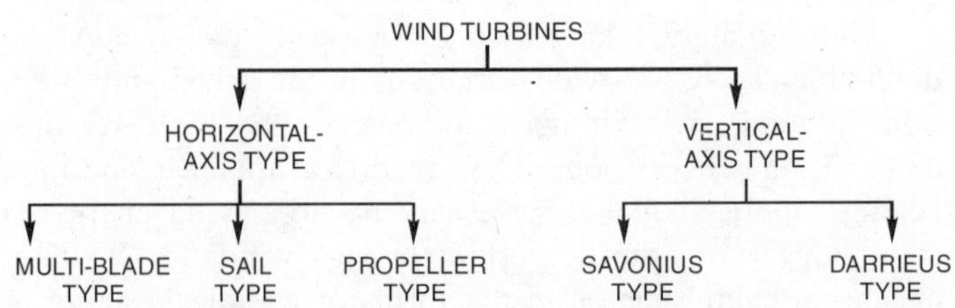

Fig. 7.40. ***Classification of Wind Turbines***

In *horizontal-axis turbines,* the axis of rotation is horizontal with respect to ground (and roughly parallel to the wind stream) while in vertical-axis turbines, the axis of rotation is verticle with respect to ground (and roughly perpendicular to the wind stream).

7.8.4. Wind Turbine Components

Wind turbines are designed to exploit the wind energy that exists at a location. Aerodynamic modeling is employed for determination of optimum tower height, control systems, number of blades and blade shape.

Conventional horizontal-axis wind turbines can be divided into three components.

The *rotor component* includes the blades for converting wind energy to low speed rotational energy and costs about 20% of the total cost of wind turbine.

The *generator component* includes the electrical generator, the control electronics and most likely a gearbox for converting the low speed incoming rotation to high speed rotation suitable for generation of electrical energy. It approximately costs about one-third of total cost of wind turbine.

The *structural support component* includes the tower and rotor yaw mechanism and costs about 15% of the wind turbine cost.

Other equipment including controls, electrical cables, ground support equipment and interconnection equipment.

7.8.5. Horizontal-Axis Wind Turbine (HAWT)

Horizontal-Axis Wind Turbine (HAWT) have the main rotor shaft and electrical generator at the top of the tower and must be pointed into the wind. Small turbines are pointed by a simple wind vane, while large turbines generally use a wind sensor coupled with a servo motor. Most have a gearbox, which turns the slow rotation of the blades into a high speed rotation that is more suitable for driving an electrical generator.

Fig. 7.42. ***Types of Wind Turbines***

Since a tower produces turbulence behind it, the turbine is usually pointed upwind of the tower. Turbine blades are made stiff to prevent the blades from being pushed into the tower and are sometimes tilted forward into the wind a small amount.

Downwind machines have been built, despite the problem of turbulence (mast wake), because they do not require an additional mechanism for keeping them in line with the wind and because in high winds the blades can be allowed to bend which reduces their swept area and thus their wind resistance. Since cyclic turbulence may lead to fatigue failure most HAWTs are upwind machines.

Fig. 7.42. ***Three-Bladed Wind Turbines***

Turbines used in wind farms for commercial production of electrical energy are usually three-bladed and pointed into the wind by computer-controlled motors. These have high tip speeds of over 320 kmph, high efficiency and low torque ripple, resulting in good reliability. The blades are usually coloured light gray to blend in with the clouds and range in length from 20 to 40 m or more. The tubular steel towers range from 60 to 90 m tall. The blades rotate at 10-22 rpm. At 22 rpm, the tip speed exceeds 90 m/s. A gearbox is commonly used for stepping up the speed of the generator, although designs may also use direct drive of an annular generator. Some models operate at constant speed, but more energy can be collected by variable-speed turbines which use a solid-state power converter to interface to the transmission system. All turbines are equiped with protective features to avoid damage at high wind speeds, by feathering the blades into the winds which ceases their rotation, supplemented by brakes.

Advantages: 1. Variable blade pitch, which provides the turbine blades the optimum angle of attack. Allowing the angle of attack to be remotely adjusted provides greater control, so the turbine collects the maximum amount of wind energy for the time of day and season.

2. The tall tower base allows access to stronger wind in sites with wind shear. In some wind shear sites, the wind speed can increase by 20% and the power output by 34% for every 10 m in elevation.

3. High efficiency due to movement of blades always perpendicular to the wind, receiving power through the whole rotation. In contrast, all vertical-axis wind turbines and most proposed airborne wind turbine designs, involving various types of reciprocating actions, need airfoil surfaces to backtrack against the wind for part of the cycle. Blacktracking against the wind leads to inherently poor efficiency.

4. The face of a horizontal axis blade is struck by the wind at a consistent angle regardless of the position in its rotation. This leads to a consistent lateral wind loading over the course of a rotation, reducing vibrations and audible noise coupled to the tower or mount.

Disadvantages: 1. Transportation difficulties in carrying tall towers and blades of length up to 45 m. Transport can now amount to 20% of equipment costs.

2. Tall HAWTs are difficult to install, requiring very tall and expensive cranes and skilled operators.

3. Requirement of massive tower construction for supporting the heavy blades, gearbox and generator.

4. Reflections from tall HAWTs may affect side lobes of radar installations producing signal clutter, although filtering can suppress it.

5. Their height makes them obtrusively visible across large areas disrupting the appearance of the landscape and sometimes causing local opposition.

6. Downwind variants suffer from fatigue and structural failure caused due to turbulence when a blade passes through the tower's wind shadow (for this reason, the majority of HAWTs use an upwind design, with the rotor facing the wind in front of the tower).

7. HAWTs need an additional yaw control mechanism to turn the blades and nacelle toward the wind.

8. In order to minimize fatigue loads due to wake turbulence, wind turbines are usually sited a distance of 5 rotor diameters from each other, but the spacing depends on the manufacturer and turbine model.

7.8.6. Vertical-Axis Wind Turbine

Vertical-axis wind turbines (VAWTs) have the main rotor shaft arranged vertically. The main advantages of such arrangement are that the turbine does not require to be pointed into the wind to be effective. This is an advantage on sites where the wind direction is highly variable. With a vertical axis, the generator and gearbox can be placed near the ground, so the tower does not require support and it is more accessible for maintenance. Drawbacks are that some designs develops pulsating torque.

Because of difficulty in mounting vertical-axis turbines on towers, they are usually installed nearer to the base on which they rests, such as the ground or a building rooftop. The wind speed is slower at a lower altitude, so less wind energy is available for a turbine of given size. Air flow near the ground and other objects can cause turbulent flow introducing issues of vibration, including noise and bearing wear that may increase the maintenance cost and shorten the service life. However, when a turbine is mounted on a rooftop, the building generally redirects wind over the roof and this can double the wind speed at the turbine. If the height of the rooftop mounted turbine tower is around 50% of tbe building height, this is near the optimum for maximum wind energy and minimum wind turbulence.

Eggbeater or Darrieus turbines have good efficiency but develop large torque ripple and cylindrical stress on the tower, that results in poor reliability. They also usually need some external power source or an adidtional Savonius rotor to start turning due to poor starting torque. The torque ripple is reduced by using 3 or more blades resulting in a higher solidity for the rotor. Newer Darrieus type turbines are not held up by guy-wires but have an external superstructure connected to the top bearing.

Giromill is subtype of Darrieus turbines with straight, as opposed to curved, blades. The cycloturbine has variable pitch in order to reduce the torque pulsation and is self starting. The advantages of variable pitch are: high starting torque; a wide, relatively flat torque curve; a lower blade speed ratio, a higher coefficient of performance; more effcient in turbulent winds; and a lower blade speed ratio resulting in lower blade bending stresses. Straight Vee or curved blades may be used.

Savonius wind turbines are drag type devices with two or more long helical scoops to provide a smooth torque.

Advantages: 1. VAWTs are usually mounted with the lower bearing mounted near the ground, therefore, massive tower structure is not necessarily required.

2. Design without yaw mechanisms are possible with fixed pitch rotor designs.

3. VAWTs may be built at locations where toller structures are prohibited.

4. VAWTs may have a lower noise figure.

5. VAWTs situated close to the ground can take advantage of locations where mesas, hilltops, ridgelines and passes funnel the wind and increase wind velocity.

6. The generator of a VAWT can be located nearer the ground, making it easier to maintain the moving parts.

7. VAWTs are lower wind startup speeds than HAWTs. Typically, they start producing electricity at 10 kmph.

Disadvantages: 1. VAWT may not develop as much energy at a given site as a HAWT with the same footprint or height. This is due to location of rotor close to the ground where wind speeds are lower because of the ground's surface drag.

2. The stress in each blade due to wind loading changes sign twice during each revolution as the apparant wind direction moves through 360 degrees. This reversal of the stress increases the likelihood of blade failures by fatigue.

3. While VAWTs components are located on the ground, they are also located under the weight of the structure above it, which can make replacement of parts very difficult if the structure is not designed properly.

4. A VAWT that uses guy-wires to hold it in place puts stress on the bottom bearing as all the weight of the rotor is on the bearing. Guy wires attached to the top bearing increase the downward thrust in wind gusts. Solution of this problem needs a superstructure to hold a top bearing in place in order to eliminate the downward thrusts of gust events in guy wired models.

7.8.7. Wind Energy Generators (WEG)

Wind energy conversion devices are commonly known as *wind turbines* because they convert the energy of the wind stream into energy of rotation; the component which rotate is called the *rotor*. The terms turbine and rotor are however regards as being synchronous. The combination of the wind turbine and generator is called the *aerogenerator.*

The fraction of free flow wind power that can be extracted by a rotor is known as power coefficient *i.e.*,

$$\text{Power coefficient, K} = \frac{\text{Power of wind rotor, } P_1}{\text{Power available in the wind, } P_2}$$

The maximum theoretical power coefficient is 0.593.

The amount of air passing per unit time,

Q = Area through which air passes, A × Velocity of air, v.

Mass of air travelling through area A swept by the rotating blades of windmill type generator

$M = \rho Q = \rho A v$ where ρ is density of air

Kinetic energy of moving air,

$$KE = \frac{1}{2}Mv^2 = \frac{1}{2}\rho Av \times v^2 = \frac{1}{2}\rho Av^3$$

$$\text{Available wind power} = KE = \frac{1}{2}\rho Av^3 \quad ...(7.11)$$

The physical conditions in a wind turbine are such that only a fraction of available wind power can be converted into useful work. The power available in the wind increases rapidly with the speed and therefore wind energy conversion (WEC) machines should be installed preferably in areas where winds are strong and persistent.

Wind turbine generators have been built having capacity in the range of a kW or so to a few thousands kWs. Wind power has been successfully used for cooling of homes, space heating, for operating irrigation pumps, navigational signals and for offshore drilling operations.

Basic Components of Wind Energy Conversion System

Various parts of a wind electric generator power plants are shown in Fig. 7.43.

Basic Components of a Wind Energy Conversion System (WECS): Aeroturbines convert the wind energy into rotary mechanical energy. A mechanical interface consisting of a step up gear and a coupling transmits the rotary mechanical energy to an electric generator. The generator output is connected either to load or power grid. The purpose of controller is to sense wind speed, wind direction, shaft speed and torques, output power and generator temperature.

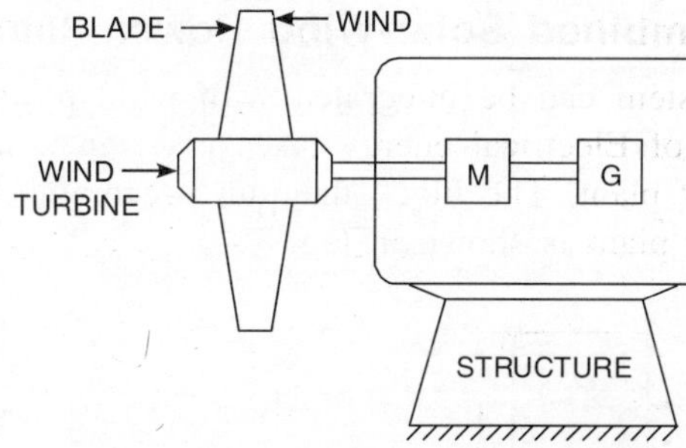

Fig. 7.43. ***Basic Components of a Wind-Electric Generator Power Plant***

Block diagram of WECS is shown in Fig. 7.44.

7.8.8. Combined Wind-Diesel Power Plants

The wind power plant can be integrated with diesel generator power plant, as shown in Fig. 7.45.

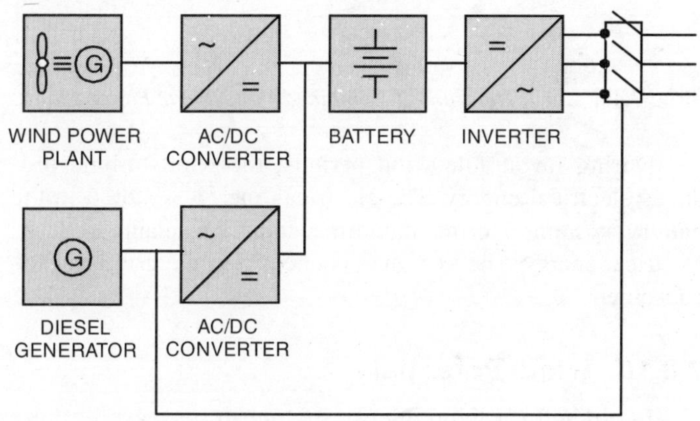

Fig. 7.45. ***Block Diagram of Combined Wind-Diesel Power Plant***

Generation of electrical energy takes place in wind/diesel power plant. If the wind is not favourable for a prolonged period or the machine is under maintenance, the diesel generator set supplies the required electrical power. A storage battery bank, inverter and power distribution electric grid make the power supply system complete. This combined power plant is useful for remote areas where wind is favourable and diesel supply is ensured.

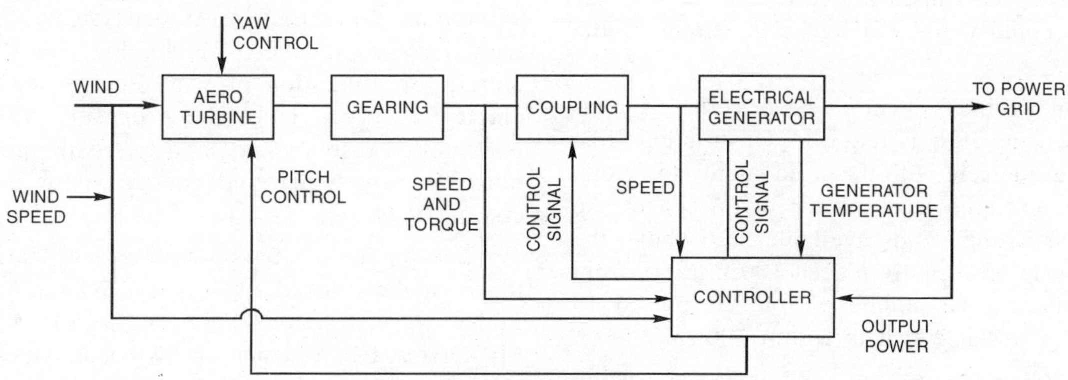

Fig. 7.44. ***Block Diagram of WECS***

7.8.9. Combined Solar-Wind Power Plant

The PV system can be integrated with wind power plant. Generation of Electrical energy takes place in a solar and wind power plant. The block diagram of combined wind-solar power plant is shown in Fig. 7.46.

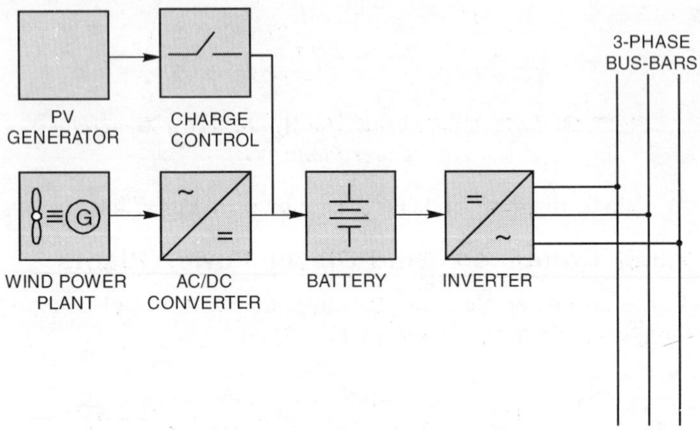

Fig. 7.46. *Block Diagram of Combined Wind-Solar Power Plant*

During favourable wind periods, the wind turbine produces electrical energy. The electrical energy is stored in the battery system. During day time solar PV plant produces electrical energy. The system is connected to the grid or to the consumer.

7.8.10. Wind Potential

In order for a wind energy system to be feasible there must be an adequate wind supply. A wind energy system usually requires an average annual wind speed of at least 15 kmph.

A wind generator develops lesser power in summer than in winter at the same wind speed as air has lower density in summer than in winter. Similarly, a wind generator produces lesser power in higher altitude—as air pressure as well as air density is lower than that at lower altitudes.

The wind speed is the most important factor influencing the amount of energy a wind turbine can develop. Increase in wind speed causes increase in amount of air passing the rotor and thereby increase in output of the wind system.

In order for a wind system to be effective, a relatively consistent wind flow is required. Obstructions like trees or hills can interfere with the wind supply to rotors. To avoid this, rotors are placed on top of towers to take advantage of the strong winds available high above the ground. The towers are usually placed 100 m away from the nearest obstacle. The middle of rotor is placed 10 metres above any obstacle that is within 100 m.

An important consideration in wind power generation is the economic flexibility of the wind power. The windmill designed to generate 1 MW at wind speeds above 48 kmph was erected in Vermount (USA) in 1941 had to be abandoned in 1945 for financial reasons as electricity could be made available at half the price from conventional sources. Another 9 MW windmill was proposed for the same place but the project was not found to be economically viable. However, wind power project may prove feasible for small power needs in isolated and remote areas. A local windmill generator may offer a cheaper alternative in case long transmission lines are required to bring power from grids to these remote and isolated areas.

7.8.11. Wind Energy in India

During the last four decades a few hundred small windmills have been installed in rural areas in south India for pumping water. An 8 meter diameter windmill was installed in village near Madurai in 1973. This windmill is used for irrigating fields and supplying water for domestic and dairy use. This windmill lifts 150 kg of water per minute to an height of 6 m when the wind speed is 16 kmph. A 220 V, 6 kW dc wind power plant supplied by Allgaier Werke Gmbh, West Germany has been installed at Central Power Research Institute, Bangalore. The annual output from this power plant has been estimated to be about 15,000 kWh.

Asia's largest wind power plant named 'Pawan Shakti' has been commissioned at Lamba near Porbandar (Gujarat) in 1990. This plant has 50 wind turbine generators, each of capacity 200 kW and has been set up as a joint venture of GEDA (Gujarat Energy Development Agency), GSEB (Gujarat State Electricity Board), and Ministry of Non-conventional Energy Resources, Govt. of India. Danish International Development Agency contributed about Rs 20 crores towards this project.

The wind energy sector has installed capacity of 11,807 MW (as on March 31, 2010). In terms of wind power installed capacity, India is ranked 5th in the world. Today India is a major player in the global wind energy market. Major wind power plants (100 MW and larger) are listed below in tabular form.

Indian Wind Energy Association has estimated that with the current level of technology, the 'on-shore' potential for utilization of wind energy for generation of electrical energy is of the order of 65,000 MW. The unexploited resource availability has the potential to sustain the growth of wind energy sector in India in the coming years.

The potential sites of wind energy harnessing in India are in the states of Tamil Nadu, Gujarat, Andhra Pradesh, Maharashtra, Karnataka, Kerala, Lakshdweep, Rajasthan, MP, Orissa, UP, Andeman and Nicobar. Among these Tamil Nadu has highest potential of wind energy.

List of Major Wind Power Plants (10 MW and Larger) in India

Power Plant	Producer	Location	State	Total Capacity (MWe)
Vankusawade Wind Park	Suzlon Energy Ltd.	Satara Dist.	Maharashtra	259
Sipla	CLP Wind Farms (India) Private Ltd.	Jaisalmer	Rajasthan	102
Samana	CLP Wind Farms (India) Private Ltd.	Jamnagar	Gujarat	101
Theni	CLP Wind Farms (India) Private Ltd.		Tamil Nadu	99
Saundatti	CLP Wind Farms (India) Private Ltd.	Belgaum	Karnataka	84
Khandke	CLP Wind Farms (India) Private Ltd.	Ahmednagar	Maharashtra	50
Narmada	CLP Wind Farms (India) Private Ltd.	Nallakonda	Andhra Pradesh	50
Harapanahalli	CLP Wind Farms (India) Private Ltd.	Davanagere	Karnataka	40
Cape Comorin	Aban Loyd Chiles Offshore Ltd.	Kanyakumari	Tamil Nadu	33
Kayathar Subhash	Subhash Ltd.	Kayathar	Tamil Nadu	30
Ramakkalmedu	Subhash Ltd.	Ramakkalmedu	Kerala	25
Muppandal Wind	Muppandal Wind Farm	Muppandal	Tamil Nadu	22
Gudimangalam	Gudimangalam Wind Farm	Gudimangalam	Tamil Nadu	21
Puthlur RCI	Wescare (India) Ltd.	Puthlur	Andhra Pradesh	20
Lamda Danida	Danida India Ltd.	Lamda	Gujarat	15
Chennai Mohan	Mohan Breweries & Distilleries Ltd.	Chennai	Tamil Nadu	15
Jamgudrani MP	MP Windfarms Ltd.	Dewas	Madhya Pradesh	14
Jogmatti BSES	BSES Ltd.	Chitradurga Dist	Karnataka	14
Perungudi Newam	Newam Power Company Ltd.	Perungudi	Tamil Nadu	12
Kethanur Wind Farm	Kethanur Wind Farm	Kethanur	Tamil Nadu	11
Hyderabad APSRTC	Andhra Pradesh State Road Transport Corp.	Hyderabad	Andhra Pradesh	10
Muppandal Madras	Madras Cements Ltd.	Muppandal	Tamil Nadu	10
Poolavadi Chettinad	Chettinad Cement Corp. Ltd.	Poolavadi	Tamil Nadu	10
Shalivahana Wind	Shalivahana Green Energy Ltd.	Tirupur	Tamil Nadu	20.4

7.8.12. Wind Turbine Size and Power Ratings

Wind turbines are availabe in a variety of sizes and therefore power ratings. The largest machine has blades that span more than the length of a football field, stands 20 building storeys high and generates enough electricity to supply 1,400 homes. A small home-sized wind machine has rotor between 2.44 and 7.62 metres in diameter and stands upwards of 9.144 m and can meet the power requriements of an all-electric home or small business. Utility scale turbines range in size from 50 to 750 kW. Single small turbines, below 50 kW, are employed for homes, telecommunication dishes or water pumping.

Small wind turbines have speed in the range of 100-300 rpm while large wind turbines have speeds in 30-600 rpm range. Large wind turbines are employed for supplying generated power into the grid.

7.8.13. Site Selection For WTGS

The site selection for a proposed WTGS is one of the most significant aspects. The main factors governing the selection of site are as under:

1. Availability of Wind. As already mentioned, an annual mean wind speed of about 18 kmph (*i.e.*, 5 m/s) or more is required for the installation of an economically viable WTGS. Wind data (such as wind speed, frequency distribution of wind speed, direction, turbulence, frequency of gusts and storms etc.) for a period of at least one year should be collected for different prospective sites.

2. Availability of Land. Large wind forms are required for installation of WTGS—about 25 acres for an installation of 1 MW. The actual land required for WTGS is only about 10% of this value and the remaining land is required for cultivation of low height vegetation.

3. Accessibility To Land. The land should be accessible for transportation of machinery and construction material.

4. Grid Stability. Grid should be free from voltage and frequency variations as far as possible.

7.8.14. Advantages of Wind-Generated Electricity

Wind energy is free, renewable resource, therefore, no matter how much is used today, there will still be the same supply in the future. Wind energy is also a source of clean, non-polluting electricity. Unlike conventional power plants, wind plants produce no air pollutants or greenhouse.

7.8.15. Environmental Concerns

Although wind power plants have relatively little impact on the environment compared to fossil fuel power plants, there is some concern over the noise produced by the rotor blades, aesthetic (visual) impacts and birds and bats having been killed (avian/bat mortality) by flying into the rotors. Most of these problems have been resolved or largely reduced through technological development or by properly siting wind plants.

7.8.16. Cost Issues

Even the cost of wind power has decreased dramatically in the last decade, the technology needs a higher initial investment than fossil-fueled generators. Roughly 80% of the cost is the machinery, with the balance being site preparation and installation. If wind generating systems are compared with fossil-fueled systems on a "life cycle" cost basis (counting fuel and operating expenses for the life of the generator), however, wind costs are much more competitive with other generating technologies because there is no fuel to purchase and minimal operating expenses.

7.8.17. Supply and Transport Issues

The major challenge to using wind as a source of power is that it is intermittent and does not always blow when electricity is required. Wind cannot be stored (although wind-generated electrical energy can be stored with the use of storage batteries) and not all winds can be harnessed to meet the timing of electricity demands. Further, good wind sites are located in remote locations far from load centers such as cities. Finally, wind resource development may compete with other uses for the land and those alternatives use may be more highly valued than electrical energy generation. However, wind turbines can be located on land that is also used for grazing of even farming. Transportation of windmill components particularly blades is difficult due to their large sizes.

7.9. TIDAL POWER GENERATION

Tidal power also represents a permanent source of energy. It can be predicted fairly accurately and is free from all types of pollutions. The rise and fall of tides nearly twice a day are associated with the rotation of earth every 24 hours in relation to both Sun and Moon. In large portion of the world tidal ranges are very low (say 1 metre or so) and, therefore, not suitable for electric power generation. However at some places the various coastal complexities give good tidal patterns owing to resonant effects of local geography. The tide amplitude may be considerable 18 to 21 m have been noted near the Magellan Strait and the shores of America respectively. There is an enormous amount of energy in waves and in tides but it is difficult to harness and control energy. The use of tides for electric power generation is practical in a few favourably situated sites where the geography of an inlet or bay favours the construction of a large scale hydroelectric power plant. For this a dam would have to be built across the mouth of the bay. It will have large gates in it and also low head water turbines installed in it. For storing water in the storage basin the gates are opened at the time of high tide and after storing water the gates are closed. After the tide has receded, there is a working water head between the basin water and open sea and the water is allowed to flow back to the ocean through water turbines installed in the dam. Thus the tidal energy stored at height can be utilised to drive a turbine coupled to an electric generator and thus generate electricity. Because of availability of low head bulb type* units are employed for such projects. The principle of operation is illustrated in Fig. 7.47.

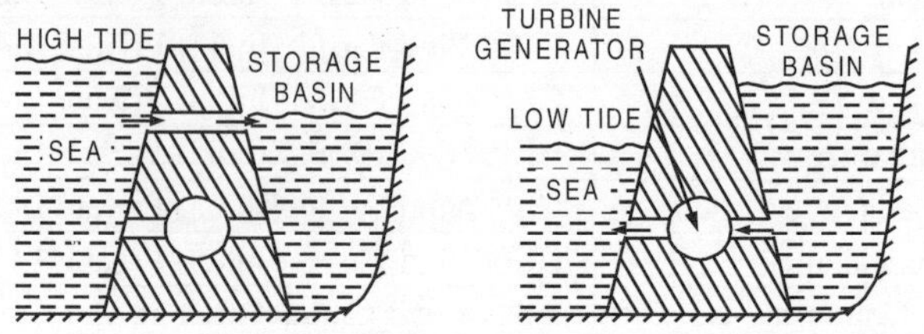

(a) (b)

Fig. 7.47

Though the arrangement explained above is simple and cheap but it is not useful due to variations in the output (the electric power generation is not continuous). By using reversible water turbine the turbine can run continuously both during high tide and low tide. The principle of operation is illustrated in Fig. 7.48. This arrangement is costlier but provides greater flexibility and continuous power output.

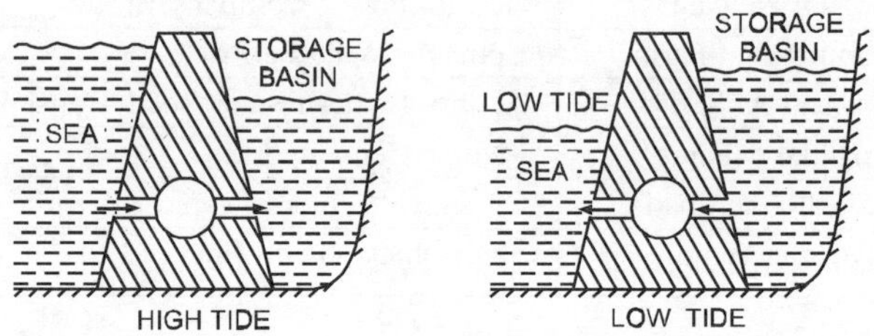

(a) (b)

Fig. 7.48

The above source for generation of electricity has been vigorously investigated in some countries like France, Germany, UK, Canada and USA. Probably the best known and the most recent tidal scheme, which utilises the tidal energy for generation of electrical energy has been constructed in France at Rance Estuary where a power station of 240 MW capacity has been in operation since 1967. This plant has 24 units of bulb type turbines, each of 10 MW capacity. The annual output of this plant is around 540 million kWh. The units are designed for two-way operation—from storage basin to sea and from sea to the storage basin. The units operate as pumps accelerating the tidal effect at small water level difference between the storage basin and the sea and they operate as turbine when one level is considerably above the other. The generators operate as motors driving the turbines as pumps during the pumping operation. The plant gates are used to accelerate filling and emptying of the storage basin at small differential water levels. This can be done with or without the operation of pumps. Operation cycle is shown in Fig. 7.49. Fully automatic operation (for starting, stopping and regulating generator sets and opening/closing valves to control power output) was introduced in 1982.

---

* A bulb type unit can be defined as a hydroelectric power unit installed in a hydraulic duct with its centre line coinciding with the flow axis.

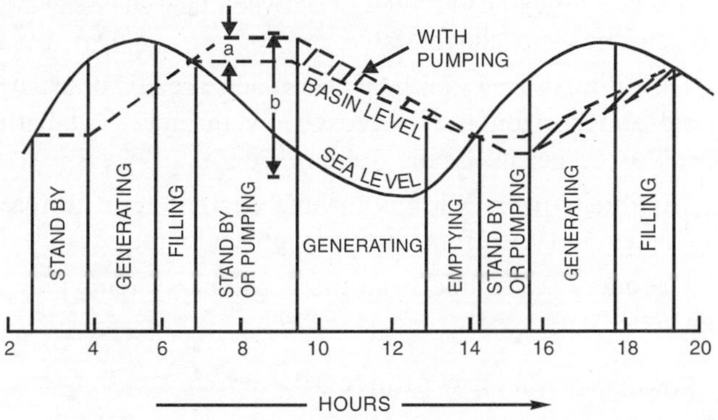

Fig. 7.49. ***Operation Cycle For a Bulb Unit***

Tidal fences can also harness the energy in the tides. A tidal fence has vertical axis turbines mounted within a fence structure, known as a caisson that completely blocks a channel, forcing all the water through it. Unlike barrage stations, tidal fences can be employed in unconfined basins, such as in a channel between the main land and a nearly offshore island or between two islands. As a result, tidal fences have much less impact on the environment, because they do not need flooding the basin. They are also comparatively cheaper to install. Tidal fences have the advantage of being able to generate electrical power once each individual module is installed. Tidal fences are not free of environmental and economic impacts, however, since the caisson can disrupt the movement of large marine animals and shipping. A 55 MW tidal fence is planned for the San Bemadino Strait in the Philippines.

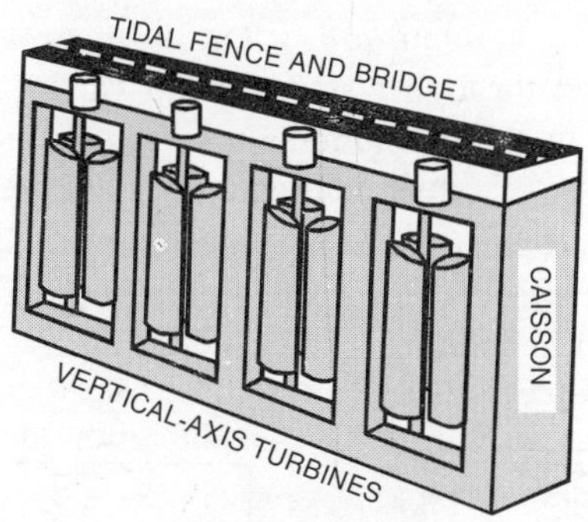

Fig. 7.50. ***Tidal Fence***

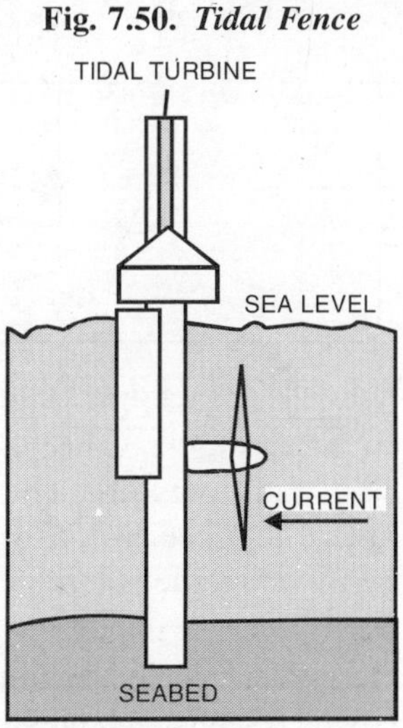

Fig. 7.51. ***Tidal Turbine***

Tidal turbines are basically wind turbines that can be installed wherever there is strong tidal flow, as well as in river estuaries. Since water is about 800 times as dense as air, tidal turbines will have to be much sturdier than wind turbines. They will be heavier and costlier to build, but will be also to capture more energy.

The advantages of tidal power are:

1. It is free from the problems of uprooting the people and disturbing the ecology balance.
2. It is everlasting and is not influenced by the changing mood of the nature such as failure of the monsoon.
3. No extra submerging of land is involved.
4. Tidal power generation is free from pollution.
5. The biggest advantage of tidal power plant is that it is inexhaustible.

Limitations of tidal power are:

(*i*) The major drawback of tidal power plants is their uneven operation. Variations in the tidal energy available through the lunar day and lunar month different from their solar counterparts prevent the tidal power from being regularly used in power systems during the periods of peak demand.

(*ii*) The other factors tending to limit the use of tidal schemes are the heavy initial capital and long construction period. The interest and depreciation component of the generation cost is significant.

(*iii*) Sea water is corrosive.

(*iv*) The conversion devices are relatively complicated in construction.

(*v*) Because of variable tidal range, the plant efficiency is affected.

(*vi*) Sedimentation and silteration of basins are the serious problems associated with tidal power plant.

However, many recent technological developments and improvements, both in design (*e.g.*, dynamic tidal power, tidal lagoons) and turbine technology (*e.g.*, new axial turbines, crossflow turbines) are suggesting that the total availability of tidal power may be much more than previously assumed, and that economic and environmental costs may be reduced to competitive levels. Such a scheme, if developed, can play a vital role in conservation of conventional resources.

Historically, tide mills have been used, both in Europe and on the Atlantic coast of North America. The earliest occurrences date from the Middle ages, or even from Roman times. France is currently the only country that has significantly harnessed tidal energy and has the tidal power plant of 240 MW in the world. The Seven Barrage is a proposed tidal power station to be built across the Bristol Channel (Seven Estuary) in UK. The River Severn has a tidal range of 14 m, making it perfect for tidal power generation. The Severn Barrage would involve the construction of a 16 km long barrage between Lavernock Point (Wales) and Brean Down (England). A total of 214 turbines each of 40 MW would be built into the barrage, making it a colossal of power plant of 8,560 MW of installed capacity with an average generation of 17 GWh/annum.

The drawback of uneven operation of tidal power plants however, can be overcome by operating the tidal power plant in conjunction with pumped-storage plant or run-of river power plant having a storage reservoir. In the tidal power-pumped storage scheme, the surplus energy generated (during the periods, the tidal power plant is generating electrical power more than requirements) by the tidal power plant can be used by the pumped storage plant to pump water into the upper reservoirs. When the tidal power plant output is at minimum, the pumped storage plant is run to deliver power to the system. This arrangement is attractive technically, but is costlier.

In the tidal power run-of river plants, the hydro plant raises its output when that of the tidal plant is declining until it shutsdown. When the tidal plant is operating at adequate capacity, water is stored in the reservoir of the hydro plant.

7.9.1. Tidal Power Generation Methods

There are three methods of tidal power generation as detailed below:

1. *Tidal stream systems* use kinetic energy of flowing water to drive power turbines as moving air is employed for driving windmills. This method is becoming popular because of the lower cost and lower ecological impact compared to barrages.

2. *Barrages* use the potential energy in the difference in height (or head) between high and low tides. Barrages are essentially dams across the full width of a tidal estuary, and involve very high civil infrastructure costs, and drawbacks of a worldwide shortage of viable sites, and environmental issues.

3. *Dynamic tidal power* exploits a combination of potential and kinetic energy by constructing long dams of 30-50 km length from the coast straight out into the sea or ocean, without enclosing an area. Both the obstruction of the tidal flow by the dam—as well as the tidal phase differences introduced by the presence of dam (which is not negligible in length compared to the tidal wavelength)—lead to hydraulic head differences along the dam. Turbines in the dam are used for conversion of power (6-15 GW per dam). In shallow coastal seas featuring strong coast-parallel oscillating tidal currents (common in UK, China and Korea), a significant water level difference (2-3 m) will appear between both sides of the dam.

Modern advances in turbine technology may eventually see large amounts of power generated from the ocean, especially tidal currents using the tidal stream designs but also from the major thermal current systems such as the Gulf stream, which is covered by the more general term marin current power. Tidal stream turbines may be arranged in high velocity areas where natural tidal current flows are concentrated such as the west and east coasts of Canada, the Strait of Gilbraltar, the Bosporus, and numerous sites in Southeast Asia and Australia. Such flows occur almost anywhere where there are entrances to bap and rivers, or between land masses where water currents are concentrated.

The turbines used in tidal power plants are: axial turbines; vertical and horizontal axis cross flow turbines; oscillating devices using aerofoils and venture effect.

Turbine Power. Various turbine designs have different efficiencies and therefore different power outputs.

The energy available from these kinetic systems may be given as

$$\text{Power generated, P (watts)} = \frac{1}{2}\xi \rho A v^3 \qquad ...(7.17)$$

where ξ is turbine efficiency, ρ is the density of sea water (1,025 kg/m^3), A is the swept area of the turbine in m^2 and v is the velocity of flow.

Relative to an open turbine in free stream, depending on the geometry of the shroud shrouded turbines are capable of as much as 3 to 4 times the power of same turbine rotor in open flow.

7.9.2. Wave Energy Conversion Devices

There are a number of wave energy conversion devices developed and experimented.

1. Hydraulic Accumulator Wave Machine. The machine consists of the following main components.

1. *Pressure Amplifier*: The waves enter the cylinder of pressure amplifier at the bottom and move the main piston. The pressure of the closed loop fluid is amplified to about 5 bar.
2. *Hydraulic Accumulator*: The high pressure fluid is conducted through a one-way up valve to a hydraulic accumulator. The accumulator has air cushions on the top which assists in maintaining a constant pressure.
3. *Pelton Turbine*: Part of high pressure fluid flows through a Pelton wheel or Francis hydraulic turbine that drives an electrical generator.

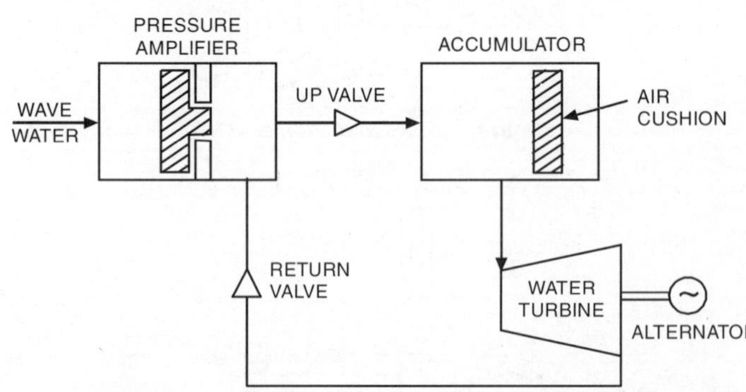

Fig. 7.52. ***Hydraulic Accumulator Wave Machine***

On the trough of the wave, the composite piston is pushed downward by the high fluid pressure. The exhaust water is sucked via return valve. The capacity of hydraulic accumulator is large enough to permit continuous turbine operation though the waves are cyclic.

2. High Level Reservoir Wave Machine. It also consists of pressure amplifier which sends high pressure water to a high level reservoir on a shore line. The potential energy of water in the reservoir is used to run a water turbine coupled to an electrical generator.

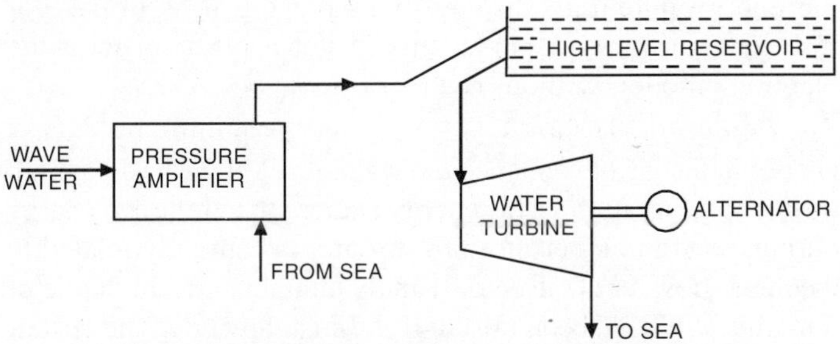

Fig. 7.53. ***High Level Reservoir Wave Machine***

3. The Dolphin Type Wave Power Machine. The major components of the system are a dolphin, a float, a connecting rod and two electrical generators (Fig. 7.54).

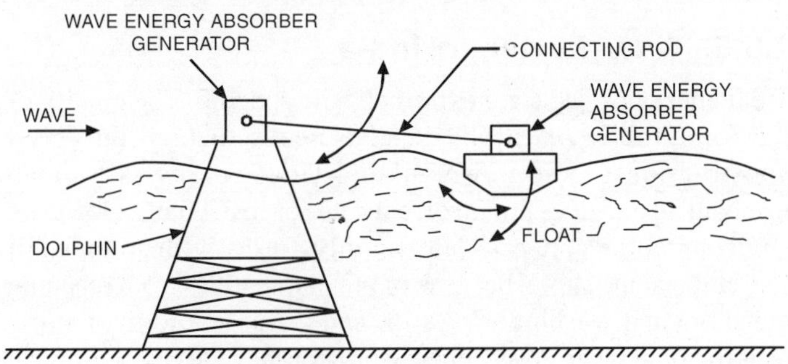

Fig. 7.54. ***Dolphin Type Wave Generator***

The float has two motions. The rolling motion about its own fulcrum with the connecting rod is amplified and converted into continuous rotary motion with the help of gears. The electrical generator is driven. The other vertical motion is also amplified and converted into rotary motion to drive the gears.

4. Dam-Atoll Wave Machine. The waves wrap themselves around the atolls (small volcanic islands) from all sides, producing a spiral in the centre. A turbine is driven when water is discharged radially outward.

7.9.3. Tidal Power Plants

For utilization of tidal energy, water must be trapped at high tide behind a dam or barrage and then made to drive turbine coupled to an electric generator as it returns to sea during low tides. The available energy is proportional to the square of the amplitude. The main component of tidal power plants are: (*i*) dam (*ii*) sluice ways from basin to sea and vice versa and (*iii*) power house.

The function of a dam is to form a barrier between the sea and the basin or between one basin to the other basin in case of multiple basins. The most suitable word for tidal power plant is *barrage*. Barrages have to resist waves whose shock can be severe and where pressure changes sides continuously.

The sluice ways are gate controlled devices. They are employed to fill the basin during the high tide or empty the basin during low tide. In existing plants, vertical lift gates have been employed. Flap gates are also used. The flap gates allow only in the direction of sea to basin. Therefore, the level of the basin rises.

Auxiliary equipments, turbines and generators are the main components of the power house. Large sized turbines are used because of low head available. Bulb types and rim type turbines are commonly used. Shaft turbines are also under steady.

7.9.4. Classification of Tidal Power Plants

Tidal power plants can be classified on the basis of basins used in power generation. There are two types of basin systems *viz.*, (*i*) single basin system and (*ii*) double basin system.

1. Single Basin System. Single basin system can generate power only intermittently. This is the simplest system of generating tidal power. The single basin scheme has only one basin. The basin is separated from the sea by a dam (barrage, Dyke). The sluice way is opened during high tide to fill the basin. The turbine-generator units are mounted within the ducts inside the barrage.

The single basin system can be further classified as (*i*) single ebb-cycle system, (*ii*) single tide-cycle system, (*iii*) double cycle system.

In *single ebb-cycle system,* when the high tide (flood side) are falling, sluices are opened to permit the sea water to enter the basin, while the turbine sets are shut. The level of the basin begins increasing. The energy is stored in the form of tidal range. Tidal range provides water head during low tides. The generation of power takes place, when the water from the basin flows over the turbine into the low level sea water. The turbines are designed for single way operation. The power output from such system is intermittent in nature and highly variable.

In *single tide cycle system,* the generation is affected when the sea is at flood tide. The sea water is admitted into the basin over the turbines. As the flood tide period is over and the sea level begins falling again, the generation is stopped. The basin is drained into the sea through the sluice ways. In this system also the power output is intermittent.

In *double cycle system,* the reversible turbines are installed and power is generated during filling and emptying of basin. Filling process occurs when the ocean is at high tide while the water in basin at low tide level, the emptying occurs when the ocean is at low tide and basin at high tide level. The flow of water in both directions is used to drive the reversible turbines. Each turbine drives the generator. In this system also continuous generation of power is not possible because of short duration. Electric power is generated during two short periods, during each tidal period of 12 hours 25 minutes or once every 6 hours and 12.5 minutes.

2. Double Basin System. There are two basins at different levels. A dam is provided between two basins. The turbines are located in the dam. The sluice gates are provided in the dam. One basin is called the upper basin, the water level is maintained above that in the other, the low basin. The high level basin gates are called the inlet gates and low level gates as outlet gates. The upper basin is filled with water. When the water level in upper basin A provides a sufficient difference of head between the two basins, the turbines are started. The water flows from basin A to basin B through the turbines and the power is generated. The power generation thus continues simultaneously along with filling up of water in basin A. When the tide attains its peak value, the water level in basin A is maximum, the inlet sluices are then closed. The water flows from the upper basin to the lower basin through the turbines. Thus, the water level in the upper basin falls and that in the lower basin rises. When the rising level in lower basin B becomes equal to the level of the falling tide, the outlet sluices are opened. When the tide reaches its lower most level, the outlet gates are closed. After some time the tide rises. When its level becomes equal to low level of the upper basin, the inlet gates are opened. Consequently the level of water in basin A starts rising. Thus, the cycle is repeated.

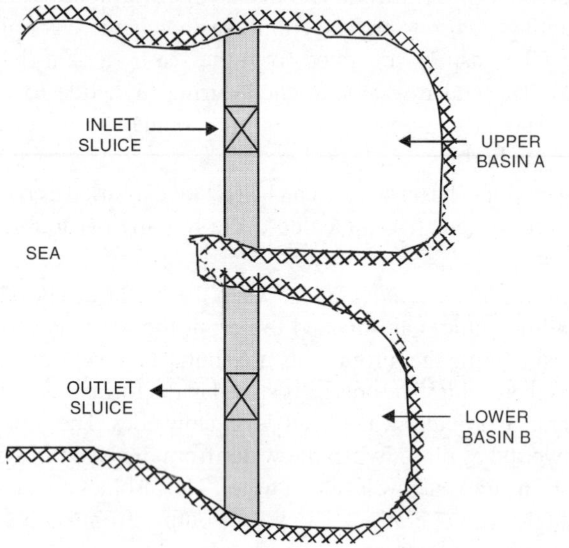

Fig. 7.55. *Double Basin System*

Two basin schemes have the advantages over normal schemes is that generation time can be adjusted with high flexibility and it is also possible to generate almost continuously. In normal estuarine situations, however, two basin schemes are very expensive to construct due to the cost of extra length of barrage. There are some favourable geographies, however, which are well suited to this type of scheme.

The tidal power scheme may be designed to operate in any one of following modes:

Ebb Generation. The basin is filled through the sluices until high tide. Then, the sluice gates are closed. At this stage, there may be pumping to raise the level further. The turbine gates are closed until the sea level falls to develop sufficient head across the barrage, and then are opened so that the turbines generate power until the head is again low. Then the sluices are opened, turbines disconnected and the basin is filled again. The cycle repeats. Ebb generation, also known as outflow generation, takes its name because generation occurs as the tide changes tidal direction.

Flood Generation. The basin is filled through the turbines, which produce at tide flood. This is usually much less efficient than Ebb generation, because the volume contained in the upper half of the basin (which is where Ebb generation operates) is greater than the volume of the lower half (filled first during flood generation). Thus, the available level difference between the basin side and the sea side of the barrage, reduces more quickly than it would in Ebb generation. Rivers flowing into the basin may further reduce the energy potential, instead of increasing it as in case of Ebb generation. Of course, this is not a problem with the "lagoon" model, without river in flow.

Pumping. Turbines are operated as pumps by excess energy in the grid to enhance the water level in the basin at high tide (Ebb generation). This energy is more than returned energy during generation, because power output is strongly related to the head. If water is raised 0.61 m by pumping on a high tide of 3 m, this will have been raised by 3.7 m at low tide. The cost of a 0.61 m rise is returned by the benefits of a 3.7 m rise. This is since the correlation between the potential energy is not a linear relationship, rather is related by the square of the tidal high variation.

Two Basin Schemes. Already described on page 147.

7.9.5. Tidal Power in India

Tidal energy projects are extremely site specific. The quality of the topography of the basin also needs to facilitate civil construction of the power plant. Tidal energy is a clean mechanism and does not involve the use of fossil fuels. However, environmental concerns are mainly to do with higher silt formation at the shore, because of preventing tides from reaching the shore and washing away silt, and disruption to marine life near the tidal basin. Wave energy projects have lesser ecological impact than tidal wave energy projects. In terms of reliability, tidal energy projects are believed to be more predictable than those harnessing solar or wind energy, since occurrences of tides are fully predictable.

Since India is surrounded by sea on three sides, its potential to harness tidal energy has been recognized by Govt. of India. Potential sites for tidal power development have already been located. The most attractive locations are the Gulf of Cambay and the Gulf of Kachchh on west coast where maximum tidal range is 11 m and 8 m with average tidal range of 6.77 m and 5.23 m respectively. The Ganges Delta in Sunderbans in West Bengal also has good locations for small scale tidal power development. The maximum tidal range in Sunderbans is about 5 m with an average tidal range of 2.97 m. The identified economic tidal power potential in India is of the order of 8,000-9,000 MW with about 7,000 MW in the Gulf of Cambay, about 1,200 MW in the Gulf of Kachchh and less than 100 MW in Sunderbans. The Kachchh tidal power project with an installed capacity of about 900 MW is estimated to cost about Rs 1,460/- crore generating electrical energy at about 90 paise per kWh. The techno-economic feasibility report is now being examined.

West Bengal 100 MW tidal power projects the first of its kind in the country, has already received clearance from the union Ministry of environment and forests. The West Bengal Renewable Energy Development Agency (WBREDA) will implement this Rs 40 crore pilot project in the Sunderbans area.

This eco-friendly will electrify several Sunderbans villages, benefiting about 200,000 people in the area. With a generating capacity of 3.6 MW, the project will comprise two barrages built across the upstream and downstream ends of the

Durgaduoni Creek which runs between the Gosaba and Bali-Bijoynagar islands and connects Bidyadhari and Gomti rivers. A bypass canal built at the downstream end will have a powerhouse and sluice gates. The plant will run for 14 hours a day (when the tide moves in or out). Electricity produced through an underwater horizontal axis turbine will be transmitted through a local grid to villages in Gosaba block. Currently, these villages are receiving 8 hours of dark power supply daily from a local biomass plant.

France is currently the only country that has significantly harnessed tidal energy and has the largest tidal power station in the world. Built in 1966, the La Rance tidal power station of Electricite de France (EdF) in Mont Saint Michel (northern France) has a generating capacity of 240 MW. It has 24 bulb-type turbines, each of 10 MW rating. The Severn Barrage is a proposed tidal power station to be built across the Bristol Channel (Severn Estuary) in U.K. The River Severn has a tidal range of 14 m, making it perfect for tidal power generation. The Severn Barrage would involve the construction of a 16-km long barrage between Lavernock Point (Wales) and Brean Down (England). A total of 214 turbines each of 40 MW would be built into the barrage, making it a colossal of power plant of 8,560 MW of installed capacity with an average annual generation of 17 GWh.

7.9.6. Economics

Tidal barrage power schemes have a capital cost and a very low running cost. As a result, a tidal power scheme may not produce returns for many years, and investors may be reluctant to participate in such projects.

7.10. GEOTHERMAL POWER GENERATION

Geothermal energy is the energy which lies embedded within the earth. Starting from the centre of the earth, it consists of the following zones.

(*i*) Solid metallic core (*ii*) A molten core (*iii*) A mantle of solid rock, 3,400 km thick, in layers (*iv*) A thin crust *i.e.*, outer shell about 35 km thick.

Temperature range from 3,000 to 4,000°C in the core and the base of the mantle. Therefore, underfoot there is tremendous heat energy which makes its presence felt in the eruption of volcanoes and in the spouting of hot springs and geysers Fig. 7.56.

Four-fifth of this heat comes from the slow decay of radioactive isotopes within the earth and about one-fifth is heat of the dust and gas clouds which coalesced to form the planet about 5 billion years before.

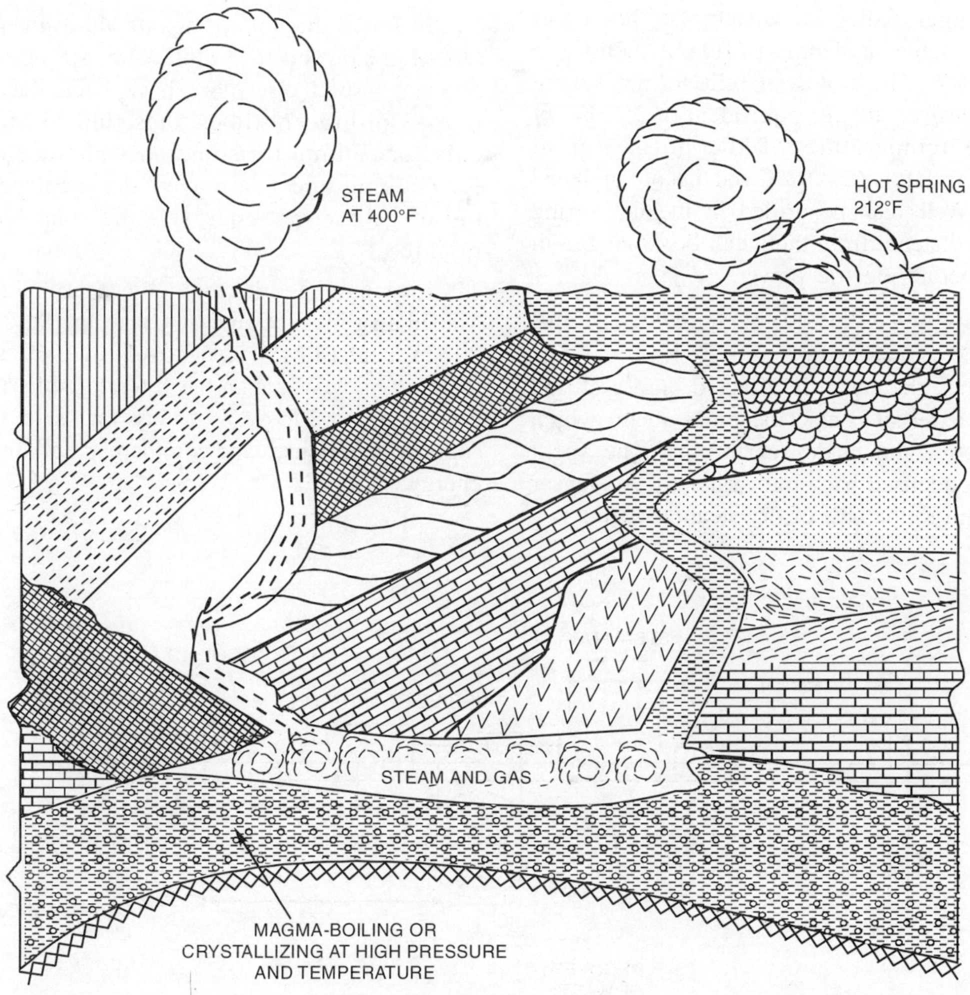

Fig. 7.56. ***Geothermal Deposit***

There are at least seven types of geothermal resources namely (*i*) dry steam fields (*ii*) wet steam fields (*iii*) hot water (*iv*) geopressure fields (*v*) Magma deposits (*vi*) hot dry rock (*vii*) volcanoes. The first three are called *hydrothermal reservoirs*, owing to involvement of water in some form, and are the best resources for production of geothermal energy at present.

Dry steam field is the most desirable form of geothermal energy. The steam is clean and easy to convert into electrical energy. Overall cost is considerably less than fossil or nuclear power. The geysers in California (USA) are good examples of dry steam fields. Steam from the well is collected, filtered to remove abrasive particles and passed through the steam-turbines coupled to electric generators. The capacity is 522 MW. Similarly dry and slightly superheated steam from wells near Lardello in Italy, has been used for generation of electrical energy since 1904. The present installed capacity of geothermal plants in Italy is more than 500 MW. Electrical energy is also being produced commercially in New Zealand, USA, Japan, Soviet Union and Mexico. However dry steam wells are rare.

There are about 300 known hot springs in India, located in the Himalayan Mobile belt, West coast region, Narmada-Sone river region and in Bihar-Bengal belt region. There is a plan to utilise the geothermal energy for generation of electrical energy in Pugga Valley in Ladakh. The hot water under pressure is available at depths of 11-32 metres, at temperatures of 50-110°C. The heat availability of this source is about 6,000 k calories/second *i.e.*, equivalent of 25 MW of electric power. The temperature of the hot water at Manikaran near Kulu in HP is 69 – 93°C and that of Surajkund (Hazaribagh) is 87°. Wells can be drilled in the hot spring region and the saturated steam from them can be expanded in steam turbines to generate electric power.

A schematic diagram of a typical power plant is shown in Fig. 7.57.

In dry steam plants, the steam is directed onto a closed water flow which on external contact becomes steam which in turn is used to move the blades of turbine. The spent water steam is then condensed back into water by condenser and then exposed again to the heat of dry steam. So, this is a closed cycle system.

Dry steam plants are already in operation in Italy, California and Japan. Residential heating systems based on geothermal resources are in use in Iceland, USA etc. However, despite these successful instances the best use of this resource lies in the generation of electrical energy.

Wet steam fields are twenty times more than the dry steam fields in the world. They give wet steam—a mixture of hot water and steam under high pressure. The steam is separated and expanded in turbines to generate electricity. Such a plant of capacity of 75 MW has been set up in Mexico. The hot water is treated for removal of its minerals and then used for agricultural or municipal purposes. Hot water can also be used for Desalination plants, air conditioning and refrigeration such as a 100 room hotel at Rotorua (NZ), heating of buildings, district heating, animal husbandry and industrial processes. In Iceland about 40% of the population live in geothermally heated houses. The hot water can itself be used to generate electric power by transferring its heat to a secondary medium having boiling point lower than that of water *e.g.*, butane and then expanding its vapour through vapour turbines.

Hot water gushing out from the earth's interior can be used to evaporate butane or Freon to run vapour turbines and thereby generate electricity, or for other purposes as mentioned above.

In India, the hot springs of Vajreshwari and Ganeshpuri, Off Mumbai, and the sulphur hot springs in foothills of the Himalayas end elsewhere have been known since generations. Until recent times, these hot springs have been used only for bathing for some possible medicinal benefits and for cooking and heating by the local people. The potential of geothermal energy is the smallest among all other resources.

In Japan, geothermal power production was begin at Matsukawa in 1966 and at Otake in 1967. The capacity was 20 MW and 13 MW respectively in 1969, and later increased to about 60 MW. Japan has a total geothermal power production of about 215 MW. Many towns in United States heat some of their houses and commercial buildings with geothermal energy.

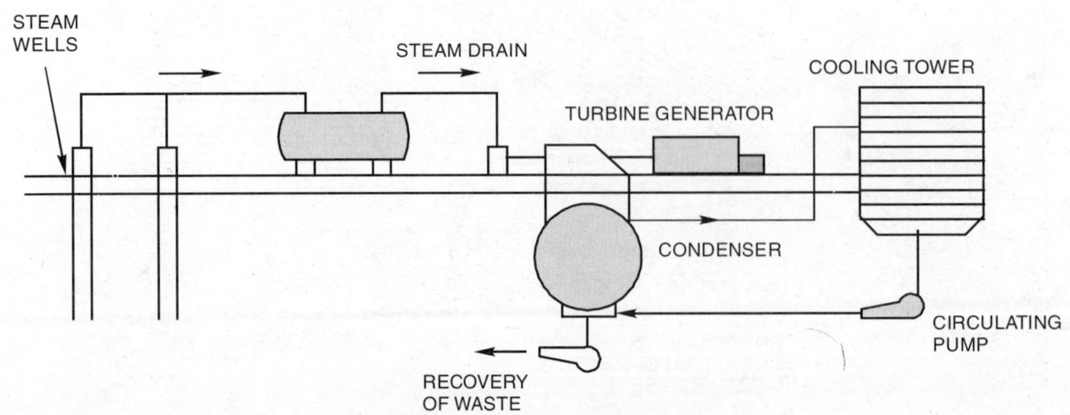

Fig. 7.57. ***Geothermal Power Plant***

7.10.1. Geothermal Electricity

According to the International Geothermal Association (IGA) sources, about 10,715 MW of geothermal power in 24 countries is online. In 2010, the United States led the world in geothermal electricity production with 3,086 MW of installed capacity from 77 power plants. The largest group of geothermal power plants in the world's is located at the Geysers, a geothermal field in California. The Philippines is the second highest producer, with 1,904 MW of capacity on line. Geothermal power makes up about 18% of the country electricity generation.

Geothermal electric power plants, were traditionally built exclusively on the edges of tectonic plates where high temperature geothermal resources are available near the surface. The development of binary cycle power plants and impovements in drilling and extraction technology enhanced geothermal systems over a much greater geographical range. Demonstration projects are operational in Landau-Pfolz, Germany and Soultz-sous-Forets, France. Other demonstration projects are under construction in Australia, UK and USA.

The thermal efficiency of geothermal electric power plants is low, about 10.23% because geothermal fluids do not attain the high temperatures of steam from boilers. The thermodynamic laws restricts the efficiency of heat engines in extracting useful energy. Exhaust heat is wasted, unless it can be used directly and locally, for example in greenhouses, timber mills, and district heating. Though system efficiency does not materially affect operational cost as it would for plants using fuels, but it does affect return on the capital cost of the plant. In order to generate more energy than the pumps consume, electricity generation needs relatively hot fields and specialized heat cycles. Because geothermal power does not rely on variable sources of energy, unlike, for example, wind or solar, its capacity factor can be quite large—up to 96% has been demonstrated. The global average was 73% in 2005.

7.10.2. Geothermal Power Plants Technology

For conversion of geothermal energy into electrical energy, heat must be extracted first for its conversion into useable form. Mite-or-more-deep wells can be drilled into under ground reservoirs to tap steam and very hot water that drive turbines coupled to electric generators.

There are basically four types of geothermal power plants which are operating nowadays. These power plants are described briefly as shown in Fig. 7.58.

1. Flashed Steam Plants. The extremely hot water from drill holes when released from deep reservoirs high pressure steam (known as flashed steam) is released. The force of steam is used to rotate turbines. The steam gets condensed and is converted into water again, which is returned to the reservoir. Flushed steam plants are widely distributed throughout the world.

Fig. 7.58. ***Geothermal Power Plant in Iceland***

2. Dry Steam Plants. Usually geysers are the main source of dry steam. The geothermal reservoirs producing mostly steam and little water are employed in electric power generation schemes. As steam from the reservoir shoots out, it is used to drive a turbine, after sending the steam through a rock-catcher. The rock-catcher protects the turbine from rocks which come along with steam.

3. Binary Power Plants. In this type of power plant, the geothermal water is passed through a heat exchanger where its heat is transferred to a secondary liquid, viz., isobutene, isopentane or ammonia-water mixture present in an adjacent, separate pipe. Because of this double-liquid heat exchanger system, it is called a *binary power plant.* The secondary liquid, also known as working fluid, should have lower boiling point than water. It turns into vapour on getting required heat from the hot water. The vapour from the working fluid is employed for rotating the turbines. The binary system is therefore useful in geothermal reservoirs which are relatively low in temperature gradient.* Since the system is a completely closed one, there is a minimum chance of heat loss. Hot water is immediately recycled back into the reservoir. The working fluid is also condensed back to the liquid and used over again and again.

---

* The temperature difference with the earth depends on

(*i*) The thermal properties of earth's interior and their radial and lateral variations.

(*ii*) Movements of fluids or solid rock materials occurring at rates of more than a few mm/annum.

Thermal (or temperature gradient) is defined as

$$\text{Thermal gradient} = \frac{\text{Heat flux}}{\text{Thermal conductivity}}$$

Thermal gradient will be high if either heat flux is high or thermal conductivity is low.

The heat energy in earth's interior is due to radioactivity. Regions of higher radioactivity have higher heat flux and are potential geothermal sites.

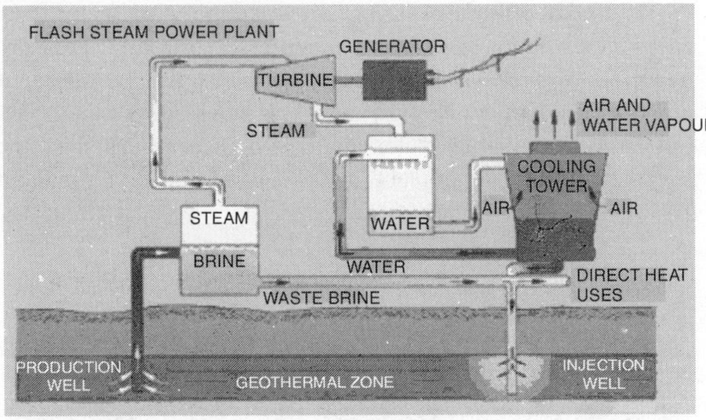

Fig. 7.59. *Electrical Power Generation Scheme Using Geothermal Energy*

4. Hybrid Power Plants. Some geothermal fields produce boiling water as well as steam, which are also employed in power generation. In this system of power generation, the flashed and binary systems are combined to make use of both steam and hot water. Efficiency of hybrid power plants is, however, less than that of dry steam plants.

Enhanced Geothermal System. The term enhanced geothermal systems (EGS), also called the engineered geothermal systems (formerly known as hot dry rock geothermal), refers to a variety of engineering techniques used to artificially create hydrothermal resources (underground steam and hot water) that can be used for generation of electricity. Traditional geothermal plants exploit naturally occurring hydrothermal reservoirs and are limited by the size and location of such natural reservoirs. EGS reduces these constraints by allowing for the creation of hydrothermal reservoirs in deep, hot but naturally dry geological formations. EGS techniques can also extend the lifespan of natural occurring hydrothermal resources. Given the costs and limited full-scale system research to date, EGS remains in its fancy, with only a few research and pilot projects existing around the world and no commercial-scale EGS plants to date. The technology is so promising, however, that a number of studies have found that EGS could quickly become wide spread.

7.10.3. Other Applications of Geothermal Energy

Low-temperature (300°F or 149°C) geothermal resources are typically used in direct-use applications like district heating, greenhouses, fisheries, mineral recovery, and industrial process heating. However, some low-temperature resources can be used for generation of electricity using binary cycle electricity generating technology. Direct heating is far more efficient than electric power generation and places less demanding temperature requirements on the heat resource. Heat may come from cogeneration via, a geothermal electrical plant, or from smaller wells or heat exchangers buried in shallow ground. Geothermal heat pumps can be used for space heating essentially anywhere.

7.10.4. History of Geothermal Energy Worldwide

In the 20th century, demand for electrical energy led to the consideration of geothermal power as a generating source. Prince Piero Ginori Conti tested the first geothermal power generator on July 4, 1904, at the same Larderello dry steam field where geothermal acid extraction began. It successfully lit four light bulbs. Later in 1911, the world's first commercial geothermal power plant was built there. It was the world's only industrial producer of geothermal electricity until New Zealand built a plant in 1958.

In 1960, Pacific gas and Electric began operation of the first successful geothermal electric power plant in the United States at the Geysers in California. The original turbine lasted for more than 30 years and produced 11 MW net power.

The binary cycle power plant was first demonstrated in 1967 in the USSR and later introduced to the US in 1981. In 2006, a binary cycle plant in Chena Hot Springs, Alaska, came on-line, producing electricity from a record low fluid temperature of 57°C (135°F).

Installed geothermal electrical power plant capacity as on 2007 was about 10,000 MW. The main countries having major electric generation installed capacities were USA (3,000 MW), Philippines (2,000 MW), Indonesia (1,000 MW), Mexico (1,000 MW), Italy (900 MW), Japan (600 MW), New Zealand (500 MW), Iceland (450 MW). The other region includes the Latin American countries, African countries and Russia.

7.10.5. India Geothermal Energy Resources in India

India's geothermal energy capacity have been estimated to produce 10,000 MW of power—a figure which is much higher than the combined power being produced from non-conventional energy sources like wind, solar and biomass. But yet geothermal power has not been explored. With the existing open economic policies of the Govt., and large incentives given to non-conventional energy sectors, the future of geothermal energy sector in India appears to be bright.

Several geothermal provinces in India characterized by heat flow (78-468 mW/m^2) and thermal gradients (47-100°C/km) discharge about 400 thermal springs. After the oil crisis in 1970s, the Geological Survey of India conducted lot of survey to explore the possibilities of geothermal power harnessing. The investigations carried out in the past have identified several sites which are suitable for power generation as well as for direct use. These provinces are capable of generating 10,000 MW of power. Though geothermal power production in Asian countries like Indonesia, Philippines has gone up, India with its around 10,000 MW geothermal power potential is yet to harness to its full capacity. However, with the growing environmental problems associated with thermal power plants, future for geothermal power in India apears to be bright.

The following Table indicates the temperatures, heat flow values and geothermal gradient of the geothermal provinces.

Table 7.3. Geothermal Provinces Data

Province	Surface T°C	Reservoir T°C	Heat Flow in mW/m^2	Thermal Gradient in °C/km
Himalaya	>90	260	468	100
Cambay	40-90	150-175	80-93	70
West Coast	46-72	102-137	75-129	47-59
SONATA	60-95	105-217	120-290	60-90
Godavari	50-60	175-215	93-104	60

From the above table it is clear that with the available technology all the above thermal provinces can be exploited for power generation as well for direct use.

7.10.6. Economics Related to Geothermal Energy Harnessing

Geothermal power requires no fuel (except for pumps), and is therefore, immune to fuel cost fluctuations, but capital costs are significant. Drilling account for over half the costs, and exploration of deep resources entails significant risks.

Unlike conventional power plants that run on fuel that is to be purchased over the life of the plant, geothermal power plants use a renewable source that is not susceptible to price fluctuations. The price of geothermal is within range of other electricity choices available today when the costs of lifetime of the plant are considered.

Most of the costs related to geothermal power plants are related to resource exploration and plant construction. Like oil and gas exploration, it is expensive and because only one in five wells yield a reservoir suitable for development. Geothermal developers must prove that they have reliable resource before they can secure millions of dollar required to develop geothermal resources.

Drilling. Although the cost of drilling geothermal has decreased during last two decades, exploration and drilling remain expensive and risky. Drilling costs alone account for as much as one-third to one-half of the total cost of a geothermal project. Location of best resources can be difficult, and developers may drill many dry wells before they discover a viable source. Because rocks in geothermal areas are usually extremely hard and hot, replacement of drilling equipment is required. Individual productive geothermal wells generally yield between 2 MW and 5 MW of electricity; each may cost \$1 million to \$5 million to drill. A few highly productive wells are capable of producing 25 MW or more power.

Transmission. Geothermal power plants are to be located near specific areas near a reservoir because it is not practical to transport steam or hot water over distances exceeding about 3 km. Since many of the best geothermal resources are located in the rural areas, developers may be limited by their ability to supply electricity to the power grid. New power lines are expensive to construct and difficult to site. Many existing transmission lines are operating near capacity and may not be able to transmit electricity without significant upgrading. Consequently, any significant increase in the number of geothermal power plants will be limited by those plants ability to connect, upgrade or build new lines to access to the power grid and whether the grid is able to deliver additional power to the market.

7.10.7. Barriers

1. Energy sources like wind, solar and hydro are more popular and better established; these factors could make developers decided against geothermal.
2. Main disadvantages of building a geothermal energy plant mainly lie in the exploration stage, which can be capital intensive and high-risk; many companies whose commission surveys are often disappointed, as quite often, the land they were interested in, cannot support a geothermal energy plant.
3. Some areas of land may have the sufficient hot rocks to supply hot water to a power station, but many of these areas are located in harsh areas of the world (near the poles), or high up in mountains.
4. Finding a suitable build location.
5. Harmful gases can escape from deep within the earth, through the holes drilled by the constructors. The plant must be able to contain any leaked gases, but disposing of gas can be very tricky to do safely.

7.10.8. Sustainability

Geothermal power is considered to be sustainable because any projected heat extraction is small compared to the Earth's heat content. The earth has an internal heat content of 10^{31} joules (3×10^{18} MWh). About 20% of this residual heat from planetary accretion, and the remainder is attributed to higher radioactive decay rates that existed in the past. Natural heat flows are not in equilibrium, and the planet is slowly cooling down on geologic timescales. Human extraction taps a minute fraction of the natural outflow, often without accelerating it.

Even though geothermal power is globally sustainable, extraction must still be monitored to avoid local depletion. Over the course of decades, individual wells draw down local temperatures and water levels until a new equilibrium is reached with natural flows. The three oldest sites, at Lorderello, Wairakei, and the Geysers have experienced reduced output because of local depletion. Heat and water, in uncertain proportions, were extracted faster than they were replenished. If production is reduced and water re-injected, these wells could their theoretically recover their full potential. Such mitigation strategies have already been implemented at some sites. The extraction of several geyser fields has also been attributed to geothermal power development.

7.10.9. Effects of Geothermal Energy on Environments

Fluids drawn from the deep earth carry a mixture of gases, notably carbon dioxide (CO_2), hydrogen sulphide (H_2S), methane (CH_4) and ammonia (NH_3). These pollutants contribute to global warming, acid, rain, and noxious smells if released. Existing geothermal electric power plants emit an average of 122 kg of CO_2 per MWh of electrical energy, a small fraction of the emission intensity of conventional fossil fuel plants. Plants that experience high level of acids and volatile chemicals are usually equiped with emission-control systems to reduce the exhaust.

Fig. 7.60. ***Geothermal Power Station in Philippines***

In addition to dissolved gases, hot water from geothermal sources may hold in solution trace amounts of toxic chemicals like mercury, arsenic, boron and antimony. These chemicals precipitate as the water cools and can cause environmental damage if released. The modern practice of injecting cooled geothermal fluids back into the earth to stimulate production has the side benefit of reducing this environmental risk.

Direct geothermal heating systems contain pumps and compressors, which may consume energy from a polluting source. This parasitic load is normally a fraction of the heat output, so it is always less polluting than electric heating. However, if the electricity is produced by burning fossil fuels, then the net emissions of geothermal heating may be comparable to directly burning the fuel for heat. For instance, a geothermal heat pump powered by electricity from a combine cycle natural gas plant would produce about as much pollution as a natural gas condensing furnace of the same size. Therefore, the environmental value of direct geothermal heating aplications is highly dependent on the emission intensity of the neighbouring electric power grid.

Plant construction can adversely affect land stability. Enhanced geothermal system can trigger earthquakes as part of hydraulic fracturing.

Geothermal has minimal land and fresh water requirements. Geothermal plant use 3.5 km^2 per GW of electric generation (not capacity) versus 32 and 12 km^2 for coal facilities and wind form respectively. They use 20 litres of fresh water per MWh versus over 1,000 litres per MWh for nuclear, coal, or oil.

7.11. OCEAN THERMAL ENERGY

The ocean can develop two types of energy viz., thermal energy from the sun's heat and mechanical energy from tides and waves.

Ocean covers more than 70% of Earth's surface, making them the world's largest solar collectors. The sun's heat warms the surface water a lot more than the deep ocean water, and this temperature difference (or temperature gradient) produces thermal energy. The ocean thermal energy is immense. In tropical regions, the surface water can be 40 or more degree warmer than the deep water. This temperature difference can be utilized in a heat generator to generate electric power. This is called ocean thermal energy conversion (OTEC). OTEC has the potential to generate more energy than tidal, wave and wind energy combined.

The OTEC systems can be open or closed. In a closed system, an evaporator turns warm surface water into steam under pressure. This steam rotates a turbine coupled to a electric generator. Water pumps bring cold deep water through pipes to a condenser on the surface. The cold water condenses the steam and the closed cycle begins again. In an open system, the steam is turned into fresh water and new surface water is added to the system. A transmission cable carries the electricity to the shore.

The OTEC systems must have a temperature difference of about 25°C to operate. This restricts OTEC's application to tropical region where the surface waters are very warm and there is deep cold water.

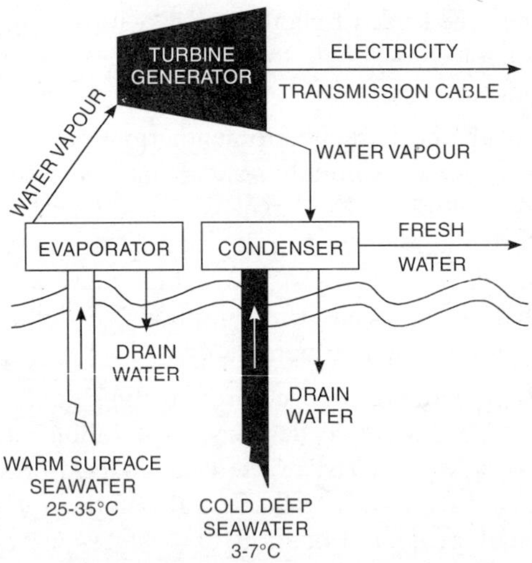

Fig. 7.61. ***Ocean Thermal Energy Conversion (OTEC) Scheme***

The temperature difference is small even in tropical regions, therefore, OTEC systems have very low efficiencies and consequently have very high capital cost.

The first plant for utilizing the Ocean Thermal Energy had a capacity of 40 kW and was tested in Cuba on 1930 by Georges Claude. The first modern OTEC power plant, which was located near Hawaii (USA), had a thermal capacity of 2.8 MW and electric power output of 50 kW (gross). However, the net power output of the plant was 18 kW. Hence, the overall efficiency was only 0.643%.

Although Japan has no potential OTEC sites, it has been a major contributor to the development of the technology, primarily for export to other countries. Beginning in 1970, the Tokyo Electric Power Company successfully built and deployed a 100 kW closed-cycle OTEC plant on the island of Nauru. The plant which became operational in 1981, produced about 120 kW; 90 kW was used to power the plant itself and the remaining was used to power a school and several places in Nauru. This set a world record for power output from an OTEC system where the power was sent to a real power grid.

India piloted a 1-MW floating OTEC plant near Tamil Nadu. Its government continues to sponsor various research in developing floating OTEC facilities.

Today there are several experimental OTEC plants, but no large operations. There are many challenges to widespread use. The OTEC sysetms are not very energy efficient. Pumping of water is a giant engineering challenge. In addition, the electricity is to be transported to land. It will be 10 to 20 years before the technology is available to generate and transmit electrical power economically from OTEC systems.

7.11.1. Ocean Thermal Energy Conversion

OTEC uses the temperature difference that exists between deep and shallow waters to run a heat engine. As with any heat engine, the greatest efficiency and power is developed with the largest temperature difference. This temperature difference usually increases with decreasing latitude, *i.e.*, near the equator, in the tropics. Historically, the main technical challenge of OTEC was to produce significant amounts of power efficiently from this very small temperature ratio. Changes in efficiency of heat exchange in modern designs allow performance approaching the theoretical maximum efficiency. The small temperature difference makes energy extraction comparatively difficult and expensive, due to low thermal efficiency. Earlier OTEC systems had an overall efficiency of only 1 to 3% (the theoretical maximum efficiency lies between 6% and 7%). Current designs under review will operate closer to the theoretical maximum efficiency. The energy carrier, sea water, is free, though it has an access cost associated with the pumping materials and pump energy costs. Although an OTEC power plant operates at a low overall efficiency, it can be configured to operate continuously as a base load power generation system.

A heat engine is a thermodynamic device placed between a high temperature reservoir and a low temperature reservoir. As heat flows from one reservoir to another reservoir, the engine converts some of heat energy to work energy. This principle is employed in steam engines and steam turbines. While in refrigeration, direction of flow of both the heat and work energy is reversed. Rather than using heat energy from the burning of fuel, OTEC power draws on temperature differences caused by the sun's warming of the ocean surface.

The most commonly used heat cycle for OTEC is the *Rankin cycle,* that uses a low-pressure turbine. OTEC systems may be either open-cycle or closed-cycle. Open-cycle engines use vapour from the sea water, itself as the working fluid while closed-cycle engines use working fluids that are typically thought of as refrigerants like ammonia or R-134a.

7.11.2. Types of OTEC Power Plants

OTEC power plants may be classified on the basis of (*i*) location and (*ii*) cycle as below:

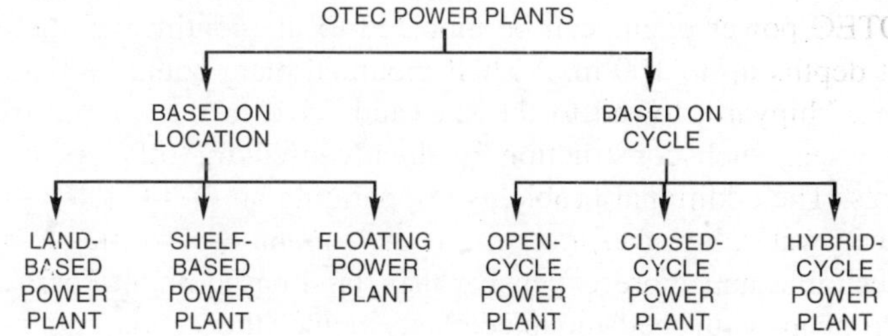

1. Power Plants Depending Upon Location

Land-Based Power Plant. Land-based and near-shore facilities offer three main advantages over those located in deep water. Power plants constructed on or near land do not require sophisticated mooring, lengthy power cables, or the more extensive maintenance associated with open-ocean environments. They can be installed in sheltered areas so that they are relatively safe from storms and heavy seas. Electricity, desalinated water, and cold, nutrient-rich sea water could be transmitted from near-shore facilities via trestle bridges or causeways. In addition, land-based, or near-shore sites allow OTEC power plants to operate with related industries such as mariculture or those that need desalinated water.

Favoured locations include those with narrow shelves (volcanic islands) steep (15-20 degrees) offshore slopes, and relatively smooth sea floors. These sites reduce the length of the cold-water in take pipe. A land-based power plant could be built well inland from the shore, offering more protection from storms, or on the beach, where the pipes would be shorter. In either case, easy access for construction and operation helps lower the costs.

Land-based or near-shore sites can also support mariculture. Mariculture tanks or lagoons built on shore allow workers to monitor and control miniature marine environments. Mariculture products can be delivered to market with relative ease via standard transport (rail roads or highways).

One disadvantage of land-based facilities arises from the turbulent wave action in the surf zone. Unless the OTEC power plants water supply and discharge pipes are buried in protective trenches, they will be subject to extreme stress during storms and prolonged periods of heavy seas. Also, the mixed discharge of cold and warm seawater may need to be carried several hundred meters offshore to reach the proper depth before it is released. This arrangement needs additional expense in construction and maintenance.

OTEC systems can avoid some of the problems and expenses of operating in a surf zone if they are built just offshore if waters ranging from 10 to 30 meters deep (Ocean Thermal Corporation 1984). This type of plant would use shorter (and therefore less costly) intake and discharge pipes, which would avoid the dangers of turbulent surf. The plant itself, however, would require protection from the marine environment, such as breakwaters and erosion-resistant foundations and the plant output would need to be transmitted to shore.

Shelf Based Power Plant. To avoid turbulent surf zone as well as to have closer access to the cold water resource, OTEC power plants can be mounted to the continental shelf at depths up to 100 m. A shelf-mounted plant could be built in a shipyard, towed to the site, and fixed to the bottom of the sea. Such construction is already used for offshore oil rigs. The additional problems of operating an OTEC plant in deep water, however, may make shelf-mounted facilities less desirable and more expensive than land-based counterparts. Problems with shelf-mounted plants include the stress of open-sea conditions and more difficult product delivery. Having to consider strong ocean currents and large waves requires additional engineering and construction expense. Platforms need extensive pilings to maintain a stable base for OTEC operation. Power delivery could also become costly due to requirement of long under water cables for reaching land. These reasons make the shelf-mounted plants less attractive.

Floating Power Plants. Floating OTEC facilities could be designed to operate off-shore. Although potentially preferred for systems with a large power capacity, floating facilities present several difficulties. This type of power plant is more difficult to stabilize, and the difficulty of mooring it in very deep water may cause problems with power delivery. Cables attached to floating platforms are more susceptible to damage, especially during storms cables at depth exceeding 1,000 m are difficult to maintain and repair. Riser cables, which span the distance between the sea bed and the plant, need to be constructed to resist entanglement.

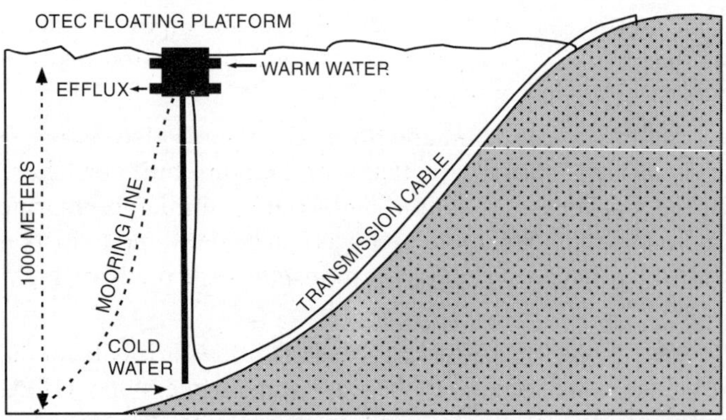

Fig. 7.62. ***Floating Type OTEC Power Plant***

As with shelf-mounted plants, floating plants require a stable base for continuous OTEC operation. Major storms and heavy seas can break the vertically suspended cold-water pipe and interrupt the intake of warm water as well. To reduce such problems pipes can be made of relatively flexible polyethylene attached to the bottom of the platform and gimballed with joints or collars. Pipes may need to be uncoupled from the plant to prevent storm damage. As an alternative to a warm-water pipe, surface water can be drawn directly into the platform; however, it is necessary to prevent the intake flow from being interrupted during violent motions caused by heavy seas.

If a floating power plant is to be connected to power delivery cables, it requires the plant to remain relatively stationary. Mooring is an acceptable method, but current mooring technology is limited to depths of about 2,000 metres. Even at shallower depths, the cost of mooring may prohibit commercial OTEC ventures.

2. Power Plants Depending on the Cycle Used. The cold seawater is an integral part of each of the three types of OTEC systems: closed-cycle, open-cycle, and hybrid. For operation, the cold seawater is to be brought to the surface. This can be done through direct pumping. A second method is of desalination of the seawater near the sea floor; this lowers its density, which will cause it to "float" up through a pipe to the surface.

The alternative to costly and massive, cold water pipes bring condensing cold water to the surface is to pump the much smaller volume of vapourised low boiling point fluid into the depths to be condensed thus overcoming a massive technical and environmental problem and lowering the cost of OTEC.

Closed-Cycle OTEC Power Plant: In a closed-cycle system, fluid with a low boiling point, such as ammonia, is used to rotate a turbine coupled to an electric generator. Warm surface seawater is pumped through a heat exchanger where the low-boiling-point fluid is vapourized. The expanding vapour rotates the turbo-generator. Then, cold, deep water pumped through a second heat exchanger condenses the vapour back into a liquid, which is then recycled through the system.

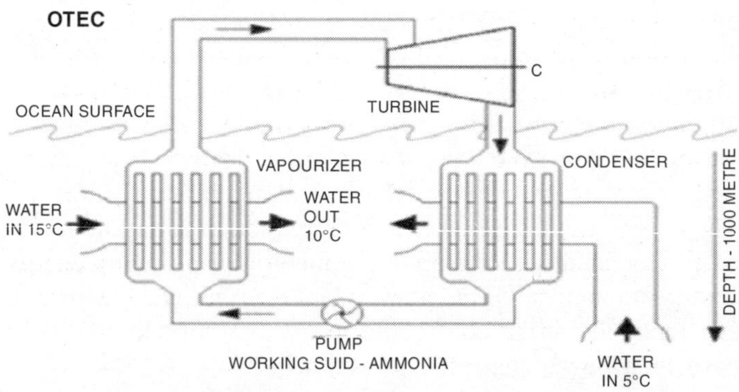

Fig. 7.63. ***Closed-Cycle OTEC Power Plant***

Open-Cycle OTEC Power Plant. In open-cycle OTEC power plant, the tropical ocean's warm surface is used to produce electricity. When warm seawater is placed in a low-pressure

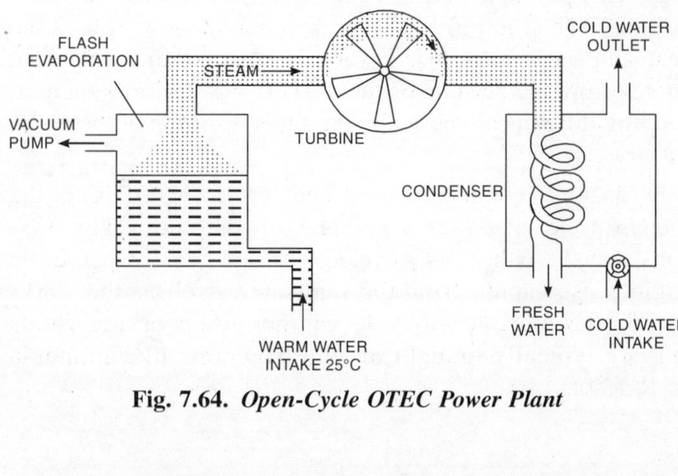

Fig. 7.64. ***Open-Cycle OTEC Power Plant***

container, it boils. The expanding steam drives a low-pressure turbine coupled to an electric generator. The steam, which has left its salt and contaminants behind the low-pressure container, is pure fresh water. It is condensed back into a liquid by exposure to cold temperatures from deep-ocean water. This method has the advantage of producing desalinized fresh water, suitable for drinking water or irrigation.

Hybrid OTEC Power Plant. A hybrid cycle combines the features of both the closed-cycle and open-cycle systems. In a hybrid OTEC system, warm seawater enters a vacuum chamber where it is flash-evaporated into steam, similar to the open-cycle process. The steam vapourises the ammonia working fluid of a closed-cycle loop on the other side of an ammonia vapourizer. The vapourised fluid then drives a turbine coupled to an electric generator producing electricity. The steam condenses within the heat exchanger and provides desalinated water.

The electrical energy generated by the system can be supplied to a utility grid or used to produce methanol, hydrogen, refined metals, ammonia, and similar products.

7.11.3. OTEC Advantages

The variety of products and services are the major advantage of OTEC plants. Ocean thermal is not relatively clean and produce more pollutants that contribute to global warming. The open cycle eliminates the need for a surface heat exchanger. The major advantage of OTEC for power generation is that the ocean energy is renewable source. Since no fuel is used, the OTEC system may become more economical in future.

7.11.4. Limitations of OTEC Power Plants

Problems facing commercial development of OTEC systems are enormous.

1. Because of low pressure, large sized turbines are required. Steam turbines capable of generating 10 MW or more using low pressure steam have yet to be developed.
2. Size of plant is limited due to large size of components.
3. The maintenance of vacuum in the pumps requires very large size vacuum pumps.
4. Low temperature differences result in very low plant efficiencies, very large plant size and huge capital cost.
5. For open-cycle systems, the problems associated with the design of evaporator, operation and maintenance are many.
6. For closed-cycle systems, heat exchangers of very large size have to be designed and built and the working fluid is expensive.
7. Pollution can be caused by closed-cycle OTEC if intermediate chemicals leak into ocean.
8. OTEC plant can alter and may damage the surrounding ecosystem.
9. Pumps handling large amounts of water have to be developed.
10. A 100 MW OTEC power plant may require 1 km long and 30 m diameter pipe.
11. The whole plant has to be stationed and moored at large depth.
12. Construction of floating power plant is difficult.
13. Cost of electrical energy generated from open-cycle OTEC is very high.
14. The development of the electrical cables required to carry the power to shore needs special care because it is subjected to severe stresses from its own weight and ocean currents and eddies.
15. An OTEC power plant must be capable of withstanding severe ocean storms over its lifetime, corrosion by seawater salts, erosion due to large volume flows, biofouling due to algae growth and encruotation by various marine life such as barnacles.

7.11.5. Applications of OTEC Plants

In addition to generation of electrical energy, open-cycle plants can be used to produce desalinated water, which can be used for irrigation and human consumption. A closed-cycle OTEC plant can also act as a chemical treatment plant. An OTEC plant can also be employed for pumping deep seawater. The cold water then can be used for air-conditioning and refrigeration if it is brought back to shore. In addition, while the sea water surrounding the plant is being moderated, the enclosing area can be used for aquaculture and mariculture.

The power generated in the OTEC plants can be used for producing hydrogen by means of electrolysis of water.

7.12. BIOMASS

Biomass, a renewable energy source, is biological material living or recently living organisms like wood, waste, hydrogen gas and alcohol fuels. Biomass is commonly plant matter grown to produce electricity or heat. In this sense, living biomass can also be included, as plants can also generate electricity while alive. The most conventional way in which biomass is used, however, still relies on direct incineration. Forest residues like dead trees, branches and tree stumps, yard clipping, wood chips and garbage are often employed for this. However, biomass also includes plant or animal matter used for producing fibres or chemicals. It excludes organic materials like fossil fuels which have been transformed by geological processes into substances such as coal or petroleum.

Industrial biomass can be grown from numerous types of plants, including miscanthus, switchgrass, hemp, corn, poplar, willow, sorghum, sugarcane and a variety of topic tree species, ranging, from eucalyptus to oil palm (palm oil). The particular plant used is usually not important to the end products, but it does affect the processing of the raw material.

Although the fossil fuels have their origin in ancient biomass, they are not considered biomass by the generally accepted definition because they contain carbon that has been "out" of the carbon cycle for a very long time. Their combustion therefore disturbs the CO_2 content in the atmosphere.

Biomass is carbon, hydrogen and oxygen based. Nitrogen and small quantities of other atoms, including alkali, alkaline earth and heavy metals can be found as well. Metals are often found in functional molecules such as the porphyrins which include chlorophyll which contains magnesium.

Plants in particular combine water and carbon dioxide to sugar building blocks. The required energy is produced from light via photosynthesis based on chlorophyll. On average, between 0.1 and 1% of the available light is stored as chemical energy in plants. The sugar building blocks are the starting point for the major fractions found in all terrestrial plants, lignin, hemicellulose and cellulose.

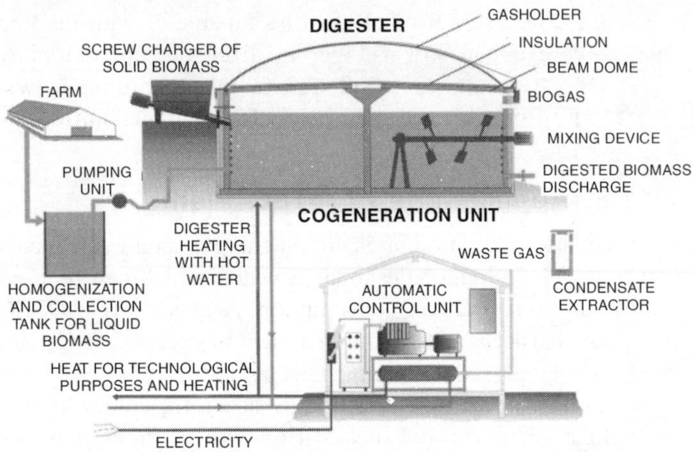

Fig. 7.65. ***Biomass Energy Conversion Scheme***

Biomass does not add CO_2 to the atmosphere as it absorbs the same amount of carbon in growing as it releases when consumed as a fuel. Its advantage is that it can be used to produce electricity with the same equipment that is now being used for burning fossil fuels. Biomass is an important source of energy and most important fuel worldwide after coal, oil and natural gas. Bio-energy, in the form of biogas, which is derived from biomass, is expected to become one of the key energy resources for global sustainable development. Biomass offers higher energy efficiency through form of biogas than by direct burning.

Bio-energy is being used for cooking, mechanical applications, pumping, power generation.

Biomass energy is derived from five distinct energy sources: (*i*) Garbage (*ii*) Wood (*iii*) Waste (*iv*) Landfill gases and (*v*) Alcohol fuel.

Wood energy is derived both from direct use of harvested wood as a fuel and from wood waste streams. The largest source of energy from wood is pulping liquor or "black liquor", a waste product from processes of the pulp, paper and paperboard industry. Wood energy is the second-largest source of biomass energy. The main contributors of waste energy are municipal solid waste (MSW), manufacturing waste and landfill gas. Biomass alcohol fuel or ethanol is derived primarily from sugarcane and corn. It can be used directly as a fuel or as an additive to gasoline.

Biomass can be converted to other usable forms of energy like methane gas or transportation fuels like ethanol or biodiesel. Methane gas is the main ingradient of natural gas. Smelly stuff, like rotting garbage and agricultural and human waste, release methane gas—also called "landfill gas" or "biogas". Crops like corn and sugar cane can be fermented to produce the transportation fuel, ethanol. Biodiesel, another transportation fuel, can be produced from left-over food products like vegetable oils and animal fats. Also, biomass to liquids (BTLs) and cellulosic ethanol are still under research.

There are a number of technological options available to make use of a wide variety of biomass types as a renewable energy source. Conversion technologies may release the energy directly, in the form of heat or electricity, or may convert it to another form, such as liquid biofuel or combustible biogas. While for some classes of biomass resource there may be a number of usage options, for others there may be only one appropriate technology.

The *practical applications* of biomass energy include biogas plants, biomass briquetting, electricity generation, bio-fuel etc.

Biogas is a clean and efficient fuel, generated from cow-dung, human waste or any kind of biological materials derived through anaerobic fermentation process. The biogas consists of 60% methane with rest mainly CO_2. Biogas is a safe fuel for cooking and lighting. By-product is usable as high-grade manure.

Biomass fuels account for about one-third of the total fuel used in the country. It is the most important fuel used in over 90% of the rural households and about 15% of the urban house holds. The types of biogas plant designs popular are: floating drum type, fixed dome-type and bag-type portable digester.

The process of densifying loose agro-waste into a solidified biomass of high density, which can be conveniently used as a fuel, is called biomass *briquetting*. It is pollution free and ecofriendly.

Electricity Generation Using Biomass. From the ancient time to the present, the most common way to capture the energy from biomass was to burn it to make heat. Since the industrial revolution this biomass fired heat has produced steam power and more recently this biomass fired steam power has been used to produce electricity. Burning biomass in conventional boilers can have numerous environmental and air-quality and advantages over burning fossil fuels.

For biomass power plants generating electricity, it is pretty much like a fossil fuel power plant.

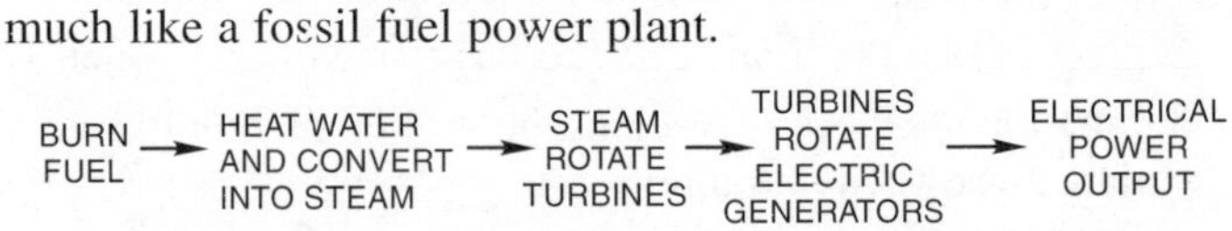

Fig. 7.66. ***Biomass Electricity Generation Process***

Advances in recent years have shown that there are even more efficient and cleaner ways to use biomass. It can be converted into liquid fuels, for example or "cooked" in a process called "gasification" to produce combustible gases, which reduces various kinds of emissions from biomass combustion, especially particulates.

Unlike other renewable energy sources, biomass can be converted directly into liquid fuels—*biofuels*—for our transportation requirements (cars, trucks, buses, aeroplanes and trains). Two most common types of biofuels are ethanol and biodiesel.

7.12.1. Processes of Converting Biomass into Electricity

There are various methods of converting biomass into electricity and some of these are given below:

1. Direct Combustion. The oldest and most common method for conversion of biomass to electricity is to burn it to produce steam and thus generate electricity. The problem with direct combustion of biomass are that much of the energy is wasted and that it can cause some pollution if it is not carefully controlled. Direct combustion can be accomplished in a plant using solely biomass (a "dedicated plant") or in a plant made to burn another fuel, usually coal.

2. Co-firing. An approach that may increase the use of biomass energy in the short term is to mix it with coal and burn it at a power plant designed for coal—a process known as "co-firing". Through gasification, biomass can also be co-fired at natural gas-powered plants.

The benefits associated with biomass co-firing can include lower operating costs, reductions of harmful emissions like sulphur and mercury, greater energy security and with the use of beneficial biomass, lower carbon emissions. Co-firing is also one of the more economically viable ways to increase biomass generation today, since it can be accomplished with modifications to existing facilities.

3. Biomass Gasification. By heating biomass in the presence of a carefully controlled amount of oxygen and under pressure, it can be converted into a mixture of hydrogen and carbon monoxide called syngas. This syngas is often refined to remove contaminants. Equipment can also be added to separate and remove the carbon dioxide in a concentrated form. The syngas can then be run directly through a gas turbine or burned and run through a steam turbine to produce electricity. Biomass gasification is generally cleaner and more efficient than direct combustion of biomass. Syngas can also be further processed to make liquid biofuels or other useful chemicals.

4. Anaerobic Digestion. Micro-organisms break down biomass to produce methane and carbon dioxide. This can occur in a carefully controlled way in anaerobic digesters used to process sewage or animal manure. Related processes happen in a less-controlled manner in landfills, as biomass in the garbage breaks down. A portion of this methane can be captured and burned for heat and power. In addition to generating biogas, which displaces natural gas from fossil fuel sources, such collection processes keep the methane from escaping to the atmosphere, reducing emissions of a powerful global warming gas. Methane can be used as an energy source in many ways. Most facilities burn it in a boiler to generate steam for generation of electrical energy or for industrial processes. Two new ways include the use of microturbines and fuel cells. Microturbines have outputs of 25-250 kW. About the size of a refrigerator, they can be used where there are space limitations for power generation. Methane can also be used as the "fuel" in a fuel cell. Fuel cells operate much like batteries but never require recharging, producing electricity as long as there is fuel.

5. Pyrolysis. In addition to gas, liquid fuels can be produced from biomass through a process called *pyrolysis.* Pyrolysis occurs when biomass is heated in the absence of oxygen. The biomass then turns into liquid known as pyrolysis oil, which can be burned like petroleum to generate electrical energy.

In addition to above mentioned processes, several biopower technologies can be used in small, modular systems. A small, modular system produces electricity at a capacity of 5 MW or less. This system is designed for use at the small town level or even at the consumer level. For example, some farmers use the waste from their live stock to provide their farms with electricity. Not only do these systems provide renewable energy, they also help farmers meet environmental regulations.

Electricity Generation Using Biomass Gasifier. Biomass gasifiers convert the solid biomass (basically wood waste, agricultural residues, etc.) into a combustible gas mixture, known as producer gas. The conversion efficiency of the gasification process is in the range of 60-70%. The producer gas consists of mainly carbon monoxide, hydrogen, nitrogen gas and methane and has a lower calorific value (1,000-1,200 kcal/Nm^3).

The "biomass gasification—electricity generation" system is a technology which converts any kind of biomass energy with low heat value (such as waste from agriculture and forest and organic waste) into combustible gas and then feeds this gas to generator for generation of electrical energy.

Discovery of method of biomass gasification for generation of electrical energy can solve both problems of effective use of renewal energy and environmental pollution from organic waste. That is why, the technology of biomass gasification for electricity generation attracts more and more research as well as applications. Thereby, this technology is being continuously optimised.

The model of biomass gasification for electrical energy generation can be realized in Fig. 7.67.

As shown in Fig. 7.67 biomass gasification for electricity generation can be realized in three ways:

1. Fuel gas produced in a biomass gasifier enters directly into a boiler to generate steam, which then drives a steam turbine coupled to an electric generator for generation of electricity.

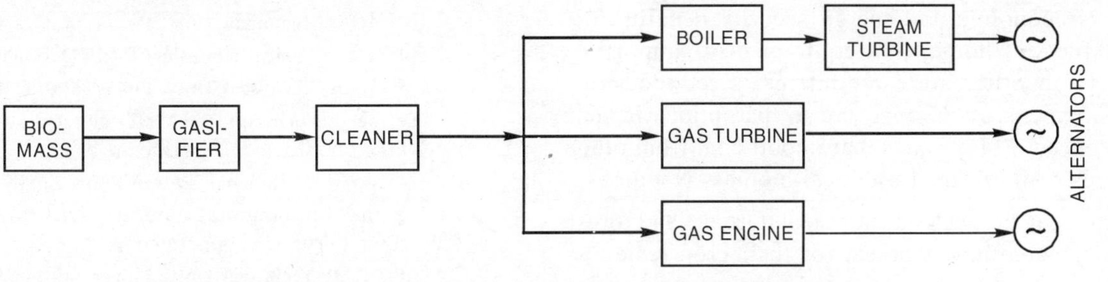

Fig. 7.67. ***Block Diagram of Biomass Electricity Generation Using Gasifier***

2. The clean gas drives a gas turbine coupled to an electric generator for generation of electricity.
3. The clean gas drives a gas engine coupled to an electric generator for generation of electricity.

Above methods of electric energy generation correspond to large-scale, medium-scale and small-scale generation respectively.

Today, commercially successful technologies for biomass gasification for electrical energy generation using gas engines get wide application because of their small system capacity, low investment, compact structure, reliable technique, low operating cost, simple operation and maintenance and their low demand for gas quality.

Electricity Generation System Equipped With a Gas Engine mainly consists of a gasifier, a gas cleaner and gas engine.

A *gasifier* is a system that converts solid biomass into combustible gas. Biomass is combusted by way of controlling the flow of air into the gasifier for conversion of solid state into gas state, producing a combustible gas consisting of H_2, CO, CH_4 and C_nH_m.

The gas temperature in the outlet of the gasifier is in the range of 350°C as 650°C depending on the type of gasifier. The gas contains impurities like dust and uncracked tar. For reliable operation of gas engine operation over a long period of time, it is necessary to clean the gas at temperatures below 40°C as well as to reduce the content of dust and tar below 50 mg/Nm^3. After cleaning the gas is supplied to the gas engine for generation of electrical energy.

In the gas engine, the gas is mixed with air, burns and drives the main shaft to rotate at a high speed. The latter then drives the electric generator to produce electrical energy.

Through above procedure, any waste can be converted into electrical energy, thereby solving the problems of pollution from waste.

7.12.2. Benefits of Biomass

If performed well, biomass energy brings numerous environmental benefits, particularly reducing many types of air pollution and net carbon emissions. Biomass can be grown and harvested in ways that protect soil quality, avoid erosion and maintain wildlife habitat. However, the environmental benefits of biomass depend on developing beneficial biomass resources and avoiding harmful resources, which having policies that can distinguish between them.

In addition to its many environmental benefits, beneficial biomass offers economic and energy security benefits. By growing our fuels at home, the requirement of importing fossil fuels from other states/countries is reduced and consequently, expenses and exposure to disruptions in that supply are reduced. Many states that import coal from other states/countries could instead use local biomass resources.

With increasing biomass development, farmers and forest owners gain valuable new markets for their crop residues, new energy crops and forest residues and we could substantially reduce our global warming emissions.

7.12.3. Biomass Energy in India

India being an agrarian country there is easy availability of agricultural based mass which can be used to produce energy. Burning of biomass is the easiest and oldest method of producing energy but also the least efficient.

Over 70% of the population of India is in villages. There electricity and steady supply of water are crucial for survival and for economic growth and social development.

Biomass exists in these villages and needs to be tapped intelligently to provide not only electricity but also water for irrigation and cultivation of fields for further increasing production of biomass (either as a main product or as a by-product), ensuring steady generation of electrical power. An added bonus is the availability of waste biomass from the biomass gasified plant to be used as fertilizer.

Most common source of biomass is wood waste and agricultural wastes. In India development of biomass gasification has received serious attention with establishment of biomass research centers and gasifier action research centres at various locations spread all over the country. These institutions have played a key role in upgradation and adaption of suitable technologies, testing, monitoring and development of biomass gasification systems. Studies reveal that the poor grade of land suitable only for scrub vegetation can be turned to advantage and form an excellent source of biomass—fast growing trees and shrubs.

In India more than 2,000 gasifiers are estimated to have been established with a capacity exceeding 22 MW and a number of villages have been electrified with biomass gasifier based generators. MNES has actively promoted R&D programmes for efficient utilization of biomass and agrowastes and further efforts are on.

Biomass gasification offers immense scope and potential for water pumping, electricity generation, heat generation for cooking gas—smokeless environment and rural electrification for better health care, better education and improved quality of life.

Exercises

1. What are the different sources of non-conventional energy ? Give one advantage and one disadvantage of each source. **[G.B. Technical Univ. Power Station Practice 2011-12]**
2. Discuss briefly, renewable and non-renewable energy sources. **[Rajasthan Technical Univ. 2011]**
3. Discuss various non-conventional energy resources with reference to present Indian Energy scenario. **[G.B. Technical Univ. Power Station Practice 2010-11]**
4. Explain the basic principle of MHD power generation. **[G.B. Technical Univ. Power Station Practice 2010-11]**
5. Describe the principle of MHD generation. How are MHD systems classified ? Describe in brief. **[M.D. Univ. Power Plant Engineering Dec. 2008]**
6. Give the fundamental operation of MHD generation. Whether the power generated is ac or dc ? **[U.P. Technical Univ. Power Station Practice 2005-06]**

7. Draw schematic diagram of an MHD power generating system having heat recovery steam generator. Explain the function of the system.
[M.D. Univ. Electrical Power Generation May 2011]
8. What is MHD power generation ? Explain basic principle and also draw schematic diagram of an MHD power generating system having heat recovery steam generation.
[U.P. Technical Univ. Power Station Practice 2009-10]
9. Write the advantages and disadvantages of MDH generation.
[U.P. Technical Univ. Power Station Practice 2008-09]
10. Give advantages of MHD generation. With block diagram explain closed cycle MHD system. Derive equation for voltage and power output of MHD generator.
[Rajasthan Univ. Generation of Electrical Power 2006]
11. Derive the output formula for MHD power generator. What are the different factors in MHD generation that influence the output power ?
[U.P. Technical Univ. Power Station Practice 2006-07]
12. What is the basic difference between thermoelectric and thermionic conversion systems ? Explain briefly the working of thermoelectric generator.
13. Describe, with the help of neat line diagram, the working of the thermionic power generation.
[M.D. Univ. Power Plant Engineering May 2008]
14. Explain the principle of operation of a solar cell.
[U.P. Technical Univ. Power Station Practice 2007-08]
15. What is photovoltaic cell ? Where they are used ?
[U.P. Technical Univ. Power Station Practice 2008-09]
16. Describe, in brief, the working principle of solar photovoltaic cell. **[U.P. Technical Univ. Power Station Practice 2004-05]**
17. Describe the principle of a fuel cell and discuss the choice of fuels required. Discuss the advantages of fuel cells.
[Pb. Univ. Direct Energy Conversion December 1992]
18. What is a fuel cell ? Describe a hydrogen oxygen cell.
[M.D. Univ. Power Plant Engineering May 2008]
19. Explain why direct energy conversion processes are becoming important as compared to conventional generation.
[Pb. Univ. Direct Energy Conversion December 1992]
20. Discuss in brief solar energy collectors.
[G.B. Technical Univ. Power Station Practice 2010-11]
21. Write in brief the methdos of generation of electrical energy from solar energy.
[M.D. Univ. Electrical Power Generation May 2011]
22. Discuss with the help of line diagram a power generating system illustrating the use of flat plate collectors as a source of energy and R-113 as a working fluid of the system.
[M.D. Univ. Power Plant Engineering May 2007; May 2009]
23. Draw a line diagram of solar power plant (high temperature) and explain its working.
[M.D. Univ. Power Plant Engineering Dec. 2009]
24. Explain the conversion of solar energy into electric energy in a solar power plant.
[Rajasthan Univ. 2006; Rajasthan Technical Univ. 2010, 2011]
25. How can solar energy be converted into electrical energy ? Give a diagram showing the elements of such a plant.
[U.P. Technical Univ. Power Station Practice 2009-10]
26. Describe the principle and applications of solar ponds and solar collectors.
[M.D. Univ. Power Plant Engineering Dec. 2008]
27. What are the difficulties associated with solar power generation ? Give its major applications.
[U.P. Technical Univ. Power Station Practice 2003-04]
28. Discuss the future prospects of solar energy use.
[U.P. Technical Univ. Power Station Practice 2007-08]
29. How can wind energy be converted into electrical energy ?
[Rajasthan Univ. 2004]
30. Draw a neat sketch of any wind electric generating power plant. Explain how does it work.
[M.D. Univ. Power Plant Engineering May 2008]
31. Explain the operation of wind power generation in an isolated power system.
[U.P. Technical Univ. Power Station Practice 2007-08]
32. Describe layout of wind power generation. What are the technical problems which may arise ?
[U.P. Technical Univ. Power Station Practice 2006-07]
33. Explain how wind turbines are used for generating electrical power. **[U.P. Technical Univ. Power Station Practice 2008-09]**
34. Explain, in brief, the scope of wind energy in India.
[A.M.I.E. Sec. B Power System Summer 1996]
35. Explain, briefly, the tidal power generation.
[U.P. Technical Univ. Power Station Practice 2004-05]
36. Discuss a tidal power scheme in brief.
[U.P. Technical Univ. Power Station Practice 2007-08]
37. Name different tidal power schemes. What is the tidal barrage ? Differentiate between tidal power generation and ocean thermal power generation.
[U.P. Technical Univ. Power Station Practice 2006-07]
38. Explain Sunderban tidal power scheme.
[Rajasthan Technical Univ. Generation of Electrical Power 2010]
39. Name the components of tidal power plant. Discuss tidal power plant installed in India after mentioning their location and capacity etc.
[M.D. Univ. Power Plant Engineering May 2008; May 2009]
40. Discuss the geothermal energy generation.
[U.P. Technical Univ. Power Station Practice 2007]
41. What is geothermal energy ? How geothermal energy can be used for generating power ?
[U.P. Technical Univ. Power Station Practice 2008-09]
42. How can geothermal energy utilised for electricity generation ?
[Rajasthan Univ. 2004; Rajasthan Technical Univ. 2011]
43. What are the different sources of geothermal energy ? Discuss any two sysetms used for generating the power using thermal energy.
[M.D. Univ. Power Plant Engineering May 2007]
44. Discuss different systems used for generating power using geothermal energy.
[M.D. Univ. Power Plant Engineering Dec. 2009]
45. What is geothermal energy ? Discuss the difficulties in extracting geothermal energy.
[U.P. Technical Univ. Power Station Practice 2007-08]
46. Describe open-cycle OTEC plant with the help of flow diagram. What modifications can be carried out to the open-cycle OTEC plant to make it economically feasible.
[M.D. Univ. Power Plant Engineering Dec. 2008]
47. What is ocean thermal energy ? How the ocean thermal energy plants are classified ? Explain the close-cycle process of ocean thermal energy conversion process.

48. Discuss the problems associated with conversion of ocean thermal energy into electrical power.

[U.P. Technical Univ. 2008-09]

49. What is biomass energy ? Explain the various processes of biomass conversion to electricity.

50. What is biomass gasification process ? What is gasifier ? How is it used to generate electricity from biomass ?

51. Draw the block diagram of biomass energy unit. Explain each in detail. Compare the merits and demerits of electrical energy with the other forms of energy.

[U.P. Technical Univ. Power Station Practice 2007]

52. Write short notes on the following:

(*i*) MHD power generation.

[Pb. Univ. December 1995; A.M.I.E. Sec. B 2000]

(*ii*) MHD power generation (open cycle).

[M.D. Univ. December 2009]

(*iii*) Closed cycle MHD system. **[M.D. Univ. May 2007]**

(*iv*) Thermoelectric power generation. **[M.D. Univ. May 2007]**

(*v*) Thermionic power generation. **[M.D. Univ. Dec. 2009]**

(*vi*) Solar cells.

[Pb. Univ. Dec. 1995; U.P. Technical Univ. 2009-10]

(*vii*) Photovoltaic power system (solar cell).

[M.D. Univ. May 2009]

(*viii*) Fuel cell. **[A.M.I.E. Sec. B. Winter 1994; Summer 1999; M.D. Univ. May 2007]**

(*ix*) Solar power generation.

(*x*) Wind power generation.

(*xi*) Tidal power generation.

[U.P. Technical Univ. 2009-10; G.B. Technical Univ. 2010-11]

(*xii*) Geothermal power generation.

[Pb. Univ. January 1993; Rajasthan Univ. 2006]

(*xiii*) Advantages, disadvantages and applications of geothermal energy. **[M.D. Univ. Dec. 2008]**

(*xiv*) Geothermal energy harnessing techniques.

[G.B. Technical Univ. 2010-11]

Short Answer Type Questions With Answers

Q. 1. What are renewable and non-renewable energy resoures ?

Ans. Renewable energy resources are those resources which can be used to produce energy again and again *e.g.*, hydro energy, solar energy, tidal energy, geothermal energy etc. Non-renewable energy resources are those sources which cannot be replenished once they are used *e.g.*, coal, oil, gas etc.

Q. 2. What are conventional and non-conventional energy resources ?

Ans. The fossil fuels (*i.e.*, coal, petroleum and natural gas), hydro power and nuclear fuels fall under the category of conventional energy resources. Conventional energy resources are used conventionally and can be stored. The alternative of conventional energy resources are known as non-conventional energy resource. The non-conventional energy resources cannot be easily stored. Solar energy, wind energy, tidal energy, geothermal energy, ocean thermal energy, bio energy fall under the category of non-conventional energy resources.

Q. 3. What is MHD power generation ?

Ans. MHD power generation is the production of electric power utilizing a high temperature conducting plasma moving through an intense magnetic field. It is a direct heat-to-electricity converion technique based on Faraday's laws of electromagnetic induction.

Q. 4. How are MHD systems classified ?

Ans. MHD systems may be classified as (*i*) open cycle systems and (*ii*) closed cycle systems. In an open cycle system, the working fluid after doing useful work (generating electrical energy) is discharged to the atmosphere through a stack while in a closed cycle system the working fluid is recycled to the heat source and thus used again and again. The operation of MHD generators directly on combustion products is an open cycle system using air as working fluid. In closed cycle systems gases used on the working fluid are helium or argon.

Q. 5. Whether power generated by MHD systems is ac or dc ?

Ans. The power generated by MHD system is dc which can be converted into ac by using an inverter.

Q. 6. What are the different factors in MHD generation that influence the output power ?

Ans. There are three parameters which govern the power output, for constant plate area and spacing and these parameters are the gas velocity, flux density and gaseous conductivity. The power output varies as (*i*) square of flux density, (*ii*) square of gas velocity and (*iii*) directly as the gaseous conductivity *i.e.*, $P \propto B^2$; $P \propto v^2$ and $P \propto \sigma$.

Q. 7. What is basic difference between thermoelectric power generator and thermionic converter ?

Ans. The thermoelectric power generator is a device which directly converts heat energy into electrical energy and its working is based upon Seebeck thermoelectric effect while a thermionic converter transforms heat energy directly into electrical energy by utilizing thermionic emission.

Q. 8. What are the main drawbacks of thermoelectric generators ?

Ans. The main drawbacks of thermoelectric generators are high cost, low efficiency and low output.

Q. 9. What is solar cell ?

Ans. Solar cell is a device which directly converts incident solar radiation into electric current. It is based on photovoltaic energy conversion phenomenon and that is why it is also called photovoltaic cell.

Q. 10. What is the principle of operation of a solar cell ?

Ans. The solar cell operate on the principle of photovoltaic effect, which is a process of generating emf as a result of absorption of ionizing radiation.

Q. 11. What are the different types of solar cells ?

Ans. According to type of crystal, the solar cells are of three types:

1. Monocrystalline silicon solar cells (band gap 1.12 eV)
2. Polycrystalline silicon solar cells (band gap 1.12 eV)
3. Thin film or amorphous silicon solar cells (band gap 1.75 eV)

Q. 12. Where are photovoltaic systems used ?

Ans. PV systems are best suited to rural areas and are used for (*i*) pumping water for drinking and irrigation, (*ii*) street lighting, (*iii*) rural telephone exchange operation, (*iv*) battery charging, (*v*) radio beacons for ship navigation at airports and in pocket calculators, watches, toys, electric fences etc.

Q. 13. What is fuel cell ?

Ans. A fuel cell is an electrochemical cell that converts chemical energy into electrical energy without combustion. In fuel cells, the chemical energy of reactants is converted into electrical energy as an isothermal process.

Q. 14. What are the advantages of fuel cells ?

Ans. The advantages of fuel cells are: (*i*) high efficiency (about 40%), (*ii*) simple and safe, (*iii*) noiseless operation and little maintenance due to absence of moving parts, (*iv*) pollution free, (*v*) compact, lighter and smaller and (*vi*) water bi-product of reaction.

Q. 15. Where are fuel cells used ?

Ans. Fuel cells can be used in vehicles, submarine, spacecrafts, locomotives and defence etc.

Q. 16. What is solar energy ?

Ans. Solar radiation in the form of electromagnetic radiation emitted by the sun is called the solar energy and this energy can be converted to useful form of energy, such as heat and electricity by different technologies. Solar energy is an essentially inexhaustible source potentially capable of meeting significant portion of the world's future energy needs with a minimum of adverse environmental consequences. Solar energy is the most promising of the unconventional energy resources. In shear size, it has the potential to supply all the energy requirements: electrical thermal process, and chemical, an even transportation fuel.

Q. 17. What is solar constant ?

Ans. The rate at which solar radiation strikes earth's upper atmosphere is expressed as the solar constant. This is the average amount of energy received in a unit time on a unit area perpendicular to the sun's direction at a mean distance of the earth from the sun (149,604,970 km).

Q. 18. What is solar collector and what are their types ?

Ans. A solar collector is a device that absorbs the incident solar radiation and converts into useful heat which is used for heating a collector fluid such as water, oil or air. The surface of a solar collector is designed for high absorption and low emission. Solar collectors can be classified as non-concentrating or flat-plate type and concentrating or focusing collectors.

In the non-concentrating type, the collector area (*i.e.*, the area that intercept the solar radiation) is the same as the absorber area (*i.e.*, the area absorbing the radiation). In these types, the whole solar panel absorbs the light. Concentrating collectors make use of optical systems in the form of reflectors or refractors to concentrate the energy of direct solar radiation on the absorbing surface. The reflectors may be flat mirrors or in the shape of a parabolic trough or paraboloidal dish concentrating solar collectors are employed to heat a working fluid for power cycles.

Q. 19. What are the important components of flat plate collectors ?

Ans. Flat plate collector consists of transparent cover, absorber plate, series of tubes and housing with thermal insulation.

Q. 20. Where are flat plate solar collectors used ?

Ans. Flat plate solar collectors are used for water heating, space heating and power generation.

Q. 21. What are the different types of solar thermal power plants ?

Ans. Solar thermal power plants are of two types: (*i*) solar distributed collector power plants and (*ii*) solar central receiving power plants.

Q. 22. What are solar power towers ?

Ans. A solar power tower is a large tower surrounded by tracking mirrors called heliostats. These mirrors align themselves and focus sunlight on the receiver at the top of tower. The receiver collects the sun's heat in a heat transfer fluid, which is used to produce steam for a conventional steam turbine located at the foot of tower for generation of electricity.

Q. 23. What are the main components of a solar chimney power plant ?

Ans. The main components of a solar chimney power plant are air heater of large area, a high chimney and air turbine.

Q. 24. What are solar ponds ?

Ans. Solar ponds are also called the solar salt ponds. A solar pond is a shallow body of water, about 1 m deep containing dissolved salts to generate a stable density gradient (fresh water on the top and denser salt water at bottom). Solar radiation passes through the upper layer to the bottom layer. The upper layer provides thermal insulation. Convection of water particles is prevented by the graded salt concentration. Temperature of water in upper layers is lower. Heat from the lower layers of water is exchanged with the working fluid in the heat exchanger.

Q. 25. What is a solar furnace ?

Ans. A solar furnace is a structure used to harness the rays of the sun in order to produce high temperatures. This is achieved by using a curved mirror (or an array of mirrors) acting as a parabolic reflector to concentrate light (Insolation) onto a focal point. The temperature of the focal point may reach up to 3,000°C, and this heat can be used to generate electricity, melt steel or make hydrogen fuel.

Q. 26. What is wind energy ? What causes wind to blow ?

Ans. Wind is air in motion. Winds are generated due to the atmospheric temperature difference caused by different areas getting different fluxes of sunlight. Air mass tends to move from hotter to the cooler regions, thereby generating winds.

Q. 27. Differentiate wind turbine and wind mill.

Ans. A wind turbine is a rotary device that extracts energy from the wind and converts into rotary mechanical energy. If the mechanical energy is used directly by machinery, such as for pumping water or grinding stones, the machine is called a *windmill*. In case the mechanical energy is instead converted into electricity, the machine is known as wind generator, *wind turbine generator (WTG), wind power unit (WPU), wind energy converter (WEC) or aerogenerator.*

Q. 28. Give the classification of wind turbines.

Ans. Basically, the wind turbines are of two types namely *horizontal axis wind turbines,* such as traditional farm windmills used for pumping water and the *vertical axis wind turbines,* such as the eggbeater-style Darrieus model, named offer its French inventer.

Most large modern wind turbines are horizontal axis turbines.

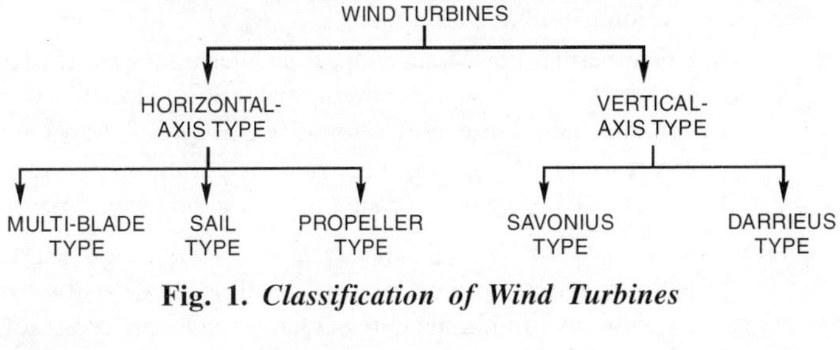

Fig. 1. ***Classification of Wind Turbines***

In *horizontal-axis turbines,* the axis of rotation is horizontal with respect to ground (and roughly parallel to the wind stream) while in vertical-axis turbines, the axis of rotation is verticle with respect to ground (and roughly perpendicular to the wind stream).

Q. 29. What factors are taken into consideration in site selection for a wind plant ?

Ans. The main factors governing the site selection for a wind power plant are: (*i*) availability of wind, (*ii*) availability of land, (*iii*) accessibility to land and (*iv*) grid stability.

Q. 30. What are the advantages of generation of electricity from winds.

Ans. Wind energy is free, renewable resource, therefore, no matter how much is used today, there will still be the same supply in the future. Wind energy is also a source of clean, non-polluting electricity. Unlike conventional power plants, wind plants produce no air pollutants or greenhouse.

Q. 31. What is tidal power ?

Ans. Tidal power, also called the *tidal energy*, is a form of hydropower that converts the energy of tides into electricity or other useful forms.

Q. 32. Enumerate the advantages of tidal wave energy.

Ans. The main advantages of tidal wave energy are:

(*i*) It is a free and renewable energy source.

(*ii*) It is a relatively pollution free.

(*iii*) Wave-power devices do not require large land masses like solar or wind.

(*iv*) After removal of power, the waves are in placid state.

Q. 33. What are the limitations of tidal wave energy ?

Ans. The limitations of total wave energy are: (*i*) corrosive sea-water, (*ii*) high capital investment, (*iii*) relative complicated energy conversion devices, (*iv*) very severe peak stresses-in storms, (*v*) relative scarcity of accessibility sites of large wave activity and (*vi*) relatively unknown problems of biological growth of marine organisms.

Q. 34. What is geothermal energy ?

Ans. Geothermal energy is the earth's natural heat available inside the earth. It originates from the original formation of the planet, from radioactive decay of minerals, from volcanic activity and from solar energy absorbed at the surface.

Q. 35. What are the various types of geothermal resources ?

Ans. Geothermal resources are of the following types:

(*i*) Hydrothermal, (*ii*) Geopressurised, (*iii*) Hard dry rock and (*iv*) Active volcanic vents and magna.

Q. 36. Classify the geothermal power plants.

Ans. There are basically four types of geothermal power plants, namely flashed steam plant, dry steam plant, binary power plant and hybrid power plant, which are in operation.

Q. 37. How do the geothermal power plants differ from conventional thermal power plants ?

Ans. The main differences between geothermal power plants and conventional power plants are:

1. Geothermal power plants have smaller unit sizes (5, 10, 15 MW), whereas the conventional steam thermal power plants have large unit sizes (200, 500, 800 MW)
2. A variety of systems are used in geothermal power plants for extraction of steam from geothermal energy resources.
3. Geothermal power plants need a large flow of geothermal fluid (due to lesser temperature and pressure).
4. Geothermal power plants are located on the geothermal field. It is not technically feasible to transport hot geothermal fluid over long distances due to the fact that there will be drop in pressure and temperature.

Q. 38. What is ocean thermal energy ?

Ans. The sun's heat warms the surface water a lot more than the deep ocean water, and this temperature difference creates thermal energy.

Q. 39. How are ocean thermal energy plants classified ?

Ans. OTEC power plants may be classified on the basis of (*i*) location and (*ii*) cycle as below:

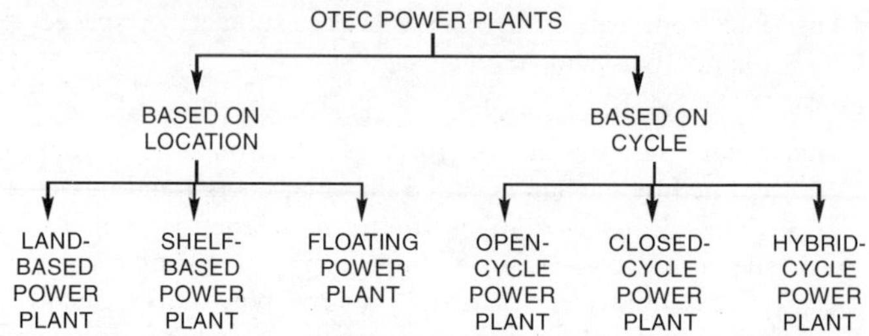

Q. 40. What are the problems that are associated with conversion of ocean thermal energy into electrical power ?

Ans. The technical difficulties associated with OTEC plants are:

1. Degradation of heat exchanger performance by dissolved gases.
2. Degradation of heat exchanger performance by microbial fouling.
3. Improper sealing.

The evaporator, turbine and condenser operate in partial vacuum ranging from 3% to 1% atmospheric pressure. This poses a number of practical concerns. First, the system must be carefully sealed to avoid in-leakage of atmospheric air that can severely degrade or shut down operation.

Q. 41. What is biomass energy ?

Ans. Organic matter derived from biological organism (plants, algae, animals) are called biomass and the energy obtained from biomass is called biomass energy.

Q. 42. What are the biomass sources?

Ans. Biomass energy is derived from five distinct energy sources: (*i*) Garbage, (*ii*) Wood, (*iii*) Waste, (*iv*) Landfill gases, (*v*) Alcohol fuel.

Q. 43. Name the biomass conversion processes.

Ans. The processes used for the biomass conversion to energy or to fuels are: (*i*) Direct combustion (*ii*) Thermochemical conversion and (*iii*) Biochemical conversion.

Q. 44. Enumerate the applications of biomass energy.

Ans. The practical applications of biomass energy include (*i*) bio-gas plants, (*ii*) biomass briquetting, (*iii*) electricity generation and (*iv*) bio fuel etc.

Q. 45. What is gasifier and what are its applications ?

Ans. A gasifier is a system that converts solid biomass energy into combustible gas. Biomass is combusted by way of controlling the flow of air into the gasifier to convert the solid state into gas state, generating a combustible gas which mainly consists of H_2, CO, CH_4 and C_nH_m.

Applications of gasifier are heat generation, water pumping and electricity generation.

❑❑❑

CHAPTER 8

Selection of Power Plant

INSIDE THIS CHAPTER

8.1. INTRODUCTION

Energy is the basic necessity for the economic development of a country. The development of electrical power and its use affect the standard of living in a country. With increased industrial development in a country there is more and more demand for electric power. In order to meet the large and ever-increasing power demand it becomes necessary to tap all the sources in the country for the development of electric power and utilise them most economically to produce and use maximum power. The power plants that can be developed are steam, hydroelectric, nuclear, diesel and gas turbine. The different types of power plants are located in places best suited to each type. Though hydroelectric power plants have numerous advantages such as operating costs and conservation of coal, the total power demand of a country or a state can never be economically met from hydro resources alone. Also, steam power plants are required to increase the firm capacity of a system mainly supplied from hydro resources. Hydroelectric power plants can be developed only at sites where huge quantity of water is available at sufficient head, and where storage of water is possible by use of dams in naturally favourable positions. The steam (coal based) power plants are adopted only when coal supply is available in plenty, large amount of power is required to be generated and financial, climatic and geographical conditions are not favourable to hydroelectric power plants. Nuclear power plants are installed only when neither water nor coal as a source of power is available. The gas turbine power plants are installed to supply peak loads and diesel-electric power plants have limited field of application as emergency power plants in private industries and other installations to supply power for short intervals in the event of failure of power supply from grid. This is because the escalating costs of fuel oil have made operation of diesel electric power plants very uneconomical.

8.2. COMPARISON OF VARIOUS TYPES OF POWER PLANTS

The comparison of hydroelectric, steam (coal based), nuclear, diesel and gas turbine power plants is given below in tabular form.

S. No.	Particulars	Hydroelectric Power Plants	Steam Power Plants	Nuclear Power Plants	Diesel Power Plants	Gas Turbine Power Plants
1.	Site	The site should be such that sufficient area is available, water in huge quantity and at sufficient head is available, the land is cheap and transportation facilities are available. Such power plants are usually	These power plants are located near the load centre but other factors such as supply of fuel and water, cost of land, available area and transportation facilities are also to be kept in view. Location of steam power plants is	These plants are located away from thickly populated areas to avoid radioactive pollution but as near the load centre as possible.	These plants can be installed any where because they require less space and small quantity of water.	These plants can be installed any where.

S. No.	Particulars	Hydroelectric Power Plants	Steam Power Plants	Nuclear Power Plants	Diesel Power Plants	Gas Turbine Power Plants
		located far from the load centre.	somewhat flexible as compared with that of hydroelectric power plants.			
2.	Space requirements	Very large space is required because of the reservoir (Maximum).	Space required is much more than diesel, gas turbine and nuclear power plants but less than that for hydroelectric power plants.	Space required is less than that for hydro and steam power plants but more than that for diesel and gas turbine power plants.	Space required is minimum.	Space required is more than that for diesel power plants but less than that for other power plants.
3.	Auxiliaries requirements	Small	Large	Less than that required in steam power plants but more than that for other power plants.	Minimum	More than that required in diesel and hydro but less than that for steam and nuclear power plants.
4.	Simplicity and cleanliness	Most simple and clean. No air pollution.	Less clean because of atmosphere pollution and ash handling. High air pollution.	Handling of equipment is very complicated. Nuclear waste disposal is, however, a major problem. No air pollution.	Simpler and cleaner than steam power plants and nuclear power plants. Air pollution-low.	Simpler and cleaner than steam power plants and nuclear power plants. Air pollution less than steam but more than hydro and nuclear.
5.	Starting time	5-10 minutes.	2-3 hours.	2-3 hours.	5-10 minutes.	10-15 minutes.
6.	Reliability	Simple, robust and more reliable.	Less reliable.	Reliable.	Less reliable.	Reliable.
7.	Plant life	40-50 years.	20-25 years.	20-30 years.	15-20 years.	15-16 years
8.	Overall efficiency	Most efficient (80-85%).	About 25%	30-40%	30-35%	20%
9.	Erection period	10-15 years.	3-5 years.	10-12 years	1-2 years	1-2 years
10.	Limit source of power	Source of power *i.e.*, water in case of hydro power plants is not dependable because it depends upon the rain fall which is at the whim of nature.	Source of power *i.e.*, coal reserves all over the world are limited and therefore coal mines are being exhausted and the time may come when all of these might finish.	The source of power is unlimited since large deposits of fissionable materials over the world are available.	Source of power *i.e.*, diesel is not available in plenty due to limited reserves.	Available in sufficient quantity.
11.	Maintenance requirements	Very low as few skilled engineers and small operating staff is required.	Quite high because large operating staff and more skilled engineers are required.	Very high because skilled and well trained staff is required for operation and maintenance.	Very low because lesser operating and supervising staff is required.	Low
12.	Capital cost	Quite high because of dam construction and excavation work.	Medium—lower than those of hydroelectric and nuclear power plants.	Very high because of huge investment on building a nuclear reactor.	Quite low	Very low
13.	Operating cost	Minimum	Medium	Quite low	Very high	Quite high
14.	Fuel transportation cost	Practically nil	Maximum because huge amount of coal is to be transported to the plant site.	Minimum because small quantity of fuel is required.	Higher than hydro and nuclear power plants.	Medium
15.	Standby losses	Nil	Maximum as the boiler remains in operation even when turbine is not in operation.	Less	Less	No standby losses

S. No.	Particulars	Hydroelectric Power Plants	Steam Power Plants	Nuclear Power Plants	Diesel Power Plants	Gas Turbine Power Plants
16.	Transmission and distribution cost	Long transmission lines are required and hence transmission and distribution cost is very high.	Short transmission lines are required and so transmission and distribution cost is moderate.	Moderate	Transmission cost is nil and distribution cost is also very low.	Transmission and distribution cost is very low.
17.	Outage rate	Very low	High	Low	Low	Low
18.	Field of application	These plants are most economical where water source is available at a sufficient head. Can be employed as peak load as well as base load power plants.	These plants are most economical near coal mines and by the side of river or canal. Can be employed as peak load as well as base load power plants.	Where neither water nor coal as a source of power is available as in Rajasthan, these plants are more feasible and adopted. Such plants are employed as base load power plants.	Such plants are installed to supply power in emergency and as standby power plants.	Such plants are installed to supply peak load and as standby power plants.

8.3. SELECTION AND LOCATION OF POWER PLANTS

The selection of type of power plant depends upon several factors such as availability of fuel and its cost; costs involved in construction and maintenance of power plant; availability of land etc.

The selection of the site for a power plant depends upon many factors such as cost of transmission of energy, cost of fuel, cost of land and taxes, requirement of space, availability of site for water power, storage space for fuel, transport facilities, availability of cooling water, nature of load, degree of reliability, pollution and noise, interest and depreciation etc.

The following discussion provides various points in detail. All these points should be considered together in taking final decision about the selection and location of power plant.

1. Cost of Transmission of Energy. A power plant should be located as near to the load centre as possible. This reduces the transmission costs and losses in transmission.

Hydroelectric, steam (coal based) and nuclear power plants cannot be located near the load centres and need transmission lines of larger, shorter and moderate length, as discussed earlier. Diesel and gas turbine power plants can be located anywhere and so no transmission line is required. However, the modern power plants are of large capacities and feed a grid which supplies power to large areas. As such other considerations become more significant than the consideration of location of plant near the load centre.

2. Cost of Fuel. The cost of fuel is an important consideration in deciding the type of power plant for a particular location. With hydroelectric power plants there is no cost of fuel as the source of power is water but water should be available in huge quantity and at sufficient head.

For steam plants, the cost of coal for the boilers and the availability of coal are important considerations. If cheap oil is available, the boilers may be oil fired and steam plants may compete other types of power plants. Steam power plants are most economical near coal mines and by the side of river or canal.

Fuel used in nuclear power plant is expensive and is difficult to recover. Diesel engines need oil as fuel and cost of this depends on the location of the power plant. For instance, if the plant is to be located near oil fields where oil is available in plenty and at cheaper rates and the transportation cost is extremely small, it may be economical to install a diesel power plant from the point of view of fuel cost.

3. Cost of Land and Taxes. The cost of land as well as taxes on the land are dependent on the situation of the proposed power plant. The cost of land near a load centre (in a big city) may be very high as compared to that at remote place. The taxes also would be higher for land in the centre of a commercial city. This is therefore, an important point to be considered in the economic selection of site.

4. Requirement of Space. The space and building requirements of the power plant is another point to be considered. If the building and space required are both large, the cost of land as well as building will be large. Thus, large power plants in the hearts of big cities and near the centre of gravity of load are not economical.

A hydroelectric power plant requires a lot of space for civil engineering construction works like dams etc., the building has to be much larger than required by other types of power plants, and a number of arrangements have to be provided in the building as special features which are quite different from those in other types of power plants. The location, however, is suitable at site where water is available at enough head and the cost of land is not excessive. Steam power plants do not need so much space, but they require space more than that required by other power plants (diesel, gas turbine, nuclear). Diesel power plant need minimum space. The space required for gas turbine power plant is more than that for diesel power plant but less than that for hydro, steam and nuclear power plants. In case of nuclear power plants, the substrata must be strong enough to support the heavy reactors which may weigh as high as 1,000,000 tonnes and impose bearing pressures around 50 tonnes/m^2. Area remote from coal fields and hydro site are preferable so as to improve the reliability of supply over the area. The site selected for nuclear power plants should have adequate space and arrangement for the disposal of radioactive waste.

5. Availability of Site For Water Power. In deciding the type of power plant for a given location, it is desirable to investigate whether any suitable source of water and sites are available for the development of hydroelectric power plant. Some of the considerations are topography of land, rainfall during the year, catchment area, suitable site for storing water behind the dam, huge quantity of water at sufficient head and cost of civil engineering works.

6. Storage Space For Fuel. A steam (coal-based) power plant needs space for storage of coal in amounts depending on the size of plant. A supply of coal for the next 2/3 weeks at least should always be available on site. The amount of reserve stock required depends on the location of power plant. If the plant is near a coal mine it may not be necessary to store a large quantity, while if the coal has to be transported from a long distance over lake or rail etc., a large stock may be required.

For oil fuel the space required is comparatively small. The quantity required as space depends to a great extent on the degree of security of oil supplies.

7. Transportation Facilities. The location of a power plant is dependent to some extent on the availability of transportation facilities. It is necessary to have a railway, line available and extended to the yard for bringing in heavy machinery for installation in the beginning and for bringing in fuel-coal or oil as well as material required for maintenance. Alternatively, the plant may be located on the bank of a river where facilities are available for transport by barges or ships. The machinery to be transported for a steam plant is rather heavy, as the plant will in general be much larger than a diesel plant. For a diesel power plant the problem of transport is not so acute, the plant is generally small, the machines and equipment to be brought in are not very heavy, and large quantities of fuels are not required. Road transport should, however, be available. For hydroelectric power plants, remote from the normal available rail and other transport facilities, roads are to be built leading to the site and arrangements made so that the equipment can be taken to site. Heavy equipment may have to be transported in parts and assembled on site. As nuclear power plant needs very little fuel, hence it does not require direct rail facilities for fuel transport. However, transportation facilities are required during the construction stage.

8. Availability of Cooling Water. A steam power plant needs larger quantities of cooling water than diesel and gas turbine power plants. Water is circulated through condenser tubes to condense the steam and to maintain a high vacuum in the turbine condenser for high efficiency. Thus, site adjacent to larger bodies of water are preferable. Alternatively, tube-wells and cooling towers have to be installed and their cost must be taken into account. Availability of reasonable pure water as make up water is another requirement. A diesel power plant requires water for cooling the cylinders of engines, and for this, water is required to be circulated through the cylinder jackets. The quantity of water required is, however, very small compared to the requirement of the steam power plant, and the diesel power plant is much smaller. The diesel power plant, therefore, can be located at a place that is suitable from other points of view. Gas turbine power plants, also need little cooling water and can be located anywhere if other requirements are met. Nuclear power plants need more than twice the water required for the steam plant of the same size for cooling etc. Hence, the site selected for nuclear power plant should be near a river, or lake or by sea side.

9. Disposal of Ash. A steam power plant produces huge quantity of ash. A site where ash can be disposed off easily will naturally be beneficial. In case of nuclear power plants, the disposal of products (radioactive in nature) is a big problem. They have either to be disposed off in a deep trench or in a sea away from the seashore.

10. Pollution and Noise. A site for a power plant near a load centre may be objectionable from the point of view of noise and pollution. Diesel engines are noisy and people in the vicinity are likely to raise complaints. Similarly gas turbine power plants also produce noise. A steam power plant in the centre of city, may also cause noise, though not so much as a diesel power plant. Smoke from the chimney of a steam power plant is undesirable in cities. There is a danger of presence of radioactivity in the atmosphere near a nuclear power plant. However, as a precautionary measure, a dome is used in the plant which does not allow the radioactivity to spread by wind or underground water ways. Noise of exhaust from diesel engines is objectionable, but may be reduced to some extent by means of silencers.

11. Nature of Load. The nature of load to be supplied affects the choice of the power plant to a certain extent. If the load varies suddenly to a great extent; so that the load factor is poor, and if the load is not very large, a diesel or a gas turbine plant will be choosen as they are more adaptable than a steam/nuclear power plants as they can be started quickly and can be put to share full load within few minutes. With a steam plant the boilers have to kept banked and ready to pick up load any time. The steam power plant also requires a longer time for starting and for taking load, as the necessary vacuum has to be created and maintained before the load can be shared. Nuclear power plants are not well suited for varying load since the reactor does not respond to the fluctuations of load efficiently. Nuclear power plants also needs a longer time (2-3 hours) for starting and taking load. If hydroelectric power plant of small or suitable capacity is available, it can be started quickly and take up load, and thus can be employed for the varying loads.

12. Reliability of Supply. In case all the big power plants are located on one side of a state, the reliability of supply in remote areas would be poor. As such power plants should be located in different areas of the state so that reliability of supply is good everywhere.

13. Equitable Growth of Different Areas. It has been seen that availability of power from a nearby source in an area encourages setting up of heavy industries in that area and consequently, the ancilliary industries are also to come up. This can improve the economy of a backward area significantly. In a welfare state committed to growth of all regions, some power plants should, therefore, be installed in backward areas. In such areas land and labour will also be available at cheaper rates.

14. Interest and Depreciation. The capital required initially depends on the total investment on the power plant, power system layout and installation. The rate of interest on capital depends only on the availability of money in the market, and not on the type of power plant. The depreciation cost, however, varies with the type of power plant. The depreciation in case of hydroelectirc, steam, nuclear, diesel and turbine power plants may be taken as 1-2%, 3½-4%, 3-4%, 5-6% and 6½% respectively.

In case of plant proposed for an industrial complex, the problems of selection of the type of the plant and its location are quite simple. If the site is near a lake, river or sea the condensing steam power plant may prove economical. If there is a scarcity of water supply, then the possibility of setting up of either a steam plant or a diesel plant using spray ponds or cooling towers may be investigated. If the site is near a river with a favourable hydro site, then the hydro plant can be set up. Any how before deciding the type of plant, the total annual cost for each option must be worked out.

The different points mentioned above are some of the considerations to be made in selecting the type of generating plant and its location. There can be no generalization in respect to the choice of the type of plant. Each case will have to be studied individually and decided on its merits by working out the economics involved. Whatever the choice, the power supply must be reliable and available at cheaper rates.

8.4. SIZE OF POWER PLANT

The **size of plant** depends on the purpose (private industry supply, regional supply or grid supply), for which the plant is to be set up. If it is to be set up for a private industry; the size would be based on power requirements of different sections of the industry and the likely increase in power demand in near future. If it is to be set up as an emergency or standby plant, the size would be governed by the load that must be supplied by the plant in the event of failure of grid supply. The size of the plant for supplying power to a given area or region will depend on the present requirements of the area and expected increase in power demand in next 5-10 years. The size of the plant for supplying power to a power grid is based on the additional power requirement of the area being supplied by the grid and the likely increase in power demand in next 10-20 years or so. The size of a proposed hydroelectric power plant for grid supply is, in addition to power demand, also governed by the rate of flow of water and height of fall (or water head). The size of steam plant to be added to the grid is, in addition to power demand, also governed by the fact whether the plant will be used as a base load plant or as a peak load plant and the extent to which it will be required to contribute to increase the firm capacity of the hydro plants.

Large sized plants have many advantages (mostly economic). Some of the costs are hardly affected by the plant size. The cost of office space, shops, docks and landscaping can be spread over more capacity. Coal handling equipment, cooling facilities and other appurtenances can also be operated at lesser cost per kWh generated in case of larger plants. Broadly speaking, a large sized plant will produce electricity at a lower cost.

However, the size of the plant is limited by some factors. Thermal power plants require lot of space for coal storage, ash disposal, cooling towers and large quantities of water. The reliability of the system is also dependent on the size of the plant. The maximum size of the plants and capability of system interconnection are also related. The environmental pollution problems are also more severe in case of larger plants.

The advantages of large power plants seem to outweigh the disadvantages and so large size plants are favoured. As a matter of fact, the increase in the power demand has forced the utilities to go in for larger size power plants.

In India six sites (Singrauli, Korba, Ramagundem, Farraka, Rihand and Vindhyachal) have been recommended for super thermal power stations of 2,000 MW each while at the time of independence, the maximum plant size was about 50 MW.

In USA, 5,000 MW size plants are being set up at present and in the coming decade the maximum plant size may be as high as 10,000 MW.

8.5. CHOICE OF SIZE AND NUMBER OF GENERATING UNITS

It will be discussed later on (Refer to Art. 14.10)

8.6. VARIOUS TYPES OF RESERVES IN POWER SYSTEM

Every system must have a certain amount of reserve generating capacity to provide for overhaul of generating equipment, forced outage of equipment etc. Various types of reserves are given below:

1. Operating Reserve. It refers to capacity in service in excess of peak load. It is provided for regulation within the hour to cover minute to minute variations, load forecasting errors, loss of equipment and maintenance of equipment.

The operating reserve is made up of the spinning reserve as well as the non-spinning or supplemental reserve.

The *spinning reserve* is the extra generating capacity that is available by increasing the power output of generators that are already connected to the power system. For most generators, this increase in power output is achieved by increasing the torque applied to the turbine rotor.

The *non-spinning or supplemental reserve* is the extra generating capacity that is not currently connected to the system but can be brought on line after a short delay. In isolated power systems, this typically equates to power available from fast-start generators. However, in interconnected power systems, this may include the power available on short notice by importing from other systems or retracing power that is currently being exported to other systems.

8.7. EFFICIENCIES OF VARIOUS POWER PLANTS

Hydro turbines, the oldest and most commonly used *renewable energy* source, have the highest efficient of all power conversion processes. The potential head of water is available right next to the turbine, so there are no energy conversion losses, only the mechanical and copper losses in the turbine and generator and the tail end loss. The efficiency is in the range of 85 to 90%.

Steam (coal fired) power plants accounts for almost 41% of the world's electricity generation. Coal fired power plants operate on the modified rankine thermodynamic cycle. The efficiency is dictated by the parameters of this thermodynamic cycle. The overall coal plant efficiency ranges from 32 to 42%. This is mainly dictated by the *superheat* and *reheat* steam temperatures and superheat pressures. Most of the large power plants operate at steam pressures of 170 bars and 570°C superheat, and 570°C reheat temperatures. The efficiencies of these plants range from 35 to 38%. Super critical power plants operating at 220 bar and 600/600°C can achieve efficiencies of 42%. Ultra super critical pressure power plants at 300 bar and 600/600°C can achieve efficiencies in the range of 45% to 48%.

The efficiency of *nuclear power plants* is little different. On the steam turbine side they use the ranking thermodynamic cycle with steam temperatures at saturated conditions. This provides a lower thermal cycle efficiency than the high temperature coal fired power plants. Thermal cycle efficiencies are in the range of 38%. Since the energy release rate in nuclear fission is extremely high, the energy transferred to steam is a very small percentage—only around 0.7%. This makes the overall plant efficiency only around 0.27%. But one does not consider the fuel efficiency in nuclear power plants, fuel availability and radiation losses take center stage.

Diesel engines, large capacity industrial engines, deliver efficiencies in the range of 35-42%.

Natural gas fired (including CNG fired) power plants account for almost 20% of the world's electricity generation. These power plants use gas turbines or gas turbine base combined cycles. Gas turbines in the simple cycle mode, only gas turbines running, have an efficiency of 32-38%. The most important parameter that dictates the efficiency is the maximum gas temperature possible. The latest gas turbines with technological advances in materials and aerodynamics has efficiencies up to 38%. In the combined cycle mode, the new "H class" gas turbines with a triple pressure HRSG and steam turbine can run at 60% efficiency at ISO conditions. This is by far the highest efficiency in the thermal power field.

Non-conventional Energy Sources

Wind turbines have an overall conversion efficiency of 30-45%.

Solar thermal systems can achieve efficiency up to 20%. The moving path of the sun and the weather conditions drastically alter the incident solar radiation. The efficiency on an annual basis, around 12%, is considerably less than on daily basis.

Geothermal systems, on the other hand, also use the Rankine cycle with steam temperatures at saturation point. Since there is no other conversion loss, this plant can achieve efficiencies around 35%.

The power industry is trying to increase the conversion efficiency of power plants to maximise electricity generation and reduce environmental impact.

8.8. ECONOMIC COMPARISON OF ALTERNATIVES

Generally, limited alternatives are available in generation system planning. However, some alternatives concerning the number and size of units, auxiliaries and steam pressure and temperature (in case of steam plants) are usually available, especially for captive plants, and are to be compared for maximum benefits. The methods available for economic comparison of alternatives of power generation are detailed below:

1. Annual Cost Method. The generation cost can broadly be divided into fixed cost and operating or running cost. In annual cost method, all costs are converted into equivalent annual figures. The annual costs of the different alternatives, are computed by using the same fixed charge rate and the same period of analysis. The alternative with minimum annual cost is considered the best. Annual cost is affected by both fixed and operating cost.

2. Present Worth Method. Here the present worth of all alternatives is determined. The present worth is the sum which if invested at the start of the project would yield an amount equal to the annual cost each year. The present worth is determined using same period of analysis and same time base for each alternative irrespective of the fact whether they have a common life or not or whether they have been started in the same year or not. The alternative with the lowest present worth will represent overall expenditure and is the most suitable alternative.

3. Capitalized Cost. This method assumes that the project once established will continue for ever, otherwise it is similar to present worth method. The capitalized cost is the amount on which the annual interest will cover the total annual costs for ever. The alternative with the lowest capitalized cost will be the optimum alternative.

4. Rate of Return. In this method, all items of annual cost except interest and depreciation are calculated as in method of annual cost. The difference between cost of purchasing power from utility and the annual cost (or the difference between the annual costs of different alternatives, excluding interest and depreciation in case power from utility is not available) represents a saving which is available for annual interest on interest and depreciation. If the difference between annual costs is A and investment is P, then rate of return r is determined from the following relation

$$\frac{A}{P} = \frac{r}{1-(1+r)^{-n}}$$

where n is the life of plant in years.

All alternatives in which value of r comes out to be less than the minimum rate specified are rejected. All the other remaining alternatives are arranged in increasing order of annual cost. The rate of return on increment investment *i.e.*, r is determined by knowing the difference ΔA between annual costs of different alternatives and diffference ΔP between investments in successive alternatives from the relation

$$\frac{\Delta A}{\Delta P} = \frac{r}{1-(1+r)^{-n}}$$

Alternative is choosen by comparing this value of r.

Exercises

1. Compare hydro and thermal power plants. **[M.D. Univ. Power Plant Engineering Dec. 2008]**

2. Write the short note on "comparison of conventional power plants". **[Rajasthan Univ. Generation of Electrical Power 2006]**

3. Write short note on site selection and location of power plants. **[Generation of Electrical Power Rajasthan Univ. 2007; Rajasthan Technical Univ. 2008, 2009, 2011]**

4. Describe in detail the procedure for selection of plant. **[Pb. Technical Univ. Generation of Electric Power May 2010]**

5. Explain the factors that govern the selection of plant size.

6. State the advantages of large power plants. **[Pb. Technical Univ. Generation of Electric Power May 2008]**

7. How can we choose the number of generating units ? **[Pb. Technical Univ. Generation of Electric Power Dec. 2009]**

8. Explain the various types of reserves in power system. **[Rajasthan Univ. Generation of Electrical Power 2006]**

9. Distinguish between operating reserve and spinning reserve. **[Pb. Technical Univ. Generation of Electric Power May 2008]**

10. Hydroelectric power plants have the highest efficiency. Comment on it.

11. Enumerate and explain the methods for economic comparison of alternatives.

12. Write short notes on the following:

(*i*) Comparison of conventional power plants.

(*ii*) Selection of plants. **[M.D. Univ. December 2008]**

(*iii*) Selection and location of power plants.

(*iv*) Types of reserves and size of plants. **[Rajasthan Univ. 2004, 2005]**

(*v*) Efficiencies of various power plants.

Short Answer Type Questions With Answers

Q. 1. Why are diesel power plants rarely used ?

Ans. Diesel power plants have limited field of application as emergency power plants in private industries and other installations to supply power for short intervals in the event of failure of power supply from grid. This is because the escalating costs of fuel oil have made operation of diesel power plants very uneconomical.

Q. 2. What are factors affecting selection of plant location ? **[Pb. Technical Univ. May 2010]**

Ans. The factors affecting selection of plant location are : availability of fuel and its cost, availability of land and its cost and local taxes, cost of transmission of energy, requirement of space, availability of site for water power, storage space for fuel, transportation facilities, availability of cooling water, degree of reliability, pollution and noise, disposal of ash, nature of load etc.

Q. 3. Is cost of fuel directly proportional to power output ? **[Pb. Technical Univ. December 2008; May 2009]**

Ans. No doubt, the cost of fuel increases with the increase in power output but does not follow directly linear law *i.e.*, the cost of fuel is not directly proportional to output. Fuel cost is directly proportional to the power output in addition to losses occurring in the conversion process.

Q. 4. Why is availability of cooling water significant for a steam power plant ?

Ans. Large quantity of water is required in a steam power plant. It is required (*i*) to raise the steam in boilers (*ii*) for cooling purposes such as in condensers (*iii*) as a carrying medium such as disposal of ash and (*iv*) for drinking purposes.

In steam power plants approximately 1.26×10^6 kcals of heat per MW per hour has to be disposed off in the condenser. In case of direct circulation from the source of water 120 m^3 of water is required per MW per hour for this purpose. In case cooling towers are used about 2.4 m^3 of water per MW per hour will be required in addition as make-up water. With pond cooling slightly large quantity of water is required to meet evaporation and drift losses etc. but cooling effect is better. The efficiency of direct cooled plant is about 0.5% higher than that of the plant cooled by cooling towers. This means a saving of about Rs 7.5 lacs per year in fuel cost for a 2,000 MW station.

Hence availability of cooling water is significant for steam (coal based) power plants.

Q. 5. Why are fuel transport facilities important in case of coal based steam power plants ?

Ans. Central coal based steam power plants need huge quantity of coal every day. The site should be such that coal can be transported easily from mines to power plant.

Q. 6. Why are nuclear power plants located away from thickly populated area ?

Ans. There should be a reasonable distance between the nuclear power plant and the nearest populated area from the point of view of safety as there is a danger of presence of radioactivity in the atmosphere near the plant.

Q. 7. Which type of plant has highest capital cost ? **[Pb. Technical Univ. December 2008]**

Ans. Nuclear power plant has highest capital cost because of heavy investment on building a nuclear reactor.

Q. 8. What is the largest size of thermal unit being used in India ? **[Pb. Technical Univ. December 2007]**

Ans. The largest size of thermal unit being used in India is 660 MW in Mundra thermal power station (Gujarat).

Q. 9. Why large sized units are preferred in steam power plants ?

Ans. Large size units have lower capital cost per kW, need less floor area, require less operating labour and have better efficiency. Thus, large sized units are economical both from the point of view of initial investment and low operating cost.

As the size of plant increases, it may be worth considering the use of larger size of generating units in the power station, if they are available and load curve permits their use.

Q. 10. What are the advantages of large size plants ?

Ans. Large sized plants have many advantages (mostly economic). Some of the costs are hardly affected by the plant size. The cost of office space, shops, docks and landscaping can be spread over more capacity. Coal handling equipment, cooling facilities and other appurtenances can also be operated at lesser cost per kWh generated in case of larger plants. Broadly speaking, a large sized plant will produce electricity at a lower cost.

Q. 11. What is spinning reserve ? **[Pb. Technical Univ. December 2008]**

Ans. The spinning reserve is the extra generating capacity that is available by increasing the power output of generators that are already connected to the power system. For most generators, this increase in power output is achieved by increasing the torque applied to the turbine rotor.

Q. 12. What is non-spinning reserve ? **[Pb. Technical Univ. May 2009]**

Ans. The non-spinning or supplemental reserve is the extra generating capacity that is not currently connected to the system but can be brought on line after a short delay. In isolated power systems, this typically equates to power available from fast-start generators. However, in interconnected power systems, this may include the power available on short notice by importing from other systems or retracing power that is currently being exported to other systems.

Q. 13. Distinguish between operating reserve and spinning reserve ? **[Pb. Technical Univ. May 2008]**

Ans. The operating reserve refers to capacity in service in excess of peak load. It is provided for regulation within the hour to cover minute to minute variations, load forecasting errors, loss of equipment and maintenance of equipment.

The operating reserve is made up of the spinning reserve as well as the non-spinning or supplemental reserve.

The spinning reserve is the extra generating capacity that is available by increasing the power output of generators that are already connected to the power system. For most generators, this increase in power output is achieved by increasing the torque applied to the turbine rotor.

Q. 14. Give the operating efficiencies of different power plants.

Ans. The operating efficiencies of hydro, coal-based steam, nuclear, diesel, gas turbine, wind turbines, power plants, solar thermal systems and geothermal systems are respectively 85-90, 32-42, 30-35, 35-42, 32-38, 30-45, 12 and 35 per cent.

❑❑❑

CHAPTER 9

Economic Operation of Power Plants

INSIDE THIS CHAPTER

9.1. INTRODUCTION

A modern power system is invariably supplied by a number of power plants. The purpose of economic operation of power system is to reduce the operating cost of generation to the minimum. The total generator operating cost includes fuel, labour and maintenance costs. For simplicity fuel cost is the only one considered to be variable. The fuel cost is meaningful in case of thermal and nuclear power plants. Hydro plants have negligible operating cost, but are required to operate under constraint of availability of water for hydro generation in a given period of time. It is, however, unrealistic to neglect transmission losses particularly when long distance transmission of power is involved.

The control engineer or the load dispatcher has to consider a number of factors while interchanging energy from one power station to another station—how much energy to interchange, the cost of supplying energy to interconnections, the cost of energy received from the interconnection, etc. It is necessary to arrange the operation of each generating unit on the system load curve every hour so as to reduce production costs during the year to a minimum. The factors directly affecting the economy of operation are many, such as variation in fuel cost, labour cost, and weather conditions, normal and emergency equipment ratings, reserve requirements, voltage limitations, characteristics of prime movers, transmission losses etc. However, the major factor controlling the most desirable load allocation between different generating units is the total operating cost. The cost of fuel forms the major portion of the variable cost of the power plant, and the aim of scheduling the operation of generating units is to obtain minimum fuel cost.

By economic load scheduling, we mean to determine the generations of different power plants such that the total operating cost is minimum, and at the same time the total demand and losses at any instant is met by the total generation.

9.2. METHODS OF LOADING TURBO-ALTERNATORS

There are several metods of loading the turbo-alternators. Of course, the *incremental loading*, the most scientific and rational method of operation of units, will always result in the maximum operating efficiencies. In this method of loading, the total system load is divided among the generating units that all the units operate at equal incremental costs. The order in which the units are brought into action depends on the heat rates. Further, loading is adjusted by incremental rates.

The other methods of loading are:

1. Base Load To Capacity. In this method, the different units are successively loaded to capacity in the order of their efficiencies. Accordingly, all but the most efficient unit are operated at minimum loads until the most efficient one is loaded to capacity, and it is assumed that arrangement would result in highest overall efficiency. But such an assumption will be correct only if the division of load is the same as that obtained by the incremental rate method.

2. Base Load To Most Efficient Load. In this method, the various units are successively loaded to their most efficient loads in the ascending order of their heat rates. After that, all the units are loaded to capacity in the same order.

3. Loading Proportional To Capacity. In this method of loading, the units are loaded in proportion to their capacities.

4. Loading Proportional To Most Efficient Load. In this method, the units are loaded in proportion to their most efficient loads. After the units are loaded up to their most efficient loads, the additional load on each unit is made proportional to the difference between the rated capacity and the most efficient load.

9.3. DISTRIBUTION OF LOAD BETWEEN GENERATING UNITS IN A POWER PLANT

To compute economic load division between units, it is necessary to express the operating costs in terms of output. For the steam turbine, steam generator or steam station as a whole, the diesel engine, or the gas turbine, the input is expressed in of kJ per hour and the output or load in MW. For hydro plant, the input will be in terms of m^3/s of water flow and output (or load) in MW. The input/output curves for a steam power station and a hydro power station are shown in Fig. 9.1.

The efficiency at any load can be measured by taking the input from the input/output curve corresponding to the load. The efficiency is then given as

$$\eta = \frac{\text{Output in MW} \times 1{,}000}{\text{Input in kJ per second}} \times 100$$

$$= \frac{\text{Output in MW} \times 1{,}000 \times 3{,}600}{\text{Input in kJ per hour}} \times 100 \text{ per cent} \quad ...(9.1)$$

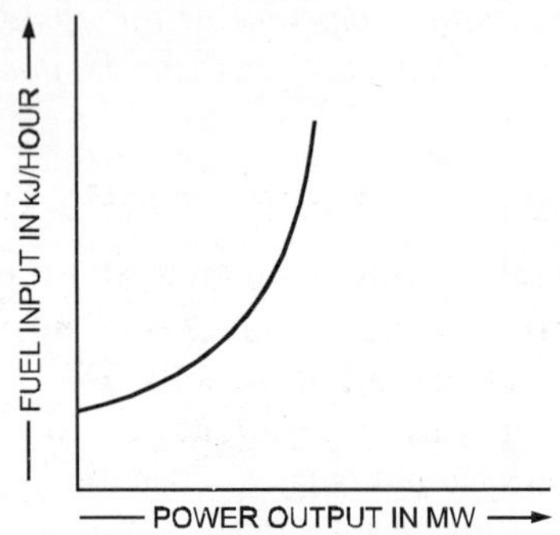

(a) For a Typical Steam Plant

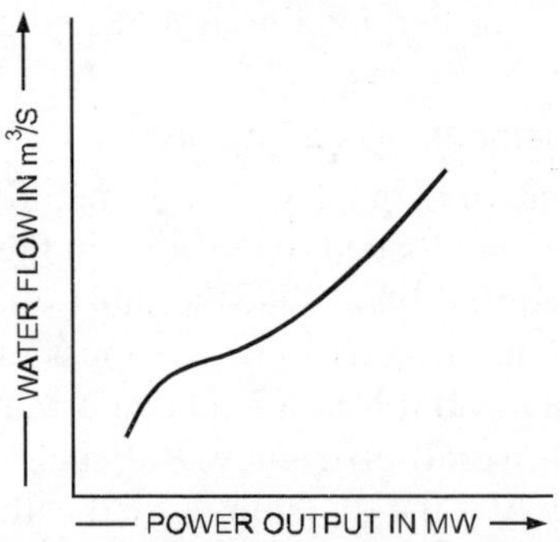

(b) For a Typical Hydro Power Plant

Fig. 9.1. ***Input-Output Curves***

Heat rate is defined as the ratio of the input to the output

$$i.e., \text{ Heat rate} = \frac{\text{Input}}{\text{Output}} \text{ in kilojoules per kWh.}$$

The incremental rate of a machine at any given output is the rate of change of the input with respect to the output *i.e.*, the ratio of small change in input to a small change in output is known as *incremental* rate of a machine. Numerically it is equal to the slope of the input/output curve at a point corresponding to that output. Mathematically, it is the first derivative of the input with respect to the output.

In the input-output characteristic if input is expressed in kJ/hour and output in MW, the slope of the curve gives *incremental fuel rate in* kJ/MWh and if the input is expressed in Rs per hour, the slope of the curve gives *incremental fuel cost* in Rs/MWh. Mostly incremental fuel cost is taken as the incremental cost (IC) and forms the basis of load division.

If the equation of the input-output curve is given in the form of an algebraic equation, then, the incremental rate can be easily determined by differentiating the equation with respect to the output.

The incremental rate can also be determined graphically, from the input/output curve. Corresponding to a given output, a point can be located on the curve. Then the slope of the tangent to the curve through this point gives the incremental rate at the given output.

Alternative method of determination of incremental rate is by taking input values corresponding to the series of output values from the input/output curve. The interval between two consecutive values of the outputs should be small and constant. The incremental rate is then calculated as the ratio of the input difference to the output difference and it is assumed to be a function of the mid-point.

The shapes of the heat rate curve and the incremental rate curve are shown in Fig. 9.2. Here, the heat rate curve and the incremental rate curve are plotted on the common coordinates. It is clear that the two curves cross each other at a point where heat rate is minimum.

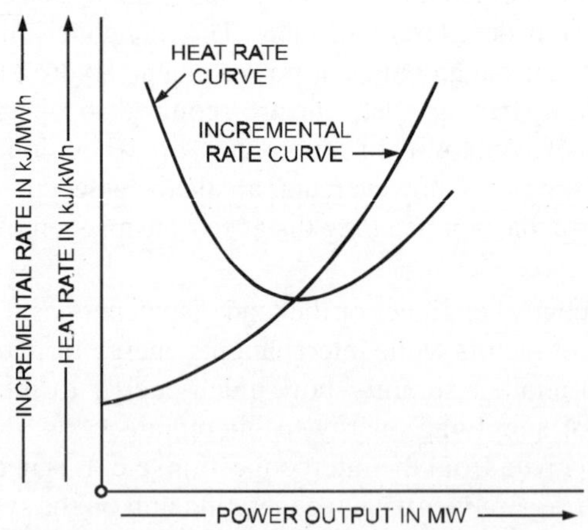

Fig. 9.2. ***Heat Rate and Incremental Rate Curves***

9.4. ECONOMIC LOAD NEGLECTING TRANSMISSION LOSSES

9.4.1. Two Generation Units

Consider two turbo alternators operating in parallel and sharing the load P_1 and P_2 and costing C_1 and C_2.

$$\text{Total output, } P = P_1 + P_2 \quad ...(9.2)$$

$$\text{Total cost, } C = C_1 + C_2 \quad ...(9.3)$$

For most economical loading of alternator, total cost C should be minimum and it will be minimum when the first differential coefficient of Eq. (9.3) with respect to P_1 is zero

$$i.e., \quad \frac{dC}{dP_1} = \frac{dC_1}{dP_1} + \frac{dC_2}{dP_1} = 0 \quad ...(9.4)$$

$$\text{or} \quad \frac{dC_1}{dP_1} = \frac{-dC_2}{dP_1} = \frac{-dC_2}{dP_2} \times \frac{dP_2}{dP_1} \quad ...(9.5)$$

From Eq. (9.2), we have

$$P_2 = P - P_1$$

Differentiating above expression with respect to P_1, we have

$$\frac{dP_2}{dP_1} = 0 - 1 = -1$$

Substituting $\frac{dP_2}{dP_1} = -1$ in Eq. (9.5), we have

$$\frac{dC_1}{dP_1} = -\frac{dC_2}{dP_2} \times -1 = \frac{dC_2}{dP_2} \quad ...(9.6)$$

But $\frac{dC_1}{dP_1}$ and $\frac{dC_2}{dP_2}$ are incremental costs of alternators 1 and 2.

Thus, the condition for minimum total cost of operation of two alternators is that incremental costs of two alternators should be equal.

The same conclusion can be arrived at through a qualitative analysis. Let the two units operate at different incremental costs. Let us transfer some load from the unit operating at higher incremental cost to the unit operating at lower incremental cost. Reduction of load on the unit with higher incremental cost will mean a greater reduction in cost than the increase in cost owing to the increase in load on the unit with lower incremental cost. In the limiting case, when the distribution of load among two units is such that both operate at the same incremental cost, maximum economy will be achieved.

9.4.2. Larger Number of Units

The above qualitative discussion about the two unit system can be extended to a system with a multiple generating units. For achieving economical loading the loads should be so allocated among the different generating units that every generating unit operates at the same incremental cost. This can be illustrated as below.

For a plant with *m* generating units.

$$P = P_1 + P_2 + ... P_m = \sum_{n=1}^{m} P_n \quad ...(9.7)$$

$$C = C_1 + C_2 + ... C_m = \sum_{n=1}^{m} C_n \quad ...(9.8)$$

Since total load is constant

$$dP = dP_1 + dP_2 + ... + dP_m = 0$$

$$\text{or} \quad dP_1 = -dP_2 - dP_3 - ... - dP_m \quad ...(9.9)$$

Since, the input or cost of any generating unit depends only on the output of that unit, the total input/cost would be minimum only when

$$dC = \sum_{n=1}^{m} \frac{dC_n}{dP_n} . dP_n = 0$$

$$\text{or} \quad \frac{dC_1}{dP_1} . dP_1 + \sum_{n=2}^{m} \frac{dC_n}{dP_n} . dP_n = 0$$

$$\text{or} \quad \frac{dC_1}{dP_1}(-dP_2 - dP_3 ... - dP_m) + \sum_{n=2}^{m} \frac{dC_n}{dP_n} . dP_n = 0$$

$$\text{or} \quad \sum_{n=2}^{m} \left[\frac{dC_n}{dP_n} - \frac{dC_1}{dP_1} \right] dP_n = 0 \quad ...(9.10)$$

The above equation will be satisfied only when each term will be zero. So for $n = 2, 3, 4, ... m$.

$$\frac{dC_1}{dP_1} = \frac{dC_n}{dP_n}$$

$$\text{or} \quad \frac{dC_1}{dP_1} = \frac{dC_2}{dP_2} = ... \frac{dC_m}{dP_m} = \lambda \quad ...(9.11)$$

i.e., for optimum economy every generating unit should operate at the same incremental rate.

The economic distribution of load among *m* generating units operating in parallel in a plant can alternatively be found by using **Lagrangian method.** The total power received by the plant bus and transferred to the load is given by Eq. (9.7) whereas the total input/cost to the *m* generating units is given by Eq. (9.8).

Since the input to any generating unit depends only on the output of that unit, the total input/cost would be minimum only when *dP* is zero.

$$i.e., \quad dC = \frac{\partial C_1}{\partial P_1} \times dP_1 + \frac{\partial C_2}{\partial P_2} \times dP_2 + ... \frac{\partial C_m}{\partial P_m} \times dP_m = 0 \quad ...(9.12)$$

Multiplying Eq. (9.9) by λ, subtracting the resulting equation from Eq. (9.12) and rearranging the terms, we have

$$\left[\frac{\partial C_1}{\partial P_1} - \lambda\right] dP_1 + \left[\frac{\partial C_2}{\partial P_2} - \lambda\right] dP_2 + ... + \left[\frac{\partial C_m}{\partial P_m} - \lambda\right] dP_m = 0 \quad ...(9.13)$$

The above equation will be satisfied only when each term equals zero. Since the input to a generating unit depends on the output of that unit only, the partial derivative becomes full derivative *i.e.*, $\frac{\partial C_m}{\partial P_m}$ becomes $\frac{dC_m}{dP_m}$.

Equation (9.13) will be satisfied only if

$$\frac{dC_1}{dP_1} = \frac{dC_2}{dP_2} = \frac{dC_3}{dP_3} = ... \frac{dC_m}{dP_m} = \lambda \quad ...(9.14)$$

The constant λ is referred to as *Lagrangian multiplier* and here it is the incremental rate of the unit. For optimum economy every generating unit should be operated at the same incremental rate.

The condition for economic loading among several generating units within a plant is represented by a graphical interpretation in Fig. 9.3.

The load allocation among several generating units can be found by preparing operating schedules for different values of λ.

The operating schedule that gives the time sum of the outputs of the generating units equal to the total load on the system will be the operating schedule for distribution of loads among the generating units for the given load.

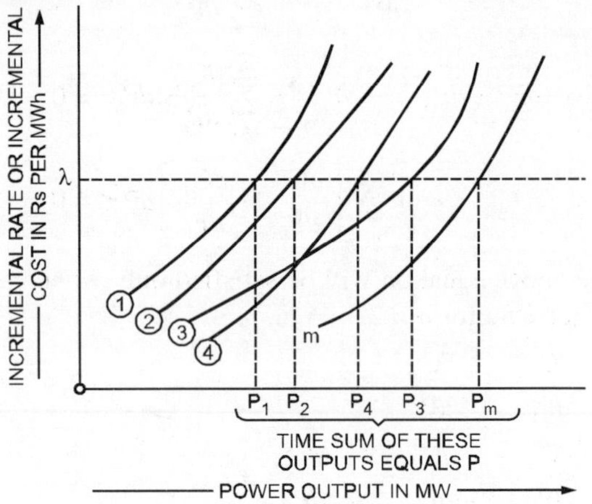

Fig. 9.3. ***Graphical Interpretation of Economic Dispatch of m Generating Units***

With the increase in load, the individual load will increase. One or several units will attain their rated capacities earlier than other units. After this happens the remaining units should be operated at equal incremental rate costs. If the minimum and maximum loads of some or all the units are specified, some units will not be able to operate at the same incremental cost and still meet the constraint of the minimum and maximum outputs.

The incremental costs are usually non-linear functions of the output. Most of the time conversion of these non-linear functions into linear functions is possible without a loss of much accuracy and the operating schedule can be had analytically. However, if this is not possible, use of a digital technique becomes essential.

Example 9.1. The input-output curve of 20 MW generating station is given by the following equation:

$$\mathbf{I = (7.5 + 0.125L + 0.16L^2) \times 10^6}$$

where I is in kcal/hour and L is in MW. Find the average heat rate of this station for a day when it was operating at a load of 20 MW for 16 hours and was kept hot at zero for the remaining 8 hours. Find the saving in the heat rate if the same energy is produced for the whole day at 100% load factor.

[M.D. Univ. Power Plant Engineering Dec. 2008]

Solution : From given datas, $I = (7.5 + 0.125L + 0.16L^2) \times 10^6$

Heat input per hour for full load of 20 MW,

$$I_1 = (7.5 + 0.125 \times 20 + 0.16 \times 20^2) \times 10^6 = 74 \times 10^6 \text{ kcals}$$

Heat input per hour for zero load,

$$I_0 = (7.5 + 0.125 \times 0 + 0.16 \times 0^2) \times 10^6 = 7.5 \times 10^6 \text{ kcals}$$

Total input per day

$$= I_1 \times 16 + I_0 \times 8 = 74 \times 10^6 \times 16 + 7.5 \times 10^6 \times 8 = 1{,}244 \times 10^6 \text{ kcals}$$

Total units generated per day

$$= 20 \times 16 + 0 \times 8 = 320 \text{ MWhr}$$

$$\text{Average heat rate} = \frac{\text{Total input}}{\text{Total output}} \text{ for a given time duration}$$

$$= \frac{1{,}244 \times 10^6}{320 \times 10^3} = 3{,}887.5 \text{ kcals/kWhr} \quad \textbf{Ans.}$$

In second case, when same energy is produced at 100% load factor in 24 hour

$$\text{Average load} = \frac{\text{Total units generated per day}}{24} = \frac{320}{24} = 13.33 \text{ MW}$$

Heat input per hour for 13.33 MW load

$$I = (7.5 + 0.125 \times 13.33 + 0.16 \times 13.33^2) \times 10^6 = 37.6 \times 10^6 \text{ kcals}$$

$$\text{Now heat rate} = \frac{37.6 \times 10^6 \times 24}{13.33 \times 10^3 \times 24} = 2{,}820 \text{ kcals/kWhr}$$

Saving in heat rate = 3,887.5 – 2,820 = 1,067.5 kcal/kWh **Ans.**

Example 9.2. An input-output curve of a 10 MW thermal station is given by an equation $I = (18 + 12L + 0.5L^2) \times 10^6$ kcal/h where I is in kcal per hour and L is load on the power plant in MW, find

(*i*) the load at which the efficiency of the plant will be maximum and

(*ii*) the increase in input required to increase the station output from 5 MW to 7 MW by using the input-output equation and by incremental rate curve.

[A.M.I.E. Sec B. Power Plant Engineering Winter 1995]

Solution : (*i*) Let the maximum efficiency be at a load of x MW. So

$$\text{Output} = x \times 1{,}000 \times 3{,}600 = 3.6 \times 10^6 x \text{ kJ/hour}$$

$$\text{Input} = (18 + 12x + 0.5x^2) \times 10^6 \times 4.18 \text{ kJ/hour}$$

$$\text{Efficiency, } \eta = \frac{\text{Output}}{\text{Input}} = \frac{3.6 \times 10^6 x}{(18 + 12x + 0.5x^2) \times 10^6 \times 4.18} = \frac{0.86x}{18 + 12x + 0.5x^2}$$

Differentiating both sides of the above equation w.r.t. x, we have

$$\frac{d\eta}{dx} = \frac{(18 + 12x + 0.5x^2) \times 0.86 - 0.86x(12 + x)}{(18 + 12x + 0.5x^2)^2} = \frac{0.86(18 - 0.5x^2)}{(18 + 12x + 0.5x^2)^2}$$

Efficiency η will be maximum when $\frac{d\eta}{dx} = 0$

or $18 - 0.5x^2 = 0$

or $x = \sqrt{36} = 6$ MW **Ans.**

(*ii*) Input required for a load of 7 MW

$$= (18 + 12 \times 7 + 0.5 \times 7^2) \times 10^6 = 126.5 \times 10^6 \text{ kcal/h}$$

Input required for a load of 5 MW

$$= (18 + 12 \times 5 + 0.5 \times 5^2) \times 10^6 = 90.5 \times 10^6 \text{ kcal/h}$$

So required increase in input

$$= 126.5 \times 10^6 - 90.5 \times 10^6 = 36 \times 10^6 \text{ kcal/hour} \quad \textbf{Ans.}$$

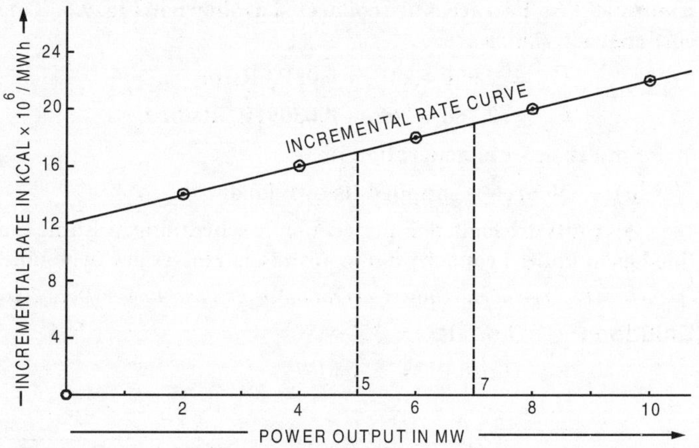

Fig. 9.4

Incremental rate of the plant is given as

$$\text{IR} = \frac{dI}{dL} = \frac{d}{dL}(18 + 12L + 0.5L^2) \times 10^6$$

$$= (12 + L) \times 10^6 \text{ kcal/MWh}$$

Data for incremental rate curve are worked out as follows :

Load in MW	Incremental Rate in kcal/MWh
2	$(12 + 2) \times 10^6 = 14 \times 10^6$
4	$(12 + 4) \times 10^6 = 16 \times 10^6$
6	$(12 + 6) \times 10^6 = 18 \times 10^6$
8	$(12 + 8) \times 10^6 = 20 \times 10^6$
10	$(12 + 10) \times 10^6 = 22 \times 10^6$

Incremental rate curve is drawn from the above data and is shown in Fig. 9.4.

Average incremental rate = 18×10^6 kcal/MWh

So required increase in input for an output from 5 MW to 7 MW

$= 18 \times 10^6 \times (7 - 5) = 36 \times 10^6$ kcal/hour **Ans.**

Example 9.3. A constant load of 300 MW is supplied by two generators, 1 and 2, for which the incremental fuel costs are $\frac{dF_1}{dP_1} = 0.10P_1 + 20$ Rs/MWh; $\frac{dF_2}{dP_2} = 0.12P_2 + 15$ Rs/MWh with power P in MW and cost F in Rs/hr. Determine (*i*) the most economical division of load between the generators and (*ii*) the saving in Rs/day thereby obtained compared the equal load sharing between generators. [*U.P.S.C. I.E.S. Electrical Engineering II, 1994*]

Solution : Total load, $P = P_1 + P_2 = 300$ MW ...(*i*)

For most economical load division,

$$\frac{dF_1}{dP_1} = \frac{dF_2}{dP_2}$$

or $0.10P_1 + 20 = 0.12P_2 + 15$ ...(*ii*)

Solving Eqs. (*i*) and (*ii*), we have

$P_1 = 140.9$ MW and $P_2 = 159.1$ MW **Ans.**

Integrating equation for $\frac{dF_1}{dP_1}$ and $\frac{dF_2}{dP_2}$, we have

$$F_1 = \frac{0.1P_1^2}{2} + 20P_1$$

and $$F_2 = \frac{0.12P_2^2}{2} + 15P_2$$

Saving on account of economic sharing in comparison to equal division,

$$\Delta F = \Delta F_1 + \Delta F_2$$

$$= \frac{0.1}{2}(150^2 - 140.9^2) + 20(150 - 140.9)$$

$$+ \frac{0.12}{2}(150^2 - 159.1^2) + 15(150 - 159.1)$$

$= 314.36 - 305.27 =$ Rs 9.09/hour

Net saving per day

$=$ Rs $9.09 \times 24 =$ Rs 218.16 **Ans.**

Example 9.4. The fuel inputs of plants 1 and 2 are given as

$$F_1 = 0.2P_1^2 + 40P_1 + 120 \text{ Rs/hour}$$

$$F_2 = 0.25P_2^2 + 30P_2 + 150 \text{ Rs/hour}$$

Determine the economic operating schedule and the corresponding cost of generation if the maximum and minimum loading of each machine is 100 MW and 25 MW, the demand is 180 MW and transmission losses are neglected. If the load is equally shared by both units, determine the saving obtained by loading the units as per incremental cost.

[*U.P.S.C. I.E.S. Electrical Engineering II, 2003*]

Solution : Total load, $P = P_1 + P_2 = 180$ MW ...(*i*)

For most economical division of load

$$\frac{dF_1}{dP_1} = \frac{dF_2}{dP_2}$$

i.e., $0.4P_1 + 40 = 0.5P_2 + 30$

or $0.4P_1 - 0.5P_2 = -10$ ...(*ii*)

Solving Eqs. (*i*) and (*ii*), we have

$P_1 = 88.889$ MW and $P_2 = 91.111$ MW **Ans.**

Cost of generation, C

$= C_1 + C_2$

$= 0.2 \times (88.889)^2 + 40 \times 88.899 + 120$

$+ 0.25 \times (91.111)^2 + 30 \times 91.111 + 150$

$= 1{,}580.25 + 3{,}555.96 + 120 + 2{,}075.30 + 2{,}733.33 + 150$

$=$ Rs 10,214.85 per hour **Ans.**

When load is equally shared

$$P_1 = P_2 = \frac{P}{2} = \frac{180}{2} = 90 \text{ MW}$$

Now, $$\Delta F_1 = \int_{88.889}^{90} (IC_1)\, dP_1 = \int_{88.889}^{90} (0.4P_1 + 40)\, dP_1$$

$$= \left[0.2P_1^2 + 40P_1\right]_{88.889}^{90} = 5{,}220 - 5{,}135.81 = \text{Rs } 84.20$$

$$\Delta F_2 = \int_{91.111}^{90} (IC_2)\, dP_2$$

$$= \int_{91.111}^{90} (0.5P_2 + 30)\, dP_2$$

$$= \left[0.25P_2^2 + 30P_2\right]_{91.111}^{90}$$

$= 4{,}725 - 4{,}808.65 = -$Rs 83.65

The significance of negative sign means decrease in cost.

Total saving $= \Delta F_1 + \Delta F_2$

$=$ Rs 84.20 – Rs 83.65 = Rs 0.55 per hour **Ans.**

Example 9.5. The incremental cost in Rs per MWh of two 250 MW units are as under :

$$\frac{dC_1}{dP_1} = 0.2P_1 + 30$$

$$\frac{dC_2}{dP_2} = 0.15P_2 + 40$$

The minimum load of each is 20 MW. Find the load division between the two units as the total load varies from 40 MW to 500 MW. [*Pb. Technical Univ. Generation of Electric Power May 2008*]

Solution : When the load is 40 MW, it will be shared equally by the two units. The incremental cost of unit 1 will be Rs 34 per MWh and that of unit 2 will be Rs 43 per MWh at the load of 20 MW. As the load is increased more load will be added to unit 1 till incremental cost of unit 1 also becomes Rs 43 per MWh. This happens when load on unit 1 is 65 MW and total load is 85 MW. After this load division is calculated by assuming succesively higher values of incremental costs and finding P_1 and P_2 and total load P. The results are given in the table below:

Table 9.1. Load Division For Example 9.5

Plant Incremental Cost Rs/MWh	Load on Unit 1 P_1 MW	Load on Unit 2 P_2 MW	Total load $P = P_1 + P_2$ MW
	20	20	40
	50	20	70
43	65	20	85
45	75	33.3	108.33
50	100	66.66	166.66
55	125	100	225
60	150	133.33	283.33
65	175	166.66	341.66
70	200	200	400
75	225	233.3	458.33
77.5	237.5	250	487.5
80	250	250	500

When the incremental cost attains a value of 77.5 unit 2 will suply its maximum load of 250 MW. At this point, unit 1 will supply 237.5 MW. After this point unit 2 will continue supplying 250 MW and unit 1 will meet the remaining requirement. This shown graphically in Fig. 9.5.

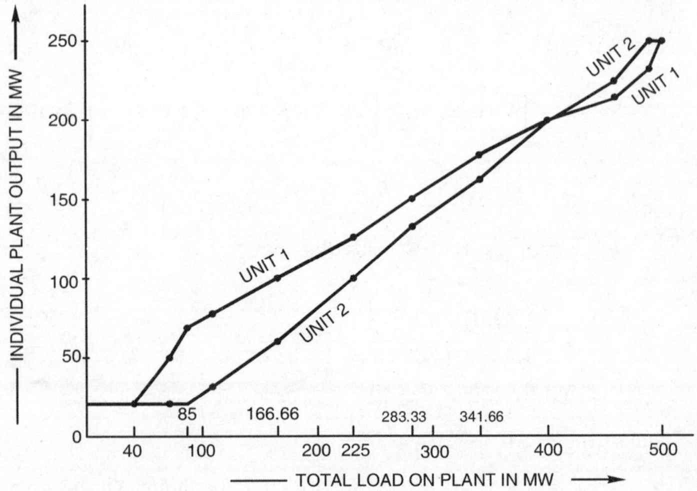

Fig. 9.5

Example 9.6. Two areas are connected as shown in Fig. 9.6. Total cost characteristics are:

$$C_1 = 2{,}000 + 20P_1 + 0.05P_1^2 \text{ Rs/hr}$$

$$C_2 = 2{,}750 + 26P_2 + 0.03091P_2^2 \text{ Rs/hr}$$

Find the cost of generation in

(*a*) Each area is supplied its own load.

(*b*) Units are loaded as per economic scheduling. Assume that line has unlimited capacity and transmission losses are negligible.

[*Pb. Technical Univ. Generation of Electric Power Dec. 2009*]

Solution : Load P_1 = 350 MW

Load P_2 = 550 MW

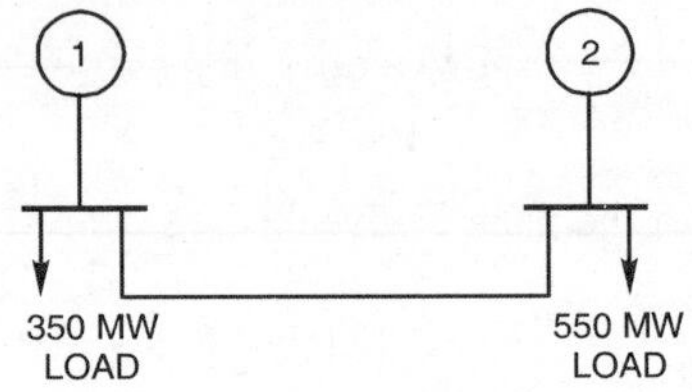

Fig. 9.6

$$C_1 = 2{,}000 + 20P_1 + 0.05P_1^2$$
$$= 2{,}000 + 20 \times 350 + 0.05 \times 350^2$$
$$= \text{Rs } 15{,}125 \text{ per hour} \quad \textbf{Ans.}$$

$$C_2 = 2{,}750 + 26P_2 + 0.03091P_2^2$$
$$= 2{,}750 + 26 \times 550 + 0.03091 \times 550^2$$
$$= \text{Rs } 26{,}400.275 \text{ per hour} \quad \textbf{Ans.}$$

$$\text{Total cost} = C_1 + C_2 = \text{Rs } (15{,}125 + 26{,}400.275) \text{ per hour}$$
$$= \text{Rs } 41{,}525.275 \text{ per hour} \quad \textbf{Ans.}$$

$$\frac{dC_1}{dP_1} = 20 + 0.1P_1$$

$$\frac{dC_2}{dP_2} = 26 + 0.06182P_2 \ldots$$

For economic loading,

$$\frac{dC_1}{dP_1} = \frac{dC_2}{dP_2}$$

i.e., $20 + 0.1P_1 = 26 + 0.06182P_2$

or $0.1P_1 - 0.06182P_2 = 6$ ...(*i*)

Total load $P_1 + P_2 = 350 + 550 = 900$ MW ...(*ii*)

Solving Eqs. (*i*) and (*ii*), we have

$$P_1 = 380.9 \text{ MW and } P_2 = 519.1 \text{ MW}$$

$$\text{Cost } C_1 = 2{,}000 + 20 \times 380.9 + 0.05 \times (380.9)^2$$
$$= \text{Rs } 16{,}872.25 \text{ per hour} \quad \textbf{Ans.}$$

$$\text{Cost } C_2 = 2{,}750 + 26P_2 + 0.03091P_2^2$$
$$= 2{,}750 + 26 \times 519.1 + 0.03091 \times (519.1)^2$$
$$= \text{Rs } 24{,}575.75 \text{ per hour} \quad \textbf{Ans.}$$

$$\text{Total cost C} = C_1 + C_2 = (\text{Rs } 16{,}872.25 + \text{Rs } 24{,}575.75) \text{ per hour}$$
$$= \text{Rs } 41{,}448 \text{ per hour} \quad \textbf{Ans.}$$

Example 9.7. Fuel cost characteristics of two thermal plants are as under :

$$C_1 = 7{,}700 + 52.8P_1 + 5.5 \times 10^{-3}P_1^2 \text{ Rs/hour}$$

$$C_2 = 2{,}500 + 15P_2 + 0.05P_2^2 \text{ Rs/hour}$$

The limits of generation of the two units are $200 \leq P \leq 800$ MW. The load curve is as shown in Fig. 9.7. Find the daily operating schedule to minimise the operating costs. The cost of taking a unit off and then putting it on is Rs 1,000.

[Pb. Technical Univ. Generaton of Electrical Power Dec. 2007]

Solution : For 1,200 MW load period it is necessary to operate both units

$$\frac{dC_1}{dP_1} = 52.8 + 0.011P_1$$

$$\frac{dC_2}{dP_2} = 15 + 0.1P_2$$

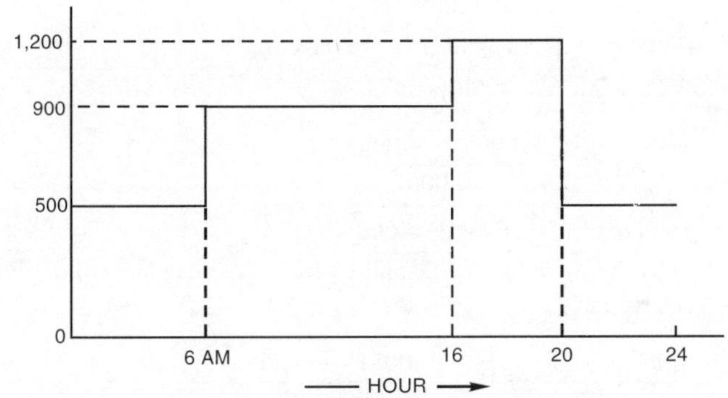

Fig. 9.7

Applying equal incremental cost concept, we have

$$\frac{dC_1}{dP_1} = \frac{dC_2}{dP_2}$$

i.e., $52.8 + 0.011P_1 = 15 + 0.1P_2$

or $0.011P_1 - 0.1P_2 = -37.8$ ...(*i*)

and total load $P_1 + P_2 = 1{,}200$ ...(*ii*)

Solving Eqs. (*i*) and (*ii*), we have

$P_1 = 740.54$ MW

$P_2 = 459.46$ MW

For load of 900 MW, it is necessary to operate both units and

$P_1 + P_2 = 900$ ...(*iii*)

Solving Eqs. (*i*) and (*iii*), we have

$P_1 = 470.27$ MW and $P_2 = 429.73$ MW

For 500 MW load, let both units be running, then

$P_1 + P_2 = 500$ ...(*iv*)

Solving Eqs. (*i*) and (*iv*), we have

$P_1 = 109.9$ MW and $P_2 = 390.1$ MW

However, it is specified that minimum load on any unit is 200 MW, therefore, $P_1 = 200$ MW and $P_2 = 300$ MW

Operating cost per hour,

$$C_1 = 7{,}700 + 52.8 \times 200 + 5.5 \times 10^{-3} \times (200)^2 = \text{Rs } 18{,}480$$

Operating cost per hour,

$$C_2 = 2{,}500 + 15 \times 300 + 0.05 \times (300)^2 = \text{Rs } 11{,}500$$

Total cost for 10 hour period of 500 MW

$$= 10(C_1 + C_2) = 10 \times (18{,}480 + 11{,}500) = \text{Rs } 299{,}800$$

Another option available to operate only unit 2 for 500 MW load.

The cost of operation of unit 2 at 500 MW for 10 hours

$$= 10[2{,}500 + 15 \times 500 + 0.05 \times (500)^2] = \text{Rs } 225{,}000$$

Shuting down and starting cost of unit one is Rs 1,000

So, total cost for second operation is

Rs 225,000 + Rs 1,000 = Rs 226,000

But when both units are running for a load of 500 MW, the cost is Rs 299,800 therefore, second option of running unit 2 for 500 MW load duration is economical. Therefore, operating schedule load is

500 MW Load: P_1 = OFF and $P_2 = 500$ MW
900 MW Load: $P_1 = 470.27$ MW and $P_2 = 429.73$ MW
1,200 MW Load: $P_1 = 740.54$ MW and $P_2 = 459.46$ MW
Ans.

Example 9.8. Incremental fuel costs in rupees per MWh for a plant consisting of two units are $\frac{dC_1}{dP_{G1}} = 0.20P_{G1} + 40$; $\frac{dC_2}{dP_{G2}} = 0.25P_{G2} + 30$. Assume that both units are operating at all times and total load varies from 40 MW to 250 MW and maximum and minimum loads on each unit are to be 125 and 20 MW respectively. How will the load be shared between the two units as the system load varies over the full range ? What are the corresponding values of the plant incremental costs ?

Find the saving in fuel cost in rupees per hour for the optimal scheduling of a total load of 130 MW as compared to the equal distribution of the same load between the two units.

[Pb. Univ. Electrical Power II (Generation) Dec. 1995]

Solution : When the total load is 40 MW, it will be shared equally by the two units. The incremental costs for units 1 and 2 will be

$$\frac{dC_1}{dP_{G1}} = 0.20 \times 20 + 40 \textit{ i.e., } 44 \text{ Rs/MWh}$$

and $$\frac{dC_2}{dP_{G2}} = 0.25 \times 20 + 30 \textit{ i.e., } 35 \text{ Rs/MWh}$$

Thus we find that at a total minimum load of 40 MW (since both the units are operating) unit 1 will operate at higher incremental fuel cost than unit 2. Therefore, additional load on unit 2 should be increased till its incremental fuel cost becomes Rs 44/MWh

So, $0.25P_{G2} + 30.0 = 44$

or $$P_{G2} = \frac{44 - 30}{0.25} = 56 \text{ MW}$$

Unit 1 is loaded at incremental cost of Rs 44/MWh and delivering 20 MW. The total load being delivered at equal incremental cost of Rs 44/MWh is 20 + 56 *i.e.*, 76 MW.

Afterwards, go on increasing the load on each unit so that both of the units operate at the same incremental costs. The load distribution is found by assuming successively higher values of incremental costs and determine P_1, P_2 and the total load. The results are given below :

Table 9.2. Load Division For Example 9.8

Plant Incremental Cost in Rs/MWh	P_{G1} in MW	P_{G2} in MW	Total Load $P = P_{G1} + P_{G2}$ in MW
	20	20	40
44	20	56	76
50	50	80	130
55	75	100	175
60	100	120	220
61.25	106.25	125	231.25
65	125	125	250

When incremental cost attains a value of 61.25, unit 2 will supply its maximum load of 125 MW. At this point unit 1 will supply 106.25 MW. After this unit 2 will continue to supply a load 125 MW and the unit 1 will meet the remaining requirement. Increase the load on the first unit till it is 125 MW so that total load on both the units together would be 250 MW.

From above table it is found that $P_{G1} = 50$ MW and $P_{G2} = 80$ MW for the corresponding total load of 130 MW. If each unit is loaded to share equal load *i.e.*, $\frac{130}{2}$ or 65 MW, Increase in cost for unit 1 is given as

$$\Delta F_1 = \int_{50}^{65} (0.20 P_{G1} + 40)\, dP_{G1} = \left| 0.10 P_{G1}^2 + 40 P_{G1} \right|_{50}^{65}$$

$$= \text{Rs } 772.5/\text{hour}$$

Increase in cost for unit 2,

$$\Delta F_2 = \int_{80}^{65} (0.25 P_{G2} + 30)\, dP_{G2} = \left| 0.125 P_{G2}^2 + 30 P_{G2} \right|_{80}^{65}$$

$$= -\text{ Rs } 721.875/\text{hour}$$

The significance of negative sign means decrease in cost.

The net increase in cost because of the departure from an economic distribution is Rs 772.5 – Rs 721.875 = Rs 50.625/hour.

So net saving in fuel cost = Rs 50.625/hour **Ans.**

Example 9.9. Consider the following three incremental cost (IC) curves :

$$\mathbf{P_{G1} = -100 + 50\, IC_1 - 2\, (IC_1)^2}$$
$$\mathbf{P_{G2} = -150 + 60\, IC_2 - 2.5\, (IC_2)^2}$$
$$\mathbf{P_{G3} = -80 + 40\, IC_3 - 1.8\, (IC_3)^2}$$

where ICs are in Rs/MWh and P_{GS} are in MW.

The total load at a certain hour of the day is 400 MW. Neglect transmission loss and find optimum values of generation *i.e.*, P_{G1}, P_{G2} and P_{G3}. [*U.P.S.C. I.E.S. Elec. Engineering II, 1995*]

Solution : The optimum values of generation are obtained by setting the three incremental costs (ICs) as equal say λ and constraining $P_{G1} + P_{G2} + P_{G3} = 400$ MW

or $(-100 + 50\lambda - 2\lambda^2) + (-150 + 60\lambda - 2.5\lambda^2) + (-80 + 40\lambda - 1.8\lambda^2) = 400$

or $6.3\lambda^2 - 150\lambda + 730 = 0$

$$\text{or } \lambda = \frac{150 \pm \sqrt{(150)^2 - 4 \times 6.3 \times 730}}{2 \times 6.3} = 6.82 \text{ or } 16.99$$

Using λ = 6.82, we have

$P_{G1} = -100 + 50 \times 6.82 - 2 \times 6.82^2 = 148$ MW **Ans.**

$P_{G2} = -150 + 60 \times 6.82 - 2.5 \times 6.82^2 = 143$ MW **Ans.**

$P_{G3} = -80 + 40 \times 6.82 - 1.8 \times 6.82^2 = 109$ MW **Ans.**

Example 9.10. The cost functions is Rs/hr for three thermal plants are:

$$\mathbf{F_1 = 350 + 7.2P_1 + 0.004P_1^2}$$
$$\mathbf{F_2 = 500 + 7.3P_2 + 0.0025P_2^2 \text{ and}}$$
$$\mathbf{F_3 = 600 + 6.74P_3 + 0.003P_3^2}$$

If P_1, P_2 and P_3 are in MW, find optimal schedule for a demand of 600 MW. [*Rajasthan Univ. Power System Engineering 2006*]

Solution :

$$\frac{dF_1}{dP_1} = 7.2 + 0.008P_1 \quad ...(i)$$

$$\frac{dF_2}{dP_2} = 7.3 + 0.005P_2 \quad ...(ii)$$

$$\frac{dF_3}{dP_3} = 6.74 + 0.006P_3 \quad ...(iii)$$

Optimum schedule is achieved when

$$\frac{dF_1}{dP_1} = \frac{dF_2}{dP_2} = \frac{dF_3}{dP_3}$$

Equating Eqs. (*i*) and (*ii*), we have

$$0.008P_1 + 7.2 = 0.005P_2 + 7.3$$

$$\text{or } P_1 = \frac{0.1 + 0.005 P_2}{0.008} \quad ...(iv)$$

Equating Eqs. (*ii*) and (*iii*), we have

$$0.005P_2 + 7.3 = 0.006P_3 + 6.74$$

$$\text{or } P_3 = \frac{0.005 P_2 + 0.56}{0.006} \quad ...(v)$$

Total load, $P_1 + P_2 + P_3 = 600$ ...(*vi*)

Substituting values of P_1 and P_3 from Eqs. (*iv*) and (*v*) in above Eq. (*vi*), we have

$$\frac{0.1 + 0.005 P_2}{0.008} + P_2 + \frac{0.005 P_2 + 0.56}{0.006} = 600$$

$$12.5 + 0.625P_2 + P_2 + 0.833P_2 + 93.33 = 600$$

$$\text{or } P_2 = \frac{494.16667}{2.45833} = 201 \text{ MW } \textbf{Ans.}$$

$$\text{Now } P_1 = \frac{0.1 + 0.005 \times 201}{0.008} = 138.15 \text{ MW } \textbf{Ans.}$$

$$\text{and } P_3 = \frac{0.005 \times 201 + 0.56}{0.006} = 260.85 \text{ MW } \textbf{Ans.}$$

Example 9.11. Determine the optimum economic operation point for the three thermal units when delivering a total of 1,000 MW.

Unit A	**P_{max} = 600 MW, P_{min} = 150 MW**
I-O Curve:	**$H_A = 500 + 7P_A + 0.0015P_A^2$**
Unit B	**P_{max} = 500 MW, P_{min} = 125 MW**
I-O curve:	**$H_B = 300 + 7.88P_B + 0.002P_B^2$**
Unit C	**P_{max} = 300 MW, P_{min} = 75 MW**
I-O Curve:	**$H_C = 80 + 7.99P_C + 0.005P_C^2$**
Fuel costs:	**Unit A : 1.1 unit of price/MBtu**
	Unit B : 1.0 unit of price/MBtu
	Unit C : 1.0 unit of price/MBtu

Find the values of P_A, P_B and P_C for optional operation.

[*U.P. Technical Univ. Power Station Practice 2008-09*]

Solution : Fuel cost characteristics are :

$$F_A = H_A \times 1.1 = 550 + 7.7P_A + 0.00165P_A^2$$
$$F_B = H_B \times 1.0 = 300 + 7.88P_B + 0.002P_B^2$$
$$F_C = H_C \times 1.0 = 80 + 7.99P_C + 0.005P_C^2$$

$$\text{Now } \frac{dF_A}{dP_A} = 7.7 + 0.0033P_A \quad ...(i)$$

$$\frac{dF_B}{dP_B} = 7.88 + 0.004P_B \quad ...(ii)$$

$$\frac{dF_C}{dP_C} = 7.99P_C + 0.01P_C \quad ...(iii)$$

Optimum schedule is achieved when

$$\frac{dF_A}{dP_A} = \frac{dF_B}{dP_B} = \frac{dF_C}{dP_C}$$

Equating Eqs. (*i*) and (*ii*) and (*i*) and (*iii*), we have respectively

$$P_B = \frac{-0.18 + 0.0033P_A}{0.004} = -45 + 0.825P_A \quad ...(iv)$$

$$\text{and } P_C = \frac{-0.29 + 0.003P_A}{0.01} = -29 + 0.33P_A \quad ...(v)$$

Total load, $P_A + P_B + P_C = 1{,}000$ MW ...(*vi*)

Substituting values of P_B and P_C from Eqs. (*iv*) and (*v*) in Eq. (*vi*), we have

$$P_A + (-45 + 0.825P_A) + (-29 + 0.33P_A) = 1{,}000$$

or $P_A = \frac{1{,}074}{2.155} = 498$ MW **Ans.**,

which is within limits of 150 MW and 600 MW

From Eq. (iv), we have

$P_B = -45 + 0.825 \times 498 = 366$ MW **Ans.**,

which is within limits of 125 MW and 500 MW

From Eq. (*v*)

$P_C = -29 + 0.33 \times 498 = 136$ MW **Ans.**,

which is within limits of 75 MW and 300 MW

9.5. OPTIMUM ECONOMIC OPERATION OF POWER SYSTEMS CONSIDERING TRANSMISSION LOSSES

To introduce the complexity of transmission losses, let us consider the simplest possible example. Let one unit (or plant) be connected to a bus that also supplies the one-system load and let another unit (or plant) feed the same bus over a transmission line. Now when the unit (or plant) connected to the line increases its generation (output) by an increment, the same increment is not delivered to the bus. The change in power at the bus is the original increment less the incremental losses in the line. In this simple case the incremental losses incurred in the transmission line are function of the power of the one generator (or plant) and the line characteristics. It is evident that the incremental cost curve at the generator (or plant) could be modified to include the cost of the incremental losses in the line. Thus the new curve would be the incremental cost curve for the unit (or plant) for supplying power over the transmission line to the bus and would include the incremental cost of transmission losses. The new curve would be used in conjunction with the curve from the other unit (or plant) to load the system.

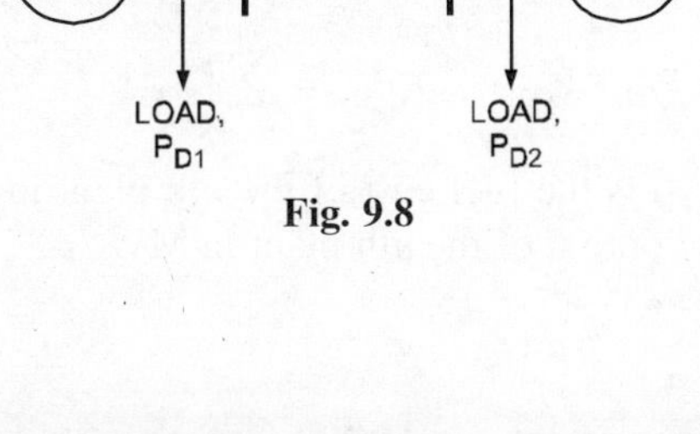

Fig. 9.8

For the sake of simplicity let us neglect shunt capacitance and shunt conductance of the line and assume impedance of the line Z (in magnitude) = $\sqrt{R^2 + X^2}$ ohms per phase

$$Y = \frac{1}{Z} = \frac{1}{R + jX}\begin{bmatrix} 1 & -1 \\ -1 & 1 \end{bmatrix} \quad ...(9.15)$$

Transmission losses are given as

$$P_L = \text{Total generation} - \text{total demand}$$
$$= (P_{G1} + P_{G2}) - (P_{D1} + P_{D2})$$
$$\text{or } P_L = (P_{G1} - P_{D1}) + (P_{G2} - P_{D2}) = P_1 + P_2 \text{ (say)}$$

Obviously, P_1 and P_2 will be the function of the bus voltages V_1 and V_2 and power angle ∂. If bus voltages are maintained to be constant, the transmission losses then become function of ∂ only.

Total cost of generation, $C = C_1(P_{G1}) + C_2(P_{G2})$

As P_{G1} and P_{G2} are dependent on only the power angle ∂,

$$\frac{dC}{d\partial} = \frac{dC_1}{d\partial} + \frac{dC_2}{d\partial} = \frac{dC_1}{dP_{G1}} \times \frac{dP_{G1}}{d\partial} + \frac{dC_2}{dP_{G2}} \times \frac{dP_{G2}}{d\partial}$$

$$\text{or } \frac{dC}{d\partial} = \text{Incremental fuel cost of plant 1} \times \frac{dP_{G1}}{d\partial}$$
$$+ \text{ incremental fuel cost of plant 2} \times \frac{dP_{G2}}{d\partial} \quad ...(9.16)$$

However, if the constancy of bus voltages V_1 and V_2 have not been assumed constant, the situation will be very complicated.

The transmission losses, P_L can be computed utilising loss formula coefficients or "B-coefficients". These B coefficients for a given system are assumed to remain constant and their evaluation depends upon the open-circuit impedance matrix of the system network.

The transmission losses are given as

$$P_L = \sum_m \sum_n P_{Gm} B_{mn} P_{Gn} \quad ...(9.17)$$

where P_L are transmission losses, P_G is plant loading, B are loss coefficients, all in per units (pu) and $\sum_m$ and $\sum_n$ indicate independent summation to include all sources

$$\text{or } P_L = [P_{G1} \dots P_{Gm}]_{1 \times m} [B_{mn}]_{m \times n} \begin{bmatrix} P_{G1} \\ \vdots \\ P_{Gn} \end{bmatrix}_{n \times 1}$$

For a two generation source system

$$P_L = P_{G1}^2 B_{11} + 2P_{G1}P_{G2}B_{12} + P_{G2}^2 B_{22} \quad ...(9.18a)$$

For a three generation source system

$$P_L = P_{G1}^2 B_{11} + P_{G2}^2 B_{22} + P_{G3}^2 B_{33} + 2P_{G1}P_{G2}B_{12} + 2P_{G2}P_{G3}B_{23} + 2P_{G3}P_{G1}B_{13} \quad ...(9.18b)$$

The matrix representation of the above loss equation is

$$P_L = P_G^T B P_G \quad ...(9.19)$$

For a total of k sources

$$P_G = \begin{bmatrix} P_{G1} \\ P_{G2} \\ \vdots \\ P_{Gk} \end{bmatrix} \text{ and } B = \begin{bmatrix} B_{11} & B_{12} & \dots & B_{1k} \\ B_{21} & B_{22} & \dots & B_{2k} \\ \vdots & \vdots & & \vdots \\ B_{k1} & B_{k2} & & B_{kk} \end{bmatrix}$$

and P_G^T is the transpose of P_G.

The B coefficients are dependent on the parameters of the system network, configuration, plant power and voltages etc. Number of assumptions are necessarily made for simplification of the otherwise cumbersome equations.

For a two area system, $m = n = 2$

$$\therefore \quad B_{mn} = \begin{bmatrix} B_{11} & B_{12} \\ B_{21} & B_{22} \end{bmatrix}$$

In a power system $B_{12} = B_{21}$ as the system network is passive and bilateral one

$$\therefore P_L = \begin{bmatrix} P_{G1} & P_{G2} \end{bmatrix} \begin{bmatrix} B_{11} & B_{12} \\ B_{21} & B_{22} \end{bmatrix} \begin{bmatrix} P_{G1} \\ P_{G2} \end{bmatrix}$$

$$= B_{11} P_{G1}^2 + 2 B_{12} P_{G1} P_{G2} + B_{22} P_{G2}^2$$

i.e., $P_L = f(P_{G1}, P_{G2}) = [P_G^T][B][P_G]$

The incremental transmission losses are then $\dfrac{dP_L}{dP_G}$.

Example 9.12. For the system given in Fig. 9.9, derive an expression for the transmission loss. Find loss coefficients in terms of voltage and power factor. Assume that I_1 and I_2 are in phase. *[Rajasthan Univ. 2005]*

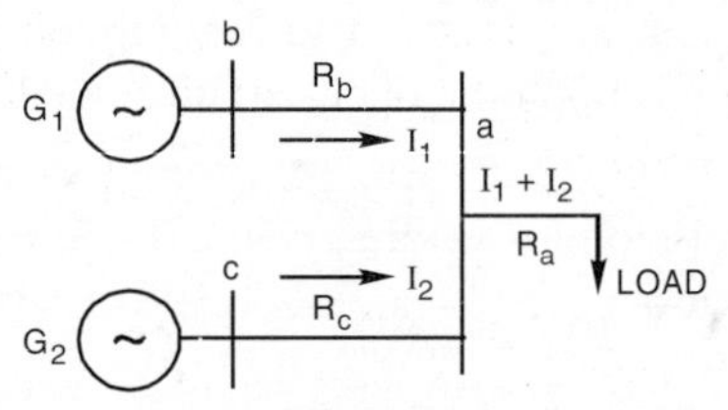

Fig. 9.9

Solution : Let R_a, R_b and R_c be the resistances of line segments a, b and c respectively. The total power loss in the 3-phase transmission system is given as

$$P_L = 3|I_1|^2 R_b + 3|I_2|^2 R_c + 3|I_1 + I_2|^2 R_a \quad \text{...(9.20)}$$

Since I_1 and I_2 are in phase

$$|I_1 + I_2| = |I_1| + |I_2| \quad \text{...(9.21)}$$

If P_{G1} and P_{G2} are the 3-phase power outputs of generating sources G_1 and G_2 at power factors $\cos\phi_1$ and $\cos\phi_2$ and V_1 and V_2 are the bus voltages of generators G_1 and G_2 respectively, then

$$|I_1| = \frac{P_{G1}}{\sqrt{3}\,|V_1|\cos\phi_1} \quad \text{...(9.22a)}$$

$$\text{and } |I_2| = \frac{P_{G2}}{\sqrt{3}\,|V_2|\cos\phi_2} \quad \text{...(9.22b)}$$

Substituting the above values of $|I_1|$ and $|I_2|$ in Eq. (9.20), we have

$$P_L = P_{G1}^2 \times \frac{R_a + R_b}{|V_1|^2 (\cos\phi_1)^2} + 2P_{G1}P_{G2} \times \frac{R_a}{|V_1||V_2|(\cos\phi_1)(\cos\phi_2)} + P_{G2}^2 \frac{R_a + R_c}{|V_2|^2 (\cos\phi_2)^2}$$

$$\text{or } P_L = P_{G1}^2 B_{11} + 2P_{G1}P_{G2}B_{12} + P_{G2}^2 B_{22} \quad \text{...(9.23)}$$

$$\text{where } B_{11} = \frac{R_a + R_b}{|V_1|^2 (\cos\phi_1)^2}$$

$$B_{12} = \frac{R_a}{|V_1||V_2|(\cos\phi_1)(\cos\phi_2)}$$

$$\text{and } B_{22} = \frac{R_a + R_c}{|V_2|^2 (\cos\phi_2)^2}$$

Equation (9.23) is called the loss formula and the terms B_{11}, B_{12} and B_{22} are called the loss coefficients or B coefficients.

If all the quantities are in pu, the loss coefficients will be in pu and if the voltages are line-to-line in kV and resistances are in ohms, then coefficients will be in reciprocal megawatts.

9.6. ECONOMIC LOAD DISTRIBUTION BETWEEN DIFFERENT PLANTS IN AN INTEGRATED SYSTEM CONSIDERING TRANSMISSION LOSSES

Since the plants are generally long distance aparts in an integrated system, it becomes imperative to consider transmission losses in deciding the load allocation to different plants. This leads to the dispatch of power in an economic way so as to make the overall cost to be the minimum.

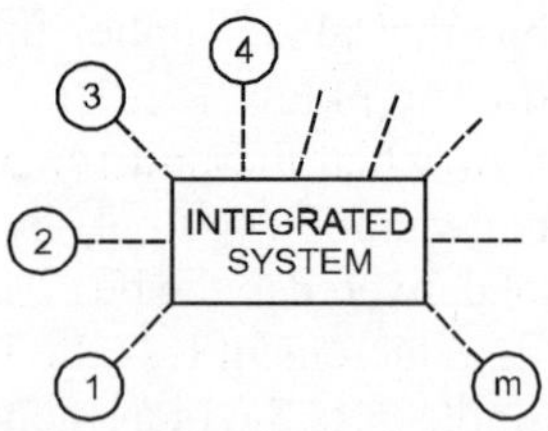

Fig. 9.10. ***An Integrated Power System Consisting of m Plants Interconnected***

Let there be m plants in a system integrated by transmission lines and tie (Fig. 9.10). Let P_{G1}, P_{G2}, ... P_{Gm} be the generation output of the plants in MW, P_D be total load in MW and P_L the losses in transmission lines.

Now we have

Total fuel cost,

$$C \text{ (Rs/hour)} = C_1(P_{G1}) + C_2(P_{G2}) + \dots + C_m(P_{Gm})$$

$$= \sum_{n=1}^{m} C_n(P_{Gn}) \quad \text{...(9.24)}$$

$$P_D + P_L = P_{G1} + P_{G2} + \dots + P_{Gm} = \sum_{n=1}^{m} P_{Gn} \quad \text{...(9.25)}$$

where $C_n(P_{Gn})$ is the fuel cost of the nth plant in Rs per hour and P_{Gn} is the output of the nth plant in MW.

Since transmission losses P_L are function of bus voltages and thus reactive power flow in the system. In our analysis let us ignore the effect of reactive power flow or we assume the magnitudes of all voltages to be constant. Thus transmission losses P_L are function of plant outputs ($P_{G1}, P_{G2}, P_{G3}, \ldots P_{Gn}$).

Equation (9.25) may be rewritten as

$$H(P_{G1}, P_{G2}, \ldots P_{Gn}) = P_D + P_L - \sum_{n=1}^{m} P_{Gn} \quad \ldots(9.26)$$

Application of Lagrangian technique, utilized for optimum true or real power dispatch gives

$$C^* = C + \lambda H(P_{G1} \ldots P_{Gn}) \quad \text{where } \lambda \text{ is the multiplier}$$

$$\text{or } C^* = \sum_{n=1}^{m} C_n(P_{Gn}) + \lambda\left(P_D + P_L - \sum_{n=1}^{m} P_{Gn}\right) \quad \ldots(9.27)$$

Differentiating the above equation with respect to P_{Gn} and equating to zero, we have

$$\frac{\partial C^*}{\partial P_{Gn}} = \frac{dC_n}{dP_{Gn}} + \lambda \frac{\partial P_L}{\partial P_{Gn}} - \lambda = 0 \quad \text{for } n = 1, 2, \ldots m \quad \ldots(9.28)$$

The above equation may be rewritten as

$$\frac{dC_n}{dP_{Gn}} = \lambda\left[1 - \frac{\partial P_L}{\partial P_{Gn}}\right] \quad \text{for } n = 1, 2, 3, \ldots m \quad \ldots(9.29)$$

The above Eq. (9.29) represents the modified economic operation criterion for the plants with transmission losses considered.

The above Eq. (9.29) may be rewritten as

$$\frac{dC_n}{dP_{Gn}}\left[\frac{1}{1 - \frac{\partial P_L}{\partial P_{Gn}}}\right] = \lambda \quad \text{or} \quad \frac{dC_n}{dP_{Gn}} \cdot L_n = \lambda \quad \ldots(9.30)$$

where $L_n = \dfrac{1}{1 - \frac{\partial P_L}{\partial P_{Gn}}}$ and is known as penalty factor for the nth plant and Eq. (9.30) is called the "exact coordination equation". $\dfrac{dC_n}{dP_{Gn}}$ being the incremental fuel cost, $\dfrac{\partial P_L}{\partial P_{Gn}}$ are known as *incremental transmission losses.*

The minimum fuel cost for a given total load is obtained by solving simultaneous equations represented by Eq. (9.29).

The n optimum loading equations along with the power balance Eq. (9.25) are sufficient to determine the $(n + 1)$ unknown. ($P_{G1}, P_{G2}, P_{G3}, \ldots P_{Gn}$ and λ). λ is thus given by

$$\lambda = \frac{\text{Incremental fuel cost}}{1 - \text{incremental transmission losses}} \quad \ldots(9.31)$$

From Eq. (9.30)

$$\frac{dC_1}{dP_{G1}} L_1 = \frac{dC_2}{dP_{G2}} L_2 = \frac{dC_3}{dP_{G3}} L_3 = \ldots \frac{dC_m}{dP_{Gm}} L_m = \lambda \quad \ldots(9.32)$$

So, minimum fuel cost is obtained when the incremental fuel cost of each plant multiplied by its penalty factor is the same for all the plants.

9.7. COORDINATION EQUATIONS

Coordination equations represent the coordination between various generating units as regards to their load allocation. If C is the total cost, P_G is the total generation for the system having n units, the coordination equations can be formulated as below:

$$C = C_1 + C_2 + \ldots + C_n \quad \ldots(9.33)$$

$$P_G = P_{G1} + P_{G2} + \ldots + P_{Gn} \quad \ldots(9.34)$$

Assuming that the total load, P_D at the receiving end is constant.

$$P_G - P_L = \text{Constant} = P_D \quad \ldots(9.35)$$

where P_L is transmission loss

$$dP_G - dP_L = 0 \quad \ldots(9.36)$$

$$\text{and } dP_G = dP_{G1} + dP_{G2} + \ldots + dP_{Gn} \quad \ldots(9.37)$$

Transmission power loss P_L is a function of plant outputs

$$dP_L = \frac{\partial P_L}{\partial P_{G1}} dP_{G1} + \frac{\partial P_L}{\partial P_{G2}} dP_{G2} + \ldots + \frac{\partial P_L}{\partial P_{Gn}} dP_{Gn} \quad \ldots(9.38)$$

From Eqs. (9.36), (9.37) and (9.38), we have

$$dP_G - dP_L = 0$$

$$= \left(1 - \frac{\partial P_L}{\partial P_{G1}}\right) dP_{G1} + \left(1 - \frac{\partial P_L}{\partial P_{G2}}\right) dP_2 + \ldots \left(1 - \frac{\partial P_L}{\partial P_{Gn}}\right) dP_{Gn} \quad \ldots(9.39)$$

$$\text{or } -dP_{G1} = \frac{1 - \frac{\partial P_L}{\partial P_{G2}}}{1 - \frac{\partial P_L}{\partial P_{G1}}} dP_{G2} + \frac{1 - \frac{\partial P_L}{\partial P_{G3}}}{1 - \frac{\partial P_L}{\partial P_{G1}}} + \ldots + \frac{1 - \frac{\partial P_L}{\partial P_{Gn}}}{1 - \frac{\partial P_L}{\partial P_{G1}}} \quad \ldots(9.40)$$

For C to be minimum

$$dC = 0 = \frac{\partial C}{\partial P_{G1}} dP_{G1} + \frac{\partial C}{\partial P_{G2}} dP_{G2} + \ldots + \frac{\partial C}{\partial P_{Gn}} dP_{Gn} \quad \ldots(9.41)$$

Substituting the value of dP_{G1} from Eq. (9.40) into Eq. (9.41), we have

$$\left[\frac{\partial C}{\partial P_{G2}} - \frac{1 - \frac{\partial P_L}{\partial P_{G2}}}{1 - \frac{\partial P_L}{\partial P_{G1}}} \cdot \frac{\partial C}{\partial P_{G1}}\right] dP_{G2} + \left[\frac{\partial C}{\partial P_{G3}} - \frac{1 - \frac{\partial P_L}{\partial P_{G3}}}{1 - \frac{\partial P_L}{\partial P_{G1}}} \cdot \frac{\partial C}{\partial P_{G1}}\right] dP_{G3}$$

$$+ \ldots + \left[\frac{\partial C}{\partial P_{Gn}} - \frac{1 - \frac{\partial P_L}{\partial P_{Gn}}}{1 - \frac{\partial P_L}{\partial P_{G1}}}\right] dP_{Gn} = 0 \quad \ldots(9.42)$$

The above equation is satisfied only if each term is zero.

Also the cost of a unit depends on the output of that unit only, the partial derivatives become full derivatives *i.e.*, $\dfrac{\partial C}{\partial P_G}$ becomes $\dfrac{\partial C_n}{\partial P_{Gn}}$.

$$\therefore \frac{dC_1}{dP_{G1}}\left[\frac{1}{1 - \frac{\partial P_L}{\partial P_{G1}}}\right] = \frac{dC_2}{dP_{G2}}\left[\frac{1}{1 - \frac{\partial P_L}{\partial P_{G2}}}\right] = \ldots = \frac{dC_n}{dP_{Gn}}\left[\frac{1}{1 - \frac{\partial P_L}{\partial P_{Gn}}}\right] = \lambda \quad \ldots(9.43)$$

i.e., $$\frac{dC_n}{dP_{Gn}} = \lambda\left(1 - \frac{\partial P_L}{\partial P_{Gn}}\right)$$

or $$\frac{dC_n}{dP_{Gn}} + \lambda\frac{\partial P_L}{\partial P_{Gn}} = 1 \qquad ...(9.44)$$

which is a coordination equation.

Also $\sum P_{Gn} - P_D - P_L = 0$ ...(9.45)

which is a power balance equation.

9.8. ITERATIVE PROCEDURE OF SOLVING COORDINATION EQUATIONS

The coordination Eq. (9.44) cannot be solved directly, because the incremental transmission loss (ITL) is dependent on power outputs from all the units. So, for solving the coordination equation an iterative procedure involving a method of successive approximations is adopted.

The coordination equation for *n*th unit is given as

$$\frac{dC_n}{dP_{Gn}} + \lambda\frac{\partial P_L}{\partial P_{Gn}} = \lambda$$

Substituting $\frac{dC_n}{dP_{Gn}} = 2C_nP_{Gn} + b_n$ in above equation, we have

$$2C_nP_{Gn} + b_n + \lambda\sum_m 2P_{Gm}B_{mn} = \lambda$$

or $$P_{Gn} = \frac{1 - \frac{b_n}{\lambda} - \sum_{m\neq n} 2P_{Gm}B_{mn}}{2\frac{C_n}{\lambda} + 2B_{mn}} \qquad ...(9.46)$$

For a simple two unit system, the power outputs P_{G1} and P_{G2}, we have

$$P_{G1} = \frac{1 - \frac{b_1}{\lambda} - 2P_{G2}B_{12}}{\frac{2C_1}{\lambda} + 2B_{11}} \qquad ...(9.47)$$

and $$P_{G2} = \frac{1 - \frac{b_2}{\lambda} - 2P_{G1}B_{12}}{\frac{2C_2}{\lambda} + 2B_{22}} \qquad ...(9.48)$$

The iterative procedure is as follows :

(*i*) Assume $P_{G2} = 0$.
(*ii*) Determine P_{G1} from Eq. (9.47) taking $P_{G2} = 0$.
(*iii*) Determine P_{G2} from Eq. (9.48) using the value of P_{G1} determined in above step (step (*ii*)).
(*iv*) Determine P_{G1} from Eq. (9.47) using the value of P_{G2} determined in above step (step (*iii*)).
(*v*) Repeat steps (*iii*) and (*iv*) one after the other till the values of P_{G1} and P_{G2} determined in any previous step match values of P_{G1} and P_{G2} determined later.

Example 9.13. A two bus system is shown in the figure. If 100 MW is transmitted from plant 1 to the load, a transmission loss of 10 MW is incurred. Find the required generation for each plant and the power received by the load when the system cost is Rs 25/MWh. The incremental fuel costs of the two plants are :

$$\mathbf{I_{C1} = \frac{dF_1}{dP_{G1}} = 0.02P_{G1} + 16 \text{ Rs/MWh}}$$

$$\mathbf{I_{C2} = \frac{dF_2}{dP_{G2}} = 0.04P_{G2} + 20 \text{ Rs/MWh.}}$$

[*U.P.S.C. I.E.S. Electrical Engineering II, 1996*]

Solution : Since there is no load at bus 1, evidently the transmission loss does not directly depend on P_{G2}. Thus B_{12} and B_{22} both are zero.

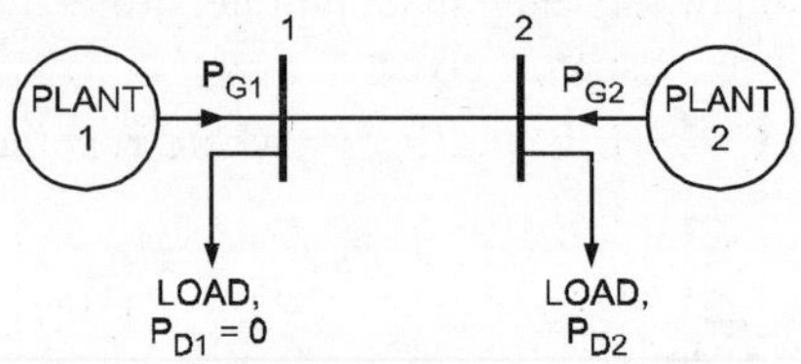

Fig. 9.11

Further a power of 100 MW flows from bus 1 to bus 2 and causes a loss of 10 kW. So

$10 \times 10^{-3} = B_{11} \times 100^2$

or $B_{11} = \frac{10^{-2}}{10^4} = 1 \times 10^{-6}\ MW^{-1}$

$P_L = 1 \times 10^{-6}\ P_{G1}^2$ and $\frac{\partial P_L}{\partial P_{G2}} = 0$

Penalty factor for plant 1,

$$L_1 = \frac{1}{1 - \frac{\partial P_L}{\partial P_{G_1}}} = \frac{1}{1 - 2\times10^{-6}\,P_{G1}}$$

Penalty factor for plant 2, $L_2 = \frac{1}{1-0} = 1$

The generation at each plant is required to be determined for λ = Rs 25/MWh

$$\frac{dF_1}{dP_{G1}}\cdot L_1 = \frac{dF_2}{dF_{G2}}\cdot L_2 = \lambda$$

So $$\frac{0.02\,P_{G1} + 16}{1 - 2\times10^{-6}\,P_{G1}} = 25$$

or $0.02P_{G1} + 16 = 25 - 50 \times 10^{-6}\,P_{G1}$

or P_{G1} = 448.9 MW **Ans.**

and $\frac{dF_2}{dF_{G2}}\cdot L_2 = 25$

or $0.04P_{G2} + 20 = 25$

or $P_{G2} = \frac{25-20}{0.04}$ = 125 MW **Ans.**

Power loss = $B_{11}P_{G1}^2 = 1 \times 10^{-6} \times (448.9)^2$ MW = 0.2015 MW

Power received by the load = 448.9 + 125 – 0.2015 = 573.7 MW **Ans.**

Example 9.14. The incremental cost characteristic of a two plant system are $IC_1 = 1.0\ P_1 + 85$ Rs/MWh; $IC_2 = 1.2\ P_2 + 72$ Rs/MWh where P_1 and P_2 are in MW. The loss coefficient matrix

in MW^{-1} is given by $\begin{bmatrix} 0.015 & -0.001 \\ -0.001 & 0.02 \end{bmatrix}$. Compute the optimum scheduling with λ = 150 Rs/MWh.

[U.P.S.C. I.E.S. Elec. Engineering II, 1992]

Solution : $IC_1 = 1.0P_1 + 85$ Rs/MWh

$IC_2 = 1.2P_2 + 72$ Rs/MWh

$$\begin{bmatrix} B_{11} & B_{12} \\ B_{21} & B_{22} \end{bmatrix} = \begin{bmatrix} 0.015 & -0.001 \\ -0.001 & 0.02 \end{bmatrix}$$

Total transmission loss is given as

$$P_L = B_{11}P_1^2 + 2B_{12}P_1P_2 + B_{22}P_2^2$$

$$\frac{\partial P_L}{\partial P_1} = 2B_{11}P_1 + 2B_{12}P_2 = 0.03P_1 - 0.002P_2$$

$$\frac{\partial P_L}{\partial P_2} = 2B_{22}P_2 + 2B_{12}P_1 = 0.04P_2 - 0.002P_1$$

Then penalty factor for plant 1,

$$L_1 = \frac{1}{1 - \frac{\partial P_L}{\partial P_1}} = \frac{1}{1 - (0.03P_1 - 0.002P_2)}$$

Penalty factor for plant 2,

$$L_2 = \frac{1}{1 - \frac{\partial P_L}{\partial P_2}} = \frac{1}{1 - (0.04P_2 - 0.002P_1)}$$

Now $IC_1L_1 = IC_2L_2 = \lambda$

or $$\frac{1.0P_1 + 85}{1 - (0.03P_1 - 0.002P_2)} = \frac{1.2P_2 + 72}{1 - (0.04P_2 - 0.002P_1)} = \lambda = 150$$

or $5.5P_1 - 0.3P_2 - 65 = 0$ ...(*i*)

and $7.2P_2 - 0.3P_1 - 78 = 0$ ...(*ii*)

Solving Eqs. (*i*) and (*ii*) we have, P_1 = 12.437 MW and P_2 = 11.35 MW **Ans.**

Example 9.15. For a simple two unit system, the loss coefficients are : B_{11} = 0.001 per MW, B_{12} = –0.0005 per MW and B_{22} = 0.0024 per MW. The incremental production costs of two units are:

$$\frac{dC_1}{dP_1} = 0.08P_1 + 16 \text{ and } \frac{dC_2}{dP_2} = 0.08P_2 + 12 \text{ Rs/MWh}$$

Find the generations P_1 and P_2 for λ = 20. Also compute the transmission loss received power and efficiency of the transmission system. *[Rajasthan Univ. Power System Engineering 2005, 2008]*

Solution : Total transmission loss is given as

$$P_L = B_{11}P_1^2 + 2B_{12}P_1P_2 + B_{22}P_2^2$$

$$= 0.001P_1^2 + 2 \times (-0.0005)P_1P_2 + 0.0024P_2^2 \quad ...(i)$$

So, $$\frac{\partial P_L}{\partial P_1} = 0.002P_1 - 0.001P_2 \quad ...(ii)$$

and $$\frac{\partial P_L}{\partial P_2} = -0.001P_1 + 0.0048P_2 \quad ...(iii)$$

Then penalty factor for plant 1,

$$L_1 = \frac{1}{1 - \frac{\partial P_L}{\partial P_1}} = \frac{1}{1 - (0.002P_1 - 0.001P_2)} \quad ...(iv)$$

Penalty factor for plant 2

$$L_2 = \frac{1}{1 - \frac{\partial P_L}{\partial P_1}} = \frac{1}{1 - (0.0048P_2 - 0.001P_1)} \quad ...(v)$$

Now $IC_1 \times L_1 = IC_2 \times L_2 = \lambda$

or $$\frac{0.08P_1 + 16}{1 - (0.002P_1 - 0.001P_2)} = \frac{0.08P_2 + 12}{1 - (0.0048P_2 - 0.001P_1)} = \lambda = 20 \quad ...(vi)$$

or $$P_1 = \frac{4 + 0.02P_2}{0.12} \quad ...(vii)$$

and $$P_2 = \frac{8 + 0.02P_1}{0.176} \quad ...(viii)$$

Solving Eqs. (*vii*) and (*viii*), we have

P_1 = 41.7 MW **Ans.**

P_2 = 50.2 MW **Ans.**

Substituting values of P_1 and P_2 in Eq. (*i*), we have

$$P_L = 0.001 \times (41.7)^2 - 0.001 \times 41.7 \times 50.2 + 0.0024 \times (50.2)^2$$
$$= 1.73889 - 2.09334 + 6.048096 = 5.694 \text{ MW}$$

Received power,

$P_D = P_1 + P_2 - P_L = 41.7 + 50.2 - 5.694 = 86.204$ MW **Ans.**

Efficiency of transmission system,

$$\eta = \frac{P_D}{P_1 + P_2} \times 100 = \frac{86.0204}{41.7 + 50.2} \times 100 = 93.6\% \text{ **Ans.**}$$

9.9. OPTIMAL UNIT COMMITMENT

It is not economical to operate all the units available all the time. The unit commitment deals with the problem of determination of a units of generating power plant which should operate for a particular load to minimize the operating cost.

The unit commitment problem is applicable to thermal power plants only because for other types of generating plants like hydro, the operating cost and start-up times are negligible, therefore, their on-off status is not significant.

A simple but an inaccurate method of solving unit commitment problem is assignment of priority to the generating units such that the most efficient unit is loaded first to be followed by the most efficient unit from the remaining units as the load increases.

Another method of solving this problem is computation of most economical operating cost for all the possible combinations of the units and then select that combination which has the least operating cost among these. This method is very cumbersome and highly time consuming. Considerable computational savings can be achieved by adopting some optimization techniques such as branch and bounce or a dynamic programming method for arriving at minimum operating cost combination as certain combinations need not be tried at all in these methods.

9.10. OPTIMAL UNIT COMMITMENT USING DYNAMIC PROGRAMMING METHOD

In practice, the unit commitment problem is solved for the complete load cycle. The total number of units available, their individual operating cost characteristics and the load cycle on the generating stations are assumed to be known in advance. Further, it is assumed that the load on each unit or combination of units changes in suitably but uniform step of size ΔP_D MW (*e.g.*, 1 MW).

Let a cost function $F_N(x)$ be defined as below:

$F_N(x)$ = The minimum cost in Rs/hour of generating x MW by N units

$F_N(y)$ = Cost of generating y MW by the Nth unit (known in advance)

$F_{N-1}(x - y)$ = the minimum cost of generating $(x - y)$ MW by the remaining (N – 1) units

[for N = 2, $F_{2-1}(x - y) = F_1(x - y) = f_1(x - y)$ known in advance]

The application of dynamic programming results in the following recursive relations:

$$F_N(x) \xrightarrow[y]{\text{Min}} \{f_N(y) + f_{N-1}(x-y)\} \quad ...(9.49)$$

Here, the value of y is varied from minimum load level to the maximum load level in the discrete steps of ΔP_D providing a set of values of the expression $\{f_N(y) + f_{N-1}(x - y)\}$. The minimum of these values is optimum value of $f_N(x)$.

By use of recursive relation given by Eq. (9.49), we can easily find the combination of units which needs minimum operating costs for load ranging in convenient steps from the minimum permissible load of the smallest unit to the sum of the capacities of all the available units. In this process the total minimum operating cost and the load shared by each unit of the optimal combination are automatically found for each load level.

Starting arbitrarily with two units, the most economical combination is found for all the discrete plant load levels using Eq. (9.49) where for N = 2, $F_{N-1}(x - y) = F_1(x - y) = f_1(x - y)$ which is known. At each plant load level the most economic solution may to run either unit or both units with a certain load sharing between the two. It may also be possible that all the combinations considered are infeasible for certain load levels. The most economical cost curve in discrete form for the two units thus obtained, can be viewed as the cost curve of a single equivalent unit. The third unit is now added and the process is repeated for determination of the cost curve of the combination of the three units considering the first two units as single equivalent unit. It may be noted that in this process the operating combinations of third and first, also third and second are not required to be worked out resulting in cosiderable saving in computational effort. The process is repeated, till all available units are exhausted. The advantage of this approach is that having obtained the optimal way of loading k units, it is quite easy to find the optimal manner of loading $(k + 1)$ units.

From the results obtained by the above procedure a unit commitment table, which indicates the units in operation requiring minimum operating cost for the different loads, is prepared. This unit commitment table is prepared once and for all for a given set of units. If the load cycle on a plant load changes, it will need to change the starting and stopping times of the units with the basic unit commitment table remaining unchanged.

Using the unit commitment table and increasing load in steps, the most economical plant operating cost is determined for the complete range of plant capacity by equal incremental cost criterion. The result is the overall plant cost characteristic in the form of set of data points. A quadratic equation (or higher order equation, if required) can then be fitted to this data for later use in economic load sharing among generating plants.

Exercises

1. Discuss the importance of optimal load allocation in power plants. What information must be available for optimum load allocation ?

[U.P. Technical Univ. Power Station Practice 2009-10]

2. State the methods of loading the turbo generators.

[Pb. Technical Univ. Generation of Electric Power Dec. 2010]

3. Draw and discuss the following curves for thermal generating unit: (*i*) Input-output curve, (*ii*) Heat rate curve, (*iii*) Efficiency curve, (*iv*) Incremental fuel cost.

[Rajasthan Technical Univ. Power System Engineering 2011]

4. What are input and output characteristics of steam and hydro plants ? Explain with a neat graph.

[G.B. Technical Univ. Power Station Practice 2011-12]

5. Why is incremental fuel cost generally taken as total incremental cost for purpose of load allocation to thermal plant ? Explain the equal incremental cost criteria for optimum scheduling.

[G.B. Technical Univ. Power Station Practice 2010-11]

6. Show that for minimum continued input to carry a combined output the incremental rate for each unit will be equal.

[M.D. Univ. Power Plant Engineering, Dec. 2009]

7. Derive the expression for incremental cost of generation between different plants without considering the transmission losses.

[Pb. Technical Univ. Generation of Electric Power, May 2010]

8. Prove that the incremental rate curve intersects the heat rate curve.

[Rajasthan Technical Univ. Power System Engineering 2011]

9. Discuss the following :
   (*i*) Equality and inequality constraints.
   (*ii*) Penalty factor
   (*iii*) Loss coefficients and
   (*iv*) Economic load scheduling of thermal plants.

[G.B. Technical Univ. Power Station Practice 2011-12]

10. What is penalty factor ? Explain the incremental costs criterion for optimum scheduling in power plants. Why is incremental fuel cost generally taken as total incremental cost for the purpose of load allocation in thermal plants ?

[U.P. Technical Univ. Power Station Practice, 2009-10]

11. Find out the relation for 'optimum generation scheduling' of power plant including transmission losses.

[Rajasthan Univ. Power System Engineering 2004, 2005, 2008]

12. Explain the optimum economic operation of power system considering transmission losses.

[U.P. Technical Univ. Power Station Practice 2008-09]

13. Derive an expression for transmission loss coefficients.

[Pb. Technical Univ. Generation of Electric Power, May 2010]

14. Write and explain coordination equations for thermal plants.

[Pb. Technical Univ. Generation of Electric Power, May 2010]

15. Describe about coordination equations.

[Pb. Technical Univ. Generation of Electric Power, Dec. 2009]

16. Write the iterative procedure to solve coordination equation.

[Pb. Technical Univ. Generation of Electric Power, Dec. 2009]

17. Explain dynamic programming solution for unit commitment.

[Rajasthan Technical Univ. Power System Engineering 2011]

18. What is dynamic programming ? How can it be used to specify the units to be operated for a given load ?

[Pb. Technical Univ. Generation of Electric Power, May 2009]

19. What do you understand by optimal unit commitment ? Explain the dynamic programming method for solution of optimal unit compliment problem.

[Rajasthan Univ. Power Station Practice 2007, 2008]

20. Write short notes on the following:

(*i*) Methods of loading of turbo-alternators.

(*ii*) Optimum economic operation of power plants neglecting transmission losses.

(*iii*) Coordination equations.

(*iv*) Iterative procedure of solving coordination equations.

Short Answer Type Questions With Answers

Q. 1. Is the cost of fuel directly proportional to power output ?

[Pb. Technical Univ. December 2008]

Ans. No, the cost of fuel does not vary directly in proportion to the power output. It is directly proportional to the power output plus the losses in the conversion process.

Q. 2. What is input-output curve ? **[Rajasthan Technical Univ. 2011]**

Ans. The graphical relation between the fuel input, in Rs per hour, (or in million kilocalories per hour) and power output in MW for any power plant is known as input-output curve. A typical input-output curve is shown in Fig. 1. The P_{min} is the minimum loading limit below which it is uneconomical or technically infeasible to operate the unit or plant and P_{max} is the maximum output limit owing to its rating.

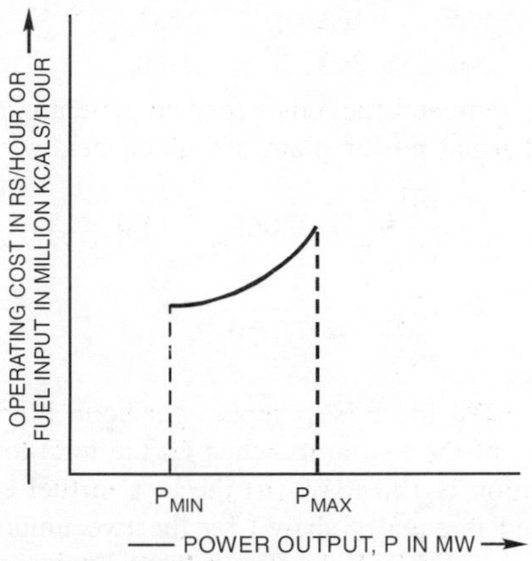

Fig. 1. *A Typical Input-Output Curve*

Q. 3. What is heat rate ?

Ans. Heat rate is defined as the ratio of input of fuel to corresponding output of power. It is found from the curve drawn by taking power output in MW on *x*-axis and fuel input in kilocalories per hour on *y*-axis. It is an indication of efficiency of the power plant.

$$\text{Heat rate} = \frac{\text{Input}}{\text{Output}} \text{ in kilocalories or kilojoules per kWh}$$

Q. 4. What is heat rate curve ?

Ans. Heat rate curve is a graphical relation between the heat rate in kilojoules or kilocalories per MWh and power output in MW.

Q. 5. What is incremental rate of a machine ?

Ans. The incremental rate of a machine at any given output is the rate of change of the input with respect to the output *i.e.,* the ratio of small change in input to a small change in output is known as *incremental* rate of a machine. Numerically it is equal to the slope of the input/output curve at a point corresponding to that output. Mathematically, it is the first derivative of the input with respect to the output.

Q. 6. What is incremental cost ?

Ans. In the input-output characteristic if input is expressed in kJ/hour and output in MW, the slope of the curve gives *incremental fuel rate in* kJ/MWh and if the input is expressed in Rs per hour, the slope of the curve gives *incremental fuel cost* in Rs/MWh. Mostly incremental fuel cost is taken as the incremental cost (IC) and forms the basis of load division.

Q. 7. How is incremental cost calculated ?

[Pb. Technical Univ. Dec. 2009]

Ans. In the input-output characteristic if the input is expressed in Rs/hour and output in MW, the slope of the curve gives incremental fuel cost in Rs/MWh. Mostly incremental fuel cost is taken as the incremental cost.

i.e., incremental cost is the ratio of change in input to corresponding change in output power.

$$\text{Mathematically, Incremental cost} = \frac{\Delta \text{ Input}}{\Delta \text{ Output}}.$$

Q. 8. What are the units of incremental cost ?

[Pb. Technical Univ., May 2009]

Ans. The units of incremental cost are Rs per MWhr.

Q. 9. What does economic load scheduling mean ?

[Pb. Technical Univ., May 2008]

Ans. By economic load scheduling, we mean to determine the generations of different power plants such that the total operating cost is minimum and at the same time the total demand and losses at any instant is met by the total generation.

Q. 10. Why is incremental fuel cost generally taken as total incremental cost for the purpose of load allocation to thermal plants ? **[U.P. Technical Univ. 2009-10; G.B. Technical Univ. 2010-11; Pb. Technical Univ. May 2009]**

Ans. The total incremental cost is made up of incremental fuel cost plus i cremental cost of other items, such as water, supplies, wages etc. It is very difficult to express costs of water, supplies, wages etc. directly as a function of output. Therefore, for simplification, these costs are taken as fixed percentage of incremental fuel cost. Hence, the incremental fuel cost is usually taken as total incremental cost.

Q. 11. What is Lagrangian multiplier ?

[Pb. Technical Univ; Dec. 2008]

Ans. Lagrangian multiplier, denoted by Greek letter λ, is the incremental cost of the unit in Rs per MWh. For optimum economy every generating unit should be operated at the same incremental cost, λ.

Q. 12. Why should transmission losses be included in incremental cost criterion ? **[Pb. Technical Univ. Dec. 2007]**

Ans. When a generating unit supplies power over a transmission line, and when generation of the unit increases, the same increment is not delivered to the load. The change in power at the load end or bus is the original incremental less the incremental loss in the line. Thus, the incremental loss in the transmission line is the function of power of the generating unit (or plant) and the line characteristic. Evidently the incremental cost curve at the generating unit or plant is to be modified to include the cost of the incremental losses in the line.

Q. 13. Write transmission loss B-matrix.

[Pb. Technical Univ. May 2010]

Ans. The transmission losses, P_L can be computed utilising loss formula coefficients or "B-coefficients". These B-coefficients for a given system are assumed to remain constant and their evaluation depends upon the open-circuit impedance matrix of the system network.

The transmission losses are given as

$$P_L = \sum_m \sum_n P_{Gm} B_{mn} P_{Gn}$$

where P_L are transmission losses, P_G is plant loading, B are loss coefficients, all in per units (pu) and $\sum_m$ and $\sum_n$ indicate independent summation to include all sources

$$\text{or } P_L = \left[P_{G1} \dots P_{Gm}\right]_{1\times m} \left[B_{mn}\right]_{m\times n} \begin{bmatrix} P_{G1} \\ \vdots \\ P_{Gn} \end{bmatrix}_{n\times 1}$$

The B-coefficient matrix is written as

$$\begin{bmatrix} B_{11} & B_{12} & \dots & B_{1n} \\ B_{21} & B_{22} & \dots & B_{2n} \\ \vdots & \vdots & \vdots & \vdots \\ B_{n1} & B_{n2} & \vdots & B_{nn} \end{bmatrix}$$

$$\text{where } B_{mn} = \frac{\cos(\partial_m - \partial_n)}{|V_m||V_n|(\cos\phi_m)(\cos\phi_n)k} \sum d_{km}\, d_{kn} R_k$$

and $d_{km} = \dfrac{i_{km}}{i_L}$, R_k = resistance, i_L = line current.

Q. 14. What are coordination equations ?

Ans. Coordination equations represent the coordination between various generating units as regards to their load allocation. These equations provide relationships between cost of energy on one hand and power output on the other.

Q. 15. Define penalty factor.

Ans. Penalty factor is defined as the ratio of small change in power of one unit to the small change in power received when only this unit supplies small change in power received.

Q. 16. What do you mean by iterative procedure ?

Ans. In an iterative procedure, the coordination equations are solved by method of successive approximations. A first guess of initial solution is assumed and error from actual value is determined. The assumed value is corrected again and again so as to reduce error to acceptable level.

Problems

1. The input-output curve of a certain 30 MW power station is given by $I = (32 + 32L + 1.68L^2) \times 10^6$ where I is the heat input in kJ/h and L is the load in MW.

   (*i*) Determine the heat input per day to the power station if it works at full load for 18 hours and at half load for the remaining period; (*ii*) Also find the savings in heat input per kWh of energy produced if the plant works at uniform load for all 24 hours generating the same amount of energy. **[A.M.I.E. Sec B. Power Plant Engineering Winter 1999]**

   [**Ans.** $50{,}412 \times 10^6$ kJ per day; 2,700 calories/kWh]

2. The incremental costs in Rs per MWh of two 200 MW units are as under :

$$\frac{dC_1}{dP_1} = 0.2P_1 + 30$$

$$\frac{dC_2}{dP_2} = 0.15P_2 + 40$$

   The minimum load of each is 18 MW. Find the load division between the two units as the total load varies from 35 MW to 450 MW.

   [Pb. Technical Univ. Generation of Electric Power, Dec. 2009]

   [**Ans.**

P_1	P_2	$P_1 + P_2$ in MW
17.5	17.5	35
⋮	⋮	⋮
63.125	17.5	80.625
100	66.6	166.66
125	100	225
150	133.33	283.33
175	166.66	341.66
200	200	400

3. The incremental fuel costs for two generating units A and B of a thermal power plant are given below:

$$\frac{dF_a}{dP_a} = 0.06P_a + 11.4$$

$$\frac{dF_b}{dP_b} = 0.07P_b + 10$$

   P is in MW and F is in rupees per hour. Find the economic loading of the two units when (*i*) the total load supplied by the station is 150 MW, (*ii*) the loss in fuel cost per hour if the load is equally shared by the two units.

   [M.D. Univ. Power Plant Engineering, Dec. 2008]

   [**Ans.** 72 MW and 78 MW; Rs 1.365 per hour]

4. The incremental costs in Rs per MW-hour of two 250 MW units are as under

$$\frac{dF_1}{dP_{G1}} = 0.2P_{G1} + 30 \; ; \; \frac{dF_2}{dP_{G2}} = 0.15P_{G2} + 40.$$

Compute the saving per year for economical load allocation compared to equal load for a total system load of 225 MW.

[Pb. Univ. Electrical Power II (Generation) January 1991]

[**Ans.** Rs 239,531 per year]

5. The respective input-output characteristics of the two thermal power plants of capacities 100 MW and 200 MW are $Q_1 = 30 + 2P_1 + 0.003P_1^2$; $Q_2 = 40 + 2.2P_2 + 0.001P_2^2$ where P_1 and P_2 and Q_1 and Q_2 are in MW and kJ × 10^6/hour (P-power and Q-heat input). For the maximum economy get the power sharing between the two power plants to meet the load of 240 MW. Taking heat value of coal as 12,000 kJ/kg, find coal consumption per day and coal consumption per kWh.

[A.M.I.E. Sec. B. Power Plant Engineering Winter 1994]

[**Ans.** 85 MW and 155 MW; 1,253.4 tonne per day; 0.0218 kg per kWh]

6. The fuel inputs per hour of plants 1 and 2 are given as

$$F_1 = 0.2P_1^2 + 40P_1 + 120 \text{ Rs per hr}$$

$$F_2 = 0.25P_2^2 + 30P_2 + 150 \text{ Rs per hr}$$

Determine the economic operating schedule and the corresponding cost of generation if the maximum and minimum loading on each unit is 100 MW and 25 MW, the demand is 180 MW and transmission losses are neglected. If the load is equally shared by both the units, determine the saving obtained by loading the units as per equal incremental production cost.

[Rajasthan Univ. Power System Engineering 2004]

[**Ans.** 88.89 MW and 91.11 MW; Rs 10,214.43/hour; Re 0.57/hour]

7. A two bus system is shown in Fig. 1. If a load of 125 MW is transmitted from plant 1 to the load, a loss of 15.625 MW is incurred. Determine the generation schedule and the load demand if the cost of received power is Rs 24/MWhr. The incremental production costs of the plant are :

$$\frac{dF_1}{dP_1} = 0.025P_1 + 15$$

$$\frac{dF_2}{dP_2} = 0.05P_2 + 20$$

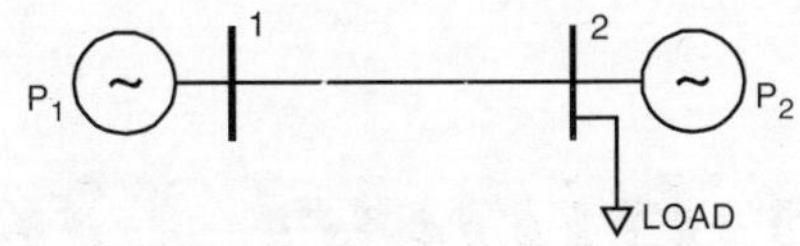

Fig. 1

[U.P. Technical Univ. Power Station Practice 2006-07]

[**Ans.** 123.28 MW, 80 MW and 188.1 MW]

8. A system consists of two plants connected by a transmission line as shown in Fig. 2. The load is at plant 2. The transmission loss calculations reveal that a transfer of 100 MW from plant 1 to plant 2 means a loss of 15 MW. Find the required generation at each plant for $\lambda = 60$. Assume that the incremental costs of the two plants are given by

$$\frac{dC_1}{dP_1} = 0.2P_1 + 22 \text{ Rs/MWh}$$

$$\frac{dC_2}{dP_2} = 0.15P_2 + 30 \text{ Rs/MWh}$$

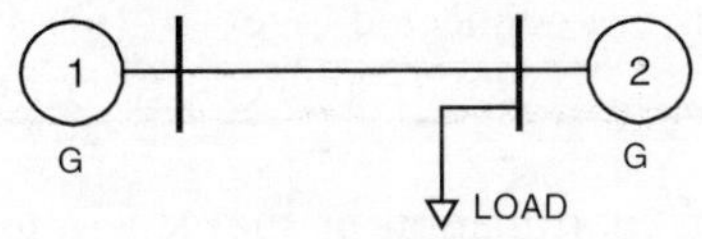

Fig. 2

[Rajasthan Technical Univ. Power System Engineering 2011]

[**Ans.** 100 MW, 200 MW]

❑❑❑

CHAPTER 10

Combined Operation of Power Plants

INSIDE THIS CHAPTER

10.1. INTRODUCTION

In order to have most economic operation and ensure optimum utilisation it becomes necessary to operate various power plants located in region in an integrated system. For this purpose all the power plants are connected together, (the system is called the *interconnected system*) and centrally supervised and controlled.

The proper coordination of the generation of electrical power in a group of power plants connected to a common grid system leads to considerable saving as compared with the same loads fed from a number of independent power plants. First of all, it enables the total generating capacity to be reduced, since less spare capacity is required, secondly, it allows the system to be planned so that the maximum size of generating unit is installed (and hence the cost per kW installed is a minimum) and, thirdly, it enables the power plant to be loaded so that the minimum amount of fuel is consumed by concentrating generation in most efficient power plants.

10.2. MERITS OF COMBINED OPERATION OF POWER PLANTS

The combined working of various types of power plants have various benefits, as discussed below.

1. *Flexibility of Operation*. A fact that has a very important bearing on the cost of electrical energy is that, not only is the demand higher in winter than that in summer, but it also varies throughout the 24 hours of each day, being lowest during the night and rising during the day with peaks at certain times. The varying load on a system calls for solution of some very complex operating problems. The system requirement is met by loading the most efficient plant in order of merit of operating cost—the power plants with the minimum operating cost are put on load for the maximum time, plants with higher operating cost are only loaded when required by the load demand on the system and those with the maximum operating cost are kept in reserve and would operate only in emergency.

The major drawback of a steam power plant is that it requires appreciable time to start up, synchronise and take up load. It is also necessary to keep the boilers hot throughout the day and night in order to keep the stresses and strains, caused by the temperature variations, to a minimum. It means a continuous unavoidable loss of energy (1 to 2.5% of full-load consumption). On the other hand, a hydroelectric power plant can be started, synchronised and loaded very quickly and has the advantage of low operating cost. Such a power plant can also meet the sudden variations of load demand easily and so it can be profitably employed both for supplying base loads as well as peak loads. Nuclear power plant is not suitable for variable load operation as the reactors cannot be easily controlled to respond quickly to load variations. Such a plant is used as base load plant only operating at high load factors of over 0.8. A gas turbine power plant having the advantage of low initial cost and low start up time is ideally suited for supplying peak loads. Diesel power plant, because of its uneconomical operating cost, is operated only when necessary *i.e.*, in emergency. Thus if various types of power plants are operated in combination, the operation will become flexible *i.e.*, the steam power plants and nuclear power plants are operated to supply base loads, the hydroelectric power plants to supply base loads or peak loads depending upon the availability of water, gas turbine power plants for meeting with peak demand of the system and existing diesel power plants in emergency only.

2. *Reliability of Supply*. The reliability of a steam power plant depends upon the coal supply, that of a hydroplant on the stream flow. Due to greater diversity, a combined operation of various types of power plants would be more reliable than the individual steam or hydro power system.

3. *Reserve Capacity.* In any electric power system, the system has to have a certain reserve capacity in order to meet any unforeseen excess power demand and to take care of the forced outages of units and schedule maintenance of units. In combined operation of several plants the reserve capacity required is reduced. In a combined system the effect of all the diversities may result in a still larger benefit from the point of view of reserve capacity.

4. *Better Utilisation of Hydro Power.* In combined power system the hydro power can be utilised in a more effective way. During periods of heavy run-off, the hydro power plants can be operated to supply base load resulting in conservation of fossil fuels and during periods of draught the steam power plants can be used as base load plants to allow water to be stored in reservoirs and thus make the capacity of tne hydro plants effective.

5. Economic Benefits. We can divide the countries of the world in three groups as regard power potential.

First group of countries (such as Norway, Sweden) is that which have bulk of hydro power resources but fossil fuel is to be imported. Second group of countries (such as UK) is that which have very limited hydro power resources and have to depend mainly on thermal generation. Third group of countries (such as India, USA) is that which have to exploit both hydro and thermal resources to meet the power requirements.

In case of countries, falling under group 1, there might be various plant sites and the comparison will have to be made in terms of cost of producing equivalent electric power and its transmission to load centre. Some sites will be suitable for base load power plants and the others might be suitable for peak load service. The cost of generation of electrical energy at various sites will have to be studied for various load factors. The load factor to be adopted will be that which will provide the lowest overall cost of generation.

There is one drawback of the above system. The maximum run-off may not occur during the periods of maximum demand and for the hydro plant to supply the whole demand throughout the year, a storage will have to be built to cover the seasonal variations of run-off. This storage may be either impracticable or very costly to provide. The storage requirements can be reduced by using thermal generation for supplying peak loads. The load factor of the thermal plant will be high only for a small period and its annual load factor will be low.

In case of countries falling in second group *i.e.*, for the countries having very limited hydro power sources and abundant coal reserves, it will be economical to develop the hydro power plants to meet peak load requirements. Such hydro power plants will run at low load factors and enable the thermal power plants to run at high load factors in an economic manner. It is noteworthy that in hydro plants, the cost of storage is the major item in the total cost and this item does not depend much on the annual load factor, therefore, the capital cost will be reduced for low load factor than for high load factor. It is found that incremental costs resulting from reduced load factor *i.e.*, enhanced amount of installed capacity is less than the capital cost of raising the installed capacity of a thermal plant. An advantage of increasing capacity of hydro plant is that the spoilage can be reduced at the time of high run-off by operating the hydro plant for longer periods and thereby saving the fuel cost in the thermal power plants.

For the countries falling in third group, proper co-ordination between hydro and thermal power plants is very essential. The hydro plants must be planned to generate enough electric power during adverse periods so that these can meet the needs of that portion of load curve to which they have been assigned. It will result in conserving the fossil fuels to the maximum extent. The thermal power plants should be set up to enhance the firm capacity of the power system. The run-off river power plants would generally meet the entire or part of the base loads. The remaining demand can be met by judicious mix of reservoir type hydro plants, coal-based thermal plants and nuclear plants.

For any particular scheme, there must be some ratio of the hydro plant to the total demand which gives the lowest annual cost of electric generation. In regions where both cheap fuel and favourable hydro sites are available, this ratio can be 25 to 40. In regions, where fuel is costly and the suitable hydro sites are available, this ratio will be much higher. But for regions where fuel is cheap and the cost of development of hydro power plant is high, the ratio may be quite low.

10.3. BASE LOAD AND PEAK LOAD

As already mentioned, the load on the power plant is seldom constant rather it varies from time to time, as shown in Fig. 10.1.

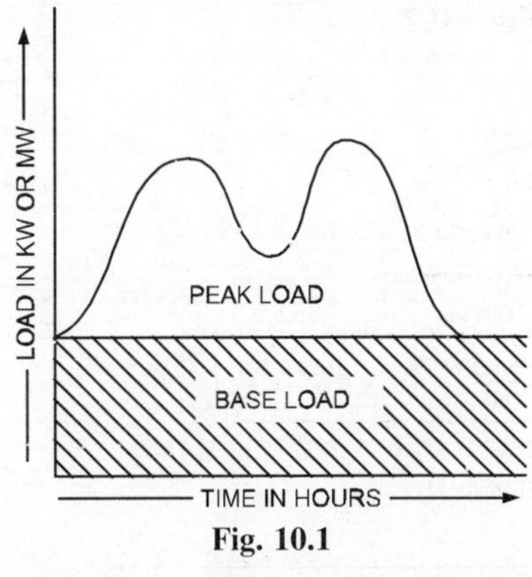

Fig. 10.1

The load on any power plant can be conveniently considered into two parts namely (*i*) base load and (*ii*) peak load.

The unvarying load, which occurs almost the whole day on the power plant is called the *base load* whereas the various peak demands of the load over and above the base load of the power plant is called the *peak load.*

Co-ordination of Base Load and Peak Load Power Plants. If the load represented by Fig. 10.1 is to be supplied from one power plant only, then the installed capacity of the plant should be equal to the peak load or more. Such a power plant would be uneconomical since the peak load occurs only for a short period in a year and, therefore, the capacity equal to the difference of the peak load and base load would remain idle for the major portion of the year. Hence such a power load should not be supplied from a single power plant. There would be some power stations supplying the base load and others, possibly of different types, supplying the peak load. Thus the co-ordination of operation of different power stations is essential.

The power plants to be employed as base power plants should have (*i*) low operating cost (*ii*) capability of working continuously for the long periods (*iii*) requirement of few operating personnel and their repair should be economical and speedy.

The power plants to be employed as peak power plants should have the capability of quick start, synchronisation and taking up of system load and quick response to load variations.

The hydro power plants are well suited for both base load and peak load operations. The hydro power plants should be employed for base load operation as far as possible because of their higher capital cost. However, during the periods of draught, the hydro power plants may be used as peak load plants.

A steam power plant gives minimum cost of generation per unit when employed as base load plant. However, in order to save fuel it may be used as peak load plant.

Nuclear power plants are suitable only for base load operation at high load factors of over 0.8.

Gas turbine power plants are suitable for supplying peak loads.

Diesel engine power plants are playing a very little role in bulk power generation because of their uneconomical operating costs.

A typical load allocation to various power stations is illustrated in Fig. 10.2.

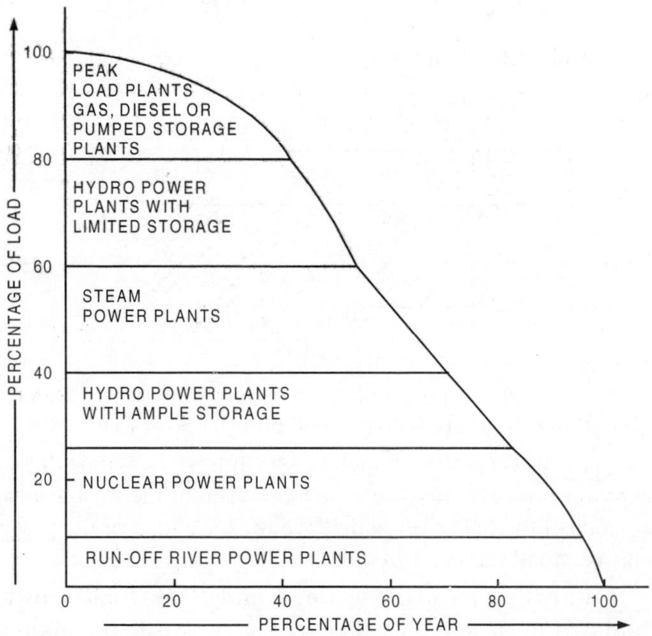

Fig. 10.2

10.4. INTERCONNECTION OF POWER STATIONS

Several generating stations connected to each other form an *interconnected grid system.* Although interconnection of power stations involves extra cost, yet considering the benefits derived from such an arrangement, it is gaining much favour these days.

Some of the advantages of interconnected power system are as follows:

1. Reduced Plant Reserved Capacity. Every power station is required to have a standby unit for emergencies. By interconnecting the large generating stations, the amount of reserve plant capacity in the system, as a whole, is reduced, the load factor and efficiency of operation are improved.

2. Increased Reliability of Supply. The reliability of supply is increased since in case of fault at one station, the power can be fed from other stations.

3. Increased Effective Capacity of Power System. The load curves of different interconnected stations are likely to differ. The result is that the maximum demand on the system is much reduced as compared to the sum of individual maximum demands on various power stations. In other words, the diversity factor of the whole system is improved, thereby increasing the effective capacity of the whole system.

4. Economical Operation. Interconnection between hydroelectric power station and thermal power station makes their operation economical. When water is available in sufficient quantity such as in rainy season, the hydroelectric power station is employed as a base load station and thermal power station is employed as peak load station because the running cost of hydroelectric plant is low. When water is not available in sufficient quantity, such as in summer season, the thermal power station takes base load and hydroelectric power station takes peak load.

Economy is also obtained by sharing the total demand among the interconnected power stations in such a way that the more efficient power stations run continuously throughout the year at high load factor and less efficient plants operate for peak load hours.

5. Use of Older Plants. The interconnected system makes it possible to employ the older and less efficient plants to carry peak loads of short durations. Although such plants may be inadequate when used alone, yet they have sufficient capacity to carry short peaks of loads when interconnected with other modern power plants. Thus, interconnected system makes use of obsolete plants possible.

6. Exchange of Peak Loads. An important benefit of interconnected system is that the peak load of the power stations can be exchanged. If load curve of a power station indicates a peak demand exceeding the rated capacity of the power station, then the excess load can be shared by other power stations interconnected to it.

7. Reduced Capital Cost. With interconnected stations, generators of larger sizes can be employed resulting in reduced capital cost per kW. In interconnected grid system large generators can be used so that in case of failure of any one, power could be supplied from other stations.

8. Savings in Operating Costs. In an interconnected system, it would be possible to allocate the total system load to the various individual power stations in the most economic manner and the load dispatching will be fully co-ordinated resulting in large savings in operating costs.

10.5. INTERCONNECTED HYDROELECTRIC STATIONS

There are large possibilities that interconnected hydro power stations are of different types such as run-off river power plants with small pondage, hydro plants with a forebay large enough to keep the turbines running at full load for hours together and high head power plants with storage capacity large enough to operate continuously at full load for weeks/ months together.

For economical operation of such an interconnected system the best way is to utilize the rainfall in the catchment area to the maximum possible and this can be done by running power stations supplied by long open channels and rivers during heavy rainfall. The noteworthy point is that during heavy rainfalls open channels and rivers get considerable amount of surface water that drains into the streams and this inflow may in certain cases exceed what is required for power generation by turbines operating on full load. The practice is, therefore, to make the forebay down to a low level just before heavy rains set in and reduce the load on stations with high head plants and ample storage capacity. This arrangement results in high water levels of storage reservoirs at the beginning of the dry season and generate maximum possible electrical energy before the rainy season commences. The interconnection of different types of hydro power stations provides more effective utilization of water for generation of electrical energy.

10.5.1. Methods of Loading of Hydro Power Plants

The performance characteristics of different units in hydro power stations may vary, and, therefore, determination of division of load among the various units, considering the incremental costs becomes essential. The required information for the economic scheduling of hydro power stations is as follows: the efficiency curve of each unit, the input-output curve indicating the total rate of flow of water against the load, and the incremental rate curve showing the change in the rate of flow required per unit change in load. The methods of determining the incremental costs of thermal plants were studied in the last chapter. Methods for very accurate determination of the incremental costs from hydro plants have not yet been well developed. Each hydro power station has a small reservoir near to it for short-term regulation of the water required under various loading conditions. The reservoirs for long-term regulation may be situated at a greater distance from the power stations and may not affect the head at the power station to an appreciable extent, even with change of load on the power stations.

The losses in the conduits and water turbines increase with the increase in required rate of flow of water, and the incremental cost of power from a hydro plant is higher during peak generation compared to that when supplying a smaller load. The incremental cost also varies with the water head at the power station, different load conditions, and the time of their occurrence. Long transmission lines are usually necessary in case of hydro plants, and the effect of transmission losses is to be included in economic schedule studies. The incremental cost of power supplied from a hydro plant may be determined by the analysis of losses owing to short-term regulation. For large power plants with large reservoirs, the incremental cost of water at power stations, with long-term regulation is required to be determined. Thus, this is very involved study and needs estimates based on probability calculations of the flow of water available from the expected rain fall etc., and details of storage in reservoirs at different times during the period under review.

When a number of identical machines having identical input-output curves are installed in a hydro power station, then the most economical method of operation can be had in the following manners:

1. One unit may carry the load variations, while all the other units operating at their points of maximum efficiency.
2. The more preferable method is loading of all the machines equally and operating at points of equal slope on their input-output curves.

If there is a fall in load equal to capacity of one unit, it is best for economy to shut-off that machine and redistribute the load among the remaining units.

If in a power plant, the machines are provided with individual penstock, the improved efficiency is obtained by adding units in addition to those already in operation. This is due to the fact that the leakage loss in the standby units is either reduced or eliminated. This is shown in Fig. 10.3.

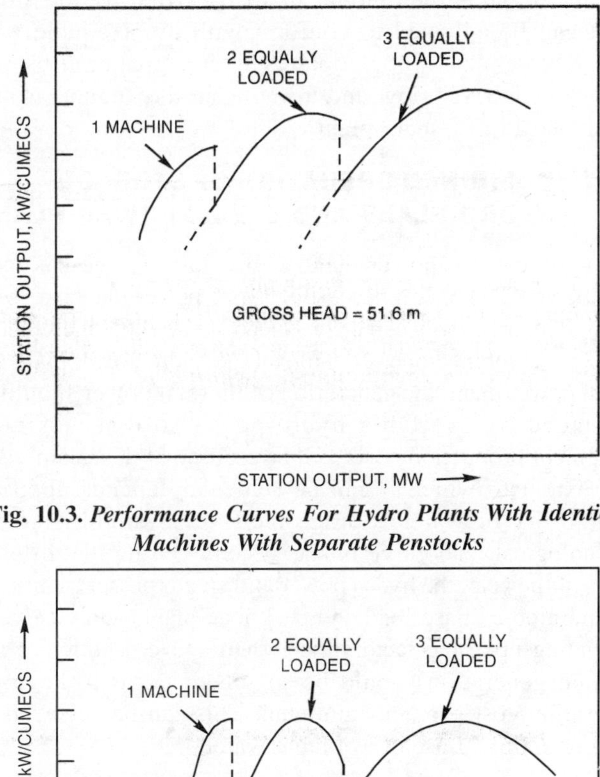

Fig. 10.3. *Performance Curves For Hydro Plants With Identical Machines With Separate Penstocks*

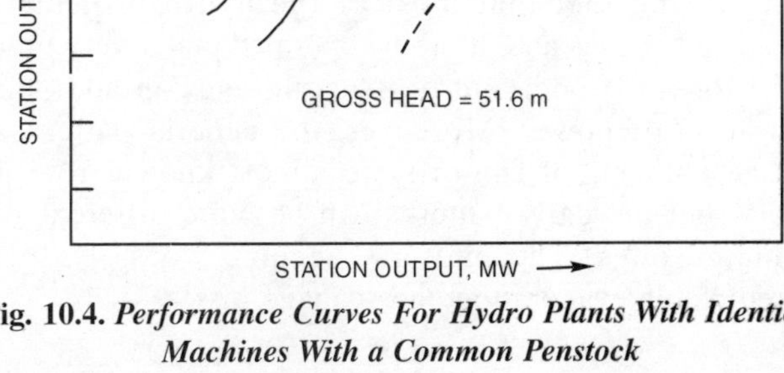

Fig. 10.4. *Performance Curves For Hydro Plants With Identical Machines With a Common Penstock*

However, when all the machines are provided with one common pipeline, then the performance will be as depicted in Fig. 10.4. As the additional units are put in operation, the leakage loss is reduced but the pipe friction loss is increased. A stage will be reached when the gain due to decrease in leakage loss is offset by the increased pipe friction loss. Thus, as obvious from Fig. 10.4, the plant will operate at the highest efficiency with only two units in operation.

10.6. COMBINED OPERATION OF RUN-OFF RIVER POWER PLANT AND STEAM POWER PLANT

As discussed in chapter 2, the run-off river power plant has a small pondage and uses water as it is available. The run-offs of river vary widely during the year—very large in rainy season and very low in dry season. As such the firm capacity of such plants is very low. The utility of such plants can be considerably increased by operating them in combination with steam power plants. The run-off river power plant can be employed as a base load power plant (up to its installed capacity) during rainy seasons (periods of high stream flows) while the steam power plant may supply the peak load. During dry season (periods of lean flow) the steam power plant is used to supply the base load and run-off river power plant for supplying peak load. Thus the amount of load carried by the steam plant can be adjusted to conform with the available river flow. Such a plan of operation results in a greater utilisation factor of the river flow and a saving in the amount of fue! consumed in the steam plant.

10.7. COMBINED OPERATION OF STORAGE HYDRO PLANT AND STEAM POWER PLANT

The hydroelectric power plant with ample storage has more firm power. The operating cost of hydro power plants are very low and hydro energy can be produced at very little incremental cost. In a combined operation of storage hydro power plant and steam power plant, the generation of thermal power should be displaced by available hydro power so that maximum decrement production costs will be realised at the steam plant. The extent to which it can be achieved depends upon the availability of water, storage facilities available and the chronological load curve of the system. During the different times of the year, the hydro power and steam power plants can be operated as base load and peak load plants and vice versa depending upon the factors mentioned above. The hydro and thermal generation should be so scheduled as to keep the generation cost per unit minimum. This can be achieved by making optimum use of available water.

The operation of a combined system is cumbersome owing to the fact that most of the hydro projects are multipurpose ones and are to be operated under very unique restrictions of flood control irrigation etc., in addition to generation of power. Moreover, the general pattern and average amounts of run-offs are almost known from past history, but the daily patterns can be quite different from estimated ones. The matter's become worse owing to unpredictable evaporation and seepage losses.

The aim of the combined operation of hydro and thermal power stations is to keep the total operating cost minimum and simultaneously to meet the total demand and losses of the power system at any instant. It means that the available water should be put to an optimum use.

Optimisation of hydro power generation involves quite a few variables like operation of various hydro plants at points of their best efficiency, optimisation of available head and reducing the spilling of water to the minimum. The operation of different hydro plants at the points of their best efficiency is concerned with hourly variations in system load demand, division of load between various plants etc. The optimisation of water head and minimisation of spilling of water are contradictory to each other. Reducing the spilling of water needs storage space in reservoir *i.e.*, the water head is less than the maximum. Spilling of water also increases the tail water elevation and thereby reduces the effective water head. With a reasonably high development of storage and where a considerable down stream is developed, reducing the spilling of water is found to be more significant.

10.7.1. Long Term Hydro-Thermal Co-ordination

When the hydro plant capacity in a system is ample and the reservoirs are full, the starting of thermal power plants should be delayed as long as possible. Normally, the thermal plants should supply part of load throughout the year and they should supply the deficit during dry seasons. They should be operated for sufficient time for the required purpose. It may also be possible as an alternative to supply all the load by hydro plants for some time and thus delay the starting of thermal power plants. When the storage in the reservoirs approaches a minimum level, the water discharge should be reduced and the thermal plants started. In determining the minimum zone of the storage reservoirs, it is necessary to have a detailed statistical analysis of their in flow, evaporation losses, etc. The minimum zone corresponds to the minimum water level in the reservoirs below which the hydro plants cannot be oeprated from the reservoir water supply.

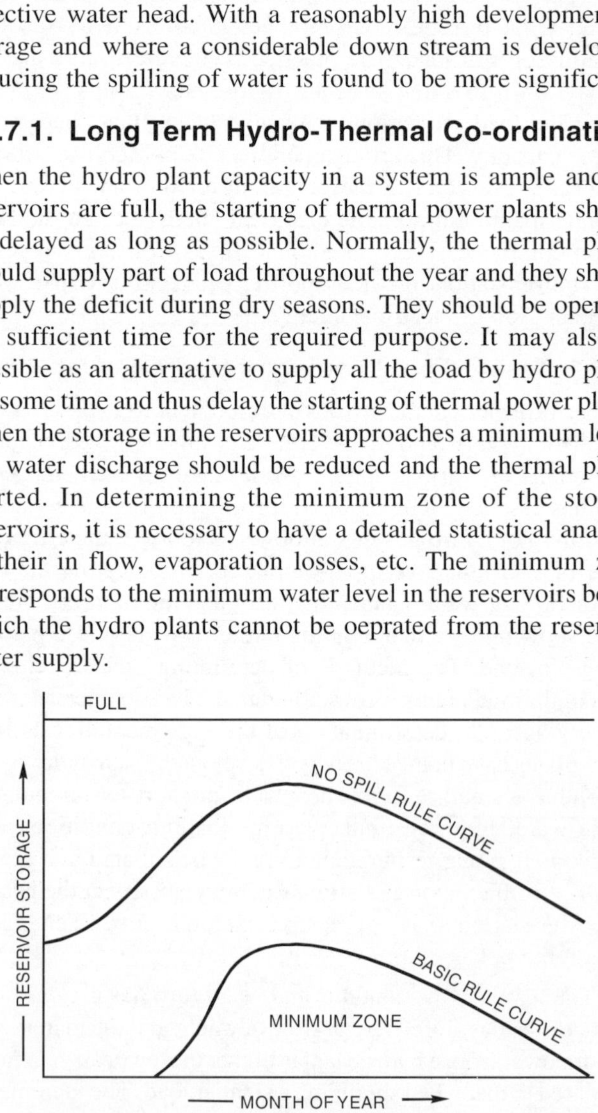

Fig. 10.5. ***Basic Rule Curve, No-Spill Rule Curve and Minimum Zone***

The minimum level is a function of the time of the year, as depicted in Fig. 10.5. This curve is usually known as *basic rule curve* and the area under this zone is called the *minimum zone.* The basic rule curve represents accumulations of water, in reverse order of time, of the deficiency between the firm load and energy available from the stream flow in critically dry periods. As soon as the reservoir storage tends to reach the basic rule curve, the thermal plants should be started and run at maximum output in order that the firm load is supplied and reservoir does not further deplete. There is another curve, known as *no-spill curve,* which represents the accumulations, in reverse order of time, of the surplus between the firm load and the energy available from stream flow in the wettest period. There is no risk of spill if the storage contents are below the 'no-spill rule curve'. Efforts should be to keep the reservoir contents up to the no-spill curve as far as possible so as to provide additional head without any risk of spill.

The economic long-term hydro-thermal scheduling using large storages is a very complicated problem. A large number of variable factors are involved, resulting in many sets of nonlinear equations. It is not possible to solve such problems by long-hand methods. However, the development of fast computers made it possible to solve a number of complicated equations quickly, as well to arrange programming of the economic working of the power plants.

In general, there are two extreme possibilities. The hydro plant is operated at maximum discharge until the allotted water is consumed, after which the hydro plant is shut down for the rest of the day and the remaining load is supplied by the thermal plant(s). An alternative is to keep the hydro plant shut down and inoperative as long as possible, and to put it in commission only when there is enough time left during the day to consume the allotted quantity of water. Another method is to reduce the total thermal cost by changing the hourly division of load between the thermal and hydro plants using incremental programming. The incremental value of the water should be determined for the purpose. A number of iterations on computer help in finally obtaining the optimum hydro plants schedule.

10.7.2. Short Term Hydro-Thermal Co-ordination

When both thermal and hydro plants are avialable in a power system they should be scheduled over a given period to achieve maximum economy by minimizing the fuel cost of the thermal-plant units when using desired amounts of water for the hydro plant units.

The following informations is required for the study—

1. Load cycle.
2. Incremental fuel costs of steam power plants.
3. Expected water inflow during the period of operation.
4. Hydrogeneration as a function of head and discharge.
5. Reservoir elevation as a function of storage that is function of inflow, water use, time, evaporation, seepage etc.
6. Incremental transmission loss characteristics.
7. Water intake losses.
8. Tail race elevation as a function of water discharge.

If it is assumed that hydro plants will operate at constant heads, and that the thermal plants will operate with the minimum cost of fuel during the year; the economical schedule can be worked out by using co-ordination equations.

$$\frac{dC_n}{dP_{sn}} + \frac{\partial P_L}{\partial P_{sn}} = \lambda \quad ...(10.1) \qquad \text{for } n = 1, 2, 3, \ldots \alpha$$

$$r_j \frac{dW_j}{dP_{Hj}} + \lambda \frac{\partial P_L}{\partial P_{Hj}} = \lambda \quad ...(10.2) \quad \text{for } j = \alpha+1, \alpha+2, \alpha+3 \ldots \beta$$

where $\frac{dC_n}{dP_{sn}}$ is incremental fuel cost of *n*th thermal plant in Rs per MWh

$\frac{dW_j}{dP_{Hj}}$ is incremental water rate of *j*th hydro plant in m^3/sec-MW

$\frac{\partial P_L}{P_{sn}}$ is incremental transmission loss of *n*th steam plant

$\frac{\partial P_L}{\partial P_{Hj}}$ is incremental transmission loss of *j*th hydro plant.

P_{sn} is output of *n*th steam plant in MW.

P_{Hj} is output of *j*th hydro plant in MW

r_j is a constant of conversion of incremental water rate of *j*th plant into an equivalent incremental cost.

α is number of thermal plants.

and $\beta - \alpha$ is number of hydro plants.

Knowing the amount of water that would be available for use at a hydro plant r_j can be chosen. Using this value, co-ordination equation is solved for the economic scheduling of plants in the system. Equations can be solved for different values of λ. If $\frac{dC_n}{dP_{sn}}$ and $\frac{dW_j}{dP_{Hj}}$ in the above equations are assumed to be linear functions, an approximate solution can be had.

Another method of solution would be equal incremental-plant cost method.

Sometimes, the practice is to operate hydro plants near the points of maximum efficiency except when surplus water is available. When hydro plants are operated in this way, the thermal plants can be scheduled either by equal incremental fuel costs or by solution of co-ordination equations.

Example 10.1. A steam station and hydro station feed an area jointly. The hydrostation is run for 14 hours daily and steam station runs for all 24 hours. The production cost characteristic of steam station is

$$C = 5 + 8P_s + 0.05P_s^2 \text{ Rs/hour}$$

If the load on steam station, when both plants are in operation is 250 MW, the incremental water rate of hydro plant is $\frac{dW}{dP_h} = 30 + 0.05\,P_h$ m^3/MW-seconds.

The total quantity of water used during 14 hours is 500 million cubic metres. Find the load on hydro plant and cost of water use. Assume that the load on hydro plant is constant for the 14 hour period.

[Pb. Technical Univ. Generation of Electric Power, Dec. 2007]

Solution : The production cost characteristic of steam plant is

$$C = 5 + 8P_s + 0.05P_s^2 \text{ Rs/hour}$$

Incremental cost for steam plant for $P_s = 250$ MW load on steam plant

$$\lambda = \frac{dC}{dP_s} = 8 + 0.1P_s = 8 + 0.1 \times 250 = \text{Rs } 33 \text{ MW hour}$$

Total quantity of water used during 14 hours

$$= P_h(30 + 0.05P_h) \times 14 \times 3{,}600 \text{ m}^3$$

But it is given as 500×10^6 m^3

So, $P_h(30 + 0.05P_h) \times 14 \times 3{,}600 = 500 \times 10^6$

or $0.05P_h^2 + 30P_h - 9{,}920.635 = 0$

or load on hydro plant, $P_h = 237$ MW **Ans.**

$$\text{Cost of water use, } r = \frac{\lambda}{\frac{dW}{dP_h}} = \frac{33}{30 + 0.05 \times 237}$$

$$= \text{Re } 0.7885\text{/hour/m}^3 \text{ second}$$ **Ans.**

10.8. COMBINED OPERATION OF WIND POWER UNIT AND OTHER POWER PLANTS

The wind power units cannot provide firm power, as already discussed in Art. 7.8. However, these units can be employed in conjunction with steam and hydro power plants.

10.8.1. Combination of Wind and Steam Power

By operating wind power units in conjunction with a steam power plant supplying major portion of the demand of the power system, there can be savings in fuel. For its economic justification the cost of wind energy may be compared with the cost of the generating additional units from the existing steam plants. Wind power would not be economical beyond the total incremental cost of generating in the steam station. If the steam power station is less efficient and operates at low load factors, it might be economical to add wind power unit(s) to the system.

10.8.2. Combination of Wind and Hydro Power

The firm capacity of an electrical power system can be increased by operating the wind power unit(s) in combination with storage hydro power plants, particularly if there is diversity between the occurrence of the rain and wind. During wind season, the wind power can be supplied to the electrical power system and thus the load on the hydro plants can be reduced. This will result in the conservation of stored water which can be used afterwards more beneficially.

Exercises

1. State the advantages from operating the hydro and thermal power plants in co-ordination.
   [Pb. Technical Univ. Generation of Electric Power, May 2008]
2. What advantages accrue from operating a hydro and a thermal plant in co-ordination ?
   [Pb. Technical Univ. Generation of Electric Power, Dec. 2009]
3. What do you understand by 'peak load plant' and 'base load plant' ?
   Which of the following plants are suitable for use as base load plants ?
   Give reasons for your answer :
   (*i*) Run-off river plant (*ii*) Storage hydro plant (*iii*) Pumped storage plant (*iv*) Steam power plant (*v*) Nuclear power plant. **[Pb Univ. Electrical Power II (Generation) Dec. 1995]**
4. Discuss the conditions with reasonings under which electric power generation by (*i*) hydro power stations (*ii*) thermal power stations (*iii*) diesel power stations and (*iv*) nuclear power stations become economical. Which of them would you recommend for base load and peak load operations ?
   [Pb. Univ. Electrical Power II, January 1993]
5. What are the advantages of interconnection of power stations ? Two power stations, one thermal station and the other hydro, how they will be connected in parallel ? Which will be used as base load station and which will be used as peak load station ? Justify your answer.
   [U.P. Technical Univ. Power Station Practice 2008-09]
6. Explain base load and peak load stations. In an interconnected hydroelectric and steam stations, which is used as peak load stations and why ? Run-off river plant, nuclear plant, hydro plant with storage, steam plant and hydro plant with limited storage are connected in the system. Show with the help of a diagram how load allocation may be done.
   [U.P. Technical Univ. Power Station Practice 2007-08]
7. What are the general considerations for interconnection and combination of different types of power generating stations.
   [A.M.I.E. Sec. B. Power Systems Winter 1997]
8. What is an interconnected grid ? What are the advantages and drawbacks of interconnected system ?
   [Pb. Univ. Electrical Power II (Generation) Dec; 1982]
9. State what is meant by base load and peak load stations. Discuss the combined operation of a run-off river plant and a steam plant in combined hydroelectric and steam station operation.
   [Agra Univ. Generation & Switchgear Practice 1981]
10. What are the advantages of combined working of thermal power plant and hydroelectric plant ? Discuss briefly the need for co-ordination of these plants in power system.
    [A.M.I.E. Sec B. Power Plant Engineering Winter 1986]
11. Why is it necessary to operate hydro and steam plants in combination ? **[A.M.I.E. Sec B. Power Systems Summer 2000]**
12. What are the factors that govern combined operation of hydro and steam plants ?
13. Discuss the importance of short-term hydrothermal co-ordination. Why it is necessary to operate steam and hydro plants in combination ? Explain.
    [G.B. Technical Univ. Power Station Practice 2010-11]
14. Why is it necessary to operate hydro and steam plants in combination ? Discuss the importance of short-term hydrothermal co-ordination ?
    [U.P. Technical Univ. Power Station Practice, 2009-10]
15. Explain hydrothermal scheduling with various power system constraints.
    [U.P. Technical Univ. Power Station Practice 2006-07]
16. Write short notes on the following:
    (*i*) Merits and demerits of combined operation of power plants.

(*ii*) Base load and peak load plants. **[Pb. Univ. Dec. 1993]**

(*iii*) Interconnected system. **[Pb. Univ. May 1994]**

(*iv*) Combined working of

(*a*) Run-off river plant and steam plant.

(*b*) Storage hydro plant and steam plant. **[Pb. Univ. April 1981]**

(*v*) Utility of steam-hydro interconnected system. **[A.M.I.E. Sec. B. Summer 1996]**

(*vi*) Load division on hydro steam system. **[Pb. Technical Univ. Dec. 2010]**

Short Answer Type Questions With Answers

Q. 1. Why is it necessary to operate hydro and steam plants in combination ? **[U.P. Technical Univ. 2009-10; G.B. Technical Univ. 2010-11]**

Ans. It is necessary to operate hydro and steam plants in combination because such a combination provides (*i*) flexibility of operation, (*ii*) reliability of power supply, (*iii*) reduced reserve capacity, (*iv*) better utilization of hydro power and (*v*) better economy.

Q. 2. What do you understand by base load and peak load ?

Ans. The unvarying load, which occurs almost the whole day on the power plant is called the *base load* whereas the various peak demands of the load over and above the base load of the power plant is called the *peak load.*

Q. 3. What do you understand by 'peak load plant' and 'base load plant'. **[Pb. Univ. Dec. 1995]**

Ans. The power plant that supplies various peak demands of the load over and above base load of the power system is known as peak load plant while the plant supplying base load (unvarying load) of the system is known as base load plant.

Q. 4. What are the requirements of plant so that it is suitable for base load operation ? **[Pb. Technical Univ. Dec. 2007]**

Ans. The power plants to be employed as base power plants should have (*i*) low operating cost (*ii*) capability of working continuously for the long periods (*iii*) requirement of few operating personnel and their repair should be economical and speedy.

Q. 5. What are the requirements of power plants that are suitable for peak load operation ?

Ans. The power plants to be employed as peak power plants should have the capability of quick start, synchronisation and taking up of system load and quick response to load variations.

Q. 6. Why the hydroelectric power plants are suited to both base load and peak load operations ?

Ans. The hydroelectric power plants are suited to base load operation because their operating costs are very low and suited to peak load operation because they can be started, synchronised and loaded very quickly and sudden variations of load demand can be met easily.

Q. 7. Why is it necessary to operate nuclear power plant as base load plant supplying load at load factor higher than 0.8 ?

Ans. The operating cost is quite low and once the installation is completed, the loading of the power plant will have no effect on the generation cost. Therefore, a nuclear power plant is always operated as a base load plant. The nuclear power plants are usually not operated at a load factor less than 0.8.

Q. 8. Why are diesel power plants rarely used ?

Ans. Diesel power plants are used in emergency only owing to their uneconomical operating costs ?

Q. 9. Run-off river plant, nuclear plant, hydro plant with storage, steam plant and hydro plant with limited storage are connected in the system, show with the help of a diagram how load allocation may be done. **[U.P. Technical Univ. 2007-08]**

Ans. A typical allocation to various power plants is shown in Fig. 10.2.

Q. 10. Explain the function of power grid. **[Pb. Technical Univ. May 2008]**

Ans. Power grid is an interconnection of major power plants by means of EHV transmission lines. As far as power sector is concerned, India is divided in five zones/regions—Northern, Southern, Eastern, Western and North-Eastern regions. Each of these regions has a grid network of its own which is managed at regional load dispatch centres. These grids are further interconnected to form *National Power Grid.* The functions of national power grid include supplying of power to power deficit areas by evacuating power from power surplus areas. It is also responsible for maintaining frequency and voltage levels. Another function is to prevent complete outage in an area in case of failure of power plants of the concerned area.

Q. 11. Why is it advantageous to interconnect various power plants ?

Ans. Some of the advantages of interconnected power system are:

(*i*) Reduced plant reserved capacity, (*ii*) Increased reliability of supply, (*iii*) Increased effective capacity of the power system, (*iv*) Economical operation, (*v*) Exchange of peak loads, (*vi*) Possibility of using obsolete plants, (*vii*) Reduced capital cost and (*viii*) Saving in operating costs.

Q. 12. What is run-off river plant ?

Ans. Some hydro power plants are so located that the water is taken from the river directly, and no pondage or storage is possible. Such plants are called the *run-off river power plants without pondage.* Such plants can use water only as and when available; these cannot be used at any time at will or fit any desired portion of the load curve.

The usefulness of run-off river power plants is increased by pondage. Pondage refers to storage at the plant which makes it possible to cope, hour to hour, with fluctuations of load throughout a week or some longer period depending on the size of pondage.

Q. 13. Why is it necessary to operate run-off river plant in combination with steam plant ? How are they operated in rainy season and dry season ? **[Pb. Technical Univ. Dec. 2008]**

Ans. The run-off river power plant has a small pondage and uses water as it is available. The run-offs of river vary widely during the year-very large in rainy season and very low in dry season. As such the firm capacity of such plants is very low. So, in order to increase utility of such plants, it becomes necessary to operate it in combination with steam plant. The run-off river power plant can be employed as a base load power plant (up to its installed capacity) during rainy seasons (periods of high stream flows) while the steam power plant may supply the peak load. During dry season (periods of lean flow) the steam power plant is used to supply the base load and run-off river power plant for supplying peak load. Thus the amount of load carried by the steam plant can be adjusted to conform with the available river flow. Such a plan of operation results in a greater utilisation factor of the river flow and a saving in the amount of fuel consumed in the steam plant.

Q. 14. What is the meaning of basic rule curve ?

[Pb. Technical Univ. Dec. 2009]

Ans. The basic rule curve is that curve which represents accumulations of water, in reverse order of time, of the deficiency between firm load and energy available from the stream flow in critically dry periods.

Q. 15. What is meant by incremental water rate ?

[Pb. Technical Univ. May 2009]

Ans. Incremental water rate is a change in output power due to corresponding change in input water flow.

Q. 16. Discuss the methods used for computing the generation schedules in a combined hydrothermal system.

[Pb. Technical Univ. Dec. 2007]

Ans. The various methods for computing the generation schedules in a combined hydro thermal system are as follows:

1. *Constant Hydro Generation.* In constant hydro generation, we keep hydro generation constant throughout the operating period at such a value that the desired quantity of water is consumed. The balance load (varying from time to time) is supplied by the thermal plants.
2. *Constant Steam Generation.* In constant steam generation, we keep steam generation constant throughout the operating period and the balanced load is supplied by the hydro plants. The thermal generation is kept such that the hydro plants would consume the desired quantity of water over the operating period.
3. *Maximum Hydro Efficiency Method.* In this method, the hydro plants are operated at the point of maximum efficiency during peak load periods. During the off-peak hours hydro generation is scheduled so as to consume desired quantity of water and operate hydro plants as near the point of maximum efficiency as possible. The balance load is supplied by thermal plants.

The scheduling of thermal plants, for supplying the balance load, is accomplished either by equal incremental cost criterion or by solution of co-ordination equations (including transmission losses).

❑❑❑

CHAPTER 11

Environmental Aspects of Electric Energy Generation

INSIDE THIS CHAPTER

11.1. INTRODUCTION

Environment is a concern for the modern civilization. The modern civilization is a product of industrial revolution which has undergone a sea change year after year. In today's concept electric power is the motive force for industrialization and this forms a nerve centre for countries continued socioeconomic development. Conversion of energy from one form to another has undesired effects and the pollutants produced in the process have to be disposed off. Figure 11.1 brings out all the associated problems at a glance. Pollution has become a nightmarish problem and strong national and international pressure groups have sprung up and are having a definite impact on the development of energy resources. Government awareness has created increasing legislation at national and international levels. The power engineers have to be fully conversant with these.

Emphasis is being laid on energy conservation issues, curtailment of transmission losses, theft, subsidized power supplies and above all on sustainable development with appropriate technology wherever feasible. In development of power plants (thermal, hydro and nuclear), we will have to

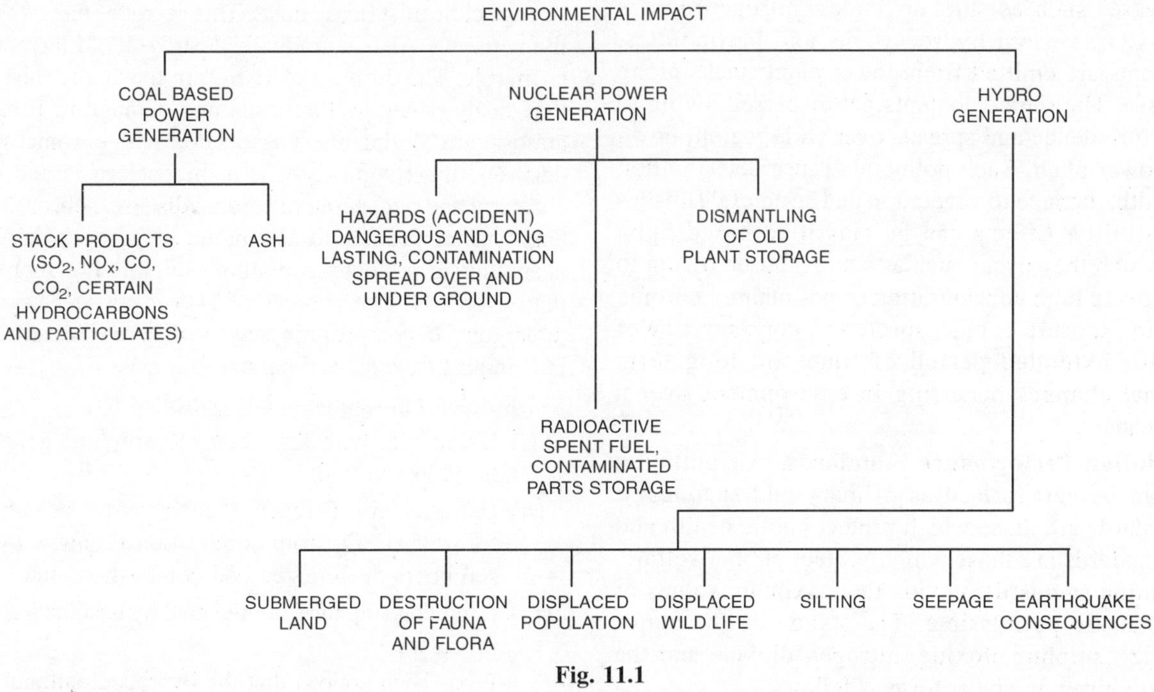

Fig. 11.1

assume that no irreversible damage is caused to the environment which would affect the living conditions of the future generations. Irreversible damages, such as, ozone layer holes and global warming caused due to increase in carbon dioxide in the atmosphere are already going up.

11.2. AIR POLLUTION

The life zone of earth is known as "*biosphere*" or "*ecosphere*" which includes lithosphere (solid), atmosphere (air) and the entire hydrosphere (the world's total water resources—rivers, lakes, oceans etc.). For the economical advancement, man has been restoring to careless and reckless exploitation of the various energy sources which is gradually affecting the ecological balance of nature and making the environment more and more unfit for supporting life. Such a deterioration of environment is termed as "*pollution*". Thus, pollution may be defined as contamination of soil, air and water with undesirable amounts of material and/or heat. The substances (including heat) causing pollution are called the *pollutants.* Pollution has always been around, but it has become acute during the last few decades owing to rapid technological advancement being made by man.

The major contribution of electric power generation is from thermal power plants. The thermal power plants definitely contribute to the economic growth, but they bring with it like air and water pollution, health hazards, displacement of popuiation, resettlement, rehabilation of project affected people and changes in land use patterns.

The air pollution by thermal plants is caused due to emission of flue gases into air. The air pollution may be divided into two groups. Stack emission and cooling tower emissions.

11.2.1. Stack Emission

Thermal power plants using fossil fuels emit large amount of harmful gases such as sulphur oxides, nitrogen oxides (NO_x), CO, CO_2, certain hydrocarbons and particulates. These pollutants are emitted from power plant stacks along with flue gases. The plume contents get dispersed owing to atmospheric turbulence and spreads over wide region, down wind from power plant. Such pollutants cause deterioration of human health, damage to vegetation and structural finishes and poor visibility. Effects can be classified based upon duration in which they occur such as acute (caused owing to short exposures to high concentration of pollutants), chronic (caused due to exposure or oganisms to low concentration of pollutants for extended period of time and long term (*i.e.*, abnormal changes occurring in environment over a period of decades).

Air Pollution Performance Standards. Air pollution standards can be categorised as primary and secondary. Primary standards are those which protect public health and secondary standards are those which protect public welfare. The air pollution standards provide the maximum values of pollutants that are permissible. The standards for major pollutants viz., sulphur dioxide, nitrogen dioxide and the particulates are given in tabular form as follows:

Table 11.1.

Pollutant	Primary Standard	Secondary Standard	Description
Sulphur Dioxide (SO_2)	80 μ gm/m^3 (0.03 ppm)	60 μ gm/m^3 (0.02 ppm)	Annual Arithmetic Mean
	—	1,300 μ gm/m^3 (0.5 ppm)	Maximum 3 Hour Level*
	365 μ gm/m^3 (0.14 ppm)	260 μ gm/m^3 (0.1 ppm)	Maximum 24 Hour Level*
Nitrogen Dioxide (NO_2)	100 μ gm/m^3 (0.05 ppm)	Same as primary	Annual Arithmetic Mean
	250 μ gm/m^3 (0.13 ppm)	-do-	Maximum 24 Hour Level*
Particulates	75 μ gm/m^3	60 μ gm/m^3	Annual Geometric Mean
	260 μ gm/m^3	150 μ gm/m^3	Maximum 24 Hour Level*

* Not To Be Exceeded More Than Once Per Year.

A brief account of various pollutants, their likely impact and methods of abatements are given below.

1. Oxides of Sulphur (SO_2). Most of sulphur present in the fossil fuel is converted into SO_2 in the combustion chamber before being emitted by the chimney. Particularly injurious to plants and animal life is the pollution of air by SO_2. The latter causes respiratory ailment when present in amounts of 20 mg/m^3 and endanger to life in amounts of 400 to 500 mg/m^3. Vegetable and plant life are most sensitive to the content of SO_2 gas in the atmosphere. Owing to toxic effect of SO_2, the surface of leaves or needles get deteriorated due to destruction of their chlorophyll. SO_2 gas is gradually oxides to sulphur trioxide (SO_3) under the influence of sunglight. This in turn interact with atmospheric moisture and forms sulphuric acid (H_2SO_4), which produces a strong acidic mist (toxic mist). This corrodes metals and destroys living tissues. Also, it gradually destroys building materials such as marble. The droplets of H_2SO_4 in the acidic mist are so fine that easily penetrate the tissue and damage it. Effect of toxic atmospheric pollutants is associated with chronic non specific diseases: most significant is a therosclerosis and the related coronary and degenerative cardiac diseases, chronic bronchitis, emphysema, bronchioil asthma etc. Hydrogen sulphide (H_2S) is another air pollutant containing sulphur. It is more poisonous than even CO. The sources of H_2S in air are: organic matter undergoing decomposition, sewage or some industrial operation. H_2S in air blackens lead paints.

Sulphur emission can be controlled by:

(*i*) Using fuel with less than 1% sulphur; generally not a feasible solution;

(*ii*) Using chemical reaction for removal of sulphur in the form of H_2SO_4 from combustion products by limestone scrubbers or fluidized bed combustion; and

(*iii*) removing sulphur from the coal by gasification or flotation processes.

It has been noticed that the byproduct sulphur could off-set, the cost of sulphur recovery plant.

2. Oxides of Nitrogen (NO_x). At high temperatures existing in the flame core of high-power boilers, the nitrogen of fuel and air may partially be oxidized to form nitrogen oxides NO and NO_2. These are toxic and produce a sharp irritating effect, especially on the mucous membrane of the eye. Nitrogen oxides are poorly soluble in liquids and therefore, can penetrate deep into lungs and cause injury to the alveolar epithelium and bronchi. People living in NO_2 contaminated areas suffer from reduced respiratory function, have a higher percentage of respiratory disease and exhibit certain changes in the peripheric blood. In large concentration, NO_2 may lead to serious lung congestion and even death. NO_2 had also got adverse effects on plant life. Nitrogen oxides absorb the natural radiation both in the ultraviolet and visible region of spectrum and thus reduce the transparency of the atmosphere and provokes the formation of photochemical mist or smoke (smoke + fog).

NO in air is converted into NO_2. At the stack exit, NO_2 constitutes only 10-15% of all nitrogen oxides present in the flue gases of thermal power plants, the remaining 85-90% being mainly NO. As the smoke plume moves in the atmosphere, the concentration of NO_2 in the flue gases increases due to the photochemical actions up to 60-70% in the zone of maximum concentration. This circumstance is very important, it results in a stronger harmful effect of combustion products on nature and living organisms, since NO_2 is roughly 3-3.5 times more toxic than NO.

Emission of NO_x can be controlled by fitting advanced technology burners which can assure more complete combustion, thereby reducing these oxides from being emitted by the stack. These can also be removed from the combustion products by absorption process by certain solvents going on to the stack.

3. Oxides of Carbon (CO, CO_2). Carbon monoxide is very toxic pollutant but it gets converted into carbon dioxide in the open atmosphere (if available) surrounding the power station. On the other hand, CO_2 has been identified as a major cause of global warming. It is not yet a serious problem in developing countries.

4. Hydrocarbons. During oxidation process in combustion chamber certain light weight hydrocarbons may be formed. The compounds are major source of photochemical reaction that adds to depletion of ozone layer.

5. Air Borne Particulates. Flue gases emitted from stacks of coal-based thermal power plants contains a substantial quantity of particulates. The amount of particulates depends upon ash content in the coal. Ash content is particularly high in the Indian coal. Particulates come out of the stack in the form of fly ash. Particulates pose a significant threat to human health. When flue gases contain both particulates and SO_2 the damage becomes significant. The hazards include bronchitis, respiratory diseases and excessive death rate. In addition particulates cause poor visibility.

Concentration of pollutants can be reduced by the dispersal over a wider area by using high stacks. Precipitation can be employed for removing particles as the flue gases rise up the stack. Precipitators have a high efficiency, up to 99% for large particles, but poor performance for particles of size less than 0.1 μm in diameter. The efficiency of precipitators is high with reasonable sulphur content in flue gases but drops for the low sulphur content coals; 99% for 3% sulphur and 83% for 0.5% sulphur. Fabric filters in the form of bag houses also have been used and are located before the flue gases enter the stack.

11.2.2. Cooling Tower Impacts

Steam from low-pressure turbine has to be liquefied in a condenser and reduced to lowest possible temperature to maximize the thermodynamic efficiency. The best efficiency of steam cycle practically achievable is about 40%. It means that 60% of the energy in steam at the end of cycle is discharged into the environment in the form of waste heat. Fossil fuel plants discharge about 15% of the waste heat directly to atmosphere (along with flue gases) and the remaining waste heat to cooling water. A nuclear power plant discharges all the waste heat to cooling water.

Cooling water is supplied from a natural source of supply such as river, canal, sea or lake or cooling towers through screens to remove the matter, that might choke the condenser tubes. It is circulated through the condenser for condensing the steam and finally discharged to the suitable position near the source of supply. During the passage its temperature rises and in the case of cooling towers the heat must be extracted before the water is again pumped to the condenser. In the former case, the rise of temperature of water discharged into supply source (such as river, canal or sea), threathens sea, canal and river life around in sea and down stream in river/canal. These are serious environmental objections and many times cannot be overruled and also there may be legislation against it.

In case of cooling towers, cool water is circulated around the condenser tube to remove heat from the exhaust steam so as to condense it. The circulating water gets hot in the process. It is pumped to the cooling towers and is sprayed through nozzles into a rising volume of air. Some of the water evaporates providing cooling. The latent heat of water is 2×10^6 J/kg and cooling can occur fast. Of course, the water evaporated must be made up in the system by adding fresh water from the source. These cooling towers are called the *wet towers.*

Cooling towers are huge hyperbolic structures measuring more than 100 m in height and employs air draft, which is produced due to shape and height of the tower, for cooling the water. The harmful effects of cooling tower are:

1. Drift Deposits. Sometimes the surrounding air cannot absorb moisture and as such small droplets of circulating water escape from the tower. These droplets, termed as *drift,* carry with them salts and chemicals. The aproximate amount of drift is 0.002% of circulating water. When deposited, the salts and chemicals damage vegetation, causes weathering and corrosion of metals and change in properties of soil.

2. Fog. Fog is formed when cooling tower water added to the atmosphere gets condensed. It is formed at ground level. The conditions of wind may result in concentration of dense fog over a small area reducing the visibility in that area.

3. Visible Plumes. Cooling towers may even produce visible plumes which are likely to pollute atmosphere and cause respiratory diseases.

11.3. AQUATIC IMPACTS

When a nearby river or estuary is available, its water can be used for cooling purposes in thermal power plants. Use of water from an open cooling system involves the withdrawal, use and discharge of water back to the source. The impacts of these processes are hazardous as given below.

1. Thermal Effects. Lot of heat is injected into biosphere from thermal power plants, through exhaust gases and waste water. The major problem is, the effect of discharge of large quantities of heated waste water into natural water source. Discharged into natural water source, hot water raises the aquatic temperature and disturbs the natural ecological balance with an unwanted effect produced on flora and fauna. The addition of heat to water reduces the water's ability to hold dissolved gases, including dissolved oxygen vital to aquatic life. When the temperature of water exceeds 35°C, the dissolved oxygen content is too low to support life. At low temperatures, however, the aquatic growth is usually increased by the warm water and the plants and fishes grow at a faster rate. In fact, the warm water discharged from a power plant is usually very popular place to fish, particularly during cold weather.

Thermal power plants are sources of the following types of waste water:

(*i*) Waste water of water treatment plants and condensate cleaners.

(*ii*) Waste water contaminated with petroleum products.

(*iii*) Cooling water which mainly causes thermal contamination of water basins or sources.

(*iv*) Water used for washing of external surfaces of fuel oil fired steam generators and peak load heating boilers.

(*v*) Used solutions after chemical cleaning and preservation of thermal equipment.

(*vi*) Water from hydraulic ash-disposal systems of solid-fuel-fired thermal power plants.

(*vii*) Used water after hydraulic cleaning of fuel conveying systems.

For manimizing thermal effects, the discharge designs should include optimal mixing and offshore, submerged heat diffusers.

2. Chemical Contamination. Water taken from rivers and lakes contains a number of substances which cause corrosion and scaling in condenser and for removal of these impurities some chemicals are used. These chemicals are discharged along with hot water into the natural water source and produces chemical contamination. Owing to increased temperature, the metabolic rate of organism is enhanced and the chemical reactions of pollutants are speeded up. Chemical residuals are harmful to aquatic life. The other chemicals may also be toxic.

3. Entrapment. Hydraulic forces in the stream prior to flow of streams through the screens cause entrapment. Such forces may restrict the movement of aquatic life and result in death of some species owing to starvation, exhaustion or asphyxiation. Entrapment effect can be reduced by proper design, location and capacity of intake structures.

4. Entrainment. Eggs and newly hatched fish are sometimes entrained into the power station along with intake water. While passing through the cooling system these organisms are subjected to thermal, mechanical and chemical stresses. These stresses may be lethal.

Miscellaneous Effects. Coal power plants discharge water polluted by chemicals and impurities back into natural water sources, causing significant environmental damage.

Coal preparation uses large quantities of water and chemicals to separate impurities from mined coal, washing away the waste in a sludge known as slurry.

A common chemical from discharged water is copper, which can reach from water condenser piping and end up in discharge water, sometimes at toxic level. In addition, water discharged from waste treatment have been shown to have high concentrations of arsenic, cadmium, chromium, lead, selenium, sulphates and boron.

Chlorine and its by-products are present in the discharged water plume and can be toxic to acquatic life, even at low concentrations. High water temperatures can magnify the damaging impacts of chlorine.

While burning in power plants, coal releases mercury into atmosphere. This mercury falls to earth in rain, running into lakes, rivers and stream. Bacteria in the water transforms this mercury into toxic methyl mercury. When fish consume this bacteria, they become contaminated. Fish that eat other contaminated fish end up with even higher levels of toxic mercury in their flesh. Humans can be contaminated with this methyl-mercury by eating contaminated fish.

11.4. HYDROELECTRIC PLANT IMPACTS

Regarded usually clean and harmless, the hydroelectric power plants have following environmental impacts.

1. Since hydro power plants require huge amount of water storage, large dams have to be constructed. This leads to displacement of inhabitants of the area. As such historical and ecological nature of area gets disturbed.
2. The construction of dams for hydroelectric plants slows down the flow of rivers and thus causes the pollution of water, the growth of deleterious blue-green algae, encourages the reproduction of epidemic-carrying bacteria, checks, the flood waters with the result that water meadows cease to exist and sometimes salinization of soil occurs.
3. In addition to environmental degradation and disturbing ecological balance, the hydro plants disturb the demographic balance also. Large number of workers required for construction are brought into the area and they disturb the very nature of local population.
4. Large acquisition means destruction of forest cover which is harmful for environment.

11.5. SOCIAL AND ECONOMIC IMPACTS

Power plants create a number of social and economic impacts, which are mostly because of competition for land and water, availability of community services and cultural threats. There may be industrial growth around the power plant and this may significantly change the features of the area and population. The social and economic impacts are summarized as below.

1. Demographic Impacts. There may be a rapid influx of large construction force in the area. A majority of these workers tend to remain there even after the completion of construction work of the power plant. This influx may result in a change in ethnic composition, age structure, sex ratio and educational status of the population. The area may get unbanised and its population density may increase.

2. Land Use Impacts. Land use pattern is affected due to plant construction and changing social structure. Agricultural land gets converted into residential and commercial property.

3. Housing Impacts. There may be changes in residency duration and mobility, enhanced housing demand, increase in housing values and rents and possibility of out migration of existing residents.

4. Public Services. Because of increase in the area population, there may be an increase in demand for water supply, waste disposal, sewerage, educational, medical and transport facilities, and police, fire protection etc.

5. Community Structure. The traditional rural community values may be replaced by urban values and preferences. Number and type of community organisations may change. An accelerated way of life may be experienced in area.

6. Psychological Impacts. There may be a change in type and number of crimes. There may be increase in social problems, such as, alcoholism, divorce and suicide.

7. Economic Impacts. Because of plant construction, job opportunities and employment rate of the area will be certainly increased. There may be a change in business and industrial composition of the community. There will be an upward trend in the family incomes, retail prices, taxes and community's gross product.

8. Recreational and Cultural Impacts. Because of increase in population of the area there may be an increase in demand of parks, playgrounds, swimming pools, theatres and other recreational and social centres.

9. Political Impacts. The strengths of local political parties and the extent of citizen participation in politics may change owing to additional large population of different social, professional and economic status. The appearance of trade union activities in the area may completely erode the base of the existing political parties of the area.

11.6. GREENHOUSE EFFECT AND GLOBAL WARMING

A greenhouse is an enclosure made of glass. The glass will allow short wavelength solar radiations to pass through it and enter the enclosure. These radiations strike the earth surface inside the enclosure and get reflected as long wave radiations. These long wave radiations are not allowed by glass to pass through it. Thus, the heat energy of the radiations gets entrapped inside the enclosure, raising its temperature. This is used to grow plants and vegetables inside the enclosure and keep them green. Such an enclosure is known as *"Greenhouse"* and the phenomenon as a *"Greenhouse Effect"*.

In the same way, the atmosphere around the earth acts like glass and results in green house effect which keeps the earth warm and green. The gases in the atmosphere which contribute to green house effect are: CO_2, water vapours, CH_4, NO_x and CFCs (Chloro Fluoro Carbons), PFC (Petro Fluorocarbon), SF_6 (Sulphur hexafluoride) *i.e.*, the environmentally harmful refrigerants and their less harmful substitutes, like HCFCs, HFCS, HCS, and are known as *Greenhouse Gases.* Increased CO_2 content of the atmosphere can affect the climate of the planet. CO_2 absorbs radiant energy of the long wave type emitted from the surface of the earth, heats up and thus prevents the heat from being lost by the earth. The increased CO_2 content of the atmosphere may lead to a rise of several degrees in the lower atmosphere temperature, which may cause the glaciers of Greenland and Antarctica to melt and submerge the portion of land inhabited now nearly by one-fourth of the population. Sources of CH_4 gas are: rice paddies, waste disposal, oil recovery operations, natural gas leakages and coal mines etc. CFCs gases (F-11, F-12, R-11 and R-12) are used in refrigeration and air-conditioning industry. These results in the depletion of Ozone (O_3) layer. Ozone gas prevents ultraviolet sun rays to reach earth. Its depletion is resulting in harmful effects on earth's atmosphere. CH_4 gas is 24 times more harmful than CO_2 as Green House Gas. Increasing percentage of these gases in the atmosphere has been leading to changing weather and climatic conditions. All this will increase the risk of natural disasters (Cyclones, floods and droughts, etc.), over the next two centuries. The devastation will be greater than the combined effect of the two World Wars. The most affected will be poor and developing countries.

There is a direct link between the rise in global temperature (by 2 to 3°C in the coming 50 years) and damage to ecosystems. Areas that would experience the worst forest loss would include Eurasia, Eastern China, Canada and the Amazon Western Africa, Southern Europe and Eastern US states are at risk from dwindling fresh water supplies and droughts. In India also climate has been the biggest casuality. Climate changes could be slowed by burying GHGs deep below the ocean under thick, cold sediment that would trap it for thousands of years. Many governments are already exploring ways to pump CO_2 under land or directly into the sea—in a process known as Carbon sequestration—to meet emission caps. The best route to contain Greenhouse disaster is to switch over to non-fossil fuel energy resources to meet the increasing energy needs.

The increasingly unstable climate can result in flooding, disease and famine, and migration on an unprecendented scale, drought and crop failure and from that intensified competition for food, water and energy in regions where resources are already stretched to the limits, economic disruption on the scale not seen since the end of World War II. This is not about narrow national security—it is about our collective security in a fragile and increasingly independent world. And tragically, once again, it will be those who are most vulnerable and least able to cope that will be hit first. There is no choice between a stable climate and the fight against poverty. Without the first, the second will fail.

According to IPCC (Intergovernmental Panel on Climate Change) predictions, sea level will rise by between 30 cm to 60 cm over the next 100 years, the temperatures are set to rise by 1.4 to 5.8°C. According to IPCC, it is the poorest of poor and this includes poor people even in prosperous societies, who are going to be worst hit by the consequences of global warming. The Arctic regions, Sub-Saharan Africa, small islands and Asian mega deltas including the Ganga and Brahampurtra, will be affected most.

Consequences of Global Warming For India. As per IPCC :

1. Climate change will make monsoons unpredictable. As a result, rain-fed wheat cultivations in South Asia will suffer in a big way. There will be a steep fall in food grain production in India. In case, winter temperature rises by 0.5°C, the per hectare wheat production will fall by 0.45 tons. Similarly, the production of cereals and many other agricultural crops will go down.
2. Total agricultural land will reduce and the land may not remain suitable to grow the present crops. All this would cause food insecurity and loss of livelihood in rural India resulting in migration to cities and more slums in the cities.
3. Huge coastal erosions by rising sea water levels (about 0.4 m) resulting from faster melting of glaciers in Himalaya Hindu-Kush ranges. This will also damage necessary areas for fisheries and cause flooding. It will also increase salinity of ground water in the Sunderbans and on sufrace water in coastal areas.
4. India's per capita water availability is expected to fall from 1,820 million m^3/year in 2001 to 1,140 million m^3/year in 2050.
5. The projected decrease in winter precipitation over the Indian Sub-continent would reduce the total seasonal precipitation during December, January and February. This means lesser storage and greater water stress.
6. In continuation of point 3 (above) a mere 0.45 m rise in sea level will destroy 75% of the mangrove forests in the Sunderbans (World Heritage). This will leave India and Bangladesh vulnerable to disturbances in the Bay of Bengal. Change in Sunderbans ecology will effect the livelihood of three crore people of India and Bangladesh.

ITT's research and ISRO data confirm the above findings of IPCC.

(*i*) ISRO data indicate that Indian seas are rising by 2.5 mm per year.

(*ii*) IITD research on the data indicates that if this trend continues, the sea would rise between 15 to 38 cm by 2050.

(*iii*) 4.6% of India's coast line could be inundated.

(*iv*) 7.1 million people of Gujarat, Mumbai, South Kerala, Deltas of the Ganga, A.P. and Orissa would be hit.

(*v*) Half of Mumbai and many coastal cities in India might be submerged when the sea level rises by 0.5 m.

(*vi*) Intensity of cyclones hitting India, especially its east coast, will also increase.

Impact of Clouds on Climate Change

Warm oceans send more water vapours into the air, and the resulting clouds accelerate global warming by trapping more of the Sun's heat in the atmosphere which further warms the oceans.

Further complicating the issues are layers of haze containing pollutants from human activity. Such pollutants include : sulphates, soot, dust and nitrates. These tend to make the atmosphere brighter, reflecting more of the Sun's heat back into the space. The IPCC has found that the net effect of the added pollutants is to cool the atmosphere. This raises a problem for Governments attempting to clean the air by removing pollutants. If this cooling effect is taken away, then the heating effect would be exacerbated. It is highly complex problems.

Exercises

1. Write about air pollution performance standards.
[Pb. Technical Univ. Generation of Electric Power, Dec. 2009]

2. Explain the impact of thermal and hydro power stations on environment.
[Rajasthan Technical Univ. Generation of Electrical Power 2011]

3. What are the acquatic impacts of pollution ?
[Pb. Technical Univ. Generation of Electric Power Dec. 2010]

4. Discuss the bad effects of using the river water as cooling water for steam plants.
[Rajasthan Technical Univ. Generation of Electrical Power 2010]

5. Explain the environmental impacts of hydro power plants.
[Pb. Technical Univ. Generation of Electric Power May 2008]

6. Write short notes on the following :

(*i*) Environmental aspects of electric power generation.
[West Bengal Technological University, 2009-10]

(*ii*) Impact of power plant on environment.
[Rajasthan Technical Univ. 2010]

(*iii*) Bad effects of using river water as cooling water in steam plants.

(*iv*) Aquatic impacts of power plants.

(*v*) Social and economic impacts of power plants.

(*vi*) Greenhouse effect and global warming.

Short Answer Type Questions With Answers

Q. 1. Name various stack emissions.
[Pb. Technical Univ. December 2009]

Ans. Thermal power plants using fossil fuels emit large amount of harmful gases such as SO_2, nitrogen oxides (NO_x), CO, CO_2, certain hydrocarbons and particulates.

Q. 2. What are particulates ? What are their harmful effects ?
[Pb. Technical Univ. December 2007]

Ans. Particulates are small particles that are emitted from stacks of coal-fired plants. Particulates affect the health of humans. They are responsible for diseases such as bronchitis, respiratory diseases and excessive death rate.

Q. 3. What are various cooling tower impacts ?
[Pb. Technical Univ. May 2008]

Ans. Various cooling tower impacts are : (*i*) Drift deposits, (*ii*) Fog and (*iii*) Visible plumes.

Q. 4. What are the aquatic impacts of pollution ?

[Pb. Technical Univ. December 2010]

Ans. Aquatic impacts of pollution are :

(*i*) Thermal effects, (*ii*) Chemical contamination, (*iii*) Entrapment, (*iv*) Entrainment, (*v*) Miscellaneous effects like decrease of oxygen in water.

Q. 5. Explain interdependence of energy and environment.

[Pb. Technical Univ. May 2010]

Ans. Energy and environment are interrelated. A healthy environment would lead to more sources of energy. On the other hand the non-polluting sources of energy lead to healthy environments.

Q. 6. Enumerate social and economic impacts of power plants.

Ans. The social and economic impacts of power plants are: (*i*) Demographic impacts, (*ii*) Land use impacts, (*iii*) Housing impacts, (*iv*) Public services, (*v*) Change in community structure, (*vi*) Psychological impacts, (*vii*) Economic impacts, (*viii*) Recreational and cultural impacts, (*ix*) Political impacts.

Q. 7. What are the environmental impacts of hydro power plants ?

Ans. Refer to Art. 11.4.

Q. 8. What is greenhouse effect ?

Ans. The *greenhouse effect* is a process by which thermal radiation from a planetary surface is absorbed by atmospheric *greenhouse gases* and is re-radiated in all directions. Since part of this re-radiation is back towards the surface, energy is transferred to the surface and the lower atmosphere. As a result, the temperature there is higher than it would be if direct heating by solar radiation were the only warming mechanism.

Solar radiation at the high frequencies of visible light passes through the atmosphere to warm the planetary surface, which then emits this energy at the lower frequencies of infrared thermal radiation. Infrared radiation is absorbed by greenhouse gases, which in turn re-radiate much of the energy to the surface and lower atmosphere. The mechanism is named after the effect of solar radiation passing through glass and warming a *greenhouse,* but the way it retains heat is fundamentally different as a *greenhouse* works by reducing air flow, isolating the warm air inside the structure so that heat is not lost by convection.

The greenhouse effect was discovered by French mathematician Joseph Fourier in 1824, first reliably experimented on by Irish physicist John Tyndall in 1858, and first reported quantitatively by the Swedish scientist Svante Arrhenius in 1896.

If an ideal thermally conductive blackbody was at the same distance from the Sun as the Earth is, it would have a temperature of about 5.3°C, However, since the earth reflects about 30% (or 28%) of the incoming sunlight, the planet's *effective temperature* (the temperature of a blackbody that would emit the same amount of radiation) is about –18°C or –19°C, about 33°C below the actual surface temperature of about 14°C or 15°C. The mechanism that produces this difference between the actual surface temperature and the effective temperature is due to the atmosphere and is known as the greenhouse effect.

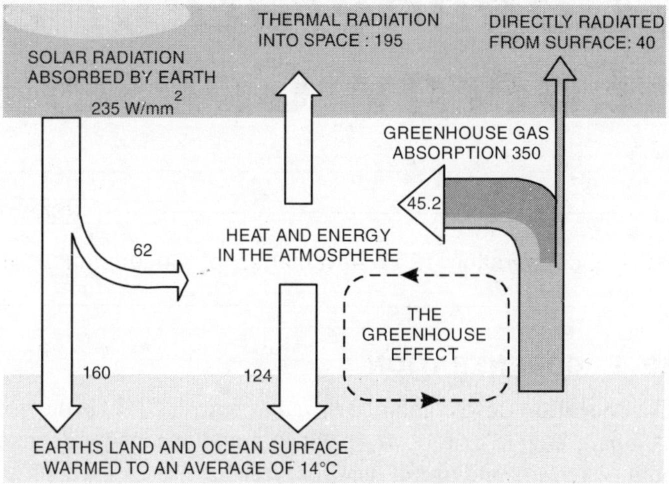

Fig. 1. ***A Representation of The Exchanges of Energy Between The Source (The Sun), The Earth's Surface. The Earth's Atmosphere, and The Ultimate Sink Outer Space***

Earth's natural greenhouse effect makes life as we know it possible. However, human activities, primarily the burning of fossil fuels and clearing of forests, have greatly intensified the natural greenhouse effect, causing *global warming.*

The ability of the atmosphere to capture and recycle energy emitted by the Earth surface is the characteristic of the greenhouse effect.

By their percentage contribution to the greenhouse effect on Earth the four major gases are : Water vapour : 36-70%; CO_2 : 9-26%; Methane : 4-9%, Ozone : 3-7%.

The major non-gas contributor to the Earth's greenhouse effect, *clouds* also absorb and emit infrared radiation and thus have an effect on radiation properties of the atmosphere.

❑❑❑

CHAPTER 12

Cogeneration and Energy Conservation

INSIDE THIS CHAPTER

12.1. COGENERATION

Cogeneration (also combined *heat and power*, CHP) is the use of a *heat engine* or a power plant to simultaneously generate both *electricity* and useful heat (or steam).

All thermal power plants emit a certain amount of heat during electricity generation. This heat can be released into the natural environment through cooling towers, flue gases or by some other means. Alternatively, this waste heat may be used for some useful purpose such as heating or producing steam. Thus, a thermal power plant may use fuel energy to generate electrical energy only or generate electrical energy alongwith the production of steam for industrial purposes. The industries in which the steam is used for different processing purposes are : (*i*) Paper mills, (*ii*) Textile mills; (*iii*) Sugar mills and refineries; (*iv*) Chemical works; (*v*) Breweries; (*vi*) Plastic manufacturing; (*vii*) Food processing; (*viii*) Launderies and hotels, (*ix*) Metallurgy, (*x*) Places that use district heating.

In generation of electrical energy alongwith production of steam for industrial purposes, known as cogeneration, steam for a certain section of the turbine is extracted for the industrial purposes and the remaining steam is allowed to expand in the turbine or the exhaust steam may be used for the purpose of process steam.

A cogeneration system can be either an inplant power generation system or a reject heat utilisation system. In implant power generation, the steam is produced at a higher temperature and pressure than required for manufacturing purposes. The steam so produced is used in a turbine-generator set to generate electrical energy and the exhaust steam from turbine is used for manufacturing purpose (refer to Fig. 12.1). The inplant power generation is employed in industries.

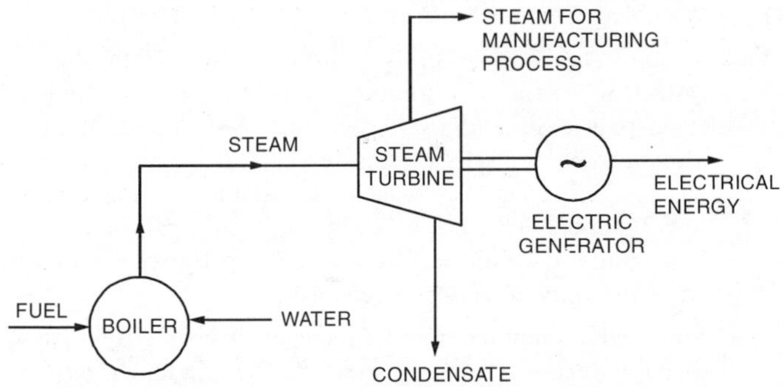

Fig. 12.1. ***Inplant Power Generation System (Cogeneration)***

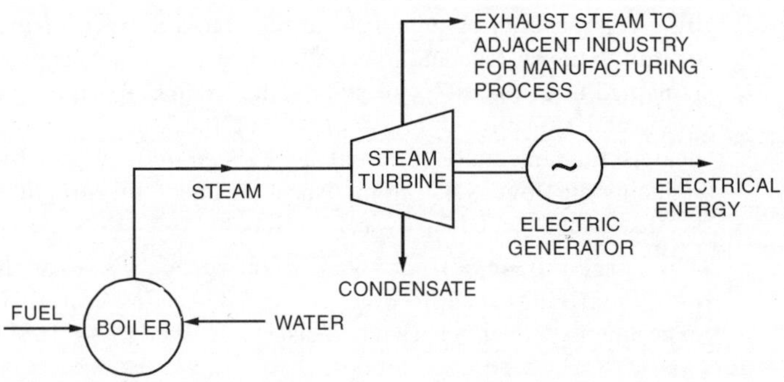

Fig. 12.2. ***Reject Heat Utilisation System***

In reject heat utilisation system, which is used in power plants, some heat is extracted from the turbine (at a suitable temperature and pressure) and supplied to an adjacent industry for manufacturing process, as shown in Fig. 12.2.

There are two main categories of cogeneration viz topping cycle and bottoming cycle.

In a *topping cycle,* the fuel energy is used to generate steam at high temperature and at high pressure and electrical energy is generated in the usual manner. Depending upon the requirements, the process steam at low temperature and pressure is either extracted from the turbine at an intermediate stage (similar to feed water heating) or taken at the turbine exhaust. In a *bottoming cycle*, the steam produced in a boiler at high temperature and pressure is directly used for the manufacturing process and the reject heat from the process is used for generation of electrial energy, obviously at low efficiency.

The combined efficiency of the bottoming cycle will most certainly be less than that for separate generation of process steam and electrical energy. Thus, the bottoming cycle is of little thermodynamic or economic interest. Generally the steam required for industrial processes is at low temperature whereas high temperature steam is required for generation of electrical energy, only the topping cycle, therefore, is suitable for cogeneration. The overall efficiency of a topping cycle is about 75% while the combined efficiency of two separate system (for generating electrical energy and steam) is about 55%.

12.2. ADVANTAGES OF COGENERATION

Advantages of cogeneration are as follows :

1. Fuel Economy. There is a considerable economy in the consumption of primary fuels (coal, oil or gas) in cogeneration. This is because of higher thermodynamic efficiency of cogeneration system as compared to separate power producing and heat producing systems. Moreover, the extra fuel required to generate electrical energy for the same quantity of steam produced for process needs is only around 10%.

2. Lower Capital Cost. Capital cost for setting up facilities reduces with cogeneration. It is observed that an industry requiring steam for processing has to invest in boilers. The additional investment required to upgrade boiler so as to generate electrical energy is very small as comapred to the cost of the boiler. It has been found that additional investment in cogeneration system is only around 50% of the investment required by an electric utility to supply the same power to the industry. Thus, there is a large saving in capital costs if cogeneration is adopted.

3. Smaller Gestation Period. For a utility to set up a generation unit, about 6 years are required. But installation of cogeneration system in an industry requires 3 years. Because of smaller gestation period, there is saving in interest, early utilization of facility, early return on investment and lesser chances of cost escalation of the project.

4. Saving To Industry From Power Cuts and Power Interruption. The power cuts and power supply interruption cause huge losses to industries. In case, an industry opts for cogeneration to meet its power needs, it will not face any power cuts or power interruption. Thus, it is an additional advantage of cogeneration.

12.3. COGENERATION TECHNOLOGIES

The commonly used cogeneration technologies are: (*i*) Steam turbine system, (*ii*) Gas turbine systems, (*iii*) Combined cycle system and (*iv*) Diesel engine sysem. All of these cogeneration technologies are presently employed for topping cycles. These cogeneration technologies are briefly described as follows.

1. Steam Turbine System. This system use boiler, steam turbine, generator and other auxiliaries, as in case of thermal power plant. The fuel used may be natural gas, oil, coal, wood or other similar material. The high pressure steam, which is generated by boiler, is used to drive a turbine coupled to an electric generator. The low pressure of the steam turbine is employed for industrial applications, such as, space heating etc. Back pressure steam turbines are usuallly preferred for this system as they exhaust low pressure steam and are more efficient. The introduction of this system leads to increase in pollution (nitrogen oxides, particulates and fly ash) and needs special pollution control measures. There is an energy saving up to 15% in this system of cogeneration.

2. Gas Turbine System. This system consists of a compressor, combustion chamber, gas turbine and electric generator. The air is sucked in by the compressor from the atmosphere through the filter and is compressed and finally supplied to the combustion chamber. Fuel is burnt in the combustion chamber and is used for heating the compressed gas. The products of combustion comprising of mixture of gases at high temperature and pressure are passed through the gas turbine. These gases expand in the turbine which drives the electric generator (or alternator). The exhaust from the gas turbine is used as process heat. If required, the exhaust from gas turbine can be used to generate steam in a heat recovery boiler. The advantage of this technology is that energy saving around 25% can be achieved. The disadvantage is that due to scarcity and high cost of petroleum products, cost of electrical energy generated by this process is high. Also, the maintenance requirements of gas turbines are as compared to those in case of steam turbines.

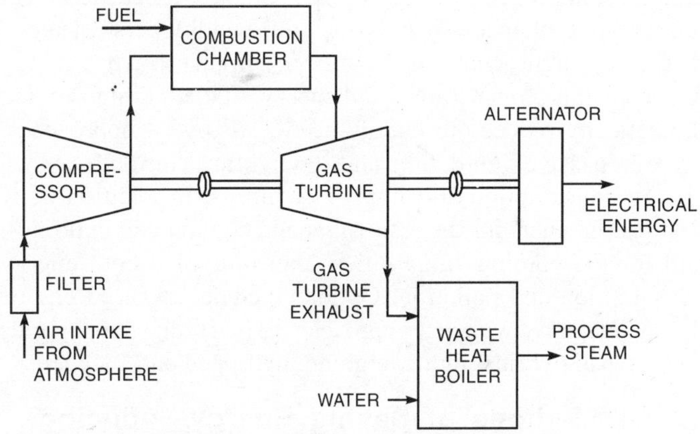

Fig. 12.3. *Gas Turbine Cogeneration System*

3. Combined Cycle System. This system is a combination of gas turbine and steam turbine system and already explained in Art. 6.10.

4. Diesel Engine System. In this system, diesel engine obtains energy from a liquid, usually called the diesel oil, and drives an electric generator (an alternator or dc generator). The engine exhaust and jacket cooling are employed to raise steam in a waste heat recovery boiler where the waste heat is transferred to water.

This system has a higher electricity to thermal ratio than that needed for most industries. So, this system can be adopted only if electrical energy can be sold. The generation cost per unit of electrical energy is higher than in other systems. The advantage of this system is that the environmental pollution is minimum. Saving is around 26%.

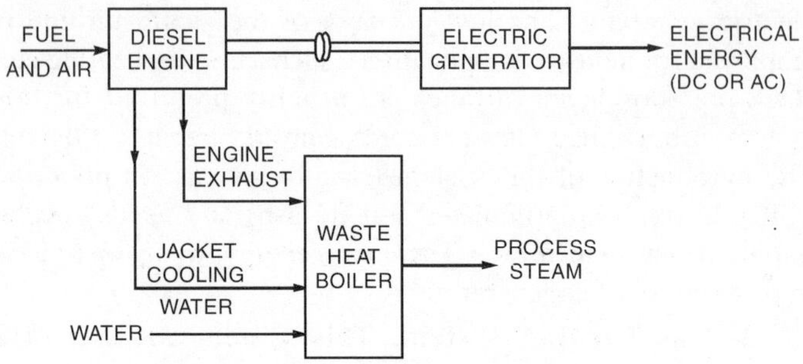

Fig. 12.4. *Diesel Engine Cogeneration System*

12.4. ENERGY CONSERVATION

Energy conservation refers to the methods of reduction in energy consumption by way of elimination of wastage and promotion of efficiency. The energy consumption can be reduced by adopting various ways of energy conservation which includes efficient use of technologies and avoiding energy wastages. Energy conservation is the key to hold off climate change/global warming, as one unit of electricity saved is equal to 3 or 4 units of electricity generated. Largely enhanced energy efficiency provides both the best short-term opportunity for addressing the major energy challenges and an essential component of a long-term strategy. The technology pathways for efficiency involve buildings vehicles and industrial processes. Around 2/3rd electrical energy is used for residential and commercial buildings. Improved lighting, HAVC appliances [energy efficient household appliances such as ACs (air-conditioners), heaters, diswashers, lights etc., and also office equipment], active energy managements, cogeneration and energy efficient designs could dramatically reduce our energy needs. Also, new approaches, such as passive ventilation and daylighting can both reduce energy consumption and improve comfort. In addition, new designs (greener) for the residential and commercial buildings, and for the coming "giga cities" can minimize both energy consumption and pollution. Options include advanced engine design integrated with new approaches to fuel utilization, hybrids and plug-in hybrids, light weighing, hydrogen and fuel cells.

12.4.1. Methods of Saving Energy/Reducing Energy Costs

1. By Using Energy Efficient Products. In India, to ensure efficient consumer appliances, Ministry of Power has set up an agency : Bureau of Energy Efficiency (BEE) to provide energy efficiency labels (Green Labels). About 90% of companies manufacturing tube lights. 60% of refrigerators and 50% of air-conditioners have voluntarily opted for this label. BEE hopes to make it mandatory soon. According to BEE calculations, the country could prevent waste of as many as 18 billion units of electricity annually just by saving this end-use energy.

Next on agenda : Green inverters, diesel pumps, transformers and consumer electronics.

Treadle water pump has been developed by the International Development Enterprises (India) (IDE-I). It uses human power to pump water from wells, streams and lakes up to the fields. A study by The Energy Resources Institute (TERI) clearly indicates that these devices can be used in place of diesel pumps in households. It reduces diesel fuel use by 0.45 litre per hour on average and saving 0.65 T of CO_2 per year.

2. For Lighting. (*i*) One of the best energy saving devices is the light switch. Turn-off lights when not required.

(*ii*) Replace 40 W tube light with 36 W tube light with electronic ballast.

(*iii*) As far as possible, use task lighting which concentrates light where it is required.

(*iv*) Clean tube lights and bulbs regularly so as to increase their luminous efficiency.

(*v*) Use Compact Fluorescent Light (CFL) in place of incandescent lamps, the General Lighting System, GLS. A CFL converts electricity to light up to 5 times more efficiently than ordinary lamps. A 15 W CFL is equivalent to 60 W ordinary bulb and saves up to 80% electricity.

India uses 18,000 MW of electric power for lighting purposes. By switching over to CFLs, there can be saving of 10,000 MW electric power and CO_2 emissions can be reduced by 55 million tonnes. Other advantages of CFLs are : Less chances of getting fused, Daylight brightness, weather friendly (disposes no heat), and no fire risk. However, biggest problem of a CFL is its safe disposal. Ministry of Energy is formulating a policy regarding its safe disposal.

Another option is use of LEDs (Light Emitting Diodes). At present, they have limited use because of their low energy intensity. They use 40% less energy than CFL. Life is 10 years. Even though, they have quite high initial cost, but they are most eco-friendly and energy saving man made lights. Energy efficient LEDs are popping up in public places around the world: from parks in U.S. to streets in China. Although LEDs can run at triple the cost of conventional light, they consume 30 to 90% less power and can last 20 to 30 years without being replaced or even wiped clean. So, these are cost savings in the long run. A comparison of LED, CFL and incandescent light is given below by "Lucifier lights".

Description	Lucifier (LED)	CFL	Incandescent
Life in Hours	50,000	10,000	1,500
Number of Units Over The Period	1	5	33.5
Cost Rs	850	5 × 230 = 1,150	33.5 × 25 = 837.5
Lux Per Watt	94	19	11
Input, W	4	40	350
Electrical Energy Used, kWh	200	2,000	17,500
Electrical Energy Cost @ Rs 5.50 Per Unit	1,100	11,000	96,250
Total cost Rs	1,950	12,150	97,087.50

∴ Total saving over a CFL = 12,150 – 1,950 = Rs 10,200

Total saving over an incandescent light

= 97,087.50 – 1,950 = Rs 95,137.50

3. For Room Air-Conditioners:

(*i*) Use ACs as sparingly as possible.

(*ii*) Trees and shrubs around your house keep the hottest sun off. Also, shade your windows by curtains. These steps can enhance the efficiency of ACs by 40%.

(*iii*) Each degree above 22°C increases efficiency of AC by about 1%. So, set the AC thermostat at 25° for best comfort at lowest cost.

(*iv*) Clean filter of the AC regularly. A clogged filter restricts air flow and can damage the unit.

4. For Refrigerators:

(*i*) Refrigerator motors and compressors produce heat, so allow enough space around the refrigerator so that there may be continuous flow of air.

(*ii*) Assure adequate air circulation inside the refrigerator:

(*iii*) When it's dark, place a lit flashlight inside the refrigerator and close the door. If light around the door is seen, get the seals replaced.

(*iv*) Keep the refrigerator away from all sources of heat, including direct sunlight, radiators and appliances like the oven and cooking range.

(*v*) Allow hot and warm foods to cool and cover them well before putting them in refrigerator. This will reduce energy consumption along with condensation.

(*vi*) Keep the frequency of opening refrigerator door to minimum.

5. For Water Heaters. (*i*) Use solar water heater for heating of water.

(*ii*) By reducing the temperature setting of water heater from 600°C to 500°C, there will be a saving of 18% of the energy consumed at higher setting.

(*iii*) Insulate hot water pipes, especially where they are run through unheated areas so as to reduce heat loss. But never insulate plastic pipes.

6. For Microwave Ovens:

(*i*) Use of microwave oven saves up to 50% on your cooking energy costs for a conventional oven, especially for small quantities of food.

(*ii*) Microwave ovens cook food from the outside edge toward the centre of the disk, so if you are cooking more than one item, place larger and thicker item on the outside.

7. For Electric Kettles:

(*i*) Electric kettle is more efficient than an electric cook top element, so prefer use of electric kettle for water heating.

(*ii*) While buying a new electric kettle, choose one with automatic shut off button and heat resistant handle.

(*iii*) Regularly clean your electric kettle by combining boiling water and vinegar so as to remove mineral deposits. It will save energy as dirty kettle draws more energy for heating.

8. For Computers:

(*i*) Turn off your home/office equipment when not in use. A computer that runs for 24 hours a day, for instance, consumes more power than an energy efficient refrigerator.

(*ii*) If your computer must be left on, turn off the monitor; this device alone consumes more than half the system's energy.

(*iii*) Set computers, monitors, and copiers to sleep-mode when not in use. This will reduce energy costs by approximately 40%.

9. Unplug Your TV Home Entertainment. These sytems are big energy hogs to begin with. But many of them have a feature that eats even more power. They remain in standby mode after you shut them off, continuing to draw electricity. Standby power is responsible for 5% to 10% of total electricity consumption in most homes and accounts for roughly 1% of global CO_2 emissions, according to International Energy Agency. "Many people don't realize that they spend more money to power DVD players when they are turned off than when they are in use."

Nstar suggests that customers connect their electronic devices to power strips and turn off the strips when devices are not in use—which guarantees the gadgets won't go into standby mode. You could also simply unplug your devices after turning them off.

10. Do not run kitchen and bathroom exhaust fan for long periods.

11. Conduct preventive maintenance checks regularly.

12. Ensure proper input voltage.

13. Use BIS approved pumps and motors.

14. Stop idle running of vehicle at traffic lights.

Above, we have discussed the various ways of energy conservation. The energy conservation is the best way toward power sector reforms and to provide energy security to the country. The power can be saved by monitoring of all the commercial buildings, industries, municipal corporations/ committees, agricultural machinery, transport etc. by conducting regular energy audits and national level awareness through media, newspaper, distributing pamphlets, conducting workshops/ seminars. The Bureau of Energy Efficiency has been established by Govt of India under Ministry of Power to coordinate, regulate and enforce the Energy Conservations Act, 2001.

Commercial and residential buildings account for more than 30% of the electricity consumption in India. We add more than 40 million m^2 of commercial and residential space annually and this results in the additional burden of around 5.5 billion units of electricity. However, it is possible to cut this figure down by 2 billion units through the adoption of energy efficiency measures. Energy efficiency can be achieved through an even mix of architectural techniques combined with technological interventions. The applications of these techniques would bring down energy consumption and demand. Renewable energy technique such as solar and wind power can be used to supplement the use of grid power.

Exercises

1. Explain the concept of "cogeneration".
   [Rajasthan Technical Univ. 2011]
2. Explain the inplant cogeneration and its applications.
   [Pb. Technical Univ. Generation of Electric Power, May 2008]
3. Discuss the benefits of cogeneration system.
   [Pb. Technical Univ. Generation of Electric Power Dec. 2009]
4. What are the advantages of cogeneration ? What do you understand by topping and bottoming cycles ?
   [Pb. Technical Univ. Generation of Electric Power May 2010]
5. Compare the different cogeneration techniques.
   [Pb. Technical Univ. Generation of Electric Power, Dec. 2007]
6. Write short notes on the following:
   (*i*) Concept of cogeneration.
   [Rajasthan Technical Univ. 2011]
   (*ii*) Benefits of cogeneration.
   (*iii*) Cogeneration technologies.
   (*iv*) Enegy conservation.

Short Answer Type Questions With Answers

Q. 1. What do you mean by word 'cogeneration' ?
[Rajasthan Univ. 1999, 2002]

OR

What do you understand by cogeneration ?
[Pb. Technical Univ. May 2010]

Ans. Cogeneration is sequential of energy contained in fuel into two or more usable forms. In one manifestation coal is burnt in the boiler to generate steam. The steam so generated provides heat for manufacturing process in addition to generation of electrical energy. In another manifestation gas is used in gas turbine to generate electricity. The remaining heat is utilized in generating steam in a heat recovery boiler. This steam may be used for generation of additional electricity or as process steam for manufacturing process.

Q. 2. What is topping cycle and why is it suitable for cogeneration ?

Ans. In a topping cycle, the fuel energy is used to generate steam at high temperature and at high pressure and electrical energy is generated in the usual manner. Depending upon the requirement, the process steam at low temperature and pressure is either extracted from the turbine at an intermediate stage (similar to feed water heating) or taken at the turbine exhaust. Generally, the steam required for industrial processes is at low temperature whereas high temperature steam is required for generation of electrical energy, only the topping cycle, therefore, is suitable for cogeneration. Also, the overall efficiency of topping cycle is about 75% while the combined efficiency of two separate systems (for generating electrical energy and steam) is about 55%.

Q. 3. What is bottoming cycle ? Why is bottoming cycle of little thermodynamic or economic interest ?

Ans. In a bottoming cycle, the steam produced in a boiler at high temperature and pressure is directly used for manufacturing process and the reject heat is used for generation of electrical energy. The bottoming cycle is of little thermodynamic or economic interest because its combined efficiency is less than that for separate generation of process steam and electricity.

Q. 4. Which type of industries are most suitable for cogeneration ?

Ans. Energy intensive industries are suitable for cogeneration. The industries which need both electricity and steam for their manufacturing processes are most suitable for cogeneration. Such industries are paper mills, textile mills, sugar mills and refineries, chemical works, breweries, plastic manufacturing, food processing, launderies and hotels etc.

Q. 5. List benefits of cogeneration.

Ans. Benefits of cogeneration are :
(*i*) Fuel economy
(*ii*) Lower capital cost
(*iii*) Smaller gestation period and
(*iv*) Saving to industry from power cuts and power supply interruption.

Q. 6. What are the cogeneration technologies ? (Give names only).

Ans. The cogeneration technologies are steam turbine system, gas turbine system, combined cycle system and diesel engine system. In addition fluidized bed combustion, MHD, fuel cells and some other conversion systems are under development.

Q. 7. What is energy conservation ?

Ans. Energy conservation refers to the methods of reduction in energy consumption by way of elimination of wastage and promotion of efficiency. The energy consumption can be reduced by adopting various ways of energy conservation which includes efficient use of technologies and avoiding energy wastages. Energy conservation is the key to hold off climate change/global warming, as one unit of electricity saved is equal to 3 or 4 units of electricity generated.

❑❑❑

CHAPTER 13

Interconnected Systems

INSIDE THIS CHAPTER

13.1. INTRODUCTION

It is necessary to provide an uninterrupted continuous supply. In case of an isolated plant, a considerable amount of reserve capacity would be required to ensure the continuity of the supply. This would naturally increase the initial cost resulting in increase in the fixed cost owing to interest and depreciation. Moreover some units are required either to be lightly loaded or floating on the lines in order to meet the unpredicted sudden increase in demand. The efficiency of the lightly loaded units is low, running cost per unit, therefore, increases. In general overall cost per unit generated will be high.

In some parts of the country there are large resources of generating power but their requirement is low whereas in other parts power requirement is large in comparison to the available resources of power.

The above difficulties are overcome by interconnecting power stations. Interconnection of power stations, as already mentioned in chapter 10, increases the reliability of the supply, reduces the reserve capacity required, improves the load factor, diversity factor and efficiency, makes use of more efficient plants as base load plants and less efficient plants as peak load plants possible and reduces the capital cost per kW and overall cost per unit generated.

The parallel operation of power stations is essentially an extension of parallel operation of alternators, with the exception of resistances and reactances of the interconnectors. We will, therefore, make thorough study of the alternators when operated in parallel.

However, the problems of load and frequency control are more difficult in large interconnected system with many power stations scattered over a wide area in comparison with a system having one or two generating stations.

13.2. PARALLEL OPERATION OF ALTERNATORS

13.2.1. Requirements For Successful Parallel Operation

The alternators to be operated in parallel should meet some requirements so as to make their operation proper. These requirements are as follows:

1. They must have the same output voltage rating.
2. The rated speeds of the machines should be such as to give the same frequency (f being equal to PN/120).
3. The alternators should be of the same type so as to generate voltages of the same waveform. They may differ in their kVA ratings.
4. The prime movers of the alternators should have same speed-load characteristics, which of course must be drooping ones, so as to load the generators in proportion to their output ratings.
5. The alternators should have reactances in their armatures, otherwise they will not operate in parallel successfully, as illustrated in Art. 13.2.6.

13.2.2. Synchronising of Alternators

The process of connecting an alternator in parallel with another alternator or with common bus-bars (bus-bars to which a number of alternators are connected) is called *synchronising*.

The following conditions must be fulfilled for proper synchronising of alternators :

1. The terminal voltage of the incoming machine must be exactly equal to that of the others, or of the bus-bars connecting them.

2. The speed of the incoming machine must be such that its frequency $\left(\text{being euqal to } \frac{PN}{120}\right)$ equals bus-bar frequency.
3. The phase of the incoming machine voltage must be the same as that of the bus-bar voltage relative to the load *i.e.,* the phase voltages of the incoming machine and the bus-bar should be in phase opposition. This implies that there will be no circulating current between the windings of the alternators already in operation (the bus-bars) and the incoming machine.

In case of 3-phase alternators, an additional requirement to be met is that the phase sequence of the incoming machine is the same as that of the bus-bars.

The condition of same phase sequence is checked by the phasing out at the time of commissioning of alternator and the remaining three conditions have to be checked each time the alternator is to be put in parallel with the bus-bars or other alternators already in operation.

13.2.3. Synchronising Current

Once a synchronous machine is synchronised, it will tend to remain in synchronism with the other alternators. Any tendency to depart from the condition of synchronism is opposed by a synchronising torque produced due to circulating current flowing through the alternators.

When two alternators are in exact synchronism, the two alternators have equal induced emfs which are in exact phase opposition, as shown in Fig. 13.1(*a*). No circulating current flows round the local circuit.

When the induced emfs of the two alternators are equal in magnitude but not in exact phase opposition, as shown in Fig. 13.1 (*b*), their resultant emf acts round the local circuit and causes flow of current called the *synchronising current*, I_{sy}.

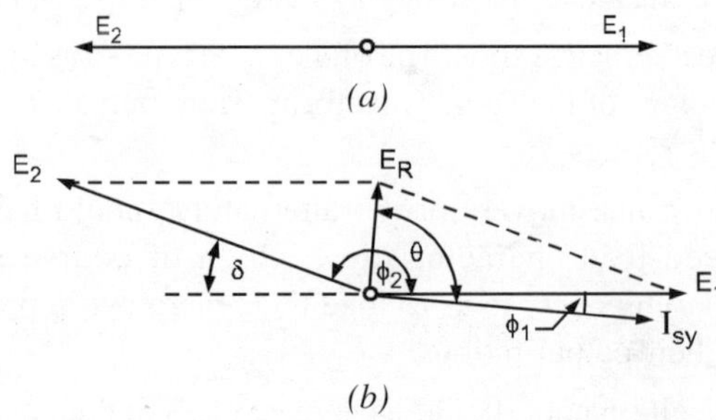

(*a*)

(*b*)

Fig. 13.1

If any alternator (say, the second alternator) due to some disturbance tends to retard, E_2 falls back by a phase angle δ electrical degrees, as shown in Fig. 13.1 (*b*). Now though their induced emfs E_1 and E_2 are equal in magnitude but have a phase difference of 180° – δ. Let each of the induced emfs E_1 and E_2 be equal to E.

Resultant emf,

$$E_R = 2E\cos\left[\frac{180° - \delta}{2}\right] = 2E\cos\left(90° - \frac{\delta}{2}\right) = 2E\sin\frac{\delta}{2} \quad ...(13.1)$$

$$= 2E \times \frac{\delta}{2} = E\delta \qquad \because \delta \text{ is very small}$$

$$\text{Synchronising current, } I_{sy} = \frac{E_R}{Z_s} = \frac{E\delta}{Z_s} \quad ...(13.2)$$

where Z_s is the combined synchronous impedance per phase of the two alternators (or of one alternator only if it is connected to infinite bus-bars).

The synchronising current I_{sy} lags behind the resultant emf E_R by an angle θ given by $\theta = \text{Tan}^{-1}\frac{X_s}{R_e}$ where X_s is the combined synchronous reactance and R_e is the effective resistance of the two alternators (or of one alternator only if it is connected to infinite bus-bars).

If resistance R_e is very small as compared to synchronous reactance X_s then

Synchronising current $I_{sy} = \frac{E\delta}{X_s}$ and lags behind E_R by 90° *i.e.,* almost in phase with E_1.

Thus I_{sy} is generating current with respect to machine no. 1 and motoring current with respect to machine no. 2 (in generator action the current flows in the direction of the induced emf while in motor action current flows in a direction opposite to that of the induced emf). This current I_{sy} sets up a synchronising torque T_{sy} which tends to accelerate the motoring machine (*i.e.,* machine no. 2) and decelerate the generating machine (*i.e.,* machine no. 1).

Similarly if the machine no. 2 tends to speed up, induced emf E_2 will tend to advance in phase and the synchronising current I_{sy} will be generating current with respect to machine no. 2 and motoring current with respect to machine no. 1. The torque developed due to synchronising current I_{sy} will now tend to retard machine no. 2 and accelerate machine no. 1.

Thus, any departure from synchronism results in development of a synchronising torque T_{sy} which tends to keep the machines in synchronism.

13.2.4. Synchronising Power

In above case (Art. 13.2.3), machine no. 1 supplies power equal to $E_1 I_{sy} \cos\phi_1$ and the machine no. 2 receives power equal to $E_2 I_{sy} \cos(180° - \phi_2)$.

Power supplied by machine no. 1

= Power supplied to machine no. 2 + copper losses.

The power supplied by machine no. 1 is called the *synchronising power* and is given by the expression

$$P_{sy} = E_1 I_{sy} \cos\phi_1 = E_1 I_{sy} = E \times \frac{E\delta}{X_s} = \frac{\delta E^2}{X_s} \quad ...(13.3)$$

$$\because E_1 = E \text{ and } \phi \text{ is very small}$$

Total synchronising power for 3-phases = $3P_{sy} = \frac{3\delta E^2}{X_s}$.

13.2.5. Synchronising Torque

If T_{sy} be the synchronising torque in N-m, then total synchronising power

$$3P_{sy} = T_{sy} \times \frac{2\pi N_s}{60}$$

or Synchronising torque, $$T_{sy} = \frac{3P_{sy} \times 60}{2\pi N_s} \quad ...(13.4)$$

where N_s is the synchronous speed in rpm and is equal to $\frac{120f}{P}$.

13.2.6. Effect of Reactance

Since with respect to local circuit, the emf of an alternator is in phase opposition to the emf of another alternator with which it is working in parallel, the machines run as synchronous motor with respect to another. Hence if due to certain reasons, the input to machine no. 2 is cut off, which is quite probable, it must receive wattful motor current from the other.

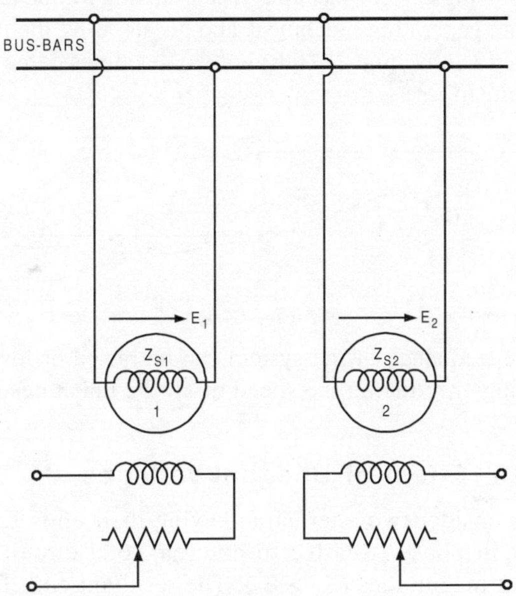

Fig. 13.2

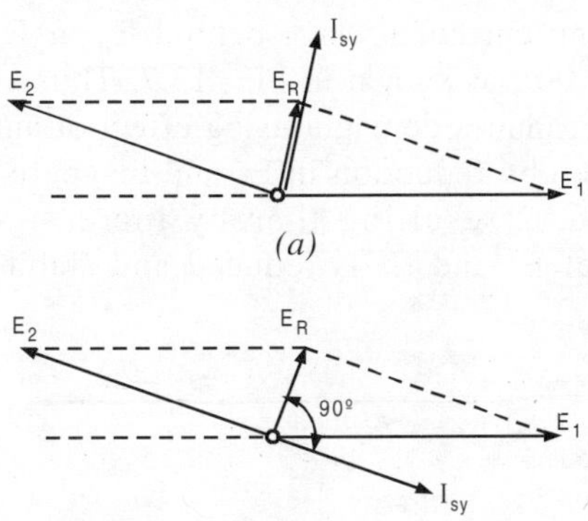

(b)

Fig. 13.3

Consider two machines having resistance but negligible reactance. Their emfs E_1 and E_2 will be practically in phase opposition, as shown in Fig. 13.3 (*a*) and their resultant emf E_R will be almost in quadrature with E_1 and E_2. The synchronising current I_{sy} given by the relation $I_{sy} = \frac{E_R}{R_1 + R_2}$ will be in phase with E_R and, therefore, in quadrature with E_1 and E_2. Hence the synchronising current I_{sy} will be wattless current and convey no power from machine no. 1 to machine no. 2, which needs help.

Now consider the two machines having the synchronous reactance, but no resistance. The synchronising current I_{sy} will be in quadrature with resultant emf E_R, therefore, almost in phase with one of the emfs, say in phase with E_1, as shown in Fig. 13.3 (*b*). Thus machine no. 1 will supply power to machine no. 2 and keep the machine no. 2 running.

From the above discussions it is obvious that without reactance the machines could not develop motor and generator power, respectively to restore synchronism; parallel operation would not be possible.

13.2.7. Effect of Increasing The Excitation of One of The Alternators

For simplicity, let us consider two identical alternators sharing equally a load whose power factor is cos φ. If both machines have exactly the same excitation it will be found that their currents I_1 and I_2 are equal in magnitude (say I each) and in phase, since the conditions are identical for both machines. The phasor diagram for the total load, for one phase is given in Fig. 13.4 (*a*). Now, if the excitation of one of the alternators is increased, it will cause flow of synchronising current I_{sy} almost

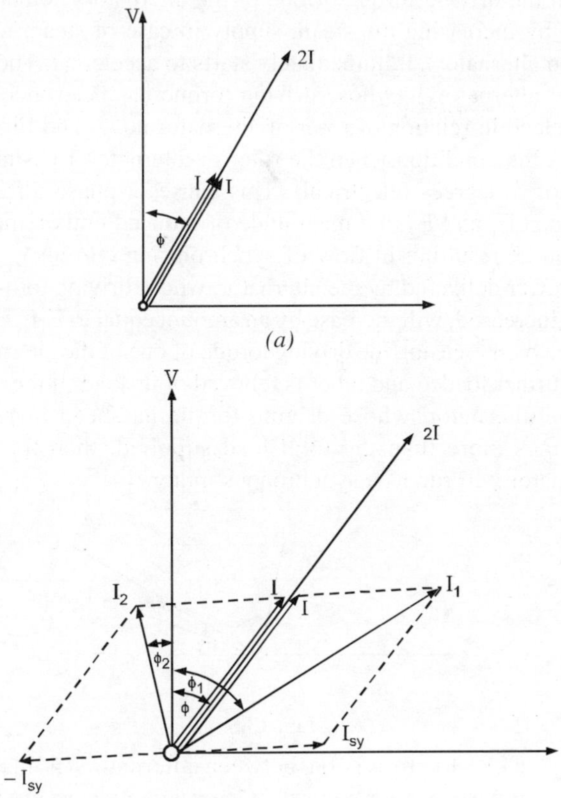

(b)

Fig. 13.4

in quadrature with supply voltage V. Therefore, the load current of alternator 1, whose excitation has been increased, will be I_1, the phasor sum of I_{sy} and I and that of alternator 2 will be I_2, the phasor difference of I_{sy} and I. Hence power factor cos ϕ_1 of alternator 1 decreases and that of the other improves. Because synchronising current I_{sy} is in quadrature with V, therefore, it does not change wattful (active) components but changes wattless (reactive) components. Hence by changing the excitation, the power factors of the alternators are changed.

If the excitation of an alternator operating in parallel with other alternators is increased above its normal value of excitation, its power factor changes in the lagging direction and its current output increases with no appreciable change in its kW load. Likewise, if the generator is underexcited its power factor becomes more leading and its current output increases with no change in kW output. This increase in current in either case is not supplied to the load but circulates between the alternators connected to the system, thereby increasing their losses and reducing their useful capacity. It is desirable in most cases, therefore, to operate each alternator at the same power factor (at or near the power factor of the load) keeping the circulating current to the minimum.

In general, the proper amount of field excitation for alternators operating in parallel is the amount of excitation each alternator would need if it were carrying its load alone at the same voltage and frequency.

13.2.8. Effect of Increasing The Driving Torque of One of the Alternators

When the driving torque of one of the alternators is increased (*e.g.,* by increasing the steam supply in case of steam turbine driven alternator), it immediately starts to accelerate. The rotor of the alternator 1 (whose driving torque has been increased) takes lead in relation of rotor of alternator no. 2, and Fig. 13.5 shows the condition when the rotor of alternator 1 has taken a lead of δ degrees (electrical). This causes a phase difference between E_1 and E_2, the magnitude of induced emfs remaining the same, resulting in flow of synchronising current I_{sy}. Thus the power delivered by the alternator, whose driving torque has been increased, will increase by an amount equal to $E_1 I_{sy} \cos \phi_1$. Hence by increasing the driving torque of one of the alternators, it is further loaded and other is relieved of its load. If the output of the alternator, whose driving torque has been increased, becomes more than the total load supplied, then the other alternator will run as a synchronous motor.

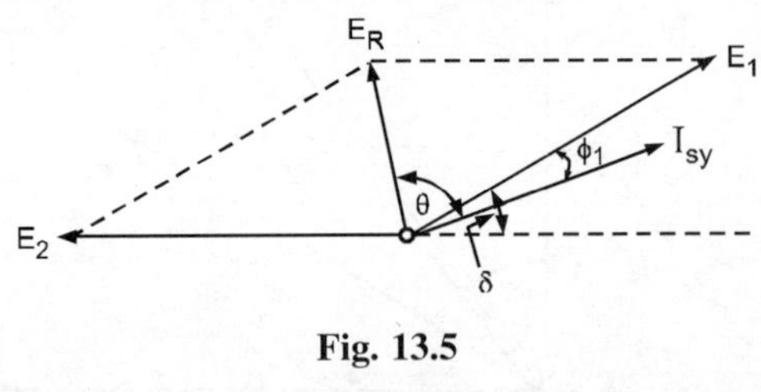

Fig. 13.5

The kW load division between alternators is made by adjusting the governor controls. One prime-mover governor is opened while the other is closed simultaneously. Thus, the system frequency is maintained constant while the load is shifted from one machine to the other. Governor control switches are mounted on the switchboard so that the operator is able to watch the switchboard instruments while making adjustments of load division.

13.2.9. Effect of Change in Speed of One of The Alternators

When the two alternators are in exact synchronism the two alternators have equal potential differences and are in exact phase opposition, as shown in Fig. 13.1 (*a*). No circulating current flows round the local circuit. But when the speed of machine 2 is reduced, E_2 falls back by a phase angle δ electrical degrees, as shown in Fig. 13.6, the magnitude of induced emfs E_1 and E_2 remaining the same. It will cause resultant emf E_R acting in the local circuit, which will cause a flow of synchronising current I_{sy} in the local circuit. The synchronising current I_{sy} will lag behind E_R by θ where $\theta = \tan^{-1} \dfrac{X_s}{R_e}$. Since R_e is negligible as compared to X_s, therefore, θ = 90°. The synchronising current I_{sy} will be generating current with respect to machine 1. This synchronising current sets up a synchronising torque tending to retard the generating machine 1 and accelerating the motoring machine *i.e.,* machine 2. Thus the two machines come again in synchronism.

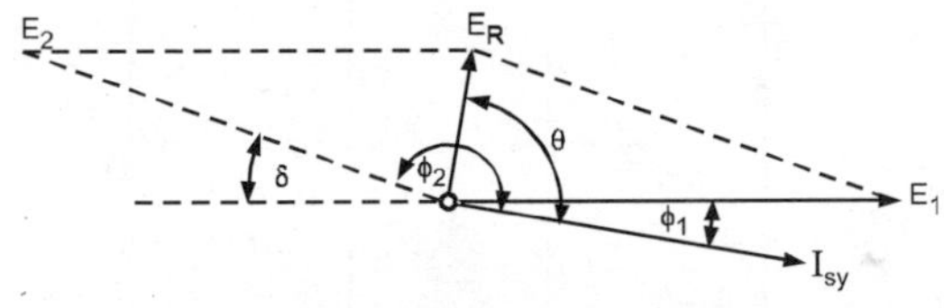

Fig. 13.6

The frequency of the system can be raised or lowered by increasing or reducing the speed of all the machines simultaneously.

13.2.10. Effect of Unequal Voltages

Let us consider two alternators having their emfs E_1 and E_2 exactly in phase (relative to external load circuit) but of different magnitudes ($E_1 > E_2$). The resultant emf, E_R being equal to $E_1 - E_2$ acts in the local circuit and causes synchronising current I_{sy} around the local circuit. This synchronising current I_{sy} lags behind E_R or E_1 by 90° and leads E_2 by 90°, as shown in Fig. 13.7. Thus synchronising current I_{sy} produces demagnetising effect on machine no. 1, resulting thereby reduction in E_1 and magnetising effect on machine no. 2 resulting thereby increase in E_2. Thus difference of E_1 and E_2 is reduced and stable condition is established.

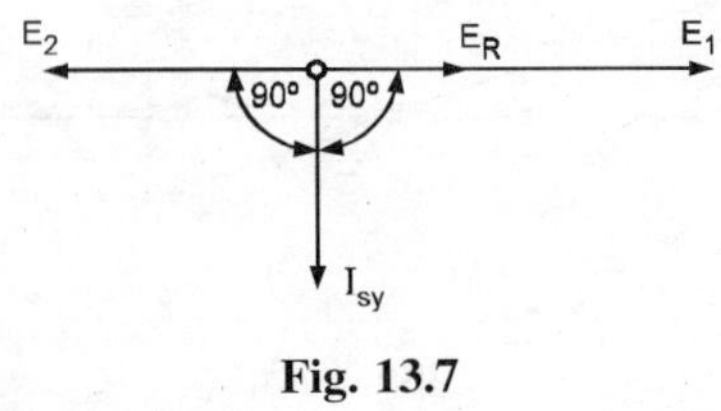

Fig. 13.7

13.2.11. Load Sharing Between Two Alternators

Consider two machines with identical speed-load characteristics running in parallel with a common terminal voltage of V volts and load impedance Z.

Let the generated emfs of the two machines 1 and 2

operating in parallel be E_1 and E_2 respectively and synchronous impedances per phase be Z_{s_1} and Z_{s_2} respectively.

Terminal voltage of machine 1, $\mathbf{V} = \mathbf{E_1} - \mathbf{I_1 Z_{s_1}}$ ...(13.5)

Similarly terminal voltage of machine 2,

$$\mathbf{V} = \mathbf{E_2} - \mathbf{I_2 Z_{s_2}} \quad ...(13.6)$$

Also $\mathbf{V} = \mathbf{IZ} = (\mathbf{I_1} + \mathbf{I_2})\,\mathbf{Z}$ ...(13.7)

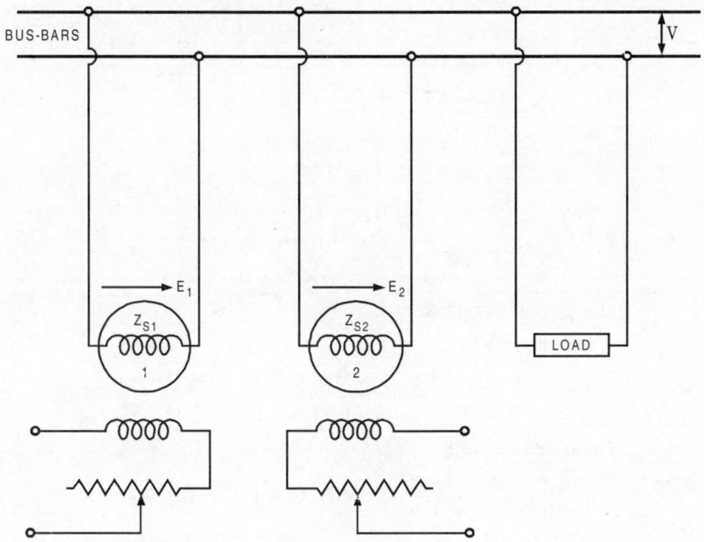

Fig. 13.8

From Eqs. (13.5) and (13.6), we have

$$\mathbf{I_1} = \frac{\mathbf{E_1} - \mathbf{V}}{\mathbf{Z_{s_1}}} \quad ...(13.8)$$

and $$\mathbf{I_2} = \frac{\mathbf{E_2} - \mathbf{V}}{\mathbf{Z_{s_2}}} \quad ...(13.9)$$

Adding Eqs. (13.8) and (13.9), we have

$$\mathbf{I_1} + \mathbf{I_2} = \frac{\mathbf{E_1} - \mathbf{V}}{\mathbf{Z_{s_1}}} + \frac{\mathbf{E_2} - \mathbf{V}}{\mathbf{Z_{s_2}}}$$

or $$\frac{\mathbf{V}}{\mathbf{Z}} = \frac{\mathbf{E_1} - \mathbf{V}}{\mathbf{Z_{s_1}}} + \frac{\mathbf{E_2} - \mathbf{V}}{\mathbf{Z_{s_2}}}$$

$\because$ from Eq. (13.7) $\mathbf{I_1} + \mathbf{I_2} = \frac{\mathbf{V}}{\mathbf{Z}}$

or $$\mathbf{V}\left(\frac{1}{\mathbf{Z_{s_1}}} + \frac{1}{\mathbf{Z_{s_2}}} + \frac{1}{\mathbf{Z}}\right) = \frac{\mathbf{E_1}}{\mathbf{Z_{s_1}}} + \frac{\mathbf{E_2}}{\mathbf{Z_{s_2}}}$$

or $$\mathbf{V} = \frac{\dfrac{\mathbf{E_1}}{\mathbf{Z_{s_1}}} + \dfrac{\mathbf{E_2}}{\mathbf{Z_{s_2}}}}{\dfrac{1}{\mathbf{Z_{s_1}}} + \dfrac{1}{\mathbf{Z_{s_2}}} + \dfrac{1}{\mathbf{Z}}} \quad ...(13.10)$$

Example 13.1. Two single-phase generators operate in parallel on a load impedance of Z ohms. Their emfs are E_1 and E_2 and their synchronous impedances Z_1 and Z_2. Deduce the terminal voltage in terms of these emfs and admittances Y, Y_1 and Y_2.

Determine the terminal voltage and kW output of each machine if $E_1 = 100$, $E_2 = 110$ volts, $Z = 3 + j4$ and $Z_1 = Z_2 = 0.2 + j1$ ohms.

[*London Univ. 1945*]

Solution : From Eq. (13.10) terminal voltage in terms of emfs and impedances Z, Z_1 and Z_2 is given as

$$\mathbf{V} = \frac{\dfrac{\mathbf{E_1}}{\mathbf{Z_1}} + \dfrac{\mathbf{E_2}}{\mathbf{Z_2}}}{\dfrac{1}{\mathbf{Z_1}} + \dfrac{1}{\mathbf{Z_2}} + \dfrac{1}{\mathbf{Z}}}$$

Substituting $\mathbf{Z_1} = \frac{1}{\mathbf{Y_1}}$, $\mathbf{Z_2} = \frac{1}{\mathbf{Y_2}}$ and $\mathbf{Z} = \frac{1}{\mathbf{Y}}$ in above equation, we have

$$\mathbf{V} = \frac{\mathbf{E_1 Y_1} + \mathbf{E_2 Y_2}}{\mathbf{Y_1} + \mathbf{Y_2} + \mathbf{Y}}$$

With the given datas in question

$$\mathbf{Y_1} = \mathbf{Y_2} = \frac{1}{0.2 + j1} = \frac{0.2 - j1}{(0.2)^2 + 1^2} = 0.1923 - j\,0.9615 \text{ S}$$

$$\mathbf{Y} = \frac{1}{\mathbf{Z}} = \frac{1}{3 + 4j} = \frac{3 - j4}{3^2 + 4^2} = 0.12 - j0.16 \text{ S}$$

Terminal voltage, $$\mathbf{V} = \frac{\mathbf{E_1 Y_1} + \mathbf{E_2 Y_2}}{\mathbf{Y_1} + \mathbf{Y_2} + \mathbf{Y}}$$

$$= \frac{100(0.1923 - j0.9615) + 110(0.1923 - j0.9615)}{2 \times (0.1923 - j0.9615) + 0.12 - j0.16}$$

$$= \frac{40.383 - j201.915}{0.5046 - j2.083} = \frac{205.914 \angle -78.69^\circ}{2.1432 \angle -76.38^\circ}$$

$$= 96.08 \angle -2.31^\circ \text{ V}$$

$$= 96 - j3.87 \text{ V } \textbf{Ans.}$$

$$\mathbf{I_1} = \frac{\mathbf{E_1} - \mathbf{V}}{\mathbf{Z_1}}$$

$$= (\mathbf{E_1} - \mathbf{V})\mathbf{Y_1}$$

$$= [100 - (96 - j\,3.87)]\,[0.1923 - j\,0.9615]$$

$$= (4 + j\,3.87)\,(0.1923 - j\,0.9615)$$

$$= 5.5657 \angle 44.05^\circ \times 0.9805 \angle -78.69^\circ$$

$$= 5.457 \angle -34.64^\circ \text{ A}$$

$$\mathbf{I_2} = \frac{\mathbf{E_2} - \mathbf{V}}{\mathbf{Z_2}}$$

$$= (\mathbf{E_2} - \mathbf{V})\mathbf{Y_2}$$

$$= [110 - (96 - j3.87)]\,[0.1923 - j\,0.9615]$$

$$= (14 + j\,3.87)\,(0.1923 - j\,0.9615)$$

$$= 14.525 \angle 15.45^\circ \times 0.9805 \angle -78.69^\circ$$

$$= 14.24 \angle -63.24^\circ \text{ A}$$

Power output of machine 1 = $\mathbf{V}_c\mathbf{I}_1$

$= 96.08 \angle{+2.31^\circ} \times 5.457 \angle{-34.64^\circ}$

= 443 watts or 0.443 kW **Ans.**

Power output of machine 2 = $\mathbf{V}_c\mathbf{I}_2$

$= 96.08 \angle{2.31^\circ} \times 14.24 \angle{-63.24^\circ}$

= 664.8 watts or 0.665 kW **Ans.**

Example 13.2. A 2,000 kVA, 3-phase, 8-pole alternator runs at 750 rpm in parallel with other machines on 6,000 V bus-bars. Find synchronising power on full load 0.8 pf lagging per mechanical degree of displacement and the corresponding synchronising torque. The synchronous reactance is 6 ohms per phase.

Solution : Voltage per phase, $V = \dfrac{6{,}000}{\sqrt{3}} = 3{,}464$ V

Full-load current, $I = \dfrac{\text{Rated kVA} \times 1{,}000}{\sqrt{3}\,V_L} = \dfrac{2{,}000 \times 1{,}000}{\sqrt{3} \times 6{,}000}$

= 192.45 A

Synchronous reactance per phase, $X_s = 6\ \Omega$

Power factor, $\cos\phi = 0.8$

Phase angle, $\phi = \text{Cos}^{-1}\, 0.8 = 36.87^\circ$ (lagging)

$\sin\phi = \sin 36.87^\circ = 0.6$

Displacement in electrical degrees

$$= 1 \times \frac{P}{2} = 1 \times \frac{8}{2} = 4^\circ = 4 \times \frac{\pi}{180} = \frac{\pi}{45} \text{ radian}$$

Induced emf, $\mathbf{E} = \mathbf{V} + j\mathbf{I}X_s$

$= 3{,}464 + j192.45\,(0.8 - j\,0.6) \times 6$

$= (4{,}156.82 + j\,923.76)$ or $4{,}258 \angle{12.53^\circ}$ V

Synchronising power for phases,

$$3P_{sy} = \frac{3\delta \mathbf{EV}}{X_s} = \frac{3 \times \frac{\pi}{45} \times 4{,}258 \cos 12.53^\circ \times 3{,}464}{6}$$

= 502,600 W or 502.6 kW **Ans.**

Synchronising torque, $T_{sy} = \dfrac{3P_{sy}}{2\pi N_s/60} = \dfrac{502{,}600}{2\pi \times 750/60}$

= 6,400 Nm **Ans.**

Example 13.3. A 750 kVA, 11 kV, 4-pole, 3-phase alternator has a resistance of 1.4 Ω per phase and a synchronous reactance of 30 Ω per phase. Calculate the synchronous power per mechanical degree displacement for full load for 0.8 power factor lagging, the terminal voltage being 11 kV.

Solution : Voltage per phase, $V = \dfrac{11 \times 1{,}000}{\sqrt{3}} = 6{,}351$ V

Full-load current, $I = \dfrac{750 \times 1{,}000}{\sqrt{3} \times 11{,}000} = 39.36$ A

Synchronous impedance per phase,

$\mathbf{Z}_s = (1.4 + j\,30)\ \Omega$

Rotor displacement,

$$\delta = 1 \times \frac{P}{2} = 1 \times \frac{4}{2} = 2 \text{ electrical degrees}$$

$$= 2 \times \frac{\pi}{180} = \frac{\pi}{90} \text{ elec. radian}$$

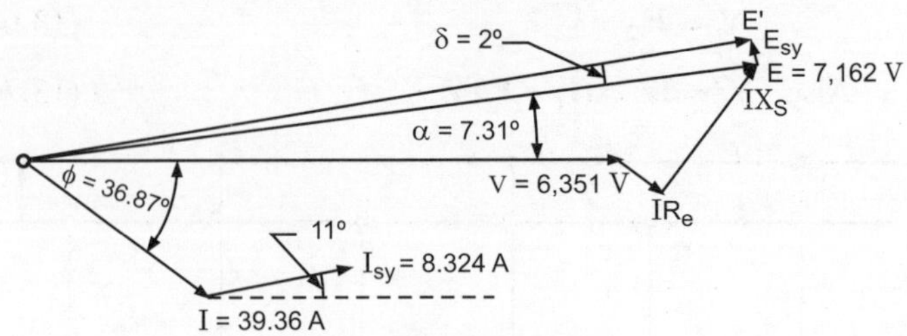

Fig. 13.9

Power factor, $\cos\phi = 0.8$

Phase angle, $\phi = \text{Cos}^{-1}\, 0.8 = 36.87^\circ$

$\sin\phi = \sin 36.87^\circ = 0.6$

Induced emf per phase,

$\mathbf{E} = \mathbf{V} + \mathbf{I}\mathbf{Z}_s$

$= (6{,}351 + j0) + 39.36\,(0.8 - j0.6)\,(1.4 + j30)$

$= 7{,}103.56 + j911.58 = 7{,}162 \angle{7.31^\circ}$ V

Synchronising emf,

$$E_{sy} = 2E \sin\frac{\delta}{2} = 2 \times 7{,}162 \sin 1^\circ = 250 \text{ V}$$

$$\mathbf{I}_{sy} = \frac{\mathbf{E}_{sy}}{\mathbf{Z}_s} = \frac{250}{(1.4 + j30)} = 8.324 \angle{87.33^\circ} \text{ A}$$

I_{sy} lags E_{sy} by 87.33° or leads V by $\left[\alpha + \dfrac{\delta}{2} + (90^\circ - 87.33^\circ)\right]$ *i.e.*, $\left(7.31^\circ + \dfrac{2}{2} + 2.67^\circ\right)$ or 11°

Synchronising power for 3-phases,

$$3P_{sy} = \frac{3VI_{sy}\cos 11^\circ}{1{,}000} \text{ kW}$$

$$= \frac{3 \times 6{,}351 \times 8.324 \cos 11^\circ}{1{,}000} = 155.68 \text{ kW } \textbf{Ans.}$$

Example 13.4. Two alternators working in parallel supply a lighting load of 3,000 kW and a motor load aggregating to 5,000 kW at a pf of 0.72. One machine is loaded up to 5,000 kW at 0.8 pf lagging. What is the load and power factor of the other machine ?

[*U.P.S.C. I.E.S. Electrical Engineering II, 2006*]

Solution : Phase angle of lighting load,

$\phi_L = \text{Cos}^{-1}\, 1.0 = 0^\circ$ as pf of lighting load is assumed to be unity

Phase angle of motor load,

$\phi_M = \text{Cos}^{-1}\, 0.72 = 43.95^\circ$ (lagging) as pf of motor load is assumed to be lagging

Total load, $P = P_L + P_M = 3{,}000 + 5{,}000 = 8{,}000$ kW

Total reactive power to be supplied,

$Q = Q_L + Q_M$

$= P_L \tan\phi_L + P_M \tan\phi_M$

$= 3{,}000 \tan 0^\circ + 5{,}000 \tan 43.95^\circ$

$= 0 + 4{,}819 = 4{,}819$ kVAR

Load supplied by first alternator,

$P_A = 5{,}000$ kW

Phase angle, $\phi_A = \text{Cos}^{-1}\, 0.8 = 36.87°$ (lagging)

Reactive power supplied by first alternator

$$Q_A = P_A \tan \phi_A = 5{,}000 \tan 36.87° = 3{,}750 \text{ kVAR}$$

Power to be supplied by second alternator

$$P_B = P - P_A = 8{,}000 - 5{,}000 = 3{,}000 \text{ kW}$$

Reactive power to be supplied by the second alternator

$$Q_B = Q - Q_A = 4{,}819 - 3{,}750 = 1{,}069 \text{ kVAR}$$

Phase angle of second alternator,

$$\phi_B = \text{Tan}^{-1} \frac{Q_B}{P_B} = \text{Tan}^{-1} \frac{1{,}069}{3{,}000} = 19.6° \text{ (lagging)}$$

Power factor of second alternator,

$$\text{Cos}\, \phi_B = \text{Cos } 19.6° = 0.942 \text{ (lagging)}$$

Thus, second machine is loaded with load of 3,000 kW at a pf of 0.942 (lagging) **Ans.**

Example 13.5. Two 200 MVA alternators operate in parallel. The frequency drops in the first machine from 50 Hz at no load to 48 Hz at full load, where as in case of the other machine, the frequency drops from 50 Hz to 47 Hz under the same conditions.

***(a)* How the two machines will share a total load of 300 MW ?**

***(b)* Determine the maximum load at unity pf which can be delivered by the two machines without overloading any of them.**

[U.P.S.C. I.E.S. Electrical Engineering II, 2003]

Solution : Speed-load characteristics are drawn for driving turbines of machines 1 and 2 in Fig. 13.10. CA represents the speed-load characteristic for turbine driving machine 1 and CB represents the speed-load characteristic of machine 2.

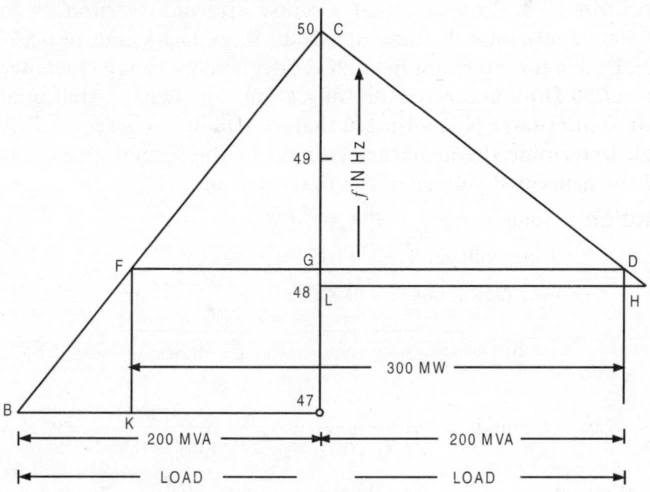

Fig. 13.10

(*a*) Since the machines are operating in parallel, therefore, their frequency must be same. Let DF be drawn through G, so that the total load supplied is 300 MW.

Load supplied by machine 1 = DG = LA – HA

$$= 200 - \frac{DH}{CL} \times AL$$

$$= 200 - \frac{OG - OL}{CL} \times AL$$

$$= 200 - \frac{OG - 1}{50 - 48} \times 200 = 300 - 100\, OG$$

Load supplied by machine 2

$$= GF = OB - BK = 200 - \frac{FK}{CO} \times OB$$

$$= 200 - \frac{OG}{50 - 47} \times 200 = 200 - 66.667\, OG$$

Total load supplied

$$= DG + GF = 300 - 100\, OG + (200 - 66.667 OG)$$

$$\text{or } 300 = 500 - 166.667 OG$$

$$\text{or } OG = \frac{200}{166.667} = 1.2$$

So, load supplied by machine 1

$$= 300 - 100\, OG = 300 - 100 \times 1.2 = 180 \text{ MW} \quad \textbf{Ans.}$$

Load supplied by machine 2

$$= 200 - 66.667 OG = 200 - 66.667 \times 1.2 = 120 \text{ MW} \quad \textbf{Ans.}$$

(*b*) Machine 1, which has better speed regulation will be loaded first to its full-load rating, so it will operate on maximum load of 200 MW (pf being unity) at frequency $f = 48$ Hz

Machine 2 will supply load of (200 – 66.667 × 1) *i.e.*, 133,333 MW at 48 frequency

Total load supplied = 200 + 133.33 = 333.333 MW **Ans.**

Example 13.6. The following two synchronous machines are operating in parallel

Machine A	**50 MW**	**6% speed regulation.**
Machine B	**50 MW**	**3% speed regulation.**

***(i)* Determine the load taken by each machine for a total load of 80 MW when the speed changers are set to give rated speed at 100% rated output.**

***(ii)* The speed changer of machine A is so adjusted that 80 MW is equally shared. Find the output of machine A for the rated speed and also its percentage speed at no load.**

[U.P.S.C. I.E.S. Elec. Engineering II, 1992]

Solution : Speed-load characteristics for machines A and B are represented by lines CA and DB respectively in Fig. 13.11 (*a*).

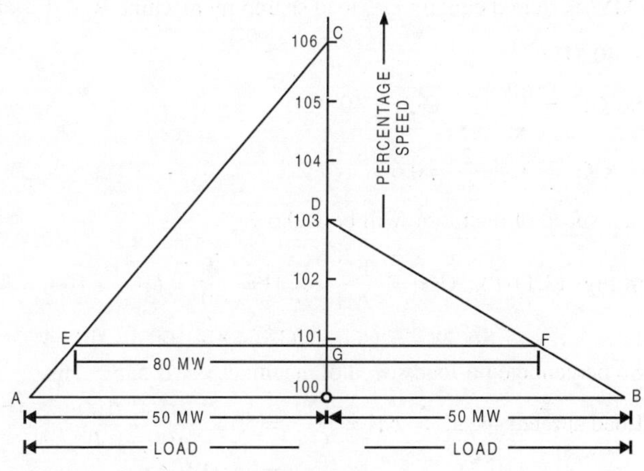

Fig. 13.11(*a*)

Since the machines are operating in parallel, therefore, their frequency or percentage speed must be the same. Let EF be drawn through G so that the total load supplied is 80 MW

Load supplied by machine A,

$$EG = AO \times \frac{CG}{CO} = 50 \times \frac{CO - OG}{CO} = 50 \times \frac{6 - OG}{6}$$

$$= \frac{25}{3}(6 - OG)$$

Load supplied by machine B,

$$GF = BO \times \frac{DG}{DO} = 50 \times \frac{3 - OG}{3} = \frac{50}{3}(3 - OG)$$

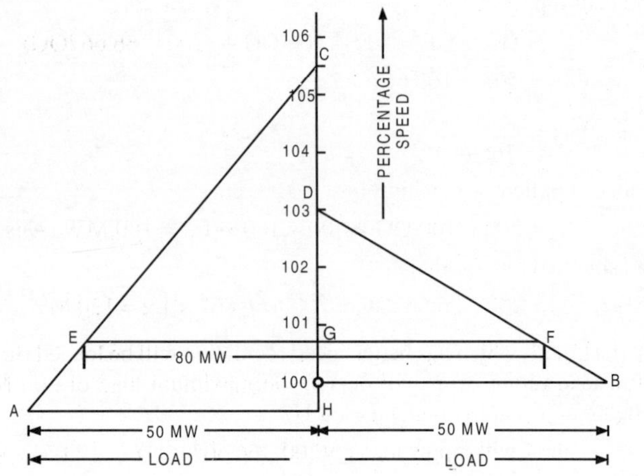

Fig. 13.11(*b*)

Since EG + GF = EF = Total load supplied

$$\frac{25}{3}(6 - OG) + \frac{50}{3}(3 - OG) = 80$$

or $OG = \frac{60}{75} = 0.8$

So load supplied by machine A, P_A

$$= \frac{25}{3}(6 - OG) = \frac{25}{3}(6 - 0.8) = 43.333 \text{ MW}$$

Load supplied by machine B,

$$P_B = \frac{50}{3}(3 - OG) = \frac{50}{3}(3 - 0.8) = 36.667 \text{ MW}$$

(*ii*) Now speed changer of machine A is so adjusted that a total load of 80 MW is shared equally *i.e.,* load shared by machine B is $\frac{1}{2} \times 80$ MW = 40 MW

So $GF = \frac{50}{3}(3 - OG) = 40$

or $OG = 3 - \frac{12}{5} = 0.6$

i.e., speed of operation will be 100.6%

From Fig. 13.11(*b*), $CG = \frac{EG}{AH} \times CH = \frac{40}{50} \times (106 - 100) = 4.8$

So OC = CG + OG = 4.8 + 100.6 = 105.4

So percentage no-load speed of machine A is 105.4% **Ans.**

$$\text{Load at rated speed} = AH \times \frac{CO}{CH} = 50 \times \frac{105.4 - 100}{106 - 100}$$

= 45 MW **Ans.**

Example 13.7. Two synchronous generators are supplying a common load. Generator 1 has no load frequency of 51.5 Hz and regulation of 1 MW/Hz. Generator 2 has no load frequency of 51.0 Hz and regulation of 1 MW/Hz. The total load is 2.5 MW at 0.8 pf lagging.

(*i*) At what frequency, are the generators supplying this load and how much power is supplied by each generator ?

(*ii*) An additional load of 1 MW is attached to this system. What will be the new frequency ?

(*iii*) How much is governor set-point of generator 2 to be adjusted to bring the system frequency at 50 Hz for 3.5 MW system load ? [*U.P.S.C. I.E.S. Elec. Engineering II, 1997*]

Solution : (*i*) Let the frequency of supply be *x* Hz at a load of 2.5 MW

Load supplied by generator 1 = $(51.5 - x) \times 1 = (51.5 - x)$ MW

Load supplied by generator 2 = $(51 - x) \times 1 = (51 - x)$ MW

Total load supplied = $(51.5 - x) + (51 - x) = 2.5$

or x = 50 Hz **Ans.**

Load supplied by generator 1 = 51.5 – 50 = 1.5 MW **Ans.**

Load supplied by generator 2 = 51 – 50 = 1.0 MW **Ans.**

(*ii*) Let the new frequency of supply be x_1 Hz

Now total load supplied = $(51.5 - x_1) \times 1 + (51 - x_1) \times 1 = 3.5$

or x_1 = 49.5 Hz **Ans.**

Load supplied by generator 1 = (51.5 – 49.5) × 1 = 2 MW **Ans.**

Load supplied by generator 2 = (51.0 – 49.5) × 1 = 1.5 MW **Ans.**

(*iii*) Load supplied by generator 1 at system frequency of 50 Hz

= (51.5 – 50) × 1 = 1.5 MW

Load to be supplied by generator 2 at system frequency of 50 Hz

= 3.5 – 1.5 = 2 MW

No load frequency of generator 2

$$= 50 + \frac{\text{Load to be supplied in MW}}{\text{Generator regulation in MW / Hz}}$$

$$= 50 + \frac{2}{1} = 52 \text{ Hz } \textbf{Ans.}$$

Example 13.8. Two identical 3-phase alternators running in parallel supply total demand of 1,500 kW at 11 kV and pf 0.867 (lag). Each alternator supplies half the demand and has a reactance (syn) of 50 Ω/ph and resistance of 4 Ω/ph. The field excitation of the first alternator is so adjusted that its armature current is 50 A (lag). Determine the armature current of the second alternator and the generated voltage of the first alternator.

Solution : Total demand, P = 1,500 kW

Line voltage, V_L = 11 kV = 11,000 V

Power factor, cos φ = 0.867

$$\sin\phi = \sqrt{1 - (\cos\phi)^2} = \sqrt{1 - (0.867)^2} = 0.4983$$

$$\text{Load current, } I_L = \frac{P}{\sqrt{3}\, V_L \cos\phi} = \frac{1{,}500 \times 1{,}000}{\sqrt{3} \times 11{,}000 \times 0.867} = 90.8 \text{ A}$$

Wattfull component of load current = 90.8 × 0.867 = 78.72 A

Wattless component of load current = 90.8 × 0.4983 = 45.24 A

In identical conditions the current supplied by each machine

$$= \frac{90.8}{2} = 45.4 \text{ A}$$

Assuming power supplied constant, the wattfull component of both machines is the same as $\frac{78.72}{2}$ = 39.36 A. But due to variation in excitation of first machine the wattless components of load current would be redivided.

The armature current of first machine after changing the excitation is 50 A (lag). The active or wattfull component of current of first machine after changing the excitation = 39.36 A as before.

Wattless component of current of first machine after changing the excitation

$$= \sqrt{50^2 - (39.36)^2} = 30.83 \text{ A}$$

Wattless component of current of second machine after changing the excitation

$$= 45.24 - 30.83 = 14.41 \text{ A}$$

Armature current of second alternator after changing the excitation

$$= \sqrt{(39.36)^2 + (14.41)^2} = 41.91 \text{ A } \textbf{Ans.}$$

Power factor of first machine,

$$\cos\phi_1 = \frac{39.36}{50} = 0.7872$$

$$\sin\phi_1 = \sqrt{1 - (\cos\phi_1)^2} = \sqrt{1 - (0.7872)^2} = 0.6166$$

Generated voltage of first machine

$$= \sqrt{(V\cos\phi_1 + IR_e)^2 + (V\sin\phi_1 + IX_s)^2}$$

$$= \sqrt{\left(\frac{11{,}000}{\sqrt{3}} \times 0.7872 + 50 \times 4\right)^2 + \left(\frac{11{,}000}{\sqrt{3}} \times 0.6166 + 50 \times 50\right)^2}$$

$$= 8{,}258.6 \text{ V}$$

Line generated voltage of first machine

$$= \sqrt{3} \times 8{,}258.6 = 14{,}304 \text{ V or } 14.3 \text{ kV } \textbf{Ans.}$$

13.3 SYNCHRONOUS MACHINES ON INFINITE BUS-BARS

A supply system with a large number of synchronous machines connected in parallel is referred to as *infinite bus-bars*. Any additional machine, whether to operate as a generator or as a motor is connected in parallel with the bus-bars.

Infinite bus-bars represent a system of large capacity whose frequency and the phase and magnitude of voltage are not affected even if there is a variation of excitation or power of a synchronous machine connected to it.

Here we will investigate the effect of varying the excitation and torque of the prime mover of a synchronous machine connected to infinite bus-bars.

13.3.1. Effect of Varying The Excitation of a Synchronous Machine Connected to Infinite Bus-bars

Consider an alternator, which has been just paralleled to infinite bus-bars and supplying no load (the torque supplied to its shaft being sufficient to meet with internal losses only). At this instant the open-circuit voltage of the incoming machine, E is equal and opposite to bus-bar voltage V, as shown in Fig. 13.12(*a*) and there is no current through the alternator circuit.

Let the excitation of the incoming machine be increased resulting increase in open-circuit voltage of the machine. The phasor sum of V and E is no longer zero and their phasor sum E_R will cause a synchronising current I_{sy} to flow in the alternator circuit making it to act as a generator supplying power to the bus-bars. This can be seen from the projections of I_{sy} on the phasors E and V of Fig. 13.12 (*b*). But the torque supplied to the shaft of the incoming machine were such that it was supplying the friction, windage and iron losses of the machine only, and the condition of Fig. 13.12 (*b*) shows an electrical output of $VI_{sy}\cos\theta$. This indicates that at this particular instant, the prime mover is not supplying enough energy to make up the internal losses and electrical output, and therefore, the incoming generator will tend to slow down. Actually it will not slow down but will get retarded momentarily to change the time phase of E to the position shown in Fig. 13.12 (*c*). The delay in the time phase of E will cause a shift in E_R in a clockwise direction and likewise a clockwise shift of I_{sy}.

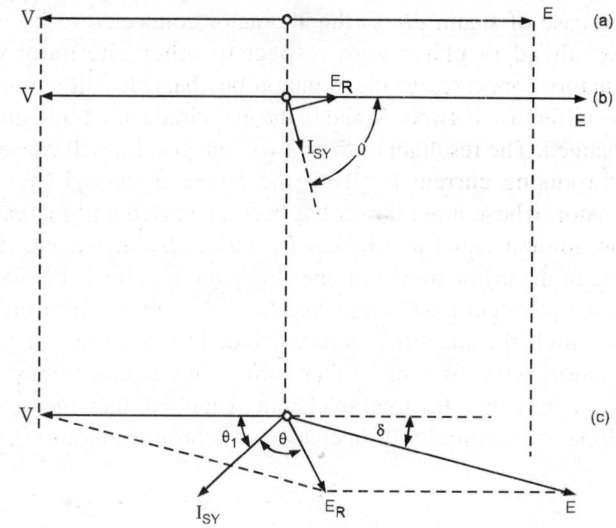

Fig. 13.12

Due to shifting of I_{sy} from the position of Fig. 13.12 (*b*) to that of 13.12 (*c*) the alternator synchronising current I_{sy} shifts the direction of its power component and it will now receive energy from the bus-bars ($VI_{sy}\cos\theta_1$) sufficient to meet the copper losses of the alternator. The amount of power transferred as a result of this change in excitation is determined by the change in the copper losses only. The magnitude of the synchronising current I_{sy} for this condition of operation is

$$\mathbf{I}_{sy} = \frac{\mathbf{E} - \mathbf{V}}{R + jX_s} = \frac{\mathbf{E}_R}{R + jX_s} \qquad \text{...(13.11)}$$

If there were no armature resistance, the synchronising current I_{sy} would always be at 90º with respect to E and V. Thus variation in excitation will cause a large change in the reactive current supplied by the alternator, and because of armature resistance, there will be a slight change in the power that it receives or delivers to the bus. *For all practical purposes it is assumed that variation in excitation causes change in power factor.*

Figure 13.12 (*c*) shows stable operation of the machine on the bus-bars. A tendency of E to advance in time would tend to make it supply power to the bus-bars and thereby experiencing a retarding torque. A tendency for E to delay in time would tend to convert it into a synchronous motor, and it would be driven by energy drawn from the bus-bars.

It may be noted that when the alternator is overexcited, it operates at lagging power factor and when underexcited it operates at leading power factor.

When the excitation of an alternator connected to infinite bus-bars and supplying load at lagging pf is increased, the alternator falls back (*i.e.,* load angle decreases), power factor further drops and armature current increases due to reduced power factor; active component being constant.

13.3.2. Effect of Varying the Torque of The Prime Mover

When the torque of the prime mover of the alternator connected to the infinite bus-bars is increased (by increasing steam supply in the case of steam drive) the alternator connected to it will swing ahead in phase with respect to other alternator or alternators connected to the common bus-bars. It will cause a phase difference between V and E, the magnitude of E remaining unchanged. The resultant of these two voltages, E_R will cause a synchronising current I_{sy}. Thus the power delivered by the alternator, whose input power has been increased will increase by an amount equal to $EI_{sy}\cos\theta_1$. Hence by increasing the torque of the prime mover of one alternator, it is further loaded and an equivalent load is removed from the other unit or units with which the machine is paralleled. If the output of the alternator, whose prime mover torque has been increased, become more than the total load being supplied, then the other machine or machines will operate as synchronous motor(s).

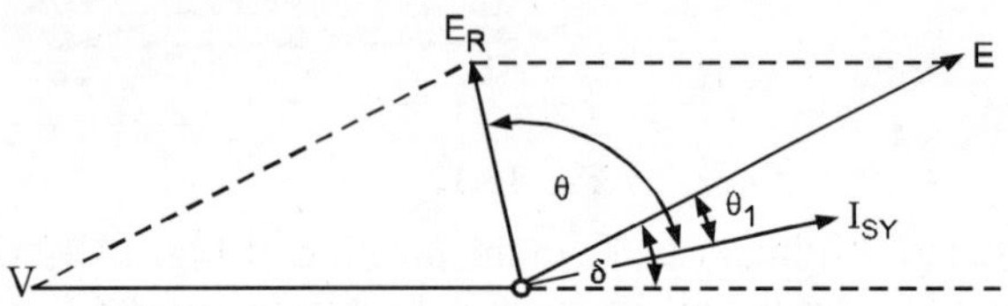

Fig. 13.13

Thus active and reactive power loading of an alternator operating on infinite bus is controlled by controlling the input power to it and excitation respectively.

13.3.3. Power Output

Let us consider a 3-phase synchronous generator having effective resistance R_e ohms per phase and synchronous reactance X_s ohms per phase. Let it be connected to the bus-bars having phase voltage of V volts and supply current I amperes at power factor $\cos\phi$. Let the phase angle between induced emf E and bus-bar voltage V be δ.

The phasor diagram is drawn with voltage phasor as reference phasor (Fig. 13.14).

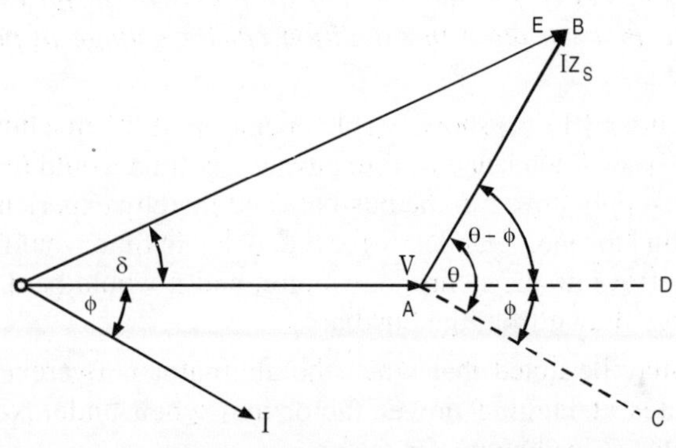

Fig. 13.14

Power output per phase,

$$P = VI\cos\phi$$
$$= VI\cos[\theta - (\theta - \phi)]$$
$$= VI\frac{Z_s}{Z_s}[\cos\theta \times \cos(\theta-\phi) + \sin\theta \times \sin(\theta-\phi)]$$
$$= \frac{V}{Z_s}[IZ_s\cos(\theta-\phi)\cos\theta + IZ_s\sin(\theta-\phi)\sin\theta] \quad ...(13.12)$$

Now $IZ_s\cos(\theta-\phi) = E\cos\delta - V$

and $IZ_s\sin(\theta-\phi) = E\sin\delta$

Substituting $IZ_s\cos(\theta-\phi) = (E\cos\delta - V)$ and $IZ_s\sin(\theta-\phi) = E\sin\delta$ in the Eq. (13.12) for power output, we have

$$P = \frac{V}{Z_s}[(E\cos\delta - V)\cos\theta + E\sin\delta\sin\theta]$$
$$= \frac{EV}{Z_s}(\cos\delta\cos\theta + \sin\delta\sin\theta) - \frac{V^2}{Z_s}\cos\theta$$
$$= \frac{EV}{Z_s}\cos(\theta-\delta) - \frac{V^2}{Z_s}\cos\theta \quad ...(13.13)$$

Since effective resistance R_e is generally negligible, therefore, synchronous impedance Z_s can be taken equal to synchronous reactance X_s and internal angle $\theta = 90°$ and the expression for power output is reduced to the following expression

$$P = \frac{EV}{X_s}\cos(90° - \delta) - \frac{V^2}{X_s}\cos 90° = \frac{EV}{X_s}\sin\delta \quad ...(13.14)$$

Equation (13.13) is for electrical output in terms of constant bus-bar voltage V, induced emf E, (or excitation as the induced emf depends upon the excitation), synchronous impedance Z_s, internal angle θ being equal to $\tan^{-1}\frac{X_s}{R_e}$ and load angle δ.

Condition for maximum power output can be determined by differentiating the Eq. (13.13) with respect to δ and equating it to zero.

$$\frac{dP}{d\delta} = \frac{EV}{Z_s}\sin(\theta-\delta) = 0$$

or $\sin(\theta - \delta) = 0$

or $\delta = \theta$

i.e., Power output for constant bus-bar voltage V and fixed excitation will be maximum when $\delta = \theta$ and the expression for maximum power output per phase will become

$$P_{max} = \frac{EV}{Z_s}\cos(\theta-\theta) - \frac{V^2}{Z_s}\cos\theta$$
$$= \frac{EV}{Z_s} - \frac{V^2}{Z_s}\cos\theta \quad ...(13.15)$$

Now from ΔAOB of phasor diagram shown in Fig. 13.14

$$IZ_s = \sqrt{V^2 + E^2 - 2VE\cos\delta}$$

$$\text{or } I = \frac{\sqrt{V^2 + E^2 - 2VE\cos\delta}}{Z_s}$$

Current corresponding to maximum power output *i.e.*, when $\delta = \theta$

$$I_{max} = \frac{\sqrt{V^2 + E^2 - 2VE\cos\theta}}{Z_s}$$

and corresponding power factor,

$$\cos\phi = \frac{P_{max}}{VI_{max}}$$

$$= \frac{\frac{EV}{Z_s} - \frac{V^2}{Z_s}\cos\theta}{\frac{V}{Z_s}\sqrt{V^2 + E^2 - 2VE\cos\theta}}$$

$$= \frac{E - V\cos\theta}{\sqrt{V^2 + E^2 - 2VE\cos\theta}} \quad ...(13.16)$$

Example 13.9. A 3-phase, 11,000 V, star-connected turbo-alternator having a synchronous reactance of 6 Ω per phase and negligible resistance has an armature current of 200 A at unity power factor when operating on constant frequency, constant voltage bus-bars. If the steam admission remains the same and the emf is raised by 25%, determine the new values of current and power factor.

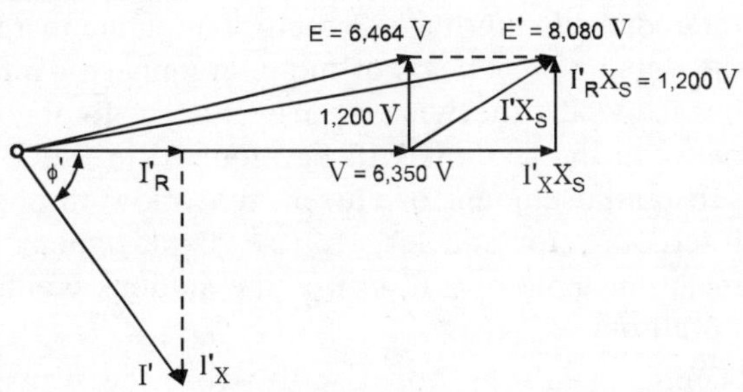

Fig. 13.15

Solution : Terminal voltage per phase, $V = \frac{11,000}{\sqrt{3}} = 6,350$ V

Reactance per phase, $X_s = 6\ \Omega$

Armature current, I = 200 A

Power factor, $\cos\phi = 1.0$

Phase angle, $\phi = \text{Cos}^{-1}\ 1.0 = 0°$

At unity power factor the reactance drop

$$IX_s = I_R X_s = 200 \times 6 = 1,200 \text{ V}$$

The phasor diagram for this condition is shown in Fig. 13.15

The emf generated per phase,

$$E = \sqrt{(6,350)^2 + (1,200)^2} = 6,464 \text{ V}$$

If the excitation is increased keeping the steam supply or power input same, the power output will not change. The result is to give the armature current a lagging reactive component which exerts a demagnetising effects on the main field and neutralise the increase in excitation. The phasor diagram is given in Fig. 13.15, which shows that the locus of the emf phasor is the dotted line perpendicular to $I'_R X_s$. I'_R is the power or active component of new armature current = 200 A as before and I'_X is the reactive component of new armature current.

With increased excitation the emf generated per phase

$$E' = 1.25E = 1.25 \times 6,464 = 8,080 \text{ V}$$

From phasor diagram shown in Fig. 13.15

$$(E')^2 = (V + I'_X X_s)^2 + (I'_R X_s)^2$$

$$\text{or } (8,080)^2 = (6,350 + 6I'_X)^2 + (200 \times 6)^2$$

$$\text{or } 6,350 + 6I'_X = \sqrt{(8,080)^2 - (1200)^2} = 7,990 \text{ V}$$

$$\text{or } I'_X = \frac{7,990 - 6350}{6} = 273.33 \text{ A}$$

New value of armature current,

$$I' = \sqrt{(I'_R)^2 + (I'_X)^2} = \sqrt{(200)^2 + (273.33)^2}$$

$$= 338.7 \text{ A } \textbf{Ans.}$$

$$\text{New power factor, } \cos\phi' = \frac{I'_R}{I'} = \frac{200}{338.7} = 0.59 \text{ (lag) } \textbf{Ans.}$$

13.4. INTERCONNECTED POWER STATIONS

When large loads are to be supplied from two power stations, the power stations are required to be interconnected so that there is no overloading and the loads are shared almost equally.

Consider two power stations S_1 and S_2 supplying load currents I_1 and I_2 to the load through transmission lines 1 and 2 in addition to their local loads. The power stations are also interconnected by an interconnector, which is carrying current I_i. The transfer of power is taking place as shown by direction of arrows (Fig. 13.16).

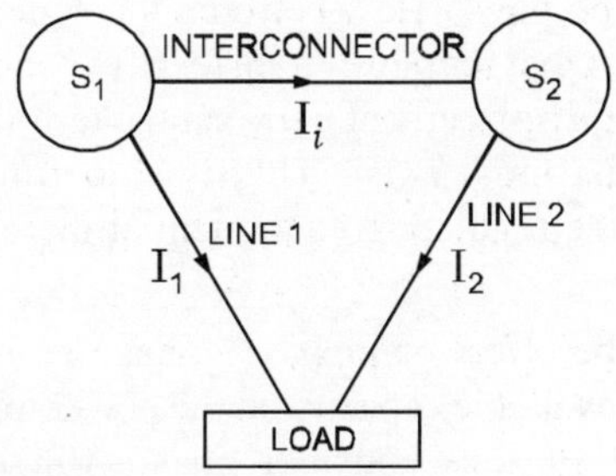

Fig. 13.16. ***Interconnected Power Stations***

Since local loads are also connected to power stations S_1 and S_2, voltages at their bus-bars have to be maintained constant at the same value as for consumers at load end.

In order that both transmission lines 1 and 2 deliver equal power and the system operates at the same terminal voltage, it is necessary that active components of the line currents I_1 and I_2 are equal. In order that the total current in lines 1 and 2 be a minimum it is necessary that they should also have reactive components equal. In order to achieve this in practice a regulating equipment is required to be installed at the sending end of each transmission line and in the interconnector. The regulating equipment installed at the sending end of each transmission line and in the interconnector will look after the impedance drops in the transmission lines and interconnector.

As in case of alternators operating in parallel, for stable operation of the interconnected generating stations, it is necessary that the two stations must be interconnected through reactor. The incorporation of interconnector introduces angular displacement between the two stations, enabling the power to flow from one station to the other as required by operating conditions. The incorporation of an impedance will cause an unnecessary loss of power.

13.4.1. Load Sharing

Consider two generating stations A and B operating in parallel and connected through an interconnector. Let the terminal voltage of operating stations A and B be V_A and V_B respectively, equal in magnitude but displaced from each other by an angle θ. The resultant of voltages V_A and V_B, V_{AB} acts along the interconnector and causes a current of I to flow through it. The current, I in the interconnector lags almost 90° behind the resultant voltage V_{AB} as most of the interconnectors are designed to have reactance and negligibly small resistance.

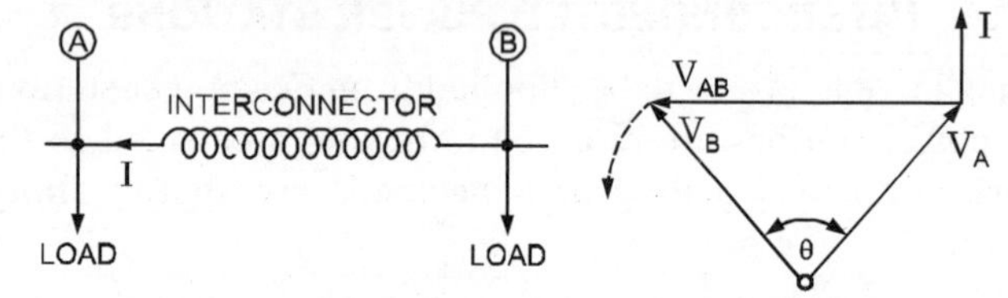

Fig. 13.17. ***Transfer of Power By Angular Displacement***

This current will be almost in phase with the voltage of the generating station, which is leading *i.e.*, V_B in this particular case and the power transfer will result from power station B to station A in the above case. When two voltages V_A and V_B are in phase, there is no power flow between the generating stations. It is thus obvious that for power transfer between interconnected stations angular displacement between the terminal voltages of the two stations must exist. This is sometimes artificially introduced by the regulating equipment at the bus-bars of one of the stations.

To study the effect of inphase voltage boost of station B having surplus available capacity on the power transfer between the generating stations consider phasor diagram shown in Fig. 13.18. The voltage of station B is boosted from V_B to V'_B, the voltage drop in interconnector is V'_{AB} and the current I through the interconnector lags almost 90° behind the V'_{AB} in case the interconnector is reactive. The current in this situation is in phase with V_A and is in quadrature with V'_B and therefore no real power can be supplied by station B to station A in this condition.

Hence it is obvious that if the interconnector had only reactance, inphase voltage boost between the generating stations would not enable the interconnector to transfer real power between them.

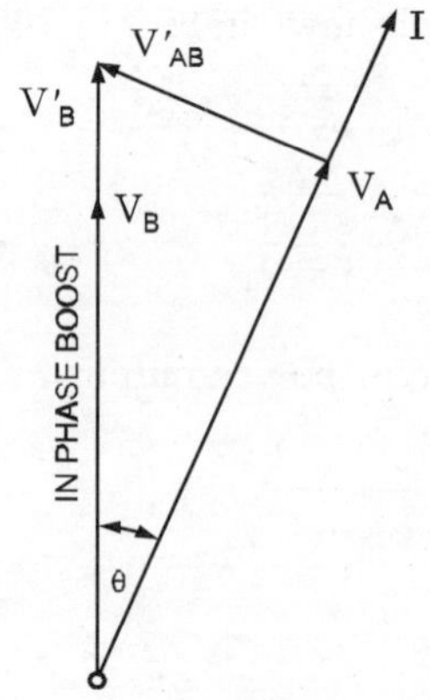

Fig. 13.18. ***Inphase Voltage Boost***

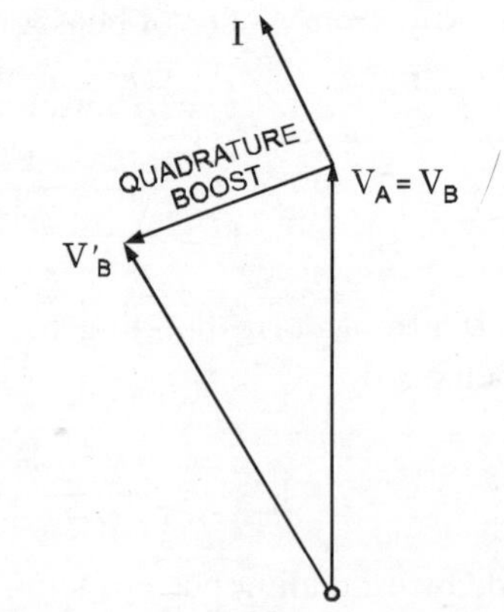

Fig. 13.19. ***Quadrature Voltage Boost***

To study the effect of quadrature voltage boost of station B having surplus available capacity on the power transfer between the generating stations consider phasor diagram shown in Fig. 13.19. The voltage of generating stations B, V_B is increased to V'_B by the quadrature boost $V_BV'_B$, the voltage drop in the interconnector is V'_BV_B and the current through the interconnector is in phase with V_B and the power is transferred from generating station B to generating station A.

Thus, we see that quadrature voltage boost enables the real power to be transferred between power stations without changing the rotor angular displacement.

From the above discussion we conclude that

(*i*) For stable operation of the interconnected power stations, it is necessary that the two stations must be interconnected through a reactor but undue increase in the value of reactance of the interconnector must not be allowed otherwise a large phase displacement between the stations will be required in order that a substantial amount of current may flow through the interconnector and large phase displacement would mean unstable operation *i.e.*, the stations would tend to fall out of steps.

(*ii*) In phase voltage boost will affect the transfer of wattless current through the interconnector and will be of little help in transfer of real power between the stations.

(*iii*) Quadrature voltage boost will affect the transfer of real power between power stations. This method results in increase in inherent stability of the system and sudden loads can be dealt with easily.

13.4.2. Power Limit of Interconnectors

We have seen that power transmitted through the interconnector depends upon the phase displacement between the voltage of two generating stations and will be maximum when the phase angle between the two voltages is 90°. If the phase displacement exceeds 90°; the increase in current is more than counterbalanced by the fall in power factor. Let the station B transmit power to station A and phase displacement between the two generating station voltages be 90°.

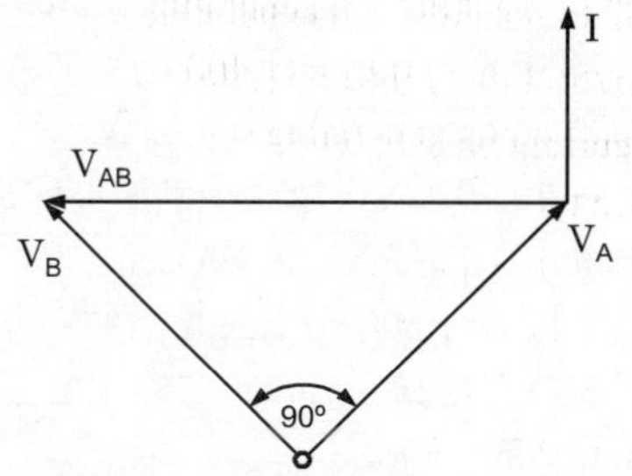

Fig. 13.20

Under these conditions current I lags behind the voltage V_B by 45°, and leads the voltage V_A by 45°, the former station being at the sending end and the latter being at the receiving end and

Maximum power transferred,

$$P_{max} = VI \cos 45^\circ$$

$$= V \times \frac{\sqrt{2}\,V}{X} \times \frac{1}{\sqrt{2}} = \frac{V^2}{X} \text{ watts/phase}$$

$$= \frac{V^2}{1{,}000\,X} \text{ kW/phase}$$

$$\because\ I = \frac{V_{AB}}{X} = \frac{\sqrt{2}\,V}{X}$$

The maximum power transferred is also referred to as the synchronous capacity of the interconnector and it is defined as the kW transmitted per radian of the displacement between the two voltages of the power stations.

13.4.3. Interconnectors in Parallel

It becomes sometimes necessary to connect two power stations S_1 and S_2 through two interconnectors in parallel running along the same or different routes. This situation arises when the power under transfer exceeds the capacity of a single interconnector. The load shared by two interconnectors will be in the inverse ratio of their impedances $\mathbf{Z}_1$ and $\mathbf{Z}_2$ as the voltages at the sending end and receiving end of the both the interconnectors are the same. The load carried by the two interconnectors can be different from this recourse if suitable voltage regulating equipment is provided in each line.

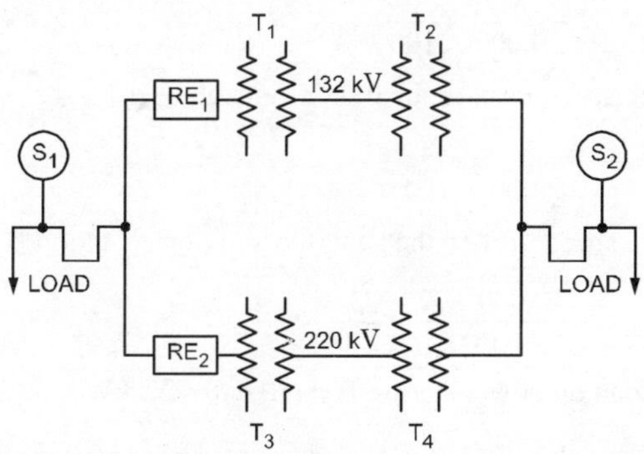

Fig. 13.21. ***Interconnection at Two Different Voltages***

The problem of correct subdivision of load becomes complicated when the two interconnectors connecting two power stations operate at different voltages, say 132 kV and 220 kV. The conditions are diagrammatically represented in Fig. 13.21 wherein RE is the voltage regulating equipment (say tap-changing transformer or induction regulator), S_1 and S_2 are the interconnected power stations and T_1, T_2, T_3, and T_4 are power transformers of suitable rating and turn-ratios.

As regard load division, the effect of transformers can be taken into account simply by adding their resistances and reactances to the impedance (resistance and reactance) of the concerned interconnector. The equivalent system to the receiving-end voltage can be represented as in Fig. 13.22, where T is a common tap changing transformer and IR_1 and IR_2 are the induction regulators. Receiving-end current I_R is fixed and its components are I_1 and I_2 so that $\mathbf{I}_R = \mathbf{I}_1 + \mathbf{I}_2$. The arithmetic sum of I_1 and I_2 will be minimum when they are in phase. I_1 and I_2 are assumed to be the currents which the two lines are designed to carry on the basis of correct division of load between them.

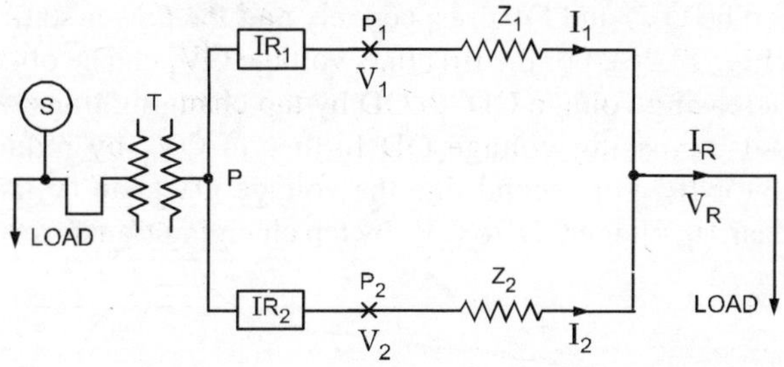

Fig. 13.22. ***Equivalent System For System Shown in Fig. 13.21***

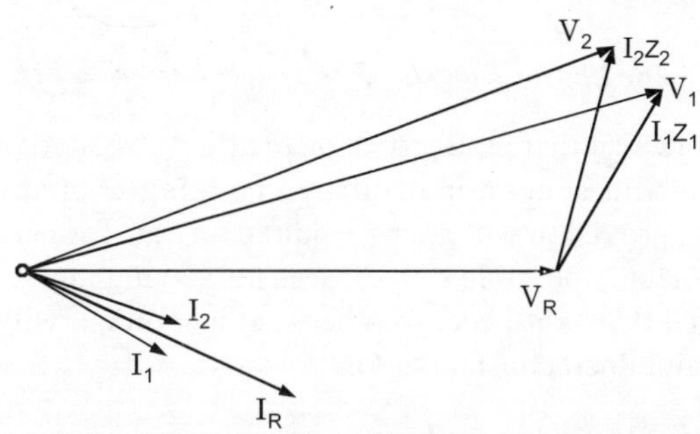

Fig. 13.23. ***Voltage Phasor Diagram at P_1 and P_2***

The voltage at P_1 and P_2 are given by phasors OV_1 (phasor sum of V_R and I_1Z_1) and OV_2 (phasor sum of V_R and I_2Z_2) respectively (Fig. 13.23), which can be possibly obtained by means of one common tap changing transformer T and a separate induction regulator for each line. The voltage at P then corresponds to OD or OD′ (Fig. 13.24) and voltage at P_1 is the phasor sum of OD (or OD′) and DV_1 (or $D'V_1$) the latter having been obtained by means of the induction regulator IR_1. Similarly it can be shown that voltage required at P_2 is OV_2.

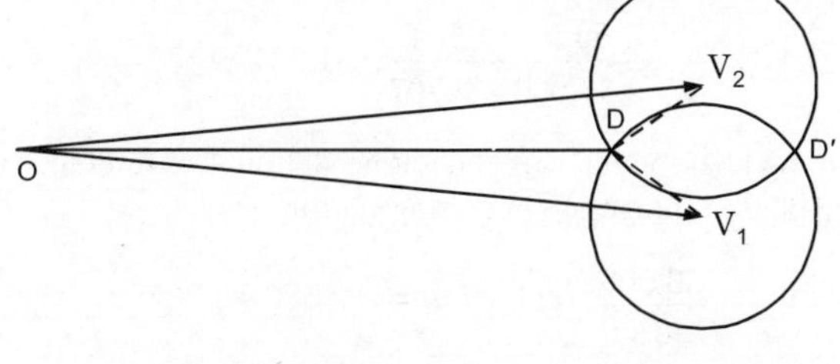

Fig. 13.24. ***Application of Induction Regulator***

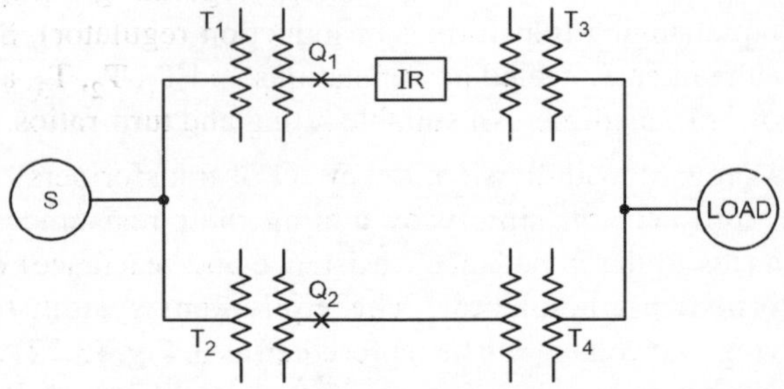

Fig. 13.25

The same results can be obtained by using two separate tap changing transformers T_1 and T_2 –one in each line and one induction regulator IR in one of the lines, as shown in Fig. 13.25. Let the desired voltages at Q_1 and Q_2 for proper division of load be OV_1 and OV_2 respectively and the power station be OD (Fig. 13.26). For the first line, voltage OV_1 can be obtained by increasing voltage OD′ to OD by tap changing transformer T_1 and increasing voltage OD further to OV_1 by induction regulator IR. For second line the voltage OV_2 can be had by increasing voltage OD′ to OV_2 by tap changing transformer T_2.

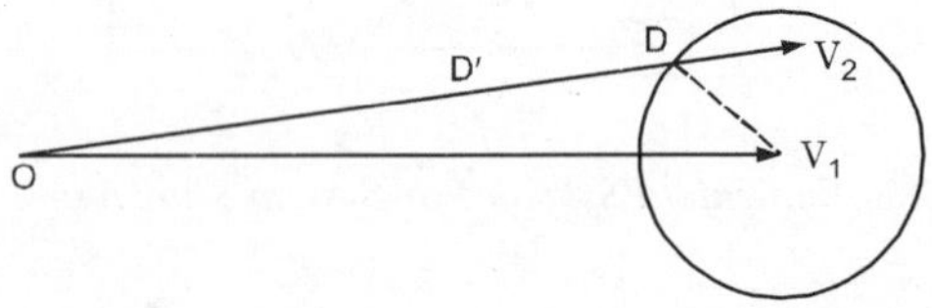

Fig. 13.26. ***Phasor Diagram For System Shown in Fig. 13.25***

If however there is a provision to affect a change from OD′ to OD in the generator itself by changing excitation, tap-changing provision will not be required on transformer T_1. This arrangement is possible only if S_1 is not having any local load connected across it because otherwise its voltage will require to be maintained constant at OD′.

Example 13.10. Two 3-phase generating stations A and B are linked together through 33 kV interconnector having a resistance and a reactance of 0.8 and 4.0 ohms respectively per phase. The load on the generators at station A is 80 MW at a pf of 0.8 lagging and the local load taken by the consumers connected directly to the bus-bars is 50 MW at a pf of 0.707 lagging. Calculate the load in kW received from station A by station B, its power factor and the phase difference between the voltages of A and B.

Solution : Load current of machines on generating station A,

$$I_A = \frac{80 \times 10^6}{\sqrt{3} \times 33{,}000 \times 0.8} = 1{,}750 \text{ A}$$

Local load current on generating station A,

$$I_C = \frac{50 \times 10^6}{\sqrt{3} \times 33{,}000 \times 0.707} = 1{,}237\text{A}$$

Taking phase voltage of generating station A as reference phasor, we have phase voltage of generating station A,

$$\mathbf{V}_A = \frac{33{,}000}{\sqrt{3}} = (1 + j\,0) = (19{,}050 + j\,0) \text{ V}$$

Load current of machines on generating station A,

$$\mathbf{I}_A = 1{,}750(0.8 - j\,0.6) = (1{,}400 - j\,1{,}050) \text{ amperes}$$

Local load current on generating station A,

$$\mathbf{I}_C = 1{,}237(0.707 - j\,0.707) = (875 - j\,875) \text{ amperes}$$

Current transferred through interconnector,

$$\mathbf{I}_B = \mathbf{I}_A - \mathbf{I}_C = (1{,}400 - j\,1{,}050) - (875 - j\,875)$$
$$= (525 - j\,175) = 553.4 \text{ A}$$

Voltage drop in interconnector

$$= \mathbf{I}_B\,(0.8 + j\,4) = (525 - j\,175)\,(0.8 + j\,4)$$
$$= (1{,}120 + j\,1{,}960) \text{ volts}$$

Voltage of generating station B,

$$\mathbf{V}_B = \mathbf{V}_A - \mathbf{I}_B\,(0.8 + j\,4) = 19{,}050 - (1{,}120 + j\,1{,}960)$$
$$= (17{,}930 - j\,1{,}960) = 18{,}040 \text{ volts}$$

Phase difference between V_B and I_B

$$= \text{Tan}^{-1}\frac{-1{,}960}{17{,}930} - \text{Tan}^{-1}\frac{-175}{525}$$

$$= -6° \,14' - (-18° \,26') = 12° \,12' \text{ with } I_B \text{ lagging}$$

Power factor of load received by station B

$= \text{Cos } 12° \,12' = 0.977$ (lagging) **Ans.**

Power received by station B

$$= 3\,V_B I_B \cos\phi$$

$$= \frac{3 \times 18{,}040 \times 553.4 \times 0.977}{10^6} = 29.26 \text{ MW} \textbf{ Ans.}$$

Phase angle between voltages of A and B

$= 6° \,14'$ with V_B (lagging) **Ans.**

Example 13.11. An induction motor-synchronous generator links two power stations A and B, the capacities of which are 20,000 kW and 10,000 kW, with speed regulation of 2% and 3% respectively. The induction motor connected to A's bus-bars has a 2,000 kW rating and a full-load slip of 4%. Find the load on the interconnector when there is load of 10,000 kW on B's bus bar due to its own consumers. A have no external load.

Solution : Let the load on the interconnector be X kW

Load sent by power station A to station B through motor-generator set = X kW

% age speed drop in the generators of power station A

$$= \frac{X}{20{,}000} \times \frac{2}{100}$$

% age speed drop in the motor-generator set

$$= \frac{X}{2{,}000} \times \frac{4}{100}$$

% age speed drop in the generators of power station B

$$= \frac{10{,}000 - X}{10{,}000} \times \frac{3}{100}$$

∵ Load on power station B = (10,000 – X) kW

The speed drop in the generators of power station A fixes the frequency of the motor of the motor-generator set. The frequency of power station B will be modified by the drop in speed of the motor-generator set. Since the synchronous generator of motor-generator set and the generators of power station B must have the same frequency, therefore drop in speed of the generators of power station B must be equal to combined drop of speeds of power station A's generators and that of the motor-generator set.

$$\therefore \quad \frac{10{,}000 - X}{10{,}000} \times \frac{3}{100} = \frac{X}{20{,}000} \times \frac{2}{100} + \frac{X}{2{,}000} \times \frac{4}{100}$$

$$\text{or } X = 1{,}250 \text{ kW } \textbf{Ans.}$$

Example 13.12. Two stations A and B are interconnected by an interconnector of reactance 4 Ω. The station A supplies a load of 10,000 kW at a pf 0.8 lagging through its own bus-bar, while station B has a load of 8,000 kW at a pf 0.9 lagging. The loads between the stations are equalised by flow of power in the interconnector. Calculate the power transmitted and the pf of each station. The bus-bar voltage at station A and B is 11 kV.

Solution : Since the load on two stations are equalised, therefore, each station will supply half of the total load *i.e.*, $\frac{10{,}000 + 8{,}000}{2}$ *i.e.*, 9,000 kW and power of 1,000 kW will be transferred from station B to station A through the interconnector.

Taking phase voltage of generating station B as reference phasor, we have phase voltage of generating station B, $\mathbf{V}_B = \frac{11{,}000}{\sqrt{3}}(1 + j0)$

$= 6{,}350 \angle 0°$ volts.

The phase voltage of generating station A will be equal to that of station B in magnitude but displaced by angle θ.

$\therefore \; \mathbf{V}_A = 6{,}350 \angle{-\theta}$ volts

The load current at stations A and B will lag behind their respective bus-bar voltages by angle $\phi_1 = \cos^{-1} 0.8 = 36.8°$ and $\phi_2 = \cos^{-1} 0.9 = 25.8°$, as shown in the phasor diagram [Fig. 13.27(*b*)].

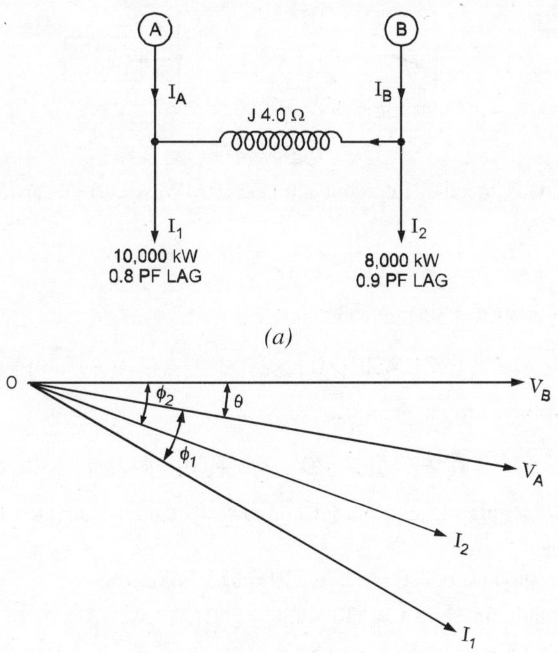

(b) Phasor Diagram

Fig. 13.27

$$\mathbf{I}_1 = \frac{10{,}000 \times 1{,}000}{\sqrt{3} \times 11{,}000 \times 0.8} = 656\angle{-(36.8 + \theta)°} \text{ A}$$

$$\mathbf{I}_2 = \frac{8{,}000 \times 1{,}000}{\sqrt{3} \times 11{,}000 \times 0.9} = 467\angle{-25.8°} \text{ A}$$

Current in interconnector,

$$\mathbf{I}_C = \frac{\mathbf{V}_B - \mathbf{V}_A}{4.0 \angle 90°}$$

$$= \frac{6{,}350 - 6{,}350\angle{-\theta}}{4 \angle 90°}$$

$$= 1{,}587.5\angle{-90°} - 1{,}587.5\angle{-(90 + \theta)°}$$

Total current supplied by station B,

$$\mathbf{I}_B = \mathbf{I}_2 + \mathbf{I}_C$$
$$= 467\angle{-25.8°} + 1{,}587.5\angle{-90°} - 1{,}587.5\angle{-(90° + \theta)}$$
$$= (420.3 - j203.5) - j1{,}587.5 + 1{,}587.5 \sin\theta + j1{,}587.5 \cos\theta$$
$$= (420.3 + 1{,}587.5 \sin\theta) + j(1{,}587.5 \cos\theta - j1{,}791)$$

Real component of current = (420.3 + 1,587.5 sin θ)

Power generated at station B

$$= \sqrt{3} \times 11{,}000\,(420.3 + 1{,}587.5 \sin\theta) = 9{,}000 \times 1{,}000$$

$$\text{or } 420.3 + 1{,}587.5 \sin\theta = \frac{9{,}000 \times 1{,}000}{\sqrt{3} \times 11{,}000} = 472.37$$

$$\text{or } \sin\theta = \frac{472.37 - 420.3}{1{,}587.5} = 0.0328$$

$$\text{or } \theta = \sin^{-1} 0.0328 = 1.9°$$

Total current supplied at station B,

$$\mathbf{I}_B = (420.3 + 1{,}587.5 \times 0.0328) + j(1{,}587.5 \cos 1.9° - 1{,}791)$$
$$= 472.37 + j\,(1{,}587.5 \times 0.9995 - 1{,}791) = 472.37 - j204.3$$

Phase angle of station B,

$$\phi_B = \tan^{-1} \frac{-204.3}{472.37} = \tan^{-1} - 0.4325 = -23.4°$$

Power factor of station B,

$$\cos\phi_B = \cos(-23.4°) = 0.9178 \text{ (lag) } \textbf{Ans.}$$

Total current supplied at station A,

$$\mathbf{I}_A = \mathbf{I}_1 - \mathbf{I}_C$$
$$= 656\angle{-(36.8 + 1.9)°} - 1{,}587.5\angle{-90°} + 1{,}587.5\angle{-(90 + 1.9)°}$$
$$= 656\angle{-38.7°} - 1{,}587.5\angle{-90°} + 1.587.5\angle{-91.9°}$$
$$= 511.6 - j410.6 + j1{,}587.5 - 52.08 - j1{,}586.7$$
$$= 459.52 - j\,409.8$$

Phase angle of station A

$$= \tan^{-1} \frac{-409.8}{459.52} = -41.73°$$

Power factor of station A,

$$\cos\phi_A = \cos(-41.73°) = 0.746 \text{ (lag) } \textbf{Ans.}$$

Example 13.13. Two generating stations A and B operate in parallel linked by an 11 kV interconnector having an impedance of (2.7 + *j* 1.5) ohms per phase. The supply voltage at both the stations are in phase. Determine the boosting voltage required at station A when a load of 2.5 MW at unity pf is taken at station B.

Solution : Since the voltage on both the stations is 11 kV

$$\therefore \quad \mathbf{V}_A = \mathbf{V}_B = \frac{11{,}000}{\sqrt{3}} = 6{,}350 \text{ volts}$$

Load current, $I = \frac{2.5 \times 10^6}{\sqrt{3} \times 11,000 \times 1.0} = 131.2$ A

Voltage drop in interconnector

$= 131.2\,(2.7 + j1.5) = 405.3\ \angle 29.05°$ V/phase

Voltage boost required at station A (line to line)

$= \sqrt{3} \times 405.3 = 702$ V **Ans.**

Example 13.14. Two 11 kV generating stations A and B generating 25 MW each at unity pf are interconnected by an interconnector which has 30% voltage drop for a power transfer of 15 MW. The load at station A is 30 MW whereas that at station B is 20 MW both at unity power factor, calculate the phase angle between the bus-bar voltages of the two generating stations.

Solution : The rating of interconnector is 30% for a power transfer of 15 MW.

Actually a power transfer of 5 MW is taking place from generating station B to A. Hence the percentage voltage drop for transfer of power of 5 MW at unity power factor $= \frac{5}{15} \times 30 = 10\%$

Voltage drop in the interconnector $= \frac{10}{100} \times 11,000 = 1,100$ volts

Let the displacement between E_A and E_B be θ.

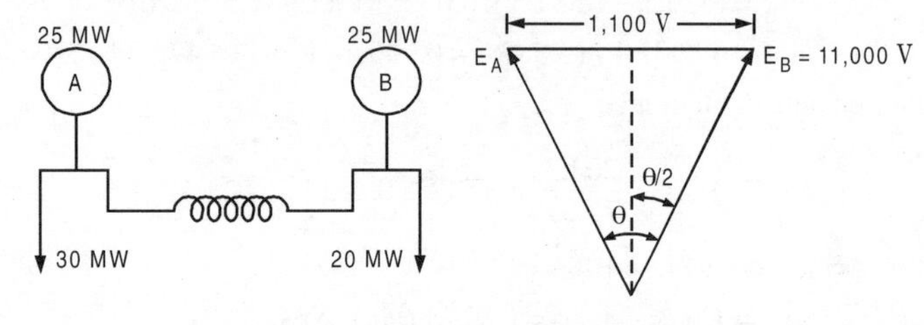

(a) (b)

Fig. 13.28

Now from Δ shown in Fig. 13.28 (*b*)

$$\text{Sin}\ \frac{\theta}{2} = \frac{1,100/2}{E_B}$$

$$\text{or Sin}\ \frac{\theta}{2} = \frac{1,100/2}{11,000} = 0.05$$

or $\theta = 2 \sin^{-1} 0.05 = 5° 44'$ **Ans.**

Example 13.15. Two power stations A and B operate in parallel and are interconnected by a short transmission line. The station capacities are 10 MW and 5 MW respectively and the generating sets have uniform speed regulation (from no load to full load) of 2% and 4% respectively. Calculate the output of each station and the load of the interconnector when the load on each station bus-bar is 6 MW. [*Pb. Univ. Elec. Generation April 1982*]

Solution : Let the load on the interconnector be X MW

Load sent by power station A to B = X MW

Load on each power station = 6 MW

Output of power station A = (6 + X) MW

Output of power station B = (6 – X) MW

Percentage speed drop in the generators of power station A

$$= \frac{6 + X}{10} \times 2$$

Percentage speed drop in the generators of power station B

$$= \frac{6 - X}{5} \times 4$$

The speed drop in the generators of both the power stations A and B should be same

$$i.e., \quad \frac{6 + X}{10} \times 2 = \frac{6 - X}{5} \times 4$$

or X = 3.6 MW **Ans.**

Output of power station A = 6 + 3.6 = 9.6 MW **Ans.**

Output of power station B = 6 – 3.6 = 2.4 MW **Ans.**

Example 13.16. A load of 10 MW at unity power factor and 33 kV is supplied from a power station by two 3-phase interconnectors A and B having impedances per phase of (3 + *j* 4) W and (5 + *j* 6) W respectively. Determine the necessary boost so that interconnectors A and B transfer load of 4 MW and 6 MW respectively.

Solution : Receiving-end voltage, $V_R = \frac{33,000}{\sqrt{3}} = 19,050$ V

Load carried by interconnector A is 4 MW

so the current carried by it, $\mathbf{I}_A = \frac{4 \times 10^6}{\sqrt{3} \times 33,000 \times 1.0}$

$= 70\ \angle 0°$ A

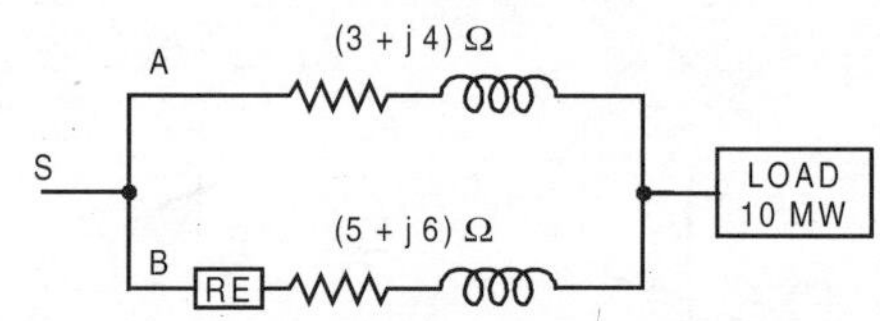

Fig. 13.29

Load carried by interconnector B is 6 MW, so current carried by it,

$$\mathbf{I}_B = \frac{6 \times 10^6}{\sqrt{3} \times 33,000 \times 1.0} = 105\ \angle 0°\ \text{A}$$

Voltage drop in connector A

$= \mathbf{I}_A\,\mathbf{Z}_A = 70\ \angle 0° \times (3 + j\,4) = (210 + j\,280)$ V/phase

Voltage drop in connector B

$= \mathbf{I}_B\,\mathbf{Z}_B = 105\ \angle 0° \times (5 + j\,6) = (525 + j\,630)$ V/phase

So regulating equipment connected in interconnector B must provide

Inphase boost = 525 – 210 = 315 V/phase

Quadrature boost = 630 – 280 = 350 V/phase

So necessary boost per phase

$= (315 + j\,350) = 470.9\ \angle 48°$ V **Ans.**

Voltage to be maintained at sending end

$= 19,050 + \mathbf{I}_A\,\mathbf{Z}_A$

$= 19,050 + 210 + j\,280$

$= 19,260 + j\,280 = 19,262\ \angle 0.833°$ V/phase

13.5. VOLTAGE AND REACTIVE POWER CONTROL

Voltage control and reactive power control are interrelated and need to be therefore considered together. One of the most troublesome features associated with the operation of overhead transmission system is the inherent variation of voltage at the receiving end, due to variations in load. These fluctuations of voltage are to be kept within the reasonable limits fixed by IE rules (Rule No. 54).

When power is supplied to a load through a transmission line keeping sending-end voltage constant, the receiving-end (or load) voltage undergoes variations depending upon the magnitude of the load and power factor of the load. The higher the load with smaller power factor the greater is the voltage variation. The voltage variation at a load is an indication of the unbalance between the reactive power generated and absorbed by that load. When the reactive power generated exceeds the reactive power absorbed, the voltage goes up and vice versa.

This problem is illustrated below:

Let P be the power to be transferred per phase in watts, Q the reactive power to be transferred per phase in VARs, V_S the sending-end voltage per phase, V_R the receiving-end voltage per phase and (R + *j*X) be the impedance of transmission line per phase in ohms.

$$\Delta V = V_S - V_R \quad \text{... drop in line voltage}$$

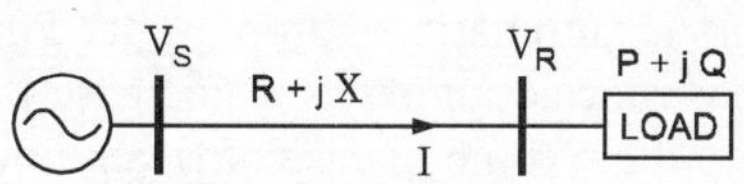

Fig. 13.30

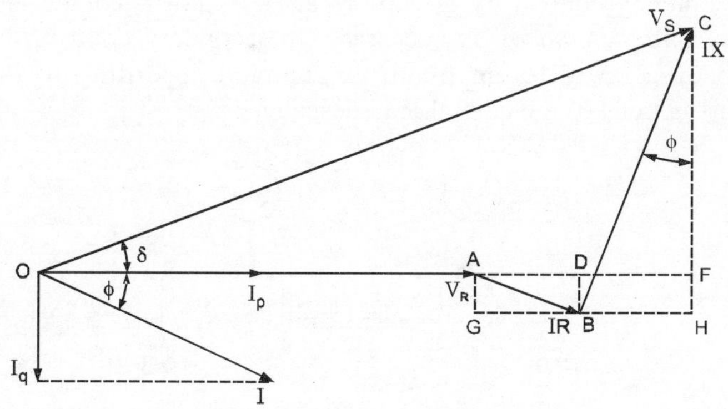

Fig. 13.31. ***Phasor Diagram***

From phasor diagram shown in Fig. 13.31, we have

$$V_S^2 = (V_R + AF)^2 + CF^2$$

$$= (V_R + AD + DF)^2 + (CH - HF)^2$$

$$= (V_R + IR\cos\phi + IX\sin\phi)^2 + (IX\cos\phi - IR\sin\phi)^2$$

$$\text{or } V_S^2 = \left(V_R + \frac{PR}{V_R} + \frac{XQ}{V_R}\right)^2 + \left(\frac{PX}{V_R} - \frac{QR}{V_R}\right)^2$$

$$\because P = V_R I\cos\phi \text{ and } Q = V_R I\sin\phi$$

Neglecting $\left(\frac{PX}{V_R} - \frac{QR}{V_R}\right)$ in comparison with $V_R + \frac{PR}{V_R} + \frac{XQ}{V_R}$ we have

$$V_S = V_R + \frac{PR}{V_R} + \frac{XQ}{V_R}$$

$$\text{or } \Delta V = V_S - V_R = \frac{PR}{V_R} + \frac{XQ}{V_R} \quad \text{...(13.17)}$$

If resistance R is negligible in comparison to X, then

$$\Delta V = V_S - V_R = \frac{XQ}{V_R} \quad \text{...(13.18)}$$

Hence line voltage drop depends mainly on the flow of reactive power Q.

The power angle δ between V_R and V_S is proportional to CF *i.e.*, $\delta \propto \frac{PX - QR}{V_R} \propto \frac{PX}{V_R}$ neglecting R in comparison with X.

The power angle δ depends mainly on the flow of true power.

Thus in order to keep the receiving-end voltage V_R constant for a given sending-end voltage V_S, the drop $\frac{Q}{V_R}X$ must be kept constant. Since in this the only variable quantity is Q, it is this reactive VARs which must be locally adjusted to keep this quantity fixed. The local generation of VARs can be obtained by connecting shunt capacitors or synchronous capacitors and or shunt inductors (for light loads or capacitive loads).

Alternatively the receiving-end voltage V_R can also be kept constant for a given sending-end voltage V_S by keeping the product QX constant. This is achieved by introducing series capacitors which will reduce the net reactance of the system. Since the voltage variation will be more for larger loads (larger reactive power), the variation could be controlled by switching-in suitable series capacitors.

13.5.1. Methods of Voltage Control

The methods usually employed for voltage control are by use of (*i*) shunt capacitors and reactors (*ii*) series capacitors (*iii*) synchronous capacitors (*iv*) tap-changing transformers and (*v*) booster transformers. The first three methods could also be categorised as reactive VAR injection methods.

Earlier the voltage control used to be affected by adjusting the excitation of the generator (alternator) at the generating (or sending) end. The larger the reactive power required by the load the more is the excitation to be provided at the generating end. This method proved useful in small isolated system where there was no local load at the generating end. Also there are limits for the excitation as well. Excitation below a certain limit may cause instability of the system in case the machine is connected to a synchronous load. The excitation above a certain level will result in overheating of the rotor. Thus the voltage regulation by excitation control is restricted by the permissible voltage rise at the sending end and by the difficulty of designing efficient generating plant for a wide range of excitation.

1. Shunt Capacitors and Reactors. Shunt capacitors are employed across an inductive load where as reactors are employed across capacitive loads or lightly loaded line. In both cases the effect is to supply the requisite reactive power to maintain receiving-end voltage constant (shunt capacitors across an inductive load supply part of the reactive VARs required by the load whereas shunt reactors connected across capacitive loads or lightly loaded lines absorb some of the lagging VARs). Capacitors are connected either directly to a bus-bar or through a tertiary winding of the main transformer and are arranged along the route to minimize the losses and voltage drops. The drawback of this method of voltage control is that as the voltage falls the correction VARs also falls *i.e.,* when it is most needed, its effectiveness falls. Similarly, on light loads when the corrective VARs requirement is comparatively less, the capacitor output is large.

2. Series Capacitors. Series capacitors are connected in series with the line conductors, as illustrated in Fig. 13.32. They reduce the effect of inductive reactance between the sending end and receiving end of the line. One drawback of this method is that high voltage is produced across the capacitor terminals when short circuit current flows through them. Therefore special protective devices (such as spark gaps with high speed contactors) are used for the protection of the capacitors under such condition.

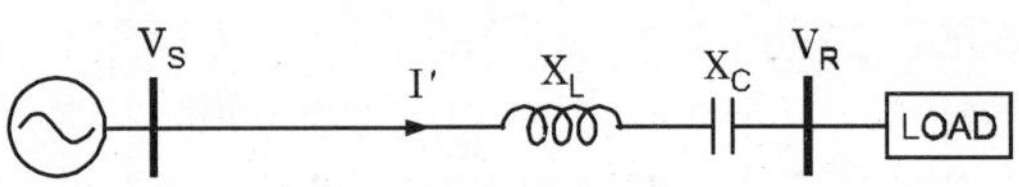

Fig. 13.32

If the load VAR requirement is small, series capacitors are of little use. With series capacitors the reduction in line current is small. Hence if thermal considerations limit the current, little advantage is obtained and shunt compensation should be employed. The shunt capacitors improves the power factor of the system whereas the series capacitors have little effect on the power factor.

For long transmission lines where the total reactance is high, series capacitors are effective in improving stability of the system.

Synchronous phase modifiers will be discussed in part II (Transmission and Distribution) of the book and tap changing transformers and booster transformers will be discussed in part III (Switchgear and Protection).

13.6. LOAD-FREQUENCY CONTROL

In a power system, both active and reactive power demands continually vary with the rising or falling trend. Power input (steam input to turbo-generators or water input to hydro-generators) must, therefore, be continuously regulated to match the active power demand, otherwise the machine speed will change with consequent change in frequency, which may be highly undesirable. Also the excitation of generators must be continuously regulated to match the reactive power demand with reactive generation, failing which the voltages at various system buses may go beyond the prescribed limits.

It is necessary to maintain the frequency of the power system constant (maximum permissible variation in supply frequency is ±0.5 Hz). The reasons for it are as follows:

1. The frequency control keeps the balance between generation and absorption of real power and thus makes the operation of power station in parallel satisfactory.
2. The speed of synchronous and induction motors, extensively used in industry as prime movers, depends upon supply frequency (the synchronous speed being equal to $120\,f/P$ where f is supply frequency and P is the number of poles) and so change in supply frequency causes variations in speed of motors of consumers– not desirable for consumers, particularly the process industries depending on constant speed drives.
3. The variation in supply frequency beyond permissible limits also affects the performance of electric motors.
4. The extensive use of synchronous clocks establishes a strong requirement for maintaining supply frequency constant in order to have correct timing from such clocks.

When the load on a generator or a group of generators increases, the rotors slow down resulting in reduction in frequency. However, the governors adjust the input so as to bring the frequency to the original level. This control of frequency by the action of governors is called the *primary control.* The action of the governors is automatic. A drop in speed due to increased load causes governor action so as to increase the input (admit more steam into the turbine in case of steam power plants) and thus output. In the event of loss of load or sudden change in load, the governor controls the speed of generators. However, frequency control by governors alone is not adequate and '*secondary control*' is necessary. In secondary control, the loading on different plants is changed according to the instructions of the central load despatcher.

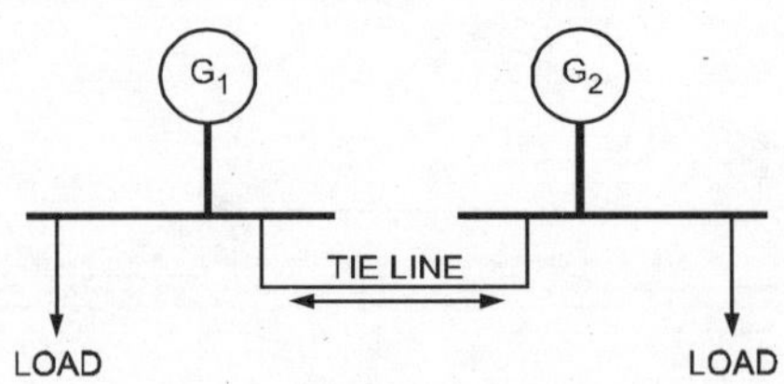

Fig. 13.33

Load and frequency control of interconnected generators introduces problems which are relatively simple in a system having one or two generating stations but are more difficult and complex in large interconnected systems with many stations scattered over a wide area.

Methods commonly employed for load- frequency control in interconnected power systems are explained below :

Very small isolated generating stations can have *manual control* of frequency. The governors adjust the input to bring the frequency within permissible limits.

1. Flat Frequency Control. Consider two stations A and B operating in parallel and interconnected by a tie-line.

The frequency of the system is maintained constant by regulating only one station A and without any regulation of station B. If the load at either station A, station B or both changes, the generation at station A is changed to maintain a balance between generation and load on the system. This type of frequency control is called *flat frequency control*.

The drawback of this method of frequency control is that station A must have enough capacity to absorb the load variations for the entire system. Secondly the tie-line between the two stations would have to absorb all load variations at station B as the generation at station B is maintained constant. These may make the operation of generator at station A uneconomical and could result in certain limitations in the operation of the system.

2. Parallel Frequency Control. In this method of frequency control both stations A and B would be regulated simultaneously to maintain frequency constant. By this method, the load swings are shared by both stations A and B and therefore, the swings on each station and on the line would be reduced.

3. Flat Tie-Line Control. In this method of frequency control the increase in load of an area is met by increasing the generation in that area and thus power flow in the tie-line is kept constant irrespective of load demands. This method is used when a small system and a large system are interconnected through a tie-line. The large system maintains the system frequency constant while the small system is controlled to keep the tie-line power constant. This method is not suitable when two or more large systems are interconnected as in such cases and with this type of control the tie-line power and frequency deviation have a tendency to swing back and forth (in addition to the swings at tie-line natural frequency) following a load change. The control equipment used consists of the frequency controller only at the larger system and tie-line power controller-recorder at the smaller system.

4. Tie-line Load Bias Control. This is the modification of (1) and (3) in that the system is allowed to follow its normal regulating characteristic directed toward holding normal frequency. This is the most widely used method on large interconnections. All power systems assist in regulating frequency and tie-line power flow regardless of where from the frequency variation orginates. The amount of assistance given by any one system is controlled by simple adjustment of the control equipment. The control equipment consists of load frequency controller and tie-line load recorder-controller. The tie-line instrument biases the frequency controller till a desired relationship between tie-line loading and system frequency is had.

Automatic Load-Frequency Control. In modern large interconnected systems, manual operation is not feasible as continuous watch on the frequency and loadings of various generators and tie-lines has to be kept by the operators and they have to continuously make adjustments. The manual controls are sluggish and involve inherent human time lags. Obviously automatic control is more efficient method of load-frequency control.

The problem of automatic load-frequency control resolves itself into (*i*) the measurement of a quantity (*ii*) interpretation of the measurement in terms of deviation from a control point, and (*iii*) the application of corrections to restore the measured quantity to its normal value. In some cases more than one measurement is needed for proper operation of the control equipment. In one of these control systems, developed by Leeds and Northrup Co, both load and frequency are measured to give various types of combined load-frequency control.

The loads on various generators, stations and systems are measured through the summation of various thermal-converter millivolt outputs and frequency is measured by a frequency-bridge type instrument. As the system frequency changes, the bridge circuit is rebalanced by the instrument movement which positions a slide wire used for transmission of a direct voltage. All these datas are fed into master controller, it is able to detect the need for more or less generation and to send impulses to the different stations calling for load increase or decrease.

By the use of area requirement, proportional load control, the equipment is able to call for changes at the several generating stations such that they each, in effect, supply the load of their respective areas, thereby causing a minimum of power flow over the tie feeders from one station to another.

Within each generating station it is possible to select the units that will be employed for regulation and for adjustment of the percentage of the required load change that is placed on each machine.

The following three types of area control may be used, one at a time, as selected by the system operator.

Flat frequency control varies the power input to the prime mover in order to correct the system frequency to the pre-determined value.

Flat tie-line control varies the power input to the prime mover so that the tie-line load is corrected to a predetermined schedule. In this method another system is required to maintain frequency constant.

Tie-line bias control is a modification of the above two types of area control and has been found to work very satisfactorily on power systems where a large number of generators and stations are required to be kept under control. If the system frequency variation is more than a predetermined value, say 1/4-1/2 Hz, the control can be made to change from automatic to "manual" automatically and sound an alarm, so the system operator can correct the faulty condition.

13.7. LOAD DISPATCHING

The operation of a modern power system has become very complex. Frequency and voltage is to be maintained within limits in addition to ensuring reliability of power supply and for maintaining the frequency and voltage within limits it is essential to match the generation of active and reactive power with the load demand. For ensuring reliability of power supply it is necessary to put additional generation capacity into the system in the event of outage of generating equipment at some station. Over and above this, it is also necessary to ensure the cost of electric supply to the minimum.

In large power systems, a central load dispatcher is necessary to assign loads to various stations and units in accordance with a predetermined schedule, modified from time to time as the actual load differs from predicted load or as emergencies arise owing to loss of generating units or tie-lines. The load dispatching may be by telephone, remote telemetering and signalling, or both.

Load assignment to a particular station varies with the type and function of the station and its relation to the system. Frequency control is sometimes assigned to one of the largest generating station, station feeding local load and a tie load carry their local load plus or minus a schedule tie load, whereas stations with no local load follow a schedule tie-line load.

In a manual load dispatch system, the control engineer at the area control centre obtains the telemetered information, calculates the station outputs for different generating stations and advises the different stations to increase or decrease generation. With systems becoming more and more complex, it is very difficult to satisfy all the system constraints with manual control, and it has become therefore, necessary to make use of on-line computing techniques. The use of on-line computers for load dispatching is increasing day by day. The on-line computer aided control system can be either open-loop or closed-loop type.

Up to date information is supplied to the control engineer by the open-loop systems so that the corrective action could be taken by the control engineer. The computer is kept up to date with information such as a basic load prediction, restrictions on plants, forecasted switching changes, synchronising time of sets, merit-order costs of plants and power outputs of individual sets. By reading the present state of the systems, the computer calculates outputs and line flows, and provides a statement of the expected load on each set in half an hour's time. The outputs of the sets are adjusted manually in accordance with the findings of the computer with respect to economy and system security *i.e.,* computer is employed simply as a tool for high speed security checking and load prediction.

The closed-loop on line computer systems performs all the above functions of data storage and security checking and in addition take corrective action. The closed-loop operation permits a great variety of emergency conditions to be catered successfully and more efficiently. Automatic control involves the provision of a machine controller connected to the governor of each turbo-generator unit. Thus outputs of the different generating units are altered in compliance with requirements (to meet the demand with maximum economy subject to system security). Examples of closed loop applications include automatic starting, loading and shut down of generators, automatic switching at substations.

The total interconnected ac network, called the *National grid,* is controlled from National load centre, called the National load dispatch centre. The National load control centre allocates (*i*) the MW generation to each regional grid depending upon the prevailing MW demand in that regional grid and required MW export/import from the regional grid (*ii*) amount of MW power through tie lines between neighbouring regional grids. The National grid operates at a common prevailing frequency (50 Hz in India).

Each regional load control centre controls load and frequency of its own by matching generation in various generating stations with total regional MW demand plus MW losses plus/minus amount of tie-line power flow.

The network controller installed in each regional load control centre is in communication with the national load control as well as with control rooms in generating stations and major substations in its zone through power line communication channels, microwave communication channels, telephone communication channel, Fax etc.

Load/generation controller installed in generating station control rooms ensures that the station frequency is within prescribed limits. The settings of the turbine input are adjusted by the station load controller automatically depending upon required generation allocated by the regional grid and the turbine governor of each generating unit operates to control the speed and frequency automatically.

The task of the load control centre is to keep the exchange of power between various zones (Electricity Boards or areas) and system frequency at desired values. Each zone may have its own load control centre for regulating the generating stations and loads in its own zone. The National load control centre regulates the exchange of power between different regional zones. The function is performed automatically by *network controller* installed in the load control centre.

Exercises

1. Discuss the advantages and limitations of interconnected system of power stations.
2. Two alternators having normal reactances in their armature circuits are connected in parallel. Will they successfully operate ? Explain your answer with the help of phasor diagrams.
3. Derive an expression for the synchronising power of two alternators operating in parallel in terms of their emf (E), synchronous reactance (X_s) and electrical angle of displacement (θ). Neglect the winding resistances and assume that alternators are not supplying any load.

   [U.P. Technical Univ. Power Station Practice 2007-08]
4. What is the effect of changing excitation of one of the machines when two alternators are running in parallel ?

   [Pb. Univ. Electrical Power II (Generation) Dec. 1993]
5. Explain, how the active and reactive power loading of an alternator working on infinite bus-bar is controlled.
6. Explain, with diagrams, the changes in the armature current and power factor of an alternator connected to constant voltage constant frequency bus-bars when the excitation is varied over a certain range and input to the prime mover is kept constant.
7. Explain the necessity of interconnecting reactor for successful parallel operation of two generating stations. Obtain an expression for the power limit of the interconnector.

   [U.P.S.C. I.E.S. Generation, Transmission & Distribution of Power 1972]

8. What are the necessary conditions under which power is transmitted through an interconnector between two power stations ?
9. Explain the method of affecting a power transfer between two stations of an interconnected system by phase angle control.
10. Describe the method of controlling (*i*) the active power (kW) and (*ii*) the reactive power (kVAR) between two interconnected power stations.
11. What are the advantages and disadvantages of interconnection of power stations? Explain the methods of affecting a power transfer between two stations. **[Agra Univ. 1979]**
12. Discuss the importance of interconnection in power system. Explain how reactive power can be controlled in such systems. **[A.M.I.E. Sec. B. Power Systems Summer 2000]**
13. Explain how the voltage of a load bus is maintained constant by injection of reactive power ? What are the sources of reactive power ? **[Pb. Univ. Power System Analysis May 1995]**
14. Explain in brief the various methods of load frequency control. **[Pb. Univ. Power System Analysis May 1995]**
15. What are the objectives of automatic generation control ? Discuss in brief how these objectives are met.

    [U.P.S.C. I.E.S. Electrical Engineering-II 1998]
16. Discuss automatic load dispatching in modern power systems.**[Pb. Univ. Power System Analysis January 1993]**
17. Describe briefly the principle of operation of load dispatch organisation coordinating different types of power plants in a power system.

    [A.M.I.E. Sec. B. Power Systems Summer 1993]
18. Write the importance of (*i*) load dispatch organisation and (*ii*) interconnection of power stations.

    [A.M.I.E. Sec. B. Power Systems Summer 1996]
19. What are the objectives of 'automatic generation control' ? Discuss in brief how these objectives are met.

    [U.P.S.C. I.E.S. Elec. Engineering II, 1998]
20. Write short notes on the following :
    (*i*) Interconnected system.
    (*ii*) Synchronous machines on infinite bus-bars.
    (*iii*) Economical loading of alternators.
    (*iv*) Load sharing and power transfer between stations.
    (*v*) Power limit of interconnectors.
    (*vi*) Load frequency control in interconnected power system. **[A.M.I.E. Sec. B. Winter 1993]**
    (*vii*) Load dispatching organisation.
    [A.M.I.E. Sec. B. Winter 1993]
    (*viii*) Automatic load dispatching. **[Pb. Univ. May 1995]**

Short Answer Type Questions With Answers

Q. 1. Why are alternators put in parallel ?

[Bharathidhasan Univ. Nov. 1996]

Ans. For high efficiency of operation, reliability, convenience and economy in maintenance and repair and possibility of additions to plant capacity with the growth in load on the power station.

Q. 2. What is meant by synchronising of alternators ?

[Manamanian Univ. Nov. 1996; Madras Univ. October 1995]

Ans. The process of connecting an alternator in parallel with another alternator or with the common bus-bars is called the synchronising.

Q. 3. What are the conditions for parallel operation of alternators ?

[Madras Univ. April 1997]

Ans. Terminal voltage, frequency and the phase of the incoming machine must be the same as that of the bus-bars (or other alternators already operating in parallel).

In case of 3-phase alternators, an additional requirement to be met is that the phase sequence of the incoming machine must be the same as that of the bus-bars.

Q. 4. What is the effect of varying excitation of an alternator running in parallel with another alternator ?

Ans. Variation in excitation of an alternator running in parallel with another alternator causes variation in wattless component and, therefore, of reactive kVA, without affecting the sharing of the kW load or power.

Q. 5. What is the effect of increasing the steam admission to the prime mover connected to an alternator ?

[Manamanian Sundaranar Univ. April 1996]

Ans. By increasing the steam admission to the prime mover of an alternator will load it further.

Q. 6. How can the distribution of load between two alternators operating in parallel be varied ?

Ans. Distribution of load between two alternators operating in parallel can be varied by adjusting the governor control. One prime mover governor is opened while the other is closed simultaneously so as to maintain the system frequency constant.

Q. 7. What is the effect of increasing the driving torque of the prime mover of one of the two alternators operating in parallel ?

Ans. By increasing the driving torque of the prime mover of one of the two alternators operating in parallel, it will be further loaded and other will be relieved of its load. If the output of the alternator, whose driving torque has been increased, becomes more than the total load supplied, then the other alternator will run as a synchronous motor.

Q. 8. What is meant by infinite bus-bars ?

[Madurai Kamarajar Univ. Nov. 1996]

Ans. Infinite bus-bars represent a system of large capacity whose frequency and the phase and magnitude of the voltage are not affected even if there is a variation of excitation or power of a synchronous machine connected to it.

Q. 9. Will an increase in the excitation of an alternator connected to an infinite bus increase its real power generation ?

[A.M.I.E. Sec. B. Summer 1993]

Ans. No.

Q. 10. How will the power angle and power factor of a 3-phase alternator supplying power to an infinite bus at constant excitation, change if the steam input is reduced ?

[A.M.I.E. Sec. B Advanced Elec. Machines Summer 1996]

Ans. Both power angle and power factor will decrease if the alternator was supplying load to the infinite bus at lagging power factor.

Q. 11. At what power angle a synchronous generator will develop maximum power ? **[Bharathidhasan Univ. Nov. 1995]**

Ans. Synchronous generator will develop maximum power when power angle $\delta = \theta$, internal angle where $\theta = \text{Tan}^{-1} \dfrac{X_s}{R_e}$.

Q. 12. What is the effect of changing the excitation of a synchronous machine connected to infinite bus-bars and supplying load at lagging pf ?

Ans. When the excitation of an alternator connected to infinite bus-bars and supplying load at lagging pf is increased, the alternator falls back (*i.e.,* load angle decreases), power factor further drops and armature current increases due to reduced power factor; active component being constant.

Q. 13. What is the effect of changing the torque of the prime mover of an alternator connected to infinite bus-bars ?

Ans. By increasing the torque of the prime mover of one alternator, it is further loaded and an equivalent load is removed from the other unit or units with which the machine is paralleled. If the output of the alternator, whose prime mover torque has been increased, become more than the total load being supplied, then the other machine or machines will operate as synchronous motor(s).

Q. 14. What is meant by the synchronous capacity of an interconnector ?

Ans. The maximum power transferred is referred to as the synchronous capacity of the interconnector and it is defined as the kW transmitted per radian of the displacement between the two voltages of the power stations.

Q. 15. What is the drawback of shunt capacitor/reactor voltage control ?

Ans. The drawback of this method of voltage control is that as the voltage falls the correction VARs also falls *i.e.*, when it is most needed, its effectiveness falls. Similarly, on light loads when the corrective VARs requirement is comparatively less, the capacitor output is large.

Q. 16. Why it is necessary to maintain the frequency of the power system constant ?

Ans. The reasons for maintaining the frequency of the power system constant are as follows :

1. The frequency control keeps the balance between generation and absorption of real power and thus makes the operation of power station in parallel satisfactory.
2. The speed of synchronous and induction motors, extensively used in industry as prime movers, depends upon supply frequency (the synchronous speed being equal to 120 f/P where f is supply frequency and P is the number of poles) and so change in supply frequency causes variations in speed of motors of consumers–not desirable for consumers, particularly the process industries depending on constant speed drives.
3. The variation in supply frequency beyond permissible limits also affects the performance of electric motors.
4. The extensive use of synchronous clocks establishes a strong requirement for maintaining supply frequency constant in order to have correct timing from such clocks.

Problems

1. A 2000 kVA, 3-phase, 8-pole alternator, runs at 750 rpm in parallel with other machines on 6,000 V bus-bars. Find synchronising power on full load at 0.8 pf (lagging) per mechanical degree of displacement and the corresponding synchronising torque. The synchronous reactance is 6 Ω/phase. [**Ans.** 500 kW; 648.6 kg-m]

2. Two single-phase generators operate in parallel on a load impedance of Z ohms. Their emfs are E_1 and E_2 and their impedances are Z_1 and Z_2. Deduce the terminal voltage in terms of these emfs and the impedances Z, Z_1 and Z_2.

   Determine the terminal voltage and the kW output of each machine if E_1 = 100 V; E_2 = 110 V; Z = 3 + j 4 and $Z_1 = Z_2$ = 0.2 + j1 ohms.

   [**Ans.** 96.10 V; 0.44 kW and 0.66 kW]

3. Two alternators running in parallel supply the following loads. 1,500 kW at pf 0.9 lagging, 1,000 kW at pf 0.8 lagging, 800 kW at pf unity and 600 kW at pf 0.8 leading.

   The load on one machine is adjusted to 2,000 kW at pf 0.95 lagging. Find the load pf of the other machine.

   [**Ans.** 1,900 kW at pf 0.98 (lagging)]

4. Two identical alternators rated 1,250 kVA at 0.8 pf supply a load in parallel. The setting of the prime mover of the first machine is such that it drops from 50 Hz on no load to 48 Hz at full load. The second machine drops from 50 Hz to 47 Hz under similar conditions. How will they share a load of 1500 kW ? What will be the frequency of the load ?

   [A.M.I.E. Sec. B. Power Systems Summer 1995]

   [**Ans.** 900 kW; 600 kW; 48.2 Hz]

5. The speed regulation of two 1,000 kW alternators running in parallel are 100% to 104% and 100% to 105% from full load to the no load respectively. How will the two alternators share a load of 1,200 kW and also find the load at which one machine ceases to supply any portion of the load. [**Ans.** 555.55 kW and 644.45 kW; 200 kW]

6. Two 1,000 kVA, 3-ϕ, alternators are operating in parallel at unity power factor. The governor of the prime mover of the first machine is so adjusted that its frequency falls uniformly from 50 Hz at no load to 48 Hz at full load. The corresponding figures for the second machine are 50 Hz and 48.5 Hz respectively. How will the alternators share a load of 1,500 kW ? [**Ans.** 642.86 kW, 857.14 kW]

7. Two 75 MW, 3-ϕ alternators are operating in parallel and are sharing equally a load of 150 MW corresponding to full load of each machine. The settings of the governors are such that the rise in speed from full load to no load is 2% in the first machine and 3% in the other machine. The speed-regulation characteristics may be assumed to be straight line in both the cases. Determine the load on each machine when the total load is 100 MW. [**Ans.** 45 MW, 55 MW]

8. Two generating stations A and B operating on 50 Hz are linked by an interconnector. The capacity of generating station A is 30 MW and it has a straight line speed regulation of 3% from no load to full load. The corresponding values for generating station B are 20 MW and 4% respectively.

   Find the loads on generators of A and B and on the interconnector when the load of generating stations A and B are 25 MW and 20 MW respectively.

   [**Ans.** 30 MW; 15 MW; 5 MW]

9. A-3 phase, star-connected turbo-alternator having a synchronous reactance of 10 Ω per phase and negligible armature resistance has an armature current of 220 A at

unity pf. The supply voltage is constant at 11 kV and at constant frequency. If the steam admission is kept unchanged and the emf is raised by 25% determine the current and power factor.

If the higher value of excitation is maintained and the steam supply is slowly increased, at what power output will the alternator break away from synchronism?

Draw phasor diagram under maximum power condition.

[**Ans.** 288.6 A; 0.762 (lagging); 16,200 kW]

10. Two 3-phase generating stations A and B are linked together through a 33 kV interconnector having a reactance of 3 Ω/phase. The load on the generators at A is 100 MW at a pf of 0.8 lagging, and the local load taken by the consumers connected to bus-bars is 60 MW at a pf of 0.707 lagging. Calculate the load in MW received from station A to station B, its power factor and phase difference between the voltages A and B.

[U.P.S.C. I.E.S. Generation, Transmission & Distribution of Power 1972]

[**Ans.** 40 MW; 0.97; 6.6°]

11. An interconnector which links two stations, A and B, has an impedance of $(3 + j2)$ ohms per phase. If the supply voltages at A and B are maintained at 33 kV and in phase with each other, suggest an arrangement for voltage control. If a booster transformer is used for this purpose, find the booster voltage per phase in magnitude and phase angle which should be supplied at station A to allow the transfer of a load of 10 MW at unity power factor from station A to station B.

[**Ans.** 631 $\angle 33.7°$ volts/phase]

12. Two generating stations are linked together by a 33 kV interconnector, having a resistance of 0.7 Ω and a reactance of 3.5 Ω per phase. One station has a load of 60 MW at a power factor of 0.8 lagging while the other has a load of 40 MW at 0.71 lagging power factor. Calculate the load transferred from one station to the other through the interconnector, and the phase difference between the voltages at the two stations. The capacity of each generating station is 50 MW.

[**Ans.** 19.72 kW; 3.6°]

13. Two stations A and B are interconnected by an interconnector having negligible resistance and reactance of 3 Ω per phase. Station A supplies a load of 5,000 kW at a pf of 0.9 lagging at a bus-bar voltage of 10 kV, while station B supplies 7,000 kW at a pf 0.8 lagging at the same voltage. If the station loads are equalised by the flow of power in the interconnector cable, determine the station power factors. [**Ans.** 0.9323 lag; 0.707 lag]

14. Stations A and B are interconnected by a tie-line of 2.8 Ω reactance. Station A supplies a load of 7 MW at 11 kV and 0.9 pf lagging to its own area and station B supplies 10 MW at 11 kV and 0.75 pf lagging to its own area. If it is desired to equalise the station loads by affecting a power transfer over the tie-line, estimate the new station power and the phase angle between the bus voltages of stations A and B.

[Agra Univ. Transmission & Distribution of Electrical Power 1979]

[**Ans.** 0.9278 lag; 0.6947 lag; 2°]

❑❑❑

CHAPTER 14

Power Plants Economics

INSIDE THIS CHAPTER

14.1. Introduction **14.2.** Classification of Costs **14.3.** Cost Analysis of Power Plants **14.4.** Interest and Depreciation **14.5.** Methods of Determination of Depreciation **14.6.** Economics of Power Generation **14.7.** Significance of Load Factor and Diversity Factor **14.8.** Capacity Scheduling **14.9.** Load Sharing Between Base Load and Peak Load Plants **14.10.** Choice of Size and Number of Generating Units **14.11.** Load Forecasting.

14.1. INTRODUCTION

In all engineering works, the question of cost is of first importance. In most cases the cost decides whether a certain project will be carried out or not, although political and other considerations may intervene sometimes. The electrical power supplier is required to supply power to a large number of consumers (whenever apply for it) to meet their requirements. While designing electrical power generating stations and other systems efforts are made to achieve overall economy so that the per unit cost of generation is the lowest possible. This will enable the supplier to supply electrical energy to its consumers at reasonable rates. The problem of determining the cost of any scheme is often difficult, because the cost varies considerably with time, tariff conditions and even with convention. The schemes involving the use of aluminium, which used to be considered very expensive a century ago, are now considered to be cheap due to lower cost. Economic problem occurs in the fields of generation, transmission, distribution and utilization of electrical power. The electrical engineer has to adopt the cheapest and most convenient scheme after having study of the relative costs of the possible schemes.

14.2. CLASSIFICATION OF COSTS

In general the cost of generating electrical energy can be divided into the following three elements :

1. ***Fixed Cost.*** This cost is independent of maximum demand and energy output. It is due to the annual cost of central organisation, interest on the capital cost of land (especially if some land is held for future development) and salaries of high officials.

The annual expenditure on central organisation and salaries of high officials is fixed since it has to be met whether the plant has high or low maximum demand or it generates less or more energy (units). Further, the capital investment on the land is fixed and hence the amount of interest is also fixed.

2. ***Semi-fixed Cost.*** This cost depends upon the maximum demand but is independent of energy output. The semi-fixed cost is due to annual interest and depreciation on the capital cost of the generating plant, transmission and distribution network, buildings and other civil works, all types of taxes and insurance charges and salaries of management and clerical staff. Also yearly compensation given to the workers is included under this head. The semi-fixed cost is approximately proportional to maximum demand.

The size and cost of the installation is governed by the maximum demand on the power plant. The greater the maximum demand on the power plant, the greater is its size and cost of installation. Further, the taxes, insurance charges and the strength of clerical staff depend upon the size of the plant and hence upon the maximum demand.

3. ***Running or Operating Cost.*** This cost depends upon the number of hours the plant is in operation or upon the number of units of electrical energy generated. The running or operating cost is due to annual cost of fuel, lubricating oil, water, maintenance and repair cost of equipment and wages and salaries of operational and maintenance staff and salaries of supervisory staff engaged in the running of the plant. The operating cost is approximately proportional to units generated.

Total annual cost incurred in the power generation is represented by the expression

$$E = a + b \text{ kW} + c \text{ kWh}$$

where a, b and c are constants. Fixed and semi-fixed cost being independent of the amount of energy generated is also called "*standing cost*".

In deciding any scheme for any given service the choice must be such that the total operating cost (sum of annual fixed cost, semi-fixed cost and operating cost) be minimum.

14.3. COST ANALYSIS OF POWER PLANTS

The generating cost per unit (kWh) of energy depends upon the cost covering the purchase, installation and erection of equipment, cost of fuel, labour, repair etc. The generation cost can, broadly be divided into *fixed cost* and the operating or *running cost*.

1. Fixed Cost. The *annual fixed cost* consists of the interest on the total investment, all types of taxes and insurance charges, salaries of high officials, management and clerical staff. The annual contribution to the fund to pay off the capital expended after useful life of the plant and yearly compensation paid to the workers is also included in the annual fixed cost. This cost varies approximately in direct proportion with the installed capacity of the plant and it does not depend upon the fact whether the plant supplies any energy or not.

The total investment or capital cost of a plant includes the preliminary cost, cost of land and other real estate, cost of design and planning, cost of building and equipment, cost of transportation, erection and installation of equipment and overheads etc. In case of hydroelectric power plants, the capital cost of the plant includes the costs of dam, earthwork, excavation, railhead, highways and other civil works and compensation to land/property owner (whose lands would be submerged in the reservoir) in addition to the costs mentioned above.

Cost of land and building will depend upon the location of the plant. If the plant is situated near a city, the land will be costlier than that in case the plant is located far from the city or residential areas.

The cost of steam plant located near a river, where adequate supply of cooling water is available, would be lesser than that of one in which cooling towers or ponds would have to be made available. In general a larger plant costs less per kW of installed capacity than a smaller one. A plant having a smaller number of large sized generating units costs less than a plant of the same total generating capacity but having a large number of small sized units. The capital costs of the plants vary widely, even in case of plants of similar types, and depend to a large extent on the market conditions.

2. Operating or Running Costs. The operating cost of a plant means the expenses which vary with the extent of operation or the amount of energy generated. This cost is due to annual cost of fuel, lubricating oil, water, maintenance and repair cost of equipment and wages and salaries of operational and maintenance staff and salaries of supervisory staff engaged on the running of the plant. The operating cost is approximately proportional to units generated. In steam, diesel and gas plants, the operating costs form a major portion of total annual cost whereas in hydro, nuclear and solar plants, the fixed costs overshadow the small operating costs.

3. Investor Profit. If the power plant is the public property, as is the case in India, then the customers will be taxpayers to share the burden of the Government. For this purpose, there is an item in the rates to cover taxes in place of the investors profit. These taxes will be paid by the consumers in the form of electric consumption bill. This amount is collected in twelve instalments per annum or six instalments per annum.

14.3.1. Fixed and Operating Costs of Hydropower Plants

The total capital outlay or the investment on a hydropower station includes the costs of preliminary surveys and investigations of the topography and geology of the proposed plant site, purchase of water rights and the land required for providing sufficient storage and pondage, compensation to the ousters, costs of preparation of detailed designs and specifications, costs of testing of materials of construction, costs of carrying out experimental work and model tests on designs for hydraulic structures, costs of construction and the purchase and installation of equipment, interest on the capital during construction, costs of new roads, railway lines, residential houses and other civil engineering works. The capital cost is very much affected by topographical and geological conditions.

The annual fixed costs include interest on the capital cost, depreciation, all types of taxes and insurance charges and salaries of high officials, management and clerical staff. The total cost of construction of a hydropower plant is invariably higher than that of a steam power plant of the same capacity. The fixed charges of a hydropower plant are about 60 to 70 per cent of the total cost of power and these are independent of the plant output.

The operating costs of hydropower plants include annual costs of lubricating oil, maintenance and repair cost of equipment and wages and salaries of operational and maintenance staff and salaries of supervisory staff engaged in the running of the plant. Because of the absence of fuel cost, the operating cost of a hydropower plant is very small.

14.3.2. Fixed and Operating Costs of Steam Power Plants

The capital cost of a steam power plant includes the costs of land, design and planning specifications, installation, powerhouse building, equipment, installation, testing, commissioning, overheads etc.

The range of cost for steam power plants is smaller than that for hydropower plants, the reason being that in case of steam plants, only about 15% of the cost is influenced by topographical and geological conditions, whereas in the case of hydropower plants, around 75% of the cost is affected by these conditions.

The annual fixed costs include interest on the capital, depreciation, all types of taxes and insurance charges and salaries of high officials, management and clerical staff.

The operating costs of a steam power plant include cost of fuel including its handling and ash handling, cost of water for boiler feed, condensers, cooling and house service, costs of oil, waste, stores and other supplies, repairs and maintenance cost of equipment, and salaries and wages of operational and maintenance staff and salaries of supervisory staff engaged in the running of the plant.

In the case of a steam power plant, the cost of coal can be split into two components—a small fixed component (required to keep the plant in readiness to meet the demand for power and independent of energy generated) and a large running component (cost of coal equivalent of the heat converted into electric power plus the heat rejected in the condenser). The latter component is directly proportional to the energy generated. The amount of fuel required per unit generated can be determined by dividing the heat rate by the calorific value of the coal. The total amount of coal consumed per unit (kWh) of energy generated and its cost depend upon the plant capacity, its efficiency, load factor, transportation charges and cost of storage etc. The fuel cost influences the total cost of steam power quite considerably—a change in fuel cost by 20% may cause about 8% variation in the total cost per kWh generated.

As already discussed, a number of environmental problems are created in case of steam power plants, and therefore, the flue gases will need cleaning *i.e.,* removal of particulates, SO_2 and NO_2 etc. so as to keep the pollutants within the safe limits. The fixed and operating costs of these air purification systems are to be included in the cost analysis of steam power plants.

14.3.3. Fixed and Operating Costs of Nuclear Power Plants

The components constituting the capital cost of a nuclear power plant are almost the same as those for steam power plants. However, the annual fixed cost in case of nuclear power plants is quite high in comparison to that of steam power plants, because of the following reasons.

1. The sophisticated technology is required for the manufacture of nuclear reactors and, therefore, the capital cost becomes pretty high.
2. The cooling water requirements of a nuclear power plant are very heavy and so the cooling tower cost becomes high as compared to that of a steam power plant of the same capacity.

In case of nuclear power plants though the quantity of fuel required is very small, problems of fuel transportation, fuel handling and ash handling are eliminated (as in case of steam power plants) but the cost of the fuel required is enormously high. Consequently the operating cost of a nuclear power plant is lesser than that of a steam power plant but higher than that of a hydropower plant. The operating efficiency of a nuclear power plant is high when it is operated at a high load factor. This is the reason that nuclear power plants are always operated as base load plants.

14.4. INTEREST AND DEPRECIATION

Usually money is borrowed from banks or insurance companies or other financial institutions for big projects such as generating plant and at the end of a year, the undertaking is required to pay the interest on the capital cost. Even if equipment is purchased out of reserve cash, the normal rate of interest must still be allowed for, since the money might otherwise be earning this interest. If a capital outlay required for a certain installation is P and the rate of interest per unit is r per annum then an amount of rP per annum must be provided as interest, and this is added to the annual operating cost. Rate of interest to be paid depends upon the money market position and the credit of the borrower and varies between 8 to 10 per cent (0.08 to 0.10 per unit) per annum.

If the installation were to last for ever, the interest is the only charge that would have to be made. In practice after a certain time (from ten to forty or fifty years) the plant and machinery have to be replaced, due to its getting old, obsolete and inadequate and it is, therefore, necessary to set aside certain amount annually to produce a sufficient sum at the end of the probable life for replacing the plant and machinery by new ones. The amount is known as *depreciation* and depends upon the type and make of machines. There are several methods of determining the yearly subscription towards this fund and some of these methods will be discussed in Art 14.5.

14.4.1. Life of Generating Plant

The depreciation charge represents the amount which is set aside from income every year and placed in depreciation reserve. For calculating this charge, it is necessary to estimate the useful life of the plant.

The expected physical life of *conventional steam power plant* is 40–45 years or even more. There is, however, the possibility that technological developments will make the plant obsolete before its physical life is over. Financial practice, therefore, is to write it off over 25–30 years. Conventional steam power plant of this age is quite serviceable, however, and it would not be economical or desirable under present conditions to scrap such plant. This, of course, was not known 30 years before nor do we know for certain what the conditions will warrant 30 years ahead. In current planning, economic appraisal is done on the basis of the capital being amortized over a life of 30 years. In the event, the plant would not necessarily be taken out of service at the end of 30 years, but in planning for the period 8 to 10 years ahead it is assumed that the old plant would be retained until it has reached 35 years of age. Even at this stage the plant would probably be still capable of further operational life and, before a decision was taken to scrap it, consideration would be given as to whether the old plant should be retained for peak load operation for a year or so longer or whether it would pay to scrap it (and save repairs, renewals, surveys, wages etc.), or to replace it by new capacity involving capital investment but with prospects of considerable saving in running costs, including fuel.

The current practice to assume for **nuclear plant** a financial life of 25 years.

Hydropower plant and pumped storage plant contain a large element of civil works (dams and tunnels, etc) having a very long physical life, say 80 years or even longer. The mechanical and electrical items (turbines, generators, etc) have, however, a life comparable to similar items in steam power stations and may be regarded for financial purposes a life of 30 years. Once a hydro scheme has been built its operating costs are so low as to make it unlikely to be displaced by technological developments.

14.5. METHODS OF DETERMINATION OF DEPRECIATION

Depreciation is the most important item in the fixed costs and it represents the reduction in the value of the equipment and other property of the plant every year due to continuous wear and tear and also due to obsolescence. The first factor can be reduced by proper maintenance of the equipment and the buildings, but the second factor is quite unpredictable. To account for the depreciation, a certain fixed amount is set aside every year so that by the time the life span of the plant is over, the total amount accumulated equals the replacement cost (original cost of the equipment less the expected salvage value). The methods commonly used for determination of annual depreciation charges are (*i*) straight line method, (*ii*) diminishing value method and (*iii*) sinking fund method.

1. *Straight Line Method.* This method assumes that certain depreciation occurs according to the straight line law and, therefore, in this method a constant depreciation charge is made every year on the basis of total depreciation (initial cost – scrap or salvage value) and useful life of the equipment/property.

i.e., Annual depreciation charge

$$= \frac{\text{Initial cost} - \text{salvage value}}{\text{Number of years of useful life}} = \frac{P - S}{n} \qquad ...(14.1)$$

Though this method is very popular because of its simplicity but it does not take into account the amount of interest earned by the amount set aside yearly. The assumption of constant depreciation every year is also not correct.

Figure 14.1 illustrates this method graphically.

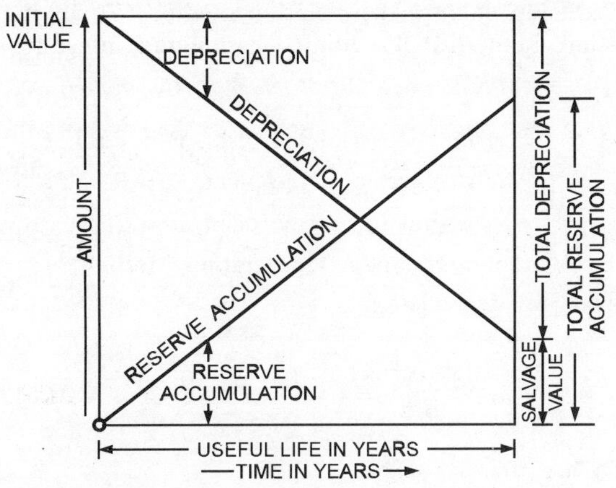

Fig. 14.1. ***Straight Line Method of Determination of Depreciation***

2. *Diminishing Value or Declining Balance Method.* In this method provision is made for setting each year a fixed rate, first applied to the original cost and then to the diminishing value; such rate depending upon the useful life of the plant. This method results in distribution of the total expenses on the plant (depreciation plus maintenance cost) over its total useful life because in this method depreciation charges are heavy in early years when maintenance charges are low and depreciation charges are low in the late years when maintenance charges are heavy. The serious drawback of this method is that an heavy burden towards depreciation charges is imposed in the early years when the plant is to develop and build up its income. In this method also the amount of interest earned by the reserve accumulation is ignored.

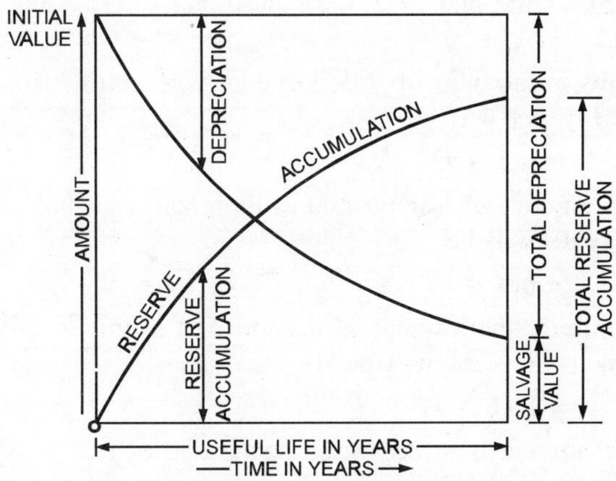

Fig. 14.2. ***Diminishing Value Method of Determination of Depreciation***

Let the capital cost of the plant be P, salvage value after useful life of equipment of n years be S and annual unit depreciation be x.

The value of plant after one years $= P(1 - x)$

The value of plant after two years $= P(1 - x)(1 - x)$

$$= P(1 - x)^2$$

The value of plant after n years, $S = P(1 - x)^n$

$$\text{or} \quad (1 - x) = \left(\frac{S}{P}\right)^{\frac{1}{n}}$$

$$\text{or} \quad x = 1 - \left(\frac{S}{P}\right)^{\frac{1}{n}} \qquad ...(14.2)$$

Deposit to be made at the end of first year of completion of the plant

$$= xP = P\left[1 - \left(\frac{S}{P}\right)^{\frac{1}{n}}\right] \qquad ...(14.3)$$

Deposit to be made at the end of second year of completion of the plant

$$= x(1 - x)P = P\left(\frac{S}{P}\right)^{\frac{1}{n}} P\left[1 - \left(\frac{S}{P}\right)^{\frac{1}{n}}\right] \text{ and so on}$$

Figure 14.2 illustrates this method graphically.

3. *Sinking Fund Method.* In this method provision is made for setting aside each year such a sum as, invested at certain interest rate compounded annually, will give the amount equal to replacement cost of the installation at the end of its useful life. As compared to straight line method, it requires smaller annual amounts and also the amounts for annuity are uniform.

Let P be the initial value of the equipment, n be the useful life of equipment in years, S be the scrap or salvage value after useful life of equipment of n years, and r is the rate of interest per annum expressed as a fraction.

If the annual deposit be q, it will earn interest rq in one year, so that it will worth $q + rq = q(1 + r)$ at the end of one year. Thus its value will be multiplied by the ratio $(1 + r)$ every year and it will worth $q(1 + r)^2$ at the end of 2 years, $q(1 + r)^3$ at the end of 3 years and $q(1 + r)^n$ at the end of n years.

Thus an amount of q deposited at the end of first year, in $n - 1$ years will be

$$= q(1 + r)^{n-1}$$

An amount of q deposited at the end of second year, in $n - 2$ years will be

$$= q(1 + r)^{n-2}$$

Similarly the amount of q deposited at the end of third year, in $n - 3$ years will be

$$= q\,(1 + r)^{n-3}$$

and the amount of q deposited at the end of $(n - 1)$th year, in one year will be $= q\,(1 + r)$

Thus after n years total sinking fund will be

$$= q[(1 + r)^{n-1} + (1 + r)^{n-2} + (1 + r)^{n-3} + \ldots (1 + r)]$$

The above terms are in geometrical progression and

$$\text{Total sum} = \frac{\frac{1}{1+r}q - q(1+r)^{n-1}}{\frac{1}{1+r} - 1}$$

$$= q\frac{1-(1+r)(1+r)^{n-1}}{1-(1+r)} = q\frac{(1+r)^n - 1}{r}$$

This must be equal to the cost of replacement Q which is equal to (P – S).

$$\text{So,} \quad Q = q\frac{(1+r)^n - 1}{r}$$

$$\text{or annual deposit, } q = \frac{Qr}{(1+r)^n - 1} \qquad \ldots(14.4)$$

This method is illustrated graphically in Fig. 14.3.

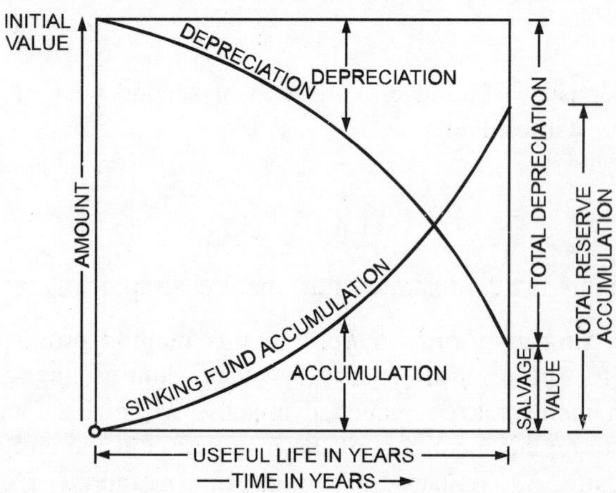

Fig. 14.3. ***Sinking Fund Method of Determination of Depreciation***

Example 14.1. A 300 kVA distribution transformer costs Rs 20,000 and has a salvage value of Rs 1,000 at the end of 20 years. Determine the depreciated value of the plant at the end of 10 years on the following methods of assessment :

(*i*) Straight line depreciation (*ii*) Sinking fund depreciation of 8% compounded annually.

[*Pb. Univ. Electrical Power II (Generation) Dec. 1995*]

Solution : Initial cost of the plant, P = Rs 20,000.00

Salvage value of the plant, S = Rs 1,000.00

Useful life of the plant, n = 20 years

(*i*) **By Straight Line Method.** Annual deposit to be made,

$$q = \frac{P - S}{n} = \text{Rs } \frac{20{,}000 - 1{,}000}{20} = \text{Rs } 950.00$$

Depreciated value of the plant at the end of 10 years

$$= p - q \times 10 = \text{Rs } 20{,}000 - \text{Rs } 950 \times 10$$

$$= \text{Rs } 10{,}500.00 \textbf{ Ans.}$$

(*ii*) **By Sinking Fund Method.** Rate of interest, $r = 8\% = 0.08$

Annual deposit in the Sinking Fund,

$$q = \text{Rs } (P - S)\left[\frac{r}{(1+r)^n - 1}\right]$$

$$= \text{Rs } \frac{(20{,}000 - 1{,}000) \times 0.08}{(1 + 0.08)^{20} - 1} = \text{Rs } 415.20$$

Total amount collected at the end of 10 years

$$= q\frac{(1+r)^{10} - 1}{r} = \text{Rs } 415.20 \times \frac{(1 + 0.08)^{10} - 1}{0.08}$$

$$= \text{Rs } 6{,}015.00$$

Depreciated value of the plant at the end of 10 years

$$= \text{Rs } 20{,}000 - \text{Rs } 6{,}015 = \text{Rs } 13{,}985 \textbf{ Ans.}$$

Example 14.2. A plant costs Rs. 7.56 × 10^5 and it is estimated that after 25 years it will have to be replaced by a new one. At that instant its salvage value will be Rs 1.56 × 10^5. Calculate (*i*) annual deposit to be made in order to replace the plant after 25 years, (*ii*) the value of plant after 10 years on the following basis :

(*i*) reducing balance depreciation method (*ii*) sinking fund depreciation method at 8% annual compound interest.

[*Pb. Univ. Electrical Power II (Generation) Dec. 1994*]

Solution : Initial cost of the plant, P = Rs 7.56 × 10^5

Salvage value of the plant, S = Rs 1.56 × 10^5

Useful life of the plant, n = 25 years

(*i*) **By Reducing Balance Depreciation Method.**

Annual unit depreciation,

$$x = 1 - \left(\frac{S}{P}\right)^{\frac{1}{n}} = 1 - \left(\frac{1.56 \times 10^5}{7.56 \times 10^5}\right)^{\frac{1}{25}} = 0.0612 \textbf{ Ans.}$$

Value of plant after 10 years

$$= \text{Rs } P\,(1 - x)^{10} = \text{Rs } 7.56 \times 10^5\,(1 - 0.0612)^{10}$$

$$= \text{Rs } 4.02 \times 10^5 \textbf{ Ans.}$$

(*ii*) **By Sinking Fund Depreciation Method.**

Rate of interest, $r = 8\% = 0.08$

Cost of replacement, Q = P – S

$$= \text{Rs } (7.56 \times 10^5 - 1.56 \times 10^5)$$

$$= \text{Rs } 6 \times 10^5$$

Annual deposit in the sinking fund,

$$q = \text{Rs } \frac{Qr}{(1+r)^n - 1} = \text{Rs } \frac{6\times10^5\times0.08}{(1+0.08)^{25}-1} = \text{Rs } 8{,}207$$

Total amount collected at the end of 10 years

$$= q\frac{(1+r)^{10}-1}{r} = \text{Rs } 8{,}207\times\frac{(1+0.08)^{10}-1}{0.08}$$

$$= \text{Rs. } 1.19 \times 10^5$$

Depreciated value of the plant at the end of 10 years

$$= \text{Rs } (7.56 \times 10^5 - 1.19 \times 10^5) = \text{Rs } 6.37 \times 10^5 \textbf{ Ans.}$$

Example 14.3. The cost of a small preheater is Rs 12,000 and its expected life is 16 years. The net annual instalment to recover the cost of equipment is Rs 425. The interest is 5%. Using sinking fund method find out the salvage value of the preheater after 16 years of service. *[A.M.I.E. Sec. B. Power Plant Engineering Winter 1996]*

Solution : Initial cost of preheater, P = Rs 12,000

Useful life of the preheater, n = 16 years

Rate of interest, r = 5% = 0.05

Annual deposit in the sinking fund, q = Rs 425

Total amount collected at the end of 16 years

$$= q\frac{(1+r)^{16}-1}{r} = \text{Rs } 425\times\frac{(1+0.05)^{16}-1}{0.05} = \text{Rs } 10{,}054$$

Salvage value,

S = Depreciated value of the plant at the end of 16 years

= Rs 12,000 – Rs. 10,054 = Rs 1,946 **Ans.**

Example 14.4. The equipment in a power station cost Rs 1,560,000 and has a salvage value of Rs 60,000 at the end of 25 years. Determine the depreciated value of the equipment at the end of 20 years by sinking fund method at 5% compound interest annually. *[Rajasthan Univ. Generation of Electrical Power 2004, 2005]*

Solution : Initial cost of equipment, P = 1,560,000

Useful life of equipment, n = 25 years

Rate of interest, r = 5% = 0.05

Salvage value of equipment S = Rs 60,000

Replacement cost, Q = P – S = Rs 1,560,000 – Rs 60,000 = 1,500,000

$$\text{Annual deposit, } q = \frac{Qr}{(1+r)^n - 1} = \frac{1{,}500{,}000\times0.05}{(1+0.05)^{25}-1} = \text{Rs } 31{,}428.69$$

The amount collected at the end of 20 years

$$= \frac{q(1+r)^{20}-1}{r} = 31{,}428.69\times\frac{(1+0.05)^{20}-1}{0.05} = \text{Rs } 1{,}039{,}219$$

Depreciated value of the equipment at the end of 20 years

= Rs (1,560,000 – 1,039,219) = Rs 520,781 **Ans.**

14.6. ECONOMICS OF POWER GENERATION

The generation of electrical energy economically is not an ordinary matter, rather it requires a long experience to decide about the type, location and the rating of generating stations.

The generating stations may be steam, hydro, nuclear, diesel or any other type. This factor mainly depends upon the natural sources available in the areas. Steam power stations are best suited near the coal fields and also adopted where coal supply is available in plenty at reasonable rates, large amount of power is required to be generated and financial, climatic and geographical conditions are not favourable to hydro and nuclear power stations. Hydropower stations are best suited in case water is available at certain height and nuclear power stations are best suited in area far away from collieries and where fuel costs are high and alternative cheap hydropower is not available as in Rajasthan. Diesel power stations are installed where supply of coal and water is not available in sufficient quantity or where power is to be generated in small quantity or where standby sets are required for continuity of supply such as in hospitals, telephone exchanges, radio-stations and cinemas.

The power station should be as near as possible to the centre of the load so that the transmission cost and losses are minimum. This factor is important when dc supply system is adopted. However, in case of ac supply system where transformation of energy from lower to higher voltage and vice versa is possible, power station can be located at places other than that of centre of load provided other conditions are favourable.

The other considerations for the design of the power station are reliability, minimum capital and operating costs. For this it is necessary that the layout should be such that the maintenance and repairs can be carried out easily; the design must be compact and well planned; equipment used must be standard one so that capital cost is reduced and replacement of worn-out parts becomes easy and equipment used should be simple so that it can be operated by semi-skilled workers. The design of power station should be such that the station could be divided into a number of sections to avoid complete shutdown of the station when the fault occurs. Another advantage of sectionalizing is that the erection can be completed and put into service in parts. The equipment must be automatic whenever possible in order to save the labour cost provided the reliability of the station is not affected. The automatic equipment should be such that it could be operated manually in case it is required. The scheme employed should be such that extension could be made to meet with the increase in demand in future without incurring further heavy expenditure.

Before a power project is taken into hand, the project engineer should have the following information with him.

(*i*) Estimate of probable load.

(*ii*) Future load conditions.

(*iii*) Location of the loads, specially in case of hydroelectric generating stations because the cost of transmission is also required to be considered.

For deciding the type and rating of generating plant it is necessary that engineer may be familiar with the following important terms.

1. *Load Curves.* The load on the power station is seldom constant; it varies from time to time. The daily variations in load on the power station from time to time—hourly or half hourly can be plotted on a graph taking load on Y-axis and time on X-axis. The curve so obtained is known as *daily load curve*. From the daily load curves of a particular month, the monthly load curve can be plotted by calculating the average value of power at a particular time of the day. Similarly if we consider such monthly load curves of a particular year, and find the average value of the power at a particular time of the day, the annual load curve can be obtained. The load curve bears a great importance in generation. The load curves supply the following informations :

(*i*) The variation of the load during different hours of the day.

(*ii*) The area under the curve represents the total number of units generated in a day.

(*iii*) The peak of the curve represents the *maximum demand* on the station on the particular day.

(*iv*) The area under the load curve divided by the number of hours represents the average load on the power station.

(*v*) The ratio of the area under the load curve to the total area of the rectangle in which it is contained gives the *load factor*.

The type of variations in load or demand on a plant depends on the domestic and industrial users (their type and habits). The load requirements of an individual consumer are different at different timings in the day. The load curve for Sunday will be different from those for other days of the week. The load curves for winter and summer will also differ. The curves for rural, suburban and urban areas will also differ.

The load curves for Sunday, other weekdays, summer and winter for a typical metropolitan area are shown in Fig. 14.4.

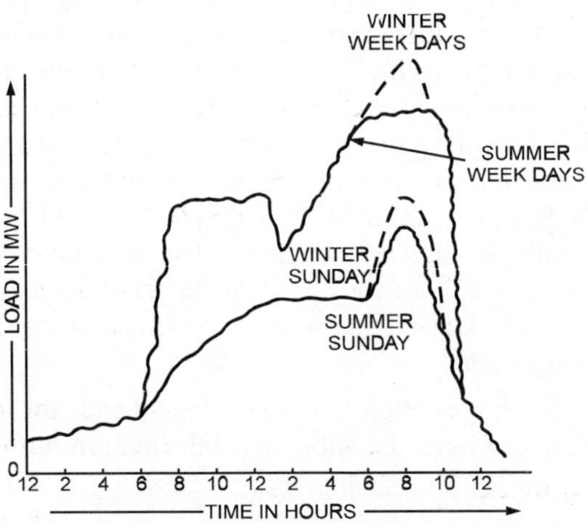

Fig. 14.4. ***Chronological Load Curves***

During the early morning hours the demand is always low. Around 5 AM the load starts increasing because of increasing industrial and traction load. Around 9 AM the load reaches a high value and remains almost constant till evening except for some dip during lunch hours. The load again starts increasing in the evening hours as the residential and traction load start coming up. The peak occurs around 7–9 PM and then the load starts decreasing. On Sundays the load demand is only due to residential consumers and so low as compared to those of other weekdays.

Thus, on the basis of above informations the load curves help in deciding the size of the units to be installed and also in preparing the schedule of operation of the generating units.

2. *Load Duration Curve.* This is another type of curve which indicates the variation of load, but with the loads arranged in descending order of magnitude *i.e.,* the greatest load on the left, lesser loads towards the right and the least load at the extreme right. This curve gives the number of hours for which a particular load lasts during the day. The area under this curve like *load curve* or *chronological load curve* gives the total number of units generated for the period considered. From this curve the load factor of the station can also be determined. From these curves the distribution of load between various generating units can also be predicted. (For illustration refer to Example 14.12)

3. *Integrated Load Duration Curve.* This curve gives the total number of units generated for the given demand. The ordinate represents the demand in kW or MW and the abscissa represents the units (kWh) generated at or below a given demand (in kW or MW). Such a curve can be obtained from the load duration curve keeping the abscissa corresponding to each ordinate equal to the area of the duration curve up to the value of that ordinate.

Such curves are useful in that with a given number of units (kWh) available, say from a river flow, the load (kW) that could be carried at the base or peak may be easily determined. (For illustration refer to Example 14.12)

4. *Mass Curve.* This curve is plotted with units (kWh) as ordinate and time as abscissa. Thus a mass curve gives the total energy consumed by the load up to a particular time in a day. This curve can be easily plotted from the chronological load curve by summing up the energy consumed up to different times starting at the zero time.

This curve is used in the study of variations between the rate of water flow and the electrical load and for the determination of the necessary storage.

5. *Connected Load.* The sum of the continuous ratings of all electrical equipment connected to the supply system is known as *connected load.*

6. *Maximum Demand.* It is not necessary that all the connected load be switched to a system at a time. The greatest of all "short time interval averaged" (15 minutes or $\frac{1}{2}$ hour or 1 hour) during a given period (a day, a month or a year), on the power station is called the *maximum demand.* It is sometimes also called as *system peak.* It is to be clarified here that the maximum demand never means the greatest instantaneous maximum demand but the greatest short time average demand occurring during a long period of time under consideration. Hence whenever the maximum demand is to be mentioned it is essential to indicate the short interval of time over which averaging has been carried out for the measurement of demand and the period of load under consideration. It is the maximum demand which determines the size and the cost of the installation.

7. *Demand Factor.* The ratio of actual maximum demand on the system to the total rated load connected to the system is called the *demand factor.* It is always less than unity.

$$\text{Mathematically demand factor} = \frac{\text{Maximum demand}}{\text{Connected load}} \quad ...(14.5)$$

The idea of a demand factor was introduced due to the fact that all the equipment connected to the system will not be worked at a time in practice and the kW or kVA maximum demand of a group of electricity consuming devices will be always less than the sum of the kW or kVA ratings or capacities of these devices.

The minimum capacity of the generating plant must be such as to meet the maximum demand, the value of demand factor, therefore, determines the capacity and hence cost of the power equipment required to serve a given load.

8. *Average Load or Demand.* The average load or demand on the power station is the average of loads occurring at the various events. It can also be stated as energy delivered in a given period divided by the number of hours in that period. Depending upon the duration of time period such as a day, a month or a year, we get daily, monthly or annual average load.

$$\text{Daily average load} = \frac{\text{kWh supplied in a day}}{24} \quad \text{...(14.6)}$$

$$\text{Monthly average load} = \frac{\text{kWh supplied in a month}}{24 \times 30} \quad \text{...(14.7)}$$

$$\text{Annual average load} = \frac{\text{kWh supplied in a year}}{24 \times 365} \quad \text{...(14.8)}$$

9. *Load Factor.* The ratio of average load to the maximum demand during a certain period of time such as a day or a month or a year is called the *load factor*. Since average load is always less than the maximum demand, load factor is therefore, always less than unity.

$$\text{Load factor} = \frac{\text{Average demand}}{\text{Maximum demand}} \quad \text{...(14.9)}$$

Load factor may also defined as the ratio of the number of units actually generated in a given period to the number of units which could have been generated had the load or demand remained at the maximum value throughout this period.

i.e., Load Factor

$$= \frac{\text{Units generated in a given period}}{\text{Maximum demand} \times \text{(number of hours of operation in the given period)}} \quad \text{...(14.10)}$$

10. *Diversity Factor.* The maximum demands of all the consumers supplied from an installation do not occur usually at the same time, maximum demand on the installation is, therefore, always less than the sum of individual maximum demands of all consumers connected to it.

The ratio of sum of the individual maximum demands of all the consumers supplied by it to the maximum demand of the power station is called the *diversity factor.*

i.e., Diversity factor

$$= \frac{\text{Sum of individual maximum demands}}{\text{Maximum demand of power station}} \quad \text{...(14.11)}$$

It is always greater than unity.

11. *Coincidence Factor.* The reciprocal of the diversity factor is called the coincidence factor.

i.e., Coincidence factor

$$= \frac{\text{Maximum demand of power station}}{\text{Sum of individual maximum demands}} \quad \text{...(14.12)}$$

It is always less than unity.

12. *Capacity Factor or Plant Factor.* Every plant has to have a reserve capacity so as to take care of the future expansion and increase in load and, therefore, total installed capacity of the plant is usually greater than that actually required (maximum demand).

Capacity (or plant) factor is defined as the ratio of the average load to the rated capacity of the power plant.

i.e., Capacity (or plant) factor

$$= \frac{\text{Average demand}}{\text{Rated capacity of power plant}} \quad \text{...(14.13)}$$

Plant capacity factor may also be defined as the actual energy generated divided by the maximum possible energy that the plant might have generated during a given period.

i.e., Plant capacity factor

$$= \frac{\text{Units generated in a given period}}{\text{Rated capacity of the power plant in kW} \times \text{(number of hours in the given period)}} \quad \text{...(14.14)}$$

The capacity factor depicts the extent of the use of the power plant. It is different from load factor and the difference is an indication of the reserve capacity.

Capacity factor

$$= \frac{\text{Maximum demand}}{\text{Rated capacity}} \times \text{load factor} \quad \text{...(14.15)}$$

13. *Utilization Factor.* It is a measure of the utility of the power plant capacity and is the ratio of maximum demand to the rated capacity of the power plant. It is always less than unity.

i.e., Utilization factor

$$= \frac{\text{Maximum demand on the power station}}{\text{Rated capacity of the power station}} \quad \text{...(14.16)}$$

14. *Plant Operating Factor or Plant Use Factor.* It is defined as the ratio of actual energy generated during a given period (say a year) to the product of capacity of the plant and the number of hours the plant has been actually in operation during the period.

i.e., Plant use factor

$$= \frac{\text{Total kWh generated}}{\text{Rated capacity of the plant} \times \text{number of operating hours}} \quad \text{...(14.17)}$$

15. *Installed Capacity.* The total of station capacities available to supply the system load is called the *installed capacity*.

16. *Firm Power.* It is the power intended to be always available (even under emergency conditions).

17. *Prime Power.* It is the power (mechanical, hydraulic or thermal) that is always available for conversion into electrical power.

18. *Dump Power.* It is power in excess of the load requirements and it is made available by surplus water. It refers to hydropower plants.

19. *Cold Reserve.* It is that reserve generating capacity which is available for service but not normally ready for immediate loading.

20. *Hot Reserve.* It usually refers to boiler excess capacity, which is kept hot and with steam pressure, ready for use.

21. *Operating Reserve.* It refers to capacity in service in excess of peak load.

22. *Spinning Reserve.* It is the generating capacity connected to the bus and ready to take load.

In order to take instantaneous peak loads, the system must have capacity in *readiness to serve.* For this purpose certain machines are kept running on no load *i.e., floating* on the system. This is called the *spinning reserve* of the *system.*

14.7. SIGNIFICANCE OF LOAD FACTOR AND DIVERSITY FACTOR

Load factor and diversity factor play an important part in the cost of the supply of electrical energy. *Higher the values of load factor and diversity factor, lower will be the overall cost per unit generated.*

Higher load factor means greater average load, resulting in greater number of units generated for a given maximum demand. Thus, the standing charges, which are proportional to maximum demand and independent of number of units generated, can be distributed over a larger number of units supplied and therefore overall cost per unit of electrical energy generated will be reduced.

The capital cost of the power station depends upon the capacity of the power station. Lower the maximum demand of the power station, the lower is the capacity required and therefore lower is the capital cost of the plant. With a given number of consumers the higher the diversity factor of their loads, the smaller will be the capacity of the plant required and consequently the fixed charges due to capital investment will be much reduced.

The suppliers should always try to improve the load factor as well as diversity factor by inducing the consumers to use the electrical energy during off-peak hours and they may be charged at lower rates for such schemes.

Typical values of demand factors and diversity factors for different types of consumers are given below :

Type of Consumers	Demand Factors	Diversity Factors
Domestic lighting up to 1 kW	0.5–0.7	3–5
Domestic power	0.5	1.5–2
Commercial loads mostly of lighting	0.5–0.7	1.5–2
Industrial power	0.5–0.8	1.5–2

14.8. CAPACITY SCHEDULING

Capacity scheduling means the order in which the various generating units should be placed in operation in order to meet the total plant or system load. Certain units will be put in operation only during peak load demand hours. To ensure continuity of power supply, sufficient generating capacity should be in operation at all times so that if one unit fails, there is no interruption in power supply. If the load on the system/plant is equal to or less than the capacity of one generating unit, then it is always advisable to run two units to supply the load so that noninterrupted power supply can be maintained. The order of putting the machines in operation depends upon their efficiencies. The machines are loaded in the descending order of the efficiency. The load is divided among the various units on the basis of incremental rates.

The capacity of the generating unit in active operation is known as the 'spinning capacity'. Spinning capacity should always be more than the load in order to avoid interruption. "Spinning reserve" is the difference between the spinning capacity and the load on the system. The minimum magnitude of the spinning reserve is usually equal to the capacity of the largest unit.

14.9. LOAD SHARING BETWEEN BASE LOAD AND PEAK LOAD PLANTS

Base load plants run throughout the year and have high load factors. The economic characteristics of base load plants should be such that they supply power at high capital costs but low cost of operation. Hydro and nuclear power plants are usually classified as base load plants. Peak load plants run for a few hours in the year and work at low load factors. The economic characteristics of peak load plants should be such that they supply power at low capital costs, although at high cost of operation. Peak load power plants should be capable of starting quickly and should be inexpensive in starting and shutting down operations.

Let the operating costs of base load and peak load plants be Rs $(a_1\text{kW} + b_1\text{kWh})$ and Rs $(a_2\,\text{kW} + b_2\,\text{kWh})$ where $a_1 > a_2$ and $b_2 > b_1$.

Let P be the maximum demand on the power plant and x be the total number of units generated.

If P_1 is the maximum demand on base load plant and x_1 is the number of units generated by the base load plant.

Peak load on peak-load plant, $P_2 = P - P_1$

Units generated by peak-load plant, $x_2 = x - x_1$

Total annual cost of operation of the system,

$$C = a_1P_1 + a_2(P - P_1) + b_1x_1 + b_2(x - x_1)$$
$$= P_1(a_1 - a_2) + a_2P + x_1(b_1 - b_2) + b_2x$$

or $$\frac{dC}{dP_1} = (a_1 - a_2) + \frac{dx_1}{dP_1}(b_1 - b_2)$$

For minimum cost, $\frac{dC}{dP_1} = 0$

$$\therefore\ (a_1 - a_2) + \frac{dx_1}{dP_1}(b_1 - b_2) = 0$$

or $$\frac{dx_1}{dP_1} = \frac{a_1 - a_2}{b_2 - b_1} \text{ hours} \quad ...(14.18)$$

i.e., for economic load sharing, peak-load power plant will operate for $\frac{a_1 - a_2}{b_2 - b_1}$ hours per year.

14.10. CHOICE OF SIZE AND NUMBER OF GENERATING UNITS

The selection of units and their operation play an important role in the working of a power station and on the economics of power generation. The number of units and the size of each unit is decided from the load curve. The following factors should be considered while deciding the number of units and preparing the operating schedule.

1. The total capacity of the generating units must be capable of meeting the peak demand of the power station.
2. Since the machines operate with maximum efficiency at the three-fourth of the rated capacity, hence the number and size of units must be so selected that they operate at maximum efficiency and better overall efficiency and load factor of the power station is had.
3. Reliability of service is a very important factor. Cheaper power without reliability of supply is of no use. There should be a spare set of capacity of that largest unit in the power station so that maintenance and repairs or overhaul of the working units may be carried out without any disturbance in power supply.
4. The growth of the demand in near future should be kept in view.
5. The capacity of the power station should be 15 or 20% more than expected maximum demand.

The minimum number of generating units chosen could be one having a capacity equal the maximum demand on the power station. Large size units have lower capital cost per kW, need less floor area, require less operating labour and have better efficiency. Thus large sized units are economical both from the point of view of initial investment and operating cost.

The size of thermal units, in India, has been progressively increasing for the past many years. Around 1955 only 30 MW units were popular while now 500 MW units are being installed. The number and size of units in hydropower stations are governed by the availability of head and water. The earliest hydro units in India installed before independence were of 5 MW capacity while the largest hydro units in operation now are 165 MW at Dehar power plant of Beas-Sutlej project in Himachal Pradesh.

However, selection of a single generating unit having a capacity equal to the maximum demand on the system or slightly more has got following drawbacks.

1. As the load on the system varies at different times of the day, there is a considerable period during the day when the load on the system would be much less than the peak load. During this light load period the generating unit may be running at 50 per cent of the rated output or even practically on no load. Hence the unit would not be running at all times under conditions best suited for its operation to give maximum efficiency. So generating cost per unit would be high.
2. There will be complete failure of supply due to breakdown or maintenance and repairs to be carried out on the generating unit for weeks together.

If the plant is an isolated one and another unit of equal capacity is installed, as standby unit, so as to ensure continuity of supply, there will be substantial increase in the initial investment. In general the system reserve requirements increase with increase in unit size.

Alternatively the size and number of generating units may be so chosen as to fit the load curve as closely as possible. Then each unit can be made to operate in such a way that it operates almost at full load or at a load to give maximum efficiency. The reserve required in this case would be only one unit of the largest size chosen and would be much smaller than the maximum reserve capacity needed in the former case. As the reserve capacity required would be lesser, it would result in better plant capacity factor and plant use factor.

If all the machines of same ratings and identical in operation are chosen, the initial cost would be less, space required would be less, requirements of spare parts would be less, the appearance would be symmetrical and good looking.

The drawbacks and difficulties involved in this alternative of selecting large number of smaller units are given below :

1. The floor area required and initial cost per kW would be increased.
2. There would be substantial increase in maintenance cost and in personnels for handling and operating the equipment.

From the above discussion it is concluded that the number of units in a power station should neither be very large nor very small. As such the size of units for a power station has to be compatible with the total capacity of the power station.

Example 14.5. A residential consumer has 10 lamps of 40 W each connected at his premises. His demand is as follows :

From 12 midnight to 5 AM – 40 W
From 5 AM to 6 PM – No load
From 6 PM to 7 PM – 320 W
From 7 PM to 9 PM – 360 W
From 9 PM to 12 midnight – 160 W

Plot the load curve. Find the average load, maximum load, load factor and electric energy consumption during the day.

[Pb. Univ. Electrical Power II (Generation) May 1995]

Solution : Load curve is drawn as shown in Fig. 14.5.

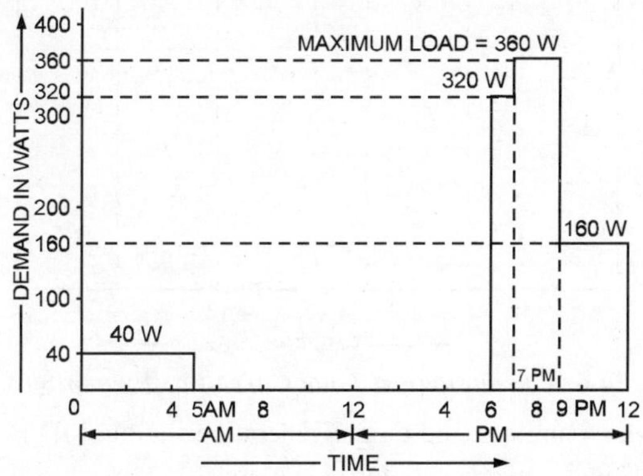

Fig. 14.5

Electrical energy consumption during the day

= Area of load curve

= 40 × 5 + 320 × 1 + 360 × 2 + 160 × 3

= 1,720 Wh or 1.72 kWh **Ans.**

$$\text{Average load} = \frac{\text{Energy consumed during 24 hours}}{24}$$

$$= \frac{1{,}720}{24} = 71.67 \text{ W } \textbf{Ans.}$$

Maximum load = 360 W

$$\text{Load factor} = \frac{\text{Average load}}{\text{Maximum demand}} = \frac{71.67}{360}$$

$$= 0.1991 \text{ or } 19.91\% \textbf{ Ans.}$$

Example 14.6. Define the terms (*i*) load factor, (*ii*) diversity factor and discuss their influence on the cost of generation.

A generating station has daily load cycle as

Time (Hrs)	0–6	6–10	10–12	12–16	16–20	20–24
Load (MW)	40	50	60	50	70	40

Draw the load curve and load duration curve and find (*i*) maximum demand (*ii*) units generated and (*iii*) load factor.

[*U.P. Technical Univ. Power Station Practice 2007*]

Solution: From the given data, the load curve is drawn as shown in Fig. 14.6.

For drawing load duration curve, the load oridinates are rearranged in descending order *i.e.,* greatest to the left, lesser loads towards the right and the least load at the extreme right, as shown in Fig. 14.7 (data given below)

Load in MW	Duration of Hours
70	4
60 and above	6
50 and above	14
40 and above	24

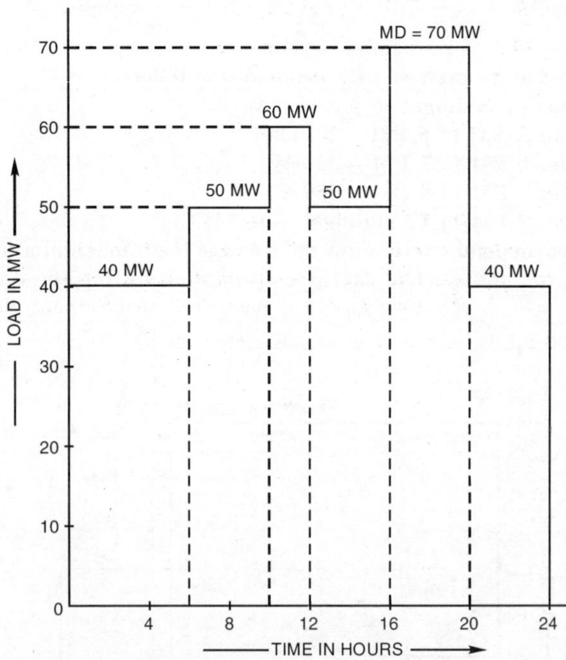

Fig. 14.6. ***Chronological Load Curve For Example 14.6.***

(*i*) Maximum demand on the generating station, MD = 70 MW

(*ii*) Units generated during 24 hours

= 70 × 4 + 60(6 – 4) + 50(14 – 6) + 40(24 – 14)

= 280 + 120 + 400 + 400

= 1,200 MWh or 12 × 10^5 kWh **Ans.**

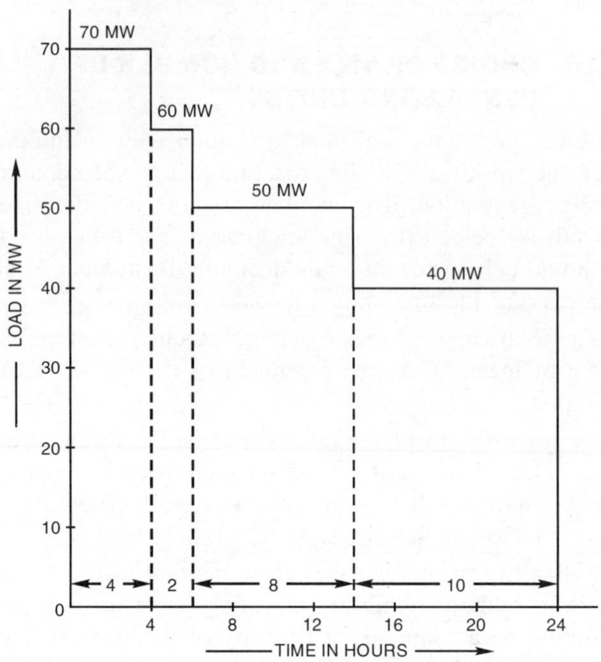

Fig. 14.7. ***Load Duration Curve For Example 14.6.***

(*iii*) Average load

$$= \frac{\text{Energy generated in MWh during the day}}{24}$$

$$= \frac{1{,}200}{24} = 50 \text{ MW } \textbf{Ans.}$$

$$\text{Load factor} = \frac{\text{Average load}}{\text{Maximum demand}} = \frac{50}{70} = 0.714 \text{ or } 71.4\% \textbf{ Ans.}$$

Example 14.7. The load on a power plant on a typical day is as under :

Time	12-5 AM	5-9 AM	9-6 PM	6-10 PM	10-12 Midnight
Load (MW)	20	40	80	100	20

Plot the chronological load curve and load duration curve. Find the load factor of the plant and the energy supplied by the plant in 24 hours.

[*Pb. Univ. Electrical Power II (Generation) January 1993*]

Solution : From the given data, the load curve is drawn, as shown in Fig. 14.8.

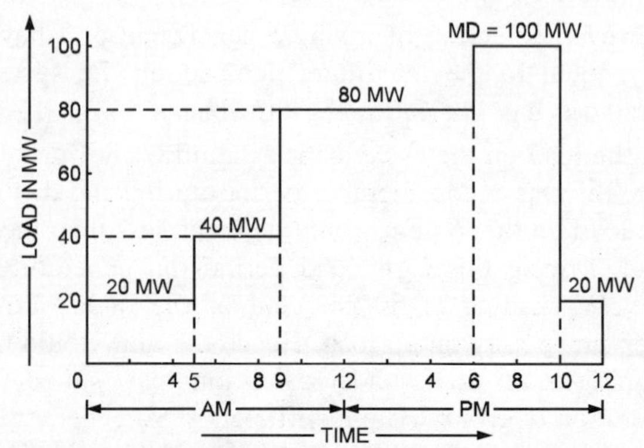

Fig. 14.8. ***Chronological Load Curve For Example 14.7***

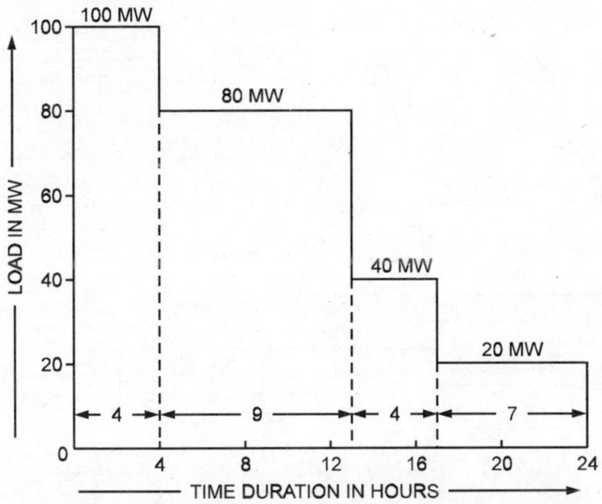

Fig. 14.9. ***Load Duration Curve For Example 14.7***

For drawing the load duration curve, the load ordinates are re-arranged in descending order *i.e.,* the greatest on the left, lesser loads towards the right and the least load at the extreme right, as illustrated in Fig. 14.9. (data given below).

Load in MW	Duration of Hours
100	4
80 and above	13
40 and above	17
20 and above	24

Maximum demand on the power plant, MD = 100 MW

Energy supplied during 24 hours

$= 100 \times 4 + 80 \times (13 - 4) + 40\,(17 - 13) + 20\,(24 - 17)$

= 1,420 MWh **Ans.**

Average load

$$= \frac{\text{Energy generated in MWh during the day}}{24}$$

$$= \frac{1{,}420}{24} = 59.167 \text{ MW}$$

Load factor

$$= \frac{\text{Average load}}{\text{Maximum demand}} = \frac{59.167}{100} = 0.59167 \text{ or } 59.167\% \textbf{ Ans.}$$

Example 14.8. Draw the integrated load duration curve and mass curve for the chronological load curve given in Example 14.7.

Solution : Integrated load duration curve is drawn between the MW demand and the total energy generated up to the MW demand and is obtained from the load duration curve. The following table gives the energy at different load levels.

Load in MW	20	40	80	100
Energy Generated in MWh	20 × 24 = 480	480 + 20 × 17 = 820	820 + 40 × 13 = 1,340	1,340 + 20 × 4 = 1,420

Integrated load duration curve (energy load curve) is drawn as illustrated in Fig. 14.10.

Mass curve is plotted between the energy generated up to a particular time in a day and the time. The energy generated up to different times of the day are obtained from the chronological load curve as given.

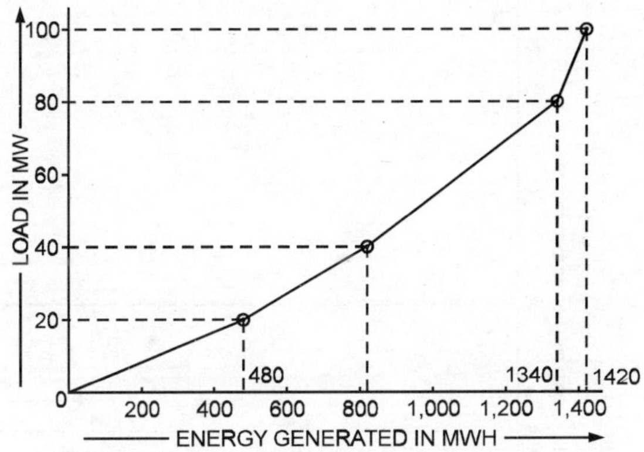

Fig. 14.10. ***Integrated Load Duration Curve***

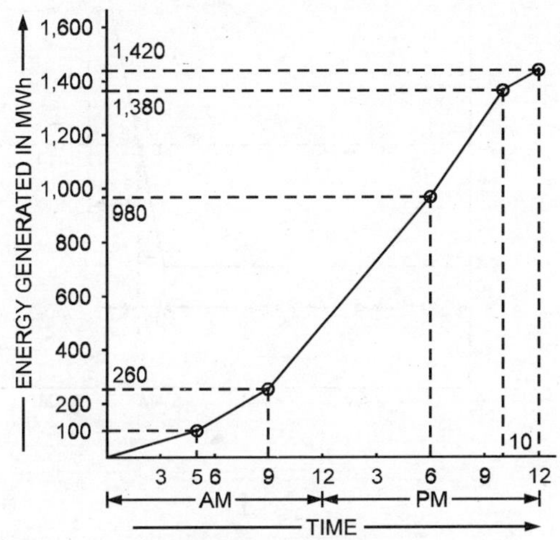

Fig. 14.11. ***Mass Curve***

Time	0	5 AM	9 AM	6 PM	10 PM	12 Mid-night
Energy Generated in MWh up to That Time	0	20 × 5 = 100	100 + 40 × 4 = 260	260 + 80 × 9 = 980	980 + 100 × 4 =1,380	1,380 + 20 × 2 = 1,420

Mass curve is drawn as illustrated in Fig. 14.11.

Example 14.9. An electrical system experiences linear changes in load such that its daily load curve is defined as follows :

Time	12 PM	2 AM	6 AM	8 AM	12 NOON	12.30	1 PM	5 PM	6 PM	12 Midnight
Load (MW)	24	12	12	60	60	48	60	60	84	24

(*i*) **Plot the chronological curve and load duration curve of the system.**

(*ii*) **Find the load factor.**

(*iii*) **What is the utilization factor of the plant serving this load if its capacity is 120 MW?**

[*M.D. Univ. Power Plant Engineering May 2007*]

Solution : (*i*) From the given data, the chronological curve is drawn as shown in Fig 14.12.

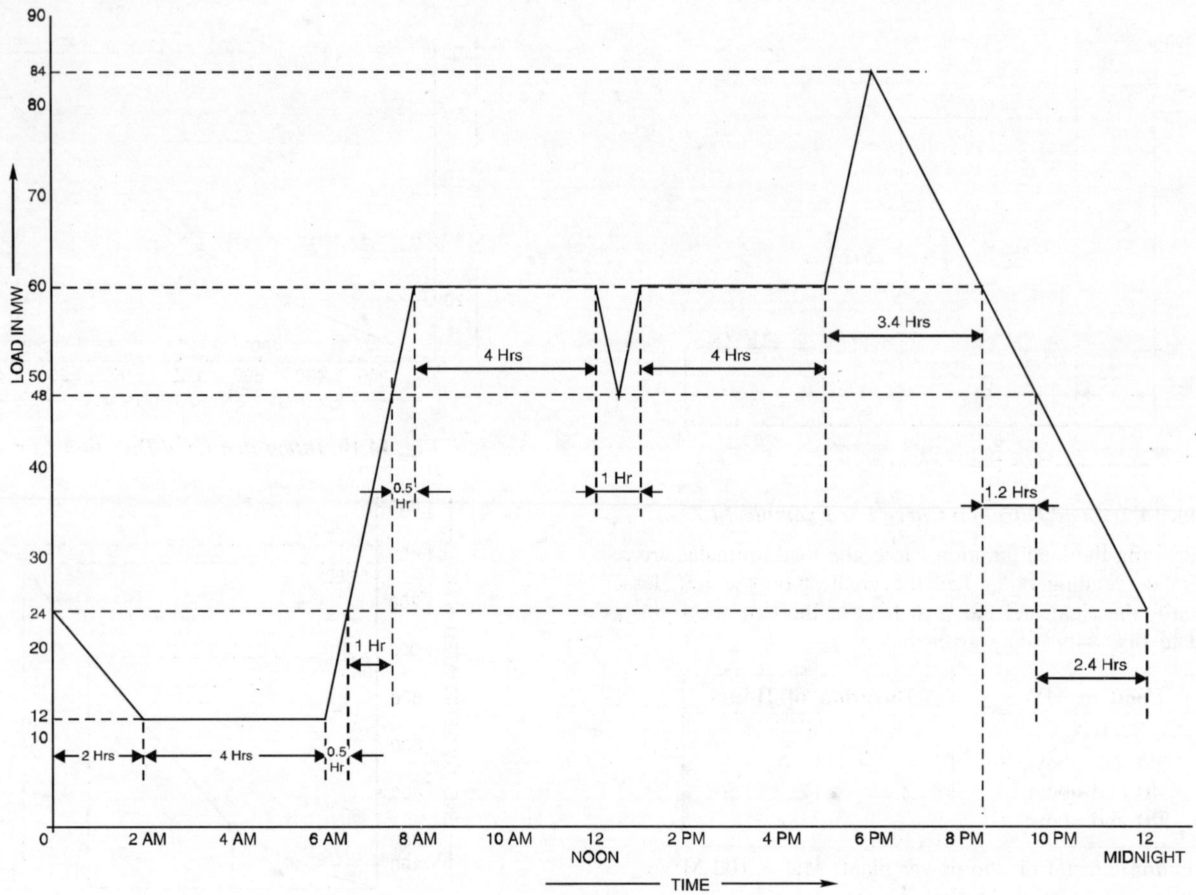

Fig. 14.12

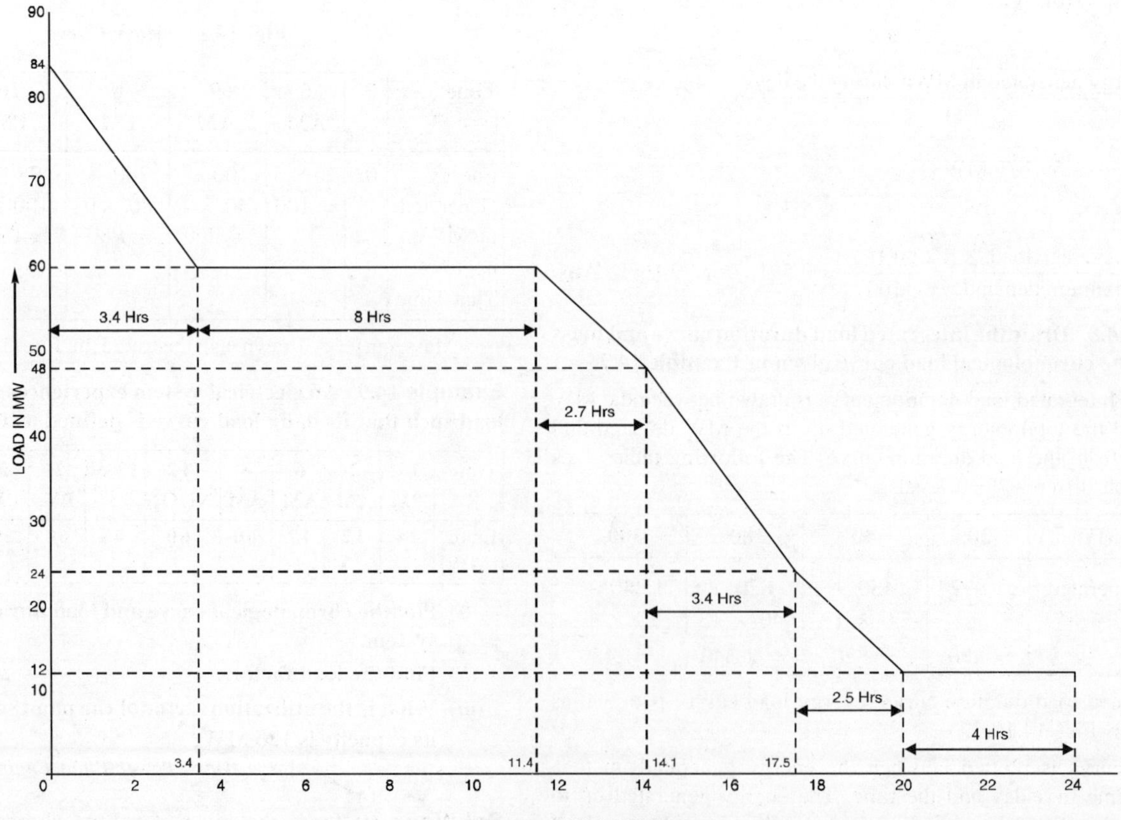

Fig. 14.13

From Fig. 14.12, it is observed that

Load falls from 84 MW to 60 MW during the time duration of 3.4 hours (5 PM to 8.24 PM)

Load remains constant at 60 MW for time duration of 8 hours 8 AM to 12 AM and 1 PM to 5 PM)

Load varies from 60 MW to 48 MW during the time duration of 0.5 + 1 + 1.2 *i.e.* 2.7 hours (7.30 to 8.00 hours, 12 to 13.00 hours and 20.24 hrs to 21.36 hrs)

Load varies from 48 MW to 24 MW during the time duration of 1 + 2.4 *i.e.* 3.4 hours (6.30 AM to 7.30 AM and 9.36 to 12.00 midnight)

Load varies from 24 MW to 12 MW during the time duration of 2 + 0.5 *i.e.* 2.5 hours (0 to 2 AM and 6 AM to 6.30 AM)

Load remains constant at 12 MW for time duration of 4 hours (2 AM to 6 AM)

From the above datas the load duration curve is drawn as shown in Fig. 14.13.

(*ii*) Energy required by the system in 24 hours

$$= \frac{24+12}{2}\times 2 + 12\times 4 + \frac{12+60}{2}\times 2 + 60\times 4$$

$$+ \frac{60+48}{2}\times\frac{1}{2} + \frac{48+60}{2}\times\frac{1}{2} + 60\times 4$$

$$+ \frac{60+84}{2}\times 1 + \frac{84+24}{2}\times 6$$

$$= 36 + 48 + 72 + 240 + 27 + 27 + 240 + 72 + 324$$

$$= 1{,}086 \text{ MWh}$$

$$\text{Average load} = \frac{\text{Energy required in MWh in 24 hours}}{24}$$

$$= \frac{1{,}086}{24} = 45.25 \text{ MW}$$

$$\text{Load factor} = \frac{\text{Average load}}{\text{Maximum demand}} = \frac{45.25}{84}$$

$$= 0.5387 \text{ or } 53.87\% \textbf{ Ans.}$$

(*iii*) Utilization factor

$$= \frac{\text{Maximum demand in MW}}{\text{Plant capacity in MW}} = \frac{84}{120} = 0.7 \textbf{ Ans.}$$

Example 14.10. The following data were collected from the daily load curves of a power system during a year :

Load kW	Duration hours
15,000	87
12,000	and over 876
10,000	and over 1,752
8,000	and over 2,628
6,000	and over 4,380
4,000	and over 7,000
2,000	and over 8,760

Construct the annual load duration curve and find the load factor of the system.

[*Pb. Univ. Electrical Power II (Generation) Dec. 1995*]

Solution : Annual load duration curve is plotted as shown in Fig. 14.14.

Units generated in 8,760 hours

$$= 15{,}000 \times 87 + 12{,}000\,(876 - 87) + 10{,}000\,(1{,}752 - 876) + 8{,}000\,(2{,}628 - 1{,}752) + 6{,}000\,(4{,}380 - 2{,}628) + 4{,}000\,(7{,}000 - 4{,}380) + 2{,}000\,(8{,}760 - 7{,}000)$$

$$= 51{,}053{,}000$$

$$\text{Average load} = \frac{\text{Units generated}}{\text{Time in hours}} = \frac{51{,}053{,}000}{8{,}760} = 5{,}828 \text{ kW}$$

$$\text{Load factor} = \frac{\text{Average load}}{\text{Maximum demand}} = \frac{5{,}828}{15{,}000}$$

$$= 0.3885 \text{ or } 38.85\% \textbf{ Ans.}$$

Example 14.11. Make two points of consideration for favouring the choice of each of the following alternatives of units for a thermal power plant with a maximum demand of 1,000 MW. (*i*) One unit of 1,000 MW (*ii*) Two units of 500 MW (*iii*) One unit of 400 MW and three units of 200 MW.

[*A.M.I.E. Sec. B. Power Systems Summer 1993*]

Solution : (*i*) One unit of 1,000 MW

Capital cost per kW would be less, requirement of floor area and operating staff will be less.

(*ii*) Two units of 500 MW Each

Reserve required in this case would be much smaller than that required in the former case (500 MW capacity against 1,000 MW capacity) and high operating efficiency during light load hours.

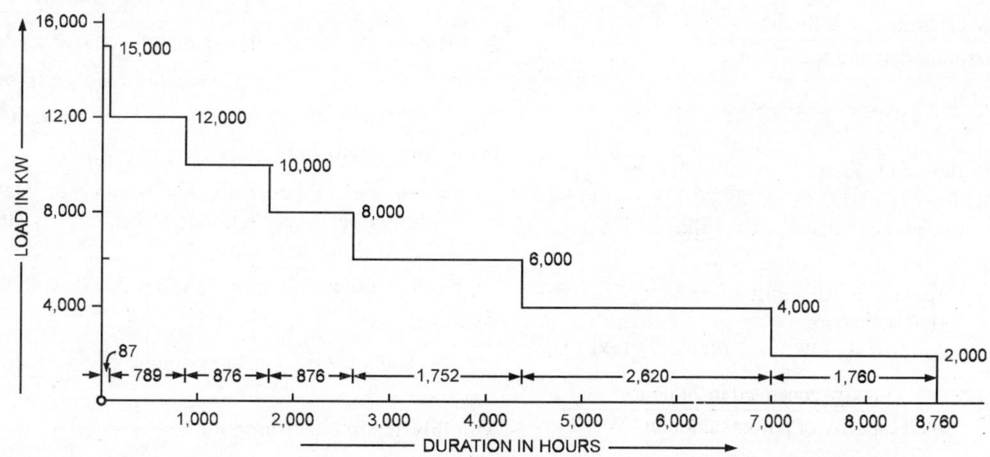

Fig. 14.14

(iii) One unit of 400 MW and Three units of 200 MW Each

There is possibility of operating each machine in such a way that it operates at full load or at a load to give maximum efficiency and reserve required would be much smaller (400 MW capacity only).

Example 14.12. A generating station a has maximum demand of 10,000 kW and the daily load on the station is as follows :

Time	kW
11 PM to 6 AM	2,000
6 AM to 8 AM	3,500
8 AM to 12 Noon	8,000
12 Noon to 1 PM	3,000
1 PM to 5 PM	7,500
5 PM to 7 PM	8,500
7 PM to 9 PM	10,000
9 PM to 11 PM	4,500

Choose the size and number of generating units. What reserve plant would be necessary ? Find the load factor and plant capacity factor. *[Pb. Univ. Electrical Power II (Generation) January 1993]*

Solution : The chronological load curve is shown in Fig. 14.15.

Maximum demand = 10,000 kW

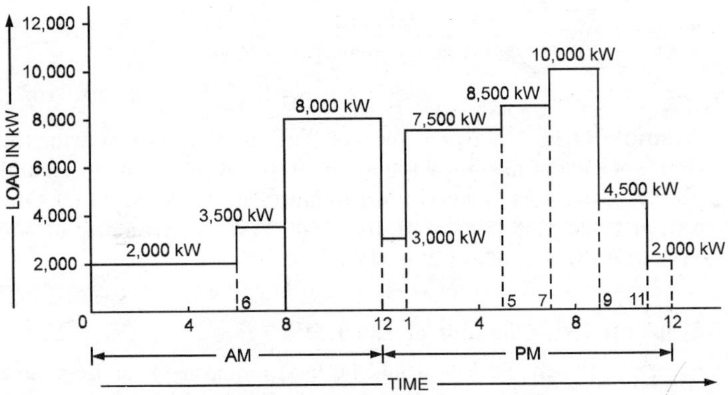

Fig. 14.15

Energy generated in 24 hours

$= 2,000 \times 7 + 3,500 \times 2 + 8,000 \times 4 + 3,000 \times 1 + 7,500 \times 4 + 8,500 \times 2 + 10,000 \times 2 + 4,500 \times 2$

$= 132,000$ kWh

$$\text{Load factor} = \frac{\text{Energy generated in 24 hours}}{\text{Maximum demand} \times 24}$$

$$= \frac{132,000}{10,000 \times 24} = 0.55 \text{ or } 55\% \text{ **Ans.**}$$

From the shape of the load curve it is obvious that three generating units of capacities 5,000, 3,000 and 2,000 kW would be the best choice to meet the given daily load on the station (see operating schedule).

Reserve plant capacity = Capacity of the unit of the largest size = 5,000 kW **Ans.**

Installed capacity = $5,000 \times 2 + 3,000 + 2,000 = 15,000$ kW

$$\text{Plant capacity factor} = \frac{\text{Energy generated in 24 hours}}{\text{Rated capacity of power station in kW} \times 24}$$

$$= \frac{132,000}{15,000 \times 24} = 0.3667 \text{ or } 36.67\% \quad \textbf{Ans.}$$

Operating schedule will be as follows :

One unit of 2,000 kW from 11 PM to 6 AM.

One unit of 2,000 kW in parallel with one unit of 3,000 kW from 6 AM to 8 AM.

One unit of 5,000 kW in parallel with one unit of 3,000 kW from 8 AM to 12 noon. (2,000 kW unit stopped)

One unit of 3,000 kW with 5,000 kW unit floating on the system from 12 noon to 1 PM.

One unit of 5,000 kW in parallel with one unit of 3,000 kW from 1 PM to 5 PM.

All three units of 5,000 kW, 3,000 kW and 2,000 kW in parallel from 5 PM to 9 PM.

One unit of 3,000 kW and one unit of 2,000 kW in parallel from 9 PM to 11 PM (5,000 kW unit stopped)

At 11 PM unit of 3,000 kW will also be stopped.

Example 14.13. A generating station has a connected load of 23 MW and a maximum demand of 20 MW, the units generated being 61.5×10^6 per annum, calculate (*a*) demand factor (*b*) Average demand (*c*) Load factor.

[Generation of Electrical Power Rajasthan Univ. 2007; Rajasthan Technical Univ. 2011]

Solution : Maximum demand = 20 MW

Connected load = 23 MW

Units generated = 61.5×10^6 kWh

$$\text{Average load} = \frac{\text{Units generated per annum in kWh}}{24 \times 365}$$

$$= \frac{61.5 \times 10^6}{24 \times 365} = 7,020 \text{ kW}$$

$$(a) \text{ Demand factor} = \frac{\text{Maximum demand}}{\text{Connected load}} = \frac{20}{23} = 0.869 \textbf{ Ans.}$$

(*b*) Average demand = 7,020 kW **Ans.**

$$(c) \text{ Load factor} = \frac{\text{Average load}}{\text{Maximum demand}} = \frac{7,020}{20 \times 1,000}$$

$= 35.1\%$ **Ans.**

Example 14.14. A consumer has the following connected loads: 10 lamps of 60 W each and two heaters of 1,000 W each. His maximum demand is 1,500 W. On the average he uses 8 lamps 5 hours a day and each heater for 3 hours a day. Find his average load, monthly energy consumption and load factor.

[A.M.I.E. Sec. B. Power Systems Winter 1993]

Solution : Maximum demand, MD = 1,500 W = 1.5 kW

Connected load = $10 \times 60 + 2 \times 1,000$ = 2,600 W

Daily energy consumption

= Number of lamps used × wattage of each lamp × working hours per day + number of heaters × wattage of each heater × working hours per day

$= 8 \times 60 \times 5 + 2 \times 1,000 \times 3 = 8,400$ Wh or 8.4 kWh

Average load

$$= \frac{\text{Daily energy consumption in Wh}}{24} = \frac{8,400}{24} = 350 \text{ W } \textbf{Ans.}$$

Monthly energy consumption

= Daily energy consumption × 30 = 8.4 × 30 = 252 kWh **Ans.**

Monthly load factor

$$= \frac{\text{Monthly energy consumption}}{\text{MD} \times 24 \times 30} = \frac{252}{1.5 \times 24 \times 30}$$

$= 0.2333$ or 23.33% **Ans.**

Example 14.15. A residential consumer has the following connected load : 8 bulbs of 100 W each; 2 fans of 60 W each, 2 light plug points of 100 W each. His use of electricity during a day is as under.

12 midnight–5 AM = One fan

5 AM–7 AM = 2 Fan + 1 light point

7 AM–9 AM – Nil

9 AM–6 PM = 2 Fans

6 PM–Midnight = 2 Fans + 4 bulbs

Find (*a*) Maximum demand (*b*) Connected load (*c*) Demand factor (*d*) Energy consumed during 24 hours.

[*Rajasthan Univ. Generation of Electrical Power 2007*]

Solution : Daily load cycle of the consumer is given below:

Time	*Watts*
12 midnight – 5 AM	1 × 60 = 60 watts
5 AM to 7 AM	2 × 60 + 1 × 100 = 220 watts
7 AM to 9 AM	Nil
9 AM to 6 PM	2 × 60 = 120 watts
6 PM to Midnight	2 × 60 + 4 × 100 = 520 watts

(*a*) Maximum demand, MD = 520 watts **Ans.**

(*b*) Connected load = Number of bulbs × wattage of each bulb + Number of fans × wattage of each fan + Number of plug points × wattage of each plug point

$= 8 \times 100 + 2 \times 60 + 2 \times 100$

$= 800 + 120 + 200 = 1{,}120$ W **Ans.**

(*c*) Demand factor $= \frac{\text{Maximum demand}}{\text{Connected load}} = \frac{520}{1{,}120} = 0.4643$ **Ans.**

(*d*) Energy consumed during 24 hours

$= 60 \times 5 + 220 \times 2 + 120 \times 9 + 520 \times 6$

$= 300 + 440 + 1{,}080 + 3{,}120 = 4{,}940$ Wh or 4.94 kWh **Ans.**

Example 14.16. A power station is to supply four regions of loads whose peak loads are 15,000 kW, 10,000 kW, 9,000 kW and 11,000 kW respectively. The diversity factor of the loads at the station is 1.5 and the average annual load factor is 50%. Calculate the maximum demand on the station and the annual energy from the station. Suggest the installed capacity and the number of units taking all aspects into account. [*A.M.I.E. Sec. B. Power Systems Summer 1995*]

Solution : Sum of individual maximum demands

$= 15{,}000 + 10{,}000 + 9{,}000 + 11{,}000 = 45{,}000$ kW

Diversity factor = 1.5

Maximum demand on the power station,

$$\text{MD} = \frac{\text{Sum of individual maximum demands}}{\text{Diversity factor}} = \frac{45{,}000}{1.5}$$

$= 30{,}000$ kW **Ans.**

Annual average load

= MD × average annual load factor = 30,000 × 0.5 = 15,000 kW

Annual energy supplied

= Annual average load × 365 × 24 = 15,000 × 8,760

$= 131.4 \times 10^6$ kWh **Ans.**

Consider that there is an increase of maximum demand by 50% in the next 5 year period. Thus maximum demand will become 30,000 × 1.5 = 45,000 kW or 45 MW. Hence three similar units of 15 MW each can be chosen to meet the present demand with one unit standby to meet emergency breakdown and carry out routine maintenance and repairs. It would necessitate the minimum number of spare parts and appearance would be symmetrical and good looking. One additional unit of 15 MW will be installed with the growth of demand, for which provision will be made now.

Example 14.17. A power station has to meet the following demand:

Group A : 200 kW between 8 AM and 6 PM

Group B : 100 kW between 6 AM and 10 AM

Group C : 50 kW between 6 AM and 10 AM

Group D : 100 kW between 10 AM and 6 PM and then 6 PM and 6 AM.

Plot daily load curve and determine (*i*) Diversity factor (*ii*) Units generated per day (*iii*) Load factor.

[*U.P. Technical Univ. Power Station Practice 2008-09*]

Solution : The given load cycle can be tabulated as under :

Time in Hours	0–6	6–8	8–10	10–18	18–24
Group A	—	—	200 kW	200 kW	—
Group B	—	100 kW	100 kW	—	—
Group C	—	50 kW	50 kW	—	—
Group D	100 kW	—	—	100 kW	100 kW
Total load on Power Station	100 kW	150 kW	350 kW	300 kW	100 kW

From the above table, it can be seen that load on the power station is as follows :

0–6 hours 100 kW

6–8 hours 150 kW

8–10 hours 350 kW

10–18 hours 300 kW

18–24 hours 100 kW

The load curve is drawn, from the given data, as shown in Fig. 14.16.

Maximum demand, MD = 350 kW

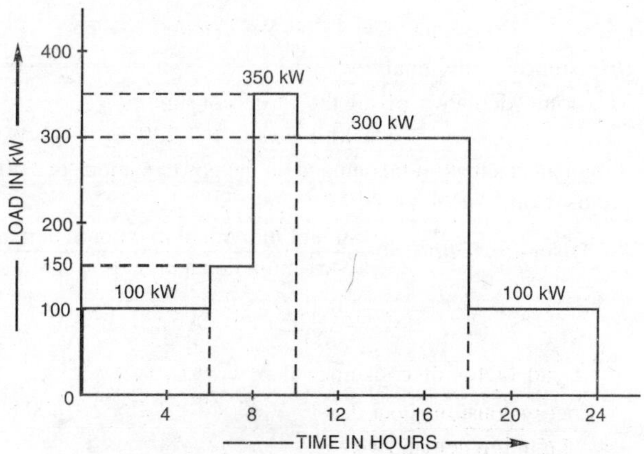

Fig. 14.16. ***Daily Load Curve For Example 14.17***

Sum of individual maximum demands of all the four groups

= 200 + 100 + 50 + 100 = 450 kW

(*i*) Diversity factor

$$= \frac{\text{Sum of individual maximum demands}}{\text{Maximum demand on power station}} = \frac{450}{350} = 1.286 \textbf{ Ans.}$$

(*ii*) Units generated per day

= Area (in kWh) under load curve

= 100 × 6 + 150 × 2 + 350 × 2 + 300 × 8 + 100 × 6

= 4,600 kWh **Ans.**

(*iii*) Average load

$$= \frac{\text{Units generated per day}}{24} = \frac{4,600}{24} = 191.7 \text{ kW}$$

Load factor

$$= \frac{\text{Average load}}{\text{Maximum demand}} = \frac{191.7}{350} = 0.548 \text{ or } 54.8\% \textbf{ Ans.}$$

Example 14.18. The daily demands of three consumers are given below :

Time	Consumer 1	Consumer 2	Consumer 3
12 Mid night to 8 AM	**No Load**	**200 W**	**No load**
8 AM to 2 PM	**600 W**	**No load**	**200 W**
2 PM to 4 PM	**200 W**	**1,000 W**	**1,200 W**
4 PM to 10 PM	**800 W**	**No load**	**No Load**
10 PM to Midnight	**No Load**	**200 W**	**200 W**

Plot the load curve and find : (*i*) maximum demand of individual consumer (*ii*) diversity factor (*iii*) load factor of individual consumer (*iv*) load factor of the station.

[*U.P. Technical Univ. Power Station Practice 2007-08*]

Solution : Daily demand of all the three consumers is as under :

0–8 hours 200 W

8–14 hours 600 + 200 = 800 W

14–16 hours 200 + 1,000 + 1,200 = 2,400 W

16–22 hours 800 W

22–24 hours 200 + 200 = 400 W

The load curve is drawn as illustrated in Fig. 14.17.

(*i*) Maximum demands of

Consumer 1 = 800 W **Ans.**

Consumer 2 = 1,000 W **Ans.**

Consumer 3 = 1,200 W **Ans.**

(*ii*) Sum of individual

Maximum demands of all the three consumers

= 800 + 1,000 + 1,200 = 3,000 W

Simultaneous maximum demand on the power stations is 2,400 W that occurs from 2 PM to 4 PM

$$\text{Diversity factor} = \frac{\text{Sum of individual maximum demands}}{\text{Maximum demand on power station}}$$

$$= \frac{3,000}{2,400} = 1.25 \textbf{ Ans.}$$

(*iii*) Load factor of consumer 1

$$= \frac{\text{Energy consumed per day}}{\text{Maximum demand} \times 24} = \frac{600 \times 6 + 200 \times 2 + 800 \times 6}{800 \times 24}$$

= 0.4583 or 45.83% **Ans.**

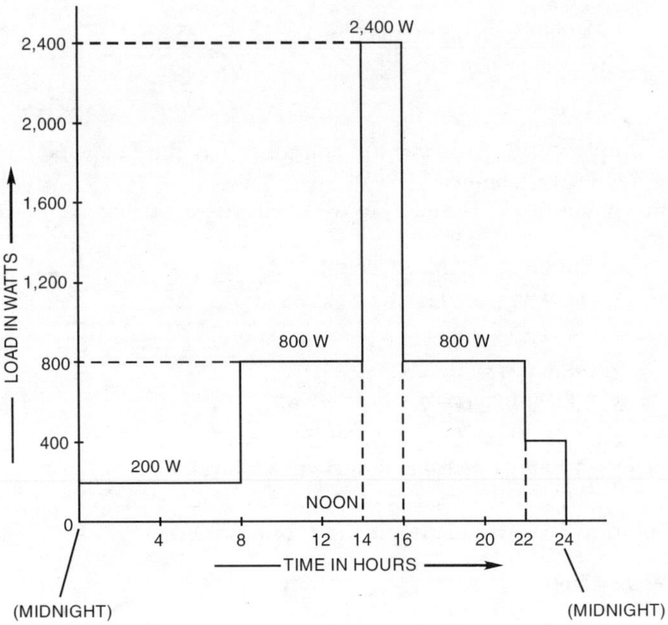

Fig. 14.17

$$\text{Load factor of consumer 2} = \frac{200 \times 8 + 1,000 \times 2 + 200 \times 2}{1,000 \times 24}$$

= 0.1667 or 16.67% **Ans.**

$$\text{Load factor of consumer 3} = \frac{200 \times 6 + 1,200 \times 2 + 200 \times 2}{1,200 \times 24}$$

= 0.1389 or 13.89% **Ans.**

(*iv*) Load factor of the power station

$$= \frac{200 \times 8 + 800 \times 6 + 2,400 \times 2 + 800 \times 6 + 400 \times 2}{2,400 \times 24}$$

= 0.2916 or 29.16% **Ans.**

Example 14.19. A power station has the following load :

Residential lighting load : Maximum load = 1,000 kW
Load factor = 20%
Diversity between consumers = 1.3

Commercial load : Maximum load = 2,000 kW
Load factor = 30%
Diversity between consumers = 1.1

Industrial load : Maximum demand = 5,000 kW

Load factor = 80%

Diversity between consumers = 1.2

Overall diversity factor may be taken as 1.4.

Find (*i*) maximum demand on system.

(*ii*) Daily energy consumption of each type of load and total energy consumption.

(*iii*) Connected load of each type, assuming the demand factor of each is 100%. [*Rajasthan Univ. Generation of Electrical Power 2000*]

Solution : The sum of maximum demands of three types of loads

= 1,000 + 2,000 + 5,000 = 8,000 kW

Group diversity factor = 1.4

(*i*) Maximum demand on the system

$$= \frac{\text{Sum of individual maximum demands}}{\text{Group diversity factor}}$$

$$= \frac{8,000}{1.4} = 5,714.28 \text{ kW} \textbf{ Ans.}$$

(*ii*) Average load = Maximum demand × load factor

So Average demand of domestic load

$$= 1,000 \times \frac{20}{100} = 200 \text{ kW}$$

Average demand of commercial load

$$= 2,000 \times \frac{30}{100} = 600 \text{ kW}$$

Average demand of industrial load

$$= 5,000 \times \frac{80}{100} = 4,000 \text{ kW}$$

∵ Daily consumption = Average load in kW × 24

Daily consumption for lighting load

= 200 × 24 = 4,800 kWh **Ans.**

Daily consumption for commercial load

= 600 × 24 = 14,400 kWh **Ans.**

Daily consumption for industrial load

= 4,000 × 24 = 96,000 kWh **Ans.**

Total energy consumption = 4,800 + 14,400 + 96,000

= 115,200 kWh **Ans.**

(*iii*) Maximum demand of domestic consumers

= Maximum domestic demands × diversity factor

= 1,000 × 1.3 = 1,300 kW

Maximum demand of commercial consumers

= 2,000 × 1.1 = 2,200 kW

Maximum demand of industrial consumers

= 5,000 × 1.2 = 6,000 kW

$$\because \text{ Connected load } = \frac{\text{Maximum demand}}{\text{Demand factor}}$$

So connected load of domestic consumers

$$= \frac{1,300}{1} = 1,300 \text{ kW} \quad \textbf{Ans.}$$

Connected load of commercial consumers

$$= \frac{2,200}{1} = 2,200 \text{ kW} \quad \textbf{Ans.}$$

Connected load of industrial consumer

$$= \frac{6,000}{1} = 6,000 \text{ kW} \quad \textbf{Ans.}$$

Example 14.20. A new housing development is to be added to the lines of public utility. There are 1,000 apartments each having a connected load of 4 kW, also stores and services are included of the characteristics shown in the following table :

Stores or Services	Connected Load (kW)	Demand Factor, Per cent
1 Laundry	20	68
2 Churches	10 each	56
1 Restaurant	60	52
1 Book store	5	66
2 Medical stores	10 each	79
2 Grocery stores	5 each	73
1 Theatre	100	48

The demand factor for the apartments is 45 per cent. The group diversity factor of the residential load for this system is 3.6 and peak diversity factor is 1.4. The commercial load group diversity factor is 1.5 and the peak diversity factor is 1.1.

Find the increase in peak demand on the total system delivery from the station bus resulting from the addition of this development on the distribution system. Assume line losses at 5% of the delivered energy.

[*A.M.I.E. Sec. B. Power Plant Engineering Summer 1994*]

Solution : Maximum demand of each apartment

= Connected load of each apartment × demand factor

= 4 × 0.45 = 1.8 kW

Maximum demand of 1,000 apartments = 1,000 × 1.8 = 1,800 kW

Demand of 1,000 apartments at the time of the system peak

$$= \frac{\text{Maximum demand of 1,000 apartments}}{\text{Group diversity factor} \times \text{peak diversity factor}} = \frac{1,800}{3.6 \times 1.4}$$

= 357 kW

Maximum demand

$$= \frac{\text{Connected load} \times \text{demand factor}}{\text{Group diversity factor}}$$

So maximum demand of all stores/services

$$= \left(\frac{20 \times 0.68 + 2 \times 10 \times 0.56 + 60 \times 0.52 + 5 \times 0.66 + 2 \times 10 \times 0.79 + 2 \times 5 \times 0.73 + 100 \times 0.48}{1.5}\right)$$

= 86.933 kW

Demand of all stores/services at the time of system peak

$$= \frac{86.933}{1.1} = 79 \text{ kW}$$

Increase in demand on the distribution system

= 357 + 79 = 436 kW

$$\text{Increase in demand on the station bus} = \frac{436}{1 - 0.05} = 459 \text{ kW} \quad \textbf{Ans.}$$

Example 14.21. The maximum demand of a power plant is 40 MW. The capacity factor is 0.5 and utilization factor is 0.8. Find

(*i*) Load factor (*ii*) Plant capacity (*iii*) Reserve capacity (*iv*) Annual energy production.

[*Rajasthan Technical Univ. Generation of Electrical Power 2008, 2009*]

Solution : Maximum demand, MD = 40 MW

Capacity factor = 0.5

Utilization factor = 0.8

Dividing Eq (14.15) by Eq (14.16), we have

$$(i) \text{ Load factor } = \frac{\text{Capacity factor}}{\text{Utilization factor}} = \frac{0.5}{0.8} = 0.625 \quad \textbf{Ans.}$$

$$(ii) \text{ Plant capacity } = \frac{\text{Maximum demand}}{\text{Utilization factor}} = \frac{40}{0.8} = 50 \text{ MW} \quad \textbf{Ans.}$$

(*iii*) Reserve capacity = Plant capacity – maximum demand

= 50 – 40 = 10 MW **Ans.**

(*iv*) Annual energy production

= Maximum demand × load factor × 365 × 24

= 40 × 0.625 × 365 × 24 kWh

= 219,000 kWh **Ans.**

Example 14.22. A generating station has a maximum demand of 20 MW, a load factor of 60%, plant capacity factor of 48% and a plant use factor of 80%. Find *(a)* the daily energy consumed *(b)* The reserve capacity of the plant *(c)* the maximum energy that could be produced daily, if the plant were running all the time and *(d)* the maximum energy that could be produced daily, if the plant were running fully loaded.

[Pb. Technical Univ. Generation of Electric Power May 2010]

Solution : Maximum demand, MD = 20 MW

Load factor, LF = 60% = 0.6

Average load = MD × LF = 20 × 0.6 = 12 MW

(*a*) Daily energy consumed = Average load × 24

= 12 × 24 = 288 MWh **Ans.**

(*b*) Generating station installed capacity

$$= \frac{\text{Average load}}{\text{Plant capacity factor}} = \frac{12}{0.48} = 25 \text{ MW}$$

Reserve capacity of the plant

= Installed capacity – maximum demand

= 25 – 20 = 5 MW **Ans.**

(*c*) The maximum energy that could be produced daily if the plant were running all the time

= Rated capacity × 24 = 25 × 24 = 600 MWh **Ans.**

(*d*) The maximum energy that could be produced daily if the plant are fully loaded

= Plant use factor × rated capacity of the plant × number of operating hours

= 0.8 × 25 × 24 = 480 MWh **Ans.**

Example 14.23. The annual load duration curve of a small hydro plant shows 438 × 10^4 kWh of energy during the year. It is a peak load plant with 20% annual load factor. Find the station capacity if plant capacity factor is 15%. Find the reserve capacity of the plant. *[Pb. Technical Univ. Generation of Electric Power May 2008]*

Solution : Average load

$$= \frac{\text{Energy generated per annum}}{24 \times 365} = \frac{438 \times 10^4}{24 \times 365} = 500 \text{ kW}$$

Maximum demand, MD

$$= \frac{\text{Average load}}{\text{Load factor}} = \frac{500}{0.2} = 2{,}500 \text{ kW or } 2.5 \text{ MW}$$

Plant capacity

$$= \frac{\text{Average load}}{\text{Capacity factor}} = \frac{500}{0.15} = 3{,}333 \text{ kW or } 3.333 \text{ MW}$$ **Ans.**

Reserve capacity

= Installed capacity – maximum demand

= 3.333 – 2.5 = 0.833 MW **Ans.**

Example 14.24. The load connected to a station is 100 MW and the demand factor of the consumers is 1.25. Estimate the capacity of the generating station when load factor is 60%, capacity factor is 50%, diversity factor of distribution is 1.33 and that of transmission is 1.5. Take the efficiency of transmission and distribution as 90%. *[Pb. Univ. Electrical Power II (Generation) Dec. 1993]*

Solution : Maximum demand on the distribution network

$$= \frac{\text{Connected load to distribution network}}{\text{Demand factor of consumers}} = \frac{100}{1.25} = 80 \text{ MW}$$

Maximum demand of the generating station

$$= \frac{\text{Maximum demand on the distribution network}}{\text{Diversity factor of distribution} \times \text{diversity factor of transmission} \times \text{efficiency of transmission and distribution}}$$

$$= \frac{80}{1.33 \times 1.5 \times 0.9} = 44.556 \text{ MW}$$

Capacity of the generating station

$$= \frac{\text{Maximum demand on the generating station}}{\text{Plant capacity factor}} \times \text{load factor}$$

$$= \frac{44.556 \times 0.6}{0.5} = 53.5 \text{ MW}$$ **Ans.**

Example 14.25. The peak load on a power station is 30 MW. The loads having maximum demands of 25 MW, 10 MW, 5 MW and 7 MW are connected to power station. The capacity of the plant is 40 MW and the annual load factor is 50%. Find (*i*) average load on the power station (*ii*) energy supplied per annum (*iii*) demand factor and (*iv*) diversity factor.

[M.D. Univ. Power Plant Engineering May 2009]

Solution : (*i*) Average load = Maximum demand × load factor

= 30 × 0.5 = 15 MW

(*ii*) Energy supplied per annum

= Average load × 24 × 365

= 15 × 24 × 365 = 131, 400 MWh or 131.4 × 10^6 kWh **Ans.**

$$(iii) \text{ Demand factor} = \frac{\text{Maximum demand}}{\text{Connected load}} = \frac{30}{25+10+5+7} = 0.6383$$ **Ans.**

$$(iv) \text{ Diversity factor} = \frac{\text{Sum of individual maximum demands}}{\text{Maximum demand of power station}}$$

$$= \frac{25+10+5+7}{30} = 1.567$$ **Ans.**

Example 14.26. The peak load on a 50 MW power station is 39 MW. It supplies power through four transformers whose connected loads are 17, 12, 9 and 10 MW. The maximum demands on these transformers are 15, 10, 8 and 9 MW respectively. If the annual load factor is 50% and the plant is operating for 65% of the period in the year, find out

(*i*) average load on the station (*ii*) energy supplied per year (*iii*) demand factor (*iv*) diversity factor (*v*) use factor for the power station. *[U.P. Technical Univ. Power Station Practice 2009-10; M.D. Univ. Power Plant Engineering May 2007]*

Solution : Power station rated capacity = 50 MW

Maximum demand on the power station = 39 MW

Sum of connected loads = 17 + 12 + 9 + 10 = 48 MW

Sum of maximum demands on transformers

= 15 + 10 + 8 + 9 = 42 MW

Annual load factor = 50% or 0.5

Plant operating period = 0.65 × 8,760 = 5,694 hours

(*i*) Average load on station = MD × LF = 39 × 0.5

= 19.5 MW **Ans.**

(*ii*) Energy supplied per year = Average load × 8,760

= 19.5 × 8,760

= 170,820 MWh **Ans.**

(*iii*) Demand factor

$$= \frac{\text{Maximum demand}}{\text{Sum of connected loads}} = \frac{39}{48} = 0.8125 \text{ or } 81.25\%$$ **Ans.**

(*iv*) Diversity factor

$$= \frac{\text{Sum of maximum demands}}{\text{Maximum deamand}} = \frac{42}{39} = 1.077$$ **Ans.**

(*v*) Use factor= $\dfrac{\text{Energy generated per year}}{\text{Rated capacity} \times \text{number of operating hours}}$

$$= \frac{170{,}820 \times 10^3}{50 \times 10^3 \times 5{,}694} = 0.6 \text{ or } 60\% \textbf{ Ans.}$$

Example 14.27. From a load duration curve, the following data are available :

The maximum demand on the system is 25 MW. The load supplied by two units is 15 MW and 12.5 MW. Unit no 1 acts as a base load unit and no 2 as a peak load unit. The base load unit works for 100% of the time and peak load unit for only 40% of the time. The energy generated by unit no 1 is 1×10^8 units and that by no 2 is 1×10^7 units. Determine the load factor, plant capacity factor, and plant use factor of each unit and load factor of the total plant.

[*Pb. Univ. Electrical Power II (Generation) May 1994*]

Solution : For Standby Unit. Assuming maximum demand on base load unit equal to its rated capacity *i.e.,* 15 MW.

Peak load on the standby unit

= Maximum demand of the system – maximum demand on base load unit

= 25 – 15 = 10 MW.

Capacity of standby unit = 12.5 MW

Annual standby unit output = 1×10^7 kWh

Hours of operation of standby unit during the year

= 0.4 × 8760 = 3,504 hours

Annual average load of standby unit

$$= \frac{\text{Output in kWh}}{24 \times 365} = \frac{1 \times 10^7}{8{,}760} = 1{,}142 \text{ kW}$$

Annual load factor

$$= \frac{\text{Annual average load}}{\text{Maximum demand}} = \frac{1{,}142}{10 \times 10^3} = 0.114 \text{ or } 11.4\% \textbf{ Ans.}$$

Plant capacity factor

$$= \frac{\text{Average Load}}{\text{Rated Capacity}} = \frac{1{,}142}{12.5 \times 10^3} = 0.0913 \text{ or } 9.13\% \textbf{ Ans.}$$

Plant use factor

$$= \frac{\text{Total units generated}}{\text{Rated capacity of unit} \times \text{number of operating hours}}$$

$$= \frac{1 \times 10^7}{12.5 \times 10^3 \times 3{,}504} = 0.2283 \text{ or } 22.83\% \textbf{ Ans.}$$

For Base Load Unit

Annual output = 1×10^8 kWh

$$\text{Annual average load} = \frac{\text{Output in kWh}}{8{,}760} = \frac{1 \times 10^8}{8{,}760} = 11{,}416 \text{ kW}$$

$$\text{Annual load factor} = \frac{\text{Annual average load}}{\text{Maximum demand}} = \frac{11{,}416}{15 \times 10^3} = 0.761 \text{ or } 76.1\% \textbf{ Ans.}$$

$$\text{Capacity factor} = \frac{\text{Annual average load}}{\text{Rated capacity}} = \frac{11{,}416}{15{,}000} = 0.761 \text{ or } 76.1\% \textbf{ Ans.}$$

$$\text{Use factor} = \frac{\text{Total kWh generated}}{\text{Rated capacity of unit} \times \text{number of operating hours}}$$

$$= \frac{1 \times 10^8}{15{,}000 \times 8{,}760} = 0.761 \text{ or } 76.1\% \textbf{ Ans.}$$

Load factor of the total plant

$$= \frac{\text{Total kWh generated by both units}}{\text{Rated capacity of the plant} \times 8{,}760}$$

$$= \frac{1 \times 10^8 + 1 \times 10^7}{(15 + 12.5) \times 10^3 \times 8{,}760} = 0.4566 \text{ or } 45.66\% \textbf{ Ans.}$$

Example 14.28. The yearly duration curve of a certain plant can be considered as a straight line from 140 MW to 30 MW. Power is supplied with one generating unit of 95 MW capacity and two units of 45 MW capacity each. Determine :

(*i*) Installed capacity (*ii*) Load factor (*iii*) Plant capacity factor (*iv*) Maximum Demand (*v*) Utilization factor.

[*Pb. Univ. Electrical Power II (Generation) Dec. 1993*]

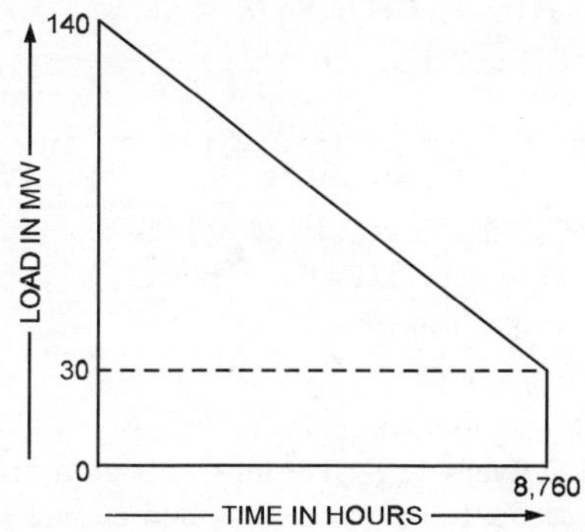

Fig. 14.18. ***Load Duration Curve For Example 14.28***

Solution : (*i*) Installed capacity = 95 + 2 × 45 = 185 MW **Ans.**

(*iv*) Maximum demand = 140 MW

$$\text{Average load} = \frac{140 + 30}{2} = 85 \text{ MW}$$

$$(ii) \text{ Load factor} = \frac{\text{Average load}}{\text{Maximum demand}} = \frac{85}{140} = 0.607 \text{ or } 60.7\% \textbf{ Ans.}$$

$$(iii) \text{ Plant capacity factor} = \frac{\text{Average load}}{\text{Rated capacity}} = \frac{85}{185} = 0.46 \text{ or } 46\% \textbf{ Ans.}$$

$$(v) \text{ Utilization factor} = \frac{\text{Maximum demand}}{\text{Rated capacity}} = \frac{140}{185} = 0.757 \text{ or } 75.7\% \textbf{ Ans.}$$

Example 14.29. For a certain industrial organisation the maximum and minimum power demands are 50 MW and 10 MW respectively. Its load duration curve can be approximated as shown in Fig. 14.19. The hydro power available at the time of minimum

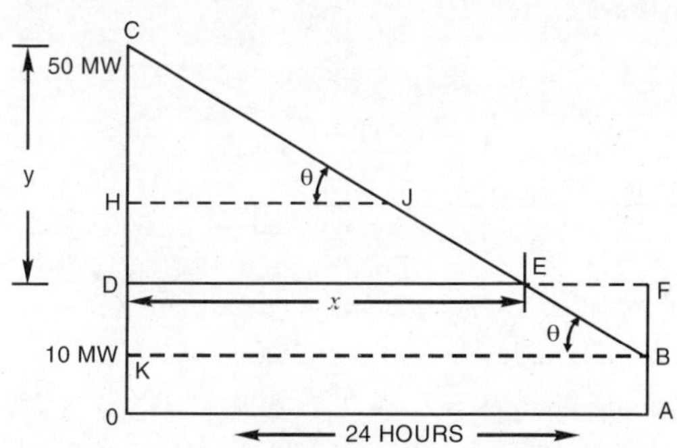

Fig. 14.19

regulated flow is just sufficient to take a peak load of 60 MWh per day. Further, it is observed that it will be economical to pump water from tail race to the reservoir tank by utilizing the steam power plant during off peak periods and thus running the station at 100 per cent load factor. If the efficiency of the steam electric conversion is 60%, calculate the maximum capacity of each type of plant.

[*U.P.S.C.I.E.S. Electrical Engineering-II 2006*]

Solution : Let the load of 60 MWh per day be represented by ΔCJH.

So $\frac{1}{2}$HJ × CH = 60 MWh

or HJ × CH = 120 ...(*i*)

From Fig. 14.19

$$HJ = \frac{CH}{\text{Tan } \theta} = \frac{CH}{CK/BK} = \frac{CH}{(50-10)/24} = \frac{24}{40} CH = 0.6\ CH$$

Substituting HJ = 0.6 CH in Eq (*i*), we have

0.6 CH × CH = 120

$$\text{or } CH = \sqrt{\frac{120}{0.6}} = 14.142 \text{ MW}$$

So Maximum capacity of hydropower plant = 14.142 MW

Maximum capacity of steam power plant

$$= \frac{50 - 14.142}{0.6} = 59.763 \text{ MW} \quad \textbf{Ans.}$$

Example 14.30. A generating station has an installed capacity of 50,000 kW and delivers 220 × 10^6 units per annum. If the annual fixed charges are Rs 160 per kW installed capacity and running charges are 4 paise per kWh, determine the cost per unit generated.

[*Pb. Technical Univ. Generation of Electric Power December 2010*]

Solution : Installed capacity = 50,000 kW

Fixed charges = Rs 160 per kW installed capacity

Total fixed charges = Rs 160 × 50,000 = Rs 8,000,000

Total number of units generated per annum

= 220 × 10^6

Running cost = 4 paise per kWh

$$\text{Total running cost} = \text{Rs } \frac{4}{100} \times 220 \times 10^6 = \text{Rs } 8,800,000$$

Total annual charges = Fixed charges + running charges

= Rs 8,000,000 + Rs 8,800,000

= 16,800,000

$$\text{Cost per unit generated} = \frac{\text{Total annual charges per annum}}{\text{Total delivered units per annum}}$$

$$= \text{Re } \frac{16,800,000}{220 \times 10^6}$$

= Re 0.0764 or paise 7.64 **Ans.**

Example 14.31. The output of generating station is 500 × 10^6 kWh per annum. The average load factor is 60%. If the annual fixed charges are Rs 60 per kW of installed capacity and the annual running charges are 5 paise per kWh, what is the cost of energy per kWh at the bus-bars ? [*Agra Univ. Generation and Switchgear Practice 1982*]

Solution : Output energy per annum = 500 × 10^6 kWh

$$\text{Average load} = \frac{500 \times 10^6}{24 \times 365} = 57,000 \text{ kW}$$

$$\text{Maximum demand} = \frac{\text{Average load}}{\text{Load factor}} = \frac{57,000}{0.6} = 95,000 \text{ kW}$$

Assuming installed capacity equal to maximum demand

Fixed charges = Rs 60 × 95,000 = Rs 5,700,000

Annual running charges = Rs 0.05 × 500 × 10^6 = Rs 25,000,000

Total annual charges = Rs 5,700,000 + Rs 25,000,000

= Rs 30,700,000

Cost of energy at the bus-bar

$$= \frac{30,700,000}{500 \times 10^6} = \text{Re } 0.0614 \text{ per kWh or } 6.14 \text{ P per kWh } \textbf{Ans.}$$

Example 14.32. A generating plant has a maximum capacity of 100 kW and costs Rs 160,000. The fixed charges are 12% consisting of 5% interest, 5% depreciation and 2% taxes. Find the fixed charges per kWh if the load factor is (i) 100% and (ii) 50%.

[*U.P. Technical Univ. Power Station Practice 2006-07*]

Solution : Assuming maximum demand equal to generating plant capacity

Maximum demand = 100 kW

Annual fixed charges = Rs 0.12 × 160,000 = Rs. 19,200

(*i*) *When load factor is 100% i.e. unity*

Units generated per annum

= Maximum demand × LF × hours in a year

= 100 × 1 × 8,760 = 876,000 kWh

$$\text{Fixed charges per kWh} = \frac{\text{Annual fixed charges}}{\text{Units generated / annum}}$$

$$= \frac{19,200}{876,000} = \text{Re } 0.0219 \text{ or } 2.19 \text{ paise } \textbf{Ans.}$$

(*ii*) *When load factor is 50% i.e., 0.5*

Units generated per annum = 100 × 0.5 × 8,760 = 438,000 kWh

$$\text{Fixed charges/kWh} = \frac{19,200}{438,000} = \text{Re. } 0.0438 \text{ or } 4.38 \text{ paise } \textbf{Ans.}$$

Example 14.33. The annual fixed and operating cost of a 500 MW station are Rs 3,000 per kW of the installed capacity per annum and Rs 0.60 per kWh respectively. Plot the variation of unit cost of energy with load factor. [*A.M.I.E. Sec. B. Power Systems Summer 2000*]

Solution : Installed capacity = 500 MW or 5 × 10^5 kW

Assuming maximum demand on the power station to be equal to installed capacity *i.e.*, 5 × 10^5 kW

Fixed cost = Rs. 3,000 × 5 × 10^5 = Rs. 15 × 10^8

Units generated per annum for unity load factor

= 5 × 10^5 × 8,760 = 438 × 10^7 kWh

Units generated per annum for load factors 0.2, 0.4, 0.6 and 0.8 will be 0.2, 0.4, 0.6 and 0.8 times of units generated per annum for unity load factor *i.e.*, 87.6 × 10^7 kWh, 175.2 × 10^7 kWh, 262.8 × 10^7 kWh 350.4 × 10^7 kWh respectively.

Total cost of electrical energy generated per unit for station load factor of unity

= Fixed cost per kWh + operating cost per kWh

$$= \frac{\text{Total fixed cost}}{\text{Total units generated}} + \text{operating cost per kWh}$$

$$= \text{Re } \frac{15 \times 10^8}{438 \times 10^7} + \text{Re } 0.60 = \text{Re } 0.34 + \text{Re } 0.60 = \text{Re } 0.94/\text{unit}$$

Similarly total cost per unit of electrical energy generated for station load factor of 0.8

$$= \text{Re } \frac{15 \times 10^8}{350.4 \times 10^7} + \text{Re } 0.60 = \text{Re } 0.43 + \text{Re } 0.60 = \text{Rs } 1.03/\text{unit}$$

Total cost per unit of electrical energy generated for station load factor of 0.6

$$= \text{Re } \frac{15 \times 10^8}{262.8 \times 10^7} + \text{Re } 0.60$$

= Re 0.57 + Re 0.60 = Rs 1.17/unit

Total cost per unit of electrical energy generated for station load factor of 0.4

$$= \text{Re}\ \frac{15 \times 10^8}{175.2 \times 10^7} + \text{Re}\ 0.60 = \text{Re}\ 0.86 + \text{Re}\ 0.60 = \text{Rs}\ 1.46/\text{unit}$$

Total cost per unit of electrical energy generated for station load factor of 0.2

$$= \text{Re}\ \frac{15 \times 10^8}{87.6 \times 10^7} + \text{Re}\ 0.60 = \text{Rs}\ 1.71 + \text{Re}\ 0.60 = \text{Rs}\ 2.31/\text{unit}$$

Variation of unit cost of energy with load factor is shown in Fig. 14.20.

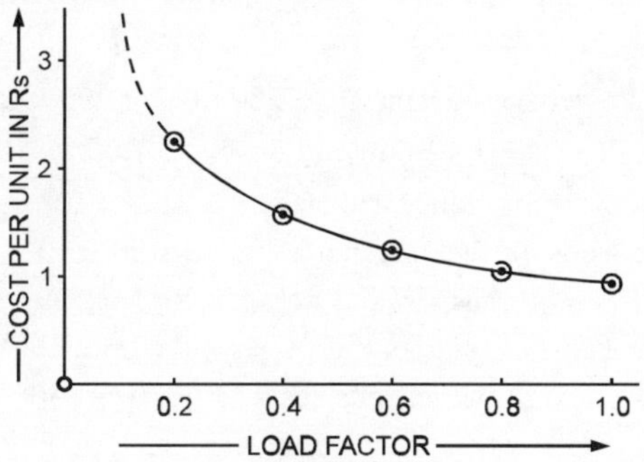

Fig. 14.20

Example 14.34. A peak load on the thermal plant of 12 MW capacity is 10 MW. The plant annual load factor is 0.7. Design a two part tariff from the following data :

Cost of plant Rs 7,000/kW installed capacity

Interest and depreciation = 10% of the capital cost

The capital cost of transmission and distribution = Rs 3 × 10^5

Interest and depreciation on it= 5%

Operating cost = Rs 3 × 10^5/year

Cost of fuel = Rs 50/ton

Plant maintenance cost = Rs 25,000/year (fixed)

Plant maintenance cost = Rs 35,000/year (running)

Coal used = 3 × 10^4 ton per year

[*M.D. Univ. Power Plant Engineering May 2009*]

Solution : Installed capacity = 12 MW

Maximum demand, MD = 10 MW

Plant annual load factor = 0.7

Total units generated/annum = MD × load factor × 8,760

= 10 × 0.7 × 8,760

= 61,320 MWh = 61.32 × 10^6 kWh

Fixed costs

Capital cost of plant = Rs 7,000 × 12 × 10^3 = Rs 84 × 10^6

Interest and depreciation (fixed cost)

$$= \text{Rs}\ \frac{10}{100} \times 84 \times 10^6 = \text{Rs}\ 8.4 \times 10^6$$

Capital cost of transmission and distribution = Rs 3 × 10^5

Interest and depreciation on transmission and distribution

$$= \text{Rs}\ \frac{5}{100} \times 3 \times 10^5 = \text{Rs}\ 0.15 \times 10^5$$

Plant maintenance cost (fixed) = Rs 25,000

Total fixed cost = Rs (8.4 × 10^6 + 15,000 + 25,000)

= Rs 8,440,000

So fixed cost per kW per annum

$$= \text{Rs}\ \frac{8{,}440{,}000}{\text{MD}} = \text{Rs}\ \frac{8{,}440{,}000}{10{,}000} = \text{Rs}\ 844\ \text{per kW/annum}$$

Running costs

Operating cost = Rs 300,000 per annum

Maintenance cost = Rs 35,000 per annum

Fuel cost = Rs 50 × 3 × 10^4 = Rs 1,500,000

Total running cost = Rs 1,500,000 + 300,000 + 35,000

= Rs 1,835,000

$$\text{Running cost per kWh} = \frac{\text{Total running cost/annum}}{\text{Units generated/annum}}$$

$$= \text{Re}\ \frac{1{,}835{,}000}{61.32 \times 10^6}$$

= Re 0.03/kWh or 3 paise per kWh

So two part tariff is Rs 844 per kW of maximum demand per annum + 3 paise per kWh **Ans.**

Example 14.35. What are the different factors related to plants and consumers for power plant economics ?

Estimate the generating cost per kWh delivered from a generating station from the following data :

Plant capacity = 50 MW

Annual load factor = 40%

Capital cost = Rs 1.2 crores, annual cost of wages, taxation etc. Rs. 4 lacs, cost of fuel, lubrication, maintenance etc. = 1.0 paise/kWh generated, interest 5% per annum, depreciation 6% per annum of initial value.

[*U.P. Technical Univ. Power Station Practice 2006-07*]

Solution : Assuming maximum demand equal to the capacity of generating plant

Maximum demand = 50 MW

Capital cost = Rs 1.2 crores = Rs 1.2 × 10^7

Interest and depreciation per annum

$$= \text{Rs}\ \frac{5+6}{100} \times 1.2 \times 10^7 = \text{Rs}\ 13.2 \times 10^5$$

Wages and taxation = Rs 4 × 10^5

Total annual fixed charges = Rs (13.2 × 10^5 + 4 × 10^5)

= Rs 17.2 × 10^5

Units generated per annum = MD × LF × 8,760 kWh

$$= 50 \times 10^3 \times \frac{40}{100} \times 8{,}760$$

= 1.752 × 10^5 kWh

Annual cost of fuel, lubrication and maintenance etc.

= Rs 0.01 × 1,752 × 10^5 = Rs 17.52 × 10^5

Total annual charges

= Total annual fixed charges + Total running charges

= Rs (17.20 × 10^5 + 17.52 × 10^5) = Rs 34.72 × 10^5

Generating cost per kWh

$$= \frac{\text{Total annual charges}}{\text{Total units generated / annum}}$$

$$= \text{Rs}\ \frac{34.72 \times 10^5}{1{,}752 \times 10^5} = \text{Re}\ 0.0198 \text{ or } 1.98 \text{ paise}$$ **Ans.**

Example 14.36. A power plant of 180 MW installed capacity has the following data :

Capital cost = Rs 2,160/kW installed

Interest and depreciation = 12%

Annual load factor = 60%

Annual capacity factor = 50%
Annual running charges = Rs 36 × 10^6
Energy consumed by power auxiliaries = 60%
Calculate (*i*) reserve capacity (*ii*) Cost of generation per kWh

[*M.D. Univ. Power Plant Engineering December 2009*]

Solution : Total installed capacity of plant = 180 MW

Annual capacity factor = 50% = 0.5

Average load = Plant installed capacity × capacity factor = 180 × 0.5 = 90 MW

Annual load factor = 60% = 0.6

$$\text{Maximum demand, MD} = \frac{\text{Average load}}{\text{Load factor}} = \frac{90}{0.6} = 150 \text{ MW}$$

(*i*) Reserve capacity = Installed capacity – MD = 180 – 150 = 30 MW **Ans.**

Total units generated per annum

= Average load × 8,760 = 90 × 10^3 × 8,760 = 788.4 × 10^6 kWh

Net energy delivered per annum

= (Total units generated – energy consumed by power auxiliaries) per annum

$$= 788.4 \times 10^6 - \frac{6}{100} \times 788.4 \times 10^6 = 741.096 \times 10^6 \text{ kWh}$$

Capital cost of plant = Rs 2,160 × 180 × 10^3 = Rs 388.8 × 10^6

Interest and depreciation on plant capital cost

$$= \text{Rs } 388.8 \times 10^6 \times \frac{12}{100} = \text{Rs } 46.656 \times 10^6$$

Annual running charges = Rs 36 × 10^6

Total annual cost = Rs 46.656 × 10^6 + Rs 36 × 10^6 = Rs 82.656 × 10^6

Cost of generation per kWh

$$= \frac{\text{Total annual cost}}{\text{Energy delivered per annum}}$$

$$= \text{Re } \frac{82.656 \times 10^6}{741.096 \times 10^6}$$

= Re 0.1115 per unit or 11.15 paise per unit **Ans.**

Example 14.37. Determine the generation cost per unit of energy from the following plant data.

Installed capacity = 120 MW
Capital cost of plant = Rs 10,000 per kW
Interest and depreciation = 15%
Fuel consumption = 0.64 kg/kWh
Fuel cost = Rs 500 per 1,000 kg
Salaries, wages, repairs and other operating costs per annum = Rs 10,000,000/-
Peak load = 100 MW
Load factor = 60%.

[*Rajasthan Technical Univ. Generation of Electrical Power 2011*]

Solution : Energy generated per annum

= Maximum demand × load factor × 8,760

$$= 100 \times 10^3 \times \frac{60}{100} \times 8{,}760 = 525.6 \times 10^6 \text{ kWh}$$

Fixed cost (annual interest and depreciation on capital cost)

$$= \text{Rs } 10{,}000 \times 120 \times 10^3 \times \frac{15}{100} = \text{Rs } 180 \times 10^6/\text{annum}$$

Fuel consumption = 0.64 × 525.6 × 10^6 = 336.384 × 10^6 kg

$$\text{Fuel cost} = \text{Rs } \frac{500}{1{,}000} \times 336.384 \times 10^6 = \text{Rs } 168.192 \times 10^6$$

Annual other cost = Rs 10 × 10^6

Total annual cost = Rs (180 × 10^6 + 168.192 × 10^6 + 10 × 10^6) = Rs 358.192 × 10^6

$$\text{Generation cost} = \frac{\text{Total annual cost}}{\text{Units generated per annum}}$$

$$= \text{Re } \frac{358.192 \times 10^6}{525.6 \times 10^6} = \text{Re } 0.68/\text{kWh}$$

or 68 paise per kWh **Ans.**

Example 14.38. A 200 MW steam power station is estimated to cost Rs 1,000 millions. Other costs are as follows :

(*a*) **Insurance and taxes Rs 20 lacs/annum.**
(*b*) **Fuel and lubricants Rs 200 lacs/annum.**
(*c*) **Transport and storage Rs 30 lacs/annum.**
(*d*) **Salaries and wages Rs 36 lacs/annum.**
(*e*) **Miscellaneous Rs 6 lacs/annum.**

Reckoning interest and depreciation at 15% per annum of capital cost, determine the cost of energy generated per unit if the power station works at an average load factor of 60%. What would it be, if the load factor be increased to 75% with a consequent increase in fuel costs by 10%, other costs remaining the same.

Solution : Power station capacity = 200 MW or 200 × 10^3 kW

Power station cost = Rs 1,000 millions = Rs 1,000 × 10^6

Annual cost of insurance and taxes = Rs 2 × 10^6

Annual cost of fuel and lubricants = Rs 20 × 10^6

Annual cost of transport and storage = Rs 3 × 10^6

Annual cost of salaries and wages = Rs 3.6 × 10^6

Annual miscellaneous costs = Rs 0.6 × 10^6

$$\text{Annual interest and depreciation} = \text{Rs } 1{,}000 \times 10^6 \times \frac{15}{100} = \text{Rs } 150 \times 10^6$$

Total annual cost

= Rs (2 + 20 + 3 + 3.6 + 0.6 + 150) = Rs 179.2 × 10^6

Total energy generated per annum

= Maximum demand × average load factor × 8,760

= 200 × 10^3 × 0.6 × 8,760 = 1,051.2 × 10^6 kWh

Generation cost

$$= \frac{\text{Total annual cost}}{\text{Total generation}} = \text{Re } \frac{179.2 \times 10^6}{1{,}051.2 \times 10^6}$$

= Re 0.17 or Paise 17/kWh **Ans.**

If the load factor is improved to 75%

Total energy generated per annum

= 200,000 × 0.75 × 8,760 = 1,314 × 10^6 kWh

Increase in operating cost

$$= 10\% \text{ of fuel cost} = \text{Rs } 20 \times 10^6 \times \frac{10}{100} = \text{Rs } 2 \times 10^6$$

Hence total operating cost

= Rs 179.2 × 10^6 + Rs 2 × 10^6 = Rs 181.2 × 10^6

Generation cost

$$= \text{Re } \frac{181.2 \times 10^6}{1{,}314 \times 10^6} = \text{Re } 0.138 \text{ or paise } 13.8/\text{kWh}$$ **Ans.**

Note : Maximum demand has been assumed equal to power station capacity.

Example 14.39. The following data refer to a 100 MW thermal power plant :

Capital cost = Rs 3 × 10^9; Annual interest and depreciation rate = 12%; Annual taxation = Rs 10^7; Annual salary and wages = Rs 2 × 10^7; Cost of coal of Rs 700/tonne; Heating value of coal = 14,000 kJ/kg; Thermal efficiency = 38% at 100% load and 30% at 50% load. Calculate unit cost in Rs/kWh at 100% and 50% load. Comment on your answer.

[*A.M.I.E. Sec. B. Power Plant Engineering Winter 1994*]

Solution : Capital cost = Rs 3 × 10^9

Annual interest and depreciation = 12% of Rs 3 × 10^9

= Rs 360 × 10^6

Annual taxation = Rs 10 × 10^6

Annual salary and wages = Rs 20 × 10^6

Units generated at 100% capacity factor

= 100 × 10^3 × 8,760 = 876 × 10^6 kWh

Coal required per annum

$$= \frac{876 \times 10^6 \times 1{,}000 \times 3{,}600}{0.38 \times 14{,}000 \times 1{,}000} = 592.782 \times 10^6 \text{ kg}$$

∵ 1 kWh = 1,000 × 3,600 watt-seconds

$$\text{Annual fuel cost} = \text{Rs } \frac{700 \times 592.78 \times 10^6}{1{,}000} = \text{Rs } 414.95 \times 10^6$$

Total annual costs = Rs (360 + 10 + 20 + 414.95) × 10^6

= Rs 804.95 × 10^6

$$\text{Generating cost} = \text{Re } \frac{804.95 \times 10^6}{876 \times 10^6} = \text{Re } 0.92/\text{kWh}$$ **Ans.**

Units generated at 50% capacity factor

= 100 × 10^3 × 0.5 × 8,760 = 438 × 10^6 kWh

Coal required per annum

$$= \frac{438 \times 10^6 \times 1{,}000 \times 3{,}600}{0.3 \times 14{,}000 \times 1{,}000} = 375.4286 \times 10^6 \text{ kg}$$

Annual fuel cost

$$= \text{Rs } \frac{700 \times 375.4286 \times 10^6}{1{,}000} = \text{Rs } 262.80 \times 10^6$$

Total annual costs

= Rs (360 + 10 + 20 + 262.80) × 10^6 = Rs 652.80 × 10^6

$$\text{Generating cost} = \text{Rs } \frac{652.80 \times 10^6}{438 \times 10^6} = \text{Rs } 1.49/\text{kWh}$$ **Ans.**

As the capacity factor or load factor falls, generating cost per unit increases due to fixed annual costs of interest and depreciation, taxation, salary and wages and poor operating efficiency of the plant at low capacity or load factor.

Example 14.40. An industrial consumer is offered electrical energy from (*a*) private diesel generating station (*b*) public supply undertaking. Calculate which scheme will be economical for the consumer having a maximum demand of 2,000 kW at 0.8 pf lagging at 40% load factor in each case. Given that

Private Plant : Capital cost Rs 1,200,000, cost of fuel oil Rs 70.00 per tonne, consumption 0.3 kg/kWh, cost of lubricating oil 0.15 paise/kWh. Taxation and insurance 0.25 paise per kWh generated, cost of repairs 0.2 paise/kWh. Wages Rs 30,000 per annum. Rate of interest and depreciation is 12%.

Public Supply Undertaking : The tariff is Rs 50.00 per kVA of maximum demand plus 2.0 paise per kWh consumed.

[Pb. Univ. Elec. Power II (Generation) December 1994]

Solution : Maximum demand, MD = 2,000 kW

Load factor, LF = 40% or 0.4

Average load = MD × LF = 2,000 × 0.4 = 800 kW

Units generated per annum

= 800 × 8,760 = 7.008 × 10^6 kWh

Maximum demand in kVA

$$= \frac{\text{kW}}{\text{Power factor}} = \frac{2{,}000}{0.8} = 2{,}500$$

(*a*) **Private Generating Plant :**

Capital cost = Rs 1.2 × 10^6

Annual interest and depreciation

= 12% of Rs 1.2 × 10^6 = Rs 14.4 × 10^4

$$\text{Annual fuel cost per kWh} = \text{Re } \frac{70}{1{,}000} \times 0.3 = \text{Re } 0.021 \text{ or paise } 2.1$$

Annual wages = Rs 30,000

Total annual charges

= Annual charges of interest and depreciation + annual wages + annual fuel cost + annual lubricating oil cost + annual taxation and insurance + annual repair cost

$$= \text{Rs } 144{,}000 + \text{Rs } 30{,}000 + \text{Rs } \frac{7.008 \times 10^6}{100}(2.1 + 0.15 + 0.25 + 0.2)$$

= Rs 144,000 + Rs 30,000 + 7.008 × 10^4 × 2.7 = Rs 363,216.00

(*b*) **Public Supply Undertaking :**

Maximum demand charges = Rs 50 × 2,500 = Rs 125,000.00

$$\text{Electrical energy charges} = \text{Rs } \frac{2}{100} \times 7.008 \times 10^6 = \text{Rs } 140{,}160.00$$

Total annual charges = Rs (125,000 + 140,160) = Rs 265,160.00

Evidently alternative of supply (*b*) is quite economical.

Example 14.41. Compare the cost of generation per kWh between (*i*) a nuclear station and (*ii*) hydroelectric station when the system requires a load of 400 MW with a load factor of 80%. The capital cost in the case of nuclear station is Rs 3,000.00 per kW of installed capacity and in case of hydroelectric station it is Rs 2,000 per kW of installed capacity. The operating cost for the nuclear unit and the hydro unit are 2 P and 1.25 P respectively. Transmission cost per kWh is negligible in case of nuclear plant, while it is 0.25 P per kWh in hydro plant. The rate of interest and depreciation may be taken as 12% in both the cases.

[U.P.S.C. I.E.S. Generation, Transmission & Distribution of Power 1972]

Solution : Maximum load = 400 MW = 400,000 kW

Load factor = 80% = 0.8

Unit generated per annum = Maximum load × load factor × 365 × 24

= 400,000 × 0.8 × 8,760 = 28 × 10^8 kWh

(*i*) **Nuclear Power Station :**

Capital cost = Rs 3,000 × 400,000 = Rs 12 × 10^8

Annual cost towards interest and depreciation

$$= \text{Rs } 12 \times 10^8 \times \frac{12}{100} = \text{Rs } 1.44 \times 10^8$$

Annual operating cost = Rs 0.02 × 28 × 10^8 = Rs 0.56 × 10^8

Total annual cost = Rs 1.44 × 10^8 + Rs 0.56 × 10^8 = Rs 2.0 × 10^8

$$\text{Cost of generation per kWh} = \frac{2.0 \times 10^8}{28 \times 10^8}$$

= Re 0.0714/kWh

= Paise 7.14/kWh **Ans.**

(*ii*) **Hydro Power Station :**

Capital cost = Rs 2,000.00 × 400,000 = Rs 8 × 10^8

Annual interest and depreciation

$$= \text{Rs } 8 \times 10^8 \times \frac{12}{100} = \text{Rs } 0.96 \times 10^8$$

Annual operating cost = Rs 0.0125 × 28 × 10^8 = Rs 0.35 × 10^8

Annual transmission cost = 0.0025 × 28 × 10^8 = Rs 0.07 × 10^8

Total annual cost = Rs 0.96 × 10^8 + Rs 0.35 × 10^8 + Rs 0.07 × 10^8

= Rs 1.38 × 10^8

Cost of generation per kWh

$$= \text{Rs } \frac{1.38 \times 10^8}{28 \times 10^8} = \text{Re } 0.0493/\text{kWh} = \text{Paise } 4.93/\text{kWh}$$ **Ans.**

Example 14.42. A power system, having a maximum demand of 100 MW, has a load factor of 30%. It can be supplied by one of the following schemes :

(*i*) A steam station capable of supplying the whole load.

(*ii*) A nuclear station capable of supplying the whole load.

(*iii*) A hydro station capable of supplying the whole load.

(*iv*) A steam power station in conjunction with a hydro station, the latter supplying 100 × 10^6 units per annum with a maximum output of 40 MW.

	Steam	Hydro	Nuclear
Capital cost per kW installed capacity	**Rs 600**	**Rs 1,500**	**Rs 2,000**
Interest and depreciation on capital	**12%**	**10%**	**10%**
Operating cost per unit	**5 P**	**1 P**	**2 P**
Transmission cost per unit	**negligible**	**0.5 P**	**negligible**

Workout total cost of the scheme per annum in each case.

[*U.P.S.C. I.E.S. Generation, Transmission and Distribution of Power 1971*]

Solution :

Maximum demand on the power system = 100 MW = 100,000 kW

Load factor = 30% = 0.3

Average load = 100,000 × 0.3 = 30,000 kW

Total number of units generated per annum

= 30,000 × 8,760 = 26.28 × 10^7 kWh

(*i*) Steam Power Station :

Capital cost = 600 × 100,000 = Rs 6 × 10^7

Annual interest and depreciation

= Rs 6 × 10^7 × 0.12 = Rs 0.72 × 10^7 = Rs 72 lacs

Annual operating cost = Rs 0.05 × 26.28 × 10^7 = Rs. 131.4 lacs

Annual transmission cost = negligible

Total annual cost = Rs 72 lacs + Rs 131.4 lacs

= Rs 203.4 lacs **Ans.**

(*ii*) Nuclear Power Station :

Capital cost = Rs 2,000 × 100,000 = Rs 20 × 10^7

Annual interest and depreciation = Rs 20 × 10^7 × 0.1 = Rs 200 lacs

Annual operating cost = Rs 0.02 × 26.28 × 10^7

= Rs 52.56 lacs

Annual transmission cost = negligible

Total annual cost = Rs 200 lacs + Rs 52.56 lacs

= Rs 252.56 lacs **Ans.**

(*iii*) Hydropower Station :

Capital cost = Rs 1,500 × 100,000 = Rs 15 × 10^7

Annual interest and depreciation

= Rs 15 × 10^7 × 0.1 = Rs 150 lacs

Annual operating cost = Rs 0.01 × 26.28 × 10^7 = Rs 26.28 lacs

Annual transmission cost = Rs 0.005 × 26.28 × 10^7 = Rs 13.14 lacs

Total annual cost = Rs (150 + 26.28 + 13.14) lacs

= Rs 189.42 lacs **Ans.**

(*iv*) Steam Power Station in conjunction with a Hydropower Station :

Maximum demand on hydropower station = 40,000 kW

Annual units generated by hydropower station = 10 × 10^7 kWh

Capital cost of hydropower station = Rs 1,500 × 40,000

= Rs 600 lacs

Annual interest and depreciation = Rs 600 × 0.1 = Rs 60 lacs

Annual operating cost = Rs 0.01 × 10 × 10^7 = Rs 10 lacs

Annual transmission cost = Rs 0.005 × 10 × 10^7 = Rs 5 lacs

Maximum demand on steam power station

= 100,000 – 40,000 = 60,000 kW

Annual units generated by steam power station

= 26.28 × 10^7 – 10 × 10^7 = 16.28 × 10^7 kWh

Capital cost of steam power station

= Rs 600 × 60,000 = Rs 360 lacs

Annual interest and depreciation

= Rs 360 × 0.12 = Rs 43.2 lacs

Annual operating cost

= Rs 0.05 × 16.28 × 10^7 = Rs 81.40 lacs

Annual transmission cost = negligible

Total annual cost = Rs (60.00 + 10.00 + 5.00 + 43.20 + 81.40) lacs

= Rs 199.60 lacs **Ans.**

Example 14.43. A load having a maximum demand of 100 MW at 30% load factor may be supplied by one of the following schemes :

(*i*) A steam plant capable of supplying the whole load.

(*ii*) A steam plant in conjunction with a pumped storage plant capable of supplying 10^6 kWh energy per year with a maximum load of 40 MW. Using the following data, find the most economic scheme among the two.

Capital cost of steam station is Rs 1,000/kW of installed capacity. Capital cost of pumped storage plant is Rs 700/kW of installed capacity. Operating cost of steam station = 2.5 P/kWh. Interest and depreciation together on capital cost is 15%. Assume no reserve capacity is required for both the schemes.

[*A.M.I.E. Sec. B. Power Plant Engineering Summer 1993*]

Solution : Maximum demand, MD = 100 MW or 100,000 kW

Load factor, LF = 30% or 0.3

Average load = MD × LF = 100,000 × 0.3 = 30,000 kW

Total number of units generated/annum

= 30,000 × 8,760 = 26.28 × 10^7 kWh

(*i*) When supplied by steam power station alone

Capital cost = Rs 1,000 × 100,000 = Rs 10 × 10^7

Annual interest and depreciation

$$= \text{Rs } 10 \times 10^7 \times \frac{15}{100} = \text{Rs } 1.5 \times 10^7$$

$$\text{Annual operating cost} = \text{Rs } \frac{2.5}{100} \times 26.28 \times 10^7 = \text{Rs } 0.657 \times 10^7$$

Total annual cost = Rs (1.5 + 0.657) × 10^7 = 2.157 × 10^7

$$\text{Cost per kWh} = \frac{\text{Total annual cost}}{\text{Total units generated}}$$

$$= \text{Re } \frac{2.157 \times 10^7}{26.28 \times 10^7} = \text{Re } 0.082 \text{ or Paise } 8.2 \textbf{ Ans.}$$

When supplied by steam power station in conjunction with a pumped storage scheme.

Maximum demand on pumped storage station

= 40 MW or 40,000 kW

Maximum demand on steam power station

= 100 – 40 = 60 MW or 60,000 kW

Energy supplied by pumped storage plant

$= 10^6$ kWh $= 0.1 \times 10^7$ kWh

Energy supplied by steam power station

$= 26.28 \times 10^7 - 0.1 \times 10^7 = 26.18 \times 10^7$ kWh

Total capital cost = Rs $1{,}000 \times 60{,}000$ + Rs $700 \times 40{,}000$

= Rs 8.8×10^7

Annual interest and depreciation

$$= \text{Rs } 8.8 \times 10^7 \times \frac{15}{100} = \text{Rs. } 1.32 \times 10^7$$

Total annual operating cost

$$= \text{Rs } \frac{2.5}{100} \times 26.18 \times 10^7 = \text{Rs } 0.6545 \times 10^7$$

neglecting operating cost of pumped storage plant

Total annual cost = Rs 1.32×10^7 + Rs 0.6545×10^7

= Rs 1.9745×10^7

$$\text{Cost per kWh} = \text{Re } \frac{1.9745 \times 10^7}{26.28 \times 10^7} = \text{Re } 0.075 \text{ or Paise } 7.5 \textbf{ Ans.}$$

Obviously scheme (*ii*) is economical.

Example 14.44. The annual electricity requirements of an industry are 20,000 MWh with a maximum demand of 50 MW.

The requirements can be met from a utility charging Rs 800 per kW of maximum demand plus 30 paise per kWh. Alternatively the industry can set up a private steam plant, which will have a capacity of 60 MW. The following different plants for setting up the steam plants are available

	Plant A	**Plant B**	**Plant C**
Total capital cost of power plant	**9,500/kW**	**10,000/kW**	**11,000/kW**
Heat rate kcal/kWh	**3,500**	**3,000**	**2,600**

Taxes 4%, insurance 0.5%, interest rate 8%, Depreciation rate may be calculated by straight line method taking life as 20 years. Heat value of coal = 5,000 kcal/kg, fuel cost = Rs 400 per 1,000 kg; Annual salaries, supply and maintenance = Rs 620,000 per year. Compare the different plants and find most economical scheme.

[*Rajasthan Univ. Generation of Electrical Power 2005*]

Solution : Depreciation rate $= \frac{P - S}{n} = \frac{P}{n}$ assuming zero salvage value

$= \frac{P}{n \times P}$ where n is the plant life in years

$= \frac{1}{n} = \frac{1}{20} = 0.05$

Total fixed charge rate

= Interest + depreciation + taxes + insurance

$= 0.08 + 0.05 + 0.04 + 0.005 = 0.175$

Capital investment of

Plant A = Rs $9{,}500 \times 60 \times 10^3$ = Rs 57×10^7

Plant B = Rs $10{,}000 \times 60 \times 10^3$ = Rs 60×10^7

Plant C = Rs $11{,}000 \times 60 \times 10^3$ = Rs 66×10^7

Annual fixed cost of

Plant A = Rs $0.175 \times 57 \times 10^7$ = Rs 99.75×10^6

Plant B = Rs $0.175 \times 60 \times 10^7$ = Rs 105×10^6

Plant C = Rs $0.175 \times 66 \times 10^7$ = Rs 115.5×10^6

Annual electricity consumption = 20,000 MWh $= 20 \times 10^6$ kWh

Annual fuel cost

$$= \text{Annual energy consumption} \times \frac{\text{Heat rate in k cal/kWh}}{\text{Heat value of coal in kcal/kg}} \times \text{fuel cost in Rs per kg}$$

Annual fuel cost of

$$\text{Plant A} = \text{Rs } 20 \times 10^6 \times \frac{3{,}500}{5{,}000} \times \frac{400}{1{,}000} = \text{Rs } 5.6 \times 10^6$$

$$\text{Plant B} = \text{Rs } 20 \times 10^6 \times \frac{3{,}000}{5{,}000} \times \frac{400}{1{,}000} = \text{Rs } 4.8 \times 10^6$$

$$\text{Plant C} = \text{Rs } 20 \times 10^6 \times \frac{2{,}600}{5{,}000} \times \frac{400}{1{,}000} = \text{Rs } 4.16 \times 10^6$$

Annual salaries, supply and maintenance charges

= Rs 620,000 = Rs 0.62×10^6

Comparison of plant annual costs :

Cost \ Plant	A	B	C
Annual fixed cost	Rs 99.75×10^6	Rs 105×10^6	Rs 115.5×10^6
Fuel cost	Rs 5.6×10^6	Rs 4.8×10^6	Rs 4.16×10^6
Salaries, supply and maintenance charges	Rs 0.62×10^6	Rs 0.62×10^6	Rs 0.62×10^6
Total	Rs 105.97×10^6	Rs 110.42×10^6	Rs 120.28×10^6

Annual cost of purchasing electricity from utility

= Demand charge in Rs/kW × MD + Energy charges in Re/kWh × energy consumption

$$= \text{Rs } 800 \times 50 \times 10^3 + \text{Rs } \frac{30}{100} \times 20 \times 10^6$$

= Rs 40×10^6 + Rs 6×10^6 = Rs 46×10^6

Thus it would be cheaper to purchase electricity from the utility as compared to set up own private plant.

In case supply from utility is not available, plant A with the lowest annual cost should be preferred.

Example 14.45. A region has a maximum demand of 500 MW at a load factor of 50%. The load-duration curve can be assumed to be a triangle. The utility has to meet this load by setting up a generating system which is partly hydro and partly thermal. The costs are as under:

Hydro plant Rs 6,000 per kW per annum and operating expenses of 30 paise per kWh.

Thermal plant Rs 3,000 per kW per annum and operating expenses of Rs 1.30 per kWh.

Determine the capacity of hydro plant, energy generated annually by each and overall generation cost per kWh.

Solution : Maximum demand of the region

= 500 MW or 500,000 kW

Load factor, LF = 50% or 0.5

Total energy generated/annum

= MD × LF × 8,760 = $5{,}00{,}000 \times 0.5 \times 8{,}760$

$= 219 \times 10^7$ kWh

The load duration curve is shown in Fig. 14.21. Hydro plant, being cheaper in operating cost, would be used as base-load power station.

Let the capacity of hydro plant be P kW and the energy generated by it be E kWh/annum

Capacity of thermal plant = (500,000 – P) kW

Energy generated by thermal plant = $(219 \times 10^7 - E)$ kWh

Annual generation cost of hydro plant = Rs (6,000 P + 0.30 E)

Annual generation cost of thermal plant

= Rs $[3{,}000\,(500{,}000 - P) + 1.3\,(219 \times 10^7 - E)]$

Total cost C

$$= 6{,}000\ P + 0.30\ E + 3{,}000\ (500{,}000 - P) + 1.3\ (219 \times 10^7 - E)$$

or $C = 434.7 \times 10^7 + 3{,}000\ P - E$

For minimum cost, $\frac{dC}{dP} = 0$

or $3{,}000 - \frac{dE}{dP} = 0$

or $\frac{dE}{dP} = 3{,}000$

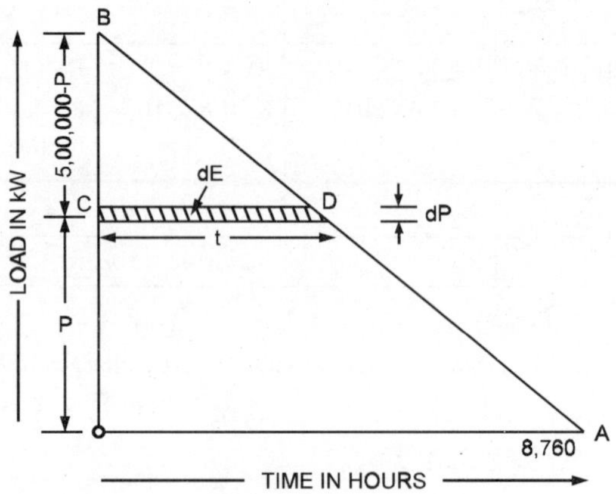

Fig. 14.21

From Fig. 14.21, $dE = dP \times t$

$$\therefore\ t = \frac{dE}{dP} = 3{,}000 \text{ hours}$$

From triangles BCD and BOA, we get

$$\frac{500{,}000 - P}{500{,}000} = \frac{3{,}000}{8{,}760}$$

or $P = 500{,}000 - \frac{3{,}000}{8{,}760} \times 500{,}000$

= 328,767 kW = 330,000 kW (say) or 330 MW

Capacity of thermal plant = 500 – 330 = 170 MW

Energy generated by thermal plant

$$= \frac{170{,}000 \times 3{,}000}{2} = 255 \times 10^6 \text{ kWh}$$ **Ans.**

Energy generated by hydro plant,

$E = (219 \times 10^7 - 255 \times 10^6) = 1{,}935 \times 10^6$ kWh **Ans.**

Total annual generation cost

= Rs $(434.7 \times 10^7 + 3{,}000 \times 330{,}000 - 1{,}935 \times 10^6)$

= Rs 340.2×10^7

Generation cost per kWh

$$= \frac{\text{Total cost}}{\text{Total generation}} = \text{Rs } \frac{340.2 \times 10^7}{219 \times 10^7} = \text{Rs } 1.553$$ **Ans.**

Example 14.46. A generating station has a maximum demand of 100 MW and a yearly load factor of 40%. Generating costs inclusive of station capital costs are Rs 900 per annum per kW demand plus 15 paise per kWh transmitted. The annual capital charge for the transmission system is Rs 4 × 10^7 and for distribution system is Rs 3 × 10^7, the respective diversity factors being 1.2 and 1.25. The efficiency of transmission system is 90% and that of distribution system inclusive of substation losses 85%. Find the yearly cost per kW and cost per kWh supplied (*i*) at the substation and (*ii*) on the consumer premises.

Solution : Maximum demand on generating station

= 100 MW = 10^5 kW

Annual capital generating cost = Rs 900 × 10^5 = Rs 9 × 10^7

Annual capital charges for transmission system = Rs 4 × 10^7

Hence total fixed charges associated with the supply of energy on the substations

= Rs 9 × 10^7 + Rs 4 × 10^7 = Rs 13 × 10^7

Aggregate of all maximum demands on the substations

= MD on generating station × diversity factor

= 100,000 × 1.2 = 1.2 × 10^5 kW

Annual cost per kW demand on the substations

$$= \text{Rs } \frac{13 \times 10^7}{1.2 \times 10^5} = \text{Rs } 1{,}083.33$$ **Ans.**

Annual capital charges for distribution system

= Rs 3 × 10^7

Total fixed charges associated with the supply of energy on the consumer premises

= Rs 13 × 10^7 + Rs 3 × 10^7 = Rs 16 × 10^7

Aggregate of maximum demands of all the consumers

= Aggregate of all maximum demands on substations × diversity factor

= 1.2 × 10^5 × 1.25 = 1.5 × 10^5 kW

Annual cost per kW demand on the consumer premises

$$= \text{Rs } \frac{16 \times 10^7}{1.5 \times 10^5} = \text{Rs } 1{,}066.67$$

Since transmission and distribution efficiencies are 90% and 85% respectively so for every kWh transmitted only 0.9 kWh reaches the substation and 0.9 × 0.85 *i.e.,* 0.765 kWh reaches the consumer premises and costs 15 paise.

∴ Cost per kWh delivered on the substations

$$= \text{Paise } \frac{15}{0.9} = \text{Paise } 16.67$$ **Ans.**

Cost per kWh delivered on the consumer premises

$$= \text{Paise } \frac{15}{0.9 \times 0.85} = \text{Paise } 19.6$$ **Ans.**

14.11. LOAD FORECASTING

When planning exploitation of the natural energy sources of a country, progress takes time and a lot of capital investment is required. So advance decisions are required to be taken for judicious and profitable investment in various projects to make them effective, economical and useful. For this purpose, the anticipated electrical energy demand should be known. The resources available in the country for electrical power generation *e.g.*, for steam power stations, hydropower stations, and nuclear power stations, can then be developed considering the electrical power and energy needs, and the locations or regions where the demand is expected.

Forecasts may be long term forecasts, medium term forecasts or short term forecasts.

Long term forecasts covering a period of 20 years provide a basis for examining the energy supply problems of the country in a broader perspective in relation to various other modes of development, including conversion and substitution of different forms of primary and secondary energy.

Construction and installation of equipment in power plants takes 4 to 6 years. Medium term forecasts of 5 or 6 years are, therefore, very significant for planning the size of power plants. Forecasts of demand and energy are required to estimate the additional installed capacity required to facilitate the plant maintenance programme and to estimate the firm capacity of the restricted hydroplants.

Short term forecasts of one or two years are mainly required for deciding operating procedures, and for preparing budget estimates.

Forecasting of load requirements is an essential part of power project design. The forecast is based upon (*i*) past development of power projects in the region (*ii*) applications received from different types of consumers and (*iii*) trends towards the development of industries and other potentialities for load development. The methods, that are generally used for forecasts or estimates of future demand of electrical energy are :

1. Load survey methods.
2. Methods of extrapolation.
3. Mathematical methods.
4. Mathematical methods using economic parameters.

1. Load Survey Methods : In load survey methods, the area of interest is visited and the existing and future load needs are estimated, taking into considerations of the communities in the region and the factors which are likely to increase the load demand of the region in the near future. Loads are grouped under different categories such as residential/domestic, commercial, industrial etc. Detailed information, categorywise and areawise is collected to estimate the demand (and its growth rate), duration and time of occurrence of load and the energy requirements. The existing data on loads in the various categories and the expected increase in each category of load yearly for the coming 5 or 6 years is then used in the preparation of load demand and energy forecasts.

2. Methods of Extrapolation : These methods are essentially statistical. These methods involve the study of data collected from different records. These methods are used to compare the trends towards increase in demand and in energy consumption during past few years comparable with the period of the forecasts. These methods are used in determination of the average rate of increase of energy consumption in the past, and also in the countries with similar economic structure. Then, these datas are used in forecasting and estimating the future needs. These methods, however, have a drawback that these do not consider unforseen developments, such as the establishment of large and major industries which may alter the future load requirements altogether. For example, in our country since independence there has been considerable emphasis on power projects under various five year plans, which could not have been predicted from pre-independence conditions.

3. Mathematical Methods : These methods use linear or exponential curve fitting techniques for extrapolation of the curves indicating energy consumption for past several years. Thee methods can be applied only when adequate statistical data over a sufficiently long period on the past consumption of electrical energy is available.

4. Mathematical Methods Using Economic Parameters : These methods are based on the assumption that per capital consumption of electrical energy in a country depends upon economic factors, such as specific gross investments, industrial production, specific GNP, steel consumption, housing transport, etc. Indices of these parameters are found and suitably used for extrapolation by linear or exponential approximations.

Exercises

1. Differentiate between fixed and operating costs of power plants. List the items which constitute the fixed and operating cost.
[Generation of Electrical Power, Pb. Technical Univ, May 2008; May 2009; Rajasthan Univ. 2000, 2007; Rajasthan Technical Univ. 2008;2009]
2. Discuss and explain the various components of (*i*) fixed cost and (*ii*) operating cost and make a comparative assessment of these in thermal, hydro and nuclear power plants.
[A.M.I.E. Sec. B. Power Systems Summer 1996]
3. What is depreciation reserve? Discuss the methods to calculate the depreciation charges.
[Pb. Technical Univ. Generation of Electric Power Dec. 2007]
4. What is depreciation reserve ? Why is it necessary to maintain it ? Discuss the methods to calculate the depreciation charges.
[Pb. Univ. Electrical Power II (Generation) October 1991; U.P. Technical Univ. Power Station Practice 2007-08; Rajasthan Technical Univ. Generation of Electrical Power 2008, 2010]
5. What is depreciation reserve?
Briefly explain the sinking fund method to find it. Prove that the depreciation charge by the sinking methods is given by :
$$q = (C - S)\left[\frac{r}{(1+r)^n - 1}\right]$$
where C = Initial cost of plant, S = Salvage cost, r = Annual rate of interest, n = Useful life of plant.
[Rajasthan Univ. Generation of Electrical Power 2000, 2004, 2005]
6. Explain the difference between chronological load curve and load duration curve.
[A.M.I.E. Sec. B. Power Systems Summer 1998]
7. Explain the load duration curves and write the various factors that may be calculated.
[Pb. Technical Univ. Generation of Electric Power May 2010]
8. Convert a chronological load curve into load duration curve.
[Rajasthan Univ. Generation of Electrical Power 2006]
9. Explain the load duration curve and mass curve with their complete information.
[Rajasthan Univ. Generation of Electrical Power 2007]
10. Explain (*i*) chronological load curve (*ii*) load duration curve and (*iii*) energy load curve.
[Pb. Univ. Electrical Power II (Generation) January 1993]
11. What are the different factors related to plants and consumer for power plant economics.
[U.P. Technical Univ. Power Station Practice 2006-07]
12. Define the following:
(*i*) Diversity factor (*ii*) Load factor (*iii*) Plant capacity factor (*iv*) Plant use factor.
[U.P.S.C. I.E.S. Electrical Engineering II, 2004]

13. Define load factor, utility factor, capacity factor, demand factor, and diversity factor.

[Pb. Univ. Electrical Power II (Generation) Dec. 1995]

14. Define installed capacity, installed reserve, cold reserve, operating reserve, and spinning reserve.

[Pb. Univ. Electrical Power II (Generation) Dec. 1995]

15. Explain in brief about plant capacitor factor.

[M.D. Univ. Power Systems-I, 2010-2011]

16. Differentiate between fixed and operating costs of power plants. List the items which constitute the fixed and operating costs. Also discuss the role of load factor on the cost of electrical energy.

[G.B. Technical Univ. Power Station Practice 2010-11]

17. Discuss the effect of load factor on the cost of generation in a power system. Explain the following terms :

Connected load, Maximum demand and Demand factor.

[U.P. Technical Univ. Power Station Practice 2006-07]

18. Explain the terms load factor and diversity factor. How do these factors influence the cost of generation?

[Rajasthan Technical Univ. Generation of Electrical Power 2008]

19. What is the effect of load factor and load diversity factor on power system economics? Describe fixed and operating costs of hydroelectric power systems.

[Rajasthan Technical Univ. Generation of Electrical Power 2011]

20. Define the terms (*i*) load factor, (*ii*) diversity factor and discuss their influence on the cost of generation.

[U.P. Technical Univ. Power Station Practice 2007]

21. Discuss the role of load factor on the cost of electrical energy. Briefly discuss the factors affecting economics of generation and distribution of power and how to reduce the power generation cost.

[U.P. Technical Univ. Power Station Practice 2009-10]

22. Discuss the factors affecting economics of generation and distribution of power and how to reduce the power generation cost.

[G.B. Technical Univ. Power Station Practice 2010-11]

23. How the number and size of units in an isolated power plant are selected ? **[Pb. Univ. June 1991]**

24. Write short notes on the following:

(*i*) Generation cost and annual fixed cost. **[Raj Univ. 2001]**

(*ii*) Load curves. **[Pb. Technical Univ. December 2010]**

(*iii*) Load division on hydro-steam system.

[Pb. Technical Univ. December 2010]

(*iv*) Load factor. **[Pb. Technical Univ. December 2010]**

(*v*) Plant capacity factor. **[Pb. Technical Univ. Dec. 2010]**

(*vi*) Utilization factor. **[Pb. Technical Univ. Dec. 2010]**

25. Write short notes on demand factor, capacity factor, utilization factor and mass curve. **[Raj Univ. 2002]**

Short Answer Type Questions With Answers

Q. 1. State factors on which generating cost depends.

[P.T.U. May 2008]

Ans. The generating cost per unit (kWh) of energy depends upon the cost covering the purchase, installation and erection of equipment, cost of fuel, labour, repair etc.

Q. 2. What items constitute the fixed and running charges of generation of electrical energy ?

Ans. The fixed charges include the annual interest and depreciation on the capital cost of generating plant, transmission and distribution network, buildings and other civil works, all types of taxes and insurance charges and salaries of high officials, management and clerical staff and compensation given to the workers.

The running or operating cost includes annual costs of fuel, lubricating oil, water, maintenance and repairs of equipment and salaries of operational and maintenance staff and salaries of supervisory staff engaged in the running of the plant.

Q. 3. Why is fixed cost independent of maximum demand and units generated? **[P.T.U. December 2010]**

Ans. Fixed cost is due to the annual cost of central organisation, interest on the capital cost of land (especially if some land is held for future development) and salaries of high officials.

The annual expenditure on central organisation and salaries of high officials is fixed since it has to be met whether the plant has high or low maximum demand or it generates less or more energy (units). Further, the capital investment on the land is fixed and hence the amount of interest is also fixed. As such fixed cost is independent of maximum demand and units generated.

Q. 4. What is depreciation reserve?

[Rajasthan Univ. 2000, 2004, 2005]

Ans. The depreciation charge represents the amount which is set aside from income every year so as to produce a sufficient sum at the end of the probable life for replacing the plant and machinery by new ones and the amount so accumulated is called the depreciation reserve.

Q. 5. What do you mean by off peak energy utilization?

[Rajasthan Univ. 2001]

Ans. Off peak energy utilization means encouraging the consumers to use electricity during off peak hours by giving a special discount. Its benefits are : 1. Improvement of plant capacity factor resulting in reduction of capital cost and better utilization of investment. 2. Reduced requirement for peaking units resulting in lower plant capacity and lower production and maintenance costs of fossil peaking units.

Q. 6. How does the load duration curve differ from chronological load curve ?

Ans. The daily variations in load on the power station from time to time—hourly or half hourly can be plotted on a graph taking load on Y-axis and time on X-axis. The curve so obtained is known as chronological load curve. The load duration curve indicates the variation of load, but with the load arranged in descending order of magnitude *i.e.*, the greatest load on the left, lesser loads towards the right hand and the least load at the extreme light.

Q. 7. What informations can be supplied by load curves ?

Ans. The informations that can be had from load curves are (*i*) variations of load during different hours of the day (*ii*) total number of units generated in a day (*iii*) maximum demand on the power station (*iv*) average load and (*v*) load factor.

Q. 8. Define energy load curve. **[Rajasthan Univ. 2001]**

Ans. Energy load curve gives the total number of units generated for the given demand. The ordinate represents the demand in kW or MW and the abscissa represents the units (kWh) generated at or below a given demand (in kW or MW). Such a curve can be obtained from the load duration curve keeping the abscissa corresponding to each ordinate equal to the area of the duration curve up to the value of that ordinate.

Such curves are useful in that with a given number of units (kWh) available, say from a river flow, the load (kW) that could be carried at the base or peak may be easily determined.

Q. 9. How does energy load curve differ from mass curve ?

Ans. The curve drawn between demand in kW or MW on Y-axis and units generated (kWh) on X-axis is known as integrated load duration curve or energy load curve while mass curve is plotted with units (kWh) generated as ordinate and time as abscissa.

Q.10. What is load factor ? **[M.D. Univ. 2010-11]**

OR

Define load factor. **[P.T.U. December 2008]**

OR

Define load factor of electric installation. **[P.T.U. May 2010]**

Ans. The ratio of average load to the maximum demand during a certain period of time such as a day or a month or a year is called the *load factor*. Since average load is always less than the maximum demand, load factor is therefore, always less than unity.

$$\text{Load factor} = \frac{\text{Average demand}}{\text{Maximum demand}}$$

Load factor may also defined as the ratio of the number of units actually generated in a given period to the number of units which could have been generated had the load or demand remained at the maximum value throughout this period.

i.e., Load Factor

$$= \frac{\text{Units generated in a given period}}{\text{Maximum demand} \times (\text{number of hours of operation in the given period})}$$

Q. 11. Can the load factor of the sysem be 100% ? **[P.T.U. May 2010]**

Ans. Since average load is always less than the maximum demand, load factor being the ratio of average load to maximum demand is always less than unity. As such load factor of the system cannot be 100%.

Q. 12. Why even if the maximum demand and load factor of the two systems are equal, their load duration curves may not be similar? **[P.T.U. December 2008]**

Ans. Load factor is the ratio of average load to maximum load. Even if load factor and maximum demands are the same, the load duration curve, which is plotted between the load in MW and time (load arranged in descending order of magnitude) and indicates the total number of units generated for the period considered, may not be same. This is because the maximum demand may not occur simultaneously. Thereby, the area under load duration curves might be same but they might differ in shape.

Q. 13. Define demand factor. **[P.T.U. December 2008]**

Ans. The ratio of actual maximum demand on the system to the total rated load connected to the system is called the *demand factor*. It is always less than unity.

$$\text{Mathematically demand factor} = \frac{\text{Maximum demand}}{\text{Connected load}}$$

The idea of a demand factor was introduced due to the fact that all the equipment connected to the system will not be worked at a time in practice and the kW or kVA maximum demand of a group of electricity consuming devices will be always less than the sum of the kW or kVA ratings or capacities of these devices.

The minimum capacity of the generating plant must be such as to meet the maximum demand, the value of demand factor, therefore, determines the capacity and hence cost of the power equipment required to serve a given load.

Q. 14. Can demand factor be more than 1? **[P.T.U. May 2009]**

Ans. Demand factor, defined as the ratio of maximum demand on the system to the total rated load connected to the system, cannot be more than unity because maximum demand on the system can never be more than rated load connected to the system.

Q. 15. What is diversity factor? **[M.D. Univ. 2010-11]**

OR

Define diversity factor. **[P.T.U. December 2009]**

Ans. The maximum demands of all the consumers supplied from an installation do not occur usually at the same time, maximum demand on the installation is, therefore, always less than the sum of individual maximum demands of all consumers connected to it.

The ratio of sum of the individual maximum demands of all the consumers supplied by it to the maximum demand of the power station is called the *diversity factor.*

$$i.e., \text{ Diversity factor} = \frac{\text{Sum of individual maximum demands}}{\text{Maximum demand of power station}}$$

It is always greater than unity.

Q. 16. Differentiate between diversity factor and load factor. **[P.T.U. May 2008]**

Ans. Diversity factor is defined as the ratio of the sum of the individual maximum demands to maximum demand of the group whereas load factor is defined as the ratio of average load to the maximum demand.

Q. 17. Discuss why, maximum demand of a group of consumers is always less than the sum of their individual maximum demands? **[P.T.U. May 2009]**

Ans. The maximum demands of all the consumers supplied from an installation do not occur usually at the same time, maximum demand on the installation is, therefore, always less than the sum of individual maximum demands of all consumers connected to it.

Q. 18. What is diversity factor and how does it influence the cost of generation ?

Ans. The ratio of sum of individual maximum demands of all the consumers supplied by it to the maximum demand of the power station is called the diversity factor. It is always greater than unity. Higher the diversity factor better it is. With a given number of consumers the higher the diversity factor of their loads, the smaller will be capacity of the power plant required and consequently the fixed charges due to capital investment will be much reduced.

Q. 19. How is diversity helpful in reducing the costs? **[P.T.U. Dec. 2007; Dec. 2008]**

Ans. Diversity helps in reducing the costs as the diversity would mean that the maximum demands of all the consumers supplied from an installation would not occur at the same time, this would reduce total maximum demand on the installation. With a given number of consumers the higher the diversity factor of their loads, the smaller will be capacity of the power plant required and consequently the fixed charges due to capital investment will be much reduced.

Q. 20. What is capacity factor? **[P.T.U. December 2007]**

Ans. Every plant has to have a reserve capacity so as to take care of the future expansion and increase in load and, therefore, total installed capacity of the plant is usually greater than that actually required (maximum demand).

Capacity (or plant) factor is defined as the ratio of the average load to the rated capacity of the power plant.

i.e., $\text{Capacity (or plant) factor} = \frac{\text{Average demand}}{\text{Rated capacity of power plant}}$

Plant capacity factor may also be defined as the actual energy generated divided by the maximum possible energy that the plant might have generated during a given period.

i.e., Plant capacity factor

$$= \frac{\text{Units generated in a given period}}{\text{Rated capacity of the power plant in kW} \times \text{(number of hours in the given period)}}$$

The capacity factor depicts the extent of the use of the power plant. It is different from load factor and the difference is an indication of the reserve capacity.

$$\text{Capacity factor} = \frac{\text{Maximum demand}}{\text{Rated capacity}} \times \text{load factor}$$

Q. 21. Discuss why utilization factor of a plant may even be greater than unity? **[Rajasthan Univ. 2001]**

Ans. Utilization factor is a measure of the utility of the power plant capacity and is defined as the ratio of maximum demand to the rated capacity of the power plant. It is always less than unity because power plant is never operated beyond its rated capacity.

Q. 22. Explain various types of reserves in power system. **[Rajasthan Univ. 2006]**

Ans. Every system must have certain amount of reserve generating capacity to provide for overhaul of generating equipment, forced outage of equipment etc. Various types of reserves are given below.

Operating reserve refers to capacity in service in excess of peak load. It is provided for regulation within the hour to cover minute to minute variations, load forecasting errors, loss of equipment and maintenance of equipment. It consists of spinning and/or non spinning reserve.

(*a*) **Spinning Reserve :** It is the generating capacity connected to the bus and ready to take load.

In order to take instantaneous peak loads, the system must have capacity in *readiness to serve.* For this purpose certain machines are kept running on no load *i.e., floating* on the system. This is called the *spinning reserve* of the *system.*

(*b*) Non Spinning Reserve is that operating reserve which is capable of being connected to the bus and loaded within a specified time.

Q. 23. Distinguish between operating reserve and spinning reserve. **[P.T.U. May 2008]**

Ans. Refer to Answer of Q.22.

Q. 24. What is no spinning reserve? **[P.T.U. 2009]**

Ans. Non Spinning Reserve is that operating reserve which is capable of being connected to the bus and loaded within a specified time.

Q. 25. What do you understand by (*i*) base load and (*ii*) peak load of a power plant?**[P.T.U. December 2010]**

Ans. The unvarying load, which occurs almost the whole day on the power plant is called the base load whereas the various peak demands of the load over and above the base load of the power plant is called the peak load.

Q. 26. Differentiate between base load plants and peak load plants. **[P.T.U. May 2010]**

Ans. Base load plants run throughout the year and have high load factors. The economic characteristics of base load plants should be such that they supply power at high capital costs but low cost of operation. Hydro and nuclear power plants are usually classified as base load plants. Peak load plants run for a few hours in the year and work at low load factors. The economic characteristics of peak load plants should be such that they supply power at low capital costs, although at high cost of operation. Peak load power plants should be capable of starting quickly and should be inexpensive in starting and shutting down operations.

Q. 27. Generally which type of generation is used for base load demand? **[P.T.U. May 2009]**

Ans. The economic characteristics of base load plants should be such they supply power at high capital cost but low operating cost. Hydro and nuclear power plants are usually used for supplying base load demand.

Q. 28. Define load forecasting.

Ans. Prediction of load on the system for future is termed as load forecasting.

Problems

1. A plant costs Rs. 80,000/- and has useful life of 15 years. If the salvage value of the equipment is Rs 5,000/-, determine the amount which should be saved annually to replace the equipment at the end of that time by (*i*) straight line method (*ii*) the sinking fund method. Assuming that the annual cost of compound interest is 5 per cent.

[Rajasthan Univ. Generation of Electric Power 2001]

[**Ans,** (*i*) Rs 5,000 per annum (*ii*) Rs. 3,476 per annum]

2. The equipment in power station costs Rs 1,560,000 and has salvage value of Rs 60,000 at the end of 25 years. Determine the depreciation value of the equipment at the end of 20 years on the following methods (*i*) Diminishing value method (*ii*) sinking fund method at 5% compound interest annually.

[Rajasthan Univ. Generation of Electric Power 2003]

[**Ans.** (*i*) Rs. 115,615 (*ii*) Rs 520,781]

3. A plant costing Rs 65,000 has a useful life of 15 years. Find the amount which should be saved annually to replace the equipment at the end of that time (*i*) by the straight line method and (*ii*) by the sinking fund method if the annual rate of compound interest is 15%. Assume that the salvage value of the equipment is Rs 5,000.

[Pb. Univ. Generation December 1982]

[**Ans.** (*i*) Rs 4,000.00 (*ii*) Rs 1,261.00]

4. A plant costs Rs 85,000 and has a useful life of 20 years. What amount should be saved annually to replace the equipment at the end of its useful life by sinking fund method, the annual rate of compound interest being 6%? Assume the salvage value of the equipment to be Rs 5,000.

[Pb. Univ. Electrical Power II (Generation) May 1994]

[**Ans.** Rs 2,175]

5. The load curve of an electrical system is linear with the following values at different times of the day :

Time	12	5	9	6	8	10	12
		AM			PM		
Load(MW)	50	50	100	100	150	80	50

Plot chronological load curve. Find the energy required by the system in one day and the system daily load factor.

[Pb. Univ. Electrical Power II (Generation) October 1991; Pb. Technical Univ. Generating Electric Power Dec. 2007]

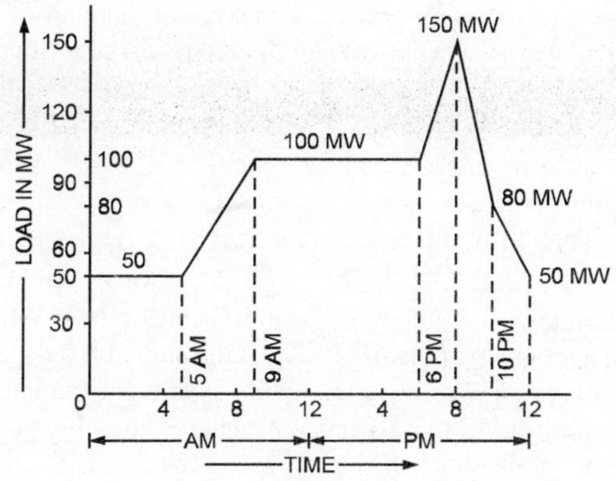

Fig. 1

Chronological load curve drawn is shown in Fig. 1.

[**Ans.** 2,060 MWh, 57.22%]

6. A generating station has the following daily loads :

0 to 6 hr	4,000 kW
6 to 8 hr	3,000 kW
8 to 12 hr	8,000 kW
12 to 14 hr	2,000 kW
14 to 18 hr	7,500 kW
18 to 20 hr	3,000 kW
20 to 24 hr	4,500 kW

Sketch load duration curve and determine the load factor and plant capacity factor assuming capacity of the plant 11,000 kW. [**Ans.** 62.5%; 45.45%]

7. A proposed station has to supply load as follows:

Time in hours	Load in kW
01 – 08	10,000
08 – 12	20,000
12 – 17	25,000
17 – 20	18,000
20 – 23	35,000
23 – 01	20,000

After drawing the load curve, find out the load factor. Also choose suitable generator units to supply this load, maintaining reliability of supply. Prepare operation schedule for the machines and calculate plant use factor.

[**Ans.** 56.43% ; Plant use factor with 4 units of 10 MW each and one unit of 5 MW – 98.75% ; Plant use factor with 2 units of 15 MW each and two units of 10 MW each –92%]

8. The daily load curve of a power plant is given by the table below :

Time	12	2	4	6	8	10	12	2	4	6	8	10	12
Load kW × 10^3	2	2.5	3	4	6	6.5	6.5	5	6	8	9	5	2

(*a*) Find the daily load factor.

(*b*) All loads in excess of 400 kW are carried by unit no 2 rated at 600 kW. Find its use factor.

[**Ans.** (*a*) 0.814 (*b*) 41.7%]

9. Define load duration curve. Draw the daily load duration curve for a power station which has the following load cycle.

Load	100 kW	5,000 kW	500 kW	3,000 kW	2,000 kW
Time	4 hrs	3 hrs	6 hrs	6 hrs	5 hrs

[Rajasthan Univ. Generation of Electrical Power 2005]

Ans.

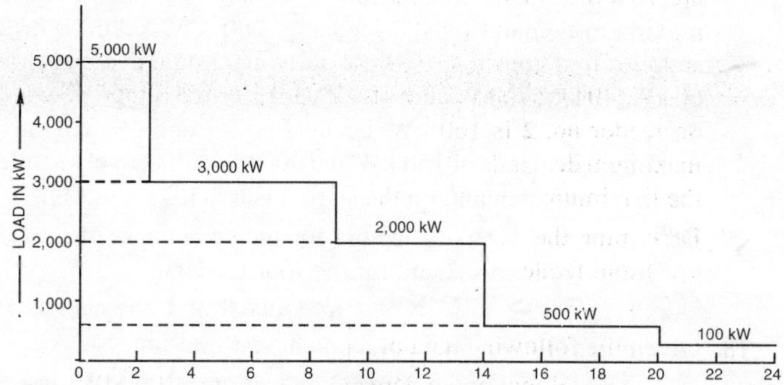

Fig. 2

10. A diesel station supplies the following loads to various consumers :

Industrial consumers = 1,500 kW
Commercial load = 750 kW
Domestic load = 100 kW
Other load = 450 kW

If the maximum demand on the station is 2,500 kW and the number of kWh generated per year is 45 × 10^5, determine: (*i*) Diversity factor (*ii*) Annual load factor.

[Rajasthan Univ. Generation of Electrical Power 2003]

[**Ans.** (*i*) 1.12 (*ii*) 0.2055]

11. A power station has a maximum demand of 15,000 kW. The annual load factor is 50% and plant capacity factor is 40%. Determine the reserve capacity of the plant.

[Pb. Technical Univ. Generation of Electric Power Dec. 2010]

[**Ans.** 3,750 kW]

12. A generating station has a maximum demand of 450 MW. The annual load factor is 45% and capacity factor is 40%. Find the reserve capacity of the plant.

[Pb. Technical Univ. Generation of Electric Power Dec. 2009]

[**Ans.** 56.25 MW]

13. The maximum demand of a power plant is 80 MW. The capacity factor is 0.5 and the utilization factor is 0.8. Find (*i*) load factor (*ii*) plant capacity (*iii*) reserve capacity (*iv*) annual energy production.

[Pb. Univ. Electrical Power II (Generation) April 1993]

[**Ans.** (*i*) 62.5% (*ii*) 100 MW (*iii*) 20 MW (*iv*) 438,000 MWh]

14. The maximum demand of a power station is 80 MW, the capacity factor is 0.6 and the utilization factor is 0.9. Find (*i*) the load factor (*ii*) plant capacity (*iii*) reserve capacity (*iv*) annual energy production.

[Rajasthan Univ. Generation of Electrical Power 2001]

[**Ans.** (*i*) 0.6667 (*ii*) 88.889 MW (*iii*) 8.889 MW (*iv*) 467,200 MWh]

15. The yearly load duration of a power plant is a straight line. The maximum load is 30 MW and the minimum load is 20 MW. The capacity of the plant is 35 MW. Find (*i*) plant capacity factor (*ii*) load factor (*iii*) utilization factor.

[Rajasthan Univ. Generation of Electrical Power 2006]

[**Ans.** (*i*) 0.71428 (*ii*) 0.833 (*iii*) 0.857]

16. A power station is to supply four regions of loads whose peak loads are 10,000 kW, 5,000 kW, 8,000 kW and 7,000 kW. The diversity factor of the load at the station is 1.5 and average annual load factor is 60%. Calculate the maximum demand on the station and annual energy supplied from the station. Suggest the installed capacity and number of units taking all aspects into account.

[Pb. Univ. Electrical Power II (Generation) Dec. 1993, Dec. 1994]

[**Ans.** 20 MW; 105.12 × 10^6 kWh ; 3 units, each of 10 MW capacity]

17. A substation supplies power by four feeders to its consumers. Feeder no. 1 supplies six consumers whose individual demands are 70 kW, 90 kW, 20 kW, 50 kW, 10 kW and 20 kW, while maximum demand on the feeder is 200 kW. Feeder no. 2 supplies four consumers whose daily maximum demands are 60 kW, 40 kW, 70 kW and 30 kW while the maximum demand on feeder no. 2 is 160 kW. Feeders nos. 3 and 4 have daily maximum demands of 150 kW and 200 kW respectively, while the maximum demand on the station is 600 kW.

Determine the diversity factors for the consumers of feeder no. 1 and feeder no. 2 and for the four feeders.

[Ans. 1.3; 1.25 and 1.183]

18. Given the following data of a power station:

(*i*) Annual maximum demand on station – 100 MW

(*ii*) Maximum demands of different types of load supplied – 40 MW, 30 MW, 25 MW and 20 MW

(*iii*) Average load factor of the station – 50% and

(*iv*) Capacity of station – 2 units of 50 MW each and one unit of 25 MW.

Determine (*a*) the number of units (kWh) supplied annually (*b*) diversity factor and (*c*) utilization factor.

[Ans. (*a*) 438 × 10^6 kWh (*b*) 1.15 (*c*) 0.83]

19. A generating station supplies the following loads :

15 MW, 12 MW, 8.5 MW, 6 MW and 0.45 MW. The annual load factor of the power station is 48% Calculate (*i*) the number of units supplied annually (*ii*) diversity factor, and (*iii*) demand factor, if the station has a maximum demand of 22 MW.

[Ans. (*i*) 92.5 × 10^6 kWh (*ii*) 1.907 (*iii*) 52.5%]

20. A power station has a maximum demand of 200 MW, a plant load factor of 0.6, a plant capacity factor of 0.5 and a plant use factor of 0.8. Find (*i*) the daily energy consumed (*ii*) the reserve capacity of the plant and (*iii*) maximum energy that can be produced daily.

[A.M.I.E. Sec. B. Power Plant Engineering Summer 1996]

[Ans. (*i*) 2,880 MWh (*ii*) 40 MW (*iii*) 5,706 MWh]

21. The maximum demand of a power supply system is 30 MW. The annual factors are as follows.

Load factor = 60% ; Capacity factor = 50% ; Use factor = 75%.

Determine (*i*) the annual energy production (*ii*) reserve capacity and (*iii*) the hours during which the plant is not in service per year.

[A.M.I.E. Sec. B. Power Plant Engineering Winter 1993]

[Ans. (*i*) 157.68 × 10^6 kWh (*ii*) 6 MW (*iii*) 2,920 hours]

22. A generating station has following data :

Installed capacity = 300 MW, capacity factor = 50%. Annual load factor = 60%. Annual fuel cost = Rs 9 × 10^7. Capital cost = Rs 10^7, annual interest and depreciation = 10%. Calculate (*i*) minimum reserve capacity of the station and (*ii*) the cost per kWhr generated.

[Rajasthan Univ. Generation of Electrical Power 2006]

[Ans. (*i*) 50 MW (*ii*) 1,314 × 10^6 kWh]

23. A base load station having a capacity of 18 MW and a standby station having a capacity of 20 MW share a common load. Find the annual load factor, use factor, and capacity factor of the two power stations from the following data :

Annual standby station output = 7.35 × 10^6 kWh

Annual base load station output = 101.35 × 10^6 kWh

Peak load on the standby station = 12 MW

Hours of use of standby station during the year = 2,190

[Ans. Standby station : 7% ; 16.78% ; 4.2%. Base load Station : 64.28% ; 64.28% ; 64.28%]

24. A generating station has got maximum demand of 50 MW. Calculate the cost per kWh delivered from the following data: (*i*) Capital cost of Rs 95 × 10^6 ; (*ii*) Annual cost of fuel and oil Rs 9 × 10^6 ; (*iii*) Taxes, wages and salaries Rs 6 × 10^6.

The rate of interest and depreciation is 10% ; Annual load factor is 50%.

[Pb. Univ. Electrical Power II (Generation) Dec. 1993]

[Ans. Paise 11.2/kWh]

25. The output of a generating station is 500 × 10^6 kWh per year and average load factor is 70%. If the annual fixed charges are Rs 500 per kW of installed plant capacity and annual running charges are 50 paise per kWh, what is the cost per kWh of energy at the bus bar ?

[Ans. paise 58.2 per kWh]

26. A small generating unit of 5,000 kW capacity supplies the following loads :

(*i*) Domestic consumers with a maximum demand of 3,000 kW at a load factor of 20 per cent :

(*ii*) Small industrial load with a maximum demand of 2,000 kW at a load factor of 50 per cent.

Find the overall energy rate for both type of consumers. Use the following data :

Capital cost of plant = Rs 2,000 per kW of installed capacity.

Total running cost = Rs 600,000 per year.

Annual rate of depreciation and interest on capital cost is 10 per cent. **[Ans.** (*i*) Paise 15.7 per unit ; Paise 8.85 per unit]

27. The fixed costs of a thermal station are Rs 262.80 per kW of installed capacity per year. The fuel and operating costs are 3 paise per kWh generated. Find the cost of the electrical energy generated per kWh. at station load factors of 100%, 75%, 50% and 25%.

Plot the curves showing the variations of the cost of energy per kWh generated.

[Agra Univ. Generation & Switchgear Practice 1976]

[Ans. Paise 6, Paise 7, Paise 9, Paise 15 per kWh respectively]

28. A 100 MW, steam power station is estimated to cost Rs 100 crores. The operating expenses are estimated as follows :

Cost of fuel and oil = Rs 700 lacs/annum ; Transportation and storage = Rs 100 lacs/annum.

Salaries and wages = Rs 100 lacs/annum ; Miscellaneous = Rs 100 lacs/annum.

Reckoning interest and depreciation 10% of capital cost, calculate the cost of generation per unit if the average load factor of the power station is 60%.

What economics would be affected if the load factor was improved to 80% and the operating expenses increasing by only 10% thereby ?

[Ans. 38 paise/kWh ; Reduction in cost/kWh by 21.25%]

29. A system with a maximum demand of 100,000 kW and a load factor of 30% is to be supplied by either (*a*) steam station alone or (*b*) a steam station in conjunction with a water storage scheme, the latter supplying 100 million units with a maximum output of 40,000 kW. The capital cost of the steam and storage stations are Rs 600 per kW and Rs 1,200 per kW respectively. The corresponding operating costs are 15 paise and 3 paise per kWh respectively. The interest on capital cost is 15% per annum. Calculate the overall generating cost per kWh and state which of the two projects will be economical.

[Ans. Paise 18.425 per kWh ; Paise 15.23 per kWh]

30. A steam power station having a capital cost of Rs 1,000/- per kW and maintenance cost of 2 P per kWh, using coal of calorific value of 6,110 k cal/kg. The overall thermal efficiency of the station is estimated at 25% or (*i*) a hydro station having a capital cost including transmission lines etc. of Rs 3,000/- per kW and a running cost of 0.5 P/kWh. Allowing 12% for the steam station and 9% for the hydro station for interest and depreciation per year calculate the price of coal above which steam generation is uneconomical, given 1 kWh = 860 kcals. [**Ans.** Rs 34.2 per tonne]

31. An electric supply undertaking having maximum load of 100 MW generates 375 million kWh per year, and supplies consumers having an aggregate maximum demand of 165 MW. The annual expenses including capital charges are Rs 32×10^5, fixed expenses connected with generation Rs 48×10^5, transmission and distribution expenses Rs 56×10^5. Assuming that 90 per cent of the fuel cost is assigned to the running charges and that the losses in transmission and distribution are 15% of the kWh generated, deduce a two part tariff to represent the actual cost of supply to the consumer.

[**Ans.** Rs 65 per kW of maximum demand plus paise 0.9 per kWh]

32. A power system, having a maximum demand of 100 MW, has a load factor 30%. It can be supplied by one of the following schemes :

(*i*) A steam station capable of supplying the whole load
(*ii*) A nuclear station capable of supplying the whole load
(*iii*) A hydro station capable of supplying the whole load.

	Steam	Hydro	Nuclear
Capital cost/kW installed capacity	Rs 600	Rs 1,500	Rs 2,000
Interest and depreciation on capital	12%	10%	10%
Operating cost per unit	5 P	1 P	2 P
Transmission cost per unit	Negligible	0.5 P	Negligible

Workout total cost of the scheme per annum in each case.

[Pb. Univ. Electrical Power II (Generation) May 1995]

[**Ans.** Steam – Rs 203.4 lacs; Hydro – Rs 189.42 lacs; Nuclear – Rs 252.56 lacs]

33. A generating station has a maximum demand of 80,000 kW and yearly load factor of 50%. Generating costs inclusive of station capital cost are Rs 70 per annum per kW demand plus 1.5 P per kWh transmitted. The annual capital charge for the transmission system is Rs 30 lakhs and for distribution system Rs 25 lakhs, respective diversity factors being 1.2 and 1.3. The efficiency of the transmission system is 90% and that of distribution system inclusive of substation losses is 83%. Find the yearly cost per kW demand and the cost per kWh supplied on the consumer premises.

[**Ans.** Rs 88.94 per annum per kW demand ; Paise 2.0 per kWh delivered at the consumer premises]

34. Compute the generation cost per kWh from the following data :
Installed capacity = 200 MW; Capital cost = Rs 3,000 per kW; Interest and depreciation = 12%; Fuel consumption = 0.9 kg/kWh; Fuel cost = Rs 70 per 1,000 kg; Other operating costs = 30% of fuel costs; Load factor = 80%; Peak Load = 170 MW.

[Pb. Univ. Electrical Power II (Generation) Oct. 1991, Jan. 1993]

[**Ans.** 14.23 paise per kWh]

❑❑❑

CHAPTER 15

Tariffs and Power Factor Improvement

15.1. TARIFFS

Tariff means the schedule of rates or charges. Tariff, in case of electric supply, means the schedule or rates framed for supply of electrical energy to different classes of consumers.

15.1.1. Objective of Tariff

The main objective of the tariff is to distribute equitably the cost of supplying energy among the various classification of use. Therefore, a tariff must cover the following items :

(*i*) Recovery of cost of capital investment in generating, transmitting and distributing equipment.

(*ii*) Recovery of cost of operation, supplies and maintenance of equipment.

(*iii*) Recovery of cost of metering equipment, billing, collection costs and miscellaneous services and

(*iv*) A satisfactory return on the total capital investment.

15.1.2. Principal Factors Affecting Framing of Tariff

The principal factors involved in fixing of a tariff are enumerated below :

1. Proper return is secured from each consumer. While fixing the tariff for different classes of consumers, it has to be taken into account whether the tariff will result in a revenue meeting all the expenditure of the supply authority. In addition, the tariff should bring forth sufficient money to enable future expansion to meet an anticipated load requirement.
2. The consumers are encouraged to make more extended use of electricity.
3. The tariff should be simple and capable of easy explanation to the public. A complicated tariff may cause an opposition from the public which is generally distrustful of the supply authorities.
4. The consumers are charged according to what the energy costs. The tariff should be such as to satisfy the consumers of all categories. A big consumer should be charged at a lower rate than a small consumer. This is because increased energy consumption spreads the fixed charges over a greater number of units, thus reducing the overall cost of producing electrical energy.
5. The consumers are encouraged to use power during off-peak hours and penalised for high loads demanded at system peak by making a provision for higher demand charges.
6. The consumers are penalised for poor power factor.

The type of service rendered by the supply authority also determines the tariff that must be charged to a consumer. This is explained below by giving some examples. Electric supply to a domestic consumer, having usually light and fan points, makes the life of an ordinary citizen convenient and pleasant. But such a class of consumers cannot be charged at higher rates. This is because in a welfare state like India, the state has the responsibility to provide some basic necessities to its citizens. Similarly, the farmer for supply of electrical energy to agricultural loads cannot be charged at a higher rate because of the national concern for agricultural output. However, the consumers using electricity for extra comforts such as for air-conditioning and for commercial purposes can be charged at higher rates.

15.2. TYPES OF TARIFFS

Various types of tariffs, which are commonly used, are described below.

1. *Flat Demand Tariff.* This is one of the earliest form of tariffs used for charging the consumers for electrical energy consumption. In this case, the total demand and the energy consumption were fixed.

In early days, when the use of electricity was mainly restricted to very few applications such as for lamps etc. and the number of hours of use of the lamps were fixed, they were charged on the basis of number of lamps installed or total load connected in kW and fixed number of hours of use in a month or in a quarter of a year or in a year. Since the total demand and energy consumption were fixed and all the consumers were of alike class, so the rate could be expressed directly as certain amount per lamp or per unit of load installed. If x is the number of lamps or load connected in kW and a is the rate per lamp or per kW of connected load then

$$\text{Energy charges} = \text{Rs } ax$$

In this type of tariff the metering equipment, meter reading, billing and accounting costs are eliminated resulting in overall decrease of total cost. The main disadvantage of this tariff is that it does not differentiate between the consumers who use same appliance or equipment but for different duration of time. It encourages the consumers to keep their appliances connected to the supply mains even when not required. Nowadays the tariff of this type is restricted to use such as in street lighting, signal systems, sign lighting and other loads where the energy consumption in quantity could be readily predicted.

2. *Simple Tariff.* This is the simplest type of tariff according to which the cost of energy is charged on the basis of units consumed. The rate can be derived as

$$\text{Cost/kWh} = \frac{\text{Annual fixed cost} + \text{annual operating cost}}{\text{Total number of units supplied to the consumers per annum}}$$

Although this method of deriving the rate is simple and also the drawback of encouraging the consumers to keep their appliances connected to the supply mains even when not required is removed as they have to pay for all the energy consumed but suffers from the following drawbacks :

(*i*) There is no discrimination among the different types of consumers (domestic, industrial, bulk) having different load factor, diversity factor and power factor.

(*ii*) The cost per kWh delivered is higher.

The simple tariff can be made suitable by having the following modifications.

(*i*) A discount be allowed to the consumer who consumes more electricity than an average consumer.

(*ii*) Special tariffs be framed for the various types of consumers just as the domestic consumers using electricity for light and fans be charged at higher rates as compared to power consumers. The reason for it is that the load factor of light and fan load is low and the demand is during peak hours of the day.

(*iii*) The consumers be encouraged to use electricity during off-peak hours by giving a suitable discount. Such a tariff is called the *off-peak tariff.*

(*iv*) The consumers be encouraged to use electricity at high load factor and power factor by framing special types of tariffs such as maximum demand tariff and power factor tariff.

3. *Flat Rate Tariff.* This type of tariff differs from the former one in the sense that the different types of consumers are charged at different rates *i.e.,* the flat rate for light and fan loads is slightly higher than that for power load. The rate for each category of consumers is arrived at by taking into account its load factor and diversity factor.

The method is usually most popular with the public since it can easily be understood by the consumers and the calculations at the supplier's end are simple. If the energy consumed during the billing period is x units and the flat rate is Rs a per unit then charges are Rs ax.

The disadvantages of this tariff are :

(*i*) Separate meters are required for different types of supply.

(*ii*) Difficulty is experienced to derive the load factor and diversity factor for various types of loads to be employed in deciding the tariff.

(*iii*) The consumer is charged for the total quantity of energy consumed at the same rate irrespective of the magnitude of energy consumed while increased generation or consumption spread the fixed charges over a greater number of units and so the overall cost per unit decreases as the consumption increases.

4. *Step Rate Tariff.* The step rate tariff is a group of flat rate tariffs of decreasing unit charges for higher range of consumption, say for example

Rs 4.0 per unit if the consumption does not exceed 50 kWh

Rs 3.50 per unit if the consumption exceeds 50 kWh but does no exceed 200 kWh

Rs 3.0 per unit if the consumption exceed 200 kWh.

This type of tariff takes into account the fact of lower generation cost owing to higher energy consumption consequent to improvement of load factor and therefore promotes the use of electricity. The drawback of this tariff is however that by increasing the energy consumption so as just to enter the next range from the final stage of previous range, the total energy cost is reduced. Thus there is tendency with the consumer, just approaching the limit of the step to, anyhow cross the step and enter the next one in order to reduce the total energy cost. This drawback is removed in *block rate tariff* explained below.

5. *Block Rate Tariff.* In this type of tariff a given block of energy is charged at higher rate and succeeding blocks of energy are charged at progressively reduced rates, say for example

The first 25 units may be charged at the rate of Rs 4.0 paise/unit.

The next 40 units may be charged at the rate of Rs 3.50 per unit.

The consumption exceeding 65 units may be charged at the rate of Rs 3.0 per unit.

The advantage of such a tariff is that the consumer gets an incentive for consuming more electrical energy. This increases the load factor of the system and hence the generation cost is reduced. Its main drawback is that it lacks a measure of the consumer's demand. This tariff is most popular nowadays among domestic, commercial and small industrial consumers.

6. *Hopkinson Demand Rate or Two Part Tariff.* The total charge to be made to the consumer is split into two components namely fixed charge and running charge. Since fixed charge is independent of energy consumed and proportional to the maximum demand so the fixed charge is made at a certain amount per kW of maximum demand, which can be assessed on the basis of the rateable value of the premises or on the number of rooms excluding bath rooms, attics etc. or on the connected load or on the total combined kW capacity of all the consuming devices owned by a particular consumer. Running charge is made at a certain amount per kWh for the total energy consumed.

This type of tariff is expressed by the expression

Total energy charges, e = Rs $a \times$ kW + Rs b kWh where Rs a is the charge per kW of maximum demand assessed and Rs b is the charge per kWh of energy consumed.

This tariff is mostly applicable to medium industrial consumers.

In this tariff charge made on maximum demand recovers the fixed charges such as interest and depreciation on the capital cost of building and equipment, taxes and insurance charges and operating cost which is independent of energy supplied by it and varies with variation of maximum demand. Charge made on total energy consumption recovers the operating cost, which varies with variation in energy supplied.

The drawback of this tariff from the point of view of consumer is that he is to pay his fixed charge irrespective of the consumption. For example, if during any month any industry remains closed, the owner will be required to pay the fixed charges unnecessarily.

7. *Maximum Demand Tariff or Wright Demand Rate.* This tariff is similar to that of two part tariff except that in this case maximum demand is actually measured by a maximum demand indicator instead of merely assessing it on the basis of rateable value. In this tariff the drawback of the two part tariff is removed. This tariff is almost applicable to all bulk supplies and large industrial consumers, who have a control over their maximum demand.

Such a tariff induces the consumer to keep his maximum demand at a low value.

8. *Power Factor Tariffs.* Since the efficiency of plant and equipment depends upon the power factor, therefore, in order to increase the utility of plant and equipment to the maximum, the plant must be operated at the most economical power factor. That is why, sometimes consumers are penalised for poor power factor by applying the following types of power factor tariffs :

(*a*) KVA Maximum Demand Tariff. It is a modified form of two part tariff. In this case maximum demand is measured in kVA instead of in kW. This type of tariff encourages the consumers to operate their machines and other equipment at improved power factor, because low power factor will cause more demand charges.

(*b*) kWh and kVARh Tariff. In this type of tariff kWh and kVARh are measured and charged separately. Since kVARh decreases with the increase in power factor, therefore, the consumer tries to improve the power factor of his installation in order to decrease the charges on account of kVARh recorded.

(*c*) Sliding Scale or Average Power Factor Tariff. In this type of tariff, an average power factor, say 0.8 lagging may be taken as reference and a surcharge for each 0.01 by which the average power factor falls below this figure may be made. Similarly a discount may be allowed for each 0.01 by which the average power factor rises above this figure. Such tariffs are rarely used.

9. *Three Part Tariff or Doherty Rate.* In this tariff total charge is split into three elements namely fixed charge, semi-fixed charge and variable charge.

Constant or fixed charge recovers the cost on account of expenses in giving a supply (of any sort or of any magnitude) service connection (incurred once only but charged for either by a capital or an annum sum or charged each billing period), office and service expenses including accounting, metering and establishment charges.

Semi-fixed charge, which are levied on the basis of maximum demand in kW or kVA, recovers the plant initial cost as well as that operating cost which is independent of total energy supplied by it and varies with the variation in maximum demand.

Variable charge, which are levied on the basis of energy consumed, recovers the cost varying with the variation in energy supplied such as on account of fuels etc.

So the general expression for the recovery of the cost split into three sections mentioned above can be written as

Total charges = Rs $a + b$ kW + c kWh

where a is a constant charge made each billing period.

b = Unit charge in Rs per kW of metered maximum demand in kW during billing period. In some cases it is also charged in Rs per kVA so that the consumers are penalised for poor power factor.

and c = Unit charge for energy in Rs per kWh of energy consumed.

This type of tariff is usually applicable to bulk supplies.

10. *Off-Peak Tariff.* The load on the power station usually has pronounced peak loads in the morning and early evening and a very low load during the night (from 10 PM to 6 AM). During the night, therefore, and other off-peak period which may occur, a large proportion of the generating and distribution equipment will be lying idle. In case the consumers are encouraged to use electricity during off peak hours by giving a special discount, the energy can be supplied without incurring an additional capital cost and should, therefore, prove very profitable.

This type of tariff is very advantageous for certain processes such as water heating by thermal storage, pumping, refrigeration etc.

15.3. SPOT PRICING

The major concern of electric utilities all over the world is very large demand during peak load hours and surplus energy during off peak hours. This put the utilities in awkward situation of capacity shortages during peak load hours and large surplus during off peak hours. In spot princing the consumer is charged based on the time at which consumption occur. This is resorted to offset the imbalance caused because of large demand during peak load conditions and off peak conditions. The price is varied for reflecting the utility cost of providing the electrical energy at a given time. The spot pricing encourages the consumer to shift the maximum demand to off peak hours when the supply is abundant and prices are low. The spot pricing rate can be expressed as

Instantaneous spot rate

= Incremental operating cost + quality of supply component.

The quality of supply component is choosen so that the resultant price reflects the marginal value of unserved energy. Spot pricing rates are economically rational as they embody the principles of marginal cost theory and result in efficient location of resources. Spot pricing is also sometimes termed as 'load adaptive pricing', 'flexible pricing', 'homeostatic control pricing', 'real time pricing' or 'response pricing'. It is also known as 'Time of Day Tariff'.

Because of recent advancement in microcomputers and telecommunications and a drastic reduction in metering and communication costs, spot pricing has become feasible and many utilities in developed countries have adopted at least partial spot pricing in their industrial and commercial rate structures.

15.4. TYPES OF CONSUMERS AND THEIR TARIFFS

The main types of load on a system are domestic, commercial agricultural, industrial, municipal, traction etc. Accordingly the consumers may be categorised as domestic consumers, commercial consumers, agricultural consumers, industrial consumers (small, medium and large), bulk consumers etc.

15.4.1. Domestic Consumers

Residential load consists of lights, fans, and appliances such as radios, TVs, heaters, electric irons, refrigerators, electric water heaters, washing machines, coolers, air-conditioners, domestic pump sets etc. Domestic consumers are given single phase supply up to a load of 5 kW and a 3-phase supply for loads exceeding 5 kW.

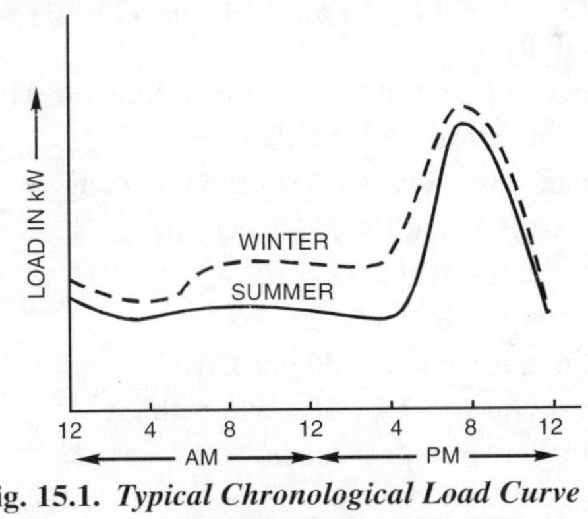

Fig. 15.1. ***Typical Chronological Load Curve For Domestic Consumers***

The small consumers generally use all the light points simultaneously so their demand factor is high (almost unity). Big domestic consumers may have demand factor low (around 0.5). During summer the major part of domestic load may consist of fan load during day and light and fan load during evenings and early mornings and some air-conditioner load. During winter, the major load is light load during evenings and early mornings and some heater load. Typical chronological load curve for domestic consumers is given in Fig. 15.1. The tariffs applicable to domestic consumers are simple tariff, flat rate tariff or block rate tariff. In addition meter rent and electricity duty are also charged from the consumers.

15.4.2. Commercial Consumers

Non-residential premises, such as shops, business-houses, cinemas, hotels, public offices, clubs etc. fall under this category. The load mainly consists of lights, fans and small electric appliances. The load remains fairly constant from around morning 10 to evening 9 hours. During night the load may consist of some lighting load. The demand factor is fairly high. Such consumers are given single phase supply for loads up to 5 kW and three phase supply for loads exceeding 5 kW. Typical chronlogical load curve for commercial consumers is shown in Fig. 15.2. The tariffs applicable to commercial consumers are also simple tariff, flat rate tariff or block rate tariff but charges per unit are higher in comparison to those in case of domestic consumers. In addition, meter rent and electricity duty are also charged from the consumers. Sometimes there is also a provision for minimum charges in case the energy consumption remains below a certain prescribed limit.

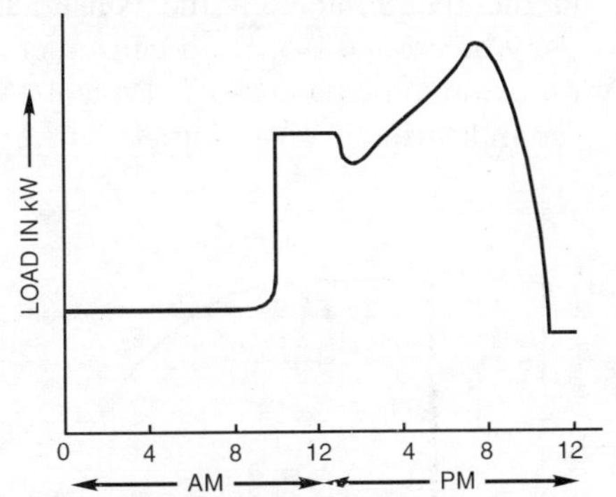

Fig. 15.2. ***Typical Chronological Load Curve For Commercial Consumers***

15.4.3. Agricultural Consumers

Consumers drawing power up to 20 kW for irrigation pumping units are categorized as agricultural consumers. Such

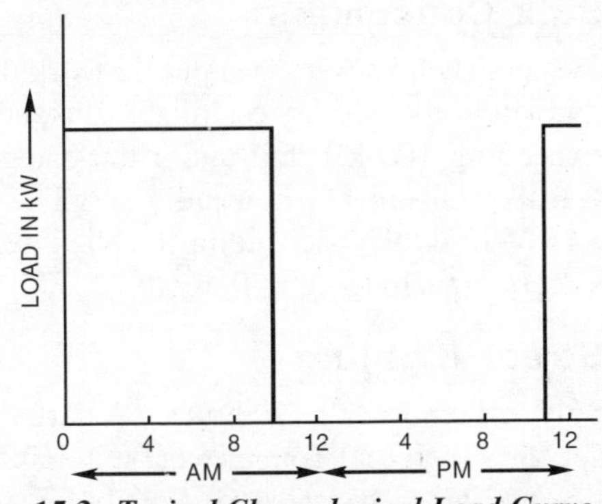

Fig. 15.3. ***Typical Chronological Load Curve For Agricultural or Irrigation***

consumers are given a three phase supply. The loads of the tubewells used for irrigation constitute a substantial portion of the system load. The demand factor and diversity factor are both almost unity. Figure 15.3 depicts the typical chronological load curve for agricultural or irrigation load.

Agricultural consumers are charged at a flat rate tariff which may be either on the basis of a fixed charge per unit consumed or a fixed charge per kW or hp of connected load.

15.4.4. Industrial Consumers

Industrial consumers may further be categorised as small industrial consumers, medium industrial consumers and large industrial consumers according to the rating of loads.

Small industrial consumers are owners of small workshops, atta chakkis, wheat threshers, saw machines and other small manufacturing and repair shops with load not exceeding 20 kW. The demand factor depends on the nature of load but usually it is high (around 0.8). Such consumers are given 3-phase supply at 415 V. A block tariff is usually offered to such consumers.

Industrial consumers with loads exceeding 20 kW but not exceeding 100 kW fall under the category of *medium industrial consumers.* They are given three phase supply at 415 V and are usually charged on two part tariff.

Industrial consumers with loads exceeding 100 kW are categorized as *large industrial consumers.* They are supplied power at 11 or 33 kV, and in rare cases, at 415 V three phase depending on the requirement of the consumers. The demand factor may be around 0.5. Such consumers are usually charged kVA maximum demand tariff. Typical chronological load curve for industrial load is depicted in Fig. 15.4.

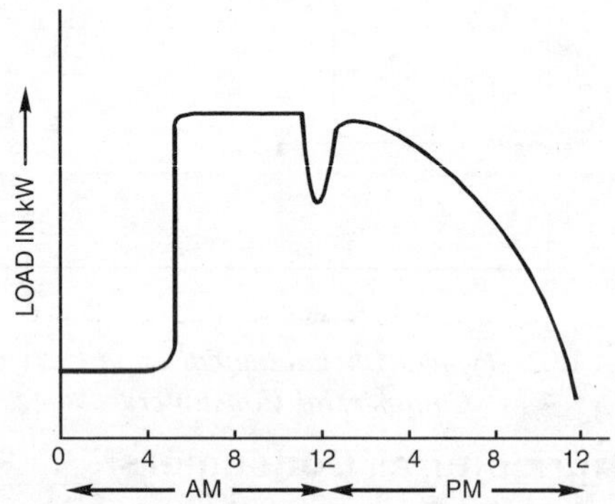

Fig. 15.4. ***Typical Chronological Load Curve For Industrial Load***

15.4.5. Bulk Consumers

Power consumers such as railways, public work departments, educational institutions, military establishments, hospitals having loads exceeding 100 kW fall under the category of bulk consumers. Bulk consumers are usually supplied by 3-phase supply at 415 V or 11 kV depending on their requirements. Such consumers are charged at flat rate.

15.4.6. Street Lighting

Power supply given for the lighting of parks, roads and streets under the municipal committees, municipal boards or panchayats comes under this category. Supply for street lighting is given at 415 V three phase or 240 V single phase. Such a load has demand factor and diversity factor of unity. The switching on the lights and their switching off is synchronised with dusk and dawn respectively. Separate distributors are run for street lighting to enable their switching simultaneously. The tariff charged for street lighting is such that it recovers the cost of the energy consumed as well as the cost of replacement of lamps. Chronological load curve for street lighting is shown in Fig. 15.5.

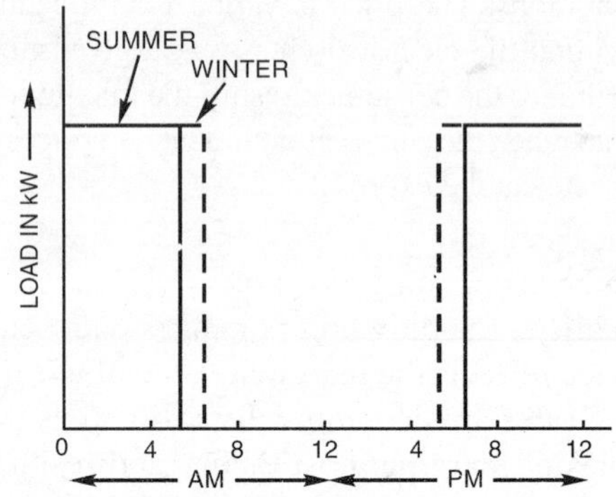

Fig. 15.5. ***Typical Chronological Load Curve For Street Lighting***

15.4.7. Water Supply

The load for water supply is for pumping water to the overhead tanks. It is generally possible to fit this load during system off peak hours, usually during night hours.

Example 15.1. The daily load of an industrial consumer is as follows :

100 kW for 9 hours, 125 kW for 6 hours, 50 kW for 7 hours and 5 kW for 2 hours.

The tariff rate is Rs 800 per kW of maximum demand per year plus Rs 2.50 per kWh. Determine the energy consumption per year (365 days) and yearly bill. [*West Bengal Univ. of Technology 2009-10*]

Solution : Daily energy consumption

$= 100 \times 9 + 12.5 \times 6 + 50 \times 7 + 5 \times 2$

$= 900 + 750 + 350 + 10 = 2{,}010$ kWh

Annual energy consumption $= 2{,}010 \times 365 = 733{,}650$ kWh **Ans.**

Maximum demand = 125 kW

Demand charges per annum = Rs 800×125 = Rs 100,000

Energy charges per annum = Rs $2.50 \times 733{,}650$ = Rs 1,834,125

Yearly bill = Demand charges per annum + energy charges per annum

= Rs 100,000 + Rs 1,834,125 = Rs 2,834,125 **Ans.**

Example 15.2. Light load : 6 tube lights 40 watts each working for 4 hours daily

Fan load : 6 fans 100 watts each working for 6 hours daily

Refrigerator load : 2 kWh daily

Miscellaneous load : 2 kW for 2 hours daily

Find the monthly bill at the following rate :

First 20 units—Rs 0.50 per kWh

Next 30 units—Rs 0.40 per kWh

Remaining units—Rs 0.30 per kWh

Constant charge—Rs 2.50 per month

Discount for prompt payment = 5 per cent.

[*Rajasthan Technical Univ. Generation of Electrical Power 2008, 2009*]

Solution : Daily energy consumption

$$= \frac{(6 \times 40 \times 4) + (6 \times 100 \times 6)}{1{,}000} + 2 + 2 \times 2 = 10.56 \text{ kWh}$$

Monthly (30 days) consumption

$= 10.56 \times 30 = 316.8$ kWh or units

Monthly bill $=$ (Energy charges + constant charge) (1 – discount)

$= \text{Rs } [20 \times 0.5 + 30 \times 0.4 + (316.8 - 50) \times 0.3 + 2.5]\,(1 - 0.05)$

$= \text{Rs } (10 + 12 + 80.04 + 2.5)\,(0.95) = \text{Rs } 99.31$ **Ans.**

Example 15.3. A consumer has an annual consumption of 70,080 kWh. The charge is Rs 100/kW of maximum demand plus 5 paise/kWh.

(*i*) Find the annual bill and the overall cost per kWh if the load factor is 40%.

(*ii*) What is the overall cost per kWh if the consumption was reduced by 25% with the same load factor ?

[*G.B. Technical Univ. Power Station Practice 2011-12*]

Solution : (*i*) Annual energy consumption = 70,080 kWh

Annual load factor, LF = 40% or 0.4

$$\text{Average load} = \frac{\text{Annual energy consumption}}{8{,}760} = \frac{70{,}080}{8{,}760} = 8 \text{ kW}$$

$$\text{Maximum demand, MD} = \frac{\text{Average load}}{\text{LF}} = \frac{8}{0.4} = 20 \text{ kW}$$

Annual energy consumption bill

$=$ Demand charges/annum + energy charges/annum

$= \text{Rs } 100 \times 20 + \text{Rs } 0.05 \times 70{,}080$

$= \text{Rs } 2{,}000 + \text{Rs } 3{,}504 = \text{Rs } 5{,}504$ **Ans.**

$$\text{Overall cost per kWh} = \frac{\text{Annual energy consumption bill}}{\text{Annual energy consumption}}$$

$$= \text{Rs } \frac{5{,}504}{70{,}080} = \text{Re } 0.0785 \text{ or } 7.85 \text{ paise}$$ **Ans.**

(*ii*) Annual energy consumption

$= (1 - 0.25) \times 70{,}080 = 52{,}560$ kWh

$$\text{Average load} = \frac{52{,}560}{8{,}760} = 6 \text{ kW}$$

$$\text{Maximum demand, MD} = \frac{\text{Average load}}{\text{Load factor}} = \frac{6}{0.4} = 15 \text{ kW}$$

Annual energy consumption bill

$= \text{Rs } 100 \times 15 + \text{Rs } 0.05 \times 52{,}560$

$= \text{Rs } 1{,}500 + \text{Rs } 2{,}628 = \text{Rs } 4{,}128$

$$\text{Overall cost per kWh} = \text{Rs } \frac{4{,}128}{52{,}560}$$

$= \text{Re } 0.0785$ or 7.85 paise **Ans.**

Note: Overall cost per kWh remains the same in the second case as in case (*i*). It is because of same load factor in both the cases, the energy consumption and maximum demands are reduced in the same ratio and therefore, the fixed and running charges.

Example 15.4. An industrial undertaking has connected load of 200 kW. The maximum demand is 150 kW. On average each machine works for 70% of time. Find yearly expenditure on electricity if the tariff is Rs 3,000 + Rs 700 per kW of maximum demand per year + Rs 0.60 per kWh.

[*Rajasthan Univ. Generation of Electrical Power 2003*]

Solution : Yearly energy consumption

$= 200 \times 0.7 \times 365 \times 24 = 1{,}226{,}400$ kWh

Fixed charges/annum $=$ Rs 3,000

Demand charges/annum $= \text{Rs } 700 \times 150 = \text{Rs } 105{,}000$

Energy charges/annum $= \text{Rs } 0.60 \times 1{,}226{,}400 = \text{Rs } 735{,}840$

Yearly expenditure on electricity

$= \text{Rs } 3{,}000 + \text{Rs } 105{,}000 + \text{Rs } 735{,}840 = \text{Rs } 843{,}840$ **Ans.**

Example 15.5. The annual consumption of domestic customer is $1{,}500 \times 10^3$ units and the annual load factor is 40%. What will be the saving in the average unit cost if the load factor improves to 100%? The two part tariff is Rs 500 per kW of maximum demand per year + 40 paise per kWh.

[*Rajasthan Univ. Generation of Electrical Power 2000*]

Solution : Annual energy consumption $= 1{,}500 \times 10^3$ kWh

$$\text{Average load} = \frac{\text{Annual energy consumption}}{365}$$

$$= \frac{1{,}500 \times 10^3}{365 \times 24} = 171.233 \text{ kW}$$

Load factor $= 40\% = 0.4$

$$\text{Maximum demand} = \frac{\text{Average load}}{\text{Load factor}} = \frac{171.233}{0.4} = 428 \text{ kW}$$

In case the consumer improves the load factor to 100 per cent, his maximum demand will be reduced to $\frac{171.233}{1}$ *i.e.* 171.233 kW

Annual energy bill with 40% load factor

$= \text{Rs } 500 \times 428 + \text{Rs } 0.40 \times 1{,}500 \times 10^3 = \text{Rs } 814{,}000$

$$\text{Unit cost} = \text{Rs } \frac{814{,}000}{1{,}500 \times 10^3} = \text{Re } 0.5427 \text{ or paise } 54.27$$

Annual energy bill with 100% load factor

$= \text{Rs } 500 \times 171.233 + \text{Rs } 0.40 \times 1{,}500 \times 10^3 = \text{Rs } 685{,}616.50$

$$\text{Unit cost} = \text{Rs } \frac{685{,}616.50}{1{,}500 \times 10^3} = \text{Re } 0.4571 \text{ or paise } 45.71$$

Saving in the average unit cost

$=$ Paise 54.27 – paise 45.71 = Paise 8.56/kWh **Ans.**

Example 15.6. An industrial consumer has single phase 230V supply. His monthly energy consumption is 2,020 kWh. A maximum demand indicator installed at his premises indicates 40 A which is charged at unity power factor for 2 hours daily at Rs 3.50 per kWh. The remaining units are charged at Rs 1.80 per kWh. Find the monthly bill for 30 days and average tariff per kWh.

[*Generation of Electrical Power, Pb. Technical Univ. May 2008; Rajasthan Univ. 2001, 2004, 2005, Rajasthan Technical Univ. 2008*]

Solution : Monthly energy consumption corresponding to maximum demand

$$= \frac{\text{Voltage} \times \text{maximum demand of current} \times \text{pf} \times \text{working hour} \times 30}{1{,}000}$$

$$= \frac{230 \times 40 \times 1 \times 2 \times 30}{1{,}000} = 552 \text{ kWh}$$

Cost of energy corresponding to maximum demand

$= \text{Rs } 3.50 \times 552 = \text{Rs } 1{,}932$

Energy cost for remaining energy consumption

$= \text{Rs } 1.80 \times (2{,}020 - 552) = \text{Rs } 2{,}642.40$

Total monthly bill $= \text{Rs } 1{,}932 + \text{Rs } 2{,}642.40 = \text{Rs } 4{,}514.40$ **Ans.**

$$\text{Average tariff} = \frac{\text{Monthly energy cost}}{\text{Monthly energy consumption}}$$

$$= \text{Rs } \frac{4{,}514.40}{2{,}020} = \text{Rs } 2.265 \text{ per kWh}$$ **Ans.**

Example 15.7. An industrial consumer having a maximum demand of 100 kW, maintains a load factor of 60%. The tariff rates are Rs 900 per kVA of maximum demand per annum plus Rs 1.80 per kWh of energy consumed. If the average power factor is 0.8 lagging, calculate the total energy consumed per annum and the annual electricity bill. Also workout the overall cost per kWh consumed.

Solution : Maximum demand = 100 kW

Average power factor = 0.8

$$\text{Maximum demand in kVA} = \frac{\text{Maximum demand in kW}}{\text{Average power factor}} = \frac{100}{0.8} = 125$$

Load factor = 60% or 0.6

Total energy consumed per annum

= Maximum demand in kW × load factor × 8,760

= 100 × 0.6 × 8,760 = 525,600 kWh **Ans.**

Annual demand charges = Rs 900 × 125 = Rs 112,500

Annual energy charges = Rs 1.80 × 525,600 = Rs 946,080

Annual electricity charges = Rs (112,500 + 946,080)

= Rs 1,058,580 **Ans.**

$$\text{Overall cost per kWh supplied} = \text{Rs } \frac{1{,}058{,}580}{525{,}600} = \text{Rs } 2.02 \textbf{ Ans.}$$

Example 15.8. A consumer takes a steady load of 250 kW at a power factor of 0.8 lagging for 10 hours per day and 300 days per annum. Estimate the annual payment under each of the following tariffs

(*i*) Rs 1.20 per kWh + Rs 1,200 per kVA per annum.

(*ii*) Rs 1.20 per kWh + Rs 1,200 per kW per annum + 25 paise per kVARh.

Solution : Maximum demand = 250 kW

Annual energy consumption

= Steady load in kW × working hours per day × working days per annum

= 250 × 10 × 300 = 750,000 kWh

$$\text{Maximum demand in kVA} = \frac{\text{kW}}{\text{PF}} = \frac{250}{0.8} = 312.5$$

kVARh consumed per annum

= kWh × tan ($\cos^{-1}$ pf)

= 750,000 × tan ($\cos^{-1}$ 0.8) = 750,000 × 0.75 = 562,500

(*i*) Annual payment under tariff (*i*)

= Rs 1.20 × 750,000 + Rs 1,200 × 312.5

= Rs 1,275,000.00 **Ans.**

(*ii*) Annual payment under tariff (*ii*)

= Rs 1.20 × 750,000 + Rs 1,200 × 250 + Rs 0.25 × 562,500

= Rs (900,000 + 300,000 + 140,625) = Rs 1,340,625.00 **Ans.**

Example 15.9. Calculate the number of units to be consumed so that the annual bill on the basis of two part tariff is same from the following data :

Maximum demand = 10 kW

Two part tariff—Rs 1,200 per annum per kW of maximum demand plus Rs 1.80 per unit consumed.

Flat rate tariff—Rs 2.40 per unit.

Solution : Maximum demand = 10 kW

Let the consumption be x units so that the annual bill on the basis of two part tariff and flat rate tariff is the same.

Annual bill under two part tariff

= Rs 1,200 × 10 + Rs 1.80 × x = Rs (1.8x + 12,000)

Annual bill under flat rate tariff

= Rs 2.40 × x = Rs 2.4x

Annual bill will be same if (1.8x + 12,000) = 2.4x

$$\text{or } x = \frac{12{,}000}{(2.4 - 1.8)} = 20{,}000 \text{ units } \textbf{Ans.}$$

Example 15.10. The monthly reading of a consumer's meter are as follows :

Maximum demand = 50 kW

Energy consumed = 36,000 kWh

Reactive energy = 23,400 kVARh

If the tariff is Rs 80 per kW of maximum demand plus 8 paise per unit plus 0.5 paise per unit for each 1% of power factor below 86%, calculate the monthly bill of the consumer.

[*Generation of Electrical Power Rajasthan Univ. 2003, 2004, 2005, 2006; Rajasthan Technical Univ. 2011*]

Solution : Average load

$$= \frac{\text{Monthly energy consumption}}{24 \times 30} = \frac{36{,}000}{24 \times 30} = 50 \text{ kW}$$

$$\text{Power factor angle } \phi = \text{Tan}^{-1} \frac{\text{kVA hr}}{\text{kWh}} = \text{Tan}^{-1} \frac{23{,}400}{36{,}000} = 33.024°$$

Power factor = Cos ϕ = Cos 33.024° = 0.838444 or 83.444%.

Power factor surcharge = Rs 0.005 × 36,000 (86 – 83.8444) = Rs 388

Monthly bill = Rs (80 × 50 + 0.08 × 36,000 + 388) = Rs 7,268 **Ans.**

Example 15.11. A factory works for 16 hours a day for 300 days in a year. The following two systems of tariff are available :

High voltage supply at Re 1 per unit plus Rs 50 per month per kVA of maximum demand. Low voltage supply at Rs 60 per month per kVA of maximum demand plus Rs 1.10 per unit.

The factory has an average load of 250 kW at 0.8 power factor and a maximum demand of 300 kW at the same power factor.

The high voltage equipment costs Rs 500 per kVA and losses can be taken as 5%. Interest and depreciation charges are 12%. Calculate the difference in the annual cost between the two systems. [*A.M.I.E. Sec. B. Power Plant Engineering Summer 1998*]

Solution : Average load = 250 kW

Working hours per annum

= Number of working hours/day × number of working days in a year

= 16 × 300 = 4,800

Number of units consumed on low voltage supply

= 250 × 4,800 = 1,200,000

$$\text{Maximum demand on low voltage supply} = \frac{300}{0.8} = 375 \text{ kVA}$$

Annual cost under low voltage supply tariff

= Rs 60 × 12 × 375 + Rs 1.10 × 1,200,000 = Rs 1,590,000.00

High voltage equipment losses = 5% or 0.05

Rating of high voltage equipment

$$= \frac{\text{Maximum demand in kW}}{\text{Power factor} \times (1 - \text{losses})} = \frac{300}{0.8 \times 0.95} = 395 \text{ kVA}$$

Number of units consumed on high voltage supply

$$= \frac{1{,}200{,}000}{0.95} = 1{,}263{,}158$$

Annual cost of high voltage equipment

= Rs 500 × 395 = Rs 197,500

Annual interest and depreciation charges

$$= \text{Rs } 197{,}500 \times \frac{12}{100} = \text{Rs } 23{,}700.00$$

Annual cost under high voltage supply tariff

= Annual interest and depreciation charges on hv equipment + annual maximum demand charges + annual energy charges

= Rs 23,700.00 + Rs 50 × 12 × 395 + Rs 1.0 × 1,263,158

= Rs 1,523,858.00 **Ans.**

Thus high voltage supply will be cheaper and saving per annum

= Rs 1,590,000.00 – Rs 1,523,858 = Rs 66,142.00

15.5. POWER FACTOR

The cosine of the angle between voltage and current in an ac circuit is known as *power factor.*

In an ac circuit, there is generally a phase difference between voltage and current. In an inductive circuit, the current lags behind the applied voltage and the power factor of the circuit is referred to as lagging. In a capacitive circuit the current leads the applied voltage and therefore, the power factor of the circuit is said to be leading.

Consider an inductive circuit, which draws a current I from the supply mains lagging behind the supply voltage V by an angle ϕ, known as phase angle, the phasor diagram is shown in Fig. 15.6.

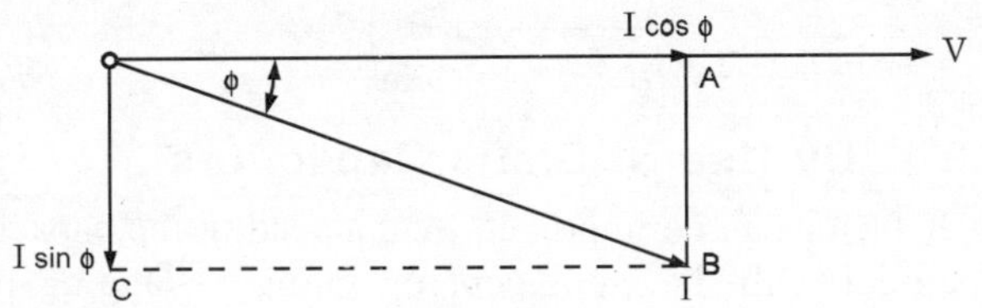

Fig. 15.6

The current I can be resolved into two components, one along the voltage phasor and the other perpendicular to it. The component along the voltage phasor, I cos ϕ is called the *in- phase* or *active* component of current, and the one perpendicular to the voltage phasor, I sin ϕ is called the *out of phase* or *wattless* or *reactive* component of current.

If all these components are multiplied by voltage V, the product of voltage V and in-phase component of current I cos ϕ *i.e.,* VI cos ϕ will represent the true power of the circuit in watts or kW, whereas the product of voltage V and the quadrature component of current I sin ϕ *i.e.,* VI sin ϕ will represent the reactive power in VARs or kVARs and the product of voltage V and current I *i.e.,* VI will represent the apparent power in volt-amperes or kVA. Thus we get a power triangle, as shown in Fig. 15.7.

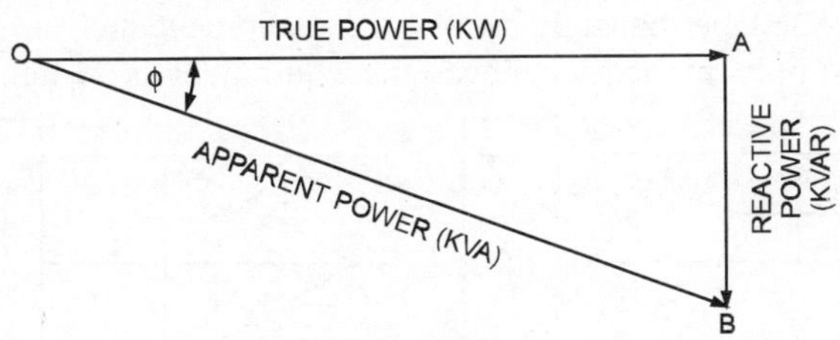

Fig. 15.7

From power triangle OAB shown in Fig. 15.7.

OA = kW component of power

AB = kVAR component of power

$$\text{Cos } \phi = \frac{OA}{OB} = \frac{kW}{kVA}, \text{ the ratio of true power and apparent power} \quad ...(15.1)$$

$$\text{Reactive power in kVAR} = \text{Apparent power} \times \sin\phi$$
$$= \text{kVA} \sin\phi$$
$$= (\text{kVA} \cos\phi) \times \frac{\sin\phi}{\cos\phi} = \text{kW} \tan\phi$$

i.e., smaller the reactive component of power, the smaller is the phase angle and the higher is the power factor.

For leading currents the triangle becomes reversed. This fact provides a key to the power factor improvement. If a device drawing leading reactive power is connected in parallel with the inductive load, then the lagging reactive power of the load will be partly neutralised, resulting in improvement of the power factor of the system.

15.6. DISADVANTAGES OF LOW POWER FACTOR

The current for a given load supplied at constant voltage will be higher at a lower power factor and lower at higher power factor. For example if load P is to be supplied at terminal voltage V and at power factor cos ϕ by a 3-phase balanced system then load current is given by

$$I_L = \frac{P}{\sqrt{3}\, V \cos\phi}$$

If P and V are constant, the load current, I_L is inversely proportional to power factor, cos ϕ *i.e.,* lower the power factor, higher the current and vice versa. The higher current due to poor power factor affects the system and results in following disadvantages.

(*i*) Rating of generators and transformers are proportional to their output current hence inversely proportional to power factor, therefore, large generators and transformers are required to deliver same load but at low power factor.

(*ii*) The cross-sectional area of the bus-bar, and the contact surface of the switchgear is required to be enlarged for the same power to be delivered but at low power factor.

(*iii*) For the same power to be transmitted but at low power factor, the transmission line or distributor or cable have to carry more current. The size of the conductor will have to be increased if current density in the line is to be kept constant. Thus more conductor material is required for transmission lines, distributors and cables to deliver the same load but at low power factor.

(*iv*) Energy losses are proportional to the square of the current hence inversely proportional to the square of the power factor *i.e.,* more energy losses incur at low power factor, which results in poor efficiency.

(*v*) Low lagging power factor results in large voltage drop in generators, transformers, transmission lines and distributors which results in poor regulation. Hence extra regulating equipment is required to keep the voltage drop within permissible limits.

(*vi*) Low lagging power factor reduces the handling capacity of all the elements of the system.

Thus we see that the low power factor leads to a high capital cost for the alternators, switchgears, transformers, transmission lines, distributors and cables etc.

Keeping in view the various drawbacks associated with the low power factor, the power suppliers insist on a power factor of 0.8 or above for industrial establishments. The power tariffs are devised to penalize the consumers with low lagging power factor and to encourage them to instal power factor correction devices or equipment.

15.7. CAUSES OF LOW POWER FACTOR

(*i*) All ac motors (except overexcited synchronous motors and certain type of commutator motors) and transformers operate at lagging power factor. The power factor falls with the decrease in load. For example an induction motor has a reasonable higher power of 0.85 at full load, 0.8 at 75% of full load, 0.7 at half-full load, 0.5 at 25% of full load and as low as 0.1 on no load.

(*ii*) Arc lamps and electric discharge lamps operate at low lagging power factor.

(*iii*) Due to increased supply mains voltage, which usually occurs during low-load periods such as lunch hours, night hours etc, the magnetizing current of inductive reactances increase and power factor of the electrical plant as a whole comes down.

(*iv*) The power factor at which motors operate falls due to improper maintenance and repairs of motors. In repaired motors, less wire is sometimes used than originally wound motors, therefore, in such motors leakage of magnetic flux increases and power factor of the motor decreases.

In case of heavily worn-out bearings, the rotor may catch at the stator.

Some metal is sometimes removed from the rotor by turning instead of replacing the defective bearings. In doing so, the length of air gap between stator and rotor increase, due to which greater magnetising current is required and, therefore, power factor drops.

(*v*) Industrial heating furnaces such as arc and induction furnaces operate on very lagging power factor.

The average power factors of some of the common appliances are given below :

Type of Load	Power Factor
Incandescent lamps	0.98 – 1.0
Fluorescent lamps	0.6 – 0.8
Neon lamps used for advertisements	0.4 – 0.5
Arc lamps used in cinemas	0.3 – 0.7
Fans	0.5 – 0.8
Induction motors	0.5 – 0.85
Fractional kW motors	0.4 – 0.75
Induction heaters	0.85
Resistance furnaces	0.6 – 0.9
Arc furnaces	0.85
Induction furnaces	0.6
Arc welders	0.3 – 0.4
Resistance welders	0.4 – 0.75

15.8. METHODS OF POWER FACTOR IMPROVEMENT

As we have already indicated, the low power factor is almost invariably due to inductive nature of load and, therefore, the logical corrective is to connect such devices across the load, which take leading reactive power such as static capacitors, synchronous machines or synchronous condensers. The leading reactive component of current drawn by power factor correcting device neutralises the lagging reactive component of current drawn by the load partly or completely. Power factor of the system will become unity when lagging reactive component of load current is completely neutralised by the leading reactive component of current drawn by power factor correcting device.

Let the current drawn by an inductive circuit be I lagging behind the applied voltage by an angle ϕ. The leading current required to neutralize the lagging reactive component of current drawn by the inductive circuit (equipment) to give unity power factor.

$$= I \sin \phi = I\sqrt{(1 - \cos^2 \phi)} = I\sqrt{1 - (pf)^2} \quad ...(15.2)$$

15.8.1. By Use of Static Capacitors

Power factor can be improved by connecting the capacitors* in parallel with the equipment operating at lagging power factor such as induction motors, fluorescent tubes. Static capacitors have the advantages of small losses (less than $\frac{1}{2}$ per cent) or higher efficiency (say 99.6%), low initial cost, little maintenance owing to absence of rotating parts, easy installation being lighter in weight and capability to operate under ordinary atmospheric conditions. However, they have drawbacks of short service life (8 to 10 years), getting damaged on overvoltages and uneconomical repair. The current drawn by induction motors or fluorescent tubes can be resolved into two components ; the active component, which is in phase with the supply voltage and the quadrature or wattless component of constant magnitude. The capacitors draw current leading the supply voltage by 90° approximately and neutralise the quadrature or wattless component of current drawn by the equipment across which these are connected. In case of 3-phase loads, capacitors can be connected either in star or delta, as shown in Figs. 15.8 and 15.9. These capacitors remain connected permanently across the equipment and are across the supply mains, whenever the equipment is switched on.

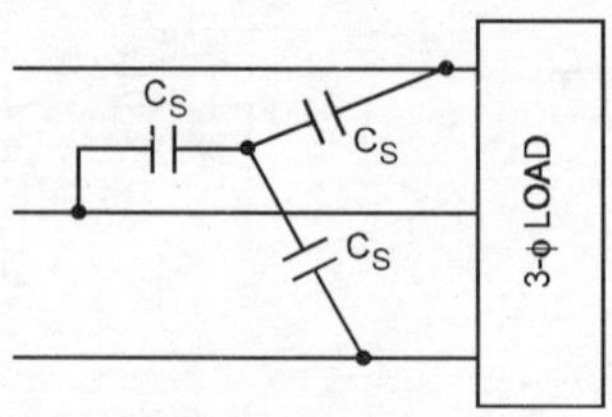

Fig. 15.8. ***Star-Connected Capacitors***

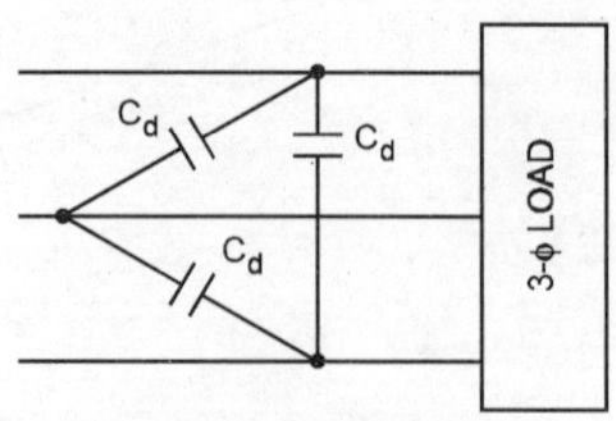

Fig. 15.9. ***Delta-Connected Capacitors***

The value of the static capacitors for the improvement of the power factor can be determined as follows :

The leading current required to neutralise the lagging reactive component of the current drawn by the equipment to give unity power factor is expressed as

$$I_C = I_L = I \sin \phi = I\sqrt{(1 - \cos^2 \phi)} = I\sqrt{1 - (pf)^2}$$

* The capacitors are generally designated as static capacitors to distinguish from the so called synchronous condensers which are synchronous motors operating under no-load condition and taking leading current.

The value of capacitance in star bank is given by

$$C_S = \frac{I_C}{2\pi f V} = \frac{I}{2\pi f V}\sqrt{1-(pf)^2} \qquad ...(15.3)$$

where V is the phase voltage, I is the phase current and f is supply frequency.

For given kVAR and line voltage the delta value will be one-third of star value.

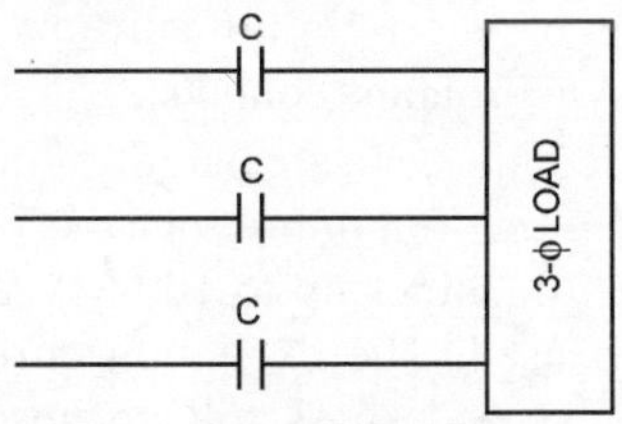

Fig. 15.10

Power factor can also be improved by connecting static capacitors in series with the line, as shown in Fig. 15.10. Capacitors connected in series with the line neutralize the line reactance. The capacitors, when connected in series with the line, are called the *series capacitors,* and when connected in parallel with the equipment, are called the *shunt capacitors.*

Shunt capacitors are used in factories, plants and also on transmission lines.

Series capacitors are used on long transmission lines as they provide automatic compensation with the variations in load. The capacity of the capacitors to neutralize the line reactance is given by

$$C = \frac{1}{(2\pi f)^2 L}$$

where f is the supply frequency and L is the inductance of the line per phase.

The value of reactance required is usually very large but reduced to reasonable value by use of a transformer, as shown in Fig. 15.11.

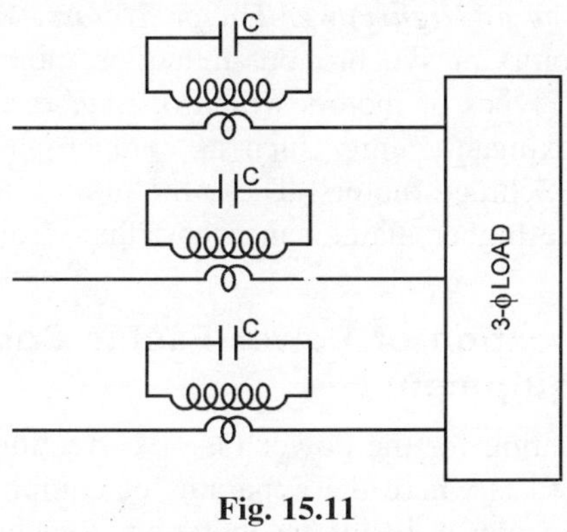

Fig. 15.11

Shunt capacitors are used in ratings from 15 kVAR to 10,000 kVAR. Small capacitors, up to a few hundred rating are used on individual distribution circuits of customers. Capacitors banks of 500 – 3,000 kVAR ratings are employed in small distribution substations and those with larger rating at big substations.

Three phase capacitor banks can be connected in star earthed, star unearthed or in delta arrangements. Ungrounded star connection is preferred because of easier protection. In this method, the fault current in case of a fault in any unit in one of the phases is restricted by the capacitors in the sound phases. This results in the use of smaller fuses and less protection materials. The capacitor must be provided with a suitable discharge device to dissipate the stored energy, and to reduce the residual voltage to a safer value within a short period (50 V or less within one minute in case of medium voltage capacitors and within five minutes in case of high voltage capacitors as per ISS 2834–1963). The discharge resistance is usually incorporated within the 'unit' itself in the case of medium voltage capacitors and in case of high voltage capacitors, potential transformers of the circuit breakers are generally utilised as a discharge device.

The reactive output of the capacitors in kVAR is given by

$\frac{\text{Voltage} \times \text{line current}}{1{,}000}$ kVAR or $2\pi f CV^2 \times 10^{-9}$ kVAR in case of single phase circuit ...(15.4)

where V is the line voltage, f the supply frequency and C is the capacitance in microfarads.

and $\frac{\sqrt{3}\ \text{voltage} \times \text{line current}}{1{,}000}$ kVAR or $2\pi f CV_L^2 \times 10^{-9}$ kVAR in case of 3-phase star-connected circuits. ...(15.5)

and $= 6\pi f CV_L^2 \times 10^{-9}$ kVAR in case of 3-phase delta-connected circuits ...(15.6)

where V_L is the line voltage, f is the supply frequency in Hz and C is the capacitance in μF between the line terminals.

Thus we see that the corrective capacity of the capacitors is a function of the line voltage and supply frequency, varying in accordance with the square of the voltage and directly with the supply frequency. The units as manufactured are designed for a variation of voltage of ±10% of normal voltage. It is, therefore, impossible to overload these units so long as normal voltage and frequency are maintained.

The characteristics of capacitors, in general, are similar to those of synchronous condensers except that the corrective kVAR of the synchronous condenser is adjustable and may be controlled automatically whereas that of the capacitor is fixed unless there is a possibility of change in the number of units connected.

15.8.2. By Use of Synchronous or High Power Factor Machines

Synchronous machines are excited by dc, and the power factor may be controlled by controlling the field excitation. The various synchronous machines available for power factor correction comprise synchronous motors, synchronous condensers, synchronous converters, synchronous phase modifiers, phase advancers, and synchronous-induction motors.

1. *By Use of Synchronous Motors.* These motors have characteristics that make them adaptable for a wide range of applications. The speed is constant, the efficiency is high and uniform from light loads up to considerable overloads, and the starting characteristics compare favourably with those of induction motors. Another desirable characteristic of the synchronous motor is its tendency to maintain a constant load voltage even if there are variations in the supply voltage. When the line voltage increases, the leading reactive kVA falls and when the line voltage falls, the leading reactive kVA increases. The usual practice is to keep the field excitation constant at a value corresponding to normal full-load rating as regards output and power factor. Synchronous motors are designed for 1.0 – 0.8 leading power factors at full load. The unity power factor motor costs less and has a higher efficiency, but if fully loaded, it cannot furnish leading reactive kVA to compensate for lagging reactive kVA in the system.

2. *By Use of Synchronous Condensers.* An overexcited synchronous motor running on no load is called the *synchronous condenser* or *synchronous phase advancer* and behaves like a capacitor, the capacitive reactance of which depends upon the motor excitation. Power factor can be improved by using synchronous condensers like shunt capacitors connected across the supply.

In phasor diagram (Fig. 15.12), phasor I_L represents the current drawn by the industrial load, lagging behind the applied voltage V by a large angle ϕ_L and phasor I_M represents the current drawn by the synchronous condenser leading the applied voltage V by the angle ϕ_M. The resultant current I is the phasor sum of I_L and I_M and now angle of lag ϕ is much smaller than ϕ_L. Thus overall power factor is improved from $\cos \phi_L$ to $\cos \phi$ by the use of the synchronous condenser. In this way the power factor can be made unity even.

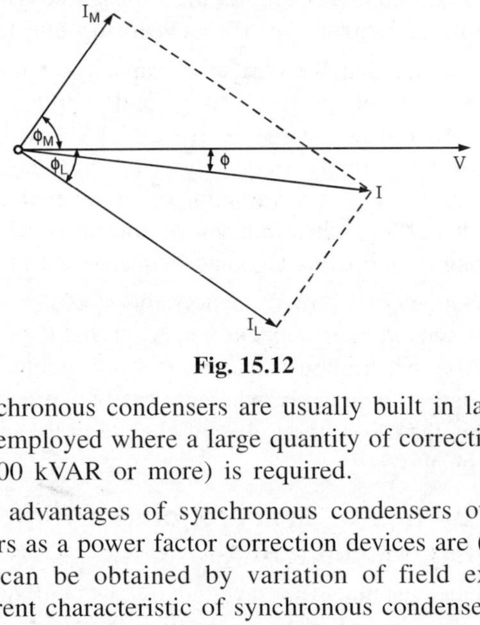

Fig. 15.12

Synchronous condensers are usually built in large units and are employed where a large quantity of corrective kVAR (say 5,000 kVAR or more) is required.

The advantages of synchronous condensers over static capacitors as a power factor correction devices are (*i*) a finer control can be obtained by variation of field excitation; (*ii*) inherent characteristic of synchronous condensers of stabilizing variations in the line voltage and thereby automatically aid in regulation, (*iii*) possibility of overloading a synchronous condenser for short periods, and (*iv*) improvement in the system stability and reduction of the effect of sudden changes in load owing to inertia of synchronous condenser. By use of synchronous condensers at intermediate stations, the voltage of the line can be kept constant at various points along its length, thereby, increasing the current carrying capacity of the line and improvement of power factor.

The disadvantages of synchronous condensers over static capacitors as power factor correcting devices are : (*i*) except in size above about 5,000 kVAR, the cost is higher than that of static capacitors of the same rating ; (*ii*) comparatively higher maintenance and operating costs ; (*iii*) comparatively lower efficiency (say 97%) due to losses in rotating parts and heat losses, (*iv*) noise is produced in operation (*v*) an auxiliary equipment is required for starting synchronous condensers ; (*vi*) possibility of synchronous condensers falling out of synchronism causing in interruption of supply ; and (*vii*) increase of short-circuit currents when the fault occurs near the synchronous condenser.

Synchronous condensers are largely employed by utilities at large substations for improving the power factor and voltage regulation. Machines up to 100 MVAR rating or even higher have been used. The excitation current is regulated automatically to give a desired voltage level.

3. *By Use of Phase Advancers.* The power factor of an induction motor falls mainly due to its exciting current drawn from the ac supply mains, because exciting current lags behind the voltage by $\pi/2$. It may be improved by equipping the set with an ac exciter or phase advancer which supplies this exciting current to the rotor circuit at slip frequency. Such an exciter may be mounted on the same shaft as the main motor or may be suitably driven from it. Use of phase advancer is not generally economical in connection with motors below 150 kW output but above this size, phase advancers are frequently employed. Shunt and series type of phase advancers are available according to whether the exciting winding of the advancer is connected in parallel or series with the rotor winding of the induction motor.

There are two main advantages of phase advancers (*i*) lagging kVAR drawn by the motor are considerably reduced due to supply of exciting ampere-turns at slip frequency and (*ii*) the phase advancers can be conveniently employed where the use of synchronous motor is inadmissible.

4. *By Use of Synchronous-Induction Motors.* These are special types of motors which operate at certain loads as synchronous motors and at other loads as induction motors.

5. *By Use of High Power Factor Motors.* Besides synchronous motors or synchronous-induction motors there are other several types of motors which operate at a power factor of approximately unity such as compensated induction motors, and Schrage motors. These motors are more expensive and have higher maintenance cost than ordinary induction motors.

15.8.3. Location of Power Factor Correction Equipment

The best location for the power factor correction equipment to be installed is where the apparatus or equipment responsible for low power factor is operating. Synchronous condensers are used at load centres where considerable corrective kVAR is required whereas static capacitors are justifiably used in smaller units and may be placed closer to the point where the load of inductive nature is installed and thereby relieving the distributors and feeders from carrying excessive currents owing to low power factor.

In case of transmission system, if synchronous condensers are to be employed for power factor improvement then these should be installed at the receiving end so that not only the generators but also the transmission lines are relieved of carrying excessive current due to poor power factor. However, if synchronous condensers are installed near the generators then only generators will be relieved from the excessive current component and the transmission lines will have to carry it.

15.9. ADVANTAGES OF POWER FACTOR IMPROVEMENT

Modern alternators are designed to have high reactance in order that alternators may not get damaged at the time of short circuit and normally reactance is 20 times of resistance. Any change in current and power factor causes the change in terminal voltage, so voltage can be kept fairly constant by power factor control.

If the power factor of the supply or power station is raised to unity, the current for the same amount of power to be supplied is reduced to minimum. This results in reduction of transmission line copper losses, and reduction of voltage drop in transmission line and in alternator windings, as copper losses are directly proportional to the square of supply current and voltage drop is directly proportional to the current.

The terminal voltage of an alternator is given by the phasor difference of induced emf and voltage drop in synchronous impedance. Neglecting the resistance as compared to synchronous reactance, the synchronous impedance can be considered purely inductive, so synchronous impedance voltage drop leads the load current by $\pi/2$.

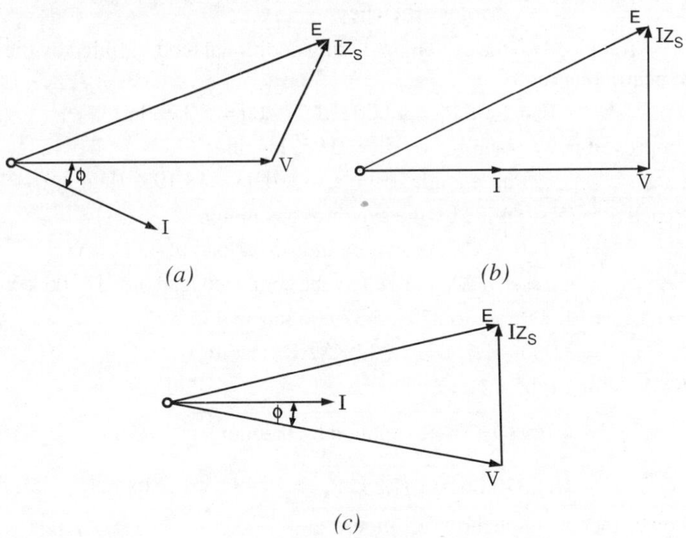

Fig. 15.13

From the phasor diagrams shown in Figs. 15.13 (*a*), 15.13 (*b*) and 15.13 (*c*) for lagging, unity and leading power factor respectively it is obvious that as the power factor is raised from lagging to unity, the difference of terminal voltage and induced emf is reduced and this difference can be reduced to zero by making the power factor, leading and less than unity. Hence to have zero regulation the power factor should be made leading one, so that no other regulating equipment is required.

It is not economical to raise the power to unity or leading one due to the following reasons :

1. In case the power factor is improved to unity for full-load conditions, it would become leading for loads less than full load (unless some capacitors are switched off which is usually difficult). At power factors lower than unity (lagging or leading) the current for the same amount of power to be supplied is increased and thereby energy losses in generators, transformers, transmission lines and distribution lines will be increased.
2. As the power factor approaches unity, the capacity of power factor correction device increases more rapidly *i.e.*, the power factor of an installation can be improved from 0.7 or 0.8 to 0.8 or 0.9 by a much smaller capacitive kVAR than that required for raising the power factor from 0.9 to unity.

The advantages of good (or improved) power factor are : (*i*) reduction in load current ; (*ii*) increase in voltage level across the load ; (*iii*) reduction in energy losses in the system (generators, transformers, transmission lines and distributors) due to reduction in load current ; (*iv*) reduction in kVA loading of the generators and transformers which may relieve an overloaded system or release capacity for additional growth of load and (*v*) reduction in kVA demand charge for large consumers.

Example 15.12. Find the power factor at an installation, supplying the following loads :

(*a*) 300 kW at unity power factor (*b*) 1,000 kW at 0.9 lagging power factor and (*c*) 1,500 kW at 0.8 lagging power factor. Also find the maximum load at unity power factor which can be supplied by the substation.

[*Rajasthan Univ. Generation of Electrical Power 1997, 2003, 2004*]

Solution : Total load on the substation

$$= 300 + 1{,}000 + 1{,}500 = 2{,}800 \text{ kW}$$

Total reactive load on the substation

$$= 300 \tan(\cos^{-1} 1.0) + 1{,}000 \tan(\cos^{-1} 0.9) + 1{,}500 (\cos^{-1} 0.8)$$

$$= 300 \tan 0° + 1{,}000 \tan 25.842° + 1{,}500 \tan 36.87°$$

$$\because \text{kVAR} = \text{kW} \tan\phi \text{ and } \phi = \cos^{-1}\text{pf}$$

$$= 0 + 484.3 + 1{,}125 = 1609.3 \text{ kVAR}$$

Total kVA load on the substation

$$= \sqrt{(\text{kW})^2 + (\text{kVAR})^2} = \sqrt{(2{,}800)^2 + (1{,}609.3)^2}$$

$$= 3{,}229.53 \text{ kVA}$$

$$\text{Power factor at the substation} = \frac{\text{kW}}{\text{kVA}} = \frac{2{,}800}{3{,}229.53} = 0.867 \text{ (lagging)} \textbf{ Ans.}$$

The maximum unity power factor load which can be supplied by the substation = Total kVA × pf = 3,229.53 × 1.0 = 3,229.53 kW **Ans.**

Example 15.13. A single phase motor connected to a 240V, 50 Hz supply takes a current of 20A at a pf of 0.75 lagging. A capacitor is shunted across the motor terminals to improve the pf to 0.9. Determine the capacitance of the capacitor to be used.

[*Rajasthan Univ. Generation of Electrical Power 2003*]

Solution : Initial motor current, $I_1 = 20\text{A}$

Initial power factor, $\cos\phi_1 = 0.75$

Lagging reactive component of initial motor current

$$= I_1 \sin\phi_1 = I_1\sqrt{1-\cos^2\phi} = 20\sqrt{1-0.75^2} = 13.229\text{A}$$

In-phase component of initial motor current

$$= I_1 \cos\phi = 20 \times 0.75 = 15\text{A}$$

Motor current after improvement of power factor

$$I_2 = \frac{\text{Active component of initial current}}{\text{Improved power factor}} = \frac{15}{0.9} = 16.667\text{A}$$

Lagging reactive component of motor current after pf improvement,

$$= I_2 \sin\phi_2 = I_2\sqrt{1-\cos\phi_2{}^2} = 16.667\sqrt{1-0.9^2}$$
$$= 7.625 \text{ A}$$

Leading reactive component drawn by the capacitor

$$I_C = I_1 \sin\phi_1 - I_2\sin\phi_2 = 13.229 - 7.265 = 5.964\text{A}$$

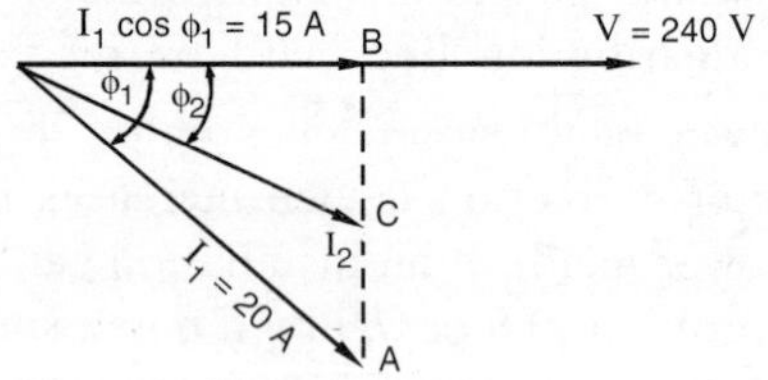

Fig. 15.14

Capacitance of the capacitor, $C = \frac{I_C}{2\pi f V}$ $\quad \because I_C = 2\pi f CV$

$$= \frac{5.964}{2\pi \times 50 \times 240} = 791\ \mu\text{F} \textbf{ Ans.}$$

Example 15.14. A 400 V, 50 Hz, 3-phase line delivers 200 kW at 0.8 pf lagging. It is desired to raise the line power factor to unity by installing shunt capacitors. Calculate the capacitance of each unit if they are connected in (*i*) star and (*ii*) delta.

[Rajasthan Technical Univ. Generation of Electrical Power 2008]

Solution : Load P = 200 kW

Power factor, $\cos\phi_1 = 0.8$ (lag) ; $\phi_1 = \text{Cos}^{-1}0.8 = 36.87°$ and Tan $36.87° = 0.75$

Power factor, $\cos\phi_2 = 1.0$; $\phi_2 = \text{Cos}^{-1}1.0 = 0°$ and Tan $0° = 0$

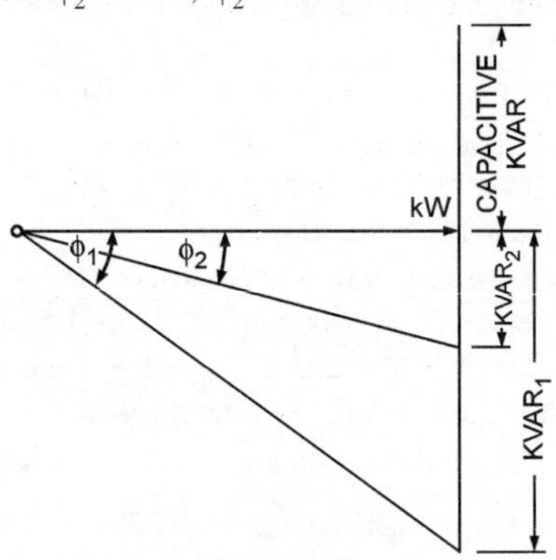

Fig. 15.15

Leading kVAR supplied by static capacitors,

Capacitive kVAR = $\text{kVAR}_1 - \text{kVAR}_2$

$$= P(\tan\phi_1 - \tan\phi_2) = 200\,(0.75 - 0) = 150$$

Since reactive kVAR output of capacitors when connected in star (Eq. 15.5) is given by $2\pi f C V_L{}^2 \times 10^{-9}$, where C is in μF

$$\therefore\ 2\pi f C V_L{}^2 \times 10^{-9} = 150$$

$$\text{or } C = \frac{150\times10^9}{2\pi f V_L^2} = \frac{150\times10^9}{2\pi\times50\times(400)^2} = 2{,}984\ \mu\text{F} \textbf{ Ans.}$$

Since reactive kVAR output of capacitors when connected in delta (Eq. 15.6) is given as $6\pi f C V^2 \times 10^{-9}$ where C is in μF

$$\therefore\ C = \frac{150\times10^9}{6\pi f V_L{}^2} = \frac{150\times10^9}{6\pi\times50\times(400)^2} = 995\ \mu\text{F} \textbf{ Ans.}$$

Example 15.15. The load on the mains of a supply system is 1,000 kW at a power factor of 0.707 lagging. What must be the kVA rating of the power factor improvement equipment which takes leading current at a power factor of 0.1 in order to raise the power factor of the entire system to 0.95?

[Pb.Technical Univ. Generation of Electric Power May 2010]

Solution : Load, P = 1,000 kW

Phase angle of load, $\phi_1 = \cos^{-1} 0.707 = 45°$ (lagging)

Improved phase angle of load, $\phi_2 = \cos^{-1} 0.95 = 18.195°$ (lagging)

Required rating of pf improvement equipment

$$= P(\tan\phi_1 - \tan\phi_2)$$
$$= 1{,}000\,(\tan 45° - \tan 18.195°)$$
$$= 1{,}000\,(1.0 - 0.32868) = 671.32 \text{ kVAR}$$

Phase angle of pf improvement equipment,

$$\phi = \cos^{-1} 0.1 = 84.26° \text{ (leading)}$$

Rating of pf improvement equipment

$$= \frac{\text{Reactive kVAR supplied by the equipment}}{\sin\phi}\text{kVA}$$

$$= \frac{671.32}{\sin 84.26°} = 674.7 \text{ kVA} \textbf{ Ans.}$$

Example 15.16. An industrial load of 4,000 kW is supplied at 11 kV, the power factor being 0.8 lagging. A synchronous motor is required to meet an additional load of 1,500 hp (1,103.25 kW) and the same time to raise the resultant power factor to 0.95 lagging. Determine the kVA capacity of the motor and the power factor at which it must operate. Take the efficiency of the motor as 80%.

[U.P.S.C. I.E.S. Electrical Engineering-II, 2011]

Solution : Industrial load, $P_1 = 4{,}000$ kW

Power factor of industrial load, $\cos\phi_1 = 0.8$ (lagging)

Phase angle, $\phi_1 = \text{Cos}^{-1}\ 0.8 = 36.87°$ (lagging)

$$\text{Motor load, } P_2 = \frac{\text{Additional load}}{\text{Motor efficiency}} = \frac{1{,}103.25}{0.8} = 1{,}379.06 \text{ kW}$$

Total load on the system including additional load supplied by the synchronous motor,

$$P = P_1 + P_2 = 4{,}000 + 1{,}379.06 = 5{,}379.06 \text{ kW}$$

Improved power factor, $\cos\phi_2 = 0.95$ (lagging)

Phase angle, $\phi_2 = \text{Cos}^{-1}\ 0.95 = 18.195°$ (lagging)

Reactive kVA supplied by the synchronous motor

= Reactive kVA drawn by industrial load of 4,000 kW

– reactive kVA drawn by the combined load of 5,379.06 kW

$$= 4{,}000 \tan 36.87° - 5{,}379.06 \tan 18.195°$$
$$= 3{,}000 - 1{,}768 = 1{,}232 \text{ KVAR (leading)}$$

KVA rating of the motor

$$= \sqrt{P_2^2 + (\text{KVAR supplied by the motor})^2}$$
$$= \sqrt{(1{,}379.06)^2 + (1{,}232)^2} = 1{,}849 \text{ kVA} \textbf{ Ans.}$$

Power factor of synchronous motor

$$= \frac{1{,}379.06}{1{,}849} = 0.7458 \text{ (leading)} \textbf{ Ans.}$$

Example 15.17. A 400 V, 3-phase installation draws a current of 50A at 0.8 lagging power factor. It is desired to install a synchronous motor to improve the overall power factor to 0.95 lagging. The synchronous motor will drive a 25 hp (metric) load at an efficiency of 0.9. Find kVA and pf of synchronous motor.

[Rajasthan Univ. Generation of Electrical Power 1998]

Solution : Load of installation,

$$P_1 = \sqrt{3}\,V_L I_L \cos\phi_1 = \sqrt{3}\times400\times50\times0.8$$
$$= 27{,}713 \text{ W or } 27.713 \text{ kW}$$

kVAR rating of installation,

$$Q_1 = P_1 \tan\phi_1 = 27.713 \tan(\cos^{-1} 0.8) = 20.785 \text{ kVAR}$$

kW input to synchronous motor, $P_2 = \frac{25 \times 735.5}{0.9 \times 1,000} = 20.43$ kW

Total load, $P = P_1 + P_2 = 27.713 + 20.43 = 48.143$ kW

Overall power factor angle, $\phi = \text{Cos}^{-1} 0.95 = 18.195°$ (lagging)

Total reactive kVA, $Q = P \tan\phi = 48.143 \tan 18.195°$

$= 48.143 \times 0.3287 = 15.82$ (lagging)

kVAR supplied by synchronous motor

$$Q_2 = Q_1 - Q = 20.785 - 15.82 = 4.965 \text{ kVAR (leading)}$$

kVA of synchronous motor $= \sqrt{P_2^2 + Q_2^2} = \sqrt{(20.43)^2 + (4.965)^2}$

$= 21.025$ kVA **Ans.**

Power factor of synchronous $= \frac{20.43}{21.025} = 0.9717$ (leading) **Ans.**

Example 15.18. A generator station is connected with the following loads :

***(i)* A lighting load of 200 kW *(ii)* an induction motor of 200 hp (metric) pf 0.7 lagging and efficiency 80% *(iii)* A synchronous motor of 100 hp (metric) efficiency 90%. Calculate the power factor of the synchronous motor so that the combined power factor becomes unity.**

[P.b. Univ. Electrical Power II (Generation) December 1995]

Solution : Input of induction motor

$$P_M = \frac{\text{Output in hp} \times 735.5}{\text{Motor efficiency}} = \frac{200 \times 735.5}{0.8}$$

$= 183,875$ W or 183.875 kW

Total load

= Lighting load + induction motor load + synchronous motor load

$= 200 + 183.875 + \frac{100 \times 735.5}{0.9 \times 1,000} = 200 + 183.875 + 81.722$

$= 465.6$ kW

Leading kVAR drawn by synchronous motor

= Lagging kVAR drawn by induction motor

$= 183.875 \tan(\cos^{-1} 0.7) = 183.875 \times 1.02 = 187.6$

∵ pf of lighting load is considered unity and pf after improvement is also unity

Power factor of synchronous motor

$$= \text{Cos} \tan^{-1} \frac{\text{Leading kVAR drawn by synchronous motor}}{\text{Power drawn by synchronous motor}}$$

$$= \text{Cos} \tan^{-1} \frac{187.6}{81.722} = 0.4 \text{ (leading)} \textbf{ Ans.}$$

Example 15.19. A 3-phase synchronous motor having a mechanical load (including losses) of 122 kW is connected in parallel with a load of 510 kW at 0.8 pf lagging. The excitation of the motor is adjusted so that the kVA input to the motor becomes 140 kVA. Determine the new power factor of the whole system.

Solution : Load, $P_1 = 510$ kW

Load power factor, $\cos\phi_1 = 0.8$ lagging

kVAR (lagging) drawn by the load, kVAR_1

$= P_1 \tan\phi_1 = 510 \tan(\cos^{-1} 0.8) = 382.5$

kVA input to the synchronous motor = 140

kW input to the synchronous motor, $P_2 = 122$ kW

kVAR (leading) drawn by the synchronous motor,

$$\text{kVAR}_2 = \sqrt{(\text{kVA})^2 - (P_2)^2} = \sqrt{(140)^2 - (122)^2} = 68.673$$

Total kVAR drawn by the whole system

$= \text{kVAR}_1 - \text{kVAR}_2 = 382.5 - 68.673 = 313.827$

Total load on the system, $P = P_1 + P_2 = 510 + 122 = 632$ kW

New power factor of the system, $\text{Cos}\,\phi = \text{Cos} \tan^{-1} \frac{313.827}{632}$

$= 0.8956$ (lagging) **Ans.**

Example 15.20. A 3 φ, 50 Hz, 400 V motor develops 74.6 kW, the power factor being 0.75 lagging and efficiency 93%. A bank of capacitors is connected in delta across the supply terminals and power factor is improved to 0.95 lagging. Each of the capacitance units is built of 4 similar 100V capacitors. Determine the capacitance of each capacitor.

[Rajasthan Univ. Generation of Electrical Power 2006]

Solution : Input to motor $= \frac{\text{Motor input}}{\eta} = \frac{74.6}{0.93} = 80.2$ kW

Initial power factor, $\cos\phi_1 = 0.75$ (lagging)

Power factor after improvement, $\cos\phi_2 = 0.95$ (lagging)

kVAR rating of capacitor bank

$= P(\tan\phi_1 - \tan\phi_2)$

$= 80.2 [\tan(\cos^{-1} 0.75) - \tan(\cos^{-1} 0.95)]$

$= 80.2 (0.881917 - 0.328684) = 44.37$ kVAR

Since reactive kVA output of the capacitors when connected in delta is given by $6\pi f C V_L^2 \times 10^{-9}$; where C is the capacitance in μF, V_l is the line voltage in volts and f is supply frequency in Hz.

$$\therefore C = \frac{44.37}{6\pi \times 50 \times 400^2 \times 10^{-9}} = 294 \ \mu F$$

The bank of capacitors used to improve the pf is connected in delta so phase voltage is 400 V, and therefore, each phase of bank will have 4 capacitors connected in series.

So capacitance of each unit = 4 × 294 = 1,176 μF **Ans.**

Example 15.21. A 35 kW induction motor has pf 0.9 and efficiency 0.9 at full load, power factor 0.6 and efficiency 0.7 at half full-load. At no load the current is 25% of full-load current and power factor 0.1. Capacitors are supplied to make the line power factor 0.8 at half full-load. With these capacitors in circuit, find the line power factor at *(i)* full load, and *(ii)* no load.

Solution : Input to motor at full load,

$$P_f = \frac{\text{Rated output}}{\eta} = \frac{35}{0.9} = 38.88 \text{ kW}$$

kVAR (lagging) drawn by the motor at full-load,

$$\text{kVAR}_{f_1} = P_{f_1} \tan\phi_{f_1} = 38.88 \tan(\cos^{-1} 0.9) = 18.835$$

Input to motor at half full-load,

$$P_{h_1} = \frac{\frac{1}{2} \times 35}{0.7} = 25 \text{ kW}$$

kVAR (lagging) drawn by the motor at half full-load

$$\text{kVAR}_{h_1} = P_{h_1} \tan(\cos^{-1} \phi_{h_1}) = 25 \tan(\cos^{-1} 0.6) = 33.33$$

Full-load current,

$$I_f = \frac{35 \times 1,000}{\sqrt{3} V_L \times 0.9 \times 0.9} = \frac{24,947}{V_L} \text{ amperes}$$

No-load current, $I_0 = 0.25\, I_f = \dfrac{24{,}947 \times 0.25}{V_L} = \dfrac{6{,}237}{V_L}$ amperes

No-load input, $P_0 = \dfrac{\sqrt{3} V_L I_0 \cos\phi_0}{1{,}000} = \dfrac{\sqrt{3} \times V_L \times 6{,}237}{1{,}000 \times V_L} \times 0.1$

$= 1.08$ kW

kVAR (lagging) drawn by the motor at no load,

$kVAR_{0_1} = 1.08 \tan(\cos^{-1} 0.1) = 10.75$

kVAR (lagging) drawn from mains at half load with capacitors,

$kVAR_{h_2} = P_h \tan\phi_{h_2} = 25 \tan(\cos^{-1} 0.8) = 18.75$

kVAR supplied by capacitors,

$kVAR_C = kVAR_{h1} - kVAR_{h2} = 33.33 - 18.75 = 14.58$

kVAR drawn from mains at full load with capacitors,

$kVAR_{f_2} = kVAR_{f_1} - kVAR_C = 18.835 - 14.58 = 4.255$

(*i*) Line power factor at full load

$$= \text{Cos}\left(\tan^{-1} \frac{kVAR_{f2}}{P_f}\right) = \text{Cos}\left(\tan^{-1} \frac{4.265}{38.88}\right)$$

$= 0.994$ lagging **Ans.**

kVAR drawn from mains at no load with capacitors,

$kVAR_{0_2} = kVAR_{0_1} - kVAR_C = 10.75 - 14.58 = -3.83$

(*ii*) Line power factor at no load

$$= \text{Cos}\left(\tan^{-1} \frac{kVAR_{0_2}}{P_0}\right) = \text{Cos}\left(\tan^{-1} \frac{-3.83}{1.08}\right)$$

$= 0.272$ leading **Ans.**

15.10. ECONOMICS OF POWER FACTOR IMPROVEMENT

When the power factor is improved it involves an expenditure on account of the power factor correcting equipment. Improvement of power factor will result in reduction of maximum demand and thus affect an annual saving over the maximum demand charge but on the other hand an expenditure is to be incurred every year in the shape of interest and depreciation on account of the investment made over the power factor correcting equipment. The limit of the power factor at which the net saving (saving in annual maximum demand charges less annual expenditure incurred on power factor correcting equipment) is maximum is known as *economical limit of power factor correcting.* It will be seen that the economical limit of power factor correction is governed by the relative costs of the supply and power factor correcting equipment.

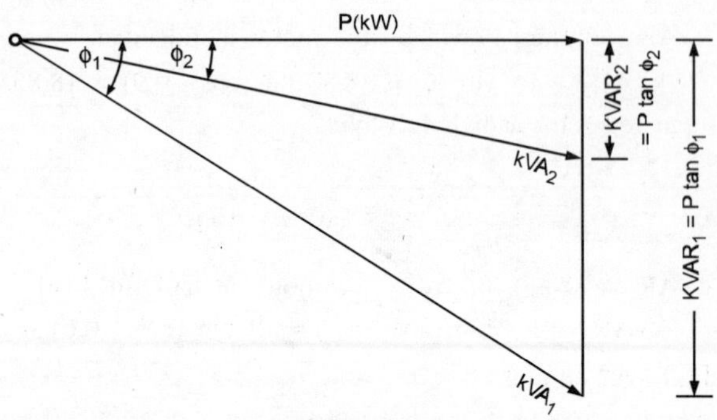

Fig. 15.16

Consider a consumer taking a peak load of P kW at a power factor of $\cos\phi_1$ and charged at the rate of Rs *x* per kVA of maximum demand per annum. Let the expenditure per kVAR per annum of the power factor correction equipment be Rs *y*.

Corresponding to maximum demand of P kW at power factor of $\cos\phi_1$

Maximum demand in kVA, $kVA_1 = \dfrac{P}{\cos\phi_1} = P \sec\phi_1$

Reactive power, $kVAR_1 = P \tan\phi_1$

Let by installing the power factor correction equipment (such as phase advancer) power factor become $\cos\phi_2$.

New maximum demand in kVA,

$$kVA_2 = \frac{P}{\cos\phi_2} = P\sec\phi_2$$

New reactive power, $kVAR_2 = P\tan\phi_2$

Annual saving in maximum demand charges

$= \text{Rs } x\,(kVA_1 - kVA_2) = \text{Rs } x(P\sec\phi_1 - P\sec\phi_2)$

$= \text{Rs } Px(\sec\phi_1 - \sec\phi_2)$.

Leading kVAR supplied by power factor correction equipment

= Difference of the reactive power drawn before and after power factor correction

$= P(\tan\phi_1 - \tan\phi_2)$

Annual cost of pf correction equipment

$= \text{Rs } y \times P(\tan\phi_1 - \tan\phi_2)$

Net annual saving,

$S = xP(\sec\phi_1 - \sec\phi_2) - yP(\tan\phi_1 - \tan\phi_2)$

This saving will be maximum if the differentiation of it with respect to ϕ_2 is zero.

i.e., $\dfrac{dS}{d\phi_2} = 0$

i.e., $\dfrac{dS}{d\phi_2}[xP(\sec\phi_1 - \sec\phi_2) - yP(\tan\phi_1 - \tan\phi_2)] = 0$

or $-xP\sec\phi_2\tan\phi_2 + yP\sec^2\phi_2 = 0$

or $\text{Tan}\,\phi_2 = \dfrac{y}{x}\sec\phi_2$

or $\text{Sin}\,\phi_2 = \dfrac{y}{x}$

$$\text{or Cos}\,\phi_2 = \sqrt{1 - \sin^2\phi_2} = \sqrt{1 - \left(\frac{y}{x}\right)^2} \quad ...(15.7)$$

From, the above expression, value of most economical power factor $\cos\phi_2$ can be determined which is independent of original factor $\cos\phi_1$ and is governed by the relative costs of supply and power factor correction equipment.

Example 15.22. Calculate the value of the new power factor when the tariff is Rs 1,350 per kVA of maximum demand plus a flat rate paise 80 per kWh. Assume additional cost of condensers etc. at Rs 1,050 per kVA of such plant. Rate of interest and depreciation together is taken as 10%.

Solution : Maximum demand charges,

x = Rs 1,350 per kVA/annum

Cost of phase advancing plant = Rs 1,050 per kVA

Expenditure on phase advancing plant,

y = Annual interest and depreciation

$$= \text{Rs } 1{,}050 \times \frac{10}{100} = \text{Rs } 105/\text{kVAR/annum}$$

Most economocial power factor,

$$\cos\phi_2 = \sqrt{1-\left(\frac{y}{x}\right)^2} = \sqrt{1-\left(\frac{105}{1{,}350}\right)^2} = 0.997 \text{ (lagging)} \quad \textbf{Ans.}$$

Example 15.23. A factory operates at 0.8 pf lagging and has a monthly demand of 750 kVA. The monthly power rate is Rs 8.50 per kVA. To improve the power factor, 250 kVA capacitors are installed in which there is negligible loss. The installed cost of equipment is Rs 20,000 and fixed charges are estimated as 10% per year. Calculate the annual savings affected by the use of capacitors.

[Pb. Technical Univ. Generation of Electric Power December 2010]

Solution : Maximum demand = 750 kVA

Load power factor, $\cos\phi$ = 0.8

Power factor angle $\phi = \cos^{-1} 0.8 = 36.87°$

kW component of load, P = kVA $\cos\phi$ = 750 × 0.8 = 600 kW

kVAR component of load, Q_1 = kVA $\sin\phi$ = 750 sin 36.87° = 450 (lagging)

Leading kVAR supplied after pf improvement

$Q = Q_1$ – kVAR supplied by capacitors = 450 – 250 = 200

kVA demand after pf improvement

$$= \sqrt{P^2+Q^2} = \sqrt{600^2+200^2} = 632.45$$

Reduction in kVA demand = 750 – 632.45 = 117.55

Annual savings on kVA charge = Rs 8.50 × 12 × 117.55 = Rs 11,990

Fixed charges per annum on the capacitors

$$= \text{Rs } 20{,}000 \times \frac{10}{100} = \text{Rs } 2{,}000$$

Annual savings = Rs 11,990 – Rs 2,000 = Rs 9,900 **Ans.**

Example 15.24. An industry consumes 80,000 units of electricity in a year at an average pf 0.707 lag. The recorded maximum demand is 500 kVA. The tariff is Rs 120/kVA maximum demand + 2.5 paise per kWhr. Calculate the annual cost of supply and find out annual saving in cost by installing phase advancing plant costing Rs 50 per kVAR which raises the power factor from 0.707 to 0.9 lagging. Allow 10% per year on the cost of phase advancing plant to cover all additional costs. *[Rajasthan Univ. Generation of Electrical Power 2006]*

Solution : Maximum demand = 500 kVA

Annual energy consumption = 80,000 units

Annual cost of supply

= Annual maximum demand charges + annual energy charges

= Rs 120 × 500 + Rs 0.025 × 80,000 = Rs 62,000 **Ans.**

Initial power factor, $\cos\phi_1$ = 0.707

Phase angle $\phi_1 = \text{Cos}^{-1}\, 0.707 = 45°$ (lagging)

Power factor after improvement, $\cos\phi_2$ = 0.9

Phase angle $\phi_2 = \text{Cos}^{-1}\, 0.9 = 25.842$ (lagging)

Load, P = 500 × 0.707 = 353.5 kW

Leading kVAR supplied by phase advancing plant

$= P(\tan\phi_1 - \tan\phi_2) = 353.5(\tan 45° - \tan 25.842°) = 182.3$ kVAR

KVA demand after improvement of power factor

$$= \frac{P}{\text{Cos}\,\phi_2} = \frac{353.5}{0.9} = 392.8 \text{ kVA}$$

Reduction in kVA demand = 500 – 392.8 = 107.2 kVA

Annual saving in kVA demand charges

= Rs 120 × 107.2 = Rs 12,864

Annual cost on phase advancing plant

$$= \text{Rs } 50 \times 182.3 \times \frac{10}{100} = \text{Rs } 911.50$$

Annual saving = Rs 12,864 – Rs 911.50 = Rs 11,952.50 **Ans.**

Example 15.25. A consumer takes a steady load of 300 kW at a lagging power factor of 0.7 for 3,000 hours a year. The tariff is Rs 1,300 per kVA of maximum demand per annum plus Re 0.80 per kWh. The annual cost of the phase advancing plant is Rs 130 per kVAR. Determine the annual saving if the pf of the load is improved.

Solution : Maximum demand charges,

x = Rs 1,300 per kVA/annum

Expenditure on phase advancing plant,

y = Rs 130 per kVAR/annum

Let the power factor be improved to most economical limit

Most economical new power factor,

$$\cos\phi_2 = \sqrt{1-\left(\frac{y}{x}\right)^2} = \sqrt{1-\left(\frac{130}{1{,}300}\right)^2} = 0.995 \text{ (lagging)}$$

Capacity of phase advancing plant

= Load in kW $(\tan\phi_1 - \tan\phi_2)$

$= 300\,[\tan(\cos^{-1} 0.7) - \tan(\cos^{-1} 0.995)] = 276$ kVAR

Annual saving in demand charges

$$= \text{Rs } x\left(\frac{P}{\cos\phi_1} - \frac{P}{\cos\phi_2}\right) = \text{Rs } 1{,}300\left(\frac{300}{0.7} - \frac{300}{0.995}\right)$$

= Rs 165,180

Annual expenditure on phase advancing plant

$= \text{Rs } y\,(\text{kVAR}_1 - \text{kVAR}_2)$ = Rs 130 × 276 = Rs 35,880

Net saving in annual cost = Rs (165,180 – 35,880)

= Rs 129,300 **Ans.**

kWh charges will be same as there will be no change in energy consumption.

Example 15.26. The load on a certain installation may be considered constant at 1,200 kVA, 0.75 lagging power factor for 3,000 hours per annum. The tariff is Rs 1,300 per kVA MD plus 80 paise per kWh.

(*i*) Determine the annual charge of energy.

(*ii*) Power factor improving apparatus is installed to improve the pf to 0.95 lagging. Determine the kVAR required and the new annual charge of energy if the power factor improving apparatus costs Rs 1,200 per kVAR, annual interest and depreciation charges are 10% of the capital cost.

Solution : Maximum demand = 1,200 kVA

Annual energy consumption

= Average load in kW × working hours/annum

$= 1{,}200 \times 0.75 \times 3{,}000 = 27 \times 10^5$ kWh

(*i*) Annual energy charges

= Maximum demand charges + unit charges

$= \text{Rs } 1{,}300 \times 1{,}200 + \text{Rs } 0.80 \times 27 \times 10^5 = \text{Rs } 3{,}720{,}000$ **Ans.**

(*ii*) kVAR required = $P(\tan\phi_1 - \tan\phi_2)$

$= 1{,}200 \times 0.75\,[\tan\cos^{-1}(0.75) - \tan\cos^{-1}(0.95)]$

= 900 (0.8819 – 0.3287) = 498 **Ans.**

Capital cost of pf improving apparatus

= Rs 1,200 × 498 = Rs 597,600

Annual interest and depreciation charges

$$= \text{Rs } 597{,}600 \times \frac{10}{100} = \text{Rs } 59{,}760$$

Maximum demand

$$= \frac{P}{\text{Cos } \phi_2} = \frac{1{,}200 \times 0.75}{0.95} = 947 \text{ kVA}$$

New annual charges

= Rs 1,300 × 947 + Rs 27 × 10^5 × 0.80 + Rs 59,760

= Rs 1,231,100 + Rs 2,160,000 + Rs 59,760

= Rs 3,450,860 **Ans.**

15.11. ECONOMICAL COMPARISON OF THE TWO METHODS OF INCREASING THE POWER SUPPLIED

The increase in power demand on the generating station can be met either by increasing the capacity of the generating plant working at the same pf or by raising the power factor of the system by installation of phase advancers. Owing to improvement of power factor in the beginning the saving in the generating and distributing plant outweighs the extra cost of the pf correction equipment in most of the cases but as the power factor is raised further its cost begins to approximate to the saving and finally any saving over the plant is obtained by incurring a greater expenditure on the pf correcting equipment. Thus there is a limit beyond which it is not economical still further to improve the power factor. The maximum value to which the power factor can be economically raised entirely depends upon the relative costs of the generating plant and phase advancing plant.

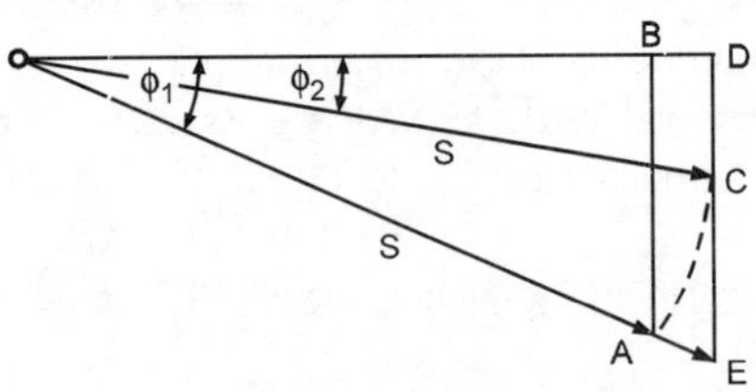

Fig. 15.17

Let the rating of generating station be S in kVA, supplying load at power factor cos ϕ_2.

Assume new demand can be met by improving the power factor from cos ϕ_1 to cos ϕ_2 with the same capacity of the plant.

Existing load in kW, $P_1 = S \cos \phi_1$

New load in kW, $P_2 = S \cos \phi_2$

(i) Cost by increasing the capacity of the generating station

Increase in load = $P_2 - P_1 = S(\cos \phi_2 - \cos \phi_1)$

Increase in the capacity of the generating plant operating at pf cos ϕ_1

$$= \frac{\text{Increase in load in kW}}{\text{Power factor of plant}} \text{ kVA}$$

$$= \frac{P_2 - P_1}{\text{Cos } \phi_1} = \frac{S(\cos \phi_2 - \cos \phi_1)}{\text{Cos } \phi_1} \text{ kVA}$$

Increase in annual cost due to increase in capacity of the plant

$$= \text{Rs } xS \frac{\cos \phi_2 - \cos \phi_1}{\cos \phi_1} \quad \text{...(15.8)}$$

where x is the annual cost per kVA of generating plant

(ii) Cost on Power Factor Correction Equipment

Reactive power drawn by load operating at the old power factor, cos ϕ_1

$= P_2 \tan \phi_1 = S \cos \phi_2 \tan \phi_1$

Reactive power supplied by the plant

= Capacity of the plant in kVA sin ϕ_2 = S sin ϕ_2

since kVA rating of the plant remains the same

kVAR rating of the pf correction equipment

= Reactive power drawn by load – reactive power supplied by the plant

$= S \cos \phi_2 \tan \phi_1 - S \sin \phi_2 = S(\cos \phi_2 \tan \phi_1 - \sin \phi_2)$

Annual cost on pf correction equipment

$$= \text{Rs } y\, S(\cos \phi_2 \tan \phi_1 - \sin \phi_2) \quad \text{...(15.9)}$$

where y is the annual cost per kVAR rating of power factor correction equipment.

Power factor correction equipment will be cheaper if annual cost on it is less than annual cost on account of increasing the generating capacity.

$$\text{or} \quad Sy(\tan \phi_1 \cos \phi_2 - \sin \phi_2) < Sx \frac{\cos \phi_2 - \cos \phi_1}{\cos \phi_1}$$

$$\text{or } y(\tan \phi_1 \cos \phi_2 - \sin \phi_2) < x \frac{\cos \phi_2 - \cos \phi_1}{\cos \phi_1}$$

In the limiting case, the maximum cost of pf correction equipment, which would justify its installation, should be equal to cost on account of increasing the generating plant capacity.

$$\textit{i.e.,} \quad y(\tan \phi_1 \cos \phi_2 - \sin \phi_2) = \frac{x(\cos \phi_2 - \cos \phi_1)}{\cos \phi_1}$$

$$\text{or } y\left(\frac{\sin \phi_1 \cos \phi_2 - \sin \phi_2 \cos \phi_1}{\cos \phi_1}\right) = x \frac{\cos \phi_2 - \cos \phi_1}{\cos \phi_1}$$

$$\text{or } y \sin(\phi_1 - \phi_2) = x(\cos \phi_2 - \cos \phi_1)$$

$$\text{or } y = \frac{x(\cos \phi_2 - \cos \phi_1)}{\sin(\phi_1 - \phi_2)} \quad \text{...(15.10)}$$

Example 15.27. A power plant is working at its maximum kVA capacity with a lagging pf of 0.7. It is now required to increase its kW capacity to meet the demand of additional load. This can be done

***(i)* By increasing the pf to 0.85 (lagging) by pf correction equipment.**

or

***(ii)* By installing additional generation plant costing Rs 800 per kVA.**

What is the maximum cost per kVA of pf correction equipment to make its use more economical than the additional plant ?

[U.P. Technical Univ. Power Station Practice 2008-09]

Solution : Let the rating of the power plant be S kVA

Increase in load

$= P_2 - P_1 = S(\cos \phi_2 - \cos \phi_1) = S(0.85 - 0.7) = 0.15$ S

Increase in capacity of the generating plant

$$= \frac{P_2 - P_1}{\text{Plant } pf} = \frac{0.15 \text{ S}}{0.7} = 0.2143 \text{ S}$$

Cost of additional plant = Rs 800 × 0.2143 S = Rs 171.43 S

Let the rate of interest and depreciation be m per cent per annum

Increase in annual cost due to increase in capacity of power plant

$$= \text{Rs } \frac{m}{100} \times 171.43 \text{ S} = \text{Rs } 1.7143 \text{ mS} \quad ...(i)$$

Reactive power drawn by load

$$= P_2 \tan \phi_1 = 0.85 \text{ S} \tan (\cos^{-1} 0.7) = 0.867 \text{ S}$$

$$\because P_2 = \text{S} \cos \phi_2 \text{ and load pf is the same}$$

Reactive power supplied by the plant

$$= \text{S} \sin \phi_2 = \text{S} \sin (\cos^{-1} 0.85) = 0.5268 \text{ S}$$

$$\because \text{new pf of the plant is } 0.85$$

kVAR rating of power factor correction equipment

= kVAR drawn by load – kVAR supplied by the plant

$$= 0.867 \text{ S} - 0.5268 \text{ S} = 0.3402 \text{ S}$$

If y is the cost per kVAR of pf correction equipment then total cost on pf correction equipment

$$= 0.3402y \text{ S}$$

Assume rate of interest and depreciation the same as in case of plant *i.e.*, m per cent per annum

The annual cost on pf correction equipment

$$= \text{Rs } \frac{m}{100} \times 0.3402 \, y \text{ S} = \text{Rs } 0.003402my \text{ S} \quad ...(ii)$$

In limiting case both the cost must be equal

i.e., $0.003402 \, m \, y \, \text{S} = 1.7143 \text{ mS}$

$$\text{or } y = \frac{1.7143}{0.003402} = \text{Rs } 503.60 \text{ per kVAR} \quad \textbf{Ans.}$$

Alternative Method. Annual cost per kVA of generating plant,

$$x = \text{Rs } 800 \times \frac{m}{100} = \text{Rs } 8 \, m$$

where m is the rate of interest and depreciation per annum

Original power factor,

$$\cos \phi_1 = 0.707 \text{ and therefore } \phi_1 = \text{Cos}^{-1} 0.7 = 45.573°$$

Improved power factor,

$$\cos \phi_2 = 0.87 \text{ and therefore } \phi_2 = \text{Cos}^{-1} 0.85 = 31.788°$$

Annual cost per kVAR of pf correcion equipment,

$$y = \frac{x(\cos \phi_2 - \cos \phi_1)}{\sin (\phi_1 - \phi_2)} = \frac{8m(0.85 - 0.7)}{\sin (45.573° - 31.788°)}$$

$$= 5.036 \, m$$

Cost per kVAR of power factor correction equipment

$$= \text{Rs } 5.036 \, m \times \frac{100}{m} = \text{Rs } 503.60 \textbf{ Ans.}$$

15.12. CHOICE OF EQUIPMENT

At the time of purchasing equipment usually several alternatives are made available, each possessing one or more special features. The choice between alternative types of electrical equipment usually depends upon the relative costs of the alternative equipment. In selecting equipment for a particular service, the choice must fall upon that which makes the sum of the fixed charges (annual interest and depreciation on the capital cost of equipment) and the operating charges minimum. It is usually found that a saving can be affected only by incurring increased fixed costs and vice versa, for instance, an electric motor of a particular type and size of lower operating efficiency will usually cost less than a similar motor of higher efficiency, although the latter will have a lower operating cost on account of its lower loss and may, therefore, have a lower operating cost.

Typical problems on choice of a transformer and induction motors are worked out below.

Example 15.28. It is necessary to choose a transformer to supply a load which varies over 24 hour period in the manner given below :

500 kVA for 4 hours, 1,000 kVA for 6 hours, 1,500 kVA for 12 hours and 2,000 kVA for the rest of the period. Two transformers each rated at 1,500 kVA have been quoted. Transformer I has iron loss of 2.7 kW and full-load copper loss of 8.1 kW, while transformer II has an iron loss and full-load copper loss of 5.4 kW each.

(*i*) Calculate the annual cost of supplying losses for each transformer, if electrical energy costs Rs 2.5 per kWh.

(*ii*) Determine which transformer should be chosen if the capital cost of the transformer I is Rs 25,000 more than that of transformer II and annual charges of interest and depreciation are 10%.

(*iii*) What difference in capital cost will reverse the decision made in (*ii*) above ?

Solution : Transformer I

Energy loss per day due to iron loss = 24 × 2.7 = 64.8 kWh

Energy loss per day due to copper loss

$$= \left(\frac{500}{1{,}500}\right)^2 \times 8.1 \times 4 + \left(\frac{1{,}000}{1{,}500}\right)^2 \times 8.1 \times 6 + \left(\frac{1{,}500}{1{,}500}\right)^2 \times 8.1 \times 12 + \left(\frac{2{,}000}{1{,}500}\right)^2 \times 8.1 \times (24 - 4 - 6 - 12)$$

$$= 3.6 + 21.6 + 97.2 + 28.8 = 151.2 \text{ kWh}$$

Total energy loss per annum

= (64.8 + 151.2) × 365 = 78,840 kWh

Annual cost of supplying losses

= Rs 2.50 × 78,840 = Rs 197,100 **Ans.**

Transformer II

Energy loss per day due to iron loss = 24 × 5.4 = 129.6 kWh

Energy loss per day due to copper loss

$$= \left(\frac{500}{1{,}500}\right)^2 \times 5.4 \times 4 + \left(\frac{1{,}000}{1{,}500}\right)^2 \times 5.4 \times 6 + \left(\frac{1{,}500}{1{,}500}\right)^2 \times 5.4 \times 12 + \left(\frac{2{,}000}{1{,}500}\right)^2 \times 5.4 \times 2$$

$$= 2.4 + 14.4 + 64.8 + 19.2 = 100.8 \text{ kWh}$$

Total energy loss per annum = (129.6 + 100.8) × 365 = 84,096 kWh

Annual cost of supplying energy losses

= Rs 2.5 × 84,096 = Rs 210,240 **Ans.**

(*ii*) The capital cost of transformer I

= Rs 25,000 more than that of transformer II

Extra annual charges in case transformer I is chosen

$$= \text{Rs } 25{,}000 \times \frac{10}{100} = \text{Rs } 2{,}500$$

Annual saving in energy cost due to losses

= Rs 210,240 – Rs 197,100 = Rs 13,140

Since annual saving in cost of supplying energy losses is much more than extra annual charges due to extra cost of transformer I, therefore, transformer I will be chosen **Ans.**

(*iii*) Decision taken in (*ii*) above will be reversed *i.e.*, transformer II will be chosen in case annual charges due to extra capital cost of transformer I exceeds the annual saving in the energy cost due to losses.

i.e., $x \times \frac{10}{100} >$ Rs 13,140

where x is the capital cost of transformer I in excess

or $x >$ Rs $13{,}140 \times \frac{100}{10}$ or x exceeds Rs 131,400 **Ans.**

Example 15.29. The motor of a 30 hp condensate pump has got burnt beyond economical repairs. Two alternatives have been proposed to replace it by :

Motor A-Cost = Rs 150,000 ; η at full load = 90% ; η at half load = 86%

Motor B-Cost = Rs 100,000 ; η at full load = 85% ; η at half load = 82%

The life of each is 20 years and its salvage value is 10% of the initial cost. The rate of interest is 5% annually. The motor operates at full load for 25% of the time and at half load for the remaining period. The annual maintenance cost of motor A is Rs 10,500 and that of motor B is Rs 6,000. The energy rate is Rs 2.50 per kWh. Which motor will you recommend and why ?

Solution : Motor A

Capital cost = Rs 150,000

Annual interest on capital cost

$$= \text{Rs } 150{,}000 \times \frac{5}{100} = \text{Rs } 7{,}500.00$$

Annual depreciation charges

$$= \frac{\text{Original cost} - \text{salvage value}}{\text{Life of motor in years}}$$

$$= \text{Rs } \frac{150{,}000 - \frac{10}{100} \times 150{,}000}{20}$$

= Rs 6,750.00

Annual maintenance cost = Rs 10,500.00

Energy consumed per annum

$$= \frac{30 \times 735.5}{0.9 \times 1{,}000} \times 0.25 \times 8{,}760 + \frac{\frac{1}{2} \times 30 \times 735.5}{0.86 \times 1{,}000} \times 0.75 \times 8{,}760$$

= 53,691 + 84,283 = 137,974 kWh

Annual energy cost = Rs 2.50 × 137,974 = Rs 344,935

Total annual cost = Rs (7,500 + 6,750 + 10,500 + 344,935)

= Rs 369,685.00

Motor B Capital cost = Rs 100,000

Annual interest on capital cost

$$= \text{Rs } 100{,}000 \times \frac{5}{100} = \text{Rs } 5{,}000.00$$

Annual depreciation charges

$$= \text{Rs } \frac{100{,}000 - \frac{10}{100} \times 100{,}000}{20} = \text{Rs } 4{,}500$$

Annual maintenance charges = Rs 6,000

$$\text{Energy input per annum} = \frac{30 \times 0.7355}{0.85} \times 0.25 \times 8{,}760 + \frac{\frac{1}{2} \times 30 \times 0.7355}{0.82} \times 0.75 \times 8{,}760$$

= 56,850 + 88,395 = 145,245 kWh

Annual energy cost = Rs 2.50 × 145,245 = Rs 363,112

Total annual cost = Rs (5,000 + 4,500 + 6,000 + 363,112)

= Rs 378,612

Since total annual cost of motor B is more by Rs (378,612 – 369,685) *i.e.*, Rs 8,927 than that of motor A, so motor A is recommended.

Example 15.30. A consumer requires an induction motor of 50 hp (metric). He is offered two motors of the following specifications :

Motor (A)-efficiency 88% and power factor 0.9 ;

Motor (B)-efficiency 90% and power factor 0.81 ;

The consumer is being charged on a two part tariff of Rs 1,400 per kVA of the maximum demand plus Re 1.0 per unit. The power factor of the motor B is to be raised to 0.89 by installing condensers. The motor B costs Rs 3,000.00 less than A.

The cost of condensers is Rs 1,200 per kVAR. Determine which motor is more economical and by how much. Assume rate of interest and depreciation 10% and working hours of motors 2,400 hours in a year.

Solution : $\text{Motor input} = \frac{\text{Motor output}}{\eta} = \frac{50 \times 0.7355}{0.88} = 41.8 \text{ kW}$

$$\text{Motor input in kVA} = \frac{41.8}{pf} = \frac{41.8}{0.9} = 46.43$$

Annual energy consumption

= Motor input in kW × number of working hours/annum

= 41.8 × 2,400 = 100,320 kWh

Annual energy charges

= Rs 1.0 × 100,320 + Rs 1,400 × 46.43 = Rs 165,320

Fixed annual charges of motor A in excess over those of motor B

$$= \text{Rs } 3{,}000 \times \frac{10}{100} = \text{Rs } 300.00$$

Motor B : $\text{Motor input} = \frac{50 \times 0.7355}{0.9} = 40.86 \text{ kW}$

$$\text{Motor input in kVA with condensers} = \frac{40.86}{0.89} = 45.91$$

Energy consumed per annum = 40.86 × 2,400 = 98,064 kWh

Annual energy charges

= Rs 1,400 × 45.91 + Rs 1.0 × 98,064 = Rs 162,338

Leading kVAR supplied by the condensers

$= 40.86 [\tan (\cos^{-1} 0.81) - \tan (\cos^{-1} 0.89)]$

= 40.86 (0.724 – 0.512) = 8.66

Capital cost of condensers = Rs 1,200 × 8.66 = Rs 10,392

Annual interest and depreciation charges

$$= \text{Rs } 10{,}392 \times \frac{10}{100} = \text{Rs } 1{,}039$$

Total annual charges

= 162,338 + 1,039 = 163,377.00

Motor B will be economical and annual saving will be

Rs (165,320 + 300 – 163,377) = Rs 2,243.00 **Ans.**

Exercises

1. What are the main points to be considered while fixing the tariff ?
2. What are the main objectives in framing a tariff ? Discuss various types of tariffs.
[U.P. Technical Univ. Power Station Practice 2002-03]
3. Define the following terms in connection with the power supply system : connected load, maximum demand, two part tariff, diversity factor.
[G.B. Technical Univ. Power Station Practice 2011-12]
4. What are the different types of tariff used by the electricity authority ? Discuss any two of them and indicate the type of consumers where such tariffs are used.
[U.P. Technical Univ. Power Station Practice 2007-08]
5. Discuss different types of tariffs used for charging the consumers of electrical energy.
[G.B. Technical Univ. Power Station Practice 2010-11]
6. Explain giving examples :
(*i*) Flat rate tariff (*ii*) Block rate tariff (*iii*) Two part tariff (*iv*) Power factor tariff.
[Pb. Univ. Electrical Power II (Generation) Dec. 1995]
7. Explain the straight meter rate tariff and block meter rate tariff.
[Pb. Technical Univ. Generation of Electric Power May 2008]
8. What do you understand by tariff? Explain briefly various types of tariffs.
[Pb. Technical Univ. Generation of Electric Power Dec. 2010]
9. Discuss the different types of tariffs used for charging the consumers electrical energy.
[Pb. Technical Univ. Generation of Electric Power May 2009]
10. What is the special feature of two part tariff ? For which category of consumers is it used ?
Discuss the importance of encouraging customers to use electricity during off-peak hours.
[Pb. Univ. Electrical Power II (Generation) Dec. 1993]
11. What is the objective of tariff ? What type of tariff is employed for domestic consumers ? Why this tariff is not employed for bulk consumers ? Suggest and explain the tariff which encourages the consumers to keep load factor and power factor high.
12. In modern LT distribution stations the VAR meters are installed. What is the mechanism of charging the tariff related to reactive power and how the above can be effectively installed ?
[U.P. Technical Univ. Power Station Practice 2005-06]
13. A distribution company wants to charge and penalize the consumers in case of excessive loads of poor power factor. What charges should be made in present existing substation so as to meet the above ?
[U.P. Technical Univ. Power Station Practice 2005-06]
14. Give the basis for expressing the service cost of an electric supply as A + B kW + C kWh and explain the factors on which A, B and C depend.
15. What are objective of tariffs? What is the general form of tariff? Describe the straight meter rate and block meter rate tariffs. What do you mean by spot pricing?
[Generation of Electrical Power Rajasthan Univ. 1997, Rajasthan Technical Univ. 2011]
16. What is spot pricing? What is its importance?
[Pb. Technical Univ. Generation of Electric Power Dec. 2008]
17. What are different types of loads and draw their chronological load curves.
[Pb. Technical Univ. Generation of Electric Power Dec. 2007]
18. What are the disadvantages of low power factor? Why do some agencies put a penalty for low power factor?
[Generation of Electrical Power Rajasthan Univ. 2004 ; Rajasthan Technical Univ. 2007]
19. What is the effect of low power factor ? Explain the different methods to improve the power factor of the system.
[U.P. Technical Univ. Power Station Practice 2003-04]
20. What are the causes and effects of poor power factor ? Discuss any one method to improve the power factor.
[U.P. Technical Univ. Power Station Practice 2007]
21. What are the advantages of power factor improvement?
[Pb. Technical Univ. Generation of Electric Power May 2009]
22. What are the advantages of power factor improvement. Explain how a synchronous capacitor can improve the power factor.
[U.P. Technical Univ. Power Station Practice 2007-08]
23. Explain the necessity of improving power factor.
[G.B. Technical Univ. Power Station Practice 2010-11]
24. Compare the advantages and disadvantages of using a synchronous condenser and a static capacitor for power factor improvement.
[Rajasthan Technical Univ. Generation of Electrical Power 2011]
25. What do you understand by power factor ? Explain the necessity of improving power factor. Also discuss the factor which determines the economical limit of power factor correction.
[U.P. Technical Univ. Power Station Practice 2009-10]
26. What factors determine the economical limit of power factor correction ? Show that the economical limit to which the power factor of a lagging power factor load can be raised is independent of the original value of the power factor if the tariff consists of a fixed charge per kVA of maximum demand plus a flat rate per kWh.
27. Calculate the most economical power factor when kW demand is constant. **[Rajasthan Technical Univ. Generation of Electrical Power 2008, 2009, 2011]**
28. Explain how the installation of a pf improvement device can augment the kW capacity of a generating station of a given kVA capacity.
Derive the conditions under which the installation of a pf improvement plant for increasing the kW capacity of a generating station is preferable to installing additional generating equipment.
29. List and explain the factors that influence the economic choice of equipment.
30. Write short notes on the following :
(*i*) Tariffs **[Pb. Univ. December 1993]**
(*ii*) Tariffs methods of electrical energy
[M.D. Univ. May 2007]
(*iii*) Hopkinson demand rate **[Rajasthan Univ. 1995, 2001]**
(*iv*) Two part tariff. **[West Bengal Univ. of Technology 2010]**
(*v*) Power factor tariff.
(*vi*) Power factor improvement. **[Pb. Univ. Janvary 1993]**
(*vii*) Power factor improvement using synchronous condensers.
[Rajasthan Univ. 2002; 2005]
(*viii*) Necessity of power factor improvement.
(*ix*) Economic limits of power factor improvement.
(*x*) Choice of equipment.

Short Answer Type Questions With Answers

Q. 1. What is meant by tariff? **[Rajasthan Univ. 2003]**

Ans. Tariff means the schedule of rates or charges. Tariff, in case of electric supply, means the schedule or rates framed for supply of electrical energy to different classes of consumers.

Q. 2. Name the different types of tariffs. **[Pb. Technical Univ. December 2008, May 2009]**

Ans. Different types of tariffs are :
(*i*) Flat demand tariff (*ii*) Simple tariff (*iii*) Flat rate tariff (*iv*) Step rate tariff (*v*) Block rate tariff (*vi*) Hopkinson demand or two part tariff (*vii*) Maximum demand tariff or Wright demand rate (*viii*) Power factor tariffs (*ix*) Three part tariff or Doherty rate tariff (*x*) Off peak tariff.

Q. 3. What is tariff? What is general form of tariff? **[Rajasthan Univ. 2005]**

Ans. Tariff means the schedule of rates or charges. Tariff, in case of electric supply, means the schedule or rates framed for supply of electrical energy to different classes of consumers.
General form of tariff is
Total energy charges, e = Rs $a + b$ kW + c kWh
where a is a constant charge made each billing period.
b = Unit charge in Rs per kW of metered maximum demand in kW during billing period. In some cases it is also charged in Rs per kVA so that the consumers are penalised for poor power factor.
and c = Unit charge for energy in Rs per kWh of energy consumed.

Q. 4. Why is tariff less for power load than lighting load? **[Rajasthan Univ. 2003]**

Ans. Power load spreads the fixed charges over greater number of units, thus reduces the overall cost of generation of electrical energy. Lighting load has low load factor and therefore cost of generation of electrical energy is higher. These are the reasons that tariff is less for power load than lighting load.

Q. 5. What type of tariff is employed for domestic consumers ?

Ans. Block rate tariff.

Q. 6. What is two part tariff ?

Ans. In the two part tariff, the total charge to be made to the consumer is divided into two components namely fixed charge (proportional to connected load or maximum demand) and the running or operating charges (proportional to the units consumed).

Q. 7. For which category of consumers two part tariff is used ?

Ans. Two part tariff is mostly applicable to medium industrial consumers.

Q. 8. How do demand factor, load factor and diversity factor in a power system affect the fixation of tariffs? **[Rajasthan Univ. 2004]**

Ans. The tariff must cover the following items :
(*i*) Recovery of cost of capital investment in generating, transmitting and distributing equipment.
(*ii*) Recovery of cost of operation, supplies and maintenance of equipment.
(*iii*) Recovery of cost of metering equipment, billing, collection costs and miscellaneous services and
(*iv*) A satisfactory return on the total capital investment.

The unit cost depends quite substantially on the load factor. The fixed cost remains constant irrespective of the load factor. At low load factor the fixed cost is shared by a smaller number of units of energy resulting in a relatively higher energy cost. At high load factor, the same fixed cost is shared by a large number of units of energy thereby reducing the unit energy cost.

The demand factor, being the ratio of maximum demand to connected load, determines the capacity of the system and hence the cost of power equipment required to serve a given load. Thus demand factor affects the fixtation of tariff.
Diversity between the loads of different consumers and different areas leads to reduction in peak demand resulting in reduction in generation, transmission and distribution costs. So the fixation of tariff is affected.

Q. 9. Explain reasons, why power factor tariff is imposed ? **[Rajasthan Univ. 2005]**

Ans. Since the efficiency of plant and equipment depends upon the power factor, therefore, in order to increase the utility of plant and equipment to the maximum, the plant must be operated at the most economical power factor. That is why, sometimes consumers are penalised for poor power factor by imposing the power factor tariff.

Q. 10. Which type of tariff encourages the consumers to keep the load factor and power factor high ?

Ans. KVA maximum demand tariff encourages the consumers to keep the load factor and power factor high.

Q. 11. Can a power factor be included in a tariff ?

Ans. Yes, power factor is included in power factor tariffs such as kVA maximum demand tariff, kWh and kVARh tariff and sliding scale or average power factor tariff.

Q. 12. What type of tariff is usually applied to bulk consumers.

Ans. Three part tariff.

Q. 13. What are the advantages of using industrial/agricultural loads during night hours ?

Ans. During night hours a large proportion of the generating and distribution equipment remains idle due to extremely low demand. In case industrial and agricultural consumers are encouraged to use electricity during night and other off-peak hours by giving a special discount, the energy can be supplied without incurring any additional capital cost.

Q. 14. What are the different types of loads ?

Ans. The main types of loads on a power system are domestic, commercial, agricultural, industrial, municipal, traction etc.

Q. 15. Draw an approximate load curve for domestic load during 24 hours. **[Pb. Technical Univ. May 2008]**

Ans. Refer to Fig. 15.1

Q. 16. Define power factor. **[Pb. Technical Univ. May 2009]**

Ans. The cosine of the angle between voltage and current in an ac circuit is known as power factor.

Q. 17. Define active, reactive and apparent power. **[Pb. Technical Univ. December 2008]**

Ans. The product of voltage V, current I and power factor cos ϕ *i.e.* VI cos ϕ is known as active power.
The product of voltage V, current I and sine of phase angle ϕ *i.e.*, VI sin ϕ is known as reactive power.
The product of voltage V and current I *i.e.*, VI is known as apparent power.
These three powers are related as follows :

$$\text{Apparent power} = \sqrt{(\text{Active power})^2 + (\text{reactive power})^2}$$

Q. 18. Why is the power factor not more than unity? **[Pb. Technical Univ. December 2010]**

Ans. Power factor being the cosine of the phase angle between voltage V and current I can never exceed unity.

Q. 19. What are the effects of low power factor. **[Pb. Technical Univ. December 2007]**

Ans. Refer to Art 15.6.

Q. 20. What is the effect of power factor on the cost of generation? **[Pb. Technical Univ. December 2010]**

Ans. Lower power factor means higher kVA for the given load, therefore, large sized generators, transformers, switchgear, bus-bars, more energy loss and poor voltage regulation thereby increasing the cost of generation.

Q. 21. What is the need of improving power factor?

[Pb. Technical Univ. December 2009]

Ans. Low power factor causes the ratings of generators and transformers, cross-sectional area of the bus-bars and the contact surface of the switchgear, the size of the feeders and distributors, energy losses, voltage drops in generators, transformers, transmission lines and distributors to increase. So power factor needs to be improved to reduce the generation, transmission and distribution costs.

Q. 22. What are the advantages of power factor improvement?

[Pb. Technical Univ. December 2007]

Ans. The advantages of good (or improved) power factor are : (*i*) reduction in load current ; (*ii*) increase in voltage level across the load ; (*iii*) reduction in energy losses in the system (generators, transformers, transmission lines and distributors) due to reduction in load current ; (*iv*) reduction in kVA loading of the generators and transformers which may relieve an overloaded system or release capacity for additional growth of load and (*v*) reduction in kVA demand charge for large consumers.

Q. 23. What is the importance of power factor?

[Pb. Technical Univ. Dec. 2010]

Ans. Low power factor leads to high capital cost for generators, transformers, switchgears, transmission lines, distributors and cables. So the power factor is very important.

Q. 24. Where are shunt capacitors located?

[Pb. Technical Univ. December 2009]

Ans. Refer to Art. 15.8.3.

Q. 25. Should the power factor to be raised to 1.

[Pb. Technical Univ. May 2009]

OR

Why it is not economical to raise the power factor to unity?

[Rajasthan Univ. 2002, 2004]

Ans. As the power factor approaches unity, the capacity of the power factor improving equipment increases more rapidly i.e. the power factor of an installation can be improved from 0.8 (lagging) to 0.9 (lagging) by a much small capacity kVAR than which will be required to raise the power factor from 0.9 to unity. So it is not economical to raise the power factor to unity.

Q. 26. What factors determine the economical limit of power factor correction ?

Ans. Economical limit of power factor correction is governed by the relative costs of the supply and power factor correcting equipment.

Problems

1. Calculate the annual energy cost for industrial consumer which takes a load of 20 kW for 1 hour per day, 150 kW for 7 hours per day and 50 kW for 8 hours per day. The tariff in force is Rs 300 per kW and Rs 1.50 per kWh. Assume 6 working days in a week.**[Ans.** Rs 732,960**]**
2. The yearly consumption of an industrial consumer is 5×10^6 kWh with a maximum demand of 1,500 kW. Compare the annual cost of electrical energy under the following two tariffs.
   (*a*) Rs 1,500 per kW plus 40 paise per kWh.
   (*b*) Flat rate of Rs 1.50 paise per kWh.
   [Ans. (*a*) Rs 4,250,000 (*b*) Rs 7,500,000**]**
3. Load factor of a consumer is 35% and monthly consumption is 504 kWh. If the rate of electricity is Rs 180 per kW of maximum demand plus Rs 2.00 per kWh, find the monthly bill and the average cost per kWh.
   [U.P. Technical Univ. Power Station Practice 2004-05]
   [Ans. Rs 1,037.60**]**
4. An industrial consumer having a maximum demand of 100 kW, maintains a load factor of 60%. The tariff rates are Rs 1,500 per kVA of maximum demand per annum plus paise 30 per kWh of energy consumed. If the average pf is 0.8 lagging, calculate, the total energy consumed per annum and annual bill. **[Ans.** 525,600 kWh ; Rs 345,180**]**
5. Compute the cost of electrical energy and average unit cost for consuming 450 kWh under block rate tariff quoted as below.
   First 50 kWh at Rs 3.80 per kWh
   Next 50 kWh at Rs 3.20 per kWh
   Next 50 kWh at Rs 2.80 per kWh
   Next 50 kWh at Rs 2.40 per kWh
   Excess over 200 kWh at Rs 2.10 per kWh.
   [Ans. Rs 1,135.00 ; Rs 2.52/kWh**]**
6. Calculate the number of units to be consumed which will justify the two part tariff over flat rate tariff from the following data : Maximum demand 5 kW, Two part tariff Rs 1,200 per annum per kW of maximum demand plus Rs 2.10 per unit consumed. Flat rate tariff is Rs 2.60 per unit. **[Ans.** 12,000 units**]**
7. A light industry has a maximum demand of 100 kW. Two alternative tariffs are as follows :
   (*i*) A fixed charge of Rs 900 per kW plus a running charge of Rs 1.50 per unit.
   (*ii*) A charge of Rs 1.80 per unit flat.
   If the factory runs for 3,000 hours with a load factor of 80% which tariff is economical ?
   [Ans. Flat rate is cheaper for Rs 18,000**]**
8. The maximum demand of a consumer is 1 MW and the units consumed per annum are 500,000 kWh. Calculate the reduction in cost if the power factor is raised from 0.5 to 0.8. The tariff is Rs 900 per annum per kW demand plus Rs 1.50 per kWh plus Re 0.30 per kVAh reactive. Calculate the flat rate tariff for a unity power factor load.
   [Ans. Rs 147,310.00; Rs 3.30 per kWh**]**
9. A supply is offered on the basis of fixed charges of Rs 900 per annum per kW plus 90 paise per unit or alternatively at the rate of Rs 1.80 per unit for the first 400 units per annum and Rs 1.50 per unit for all the additional units. The maximum demand is 1 kW. Find the number of units consumed per annum for which the cost under these two tariffs becomes the same.
   [Ans. 1,300 units**]**
10. A consumer, whose connected load is 400 kVA, consumes 1×10^6 kWh per annum and the maximum demand registered is 250 kVA. He is offered the following two alternative tariffs :
    Tariff A : Rs 1,000 per annum per kVA of connected load + first 150 kWh per kVA of connected load at Rs 3.0/kWh, next 50,000 kWh at Rs 2.00 per kWh and the rest at Re 1.0/kWh.
    Tariff B : Rs 1,200 per annum per kVA of maximum demand + first 150 kWh per kVA of maximum demand at Rs 3.0 per kWh, next 50,000 kWh at Rs 2.0/kWh and the rest at Re 1.0/ kWh.
    Which tariff is economical and what is the annual gain ?
    [Ans. Tariff B is cheaper ; Annual gain = Rs 145,000.00**]**
11. A consumer has a yearly consumption of 200,000 units. The tariff is Rs 1,000.00 per annum per kW of maximum demand + Rs 2.0 per kWh. Find out the total annual cost and overall cost per unit (*i*) consumption were 200,000 units per annum with the load factor of 25% (*ii*) consumption were reduced by 25% with the same load factor.
    [Ans. (*i*) Rs 491,324.00 and Rs 2.46/kWh
    (*ii*) Rs 368,493.00 and Rs 2.46/kWh**]**
12. Calculate the minimum two part tariff to be charged from the following data :
    Generating cost per unit – Re 1.0

Annual generating cost per kW of maximum demand – Rs 1,250.00
Units supplied annually – $4,380 \times 10^4$
Annual load factor of the station – 50%
Annual charges for distribution – Rs 25×10^5
Diversity factor for distribution work – 1.25 and the distribution losses are 10%.
[**Ans.** Rs 1,250 per kW of maximum demand plus Rs 1.10/kWh]

13. Two systems of tariff are available for a factory, working 8 hours a day for 300 working days in a year.
(*i*) HV supply at Rs 1.25 per unit plus Rs 1,350 per annum per kVA of maximum demand.
(*ii*) LV supply at Rs 1,500 per annum per kVA of maximum demand plus Rs 1.375 per unit.
The factory has an average load of 200 kW at 0.8 pf and a maximum demand of 250 kW at the same power factor. The hv equipment costs Rs 1,250.00 per kVA and losses can be taken as 4%. Interest and depreciation charges are 12%. Calculate the difference in annual costs between the two systems.
[**Ans.** HV supply is cheaper for Rs 15,470.00]

14. An industrial load requiring 500 kW at 0.8 pf can be supplied on the following tariff rates :
(*i*) HT supply at Rs 1,500 per kVA per annum + 60 paise per unit.
(*ii*) LT supply at Rs 1,650 per kVA per annum + 75 paise per unit.
Transformer and switchgear for ht supply costs Rs 1,125 per kVA, the full load losses being 2.5%. The fixed charges are 25 per cent per annum of the capital cost of ht plant. If the installation works on full load, find the number of working hours per week above which ht supply is cheaper. There are 48 weeks in a year. [**Ans.** 34.2 hours per week]

15. A synchronous motor having a power consumption of 50 kW is connected in parallel with a load of 200 kW having pf 0.8 lagging. The excitation of motor is adjusted until the combined pf becomes 0.9 lagging. Calculate the kVA input to the motor.
[**Ans.** 57.76 kVA]

16. Describe the objectives and different forms of the tariff. The monthly readings of a consumer meter are as :
Maximum demand 50 kW. Energy consumed = 36,000 kWh and reactive energy = 23,400 kVAh.
If the tariff is Rs 80 per kW of maximum demand plus 8 paise per unit plus 0.5 paise per unit for each 1% of pf below 87%, calculate the monthly bill of the consumer.
[U.P. Technical Univ. Power Station Practice 2007]
[**Ans.** Rs 7,448]

17. It is desired to correct the power factor to 0.95 by means of static condensers connected across each phase of a 3-phase, 400 V, 50 Hz motor installation having a maximum load of 50 kVA at pf of 0.75 lagging. What must be the capacity of each delta-connected capacitor ? [**Ans.** 244 μF]

18. A 340 kW, 3,300 V, 50 Hz, 3-phase star-connected induction motor has full load efficiency of 85% and pf of 0.8 lagging. It is desired to improve the power factor to 0.96 lagging by using a bank of three capacitors. Calculate (*i*) The kVA rating of the condenser bank (*ii*) The capacitance of each limb of the condenser bank connected in delta (*iii*) The capacitance of each capacitor, if one each of the limbs of the delta-connected condenser bank is formed by using 6 similar 3,300 V capacitors.
[**Ans.** (*i*) 183.333 (*ii*) 17.86 μF (*iii*) 2.977 μF]

19. Derive the most economic angle of lag at which a consumer should take power when the charge is based on a fixed sum per kVA of maximum demand plus a flat rate per kWh.
Calculate this value when the tariff is Rs 1,000 per kVA of maximum demand plus a flat rate per kWh. The annual cost of condenser and additional equipment for power factor improvement is 10% of the capital cost of Rs 800 per kVA of such plant. [**Ans.** 0.9968 lagging]

20. A consumer is taking a load of 20 kW at 0.8 pf (lagging).
(*i*) Find the rating of the capacitor to raise the pf to 0.95 lagging.
(*ii*) If a phase advancing device is used which takes current at leading power factor of 0.1, find the rating of the device.
[U.P. Technical Univ. Power Station Practice 2002-03]
[**Ans.** (*i*) 8.426 kVA, (*ii*) 8.468 kVA]

21. A consumer has an annual consumption of 1 million kWh. The tariff is Rs 1,000 per year per kVA of maximum demand plus Re 1.0 per kWh. Determine the annual power cost if the power factor is 0.75 and the load factor is 50%. What would be most economical value of power factor, if the loss free phase advancing apparatus costs Rs 1,000 per kVAR, the rate of interest and depreciation 20% per annum.
[**Ans.** Rs 1,304,400; 0.98]

22. The power factor of an industrial three-phase load of 490 kW is to be improved from 0.7 lagging to 0.97 lagging by connecting loss free delta-connected capacitors across the 6.6 kV, 50 Hz supply. The cost of suitable capacitors and control gear is Rs 200 per kVAR and the annual tariff charge is Rs 120 per kVA maximum demand. The annual interest and depreciation charges are 15%. Calculate (*a*) the total kVAR rating of capacitors required. (*b*) the required value of capacitance per phase (*c*) the net annual saving. **[GATE 1992]**
[**Ans.** (*a*) 377 kVAR (*b*) 9.183 μF (*c*) Rs 12,072]

23. For increasing the kW capacity of a power plant working at 0.7 lagging power factor the necessary increase in power can be obtained by raising the power factor to 0.9 or by installing additional plant. What is the maximum cost per kVA of power factor correction apparatus to make its use more economical than addtional plant at Rs 2,400 per kVA? [**Ans.** 1,422/kVA]

24. A system is working at its maximum kVA capacity with a lagging pf of 0.7. An anticipated increase of load can be met by one of the two following methods.
(*i*) By raising the pf of the system to 0.866 by installing phase advancing equipment.
(*ii*) By installing extra generating plant. If the total cost of generating plant is Rs 100 per kVA, estimate the limiting cost per kVA of phase advancing equipment to make its use more economical than the additional generating plant. Interest and depreciation charges may be assumed 10% in each case.
[U.P. Technical Univ. Power Station Practice 2007-08]
[**Ans.** Rs 61.88]

25. Tender A and B for a 1,000 kVA 0.8 pf transformer are :
A—Full-load efficiency = 98.3% and core loss = 7 kW at rated voltage.
B—Full-load efficiency = 98.7% and core loss = 4 kW but costs Rs 5,000.00 more than A.
The service needed is 1,800 hours per annum at 1,000 kVA, 600 hours at 600 kVA, 400 hours at 15 kVA and remaining period shutdown. Take the annual charges for interest and depreciation at 12½% of capital costs and energy costs at 8 paise per kWh. Which is better tender and what would be the annual savings? **[Allahabad Univ., 1967]**
[**Ans.** Tender B is cheaper; Rs 211.00]

❑❑❑

MULTIPLE CHOICE TYPE QUESTIONS

Note: Each of the following questions comprises multiple responses (two or more than two answers). Select the correct answers. In case you feel that there is more than one correct answer, select the answer which you consider the best. In any case choose ONLY ONE answer or response for each question.

1. The commercial sources of energy are
(a) fossile fuels, water and radioactive substances.
(b) solar, wind, biomass.
(c) wood, animal wastes and agricultural wastes.
(d) none of the above.

2. Which of the following power plants is the least reliable ?
(a) Wind. (b) Tidal.
(c) Geothermal. (d) Solar.

3. Which of the following methods of generating electric power from the sea water is more advantageous ?
(a) Ocean currents. (b) Wave power.
(c) Tidal power. (d) None of the above.

4. What is the total power installed capacity (approximate) in India?
(a) 125,000 MW (b) 175,000 MW
(c) 200,000 MW (d) 50,000 MW
[U.P.S.C. I.E.S. E.E.-II, 2007]

5. The power development programmes that are pssible in Tamil Nadu are largely based on
(a) steam power plants. (b) hydroelectric power plants.
(c) nuclear power plants. (d) diesel power plants.

6. The power output from an hydroelectric power plant depends on
(a) type of dam, type of catchment area and discharge.
(b) type of dam, head and system efficiency.
(c) discharge, head and system efficiency.
(d) type of turbine, type of dam and type of catchment area.

7. What is the correct expression for the electrical power developed by a hydroelectric plant in kW?
(a) $\frac{0.736}{75}$ wQHη (b) $\frac{75}{0.736}$ wQHη
(c) 75 × 0.736 wQHη (d) $\frac{7}{75 \times 0.736}$ wQHη
(Symbols have their usual meaning) **[U.P.S.C. I.E.S. E.E.-II, 2007]**

8. Taking the density of water to be 1,000 kg/m^3, how much power would be developed by a hydroelectric generator unit, assuming 100% efficiency, with 1.0 m head and 1.0 m^3/s discharge?
(a) 2.90 kW (b) 4.45 kW (c) 9.80 kW (d) 19.60 kW
[U.P.S.C. I.E.S. E.E.-II, 2008]

9. One million cubic meters of water is stored in a reservoir feeding a water turbine. The density of water is 993 kg/m^3. If the centre of mass of water is 50 meters above the turbine and the losses are negligible, the energy produced by that volume of water is:
(a) 135.3 MWhr. (b) 130 MWhr.
(c) 120 MWhr. (d) 140 MWhr.
[U.P.S.C. I.E.S. E.E.-II, 2011]

10. If power P available from a hydro-scheme is given by the formula P = 9.81 QH, where Q is the flow rate through the turbine in *l/s* and H is the head in metres, then P will be in units of
(a) W (b) HP (c) kJ/s (d) kWh
[U.P.S.C. I.E.S. E.E.-II, 2000]

11. If the discharge is 1 m^3/s and head of the water is 1 m, then the power generated by the alternator in one hour (assume 100% efficiency of generator and turbine) will be
(a) 10 kW (b) $\frac{73}{75}$ kW (c) $\frac{746}{75}$ kW (d) 100 kW
[U.P.S.C. I.E.S. E.E.-II, 1997]

12. An hydroelectric power station is supplied from a reservoir of capacity 3×10^7 m^3 at a head of 150 m. The overall efficiency of the power plant is 70% . Energy available from the plant will be
(a) 12.2625×10^6 kWh (b) 8.58375×10^6 kWh
(c) 1.25×10^6 kWh (d) 0.875×10^6 kWh

13. A hydroelectric power station is commonly found in
(a) desert areas. (b) hilly areas.
(c) swamps. (d) grasslands.
[A.M.I.E. Sec B. Summer 1993]

14. In hydroelectric power plants
(a) operating cost is low and initial cost is high.
(b) operating cost is high and initial cost is low.
(c) both operating cost as well initial cost are high.
(d) both operating cost as well as initial cost are low.

15. The advantage(s) of hydro plants is/are
(a) low operating cost.
(b) they can be started and loaded very quickly.
(c) they can be used as base load and peak load plants as well.
(d) all the above. **[A.M.I.E. Sec B. Power Systems, Summer 1998]**

16. The largest size of hydroelectric unit in India is
(a) 350 MW (b) 500 MW
(c) 165 MW (d) 90 MW
[A.M.I.E. Sec B. Power Systems, Summer 1996]

17. Gross head of an hydroelectric power station is
(a) the difference of water level between the level in the storage and tail race.
(b) the height of water level in the river where the tail race is provided.
(c) the height of water level in the river where the storage is provided.
(d) all of the above.

18. With reference to hydropower station, the graphical representation of the discharge as a function of time is known as:
(a) Monograph. (b) Hectrograph.
(c) Load duration curve. (d) Hydrograph.
[U.P.S.C. I.E.S. E.E.-II, 2011]

19. A graphical representation of the discharge and time is known as
(a) load curve. (b) load duration curve.
(c) monograph. (d) hectograph.
(e) hydrograph. **[A.M.I.E. Sec B. Power Plant Engineering Summer 1994]**.

20. An hydrograph indicates
(a) the discharge at any time during the period under consideration.

Answers: 1. (a), **2.** (a), **3.** (c), **4.** (a), **5.** (c), **6.** (c), **7.** (a), **8.** (c), **9.** (a), **10.** (a), **11.** (c), **12.** (b), **13.** (b), **14.** (a), **15.** (d), **16.** (c), **17.** (a), **18.** (d), **19.** (e), **20.** (d),

(b) the maximum and minimum run-off during the period.
(c) the average run-off during the period.
(d) all of the above.

21. Hydrograph is similar to
(a) load duration curve. (b) mass curve.
(c) energy load curve. (d) chronological load curve.
[A.M.I.E. Sec B. Power Systems, Summer 2000]

22. The factors affecting the run-off are
(a) rain fall pattern, shape and size of catchment area.
(b) the topography and nature of soil in the catchment area.
(c) amount of vegetation and weather condition in the catchment area.
(d) all of the above.

23. The flow duration curve at a given head of a hydroelectric plant is used to determine
(a) total power available at the site.
(b) total units of energy available.
(c) load-factor at the plant. (d) diversity factor for the plant.
[U.P.S.C. I.E.S. E.E.-II,1994]

24. The area under a flow duration curve represents.
(a) total units of energy available.
(b) total power available at site.
(c) total quantity of run-off during that period.
(d) maximum rate of run-off during that period.

25. A mass curve can be plotted from
(a) load duration curve.
(b) chronological load curve.
(c) energy load curve.
(d) both load duration curve and chronological load curve.
[A.M.I.E. Sec B. Design of Electrical Systems Summer 2005]

26. Storage requirement can be determined from
(a) hydrograph. (b) flow-duration curve.
(c) mass curve.
(d) either by hydrograph or by mass curve.

27. In hydropower stations what is an enlarged body of water just above the intake and used as a regulating reservoir, called?
(a) Spillways. (b) Forebay.
(c) Reservoir. (d) Penstock.
[U.P.S.C. I.E.S. E.E.-II, 2006]

28. A penstock is used as a conduit between
(a) the steam chest and the turbine in a thermal station.
(b) the dam and the turbine in a hydrostation.
(c) the turbine and the discharge drain.
(d) the heat exchanger and the turbine in a nuclear power plant.
(e) None of these.
[A.M.I.E. Sec B. Power Plant Engineering, Summer 1998]

29. In high head hydroelectric power plant, the velocity of water flow in penstock is around
(a) 2 m/s (b) 4 m/s (c) 7 m/s (d) 10 m/s

30. Location of a surge tank, in an hydroelectric power station, is near
(a) turbine. (b) tail race.
(c) reservoir. (d) dam.

31. The function of a surge tank is to
(a) supply water at constant pressure.
(b) relieve water hammer pressures in the penstock pipe.
(c) produce surge in the pipeline.
(d) none of the above.

32. Water hammer occurs in
(a) surge tank. (b) penstock.
(c) turbine casing. (d) draft tube.
[A.M.I.E. Sec B. Power Plant Engineering, Winter 1999]

33. In a medium or high head hydroelectric power station, a surge tank is provided to
(a) reduce the length of the penstock pipes.
(b) augment water at the forebay.
(c) control the pressure variations in the penstock pipes due to sudden load changes.
(d) control the water flow through the turbine.
[U.P.S.C. I.E.S. E.E.-II, 1996]

34. Where is the draft tube of a hydropower station that is an airtight pipe located?
(a) Near the surge tank.
(b) In between the penstock and the runner.
(c) In between the runner exhaust and the tailrace.
(d) At the beginning of penstock. **[U.P.S.C. I.E.S. E.E.-II, 2009]**

35. The draft tube is provided to
(a) raise the water surface of the stream to create an artificial head.
(b) reduce the effect of water hammer.
(c) increase the acting head on the water wheel.
(d) none of the above.

36. For variable heads of near about but less than 30 metres, which type of turbines is used in hydropower stations?
(a) Pelton. (b) Kaplan.
(c) Francis. (d) None of the above.
[U.P.S.C. I.E.S. E.E.-II, 2006]

37. For harnessing lower variable water heads, the suitable hydraulic turbine with high percentage of reaction and runner adjustable vanes is
(a) Kaplan (b) Francis (c) Pelton (d) Impeller
[GATE E.E.-2004]

38. For harnessing low variable water heads, the suitable hydraulic turbine with reaction and adjustable vanes is:
(a) Francis (b) Impeller.
(c) Kaplan. (d) Pelton.
[U.P.S.C. I.E.S. E.E.-II, 2011]

39. For high head and low discharge the water turbine used is
(a) Pelton wheel. (b) Kaplan turbine.
(c) Francis turbine. (d) propeller turbine.

40. The specific speed (N_s) of a turbine is given by the equation
(a) $N_s = \dfrac{N\sqrt{P}}{H^{0.75}}$ (b) $N_s = \sqrt{N}.P/H^{3/2}$
(c) $N_s = \dfrac{N\sqrt{P}}{H^{1.25}}$ (d) $N_s = N\sqrt{P}/H^{2/3}$

41. The "specific speed" of a water turbine is the speed at which the turbine develops
(a) maximum horse power. (b) unit horse power at all heads.
(c) unit horse power at unit head.
(d) minimum horse power. **[U.P.S.C. I.E.S. E.E.-II, 1993]**

42. The specific speeds of Kaplan, Francis and Pelton turbines are in
(a) the increasing order. (b) the decreasing order.
(c) neither increasing nor decreasing order.
[A.M.I.E. Sec B. Summer 1993]

Answers: 21. (d), **22.** (d), **23.** (a), **24.** (c), **25.** (b), **26.** (d), **27.** (b), **28.** (b), **29.** (c), **30.** (a), **31.** (b), **32.** (b), **33.** (c), **34.** (b), **35.** (c), **36.** (b), **37.** (a), **38.** (c), **39.** (a), **40.** (c), **41.** (c), **42.** (b),

43. An impulse turbine
(a) is most suited for low head and high discharge power plants.
(b) operates by initial complete conversion to kinetic energy.
(c) makes use of a draft tube.
(d) always operates submerged.

44. Turbines installed at Bhakra Nangal are
(a) Pelton wheels. (b) Francis turbines.
(c) Kaplan turbines. (d) Propeller turbines.

45. In water turbines, the runaway speed of Pelton wheel is
(a) 1.8 times rated speed. (b) 2 – 2.2 times rated speed.
(c) 2.5–3 times rated speed. (d) full load speed.

46. The number of buckets (Z) for a Pelton wheel in terms of jet ratio(m) is given by empirical relation
(a) Z = 15 m + 0.5 (b) Z = 0.5 m + 15
(c) Z = 2 m + 15 (d) Z = 2 + 15 m

47. In reaction turbine, the function of draft tube is to
(a) reduce water hammer. (b) increase flow rate of water.
(c) convert kinetic energy of water to potential energy by a gradual expansion in divergent part.
(d) none of the above.

48. In Francis turbine runner, the number of blades is usually of the order of
(a) 16–24 (b) 8–10 (c) 6–8 (d) 3–6

49. The maximum head of a Kaplan turbine is limited to
(a) 25 m (b) 70 m (c) 125 m
[A.M.I.E. Sec B. Power Plant Engineering, Winter 1996]

50. Which of the following statements is/are correct?
In pumped storage power plants,
(a) water is recirculated through water turbines.
(b) reversible turbines are used which operate as turbines for power generation during peak load hours and as pumps for pumping water during peak-off hours.
(c) plain Francis turbines are used.
(d) both (a) and (c) above are employed. **[U.P.S.C. I.E.S. E.E.-II, 2007]**

51. Consider the following statements regarding pumped storage plants:
1. A pumped storage plant is a base load plant.
2. The starting time of a pumped storage plant is very short.
3. Reversible turbines and pumps are highly suitable for pumped storage plants.

Which of the statements given above are correct?
(a) 1 and 2. (b) 1 and 3.
(c) 2 and 3. (d) 1, 2 and 3.
[U.P.S.C. I.E.S. E.E.-II, 2004]

52. In pump storage hydropower plant, the electrical machine is made to work alternately as generator and motor. The efficiency of the generator working at the same electric power level is
(a) greater than that as motor.
(b) equal to that as motor. (c) less than that as motor.
(d) greater or less than that as motor depending on the type of the machine. **[U.P.S.C. I.E.S. E.E.-II, 2001]**

53. A 'Pumped storage hydroelectric plant' consists of:
(a) A synchronous machine and a multistage centrifugal pump in one shaft.
(b) A synchronous machine, a reaction turbine, and a multistage centrifugal pump all in one shaft.
(c) An induction generator, a synchronous machine and a reaction turbine, all in one shaft.
(d) An induction generator, a synchronous machine, and a multistage centrifugal pump, all in one shaft.
[U.P.S.C. I.E.S. E.E.-II, 2011]

54. Consider the following statements:
Pumped storage plants when operated in interconnected power systems serve to
1. increase load factor of steam plant.
2. provide added capacity to meet peak loads.
3. decrease load factor of steam plant.
4. provide added capacity to meet base loads.

Which of the above statements are correct?
(a) 1, 2, 3 and 4. (b) 1 and 3 only.
(c) 1 and 2 only. (d) 3 and 4 only.
[U.P.S.C. I.E.S. E.E.-II, 2010]

55. Consider the following statements regarding the pumped storage plants:
1. A pumped storage plant is a peak load plant.
2. The starting time of a pumped storage plant is very long.
3. Reversible turbines and pumps are very suitable for pumped storage plants.
4. Pumped storage plants can be used for load frequency control.

Which of the above statements is/are correct?
(a) 1 only. (b) 1 and 2 only.
(c) 1, 3 and 4. (d) 3 and 4 only.
[U.P.S.C. I.E.S. E.E.-II, 2009]

56. In India, pumped storage power plant
(a) does not exist at all. (b) exists is Rajasthan alone.
(c) exists in Kadampurai (TN).
(d) exists in Haryana.

57. The pumped storage scheme is employed to supply
(a) during the peak hours. (b) during the off-peak hours.
(c) system base load. **[A.M.I.E. Sec B. Power Systems, Summer 1999]**

58. In pumped storage scheme, the generator is also used as
(a) induction generator or synchronous condenser.
(b) induction generator or synchronous motor.
(c) synchronous generator or induction generator.
(d) synchronous motor or synchronous condenser.
[U.P.S.C. I.E.S. E.E.-II, 1993]

59. Pump storage schemes are used to improve
(a) the power factor. (b) the diversity factor.
(c) the load factor. (d) the quality of voltage.
(e) the reactive power capacity.
(f) the plant capacity factor as well as the load factor of the power system. **[A.M.I.E. Sec B. Power Plant Engineering, Summer 1998]**

60. A hydel power plant of run-off-river type should be provided with pondage so that the
(a) firm capacity of the plant is increased.
(b) operating head is controlled.
(c) pressure inside the turbine casing remains constant.
(d) kinetic energy of the running water is fully utilized.
[U.P.S.C. I.E.S. E.E.-II, 1998]

61. Bulb turbines are
(a) low head axial flow turbines.
(b) high speed turbines. (c) high head turbines.
(d) high pressure turbines.

62. Bulb turbines
(a) can be used at very low and widely varying heads.
(b) are smaller, faster and easier to build than other types and are cheaper in capital cost than Kaplan turbines.
(c) are universally used for small hydro schemes.
(d) all of the above.

63. In turbulent flow
(a) cohesion is more effective than momentum transfer in causing shear stress.

Answers: 43. (b), **44.** (a), **45.** (a), **46.** (b), **47.** (c), **48.** (a), **49.** (a), **50.** (b), **51.** (c), **52.** (c), **53.** (b), **54.** (c), **55.** (c), **56.** (c), **57.** (a), **58.** (d), **59.** (f), **60.** (a), **61.** (a), **62.** (d), **63.** (b),

(b) shear stresses are usually larger than those in laminar flow.
(c) the fluid particles move in an orderly manner.
(d) momentum transfer is on molecular scale only.

64. Rotameter is used for measuring
(a) fluid discharge. (b) fluid density.
(c) fluid viscosity. (d) all of the above.

65. A hydraulic turbine having rated speed of 250 rpm is connected to a synchronous generator. In order to produce power at 50 Hz, the number of poles required in the generator are
(a) 6 (b) 12 (c) 16 (d) 24
[GATE E.E.-I, 2004]

66. A steam power station needs space
(a) less than that required by the diesel power station of the same output.
(b) less than that required by atomic power station of the same output.
(c) less than that required by hydropower station of the same capacity.
(d) less than that required by a gas turbine power station of the same output.

67. Consider the following statements concerning steam power plants:
1. Maintenance and operating costs are low.
2. Water is required in huge quantity.
3. Requires long time for starting.
4. Handling of coal and disposal of ash can be done easily.
Which of the above statements is/are correct?
(a) 1 only. (b) 2 and 3.
(c) 3 only. (d) 3 and 4.
[U.P.S.C. I.E.S. E.E.-II, 2009]

68. Operating cost of steam power station is
(a) less than that of nuclear power plants.
(b) less than that of diesel/gas turbine power plants.
(c) less than that of hydroelectric power plants.
(d) is the same as that of nuclear power plant.

69. The public sector unit associated with the manufacturing of steam power plant equipment in India is
(a) Bharat Heavy Electricals Ltd.
(b) N.T.P.C.
(c) Heavy Engineering Corporation.
(d) Neyveli Lignite Corporation.

70. Public sector undertaking associated with erection and commissioning of steam power plants in India is
(a) Indian Electricity Authority.
(b) National Thermal Power Corporation Ltd.
(c) Bharat Coking Coal Ltd. (d) Coal India Ltd.

71. On which one of the following cycles does a modern steam power plant work?
(a) Carnot cycle (b) Rankine cycle
(c) Otto cycle (d) Bell-Coleman cycle
[U.P.S.C. I.E.S. E.E.-II, 2005]

72. Rankine cycle efficiency of a steam power plant is in the range of
(a) 20–30%. (b) 30–45%.
(c) 45–60%. (d) 60–80%.

73. In Rankine cycle, the work output from a turbine is given by
(a) change of enthalpy between inlet and outlet.
(b) change in internal energy between inlet and outlet.
(c) change of entropy between inlet and outlet.
(d) change in temperature between inlet and outlet.

74. Binary vapour cycles are used for
(a) increasing the turbine efficiency.
(b) improving the condenser performance.
(c) improving the plant efficiency.
(d) all of the above.

75. Binary vapour cycles are not being adopted because
(a) initial cost of such plants is high.
(b) such plants are suitable for high load factors only.
(c) pipes of thicker sections are required.
(d) ideal vapours are not available.

76. The overall efficiency of a thermal power station is equal to
(a) that of Rankine cycle. (b) that of regenerative cycle.
(c) that of Binary vapour cycle.
(d) product of boiler efficiency, turbine efficiency and generator efficiency.

77. Rankine cycle operating between pressure limits of P_1 and P_2 has thermal efficiency
(a) lower than that of Carnot cycle operating between same pressure limits.
(b) higher than that of Carnot cycle operating between same pressure limits.
(c) same as that of Carnot cycle operating between same pressure limits.
(d) none of the above.

78. Reheat cycle in a steam power station is employed for
(a) improving thermal efficiency.
(b) improving condenser performance.
(c) reducing heat losses. (d) utilising heat of flue gases.

79. In a regenerative cycle, feed water is heated by
(a) heaters.
(b) drained steam from the turbine.
(c) exhaust gases. (d) all of the above.

80. In a regenerative feed heating cycle,
(a) the optimum value of the fraction of steam extracted for feed water heating increases with the increase in Rankine cycle efficiency.
(b) the greatest economy is affected when steam is extracted from several places in different stages of the steam turbine.
(c) thermal efficiency is reduced.
(d) both (a) and (b).

81. What is the approximate efficiency of a normal thermal power station?
(a) 30-40% (b) 45-55% (c) 20-25% (d) 60-70%
[U.P.S.C. I.E.S. E.E.-II, 2007]

82. In a thermal power station, a typical heat balance sheet, for a large turbine and surface condenser taken together, is the percentage distribution of heat energy in:
1. work done or thermal efficiency.
2. friction and windage loss.
3. heat to circulating water.
4. heat in condensate to be retained to the boiler.
The percentage amount of heat in the heads stated above, in the descending order is:
(a) 3, 4, 1 and 2. (b) 2, 1, 4 and 3.
(c) 3, 1, 4 and 2. (d) 2, 4, 1 and 3.
[U.P.S.C. I.E.S. E.E.-II, 2011]

83. Maximum efficiency of modern coal fired steam-raising thermal power plants is restricted to about 0.35 (a low value), mainly because of

Answers: **64.** (a), **65.** (d), **66.** (c), **67.** (b), **68.** (b), **69.** (a), **70.** (b), **71.** (b), **72.** (b), **73.** (a), **74.** (c), **75.** (a), **76.** (d), **77.** (b), **78.** (a), **79.** (b), **80.** (d), **81.** (a), **82.** (c), **83.** (d),

(a) low alternator efficiency. (b) high energy loss in boilers.
(c) low steam turbine mechanical efficiency.
(d) high energy loss from turbine exhaust to condenser.
[U.P.S.C. I.E.S. E.E.-II, 2008]

84. Which of the following contributes to the Rankine cycle efficiency improvement in a steam power plant ?
(a) Use of high pressure.
(b) Reheat of steam at intermediate stages.
(c) Use of steam for heating of boiler feed water.
(d) All of the above.

85. The steam power plant efficiency can be improved by
(a) using large quantity of water.
(b) burning large quantity of coal.
(c) using high pressure and high temperature steam.
(d) decreasing the load on the plant.

86. The average load factor of thermal power plants in India is
(a) 100% (b) 80-95% (c) 50-60% (d) 35-45%
(e) 20-30% [A.M.I.E. Sec B. Summer 1994]

87. As the size of a thermal generating unit increases, the capital cost per kW of installed capacity
(a) increases. (b) decreases.
(c) remains the same. (d) may increase or decrease.
[A.M.I.E. Sec B. Power Systems, Summer 1996]

88. The modern trend in electric power generation is
(a) to have a large number of small size thermal plants located at different places.
(b) to have large size thermal plants near load centres.
(c) to have large size thermal plants located near coal fields.
(d) none of the above. [A.M.I.E. Sec B. Power Systems, Summer 1998]

89. In the feed water and steam circuit of a thermal power plant, the condensate flows from the condensate extraction pump to the boiler feed pump and, then to the boiler. The HP and LP heaters are located in this circuit in relation to the boiler feed pump (BFP) as
(a) both HP and LP heaters after BFP.
(b) HP heater before and, LP heater after BFP.
(c) both HP and LP heaters before BFP.
(d) LP heater before and, HP heater after BFP.
[U.P.S.C. I.E.S. E.E.-II, 1996]

90. In thermal power plants, the pressure in the working fluid cycle is developed by
(a) condenser. (b) superheater.
(c) feed water pump. (d) turbine. [GATE E.E. E.E.-2004]

91. Deaerative heating is done to
(a) heat the steam. (b) heat the feed water.
(c) remove dissolved gases in water.
(d) remove dissolved solid impurities in water.
[A.M.I.E. Sec B. Power Plant Engineering, Winter 1999]

92. In a thermal power plant, the feed water coming to the economiser is heated using
(a) HP steam. (b) LP steam.
(c) direct heat in the furnace. (d) flue gases. [GATE E.E. 2000]

93. Arrange the following in the correct sequence in which the flue gas passes through them after coming out of the boiler in a thermal power station:
(1) ID Fan. (2) Air preheater.
(3) Economiser. (4) Electrostatic precipitator.
Select the answer using the codes given below Codes
(a) 4, 3, 2, 1 (b) 3, 2, 4, 1 (c) 2, 1, 4, 3 (d) 1, 4, 3, 2
[U.P.S.C. I.E.S. E.E.-II, 1994]

94. In bleeder turbines, part of the steam is extracted for
(a) condensation. (b) reheating.
(c) feed water heating. (d) secondary air heating.

95. Which of the following is not a fire tube boiler ?
(a) Babcock and Wilcox. (b) Lancashire.
(c) Locomotive. (d) Cochron.

96. Water is supplied to the boiler at
(a) atmospheric pressure.
(b) more than steam pressure in the boiler.
(c) slightly more than atmospheric pressure.
(d) any pressure.

97. The efficiency of a boiler used in a modern steam power plant is of the order of
(a) 30% (b) 50% (c) 85% (d) 100%

98. Steam pressure used in steam power plants in India is in the range of
(a) 110–170 kg/cm^2 (b) 40–80 kg/cm^2
(c) 20–40 kg/cm^2 (d) 250 kg/cm^2

99. Critical pressure for steam is in the range of
(a) 50 – 100 kg/cm^2 (b) 100 – 200 kg/cm^2
(c) 200 – 250 kg/cm^2 (d) above 300 kg/cm^2

100. A super critical boiler is one that operates above the pressure and temperature of the following values
(a) 100 kg/cm^{2}and 540° C (b) 218 kg/cm^2 and 540° C
(c) 100 kg/cm^2 and 373° C (d 218 kg/cm^2 and 373° C
[A.M.I.E. Sec B. Summer 1993]

101. In a boiler, the carry over of slugs of water into the piping because of dirty water is called the
(a) foaming. (b) pitting. (c) scalping. (d) beating.

102. Which of the following steam will have least enthalpy ?
(a) Dry and saturated steam at 25 kg /cm^2.
(b) Wet steam at 25 kg/cm^2.
(c) Superheated steam at 25 kg/cm^2.
(d) All of the above will have the same enthalpy.

103. The boilers using lignite as fuel do not use
(a) spreader stoker. (b) travelling grate stoker.
(c) underfeed stoker. (d) all of the above.

104. Pipes carrying steam are generally of
(a) steel. (b) cast iron.
(c) aluminum. (d) cobalt.

105. For given power a water tube boiler occupies
(a) less floor area and less height.
(b) more floor area and less height.
(c) less floor area and more height.
(d) more floor area and more height than a fire tube boiler.
[A.M.I.E. Sec B. Power Plant Engineering, Winter 1995]

106. The selection of the type and size of a steam boiler depends upon
(a) the power required and working pressure.
(b) the geographical position of the powerhouse.
(c) the fuel and water availability.
(d) all of the above factors.
[A.M.I.E. Sec B. Power Plant Engineering, Winter 1995]

107. In a superheater
(a) pressure rises and temperature drops .
(b) temperature rises and pressure drops .
(c) temperature rises and pressure remains unchanged.
(d) pressure rises and temperature remains the same.

Answers: 84. (d), **85.** (c), **86.** (c), **87.** (b), **88.** (c), **89.** (d), **90.** (c), **91.** (c), **92.** (d), **93.** (b), **94.** (c), **95.** (a), **96.** (b), **97.** (c), **98.** (a), **99.** (c), **100.** (b), **101.** (a), **102.** (b), **103.** (c), **104.** (a), **105.** (c), **106.** (a), **107.** (c),

108. Within the boiler, the steam has highest temperature in
(a) superheater. (b) water tubes.
(c) water walls. (d) water drum.

109. Which of the following enters the superheater ?
(a) Superheated steam. (b) Wet steam.
(c) Hot water. (d) Cold water.

110. Superheated steam is always
(a) at a pressure more than that of the boiler steam.
(b) at a pressure lower than the maximum cycle pressure.
(c) at a temperature higher than the temperature of saturation corresponding to the steam pressure.
(d) none of the above.

111. Steam is superheated in order to
(a) improve Rankine cycle efficiency.
(b) reduce initial condensation losses.
(c) avoid too high temperature in the last stage of the turbine.
(d) all of the above.

112. The function of the economiser is to
(a) heat up the incoming water with exhaust steam.
(b) heat up the pulverised fuel by exhaust gases.
(c) heat up the incoming air by exhaust gases.
(d) heat up the incoming water by exhaust gases.
[A.M.I.E. Sec B. Winter 1994]

113. The main function of economiser of a boiler plant is to
(a) increase steam production.
(b) reduce fuel consumption.
(c) increase steam pressure.
(d) increase life of the boiler. **[A.M.I.E. Sec B. Winter 1994]**

114. In a steam power plant heat from the flue gases is recovered in
(a) a condenser. (b) a chimney.
(c) economiser and air preheater. (d) a desuperheater.

115. Economiser in a steam power plant
(a) improves the boiler efficiency by 10-12%.
(b) saves fuel consumption by 5-15%.
(c) becomes a necessity for pressure exceeding 70 kg/cm^2.
(d) all of the above.

116. The advantage of reheating of steam in a turbine is that
(a) it increases the efficiency of the turbine.
(b) it reduces the wears on the blades.
(c) it increases the workdone through the turbine.
(d) all of the above advantages.
[A.M.I.E. Sec B. Power Plant Engineering, Winter 1995]

117. Air preheater in a steam power plant
(a) recovers the heat from the flue gases leaving the economiser.
(b) improves combustion rate.
(c) raises the temperature of the furnace gases.
(d) all of the above.

118. Condensers in thermal power plants are for condensing
(a) steam to water. (b) water to ice.
(c) hydrogen gas to liquid gas.
(d) carbon dioxide to dry ice.
[A.M.I.E. Sec B. Power Systems, Summer 1993]

119. A condenser in a steam power plant condenses steam coming out of
(a) turbine. (b) boiler.
(c) economiser. (d) superheater.

120. In a steam power plant water is used for cooling purposes in
(a) economiser. (b) condenser.
(c) superheater. (d) electrostatic precipitator.

121. In a steam turbine cycle, the lowest pressure occurs in
(a) condenser. (b) turbine inlet.
(c) boiler. (d) superheater.

122. Increase in condenser back pressure will lead to a
(a) loss. (b) gain. (c) none of these.
[A.M.I.E. Sec B. Power Plant Engineering, Summer 1996]

123. In which part of the thermal power plant, the steam pressure is less than that of atmosphere ?
(a) Boiler. (b) Turbine.
(c) Superheater. (d) Condenser.

124. The major function of the condenser is to
(a) remove the condensate for boiler feed water.
(b) condense steam.
(c) reduce the back pressure so that maximum heat energy can be extracted from steam.
(d) provide a closed cycle. **[A.M.I.E. Sec B. Summer 1993]**

125. The function of a condenser in a steam power plant is to
(a) condense the large volume of steam to water to be used as boiler feed water.
(b) receive the large volume of steam exhausted from the steam turbine.
(c) maintain pressure below atmospheric so that maximum heat energy can be extracted from steam.
(d) all of the above.

126. Evaporative type of condenser has
(a) steam in pipes surrounded by water.
(b) water in pipes surrounded by steam outside.
(c) steam and cooling water mixed to give condensate.
(d) none of the above.

127. In a jet type condenser
(a) steam passes through tubes and cooling water surrounds them.
(b) water passes through tubes and steam surrounds them.
(c) steam and cooling water mix.
(d) none of the above.

128. In a shell and tube surface condenser
(a) steam passes through the tubes and cooling water surrounds them.
(b) cooling water passes through the tubes and steam surrounds them.
(c) steam and water mix to give condensate.
(d) none of the above.

129. In a regenerative surface condenser
(a) there is no pump to remove condensate and the condensate gets removed by gravity.
(b) there is only one pump for removing air and condensate.
(c) there are two pumps to remove air and condensate.
(d) there are three pumps to remove air, condensate and vapour.

130. In a surface condenser, on removal of air
(a) absolute pressure of condenser is reduced.
(b) absolute pressure of condenser is increased.
(c) absolute pressure of condenser remains unaffected.
(d) temperature of condensed steam is increased.

131. The purpose of the boiler feed pump is to
(a) pump hot air into the boiler.
(b) pump pulverised coal into the boiler.
(c) pump out steam from the boiler.
(d) pump water into the boiler.
(e) none of the above. **[A.M.I.E. Sec B. Winter 1994]**

Answers: **108.** (a), **109.** (b), **110.** (b), **111.** (d), **112.** (d), **113.** (b), **114.** (c), **115.** (d), **116.** (d), **117.** (d), **118.** (a), **119.** (a), **120.** (b), **121.** (a), **122.** (a), **123.** (d), **124.** (c), **125.** (d), **126.** (a), **127.** (c), **128.** (b), **129.** (c), **130.** (a), **131.** (d),

132. The function of air pumps in a condenser is to
(a) remove water.
(b) air leaking into the condenser and maintain vacuum.
(c) maintain atmospheric pressure in the condenser.
(d) both (a) and (b).

133. Wet air pump is to remove
(a) condensate only. (b) air only.
(c) both air and condensate. (d) vapour from the condenser.

134. In a steam power plant cooling towers are used for
(a) cooling condenser outlet water.
(b) cooling exhaust steam. (c) cooling feed water.
(d) all of the above.

135. Spray ponds are used for cooling warm water coming out of condenser in
(a) large power plants. (b) medium power plants.
(c) small power plants.
(d) both in medium and large power plants.

136. The drawback (s) of spray ponds is /are
(a) large ground area is required especially in case of large sized power plants.
(b) loss of water due to evaporation.
(c) considerable quantity of water is carried away in air when its velocity is high.
(d) all of the above.

137. In which of the following steam turbines the back pressure will be below the atmospheric one ?
(a) Non-condensing turbine. (b) Condensing turbine.
(c) Topping turbine. (d) None of the above.

138. The modern steam turbines are
(a) reaction turbines. (b) impulse turbines.
(c) impulse reaction turbines.
(d) none of the above.

139. In case of reaction steam turbine
(a) there is a enthalpy drop both in fixed and moving blades.
(b) there is enthalpy drop only in moving blades.
(c) there is enthalpy drop only in fixed blades.
(d) none of the above.

140. The largest size of steam turbine installed in India is
(a) 100 MW (b) 250 MW
(c) 500 MW (d) 1,000 MW

141. As compared to steam at entry to the turbine which of the following will be larger at exit ?
(a) Flow rate. (b) Pressure.
(c) Specific volume. (d) Specific enthalpy.

142. For the same power, the size of a turbine
(a) increases with speed. (b) remains with speed.
(c) decrease with speed.
[A.M.I.E. Sec B. Power Plant Engineering, Summer 1993]

143. The pressure on the two sides of the impulse wheel of a steam turbine
(a) is different. (b) is the same.
(c) reduces from one side to the other side.
(d) increases form one side to the other side.

144. Reheat factor in steam turbines depends on
(a) stage efficiency. (b) exit pressure.
(c) initial pressure and temperature.
(d) all of the above.

145. In steam turbines, the reheat factor
(a) increases with the increase in number of stages.
(b) reduces with the increase in number of stages.
(c) remains the same irrespective of number of stages.
(d) none of the above.

146. The value of reheat factor for a multistage steam turbine lies in the range of
(a) 1.005 to 1.03 (b) 1.01 to 1.06
(c) 1.02 to 1.1 (d) 1.10 to 1.2
[A.M.I.E. Sec B. Winter 1994]

147. Steam turbines are governed by
(a) nozzle control governing.(b) throttle governing.
(c) bypass governing. (d) all of the above.

148. Topping turbines are
(a) low pressure condensing ones.
(b) high pressure condensing ones.
(c) high pressure non-condensing ones.
(d) low pressure non-condensing ones.

149. Compounding of steam turbines is done in order to
(a) reduce the rotor speed. (b) increase the rotor speed.
(c) balance the turbine. (d) none of the above.

150. The governing employed for medium and larger sized steam turbines is
(a) bypass. (b) nozzle.
(c) throttle. (d) combination of (b) and (c).

151. The ratio of exit pressure to inlet pressure of maximum mass flow rate per unit area of steam through a nozzle, when the steam is initially super heated, is
(a) 0.65 (b) 0.578 (c) 0.5457 (d) 0.5325

152. The ratio of exit pressure to inlet pressure of maximum mass flow rate per unit area of steam through a nozzle, when the steam is initially dry saturated, is
(a) 0.5325 (b) 0.5457 (c) 0.578 (d) 0.65

153. The effect of considering friction in steam nozzles for the same pressure ratio leads to
(a) increase in dryness fraction of exit steam.
(b) decrease in exit velocity from the nozzle.
(c) no change in exit velocity from the nozzle and quality of exit steam.
(d) both (a) and (b).

154. The steam is superheated to the highest economical temperature in order to
(a) reduce the requirement of steam for a given output of energy owing to its high internal energy and thus reduce the turbine size.
(b) reduce the mechanical resistance to the flow of steam over turbine blades and increase the efficiency as superheated steam is dry and so turbine blades remain dry.
(c) avoid corrosion and pitting at the turbine blades occurring owing to steam dryness.
(d) all of the above.

155. Any leakage of air into the condenser destroys the vacuum and causes
(a) an increase in air pressure in the condenser limiting the useful heat drop in the steam turbine.
(b) lowering of the partial pressure of the steam and of the saturation temperature along with it.
(c) undercooling of the condensate more severe.
(d) all of the above.

Answers: **132.** (b), **133.** (c), **134.** (a), **135.** (c), **136.** (d), **137.** (b), **138.** (c), **139.** (a), **140.** (c), **141.** (c), **142.** (c), **143.** (b), **144.** (d), **145.** (a), **146.** (b), **147.** (d), **148.** (c), **149.** (a), **150.** (b), **151.** (c), **152.** (c), **153.** (d), **154.** (d), **155.** (d),

156. A 200 MW steam power plant will consume nearly _______ tonnes of coal per day.
(a) 1,000 (b) 2,000 (c) 4,000 (d) 7,500

157. Heating value of coal is roughly
(a) 1,000-1,500 kcal /kg (b) 3,000-4,500 kcal /kg
(c) 5,000-6,500 kcal /kg (d) 7,500-10,000 kcal /kg

158. Calorific value of coal largely depends upon
(a) ash content. (b) size of coal particles.
(c) moisture content. (d) volatile matter.

159. The coal of the lowest calorific value is
(a) anthracite. (b) bituminous coal.
(c) lignite. (d) steam coal.

160. The coal having highest calorific value is
(a) anthracite. (b) bituminous.
(c) lignite. (d) peat.

161. Coal used in thermal power plants is also known as
(a) soft coal. (b) steam coal.
(c) charcoal. (d) coke.

162. Coking cokes
(a) do not form ash. (b) burn completely.
(c) form lumps or masses of coke.
(d) burn freely.

163. Which of the following is considered to be superior quality of coal ?
(a) Bituminous. (b) Coke.
(c) Lignite. (d) Peat.

164. Low grade coals have
(a) low carbon content. (b) low calorific value.
(c) low moisture content. (d) low ash content.

165. Coal is usually considered of
(a) lava origin. (b) animal origin.
(c) vegetable origin. (d) none of the above.

166. Coal rank classifies coal as per its
(a) specific gravity. (b) degree of metamorphism.
(c) carbon percentage. (d) ash content.

167. The average ash content in Indian coals is around
(a) 5%. (b) 10%. (c) 20%. (d) 30%.

168. The coal that has highest ash content is
(a) lignite. (b) coking coal.
(c) bituminous coal. (d) steam coal.

169. Ash content of coal can be reduced by
(a) pulverising. (b) washing.
(c) slow burning. (d) mixing with high grade coal.

170. Combustible elements in the fuel are
(a) carbon and hydrogen. (b) carbon, hydrogen and ash.
(c) carbon, hydrogen and sulphur.
(d) carbon, nitrogen and ash.

171. The percentage of carbon in anthracites is usually
(a) more than 90%. (b) about 70%.
(c) about 50%. (d) below 40%.

172. Solvent refined coal has low percentage of
(a) impurities. (b) sulphur.
(c) ash. (d) all of the above.

173. Presence of sulphur content in coal will cause
(a) spontaneous combustion during coal storage and air heater corroding.
(b) clinkering and slagging. (c) facilitating ash precipitation.
(d) all of the above.

174. Sulphur content of liquid fuels assumes importance from the point of view of
(a) firing rate. (b) heating rate.
(c) corrosion. (d) efficiency.

175. Chemical composition of coal is given by
(a) ultimate analysis. (b) proximate analysis.
(c) gross analysis. (d) any of the above.

176. Ultimate analysis of fuel determines the percentage of
(a) ash, volatile matter and moisture.
(b) total carbon by weight.
(c) total carbon by weight—unit weight of H_2, O_2,N_2, sulphur and ash.
(d) none of the above.

177. Complete combustion of pulverized coal in a steam raising thermal power plant is ensured by what type of an analysis of flue gas going out by the chimney?
(a) O_2 content for given air intake.
(b) CO_2 content for given fuel rate feed.
(c) CO content.
(d) All of these. **[U.P.S.C. I.E.S. E.E.-II, 2009]**

178. The proximate analysis of coal gives percentage by weight of
(a) moisture, and volatile matter.
(b) moisture, volatile matter, fixed carbon and ash.
(c) carbon, hydrogen, oxygen, nitrogen, sulphur and ash.
(d) carbon, hydrogen, and oxygen. **[A.M.I.E. Sec B. Winter 1994]**

179. The ash content of bituminous coal is
(a) 2 to 3% (b) 6 to 12%
(c) 15 to 20% **[A.M.I.E. Sec B. Power Plant Engineering, Winter 1996]**

180. The most important factors to be considered in the selection of fuel for power plants include
(a) cost of fuel. (b) calorific value of fuel.
(c) none of the above. (d) both (a) and (b).

181. Live storage of coal in power plant means
(a) a covered storage near the boiler furnace having coal sufficient to meet 24 hour demand of the power plant.
(b) coal in transit. (c) coal ready for combustion.
(d) storage of coal sufficient to meet one week demand of the plant readily.

182. In a steam power plant, coal is carried from storage to boiler generally by means of
(a) trolleys. (b) V-belts. (c) buckets. (d) manually.

183. Belt conveyors can be employed for transporting coal at inclination upto
(a) 75° (b) 60° (c) 30° (d) 15°

184. The maximum length of a screw conveyor is around
(a) 85-90% (b) 60-75%
(c) 40-60% (d) 15-30%

185. In small power plants, coal is unloaded by means of
(a) belt conveyors. (b) coal accelerators.
(c) lift trucks. (d) all of the above.

186. Load carrying capacity of a belt conveyor is around
(a) 10–20 tonnes per hour. (b) 20–40 tonnes per hour.
(c) 50–100 tonnes per hour. (d) 100–160 tonnes per hour.

Answers: **156.** (b), **157.** (c), **158.** (a), **159.** (c), **160.** (a), **161.** (b), **162.** (c), **163.** (a), **164.** (b), **165.** (c), **166.** (b), **167.** (c), **168.** (a), **169.** (b), **170.** (c), **171.** (a), **172.** (d), **173.** (d), **174.** (c), **175.** (a), **176.** (c), **177.** (d), **178.** (b), **179.** (b), **180.** (d), **181.** (a), **182.** (b), **183.** (c), **184.** (a), **185.** (b), **186.** (c),

187. Bucket elevators are employed for
(a) carrying coal in vertical direction.
(b) carrying coal in horizontal direction.
(c) carrying coal in any direction.
(d) none of the above.

188. In coal preparation plants, the magnetic separation are used for removing
(a) dust. (b) iron particles.
(c) clinkers. (d) all of the above.

189. Burning of low grade fuel can be improved by
(a) pulverising.
(b) blending with better quality coals.
(c) oil assisted ignition. (d) any of the above.

190. Which one is essential for combustion of fuel ?
(a) Oxygen. (b) Correct fuel-air ratio.
(c) Proper ignition temperature.
(d) All of the above three.

191. The percentage of O_2 in atmospheric air is
(a) 21 % by volume. (b) 23 % by weight.
(c) 79 % by volume and 77% by weight.
(d) both (a) and (b).

192. The proper indication of incomplete combustion is
(a) the smoking exhaust from chimney.
(b) high temperature of flue gas.
(c) high CO content in flue gases at exit.
(d) high CO_2 content in flue gases at exit.

193. The stoker used in large capacity boilers, when unpulverised coal is used, is
(a) overfeed stoker. (b) underfeed stoker.
(c) any of the above. (d) none of the above.

194. Overfeed stoker includes
(a) travelling grate. (b) chain grate.
(c) spreader. (d) all of the above.

195. The equipment used for supplying coal to the boiler, when unpulverised coal is being used, is
(a) skip hoist. (b) stoker.
(c) any of the above. (d) none of the above.

196. Capacity of the underfeed stoker is in the range of ________ of coal burned per hour.
(a) 100 – 250 kg (b) 100 – 500 kg
(c) 100 – 2,000 kg (d) 100 – 5,000 kg

197. Travelling grate stoker can burn coal at the rate of
(a) 50 to 75 kg/m^2h (b) 75 to 100 kg/m^2h
(c) 100 to 150 kg/m^2h (d) 150 to 200 kg/m^2h
[A.M.I.E. Sec B. Winter 1994]

198. Pulverised coal is
(a) non-smoking coal. (b) coal free from ash.
(c) coal broken in fine particles.
(d) coal which burns for long time.

199. Equipment used for pulverising the coal is the
(a) hopper. (b) stoker. (c) ball mill. (d) burner.

200. Pulverised fuel is used for
(a) better burning.
(b) increased calorific value of coal.
(c) less radiation loss. (d) medium size units.
[A.M.I.E. Sec B. Summer 1993]

201. The advantages of using pulverised fuel include
(a) higher boiler efficiency, low air requirement and low fan power.
(b) easy and complete combustion.
(c) fast response to load changes and low banking losses.
(d) all of the above.

202. The drawbacks of using pulverised fuel include
(a) additional investment in pulverising plant and stack flyash removal equipment.
(b) extra power requirements for the pulverising plant.
(c) reduced power plant efficiency.
(d) both (a) and (b).

203. Ash is the main waste product of steam power plants with low grade coal (may be in tonnes per day). This ash
(a) can be used in building construction.
(b) can be used in brick making near the plant side.
(c) it can be dumped in disused mines, river or sea or in an area excavated in the waste land, as the case may be.
(d) all of the above.

204. Flyash generally results from
(a) fluidized bed boilers. (b) pulverised coal boilers.
(c) diesel engines. (d) gas turbines.

205. Dust can be separated from gas by
(a) impingement upon small baffles.
(b) sudden velocity decrease owing to enlargement.
(c) abrupt changes of flow direction.
(d) any of the above.

206. Which of the following equipment is installed in steam power plants to reduce air pollution ?
(a) Desuper heater. (b) Air filter.
(c) Air electrostatic precipitator.
(d) Stock. (e) None of them.
[A.M.I.E. Sec B. Winter 1993]

207. In coal-fired thermal power stations, what are the electrostatic precipitators used for?
(a) To remove dust particles settling on the bus bar conductors in the station switchyard.
(b) To condense steam by electrostatic means.
(c) To keep the air heaters clean.
(d) To collect the dust particles from the flue gases.
[U.P.S.C. I.E.S. E.E.-II, 2007]

208. Electrostatic precipitator is installed between
(a) induced fan and chimney.
(b) air preheater and induced fan.
(c) economiser and air preheater.
(d) boiler furnace and economiser.

209. The draught produced by a chimney is called the
(a) forced draught. (b) induced draught.
(c) natural draught. (d) balanced draught.

210. The draught produced by steel chimney in comparison to that produced by brick chimney, for the same height, is
(a) more. (b) less.
(c) the same. (d) unpredictable.

211. Artificial draught is produced by
(a) a forced fan. (b) an induced fan.
(c) induced and forced fan. (d) all of the above.

212. For the same draught produced the power of forced draught fan, in comparison to that of induced draught fan, is
(a) more. (b) less. (c) the same.
(d) may be more or less.

213. The artificial draught is usually designed to produce
(a) more draught. (b) less smoke.

Answers: 187. (a), **188.** (b), **189.** (d), **190.** (d), **191.** (d), **192.** (c), **193.** (a), **194.** (d), **195.** (b), **196.** (d), **197.** (d), **198.** (c), **199.** (c), **200.** (a), **201.** (d), **202.** (d), **203.** (d), **204.** (b), **205.** (d), **206.** (c), **207.** (d), **208.** (b), **209.** (c), **210.** (a), **211.** (d), **212.** (b), **213.** (d),

(c) low temperature of chimney gas.
(d) all of the above.

214. The draught produced by a chimney of a given height at a given outside temperature
(a) decreases with the increase in temperature of chimney gases.
(b) increases with the increase in temperature of chimney gases.
(c) remains unchanged irrespective of temperature of chimney gases.
(d) may increase or decrease.

215. The natural draught
(a) is provided by a chimney.
(b) is used in small steam boilers.
(c) can also be used in large steam boilers if economisers and preheaters have been used.
(d) both (a) and (b).

216. The magnitude of natural draught depends upon
(a) average temperature difference between flue gases within the chimney and the outside air.
(b) height of the chimney above the furnace grate level.
(c) weather and operating conditions.
(d) all of the above.

217. For forced draught the blower is located
(a) at the top of the chimney.
(b) near the base of the chimney.
(c) near the base of the boiler.
(d) any where permissible.

218. The height of chimney, in a steam power stations, is governed by
(a) the draught to be produced.
(b) the quantity of flue gases being produced.
(c) pollution control. (d) all of the above.

219. The draught produced in locomotive boilers is by
(a) forced fan. (b) steam jet.
(c) chimney. (d) motion of locomotive.

220. The pressure at the furnace is minimum in case of
(a) induced draught. (b) forced draught.
(c) balanced draught. (d) natural draught.

221. In forced draught the main function of the chimney is to
(a) discharge gases high up in the atmosphere from the point of view of air pollution.
(b) accelerate the fuel combustion.
(c) reduce the temperature of the discharged hot gases.
(d) all of the above.

222. Induced draught fans are used to
(a) cool the steam let out by the turbine in thermal station.
(b) cool the hot gases coming out of the boiler.
(c) force the air inside the coal furnace.
(d) control the heat generated in a nuclear reactor.
(e) pull the gas out of the chimney.
[A.M.I.E. Sec B. Power Plant Engineering, Summer 1998]

223. What is the nominal pH value of water that is to be maintained in a steam raising thermal power station?
(a) 0.0 (b) 7.0
(c) 8.5 (d) 14.3 **[U.P.S.C. I.E.S. E.E.-II, 2009]**

224. The feed water treatment is done mainly to avoid
(a) embrittlement problem. (b) carry over problem.
(c) corrosion and scale formation problems.
(d) all of the above problems.

225. Boiling down of boiler water is done so as to
(a) control the solid concentration in the boiler water by removing some of the concentrated saline water.
(b) remove dissolved gases in the water.
(c) reduce boiler pressure.
(d) increase steam temperature.

226. It is necessary to heat water before feeding it back to the boiler as because of this
(a) the dissolved oxygen and carbon dioxide which would otherwise corrode boiler are removed in the feed water heater.
(b) thermal stresses due to cold water entering the boiler drum are avoided.
(c) some other impurities carried by steam and condensate due to corrosion in boiler and condenser are precipitated outside the boiler.
(d) all of the above.

227. Huge quantity of water is required in a steam power plant. It is required
(a) to raise the steam in boilers.
(b) for cooling purposes such as in condensers.
(c) as a carrying medium such as in disposal of ash.
(d) all of the above .

228. For a 3-element feed water control in a coal-fired thermal power station, measurements of level of water in the boiler drums is made so that the water level does not
(a) exceed a specified upper limit.
(b) fall below a specified lower limit.
(c) violate specified upper and lower limits.
(d) restrict to a specified limit. **[U.P.S.C. I.E.S. E.E.-II, 2010]**

229. The auxiliary consumption in a thermal power (steam) station is
(a) 2–5% (b) 8–10% (c) 15–20% (d) 20–25%
[A.M.I.E. Sec B. Power Plant Engineering, Summer 1996]

230. Turbo-alternators run at
(a) a variable speed around 2,000 rpm.
(b) a constant speed of 1,000 rpm.
(c) a constant speed of 3,000 rpm.
(d) a variable speed above 1,000 rpm.
(e) none of the above.
[A.M.I.E. Sec B. Power Plant Engineering, Summer 1998]

231. A coal-fired steam power station working at a plant load factor of 80% has one 500 MW generating unit. If the heat content of coal is 2 kWh/kg, the overall plant efficiency is 40% and a train load of coal is 2,000 metric tons, then the number of trains required daily for the plant is
(a) 1 (b) 6 (c) 10 (d) 24
[U.P.S.C. I.E.S. E.E.-II, 2000]

232. Large size steam plants and nuclear plants are suitable for
(a) peak loads. (b) intermediate loads.
(c) base loads. (d) both base and peak loads
[A.M.I.E. Sec B. Power Systems, Summer 1996; Summer 1998]

233. The efficiency of a nuclear power plant is less than that of a coventional fuel fired thermal plant because of
(a) less rejection of heat in the condenser.
(b) higher temperature conditions.
(c) higher pressure conditions
(d) low temperature and pressure conditions.
[A.M.I.E. Sec B. Power Plant Engineering, Winter 1995]

Answers: **214.** (b), **215.** (d), **216.** (d), **217.** (c), **218.** (a), **219.** (b), **220.** (c), **221.** (a), **222.** (c), **223.** (c), **224.** (d), **225.** (a), **226.** (d), **227.** (d), **228.** (c), **229.** (a), **230.** (c), **231.** (b), **232.** (c), **233.** (d),

234. The first nuclear power plant installed in India is
(a) Kota (Rajasthan). (b) Tarapur (Maharashtra).
(c) Kalpakkam (TN). (d) Narora (U.P.).

235. Western U.P, Rajasthan, Punjab, Haryana, Tamil Nadu are quite suitable areas for
(a) hydroelectric power plants.
(b) steam power plants. (c) nuclear power plants.
(d) all of the above.

236. One atomic mass unit (amu) is approximately equal to
(a) 1.66×10^{-24} kg (b) 1.66×10^{-27} kg
(c) 1.6×10^{-19} kg (d) 1.6×10^{-13} kg

237. One atomic mass unit is equaivalent to
(a) 931.4 Mev. (b) 251.2 Mev.
(c) 120.4 Mev. **[A.M.I.E. Sec B. Power Plant Engineering, Winter 1996]**

238. One kilogram of natural uranium gives energy equivalent to about
(a) 100 kg of coal. (b) 1,000 kg of coal
(c) 5,000 kg of coal. (d) 10,000 kg of coal.
[A.M.I.E. Sec. B. Power Systems, Summer 1996; Summer 2000]

239. Generation of one watt needs about ______ fissions per second.
(a) 3.1 (b) 3.1×10^6
(c) 3.1×10^7 (d) 3.1×10^{10}

240. Particles having the same atomic numbers but different mass numbers are called the
(a) beta particles. (b) decayed particles.
(c) isotopes. (d) neutrons.

241. In a nuclear reactor thermal energy is obtained from
(a) fission of radioactive materials.
(b) fussion of radioactive materials.
(c) Burning of the fuel rods in oxygen.
[A.M.I.E. Sec. B. Power Systems, Summer 1999]

242. Nuclear reactors usually employ
(a) fission. (b) fusion.
(c) both fission and fusion. (d) none of the above.

243. In nuclear chain fission reaction each neutron causing fission produces
(a) no new neutron. (b) less than one new neutron.
(c) one new neutron. (d) at least one new neutron.

244. Which of the following is not a secondary nuclear fuel ?
(a) Pu_{239} (b) U_{233}
(c) U_{235} (d) none of above.

245. Which of the following are the fertile materials ?
(a) U_{238} and Th_{232} (b) U_{238} and Th_{239}
(c) U_{233} and Pu_{239} (d) U_{238} and Pu_{239}

246. Which of the following are the fissile materials ?
(a) U_{238} and Th_{232} (b) U_{235}, U_{233} and Pu_{239}
(c) U_{235} and Th_{239} (d) None of the above.

247. In the nuclear fission reactions, isotope of uranium used is
(a) U_{234} (b) U_{235} (c) U_{233} (d) U_{238}

248. When a nuclear reactor is operating at constant power the multiplication factor
(a) is equal to unity. (b) is less than unity.
(c) is greater than unity. (d) may be any of the above.

249. One fission of U_{235} releases energy of
(a) 335 MeV (b) 200 MeV
(c) 430 MeV (d) 525 MeV

250. Which of the following statements are correct ?
1. Nuclear fission occurs whenever a uranium nucleus reacts with a neutron.
2. Nuclear fission is accompanied by the release of neutrons and gamma rays.
3. About 200 MeV of energy is released in the fission of a uranium nulceus.
4. Energy from the fission of uranium nucleus is released mainly as the kinetic energy of the neutrons and the energy of gamma radiations.

Select the correct answer using the codes given below.
Codes:
(a) 1, 2, 3 and 4. (b) 2 and 3.
(c) 2, 3 and 4. (d) 1 and 4.
[U.P.S.C. I.E.S. E.E.-II, 1993]

251. Consider the following statements:
1. Nuclear fission occurs whenever Uranium reacts with a neutron.
2. Nuclear fission is accompanied by the release of neutrons and gamma rays.
3. About 200 MeV of energy is released in the fission of a U^{235} nucleus.
4. Energy from the fission of U^{235} nucleus is released mainly as kinetic energy of the neutrons and the energy of gamma radiations.

Which of the above statements are correct?
(a) 1, 2, 3 and 4. (b) 3 and 4 only.
(c) 2, 3 and 4 only. (d) 1, 2 and 4 only.
[U.P.S.C. I.E.S. E.E.-II, 2010]

252. In what form the initial energy will be released for the 200 MeV per fission by a neutron in a slow thermal nuclear reactor?
(a) Heat energy. (b) Electromagnetic radiation.
(c) Kinetic energy of particles and electromagnetic radiation.
(d) Kinetic and sound energy. **[U.P.S.C. I.E.S. E.E.-II, 2009]**

253. The operation of a nuclear reactor is controlled by controlling the multiplication factor (k), defined as

$$K = \frac{\text{Number of neutrons of any one generation}}{\text{Number of neutrons of immediately preceding generation}}$$

The power level of the reactor can be increased by
(a) raising the value of k above 1 and, keeping it at that raised value.
(b) raising the value of k above 1, but later bringing it back to k = 1.
(c) lowering the value of k below 1 and keeping at that lowered value.
(d) lowering the value of k below 1, but later bringing it back to k = 1. **[U.P.S.C. I.E.S. E.E.-II, 1994]**

254. Critical mass of the fuel is the amount required to make the multiplication factor
(a) less than unity. (b) more than unity.
(c) equal to unity. (d) 0.5

255. The power output of a nuclear power station is proportional to
(a) the rate at which fission reaction occurs
(b) square root of the rate at which fission reaction occurs.
(c) square of the rate at which fission reaction occurs.
(d) none of the above.
[A.M.I.E. Sec. B. Power Plant Engineering, Winter 1999]

256. The function of moderator in a nuclear reactor is
(a) to absorb the excess neutrons.
(b) to increase the energy of the neutrons.
(c) to slow down the neutrons.
[A.M.I.E. Sec B. Power Systems, Summer 1999]

Answers: 234. (b), **235.** (c), **236.** (b), **237.** (a), **238.** (d), **239.** (d), **240.** (c), **241.** (a), **242.** (a), **243.** (d), **244.** (c), **245.** (a), **246.** (b), **247.** (c), **248.** (a), **249.** (b), **250.** (a), **251.** (a), **252.** (c), **253.** (b), **254.** (c), **255.** (a), **256.** (c),

257. A moderator material should have
(a) small atomic mass. (b) large atomic mass.
(c) none of the above. **[A.M.I.E. Sec B. Power Systems, Summer 2000]**

258. Graphite is used in nuclear power plant as a
(a) fuel. (b) coolant. (c) moderator. (d) electrode.
[A.M.I.E. Sec B. Summer 1993]

259. Which of the following materials cannot be used as a moderator ?
(a) Deuterium. (b) Graphite.
(c) Heavy water. (d) Beryllium.

260. Consider the following moderators used in nuclear reactors:
1. Graphite. 2. Beryllium.
3. Heavy water.
Their correct sequence in increasing order of their neutron absorption cross section is
(a) 1, 3, 2. (b) 2, 1, 3.
(c) 3, 1, 2. (d) 3, 2, 1. **[U.P.S.C. I.E.S. E.E.-II, 1999]**

261. Which material is used in controlling chain reaction in a nuclear reactor ?
(a) Thorium. (b) Heavy water.
(c) Boron. (d) Beryllium. **[GATE E.E. 1996]**

262. Control rods used in nuclear reactors are made of
(a) zirconium. (b) boron.
(c) beryllium. (d) lead. **[U.P.S.C. I.E.S. E.E.-II, 1997]**

263. In a nuclear reactor, chain reaction is controlled by introducing
(a) iron rods. (b) cadmium rods.
(c) graphite rods. (d) brass rods.
[U.P.S.C. I.E.S. E.E.-II, 2000]

264. Heavy water is used in nuclear power plant as
(a) fuel. (b) coolant.
(c) reflector. (d) moderator.
(e) shield. (f) both (b) and (d)

265. Which of the following materials can be used as a coolant ?
(a) Heavy water. (b) CO_2.
(c) Helium. (d) Any of the above.

266. The function of reflector in a nuclear reactor is to
(a) bounce back most of the neutrons that escape from the fuel core.
(b) reduce the speed of the neutrons.
(c) stop the chain reaction. (d) all of the above.

267. Reflectors of a nuclear reactor are made of
(a) cast iron. (b) beryllium.
(c) steel. (d) boron.

268. Thermal shielding is provided to
(a) absorb the fast neutrons.
(b) protect the operating personnel from exposure to radiations.
(c) prevent the reactor wall from getting heated.
(d) all of the above.

269. The radiation shield for a nuclear power reactor for biological safety is provided by having the reactor:
(a) immersed in water pool. (b) encased by thick metal walls.
(c) encased by thick concrete wall.
(d) isolated from outside word with strong magnetic fields.
[U.P.S.C. I.E.S. E.E.-II, 2011]

270. In boiler water reactor (BWR)
(a) the feed water acts both coolant and a moderator.
(b) enriched uranium is used as fuel.
(c) there is a danger of radioactive contamination of steam.
(d) all of the above.

271. Tarapur atomic power station has
(a) CANDU type reactors. (b) boiling water reactors.
(c) pressurised water reactors.
(d) gas cooled reactors.

272. The thermal efficiency of a boiler water reactor is
(a) greater than. (b) equal to.
(c) less than that of a pressurized water reactor.
[A.M.I.E. Sec B. Power Plant Engineering, Winter 1996]

273. In a pressurized water reactor (PWR)
(a) light or heavy water is used as both coolant and moderator.
(b) water is used at high pressure (as high as 140 kg/cm^2) so that it is available in liquid form at about 280ºC.
(c) the coolant water boils in the reactor core.
(d) both (a) and (b).

274. A CANDU reactor uses
(a) highly enriched uranium as fuel and light water as moderator and coolant.
(b) natural uranium as fuel and heavy water as moderator and coolant.
(c) enriched uranium as fuel and ordinary water as moderator and cooler.
(d) only fertile material.

275. Which one of the following is employed as a moderator by CANDU type of slow thermal nuclear reactors?
(a) Water. (b) Heavy water.
(c) Graphite. (d) Beryllium.
[U.P.S.C. I.E.S. E.E.-II, 2006]

276. Which one of the following fuels is used by the slow thermal nuclear reactors for power generation?
(a) U^{235} (b) U^{238} (c) Th^{232} (d) Pu^{239}
[U.P.S.C. I.E.S. E.E.-II, 2007]

277. Consider the following statements regarding the nuclear power plants:
1. A thermal reactor needs a moderator material.
2. In a nuclear reactor, multiplication factor is kept almost equal to one.
3. Nuclear power plants are used as peak load plants only.
Which of the statements given above are correct?
(a) 1, 2 and 3. (b) 1 and 2 only.
(c) 2 and 3 only. (d) 1 and 3 only.
[U.P.S.C. I.E.S. E.E.-II, 2008]

278. In a thermal nuclear reactor
1. the purpose of moderator is to slowdown fast neutrons produced due to fission.
2. the moderator material must have low molecular weight.
3. ordinary water can be used as moderator with natural uranium as fuel.
4. the multiplication factor is kept slightly greater than unity during its normal functioning.
Of these statements
(a) 1 and 3 are correct. (b) 3 and 4 are correct.
(c) 1, 2 and 3 are correct. (d) 1, 2 and 4 are correct.
[U.P.S.C. I.E.S. E.E.-II, 2001]

279. A heavy water reactor
(a) has higher neutron flux which can be produced at a given power level.
(b) requires less time to shutdown than in the case of graphite because of photo-neutron production.
(c) is not stable.
(d) cannot use natural uranium as fuel.
[A.M.I.E. Sec B. Power Plant Engineering, Winter 1995]

Answers: **257.** (a), **258.** (c), **259.** (a), **260.** (d), **261.** (c), **262.** (b), **263.** (b), **264.** (f), **265.** (d), **266.** (a), **267.** (b), **268.** (d), **269.** (c), **270.** (d), **271.** (b), **272.** (a), **273.** (d), **274.** (b), **275.** (b), **276.** (a), **277.** (b), **278.** (d), **279.** (a),

280. The conversion ratio of a breeder reactor is
(a) less than unity. (b) more than unity.
(c) equal to unity. (d) none of the above.

281. Fast breed reactors are best suited for India owing to
(a) large thorium deposits. (b) large plutonium deposits.
(c) large uranium deposits. (d) none of the above.

282. Gas cooled reactor
(a) uses CO_2 or helium as coolant and graphite as the moderator.
(b) uses coolant at pressure of 14-28 kg/cm^2 and at temperature of 700- 800ºC.
(c) uses heavy water as coolant as well as moderator.
(d) both (a) and (b).

283. A sodium graphite reactor uses
(a) sodium as moderator and graphite as coolant.
(b) sodium as coolant and graphite as moderator.
(c) a mixture of sodium and graphite as coolant.
(d) a mixture of sodium and graphite as moderator.

284. The economic prospects of nuclear power generation in India to a great extent depends on successful development of
(a) fast reactors using enriched uranium as the fuel thus avoiding import of heavy water.
(b) thermal reactors using natural uranium as the fuel and avoiding import of enriched uranium.
(c) breeder reactors using thorium as blanket material and U_{233} as the fuel.
(d) reactor using plutonium as the fuel, natural uranium as blanket and liquid sodium as the coolant. **[U.P.S.C. I.E.S. E.E.-II, 1993]**

285. Compared to steam engines, the internal combustion engines have
(a) much higher thermal efficiency.
(b) almost same thermal efficiency as that of steam engines.
(c) much lower thermal efficiency.
(d) can have lower or higher thermal efficiency.
[A.M.I.E. Sec B. Power Plant Engineering, Winter 1995]

286. The lubrication cost in a diesel power plant is
(a) high. (b) moderate. (c) low.
[A.M.I.E. Sec B. Power Plant Engineering, Winter 1996]

287. Which of the following statements is True ?
(a) Diesel power plants employ only four stroke engines.
(b) Space requirement for diesel power plants in more than that for a conventional steam power plant.
(c) A two stroke engine develops more power for the same speed and piston displacement.
(d) The specific fuel consumption of a diesel power plant decreases with the decrease in load.

288. The diesel engines are available in sizes from
(a) 0–5 kW (b) 5–50 kW
(c) 75–3,750 kW (d) above 10,000 kW

289. The compression ratio in diesel engine may be in the range of
(a) 15–20 (b) 10–15
(c) 5–10 (d) 3–5

290. Air standard efficiency of a diesel engine depends upon
(a) expansion ratio. (b) compression ratio.
(c) fuel used. (d) none of the above.

291. In case of diesel engines, the thermal efficiency is in the range of
(a) 10–20 per cent. (b) 20–30 per cent.
(c) 30–40 per cent. (d) 40–50 per cent.

292. In ideal diesel cycle the working substance is
(a) air. (b) diesel.
(c) mixture of air and diesel. (d) any combustible gas.

293. Air-fuel ratio required for combustion in diesel engine is around
(a) 20 : 1 (b) 15 : 1 (c) 10 : 1 (d) 5 : 1

294. In a diesel engine fuel is injected at a pressure of
(a) 20–25 kg/ cm^2 (b) 50–75 kg/cm^2
(c) 90–130 kg/cm^2 (d) 130–160 kg/cm^2

295. Maximum temperature developed in a diesel engine cylinder is in the range of
(a) 2,000–2,500ºC (b) 1,500–2,000ºC
(c) 1,000–1,500ºC (d) 800–1,000ºC

296. In a diesel engine, the heat lost to the cooling water is about
(a) 50% (b) 30% (c) 20% (d) 15%

297. The highest proportion of total heat supplied to a diesel engine goes to
(a) useful output. (b) heat lost to exhaust gases.
(c) heat lost to cooling water.
(d) heat lost to friction, windage and radiation losses.

298. Which of the following diesel engines have minimum air consumption per bhp ?
(a) 2 stroke, air injection. (b) 4 stroke, air injection.
(c) 4 stroke, mechanical injection.
(d) All of the above almost equal air consumption.

299. Heating value of diesel oil is around
(a) 4,000 kcals/kg. (b) 10,000 kcals/kg.
(c) 15,000 kcals/kg. (d) 20,000 kcals/kg.

300. In a four stroke petrol engine during suction stroke
(a) only petrol is sucked in. (b) only air is sucked in.
(c) mixture of petrol and air is sucked in.
(d) any combustible gas is sucked in.

301. In a four stroke engine, the four operations (suction, compression, expansion and exhaust) are completed in _______ revolution(s) of crankshaft.
(a) eight (b) four (c) two (d) one

302. In a two stroke engine, the four operations (suction, compression, expansion and exhaust) are completed in _______ revolution(s) of crankshaft.
(a) one (b) two (c) four (d) eight

303. Most of the heat generated in IC engines goes to
(a) cooling water. (b) exhaust gases.
(c) friction, windage and radiation loses.
(d) lubricating oil.

304. Reciprocating motion of piston is converted into rotary one by
(a) crank shaft. (b) connecting rod.
(c) gudgeon pin. (d) gear box.

305. A two stroke engine may be identified by
(a) cooling system. (b) lubrication system.
(c) absence of valves. (d) any of the above.

306. The performance of engines of different outputs, speeds and sizes can be compared on the basis of
(a) specific fuel consumption. (b) cylinder volume.
(c) compression ratio. (d) piston size.

307. The thermal efficiency of a diesel engine in comparison to that of a petrol engine is
(a) lower. (b) higher.
(c) the same. (d) may be lower or higher.

Answers: **280.** (b), **281.** (a), **282.** (d), **283.** (b), **284.** (c), **285.** (a), **286.** (a), **287.** (c), **288.** (c), **289.** (a), **290.** (b), **291.** (c), **292.** (a), **293.** (b), **294.** (c), **295.** (a), **296.** (b), **297.** (a), **298.** (c), **299.** (b), **300.** (c), **301.** (c), **302.** (a), **303.** (a), **304.** (b), **305.** (c), **306.** (a), **307.** (b),

308. In a multicylinder engine, the firing order should be in particular sequence because it
(a) will operate ignition system smoothly.
(b) will give smooth turning moment.
(c) will provide the best performance of the engine.
(d) all of the above.

309. High output diesel engines are started by
(a) self starter. (b) compressed air.
(c) battery. (d) cranking.

310. The temperature of cooling water leaving the diesel engine should not exceed
(a) 25°C (b) 40°C (c) 70°C (d) 85°C

311. Air filter is used in
(a) steam power plants. (b) nuclear power plants.
(c) hydropower plants. (d) diesel power plants.

312. For supercharging of diesel engines, the air is supplied through
(a) injector. (b) reciprocating compressor.
(c) centrifugal blower. (d) any of the above.

313. Supercharging of diesel engines
(a) supply pressurized air during suction.
(b) increases the power output of the engine.
(c) reduces the specific fuel consumption.
(d) both (a) and (b).

314. The temperature at which a lubricating oil will give sufficient vapours to form a combustible mixture with air is called the
(a) flash point. (b) pour point.
(c) fire point. (d) combustion point.

315. Which of the following will be the lowest for a lubricating oil ?
(a) Flash point. (b) Pour point.
(c) Fire point. (d) Combustion point.

316. Fire and flash points of oils are important from the point of view of
(a) heating value. (b) viscosity.
(c) ignition and storage hazards.
(d) tendency of freezing during winter.

317. For larger output diesel electric power plants the engines used are
(a) horizontal type.
(b) vertical type, multicylinder and supercharged.
(c) air cooled. (d) low rpm.

318. The gas turbine power plants have the advantages of
(a) simplicity of design and installation, high reliability, compactness, clean exhaust and low initial cost.
(b) quick starting and capability of putting to share full load in few minutes.
(c) high overall efficiency, low specific consumption and long life in comparsion to that of steam power plants.
(d) both (a) and (b).

319. The main drawbacks of gas turbine power plants are
(a) low overall efficiency, noisy operation and limited unit capacity.
(b) inability of using coal or heavy residual petroleum as fuels.
(c) high initial cost, poor reliability and large space requirements.
(d) both (a) and (b).

320. A gas turbine works on
(a) Carnot cycle. (b) Brayton cycle.
(c) dual cycle. (d) Rankine cycle.
(e) regenerative cycle.
[A.M.I.E. Sec B. Summer 1994]

321. Gas turbine is widely used in
(a) locomotives. (b) automobiles.
(c) aircrafts. (d) pumping stations.

322. Gas turbine power plants are not widely used
(a) as peak load plants. (b) as standby power plants.
(c) as base load plants.
(d) in combination with steam power plants.

323. In comparsion to steam power plants, the gas power plants have
(a) low initial cost but higher operating cost.
(b) high initial cost but low operating cost.
(c) both initial as well operating costs high.
(d) both initial as well as operating costs low.

324. In which of the following places 70 MW gas turbine power plant being operated as base load plant ?
(a) Nangal (H.P.). (b) Namrup (Assam).
(c) Jharia (Bihar). (d) Kokkata (W.B.).

325. The gas turbine power plants mainly use
(a) coal and peat.
(b) kerosine oil, diesel oil, residual oil.
(c) gas oil.
(d) natural gas and liquid petroleum fuels.

326. Compressor used in gas turbine is
(a) reciprocating compressor.
(b) screw compressor.
(c) multistage axial flow compressor.
(d) either (a) or (b).

327. The blades of gas turbines are made of
(a) stainless steel. (b) carbon steel.
(c) high alloy steel. (d) high nickel alloy.

328. The pressure ratio in a gas turbine power plant is
(a) 5 to 6 (b) 9 to 12 (c) 2 to 3
[A.M.I.E. Sec B. Power Plant E., Winter 1996]

329. The highest temperature ever used for gas turbine power plants is about
(a) 1,000°C (b) 500°C (c) 2,000°C
[A.M.I.E. Sec B. Power Plant Engineering, Winter 1996]

330. The heating value of gaseous fuels is about
(a) 300 kJ/litre. (b) 100 kJ/litre.
(c) 30 kJ/litre. (d) 10 kJ/litre.

331. Which of the following statements is not correct ?
(a) The efficiency of modern gas turbine power plants is more than 50%.
(b) One advanced feature of modern gas turbine power plants is higher firing temperature.
(c) The installed capacity of gas turbine power plants in India exceeds 20% of total installed generating capacity.
(d) The gas turbine power plants have the advantages of quick starting and capability of putting to share full load in few minutes.

332. Which of the following statements is correct ?
(a) In a gas turbine plant, the combustion takes place in the compressor.
(b) Aeroderivative turbines use clean fuel.
(c) Capital cost of a gas turbine power plant is higher than that of a steam power plant.
(d) Gas turbine power plants have the drawbacks of poor reliability and large space requirements.

333. The compressor has to be started
(a) before starting of gas turbine.
(b) after starting of gas turbine.
(c) simultaneously with starting of gas turbine.
(d) at any time (*i.e.* before or after starting of gas turbine).

Answers: **308.** (c), **309.** (b), **310.** (c), **311.** (d), **312.** (c), **313.** (d), **314.** (a), **315.** (b), **316.** (c), **317.** (b), **318.** (d), **319.** (d), **320.** (b), **321.** (c), **322.** (c), **323.** (a), **324.** (b), **325.** (d), **326.** (c), **327.** (d), **328.** (a), **329.** (a), **330.** (c), **331.** (c), **332.** (b), **333.** (a),

334. For starting gas turbine, its rotor is usually motored up to
(a) 10 per cent of turbine rated speed.
(b) 25 per cent of turbine rated speed.
(c) half of turbine rated speed. (d) turbine rated speed.

335. In comparsion to diesel engines the gas turbines have maximum combustion pressure
(a) less. (b) same.
(c) more. (d) unpredictable.

336. For the same cylinder size and speed which of the following engines will develop more power ?
(a) Petrol engine. (b) Diesel engine.
(c) Supercharged engine. (d) Gas engine.

337. Greater part of the power developed by the turbine is used in driving the compressor. It is around
(a) 75% (b) 65% (c) 45% (d) 35%

338. The work ratio of closed cycle gas turbine power plant depends upon
(a) only on pressure ratio.
(b) temperature ratio of the cycle and specific heat ratio.
(c) pressure ratio, temperature ratio and specific heat ratio.
(d) pressure ratio and specific heat ratio.

339. Overall efficiency of gas turbine is
(a) less than that of diesel cycle.
(b) more than that of diesel cycle.
(c) more than that of otto cycle.
(d) equal to that of Rankine or Carnot cycle.

340. In comparison to diesel power plant the gas turbine power plant has
(a) lower overall thermal efficiency.
(b) higher overall thermal efficiency.
(c) the same overall thermal efficiency.
(d) unpredictable.

341. Mechanical efficiency of a gas turbine in comparison to that of reciprocating IC engine is
(a) lower. (b) higher.
(c) the same. (d) unpredictable.

342. Thermal efficiency of a gas turbine improves because of
(a) reheating of gas after partial expansion.
(b) removing the heat from compressed air between the stages of compression.
(c) utilizing the heat of exhaust gases to heat the compressed air before it is sent to combustion chamber.
(d) all of the above.

343. The thermal efficiency of a gas turbine cycle with ideal regenerative heat exchanger is
(a) less than work ratio. (b) more than work ratio.
(c) equal to work ratio. (d) unpredictable.

344. With the increase in pressure ratio thermal efficiency of a simple gas turbine power plant with fixed turbine inlet temperature
(a) increases. (b) reduces.
(c) first increases and then falls.
(d) first falls and then increases.

345. The combined cycle power plants are more appropriate for
(a) base loads. (b) peak loads.
(c) intermediate loads. (d) for both base and peak loads.

346. The combined cycle power plants are more appropriate for capacity factors of
(a) less than 20% (b) 20 to 50%
(c) 50 to 70% (d) 100%

347. In a two stage gas turbine plant reheating after first stage
(a) improves thermal efficiency but reduces the work ratio.
(b) increases work ratio but reduces thermal efficiency.
(c) reduces both of the work ratio and thermal efficiency.
(d) increases both of the work ratio and thermal efficiency.

348. In a two stage gas turbine power plant intercooling and reheating
(a) improves the work ratio but reduces the thermal efficiency.
(b) reduces the work ratio but improves the thermal efficiency.
(c) improves both of the work ratio and thermal efficiency.
(d) reduces both of the work ratio and thermal efficiency.

349. A jet aircraft is powered by
(a) petrol engine. (b) diesel engine.
(c) gas turbine. (d) solar cells.
(e) batteries. **[A.M.I.E. Sec B. Summer 1994]**

350. For a jet propulsion unit, ideally the work of compressor and that of gas turbine are
(a) different. (b) equal.
(c) not related to each other. (d) unpredictable.

351. In a closed cycle operation of gas turbines
(a) load variation is affected by controlling the absolute pressure and the mass flow of the circulating air.
(b) high efficiency is maintained over the complete range of operating loads.
(c) maximum unit capacity is increased.
(d) all of the above.

352. Coal and peat is not generally used in gas turbines as fuel because of
(a) coal handling and ash handling problems.
(b) their higher costs. (c) their non-availability.
(d) all of the above.

353. The gas turbine power plant can be used in combination with steam power plants by
(a) using exhaust gases from gas turbine power plant in heating of feed water coming from the steam turbine condenser.
(b) using exhaust gases from gas turbine power plant as combustion air in steam boiler.
(c) using gas from a supercharged boiler for expansion in gas turbine.
(d) any of the above.

354. In an MHD generator the conductor is made of
(a) copper or aluminium. (b) liquid metal.
(c) gas. (d) liquid metal or gas.

355. The nature of the current developed in MHD generator is
(a) ac. (b) dc. (c) either dc or ac.

356. Power output per unit volume of an MHD generator is proportional to
(a) specific electrical conductivity of gas.
(b) square of the magnetic field strength.
(c) square of the fluid velocity.
(d) all of the above.

357. Seeding material is injected into the working fluid, in MHD generator, in order to
(a) increase the conductivity of the gas.
(b) reduce the conductivity of the gas.
(c) reduce the temperature of the gas.
(d) none of the above.

Answers: 334. (c), **335.** (a), **336.** (c), **337.** (b), **338.** (c), **339.** (a), **340.** (a), **341.** (b), **342.** (d), **343.** (c), **344.** (c), **345.** (d), **346.** (b), **347.** (b), **348.** (a), **349.** (c), **350.** (b), **351.** (d), **352.** (a), **353.** (d), **354.** (d), **355.** (b), **356.** (d), **357.** (a),

358. Seeding material commonly used in MHD generator is
(a) sodium. (b) aluminium.
(c) potassium. (d) cesium.

359. Which of the following liquid metal is not used as an MHD working fluid ?
(a) Potassium. (b) Sodium
(c) Lithium (d) all of the above

360. The exhaust from an MHD generator is at a temperature of about
(a) 800°C (b) 2,200°C (c) 3,500°C (d) 1,200°C

361. In an MHD generator the ionization potential of the seeding material should be
(a) very high. (b) the lowest possible.
(c) neither high or low. (d) either high or low.

362. An MHD system for efficient operation needs
(a) a magnet capable of producing high magnetic flux density.
(b) injection and recovery of seed material
(c) arrangement for superheating of air.
(d) all of the above.

363. The advantage(s) of an MHD system is/are
(a) low pollution. (b) good efficiency.
(c) conservation of fuel. (d) absence of moving parts.
(e) all of the above.

364. The concept of MHD generation depends much more on the _____ of the gas in the duct.
(a) permeability (b) permittivity
(c) conductivity (d) none of these

365. The materials used for electrodes in thermoelectric generators are
(a) metal conductors. (b) carbon.
(c) semiconductors. (d) insulators.

366. Thermionic convertor utilizes
(a) Seebeck effect. (b) thermionic emission effect.
(c) Peltier effect. (d) Thomson effect.

367. The solar or photovoltaic cell converts
(a) chemical energy into electrical energy.
(b) solar radiations into electrical energy.
(c) solar radiations into thermal energy.
(d) thermal energy into electrical energy.

368. Photovoltaic solar energy conversion system makes use of
(a) solar pond. (b) fuel cell.
(c) Edison cell. (d) none of the above.

369. The efficiency of solar cell is about
(a) 25% (b) 15% (c) 40% (d) 60%

370. Solar cells are made of
(a) aluminium. (b) germanium.
(c) silicon. (d) cadmium.

371. The voltage of a solar cell is
(a) 0.5–1 V (b) 1–2 V (c) 2–3 V (d) 4–5 V

372. The output of a solar cell is of the order of
(a) 0.5 W (b) 1.0 W (c) 5.0 W (d) 10.0 W

373. Solar cells, for power generation, have the mains drawback(s) of
(a) low efficiency. (b) lack of availability.
(c) high cost and maintenance problems.
(d) all of the above.

374. A module is a
(a) series-arrangement of solar cells.
(b) parallel arrangement of solar cells.
(c) series-parallel arrangement of solar cells.
(d) none of the above.

375. For satellites the source of energy is
(a) solar cell. (b) fuel cell.
(c) Edison cell. (d) cryogenic storage.

376. Fuel cells have conversion efficiencies of about
(a) 25% (b) 40% (c) 65% (d) 85%

377. Fuel cell converts chemical energy into
(a) heat energy.
(b) low-voltage direct current electrical energy.
(c) low-voltage alternating current electrical energy.
(d) mechanical energy. **[A.M.I.E. Sec B. Winter 1993]**

378. Fuel cells for power generation have the drawbacks of
(a) very high development costs.
(b) low service life and low voltage.
(c) noise, pollution and maintenance problems.
(d) both (a) or (b).

379. The disadvantage(s) of renewable sources of energy is/are
(a) intermittency. (b) lack of dependability.
(c) availability in low energy densities.
(d) all of the above.

380. Biogas consists of
(a) only methane.
(b) methane and carbon dioxide with some impurities.
(c) only ethane. (d) a special organic gas.

381. The main source(s) of production of biogas is /are
(a) wet cow dung. (b) human waste.
(c) wet livestock waste. (d) all of the above.

382. The main by-product of the biogas plant is
(a) biomass. (b) biogas.
(c) organic manure. (d) none of the above.

383. Biogas plants are suitable for
(a) metallurgical industries. (b) commercial complexes.
(c) rural areas. (d) coal mines.
[A.M.I.E. Sec B. Summer 1993]

384. The energy radiated by Sun on a bright sunny day is about
(a) 2.5 kW/m^2 (b) 1.0 kW/m^2
(c) 500 W/m^2 (d) 200 W/m^2

385. Most of solar radiations received on earth surface is within the range of
(a) 0.25-0.4 micron. (b) 0.4-0.8 micron.
(c) 0.6-0.95 micron. (d) 0.1-0.25 micron.

386. Which of the following area is preferred for solar plants ?
(a) Coastal areas. (b) Hot arid zones.
(c) Mountain tops. (d) High rainfall zones.

387. In solar power plants the solar heat is transferred to
(a) molten salts. (b) liquid metals.
(c) water steam. (d) any of the above.

388. Main applications of solar energy may be considered as
(a) direct thermal applications.
(b) fuel from biomass. (c) solar-electric applications.
(d) all of the above.

389. Insulation is referred to as
(a) diffuse radiations received at any time.
(b) direct radiations received at any time.
(c) both of the above. (d) none of the above.

Answers: **358.** (c), **359.** (b), **360.** (b), **361.** (b), **362.** (d), **363.** (e), **364.** (c), **365.** (c), **366.** (b), **367.** (b), **368.** (a), **369.** (b), **370.** (c), **371.** (a), **372.** (b), **373.** (c), **374.** (c), **375.** (a), **376.** (c), **377.** (b), **378.** (d), **379.** (d), **380.** (b), **381.** (d), **382.** (c), **383.** (c), **384.** (b), **385.** (a), **386.** (b), **387.** (d), **388.** (d), **389.** (c),

390. Solar thermal power generation can be had by using
(a) flat plate collectors.
(b) focussing or concentrating collectors.
(c) solar ponds. (d) any of the above.

391. The function of a solar collector is of converting solar energy into
(a) radiations. (b) electrical energy directly.
(c) thermal energy. (d) any of the above.

392. In a solar collector, the transparent cover is provided to
(a) protect the collector from dust.
(b) reduce the heat losses from collector beneath to atmosphere.
(c) transmit solar radiation only .
(d) all of the above.

393. Flat plate collector
(a) has plate blackened in order to absorb maximum amount of direct radiations.
(b) is placed at a proper inclination to the Sun so as to absorb solar radiations.
(c) is mainly used for water heating,space-heating, and drying.
(d) all of the above.

394. The flat plate collector gives a
(a) temperature of about 90°C with an efficiency of 30–35%.
(b) temperature of about 120°C with an efficiency of 45%.
(c) temperature of about 150°C with an efficiency of 52% .
(d) none of the above.

395. Temperature attained by cylindrical parabolic collector is of the order of
(a) 50–100°C (b) 100–150°C
(c) 150–200°C (d) 200–300°C and above.

396. Reflector mirrors employed for exploiting solar energy are called the
(a) mantle. (b) heliostats. (c) diffusers. (d) ponds.

397. Sun tracking is required in case of
(a) cylindrical parabolic and paraboloid.
(b) flat plate collector. (c) both (*a*) and (*b*).
(d) none of the above.

398. A pyrometer can be used for measurement of
(a) diffuse radiations only. (b) direct radiations only.
(c) both, direct as well as diffuse radiations.
(d) none of the above.

399. Wind as a source of power
(a) is non-steady and unreliable.
(b) does not possess the basic requirements of any energy source.
(c) is plentiful, inexhaustible, renewable. and non-polluting.
(d) all of the above.

400. The drawback(s) of wind power plant is/are
(a) unreliability and non-steadiness.
(b) unwieldy size, high structural area and quite large finance requirements.
(c) output voltage and frequency fluctuations.
(d) all of the above.

401. Maximum wind energy available is proportional to
(a) air density. (b) cube of the wind velocity.
(c) square of the rotor diameter.
(d) all of the above.

402. Under favourable conditions, use of wind power for _______ is possible.
(a) small sized plants (b) medium sized plants
(c) large scale generation (d) (a) or (b).

403. Tidal power schemes, could not be found economically justified because of
(a) high cost of civil engineering works.
(b) non-availability of tidal energy in India.
(c) both (a) and (b).

404. Tidal energy utilizes
(a) kinetic energy of water. (b) potential energy of water.
(c) both potential as well as kinetic energy of water.
(d) none of the above.

405. The turbine normally employed in tidal power plants is
(a) simple impulse type. (b) propeller type.
(c) reaction type. (d) reversible type.

406. A tidal power plant is installed in India near
(a) Bay of Bengal. (b) Visakhapatnam.
(c) Goa. (d) Gulf of Cambay.
[A.M.I.E. Sec B. Summer 1994]

407. Geothermal energy is
(a) inexhaustible energy source. (b) a renewable energy source.
(c) alternative energy source. (d) all of the above.

408. A geothermal field may yield
(a) dry steam. (b) wet steam.
(c) hot water. (d) all of the above.

409. Geothermal steam and hot water may contain
(a) NH_3 (b) CO_2 (c) H_2S
(d) H_2S, CO_2, NH_3 and random gas.

410. When geothermal energy is available in the form of saline water, power is developed using
(a) binary cycle system. (b) flashed-steam system.
(c) total flow system. (d) any of the above.

411. In hydrothermal source of geothermal energy
(a) hot water or steam is available.
(b) molten leva is available. (c) hot gases are available.
(d) all of the above.

412. A potential geothermal source region should have
(a) low thermal gradient. (b) high thermal gradient.
(c) high thermal conductivity.
(d) none of the above.

413. In hydrothermal systems when steam, water and dissolved solids are available as source of energy, the entrained solid are removed by
(a) strainers. (b) centrifugal separators.
(c) filters. (d) any of the above.

414. In comparison to conventional steam power plants geothermal power plants have
(a) lower load factor. (b) the same load factor.
(c) higher load factor. (d) unpredictable.

415. Presence of sand in geopressured water is likely to cause
(a) erosion problem. (b) water circulation problem.
(c) heat exchange problem. (d) none of the above.

416. Presence of non-condensable gases in geo-pressured water causes
(a) pollution problem. (b) corrosion problem.
(c) flow problem. (d) all of the above.

417. Petrothermal systems are composed of hot dry rock with
(a) petrochemicals. (b) no underground water.
(c) large underground water.(d) dense gases.

418. In petrothermal systems of geothermal energy having hot dry rock but no underground water, the energy is obtained by

---

Answers: **390.** (d), **391.** (c), **392.** (d), **393.** (d), **394.** (a), **395.** (d), **396.** (b), **397.** (a), **398.** (c), **399.** (d), **400.** (d), **401.** (d), **402.** (d), **403.** (a), **404.** (b), **405.** (d), **406.** (d), **407.** (d), **408.** (d), **409.** (d), **410.** (d), **411.** (a), **412.** (b), **413.** (b), **414.** (c), **415.** (a), **416.** (d), **417.** (b), **418.** (c),

(a) creating water wells. (b) circulating compressed air.
(c) pumping water. (d) any of the above.

419. In geothermal power plants waste water is
(a) discharged back to earth. (b) discharged into sea.
(c) recirculated after cooling in cooling towers.
(d) evaporated in ponds.

420. In a fuel cell cathode is of
(a) oxygen. (b) ammonia.
(c) hydrogen. (d) carbon monoxide.

421. In a fuel cell electrical energy is produced by
(a) reaction of hydrogen with oxygen.
(b) thermionic action.
(c) combustion of fuel in absence of oxygen.
(d) any of the above.

422. Direct conversion of heat energy into electrical energy is possible through
(a) MHD generators. (b) thermionic converters.
(c) fuel or solar cell. (d) both (a) and (b).

423. Which of the following power plants can generate power at unpredictable or uncontrollable time ?
(a) Tidal power plant. (b) Wind power plant.
(c) Solar power plant. (d) Any of the above.

424. For under water movement of a submarine the power is provided by
(a) tidal energy. (b) MHD generator.
(c) batteries. (d) diesel engine.
(e) thermionic generator. **[A.M.I.E. Sec B. Winter 1993]**

425. The power station that does not require any moving part is
(a) Thermionic convertor. (b) Wind power generation.
(c) Tidal power plant. (d) Solar power generation.

426. A generating station which has a high investment cost and low operating cost is usually operated as a
(a) peak load station. (b) base load station.
(c) medium load station. (d) None of the above.
[A.M.I.E. Sec B. Summer 1994]

427. Which of the following power plants normally operates at high speeds ?
(a) Hydroelectric. (b) Steam-turbine.
(c) Petrol engine. (d) Diesel engine.

428. Which of the following power plants is free from environmental problems ?
(a) Diesel engine. (b) Nuclear.
(c) Hydroelectric. (d) Steam.

429. Which power plant cannot have single unit of 100 MW capacity ?
(a) Steam power plant. (b) Nuclear power plant.
(c) Hydroelectric power plant.
(d) Diesel engine power plant.
[A.M.I.E. Sec B. Power Systems, Winter 1999]

430. Which of the following plants will take least time in starting from cold conditions to full-load operation ?
(a) Nuclear power plant. (b) Steam power plant.
(c) Hydroelectric plant. (d) Gas turbine plant
[A.M.I.E. Sec B. Power Systems, Winter 1999]

431. Which of the following generating plants has the minimum operating cost ?
(a) Steam plant. (b) Hydroelectric plant.
(c) Nuclear plant. (d) Diesel plant.
[A.M.I.E. Sec B. Power Systems, Winter 1998]

432. Which of the following plants has the maximum capital cost ?
(a) Steam plants. (b) Hydro-plants.
(c) Diesel plants. (d) Nuclear plants.
[A.M.I.E. Sec B. Power Systems, Winter 1996]

433. In a steam power plant, which of its following components needs maximum maintenance attention ?
(a) Condenser. (b) Boiler.
(c) Steam turbine. (d) Superheater.

434. A steam power plant is being supplied with coal having much more ash content than that for which it was designed. Which of its following units needs major modifications ?
(a) Ash handling unit. (b) Pulverising unit.
(c) Condenser. (d) Cooling towers.

435. A diesel power plant is generally used as a /an
(a) base load plant. (b) peak load plant.
(c) standby plant/emergency plant.
(d) any of the above.

436. A gas turbine power plant is best suited for
(a) base loads. (b) peak loads.
(c) emergency. (d) none of the above.

437. For the same cylinder dimensions and same rpm, which of the following engines will produce least power ?
(a) Petrol engine. (b) Diesel engine.
(c) Supercharged engine.

438. Ships are usually supplied power by
(a) steam accumulators. (b) hydraulic turbine.
(c) diesel engines. (d) none of the above.

439. Gas turbine power plants for maximum efficiency may have
(a) heat exchangers. (b) multistage compression.
(c) reheating. (d) all of the above.

440. Air will not be the working substance in a
(a) closed cycle gas turbine. (b) open cycle gas turbine.
(c) petrol engine. (d) diesel engine.

441. Which of the following plants can never have 100 percent load factor ?
(a) Nuclear power plant. (b) Hydroelectric power plant.
(c) Base load plant. (d) Peak load plant.

442. Which of the following power plants needs highly skilled/qualified engineers for its operation ?
(a) Hydro. (b) Steam. (c) Nuclear. (d) Gas turbine.

443. Which of the following power plants will take long period in erection and installation ?
(a) Steam. (b) Hydro. (c) Nuclear. (d) Gas-turbine.

444. Major share of power generated in India is through
(a) hydroelectric power plants.
(b) nuclear power plants.
(c) thermal power plants. (d) gas turbine power plants.

445. When a 'pumped storage' power plant is operated in conjunction with a steam power plant
(a) The operating cost of the steam plant becomes optimum.
(b) Load factor of the steam plant is increased.
(c) Chances of the tripping of the system decrease because of the use of two heterogeneous types of plants.
(d) Operation of the storage plant in the pumping-mode during the low-load period improves the steam-plant stability.
[U.P.S.C. I.E.S. E.E.-II, 2003]

Answers: 419. (a), **420.** (c), **421.** (a), **422.** (d), **423.** (d), **424.** (c), **425.** (a), **426.** (b), **427.** (b), **428.** (c), **429.** (d), **430.** (c), **431.** (b), **432.** (d), **433.** (b), **434.** (a), **435.** (c), **436.** (b), **437.** (a), **438.** (c), **439.** (d), **440.** (a), **441.** (d), **442.** (c), **443.** (b), **444.** (c), **445.** (b),

446. Out of the following plant categories
(i) Nuclear (ii) Run-off-river
(iii) Pump Storage (iv) Diesel
The base load power plants are
(a) (i) and (ii). (b) (ii) and (iii).
(c) (i), (ii) and (iv). (d) (i), (iii) and (iv).
[GATE E.E. 2009]

447. The essential requirement(s) for power plants to be operated as base load plants is /are
(a) low operating cost.
(b) capability of operating continuously for long periods.
(c) requirement of few operating personnel and economical repair.
(d) all of the above.

448. The generating station suitable to operate as peak load plant is
(a) Thermal power station. (b) Nuclear power station.
(c) Pumped storage power station.
(d) none of these. **[W.B. Univ. of Technology Power System III, 2009]**

449. Which of the following power stations is mainly used to cover peak load on the system?
(a) Coal based thermal power plant.
(b) Nuclear power plant.
(c) Gas based thermal power plant.
(d) Pumped storage hydropower plant. **[U.P.S.C. I.E.S. E.E.-II, 2010]**

450. The essential requirement(s) for power plants to be operated as peak load plants are
(a) capability of quick start, synchronisation and taking up of system load.
(b) quick response to load variations.
(c) low capital cost. (d) all of the above.

451. Efficiency is of little importance in case of
(a) base load power plants.
(b) peak load plants.
(c) standby plants. (d) both (b) and (c).

452. The annual cost of generating station can be expressed in the form of Rs (A + B × kW + C × kWh) where A, B, C are constants and kW and kWh represent capacity of the station and energy generated per year respectively. The choice between the "base-load station" and the "peak-load station" basically depends on the fact that
(a) factor B should be less for the peak-load station and factor C should be less for the base-load station.
(b) factor C should be less for the peak-load station and factor B should be less for the base load station.
(c) the factors B and C should both be less for the base-load station.
(d) the factors B and C should both be less for the peak-load station. **[U.P.S.C. I.E.S. E.E.-II, 1993]**

453. The modern trend in electric power generation is
(a) to have a large number of small size thermal plants located at different places.
(b) to have large size thermal plants near load centre.
(c) to have large size thermal plants located near coal fields.
(d) none of the above. **[A,M.I.E. Sec B. Power Systems, Winter 1998]**

454. Which of the following are the advantages of interconnected operation of power systems ?
1. Less reserve capacity requirement.
2. More reliability.
3. High power factor.
4. Reduction in short-circuit level.
Select the correct answer using the codes given below:
(a) 1 and 2. (b) 2 and 3.
(c) 3 and 4. (d) 1 and 4.
[U.P.S.C. I.E.S. E.E.-II, 1998]

455. Interconnected systems have the advantage(s) of
(a) reduced reserve plant capacity, capital cost per kW and economy in operation.
(b) improved load factor, diversity factor and operation efficiency and increased reliability of supply.
(c) all of the above. (d) none of the above.

456. An interconnected system has the following power plants
1. Nuclear, 2. Steam, 3. Diesel, 4. Gas turbine, 5. Hydro with storage, 6. Run-off the river, 7. Pumped storage.
Which of the above three power plants are used exclusively for supplying peak loads ?
(a) 1, 2, 3 (b) 3, 4, 7
(c) 4, 5, 6 (d) 2, 3, 5

457. Which two of the power plants mentioned in Q. 440 are used exclusively for supplying base load ?
(a) 1, 5 (b) 1, 2 (c) 1, 6 (d) 4, 6

458. In a moderate size steam power station, electrical power is generated at
(a) 440 V (b) 1.1 kV (c) 11 kV (d) 33 kV

459. If two synchronous generators are connected, loss of synchronism will result in
(a) stalling of generators. (b) wild fluctuations in current.
(c) wild fluctuations in current and voltage.
(d) none of the above. **[A.M.I.E. Sec B. Summer 1995]**

460. A large synchronous generator is feeding power into an infinite bus at slightly lagging power factor. If a total loss of field occurs and the system can supply sufficient reactive power without a large terminal voltage drop, the unit will
(a) continue to run as a synchronous generator and no tripping is necessary.
(b) get short-circuited and it should be tripped instantaneously.
(c) run as an induction generator and it should be tripped after a time delay.
(d) run as a synchronous motor and it should be tripped after a time delay. **[U.P.S.C. I.E.S. E.E.-II, 1993]**

461. Unit of regulation of speed governor is
(a) Hz/MW (b) MW/Hz
(c) Unitless (d) km/sec
[W.B. Univ. of Technology Power System III 2009-10]

462. A turbine-generator set has a regulation constant of 6% on the generator rating of 120 MVA, 50 Hz. The generator frequency decreases by 0.01 Hz. The increase in turbine output for steady-state operation is:
(a) 0.8 MW (b) 0.4 MW
(c) 0.2 MW (d) 40 MW
[U.P.S.C. I.E.S. E.E.-II, 2011]

463. Three identical generators supply power in a system having lossless transmission lines. Generator 1 is equipped with a speed governor that maintains its speed constant at the rated value while generators 2 and 3 have governors with droops of 5% and 4% respectively. For a given increase in system load in the steady state
(a) Generators 1, 2, 3 will share the increased load in the ratio of 0 : 5 : 4.
(b) Generators 1, 2, 3 will share the increased load equally.
(c) Generators 1, 2, 3 will share the increased load in the ratio of 0 : 4 : 5.

Answers: **446.** (a), **447.** (d), **448.** (c), **449.** (d), **450.** (d), **451.** (d), **452.** (a), **453.** (c), **454.** (a), **455.** (c), **456.** (b), **457.** (c), **458.** (c), **459.** (a), **460.** (c), **461.** (a), **462.** (b), **463.** (d),

(d) Generator 1 will alone take the entire increased load and the output of generators 2 and 3 will remain unchanged. [U.P.S.C. I.E.S. E.E.-II, 2010]

464. Three generators are feeding a load of 100 MW. The details of the generators are

	Rating (MW)	Efficiency(%)	Regulation (pu) on 100 MVA base
Generator-1	100	20	0.02
Generator-2	100	30	0.04
Generator-3	100	40	0.03

In the event of increased load power demand, which of the following will happen?
(a) All the generators will share equal power
(b) Generator-3 will share more power compared to Generator-1
(c) Generator-1 will share more power compared to Generator-2
(d) Generator-2 will share more power compared to Generator-3 [GATE E.E. 2009]

465. Two generators rated at 200 MW and 400 MW are operating in parallel. Both the governors have a drop of 4% when the total load is 300 MW. They share the load as (suffix 1 is used for generator 200 MW and suffix 2 is used for generator 400 MW)
(a) P_1 = 100 MW and P_2 = 200 MW
(b) P_1 = 150 MW and P_2 = 150 MW
(c) P_1 = 200 MW and P_2 = 100 MW
(d) P_1 = 200 MW and P_2 = 400 MW [U.P.S.C. I.E.S. E.E.-II, 1999]

466. For stable operation of interconnected system, the passive element that can be used as the interconnecting element is
(a) reactor. (b) resistor. (c) capacitor.
(d) resistor and capacitor. [A.M.I.E. Sec B. Winter 1993, Summer 1995]

467. To meet the reactive power requirements at load centres usually
(a) shunt capacitors are used.
(b) series capacitors are used.
(c) tap changing transformers are used.
(d) shunt reactors are used. [A.M.I.E. Sec B. Summer 1994]

468. A power system needs injection of VARs at
(a) off-peak load. (b) peak load.
(c) both peak and off-peak load.
(d) full load.

469. The injection of VARs is required to
(a) compensate for line losses.
(b) get a good voltage profile.
(c) increase the voltage at the receiving end.
(d) all of the above.

470. Shunt capacitors in a substation
(a) consume lagging VARs. (b) deliver lagging VARs.
(c) consume active power. (d) deliver active power.
[A.M.I.E. Sec B. Power Systems, Summer 1993]

471. In a 400 kV power network, 360 kV is recorded at a 400 kV bus. The reactive power absorbed by a shunt reactor rated for 50 MVAR, 400 kV connected at the bus is
(a) 61.73 MVAR (b) 55.56 MVAR
(c) 45.0 MVAR (d) 40.5 MVAR
[U.P.S.C. I.E.S. E.E.-II, 1995]

472. The principle of incremental costs is employed for deciding the
(a) sequence of adding units.
(b) load allocation between units in operation.
(c) total plant capacity to be operated.
(d) all of the above.

473. For a thermal power plant the input output characteristic is given by
$Q = 8 + P + 0.08\,P^2$
where Q and P are heat input and power output, respectively in MW. The maximum thermal efficiency would be
(a) 26.5% (b) 30.5% (c) 34.5% (d) 38.5%
[A.M.I.E. Sec B. Power Plant Engineering, Winter 1994]

474. For economy measure the generators at a power plant operate at
(a) equal loads. (b) load proportional to the ratings.
(c) equal incremental cost. [A.M.I.E. Sec B. Power Systems, Summer 1999]

475. Two generating plants feed a load centre through a transmission network. For maximum economy
(a) the incremental cost of power supplied at the load centre should be the same for both the plants.
(b) the incremental fuel cost should be the same for both the plants.
(c) the more efficient plant should share more load.
(d) none of the above.

476. In the optimum generator scheduling different power plants, the minimum fuel cost is obtained when
(a) only the incremental fuel cost of each plant is the same.
(b) the penalty factor of each plant is the same.
(c) the ratio of the incremental fuel cost to the penalty factor of each plant is the same.
(d) the incremental fuel cost of each plant multiplied by its penalty factor is the same. [U.P.S.C. I.E.S. E.E.-II, 1994]

477. In terms of power generation and B_{mn} coefficients, the transmission loss for a two-plant system is (notations having their usual meanings)
(a) $P_1^2 B_{11} + 2P_1P_2 B_{12} + P_2^2 B_{22}$
(b) $P_1^2 B_{11} - 2P_1P_2 B_{12} + P_2^2 B_{22}$
(c) $P_2^2 B_{11} + 2P_1P_2 B_{12} + P_1^2 B_{22}$
(d) $P_1^2 B_{11} + P_1P_2 B_{12} + P_2^2 B_{22}$ [U.P.S.C. I.E.S. E.E.-II, 2000]

478. If the generating station is situated very close to the load centre, the penalty factor for this unit is
(a) zero. (b) almost unity.
(c) negative. (d) very high.

479. The cost function of a 50 MW generator is given by (P_1 is the generating loading)
$F(P_1) = 225 + 53\,P_1 + 0.02\,P_1^2$
When 100% loading is applied, the incremental fuel cost (IFC) will be
(a) Rs 55 per MWh (b) Rs 55 per MW
(c) Rs 33 per MWh (d) Rs 33 per MW
[U.P.S.C. I.E.S. E.E.-II, 2000]

480. Two generating stations connected to a load centre having capacity of 50 MVA and 75 MVA deliver 100 MW to the load. The incremental cost of plant 1 is $15 + 0.15P_1$ and that of the plant 2 is $18 + 0.15\,P_2$. What are the values of P_1 and P_2, respectively?
(a) 60 MW and 40 MW (b) 50 MW each.
(c) 72 MW and 28 MW (d) 30 MW and 70 MW
[U.P.S.C. I.E.S. E.E.-II, 2008]

481. The incremental cost curves in Rs/MWhr for two generators supplying a common load of 700 MW are shown in the figures. The maximum and minimum generation limits are also indicated. The optimum generation schedule is:

Answers: 464. (c), **465.** (a), **466.** (a), **467.** (a), **468.** (c), **469.** (b), **470.** (b), **471.** (d), **472.** (b), **473.** (b), **474.** (c), **475.** (a), **476.** (d), **477.** (a), **478.** (b), **479.** (a), **480.** (b), **481.** (c),

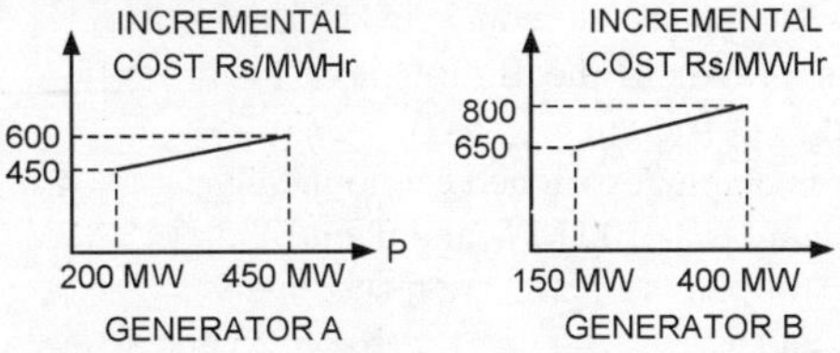

(a) Generator A : 400 MW, Generator B : 300 MW
(b) Generator A : 350 MW, Generator B : 350 MW
(c) Generator A : 450 MW, Generator B : 250 MW
(d) Generator A : 425 MW, Generator B : 275 MW

[GATE E.E. 2007]

482. A lossless power system has to serve a load of 250 MW. There are two generators (G_1 and G_2) in the system with cost curves C_1 and C_2 respectively defined as follows:

$$C_1(P_{G1}) = P_{G1} + 0.055 \times P_{G1}^2$$

$$C_2(P_{G2}) = 3P_{G2} + 0.03 \times P_{G2}^2$$

where P_{G1} and P_{G2} are the MW injections from generator G_1 and G_2 respectively. Thus, the minimum cost dispatch will be
(a) $P_{G1} = 250$ MW; $P_{G2} = 0$ MW
(b) $P_{G1} = 150$ MW; $P_{G2} = 100$ MW
(c) $P_{G1} = 100$ MW; $P_{G2} = 150$ MW
(d) $P_{G1} = 0$ MW; $P_{G2} = 250$ MW **[GATE E.E. 2008]**

483. The incremental cost characteristics of two generators delivering a total load of 200 MW are as follows :

$$\frac{dF_1}{dP_1} = 2.0 + 0.01P_1 \qquad \frac{dF_2}{dP_2} = 1.6 + 0.02P_2$$

What should be the values of P_1 and P_2, for economic operation?
(a) $P_1 = P_2 = 100$ MW (b) $P_1 = 80$ MW, $P_2 = 120$ MW
(c) $P_1 = 200$ MW, $P_2 = 0$ (d) $P_1 = 120$ MW, $P_2 = 80$ MW

[U.P.S.C. I.E.S. E.E.-II, 2007]

484. The incremental cost characteristics of two generators delivering a load of 200 MW are as follows:

$$\frac{dF_1}{dP_1} = 2.0 + 0.01\ P_1$$

$$\frac{dF_2}{dP_2} = 1.6 + 0.02\ P_2$$

For economic operation the generators P_1 and P_2 should be
(a) 120 MW and 80 MW (b) 80 MW and 120 MW
(c) 120 MW and 120 MW (d) 80 MW and 80 MW

[U.P.S.C. I.E.S. E.E.-II, 2011]

485. The incremental fuel costs for two generating units G_1 and G_2 are given by:

$IC_1 = 25 + 0.2\ PG_1$ and $IC_2 = 32 + 0.2\ PG_2$ where PG_1 and PG_2 are real powers generated by the units. The economic allocation for a total load of 250 MW, neglecting transmission loss, is given by
(a) $PG_1 = 142.5$ MW and $PG_2 = 107.5$ MW
(b) $PG_1 = 109.75$ MW and $PG_2 = 140.25$ MW
(c) $PG_1 = 125$ MW and $PG_2 = 125$ MW
(d) $PG_1 = 100$ MW and $PG_2 = 150$ MW

[U.P.S.C. I.E.S. E.E.-II, 2011]

486. A load centre is at an equidistant from the two thermal generating stations G_1 and G_2 as shown in the figure. The fuel cost characteristics of the generating stations are given by

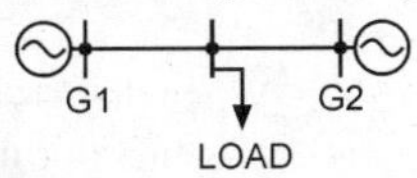

$F_1 = a + bP_1 + cP_1^2$ Rs/hour
$F_2 = a + bP_2 + cP_2^2$ Rs/hour

where P_1 and P_2 are the generation in MW of G_1 and G_2, respectively. For most economic generation to meet 300 MW of load, P_1 and P_2, respectively, are
(a) 150, 150 (b) 100, 200
(c) 200, 100 (d) 175, 125 **[GATE E.E. 2005]**

487. A load center of 120 MW derives power from two power stations connected by 220 kV transmission lines of 25 km and 75 km as shown in the figure below. The three generators G1, G2 and G3 are of 100 MW capacity each and have identical fuel cost characteristics. The *minimum loss generation schedule* for supplying the 120 MW load is

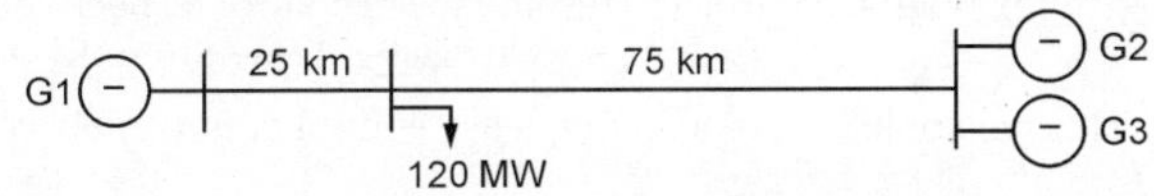

(a) P1 = 80 MW + losses, P2 = 20 MW, P3 = 20 MW
(b) P1 = 60 MW, P2 = 30 MW + losses, P3 = 30 MW
(c) P1 = 40 MW, P2 = 40 MW, P3 = 40 MW + losses
(d) P1 = 30 MW + losses, P2 = 45 MW, P3 = 45 MW

[GATE E.E., 2011]

488. Incremental fuel costs (in some appropriate unit) for a power plant consisting of three generating units are

$$IC_1 = 20 + 0.3\ P_1,\ IC_2 = 30 + 0.4\ P_2,\ IC_3 = 30$$

where P_i is the power in MW generated by unit *i*, for *i* = 1, 2 and 3. Assume that all the three units are operating all the time. Minimum and maximum loads on each unit are 50 MW and 300 MW respectively. If the plant is operating on economic load dispatch to supply the total power demand of 700 MW, the power generated by each unit is
(a) $P_1 = 242.86$ MW; $P_2 = 157.14$ MW; and $P_3 = 300$ MW
(b) $P_1 = 157.14$ MW; $P_2 = 242.86$ MW; and $P_3 = 300$ MW
(c) $P_1 = 300.0$ MW; $P_2 = 300.0$ MW; and $P_3 = 100$ MW
(d) $P_1 = 233.3$ MW; $P_2 = 233.3$ MW; and $P_3 = 233.4$ MW

[GATE E.E. 2003]

489. Consider a power system with three identical generators. The transmission losses are negligible. One generator (G_1) has a speed governor which maintains its speed constant at the rated value, while the other generators (G_2 and G_3) have governors with a droop of 5%. If the load of the system is increased, then in steady state.
(a) generation of G_2 & G_3 is increased equally while generation of G_1 is unchanged.
(b) generation of G_1 alone is increased while generation of G_2 and G_3 is unchanged.
(c) generation of G_1, G_2 and G_3 is increased equally.
(d) generation of G_1, G_2 and G_3 is increased in the ratio 0.5 : 0.25 : 0.25. **[GATE E.E. 2002]**

490. The unit of transmission loss coefficient is
(a) MW (b) $(MW)^{-1}$
(c) $(MW)^{-2}$ (d) unitless.
(e) none of these. **[W.B. Univ. of Technology Power System 2009-10]**

491. If penalty factor of a plant is unity, its incremental transmission loss is
(a) 1.0 (b) –1.0
(c) 0.0 (d) 2.0

[W.B. Univ. of Technology Power System 2009-10]

Answers: 482. (c), **483.** (d), **484.** (a), **485.** (a), **486.** (c), **487.** (a), **488.** (a), **489.** (b), **490.** (b), **491.** (c),

492. The main objectives of load frequency control in a power system are:
1. to bring the steady state error to zero after load change.
2. to maintain the net tie-line flow.
3. to maintain voltages on all buses.
4. to economize the cost of generation.

(a) 1 and 2. (b) 2 and 3.
(c) 3 and 4. (d) 1, 2, 3 and 4.
[U.P.S.C. I.E.S. E.E.-II, 2011]

493. In AGC, the voltage and frequency is controlled by
(a) excitation control. (b) turbine control.
(c) turbine speed control and excitation control respectively.
(d) excitation control and turbine speed control respectively.
[W.B. Univ. of Technology Power System III 2009-10]

494. The main objective of load frequency controller is to apply control of:
(a) frequency alone.
(b) frequency and at the same time of real power exchange via the outgoing lines.
(c) frequency and at the same time of reactive power exchange via the outgoing lines.
(d) frequency and bus voltages. **[U.P.S.C. I.E.S. E.E.-II, 2011]**

495. Load frequency control uses
(a) proportional controllers alone.
(b) integral controllers alone.
(c) both proportional and integral controllers.
(d) either proportional or integral controllers.
[U.P.S.C. I.E.S. E.E.-II, 1999]

496. Load frequency controls are carried out with
(a) P controllers only. (b) I controllers only.
(c) D controllers only. (d) PID controllers.
[U.P.S.C. I.E.S. E.E.-II, 2010]

497. Consider the following statements regarding load frequency control:
1. Time constant of automatic load frequency control is about 15 seconds.
2. Integral control eliminates static frequency drop.
3. In tie-line load bias control, the control signal for each area is proportional to change in frequency as well as change in tie-line power.

Which of the statements given above are correct?
(a) 1, 2 and 3. (b) 1 and 2.
(c) 1 and 3. (d) 2 and 3.
[U.P.S.C. I.E.S. E.E.-II, 2004]

498. Consider a power system with two plants S_1 and S_2 connected through a tie line as shown in the figure.

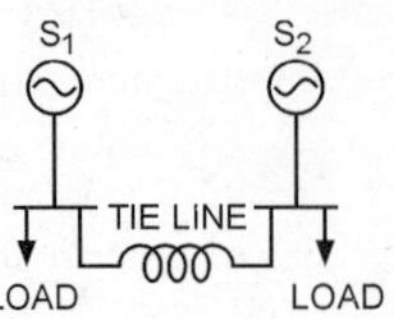

When the load-frequency control of the system is considered, the Flat tie line control system is preferred over the 'Flat frequency regulation system', because
(a) It is advantageous to control the frequency from any one particular plant without disturbing the other one during load-swings on either S_1 or S_2 areas.
(b) This ensures that only the more efficient plant's input is controlled for load variation in any area.
(c) Only the tie line is required to absorb the load-swings.
(d) The load-change in a particular area is taken care of by the generator in that area resulting in the tie-line loading to remain constant. **[U.P.S.C. I.E.S. E.E.-II, 2003]**

499. Two power plants interconnected by a tie-line as shown in the figure have loss formula coefficient $B_{11} = 10^{-3}$ MW^{-1}. Power is being despatched economically with plant '1' as 100 MW and plant '2' as 125 MW. The penalty factors for plants 1 and 2 are respectively

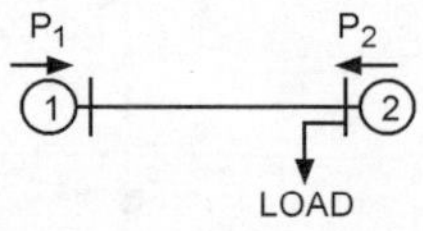

(a) 1 and 1.25 (b) 1.25 and 1
(c) 1 and zero (d) zero and 1
[U.P.S.C. I.E.S. E.E.-II, 2001]

500. An isolated 50 Hz synchronous generator is rated at 15 MW which is also the maximum continuous power limit of its prime mover. It is equipped with a speed governor with 5% droop. Initially, the generator is feeding three loads of 4 MW each at 50 Hz. One of these loads is programmed to trip permanently if the frequency falls below 48 Hz. If an additional load of 3.5 MW is connected then the frequency will settle down to
(a) 49.417 Hz (b) 49.917 Hz
(c) 50.083 Hz (d) 50.583 Hz **[GATE E.E. 2007]**

501. A power system has two synchronous generators. The Governor-turbine characteristics corresponding to the generators are
$P_1 = 50\,(50 - f)$, $P_2 = 100\,(51 - f)$
where f denotes the system frequency in Hz, and P_1 and P_2 are, respectively, the power outputs (in MW) of turbines 1 and 2. Assuming the generators and transmission network to be lossless, the system frequency for a total load of 400 MW is
(a) 47.5 Hz (b) 48.0 Hz
(c) 48.5 Hz (d) 49.0 Hz **[GATE E.E. 2001]**

502. For a synchronous generator connected to an infinite bus through a transmission line, how are the change of voltage (ΔV) and the change of frequency (Δf) related to the active power (P) and the reactive power (Q)?
(a) ΔV is proportional to P and Δf to Q.
(b) ΔV is proportional to Q and Δf to P.
(c) Both ΔV and Δf are proportional to P.
(d) Both ΔV and Δf are proportional to Q.
[U.P.S.C. I.E.S. E.E.-II, 2006]

503. The speed regulation parameter R of a control area is 0.025 Hz/MW and load frequency constant D is 2 MW/Hz. The area frequency response characteristic (AFRC) is
(a) 42.0 MW/Hz (b) 40.0 MW/Hz
(c) 20 MW/Hz (d) 2 MW/Hz
[U.P.S.C. I.E.S. E.E.-II, 2001]

504. What is the correct expression for Area Control Error (ACE) for an interconnected 2-area power system, if ΔP_t is incremental tie line power deviation, Δf is frequency deviation and b is frequency bias coefficient?
(a) $ACE = b\Delta P_t + \Delta f$ (b) $ACE = \Delta P_t + b\Delta f$
(c) $ACE = \frac{1}{b}\Delta P_t + \Delta f$ (d) $ACE = \Delta P_t + \frac{1}{b}\Delta f$
[U.P.S.C. I.E.S. E.E.-II, 2007]

505. The voltage of a bus can be controlled by controlling the
(a) phase angle. (b) reactive power of the bus.
(c) active power of the bus.
(d) phase angle and the reactive power. **[A.M.I.E. Sec B. Summer 1995]**

506. The permissible variation of frequency in power system is
(a) ± 1% (b) ± 3% (c) ± 5% (d) ± 10%
[A.M.I.E. Sec B. Winter 1993; Summer 1999]

507. When there is a change in load in a power station having a number of generator units operating in parallel, the system frequency is controlled by

Answers: 492. (a), **493.** (c), **494.** (b), **495.** (c), **496.** (d), **497.** (d), **498.** (b), **499.** (b), **500.** (a), **501.** (b), **502.** (b), **503.** (a), **504.** (b), **505.** (b), **506.** (b), **507.** (a),

(a) adjusting the steam input to the units.
(b) adjusting the field excitation of the generators.
(c) changing the load divisions between the units.
(d) injecting reactive power at the station bus-bar.

[U.P.S.C. I.E.S. E.E.-II, 1994]

508. Load frequency control is achieved by properly matching the individual machine's
(a) reactive powers. (b) generated voltages.
(c) turbine inputs. (d) turbine and generator ratings.

[U.P.S.C. I.E.S. E.E.-II, 1998]

509. In the load-frequency control system with free governor action, the increase in load-demand under steady conditions is met
(a) only by increased generation due to opening of steam valve.
(b) only by decrease of load-demand due to drop in system frequency.
(c) partly by increased generation and partly by decrease of load demand.
(d) partly by increase generation and partly by increased excitation.

[U.P.S.C. I.E.S. E.E.-II, 1996]

510. During load shedding
(a) system voltage is reduced.
(b) system frequency is reduced.
(c) system loads are switched off.
(d) system power factor is changed.

[A.M.I.E. Sec B. Power Systems, Winter 1999]

511. When the power system is not in a position to meet the load it will resort to
(a) power factor improvement at the generators.
(b) load shedding. (c) efficient plant operation.
(d) penalising high load consumers by increasing the charges.

512. Load shedding is done for
(a) reducing peak demand on the system.
(b) repairing of machines. (c) power factor improvement.
(d) efficient operation of equipment.

513. When load shedding is resorted to, it can be concluded that
(a) plant is under repair.
(b) load on the system is more than the installed capacity.
(c) both of the above (a) and (b).
(d) none of the above.

514. Generators for base load plants are usually designed for maximum efficiency around
(a) 20% over-load. (b) full-load.
(c) 75% full-load. **[A.M.I.E. Sec B. Power Systems, Winter 1998]**

515. Generators for power plants to supply exclusively peak loads are usually designed for maximum efficiency to occur at
(a) full load. (b) 50 – 75% full load.
(c) 25 – 50% full load. (d) 10% full load.

516. A pilot exciter is provided on synchronous generator to
(a) provide starting torque to the generator.
(b) to supplement the power generated by the main generator.
(c) excite the main exciter. (d) none of the above.

517. Annual operating cost of a generating plant consists of
(a) fixed charges. (b) semifixed charges.
(c) operating or running charges.
(d) all of the above.

518. For a power plant the expenditure on which of the following items is expected to be almost negligible ?
(a) Publicity. (b) Taxes.
(c) Insurance. (d) Maintenance.

519. In a steam power station which of the following is not a fixed cost ?
(a) Interest on capital cost of land and buildings.
(b) Salaries of high officials.
(c) Fuel and lubricating oil cost.
(d) Insurance charges.

520. In a power plant which of the following items does not fall in the category of operating cost ?
(a) Salaries of operational and maintenance staff.
(b) Maintenance and repair cost.
(c) Salaries of supervisory staff engaged on the running of the plant.
(d) Salaries of management and clerical staff.

521. In a power plant, which of the following items fall in the category of semifixed charges ?
(a) Annual interest and depreciation on capital cost.
(b) All types of taxes and insurance charges.
(c) Salaries of management and clerical staff.
(d) All of the above.

522. For a nuclear power plant, the useful life is expected to be about
(a) 10 years. (b) 30 years.
(c) 60 years. (d) 80 years.

523. The interest on the capital cost is included in
(a) annual fixed cost. (b) annual operating cost.
(c) both (a) and (b). (d) neither (a) nor (b).

524. The capital cost of a power plant depends on
(a) total installed capacity only. (b) total number of units only.
(c) total installed capacity and number of units as well.
(d) neither the installed capacity nor number of units.

[A.M.I.E. Sec B. Power Systems Summer 1998; Summer 2000]

525. The long term load forecast is required for
(a) operation of plant. (b) economic operation of plant.
(c) planning the addition in generation capacity.
(d) both (b) and (c).

526. The annual depreciation reserve depends on
(a) capital cost only. (b) salvage value only.
(c) on the method of calculation depreciation reserve.
(d) all of the above.

527. Salvage value of a plant
(a) is always positive. (b) is always zero.
(c) is always negative. (d) may be any of the above.

528. An equipment purchased for Rs 25 lakhs, 3 years before, has now market value of 32 lakhs, It shows that
(a) the value has depreciated as per straight line method.
(b) the value has depreciated as per diminishing value method.
(c) the value has depreciated as per sinking fund method.
(d) the value has appreciated with the time.

529. Ideally depreciated value of the plant plus the accumulation in the depreciation fund should be equal to
(a) obsolescence rate. (b) sinking fund.
(c) original invested value. (d) salvage value.

530. In a straight line method, annual depreciation charges are calculated by
(a) the capital cost minus salvage value divided by the number of years of life.
(b) the capital cost divided by the number of years of life.
(c) making a provision for setting each year a fixed rate, first applied to the original cost and then to the diminishing value, depending upon the useful life of the plant.
(d) none of the above.

Answers: **508.** (c), **509.** (c), **510.** (c), **511.** (b), **512.** (a), **513.** (b), **514.** (b), **515.** (b), **516.** (c), **517.** (d), **518.** (a), **519.** (c), **520.** (d), **521.** (d), **522.** (b), **523.** (a), **524.** (c), **525.** (c), **526.** (d), **527.** (d), **528.** (d), **529.** (c), **530.** (a),

531. Annual instalment towards depreciation increases with the decrease in interest rate in case of
(a) straight line depreciation. (b) sink fund depreciation.
(c) reducing balance depreciation.
(d) all of the above.

532. Annual depreciation of the plant is proportional to the earning capacity of the plant in case of
(a) straight line depreciation. (b) sinking fund depreciation.
(c) reducing balance depreciation.
(d) none of the above.

533. Annual estimated depreciation charges for a plant are heavy during early years in case of
(a) diminishing value method.
(b) sinking fund method. (c) straight line method.
(d) none of the above.

534. In a power plant insurance cover is provided for
(a) equipment only. (b) unskilled workers only.
(c) unskilled and skilled workers.
(d) equipment and skilled and unskilled workers.

535. A company can raise funds through
(a) fixed deposits. (b) shares.
(c) bonds. (d) any of the above.

536. All of the following are the sources of borrowing money except
(a) shares. (b) debentures.
(c) bonds. (d) fixes deposits.

537. Which of the following are not repayable after a stipulated period ?
(a) Fixed deposits. (b) Shares.
(c) Bonds. (d) Cash certificates.

538. During summer season load on the system increases, this is due to
(a) increase in demand of water supply.
(b) increased business activity.
(c) increased use of fans, air coolers and air conditioners.
(d) vacations in schools and institutions.

539. The connected load of a domestic consumer (medium income group) is usually about
(a) 2–5 kW (b) 10–20 kW
(c) 20–30 kW (c) below 1 kW

540. Power demand can be estimated approximately by
(a) load survey method. (b) mathematical method.
(c) statistical method. (d) economic parameters.
(e) all of the above.

541. A load curve is a plot of
(a) load versus duration of time.
(b) load versus current. (c) load versus time.
(d) total number of units generated versus time.

542. The chronological load curve indicates
(a) variation in demand factor during 24 hours.
(b) variation in demand from instant to instant during 24 hours.
(c) the total energy consumed upto different times of the day.
(d) the total number of hours for which a particular load lasts during the day.

543. The area under the load curve represents
(a) system voltage. (b) current.
(c) energy consumed. (d) maximum demand.
(e) average demand. **[A.M.I.E. Sec B. Summer 1994]**

544. The area under daily load curve divided by 24 gives
(a) average load for the day. (b) maximum demand.
(c) connected load. (d) demand factor.

545. Which of the following statements regarding load curve is true ?
(a) The area under the curve gives average demand.
(b) The ratio of area under the load curve to the total area of the rectangle in which it is contained gives the load factor for the day.
(c) The area of the load curve divided by the number of hours gives the load factor.
(d) The area of the load curve divided by the number of hours gives the demand factor.

546. Load curve of a power generating station is of always
(a) negative slope. (b) zero slope.
(c) positive slope.
(d) any combination of (a), (b) and (c).

547. Load curve helps in deciding the
(a) total installed capacity of the plant.
(b) size of the generating units.
(c) operating schedule of the generating units.
(d) all of the above.

548. Load duration curve gives
(a) the variations of load during different hours of the day.
(b) average load.
(c) the number of hours for which a particular load lasts during the day.
(d) none of the above.

549. The load duration curve for unity load factor will be of
(a) rectangular shape. (b) triangular shape.
(c) L-shape. (d) I-shape.

550. The mass curve represents
(a) average load.
(b) the total energy consumed by the load upto a particular time in a day.
(c) the number of hours for which a particular load lasts during the day.
(d) the variation of load during different hours of the day.

551. A mass curve can be plotted from
(a) load duration curve. (b) chronological load curve.
(c) energy load curve.
(d) both load duration curve and chronological load curve.
[A.M.I.E. Sec B. Summer 1998, 2000]

552. Integrated load duration curve representing total number of units (kWh) generated for a given demand in kW can be plotted directly from
(a) load curve. (b) load duration curve.
(c) mass curve. (d) any of these.

553. In a load duration curve for an integrated power system, the uppermost crest represents the energy contributed by which one of the following?
(a) Base power stations. (b) Major thermal stations.
(c) Peaking hydro or gas turbine stations.
(d) Non-conventional power stations. **[U.P.S.C. I.E.S. E.E.-II, 2007]**

554. During which season the load on a power system is maximum ?
(a) Autumn. (b) Rainy. (c) Summer. (d) Winter.

555. Maximum demand on the power system is
(a) the greatest of all 'short time interval averaged' demand during a period.
(b) instantaneous maximum value of kW supplied during a period.
(c) instantaneous maximum value of kVA supplied during a period.
(d) maximum value of units (kWh) supplied during a period.

Answers: **531.** (b), **532.** (c), **533.** (a), **534.** (d), **535.** (d), **536.** (a), **537.** (b), **538.** (c), **539.** (a), **540.** (c), **541.** (c), **542.** (b), **543.** (c), **544.** (a), **545.** (b), **546.** (d), **547.** (d), **548.** (c), **549.** (a), **550.** (b), **551.** (b), **552.** (b), **553.** (d), **554.** (d), **555.** (a),

556. The power system experiences peak demand from
(a) midnight to 8 A.M. (b) 8 A.M. to 2 P.M.
(c) 2 P.M. to 6 P.M. (d) 6 P.M. to 10 P.M.

557. Connected load means
(a) installed electrical load in the premises of a consumer.
(b) maximum load a consumer draws.
(c) load drawn by consumer at any instant.
(d) none of the above.

558. The maximum demand of a power station is
(a) sum of the maximum demands of all its consumers.
(b) greatest average load in a specified time.
(c) peak value of load in a specified time.
(d) all of the above.

559. Demand factor is defined as the ratio of
(a) average load to maximum demand.
(b) maximum demand to connected load.
(c) connected load to maximum demand.
(d) maximum demand to average load.

560. Demand factor on a power system is
(a) always greater than unity. (b) normally greater than unity.
(c) always lesser than unity. (d) normally lesser than unity.

561. The load factor is equal to
(a) average load/peak load. (b) peak load/average load.
(c) average load/connected load. (d) average load/base load.
[A.M.I.E Sec B Design of Electrical Systems Summer 2005]

562. A power station's plant load factor is defined as the ratio of
(a) the energy generated to that of maximum energy that could have been generated.
(b) average load to peak load.
(c) minimum load to peak load.
(d) minimum load to average load. [U.P.S.C. I.E.S. E.E.-II, 2006]

563. Load factor of a power plant is
(a) generally equal to unity. (b) always less than unity.
(c) always more than unity. (d) normally more than unity.

564. The load factor for domestic loads may be taken as
(a) about 85% (b) 50 – 60%
(c) 25–50% (d) 10 – 15%

565. Load factor for heavy industries may be taken as
(a) 70–80% (b) 40–50%
(c) 25–40% (d) 20–25%

566. The annual load duration curve of a power supply system may be considered as a straight line from 40 MW to 8 MW. The load factor of the system is
(a) 20% (b) 40% (c) 60% (d) 83.33%
(e) none of the above. [A.M.I.E. Sec B. Winter 1993]

567. An industrial consumer has a load pattern of 2,000 kW 0.8 lag for 12 hours and 1,000 kW unity power factor for 12 hours. The load factor is
(a) 0.5 (b) 0.75 (c) 0.6 (d) 2.0
[GATE E.E. 1999]

568. The maximum demand of a consumer is 2 kW and the corresponding daily energy consumption is 30 units. What is the corresponding load factor?
(a) 25% (b) 50% (c) 62.5% (d) 75%
[U.P.S.C. I.E.S. E.E.-II, 2009]

569. The load curve of a system is shown in the figure. The load factor of the system is
(a) 1.66%
(b) 6.013%
(c) 16.6%
(d) 60.13%

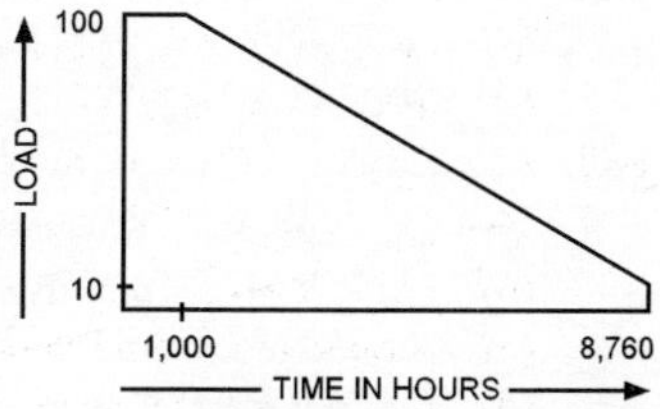

[U.P.S.C. I.E.S. E.E.-II, 1997]

570. Diversity factor is the ratio of
(a) sum of maximum demands of consumers/system maximum demand.
(b) maximum demand of consumers/average demand.
(c) demand of all consumers/average demand.
(d) none of the above. [A.M.I.E. Sec B. Power Systems, Winter 1998]

571. The curve shown in the figure is
(a) chronological load curve.
(b) annual load duration curve.
(c) mass curve.
(d) energy load curve.

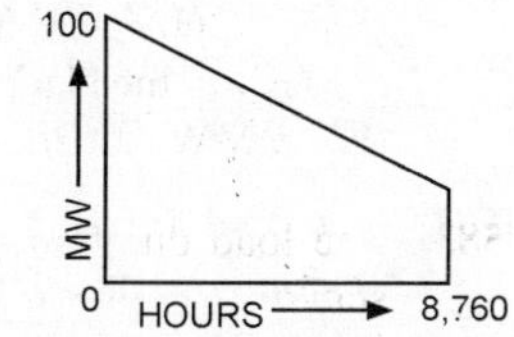

[A.M.I.E Sec B Design of Electrical Systems Summer 2005]

572. The daily energy produced in thermal power station is 720 MWh at a load factor of 0.6. What is the maximum demand of the station?
(a) 50 MW (b) 30 MW (c) 72 MW (d) 720 MW
[U.P.S.C. I.E.S. E.E.-II, 2008]

573. Diversity factor in a power system is
(a) always less than unity. (b) normally less than unity.
(c) always more than unity. (d) normally more than unity.
[A.M.I.E. Sec B. Summer 1993]

574. Diversity factor × maximum demand is
(a) average demand.
(b) sum of consumer's maximum demands.
(c) installed capacity. (d) generated power.
[A.M.I.E. Sec B. Summer 1994]

575. Diversity factor has direct effect on
(a) fixed cost per unit generated.
(b) operating cost per unit generated.
(c) both (a) and (b). (d) neither (a) nor (b).

576. The knowledge of diversity factor helps in computing
(a) plant capacity. (b) average load.
(c) units (kWh) generated. (d) peak demand.

577. Plant capacity factor is
(a) actual energy produced/maximum possible energy that could have been produced (based on installed capacity).
(b) actual energy produced /plant capacity hours for which the plant has been in operation.
(c) energy generated in a given period/ (maximum demand x hours of operation in given period).
(d) none of the above. [A.M.I.E. Sec B. Winter 1993]

578. Which of the following statement is correct ?
(a) Load factor = Capacity factor × utilisation factor.
(b) Utilisation factor = Capacity factor × load factor.
(c) Capacity factor = $\frac{\text{Load factor}}{\text{Utilisation factor}}$
(d) Capacity factor = Load factor × utilisation factor.
[A.M.I.E. Sec B. Power Systems, Summer 2000]

579. Utilisation factor is defined as the ratio of
(a) average demand to rated capacity of the power plant.
(b) maximum demand on the power plant to the rated capacity of the power plant.

Answers: **556.** (d), **557.** (a), **558.** (b), **559.** (b), **560.** (c), **561.** (a), **562.** (b), **563.** (b), **564.** (d), **565.** (a), **566.** (c), **567.** (b), **568.** (c), **569.** (d), **570.** (a), **571.** (b), **572.** (a), **573.** (c), **574.** (b), **575.** (a), **576.** (a), **577.** (a), **578.** (d), **579.** (b),

(c) rated capacity of the power plant to the maximum demand.
(d) none of the above.

580. Capacity factor will be very low when the power plant
(a) is operated as base load plant.
(b) is operated for supplying base loads as well as peak loads.
(c) is operated in emergency only.
(d) is under maintenance.

581. A thermal generating station has an installed capacity of 15 MW and supplies a daily load of 10 MW for 12 hours and 5 MW of remaining 12 hours. The plant capacity factor for this station is
(a) 1 (b) 0.75 (c) 0.67 (d) 0.5
[U.P.S.C. I.E.S. E.E.-II, 1999]

582. A generating station has a maximum demand of 30 MW, a load factor of 60% and a plant capacity factor of 50%. The reserve capacity of the plant is
(a) 5 MW (b) 4 MW (c) 6 MW (d) 10 MW
[U.P.S.C. I.E.S. E.E.-II, 2003]

583. The load duration curve for a power station is as shown in the given figure. The reserve capacity in the plant at 70% capacity factor is

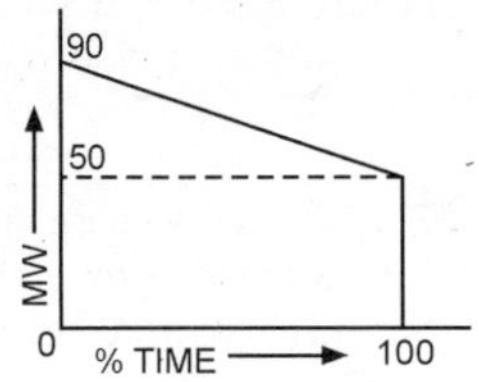

(a) zero.
(b) 10 MW
(c) 30 MW
(d) 50 MW **[U.P.S.C. I.E.S. E.E.-II, 2001]**

584. If the rated plant capacity and maximum load of a power station are equal then
(a) load factor is unity. (b) capacity factor is unity .
(c) utilisation factor is unity.
(d) load factor and capacity factor are equal.
(e) both (c) and (d).

585. Spinning reserve is
(a) the reserve generating capacity which is available for service but not in operation.
(b) the reserve generating capacity which is connected to the bus and is ready to take load.
(c) the reserve generating capacity which is in operation but is not in service.
(d) none of the above. **[A.M.I.E. Sec B. Winter 1993]**

586. The power which must be available ever under emergency conditions is known as
(a) spinning reserve. (b) cold reserve.
(c) firm reserve. (d) hot reserve.
[A.M.I.E. Sec B. Power Systems, Summer 2000]

587. In a power plant, a reserve generating capacity which is in operation but not in service, in called the
(a) cold reserve. (b) hot reserve.
(c) spinning reserve. (d) firm reserve.

588. In a power plant, a reserve generating capacity, which is in operable condition and available for service, but not in operation is called the
(a) spinning reserve. (b) firm reserve.
(c) cold reserve. (d) hot reserve.

589. A power plant has a maximum demand of 15 MW. The load factor is 50% and the plant factor is 40%. The operating reserve is
(a) 3 MW (b) 3.75 MW
(c) 6 MW (d) 7.5 MW
[U.P.S.C. I.E.S. E.E.-II, 1996]

590. Which of the following relations holds good ?
(a) Maximum demand = Connected load × demand factor.
(b) Average load = Maximum load × load factor.
(c) Capacity factor = Utilisation factor × load factor.
(d) All of the above.

591. Two areas A and B have equal connected loads. However, load diversity in area A is more than in area B. Then
(a) maximum demands of the two areas would be equal.
(b) maximum demand of A would be more than that of B.
(c) maximum demand of B would be more than that of A.
(d) maximum demand of A may be more or less than that of B.
[A.M.I.E. Sec B. Power Systems, Summer 1998]

592. A large diversity factor of the load in a power system
(a) reduces the installation cost.
(b) increases the installation cost.
(c) does not affect the installation cost.
[A.M.I.E. Sec B. Power Systems, Summer 1999]

593. As the load factor of a generating plant increases, the generation cost per kWh generated
(a) decreases. (b) increases.
(c) remains same. (d) none of these.
[W.B. Univ. of Technology Power System I 2009-10]

594. In order to have a lower cost of electrical energy generation
(a) the load factor and diversity factor should be low.
(b) the load factor should be low but diversity factor should be high.
(c) the load factor should be high but the diversity factor should be low.
(d) the load factor and diversity factor should be high.
[GATE E.E. 1995]

595. In a power station, the cost of generation of power reduces most effectively when
(a) diversity factor alone increases.
(b) both diversity factor and load factor increase.
(c) load factor alone increases.
(d) both diversity factor and load factor decrease.
[U.P.S.C. I.E.S. E.E.-II, 1994)

596. The economics of power plant is greatly influenced by:
1. load factor 2. utilization factor
3. unit capacity 4. type of load
(a) 1, 2, 3 and 4. (b) 1, 3 and 4.
(c) 1, 2 and 3. (d) 2, 3 and 4.
[U.P.S.C. I.E.S. E.E.-II, 2011]

Questions 597 and 598 refer to the following data:

A region has a maximum demand of 500 MW at a load factor of 50 per cent. The load-duration curve can be assumed to be a triangle. The utility has to meet this load by setting up a generating system which has partly hydro and partly thermal. The costs are as under:

Hydro plant-Rs. 600 per kW per annum plus 3 p per kWh.
Thermal plant-Rs. 300 per kW per annum plus 13 p per kWh.

597. The respective capacity of hydro and thermal power plants, for minimum cost, are
(a) 330 MW and 170 MW (b) 170 MW and 330 MW
(c) 93.75 MW and 406.25 MW
(d) 406.25 MW and 93.75 MW

598. The energy generated annually by hydro and thermal power plant, for minimum cost, are
(a) 1,935 × 10^6 kWh and 255 × 10^6 kWh.
(b) 255 × 10^6 kWh and 1,935 × 10^6 kWh.
(c) 187.5 ×10^6 kWh and 2,002.5 × 10^6 kWh.
(d) 2,002.5 ×10^6 kWh and 187.5 × 10^6 kWh.

Answers: 580. (c), **581.** (d), **582.** (c), **583.** (b), **584.** (e), **585.** (b), **586.** (c), **587.** (b), **588.** (c), **589.** (b), **590.** (d), **591.** (c), **592.** (a), **593.** (a), **594.** (d), **595.** (b), **596.** (a), **597.** (a), **598.** (a),

599. Flat rate tariff can be charged on the basis of
(a) connected load. (b) units consumed.
(c) maximum demand. (d) both (a) and (b).

600. Domestic consumers are usally charged
(a) flat demand tariff. (b) block rate tariff.
(c) flat rate tariff. (d) off peak tariff.

601. If the tariff for electrical energy charges provides incentive by way of reduced charges for higher consumption, then it may be concluded that the
(a) load factor is high. (b) power factor is high.
(c) plant has sufficient reserve capacity.
(d) power is generated through hydropower plant.

602. Two part tariff is charged on the basis of
(a) connected load. (b) units consumed.
(c) maximum demand. (d) both (a) and (b).

603. Block rate tariff, where energy charge decreases with the increase in energy consumption,
(a) encourages the consumers for more consumption.
(b) discourages the consumers for more consumption.
(c) encourages the consumers to restrict their demands.
(d) encourages the consumers to improve the power factor.

604. The penalty for low power factor is imposed on
(a) residential and commercial consumers.
(b) industrial consumers. (c) agricultural consumers.
(d) all of the above.

605. Doherty rate tariff is applied to
(a) domestic consumers. (b) medium industrial consumers.
(c) bulk supplies. (d) municipal loads.

606. Maximum demand tariff is generally not applied to the domestic consumers owing to their
(a) low maximum demand. (b) low load factor.
(c) low power factor. (d) low energy consumption.

607. Low power factor is usually not due to
(a) discharge lamps. (b) incandescent lamps.
(c) arc lamps. (d) induction furnaces.

608. The primary reason for low power factor is owing to installation of
(a) synchronous motors. (b) dc motors.
(c) induction motors. (d) commutator motors.

609. Low power factor has the drawback(s) of
(a) increased transmission and distribution losses.
(b) poor voltage regulation.
(c) high cost of equipment for a given load.
(d) all of the above.

610. The low power factor of an industrial plant is uneconomical for
(a) electric supply utility only.
(b) owner of the plant only.
(c) both the owner of the plant and the electric supply utility.
(d) either (a) or (b).

611. For a consumer the most economical power factor is usually
(a) 0.25–0.5 lagging. (b) 0.25–0.5 leading.
(c) 0.85–0.95 lagging. (d) 0.85–0.95 leading.

612. Power factor can be improved by using
(a) static capacitors. (b) synchronous condensers.
(c) phase advancers. (d) all of the above.

613. Static capacitors are rated in terms of
(a) kVAR (b) kW (c) kVA (d) kWh

614. For power factor improvement static capacitors have the advantage(s) of
(a) small losses. (b) easy installation.
(c) low initial cost and little maintenance.
(d) all of the above.

615. For power factor improvement static capacitors have the drawback(s) of
(a) short-service life.
(b) getting damaged by high voltage.
(c) not repairable. (d) all of the above.

616. For power factor improvement synchronous condensers have the drawback(s) of
(a) comparatively higher maintenance and operating costs.
(b) requirement of auxiliary equipment for their starting.
(c) possibility of falling out of synchronism causing in interruption of supply.
(d) all of the above.

617. Advantages of the improved power factor are
(a) increase in operating efficiency of the power system.
(b) improvement in voltage regulation.
(c) reduction in overall cost per unit.
(d) better utilization of kW capacities of prime movers, transformers, switchgear and the lines.

618. The most suitable location for a power factor improvement device is
(a) near the electrical appliance which is responsible for the poor power factor.
(b) at the receiving end, in case of a transmission line.
(c) at the sending end. (d) any where in the circuit.
(e) both (a) and (b).

619. An industrial installation has a power factor of 0.8 lagging. It would be economical to improve pf to
(a) unity. (b) about 0.8 leading.
(c) about 0.95 lagging. (d) about 0.95 leading.
[A.M.I.E. Sec B. Power Systems, Summer 2000]

620. A 3-phase, 11 kV, 50 Hz, 200 kW load has a power factor of 0.8 lag. A delta-connected 3-phase capacitor is used to improve the pf to unity. The capacitance per phase of the capacitor in micro-farads is
(a) 3.948 (b) 1.316 (c) 0.439 (d) 11.844
[GATE E.E. 1999]

621. The most economical limit of power factor correcting is governed by
(a) original power factor.
(b) relative costs of the supply and power factor correction equipment.
(c) both (a) and (b). (d) none of the above.

622. There is a limit beyond which it is not economical still further to improve the power factor in order to meet the increased demand on the generating station. The maximum value to which the pf can be economically raised entirely depends upon
(a) the cost of the power factor correction equipment.
(b) the cost of the generating plant.
(c) the relative costs of the generating plant and phase advancing plant.
(d) none of the above.

623. A consumer requires an induction motor of 50 metric bhp. He is offered two motors of the following specifications:
Motor A – efficiency 88% and pf 0.9(lag). Motor B – efficiency 90% and pf 0.81.

Answers: 599. (b), **600.** (b), **601.** (c), **602.** (d), **603.** (a), **604.** (b), **605.** (c), **606.** (a), **607.** (b), **608.** (c), **609.** (d), **610.** (c), **611.** (c), **612.** (d), **613.** (a), **614.** (d), **615.** (d), **616.** (d), **617.** (d), **618.** (e), **619.** (c), **620.** (b), **621.** (b), **622.** (c), **623.** (a),

The consumer is being charged on a two part tariff of Rs 1,050 per kVA of maximum demand plus 75 paise per unit. The pf of the motor B is raised to 0.9 by installing condensers. The motor B costs Rs 2,250.00 less than A. The cost of condenser is Rs 900 per kVAR. Assume rate of interest and depreciation as 10% and working hours of motor as 2,400 hours in a year. Which of the two motors A and B is cheaper and by how much ?

(a) Motor B is more economical with annual saving of Rs 1,666.50.
(b) Motor A is more economical with annual saving of Rs 163.00.
(c) Overall annual operating cost are almost same in both cases.
(d) None of the above.

624. The medium sized induction motor will operate at maximum power factor while operating at
(a) full load. (b) 50 per cent full load.
(c) 25 per cent full load. (d) any load.

625. A synchronous condenser is virtually a/an
(a) induction motor.
(b) over-excited synchronous motor.
(c) under-excited synchronous motor.
(d) commutator motor.

Answers: 624. (a), **625.** (b).

ASSERTION (A)/REASON (R) TYPE QUESTIONS

Note : Each of the following questions consists of two statements, one labelled the 'Assertion (A)' and the other labelled the 'Reason (R)'. Examine the two statements carefully and decide if the Assertion (A) and Reason (R) are individually true and if so, whether the reason (R) is correct explanation of the Assertion (A). Select your answers to these questions using the codes given below:
Codes: (a) Both A and R are true and R is a correct explanation of A.
(b) Both A and B are true and R is not a correct explanation of A.
(c) A is true but R is false. (d) A is false but R is true.

1. (A): The major power development programmes in Bihar, should be primarily based on conventional steam power plants.
(R): Bihar has abundant coal reserves.

2. (A): The operating cost of hydroelectric power plants are the lowest.
(R): There are no standby losses in hydroplants.

3. (A): The hydroplants have constant speed and constant frequency.
(R): Very acute governing is possible with water turbines.

4. (A): Availablility of storage may convert an uneconomical hydro-project into an economical one.
(R): Storage increases the firm capacity of the power plant and increase in firm capacity of hydroplant means dispensing with the alternative steam capacity.

5. (A): Pelton turbine is not suitable for heads below 200 m.
(R): The diameter of a pelton wheel runner, for the same output but low head becomes large and unwieldy.

6. (A): The draft tube in hydropower stations permits a negative of suction head to be established at the runner exit.
(R): The turbine can be installed above the tail-race level without loss of head. [U.P.S.C. I.E.S. E.E..-II, 1995]

7. (A): In a high or medium head hydroelectric power station having a long penstock, a surge tank is provided near the turbine.
(R): A surge tank stores additional water to be released during the peak load period. [U.P.S.C. I.E.S. E.E., 2000]

8. (A): Hydroelectric plants with large storage can be best used as base load plants with high load factor.
(R): For a plant to be used as a base load plant, the unit cost of energy generated by the plant should be low. [U.P.S.C. I.E.S. E.E.-II, 2003]

9. (A): A pumped storage plant is very suitable for supplying peak loads. During off peak period, water is pumped back from tail race pond to head water pond.
(R): The starting time of pumped storage plant is very short. [U.P.S.C. I.E.S. E.E.-II, 2005]

10. (A): Huge amount of coal is required for raising the steam (about 2,000 tonnes per day for a 200 MW station) in a steam power station.
(R): As per government policy only low grade coal with 30-40 percent ash content can be used in steam power plants for generation purposes.

11. (A): Fire tube boilers are commonly used in central steam power plants.
(R): Fire tube boilers are compact in size, have low initial cost and have the ability to raise rapidly large quantities of steam per unit area of fire grate.

12. (A): Regenerative air preheaters are widely used in steam power plants.
(R): The rotor elements, in regenerative air preheaters, provide large surface area to the flue gases.

13. (A): Commercial steam turbines use series combination of impulse and reaction types of turbines.
(R): Impulse-reaction turbines are cheaper in initial cost.

14. (A): For the optimum loading of a number generating units in a power plant, the units must operate at the same incremental fuel cost.
(R): When units operate at the same incremental fuel cost, the active and reactive load diversions are equal. [U.P.S.C. I.E.S. E.E.-II, 1993]

15. (A): Nuclear power plants are not well suited for varying loads.
(R): The nuclear reactor does not respond to the load fluctuations efficiently.

16. (A): Diesel electric power plants are usually used as base load plants.
(R): Such plants can be located any where and there are no standby losses.

Answers: 1. (a), **2.** (b), **3.** (a), **4.** (a), **5.** (a), **6.** (b), **7.** (d), **8.** (d), **9.** (a), **10.** (a), **11.** (d), **12.** (a), **13.** (c), **14.** (c), **15.** (a), **16.** (d),

17. (A): Gas turbine power plants are generally used as base load power plants.

(R): The gas turbine power plants have some what higher fuel costs and low initial cost.

18. (A): Petroleum fuels such as kerosine, gas oil, diesel oil, residual oil have limited use, in gas turbines, as fuels.

(R): Petroleum fuels like gas oil, diesel oil, residual oil and kerosine have low calorific value.

19. (A): Intercooling in gas turbines reduces the size of the turbine and compressor for the same output.

(R): Intercooling removes the heat from the compressed air before it is sent to the combustion chamber.

20. (A): In almost all the fields open cycle turbine power plants are employed.

(R): Closed cycle gas turbine plants have the additional complexity of the plant including externally fired heater, gas cooler and various heat exchangers.

21. (A): The seeding material ionizes more readily than the gas and increases the conductivity of the gas.

(R): Seeding material have comparatively higher ionization potential.

22. (A): Now great emphasis is being given for the development of new and alternative (non-conventional) methods of power generation.

(R): Non-conventional methods of power generation give low cost of production of electrical energy and also these methods are more reliable.

23. (A): Optimum loading of the power stations interconnected by transmission line is obtained when the incremental fuel cost of each station divided by its penalty factor is the same for all the stations in the system.

(R): The total load demand and transmission losses at any instant must be met by the total generation at that instant and the penalty factor is a function of the incremental transmission loss. **[U.P.S.C. I.E.S. E.E.-II, 1993]**

24. (A): During winter season the load on the system increases.

(R): There is an increase of load on the system during winter season due to heaters and geysers.

25. (A): As the load factor of the power plant increases, the cost per kWh of the energy generated decreases.

(R): Higher load factor means greater average load, resulting in larger number of units generated for a given maximum demand or capacity, and thus there is a distribution of standing charges over a large number of units supplied and therefore reduction in unit cost.

26. (A): Synchronous condensers are used for pf improvement only when kVAR requirement exceeds 5,000.

(R): Initial as well as operating costs of synchronous condensers are much higher than those of static capacitors and do not justify their installation for low ratings.

27. (A): The best location for the pf correction equipment to be installed is the generating end.

(R): Generators are relieved of carrying excessive current due to poor power factor.

28. (A): It is not economical to raise the power factor to unity or leading one .

(R) As the pf approaches unity, the capacity of pf correction device increases more rapidly.

Answers: 17. (d), **18.** (c), **19.** (c), **20.** (a), **21.** (c), **22.** (c), **23.** (d), **24.** (a), **25.** (a), **26.** (a), **27.** (d), **28.** (a).

MATCHING TYPE QUESTIONS

Note: Each of the following questions consists of two lists. Match the items given in list 1 and those in list 2. Select your answers to these questions using the codes given in each of the questions.

1.

List 1 (Classification of Head)	**List 2 (Type of Turbines)**
A. Low head, 2–15 m	1. Propeller or Kaplan.
B. Medium head, 15–17 m	2. Kaplan or Francis.
C. High head, 70–500 m	3. Pelton.
D. Very high head > 500 m	4. Francies or Pelton.

Codes:	A	B	C	D
(a)	1	3	4	2
(b)	4	2	1	3
(c)	1	2	4	3
(d)	4	3	1	2

[U.P.S.C. I.E.S. E.E.-II, 2005]

2.

List 1 (Classification Based on Head)	**List 2 (Types of Turbine)**
A. Low head, 2 – 15 m	1. Pelton.
B. Medium head, 16 – 70 m	2. Francis or Pelton.
C. High head, 71 – 500 m	3. Kaplan or Francis.
D. Very high head, > 500 m	4. Propeller or Kaplan.

Codes:	A	B	C	D
(a)	4	3	2	1
(b)	3	4	1	2
(c)	4	3	1	2
(d)	3	4	2	1

[U.P.S.C. I.E.S. E.E.-II, 2004]

3.

List 1 (Type of Fuel)	**List 2 (Nuclear Fuel)**
A. Fertile	1. U-238
B. Fissile	2. Pu-239
	3. U-235
	4. Th-232

Codes:	A	B
(a)	1, 2	3, 4
(b)	2, 3	1, 4
(c)	3, 4	1, 2
(d)	1, 4	2, 3

[U.P.S.C. I.E.S. E.E.-II, 1993]

Answers: 1. (c), **2.** (a), **3.** (d),

4.

List 1 (Material)	List 2 (Use)
A. Deuterium	1. A fissile fuel of nuclear reactor.
B. Heavy water	2. A fertile fuel of nuclear reactor.
C. Uranium-235	3. Used as fuel in fusion reactor.
D. Thorium-232	4. Used as a mod erator in nuclear reactor.

Codes:	A	B	C	D
(a)	1	2	3	4
(b)	3	2	4	1
(c)	3	4	1	2
(d)	4	1	2	3

[U.P.S.C. I.E.S. E.E.-II, 1995]

5.

List 1 (Components)	List 2 (Materials)
A. Moderator	1. Boron
B. Control Rod	2. Concrete
C. Coolant	3. Graphite
D. Shield	4. Sodium

Codes:	A	B	C	D
(a)	3	1	4	2
(b)	1	3	2	4
(c)	3	1	2	4
(d)	1	3	4	2

[U.P.S.C. I.E.S. E.E.-II, 2003]

6. Nuclear reactors are classified on different basis, namely (A) Neutron energy (B) Fuel used (C) Moderator used (D) Coolant used (E) Type of core used.

The correct combination of reactors corresponding to their respective basis of classification is

Codes:	Thermal reactors	Beryllium reactor	Homogeneus reactors
(a)	(A)	(C)	(E)
(b)	(B)	(D)	(C)
(c)	(A)	(C)	(D)
(d)	(B)	(D)	(E)

[U.P.S.C. I.E.S. E.E.-II, 1996]

7.

List 1 (Power Station)	List 2 (Component)
A. Hydroelectric power	1. Grate and stokers station.
B. Nuclear power station	2. Forebay.
C. Steam power station	3. Pressurized water reactor.
D. Pressure gasification combined cycle power plant.	4. Fluidized bed gasifier.

Codes:	A	B	C	D
(a)	1	4	2	3
(b)	2	3	1	4
(c)	1	3	2	4
(d)	2	4	1	3

[U.P.S.C. I.E.S. E.E.-II, 2006]

8.

List 1 (Name of Power Plant)	List 2 (Plant Features)
A. Thermal	1. High operating cost.
B. Nuclear	2. High capital cost.
C. Hydro	3. High plant life.
D. Diesel	4. High fuel transportation cost.

Codes:	A	B	C	D
(a)	3	1	4	2
(b)	4	2	3	1
(c)	3	2	4	1
(d)	4	1	3	2

[U.P.S.C. I.E.S. E.E.-II, 2004]

9.

List 1 (Name of Power Plant)	List 2 (Feature)
A. Hydro	1. High maintenance cost.
B. Nuclear	2. High transmission and distribution cost.
C. Steam	3. High overall generation cost.
D. Diesel	4. Less efficient.

Codes:	A	B	C	D
(a)	2	1	4	3
(b)	1	2	3	4
(c)	2	3	4	1
(d)	3	4	2	1

10.

List 1 (Power Plant)	List 2 (Application)
A. Nuclear	1. Base load
B. Diesel	2. Standby
C. Gas turbine	3. Base or peak load
D. Hydro	4. Peak load

Codes:	A	B	C	D
(a)	1	2	3	4
(b)	2	3	4	1
(c)	3	4	2	1
(d)	1	2	4	3

11.

List 1 (Power Generating Unit)	List 2 (Efficiency)
A. Hydraulic turbine	1. 10–12%
B. Solar cell	2. 60–70%
C. Diesel engine	3. 85–90%
D. Fuel cell	4. 25%

Codes:	A	B	C	D
(a)	1	2	3	4
(b)	2	3	4	1
(c)	3	1	4	2
(d)	3	4	1	2

[A.M.I.E. Sec B. Winter 1993]

12.

List 1 (Type of Tariff)	List 2 (Type of Consumers)
A. Block rate	1. Large industrial
B. Flat rate	2. Medium industrial
C. Two part tariff	3. Domestic and commercial
D. Power factor tariff	4. Agricultural and street lighting

Codes:	A	B	C	D
(a)	3	4	2	1
(b)	1	2	3	4
(c)	2	3	4	1
(d)	4	1	2	3

Answers: 4. (c), **5.** (a), **6.** (c), **7.** (b), **8.** (b), **9.** (a), **10.** (d), **11.** (c), **12.** (a).

TRUE(T)/FALSE(F) TYPE QUESTIONS

Note : Identify whether the following statements are TRUE (T) or FALSE (F).

1. Kerala, Karnataka and J&K are very rich in water power resources and so power development programmes should be based on this source.
2. The power plant should be set up near the load centre as far as possible.
3. In India per capita consumption of electricity is above 1,000 kWh.
4. The present trend, all over the world, is to encourage private sector in energy sector.
5. Generally hydro projects are multipurpose.
6. Hydroelectric power plants are usually located at the load centres.
7. Hydropower plants have low operating cost but construction cost is enormously high.
8. Availability of head of water has considerable effect on the cost and economics of generation.
9. The cost of hydro energy is almost independent of load factor.
10. The firm capacity of hydroelectric power plants is low and so needs to be backed up by steam plants.
11. Availability of storage facility increases the firm capacity of the hydroelectric power plant.
12. The whole area behind the dam draining into a river stream across which the dam has been constructed at a suitable place is called the catchment area.
13. Hydrology is a science which deals with the depletion and replenishment of water resources.
14. The precipitation includes all the water that falls from atmosphere to earth.
15. Total volume of discharge upto any time is given by the area under the hydrograph up to that time.
16. "Flow duration curve" gives the relation between flows and lengths of time during which they are available.
17. Flow duration curve cannot be used for preliminary studies.
18. The basic concept of unit hydrograph is that the hydrographs of run-off from two identical storms would not be the same.
19. Mass curves are used in determination of capacity of storage reservoir in hydro-projects.
20. The "firm power" is also called the primary power.
21. Water held in upstream reservoir is called the pondage.
22. The water behind the dam at the power station is called the storage.
23. Necessity of pondage arises because of plant operation at a load factor less than 100%.
24. A weir is a low overflow dam across a stream to measure flow or maintain water level, at a lake outlet.
25. An emergency spillway comes into action when the occurring flood discharge exceeds the designed flood discharge.
26. A head race is a channel that conducts water from the wheels.
27. A flume is a closed channel erected on the surface or supported above ground on a trestle.
28. A tunnel is an open channel excavated through a natural obstruction.
29. The tunnels are generally the least costly type of conduit for a given length.
30. Exposed penstocks last longer and are easily accessible for inspection and maintenance.
31. Over flow surge tank is very satisfactory and economical.
32. Steel penstocks can be designed for any head with thickness varying with pressure and diameter.
33. Inclined surge tank is cheaper than ordinary one.
34. Hydro turbines can be classified as Pelton, Francis and propeller turbines.
35. The cavitation effect cannot be reduced by polishing the surfaces.
36. The ratio of the volume of water actually striking the runner to the volume of water supplied to the turbine is called the volumetric efficiency.
37. Specific speed of a turbine is the speed of a scale model of turbine which develops 1 metric horse power under a head of 1.0 m.
38. The runner of too low specific speed turbine with low available head increases the cost of the generator because of low turbine speed.
39. An increase in specific speed of turbine is accompanied by lower maximum efficiency and greater depth of excavation of the draft tube.
40. Turbines with low specific speeds operate under a low head and high discharge condition.
41. Specific speed of a hydraulic turbine is directly proportional to power output.
42. Most of the hydro-machines have horizontal configuration.
43. In a Pelton wheel the angle of deflection of the jet through the buckets is taken as 120° if it is not given.
44. Pelton turbines are essentially low head high capacity machines. **[A.M.I.E. Sec B. Power Plant Engineering, Winter 1998]**
45. In case of large sized impulse turbines, vertical shaft arrangement is commonly used.
46. Compared to Francis turbine the efficiency of a Pelton turbine is less affected by variation of load. **[A.M.I.E. Sec B. Power Plant Engineering, Summer 2000]**
47. In a reaction turbine the pressure energy of water is converted into kinetic energy.
48. In reaction turbiness the runner utilises both potential and kinetic energies.

Answers: **1.** (T), **2.** (T), **3.** (F), **4.** (T), **5.** (T), **6.** (F), **7.** (T), **8.** (T), **9.** (T), **10.** (T), **11.** (T), **12.** (T), **13.** (T), **14.** (T), **15.** (T), **16.** (T), **17.** (F), **18.** (F), **19.** (T), **20.** (T), **21.** (F), **22.** (F), **23.** (T), **24.** (T), **25.** (T), **26.** (F), **27.** (F), **28.** (F), **29.** (F), **30.** (T), **31.** (F), **32.** (T), **33.** (F), **34.** (T), **35.** (F), **36.** (T), **37.** (T), **38.** (T), **39.** (T), **40.** (F), **41.** (F), **42.** (F), **43.** (F), **44.** (F), **45.** (F), **46.** (T), **47.** (F), **48.** (T),

49. In a propeller turbine the blades are adjustable.
50. Fixed blades and moving blades are provided on reaction turbines.
51. Francis turbine is used for high head low discharge hydroelectric plants.
52. Cavitation is a serious problem in a Francis turbine as in all reaction turbines.
53. A Kaplan runner is also capable of reverse operation and is ideally suitable for a pumped storage scheme.
54. For low heads, propeller turbines are favoured.
55. Kaplan turbines have variable angle blades for giving high efficiency.
56. Bulb turbines are suitable for small hydroelectric power plants.
57. Standardised bulb sets with unit output ranging from 100 to 1,000 kW operating under heads between 1.5 to 10 m are used in micro-hydel plants.
58. The starting time of a pumped storage plant is very short.
59. Steam power plants have low maintenance and operating costs.
60. Steam power plants are costlier in initial cost in comparison with other types of power plants of same capacity.
61. Regenerative heating improves the thermal efficiency of the cycle.
62. Regenerative cycle thermal efficiency is always lesser than simple Rankine cycle thermal efficiency.
63. The boilers producing steam at pressures of 20 bar and above are called the high pressure boilers.
64. Fire tube boilers use is limited to low cost, small size and low pressure (to about 10 kg/cm^2) plants.
65. Boilers used in central steam power plants of large capacity are invariably water tube boilers.
66. Super-critical pressure boilers with pressures as high as 350 atm, temperature 600° C and capacity 10,00,000 kg/hour are available.
67. In case of oil fired boilers, the excess air is usually 5-12%.
68. Steam generated in boilers for power plants is invariably dry and saturated.
69. Feed water heating with steam at low pressure than boiler pressure usually reduces the overall plant efficiency.
70. For good performance feed water heaters must be drained and vented.
71. Water solids carried over in the steam leaving a boiler drum are called "Carry over".
72. The steam pipes for high temperature applications are made of mild steel.
73. The plain refractory walls are suitable for small plants where the furnace temperature may not be high.
74. Heat recovery equipment does not include superheater and economiser.
75. The superheater tubes are heated by the heat of the combustion gases during their passage from the furnace to the chimney.
76. Superheaters consist of groups of tubes made of carbon steel for steam temperature of 650°F.
77. Superheaters are mainly of two types viz. radiant superheater and convection superheater. Radiant superheaters are more commonly used.
78. Modern steam plants have reheaters as well as superheaters in the same gas passage of the boiler in order to have steam dry as far as possible through the last stage of the turbine.
79. The economiser utilizes the waste heat of the flue gases to reheat the steam.
80. In an economiser, the flue gases flow through the tubes and water outside the tubes across it.
81. Preheating of feed water removes dissolved oxygen and carbon dioxide which corrode boiler.
82. Air from air preheater is sent to the forced draught fans.
83. Corrosion of air preheater tubes can be minimised by limiting the cooling of flue gases to a temperature above the dew point.
84. The recuperative type air preheaters are compact, lighter in weight, occupy less space and provide enormous amount of heat transfer surface and therefore widely used in steam plants.
85. In a jet condenser the exhaust steam and cooling water do not come in direct contact with each other and so do not mix.
86. In counter-flow jet condenser the steam and cooling water enter the condenser from opposite directions.
87. In a surface condenser cooling water flows through the tubes, the steam enters from the top and get condensed due to contact with the tube surfaces and the condensate leaves from the bottom.
88. Maintenance of high vacuum in the condenser is essential for efficient operation.
89. An air pump that removes the moist alone is called a dry air pump.
90. The cooling water supply to the condenser helps in maintaining a low pressure in it.
91. The function of a steam separator is to remove the entrained water particles from the steam conveyed to the steam turbines.
92. In a cooling pond some spray or cooling devices are used.
93. In a cooling tower water is made to trickle down drop by drop so as to come in contact with air moving in the opposite direction.
94. The steam turbine efficiency is considerably reduced when throttle governing is done at low loads.
95. A steam nozzle may be defined as a passage of varying x-section, through which heat energy of steam is converted into potential energy.
96. Fixed and moving blades are provided on impulse turbine.
97. Nozzles are provided on reaction turbines.
98. In an impulse turbine there is a gradual pressure drop and occur continuously on fixed and moving blades.
99. The velocity compounded impulse turbine has high steam consumption and poor efficiency.
100. The blades of impulse steam turbines are of symmetrical profile.
101. In steam turbines, governing may be centrifugal or hydraulic.
102. Steam turbines are manufactured in different sizes up to 1,000,000 hp or even more and have horizontal configuration.
103. For the same turbine output, the coal feed rate to the boiler decreases on implementing regenerative feed heating in a steam power cycle. **[A.M.I.E. Sec B. Winter 1998]**

Answers: **49.** (F), **50.** (T), **51.** (F), **52.** (T), **53.** (T), **54.** (T), **55.** (T), **56.** (T), **57.** (T), **58.** (T), **59.** (F), **60.** (F), **61.** (T), **62.** (F), **63.** (F), **64.** (T), **65.** (T), **66.** (T), **67.** (T), **68.** (F), **69.** (F), **70.** (T), **71.** (T), **72.** (F), **73.** (T), **74.** (F), **75.** (T), **76.** (F), **77.** (F), **78.** (T), **79.** (F), **80.** (F), **81.** (T), **82.** (F), **83.** (T), **84.** (F), **85.** (F), **86.** (T), **87.** (T), **88.** (T), **89.** (T), **90.** (T), **91.** (T), **92.** (F), **93.** (T), **94.** (T), **95.** (T), **96.** (F), **97.** (F), **98.** (F), **99.** (T), **100.** (T), **101.** (T), **102.** (T), **103.** (T),

104. Bituminous coal has low percentage of volatile matter.

105. Anthracite is very soft coal.

106. Peat is the first stage in the formation of coal from wood.

107. An orsat analysis gives volumetric analysis of exhaust gases.

108. A belt conveyor is very suitable means for transporting small quantities of coal over small distances.

109. Belt conveyor is not suitable for greater heights and shorter distances.

110. Chain grate stokers are quite suitable for high ash content coal feeding.

111. In stoker firing fuel of cheaper grade cannot be used.

112. Spreader stokers can be used for burning any type coal from lignite to semi-bituminous.

113. Overfeed stoker feeds the fuel into the grate below the point of air admission.

114. Chain grate and travelling grate stokers are similar in appearance but differ in the grate construction.

115. The underfeed stoker is suitable for burning semi-bituminous and bituminous coals.

116. Bucket elevators are used for carrying coal in horizontal direction.

117. Overfeed stoker is used when unpulverised coal is to be used in large capacity boiler.

118. In burners, too much secondary air may cool the mixture and prevent its heating to the ignition temperature.

119. Excess air requirements in case of pulverised fuels are high in comparison to those in stoker firing.

120. Pulverised coal is fed to the boiler through burners.

121. Coal is pulverised so as to increase its surface exposure, thus resulting in fast combustion without using large quantities of excess air.

122. Most of the steam power plants use pulverised coal in India because of low ash content of coal used.

123. The air supplied for complete combustion is called the primary air.

124. In an electrostatic precipitator dust particles are ionised.

125. Forced draught is a negative pressure drop.

126. Steam jet draught is simple and easy method of producing artificial draught.

127. A balanced draught combines the advantages of both induced and forced draughts.

128. Presence of sulphur in the fuel is considered desirable.

129. For minimizing pollution occurring because of gases discharged from the chimney the chimney height is increased.

130. Dissolved oxygen from water can be removed by adding sodium sulphite.

131. Foaming and priming of the boiler water forms the carry over.

132. Blow down of boiler reduces concentration of solids within the drum.

133. Chromium improves corrosion and oxidation resistance.

134. Zeolite removes hardness almost completely but cannot remove alkalinity.

135. Sodium carbonate is essentially responsible for the scale formation.

136. The scale formation reduces heat transfer and simultaneously increases the metal wall temperature.

137. "Foaming" is the weakening of boiler steel owing to inner crystalline cracks.

138. Demineralisation is usually the most costly method of producing make-up water for high pressure boilers.

139. Nuclear power plants have higher capital cost but lower fuel cost in comparison to those of a conventional coal-fired power plants.

140. A nuclear power plant needs less cooling water as compared to conventional thermal power plants.

141. There is maximum flexibility in the site selection for a nuclear power station.

142. Slow neutrons are very efficient in causing fission of U_{235}.

143. A moderator is used to absorb gamma rays in nuclear power plants.

144. The function of moderator in a nuclear reactor is to slow down the neutron from high velocities.

145. Two principal natural isotopes of uranium are U_{238} and U_{231}.

146. Radioactivity is the phenomenon of spontaneous emission of powerful radiations exhibited by light elements.

147. Radioactive radiations are less penetrating.

148. During a nuclear reaction, the change in mass of the particle represents the release or absorption of energy.

149. 'Nuclear fusion' is the process of combining or fusing two lighter nuclei into a stable and heavier nuclide.

150. Nuclear fission is accompanied by the emission of neutrons and gamma rays.

151. The function of reflector in a nuclear reactor is to stop the chain reaction.

152. In nuclear reactors heavy water is used to slow down neutrons.

153. Graphite, heavy water and light water are the most commonly used moderators in nuclear reactors.

154. The control rods of the reactor absorb the excess electrons produced by fissioning in a nuclear reactor.

155. Half life represents the rate of decay of the radioactive isotopes.

156. The average life is twice of half life.

157. The proportion of fissile material in the fuel is considerably important in determining the critical size of the reactor.

158. Chain reaction is that process in which the number of neutrons goes on increasing in geometrical progression during fission till whole of the fissionable material is disinte- grated.

159. In light water reactor water acts as coolant as well as moderator.

160. A fast breeder reactor produces less fissionable material than that consumed by it.

161. A breeder reactor converts fertile materials into fissionable materials.

162. In a gas cooled reactor there is ample corrosion problem.

163. In a boiling water reactor the feed water acts both as a coolant and a moderator and an enriched uranium is used as fuel.

Answers: **104.** (F), **105.** (F), **106.** (T), **107.** (T), **108.** (F), **109.** (T), **110.** (F), **111.** (F), **112.** (T), **113.** (F), **114.** (T), **115.** (T), **116.** (F), **117.** (T), **118.** (T), **119.** (F), **120.** (T), **121.** (T), **122.** (F), **123.** (F), **124.** (T), **125.** (F), **126.** (T), **127.** (T), **128.** (F), **129.** (T), **130.** (T), **131.** (T), **132.** (T), **133.** (T), **134.** (T), **135.** (F), **136.** (T), **137.** (F), **138.** (F), **139.** (T), **140.** (F), **141.** (T), **142.** (T), **143.** (F), **144.** (T), **145.** (F), **146.** (F), **147.** (F), **148.** (T), **149.** (T), **150.** (T), **151.** (F), **152.** (T), **153.** (T), **154.** (T), **155.** (T), **156.** (F), **157.** (T), **158.** (T), **159.** (T), **160.** (F), **161.** (T), **162.** (F), **163.** (T),

164. Zirconium clad enriched uranium is used as fuel in pressurized water reactor.

165. In an heterogeneous reactor the fuel and moderator are mixed to give a homogeneous material.

166. Water can be used as a coolant in a fast nuclear reactor.
[A.M.I.E. Sec B. Power Plant Engineering, Summer 2000]

167. For the same heating value liquid fuels occupy more space in comparison to solid fuels.

168. A chemical fuel does not release heat energy on combustion.

169. The amount of excess air supplied varies with the type of fuel and the firing conditions.

170. Diesel power plants cannot respond to varying loads easily.

171. Standby losses in diesel electric power plants are enormously high.

172. Maintenance and lubrication cost in case of diesel power plants is quite low in comparison to that in other plants.

173. There is less fire hazard in case of diesel electric power plants.

174. Engines driving electrical generators are of lower speed and simple combustion chamber.

175. No spark plug is used in diesel engines .

176. In an IC engine the combustion of fuel takes place outside the engine cylinder.

177. The top position of the piston towards cover end side of the cylinder is known as bottom dead centre.

178. In a petrol engine the power is produced by compression ignition.

179. Depreciation grows faster in diesel power plant.
[A.M.I.E. Sec B. Power Plant Engineering, Winter 1996]

180. Dry sump lubrication is usually used for high capacity engines.

181. Semi-pressure system of lubrication is combination of splash and pressure systems.

182. The gas turbine is different from the diesel engine in the respect of combustion region being external to the prime mover.

183. Nowadays the gas turbine power plants use mainly gas oil and residual oil.

184. In gas turbines thermal efficiency is lower in case of closed cycle in comparison to that of open cycle.

185. The efficiency of coal fired gas turbine power plant is higher than that of oil fired plant.

186. The problem of ash disposal does not arise in case of gas turbine plants.

187. A combined cycle power generating plant utilizes the waste heat from gas turbines.

188. Gas turbines are self starting.

189. In almost all the fields closed cycle gas turbine plants are used.

190. A heat exchanger is usually employed in large gas turbine units for industrial power.

191. Minerals such as sodium and calcium prove very harmful for turbine blades.

192. Combined cycle power plants are less efficient than single cycle power plants.

193. The MHD power generator consists of a pipe whose diameter increases gradually from beginning to end.

194. If a liquid metal is used as a working fluid in an MHD system the conductivity can be increased many times, and can also be used at low temperatures under low magnetic fields.

195. MHD power generator operates on the principle that when a hot, partially ionized and compressed gas is expanded in a duct, and forced through a strong magnetic field, electric potential is generated in the gas.

196. An MHD generator cannot be used in combination with a conventional thermal power plant because its overall efficiency will decrease.

197. The process of adding a material of low ionization potential to the gas to increase its conductivity is called regenerating.

198. MHD power plants are less efficient in comparison to conventional steam power plants.

199. An ionic gas is capable of carrying current.

200. Solar cells convert solar energy directly into electrical energy.

201. In a fuel cell hydrogen reacts with oxygen to produce electricity.

202. Operating cost of a wind mill is negligible.

203. The winds as source of power generation are quite steady and dependable.

204. Fuel cells are particularly suited for low voltage and high current applications.

205. India has geothermal resources such as hot spring water of base temperatures of 180 to 200ºC in the Himalyan ranges in Himachal Pradesh.

206. The run-off river power plants have very high firm capacity.

207. Bengal and Bihar are ideal for nuclear power plants.

208. A super thermal power plant has units of 500 MW and above.

209. Among the operating costs of a steam power plant fuel has the largest share.

210. Maximum potential for hydropower plants is in northern India.

211. Plant heat rate can be improved by using fuel of better quality or by adopting better thermodynamic conditions in the plant design.

212. Diesel power plants have low operation efficiency than that of steam power plants.

213. In a generating station it is always desirable to have two or more units.

214. Single peak in load curve means high load factor.

215. High load factor can be expected in peak load plants.

216. Major part of electrical power generated in India is shared by nuclear power plants.

217. In an interconnected power system the peak load is supplied by hydropower plants when the maximum flow demand is less than stream flow while the base load is supplied by thermal power plants.

218. The penalty factor is always less than 1.

219. Generally the alternators employed in hydropower stations have vertical configuration.

Answers: **164.** (T), **165.** (F), **166.** (F), **167.** (F), **168.** (F), **169.** (T), **170.** (F), **171.** (F), **172.** (F), **173.** (T), **174.** (T), **175.** (T), **176.** (F), **177.** (F), **178.** (F), **179.** (T), **180.** (T), **181.** (T), **182.** (T), **183.** (F), **184.** (F), **185.** (F), **186.** (T), **187.** (T), **188.** (F), **189.** (F), **190.** (T), **191.** (T), **192.** (F), **193.** (T), **194.** (T), **195.** (T), **196.** (F), **197.** (F), **198.** (F), **199.** (T), **200.** (T), **201.** (T), **202.** (T), **203.** (F), **204.** (T), **205.** (T), **206.** (F), **207.** (F), **208.** (T), **209.** (T), **210.** (T), **211.** (T), **212.** (F), **213.** (T), **214.** (F), **215.** (F), **216.** (F), **217.** (F), **218.** (F), **219.** (T),

220. In modern power stations each generator has its own transformer.

221. A plant layout is the first step in installing a power plant. [A.M.I.E. Sec B. Winter 1996]

222. The annual depreciation of a hydropower plant is about 5 per cent.

223. The power plant capacity can be determined by studying the load duration curve and anticipated future demands.

224. Maximum demand on a power system is the instantaneous maximum value of kVA supplied during a given period.

225. Higher the maximum demand, better is the utilisation of total plant capacity.

226. The main purpose of design and operation of the power plant is to generate the electrical energy at minimum possible cost.

227. Transmission lines also form a part of capital cost of the power plant.

228. Depreciation is a fixed cost for a power plant.

229. The salaries of high officials is accounted for as running charges.

230. Fuel consumption depends upon the installed capacity of the power plant.

231. Cost of labour is considered as operating cost where as the depreciation is taken as fixed cost.

232. Obsolescence is because of technological advancements.

233. Salvage value of a plant is always zero.

234. Depreciation accounts for the wear and tear with use.

235. Operating cost of a nuclear plant is more than that of a hydroplant.

236. Operating cost of a nuclear plant is more than that of a steam power plant.

237. Operating cost of a diesel power plant is less than that of a gas turbine plant.

238. The peak of the load curve represents the maximum demand on the power station on the particular day.

239. Load curve is derived from the load duration curve.

240. Higher the value of load factor, lower is the cost of generation.

241. Higher the value of the diversity factor lesser will be the value of fixed charges per unit generated.

242. Domestic consumers have maximum demand during early night hours.

243. There is an inherent reserve in most electrical supply systems in that the load can be reduced by reducing the voltage.

244. A power plant having maximum demand more than installed capacity will have utilisation factor of lesser than unity.

245. While selecting units continuous rating of generators is taken into account.

246. In a generating station it is always desirable to have two or more generating units.

247. For a generator at a particular load, when incremental heat rate is equal to heat rate, it will operate at maximum efficiency.

248. In two part tariff, the consumer with low power factor will be charged more on account of demand charges.

249. In a two part tariff energy charges do not depend on power factor.

250. In a power factor tariff, charge per unit to be recovered from a consumer having high power factor is less.

251. Hopkinson demand rate is a three part tariff.

252. Doherty rate tariff is a two part tariff.

253. Industrial plants usually operate on leading power factor .

254. Power factor tariff is charged on the basis of maximum demand in kVA.

255. Low power factor leads to higher energy loss.

256. Low power factor results in poor voltage regulation.

257. It is advisable to improve power factor to unity.

258. Improvement in power factor of an installation results in increase in kVA demand of the installation.

259. Capacitor banks are installed for improvement of power factor.

260. For smaller kVAR requirements it is economical to use static capacitors rather than synchronous capacitors.

Answers: **220.** (T), **221.** (T), **222.** (F), **223.** (T), **224.** (F), **225.** (T), **226.** (T), **227.** (T), **228.** (T), **229.** (F), **230.** (F), **231.** (T), **232.** (T), **233.** (F), **234.** (T), **235.** (T), **236.** (F), **237.** (F), **238.** (T), **239.** (F), **240.** (T), **241.** (T), **242.** (T), **243.** (T), **244.** (F), **245.** (T), **246.** (T), **247.** (T), **248.** (F), **249.** (T), **250.** (T), **251.** (F), **252.** (F), **253.** (F), **254.** (T), **255.** (T), **256.** (T), **257.** (F), **258.** (F), **259.** (T), **260.** (T).

FILL IN THE BLANKS TYPE QUESTIONS

Note : Fill in the blanks with appropriate word or words.

1. In India more than 60% of electric power is generated by ______ power plants.

2. The installed capacity as on 31-3-99 was ______ MW while total electrical energy generation as on 31-3-99 was ______ million kWh per year.

3. An hydroelectric power plant operates under an effective head of 50 m and discharge of 94 m^3/s. The power developed will be ______ MW.

4. The ______ power plant will give least running cost of production of electricity. [A.M.I.E. Sec B. Summer 1994]

5. Useful life of hydroelectric power plants is about ______ years.

6. A hydrograph is a plot of ______ vs ______.

[A.M.I.E. Sec B. Power Plant Engineering, Winter 1998]

7. One day-second metre of run-off = ______ m^3/day.

8. ______ is a curve which indicates the total volume of run-off in m^3 up to a certain time.

9. Storage capacity should be determined with the help of ______ curve.

10. ______ is a closed conduit for supplying water under pressure to the turbine.

11. ______ are employed where the slope is too great for a canal.

12. The main purpose of ______ in a pipeline is to relieve the pressure due to water hammer. [A.M.I.E. Sec B. Winter 1993]

13. ______ is provided to absorb sudden changes in water requirements.

14. An earth dam has a very _______ base as compared to its height.

15. In a _______ turbine the ratio of width of the wheel to its diameter varies from 0.1 to 0.4.

16. In a _______ turbine the flow ratio varies from 0.4 to 0.6.

17. Pelton turbine uses _______ energy of water and is suitable for _______ heads. **[A.M.I.E. Sec B. Power Plant Engineering, Winter 1999]**

18. In case of reaction turbines _______ shaft arrangement is universally adopted.

19. The efficiency of Pelton turbine is _______ than that of a Kaplan turbine. **[A.M.I.E. Sec B. Summer 1993]**

20. High specific speed turbines work under _______ head and _______ discharge conditions. **[A.M.I.E. Sec B. Power Plant Engineering, Winter 1999]**

21. The _______ turbine can be operated both in the turbine mode and pump mode. **[A.M.I.E. Sec B. Power Plant Engineering, Summer 2000]**

22. Bulb turbines are used for _______ head plants. **[A.M.I.E. Sec B. Summer 1993.]**

23. The efficiency of a pumped storage plant is in the range of _______ per cent.

24. Hydro alternators are of a _______ type. **[A.M.I.E. Sec B. Summer 1996]**

25. Steam power plants are _______ in operating cost in comparison to hydro and nuclear power plants.

26. Industrial power plants or captive power plants are normally _______ .

27. In water tube boilers water flows _______ the tube and hot gases _______ .

28. All modern high pressure boilers use _______ circulation of water.

29. _______ is the largest heat exchanger in a steam power plant. **[A.M.I.E. Sec B. Summer 1998]**

30. The furnace wall containing water circulating pipes is called _______. **[A.M.I. E. Sec B. Winter 1994]**

31. The function of superheater is to _______ the steam temperature.

32. A _______ is a device which removes the last traces of moisture from the saturated steam leaving the boiler tubes and also increases the temperature above the saturation temperature.

33. The function of the _______ is to superheat the partly expanded steam from the turbine.

34. An _______ is a device that utilises the waste heat of the flue gases to heat the feed water.

35. Economiser and _______ are such devices which recover the heat from the flue gases on their way to chimney and raise the temperature of feed water and air supplied for combustion respectively.

36. Generally economisers improve boiler efficiency by _______ % to _______ %. **[A.M.I.E. Sec B. Summer 1998]**

37. Air preheaters utilises the _______ left in the _______ gases before they are discharged to the atmophere. **[A.M.I.E. Sec B. Summer 1998]**

38. Air preheater improves the _______ of a steam power plant. **[A.M.I.E. Sec B Summer 1994]**

39. A feed water heater is basically a _______.

40. The make-up water in a modern steam power plant is about _______ to _______ per cent.

41. The removal of the mud and other impurities of water from the lowest part of the boiler is called the _______.

42. With the addition of heat recovery apparatus, the efficiency of modern boilers can be increased to between _______ % and _______ %. **[A.M.I.E. Sec B. Summer 1994]**

43. In a steam power plant the upper limit of feed water heaters may be _______ to _______. **[A.M.I.E. Sec B. Winter 1994]**

44. Regenerative feed water heating improves _______ of the cycle. **[A.M.I.E. Sec B. Summer 1996]**

45. _______ is a device in which the steam condenses and heat so released is absorbed by the cooling water.

46. _______ consists of a water tank in which hot water is distributed by pipes throughout and is sprayed in air through nozzles at a suitable pressure.

47. The blades of steam reaction type turbines are of _______ section.

48. There are two material problems of steam turbine blades, namely _______ and stress corrosion cracking. **[A.M.I.E. Sec B. Winter 1993]**

49. The percentage of ash content in the coal varies from about _______ per cent in good quality coals to _______ per cent in poor quality coals.

50. Lignite is _______ calorific value coal. **[A.M.I.E. Sec B. Summer 1994]**

51. Any natural fuel of any organic nature such as coal, crude oil and natural gas is called the _______ fuel.

52. Coaking coal is a form of _______ coals.

53. Hydrogeneous oil contains excess of _______.

54. The average coal consumption varies from _______ kg per kWh generated for plants using high grade coal and operating at high load factor to _______ kg per kWh generated for plants using low grade coal and operating at low load factor.

55. A _______ is power operated fuel feeding mechanism and grate.

56. Travelling grate stokers are of two types namely bar grate and _______ grate.

57. The equipment used for pulverising the coal is _______. **[A.M.I.E. Sec B. Summer 1994]**

58. _______ removes particulate matter by giving electrical charge to the particles.

59. The electrostatic precipitators are most effective on _______. **[A.M.I.E. Sec B. Summer 1994]**

60. In electrostatic precipitator the applied voltage is of the order of _______ to _______ V dc.

61. The small pressure difference that causes a flow of gas to take place is called the _______.

62. In _______ the water is allowed to stand at stand-still in large tanks inorder to allow solid matter to settle down.

63. The efficiency of large turbo-alternator can be taken as _______% with hydrogen cooling. **[A.M.I.E. Sec B. Summer1996]**

64. _______ is the basic unit for describing the intensity of radioactivity.

65. In a nuclear power plant, the power is produced by nuclear fission _______ . **[A.M.I.E. Sec B. Summer 1994]**

66. _______ or _______ is used to suppress the rate of nuclear reaction. **[A.M.I.E. Sec B. Winter 1994]**

67. Natural uranium mainly consists of _______, however, it is only _______ isotope that fissions readily. **[A.M.I.E. Sec B. Winter 1999]**

68. CANDU PHW stands for _______. **[A.M.I.E. Sec B. Winter 1999]**

69. _______ type reactors do not require any moderator. **[A.M.I.E. Sec B. Winter 1998]**

70. The fuel used in diesel engine power plant is ________. [A.M.I.E. Sec B. Winter 1993]

71. The overall efficiency of diesel engine is from ________ % to ________ %. [A.M.I.E. Sec B. Power Plant Engineering, Summer 1996]

72. A turbocharger is a part of ________. [A.M.I.E. Sec B. Summer 1994]

73. In diesel engines ________ are provided to remove dust and other suspended impurities from the air to be supplied to the engines.

74. The efficiency of open cycle gas turbine plant can be improved to ________ per cent by using regenerator, intercooler and reheater.

75. The thermal efficiency of diesel plant is ________ than the gas turbine plant. [A.M.I.E. Sec B. Summer 1993]

76. Combined cycle plants are ________ efficient than single cycle plants.

77. Aero-derivative turbine has ________ fan(s).

78. Aero-derivative turbine uses ________ fuel.

79. The main constituents of natural gas are methane and ________.

80. The process of adding a material of low ionization potential to the gas to increase its conductivity is called ________.

81. ________ and ________ salts are used as seed materials for MHD generation. [A.M.I.E. Sec B. Summer 1998]

82. Thermoelectric converters operate on the principle of ________.

83. Modern solar cells are made of ________. [A.M.I.E. Sec B. Winter 1993]

84. An ordinary fuel cell uses ________ as fuel and ________ as oxidizer. [A.M.I.E. Sec B. Winter 1998]

85. The maximum possible theoretical efficiency of a wind energy convertor is ________ %. [A.M.I.E. Sec B. Winter 1998]

86. Tidal plants can give only ________ supply of power. [A.M.I.E. Sec B. Summer 1993]

87. Peak load plants operate at ________ load factor.

88. The unit of heat rate is ________.

89. The penalty factor is always ________.

90. The unit of incremental cost is ________.

91. For most economic generation of power in a thermal station having several units, ________ should be same for all units.

92. When the power system is not in a position to meet the demand it will resort to ________.

93. Standard frequency for power supply in India is ________ Hz.

94. H_2 cooled generators are ________ than the air cooled generators. [A.M.I.E. Sec B. Summer 1993]

95. Modern alternators use ________ regulators.

96. The value of the plant at the end of its useful life is known as ________ value.

97. A load curve is the relationship between ________. [A.M.I.E. Sec B. Summer 1994]

98. Load curve is a graph between ________ on X-axis and ________ on Y-axis. [A.M.I.E. Sec B. Summer 1999]

99. ________ curve gives the number of hours for which a particular load lasts during the day.

100. The sum of continuous ratings of all the electrical equipment connected to the supply system is known as ________ load.

101. The greatest of all 'short time interval averaged' during a given period on the power station is called the ________.

102. Load factor is the ratio of ________ to ________. [A.M.I.E. Sec B. Summer 1999]

103. Demand factor is the ratio of ________ to ________. [A.M.I.E. Sec B. Summer 1999]

104. Diversity factor is the ratio of ________ to ________. [A.M.I.E. Sec B. Summer 1999]

105. The peak load of a 40 MW capacity power station is 30 MW. The loads having maximum demands of 25 MW, 10 MW, 5 MW and 7 MW are connected to the system. The diversity factor of the plant is ________. [A.M.I.E. Sec B. Summer 1993]

106. In Q. 105, if the load factor for the year is 50%, the energy supplied by the system is ________ kWh/year. [A.M.I.E. Sec B. Summer 1993]

107. Plant capacity factor is the ratio of ________ to ________. [A.M.I.E. Sec B. Summer 1999]

108. The utilisation factor is defined as the ratio of ________ to ________. [A.M.I.E. Sec B. Summer 1999]

109. Firm power is the power intended to be available ________ even under ________ condition. [A.M.I.E. Sec B. Summer 1999]

110. Cold reserve is the reserve generating capacity which is available for ________ but is not in ________. [A.M.I.E. Sec B. Summer 1999]

111. Hot reserve is the reserve generating capacity which is in ________ but not in ________. [A.M.I.E. Sec B. Summer1999]

112. Spining reserve is that generating capacity which ________ and is ready to ________. [A.M.I.E. Sec B. Summer 1999]

113. Most of the power generating units are conservatively rated. They can be easily overloaded by about ________ percent for short durations.

114. ________ load includes domestic lights, power for domestic appliances and small motors for pumping water.

115. ________ load includes lighting for commercial establishments, advertising, and electrical appliances used in shops and restaurants.

116. ________ load consists of street lighting, power required for water supply and drainage purposes.

117. ________ load includes electrical power required for driving water pumps to supply water to fields.

118. ________ are the different methods of charging the consumers for electrical energy consumed by them.

119. ________ is applied to medium industrial consumers.

120. ________ is applied to bulk supplies.

ANSWERS

1. steam
2. 94,055; 448,000
3. 46.1
4. hydroelectric
5. 40
6. discharge or flow; time
7. 86,400
8. Mass curve
9. mass curve
10. Penstock
11. Penstocks
12. a surge tank
13. Surge tank
14. wide
15. Francis
16. Francis
17. kinetic; high
18. vertical
19. lower
20. low; large
21. Kaplan
22. low
23. 60-70
24. salient pole
25. costlier
26. non-condensing
27. inside; outside
28. forced
29. Boiler
30. water wall
31. raise
32. superheater
33. reheaters
34. economiser
35. air preheaters
36. 4; 6
37. heat; flue
38. efficiency
39. heat exchanger
40. 2; 5
41. blowing off
42. 10; 13
43. 5; 9
44. thermal efficiency
45. Condenser
46. Spray pond
47. aerofoil
48. pitting
49. 5; 40
50. low
51. fossil
52. bituminous
53. moisture
54. 0.16; 1.6
55. stoker
56. chain
57. ball mill
58. Electrostatic precipitator
59. fine dust particles
60. 25,000; 100,000
61. draught
62. coagulation
63. 98–99
64. Curie
65. of heavy atom like U_{235}
66. Moderator; control rod
67. U_{238} ; U_{235}
68. Canadian deuterium uranium pressurised heavy water.
69. Fast breeder
70. light diesel oil
71. 28; 35
72. diesel engine
73. air filters
74. 30
75. more
76. more
77. no
78. clean
79. ethane
80. seeding
81. Cesium; potassium
82. Seebeck effect
83. silicon
84. hydrogen; oxygen
85. 59
86. uneven
87. low
88. kcal/MWh;
89. more than 1
90. Rs per MWh
91. incremental costs
92. load shedding
93. 50
94. of large size
95. electronic
96. salvage
97. load on the power system with the time.
98. time; kW
99. Load duration
100. connected
101. maximum demand
102. average load; maximum demand
103. maximum demand; connected load
104. sum of individual maximum demand; maximum demand of power plant.
105. 1.5667
106. 131.4×10^6
107. average load; rated capacity
108. maximum demand; rated plant capacity
109. always; emergency
110. service; operation
111. operation; service
112. is connected to bus; take load
113. 10–15
114. Domestic
115. Commercial
116. Municipal
117. Irrigation
118. Tariffs
119. Hopkinson demand rate
120. Doherty rate.

❑❑❑

PART II

TRANSMISSION AND DISTRIBUTION OF ELECTRICAL POWER

CHAPTER 1

Supply Systems

INSIDE THIS CHAPTER

1.1. POWER SYSTEM STRUCTURE

Power is the basic need for the economic development of any country. Availability of electricity has been the most powerful vehicle of introducing economic development and social change throughout the world. The process of modernisation, increase in productivity in industry and agriculture and the improvement in the standard of living of the people basically depend upon the adequate supply of electrical energy. Appropriately the programmes relating to the generation, transmission and distribution of electrical energy have been the highest priority in the national planning process.

Cenerating stations, transmission lines and distribution systems are the main components of an electrical power system. Generating stations and distribution systems are connected through transmission lines, which also connect one power system to another. A distribution system connects all the loads in a particular area to the transmission line.

The transmission system of an area (or state) is called the *grid.* The different grids are interconnected through tie lines to form a regional grid and the different regional grids are further interconnected to form a national grid. Each grid operates independently. However, power can be transmitted from one grid to another, via the tie lines under conditions of sudden loss of generation or increase in load.

The distribution system in India is quite extensive. The voltages used in distribution systems are 11, 6.6 and 3.3 kV at ht level (or for primary distribution) and 415 volts at lt level (or for secondary distribution).

The country has been divided into five regions (Northern, Western, Eastern, Southern and Northern-Eastern), each with a regional electricity board, so as to promote integrated operation of the constituent power systems. Each State has a State Electricity Board, responsible for generation, transmission and distribution of electrical power in their respective states. Central Government has also control over many generating plants, transmission lines and substations through central organisations like NTPC (National Thermal Power Corporation), NHPC (National Hydroelectric Power Corporation), PGCIL (Power Grid Corporation of India Ltd.) etc. Regional load despatch centres have been established for coordinating the activities of state load despatch centres so as to ensure optimum utilization of generation and transmission facilities in the regions. Setting-up of a National Load Despatch centre is under active consideration.

Electrical energy is normally generated at the power stations far away from the urban areas where consumers are located. The problem, therefore, is to transport (or more usually called transmit) the large blocks of power over long distances economically. In this volume of the book we will deal with the type of equipment, techniques and problems involved in transmitting large blocks of power over long distances from the generating stations to the load centres.

1.2. ELECTRICAL SUPPLY SYSTEM

Electrical energy is generated at the power stations (hydro-electric, thermal or nuclear) which are usually situated far away from the load centres. Hence an extensive network of conductors between the power stations and the consumers is required. This network of conductors may be divided into two main components, called the *transmission system and the distribution system.* The generation, transmission and distribution system of electrical power is called the *electrical power supply system.* The pictorial diagram of a typical electrical supply system is shown in Fig. 1.1.

The *transmission system is* to deliver bulk power from power stations to the load centres and large industrial consumers beyond the economical service range of the regular primary distribution lines where as *distribution system* is to deliver power from power stations or substations to the various consumers.

Although electrical power can be transmitted and distributed by either ac or dc but in practice 3-phase 3-wire ac system is universally adopted for transmission of large blocks of power and 3-phase 4-wire ac system is usually adopted for distribution of electrical power.

The transmission system may be further divided into primary and secondary (or sub) transmission. Similarly the distribution system may be divided into primary and secondary distribution.

The maximum generation voltage in advanced countries is 33 kV while that in India is 11 kV. The primary transmission voltages are 220-765 kV (220, 400 or 765 kV) depending upon the distance, the amount of power to be transmitted and the system stability. Secondary transmission voltages are 33-132 kV (33, 66, or 132 kV). The voltages for primary distribution are 3.3, 6.6 or 11 kV, most usual value adopted in practice is 11 kV. The usable voltage for secondary distribution is 415/240 V.

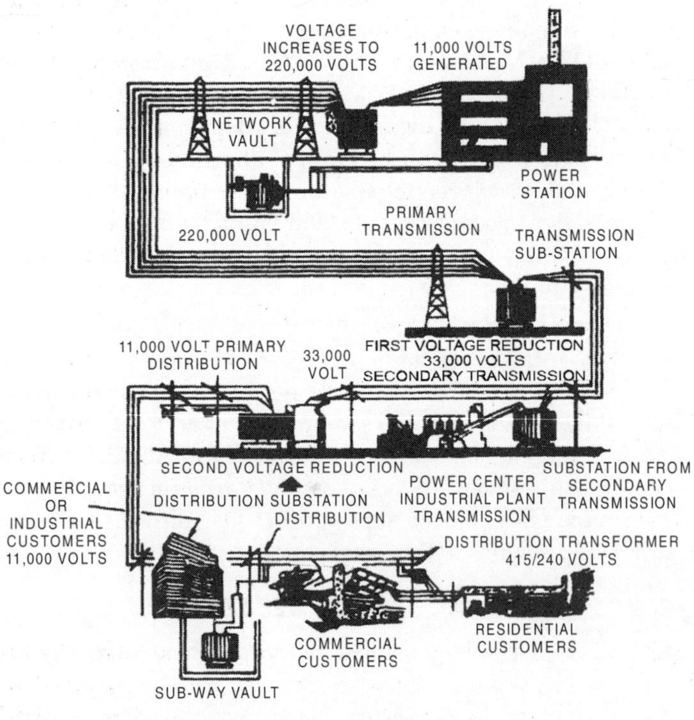

Fig. 1.1. ***Pictorial Diagram of a Typical Electrical Power Supply System***

The distribution system may further be divided into feeders, distributors and service mains.

(*i*) *Feeders.* Feeders are the conductors which connect the stations (in some cases generating stations) to the areas to be fed by those stations. Generally from feeders no tapping is taken to the consumers, therefore, current loading of a feeder remains the same along its length. It is designed mainly from the point of view of its current carrying capacity.

(*ii*) *Distributors.* Distributors are the conductors from which numerous tappings for the supply to the consumers are taken. The current loading of a distributor varies along its length. Distributors are designed from the point of view of the voltage drop in them.

(*iii*) *Service Mains.* Service mains are the conductors, which connect the consumer's terminals to the distributor. Elements of a distribution system are shown in Fig. 1.2.

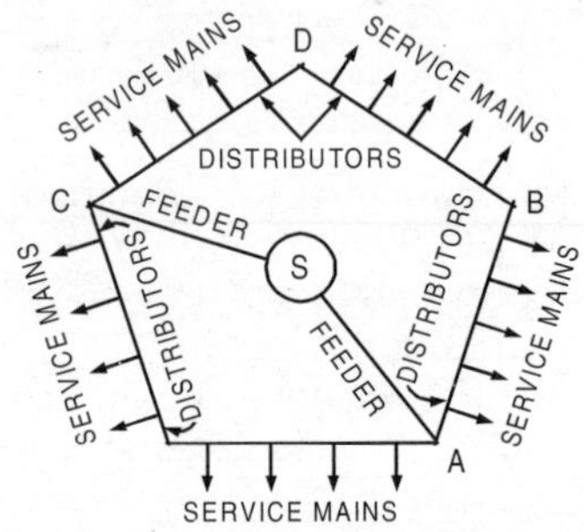

Fig. 1.2. ***Elements of a Distribution System***

A distributor is subject to the legal requirement that the voltage at the consumer's terminals should be maintained within ±6% of the declared (or rated) voltage. However, there is no such legal restriction on a transmission line and the voltage can vary as much as 10% or even 15% due to variations in loads.

Modern ac electrical power supply system usually consists of the following elements:

(*i*) generating stations (*ii*) step-up transformer substations (*iii*) primary transmission lines (*iv*) primary transmission substations (*v*) secondary (or sub) transmission lines (*vi*) secondary transmission substations (*vii*) primary (high voltage) distribution lines (or networks) (*viii*) service transformer banks (*ix*) secondary (or low voltage) distribution lines (or networks). Essentially elements (*ii*), (*iii*), (*iv*), (*v*) and (*vi*) fall in the transmission system while the elements (*vii*), (*viii*) and (*ix*) fall in the distribution system.

All systems of power transmission and distribution may or may not include all elements enumerated above; for example, some systems may have no primary transmission, some may not have secondary transmission and the others may not have transmission at all, being very small and so on.

1.3. TYPICAL AC ELECTRICAL POWER SUPPLY SYSTEM

Block diagram of a typical ac electrical power supply system is shown is Fig. 1.3. Electrical power is generated by 3-phase alternators operating in parallel at the generating stations. The generation voltage in India is 11 kV. For economy reasons this voltage is stepped up from 11 kV to 220 kV* at the generating stations by means of 3-phase transformers.

* This voltage may be 400 or 765 kV depending upon the length of transmission line and the amount of electrical power to be transmitted.

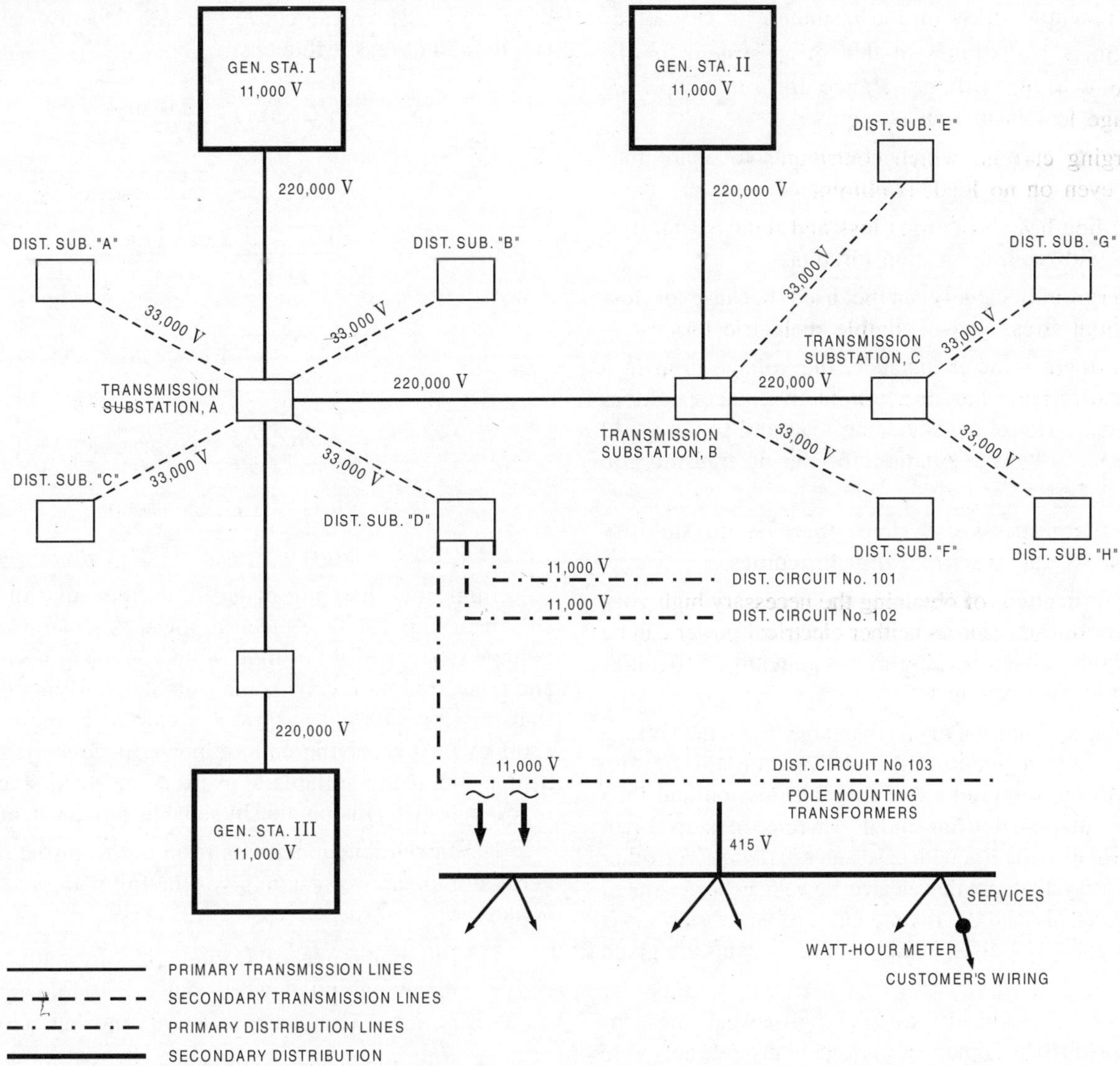

Fig. 1.3. *Block Diagram of a Typical AC Electrical Power Supply System*

The primary (or high voltage) transmission lines transmit electrical power from the generating stations to primary transmission substations. At these primary substations the voltage is stepped down to 33-132 kV level by means of stepdown transformers. From the primary substations electrical power is transmitted through secondary (or sub) transmission lines at 33 kV (66 or 132 kV in some cases) to various secondary transmission substations. At the secondary substations the voltage is further stepped down to 11 kV by means of 3-phase transformers and the power is supplied to distribution substations through primary (or high voltage) distribution lines at 11 kV. The distribution substations are located at suitable places in the area in which power is to be supplied. At these substations the voltage is stepped down to 415 V. The secondary (or low voltage) distribution lines, called the *distributors,* are laid along the roads and service connections to the consumers are tapped off from the distributors.

Most of the domestic, commercial and small scale industrial consumers are supplied power at low voltage (at 415 volts for 3-phase loads and 240 volts for single phase loads). However, large commercial organisations, public buildings and industries having load in excess of 100 kW prefer to have bulk power supply at 11 kV (or even at 33 kV in case of large industries). This is accomplished by taking 11 kV line directly from secondary transmission substations right up to the premises of such consumers and connecting it to an 11 kV/415 V stepdown transformer.

1.4. COMPARISON BETWEEN DC AND AC SYSTEMS FOR TRANSMISSION AND DISTRIBUTION

Transmission of electric power by high voltage dc system is superior to that by high voltage ac system in many respects.

(*i*) It requires only two conductors for transmission and it is also possible to transmit the power through only one conductor by using earth as returning conductor, hence much copper is saved.

(*ii*) No inductance, capacitance, phase displacement and surge problem in dc transmission.

(*iii*) There is no skin effect in dc, cross section of line conductor is, therefore, fully utilised.

(*iv*) The potential stress on the insulation in case of dc system is $1/\sqrt{2}$ times of that in ac system for the same working voltage. Hence for same working voltage less insulation is required.

(*v*) Charging current, which contributes to continuous loss even on no load, is eliminated.

(*vi*) A dc line has less corona loss and reduced interference with communication circuits.

(*vii*) Underground cables can be used because of less potential stress and negligible dielectric loss.

(*viii*) Since there is no inductance, the voltage drop in a dc transmission line due to inductive reactance does not exist. Hence for the same load and sending-end voltage, voltage regulation for the dc transmission line is better.

(*ix*) In dc transmission system, there is no stability problems and synchronising difficulties.

The only difficulty is of obtaining the necessary high voltage required for transmission as neither electrical power can be generated at high voltage because of commutating difficulties nor voltage can be stepped up.

In case of ac systems the main advantage is that the voltage can be stepped up at generating end by means of step-up transformers to the required value for transmission and then stepped down at distributing end by means of step-down transformers for distribution with ease and efficiency. The other advantages of ac systems are that electric power can be generated at high voltage easily and the maintenance of ac substations is easier and cheaper. The disadvantages of ac systems are given below:

(*i*) In case of overhead lines spacing between the conductors is to be kept more in order to provide adequate insulation and to avoid corona loss.

(*ii*) More copper is required.

(*iii*) Greater the length of transmission line more is the capacitance of the line. There is a continuous loss on account of charging current even though the line is open.

(*iv*) The construction of transmission lines for ac systems is not so easy as for dc systems.

(*v*) The alternators are to be synchronised before putting them into parallel.

(*vi*) The variation in speeds of alternators are to be controlled within very low limits.

Distribution by ac system is undoubtedly superior to that by dc system as in ac system voltage control is easy by means of transformers.

The best method is to employ ac system for generation and distribution and dc system for transmission.

Transmission of electric power by dc system became possible because of introduction of mercury-arc rectifiers and thyratrons*, which can convert ac into dc and vice versa efficiently and at reasonable cost.

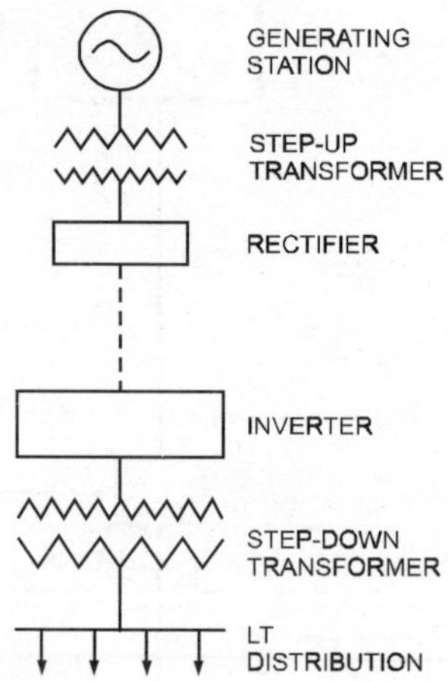

Fig. 1.4. ***Line Diagram of Typical HVDC Transmission System***

Single line diagram of an hv dc transmission system is shown in Fig. 1.4. The electric power is generated as ac, its voltage is stepped up to high voltage by step-up transformer and converted into dc by some suitable rectifying device. The transmission of electric power is carried at high dc voltage (500 kV). At receiving end, dc power is converted back into 3-phase ac using suitable converters and then stepped down to low voltage for distribution by suitable step-down transformers.

The favourable conditions resulting from the use of converters in the above system gives the following additional advantages:

1. Independence of frequencies at sending-end and receiving-end networks makes possible generation of power at a frequency different from that of load system.
2. With the use of suitable converters, power can be re versed, increased or decreased in the system at will and with rapidity *i.e.*, the control is easy and rapid.
3. Because of absence of charging and stability problems, there is no limit of distance over which dc power can be transmitted.
4. Keeping in view the circuit breaker capacities etc., the short-circuit currents are not transmitted through lines and hence a tremendous amounts of saving is affected by installing smaller circuit breakers.
5. The smaller amount of power can be transmitted much more economically.

1.5. COMPARISON BETWEEN OVERHEAD AND UNDERGROUND SYSTEMS

Transmission and distribution of electric power can be carried out by overhead as well as underground systems. Comparison between the two is given below:

(*i*) *Public Safety*: Underground system is more safer than overhead system.

* Since the early 1970s, all new HVDC schemes use thyristors instead of earlier used mercury-arc valves.

(*ii*) *Initial Cost*: Underground system is more expensive. The approximate cost ratios are as below :

System Voltage in kV	0.4	11	33	66	132	220	400
Cost Ratio Cable/OH Line	2	3	5	7	9	13	24

(*iii*) *Flexibility*: Overhead system is more flexible than underground system. In overhead system new conductors can be laid along the existing ones for load expansion. In case of underground system new conductors are to be laid in new channels.

(*iv*) *Working Voltage*: The underground system cannot be operated above 66 kV because of insulation difficulties but overhead system can be designed for operation up to 400 kV or higher even.

(*v*) *Maintenance Cost*: Maintenance cost of underground system is very low in comparison with that of overhead system.

(*vi*) *Frequency of Faults or Failures*: As the cables are laid underground, so these are not easily accessible. The insulation is also better, so there are very few chances of power failures or fault as compared to overhead system.

(*vii*) *Frequency of Accidents*: The chances of accidents in underground system are very low as compared to overhead system.

(*viii*) *Voltage Drop*: In underground system because of less spacing between the conductors inductance is very low as compared to overhead lines, therefore, voltage drop is low in underground system.

(*ix*) *Appearance*: Underground system of distribution or transmission is good looking because no wiring is visible. Due to its good looking, inspite of its higher cost it is adopted in modern cities like Chandigarh, Faridabad.

(*x*) *Fault Location and Repairs*: Though there are very rare chances of occurring fault in underground system, but if occurs it is very difficult to locate that fault and its repair is difficult and expensive.

(*xi*) *Charging Current*: On account of less spacing between the conductors the cables have much capacitance, so draw higher charging current.

(*xii*) *Jointing*: Jointing of underground cables is difficult so tapping for loads and service mains is not conveniently possible in underground system.

(*xiii*) *Damage Due to Lightning and Thunderstorm*: Underground system is free from interruption of service on account of thunderstorm, lightning and objects falling across the wires.

(*xiv*) *Surge Effect*: In underground system surge effect is smoothened down as surge energy is absorbed by the sheath.

(*xv*) *Interference to Communication Circuits*: In underground system there is no interference to communication circuits.

Mostly the high voltage transmission is carried out by overhead system due to low cost. However, distribution in conjested areas and in modern cities, like Chandigarh, is carried out by underground cables.

The overhead line as a mean of transmitting electrical power over long distances is cheap and efficient (a 400 kV double-circuit line may transmit 2,000 MW over a distance of 160 km in either direction with an efficiency of 98%). It is reliable—there are only one or two breakdowns due to lightning, and one or two due to other reasons, such as frost and fog, per 160 km of a line a year. Most of the breakdowns which do occur are transient and none out of them can be rendered innocuous, as far as system is concerned, by using high-speed automatic-reclosing circuit breakers. Repairs, where necessary, can easily and quickly be carried on site. The only real disadvantage of the overhead line is that it needs large electrical clearances and it is sometimes not possible to get a number of circuits into congested areas.

EHV power cables are used in congested or built-up areas, in the vicinity of airfields, line terminations and transformer banking connections where the use of overhead lines is not permitted because of space limitations, safety requirements, amenities, etc. The major factor against a greater use of insulated cables is the high price of the equipment.

1.6. REPRESENTATION OF POWER SYSTEMS

A complete diagram of power system representing all the three phases becomes too complicated and cumbersome for a system of practical size, so much so that it may no longer convey the information it is intended to convey. It is much, more practical to represent a power system by means of simple symbols for each component resulting in what is called a single line diagram.*

1.6.1. Single Line Diagram

The single line diagram of a power system network shows the *main connections* and *arrangements* of the system components along with their data (such as output rating, voltage, resistance and reactance etc.). In case of transmission lines sometimes the conductor size and spacings are given. It is not necessary to show all the components of the system on a single line diagram, *e.g.*, circuit breakers need not be shown in a load flow study but are must for a protection study. In a *single line diagram*, the system components are usually drawn in the form of their symbols. Generators and transformer connections—star, delta and neutral earthing are indicated by symbols drawn by the side of the representation of these elements. Circuit breakers are represented by rectangular blocks. Figure 1.5 represents the single line diagram of a typical power system. The ratings of generator, motor and transformers are given in the diagram.

1.7. POWER SYSTEM COMPONENTS

Simplified single line diagram for a typical power system is shown in Fig. 1.6. Major components of a power system are : synchronous generators, synchronising equipment, circuit

* Any 3-phase network consisting of generators, transformers, transmission lines etc., can be solved as a single phase network composed of one of the three phases and a neutral return, if it is balanced under normal operation. Many times, the components of the system are shown in a single line diagram omitting the neutral also.

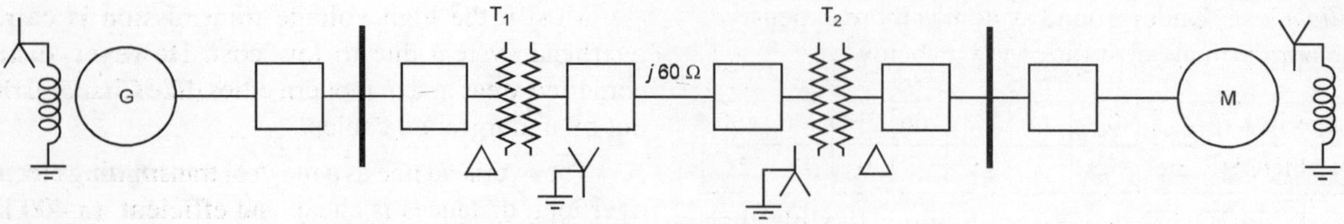

Generator	:	40 MVA, 11 kV, X" = 20 %
Motor	:	30 MVA, 11 kV, X" = 30 %
Transformer T_1	:	40 MVA, 11/220 kV, X = 15 %
Transformer T_2	:	40 MVA, 220/11 kV, X = 15 %

Fig. 1.5. *Single Line Representation of a Typical Power System*

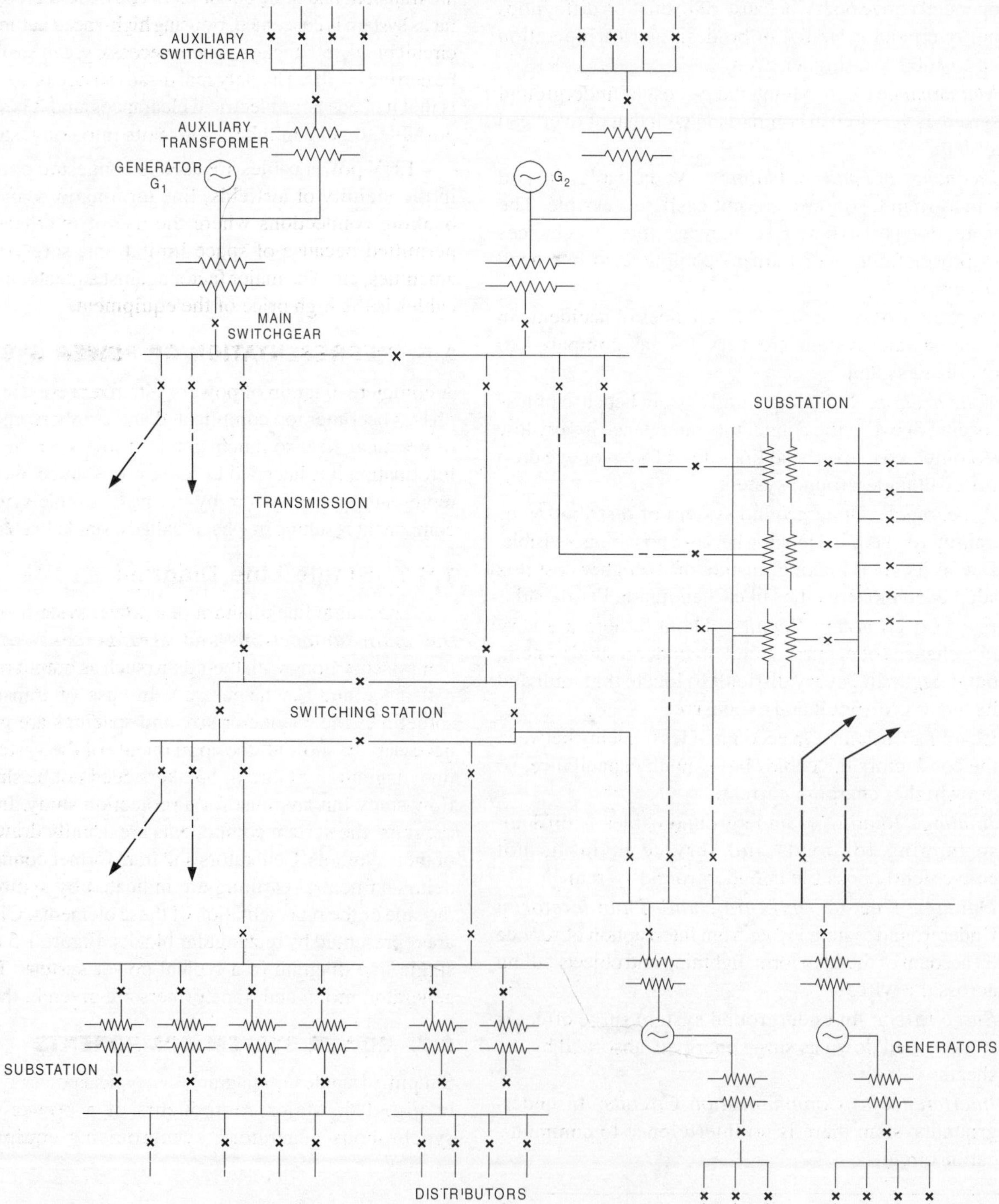

Fig. 1.6. *Simplified Single Line Diagram For a Typical Power System*

breakers, isolators, earthing switches, bus-bars, transformers, transmission lines, current transformers, potential transformers, relay and protection equipment, lightning arresters, station transformer, motors for driving auxiliaries in power station.

1.8. INFLUENCE OF WORKING VOLTAGE ON SIZE OF FEEDERS AND DISTRIBUTORS IN DC SYSTEM

There are following advantages of adopting high voltage for transmission and distribution of electrical energy in dc system.

(*i*) *It reduces the size (area of cross section of the core carrying the current) of the feeders and distributors.*

Consider first of all a feeder. As no consumer is taped off from a feeder so voltage drop in feeder will not be so important provided that it is not outside the range of compound generators in the generating stations. So the fundamental basis of designing of the size of a feeder is the current-carrying capacity.

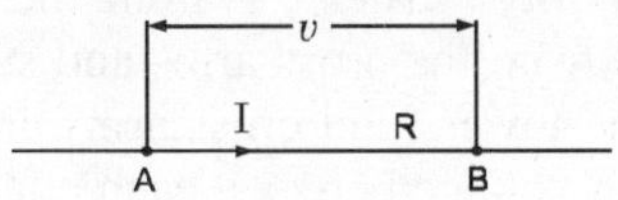

Fig. 1.7

Though the permissible current density for a given type of cable is not constant but decreases somewhat with the increase in size due to poor cooling facilities with higher sizes but for approximation we assume it as constant.

Now if supply voltage is increased n times, then for a given power to be delivered the current is reduced to $\frac{1}{n}$th so the size of the cable is also reduced approximately to $\frac{1}{n}$th.

Consider a distributor in which voltage drop is to be considered mainly while deciding its size. This is because the voltage drop in secondary distribution lines is to be limited to a specified value. Let distributor AB carry a current of I amperes and let permissible voltage drop be of v volts. Then the resistance of the distributor is given by $R = \frac{v}{I}$. Now if the supply voltage is increased n times, then for a given power to be delivered the current in the distributor becomes $\frac{I}{n}$ and permissible voltage drop increases for a given percentage of permissible voltage drop in the distributor and becomes nv volts. The resistance of the distributor with high voltage of distribution is given by

$$R' = \frac{nv}{1/n} = n^2R.$$

Since for a given length, the area of cross section of a cable is inversely proportional to its resistance, so by increasing the supply voltage to n times, the area of cross section can be reduced to $\frac{1}{n^2}$ times for supplying a given power with same per cent of voltage drop in feeder.

(*ii*) *It increases the efficiency of transmission.*

Let voltage at feeding point = V volts

Current supplied by the feeder = I amps

Resistance of each conductor of feeder = R ohms

Total voltage drop in feeder = 2IR

Voltage at supply end = V + 2IR

Efficiency of transmission,

$$\eta_1 = \frac{VI}{(V+2IR)I} = \frac{V}{(V+2IR)} = \left(1 - \frac{2IR}{V+2IR}\right) \quad ...(1.1)$$

Now let the voltage at feeding point be raised to nV volts then for transmission of same power

$$\text{Current in feeder} = \frac{VI}{nV} = \frac{I}{n}$$

Assuming current density constant, the size of the cable of feeder is reduced to $\frac{1}{n}$ th, hence resistance of each cable of feeder becomes nR

Voltage drop in feeder = Current × resistance

$$= \frac{I}{n} \times 2nR = 2IR \text{ (same as before)}$$

Voltage at supply end = nV + 2IR

Efficiency of transmission,

$$\eta_2 = \frac{\text{Output}}{\text{Input}} = \frac{nV \times \frac{I}{n}}{\frac{I}{n}(nV+2IR)}$$

$$= \frac{V}{V + \frac{2IR}{n}} = 1 - \frac{2IR}{nV+2IR} \quad ...(1.2)$$

Comparing Eqs. (1.1) and (1.2) we conclude that transmission efficiency is sufficiently increased. For example if I = 50 A; R = 0.4 Ω; V = 100 V and n = 10 then

$$\eta_1 = 1 - \frac{2IR}{V+2IR} = 1 - \frac{2 \times 50 \times 0.4}{100 + 2 \times 50 \times 0.4}$$

$$= 0.714 \text{ or } 71.4\%$$

$$\eta_2 = 1 - \frac{2IR}{nV+2IR} = 1 - \frac{2 \times 50 \times 0.4}{10 \times 100 + 2 \times 50 \times 0.4}$$

$$= 0.9615 \text{ or } 96.15\%$$

Thus with the increase in transmission voltage, efficiency of transmission increases and cost of copper in cable decreases. For this reason it is advantageous to transmit power at high voltage.

Example 1.1. What is the percentage saving in feeder copper if the line voltage in a 2-wire dc system be raised from 220 volts to 400 volts for the same power transmitted over same distance and having the same power loss.

[*U.P. Technical Univ. Elements of Power System 2006-07*]

Solution : Let the power delivered be P watts

Load current, $I_1 = \frac{P}{220}$ A when line voltage is 220 V

Load current, $I_2 = \frac{P}{400}$ A when line voltage is 400 V

Let the power be transmitted over length l and power loss be W watts.

When line voltage is 220 V

$$\text{Power losses in line, W} = I_1^2 \times 2R_1 = \left(\frac{P}{220}\right)^2 \times 2\rho\frac{l}{a_1}$$

where ρ is the resistivity of copper, and a_1 is the area of cross section of copper required

$$\text{or } a_1 = \frac{P^2 \rho l}{24{,}200\ W}$$

Volume of copper required,

$$v_1 = 2a_1 l = \frac{P^2 \rho l^2}{12{,}100\ W} \qquad ...(i)$$

When line voltage is 400 V

$$\text{Power losses in line, W} = I_2^2 \times 2R_2 = \left(\frac{P}{400}\right)^2 \times 2\rho\frac{l}{a_2} = \frac{P^2 \rho l}{80{,}000 a_2}$$

$$\text{or } a_2 = \frac{P^2 \rho l}{80{,}000\ W}$$

Volume of copper required,

$$v_2 = 2a_2 l = \frac{P^2 \rho l^2}{40{,}000\ W} \qquad ...(ii)$$

Percentage saving in copper

$$= \frac{v_1 - v_2}{v_1} \times 100 = \frac{\dfrac{P^2 \rho l^2}{12{,}100\ W} - \dfrac{P^2 \rho l^2}{40{,}000\ W}}{\dfrac{P^2 \rho l^2}{12{,}100\ W}} \times 100$$

$$= \left[1 - \frac{12{,}100}{40{,}000}\right] \times 100$$

$$= (1 - 0.3025) \times 100 = 69.75\% \quad \textbf{Ans.}$$

1.9. VARIOUS SYSTEMS OF TRANSMISSION OF ELECTRICAL POWER

Though in practice 3-phase 3-wire ac system is universally used for transmission and 3-phase 4-wire ac system is used for distribution of electric power but for special purposes other systems may also be used. The various systems of power transmission are :

(*a*) *DC Systems :*

(*i*) DC two-wire system.

(*ii*) DC two-wire system with midpoint earthed.

(*iii*) DC three-wire system.

(*b*) *Single-Phase AC Systems:*

(*i*) Single-phase two-wire system.

(*ii*) Single-phase two-wire system with midpoint earthed.

(*iii*) Single-phase three-wire system.

(*c*) *Two-Phase AC Systems:*

(*i*) Two-phase four-wire system.

(*ii*) Two-phase three-wire system.

(*d*) *Three-Phase AC Systems:*

(*i*) Three-phase 3-wire system.

(*ii*) Three-phase 4-wire system.

From such a big list of possible systems of transmission it is difficult to decide the best system without making comparison. The basis for comparison between the various systems of power transmission is usually economy. Since in a transmission system, the cost of the conductor material accounts for a major part of the total cost, the best system for transmission of electrical power is that for which the volume of conductor material required is minimum. Thus the requirement of volume of conductor material forms the basis of comparison between various systems.

In making comparison of the volume of conductor material required for various transmission systems, the basis will be the *equal maximum stress on the dielectric.* This is because the voltage is only limited by the problem of insulating the conductors against disruptive discharge.

For comparing the amount of conductor material required for different systems two cases arise:

(*i*) *When overhead system is employed for transmission of power.* In overhead system the conductors are insulated from the cross arms and supporting towers and as the towers and cross arms are earthed so the maximum voltage between each conductor and earth forms the basis of comparison of volume of conductor material required.

(*ii*) *When underground cables are employed for transmission of power.* In underground cables the maximum disruptive stress is between the two conductors of the cable; so the maximum voltage between conductors forms the basis of comparison of volume of conductor material required.

1.10. COMPARISON OF COST OF CONDUCTORS—OVERHEAD SYSTEMS

The assumptions made for comparison are:

(*i*) In all cases power to be transmitted is same (say, P watts).

(*ii*) The distance over which the power is to be transmitted is same (say, *l*).

(*iii*) The line losses are same (say, W watts).

(*iv*) The maximum voltage to earth is same (say, V_m volts).

(*a*) **DC Systems**

(*i*) *DC 2-Wire System With One Conductor Earthed* [Fig. 1.8(*i*)].

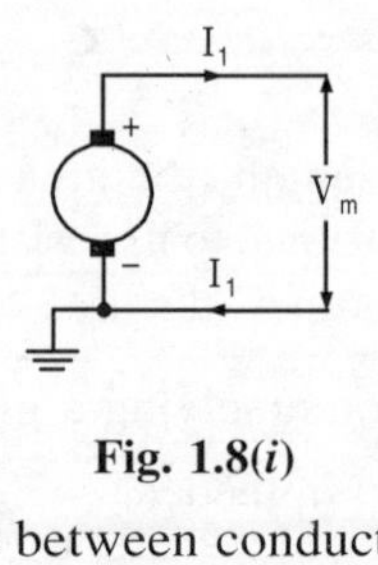

Fig. 1.8(*i*)

Maximum voltage between conductors $= V_m$ volts

Power to be transmitted = P watts

$$\text{Load current, } I_1 = \frac{P}{V_m}$$

Line losses, $W = 2I_1^2 R_1 = 2\left(\frac{P}{V_m}\right)^2 \rho\frac{l}{a_1}$ Since $R_1 = \rho\frac{l}{a_1}$

or Area of cross section of conductor, $a_1 = \frac{2\rho P^2 l}{W{V_m}^2}$

Volume of conductor material required $= 2a_1 l = \frac{4\rho P^2 l^2}{W{V_m}^2}$ ...(1.3)

This system is usually formed the basis for comparison with other systems. Thus the volume of conductor material required in this system shall be taken as the basic quantity *i.e.*,

$$\frac{4\rho P^2 l^2}{W{V_m}^2} = K \text{ (say)}$$

(*ii*) *DC Two-Wire System With Midpoint Earthed.* This system is shown in Fig. 1.8 (*ii*). The maximum voltage between any conductor and earth is V_m volts so

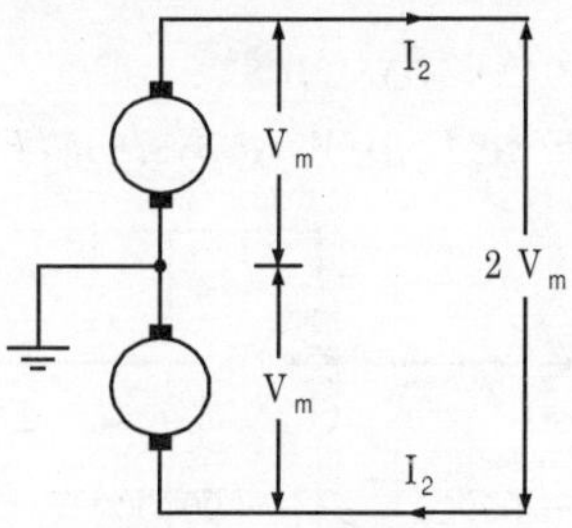

Fig. 1.8(*ii*)

Maximum voltage between conductors = $2V_m$ volts

Load current, $I_2 = \frac{P}{2V_m}$

Line losses, $W = 2I_2^2 R_2 = 2\left(\frac{P}{2\,V_m}\right)^2 \rho\frac{l}{a_2} = \frac{P^2\rho l}{2a_2 V_m^2}$

or $a_2 = \frac{P^2\rho l}{2WV_m^2}$

Volume of conductor material required

$$= 2a_2 l = \frac{P^2\rho l^2}{WV_m^2} = 0.25\text{ K} \quad ...(1.4)$$

Hence volume of conductor material required is one-fourth of that required in two-wire dc system with one conductor earthed.

(*iii*) *DC Three-Wire System.* In a 3-wire dc system, there are two outers and one middle wire, called the *neutral wire* which is grounded at the generator end, as illustrated in Fig. 1.8 (*iii*). For balanced load on the system, the middle wire carries no current.

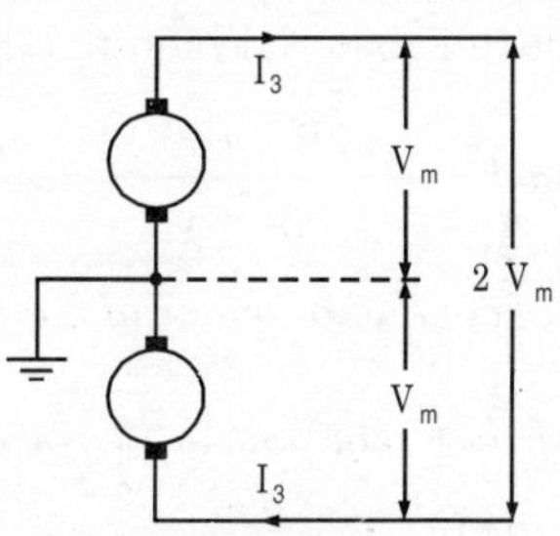

Fig. 1.8(*iii*)

Assuming load balanced

Maximum voltage between the outer and earth = V_m volts

Load current, $I_3 = \frac{P/2}{V_m} = \frac{P}{2V_m}$

Line losses, $W = 2I_3^2 R_3 = 2\left(\frac{P}{2V_m}\right)^2 \frac{\rho l}{a_3} = \frac{P^2\rho l}{2a_3 V_m^2}$

or $a_3 = \frac{P^2\rho l}{2WV_m^2}$

Assuming area of cross section of neutral wire as half of that of any of the outers

Volume of conductor material required

$$= 2.5a_3 l = \frac{2.5P^2\rho l^2}{2WV_m^2} = 0.3125\text{ K} \quad ...(1.5)$$

Hence volume of conductor material required is 0.3125 times of that required in two-wire dc system with one conductor earthed.

(*b*) Single-Phase AC Systems

(*i*) *AC Single-Phase Two-Wire System With One Conductor Earthed* Fig. [1.8 (*iv*)].

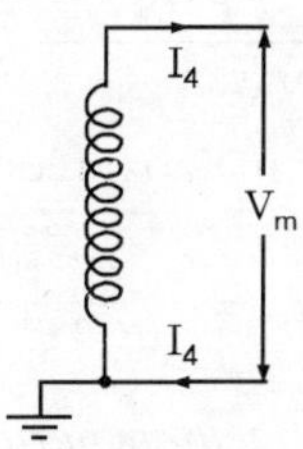

Fig. 1.8(*iv*)

Peak value of voltage between conductors = V_m volts

RMS value of voltage between conductors $= \frac{V_m}{\sqrt{2}}$ volts

Load current, $I_4 = \frac{P}{\frac{V_m}{\sqrt{2}}\cos\phi} = \frac{\sqrt{2}P}{V_m\cos\phi}$

where $\cos\phi$ is the power factor of the load

Line losses, $W = 2I_4^2 R_4$

$$= 2\left(\frac{\sqrt{2}P}{V_m\cos\phi}\right)^2 \frac{\rho l}{a_4} = \frac{4P^2\rho l}{a_4 V_m^2\cos^2\phi}$$

or $a_4 = \frac{4\,P^2\rho l}{WV_m^2\cos^2\phi}$

Volume of conductor material required

$$= 2a_4 l = \frac{8P^2\rho l^2}{WV_m^2\cos^2\phi} = \frac{2}{\cos^2\phi}\text{ K} \quad ...(1.6)$$

Hence volume of conductor material required in this system is $\frac{2}{\cos^2\phi}$ *times of that required in 2-wire dc system with one conductor earthed.*

(ii) AC Single-Phase Two-Wire System With Midpoint Earthed [Fig. 1.8 (*v*)].

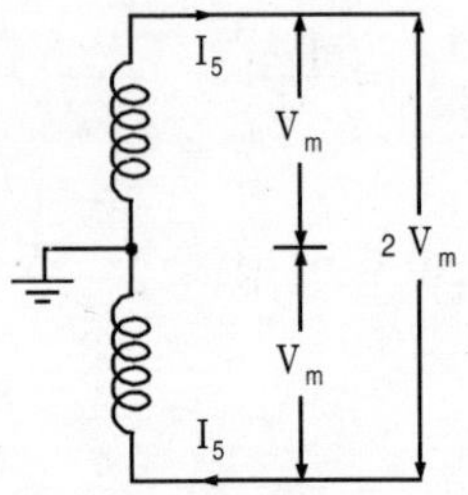

Fig. 1.8(*v*)

Peak value of voltage between conductors $= 2V_m$ volts

RMS value of voltage between conductors

$$= \frac{2V_m}{\sqrt{2}} = \sqrt{2}\,V_m \text{ volts}$$

$$\text{Load current, } I_5 = \frac{P}{\sqrt{2}V_m \cos\phi}$$

$$\text{Line losses, } W = 2I_5^2 R_5$$

$$= 2\left(\frac{P}{\sqrt{2}V_m \cos\phi}\right)^2 \frac{\rho l}{a_5} = \frac{P^2 \rho l}{a_5 V_m^2 \cos^2\phi}$$

$$\text{or } a_5 = \frac{P^2 \rho l}{W V_m^2 \cos^2\phi}$$

Volume of conductor material required

$$= 2a_5 l = \frac{2P^2 \rho l^2}{W V_m^2 \cos^2\phi} = \frac{0.5\,K}{\cos^2\phi} \quad \text{...(1.7)}$$

Hence volume of conductor material required in this system is $\frac{0.5}{\cos^2\phi}$ *times of that required in 2-wire dc system with one conductor earthed.*

(iii) AC Single-Phase Three-Wire System [Fig. 1.8 (*vi*)].

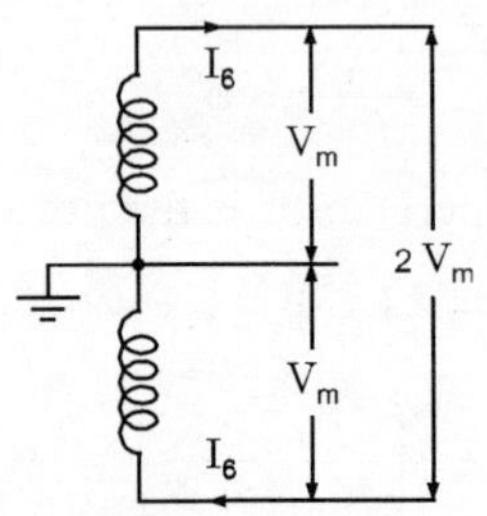

Fig. 1.8(*vi*)

Maximum voltage between outer conductors $= 2V_m$ volts

$$\text{RMS value of voltage between outer and earth} = \frac{V_m}{\sqrt{2}} \text{ volts}$$

$$\text{Load current, } I_6 = \frac{P/2}{\frac{V_m}{\sqrt{2}}\cos\phi}$$

$$= \frac{P}{\sqrt{2}V_m \cos\phi} \text{ load being assumed balanced}$$

$$\text{Line losses, } W = 2I_6^2 R_6$$

$$= 2\left(\frac{P}{\sqrt{2}V_m \cos\phi}\right)^2 \frac{\rho l}{a_6} = \frac{P^2 \rho l}{V_m^2 \cos^2\phi\, a_6}$$

$$\text{or } a_6 = \frac{P^2 \rho l}{W V_m^2 \cos^2\phi}$$

Assuming cross section of neutral wire half of that of any of the outers

Volume of conductor material required

$$= 2.5 a_6 l = \frac{2.5 P^2 \rho l^2}{W V_m^2 \cos^2\phi} = \frac{0.625\,K}{\cos^2\phi} \quad \text{...(1.8)}$$

Hence volume of conductor material required in this system is $\frac{0.625}{\cos^2\phi}$ *times of that required in 2-wire dc system with one conductor earthed.*

(*c*) Two-Phase AC Systems

(i) AC Two-Phase Four-Wire System [Fig. 1.8 (vii)].

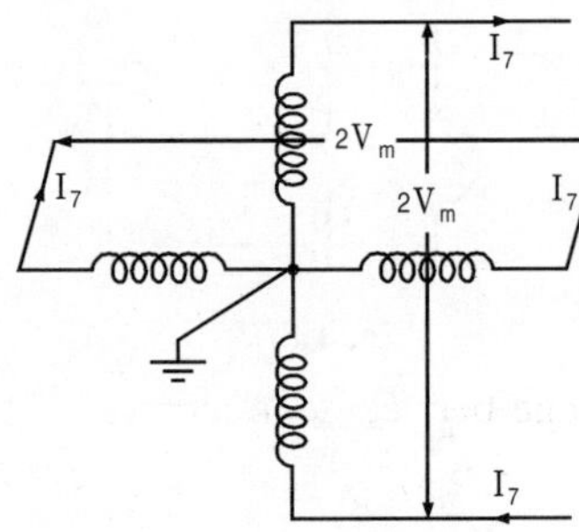

Fig. 1.8(*vii*)

$$\text{RMS value of voltage between the outers} = \frac{2V_m}{\sqrt{2}} = \sqrt{2}\,V_m$$

$$\text{Load supplied by each phase} = \frac{P}{2}$$

$$\text{Load current per phase, } I_7 = \frac{P}{2\sqrt{2}V_m \cos\phi}$$

$$\text{Line losses, } W = 4I_7^2 R_7$$

$$= 4\left(\frac{P}{2\sqrt{2}V_m \cos\phi}\right)^2 \rho \frac{l}{a_7} = \frac{P^2 \rho l}{2V_m^2 a_7 \cos^2\phi}$$

$$\text{or } a_7 = \frac{P^2 \rho l}{2V_m^2 W \cos^2\phi}$$

Volume of conductor material required

$$= 4a_7 l = \frac{2P^2 \rho l^2}{V_m^2 \cos^2\phi\, W} = \frac{0.5\,K}{\cos^2\phi} \quad \text{...(1.9)}$$

Hence volume of conductor material required in this system is $\frac{0.5}{\cos^2\phi}$ *times of that required in two-wire dc system with one conductor earthed.*

(ii) AC Two-Phase Three-Wire System [Fig. 1.8 (viii)].

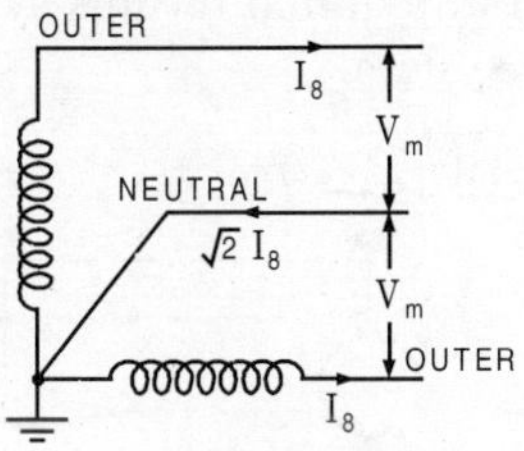

Fig. 1.8(*viii*)

Considering balanced load, RMS value of voltage between any outer and neutral

$$= \frac{V_m}{\sqrt{2}}$$

Current in each of outer conductors,

$$I_8 = \frac{P/2}{\frac{V_m}{\sqrt{2}}\cos\phi} = \frac{P}{\sqrt{2}V_m\cos\phi}$$

Current in neutral wire

$$= \sqrt{(I_8^2 + I_8^2)} = \sqrt{2}\,I_8$$

Assuming current density constant, area of cross section of neutral wire is $\sqrt{2}$ times of that of either of the outers.

So resistance of neutral wire

$$= \frac{R_8}{\sqrt{2}} = \frac{\rho l}{\sqrt{2}a_8} \qquad \text{where } R_8 = \frac{\rho l}{a_8}$$

$$\text{Line losses, W} = 2I_8^2R_8 + (\sqrt{2}I_8)^2\frac{R_8}{\sqrt{2}}$$

$$= I_8^2\frac{\rho l}{a_8}(2+\sqrt{2}) = \frac{P^2\rho l}{2V_m^2\cos^2\phi a_8}(2+\sqrt{2})$$

$$\text{or } a_8 = \frac{P^2\rho l}{2V_m^2\cos^2\phi W}\left(2+\sqrt{2}\right)$$

Volume of conductor material required

$$= 2a_8l + \sqrt{2}a_8l = a_8l(2+\sqrt{2})$$

$$= \frac{P^2\rho l^2}{2V_m^2\cos^2\phi W}(2+\sqrt{2})^2$$

$$= \frac{P^2\rho l^2}{V_m^2\cos^2\phi W}(\sqrt{2}+1)^2$$

$$= \frac{1.457}{\cos^2\phi}K \qquad \text{...(1.10)}$$

Hence volume of conductor material required in this system is $\frac{1.457}{\cos^2\phi}$ *times of that required in 2-wire dc system with one conductor earthed.*

***(d)* 3-Phase AC Systems**

(i) AC 3-Phase 3-Wire System *[Fig. 1.8 (ix)].*

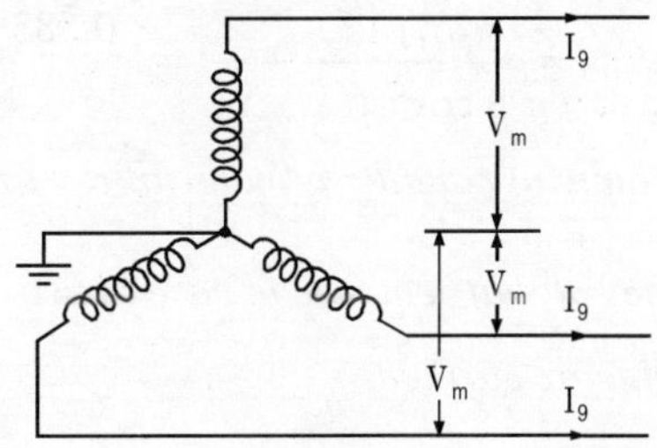

Fig. 1.8(*ix*)

$$\text{RMS value of voltage per phase} = \frac{V_m}{\sqrt{2}}$$

$$\text{Load current per phase, } I_9 = \frac{P/3}{\frac{V_m}{\sqrt{2}}\cos\phi} = \frac{\sqrt{2}P}{3V_m\cos\phi}$$

$$\text{Line losses, W} = 3I_9^2R_9$$

$$= 3\left(\frac{\sqrt{2}P}{3V_m\cos\phi}\right)^2\frac{\rho l}{a_9} = \frac{2P^2\rho l}{3V_m^2\cos^2\phi\, a_9}$$

$$\text{or } a_9 = \frac{2P^2\rho l}{3V_m^2\cos^2\phi W}$$

Volume of conductor material required

$$= 3a_9l = \frac{2P^2\rho l^2}{V_m^2\cos^2\phi W} = \frac{0.5\text{ K}}{\cos^2\phi} \qquad \text{...(1.11)}$$

Hence volume of conductor material required in this system is $\frac{0.5}{\cos^2\phi}$ *times of that required in two-wire dc system with one conductor earthed.*

(ii) AC 3-Phase 4-Wire System. [Fig. 1.8(x)].

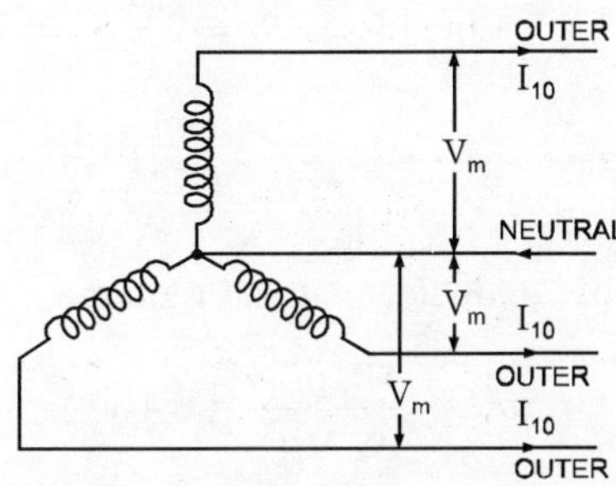

Fig. 1.8(*x*)

Assuming balanced load, there will be no current in neutral wire and copper losses will be same as in 3-phase 3-wire system

$$\text{i.e., } W = \frac{2P^2\rho l}{3V_m^2\cos^2\phi\, a_{10}}$$

$$\text{or } a_{10} = \frac{2P^2\rho l}{3V_m^2\cos^2\phi W}$$

Taking cross section of neutral wire as half of either outer

Volume of conductor material required

$$= 3.5a_{10}\, l = \frac{7\rho P^2 l^2}{3\cos^2\phi\, V_m^2\, W} = \frac{0.583}{\cos^2\phi} K \qquad ...(1.12)$$

Hence volume of conductor material required in this case is $\dfrac{0.583}{\cos^2\phi}$ *times of that required in case of two-wire dc system with one conductor earthed.*

1.11. COMPARISON OF COST OF CONDUCTORS—UNDERGROUND SYSTEMS

The assumptions made for comparison are:

(*i*) In all cases the power to be transmitted is same (say, P watts).

(*ii*) The distance over which the power is to be transmitted is same (say, l).

(*iii*) The line losses are same (say, W watts).

(*iv*) The maximum voltage between two conductors is same (say, V_m volts).

(*a*) DC Systems

(*i*) *DC Two-Wire System [Fig. 1.9 (i)].*

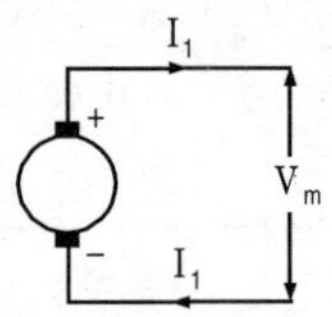

Fig. 1.9(*i*)

Voltage between two conductors $= V_m$ volts

$$\text{Load current, } I_1 = \frac{P}{V_m}$$

$$\text{Line losses, } W = 2I_1^2 R_1 = \frac{2P^2\rho l}{V_m^2 a_1}$$

$$\text{or } a_1 = \frac{2P^2\rho l}{W\, V_m^2}$$

Volume of conductor material required

$$= 2a_1 l = \frac{4P^2\rho l^2}{W\, V_m^2} = K \text{ (say)} \qquad ...(1.13)$$

(*ii*) *DC Two-Wire System With Midpoint Earthed [Fig. 1.9 (ii)].*

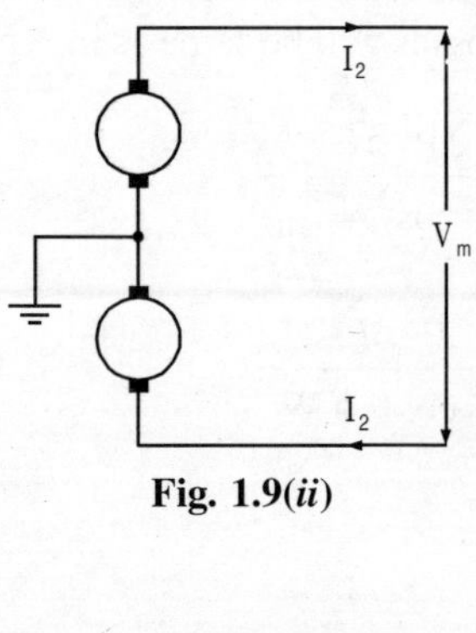

Fig. 1.9(*ii*)

This system is the same as a 2-wire dc system, so volume of conductor material required for this system is the same as that in a 2-wire dc system.

(*iii*) *DC Three-Wire System [Fig. 1.9 (iii)].*

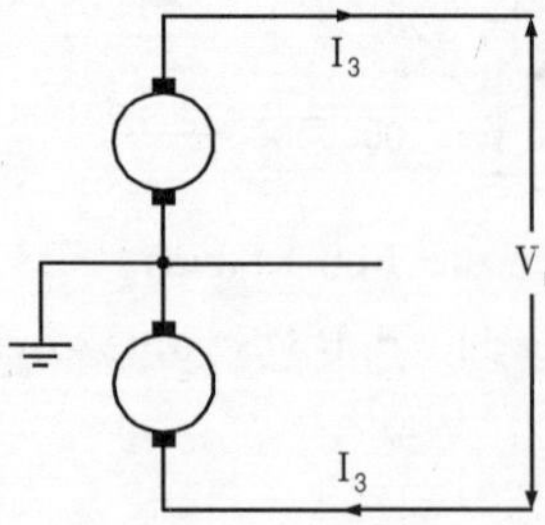

Fig. 1.9(*iii*)

Assuming balanced load, the current through the neutral or mid wire is zero.

Maximum voltage between outers $= V_m$ volts

$$\text{Load current, } I_3 = \frac{P}{V_m}$$

$$\text{Line losses, } W = 2I_3^2 R_3 = \frac{2P^2\rho l}{V_m^2 a_3}$$

$$\text{or } a_3 = \frac{2P^2\rho l}{V_m^2\, W}$$

Assuming cross section of neutral as half of that of either outer

Volume of conductor material required

$$= 2.5a_3 l = \frac{5P^2\rho l^2}{V_m^2\, W} = 1.25\text{ K} \qquad ...(1.14)$$

Hence volume of conductor material required in this system is 1.25 times of that required in a two-wire dc system.

(*b*) Single-Phase AC Systems

(*i*) *AC Single-Phase Two-Wire System [Fig. 1.9 (iv)].*

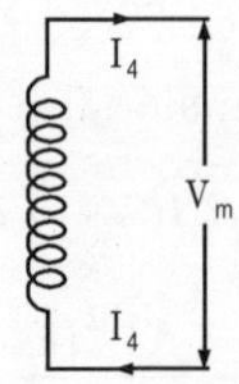

Fig. 1.9(*iv*)

Maximum value of voltage between outers $= V_m$ volts

$$\text{RMS value of voltage between outers} = \frac{V_m}{\sqrt{2}} \text{ volts}$$

$$\text{Load current, } I_4 = \frac{P}{\dfrac{V_m}{\sqrt{2}}\cos\phi} = \frac{\sqrt{2}P}{V_m\cos\phi}$$

where $\cos\phi$ is the power factor of the load

$$\text{Line losses, } W = 2I_4^2 R_4 = 2I_4^2\frac{\rho l}{a_4} = \frac{4P^2\rho l}{V_m^2\cos^2\phi\, a_4}$$

$$\text{or } a_4 = \frac{4P^2 \rho l}{V_m^2 \cos^2 \phi \, W}$$

Volume of conductor material required

$$= 2a_4 l = \frac{8P^2 \rho l^2}{V_m^2 \cos^2 \phi \, W} = \frac{2K}{\cos^2 \phi} \quad ...(1.15)$$

Hence volume of conductor material required in this system is $\frac{2}{\cos^2 \phi}$ *times of that required for a two-wire dc system.*

(ii) AC Single-Phase Two-Wire System With Midpoint Earthed [Fig. 1.9 (v)].

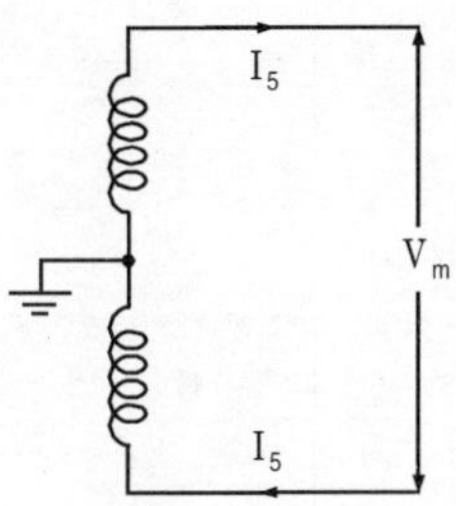

Fig. 1.9(*v*)

This system is the same as a 2-wire single-phase ac system, so volume of conductor material required in this case is also $\frac{2}{\cos^2 \phi}$ times of that required in a two-wire dc system.

(iii) AC Single-Phase Three-Wire System [Fig. 1.9 (vi)]. Assuming balanced load, the system reduces to a single-phase, 2-wire ac system except that a neutral conductor of half the cross section is provided in addition.

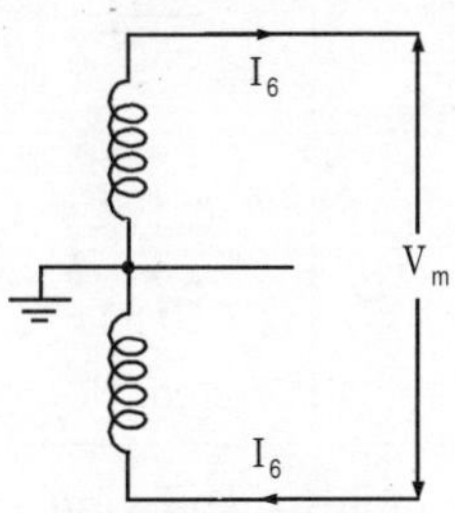

Fig. 1.9(*vi*)

So volume of conductor material required

$$= 2.5a_4 l = \frac{10\rho P^2 l^2}{V_m^2 \cos^2 \phi \, W} = \frac{2.5}{\cos^2 \phi} K \quad ...(1.16)$$

Hence volume of conductor material required in this case is $\frac{2.5}{\cos^2 \phi}$ *times of that required in a two-wire dc system.*

(*c*) Two-Phase AC Systems

(i) AC Two-Phase 4-Wire System [Fig. 1.9 (vii)].

In this system each phase shares the half of the total load. This system is equivalent to two-wire ac system. In this case cross section area of each conductor is taken half of that of single phase two-wire ac system but four wires are required in place of two wires, so the same volume of conductor material is required *i.e.,* $\frac{2}{\cos^2 \phi}$ times of that required in case of two-wire dc system.

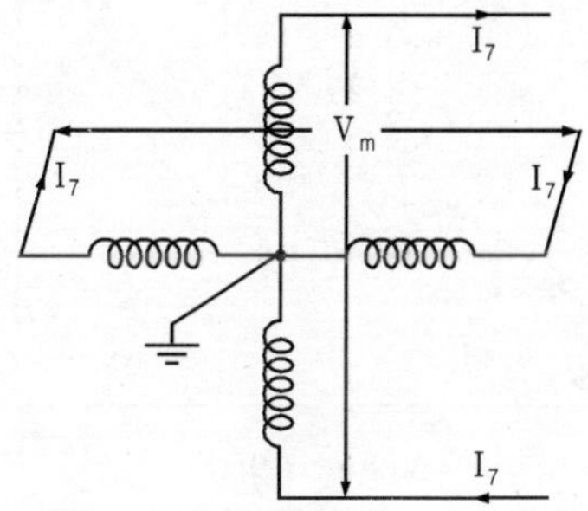

Fig. 1.9(*vii*)

(ii) AC Two-Phase Three-Wire System [Fig. 1.9 (viii)].

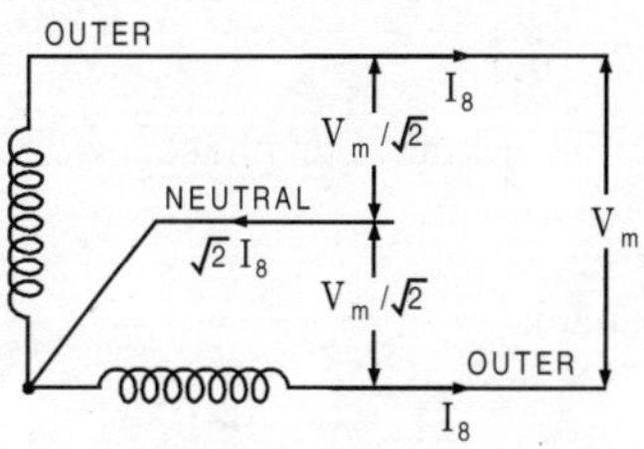

Fig. 1.9(*viii*)

$$\text{RMS value of voltage between outers} = \frac{V_m}{\sqrt{2}}$$

$$\text{RMS value of voltage per phase} = \frac{V_m}{\sqrt{2}\,\sqrt{2}} = \frac{V_m}{2}$$

$$\text{Load current per phase, } I_8 = \frac{P}{2 \times \frac{V_m}{2} \cos \phi} = \frac{P}{V_m \cos \phi}$$

The current in middle wire $= \sqrt{2} I_8$

The cross section of middle wire is taken $\sqrt{2}$ times of either outer so that current density may remain the same.

$$\text{Line losses, W} = 2I_8^2 \frac{\rho l}{a_8} + (\sqrt{2} I_8)^2 \frac{\rho l}{\sqrt{2}\, a_8}$$

$$= \frac{P^2 \rho l}{V_m^2 \cos^2 \phi \, a_8} (2 + \sqrt{2})$$

$$\text{or } a_8 = \frac{P^2 \rho l}{V_m^2 \cos^2 \phi \, W} (2 + \sqrt{2})$$

Volume of conductor material required

$$= 2a_8 l + \sqrt{2} a_8 l$$

$$= a_8 l (2 + \sqrt{2}) = \frac{P^2 \rho l^2}{V_m^2 \cos^2 \phi \, W} (2 + \sqrt{2})^2$$

$$= \frac{2.914 \text{ K}}{\cos^2 \phi} \quad ...(1.17)$$

Hence conductor material required in this system is $\frac{2.914}{\cos^2 \phi}$ *times of that required in a 2-wire dc system.*

(*d*) 3-Phase AC Systems

(*i*) *AC Three-Phase Three-Wire System [Fig. 1.9 (ix)].*

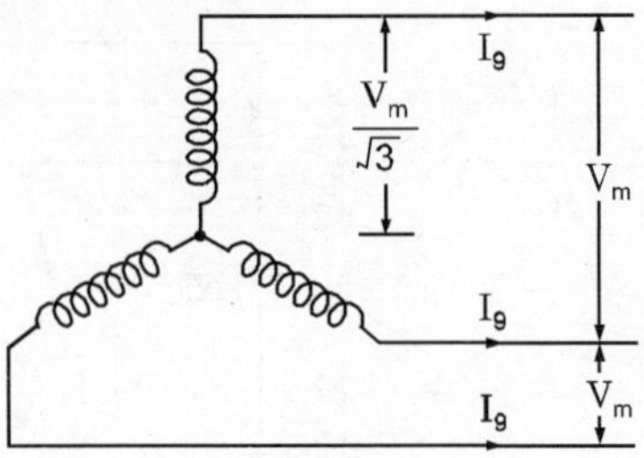

Fig. 1.9(*ix*)

Maximum value of voltage between the conductors = V_m

$$\text{RMS value of voltage between the conductors} = \frac{V_m}{\sqrt{2}}$$

$$\text{RMS value of voltage per phase} = \frac{V_m}{\sqrt{2}\sqrt{3}} = \frac{V_m}{\sqrt{6}}$$

$$\text{Load current, } I_9 = \frac{P}{3\dfrac{V_m}{\sqrt{6}}\cos\phi} = \frac{\sqrt{6}P}{3V_m\cos\phi}$$

$$\text{Line losses, } W = 3I_9^2\frac{\rho l}{a_9} = \frac{3\times 6\times P^2\rho l}{9V_m^2\cos^2\phi\, a_9}$$

$$\text{or } a_9 = \frac{2P^2\rho l}{V_m^2\cos^2\phi\, W}$$

Volume of conductor material required

$$= 3a_9 l = \frac{6P^2\rho l^2}{V_m^2\cos^2\phi\, W} = \frac{1.5}{\cos^2\phi}K \quad ...(1.18)$$

Hence volume of conductor material required in this case is $\frac{1.5}{\cos^2\phi}$ *times of that required in case of a two-wire dc system.*

(*ii*) *AC Three-Phase Four-Wire System [Fig. 1.9 (x)].*

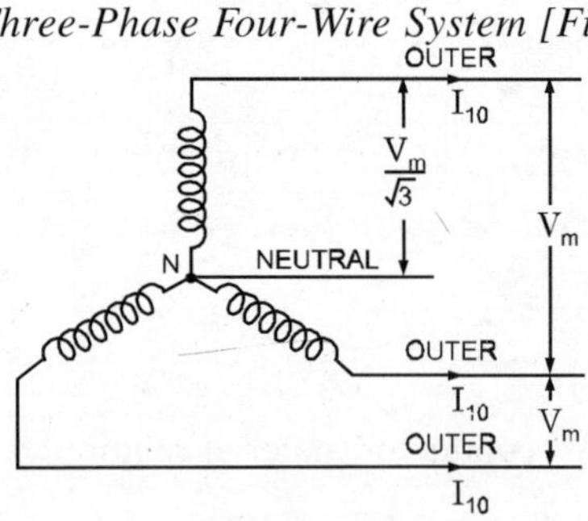

Fig. 1.9(*x*)

Assuming balanced load, this system is reduced to a 3-ϕ, ac system except that an additional wire, called the *neutral wire,* is provided of the half the cross section of that either outer

So volume of conductor material required

$$= 3.5a_9 l = \frac{7P^2\rho l^2}{V_m^2\cos^2\phi\, W} = \frac{1.75\,K}{\cos^2\phi} \quad ...(1.19)$$

Hence volume of conductor material required in this system is $\frac{1.75}{\cos^2\phi}$ *times of that required in case of a 2-wire dc system.*

1.12. RESULTS OF COMPARISON OF TRANSMISSION SYSTEMS

The results obtained above are summarised as below :

Systems	Volume of conductor material required on the basis of	
	Same maximum voltage between conductor and earth	Same maximum voltage between any two conductors
(*a*) *DC systems*		
(*i*) Two-wire	1	1
(*ii*) Two-wire with mid-point earthed	0.25	1
(*iii*) Three-wire	0.3125	1.25
(*b*) *AC Single-Phase Systems*		
(*i*) Two-wire	$\frac{2}{\cos^2\phi}$	$\frac{2}{\cos^2\phi}$
(*ii*) Two-wire with mid-point earthed	$\frac{0.5}{\cos^2\phi}$	$\frac{2}{\cos^2\phi}$
(*iii*) Three-wire	$\frac{0.625}{\cos^2\phi}$	$\frac{2.5}{\cos^2\phi}$
(*c*) *AC Two-phase Systems*		
(*i*) Two-phase four-wire	$\frac{0.5}{\cos^2\phi}$	$\frac{2}{\cos^2\phi}$
(*ii*) Two-phase three-wire	$\frac{1.457}{\cos^2\phi}$	$\frac{2.914}{\cos^2\phi}$
(*d*) *AC Three-phase Systems*		
(*i*) Three-phase three-wire	$\frac{0.5}{\cos^2\phi}$	$\frac{1.5}{\cos^2\phi}$
(*ii*) Three-phase four-wire	$\frac{0.583}{\cos^2\phi}$	$\frac{1.75}{\cos^2\phi}$

From the above table it is obvious that for transmission, dc system is an ideal one from economic point of view, particularly when it is remembered that the power factor of an ac system is usually considerably less than unity.

Two-phase three-wire system is obviously quite unsuitable for long distance transmission and needs no further consideration.

Among the remaining ac systems for overhead systems there is no difference in single-phase, 2-wire and three-phase 3-wire systems and for underground systems there is a decided saving in conductor material with the three phase system. Considering other factors such as efficiency of operation and convenience three-phase three-wire system is usually adopted.

Example 1.2. A 50 km long transmission line supplies a load of 5 MVA at 0.8 pf lagging at 33 kV. The efficiency of transmission is 90 per cent. Calculate the volume of conductor required for the line when

(*a*) single-phase two-wire system is used

(*b*) three-phase three-wire system is used.

Take resistivity of aluminium as 2.85×10^{-8} Ω-m.

[Rajasthan Univ. Transmission & Distribution 2002]

Solution : Power transmitted, $P = 5 \times 0.8$ MW or 4×10^6 W

Transmission efficiency, $\eta = 90\%$

$$\text{Line losses, } W = 10 \text{ per cent of power transmitted}$$

$$= \frac{10}{100} \times 4 \times 10^6 = 4 \times 10^5 \text{ W}$$

(a) Single-Phase Two-Wire System

$$\text{Load current, } I_1 = \frac{\text{MVA}}{V} = \frac{5 \times 10^6}{33 \times 10^3} = 151.52 \text{ A}$$

$$\text{Line loss, } W = 2I_1^2 R_1 = 2I_1^2 \times \rho \frac{l}{a_1}$$

$$\text{or } 4 \times 10^5 = \frac{2 \times (151.52)^2 \times 2.85 \times 10^{-8} \times 50 \times 1{,}000}{a_1}$$

$$\text{or } a_1 = \frac{2 \times (151.52)^2 \times 2.85 \times 10^{-8} \times 50 \times 1{,}000}{4 \times 10^5}$$

$$= 1.636 \times 10^{-4} \text{ m}^2$$

Volume of conductor required,

$$v_1 = 2a_1 l = 2 \times 1.636 \times 10^{-4} \times 50 \times 1{,}000$$

$$= 16.36 \text{ m}^3 \quad \textbf{Ans.}$$

(b) Three-Phase Three-Wire System

$$\text{Load current, } I_L = \frac{\text{MVA} \times 10^6}{\sqrt{3} \times V_L} = \frac{5 \times 10^6}{\sqrt{3} \times 33 \times 1{,}000} = 87.48 \text{ A}$$

$$\text{Line loss, } W = 3I_2^2 R_2 = 3 \times I_2^2 \times \rho \frac{l}{a_2}$$

$$\text{or } a_2 = \frac{3 \times I_2^2 \times \rho \times l}{W}$$

$$= \frac{3 \times (87.48)^2 \times 2.85 \times 10^{-8} \times 50{,}000}{4 \times 10^5}$$

$$= 0.818 \times 10^{-4} \text{ m}^2$$

Volume of conductor required,

$$v_2 = 3a_2 l = 3 \times 0.818 \times 50{,}000 = 12.27 \text{ m}^3$$

Example 1.3. A 3-phase 4-wire system is used for lighting. Compare the amount of conductor material required with that needed for a 2-wire direct current system with the same lamp voltage. Assume the same losses and balanced load. The neutral is one-half the cross section of one of the respective outers.

Solution : Let the lighting load to be supplied be P watts at voltage V volts (between line and neutral in case of 3-phase, 4-wire system) and line losses be W watts in both of the cases:

(i) Two-wire dc system

$$\text{Load current, } I_1 = \frac{P}{V} \text{ amperes}$$

$$\text{Line losses, } W = 2I_1^2 R_1 = 2\frac{P^2}{V^2} \rho \frac{l}{a_1} = \frac{2P^2 \rho l}{V^2 a_1}$$

where ρ is the resistivity of conductor material, l is the length of line, a_1 is the cross section of conductor in case of 2-wire dc system.

$$\text{or Area of cross section of conductor, } a_1 = \frac{2P^2 \rho l}{V^2 W}$$

$$\text{Volume of conductor material required} = 2a_1 l = \frac{4P^2 \rho l^2}{V^2 W} \quad ...(i)$$

(ii) Three-Phase Four-Wire System

$$\text{Power supplied by each phase} = \frac{P}{3} \quad (\because \text{ load is balanced})$$

RMS voltage between line and neutral = V volts

Power factor of lighting load be assumed unity

$$\text{Load current, } I_2 = \frac{P}{3V} \text{ amperes}$$

$$\text{Line losses } W = 3I_2^2 R_2 = 3\left(\frac{P}{3V}\right)^2 \rho \frac{l}{a_2} = \frac{P^2 \rho l}{3V^2 a_2}$$

where a_2 is the cross section of conductor in case of 3-phase 4-wire system

$$\text{or } a_2 = \frac{P^2 \rho l}{3V^2 W}$$

Since the neutral is one-half of the cross section of the outer.

Volume of conductor material required

$$= 3.5 a_2 l = \frac{3.5 P^2 \rho l^2}{3V^2 W} = \frac{3.5}{12} \times \frac{4P^2 \rho l^2}{V^2 W}$$

$$= 0.292 \times \frac{4P^2 \rho l^2}{V^2 W}$$

i.e., In case of 3-phase 4-wire system conductor material required will be 0.292 times of that required in a 2-wire dc system. **Ans.**

Example 1.4. An existing single-phase ac system comprising of two overhead conductors is to be converted into a 3-phase, 3-wire system by providing an additional conductor of same size. Calculate the percentage of additional load that can be transmitted by the three phase system if the operating line voltage and percentage line losses remain the same in both the systems.

[A.M.I.E. Sec. B. Power Systems Summer 1998]

Solution : Let the operating line voltage and power factor in both the systems be V volts and cos ϕ respectively. If I_1 is single phase current, I_2 is the three phase current and R is the resistance of each conductor, then

in single-phase ac system

$$\text{Power transmitted, } P_1 = VI_1 \cos\phi \text{ watts}$$

$$\text{Line losses, } W_1 = 2I_1^2 R \text{ watts}$$

$$\text{Percentage line losses} = \frac{W_1}{P_1} \times 100 = \frac{2I_1^2 R}{VI_1 \cos\phi} \times 100 \quad ...(i)$$

and in three-phase ac system

$$\text{Power transmitted, } P_2 = \sqrt{3} VI_2 \cos\phi$$

$$\text{Lines losses, } W_2 = 3I_2^2 R$$

$$\text{Percentage line losses} = \frac{W_2}{P_2} \times 100 = \frac{3I_2^2 R}{\sqrt{3} VI_2 \cos\phi} \times 100 \quad ...(ii)$$

For the same percentage line losses in both the cases, we have

$$\frac{2I_1^2 R}{VI_1 \cos\phi} \times 100 = \frac{3I_2^2 R}{\sqrt{3} VI_2 \cos\phi} \times 100$$

$$\text{or } I_2 = \frac{2}{\sqrt{3}} I_1$$

So, power transmitted in case of 3-phase, 3-wire system,

$$P_2 = \sqrt{3}\,V \times \frac{2}{\sqrt{3}} I_1 \cos\phi = 2VI_1 \cos\phi = 2P_1$$

Additional load transmitted $= P_2 - P_1 = 2P_1 - P_1 = P_1$

Percentage of additional load $= \dfrac{P_2 - P_1}{P_1} \times 100 = \dfrac{P_1}{P_1} \times 100 = 100\%$

Thus, 100 per cent additional load can be transmitted by converting single phase line to three-phase line. **Ans.**

Example 1.5. An existing dc three-wire system is to be converted into a 3-phase 4-wire system by adding a fourth wire equal in cross section to each outer of the dc system. Assuming the same supply and load voltages to neutral and balanced conditions find the extra power at unity power factor that can be supplied by the ac system.

Solution : Let the supply and load voltages to neutral be V_S and V_L volts respectively and resistance of each conductor be R ohms.

Line current in 3-wire dc system,

$$I_1 = \frac{V_S - V_L}{R}$$

Power supplied, $P_1 = 2V_LI_1 = \dfrac{2V_L}{R}(V_S - V_L)$ ...(*i*)

Line current in 3-phase 4-wire ac system,

$$I_2 = \frac{V_S - V_L}{R}$$

Power supplied, $P_2 = 3V_LI_2 \cos\phi = \dfrac{3V_L}{R}(V_S - V_L) \times 1$ ...(*ii*)

Comparing Eqs. (*i*) and (*ii*), we get

$$P_2 = 1.5\,P_1$$

i.e., 50 per cent extra power can be supplied at unity pf by 3-phase, 4-wire ac system **Ans.**

Example 1.6. A 3-wire dc system is converted to a 4-wire 3-phase ac system by the addition of another wire equal in section to one of the outers. For the same effective voltage between outers and neutral at the consumer's terminals and the same percentage loss, find the percentage additional load that can be supplied. Assume balanced load and in the ac system a power factor of 0.9.

[*M.D. Univ. Transmission and Distribution, May 2011*]

Solution : Let the voltage between outers and neutral be V volts in both of the cases and currents in 3-wire dc system and 3-phase 4-wire system be I_1 and I_2 respectively.

3-Wire dc system

Power supplied, $P_1 = 2I_1V$

Power loss $= 2I_1^2R$

Percentage power loss $= \dfrac{2I_1^2R}{2I_1V} \times 100 = \dfrac{100I_1R}{V}$ ...(*i*)

3-phase 4-wire ac system

Power supplied, $P_2 = 3I_2V \cos\phi = 3I_2V \times 0.9 = 2.7I_2V$

Power loss $= 3I_2^2R$

Percentage power loss $= \dfrac{3I_2^2R}{2.7I_2V} \times 100 = \dfrac{100I_2R}{0.9V}$ ...(*ii*)

Since, load is balanced, therefore neutral wire will carry no current.

Since, percentage loss is same, therefore comparing Eqs. (*i*) and (*ii*), we get

$$\frac{100I_1R}{V} = \frac{100I_2R}{0.9\,V}$$

or $I_2 = 0.9I_1$

Hence power supplied in 3-phase 4-wire system,

$$P_2 = 2.7 \times 0.9I_1V = 2.43I_1V$$

Percentage increase in load supplied

$$= \frac{P_2 - P_1}{P_1} \times 100 = \frac{2.43I_1V - 2I_1V}{2I_1V} \times 100 = 21.5\% \text{ Ans.}$$

1.13. CHOICE OF WORKING VOLTAGE FOR TRANSMISSION

Before discussing the method of determination of the most economical working voltage for transmission we will discuss here the advantages and limitations of high voltage transmission.

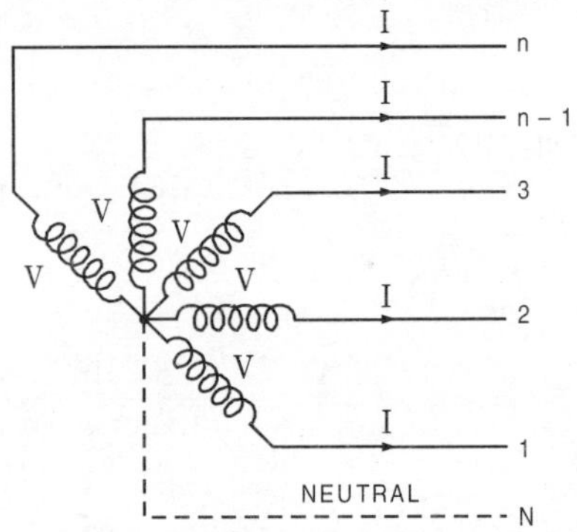

Fig. 1.10. *Polyphase Transmission System*

The advantages of adopting high voltage for transmission are given below:

(*i*) With the increase in transmission voltage the size of the conductor (area of cross section of the core carrying the current) is reduced. This further reduces the cost of the supporting structure materials.

(*ii*) With the increase in transmission voltage, line current is reduced, which results in reduction of line losses.

(*iii*) With the increase in transmission voltage reduction in line losses results in higher efficiency.

(*iv*) Due to low current at high transmission voltage, the voltage drop in the lines is low. This leads to better voltage regulation.

The above facts can be shown very clearly as follows:

Let the power delivered by *n*-phase system be P watts per phase at a phase voltage of V volts and power factor cos ϕ.

Load current per phase, $I = \dfrac{P}{V\cos\phi}$

Resistance of conductor, $R = \rho\dfrac{l}{a}$

where l is the length of line, a is the cross section of conductor and ρ is the resistivity of conductor material.

Power loss in the line (per phase),

$$W = I^2R = \frac{P^2}{V^2 \cos^2 \phi} \frac{\rho l}{a} \quad ...(1.20)$$

Also cross section of conductor,

$$a = \frac{I}{\delta} = \frac{P}{V \cos \phi \delta} \quad \text{where } \delta \text{ is the current density}$$

So substituting $a = \frac{P}{V \cos \phi \delta}$ in Eq. (1.20), we get

$$W = \frac{P^2}{V^2 \cos^2 \phi} \times \rho l \times \frac{V \cos \phi \delta}{P} = \frac{P \rho l \delta}{V \cos \phi} \quad ...(1.21)$$

Transmission efficiency,

$$\eta = \frac{\text{Output}}{\text{Input}} = \frac{\text{Output}}{\text{Output + line losses}}$$

$$\text{or } \eta = \frac{P}{P + \frac{P l \delta \rho}{V \cos \phi}} = 1 - \frac{\rho \delta l}{V \cos \phi} \quad ...(1.22)$$

Resistance drop per line

$$= IR = \frac{P \rho l \times V \cos \phi \delta}{V \cos \phi \times P} = \rho l \delta \quad ...(1.23)$$

Volume of conductor material required per phase

$$= al = \frac{Pl}{\delta V \cos \phi} \quad ...(1.24)$$

Substituting $\delta = \frac{WV \cos \phi}{P \rho l}$ from Eq. (1.21) in Eq. (1.24), we get

Volume of conductor material required per phase

$$= \frac{Pl}{V \cos \phi} \times \frac{\rho P l}{WV \cos \phi} = \frac{P^2 l^2 \rho}{WV^2 \cos^2 \phi} \quad ...(1.25)$$

Hence from Eqs. (1.21), (1.22), (1.23), and (1.24) it is obvious that for constant values of P, l, δ, $\cos \phi$ and ρ

(*a*) With the increase in transmission voltage line losses are reduced since line losses are inversely proportional to supply voltage, V.

(*b*) The transmission efficiency increases with the increase in transmission voltage, V.

(*c*) Resistance drop per line is constant. Hence percentage resistance drop decreases with the increase in transmission voltage V.

(*d*) The volume of conductor material required is inversely proportional to the supply voltage.

(*e*) From Eq. (1.25) it is obvious that for the same line losses and for transmission of given amount of power over a given distance through the conductors of given material and at a given power factor *i.e.*, for constant values of W, P, l, ρ and $\cos \phi$ conductor material required is inversely proportional to the square of supply voltage.

The above formula, besides showing the great saving of conductor material by employing a high transmission voltage, indicates also the importance of working as near unity power factor as possible. The effect of operating at, say, 0.707 power factor instead of unity, is to increase the amount of conductor material used in the ratio of 1: $\frac{1}{(0.707)^2}$ or an increase of 100%. A very considerable economy can be affected in the conductor material if the transmission line is operated at high power factor.

Where there are several advantages of high voltage for transmission, there are some limitations also, which are given below:

(*a*) With the increase in voltage of transmission, the insulation required between the conductors and the earthed tower increases. This increases the cost of line supports.

(*b*) With the increase in voltage of transmission, more clearance is required between conductors and ground. Hence higher towers are required.

(*c*) With the increase in voltage of transmission, more distance is required between the conductors. Therefore, longer cross arms are required.

For every transmission line there is a superior limit fixed for the voltage to be employed, beyond which nothing is gained in the matter of economy. This limit is reached when the cost of conductors, cost of insulators, supports, transformers, switchgear, lightning arrestors and also the cost of erection is minimum. The method of finding the optimum voltage is to choose a certain standard voltage and calculate the cost of (*i*) transformers to be used at the generating and receiving stations, (*ii*) switchgear consisting of isolators, bus-bars, circuit breakers, relays etc, (*iii*) lightning arresters, (*iv*) insulators, (*v*) supports including the cross arms and other fittings, (*vi*) conductors and cost of erection for the given voltage of generation, given power to be transmitted and given length of the transmission system. The total sum of all the costs gives the total capital cost of transmission. Thus for various standard voltages the total capital cost is worked out and a curve of capital cost against the transmission voltage is plotted. The curve plotted is of the form shown in Fig. 1.11. From the curve it is observed that the optimum

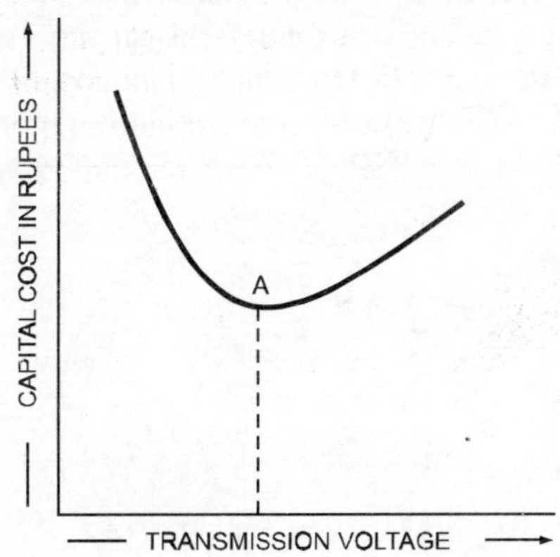

Fig. 1.11

voltage from economy point of view will be voltage corresponding to point A on the curve. In fact higher voltage would be choosen as with large working voltage, power loss and voltage drop are reduced, load can be increased if required and control becomes easier.

In practice it is not possible to determine the economic voltage by the method explained above. To avoid the complications and labour *Cable Research Hand Book* gives an empirical formula for determining the optimum voltage for lines more than 30 km long. The formula is

Approximate economical voltage for transmission in kV

$$= 5.5\sqrt{\frac{\text{km}}{1.6} + \frac{\text{load in kVA}}{150}} \quad ...(1.26)$$

The noteworthy point is that power to be transmitted and distance of transmission have been taken into account in the above formula, because both of these factors have considerable influence in arriving at economic voltage for transmission, as explained below.

With the increase in distance of transmission, the cost of terminal apparatus is reduced resulting in higher economic transmission voltage. Similarly if the power to be transmitted becomes large, the cost per kW of the terminal station equipment is reduced.

As a rough guide the voltage for transmission is chosen as 625 volts per km though in practice the voltage per km varies from about 400 to 900 volts for longer to shorter distances. The choice is usually limited in practice by the requirement of standardisation and for satisfactory regulation without excessive equipment cost.

The voltages normally adopted for transmission are given below.

Distance in km	15-30	30-60	60-100	100-200	200-300	300-400
Transmission voltage in kV	11	33	66	132	220	400

As already mentioned, the most common transmission voltages employed in India are 33 kV, 66 kV, 132 kV, 220 kV, 400 kV and 765 kV.

Example 1.7. Estimate the weight of copper required to supply a load of 100 MW at UPF by a 3-phase, 380 kV system over a distance of 100 km. The neutral point is earthed. The resistance of the conductor is 0.045 Ω/cm²/km. The weight of copper is 0.01 kg/cm³. The efficiency of transmission can be assumed to be 90 per cent. *[A.M.I.E. Sec. B. Power Systems Summer 1994]*

Solution : Line current,

$$I_L = \frac{P \text{ in MW} \times 10^6}{\sqrt{3} \times V_L \text{ in kV} \times 10^3 \times \cos\phi} = \frac{100 \times 10^6}{\sqrt{3} \times 380 \times 10^3 \times 1.0}$$

$$= 152 \text{ A}$$

Power losses in line,

$$W = P(1 - \eta) = 100(1 - 0.9) = 10 \text{ MW}$$

Line loss per conductor

$$= I_L^2 R = \frac{W}{3} = \frac{10 \times 10^6}{3} \text{ watts}$$

$$\text{Resistance per conductor, } R = \frac{10 \times 10^6}{3 \times (152)^2} = 144.4\ \Omega$$

$$\text{or Resistance per conductor per km} = \frac{144.4}{100} = 1.444\ \Omega$$

$$\text{Conductor area, } a = \frac{0.045}{1.444} = 0.03 \text{ cm}^2$$

Volume of copper required per km run

$$= 0.03 \times 100 \times 1{,}000 = 3{,}000 \text{ cm}^3$$

Weight of copper required per km run

$$= 0.01 \times 3{,}000 = 30 \text{ kg}$$

Weight of copper required for 3 conductors, each of 100 km length

$$= 3 \times 100 \times 30 = 9{,}000 \text{ kg}$$ **Ans.**

1.14. ECONOMIC SIZE OF LINE CONDUCTOR–KELVIN'S LAW

In case of a transmission system carrying bulk power over a long distance, the question of voltage regulation is insignificant; in some very long lines a regulation of 40 per cent is considered satisfactory. The only consideration is the economy. Hence the cross section of the conductor of a transmission line (or feeder) is decided on the basis of its current carrying capacity and, where practicable, of minimum cost, both fixed and operating.

The fixed cost is on account of annual interest and depreciation on capital cost of conductors, supports and insulators and the cost of their erection in case of overhead systems. For underground systems the fixed cost is on account of annual interest and depreciation on cost of conductors, insulation and the cost of laying the cables. For a particular voltage cost of insulation is practically constant and does not change with the cross section of the conductor, but the cost of conductors varies directly as the cross section of the conductor irrespective of the system of transmission. In case of overhead system the cost of supports and their erection partly varies as the cross section of the conductors and partly constant. Thus total annual fixed cost may be represented as Rs $(P_1 + P_2a)$ where P_1 and P_2 are constants and a is the area of cross section of conductor.

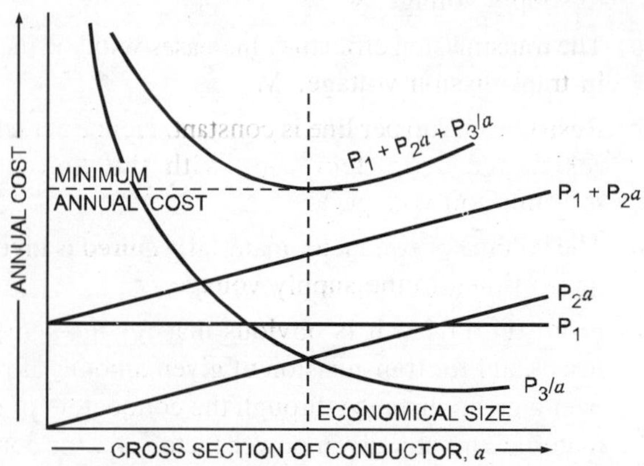

Fig. 1.12. ***Illustration of Kelvin's Law***

The annual running cost is on account of energy lost in the conductor due to its ohmic resistance *i.e.*, I^2R losses; losses in insulating material and in metallic sheaths (for insulated cables). The losses in insulating material and in metallic sheaths are very small and in comparison to ohmic losses may be neglected for the low voltage, but at higher voltages these are considerable. The ohmic resistance R is inversely proportional to area of cross section of conductor, and therefore for a given curve of demand for current through out the year *i.e.*, for a given annual load curve, the energy lost in the conductor will be proportional to the resistance and therefore, inversely proportional to area of cross section. Hence annual running cost on account of energy lost in transmission line may be represented as $\frac{P_3}{a}$ where P_3 is a constant.

$$\therefore \text{ Total annual cost, } C = P_1 + P_2 a + \frac{P_3}{a} \quad ...(1.27)$$

This total annual cost will be minimum if the differentiation of it with respect to a is zero

$$i.e., \text{ if } \frac{dC}{da} = 0$$

$$\text{or } \frac{d}{da}\left(P_1 + P_2 a + \frac{P_3}{a}\right) = 0$$

$$\text{or } P_2 - \frac{P_3}{a^2} = 0$$

$$\text{or } P_2 a = \frac{P_3}{a} \quad ...(1.28)$$

i.e., if variable part of annual cost on account of interest and depreciation on the capital outlay is equal to the annual cost of electrical energy wasted in the conductors, the total annual cost will be minimum and the corresponding size of conductor will be most economical. This statement is modified version of *Kelvin's law**. Though theoretically it holds good but in actual working condition number of difficulties are faced in its application.

1.14.1. Practical Limitations To Application of Kelvin's Law

Though theoretically Kelvin's law is true but in actual working, an economical cross section of the conductor determined by Kelvin's law may not suit because of the following factors:

(*i*) The assumption that the total annual cost on account of interest and depreciation on the capital outlay is of the form $P_1 + P_2 a$ is strictly speaking not true. For example, neither the cost of cable dielectric and sheath in underground cables vary according to this simple law, nor the cost of laying.

(*ii*) The size of the conductor determined may be of such a small cross section that it may cause too much voltage drop in the line.

(*iii*) The size of the conductor determined may be too weak or too strong from mechanical consideration. In case the size determined is too weak, it is advisable to go to the higher size irrespective to economy.

(*iv*) In case of cables the sizes of conductors determined by Kelvin's law usually gives higher current density thereby giving excessive heating. The only remedy is to decide the size of conductors in case of cables on the basis of current carrying capacity.

(*v*) The diameter of the conductor may be so small as to cause high corona loss.

(*vi*) Interest and depreciation on the capital outlay depends upon commercial conditions and estimated life of conductors and also the probable scrap value at the end of estimated life, which are not known with a fair degree of accuracy.

(*vii*) It is difficult to estimate the energy loss in the line without actual load curves, which are definitely not available at the times of estimation.

(*viii*) It is also difficult to estimate the cost per unit of the energy wasted in the line. The cost per unit of the energy wasted is not the same as that of cost of generation per unit since their cost per unit depends upon load factors which are different for the line losses and the generation.

(*ix*) In the case of cables there are sheath losses and with high voltages dielectric losses also. Dielectric loss occurs continuously therefore load factor of dielectric loss is 100%. Hence the cost per unit of energy lost as dielectric loss is less than cost per unit supplying the line loss.

It is impossible to derive an exact formula for the conductor cross section which will take into account all the above factors.

In case of overhead lines the question of temperature rise is of less importance and there are no losses other than I^2R loss so cross sections derived from Kelvin's law have more chances of being acceptable. In the case of underground systems it is advisable to examine the available cross sections above and below the calculated section and to ensure that temperature rise will not be excessive.

Example 1.8. A 2-conductor cable one km in length is required to supply a constant load of 200 A throughout the year. The cost of cable is Rs (100*a* + 40) per metre where *a* is the area of cross section of the conductor in cm². The cost of energy is 10 P per kWh and interest and depreciation charges amount to 10%. Specific resistivity of copper is 1.85 μΩ-cm. Find the cross section area '*a*'.

[*Rajasthan Technical Univ. Transmission & Distribution of Electrical Power 2011*]

Solution : Resistance of each conductor of area a cm² and one km in length

* According to Kelvin's law the most economical cross section of a conductor is that which makes the annual cost of electrical energy wasted in the conductor equal to the annual cost of interest and depreciation on the capital cost of the conductor. This law was based on the assumption that the cost of the supporting structures *e.g.,* towers, insulators and the cost of erection-cum-commisioning *i.e.*, stringing of the conductors etc are fixed and independent of the area of cross section of the conductor. But this is not true in practice. So the Kelvin's law was modified in order to take the above fact into account by assuming that the initial capital cost on the complete line installation can be split-up into two parts, one independent of the area of cross section of the conductor and the other dependent upon it, directly. Thus total annual fixed cost can be represented as Rs $(P_1 + P_2 a)$.

$$R = \frac{\rho l}{a} = \frac{1.85 \times 10^{-6} \times 1{,}000 \times 100}{a} = \frac{0.185}{a}\ \Omega$$

Energy loss per annum

$$= \frac{2I^2 R \times 8{,}760}{1{,}000}$$

$$= \frac{2 \times (200)^2 \times 0.185 \times 8{,}760}{1{,}000 \times a} = \frac{129{,}648}{a}\ \text{kWh}$$

Annual cost of energy loss,

$$\frac{P_3}{a} = \text{Rs}\ \frac{10}{100} \times \frac{129{,}648}{a} = \text{Rs}\ \frac{12{,}964.8}{a}$$

Capital cost of feeder cable (variable cost)

$= \text{Rs } 100 \times a \times 1{,}000 = \text{Rs } 100{,}000a$

Annual charges on account of interest and depreciation on variable cost of feeder cable,

$$P_2 a = \text{Rs } 100{,}000a \times \frac{10}{100} = \text{Rs } 10{,}000a$$

For most economical cross section of conductor

$$\frac{12{,}964.8}{a} = 10{,}000a$$

$$\text{or } a = \sqrt{\frac{12{,}964.8}{10{,}000}} = 1.14 \text{ cm}^2 \textbf{ Ans.}$$

Example 1.9. Determine the best current density in amperes/mm² for a three phase overhead line. The line is in use for 2,600 hours per year and if the conductor costs Rs. 3.0/kg. It has a specific resistance of 1.73 × 10⁻⁸ Ω-m and weighs 6,200 kg/m³. Cost of energy is 10 paise/unit. Interest and depreciation is 12% of conductor costs. [*U.P. Technical Univ. Elements of Power System 2006-07*]

Solution : Let I be the line current and 'a' be the cross sectional area of conductors in mm².

Weight of conductor of 1 km length

= Volume of conductor × specific gravity of material

$$= \frac{a \times 1{,}000 \times 6{,}200}{1{,}000 \times 1{,}000} = 6.2a \text{ kg}$$

Cost of conductor of 1 km length

$= \text{Rs } 3 \times 6.2a = \text{Rs } 18.6a$

Annual interest and depreciation per conductor per km

$$= \text{Rs } 18.6a \times \frac{12}{100} = 2.232a$$

Energy lost per annum per conductor per km

$$= I^2 R \times \frac{\text{Time in hours}}{1{,}000}$$

$$= \frac{I^2 \times 1.73 \times 10^{-8} \times 1{,}000}{a \times 10^{-6}} \times \frac{2{,}600}{1{,}000} = 44.98 \frac{I^2}{a}$$

Annual cost due to energy losses

$$= \text{Rs } \frac{10}{100} \times 44.98 \frac{I^2}{a} = \text{Rs } 4.498 \frac{I^2}{a}$$

For most economical cross section of conductor

$$2.232a = 4.498 \frac{I^2}{a}$$

$$\text{or } \frac{I}{a} = \sqrt{\frac{2.232}{4.498}} = 0.705 \text{ A/mm}^2$$

∴ The best current density = 0.705 A/mm² **Ans.**

Example 1.10. A 3-core, 11 kV cable supplies a load of 1,500 kW at 0.8 pf lagging for 300 days in a year at an average of 8 hours per day. The capital cost per km of cable is Rs 8,000 + 20,000*a* (*a* is the cross-sectional area of cable). The resistance per km of a cable of cross-sectional area 1 cm² is 0.173 Ω. If the energy loss cost is 2 paisa per unit, and the rate of interest and depreciation is 12%, calculate the most economical current density and diameter of the conductor. [*A.M.I.E. Sec. B. Power Systems Winter 1998*]

Solution : Capital cost (variable) of cable = Rs 20,000a per km length.

Annual charges on account of interest and depreciation on variable cost of cable,

$$P_2 a = \text{Rs } 20{,}000a \times \frac{12}{100} = \text{Rs } 2{,}400a \text{ per km length}$$

Resistance of each conductor of cross sectional area a cm²,

$$R = \frac{0.173\ \Omega}{a} \text{ per km length}$$

Current in each conductor,

$$I = \frac{\text{Power supplied in kW}}{\sqrt{3} \times \text{line voltage in kV} \times \text{pf}} = \frac{1{,}500}{\sqrt{3} \times 11 \times 0.8}$$

$= 98.412$ A

Total power loss in all the three conductors for length of 1 km,

$$W = 3I^2R = 3 \times (98.412)^2 \times \frac{0.173}{a} = \frac{5{,}026.47}{a} \text{ watts}$$

Energy loss per annum

= W in kW × working hours/annum

$$= \frac{5{,}026.47}{a \times 1{,}000} \times 300 \times 8 = \frac{12{,}064}{a} \text{ kWh}$$

Annual cost of energy loss,

$$\frac{P_3}{a} = \text{Rs } 0.02 \times \frac{12{,}064}{a} = \frac{241.28}{a}$$

For most economical cross section of conductor,

$$2{,}400a = \frac{241.28}{a}$$

$$\text{or } a = \sqrt{\frac{241.28}{2{,}400}} = 0.317 \text{ cm}$$

and diameter of conductor,

$$d = \sqrt{\frac{4a}{\pi}} = \sqrt{\frac{4 \times 0.317}{\pi}} = 0.635 \text{ cm} \quad \therefore a = \frac{\pi}{4}d^2 \textbf{ Ans.}$$

Current density,

$$\delta = \frac{I}{a} = \frac{98.412}{0.317} = 310.4 \text{ A/cm}^2 \textbf{ Ans.}$$

Example 1.11. The cost per km for each of the copper conductor of a section *a* cm² for a transmission line is Rs (2,800*a* + 1,300). The load factor of the load current is 80% and the load factor of the losses is 65%. The rate of interest and depreciation is 10% and the cost of energy is 50 paisa per kWh. Find the most economical current density for the transmission line by the use of Kelvin's law. Given ρ = 1.78 × 10⁻⁸ Ω-m.

[*A.M.I.E. Sec. B. Power Systems Winter 1999; Pb. Technical Univ. Power System-I (T&D) May 2010*]

Solution : Capital cost (variable) of transmission line

= Rs 2,800 a per conductor per km length.

Annual charges on account of interest and depreciation on variable cost of transmission line,

$$P_2a = \text{Rs } 2{,}800a \times \frac{10}{100}$$

$= \text{Rs } 280a$ per conductor per km length

Resistance of each conductor of cross-sectional area a cm^2,

$$R = \frac{1.78 \times 10^{-8} \times 1{,}000}{a \times 10^{-4}} = \frac{0.178}{a} \ \Omega \text{ per km length}$$

If I is the current in amperes in each conductor, then power losses in each conductor for a length of 1 km,

$$W = I^2R = \frac{0.178I^2}{a} \text{ watts}$$

Energy loss per annum per conductor for a length of 1 km,

= W in kW × load factor of losses × 8,760 kWh

$$= \frac{0.178}{a} \frac{I^2 \times 0.65 \times 8{,}760}{1{,}000} = \frac{1.0135321I^2}{a} \text{ kWh}$$

Annual cost of energy loss for a length of 1 km,

$$\frac{P_3}{a} = \text{Rs } \frac{50}{100} \times \frac{1.0135321I^2}{a} = \text{Rs } \frac{0.5067661I^2}{a}$$

For most economical cross section of conductor,

$$P_2a = P_3/a$$

i.e., $280\,a = \dfrac{0.5067661I^2}{a}$ or $\dfrac{I}{a} = \sqrt{\dfrac{280}{0.506766}} = 23.5 \text{ A/cm}^2$

i.e., most economical current density for the transmission line,

$$\delta = \frac{I}{a} = 23.5 \text{ A/cm}^2 \textbf{ Ans.}$$

Example 1.12. The cost of a 3-ϕ overhead line having cross-sectional area '*a*' cm² is Rs (500 + 2,000*a*) per km. Calculate the most economical current density for the conductor if the rate of interest and depreciation is 12% per annum and the cost of energy wasted is 5 paise per kWh. The resistivity of conductor may be taken as 1.78 μΩ-cm. Take load factor of 12%.

[*U.P. Technical Univ. Elements of Power System 2008-09*]

Solution : Capital cost (variable) of transmission line

= Rs 2,000a per conductor per km length

Annual charges on account of interest and depreciation on variable cost of transmission line,

$$P_2a = \text{Rs } 2{,}000a \times \frac{12}{100}$$

= Rs 240a per conductor per km length.

Resistance of each conductor of cross-sectional area a cm^2

$$R = \frac{\rho l}{a} = \frac{1.78 \times 10^{-8}}{a \times 10^{-4}} \times 1{,}000 = \frac{0.178}{a} \ \Omega \text{ per km length}$$

If I is the current in amperes in each conductor, then power loss in each conductor for a length of 1 km

$$W = I^2R = \frac{0.178}{a} I^2 \text{ watts}$$

Energy loss per annum per conductor for a length of 1 km,

= W in kW × Load factor of losses × 8,760 kWh

$$= \frac{0.178}{a} I^2 \times \frac{0.12 \times 8{,}760}{1{,}000} = \frac{0.18711361I^2}{a} \text{ kWh}$$

Annual cost of energy loss for a length of 1 km,

$$\frac{P_3}{a} = \text{Rs } \frac{5}{100} \times \frac{0.18711361I^2}{a} = \text{Rs } \frac{0.009355681I^2}{a}$$

For most economical cross section of conductor,

$$P_2a = \frac{P_3}{a}$$

$$\text{or } 240a = \frac{0.009355681I^2}{a}$$

$$\text{or } \frac{I}{a} = \sqrt{\frac{240}{0.00935568}} = 160.2 \text{ A/cm}^2 \textbf{ Ans.}$$

Example 1.13. Describe the main limitations of Kelvin's law. The cost of a 3-phase overhead transmission line having cross-sectional area A cm² is Rs (500 + 2,000 A) per km. Calculate the most economical current density for the conductor if the rate of interest and depreciation is 12% per annum. The cost of energy wasted is Rs 0.05/kWh. The resistance of each conductor is $\frac{0.17}{A}$ ohm/km. Take load factor for losses = 12%.

[*U.P. Technical Univ. Elements of Power System 2007-08*]

Solution : Annual charges on account of interest and depreciation on variable cost of transmission line is

$$P_2A = \text{Rs } 2{,}000A \times \frac{12}{100}$$

= Rs. 240 A per conductor per km length

Resistance of each conductor,

$$R = \frac{0.17}{A} \ \Omega \text{ per km length}$$

Power loss in each conductor for a length of 1 km,

$$W = I^2R = \frac{0.17I^2}{A} \text{ watts}$$

where I is the current in amperes in each conductor.

Energy loss per annum per conductor for a length of 1 km

$$= \frac{0.17}{A} \frac{I^2 \times 0.12 \times 8{,}760}{1{,}000} = \frac{0.1787I^2}{A} \text{ kWh}$$

Annual cost of energy loss for a length of 1 km

$$\frac{P_3}{A} = \frac{\text{Rs } 0.05 \times 0.1787I^2}{A} = \text{Rs } \frac{0.008935I^2}{A}$$

For most economical cross section of conductor

$$P_2A = \frac{P_3}{A}$$

$$\text{or } 240\text{ A} = \frac{0.008935I^2}{A}$$

Most economical current density

$$= \frac{I}{A} = \sqrt{\frac{240}{0.008935}} = 163.89 \text{ A/cm}^2 \textbf{ Ans.}$$

Example 1.14. Determine the most economical size of a 3-phase line which supplies the following loads at 10 kV.

(*i*) 1,000 kW at 0.8 pf (lag) for 10 hours,

(*ii*) 500 kW at 0.9 pf (lag) for 8 hours and

(*iii*) 100 kW at upf for 6 hours.

The above gives the daily load cycle. The cost per km of the completely erected line is Rs (8,000*a* + 1,500) where *a* is the area of cross section of each conductor in cm². The combined interest and depreciation is 10 per cent per annum of capital cost. Cost of energy losses is 80 paise per kWh. Resistivity of conductor material = 1.72 × 10^{-6} Ω-cm.

Solution : Let the length of the line be L km

Capital cost of line

$$= \text{Rs } (8{,}000a + 1{,}500) \times L$$

Annual cost on account of interest and depreciation on capital (variable) cost

$$= \text{Rs } \frac{10}{100} \times 8{,}000aL = \text{Rs } 800aL$$

The load currents

(*i*) at 1,000 kW, 0.8 pf (lag),

$$I = \frac{1{,}000 \times 1{,}000}{\sqrt{3} \times 10{,}000 \times 0.8} = \frac{125}{\sqrt{3}} \text{ A}$$

(*ii*) at 500 kW, 0.9 pf (lag),

$$I = \frac{500 \times 1{,}000}{\sqrt{3} \times 10{,}000 \times 0.9} = \frac{500}{9\sqrt{3}} \text{ A}$$

(*iii*) at 100 kW, upf,

$$I = \frac{100 \times 1{,}000}{\sqrt{3} \times 10{,}000 \times 1} = \frac{10}{\sqrt{3}} \text{ A}$$

Resistance of line conductor,

$$R = \frac{\rho l}{a} = \frac{1.72 \times 10^{-6} \times 1{,}000 \times 100 \times L}{a} \text{ ohms} = \frac{0.172\text{ L}}{a} \text{ ohms}$$

Energy lost per annum

$$= \frac{3I^2Rt}{1{,}000} \text{ kWh}$$

$$= \frac{3 \times 0.172L}{a} \times \frac{365}{1{,}000}\left[\left(\frac{125}{\sqrt{3}}\right)^2 \times 10 + \left(\frac{500}{9\sqrt{3}}\right)^2 \times 8 + \left(\frac{10}{\sqrt{3}}\right)^2 \times 6\right]$$

$$= \frac{11{,}397L}{a} \text{ kWh}$$

Cost of energy wasted per annum

$$= \text{Rs } 0.80 \times \frac{11{,}397L}{a} = \text{Rs } \frac{9{,}117.6L}{a}$$

For most economical cross section of conductor

$$800aL = \frac{9{,}117.6L}{a}$$

$$\text{or } a = \sqrt{\frac{9{,}117.6}{800}} = 3.376 \text{ cm}^2 \textbf{ Ans.}$$

Exercises

1. Draw single line diagram of a typical power system. Locate various sections of it and mention there on typical voltages of generation, transmission and distribution.
2. Give a typical layout of a power system between generation and use of electrical power. **[A.M.I.E. Sec. B. Power Systems Winter 1998]**
3. Explain the terms (*i*) Feeder (*ii*) Distributor and (*iii*) Service mains. **[M.D. Univ. Power System-I, Dec. 2008]**
4. Distinguish between a feeder and a distributor. Draw single line diagram representing elements of a transmission system. **[Rajasthan Technical Univ. Transmission and Distribution of Electrical Power 2008]**
5. What is meant by the terms feeder, distributor and service mains ? Why the distribution by ac is considered superior to that by dc. **[Rajasthan Technical Univ. Transmission and Distribution of Electrical Power 2011]**
6. Compare between ac and dc systems for transmission and distribution. **[M.D. Univ. Power System-I, May 2011]**
7. Compare the advantages of using ac transmission system over dc system. **[U.P. Technical Univ. Elements of Power System 2002-03]**
8. Give comparison between underground cables with overhead transmission lines. **[Pb. Technical Univ. Power System-I (T&D), Dec. 2006; Dec. 2008]**
9. Discuss the advantages of using overhead lines compared to underground cables. **[U.P. Technical Univ. Elements of Power System 2006-07]**
10. What do you mean by single line diagram of power system ? What is the usefulness of drawing single line diagram ? **[U.P. Technical Univ. Elements of Power System 2007-08]**
11. What do you understand by single line diagram of a power system ? What is the need of this diagram ? **[U.P. Technical Univ. Elements of Power System 2009-10]**
12. Explain the influence of working voltage on the size of feeders and distributors in dc system. **[Pb. Technical Univ. Power System-I (T&D) Dec. 2008]**
13. Discuss the effect of transmission voltage on line performance. Derive any formula to support your answer. **[U.P. Technical Univ. Elements of Power System 2005-06]**
14. Distinguish between a feeder, distributor and service main in distribution scheme. Show that the cross section of a feeder and distributor would be reduced to 1/k and 1/k² of their respective values with an increase in working voltage by k times. **[U.P.S.C. 1972]**
15. Discuss the effect of increase in voltage on the transmission line efficiency. **[U.P. Technical Univ. Elements of Power System 2006-07]**
16. Compare the cross-sectional areas of conductors for dc two-wire system and single phase ac system assuming equal length, equal power, and equal losses. **[Pb. Technical Univ. Power System-I (T&D) May 2010]**
17. Find the ratio of volume of copper required to transmit a given power over a given distance by overhead system using (*i*) dc two-wire system, (*ii*) 3-ϕ, 3-wire ac system. **[Rajasthan Univ. Transmission & Distribution of Electrical Power 2007]**
18. Compare the conductor requirement for 1-ϕ ac system with 3-ϕ ac system. Assume same maximum potential difference between conductor and earth. **[U.P. Technical Univ. Elements of Power System 2008-09]**

19. Show that for overhead system, the ratio of the volume of conductors in two wire dc (mid point earthed) single phase ac (mid point earthed) and three-phase, three-wire ac are given by :

$$v_1 : v_2 : v_3 :: 1 : \frac{2}{\cos^2 \phi} : \frac{2}{\cos^2 \phi}$$

where cos ϕ is the power factor of the load. Assume equal power transmitted over equal length with equal losses and maximum voltage to earth to be same in all cases.

[Transmission & Distribution of Electrical Power, Rajasthan Univ. 2001, 2006; Rajasthan Technical Univ. 2008, 2009]

20. Compare the copper volume of single-phase and three-phase three-wire supply system assuming equal power loss in transmitting same power over given distance.

[Pb. Technical Univ. Power System-I (T&D), May 2009]

21. A 3-phase 4-wire system is used for lighting. Compare the amount of copper required with that needed for a 2-wire dc system with the same maximum voltage. Assume the same losses and balanced load. The neutral is one half of the respective outers.

[Rajasthan Technical Univ. Transmission & Distribution of Electrical Power 2011]

22. Show that at higher voltages and higher power factor, the efficiency of transmission line will be increased. What are the limiting factors of high voltage transmission lines ?

[U.P. Technical Univ. Elements of Power System 2007-08]

23. Discuss about the factors affecting choice of transmission voltage level.

[U.P. Technical Univ. Elements of Power System 2009-10]

24. What are the advantages of high voltage transmission ? Give the limitations also.

[Pb. Technical Univ. Power System-I (T&D) May 2009]

25. Prove that the volume of conductor required in a transmission system is inversely proportional to the square of voltage as well as power factor of the load. State the conditions under which this statement is true ?

[Rajasthan Technical Univ. Transmission & Distribution of Electrical Power 2011]

26. What is the most economical cross-sectional area ? Deduce its expression using Kelvin's law.

[Pb. Technical Univ. Power System-I (T&D), May 2010]

27. Explain the practical limitations to application of Kelvin's law for economical size of line conductor.

[Pb. Technical Univ. Power System-I (T&D) Dec. 2008]

28. Write short notes on the following :

(*i*) DC system versus ac system for transmission and distribution of electrical power.

(*ii*) Underground versus overhead system.

(*iii*) Power system elements.

[G.B. Tecnical Univ. 2011-12]

(*iv*) Choice of voltage for transmission line.

[W.B. Technological Univ. 2009-10]

(*v*) Kelvin's law and its limitations.

[Rajasthan Technical Univ. 2011]

Short Answer Type Questions With Answers

Q. 1. What is the voltage that has been selected for HVDC transmission ?

Ans. ± 500 kV.

Q. 2. What is an electric power supply system ?

Ans. The system of generation, transmission and distribution of electrical power is called the electrical power supply system.

Q. 3. What is the usable voltage for secondary distribution ?

Ans. 415/240 V (415 volts for 3-phase loads and 240 V for single phase loads).

Q. 4. What is the difference between a feeder and distributor ?

[M.D. Univ. May 2008; Pb. Technical Univ. May 2009]

Ans. Feeders are the conductors which connect the stations (in some cases generating stations) to the areas to be fed by those stations. Generally from feeders no tapping is taken to the consumers, therefore, current loading of a feeder remains the same along its length. It is designed mainly from the point of view of its current carrying capacity.

Distributors are the conductors from which numerous tappings for the supply to the consumers are taken. The current loading of a distributor varies along its length. Distributors are designed from the point of view of the voltage drop in them.

Q. 5. What is the legal requirement to which a distributor is subjected ?

Ans. A distributor is subject to the legal requirement that the voltage at the consumer's terminals should be maintained within ±6% of the declared (or rated) voltage.

Q. 6. Why is ac distribution considered superior to dc ?

[Pb. Technical Univ. Dec. 2008; Rajasthan Technical Univ. 2011]

Ans. Distribution by ac system is undoubtedly superior to that by dc system as in ac system voltage control is easy by means of transformers.

Q. 7. List advantages of dc system over ac system.

[Pb. Technical Univ. May 2008]

Ans. DC system has the following advantages over dc system:

(*i*) Saving in conductor material, (*ii*) No inductance, capacitance, phase displacement and surge problem, (*iii*) No skin effect, (*iv*) Less insulation requirement, (*v*) No charging current, (*vi*) Less corona loss and reduced interference with communication circuits, (*vii*) Better voltage regulation, (*viii*) No stability problems and synchronising difficulties.

Q. 8. Give limitations of ac transmission over dc transmission.

[Pb. Technical Univ. Dec. 2007]

Ans. AC transmission has got the following limitations over dc transmission.

(*i*) In case of overhead lines spacing between the conductors is to be kept more in order to provide adequate insulation and to avoid corona loss.

(*ii*) More copper is required.

(*iii*) Greater the length of transmission line more is the capacitance of the line. There is a continuous loss on account of charging current even though the line is open.

(*iv*) The construction of transmission lines for ac systems is not so easy as for dc systems.

(*v*) The alternators are to be synchronised before putting them into parallel.

(*vi*) The variation in speeds of alternators are to be controlled within very low limits.

Q. 9. What is the percentage saving in feeder conductor if the line voltage in a 2-wire dc system be raised from 200 V to 500 V for the same power to be transmitted ?

Ans. 84%.

Q. 10. List various systems of transmission of electrical power. **[Pb. Technical Univ. Dec. 2008]**

OR

What are possible system used in India for power transmission ? **[Pb. Technical Univ. Dec. 2006]**

Ans. Though in practice 3-phase 3-wire ac system is universally used for transmission and 3-phase 4-wire ac system is used for distribution of electric power but for special purposes other systems may also be used. The various systems of power transmission are :

(*a*) *DC Systems :*

(*i*) DC two-wire system.

(*ii*) DC two-wire system with midpoint earthed.

(*iii*) DC three-wire system.

(*b*) *Single-Phase AC Systems:*

(*i*) Single-phase two-wire system.

(*ii*) Single-phase two-wire system with midpoint earthed.

(*iii*) Single-phase three-wire system.

(*c*) *Two-Phase AC Systems:*

(*i*) Two-phase four-wire system.

(*ii*) Two-phase three-wire system.

(*d*) *Three-Phase AC Systems:*

(*i*) Three-phase 3-wire system.

(*ii*) Three-phase 4-wire system.

Q. 11. Why does the maximum voltage between each conductor and earth form the basis of comparison of volume of conductor material required in overhead system ?

Ans. In overhead system the conductors are insulated from the cross arms and supporting towers and as the towers and cross arms are earthed so the maximum voltage between each conductor and earth forms the basis of comparison of volume of conductor material required in overhead system.

Q. 12. Why does the maximum voltage between conductors form the basis of comparison of volume of conductor material required in underground system ?

Ans. In underground cables the maximum disruptive stress is between the two conductors of the cable; so the maximum voltage between conductors forms the basis of comparison of volume of conductor material required in underground system.

Q. 13. What is the ratio of volume of conductor material required for transmission of a given amount of power over a given distance by overhead system using

(*i*) dc 2-wire system and (*ii*) 3-ϕ, 3-wire ac system ?

Ans. $0.5/\cos^2\phi$.

Q. 14. What is the ratio of volume of conductor material required for transmission of a given amount of power over a given distance by underground system using (*i*) Two-phase 4-wire system; (*ii*) Three-phase three-wire system.

Ans. Three-fourth.

Q. 15. What is the effect of voltage on transmission efficiency ? **[Pb. Technical Univ. Dec. 2006]**

Ans. The transmission efficiency increases with the increase in transmission voltage.

Q. 16. State the empirical formula for determining the system voltage of transmission line. **[B.T.E. A.P. June 1996]**

Ans. Approximate economical voltage for transmission in kV

$$= 6.5\sqrt{\frac{\text{km}}{1.6} + \frac{\text{Load in kVA}}{150}}\,.$$

Q. 17. What is modified version of kelvin's law ?

Ans. According to Kelvin's law the most economical cross section of a conductor is that which makes the annual cost of electrical energy wasted in the conductor equal to the annual cost of interest and depreciation on the capital cost of the conductor. This law was based on the assumption that the cost of the supporting structures *e.g.,* towers, insulators and the cost of erection-cum-commisioning *i.e.,* stringing of the conductors etc are fixed and independent of the area of cross section of the conductor. But this is not true in practice. So the Kelvin's law was modified in order to take the above fact into account by assuming that the initial capital cost on the complete line installation can be split-up into two parts, one independent of the area of cross section of the conductor and the other dependent upon it, directly. Thus total annual fixed cost can be represented as Rs $(P_1 + P_2a)$.

Problems

1. What is the percentage saving in feeder copper if the line voltage in a 2-wire dc system be raised from 220 volts to 400 volts for the same power transmitted over same distance and having the same power loss ? **[Ans.** 69.75%]

2. A single-phase load of 10,000 kW is transmitted by a pair of overhead conductors. If a third conductor of the same cross section be added and a 3-phase supply be thus substituted for the original single phase, calculate the 3-phase load which can be now transmitted if voltage between the wires and percentage loss in the line remain unchanged. **[Ans.** 20,000 kW]

3. A dc 2-wire overhead distribution system is to be converted into a 3-ϕ, 3-wire ac system by installing a third conductor of the same section as the other two. Calculate the additional percentage load which can be distributed by the ac system, assuming that the voltage between the conductors and the percentage power loss remain the same. Assume a balanced 3-ϕ load at (*i*) unity pf and (*ii*) 0.8 lagging pf. **[Ans.** (*i*) 100% (*ii*) 28%]

4. A 3-wire dc transmission system is converted into a 4-wire, 3-phase balanced ac transmission system by addition of another conductor, equal in cross section to other conductors. Assume that the middle (grounded) conductor in dc system carries negligible current and the voltage between the outer conductors is 2 V. Assume that the rms voltage to ground for ac system is V. For the same

percentage power loss find the additional power that can be transmitted by the ac system at 0.9 pf. [**Ans.** 21.5%]

5. A 3-wire dc supply system is converted to a 3-phase, 4-wire ac system by the addition of another conductor identical to the outers. For the same voltage at consumer's terminals and the same percentage copper loss, find the percentage additional load that can be supplied. Assume balanced loads at unity power factor.

[A.M.I.E. Sec. B. Power Systems Winter 1996] [**Ans.** 50%]

6. The cost per km for each of the copper conductor of a section '*r*' square cm for a transmission line is Rs (2,000*r* + 1,300). The load factor of the load current is 80% and the load for the losses is 65%. The rate of interest and depreciation is 10% and the cost of energy is 5 paise/kWh. Find the most economical current density for the transmission line by the use of Kelvin's law. Given $\rho = 1.78 \times 10^{-8}$ Ω-m.

[Rajasthan Univ. Transmission and Distribution of Electrical Power 2002, 2006]

[**Ans.** 74 A/cm^2]

7. A 500 V, 2-core feeder 0.8 km long is required to supply a constant load of 100 kW. The cost of the cable including installation charges is Rs (6*a* + 1.3) per metre where *a* is the cross-sectional area of each conductor in cm^2. Interest and depreciation total 10%. Determine the most economical size. Cost of energy is 2P per unit. Specific resistance of copper is 1.75×10^{-6} Ω per cm^2 cross-sectional area and 1 cm long. [**Ans.** 2.022 cm^2]

8. Find the most economical cross section of a 3-core feeder cable 250 m long supplying a load of 80 kW, 0.8 pf (lagging) at 400 V for 4,000 hrs per year. The cost of the cable including installation is Rs (150 *a* + 250) per metre where *a* is the cross section of each conductor in cm^2. Interest and depreciation total 10%. Cost of energy wasted is Rs 1 per unit. The resistance per km of conductor 1 cm^2 cross section is 0.173 Ω.

[A.M.I.E. Sec. B. Power Systems Winter 1996]

[**Ans.** 1.7 cm^2]

9. An industrial load is supplied by a 3-phase cable from a substation at a distance of 6 km. The voltage at the load is 11 kV. The daily load cycle for six days in a week for the entire year is as given below:

(*i*) 700 kW at 0.8 pf for 7 hours.

(*ii*) 400 kW at 0.9 pf for 3 hours.

(*iii*) 88 kW at unity pf for 14 hours.

Compute the most economical cross section of conductors for a cable whose cost is Rs. (5,000 A + 1,500) per km (including the cost of laying etc.). The tariff for the energy consumed at the load is Rs. 150 per annum per kVA of MD plus 5 paise per unit. Assume the rate of interest and depreciation as 15%. The resistance per km of conductor is (0.173/A) ohms. [**Ans.** 0.63 cm^2]

10. A 500-V, 2-core feeder cable 4 km long supplies a maximum current of 200 A and the demand is such that the copper loss per annum is such as would be produced by the full load current flowing for 6 months. The resistance of the conductor 1 km long and 1 sq cm. cross-sectional area is 0.17 Ω. The cost of cable including installation is Rs. (120 A + 24) per metre where A is the area of cross section in cm^2 and interest and depreciation charges are 10%. The cost of energy is 4 paise per kWh. Determine the most economical cross section. [**Ans.** 44.6 mm^2]

❑❑❑

CHAPTER 2

Mechanical Design of Overhead Lines

INSIDE THIS CHAPTER

2.1. INTRODUCTION

Transmission and distribution lines are vital links between generating stations and consumers as power from generating stations is transmitted at high voltage (such as 132, 220, 400 or 765 kV) over long distances to the major load centres and then the power is distributed to various substations located at various places and localities through distribution lines. Because of tremendous industrial growth, requirement of power has increased manifold. Hence it becomes imperative that transmission and distribution of power from the generating stations to the various consumers is carried out with minimum possible loss and disturbance. This objective can be achieved only if the transmission and distribution system is so designed and constructed that it is an efficient, technically sound and reliable system. The line should have sufficient current carrying capacity so as to transmit the required power over a given distance without an excessive voltage drop and overheating. The line losses should be small and insulation of the line should be adequate to cope with the system voltage. The line should have sufficient mechanical strength to cope with the worst probable (not worst possible)* weather conditions and provide satisfactory service over a long period of time without the necessity of too much maintenance.

2.2. MAIN COMPONENTS OF OVERHEAD LINES

The main components of an overhead line are enlisted, as below.

1. **Supports.** Poles or towers depending upon the working voltage and the region where these are used. The function of the line support is obviously to support the conductors so as to keep them at a suitable level above the ground.
2. **Cross arms and Clamps.** These are either of wood or steel angle section and are used on pole structures to support the insulators and conductors.
3. **Insulators.** Pin, strain or suspension types, as the case may be, for supporting the conductors and taking strain or suspending the conductors respectively.
4. **Conductors.** Copper, aluminium or ACSR or of any other composition depending upon the current to be carried and the span of the line.
5. **Guys and Stays.** Braces or cables are fastened to the pole at the termination or angle poles to resist lateral forces.
6. **Lightning Arrestors** to discharge excessive voltages built upon the line, to earth, due to lightning.
7. **Fuses and Isolating Switches** to isolate different parts of the overhead system.
8. **Continuous Earth Wire** is run on the top of the towers to protect the line against lightning discharges.
9. **Vee Guards** are often provided below bare overhead lines running along or across public streets to make the line safe if it should break.
10. **Guard Wires** are provided above or below power lines while crossing telephone or telegraph lines. The guard wires and steel structures are solidly connected to earth.
11. **Phase Plates** in order to distinguish the various phases.

* The cost of the system which would be capable of withstanding a worst possible weather condition would be extremely large and from economical aspects it is better to bear the risk of failure under such extreme worst conditions.

12. Bird Guards. A stick of ebonite with rounded top is fixed near the insulator on the cross arm to prevent flash-over due to birds pecking on the conductors (on lines with pin insulators).

13. Danger Plate. It is provided on each pole, as a warning measure indicating the working voltage of the line and the word "danger". It is provided at a height of 2.5 m from the ground.

14. Barbed Wire. Barbed wire is wrapped on a pole at a height of about 2.5 m from the ground for atleast 1 metre. This prevents climbing by unauthorised persons.

15. Miscellaneous Items such as vibration dampers, top hampers, beads for jumpers etc.

2.3. LINE SUPPORTS

The function of line support is obviously to support the conductors. Line support must be capable of carrying the load due to insulators and conductors including the ice and wind loads on the conductors along with the wind load on the support itself. The line supports are of various types including wood, steel and reinforced concrete poles, and steel towers either of the rigid or flexible type.

A distinction is drawn between straight poles which lie in direct line of transmission and normally only support the conductors and the special poles which may carry some load due to conductor tension. These latter supports may be angle towers, terminal towers, towers at tee-off parts, anchor towers, or towers for some special purpose such as for crossing the rivers. The dimensions of these line supports are governed by ISI regulations. In case of telegraph or railway line crossing special requirements are to be met with.

The main requirements of the line supports are:

1. *High mechanical strength* to withstand the weight of conductors and wind load etc.
2. *Light in weight* without the loss of mechanical strength.
3. *Cheaper in cost.*
4. *Low maintenance cost.*
5. *Longer life.*
6. *Good looking.*
7. *Easy accessibility* for painting and erection of line conductors.

The choice of line supports for a particular situation depends upon the line span, cross-sectional area, line voltage, cost and local conditions.

The design of an overhead line support depends upon the fact whether the support is rigid or has a certain amount of flexibility in the direction of the line. Wooden poles and some special types of steel structures are of latter type and only the transverse wind pressure occurring upon the conductors and upon the support itself is usually considered in their design. The longitudinal pull of the conductors is normally balanced on either side of the support but in the event of breaking of one or more conductors on one side, there will be an unbalanced load which may be far in excess than transverse wind pressure. With flexible supports this unbalanced load is quickly absorbed by an increase in sag in undamaged span on account of bending of the supports on each side of the wrecked span towards the adjoining spans. After three or four spans the longitudinal pull becomes negligible. A certain general rigidity in the longitudinal direction is provided for in practice by using rigid anchoring towers at an interval of 1.5 km or so. These anchoring towers are designed to withstand the breaking of one wire in three on one side as well as the transverse load.

In case of rigid supports such as the lattice-steel broad-base structures equal strength is usually provided in both the longitudinal and transverse directions and every tower is designed to withstand the unbalanced load on account of breaking of one wire in three on the same side. In addition, anchor towers are often provided, in which the support is capable of withstanding the failure of two conductors out of three, or even of all conductors on one side.

1. Wooden Poles. These supports are cheapest, easily available, provide insulating properties and therefore, are

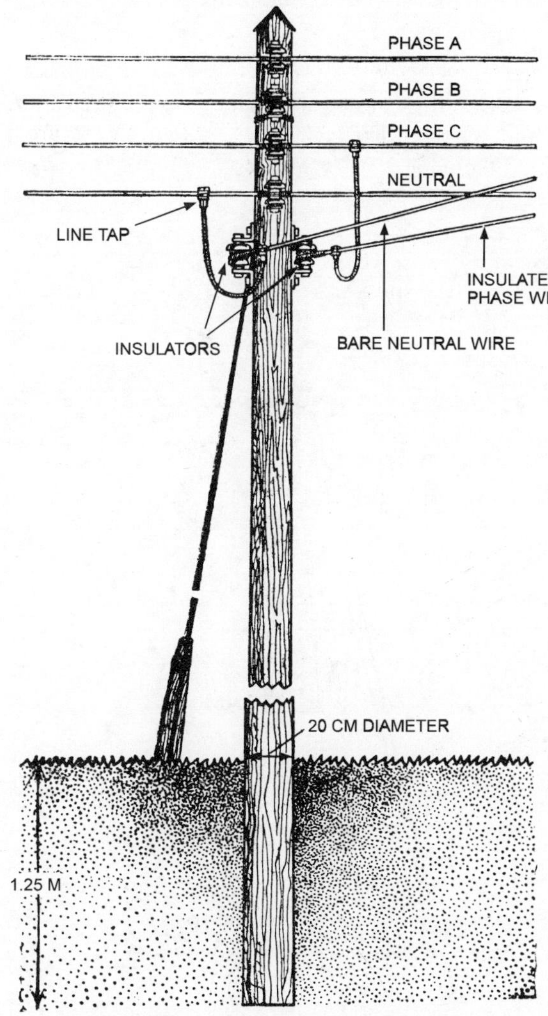

Fig. 2.1. ***Single Pole Construction Showing Installation of Three-Phase Four Wire Cable in Vertical Formation. A Two-Wire Service Tapping Supplies Consumers Homes***

extensively used for the distribution purposes specially in rural electrification keeping the cost low. Their use is usually limited to low pressures (up to 22 kV) and for short spans (up to 60 metres). In districts having a plentiful supply of suitable timber and where the cost of transporting steel towers is high single and 'H' poles have been used for overhead lines operating at voltages up to 130 kV and average span lengths of 150 metres. Sal or chir wooden poles up to 11 metre length with minimum circumference of 38 cm at the top and 66 cm at the bottom are used. These poles must be straight, strong with gradual taper and free from large knots. The poles should be properly seasoned to prevent rapid decay due to opening of cracks. To prevent decay owing to snow and rain, the wooden poles are protected by an aluminium, zinc or cement cap at the top. The portion of the pole *i.e.,* 2 metres from bottom which is buried in the ground must be treated with creosite oil or any preservative compound. The wooden poles well impregnated with creosite oil or any preservative have life from 25 to 30 years.

Wooden poles are very elastic and lines employing wooden supports are often designed throughout for the transverse load. Longitudinal strength at terminals and for anchor support is provided by means of guys. Double pole structures of A or H types are often employed for obtaining a higher transverse strength than that could be economically provided by means of single poles.

Single member poles, shown in Fig. 2.1, are ordinary poles and are used in all positions where there is no undue stress or tension and where no transformer or switchgear are to be mounted on them.

Fig. 2.2. ***"A" Type Pole Utilized For 3-Phase 3-Wire HT Transmission Where Route of Cables Turns at an Angle***

"A" poles shown in Fig. 2.2 are used mainly where bends in the lines cause strain and single poles are not suitable. "A" poles consist of two member poles spaced at the base and joined at the top, held together by cross-bars in the form of the letter A.

"H" poles shown in Fig. 2.3 comprising two single poles strapped together by steel or wooden cross-pieces are used mainly where transformers and switchgear are to be mounted on them.

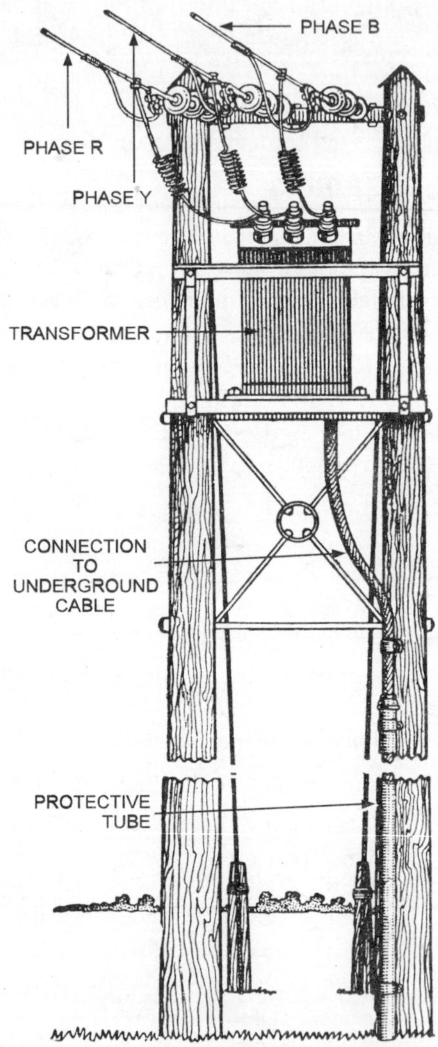

Fig. 2.3. ***"H" Pole Supporting Transformer***

Four member poles, shown in Fig. 2.4, comprise two 'H' units in the form of a square joined by cross-bars. They are used where extra-heavy transformers and switchgear are required, usually at the junction of number of circuits.

The height of a wooden pole depends upon clearance above the ground surface and secondly, the number of cross-arms and other equipment to be attached. Normally, the height of wooden pole is 10 to 12 m.

The main drawbacks of wooden supports are : tendency to rot below the ground level, comparatively smaller life, less mechanical strength and requirement of periodical inspection.

Wooden poles are covered under IS: 876-1961.

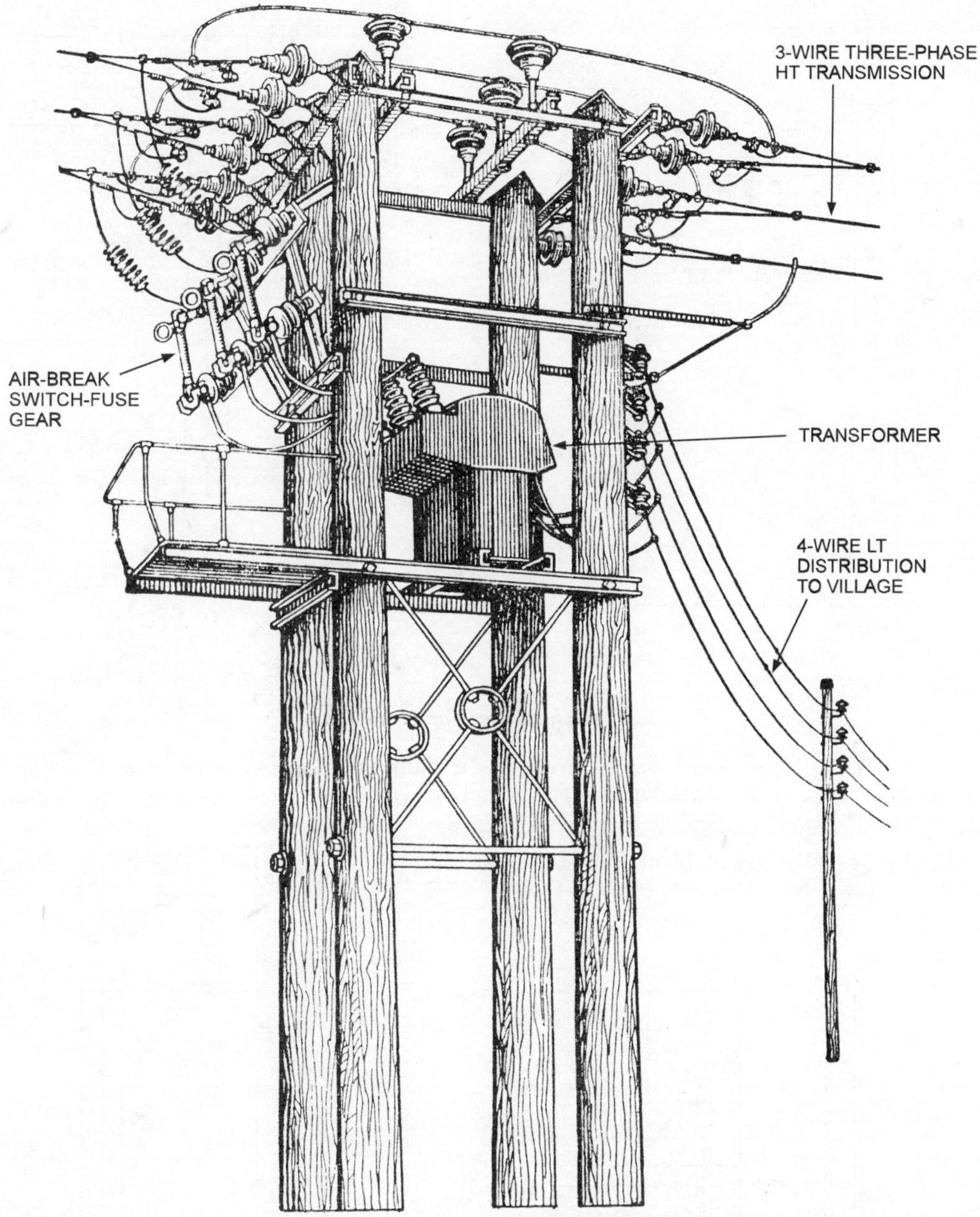

Fig. 2.4. ***4-Member Pole Construction With Transformer, Switchgear and Fuses. LT Four-Wire 3-Phase Distributor is Tapped Off Transformer Secondary To Supply a Village***

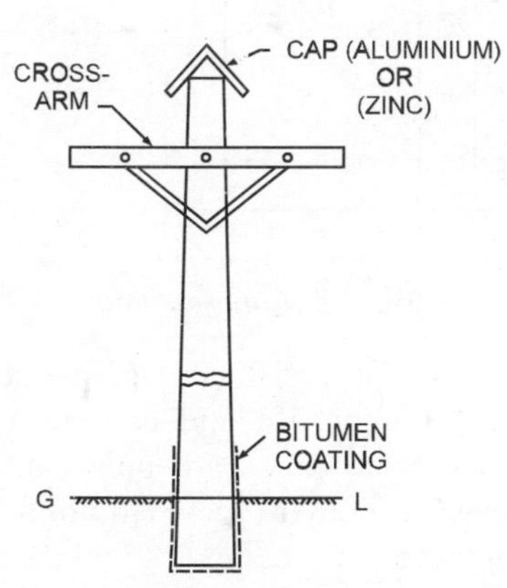

Fig. 2.5. ***Wood Pole With Protection Cap***

2. Steel Poles. The steel poles are of three types (*i*) tubular poles (*ii*) rail poles and (*iii*) rolled steel joists. The tubular poles are of round cross section, the rail poles are of the shape of track used for railways and rolled steel joists are of I cross section.

Such poles possess greater mechanical strength and permit use of longer spans (50-80 m) but cost is higher. Their life is longer than that of wooden poles and life is increased by regular painting. These poles are set in concrete muffs at the foundation inorder to protect them from chemical action. The average life of steel poles is more than 40 years.

The advantages of tubular poles are that these are lighter in weight and easy to install though initial cost is little more as compared to wooden poles. It does not require special

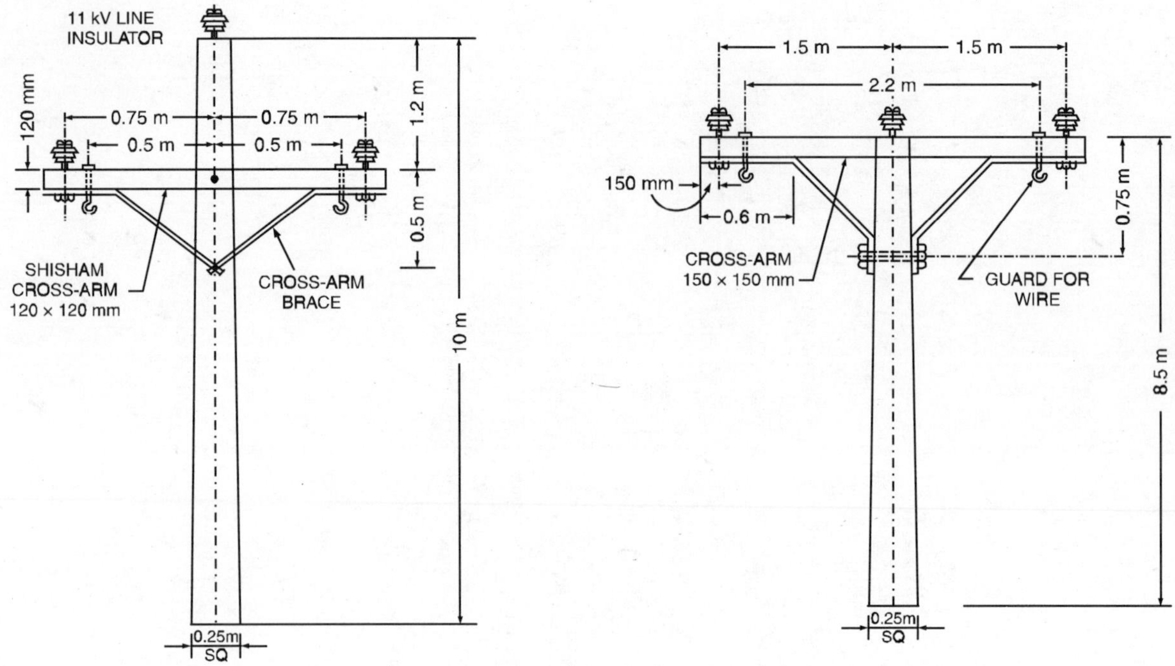

(a) Triangular Formation *(b) Horizontal Formation*

Fig. 2.6. ***Wooden Poles For 11 kV Line***

equipment for its erection. The tubular poles are covered under IS: 2713-1969. Tubular poles in height of 9 to 11 m are generally used for distribution purposes in cities to give good appearance. Steel rail poles in height of 11 m and 13 m are used for transmission purposes at 11 kV and 33 kV respectively.

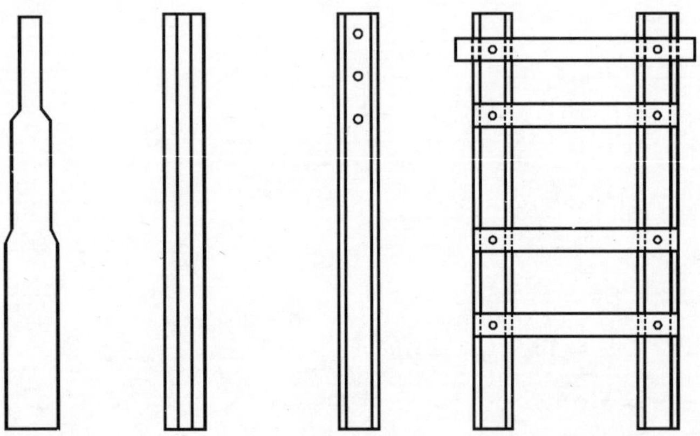

(a) Tubular Pole *(b) Rail Pole* *(c) Rolled Steel Joist Pole* *(d) H-Type RS Joist Pole*

Fig. 2.7. ***Steel Poles***

3. RCC Poles. Poles made of *reinforced cement concrete* (RCC), usually called the *concrete poles,* are extensively used for low voltage and high voltage distribution lines up to 33 kV. Their construction should conform to the standard specification for RCC work, but in no case the dimensions shall be less 25 cm × 25 cm at the bottom and 13 cm × 13 cm at the top. RCC poles must conform to IS: 785-1964. These poles are of two types in shape. One type is of square cross section from bottom to top. The other type has rectangular bottom and square top with rectangular holes in it to facilitate the climbing of poles and reduce the weight of poles. These give good appearance, require no maintenance, have got

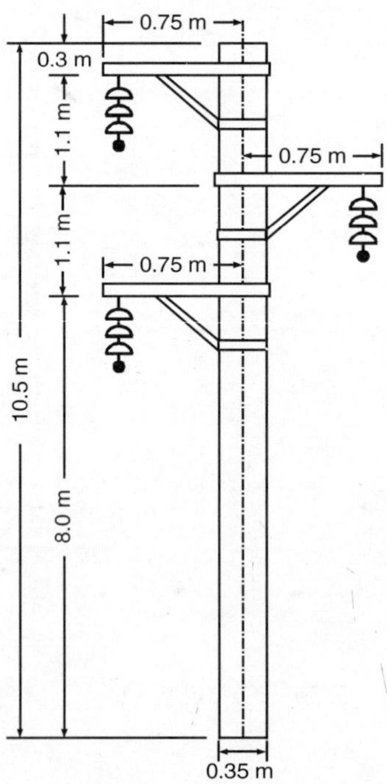

Fig. 2.8. ***Concrete Pole***

insulating properties and resistance against chemical action, very strong, have longer life and can be used for longer spans (80-200 m). Such poles are most suitable for water logged situations where other types will not be at all suitable, as due to standing water wooden poles will decay very rapidly, and steel construction will be having deposit of rust. Since these poles are very bulky and heavy, therefore, transportation cost is heavy and need care in handling and erection. They may be manufactured at site itself to avoid heavy cost of transportation. However, prestressed concrete poles, called the PCC poles, are less bulky and lighter than RCC poles. PCC poles are extensively used on 11 kV and lt lines. PCC poles must conform to IS: 1678-1960.

4. Lattice Steel Towers. Though there is no hard and fast rule but wooden poles are generally used for distribution purposes in rural area, the steel tubular poles and concrete poles are usually used for distribution in urban area to give good appearance and steel rails or narrow-base, lattice-steel towers are used for transmission at 11 kV and 33 kV and broad-base lattice-steel towers are used for transmission purposes at 66 kV and above. The broad-base, lattice-steel towers are mechanically stronger and have got longer life. Due to their robust construction long spans (300 m or above) can be used and are much useful for crossing fields, valleys, railway lines, river etc. Even though these are two to four times costlier than wooden poles, yet for tall supports and longer spans these prove more economical. Furthermore their reliability is of a high degree. The substantial construction of the towers renders them capable of withstanding the most severe climatic conditions, and immune from destruction by forest fires. The risk of service interruptions, due to broken or punctured insulators, is considerably reduced owing to use of large spans. Lightning troubles are also minimised as each tower is a lightning conductor, whereas on wooden pole lines shattered poles and wrecked line sections are not infrequent.

Steel towers are fabricated from painted or galvanised angle sections which can be transported separately and the erection done on site. At a moderate cost these can be designed for double circuit giving a further insurance against

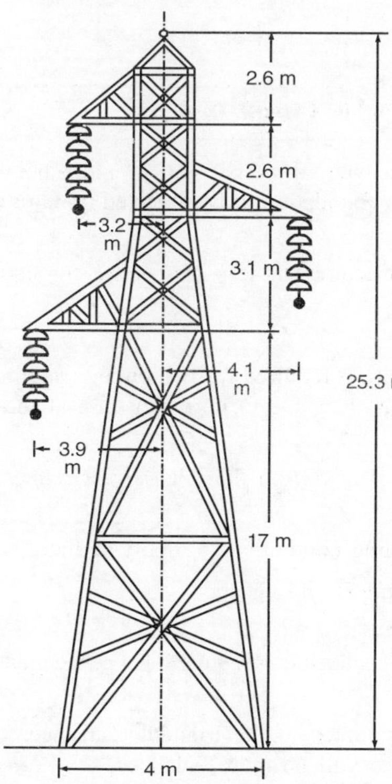

(a) Single Circuit Tower

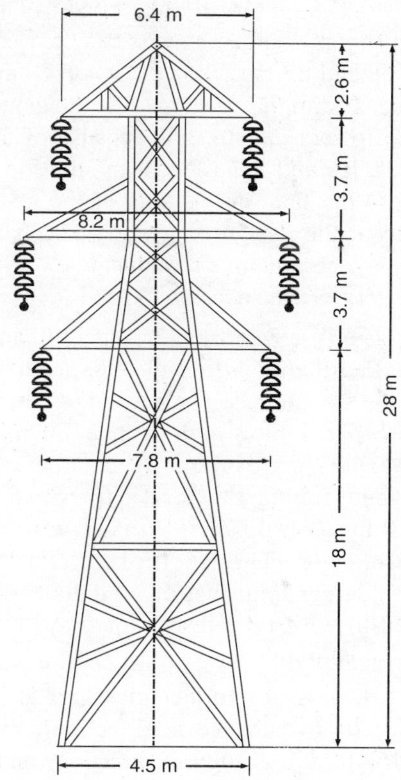

(b) Double Circuit Tower

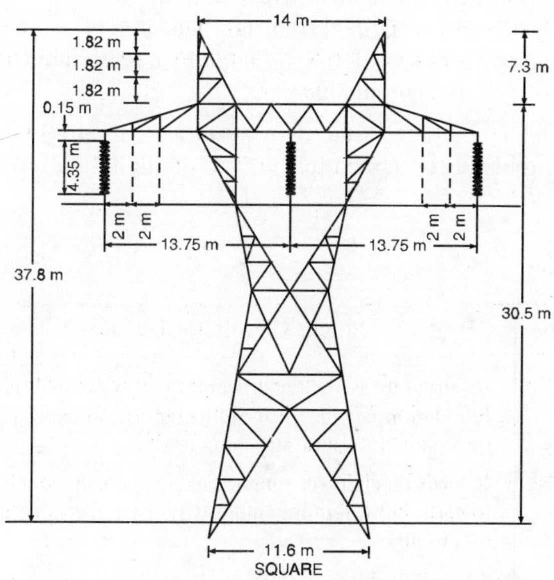

(c) Steel Tower

Fig. 2.9

discontinuity of supply. In case of breakdown to one circuit it is possible to carry out repairs while maintaining the continuity of supply on the other circuit. The height of the tower depends on the line voltage and length of span. The legs of the towers are set in special concrete foundations. The forces to be taken into account in the design of a tower are vertical loads of conductors, insulators, fittings and tower itself, wind pressure on conductors and wind pressure on tower itself.

Steel towers can be broadly classified as tangent towers and deviation towers. Tangent towers can be used for straight runs of the line and up to 2° line deviation from the straight run. The line is straight or along the tangent to the line route. In such towers the stress is because of the weight of the conductors, ice and wind loads. In addition extra forces due to break in the line on one side of the tower is also to be considered in the design of towers. The base of such a steel tower may be square or rectangular. Insulators used with such towers are suspension types.

For deviations exceeding 2°, special angle towers, sometimes called the *deviation towers,* are used. They are used where the transmission line changes direction. Such towers have broader base and stronger members as they are to withstand the resultant force due to change in direction in addition to the forces to which the tangent towers are subjected. Insulators used with such towers are of strain type. The cost of deviation tower is comparatively larger than that of a tangent tower because it is designed to withstand heavy loading as compared to standard or tangent tower.

Deviation towers are further classified as *small angle towers* (for 2° to 15° change in direction), medium angle towers (for 15° to 30° change in direction) and large angle towers (for 30° to 60° change in direction and dead end).

For protection against corrosion the steel towers are periodically painted or galvanised. The life of steel towers can be made almost indefinitely large by a reasonable amount of attention to their maintenance.

The advantages and disadvantages of single circuit and double circuit designs are given in tabular form in Table 2.1.

2.4. CONDUCTOR MATERIALS

The conductor is one of the important items as in a transmission and distribution system of electric power, the cost of the conductor material accounts for a major part of the total cost. So proper choice of conductor material and size of the conductor is of utmost importance. The conductor material used for transmission and distribution of electric power must have the following characteristics:

(*i*) High electric conductivity *i.e.*, low specific resistance.

(*ii*) High tensile strength in order to withstand the mechanical stresses.

(*iii*) Low specific gravity in order to give low weight per unit volume.

(*iv*) Low cost in order to be used over long distances usually involved in transmission lines.

(*v*) Easy availability.

(*vi*) Should not be brittle.

No single conductor material meets all of the above requirements. Hence a compromise will have to be made between the cost and the desired electrical and mechanical properties in the selection of a conductor material for a given situation.

The most commonly used conductor materials for overhead lines are copper, aluminium, steel-cored aluminium, galvanized steel and cadmium copper.

All conductors used for overhead lines are preferably stranded in order to increase the flexibility. Solid wires, except of smaller sizes, are difficult to handle and when employed for long spans tend to crystallise at the points of support because of swinging in winds.

Table 2.1

S.No.	Single Circuit Design	Double Circuit Design
1.	Its structure is lighter in weight and requires less strong foundation because it is subjected to low wind pressure on conductors and structure itself.	Its structure is heavier in weight and of more height. It requires relatively stronger foundation. It is subjected to more wind pressure.
2.	It needs much lower support for equal conductor clearance to earth but it requires more way leave for same number of circuits.	It needs taller structure but less way leave for equal number of circuits.
3.	Two earth wires are required for single circuit as these cannot be disposed at the top.	Only one earth wire is required for two circuits and more protection against lightning is had due to its disposition at the top.
4.	Danger of flashover is most unlikely and repairs can be carried out without danger to workmen from other circuits.	There is always danger from the other live circuit.
5.	Reliability regarding continuity of supply is less.	Reliability regarding continuity of supply is more.
6.	It is more expensive for two circuits than the double circuit design.	It is most economical and cheaper.
7.	Greater spacing of conductors is required resulting in greater inductive reactance.	Lesser spacing of conductors is required, hence the inductive reactance is less.
8.	The phase performance along the line is unbalanced as the central conductor passes at the top of the support, which is an obvious drawback.	It gives better approach to the triangular arrangement, hence the phase performance will be more balanced.

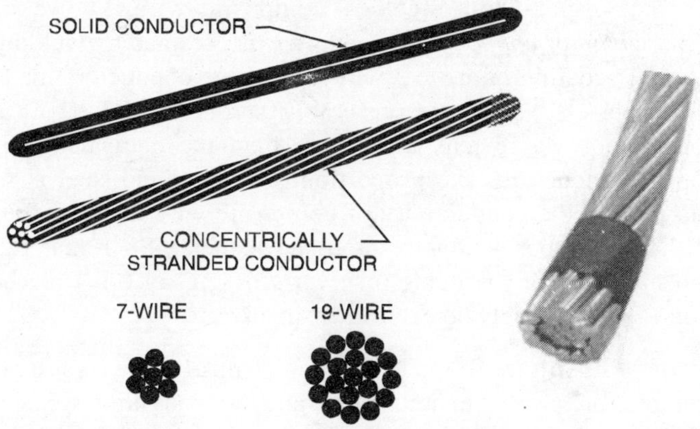

Fig. 2.10. ***Types of Conductors***

Stranded conductors usually have a central wire around which there are successive layers of 6, 12, 18, 24 wires. For n layers, the total number of individual wire is $3n(n+1)+1$. If the diameter of each strand is d then diameter of the stranded conductor will be $(2n+1)d$. In the process of manufacture adjacent layers are spiralled in opposite directions so that the layers are bound together. The method of construction is called as '*concrete lay*'.

With conductors of large cross section, however, another method known as '*rope lay*' is sometimes employed in order to give more flexibility.

1. Stranded Hard Drawn Copper. Hard drawn copper conductor is the best conductor owing to its high electrical conductivity and great tensile strength for all types of transmission. Though hard drawing (cold rolling and drawing) reduces the conductivity slightly but increases the tensile strength considerably. It does not corrode in normal atmosphere and is not subjected to electrolytic troubles. It has higher current density so lesser cross-sectional area of conductor is required and so lesser area is subjected to wind loads. It is quite homogeneous, durable and of high scrap value. The other advantages of hard drawn copper are long life and ease of jointing.

However, medium hard-drawn copper conductor is suitable for distribution lines and soft-drawn copper wires are suitable for secondary distribution circuits, and for service connections to buildings.

Though copper is ideally suitable for the transmission and distribution, but owing to the difficulty of importing it due to lack of foreign exchange in our country, the trend nowadays is to use aluminium in place of copper. The use of copper is being restricted for the manufacturing of the machines only.

2. Aluminium. Aluminium is cheaper in cost and lighter in weight but is poor in conductivity and tensile strength as compared to copper. Its conductivity is 60% to that of copper and density 0.303 time that of copper. An aluminium conductor has a diameter about 1.26 times that of copper conductor of equal resistance but due to its low density only half weight of aluminium is required to that of copper. The tensile strength is much lower than that of copper (45 per cent to that of copper), but the larger cross-sectional area of metal neutralizes the difference to some extent, and an aluminium conductor has about 75% of the ultimate strength of equivalent copper conductor. For the same conductivity aluminium conductor having 1.66 times the cross section of copper is required thus causes a greater surface for wind pressure and supporting structures are required to be designed for a greater transverse strength. In many cases also, high towers must be employed with aluminium conductors than would be required with copper conductors having the same length of span. As a result of the reduced working stress permissible and also owing to the fact that the linear coefficient of expansion of aluminium is 1.4 times that of copper, the sag is greater in aluminium wires. The aluminium conductor being liable to swing, requires larger cross arms. Due to low melting point of aluminium it cannot withstand short circuits etc. Jointing of aluminium is also difficult as compared to that of copper.

These days there is a great trend towards the aluminium as conductor material, because of its greater availability and cheapness in comparison to copper. All aluminium stranded conductor (AAC) weighs only half as the equivalent copper conductor and cost per unit length at prevailing market rates, is considerably less. All aluminium stranded conductors are mainly used for low voltage distribution overhead lines having short spans of up to 65 m.

3. Steel-Cored Aluminium. Steel-cored aluminium conductor, abbreviated as **ACSR** (*aluminium conductor steel reinforced*), consists of a core of galvanised steel strand surrounded by a number of aluminium strands. The steel conductors used are galvanised in order to prevent rusting and electrolytic corrosion, the cross section of two metals are in the ratio 1:6 but in case of high strength conductors their ratio may be 1:4. The steel core takes a greater percentage of mechanical stresses while the aluminium carries the bulk of current. ACSR conductor being of high tensile strength and lighter in weight produces small sag and therefore longer spans can be used. The ACSR conductor has a largest diameter than any other type of conductor of same resistance, so corona losses are reduced, but stronger supports are required for a given span. The ACSR conductor gets deteriorated in service to the atmospheric corrosion due to bimetallic action of zinc and the aluminium, electrochemical in nature. The skin effect is very predominating in ACSR conductors hence the resistance of the composite conductor is taken equal to that of aluminium covering alone. The reactance of ACSR conductor is taken to be equal to that of a non-magnetic conductor having diameter equal to that outside of the ACSR conductor.

Another advantage with ACSR conductor is that, because of reduced corona losses, critical voltage limit of the conductor can be raised by 30 to 50% as compared to copper conductors.

Because of use of larger span, the number of line supports may be reduced by about 25%. Thus the overall cost of supports, foundations, insulators and erection is considerably reduced. In addition, there is a substantial saving in maintenance cost. Further, since most faults occur at the line supports, so the frequency of occurrence of faults is automatically reduced.

Composite type of conductor, being costlier than the all-aluminium type, is chiefly used in those cases where an all aluminium conductor would require a much higher and more costly support than copper. This occurs principally where spans are 100 metres and over are involved.

4. Galvanised Steel. Galvanised steel conductors have been used to advantage for extremely long spans, or for short line sections exposed to normally high stresses due to climatic conditions. These conductors are found most suitable for lines supplying rural areas and operating at voltages of about 11 kV, where cheapness is the main consideration. Iron or steel wire use is most advantageous for transmission of small power over a short distance, where the size of copper conductor desirable from economical considerations comes out to be smaller than 8 SWG, which cannot be used because of poor mechanical strength. This conductor is not suitable for EHT lines for the purpose of transmitting large amounts of power over a long distance due to its following properties:

(*i*) Poor conductivity, 13% that of copper.

(*ii*) High internal reactance.

(*iii*) It is subjected to eddy current and hysteresis.

Nowadays use of galvanised steel wire is limited to telecommunication lines, stay wires, earth wires and guard wires. Galvanised steel wires in small sizes are employed as binding wires. Stranded galvanised steel wires are used as guy wires and earth/ground wires. In *lt*, 11 kV and 33 kV lines, the earth wire is provided below the line conductors for the safety of personnals, cattles and material moving under the line as a precaution in case a line conductor snaps. In case of extra-high voltage lines of 66, 132 or 220 kV, the ground wire is provided above the line conductors for protection against lightning.

5. Cadmium Copper. The conductor being used in certain cases is copper alloyed with cadmium. Addition of 1 or 2% of cadmium in copper increases the tensile strength by about 40% and reduces the conductivity only by 17% below that of pure copper. However, cadmium copper is costlier than pure copper. Use of cadmium copper will be economical for a line with long spans and small cross section *i.e.*, where the cost of conductor material is comparatively small in comparison to that of supports etc. Cadmium copper conductors are also employed for telephone and telegraph lines where currents involved are quite small. However, owing to scarcity of copper, cadmium copper conductors on communication lines are being replaced by ACSR conductors.

6. Copper-Clad Steel. A composite wire, known as *copper-clad* or *copper-weld steel wire,* is obtained by welding a copper coating on a steel wire core. Line conductors made of copper-clad steel are preferably stranded, and have a considerably large tensile strength than the equivalent all-copper conductors. The proportion of copper and steel is so chosen that the conductivity of composite wire is 30 to 40% of that of copper conductor of equai diameter. Such material appears to be very suitable for river-crossings or other places where an extremely long span is involved.

7. Phosphor Bronze. When harmful gases such as ammonia are present in atmosphere and the spans are extremely long, phosphor bronze is most suitable material for an overhead line conductor. In this conductor some strands of phosphor bronze are added to the cadmium copper.

The properties of the conductor materials of various types are given below in tabular form in Table 2.2.

Example 2.1. A given amount of power is to be transmitted by an overhead line. Compare the diameter and weight of aluminium conductor with those of copper for the same power loss in the line. The following data may be assumed:

Specific resistance of aluminium = 2.85 μΩ-cm

Specific resistance of copper = 1.70 μΩ-cm

Specific gravity of aluminium = 2.71

Specific gravity of copper = 8.89

[*Rajasthan Univ. Transmission and Distribution of Electrical Power 2003*]

Solution : For the same power loss in the line, the resistances of aluminium and copper conductors must be same *i.e.*,

$$R_{copper} = R_{aluminium}$$

$$\text{or} \quad \rho_{copper} \times \frac{l}{a_{copper}} = \rho_{aluminium} \times \frac{l}{a_{aluminium}}$$

∵ Length of the line is the same

$$\text{or} \quad \frac{a_{copper}}{a_{aluminium}} = \frac{\rho_{copper}}{\rho_{aluminium}}$$

$$\text{or} \quad \frac{\frac{\pi}{4} d^2_{copper}}{\frac{\pi}{4} d^2_{aluminium}} = \frac{1.7}{2.85}$$

$$\text{or} \quad \frac{d_{aluminium}}{d_{copper}} = \sqrt{\frac{2.85}{1.7}} = 1.295 \textbf{ Ans.}$$

Table 2.2

Mechanical Properties	Copper	Aluminium	Steel	Aluminium and Steel 1 : 3	Aluminium and Steel 1 : 4
Specific Weight (kg/m³)	8,900	2,700	7,860	3,450	3,700
Young Modulus (kg/mm²)	13,000	5,600	20,700	7,500	8,300
Ultimate Tensile Strength (kg/mm²)	40	18	40 to 320	120	120
Specific resistance (Ω-m/mm²) at 20°C	0.01786	0.0287	0.178		
Resistance temperature coefficient	0.0038	0.004	0.00496		

$$\frac{\text{Weight of aluminium}}{\text{Weight of copper}} = \frac{\text{Specific graviy of aluminium, } \sigma_{\text{aluminimum}}}{\text{Specific gravity of copper, } \sigma_{\text{copper}}} \times \frac{\text{Volume of aluminium conductor}}{\text{Volume of copper conductor}}$$

$$= \frac{\sigma_{\text{aluminium}}}{\sigma_{\text{copper}}} \times \frac{a_{\text{aluminium}}}{a_{\text{copper}}} \times \frac{l_{\text{aluminium}}}{l_{\text{copper}}}$$

$$= \frac{2.71}{8.89} \times \frac{2.85}{1.7}$$

because length of line is same in both cases

$= 0.511$ **Ans.**

Example 2.2. Electric power of 30 MW at a pf of 0.8 lagging is to be transmitted by a 132 kV, 3-phase, 3-wire transmission line over a distance of 120 km. The efficiency of transmission is 90 per cent. Calculate:

(*i*) **Weight of material required in case of copper having resistivity of 1.78×10^{-8} Ω-m and specific gravity of 8.9.**

(*ii*) **Weight of material required in case of aluminium having resistivity of 2.6×10^{-8} Ω-m and specific gravity of 2.**

Solution : Power to be transmitted, $P = 30 \text{ MW} = 30 \times 10^6$ watts

Line voltage, $V_L = 132 \text{ kV} = 132 \times 10^3$ volts

Length of 3-phase, 3-wire line, $l = 120 \text{ km} = 120 \times 10^3$ metres.

Line current, $I_L = \dfrac{P}{\sqrt{3} V_L \cos\phi}$

Power loss,

$$W = 3 I_L^2 R = 3 \times \left(\frac{P}{\sqrt{3} V_L \cos\phi}\right)^2 \times \rho \frac{l}{a} = \frac{P^2 \rho l}{V_L^2 \cos^2 \phi\, a}$$

$$\text{or } a = \frac{P^2 \rho l}{V_L^2 \cos^2 \phi\, W} = \frac{P^2 \rho l}{V_L^2 \cos^2 \phi (1-\eta) P}$$

$\because$ Power loss, $W = (1 - \eta)P$

Volume of conductor material required

$$= 3al = \frac{3 \rho l^2 P}{(1-\eta) V_L^2 \cos^2 \phi} \text{ m}^3$$

Weight of conductor material required

$$= \frac{3 \rho l^2 P}{(1-\eta) V_L^2 \cos^2 \phi} \times D$$

Weight of copper required

$$= \frac{3 \rho l^2 P}{(1-\eta) V_L^2 \cos^2 \phi} \times 8{,}900$$

$$\because \text{specific gravity of copper} = \frac{8.9 \times 10^6}{1{,}000} = 8{,}900 \text{ kg/m}^3$$

$$= \frac{3 \times 1.78 \times 10^{-8} \times (120 \times 10^3)^2 \times 30 \times 10^6}{(1-0.9) \times (132 \times 10^3)^2 \times (0.8)^2} \times 8{,}900$$

$= 184{,}000$ kg **Ans.**

Weight of aluminium required

$$= \frac{3 \times 2.6 \times 10^{-8} \times (120 \times 10^3)^2 \times 30 \times 10^6}{(1-0.9) \times (132 \times 10^3)^2 \times (0.8)^2} \times 2{,}000$$

$= 60{,}400$ kg **Ans.**

2.5. CROSS ARMS

The function of a line support (pole) is to support the line conductors at a safe distance from ground whereas the function of cross arms is to keep the conductors at a safe distance from each other and from the pole. Cross arm is a *cross-piece* fitted to the pole top end portion by means of brackets, known as *pole brackets,* for supporting insulators.

Cross arms are of various types such as MS channel, angle iron or wooden. These may be straight, U-shaped, V-shaped or zig-zag shaped, as shown in Fig. 2.11. Wooden cross arms are commonly employed on 11 kV and 33 kV lines. These are made of sal wood, seasoned sheesam wood or creosoted fire wood. Grains of wood cross arms should be lengthwise and completely free from knots. Wooden cross arms are preferred owing to their insulating property which provides safety to line staff and minimizes flash-over due to birdage. The usual lengths and cross sections of wooden cross arms in use are : 1.5 m × 125 mm × 125 mm for 11 kV lines and 2.1 m × 125 m × 125 mm for 33 kV lines. Wooden cross arms need replacement owing to decay after every 5-7 years depending upon weather conditions.

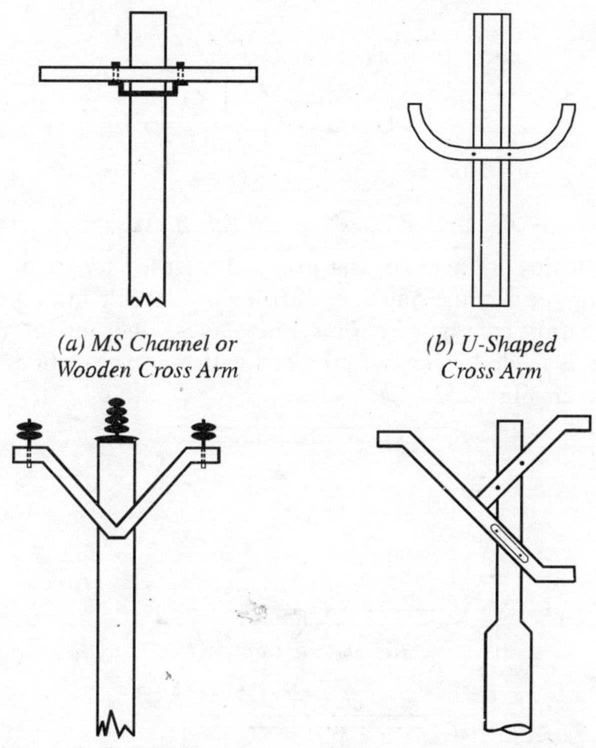

(*a*) *MS Channel or Wooden Cross Arm* (*b*) *U-Shaped Cross Arm*

(*c*) *V-Shaped Cross Arm* (*d*) *Zig-zag Cross Arm*

Fig. 2.11

Steel cross arms are stronger and are generally used on steel poles. For *lv* distribution, the angle iron or channel iron cross arms shall of size not less than 50 mm × 50 mm × 6.4 mm and 76 mm × 38 mm. The length of the cross arms shall be suitable for the spacing of the conductors. Cross arms shall be suitable and strong enough to withstand the resultant forces caused by insulators, their pins and dead weight of insulator attachments etc. To avoid birdage on *ht* lines, V-shaped cross arms are used with pin insulators while straight cross arms are used with disc insulators. The cross arm is fixed to the pole in such a manner that the load of the conductors is taken by the cross arm and not the clamp or bolt that fixes the cross arm to the pole.

In order to prevent arcing, the construction of the cross arms should be such that under the worst conditions, the spacing between conductors, when swinging, would never be less than that given below :

Working Voltage	Spacing
6.6 kV	76 mm
11 kV	101 mm
33 kV	190 mm and so on

2.6. POLE BRACKETS AND CLAMPS

Pole brackets of different types are shown in Fig. 2.12.

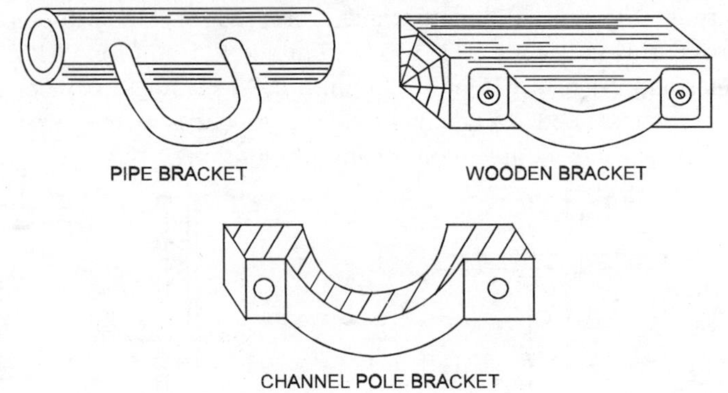

Fig. 2.12. ***Various Types of Pole Brackets***

Clamps are made of flat iron and are used for fixing or holding service line stay wires, earth wire, shackle insulators, cross arms etc. In case of service lines, one end of the clamp is made longer and provided with an 'eye' section, as shown in Fig. 2.13 (*d*).

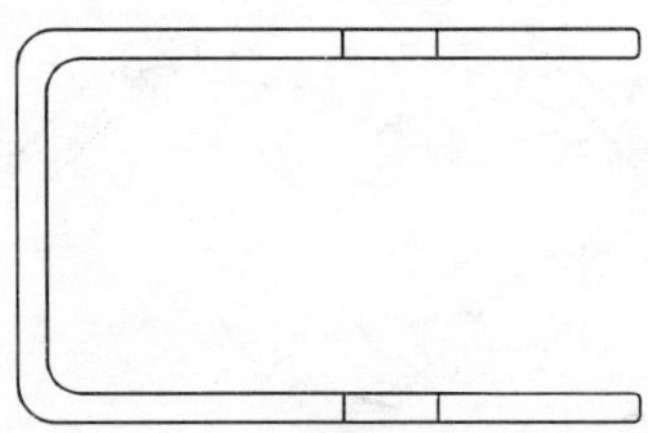

(a) Shackle Clamp

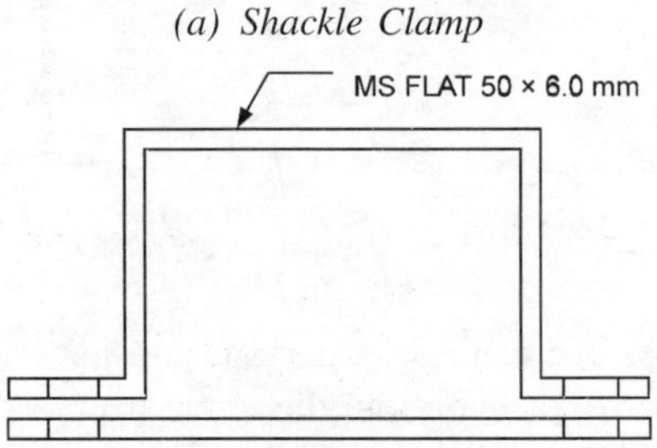

(b) Cross-arm Clamp

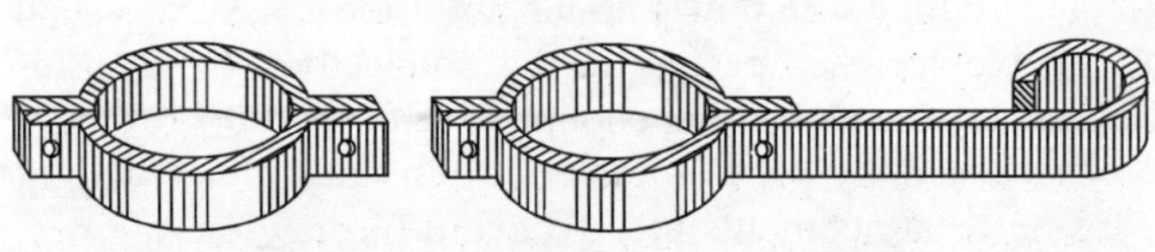

(c) Stay Clamp *(d) Eye Clamp*

Fig. 2.13. ***Various Types of Clamps***

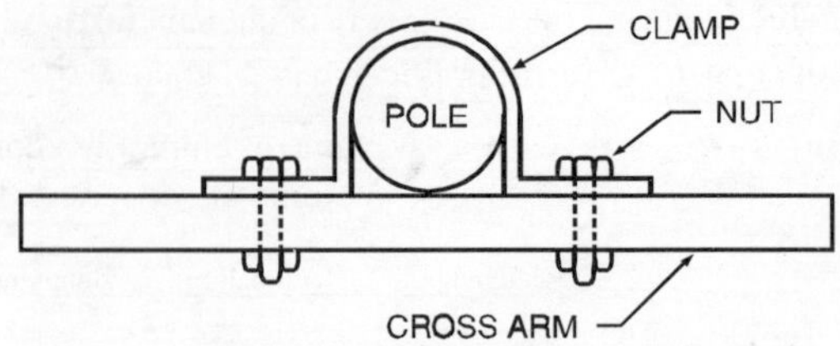

Fig. 2.14. ***Method of Fixing Cross Arm To Pole***

2.7. GUYS AND STAYS

It becomes essential to stay overhead line supports at angle and terminal positions as the poles take the pull due to conductors. The theoretical angle between the pole and stay should be 45°, but in general practice it is not always possible to obtain this, and so stay design is generally based on a minimum angle of 30° between stay and pole.

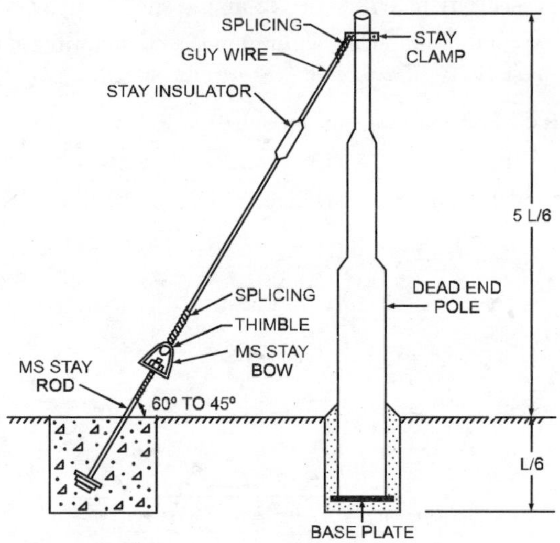

Fig. 2.15

Stay set consists of MS rod of 19 mm diameter, stay bow, checknut, thimbles, stay wire (either $\frac{7}{8}$ or $\frac{7}{10}$ SWG GI wire), stay clamp and CI anchor plate 450 × 450 mm having 4.8 mm hole in the centre. Except the plate all other parts are galvanised. Stay rod is embedded in cement concrete 1:3:6 to a depth of not less than 1.67 m keeping length of 46 cm of the rod projected above the ground level. One end of stay wire is fixed to the stay rod at the bottom and to the stay clamp to the pole by means of well spliced joints with a strain insulator and turn buckle inserted in the middle and near the top respectively. The stay or guy is tightened by means of a stay bow and anchor rod to the required tension. Mild steel thimbles are used at both ends of the stay wire, otherwise strands of stay wire may get damaged. The anchor rod is projected far enough above the ground level so that the guy wire does not come into contact with the soil of the ground.

An egg type strain insulator is inserted in the guy wire for safety. It isolates stay wire electrically from metal support. The two ends of guy are threaded through the insulator in such a way that the porcelain of the insulator is under compression, making it possible to withstand a large pull. In case of breakage of porcelain, the guy will still be effective due to linking of the two ends. Stay wires are galvanized steel wires having tensile strength of 7 tons/cm^2 and are usually of stranded section.

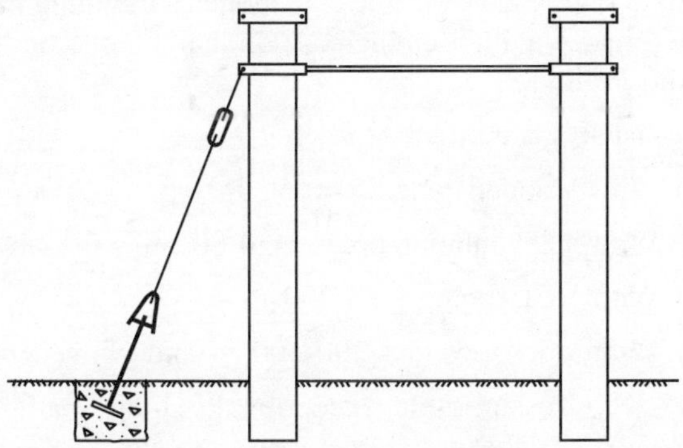

(a) Single Stay Arrangement Where Space Available is Limited

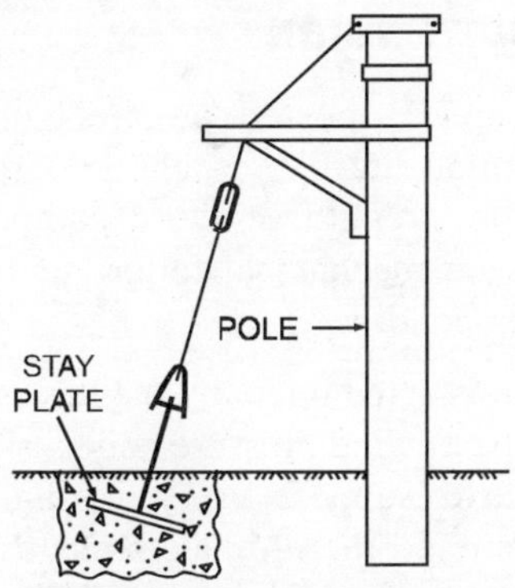

(b) Stay To a Pole With The Help of Stay Pole and Stay Wire (For Road-Crossings)

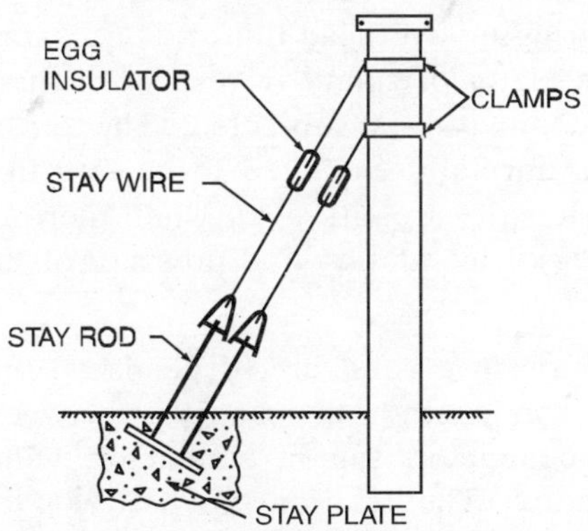

(c) Single Stay Arrangement Where No Space is Available For Laying Stay Rod

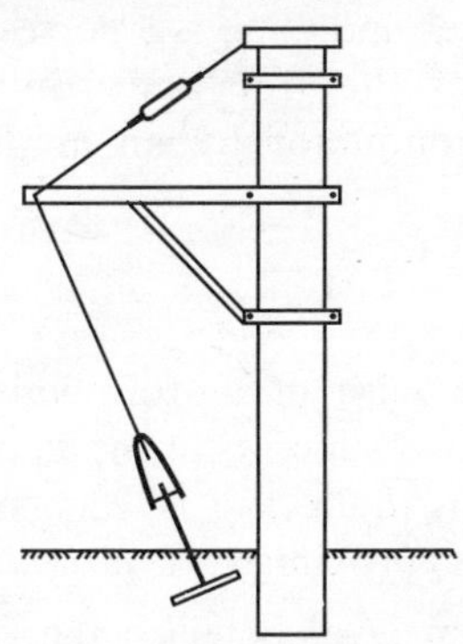

(d) Double Stay Arrangement

Fig. 2.16. ***Stay Arrangements***

The stays are provided to angle poles, terminal poles, H-pole structures, where the span on the two sides of the pole are very much different and to supports subjected to uprooting pressure. Stay wires are placed on the poles and stay rods fixed before stretching of line conductors on the poles. This is very important, otherwise, in case the line conductors are placed first, the poles would be pulled out of the position. The stay wire is held up on the pole either by means of guy hook or through the bolt or may be fixed to the clamp. The stay wire is drawn up until the pole is pulled over slightly towards the stay in order to keep the proper tension.

Different stay arrangements are shown in Fig. 2.16.

2.8. CONDUCTORS CONFIGURATION, SPACING AND CLEARANCES

2.8.1. Conductor Configuration

Several conductor configurations are possible, but three configurations are the most common *i.e.*, horizontal configuration (or horizontal disposition of conductors), vertical configuration and triangular configuration.

There is no special advantage in using the symmetrical *delta* or *triangular configuration* [Fig. 2.6(*a*)] and in most cases flat horizontal or vertical configurations are employed from mechanical considerations, particularly when suspension insulators are used. In *horizontal configuration,* all the conductors are mounted over one cross arm, as shown in Fig. 2.6(*b*). Though such an arrangement of conductors needs supports of smaller height but needs a wider right of way. In certain congested areas where it is not possible to have horizontal arrangement of conductors, the conductors are placed in vertical formation (along the length of pole one below the other). The drawbacks of vertical formations are taller towers and more lightning hazards. There are places where both horizontal and vertical formations are applied.

In unsymmetrical arrangement of conductors, the conductors are usually transposed at regular intervals in order to balance the electrical characteristics of various phases, and prevent inductive interference with neighbouring communication circuits.

Experience shows that a vertical configuration is the most economical for double circuit lines and horizontal or L-type configuration for single circuit lines.

2.8.2. Conductor Spacings

The spacing of conductors is determined by considerations partly electrical and partly mechanical. Larger spacing causes increase in inductance of the line and hence the voltage drop, so that to keep the latter within a reasonable value the conductors should be as close together as is consistent with the prevention of corona.

The basic consideration regarding the minimum spacing between conductors is that the electrical clearances between conductors under the worst condition *i.e.*, maximum temperature and wind pressure, shall not be less than the limits for safety, particularly at the mid spans. Owing to the action of the gusts of wind, conductor has got tendency to move about in an elliptical path, therefore, in case of suspension insulators, the minimum clearance to supporting structures should be calculated with a 45° swing of the suspension string towards the structure.

An empirical formula commonly employed for determination of spacing of conductors for an aluminium conductor line is given below.

$$\text{Spacing} = \sqrt{S} + \frac{V}{150} \text{ metres} \qquad \ldots(2.1)$$

where S is sag in metres and V is line voltage in kV.

Some typical spacing values are given below:

Table 2.3

Line voltage in kV	0.4	11	33	66	132	220	400	765
Spacing in metres	0.2	1.2	2.0	2.5	3.5	6.0	11.5	14

2.8.3. Conductor Clearances

The minimum vertical clearances between the ground and the conductor are recommended vide IE rule no. 77 (see at page 51).

These values are given below:

Table 2.4

Line Voltage in kV / Clearance To Ground in metres	0.4	11	33	66	132	220	400
Across Street	5.8	5.8	6.1	6.1	6.1	7.0	8.4
Along Street	5.5	5.5	5.8	6.1	6.1	7.0	8.4
Other Areas	4.6	4.6	5.2	5.5	6.1	7.0	8.4

2.9. SPAN LENGTHS

Neglecting the deciding influence upon the span length of such local conditions as the necessity for following the configurity of roads, canals or railways, it is interesting to note that there is one definite value for the span length which will give the minimum overall cost of the line. As the length of span increases, the number of insulators and supports decreases resulting in decrease in cost but at the same time the height of the support will go up to allow for more sag and also the length of the cross arms will have to be increased to take up increased spacing, this will cause increase in cost. Moreover the length of the span depends upon the working voltage, higher the working voltage of the system, the greater will be the economical length of span owing to the higher relative cost of insulators to supports. Moreover the insulators constitute the weakest part of transmission line and reduction in number of towers per km with the use of longer span increases the reliability of the line. Thus it is not possible to give any hard and fast rule as to the best span length to be adopted, and the only way to determine it is to calculate the total cost per km for a number of different span lengths, and plot the results to get the most economical span length.

Many a time it happens that the conductor size determined from electrical calculations comes out rather small, it is possible to reduce the total cost of line by using a thicker and stronger conductor, and increasing the span length. Sometimes it is not feasible to determine the height of the line support and span length on the basis of line cost alone as lightning hazards increase greatly with the increase in height of conductors above the ground.

The usual spans are :

(*a*) With wooden poles : 40-50 m.

(*b*) With steel tubular poles : 50-80 m.

(*c*) With RCC poles : 80-200 m

(*d*) With steel towers : 200-400 m and above.

For river-crossings etc exceptionally long spans up to 800 m or so have been satisfactorily employed.

2.10. SAG AND TENSION

Overhead lines are supported on mechanical structures consisting of components like insulators, cross arms, poles or towers etc. The strength of these components must be such that there is no mechanical failure of line, even under the worst weather conditions.

The conductor is acted upon by the forces such as weight of the conductor itself, wind pressure and tension. A conductor stretched between two supports will have an ultimate strength at which it will fail and the ultimate strength of a conductor depends upon the type of conductor material used for overhead line.

While stringing overhead lines it is necessary to allow a reasonable factor of safety in respect of the tension to which the conductor is subjected. The tension in the conductor is normally expected to be less than 50% of its ultimate tensile strength even when there is 12.7 mm radial coating of ice and a wind pressure of the order of 380 N/m^2.

The tension in a conductor depends on the diameter of the conductor, length of the conductor between supports, material of conductor, sag in conductor, wind pressure and temperature. The relationship between tension and sag is dependent on the loading conditions and temperature variations. For instance, the tension increases when temperature decreases and there is a corresponding decrease in sag. Icing up of the line and wind loading causes stretching of the conductor by an amount governed by the line tension.

2.10.1. Sag

A perfectly flexible wire of uniform cross section, when strung between the two supports at the same level, will form a *catenary.* However, if the sag is very small compared to the span, its shape approximates a *parabola.*

The difference in level between the points of supports and the lowest point on the conductor is known as *sag.*

The factors affecting the sag in an overhead line are given below :

(*i*) *Weight of the Conductor.* This affects the sag directly. Heavier the conductor, greater will be the sag. In locations where ice formation takes place on the conductor, this will also cause increase in the sag.

(*ii*) *Length of the Span.* This also affects the sag. Sag is directly proportional to the square of the span length. Hence other conditions, such as type of conductor, working tension, temperature etc, remaining the same a section with longer span will have much greater sag.

(*iii*) *Working Tensile Strength.* The sag is inversely proportional to the working tensile strength of conductor if other conditions such as temperature, length of span etc remain the same. Working tensile strength of the conductor is determined by multiplying the ultimate stress and area of cross section and dividing by a factor of safety.

(*iv*) *Temperature.* All metallic bodies expand with the rise in temperature and, therefore, the length of the conductor increases with the rise in temperature, and so does the sag.

The sag plays an important role in the design of overhead line. It is disadvantageous to provide either too high sag or too low sag. In case the sag is too high, more conductor material is required, more weight on the supports is to be supported, higher supports are necessary and there is a chance of greater swing-amplitude due to wind load. On the other hand in the case of too low sag, there is more tension in the conductor and thus the conductor is liable to break if any additional stress is to be taken, such as due to vibration of line or due to fall in temperature.

Normally two conditions should be investigated, when making sag-tension calculations:

(*a*) *At Minimum Temperature.* The lowest sag and maximum tension in conductor section occurs when the temperature is minimum and wind maximum. Under these conditions, tension on the conductor should not exceed the breaking strength of the conductor divided by a factor of safety of 2.5, as prescribed by IE Rule No. 76.

(*b*) *At Maximum Temperature.* On the other hand maximum sag occurs when temperature is maximum and there is no wind pressure. Under these conditions, electrical clearances should be above the minimum values prescribed under IE Rule No. 77.

Approximate formulae for the calculation of sag is given below

$$S = \frac{wL^2}{T}$$

where w = weight of conductor in kg per metre length, L = length of the span in metres, and T = tension in the conductor in kg.

2.10.2. Calculation of Sag and Tension

As mentioned earlier, a conductor strung between two supports at the same level forms a catenary. For extra high voltage lines, when sag exceeds 10% of the span length, formulae based on catenary should be used.

Consider a portion OP of a curved length l of a wire hanging in still air with O as the lowest point on the wire [Fig. 2.17 (*a*)]. Let the weight of conductor per metre length be w kg.

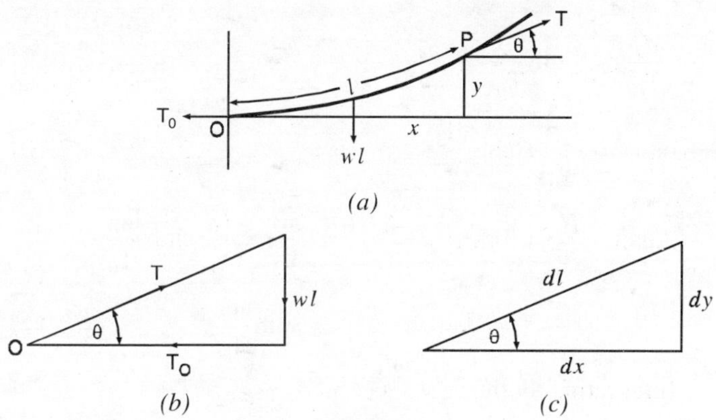

Fig. 2.17

If T_0 is the tension at point O (lowest point on the wire and where the curve is horizontal) and T the tension at point P (distant l from the lowest point O), the portion OP (length l) is in equilibrium under the action of three forces, namely T_0, T and the weight of the wire of length l acting vertically downward through centre of gravity, wl.

The above three forces can be represented by a triangle shown in Fig. 2.17 (*b*) and from this triangle

$$\text{Tan } \theta = \frac{wl}{T_0} \qquad ...(2.2)$$

The horizontal and vertical distances of the length OP of the wire are x and y respectively, since the coordinates of the point P are (x, y).

Now from triangle shown in Fig. 2.17 (*c*),

$$\text{Tan } \theta = \frac{dy}{dx} \qquad ...(2.3)$$

and also $dl^2 = dx^2 + dy^2$ ...(2.4)

or
$$\left(\frac{dl}{dx}\right)^2 = 1 + \left(\frac{dy}{dx}\right)^2 \qquad ...(2.5)$$

$$= 1 + \tan^2 \theta \quad \because \text{ from Eq. (2.3) } \frac{dy}{dx} = \text{Tan } \theta$$

$$= 1 + \left(\frac{wl}{T_0}\right)^2 \quad \because \text{ from Eq. (2.2) Tan } \theta = \frac{wl}{T_0}$$

or
$$\frac{dl}{dx} = \sqrt{1 + \left(\frac{wl}{T_0}\right)^2}$$

or
$$dx = \frac{dl}{\sqrt{1 + \dfrac{w^2 l^2}{T_0^2}}} \qquad ...(2.6)$$

Integrating both sides, we get

$$x = \frac{T_0}{w} \sinh^{-1} \frac{wl}{T_0} + C \quad ...(2.7)$$

where C is a constant of integration

From initial conditions *i.e.*, when $x = 0$, $l = 0$, we get $C = 0$ and Eq. (2.7) becomes

$$x = \frac{T_0}{w} \sinh^{-1} \frac{wl}{T_0} \quad ...(2.8)$$

$$\text{or} \quad \frac{wl}{T_0} = \text{Sinh} \frac{wx}{T_0} \quad ...(2.9)$$

$$\text{or } l = \frac{T_0}{w} \sinh \frac{wx}{T_0} \quad ...(2.10)$$

$$\text{Again } \frac{dy}{dx} = \text{Tan } \theta = \frac{wl}{T_0} = \frac{w}{T_0} \times \frac{T_0}{w} \sinh \frac{wx}{T_0}$$

$$= \text{Sinh} \frac{wx}{T_0} \quad ...(2.11)$$

Integrating both sides, we get

$$y = \frac{T_0}{w} \cosh \frac{wx}{T_0} + D \quad ...(2.12)$$

where D is another constant of integration

From initial conditions *i.e.*, when $x = 0$, $y = 0$, we get $D = \frac{-T_0}{w}$ and Eq. (2.12) becomes

$$y = \frac{T_0}{w} \cosh \frac{wx}{T_0} - \frac{T_0}{w} = \frac{T_0}{w}\left(\cosh \frac{wx}{T_0} - 1\right) \quad ...(2.13)$$

Equation (2.13) is the equation of the curve, called the *catenary*.

The function cosh is the hyperbolic cosine and is such that

$$\text{Cosh} \frac{wx}{T_0} = 1 + \frac{\left(\frac{wx}{T_0}\right)^2}{\lfloor 2} + \frac{\left(\frac{wx}{T_0}\right)^4}{\lfloor 4} + \frac{\left(\frac{wx}{T_0}\right)^6}{\lfloor 6} + \ldots$$

$$\text{So } y = \frac{T_0}{w}\left[1 + \frac{\left(\frac{wx}{T_0}\right)^2}{\lfloor 2} + \frac{\left(\frac{wx}{T_0}\right)^4}{\lfloor 4} + \frac{\left(\frac{wx}{T_0}\right)^2}{\lfloor 6} - 1\right]$$

If the fourth and higher order terms are neglected,

$$y \simeq \frac{T_0}{w} \cdot \left(\frac{wx}{T_0}\right)^2 \cdot \frac{1}{\lfloor 2} = \frac{wx^2}{2T_0} \quad ...(2.14)$$

From Fig. 2.17(*b*) the tension T at point P is given by

$$T = \sqrt{(T_0)^2 + (wl)^2}$$

$$= \sqrt{(T_0)^2 + \left(T_0 \sinh \frac{wx}{T_0}\right)^2}$$

$$= T_0 \sqrt{1 + \left(\sinh \frac{wx}{T_0}\right)^2} = T_0 \cosh \frac{wx}{T_0} \quad ...(2.15)$$

If the line is supported between two points A and B at the same level and the length of the span is L (Fig. 2.18), then at the supports $x = \pm\frac{L}{2}$ and

$$T = T_0 \cosh \frac{wL}{2T_0} \quad ...(2.16)$$

The sag S is the value of y at A or B and is given by

$$S \simeq \frac{w}{2T_0}\left(\frac{L}{2}\right)^2 = \frac{wL^2}{8T_0} \quad ...(2.17) \text{ refer to Eq. (2.14)}$$

Length of line in a half span

$$= \frac{T_0}{w} \sinh \frac{wL}{2T_0} \quad ...(2.18) \text{ refer to Eq. (2.10)}$$

$$= \frac{T_0}{w}\left[\frac{wL}{2T_0} + \frac{w^3L^3}{48T_0^3} + \ldots\right]$$

$$\because \sinh \frac{wL}{2T_0} = \frac{wL}{2T_0} + \left(\frac{wL}{2T_0}\right)^3 \times \frac{1}{\lfloor 3} + \ldots$$

Neglecting terms of order exceeding cube, we have Length of line in a half span

$$\simeq \frac{T_0}{w} \times \frac{wL}{2T_0} + \frac{T_0}{w} \times \frac{w^3L^3}{48T_0^3} \simeq \frac{L}{2} + \frac{w^2L^3}{48T_0^2} \quad ...(2.19)$$

or Length of line in a full span length

$$\simeq L + \frac{w^2L^3}{24T_0^2} \quad ...(2.20)$$

$$\text{Tension, } T = T_0 \cosh \frac{wx}{T_0} \quad \text{refer to Eq. (2.15)}$$

$$= T_0\left(1 + \frac{w^2x^2}{2T_0^2} + \frac{w^4x^4}{24T_0^4} + \ldots\right) \simeq T_0\left(1 + \frac{w^2x^2}{2T_0^2}\right)$$

Substituting $x = \frac{L}{2}$ in above equation, we have

$$T = T_0\left(1 + \frac{w^2L^2}{8T_0^2}\right) = T_0 + \frac{w^2L^2}{8T_0} = T_0 + wS$$

$$\because \text{ from Eq. (2.17) } \frac{wL^2}{8T_0} = S$$

The maximum tension occurs when $x = \frac{L}{2}$ so maximum tension is given as

$$T_{max} = T_0 + wS \quad ....(2.21)$$

For very small sags the term wS may be neglected in comparison with T_0, and the tension is considered approximately uniform through out the conductor.

Approximate Formulae. Neglecting $\frac{w^2L^2}{8T_0}$, being very small compared with unity, T becomes nearly equal to T_0 and

$$\text{Sag, } S = \frac{wL^2}{8T} \qquad ...(2.22)$$

$$\text{Total length of wire} = L + \frac{w^2L^3}{24T^2} \qquad ...(2.23)$$

The sag could also be very simply deduced as follows:

Consider a conductor suspended between two, equal level supports A and B. The conductor is assumed to be flexible and sags below the level AB due to its own weight. Though the exact shape of the conductor is that of a catenary but except for lines with very long span and large sag, it is sufficiently accurate to assume that the shape of the hanging conductor is that of a parabola $y = ax^2$ where a is a constant for a given conductor and O is the origin. The curve at point O, being the lowest one, will be horizontal.

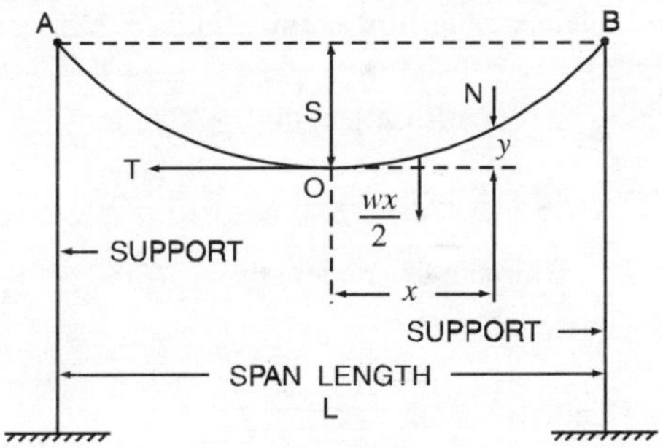

Fig. 2.18

Let the length of span (*i.e.*, horizontal distance between supports) be L metres, weight of conductor per metre length be w kg and tension in the conductor be T kg.

Assuming curvature very small, the length of conductor hanging between two supports may be taken equal to the length of span and maximum sag may be considered at half span length with equi-level supports.

Consider x metre length of conductor between mid-point O and point N. The two external forces acting on the portion ON of the conductor are

(*i*) the tension T acting at point O and

(*ii*) the weight of the conductor of x metres length *i.e.*, wx acting at $\frac{x}{2}$ metres from point N.

Equating the moments of the above two forces about point N, we get

$$Ty = wx \times \frac{x}{2}$$

$$\text{or } y = \frac{wx^2}{2T} \qquad ...(2.24)$$

At point B, $y = S$ and $x = \frac{L}{2}$

$$\text{or Sag, } S = \frac{w(L/2)^2}{2T} = \frac{wL^2}{8T} \qquad ...(2.25)$$

Supports at Different Levels. When transmission lines are run on steep inclines, as in case of hilly areas, the two supports A and B will be at different levels. The shape of the conductor strung between the supports may be assumed to be a part of the parabola. In this case, the lowest point of the conductor will not lie in the middle of the span.

Consider an overhead line conductor AOB, supported over the supports A and B, as illustrated in Fig. 2.19.

Let the difference in levels between the two supports be h and the lowest point O of the conductor be at a distance of x metres from the support at low level.

So, distance of O from higher level support

$= (L - x)$ metres

From Eq. (2.24)

$$S_1 = \frac{wx^2}{2T} \qquad ...(2.26)$$

$$\text{and } S_2 = \frac{w(L-x)^2}{2T} \qquad ...(2.27)$$

Difference in two level of supports,

$$h = S_2 - S_1 = \frac{w(L-x)^2}{2T} - \frac{wx^2}{2T}$$

$$= \frac{wL^2 + wx^2 - 2wLx - wx^2}{2T}$$

$$= \frac{wL^2 - 2wLx}{2T}$$

or $2Th = wL^2 - 2wLx$

or $2wLx = wL^2 - 2Th$

$$\text{or } x = \frac{L}{2} - \frac{Th}{wL} \qquad ...(2.28)$$

From the above Eqs. (2.26), (2.27) and (2.28) location of point O and thus the values of S_1 and S_2 can be determined.

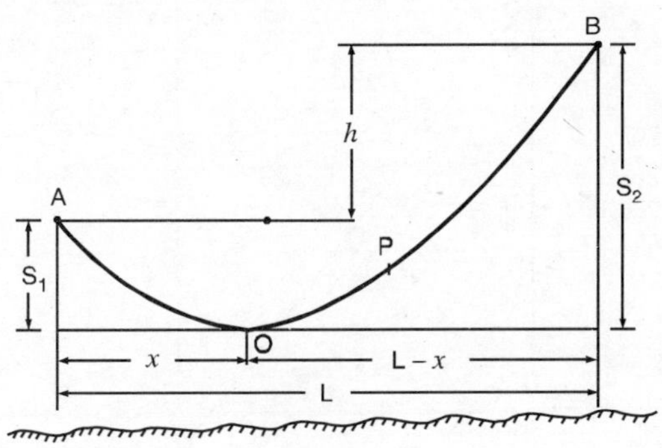

Fig. 2.19

Effect of Ice. In areas where it becomes too cold in winter, there is a possibility of formation of an ice coating on the line conductors. The formation of an ice coating on a line conductor has a two fold effect—increase in weight and effective diameter of the conductor. In this condition the weight of conductor, together with weight of ice acts vertically downwards. Thus, the total vertical weight acting on the conductor per metre length is $w_c + w_i$ where w_c is the weight of conductor in kg per metre length and w_i is the weight of ice coating per metre length, w_c is known and w_i is determined as follows:

Let the diameter of conductor be D metres and radial thickness of ice coating be r metre, as illustrated in Fig. 2.20. The overall diameter of ice covered conductor, as obvious from Fig. 2.20, becomes equal to (D + 2r) metre.

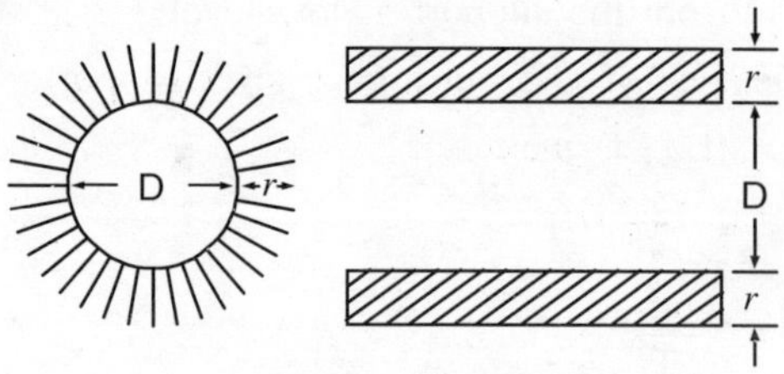

Fig. 2.20. ***Ice Coating***

Volume of ice coating per metre length of conductor

$$= \frac{\pi}{4}[(D + 2r)^2 - D^2]$$

$$= \frac{\pi}{4}[4Dr + 4r^2] = \pi r(D + r)\ m^3$$

The density of ice is approximately 920 kg per m^3, so

The weight of ice coating per metre length,

$$w_i = 920 \times \pi r(D + r)\ kg$$
$$= 2{,}890.3r(D + r)\ kg/m \qquad ...(2.29)$$

Combined Effect of Wind and Ice. Due to weight of ice deposits on the line, and the wind pressure, the mechanical stress increases in the conductor and, therefore, the line must be designed to withstand these stresses and tensions. Under this condition, the weight of the conductor, together with weight of ice acts vertically downwards while the wind loading w_w acts horizontally, as shown in Fig. 2.21.

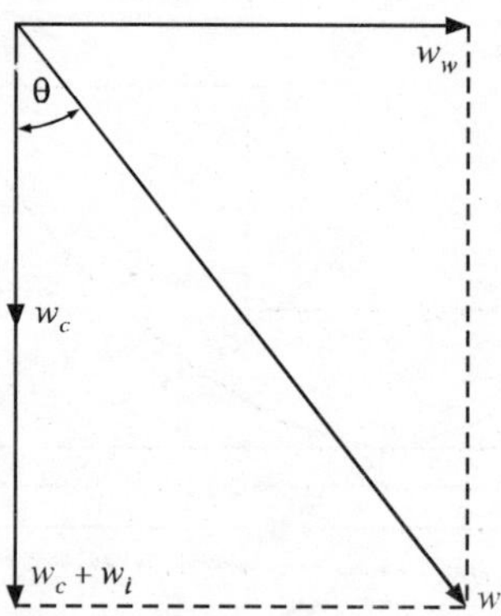

Fig. 2.21

Resultant weight per metre length of conductor including ice coating and wind force,

$$w_r = \sqrt{(w_c + w_i)^2 + w_w^2} \qquad ...(2.30)$$

where w_w = Wind force in kg per metre length

= Wind pressure per m^2 of projected area × projected area per metre length

$= p \times (D + 2r)$ ...(2.31)

$$\therefore \quad \text{Maximum sag} = \frac{w_r L^2}{8T} \qquad ...(2.32)$$

When the ice and wind are acting simultaneously, the lowest point of the conductor does not remain vertically down but away from it at an angle θ given by the expression

$$\theta = \text{Cos}^{-1}\frac{w_c + w_i}{w_r} \qquad ...(2.33)$$

The sag calculated from Eq. (2.32) will not be vertical sag but will be *slant sag* and vertical sag will be obtained by multiplying the slant sag with cos θ

$$i.e., \quad \text{Vertical sag} = \frac{w_r L^2}{8T}\cos\theta \qquad ...(2.34)$$

Example 2.3. An overhead transmission line has a span of 220 m the conductor weighing 604 kg/km. Calculate the maximum sag if the ultimate tensile strength of conductor is 5,758 kg. Assume a factor of safety = 2. *[M.D. Univ. Transmission and Distribution, 2011]*

Solution : Span length, L = 220 m

Weight of conductor per metre length, $w = \frac{604}{1{,}000}$ kg = 0.604 kg

Ultimate tensile length = 5,758

Factory of safety = 2

$$\text{Working tension, T} = \frac{\text{Ultimate tensile strength}}{\text{Factor of safety}} = \frac{5{,}758}{2} = 2{,}879\ kg$$

$$\text{Maximum sag, S} = \frac{\omega L^2}{8T} = \frac{0.604 \times 220^2}{8 \times 2{,}879} = 1.269\ m \quad \textbf{Ans.}$$

Example 2.4. A 132 kV transmission line has the following data:

Wt of conductor = 680 kg/km, Length of span = 260 m, Ultimate strength = 3,100 kg ; Safety factor = 2.

Calculate the height above ground at which the conductor should be supported. Ground clearance required is 10 m.

Solution : Working tension,

$$T = \frac{\text{Ultimate strength}}{\text{Safety factor}} = \frac{3{,}100}{2} = 1{,}550\ kg$$

Weight of conductor per metre length,

$$w = \frac{680}{1{,}000} = 0.68\ kg$$

$$\text{Maximum sag, S} = \frac{wL^2}{8T} = \frac{0.68 \times (260)^2}{8 \times 1{,}550} = 3.7\ \text{metres}$$

The height above ground at which the conductor should be supported

= Clearance + sag = 10 + 3.7 = 13.7 metres **Ans.**

Example 2.5. Calculate the horizontal component of tension and maximum sag for a span of 300 m if the maximum tension in the conductor be 3,500 kg and weight of conductor is 700 kg/km. Determine also the location of the points on the conductor at which the sag will be half of the above value.

[A.M.I.E. Sec. B. Power Systems Winter 1996; Pb. Technical Univ. Power System-I (T&D) Dec. 2008]

Solution : Weight of conductor per metre length,

$$w = \frac{700}{1,000} = 0.7 \text{ kg}$$

From Eqs. (2.17) and (2.21), we have

$$\text{Maximum sag, S} = \frac{wL^2}{8T_0} = \frac{0.7 \times (300)^2}{8T_0} = \frac{7,875}{T_0} \quad ...(i)$$

and maximum tension,

$$T_{max} = T_0 + wS = T_0 + 0.7 \times \frac{7,875}{T_0}$$

$$= T_0 + \frac{5,512.5}{T_0} \quad ...(ii)$$

where T_0 is the horizontal component of tension in kg

Substituting the value T_{max} = 3,500 kg in Eq. (*ii*), we have

$$3,500 = T_0 + \frac{5,512.5}{T_0}$$

or $T_0^2 - 3,500\, T_0 + 5,512.5 = 0$

$$\text{or } T_0 = \frac{3,500 \pm \sqrt{(3,500)^2 - 4 \times 5,512.5}}{2}$$

$$= (1750 \pm 1,748) \text{ kg}$$

As $T_0 \simeq T_{max}$ hence +ve sign will be taken, so

T_0 = 1750 + 1,748.42 = 3,498.42 kg **Ans.**

$$\text{Maximum sag, S} = \frac{wL^2}{8T_0} = \frac{0.7 \times (300)^2}{8 \times 3,498.42} = 2.251 \text{ metres}$$ **Ans.**

$$\text{Now for sag, } y = \frac{S}{2} = \frac{2.251}{2} = 1.1255 \text{ m}$$

Since from Eq. (2.14),

$$y = \frac{wx^2}{2T_0}$$

$$\therefore\ x = \sqrt{\frac{2T_0 \times y}{w}} = \sqrt{\frac{2 \times 3,498.42 \times 1.1255}{0.7}}$$

$$= 106 \text{ metres}$$

Hence, for half of maximum value of sag the points on the conductor will be located having value of *x* coordinate = 106 m on both sides of point O.

Example 2.6. A transmission line has a span of 150 m between level supports. The cross-sectional area of the conductor is 1.25 cm^2 and weighs 100 kg/100 m. The breaking stress is 4,220 kg/cm^2, calculate the factor of safety if the sag of the line is 3.5 m. Assume a maximum wind pressure of 100 kg/m^2.

[*Rajasthan Technical Univ. Transmission & Distribution of Electrical Power 2011*]

Solution : Span length, L = 150 m

$$\text{Weight of conductor per metre length, } w_c = \frac{100}{100} = 1.0 \text{ kg}$$

Conductor diameter,

$$d = \sqrt{\frac{\text{cross-sectional area}}{\pi/4}} = \sqrt{\frac{1.25 \times 4}{\pi}} = 1.2616 \text{ cm}$$

Wind force per metre length of conductor,

w_w = Wind pressure × conductor diameter

$= 100 \times 1.2616 \times 10^{-2} = 1.2616$ kg

Resultant force per metre length,

$$w_r = \sqrt{w_c^2 + w_m^2} = \sqrt{1.0^2 + 1.2616^2} = 1.61 \text{ kg}$$

Working tension,

$$T = \frac{\omega_r L^2}{8\,S} \qquad \because\ S = \frac{\omega_r L^2}{8T}$$

$$= \frac{1.61 \times 150^2}{8 \times 3.5} = 1,294 \text{ kg}$$

$$\text{Factor of safety} = \frac{\text{Breaking stress} \times \text{area of cross section}}{\text{Working tension, T}}$$

$$= \frac{4,220 \times 1.25}{1,294} = 4$$ **Ans.**

Example 2.7. An overhead line has the following data :

Span length = 160 m

Conductor diameter = 0.95 cm

Weight per unit length of the conductor = 0.65 kg/m

Ultimate stress = 4,250 kg/cm^2

Wind pressure = 40 kg/m^2 of the projected area

Factor of safety = 5

Calculate the sag.

[*U.P. Technical Univ. Elements of Power System 2002-03, 2006-07*]

Solution :

Span length = 160 m

Weight of conductor per metre length, w_c = 0.65 kg

$$\text{Working stress} = \frac{\text{Ultimate stress}}{\text{Factor of safety}} = \frac{4,250}{5} = 850 \text{ kg/cm}^2$$

Working tension, T = Working stress × cross-sectional area

$$= 850 \times \frac{\pi}{4}(0.95)^2 = 602.5 \text{ kg}$$

Wind force per metre length of conductor,

w_w = Wind pressure × conductor diameter

$= 40 \times 0.95 \times 10^{-2} = 0.38$ kg/m

Resultant force per metre length,

$$w_r = \sqrt{w_c^2 + w_w^2} = \sqrt{0.65^2 + 0.38^2} = 0.7529 \text{ kg}$$

$$\text{Sag, S} = \frac{w_r L^2}{8T} = \frac{0.7529 \times 160^2}{8 \times 602.5} = 4 \text{ m}$$ **Ans.**

Example 2.8. The line supports of an overhead line are 180 m apart and are at the same level. The line conductor is of hard drawn copper having a diameter of 1.27 cm. Calculate the maximum sag (*a*) in still air; and (*b*) in a wind pressure of 33.7 kg per square metre of the projected area of the ice covered line. The radial thickness of the ice is 1.25 cm. Assume that the maximum stress in the line is not to increase one-fifth of the ultimate strength of 4,220 kg per square cm. The weight of conductor per metre run is 1.13 kg.

Solution : Working stress,

$$f = \frac{\text{Ultimate stress}}{\text{Factor of safety}} = \frac{4{,}220}{5} = 844 \text{ kg/cm}^2$$

Area of cross section of conductor,

$$a = \frac{\pi}{4} D^2 = \frac{\pi}{4} \times (1.27)^2 = 1.267 \text{ cm}^2$$

Working tension, $T = f \times a = 844 \times 1.267 = 1{,}069.15$ kg

(*a*) Sag in still air $= \frac{w_c L^2}{8T} = \frac{1.13 \times (180)^2}{8 \times 1{,}069.15} = 4.28$ m **Ans.**

(*b*) Weight of ice coating per metre length,

$w_1 = 2{,}890.3r(D + r)$ kg/m refer to Eq. (2.29)

$= 2{,}890.3 \times 1.25 \times 10^{-2}(1.27 + 1.25) \times 10^{-2} = 0.91$ kg

Wind force per metre length of conductor,

w_w = Wind pressure $p \times (D + 2r)$

$= 33.7 \times (1.27 + 2 \times 1.25) \times 10^{-2} = 1.2705$ kg

Resultant force per metre length of conductor,

$$w_r = \sqrt{(w_c + w_i)^2 + w_w^2} = \sqrt{(1.13 + 0.91)^2 + (1.2705)^2}$$

$$= 2.403 \text{ kg}$$

Maximum sag $= \frac{w_r L^2}{8T} = \frac{2.403 \times (180)^2}{8 \times 1{,}069.15} = 9.104$ m **Ans.**

Example 2.9. A transmission line conductor having a diameter of 19.5 mm weighs 0.85 kg/m. The span is 275 metres. The wind pressure is 39 kg/m^2 of projected area with ice coating of 13 mm. The ultimate strength of the conductor is 8,000 kg. Calculate the maximum sag if the factor of safety is 2 and ice weighs 910 kg/m^3.

[*U.P. Technical Univ. Elements of Power System 2005-06; Pb. Technical Univ. Power System (T&D), May 2010*]

Solution : Working tension,

$$T = \frac{\text{Ultimate strength}}{\text{Factor of safety}} = \frac{8{,}000}{2} = 4{,}000 \text{ kg}$$

Weight of ice coating per metre length,

w_i = Density of ice $\times \pi r(D + r)$

$= 910 \times \pi \times 13 \times 10^{-3} \times (19.5 + 13) \times 10^{-3} = 1.208$ kg

Wind force per metre length of conductor,

$w_w = p \times (D + 2r) = 39(19.5 + 2 \times 13) \times 10^{-3} = 1.7745$ kg

Resultant force per metre length of conductor,

$$w_r = \sqrt{(w_c + w_i)^2 + w_w^2} = \sqrt{(0.85 + 1.208)^2 + (1.7745)^2}$$

$$= 2.717 \text{ kg}$$

Maximum sag,

$$S = \frac{w_r L^2}{8T} = \frac{2.717 \times 275^2}{8 \times 4{,}000} = 6.422 \text{ m } \textbf{Ans.}$$

Example 2.10. A transmission line conductor is having a diameter of 20 mm and weighs 1.0 kg/m. The span is 280 m. The wind pressure is 40 kg/m^2 of projected area with ice coating of 10 mm. The ultimate strength of conductor is 10,000 kg. Calculate the maximum sag if the factor of safety is 2 and ice weighs 910 kg/m^3.

[*U.P. Technical Univ. Elements of Power System 2004-05*]

Solution : Weight of conductor per metre length, $w_c = 1.0$ kg

Weight of ice per metre length,

w_i = Density of ice $\times \pi r(D + r)$

$= 910 \times \pi \times 10 \times 10^{-3}(20 + 10) \times 10^{-3} = 0.85765$ kg

Wind force per metre length of conductor,

$w_w = p \times (D + 2r) = 40 \times (20 + 2 \times 10) \times 10^{-3} = 1.6$ kg

Resultant force per metre length of conductor

$$= \sqrt{(w_c + w_i) + w_w^2} = \sqrt{(1 + 0.85765)^2 + (1.6)^2}$$

$$= 2.4517 \text{ kg}$$

Maximum sag,

$$S = \frac{w_r L^2}{8T} = \frac{2.4517 \times 280^2}{8 \times \frac{10{,}000}{2}} \qquad \because T = \frac{\text{Ultimate strength}}{\text{Factor of safety}}$$

$= 4.8$ metres **Ans.**

Example 2.11. An overhead line is erected across a span of 250 metres on level supports. The conductor has a diameter of 1.42 cm and has a dead weight of 1.09 kg/m. The line is subjected to a wind pressure of 37.8 kg per square metre of the projected area. The radial thickness of ice is 1.25 cm. The line is carried by insulator string 1.43 m long. Calculate (*i*) sag in an inclined direction (*ii*) sag in a vertical direction and (*iii*) the height of lowest cross-arm to give a minimum ground clearance of 7.62 m under bad weather conditions. Assume one cubic metre of ice weight 913.5 kg and a maximum stress of 1,050 kg/cm^2.

[*A.M.I.E. Sec. B. Power Systems Summer 1995; W.B. Technical Univ. Power System-I, 2009-10*]

Solution : Working tension,

$$T = f \times \frac{\pi}{4}(D^2) = 1{,}050 \times \frac{\pi}{4}(1.42)^2 = 1{,}662.86 \text{ kg}$$

Weight of ice coating per metre length,

$w_i = 913.5 \times \pi r(D + r)$

$= 913.5 \times \pi \times 1.25 \times 10^{-2} \times (1.42 + 1.25) \times 10^{-2} = 0.9578$ kg

Wind force per metre length of conductor,

$w_w = p \times (D + 2r) = 37.8\ (1.42 + 2 \times 1.25) \times 10^{-2}$

$= 1.48176$ kg/m

Resultant force per metre length of conductor,

$$w_r = \sqrt{(w_c + w_i)^2 + w_w^2}$$

$$= \sqrt{(1.09 + 0.9578)^2 + (1.48176)^2} = 2.5277 \text{ kg}$$

(*i*) Sag in an inclined direction,

$$S = \frac{w_r L^2}{8T} = \frac{2.5277 \times 250^2}{8 \times 1{,}662.86} = 11.8756 \text{ m } \textbf{Ans.}$$

(*ii*) Sag in vertical direction,

$S' = S \cos \theta$

$$= \frac{w_r L^2}{8T} \times \frac{w_c + w_i}{w_r}$$

$$= \frac{(w_c + w_i)L^2}{8T} = \frac{(1.09 + 0.9578) \times 250^2}{8 \times 1662.86} = 9.62 \text{ m } \textbf{Ans.}$$

(*iii*) Height of lowest cross arm

= Minimum ground clearance + vertical sag + insulator string length

$= 7.62 + 9.62 + 1.43 + 18.67$ m **Ans.**

Example 2.12. Derive the expression for the lowest point of the catenary curve if the supports are at different levels. Given that weight of conductor = 0.35 kg/m, Maximum allowable strength = 800 kg. Span length = 160 m and safety factor = 2, find the minimum point of catenary if supports are at 70 m and 65 m.

[*U.P. Technical Univ. Elements of Power System 2007-08*]

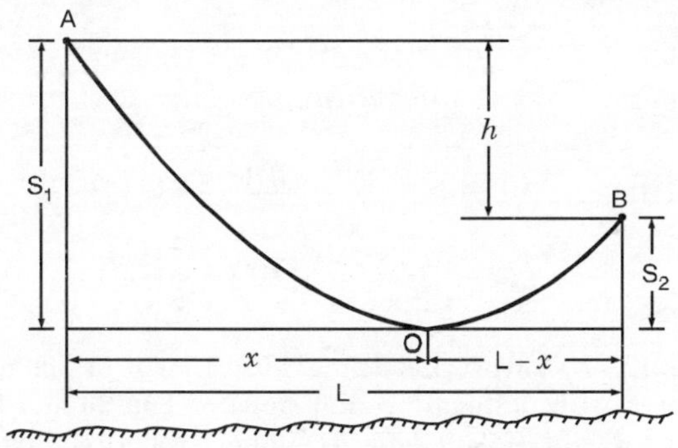

Fig. 2.22

Solution : Weight of conductor per metre length,

$$w = 0.35 \text{ kg/m}$$

Maximum allowable stress = 800 kg

Safety factor = 2

$$\text{Allowable tension, } T = \frac{800}{2} = 400 \text{ kg}$$

Difference in level of two supports,

$$h = 70 - 65 = 5 \text{ m}$$

Span length, L = 160 m

Distance of lowest point of conductor O from the taller support,

$$x = \frac{L}{2} + \frac{Th}{wL} = \frac{160}{2} + \frac{400 \times 5}{0.35 \times 160} = 115.71 \text{ m } \textbf{Ans.}$$

$$\text{Sag } S_1 = \frac{wx^2}{2T} = \frac{0.35 \times 115.71^2}{2 \times 400} = 5.8576 \text{ m}$$

The minimum point of catenary is at 70 – 5.8576, *i.e.*, 64.1424 m above ground. **Ans.**

Example 2.13. An overhead line with stranded copper conductors is supported on two poles 200 m apart having a difference in level of 10 metres. The conductor diameter is 2 cm and weighs 2.3 kg per metre length. Calculate the sag at the lower support under the conditions if wind provides a pressure of 57.5 kg per square metre of the projected area and a factor of safety is 4. The maximum tensile strength of copper is 4,220 kg per square cm.

Solution : Horizontal wind force per metre length,

$$w_w = \text{Wind pressure} \times D = 57.5 \times 2 \times 10^{-2} = 1.15 \text{ kg}$$

Resultant force per metre length,

$$w_r = \sqrt{w_c^2 + w_w^2} = \sqrt{2.3^2 + 1.15^2} = 2.5715 \text{ kg}$$

Difference in level of two supports,

$$h = 10 \text{ metres}$$

Working stress,

$$f = \frac{\text{Maximum tensile strength}}{\text{Factor of safety}} = \frac{4{,}220}{4} = 1{,}055 \text{ kg/cm}^2$$

Allowable tension,

$$T = f \times \frac{\pi}{4}(D^2) = 1{,}055 \times \frac{\pi}{4}(2)^2 = 3{,}314.38 \text{ kg}$$

Distance of lowest point of conductor O (Fig. 2.19) from the support at low level,

$$x = \frac{L}{2} - \frac{Th}{w_r L} = \frac{200}{2} - \frac{3{,}314.38 \times 10}{2.5715 \times 200} = 35.556 \text{ m}$$

$$\text{Slant sag, } S = \frac{w_r x^2}{2T} = \frac{2.5715 \times (35.556)^2}{2 \times 3{,}314.38} = 0.4904 \text{ m } \textbf{Ans.}$$

$$\text{Vertical sag, } S' = \frac{w_c x^2}{2T} = \frac{2.3 \times (35.556)^2}{2 \times 3{,}314.38} = 0.4386 \text{ m } \textbf{Ans.}$$

Example 2.14. An overhead transmission line conductor having parabolic configuration weighs 1.925 kg per metre length, area of cross section 2.2 cm² and an ultimate strength of 8,000 kg/cm². When erected between supports 600 metre apart and having 15 metre difference in height, determine the vertical sag from the taller of the two supports which must be allowed so that the factor of safety shall be 5 with the wire loaded due to 1 kg of ice per metre and no wind pressure.

Solution : Effective weight of conductor per metre length,

$$w = 1.925 + 1 = 2.925 \text{ kg}$$

Difference in level of two supports,

$$h = 15 \text{ metres}$$

$$\text{Allowable tension, } T = \frac{8{,}000 \times 2.2}{5} = 3{,}520 \text{ kg}$$

Distance of lowest point of the conductor, O (Fig. 2.19) from the support at low level

$$x = \frac{L}{2} - \frac{Th}{wL} = \frac{600}{2} - \frac{3{,}520 \times 15}{2.925 \times 600} = 270 \text{ metres}$$

Distance of lowest point of the conductor from the taller support

$$= L - x = 600 - 270 = 330 \text{ metres}$$

Vertical sag from the taller of the two supports

$$S_2 = \frac{w(L-x)^2}{2T} = \frac{2.925 \times (330)^2}{2 \times 3{,}520} = 45.25 \text{ m } \textbf{Ans.}$$

Example 2.15. A transmission line conductor at a river crossing is supported from two towers at heights of 50 m and 80 m above water level. The horizontal distance between the towers is 300 metres. If the tension in the conductor is 2,000 kg, find the clearance between the conductor and water level at a point midway between the towers. Weight of conductor per metre = 0.844 kg. Assume that the conductor takes the shape of parabolic curve. [*Rajasthan Technical Univ. Transmission & Distribution of Electrical Power 2008*]

Solution : Difference in level between the two supports,

$$h = 80 - 50 = 30 \text{ m}$$

Distance of lowest point of conductor, O from the support at low level

$$x = \frac{L}{2} - \frac{Th}{wL} = \frac{300}{2} - \frac{2{,}000 \times 30}{0.844 \times 300} = -86.967 \text{ m}$$

The negative value of x shows that support A is on the same side of O as support B (Fig. 2.23)

Distance of midpoint P from O

$$= \frac{L}{2} - x = \frac{300}{2} - (-86.967) = 236.967 \text{ metres}$$

Distance of point B from O

$$= L - x = 300 + 86.967 = 386.967 \text{ metres}$$

Height of midpoint P above O,

$$S_{mid} = \frac{w\left(\frac{L}{2}-x\right)^2}{2T} = \frac{0.844 \times (236.967)^2}{2 \times 2{,}000} = 11.848 \text{ m}$$

Height of point B from O,

$$S_2 = \frac{w \times (L-x)^2}{2T} = \frac{0.844 \times (386.967)^2}{2 \times 2{,}000} = 31.596 \text{ m}$$

Hence mid point P is (31.596 – 11.848) *i.e.*, 19.748 metres below point B or (80 – 19.748) *i.e.*, 60.252 m above the water level **Ans.**

Example 2.16. An overhead transmission line at a river crossing is supported from two towers at heights of 25 m and 75 m above the water level. The horizontal distance between the towers is 250 m. If the required clearance between the conductor and the water mid way between the towers is 45 metres and if both the towers are on the same side of the point of maximum sag of the parabolic configuration, find the stringing tension in the conductor. The weight of the conductor is 0.7 kg/m.

[Rajasthan Technical Univ. Transmission & Distribution of Electrical Power 2011]

Solution : Let T be the stringing tension and x be the distance between the lowest point O of the conductor and the support at low level, as shown in Fig. 2.23

Since $h = S_2 - S_1$

$$\text{or } h = \frac{w(L-x)^2}{2T} - \frac{wx^2}{2T} = \frac{wL^2 - 2wLx}{2T}$$

$$\text{or } h = \frac{wL}{2T}(L-2x) \quad \ldots(i)$$

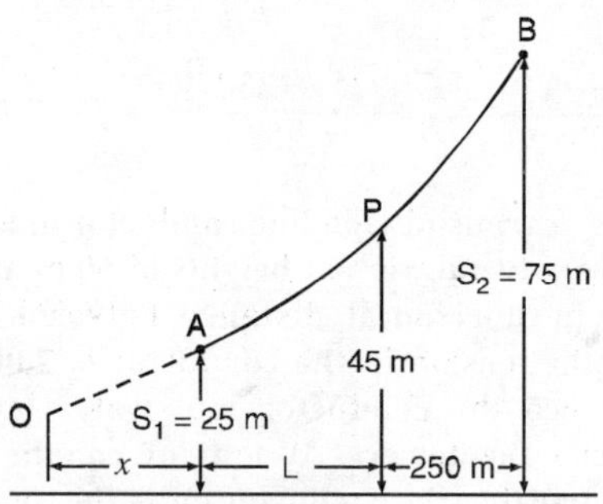

Fig. 2.23

For points A and B,

$h = 75 - 25 = 50$ metres

$L = 250$ metres

and For points A and P (mid way point between the towers)

$h = 45 - 25 = 20$ metres

and $L = 125$ metres

Substituting the above values of h and L in Eq. (*i*), we get

$$50 = \frac{0.7 \times 250}{2T}[250 - 2x] \quad \ldots(ii)$$

$$20 = \frac{0.7 \times 125}{2T}[125 - 2x] \quad \ldots(iii)$$

Dividing Eq. (*ii*) by Eq. (*iii*), we get

$$2.5 = 2 \times \frac{250 - 2x}{125 - 2x}$$

or $2.5\,(125 - 2x) = 500 - 4x$

or $x = -500 + 312.5 = -187.5$ metres

Substituting $x = -187.5$ in Eq. (*ii*), we get

$$50 = \frac{0.7 \times 250}{2T}[250 - 2 \times (-187.5)]$$

$$\text{or } T = \frac{0.7 \times 250}{2 \times 250}[250 + 375] = 1{,}093.75 \text{ kg}$$ **Ans.**

Example 2.17. An overhead line with a span of 300 m and in a hilly area with uniform ground slope of 1 in 20 and having a conductor weight of 0.80 kg/m, is supported between two towers at equal height of 30 m above the ground level. Assuming the horizontal component of tension in the line to be 1,500 kg; find (*i*) the clearance at the point where conductor has its lowest elevation : and (*ii*) the minimum clearance of the line from ground.

[A.M.I.E. Sec. B. Power Systems Summer 1999]

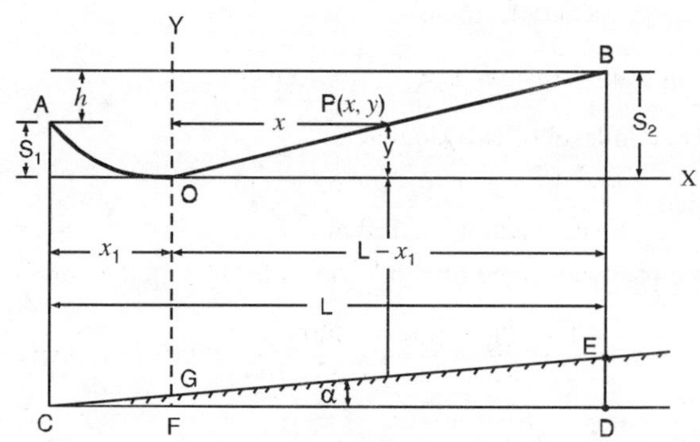

Fig. 2.24

Solution : The lowest conductor AOB and the profile of the ground CE are shown in Fig. 2.24. Take origin O at the lowest point of conductor. Let OX and OY be the horizontal and vertical reference axis.

$$\text{Slope} = \tan \alpha = \frac{1}{20}$$

$$ED = CD \tan \alpha = L \tan \alpha = 300 \times \frac{1}{20} = 15 \text{ m}$$

Height of the lowest conductor from the ground

= AC = BE = 30 m

Height BD = BE + ED = 30 + 15 = 45 m

The equation of parabola AOB is given by

$$y = \frac{wx^2}{2T_0}$$

$$\text{Thus } S_1 = \frac{wx_1^2}{2T_0}$$

$$\text{and } S_2 = \frac{w(L-x_1)^2}{2T_0}$$

Vertical distance between the supports

$$h = ED = 15 \text{ m}$$

$$\text{Also } h = S_2 - S_1 = \frac{w(L - x_1)^2}{2T_0} - \frac{wx_1^2}{2T_0} = \frac{w}{2T_0}(L^2 - 2Lx_1)$$

$$\text{or } x_1 = \frac{L}{2} - \frac{T_0 h}{wL} = \frac{300}{2} - \frac{1{,}500 \times 15}{0.8 \times 300} = 56.25 \text{ m}$$

$$S_1 = \frac{wx_1^2}{2T_0} = \frac{0.8 \times (56.25)^2}{2 \times 1{,}500} = 0.84375 \text{ m}$$

$$\text{and } S_2 = \frac{w(L - x_1)^2}{2T_0} = \frac{0.8\,(300 - 56.25)^2}{2 \times 1{,}500} = 15.84375 \text{ m}$$

Clearance of the lowest point of the lowest conductor from the ground

$$= OG = AC - S_1 - GF$$

$$= 30 - 0.84375 - CF \tan \alpha \text{ (or } x_1 \tan \alpha)$$

$$= 30 - 0.84375 - 56.25 \times \frac{1}{20} = 26.34375 \text{ m} \textbf{ Ans.}$$

Equation of the slope of the ground is given as

$$y = mx + C = x \tan \alpha - OG = \frac{x}{20} - 26.34375$$

Clearance from ground at any point P (x, y)

$$C_1 = y - \left(\frac{x}{20} - 26.34375\right) = \frac{wx^2}{2T_0} - \frac{x}{20} + 26.34375$$

For minimum value of C_1,

$$\frac{dC_1}{dx} = 0$$

$$\text{or } \frac{2wx}{2T_0} - \frac{1}{20} = 0$$

$$\text{or } x = \frac{T_0}{20w} = \frac{1{,}500}{20 \times 0.8} = 93.75 \text{ m}$$

Hence minimum clearance of the line from ground will be for value of $x = 93.75$ m

and minimum clearance

$$= \frac{wx^2}{2T_0} - \frac{x}{20} + 26.34375$$

$$= \frac{0.8(93.75)^2}{2 \times 1{,}500} - \frac{93.75}{20} + 26.34375 = 24 \text{ m} \textbf{ Ans.}$$

2.11. STRINGING CHART

Under statutory conditions, the sag is required to be determined for worst probable conditions and the minimum ground clearance is to be maintained for these conditions. At the time of erection the severe conditions do not prevail, the temperature is usually higher, the designer, therefore, should know the sag to be allowed and the tension in the line to be allowed, so that under no condition there should be any danger to the line.

Stringing chart is helpful in knowing the sag and tension at any temperature. This chart gives the data for sag to be allowed and the tension to be allowed at a particular temperature.

For preparation of stringing or sag chart first of all calculate the sag and tension on the conductor under the worst conditions *i.e.*, maximum wind pressure and minimum temperature, assuming a suitable factor of safety in fixing the maximum working tension for the conductor. Now evaluate the sag and tension for a series of temperatures in steps within the working range of temperatures.

The equation for determining stringing chart of a line is derived as below:

We know that $T = T_0$; $y = \dfrac{wx^2}{2T}$; $S = \dfrac{wL^2}{8T}$ and span length*, $l = L + \dfrac{w^2L^3}{24T^2}$

Let w_1, f_1, l_1, S_1 and t_1 be the load per unit length, the stress, the span length, sag and temperature at the maximum load conditions (with the ice and wind and low temperature usually – 5.5°); w_2, f_2, l_2, S_2 and t_2 be the values under stringing conditions, a is the area of cross section of the conductor, α is the coefficient of linear expansion and E is the modulus of elasticity.

$$\text{Span length, } l = L + \frac{w^2L^3}{24T^2} = L + \frac{w^2L^3}{24a^2f^2}$$

$$= L + \frac{8S^2}{3L} \qquad ...(2.35)$$

The span length at maximum load condition is

$$l_1 = L + \frac{w_1^2L^3}{24a^2f_1^2} \qquad ...(2.36)$$

The temperature rise from t_1 to t_2 causes an increase in the span length of $l_1\alpha\,(t_2 - t_1)$ which is practically equal to $L\,\alpha\,(t_2 - t_1)$. The fall in stress from f_1 to f_2 causes a decrease in the length of $\left(\dfrac{f_2 - f_1}{E}\right) l_1 \simeq \left(\dfrac{f_1 - f_2}{E}\right) L$

The new length l_2 is thus given by

$$l_2 = l_1 - \frac{f_1 - f_2}{E} L + (t_2 - t_1)\,\alpha\,L \qquad ...(2.37)$$

$$\text{But } l_2 = L + \frac{w_2^2L^3}{24a^2f_2^2} \qquad ...(2.38)$$

$$\therefore\ L + \frac{w_2^2L^3}{24a^2f_2^2} = L + \frac{w_1^2L^3}{24a^2f_1^2} - \frac{f_1 - f_2}{E} L + (t_2 - t_1)\,\alpha\,L$$

$$\text{or } f_2^2\left[(f_2 - f_1) + \frac{w_1^2L^2E}{24f_1^2a^2} + (t_2 - t_1)\,\alpha\,E\right] = \frac{w_2^2L^2E}{24a^2} \qquad ...(2.39)$$

* Span length means length of conductor between two supports.

Equation (2.39) is a cubic equation and can be solved graphically or analytically. From this equation erection tension $T_1 = f_1 a$ can be determined such that tension $T_2 = f_2 a$ under worst probable conditions will not exceed the safe limit of tension. After determining f_2, the corresponding sag can be determined from the equation

$$S_2 = \frac{w_2 L^2}{8 f_2 a} \qquad \text{...(2.40)}$$

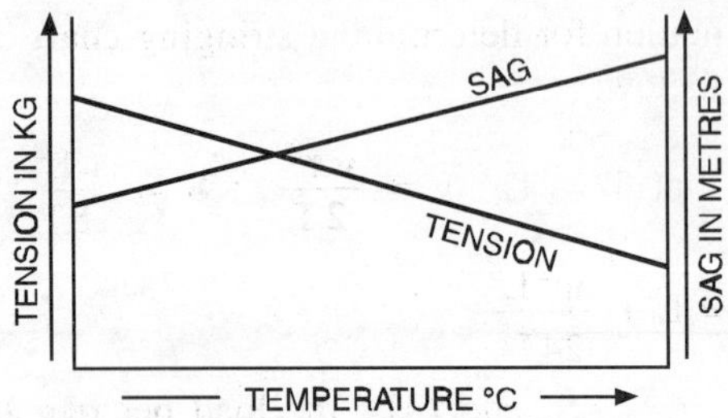

Fig. 2.25

Various values of f_2 and S_2 are calculated using Eqs. (2.39) and (2.40) repeatedly for different temperatures. Now the graph of tension vs temperature and sag vs temperature can be plotted, as shown in Fig. 2.25. This graph is plotted for a fixed span and is called the stringing chart. The stringing chart is very useful while erecting the transmission line conductors for adjusting the sag and tension properly.

Example 2.18. An overhead line having a span of 250 m is to be erected at a temperature of 40°C in still air conditions. It is desired that a factor of safety of 2 should be maintained under bad weather conditions when the temperature is 10°C and the wind load is 378 N/m² of projected area. The data for the ACSR conductor used for the line is as under :

Diameter 1.95 cm, area 2.25 cm², weight = 8.31 N/m, breaking load is 77,900 N, Coefficient of linear coefficient is 18.44 × 10⁻⁶ per °C, Young's modulus 91.4 × 10³ N/mm². Find the sag and tension under erection conditions.

[*Rajasthan Univ. Transmission and Distribution of Electrical Power 2001*]

Solution : *Under worst weather conditions:*

Temperature, $t_1 = 10°C$

$$\text{Stress, } f_1 = \frac{T_1}{a}$$

$$= \frac{\text{Breaking load}}{\text{Factor of safety}} \times \frac{1}{\text{Area of cross section}}$$

$$= \frac{77{,}900}{2 \times 2.25 \times 10^{-4}} = 1.558 \times 10^8 \text{ N/m}^2$$

Wind force per unit length,

$$w_w = \text{Wind pressure} \times D = 378 \times 1.95 \times 10^{-2} = 7.371 \text{ N}$$

Effective load per unit length,

$$w_1 = \sqrt{w_c^2 + w_w^2} = \sqrt{8.31^2 + 7.371^2} = 11.108 \text{ N}$$

Under Erection conditions:

Temperature, $t_2 = 40°C$

$w_2 = 8.31$ N/m

Now from Eq. (2.39)

$$f_2^2 \left[(f_2 - f_1) + \frac{w_1^2 L^2 E}{24 f_1^2 a^2} + (t_2 - t_1) \times \alpha E \right] = \frac{w_2^2 L^2 E}{24 a^2}$$

$$\text{or } f_2^2 \left[(f_2 - 1.558 \times 10^8) + \frac{(11.108)^2 \times (250)^2 \times 91.4 \times 10^9}{24 \times (1.558 \times 10^8)^2 \times (2.25) \times 10^{-4})^2} + (40 - 10) \times 18.44 \times 10^{-6} \times 91.4 \times 10^9 \right]$$

$$= \frac{(8.31)^2 \times (250)^2 \times 91.4 \times 10^9}{24 \times (2.25 \times 10^{-4})^2}$$

$$\text{or } f_2^2 \left[(f_2 - 1.558 \times 10^8) + 0.239 \times 10^8 + 0.5056 \times 10^8 \right]$$

$$= 3{,}246{,}773.426 \times 10^{17}$$

$$\text{or } f_2^3 - 0.8134 \times 10^8 f_2^2 - 3{,}246{,}773.4 \times 10^{17} = 0$$

By hit and trial

$$f_2 = 1.0578 \times 10^8 \text{ N/m}^2$$

Tension, $T_2 = f_2 \times a = 1.0578 \times 10^8 \times 2.25 \times 10^{-4} = 23{,}800$ N

$$\text{Sag of erection} = \frac{w_2 l^2}{8T} = \frac{8.31 \times 250^2}{8 \times 23{,}800} = 2.73 \text{ m} \quad \textbf{Ans.}$$

2.12. SAG TEMPLATE

In the initial planning stages a survey of the proposed route enables an estimated line profile to be drawn, as illustrated in Fig. 2.26. Such a profile is constructed with the horizontal scale much more reduced (say to 1/10) in comparison to vertical scale. This profile should meet the minimum clearance requirements, and the location of the supports should be such that some horizontal adjustment is possible without departing from the standard or tangent tower lengths.

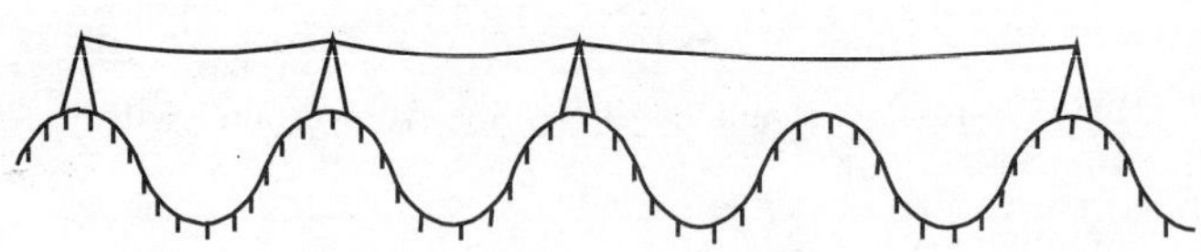

Fig. 2.26. ***Line Profile***

Use is made of *sag templates* drawn to the same scale as the line profile, to ensure the required clearance. The sag template is usually made on celluloid or tracing cloth. In sag template, shown in Fig. 2.27, the upper curve I represents the conductor line. The middle curve (*i.e.*, curve II) is below the upper curve I by a uniform vertical distance equal to the desired minimum vertical clearance to ground. This clearance to ground is governed by the operating voltage and is given, according to IE rules, in Table 2.4. The lower curve (curve III) is below the upper curve by a uniform vertical distance equal to the height of a standard tower measured to the point of support of the conductor. If the location of the left tower has been decided, the location of the right hand tower can be determined by adjusting the sag template so that the conductor line passes through the point of support on the left hand tower and the clearance line is tangent to ground at one or more points.

In the above particular case, illustrated the points of conductor support are upon the same horizontal level, but the same process applies when the route is a steeply sloping one. However, the shape and position of the conductor will always be represented by curve I, as shown in Fig. 2.27.

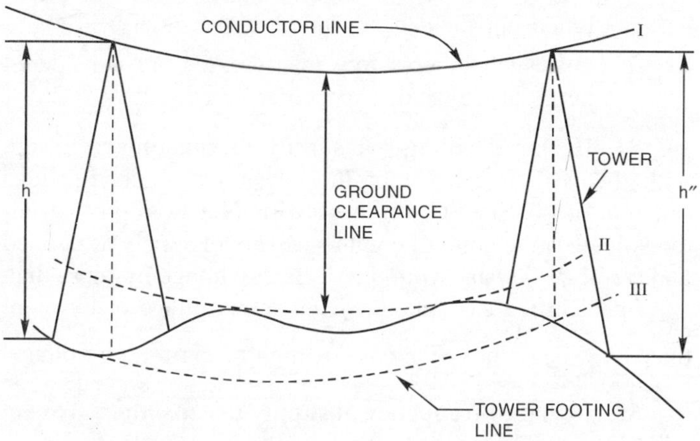

Fig. 2.27. *Sag Template Used For Locating Towers*

Sag template is a very convenient method for allocating the positions and height of the towers/supports correctly on the profile.

2.13. EQUIVALENT SPAN

It may not be possible to have a section of transmission line consisting of successive spans of equal lengths because the location of the towers depends upon the profile of land along which the transmission line is to be laid. Sometimes also, the towers are forced to be located to give spans of different lengths so that minimum interference is caused with the use of land. When the successive spans are of unequal lengths changes in tension in load or temperature will cause unequal changes in tension in the different spans.

It is very tedious to make calculations of sag and tension for each and every span individually and then to make adjustment while erecting the transmission line. In the erection of a transmission line the conductors are run out through snatch blocks attached to the support arms equally tensioned at each end of a section of five or six blocks. When the conductors are clamped to suspension insulator strings, the equal tension is maintained by insulator swing. When the conductors are bound to pin type insulators, the flexibility of the supports ensures equal tension.

It is often convenient to make sag and tension calculations in terms of a *hypothetical equivalent span,* this tension being applied to each span within the section of overhead line between the tensioning points. If there are n spans of length L_1, L_2, L_3 etc which are to be given an equivalent span L_e, then the strung length of the equivalent line must be the same as that of the individual spans. This may be expressed as

$$n\left(L_e + \frac{w^2 L_e^3}{24 T_0^2}\right) = \sum L + \frac{w^2 (\sum L^3)}{24\, T_0^2} \qquad ...(2.41)$$

Since $\Sigma L \equiv nL_e$

$$\text{So } L_e^2 = \frac{\sum L^3}{nL_e} = \frac{\sum L^3}{\sum L}$$

$$\text{or } L_e = \sqrt{\frac{L_1^3 + L_2^3 + L_3^3 + ...}{L_1 + L_2 + L_3 + ...}} \qquad ...(2.42)$$

If the line tension T_0 is determined for the equivalent span, the sag for the individual spans may be calculated from Eqs. (2.17) using the approximate value of span.

It is to be noted that the method of sag template for locating the towers should not be used for long spans as well as where the slope of the profile is very steep. In such cases, actual calculations for sag and tension should be made.

2.14. VIBRATIONS AND DAMPERS

The overhead transmission line experiences vibrations in the vertical plane and there are two types of such vibrations, in addition to normal swinging in wind, called the *aeoline vibrations* or *resonant vibrations* or *high frequency oscillations* and *galloping* or *dancing* or *low frequency vibrations.*

Simple swinging of conductors is harmless provided the clearance is sufficiently large so that the conductors do not approach within the sparking distance of each other.

Aeoline vibrations are high frequency (5–100 Hz) and low amplitude (20 mm to 50 mm) vibrations. They are caused by vortex phenomenon in light winds (5-20 km/hour). The line conductor vibrates in a number of loops. The length of the loop (half-wave length) depends on tension T and weight of conductor w per metre length and is given by $\frac{1}{2f}\sqrt{\frac{T}{w}}$. The loop length varies from 1 to 10 metres depending upon T, w and f. These vibrations are common to all conductors and are more or less always present. The harmful effects of such vibrations take place at clamps or supports where the conductor suffers fatigue and breaks eventually.

Low frequency vibrations (about one Hz) occur during sleet storms with a strong wind. The amplitude is very large, about 6 metres or more, and the conductors are said to 'dance'. Operation is almost impossible for the conductors touch one another, since the 'dancing' takes place horizontally and vertically. It is considered that the phenomenon takes place due to the fact that the line conductor receives an irregular coating of sleet. If the cross section of the coating is considered to be an ellipse, the line will experience a drag; further more, if the major axis of an ellipse makes an angle of 45° to the wind, there will be a considerable aerodynamic lift (or downward thrust) such as is experienced by an aerofoil. The conductor will thus dance horizontally and vertically with a large amplitude and in an irregular manner owing to the irregular deposition of sleet. The stranding of conductors also contributes to these vibrations. The travel of conductor follows the path of an ellipse. There is no method for prevention of these (low-frequency) vibrations. However, danger due to such vibrations can be reduced if horizontal conductor configuration is used.

The conductors are protected by dampers which prevent the resonant vibrations from reaching the conductors at the clamps or supports.

The stock bridge damper, shown in Fig. 2.28, consists of two weights attached to a piece of stranded cable 0.3 or 0.5 m long, which is clamped to the line conductor. The energy of vibration is absorbed by the stranded cable, and the vibration is rapidly damped out.

Fig. 2.28 *Stock Bridge Damper*

Another successful damper consists of a box containing a weight resting on a spring. In this case the spring absorbs the energy of vibration. The fatigue may be reduced by reinforcing the conductor for a few metres on either side of the clamp by binding metal rods or a length of the same conductor to the main conductor outside the clamp.

2.15. STATUTORY RULES AND INDIAN ELECTRICITY RULES 1956

1. Statutory Rules: The Indian Electricity Act, 1910, and Rules regulate the generation, transmission, supply and use of electric energy. In addition, there are several Acts and Rules such as the Factory Act and Rules, The Mines Act, The Lifts Act etc. which contains important clauses governing the installation and use of electrical apparatus in particular locations. All these laws and rules have statutory validity and it is the duty of every one engaged in the electrical work to know them.

2. Indian Electricity Rules 1956: These rules were framed by Central Electricity Board in accordance with the power conferred on it by section 37 of the Indian Electricity Act 1910. These rules and regulations have been framed in order to

(*i*) safeguard consumers of electrical energy from electric shock

(*ii*) minimise fire risk and

(*iii*) ensure, as far as possible, satisfactory operation of equipment and apparatus used.

Important extracts from these rules are given below:

Rule 29. Construction, Installation, Protection, Operation and Maintenance of Electric Supply Lines and Apparatus.

All electric supply lines and apparatus shall be sufficient in power and size and of sufficient mechanical strength for the work they may be required to do, and, so far as is practicable, shall be constructed, installed, protected, worked and maintained in accordance with standards for the Indian Standards Institution so as to prevent danger.

34. Accessibility of bare conductor: Where bare conductors are used in a building the owner of such conductor shall:

(*a*) ensure that they are inaccessible;

(*b*) provide in readily accessible position switches for rendering them dead whenever necessary;

(*c*) take such other safety measures as are considered necessary by the Inspector.

54. Declared voltage of supply to consumer: Except with the written consent of the consumer or the previous sanction of the State Government a supplier shall not permit the voltage at the point of commencement of supply as defined under rule 58 to vary from the declared voltage by more than 5 per cent in the case of low or medium voltage or by more than $12\frac{1}{2}$ per cent in the case of high or extra-high voltage.

55. Declared frequency of supply to consumer: Except with the written consent of the consumer or with the previous sanction of the State Government a supplier shall not permit the frequency of an alternating current supply to vary from the declared frequency by more than 3 per cent.

74. Material and strength: (1) All conductors of overhead lines other than those specified in sub-rule (1) of rule 86 shall have a breaking strength of not less than 317.5 kg. (700 lbs).

(2) Where the voltage is low and the span is of less than 50 ft. (15.24 metres) and is on the owner's or consumer premises, a conductor having an actual breaking strength of not less than 300 lbs (136.08 kg) may be used.

76. Maximum Stresses : Factor of Safety: (1) (*a*) The owner of every overhead line shall ensure that it has the following minimum factors of safety. The minimum factors of safety for supports based on crippling load shall be as follows :

(*i*) for metal supports 2.0

(*ii*) for mechanically processed concrete supports 2.5

(*iii*) for hand moulded concrete supports 3.0

(*iv*) for wood supports 3.5

The said owner shall also ensure that the strength of the supports in the direction of the line is not less than one-fourth of the strength required in the direction transverse to the line.

Provided that in the case of latticed steel or other compound structure the factors of safety shall not be less than 1.5 under such broken wire conditions as may be specified by the state government in this behalf.

(*b*) The minimum factor of safety for stay wires, guard wires or bearer wire shall be 2.5 based on the ultimate tensile strength of the wire.

(*c*) The minimum factor of safety for conductors shall be 2 based on their ultimate tensile strength. In addition the conductor tension at 90°F without external load shall not exceed the following percentages of the ultimate strength of such conductor;

Initial unloaded tension ... ... 35 per cent

Final unloaded tension ... ... 25 per cent

Provided that in the case of conductors having cross section of a generally triangular shape, such as conductors composed of 3 wires, the final unloaded tension at 90°F shall not exceed 30 per cent of the ultimate tensile strength of such conductor.

(2) For the purpose of calculation the factors of safety prescribed in sub-rule (1) :

(*a*) the maximum wind pressure shall be such as the State Government may specify in each case;

(*b*) for the cylindrical bodies the effective area shall be taken as two-thirds of the projected area exposed to wind pressure;

(*c*) for latticed steel or other compound structure the wind pressure on the lee-side members, shall be taken as one-half of the wind pressure on the windward side members and the factors of safety shall be calculated on the crippling load of struts and upon the elastic limit of tension members;

(*d*) the maximum and minimum temperatures shall be as the State Government may specify in each case.

(3) Notwithstanding anything contained in sub-rules (1) and (2) in localities where overhead lines are liable to accumulations of ice or snow the State Government may, by order in writing, specify the loading conditions for the purpose of calculating the factor of safety.

77. Clearances above ground of the lowest conductor. (1) No conductor of an overhead line, including service lines, erected across a street shall at any part thereof be at height less than:

(*a*) for low and medium voltage line ...19 ft. (5.791 m)

(*b*) for high voltage lines ...20 ft. (6.096 m)

(2) No conductor of an overhead line including service lines erected along any street shall at any part thereof be at a height less than:

(*a*) for low and medium voltage lines ...18 ft. (5.486 m)

(*b*) for high voltage lines ...19 ft. (5.791 m)

(3) No conductor of an overhead line including service lines, erected else where than along or across any street shall be at a height less than:

(*a*) for low, medium and high voltage lines up to and including 11,000 volts, if bare ...15 ft. (4.572 m)

(*b*) for low, medium and high voltage lines up to and including 11,000 volts if insulated ...13 ft. (3.963 m)

(*c*) for high voltage lines above 11,000 volts ...17 ft. (5.182 m)

(4) For extra-high voltage lines the clearance above ground shall not be less than 17 ft, (5.182 m) plus 1 foot (0.3048 m) for every 33,000 volts.

Provided that the minimum clearance along or across any street shall not be less than 20 ft. (6.096 metres).

79. Clearances from buildings of low and medium voltage lines and service lines:

(1) Where a low or medium voltage overhead line passes above or adjacent to or terminates on any building, the following minimum clearances from any accessible point, on the basis of maximum sag, shall be observed:

(*a*) For any flat roof, open balcony, verandah roof and lean-to-roof---

(*i*) When the line passes above the building, a vertical clearance of 8 feet (2.439 m) from the highest point, and

(*ii*) When the line passes adjacent to the building, a horizontal clearance of 4 feet (1.219 m) from the nearest point, and

(*b*) For pitched roof,

(*i*) When the line passes above the building, a vertical clearance of 8 feet (2.439 m) immediately under the lines, and

(*ii*) When the line passes adjacent to the building, a horizontal clearance of 4 feet (1.219 m).

(2) Any conductor so situated as to have a clearance less than that specified in sub-rule (*i*) shall be adequately insulated and shall be attached by means of metal clips at suitable intervals to a bare earthed bearer wire having a breaking strength of not less than 700 lbs. (317.5 kg).

(3) The horizontal clearance shall be measured when the line is at maximum deflection from the vertical due to wind pressure.

80. Clearance from buildings of high and extra-high voltage lines: (1) Where a high or extra-high voltage overhead line passes above or adjacent to any building or part of a building it shall have on the basis of maximum sag a vertical clearance above the highest part of the building immediately under such lines, of not less than

(*a*) for high voltage lines up to and including 33,000 volts ... ... 12 ft. (3.658 m)

(*b*) for extra-high voltage line 12 ft. (3.658 m) plus 1 foot (0.3048 m) for every additional 33,000 volts or part thereof.

(2) The horizontal clearance between the nearest conductor and any part of such building shall, on the basis of maximum deflection due to wind pressure be not less than

(*a*) for high voltage lines up to and including 11,000 volts ... ... 4ft. (1.219 m)

(*b*) for high voltage lines above 11,000 volts and up to and including 33,000 volts ... ... 6 ft.(1.829 m) (*c*) for extra-high voltage lines 6 ft. (1.829 m) plus 1 foot (0.3048 m) for every additional 33,000 volts, or part thereof.

81. Conductors at different voltages on same supports: Where conductors forming parts of systems at different voltages are erected on the same supports the owner shall make adequate provision to guard against danger to line men and others from the lower voltage system being charged above its normal working voltage by leakage from or contact with the higher voltage system and the methods of construction and the clearance between the conductors of the two systems shall be subject to the prior approval of the inspector.

85. Maximum interval between supports: All the conductors shall be attached to support at intervals not exceeding the safe limit based on the ultimate tensile strength of the conductor and the factor of safety prescribed in rule 76.

Provided that in the case of overhead lines carrying low or medium voltage conductors when erected in, over, along or across any street, the interval shall not, without the consent in writing of the inspector, exceed 220 ft. (67.05 metres).

89. Service lines from overhead lines: No service line or tapping shall be taken off from an overhead line except at a point of support.

90. Earthing: (1) All metal supports of overhead lines and metallic fittings attached there to, shall be permanently and efficiently earthed. For the purpose a continuous earth wire shall be provided and securely fastened to each pole and connected with earth ordinarily at four points in every 1.609 km (mile), the spacing between the points being as nearly equidistant as possible. Alternatively, each support and the metallic fitting attached thereto shall be efficiently earthed.

(2) Each stay wire shall be similarly earthed unless an insulator has been placed in it at a height not less than 3.048 metres (10 ft.) from the ground.

Exercises

1. Name different components of an overhead line. Also explain the purpose for which they are used.
2. Classify the various supports used for overhead lines.
[Pb. Technical Univ. Power System-I (T&D) Dec. 2006]
3. Briefly discuss the different types of supporting structures, which are used for overhead lines.
[Pb. Technical Univ. Power System-I (T&D) Dec. 2007]
4. Discuss the various types of line supports.
[Rajasthan Technical Univ. Transmission & Distribution of Electrical Power 2009]
5. What are the various types of line supports? Discuss the suitability of each with reference to system voltage and span.
6. Why steel towers are used ? Sketch and explain any one such tower.
7. Which material is used for overhead line conductor ? Discuss the reasons for selecting that material.
[U.P. Technical Univ. Elements of Power System 2009-10]
8. What advantage you get by using stranded conductors ?
[U.P. Technical Univ. Elements of Power System 2008-09]
9. What are the advantages and disadvantages of using aluminium over copper in power transmission.
[U.P. Technical Univ. Elements of Power System 2002-03]
10. What are the factors considered for the choice of the conductor materials for overhead lines ? Compare the properties of materials hard drawn aluminium, aluminium alloy for overhead line conductors.
[Rajasthan Univ. Transmission & Distribution of Electrical Power 2003]
11. What are ACSR conductors ? Give their advantages.
[Pb. Technical Univ. Power System-I, May 2008]
12. Give reasons for the following :
ACSR conductors are preferred for transmission and distribution lines.
[G.B. Technical Univ. Elements of Power System 2011-12]
13. Is sag desirable or an evil for a transmission line ? What are the disadvantages of providing too much or too small sag in a transmission line ?
14. What is sag ? What factors affects sag ?
[W.B. Technological Univ. Power System-I, 2009-10]
15. What is sag in overhead line ? Discuss the disadvantages of too small or too large sag on a line ?
[M.D. Univ. Power System-I, Dec. 2009]
16. Prove that a transmission line conductor between two supports at equal heights takes the form of a catenary.
[Pb. Technical Univ. Power System-I (T&D), May 2010]
17. Derive an expression for sag of a line supported between two supports at the same height.
[Pb. Technical Univ. Power System-I (T&D), May 2009]
18. Prove that a transmission line conductor between two supports at equal heights takes the form of a catenary. Derive expressions for sag and length of conductor.
[Rajasthan Technical Univ. Transmission and Distribution of Electrical Power 2008]
19. Show how the sag of an overhead line can be calculated in case of supports at different levels.
[M.D. Univ. Transmission & Distribution, May 2011]
20. Discuss the effect of wind and ice on sag.
[W.B. Technological Univ. Power System-I, 2009-10]
21. Deduce an approximate expression for calculating sag in overhead line with conductors suspended between level supports. Show also how the effects of wind and ice can be taken into account when making calculations for sag.
[Rajasthan Technical Univ. Transmission & Distribution of Electrical Power 2011]
22. Derive an expression for sag and tension in a power conductor strung between two supports at equal heights taking into account the wind and ice loadings also.
[G.B. Technical Univ. Elements of Power System, 2011-12]
23. What is a stringing chart ? What is its utility ?
[W.B. Technological Univ. Power System-I, 2009-10]

24. Explain the necessity of a stringing chart for a transmission line and show how such chart can be constructed.

[M.D. Univ. Transmission & Distribution, May 2011]

25. Explain the necessity and the method of preparing "stringing charts" for overhead transmission lines.

[Rajasthan Technical Univ. Transmission & Distribution of Electrical Power 2011]

26. Derive an equation to calculate the conductor tension under erection condition if the conductor tension and loading under bad weather conditions are known.

[Rajasthan Technical Univ. Transmission & Distribution of Electrical Power 2001]

27. What is the use of a sag template ? Describe it in detail.

[U.P. Technical Univ. Elements of Power System 2009-10]

28. Why do the vibrations get generated in conductors ? How are they damped ?

[U.P. Technical Univ. Elements of Power System 2009-10]

29. Explain various vibration dampers.

[U.P. Technical Univ. Elements of Power System 2004-05]

30. Write short notes on the following :

(*i*) Mechanical features of overhead line.

[Rajasthan Univ. 2002]

(*ii*) Conductor materials of transmission lines.

[Rajasthan Univ. 2001]

(*iii*) Sag templates and their applications.

[Rajasthan Univ. 2001]

(*iv*) Types of vibrations in conductors.

[Rajasthan Univ. 2003]

(*v*) Vibration and vibration dampers.

[M.D. Univ. May 2008]

Short Answer Type Questions With Answers

Q. 1. What are the various components of an overhead line ?

Ans. The main components of an overhead line are supports/towers, cross arms and clamps, insulators, conductors, guys and stays, lightning arresters, earth wire, V-guards, guard wires, phase plates, bird guards, danger plates, anti-climbing devices, vibration dampers etc.

Q. 2. What is the function of line support ?

Ans. The function of a line support is to support the line conductors and keep them at a safe distance from the ground.

Q. 3. What are the main requirements of line supports ?

Ans. The main requirements of line supports are high mechanical strength, lighter in weight, low maintenance cost, longer life, cheaper in cost, good appearance, easy accessibility for painting and erection of line conductors and immune to damage from lightning and fire.

Q. 4. What are four types of line supports ?

Ans. The four types of line supports are wooden poles, RCC or PCC poles, rail poles and steel structures or towers.

Q. 5. What are the drawbacks of wooden poles ?

Ans. The main drawbacks of wooden poles are comparatively smaller life, poor mechanical strength, tendency to rot below the ground level and requirement of periodical inspection.

Q. 6. How do tangent angle tower and deviation tower differ from each other ? **[Pb. Technical Univ. May 2010]**

Ans. Steel towers can be broadly classified as tangent towers and deviation towers. Tangent towers can be used for straight runs of the line and up to 2° line deviation from the straight run. The line is straight or along the tangent to the line route. In such towers the stress is because of the weight of the conductors, ice and wind loads. In addition extra forces due to break in the line on one side of the tower is also to be considered in the design of towers. The base of such a steel tower may be square or rectangular. Insulators used with such towers are suspension types.

For deviations exceeding 2°, special angle towers, sometimes called the *deviation towers,* are used. They are used where the transmission line changes direction. Such towers have broader base and stronger members as they are to withstand the resultant force due to change in direction in addition to the forces to which the tangent towers are subjected. Insulators used with such towers are of strain type. The cost of deviation tower is comparatively larger than that of a tangent tower because it is designed to withstand heavy loading as compared to standard or tangent tower.

Q. 7. What are the main requirements of good conductor material used for overhead lines ?

Ans. Main requirements of good conductor material used for overhead lines are high electrical conductivity, high tensile strength, low specific gravity, low cost, easy availability, free from brittleness and the material should not react with atmospheric condition.

Q. 8. What for ACSR conductor stands ?

Ans. ACSR conductor stands for aluminium conductor steel reinforced conductor.

Q. 9. Why are conductors for transmission line stranded ?

Ans. All conductors used for overhead transmission lines are stranded so as to increase the flexibility. Solid wires, except of smaller sizes, are difficult to handle when used for long spans and tend to crystallise at the points of supports because of swinging in winds.

Q. 10. Define term 'ACSR'. **[Pb. Technical Univ. Dec. 2006]**

Ans. ACSR conductor consists of a core of galvanised steel strand surrounded by a number of aluminium strands.

Q. 11. Why are ACSR conductors preferred for transmission and distribution lines ?

Ans. ACSR conductors are preferred for transmission and distribution of lines because of the following facts:

(*i*) ACSR conductor being of high tensile strength and lighter in weight produces small sag and therefore longer spans can be used.

Because of use of larger span, the number of line supports may be reduced by about 25%. Thus the overall cost of supports, foundations, insulators and erection is considerably reduced. In addition, there is a substantial saving in maintenance cost. Further, since most faults occur at the line supports, so the frequency of occurrence of faults is automatically reduced.

(*ii*) The ACSR conductor has a largest diameter than any other type of conductor of same resistance, so corona losses are reduced.

(*iii*) Another advantage with ACSR conductor is that, because of reduced corona losses, critical voltage limit of the conductor can be raised by 30 to 50% as compared to copper conductors.

Q. 12. Why is galvanised steel wire not suitable for EHT lines for the purpose of transmitting large amounts of power over a long distance ?

Ans. The galvanised steel wire is not suitable for EHT lines for the purpose of transmitting large amounts of power over a long distance because of poor conductivity, high internal reactance and eddy current and hysteresis.

Q. 13. What is the function of cross arms ?

Ans. The function of cross arms is to keep the line conductors at a safe distance from each other and from the pole.

Q. 14. What are the factors on which conductor spacings and ground clearance depend ?

Ans. Factor affecting conductor spacings is voltage while factors affecting ground clearance are voltage and area (living or non-living, urban or suburban, along the streets or across the streets, river crossings/hill crossings, railway track/bridge etc).

Q. 15. Define term 'sag' in a transmission line.

[Pb. Technical Univ. Dec. 2006]

Ans. The difference in level between the points of line supports and the lowest point on the conductor is known as sag.

Q. 16. What are the factors affecting the sag in a transmission line ?

Ans. Factors affecting the sag in a transmission line are weight of conductor, span© length (distance between the poles), working tensile strength, ice coating, wind pressure and temperature.

Q. 17. Why is it disadvantageous to provide either too high sag or too low sag ?

Ans. It is disadvantageous to provide either too high sag or too low sag owing to the following facts.

In case the sag is too high, more conductor material is required, more weight on the supports is to be supported, higher supports are necessary and there is a chance of greater swing-amplitude due to wind load. On the other hand in the case of too low sag, there is more tension in the conductor and thus the conductor is liable to break if any additional stress is to be taken, such as due to vibration of line or due to fall in temperature.

Q. 18. What is the effect of wind and ice on sag ?

Ans. *Combined Effect of Wind and Ice.* Due to weight of ice deposits on the line, and the wind pressure, the mechanical stress increases in the conductor and, therefore, the line must be designed to withstand these stresses and tensions. Under this condition, the weight of the conductor, together with weight of ice acts vertically downwards while the wind loading w_w acts horizontally, as shown in Fig. 2.21.

Q. 19. What is stringing chart ? **[Pb. Technical Univ. May 2009]**

OR

State significance of stringing chart.

[Pb. Technical Univ. Dec. 2007]

Ans. Stringing chart gives the data for sag and tension to be allowed at a particular temperature. It is very useful while erecting the transmission line conductors for adjusting the sag and tension properly.

Problems

1. An overload line has a span of 200 m, the line conductor weights 0.7 kg/m. Calculate the maximum sag if the maximum allowable tension in the line is 1,400 kg. Prove formula used.

[U.P. Technical Univ. Elements of Power System 2006-07]

[**Ans.** 2.5 m]

2. Find the sag of a transmission line with following particulars :

Span-500 m, Conductor weight per m = 1 kg, Tension 1,500 kg. [**Ans.** 20.833 m]

3. An OH line has a span of 200 metres between level supports. The conductor diameter is 1 cm and weighs 0.65 kg/metre length. The allowable tension is 550 kg. Calculate the maximum sag. [**Ans.** 5.91 m]

4. An overhead line has a span of 150 m between level supports. The conductor diameter is 0.94 cm and weighs 0.62 kg/m length. The allowable tension is 586 kg. Calculate the sag if the wind pressure is 39.2 kg/m^2 of a projected area. [**Ans.** 3.462 m]

5. A 132 kV OH line uses ACSR conductor 19.53 mm diameter, weight 844 kg per km, ultimate strength 7,950 kg. Calculate the height above ground at which the conductors with a span of 300 metres shall be supported, the factor of safety being 2, wind pressure 75 kg per square metre of projected area. Ground clearance required is 7 metres.

[**Ans.** 4.784 m]

6. A transmission line has a span of 200 m between level supports. The conductor has cross-sectional area of 1.29 cm^2, weight 1,170 kg/m and has a breaking stress of 4,218 kg/cm^2. Calculate sag for a safety factor of 5 allowing a wind pressure of 122 kg/m^2 of projected area. What is vertical sag ? **[M.D. Univ. Power System-I May 2009]**

[**Ans.** 8.92 m; 5.38 m]

7. A transmission line conductor having a diameter of 19 mm weighs 0.82 kg/m. The span is 250 m. The wind pressure is 39 kg/m^2 of projected area with ice coating of 12 mm. The ultimate strength of the conductor is 8,000 kg. Calculate the maximum sag if the factor of safety is 2.5 and ice weighs 900 kg/m^3.

[A.M.I.E. Sec. B. Power Systems Winter 1994]

[**Ans.** 6.135 m]

8. An overhead line having a span of 300 m has ACSR conductors of 1.95 cm diameter. The allowable tension is 35,600 N. Find (*i*) sag in still air condition with no ice covering. (*ii*) vertical sag when there is an ice coating of 0.8 cm and a horizontal wind pressure of 400 N/m^2 of projected area. Ice density 8,900 N/m^3. (*iii*) What should be the height of the lowest cross-arm to give a minimum ground clearance of 7.8 m under bad weather conditions ?

Line is carried by insulator string 1.6 m long. Weight of conductor = 0.85 kg/m

[A.M.I.E. Sec. B. Power Systems Summer 1996]

[**Ans.** (*i*) 2.635 m (*ii*) 4.58 m (*iii*) 13.98 m]

9. A transmission line conductor consists of hard drawn copper 240 mm^2 in cross section and has a span of 160 metres, the supporting structures being level. The conductor has an ultimate strength of 42.2 kg/mm^2 and the desired factor of safety is 5. Find the vertical sag with a wind pressure of 1.35 kg/metre length and an ice coating of 1.25 cm. Take the density of hard drawn copper as 8.9 gm/cc and weight of ice as 915 kg/m^3.

[**Agra Univ. Transmission & Distribution of Electrical Power 1977 Supp**]

[**Ans.** 4.22 metres]

10. An overhead conductor consists of 7 strands of phosphor bronze having an ultimate tensile strength of 8,000 kg/cm^2 and an area of 2.2 cm^2 when erected between supports 600 metres apart having 15 metre difference in level. Find the vertical sag which must be allowed so that the factor of safety is 5. Assume the wire to be loaded with 1 kg/m of ice. The wire weighs 2.03 kg/m.

[**Pb. Univ. Elec. Power II Nov., 1974**]

[**Ans.** 16.2 metres from the lower support]

11. An overhead line at a river crossing is supported from two towers at heights of 40 m and 90 m above water level, the horizontal distance between towers is 400 m. If maximum allowable tension is 2,000 kg, find the clearance between conductor and water at point midway between towers. Weight of conductor is 1 kg/m.

[**M.D. Univ. Power System-I, Dec. 2008; Dec. 2009**]

[**Ans.** 55 m]

12. An overhead transmission line at a river crossing is supported, from two towers 600 metres apart and having a level difference of 60 metres. The conductors consist of 7 strands of silicon bronze wire having an ultimate strength of 8,000 kg/cm^2 and an area of 2.2 cm^2. Find the vertical sag which must be allowed so that factor of safety shall be 4 with the wires loaded with 1.0 kg of ice per metre and a wind pressure of 1.75 kg per metre. Specific gravity of silicon bronze is 8.735. Assume the curve to be a parabola.

[**Kurukshetra Univ. Power System-I, April 1977**]

[**Ans.** 69.65 metres]

13. A transmission line conductor at a river crossing is supported from two towers at heights of 40 m and 90 m above water level. The horizontal distance between towers is 275 m. If the tension in the conductors is 1,362 kg. fing the clearance between the conductor and water at a point mid-way between the towers. Weight of conductor per metre run is 0.8 kg. [**Allahabad Univ. Power System-I, 1976 Supp.**]

[**Ans.** 59.46 m]

14. An overhead transmission line spanning over a river is supported at two ends by towers at 45 m and 75 m. Find the clearance between the conductor and water at a point midway between the towers. The tension is to be limited to 2,500 kg. The weight of the conductor is 0.9 kg/m and the distance between the towers is 300 metres.

[**A.M.I.E. Sec. B. Power Systems Summer 1994**]

[**Ans.** 55.95 m]

15. A transmission line conductor at a river crossing is supported from two towers at heights of 30 m and 90 m above water level. The horizontal distance between the towers is 270 m. If the tension in the conductor is 1,800 kgf and the conductor weighs 1.0 kgf per metre, find the clerance between the conductor and the water at a point midway between the towers. Assume parabolic configurations.

[**Rajasthan Technical Univ. Transmission & Distribution of Electrical Power, 2009**]

[**Ans.** 54.95 m]

16. A hilly terrain has a gradient of 1 in 20 and two towers of 36 m height are considered with a horizontal distance 300 m. An overhead line is supported by these towers wherein the lowest conductor is fixed 6 m below the top of each tower. The weight of each line conductor is 1.5 kgf/m and the tension 2,200 kgf. Assume parabolic configuration of sag and deduce and develop a method to determine the clearance of the lowest point of the conductor from the ground and the minimum clearance.

[**A.M.I.E. Sec. B. Power Systems Winter 1997**]

[**Ans.** 24.17 m, 22.337 m]

❑❑❑

CHAPTER 3

Overhead Line Insulators

INSIDE THIS CHAPTER

3.1. INTRODUCTION

The overhead line conductors are bare and not covered with any insulating covering/coating. The line conductors are, therefore, secured to the supporting structures by means of insulating fixtures, called the *insulators*, in order that there is no current leakage to the earth through the supports. Insulators are mounted on the cross arms and the line conductors are attached to the insulators so as to provide the conductors proper insulation and also provide necessary clearances between conductors and metalwork. The insulators must provide proper insulation and necessary clearances against the highest voltage in worst atmospheric conditions to which the line is likely to be subjected. The insulators also prevent short circuiting between the different phase conductors and provide necessary mechanical support for the line conductors. Thus insulator is undoubtedly one of the most important and vulnerable links in transmission and distribution practice and, therefore, proper selection is of utmost importance for the successful operation of overhead transmission and distribution system.

The important properties that an overhead line insulator must posses are:

1. *High mechanical strength* so as to bear the load due to the weight of line conductors, wind force and ice loading if any.
2. *High relative permittivity* so as to provide high dielectric strength.
3. *High insulation resistance* in order to prevent leakage of currents to earth.
4. *High ratio of rupture strength to flash-over voltage.*
5. *Ability to withstand large temperature variations i.e.*, it should not crack when subjected to high temperatures during summer and low temperatures during winter. The dielectric strength should remain unaffected under different conditions of temperature and pressure.

The material used should not be porous and should be imprevious to the fluids and gases in the atmosphere. Simultaneously it should be free from internal impurities and cracks etc as these lowers the value of dielectric strength to a great extent.

3.2. INSULATOR MATERIALS

The material most commonly used for overhead line insulators is porcelain but toughened glass, steatite and special composition materials are also used to a limited extent.

Porcelain is produced by firing at a controlled temperature a mixture of kaolin, feldspar and quartz. It is mechanically stronger than glass. It gives less trouble from leakage, and is less susceptible to temperature variations and its surface is not affected by dirt deposits. On the other hand, it is not so homogeneous as glass, owing to the fact that each component shell of a porcelain insulator is glazed during manufacturing process and its satisfactory performance in service depends to a considerable extent on the preservation of this glaze which is only of the order of 25 microns in thickness. Also fault cannot be detected easily as it is not transparent. In tension this material is usually weak and does not withstand tensile stresses exceeding 5 kg/mm^2. The dielectric strength and compressive strength of a mechanically sound porcelain insulators are about 6.5 kV/mm of its thickness and 700 kg/mm^2 respectively. Normally, it is difficult to manufacture homogeneous porcelain in the thickness required for some types of insulators and, therefore, for a particular operating voltage, a two or more piece construction is adopted in which each piece is fired and glazed separately and then they are cemented together.

If this insulating material is manufactured at lower temperature, its mechanical properties improve but the material remains porous and on putting it in service it may deteriorate. If this material is manufactured at high temperature, its porosity is reduced but the material becomes brittle. The porosity of this insulating material also reduces its dielectric strength, also any impurity or air bubble left within the material results in a lower dielectric strength. So a compromise is made between the mechanical strength and the porosity of the material and a suitable temperature of the kiln is designed.

Glass is cheaper than porcelain in the simpler shapes and if properly toughened and annealed gives high resistivity and dielectric strength (14 kV per mm of thickness of the material). Owing to high dielectric strength, the glass insulators have simpler design and even one piece design can be used. Glass is quite homogeneous material and can withstand higher compressive stresses as compared to porcelain. It has also a lower coefficient of thermal expansion which minimises the strains due to temperature changes and owing to its transparent nature flaws in the material can be readily detected by visual examination. The main disadvantage of glass is that moisture more readily condenses on its surface and facilitates the accumulation of dirt deposits, thus giving a high surface leakage. Also in large sizes the great mass of material combined with the irregular shape, may result in internal strains after cooling. Glass insulator however, can be used up to 25 kV under ordinary atmospheric conditions and well up to 50 kV in dry atmosphere.

Steatite is a naturally occurring magnesium silicate, usually found combined with oxides in varying proportions. It has a much higher tensile and bending stress than porcelain and can advantageously be used at tension towers or when the transmission line takes a sharp turn.

The **special artificial material** is used in insulators for low voltages and have an important advantage that these can be easily moulded into any shape without any internal stress. Metallic fittings can also be firmly embedded in the material without any difficulty. The disadvantage of insulators made from special artificial material is that they deteriorate rapidly in bad climatic conditions and on being subjected to flash-over their carbonised surface forms a conducting path.

3.3. FACTORS INVOLVED IN INSULATOR DESIGN

Insulators are required to withstand both electrical and mechanical stresses. In addition, the surface leakage path must have sufficiently high resistance so as to avoid any current leakage to earth. The design of the insulators must be such that the stress developed owing to contraction and expansion in any part of the insulator does not lead to any defect.

Electrical breakdown may occur either by flash-over or puncture. In a flash-over, an arc occurs between the line conductor and earth (*i.e.,* supporting pin of the insulator) and the discharge jumps across the air gaps in its path. In a puncture, the discharge occurs from conductor to pin through the body of the insulator. In case of electrical breakdown due to flash-over, the insulator continues to act in its proper capacity after the event unless fractured by the heat of the arc, but after a puncture, it is permanently damaged due to excessive heat. Thus it is very important to provide sufficient thickness of porcelain in the insulator to resist puncture by the combined effect of the line voltage and any probable transient voltage rise on the line. The ratio of puncture strength to flash-over voltage, termed as the *factor of safety*, must be high so as to provide a good margin for the protection of insulators from complete failure.

It is desirable that porcelain may not come in direct contact with a hard metal screw thread. Normally cement is used between metal and the porcelain. The cement so used must not cause any fracture by expansion or contraction.

3.4. TYPES OF INSULATORS

Various types of insulators used for overhead transmission and distribution lines are described below.

1. Pin Type Insulators. This type was amongst the earliest designs, and used for supporting line conductors. It provides the most economic, simple and efficient method of conductor and bus-bar support for voltages up to and including 33 kV. Modern pin type insulators are very reliable and inherent cracks in porcelain are very rare and never occur with toughened glass insulators. The life of modern porcelain insulators is relatively long (expected to be about 50 years).

The pin type insulator is designed to be mounted on a pin which in turn is secured to the cross arm of the pole. The insulator is screwed on the pin and the line conductor is placed in the groove at the top of the insulator and is tied down with soft copper or soft aluminium binding wire according to the conductor material. To avoid a direct contact between the porcelain and the metal pin, a soft metal (usually lead) thimble is employed. The insulator and its pin should be sufficiently mechanically stronger to withstand the resultant force due to combined effect of the weight of the conductor, wind pressure and ice loading, if any, per span length.

Fig. 3.1. *Pin Type Insulator*

For lower voltages generally one piece type of insulators are used. For high-voltage transmission lines, stronger pin type insulators are used. The high-voltage pin type insulators differ in construction from low-voltage type in that they consist of two or three pieces of porcelain cemented together. These pieces form what we call *petticoats* or *rain sheds*. The multiple shells or petticoats are provided in order to have an adequate length of leakage path so that the flash-over voltage between the line conductor and the insulator pin is increased. The petticoats or rain sheds are so designed, that even when the outer surface is wet due to rain, sufficient leakage resistance is still provided by the inner dry surfaces. It is desirable that the horizontal distance between the tip of the lower most shell is less in comparison to the vertical distance between the same tip and the cross arm, otherwise in case of an arc-over, the discharge will take place between the line conductor and cross arm rather than line conductor and the pin of the insulator, thereby, the cross arm will have to be replaced rather than the insulator.

Many engineers have a preference for one piece insulators up to and including 11 kV and use the multi-piece type for high voltages while others prefer multi-piece for all voltages. One-piece insulators are more prone to cracking, because of rain, and, therefore, their use may cause interruption of supply during the very first rain after the lines is commissioned. Multi-piece insulators have an advantage that a defect on one shed does not seriously affect the mechanical strength of the insulator, and it generally functions electrically at normal voltage until the defective unit is traced and replaced. For rural 11 kV lines, carrying less load and therefore, using light conductors, single piece insulators are employed for reasons of economy. For urban 11 kV feeders, specially feeding industrial loads, multi-piece insulators are preferred, being more reliable.

Two one-piece insulators are shown in Figures 3.2 (*b*) and 3.2 (*c*). Leakage paths are shown by dotted lines in the figure.

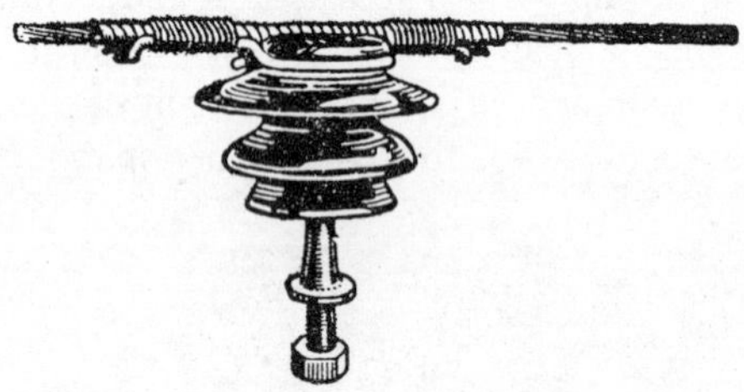

(a) Pin Type Insulator, With Conductor Made Fast By Binding

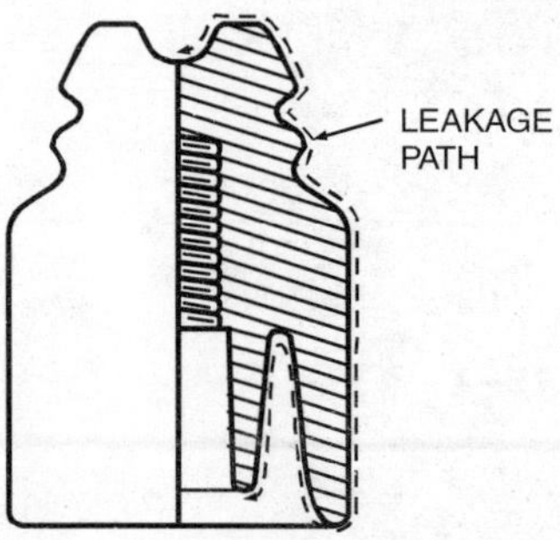

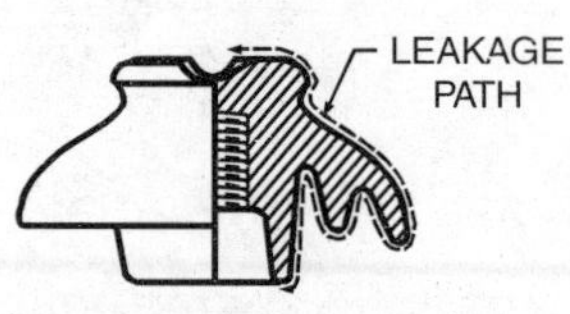

(b) For 11,000 V *(c) For 25,000 V*

Fig. 3.2. ***One-piece Pin Type Insulator***

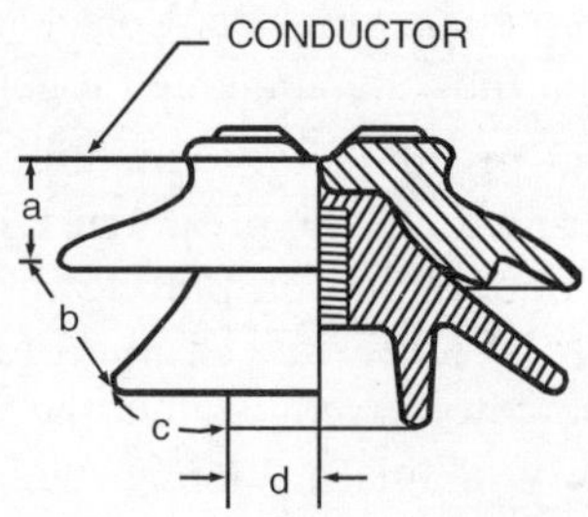

Fig. 3.3(*a*). ***33 kV Pin Type Insulator***

Two multi-part insulators for use on 33 kV and 66 kV are shown in Figs. 3.3 (*a*) and 3.3 (*b*) respectively. The flash-over distances are also shown when the insulators are wet and dry in Figs. 3.3 (*a*) and 3.3 (*b*) respectively.

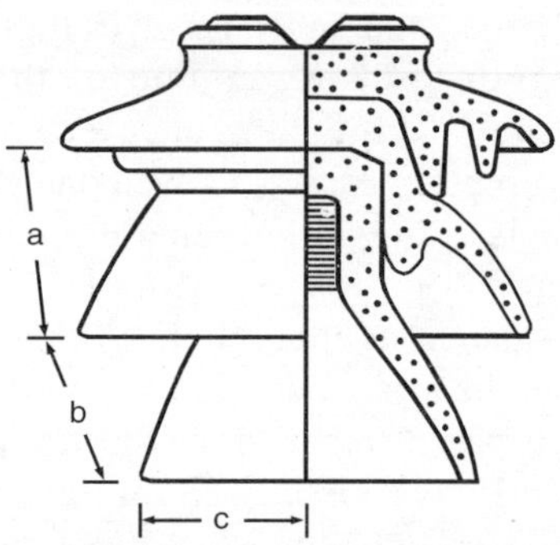

Fig. 3.3(*b*). ***66 kV Pin Type Insulator***

Flash-over distances for dry insulators, as shown in Fig. 3.3 (*a*),

$$= a + b + c + d$$

Flash-over distances for wet insulators, as shown in Fig. 3.3 (*b*),

$$= a + b + c$$

Thus it is seen that flash-over distance in case of wet insulators is less as compared to that in case of dry insulators. The rain sheds are, therefore, made to keep the inner side of the insulator dry.

The main advantage of the pin type insulator is that it is cheaper. In many cases one pin insulator can do the work of two suspension insulators. Secondly the pin insulator requires a shorter pole to give the same conductor clearance above the ground since the pin insulator raises the conductor above the cross arm while the suspension insulator suspends it below the cross arm. However, its use beyond operating voltage of 50,000 volts become uneconomical. Pin type insulators become very bulky and cumbersome when designed for the higher voltages, and owing to the larger length of the supporting pin the bending moment near the point of attachment to the cross-arm tends to become excessive. The modern practice is not to use pin type insulators beyond 33 kV.

For these the ratio of the spark-over voltage to the working voltage (usually known as safety factor) must be high for low voltage than for high voltage. This is about 10, but these are so designed that spark-over takes place before these get punctured. Surface leakage current is due to the accumulation of dirt and inorder to reduce its value the insulators are given a long leakage path by providing two or three petticoats or sheds. Insulators of this type are used on intermediate poles on straight run. These do not take tension.

2. Suspension Type Insulators. With the increase in operating voltage, the insulation required increases. Transmission lines use extremely high voltages, 765 kV, for example. At these voltages the pin type insulators become bulky, cumbersome and costly. Besides, the pin which must hold it would have to be inordinately long and large. In order to meet the problem of insulators for these high voltages, the suspension insulator was developed.

Suspension type insulators consist of a number of porcelain discs flexibly connected in series by metal links in the form of a string. The suspension insulator hangs from the cross arm of the supporting structure and the line conductor is attached to its lower end. Because there is no pin problem, we can put any distance between the cross arm and the conductor just by adding more insulators to the "string". The entire unit of suspension insulators is called a *string*. How many insulators the string consists of depends upon the working voltage, the weather conditions, the type of transmission construction, and the size of insulator used. It is noteworthy that in a string of suspension insulators one or more insulators can be replaced without replacing the whole string.

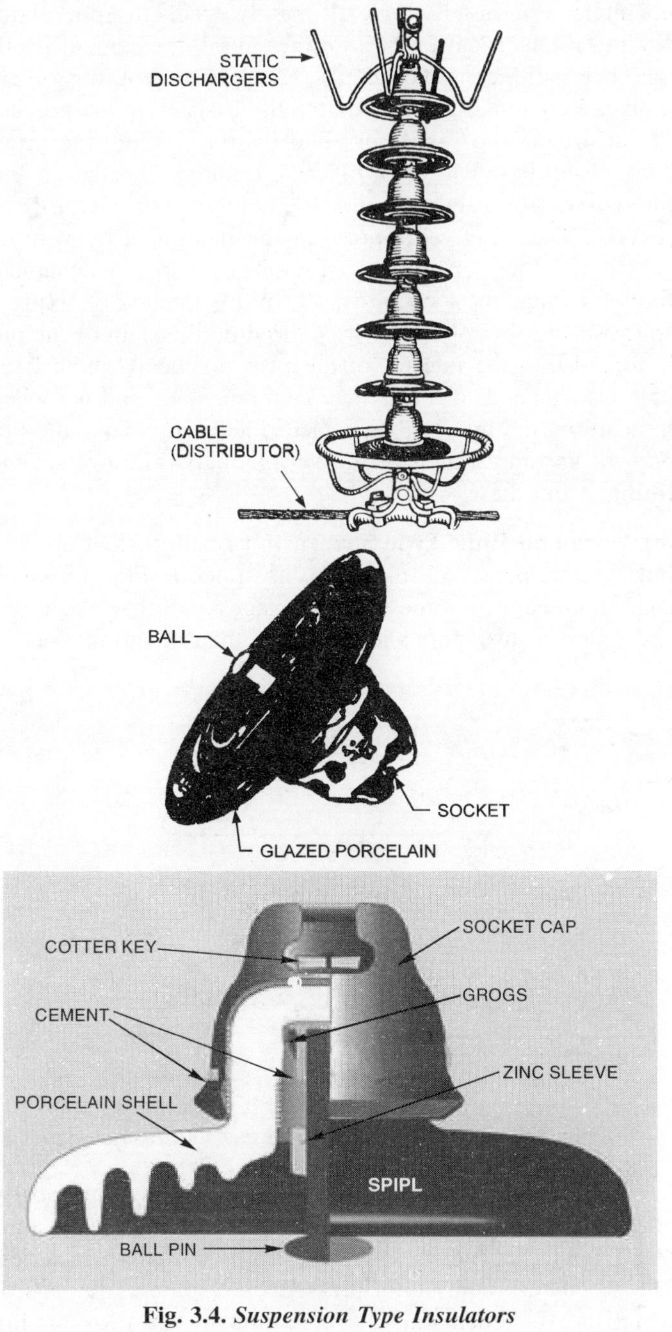

Fig. 3.4. ***Suspension Type Insulators***

The main advantages of suspension type insulators over pin type ones are enumerated below:

(*i*) Suspension type insulators are usually cheaper in cost for operating voltages above 50,000 volts.

(*ii*) Each unit of suspension type insulators is designed for comparatively low voltage (about 11,000 volts) and can be used by connecting them in series, the number depending upon the working voltage.

(*iii*) In the event of failure of an insulator, one unit, instead of the whole string, can be replaced.

(*iv*) Suspension type insulators give more flexibility to the line and mechanical stresses are reduced in this arrangement. The connection at the cross arm is such that the insulator string is free to swing in any direction, and thus takes up a position where it experiences only a pure tensile stress.

(*v*) The suspension type insulators, when used in conjunction with steel supporting structures, has the advantage of rendering the conductor less liable to be affected by lightning disturbances. At every point of support the wire is hung below the earthed cross arm, thus enabling the tower to function as a lightning rod.

(*vi*) In case of rapid increase in load on transmission line, the increased demand can be met by raising the line voltage than to provide another set of conductors. With suspension type insulators, additional line insulation required can be obtained easily by adding one or more discs to the string.

(*vii*) In case of long spans (river or valley crossings) where heavy conductor load is to be sustained, two disc insulator strings can be yoked (Fig. 3.5). Such an arrangement is not possible with pin type insulators.

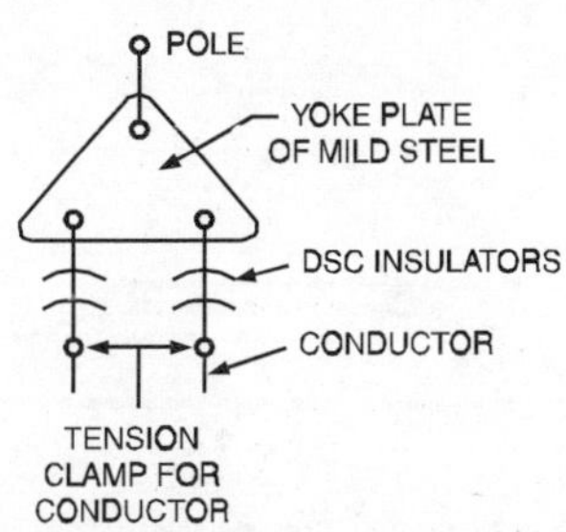

Fig. 3.5. ***Yoked Insulator Strings (Symbolic Figure)***

The disadvantage of suspension type insulators is that large spacings between conductors are required than with pin type insulators due to large amplitude of the swing of the conductors, but this is not a serious disadvantage.

There are three types of suspension insulators generally used namely, Hewlett or inter linking type, cemented-cap type and core and link type.

(*a*) Hewlett or Interlinking Type Suspension Insulators. The Hewlett suspension insulator is shown in Fig. 3.6 and is one of the earliest design. Each disc consists only of one piece of porcelain, the central bulbous portion of which is provided with two curved tunnels lying in planes at right angles to each other. The short steel strips forming the connection between individual discs are threaded through these tunnels and thus loop through each other, being separated by a layer of porcelain which is totally in compression.

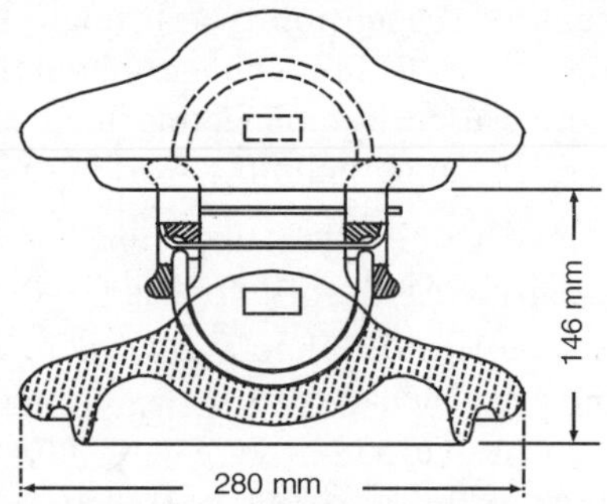

Fig. 3.6. ***Hewlett Type Suspension Insulator***

The main advantages of Hewlett type suspension insulators are:

(*i*) Simple in design.

(*ii*) High mechanical strength since the porcelain in between the two tunnels is under compression only.

(*iii*) No risk of breakage owing to the difference in expansion or contraction of the connecting links and the insulating material.

(*iv*) No risk of interruption to the service in case the porcelain between the links gets accidently broken, since the links keep the other units held together.

However, the Hewlett insulator appears to be rather more liable to puncture than other types of suspension insulators, owing to the high electrostatic stress in the material between the links.

(*b*) Cemented-Cap Type Suspension Insulators. Cemented-cap type suspension insulator shown in Fig. 3.7 is the most commonly used type and consists of a single disc-shaped piece of porcelain grooved on the under surface to increase the surface leakage path, and to a metal cap at the top, and to a metal pin underneath. The cap is recessed so as to take the pin of another unit and thus build up a string of any number of units. The cap is secured to the insulator by means of cement. A very uniform distribution of the electrostatic stress in the material between the connecting links is obtained in such insulators. The main drawback of this type of insulators has been that coefficients of cubical expansion of the three materials–porcelain, cement and steel are different and no provision is made for their expansion and the sudden temperature changes occurring in service are sufficient to set up internal stresses which ultimately crack the porcelain, leading to electrical failure. Further more, the cement itself, which is subject to volumetric changes depending on its moisture content, has often materially assisted in the process of failure of the insulator. The insulator manufacturers recognised these causes of ultimate failure of insulators and improved their designs from the point of view of reliability of service. One way of improving the design is by way of substituting the cementing of the pin by purely mechanical fixing, such as the "spring-ring". In this method of fixing a spiral spring ring of steel wire carried on the stem of the pin is forced into the interior of the insulator head which is of bulbous shape. The ring immediately expands and is locked in position, the interior being then filled with a lead alloy to prevent any movement of the various parts and protect the fittings from the weather.

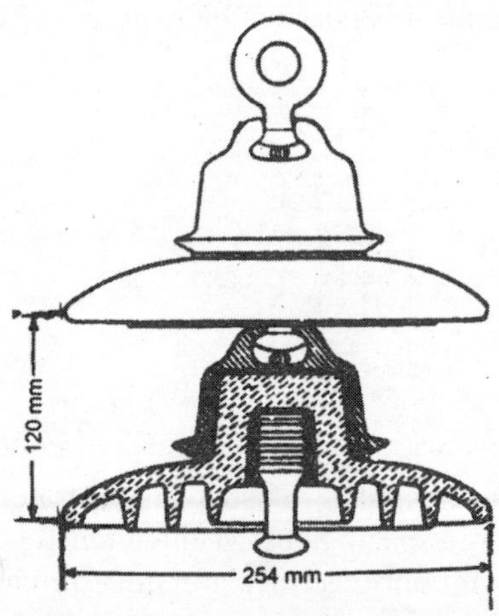

Fig. 3.7. ***Cemented Cap Type Suspension Insulator***

(*c*) Core and Link Type Suspension Insulators. Core and link type suspension insulator is illustrated in Fig. 3.8. Such an arrangement combines the advantages of both of the above two types of insulators and overcome their disadvantages.

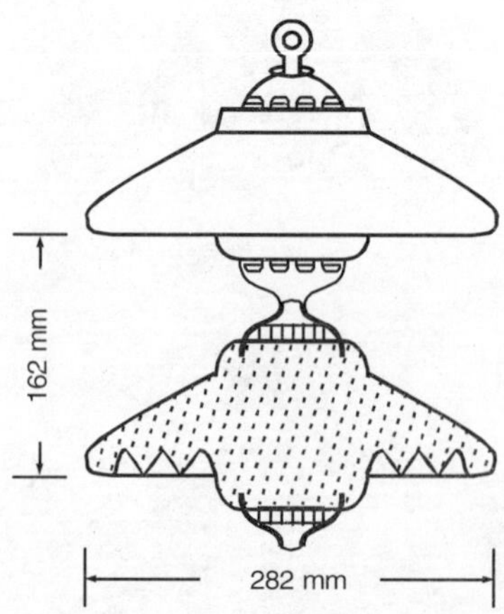

Fig. 3.8. ***Core and Link Type Suspension Insulator***

In such a construction, each insulator disc is symmetrically placed and it conforms to the electrostatic lines of force, thus avoiding placing materials of different permittivities in series. The metalwork consists of pressed steel spiders, the legs of which are fastened into the porcelain by an alloy having approximately the same coefficient of cubical expansion as the porcelain. Thus, high mechanical stresses on the porcelain, whether due to sudden temperature variations or to the employment of cement are completely eliminated. This is the recent type of construction and it allows discs to be formed out of quite thick porcelain thereby allowing the disc to be of one piece only. Added to the other advantages such an insulator has a high puncture strength.

3. Strain Insulators. Where there is a dead end of the line, or there is a corner or a sharp curve, or the line crosses river etc, the line is subjected to greater tension. Pin type insulators cannot be used in such situations because they cannot take conductor load in tension which often occurs in such situations. For low-voltage lines (say, up to 11,000 V), shackle insulators can be used, but for higher voltage transmission lines strain insulators consisting of an assembly of suspension type insulators are used, as illustrated in Figs. 3.9(*b*) and 3.10. Because of its peculiarly important job, a strain insulator must have considerable strength as well as necessary dielectric properties. Where the tension is exceedingly high, as at long river spans, two, three, or even four strings of insulators in parallel are used. The discs of strain insulators are employed in vertical plane where as the suspension insulators are used in horizontal plane.

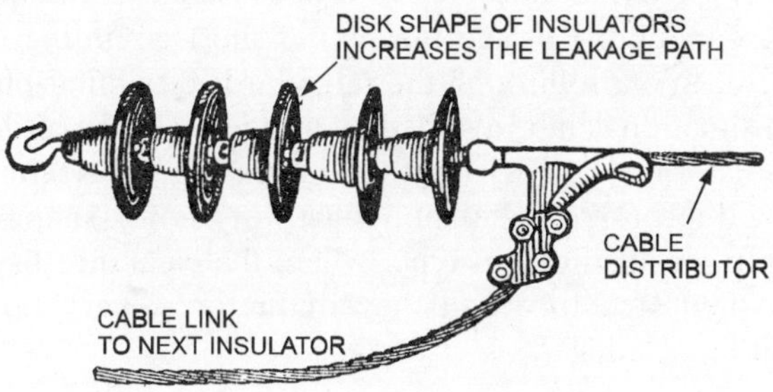

(*a*)

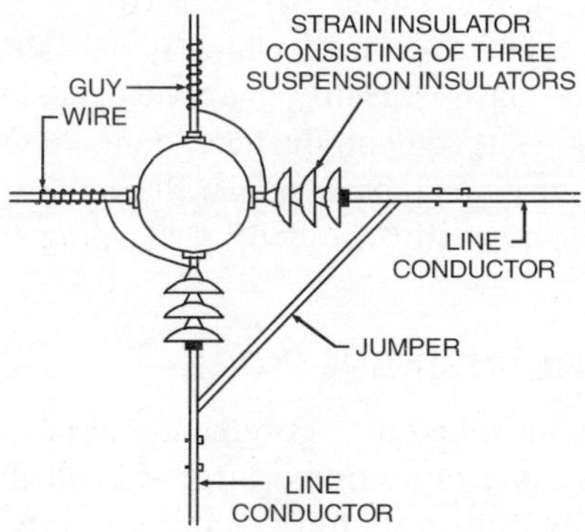

(*b*)

Fig. 3.9

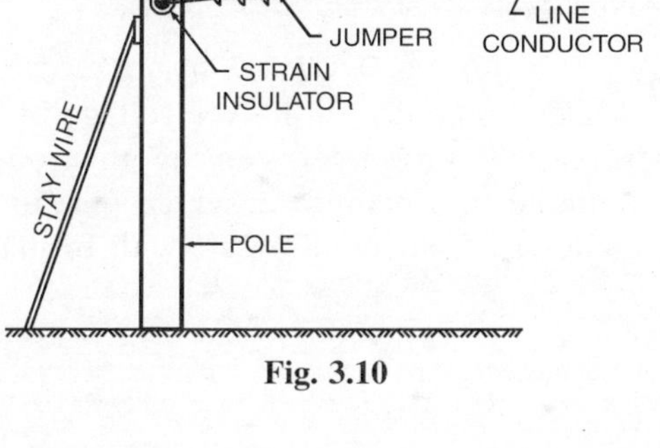

Fig. 3.10

4. Post Insulators. Such insulators are employed for supporting bus-bars and isolating switches etc. A post insulator is similar to a pin type insulator but has a metal base with a metal cap so that more than one unit can be mounted in series. This is mostly a solid core insulator made of resin. Porcelain is also used.

5. Shackle Insulators. The shackle insulators or spool insulators are almost universally used on low voltage lines and provide a very neat, efficient and economical arrangement. Its construction is shown is Fig. 3.11. Every insulator is coated with an extremely hard, smooth glaze, that reduces accumulation of surface deposits. The surface can be easily cleaned and it will not crack when subjected to temperature changes. The wet flash-over and dry flash-over voltages for shackle insulators are 10 kV and 25 kV respectively while the puncture voltage is about 35 kV. Its operating voltage is 1,000 V. Its weight, transverse mechanical load and total creepage distance are 0.5 kg, 1,150 kg and 63 mm respectively. The tapered hole in the shackle insulator distributes the load more evenly and reduces the possibility of breakage when heavily loaded.

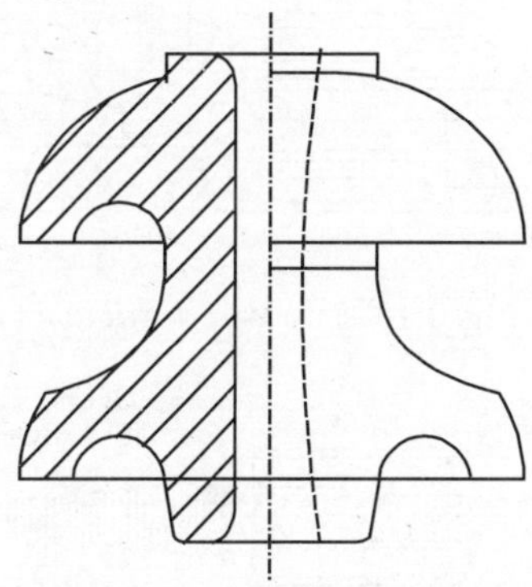

Fig. 3.11. ***Shackle Insulator***

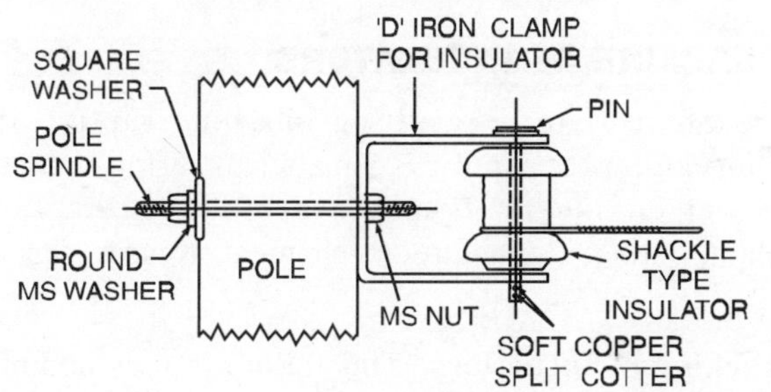

(*a*)

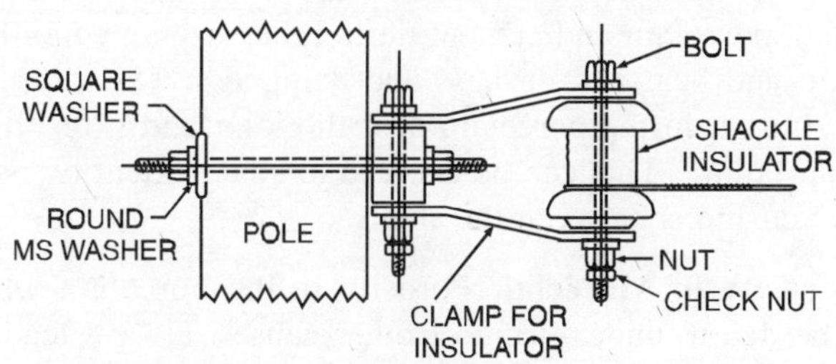

(*b*)

Fig. 3.12. ***Shackle Type Insulators Fitted To Poles***

Shackle insulators may either be mounted horizontally or vertically, and the conductors are fixed in the grooves by means of soft copper or aluminium binding wire according to the conductor material. They can be directly fixed to the pole with a bolt or to the cross arm. The insulators are bellmounted to prevent water being held in contact with the spindle. This type of insulator is used at all positions, either intermediate, terminal or angle. Where the angle exceeds 60° deviation they are generally used in conjunction with shackle straps.

6. Stay Insulators. Such insulators are of egg shape, also called *strain* or *guy* insulators, and are used in guy cables, where it is necessary to insulate the lower part of the guy cable from the pole for the safety of people and animals on the ground. This type of insulator consists of a porcelain piece pierced with two holes at right angles to each other through which two ends of the guy wires are looped. This is illustrated in Fig. 3.14. This keeps the porcelain between the two loops in compression and the guy wire remains in position even if the insulator breaks due to any reason. These insulators are provided at a height of about 3 m from the ground level. The size of insulator (small or large) used depends upon the tensile strength of stay wire.

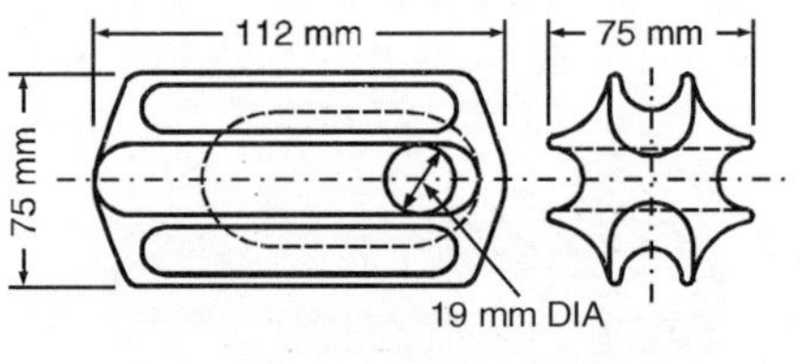

Fig. 3.13. *Stay Wire Insulator*

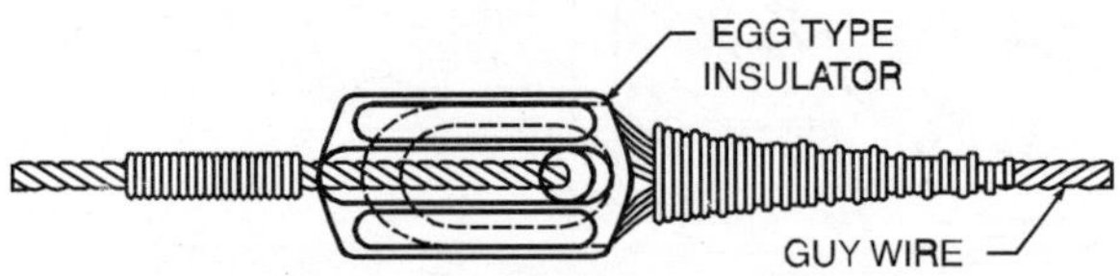

Fig. 3.14

3.5. FAILURE OF INSULATORS

Failure of insulators, resulting in a interruption in the continuity of power supply, is a matter of serious concern to the power engineers. It may occur due to cracking of porcelain, porosity, puncture, mechanical stresses, flash-over etc.

1. Cracking of Insulators. This failure is very common in case of pin type insulators and the cemented-cap type suspension insulators. This occurs due to unequal expansion of steel, porcelain and cement during the varying conditions of cold and heat and dryness and dampness. This develops high stresses in the porcelain near the joint and it results in tension failure. This can be avoided to some extent by using elastic cushions between the shells.

2. Porosity of Material. Porosity in the porcelain which may be due to under-firing or other causes, always leads to failure after a comparatively short period of service. The pores usually absorb moisture from the atmosphere or the cement, thereby decreasing the insulation resistivity of the material. This gives rise to leakage current flowing through the porcelain, resulting in a gradual rise in the temperature until the porcelain is punctured. Such a failure can be avoided by glazing the insulator, to some extent.

3. Improper Vitrification. This is another cause of the puncture of the material and it can be avoided by carrying out suitable routine tests during the course of manufacture.

4. Flash-Over. The most common cause of the insulator failure is the flash-over, that causes unequal expansion of the porcelain thereby shattering the insulator with big cracks and causing interruption of the supply. This can be avoided by providing arcing horns or rings which take up the arc and divert it away from the insulators.

5. Mechanical Stresses. Although the compressive strength of the porcelain is quite high, yet its tensile strength is not adequate and the insulator is always weak in tension and usually fails in that fashion. Such a failure is very rare because defective pieces are weeded out in the routine factory tests.

6. Short Circuits. In the case of pin insulators, the birdage is very common. Birdage means the short circuiting of the conductor to earth through the large birds or similar objects. Such a situation can be avoided by providing bird gaurds near the insulator on the cross arm, by increasing the clearance of the conductor from earthed parts or using suspension insulators, instead of pin type, where the clearance between the earthed structure and the conductor is very large to warrant any birdage.

7. Deposition of Dust. If the insulator material is not properly glazed, the water will stick over it resulting into deposition of dust etc. over it which is partially conducting and reduces the flash-over distance. The deposits of dust and like matter *e.g.* salt, cement, dust etc. on the interior surfaces can actually cause much havoc under the condition of fog and mist. This can be easily avoided by cleaning the insulators periodically.

3.6. TESTING OF INSULATORS

An insulator should have good mechanical strength to withstand the load conditions, large dielectric strength to withstand the normal operating voltage and usual overvoltages*, but it should be free from any pores or voids that may lead to breakdown.

According to ISS three distinct tests namely flash-over tests, performance tests and routine tests, are to be conducted.

1. Flash-Over Tests

(a) Power Frequency Dry Flash-over Test. This test is performed to determine dry flash-over voltage by applying gradually increasing voltage between the electrodes of the insulator mounted in a manner in which it is to be used, till the breakdown of surrounding air with formation of a

* The insulator should flash-over at such voltages at which other equipment is likely to get damaged.

sustained arc occurs. The test is repeated 5 or 10 times. For 5 readings, the means is taken as the correct flash-over voltage. In case of 10 readings, commonly used method, the median can be had by arranging 10 values in ascending order and taking the mean of 5th and 6th reading. The voltage so obtained is known as *50% dry flash-over* voltage.

The above test can also be performed by applying the same voltage to 10 insulators together. The spacing between the insulators should be at least 4 times the length of each. All the insulators may not flash-over simultaneously. The voltage till which five insulators have flashed-over is called the *50% power frequency* dry flash-over voltage. The flash-over voltage so obtained should not be less than that given in IS : 731 – 1971.

The voltage may be applied constantly and the voltage which causes flash-over after one minute is called *one minute dry flash-over voltage.*

(b) Power Frequency Wet Flash-Over Test. In this test also the insulator is mounted in a manner in which it is to be used and the voltage is applied gradually but in addition to this the insulator is sprayed with water. The standard precipitation conditions are 3 mm per minute at an angle of 45°, the water having a resistance of 10,000 Ω per cm cube at normal atmospheric temperature and pressure. If the resistivity of pure water is more, it may be reduced by adding small quantities of a mineral salt or an inorganic acid. This voltage should not be less than that given in IS : 731 – 1971.

(c) Impulse-Frequency Flash-Over Test. For this test the generator developing lightning voltage is used. It develops very high voltages at frequency of several hundred kHz. Such a voltage is applied to the insulator and the spark-over voltage is noted. The ratio of impulse spark-over voltage to spark-over voltage at power frequency is called the *impulse ratio i.e.,*

$$\text{Impulse ratio} = \frac{\text{Impulse spark-over voltage}}{\text{Spark-over voltage at power frequency}}$$

Minimum acceptable flash-over voltages are given in standards.

2. Performance Tests.

(a) Puncture Test. This test is performed to determine puncture voltage. In this test the insulator is totally immersed in an insulating oil and is subjected to the voltage being gradually increased till the puncture takes place. The voltage at which conduction begins is called the *puncture voltage.* The puncture voltage for strings should be at least 30% higher than the power frequency wet flash-over voltage and for pin insulators it should not be less than that given in IS : 731 – 1971.

As an acceptance test the voltage can be rapidly increased to the value given in standards for pin insulators and 1.3 times the wet flash-over voltage for string insulators. No conduction should take place.

(b) Porosity Test. For performance of porosity test, sample from every batch of freshly fired insulators are taken. The specimens are broken into pieces and immersed in a 1% solution of fuchsine dye in alcohal under pressure of 150 kg/cm^2. After sufficient time (the product of test duration in hours and pressure in kg/cm^2 should not be less than 1,800) the pieces are removed from the testing pot and examined. The slightest degree of porosity is indicated by a deep penetration of the dye into the body of specimen.

(c) Temperature Cycle Test. For conducting this test the insulators are quickly and completely immersed in a water-bath maintained at a temperature of 70ºC above that of the cold water and left submerged for a period of T* minutes. They then are withdrawn and quickly and completely immersed without being placed in an intermediate container, in a bath of cold water for the same period of T minutes. This heating and cooling cycle is repeated 4-5 times and the insulators are then dried to observe any cracks in the glazing.

(d) Electromechanical Test. This test is performed with suspension type insulators only and in this test the insulators are subjected to a power-frequency voltage equal to 75 per cent of their dry flash-over voltage and simultaneously to a tensile load equal to one half of the specified minimum failing load.

The tensile load is then steadily increased at a rate per minute of about 20 per cent of the specified minimum failing load, until puncture or breakage occurs. Puncture or breakage should not occur at less than the specified minimum failing load.

(e) Mechanical Test. This test is performed on pin type insulators to determine their ultimate mechanical strength. During performance of this test the pin type insulator is mounted on a rigidly fixed pin capable of withstanding, without appreciable deformation, the loads to which it is subjected during test. The insulator is subjected to a load equal to one half of the specified minimum failing load applied perpendicular to its axis in the plane of the side groove by means of a wire rope encircling the groove. The load is then increased at a rate per minute of about 20% of the specified minimum failing load until breakage occurs. Breakage should not occur at a load less than the specified minimum failing load.

3. Routine Tests

(a) Proof Load or Mechanical Test. In this test all types of insulators (pin type or suspension type) are assembled and then subjected to tensile load which is 20-25% in excess of the normal tensile load which the insulators are likely to encounter in actual operation. This load is applied for a duration of 1 minute only.

(b) Corrosion Test. In this test the insulator with its fittings is suspended in a copper sulphate solution at 15.6ºC for one minute. Then the insulator is removed, wiped, cleaned and again put in copper sulphate solution. This is repeated

* The period T in minutes is given as T = (15 + 0.7 M), where M is the mass of the insulator in kg.

four times. Now on examination there should not be any deposition of metal over it.

(*c*) *High Voltage Test.* For this test the pin insulators are inverted and are put into water up to the neck. The water is also put into the spindle hole. Then high voltage is applied for 5 minutes. There should be no damage to the insulator after carrying out this test.

For detailed description of the tests mentioned above ISS may be referred too.

3.7. DISTRIBUTION OF POTENTIAL OVER A STRING OF SUSPENSION INSULATORS

For overhead lines operating at high voltages (33 kV and above) use of number of discs connected in series, through metal links, is made. The whole unit formed by connecting a number of discs in series is known as *string of insulators.* The line conductor is secured to the bottom disc of the string and the top disc is connected to the cross arm of the pole or tower, as illustrated in Fig. 3.15. The number of discs connected in series in an insulator string depends upon the line operating voltage (higher the line operating voltage, the larger is the number of discs required for the insulator string, as given below in tabular form).

Table 3.1. ***Minimum Number of Insulator Discs Required For Transmission Lines***

Operating Voltage in kV	Number of Discs in	
	Suspension Assembly	Tension or dead-end Assembly
11	1	1
33	2	3
66	5	6
132	9	10
220	14	15
400	21	22

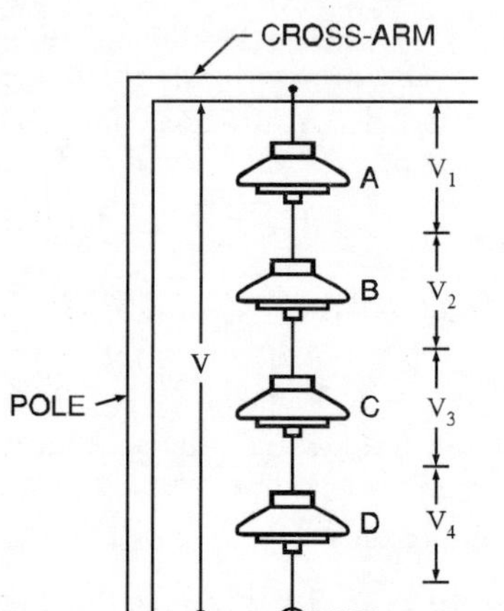

Fig. 3.15. ***Arrangement of Insulator String***

The number of discs indicated in above Table 3.1 is actually the usual number used. However, in the case of transmission lines operating at 66 kV or more, one disc less than the number indicated in Table 3.1 is used on about eight suspension structures near the substation. This is accomplished so that in the event of a lightning surge appearing on the line, the insulator string will flash-over and prevent the surge from travelling to the substation thus safeguarding the equipment there.

It is found that the voltage impressed on a string of suspension insulators (the voltage applied between the line conductor and earth) does not distribute itself uniformly across the individual discs. The line unit (unit nearest the line conductor) has the maximum value across it, the figure progressively decreasing as the unit nearest the cross arm is approached. The inequality of voltage distribution between individual units is all the more pronounced with a larger number of insulator units. This fact may be explained with the help of equivalent circuit of an insulator string (Fig. 3.16)

Each string insulator unit behaves like a capacitor having a dielectric medium between the two metallic parts (viz pin and cap), The capacitance due to two metal fittings on either side of an insulator is known as *mutual capacitance.* Further there is also a capacitance between metal fitting of each unit and the earthed pole or tower. The capacitance so formed is known as *shunt capacitance.* If a string of similar suspension insulators could be situated so far from neighbouring metalwork that the capacitance between this metalwork and the metal fitting of the insulators (*i.e.,* shunt capacitance) would be negligibly small in comparison with the capacitance of each unit (*i.e.,* mutual capacitance), then the charging current would have been the same through all of the discs, the discs being connected in series, and consequently the voltage across individual units would have been the same *i.e*., applied voltage V divided by the number of units in the string. However, in practice this condition is not fulfilled because of nearness of the tower, the cross arm, and the line. These shunt capacitances, sometimes called the *stray capacitances,* have an important effect on the voltage distribution between the units. Due to shunt capacitance, charging current is not the same through all the discs of the string (Fig. 3.16). So voltage across individual units, being directly proportional to the current flowing through them, will be different. This unequal potential distribution is undesirable and is usually expressed in terms of string efficiency.

The ratio of voltage across the whole string and the product of the number of units and voltage across the unit nearest to the line conductor is known as the string efficiency *i.e.*,

$$\text{String efficiency} = \frac{\text{Voltage across the string}}{n \times \text{voltage across the lower most unit}} \quad ...(3.1)$$

where n is the number of units in the string.

String efficiency may also be defined as

$$\text{String efficiency} = \frac{\text{Flash-over voltage of the string}}{n \times \text{flash-over voltage of one unit}}$$

The voltage distribution across different units of an insulator string and string efficiency can be mathematically determined with the help of an equivalent circuit of the insulator string (Fig. 3.16) as below. Figure 3.16 shows the equivalent circuit of a string of suspension insulators containing 4 units.

Let the mutual capacitance between the links be C and shunt capacitance between links and earth be C_1, voltage across the first unit (nearest the cross-arm) be V_1, voltage across the second unit be V_2, voltage across the third unit be V_3, voltage across the fourth unit (nearest the line conductor) be V_4 and voltage between conductor and earth be V volts

Let $\frac{C_1}{C} = K$ or $C_1 = KC$

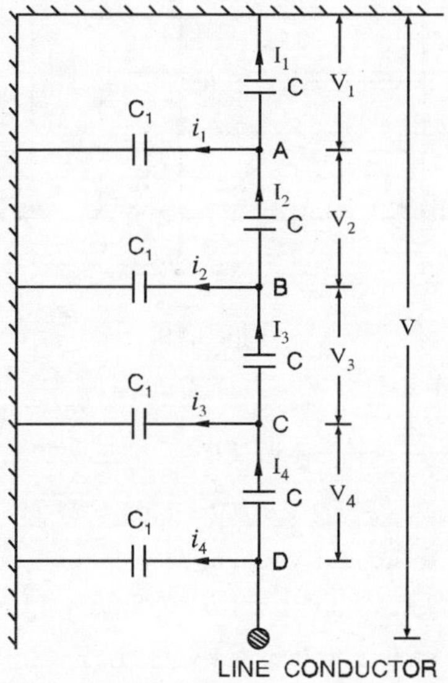

Fig. 3.16. ***Equivalent Circuit For a 4-Unit String***

Applying Kirchhoff's first law to node A, we get

$$I_2 = I_1 + i_1$$

$$\text{or } \omega CV_2 = \omega CV_1 + \omega C_1V_1$$

$$\text{or } \omega CV_2 = \omega CV_1 + \omega KCV_1$$

$$\text{or } V_2 = V_1(1 + K) \quad ...(3.2)$$

Applying Kirchhoff's first law to node B, we get

$$I_3 = I_2 + i_2$$

$$\text{or } \omega CV_3 = \omega CV_2 + \omega C_1(V_1 + V_2)$$

∵ Voltage across the second shunt capacitance C_1 from the top $= V_1 + V_2$

$$\text{or } \omega CV_3 = \omega CV_2 + \omega KC(V_1 + V_2)$$

$$\text{or } V_3 = V_2 + K(V_1 + V_2) = KV_1 + V_2(1 + K)$$

$$\text{or } V_3 = KV_1 + V_1 (1 + K) (1 + K)$$

∵ From Eq. (3.2) $V_2 = V_1(1 + K)$

$$\text{or } V_3 = V_1(1 + 3K + K^2) \quad ...(3.3)$$

Applying Kirchhoff's first law to node C, we get

$$I_4 = I_3 + i_3$$

$$\text{or } \omega CV_4 = \omega CV_3 + \omega C_1(V_1 + V_2 + V_3)$$

∵ Voltage across the third shunt capacitance C_1 from the top $= V_1 + V_2 + V_3$

$$\text{or } \omega CV_4 = \omega CV_1(1 + 3K + K^2) + \omega KC[V_1 + V_1 (1 + K) + V_1(1 + 3K + K^2)]$$

$$\text{or } V_4 = V_1(1 + 6K + 5K^2 + K^3) \quad ...(3.4)$$

Finally voltage between line conductor and earth

$$V = V_1 + V_2 + V_3 + V_4$$
$$= V_1 + V_1(1 + K) + V_1(1 + 3 K + K^2) + V_1 (1 + 6 K + 5 K^2 + K^3)$$
$$= V_1(4 + 10 K + 6 K^2 + K^3) \quad ...(3.5)$$

$$\text{Thus } \frac{V_1}{1} = \frac{V_2}{1+K} + \frac{V_3}{1+3K+K^2}$$
$$= \frac{V_4}{1+6K+5K^2+K^3}$$
$$= \frac{V}{4+10K+6K^2+K^3} \quad ...(3.6)$$

The greatest voltage will be obviously V_4 which is given as

$$V_4 = \frac{1+6K+5K^2+K^3}{4+10K+6K^2+K^3} V \quad ...(3.7)$$

and Percentage string efficiency

$$= \frac{V}{nV_4} \times 100 = \frac{V}{4 \times \frac{1+6K+5K^2+K^3}{4+10K+6K^2+K^3}} \times 100$$

Since $n = 4$, and flash-over voltage of one unit = Greatest voltage across any unit *i.e.*, V_4

or Percentage string efficiency

$$= \frac{4+10K+6K^2+K^3}{4(1+6K+5K^2+K^3)} \times 100 \quad ...(3.8)$$

Similarly derivation can be had for a string of insulators consisting of any number of units.

When the number of insulators in the string is large it becomes laborious to work out the voltage distribution across each unit, for such cases standard formula may be used.

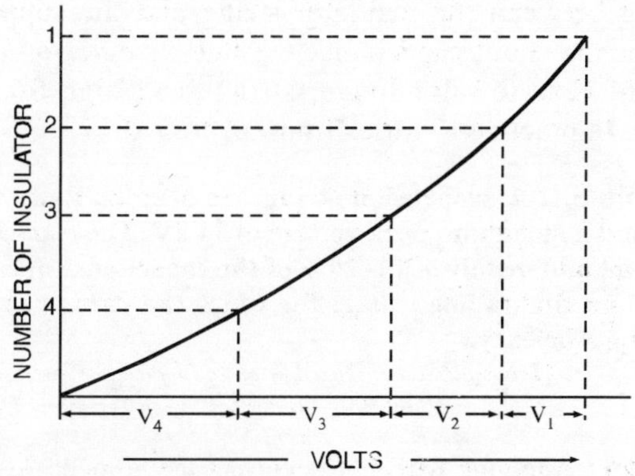

Fig. 3.17. ***Voltage Distribution Across Suspension Insulator String***

In general case if there are n units in the string, V is the maximum voltage across the string, $V_1, V_2, V_3 .....V_n$ denote the voltages across the insulator units starting from top, C is the capacitance between the links and KC be the shunt capacitance between the links and earth, the voltage distribution across the mth unit (counted from top) is given as

$$V_m = V \frac{2 \sinh\left(\frac{1}{2}\sqrt{K}\right) \cosh\left[\left(m - \frac{1}{2}\right)\sqrt{K}\right]}{\sinh\left(n\sqrt{K}\right)} \quad ...(3.9)$$

and potential adjacent to the line conductor

$$V_n = V \frac{2 \sinh\left(\frac{1}{2}\sqrt{K}\right) \cosh\left[\left(n - \frac{1}{2}\right)\sqrt{K}\right]}{\sinh\left(n\sqrt{K}\right)} \quad ...(3.10)$$

Graphical plot of how voltages are distributed across the units of an insulator string is shown in Fig. 3.17.

The following points may be noted.

1. The unit nearest to the line conductor is under maximum electrical stress and is likely to be punctured while the one nearest to the cross arm is under minimum electrical stress.
2. The voltage distribution across various units depends upon the value of K and number of discs contained in the string. The greater the value of K, the more non-uniform is the voltage distribution across the discs and lesser is string efficiency. The inequality in voltage distribution increases with the number of discs in the string. Thus a shorter string has higher efficiency than the longer one.

When the insulators are wet the value of mutual capacitance C increases while C_1 remains constant (except for the unit nearest the cross arm) so the value of K decreases, more uniform potential distribution is obtained and the string efficiency increases.

The value of K (the ratio of shunt capacitance C_1 to mutual capacitance C) varies and depends upon the length of the insulator string. The larger the number of insulator discs in a string, the longer will be the string. The longer the string, the greater must be the horizontal spacing between the insulator disc and the support (pole or tower) to make an allowance for conductor swing. The greater the horizontal spacing between the insulator string and the support, the lesser is the shunt capacitance C_1 and vice versa. Thus the value of K is low for longer strings and high for shorter strings. In practice K varies from 0.1 to 0.1667.

Example 3.1. A suspension string has 3 units. Each unit can withstand a maximum peak voltage of 11 kV. The capacitance of each joint and metalwork is 20% of the capacitance of each disc. Find (*i*) maximum line voltage for which the string can be used (*ii*) string efficiency.

[Transmission & Distribution of Electrical Power; Rajasthan Univ. 1997, 2006; Rajasthan Technical Univ. 2009]

Solution : Applying Kirchhoff's current law to node A, we have

$$I_2 = I_1 + i_1$$

or $\omega CV_2 = \omega CV_1 + \omega(0.2C)V_1$

or $V_2 = 1.2V_1$

Applying KCL to node B, we have

$$I_3 = I_2 + i_2$$

or $\omega CV_3 = \omega CV_2 + \omega(0.2C)(V_1 + V_2)$

$\because$ Voltage across the second shunt capacitance C_1 from the top $= V_1 + V_2$

or $V_3 = 0.2V_1 + 1.2V_2$

$= 0.2V_1 + 1.2 \times 1.2V_1 = 1.64V_1 \quad \because V_2 = 1.2V_1$

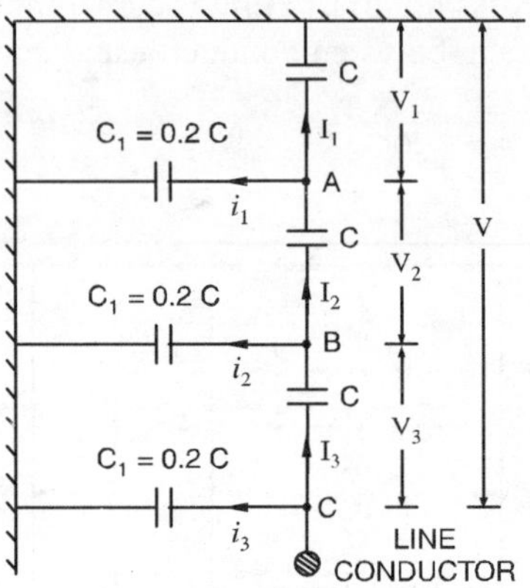

Fig. 3.18

Since $V_3 = 11$ kV

$$V_1 = \frac{V_3}{1.64} = \frac{11}{1.64} = 6.71 \text{ kV}$$

$$V_2 = 1.2V_1 = 1.2 \times 6.71 = 8.05 \text{ kV}$$

Maximum line to neutral voltage,

$$V = V_1 + V_2 + V_3 = 6.71 + 8.05 + 11 = 25.76 \text{ V}$$

Maximum line to line volt $= \sqrt{3} \times 25.76 = 44.62$ kV **Ans.**

String efficiency, $\eta = \frac{V}{3 \times V_3} \times 100 = \frac{25.76}{3 \times 11} \times 100 = 78\%$ **Ans.**

Example 3.2. In a string insulator having 5 units, the self capacitance of insulator units is 6 times to that between a unit and earth. Obtain the potential distribution across each unit in the string as percentage of potential of conductor with respect to earth. Also find the string efficiency.

[U.P. Technical Univ. Elements of Power System 2009-10]

Solution : The equivalent circuit of a 5-unit insulator string is shown in Fig. 3.19. Ratio of shunt capacitance to mutual capacitance, $K = \frac{C_1}{C} = \frac{1}{6}$.

Let the voltage of conductor to earth be V volts

From Eqs. (3.2), (3.3) and (3.4) respectively, we have

$$V_2 = V_1(1 + K) = V_1\left(1 + \frac{1}{6}\right) = \frac{7}{6}V_1 = 1.16667V_1 \quad ...(i)$$

$$V_3 = V_1(1 + 3K + K^2) = V_1\left(1 + \frac{3}{6} + \frac{1}{6^2}\right) = 1.5278V_1 \quad ...(ii)$$

$$V_4 = V_1(1 + 6K + 5K^2 + K^3) = 2.1435 V_1 \quad ...(iii)$$

If we apply Kirchhoff's first law to node D, we get

$$I_5 = I_4 + i_4$$

or $\omega CV_5 = \omega CV_4 + \omega C_1(V_1 + V_2 + V_3 + V_4)$

$\because$ voltage across the fourth capacitance from the top $= V_1 + V_2 + V_3 + V_4$

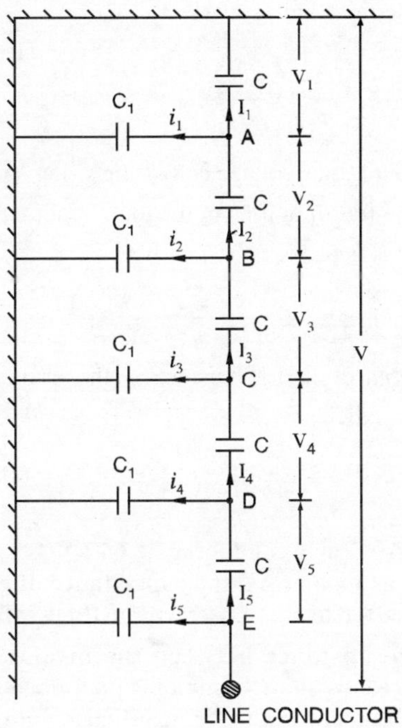

Fig. 3.19

$$\text{or } \omega CV_5 = \omega CV_1(1 + 6K + 5K^2 + K^3) + \omega KC[V_1 + V_1(1 + K) + V_1(1 + 3K + K^2) + V_1(1 + 6K + 5K^2 + K^3)]$$

$$\text{or } V_5 = V_1(1 + 10K + 15K^2 + 7K^3 + K^4) = 3.1165\ V_1 \quad ...(iv)$$

and voltage between line conductor and earth,

$$V = V_1 + V_2 + V_3 + V_4 + V_5$$
$$= V_1 + V_1(1 + K) + V_1\ (1 + 3K + K^2) + V_1(1 + 6K + 5K^2 + K^3) + (1 + 10K + 15K^2 + 7K^3 + K^4]$$
$$= [5 + 20K + 21K^2 + 8K^3 + K^4]V_1 \quad ...(v)$$

So voltage across the first unit (nearest the cross-arm),

$$V_1 = \frac{V}{5 + 20K + 21K^2 + 8K^3 + K^4}$$

$$= \frac{V}{5 + 20 \times \frac{1}{6} + 21 \times \left(\frac{1}{6}\right)^2 + 8\left(\frac{1}{6}\right)^3 + \left(\frac{1}{6}\right)^4}$$

$$= \frac{V}{5 + 3.33 + 0.583 + 0.037 + 0.0008}$$

$$= \frac{V}{8.954} \text{ or } 11.168\% \text{ of V } \textbf{Ans.}$$

Voltage across the second unit from top,

$$V_2 = 1.16667V_1 = 1.16667 \times 11.168\% \text{ of V} = 13.029\% \text{ of V } \textbf{Ans.}$$

Voltage across the third unit from top,

$$V_3 = 1.5278\ V_1 = 1.5278 \times 11.168\% \text{ of V} = 17.062\% \text{ of V } \textbf{Ans.}$$

Voltage across the fourth unit from top,

$$V_4 = 2.1435\ V_1 = 2.1435 \times 11.168\% \text{ of V} = 23.938\% \text{ of V } \textbf{Ans.}$$

Voltage across the bottom most unit,

$$V_5 = 3.1165\ V_1 = 3.1165 \times 11.168\% \text{ of V} = 34.804\% \text{ of V } \textbf{Ans.}$$

String efficiency

$$= \frac{V}{nV_n} \times 100 = \frac{V}{\frac{5 \times 34.804\ V}{100}} \times 100 = 57.46\ \% \textbf{ Ans.}$$

String efficiency will be improved in rain due to decrease in value of K.

Example 3.3. Define string efficiency :

A string of 6 insulator units has a self-capacitance equal to 10 times the pin-to-earth capacitance. Find (*i*) the voltage distribution across various units as a percentage of total voltage across the string, (*ii*) the string efficiency.

[U.P. Technical Univ. Elements of Power System 2005-06]

Solution : The equivalent circuit of a 6-unit insulator string is shown in Fig. 3.20.

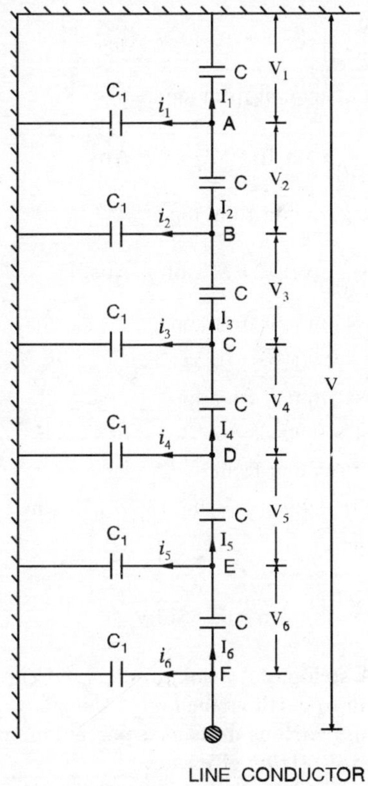

Fig. 3.20

Ratio of shunt capacitance to mutual capacitance,

$$K = \frac{C_1}{C} = \frac{1}{10} = 0.1$$

Let the voltage of conductor to earth be V volts.

From Eqs. (3.2), (3.3) and (3.4) respectively, we have

$$V_2 = V_1(1 + K) = V_1(1 + 0.1) = 1.1V_1 \quad ...(i)$$

$$V_3 = V_1(1 + 3K + K^2) = V_1(1 + 3 \times 0.1 + 0.1^2) = 1.31\ V_1 \quad ...(ii)$$

$$V_4 = V_1(1 + 6K + 5K^2 + K^3) = V_1(1 + 6 \times 0.1 + 5 \times 0.1^2 + 0.1^3) = 1.651\ V_1 \quad ...(iii)$$

From Eq. (*iv*) of Example 3.1, we have

$$V_5 = V_1(1 + 10K + 15K^2 + 7K^3 + K^4) = V_1(1 + 10 \times 0.1 + 15 \times 0.1^2 + 7 \times 0.1^3 + 0.1^4) = 2.1571\ V_1 \quad ...(iv)$$

Applying Kirchhoff's first law to node E, we have

$$I_6 = I_5 + i_5$$

or $\omega CV_6 = \omega CV_5 + \omega C_1 (V_1 + V_2 + V_3 + V_4 + V_5)$

$\because$ Voltage across the fifth capacitance from the top

$$= V_1 + V_2 + V_3 + V_4 + V_5$$

$$\text{or } V_6 = V_5 + K(V_1 + V_2 + V_3 + V_4 + V_5)$$
$$= 2.1571V_1 + 0.1(V_1 + 1.1V_1 + 1.31V_1 + 1.651V_1 + 2.1571V_1)$$
$$= 2.1571V_1 + 0.1 \times 7.2181V_1 = 2.87891V_1$$

Voltage between line conductor and earth

$$V = V_1 + V_2 + V_3 + V_4 + V_5 + V_6$$
$$= V_1 + 1.1V_1 + 1.31V_1 + 1.651V_1 + 2.1571V_1 + 2.87891V_1$$
$$= 10.09701V_1$$

$(\therefore)$ So voltage across the first unit (nearest the cross-arm)

$$V_1 = \frac{V}{10.09701} \text{ or } 9.904\% \text{ of V} \quad \textbf{Ans.}$$

Voltage across 2nd unit from top,

$$V_2 = \frac{1.1V}{10.09701} \text{ or } 10.894\% \text{ of V} \quad \textbf{Ans.}$$

Voltage across 3rd unit from top,

$$V_3 = \frac{1.31\,V}{10.09701} \text{ or } 12.974\% \text{ of V} \quad \textbf{Ans.}$$

Voltage across 4th unit from top,

$V_4 = 1.651 \times 9.904\% = 16.351\%$ of V **Ans.**

Voltage across 5th unit from top

$V_5 = 2.1571 \times 9.904\% = 21.364\%$ of V **Ans.**

Voltate across 6th unit from top

$V_6 = 2.87891 \times 9.904\% = 28.513\%$ of V **Ans.**

(*ii*) String efficiency,

$$\eta = \frac{V}{nV_n} \times 100 = \frac{V}{6 \times 0.28513\,V} \times 100 = 58.45\% \quad \textbf{Ans.}$$

Example 3.4. A string of 4 insulators has self capacitance equal to 4 times the pin to earth capacitance. Calculate: (*i*) the voltage distribution across various units as a percentage of total voltage across the string, (*ii*) string efficiency.

[*M.D. Univ. Power System-I, Dec. 2009*]

Solution : Ratio of shunt to mutual capacitance,

$$K = \frac{C_1}{C} = \frac{1}{4} = 0.25$$

Voltage across top unit,

$$V_1 = \frac{V}{4 + 10K + 6K^2 + K^3} \quad \text{... Refer to Eq. (3.6)}$$

$$= \frac{V}{4 + 10 \times 0.25 + 6 \times (0.25)^2 + (0.25)^3}$$

$= \frac{V}{6.89}$ or $\frac{100}{6.89}$ *i.e.*, 14.51% of total voltage across the string **Ans.**

Voltage across the second unit from the top

$$V_2 = (1 + K)V_1 \quad \text{...Refer to Eq. (3.2)}$$
$$= 1.25 \times 14.51\% \text{ of V}$$

= 18.14% total voltage across the string **Ans.**

Voltage across the third unit from the top

$$V_3 = (1 + 3K + K^2)V_1 \quad \text{...Refer to Eq. (3.3)}$$
$$= (1 + 3 \times 0.25 + 0.25^2)V_1$$
$$= 1.8125 \times 14.51\% \text{ of V}$$

= 26.3% of total voltage across the string **Ans.**

Voltage across the fourth unit from the top

$$V_4 = (1 + 6K + 5K^2 + K^3)V_1$$
$$= (1 + 6 \times 0.25 + 5 \times 0.25^2 + 0.25^3)\,V_1$$
$$= 2.828 \times 14.51\% \text{ of V}$$

= 41.036% of total voltage across the string. **Ans.**

Percentage string efficiency,

$$\eta = \frac{V}{4V_4} \times 100 = \frac{V}{4 \times 0.41036\,V} \times 100 = 60.92\% \quad \textbf{Ans.}$$

Example 3.5. (*a*) Each conductor at 66 kV is suspended by a string of 5 similar insulators, the capacitance of each disc being 5 times the capacitor to earth. Determine the % string efficiency.

If in (*a*) the distance between the insulators and tower structure is increased so as to make capacitance of disc 6 times the earth capacitance, will the string efficiency improve and how much ? [*A.M.I.E. Sec. B. Power Systems Summer 1997*]

Solution : Voltage between line conductor and earth, V = 66 kV

Number of units in insulator string, $n = 5$

(*a*) Shunt capacitance to mutual capacitance ratio, $K = \frac{1}{5} = 0.2$

Voltage across the top unit,

$$V_1 = \frac{V}{5 + 20K + 21K^2 + 8K^3 + K^4} \quad \text{Refer to Eq. (v) \& Example 3.2}$$

$$= \frac{66 \text{ kV}}{5 + 20 \times 0.2 + 21 \times (0.2)^2 + 8 \times (0.2)^3 + (0.2)^4}$$

$$= \frac{66}{9.9056} \text{ kV} = 6.66 \text{ kV}$$

Voltage across the bottom most or line unit,

$$V_5 = V_1(1 + 10K + 15K^2 + 7K^3 + K^4) \quad \text{Refer to Eq. (iv) \& Example 3.2}$$
$$= V_1(1 + 10 \times 0.2 + 15 \times (0.2)^2 + 7(0.2)^3 + (0.2)^4]$$
$$= 3.6576\,V_1 = 3.6576 \times 6.66 = 24.37 \text{ kV}$$

Percentage string efficiency

$$= \frac{V}{nV_n} \times 100 = \frac{66}{5 \times 24.37} \times 100 = 54.16\% \quad \textbf{Ans.}$$

(*b*) Shunt capacitance to mutual capacitance ratio, $K = \frac{1}{6}$

Voltage across the top unit, V_1

$$= \frac{V}{8.954} = \frac{66}{8.954} = 7.37 \text{ kV} \quad \text{Refer to Example 3.2}$$

Voltage across the bottom unit,

$$V_5 = 3.1165V_1 = 3.1165 \times 7.37 = 22.97 \text{ kV} \quad \text{Refer to Example 3.2}$$

Percentage string efficiency

$$= \frac{V}{nV_n} \times 100 = \frac{66}{5 \times 22.97} \times 100 = 57.46\%$$

i.e., string efficiency is improved and becomes 57.46% **Ans.**

Example 3.6. In a string of three identical suspension insulator units supporting a transmission line conductor, if the self capacitance of each unit is denoted as C farads, the capacitance of each connector pin to ground can be taken as 0.1 C farads. Determine the voltage distribution across the string if the m. ximum permissible voltage per unit is given as 20 kV. Also determine the string efficiency. [*GATE 1995; Rajasthan Univ. 2007*]

Solution : Number of units, $n = 3$

Ratio of shunt capacitance to mutual capacitance,

$$K = \frac{0.1C}{C} = 0.1$$

Voltage across bottom most unit,

V_3 = Safe working voltage of the unit = 20 kV

So voltage across top most unit,

$$V_1 = \frac{V_3}{1+3K+K^2} = \frac{20}{1+0.3+0.01} = 15.267 \text{ kV } \textbf{Ans.}$$

Voltage across middle unit,

$$V_2 = V_1(1 + K) = 15.267 \times 1.1 = 16.794 \text{ kV } \textbf{Ans.}$$

Maximum safe working voltage of the string,

$$V = V_1 + V_2 + V_3 = 15.267 + 16.794 + 20 = 52 \text{ kV}$$

String efficiency

$$= \frac{V}{nV_n} \times 100 = \frac{52}{3 \times 20} \times 100 = 86.667\% \textbf{ Ans.}$$

Example 3.7. A 3-phase transmission line is supported by a 3-unit suspension insulator string. The voltage across the unit nearest to the line is 20 kV and that across the adjacent unit is 15 kV. Determine (*i*) ratio of mutual to ground capacitance (*ii*) system line voltage. Determine the string efficiency.

[*Rajasthan Technical Univ. Transmission & Distribution of Electrical Power 2011*]

Solution : Number of units, $n = 3$

Voltage across the second unit, $V_2 = 15$ kV

Voltage across the third unit, $V_3 = 20$ kV

From Eqs. (3.2) and (3.3), we have

$$\frac{V_3}{V_2} = \frac{1+3K+K^2}{1+K}$$

$$\text{or} \quad \frac{20}{15} = \frac{1+3K+K^2}{1+K}$$

or $20 + 20K = 15 + 45K + 15K^2$

or $3K^2 + 5K - 1 = 0$

$$\text{or } K = \frac{-5 \pm \sqrt{25+12}}{6} = 0.18 \quad \text{ignoring –ve value}$$

(*i*) Ratio of mutual to ground capacitance, K = 0.18 **Ans.**

$$\text{Voltage across the top unit, } V_1 = \frac{V_2}{1+K} \qquad \text{Refer to Eq. (3.2)}$$

$$= \frac{15}{1+0.18} = 12.7 \text{ kV}$$

Line to neutral voltage,

$$V = V_1 + V_2 + V_3 = 12.7 + 15 + 20 = 47.7 \text{ kV}$$

Line to line voltage

$$= \sqrt{3} \times 47.7 = 82.62 \text{ kV } \textbf{Ans.}$$

(*ii*) String efficiency, $\eta = \frac{V}{3 \times V_3} \times 100$

$$= \frac{47.7}{3 \times 20} \times 100 = 79.5\% \textbf{ Ans.}$$

Example 3.8. A 3-phase overhead line is supported on 4-disc suspension insulators. The voltage across second and third disc are 13.2 kV and 18.2 kV respectively. Calculate the line voltage and mention the nearest standard voltage.

[*M.D. Univ. Power System-I, Dec. 2010*]

Solution : Number of units = 4

Voltage across the second unit, V_2 = 13.2 kV

Voltage across the third unit, V_3 = 18.2 kV

From Eqs. (3.2) and (3.3), we have

$$\frac{V_3}{V_2} = \frac{(1+3K+K^2)V_1}{(1+K)V_1}$$

$$\text{or} \quad \frac{18.2}{13.2} = \frac{1+3K+K^2}{1+K}$$

or $18.2 + 18.2K = 13.2 + 39.6K + 13.2K^2$

or $13.2K^2 + 21.4K - 5 = 0$

$$\text{or} \quad K = \frac{-21.4 \pm \sqrt{(21.4)^2 + 4 \times 5 \times 13.2}}{2 \times 13.2}$$

= 0.207 neglecting –ve value

Voltage across top unit,

$$V_1 = \frac{V_2}{1+K} = \frac{13.2}{1+0.207} = 10.936 \text{ kV}$$

Voltage across the bottom most unit,

$$V_4 = (1 + 6K + 5K^2 + K^3)V_1$$
$$= (1 + 6 \times 0.207 + 5 \times 0.207^2 + 0.207^3) \times 10.936$$
$$= 26.96 \text{ kV}$$

Voltage between conductor and earth

$$V = V_1 + V_2 + V_3 + V_4$$
$$= 10.936 + 13.2 + 18.2 + 26.96 = 69.294 \text{ kV}$$

Voltage between conductors

$$= \sqrt{3} \times 69.294 = 120 \text{ kV } \textbf{Ans.}$$

Nearest standard voltage is 110 kV **Ans.**

3.8. METHODS OF IMPROVING STRING EFFICIENCY

It has been seen that the voltage distribution across an insulator string is not uniform. The insulator unit nearest to the line conductor is highly stressed while that nearest to the tower is lightly stressed. The drawbacks of this non-uniform voltage distribution across the units of an insulator string are two-fold : if the voltage across the insulator unit nearest to the line conductor reaches beyond a prescribed safe value it may breakdown and the breakdown of other insulator units will take place in succession; at the same time if the voltage across the insulator unit farthest from the line conductor (or nearest to the cross arm) is too low, that unit is not utilized to its optimum capacity resulting in poor efficiency of the insulator string. This necessitates the equalisation of voltage distribution across the various insulator units of the string and thereby improving the string efficiency.

The string efficiency indicates the extent of the wastage. Though string efficiency can never be made 100 per cent, an improvement in its value is necessary so as to reduce the wastage to the minimum. Some methods used for this purpose are given below.

1. By Using Insulators With Larger Discs or By Providing Each Insulator Unit With a Metal Cap. It is clear from the expression of string efficiency that the string efficiency increases with the decrease in value of K (*i.e.*, the ratio of shunt capacitance to mutual capacitance). One method is to design the units such that the mutual capacitance (capacitance of each unit) is much greater than the shunt capacitance (capacitance to earth). This can be achieved by using insulators with larger discs or providing each insulator unit with a metal cap. The ratio K can be made $\frac{1}{6}$ to $\frac{1}{10}$ by this method

2. By Using Longer Cross Arms. The ratio of shunt capacitance to mutual capacitance, K can alternatively be reduced by using longer cross arms so that the horizontal distance from line support (pole or tower) is increased thereby decreasing the shunt capacitance. But the limitations of cost and mechanical strength of line supports do not allow the cross arms to be too long and it has been found that in practice it is not possible to obtain the value of K less than 0.1.

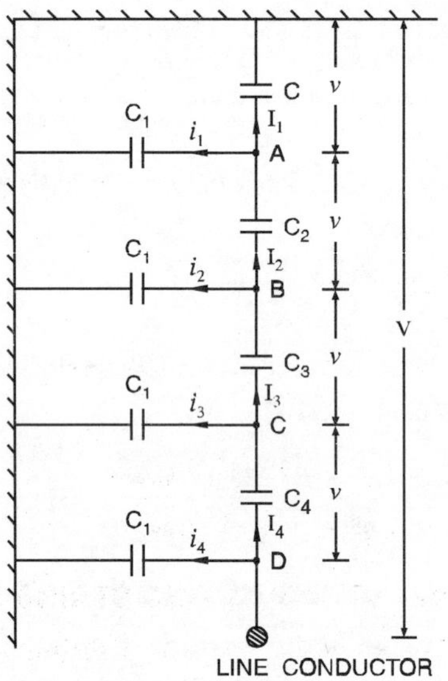

Fig. 3.21

3. By Capacitance Grading. It is seen that non-uniform distribution of voltage across an insulator string is due to leakage current from the insulator pin to the supporting structure. This current cannot be eliminated. However, it is possible that discs of different capacities are used such that the product of their capacitive reactance and the current flowing through the respective unit is same. This can be achieved by grading the mutual capacitance of the insulator units *i.e.*, by having lower units of more capacitance—maximum at the line unit and minimum at the top unit, nearest to the cross arm. It can be shown that by this method complete equality of voltage across the units of an insulator string can be obtained but this method needs a large number of different-sized insulator units. This involves maintaining spares of all varieties of insulator discs which is contrary to the tendency of standardization. So this method is not used in practice below 200 kV.

Consider a 4-unit string. Let C be the capacitance of the top unit and let the capacitances of others units be C_2, C_3 and C_4, as shown in Fig. 3.21.

Assume $C_1 = KC$

Applying Kirchhoff's first law to node A, we get

$$I_2 = I_1 + i_1$$

$$\text{or } \omega C_2 v = \omega C v + \omega C_1 v$$

$$\text{or } C_2 = C + KC = C(1 + K) \quad ...(3.11)$$

Applying Kirchhoff's first law to node B, we get

$$I_3 = I_2 + i_2$$

$$\text{or } \omega C_3 v = \omega C_2 v + \omega C_1 \times 2v$$

$$\text{or } C_3 = C_2 + 2KC$$

$$= C(1 + K) + 2KC$$

$$= C(1 + 3K) \quad ...(3.12)$$

Applying Kirchhoff's first law to node C, we get

$$I_4 = I_3 + i_3$$

$$\text{or } \omega C_4 v = \omega C_3 v + \omega C_1 \times 3v$$

$$\text{or } C_4 = C_3 + 3KC$$

$$= C(1 + 3K) + 3KC$$

$$= C(1 + 6K) \quad ...(3.13)$$

Thus, it will be possible to equalise the potential across all the units, if their capacitances are in the ratio of 1 : (1 + K) : (1 + 3 K) : (1 + 6 K) and so on.

But in practice it is impossible to obtain such units which will have their capacitances in above ratio, although nearby results can be obtained by employing standard insulators for most of the units and employing larger units adjacent to the line.

4. By Static Shielding. In case of capacitance grading, insulator units of different capacitances are used so that the flow of different currents through the respective units produce equal voltage drop. In static shielding, pin to supporting structure charging currents are exactly cancelled so that the same current flows through the identical insulator units and produce equal voltage drops across each insulator unit. This arrangement is given is Fig. 3.22. In this method a *guard* or *grading ring,* which usually takes the form of a large metal ring surrounding the bottom unit and electrically connected to the metalwork at the bottom of this unit, and therefore to the line conductor. The guard ring screens the lower units, reduce their earth capacitance C_1 and introduces a number of capacitances between the line conductor and the various insulator unit caps. These capacitances are greater for lower units and thus the voltages across them are reduced. With this method also it is impossible to obtain in practice an equal distribution of voltage but considerable improvements are possible.

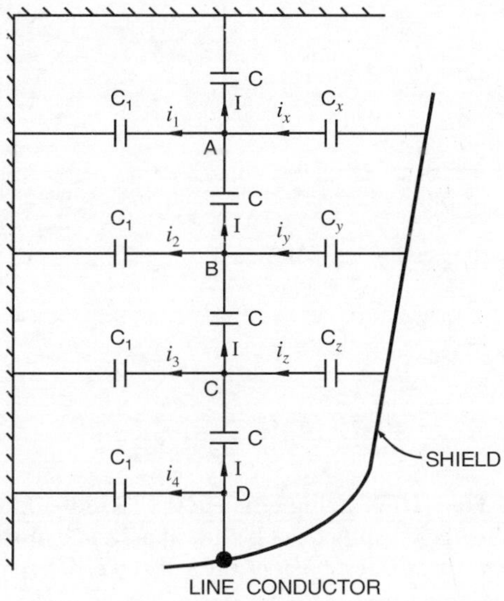

Fig. 3.22

Let the capacitances between the links and the shield be C_x, C_y and C_z respectively as shown in Fig. 3.22, and let v be the potential across each unit.

Since the capacitance of each unit is same, therefore, their charging currents I_1, I_2, I_3, and I_4 would be same, let it be I. If $C_1 = KC$

Applying Kirchhoff's first law to node A, we get

$$I + i_x = I + i_1$$

$$\text{or } i_x = i_1 \quad ...(3.14)$$

$$\text{Similarly } i_y = i_2 \quad ...(3.15)$$

$$\text{and } i_z = i_3 \quad ...(3.16)$$

$$\text{Also } i_1 = \omega C_1 v = \omega KCv \quad ...(3.17)$$

$$i_2 = 2\omega C_1 v = 2\omega KCv \quad ...(3.18)$$

$$i_3 = 3\omega C_1 v = 3\omega KCv \quad ...(3.19)$$

The potential causing current i_x is $3v$ (voltage across three units leaving the top one)

$$\text{So } i_x = \omega C_x \times 3v = 3\omega C_x v \quad ...(3.20)$$

Comparing Eqs. (3.14), (3.17) and (3.20), we have

$$3\omega C_x v = \omega KCv \text{ or } C_x = \frac{KC}{3} \quad ...(3.21)$$

The potential causing current y is $2\,v$ and therefore,

$$i_y = 2\omega C_y v \quad ...(3.22)$$

Comparing Eqs. (3.15), (3.18) and (3.22), we have

$$2\omega C_y v = 2\omega KCv \text{ or } C_y = KC \quad ...(3.23)$$

The potential causing current i_z is v and therefore,

$$i_z = \omega C_z v \quad ...(3.24)$$

Comparing Eqs. (3.16), (3.19) and (3.24), we have

$$\omega C_z v = 3\omega KCv$$

$$\text{or } C_z = 3KC \quad ...(3.25)$$

In general if there are n units

$$i_1 = \omega KCv \text{ and } i_x = (n-1)\omega C_x v$$

$$\text{or } C_x = \frac{KC}{n-1} \quad ...(3.26)$$

$$\text{Similarly } C_y = \frac{2KC}{n-2} \quad ...(3.27)$$

$$\text{and } C_z = \frac{3KC}{n-3} \quad ...(3.28)$$

or The capacitance of the pth metal link to the line is given as

$$C_p = \frac{pKC}{n-p} \quad ...(3.29)$$

Example 3.9. A string of 6 suspension insulators is to be graded to obtain uniform distribution of voltage across the string. If the pin to earth capacitance are all equal to C and the mutual capacitance of the top insulator is 10 C, find the mutual capacitance of each unit in terms of C.

[Pb. Technical Univ. Power System-I (T&D) May 2007]

Solution : Number of insulator units, $n = 6$

Capacitance to ground to top unit capacitance ratio,

$$K = 0.1 \text{ or } C = 10\,C_1$$

From equations (3.11), (3.12) and (3.13), we have

$$C_2 = (1 + K)C = (1 + 0.1)C = 1.1 \times 10C_1 = 11C_1 \textbf{ Ans.}$$

$$C_3 = (1 + 3K)C = 1.3C = 1.3 \times 10C_1 = 13C_1 \textbf{ Ans.}$$

$$C_4 = (1 + 6K)\,C = 1.6C = 1.6 \times 10C_1 = 16C_1 \textbf{ Ans.}$$

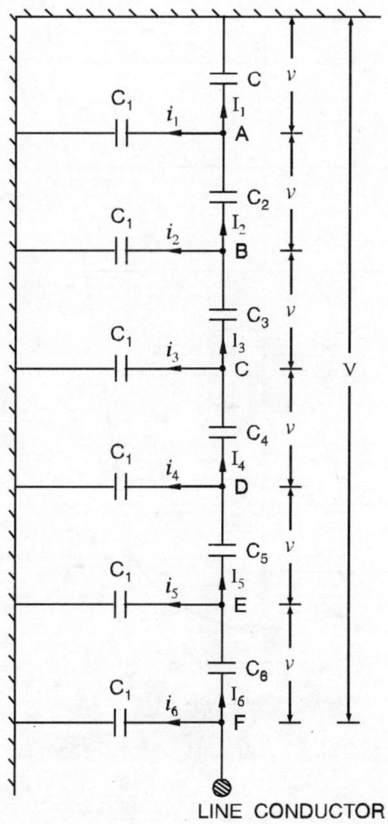

Fig. 3.23

Applying Kirchhoff's first law to node D, we have

$$I_5 = I_4 + i_4$$

$$\text{or } \omega C_5 v = \omega C_4 v + \omega C_1 \times 4v$$

$$\text{or } C_5 = C_4 + 4KC = (1 + 6K)C + 4KC$$

$$= (1 + 10K)C$$

$$= (1 + 10 \times 0.1)C = 2 \times 10C_1 = 20C_1 \textbf{ Ans.}$$

Applying Kirchhoff's first law to node E, we have

$$I_6 = I_5 + i_5$$

$$\text{or } \omega C_6 v = \omega C_5 v + \omega C_1 \times 5v$$

$$\text{or } C_6 = C_5 + 5KC = (1 + 10K)C + 5KC$$

$$= (1 + 15K)C = (1 + 15 \times 0.1) \times 10C_1 = 25C_1 \textbf{ Ans.}$$

Example 3.10. A string of 8 suspension insulators is to be fitted with a grading ring. If the pin to earth capacitances are all equal to C, find the values of line to pin capacitances that would give a uniform voltage distribution over the string.

[*Rajasthan Technical Univ. Transmission and Distribution of Electrical Power 2011; U.P. Technical Univ. Elements of Power System 2006-07*]

Solution : Number of insulator units, $n = 8$

Pin-to-either capacitance, $C' = C$

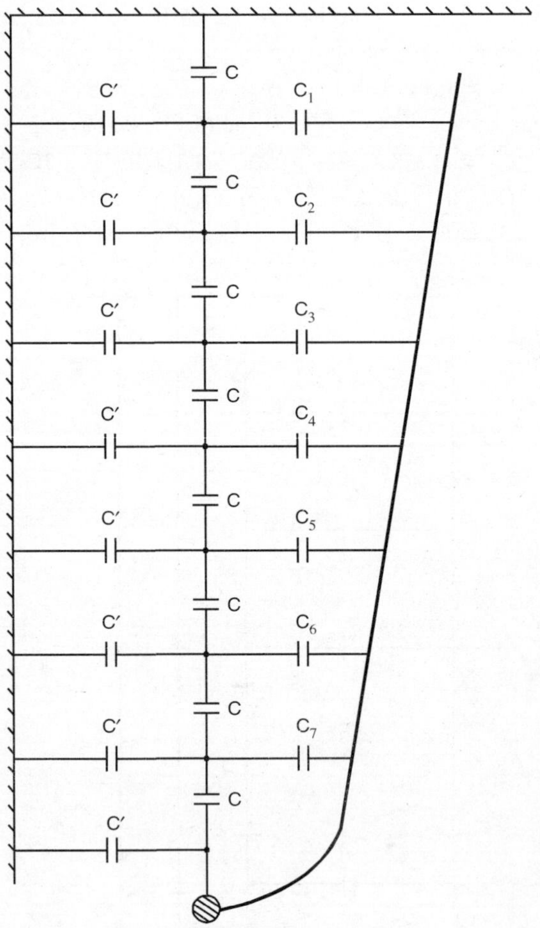

Fig. 3.24

From Eq. (3.29), the capacitance of the pth metal link to the line is given as

$$C_p = \frac{p \times \text{pin-to-earth capacitance}}{n - p}$$

$$\text{Thus } C_1 = \frac{1 \times C}{8-1} = \frac{C}{7} \textbf{ Ans.}$$

$$C_2 = \frac{2 \times C}{8-2} = \frac{C}{3} = \frac{2}{3} C \textbf{ Ans.}$$

$$C_3 = \frac{3 \times C}{8-3} = 0.6\text{ C } \textbf{Ans.}$$

$$C_4 = \frac{4 \times C}{8-4} = \text{C } \textbf{Ans.}$$

$$C_5 = \frac{5 \times C}{8-5} = \frac{5}{3}\text{C } \textbf{Ans.}$$

$$C_6 = \frac{6 \times C}{8-6} = 3\text{C } \textbf{Ans.}$$

$$C_7 = \frac{7 \times C}{8-7} = 7\text{C } \textbf{Ans.}$$

Example 3.11. Define string efficiency. Calculate its value for a string of 3 insulator units used if the capacitance of each unit to earth and line be 20% and 5% of the self capacitance of the unit. Derive the formula used. [*M.D. Univ. Power System-I; May 2008*]

Solution : Equivalent circuit of a string of insulators is shown in Fig. 3.25.

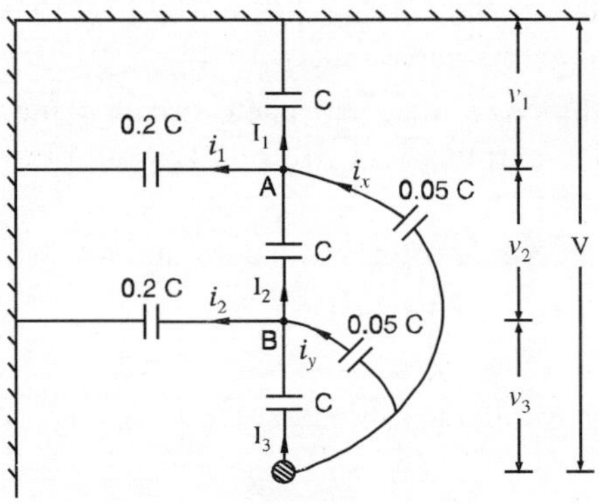

Fig. 3.25

Applying Kirchhoff's first law to node points A and B, we have

$$I_1 + i_1 = I_2 + i_x$$

$$\text{and } I_2 + i_2 = I_3 + i_y$$

$$\text{or } \omega C v_1 + 0.2\omega C v_1 = \omega C v_2 + 0.05\omega C(v_2 + v_3)$$

$$\text{and } \omega C v_2 + 0.2\omega C(v_1 + v_2) = \omega C v_3 + 0.05\omega C v_3$$

$$\text{or } 1.2v_1 = 1.05v_2 + 0.05v_3 \quad ...(i)$$

$$\text{and } 0.2v_1 = -1.2v_2 + 1.05v_3 \quad ...(ii)$$

Solving Eqs. (*i*) and (*ii*) in terms of v_1, we have

$$v_2 = \frac{25}{23.25} v_1 \text{ and } v_3 = \frac{1.65}{1.1625} v_1$$

Finally voltage between line conductor and earth

$$V = v_1 + v_2 + v_3$$

$$= v_1 + \frac{25}{23.25} v_1 + \frac{1.65}{1.1625} v_1$$

$$= 3.49462\, v_1$$

$$\text{So } v_1 = \frac{V}{3.49462} = 0.286\text{ V}$$

$$v_2 = \frac{25}{23.25} \times 0.286 = 0.308\text{ V}$$

$$\text{and } v_3 = \frac{1.65}{1.1625} \times 0.286 = 0.406 \text{ V}$$

$$\text{String efficiency} = \frac{V}{3v_3} \times 100$$

$$= \frac{V}{3 \times 0.406 \text{ V}} \times 100 = 82.1\% \textbf{ Ans.}$$

Example 3.12. In a transmission line each conductor is at 20 kV and is supported by a string of 3 suspension insulators. The air capacitance between each cap-pin junction and tower is one-fifth of the capacitance of each insulator unit. A guard ring, effective only over the line-end insulator unit is fitted so that the voltages on the two units nearest the line-end are equal. (*a*) Calculate the voltage on the line-end unit, (*b*) Calculate the value of capacitance C_x required. *[GATE 1999]*

Solution : Voltage between line conductor and earth,

$$V = 20 \text{ kV} \quad ...(i)$$

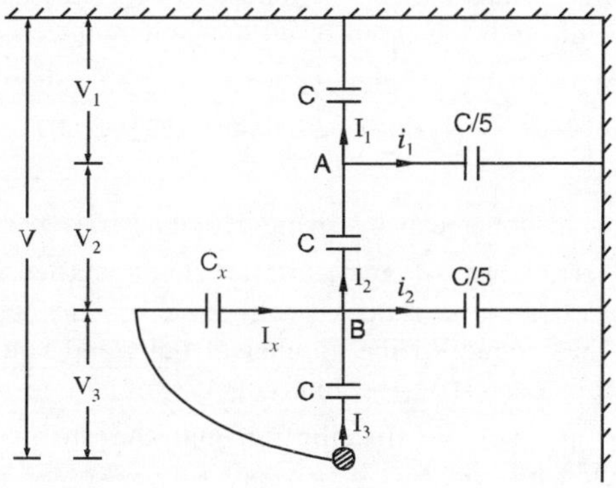

Fig. 3.26

$$\text{Mutual capacitance} = C$$

$$\text{Shunt capacitance} = \frac{C}{5} = 0.2 \text{ C}$$

Applying Kirchhoff's first law to node A

$$I_2 = I_1 + i_1$$

$$\omega CV_2 = \omega CV_1 + \omega \times 0.2CV_1$$

$$\text{or } V_2 = 1.2V_1 \quad ...(ii)$$

$$\text{Also } V_3 = V_2 \quad ...(iii)$$

$$\text{and } V = V_1 + V_2 + V_3$$

$$\text{or } 20 = V_1 + 1.2\, V_1 + 1.2\, V_1$$

$$\text{or } V_1 = \frac{20}{1 + 1.2 + 1.2} = 5.882 \text{ kV}$$

(*a*) Voltage on the line-end unit

$$= V_3 = 1.2\, V_1 = 1.2 \times 5.882 = 7.06 \text{ kV } \textbf{Ans.}$$

Applying Kirchhoff's first law to node B, we have

$$I_3 + i_x = I_2 + i_2$$

$$\omega CV_3 + \omega C_x V_3 = \omega CV_2 + \omega \frac{C}{5}(V_1 + V_2)$$

$$\text{or } C \times 1.2V_1 + C_x \times 1.2V_1 = C \times 1.2V_1 + \frac{C}{5} \times 2.2V_1$$

$$\text{or } C_x = \frac{2.2}{5 \times 1.2} C = 0.3667 \text{ C farads } \textbf{Ans.}$$

3.9. ARCING HORN AND GRADING RING

In the event of flash-over the insulator is cracked or broken up due to the heat of the arc. Grading ring, in addition to equalisation of voltage distribution across the insulator units, when used in conjunction with arcing horn fixed at the top end of the string serves the purpose of arcing shield and protects the insulator string from flash-over whenever overvoltage (under normal or abnormal condition) appears between the tower structure and the line conductor. They are designed to keep the arc away from the insulator string until it is interrupted by the device protecting the line. The arrangement of the arcing horns on a 7-unit string of suspension insulators is shown in Fig. 3.27. The combination of the arcing horn and grading ring provides path through the air medium and discharges the energy contained in the abnormal voltage and thus saves the insulator string.

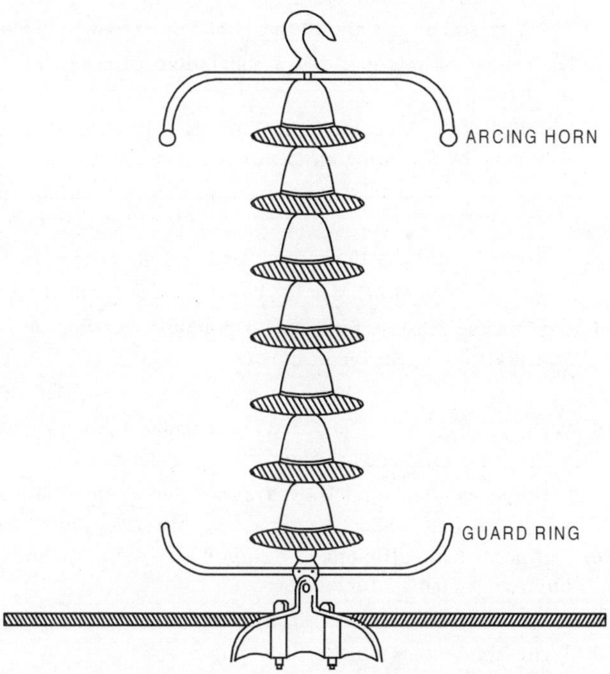

Fig. 3.27. ***Insulator String With Arcing Horn and Grading Ring***

Exercises

1. What electrical and mechanical characteristics are required for a good insulator for use in *hv* transmission lines ?
2. Name the various materials that are commonly used for overhead line insulators. Why glass insulators cannot be used above 50 kV ?
3. Name the different types of insulators used in power systems along with the uses of each insulator and explain pin type insulator in detail.
[Rajasthan Technical Univ. Transmission and Distribution of Electrical Power 2008, 2009]
4. Explain different types of insulators used in power system.
[M.D. Univ. Power System-I, May 2009]
5. Explain, with neat sketches, the constructional features of pin type and suspension type insulators. List the advantages of the latter type over the former for high voltage transmission lines. **[U.P.S.C. Electric Power Systems, 1975]**

6. Explain the suspension type insulators with neat diagram.

[Pb. Technical Univ. Power System-I (T&D) Dec. 2006]

7. Describe the following briefly:

(*a*) Pin type insulator.

(*b*) Suspension type insulator.

(*c*) Post type insulator.

[W.B. Technological Univ. Power System-I, 2009-10]

8. Explain strain and shackle insulators with the help of neat sketches.

9. Mention appropriate locations where you would use (*i*) pin type (*ii*) suspension type (*iii*) strain (*iv*) shackle and (*v*) egg type insulators.

10. Describe about the following in brief :

(*i*) String insulators.

(*ii*) Pin insulators.

(*iii*) Strain insulators.

(*iv*) Shackle insulators.

[U.P. Technical Univ. Elements of Power System 2009-10]

11. Name the various causes for the failure of overhead line insulators.

12. Describe the reasons for the failure of an insulator. What do you mean by the string efficiency of an insulator of *n* units ?

[Rajasthan Univ. Transmission & Distribution of Electrical Power, 2003]

13. List the basic tests that are carried out on insulators.

[A.M.I.E. Sec. B. Power Systems Summer 1994]

14. Explain the reasons for unequal potential distribution over a string of suspension insulators.

[M.D. Univ. Power System-I, Dec. 2010]

15. What is string efficiency ? Why is it necessary to have a high string efficiency ? How can it be achieved ?

[Rajasthan Technical Univ. Transmission & Distribution of Electrical Power 2009]

16. Define 'string efficiency'. Explain different methods of improving string efficiency.

[G.B. Technical Univ. Elements of Power System 2011-12]

17. Write short notes on the following :

(*i*) Insulator materials. **[M.D. Univ. Dec. 2007]**

(*ii*) Overhead line insulators.

[Pb. Technical Univ. May 2008]

(*iii*) String efficiency. **[Pb. Technical Univ. Dec. 2009]**

(*iv*) Use of guard rings and arcing horns.

[M.D. Univ. Dec. 2007]

Short Answer Type Questions With Answers

Q. 1. What are the important properties that an overhead line insulator must possess ?

Ans. The most important properties that an overhead line insulator must possess are : High mechanical strength, high relative permittivity, high insulation resistance, high ratio of rupturing strength to flash-over voltage and capability of withstanding temperature variations.

Q. 2. What types of insulators are used for overhead transmission and distribution lines ?

Ans. The insulators used for overhead line are pin type, suspension type, strain type and shackle insulators.

Q. 3. What are the causes of failure of insulators ?

Ans. The insulator failure may occur due to cracking of porcelain, porosity, puncture, mechanical stresses, flash-over etc.

Q. 4. How do the insulators fail ? **[A.M.I.E. Sec. B. Summer 1993]**

Ans. Electric failure of insulators may occur either by puncture or flash-over. In the case of puncture the arc passes through the insulator body while flash-over is caused by an arc discharge between the line conductor and earth through air surrounding the insulator.

Q. 5. How are insulators tested to determine flash-over voltages ?

Ans. For determination of flash-over voltage of a given insulator high voltage supply terminals are connected to the pin and insulator and the test voltage is gradually raised to a specified value in approximate 10 seconds and is maintained for one minute. The voltage is then gradually raised till a flash-over takes place. The flash-over voltage depends upon barometric pressure, temperature, humidity etc.

Q. 6. What is meant by birdage ?

Ans. Birdage means the short circuiting of the line conductor to earth through the large birds or similar objects.

Q. 7. What is string efficiency ?

[U.P.T.U. 2005-06, 2007-08; PTU Dec. 2008]

OR

What do you mean by string efficiency ? **[UPTU 2004-05]**

Ans. The efficiency of an insulator string, called the *string efficiency*, is the ratio of voltage across the whole string and the product of the number of units and voltage across the unit nearest to the line conductor.

Q. 8. How are voltage distribution and the string efficiency affected by rain ?

Ans. More uniform voltage distribution is obtained and the string efficiency is improved because of reduction in value of K [when insulators are wet, mutual capacitance C increases while the shunt capacitance C_1 (except for the unit nearest the cross arm) remains constant].

Q. 9. What are the methods, that are used for improving string efficiency ?

Ans. Various methods used for improving string efficiency are by using insulators with larger discs or by providing each insulator unit with a metal cap, by using longer cross arms, by capacitance grading or by static shielding.

Q. 10. What are the two functions that are performed by grading ring ?

Ans. The grading ring performs two functions viz equalizes the voltage distribution across the insulator units and when used in conjunction with arcing horn fixed at the top end of the string serves the purpose of arcing shield and protects the insulator string from flash-over.

Problems

1. Determine the maximum voltage that the string of suspension insulators in the figure can withstand if the maximum voltage per unit is 17.5 kV.

[U.P. Technical Univ. Elements of Power System 2008-09]

[**Ans.** 44.24 kV]

2. A string insulator has 4 units and each unit is rated for 12 kV. Find the maximum line voltage on which it can be operated safely. The mutual capacitance of unit is 10 times the capacitance between pin to earth. Find the string efficiency.

[U.P. Technical Univ. Elements of Power System 2007-08]

[**Ans.** 36.78 kV, 76.625%]

3. A string insulator has five units each rated for 11 kV. Find the maximum line voltage on which it can be operated safety. The mutual capacitance of unit is 10 times the capacitance between pin-to-earth.

[U.P. Technical Univ. Elements of Power System 2004-05]

[**Ans.** 36.8 kV]

4. A string insulator is connected across a 100 kV line. If the capacitance of each disc to earth is 0.1 of the capacitance of the insulator, calculate (*i*) the distribution of voltage on the insulator discs and (*ii*) the string efficiency.

[G.B. Technical Univ. Elements of Power System 2011-12]

[**Ans.** 7.96 kV, 8.77 kV, 10.5 kV, 13.16 kV, 17.3 kV; 66.7%]

5. A 33 kV, 3-phase, OH line is supported on a string of 3 similar insulators, the mutual capacitance of which is 10 times the shunt capacitance between the unit and earthed frame work. Calculate the voltage across each insulator and string efficiency.

[**Ans.** V_1 = 5.587 kV, V_2 = 6.146 kV, V_3 = 7.319 kV, String efficiency = 86.77%]

6. For a string insulator with four discs, the capacitance of the disc is 10 times the capacitance between the pin and earth. Calculate the voltage across each disc when used on 66 kV line. Also calculate the string efficiency.

[A.M.I.E. Sec. B. Power Systems Summer 1994]

[**Ans.** V_1 = 7.529 kV, V_2 = 8.282 kV, V_3 = 9.863 kV, V_4 = 12.43 kV, string efficiency = 76.64%]

7. A string of six insulator units has self capacitances equal to 10 times the pin to earth capacitances. Find (*a*) the voltage distribution from top to bottom insulator as a percentage of the total voltage, (*b*) the string efficiency.

[**Ans.** 9.904%, 10.894%, 12.974%, 16.351%, 21.346%, 28.512% of V, 58.45%]

8. Each line of a 3-phase system is supported by a string of 3 similar insulators. If the voltage across the line unit is 17.5 kV, calculate the line-to-neutral voltage. Assume that the shunt capacitance between each insulator and earth is $\frac{1}{8}$th of the capacitance of the insulator itself. Also find the string efficiency. [**Ans.** 44.24 kV, 84.26%]

9. A string insulator has three units. The capacitance from each joint is 12.5% of the capacitance of each unit. The voltage across any unit should not exceed 11 kV. Find the maximum voltage for the string application.

[A.M.I.E. Sec. B. Power Systems Winter 1997]

[**Ans.** 27.8 kV]

10. Each conductor of a 3-phase high voltage transmission line is suspended from cross-arm of steel tower by a string of 4 suspension type disc insulators. If the voltage across the second unit is 13.2 kV and that across the third unit is 18 kV, calculate the voltage between the conductors.

[Agra Univ 1977] [**Ans.** 70 kV]

11. Each of the three insulators forming a string has a self capacitance of C farads. The shunting capacitance of the connecting metal work of each insulator is 0.2 C to earth and 0.1 C to the line. Calculate the voltage across each insulator as a percentage of the line voltage to earth and also the string efficiency.

[A.M.I.E. Sec. B. Power Systems Winter 1998; M.D. Univ. Power System-I, May 2009]

[**Ans.** 31%; 30.3% and 38.7% of V and 86.13%]

12. Each of the three insulators forming a string has a self capacitance of 'C' Farads. The shunting capacitance of the connecting metal work of each insulator is 0.2 C to earth and 0.1 C to the line. A guard ring increases the capacitance to the line of the metal work of the lowest insulator to 0.3 C. Calculate the string efficiency of this arrangement with the guard ring. **[Agra Univ. 1979]**

[**Ans.** 95.24%]

13. Each line of a 3-phase system is suspended by a string of 3-identical insulators of self-capacitance C Farads. The shunt capacitance of metalwork of each insulator is 0.26 C to each and 0.15 C to line. Find the string efficiency if a guard ring increases the shunt capacitance to line of metalwork of the lowest insulator to 0.35C.

[Rajasthan Univ. Transmission & Distribution of Electrical Power 2005]

[**Ans.** 93.3%]

14. In a 3-unit string insulator, the joint to lower capacitance is 20% of the capacitance of each unit. By how much should the capacitance of the lowest unit be increased to get a string efficiency of 90% ? The remaining two units are left unchanged. **[Rajasthan Univ. Transmission & Distribution of Electrical Power 2003]**

[**Ans.** 26.72%]

❑❑❑

CHAPTER 4

Corona

INSIDE THIS CHAPTER

4.1. INTRODUCTION

When an alternating potential difference is applied across two conductors, whose spacing is large in comparison with their diameters, then the atmospheric air surrounding the conductors is subjected to electrostatic stresses. At low voltage there is no change in the condition of atmospheric air around the conductors. However, when the potential difference is gradually increased, a stage arrives when a faint luminous glow of violet colour appears together with a hissing noise. This phenomenon is called "*visual corona*" and is accompanied by the production of ozone which is readily detected because of its characteristic odour. The glow is due to the fact that the atmospheric air around the conductor becomes conducting due to electrostatic stresses. If the potential difference is raised still further, the glow and the noise will increase in intensity until eventually a spark-over owing to breakdown of air insulation will take place. The whole phenomenon—the hissing noise, the violet glow, and production of ozone gas is known as the *corona*. This phenomenon is very much evident in transmission lines of 100 kV and above.

If the conductors are perfectly uniform and smooth the glow will be uniform throughout the length of the conductor, otherwise the rough points will appear brighter. If the spacing between the conductors is not very large as compared with their diameters (interaxial difference is less than 15 times the diameter of the conductors) corona may bridge the conductors and cause flash-over before any luminous glow is observed. It happens due to the fact that the distance between the conductors being smaller, there is no time for the glow to occur.

In EHV ac transmission lines the visible glow of corona is mostly uniform about both the conductors but in case of HVDC transmission the glow is uniform and brighter about the positive conductor but spotty about the negative conductor.

Corona is accompanied by a loss of energy which increases very rapidly once the visual critical voltage is exceeded. Power loss, due to corona, heavily depends upon weather condition—during humid and moist climate corona loss is much increased. The energy loss accompanied by the phenomenon, called the corona, is dissipated in the form of light, heat, sound and chemical action. In case of ac system the current due to corona is non-sinusoidal. In practice this non-sinusoidal current and the nonsinusoidal voltage drop caused by it may be more important than the power loss.

The effects of corona are summarized as below:

(*i*) A violet glow is observed around the conductor.

(*ii*) It produces a hissing noise.

(*iii*) It produces ozone which can be readily detected by its characteristic odour.

(*iv*) The glow is maximum over rough and dirty surfaces of the conductor.

(*v*) It is accompanied by a power loss—the wattmeter connected in the electric circuit will show a reading.

(*vi*) The charging current under corona condition increases because the corona induces harmonic currents.

4.2. FACTORS AFFECTING CORONA

Since corona occurs due to the ionization of the air surrounding the line conductors, it is affected by the physical state of the atmosphere as well as by the condition of the line. The corona is affected by the following factors.

(*i*) *Atmosphere*: Corona is caused by the bombardment of molecules, with subsequent dislodging of electrons, by the ionised particles. Corona will thus be affected by the physical state of the atmosphere. The voltage gradient for the breakdown of the air is proportional to its density. In the stormy weather the number of ions may be more than normal, and as such the corona may occur at much less voltage compared with fair weather.

(*ii*) *Conductor*: The corona is considerably affected by the size (diameter), shape (stranded or smooth) and surface condition (dirty or clean) of the conductor. The corona decreases with the increase in diameter of conductor. A stranded conductor gives rise to more corona than a solid conductor. For stranded conductors, the shape of the cross section is a series of arcs of circles each of much smaller diameter than the conductor as a whole. The potential gradient for such a conductor will, therefore, be greater than for the equivalent smooth conductor so the breakdown voltage for such a conductor will be somewhat less than for a smooth conductor, the ratio being between 0.88 and unity. The effect of dirt on the surface is to increase the irregularity and thereby decreasing the breakdown voltage further, an average value for the ratio when both stranding and dirt are taken into account lies between 0.85 and unity.

(*iii*) *Spacing Between Conductors*: With the increase in spacing between the conductors the electrostatic stresses are reduced and therefore, the corona effect is reduced. If the spacing between the conductors is made very large as compared with their diameters, there may not be any corona effect.

(*iv*) *Line Voltage*: Line voltage largely affects the corona. At low voltage, there is no corona effect, but when the line voltage is increased to such a value that electrostatic stresses developed at the conductor surface make the atmospheric air surrounding the conductor conducting, corona effect appears.

4.3. TRIPLE-HARMONIC CURRENTS DUE TO CORONA

Corona forms when the voltage of a conductor passes the disruptive critical voltage and disappears when the voltage descends through the same value. This occurs on each conductor every half cycle, and develops a pulsation in the voltage wave having three times the generator frequency. In an earthed system this triple-frequency voltage causes a triple-frequency current to flow through the capacitance of the system to earth and back through the earthed neutral. This effect is accentuated by the fact that the effective capacitance of the conductors pulsates at triple frequency due to the increase and decrease of effective diameter of conductors caused by the corona. Because of corona triple frequency currents flow through the ground in case of a non-earthed system, the neutral has a voltage to earth of triple frequency.

4.4. ADVANTAGES AND DISADVANTAGES OF CORONA

In the past, the corona was considered as something to be avoided because of the energy loss associated with it and the distortion of the waveform. But now the corona is considered beneficial because of the following advantages.

1. On the formation of corona the sheath of air surrounding the conductor becomes conductive and there is a virtual increase in conductor diameter and due to this virtual increase in conductor diameter the maximum potential gradient or maximum electrostatic stress is reduced. Thus probability of flash-over is reduced and system performance is improved.
2. Effects of transients produced by lightning and other causes are reduced, since charges induced on the line by lightning or other causes will be partially dissipated as a corona loss. In this way it acts as a safety valve and sometimes lines have been purposely designed to have an operating voltage near to the critical voltage in order to do away with the necessity for, and expense of lightning arrestor gear. An objection to this scheme is that the critical voltage is not fixed for a given line, but may vary considerably with changes in the weather.

Disadvantages of corona are given below.

1. There is a definite loss of power, although this is not important except under abnormal weather conditions.
2. There is a non-sinusoidal voltage drop due to non-sinusoidal corona current and these may cause some interference with neighbouring communication circuits due to electromagnetic and electrostatic induction effects.
3. Owing to the formation of corona, ozone gas is produced which chemically reacts with the conductor and causes corrosion.

4.5. DIELECTRIC STRENGTH OF AIR

The value of potential gradient at which complete disruption of air occurs, is called the *disruptive strength* or *dielectric strength* of air. The breakdown strength of air at a barometric pressure of 760 mm, and temperature of 25°C is 30 kV/cm (maximum) or 21.1.kV/cm (rms) and is denoted by g_0. The value of dielectric strength of air (*i.e.*, g_0) depends upon the density of air—g_0 is proportional to density of air over a wide range and thus directly proportional to the barometric pressure, and inversely proportional to the absolute temperature. Thus breakdown strength of air at a barometric pressure of b cm of mercury and temperature of t°C becomes δg_o where

$$\delta = \frac{b \times 10}{760} \times \frac{273+25}{273+t} = \frac{3.92b}{273+t} \qquad ...(4.1)$$

4.6. DISRUPTIVE CRITICAL VOLTAGE

Let us consider the two-wire line shown in Fig. 4.1, where r is the radius of the line conductors and d is the distance between their centres, where d is very large as compared to r. If a positive charge of q coulombs per metre of conductor length is given to conductor A. then a negative charge of q coulombs per metre of conductor length will be induced on conductor B.

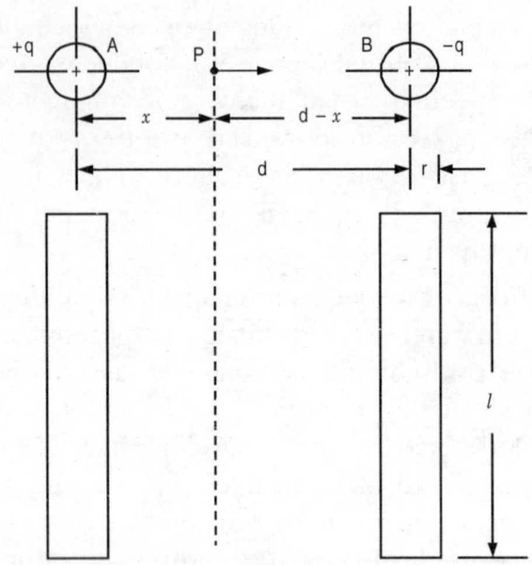

Fig. 4.1

Consider point P at a distance of x metres from conductor A.

Electric field intensity at point P due to charge on conductor A $= \frac{q}{2\pi\varepsilon_0 x}$ acting towards B as shown.

Electric field intensity at point P due to induced charge on conductor B $= \frac{q}{2\pi\varepsilon_0 (d-x)}$ acting towards B as shown

Resultant electric field intensity at point P,

$$E_x = \frac{q}{2\pi\varepsilon_0 x} + \frac{q}{2\pi\varepsilon_0 (d-x)} = \frac{q}{2\pi\varepsilon_0}\left(\frac{1}{x} + \frac{1}{d-x}\right)$$

PD between conductors A and B,

$$V = \int_r^{d-r} E_x\, dx = \int_r^{d-r} \frac{q}{2\pi\varepsilon_0}\left(\frac{1}{x} + \frac{1}{d-x}\right) dx$$

$$= \frac{q}{2\pi\varepsilon_0}\left[\log_e x - \log_e (d-x)\right]_r^{d-r} = \frac{q}{\pi\varepsilon_o} \log_e \frac{d-r}{r}$$

Now since r is very small as compared to d, $d - r \simeq d$ and, therefore,

$$V = \frac{q}{\pi\varepsilon_0} \log_e \frac{d}{r} \qquad \text{...(4.2)}$$

Now gradient at any point x from the centre of the conductor A is given by

$$E_x = \frac{q}{2\pi\varepsilon_0}\left(\frac{1}{x} + \frac{1}{d-x}\right) = \frac{q}{2\pi\varepsilon_0} \cdot \frac{d}{x(d-x)}$$

Substituting for q from Eq. (4.2) in above equation, we have

$$E_x = \frac{\pi\varepsilon_0 V}{\log_e \frac{d}{r}} \cdot \frac{1}{2\pi\varepsilon_0} \cdot \frac{d}{x(d-x)} = \frac{V}{2\log_e \frac{d}{r}} \cdot \frac{d}{x(d-x)}$$

where V is voltage between two conductors.

and

$$E_x = \frac{V'}{\log_e \frac{d}{r}} \cdot \frac{d}{x(d-x)} \qquad \text{...(4.3)}$$

where V′ is the line to neutral voltage of the system.

In case of 3-phase system,

$$V' = \frac{\text{Line-to-line voltage}}{\sqrt{3}} = \frac{V_L}{\sqrt{3}}.$$

From the expression for the potential gradient it is clear that for a given transmission system, the potential gradient increases as x decreases *i.e.*, the potential gradient is maximum when $x = r$, the surface of the conductor, and this value is given by

$$g_{max} = E_r = E_{max} = \frac{V'd}{r(d-r)\log_e \frac{d}{r}} \simeq \frac{V'}{r \log_e \frac{d}{r}}. \qquad \text{...(4.4)}$$

where r is the radius of conductors in cm, d is the spacing in cm and V′ is the value of voltage of the conductors to neutral, both g_{max} and V being expressed in rms values.

When the disruptive gradient of air is reached at the conductor surface

$$g_0 = \frac{V_{d_0}}{r \log_e \frac{d}{r}}$$

or disruptive critical voltage,

$$V_{d_0} = g_0\, r \log_e \frac{d}{r} \qquad \text{...(4.5)}$$

In practice, corrections have to be applied to the above formula for air density and surface conditions of the conductor and thus the complete formula becomes

$$V_{d_0} = g_0\, \delta m_0 r \log_e \frac{d}{r} \qquad \text{...(4.6)}$$

where δ is the air density factor [Refer to Eq. (4.1)] and m_0 is the conductor surface condition factor.

The value of m_0 may be taken as unity, 0.98 – 0.93 and 0.87 – 0.80 for polished, roughened (or weathered) and stranded conductors respectively.

In the case of stranded conductors the variations in m_0 is duc to the number of strands employed, and the varying ratio between the strand diameter and overall diameter in different sizes of conductors. It also depends on the surface conditions of the conductor, often improving after a conductor has been in service for a long time, and become weathered and the roughness and abrasions oxidised away. Any mutilation of the conductor during construction or stringing lowers m_0 considerably.

The value of V_{d_0} given by Eq. (4.6) is known as *disruptive critical voltage*, and is defined as the minimum phase to neutral voltage at which corona occurs and holds good for fair weather conditions. Its value is considerably reduced during bad atmospheric conditions, such as fog, sleet, rain and snowstorms.

Example 4.1. A single phase overhead line has two conductors of diameter 1 cm with a spacing of 1 m between centres. If the disruptive critical voltage for air is 21.1 kV/cm, for what value of the line voltage will corona commence.

[M.D. Univ. Transmission & Distribution, May 2011]

Solution : Radius of conductors, $r = \frac{1}{2}$ cm = 0.5 cm

Spacing of conductors, d = 1 m = 100 cm

Dielectric strength of air, g_0 = 21.1 kV/cm

Line voltage for commencing of corona

$$= V_{d_0} = g_0 r \log_e \frac{d}{r}$$

$$= 21.1 \times 0.5 \log_e \frac{100}{0.5} = 55.9 \text{ kV} \quad \textbf{Ans.}$$

Example 4.2. Define disruptive critical voltage. A 3-phase, 220 kV, 50 Hz transmission line consists of 1.2 cm radius conductors spaced 2 m at the corners of an equilateral triangle. Calculate the disruptive critical voltage between the lines. Irregularity factor = 0.96; temperature 20°C. Barometric pressure 72.2 cm of mercury. Dielectric strength of air = 21.1 kV (rms)/cm.

[U.P. Technical Univ. Elements of Power System 2005-06]

Solution : Air density factor, $\delta = \frac{3.92b}{273+t} = \frac{3.92 \times 72.2}{273+20} = 0.966$

Disruptive critical voltage to neutral,

$$V_{d_0} = g_0 \delta m_0 r \log_e \frac{d}{r}$$

$$= 21.1 \times 0.966 \times 0.96 \times 1.2 \log_e \frac{200}{1.2} = 120.1 \text{ kV (rms)}$$

Disruptive critical voltage from line-to-line

$$= \sqrt{3}\, V_{d_0} = \sqrt{3} \times 120.1 = 208 \text{ kV (rms)} \quad \textbf{Ans.}$$

Example 4.3. A 132 kV line with 2 cm diameter conductors is built so that corona takes place if the line voltage exceeds 210 kV (rms). If the value of the potential gradient at which ionisation occurs can be taken as 30,000 volts/cm, find the spacing between the conductors.

[Rajasthan Univ. Transmission & Distribution of Electrical Power 2007]

Solution : Conductor radius,

$$r = \frac{\text{Conductor diameter}}{2} = \frac{2}{2} \text{ cm} = 1 \text{ cm}$$

Dielectric strength of air,

$$g_0 = \frac{30{,}000}{\sqrt{2}} = 21{,}210 \text{ V/cm} = 21.21 \text{ kV/cm}$$

Disruptive critical voltage to neutral,

$$V_{d_0} = \frac{210}{\sqrt{3}} = 70\sqrt{3} \text{ kV assuming 3-phase line}$$

Assuming smooth conductors *i.e.*, irregularity factor $m_0 = 1$, standard pressure and temperature for which air density factor, $\delta = 1$ and spacing of conductors d cm, we have

$$70\sqrt{3} = 21.21 \times 1 \times 1 \times 1 \times \log_e \frac{d}{1} \quad \therefore \; V_{d_0} = g_0 m_0 \delta r \log_e \frac{d}{r}$$

or $\log_e d = \frac{70\sqrt{3}}{21.21} = 5.715$

or Spacing between conductors,

$$d = e^{5.715} = 304 \text{ cm or } 3.04 \text{ m} \quad \textbf{Ans.}$$

Example 4.4. A 3-phase overhead transmission line operates at 132 kV, 50 Hz. The conductors are arranged in a 3.0 m delta configuration. What is the minimum diameter of conductor that can be used for no corona under fair weather condition ? Assume air density factor of 0.95 and irregularity factor of 0.85.

[U.P. Technical Univ. Elements of Power System 2004-05]

Solution : Spacing of conductors, d = 3.0 m = 300 cm

Dielectric strength of air, g_0 = 21.21 kV/cm (rms), as assumed

Irregularity factor, m_0 = 0.85

Air density factor, δ = 0.95

Disruptive critical voltage to neutral,

$$V_{d_0} = \frac{132}{\sqrt{3}} = 76.21 \text{ kV}$$

$$76.21 = 21.21 \times 0.85 \times 0.95 \times r \times \log_e \frac{300}{r}$$

$$\because \; V_{d_0} = g_0 m_0 \delta r \log_e \frac{d}{r}$$

or $$r \log_e \frac{300}{r} = \frac{76.21}{21.21 \times 0.85 \times 0.95} = 4.45$$

Solving above equation by trial and error method, we have

$$r = 0.75 \text{ cm}$$

or Diameter of conductors (minimum) = 2 × 0.75 = 1.5 cm **Ans.**

Example 4.5. A conductor with 2.5 cm diameter is passed through a porcelain bushing having ε_r = 4 and internal and external diameters of 3 cm and 9 cm respectively. The voltage between the conductor and an earthed clamp surrounding the porcelain is 20 kV rms. Determine whether corona will be present in the air space round the conductor.

[U.P. Technical Univ. Elements of Power System 2008-09]

Solution : Maximum gradient on the conductor surface,

$$g_{1\,max} = \frac{q}{2\pi\varepsilon_0 r}$$

Maximum gradient on the inner side of the porcelain,

$$g_{2\,max} = \frac{q}{2\pi\varepsilon_0 \varepsilon_r r_1}$$

So, $$g_{2\,max} = \frac{r}{\varepsilon_r r_1} g_{1max} = \frac{1.25}{4 \times 1.5} g_{1max} = 0.2083 g_{1\,max}$$

Now, $$20 = g_{1max}\, r \log_e \frac{1.5}{1.25} + g_{2max} \times 1.5 \log_e \frac{4.5}{1.5}$$

$$= 1.25\, g_{1max} \log_e \frac{1.5}{1.25} + 0.2083\, g_{1max} \times 1.5 \log_e 3$$

$$= 0.2279\, g_{1max} + 0.3433\, g_{1max} = 0.5712\, g_{1max}$$

or $$g_{1\,max} = \frac{20}{0.5712} = 35 \text{ kV/cm.}$$

Since, the gradient exceeds the dielectric strength of air (21.1 kV/cm), corona will be present.

4.7. VISUAL CRITICAL VOLTAGE

In the case of parallel wires it is found that visual corona does not begin at the voltage V_{d_0} at which the disruptive gradient of air g_0 is reached, but at a higher voltage V_{v0}, called the *visual critical voltage*. Visual critical voltage is defined as the minimum phase to neutral voltage at which glow appears all along the line conductors. Thus when corona begins, the potential gradient g_v at the conductor surface is higher than the disruptive gradient g_0. Contrary to what might be expected, g_v, the apparent strength of air is not constant but depends on the size of the conductors, air being apparently stronger at the surface of smaller conductors than larger ones.

Peek states that the disruptive critical voltage must be so exceeded that the stress is greater than the breakdown value up to a distance of $0.3\sqrt{\delta r}$ cm from the conductor. Thus visual corona will occur when the breakdown value is attained at the distance $r + 0.3\sqrt{\delta r}$ from the axis, instead of at the distance r. This requires that the voltage to neutral be $1 + \frac{(0.3)}{\sqrt{\delta r}}$ times the disruptive critical voltage. Thus the visual critical voltage is

$$V_{v0} = g_0 \delta m_v r \left(1 + \frac{0.3}{\sqrt{\delta r}}\right) \log_e \frac{d}{r} \text{ kV (rms) to neutral} \quad ...(4.7)$$

where m_v is a roughness factor, which is unity for smooth conductors. When the wire is stranded or rough, m_v is less than unity. m_v may be taken as 0.98 – 0.93 for rough conductor exposed to atmospheric severities and 0.72 for local corona on stranded conductors.

Because of irregular surface of the conductor, the corona does not start simultaneously on the whole surface but it takes place at different points of the conductor which are pointed and this is known as *local corona*. For this $m_v = 0.72$. For decided or general corona along the length of the conductor m_v may be taken as 0.82.

Example 4.6. Calculate the critical disruptive voltage and critical voltages for local and general corona on a three phase overhead transmission line, consisting of three stranded copper conductors spaced 2.5 m apart at the corners of an equilateral triangle. Air temperature and pressure are 21°C and 73.6 cm Hg respectively. The conductor diameter, irregularity factor and surface factors are 10.4 mm, 0.85 and 0.7, 0.8 respectively.

[*U.P. Technical Univ. Elements of Power System 2002-03*]

Solution : Conductor radius, $r = \frac{10.4}{2}$ mm = 5.2 mm or 0.52 cm

Spacing of conductors, d = 2.5 m = 250 cm

Dielectric strength of air, g_0 = 21.1 kV/cm (rms) as assumed

$$\text{Air density factor, } \delta = \frac{3.92b}{273+t} = \frac{3.92 \times 73.6}{273+21} = 0.9813$$

Critical disruptive voltage to neutral,

$$V_{d_0} = g_0 \delta m_0 r \log_e \frac{d}{r}$$

$$= 21.1 \times 0.9813 \times 0.85 \times 0.52 \log_e \frac{250}{0.52} = 56.52 \text{ kV (rms)}$$

Critical disruptive line-to-line voltage

$$= \sqrt{3} \times 56.52 = 97.89 \text{ kV (rms)}$$

Visual critical voltage to neutral,

$$V_{vo} = g_0 \delta m_v r \left(1 + \frac{0.3}{\sqrt{\delta r}}\right) \log_e \frac{d}{r} \text{ kV (rms)}$$

Here m_v = 0.7 for local corona

and m_v = 0.8 for decided or general corona

Hence V_{vo} for local corona

$$= 21.1 \times 0.9813 \times 0.7 \times 0.52 \left(1 + \frac{0.3}{\sqrt{0.9813 \times 0.52}}\right) \log_e \frac{250}{0.52}$$

$$= 66 \text{ kV (rms)}$$

Line-to-line visual critical voltage for local corona

$$= \sqrt{3} \times 66 = 114.3 \text{ kV (rms)} \quad \textbf{Ans.}$$

Line-to-line visual critical voltage for general corona

$$= \frac{0.8}{0.7} \times 114.3 = 130 \text{ kV (rms)} \quad \textbf{Ans.}$$

Example 4.7. Find the disruptive critical voltage and visual corona voltage (local corona as well as general corona) for a 3-phase 220 V kV line consisting of 22.26 mm diameter conductors spaced in a 6 m delta configuration. The following data can be assumed: Temperature 25°C, pressure 73 cm of mercury, surface factor 0.84, irregularity factor for local corona 0.72, irregularity factor for general (decided) corona 0.82.

[*Rajasthan Univ. Transmission & Distribution of Electrical Power 2005*]

Solution : Conductor radius, $r = \frac{22.26}{2}$ mm

= 11.13 mm or 1.113 cm

Spacing of conductors, d = 6 m = 600 cm

Dielectric strength of air, g_0 = 21.1 kV/cm (rms) as assumed

$$\text{Air density factor, } \delta = \frac{3.92b}{273+t} = \frac{3.92 \times 73}{273+25} = 0.96$$

Critical disruptive voltage to neutral,

$$V_{d_0} = g_0 \delta m_0 r \log_e \frac{d}{r}$$

$$= 21.1 \times 0.96 \times 0.84 \times 1.113 \log_e \frac{600}{1.113}$$

= 119 kV per phase

Critical disruptive line-to-line voltage

$$= \sqrt{3} \times 119 = 206 \text{ kV} \quad \textbf{Ans.}$$

Visual critical voltage to neutral,

$$V_{v0} = g_0 \delta m_v r \left(1 + \frac{0.3}{\sqrt{\delta r}}\right) \log_e \frac{d}{r} \text{ kV (rms)}$$

Here m_v = 0.72 for local corona

and m_v = 0.82 for decided or general corona

$$= 21.1 \times 0.96 \times 0.72 \times 1.113 \left(1 + \frac{0.3}{\sqrt{0.96 \times 1.113}}\right) \log_e \frac{600}{1.113}$$

= 131.73 kV/phase

Line-to-line visual critical voltage for local corona

$= \sqrt{3} \times 131.73 = 228$ kV **Ans.**

Line-to-line visual critical voltage for general corona

$= \frac{0.82}{0.72} \times 228 = 134.6$ kV **Ans.**

Example 4.8. Find the disruptive critical and visual corona voltages of a grid line operating at 132 kV. The following data is given:

Conductor diameter 1.9 cm, conductor spacing = 3.81 cm, temperature = 44°C, barometric pressure = 73.7 cm, conductor surface factor: fine weather = 0.8, rough weather = 0.66.

[W.B. Technonological Univ. Power System-I, 2009-10]

Solution : Conductor radius $= \frac{1.9}{2}$ cm = 0.95 cm

Spacing of conductors, d = 3.81 m = 381 cm

Dielectric strength of air, g_0 = 21.1 kV/cm (rms) as assumed

Air density factor, $\delta = \frac{3.92b}{273+t} = \frac{3.92 \times 73.7}{273+44} = 0.91$

Irregularity factor, m_0 = 0.85 (assumed)

Critical disruptive voltage to neutral,

$$V_{d_0} = g_0 \delta m_0 r \log_e \frac{d}{r}$$

$$= 21.1 \times 0.91 \times 0.85 \times 0.95 \log_e \frac{381}{0.95}$$

= 92.94 kV (rms)

Line-to-line disruptive critical voltage

$= \sqrt{3} \times 92.94 = 160.97$ kV **Ans.**

Visual critical voltage to neutral (fine weather)

$$V_{v_0} = g_0 \delta m_v r \left[1 + \frac{0.3}{\sqrt{\delta r}}\right] \log_e \frac{d}{r}$$

$$= 21.1 \times 0.91 \times 0.8 \times 0.95 \left(1 + \frac{0.3}{\sqrt{0.91 \times 0.95}}\right) \log_e \frac{381}{0.95}$$

= 115.7 kV

Line-to-line visual critical voltage (fine weather)

$= \sqrt{3} \times 115.7 = 200$ kV **Ans.**

Line-to-line visual critical voltage (rough weather)

$= \frac{0.66}{0.8} \times 200 = 165$ kV **Ans.**

4.8. CORONA POWER LOSS

The formation of corona is associated with a loss of power, which will have some effect on the efficiency of the transmission line, but will not be of much importance to have any appreciable effect on the voltage regulation.

As in the case of the critical voltage for corona, the power loss is affected both by atmospheric and line conditions.

Under the fair weather conditions, the following empirical formula for the corona power loss has been ascertained due to Peek

$$P_c = \frac{244}{\delta}(f+25)\sqrt{\frac{r}{d}}\left(V_{ph} - V_{d_0}\right)^2 \times 10^{-5} \text{ kW/km/phase} \quad ...(4.8)$$

where V_{ph} is voltage to neutral in kV, V_{d_0} is the disruptive critical voltage to neutral in kV and f is the supply frequency in Hz.

Under the stormy weather conditions, V_{d_0} is taken to be 0.8 times its fair weather value and power loss due to corona is given by

$$P'_c = \frac{244}{\delta}(f+25)\sqrt{\frac{r}{d}}\left(V_{ph} - 0.8V_{d_0}\right)^2 \times 10^{-5} \text{ kW/km/phase} \quad ...(4.9)$$

As a matter of fact, with perfectly smooth and cylindrical conductors no loss occurs until the visual critical voltage is reached, when the loss suddenly takes a definite value as calculated by the Eq. (4.8) or Eq. (4.9). It then follows the quadratic law for higher voltages. The empirical relation as derived by Peek has certain limitations and holds good only under certain conditions (*i.e.,* the supply frequency lies between 25 and 120 Hz, the conductor radius is greater than 2.5 mm and ratio $\frac{V_{ph}}{V_{d_0}}$ exceeds 1.8). Also a small error in m_0, the irregularity factor, will lead to wrong results when using the above formula.

When the ratio $\frac{V_{ph}}{V_{d_0}}$ is less than 1.8 Peterson's formula is to be applied for determining the corona losses and is given as

$$P_c = \frac{21 \times 10^{-6} f (V_{ph})^2}{\left(\log_{10} \frac{d}{r}\right)^2} \times K \text{ kW/km/phase} \quad ...(4.10)$$

where K is a factor which varies with the ratio of $\frac{V_{ph}}{V_{d_0}}$ as given below :

V_{ph}/V_{d_0}	0.6	0.8	1.0	1.2	1.4	1.6	1.8	2.0	2.2
K	0.012	0.018	0.05	0.08	0.30	1.0	3.5	6.0	8.0

Example 4.9. A 3-phase 220 kV 50 Hz transmission line consists of 2 cm radius conductor spaced 2.5 m apart in equilateral triangular formulation. If the temperature is 20°C and atmospheric pressure 75 cm, m_0 = 0.80, determine the corona loss per km of line. *[Rajasthan Technical Univ. Transmission & Distribution of Electrical Power 2008]*

Solution : Conductor radius = 2.0 cm

Spacing of conductors, d = 2.5 m = 250 cm

Dielectric strength of air = 21.1 kV/cm (rms), as assumed

Irregularity factor, m_0 = 0.80

Air density factor, $\delta = \frac{3.92b}{273+t} = \frac{3.92 \times 75}{273+20} = 1.00$

Critical disruptive critical voltage to neutral,

$$V_{d_0} = g_0 \delta m_0 r \log_e \frac{d}{r}$$

$$= 21.1 \times 1.0 \times 0.8 \times 2 \log_e \frac{250}{2} = 163 \text{ kV}$$

Supply voltage/phase $= \frac{220}{\sqrt{3}} = 127$ kV

Corona loss,

$$P_c = \frac{244}{\delta}(f+25)\sqrt{\frac{r}{d}}(V_{ph} - V_{d_0})^2 \times 10^{-5} \text{ kW/km/phase}$$

$$= \frac{244}{1.0}(50+25)\sqrt{\frac{2}{250}}(127-163)^2 \times 10^{-5} \text{ kW/km/phase}$$

$$= 21.2 \text{ kW/km/phase}$$

Total corona loss per km for 3-phase line

= 3 × 21.2 = 63.6 kW **Ans.**

Example 4.10. A certain 3-phase equilateral transmission line has a total corona loss of 53 kW at 106 kV and a loss of 98 kW at 110.9 kV. What is the disruptive critical voltage between lines ? What is the corona loss at 113 kV ?

[Elements of Power System U.P. Technical Univ. 2004-05; G.B. Technical Univ. 2011-12]

Solution : Power loss due to corona for three phases is given by

$$P_c = 3 \times \frac{244}{\delta}(f+25)\sqrt{\frac{r}{d}}(V_{ph} - V_{d_0})^2 \times 10^{-5} \text{ kW/km}$$

Taking air density factor δ, supply frequency *f*, radius of conductor *r* and spacing of conductors *d* constant

$$P_c \,\alpha\, (V_{ph} - V_{d_0})^2$$

$$\therefore \quad 53 \,\alpha\, \left(\frac{106}{\sqrt{3}} - V_{d_0}\right) \alpha \left(61.2 - V_{d_0}\right)^2 \qquad ...(i)$$

$$\text{and} \quad 98 \,\alpha\, \left(\frac{110.9}{\sqrt{3}} - V_{d_0}\right)^2 \alpha \left(64 - V_{d_0}\right)^2 \qquad ...(ii)$$

Dividing Eq. (*ii*) by Eq. (*i*), we get

$$\frac{98}{53} = \frac{(64 - V_{d_0})^2}{(61.2 - V_{d_0})^2}$$

or $V_{d_0} = 53.4$ kV

and total corona loss at 113 kV

$$P_c \,\alpha\, \left(\frac{113}{\sqrt{3}} - V_{d_0}\right)^2 \alpha \left(65.2 - V_{d_0}\right)^2 \qquad ...(iii)$$

Dividing Eq. (*iii*) by Eq. (*i*), we get

$$\frac{P_c}{53} = \frac{(65.2 - V_{d_0})^2}{(61.2 - V_{d_0})^2} = \frac{(65.2 - 53.4)^2}{(61.2 - 53.4)^2}$$

or Corona loss,

$$P_c = 53 \times \left(\frac{11.8}{7.8}\right)^2 = 121.3 \text{ kW } \textbf{Ans.}$$

Disruptive critical voltage between lines

$$= \sqrt{3} \times V_{d_0} = \sqrt{3} \times 53.4 = 92.5 \text{ kV } \textbf{Ans.}$$

Example 4.11. Determine (*i*) the critical disruptive voltage (*ii*) the visual critical voltage; and (*iii*) the corona loss under foul weather condition for 3-phase line 160 km long, conductor dia. 1.036 cm; 2.44 m delta spacing. Air temperature 26.6°C, corresponding to an approximate barometric pressure of 73.15 cm of mercury, operating voltage 110 kV at 50 Hz, surface irregularity factor 0.85. Assume a value of $m_v = 0.72$.

Disruptive voltage under foul weather = 0.8 × fair weather value.

[A.M.I.E. Sec. B. Power Systems Winter 1998]

Solution : Conductor radius, $r = \frac{1.036}{2} = 0.518$ cm

Spacings of conductors, d = 2.44 m = 244 cm

Dielectric strength of air, g_0 = 21.1 kV/cm (rms), as assumed

Irregularity factor, m_0 = 0.85

Roughness factor, m_v = 0.72

$$\text{Air density factor, } \delta = \frac{3.92b}{273+t} = \frac{3.92 \times 73.15}{273+26.6} = 0.957$$

Critical disruptive voltage to neutral,

$$V_{d_0} = g_0 \delta m_0 r \log_e \frac{d}{r}$$

$$= 21.1 \times 0.957 \times 0.85 \times 0.518 \log_e \frac{244}{0.518}$$

$$= 54.72 \text{ kV (rms) } \textbf{Ans.} \qquad ...(i)$$

Visual critical voltage to neutral,

$$V_{v0} = g_0 \delta m_v r \left(1 + \frac{0.3}{\sqrt{\delta r}}\right) \log_e \frac{d}{r}$$

$$= 21.1 \times 0.957 \times 0.72 \times 0.518 \left(1 + \frac{0.3}{\sqrt{0.957 \times 0.518}}\right) \log_e \frac{244}{0.518}$$

$$= 66.1 \text{ kV (rms) } \textbf{Ans.} \qquad ...(ii)$$

Voltage to which the conductors are subjected,

$$V_{ph} = \frac{110}{\sqrt{3}} = 63.51 \text{ kV} \qquad ...(iii)$$

Hence from Eqs. (*i*), (*ii*) and (*iii*) it is obvious that corona will take place but it will not be visual.

Corona loss under foul weather condition (using Peek's formula under stormy weather conditions),

$$P'_c = \frac{244}{\delta}(f+25)\sqrt{\frac{r}{d}}\left(V_{ph} - 0.8V_{d_0}\right)^2 \times 10^{-5} \text{ kW/km/phase}$$

$$= \frac{244}{0.957}(50+25)\sqrt{\frac{0.518}{244}}(63.51 - 0.8 \times 54.72)^2 \times 10^{-5} \text{ kW/km/phase}$$

$$= 3.43 \text{ kW/km/phase}$$

Total corona loss for 160 km, 3-phase line

= 160 × 3 × 3.43 = 1,647 kW **Ans.**

Corona loss under foul weather condition (using Peterson formula for stormy weather conditions)

$$\text{In this case } \frac{V_{ph}}{V_{d_0}} = \frac{63.51}{0.8 \times 54.72} = 1.45$$

For this value of $\frac{V_{ph}}{V_{d_0}}$, K is 0.46 (see Table given in the text on page 81)

$$\text{Corona loss} = \frac{21\times10^{-6} f(V_{ph})^2}{\left(\log_{10}\frac{d}{r}\right)^2} \times K \text{ kW/km/phase}$$

$$= \frac{21\times10^{-6}\times50\times(63.51)^2\times0.46}{\left(\log_{10}\frac{244}{0.518}\right)^2}$$

$$= 2.726 \text{ kW/km/phase}$$

Total corona loss for 160 km, 3-phase line

= 160 × 3 × 2.726 = 1,308.5 kW **Ans.**

Example 4.12. Determine the corona characteristics of a 3-phase line 160 km long. Conductor diameter 1.036 cm, 2.44 m delta spacing air temperature 26.67°, altitude 2,440 m, corresponding to an approximate barometric pressure of 73.15 cm, operating voltage of 110 kV at 50 Hz.

[*Rajasthan Technical Univ. Transmission & Distribution of Electrical Power 2011*]

Solution : Conductor radius, $r = \frac{1.036}{2} = 0.518$ cm

Spacing of conductors, d = 2.44 m = 244 cm

Dielectric strength of air = 21.1 kV/cm (rms), as assumed

Irregularity factor, m_0 = 0.85, as assumed

Roughness factor, m_v = 0.72 for local corona, as assumed

Roughness factor, m'_v = 0.82 for general corona, as assumed

Voltage to which the conductors are subjected,

$$V_{ph} = \frac{110}{\sqrt{3}} = 63.51 \text{ kV} \quad ...(i)$$

$$\text{Air density factor, } \delta = \frac{3.92b}{273+t} = \frac{3.92\times73.15}{273+26.67} = 0.957$$

Critical disruptive voltage to neutral,

$$V_{d_0} = g_0\delta m_0 r \log_e\frac{d}{r}$$

$$= 21.1 \times 0.957 \times 0.85 \times 0.518 \log_e\frac{244}{0.518}$$

$$= 54.72 \text{ kV (rms)} \quad ...(ii)$$

Visual critical voltage to neutral in case of local corona,

$$V_{v0} = g_0\delta m_v r\left(1+\frac{0.3}{\sqrt{\delta r}}\right)\log_e\frac{d}{r}$$

$$= 21.1\times0.957\times0.72\times0.518\left(1+\frac{0.3}{\sqrt{0.957\times0.518}}\right)\log_e\frac{244}{0.518}$$

$$= 66 \text{ kV (rms)} \quad ...(iii)$$

Visual critical voltage to neutral in case of general corona,

$$V_{v0} = 66\times\frac{m'_v}{m_v} = 66\times\frac{0.82}{0.72} = 75.29 \text{ kV (rms)} \quad ...(iv)$$

From Eqs. (*i*), (*ii*), (*iii*) and (*iv*) it is obvious that the corona will take place but it will not be visual.

Power loss due to corona for fair weather conditions

According to Peek's formula,

$$P_c = \frac{244}{\delta}(f+25)\sqrt{\frac{r}{d}}\left(V_{ph}-V_{d0}\right)^2\times10^{-5} \text{ kW/km/phase}$$

$$= \frac{244}{0.957}(50+25)\sqrt{\frac{0.518}{244}}(63.51-54.72)^2\times10^{-5}$$

$$= 0.68 \text{ kW/km/phase}$$

Total corona loss for 160 km, 3-phase line

= 160 × 3 × 0.68 = 326.8 kW **Ans.**

According to Peterson formula

$$\frac{V_{ph}}{V_{d0}} = \frac{63.51}{54.72} = 1.16$$

For this value of $\frac{V_{ph}}{V_{d0}}$,

K = 0.0758 (see Table given in the text on page 81)

$$\text{and } P_c = \frac{21\times10^{-6}\times f(V_{ph})^2}{\left(\log_{10}\frac{d}{r}\right)^2}\times K \text{ kW/km/phase}$$

$$= \frac{21\times10^{-6}\times50\times(63.51)^2}{\left(\log_{10}\frac{244}{0.518}\right)^2}\times0.0758 = 0.045 \text{ kW/km/phase}$$

Total corona loss for 160 km, 3-phase line

= 160 × 3 × 0.045 = 216 kW **Ans.**

Power loss due to corona for stormy weather conditions

Disruptive critical voltage,

V_{d0} = 0.8 = 0.8 × 54.72 = 43.776 kV

Total corona loss for 160 km, 3-phase line, according to Peek's formula,

$$P'_c = 160\times3\times\frac{244}{\delta}(f+25)\sqrt{\frac{r}{d}}(V_{ph}-V_{d0})^2\times10^{-5} \text{ kW}$$

$$= 160\times3\times\frac{244}{0.957}(50+25)\times\sqrt{\frac{0.518}{244}}\times(63.51-43.776)^2\times10^{-5} \text{ kW}$$

= 1,647 kW **Ans.**

According to Peterson formula

$$\frac{V_{ph}}{V_{d0}} = \frac{63.51}{43.776} = 1.45$$

For this value of $\frac{V_{ph}}{V_{d0}}$,

K = 0.4575 (see Table given in the text on page 81)

and Corona loss,

$$P'_c = \frac{21 \times 10^{-6} \times f \times (V_{ph})^2}{\left(\log_{10} \frac{d}{r}\right)^2} \times K \text{ kW/km/phase}$$

$$= \frac{21 \times 10^{-6} \times 50 \times (63.51)^2}{\left(\log_{10} \frac{244}{0.518}\right)^2} \times 0.4575 = 2.71 \text{ kW/km/phase}$$

Total corona loss

= 160 × 3 × 2.71 = 1,300 kW **Ans.**

4.9. FACTORS AFFECTING CORONA LOSS

There are numerous factors such as system frequency, system voltage, air conductivity, air density, conductor radius, surface of conductor, load current and atmospheric conditions, on which the corona loss depends.

1. Effect of System Frequency. From Eqs. (4.8), (4.9) and (4.10) it is obvious that corona loss varies directly as the system frequency.

2. Effect of System Voltage. The electric field in the space around the conductors depends mainly on the potential difference between the conductors. Greater the potential difference, greater is the electric field and, therefore, greater is the power loss due to corona. In the region near the disruptive critical voltage, the rate of increase of corona power loss with the increase in system voltage is small but when V_{ph} is large as compared with V_{d_0} corona loss increases at a very fast rate with the increase in system voltage. This fact is quite evident from Eqs. (4.8), (4.9) and (4.10).

3. Effect of Conductivity of Air. The conductivity of air depends on the number of ions per unit volume of air, the size and charge per ion. Both of these factors vary with altitude and atmospheric conditions. During rain and thunderstorms, ion content increases and, therefore, atmosphere becomes more conducting. Higher conductivity leads to greater corona loss.

4. Effect of Density of Air. Corona loss increases with the decrease in the density of air. The corona loss of a transmission line passing through a hilly area is higher than that of a similar line in plains due to reduced value of δ at high altitudes.

5. Effect of Conductor Radius. The electric field intensity decreases with the increase in radius of conductor. Hence with conductors of large radius, electric field intensity decreases resulting in lower corona power loss.

6. Effect of Conductor Surface. The potential gradient at the surface of a stranded conductor is greater than that for the equivalent solid conductor. So breakdown voltage is low and corona loss is more for stranded conductors.

Roughness of the surface of the conductor causes a field distortion and gives rise to high potential gradient causing higher corona loss.

7. Effect of Atmospheric Conditions. Corona loss, particularly local corona discharges, increases in rain and bad atmospheric conditions, such as fog, sleet and snowstorms. The effect is more pronounced if the conductor is of larger diameter because on smaller diameter conductors the radius of a rain drop may not be very much smaller than the radius of the conductor itself.

8. Effect of Load Current. The heating of the conductor due to flow of load current through it has an indirect reducing effect on the corona loss. Heating of conductor prevents deposition of dew or snow on the surface of the conductor and reduces corona loss.

During rains, the heating of conductor has no effect on the corona loss but, after the rain it accelerates the drying of the conductor surface. The time duration for which the drops remain on the surface is reduced and so the corona loss is reduced.

For long transmission lines passing through routes of varying altitudes, the average value of corona loss is determined by determining the corona loss per km at a number of points and then taking out an average.

9. Bundling of Conductors. A bundled conductor consists of two or more parallel sub-conductors at a spacing of several diameters.These group of conductors form the phase conductors.Thus the effective diameter of the bundled conductor is much larger than that of the equivalent single conductor. The value of large diameter, as obvious from the equation for corona loss, will reduce the corona loss.

4.10. METHODS OF REDUCING CORONA EFFECTS

The critical voltage can be raised either by increasing the spacing between the conductors or the diameter of conductors. By increasing the spacing between the conductors, the voltage at which corona occurs is raised and hence corona effects can be avoided. However, the spacing cannot be increased to a large extent because it will result in heavier supporting structures (*e.g.*, the bigger cross arms and supports) and therefore high cost. The diameter of the conductors can be increased by using hollow conductors with a hemp core. Steel-cored aluminium conductors have a large diameter for a given conductivity and weight, and are thus good from the point of view of corona.

By using bundled conductors also corona effects can be avoided because effective diameter of the bundled conductor is much larger than that of the equivalent single conductor.

4.11. VOLTAGE LIMITATIONS OF LINES

The basis for the design of a transmission line is essentially financial, the most economical line being the most acceptable.

It is because power loss due to corona is of no great importance. It is considered satisfactory to design a line for operation at a voltage just below the disruptive critical voltage for fair weather (taking $\delta = 1$). It is economical to have a small corona loss in bad weather *i.e.,* for a fraction of the year (storms are experienced at intervals) rather than have large conductors and heavy supporting structures to avoid corona entirely.

Moreover, corona is considered beneficial because of its inherent advantages, mentioned in Art. 4.4.

Exercises

1. Explain the phenomenon of corona. What are the factors affecting corona ?
[Transmission & Distribution of Electrical Power Rajasthan Univ. 1998, 1999, 2002, 2005, 2007; Rajasthan Technical Univ. 2008, 2009]
2. What is corona ? Explain briefly the factors which affect corona. **[W.B. Technological Univ. Power System-I, 2009-10]**
3. Explain the situation under which corona phenomenon starts in high voltage power transmission lines. Identify the factors that affect the corona alongwith reasoning.
[U.P. Technical Univ. Elements of Power System 2005-06]
4. Explain corona and discuss its effects on performance of transmission lines.
[U.P. Technical Univ. Elements of Power System 2006-07]
5. Explain various factors which affect corona loss.
[M.D. Univ. Power System-I, Dec. 2009]
6. Explain the factors which determine the formation of corona on overhead lines. Derive an expression for the disruptive critical voltage between two smooth circular conductors, assuming the breakdown strength of air to be 30 kV/cm (peak).
[U.P. Technical Univ. Elements of Power System 2003-04]
7. Deduce from first principles expressions for the disruptive critical voltage between two smooth circular wires. Find the ratio of critical voltages when the distance between the conductors is halved, other data remaining the same.
[U.P. Technical Univ. Elements of Power System 2007-08]
8. What is corona ? Show that the maximum critical disruptive voltage occurs when the radius of conductor is $\frac{d}{e}$ where d is the distance between conductors.
[U.P. Technical Univ. Elements of Power System 2007-08]
9. Define with reference to corona the disruptive critical and visual critical voltages. Discuss the factors which affect these voltages.
[Rajasthan Technical Univ. Transmission & Distribution of Electrical Power 2009]
10. State and explain the Peek's empirical formula for calculation of corona power loss for a power transmission line. **[U.P. Technical Univ. Elements of Power System 2005-06]**
11. Write and explain the expression for power loss due to corona. What factors affect the corona loss ?
[U.P. Technical Univ. Elements of Power System 2009-10]
12. Discuss advantages and disadvantages of corona. How can corona loss be minimized ?
[U.P. Technical Univ. Elements of Power System 2008-09]
13. Discuss the factors on which corona loss depends. Suggest different means of reducing corona loss.
[U.P. Technical Univ. Elements of Power System 2007-08]
14. How is corona affected by line voltage, atmospheric conditions and conductor size, shape and condition.
[M.D. Univ. Power System-I, Dec. 2007]
15. Explain the role of bundled conductors in reducing corona loss.
[U.P. Technical Univ. Elements of Power System 2004-05]
16. Explain 'corona loss' at extra high voltages and its disadvantages. How is disruptive critical voltage estimated ? What are the effects of 'air density', 'temperature' and 'bundling' of conductors on corona loss ?
17. Explain various methods of reducing corona loss.
[M.D. Univ. Power System-I, May 2009]
18. Write short notes on the following:
(*i*) Phenomenon of corona. **[M.D. Univ. May 2008]**
(*ii*) Corona. **[M.D. Univ. Dec. 2008]**
(*iii*) Factors affecting corona loss.
[Rajasthan Univ. 2002, 2003, 2004]
(*iv*) Corona loss. **[M.D. Univ. Dec. 2010]**
(*v*) Methods of reducing corona loss.
[M.D. Univ. May 2008]

Short Answer Type Questions With Answers

Q. 1. What is corona ?

Ans. When the voltage between conductors of an overhead line exceeds the disruptive critical voltage value, a hissing noise accompanied by a violet glow appears. This phenomenon is called the corona.

Q. 2. What are the various factors which affect corona ?

Ans. Factors affecting corona are (*i*) number of ions, size and charge per ion, mean free path depending on atmospheric conditions, (*ii*) line voltage, (*iii*) ratio of conductor spacing to conductor radius, (*iv*) contour of the conductor surface, and (*v*) state of the conductor surface.

Q. 3. What are the adverse effects of corona ?

Ans. Adverse effects of corona are (*i*) power loss but not very important except under abnormal weather conditions, (*ii*) corrosion due to production of ozone gas, and (*iii*) interference with neighbouring communication circuits.

Q. 4. What do you mean by dielectric strength of air ?

Ans. The value of potential gradient at which complete disruption of air occurs, is called the dielectric strength of air.

Q. 5. What is meant by disruptive critical voltage ?

Ans. The disruptive critical voltage is defined as the minimum phase to neutral voltage at which corona occurs.

Q. 6. What is visual critical voltage ?

Ans. Visual critical voltage is defined as the minimum phase to neutral voltage at which glow appears all along the line conductors.

Q. 7. What are the various factors which affect corona loss ?

Ans. The factors affecting corona loss are system frequency, system voltage, air conductivity, air density, conductor diameter, conductor surface condition, atmospheric conditions and load current.

Q. 8. How can the corona effects be minimised ?

Ans. Corona effects can be minimised in a transmission line by using large diameter conductors which may be accomplished by using hollow conductors with a hemp core, ACSR conductors or bundled conductors. Corona effects can also be reduced but to a limited extent, by increasing the spacing between the conductors.

Problems

1. A 3-phase line has conductors of radius 0.8 cm, spaced equilaterally 2.5 m apart. If the dielectric strength of air is 30 kV/cm, determine the disruptive critical voltage at which corona will occur. Take relative air density factor δ = 0.96 and irregularity factor m_0 = 0.94. [**Ans.** 152 kV (rms)]

2. Find the disruptive critical voltage for a transmission line having :

Conductor spacing = 1 m; conductor (stranded) radius = 1 cm;

barometric pressure = 76 cm of Hg; temperature = 40 °C;

Air breakdown potential gradient (at 76 cm of Hg and at 25°C)= 21.1 kV (rms)/cm.

[Gujarat Univ. Electrical Power System-I April, 1970]

[**Ans.** 136 kV (rms)]

3. Find the disruptive critical and visual corona voltage of a grid line operating at 132 kV.

Conductor diameter = 1.9 cm; conductor spacing = 3.81 m; temperature = 44°C; barometric pressure = 73.7 cm;

Conductor surface factor: fine weather = 0.8; rough weather = 0.66. **[Gujarat Univ. Elec. Power System-I Oct., 1970]**

[**Ans.** 155.4 kV/phase, 157.6 kV/phase]

4. Find the critical disruptive voltage and the critical voltages for local and decided corona on a 3-phase line consisting of three stranded conductors spaced 2.5 m apart at the corners of an equilateral triangle. Air temperature and pressure are 21°C and 73.6 cm of Hg respectively. The conductor diameter, irregularity factor, surface factor for local and decided corona are 10.4 mm, 0.85, 0.7 and 0.8 respectively.

[A.M.I.E. Sec. B. Power Systems Summer 1995]

[**Ans.** 97.89 kV (rms); 114.46 kV (rms); 130.82 kV (rms)]

5. Determine the corona loss for 3-phase line operating at 110 kV which has conductors of 1.25 cm dia arranged in a 3.05 m delta configuration. Assume polished wires and air density factor of 1.07.

[A.M.I.E Sec. B. Power Systems Summer 1997]

[**Ans.** No corona loss]

6. A 3-phase, 220 kV, 50 Hz transmission line consists of 3.0 cm diameter conductors spaced 2 metres apart in equilateral triangle formation. If the temperature is 20 °C and atmospheric pressure 75 cm, determine the corona loss per km of the line. Take irregularity factor as 0.80.

[**Ans.** 0.425 kW/km]

❑❑❑

CHAPTER 5

Transmission Line Constants

INSIDE THIS CHAPTER

5.1. INTRODUCTION

A transmission line consists of a set of conductors carrying electrical energy in bulk and transmitting it from power stations to primary substations. The conductors are parallel to each other and are carried on supports which provide insulation between the different conductors and between each conductor and earth. Figs. 5.1(*a*), (*b*) illustrate the transmission line in which line conductors are strung on insulators fixed to the cross arms. Single circuit tarnsmission line [Fig. 5.1(*a*)] carries 3 line conductors pertaining to three phases R, Y and B and one ground wire running over the top of the towers. Double circuit transmission line [Fig. 5.1(*b*)] carries 6 line conductors. These 6 line conductors constitute two separate circuit, each consists of three line conductors pertaining to phases R, Y and B. In case of double circuit transmission line two wires, known as ground wires, run over the top of the towers.

Transmission lines are basically electrical circuits having distributed constants—resistance, inductance, capacitance and shunt conductance. The shunt conductance is mostly due to leakage over the insulators and is so small that it can be neglected. These line constants, also called the *line parameters,* are uniformly distributed along the entire length of the line and are usually expressed as resistance, inductance and capacitance per unit length. Concentration of all such parameters for the complete length of line at a single point is not possible. The performance of a transmission line depends to a considerable extent upon these parameters–series resistance causes a power loss in the conductor and so affects the transmission efficiency,

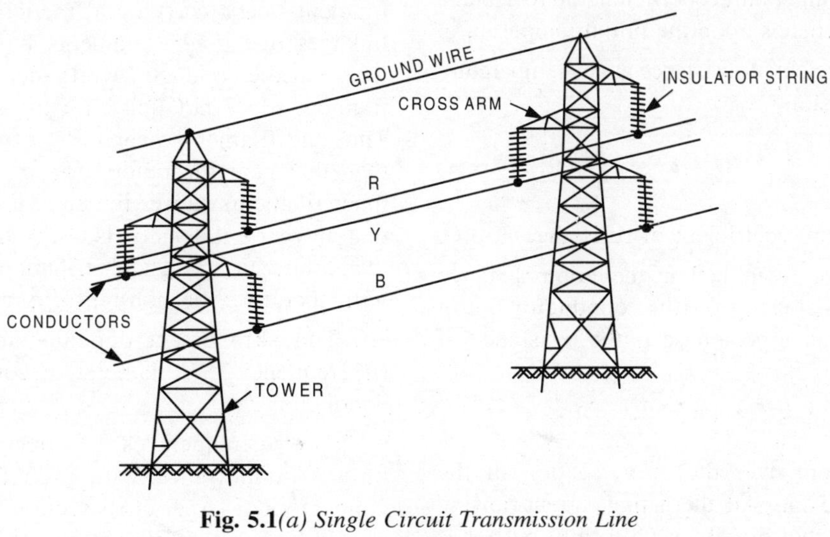

Fig. 5.1(*a*) *Single Circuit Transmission Line*

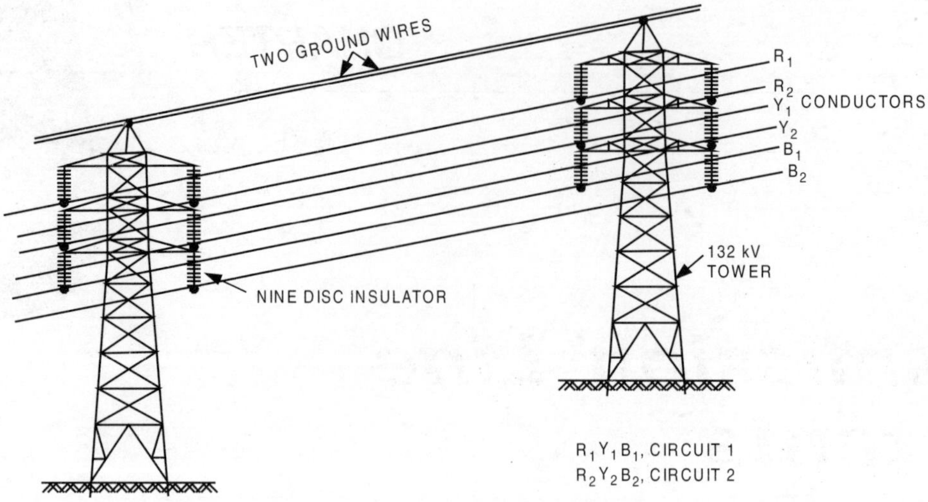

(b) Double Circuit Transmission Line

Fig. 5.2

series inductance mainly governs the power transmission capacity of the line, and shunt capacitance causes a charging current to flow in the line.

5.2. LINE RESISTANCE

Every electric conductor offers opposition to the flow of current and this opposition is called the *resistance*. The resistance of transmission line conductors is the most important cause of power loss (I^2R) in a transmission line. The ohmic resistance R of a conductor of length l and uniform cross section a is given by the expression

$$R = \rho \frac{l}{a} \qquad ...(5.1)$$

where the resistivity or specific resistance of the conductor (ρ) depends not only on the conductor material but also on its temperature.

If ρ_1 and ρ_2 are the values of resistivity corresponding to two different temperatures t_1 and t_2, then

$$\rho_2 = \rho_1[1 + \alpha\,(t_2 - t_1)] \qquad ...(5.2)$$

where α is the temperature coefficient of resistance of the material. The value of temperature coefficient of resistance is also not constant but depends upon the initial temperature.

The temperature coefficient of resistance at any temperature t_1 is given by the expression

$$\alpha_1 = \frac{\alpha_0}{1 + \alpha_0 t_1} \qquad ...(5.3)$$

where α_0 is the temperature coefficient of resistance at 0ºC.

In a single phase line or 2-wire dc line, the loop resistance is taken as double of resistance of either conductor but in 3-phase system the resistance per phase is the resistance of either conductor.

5.3. SKIN EFFECT

The distribution of current over the cross section of the conductor is uniform for dc only. An alternating current flowing through a conductor does not distribute uniformly but tends to concentrate near the surface of the conductor. In fact, in ac system no current flows through the core and the entire current is concentrated at the surface regions. The result is that the effective area of the conductor is reduced causing an increase in its ac resistance. The effective ac resistance is usually referred as the *effective resistance* of the conductor. The phenomenon is called the *skin effect* as it causes concentration of current at the skin of the conductor.

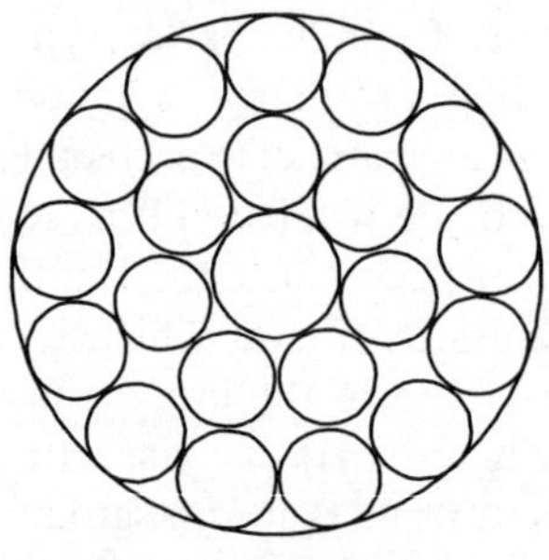

Fig. 5.2(*a*)

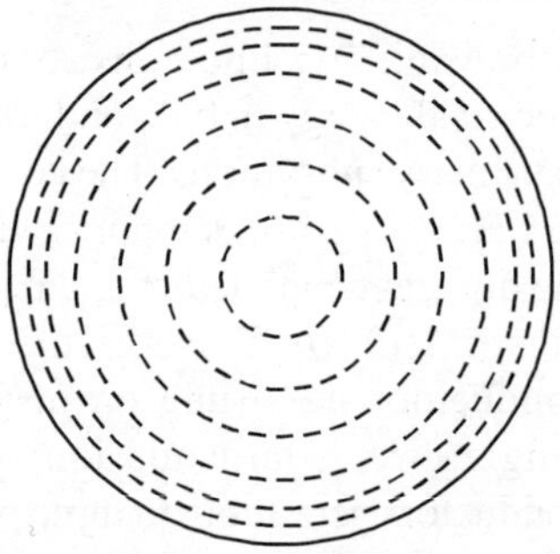

Fig. 5.2(*b*)

The skin effect can be easily explained by considering a solid conductor to be composed of a large number of annular filaments, each carrying a fraction of total current. The flux linkages due to the filaments lying at the surface link the whole of the conductor while the flux set up due to the inner filaments does not link with the surface or outer filaments. Thus the filaments near the centre are of larger inductance than that near the outer surface. The high reactance of the inner filaments causes the current to distribute in such a way that the current density is less in the interior of the conductor than at the surface. The distribution of current over the section will, therefore, be non-uniform, as shown in Fig. 5.2(*b*).

The skin effect depends upon (*i*) type of material (*ii*) frequency (*iii*) diameter of conductor, and (*iv*) shape of conductor.

At low frequencies the effect is very small, in fact it is only of importance with high frequencies or with solid conductors of larger cross section. For commercial frequency of 50 Hz or less the increase in effective resistance is inappreciable for solid copper conductors up to 1 cm in diameter; about 2.5 per cent for 2 cm diameter and 8 per cent for 2.5 cm diameter. In an aluminium wire the effect is the same as in a copper wire of equal conductivity. Thus since the resistivity of copper is 0.6 times that of aluminium, the increased resistance due to skin effect on an aluminium wire of a square mm in cross section will be of the same percentage as on a 0.6 a square mm in copper wire.

The skin effect is much smaller with stranded conductors than with solid conductors. It increases with the increase of cross section, permeability and supply frequency.

In practice stranded conductors are invariably used for transmission and distribution lines and hollow conductors for solid bus-bar. This is done in order to overcome the adverse effects of skin effect.

5.4. PROXIMITY EFFECT

The inductance and, therefore, the current distribution in a conductor is also affected by the presence of other conductors in its vicinity. This effect is known as the *proximity effect.*

While considering the skin effect it was assumed that there is no other current carrying conductor nearby but when a current carrying conductor is nearby, its flux will link with the conductor under consideration and its effect to the nearer half of the conductor will be more than with the farther half.

If conductors carry currents in the opposite directions the magnetic fields set up will tend to cause an increase in the current density in the adjacent portions of the conductors while if the currents are in the same direction, the current density is increased in remote parts of the conductor.

So like skin effect, proximity effect affects the current distribution and results in an increase in the resistance of the conductor and decrease of self reactance. Like skin effect, the proximity effect depends on the conductor size, frequency of the supply, resistivity and relative permeability of the material. This phenomenon (*i.e.*, proximity effect) is more pronounced for large conductors, high frequencies and close proximity. The magnitude of the effect, at normal supply frequencies, in the case of the wide spacing of conductors required for overhead transmission lines, is so small that it can be ignored. However, the proximity effect is pronounced in case of cables where the spacing between the conductors is small.

In the case of stranded conductors, each wire traverses alternately weaker and stronger portions of the magnetic field caused by the external current carrying conductor. Thus the average value of the field along the path of any wire remains the same, and assuming that the currents in the conductor follow to a large extent the paths of the individual wires, the proximity effect is substantially eliminated.

5.5. LINE INDUCTANCE

Inductance is the property by virtue of which a circuit opposes changes in the value of a varying current flowing through it. While the resistance of a circuit opposes the flow of both steady as well as varying currents, the inductance causes opposition only to varying currents. Inductance does not cause any opposition to steady or direct current. In case of transmission and distribution lines, the current flowing is varying or alternating current, the effect of inductance, in addition to that of resistance, is therefore to be considered. The opposition to the flow of varying current owing to inductance is viewed as a voltage drop.

It is well known fact that a current carrying conductor is surrounded by concentric circles of magnetic lines. In case of ac system this field set up around the conductor is not constant but changing and links with the same conductor as well as with other conductors. Due to these flux linkages the line possesses inductance, defined as the flux linkages per unit current. Thus for determination of inductance of a circuit, determination of flux linkages is essential.

5.5.1. Flux Linkages of a Conductor

A long straight cylindrical conductor carrying a current is surrounded by a magnetic field. The magnetic lines of force will exist inside the conductor as well as outside the conductor. Both of these fluxes contribute to the inductance of the conductor.

Flux Linkages of a Conductor Due to Internal Flux. Consider a long straight cylindrical conductor of radius *r* metres and carrying a current of I amperes, as shown in Fig. 5.3.

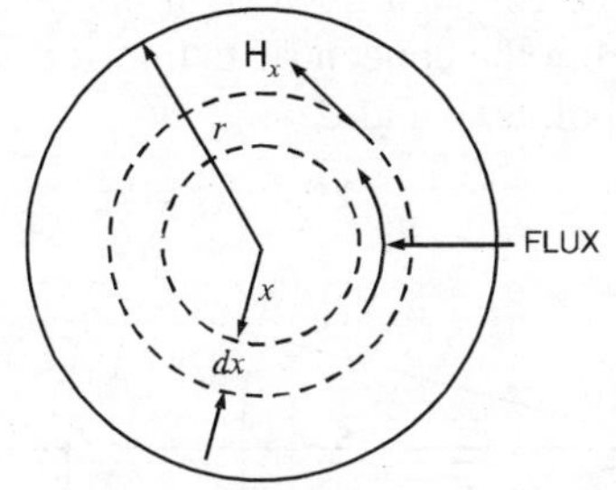

Fig. 5.3

In overhead lines it may be assumed without appreciable error that the current is uniformly distributed.

The current inside a line of force of radius *x*,

$$I_x = \frac{I}{\pi r^2} \times \pi x^2 = \frac{Ix^2}{r^2}$$

Field strength inside the conductor at a distance *x* from the centre,

$$H_x = \frac{\text{Current}}{2\pi \times \text{distance}} = \frac{Ix^2}{r^2} \times \frac{1}{2\pi x} = \frac{Ix}{2\pi r^2} \text{ AT/m}$$

Flux density,

$$B_x = \mu_0 \mu H_x \text{ Wb/m}^2 \text{ or T where } \mu_0 = 4\pi \times 10^{-7}$$

and μ is the relative permeability of the medium and for non-magnetic materials $\mu = 1$.

$$\therefore\ B_x = \frac{\mu_0 Ix}{2\pi r^2} \text{ Wb/m}^2$$

Now the flux through a cylindrical shell of radial thickness *dx* and axial length one metre,

$$d\phi = B_x \times 1 \times dx = \frac{\mu_0 x}{2\pi r^2} \cdot I \cdot dx \text{ webers}$$

But this flux links with only the current lying within the circle of radius x, *i.e.*, with I_x or $\frac{x^2}{r^2} I$

∴ Linkages of the shell,

$$d\psi = \frac{\mu_0 x}{2\pi r^2} \cdot dx \times \frac{x^2}{r^2} \times I = \frac{\mu_0 I x^3}{2\pi r^4} dx \text{ weber-turns}$$

Total flux linkages from centre of the conductor and up to the surface of the conductor,

$$\psi_{int} = \int_0^r \frac{\mu_0 I x^3}{2\pi r^4} dx \text{ Wb-turns/m}$$

$$= \frac{\mu_0 I}{8\pi} \text{ Wb-turns/m}$$

$$= \frac{4\pi \times 10^{-7} \times I}{8\pi} \text{ Wb-turns/m}$$

$$= \frac{1}{2} I \times 10^{-7} \text{ Wb-turns/m} \quad ...(5.4)$$

Flux Linkages of a Conductor Due to External Flux. Consider two points 1 and 2 distant d_1 and d_2 from the centre of the conductor. Since the flux paths are concentric circles around the conductor, whole of the flux between points 1 and 2 lies within the concentric cylindrical surfaces passing through these points 1 and 2.

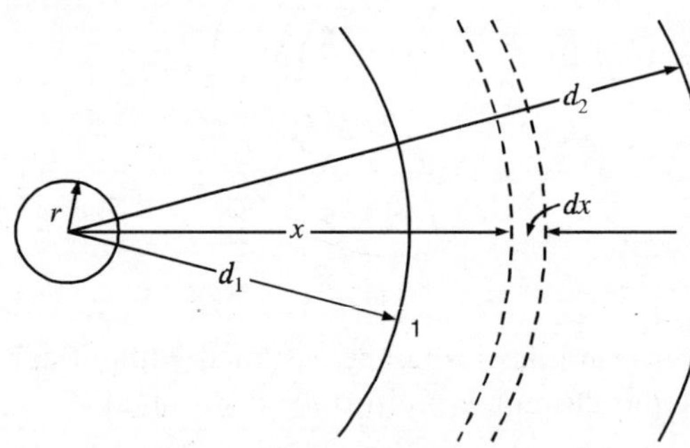

Fig. 5.4

The field strength at any distance x from the centre of the conductor ($x > r$),

$$H_x = \frac{I}{2\pi x} \text{ AT/m}$$

Flux density,

$$B_x = \mu_0 H_x = \frac{\mu_0 I}{2\pi x} \text{ Wb/m}^2$$

So the flux through a cylindrical shell of radial thickness dx and axial length one metre

$$d\phi = \frac{\mu_0 I}{2\pi x} dx \text{ webers per metre}$$

Now flux linkages per metre is equal to $d\phi$ since flux external to conductor links all the current in the conductor once and only once

$$\therefore \; d\psi = \frac{\mu_0 I}{2\pi x} dx \text{ Wb-turns per metre}$$

Total flux linkages between points 1 and 2

$$\psi_{ext} = \int_{d_1}^{d_2} \frac{\mu_0 I}{2\pi x} dx = \frac{4\pi \times 10^{-7}}{2\pi} I [\log_e x]_{d_1}^{d_2}$$

$$= 2 \times 10^{-7} I \log_e \frac{d_2}{d_1} \text{ Wb-turns/m} \quad ...(5.5)$$

5.5.2. Inductance of a Single-Phase Two-Wire Line

Consider a single-phase line consisting of two parallel conductors A and B of radii r_1 and r_2 spaced d metres apart (d being very large as compared to r_1 and r_2). Conductors A and B carry the same current (*i.e.*, $I_A = I_B$) in magnitude but opposite in directions, as one forms the return path for the other. The inductance of each conductor is due to internal flux linkages and external flux linkages and the following points are to be noted regarding external flux linkages :

1. A line of flux produced due to current in conductor A at a distance equal to or greater than $(d + r_2)$ from the centre of conductor A links with a zero net current, as the current flowing in the two conductors A and B are equal in magnitude but opposite in directions.

2. Flux lines at a distance $(d - r_2)$ link with a current I and those between $(d - r_2)$ and $(d + r_2)$ link with a current varying from I to zero.

As a simplifying assumption, it can be assumed that all the flux produced by current in conductor A links all the current up to the centre of conductor B and that the flux beyond the centre of conductor B does not link any current.

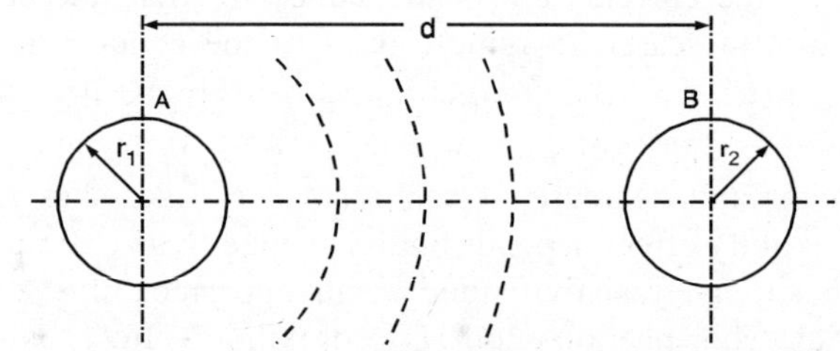

Fig. 5.5

The above assumption simplifies the calculations and results obtained are quite accurate specially when d is much greater than r_1 and r_2, as is usually the case in overhead lines.

Based on the above assumption flux linkages of conductor A due to external flux can be determined from Eq. (5.5) by substituting $d_2 = d$ and $d_1 = r_1$. Thus flux linkages of conductors A due to external flux only,

$$\psi_{A\,ext} = 2 \times 10^{-7} I \log_e \frac{d}{r_1} \text{ Wb-turns/metre} \quad ...(5.6)$$

Flux linkages of conductor A due to internal flux only,

$$\psi_{A\,int} = \frac{1}{2} I \times 10^{-7} \text{ Wb-turns/metre} \quad ...(5.7)$$

Total flux linkages of conductor A

$$\psi_A = \psi_{A\,ext} + \psi_{A\,int}$$

$$= \left(2 \times 10^{-7} I \log_e \frac{d}{r_1} + \frac{1}{2} I \times 10^{-7}\right)$$

$$= \left(0.5 + 2 \log_e \frac{d}{r_1}\right) I \times 10^{-7} \text{ Wb-turns/m}$$

Total inductance of conductor A,

$$L_A = \frac{\psi_A}{I} = \frac{\left(0.5 + 2\log_e \frac{d}{r_1}\right)}{I} \times I \times 10^{-7} \text{ H/m}$$

$$= 2 \times 10^{-7}\left(0.25 + \log_e \frac{d}{r_1}\right) \text{ H/m}$$

$$= 2 \times 10^{-7}\left(\log_e e^{1/4} + \log_e \frac{d}{r_1}\right) \text{ H/m}$$

$$= 2\times 10^{-7} \log_e \frac{d}{r_1 e^{-1/4}} \text{ H/m}$$

The product $(r_1 e^{-1/4})$ is known as *Geometric Mean Radius* (GMR) of the conductor and is equal to 0.7788 times the radius of the conductor. Let it be represented by r'_1 where

$$r'_1 = 0.7788 r_1$$

$$\therefore\ L_A = 2\times 10^{-7} \log_e \frac{d}{r'_1} \text{ H/m} \qquad ...(5.8)$$

Similarly inductance of conductor B,

$$L_B = 2\times 10^{-7} \log_e \frac{d}{r'_2} \text{ H/m}$$

Loop inductance of the line,

$$L = L_A + L_B = 2 \times 10^{-7}\left(\log_e \frac{d}{r'_1} + \log_e \frac{d}{r'_2}\right)$$

If $r'_1 = r'_2 = r'$, the loop inductance of the line is given as

$$L = 4 \times 10^{-7} \log_e \frac{d}{r'} \text{ H/m}$$

$$= 0.4 \log_e \frac{d}{r'} \text{ mH/km} \qquad ...(5.9)$$

The idea of replacing the original conductor of radii r by a fictitious conductor of radii r' is quite attractive because streamlined equations for inductance can be developed without bogging down in accounting for the internal flux.

5.5.3. Flux Linkages of One Conductor in a Group of Conductors

Consider a group of parallel conductors 1, 2, 3........ n carrying currents I_1, I_2, I_3 ..... I_n respectively, as illustrated in Fig. 5.6. Let it be assumed that the sum of the currents in various conductors is zero *i.e.*, $I_1 + I_2 + I_3 + ... I_n = 0$.

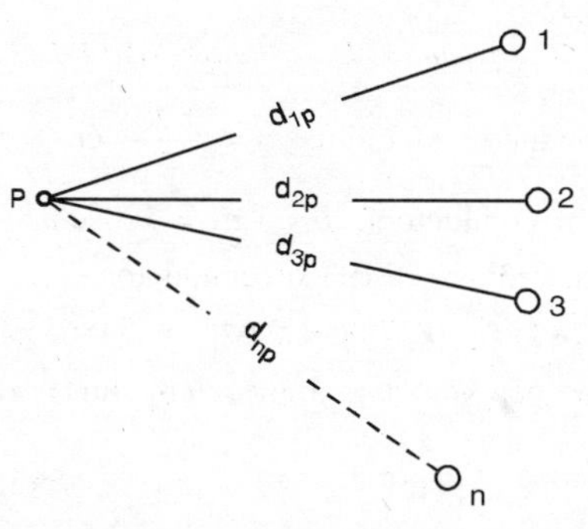

Fig. 5.6. *Cross-sectional View of a Group of n Conductors and Distances of Conductors From a Remote Point P*

Theoritically, the flux due to a conductor extends from the centre of the conductor right up to infinity but let us assume that the flux linkages extend up to a remote point P and the respective distances are as marked in Fig. 5.6.

The current in each conductor sets up a certain flux due its own current. The sum of all these fluxes is the total flux of the system and the total flux linkages of any one conductor is the sum of its linkages with all the individual fluxes set up by the conductors of the system.

Now let us determine the flux linkages of conductor 1 due to current I_1 carried by the conductor itself and flux linkages to conductor 1 due to currents carried by other conductors (2, 3, ..., n)

The flux linkages of conductor 1 due to its own current I_1 (internal and external), up to point P

$$\psi_{1p1} = 2 \times 10^{-7}\, I_1 \log_e \frac{d_{1p}}{r'_1} \text{ Wb-turns/metre}$$

The flux linkages of conductor 1 due to current in conductor 2

$$\psi_{1p2} = 2 \times 10^{-7}\, I_2 \log_e \frac{d_{2p}}{d_{12}}$$

Flux due to conductor 2 that lies between conductors 2 and 1 does not link conductor 1 and therefore the distances involved are d_{2p} and d_{12}.

Thus, the expression for flux linkages of conductor 1 due to currents in all conductors can be written as

$$\psi_{1p} = 2 \times 10^{-7}\left[I_1 \log_e \frac{d_{1p}}{r'_1} + I_2 \log_e \frac{d_{2p}}{d_{12}} + I_3 \log_e \frac{d_{3p}}{d_{13}} + ... I_n \log_e \frac{d_{np}}{d_{1n}}\right]$$

Wb-turns/m ...(5.10)

The above equation may be written as

$$\psi_{1p} = 2 \times 10^{-7}\left[I_1 \log_e \frac{1}{r'_1} + I_2 \log_e \frac{1}{d_{12}} + I_3 \log_e \frac{1}{d_{13}} + ... I_n \log_e \frac{1}{d_{1n}}\right]$$

$$+ 2 \times 10^{-7}\,[I_1 \log_e d_{1p} + I_2 \log_e d_{2p} + I_3 \log_e d_{3p} + ... I_n \log_e d_{np}] \text{ Wb-turns/m} \qquad ...(5.11)$$

To account for the total flux linkages to conductor 1, the point P must approach infinity and in this condition

$$d_{1p} \simeq d_{2p} \simeq d_{3p} \simeq d_{4p} .... d_{np} \simeq d \text{ (say)}$$

Then, $\lim_{d \to \infty} (I_1 + I_2 + I_3 + ... + I_n) \log_e d = 0$

$$\because\ I_1 + I_2 + I_3 + ... I_n = 0$$

This simplifies the Eq. (5.11) and the equation for the flux linkages to conductor 1 becomes

$$\psi_1 = 2 \times 10^{-7}\left[I_1 \log_e \frac{1}{r'_1} + I_2 \log_e \frac{1}{d_{12}} + I_3 \log_e \frac{1}{d_{13}} + ... I_n \log_e \frac{1}{d_{1n}}\right]$$

Wb-turns/m ...(5.12)

5.5.4. Inductance of Composite Conductor Lines—Self and Mutual GMDs

Consider a single phase line consisting of two parallel conductors A and B, conductor A consisting of x and conductor B of y strands, as illustrated in Fig. 5.7.

Let the conductors A and B carry currents I and –I respectively (since conductors of a 2-wire line carry the same current but in opposite directions).

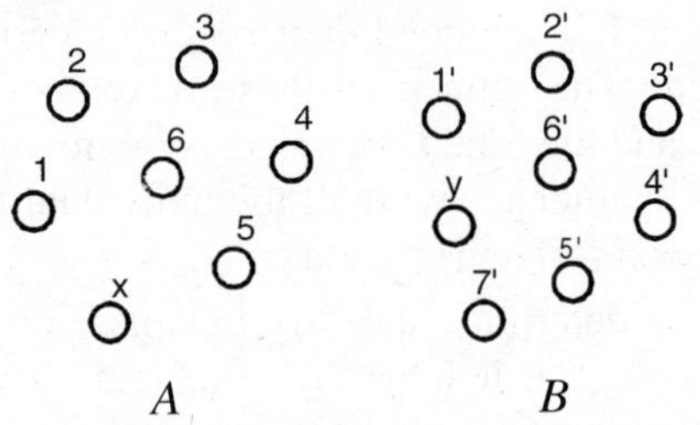

Fig. 5.7

Assuming uniform current density in both the conductors the current carried by each strand of conductor A will be $\frac{I}{x}$ while that carried by each strand of conductor B will be $\frac{-I}{y}$

Using Eq. (5.12) the flux linkages of strand 1 in conductor A can be written as

$$\psi_1 = 2 \times 10^{-7} \frac{I}{x}\left[\log_e \frac{1}{r_1'} + \log_e \frac{1}{d_{12}} + \log_e \frac{1}{d_{13}} + \ldots + \log_e \frac{1}{d_{1x}}\right]$$

$$- 2 \times 10^{-7} \frac{I}{y}\left[\log_e \frac{1}{d11'} + \log_e \frac{1}{d12'} + \log_e \frac{1}{d13'} + \ldots + \log_e \frac{1}{d_{1y}}\right]$$

$$\text{or } \psi_1 = 2 \times 10^{-7}\, I \log_e \frac{\sqrt[y]{(d_{11'}\; d_{12'}\; d_{13'} \ldots d_{1y})}}{\sqrt[x]{r_1'\, d_{12}\; d_{13} \ldots d_{1x}}} \text{ Wb-turns/m}$$

Inductance of strand 1 of conductor A

$$L_1 = \frac{\psi_1}{\frac{I}{x}} = 2 \times 10^{-7} x \log_e \frac{\sqrt[y]{(d_{11'}\; d_{12'}\; d_{13'}\; d_{14'} \ldots d_{1y})}}{\sqrt[x]{r'\, d_{12}\; d_{13}\; d_{14} \ldots d_{1x}}} \quad \ldots(5.13)$$

In above expression numerator of argument of $\log_e$ is written yth root of distances $d_{11'}, d_{12'}, d_{13'}, d_{14'} \ldots, d_1y$ multiplied together where distances $d_{11'}, d_{12'}, d_{13'}, d_{14'} \ldots, d_{1y}$ are the distances of strands 1′, 2′, 3′, 4′ ..., y (all segments of conductor B) from segment 1 under consideration.

The denominator of argument of $\log_e$ is xth root of distances $d_{12}, d_{13}, d_{14} \ldots d_{1x}$ and r' multiplied together where distances $d_{12}, d_{13}, d_{14} \ldots d_1x$ are the distances of strands 2, 3, 4, ..., n (all strands of conductor A) from strand 1. r' can also be represented by distance d_{11} and the expression for inductance for conductor A becomes

$$L_1 = 2 \times 10^{-7} x \log_e \frac{\sqrt[y]{d_{11'}\; d_{12'}\; d_{13'}\; d_{14'} \ldots . d_{1y}}}{\sqrt[x]{d_{11}\; d_{12}\; d_{13}\; d_{14} \ldots . d_{1x}}} \text{ H/m} \quad \ldots(5.14)$$

Similarly the expression for inductance for strand 2 can be written as

$$L_2 = 2 \times 10^{-7} x \log_e \frac{y\sqrt{d_{21'}\; d_{22'}\; d_{23'}\; d_{24'} \ldots . d_{2y}}}{x\sqrt{d_{21}\; d_{22}\; d_{23}\; d_{24} \ldots . d_{2x}}} \text{ H/m} \ldots(5.15)$$

Thus, we see that the different strands of a conductor have different inductances.

∴ Average inductance of strands of conductor A,

$$L_{av} = \frac{L_1 + L_2 + L_3 + \ldots . L_x}{x}$$

Since x such strands of conductor A are electrically in parallel,

Therefore inductance of conductor A,

$$L_A = \frac{L_{av}}{x}$$

$$= \frac{L_1 + L_2 + L_3 + \ldots L_x}{x^2}$$

$$= 2 \times 10^{-7} \log_e \left[\frac{\sqrt[xy]{(d_{11'}\, d_{12'}\, d_{13'}\, d_{14'} \ldots d_{1y})(d_{21'}\, d_{22'}\, d_{23'} \ldots d_{2y}) \ldots (d_{x1'}\, d_{x2'}\, d_{x3'} \ldots d_{xy})}}{\sqrt[x^2]{(d_{11}\, d_{12}\, d_{13}\, d_{14} \ldots d_{1x})(d_{21}\, d_{22}\, d_{23} \ldots d_{2x})(d_{x1}\, d_{x2}\, d_{x3} \ldots dxx)}}\right] \quad \ldots(5.16)$$

In the above expression the numerator of argument of $\log_e$ is called the GMD (often called the mutual GMD) between conductors A and B and the denominator of argument $\log_e$ is called GMR (often called self GMD). GMD (Geometric Mean Distance) and GMR (Geometric Mean Radius) are denoted by D_m and D_s respectively.

$$\therefore \; L_A = 2 \times 10^{-7} \log_e \frac{D_m}{D_{sA}} \text{ H/m}$$

$$\text{Similarly } L_B = 2 \times 10^{-7} \log_e \frac{D_m}{D_{sB}} \text{ H/m}$$

If conductors A and B are identical

$$\therefore \; D_{sA} = D_{sB} = D_s \text{ (say)}$$

Loop inductance,

$$L = L_A + L_B$$

$$= 2 \times 10^{-7} \log_e \frac{D_m}{D_s} + 2 \times 10^{-7} \log_e \frac{D_m}{D_s}$$

$$= 4 \times 10^{-7} \log_e \frac{D_m}{D_s} \text{ H/m or } 0.4 \log_e \frac{D_m}{D_s} \text{ mH/km} \ldots(5.17)$$

Example 5.1. Find the loop inductance and reactance per km of a single-phase overhead line consisting of two conductors, each 1.25 cm diameter. The spacing between conductors is 1 m and frequency is 50 Hz.

[Pb. Technical Univ. Power System-I (T&D), Dec. 2009]

Solution : Conductor of radius, $r = \frac{1.25}{2}$ cm = 0.625 cm

Spacing of conductors, d = 1 m = 100 cm

Geometric mean radius (GMR) of conductor,

r' = 0.7788 r = 0.7788 × 0.625 = 0.48675 cm

Loop inductance per km of a single phase line

$$L = 0.4 \log_e \frac{d}{r'} \text{ mH} \qquad \text{Refer to Eq. (5.9)}$$

$$= 0.4 \log_e \frac{100}{0.48675} = 2.13 \text{ mH } \textbf{Ans.}$$

Reactance of transmission line,

$$X_L = 2\pi f L = 2\pi \times 50 \times 2.13 \times 10^{-3} = 0.669\ \Omega/\text{km}\ \textbf{Ans.}$$

Example 5.2. Calculate inductance of a 500 m long, 1-ϕ, 2-wire transmission line. Each wire has identical area of cross section of 10 cm² and conductors are separated by a distance of 5 m.

[U.P. Technical Univ. Elements of Power System 2004-05]

Solution : *Conductor radius,* $r = \sqrt{\dfrac{a}{\pi}} = \sqrt{\dfrac{10}{\pi}} = 1.784\ \text{cm} \quad \because a = \pi r^2$

Spacing of conductors, $d = 5$ m

Geometric mean radius (GMR) of conductor,

$$r' = 0.7788r = 0.7788 \times 1.784 = 1.39 \text{ cm}$$

Loop inductance of 500 m long single phase line,

$$L = 0.4 \log_e \frac{d}{r'} \times l = 0.4 \log_e \frac{500}{1.39} \times \frac{1}{2} = 1.177 \text{ mH}\ \textbf{Ans.}$$

Example 5.3. Calculate the loop inductance per km of a single-phase line comprising of 2 parallel conductors 1 metre apart and 1 cm in diameter, when the material of conductor is (*i*) copper and (*ii*) steel of relative permeability 50.

[U.P. Technical Univ. Elements of Power System 2006-07]

Solution : Conductor radius, $r = 0.5$ cm

Spacing of conductors, d = 1 metre = 100 cm

(*i*) Loop inductance per km of line with copper conductor

$$= \left(1 + 4\log_e \frac{d}{r}\right) \times 10^{-7} \times 1{,}000 \text{ H}$$

$$= \left(0.1 + 0.4\log_e \frac{100}{0.5}\right) \times 10^{-3} \text{ H} = 2.22 \text{ mH}\ \textbf{Ans.}$$

(*ii*) Loop inductance per km of line with steel conductor

$$= \left(\mu + 4\log_e \frac{d}{r}\right) \times 10^{-7} \times 1{,}000 \text{ H}$$

$$= \left(50 + 4\log_e \frac{100}{0.5}\right) \times 10^{-4} \text{ H} = 7.12 \text{ mH}\ \textbf{Ans.}$$

Example 5.4. A single-phase transmission line has two parallel conductors, each of 1.2 cm diameter and 2.5 m apart. Calculate the loop inductance per km length of line if the material of the conductor is steel with relative permeability of 200.

[G.B. Technical Univ. Elements of Power System 2011-12]

Solution : Loop inductance per km of line with steel conductor

$$= \left(\mu + 4\log_e \frac{d}{r}\right) \times 10^{-7} \times 1{,}000 \text{ H}$$

$$= \left(200 + 4\log_e \frac{250}{0.6}\right) \times 10^{-4} \text{ H} = 22.4 \text{ mH}\ \textbf{Ans.}$$

Example 5.5. Calculate the GMR of a 6/3 mm Al, 1/3 mm steel ACSR conductor. *[A.M.I.E. Sec. B. Power Systems Winter 1996]*

Solution : Figure 5.8 shows an ACSR conductor with 6 outer conductors of aluminium each of radius r = 3 mm and one central conductor of steel also of radius r = 3 mm.

From Fig. 5.8

$$d_{11} = r;\ d_{12} = d_{16} = d_{17} = 2r;\ d_{14} = 4r \text{ and}$$

$$d_{13} = d_{15} = \sqrt{D_{14}^2 - D_{34}^2} = \sqrt{(4r)^2 - (2r)^2} = 2\sqrt{3}\, r$$

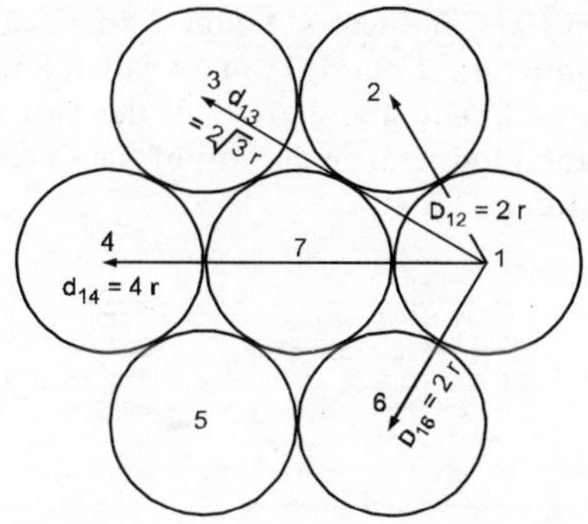

Fig. 5.8. ***Cross-Section of a 7-Strand Conductor***

$$D_{s1} = D_{s2} = D_{s3} = D_{s4} = D_{s5} = D_{s6}$$

$$= \sqrt[7]{0.7788r \times 2r \times 2\sqrt{3}r \times 4r \times 2\sqrt{3}r \times 2r \times 2r} = \sqrt[7]{299}\, r$$

$$D_{s7} = \sqrt[7]{2r \times 0.7788r \times 2r \times 2r \times 2r \times 2r \times 2r} = \sqrt[7]{49.8432}\, r$$

Geometric mean radius,

$$D_s = \sqrt[7]{D_{s1} \cdot D_{s2} \cdot D_{s3} \cdot D_{s4} \cdot D_{s5} \cdot D_{s6} \cdot D_{s7}}$$

$$= r \sqrt[49]{299^6 \times 49.8432}$$

$$= 2.176r = 2.176 \times 3 \text{ mm} = 6.528 \text{ mm}\ \textbf{Ans.}$$

Note: The overall conductor radius is $3r$. So the ratio of GMR to overall conductor radius is $2.176\ r/3r$ *i.e.*, 0.7253. With the increase in number of strands, this ratio approaches 0.7788 which is that for a solid conductor. This ratio comes out to be 0.678 and 0.777 for 3 strands and 169 strands respectively.

Example 5.6. A split-phase, single phase transmission line is shown in Fig. 5.9. Conductors 1 and 2 in parallel form one path while conductors 1′ and 2′ in parallel form the return path. The current is equally shared by the two parallel conductors. Determine the total inductance per km of the line. The radius of each conductor is 1.2 cm.

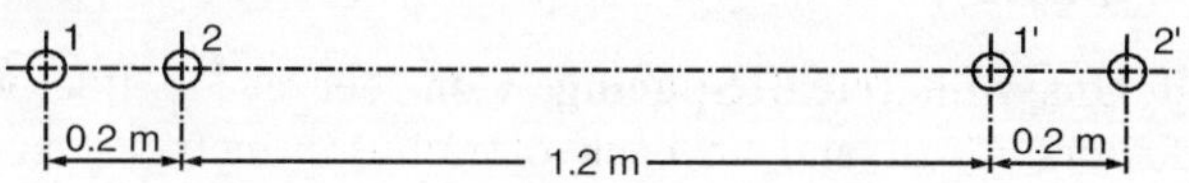

Fig. 5.9

Solution : Let the current flowing in each conductor be I amperes

Radius of each conductor, r = 1.2 cm

Geometric mean radius, $r' = 0.7788 \times 1.2 = 0.93456$ cm

Spacings of conductors, d_{12} = 0.2 m = 20 cm

$d_{11'}$ = (0.2 + 1.2) m = 140 cm ; $d_{22'}$ = (0.2 + 1.2) m = 140 cm

$d_{12'}$ = (0.2 + 1.2 + 0.2) m = 160 cm ; $d_{21'}$ = 1.2 m = 120 cm

Mutual GMD,

$$D_m = \sqrt[4]{d_{11'}\, d_{12'}\, d_{21'}\, d_{22'}} = \sqrt[4]{140 \times 160 \times 120 \times 140} = 139.28 \text{ cm}$$

Self GMD,

$$D_s = \sqrt[4]{d_{11}\, d_{12}\, d_{21}\, d_{22}} = \sqrt[4]{0.93456 \times 20 \times 20 \times 0.93456} = 4.32 \text{ cm}$$

Loop inductance of line,

$$L = 0.4 \log_e \frac{D_m}{D_s} \text{ mH/km}$$

$$= 0.4 \log_e \frac{139.28}{4.32} = 1.389 \text{ mH/km}\ \textbf{Ans.}$$

Example 5.7. A single phase double circuit transmission line is shown in Fig. 5.10. Conductors 1 and 2 in parallel form one path while conductors 1′ and 2′ in parallel form the return path. The current is equally shared by the two parallel lines. Determine the total inductance per km of the line. The diameter of each conductor is 2.0 cm.

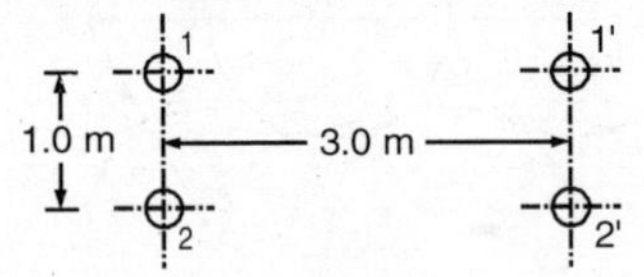

Fig. 5.10

Solution : Radius of each conductor, $r = \frac{2.0}{2} = 1.0$ cm

Geometric mean radius, $r' = 0.7788r = 0.7788$ cm

Spacing of conductors, $d_{11'} = 300$ cm

$$d_{12'} = \sqrt{(300)^2 + (100)^2} = 316.23 \text{ cm;}$$

$$d_{21'} = d_{12'} = 316.23 \text{ cm;}$$

$$d_{22'} = 300 \text{ cm}$$

$$d_{11} = d_{22} = r' = 0.7788 \text{ cm; } d_{12} = d_{21} = 100 \text{ cm}$$

Mutual GMD, $D_m = \sqrt[4]{d_{11'}\, d_{12'}\, d_{21'}\, d_{22'}}$ cm

$$= \sqrt[4]{300 \times 316.23 \times 316.23 \times 300} = 308.0 \text{ cm}$$

Self GMD, $D_s = \sqrt[4]{d_{11}\, d_{12}\, d_{21}\, d_{22}}$ cm

$$= \sqrt[4]{0.7788 \times 100 \times 100 \times 0.7788} = 8.825 \text{ cm}$$

Loop inductance $= 0.4 \log_e \frac{D_m}{D_s}$ mH/km $= 0.4 \log_e \frac{308}{8.825}$

$= 1.421$ mH/km **Ans.**

5.6. INDUCTANCE OF THREE-PHASE OVERHEAD LINES

With Unsymmetrical Spacing. Consider a 3-ϕ line with conductors A, B, and C; each of radius r metres. Let the spacing between them be d_1, d_2 and d_3 and the current flowing through them be I_A, I_B and I_C respectively.

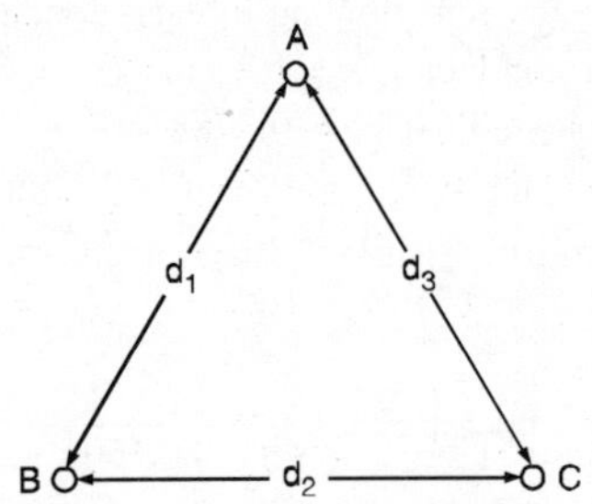

Fig. 5.11. 3-ϕ *Transmission Line With Unsymmetrical Spacing*

The flux linkages of conductor A due to its own current I_A and other conductor currents I_B and I_C

$$\psi_A = 2 \times 10^{-7} \left[I_A \log_e \frac{1}{r'} + I_B \log_e \frac{1}{d_1} + I_C \log_e \frac{1}{d_3} \right] \text{ Wb-turns/m}$$

Similarly,

$$\psi_B = 2 \times 10^{-7} \left[I_B \log_e \frac{1}{r'} + I_A \log_e \frac{1}{d_1} + I_C \log_e \frac{1}{d_2} \right] \text{ Wb-turns/m}$$

and

$$\psi_C = 2 \times 10^{-7} \left[I_C \log_e \frac{1}{r'} + I_A \log_e \frac{1}{d_3} + I_B \log_e \frac{1}{d_2} \right] \text{ Wb-turns/m}$$

...Refer to Eq. (5.12)

If the system is balanced,

$I_A = I_B = I_C = I$ (say) in magnitude

Taking I_A as a reference phasor, the currents are represented, in symbolic form as

$$\mathbf{I}_A = I;\ \mathbf{I}_B = I(-0.5 - j0.866) \text{ and } \mathbf{I}_C = I(-0.5 + j0.866)$$

Substituting these values of I_B and I_C in the expression for ψ_A we get

$$\psi_A = 2 \times 10^{-7} \left[I \log_e \frac{1}{r'} + I(-0.5 - j0.866) \log_e \frac{1}{d_1} + I(-0.5 + j0.866) \log_e \frac{1}{d_3} \right]$$

$$= 2 \times 10^{-7}\, I \left[\log_e \frac{1}{r'} + \log_e \sqrt{d_1 d_3} + j\sqrt{3} \log_e \sqrt{\frac{d_1}{d_3}} \right]$$

and $L_A = \frac{\psi_A}{I_A}$

$$= 2 \times 10^{-7} \left[\log_e \frac{1}{r'} + \log_e \sqrt{d_1 d_3} + j\sqrt{3} \log_e \sqrt{\frac{d_1}{d_3}} \right] \text{ H/m} \quad ...(5.18)$$

Similarly,

$$L_B = 2 \times 10^{-7} \left[\log_e \frac{1}{r'} + \log_e \sqrt{d_1 d_2} + j\sqrt{3} \log_e \sqrt{\frac{d_2}{d_1}} \right] \text{ H/m} \quad ...(5.19)$$

and

$$L_C = 2 \times 10^{-7} \left[\log_e \frac{1}{r'} + \log_e \sqrt{d_2 d_3} + j\sqrt{3} \log_e \sqrt{\frac{d_3}{d_2}} \right] \text{ H/m} \quad ...(5.20)$$

Thus, we see that when the conductors of a 3-phase transmission line are not equidistant from each other, *i.e.*, *unsymmetrically spaced*, the flux linkages and inductances of various phases are different which cause unequal voltage drops in the three phases and transfer of power between phases (represented by imaginary terms of the expression for inductances) due to mutual inductances even if the currents in the conductors are balanced.

The unbalancing effect on account of irregular spacing of conductors is avoided by transposition of conductors, as shown in Fig. 5.12. In practice the conductors are so transposed that each of the three possible arrangements of conductor exist for one-third of the total length of the line.

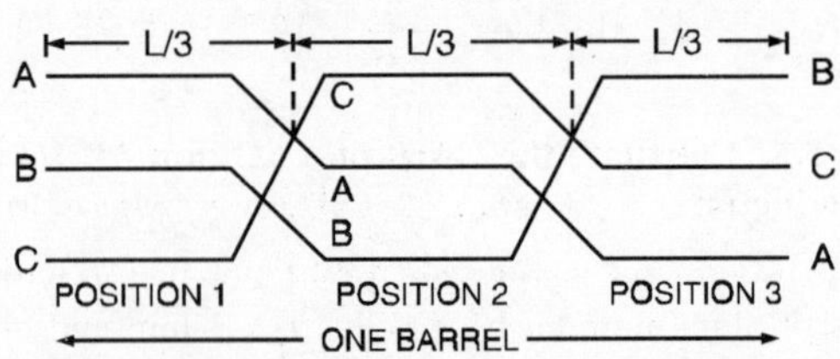

Fig. 5.12. *Transposition of a 3-Phase Line*

(i) The effect of transposition is that each conductor has the same average inductance, which is given as

$$L = \frac{L_A + L_B + L_C}{3}$$

$$= 2 \times 10^{-7}\left[\log_e \frac{1}{r'} + \frac{1}{3}\left(\log_e \sqrt{d_1 d_3} + \log_e \sqrt{d_1 d_2} + \log_e \sqrt{d_2 d_3}\right)\right.$$

$$\left. + \frac{j\sqrt{3}}{3}\left(\log_e \sqrt{\frac{d_1}{d_3}} + \log_e \sqrt{\frac{d_2}{d_1}} + \log_e \sqrt{\frac{d_3}{d_2}}\right)\right] \text{ H/m}$$

$$= 2 \times 10^{-7}\left[\log_e \frac{1}{r'} + \log_e \sqrt[3]{d_1\, d_2\, d_3} + 0\right]$$

$$= 2 \times 10^{-7} \log_e \frac{\sqrt[3]{d_1\, d_2\, d_3}}{r'} \text{ H/m} \qquad ...(5.21)$$

(ii) *If the conductors are equispaced* (let the spacing be equal to d), as shown in Fig. 5.13, the inductance of each conductor will be same and can be obtained by substituting $d_1 = d_2 = d_3$ in above expression. So inductance of each conductor

$$L = 2 \times 10^{-7} \log_e \frac{\sqrt[3]{d \times d \times d}}{r'} \text{ H/m}$$

$$= 2 \times 10^{-7} \log_e \frac{d}{r'} \text{ H/m} \qquad ...(5.22)$$

For stranded conductor r' will be replaced by D_s (self GMD)

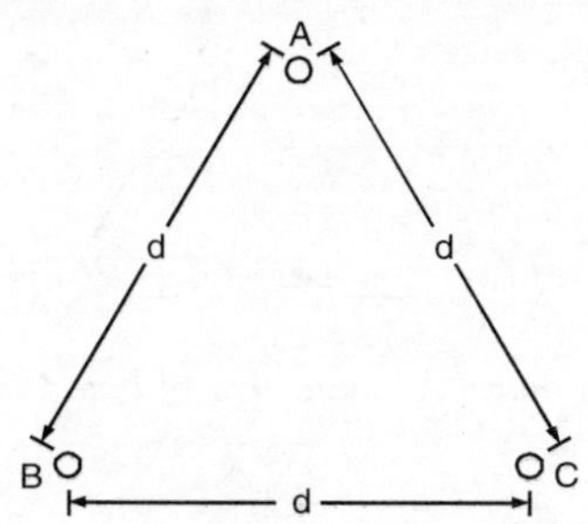

Fig. 5.13

(iii) *When the conductors of three-phase transmission line are in the same plane, as shown in Fig. 5.14.*

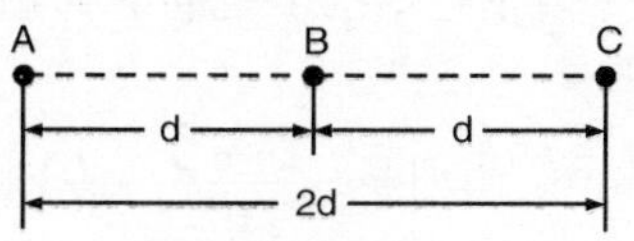

Fig. 5.14

In this position $d_1 = d$, $d_2 = d$ and $d_3 = 2d$

Substituting, $d_1 = d_2 = d$ and $d_3 = 2d$

in general Eqs. (5.18), (5.19), and (5.20) for L_A, L_B and L_C respectively, we get

$$L_A = 2 \times 10^{-7}\left[\log_e \frac{1}{r'} + \log_e \sqrt{2d \times d} + j\sqrt{3}\,\log_e \sqrt{\frac{d}{2d}}\right]$$

$$= 2 \times 10^{-7}\left[\log_e \frac{d}{r'} + \frac{1}{2}\log_e 2 - j\,0.866 \log_e 2\right] \text{ H/m} \qquad ...(5.23)$$

$$L_B = 2 \times 10^{-7}\left[\log_e \frac{1}{r'} + \log_e \sqrt{d \times d} + j\sqrt{3}\,\log_e \sqrt{\frac{d}{d}}\right] \text{ H/m}$$

$$= 2 \times 10^{-7} \log_e \frac{d}{r'} \text{ H/m} \qquad ...(5.24)$$

and

$$L_C = 2 \times 10^{-7}\left[\log_e \frac{1}{r'} + \log_e \sqrt{2d \times d} + j\sqrt{3}\,\log_e \sqrt{\frac{2d}{d}}\right] \text{ H/m}$$

$$= 2 \times 10^{-7}\left[\log_e \frac{d}{r'} + \frac{1}{2}\log_e 2 + j0.866 \log_e 2\right] \text{ H/m} \qquad ...(5.25)$$

(iv) *When the conductors are at the corners of a right-angled triangled, as shown in Fig. 5.15.* In this position

$d_1 = d_2 = d$ and $d_3 = \sqrt{2}\, d$

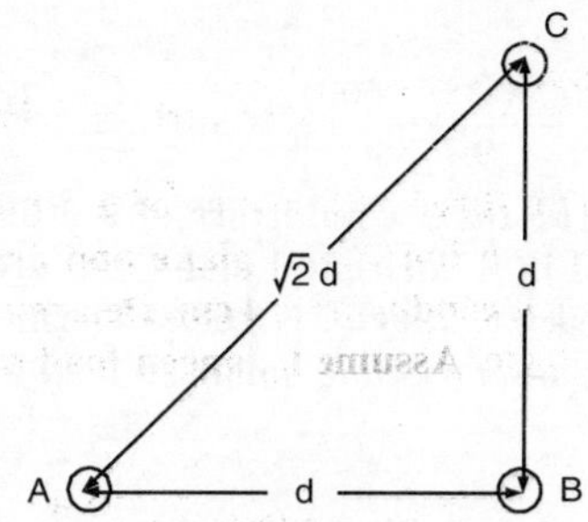

Fig. 5.15

Substituting $d_1 = d_2 = d$ and $d_3 = \sqrt{2}\, d$ in general Eqs. (5.18), (5.19), and (5.20) for L_A, L_B and L_C respectively, we get

$$L_A = 2 \times 10^{-7}\left[\log_e \frac{1}{r'} + \log_e \sqrt{\sqrt{2}d \times d} + j\sqrt{3}\,\log_e \sqrt{\frac{d}{\sqrt{2}d}}\right] \text{ H/m}$$

$$= 2 \times 10^{-7}\left[\log_e \frac{d}{r'} + \frac{1}{2}\log_e \sqrt{2} - j0.866 \log_e \sqrt{2}\right] \text{ H/m} \qquad ...(5.26)$$

$$L_B = 2 \times 10^{-7}\left[\log_e \frac{1}{r'} + \log_e \sqrt{d \times d} + \sqrt{3}\,\log_e \sqrt{\frac{d}{d}}\right] \text{ H/m}$$

$$= 2 \times 10^{-7} \log_e \frac{d}{r'} \text{ H/m} \qquad ...(5.27)$$

and

$$L_C = 2 \times 10^{-7}\left[\log_e \frac{1}{r'} + \log_e \sqrt{\sqrt{2}d \times d} + j\sqrt{3}\,\log \sqrt{\frac{\sqrt{2}d}{d}}\right] \text{ H/m}$$

$$= 2 \times 10^{-7}\left[\log_e \frac{d}{r'} + \frac{1}{2}\log_e \sqrt{2} + j\,0.866 \log_e \sqrt{2}\right] \text{ H/m} \qquad ...(5.28)$$

Example 5.8. Calculate the inductance per km of a 3-phase overhead transmission line using 1.24 cm diameter conductors when these are placed at the corners of an equilateral triangle of each side 2 metres.

[U.P. Technical Univ. Elements of Power System 2003-04]

Solution : Conductor radius, $r = \frac{1.24}{2}$ cm $= 0.62$ cm

Geometric mean radius $r' = 0.7788r = 0.7788 \times 0.62$ cm $= 0.4829$ cm

Spacing of conductors, $d = 2$ m $= 200$ cm

Inductance per phase per km $= 0.2 \log_e \frac{d}{r'}$ mH

$= 0.2 \log_e \frac{200}{0.4829}$ mH

$= 1.205$ mH **Ans.**

Example 5.9. Determine the inductance per km of a 3-ϕ transmission line using 20 mm diameter conductors when conductors are situated at the corners of a triangle with spacing of 4, 5 and 6 metres. Conductors are regularly transposed.

Solution : Conductor radius, $r = \frac{20}{2} = 10 \text{ mm} = 1.0 \text{ cm}$

Geometric mean radius, $r' = 0.7788 \times 1.0 = 0.7788$ cm

Inductance per phase,

$$L = 0.2 \log_e \frac{(d_1 d_2 d_3)^{1/3}}{r'} \text{ mH/km}$$

$$= 0.2 \log_e \frac{(400 \times 500 \times 600)^{1/3}}{0.7788} \text{ mH} = 1.29 \text{ mH} \quad \textbf{Ans.}$$

Example 5.10. The three conductors of a 3-phase transmission line are arranged in a horizontal plane and are 3 metres apart. The diameter of each conductor is 4 cm. Determine the inductance per km of each phase. Assume balanced load and R, Y, B phase sequence.

Solution : Radius of each conductor, $r = \frac{4}{2}$ cm = 2 cm

GMR, $r' = 0.7788 \times 2$ cm = 1.5576 cm

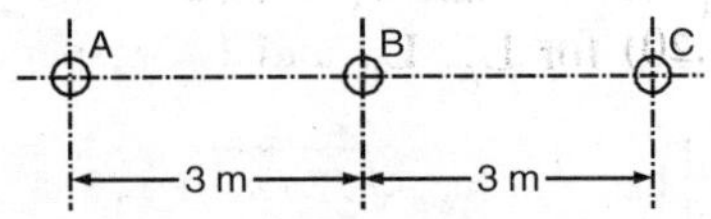

Fig. 5.16

From Eqs. (5.23), (5.24) and (5.25), respectively

$$L_A = 0.2\left[\log_e \frac{d}{r'} + \frac{1}{2}\log_e 2 - j0.866 \log_e 2\right] \text{ mH/km}$$

$$= 0.2\left[\log_e \frac{300}{1.5576} + \frac{1}{2}\log_e 2 - j\,0.866 \log_e 2\right] \text{ mH/km}$$

$$= 0.2\,[5.26 + 0.346 - j0.866 \times 0.6931]$$

$$= (1.1212 - j0.12) \text{ mH} \quad \textbf{Ans.}$$

$$L_B = 0.2 \log_e \frac{d}{r'} \text{ mH/km} = 0.2 \log_e \frac{300}{1.5576} = 1.052 \text{ mH} \quad \textbf{Ans.}$$

$$L_C = 0.2\left[\log_e \frac{d}{r'} + \frac{1}{2}\log_e 2 + j0.866 \log_e 2\right] \text{ mH}$$

$$= 0.2\left[\log_e \frac{300}{1.5576} + \frac{1}{2}\log_e 2 + j0.866 \log_e 2\right] \text{ mH}$$

$$= (1.1212 + j0.12) \text{ mH} \quad \textbf{Ans.}$$

Example 5.11. A three-phase overhead line is designed with an equilateral spacing of 3.5 m with a conductor diameter of 1.2 cm. If the line is constructed with horizontal spacing with suitably transposed conductors, find spacing between adjacent conductors which would give the same value of inductance as in the equilateral arrangement. *[A.M.I.E. Sec. B. Power Systems Summer 2000]*

Solution : Conductor radius, $r = \frac{1.2}{2}$ cm = 0.6 cm or 6 mm

Geometric mean radius, $r' = 0.7788 \times 6$ cm = 4.6728 mm

In equilateral arrangement of line conductors [Fig. 5.17 (a)]

Spacing of conductors, d = 3.5 m = 3,500 mm

Inductance per phase,

$$L = 2 \times 10^{-7} \log_e \frac{d}{r'} \text{ H/m} \qquad \text{...Refer to Eq. (5.22)}$$

$$= 2 \times 10^{-7} \log_e \frac{3{,}500}{4.6728} = 1.324 \times 10^{-6} \text{ H/m}$$

In the arrangement of conductors horizontally spaced [Fig. 5.17 (*b*)]

Let the spacing between adjacent conductors be d mm.

The effect of transposition of line is that each conductor has the same average inductance which in given as

$$L_{av} = \frac{L_A + L_B + L_C}{3}$$

Substituting the values of L_A, L_B and L_C from Eqs. (5.23), (5.24) and (5.25) respectively

$$L_{av} = \frac{1}{3} \times 2 \times 10^{-7}\left[\log_e \frac{d}{r'} + \frac{1}{2}\log_e 2 - j0.866\log_e 2 + \log_e \frac{d}{r'} + \log_e \frac{d}{r'} + \frac{1}{2}\log_e 2 + j0.866\log_e 2\right] \text{H/m}$$

$$= 2 \times 10^{-7}\left[\log_e \frac{d}{r'} + \frac{1}{3}\log_e 2\right] \text{H/m}$$

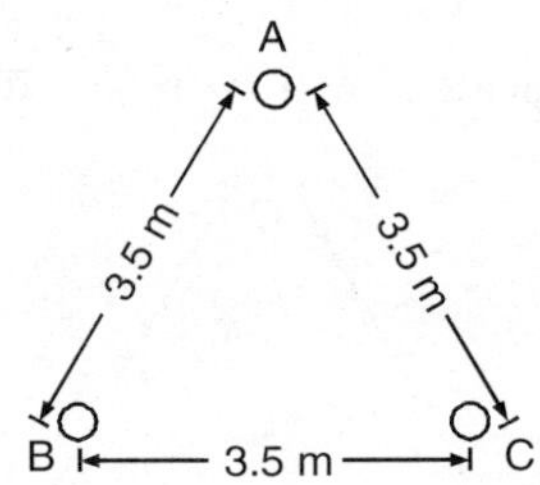

(a) Equilateral Arrangement of Line Conductors

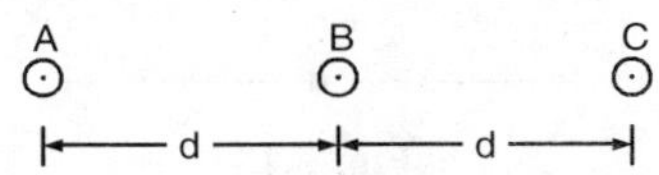

(b) Horizontally Spaced Line Conductors

Fig. 5.17

Subituting $L_{av} = 1.324 \times 10^{-6}$ and $r' = 4.6728$ mm in above equation, we have

$$1.324 \times 10^{-6} = 2 \times 10^{-7}\left[\log_e \frac{d}{4.6728} + \frac{1}{3}\log_e 2\right]$$

or $d = 4.6728\, e^{6.389}$ = 2,781.5 mm or 2.781 m **Ans.**

5.7. INDUCTANCE OF THREE-PHASE LINES WITH MORE THAN ONE CIRCUIT

It is usual practice to run 3-phase transmission lines with more than one circuit in parallel on the same towers, because it gives greater reliability and a higher transmission capacity. If such circuits are so widely separated that the mutual inductance between them becomes negligible, the inductance of the equivalent single circuit would be half of each of the individual circuits considered alone. But in actual practice the separation is not very wide and the mutual inductance is not negligible. GMD method is used for determination of inductance per phase by considering the various conductors connected in parallel as strands of one composite conductor.

It is desirable to have a configuration that provides minimum inductance so as to have maximum transmission capacity. This is possible only with low GMD and high GMR. Therefore, the practice is to have the individual conductors of a phase widely separated to provide high GMR and the distance between the phases small to give low GMD. Thus in the case of a double circuit in vertical formation the arrangement of conductors would be as illustrated in Fig. 5.18 (*a*) and not as illustrated in Fig. 5.18 (*b*) because the arrangement of conductors given in Fig. 5.18 (*a*) results in low inductance in comparison to that given by the arrangement illustrated in Fig. 5.18 (*b*).

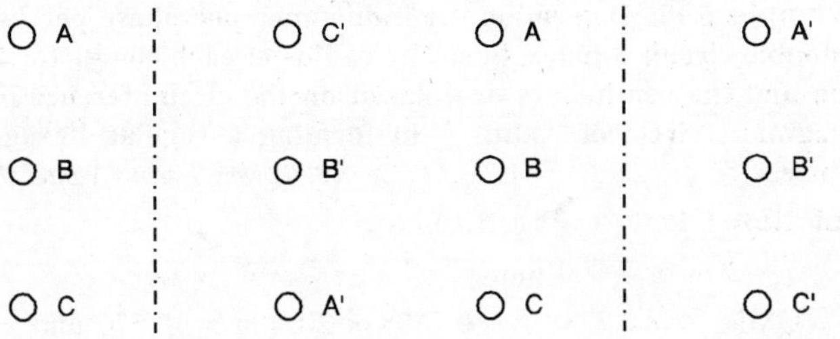

(*a*) (*b*)

Fig. 5.18. ***Arrangements of Conductors in a Double Circuit 3-Phase Line***

5.7.1. Inductance of a 3-Phase Double Circuit Line With Symmetrical Spacing

Consider a 3-phase double circuit connected in parallel-conductors A, B, C forming one circuit and conductors A', B', C' forming the other one, as illustrated in Fig. 5.19 (conductors symmetrically spaced).

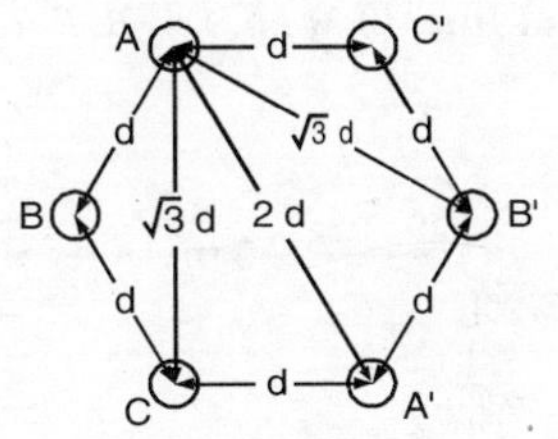

Fig. 5.19

Flux-linkages of phase A conductors

$$\psi_A = 2 \times 10^{-7}\left[I_A\left(\log_e \frac{1}{r'} + \log_e \frac{1}{2d}\right) + I_B\left(\log_e \frac{1}{d} + \log_e \frac{1}{\sqrt{3}d}\right) + I_C\left(\log_e \frac{1}{\sqrt{3}d} + \log_e \frac{1}{d}\right)\right]$$

$$= 2 \times 10^{-7}\left[I_A \log_e \frac{1}{2dr'} + (I_B + I_C)\log_e \frac{1}{\sqrt{3}d^2}\right]$$

$$= 2 \times 10^{-7}\left[I_A \log_e \frac{1}{2dr'} - I_A \log_e \frac{1}{\sqrt{3}d^2}\right] \qquad \because I_A + I_B + I_C = 0$$

$$= 2 \times 10^{-7} I_A \log_e \frac{\sqrt{3}d}{2r'} \text{ Wb-turns/m}$$

Inductance of conductor A,

$$L_A = \frac{\psi_A}{I_A} = 2 \times 10^{-7} \log_e \frac{\sqrt{3}d}{2r'} \text{ H/m} \qquad ...(5.29)$$

Similarly inductance of remaining conductors can be worked out, which will be the same as L_A. This is due to the fact that the conductors of different phases are symmetrically placed.

Since conductors are electrically in parallel, inductance of each phase

$$= \frac{1}{2} L_A = 1 \times 10^{-7} \log_e \frac{\sqrt{3}d}{2r'} \text{ H/m} \qquad ...(5.30)$$

5.7.2. Inductance of a 3-Phase Double Circuit with Unsymmetrical Spacing but Transposed

Now consider a 3-phase double circuit connected in parallel—conductors A, B and C forming one circuit and conductors A′, B′ and C′ forming the other one, as illustrated in Fig. 5.20. (conductors unsymmetrically spaced and transposed).

Since the conductors are thoroughly transposed, the conductor situations in the transposition cycle would be, as illustrated in Figs. 5.20(1), 5.20(2) and 5.20(3).

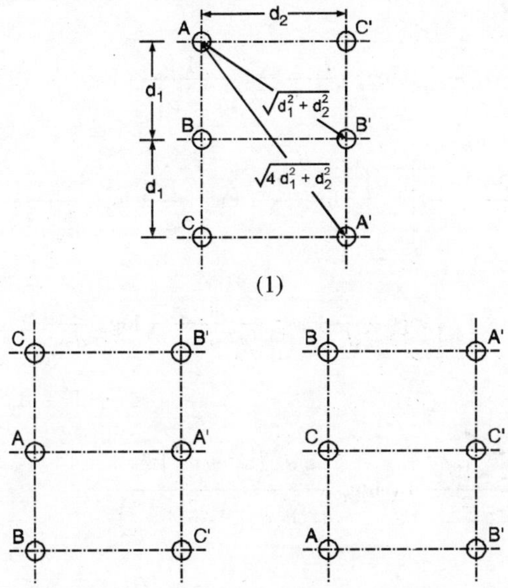

(2) (3)

Fig. 5.20

Flux linkages with conductor A in position (1)

$$\psi_{A1} = 2 \times 10^{-7}\left[I_A\left(\log_e \frac{1}{r'} + \log_e \frac{1}{\sqrt{4d_1^2 + d_2^2}}\right) + I_B\left(\log_e \frac{1}{d_1} + \log_e \frac{1}{\sqrt{d_1^2 + d_2^2}}\right) + I_C\left(\log_e \frac{1}{2d_1} + \log_e \frac{1}{d_2}\right)\right]$$

Similarly flux linkages with conductor A in position (2) and (3)

$$\psi_{A2} = 2 \times 10^{-7}\left[I_A\left(\log_e \frac{1}{r'} + \log_e \frac{1}{d_2}\right) + I_B\left(\log_e \frac{1}{d_1} + \log_e \frac{1}{\sqrt{d_1^2 + d_2^2}}\right) + I_C\left(\log_e \frac{1}{d_1} + \log_e \frac{1}{\sqrt{d_1^2 + d_2^2}}\right)\right]$$

and

$$\psi_{A3} = 2 \times 10^{-7} \left[I_A \left(\log_e \frac{1}{r'} + \log_e \frac{1}{\sqrt{4\, d_1^2 + d_2^2}} \right) + I_B \left(\log_e \frac{1}{2 d_1} + \log_e \frac{1}{d_2} \right) + I_C \left(\log_e \frac{1}{d_1} + \log_e \frac{1}{\sqrt{d_1^2 + d_2^2}} \right) \right]$$

Average flux linkages with conductor A

$$\psi_A = \frac{\Psi_{A1} + \Psi_{A2} + \Psi_{A3}}{3}$$

$$= \frac{2 \times 10^{-7}}{3} \left[I_A \left(3 \log_e \frac{1}{r'} + \log_e \frac{1}{d_2} + \log_e \frac{1}{4 d_1^2 + d_2^2} \right) + I_B \left(2 \log_e \frac{1}{d_1} + \log_e \frac{1}{2\, d_1} + \log_e \frac{1}{d_1^2 + d_2^2} + \log_e \frac{1}{d_2} \right) + I_C \left(\log_e \frac{1}{2\, d_1} + \log_e \frac{1}{d_1^2 + d_2^2} + \log_e \frac{1}{d_2} + 2 \log_e \frac{1}{d_1} \right) \right]$$

$$= \frac{2 \times 10^{-7}}{3} \left[I_A \left(3 \log_e \frac{1}{r'} + \log_e \frac{1}{d_2} + \log_e \frac{1}{4\, d_1^2 + d_2^2} \right) + (I_B + I_C) \left(\log_e \frac{1}{2 d_1} + \log_e \frac{1}{d_1^2 + d_2^2} + \log_e \frac{1}{d_2} + 2 \log_e \frac{1}{d_1} \right) \right]$$

$$= \frac{2 \times 10^{-7}}{3} \left[I_A \left(3 \log_e \frac{1}{r'} + \log_e \frac{1}{d_2} + \log_e \frac{1}{4\, d_1^2 + d_2^2} \right) - I_A \left(\log_e \frac{1}{2 d_1} + \log_e \frac{1}{d_1^2 + d_2^2} + \log_e \frac{1}{d_2} + 2 \log_e \frac{1}{d_1} \right) \right]$$

$$\because I_A + I_B + I_C = 0$$

$$\text{or } \psi_A = \frac{2 \times 10^{-7}}{3} I_A \log_e \frac{2\, d_1^3 \left(d_1^2 + d_2^2 \right) d_2}{(r')^3\, d_2 \left(4\, d_1^2 + d_2^2 \right)}$$

$$\text{or } \psi_A = 2 \times 10^{-7}\, I_A \log_e \frac{2^{1/3}\, d_1 \left(d_1^2 + d_2^2 \right)^{1/3}}{r' \left(4\, d_1^2 + d_2^2 \right)^{1/3}} \quad \text{wb-turns per metre}$$

and inductance,

$$L_A = \frac{\Psi_A}{I_A} = 2 \times 10^{-7} \log_e \frac{2^{1/3}\, d_1 \left(d_1^2 + d_2^2 \right)^{1/3}}{r' \left(4\, d_1^2 + d_2^2 \right)^{1/3}} \quad \text{H/m}$$

$$= 2 \times 10^{-7} \log_e 2^{1/3} \frac{d_1}{r'} \left(\frac{d_1^2 + d_2^2}{4\, d_1^2 + d_2^2} \right)^{1/3} \quad \text{H/m} \quad ...(5.31)$$

Inductance of each phase,

$$L = \frac{1}{2} \times \text{inductance per conductor}$$

$$= \frac{1}{2} L_A$$

$$= 2 \times 10^{-7} \log_e 2^{1/6} \left(\frac{d_1}{r'} \right)^{1/2} \left(\frac{d_1^2 + d_2^2}{4\, d_1^2 + d_2^2} \right)^{1/6} \quad \text{H/m} \quad ...(5.32)$$

If the distance d_2 is too large as compared to d_1, $\frac{d_1^2 + d_2^2}{4 d_1^2 + d_2^2}$ would tend to be unity and inductance per phase,

$$L = 2 \times 10^{-7} \log_e 2^{1/6} \left(\frac{d_1}{r'} \right)^{1/2} \quad \text{H/m}$$

$$\text{or } L = 0.2 \log_e 2^{1/6} \left(\frac{d_1}{r'} \right)^{1/2} \quad \text{mH/km} \quad ...(5.33)$$

Example 5.12. Determine the inductance per phase per km of a double circuit 3-phase line. The radius of each conductor is 20 mm and the conductors are placed on the circumference of an imaginary circle of radius 7 m forming a regular hexagonal figure. [*A.M.I.E. Sec. B. Power Systems Winter 1994*]

Solution : Refer to Fig. 5.19.

d = 7 m or 7,000 mm

GMR, $r' = 0.7788\ r = 0.7788 \times 20$ mm = 15.576 mm

From Eq. (5.30), inductance per phase

$$L = 1 \times 10^{-7} \log_e \frac{\sqrt{3}}{2} \frac{d}{r'} \quad \text{H/m}$$

$$= 0.1 \log_e \frac{\sqrt{3}}{2} \frac{d}{r'} \quad \text{mH/km}$$

$$= 0.1 \log_e 0.866 \times \frac{7{,}000}{15.576} = 0.5964 \text{ mH} \quad \textbf{Ans.}$$

Example 5.13. A part of the transposition cycle of a 3-phase double circuit line is shown in Fig. 5.21. Radius of each conductor is 0.9 cm. The conductors are solid copper. Find the inductance per phase per km of line. [*A.M.I.E. Sec. B. Power Systems Winter 1996*]

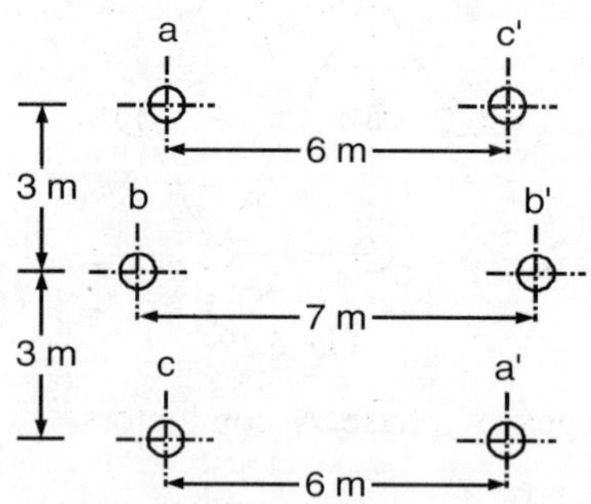

Fig. 5.21

Solution : GMR of each conductor

$r' = 0.7788 \times 0.9$ cm = 0.7 cm = 0.7×10^{-2} m

Distance between conductors a and a',

$$D_{aa}' = \sqrt{6^2 + 6^2} = 8.485 \text{ m}$$

d_{bb}' = 7 m and $d_{cc}' = d_{aa}'$ = 8.485 m

GMR of conductors of phase a,

$$D_{sa} = 4 \sqrt{d_{aa}\, d_{aa'}\, d_{a'a'}\, d_{a'a}}$$

$$= \sqrt{0.7 \times 10^{-2} \times 8.485 \times 0.7 \times 10^{-2} \times 8.485}$$

$$= 0.2437 \text{ m}$$

Similarly,

$$D_{sb} = \sqrt{0.7 \times 10^{-2} \times 7} = 0.2214 \text{ m}$$

and

$$D_{sc} = \sqrt{0.7 \times 10^{-2} \times 8.485} = 0.2437 \text{ m}$$

Self GMD,

$$D_s = \sqrt[3]{D_{sa}\, D_{sb}\, D_{sc}} = \sqrt[3]{0.2437 \times 0.2214 \times 0.2437} = 0.236 \text{ m}$$

Distance between conductors a and b,

$$d_{ab} = \sqrt{3^2 + 0.5^2} = 3.041 \text{ m}$$

Distance between conductors a and b',

$$d_{ab}' = \sqrt{3^2 + 6.5^2} = 7.159 \text{ m}$$

$$D_{ab} = \sqrt[4]{d_{ab}\, d_{ab'}\, d_{a'b}\, d_{a'b'}}$$

$$= \sqrt[4]{3.041 \times 7.159 \times 7.159 \times 3.041} = 4.666 \text{ m}$$

Similarly,

$$D_{bc} = \sqrt[4]{(3.041 \times 7.159)^2} = 4.666 \text{ m and}$$

$$D_{ca} = \sqrt[4]{(6 \times 6)^2} = 6 \text{ m}$$

$$D_{eq} \text{ or } D_m = \sqrt[3]{D_{ab}\, D_{bc}\, D_{ca}} = \sqrt[3]{4.666 \times 4.666 \times 6} = 5.074 \text{ m}$$

Inductance per phase,

$$L = 0.2 \log_e \frac{D_m}{D_s} \text{ mH/km} = 0.2 \times \log_e \frac{5.074}{0.236}$$

$$= 0.6136 \text{ mH/km } \textbf{Ans.}$$

Example 5.14. A 100 km double circuit transmission line with 7 strand copper conductors each strand having 5 mm diameter, has 6 conductors arranged in the form shown in Fig. 5.22. The line is transposed at regular intervals. Evaluate the GMD and GMR and calculate the overall inductance per phase of the line.

Solution : Radius of each strand of conductor = $\frac{5}{2}$ mm = 2.5 mm

GMR of conductor, $r' = 2.176 \times 2.5$ mm $= 5.44 \times 10^{-3}$ m

[See Example 5.5]

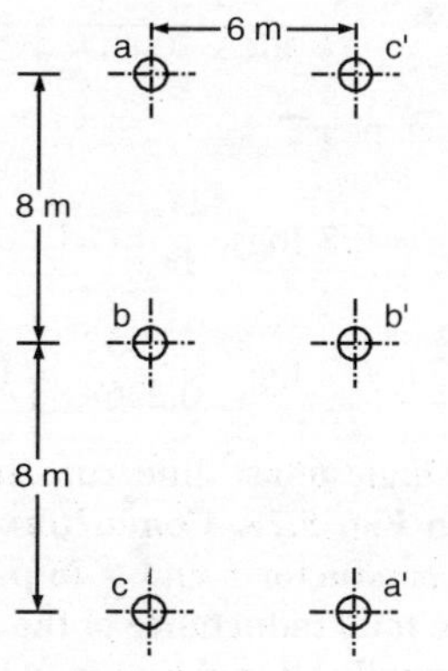

Fig. 5.22

Distance between conductors

$$d_{aa'} = \sqrt{6^2 + 16^2} = 17.089 \text{ m}; \; d_{bb'} = 6 \text{ m}$$

$$d_{cc'} = d_{aa'} = 17.089 \text{ m}; \quad d_{ab} = 8 \text{ m}$$

$$d_{ab'} = \sqrt{6^2 + 8^2} = 10 \text{ m}$$

GMR of phase a,

$$D_{sa} = \sqrt[4]{d_{aa}\, d_{aa'}\, d_{a'a'}\, d_{a'a}}$$

$$= \sqrt{r' \times d_{aa'}} = \sqrt{5.44 \times 10^{-3} \times 17.089} = 0.305 \text{ m}$$

Similarly,

$$D_{sb} = \sqrt{r' \times d_{bb'}} = \sqrt{5.44 \times 10^{-3} \times 6} = 0.1807 \text{ m}$$

$$\text{and } D_{sc} = \sqrt{r' \times d_{cc'}} = \sqrt{5.44 \times 10^{-3} \times 17.089} = 0.305 \text{ m}$$

Self GMD or GMR,

$$D_s = \sqrt[3]{D_{sa} D_{sb} D_{sc}} = \sqrt[3]{0.305 \times 0.1807 \times 0.305}$$

$$= 0.2562 \text{ m } \textbf{Ans.}$$

$$D_{ab} = \sqrt[4]{d_{ab} d_{ab'} d_{a'b} d_{a'b'}} = \sqrt{d_{ab} d_{ab'}} = \sqrt{8 \times 10} = 8.944 \text{ m}$$

Similarly

$$D_{bc} = \sqrt{d_{bc} d_{bc'}} = \sqrt{8 \times 10} = 8.944 \text{ m}$$

$$\text{and } D_{ca} = \sqrt{d_{ca} d_{ca'}} = \sqrt{16 \times 6} = 9.8 \text{ m}$$

$$D_{eq} \text{ or } D_m = \sqrt[3]{D_{ab} D_{bc} D_{ca}} = \sqrt[3]{8.944 \times 8.944 \times 9.8} = 9.22 \text{ m } \textbf{Ans.}$$

Inductance per phase,

$$L = 0.2 \log_e \frac{D_m}{D_s} \text{ mH/km} = 0.2 \log_e \frac{9.22}{0.2562}$$

$$= 0.717 \text{ mH/km}$$

Inductance of 100 km line

$$= \frac{0.717 \times 100}{1{,}000} = 0.0717 \text{ H } \textbf{Ans.}$$

Inductance per phase per km can also be determined from Eq. (5.32) *i.e.,*

$$L = 0.2 \log_e 2^{1/6} \left(\frac{d_1}{r'}\right)^{1/2} \left(\frac{d_1^2 + d_2^2}{4\, d_1^2 + d_2^2}\right)^{1/6} \text{ mH}$$

$$\text{or } L = 0.1 \log_e 2^{1/3} \left(\frac{d_1}{r'}\right) \left(\frac{d_1^2 + d_2^2}{4\, d_1^2 + d_2^2}\right)^{1/3} \text{ mH}$$

where $d_1 = 8$ m, $r' = 5.44 \times 10^{-3}$ m; $d_1^2 + d_2^2 = 8^2 + 6^2 = 100 \text{ m}^2$ and $4\, d_1^2 + d_2^2 = 4 \times 8^2 + 6^2 = 292 \text{ m}^2$

$$\text{So } L = 0.1 \log_e 2^{1/3} \left(\frac{8}{5.44 \times 10^{-3}}\right) \left(\frac{100}{292}\right)^{1/3}$$

$$= 0.1 \log_e 1.26 \times 1{,}470.6 \times 0.6996 = 0.717 \text{ mH}$$

Inductance of 100 km line

$$= 0.717 \times 10^{-3} \times 100 = 0.0717 \text{ H } \textbf{Ans.}$$

Same as before and as expected.

Example 5.15. What is transposition and why is it done ? Find the inductance per phase per km of double circuit 3-phase line shown in Fig. 5.23. The conductors are transposed and are of radius 0.9 cm each. The phase sequence is ABC.

[Pb. Technical Univ. Power System-I, Dec. 2009]

Solution : Conductor resistance, $r = 0.9$ cm

GMR of each conductor,

$$r' = 0.7788 \times 0.9 \text{ cm} = 0.7 \text{ cm} = 0.7 \times 10^{-2} \text{ m}$$

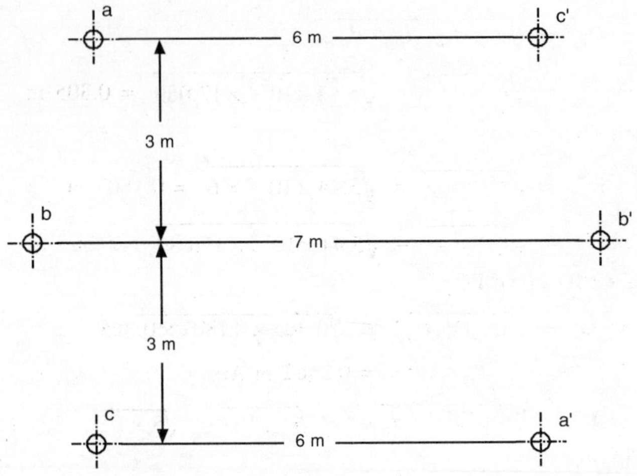

Fig. 5.23

Distance between conductors,

$$d_{aa'} = \sqrt{6^2 + 6^2} = 8.485 \text{ m}$$

$$d_{cc'} = d_{aa'} = 8.485 \text{ m}$$

$$d_{bb'} = 7 \text{ m}$$

GMR of conductors of phase a

$$D_{sa} = \sqrt[4]{d_{aa} d_{aa'} \times d_{a'a'} \times d_{a'a}}$$

$$= \sqrt{0.7 \times 10^{-2} \times (8.485) \times 0.7 \times 10^{-2} \times 8.485}$$

$$= 0.2437 \text{ m}$$

Similarly, $D_{sb} = \sqrt{0.7 \times 10^{-12} \times 7} = 0.2214$ m

$$D_{sc} = \sqrt{0.7 \times 10^{-12} \times 8.485} = 0.2437 \text{ m}$$

Self GMD, $D_s = \sqrt[3]{D_{sa} \times D_{sb} \times D_{sc}}$

$$= \sqrt[3]{0.2437 \times 0.2214 \times 0.2437} = 0.236 \text{ m}$$

Distance between conductors a and b,

$$d_{ab} = \sqrt{3^2 + 0.5^2} = 3.041 \text{ m}$$

Distance between conductors a and b'

$$d_{ab'} = \sqrt{3^2 + 6.5^2} = 7.159 \text{ m}$$

$$D_{ab} = \sqrt[4]{d_{ab} d_{ab'} d_{a'b} d_{a'b'}}$$

$$= \sqrt{d_{ab} d_{ab'}} = \sqrt{3.041 \times 7.159} = 4.666 \text{ m}$$

Similarly, $D_{bc} = \sqrt{3.04 \times 7.159} = 4.666$ m

$$D_{ca} = \sqrt{6 \times 6} = 6 \text{ m}$$

$$D_{eq} \text{ or } D_m = \sqrt[3]{D_{ab} D_{bc} D_{ca}} = \sqrt[3]{4.666 \times 4.666 \times 6} = 5.074 \text{ m}$$

Inductance per phase per km

$$= 0.2 \log_e \frac{D_m}{D_s} \text{ mH} = 0.2 \log_e \frac{5.074}{0.236}$$

$$= 0.6136 \text{ mH/km} \quad \textbf{Ans.}$$

Example 5.16. There are 6 conductors in a double circuit line. Each conductor has a radius of 12 mm. The six conductors are arranged horizontally. The centre to centre distance between the conductors is 2 m. The sequence of conductors are from left to right as follow : *a, b, c, a′, b′, c′*. Calculate the inductance per km per phase of the system.

[U.P.S.C. I.E.S. Electrical Engineering-II, 2007]

Solution :

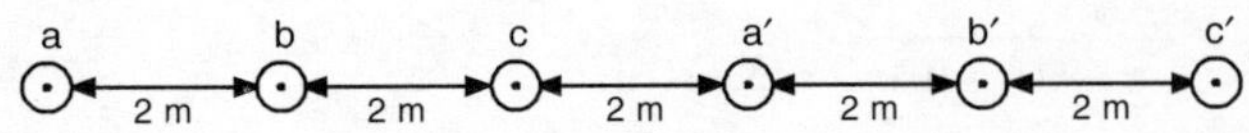

Fig. 5.24

GMR of each conductor, $r' = 0.7788 \times 12 \times 10^{-3}$ m $= 0.0093456$ m

GMR of phase a, $D_{sa} = \sqrt[4]{d_{aa} d_{aa'} d_{a'a'} d_{a'a}}$

$$= \sqrt{r' \times d_{aa'}} = \sqrt{0.0093456 \times 6} = 0.2368 \text{ m}$$

Similarly, $D_{sb} = \sqrt{0.0093456 \times 6} = 0.2368$ m

and $D_{sc} = \sqrt{0.0093456 \times 6} = 0.2368$ m

Self GMD or GMR,

$$D_s = \sqrt[3]{D_{sa} \times D_{sb} \times D_{sc}}$$

$$= \sqrt[3]{0.2368 \times 0.2368 \times 0.2368}$$

$$= 0.2368 \text{ m}$$

$$D_{ab} = \sqrt[4]{d_{ab} d_{ab'} d_{a'b} d_{a'b'}} = \sqrt[4]{2 \times 8 \times 4 \times 2}$$

$$= 3.364 \text{ m}$$

Similarly, $D_{bc} = \sqrt[4]{d_{bc} d_{bc'} d_{b'c} d_{b'c'}} = \sqrt[4]{2 \times 8 \times 4 \times 2}$

$$= 3.364 \text{ m}$$

and $D_{ca} = \sqrt[4]{d_{ca} d_{ca'} d_{c'a} d_{c'a'}} = \sqrt[4]{4 \times 2 \times 10 \times 4}$

$$= 4.229 \text{ m}$$

$$D_{eq} \text{ or } D_m = \sqrt[4]{D_{sa} D_{sb} D_{sc}}$$

$$= \sqrt{3.364 \times 3.364 \times 4.229} = 3.63 \text{ m}$$

Inductance per km per phase

$$= 0.2 \log_e \frac{D_m}{D_s} \text{ mH}$$

$$= 0.2 \log_e \frac{3.63}{0.2368} = 0.546 \text{ mH} \quad \textbf{Ans.}$$

Example 5.17. A single-phase line consists of two circuits in parallel as shown in Fig. 5.25. Conductors a and a′ in parallel form the lead while conductor *b* and *b′* in parallel form of return circuit. Calculate the total inductance of the line per km assuming that the current is equally shared by the two parallel conductors. The diameter of each conductor is 20 mm.

[Rajasthan Univ. Transmission & Distribution of Electrical Power 2003]

Solution :

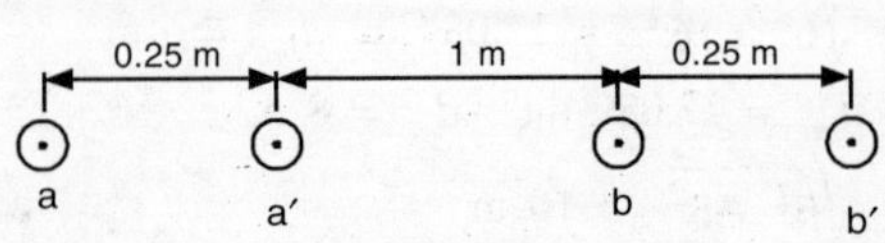

Fig. 5.25

Mutual GMD, $D_m = \sqrt[4]{d_{ab}d_{ab'}d_{a'b}d_{a'b'}}$

$= \sqrt[4]{1.25 \times 1.5 \times 1 \times 1.25} = 1.2373$ m

and self GMD,

$D_s = \sqrt[4]{d_{aa}d_{aa'}d_{a'a'}d_{a'a}}$

$= \sqrt{0.778 \times 25 \times 0.7788 \times 25}$

$\because \; d_{aa} = d_{a'a} = 0.7788$ m and $d_{aa'} = d_{a'a} = 25$ cm

$= 4.4125$ cm $= 4.4125 \times 10^{-2}$ m

Total inductance of line per km

$$= 0.4 \log_e \frac{D_m}{D_s} \text{ mH/km} = 0.4 \log_e \frac{1.2373}{4.4125 \times 10^{-2}}$$

$= 1.3335$ mH/km **Ans.**

5.8. BUNDLED CONDUCTORS

The demand of electric power is increasing throughout the world and in many countries it is doubling every five to eight years. The power stations (hydro, thermal or nuclear) are usually located far away from the load centres. Thus transmission of large amounts of power over long distances, which can be accomplished most economically only by using extra high voltages (or simply EHV), is necessary. Voltages in excess of 230 kV fall in this category. An increase in transmission voltage results in reduction of electrical losses, increase in transmission efficiency, improvement of voltage regulation and reduction in conductor material requirement. At voltages above 300 kV, corona causes a significant power loss and interference with communication circuits, if a round single conductor per phase is used. Instead of going for a hollow conductor it is preferable to use more than one conductor per phase which is called the *bundling* of conductors. Lines of 400 kV and higher voltages invariably use *bundled conductors.*

A bundled conductor is a conductor made up of two or more conductors, called the *sub-conductors*, per phase in close proximity compared with the spacing between phases (Fig. 5.26). The basic difference between a composite conductor and *a* bundled conductor is that the sub-conductors of a bundled conductor are separated from each other by a constant distance varying from 0.2 m to 0.6 m depending upon designed voltage and surrounding conditions throughout the length of the line with the help of spacers whereas the wires of a composite conductor touch each other. The bundled conductors have filter material or air space inside so that the overall diameter is increased.

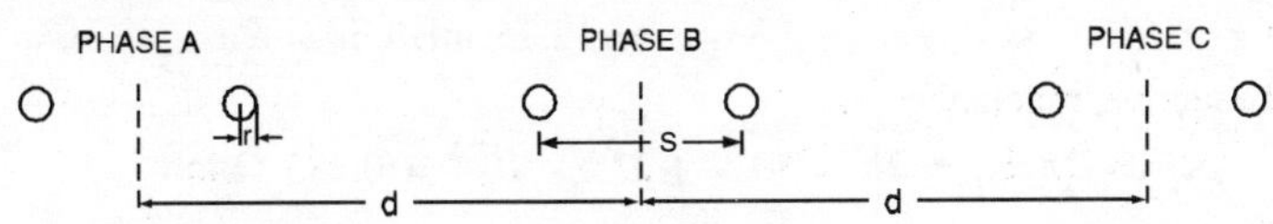

Fig. 5.26

The use of bundled conductors per phase reduces the voltage gradient in the vicinity of the line and thus, reduces the possibilities of the corona discharge.

Although the bundled conductors are used on EHV transmission lines primarily to reduce corona loss and radio interference, but they have several other advantages over single conductors such as given below :

1. The bundle conductor lines transmit bulk power with reduced losses, thereby giving increased transmission efficiency.
2. Since the bundle conductor lines have a higher capacitance to neutral in comparison with single conductor lines, therefore, they have higher charging current, which helps in improving the power factor.
3. Since by bundling, the self GMD or GMR is increased, the inductance per phase, in comparison with single conductor lines, is reduced. As a result reactance per phase is reduced.
4. Since surge impedance of a line is given by $Z_0 = \sqrt{\frac{L}{C}}$ and the bundle conductor lines have higher capacitance and lower inductance in comparison with single conductor lines, therefore, bundle conductor lines have comparatively lower surge impedance with a corresponding increase in the maximum power transfer capability. The following table represents relative power transfer with number of sub-conductors forming a bundle conductor.

Number of sub-conductors	1	2	4	8
Relative power transfer	1.0	1.3	1.6	1.7

It is to be noted that there is a little to be gained by using more than four sub-conductors per phase, two or three sub-conductors per phase are sufficient for most of the EHV lines.

(*a*)

(*b*)

(*c*)

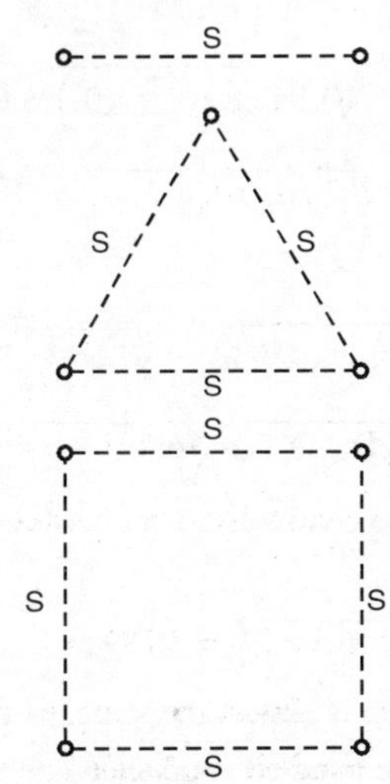

Fig. 5.27

The geometric mean radius of a bundled conductor for a two-conductor (duplex), a three-conductor (triplex) and four conductor (quadruplex) arrangements shown in Figs. 5.27 (*a*), 5.27 (*b*) and 5.27 (*c*) respectively can be determined from the following relations

$D_s = \sqrt{r's}$ for duplex arrangement ...(5.34)

$$D_s = \sqrt[3]{r's^2} \text{ for triplex arrangement} \quad ...(5.35)$$

$$D_s = 1.09\sqrt[4]{r's^3} \text{ for quadruplex arrangement} \quad ...(5.36)$$

where r' is the geometric mean radius of each subconductor of bundle and s is the spacing between subconductors of a bundle.

GMD of a bundled conductor line can be determined by taking the root of the product of distances from each conductor of a bundle to every other conductor of the other bundles. However, if the distances from centre of one bundle to the centres of other bundles is taken as the distances d_{ab}, d_{bc}, d_{ca}, accuracy will not be much affected.

Example 5.18. Determine the inductance per km per phase of a single circuit 460 kV line using two bundle conductors per phase as shown in the figure. The diameter of each conductor is 5.0 cm.

[U.P. Technical Univ. Elements of Power System 2007-08]

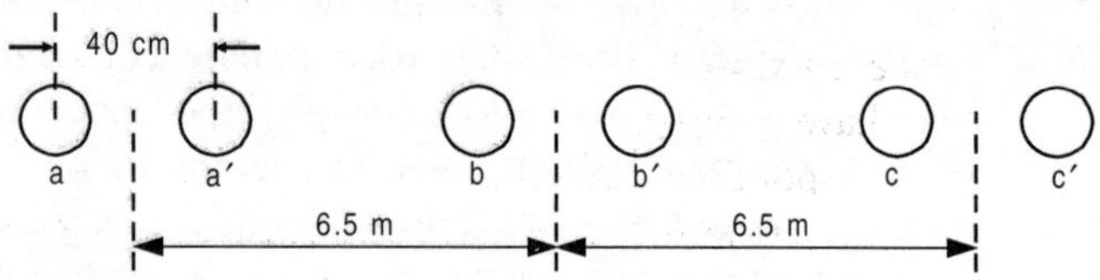

Fig. 5.28

Solution : Radius of each sub-conductor, $r = \dfrac{5.0}{2}$ cm = 2.5 cm

GMR of each sub-conductor,

$$r' = 0.7788r = 0.7788 \times 2.5 \text{ cm} = 1.947 \text{ cm} = 1.947 \times 10^{-2} \text{ m}$$

Phase-to-phase separation,

$$d = 6.5 \text{ m}$$

Spacing between sub-conductors of one phase,

$$s = 0.4 \text{ m}$$

$$\text{GMR, } D_s = \sqrt{r's} = \sqrt{1.947 \times 10^{-2} \times 0.4} = 0.088 \text{ m}$$

$$D_{ab} = D_{bc} = \sqrt[4]{d_{ab}d_{ab'}d_{a'b}d_{a'b'}} = \sqrt[4]{6.5 \times 6.9 \times 6.1 \times 6.5} = 6.494 \text{ m}$$

$$D_{ca} = \sqrt[4]{d_{ca}d_{ca'}d_{c'a}d_{c'a'}} = \sqrt[4]{13 \times 12.6 \times 13.4 \times 13} = 12.997 \text{ m}$$

$$D_{eq} \text{ or } D_m = \sqrt[3]{D_{ab}D_{bc}D_{ca}} = \sqrt[3]{6.494 \times 6.494 \times 12.997} = 8.184 \text{ m}$$

Taking the centre to centre distances between the bundles as D_{ab}, D_{bc} and D_{ca}

$$D_{eq} = \sqrt[3]{6.5 \times 6.5 \times 13} = 8.189 \text{ m}$$

which is almost the same, as the accurate value of D_{eq}.

Inductance of the bundled conductor line per phase,

$$L = 0.2 \log_e \frac{D_m}{D_s} \text{ mH/km} = 0.2 \log_e \frac{8.184}{0.088} = 0.907 \text{ mH/km } \textbf{Ans.}$$

Example 5.19. Find the inductive reactance of a 3-phase bundled conductor line with 2 conductors per phase with spacing of 40 cm. Phase to phase separation is 7 m in horizontal configuration. All conductors are ACSR with dia of 3.5 cm.

Compare the above value with that of an 'equivalent' single-conductor line. *[A.M.I.E. Sec. B. Power Systems Summer 1997]*

Solution : Radius of each sub-conductor, $r = \dfrac{3.5}{2}$ cm = 1.75 cm

GMR of each sub-conductor,

$$r' = 0.7788\, r = 0.7788 \times 1.75 \times 10^{-2} \text{ cm} = 1.363 \times 10^{-2} \text{ m}$$

Phase to phase separation,

$$d = 7 \text{ m}$$

Spacing between sub-conductors of one phase,

$$s = 40 \text{ cm} = 0.4 \text{ m}$$

$$\text{GMR, } D_s = \sqrt{r's} = \sqrt{1.363 \times 10^{-2} \times 0.4} = 0.0738 \text{ m}$$

$$D_{ab} = D_{bc} = \sqrt[4]{d_{ab}d_{ab'}d_{a'b}d_{a'b'}} = \sqrt[4]{7 \times 7.4 \times 6.6 \times 7} = 6.994 \text{ m}$$

$$D_{ca} = \sqrt[4]{d_{ca}d_{ca'}d_{c'a}d_{c'a'}} = \sqrt[4]{14 \times 13.6 \times 14.4 \times 14} = 13.997 \text{ m}$$

$$D_{eq} \text{ or } D_m = \sqrt[3]{D_{ab}D_{bc}D_{ca}} = \sqrt[3]{6.994 \times 6.994 \times 13.997} = 8.814 \text{ m}$$

Taking the centre to centre distances between the bundles as D_{ab}, D_{bc} and D_{ca},

$D_{eq} = \sqrt[3]{7 \times 7 \times 14} = 8.819$ m which is almost the same, as the accurate value of D_{eq}.

Inductance of the bundled conductor line,

$$L = 0.2 \log_e \frac{D_m}{D_s} \text{ mH/km} = 0.2 \log_e \frac{8.814}{0.0738} = 0.9565 \text{ mH/km}$$

Inductive reactance of the bundled conductor line per phase

$$X_L = 2\pi f L = 2\pi \times 50 \times 0.9565 \times 10^{-3} = 0.3 \ \Omega\text{/km } \textbf{Ans.}$$

Radius of conductor in the equivalent line with single conductor per phase,

$$r_1 = \sqrt{\frac{n\pi r^2}{\pi}} \text{ where } n \text{ is the number of sub-conductors in a bundle}$$

$$= \sqrt{2 \times (1.75)^2} = 2.4745 \times 10^{-2} \text{ m} \because n = 2 \text{ and } r = 1.75 \text{ cm}$$

GMR of each conductor,

$$r_1' = 0.7788 r_1 = 0.7788 \times 2.4745 \times 10^{-2} = 1.927 \times 10^{-2} \text{ m}$$

$$D_{m_1} = \sqrt[3]{D_{ab}D_{bc}D_{ca}} = \sqrt{7 \times 7 \times 14} = 8.819 \text{ m}$$

Inductance,

$$L_1 = 0.2 \log_e \frac{D_{m_1}}{r_1'} \text{ mH/km} = 0.2 \log_e \frac{8.819}{1.927 \times 10^{-2}} = 1.225 \text{ mH/km}$$

Inductive reactance,

$$X_{L_1} = 2\pi f L_1 = 2\pi \times 50 \times 1.225 \times 10^{-3} = 0.385 \ \Omega\text{/km}$$

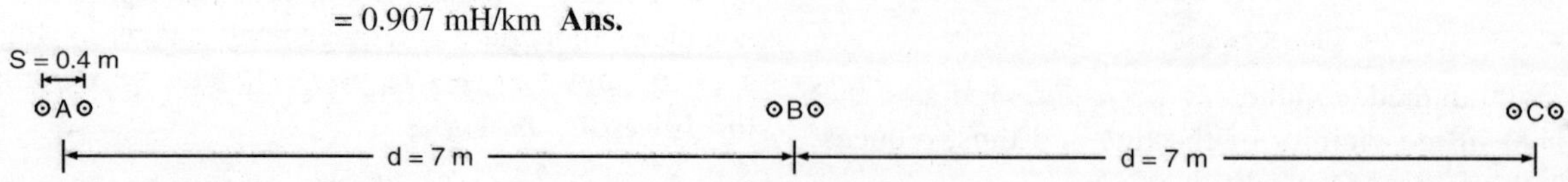

Fig. 5.29

Comments. The inductance of the bundled conductor line is significantly less than that of the line with one solid conductor per phase. The lower inductance leads to a greater transmission capacity of the bundled conductor line.

5.9. CAPACITANCE OF A TRANSMISSION LINE

We know that any two conductors separated by an insulating medium constitute a *condenser* or a *capacitor*. In case of an overhead line two conductors form the two plates of a capacitor and the air between the conductors behaves as the dieletric medium. Thus an overhead line can be assumed to have capacitance between conductors throughout the length of the line. The capacitance is uniformly distributed over the total length of the line and may be regarded as a uniform series of condensers connected between the conductors, as illustrated in Fig. 5.30.

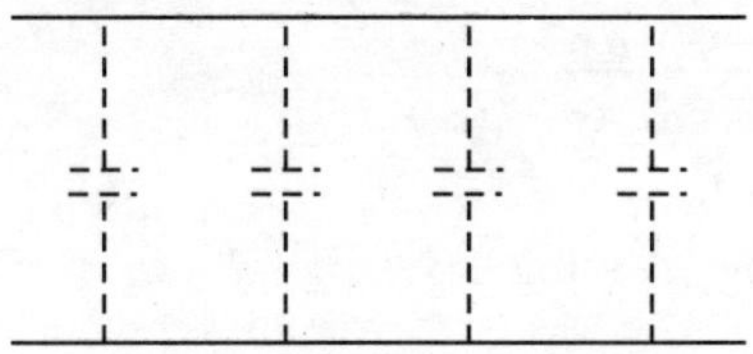

Fig. 5.30

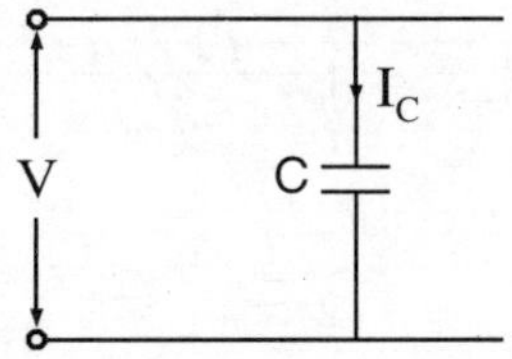

Fig. 5.31

When an alternating *pd* is applied across a transmission line, as shown in Fig. 5.31, it draws a leading current, even when supplying no load. This leading current is in quadrature with the applied voltage and is termed as the *charging current*. It must be noted that the charging current is due to the capacitive effect between the conductors of the line and is not in any way dependent on the load. The strength of the charging current depends upon the voltage of transmission, the capacitance of the line and the frequency of ac supply and is given by the expression.

$$\text{Charging current, } I_C = 2\pi f CV \qquad ...(5.37)$$

where f is the frequency of supply, C is capacitance of the line and V is the voltage.

If the capacitance of an overhead line is high, the line draws more charging current, which compensates or cancels the lagging component (reactive component) of load current (under normal load conditions the load is inductive one). Hence the resultant current flowing in the line is reduced. The reduction in resultant current flowing in the line results in (*i*) reduction of line losses and so increase of transmission efficiency, (*ii*) reduction in voltage drop or improvement of voltage regulation. The other advantages of a transmission line having high capacitance are increased load capacity and improved power factor.

5.10. ELECTRIC FIELD AND POTENTIAL DIFFERENCE

As we already know, an electric charge causes an electric field around it which theoretically extends up to infinity. If any charge is introduced in this electric field, it will be attracted or repelled according to the nature of the charge and when it will be moved, the work will be done against or by the force acting on the charge due to electric field. Hence potential in an electric field is exactly the same as potential in the gravitational field.

In general electric potential or potential at any point in an electric field is defined as the work done in moving a unit positive charge from infinity to that point.

The concept of electric potential is very important for the determination of capacitance as the latter is defined as the charge per unit potential. Now we will discuss in detail the electric potential owing to various important conductor arrangements.

5.10.1. Potential at a Charged Single Conductor

Consider a long straight cylindrical conductor A of radius r metres and having a charge of q coulombs per metre of its length.

The electric field intensity at a distance x from the centre of conductor,

$$E = \frac{q}{2\pi \epsilon_0 \epsilon_r x} \text{ volts/m}$$

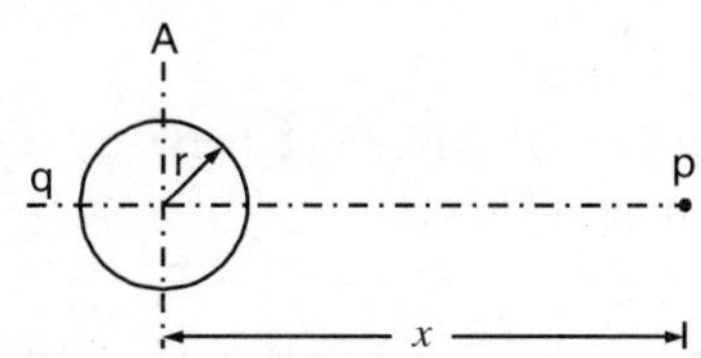

Fig. 5.32

Taking air as medium *i.e.*, $e_r = 1$

$$E = \frac{q}{2\pi \epsilon_0 x} \text{ volts/m}$$

The potential difference between conductor A and infinity distant neutral plane (a plane where E and therefore, potential is zero) will be equal to work done in bringing a unit positive charge against E from infinity to conductor surface and is given as

$$V_A = \int_r^\infty \frac{q}{2\pi \epsilon_0 x} dx = \frac{q}{2\pi \epsilon_0} \int_r^\infty \frac{dx}{x} \qquad ...(5.38)$$

5.10.2. Potential at a Charged Conductor in a Group of Charged Conductors

Consider a group of long straight conductors A, B, C, D, E, ..., N having charges of $q_1, q_2, q_3, q_4, q_5, ..., q_n$ coulombs per metre length respectively (Fig. 5.33).

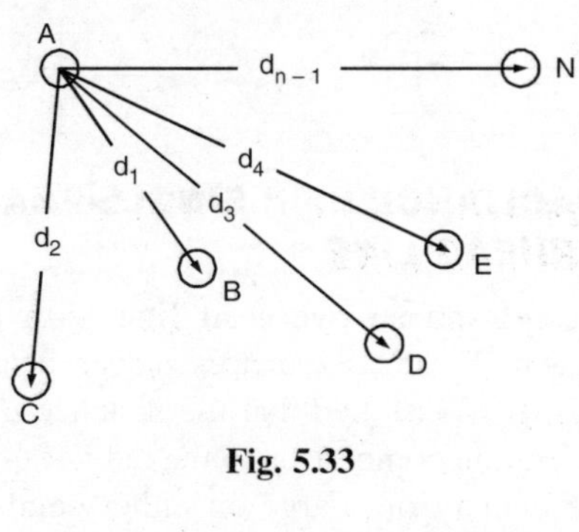

Fig. 5.33

Potential of conductor A due to its own charge q_1

$$= \int_{r}^{\infty} \frac{q_1}{2\pi \in_0} \frac{dx}{x} \text{ volts}$$

Potential of conductor A due to charge q_2

$$= \int_{d_1}^{\infty} \frac{q_2}{2\pi \in_0} \frac{dx}{x} \text{ volts}$$

since the field due to the charge q_2 extends from infinity up to a distance d_1 from conductor A.

Similarly, potential of conductor A, due to charge q_3 of conductor C placed at a distance of d_2 metres from conductor A

$$= \int_{d_2}^{\infty} \frac{q_3}{2\pi \in_0} \frac{dx}{x} \text{ volts}$$

Potential of conductor A due to charge q_n of conductor N placed at a distance of d_{n-1} metres from conductor A

$$= \int_{d_{n-1}}^{\infty} \frac{q_n}{2\pi \in_0} \frac{dx}{x} \text{ volts}$$

So overall potential difference between conductor A and infinite distant neutral plane

$$V_{AN} = \int_{r}^{\infty} \frac{q_1}{2\pi \in_0} \frac{dx}{x} + \int_{d_1}^{\infty} \frac{q_2}{2\pi \in_0} \frac{dx}{x} + \int_{d_2}^{\infty} \frac{q_3}{2\pi \in_0} \frac{dx}{x} + \dots + \int_{d_{n-1}}^{\infty} \frac{q_n}{2\pi \in_0} \frac{dx}{x}$$

$$= \frac{1}{2\pi \in_0} \big[q_1 (\log_e \infty - \log_e r) + q_2 (\log_e \infty - \log_e d_1) + q_3 (\log_e \infty - \log_e d_3) + \dots + q_n (\log_e \infty - \log_e d_{n-1}) \big]$$

$$= \frac{1}{2\pi \in_0} \left[q_1 \log_e \frac{1}{r} + q_2 \log_e \frac{1}{d_1} + q_3 \log_e \frac{1}{d_2} + \dots + q_n \log_e \frac{1}{d_{n-1}} + \log_e \infty (q_1 + q_2 + q_3 + \dots + q_n) \right]$$

Assuming balanced load *i.e.*, $q_1 + q_2 + q_3 \dots q_n = 0$

or

$$V_{AN} = \frac{1}{2\pi \in_0} \left[q_1 \log_e \frac{1}{r} + q_2 \log_e \frac{1}{d_1} + q_3 \log_e \frac{1}{d_2} + \dots + q_n \log_e \frac{1}{d_{n-1}} \right] \text{ volts} \quad \dots(5.39)$$

5.11. CAPACITANCE OF A SINGLE-PHASE OVERHEAD LINE

Consider a single-phase overhead line with two parallel conductors, each of radius r metres placed at a distance of d metres in air. It is assumed that the distance d between the conductors is large in comparison to the radii of the conductors. Therefore the density of charge on either conductor will be practically unaffected by the charge on the other conductor and will, therefore, be uniform throughout the length. A uniformly distributed charge on a conductor acts as though it is concentrated on the conductor axis. Therefore, for the purpose of our present analysis it is assumed that the charge $+q$ coulombs on conductor A and $-q$ coulombs on conductor B are concentrated at the centres of the two conductors which are separated from each other by d metres.

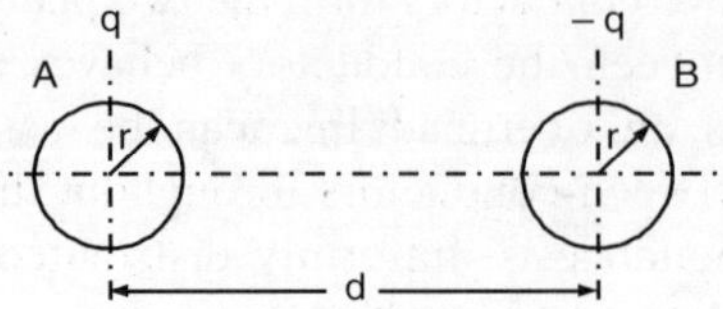

Fig. 5.34

PD between conductor A and neutral 'infinite' plane,

$$V_{AN} = \int_{r}^{\infty} \frac{q}{2\pi \in_0} \cdot \frac{d}{dx} + \int_{d}^{\infty} \frac{-q}{2\pi \in_0} \frac{dx}{x}$$

$$= \frac{q}{2\pi \in_0} \left[\log_e \infty + \log_e \frac{1}{r} - \log_e \infty + \log_e d \right]$$

$$= \frac{q}{2\pi \in_0} \log_e \frac{d}{r}$$

Similarly *pd* between conductor B and neutral 'infinite' plane,

$$V_{BN} = \int_{r}^{\infty} \frac{-q}{2\pi \in_0} \frac{dx}{x} + \int_{d}^{\infty} \frac{q}{2\pi \in_0} \frac{dx}{x} = \frac{-q}{2\pi \in_0} \log_e \frac{d}{r} \text{ volts}$$

PD between conductors A and B,

$$V_{AB} = V_{AN} - V_{BN}$$

$$= \frac{q}{2\pi \in_0} \log_e \frac{d}{r} - \frac{-q}{2\pi \in_0} \log_e \frac{d}{r} = \frac{q}{\pi \in_0} \log_e \frac{d}{r} \text{ volts}$$

Capacitance of the line,

$$C = \frac{q}{V_{AB}} = \frac{q}{\frac{q}{\pi \in_0} \log_e \frac{d}{r}} = \frac{\pi \in_0}{\log_e \frac{d}{r}} \text{ F/m} \quad \dots(5.40)$$

The above Eq. (5.40) is for capacitance between two conductors. The capacitance for each conductor, C_n (or phase to neutral) will be double of this value.

Example 5.20. Find out capacitance of a single phase line 30 km long consisting of two parallel wires each 15 mm diameter and 1.5 m apart.

Solution : Radius of each conductor, $r = \frac{15}{2}$ mm = 7.5 mm

Spacing between conductors, d = 1.5 m = 1,500 mm

Capacitance of line,

$$C = \frac{\pi \in_0}{\log_e \frac{d}{r}} \times \text{length of line in metres}$$

$$= \frac{\pi \times 8.854 \times 10^{-12}}{\log_e \frac{1{,}500}{7.5}} \times 30{,}000 = 0.1575 \ \mu\text{F} \ \textbf{Ans.}$$

5.12. CAPACITANCE OF 3-PHASE OVERHEAD LINES

5.12.1. Unsymmetrically Spaced Line

For an untransposed unsymmetrical 3-phase line the capacitances between conductor to neutral of the three conductors are different. Suppose that the line is, as shown in Fig. 5.11, and that voltages V_A, V_B, V_C are applied to the conductors with the result that the charges per metre length are q_1, q_2, q_3 respectively.

Potential of conductor A, (*w.r.t.* to neutral infinite plane)

$$V_{AN} = \frac{q_1}{2\pi\epsilon_0}\int_r^\infty \frac{dx}{x} + \frac{q_2}{2\pi\epsilon_0}\int_{d_1}^\infty \frac{dx}{x} + \frac{q_3}{2\pi\epsilon_0}\int_{d_3}^\infty \frac{dx}{x}$$

$$= \frac{1}{2\pi\epsilon_0}[q_1 \log_e \infty - q_1 \log_e r + q_2 \log_e \infty - q_2 \log_e d_1 + q_3 \log_e \infty - q_3 \log_e d_3]$$

$$= \frac{1}{2\pi\epsilon_0}\left[q_1 \log_e \frac{1}{r} + q_2 \log_e \frac{1}{d_1} + q_3 \log_e \frac{1}{d_3} + (q_1+q_2+q_3)\log_e \infty\right]$$

$$= \frac{1}{2\pi\epsilon_0}\left[q_1 \log_e \frac{1}{r} + q_2 \log_e \frac{1}{d_1} + q_3 \log_e \frac{1}{d_3}\right] \text{ volts} \quad \because q_1+q_2+q_3 = 0 \quad ...(5.41)$$

Similarly,

$$V_{BN} = \frac{1}{2\pi\epsilon_0}\left[q_2 \log_e \frac{1}{r} + q_1 \log_e \frac{1}{d_1} + q_3 \log_e \frac{1}{d_2}\right] \text{ volts} \quad ...(5.42)$$

and

$$V_{CN} = \frac{1}{2\pi\epsilon_0}\left[q_3 \log \frac{1}{r} + q_1 \log_e \frac{1}{d_3} + q_2 \log_e \frac{1}{d_2}\right] \text{ volts} \quad ...(5.43)$$

Substituting $q_3 = -(q_1 + q_2)$ in Eqs. (5.41) and (5.42), we have

$$V_{AN} = \frac{1}{2\pi\epsilon_0}\left[q_1 \log_e \frac{1}{r} + q_2 \log_e \frac{1}{d_1} - (q_1+q_2)\log_e \frac{1}{d_3}\right]$$

$$= \frac{1}{2\pi\epsilon_0}\left[q_1 \log_e \frac{d_3}{r} + q_2 \log_e \frac{d_3}{d_1}\right] \text{ volts} \quad ...(5.44)$$

and

$$V_{BN} = \frac{1}{2\pi\epsilon_0}\left[q_2 \log_e \frac{1}{r} + q_1 \log_e \frac{1}{d_1} - (q_1+q_2)\log_e \frac{1}{d_2}\right] \text{ volts}$$

$$= \frac{1}{2\pi\epsilon_0}\left[q_1 \log_e \frac{d_2}{d_1} + q_2 \log_e \frac{d_2}{r}\right] \text{ volts} \quad ...(5.45)$$

Multiplying Eq. (5.44) by $\log_e \frac{d_2}{r}$ and Eq. (5.45) by $\log_e \frac{d_3}{d_1}$, we have

$$V_{AN} \log_e \frac{d_2}{r} = \frac{1}{2\pi\epsilon_0}\left[q_1 \log_e \frac{d_3}{r}\cdot\log_e \frac{d_2}{r} + q_2 \log_e \frac{d_3}{d_1} \log_e \frac{d_2}{r}\right] \quad ...(5.46)$$

and

$$V_{BN} \log_e \frac{d_3}{d_1} = \frac{1}{2\pi\epsilon_0}\left[q_1 \log_e \frac{d_2}{d_1}\cdot\log_e \frac{d_3}{d_1} + q_2 \log_e \frac{d_2}{r}\cdot\log_e \frac{d_3}{d_1}\right] \quad ...(5.47)$$

Subtracting Eq. (5.47) from Eq. (5.46), we have

$$V_{AN} \log_e \frac{d_2}{r} - V_{BN} \log_e \frac{d_3}{d_1} = \frac{q_1}{2\pi\epsilon_0}\left[\log_e \frac{d_3}{r}\cdot\log_e \frac{d_2}{r} - \log_e \frac{d_2}{d_1}\cdot\log_e \frac{d_3}{d_1}\right]$$

$$\text{or } q_1 = 2\pi\epsilon_0 \frac{V_{AN} \log_e \frac{d_2}{r} - V_{BN} \log_e \frac{d_3}{d_1}}{\log_e \frac{d_3}{r}\log_e \frac{d_2}{r} - \log_e \frac{d_2}{d_1}\cdot\log_e \frac{d_3}{d_1}} \quad ...(5.48)$$

Capacitance of conductor A to neutral,

$$C_{AN} = \frac{q_1}{V_{AN}}$$

$$= 2\pi\epsilon_0 \frac{\log_e \frac{d_2}{r} - \frac{V_{BN}}{V_{AN}}\log_e \frac{d_3}{d_1}}{\log_e \frac{d_3}{r}\log_e \frac{d_2}{r} - \log_e \frac{d_2}{d_1}\cdot\log_e \frac{d_3}{d_1}} \text{ F/m} \quad ...(5.49)$$

Similarly capacitance of conductor B to neutral,

$$C_{BN} = 2\pi\epsilon_0 \frac{\log_e \frac{d_3}{r} - \frac{V_{CN}}{V_{BN}}\log_e \frac{d_1}{d_2}}{\log_e \frac{d_1}{r}\cdot\log_e \frac{d_3}{r} - \log_e \frac{d_3}{d_2}\cdot\log_e \frac{d_1}{d_2}} \text{ F/m} \quad ...(5.50)$$

and

$$C_{CN} = 2\pi\epsilon_0 \frac{\log_e \frac{d_1}{r} - \frac{V_{AN}}{V_{CN}}\log_e \frac{d_2}{d_3}}{\log_e \frac{d_2}{r}\cdot\log_e \frac{d_1}{r} - \log_e \frac{d_1}{d_3}\cdot\log_e \frac{d_2}{d_3}} \text{ F/m} \quad ...(5.51)$$

5.12.2. Unsymmetrical Line With Transposed Conductors

Assuming that the charge per unit length is same in every part of the transposed cycle [Figs. 5.35 (1), (2) and (3)], average value of voltage of conductor A, will be

$$\frac{V_{AN_1} + V_{AN_2} + V_{AN_3}}{3} \quad ...(5.52)$$

Voltage of conductor A in positions (1), (2), and (3) we have

$$V_{AN_1} = \frac{1}{2\pi\epsilon_0}\left[q_1 \log_e \frac{1}{r} + q_2 \log_e \frac{1}{d_1} + q_3 \log_e \frac{1}{d_3}\right] \text{ volts}$$

$$V_{AN_2} = \frac{1}{2\pi\epsilon_0}\left[q_1 \log_e \frac{1}{r} + q_2 \log_e \frac{1}{d_2} + q_3 \log_e \frac{1}{d_1}\right] \text{ volts}$$

and

$$V_{AN_3} = \frac{1}{2\pi\epsilon_0}\left[q_1 \log_e \frac{1}{r} + q_2 \log_e \frac{1}{d_3} + q_3 \log_e \frac{1}{d_2}\right] \text{ volts}$$

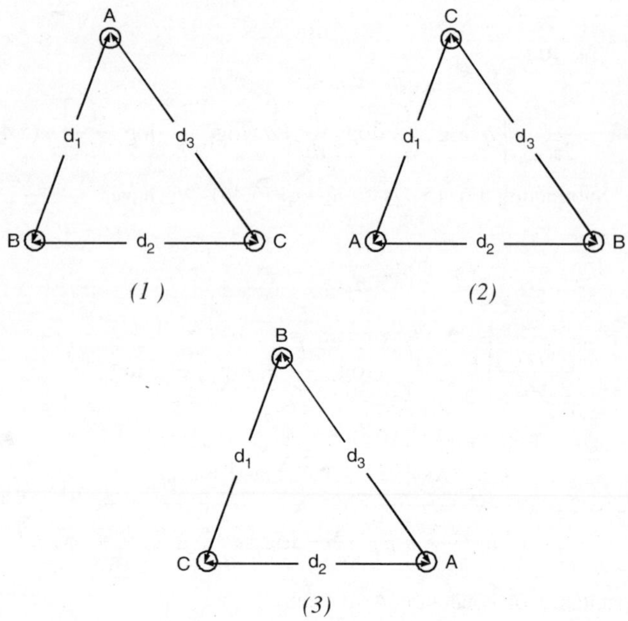

Fig. 5.35

So average value of voltage of conductor A,

$$V_{AN} = \frac{V_{AN_1} + V_{AN_2} + V_{AN_3}}{3}$$

$$= \frac{1}{6\pi \in_0}\left[q_1 \log_e \frac{1}{r^3} + q_2 \log_e \frac{1}{d_1 d_2 d_3} + q_3 \log_e \frac{1}{d_1 d_2 d_3}\right]$$

$$= \frac{1}{6\pi \in_0}\left[q_1 \log_e \frac{1}{r^3} + (q_2 + q_3)\log_e \frac{1}{d_1\ d_2\ d_3}\right] \text{ volts}$$

Substituting $q_2 + q_3 = -\ q_1$ in above equation, we have

$$V_{AN} = \frac{1}{6\pi \in_0}\left[q_1 \log_e \frac{1}{r^3} - \log_e \frac{1}{d_1\ d_2\ d_3}\right] \text{ volts}$$

$$\text{or } V_{AN} = \frac{q_1}{2\pi \in_0}\left[\log_e \frac{1}{r} + \log_e \sqrt[3]{d_1\ d_2\ d_3}\right]$$

$$= \frac{q_1}{2\pi \in_0} \log_e \frac{\sqrt[3]{d_1\ d_2\ d_3}}{r} \text{ volts}$$

Capacitance of conductor A to neutral,

$$C_{AN} = \frac{q_1}{V_{AN}} = \frac{q_1}{\dfrac{q_1}{2\pi \in_0}\log_e \dfrac{\sqrt[3]{d_1 d_2 d_3}}{r}} = \frac{2\pi \in_0}{\log_e \dfrac{\sqrt[3]{d_1 d_2 d_3}}{r}} \text{ F/m}$$

Similarly expressions for C_{BN} and C_{CN} can be obtained and we have

$$C_{AN} = C_{BN} = C_{CN} = C_N = \frac{2\pi \in_0}{\log_e \dfrac{\sqrt[3]{d_1 d_2 d_3}}{r}} \text{ F/m} \quad ...(5.53)$$

5.12.3. Equilaterally Spaced Line

For the equilateral spacing $d_1 = d_2 = d_3 = d$ (say)

$$\text{and } C_{AN} = C_{BN} = C_{CN} = C_N = \frac{2\pi \in_0}{\log_e \dfrac{\sqrt[3]{d.d.d}}{r}} = \frac{2\pi \in_0}{\log_e \dfrac{d}{r}} \quad ...(5.54)$$

Example 5.21. Calculate the capacitance of a 100 km long 3-phase, 50 Hz overhead transmission line consisting of 3-conductors, each of diameter 2 cm and spaced 2.5 m at the corners of equilateral triangle.

[U.P. Technical Univ. Elements of Power System 2006-07]

Solution : Spacing of conductors, $d = 2.5$ m $= 250$ cm

Radius of each conductor, $r = \dfrac{2.0}{2} = 1.0$ cm

$$\text{Capacitance of the line} = \frac{2\pi \in_0}{\log_e \dfrac{d}{r}} \text{ F/m}$$

$$= \frac{2\pi \times 8.854 \times 10^{-12}}{\log_e \dfrac{250}{1}}$$

$$= 10.075 \times 10^{-12} \text{ F/m}$$

Capacitance of 100 km long line

$$= 10.075 \times 100 \times 10^3 \times 10^{-12}$$

$= 1.0075$ μF **Ans.**

Example 5.22. A single 3-phase line operated at 50 Hz is arranged as follows : (Fig. 5.36). The conductor diameter is 0.6 cm. Find the inductance and capacitance per km. The line is regularly transposed. *[Rajasthan Technical Univ. Transmission & Distribution of Electrical Power 2011]*

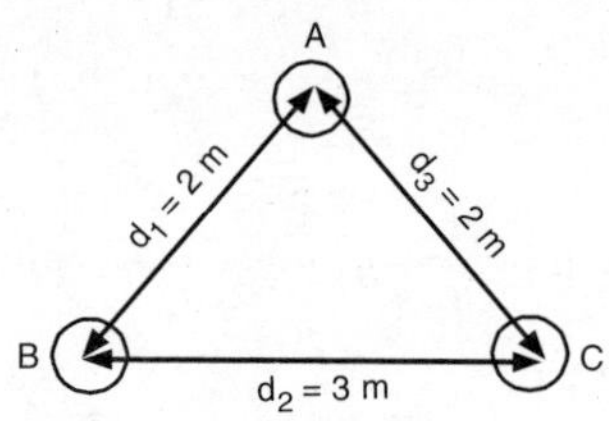

Fig. 5.36

Solution : Conductor radius, $r = \dfrac{0.6}{2}$ cm $= 0.3$ cm

Geometric mean radius, $r' = 0.7788 \times 0.3$ cm $= 0.23364$ cm

Spacing $d_1 = d_3 = 2.0$ m $= 200$ cm

$d_2 = 3$ m $= 300$ cm

Inductance per km per phase

$$= 0.2 \log_e \frac{\sqrt[3]{d_1 d_2 d_3}}{r'} \text{ mH}$$

$$= 0.2 \log_e \frac{\sqrt{200 \times 300 \times 200}}{0.23364} \text{ mH} = 1.378 \text{ mH} \quad \textbf{Ans.}$$

Capacitance per phase per metre

$$= \frac{2\pi \in_0}{\log_e \dfrac{\sqrt[3]{d_1 d_2 d_3}}{r}} = \frac{2\pi \times 8.854 \times 10^{-12}}{\log_e \dfrac{\sqrt[3]{200 \times 300 \times 200}}{0.3}}$$

$$= 8.3814 \times 10^{-12} \text{ F}$$

Capacitance per phase per km

$= 8.3814 \times 10^{-12} \times 1{,}000 = 8.3814 \times 10^{-9}$ F **Ans.**

Example 5.23. Determine the capacitance and charging current per unit length of the line for the following arrangement of conductors of diameter 20 mm shown in Fig. 5.37.

[A.M.I.E. Sec. B. Power Systems Winter 1994]

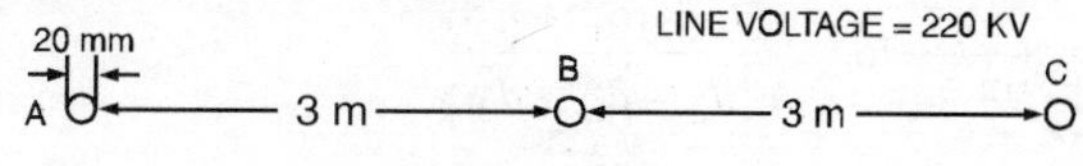

Fig. 5.37

Solution : $r = \frac{20}{2} = 10 \text{ mm} = 0.01 \text{ m}$

Spacing between conductors

$$d_1 = AB = 3 \text{ m}; \ d_2 = BC = 3 \text{ m}; \ d_3 = CA = 6 \text{ m}$$

Capacitance per phase per m length,

$$C_N = \frac{2\pi \epsilon_0}{\log_e \frac{\sqrt[3]{d_1 d_2 d_3}}{r}} = \frac{2\pi \times 8.854 \times 10^{-12}}{\log_e \frac{\sqrt[3]{3 \times 3 \times 6}}{0.01}}$$

$$= 9.3737 \times 10^{-12} \text{ F } \textbf{Ans.}$$

Charging current per phase,

$$I_C = 2\pi f C_N V_{ph}$$

$$= 2\pi \times 50 \times 9.3737 \times 10^{-12} \times \frac{220 \times 1{,}000}{\sqrt{3}} = 0.374 \text{ mA } \textbf{Ans.}$$

Example 5.24. A 3-phase, 50 Hz transmission line has flat horizontal configuration with 3.5 m between adjacent conductors. The conductors are no. 2/0 hard-drawn seven strand copper (outside conductor diameter = 1.05 cm). The voltage of the line is 110 kV. Find the capacitance to neutral and the charging current per km.

[*U.P. Technical Univ. Elements of Power System 2005-06*]

Solution : Conductor radius, $r = \frac{1.05}{2} \text{ cm} = 0.5025 \text{ cm}$

Spacings, $d_1 = 3.5$ m; $d_2 = 3.5$ m and $d_3 = 7$ m

Capacitance to neutral,

$$C_N = \frac{2\pi \epsilon_0}{\log_e \frac{\sqrt[3]{d_1 d_2 d_3}}{r}} = \frac{2\pi \times 8.854 \times 10^{-12}}{\log_e \frac{\sqrt[3]{3.5 \times 3.5 \times 7 \times 100}}{0.5025}}$$

$$= 8.21 \times 10^{-12} \text{ F/m or } 8.21 \times 10^{-12} \times 1{,}000$$

$$= 0.00821 \ \mu\text{F/km } \textbf{Ans.}$$

Charging current per phase,

$$I_C = 2\pi f C_N V_{ph}$$

$$= 2\pi \times 50 \times 0.00821 \times 10^{-6} \times \frac{110 \times 1000}{\sqrt{3}} = 0.164 \text{ A/km } \textbf{Ans.}$$

5.13. CAPACITANCE OF DOUBLE CIRCUIT 3-PHASE OVERHEAD LINES

Normally used conductor configurations are of hexagonal spacing (Fig. 5.19) and flat vertical spacing (Fig. 5.20). It has been found that modified GMD method holds good for determination of capacitance of transposed double circuit 3-phase overhead lines with equilateral spacing (conductors at the vertices of a regular hexagon) and with flat vertical spacing. It is reasonable to assume that the modified GMD method can be used for determination of capacitance of a line with any configuration intermediate between these two configurations.

In the case of calculations of inductance, determination of self GMD (or GMR) of conductor is necessary because of internal flux linkages of the conductor. But in case of calculations of capacitance, since all charges reside on the surface of the conductor, actual radius of the conductor is used.

5.13.1. Symmetrically Spaced Line

Consider a 3-phase double circuit connected in parallel—conductors A, B and C forming one circuit and conductors A′, B′ and C′ forming another circuit (conductors symmetrically spaced).

Let the charge over conductors A, B and C be q_1, q_2 and q_3 coulombs per metre length. Then charge over conductors A′, B′ and C′ will obviously be q_1, q_2 and q_3 coulombs per metre length and $q_1 + q_2 + q_3 = 0$.

Potential of conductor A *w.r.t.* neutral infinite plane (Fig. 5.19),

$$V_{AN} = \frac{1}{2\pi \epsilon_0}\left[q_1 \int_r^\infty \frac{dx}{x} + q_2 \int_d^\infty \frac{dx}{x} + q_3 \int_{\sqrt{3}d}^\infty \frac{dx}{x} + q_1 \int_{2d}^\infty \frac{dx}{x} + q_2 \int_{\sqrt{3}d}^\infty \frac{dx}{x} + q_3 \int_d^\infty \frac{dx}{x}\right]$$

$$= \frac{1}{2\pi \epsilon_0}\left[q_1 \log_e \infty - q_1 \log_e r + q_2 \log_e \infty - q_2 \log_e d - q_3 \log_e \sqrt{3}d + q_3 \log_e \infty + q_1 \log_e \infty - q_1 \log_e 2d + q_2 \log_e \infty - q_2 \log_e \sqrt{3}\, d + q_3 \log_e \infty - q_3 \log_e d\right]$$

$$= \frac{1}{2\pi \epsilon_0}\left[2(q_1 + q_2 + q_3) \log_e \infty + q_1 \log_e \frac{1}{2\,dr} + q_2 \log_e \frac{1}{\sqrt{3}\, d^2} + q_3 \log_e \frac{1}{\sqrt{3}\, d^2}\right]$$

$$= \frac{1}{2\pi \epsilon_0}\left[q_1 \log_e \frac{1}{2dr} + q_2 \log_e \frac{1}{\sqrt{3}\, d^2} + q_3 \log_e \frac{1}{\sqrt{3}\, d^2}\right]$$

$$\because q_1 + q_2 + q_3 = 0$$

$$= \frac{1}{2\pi \epsilon_0}\left[q_1 \log_e \frac{1}{2dr} + (q_2 + q_3) \log_e \frac{1}{\sqrt{3}\, d^2}\right]$$

$$= \frac{1}{2\pi \epsilon_0}\left[q_1 \log_e \frac{1}{2dr} - q_1 \log_e \frac{1}{\sqrt{3}\, d^2}\right] \quad \because q_2 + q_3 = -q_1$$

$$= \frac{q_1}{2\pi \epsilon_0} \log_e \frac{\sqrt{3}d^2}{2dr} = \frac{q_1}{2\pi \epsilon_0} \log_e \frac{\sqrt{3}d}{2r} \text{ volts}$$

Capacitance of conductor A to neutral,

$$C_{AN} = \frac{q_1}{V_{AN}} = \frac{q_1}{\frac{q_1}{2\pi \epsilon_0} \log_e \frac{\sqrt{3}\, d}{2r}} = \frac{2\pi \epsilon_0}{\log_e \frac{\sqrt{3}\, d}{2r}} \text{ F/m}$$

Similarly expressions for C_{BN} and C_{CN} can be obtained and we have

$$C_{AN} = C_{BN} = C_{CN} = C_N = \frac{2\pi \epsilon_0}{\log_e \frac{\sqrt{3}d}{2r}} \text{ F/m} \qquad ...(5.55)$$

This is because the conductors of different phases are symmetrically placed.

Equation (5.55) gives the capacitance of conductor A alone, whereas there are two conductors per phase A and A'. Therefore the capacitance of the system per phase will be twice of the capacitance of one conductor to neutral, *i.e.*,

$$\text{Capacitance per phase, } C = 2C_N = \frac{4\pi \in_0}{\log_e \frac{\sqrt{3}d}{2r}} \text{ F/m} \quad ...(5.56)$$

5.13.2. Flat Vertically Spaced Line

Consider conductors arranged, as shown in Figs. 5.20 (1), (2) and (3), corresponding to different positions in the transposition positions.

Potential of conductor A w.r.t. 'infinite' neutral plane [Fig. 5.20 (1)]

$$V_{AN_1} = \frac{1}{2\pi \in_0}\left[q_1 \int_r^\infty \frac{dx}{x} + q_2 \int_{d_1}^\infty \frac{dx}{x} + q_3 \int_{2d_1}^\infty \frac{dx}{x} + q_1 \int_{\sqrt{4d_1^2+d_2^2}}^\infty \frac{dx}{x} + q_2 \int_{\sqrt{d_1^2+d_2^2}}^\infty \frac{dx}{x} + q_3 \int_{d_2}^\infty \frac{dx}{x}\right] \text{ volts}$$

$$= \frac{1}{2\pi \in_0}\left[q_1 \log_e \frac{1}{r\sqrt{4d_1^2+d_2^2}} + q_2 \log_e \frac{1}{d_1\sqrt{d_1^2+d_2^2}} + q_3 \log_e \frac{1}{2d_1 d_2}\right] \text{ volts}$$

Similarly,

$$V_{AN_2} = \frac{1}{2\pi \in_0}\left[q_1 \log_e \frac{1}{rd_2} + q_2 \log_e \frac{1}{d_1\sqrt{d_1^2+d_2^2}} + q_3 \log_e \frac{1}{d_1\sqrt{d_1^2+d_2^2}}\right] \text{ volts and}$$

$$V_{AN_3} = \frac{1}{2\pi \in_0}\left[q_1 \log_e \frac{1}{r\sqrt{4d_1^2+d_2^2}} + q_2 \log_e \frac{1}{2d_1 d_2} + q_3 \log_e \frac{1}{d_1\sqrt{d_1^2+d_2^2}}\right]$$

$$V_{AN} = \frac{V_{AN1}+V_{AN2}+V_{AN3}}{3}$$

$$= \frac{1}{6\pi \in_0}\left[q_1 \log_e \frac{1}{r^3 d_2(4d_1^2+d_2^2)} + q_2 \log_e \frac{1}{2d_1^3 d_2 (d_1^2+d_2^2)} + q_3 \log_e \frac{1}{2d_1^3 d_2 (d_1^2+d_2^2)}\right]$$

$$= \frac{1}{6\pi \in_0}\left[q_1 \log_e \frac{1}{r^3 d_2(4d_1^2+d_2^2)} + (q_2+q_3)\log_e \frac{1}{2d_1^3 d_2 (d_1^2+d_2^2)}\right]$$

$$= \frac{1}{6\pi \in_0}\left[q_1 \log_e \frac{1}{r^3 d_2(4d_1^2+d_2^2)} - q_1 \log_e \frac{1}{2d_1^3 d_2 (d_1^2+d_2^2)}\right]$$

$$\because q_2 + q_3 = -q_1$$

$$= \frac{q_1}{6\pi \in_0} \log_e \frac{2d_1^3 d_2 (d_1^2+d_2^2)}{r^3 d_2 (4d_1^2+d_2^2)}$$

$$= \frac{q_1}{6\pi \in_0} \log_e \frac{2d_1^3}{r^3} \cdot \frac{d_1^2+d_2^2}{4d_1^2+d_2^2}$$

$$= \frac{q_1}{2\pi \in_0} \log_e \frac{2^{1/3} d_1}{r}\left(\frac{d_1^2+d_2^2}{4d_1^2+d_2^2}\right)^{1/3} \text{ volts}$$

Capacitance of conductor A,

$$C_{AN} = \frac{q_1}{V_{AN}} = \frac{q}{\frac{q}{2\pi\in_0}\log_e 2^{1/3}\frac{d_1}{r}\left(\frac{d_1^2+d_2^2}{4d_1^2+d_2^2}\right)^{1/3}}$$

$$= \frac{2\pi \in_0}{\log_e 2^{1/3}\frac{d_1}{r}\left(\frac{d_1^2+d_2^2}{4d_1^2+d^2}\right)^{1/3}} \text{ F/m} \quad ...(5.57)$$

Similarly expressions for capacitances C_{BN} and C_{CN} can be obtained which are same as above

Capacitance per phase will be double of C_{AN}, *i.e.*,

Capacitance per phase,

$$C = \frac{4\pi \in_0}{\log_e 2^{1/3}\frac{d_1}{r}\left(\frac{d_1^2+d_2^2}{4d_1^2+d_2^2}\right)^{1/3}} \text{ F/m} \quad ...(5.58)$$

Example 5.25. Six conductors of a double-circuit transmission line are arranged as shown below : (Fig. 5.38)

The diameter of each conductor is 2.5 cm. Find the capacitive reactance to neutral and the charging current per km per phase at 132 kV and 50 Hz, assuming that the line is regularly transposed. Neglect the effect of earth.

[*Agra Univ. Transmission and Distribution of Electrical Power 1979 Supp.*]

Solution : Radius of conductor, $r = \frac{2.5}{5}$ cm = 1.25 cm

The mutual GMD of the circuit is calculated as in case of determination of inductance. The self GMD of the circuit is also determined as in case of determination of inductance except that for r', r is used, as the electric charge, resides on the surface of the conductor unlike the magnetic flux which lies inside the conductor.

Distance between conductors R and R',

$$d_{RR'} = \sqrt{7^2 + (4+4)^2} = 10.63 \text{ m}$$

$$d_{BB'} = d_{RR'} = 10.63 \text{ m}$$

GMR of conductors of phase R,

$$D_{SR} = \sqrt[4]{d_{RR}\, d_{RR'}\, d_{R'R'}\, d_{R'R}}$$

$$= \sqrt{r \times d_{RR'}} = \sqrt{1.25 \times 10^{-2} \times 10.63} = 0.3645 \text{ m}$$

Similarly, $D_{SY} = \sqrt{r \times d_{YY'}} = \sqrt{1.25 \times 10^{-2} \times 9} = 0.3354$ m

and $D_{SB} = \sqrt{r \times d_{BB'}} = \sqrt{1.25 \times 10^{-2} \times 10.63} = 0.3645$ m

Self, GMD, $D_S = \sqrt[3]{D_{SR}\, D_{SY}\, D_{SB}}$

$$= \sqrt[3]{0.3645 \times 0.3354 \times 0.3645} = 0.3545 \text{ m}$$

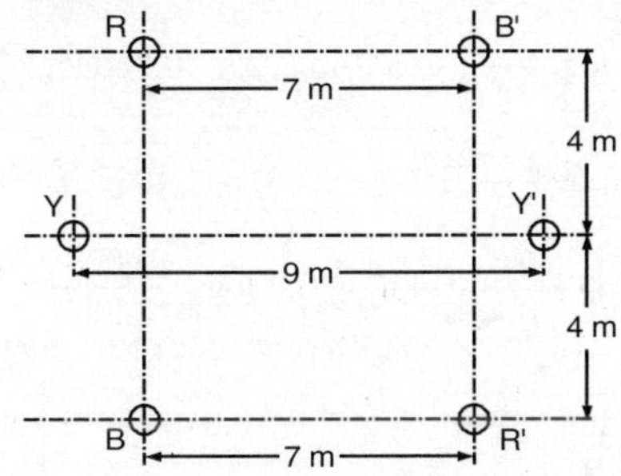

Fig. 5.38

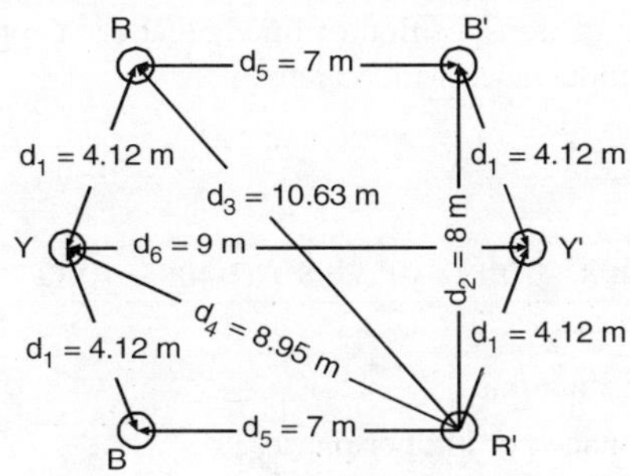

Fig. 5.39

Distance between conductors R and Y,

$$D_{RY} = \sqrt{4^2 + (4.5 - 3.5)^2} = 4.12 \text{ m}$$

Distance between conductors R and Y′,

$$D_{RY'} = \sqrt{(9-1)^2 + 4^2} = 8.95 \text{ m}$$

$$\text{So } D_{RY} = \sqrt[4]{d_{RY}\, d_{RY'}\, d_{R'Y}\, d_{R'Y'}}$$

$$= \sqrt[4]{4.12 \times 8.95 \times 8.95 \times 4.12} = 6.072 \text{ m}$$

$$\text{Similarly, } D_{YB} = \sqrt[4]{(4.12 \times 8.95)^2} = 6.072 \text{ m}$$

$$\text{and } D_{BR} = \sqrt[4]{(8 \times 7)^2} = 7.483 \text{ m}$$

$$D_{eq} \text{ or } D_m = \sqrt[3]{D_{RY} D_{YB} D_{BR}}$$

$$= \sqrt[3]{6.072 \times 6.072 \times 7.483} = 6.51 \text{ m}$$

Capacitance per phase,

$$C = \frac{2\pi \epsilon_0}{\log_e \dfrac{D_m}{D_s}}$$

$$= \frac{2\pi \times 8.854 \times 10^{-12}}{\log_e \dfrac{6.51}{0.3545}}$$

$$= 19.11 \times 10^{-12} \text{ F/m or } 19.11 \times 10^{-9} \text{ F/km}$$

Capacitance reactance to neutral

$$= \frac{1}{2\pi f C} = \frac{1}{2\pi \times 50 \times 19.11 \times 10^{-9}}$$

$$= 166.526 \text{ k}\Omega \textbf{ Ans.}$$

Charging current,

$$I_C = 2\pi f C V_{ph}$$

$$= 2\pi \times 50 \times 19.11 \times 10^{-9} \times \frac{132 \times 1{,}000}{\sqrt{3}}$$

$$= 0.4575 \text{ A/km } \textbf{Ans.}$$

Example 5.26. Determine the capacitance per phase of the line of Example 5.14.

Solution : Distances $d_1 = 8$ m and $d_2 = 6$ m

Radius of conductor, $r = 3 \times$ radius of each strand of conductor

$$= 3 \times \frac{5}{2} \text{ mm} = 7.5 \text{ mm or } 7.5 \times 10^{-3} \text{ m}$$

Capacitance per phase, from Eq. (5.58),

$$C = \frac{4\pi \epsilon_0}{\log_e 2^{1/3} \dfrac{d_1}{r} \left(\dfrac{d_1^2 + d_2^2}{4d_1^2 + d_2^2} \right)^{1/3}} \text{ F/m}$$

$$= \frac{4\pi \times 8.854 \times 10^{-12}}{\log_e 2^{1/3} \left(\dfrac{8}{7.5 \times 10^{-3}} \right) \left(\dfrac{8^2 + 6^2}{4 \times 8^2 + 6^2} \right)^{1/3}}$$

$$= 16.25 \times 10^{-12} \text{ F/m}$$

Capacitance of 100 km line

$$= 16.25 \times 10^{-12} \times 100 \times 1{,}000 = 1.625 \ \mu\text{F } \textbf{Ans.}$$

5.14. CAPACITANCE OF BUNDLED CONDUCTOR LINE

A bundled conductor line is shown in Fig. 5.40. The conductors of any one bundle are in parallel and it is assumed that the charge per bundle divides equally among the conductors of the bundle as $D_{ab} >> s$. Also $D_{ab} - s \simeq D_{ab} + s \simeq D_{ab}$ for the same reason. The results obtained with these assumptions are fairly accurate for usual spacings. Thus if charge on phase A is q_A, then a and a' have a charge of $q_A/2$ each. Similarly the charge is equally divided for phases B and C.

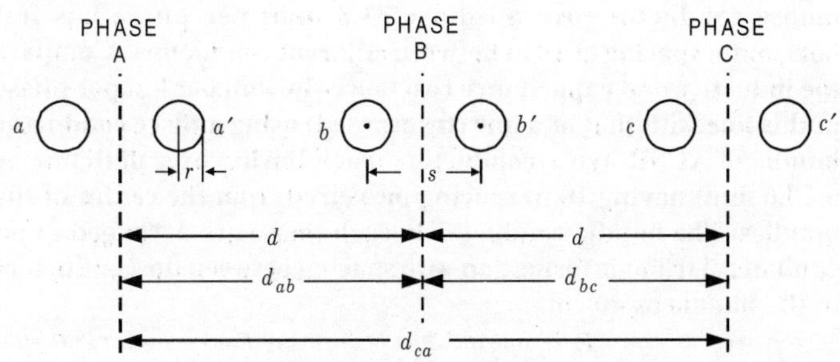

Fig. 5.40. ***Cross Section of a Bundled Conductor Three-Phase Transmission Line***

Now, writing an equation for the voltage from conductor a to conductor b, we have

$$V_{ab} = \frac{1}{2\pi\epsilon_0}\left[0.5q_A\left(\log_e \frac{d_{ab}}{r} + \log_e \frac{d_{ab}}{s}\right) + 0.5q_B\left(\log_e \frac{r}{d_{ab}} + \log_e \frac{s}{d_{ab}}\right) + 0.5q_c\left(\log_e \frac{d_{bc}}{d_{ca}} + \log_e \frac{d_{bc}}{d_{ca}}\right)\right]$$

$$= \frac{1}{2\pi\epsilon_0}\left(q_A \log_e \frac{d_{ab}}{\sqrt{rs}} + q_B \log_e \frac{\sqrt{rs}}{d_{ab}} + q_C \log_e \frac{d_{bc}}{d_{ca}}\right) \quad ...(5.59)$$

Considering the line to be transposed and proceeding in the usual manner, the final result will be

$$C_N = \frac{2\pi\varepsilon_0}{\log_e \frac{\sqrt[3]{d_{ab}d_{bc}d_{ca}}}{\sqrt{rs}}} = \frac{2\pi\varepsilon_0}{\log_e \frac{D_m}{D_s}} \quad ...(5.60)$$

Mutual GMD of the circuit will be determined as mentioned in Art. 5.8.

i.e., $D_m = \sqrt[3]{d_{ab}d_{bc}d_{ca}}$

Self GMD, D_s is determined in the same way as mentioned in Art. 5.8 except that for r', r is used.

Example 5.27. Determine the capacitance and charging current per km of the line of Example 5.18 if the line operates at 132 kV.

Solution : Self GMD, $D_s = \sqrt{rs} = \sqrt{2.5 \times 10^{-2} \times 0.4} = 0.1$ m

Mutual GMD, D_m = 8.189 m

Capacitance per km per phase, C

$$= \frac{2\pi\varepsilon_0}{\log_e \frac{D_m}{D_s}}$$

$$= \frac{2\pi \times 8.854 \times 10^{-12}}{\log_e \frac{8.189}{0.1}} \text{ F/m}$$

$= 12.628 \times 10^{-12}$ F/m or 12.628×10^{-9} F/km

Charging current per km

$$= 2\pi f C V_{ph} = 2\pi \times 50 \times 12.628 \times 10^{-9} \times \frac{132{,}000}{\sqrt{3}}$$

assuming supply frequency f = 50 Hz

= 0.3 A/km **Ans.**

Example 5.28. A 50 Hz, 3-phase line (transposed) of an ACSR moose conductor (overall dia = 31.8 mm) per phase has flat horizontal spacing of 10 m between adjacent conductors. Compare the inductive and capacitance reactances in ohm per km per phase of this line with that of a line (transposed) using a three conductor bundle of ACSR hyn *x* conductors (each having overall diameter = 19.6 mm) having 10 m spacing measured from the centre of the bundles. The bundle conductor in each phase are arranged in an equilateral triangle formation with spacing between the conductors in the bundle as 40 cm.

[*U.P. Technical Univ. Elements of Power System 2007-08*]

Solution : One Conductor ACSR Moose Conductor Line :

Radius of each conductor, $r = \frac{31.8}{2}$ mm = 15.9 mm

GMR, r' = 0.7788 × 15.9 = 12.383 mm

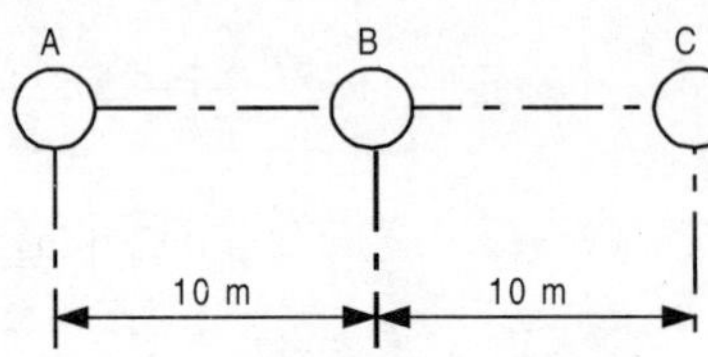

Fig. 5.41

From Eqs. (5.23), (5.24) and (5.25) respectively, we have

$$L_A = 0.2\left[\log_e \frac{d}{r'} + \frac{1}{2}\log_e 2 - j0.866 \log_e 2\right] \text{ mH/km}$$

$$= 0.2\left[\log_e \frac{10{,}000}{12.383} + \frac{1}{2}\log_e 2 + j0.866 \log_e 2\right]$$

$= 0.2[6.694 + 0.3466 - j0.6] = (1.408 - j0.12)$ mH/km

$$L_B = 0.2 \log_e \frac{d}{r'} \text{ mH/km} = 0.2 \log_e \frac{10{,}000}{12.383} = 1.3388 \text{ mH/km}$$

$$L_C = 0.2\left[\log_e \frac{d}{r'} + \frac{1}{2}\log_e 2 + j0.866 \log_e 2\right] \text{ mH/km}$$

$= (1.408 + j0.12)$ mH/km

The effect of transposition of line is that each conductor has the same average inductance which is given as

$$L_{av} = \frac{L_A + L_B + L_C}{3}$$

$$= \frac{(1.408 - j0.12) + 1.3388 + (1.408 - j0.12)}{3} \text{ mH/km}$$

= 1.385 mH/km

Inductive reactance per km per phase,

$X_L = 2\pi f L_{av} = 2\pi \times 50 \times 1.385 \times 10^{-3} = 0.435\ \Omega$ **Ans.**

Capacitance per phase,

$$C_N = \frac{2\pi\varepsilon_0}{\log_e \frac{\sqrt[3]{d_1 d_2 d_3}}{r}}$$

$$= \frac{2\pi \times 8.854 \times 10^{-12}}{\log_e \frac{\sqrt[3]{10 \times 10 \times 20}}{12.383 \times 10^{-3}}}$$

$= 8.0333 \times 10^{-12}$ F/m or 8.0333×10^{-9} F/km

Capacitive reactance per phase per km,

$$X_C = \frac{1}{2\pi f C} = \frac{1}{2\pi \times 50 \times 8.0333 \times 10^{-9}\ \Omega} = 0.396\ \text{M}\Omega$$

Three Conductor Bundled Line

GMR, $D_s = \sqrt[3]{r' \times s^2}$

$$= \sqrt[3]{0.7788 \times r \times s^2} = \sqrt[3]{0.7788 \times 19.6 \times 10^{-3} \times 0.4^2}$$

= 0.1347 m

Taking the centre to centre distances between the bundles as D_{ab}, D_{bc} and D_{ca}.

$$D_{eq} = \sqrt[3]{10 \times 10 \times 20} = \sqrt[3]{2{,}000} = 12.6 \text{ m}$$

Inductance of bundled conductor line,

$$L' = 0.2 \log_e \frac{D_m}{D_s} \text{ mH/km}$$

$$= 0.2 \log_e \frac{12.6}{0.1347} \text{ mH/km} = 0.908 \text{ mH/km}$$

Inductive reactance per km per phase,

$X'_L = 2\pi f L' = 2\pi \times 50 \times 0.908 \times 10^{-3} = 0.285\ \Omega$ **Ans.**

GMR, $D_s = \sqrt[3]{r \times s^2} = \sqrt[3]{19.6 \times 10^{-3} \times 0.4^2} = 0.1464$ m

Capacitance per km per phase,

$$C' = \frac{2\pi\varepsilon_0}{\log_e \frac{D_{eq}}{D_s}} = \frac{2\pi \times 8.854 \times 10^{-9}}{\log_e \frac{12.6}{0.1464}} = 0.01249\ \mu F$$

Capacitive reactance per km per phase

$$= \frac{1}{2\pi \times 50 \times 0.01249 \times 10^{-6}} = 0.255\ M\Omega$$

Inductive reactance of three conductor bundled line is 0.285 Ω/km/phase, while the inductive reactance of one conductor ACSR Moose conductor line is 0.435 Ω per km per phase **Ans.**

Capacitive reactance of three conductor bundled line is 0.255 MΩ per km per phase while for one conductor ACSR Moose conductor line it is 0.396 MΩ per km per phase.

5.15. EFFECT OF EARTH ON CAPACITANCE OF TRANSMISSION LINE

So far, in determination of capacitance of transmission lines, the presence of earth was ignored. But it is not true; the presence of earth affects the electric field of a line and so the capacitance. The effect of earth on capacitance can be modelled by the method of images.

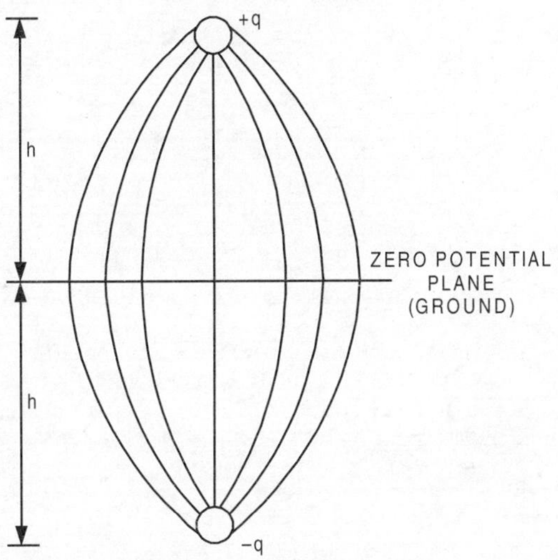

Fig. 5.42. ***Electric Field of Two Long Parallel, Oppositely Charged Conductors***

The electric field of too long, parallel conductors having charge $+q$ and $-q$ per unit is such that it has a zero potential between the conductors, as depicted in Fig. 5.42. If a conducting sheet of infinite dimensions is placed at the zero potential plane, the electric field is not disturbed. Further, if the conductor carrying charge $-q$ is now removed, the electric field above the conducting sheet stays intact, while that below it disappears. Using these well known facts in reverse, the presence of ground below a charged conductor can be replaced by a fictitious conductor having equal and opposite charge and located as far below the ground surface as the overhead conductor above it—such a fictitious conductor is the *mirror image* of the overhead conductor. This method of producing the same electric field as in the presence of earth is called the *method of images* originally suggested by Lord Kelvin.

5.15.1. Effect of Earth on the Capacitance of a Single-Phase Transmission Line

Considering the case of a single-phase overhead line, assume conductors A′ and B′ as image conductors of conductors A and B respectively, as shown in Fig. 5.43. Let the height of conductors be h metres above the earth and charge of $+q$ coulombs per metre length and $-q$ coulombs per metre length on the conductors A and B respectively.

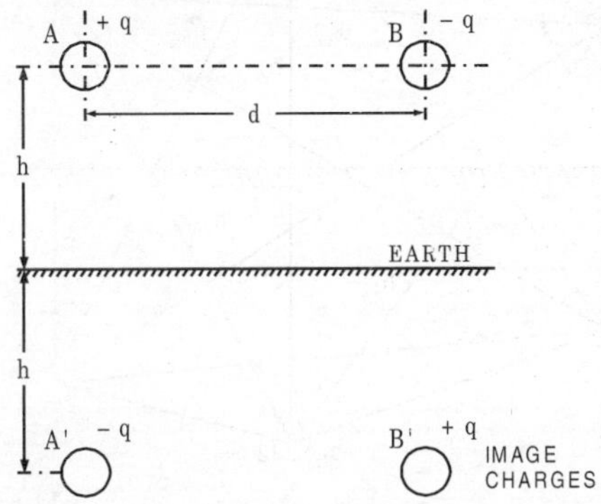

Fig. 5.43. ***Single-Phase Transmission Lines With Images***

The equation for the voltage drop V_{AB} as determined by two charged conductors A and B, and their images A′ and B′ may be written as follows :

$$V_{AB} = \frac{1}{2\pi\varepsilon_0}\left[q_A \log_e \frac{d}{r} + q_B \log_e \frac{r}{d} + q'_A \log_e \frac{\sqrt{4h^2 + d^2}}{2h} + q'_B \log_e \frac{2h}{\sqrt{4h^2 + d^2}}\right] \quad ...(5.61)$$

Substituting $q_A = q'_B = +q$ and $q'_A = q_B = -q$, we have

$$V_{AB} = \frac{q}{\pi\varepsilon_0} \log_e \frac{2hd}{r\sqrt{(4h^2 + d^2)}}$$

Capacitance between conductors A and B

$$C_{AB} = \frac{q}{V_{AB}} = \frac{\pi\varepsilon_0}{\log_e \frac{d}{r\sqrt{1 + \frac{d^2}{4h^2}}}} \quad ...(5.62)$$

The above expression for capacitance reveals that the presence of earth modifies the radius of conductor r to

$r\sqrt{\left(1 + \frac{d^2}{4h^2}\right)}$. The effect of earth on the capacitance of the system is to increase it. However, normally the distance of separation between the conductors is much smaller than the height of the conductor from the ground, therefore,

$r\sqrt{1 + \frac{d^2}{4h^2}} \simeq r$ and for all practical purposes the effect of earth on line capacitance can be neglected.

5.15.2. Effect of Earth on the Capacitance of a Three-Phase Line

Figure 5.44 shows the conductors of a 3-phase line along with image conductors. The line is assumed to be transposed and in the first part of the transposition cycle conductor A is in position 1, B is in position 2 and C is in position 3. Let the charges on line conductors be q_A, q_B and q_C while those on image conductors be $-q_A$, $-q_B$ and $-q_C$ respectively, as shown in the figure.

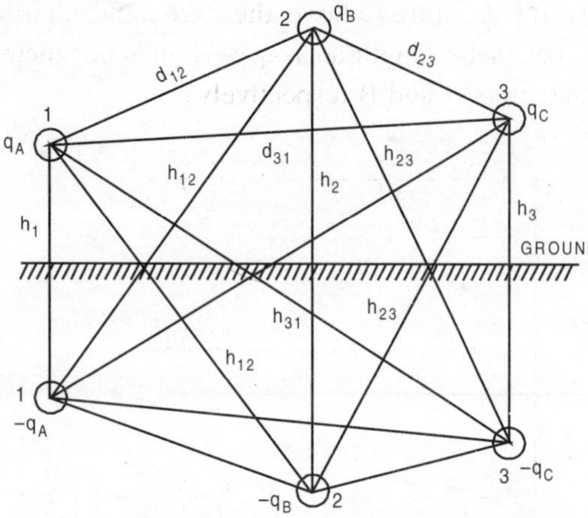

Fig. 5.44. ***Conductors of a 3-Phase Line With Image Charges***

The equation for the three sections of transposition cycle can be written for the voltage drop V_{AB} as determined by three charged conductors and their images. With conductor in position 1, B in position 2 and C in position 3, we have

$$V_{AB} = \frac{1}{2\pi\varepsilon_0}\left[q_A\left(\log_e\frac{d_{12}}{r} - \log_e\frac{h_{12}}{h_1}\right) + q_B\left(\log_e\frac{r}{d_{12}} - \log_e\frac{h_2}{h_{12}}\right)\right.$$

$$\left. + q_C\left(\log_e\frac{d_{23}}{d_{31}} - \log\frac{h_{23}}{h_{31}}\right)\right] \quad ...(5.63)$$

Similarly, equations for V_{AB} can be written for the second and third sections of the transposition cycle. If the fairly accurate assumption of constant charge per unit length of the conductor throughout the transmission cycle is made, the average value of three sections of the cycle is given by

$$V_{AB} = \frac{1}{2\pi\varepsilon_0}\left[q_A\left(\log_e\frac{D_{eq}}{r}\right) - \log_e\left(\frac{h_{12}h_{23}h_{31}}{h_1h_2h_3}\right)^{1/3}\right.$$

$$\left. + q_B\left(\log_e\frac{r}{D_{eq}} - \log_e\left(\frac{h_1h_2h_3}{h_{12}h_{23}h_{31}}\right)^{1/3}\right)\right] \quad ...(5.64)$$

where $D_{eq} = \sqrt[3]{d_{12}d_{23}d_{31}}$

The equation for the average value of the voltage V_{AC} can be determined in the same way. Proceeding on the lines of Sec. 5.12 and using $V_{AB} + V_{AC} = 3V_{AN}$ and $q_A + q_B + q_C = 0$, we ultimately have the following expression for the capacitance to neutral

$$C_N = \frac{2\pi\varepsilon_0}{\log_e\frac{D_{eq}}{r} - \log_e\sqrt[3]{\frac{h_{12}h_{23}h_{31}}{h_1h_2h_3}}} \text{ F/m to neutral} \quad ...(5.65)$$

A comparison of above equation with Eq. (5.54) reveals that the presence of earth increases the line capacitance by a small amount. However, the effect is very-very small because the heights of conductors are large as compared to the distance between them.

Example 5.29. Determine taking the effect of ground into consideration the capacitance of a single phase 50 Hz overhead line 50 km long consisting of two parallel wires each 0.5 cm in diameter and 1.5 m apart. The height of conductors above the ground is 7 m. *[Pb. Technical Univ. Power System-I (T&D), May 2010]*

Solution : Radius of each conductor, $r = \frac{0.5}{2}$ cm = 0.25 cm

Spacing between conductors, d = 1.5 m = 150 cm

Height of conductors above earth, h = 7 m = 700 cm

Capacitance per metre of line,

$$C = \frac{\pi\varepsilon_0}{\log_e\frac{d}{r\sqrt{1+\frac{d^2}{4h^2}}}} = \frac{\pi\times 8.854\times 10^{-12}}{\log_e\frac{150}{0.25\sqrt{1+\left(\frac{150}{1,400}\right)^2}}}$$

$$= 4.5 \times 10^{-12} \text{ F/m} \quad \textbf{Ans.}$$

Capacitance of 50 km long,

$$C = 4.5\times 10^{-12} \times 50 \times 1,000 \text{ F} = 0.225\ \mu\text{F} \quad \textbf{Ans.}$$

Example 5.30. Determine the capacitance of the arrangement shown in the figure when (*i*) the effect of earth is neglected and (*ii*) the effect of earth is considered. The height of the conductors from ground is 10 m and radius of each of the conductor is 2 cm. *[U.P. Technical Univ. Elements of Power System 2008-09]*

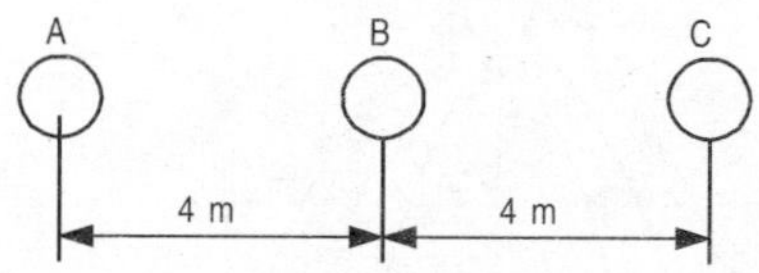

Fig. 5.45

Solution : Conductor radius, r = 2 cm = 0.02 m

Spacings d_1 = AB = 4 m;

d_2 = BC = 4 m; d_3 = CA = 8 m

Capacitance per phase per metre length (neglecting earth effect)

$$C_N = \frac{2\pi\varepsilon_0}{\log_e\frac{\sqrt[3]{d_1d_2d_3}}{r}} = \frac{2\pi\times 8.854\times 10^{-12}}{\log_e\frac{\sqrt[3]{4\times 4\times 8}}{0.02}} = 10.06\times 10^{-12} \text{ F}$$

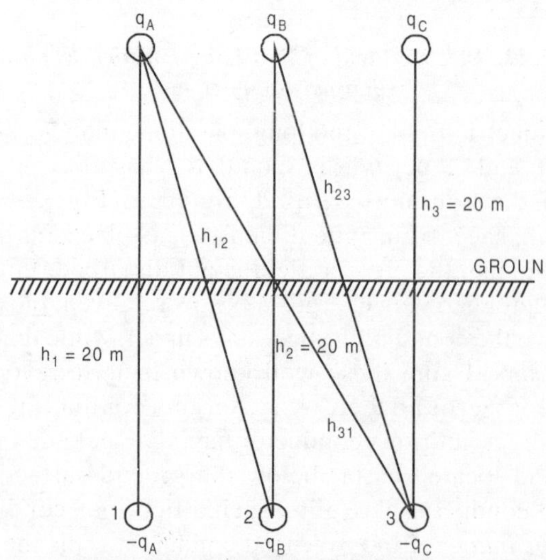

Fig. 5.46

From Fig. 5.46, we have

$h_1 = h_2 = h_3$ = 20 m

$$h_{12} = h_{23} = \sqrt{20^2 + 4^2} = 20.396 \text{ m}$$

$$h_{31} = \sqrt{20^2 + 8^2} = 21.54 \text{ m}$$

$$r = 0.02 \text{ m}$$

$$D_{eq} = \sqrt[3]{d_1 d_2 d_3} = \sqrt[3]{4 \times 4 \times 8} = 5.0397 \text{ m}$$

Capacitance per phase per metre length (considering effect of ground)

$$C_N = \frac{2\pi\varepsilon_0}{\log_e \frac{D_{eq}}{r} - \log_e \sqrt[3]{\frac{h_{12}h_{23}h_{31}}{h_1 h_2 h_3}}} \text{ F}$$

$$= \frac{2\pi \times 8.854 \times 10^{-12}}{\log_e \frac{5.0397}{0.02} - \log_e \sqrt[3]{\frac{20.396 \times 20.396 \times 21.54}{20 \times 20 \times 20}}}$$

$$= \frac{55.63 \times 10^{-12}}{5.529 - 0.0378} \text{ F} = 10.13 \times 10^{-12} \text{ F} \textbf{ Ans.}$$

5.16. TRANSPOSITION OF 3-PHASE LINES

As already mentioned, the inductance and capacitance of each phase will be different in case of conductors of three-phase line being spaced irregularly. The apparent resistance of the conductors is also affected on account of transfer of power between the phases, which occur due to mutual inductance. Thus all the three line constants are affected by irregular spacing of the conductors in a 3-phase line.

Also, due to unsymmetrical spacing, the magnetic field external to the conductors is not zero, thereby causing induced voltages in adjacent electrical circuits, particularly the telephone lines, that may cause disturbances in the telephone lines.

The unbalancing effect on account of irregular spacing of line conductors can be avoided by transposition of line conductors. Transposition of line conductors means changing the positions of the three phases on the line supports twice over the total length of the line. In practice the line conductors should be so transposed that each of the three possible arrangements of conductors exist for one-third of the total length of line. This is illustrated in Figs. 5.12 and 5.47.

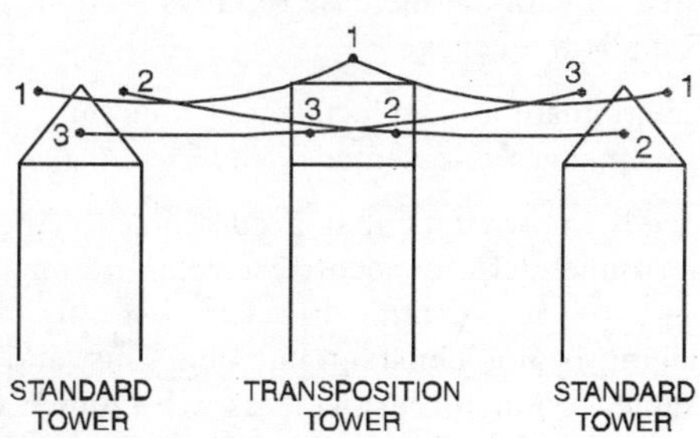

Fig. 5.47. ***Transposition of a 3-Phase Line***

An unbalanced system, may be of conductors symmetrically placed, cannot be represented by three simple inductances, without making transposition. The effect of unbalanced currents is neutralised in case of conductors being transposed regularly at intervals.

The transposition of conductors also reduces the disturbances to the nearby communication circuits. In composite line, the line carrying telephone line conductors below the power line conductors, it is also necessary to transpose the telephone line conductors in order to keep down the disturbances.

Exercises

1. Explain with justification whether the following statement is TRUE or FALSE.

   "Skin effect increases effective resistance and reactance of conductor".

   [U.P. Technical Univ. Elements of Power System 2008-09]
2. Explain difference between skin effect and proximity effect.

   [U.P. Technical Univ. Elements of Power System 2009-10]
3. Compare skin effect with proximity effect.

   [U.P. Technical Univ. Elements of Power System 2004-05]
4. Distinguish between ac and dc resistances of a conductor. Why the two differ ? Derive an expression for internal flux leakage due to a single current carrying conductor.

   [G.B. Technical Univ. Elements of Power System 2011-12]
5. Find out the flux linkage of single phase two-wire line and derive an expression for inductance per unit length.

   [Pb. Technical Univ. Power System-I (T&D), Dec. 2007]
6. Show that the inductance per unit length of an overhead line due to internal flux leakage is constant and is independent of size of conductor.

   [Pb. Technical Univ. Power System-I (T&D), May 2008]
7. Show that the inductance per loop metre of a 2-wire transmission line using solid round conductor is given by

   $$L = 4 \times 10^{-7} \log_e \left(\frac{D}{r'}\right) \text{ heneries}$$

   [Rajasthan Technical Univ. Transmission & Distribution of Electrical Power 2008]
8. Derive an expression for inductance of a single-phase transmission line.

   [U.P. Technical Univ. Elements of Power System 2005-06]
9. Define GMD and GMR. Explain the role of GMD and GMR in line inductance calculation.

   [U.P. Technical Univ. Elements of Power System 2007-08]
10. Give the concept of 'self GMD'. Derive an expression for inductance of an unsymmetrical, transposed line using solid round conductors.

    [G.B. Technical Univ. Elements of Power System, 2011-12]
11. Discuss the concept of geometric mean distance. How is this concept used to find the inductance of composite conductor line ?

    [Transmission & Distribution of Electrical Power, Rajasthan Univ. 1997, 2005, 2007; Rajasthan Technical Univ. 2008]
12. Explain the terms geometrical mean distance (GMD) and 'self GMD' in the inductance calculation of single phase transmission lines with composite conductors. Find the self GMD of a conductor consisting of seven identical strands each having a radius r.
13. Starting from first principles, derive the expression for inductance of a 3-phase unsymmetrically spaced transposed transmission line.

    [U.P. Technical Univ. Elements of Power System 2009-10]

14. Derive the expression of inductance for a 3-phase overhead line with symmetrical spacing.

[M.D. Univ. Power System-I, Dec. 2008, Dec. 2010]

15. Derive the expression for the inductance of a 3-phase line which is completely transposed.

[Pb. Technical Univ. Power System-I (T&D), May 2010]

16. Derive the expression for inductance of a 3ϕ overhead line with unsymmetrical spacing.

[M.D. Univ. Power System, Dec. 2009]

17. Derive the expression of inductance of double circuit 3-phase line with conductors completely transposed.

[Rajasthan Technical Univ. Transmission & Distribution of Electrical Power 2009]

18. Starting with flux linkage concept, find the inductance of 3-phase double circuit line with unsymmetrical spacing.

[U.P. Technical Univ. Elements of Power System 2007-08]

19. Explain the advantages of using bundled conductors in extra high voltage transmission lines and the different configurations.

[U.P. Technical Univ. Elements of Power System 2003-04]

20. Derive expressions for line-to-line capacitance and line to neutral capacitance of a single-phase line.

[U.P. Technical Univ. Elements of Power System 2003-04]

21. Derive for the capacitance of 3-phase transmission line for symmetrical and unsymmetrical spacing.

[Pb. Technical Univ. Power System-I (T&D) Dec. 2006]

22. Derive the expression for capacitance of a 3-phase overhead line with unsymmetrical spacing.

[M.D. Univ. Power System-I, May 2009]

23. Derive formula to calculate the capacitance of a double circuit line. All the conductors are of equal diameter and are spaced hexagonally.

[U.P. Technical Univ. Elements of Power System 2002-03]

24. Explain with reasons :

(*i*) In double circuit lines the conductors constituting the same phase should be spaced as far away as possible and conductors of different phases should be spaced as close as possible.

(*ii*) There is an increase in capacitance due to bundling of conductors.

[Rajasthan Univ. Transmission & Distribution of Electrical Power 2003]

25. Derive an expression for the capacitance-per unit length of a 3-phase line completely transposed. What is the effect of earth on the capacitance of the line ?

[Pb. Technical Univ. Power System-I (T&D), May 2007]

26. Explain the method of images for the calculation of capacitance of a three phase transmission line.

[U.P. Technical Univ. Elements of Power System 2002-03]

27. Show that presence of earth increases capacitance of transmission lines.

[U.P. Technical Univ. Elements of Power System 2004-05]

28. What are the advantages of transposition of conductors ?

[U.P. Technical Univ. Elements of Power System 2004-05]

29. Discuss the transposition of conductors of a three-phase transmission line.

[U.P. Technical Univ. Elements of Power System 2005-06]

30. Explain briefly about the following :

(*i*) GMD.

(*ii*) Bundled conductors.

(*iii*) Double circuit lines.

(*iv*) Transposition.

[U.P. Technical Univ. Elements of Power System 2009-10]

31. What is meant by transposition in an overhead line ? How and why it is carried out ?

[Rajasthan Univ. Transmission & Distribution of Electrical Power 2004]

32. Write short notes on the following :

(*i*) Skin effect. **[M.D. Univ. Dec. 2007, May 2008, May 2009]**

(*ii*) Proximity effect.

[M.D. Univ. May 2009, G.B. Technical Univ. 2011-12]

(*iii*) Skin, proximity effects.

[Rajasthan Univ. 1999, 2000, 2002, 2005]

(*iv*) GMR. **[Rajasthan Univ. 2007]**

(*v*) Bundle conductors. **[M.D. Univ. May 2008]**

(*vi*) Line transposition. **[Rajasthan Univ. 2005]**

(*vii*) Transposition of conductors. **[M.D. Univ. May 2008]**

Short Answer Type Questions With Answers

Q. 1. Distinguish between ac and dc resistances of a conductor. Why the two differ ? **[G.B. Technical Univ. 2011-12]**

Ans. The resistance offered to dc, by a conductor is known as dc resistance while the resistance offered to ac is known as ac resistance or effective resistance. For a given conductor ac resistance is more than its dc resistance. This is because an alternating current flowing through a conductor does not distribute uniformly but tends to concentrate near the surface of the conductor. The result is that the effective area of the conductor is reduced causing an increase in ac resistance.

Q. 2. What is skin effect ? **[Pb. Technical Univ. Dec. 2008]**

Ans. The phenomenon of concentration of an electric current near the surface of the conductor is known as skin effect.

Q. 3. On what factors does the skin effect depend ?

Ans. The skin effect depends upon the type of material, frequency of current, diameter of conductor and shape of conductor. It increases with the increase of cross section, permeability and supply frequency.

Q. 4. For determination of inductance of a circuit, determination of flux linkages is essential. Why ?

Ans. It is well known fact that a current carrying conductor is surrounded by concentric circles of magnetic lines. In case of ac system this field set up around the conductor is not constant but changing and links with the same conductor as well as with other conductors. Due to these flux linkages the line possesses inductance, defined as the flux linkages per unit current. Thus for determination of inductance of a circuit, determination of flux linkages is essential.

Q. 5. What is meant by GMR of a conductor ?

Ans. The product $r\,e^{-1/4}$ is known as GMR of a conductor and is equal to 0.7788 times the conductor radius.

Q. 6. What is self GMD of a conductor consisting of seven identical strands, each having a radius r ?

Ans. $2.176r$

Q. 7. What is a bundle conductor and why is it used ?

[U.P. Technical Univ. 2006-07]

Ans. A bundled conductor is a conductor made up of two or more conductors, called the sub-conductors, per phase in close proximity compared with the spacing of phases (Fig. 5.26).

The use of bundled conductors per phase reduces the voltage gradient in the vicinity of the line and thus, reduces the possibilities of the corona discharge.

Although the bundled conductors are used on EHV transmission lines primarily to reduce corona loss and radio interference, but they have several other advantages over single conductors such as given below :

1. The bundle conductor lines transmit bulk power with reduced losses, thereby giving increased transmission efficiency.
2. Since the bundle conductor lines have a higher capacitance to neutral in comparison with single conductor lines, therefore, they have higher charging current, which helps in improving the power factor.
3. Since by bundling, the self GMD or GMR is increased, the inductance per phase, in comparison with single conductor lines, is reduced. As a result reactance per phase is reduced.
4. Since surge impedance of a line is given by $Z_0 = \sqrt{\frac{L}{C}}$

   and the bundle conductor lines have higher capacitance and lower inductance in comparison with single conductor lines, therefore, bundle conductor lines have comparatively lower surge impedance with a corresponding increase in the maximum power transfer capability. The following table represents relative power transfer with number of sub-conductors forming a bundle conductor.

Number of sub-conductors	1	2	4	8
Relative power transfer	1.0	1.3	1.6	1.7

Q. 8. What is bundling of conductors ?

Ans. The use of more than one conductor per phase is called the bundling of conductors.

Q. 9. Differentiate between bundled and composite conductors.

[Pb. Technical Univ. Dec. 2007]

Ans. The basic difference between a composite conductor and a bundled conductor is that the sub-conductors of a bundled conductor are separated from each other by a constant distance varying from 0.2 m to 0.6 m depending upon designed voltage and surrounding conditions throughout the length of the line with the help of spacers whereas the wires of a composite conductor touch each other. The bundled conductors have filter material or air space inside so that the overall diameter is increased.

Q. 10. For what purposes bundled conductors are primarily used in transmission lines ?

Ans. The bundled conductors are used in EHV transmission lines primarily to reduce corona loss and radio interference.

Q. 11. What are the advantages of bundled conductors.

[U.P.T.U. 2007-08]

Ans. The advantages of using bundled conductors in EHT transmission lines are : Reduced possibilities of the corona discharge; reduced corona loss; reduced radio interference; reduced losses resulting in increased transmission efficiency; higher charging current resulting in improved pf; reduced inductive reactance per phase; lower surge impedance resulting in increased power transfer capability.

Q. 12. What is the effect of ground on the capacitance of a transmission line ?

Ans. The presence of ground increases the capacitance of a transmission line by a small amount.

Q. 13. Why the effect of ground on the line capacitance can be neglected ?

Ans. The effect of ground on the line capacitance is very-very small because the height of conductors is large as compared to the distance between them.

Q. 14. What is meant by transposition of line conductors ?

Ans. Transposition of line conductors means changing the positions of the three phases on the line supports twice over the total length of the line. The line conductors in practice, are so transposed that each of the three possible arrangements of conductors exist for one-third of the total length of the line.

Q. 15. Why transposition of conductors in a three-phase transmission lines is essential ?

Ans. The inductance and capacitance of each phase will be different in case of conductors of three-phase line being spaced irregularly. The apparent resistance of the conductors is also affected on account of transfer of power between the phases, which occur due to mutual inductance. Thus all the three line constants are affected by irregular spacing of the conductors in a 3-phase line.

Also, due to unsymmetrical spacing, the magnetic field external to the conductors is not zero, thereby causing induced voltages in adjacent electrical circuits, particularly the telephone lines, that may cause disturbances in the telephone lines.

Transposition of line conductors in transmission lines is essential to counter the unbalancing effect, mentioned above, of irregular spacing of line conductors.

Problems

1. A 16 km long transmission line consists of a pair of conductors 1.5 cm in diameter spaced 2 m apart. Find the loop inductance of the line.

   [Ans. 0.03735 H]
2. A single-phase line of 230 V has conductor spacing of 1.5 m. The diameter of each conductor is 1.5 cm. Calculate the loop inductance in mH of the line per km.

   [Ans. 2.22 mH]
3. A conductor is composed of 7 identical copper strands, each having a radius r. Find the self GMD of the conductor.

   [A.M.I.E. Sec B. Power Systems Summer 1997]

   [Ans. $2.176r$]

4. A 3-phase transmission line 100 km long has its conductors of 0.5 cm diameter spaced at the corners of the equilateral triangle of 120 cm side. Find the inductance per phase of the system. **[Pb Univ. April 1980]** [**Ans.** 0.129 H]

5. A single circuit 3-phase line operating at 50 Hz has a conductor diameter of 5.18 mm. The spacing between the line conductors is given in metres. D_{ry} = 3.2; D_{yb} = 4.0; D_{br} = 5.0. Calculate the inductance and inductive reactance per phase per km. **[U.P.S.C. I.E.S. 1972]**

[**Ans.** 1.52 mH ; 0.477 Ω]

6. The three conductors of a 3-phase transmission line are arranged in a horizontal plane and are 4 metres apart. The diameter of each conductor is 25 mm. Determine the inductance per km of each conductor (line-to-neutral). Assume balanced load and R, Y, B phase sequence.

Determine also the average inductance per phase for regularly transposed line.

[**Ans.** (1.273 – j 0.12) mH; 1.2 mH; (1.273 + j 0.12) mH; 1.249 mH]

7. A 3-phase transmission line has conductor diameter of 2.0 cm each and they are spaced as in the following figure. Find the inductance of the line per km per phase if the transmission line is transposed.

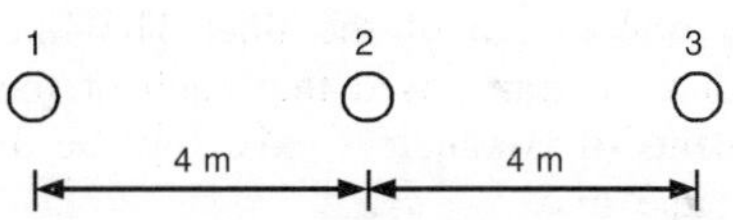

Fig. 1

[Rajasthan Univ. Transmission & Distribution of Electrical Power 2004]

[**Ans.** 11.55 × 10^{-7} H/m]

8. A 3-phase double circuit line is arranged as shown below (Fig. 2). The conductors are transposed. The radius of each conductor is 0.75 cm. Phase sequence is ABC. Find the inductance per phase per km.

[U.P.S.C. I.E.S. Electrical Engineering-II, 2011]

[**Ans.** 0.7 mH/km/phase]

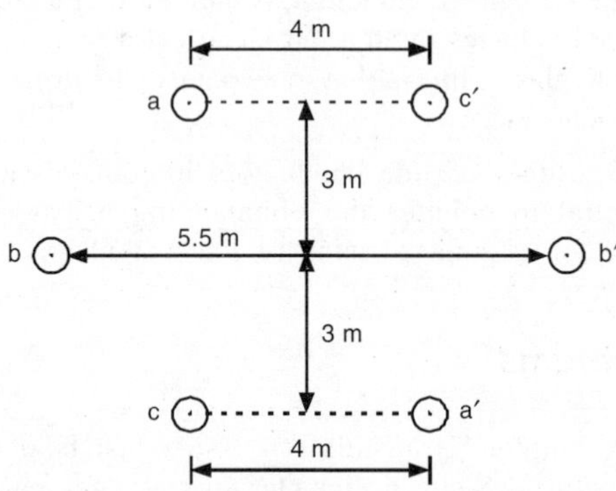

Fig. 2

9. A part of the transposition cycle of a 3-phase double circuit line is shown in Fig. 3. Radius of each conductor is 12.5 mm. The conductors are solid copper. Determine GMD and GMR and calculate the inductance per km per phase of the line. [**Ans.** GMD = 6.61 m ; GMR = 0.318 m and H = 0.606 mH/km/phase]

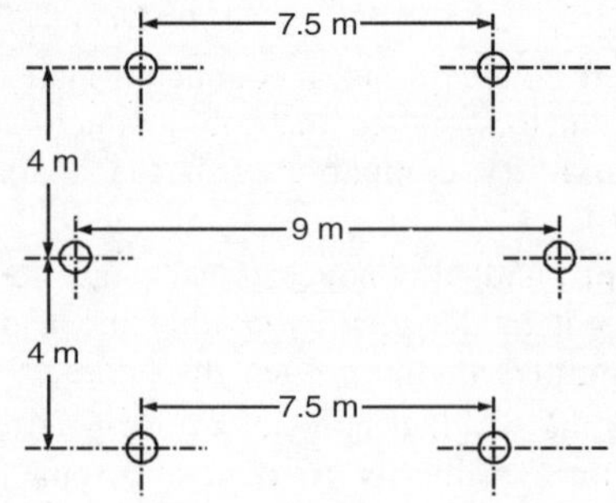

Fig. 3

10. A symmetrical double circuit, 3-phase line is to have its 6 conductors arranged, as shown in Fig. 4. Calculate the inductive reactance per phase in ohms/km. The line is completely transposed and the radius of each conductor is 1.25 cm. **[Agra Univ. 1976]** [**Ans.** 0.16 W/km]

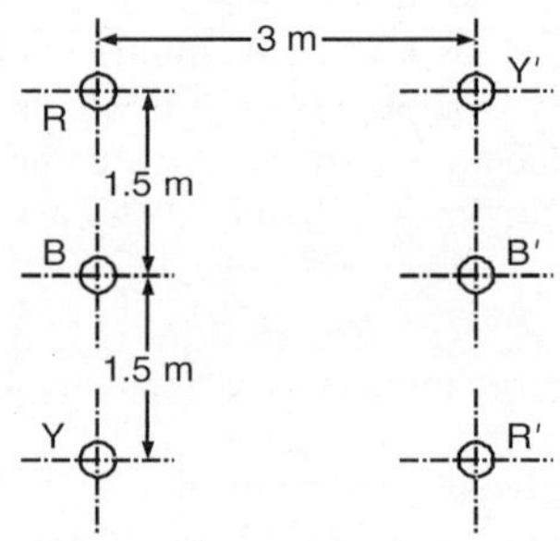

Fig. 4

11. A 460 kV bundled conductor transmission line has the conductor spacing as shown below. ACSR conductors are used. Determine the inductance per phase per km if the self GMD of each conductor is 0.9 cm. Assume line regularly transposed. (Fig. 5) [**Ans.** 1.023 mH]

12. Calculate the capacitance of a single-phase overhead line 40 km long consisting of two parallel conductors 1.5 m apart. The conductors are 1.5 cm in diameter.

[**Ans.** 0.21 μF]

13. Find out the capacitance of a single-phase line 30 km long consisting of two parallel wires each 15 mm diameter and 1.5 m apart.

[Pb. Technical Univ. Power System-I (T&D) May 2009]

[**Ans.** 0.1575 μF]

14. A single-phase, 30 km long overhead transmission line consists of two parallel long conductors each 5 mm in diameter and spaced 1.5 m apart. If the line voltage is 50 kV, 50 Hz, determine the charging current of the open-circuited line. [**Ans.** 2.04 A]

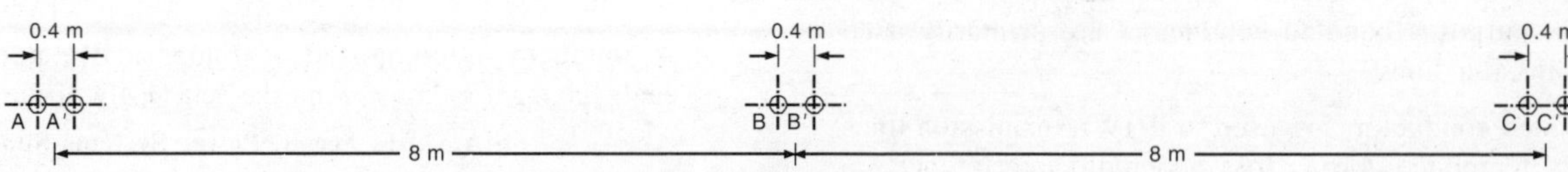

Fig. 5

15. Calculate the capacitance of a single-phase 50 Hz overhead line 40 km long consisting of two parallel wires each 5 mm in diameter, 1.5 metres apart, taking the effect of ground into consideration. The height of the conductors above ground is 7 metres. Derive any formula used.

[A.P.S. Univ. Electric Power Generation, Transmission and Distribution 1978]

[**Ans.** 0.174 μF]

16. A three-phase 50 Hz fully transposed line has conductors equilaterally spaced at 3 metres. The conductor diameter is 3 cm. Determine the capacitance to neutral per km per phase. **[Gorakhpur Univ. Transmission of Electric Power 1975]**

[**Ans.** 0.0105 mF]

17. The three conductors of a 3-phase, 3-wire line are arranged at the corners of a triangle, the sides of which are 1 m, 1.3 m and 2 m. Calculate the inductance and capacitance per km of the line, when the conductors are transposed. The conductors are 1.5 cm diameter.

[**Ans.** 1.0923 mH ; 0.0107 μF]

18. A 220 kV, 50 Hz, 200 km line has its conductors on the corners of a triangle with sides 6 m, 6 m and 12 m. The conductor radius is 1.81 cm. Find the capacitance per phase per km. [**Ans.** 1.844 μF]

19. A 3-phase, 132 kV, 50 Hz overhead transmission line has steel cored aluminium conductors of equivalent copper area of 1.5 cm^2 and effective diameter of 39.8 mm spaced equilaterally 8 m apart. Calculate the line constants. The resistivity of copper is 1.73 mΩ-cm

[A.M.I.E. Sec. B. Power Systems Summer 1994]

[**Ans.** R = 0.11533 Ω/km/conductor; L = 1.25 mH/km/phase; C = 0.00928 μF/km/phase]

20. A 220 kV, 50 Hz, 200 km long 3-phase line has its conductors on the corners of a triangle with sides 6 m, 6 m and 12 m. The conductor radius is 1.81 cm. Find the capacitance per phase per km, capacitive reactance per phase, charging current and total charging MVA.

[Rajasthan Univ. Transmission and Distribution of Electrical Power 2005]

[**Ans.** 1.84 μF/phase, 1,730 Ω/phase, 73.42 A/phase; 27.97 MVAR]

21. A 3 phase, 50 Hz, 132 kV overhead line has conductors placed in a horizontal plane 4.56 m apart. Conductor diameter is 22.4 mm. Line length is 100 km. Calculate the charging current per phase assuming complete transposition.

[M.D. Univ. Power System-I; Dec. 2008]

[**Ans.** 0.329 A/km]

22. The conductors in a single-phase line are 6 m above the ground. Each conductor is of 1.5 cm diameter and they are spaced 3 m apart. Taking the effect of earth into account, calculate the capacitance per km.

[U.P. Technical Univ. Elements of Power System 2002-03]

[**Ans.** 4.666×10^{-9} F]

❑❑❑

CHAPTER 6

Performance of Short and Medium Transmission Lines

INSIDE THIS CHAPTER

6.1. INTRODUCTION

The performance of a power system under normal balanced steady-state conditions is of primary importance in power system engineering. The transmission line is the main energy corridor in a power system. The performance of a power system depends mainly on the performance of transmission lines in the system. The important considerations in the operation of the transmission lines are voltage drop and power losses occurring in the line and efficiency of transmission. The performance of a transmission line is governed by its four parameters—series resistance R and inductance L, shunt capacitance C and conductance G. The resistance R is due to the fact that every conductor offers opposition to the flow of current. The inductance L is due to the fact that the current carrying conductor is surrounded by the magnetic lines of force. The capacitance of the line is due to the fact that the conductor carrying current forms a capacitor with the earth which is always at lower potential than the conductor and the air between them forms a dielectric medium. The shunt conductance is mainly due to flow of leakage currents over the surface of the insulators especially during bad weather. The effect of line resistance is to cause voltage drop (IR volts) and power loss (I^2R watts) in the line. The effect of line inductance is to cause voltage drop in the line in quadrature with the current flowing in the conductor, which is equal to $2\pi fLI$ volts where f is supply frequency in Hz, L is the inductance per conductor in Henries and I is the current flowing through the conductor in amperes. The effect of line capacitance is to produce a current; called the *charging current,* which is in quadrature with the voltage. This component of current in the conductor has its maximum value at the sending end of the line and decreases continuously as the receiving end of the line is approached at which its value is zero. The magnitude of charging current I_C at the sending end is given by $2\pi fCV_S$ where f is the supply frequency in Hz, C is the capacitor per conductor in farads and V_S is the voltage at the sending end to neutral in volts. The shunt conductance is in parallel with the system. Generally the leakage currents are small and the shunt conductance G is ignored in calculations.

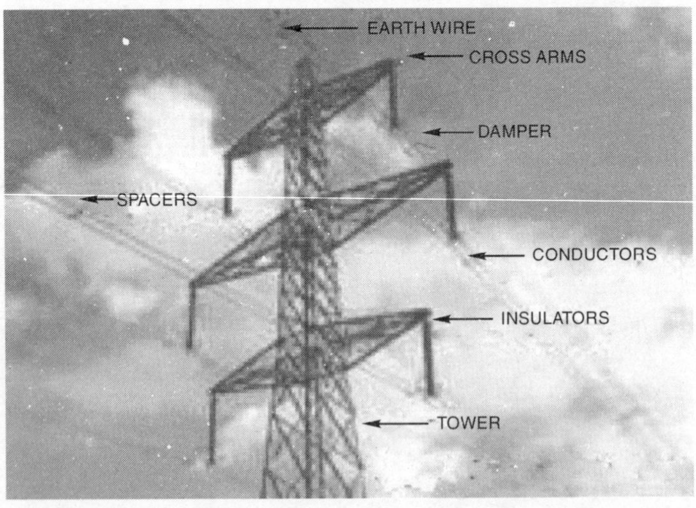

Fig. 6.1. *Typical Transmission Line*

It is noteworthy that both the reactive drop ($2\pi fLI$) and the charging current ($2\pi fCV_S$) of the line are proportional to the supply frequency and thus have a far greater influence on the performance of a 50 Hz line than on a 25 Hz line. Also that with overhead lines reactance effects are relatively more important owing to wide spacing of conductors which must be employed, whilst in the case of transmission by underground cables reactance effects are small and the capacitance effect predominates.

We are aware that in 3-phase circuit problems it is sufficient to compute results for one phase and subsequently predict results for the remaining two phases by exploiting the 3-phase symmetry. Although the transmission lines are not spaced equilaterally and not transposed, the resulting asymmetry is slight and the phases are considered to be balanced. A 3-phase transmission line carrying an equal load on each phase can, therefore be represented by its single phase equivalent in which inductance is computed for one phase of a balanced 3-phase line and capacitance is computed from line to neutral. While working with a 3-phase transmission line carrying an equal load on each phase, it is usual to assume that all the given voltages are line-to-line values, that all the currents are line currents, that all volt-amperes, power and reactive volt-amperes are 3-phase and that all the power factors are phase power factors (cosine of the angle between the phase voltage and the corresponding phase current). No doubt, all calculations are done in per phase values.

6.2. CLASSIFICATION OF TRANSMISSION LINES

For discussing the performance of the overhead transmission lines they are classified as short, medium and long transmission lines. The overhead transmission lines are classified depending upon the manner in which capacitance is taken into account.

Transmission lines having length lesser than 80 km and operating voltage lower than 20 kV fall in the first category *i.e., short transmission lines*. Due to smaller distance and lower line voltage, the capacitance effects are extremely small and, therefore, can be neglected. Hence the performance of short transmission lines depend upon the resistance and inductance of the line. Though in an actual line, the resistance and inductance are distributed over the whole length, but in case of short lines, the total resistance and inductance are assumed to be lumped at one place. In case of single phase circuits, the total loop resistance and inductance is to be taken into account, whereas, in case of 3-phase circuits only resistance and inductance to neutral *i.e.,* per phase is required to be taken into account. The effect of generators and transformers can be taken into account by adding their impedances to the impedance of the line.

Transmission lines having length between 80 km and 200 km and line voltage between 20 kV and 100 kV fall into second category *i.e., medium transmission lines*. Owing to appreciable length and voltage of the line the charging current is appreciable and, therefore, capacitance effect cannot be ignored. Though the capacitance is uniformly distributed over the entire length of the line, yet the capacitance may be assumed to be concentrated at one or more points.

The transmission lines having length above 200 km and line voltage above 100 kV fall into third class *i.e., long transmission lines*. In these lines impedance and admittance are to be considered uniformly distributed (not lumped) and, therefore, for their solution rigorous methods are employed.

It is to be emphasised here that exact solution of any transmission line must be based on the fact that the parameters or constants of any line are not lumped but are distributed uniformly throughout the length of the line. However, results obtained by assuming the constants as lumped for short and medium transmission lines are reasonably accurate.

6.3. REGULATION AND EFFICIENCY OF A TRANSMISSION LINE

By performance of a transmission line is meant the determination of regulation and efficiency of the line.

1. *Regulation.* When the load is supplied, there is a voltage drop in the line due to resistance and inductance of the line and, therefore, receiving-end voltage V_R is usually less than sending-end voltage V_S. The voltage drop *i.e.,* difference of sending-end voltage and receiving-end voltage expressed as a percentage of receiving-end voltage is called the *regulation*.

When the load is thrown off *i.e.,* when the line is supplying no load, the receiving-end voltage becomes equal to sending-end voltage and therefore, regulation can be defined as below :

Regulation is defined as the change in voltage at the receiving (or load) end when the full load is thrown off, the sending-end (or supply) voltage and supply frequency remaining unchanged. It is usually expressed as a percentage of receiving-end voltage.

Mathematically percentage voltage regulation of a transmission line is given by the expression

$$\% \text{ voltage regulation} = \frac{V_S - V_R}{V_R} \times 100 \qquad ...(6.1)$$

where V_S is the voltage at sending or supply end and V_R is the voltage at the receiving or load end. *It is to be noted here that V_S and V_R are the magnitudes of voltages.*

The *lower the voltage regulation, better it is* because low voltage regulation means little variation in receiving-end voltage due to variation in load current.

Knowledge of voltage regulation helps in maintaining the voltage at the load terminals within prescribed limits* by employing suitable voltage control equipment.

2. *Efficiency.* When the load is supplied there are line losses due to ohmic resistance of line conductors and power delivered at the load end of a transmission line is less than the power supplied at the sending end. Efficiency of a transmission line is defined as below :

Efficiency of a transmission line is defined as the ratio of power delivered at the receiving end to the power sent from the sending end.

---

* According to IE Rule 54, the voltage at the point of commencement of supply shall not vary by more than

1. ± 6% in the case of low voltage (*i.e.,* up to 250 V) and medium voltage (*i.e.,* up to 650 V)
2. + 6% to – 9% in the case of high voltage (*i.e.,* up to 33 kV)
3. $\pm 12\frac{1}{2}$ % in the case of extra high voltage (*i.e.,* exceeding 33 kV)

Mathematically transmission efficiency is given by the expression

$$\eta_T = \frac{V_R I_R \cos\phi_R}{V_S I_S \cos\phi_S} \times 100 \quad ...(6.2)$$

$$\text{or } \eta_T = \frac{V_R I_R \cos\phi_R}{V_R I_R \cos\phi_R + \text{line losses}} \times 100 \quad ...(6.3)$$

where V_R, I_R and $\cos\phi_R$ are the receiving-end voltage, current and power factor (all phase values) while V_S, I_S and $\cos\phi_S$ are the sending-end voltage, current and power factor (all phase values).

6.4. SHORT TRANSMISSION LINES

In a short transmission line the shunt conductance and shunt capacitance are neglected and so only the series resistance and inductive reactance are to be considered. The equivalent circuit of a short transmission line is shown in Fig. 6.2 (*a*) where R and X represent the total resistance and inductive reactance of both the conductors (*i.e.,* go and return) of a single phase transmission line. Since the effect of shunt capacitance is neglected, the series resistance and reactance are taken as lumped. Thus the equivalent circuit of a short transmission line is a simple ac series circuit.

The equivalent circuit shown in Fig. 6.2 (*a*), will represent a 3-phase short transmission line if R and X represent the resistance and inductive reactance to neutral, I is the current in one conductor, V_R is the receiving-end voltage to neutral, V_S is the sending-end voltage to neutral, and the load is that in one phase; the voltage drop found in this way is the drop between a phase and neutral.

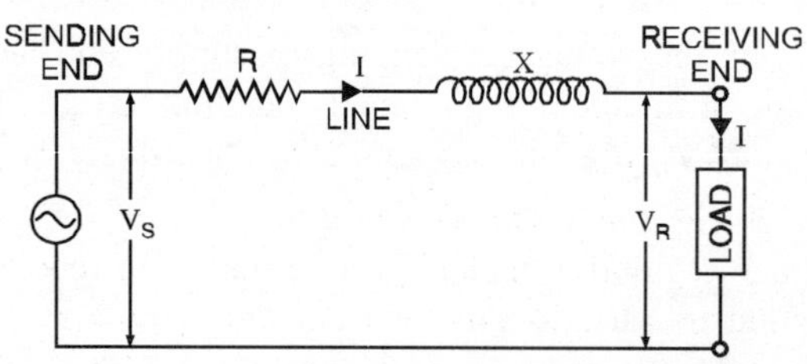

(a) Equivalent Circuit Diagram

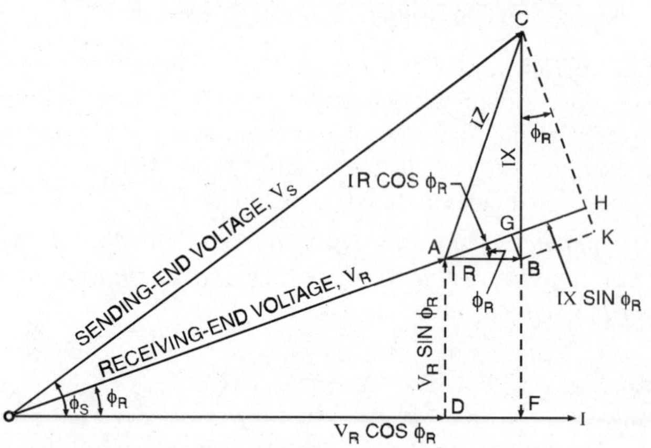

(b) Phasor Diagram

Fig. 6.2. ***Short Transmission Line***

From the equivalent circuit shown in Fig. 6.2 (*a*) it is obvious that

Receiving-end voltage,

$$\mathbf{V_R} = \mathbf{V_S} - \mathbf{I}\,(R + jX) = \mathbf{V_S} - \mathbf{IZ} \quad ...(6.4)$$

The quantity IZ is the voltage drop along the line.

Phasor diagram, taking phasor I as the reference phasor, for an inductive load (lagging power factor load) is shown in Fig. 6.2 (*b*) where phasor OI, OA, AB, BC, AC and OC represent the load current I, receiving-end voltage V_R, resistive drop IR in line, reactive drop IX in line, line impedance drop IZ and sending-end voltage V_S respectively.

From phasor diagram shown in Fig. 6.2 (*b*),

Sending-end voltage,

$$V_S = OC$$
$$= \sqrt{OF^2 + FC^2}$$
$$= \sqrt{(OD + DF)^2 + (FB + BC)^2}$$
$$= \sqrt{(V_R \cos\phi_R + IR)^2 + (V_R \sin\phi_R + IX)^2} \quad ...(6.5)$$

Sending-end phase angle,

$$\phi_S = \text{Tan}^{-1}\frac{FC}{OF} = \frac{V_R \sin\phi_R + IX}{V_R \cos\phi_R + IR} \quad ...(6.6)$$

and sending-end power factor,

$$\cos\phi_S = \frac{OF}{OC} = \frac{V_R \cos\phi_R + IR}{V_S} \quad ...(6.7)$$

Hence, knowing the conditions existing at one end of the line, those at the other end can easily be determined by means of above relations. Problems may also be solved graphically by drawing the phasor diagram to scale, but owing to very small size of impedance triangle ABC compared with rest of the diagram, accurate results cannot be obtained. So it is advisable to calculate all the quantities trigonometrically, using the graphical method as a rough check,

Percentage voltage regulation

$$= \frac{V_S - V_R}{V_R} \times 100$$

$$= \frac{\sqrt{(V_R \cos\phi_R + IR)^2 + (V_R \sin\phi_R + IX)^2} - V_R}{V_R} \times 100 \quad ...(6.8)$$

For determination of an approximate value of sending-end voltage, V_S may be taken equal to its component along V_R.

So sending-end voltage,

$$V_S \simeq OA + AG + GH$$
$$\simeq V_R + IR\cos\phi_R + IX\sin\phi_R \quad ...(6.9)$$

Percentage voltage regulation

$$\simeq \frac{V_S - V_R}{V_R} \times 100$$

$$\simeq \frac{V_R + IR\cos\phi_R + IX\sin\phi_R - V_R}{V_R} \times 100$$

$$\simeq \frac{IR\cos\phi_R + IX\sin\phi_R}{V_R} \times 100 \quad ...(6.10)$$

From the above expression it is obvious that voltage regulation of the line depends upon the resistance and reactance of the line for supplying a given load at a given voltage—larger the value of line resistance and line reactance, greater the percentage regulation.

Regulation per unit

$$= \frac{IR \cos \phi_R}{V_R} + \frac{IX \sin \phi_R}{V_R} = v_r \cos \phi_R + v_x \sin \phi_R \quad ...(6.11)$$

where v_r and v_x are the per unit values of resistance and reactance of the line.

$$\text{Power delivered to load} = V_R I \cos \phi_R$$

$$\text{Line losses} = I^2 R$$

$$\text{Power sent from supply end} = V_R I \cos \phi_R + I^2 R$$

Efficiency of transmission line,

$$\eta_T = \frac{\text{Power delivered to load}}{\text{Power supplied from supply end}} \times 100$$

$$= \frac{V_R I \cos \phi_R}{V_R I \cos \phi_R + I^2 R} \quad ...(6.12)$$

Note: The approximate formula for sending-end voltage V_S [Eq. (6.9)] gives fairly accurate results for lagging power factors (*i.e.,* inductive loads) but an appreciable error is caused for leading power factors (*i.e.,* capacitive loads).

6.4.1. Effect of Load Power Factor on Transmission Efficiency and Voltage Regulation of Line

To deliver given power at a given voltage at consumer end, line current is inversely proportional to the power factor of the load. The current, hence, the power loss in the line goes on increasing as the power factor of the load goes on decreasing. Thus efficiency of the transmission line decreases with the fall in power factor and vice versa.

Since regulation of the line is given by

$$= IR \cos \phi_R + IX \sin \phi_R$$

$$= IZ \left[\frac{R}{Z} \cos \phi_R + \frac{X}{Z} \sin \phi_R\right]$$

$$= IZ [\cos \theta \, . \, \cos \phi_R + \sin \theta \, . \, \sin \phi_R]$$

$$= IZ \cos (\phi_R - \theta) \quad ...(6.13)$$

where Z is line impedance and is equal to $\sqrt{R^2 + X^2}$ and θ is the line impedance angle and is equal to $\tan^{-1} \frac{X}{R}$.

The regulation is zero when $(\phi_R - \theta) = \frac{-\pi}{2}$ and then $\cos \phi_R = \sin \theta = \frac{X}{Z}$. If $(R + jX)$ is the impedance of the line, generator or transformer, θ is positive since X is positive. It is then necessary that ϕ_R be negative for zero regulation. If it is the impedance of a synchronous condenser, θ may be negative, so that ϕ_R may be positive for zero regulation.

The formula given above for zero regulation *viz.* $\phi_R = \theta - \frac{\pi}{2}$ is approximate. The exact relation is determined as follows :

$$V_S^2 = (V_R \cos \phi_R + IR)^2 + (V_R \sin \phi_R + IX)^2$$

For zero regulation $V_S = V_R$ [Refer to Eq. (6.5)]

$$\therefore V_R^2 = V_R^2 + 2IV_R (R \cos \phi_R + X \sin \phi_R) + I^2 (R^2 + X^2)$$

$$\therefore R \cos \phi_R + X \sin \phi_R = \frac{-IZ^2}{2V_R} \qquad \because R^2 + X^2 = Z^2$$

$$\text{or} \quad \cos (\phi_R - \theta) = \frac{-IZ}{2V_R}$$

$$\text{or} \quad \phi_R = \theta - \frac{\pi}{2} - \sin^{-1} \frac{IZ}{2V_R} \quad ...(6.14)$$

The regulation is negative *i.e.,* the receiving-end voltage is greater than the sending-end voltage, if ϕ_R is negative and numerically greater than the value given in this equation

From the expression for regulation *viz.* IR cos ϕ_R + IX sin ϕ_R the following conclusions can be drawn :

(*i*) When the load power factor, cos ϕ_R is lagging or unity or such leading that IR cos ϕ_R > IX sin ϕ_R, the voltage regulation is positive (receiving-end voltage is lesser than the sending-end voltage) and increases with the decrease in power factor for a given load current to be delivered at a given voltage at receiving end.

(*ii*) When the load power factor, cos ϕ_R is leading to the extent that IX sin ϕ_R > IR cos ϕ_R, the voltage regulation is negative (the receiving-end voltage is more than sending-end voltage) and decreases with the decrease in power factor for a given load current to be delivered at a given voltage at receiving end.

Note: For leading power factor loads Eq. (6.10) for voltage regulation of a transmission line is slightly modified. This modification is due to the fact that phase angle ϕ_R is considered positive for lagging power factor loads and negative for leading power factor loads and $\sin(-\phi_R) = -\sin \phi_R$. The modified expression for voltage regulation for leading power factor loads is given below :

$$\% \text{ voltage regulation} = \frac{IR \cos \phi_R - IX \sin \phi_R}{V_R} \times 100$$

6.4.2. Mixed Sending- and Receiving-end Conditions

Sometimes the mixed conditions arise for example when the sending-end voltage and the receiving-end power and power factor are given and the sending-end power and power factor and the receiving-end voltage are to be determined. In such circumstances the solution becomes more complicated. Such problems can be dealt with as follows:

We know that, Power delivered,

$$P = V_R I \cos \phi_R \text{ watts} \quad ...(6.15)$$

Reactive volt-amperes delivered,

$$Q = V_R I \sin \phi_R \quad ...(6.16)$$

Sending-end voltage,

$$V_S = \sqrt{(V_R \cos\phi_R + IR)^2 + (V_R \sin\phi_R + IX)^2}$$

$$\text{or } V_S^2 = V_R^2 \cos^2\phi_R + 2V_R IR \cos\phi_R + I^2R^2 + V_R^2 \sin^2\phi_R + 2V_R IX \sin\phi_R + I^2X^2$$

$$\text{or } V_S^2 = V_R^2 (\cos^2\phi_R + \sin^2\phi_R) + 2V_R (IR \cos\phi_R + IX \sin\phi_R) + I^2(R^2 + X^2)$$

$$\text{or } V_S^2 = V_R^2 + 2V_R I(R \cos\phi_R + X \sin\phi_R) + I^2(R^2 + X^2) \quad ...(6.17)$$

Substituting $V_R I \cos\phi_R = P$, $V_R I \sin\phi_R = Q$ and $I = \frac{P}{V_R \cos\phi_R}$ from Eqs. (6.15), (6.16) and (6.15) respectively in above Eq. (6.17), we have

$$V_S^2 = V_R^2 + 2(PR + QX) + \frac{(R^2 + X^2)P^2}{V_R^2 \cos^2\phi_R}$$

$$\text{or } V_S^2 V_R^2 = V_R^4 + 2V_R^2 (PR + QX) + \frac{(R^2 + X^2)P^2}{\cos^2\phi_R}$$

$$\text{or } V_R^4 - V_R^2[V_S^2 - 2(PR + QX)] + \frac{(R^2 + X^2)P^2}{\cos^2\phi_R} = 0$$

$$\text{or } V_R^4 - AV_R^2 + B = 0$$

where $A = V_S^2 - 2(PR + QX)$ and $B = \frac{(R^2 + X^2)P^2}{\cos^2\phi_R}$

$$\text{or } V_R = \sqrt{\frac{A \pm \sqrt{(A^2 - 4B)}}{2}} \quad ...(6.18)$$

From the above expression we will get two values for V_R depending on whether the +ve or –ve sign is used. The solution giving the higher value of V_R and consequently higher transmission efficiency is the one required in practice, and therefore the +ve sign is usually taken.

After determining the value of V_R, other quantities can be evaluated.

Graphical Solution. The value of receiving-end voltage V_R can also be determined graphically as explained below:

OA is drawn representing the voltage to neutral at sending end of the line and AO′ is drawn making an angle $\frac{\pi}{2} + \phi_R - \theta$ with AO where $\theta = \tan^{-1}\frac{X}{R}$ (Fig. 6.3).

BO′ is drawn perpendicular to OA and an arc of circle, OCA is drawn with O′ as centre and O′A as radius.

Now with B, midpoint of OA, as centre and BC as radius an arc cutting the arc OCA at C is drawn where $BC = \sqrt{\frac{V_S^2}{4} - (PR + QX)}$. Then OC represents the unknown voltage at the receiving end, V_R.

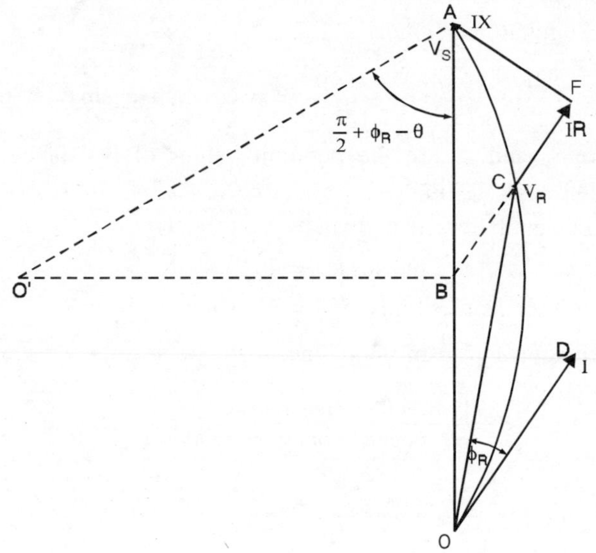

Fig. 6.3

6.4.3. Maximum Power With a Given Regulation

The maximum power, that can be transmitted with a given regulation, can be determined as follows:

From Eq. (6.17)

$$V_S^2 = V_R^2 + 2V_R I(R \cos\phi_R + X \sin\phi_R) + I^2(R^2 + X^2)$$

Substituting $V_R I \cos\phi_R = P$, $V_R I \sin\phi_R = Q$ and $I^2 = \frac{P^2 + Q^2}{V_R^2}$ from Eqs. (6.15) and (6.16) in above equation, we have

$$-V_S^2 + V_R^2 + 2PR + 2QX + \frac{(P^2 + Q^2)(R^2 + X^2)}{V_R^2} = 0 \quad ...(6.19)$$

For a given line, *i.e.*, given values of R and X, and for given receiving- and sending-end voltages, the above Eq. (6.19) gives relation between P and Q. Active power P delivered, varies with the variation in reactive power Q delivered *i.e.*, the above equation gives P as an implicit function of Q. For determining the condition for a maximum value of power P, let us differentiate the above Eq. (6.19) w.r.t. Q and put $\frac{dP}{dQ} = 0$

$$\text{So } 2X + \frac{2Q(R^2 + X^2)}{V_R^2} = 0$$

$$\text{or } Q = \frac{-V_R^2 X}{R^2 + X^2} \quad ...(6.20)$$

i.e., for maximum value of P, the value of Q is a constant negative value.

The maximum value of P is determined by substituting this value of Q in Eq. (6.19) which then becomes

$$-V_S^2 + V_R^2 + 2PR - \frac{2V_R^2X^2}{R^2+X^2} + \frac{P^2(R^2+X^2)}{V_R^2}$$

$$+ \frac{V_R^4X^2}{(R^2+X^2)^2} \times \frac{(R^2+X^2)}{V_R^2} = 0$$

or $$-V_S^2 + V_R^2 + 2PR - \frac{2V_R^2X^2}{Z^2} + \frac{P^2Z^2}{V_R^2} + \frac{V_R^2X^2}{Z^2} = 0$$

$$\because \; R^2 + X^2 = Z^2$$

or $$\frac{P^2Z^2}{V_R^2} + 2PR + \frac{V_R^2R^2}{Z^2} - V_S^2 = 0$$

So we find

$$P_{max} = \frac{V_R^2}{Z^2}\left[-R \pm \sqrt{R^2 - R^2 + \frac{V_S^2Z^2}{V_R^2}}\right]$$

$$= \frac{V_R^2}{Z^2}\left[\frac{ZV_S}{V_R} - R\right] \quad ...(6.21)$$

taking +ve sign of the square root.

If the reactance of the line can be varied, P_{max} will vary, and will have a final maximum when $\frac{dP_{max}}{dX} = 0$.

$$\text{As } P_{max} = \frac{V_R^2}{Z^2}\left[\frac{ZV_S}{V_R} - R\right]$$

$$= \frac{V_RV_S}{Z} - \frac{V_R^2R}{Z^2} = \frac{V_RV_S}{\sqrt{R^2+X^2}} - \frac{V_R^2R}{R^2+X^2}$$

the condition is

$$\frac{V_RV_SX}{(R^2+X^2)^{3/2}} = \frac{2V_R^2RX}{(R^2+X^2)^2} \quad \text{which yields } X = 0$$

or $$(R^2+X^2)^{1/2} = \frac{2RV_R}{V_S}$$

i.e., $$X = R\sqrt{\left(\frac{4V_R'^2}{V_S^2} - 1\right)}$$

The value of P_{max} is then $$P_{max} = \frac{V_S^2}{4R} \quad ...(6.22)$$

Example 6.1. A single-phase line is transmitting 1,100 kW power to a factory at 11 kV and at 0.8 pf lagging. It has a total resistance of 2 Ω and a loop reactance of 3 Ω. Determine (*i*) the voltage at the sending end, (*ii*) percentage regulation and (*iii*) transmission efficiency.

Solution : Load current,

$$I = \frac{P \times 1{,}000}{V_R \cos\phi_R} = \frac{1{,}100 \times 1{,}000}{11{,}000 \times 0.8} = 125 \text{ A}$$

(*i*) Voltage at sending end,

$$V_S = \sqrt{(V_R\cos\phi_R + IR)^2 + (V_R\sin\phi_R + IX)^2}$$

$$= \sqrt{(11{,}000 \times 0.8 + 125 \times 2)^2 + (11{,}000 \times 0.6 + 125 \times 3)^2}$$

$$= \sqrt{(9{,}050)^2 + (6{,}975)^2} = 11{,}426 \text{ V } \textbf{Ans.}$$

(*ii*) Percentage regulation

$$= \frac{V_S - V_R}{V_R} \times 100 = \frac{11{,}426 - 11{,}000}{11{,}000} \times 100 = 3.873\% \textbf{ Ans.}$$

Line losses $= I^2R = (125)^2 \times 2 = 31{,}250$ W or 31.25 kW

(*iii*) Transmission efficiency,

$$\eta_T = \frac{\text{Power delivered}}{\text{Power delivered + line loss}} \times 100$$

$$= \frac{1{,}100}{1{,}100 + 31.25} \times 100 = 97.24\% \textbf{ Ans.}$$

Example 6.2. A short three-phase transmission line of parameters R = 0.4 Ω and X = 0.4 Ω is delivering 2,000 kVA to a load at a pf of 0.8 lagging at the receiving end of the line. If the load voltage is 3,000 V, determine the voltage regulation and efficiency of the line.

[U.P. Technical Univ. Elements of Power System 2004-05]

Solution : Phase voltage at sending end, $V_R = \frac{3{,}000}{\sqrt{3}} = 1{,}732$ V

Load delivered, P = 2,000 kVA

Load power factor, $\cos\phi_L = 0.8$ lagging

$$\text{Line current, } I = \frac{P \text{ in VA}}{3 \times V_R} = \frac{2{,}000 \times 1{,}000}{3 \times 1{,}732} = 384.9 \text{ A}$$

Sending end voltage per phase,

$$V_S = V_R + IR\cos\phi_R + IX\sin\phi_R$$

$$= V_R + I(R\cos\phi_R + X\sin\phi_R)$$

$$= 1{,}732 + 384.9(0.4 \times 0.8 + 0.4 \times 0.6)$$

$$= 1{,}732 + 384.9 \times 0.56 = 1{,}947.5 \text{ V}$$

$$\text{Percentage regulation} = \frac{V_S - V_R}{V_R} \times 100 = \frac{1{,}947.5 - 1{,}732}{1{,}732} \times 100$$

$$= 12.44\% \textbf{ Ans.}$$

$$\text{Total line losses} = \frac{3I^2R}{1{,}000} = \frac{3 \times (384.9)^2 \times 0.4}{1{,}000} = 177.8 \text{ kW}$$

Power output = KVA × cos ϕ = 2,000 × 0.8 = 1,600 kW

Transmission efficiency,

$$\eta_T = \frac{\text{Output power}}{\text{Output power + line losses}} \times 100$$

$$= \frac{1{,}600}{1{,}600 + 177.8} \times 100 = 90\% \textbf{ Ans.}$$

Example 6.3. A 220 kV, 3-phase transmission line is 40 km long. The resistance per phase is 0.15 Ω per km and the inductance per phase is 1.5923 mH per km. The line is supplying a 3-phase load of 381 MVA at 0.8 pf lagging. Find: (*i*) the voltage and power factor at the sending end of the line, (*ii*) the voltage regulation and efficiency of the line.

[U.P.S.C. I.E.S. Electrical Engineering-II, 2010]

Solution : Phase voltage at receiving end, $V_R = \frac{220}{\sqrt{3}}$ kV = 127 kV

Load delivered, S = 381 MVA = 381×10^6 V

Load power factor, Cos ϕ = 0.8 lagging

$$\phi = \text{Cos}^{-1}\ 0.8 = (-36.87°)$$

$$\text{Load current, I} = \frac{S}{\sqrt{3}\ V_{RL}} = \frac{381 \times 10^6}{\sqrt{3} \times 220 \times 10^3} = 1{,}000 \text{ A}$$

Line resistance per phase, R = 0.15 × 40 = 6

Line reactance per phase,

$$X = 2\pi f L \times \text{length of line} = 2\pi \times 50 \times 1.5923 \times 10^{-3} \times 40$$

Assuming f = 50 Hz

$$= 20\ \Omega$$

Phase impedance,

$$\mathbf{Z} = R + jX = (6 + j20) = 20.88 \angle 73.3°$$

Sending-end voltage per phase,

$$\mathbf{V}_S = 127 + 1{,}000 \angle -36.87° \times 20.88 \angle 73.3° \times 10^{-3}$$

$$= 127 + 20.880 \angle 36.43°$$

$$= (143.8 + j12.34) \text{ kV} = 144.33 \angle 4.9° \text{ kV}$$

Sending-end current,

$$\mathbf{I}_S = \mathbf{I}_R = 1{,}000 \angle -36.87°$$

(*i*) Sending-end line voltage,

$$V_{SL} = \sqrt{3}\ V_S = \sqrt{3} \times 144.33 \text{ kV} = 250 \text{ kV (line-to-line)} \textbf{ Ans.}$$

Sending-end power factor

$$= \text{Cos}\ \phi_S = \text{Cos}\ [4.9° - (-36.87°)] = \text{Cos}\ 41.77°$$

$$= 0.746 \text{ (lagging)} \textbf{ Ans.}$$

(*ii*) Voltage regulation

$$= \frac{V_{SL} - V_{RL}}{V_{RL}} \times 100 = \frac{250 - 220}{220} \times 100 = 13.64\% \textbf{ Ans.}$$

Receiving-end power,

$$P_R = 381 \times 0.8 = 304.8 \text{ MW}$$

$$\text{Line losses} = \frac{3I^2R}{10^6} = \frac{3 \times (1{,}000)^2}{10^6} \times 6 = 18 \text{ MW}$$

Sending-end power,

$$P_S = P_R + \text{line losses} = 304.8 + 18 = 322.8 \text{ MW}$$

Transmission efficiency,

$$\eta_T = \frac{P_R}{P_S} \times 100 = \frac{304.8}{322.8} \times 100 = 94.42\% \textbf{ Ans.}$$

Example 6.4. A 15 km long 3-phase overhead line delivers 5 MW at 11 kV at 0.8 lagging pf. Line loss is 12% of power delivered. Line inductance is 1.1 mH per km per phase. Find sending-end voltage and voltage regulation.

Solution : Power delivered, P = 5 MW = 5,000 kW

$$\text{Phase voltage at receiving end, } V_R = \frac{V_{RL}}{\sqrt{3}} = \frac{11{,}000}{\sqrt{3}} = 6{,}351 \text{ V}$$

$$\text{Line current, I} = \frac{P \times 1{,}000}{\sqrt{3} V_{RL} \cos\phi} = \frac{5{,}000 \times 1{,}000}{\sqrt{3} \times 11{,}000 \times 0.8} = 328 \text{ A}$$

$$\text{Line losses, } 3I^2R = 12\% \text{ of P} = \frac{12}{100} \times 5{,}000 \times 1{,}000 = 600{,}000 \text{ W}$$

$$\text{Line resistance, R} = \frac{\text{Line losses}}{3I^2} = \frac{600{,}000}{3 \times 328^2} = 1.859\ \Omega/\text{phase}$$

Inductive reactance per phase,

$$X = 2\pi f L = 2\pi \times 50 \times 15 \times 1.1 \times 10^{-3} = 5.1836\ \Omega/\text{phase}$$

Phase voltage at sending end,

$$V_S = V_R + IR \cos\phi_R + IX \sin\phi_R$$

$$= 6{,}351 + 328 \times 1.859 \times 0.8 + 328 \times 5.1836 \times 0.6$$

$$= 7{,}858.93 \text{ V}$$

Line voltage at sending end,

$$V_{SL} = \sqrt{3}\ V_S = \sqrt{3} \times 7{,}858.93 = 13{,}612 \text{ V or } 13.612 \text{ kV} \textbf{ Ans.}$$

Voltage regulation

$$= \frac{V_{SL} - V_{RL}}{V_{RL}} \times 100 = \frac{13.612 - 11}{11} \times 100 = 23.746\% \textbf{ Ans.}$$

Example 6.5. A 3-phase transmission line 50 km long consists of three hard drawn copper conductors in a 1.2 m delta. Load conditions at receiving end are: 10,000 kVA at 0.8 pf lagging 33,000 volts, 50 Hz. Line is designed so that transmission loss is approximately 10%. Find (*i*) the sending-end voltage and power factor; (*ii*) efficiency; and (*iii*) regulation.

Solution : Power supplied,

$$S = 10{,}000 \text{ kVA}$$

$$\text{Line current, I} = \frac{\text{S in kVA} \times 1{,}000}{\sqrt{3} \times V_{RL}} = \frac{10{,}000 \times 1{,}000}{\sqrt{3} \times 33{,}000} = 174.95 \text{ A}$$

Line losses, $3I^2R$ = 10% of power delivered

$$= \frac{10}{100} \times 10{,}000 \times 1{,}000 \times 0.8 = 800{,}000 \text{ watts}$$

$$\text{Line resistance, R} = \frac{800{,}000}{3 \times (174.95)^2} = 8.7125\ \Omega$$

Area of cross section of conductor,

$$a = \frac{\rho l}{R} = \frac{1.73 \times 10^{-6} \times 50 \times 1{,}000 \times 100}{8.7125} = 0.9928 \text{ cm}^2$$

Radius of conductor,

$$r = \sqrt{\frac{a}{\pi}} = \sqrt{\frac{0.9928}{\pi}} = 0.562 \text{ cm}$$

Inductance of line,

$$L = 0.2 \log_e \frac{d}{r'} \text{ mH/km} \times \text{length of line in km}$$

$$= \left[0.2 \log_e \frac{120}{0.7788 \times 0.562}\right] \times 10^{-3} \times 50 = 56.14 \text{ mH}$$

Inductive reactance,

$$X = 2\pi f L = 2\pi \times 50 \times 56.14 \times 10^{-3} = 17.637\ \Omega$$

Sending-end phase voltage,

$$V_S = V_R + IR \cos\phi_R + IX \sin\phi_R$$

$$= \frac{33{,}000}{\sqrt{3}} + 174.95 \times 8.7125 \times 0.8 + 174.95 \times 17.637 \times 0.6$$

$$= 19{,}052.6 + 3{,}070.8 = 22{,}123.36 \text{ V}$$

Sending-end line voltage,

$$V_{SL} = \sqrt{3} \times 22{,}123.36 = 38{,}320 \text{ V or } 38.32 \text{ kV} \textbf{ Ans.}$$

Sending-end pf,

$$\cos\phi_R = \frac{V_R \cos\phi_R + IR}{V_S}$$

$$= \frac{19{,}052.6 \times 0.8 + 174.95 \times 8.7125}{22{,}123.36} = 0.7578 \text{ lagging } \textbf{Ans.}$$

Transmission efficiency,

$$\eta_T = \frac{\text{Output}}{\text{Output + losses}} \times 100$$

$$= \frac{10{,}000 \times 0.8}{10{,}000 \times 0.8 + 800} \times 100 = 90.91\% \textbf{ Ans.}$$

Voltage regulation

$$= \frac{V_{SL} - V_{RL}}{V_{RL}} \times 100 = \frac{38.32 - 33}{33} \times 100 = 16.12\%.$$

Example 6.6. A short 3-phase transmission line connected to a 33 kV, 50 Hz generating station at the sending end is required to supply a load of 10 MW at 0.8 lagging power factor at 30 kV at the receiving end. If the minimum transmission efficiency is to be limited to 96% estimate the per phase values of resistance and inductance of the line.

Solution : Power delivered, P = 10 MW = 10,000 kW

Phase voltage at receiving end,

$$V_R = \frac{V_{RL}}{\sqrt{3}} = \frac{30{,}000}{\sqrt{3}} = 17{,}320 \text{ V}$$

$$\text{Line current, } I = \frac{P \times 1{,}000}{\sqrt{3} V_{RL} \cos\phi_R} = \frac{10{,}000 \times 1{,}000}{\sqrt{3} \times 30{,}000 \times 0.8} = 240.56 \text{ A}$$

$$\text{Line losses, } \frac{3I^2R}{1{,}000} = \text{Power delivered} \times \left(\frac{1}{\eta_T} - 1\right)$$

$$= 10{,}000\left(\frac{1}{0.96} - 1\right) = 416.67 \text{ kW}$$

$$\text{or } R = \frac{416.67 \times 1{,}000}{3 \times (240.56)^2} = 2.4 \ \Omega/\text{phase} \textbf{ Ans.}$$

Phase voltage at sending-end,

$$V_S = \frac{V_{SL}}{\sqrt{3}} = \frac{33{,}000}{\sqrt{3}} = 19.052 \text{ V}$$

Since $V_S = V_R + IR\cos\phi_R + IX\sin\phi_R$

$$\therefore 19{,}052 = 17{,}320 + 240.56 \times 2.4 \times 0.8 + 240.56 \times X \times 0.6$$

$$\text{or } X = \frac{19{,}052 - 17{,}320 - 240.56 \times 2.4 \times 0.8}{240.56 \times 0.6}$$

$$= 8.8 \ \Omega/\text{phase}$$

Inductance per phase,

$$L = \frac{X}{2\pi f} = \frac{8.8}{2\pi \times} = 28 \text{ mH/phase} \textbf{ Ans.}$$

Example 6.7. A 3-phase line, 3 km long delivers 3,000 kW at a power factor of 0.8 (lagging) to a load. If the voltage at supply end is 11 kV, determine the voltage at the load end and the efficiency of transmission. The resistance and reactance per km of each conductor are 0.4 Ω and 0.8 respectively.

[*U.P. Technical Univ. Elements of Power System 2006-07*]

Solution : Resistance of each conductor, R = 3 × 0.4 = 1.2 Ω

Reactance of each conductor, X = 3 × 0.8 = 2.4 Ω

Let the phase voltage at the load end be V_R and load current be I amperes

Phase voltage at supply end,

$$V_S = \frac{V_{SL}}{\sqrt{3}} = \frac{11{,}000}{\sqrt{3}} = 6{,}351 \text{ V}$$

Now phase voltage at sending end,

$$V_S = V_R + I(R\cos\phi_R + X\sin\phi_R)$$

$$\text{or } 6{,}351 = V_R + I(1.2 \times 0.8 + 2.4 \times 0.6)$$

$$\text{or } V_R = 6{,}351 - 2.4I \quad \ldots(i)$$

$$\text{also } I = \frac{\text{kW output} \times 1{,}000}{3 \times V_R \times \cos\phi_R} = \frac{3{,}000 \times 1{,}000}{3 \times V_R \times 0.8}$$

$$= \frac{1{,}250{,}000}{V_R} \quad \ldots(ii)$$

Solving Eqs. (*i*) and (*ii*), we get V_R = 5,837 V and I = 214 A

Line voltage at load end,

$$V_{RL} = \sqrt{3} \times V_R = \sqrt{3} \times 5{,}837 = 10{,}110 \text{ V} \textbf{ Ans.}$$

Transmission efficiency,

$$\eta_T = \frac{\text{Output}}{\text{Output + line losses}} \times 100$$

$$= \frac{3{,}000 \times 1{,}000}{3{,}000 \times 1{,}000 + 3 \times (214)^2 \times 1.2} \times 100$$

$$= 94.8\% \textbf{ Ans.}$$

ALTERNATIVE METHOD

Power transmitted per phase,

$$P = \frac{1}{3} \times 3{,}000 \times 1{,}000 = 1 \times 10^6 \text{ watts}$$

Reactive volt-amperes transmitted per phase,

$$Q = P\tan\phi_R = 1 \times 10^6 \times \frac{0.6}{0.8} = 0.75 \times 10^6 \quad \because \cos\phi_R = 0.8$$

$$A = V_S^2 - 2(PR + QX)$$

$$= \left(\frac{11{,}000}{\sqrt{3}}\right)^2 - 2(1 \times 10^6 \times 1.2 + 0.75 \times 10^6 \times 2.4)$$

$$= 34.33 \times 10^6$$

$$B = \frac{(R^2 + X^2)P^2}{\cos^2\phi_R} = \frac{[(1.2)^2 + (2.4)^2][1 \times 10^6]^2}{(0.8)^2}$$

$$= 11.25 \times 10^{12}$$

From Eq. (6.18)

$$V_R = \sqrt{\frac{A \pm \sqrt{(A^2 - 4B)}}{2}}$$

$$= \sqrt{\frac{34.33 \times 10^6 + \sqrt{(34.33 \times 10^6)^2 - 4 \times 11.25 \times 10^{12})}}{2}}$$

$$= 5{,}831 \text{ V} \quad \text{taking + ve sign}$$

Line voltage at load end,

$$V_{RL} = \sqrt{3} \times 5{,}831 = 10{,}100 \text{ V} \textbf{ Ans.}$$

Line current,

$$I = \frac{1,000 \times 1,000}{5,831 \times 0.8} = 214.37 \text{ A}$$

Transmission efficiency,

$$\eta_T = \frac{3,000 \times 1,000}{3,000 \times 1,000 + 3(214.37)^2 \times 1.2} \times 100 = 94.8\% \textbf{ Ans.}$$

Graphical Method. Draw OA = 6,351 V (on scale 1 cm = 500 V) to represent phase voltage at supply end and draw BO′ perpendicular to OA and AO′ making an angle $\left(\frac{\pi}{2} + \cos^{-1} 0.8 - \tan^{-1} \frac{2.4}{1.2}\right)$ *i.e.,* (90° + 36.87° – 63.435°) with OA. Draw an arc of circle with O′ as centre and O′A as radius.

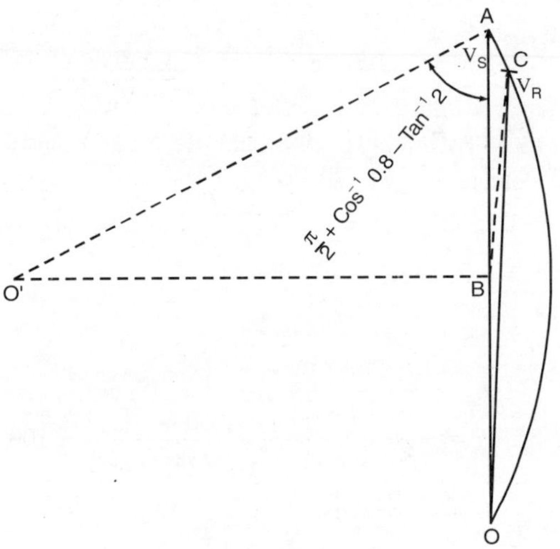

Fig. 6.4

$$BC = \sqrt{\frac{V_S^2}{4} - (PR + QX)}$$

$$= \sqrt{\frac{1}{4}\left(\frac{11,000}{\sqrt{3}}\right)^2 - 10^6 \times 1.2 + 0.75 \times 10^6 \times 2.4} = 2,661 \text{ V}$$

Now with B as centre and BC representing 2,661 V as radius draw an arc cutting the arc OCA at C.

Receiving-end phase voltage,

V_R = OC = 11.7 × 500 = 5,850 V

Line voltage at load end,

$V_{RL} = \sqrt{3} \times 5,850$ = 10,132 V **Ans.**

Line current,

$$I = \frac{1,000 \times 1,000}{5,850 \times 0.8} = 213.7 \text{ A}$$

Transmission efficiency,

$$\eta_T = \frac{3,000 \times 1,000}{3,000 \times 1,000 + 3(213.7)^2 \times 1.2} = 94.8\% \textbf{ Ans.}$$

Example 6.8. The sending-end voltage of a 3-phase overhead line with a resistance and reactance per phase of 5 Ω and 20 Ω respectively is 46.85 kV while the receiving-end voltage is 33 kV. Determine the power output at 0.8 pf (lag) and the sending-end power factor.

Solution : Phase voltage at receiving end,

$$V_R = \frac{33,000}{\sqrt{3}} = 19,050 \text{ V}$$

Let load current be I amperes

Now phase voltage at sending end,

$$V_S = V_{RL} + I\,(R \cos \phi_R + X \sin \phi_R)$$

or $$\frac{46.85 \times 1,000}{\sqrt{3}} = 19,050 + I\,(5 \times 0.8 + 20 \times 0.6)$$

or I = 500 amperes

$$\text{Power output} = \frac{\sqrt{3}\, V_{RL}\, I \cos\phi_R}{1,000} \text{ kW}$$

$$= \frac{\sqrt{3} \times 33,000 \times 500 \times 0.8}{1,000} = 22,853 \text{ kW } \textbf{Ans.}$$

Power factor at sending end,

$$\cos \phi_S = \frac{V_R \cos\phi_R + IR}{V_S} = \frac{19,050 \times 0.8 + 500 \times 5}{\frac{46,850}{\sqrt{3}}}$$

$$= 0.656 \text{ (lag) } \textbf{Ans.}$$

Graphical Method. Draw OA = $\frac{46,850}{\sqrt{3}}$ *i.e.,* 27,048 V (on scale 1 cm = 2,000 V) to represent phase voltage at supply end and BO′ perpendicular to OA and AO' making an angle $\left(\frac{\pi}{2} + \cos^{-1} 0.8 - \tan^{-1} \frac{20}{5}\right)$ *i.e.,* (90° + 36.87° – 75.97° = 50.9°) with OA. Draw an arc of circle with O′ as centre and O′A as radius. Now with O as centre and OC representing 19,050 V, the phase voltage at load end as radius draw an arc cutting the arc ACO at C.

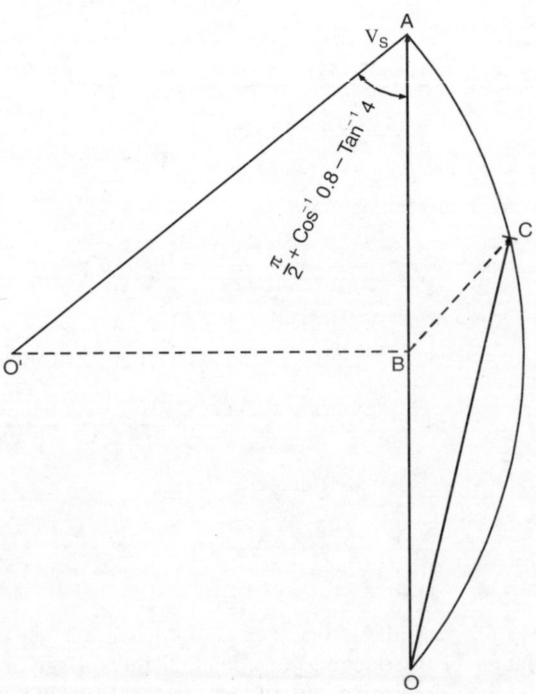

Fig. 6.5

On measurement BC = 3.3 × 2,000 = 6,600 V

$$BC^2 = \frac{V_S^2}{4} - (PR + QX)$$

$$(6{,}600)^2 = \frac{1}{4}\left(\frac{46{,}850}{\sqrt{3}}\right)^2 - [P \times 5 + P \times 0.75 \times 20]$$

$$\because \; Q = P \tan \phi_R = 0.75\, P$$

$$\text{or } P = \left[\frac{(46{,}850)^2}{12} - (6{,}600)^2\right] \times \frac{1}{20}$$

$$= 6{,}967{,}510 \text{ W or } 6{,}967.51 \text{ kW}$$

Total power output = 3P = 3 × 6,967.51 = 20,902.5 kW **Ans.**

$$\text{Load current, } I = \frac{20{,}902.5 \times 1{,}000}{\sqrt{3} \times 33{,}000 \times 0.8} = 457 \text{ A}$$

Power factor at sending end,

$$\cos \phi_S = \frac{V_R \cos\phi_R + IR}{V_S} = \frac{19{,}050 \times 0.8 + 457 \times 5}{\frac{46{,}850}{\sqrt{3}}}$$

$$= 0.648 \text{ (lag)} \textbf{ Ans.}$$

Example 6.9. What is the maximum power that can be transmitted over a 3-phase short line having per phase impedance of (0.3 + *j*0.4) ohm if the receiving-end voltage is 6,351 volts per phase and voltage regulation not to exceed 5%. Also calculate the receiving-end power factor and the total line loss when supplying maximum power. [*M.D. Univ. Transmission & Distribution May 2011*]

Solution : Receiving-end voltage per phase, V_R = 6,351 V

Voltage regulation = 5% = 0.05

Sending-end voltage per phase,

$$V_S = V_R \,(1 + \text{voltage regulation})$$

$$= 6{,}351\,(1 + 0.05) = 6{,}668.55 \text{ V}$$

Impedance per phase,

$$\mathbf{Z} = (0.3 + j0.4) = 0.5 \angle 53.13° \,\Omega$$

Maximum power per phase

$$P_{max} = \frac{V_R^2}{Z^2}\left[\frac{ZV_S}{V_R} - R\right] = \frac{6{,}351^2}{0.5^2}\left[\frac{0.5 \times 6{,}668.55}{6{,}351} - 0.3\right]$$

$$= 36{,}294{,}695 \text{ W or } 36.295 \text{ MW}$$

Maximum power transferred (total)

$$= 3P_{max} = 3 \times 36.295 = 108.88 \text{ MW } \textbf{Ans.}$$

Reactive power transferred (total), 3Q

$$= \frac{-3V_R^2 X}{R^2 + X^2} = \frac{-3 \times 6{,}351^2 \times 0.4}{0.5} \times 10^{-6} \text{ MVAR}$$

$$= 96.804 \text{ MVAR (lagging)}$$

Receiving-end power factor,

$$\text{Cos } \phi_R = \text{Cos}\left(\tan^{-1}\frac{Q}{P}\right) = \text{Cos}\left(\tan^{-1}\frac{96.804}{108.88}\right)$$

$$= \text{Cos } 41.64° = 0.7473 \textbf{ Ans.}$$

Line current,

$$I = \frac{3P_{max}}{3 \times V_R \times \cos\phi} = \frac{P_{max}}{V_R \cos\phi} = \frac{36.295 \times 10^6}{6{,}351 \times 0.7473} = 7{,}647 \text{ A}$$

Total line losses

$$= 3I^2R = \frac{3 \times 7{,}647^2 \times 0.3}{10^6} \text{ MW} = 52.63 \text{ MW } \textbf{Ans.}$$

6.5. KAPP's REGULATION DIAGRAM

As discussed, the receiving-end voltage falls with the increase in load in case of inductive loads (lagging power factor) and increases with the increase in load in case of capacitive loads (leading power factor) *i.e.,* receiving-end voltage not only depends on the load but on power factor also.

As given in Art. 6.3, regulation is defined as the change in voltage at the receiving end when full load is thrown off, the sending-end voltage remaining the same.

When the load is thrown off, the line takes only a charging current and if this current is negligibly small the receiving-end voltage V_R will be equal to sending-end voltage V_S. Hence in the case of short lines, regulation is equal to $V_S - V_R$.

The regulation for any given load power factor can be graphically determined by the Kapp's regulation diagram, which is shown in Fig. 6.6.

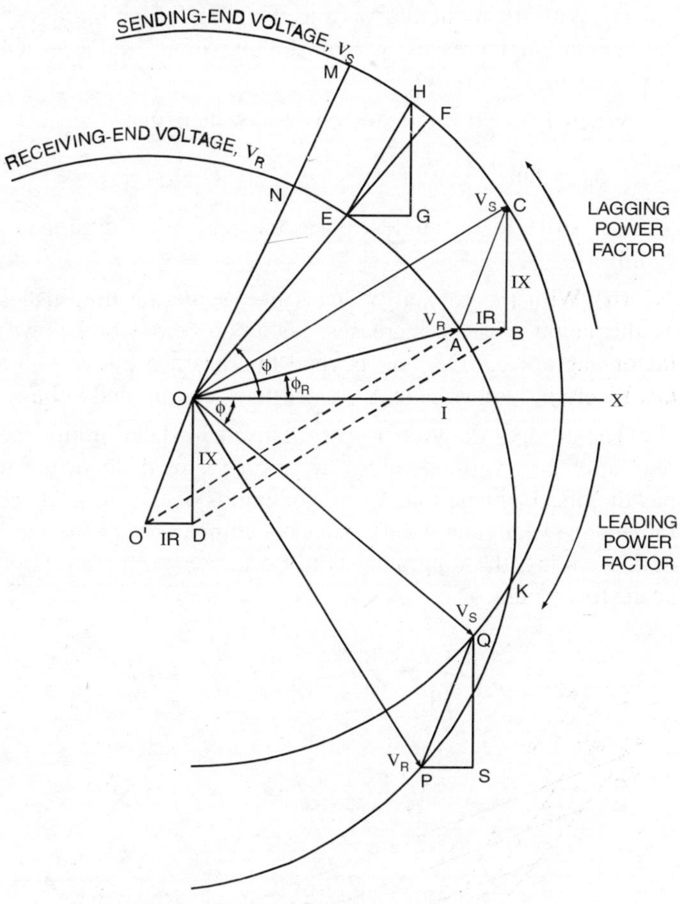

Fig. 6.6. ***Kapp's Regulation Diagram***

A phasor OA is drawn to represent the receiving-end phase voltage V_R on load and OX is drawn inclined at an angle ϕ_R to OA where cos ϕ_R is the power factor of the load. Then OX represents the phase of the load current I. AB is drawn parallel to OX and equal to resistance drop IR and BC perpendicular to OX and equal to reactance drop IX. Then ABC is the impedance triangle and CA is the total line impedance drop IZ. Hence OC represents the sending-end phase voltage, V_S and the difference of V_S and V_R

(*i.e.*, OC – OA) is the voltage drop in the line and also the regulation.

Now if the triangle CAB is transferred to OO′D as shown then O′A = OC = V_S. Hence for a given current, the locus of C is a circle with centre O and radius V_S while the locus of A is a circle of radius V_S with centre O′.

In order to determine the regulation on any power factor cos ϕ, a radius OEF is drawn inclined at an angle ϕ to OX. If the impedance triangle is drawn in position HEG, then OH = OF, the sending-end phase voltage V_S and OE = the receiving-end phase voltage V_R. Hence length EF represents the line regulation. It is obvious that the triangle HEG need not be drawn but all that is necessary, in order to determine the regulation at any power factor cos ϕ is to draw a line making an angle ϕ with OX, and measure the length of the intercept between the two circles.

From the Kapp's regulation diagram we observe that

(*i*) With the inductive loads as the angle of lag increases, the regulation increases, reaching a maximum value when OA is in line with OO′ *i.e.*,

When load power factor cos ϕ_R is such that

$$\phi_R = \tan^{-1}\frac{X}{R}$$

For still greater angles of lag the regulation diminishes again.

(*ii*) With the capacitive loads the regulation diminishes as the angle of lead increases, becomes zero when power factor angle = ∠XOK and beyond this has a negative value *i.e.*, receiving-end voltage is greater than sending-end voltage.

The Kapp's diagram is very helpful in determining the regulation for various values of power factor of load, but it has the disadvantage that since the lengths of the sides of the impedance triangle are very small in comparison to the radii of the circles, the diagram has to be drawn on a very large scale for accuracy.

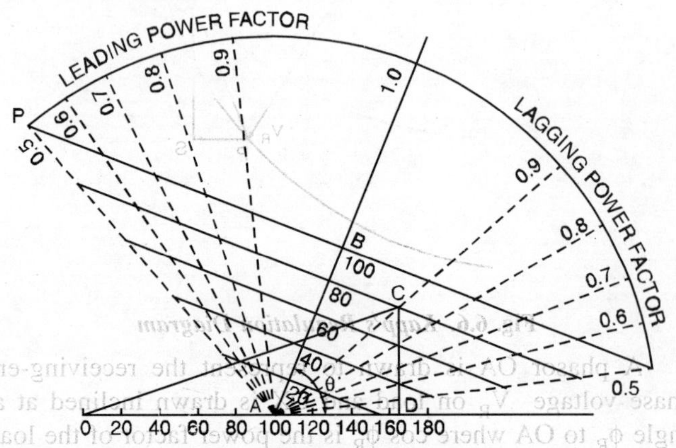

Fig. 6.7. *Perrine Baum Diagram*

Another disadvantage of Kapp's diagram is that it gives the performance of the short line only at one particular loading; but in order to determine the performance of the short line at various loads and on various power factors use of Perrine Baum diagram can be made, as shown in Fig. 6.7.

A phasor OA is drawn to represent the 100% applied voltage. AB is drawn to represent full-load impedance drop making an angle θ with OA where $\theta = \tan^{-1}\frac{X}{R}$. The line AB is divided into a number of parts, representing drop for load from 0% to 100 per cent full-load at unity power factor. Radial lines are drawn on both sides of line AB for leading and lagging power factors. Now regulation at any load and at any power factor can be determined from this diagram.

Suppose regulation at 80 per cent load and 0.9 power factor (lag) is to be determined. The interpolation of the point corresponding to this load and power factor is C and projection of this point C is D. OC represents the sending-end voltage V_S and AD represents the regulation.

Example 6.10. A single-phase transmission line is delivering 500 kVA load at 2,000 V. Its resistance is 0.4 Ω and inductive reactance is 0.6 Ω. Determine the voltage regulation by Kapp's regulation diagram if the load power factor is (*i*) 0.707 lagging and (*ii*) 0.707 leading.

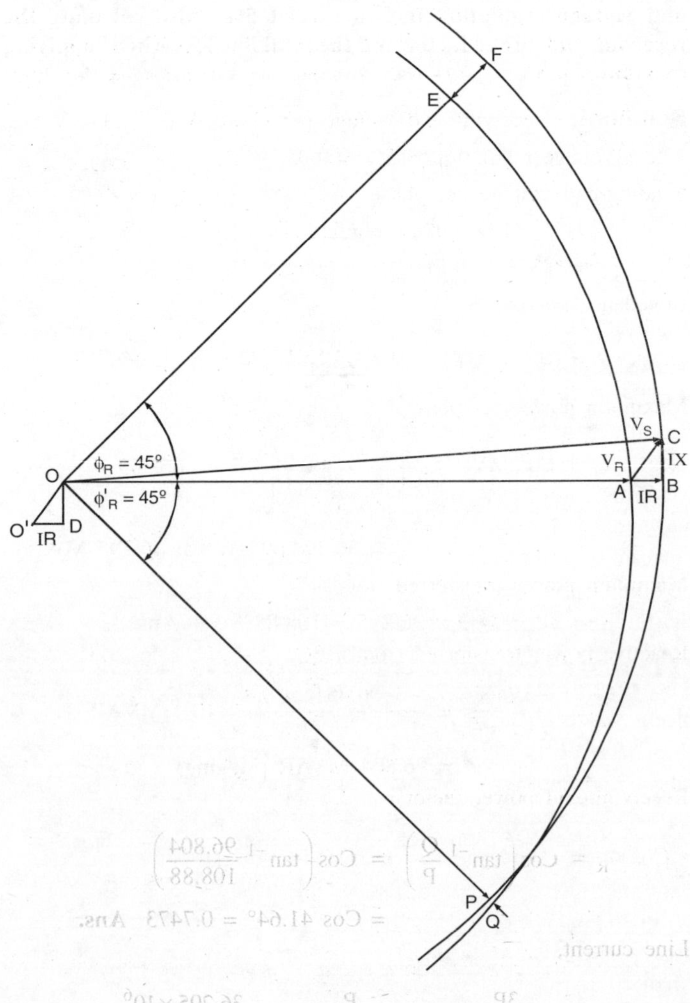

Fig. 6.8

Solution : Line resistance, R = 0.4 Ω

Line reactance, X = 0.6 Ω

Receiving-end voltage, V_R = 2,000 V

Power transmitted = 500 kVA

$$\text{Load current, I} = \frac{500 \times 1{,}000}{2{,}000} = 250 \text{ A}$$

$$IR = 250 \times 0.4 = 100 \text{ V}$$

$$IX = 250 \times 0.6 = 150 \text{ V}$$

Draw Kapp's regulation diagram on scale 1 cm = 200 V (as discussed in Art. 6.5) so that

$$OA = 2{,}000 \text{ V}$$

$$AB = 100 \text{ V}$$

$$BC = 150 \text{ V}$$

$$OD = BC = 150 \text{ V}$$

$$DO' = AB = 100 \text{ V}$$

Draw OEF and OPQ making angles of $\cos^{-1} 0.707$ *i.e.*, 45° and $-\cos^{-1} 0.707$ *i.e.*, – 45° respectively from OX to determine regulation at (*i*) 0.707 lagging and (*ii*) 0.707 leading power factors.

Now regulation at 0.707 lagging power factor

$$= EF \times 200 = 0.9 \times 200 = 180 \text{ V}$$

$$\% \text{ regulation} = \frac{\text{Voltage drop}}{V_R} \times 100 = \frac{180}{2{,}000} \times 100$$

$$= 9\% \textbf{ Ans.}$$

Regulation at 0.707 leading power factor

$$= PQ \times 200 = -0.15 \times 200 = -30 \text{ V}$$

$$\% \text{ regulation} = \frac{-30}{2{,}000} \times 100 = -1.5\% \textbf{ Ans.}$$

6.6. MEDIUM TRANSMISSION LINES

As already discussed, for transmission lines of length up to 80 km and transmitting power at relatively low voltage; (voltage below 20 kV), the capacitance is too small that its effects can be neglected. However, the effects of shunt capacitance become more and more pronounced with the increase in length and operating voltage of the line. Since medium transmission lines have sufficient length exceeding 80 km and usually operate at voltages exceeding 20 kV, the capacitive current is appreciable and, therefore, line capacitance is to be taken into account. The capacitive current is always flowing in the line while the supply end switches are closed even though the receiving end of the line may be open circuited. The magnitude of capacitive current flowing at any point along the line is that required to charge the section of the line between the given point and the receiving end, hence it has a maximum value at the sending end and diminishes at a practically uniform rate down to zero at the receiving end.

Actually the capacitance of the line is uniformly distributed over its entire length, as illustrated in Fig. 6.9. However, in order to make the calculations simple, the capacitance of the system is assumed to be divided up, and 'lumped' in the form of capacitors shunted across the line at one or more points—the more the points the closer the approximation.

The most common methods of representation, called the *localised capacitance methods*, are: (*i*) end condenser method, nominal-T method (or middle condenser method) and nominal-π method (or split-condenser method).

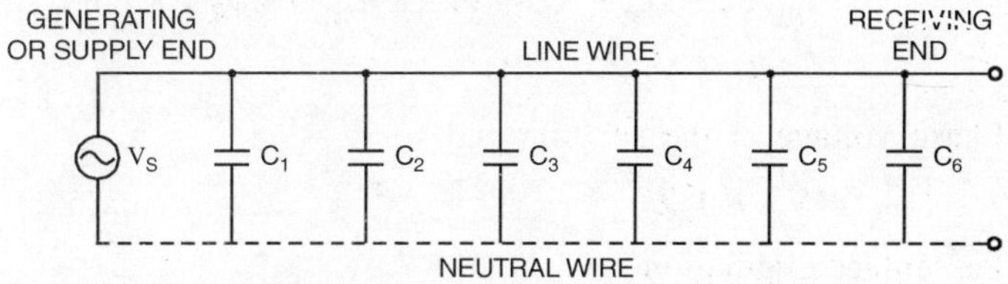

Fig. 6.9

The above methods of solution of transmission lines are only of academic interest as far as long lines are concerned because these involve considerable error owing to the actual distributed capacitance having been assumed lumped. However, for lines up to 200 km at 50 Hz, the error involved by the use of these methods will be of the order of 10 per cent for sending-end voltage calculations which is an acceptable approximation. It is interesting to note that the end condenser method and the nominal T- method over compensates (giving sending-end voltages which are too low), while the nominal-π method under compensates. Because of the approximations involved in T and π-methods, they are often referred to as nominal-T and nominal-π methods.

6.6.1. End Condenser Method

In this method the capacitance of the line is assumed to be lumped at the load-end, as shown in Fig. 6.10. This method overestimates the effect of capacitance.

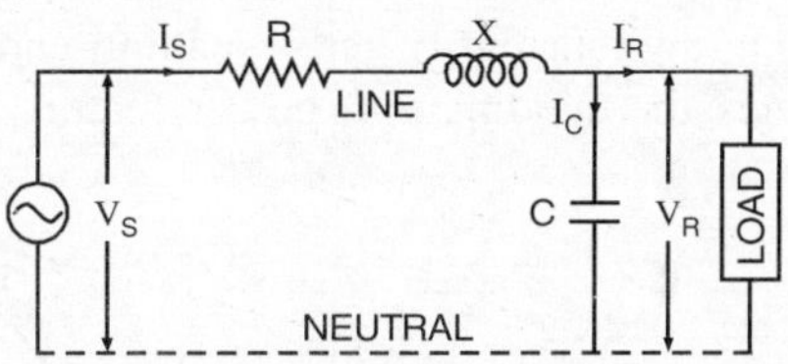

Fig. 6.10. ***Equivalent Circuit For End Condenser Method***

Let the load current per phase be $\mathbf{I_R}$ lagging behind the load end phase voltage $\mathbf{V_R}$ by and angie ϕ_R. Let resistance, and reactance per phase of the line be R and X ohms respectively. Capacitance per phase of line be C farads.

Taking receiving-end phase voltage as reference phasor, we have $\mathbf{V_R} = V_R(1 + j0)$

Receiving-end or load-end current,

$$\mathbf{I_R} = I_R(\cos \phi_R - j \sin \phi_R)$$

Capacitive current,

$$\mathbf{I_C} = j\mathbf{V_R}\omega C = j2\pi f\, C\mathbf{V_R}$$

Sending-end current,

$$\mathbf{I_S} = \mathbf{I_R} + \mathbf{I_C} = I_R (\cos \phi_R - j \sin \phi_R) + j2\pi f C\mathbf{V_R}$$

Voltage drop in line (per phase)

$$= \mathbf{I_S Z}$$

$$= \mathbf{I_S} (R + jX)$$

$$= [I_R(\cos \phi_R - j \sin \phi_R) + j2\pi f C\mathbf{V_R}] (R + jX)$$

$$= I_R R \cos \phi_R + I_R X \sin \phi_R - 2\pi f C V_R X$$

$$+ j [I_R X \cos \phi_R - I_R R \sin \phi_R + 2\pi f C V_R R]$$

$$= I_R[R \cos \phi_R + X \sin \phi_R) + j(X \cos \phi_R - R \sin \phi_R)] - 2\pi f C V_R (X - jR)$$

Phase voltage at the sending end,

$$\mathbf{V_S} = \mathbf{V_R} + \mathbf{I_S Z} \quad ...(6.23)$$

Percentage regulation

$$= \frac{V_S - V_R}{V_S} \times 100$$

Transmission efficiency,

$$\eta_T = \frac{\text{Power delivered per phase}}{\text{Power delivered per phase + line losses / phase}} \times 100$$

$$= \frac{V_R I_R \cos \phi_R}{V_R I_R \cos \phi_R + I_S^2 R} \times 100 \quad ...(6.24)$$

Phasor diagram is shown in Fig. 6.11 where horizontal line OA is the voltage to neutral at the receiving end; OD is the load current I_R that lags behind V_R by ϕ_R; DF is the capacitive current I_C that leads the receiving-end voltage V_R by 90°; OF is the sending-end current I_S that flows through the line and is the phasor sum of receiving-end current I_R and capacitive current I_C; AB is the resistive voltage drop $I_S R$ that is in phase with I_S; BC is the reactive voltage drop $I_S X$ that leads the I_S by 90°; OC is the sending-end voltage V_S that is the phasor sum of V_R, $I_S R$ and $I_S X$ (or V_R and $I_S Z$); and ϕ_S, the phase angle between the sending-end voltage V_S and sending-end current I_S is the sending-end phase angle that determines the sending-end power factor.

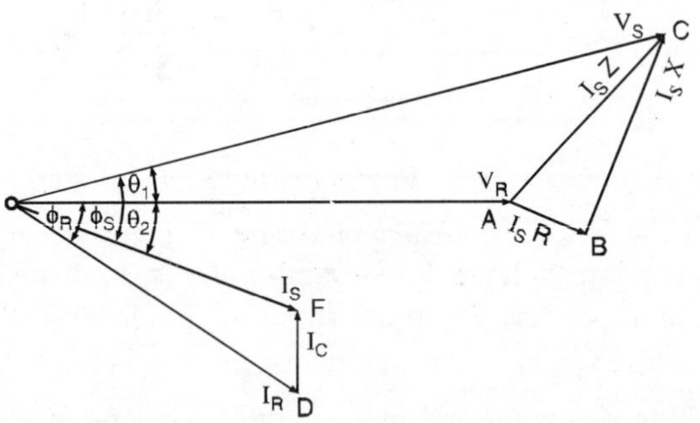

Fig. 6.11. *Phasor Diagram For End Condenser Method*

6.6.2. Nominal-T Method (or Middle Condenser Method)

In this method the whole of the line capacitance is assumed to be concentrated at the middle point of the line and half the line resistance and reactance (*i.e.*, $\frac{R}{2}$ and $\frac{X}{2}$) to be lumped on either side, as shown in Fig. 6.12. Thus in this arrangement, at the terminals of the condenser the voltage has a value V' intermediate between the sending-end voltage V_S and the receiving-end voltage V_R. A capacitive current I_C leading V' by 90° accordingly flows over first half of the line from the sending-end. The current in the receiving-end half of the line is I_R, and in the sending-end half of the line is I_S which is the phasor sum of the receiving-end (or load) current I_R and charging or capacitance current I_C.

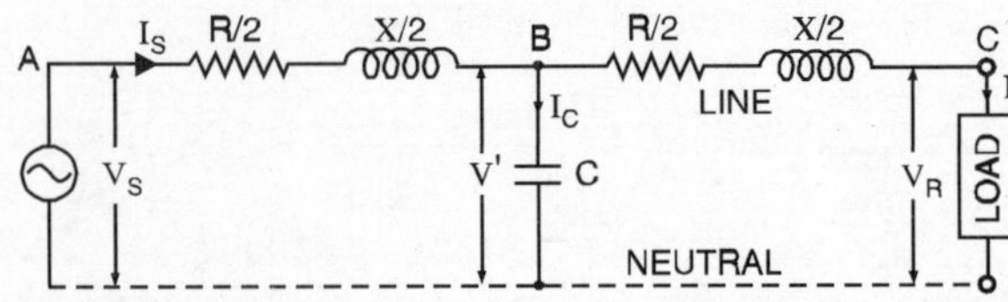

Fig. 6.12. *Equivalent Circuit For Nominal-T Method*

Taking receiving-end phase voltage (voltage to neutral) as the reference phasor we have

$$\mathbf{V_R} = V_R(1 + j0)$$

Receiving-end or load current,

$$\mathbf{I_R} = I_R(\cos \phi_R - j \sin \phi_R)$$

Voltage across the condenser,

$$\mathbf{V} = \mathbf{V_R} + \mathbf{I_R} \frac{\mathbf{Z}}{2}$$

$$= \mathbf{V_R} + \frac{\mathbf{I_R}}{2}(R + jX)$$

$$= V_R(1 + j0) + \frac{1}{2} I_R(\cos \phi_R - j \sin \phi_R)(R + jX) \quad ...(6.25)$$

Capacitive current,

$$I_C = j\omega C\mathbf{V}' = j2\pi f C\mathbf{V}' \quad ...(6.26)$$

Sending-end current,

$$\mathbf{I_S} = \mathbf{I_R} + \mathbf{I_C}$$

$$= I_R(\cos \phi_R - j \sin \phi_R) + 2\pi f C\mathbf{V}' \quad ...(6.27)$$

Sending-end voltage,

$$\mathbf{V_S} = \mathbf{V}' + I_S \frac{(R + jX)}{2} \quad ...(6.28)$$

Transmission efficiency,

$$\eta_T = \frac{V_R I_R \cos\phi_R}{V_R I_R \cos\phi_R + I_R^2 \frac{R}{2} + I_S^2 \frac{R}{2}} \quad ...(6.29)$$

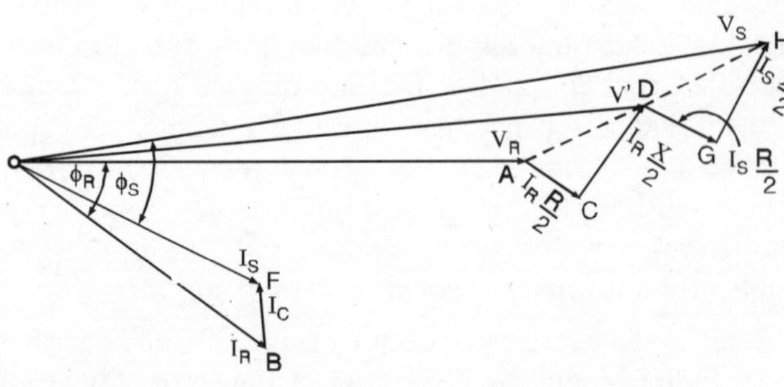

Fig. 6.13. *Phasor Diagram For Nominal-T Method*

In the phasor diagram shown in Fig. 6.13, horizontal line OA is the voltage to neutral at the receiving end; OB is the current in the receiving-end half of the line (*i.e.*, I_R) drawn ϕ_R degrees behind OA; ϕ_R is the load power factor angle; AC drawn parallel to OB is the ohmic voltage drop in the receiving-end half of the line, CD drawn 90° ahead of OB is the reactive voltage drop in the receiving-end half of the line; OD is the voltage across the capacitor; BF drawn 90° ahead of OD is the charging or capacitive current I_C; OF (the phasor sum of OB and BF) is the sending-end current I_S; DG drawn parallel to OF is the ohmic voltage drop in the sending-end half of the line; GH drawn 90° ahead of OF is the reactive voltage drop in the sending end half of the line; OH is the sending-end voltage; and ϕ_S (the angle between OH and OF) is the phase angle at the sending end.

6.6.3. Nominal-π Method (or Split-Condenser Method)

In this method the capacitance of each line conductor is assumed to be divided into two halves, one half being shunted between line conductor and neutral at the receiving end and the other half at the sending end, as shown in Fig. 6.14. The current flowing in the line at any point in between the two capacitors is I_L which is phasor sum of the load current I_R and the current I_{CR} drawn by the receiving end capacitor. Knowing the line current I_L, the voltage at the sending end can easily be determined by the fundamental impedance method used for short lines. The sending-end voltage is, therefore, more readily calculated by this method than by nominal-T method which needs the calculation of two separate halves of the line. Though the capacitance at the sending end has no effect on the voltage drop in the line but if the sending-end current I_S and power factor cos ϕ_S are to be determined, the current drawn by the sending-end capacitor, I_{CS} is to be added vectorially to that in the line.

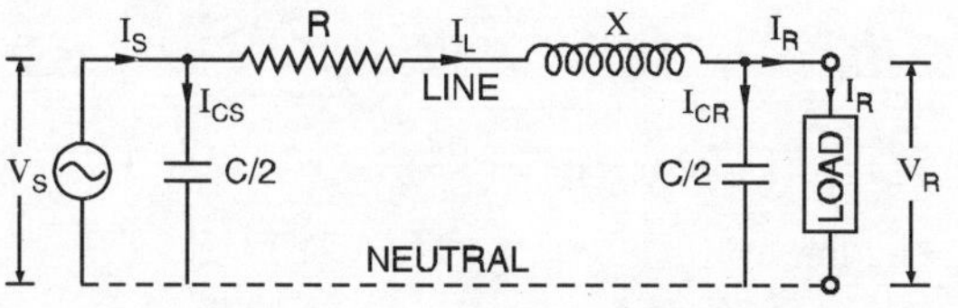

Fig. 6.14. ***Equivalent Circuit For Nominal-π Method***

Taking receiving-end phase voltage (voltage to neutral) as the reference phasor we have

$$\mathbf{V_R} = V_R(1 + j0)$$

Load current, $\mathbf{I_R} = I_R(\cos\phi_R - j\sin\phi_R)$

Charging current at the receiving end,

$$\mathbf{I_{CR}} = j\omega\frac{C}{2}\mathbf{V_R}$$

Line current, $\mathbf{I_L} = \mathbf{I_R} + \mathbf{I_{CR}}$

Sending-end voltage,

$$\mathbf{V_S} = \mathbf{V_R} + \mathbf{I_L Z} = \mathbf{V_R} + \mathbf{I_L}(R + jX) \quad ...(6.30)$$

Charging current at the sending end,

$$\mathbf{I_{CS}} = j\omega\frac{C}{2}\mathbf{V_S}$$

Sending-end current,

$$\mathbf{I_S} = \mathbf{I_L} + \mathbf{I_{CS}} \quad ...(6.31)$$

In phasor diagram shown in Fig. 6.15 horizontal line OA is the voltage to neutral at the receiving end V_R; OB is the load current I_R drawn ϕ_R degrees behind OA; ϕ_R is the load power factor angle. The capacitive current due to capacitance localised at the receiving end, I_{CR} will lead the phasor V_R by 90° and will be equal to $\frac{1}{2}\omega CV_R$ in magnitude. Let it be represented by phasor BC. The line current I_L will be the phasor sum of I_R and I_{CR} and is represented by phasor OC. The voltage drop in the line is equal to phasor sum of resistive drop I_LR in phase with phasor I_L represented by AD and reactive drop I_LX in quadrature with I_L represented by DF. The phase voltage at the sending end V_S, the phasor sum of receiving-end phase voltage V_R and the line drop is represented by phasor OF. The sending-end current I_S, which is phasor sum of I_L and I_{CS} (I_{CS} is equal to $\frac{1}{2}\omega CV_S$, in magnitude and leads V_S by 90°), is represented by phasor OG.

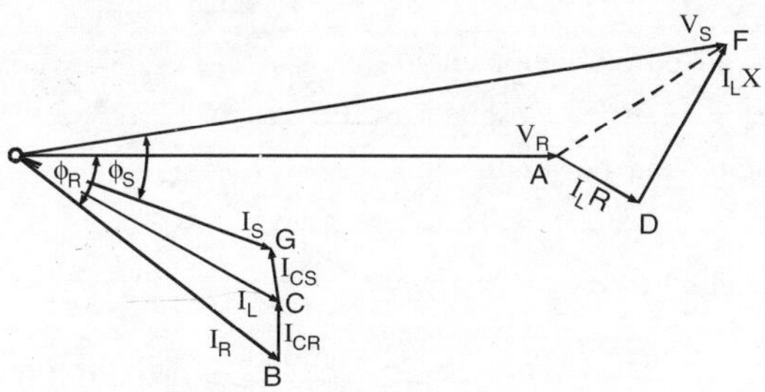

Fig. 6.15

Example 6.11. A single-phase medium transmission line 80 km long has the following constants.

Resistance per km = 0.3125 Ω; Reactance per km = 1.0 Ω; Susceptance per km = 17.5 × 10^{-6} S. Receiving-end voltage = 66 kV.

Assuming that the total capacitance of the line is localized at the receiving end alone, determine (*i*) the sending-end current (*ii*) the sending-end voltage (*iii*) the sending-end power factor (*iv*) the regulation (*v*) the line losses and (*vi*) transmission efficiency. The line is delivering 15.0 MW at 0.8 power factor lagging.

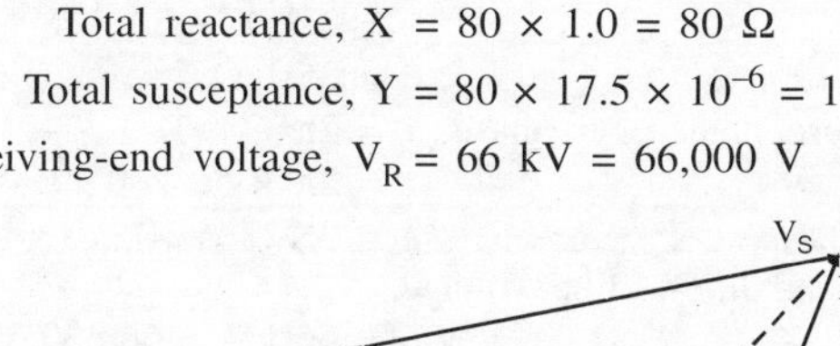

Fig. 6.16

Solution : Total resistance, R = 80 × 0.3125 = 25 Ω

Total reactance, X = 80 × 1.0 = 80 Ω

Total susceptance, Y = 80 × 17.5 × 10^{-6} = 14 × 10^{-4} S

Receiving-end voltage, V_R = 66 kV = 66,000 V

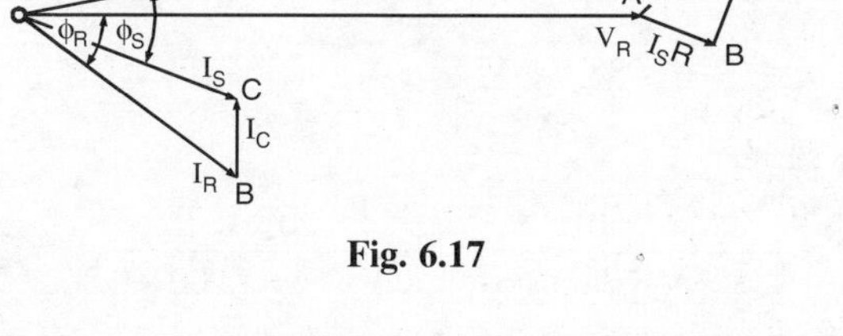

Fig. 6.17

$$\text{Load current, } I_R = \frac{15 \times 10^6}{66{,}000 \times 0.8} = 284 \text{ A}$$

Power factor at receiving end, $\cos \phi_R = 0.8$ and $\sin \phi_R = 0.6$

Taking receiving-end voltage V_R as reference phasor we have

$$\mathbf{V_R} = 66{,}000\ (1 + j0) \text{ volts}$$

$$\text{Load current, } \mathbf{I_R} = 284(0.8 - j0.6) \text{ A} = (227.2 - j170.4) \text{ A}$$

$$\text{Charging current, } \mathbf{I_C} = jYV_R = j\,0.0014 \times 66{,}000 = j\,92.4 \text{ A}$$

(*i*) Sending-end current,

$$\mathbf{I_S} = \mathbf{I_R} + \mathbf{I_C} = (227.2 - j\,170.4) + j\,92.4 = (227.2 - j\,78)$$
$$= 240.2 \angle -18.94^\circ \text{ A} \quad \textbf{Ans.}$$

(*ii*) Sending-end voltage,

$$\mathbf{V_S} = \mathbf{V_R} + \mathbf{I_S}(R + jX)$$
$$= (66{,}000 + j0) + (227.2 - j78)\ (25 + j80)$$
$$= (77{,}920 + j16{,}226) = 79{,}592 \angle 11.76^\circ \text{ volts} \quad \textbf{Ans.}$$

Sending-end power factor angle,

$$\phi_S = 11.76^\circ - (-18.94^\circ) = 30.7^\circ$$

(*iii*) Sending-end power factor,

$$\cos \phi_S = \text{Cos } 30.7^\circ = 0.86 \text{ (lag)} \quad \textbf{Ans.}$$

(*iv*) Regulation

$$= \frac{V_S - V_R}{V_{R_2}} = \frac{79{,}592 - 66{,}000}{66{,}000} = 0.206 \text{ or } 20.6\% \quad \textbf{Ans.}$$

(*v*) Line losses

$$= \frac{I_S^2 R}{1{,}000} = \frac{(240.2)^2 \times 25}{1{,}000} \text{ kW} = 1{,}442.4 \text{ kW} \quad \textbf{Ans.}$$

(*vi*) Transmission efficiency,

$$\eta_T = \frac{\text{Power output}}{\text{Power output + line losses}} \times 100$$

$$= \frac{15{,}000}{15{,}000 + 1{,}442.4} \times 100 = 91.23\% \quad \textbf{Ans.}$$

Example 6.12. Calculate the sending-end voltage, efficiency and regulation of a three-phase, 50 Hz, 100 km long transmission line, supplying a load of 20 MW at 0.8 power factor lag, at 66 kV. The resistance per phase is 10 ohms; the inductance per phase is 111.7 mH; and the capacitance per phase is 0.9954 μF. Use Nominal-T Method. [*U.P. Technical Univ. Elements of Power System 2002-03*]

Solution : Line resistance per phase,

$$R = 10\ \Omega$$

$$\text{Line reactance per phase, } X = 2\pi f L$$
$$= 2 \times \pi \times 50 \times 111.7 \times 10^{-3}$$
$$= 35.1\ \Omega$$

$$\text{Line susceptance per phase, } Y = 2\pi f C$$
$$= 2\pi \times 50 \times 0.9954 \times 10^{-6}$$
$$= 3.127 \times 10^{-4} \text{ S}$$

Receiving-end line-to-line voltage,

$$V_{RL} = 66 \text{ kV} = 66{,}000 \text{ V}$$

Receiving-end line voltage per phase,

$$V_R = \frac{66{,}000}{\sqrt{3}} = 38{,}105 \text{ V}$$

Load current,

$$I_R = \frac{\text{MW} \times 10^6}{\sqrt{3} \times V_{RL} \times \cos\phi} = \frac{20 \times 10^6}{\sqrt{3} \times 66 \times 1{,}000 \times 0.8} = 218.7 \text{ A}$$

Taking receiving-end phase voltage as reference phasor, we have

$$\mathbf{V_R} = 38{,}105 \angle 0^\circ \text{ V} = (38{,}105 + j0) \text{ V}$$

Load current,

$$\mathbf{I_R} = 218.7\ (\cos\phi - j\sin\phi)$$
$$= 218.7(0.8 - j0.6) = (175 - j131.22) \text{ A} = 218.7 \angle -36.87^\circ \text{ A}$$

Line impedance per phase,

$$\mathbf{Z} = (10 + j35.1) = 36.5 \angle 74.1^\circ\ \Omega$$

Phase voltage across condenser,

$$\mathbf{V'} = \mathbf{V_R} + \frac{1}{2}\mathbf{I_R Z_R}$$
$$= (38{,}105) + \frac{1}{2} \times 218.7 \angle -36.87^\circ \times 36.5 \angle 74.1^\circ$$
$$= (41{,}283 + j2{,}415) \text{ V}$$

Charging current,

$$\mathbf{I_C} = \mathbf{V'} \times \mathbf{Y} = (41{,}283 + j2{,}415) \times j3.127 \times 10^{-4}$$
$$= (-0.75 + j12.91) \text{ A}$$

Sending-end current,

$$\mathbf{I_S} = \mathbf{I_R} + \mathbf{I_C} = (175 - j131.22) + (-0.75 + j12.91)$$
$$= (174.25 - 118.31) \text{ A} = 210.62 \angle -34.175^\circ \text{ A}$$

Sending-end voltage per phase,

$$\mathbf{V_S} = \mathbf{V} + \frac{1}{2}\mathbf{I_S Z}$$
$$= (41{,}283 + j2{,}415) + \frac{1}{2} \times 210.62 \angle -34.175^\circ \times 36.5 \angle 74.1^\circ$$
$$= (44{,}231 + j4{,}882) = 44{,}500 \text{ V} \angle 6.3^\circ \text{ V}$$

Sending-end line volts,

$$\mathbf{V_{SL}} = \sqrt{3}V_S = \sqrt{3} \times 44{,}500 = 77{,}075 \text{ V} \quad \textbf{Ans.}$$

Percentage regulation

$$= \frac{V_{SL} - V_{RL}}{V_{RL}} = \frac{77{,}075 - 66{,}000}{66{,}000} \times 100 = 16.78\% \quad \textbf{Ans.}$$

Transmission efficiency,

$$\eta = \frac{\sqrt{3} V_{RL} I_{RL} \cos\phi_R}{\sqrt{3} V_{SL} \times I_{SL} \times \cos\phi_S} \times 100$$

$$= \frac{\sqrt{3} \times 66{,}000 \times 218.7 \times 0.8}{\sqrt{3} \times 77{,}075 \times 210.62 \cos \angle 6.3 - (-34.175^\circ)} \times 100$$

$$= \frac{66{,}000 \times 218.7 \times 0.8}{77{,}075 \times 210.62 \times 0.7607} \times 100 = 93.5\% \quad \textbf{Ans.}$$

Example 6.13. A 3-phase 50 Hz transmission line has resistance, inductance and capacitance per phase of 10 Ω, 0.1 H and 0.9 μF and delivers a load of 35 MW at 132 kV and 0.8 pf lag. Determine the efficiency and regulation of the line using nominal-T method.

[*U.P.S.C. I.E.S. Electrical Engineering-II, 2008*]

Solution : Line resistance per phase, $R = 10\ \Omega$

Line reactance per phase

$$X = 2\pi f L = 2\pi \times 50 \times 0.1 = 31.4\ \Omega$$

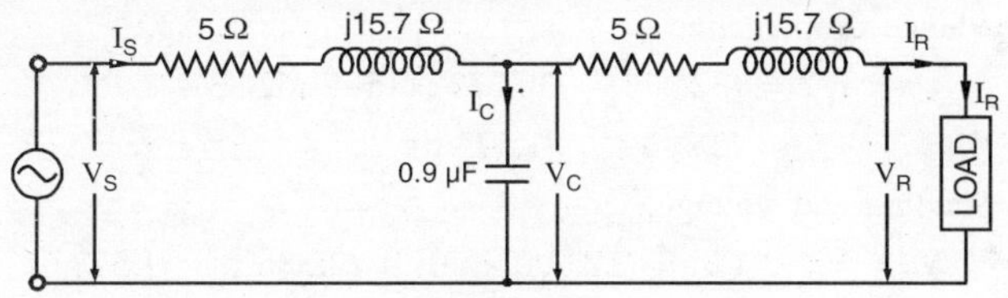

Fig. 6.18

Line susceptance per phase,

$$Y = 2\pi fC = 2\pi \times 50 \times 0.9 \times 10^{-6} = 0.282 \times 10^{-3} \text{ S}$$

Receiving-end power factor,

$$\text{Cos } \phi_R = 0.8 \text{ (lagging)}$$

$$\text{Sin } \phi_R = \sqrt{1-0.8^2} = -0.6$$

Receiving-end voltage, V_{RL} = 132 kV or 132,000 V

$$\text{Receiving-end phase voltage, } V_P = \frac{132{,}000}{\sqrt{3}} = 76{,}210 \text{ V}$$

$$\text{Load current, } I_R = \frac{35 \times 10^6}{\sqrt{3} \times 132{,}000 \times 0.8} = 191.36 \text{ A}$$

Line impedance, $Z = R + jX = (10 + j31.4)\ \Omega = 32.954 \angle 72.335°$

Taking receiving-end phase voltage as reference phasor, we have

$$\mathbf{V}_R = (76{,}210 + j0) \text{ volts or } 76{,}210 \angle 0° \text{ V}$$

Receiving-end current,

$$\mathbf{I}_R = 191.36 \angle -36.87° = (153 - j114.8) \text{ A}$$

Phase voltage across capacitor,

$$\mathbf{V}_C = \mathbf{V}_R + \mathbf{I}_R \times \frac{\mathbf{Z}}{2}$$

$$= 76{,}210 \angle 0° + 191.36 \angle -36.87° \times \frac{32.954}{2} \angle 72.335°$$

$$= (78{,}778 + j1{,}829.4) = 78{,}800 \angle 1.33° \text{ V}$$

Charging current,

$$\mathbf{I}_C = \mathbf{V}_C \times \mathbf{Y} = 78{,}800 \angle 1.33° \times 0.282 \times 10^{-3} \angle 90°$$

$$= 22.22 \angle 91.33° \text{ A} = (-0.52 + j22.2) \text{ A}$$

Sending-end current,

$$\mathbf{I}_S = \mathbf{I}_R + \mathbf{I}_C$$

$$= [(153 - j114.8) + (-0.52 + j22.2) \text{ A}$$

$$= (152.48 - j92.6) \text{ A} = 178.4 \angle -31.27° \text{ A}$$

Sending-end voltage per phase,

$$\mathbf{V}_S = \mathbf{V}_C + \mathbf{I}_S \frac{\mathbf{Z}}{2}$$

$$= 78{,}800 \angle 1.33° + 178.4 \angle -31.27° \times \frac{32.954}{2} \angle 72.335°$$

$$= 78{,}000 \angle 1.33° + 2{,}935.5 \angle 41.065°$$

$$= (78{,}778 + j1{,}829.4) + (2{,}213.6 + j1{,}928.4)$$

$$= (80{,}991.6 + j3{,}757.8) = 81{,}079 \angle 2.66° \text{ V}$$

Line loss

$$= 3\left[I_R^2 \times \frac{R}{2} + I_S^2 \times \frac{R}{2}\right]$$

$$= 3[191.36^2 \times 5 + 178.4^2 \times 5]$$

$$= 1{,}026{,}678 \text{ W or } 1.0267 \text{ MW}$$

$$\text{Transmission efficiency, } \eta = \frac{\text{Load in MW}}{\text{Load in MW + line loss}} \times 100$$

$$= \frac{35}{35 + 1.0267} \times 100 = 97.15\% \quad \textbf{Ans.}$$

$$\text{Regulation of the line} = \frac{V_S - V_R}{V_R} \times 100$$

$$= \frac{81{,}079 - 76{,}210}{76{,}210} \times 100 = 6.39\% \quad \textbf{Ans.}$$

Example 6.14. A 3-phase, 50 Hz, overhead transmission line delivers 10 MW at 0.8 pf lagging and at 66 kV. The resistance and inductive reactance of line per phase are 10 Ω and 20 Ω while capacitive admittance is 4 × 10⁻⁴ S. Calculate (*i*) sending-end voltage, (*ii*) sending-end current and power factor, (*iii*) transmission efficiency using nominal T-method.

[*M.D. Univ. Power System-I, Dec. 2010*]

Solution : Receiving-end voltage (line-to-line), V_{RL} = 66,000 V

$$\text{Receiving-end phase voltage, } V_R = \frac{66{,}000}{\sqrt{3}} = 38{,}105 \text{ V}$$

Receiving-end power factor, cos ϕ_R = 0.8 (lagging)

$$\phi_R = \text{Cos}^{-1}\ 0.8 = -36.87°$$

$$\text{Sin } \phi_R = \sin(-36.87°) = -0.6$$

$$\text{Receiving-end current, } I_R = \frac{10 \times 10^6}{\sqrt{3} \times 66 \times 10^3 \times 0.8} = 109.35 \text{ A}$$

Line impedance per phase $\mathbf{Z} = (10 + j20) = 22.36 \angle 63.435°\ \Omega$

Line susceptance per phase, $\mathbf{Y} = 0.4 \times 10^{-3}$ S

Taking receiving-end phase voltage as reference phasor, we have

$$\mathbf{V}_R = (38{,}105 + j0) \text{ or } 38{,}105 \angle 0° \text{ V}$$

Receiving-end current,

$$I_R = 109.35 \angle -36.87° \text{ A or } (86.68 - j65.01) \text{ A}$$

Phase voltage across capacitor,

$$\mathbf{V}_C \text{ or } \mathbf{V}'_R = \mathbf{V}_R + \mathbf{I}_R \times \frac{\mathbf{Z}}{2}$$

$$= 38{,}105 \angle 0° + 109.35 \angle -36.87° \times 11.18 \angle 63.435°$$

$$= 39{,}198.4 + j546.72 = 39{,}202 \angle 0.8° \text{ V}$$

Charging current,

$$\mathbf{I}_C = \mathbf{V}_C \times \mathbf{Y} = 39{,}202 \angle 0.8° \times 0.4 \times 10^{-3} \angle 90°$$

$$= 15.68 \angle 90.8° \text{ A} = (-0.22 + j15.68) \text{ A}$$

Sending-end current,

$$\mathbf{I}_S = \mathbf{I}_R + \mathbf{I}_C = (86.68 - j65.01) + (-0.22 + j15.68)$$

$$= (86.46 - j49.33) \text{ A or } 99.54 \angle -29.7° \text{ A}$$

Sending-end voltage per phase,

$$\mathbf{V}_S = \mathbf{V}_C + \mathbf{I}_S \times \frac{\mathbf{Z}}{2}$$

$$= (39{,}198.4 + j546.72) + 99.54 \angle -29.7° \times 11.18 \angle 63.435°$$

$$= (39{,}198.4 + j546.72) + 1{,}112.86 \angle 33.735^\circ \text{ V}$$

$$= (40{,}123.87 + j1{,}164.75) = 40{,}140.8 \angle 1.66^\circ \text{ V}$$

(*i*) Sending-end line-to-line voltage,

$$V_{SL} = \sqrt{3} \times 40{,}140.8 = 69{,}526 \text{ V or } 69.526 \text{ kV } \textbf{Ans.}$$

(*ii*) Sending-end current, $I_S = 99.54$ A **Ans.**

Sending-end power factor angle, $\phi_S = 1.66 - (-29.7^\circ) = 31.36^\circ$

Sending-end power factor,

$$= \cos \phi_S = \cos 31.36^\circ = 0.854 \text{ (lagging) } \textbf{Ans.}$$

$$\text{Line loss } = 3\left[I_R^2 + \frac{R}{2} + I_S^2 \times \frac{R}{2}\right]$$

$$= 3[109.35^2 \times 5 + 99.54^2 \times 5]$$

$$= 0.328 \times 10^6 \text{ W or } 0.328 \text{ MW}$$

(*iii*) Transmission efficiency,

$$\eta_T = \frac{10}{10 + 0.328} \times 100 = 96.8\% \ \textbf{Ans.}$$

Example 6.15. A single-phase transmission line delivers 1 MVA at a power factor of 0.71 lagging, 22 kV, 50 Hz. The loop resistance is 15 Ω, the loop inductance 0.2 H and the capacitance 0.5 μF. Find (*a*) the voltage (*b*) the current (*c*) the power factor at the sending end. Using nominal-π method. (*d*) If the sending-end voltage be maintained unaltered, to what value will the receiving-end voltage rise on no load ?

Solution : Load delivered, S = 1 MVA = 10^3 kVA

$$\text{Receiving-end current, } I_R = \frac{S}{V_R} = \frac{10^3 \times 1{,}000}{22 \times 1{,}000} = 45.455 \text{ A}$$

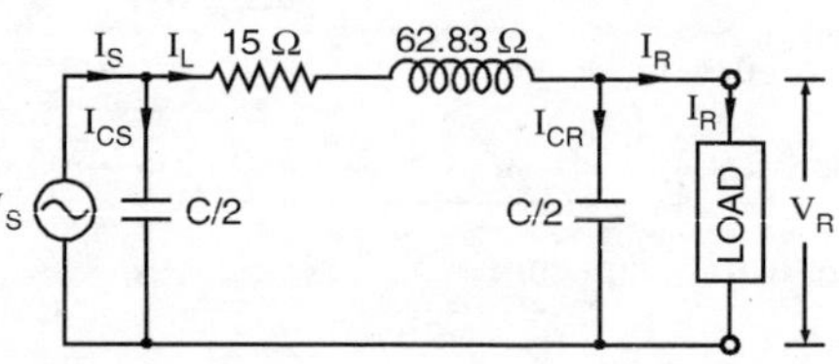

Fig. 6.19

Taking receiving-end voltage V_R as reference phasor, we have

$$\mathbf{V_R} = 22{,}000\ (1 + j0) \text{ V}$$

$$\mathbf{I_R} = 45.455(0.71 - j0.7042) = (32.273 - j32) \text{ A}$$

Inductive reactance of line,

$$X = 2\pi f L = 2\pi \times 50 \times 0.2 = 62.83\ \Omega$$

Line impedance,

$$Z = \sqrt{R^2 + X^2} = \sqrt{15^2 + 62.83^2} = 64.6^\circ \angle 76.57^\circ\ \Omega$$

Capacitive susceptance of line,

$$Y = 2\pi f C = 2\pi \times 50 \times 0.5 \times 10^{-6} = 1.5708 \times 10^{-4} \text{ S}$$

Charging current at the receiving end,

$$\mathbf{I_{CR}} = \frac{1}{2}\mathbf{Y}\mathbf{V_R} = \frac{1}{2}(0 + j1.5708 \times 10^{-4}) \times (22{,}000 + j0)$$

$$= j\,1.728 \text{ A}$$

Line current,

$$\mathbf{I_L} = \mathbf{I_R} + \mathbf{I_{CR}} = (32.273 - j32) + j1.728 = (32.273 - j30.272) \text{ A}$$

Impedance drop in line

$$= \mathbf{I_L Z} = (32.273 - j30.272)\ (15 + j62.83)$$

$$= (2{,}386.1 + j1{,}573.6) \text{ V}$$

(*a*) Sending-end voltage,

$$\mathbf{V_S} = \mathbf{V_R} + \mathbf{I_L}Z = (22{,}000 + j0) + (2{,}386.1 + j1{,}573.6)$$

$$= (24{,}386.1 + j1{,}573.6)$$

$$= 24{,}437 \angle 3.69^\circ \text{ V } \textbf{Ans.}$$

Charging current at the sending end,

$$\mathbf{I_{CS}} = \frac{1}{2}\mathbf{YV_S} = \frac{1}{2}(0 + j1.5708 \times 10^{-4})\ (24{,}386.1 + j1{,}573.6)$$

$$= (-0.124 + j1.915) \text{ A}$$

(*b*) Sending-end current,

$$\mathbf{I_S} = \mathbf{I_L} + \mathbf{I_{CS}}$$

$$= (32.273 - j30.272) + (-0.124 + j1.915)$$

$$= (32.149 - j28.357) = 42.868 \angle -41.4^\circ \text{ A } \textbf{Ans.}$$

Sending-end power factor angle,

$$\phi_S = 3.69^\circ - (-41.4^\circ) = 45.09^\circ$$

(*c*) Sending-end power factor

$$= \cos \phi_S = \cos 45.09^\circ = 0.706 \text{ (lag) } \textbf{Ans.}$$

(*d*) At no load, receiving-end voltage,

$$V_{RO} = \frac{24{,}437}{1 + \frac{1}{2}\mathbf{YZ}}$$

$$= \frac{24.437}{1 + \frac{1}{2} \times 1.5708 \times 10^{-4} \angle 90^\circ \times 64.6 \angle 76.57^\circ}$$

$$= 24{,}558 \text{ V or } 24.558 \text{ kV } \textbf{Ans.}$$

Example 6.16. A single-phase transmission line delivers 1 MVA at a power factor of 0.71 lagging, 22 kV, 50 Hz. The loop resistance is 15 Ω, the loop inductance 0.2 H and the capacitance 0.5 μF. Using nominal-π method find the sending-end voltage, current and power factor. [*U.P. Technical Univ. Elements of Power System 2005-06*]

Solution : Refer to Solution of Example 6.15.

Example 6.17. Using the nominal-π method, find the sending-end voltage and voltage regulation of 250 km, 3-phase, 50 Hz transmission line delivering 25 MVA at 0.8 power factor lagging to a balanced load at 132 kV. The line has a series impedance of 27.5 + *j*97.4 ohms and shunt admittance 7.38 × 10^{-4} mho. Neglect leakage. [*Pb. Technical Univ. Power System-I (T&D), May 2010*]

Solution : Phase voltage at the receiving end,

$$V_R = \frac{132}{\sqrt{3}} = 76.21 \text{ kV or } 76{,}210 \text{ V}$$

Power factor,

$$\cos \phi = 0.8 \text{ (lag) ; } \sin \phi = \sqrt{1 - 0.8^2} = 0.6$$

Receiving-end current,

$$I_R = \frac{\text{Load supplied in MVA} \times 10^6}{\sqrt{3}\, V_{RL}} = \frac{25 \times 10^6}{\sqrt{3} \times 132 \times 1{,}000}$$

$$= 109.35 \text{ A}$$

Taking phase voltage at receiving-end, V_R as reference phasor, we have

$$\mathbf{V_R} = (76{,}210 + j0) \text{ volts}$$

$$\mathbf{I}_R = 109.35\,(0.8 - j0.6) = (87.48 - j65.6)\text{ A}$$

Charging current at receiving-end,

$$\mathbf{I}_{CR} = \frac{1}{2}\mathbf{YV}_R = \frac{1}{2}(0 + j7.38 \times 10^{-4})\,(76{,}210 - j0)$$
$$= (0 + j\,28.12)\text{ A}$$

Line current,

$$\mathbf{I}_L = \mathbf{I}_R + \mathbf{I}_{CR} = (87.48 - j65.6) + (0 + j28.12)$$
$$= (87.48 - j\,37.48)\text{ A}$$

Sending-end voltage,

$$\mathbf{V}_S = \mathbf{V}_R + \mathbf{I}_L\mathbf{Z} = (76{,}210 + j0) + (87.48 - j37.48)\,(27.5 + j97.4)$$
$$= (82{,}266.3 + j7{,}590) = 82{,}615\ \angle 5.27^\circ\text{ V}$$

Line voltage at sending-end,

$$V_{SL} = \sqrt{3} \times 82{,}615 = 1{,}43{,}093\text{ V or } 143.1\text{ kV} \quad \textbf{Ans.}$$

$$1 + \frac{1}{2}\mathbf{YZ} = 1 + \frac{1}{2}(0 + j7.38 \times 10^{-4})\,(27.5 + j97.4)$$
$$= (0.964 + j\,0.01) = 0.964\ \angle 0.6^\circ$$

Line voltage at receiving end on no load,

$$V_{RLO} = \frac{V_{SL}}{1 + \frac{1}{2}\mathbf{YZ}} = \frac{143.1}{0.964} = 148.4\text{ kV}$$

Voltage regulation

$$= \frac{V_{RLO} - V_R}{V_R} \times 100 = \frac{148.4 - 132}{132} \times 100 = 12.42\% \quad \textbf{Ans.}$$

Example 6.18. A 50 Hz, 3-phase transmission line is 200 km long. It has a total series impedance of (35 + *j*140) Ω and shunt admittance of 930 × 10⁻⁶ ∠90° S. It delivers 40 MW power at 220 kV with 0.9 pf (lagging). Find the magnitude of the sending-end voltage. Consider nominal-π model of the line.

[U.P.S.C. I.E.S. Electrical Engineering-II, 2009]

Solution : Phase voltage at the receiving end,

$$V_R = \frac{220}{\sqrt{3}} = 127\text{ kV or } 127{,}000\text{ V}$$

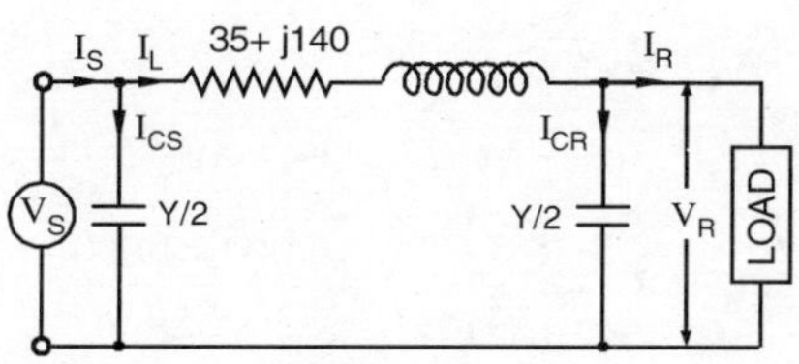

Fig. 6.20

Receiving end power factor,

$$\cos\phi_R = 0.9\text{ (lag)}$$
$$\phi_R = \cos^{-1} 0.9 = 25.84^\circ\text{ (lag)}$$
$$\sin\phi_R = \sin(-25.84^\circ) = -0.436$$

Receiving-end current,

$$I_R = \frac{\text{Load supplied in MW} \times 10^6}{\sqrt{3} \times V_{RL}\text{ in volts} \times \cos\phi_R}$$

$$= \frac{40 \times 10^6}{\sqrt{3} \times 220 \times 10^3 \times 0.9} = 116.64\text{ A}$$

Series impedance of line, $\mathbf{Z} = (35 + j140) = 144.3\ \angle 75.96^\circ\ \Omega$

Shunt admittance, $\mathbf{Y} = 930 \times 10^{-6}\ \angle 90^\circ\text{ S}$

Taking phase voltage at receiving end, $\mathbf{V}_R$ as reference phasor, we have

$$\mathbf{V}_R = 127{,}000\ \angle 0^\circ\text{ Volts}$$

Receiving-end current,

$$\mathbf{I}_R = 116.64\ \angle -25.84^\circ\text{ A} = (104.976 - j50.84)\text{ A}$$

Charging current at receiving end,

$$\mathbf{I}_{CR} = \frac{1}{2}\mathbf{Y}\,\mathbf{V}_R = \frac{1}{2} \times 930 \times 10^{-6}\ \angle 90^\circ \times 127{,}000\ \angle 0^\circ$$

$$= 59.055\ \angle 90^\circ\text{ A} = (0 + j59.055)\text{ A}$$

Line current,

$$\mathbf{I}_L = \mathbf{I}_R + \mathbf{I}_{CR}$$
$$= (104.976 - j50.84) + (0 + j59.055)$$
$$= (104.976 + j8.215) = 105.3\ \angle 4.48^\circ\text{ A}$$

Sending-end phase voltage,

$$\mathbf{V}_S = \mathbf{V}_R + \mathbf{I}_L Z$$
$$= 127{,}000\ \angle 0^\circ + 105.3\ \angle 4.48^\circ \times 144.3\ \angle 79.96^\circ$$
$$= 127{,}000\ \angle 0^\circ + 15{,}194.79\ \angle 84.44^\circ = 129{,}359\text{ V}$$

Sending-end line voltage,

$$V_{SL} = \sqrt{3} \times 129{,}359 = 224{,}000\text{ V or } 224\text{ kV} \quad \textbf{Ans.}$$

Example 6.19. A 3-phase transmission line, 100 km long has following constants.

Resistance per km per phase = 0.28 Ω; inductive reactance per km per phase 0.63 Ω; capacitive susceptance per km per phase 4 × 10⁻⁶ siemens. If the load at the receiving end is 75 MVA at pf 0.8 lagging with 132 kV between lines, calculate sending-end voltage, current and pf. Use nominal-π method.

[Pb. Technical Univ. Power System-I (T&D), Dec. 2006, Dec. 2008, May 2009]

Solution : Phase voltage at receiving-end,

$$V_R = \frac{132}{\sqrt{3}} = 76.21\text{ kV or } 76{,}210\text{ V}$$

Receiving-end power factor,

$$\cos\phi_R = 0.8\text{ lagging}$$
$$\phi_R = \cos^{-1} 0.8 = -36.87^\circ$$
$$\sin\phi_R = \sin(-36.87^\circ) = -0.6$$

Receiving-end current,

$$I_R = \frac{75 \times 10^6}{\sqrt{3} \times 132 \times 10^3 \times 0.8} = 410\ \angle -36.87^\circ\text{ A}$$

Resistance per phase,

$$R = 0.28 \times 100 = 28\ \Omega$$

Reactance per phase,

$$X = 0.63 \times 100 = 63\ \Omega$$

Impedance per phase,

$$\mathbf{Z} = (R + jX) = (28 + j63)\ \Omega = 68.942\ \angle 66°\ \Omega$$

Capacitive susceptance per phase,

$$\mathbf{Y} = 4 \times 10^{-6} \times 100 = 4 \times 10^{-4}\ S$$

Taking phase voltage at receiving end, V_R as reference phasor, we have

$$\mathbf{V_R} = 76{,}210\ \angle 0° \text{ volts or } (76{,}210 + j0) \text{ volts}$$

$$\mathbf{I_R} = 410\ \angle -36.87°\ A = (328 - j246)\ A$$

Charging current at receiving end,

$$\mathbf{I_{CR}} = \frac{1}{2}\mathbf{Y}\,\mathbf{V_R} = \frac{1}{2} \times 4 \times 10^{-4}\ \angle 90° \times 76{,}210\ \angle 0°$$

$$= 15.242\ \angle 90°\ A = (0 + j15.242)\ A$$

Line current, $\mathbf{I_L} = \mathbf{I_R} + \mathbf{I_{CR}}$

$$= (328 - j246) + (0 + j15.242)$$

$$= (328 - j230.758)\ A$$

$$= 160.83\ \angle -35.13°\ A$$

Sending-end phase voltage,

$$\mathbf{V_S} = \mathbf{V_R} + \mathbf{I_L Z}$$

$$= (76{,}210 + j0) + 160.83\ \angle -35.13° \times 68.942\ \angle 66°$$

$$= (76{,}210 + j0) + (9{,}517.2 + j5{,}689)$$

$$= 86{,}115\ \angle 3.79°\ V$$

Sending-end line-to-line voltage

$$= \sqrt{3} \times 86{,}115 = 149{,}000\ V \text{ or } 149\ kV$$ **Ans.**

Sending-end charging current,

$$\mathbf{I_{CS}} = \frac{1}{2}\mathbf{Y}\,\mathbf{V_S}$$

$$= \frac{1}{2} \times 4 \times 10^{-4}\ \angle 90° \times 86{,}115\ \angle 3.79°$$

$$= 17.223\ \angle 93.79°\ A = (-0.988 + j14.92)\ A$$

Sending-end current,

$$\mathbf{I_S} = \mathbf{I_L} + \mathbf{I_{CS}}$$

$$= (328 - j230.758) + (-0.988 + j14.92)$$

$$= (327.012 - j215.838)\ A = 391.82\ \angle -33.43°\ A$$

Sending-end current,

$$I_S = 391.82\ A$$ **Ans.**

Sending-end power factor

$$= \text{Cos}\ \phi_S = \text{Cos}\ [3.79 - (-33.43°)]$$

$$= \text{Cos}\ 37.22° = 0.7964 \text{ (lagging)}$$ **Ans.**

Example 6.20. A 20 km 3-phase transmission line delivers at the farthest end a 5,000 kVA load at 33 kV and 0.8 pf lagging and at the midpoint, a 10,000 kVA load at 0.7071 pf lagging. Neglecting the changes in line power factor and using approx formula, find the voltage at the midpoint and at the sending end and also the total current. The resistance and reactance per phase per km are 0.6 Ω and 1.5 Ω respectively.

Solution : Phase voltage at receiving end,

$$V_R = \frac{33}{\sqrt{3}} \times 1{,}000 = 19{,}052\ V$$

Power factor, $\cos\phi = 0.8$

$$\text{Sin}\ \phi = \sqrt{1 - 0.8^2} = 0.6$$

Receiving-end current,

$$I_1 = \frac{S_1}{\sqrt{3}\,V_{RL}} = \frac{5{,}000 \times 1{,}000}{\sqrt{3} \times 33 \times 1{,}000} = 87.5\ A$$

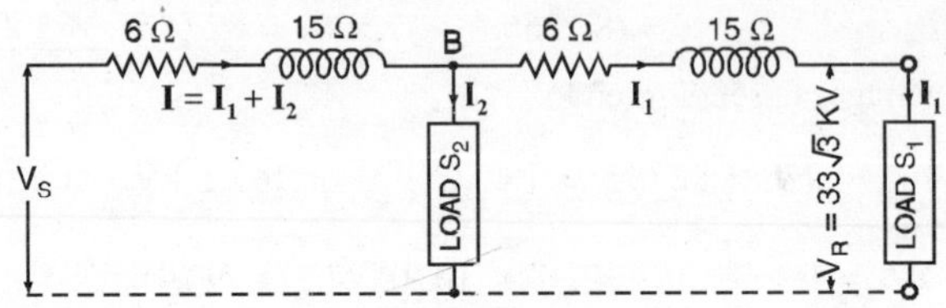

Fig. 6.21

Total resistance of line from receiving end to midpoint (*i.e.*, of 10 km length)

$$R_1 = 0.6 \times 10 = 6\ \Omega$$

Total reactance of line from receiving end to midpoint,

$$X_1 = 1.5 \times 10 = 15\ \Omega$$

Taking phase voltage at receiving end A, V_R as the reference phasor, we have

$$\mathbf{V_R} = (19{,}052 + j0) \text{ volts}$$

$$\mathbf{I_1} = 87.5\,(0.8 - j0.6) = (70 - j52.5)\ A$$

Voltage at the midpoint B,

$$\mathbf{V_B} = \mathbf{V_R} + \mathbf{I_1 Z_1}$$

$$= (19{,}052 + j0) + (70 - j52.5)\,(6 + j15)$$

$$= (20{,}259.5 + j735) = 20{,}273\ \angle 2.1° \text{ volts}$$

Line voltage at midpoint,

$$V_{BL} = \sqrt{3}\,V_B = \sqrt{3} \times 20{,}273$$

$$= 35{,}114\ V \text{ or } 35.114\ kV$$ **Ans.**

$$\text{Current, } I_2 = \frac{S_2}{\sqrt{3}\,V_{BL}} = \frac{10{,}000 \times 1{,}000}{\sqrt{3} \times 35{,}114} = 164.4\ A$$

Neglecting changes in line power factor,

$$\mathbf{I_2} = 164\,(\cos\phi_2 - j\sin\phi_2) = 164.4\,(0.7071 - j0.7071)$$

$$= (116.26 - j116.26)\ A$$

Total current flowing in the line from midpoint to sending end,

$$\mathbf{I} = \mathbf{I_1} + \mathbf{I_2} = (70 - j52.5) + (116.26 - j116.26)$$

$$= (186.26 - j168.76)\ A = 251.35\ \angle -42.2°\ A$$ **Ans.**

Impedance of line from midpoint to sending end,

$$\mathbf{Z_2} = (R_2 + jX_2) = (10 \times 0.6 + j\,10 \times 1.5) = (6 + j15)\ \Omega$$

Voltage at the sending end,

$$\mathbf{V_S} = \mathbf{V_B} + \mathbf{IZ_2}$$

$$= (20{,}259.5 + j735) + (186.26 - j168.76)\ (6 + j15)$$

$$= 23{,}908.5 + j2{,}516.3 = 24{,}040\ \angle 6^\circ\ \text{V}$$

Line voltage at sending end,

$$V_{SL} = \sqrt{3}\ V_S = \sqrt{3} \times 24{,}040 = 41{,}640 \text{ V or } 41.64 \text{ kV} \textbf{ Ans.}$$

Example 6.21. Two overhead lines are connected in parallel to supply a load of 10 MW at 0.8 pf (lagging) and 30 kV. The resistance and reactance of one line (A) are 5.5 Ω and 13.5 Ω, respectively; those of the other line (B) are 6 Ω and 11 Ω respectively. Calculate *(a)* the kVA supplied by each line *(b)* the power supplied by each line. *[I.E.E. Pt. II May, 1934]*

Solution : Impedance of line A,

$$\mathbf{Z}_A = 5.5 + j13.5 = 14.58\ \angle 67^\circ 50' \text{ ohms}$$

Impedance of line B,

$$\mathbf{Z}_B = 6 + j11.0 = 12.53\ \angle 61^\circ 24' \text{ ohms}$$

$$\mathbf{Z}_A + \mathbf{Z}_B = (5.5 + j13.5) + (6 + j11.0)$$

$$= (11.5 + j24.5) = 27.06\ \angle 64^\circ 51' \text{ ohms}$$

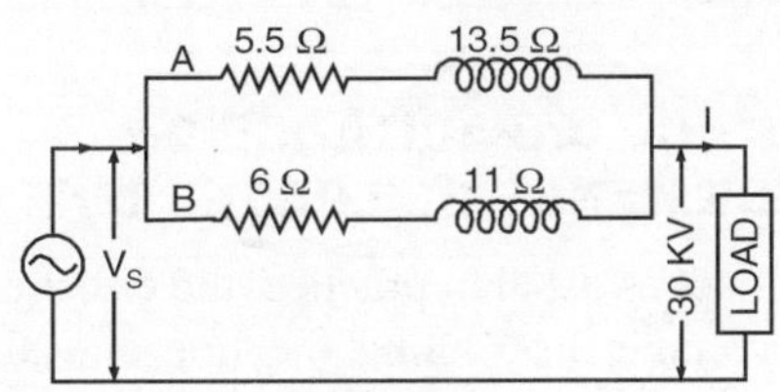

Fig. 6.22

Total load supplied in kVA,

$$\mathbf{S} = \frac{10{,}000}{0.8} \angle -36^\circ 52' = 12{,}500 \angle -36^\circ 52' \text{ KVA}$$

(a) Load supplied by line **A**,

$$\mathbf{S}_A = \mathbf{S} \cdot \frac{\mathbf{Z}_B}{\mathbf{Z}_A + \mathbf{Z}_B}$$

$$= 12{,}500 \angle -36^\circ 52' \times \frac{12.53 \angle 61^\circ 24'}{27.06 \angle 64^\circ 51'}$$

$$= \frac{12{,}500 \times 12.53}{27.06} \angle -36^\circ 52' + 61^\circ 24' - 64^\circ 51'$$

$$= 5{,}790 \angle -40^\circ 19' \text{ kVA}$$

$$= 5{,}790 \text{ kVA at pf } 0.763 \text{ (lag)} \textbf{ Ans.}$$

Load supplied by line **B**,

$$\mathbf{S}_B = \mathbf{S} \cdot \frac{\mathbf{Z}_A}{\mathbf{Z}_A + \mathbf{Z}_B}$$

$$= 12{,}500 \angle -36^\circ 52' \times \frac{14.58 \angle 67^\circ 50'}{27.06 \angle 64^\circ 51'}$$

$$= \frac{12{,}500 \times 14.58}{27.06} \angle -36^\circ 52' + 67^\circ 50' - 64^\circ 51'$$

$$= 6{,}730 \angle -33^\circ 53' \text{ kVA} = 6{,}730 \text{ kVA at pf } 0.83 \text{ (lag)} \textbf{ Ans.}$$

(b) Power supplied by line **A**

$$= 5{,}790 \times 0.763 = 4{,}415 \text{ kW} \textbf{ Ans.}$$

Power supplied by line **B**

$$= 6{,}730 \times 0.83 = 5{,}585 \text{ kW} \textbf{ Ans.}$$

6.7. FERRANTI EFFECT

In a medium or long transmission line when open-circuited or lightly loaded the receiving-end voltage is found more than the sending-end voltage. *This phenomenon of rise in voltage at the receiving end of the open-circuited or lightly loaded line is called the Ferranti effect* owing to its first being observed on the Deptford mains laid down by S.Z. de Ferranti in 1890. This effect is due to voltage drop across the line inductance, due to charging current, being in phase with the applied voltage at the sending end of the line. Thus both capacitance and inductance are necessary to cause this phenomenon. The capacitance, and therefore, charging current is negligible in short lines but significant in medium and long transmission lines. On a 300 km line operating on 50 Hz supply, the receiving-end voltage on open circuit is usually found to be roughly 5% higher than the sending-end voltage.

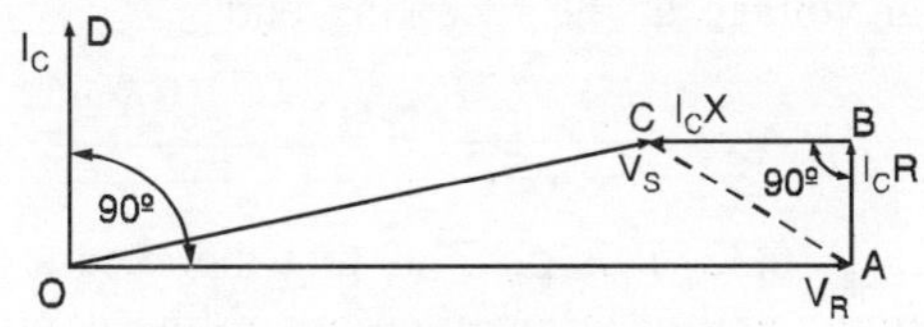

Fig. 6.23. ***Ferranti Effect***

Phasor diagram given in Fig. 6.23 illustrates the Ferranti effect.

Taking receiving-end phase voltage as reference phasor, we have

$$\mathbf{V}_R = V_R(1 + j0) \text{ represented by phasor OA}$$

Charging current,

$$\mathbf{I}_C = j\omega C V_R \text{ represented by phasor OD.}$$

Voltage at sending end,

$$\mathbf{V}_S = \mathbf{V}_R + \text{resistive drop in line} + \text{reactive drop in line}$$

$$= \mathbf{V}_R + \mathbf{I}_C R \text{ (represented by phasor AB)}$$

$$+ j\mathbf{I}_C X \text{ (represented by phasor BC)}$$

$$= \mathbf{V}_R + \mathbf{I}_C(R + jX) = \mathbf{V}_R + j\omega C V_R(R + j\omega L)$$

$$\text{or } \mathbf{V}_S = \mathbf{V}_R - \omega^2 C L V_R + j\omega C R \mathbf{V}_R \text{ represented by phasor OC.}$$

The overhead transmission lines having long length to bring the effect into prominence usually have a fairly large reactance compared with resistance, and by taking advantage of this fact a simple approximate formula for determining the magnitude of the effect can be derived.

Neglecting resistive drop $\mathbf{I}_C\mathbf{R}$, we get

Rise in voltage = **OA** – **OC**

= Inductive drop in line (refer to Fig. 6.23)

If $\mathbf{C}_0$ and $\mathbf{L}_0$ are the capacitance and inductance of the transmission line per km, length and l is the length of line in km, the capacitive reactance, $\mathbf{X}_C = \frac{1}{\omega l C_0}$.

As the capacitance of the line is uniformly distributed over the entire length of the line, therefore, the average current flowing throughout the line

$$I_C = \frac{1}{2}\frac{V_R}{X_C} = \frac{1}{2}V_R C_0 \omega l$$

The reactance of the transmission line,

$$X = \omega L_0 l$$

Rise in voltage

$$= I_C X = \frac{1}{2}V_R C_0 \omega l \times \omega L_0 l = \frac{1}{2}\omega^2 l^2 V_R C_0 L_0 \text{ volts}$$

The quantity $\frac{1}{\sqrt{L_0 C_0}}$ is constant for all overhead lines and is equal to the velocity of propagation of electromagnetic waves. It is nearly equal to the speed of light *i.e.*, 3×10^5 km/s. So

Rise in voltage at the receiving end

$$= \frac{1}{2}\omega^2 l^2 V_R \times \frac{1}{(3 \times 10^5)^2} = \frac{\omega^2 l^2 V_R \times 10^{-10}}{18} \text{ V} \quad ...(6.32)$$

From Eq. (6.32) it is obvious that the excess of voltage at the receiving end of an open-circuited line is proportional to the square of the length of the line. It is found that quite a small load eliminates this effect, and where there are transformers in the circuit at the receiving end, the magnetising current acts in such a way.

For long high voltage and extra high voltage lines shunt reactors are provided to absorb part of charging current or shunt capacitive VA, of the line on no load and light load, in order to prevent overvoltages on the line.

Example 6.22. Determine the percentage rise in voltage at the receiving end of a transmission line of length 200 km operating at 50 Hz.

Solution : Rise in voltage $= \frac{\omega^2 l^2 V_R \times 10^{-10}}{18}$ volts

Percentage rise in voltage

$$= \frac{\omega^2 l^2 V_R \times 10^{-10}}{18 \times V_R} \times 100\%$$

$$= \frac{\omega^2 l^2 \times 10^{-8}}{18} = \frac{(2\pi \times 50)^2 \times (200)^2 \times 10^{-8}}{18} = 2.19\% \textbf{ Ans.}$$

6.8. CHARGING CURRENT AND LINE LOSSES IN OPEN-CIRCUITED LINE

In transmission line operation the voltage at the receiving end is usually maintained constant irrespective of the magnitude of the load of the system. The charging current of an open-circuited line is the amount of current flowing into the line at the sending end with normal voltage and no load at the receiving end. The calculation of charging current becomes much simplified if it is assumed that voltage throughout the length of the line is same *i.e.*, there is no voltage drop in the line.

Making the assumption of voltage equality along the line, the actual value of charging current will fall uniformly from its maximum value at the sending end to zero at the receiving end.

Charging current at a distance x km from the sending-end of a l km long line is given as $I_C \frac{l-x}{l}$ where I_C is the maximum value of charging current.

The average value of the square of this quantity

$$= \frac{1}{l}\int_0^l \frac{I_C^2 (l-x)^2}{l^2} = \frac{I_C^2}{l^3}\int_0^l (l-x)^2\, dx = \frac{I_C^2}{l^3}\left[\frac{(l-x)^3}{3\times(-1)}\right]_0^l = \frac{I_C^2}{3} \text{ A}^2$$

RMS value of charging current,

$$I_{C_{rms}} = \sqrt{\frac{I_C^2}{3}} = \frac{I_C}{\sqrt{3}} \text{ A} \quad ...(6.33)$$

Power loss in each of the line conductor

$$= \left(\frac{I_C}{\sqrt{3}}\right)^2 R = \frac{1}{3} I_C^2 R \text{ watts} \quad ...(6.34)$$

6.9. EFFECT OF CAPACITANCE ON PERFORMANCE OF LOADED LINE

On inductive load, as usual in practice, the effect of capacitance is to reduce voltage drop along the line and also to reduce the copper losses of the line. The magnitude of this effect depends on the phase angle at the receiving end.

In any line, the current flowing through it can be split up into two components, the load current and the charging current. The effect of load current with lagging or unity power factor is to make the receiving-end voltage lower than sending-end voltage whereas the charging current tends to raise the receiving-end voltage. As a result, the voltage drop along the line is diminished by the amount given by the Eq. (6.32).

In case of short transmission lines the regulation and voltage drop are numerically equal. But in medium and long transmission lines, where capacitance effects have to be taken into account, the regulation is greater than voltage drop by the rise in voltage due to Ferranti effect.

The presence of capacitance also influences the power losses of the system by an amount depending on the relative values of the charging current and the quadrature component of the load current. The charging current tends to neutralize the quadrature component (lagging) of the load current, consequently the line current is reduced and, therefore, line losses are reduced. It can be easily seen that reduction in line current and line losses takes place so long as the receiving-end charging current is less than twice the quadrature component of the load current.

At unity power factor of the load there will be no quadrature component of the load current and so the effect of charging current in a line supplying a non-inductive load is always to increase the line losses. With small highly inductive load, the charging current may exceed twice the quadrature component of load current and give increased line losses.

6.10. GENERALISED CIRCUIT CONSTANTS

In any passive, bilateral and linear network with two input and two output terminals, the input voltage and the input current can be expressed in terms of output voltage and output current. Incidently, a transmission line is a 4-terminal network; two input terminals where power enters the network, and two output terminals where the power leaves the network. Such a circuit is passive as it does not contain any source of emf, linear as its impedance is independent of current flowing and bilateral as its impedance is independent of direction of current flowing.

The input voltage per phase (voltage at sending end) and the input current (current at sending end) of a transmission line can be expressed as

$$\mathbf{V}_S = \mathbf{AV}_R + \mathbf{BI}_R \quad ...(6.35)$$

$$\mathbf{I}_S = \mathbf{CV}_R + \mathbf{DI}_R \quad ...(6.36)$$

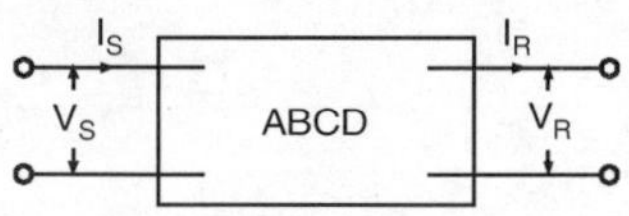

Fig. 6.24

where $\mathbf{V}_S$ is the sending-end voltage per phase, $\mathbf{I}_S$ is the sending-end current, $\mathbf{V}_R$ is the receiving-end voltage per phase, $\mathbf{I}_R$ is the receiving-end current, and **A**, **B**, **C** and **D** (generally complex numbers) are the constants, called the *generalised circuit constants* of a transmission line. The values of these constants A, B, C and D depend upon the particular method of solution employed. Once the values of these constants for a particular transmission line are known, the performance of the line can be determined easily.

The values of A, B, C and D can be determined as follows :

With receiving end open circuited

Receiving-end current $\mathbf{I}_R$ is zero, therefore, from Eqs. (6.35) and (6.36)

$$\mathbf{A} = \mathbf{V}_S/\mathbf{V}_R \quad ...(6.37)$$

and $$\mathbf{C} = \mathbf{I}_S/\mathbf{V}_R \quad ...(6.38)$$

With receiving end short circuited

Receiving-end voltage $\mathbf{V}_R$ is zero, therefore, from Eqs. (6.35) and (6.36)

$$\mathbf{B} = \mathbf{V}_S/\mathbf{I}_R \quad ...(6.39)$$

and $$\mathbf{D} = \mathbf{I}_S/\mathbf{I}_R \quad ...(6.40)$$

From the above expressions **A**, **B**, **C** and **D** can be defined as follows :

A can be defined as the ratio of sending-end voltage $\mathbf{V}_S$ to receiving-end voltage $\mathbf{V}_R$ when the line is open circuited on receiving end. **A** is dimensionless and usually a complex number.

B is defined as the ratio of sending-end voltage $\mathbf{V}_S$ to receiving-end current $\mathbf{I}_R$ when the line is short circuited on receiving end. **B** is also the complex number and has the unit of ohm.

C is defined as the ratio of sending-end current $\mathbf{I}_S$ to receiving-end voltage $\mathbf{V}_R$ when the line is open circuited on the receiving end. **C** is also the complex number and has the unit of *siemen*.

D is defined as the ratio of sending-end current $\mathbf{I}_S$ to receiving-end current $\mathbf{I}_R$ when the line is short circuited on the receiving end. **D** is dimensionless and usually a complex number

For passive network $$\mathbf{AD} - \mathbf{BC} = 1 \quad ...(6.41)$$

and in symmetrical network $$\mathbf{A} = \mathbf{D} \quad ...(6.42)$$

The values of generalised constants **A**, **B**, **C** and **D** for various methods of solution adopted are determined as follows :

(*i*) *Short Lines (Simple Series Impedance Method).* In this method the effect of the line capacitance is neglected and, therefore, the line can be considered to have series impedance. The circuit diagram is shown in Fig. 6.2 (*a*). Here we have

$$\mathbf{I}_S = \mathbf{I}_R \quad ...(6.43)$$

and $$\mathbf{V}_S = \mathbf{V}_R + \mathbf{I}_R\mathbf{Z} \quad ...(6.44)$$

Comparing Eq. (6.44) with Eq. (6.35) and Eq. (6.43) with Eq. (6.36), we have

$$\mathbf{A} = 1;\ \mathbf{B} = \mathbf{Z};\ \mathbf{C} = 0 \text{ and } \mathbf{D} = 1$$

The phasor diagram is shown in Fig. 6.2 (*b*)

Incidently $\mathbf{A} = \mathbf{D}$ and $\mathbf{AD} - \mathbf{BC} = 1 \times 1 - \mathbf{Z} \times 0 = 1$

(*ii*) *End Condenser Method*. In this method the capacitance of the line is assumed to be lumped at the load end, as shown in Fig. 6.10. This method overestimates the effect of capacitance.

Here we have $$\mathbf{I}_S = \mathbf{I}_C + \mathbf{I}_R = \mathbf{Y}\,\mathbf{V}_R + \mathbf{I}_R \quad ...(6.45)$$

and $$\mathbf{V}_S = \mathbf{V}_R + \mathbf{Z}\,(\mathbf{I}_C + \mathbf{I}_R)$$

$$= \mathbf{V}_R + \mathbf{ZI}_R + \mathbf{ZYV}_R$$

$$= \mathbf{V}_R(1 + \mathbf{ZY}) + \mathbf{ZI}_R \quad ...(6.46)$$

Comparing Eq. (6.46) with Eq. (6.35) and Eq. (6.45) with Eq. (6.36), we have

$$\mathbf{A} = 1 + \mathbf{ZY};\ \mathbf{B} = \mathbf{Z};\ \mathbf{C} = \mathbf{Y} \text{ and } \mathbf{D} = 1$$

The phasor diagram is shown in Fig. 6.11.

Incidently $\mathbf{AD} - \mathbf{BC} = (1 + \mathbf{ZY}) \times 1 - \mathbf{ZY} = 1$

(*iii*) *Nominal-T Method.* In this method the capacitance of each conductor is assumed to be concentrated at the mid-point, as shown in Fig. 6.12. The phasor diagram is shown in Fig. 6.13.

Here we have

$$\mathbf{V}_S = \mathbf{V}' + \mathbf{I}_S\frac{\mathbf{Z}}{2} \quad ...(6.47)$$

and $$\mathbf{V}' = \mathbf{V}_R + \mathbf{I}_R\frac{\mathbf{Z}}{2} \quad ...(6.48)$$

and $$\mathbf{I}_S = \mathbf{I}_R + \mathbf{I}_C$$

$$= \mathbf{I}_R + \mathbf{YV}'$$

$$= \mathbf{I}_R + \mathbf{Y}\left(\mathbf{V}_R + \mathbf{I}_R\frac{\mathbf{Z}}{2}\right)$$

$$= YV_R + I_R\left(1+\frac{YZ}{2}\right) \quad \text{...(6.49)}$$

Substituting the value of **V′** from Eq. (6.48) in Eq. (6.47), we have

$$V_S = V_R + I_R\frac{Z}{2} + I_S\frac{Z}{2}$$

$$= V_R + I_R\frac{Z}{2} + \frac{Z}{2}\left[YV_R + I_R\left(1+\frac{YZ}{2}\right)\right]$$

$$= \left(1+\frac{YZ}{2}\right)V_R + \left(Z+\frac{YZ^2}{4}\right)I_R \quad \text{...(6.50)}$$

Comparing Eq. (6.50) with Eq. (6.35) and Eq. (6.49) with Eq. (6.36), we have

$$A = D = 1+\frac{YZ}{2};\ B = Z\left(1+\frac{YZ}{4}\right) \text{ and } C = Y$$

Incidently $AD - BC = \left(1+\frac{YZ}{2}\right)^2 - ZY\left(1+\frac{YZ}{4}\right) = 1$

(*iv*) *Nominal-π Method.* In this method the capacitance of each conductor is assumed to be divided into two halves, one half being shunted between conductor and neutral at the receiving end and the other half at the sending end, as shown in Fig. 6.14. The phasor diagram is shown in Fig. 6.15.

Here we have

$$I_L = I_R + \frac{Y}{2}V_R \quad \text{...(6.51)}$$

and $I_S = I_L + \frac{Y}{2}V_S$ ...(6.52)

and $V_S = V_R + I_L Z$ ...(6.53)

Substituting $I_L = I_R + \frac{1}{2}YV_R$ from Eq. (6.51), we have

$$V_S = V_R + Z\left(I_R + \frac{1}{2}YV_R\right)$$

$$= V_R\left(1+\frac{1}{2}YZ\right) + I_R Z \quad \text{...(6.54)}$$

From Eqs. (6.51) and (6.52), we have

$$I_S = I_R + \frac{1}{2}YV_R + \frac{1}{2}YV_S \quad \text{...(6.55)}$$

Substituting the value of V_S from Eq. (6.54) in Eq. (6.55), we have

$$I_S = I_R + \frac{1}{2}YV_R + \frac{1}{2}Y\left[\left(1+\frac{1}{2}YZ\right)V_R + I_R Z\right]$$

$$= V_R Y\left(1+\frac{1}{4}YZ\right) + I_R\left(1+\frac{1}{2}YZ\right) \quad \text{...(6.56)}$$

Comparing Eq. (6.54) with Eq. (6.35) and Eq. (6.56) with Eq. (6.36), we have

$$A = D = 1 + \frac{1}{2}YZ;\ B = Z \text{ and } C = Y\left(1+\frac{1}{4}YZ\right)$$

Incidently **AD – BC**

$$= \left(1+\frac{1}{2}YZ\right)^2 - ZY\left(1+\frac{1}{4}YZ\right) = 1$$

6.10.1. Proof For Relation AD – BC = 1

Consider a two terminal pair network with parameters **A**, **B**, **C** and **D** connected to an ideal voltage source* at one end and short circuited at other end (Fig. 6.25).

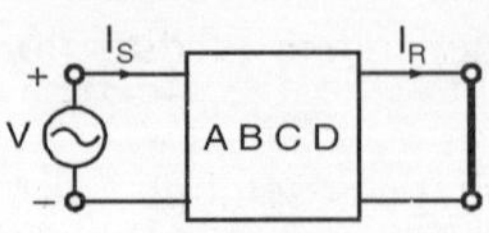

Fig. 6.25

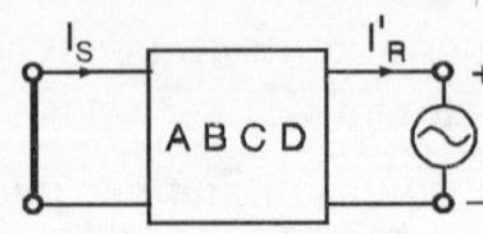

Fig. 6.26

Receiving-end voltage, $V_R = 0$

Now from Eq. (6.35), we have

$$V = A \times 0 + BI_R$$

or $I_R = \frac{V}{B}$ ...(6.57)

Now let the sending-end terminals be short circuited and the voltage source be connected to receiving-end terminals, as shown in Fig. 6.26.

Now from Eq. (6.35), we have

$$0 = AV + BI'_R \quad \text{...(6.58)}$$

and from Eq. (6.36)

$$I_S = CV + DI'_R$$

or $-I_R = CV + DI'_R$ ...(6.59)

because $I_S = -I_R$ as transmission line is a linear bilateral network.

Substituting the values of I_R and I'_R from Eqs. (6.57) and (6.58) respectively in above Eq. (6.59), we have

$$\frac{-V}{B} = CV + D\frac{(-AV)}{B}$$

or $-1 = BC - AD$

or $AD - BC = 1$ ...(6.60)

6.10.2. Two Networks in Tandem

Two networks** are said to be connected in tandem when the output of one network is connected to the input of the other network, as illustrated in Fig. 6.27. Let the two networks having constants A_1, B_1, C_1, D_1 and A_2, B_2, C_2, D_2 be connected in series

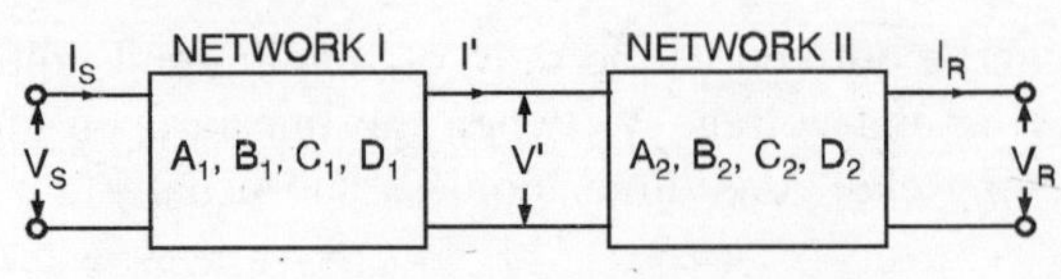

Fig. 6.27. *Two Networks in Tandem*

* Voltage source with zero internal impedance is called the ideal voltage source.

** Two networks can be two transmission lines or a transformer connected to a transmission line etc.

For network I, $\mathbf{V_S = A_1V' + B_1I'}$ ...(6.61)

and $\mathbf{I_S = C_1V' + D_1I'}$ ...(6.62)

For network II, $\mathbf{V' = A_2V_R + B_2I_R}$ ...(6.63)

and $\mathbf{I' = C_2V_R + D_2I_R}$ ...(6.64)

Substituting the values of **V′** and **I′** from Eqs. (6.63) and (6.64) in Eqs. (6.61) and (6.62) respectively, we have

$$\mathbf{V_S = A_1(A_2V_R + B_2I_R) + B_1(C_2V_R + D_2I_R)}$$
$$\mathbf{= (A_1A_2 + B_1C_2)V_R + (A_1B_2 + B_1D_2)I_R} \quad ...(6.65)$$

and
$$\mathbf{I_S = C_1(A_2V_R + B_2I_R) + D_1(C_2V_R + D_2I_R)}$$
$$\mathbf{= (C_1A_2 + D_1C_2)V_R + (C_1B_2 + D_1D_2)I_R} \quad ...(6.66)$$

Therefore, for the two networks in tandem, the generalised constants of the equivalent network are

$$\left.\begin{aligned} \mathbf{A} &= \mathbf{A_1A_2 + B_1C_2} \\ \mathbf{B} &= \mathbf{A_1B_2 + B_1D_2} \\ \mathbf{C} &= \mathbf{A_2C_1 + C_2D_1} \\ \mathbf{D} &= \mathbf{B_2C_1 + D_1D_2} \end{aligned}\right\} \quad ...(6.67)$$

6.10.3. Two Networks in Parallel

Let the two networks be connected in parallel, as illustrated in Fig. 6.28, and the currents at the sending end and receiving end be $\mathbf{I_{S_1}}$ and $\mathbf{I_{R_1}}$ for network **I** and $\mathbf{I_{S_2}}$ and $\mathbf{I_{R_2}}$ for network **II**

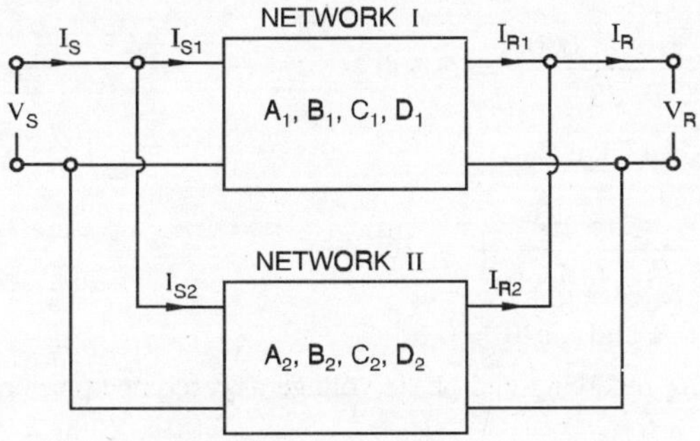

Fig. 6.28. *Two Networks in Parallel*

Here $\mathbf{I_S = I_{S_1} + I_{S_2}}$ ...(6.68)

$\mathbf{I_R = I_{R_1} + I_{R_2}}$ ...(6.69)

For network I, $\mathbf{V_S = A_1V_R + B_1I_{R_1}}$ ...(6.70)

and $\mathbf{I_{S_1} = C_1V_R + D_1I_{R_1}}$ ...(6.71)

For network II, $\mathbf{V_S = A_2V_R + BI_{R_2}}$ ...(6.72)

$\mathbf{I_{S_2} = C_2V_R + D_2I_{R_2}}$ ...(6.73)

Comparing Eqs. (6.70) and (6.72), we have

$$\mathbf{A_1V_R + B_1I_{R_1} = A_2V_R + B_2I_{R_2}}$$

or $\mathbf{(A_1 - A_2)V_R = B_2I_{R_2} - B_1I_{R_1}}$

or $\mathbf{(A_1 - A_2)V_R = B_2(I_R - I_{R_1}) - B_1I_{R_1}} \quad \because \mathbf{I_{R_1} + I_{R_2} = I_R}$

or $\mathbf{(A_1 - A_2)V_R = B_2I_R - I_{R_1}(B_1 + B_2)}$

or
$$\mathbf{I_{R_1} = \frac{B_2I_R - (A_1 - A_2)V_R}{B_1 + B_2}} \quad ...(6.74)$$

Substituting the value of $\mathbf{I_{R_1}}$ from above Eq. (6.74) in Eq. (6.70), we have

$$\mathbf{V_S = A_1V_R + \frac{B_1}{B_1 + B_2}[B_2I_R - (A_1 - A_2)V_R]}$$
$$\mathbf{= \left[A_1 - \frac{B_1(A_1 - A_2)}{B_1 + B_2}\right]V_R + \frac{B_1B_2}{B_1 + B_2}I_R}$$
$$\mathbf{= \frac{A_1B_2 + B_1A_2}{B_1 + B_2}V_R + \frac{B_1B_2}{B_1 + B_2}I_R} \quad ...(6.75)$$

Substituting the values of $\mathbf{I_{S_1}}$ and $\mathbf{I_{S_2}}$ from Eqs. (6.71) and (6.73) in Eq. (6.68), we have

$$\mathbf{I_S = C_1V_R + D_1I_{R_1} + C_2V_R + D_2I_{R_2}}$$
$$\mathbf{= (C_1 + C_2)V_R + D_1I_{R_1} + D_2(I_R - I_{R_1})}$$
$$\mathbf{= (C_1 + C_2)V_R + (D_1 - D_2)I_{R_1} + D_2I_R}$$

Substituting the value of $\mathbf{I_{R_1}}$ from Eq. (6.74), we have

$$\mathbf{I_S = (C_1 + C_2)V_R + (D_1 - D_2)\frac{B_2I_R - (A_1 - A_2)V_R}{B_1 + B_2} + D_2I_R}$$
$$\mathbf{= \left[C_1 + C_2 - \frac{(A_1 - A_2)(D_1 - D_2)}{B_1 + B_2}\right]V_R + \frac{B_1D_2 + D_1B_2}{B_1 + B_2}I_R} \quad ...(6.76)$$

Therefore, for the two networks connected in parallel, the generalised constants of the equivalent network are

$$\left.\begin{aligned} \mathbf{A} &= \mathbf{\frac{A_1B_2 + A_2B_1}{B_1 + B_2}} \\ \mathbf{B} &= \mathbf{\frac{B_1B_2}{B_1 + B_2}} \\ \mathbf{C} &= \mathbf{C_1 + C_2 - \frac{(A_1 - A_2)(D_1 - D_2)}{B_1 + B_2}} \\ \mathbf{D} &= \mathbf{\frac{B_1D_2 + B_2D_1}{B_1 + B_2}} \end{aligned}\right\} \quad ...(6.77)$$

Example 6.23. Find the A, B, C, D parameters of a 3-phase 80 km 50 Hz transmission line with series impedance of 0.15 + j 0.78 ohm per km and a shunt capacitance of $j5 \times 10^{-6}$ mho per km. [*Pb. Technical Univ. Power System-I (T&D) May 2008*]

Solution : Total series impedance of the line,

$$\mathbf{Z} = (0.15 + j0.78) \times 80 = (12 + j62.4)\ \Omega$$

Total shunt capacitance of the line

$$\mathbf{Y} = j5 \times 10^{-6} \times 80 = j4 \times 10^{-4} \text{ mho}$$

In nominal-T configuration

$$\mathbf{A = D} = 1 + \frac{1}{2}\mathbf{YZ}$$
$$= 1 + \frac{1}{2}(12 + j62.4) \times j4 \times 10^{-4}$$
$$= (0.98752 + j\,0.0024) \quad \textbf{Ans.}$$

$$\mathbf{B} = \mathbf{Z} \times \left(1 + \frac{\mathbf{YZ}}{4}\right)$$

$$= (12 + j62.4)\left[1 + \frac{1}{4} \times j4 \times 10^{-4} \times (12 + j62.4)\right]$$

$$= (12 + j62.4)\,(0.99376 + j0.0012)$$

$$= (11.85024 + j11.91072) \text{ ohms } \textbf{Ans.}$$

$$\mathbf{C} = \mathbf{Y} = j4 \times 10^{-4} = 0.0004 \text{ S } \textbf{Ans.}$$

Example 6.24. The A, B, C, D constants of a three-phase transmission line are A = D = 0.936 + *j*0.016, B = 33.5 + *j*138 Ω and C = (–0.9280 + *j*901.223) × 10^{-6} mho.

The load at the receiving end is 40 MW at 200 kV with pf of 0.86 lagging. Find the magnitude of the sending-end voltage, current, power, and voltage regulation. Assume that the magnitude of the sending-end voltage remains constant.

[*U.P. Technical Univ. Elements of Power System 2008-09*]

Solution : Receiving-end phase voltage,

$$V_R = \frac{200 \times 1{,}000}{\sqrt{3}} = 115{,}470 \text{ V}$$

Receiving-end current,

$$I_R = \frac{40 \times 10^6}{\sqrt{3} \times 200 \times 10^3 \times 0.86} = 134.27 \text{ A}$$

Power factor, $\cos \phi = 0.86$

Phase angle, $\phi = \text{Cos}^{-1}\, 0.86 = -30.68°$

$\sin \phi = 0.51$

$$\mathbf{A} = \mathbf{D} = 0.936 + j0.016 = 0.93614 \angle 0.98°$$

$$\mathbf{B} = 33.5 + j138 = 142 \angle 76.355° \; \Omega$$

$$\mathbf{C} = (-0.9280 + j901.223) \times 10^{-6}$$

$$= 901.2235 \angle 90.06° \times 10^{-6} \; \mho$$

Taking receiving-end phase voltage as reference phasor, we have

$$\mathbf{V}_R = 115{,}470(1 + j0) = 115{,}470 \angle 0° \text{ V}$$

$$\mathbf{I}_R = 134.27 \angle -30.68° \text{ A}$$

Sending-end phase voltage,

$$\mathbf{V}_S = \mathbf{AV}_R + \mathbf{BI}_R$$

$$= 0.93614 \angle 0.98° \times 115{,}470 \angle 0° + 142 \angle 76.355° \times 134.27 \angle -30.68°$$

$$= 108{,}096 \angle 0.98° + 19{,}066 \angle 45.675°$$

$$= (108{,}080 + j1848.8) + (13{,}322 + j13{,}639.6)$$

$$= (121{,}402 + j15{,}488.4) = 122{,}386 \angle 7.27° \text{ V}$$

Sending-end line voltage,

$$V_{SL} = \sqrt{3} \times V_S = \sqrt{3} \times 122{,}386$$

$$= 211{,}979 \text{ V or } 211.98 \text{ kV } \textbf{Ans.}$$

Sending-end current,

$$\mathbf{I}_S = \mathbf{CV}_R + \mathbf{DI}_R$$

$$= 901.2235 \times 10^{-6} \angle 90.06° \times 115{,}470 \angle 0° + 0.93614 \angle 0.98° \times 134.27 \angle -30.68°$$

$$= 104.064 \angle 90.06° + 125.696 \angle -29.7°$$

$$= (-0.109 + j104.064) + (109.18 - j62.28)$$

$$= (109.071 + j41.784) = 116.8 \angle 20.96° \text{ A } \textbf{Ans.}$$

Sending-end phase angle,

$$\phi_S = 20.96° - 7.27° = 13.69° = 0.9716 \text{ (leading) } \textbf{Ans.}$$

Sending-end power

$$= \sqrt{3} V_{SL} I_S \times \cos \phi_S \times 10^{-6} \text{ MW}$$

$$= \sqrt{3} \times 211.98 \times 10^3 \times 116.8 \times 0.9716 \times 10^{-6}$$

$$= 41.666 \text{ MW } \textbf{Ans.}$$

Voltage regulation

$$= \frac{V_{SL} - V_{RL}}{V_{RL}} \times 100 = \frac{211.98 - 200}{200} \times 100 = 5.99\% \textbf{ Ans.}$$

Example 6.25. Find the following for a single circuit transmission line delivering a load of 50 MVA at 110 kV and pf 0.8 lagging.

(*i*) Sending-end voltage (*ii*) sending-end current (*iii*) sending-end power (*iv*) efficiency of transmission.

Given A = D = 0.98 $\angle 3°$; B = 110 $\angle 75°$ Ω ; C = 0.0005 $\angle 80°$ ℧. [*A.M.I.E. Sec. B. Power Systems Winter 1993*]

Solution : Receiving-end phase voltage,

$$V_R = \frac{110 \times 1{,}000}{\sqrt{3}} = 63{,}500 \text{ V}$$

Receiving-end current,

$$I_R = \frac{50 \times 10^6}{\sqrt{3} \times 110 \times 10^3} = 262.43 \text{ A}$$

$\text{Cos}\, \phi = 0.8$ and $\sin \theta = 0.6$

Taking receiving-end phase voltage as reference phasor, we have

$$\mathbf{V}_R = 63{,}500(1 + j0) = 63{,}500 \angle 0° \text{ V}$$

$$\mathbf{I}_R = 262.43\,(0.8 - j\,0.6) = 262.43 \angle -36.87° \text{ A}$$

$$\mathbf{A} = \mathbf{D} = 0.98 \angle 3°;\ \mathbf{B} = 110 \angle 75° \; \Omega;\ \mathbf{C} = 0.0005 \angle 80° \text{ siemen}$$

Sending-end phase voltage,

$$\mathbf{V}_S = \mathbf{AV}_R + \mathbf{BI}_R$$

$$= 0.98 \angle 3° \times 63{,}500 \angle 0° + 110 \angle 75° \times 262.43 \angle -36.87°$$

$$= 62{,}230 \angle 3° + 28{,}867 \angle -38.13°$$

$$= (62{,}144 + j\,3{,}257) + (22{,}707 + j\,17{,}824)$$

$$= 84{,}851 + j\,21{,}081 = 87{,}430 \angle 13.95° \text{ V}$$

(*i*) Sending-end line voltage,

$$V_{SL} = \sqrt{3} \times 87{,}430 = 1{,}51{,}433 \text{ V or } 151.433 \text{ kV } \textbf{Ans.}$$

(*ii*) Sending-end current,

$$\mathbf{I}_S = \mathbf{CV}_R + \mathbf{DI}_R$$

$= 0.0005 \angle 80° \times 63{,}500 \angle 0° + 0.98 \angle 3° \times 262.43 \angle -36.87°$

$= 31.75 \angle 80° + 257.18 \angle -33.87°$

$= 5.513 + j31.27 + 213.53 - j143.33$

$= 219.043 - j112.06$

$= 246 \angle -27.1°$ A **Ans.**

Sending-end phase angle, $\phi_S = 13.95° - (-27.1°) = 41.05°$

Sending-end power factor $= \cos \phi_S = \cos 41.05° = 0.7541$

(*iii*) Sending-end power

$= \sqrt{3}\, V_{SL} I_S \cos \phi_S \times 10^{-6}$

$= \sqrt{3} \times 151{,}433 \times 246 \times 0.7541 \times 10^{-6} = 48.66$ MW **Ans.**

(*iv*) Transmission efficiency,

$$\eta = \frac{\text{Power output}}{\text{Power input}} \times 100 = \frac{50 \times 0.8}{48.66} \times 100 = 82.2\% \text{ **Ans.**}$$

Example 6.26. A 3-phase, 50 Hz, 300 km long transmission line has the following parameters: $r = 0.15\ \Omega$/km ; $x = 0.5\ \Omega$/km; $y = 3 \times 10^{-6}$ ℧/km.

If the line is represented by nominal-π, determine

(*i*) the ABCD constants of the line (derive the formula you might have used);

(*ii*) the power at unity power factor that can be received if the voltage at each end is maintained at 220 kV;

(*iii*) the sending-end voltage required if a load of 200 MW at 0.85 lagging power factor is to be delivered with the receiving-end voltage remaining at 220 kV.

[*M.D. Univ. Power System-I, May 2008*]

Solution : Total line parameters are:

Resistance, $R = r \times l = 0.15 \times 300 = 45\ \Omega$

Reactance, $X = x \times l = 0.5 \times 300 = 150\ \Omega$

Susceptance, $Y = y \times l = 3 \times 10^{-6} \times 300 = 9 \times 10^{-4}$ S

For nominal-π equivalent

(*i*) $\mathbf{A} = \mathbf{D} = 1 + \frac{1}{2}\mathbf{YZ}$

$= 1 + \frac{1}{2} \times j9 \times 10^{-4} \times (45 + j150)$

$= (0.9325 + j0.02025)$ or $0.9327 \angle 1.24°$ **Ans.**

$\mathbf{B} = \mathbf{Z} = (45 + j150)\ \Omega = 156.6 \angle 73.3°\ \Omega$ **Ans.**

$\mathbf{C} = \mathbf{Y}\left(1 + \frac{1}{4}\mathbf{YZ}\right)$

$= j9 \times 10^{-4}\left[1 + \frac{1}{4} \times j9 \times 10^{-4}(45 + j150)\right]$

$= j9 \times 10^{-4}(0.96625 + j0.010125)$

$= 8.67 \times 10^{-4} \angle 90.6°$ S **Ans.**

(*ii*) Taking receiving-end phase voltage as the reference phasor we have

$$\mathbf{V_R} = \frac{220}{\sqrt{3}} \times 1{,}000\,(1 + j0) = 127{,}000 \angle 0° \text{ V}$$

Load current, $\mathbf{I_R} = I_R \angle 0°$ A $\quad \because$ power factor is unity

Sending-end phase voltage,

$\mathbf{V_S} = \mathbf{AV_R} + \mathbf{BI_R}$

$= 0.9327 \angle 1.24° \times 127{,}000 \angle 0° + 156.6 \angle 73.3° \times I_R \angle 0°$

$= 118{,}425 + j2{,}563 + 45I_R + j150I_R$

$= (118{,}425 + 45I_R) + j(2{,}563 + 150I_R)$

or $V_S = \sqrt{(118{,}425 + 45I_R)^2 + (2{,}563 + 150I_R)^2}$

$16{,}133 \times 10^6 = 14{,}024.5 \times 10^6 + 6.57 \times 10^6 + 10{,}658 \times 10^6 I_R$

$+ 0.7689 \times 10^6 I_R + 0.024525 \times 10^6 I_R^2$

or $0.024525 I_R^2 + 11.427 I_R - 21{,}02 = 0$

$$\text{or } I_R = \frac{-11.427 \pm \sqrt{(11.427)^2 + 4 \times 0.024525 \times 2{,}102}}{2 \times 0.024525}$$

$= 141.175$ A

Power received,

$P = \sqrt{3}\, V_{RL} I_R \cos\theta \times 10^{-6}$ MW

$= \sqrt{3} \times 220{,}000 \times 141.175 \times 1.0 \times 10^{-6} = 53.8$ MW **Ans.**

(*iii*) $V_R = 127{,}000$ V

$$I_R = \frac{200 \times 10^6}{\sqrt{3} \times 2{,}20{,}000 \times 0.85} = 617.5 \text{ A}$$

Phase angle, $\phi = \cos^{-1} 0.85 = 31.79°$ (lagging)

Taking receiving-end phase voltage as the reference phasor we have

$\mathbf{V_R} = 127{,}000 \angle 0°$ V and $\mathbf{I_R} = 617.5 \angle -31.79°$ A

Sending-end phase voltage,

$\mathbf{V_S} = \mathbf{AV_R} + \mathbf{BI_R}$

$= 0.9327 \angle 1.24° \times 127{,}000 \angle 0° + 156.6 \angle 73.3° \times 617.5 \angle -31.79°$

$= 118{,}469 \angle 1.24° + 96{,}700 \angle 41.51°$

$= 118{,}441 + j2{,}563.72 + 72{,}413 + j64{,}088$

$= 190{,}854 + j66{,}652 = 202{,}158 \angle 19.25°$ V

Sending-end line voltage,

$V_{SL} = \sqrt{3} \times 202{,}158 = 350{,}150$ V or 350.15 kV **Ans.**

Example 6.27. Two 3-phase transmission lines have the generalised constants :

$\mathbf{A_1 = D_1 = 0.98 \angle 2°\ ;\ B_1 = 28 \angle 69°\ \Omega\ ;\ C_1 = 0.0002 \angle 88°\ S}$

$\mathbf{A_2 = D_2 = 0.95 \angle 3°\ ;\ B_2 = 40 \angle 85°\ \Omega\ ;\ C_2 = 0.0004 \angle 90°\ S}$

They are connected in series and delivers a load current of 200 A at 0.95 pf (lagging) at 110 kV. Determine the sending-end voltage and sending-end current.

Solution : Generalised constants of two lines connected in series

$\mathbf{A} = \mathbf{A_1 A_2} + \mathbf{B_1 C_2}$

$= 0.98 \angle 2° \times 0.95 \angle 3° + 28 \angle 69° \times 0.0004 \angle 90°$

$= 0.931 \angle 5° + 0.0112 \angle 159°$

$= 0.9275 + j0.08114 + 0.0105 + j0.004$

$= 0.917 + j0.08514 = 0.92 \angle 5.3°$

$\mathbf{B} = \mathbf{A_1B_2} + \mathbf{B_1D_2}$

$= 0.98 \angle 2° \times 40 \angle 85° + 28 \angle 69° \times 0.95 \angle 3°$

$= 39.2 \angle 87° + 26.6 \angle 72°$

$= 2.0516 + j39.1463 + 8.2199 + j\,25.2981$

$= 10.2715 + j64.4444 = 65.26 \angle 80.94°\ \Omega$

$\mathbf{C} = \mathbf{C_1A_2} + \mathbf{D_1C_2}$

$= 0.0002 \angle 88° \times 0.95 \angle 3° + 0.98 \angle 2° \times 0.0004 \angle 90°$

$= 0.00019 \angle 91° + 0.000392 \angle 92°$

$= -0.000003 + j0.00019 - 0.00001368 + j0.000392$

$= -0.00001668 + j0.000582 = 0.000582 \angle 91.67°$ S

$\mathbf{D} = \mathbf{C_1B_2} + \mathbf{D_1D_2}$

$= 0.0002 \angle 88° \times 40 \angle 85° + 0.98 \angle 2° + 0.95 \angle 3°$

$= -00794 + j\,0.000975 + 0.9275 + j\,0.08114$

$= 0.91956 + j\,0.082115 = 0.923 \angle 5.1°$

Receiving-end phase voltage,

$$V_R = \frac{110{,}000}{\sqrt{3}} = 63{,}500 \text{ V}$$

Receiving-end current, $I_R = 200$ **A**

Receiving-end power factor angle,

$$\phi_R = \text{Cos}^{-1}\, 0.95\ C = 18.195° \text{ (lagging)}$$

Taking receiving-end phase voltage as reference phasor we have

$$\mathbf{V_R} = 63{,}500 \angle 0° \text{ V}$$

and $\mathbf{I_R} = 200 \angle -18.195°$ A

Sending-end phase voltage,

$\mathbf{V_S} = \mathbf{AV_R} + \mathbf{BI_R}$

$= 0.92 \angle 5.3° \times 63{,}500 \angle 0° + 65.26 \angle 80.94° \times 200 \angle -18.195°$

$= 58{,}420 \angle 5.3° + 13{,}052 \angle 62.745°$

$= 58{,}170 + j5{,}396 + 5{,}977 + j11{,}602$

$= 64{,}147 + j16{,}998 = 66{,}361 \angle 14.84°$ V

Sending-end line voltage,

$V_{SL} = \sqrt{3} \times 66{,}361 = 114{,}940$ V or 114.94 kV **Ans.**

Sending-end current,

$\mathbf{I_S} = \mathbf{CV_R} + \mathbf{DI_R}$

$= 0.000582\angle 91.67° \times 63{,}500\angle 0° + 0.923\angle 5.1° \times 200\angle -18.195°$

$= 36.957\angle 91.67° + 184.6\angle -13.095°$

$= -1.077 + j36.9413 + 179.8 - j41.8241$

$= 178.723 - j4.8828 = 178.8\angle -1.56°$ A **Ans.**

Example 6.28. Two identical three-phase transmission lines are connected in parallel to supply a load of 100 MW at 132 kV and 0.8 pf lagging at the receiving-end. The ABCD constants of each transmission line are as follows :

$$\mathbf{A = D = 0.98 \angle 1°\ ;\ B = 100 \angle 75°\ \Omega \text{ per phase}\ ;}$$
$$\mathbf{C = 0.0005 \angle 90° \text{ siemens per phase.}}$$

Determine (*a*) the ABCD constants of the combined network (*b*) the sending-end power factor. *[GATE 1995]*

Solution : Generalised constants of two lines combined in parallel are :

$$\mathbf{A} = \frac{\mathbf{A_1B_2 + A_2B_1}}{\mathbf{B_1 + B_2}}$$

$$= \frac{0.98 \angle 1° \times 100 \angle 75° + 0.98 \angle 1° \times 100 \angle 75°}{100 \angle 75° + 100 \angle 75°}$$

$= 0.98 \angle 1°$ **Ans.**

$$\mathbf{B} = \frac{\mathbf{B_1 \cdot B_2}}{\mathbf{B_1 + B_2}} = \frac{100 \angle 75° \times 100 \angle 75°}{100 \angle 75° + 100 \angle 75°} = 50 \angle 75°\ \Omega \textbf{ Ans.}$$

$$\mathbf{C} = \mathbf{C_1} + \mathbf{C_2} - \frac{(\mathbf{A_1} - \mathbf{A_2})(\mathbf{D_1} - \mathbf{D_2})}{\mathbf{B_1 + B_2}}$$

$= \mathbf{C_1} + \mathbf{C_2}$ $\quad\because \mathbf{A_1} = \mathbf{A_2}$ and $\mathbf{D_1} = \mathbf{D_2}$ So $(\mathbf{A_1} - \mathbf{A_2}) = 0$ and $(\mathbf{D_1} - \mathbf{D_2}) = 0$

$= 0.0005 \angle 90° + 0.0005 \angle 90° = 0.001 \angle 90°$ S **Ans.**

$$\mathbf{D} = \frac{\mathbf{B_1D_2 + B_2D_1}}{\mathbf{B_1 + B_2}} = \mathbf{A} = 0.98 \angle 1° \textbf{ Ans.}$$

Receiving-end phase voltage,

$$\mathbf{V_R} = \frac{132 \times 1{,}000}{\sqrt{3}} = 76{,}210 \text{ V}$$

Receiving-end current,

$$\mathbf{I_R} = \frac{100 \times 10^6}{\sqrt{3} \times 132 \times 10^3 \times 0.8} = 546.7 \text{ A}$$

Taking receiving-end phase voltage as the reference phasor we have

$\mathbf{V_R} = 76{,}210 \angle 0°$ V and $\mathbf{I_R} = 546.7 \angle -36.87°$ A

$\because$ Phase angle, $\phi_R = \text{Cos}^{-1}\ 0.8 = -36.87°$

Sending-end phase voltage,

$\mathbf{V_S} = \mathbf{AV_R} + \mathbf{BI_R}$

$= 0.98 \angle 1° \times 76{,}210 \angle 0° + 50 \angle 75° \times 546.7 \angle -36.87°$

$= 74{,}685.8 \angle 1° + 27{,}335 \angle 38.13°$

$= 74{,}674.4 + j1{,}303.4 + 21{,}502 + j16{,}877.9$

$= 96{,}176.4 + j18{,}181.3 = 97{,}879.8 \angle 10.7°$ V

Sending-end current,

$\mathbf{I_S} = \mathbf{CV_R} + \mathbf{DI_R}$

$= 0.001 \angle 90° \times 76{,}210 \angle 0° + 0.98 \angle 1° \times 546.7 \angle -36.87°$

$= 76.21 \angle 90° + 535.766 \angle -35.87°$

$= j76.21 + 434.15 - j313.93$

$= 434.15 - j237.72 = 495 \angle -28.7°$ A

Sending-end phase angle,

ϕ_S = Phase angle between $\mathbf{V_S}$ and $\mathbf{I_S}$

$= 10.7° - (-28.7°) = 39.4°$ (lagging)

Sending-end power factor

$= \text{Cos}\ \phi_S = \text{Cos}\ 39.4° = 0.7727$ (lagging) **Ans.**

Exercises

1. Along with circuit representations, explain the basis of categorization of short, medium and long transmission lines.
[U.P. Technical Univ. Elements of Power System 2003-04]
2. Identify the criterion for classification of transmission lines. Which transmission line representation is close to actual system dynamics ?
[U.P. Technical Univ. Elements of Power System 2005-06]
3. Distinguish between short, medium and long transmission lines. **[Pb. Technical Univ. Power System-I (T&D), May 2008]**
4. Derive an expression for voltage regulation of a short transmission line with vector diagram.
[M.D. Univ. Power System-I, Dec. 2010]
5. Describe three-phase short line calculations.
[Pb. Technical Univ. Power System-I (T&D), Dec. 2006]
6. Draw the phasor diagram of a short transmission line and derive an expression for voltage regulation.
[U.P. Technical Univ. Elements of Power System 2006-07]
7. Obtain the expression of 'voltage regulation' and 'efficiency' of a short transmission line in terms of line parameters.
[U.P. Technical Univ. Elements of Power System 2007-08]
8. Define regulation, efficiency, losses and power factor at each end of the line and explain how these characteristics are affected by the constants of the line.
9. Explain the effect of load power factor on the transmission efficiency and voltage regulation of transmission line.
[Pb. Technical Univ. Power System-I (T&D), May 2009]
10. Explain the effect of load power factor on regulation and efficiency of a transmission line.
[M.D. Univ. Power System-I, Dec. 2009]
11. Define precisely the 'regulation' of a transmission line and discuss qualitatively its dependence on the load power factor. Show that the regulation is zero when $\phi_R = \theta - \pi/2$ where ϕ_R is the load power factor angle and θ is the line impedance angle. Can the regulation be negative ? If so, under what conditions ? State clearly the assumptions made.
12. Show a diagrammatic representation of a medium line in terms of its nominal-π model. Indicate also the values of series impedance and shunt admittance of the model in terms of the resistance, inductance and capacitance per unit length of the line.
13. How are transmission lines classified ? Obtain the relation between the sending-end and receiving-end voltages and currents of a medium line using nominal-π representation.
[Rajasthan Univ. Transmission & Distribution of Electrical Power 2002, 2004, 2006, 2007]
14. Draw and explain the equivalent circuits and phasor diagrams for the nominal π and T methods in detail.
[Pb. Technical Univ. Power System-I, Dec. 2006, Dec. 2008]
15. Draw phasor diagrams for a nominal π circuit and nominal T circuit of transmission line. Derive expressions for sending-end voltage and current in each case.
[Transmission & Distribution of Electrical Power Rajasthan Univ. 2005; Rajasthan Technical Univ. 2009]
16. Explain Ferranti effect and proximity effect.
[M.D. Univ. Power System-I, Dec. 2010]
17. What is Ferranti effect ? Explain it with the help of phasor diagram.
18. Explain by means of phasor diagram the 'Ferranti effect' in transmission lines.
[U.P. Technical Univ. Elements of Power System 2007-08]
19. What is Ferranti Effect ? Derive an expression for the voltage rise of an unloaded line.
20. A long transmission line is not loaded ? What will be the voltage at the receiving end compared to the sending-end voltage ? Give reasons.
21. Explain the Ferranti effect and derive an expression for power loss in an open-circuited line.
22. Explain the influence of capacitance on performance of loaded line.
[Pb. Technical Univ. Power System-I (T&D), May 2009]
23. Define the general network for localized capacitance method to represent the transmission lines over 100 km long.
[Rajasthan Univ. Transmission & Distribution of Electrical Power, 2003]
24. Explain the physical significance of the generalised circuit constants A, B, C and D of a transmission line. Find the values of A, B, C and D in the following approximate methods in terms of Z and Y.
(*i*) Simple series impedance method.
(*ii*) Nominal-π method.
(*iii*) Nominal-T method.
25. Obtain A, B, C, D constants for a short transmission line.
[U.P. Technical Univ. Elements of Power System 2007-08]
26. Derive the A, B, C, D constants for the transmission line represented by nominal-T section.
[U.P. Technical Univ. Elements of Power System 2005-06]
27. Derive A, B, C, D constants of a medium length transmission line and hence prove that AD – BC = 1.
[U.P. Technical Univ. Elements of Power System 2005-06]
28. Evaluate the generalized circuit constants for (*i*) short transmission line, (*ii*) medium transmission line nominal-T method.
[G.B. Technical Univ. Elements of Power System 2011-12]
29. Evaluate the generalized circuit constants for medium line using nominal-π method.
[M.D. Univ. Power System-I, Dec. 2008]
30. Two transmission lines having generalized circuit constants A_1, B_1, C_1, D_1 and A_2, B_2, C_2, D_2 are connected in parallel. Develop expression for overall A, B, C, D constants of the combination.
[Pb. Technical Univ. Power System-I (T&D), Dec. 2007]
31. The ABCD parameters of two transmission networks are A_1, B_1, C_1, D_1 and A_2, B_2, C_2, D_2 respectively. Determine the overall ABCD parameters (*a*) when the two networks are connected in series and (*b*) in parallel.
32. Write short notes on the following :
(*i*) Classification of transmission line.
(*ii*) Voltage regulation of transmission lines.
[Pb. Technical Univ. Dec. 2009]
(*iii*) Effect of load power factor on the transmission efficiency and voltage regulation of a transmission line.
(*iv*) Ferranti effect.
[Rajasthan Technical Univ. 2008, 2009; M.D. Univ. May 2009; W.B. Technological Univ. 2009-10]

Short Answer Type Questions With Answers

Q. 1. What are the factors which govern the performance of a transmission line ?

Ans. The factors governing the performance of a transmission line are series resistance R and inductance L, shunt capacitance C and conductance G. However, the effect of conductance is negligible.

Q. 2. How do you classify transmission lines ?

Ans. Transmission lines are classified depending upon the manner in which capacitance is taken into account.

Q. 3. What are the causes of voltage drop and line loss in a transmission line ?

Ans. The voltage drop in a transmission line is due to its resistance and inductance whereas the line loss is due to its resistance.

Q. 4. What is voltage regulation of line ?

[Pb. Technical Univ. Dec. 2008]

OR

What is regulation and its impact on transmission lines.

[Pb. Technical Univ. Dec. 2007]

Ans. ***Regulation.*** When the load is supplied, there is a voltage drop in the line due to resistance and inductance of the line and, therefore, receiving-end voltage V_R is usually less than sending-end voltage V_S. The voltage drop *i.e.,* difference of sending-end voltage and receiving-end voltage expressed as a percentage of receiving-end voltage is called the *regulation.*

When the load is thrown off *i.e.,* when the line is supplying no load, the receiving-end voltage becomes equal to sending-end voltage and therefore, regulation can be defined as below :

Regulation is defined as the change in voltage at the receiving (or load) end when the full load is thrown off, the sending-end (or supply) voltage and supply frequency remaining unchanged. It is usually expressed as a percentage of receiving-end voltage.

Mathematically percentage voltage regulation of a transmission line is given by the expression

$$\% \text{ voltage regulation } = \frac{V_S - V_R}{V_R} \times 100$$

where V_S is the voltage at sending or supply end and V_R is the voltage at the receiving or load end. *It is to be noted here that* V_S *and* V_R *are the magnitudes of voltages.*

The *lower the voltage regulation, better it is,* because low voltage regulation means little variation in receiving-end voltage due to variation in load current.

Knowledge of voltage regulation helps in maintaining the voltage at the load terminals within prescribed limits by employing suitable voltage control equipment.

Q. 5. Explain the effect of load power factor on the transmission efficiency.

Ans. To deliver given power at a given voltage at consumer end, line current is inversely proportional to the power factor of the load. The current, hence, the power loss in the line goes on increasing as the power factor of the load goes on decreasing. Thus efficiency of the transmission line decreases with the fall in power factor and vice versa.

Q. 6. What do you understand by Ferranti effect ?

[U.P.S.C. I.E.S. E.E.-II, 2006]

OR

What is Ferranti effect ? Define.

[Pb. Technical Univ. Dec. 2006]

OR

What is Ferranti effect ? **[Pb. Technical Univ. May 2009]**

Ans. The phenomenon of rise in voltage at the receiving end of the open-circuited or lightly loaded line is called the Ferranti effect.

Q. 7. How does rise in voltage at the receiving end of an open-circuited line depend upon the length of the line and system operating voltage ?

Ans. Rise in voltage at the receiving end of an open-circuited line is proportional to the square of the length of the line and varies directly in proportion to the system operating voltage.

Q. 8. How do the terms 'impedance drop', 'voltage drop' and 'voltage regulation' , in connection with transmission lines differ ?

Ans. The impedance drop in a line is the voltage drop in the line impedance and is equal to the phase difference between the sending-ending voltage and the receiving-end voltage.

The voltage drop in a line is the arithmetical difference between the sending-end and receiving-end voltages.

The voltage regulation is defined as the change in voltage at the receiving-end when full-load is thrown off, the sending-end voltage being held constant.

Q. 9. Why in medium and long transmission lines, regulation is greater than voltage drop ?

Ans. The regulation is greater than voltage drop in a medium/long transmission line due to rise in voltage owing to Ferranti effect.

Q. 10. What are the units of generalised constants of a transmission line ?

Ans. Generalised constants A and D are dimensionless whereas the units of generalised constants B and C are ohms and siemens (mho) respectively.

Problems

1. A single-phase transmission line has a resistance of 0.22 Ω and inductive reactance of 0.36 Ω. Find the voltage at the sending end to give 500 kVA at 2,000 V at the receiving end at a load power factor of 0.707 lagging. Also find the efficiency of transmission at this load.

[**Ans.** 2,102.37 V; 96.26%]

2. A 1-phase transmission line has a resistance of 0.2 Ω and an inductive reactance of 0.4 Ω. Find the voltage at the sending end to give 500 kVA at 2 kV at receiving end at load power factors of (*i*) unity (*ii*) 0.707 lagging. Illustrate with suitable phasor diagrams.

[**Ans.** (*i*) 2,052.44 V (*ii*) 2,106 V]

3. A power of 250 kW is delivered at the end of a single-phase line 15 km long at 1,000 V, and 0.8 pf lag, 50 Hz. The resistance of the conductor is 0.03 Ω/km and the inductance/conductor/km is 0.15 mH. Find the sending-end voltage, sending-end power factor and % regulation. Draw the phasor diagram. [**Ans.** 1,501.5 V; 0.72 (lag); 73.99%]

4. An overhead, single-phase transmission line delivers 1,100 kW at 33 kV at 0.8 pf lagging. The total resistance of the line is 10 Ω and the total inductive reactance is 15 Ω. Determine (*i*) sending-end voltage (*ii*) sending-end power factor and (*iii*) transmission efficiency.

[Bombay Univ. 1970]

[**Ans.** 33,710 V (*ii*) 0.794 (lag) (*iii*) 98.5%]

5. A single-phase line has an impedance of 5 $\angle 60°$ ohm. It is supplying a load of 120 A, 3.3 kV at 0.8 pf lagging. Calculate its regulation. [**Ans.** 16.94%]

6. What is the maximum length in km for a single-phase transmission line having copper conductors of 0.775 cm^2 cross section over which 200 kW at unity power factor and at 3,300 V are to be delivered ? The efficiency of transmission is 90 per cent. Take specific resistance as 1.725 mΩ-cm.

[Bombay Univ. March 1971]

[**Ans.** 13.6 km]

7. A short three-phase transmission line having resistance and reactance 5 Ω and 12 Ω respectively supplies 10 MVA load at 0.8 pf (lagging). When receiving-end voltage is maintained at 33 kV, find the sending-end voltage. Draw a vector diagram to explain. [**Ans.** 36.394 kV]

8. A 3-phase transmission line delivers 10,000 kW at 33 kV and 0.8 lagging power factor. The resistance and inductive reactance per conductor is 3 Ω and 6 Ω respectively. Calculate the sending-end voltage and the efficiency of transmission. **[U.P.S.C. I.E.S. (Elec. Engg.) 1972]**

[**Ans.** 35.2 kV; 96.5%]

9. Find the sending-end voltage, percentage voltage drop and efficiency of a 20 km, 20 kV, 50 Hz, 3-phase transmission line delivering 5 MW at a power factor of 0.8. The conductors have an effective diameter of 12 mm and a cross-sectional area of 95 mm^2. These are arranged in an equilateral triangular formation, the distance between their centres being one metre. [**Ans.** 22.2 kV : 12 % : 93.3%]

10. A 3-phase transmission line delivers a load of 5 MW at 0.8 pf (lagging). Resistance of each conductor is 0.5 Ω/km. Receiving-end voltage is 33 kV. If the line loss is not to exceed 10%, determine the length of the line. [**Ans.** 27.9 km]

11. The following data refers to a 50 Hz, 3-phase transmission line :

Length.......10 km; Sending-end voltage.........11 kV

Load delivered at receiving end........1,000 kW at 0.8 pf (lag)

Resistance of each conductor........0.5 Ω/km

Reactance of conductor...........0.56 Ω/km

Calculate (*a*) line current (*b*) receiving-end voltage (*c*) efficiency of transmission.

[Pb. Technical Univ. Power System-I (T&D), Dec. 2009]

[**Ans.** (*a*) 80.7 A, (*b*) 8,943 V, (*c*) 83.66%]

12. A 3-phase line has a resistance of 5.31 Ω and an inductance of 0.0176 H. Power is transmitted at 33 kV, 50 Hz from one end and the load at the receiving end is 3,600 kW at 80 per cent power factor. Find the line current, receiving-end voltage, sending-end power factor and efficiency of transmission. **[A.P.S. Univ. 1978]**

[**Ans.** 81.5 A, 31.9 kV, 0.796, 97.2%]

13. A 3-phase 20 km long line delivers 3 MW at 11 kV. Line loss is equal to 10% of power delivered. Line inductance is 0.9156 mH/phase/km. Fi..d sending-end voltage and regulation if the load power factor is 0.8 lagging.

[**Ans.** 12,863.66 V; 16.94%]

14. An 11 kV, 3-phase, 50 Hz ac power line supplies power to a load of 1 MW at 0.8 pf lag situated at a distance of 10 km. The resistance of each conductor is 0.5 Ω per km and reactance is 0.56 Ω per km. Determine (*i*) line current (*ii*) line losses (*iii*) receiving-end voltage and (*iv*) efficiency of transmission.

[**Ans.** (*i*) 71.5 A (*ii*) 76.68 kW (*iii*) 10.09 kV (*iv*) 92.88%]

15. A 3-phase transmission line has the following constants (line to neutral) :

R = 10 Ω, inductive reactance = 20 Ω, capacitive susceptance = 4 × 10^{-4} S

Using the nominal-T method, calculate the sending-end voltage, line current, power factor and the efficiency of transmission when supplying a balance load of 10 MW at 66 kV, power factor 0.8 lagging.

[**Ans.** 69.544 kV; 100.53 A; 0.853 (lag); 96.79%]

16. A 3-phase transmission line has the following data :

Resistance of each conductor = 20 Ω

Inductive reactance of each conductor = 40 Ω

Capacitive susceptance to neutral = 0.0004 ℧

Load at the receiving-end = 12,500 kVA at 0.8 pf lagging

Voltage at the receiving end = 66 kV

Find (*a*) sending-end voltage and current (*b*) sending-end power factor (*c*) efficiency (*d*) regulation of the line. Use nominal-T method and draw the phasor diagram.

[A.P.S. Univ. 1977]

[**Ans.** (*a*) 73.122 kV; 100.125 A (*b*) 0.84 (lag) (*c*) 93.88% (*d*) 10.8%]

17. A 3-phase, 50 Hz transmission line 100 km long delivers 200 MW at 0.9 pf lagging at 110 kV. The resistance and reactance of the line per phase are 0.2 Ω and 0.4 Ω while capacitive admittance is 2.5 × 10^{-6} siemen/km/phase. Calculate (*i*) current and voltage at sending-end, (*ii*) Efficiency of transmission line using nominal-T method.

[M.D. Univ. Power System-I, May 2009]

[**Ans.** (*i*) 110 A, 116.72 kV, (*ii*) 96.29%]

18. A 3-phase transmission line has the following constants (line to neutral), R = 10 Ω, inductive reactance = 20 Ω, capacitive susceptance = 4 × 10^{-4} S. Using the nominal-T method, calculate the sending-end voltage, line current, power factor and efficiency of transmission when supplying a balanced load of 10 MW at 66 kV, power factor 0.8 (lagging). Draw vector diagram.

[Rajasthan Technical Univ. Transmission and Distribution of Electrical Power 2008]

[**Ans.** 69.53 kV, 100 A, 0.853 (lagging), 97.34%]

19. A 3-phase, 50 Hz, 150 km long line has resistance, inductive reactance and capacitive shunt admittance of 0.1 Ω, 0.5 Ω and 3 × 10^{-6} ℧ per km per phase. If the line delivers 50 MW at 110 kV at 0.8 pf lagging, determine the sending-end voltage. Assume a nominal-π circuit for the line.

[U. P. Technical Univ. Elements of Power System 2003-04]

[**Ans.** 143.556 kV]

20. A 50 Hz, 3-phase transmission line of length 30 km has a total series impedance of $(40 + j125)$ ohms and shunt admittance of 10^{-3} mho. The load is 50 MW at 220 kV with 0.8 lagging power. Find the sending-end voltage. (Use Nominal-π representation).

[U.P. Technical Univ. Elements of Power System 2005-06]

[**Ans.** 238.04 kV]

21. A 3-phase, 100 km transmission line is delivering 50 MW, 0.8 pf lagging at 132 kV. Each conductor is having resistance of 0.1 Ω/km, reactance of 0.3 Ω/km and shunt admittance of 3×10^{-6} ℧/km. If the load is balanced and leakage is neglected, calculate the sending end voltage, efficiency and sending-end power factor of line using nominal-π representation.

[U.P. Technical Univ. Elements of Power System 2005-06]

[**Ans.** 143.98 kV, 95.893%, 0.8069 (lagging)]

22. A 50 Hz, 3-ϕ transmission line is 280 km long. It has a total series impedance of 35 + *j*140 Ω and shunt admittance of 930×10^{-4} ℧. It delivers 40,000 kW at 220 kV. With 90% power factor lagging, find the sending-end voltage, regulation, transmission efficiency and A, B, C, D constants by the nominal-T method.

[Rajasthan Univ. Transmission & Distribution of Electrical Power 2002]

[**Ans.** 255.6 kV; 2.545%; 96.0%; $0.935 \angle 0°59'$; $139.5 \angle 76.5°$ Ω ; 930×10^{-6} mho]

23. A 1,000 km long, 3-phase, 50 Hz transmission line has resistance per phase per km = 0.1 Ω, reactance per phase per km = 0.5 Ω; susceptance per phase per km = 10×10^{-6} ℧. If the line supplies a load of 20 MW at 0.9 pf lagging at 65 kV at the receiving end, calculate by the nominal-π method the regulation and efficiency of the line. Neglect leakance. **[Bombay Univ. Elec. Power-I, March 1971]**

[**Ans.** 15.15%; 94.56%]

24. A 3-ϕ, 50 Hz, 150 km long overhead line has the following constants :

Resistance/phase/km = 0.15 Ω

Inductance/phase/km = 1.2 mH

Capacitance/phase/km = 0.0096 μF

The line supplies a load of 30 MW at 0.9 power factor lagging at a line voltage of 132 kV at the receiving end. Determine voltage, current and power factor at the sending end and also percentage regulation of the line.

[A.M.I.E. Sec. B. Power Systems Summer 1996]

[**Ans.** 142.08 kV; 132.73 A; 0.9588 (lag); 9.037%]

25. A total load of 4 MW at 0.8 pf (lagging) is transmitted to a 33/11 kV substation by two 33 kV OH lines drawn in parallel. One line has a conductor resistance 2 Ω per phase and reactance 15 Ω per phase and corresponding values for the other line being 1.5 Ω and 12 Ω respectively. Calculate power transmitted by each OH line.

[**Ans.** 1.784 MW; 2.216 MW]

26. A 150 km, 3-phase, 110 kV, 50 Hz transmission line transmits a load of 40,000 kW at 0.8 pf lag at receiving end. Resistance/km/phase = 0.15 Ω, reactance/km/phase = 0.6 Ω; susceptance/km/phase = 10^{-5} S. Determine (*a*) A, B, C and D constants of the line, (*b*) regulation of the line.

[Pb. Technical Univ. Power System-I (T&D) Dec. 2006; Dec. 2008]

[**Ans.** (*a*) $0.98 \angle 10°$, $92.8 \angle 75°$ Ω; $0.00145 \angle 90.5°$ S; (*b*) 33.5%]

27. A 220 kV, 50 Hz, 3-phase overhead transmission line delivers a load of 75,000 kW at 0.8 pf lagging at the receiving end and has the following constants : **A** = **D** = $0.90 \angle 0.6°$; **B** = $153.2 \angle 84.6°$ Ω; **C** = $0.0012 \angle 90°$ S.

Calculate the sending-end voltage and current.

[Rajasthan Univ. Transmission & Distribution of Electrical Power 2004]

[**Ans.** 247 kV; 179.76 A]

28. The parameters of a 132 kV, 50 Hz, 3-ϕ transmission line are:

R = 12.84 Ω; L = 0.162 H ; C = 1.014×10^{-6} **F**

Calculate the line constants, **A**, **B**, **C** and **D** for (*i*) nominal-T circuit and (*ii*) nominal-π circuit.

[**Ans.** (*i*) **A** = **D** = $0.992 \angle 0.1°$; **B** = $52.2 \angle 75.95°$; **Ω** ; **C** = $0.00032 \angle 90°$ S; (*ii*) **A** = **D** = $0.992 \angle 0.1°$; **B** = $52.4 \angle 75.9°$ Ω ; **C** = $0.000318 \angle 90.05°$ S]

29. A three-phase, 220 kV, 50 Hz transmission line 200 km long transmits load of 75 MW at 0.8 pf lagging at the receiving end. Resistance of the line per phase per km is 0.8 Ω. Susceptance per phase per km = 6×10^{-6} ℧. Reactance per phase per km = 0.8 Ω. Determine (*a*) A, B, C, D constants of the line. (*b*) the sending-end voltage and current (*c*) sending-end power factor, (*d*) sending-end active and reactive powers (*e*) regulation and efficiency of the line. **[Pb. Univ. April 1982]**

[**Ans.** (*a*) **A** = **D** = $0.909 \angle 6.06°$, **B** = $215.66 \angle 47.89°$ Ω, **C** = $1.2 \times 10^{-3} \angle 90°$ S (by nominal-T method) (*b*) 291.7 kV, 195.78 A (*c*) 0.9981 (leading) (*d*) 98.71 MW; 6.094 MVAR (*e*) 32.57%, 75.98%]

30. A 220 kV, 50 Hz, 3-phase overhead transmission line delivers a load of 75,000 kW at 0.8 pf lagging at the receiving end and has the following constants :

A = **D** = $0.9 \angle 0.6°$; **B** = $153.2 \angle 84.6°$ Ω and **C** = $0.0012 \angle 90°$ S

Calculate the sending-end parameters.

[Pb. Univ. December 1982]

[**Ans.** 247 kV; 180 A; 4.92°; 0.996 (lag); 76,700 kW]

31. A 132 kV overhead transmission line has the following line constants :

A = **D** = $0.99 \angle 0.3°$; **B** = $70 \angle 69°$ Ω; **C** = $4 \times 10^{-4} \angle 90°$ ℧

Calculate the sending-end voltage of transmission line if the load delivered at the receiving end is 45 MVA at 0.8 pf lag. **[Pb. Univ. April 1981]**

[**Ans.** 167.215 kV]

32. A transmission line consists of two lines I and II connected in series, the line I being at the sending end and line II at the receiving end. The circuits have the following constants :

Constants	Line I	Line II
A = D	$0.895 \angle 1.4°$	$0.987 \angle 0.3°$
B	$182.5 \angle 78.6°$ Ω	$76 \angle 67.1°$ Ω
C	$0.00111 \angle 90.47°$ ℧	$0.00037 \angle 90.0°$ ℧

Determine the **A**, **B**, **C** and **D** constants of the whole line which is equivalent to the two lines in series.

[**Ans.** **A** = $0.816 \angle 2.7°$; **B** = $247 \angle 76.1°$ Ω; **C** = $0.00142 \angle 90.97°$ S ; **D** = $0.8071 \angle 47.0°$]

❑❑❑

CHAPTER 7

Performance of Long Transmission Lines

INSIDE THIS CHAPTER

7.1. INTRODUCTION

It is well known that the line constants (resistance, inductance, capacitance and conductance) of a transmission line are uniformly distributed over the entire length of the line. So far it has been assumed that the line has lumped constants and the line calculations made with such assumptions gave results with reasonable accuracy. If such an assumption of lumped line constants is applied to long transmission lines (lines having length more than 200 km), serious errors are introduced in the line performance calculations. Therefore, the performance calculations of long transmission lines are made with line constants uniformly distributed over the entire length of the line so that the results with fair degree of accuracy are obtained. Rigorous mathematical treatment is required for the solution of such transmission lines. The equivalent circuit of a 3-phase long transmission line is represented schematically by Fig. 7.1.

The whole line is divided into n-sections, each section having line constants $\frac{1}{n}$ th of those for the whole line.

The following points are noteworthy:

(*i*) The line constants (resistance, inductive reactance, capacitive susceptance and conductance) are uniformly distributed over the entire length of the line as is actually the case.

(*ii*) The resistance (R) and inductive reactance (X) are the series elements.

(*iii*) The capacitive susceptance (B) and leakage conductance (G) are shunt elements.

The leakage susceptance is due to the fact that the capacitance exists between line and neutral. The leakage conductance (G) takes into account the energy losses occurring through leakage over the insulators, or due to corona effect between the conductors.

$$\text{Thus shunt admittance, } Y = \sqrt{G^2 + B^2}$$

(*iv*) The leakage current through the shunt admittance is maximum at the sending end of the transmission line and decreases continuously as we move towards the receiving end of the line and becomes zero at the receiving end.

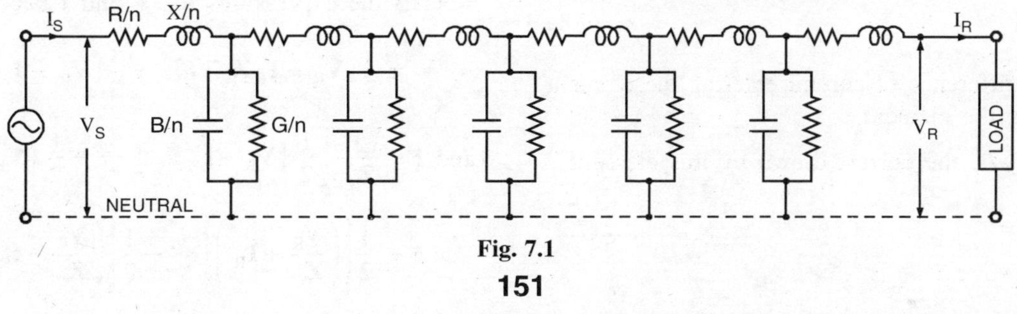

Fig. 7.1

7.2. ANALYSIS OF LONG TRANSMISSION LINES (RIGOROUS METHOD OF SOLUTION OF LONG TRANSMISSION LINES)

The exact equivalent circuit of a transmission line is shown in Fig. 7.2.

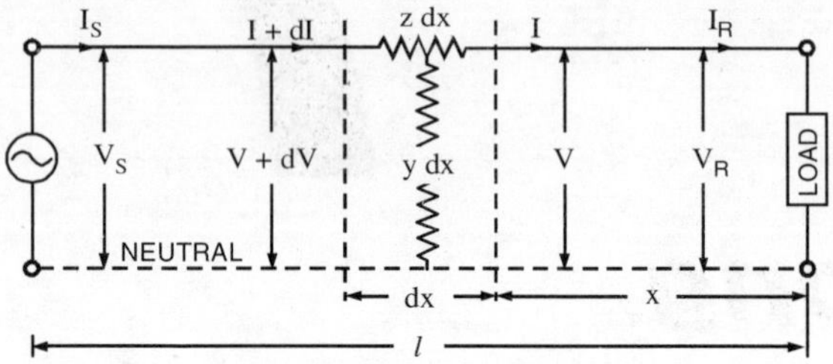

Fig. 7.2

Consider an infinitely small length dx of the line at a distance x from the receiving end.

Let r = Resistance per unit length of line

x = Reactance per unit length of line

b = Susceptance per unit length of line

g = Conductance per unit length of line

$z = \sqrt{r^2 + x^2}$, impedance per unit length of line

$y = \sqrt{g^2 + b^2}$, admittance per unit length of line

$\mathbf{V}$ = Voltage per phase at the end of element towards receiving end

$\mathbf{V} + d\mathbf{V}$ = Voltage per phase at the end of element towards sending end

$\mathbf{V}_R$ = Voltage per phase at receiving end

$\mathbf{V}_S$ = Voltage per phase at sending end

$\mathbf{I}_R$ = Current per phase at receiving end

$\mathbf{I}_S$ = Current per phase at sending end

$\mathbf{I} + d\mathbf{I}$ = Current entering the element dx

and $\mathbf{I}$ = Current leaving the element dx

Now the series impedance of element dx of the line = $z\,dx$

The shunt admittance of element dx of the line = $y\,dx$

Obviously the rise in voltage over the element length in the direction of increasing x

$d\mathbf{V} = \mathbf{I}z\,dx$, voltage drop in the element dx

$$\text{or} \quad \frac{d\mathbf{V}}{dx} = z\mathbf{I} \qquad \text{...(7.1)}$$

Similarly the difference of current entering the element and that of leaving the element,

$d\mathbf{I} = \mathbf{V}y dx$, the current drawn by this element

$$\text{or} \quad \frac{d\mathbf{I}}{dx} = \mathbf{V}y \qquad \text{...(7.2)}$$

Differentiating Eq. (7.1) w.r.t. x, we get

$$\frac{d^2\mathbf{V}}{dx^2} = z\frac{d\mathbf{I}}{dx} = z\mathbf{V}y \qquad \text{...(7.3)}$$

$$\because \text{ From Eq. (7.2) } \frac{d\mathbf{I}}{dx} = \mathbf{V}y$$

The solution of above differential equation is

$$\mathbf{V} = A_1 e^{\sqrt{yz}\,x} + A_2 e^{-\sqrt{yz}\,x} \qquad \text{...(7.4)}$$

where A_1 and A_2 are unknown constants. Differentiating Eq. (7.4) w.r.t. x, we get

$$\frac{d\mathbf{V}}{dx} = \sqrt{yz}\,\{A_1 e^{\sqrt{yz}x} - A_2 e^{-\sqrt{yz}x}\} \qquad \text{...(7.5)}$$

and from Eq. (7.1), we have

$$\mathbf{I} = \frac{1}{z}\frac{d\mathbf{V}}{dx}$$

$$\therefore \mathbf{I} = \frac{1}{z}\sqrt{yz}\,\{A_1 e^{\sqrt{yz}x} - A_2 e^{-\sqrt{yz}x}\}$$

$$= \sqrt{y/z}\,\{A_1 e^{\sqrt{yz}x} - A_2 e^{-\sqrt{yz}x}\} \qquad \text{...(7.6)}$$

Equations (7.4) and (7.6) thus give the expressions for **V** and **I** in the form of unknown constants A_1 and A_2. The values of A_1 and A_2 can be determined by applying receiving-end conditions as under

At receiving-end $x = 0$, $\mathbf{V} = \mathbf{V}_R$ and $\mathbf{I} = \mathbf{I}_R$

Substituting these values in Eqs. (7.4) and (7.6), we get

$$\mathbf{V}_R = A_1 + A_2 \qquad \text{...(7.7)}$$

$$\text{and } \mathbf{I}_R = \sqrt{\frac{y}{z}}(A_1 - A_2) \qquad \text{...(7.8)}$$

For a transmission line $\sqrt{\frac{z}{y}}$ is a constant, called the *characteristic constant,* Z_C. and $\sqrt{yz}$ is another constant, called the *propagation constant,* γ, both complex quantities.

From Eqs. (7.7) and (7.8), we have

$$A_1 = \frac{1}{2}\mathbf{V}_R + \frac{1}{2}\sqrt{\frac{z}{y}}\mathbf{I}_R = \frac{1}{2}[\mathbf{V}_R + \mathbf{I}_R\mathbf{Z}_C]$$

$$\text{and } A_2 = \frac{1}{2}\mathbf{V}_R - \frac{1}{2}\sqrt{\frac{z}{y}}\mathbf{I}_R = \frac{1}{2}[\mathbf{V}_R - \mathbf{I}_R\mathbf{Z}_C]$$

Thus the expressions for **V** and **I** become

$$\mathbf{V} = \frac{1}{2}[\mathbf{V}_R + \mathbf{I}_R\mathbf{Z}_C]\,e^{\gamma x} + \frac{1}{2}[\mathbf{V}_R - \mathbf{I}_R\mathbf{Z}_C]\,e^{-\gamma x} \qquad \text{...(7.9)}$$

$$\text{and } \mathbf{I} = \frac{1}{\mathbf{Z}_C}\left[\frac{1}{2}\{\mathbf{V}_R + \mathbf{I}_R\mathbf{Z}_C\}e^{\gamma x} - \frac{1}{2}\{\mathbf{V}_R - \mathbf{I}_R\mathbf{Z}_C\}e^{-\gamma x}\right]$$

$$\text{or } \mathbf{I} = \frac{1}{2}\left[\left(\frac{\mathbf{V}_R}{\mathbf{Z}_C} + \mathbf{I}_R\right)\right]e^{\gamma x} - \frac{1}{2}\left[\left(\frac{\mathbf{V}_R}{\mathbf{Z}_C} - \mathbf{I}_R\right)\right]e^{-\gamma x} \qquad \text{...(7.10)}$$

Expanding and rearranging Eqs. (7.9) and (7.10), we get

$$\mathbf{V} = \mathbf{V}_R \cosh \gamma x + \mathbf{I}_R \mathbf{Z}_C \sinh \gamma x \quad ...(7.11)$$

and $$\mathbf{I} = \mathbf{I}_R \cosh \gamma x + \frac{\mathbf{V}_R}{\mathbf{Z}_C} \sinh \gamma x \quad ...(7.12)$$

The sending-end voltage, $\mathbf{V}_S$ and sending-end current, $\mathbf{I}_S$ can be obtained by substituting $x = l$ in the above Eqs. (7.11) and (7.12)

$$\mathbf{V}_S = \mathbf{V}_R \cosh \gamma l + \mathbf{I}_R \mathbf{Z}_C \sinh \gamma l \quad ...(7.13)$$

and $$\mathbf{I}_S = \mathbf{I}_R \cosh \gamma l + \frac{\mathbf{V}_R}{\mathbf{Z}_C} \sinh \gamma l \quad ...(7.14)$$

Now $\gamma l = \sqrt{yz}\, l = \sqrt{yl \times zl} = \sqrt{\mathbf{YZ}}$

where **Z** is the total impedance of the line and **Y** is the total admittance of the line.

The expressions for sending-end voltage and sending-end currents are

$$\mathbf{V}_S = \mathbf{V}_R \cosh \sqrt{\mathbf{YZ}} + \mathbf{I}_R \mathbf{Z}_C \sinh \sqrt{\mathbf{YZ}} \quad ...(7.15)$$

and $$\mathbf{I}_S = \mathbf{I}_R \cosh \sqrt{\mathbf{YZ}} + \frac{\mathbf{V}_R}{\mathbf{Z}_C} \sinh \sqrt{\mathbf{YZ}} \quad ...(7.16)$$

Comparing above equations with the general voltage and current equations of the line, we have

$$\mathbf{A} = \mathbf{D} = \text{Cosh} \sqrt{\mathbf{YZ}}$$

$$\mathbf{B} = \mathbf{Z}_C \sinh \sqrt{\mathbf{YZ}}$$

and $$\mathbf{C} = \frac{1}{\mathbf{Z}_C} \sinh \sqrt{\mathbf{YZ}}$$

The relation **AD** – **BC** = 1 holds for

$$\mathbf{AD} - \mathbf{BC} = \cosh^2 \sqrt{\mathbf{YZ}} - \sinh^2 \sqrt{\mathbf{YZ}} = 1$$

Equations (7.13) and (7.14) can be written in matrix form as

$$\begin{bmatrix} \mathbf{V}_S \\ \mathbf{I}_S \end{bmatrix} = \begin{bmatrix} \cosh \gamma l & \mathbf{Z}_C \sinh \gamma l \\ \frac{1}{\mathbf{Z}_C} \sinh \gamma l & \cosh \gamma l \end{bmatrix} \begin{bmatrix} \mathbf{V}_R \\ \mathbf{I}_R \end{bmatrix} \quad ...(7.17)$$

Thus sending-end values can be obtained if $\mathbf{V}_R$, $\mathbf{I}_R$ and line parameters are known.

In case $\mathbf{V}_S$, $\mathbf{I}_S$ are known, $\mathbf{V}_R$ and $\mathbf{I}_R$ can be easily found by inverting the above Eq. (7.17)

$$\begin{bmatrix} \mathbf{V}_R \\ \mathbf{I}_R \end{bmatrix} = \begin{bmatrix} \cosh \gamma l & -\mathbf{Z}_C \sinh \gamma l \\ \frac{-1}{\mathbf{Z}_C} \sinh \gamma l & \cosh \gamma l \end{bmatrix} \begin{bmatrix} \mathbf{V}_S \\ \mathbf{I}_S \end{bmatrix} \quad ...(7.18)$$

7.3. EVALUATION OF CONSTANTS ABCD

ABCD constants can be evaluated by three methods given below:

1. *Convergent Series (Complex Angle) Method.*

In this method the hyperbolic sine and cosine are expressed in terms of their power series. The expansions are

$$\text{Cosh}\, x = 1 + \frac{x^2}{2!} + \frac{x^4}{4!} + \frac{x^6}{6!} + \ldots$$

and $$\sinh x = x + \frac{x^3}{3!} + \frac{x^5}{5!} + \frac{x^7}{7!} + \ldots$$

It is never worthwhile going beyond the seventh power. We then get

$$\mathbf{A} = \mathbf{D} = \text{Cosh} \sqrt{\mathbf{YZ}}$$

$$= 1 + \frac{\mathbf{YZ}}{2} + \frac{\mathbf{Y}^2\mathbf{Z}^2}{24} + \frac{\mathbf{Y}^3\mathbf{Z}^3}{720} + \ldots \quad ...(7.19)$$

$$\mathbf{B} = \mathbf{Z}_C \sinh \sqrt{\mathbf{YZ}}$$

$$= \mathbf{Z}_C \sqrt{\mathbf{YZ}} \left\{1 + \frac{\mathbf{YZ}}{6} + \frac{\mathbf{Y}^2\mathbf{Z}^2}{120} + \frac{\mathbf{Y}^3\mathbf{Z}^3}{5{,}040} + \ldots\right\}$$

$$= \sqrt{\frac{\mathbf{Z}}{\mathbf{Y}}} \times \sqrt{\mathbf{YZ}} \left\{1 + \frac{\mathbf{YZ}}{6} + \frac{\mathbf{Y}^2\mathbf{Z}^2}{120} + \frac{\mathbf{Y}^3\mathbf{Z}^3}{5{,}040} + \ldots\right\}$$

$$= \mathbf{Z} \left\{1 + \frac{\mathbf{YZ}}{6} + \frac{\mathbf{Y}^2\mathbf{Z}^2}{120} + \frac{\mathbf{Y}^3\mathbf{Z}^3}{5{,}040} + \ldots\right\} \quad ...(7.20)$$

and $$\mathbf{C} = \frac{1}{\mathbf{Z}_C} \sinh \sqrt{\mathbf{YZ}}$$

$$= \sqrt{\frac{\mathbf{Y}}{\mathbf{Z}}} \times \sqrt{\mathbf{YZ}} \left\{1 + \frac{\mathbf{YZ}}{6} + \frac{\mathbf{Y}^2\mathbf{Z}^2}{120} + \frac{\mathbf{Y}^3\mathbf{Z}^3}{5{,}040} + \ldots\right\}$$

$$= \mathbf{Y} \left[1 + \frac{\mathbf{YZ}}{6} + \frac{\mathbf{Y}^2\mathbf{Z}^2}{120} + \frac{\mathbf{Y}^3\mathbf{Z}^3}{5{,}040} + \ldots\right] \quad ...(7.21)$$

2. *Convergent Series (Real Angle) Method.* In this method the hyperbolic sines and cosines are expanded by the trigonometrical formulae given below, and tables of the trigonometric functions and hyperbolic functions of real numbers are used. Thus we express $\sqrt{\mathbf{YZ}}$ in the form

$\sqrt{\mathbf{YZ}} = \alpha + j\beta$ where both α and β are real

$$\text{Then } \cosh \sqrt{\mathbf{YZ}} = \text{Cosh}\,(\alpha + j\beta)$$

$$= \text{Cosh}\, \alpha \cosh j\beta + \sinh \alpha \sinh j\beta$$

$$= \text{Cosh}\, \alpha \cos \beta + j \sinh \alpha \sin \beta$$

$$\text{and } \sinh \sqrt{\mathbf{YZ}} = \text{Sinh}\, \alpha \cosh j\beta + \cosh \alpha \sinh j\beta$$

$$= \text{Sinh}\, \alpha \cos \beta + j \cosh \alpha \sin \beta.$$

3. *By Use of Hyperbolic Charts.* Woodruff has prepared a convenient set of charts for evaluating the hyperbolic functions of complex variables, usually required in transmission line calculations. But such charts are usually not available.

In general, method 1 is less laborious to use than method 2; moreover the tables of the sinh and cosh of complex numbers are not easy to use, and are not given for small differences.

7.4. SURGE IMPEDANCE

The square root of the ratio of line impedance (**Z**) and shunt admittance (**Y**) is called the *'surge impedance'* (Z_0) of the line

$$i.e., \quad \mathbf{Z}_0 = \sqrt{\frac{\mathbf{Z}}{\mathbf{Y}}}$$

where $\mathbf{Z} = R + jX$ and $\mathbf{Y} = G + jB$

Characteristic impedance ($\mathbf{Z}_C$) has also been defined in the same way. Actually characteristic impedance is a more correct nomenclature. Surge impedance is the characteristic impedance of a loss-free line.

For a heavy copper conductor and well insulated line the resistance (R) and leakage conductance (G) can be taken as zero and therefore, surge impedance, $Z_0 = \sqrt{\frac{L}{C}}$ which is a pure resistance. Its value varies between 400 Ω and 600 Ω in case of overhead transmission lines and between 40 Ω and 60 Ω incase of underground cables.

The surge impedance can also be evaluated by measuring the line impedance at the sending end when

(*i*) the line at the receiving end is open circuited.

(*ii*) the line at the receiving end is short circuited.

When the line is open circuited ($\mathbf{I}_R$) is zero so, the sending-end voltage ($\mathbf{V}_S$) and sending-end current ($\mathbf{I}_S$) are given as :

$$\mathbf{V}_S = \mathbf{A}\mathbf{V}_R \text{ and } \mathbf{I}_S = \mathbf{C}\mathbf{V}_R$$

$$\therefore \mathbf{Z}_{OC} = \frac{\mathbf{V}_S}{\mathbf{I}_S} = \frac{\mathbf{A}\mathbf{V}_R}{\mathbf{C}\mathbf{V}_R} = \frac{\mathbf{A}}{\mathbf{C}} \quad ...(7.22)$$

Similarly when the line is short circuited at the receiving end, the receiving-end voltage ($\mathbf{V}_R$) reduces to zero and the sending-end voltage ($\mathbf{V}_S$) and the sending-end current ($\mathbf{I}_S$) are given as:

$$\mathbf{V}_S = \mathbf{B}\mathbf{I}_R \text{ and } \mathbf{I}_S = \mathbf{D}\mathbf{I}_R$$

$$\therefore \quad \mathbf{Z}_{SC} = \frac{\mathbf{V}_S}{\mathbf{I}_S} = \frac{\mathbf{B}\mathbf{I}_R}{\mathbf{D}\mathbf{I}_R} = \frac{\mathbf{B}}{\mathbf{D}} \quad ...(7.23)$$

Multiplying Eqs. (7.22) and (7.23), we have

$$\mathbf{Z}_{OC} \cdot \mathbf{Z}_{SC} = \frac{\mathbf{A}}{\mathbf{C}} \times \frac{\mathbf{B}}{\mathbf{D}} = \frac{\mathbf{A}}{\mathbf{C}} \times \frac{\mathbf{B}}{\mathbf{A}} = \frac{\mathbf{B}}{\mathbf{C}} \quad ...(7.24)$$

∵ For a transmission line (which happens to be bilateral network) $\mathbf{D} = \mathbf{A}$

Substituting $\mathbf{B} = \sqrt{\frac{\mathbf{Z}}{\mathbf{Y}}} \sinh \sqrt{\mathbf{YZ}}$ and $\mathbf{C} = \sqrt{\frac{\mathbf{Y}}{\mathbf{Z}}} \sinh \sqrt{\mathbf{YZ}}$

in Eq. (7.24), we have

$$\mathbf{Z}_{OC} \cdot \mathbf{Z}_{SC} = \frac{\mathbf{B}}{\mathbf{C}} = \frac{\sqrt{\mathbf{Z}/\mathbf{Y}} \sinh \sqrt{\mathbf{YZ}}}{\sqrt{\mathbf{Y}/\mathbf{Z}} \sinh \sqrt{\mathbf{YZ}}} = \frac{\mathbf{Z}}{\mathbf{Y}} = Z_0^2 \quad ...(7.25)$$

$$\therefore \text{ Surge impedance } Z_0 = \sqrt{\mathbf{Z}_{OC}\,\mathbf{Z}_{SC}}$$

7.5. SURGE IMPEDANCE LOADING

This is defined as the load (of unity power factor) that can be delivered by the line of negligible resistance.

$$\therefore \text{ Power transmitted, } P_R = \frac{V_{RL}^2}{Z_0} \text{ MW} \quad ...(7.26)$$

where V_{RL} is the receiving-end voltage in kV and Z_0 is the surge impedance in ohms.

P_R is called the '*surge impedance loading*' or '*natural power*' of the line.

The above expression gives a limit of the maximum power* that can be delivered by a line and is useful in the design of transmission lines. This can be used for the comparison of loads that can be carried on the transmission lines at different voltages.

The power transmitted through a long transmission line can be increased either by increasing the value of receiving-end line voltage (V_{RL}) or by reducing surge impedance (Z_0).

Nowadays the trend is of employing higher and higher voltages for transmission, therefore, this is the most commonly adopted method for increasing the power limit of heavily loaded long transmission lines. But there is a limit beyond which it is neither economical nor practicable to increase the value of receiving-end line voltage.

Since the spacing between the conductors, which depends upon the line voltage employed, cannot be reduced much, so the value of surge impedance (Z_0) cannot be varied as such. However, some artificial means, such as series capacitors (capacitors in series with the transmission lines) or shunt capacitors (capacitors in parallel with the transmission lines), can be used to reduce the value of surge impedance (Z_0).

For a loss-free transmission line surge impedance, $Z_0 = \sqrt{\frac{L}{C}}$ and propagation constant, $\gamma = j\omega\sqrt{LC} = j\beta$, where β is the phase shift, which determines the torque angle δ between V_S and V_R and hence the system stability.

By use of series capacitors surge impedance (Z_0) and the phase shift (β) get reduced due to decrease of line inductance (L). This improves the system stability limit also. But it causes difficulties under short-circuit conditions of the system as no satisfactory protection of capacitors has yet been devised. These capacitors are also helpful in reducing line drops and so voltage variations.

By use of shunt capacitors though the surge impedance (Z_0) is reduced but the phase shift β is increased due to increase in the value of C. Hence stability conditions become worse specially when synchronous machines are used as loads. So this method is not used in case of long transmission lines specially where stability limits are present.

Example 7.1. A 3-phase, 50 Hz, 1,000 km long transmission line has the following line constants per phase per km uniformly distributed :

$$r = 0.22\ \Omega;\ x = 0.45\ \Omega;\ g = 4 \times 10^{-9}\ S \text{ and } b = 2.53 \times 10^{-6}\ S$$

Determine the auxiliary constants (*i*) by using convergent series of complex angles (*ii*) by using convergent series of real angles.

Solution : Impedance per km per phase,

$$\mathbf{z} = (r + jx) = (0.22 + j0.45)\ \Omega$$

* This may not necessarily be the maximum loading on the transmission lines as discussed elsewhere.

Admittance per km per phase, $\mathbf{y} = (0.004 + j2.53) \times 10^{-6}$ S

Total series impedance,

$$\mathbf{Z} = (0.22 + j0.45) \times 1{,}000$$
$$= (220 + j450)\ \Omega = 500.9\ \angle 63.95^\circ\ \Omega$$

Total shunt admittance,

$$\mathbf{Y} = (0.004 + j2.53) \times 10^{-6} \times 1{,}000$$
$$= (0.004 + j2.53) \times 10^{-3}\ \text{S} = 2.53 \times 10^{-3}\ \angle 89.91^\circ\ \text{S}$$

Hence

$$\mathbf{YZ} = 2.53 \times 10^{-3} \angle 89.91^\circ \times 500.9 \angle 63.95^\circ$$
$$= 1.267277\ \angle 153.86^\circ = (-1.1376 + j0.5583)$$
$$\mathbf{Y}^2\mathbf{Z}^2 = (1.267277)^2 \angle 307.72^\circ$$
$$= 1.606 \angle 307.72^\circ = (0.9826 - j1.27)$$
$$\mathbf{Y}^3\mathbf{Z}^3 = (1.267277)^3\ \angle 461.58^\circ = 2.0352354 \angle 101.58^\circ$$
$$= (-0.408 + j1.9938)$$

(*i*) Now auxiliary line constants

$$\mathbf{A} = \mathbf{D} = 1 + \frac{\mathbf{YZ}}{2} + \frac{\mathbf{Y}^2\mathbf{Z}^2}{24} + \frac{\mathbf{Y}^3\mathbf{Z}^3}{720} + \ldots$$
$$= 1 + \frac{-1.1376 + j0.5583}{2} + \frac{0.9826 - j1.27}{24} + \frac{-0.408 + j1.9938}{720} + \ldots$$
$$= (1 - 0.5688 + 0.0409 - 0.0005) + j(0.2792 - 0.0529 + 0.0028)$$
$$= 0.4716 + j0.2291$$
$$\mathbf{B} = \mathbf{Z}\left\{1 + \frac{\mathbf{YZ}}{6} + \frac{\mathbf{Y}^2\mathbf{Z}^2}{120} + \frac{\mathbf{Y}^3\mathbf{Z}^3}{5{,}040} + \ldots\right\}\ \Omega$$
$$= (220 + j450)\left\{1 + \frac{-1.1376 + j0.5583}{6} + \frac{0.9826 - j1.27}{120} + \frac{-0.408 + j1.9938}{5{,}040} + \ldots\right\}$$
$$= (220 + j450)\ [(1 - 0.1896 + 0.0081 - 0.00008) + j(0.09305 - 0.01058 + 0.0004)]\ \Omega$$
$$= (220 + j450)\ (0.81842 + j0.08287)\ \Omega$$
$$= 500.9 \angle 63.95^\circ \times 0.8226 \angle 5.78^\circ\ \Omega$$
$$= 412.04 \angle 69.73^\circ = (142.75 + j386.52)\ \Omega$$
$$\mathbf{C} = \mathbf{Y}\left\{1 + \frac{\mathbf{YZ}}{6} + \frac{\mathbf{Y}^2\mathbf{Z}^2}{120} + \frac{\mathbf{Y}^3\mathbf{Z}^3}{5{,}040} + \ldots\right\}\ \text{S}$$
$$= 2.53 \times 10^{-3} \angle 89.91^\circ \times 0.8226\ \angle 5.78^\circ\ \text{S}$$
$$= 2.08 \times 10^{-3} \angle 95.69^\circ = (-0.0002 + j0.00207)\ \text{S}$$

(*ii*) Using convergent series of real angles

We have $\mathbf{Z} = 500.9 \angle 63.59^\circ\ \Omega$ and $\mathbf{Y} = 2.53 \times 10^{-3} \angle 89.91^\circ$ S

$$\therefore\ \mathbf{YZ} = 2.53 \times 10^{-3} \angle 89.91^\circ \times 500.9 \angle 63.95^\circ = 1.26728\ \angle 153.86^\circ$$

and $\sqrt{\mathbf{YZ}} = 1.125735 \angle 76.93^\circ = 0.25458 + j1.09658$

Since $\alpha + j\beta = 0.25458 + j1.09658$

$\therefore\ \alpha = 0.25458$ radians or 14.586° and $\beta = 1.09658$ radians or 62.83°

From tables

Cosh α = Cosh $14.586^\circ = 1.03258$ and sinh α = sinh $14.586^\circ = 0.25734$

Cos β = Cos $62.83^\circ = 0.4566$ and sin β = Sin $62.83^\circ = 0.8896$

$$\text{Cosh}\ \sqrt{\mathbf{YZ}} = \text{Cosh } 14.586^\circ \cos 62.83^\circ + j \sinh 14.586^\circ \sin 62.53^\circ$$
$$= 1.03258 \times 0.4566 + j0.25734 \times 0.8896$$
$$= 0.4715 + j0.2289 = 0.524 \angle 25.9^\circ$$
$$\text{Sinh}\ \sqrt{\mathbf{YZ}} = \text{Sinh } \alpha \cos \beta + j \cosh \alpha \sin \beta$$
$$= \text{Sinh } 14.586^\circ \cos 62.83^\circ + j \cosh 14.586^\circ \sin 62.83^\circ$$
$$= 0.25734 \times 0.4566 + j1.03258 \times 0.8896$$
$$= 0.1175 + j0.9186 = 0.9261 \angle 82.71^\circ$$

Auxiliary line constants

$$\mathbf{A} = \mathbf{D} = \text{Cosh}\ \sqrt{\mathbf{YZ}} = 0.4715 + j0.2289 = 0.524 \angle 25.9^\circ$$
$$\mathbf{B} = \mathbf{Z}_C \sinh \sqrt{\mathbf{YZ}} = \sqrt{\frac{\mathbf{Z}}{\mathbf{Y}}} \times 0.9261 \angle 82.71^\circ$$
$$= \sqrt{\frac{500.9 \angle 63.95^\circ}{2.53 \times 10^{-3} \angle 89.91^\circ}} \times 0.9261 \angle 82.71^\circ$$
$$= 412.07 \angle 69.73^\circ\ \Omega$$
$$\mathbf{C} = \frac{1}{\mathbf{Z}_C} \sinh \sqrt{\mathbf{YZ}}$$
$$= \sqrt{\frac{2.53 \times 10^{-3} \angle 89.91^\circ}{500.9 \angle 63.95^\circ}} \times 0.9261\ \angle 82.71^\circ$$
$$= 2.08 \times 10^{-3} \angle 95.69^\circ\ \text{S}$$

Note: *Results obtained by both methods are almost same.*

Example 7.2. A 3-phase overhead transmission line has a total series impedance per phase of 200 $\angle 80^\circ$ ohms and a total shunt admittance of 0.0013 $\angle 90^\circ$ siemen per phase. The line delivers a load of 80 MW at 0.8 pf lagging and 220 kV between the lines. Determine the sending-end voltage and current by rigorous method.

Solution : Receiving-end phase voltage,

$$V_R = \frac{220{,}000}{\sqrt{3}} = 127{,}000\ \text{V}$$

$$\text{Receiving-end current, } I_R = \frac{80 \times 10^6}{\sqrt{3} \times 220{,}000 \times 0.8} = 262.43\ \text{A}$$

Power factor angle, $\phi = \text{Cos}^{-1}\ 0.8 = 36.87^\circ$ lagging

Taking receiving-end phase voltage as reference phasor, we have

$$\mathbf{V}_R = 127{,}000\ \angle 0^\circ\ \text{V and } \mathbf{I}_R = 262.43\ \angle -36.87^\circ\ \text{A}$$

Series impedance, $\mathbf{Z} = 200 \angle 80^\circ\ \Omega$; Shunt admittance, $\mathbf{Y} = 0.0013 \angle 90^\circ$ S

$$\mathbf{YZ} = 0.0013 \angle 90^\circ \times 200 \angle 80^\circ = 0.26 \angle 170^\circ$$
$$= -0.25605 + j0.04515$$

$Y^2Z^2 = 0.26^2 \angle 340° = 0.0676 \angle 340° = 0.06352 - j0.02312$

$Y^3Z^3 = 0.26^3 \angle 510° = 0.017576 \angle 510°$
$= -0.01522 + j0.008788$

Now auxiliary line constants,

$$A = D = 1 + \frac{YZ}{2} + \frac{Y^2Z^2}{24} + \frac{Y^3Z^3}{720} + ...$$

$$= 1 + \frac{-0.25605 + j\,0.04515}{2} + \frac{0.06352 - j\,0.02312}{24} + \frac{-0.01522 + j\,0.008788}{720} + ...$$

$$= (1 - 0.12802 + 0.00265 - 0.00002) + j(0.02257 - 0.00096 + 0.00001)$$

$$= (0.8746 + j0.02162) = 0.8749 \angle 1.42°$$

$$B = Z\left\{1 + \frac{YZ}{6} + \frac{Y^2Z^2}{120} + \frac{Y^3Z^3}{5{,}040} + ...\right\} \Omega$$

$$= 200 \angle 80° + \left\{1 + \frac{-0.25605 + j\,0.04515}{6} + \frac{0.06352 - j\,0.02312}{120} + \frac{-0.01522 + j0.008788}{5{,}040}\right\} \Omega$$

$$= 200 \angle 80° \{(1 - 0.042675 + 0.00053 - 0.000003) + j\,(0.007525 - 0.00019 + 0.000002)\}$$

$$= 200 \angle 80° (0.957852 + j0.007337)$$

$$= 200 \angle 80° \times 0.95788 \angle 0.43° \ \Omega = 191.57 \angle 80.43° \ \Omega$$

$$C = Y\left\{1 + \frac{YZ}{6} + \frac{Y^2Z^2}{120} + \frac{Y^3Z^3}{5{,}040} + ...\right\} S$$

$$= 0.0013 \angle 90° \times 0.95788 \angle 0.43° = 0.001245 \angle 90.43° \ S$$

Sending-end phase voltage,

$$V_S = AV_R + BI_R$$

$$= 0.8749 \angle 1.42° \times 127{,}000 \angle 0° + 191.57 \angle 80.43° \times 262.43 \angle -36.87°$$

$$= 111{,}112 \angle 1.42° + 50{,}273 \angle 43.56°$$

$$= 111{,}078 + j2{,}753 + 36{,}430 + j34{,}644$$

$$= 147{,}508 + j37{,}397 = 152{,}174 \angle 14.23° \text{ volts}$$

Sending-end line voltage,

$V_{SL} = \sqrt{3}\, V_S = \sqrt{3} \times 152{,}174 = 263{,}574$ volts $= 263.574$ kV **Ans.**

Sending-end current,

$$I_S = CV_R + DI_R$$

$$= 0.001245 \angle 90.43° \times 127{,}000 \angle 0° + 0.8749 \angle 1.42° \times 262.43 \angle -36.87°$$

$$= 158.12 \angle 90.43° + 229.6 \angle -35.45°$$

$$= -1.2115 + j158.1154 + 187.0372 - j133.1662$$

$= 185.8257 + j24.95 = 187.5 \angle 7.65°$ A **Ans.**

7.6. TRANSMISSION LINE WITH SERIES IMPEDANCE AT THE RECEIVING END

Line auxiliary constants, as determined before, are changed due to insertion of series impedance at the receiving end of a transmission line. Normally an impedance is supposed to consist of resistance and inductance and the current flowing through it is the same as the receiving-end current, I_R. The circuit is shown in Fig. 7.3.

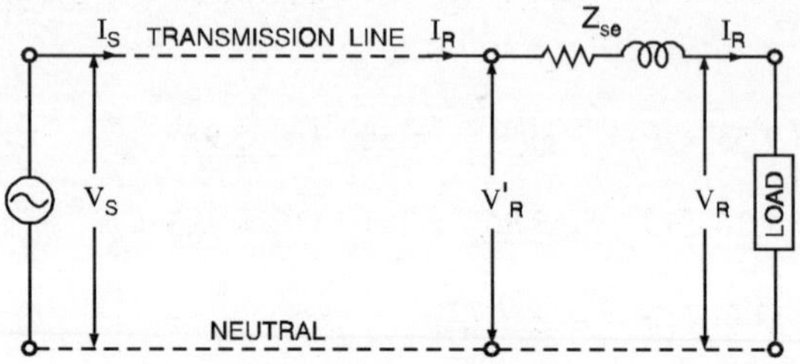

Fig. 7.3. *Transmission Line With Series Impedance At The Receiving End*

Let the auxiliary constants of the transmission line be **A**, **B**, **C** and **D** and Z_{se} by the value of series impedance of the line at the receiving-end. If V_R is the phase voltage at the load end and I_R is the load current, then the phase voltage at the end of transmission line will be

$$V'_R = V_R + I_R Z_{se} \qquad ...(7.27)$$

and $V_S = AV'_R + BI_R$

$$= A(V_R + I_R Z_{se}) + BI_R$$

$$= AV_R + (AZ_{se} + B)I_R \qquad ...(7.28)$$

and $I_S = CV'_R + DI_R$

Substituting the value of V'_R from Eq. (7.27), we have

$$I_S = C(V_R + I_R Z_{se}) + DI_R$$

$$= CV_R + (CZ_{se} + D)I_R \qquad ...(7.29)$$

Comparing with the general voltage and current expressions of the transmission line, we have

$$A_0 = A;\ B_0 = AZ_{se} + B;\ C_0 = C \text{ and } D_0 = CZ_{se} + D \qquad ...(7.30)$$

7.7. TRANSMISSION LINE WITH SERIES IMPEDANCE AT THE SENDING END

Figure 7.4. represents a transmission line with a series impedance Z_{se} at the sending end. In this case the current

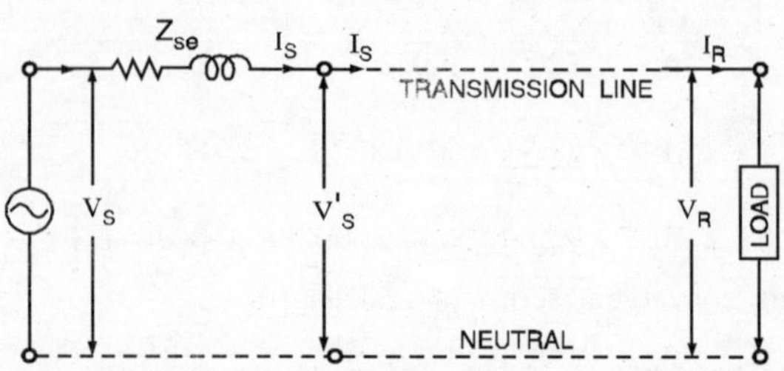

Fig. 7.4. *Transmission Line With Series Impedance At The Sending End*

passing through the series impedance Z_{se} is the same as the sending-end current I_S.

If **A, B, C** and **D** are the auxiliary constants of the transmission line, $\mathbf{V}_R$ and $\mathbf{V}_S$ are the phase voltages at the load and supply ends respectively, $\mathbf{I}_R$ is the load current and $\mathbf{V'}_S$ is the phase voltage at the starting end of the transmission line.

$$\text{then } \mathbf{V'}_S = \mathbf{AV}_R + \mathbf{BI}_R \quad ...(7.31)$$

$$\text{and } \mathbf{I}_S = \mathbf{CV}_R + \mathbf{DI}_R \quad ...(7.32)$$

$$\text{Also } \mathbf{V}_S = \mathbf{V'}_S + \mathbf{I}_S\mathbf{Z}_{se} \quad ...(7.33)$$

Substituting the values of $\mathbf{V'}_S$ and $\mathbf{I}_S$ from Eqs. (7.31) and (7.32) in Eq. (7.33), we have

$$\mathbf{V}_S = \mathbf{AV}_R + \mathbf{BI}_R + (\mathbf{CV}_R + \mathbf{DI}_R)\mathbf{Z}_{se}$$
$$= (\mathbf{A} + \mathbf{CZ}_{se})\mathbf{V}_R + (\mathbf{B} + \mathbf{DZ}_{se})\mathbf{I}_R \quad ...(7.34)$$

Comparing Eqs. (7.34) and (7.32) with the general expressions for voltage and current, we have

$$\mathbf{A}_0 = \mathbf{A} + \mathbf{CZ}_{se};\ \mathbf{B}_0 = \mathbf{B} + \mathbf{DZ}_{se};\ \mathbf{C}_0 = \mathbf{C} \text{ and } \mathbf{D}_0 = \mathbf{D} = \mathbf{A} \quad ...(7.35)$$

7.8. TRANSMISSION LINE WITH SERIES IMPEDANCES AT BOTH ENDS

In this case there are two series impedances, one at the receiving end and the other at the sending end, as shown in Fig. 7.5.

Now we have

$$\mathbf{V'}_R = \mathbf{V}_R + \mathbf{I}_R\mathbf{Z}_{seR} \quad ...(7.36)$$

$$\text{and } \mathbf{V'}_S = \mathbf{AV'}_R + \mathbf{BI}_R$$
$$= \mathbf{A}(\mathbf{V}_R + \mathbf{I}_R\mathbf{Z}_{seR}) + \mathbf{BI}_R$$
$$= \mathbf{AV}_R + (\mathbf{B} + \mathbf{A}\,\mathbf{Z}_{seR})\mathbf{I}_R \quad ...(7.37)$$

$$\mathbf{I}_S = \mathbf{CV'}_R + \mathbf{DI}_R$$
$$= \mathbf{C}(\mathbf{V}_R + \mathbf{I}_R\mathbf{Z}_{seR}) + \mathbf{DI}_R$$
$$= \mathbf{CV}_R + (\mathbf{D} + \mathbf{C}\,\mathbf{Z}_{seR})\mathbf{I}_R \quad ...(7.38)$$

$$\text{and } \mathbf{V}_S = \mathbf{V'}_S + \mathbf{I}_S\mathbf{Z}_{seS}$$
$$= \mathbf{AV}_R + (\mathbf{B} + \mathbf{AZ}_{seR})\mathbf{I}_R + \{\mathbf{CV}_R + (\mathbf{D} + \mathbf{CZ}_{seR})\,\mathbf{I}_R\}\mathbf{Z}_{seS}$$
$$= (\mathbf{A} + \mathbf{CZ}_{seS})\mathbf{V}_R + \{\mathbf{B} + \mathbf{AZ}_{seR} + (\mathbf{D} + \mathbf{CZ}_{seR})\mathbf{Z}_{seS}\}\mathbf{I}_R \quad ...(7.39)$$

Comparing Eqs. (7.39) and (7.38) with the general expressions for voltage and current of the transmission line, we have

$$\mathbf{A}_0 = \mathbf{A} + \mathbf{CZ}_{seS}$$
$$\mathbf{B}_0 = \mathbf{B} + \mathbf{AZ}_{seR} + (\mathbf{D} + \mathbf{CZ}_{seR})\mathbf{Z}_{seS} \quad ...[7.40(a)]$$
$$= \mathbf{B} + \mathbf{AZ}_{seR} + (\mathbf{A} + \mathbf{CZ}_{seR})\mathbf{Z}_{seS} \quad \because \mathbf{A} = \mathbf{D}$$
$$= \mathbf{B} + \mathbf{A}(\mathbf{Z}_{seR} + \mathbf{Z}_{seS}) + \mathbf{CZ}_{seR}\mathbf{Z}_{seS} \quad ...[7.40(b)]$$
$$\mathbf{C}_0 = \mathbf{C} \quad ...[7.40(c)]$$
$$\mathbf{D}_0 = \mathbf{D} + \mathbf{C}\,\mathbf{Z}_{seR} \quad ...[7.40(d)]$$

7.9. TRANSMISSION LINE WITH TRANSFORMERS AT BOTH ENDS

The transmission line may have a step-up transformer at the sending end and step-down transformer at the receiving end. The transformer has series impedance $\mathbf{Z}_T$ and shunt admittance $\mathbf{Y}_T$. The shunt admittance of transformer is usually so high that in most of the cases it may be taken as infinite without much loss of accuracy in which case transformers are just equivalent to series impedances and the equivalent circuit is, as shown in Fig. 7.5.

In case the shunt admittance of transformers are to be taken into account, the equivalent circuit will be as shown in Fig. 7.6.

In this case the sending-end voltage and current for the transmission line are $\mathbf{V'}_S$ and $\mathbf{I'}_S$ respectively.

Also the receiving-end voltage and current for the transmission line are $\mathbf{V'}_R$ and $\mathbf{I'}_R$ respectively.

$\mathbf{I}_{TS}$ and $\mathbf{I}_{TR}$ are the magnetising currents taken by the transformers at the sending end and receiving end of the transmission line respectively.

Now applying the transmission line treatment to the line, we have

$$\mathbf{V'}_S = \mathbf{AV'}_R + \mathbf{BI'}_R \quad ...(7.41)$$

$$\text{and } \mathbf{I}_{TR} = \mathbf{Y}_{TR}\mathbf{V}_R$$

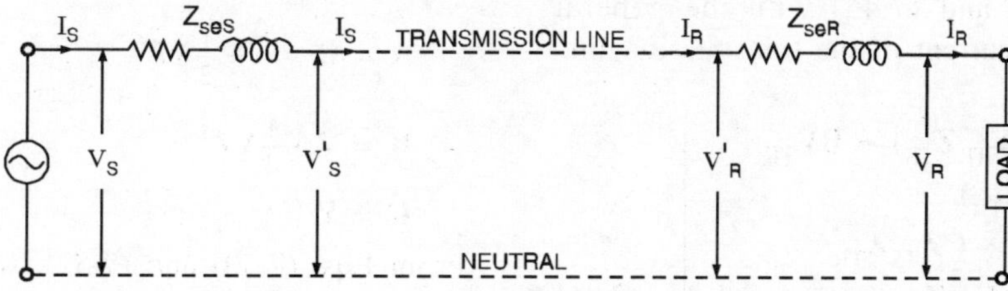

Fig. 7.5. *Transmission Line With Series Impedances At Both Ends*

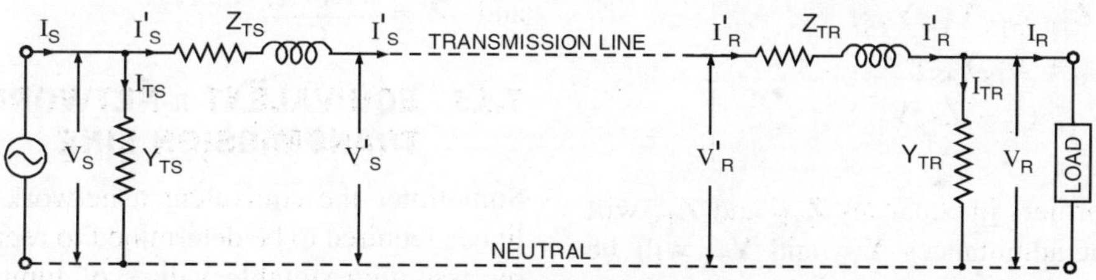

Fig. 7.6. *Transmission Line With Transformers At Both Ends*

$$I'_R = I_R + I_{TR} = I_R + Y_{TR}V_R \quad ...(7.42)$$

$$\therefore V'_R = V_R + I'_R Z_{TR}$$
$$= V_R + (I_R + I_{TR})\,Z_{TR}$$
$$= V_R + (I_R + Y_{TR}V_R)\,Z_{TR} \quad ...(7.43)$$

Substituting the value of V'_R from Eq. (7.43) and value of I'_R from Eq. (7.42) in Eq. (7.41), we get

$$V'_S = A\{V_R + (I_R + Y_{TR}V_R)Z_{TR}\} + B\,(I_R + Y_{TR}V_R)$$

Considering the sending-end current for the line

$$I'_S = CV'_R + DI'_R$$
$$= C\{V_R + (I_R + Y_{TR}V_R)Z_{TR}\} + D(I_R + Y_{TR}V_R)$$

Also

$$V_S = V'_S + I'_S Z_{TS}$$
$$= A\{V_R + (I_R + Y_{TR}V_R)Z_{TR}\} + B(I_R + Y_{TR}V_R) + CZ_{TS}\{V_R + (I_R + Y_{TR}V_R)Z_{TR}\} + DZ_{TS}(I_R + Y_{TR}V_R)$$
$$= [A(1 + Y_{TR}Z_{TR} + Z_{TS}Y_{TR}) + BY_{TR} + CZ_{TS}(1 + Y_{TR}Z_{TR})]V_R + [AZ_{TR} + B + CZ_{TS}Z_{TR} + AZ_{TS}]I_R \quad ...(7.44)$$

and

$$I_S = I'_S + I_{TS} \qquad \because A = D$$
$$= C\{V_R + (I_R + Y_{TR}V_R)Z_{TR}\} + D(I_R + Y_{TR}V_R) + Y_{TS}V_S$$
$$= C\{V_R + (I_R + Y_{TR}V_R)Z_{TR}\} + A(I_R + Y_{TR}V_R) + [\{A(1 + Y_{TR}Z_{TR} + Y_{TR}Z_{TS}) + BY_{TR} + CZ_{TS}(1 + Y_{TR}Z_{TR})\}V_R + \{AZ_{TR} + B + CZ_{TS}Z_{TR} + AZ_{TS}\}I_R]Y_{TS}$$
$$= [C + CY_{TR}Z_{TR} + AY_{TR} + AY_{TS} + AY_{TR}Z_{TR}Y_{TS} + AY_{TR}Z_{TS}Y_{TS} + BY_{TR}Y_{TS} + CZ_{TS}Y_{TS} + CZ_{TS}Y_{TR}Z_{TR}Y_{TS}]V_R + I_R[CZ_{TR} + A + AZ_{TR}Y_{TS} + BY_{TS} + CZ_{TS}Z_{TR}Y_{TS} + AZ_{TS}Y_{TS}]$$
$$= [C(1 + Y_{TR}Z_{TR} + Z_{TS}Y_{TS} + Z_{TS}Z_{TR}Y_{TR}Y_{TS}) + BY_{TR}Y_{TS} + A(Y_{TR} + Y_{TS} + Y_{TR}Y_{TS}Z_{TR} + Y_{TR}Y_{TS}Z_{TS})]V_R + [A(1 + Y_{TS}Z_{TR} + Z_{TS}Y_{TS}) + BY_{TS} + CZ_{TR}(1 + Z_{TS}Y_{TS})]I_R \quad ...(7.45)$$

Comparing Eqs. (7.44) and (7.45) with the general expressions for voltage and current of the transmission line we have

$$\left.\begin{aligned}
A_0 &= A(1 + Y_{TR}Z_{TR} + Y_{TR}Z_{TS}) + BY_{TR} + CZ_{TS}(1 + Y_{TR}Z_{TR})\\
B_0 &= A(Z_{TR} + Z_{TS}) + B + CZ_{TS}Z_{TR}\\
C_0 &= A(Y_{TR} + Y_{TS} + Y_{TR}Y_{TS}Z_{TR} + Y_{TR}Y_{TS}Z_{TS}) + BY_{TR}Y_{TS} + C(1 + Y_{TR}Z_{TR} + Z_{TS}Y_{TS} + Z_{TR}Z_{TS}Y_{TR}Y_{TS})\\
D_0 &= A(1 + Y_{TS}Z_{TR} + Z_{TS}Y_{TS}) + BY_{TS} + CZ_{TR}(1 + Z_{TS}Y_{TS})
\end{aligned}\right\} \quad ...(7.46)$$

For identical transformers impedances Z_{TR} and Z_{TS} will be equal and also shunt admittances Y_{TR} and Y_{TS} will be equal.

Let $Z_{TR} = Z_{TS} = Z_T$ (say) and $Y_{TR} = Y_{TS} = Y_T$ (say)

Substituting the above values of Z_{TR}, Z_{TS}, Y_{TR} and Y_{TS} in Eq. (7.46), we get

$$\left.\begin{aligned}
A_0 &= A(1 + 2Y_TZ_T) + BY_T + CZ_T(1 + Y_TZ_T)\\
B_0 &= 2AZ_T + B + CZ_T^2\\
C_0 &= 2AY_T(1 + Y_TZ_T) + BY_T^2 + C(1 + Y_TZ_T)^2\\
D_0 &= A(1 + 2Y_TZ_T) + BY_T + CZ_T(1 + Z_TY_T)
\end{aligned}\right\} \quad ...(7.47)$$
$$= A_0$$

7.10. EQUIVALENT T-NETWORK OF A LONG TRANSMISSION LINE

The equivalent T-network of a long transmission line is determined to represent the line accurately by assuming suitable values of lumped constants.

Let the transmission line having auxiliary constants A, B, C and D be represented by a T-network shown in Fig. 7.7.

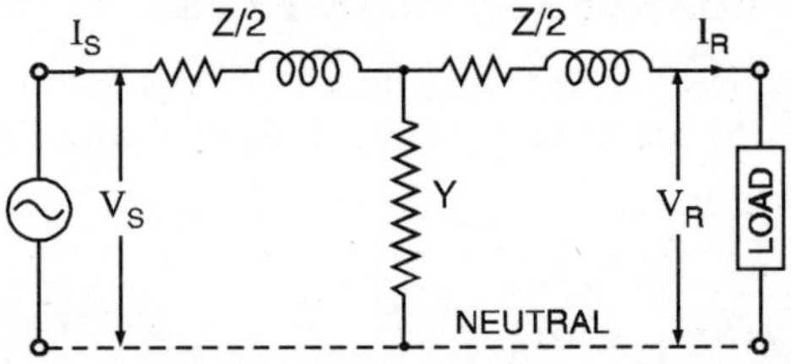

Fig. 7.7. *Equivalent T-Network of a Long Transmission Line*

Now we have

$$I_S = I_R + \left(V_R + \frac{1}{2}I_RZ\right)Y = YV_R + \left(1 + \frac{1}{2}YZ\right)I_R \quad ...(7.48)$$

and

$$V_S = V_R + \frac{1}{2}I_RZ + \frac{1}{2}I_SZ$$
$$= V_R + \frac{1}{2}I_RZ + \frac{1}{2}Z\left\{YV_R + \left(1 + \frac{1}{2}YZ\right)I_R\right\}$$
$$= \left(1 + \frac{1}{2}YZ\right)V_R + \left(Z + \frac{1}{4}YZ^2\right)I_R \quad ...(7.49)$$

Comparing Eqs. (7.49) and (7.48) with the general expressions for voltage and current of the transmission line, we have

$$A = D = 1 + \frac{1}{2}YZ \quad ...(7.50)$$

$$B = Z + \frac{1}{4}YZ^2 \quad ...(7.51)$$

$$C = Y \quad ...(7.52)$$

From Eqs. (7.50) and (7.52), we have

$$Y = C \quad ...(7.53)$$

and

$$Z = \frac{2(A-1)}{C} \quad ...(7.54)$$

7.11. EQUIVALENT π-NETWORK OF A LONG TRANSMISSION LINE

Sometimes the equivalent π-network of a long transmission line is required to be determined to represent the line accurately by assuming suitable values of lumped constants.

Let the transmission line having auxiliary constants A, B, C and D be represented by a π-network shown in Fig. 7.8.

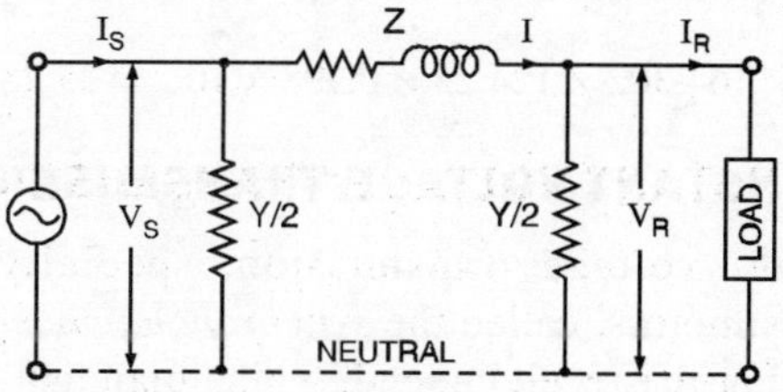

Fig. 7.8. ***Equivalent π-Network of a Long Transmission Line***

The sending-end voltage and current are given as

$$\mathbf{V_S} = \mathbf{V_R} + \mathbf{IZ} = \mathbf{V_R} + \left(\mathbf{I_R} + \frac{1}{2}\mathbf{YV_R}\right)\mathbf{Z}$$

$$= \left(1 + \frac{1}{2}\mathbf{YZ}\right)\mathbf{V_R} + \mathbf{ZI_R} \quad ...(7.55)$$

and $\mathbf{I_S} = \mathbf{I_R} + \frac{1}{2}\mathbf{YV_R} + \frac{1}{2}\mathbf{YV_S}$

$$= \mathbf{I_R} + \frac{1}{2}\mathbf{YV_R} + \frac{1}{2}\mathbf{Y}\left\{\left(1 + \frac{1}{2}\mathbf{YZ}\right)\mathbf{V_R} + \mathbf{ZI_R}\right\}$$

$$= \left(\mathbf{Y} + \frac{1}{4}\mathbf{Y^2Z}\right)\mathbf{V_R} + \left(1 + \frac{1}{2}\mathbf{YZ}\right)\mathbf{I_R} \quad ...(7.56)$$

Comparing Eqs. (7.55) and (7.56) with the general expressions for voltage and current of the transmissions line, we have

$$\mathbf{A} = \mathbf{D} = 1 + \frac{1}{2}\mathbf{YZ} \quad ...(7.57)$$

$$\mathbf{B} = \mathbf{Z} \quad ...(7.58)$$

$$\mathbf{C} = \mathbf{Y} + \frac{1}{4}\mathbf{Y^2Z} \quad ...(7.59)$$

From Eqs. (7.57) and (7.58) we get

$$\mathbf{Z} = \mathbf{B} \quad ...(7.60)$$

and $$\mathbf{Y} = \frac{2(\mathbf{A} - 1)}{\mathbf{B}} \quad ...(7.61)$$

Example 7.3. A 300 km, 132 kV 3-phase overhead line has a total series impedance of 52 + *j*200 ohms per phase and a total shunt admittance of *j*1.5 × 10^{-3} siemens per phase to neutral. The line is supplying 40 MVA at 0.8 pf lagging at 132 kV. Find the sending-end voltage, current, power factor and power.

Use (*a*) nominal-π circuit (*b*) long line equations.

[*Rajasthan Univ. Transmission & Distribution of Electrical Power 2000*]

Solution : Receiving-end phase voltage,

$$V_R = \frac{132{,}000}{\sqrt{3}} = 76{,}210 \text{ V}$$

Receiving-end current,

$$I_R = \frac{40 \times 10^6}{\sqrt{3} \times 132{,}000} = 174.95 \text{ A}$$

Power factor angle,

$$\phi = \text{Cos}^{-1}\ 0.8 = 36.87° \text{ (lagging)}$$

Taking receiving-end phase voltage as reference phasor, we have

$$\mathbf{V_R} = 76{,}210 \angle 0° \text{ V}$$

$$\mathbf{I_R} = 174.95 \angle -36.87° \text{ A}$$

Total series impedance, $\mathbf{Z} = 52 + j200 = 206.65 \angle 75.43°\ \Omega$

Total shunt admittance, $\mathbf{Y} = 1.5 \times 10^{-3} \angle 90°$ siemens

(*a*) *For the nominal circuit*

$$\mathbf{A} = \mathbf{D} = 1 + \frac{\mathbf{YZ}}{2}$$

$$= 1 + \frac{1.5 \times 10^{-3} \angle 9° \times 206.65 \angle 75.43°}{2}$$

$$= 0.85 + j0.039 = 0.85 \angle 2.63°$$

$$\mathbf{B} = \mathbf{Z} = 206.65 \angle 75.43°\ \Omega$$

$$\mathbf{C} = \mathbf{Y}\left(1 + \frac{\mathbf{YZ}}{4}\right)$$

$$= 1.5 \times 10^{-3} \angle 90° \left[1 + \frac{1.5 \times 10^{-3} \angle 90° \times 206.65 \angle 75.43°}{4}\right]$$

$$= 1.3875 \times 10^{-3} \angle 91.2° \text{ siemens}$$

Sending-end voltage per phase,

$$\mathbf{V_S} = \mathbf{AV_R} + \mathbf{BI_R}$$

$$= 0.85 \angle 2.63° \times 76{,}210 \angle 0° + 206.65 \angle 75.43° \times 174.95 \angle -36.87°$$

$$= 64{,}778.5 \angle 2.63° \times 36{,}153.4 \angle 38.56°$$

$$= 92{,}920 + j25{,}508 = 96{,}358 \angle 15.35° \text{ V}$$

Sending-end line voltage,

$$V_{SL} = \sqrt{3}V_S = \sqrt{3} \times 96{,}358 = 166{,}897 \text{ V or } 166.897 \text{ kV} \quad \textbf{Ans.}$$

Sending-end current,

$$\mathbf{I_S} = \mathbf{CV_R} + \mathbf{DI_R}$$

$$= 1.3875 \times 10^{-3} \angle 91.2° \times 76{,}210 \angle 0°$$

$$+ 0.85 \angle 2.63° \times 174.95 \angle -36.87°$$

$$= 105.74 \angle 91.2° + 148.71 \angle -34.24°$$

$$= 120.72 + j22.0 = 122.7° \angle 10.33° \text{ A}$$

Sending-end current,

$I_S = 122.7$ A **Ans.**

Sending-end power factor,

$$\text{Cos}\ \phi_S = \text{Cos}\ (15.35° - 10.33°) = 0.9962 \text{ (lagging)} \quad \textbf{Ans.}$$

Sending-end power,

$$P_S = \sqrt{3}\ V_{SL} I_{SL} \cos\phi_S \times 10^{-6} \text{ MW}$$

$$= \sqrt{3} \times 166{,}897 \times 122.7 \times 0.9962 \times 10^{-6}$$

$$= 36.335 \text{ MW} \quad \textbf{Ans.}$$

(*b*) $$\mathbf{YZ} = 1.5 \times 10^{-3} \angle 90° \times 206.65 \angle 75.43°$$

$$= 0.31 \angle 165.43° = -0.3 + j0.078$$

$$Y^2Z^2 = 0.31^2 \angle 330.86° = 0.084 - j0.0468$$

$$Y^3Z^3 = 0.31^3 \angle 496.3° = 0.0298 \angle 136.3° = -0.0215 + j0.0206$$

Now auxiliary line constants

$$A = D = 1 + \frac{YZ}{2} + \frac{Y^2Z^2}{24} + \frac{Y^3Z^3}{720}$$

$$= 1 + \frac{-0.3 + j0.078}{2} + \frac{0.084 - j0.0468}{24} + \frac{-0.0215 + j0.0206}{720}$$

$$= 1 - 0.15 + j0.039 + 0.0035 - j0.00195 - 0.00003 + j0.000029$$

$$= 0.85347 + j0.037 = 0.854 \angle 2.48°$$

$$B = \left[1 + \frac{YZ}{6} + \frac{Y^2Z^2}{120} + \frac{Y^3Z^3}{5{,}040}\right] \times Z$$

$$= [1 - 0.05 + j0.013 + 0.0007 - j0.00039 - 0.0000043 + j0.000004] \times 206.65 \angle 75.43°$$

$$= (0.9507 + j0.012614) \times 206.65 \angle 75.43°$$

$$= 0.95078 \angle 0.76° \times 206.65 \angle 75.43°$$

$$= 196.48 \angle 76.19° \, \Omega$$

$$C = Y\left[1 + \frac{YZ}{6} + \frac{Y^2Z^2}{120} + \frac{Y^3Z^3}{5{,}040}\right]$$

$$= 1.5 \times 10^{-3} \angle 90° \times 0.95078 \angle 0.76°$$

$$= 1.426 \times 10^{-3} \angle 90.76° \text{ siemens}$$

Sending-end voltage,

$$V_S = AV_R + BI_R$$

$$= 0.854 \angle 2.48° \times 76{,}210 \angle 0° + 196.48 \angle 76.19° \times 174.95 \angle -36.87°$$

$$= 65{,}083.34 \angle 2.48° + 34{,}374.18 \angle 39.32°$$

$$= 91{,}614.9 + j24{,}597.4 = 94{,}860 \angle 15.03° \text{ V}$$

Sending-end line voltage,

$$V_{SL} = \sqrt{3} \times 94{,}860 = 164.302 \text{ V or } 164.3 \text{ kV}$$ **Ans.**

Sending-end current,

$$I_S = CV_R + DI_R$$

$$= 1.426 \times 10^{-3} \angle 90.76° \times 76{,}210 \angle 0° + 0.854 \angle 2.48° \times 174.95 \angle -36.87°$$

$$= 108.675 \angle 90.76° + 149.4073 \angle -34.39°$$

$$= 121.85 + j24.277 = 124.25 \angle 11.27° \text{ A}$$

Sending-end current,

$I_S = 124.25$ A **Ans.**

Sending-end power factor,

$\cos \phi_S = \cos (15.03° - 11.27°) = 0.998$ (lagging)

Sending-end power,

$$P_S = \sqrt{3}\, V_S I_S \cos \phi_S \times 10^{-6}$$

$$= \sqrt{3} \times 164{,}302 \times 124.25 \times 0.998 \times 10^{-6} = 35.28 \text{ MVA}$$ **Ans.**

7.12. CONSTANT VOLTAGE TRANSMISSION

For constant voltage transmission, specially designed synchronous motors, called the *synchronous phase modifiers,* are installed at the receiving end, which maintain the voltage drop along the line constant. With the change in load, power factor of the system is changed by the synchronous motors and thus voltage drop along the line remains constant. The merits and demerits of the system are given below :

Merits. (*i*) Possibility of carrying increased power for a given conductor size in case of long distance heavy power transmission.

(*ii*) Improvement of power factor at times of moderate and heavy loads.

(*iii*) Possibility of better protection for the line due to possible use of higher terminal reactances.

(*iv*) Availability of steady voltage at all loads at the line terminals.

(*v*) Improvement in system stability due to inertia effect of synchronous phase modifier and reduction in effect of sudden changes in load.

Demerits. (*i*) Increase of short-circuit current of the system and, therefore, increase in the circuit breaker ratings.

(*ii*) Increased risk of interruption of supply due to falling of synchronous motors, called the *synchronous phase modifiers,* out of synchronism.

(*iii*) Lower reserve of lines in case of line trouble.

7.13. SYNCHRONOUS PHASE MODIFIERS OR COMPENSATORS

It is well known that a synchronous motor can be made to take either a lagging or leading current from the line by altering its excitation. Idle-running synchronous motors were first employed in connection with power plants to correct for low power factor of the load, and thus reduce the current and power losses in the feeders and generators. The synchronous machines used for power factor improvement are usually referred to as '*synchronous condensers*' as they are always required to take a leading current.

However, an idle-running synchronous motor can be used for voltage regulating purposes by connecting it in parallel with the load at the receiving end of a line, as shown in Fig. 7.9 (*a*). Since the machines used for voltage regulation are required to run part of the time with leading current and part of the time with lagging current depending upon the load conditions, therefore, it is more appropriate to refer to these machines as '*synchronous phase modifiers*'. They are usually of salient pole design with 6 or 8 poles with ratings up to 60 MVA, 11 kV and connected to high-voltage system through transformers.

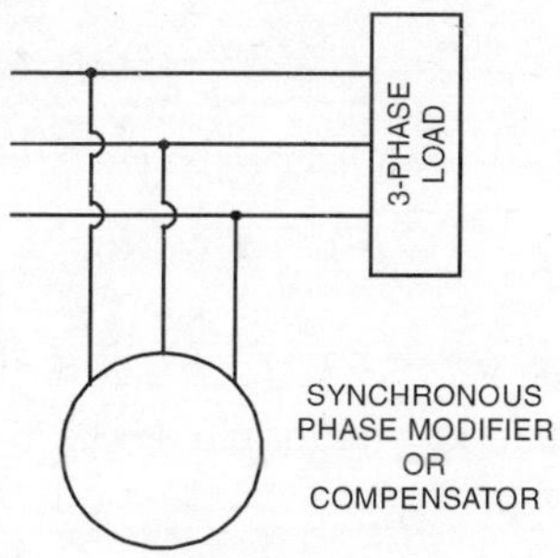

Fig. 7.9(*a*)

Synchronous phase modifiers differ from the ordinary synchronous motors in as much as they are built for the highest economical speeds, and provided with smaller shafts and bearings and special attention is paid for securing a high overall efficiency. Standard machines for this purpose are designed to give their full-load output at leading power factor, and can carry about 50% of their rated capacity at lagging power factor. Machines can, however, be designed to operate at full rating on both leading and lagging power factors but they are larger in size, have poor efficiency and are more expensive than standard machines.

A transmission line needs lagging VARs (reactive volt-amperes) at the receiving end during peak load conditions to avoid excessive voltage drop and these lagging VARs are supplied by the synchronous phase modifier when it is overexcited as it absorbs leading VARs when overexcited. During off-peak hours the transmission line needs leading VARs at the receiving end to prevent any voltage rise. These leading VARs are supplied by the synchronous phase modifier when it is underexcited as it absorbs lagging VARs when underexcited. Thus synchronous phase modifier or compensator is a very flexible source of supply of reactive VARs (leading or lagging as per need) and is well suited for voltage regulation (or compensation).

7.13.1. Determination of Synchronous Phase Modifier Capacity

The generalized equation for a transmission line is given as

$$\mathbf{V_S} = \mathbf{AV_R} + \mathbf{BI_R} \quad ...(7.62)$$

where V_S and V_R are sending-end and receiving-end voltages, I_R is the receiving end current and A and B are generalized constants.

Taking receiving-end phase voltage V_R as reference phasor, [Fig. 7.9(*b*)] the above equation may be rewritten as

$$V_S \angle\delta = AV_R(\alpha) + BI_R(\beta - \phi_R)$$
$$= AV_R \cos\alpha + jAV_R \sin\alpha + BI_R \cos(\beta - \phi_R) + jBI_R \sin(\beta - \phi_R)$$
$$V_S^2 = A^2V_R^2 + B^2I_R^2 + 2ABV_RI_R \cos\alpha \cos(\beta - \phi_R) + 2ABV_RI_R \sin\alpha \sin(\beta - \phi_R)$$
$$= A^2V_R^2 + B^2I_R^2 + 2ABV_RI_R \cos(\alpha - \beta + \phi_R)$$
$$= A^2V_R^2 + B^2I_R^2 + 2ABV_RI_R [\cos(\alpha - \beta)\cos\phi_R - \sin(\alpha - \beta)\sin\phi_R]$$

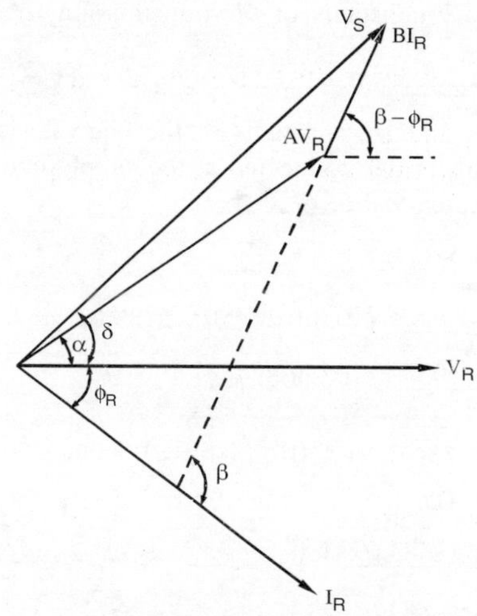

Fig. 7.9(*b*) ***Phasor Diagram Representing General Equation (7.62)***

Substituting $V_RI_R \cos\phi_R = P_R$ and $V_RI_R \sin\phi_R = Q_R$ in above equation, we have

$$V_S^2 = A^2V_R^2 + B^2I_R^2 + 2ABP_R \cos(\alpha - \beta) - 2ABQ_R \sin(\alpha - \beta) \quad ...(7.63)$$

Also $I_R = I_p - jI_q$

or $I_R^2 = I_p^2 + I_q^2$

and $I_p = \dfrac{P_R}{V_R}$ and $I_q = \dfrac{Q_R}{V_R}$

$$\text{So, } V_S^2 = A^2V_R^2 + B^2\left(\frac{P_R^2}{V_R^2} + \frac{Q_R^2}{V_R^2}\right) + 2ABP_R \cos(\alpha - \beta) - 2AB\,Q_R \sin(\alpha - \beta) \quad ...(7.64)$$

Sending-end voltage can be determined from the above expression if A, B, α, β, V_R and P_R and Q_R are given, as in normal case.

But if V_S, V_R are fixed and A, B, α, β, P_R and Q_R (load) are given and the capacity of the phase modifier is to be determined, the required quantity is Q_R. In this case Q_R is the net reactive power at the receiving end and not the reactive power for the load as in the first type of problem as mentioned above. So, if the net reactive power required to maintain certain voltages at the two ends is known, the capacity of the phase modifier can be determined.

Example 7.4. A 220 kV, 50 Hz, 3-phase transmission line is connected to station bus-bars at the sending end through a 10 MVA; 11/220 kV transformer having impedance of (1 + *j*8)%. The auxiliary constants of the transmission line are :

$A = D = 0.9 \angle 0.6°$; $B = 153.2 \angle 84.6°$ Ω and $C = 0.0012 \angle 90°$ S.

Determine the A_0, B_0, C_0 and D_0 constants of the equivalent circuit of the line and the transformer connected at the supply end. *[Pb. Technical Univ. Power System-I (T&D), May 2010]*

Solution : Transformer impedance = (1 + *j*8)%

Impedance of transformers in ohms referred to HV side,

$$\mathbf{Z}_{se} = \frac{(1 + j8)}{100} \times \frac{(220)^2}{100}$$
$$= (4.84 + j\,38.72)\ \Omega/\text{phase} = 39.02 \angle 82.87°\ \Omega/\text{phase}$$

The auxiliary constants of the transmission line are :

$\mathbf{A} = \mathbf{D} = 0.9 \angle 0.6°$, $\mathbf{B} = 153.2 \angle 84.6°$ Ω and $\mathbf{C} = 0.0012 \angle 90°$ S

$\mathbf{A}_0$, $\mathbf{B}_0$, $\mathbf{C}_0$ and $\mathbf{D}_0$ constants of the equivalent circuit of the line and the transformer connected at the supply end (Fig. 7.4) are given below [Refer to Eq. (7.35)]

$$\mathbf{A}_0 = \mathbf{A} + \mathbf{C}\mathbf{Z}_{se}$$

$$= 0.9 \angle 0.6° + 0.0012 \angle 90° \times 39.02 \angle 82.87°$$

$$= 0.9 \angle 0.6° + 0.046824 \angle 172.87°$$

$$= (0.8535166 + j0.0152346) = 0.8536 \angle 1.02° \text{ Ans.}$$

$$\mathbf{B}_0 = \mathbf{B} + \mathbf{D}\mathbf{Z}_{se}$$

$$= 153.2 \angle 84.6° + 0.9 \angle 0.6° \times 39.02 \angle 82.87°$$

$$= 153.2 \angle 84.6° + 35.118 \angle 83.43°$$

$$= 18.43549 + j187.40743 = 188 \angle 84.38° \ \Omega \text{ Ans.}$$

$$\mathbf{C}_0 = \mathbf{C} = 0.0012 \angle 90° \text{ S Ans.}$$

$$\mathbf{D}_0 = \mathbf{A} = 0.9 \angle 0.6° \text{ Ans.}$$

Example 7.5. A long transmission line has the following auxiliary constants:

$$\mathbf{A} = \mathbf{D} = \mathbf{0.98} \angle \mathbf{2°} \text{ and } \mathbf{B} = \mathbf{28} \angle \mathbf{69°} \ \Omega$$

Two similar transformers, each of impedance 12 ∠ 80° ohms are interposed at the two ends of the above transmission line. Determine the values of $\mathbf{A}_0$, $\mathbf{B}_0$, $\mathbf{C}_0$ and $\mathbf{D}_0$ to represent the line with the transformers.

Solution : The auxiliary constants of the transmission line are

$$\mathbf{A} = \mathbf{D} = 0.98 \angle 2° \text{ and } \mathbf{B} = 28 \angle 69° \ \Omega$$

$$\mathbf{C} = \frac{\mathbf{AD} - 1}{\mathbf{B}} \qquad \because \mathbf{AD} - \mathbf{BC} = 1$$

$$= \frac{0.98 \angle 2° \cdot 0.98 \angle 2° - 1}{28 \angle 69°}$$

$$= \frac{0.9604 \angle 4° - 1}{28 \angle 69°}$$

$$= \frac{0.9580605 + j0.067 - 1}{28 \angle 69°}$$

$$= \frac{0.079 \angle 122°}{28 \angle 69°} = 0.00282 \angle 53° \text{ S}$$

Series impedance connected at both ends of the transmission line

$$\mathbf{Z}_{se} = 12 \angle 80° \ \Omega$$

$\mathbf{A}_0$, $\mathbf{B}_0$, $\mathbf{C}_0$ and $\mathbf{D}_0$ constants of the equivalent circuit of the line and similar transformers connected at both ends (Fig. 7.5) are given below [Refer to Eq. (7.40)]

$$\mathbf{A}_0 = \mathbf{A} + \mathbf{C}\mathbf{Z}_{se}$$

$$= 0.98 \angle 2° + 0.00282 \angle 53° \times 12 \angle 80°$$

$$= 0.98 \angle 2° + 0.03384 \angle 133°$$

$$= (0.9794 + j0.0342) + (-0.0231 + j0.0248)$$

$$= (0.9563 + j0.059) = 0.958 \angle 3.53° \text{ Ans.}$$

$$\mathbf{B}_0 = \mathbf{B} + 2\mathbf{A}\mathbf{Z}_{se} + \mathbf{C}\mathbf{Z}_{se}^2$$

$$= 28 \angle 69° + 2 \times 0.98 \angle 2° \times 12 \angle 80°$$

$$+ 0.00282 \angle 53° \cdot (12 \angle 80°)^2$$

$$= 28 \angle 69° + 23.52 \angle 82° + 0.40608 \angle 213°$$

$$= (10.0343 + j26.1403) + (3.2734 + j23.2911) + (-0.3406 - j0.2212)$$

$$= 12.9671 + j49.2102 = 50.89 \angle 75.24° \ \Omega \text{ Ans.}$$

$$\mathbf{C}_0 = \mathbf{C} = 0.00282 \angle 53° \text{ S Ans.}$$

$$\mathbf{D}_0 = \mathbf{A}_0 = 0.958 \angle 3.53° \text{ Ans.}$$

Example 7.6. A long transmission line has the following auxiliary constants :

$$\mathbf{A} = \mathbf{D} = \mathbf{0.92} \angle \mathbf{53°} \ ; \ \mathbf{B} = \mathbf{65.3} \angle \mathbf{81°} \ \Omega$$

Two similar transformers, each of series impedance 100 ∠ 70° Ω and admittance 0.0002 ∠ – 75° S are connected at the two ends of the above transmission line. Determine $\mathbf{A}_0$, $\mathbf{B}_0$, $\mathbf{C}_0$ and $\mathbf{D}_0$ for the combination.

Solution : The auxiliary constants of the transmission lines are

$$\mathbf{A} = \mathbf{D} = 0.92 \angle 5.3° \text{ and } \mathbf{B} = 65.3 \angle 81° \ \Omega$$

$$\mathbf{C} = \frac{\mathbf{AD} - 1}{\mathbf{B}} = \frac{(0.92 \angle 5.3°)^2 - 1}{65.3 \angle 81°} = \frac{0.8464 \angle 10.6° - 1}{65.3 \angle 81°}$$

$$\because \mathbf{AD} - \mathbf{BC} = 1$$

$$= \frac{-0.1681 + j0.1557}{65.3 \angle 81°} = \frac{0.229 \angle 137.2°}{65.3 \angle 81°} = 0.0035 \angle 56.2° \text{ S}$$

Impedance of transformers connected at both ends of transmission line,

$$\mathbf{Z}_T = 100 \angle 70° \ \Omega$$

Shunt admittance of transformers connected at both ends of transmission line,

$$\mathbf{Y}_T = 0.0002 \angle -75° \text{ S}$$

$\mathbf{A}_0$, $\mathbf{B}_0$, $\mathbf{C}_0$ and $\mathbf{D}_0$ constants of the equivalent circuit of the transmission line and similar transformers connected at both ends (Fig. 7.6) are given below [Refer to Eq. (7.47)]

$$\mathbf{A}_0 = \mathbf{D}_0 = \mathbf{A}(1 + 2\mathbf{Y}_T\mathbf{Z}_T) + \mathbf{B}\mathbf{Y}_T + \mathbf{C}\mathbf{Z}_T(1 + \mathbf{Y}_T\mathbf{Z}_T)$$

$$= 0.92 \angle 5.3° (1 + 2 \times 0.0002 \angle -75° \times 100 \angle 70°)$$

$$+ 65.3 \angle 81° \times 0.0002 \angle -75° + 0.0035 \angle 56.2°$$

$$\times 100 \angle 70° (1 + 0.0002 \angle -75° \times 100 \angle 70°)$$

$$= 0.92 \angle 5.3° (1 + 0.04 \angle -5°) + 0.01306 \angle 6°$$

$$+ 0.35 \angle 126.2° (1 + 0.02 \angle -5°)$$

$$= 0.92 \angle 5.3° (1 + 0.03985 - j0.0034) + 0.01306 \angle 6°$$

$$+ 0.35 \angle 126.2° (1.02 - j0.0017)$$

$$= 0.92 \angle 5.3° \times 1.04 \angle -0.2° + 0.01306 \angle 6°$$

$$+ 0.35 \angle 126.2° \times 1.02 \angle -0.1°$$

$= 0.9568 \angle 5.1^\circ \times 0.01306 \angle 6^\circ + 0.357 \angle 126.1^\circ$

$= (0.9530 + j0.0850) + (0.013 + j0.0014) + (-0.2103 + j0.2884)$

$= 0.7557 + j0.3748 = 0.84354 \angle 26.38^\circ$ **Ans.**

$\mathbf{B_0} = 2\mathbf{AZ_T} + \mathbf{B} + \mathbf{C}\,\mathbf{Z_T^2}$

$= 2 \times 0.92 \angle 5.3^\circ \times 100 \angle 70^\circ + 65.3 \angle 81^\circ$

$+ 0.0035 \angle 56.2^\circ \times (100 \angle 70^\circ)^2$

$= 184 \angle 75.3^\circ + 65.3 \angle 81^\circ + 35 \angle 196.2^\circ$

$= (46.6915 + j\,177.9773) + (10.2152 + j\,64.496)$

$+ (-33.6103 - j\,9.7647)$

$= 23.2964 + 232.7086 = 233.87 \angle 84.28^\circ\ \Omega$ **Ans.**

$\mathbf{C_0} = 2\mathbf{AY_T}(1 + \mathbf{Y_T Z_T}) + \mathbf{B}\,\mathbf{Y_T^2} + \mathbf{C}(1 + \mathbf{Y_T Z_T})^2$

$= 2 \times 0.92 \angle 5.3^\circ \times 0.0002 \angle -75^\circ (1 + 0.0002 \angle -75^\circ$

$\times (100 \angle 70^\circ) + 65.3 \angle 81^\circ \times (0.0002 \angle -75^\circ)^2$

$+ 0.0035 \angle 56.2^\circ (1 + 0.0002 \angle -75^\circ \times 100 \angle 70^\circ)^2$

$= 0.000368 \angle -69.7^\circ (1 + 0.02 \angle -5^\circ) + 0.0000026 \angle -69^\circ$

$+ 0.0035 \angle 56.2^\circ (1 + 0.02 \angle -5^\circ)^2$

$= 0.000368 \angle -69.7^\circ (1.01992 - j0.00174) +$ zero

$+ 0.0035 \angle 56.2^\circ (1.01992 - j0.00174)^2$

$= 0.000368 \angle -69.7^\circ \times 1.02 \angle -0.1^\circ + 0.0035 \angle 56.2^\circ$
$\times 1.0404 \angle -0.2^\circ$

$= 0.000375 + \angle -69.8^\circ + 0.0036414 \angle 56^\circ$

$= (0.00013 - j0.000352) + (0.00204 + j0.003019)$

$= (0.00217 + j0.002667)$ S $= 0.003438 \angle 50.9^\circ$ S **Ans.**

Example 7.7. A long transmission line has the following auxiliary constants:

$\mathbf{A} = \mathbf{D} = \mathbf{0.945} \angle \mathbf{1.02^\circ}$, $\mathbf{B} = \mathbf{82.3} \angle \mathbf{73.03^\circ}\ \Omega$ **and** $\mathbf{C} = \mathbf{0.001376} \angle \mathbf{90.4^\circ}$ **S**

Determine (*i*) equivalent T-network and (*ii*) π-network.

Solution : (*i*) Equivalent T-network is shown in Fig. 7.7 in which Shunt admittance

$\mathbf{Y} = \mathbf{C} = 0.001376 \angle 90.4^\circ$ S **Ans.**

and impedance,

$$\mathbf{Z} = \frac{2(\mathbf{A} - 1)}{\mathbf{C}}$$

$$= \frac{2\,(0.945 \angle 1.02^\circ - 1)}{0.001376 \angle 90.4^\circ}$$

$$= 2 \times \frac{-0.0552 + j\,0.01682}{0.000688 \angle 90.4^\circ}$$

$$= 2 \times \frac{0.05766 \angle 163^\circ}{0.000688 \angle 90.4^\circ} = 167.75 \angle 72.6^\circ\ \Omega \textbf{ Ans.}$$

(*ii*) Equivalent π-network is shown in Fig. 7.8 in which Series impedance,

$\mathbf{Z} = \mathbf{B} = 82.3 \angle 73.03^\circ\ \Omega$ **Ans.**

and shunt admittance,

$$\mathbf{Y} = \frac{2(\mathbf{A} - 1)}{\mathbf{B}}$$

$$= \frac{2(0.945 \angle 1.02^\circ - 1)}{82.3 \angle 73.03^\circ}$$

$$= \frac{2 \times 0.05766 \angle 163^\circ}{82.3 \angle 73.03^\circ} = 0.001401 \angle 89.97^\circ \text{ S } \textbf{Ans.}$$

7.14. TUNED POWER LINES

The behaviour of a transmission line is described by the Eqs. (7.15) and (7.16), as given below :

$$\mathbf{V_S} = \mathbf{V_R} \cosh \sqrt{\mathbf{YZ}} + \mathbf{I_R}\,\mathbf{Z_C} \sinh \sqrt{\mathbf{YZ}}$$

$$\text{and } \mathbf{I_S} = \mathbf{I_R} \cosh \sqrt{\mathbf{YZ}} + \frac{\mathbf{V_R}}{\mathbf{Z_C}} \sinh \sqrt{\mathbf{YZ}}$$

For an overhead line shunt conductance G is always negligible and it is sufficiently accurate to neglect line resistance R as well. So with this approximation

$$\mathbf{Y} = j\omega Cl \text{ and } \mathbf{Z} = j\omega Ll$$

where L and C are the inductance and capacitance per unit length and l the length. Thus

$$\sqrt{\frac{\mathbf{Z}}{\mathbf{Y}}} = \sqrt{\frac{j\omega Ll}{j\omega Cl}} = \sqrt{\frac{L}{C}}$$

$$\text{and } \sqrt{\mathbf{YZ}} = \sqrt{j\omega Cl \times j\omega Ll} = j\omega l\sqrt{LC}$$

So the characteristic equations become

$$\mathbf{V_S} = \mathbf{V_R} \cosh (j\omega l\sqrt{LC}) + \mathbf{I_R Z_C} \sinh (j\omega l\sqrt{LC})$$

$$\text{and } \mathbf{I_S} = \mathbf{I_R} \cosh (j\omega l\sqrt{LC}) + \mathbf{V_R} \times \frac{1}{\mathbf{Z_C}} \sinh (j\omega l\sqrt{LC})$$

$$\text{or } \mathbf{V_S} = \mathbf{V_R} \cos (\omega l\sqrt{LC}) + j\mathbf{Z_C I_R} \sin (\omega l\sqrt{LC}) \quad ...(7.65a)$$

$$\text{and } \mathbf{I_S} = \mathbf{I_R} \cos (\omega l\sqrt{LC}) + \frac{j}{\mathbf{Z_C}} \mathbf{V_R} \sin (\omega l\sqrt{LC}) \quad ...(7.65b)$$

$$\because \cosh (j\omega l\sqrt{LC}) = \cos (\omega l\sqrt{LC})$$

$$\text{and } \sinh (j\omega l\sqrt{LC}) = j \sin (\omega l\sqrt{LC})$$

If there is no load, the receiving-end current I_R is zero. The charging current becomes

$$\mathbf{I_S} = \frac{j}{\mathbf{Z_C}} \mathbf{V_R} \sin (\omega l\sqrt{LC})$$

while $\mathbf{V_S} = \mathbf{V_R} \cos (\omega l\sqrt{LC})$

Thus, sending-end voltage V_S and sending-end current I_S are in quadrature. If the relation

$$\omega l\sqrt{LC} = n\pi \text{ where } n = 1, 2, 3, ... \quad ...(7.66)$$

is satisfied, there is no charging current on zero load. If the expression $\omega l\sqrt{LC} = \pi$, the $\cos (\omega l\sqrt{LC}) = -1$ and then for load conditions, the Eqs. (7.65 (*a*)) and (7.65 (*b*)) become

$$\mathbf{V_S} = -\mathbf{V_R} \text{ and } \mathbf{I_S} = -\mathbf{I_R}$$

and if $(\omega l\sqrt{LC}) = 2\pi$, the $\cos \omega l(\sqrt{LC}) = 1$ and Eqs. 7.65 (*a*) and 7.65 (*b*) become

$$\mathbf{V}_S = \mathbf{V}_R \quad \text{and} \quad \mathbf{I}_S = \mathbf{I}_R$$

In both the cases the receiving-end voltage and current are numerically equal to the sending-end values, so that there is no voltage drop on load. Such a line is called *a tuned line.*

For 50 Hz, the length of line for tuning is

$$l = \frac{n\pi}{2\pi f\sqrt{LC}} = \frac{n}{2 \times 50\sqrt{LC}} = \frac{n}{100} \times \text{velocity of light}$$

$$\because \quad \frac{1}{\sqrt{LC}} = v, \text{ the velocity of light} = 3 \times 10^8 \text{ m/s}$$

$$\text{or } 3 \times 10^5 \text{ km/s}$$

$$\text{or } l = \frac{n}{100} \times 3 \times 10^5 \text{ km} = 3{,}000 \text{ km; } 6{,}000 \text{ km } .....$$

It is too long a distance of transmission from the point of view of cost and efficiency. For a given line (length of line and frequency being fixed), the line may be tuned by increasing L or C *i.e.,* by adding series inductances or shunt capacitances at intervals along the line length, as illustrated in Figs. 7.10 (*a*) and 7.10 (*b*). This method is impractical and uneconomical for power transmission and is used in telephony where higher frequencies are employed.

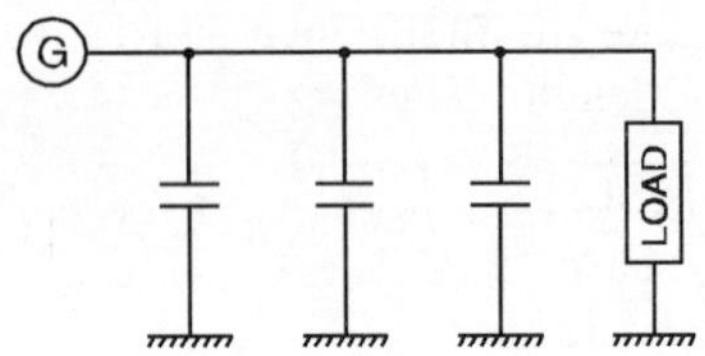

(*a*) *By Shunt Capacitance*

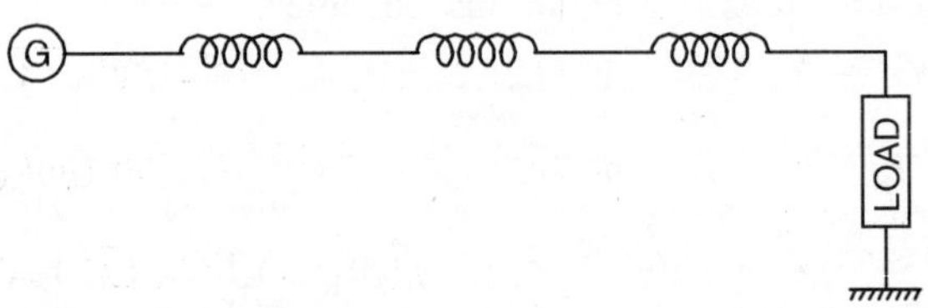

(*b*) *By Series Inductance*

Fig. 7.10. ***Tuning of Transmission Line***

Another method of tuning power lines, which is being presently experimented with, is shown in Fig. 7.11. The capacitors in series with the line neutralize the voltage drop due to the line inductance, and the inductors across the line neutralize the charging current due to line capacitance. This method is known as the *method of tuning by the use of compensating sections.*

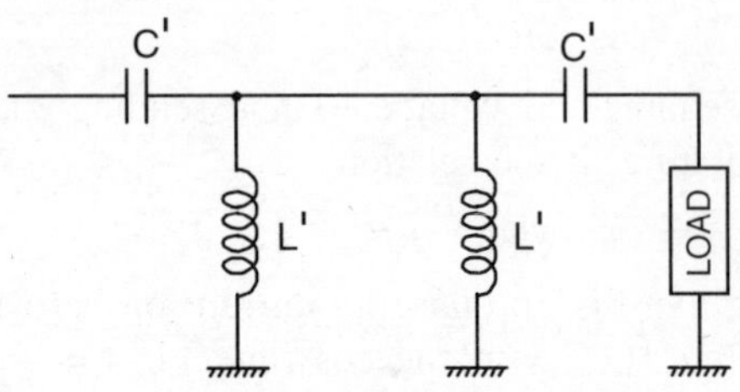

Fig. 7.11. ***Tuning of Power Transmission Line by Compensating Sections***

The values of L′ and C′ may be found with fair accuracy by replacing the line by its nominal-π equivalent (Fig. 6.14). The two capacitances C′ are equivalent to a single capacitance $\frac{1}{2}C'$ which will neutralize the inductance L of the line impedance Z if $\omega^2 L\left(\frac{1}{2}C'\right) = 1$. The inductance L′ will neutralize the charging current provided $\omega^2 L'\left(\frac{1}{2}C\right) = 1$. The exact values required in long lines, can be obtained from the π-equivalent shown in Fig. 7.8. Thus $\frac{1}{2}C'$ must neutralize Z, which is given as

$$Z = \left(\sqrt{\frac{L}{C}}\right) \sinh (j\omega\sqrt{LC}) = j\sqrt{\frac{L}{C}} \sin (\omega\sqrt{LC}) \text{ so that}$$

$$\frac{1}{j\omega C'} + Z = 0 \text{ giving } C' = \frac{2}{\omega\sqrt{\frac{L}{C}} \sin \omega\sqrt{LC}} \qquad ...(7.67)$$

Similarly L′ must neutralize, Y/2 giving

$$L' = \frac{1}{\omega\sqrt{C/L} \tan\left(\frac{1}{2}\omega\sqrt{LC}\right)} \qquad ...(7.68)$$

7.15. POWER FLOW THROUGH A TRANSMISSION LINE

So far we have presented the performance equations of a transmission line in the form of voltage and current relationships between sending- and receiving-ends. But the loads are usually expressed in terms of real, active or true power (watts/kW/MW) and reactive power (VARS/kVAR/MVAR), so it will be convenient to deal with transmission line performance equations in the form of sending- and receiving-end complex power and voltages.

Line diagram of a 3-phase transmission line is shown in Fig. 7.12. The transmission line may be supplied power from a generating station, grid switching station or grid supply point. The ends of a transmission line are designated as buses. The concept of a bus in a one line diagram is essentially the same as that of a node in a circuit diagram.

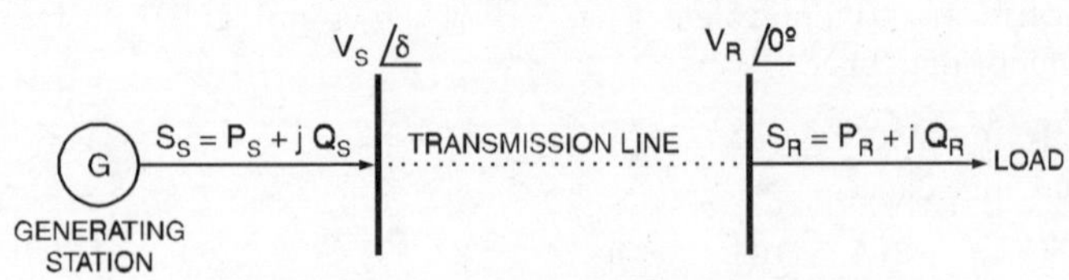

Fig. 7.12. ***A Two-Bus Power System***

Let the receiving-end voltage be $V_R \angle 0°$ (taking V_R as reference phasor) and the sending-end voltage $V_S \angle\delta$ where δ is the phase angle between sending- and receiving-end voltages. Let the generalised line constants be expressed as $\mathbf{A} = A \angle\alpha$; $\mathbf{B} = B \angle\beta$; $\mathbf{D} = D \angle\Delta$. The complex power at the receiving-end is given by the expression

$$\mathbf{S}_R = P_R + jQ_R = \mathbf{V}_R \mathbf{I}_{RC} \qquad ...(7.69)$$

where $\mathbf{I}_{RC}$ is the conjugate of receiving-end current $\mathbf{I}_R$.

Receiving-end current can, however, be expressed in terms of receiving- and sending end voltages [using Eq. (6.35)]

$$\mathbf{I}_R = \frac{\mathbf{V}_S - \mathbf{AV}_R}{\mathbf{B}}$$

$$= \frac{V_S \angle\delta - A\angle\alpha \cdot V_R \angle 0°}{B\angle\beta}$$

$$= \frac{V_S}{B} \angle\delta - \beta - \frac{A}{B} V_R \angle\alpha - \beta \qquad ...(7.70)$$

So conjugate of $\mathbf{I}_R$, $\mathbf{I}_{RC} = \frac{V_S}{B} \angle\beta - \delta - \frac{A}{B} V_R \angle\beta - a$

Substituting the value of $\mathbf{I}_{RC}$ from above equation in Eq. (7.69) we have

$$\mathbf{S}_R = P_R + jQ_R$$

$$= \frac{V_S V_R}{B} \angle\beta - \delta - \frac{A}{B} V_R^2 \angle\beta - a \qquad ...(7.71)$$

Separating real and reactive components we have

Receiving-end true power,

$$P_R = \frac{V_S V_R}{B} \cos(\angle\beta - \delta) - \frac{A}{B} V_R^2 \cos(\beta - \alpha) \quad ...(7.72)$$

and receiving-end reactive power,

$$Q_R = \frac{V_S V_R}{B} \sin(\angle\beta - \delta) - \frac{A}{B} V_R^2 \sin(\beta - \alpha) \quad ...(7.73)$$

For fixed values of V_S and V_R the power received will be maximum when cos (β – δ) = 1, or when δ = β.

$$\therefore P_{R\,max} = \frac{V_S V_R}{B} - \frac{A}{B} V_R^2 \cos(\beta - \alpha) \quad ...(7.74)$$

The corresponding value of Q_R at this power limit

$$Q_R = -\frac{A}{B} V_R^2 \sin(\beta - \alpha) \qquad ...(7.75)$$

From Eq. (6.36),

$$\mathbf{I}_R = \frac{\mathbf{I}_S - \mathbf{CV}_R}{\mathbf{D}}$$

Substituting the value of $\mathbf{I}_R$ from above equation in Eq. (6.35) we have

$$\mathbf{V}_S = \mathbf{AV}_R + \mathbf{B}\frac{\mathbf{I}_S - \mathbf{CV}_R}{\mathbf{D}}$$

$$\text{or } \mathbf{DV}_S = \mathbf{ADV}_R + \mathbf{BI}_S - \mathbf{BCV}_R$$

$$\text{or } \mathbf{V}_R(\mathbf{AD} - \mathbf{BC}) = \mathbf{DV}_S - \mathbf{BI}_S$$

$$\text{or } \mathbf{V}_R = \mathbf{DV}_S - \mathbf{BI}_S$$

Since for a four-terminal network **AD** – **BC** = 1

and therefore, $\mathbf{I}_S = \frac{\mathbf{DV}_S - \mathbf{V}_R}{\mathbf{B}}$

$$= \frac{D\angle\Delta\, V_S \angle\delta - V_R \angle 0}{B\angle\beta}$$

$$= \frac{D}{B} V_S \angle\Delta + \delta - \beta - \frac{V_R}{B} \angle -\beta \quad ...(7.76)$$

Conjugate of sending-end current,

$$\mathbf{I}_{SC} = \frac{D}{B} V_S \angle\beta - \Delta - \delta - \frac{V_R}{B} \angle\beta \quad ...(7.77)$$

Sending-end power,

$$\mathbf{S}_S = P_S + jQ_S$$

$$= \mathbf{V}_S \mathbf{I}_{SC}$$

$$= \frac{D}{B} V_S^2 \angle\beta - \Delta - \frac{V_R V_S}{B} \angle\beta + \delta \quad ...(7.78)$$

Separating real and reactive components, we have

Sending-end true power,

$$P_S = \frac{D}{B} V_S^2 \cos(\beta - \Delta) - \frac{V_R V_S}{B} \cos(\beta + \delta) \quad ...(7.79)$$

and sending-end reactive power,

$$Q_S = \frac{D}{B} V_S^2 \sin(\beta - \Delta) - \frac{V_R V_S}{B} \sin(\beta + \delta) \quad ...(7.80)$$

For fixed values of V_S and V_R, the power sent out will be maximum when

cos (β + δ) = – 1 or when β + δ = 180°

$$\therefore P_{S\,max} = \frac{D}{B} V_S^2 \cos(\beta - \Delta) + \frac{V_R V_S}{B} \quad ...(7.81)$$

The noteworthy points about Eqs. (7.71) and (7.78) are :

1. These equations give phase values of S_R and S_S in volt-amperes if V_R and V_S are phase values in volts (*i.e.,* line to neutral voltage).

2. The total 3-phase power is 3 times the power per phase. Since the line voltage is $\sqrt{3}$ times the phase voltage, these equations yield 3-phase power directly if V_R and V_S are expressed in line volts (line-to-line voltage).

3. If the voltages V_R and V_S are expressed in kV, the complex power is expressed in mega volt-amperes.

Now consider the special case of a short transmission line with a series impedance. In case of a **short transmission line.**

$$\mathbf{A} = \mathbf{D} = 1 \angle 0°, \; \mathbf{B} = \mathbf{Z} = Z\angle\theta$$

Substituting these values in Eqs. (7.72), (7.73), (7.79) and (7.80), we have

$$P_R = \frac{V_S V_R}{Z} \cos(\theta - \delta) - \frac{V_R^2}{Z} \cos\theta \qquad ...(7.82)$$

$$Q_R = \frac{V_S V_R}{Z} \sin(\theta - \delta) - \frac{V_R^2}{Z} \sin\theta \qquad ...(7.83)$$

$$P_S = \frac{V_S^2}{Z} \cos\theta - \frac{V_S V_R}{Z} \cos(\theta + \delta) \qquad ...(7.84)$$

$$\text{and } Q_S = \frac{V_S^2}{Z} \sin\theta - \frac{V_S V_R}{Z} \sin(\theta + \delta) \qquad ...(7.85)$$

The resistance of a transmission line is usually very small as compared to inductive reactance. Then Z = X and θ = 90°. Substituting these values in above equations, we have

$$P_R = \frac{V_S V_R}{X} \sin \delta \quad ...(7.86)$$

$$Q_R = \frac{V_S V_R}{X} \cos \delta - \frac{V_R^2}{X} \quad ...(7.87)$$

The above Eq. (7.87) can be further simplified by assuming $\cos \delta \simeq 1$, since δ, the power or load angle, is usually very small from considerations of system stability. Thus

$$Q_R = \frac{V_S V_R}{X} - \frac{V_R^2}{X} = \frac{V_R}{X} [V_S - V_R] \quad ...(7.88)$$

The above Eqs. [7.82 to 7.88] are for short transmission lines. These equations will also apply for a long transmission line when the line is replaced by its equivalent π-network and the shunt admittances are lumped with the receiving-end load and sending-end generating network.

A study of above Eqs. (7.86), (7.87) and (7.88) leads to some very important conclusions, such as

1. For fixed values of V_S, V_R and X, the true or real power transferred to the receiving end is proportional to sin δ while the reactive power is proportional to the magnitude of the voltage drop across the line $(V_S - V_R)$.

2. The power transferred over a transmission line will be maximum for $\delta = 90°$ and has a value of $\frac{V_S V_R}{X}$. However, δ has to be kept to values well below 90° from considerations of stability.

3. The maximum true power transferred over a given line increases with the increase in V_S and V_R. This is the reason that high and extra high voltages are adopted for the transmission of bulk power over long distances.

4. The maximum true power transferred also depends on the line reactance (or line inductance L). The line capacity can, therefore, be increased by reducing line inductance which can be accomplished by employing double circuit lines and by using bundled conductors. The line inductance can also be reduced by series compensation (by inserting series capacitors in the line). The series capacitors are usually installed at the mid point of the line.

5. As said above, the reactive power transferred over a transmission line is proportional to the line voltage drop $(V_S - V_R)$ and is independent of load or power angle δ. This means that the transfer of reactive power over a line is the major cause of voltage drop across the line. Thus the control of reactive power is necessary for maintaining a desired voltage profile.

Example 7.8. A 33 kV single circuit, 3-phase transmission line has the ABCD parameters A = D = 1 $\angle 0°$, B = 11.18 $\angle 63.43°$ Ω. The line is to deliver 7.5 MVA at 0.85 pf lagging at the load end. The receiving-end voltage is 32 kV (line to line). How much active and reactive power is to be despatched from the sending-end ? *[GATE 1993]*

Solution : Taking receiving-end phase voltage as reference phasor we have

$$V_R = \frac{32{,}000}{\sqrt{3}} \angle 0° = 18{,}475 \angle 0° \text{ V}$$

Receiving-end currrent,

$$I_R = \frac{S_R \text{ in MVA} \times 10^6}{\sqrt{3}\, V_{RL}} = \frac{7.5 \times 10^6}{\sqrt{3} \times 32{,}000} = 135.32 \text{ A}$$

Receiving-end power factor,

$\cos \phi_R = 0.85$ (lagging) or $\phi_R = \text{Cos}^{-1}\ 0.85 = 31.8°$

So $\mathbf{I}_R = 135.32 \angle -31.8°$ A

Line parameters, $\mathbf{A} = \mathbf{D} = 1 \angle 0°$ and $\mathbf{B} = 11.18 \angle 63.43°$ Ω

$$\mathbf{C} = \frac{\mathbf{AD} - 1}{\mathbf{B}} = \frac{1-1}{\mathbf{B}} = 0 \quad \because \mathbf{AD} - \mathbf{BC} = 1$$

Sending-end voltage,

$$\mathbf{V}_S = \mathbf{AV}_R + \mathbf{BI}_R$$

$$= 1 \angle 0° \times 18{,}475 \angle 0° + 11.18 \angle 63.43° \times 135.32 \angle -31.8°$$

$$= 18{,}475 \angle 0° + 1{,}512.88 \angle 31.63°$$

$$= 19{,}763.2 + j793.4 = 19{,}779 \angle 2.3° \text{ V}$$

Sending-end current,

$$\mathbf{I}_S = \mathbf{CV}_R + \mathbf{DI}_R$$

$$= 0 + 1 \angle 0° \times 135.32 \angle -31.8°$$

$$= 135.32 \angle -31.8° \text{ A}$$

Phase angle,

$$\phi_S = 2.3° - (-31.8°) = 34.1°$$

Active power despatched

$$= \frac{3\, V_S\, I_S \cos \phi_S}{10^6} = \frac{3 \times 19{,}779 \times 135.32 \cos 34.1°}{10^6} \text{ MW}$$

= 6.65 MW **Ans.**

Reactive power despatched

$$= \frac{3\, V_S\, I_S \sin \phi_S}{10^6}$$

$= 3 \times 19{,}779 \times 135.32 \sin 34.1° \times 10^{-6}$ MVAR

= 4.5 MVAR **Ans.**

Example 7.9. A typical line has the following parameters ;

A = D = 0.96 $\angle 1.0°$ and B = 100 $\angle 80°$ Ω

(*a*) If the line supplies a load of 30 MW at 0.8 pf lag, 110 kV, find the sending-end voltage and hence the regulation of the line.

(*b*) For a load of 50 MW at 0.8 pf lag, 110 kV, find the reactive power supplied by the line and by synchronous capacitors if the sending-end voltage is 120 kV. Also determine the power factor of the line at the receiving end.

(*c*) Find the maximum power that can be transmitted if the sending- and receiving-end voltages are as in (*b*).

(*d*) Find the power and pf of the load if the voltages at the two ends are 110 kV and having a phase difference of 20°.

Solution : (*a*) Receiving-end phase voltage,

$$V_R = \frac{110 \times 1{,}000}{\sqrt{3}} = 63{,}508 \text{ V}$$

Receiving-end current,

$$I_R = \frac{\text{Load in MW} \times 10^6}{\sqrt{3}\, V_{RL} \cos\phi_R} = \frac{30 \times 10^6}{\sqrt{3} \times 110 \times 10^3 \times 0.8}$$

$$= 196.82 \text{ A}$$

Taking recceiving-end phase voltage as reference phasor we have

$$\mathbf{V}_R = 63{,}508 \angle 0^\circ \text{ V}$$

Receiving-end current,

$$\mathbf{I}_R = 196.82 \angle -36.87^\circ \text{ A}$$

Sending-end phase voltage,

$$\mathbf{V}_S = \mathbf{AV}_R + \mathbf{BI}_R$$

$$= 0.96 \angle 1.0^\circ \times 63{,}508 \angle 0^\circ + 100 \angle 80^\circ \times 196.82 \angle -36.87^\circ$$

$$= 60{,}968 \angle 1.0^\circ + 19{,}682 \angle 43.13^\circ$$

$$= 60{,}958.9 + j1{,}064 + 14{,}364 + j13{,}455.7$$

$$= 75{,}322.9 + j14{,}519.7 = 76{,}709.6 \angle 10.91^\circ \text{ V}$$

Sending-end line voltage,

$$V_{SL} = \sqrt{3} \times 76{,}709.6 = 132{,}865 \text{ V or } 132.865 \text{ kV} \textbf{ Ans.}$$

Voltage regulation

$$= \frac{V_{SL} - V_{RL}}{V_{RL}} \times 100 = \frac{132.865 - 110}{110} \times 100$$

$$= 20.786\% \textbf{ Ans.}$$

(*b*) From Eq. (7.72)

Receiving-end true power,

$$P_R = \frac{V_{SL}\, V_{RL}}{B} \cos(\beta - \delta) - \frac{A}{B} V_{RL}^2 \cos(\beta - \alpha) \text{ MW}$$

where V_{SL} is the sending-end line voltage in kV and is equal to 120 kV; V_{RL} is the receiving-end line voltage in kV and is equal to 110 kV, A = 0.96, B = 100, $\alpha = 1^\circ$ and $\beta = 80^\circ$ and δ is the phase angle between sending-and receiving-end voltages. Thus, we have

$$\frac{120 \times 110}{100} \cos(\beta - \delta) - \frac{0.96}{100} \times (110)^2 \cos(80^\circ - 1^\circ) = 50$$

$$\text{or } \cos(\beta - \delta) = \frac{50 + 22.164}{132}$$

$$= 0.5467$$

$$\text{or } (\beta - \delta) = \text{Cos}^{-1}\, 0.5467 = 56.86^\circ$$

Receiving-end reactive,

$$Q_R = \frac{V_{SL}\, V_{RL}}{B} \sin(\beta - \delta) - \frac{A}{B} V_{RL}^2 \sin(\beta - \alpha)$$

$$= \frac{120 \times 110}{100} \sin(56.86^\circ) - \frac{0.96}{100} \times (110)^2 \sin(80^\circ - 1^\circ) \text{ MVAR}$$

$$= 110.53 - 114 = -3.47 \text{ MVAR} \textbf{ Ans.}$$

Thus for V_{SL} = 120 kV, V_{RL} = 110 kV and P_R = 50 MW, a leading MVAR of 3.47 must be drawn from the line along with the true power of 50 MW. Since the load requires lagging MVAR of 50 tan ($\cos^{-1}$ 0.8) *i.e.,* 37.5, the synchronous capacitors must deliver 37.5 + 3.47 *i.e.,* 40.97 MVAR **Ans.**

Power factor of the line at the receiving-end,

$$\text{Cos}\,\phi_R = \text{Cos}\left(\tan^{-1}\frac{Q_R}{P_R}\right) = \text{Cos}\left(\tan^{-1}\frac{3.47}{50}\right)$$

$$= 0.9976 \text{ (lead)} \textbf{ Ans.}$$

(*c*) For the given values of sending- and receiving-end voltages, the power transfer will be maximum for $\delta = \beta$. Thus maximum power transmitted is given as

$$P_{Rmax} = \frac{V_{SL} V_{RL}}{B} - \frac{A}{B} \times V_{RL}^2 \cos(\beta - \alpha)$$

Refer to Eq. (7.74)

$$= \frac{120 \times 110}{100} - \frac{0.96}{100} \times (110)^2 \cos 79^\circ$$

$$= 132 - 22.164 = 109.836 \text{ MW} \textbf{ Ans.}$$

(*d*) Power supplied for $V_{SL} = V_{RL}$ = 110 kV, A = 0.96, B = 100, $\alpha = 1^\circ$, $\beta = 80^\circ$ and δ, the phase difference between sending- and receiving-end voltages = 20°, is given as

$$P_R = \frac{V_{SL} V_{RL}}{B} \cos(\beta - \delta) - \frac{A}{B} V_{RL}^2 \cos(\beta - \alpha)$$

$$= \frac{110 \times 110}{100} \cos(80^\circ - 20^\circ) - \frac{0.96}{100} \times (110)^2 \cos(80^\circ - 1^\circ)$$

$$= 60.5 - 22.164 = 38.336 \text{ MW} \textbf{ Ans.}$$

and

$$Q_R = \frac{V_{SL}\, V_{RL}}{B} \sin(\beta - \delta) - \frac{A}{B} V_{RL}^2 \sin(\beta - \alpha)$$

$$= \frac{110 \times 110}{100} \sin(80^\circ - 20^\circ) - \frac{0.96}{100} \times (110)^2 \sin 79^\circ$$

$$= 104.789 - 114.026 = -9.237 \text{ MVAR}$$

$$\text{Load angle, } \phi_R = \text{Tan}^{-1}\frac{Q_R}{P_R} = \text{Tan}^{-1}\frac{-9.237}{38.336} = -13.55^\circ$$

Load power factor, cos ϕ_R = Cos (– 13.55°) = 0.9722 (leading) **Ans.**

Example 7.10. A 100 km, 50 Hz, 3-phase line with $r = 0.03\ \Omega$, $x = 0.10\ \Omega$ and $c = 0.01\ \mu F$ per km of line length has a load of 2 MW at 0.8 power factor lag at the receiving end. The receiving end is maintained at 11 kV by connecting at receiving end a bank of static capacitors rated at 1 MVAR at 11 kV. Calculate using nominal π-method, the voltage and current at the sending end of the line.

Solution : Total series impedance per phase,

$$\mathbf{Z} = (0.03 + 0.10) \times 100 = (3 + j10)\ \Omega$$

Total admittance per phase,

$$\mathbf{Y} = j2\pi fC = j2\pi \times 50 \times 0.01 \times 10^{-6} \times 100 = j314 \times 10^{-6} \text{ S}$$

For a nominal π-network (Fig. 7.8), the generalised circuit constants are :

$$\mathbf{A} = \mathbf{D} = 1 + \frac{1}{2}\mathbf{YZ}$$

$$= 1 + \frac{1}{2} j\,314 \times 10^{-6} \times (3 + j\,10)$$

$$= 0.99843 + j\,0.000471 = 0.99843 \angle 0.027^\circ$$

$$\mathbf{B} = \mathbf{Z} = (3 + j\,10) = 10.44 \angle 73.3^\circ\ \Omega$$

$$\mathbf{C} = \mathbf{Y} + \frac{1}{4}\mathbf{Y}^2\mathbf{Z}$$

$$= j314 \times 10^{-6} + \frac{1}{4} \times (j\,314 \times 10^{-6})^2 \times (3 + j\,10)$$

$$= j314 \times 10^{-6}\left[1 + \frac{(j\,314\times10^{-6})(3 + j\,10)}{4}\right]$$

$$= j314 \times 10^{-6}\ [0.999215 + j0.0002355]$$

$$= 313.75 \times 10^{-6} \angle 90^\circ \text{ S}$$

Taking receiving-end phase voltage as reference phasor we have

$$\mathbf{V_R} = \frac{11{,}000}{\sqrt{3}} \angle 0^\circ = 6{,}351 \angle 0^\circ \text{ V}$$

Load-current,

$$\mathbf{I_L} = I_L \angle -36.87^\circ$$

$$= \frac{2\times10^6}{\sqrt{3}\times11{,}000\times0.8} \angle -36.87^\circ = 131.22 \angle -36.87^\circ \text{ A}$$

Capacitor current,

$$\mathbf{I_C} = j\frac{\text{MVAR}}{\sqrt{3}\,V_C} = j\frac{10^6}{\sqrt{3}\times11{,}000} = j\,52.49 \text{ A}$$

Receiving-end current,

$$\mathbf{I_R} = \mathbf{I_L} + \mathbf{I_C}$$

$$= 131.22 \angle -36.87^\circ + 52.49 \angle 90^\circ$$

$$= (104.976 - j26.242) \text{ A or } 108.21 \angle -14.04^\circ \text{ A}$$

Phase voltage at sending end,

$$\mathbf{V_S} = \mathbf{AV_R} + \mathbf{BI_R}$$

$$= 0.99843 \angle 0.027^\circ \times 6{,}351 \angle 0^\circ$$

$$+ 10.44 \angle 73.3^\circ \times 108.21 \angle -14.04^\circ \text{ A}$$

$$= 6{,}341 \angle 0.027^\circ + 1{,}129.71 \angle 59.26^\circ = 6{,}986.6 \angle 8^\circ \text{ V}$$

Sending-end line voltage,

$$V_{SL} = \sqrt{3}\,V_S = \sqrt{3} \times 6{,}986.6 = 12{,}100 \text{ V or } 12.1 \text{ kV}$$ **Ans.**

Sending-end current,

$$\mathbf{I_S} = \mathbf{CV_R} + \mathbf{DI_R}$$

$$= 313.75 \times 10^{-6} \angle 90^\circ \times 6{,}351 \angle 0^\circ$$

$$+ 0.99843 \angle 0.027^\circ \times 108.21 \angle -14.04^\circ$$

$$= 1.993 \angle 90^\circ + 108.04 \angle -14.013^\circ$$

$$= (104.82 - j\,24.17) \text{ A or } 107.57 \angle -12.98^\circ \text{ A}$$ **Ans.**

Example 7.11. A 200 km, 3-phase, 50 Hz transmission line has the following data :

A = D = 0.938 $\angle 1.2^\circ$, B = 131.2 $\angle 72.3^\circ$ ohms/phase; C = 0.001 $\angle 90^\circ$ siemens/phase. The sending-end voltage is 230 kV. Determine (*a*) the receiving-end voltage when the load is disconnected (*b*) the line charging current (*c*) the maximum power that can be transmitted at a receving-end voltage of 220 kV and the corresponding load reactive power required at the receiving end. *[GATE 1992]*

Solution : Sending-end phase voltage,

$$V_S = \frac{230\times1{,}000}{\sqrt{3}} = 132{,}790 \text{ V}$$

When the load is disconnected,

$$I_R = 0$$

Sending-end phase voltage is given as

$$\mathbf{V_S} = \mathbf{AV_R} + \mathbf{BI_R} = \mathbf{AV_R} + \mathbf{B} \times 0$$

or Receiving-end phase voltage,

$$\mathbf{V_R} = \frac{\mathbf{V_S}}{\mathbf{A}} = \frac{132{,}790 \angle 0^\circ}{0.938 \angle 1.2^\circ} = 141{,}570 \angle -1.2^\circ \text{ V}$$

Receiving-end line voltage,

$$V_{RL} = \sqrt{3}\,V_R = \sqrt{3} \times 141{,}570 = 245{,}200 \text{ V or } 245.2 \text{ kV}$$ **Ans.**

Line charging current,

$$\mathbf{I_C} = \mathbf{CV_R} = 0.001 \angle 90^\circ \times 141{,}570 \angle -1.2^\circ$$

$$= 141.57 \angle 88.8^\circ \text{ A}$$ **Ans.**

Maximum power that can be transmitted for V_{RL} = 220 kV and V_S = 230 kV,

$$A = 0.938,\ B = 131.2,\ \alpha = 1.2^\circ,\ \beta = 72.3^\circ$$

$$P_{Rmax} = \frac{V_{SL}\,V_{RL}}{B} - \frac{A}{B} \times V_{RL}^2 \cos(\beta - \alpha) \quad \text{(Refer to Eq. 7.74)}$$

$$= \frac{230\times220}{131.2} - \frac{-0.938}{131.2} \times (220)^2 \cos(72.3^\circ - 1.2^\circ)$$

$$= 385.67 - 112.085 = 272.58 \text{ MW}$$ **Ans.**

Corresponding reactive power required at the receiving end,

$$Q_R = -\frac{A}{B} \times V_{RL}^2 \sin(\beta - \alpha)$$

$$= \frac{-0.938}{131.2} \times (220)^2 \sin(72.3^\circ - 1.2^\circ)$$

$$= 327.37 \text{ MVAR}$$ **Ans.**

7.16. POWER CIRCLE DIAGRAM

Generally it is customary to determine the sending-end voltage V_S and current I_S, while receiving-end voltage V_R and receiving-end current I_R are known. One of the most interesting feature of such circuits when operated on a fixed frequency, is that they behave according to a circular function.

The circle diagram for a short line was described by Philip and later on extended by Dwight to take into account the effect of capacitance, that is a diagram for a long line.

The data required for drawing the circle diagram for a long transmission line is obtained as follows :

The complex power at the receiving end and sending end of a transmission line are given by Eqs. (7.71) and (7.78) respectively which can be rewritten as

$$S_R = P_R + jQ_R$$

$$= -\frac{A}{B} V_R^2 \angle \beta - \alpha + \frac{V_S\,V_R}{B} \angle \beta - \delta \qquad \text{...(7.89)}$$

and

$$S_S = P_S + jQ_S$$

$$= \frac{D}{B} V_S^2 \angle \beta - \Delta - \frac{V_S\,V_R}{B} \angle \beta + \delta \qquad \text{...(7.90)}$$

From above Eqs. (7.89) and (7.90) it is obvious that each of the powers is the sum of two phasors. Since the real components of these phasors represent true power P and the imaginary components represent the reactive power Q, it is possible to plot S_R and S_S in the X-Y plane whose horizontal and vertical coordinates represent the true and reactive powers respectively. The transmission lines are usually operated with constant sending-and receiving-end voltages, so one component of each of receiving- and sending-end powers is a fixed phasor (phasor of constant magnitude and phase) while the other component is a phasor of constant magnitude but variable angle. The loci of S_R and S_S are therefore, circles drawn from the tip of constant phasors or radius.

7.16.1. Receiving-End Power Circle Diagram

The circle drawn with receiving-end true and reactive power components as the horizontal and vertical coordinates is called the "receiving-end power circle diagram".

The Eqs. (7.72) and (7.73) for receiving-end true and reactive power components can be rewritten as

$$P_R + \frac{A}{B} V_R^2 \cos(\beta - \alpha) = \frac{V_S V_R}{B} \cos(\beta - \delta) \quad ...(7.91)$$

and $$Q_R + \frac{A}{B} V_R^2 \sin(\beta - \alpha) = \frac{V_S V_R}{B} \sin(\beta - \delta) \quad ...(7.92)$$

Squaring and adding Eqs. (7.91) and (7.92) we have

$$\left[P_R + \frac{A}{B} V_R^2 \cos(\beta - \alpha)\right]^2 + \left[Q_R + \frac{A}{B} V_R^2 \sin(\beta - \alpha)\right]^2$$

$$= \frac{V_S^2 V_R^2}{B^2} [\cos^2(\beta - \delta) + \sin^2(\beta - \delta)]$$

or $$\left[P_R + \frac{A}{B} V_R^2 \cos(\beta - \alpha)\right]^2 + \left[Q_R + \frac{A}{B} V_R^2 \sin(\beta - \alpha)\right]^2 = \left[\frac{V_S V_R}{B}\right]^2 \quad ...(7.93)$$

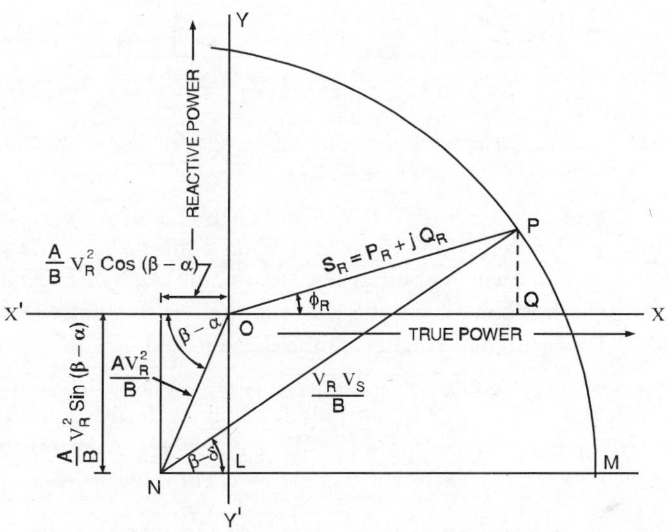

Fig. 7.13. ***Receiving-End Power Circle Diagram***

The coordinates of the centre of circle and radius of the circle given by the above equation are

x-coordinate of the centre of circle $= -\frac{A}{B} V_R^2 \cos(\beta - \alpha)$

y-coordinate of the centre of circle $= -\frac{A}{B} V_R^2 \sin(\beta - \alpha)$

Radius of the circle $= \frac{V_S V_R}{B}$

Construction of Power Circle Diagram. At first the centre of the circle is plotted on a suitable scale. From the centre an arc of circle is drawn with the calculated radius. The load line OP is then drawn from the origin inclined at an angle ϕ_R (load power factor angle) with the horizontal. Let it cut the circle at P then the receiving-end true power and reactive power will be represented by OQ and PQ respectively.

If the voltages V_S and V_R are taken phase voltages (line to neutral voltages) in volts then the powers indicated on X-axis and Y-axis will be in watts and VARs per phase respectively.

If the voltages V_S and V_R are taken line voltages (line-to-line voltages) in volts then the power indicated on X-axis and Y-axis will be in watts and VARs for all the three phases respectively.

In case the voltages V_S and V_R are taken from line-to-line and in kV then the power indicated will be in MW and MVAR and for all the three phases.

To determine the maximum power a horizontal line is drawn from the centre of the circle intersecting vertical axis at the point L and the circle at the point M. Distance LM represents the maximum power for the receiving end.

7.16.2. Sending-End Power Circle Diagram

The circle drawn with sending-end true and reactive power components as the horizontal and vertical coordinates is known as the *sending-end power circle diagram.*

Similar procedure can be followed for determining the data required for drawing the sending-end power circle diagram for long transmission line, as described below :

The Eqs. (7.79) and (7.80) for sending-end true and reactive power components can be rewritten as

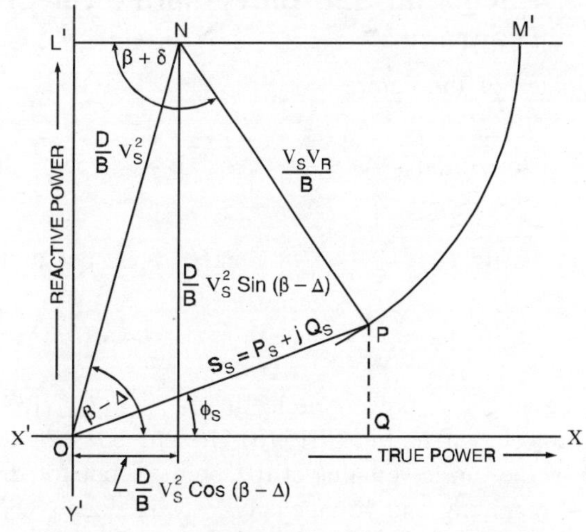

Fig. 7.14. ***Sending-End Power Circle Diagram***

$$P_S - \frac{D}{B} V_S^2 \cos(\beta - \Delta) = - \frac{V_R V_S}{B} \cos(\beta + \delta) \quad ...(7.94)$$

and $$Q_S - \frac{D}{B} V_S^2 \sin(\beta - \Delta) = - \frac{V_R V_S}{B} \sin(\beta + \delta) \quad ...(7.95)$$

Squaring-and adding Eqs. (7.94) and (7.95) we have

$$\left[P_S - \frac{D}{B} V_S^2 \cos(\beta - \Delta)\right]^2 + \left[Q_S - \frac{D}{B} V_S^2 \sin(\beta - \Delta)\right]^2 = \left[\frac{-V_R V_S}{B}\right]^2 \quad ...(7.96)$$

The coordinates of circle and radius of the circle given by the above equation are

x-coordinate of the centre of the circle = $\frac{D}{B} V_S^2 \cos(\beta - \Delta)$

y-coordinate of the centre of the circle = $\frac{D}{B} V_S^2 \sin(\beta - \Delta)$

Radius of the circle = $\frac{V_R V_S}{B}$

The sending-end power circle diagram is shown in Fig. 7.14. In the sending-end power circle diagram the maximum power for sending end is represented by L′ M′, as shown in Fig. 7.14. Sending-end true power and reactive power components are represented by OQ and PQ respectively.

7.16.3. Universal Power Circle Diagram

In the ordinary receiving-end power circle diagram described above, if the receiving-end voltage is kept constant, the centre of the circles is fixed and for different sending-end voltages circles are drawn; these will be concentric but if the receiving-end voltage is not constant the circles will not be concentric. Similar is the case with the sending-end power circle diagram. When sending-end voltage is not constant, its centre will also vary.

To overcome the above limitation the universal power circle diagram is drawn. In this diagram each distance on the original diagram is divided by $\frac{V^2}{B}$ where V is the "reference or base voltage" and B is the generalised circuit constant. This gives dimensionless units.

7.16.4. Receiving-End Universal Power Circle Diagram

Coordinates of the centre

Horizontal $-A\left(\frac{V_R}{V}\right)^2 \cos(\beta - \alpha)$ per unit

Vertical $-A\left(\frac{V_R}{V}\right)^2 \sin(\beta - \alpha)$ per unit

Radius = $\frac{V_S V_R}{V^2}$ per unit

Receiving-end power circle diagram (Fig. 7.13) is re-drawn as, shown in Fig. 7.15 (*a*). It will be seen that the centre of the circle remains at the point N but for different value of receiving-end voltage (say V'_R) distance NO_R varies as the square of receiving-end voltage. In order to have concentric circles for different values of $V_S V_R$ point N is retained as the centre of circles but the measurements of true and reactive power are made from O'_R instead of O_R where $O'_R O_R$ = NN′. Useful portions of the circle lie in the first quadrant with respect to point N and is shown in Fig. 7.15 (*b*).

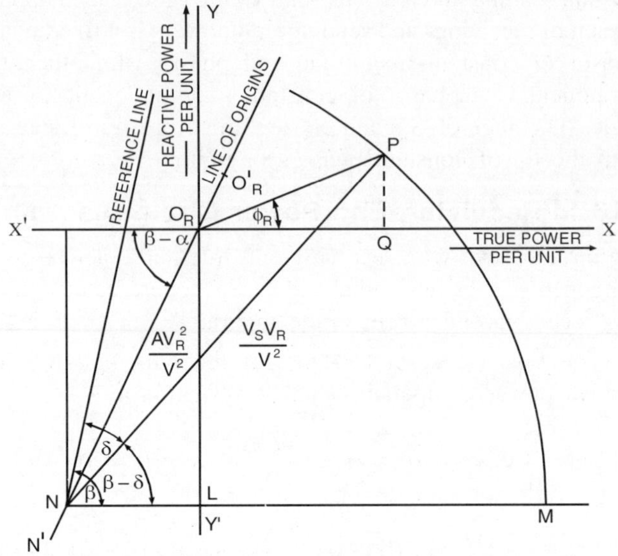

(a) Universal Power Circle Diagram For Receiving End

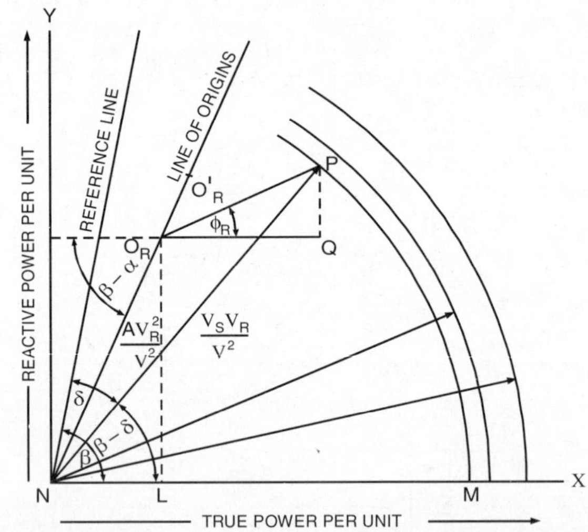

(b) Modified Universal Power Circle Diagram For Receiving End

Fig. 7.15

From O_R load line $O_R P$ is drawn inclined at angle ϕ_R (load power factor angle) to horizontal cutting the circle at P. The true power measured from the point O_R will be $O_R Q$ along the horizontal axis per unit and the reactive power will be PQ measured vertically above the point Q.

For a new value of V_R, (say V'_R) as already mentioned above the origin O_R will shift along the line of origins to O'_R but the centre of the circle is still at point N. Power will be measured from a different origin for each value of V_R.

7.16.5. Sending-End Universal Power Circle Diagram

The centre of circle is taken at point N but the point O_S (the origin for measurement of power) now lies in the second quadrant, the coordinates of which are given by

Horizontal – $D \cdot \frac{V_S^2}{V^2} \cos(\beta - \Delta)$ per unit

Vertical $D \cdot \frac{V_S^2}{V^2} \sin(\beta - \Delta)$ per unit

Radius = $\frac{V_S V_R}{V^2}$ per unit

Line drawn through N and O_S is called the *'line of origins'*. With the point N as centre an arc of circle of radius equal to $\frac{V_S V_R}{V^2}$ is drawn. From O_S the load line inclined at power factor angle ϕ_S, to the horizontal cutting the arc of the circle at P' is drawn. The true power measured from the point O_S along the horizontal axis in per unit will be O_SQ' and the reactive power measured downwards from point Q' will be Q'P'. The circle diagram is shown in Fig. 7.16.

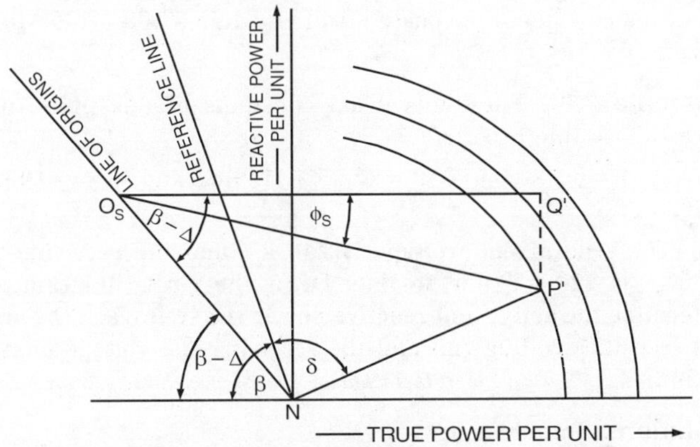

Fig. 7.16. ***Universal Power Circle Diagram For Sending End***

By coinciding the centre points of the Figs. 7.15 (*b*) and 7.16 we have a universal power circle diagram.

A universal power circle diagram is helpful in giving entire line performance readily.

Torque Angle. The angle between phasors V_S and V_R is called the "*torque angle*" and is denoted by δ. With the change in load the torque angle also changes. In the receiving-end universal power circle diagram a reference line is drawn from the point N inclined at an angle β to the horizontal, the torque angle δ will be the angle measured between the line NP and the reference line (Fig. 7.15). Similarly in the sending-end universal power circle diagram if a reference line is drawn from the point N inclined at angle β to the horizontal then the torque angle δ will be the angle measured between the line NP' and the reference line (Fig. 7.16).

Example 7.12. A 3-phase transmission line has a resistance per phase of 5 Ω and an inductive reactance of 15 Ω per phase. Determine the load which the line will be supplying at 132 kV and 0.8 pf lagging. The sending-end voltage is 140 kV.

If a synchronous phase modifier is connected in parallel with the load to improve the power factor to 0.95 lagging, determine the leading MVAR supplied by the synchronous phase modifier. Receiving-end voltage and load remain unchanged.

Solution : Power factor of the load, $\cos \phi_R = 0.8$

Load power factor angle, $\phi_R = \text{Cos}^{-1} 0.8 = 36.9°$

For short transmission line, A = 1 and α = 0°;

$$B = Z = \sqrt{5^2 + 15^2} = 15.81\ \Omega$$

$$\beta = \text{Tan}^{-1} \frac{X}{R} = \text{Tan}^{-1} \frac{15}{5} = \text{Tan}^{-1} 3 = 71.6°$$

For receiving-end power circle diagram

$$\text{Radius of circle} = \frac{V_{SL} V_{RL}}{B} = \frac{132 \times 140}{15.81} = 1{,}168.8 \text{ MVA}$$

Coordinates of the centre of the power circle are :

$$\text{Horizontal coordinate} = -\frac{A \times V_{RL}^2}{B} \cos(\beta - \alpha)$$

$$= -\frac{1 \times (132)^2}{15.81} \cos 71.6° = -347.8 \text{ MW}$$

$$\text{Vertical coordinate} = -\frac{A \times V_{RL}^2}{B} \sin(\beta - \alpha)$$

$$= -\frac{1 \times (132)^2}{15.81} \sin 71.6° = -1{,}045.6 \text{ MVAR}$$

Take scale 1 cm = 200 MW horizontally and 1 cm = 200 MVAR vertically.

$$\text{Radius of the circle is } \frac{1{,}168.8}{200} = 5.85 \text{ cm}$$

$$\text{Horizontal coordinate} = \frac{-347.8}{200} = -1.74 \text{ cm}$$

$$\text{Vertical coordinate} = \frac{-1{,}045.6}{200} = -5.23 \text{ cm}$$

and draw circle diagram (Fig. 7.17)

From origin O draw a line OP inclined at an angle $\cos^{-1} 0.8 = 36.9°$ with the horizontal cutting the circle at point *b*. From point *b* draw a perpendicular *ba* on the horizontal axis.

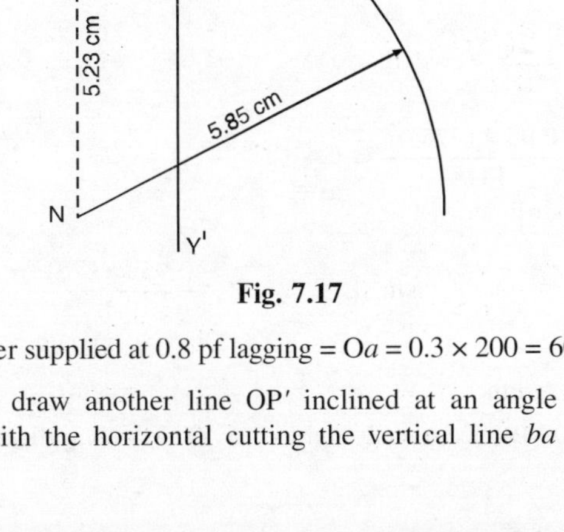

Fig. 7.17

Power supplied at 0.8 pf lagging = O*a* = 0.3 × 200 = 60 MW **Ans.**

Now draw another line OP' inclined at an angle $\cos^{-1} 0.95 = 18.2°$ with the horizontal cutting the vertical line *ba* at *c*.

Total leading MVAR supplied by the synchronous phase modifier

$= bc = 0.1 \times 200 = 20$ MVAR **Ans.**

Example 7.13. A 3-phase transmission has constants B = 141.0 $\angle 82°$ Ω and C = 0.00073 $\angle 80.19°$. A synchronous phase modifier, connected at the receiving end of the line maintains the receiving end at 275 kV at all loads. The sending-end voltage of the line is also 275 kV. Compute graphically, the output of the phase modifier (*i*) at no load and (*ii*) for a receiving-end load of 200 MVA at 0.85 power factor lag.

[*A.M.I.E. Sec. B. Power Systems Summer 1999; Pb. Technical Univ. Power System-I (T&D); May 2010*]

Solution : Generalised constants of the given transmission line

$\mathbf{B} = 141.0 \angle 82°\ \Omega$ and $\mathbf{C} = 0.00073 \angle 80.19°$ S

For a symmetrical network **AD** – **BC** = 1 and **A** = **D**

So $A^2 = 1 + \mathbf{BC}$

$= 1 + 141.0 \angle 82° \times 0.00073 \angle 80.19°$

$= 1 + 0.10293 \angle 162.19°$

$= 0.902 + j0.0315 = 0.9025 \angle 2.0°$

or $\mathbf{A} = \mathbf{D} = (0.9025 \angle 2.0°)^{1/2} = 0.95 \angle 1°$

For receiving-end power circle

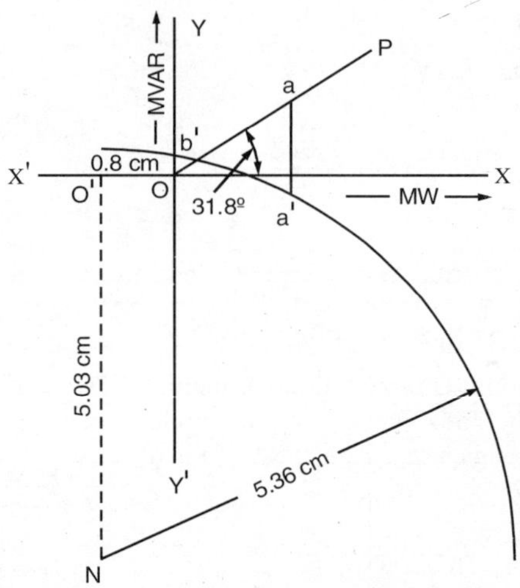

Fig. 7.18

Radius of circle

$$= \frac{V_{SL}\ V_{RL}}{B} = \frac{275 \times 275}{141.0} = 536.35 \text{ MVA}$$

Horizontal coordinate

$$= -\frac{A\ V_{RL}^2}{B} \cos(\beta - \alpha)$$

$$= \frac{-0.95 \times (275)^2}{141} \times \cos(82° - 1°) = -79.71 \text{ MW.}$$

Vertical coordinate

$$= -\frac{A \times V_{RL}^2}{B} \sin(\beta - \alpha) = \frac{-0.95 \times (275)^2}{141} \sin 81°$$

$$= -503.26 \text{ MVAR}$$

Take scale 1 cm

= 100 MW horizontally and 1 cm = 100 MVAR vertically, we have

$$\text{Horizontal coordinate} = \frac{-79.71}{100} = -0.8 \text{ cm}$$

$$\text{Vertical coordinate} = \frac{-503.26}{100} = -5.03 \text{ cm}$$

Now draw circle diagram (Fig. 7.18), taking N as centre and radius equal to $\frac{536.35}{100} = 5.36$ cm.

The operating point for zero load is *b'*. So O*b'* represents the amount of shunt compensation to maintain the specified voltages under no-load condition. Thus at no load the synchronous phase modifier takes lagging MVAR given by

O*b'* = 0.27 cm or 0.27 × 100 = 27 MVAR **Ans.**

For a load of 200 MVA at 0.85 pf lagging draw load line OP making an angle of $\cos^{-1} 0.85 = 31.8°$ with the horizontal and make O*a* $= \frac{200}{100} = 2$ cm. Drop perpendicular from point *a* meeting the receiving-end power circle at point *a'*. Now *aa'* represents the output of the phase-modifier for receiving-end load of 200 MVA at 0.85 pf lagging. On measurement *aa'* = 1.35 cm.

So the output of the phase modifier = 1.35 × 100 = 135 MVAR (leading) **Ans.**

Example 7.14. The generalised constants of one phase of a three-phase line are :

A = D = 0.9 + *j*0.012, B = (22.5 + *j*150) Ω and C = (–0.00004 + *j*0.001) S

The sending-end voltage is 240 kV and the receiving-end voltage is 220 kV line to line. Draw the circle diagram and determine the active and reactive power received when the angle between the sending-end and the receiving-end voltage phasors is 30°. [*Pb. Technical Univ. Power System-I, May 2009*]

Solution : Torque angle, $\delta = 30°$

Line constants, $\mathbf{A} = 0.9 + j0.012 = 0.9 \angle 0.76°$

$\mathbf{B} = (22.5 + j150)\ \Omega = 151.68 \angle 81.47°\ \Omega$

$\therefore$ A = 0.9 ; α = 0.76° ; B = 151.68 Ω ; β = 81.47°

For receiving-end power circle diagram

$$\text{Radius of circle} = \frac{V_{SL}\ V_{RL}}{B} = \frac{220 \times 240}{151.68} = 348 \text{ MVA}$$

Coordinates of the centre of the power circle are :

Horizontal coordinate

$$= -\frac{A\ V_{RL}^2}{B} \cos(\beta - \alpha)$$

$$= \frac{-0.9 \times (220)^2}{151.68} \times \cos(81.47° - 0.76°) = -46.36 \text{ MW.}$$

Vertical coordinate

$$= -\frac{A \times V_{RL}^2}{B} \sin(\beta - \alpha)$$

$$= \frac{-0.9 \times (220)^2 \sin(81.47° - 0.76°)}{151.68} = -283.4 \text{ MVAR}$$

Take scale 1 cm = 50 MW horizontally and

1 cm = 50 MVAR vertically

$$\text{Radius of the circle} = \frac{348}{50} = 6.96 \text{ cm}$$

$$\text{Horizontal coordinate} = \frac{-46.36}{50} = -0.93 \text{ cm}$$

$$\text{Vertical coordinate} = \frac{-283.4}{50} = -5.67 \text{ cm}$$

and draw circle diagram (Fig. 7.19).

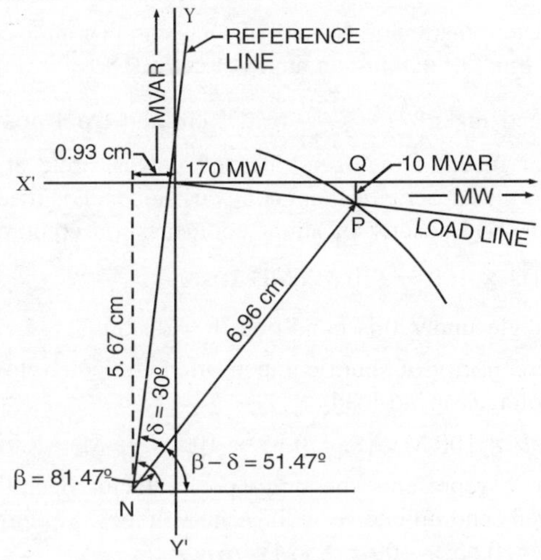

Fig. 7.19

Draw reference line making an angle $\beta = 81.47°$ with the horizontal and another line NP making an angle $\delta = 30°$ with the reference line or $81.47° - 30° = 51.47°$ with the horizontal axis cutting the arc of the circle at point P. Draw perpendicular from point P on the X-axis cutting the X-axis on the point Q.

Active power received = OQ = 3.4 × 50 = 170 MW **Ans.**

Reactive power supplied = PQ = 0.2 × 50 = 10 MVAR (leading) **Ans.**

Example 7.15. A 3-phase transmission line has a resistance per phase of 5 Ω and an inductive reactance per phase of 12 Ω and the line voltage at the receiving end is 33 kV.

Determine the voltage at the sending end when the load at the receiving end is 20 MVA at 0.8 power factor lagging.

The voltage at the sending end is maintained constant by means of a synchronous phase modifier at the receiving end which has the same rating at zero load at the receiving end as for the full-load of 16 MW load. Determine the power factor for the full-load output and the rating of the synchronous phase modifier.

Solution : Impedance of the line per phase,

$$\mathbf{Z} = 5 + j12 = 13 \angle 67.38° \ \Omega$$

Load power factor angle,

$$\phi_R = \text{Cos}^{-1}\ 0.8 = 36.87° \text{ (lagging)}$$

Load current,

$$I_R = \frac{\text{Load in MVA} \times 10^6}{\sqrt{3} \times V_{RL}} = \frac{20 \times 10^6}{\sqrt{3} \times 33{,}000} = 350 \text{ A}$$

Sending-end phase voltage,

$$\mathbf{V_S} = \mathbf{V_R} + \mathbf{I_R}\mathbf{Z}$$

$$= \frac{33{,}000}{\sqrt{3}} + 350 \angle -36.87° \times 13 \angle 67.38°$$

$$= (19{,}052.6 + j0) + (3{,}919.0 + j2309.4)$$

$$= 23{,}087 \angle 5.74° \text{ volts}$$

Sending-end line voltage,

$$V_{SL} = \sqrt{3} \times 23{,}087 \text{ V} = 40{,}000 \text{ V or } 40 \text{ kV } \textbf{Ans.}$$

For short transmission line

$$A = 1,\ \alpha = 0,\ B = Z = 13\ \Omega \text{ and } \beta = 67.38°$$

For receiving-end power circle diagram

$$\text{Radius of circle} = \frac{V_{SL} V_{RL}}{B} = \frac{33 \times 40}{13} = 101.54 \text{ MVA}$$

Coordinates of the centre of the power circle are :

Horizontal coordinate

$$= -\frac{A \times V_{RL}^2}{B} \cos(\beta - \alpha) = -\frac{1 \times (33)^2}{13} \cos 67.38°$$

$$= -32.22 \text{ MW}$$

Vertical coordinate

$$= -\frac{A \times V_{RL}^2}{B} \sin(\beta - \alpha) = -\frac{1 \times (33)^2}{13} \sin 67.38°$$

$$= -77.32 \text{ MVAR}$$

Take scale 1 cm = 20 MW horizontally and 1 cm = 20 MVAR vertically

$$\text{Radius of the circle} = -\frac{101.54}{20} = -5.1 \text{ cm}$$

$$\text{Horizontal coordinate} = -\frac{32.22}{20} = -1.6 \text{ cm}$$

$$\text{Vertical coordinate} = -\frac{77.32}{20} = -3.86 \text{ cm}$$

and draw circle diagram (Fig. 7.20)

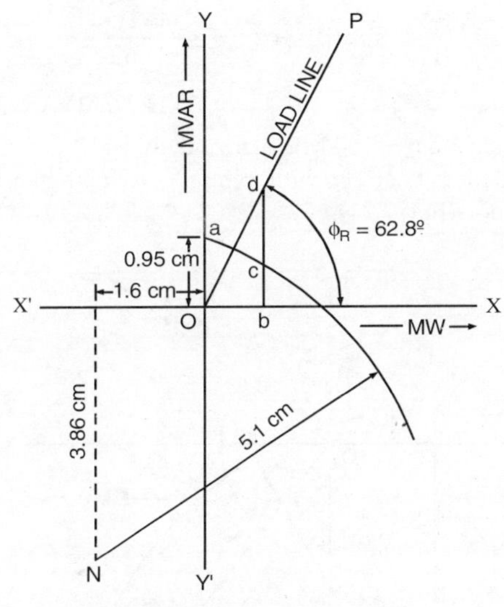

Fig. 7.20

At no load the synchronous phase modifier will be taking lagging MVAR given by

Oa = 0.95 cm = 20 × 0.95 MVAR = 19 MVAR **Ans.**

Take Ob = 16 MW, the full load supplied. At point b draw a perpendicular bcd cutting the arc of the circle at the point c. The MVAR supplied is given by the line bc. The synchronous phase modifier is supplying the same MVAR at full-load of 16 MW as at no load *i.e.,* 19 MVAR. Take cd = 19 MVAR. Join point O with the point d to give the load line.

The receiving end power factor angle at full load = Cos ϕ_R = Cos 62.8° = 0.4586 (lagging) **Ans.**

Example 7.16. A 3-phase overhead transmission line has a series impedance of (10 + *j* 30) Ω per phase. For receiving-and sending-end voltages of 132 kV and 140 kV respectively draw the receiving-end power circle diagram and determine (*i*) the maximum real power delivered by the line and the load power factor under that condition, (*ii*) the capacity of shunt compensation equipment for supplying a load of 150 MVA at 0.8 pf lagging and the power angle under that condition, (*iii*) the capacity of shunt compensation equipment to maintain the above voltage under no-load condition, and (*iv*) the unity pf load that the line can supply with voltages at above values.

[*A.M.I.E. Sec. B. Power Systems Summer 1995*]

Solution : The generalised constants of the transmission line (neglecting shunt capacitance of the line)

$\mathbf{A} = \mathbf{D} = 1 \angle 0^\circ$; $\mathbf{B} = \mathbf{Z} = (10 + j\,30) = 31.62 \angle 71.6^\circ$ Ω and $\mathbf{C} = 0$

For receiving-end power circle diagram

Radius of circle

$$= \frac{V_{SL}\,V_{RL}}{B} = \frac{132 \times 140}{31.62} = 584.4 \text{ MVAR}$$

Horizontal coordinate of centre

$$= -\frac{AV_{RL}^2}{B}\cos(\beta - \alpha) = -\frac{(132)^2}{31.62}\cos 71.6^\circ$$

$$= -174.26 \text{ MW}$$

Vertical coordinate of centre

$$= -\frac{AV_{RL}^2}{B}\sin(\beta - \alpha) = -\frac{(132)^2}{31.62}\sin 71.6^\circ$$

$$= -522.77 \text{ MVAR}$$

Take scale 1 cm = 100 MW horizontally and 1 cm = 100 MVAR vertically and draw circle with centre (–1.74 cm, –5.23 cm) and radius 5.84 cm (Fig. 7.21).

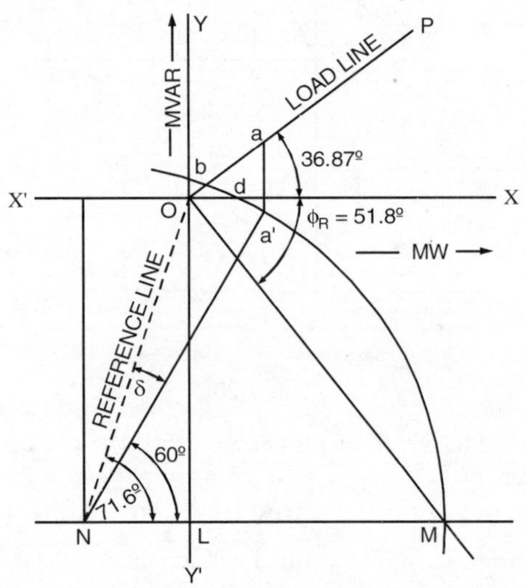

Fig. 7.21

(*i*) For determination of maximum power draw a horizontal line from the centre of the circle (N point) intersecting vertical axis at the point L and the circle at the point M. Distance LM represents the maximum power for the receiving end. On measurement LM = 4.10 cm.

So maximum real power delivered by the line

= 4.1 × 100 = 410 MW **Ans.**

Power factor of load under this condition

= Cos (XOM) = Cos 51.8° = 0.618 (lead) **Ans.**

(*ii*) For determination of capacity of shunt compensation equipment draw a load line OP making an angle of $\cos^{-1}$ 0.8 (*i.e.*, 36.87°) with the X-axis and make $oa = \frac{150}{100} = 1.5$ cm and from point *a* drop perpendicular meeting the receiving-end power circle at *a'*. Now *aa'* represents the capacity of shunt compensation and on measurement *aa'* = 1.2 cm. So capacity of shunt compensation equipment

= 1.2 × 100 = 120 MVAR **Ans.**

Power angle under this condition, δ = 11.6°

(*iii*) The capacity of shunt compensation equipment to maintain the given voltages at no-load

= *ob* × 100 MVAR = 0.35 × 100 = 35 MVAR **Ans.**

(*iv*) Point *d* represents the unity power factor operating point with the given sending-end receiving-end voltages. So unity *pf* load = *od* × 100 = 0.85 × 100 = 85 MW **Ans.**

Example 7.17. Draw the receiving-end and sending-end power circle diagrams for a 300 km line having R = 0.08 Ω/km; X = 0.4 Ω/km and Y = 5.15 × 10^{-6} $\angle 90^\circ$ S/km. From the diagram find the sending-end voltage, current and power factor angle when the line is delivering a load of 192 MW at 0.8 pf (lagging) and 275 kV. Assume π-configuration.

Solution : Resistance of line per phase,

$$R = 300 \times 0.08 = 24\ \Omega$$

Reactance of line per phase,

$$X = 300 \times 0.4 = 120\ \Omega$$

Impedance of line per phase,

$$\mathbf{Z} = (24 + j120) = 122.38 \angle 78.69^\circ\ \Omega$$

Admittance of line per phase,

$$\mathbf{Y} = 300 \times 5.15 \times 10^{-6} \angle 90^\circ \text{ S} = 1.545 \times 10^{-3} \angle 90^\circ \text{ S}$$

Representing the line by π-configuration

Line constants

$$\mathbf{A} = \mathbf{D} = 1 + \frac{\mathbf{ZY}}{2}$$

$$= 1 + \frac{1}{2} \times 122.38 \angle 78.69^\circ \times 1.545 \times 10^{-3} \angle 90^\circ$$

$$= 1 + 0.094539 \angle 168.69^\circ$$

$$= 0.9073 + j\,0.01854 = 0.9075 \angle 1.17^\circ$$

$$\mathbf{B} = \mathbf{Z} = 122.38 \angle 78.69^\circ\ \Omega$$

Load power factor angle,

$$\phi_R = \cos^{-1} 0.8 = \angle 36.9^\circ$$

Load current,

$$I_R = \frac{\text{Load in MW} \times 10^6}{\sqrt{3} \times V_{RL} \times \cos\phi_R} = \frac{192 \times 10^6}{\sqrt{3} \times 275{,}000 \times 0.8}$$

$$= 504.0 \text{ A}$$

For given line A = 0.9075 ; α = 1.17° ; B = 122.38 Ω and β = 78.69°

For receiving-end power circle diagram

Horizontal coordinate

$$= -\frac{A}{B} V_{RL}^2 \cos(\beta - \alpha)$$

$$= -\frac{0.9075}{122.38} \times (275)^2 \cos(78.69° - 1.17°) = -121 \text{ MW}$$

Vertical coordinate

$$= -\frac{A}{B} V_{RL}^2 \sin(\beta - \alpha)$$

$$= -\frac{0.9075}{122.38} \times (275)^2 \sin(78.69° - 1.17°) = -548 \text{ MVAR}$$

Take scale 1 cm = 100 MW horizontally and 1 cm = 100 MVAR vertically

Locate point N having coordinates – 1.21 cm and – 5.48 cm and draw load line OP at an angle $\cos^{-1}$ 0.8 *i.e.,* 36.9° inclined to the horizontal and to represent $\frac{\sqrt{3} V_{RL} I_R}{1,000}$ MVAR

i.e., $\frac{\sqrt{3} \times 275 \times 504}{1,000 \times 100}$ cm = 2.4 cm

Draw circle diagram with N (– 1.21 cm, – 5.48 cm) as centre and NP as radius.

From circle diagram (Fig. 7.22 *a*) NP = 7.6 cm .... (From measurement)

$$\therefore \text{NP} = \frac{V_{SL} V_{RL}}{B} = 100 \times 7.6 = 760 \text{ MVA}$$

$$\therefore V_{SL} = \frac{760 \times B}{V_{RL}} = \frac{760 \times 122.38}{275} = 338 \text{ kV}$$ **Ans.**

$\beta - \delta = 64°$

$\delta = \beta - 64° = 78.69° - 64 = 14.69°$

and $\beta + \delta = 78.69° + 14.69 = 93.38°$

For sending-end power circle diagram (Fig. 7.22*b*)

Horizontal coordinate

$$= \frac{D}{B} V_{SL}^2 \cos(\beta - \Delta)$$

$$= \frac{0.9075}{122.38} \times (338)^2 \cos(78.69° - 1.17°) = 183 \text{ MW}$$

Vertical coordinate

$$= \frac{D}{B} V_{SL}^2 \cos(\beta - \Delta) = \frac{0.9075}{122.38} \times (338)^2 \sin(78.69° - 1.17°)$$

$$= 827 \text{ MW}$$

$$\text{Radius of circle} = \frac{V_{SL} V_{RL}}{B} = 7.6 \text{ cm}$$

Draw a circle diagram with N (1.83 cm, 8.27 cm) as centre and 7.6 cm as radius. Draw NP inclined at angle $\beta + \delta$ *i.e.,* 78.69° + 14.69 = 93.38° to the horizontal cutting the arc of sending-end power circle at P. Join OP.

Sending-end power factor angle = ϕ_S = 15° (From measurement)

Sending-end power factor = Cos ϕ_S = Cos 15° = 0.966 **Ans.**

OP = 2.4 cm = 2.4 × 100 = 240 MVA ...(From measurement)

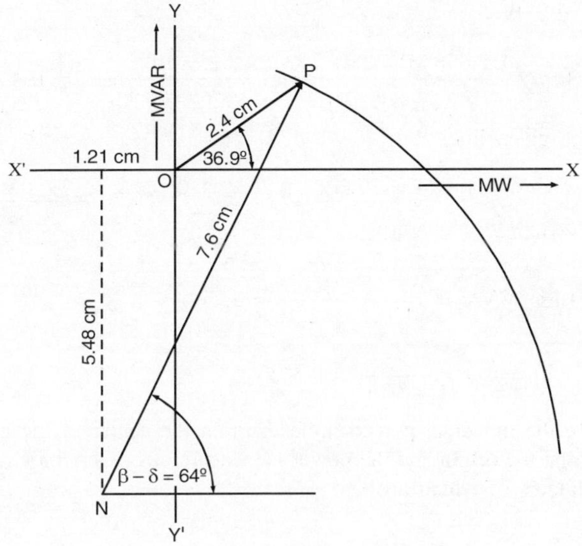

(*a*)

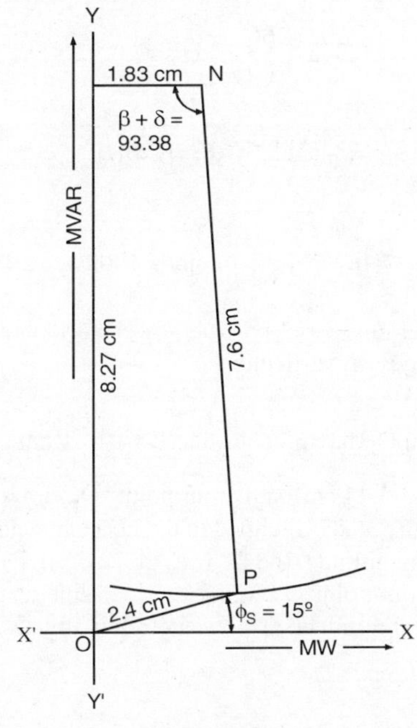

(*b*)

Fig. 7.22

$$\therefore \text{OP} = \frac{\sqrt{3}\, V_{SL}\, I_S}{1,000} \text{ MVAR}$$

$$\therefore I_S = \frac{\text{OP} \times 1,000}{\sqrt{3} \times V_{SL}} = \frac{240 \times 1,000}{\sqrt{3} \times 338} = 410 \text{ A}$$ **Ans.**

Example 7.18. A 3-ϕ transmission line has the following constants :

A = D = 0.945 $\angle 1° 01'$; B = 82.3 $\angle 73° 02'$ Ω and C = 0.001376 $\angle 90° 23'$ S

Construct a universal power circle diagram for the line and find the sending-end voltage, current and power factor when the line is supplying a load of 15 MW at 0.8 pf lagging, the receiving-end voltage being 66 kV. Choose base voltage as 60 kV.

Solution : Load power factor angle,

$$\phi_R = \text{Cos}^{-1}\ 0.8 = 36.87°$$

Load current,

$$I_R = \frac{\text{Load in MW} \times 10^6}{\sqrt{3} \times V_{RL} \times \cos\phi_R} = \frac{15 \times 10^6}{\sqrt{3} \times 66{,}000 \times 0.8} = 164 \text{ A}$$

For the given line

$$A = 0.945;\ \alpha = \angle 1°\,01';\ B = 82.3\ \Omega;\ \beta = \angle 73°\,02'$$

Base voltage, V = 60 kV

$$\text{Base MVA} = \frac{V^2}{B} = \frac{(60)^2}{82.3} = 43.74$$

$$\text{Per unit power} = \frac{15}{43.74} = 0.343$$

In the universal power circle diagram the centre of the circle is kept at the origin and the power is measured from a point O_R, the coordinates of which are

$$\text{Horizontal} = A\left(\frac{V_{RL}}{V}\right)^2 \cos(\beta - \alpha) \text{ per unit}$$

$$= 0.945\left(\frac{66}{60}\right)^2 \cos(73°\,02' - 1°\,01') = 0.353 \text{ pu}$$

$$\text{Vertical} = A\left(\frac{V_{RL}}{V}\right)^2 \sin(\beta - \alpha)$$

$$= 0.945\left(\frac{66}{60}\right)^2 \sin(73°\,02' - 1°\,01') = 1.0876 \text{ pu}$$

Take scale 1 cm = 0.2 pu true power horizontally and 1 cm = 0.2 pu reactive power vertically

Locate point O_R having coordinates $\frac{0.353}{0.2}$ cm and $\frac{1.0876}{0.2}$ cm *i.e.,* (1.76 cm and 5.44 cm) and from point O_R draw load line at an angle $\cos^{-1} 0.8$ *i.e.,* 36.87° inclined to the horizontal and an horizontal line O_R Q of magnitude 0.343 (*i.e.,* 1.72 cm). Draw a line QP perpendicular from point Q cutting the load line at the point P. NP is the radius of the universal power circle. (Fig. 7.23)

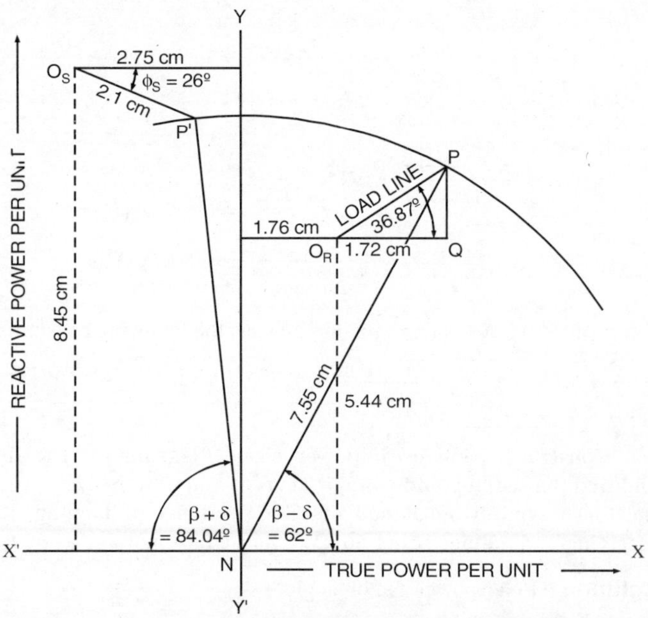

Fig. 7.23

From circle diagram

NP = 7.55 cm = = 7.55 × 0.2 = 1.51 pu ...(From measurement)

$$\therefore\ NP = \frac{V_{SL}\, V_{RL}}{V^2} \text{ per unit}$$

$$\therefore\ V_{SL} = \frac{NP \times V^2}{V_{RL}} = \frac{1.51 \times 60^2}{66} = 82.36 \text{ kV } \textbf{Ans.}$$

$\beta - \delta = 62°$ ...(From measurement)

$\therefore\ \delta = \beta - 62° = 73°\,02' - 62° = 11°\,02'$

$\therefore\ \beta + \delta = 73°\,02' + 11°\,02' = 84°\,04'$

The coordinates of the point O_S (the point of sending-end power measurement) with respect to the origin (the centre of the circle)

$$\text{Horizontal} = -D\frac{V_{SL}^2}{V^2} \cos(\beta - \Delta)$$

$$= -0.945 \times \left(\frac{82.36}{60}\right)^2 \cos(73°\,02' - 1°\,01')$$

$$= -0.55 \text{ pu}$$

$$\text{Vertical} = D\frac{V_{SL}^2}{V^2} \sin(\beta - \Delta)$$

$$= 0.945 \times \left(\frac{82.36}{60}\right)^2 \sin(73°\,02' - 1°\,01') = 1.69 \text{ pu}$$

Locate point O_S having coordinates $\frac{-0.55}{0.2}$ cm, $\frac{1.69}{0.2}$ cm, *i.e.,* (–2.75 cm, 8.45 cm) and from point N (the centre of universal circle) draw line NP′ making an angle β + δ *i.e.,* 84° 04' with the horizontal so as to cut the universal power circle at P′. Join O_S P′.

Sending-end power factor angle ϕ_S is the angle made by the line O_S P′ with the horizontal and the length of the line O_S P′ represents sending-end MVA. On measurement, O_S P' = 2.1 cm, so Sending-end MVA = 2.1 × 0.2 × 43.74 (Base MVA) = 18.37 MVA

∴ Sending-end current,

$$I_S = \frac{\text{MVA} \times 1{,}000}{\sqrt{3}\, V_{SL}} = \frac{18.37 \times 1{,}000}{\sqrt{3} \times 82.36} = 128.8 \text{ A } \textbf{Ans.}$$

Sending-end power factor angle, $\phi_S = 26°$

Sending-end power factor = $\cos\phi_S = \cos 26° = 0.899$ **Ans.**

Example 7.19. A 3-ϕ, 50 Hz transmission line has the following constants :

$$\mathbf{A = D = 0.997° \angle 0°\,6';\ B = 33.5 \angle 68°\,20'\ \Omega \text{ and } C = 0.218 \times 10^{-3} \angle 90°\,2'\ S}$$

Construct the universal power circle diagram for the line and compute the following :

(*a*) If the load supplied at the receiving end is 64 MW at 0.8 pf lag and 132 kV, find the sending-end voltage and power factor.

(*b*) If the sending-end voltage is maintained at 140 kV by a synchronous phase modifier at the receiving end, find the rating of the phase modifier when supplying load as in (*a*).

Solution : Load power factor angle, $\phi_R = \cos^{-1} 0.8 = 36.87°$

Load current,

$$I_R = \frac{\text{Load in MW} \times 10^6}{\sqrt{3} \times V_{RL} \times \cos\phi_R} = \frac{64 \times 10^6}{\sqrt{3} \times 132{,}000 \times 0.8} = 350 \text{ A}$$

For the given line A = 0.997 ; $\alpha = \angle 0^\circ 6'$; B = 33.5 Ω; $\beta = \angle 68^\circ 20'$

Let the base voltage V be 100 kV

$$\text{Base MVA} = \frac{V^2}{B} = \frac{(100)^2}{33.5} = 298.5$$

$$\text{Per unit power} = \frac{64}{298.5} = 0.2144$$

In the universal power circle diagram the centre of the circle is kept at the origin and the power is measured from point O_R, the coordinates of which are

$$\text{Horizontal} = A\left(\frac{V_{RL}}{V}\right)^2 \cos(\beta - \alpha) \text{ per unit}$$

$$= 0.997\left(\frac{132}{100}\right)^2 \cos(68^\circ\ 20' - 0^\circ\ 6') = 0.644 \text{ pu}$$

$$\text{Vertical} = 0.997\left(\frac{132}{100}\right)^2 \sin(68^\circ\ 20' - 0^\circ\ 6') = 1.614 \text{ pu}$$

Take scale 1 cm = 0.2 pu true power horizontally and 1 cm = 0.2 pu reactive power vertically

Locate point O_R having coordinates $\left(\frac{0.644}{0.2}\text{cm}, \frac{1.614}{0.2}\text{cm}\right)$ *i.e.*, (3.22 cm, 8.07 cm) and from point O_R draw load line at an angle $\cos^{-1} 0.8$ *i.e.*, 36.87° inclined to the horizontal and an horizontal line O_RQ of magnitude 0.2144 *i.e.*, (1.072 cm). Draw a line QP perpendicular from the point Q cutting the load line at the point P. NP is the radius of the universal power circle (Fig. 7.24)

From circle diagram NP = 9.9 cm ... (from measurement) = 9.9 × 0.2 = 1.98 pu

$$\because \text{NP} = \frac{V_{RL} V_{SL}}{V^2} \text{ pu}$$

$$\therefore V_{SL} = \frac{\text{NP} \times V^2}{V_{RL}} = \frac{1.98 \times 100^2}{132} = 150 \text{ kV } \textbf{Ans.}$$

$\because \beta - \delta = 63^\circ$ ...(from measurement)

$\because \delta = \beta - 63^\circ = (68^\circ\ 20' - 63^\circ) = 5^\circ\ 20'$

and $\beta + \delta = (68^\circ\ 20' + 5^\circ\ 20') = 73^\circ\ 40'$

The coordinates of the point O_S (the point of sending-end power measurement) with respect to the origin (the centre of the circle)

$$\text{Horizontal} = -D\frac{V_{SL}^2}{V^2}\cos(\beta - \Delta)$$

$$= -0.997 \times \frac{150^2}{100^2}\cos(\angle 68^\circ 20' + \angle 0^\circ 6')$$

$$= -0.83 \text{ pu}$$

$$\text{Vertical} = D\frac{V_{SL}^2}{V^2}\sin(\beta - \Delta)$$

$$= 0.997 \times \frac{150^2}{100^2}\sin(\angle 68^\circ 20' + \angle 0^\circ 6')$$

$$= 2.08 \text{ pu}$$

Locate point O_S having coordinates $\left(\frac{-0.83}{0.2}\text{cm}, \frac{2.08}{0.2}\text{cm}\right)$ *i.e.*, (–4.15 cm, 10.4 cm) and from point N (the centre of universal circle) draw line NP' making an angle β + δ *i.e.*, 73° 40' with the horizontal so as to cut the universal power circle at P'. Join O_SP'.

Sending-end power factor angle ϕ_S is the angle made by the line O_SP' with the horizontal $\phi_S = 39^\circ$ (by measurement)

Sending-end power factor = Cos ϕ_S = Cos 39° = 0.777 **Ans.**

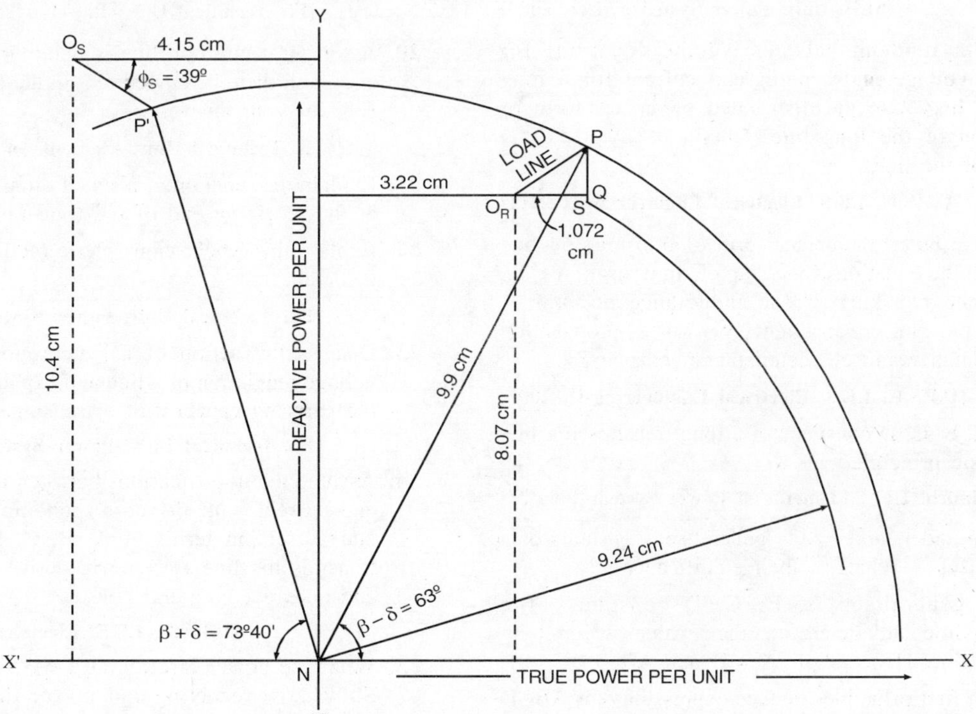

Fig. 7.24

(*b*) When sending-end voltage is 140 kV the radius of the universal power circle

$$= \frac{V_{RL} V_{SL}}{V^2} = \frac{132 \times 140}{(100)^2} = 1.848 \text{ pu}$$

Draw another circle arc with point N as centre and $\frac{1.848}{0.2} = 9.24$ cm as radius. Drop perpendicular PQS so as to cut the circle at point S.

MVAR supplied by the synchronous phase modifier

= PS × 0.2 × Base MVA

= 0.9 × 0.2 × 298.5 = 53.73 MVAR (leading) **Ans.**

Exercises

1. Derive the expression for sending-end voltage and current in terms of receiving-end voltage and current for long transmission lines. What do you understand by the "characteristic impedance" and "propagation constant" in it.
[U.P. Technical Univ. Elements of Power System 2009-10]
2. A long transmission line is terminated by an impedance equal to its characteristic impedance. Show that

$$\frac{V_R}{V_S} = e^{-\text{Y}l}$$

where V_R = Receiving-end voltage
V_S = Sending-end voltage
Y = Propagation constant
l = Length of line
[U.P. Technical Univ. Elements of Power System 2004-05]
3. Using rigorous method derive expressions for sending-end voltage and current for long transmission line.
[Rajasthan Univ. T&D of Electrical Power, 1999, 2002; M.D. Univ. Power System-I, Dec. 2009]
4. Starting from fundamentals, derive the equations for sending-end voltage and sending-end current for a long transmission line. Use the distributed parameter form of representation of the long line. Obtain the A, B, C, D parameters of the line.
[U.P.S.C. I.E.S. Electrical Engineering-II 2011]
5. Using the distributed parameter form of transmission line model, obtain the equations for V_S and I_S in terms of V_R, I_R, line resistance per km, *r*, line inductance per km, *l*, conductance per km, *g*, capacitance per km, *c*, propagation constant, gamma γ and characteristic impedance Z_C.
[U.P.S.C. I.E.S. Electrical Engineering-II, 2007]
6. Derive the A, B, C, D constants of a long transmission line from the basic principle.
[U.P. Technical Univ. Elements of Power System 2004-05]
7. What do you understand by the generalised constants of a transmission line ? What is their significance ?
8. Define the generalised A, B, C, D constants of a transmission line and determine their values for a long transmission line. Hence show **A** = **D** and **AD** – **BC** = 1.
9. Starting from first principles deduce expressions for ABCD constants of a long line in terms of its parameters. Define propagation constant and characteristic impedance.
10. Develop the long line exact equations in hyperbolic terms for sending-end voltage and current.
[U.P.S.C. I.E.S. Elec. Engineering II, 1999]
11. With reference to long transmission lines, give physical interpretation of the terms 'characteristic empedance' and 'propagation constant.' What is meant by surge impedance ?
[U.P.S.C. I.E.S. Elec. Engineering II, 1998]
12. Explain surge impedance loading of a transmission line.
[Pb. Technical Univ. Power System-I (T&D), Dec. 2009; U.P.S.C. I.E.S. Electrical Engineering-II, 2005]
13. Explain surge impedance loading with respect to an overhead transmission line. Show that

$$Z_C = \sqrt{Z_{OC} Z_{SC}}$$

[U.P. Technical Univ. Elements of Power System 2006-07]
14. Explain briefly about "SIL".
[U.P. Technical Univ. Elements of Power System 2009-10]
15. Obtain equivalent-π model of a long transmission line.
[U.P. Technical Univ. Elements of Power System 2008-09]
16. Explain the concept of regulation in long transmission line. If regulation is very high then what can be seen adversely in power network.
[U.P. Technical Univ. Elements of Power System 2005-06]
17. Draw the equivalent circuit of a long transmission line. Derive from fundamentals the following relationships between sending- and receiving-end voltages and currents.

$$\mathbf{V_S = AV_R + BI_R} \quad \text{and} \quad \mathbf{I_S = CV_R + DI_R}$$

18. What are A, B, C and D constants of a long transmission line and derive these constants for nominal T and π-network of a long transmission line.
[Pb. Technical Univ. Power System-I (T&D) May 2009]
19. Discuss the equivalent 'T' circuit for long length line.
[Pb. Technical Univ. Power System-I (T&D) Dec. 2009]
20. A long transmission line is open circuited at the receiving end. Will there be any current in the line at the sending end ? Explain your answer.
[G.B. Technical Univ. Elements of Power System 2011-12]
21. Explain the function of a synchronous phase modifier placed at the receiving end of the transmission line.
22. Explain the synchronous phase modifiers with their ratings in detail.
[Pb. Technical Univ. Power System-I (T&D), Dec. 2009]
23. Discuss the action of a synchronous phase modifier for voltage regulation of a line and explain how its use increases the carrying capacity of a transmission line.
[Pb. Technical Univ. Power System-I (T&D), May 2008]
24. Assume that the equations between the sending-end voltage and current with the corresponding receiving-end voltage and current in terms of A, B, C, D constants of a long transmission line are known, derive equations for sending-end and receiving-end power.
[U.P.S.C. I.E.S. Electrical Engineering-II, 2005]
25. What are power circle diagrams ? How are they useful ? Show how receiving-end power diagram may be drawn for a transmission line.
[M.D. Univ. Power System-I, Dec. 2007]

26. Explain under what conditions, the locus of receiving-end complex power would be a circle. Obtain the expression for the radii and cordinates of the centre of circle.

[Pb. Technical Univ. Power System-I (T&D), May 2010]

27. Show how receiving-end power circle diagram of a transmission line based on generalized (A, B, C, D) constants can be drawn.

[Pb. Technical Univ. Power System-I (T&D) Dec. 2007]

28. Show how sending-end power circle diagram of a transmission line based on generalized (A, B, C, D) constants can be drawn.

[Pb. Technical Univ. Power System-I (T&D), May 2008]

29. What is a universal power circle diagram of a transmission line ? Explain how it can be constructed and what useful informations can be obtained from it.

[M.D. Univ. Transmission & Distribution May 2011]

30. Write short notes on the following :

(*i*) Rigorous method of solution for a long transmission line. **[Rajasthan Univ. 2000, 2001]**

(*ii*) Surge impedance of a transmission line.

(*iii*) Tuned power lines.

(*iv*) Synchronous phase modifiers.

[Pb. Technical Univ. Dec. 2007]

(*v*) Power circle diagram of a transmission line.

[Pb. Technical Univ. Dec. 2009]

(*vi*) Power circle diagram.

[W.B. Univ. of Technology 2009-10]

(*vii*) Universal power circle diagram.

Short Answer Type Questions With Answers

Q. 1. What is the range of surge impedance for an overhead transmission line ?

Ans. 400–600 Ω.

Q. 2. What is the range of surge impedance in case of underground cables ?

Ans. 40–60 Ω.

Q. 3. What is surge impedance loading with respect to an overhead transmission line ? **[Pb. Technical Univ. May 2010]**

OR

What is surge impedance loading ?

[Pb. Technical Univ. May 2009]

OR

Define surge impedance loading (SIL).

[Pb. Technical Univ. Dec. 2007]

OR

Explain surge impedance loading of a transmission line.

[U.P.S.C. I.E.S. E.E.-II, 2005]

OR

What is meant by natural loading of lines ?

[Pb. Technical Univ. May 2008]

Ans. Surge impedance loading (SIL) or natural load is the load that can be delivered by a line having no resistance and the loading being driven at unity power factor.

The 'natural power' or 'surge impedance loading' of a transmission line is given as

$$P_R = \frac{V_{RL}^2}{Z_0} \text{ MW}$$

where V_{RL} is the receiving-end voltage in kV and Z_0 is the surge impedance in ohm.

The above expression gives a limit of the maximum power that can be delivered by a line and is useful in the design of transmission lines. This can be used for the comparison of loads that can be carried on the transmission lines at different voltages.

Q. 4. How does synchronous phase modifier differ from a synchronous motor ?

Ans. Synchronous phase modifier differs from an ordinary synchronous motor in as much as they are built for the highest economical speeds, and provided with smaller shafts and bearings and special attention is paid for securing a high overall efficiency.

Q. 5. How does the reactive power of a transmission line vary for given values of V_S, V_R and X ?

Ans. For fixed values of V_S, V_R and X, the reactive power is proportional to the magnitude of voltage drop in the line (*i.e.*, $V_S - V_R$).

Q. 6. Why high and extra voltages are adopted for transmission of bulk power over long distances ?

Ans. This is because the maximum power transferred over a given line increases with the increase in V_S and V_R.

Q. 7. What is meant by receiving-end power circle diagram ?

Ans. The circle diagram drawn with receiving-end true and reactive power components as the horizontal and vertical ordinates is called the receiving-end power circle diagram.

Q. 8. What is meant by sending-end power circle diagram ?

Ans. The circle diagram drawn with sending-end true and reactive power components as the horizontal and vertical ordinates is called the sending-end power circle diagram.

Q. 9. What is universal power circle diagram ?

Ans. In the ordinary receiving-end power circle diagram, if the receiving-end voltage is kept constant, the centre of the circles is fixed and for different sending-end voltages circles are drawn; these will be concentric but if the receiving-end voltage is not constant the circles will not be concentric. Similar is the case with the sending-end power circle diagram. When sending-end voltage is not constant, its centre will also vary.

To overcome the above limitation the universal power circle diagram is drawn. In this diagram each distance on the original diagram is divided by $\frac{V^2}{B}$ where V is the "reference or base voltage" and B is the generalised circuit constant. This gives dimensionless units.

Problems

1. A 3-phase transmission line, 160 km long, has the following constants :

Resistance per phase per km = 0.2 Ω

Reactance per phase per km = 0.3127 Ω

Shunt admittance per phase per km = 1.875×10^{-6} S

Determine the sending-end voltage and current by rigorous method when the line is delivering a load of 25 MVA at 0.8 power factor lagging. The receiving-end voltage is kept constant at 110 kV. **[Ans.** 116.67 kV; 131.1 A]

2. Determine the auxiliary constants of a 3-ϕ, 50 Hz, 150 km long transmission line having resistance, inductance and capacitance per phase per km of 0.2 Ω, 1.5 mH and 0.008 μF respectively. Neglect the conductance of the line.

[Ans. A = D = 0.987 $\angle 0.3°$; **B** = 76 $\angle 67.1°$ Ω ;

C = 0.37 × 10^{-3} $\angle 90.1°$ S]

3. The constants for a 132 kV line are as under :

A = D = 0.98 $\angle 3°$; **B** = 110 $\angle 75°$ Ω; **C** = 0.0034 $\angle 80°$ S

The load at the receiving end is 40 MVA at 132 kV and power factor 0.8 lagging. Draw the circle diagram and find out (*i*) sending-end voltage (*ii*) leading MVAR for this load is at the receiving end, if the sending-end voltage is 140 kV.

[Pb. Univ. Transmission and Distribution of Electrical Energy Nov. 1981]

[Ans. (*i*) 175.8 kV (*ii*) 23 MVAR]

4. A single circuit, 50 Hz transmission line is 500 km long and has the following ABCD constants:

A = D = 0.895 $\angle 1.5°$; **B** = 182.5 $\angle 78.5°$ Ω;

C = 0.00111 $\angle 90.5°$ S

Construct a Universal power circle diagram and from the diagram determine :

(*a*) The sending-end voltage for a load of 80 MVA at 215 kV and 0.8 pf lagging at the receiving end.

(*b*) The voltage regulation for the loading as in (*a*) and

(*c*) The reactive power supplied by the line and by a synchronous phase modifier in parallel with the load, when the loading is 80 MVA at 215 kV and 0.9 pf lagging. Given that the sending-end voltage is 236 kV, also determine the power factor at the receiving end of the line.

[Agra Univ. Transmission and Distribution of Electrical Power 1976]

[Ans. (*a*) 263 kV (*b*) 22.33% (*c*) 24.45 MVAR (line); 14.3 MVAR lead (SPM); 0.956 lagging]

5. A 3-phase overhead line has resistance and reactance per phase of 25 Ω and 90 Ω respectively. The supply voltage is 145 kV while the load-end voltage is maintained at 132 kV for all loads by an automatically controlled synchronous phase modifier. If the KVAR of the modifier has the same value for zero load as for a load of 50 MW, find the rating of the modifier and the power factor of the load.

[U.P.S.C I.E.S. Power Systems 1974]

[Ans. 19 MVAR; 0.95 lagging]

6. A 3-phase overhead line has resistance and reactance per phase of 5 Ω and 25 Ω. The load at the receiving end is 15 MW, 33 kV, 0.8 pf lagging. Estimate the capacity of the compensation equipment needed to supply this load with a sending-end voltage of 33 kV.

[A.M.I.E. Sec. B. Power Systems Summer 1997]

[Ans. 17.38 MVAR]

❑❑❑

CHAPTER 8

Underground Cables

INSIDE THIS CHAPTER

8.1. INTRODUCTION

As already discussed in chapter 1, the merits and demerits of transmission and distribution of electric power by underground cables in comparison with that by overhead system, the initial heavy cost* is the only factor which discouraged the use of underground cables for the purpose of transmission and distribution of electric power inspite of its numerous advantages. Underground cables are used for transmission and distribution of power where it becomes impracticable to make use of overhead construction. Such location may be congested urban area where right of way cost would be excessive or local ordinances prohibit overhead lines for reasons of safety, or around plants and substations; or crossings of wide bodies of water which for various reasons would not permit the overhead crossings. The type of cable used will depend upon the voltage and service requirements. Recent improvements in design and manufacture have led to the development of cables suitable for use at high voltages. This has made it possible to use underground cables for transmission of electric power for short or moderate distances.

An electric cable may be defined as a single conductor insulated through its full length; or two or more conductors each provided with its own insulation and laid up together under one outer protective covering.

8.2. GENERAL CONSTRUCTION OF CABLE

The general construction of an underground cable is given below:

The underground cable employed for transmission of power at high voltage consists of one central core or a number of cores (two, three or four) of tinned stranded copper or aluminium conductors insulated from each other by paper or varnished cambric or vulcanized bitumen or impregnated paper. A metallic sheath of lead or alloy or of aluminium is provided around the insulation to protect it against ingress of moisture, gases or other damaging liquids (acid or alkalies) in the soil and atmosphere. A metal is essential for sheath because no organic material is sufficiently impervious to moisture. The main advantages of lead sheaths are the comparative ease with which they are made in lead presses, their flexibility and high corrosion resistance while they have got drawbacks of large specific gravity, low mechanical strength, fluidity and small resistance to vibrations. The hardness, mechanical strength and resistance to vibrations of lead sheaths can be considerably improved by adding alloying mixtures (about 1 per cent of copper, tin, bismuth, etc). Aluminium sheaths being cheaper (costing 3.5 times less than the lead sheaths) and having much greater mechanical strength than lead sheaths, low weight and low fluidity are now being increasingly employed.

* The approximate cost ratios are as below :

System Operating Voltage in kV	0.4	11	33	66	132	220	400
Cost Ratio of UG/OH System	2	3	5	7	9	13	24

For the protection of metallic sheath against corrosion and from mechanical injury a layer of *bedding* consisting of paper tape compounded with a fibrous material (such as jute, cotton or hessian tape) is provided over the metallic sheath. The compound used should be of such a nature that it does not react with the material of the armouring and the lead sheath but simultaneously it is adhesive enough so as to stick both to the lead sheath and the armouring. Bedding is employed in paper insulated lead covered cables. It is not required in PVC cables.

Over the layer of bedding *armouring* consisting of one or two layers of galvanized steel wire or steel tape is provided to save the cable from mechanical injury and over the armouring a layer of fibrous material similar to that of bedding, known as *serving,* is provided in order to protect the armouring. Steel tape armour consists of two steel tapes coated with preservative compound, applied hellically in the same direction over a fibrous bedding, the outer layer covering the spaces between the turns of the inner layer. This provides sufficient protection against mechanical damage but steel wire armour is recommended where additional longitudinal stresses may occur during installation or in service. Double wire armours are employed for high degree of mechanical protection.

Single core cables for ac systems are not provided with armouring because eddy currents induced in the steel armour causes additional power losses. In such cables mechanical protection is provided by plastic wrap. Aluminium sheathed cables, owing to comparative hardness of sheath, need not to be armoured in the same way as the lead sheathed cables but for special conditions armouring is provided for these cables too.

Figure 8.1 illustrates the various parts of an underground cable.

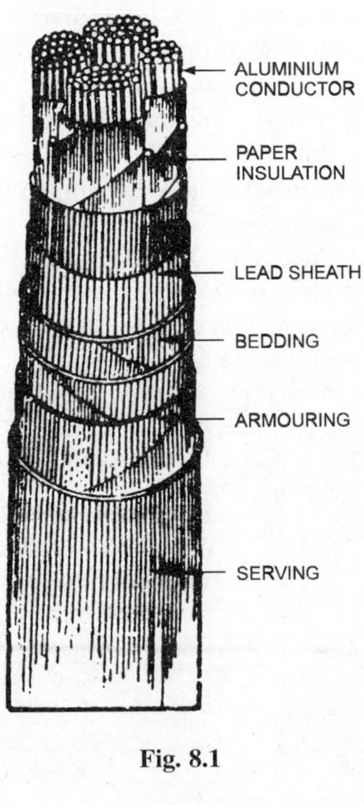

Fig. 8.1

8.3. REQUIREMENTS OF CABLE

(*i*) The conductor (usually annealed copper with about 99% purity or aluminium) used in cables should be stranded one in order to provide flexibility to the cables and should be of such x-sectional area that it may carry the desired load current without overheating and causing excessive voltage drop.

(*ii*) The insulation provided should be of such thickness that it may give high degree of safety and reliability at the working voltage for which it is designed.

(*iii*) The cables should be provided with a mechanical protection so that it may withstand the rough usage in laying it.

(*iv*) Materials used in manufacture of cables should be such as to give complete chemical and physical stability throughout.

8.4. CABLE CONDUCTORS

For all cables the conductor was, at one time, almost universally stranded copper but because of scarcity of copper, only aluminium is allowed to be used in power cables in India. Copper conductors are only used for cables used in control circuits.

The stranding is necessary to provide flexibility to cables and the conductor numbers are 3, 7, 19, 37, and so on, all numbers except 3 having a centrally disposed conductor with all the others around it. Thus a 7-strand has a central conductor with 6 around it. The various conductors are spiralled round the central conductor, and when there is more than one layer alternate layers are spiralled in opposite directions. This is to prevent 'birdcaging' when the conductor is bent.

The spiralling of the conductors obviously causes the increase in resistance, since the length of each spiralled conductor is greater than the central strand, and the flow of current is along the various conductors. It is therefore necessary to be particular in the designation of the cross section.

8.5. INSULATING MATERIALS FOR CABLES

The satisfactory operation of a cable depends to a great extent upon the characteristics of insulation employed. So the proper choice of insulating material for cables is considerably important. The main requirements of the insulating materials used for cables are :

1. *High insulation resistance* to avoid leakage current.
2. *High dielectric strength* to avoid electrical breakdown of the cable.
3. *Good mechanical properties i.e.,* tenacity and elasticity. Good tenacity is required in the material to withstand the mechanical handling of cables.
4. *Immune to attacks by acids and alkalies,* over a range of temperature of about – 18°C to 94°C
5. *Nonhygroscopic* because the dielectric strength of any material goes very much down with moisture

content. In case the insulating material is hygroscopic, it must be enclosed in a watertight covering like lead sheath etc.

6. *Noninflammable.*
7. *Low coefficient of thermal expansion.*
8. *Low permittivity.*
9 *Capability of withstanding high rupturing voltages.*

The last point but not the least is that a dielectric should not be *expensive* and it should be possible to handle it from the point of view of the manufacturer.

No one insulating material possesses all the above mentioned qualities, so the type of insulating material used in a cable depends upon the service for which the cable is required. The various insulating materials used in manufacture of cables are rubber, VIR, paper, polyvinyl chloride, varnished cambric, polyethylene, vulcanized bitumen, gutta-percha, silk, cotton, enamel etc.

1. *Rubber.* Rubber may be natural and synthetic. Natural rubber is obtained from the milky sap of tropical trees. Synthetic rubber is produced from alcohol or oil products. Its relative permittivity is between 2 and 3 and dielectric strength is 30 kV/mm. Though it possesses high insulating qualities but it absorbs moisture readily, softens when heated to a temperature of 60 or 70ºC, swells under the action of mineral oils and ages when exposed to light. Hence pure rubber cannot be used as insulating material.

2. *Vulcanized India Rubber (VIR).* It is prepared by mixing India rubber with mineral matter such as sulphur, zinc oxide, red lead etc. Though vulcanization process improves the nonhygroscopic and other properties like strength, durability, resilience, elasticity etc; but it slightly reduces its specific insulation resistance. It has dielectric strength of about 10-20 kV/mm and dielectric constant of about 2.5. The main drawback of VIR is that owing to sulphur content it attacks copper and therefore in cables using VIR insulation the copper conductor is tinned before providing this insulation. Sometimes a layer of pure rubber is also given on the conductor to protect it from sulphur.

As compared to impregnated paper, the performance of VIR is poor because VIR cables are expensive, less durable and have poor mechanical properties such as electrostatic capacity, elasticity and tenacity. VIR is used as insulating covering for low and medium voltage cables. Such type of cables are normally covered with cotton tape and jute braiding soaked in the compound in order to protect the cables against ingress of moisture.

3. *Impregnated Paper.* It is prepared from wood pulp rags or plant fibres by a suitable chemical process. It is quite cheap, has low capacitance, high dielectric strength (30 kV/mm), high insulation resistivity (of the order 10^7 Ω–m), high thermal conductivity, better thermal withstand capability and high durability. The only disadvantage is that it is hygroscopic, and even if it is impregnated with compound, it absorbs moisture, which lowers its insulation resistance. That is why paper insulated cables are always provided with some sort of protective covering and are never left unsealed. Its ends should be temporarily covered with wax or tar, even if it is required to be left unused on site while being laid. The maximum safe temperature of paper insulated cable is about 95ºC. The compound usually employed for impregnation purposes is a heavy-grade mineral oil mixed with resin, the actual proportions of oil and resin depending on the particular manufacturer.

For providing insulation over the conductor, the paper tape is lapped on to the conductor until the required thickness* is obtained, it is then dried and impregnated with insulating compound. The insulating compounds are required to have, besides high electrical insulation properties, a high viscosity at operating temperatures (50–80ºC), a low viscosity at impregnation temperatures of 50–120ºC, a smooth change in viscosity with temperature and very low coefficient of thermal expansion. The impregnation of paper insulation increases its dielectric strength from about 3–5 to 10–80 kV/mm. The dielectric constant of impregnated paper insulation is about 3.5.

Paper insulated cables are used for conveying large blocks of power, in transmission and distribution, and particularly for distribution at low voltage in congested areas where the joints are to be provided only at the terminal apparatus or where the joints are rare, owing to cheapness and durability over VIR cables.

4. *Polyvinyl Chloride.* This is a synthetic compound. For obtaining this material as a cable dielectric or sheathing it is processed with certain materials known as *plasticizer* and its type will depend upon the use of the finished product *i.e.,* PVC. It is inert to oxygen and almost inert to oils and to many alkalies and acids and therefore its use is preferred over VIR in extreme environments such as in cement factory or chemical factory. PVC has dielectric strength of 17 kV/mm and dielectric constant of 5. Maximum continuous temperature rating is 75ºC. The mechanical properties (*i.e.,* elasticity and and recovery from stretching) of PVC are not so good as those of rubber. PVC insulated cables are usually employed for low and medium voltage domestic and industrial light and power installations.

5. *Polythene.* Polythene is a straight chain polymer derived from ethylene. Its electrical properties are very good—low permittivity (dielectric constant), high insulation resistivity and very low power factor at all frequencies. The thermal dissipation properties are better than those of impregnated paper and the impulse strength compares favourably with an impregnated paper insulated cable. The maximum operating temperature under short circuits is 100ºC. It has limited use in power cables. However, a 3-core 11 kV self supporting aerial cable, which can be employed in rural distribution, has been developed.

* The thickness of insulation depends on the mechanical and electrical requirements which it has to meet in the service.

6. *Cross-Linked Polythene.* Low density polythene, when vulcanized under controlled conditions, results in cross-linking of carbon atoms and the compound produced is a new material having extremely high melting point with light weight, small dimensions, low dielectric constant and good mechanical strength. By using chemical process it has been made technically possible to cross-link polythene in conventional equipment for the manufacture of rubber. This is why the product is said to be "vulcanized" 'or cross-linked' polythene.

Cross-linked polythene (XLPE) cables have high maximum continuous temperature rating of 90° with dielectric strength of about 20 kV/mm. Because of high temperature withstand capability and very low water absorption, such cables can be buried directly in soil. These cables have been found very suitable for all voltages up to 33 kV.

7. *Varnished Cambric (or Empire Tape).* This is a cotton cloth impregnated and coated with varnish. It has very smooth surface. The cambric is lapped in the form of a tap on to the conductor and its surfaces are coated with petroleum jelly compound to give easy sliding of the surfaces. Such cables require protective covering like lead sheath because this insulating material is hygroscopic. Its dielectric strength is 4 kV/mm and its dielectric constant is 2.5 to 3.8. Such cables do not require sealing.

8. *Gutta-percha.* It is similar to rubber but it becomes soft at about 65°C. It is nonhygroscopic but cannot withstand even medium voltages. It is mostly used for submarine cables for telephone and telegraph purposes.

9. *Silk and Cotton.* This is used in low voltage cables. The conductor may have a single layer or double layer covering depending upon the requirements of service. Silk or cotton covered wires are usually used for instrument and motor winding.

10. *Enamel Insulation.* *Enamelled* wires are also used for instruments and motor winding. The wires are cheaper than silk and cotton covered wires and therefore for low voltage machines and instruments enamelled wires are used. Enamel insulation is liable to crack.

8.6. CLASSIFICATION OF CABLES

The type of cable to be used at a particular location is determined by the mechanical considerations and the voltage at which it is required to operate. Usually the operating voltage determines the type of insulation and the cables are placed in various categories depending upon the voltage for which they are designed. According to voltage, the cables are classified as :

(*i*) Low voltage (LT) cables for operating voltage up to 1,000 V.

(*ii*) High voltage (HT) cables for operating voltage up to 11,000 V.

(*iii*) Super-tension (ST) cables for operating voltage up to 33,000 V.

(*iv*) Extra high tension (EHT) cables for operating voltage up to 66,000 V.

(*v*) Extra super-voltage cables for operating voltage beyond 1,32,000 V.

A cable may be a single core or multi-core depending upon the type of service for which it is intended to be used. A multi-core cable may be two-core, three-core or four-core. For 3-phase, 3-wire systems either three single-core cables or three-core cables can be used depending upon the operating voltage and load demand. For 3-phase, 4-wire systems 4-core cables may be used.

1. Low Tension (or LT) Cables. These cables are meant for use up to 1,000 V. For voltages up to 6,600 V, the electrostatic stresses developed in cables are very small and thermal conductivity is also of not much importance so no special construction is required. The insulating materials used may be impregnated paper, varnished cambric, vulcanized rubber, or vulcanized bitumen.

The *lt* cables are of two types *viz* single-core and multi-core cables. The former type has the advantages of simplicity of construction and availability of larger conductor section. The various sizes of *lt* cables are given below:

Single core cable with aluminium conductors : 1.5 – 625 mm^2

Two-, three-, three-and-a-half and four-core cables with aluminium conductors : 1.5 – 625 mm^2

Control cables up to 61 cores with copper conductors : 1.5 and 2.5 mm^2.

Single-core *lt* cable consists of one circular core of tinned stranded copper (or aluminium) insulated by layers of impregnated paper or varnished cambric over it and a lead sheath over the insulation, as shown in Fig. 8.2. The lead sheath protects the cable against ingress of moisture and mechanical handling. If the cables are to be buried directly then an overall serving of compounded fibrous material or hessian tape is provided over the sheath in order to protect the metallic sheath against corrosion. Single core cables are usually not provided with armouring in order to avoid excessive loss in the armour.

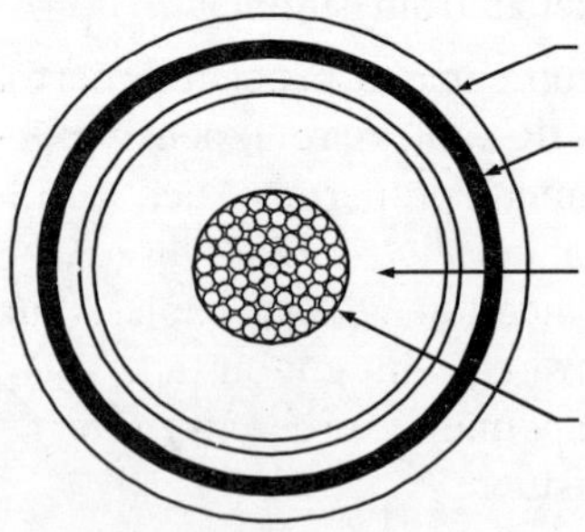

Fig. 8.2. *Single-Core LT Cable*

The multi-core cables for use up to 11,000 volts are of belted type. The belted type cable consists of either circular shaped or oval or sector shaped (oval or sector shaped core improve the copper space factor) cores of stranded copper (or aluminium) conductors wrapped around by impregnated paper and then wormed together. The interstices are filled with packing to make the cable of circular cross section. Insulating belt of impregnated paper is provided surrounding the three cores. A lead sheath enclosing the whole is provided for its protection against the entry of moisture. The cable is provided with two layers of steel tape pre-dipped in compound. The serving over the armouring consists of a compound and then one layer of impregnated hessian tape. In order to prevent adhesion, a coating of lime wash is applied to the outside of the cable.

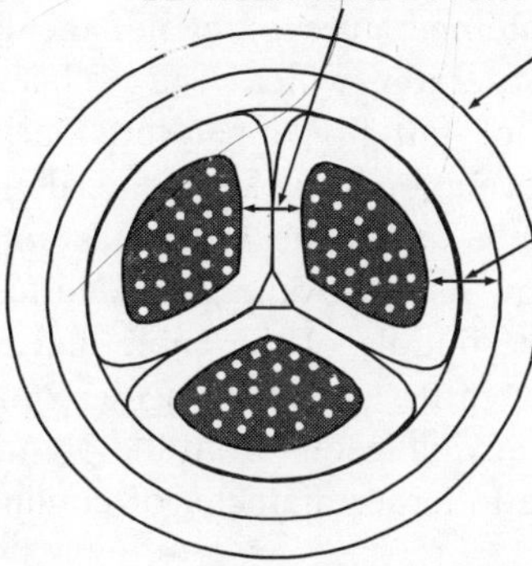

Fig. 8.3. ***Three-Core Belted Cable (Up to Lead Sheath)***

The belted type construction is not suitable for the cables used for voltages above 22 kV because of development of both the radial and tangential stresses. The tangential stresses act along the layers of insulation. The electric resistance, therefore, dielectric strength of the impregnated paper is much higher across the layers than along the layers. The leakage current on account of tangential stresses along the impregnated paper insulation causes power loss at the centre filling and local heating resulting in breakdown at any moment. Further owing to non-homogenity of dielectric in belted construction, when cables are loaded and unloaded some portions of the dielectric are less stressed whereas some portions are overstressed resulting in formation of vacuous spaces and voids. These vacuous spaces are ionised when voltage is applied and ultimately deteriorate the cable insulation.

The above difficulties have been overcome in the screened cables where leakage currents are conducted to earth through metallic sheaths.

2. Super-Tension Cables. For cables above 11,000 V a special construction is adopted. For use up to 33 kV, the cables used are screened cables where leakage currents are conducted to earth through metallic sheaths. Screened cables are of two types *viz* H-type and SL type.

(*a*) *H-Type Cables.* In the "H" type cable, invented by Hochstadter, no belt insulation is used, but each of the core is insulated with paper to the desired thickness and over this is provided a layer of metallized paper perforated to facilitate the process of impregnation, the coefficient of expansion and contraction is same as that of dielectric. Additional layer of the cotton tape with fine wires of copper is wrapped round all the three cores. Over this tape, no belt insulation is provided, but the lead sheath and armouring are provided as usual. The perforations in the metallized paper sheaths assist in the complete impregnation of cable with the compound and thus the possibility of air-duct formation is eliminated. All four screens and the lead sheath are at earth potential, with the result that the electric stresses are entirely radial and the dielectric losses are therefore reduced. Another advantage of metal foil is that it increases the heat dissipating power and there are no sheath losses. All of these effects increase the current carrying capacity of the cable. These cables have been used up to 66 kV but are usually used up to 33 kV because pressurized cables have better performance than H type cables for voltages exceeding 33 kV.

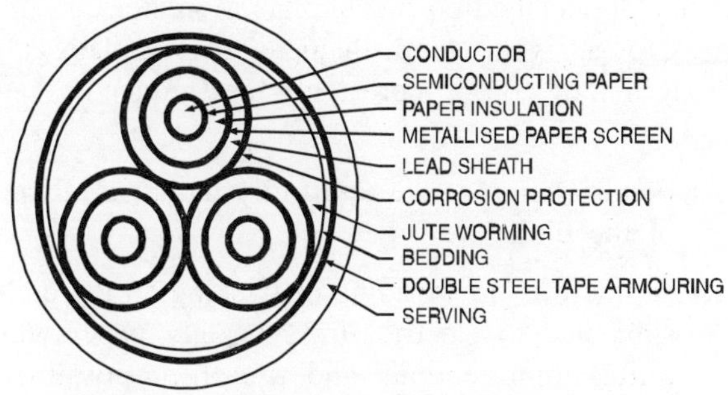

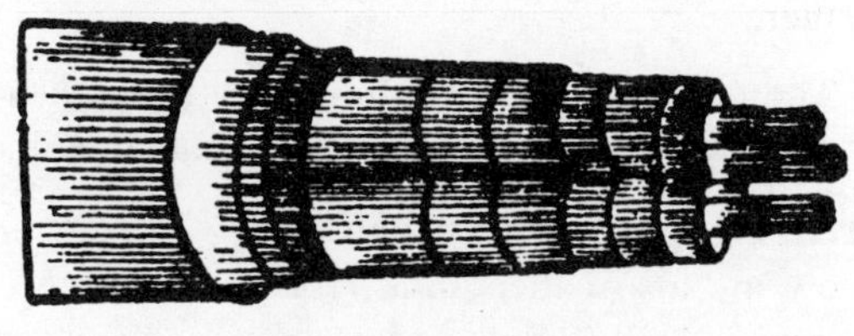

Fig. 8.4. ***H-Type Cable***

(*b*) *SL Type Cables.* In SL type cables each core is first insulated with an impregnated paper and then each of them, is separately lead sheathed. Now the three cores are just equivalent to three separate cables, each having its own lead

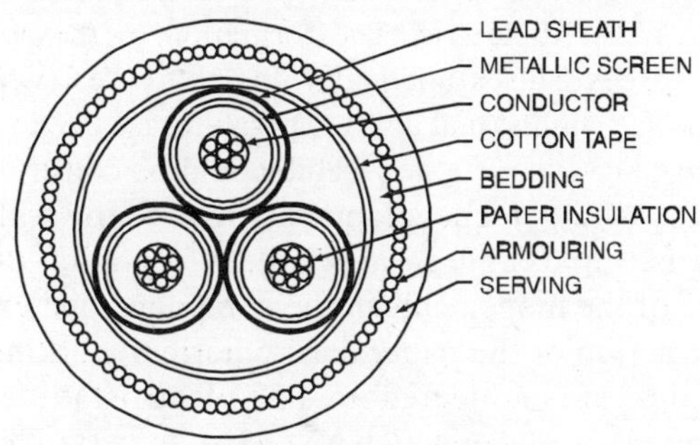

Fig. 8.5. ***SL Type Cable***

sheath. The three cables are laid up with fillers in the ordinary way, armoured and served overall with impregnated hessian tape as usual. No lead covering is provided surrounding all the three cores in addition to their individual lead sheaths. The advantages of SL type cables over H-type cables are :

(*i*) Bending of cable becomes possible owing to no overall lead sheath.

* It should not be confused with charging current.

(*ii*) Less tendency for oil drainage on hilly routes owing to elimination of filler spaces containing compound.

The disadvantage of SL type cable is that the manufacturing is difficult because of thinner lead sheaths.

SL type cables can be used up to 66 kV.

(*c*) *HSL Type Cables.* Such a cable is a combination of H-type and SL type cables. In HSL type cables each conductor is insulated, sheathed with metalized paper and in then provided with lead sheath. The three cores are then laid up and provided with filler, braided, armoured and finally served.

Advantages of Screened Type Cables Over Belted Types. (*i*) The metal sheathed core cable has a greater core to core thickness for a given overall diameter than a plain cable so the possibility of core-to-core faults are reduced to a large extent.

(*ii*) Electric stresses are uniformly radial in all sections of the dielectric.

(*iii*) As there is no worming or packing so the dielectric subjected to electric stress is only paper which is quite homogeneous and therefore possibility of formation of voids within the electric field is not there.

(*iv*) The metal sheaths help in dissipation of heat, so the current carrying capacity of the cables is increased.

3. Extra High Tension Cables. The cables considered up till now are also called *solid type* and in such cables it was assumed that the dielectric is homogeneous and there are no voids in the layers. With this assumption it is necessary to stick to maximum safe dielectric stress of about 4 or 5 kV per mm and also to maximum operating temperature of 50 or 60 °C in order to ensure safety from breakdown because the dielectric is far from homogeneous and with the normal manufacturing methods, it is almost impossible to avoid voids in the layers of dielectric. The formation of these voids or vacuous spaces causes unequal voltage stresses (which may exceed the safe limits) and also temperature rise due to leakance current. Again, even if a new cable could be constructed free from voids owing to development of modern techniques, it would be impossible to avoid their formation under the influence of the movements of the compound and expansion and contraction of the paper during normal working. When these voids are subjected to the electrostatic stresses, ionization takes place, which gives rise to certain chemical action, resulting in the deterioration of the chemical compound. Thus, these voids are sometimes the primary cause of a breakdown and therefore are required to be avoided. In order to meet the increased voltage demand the extra high tension and extra super-voltage power cables, useful for 132 kV and above, have been developed. In all such cables, the voids have been eliminated by increasing the pressure of the compound and that is why such cables are also called *pressure cables.*

Pressure cables are of two types viz., (*a*) oil-filled cables, and (*b*) gas-pressure cables.

(*a*) *Oil-filled Cables.* A single-core oil-filled cable is shown in Fig. 8.6. In this type of cable a channel is formed at the centre of the core by stranding the conductor wires around a hollow cylindrical steel spiral tape. This channel formed at the centre of core is filled with thin oil by means of oil reservoir and feeding tanks along its length and maintained at a pressure, not below atmospheric one at any point along the cable. This oil is the same light mineral oil of very low viscosity as used for initial impregnation. The system is so designed that when the oil gets expanded on account of increase in temperature of cable, the extra oil collects in the external reservoir, which sends it back during contraction on account of fall in temperature during light load conditions. Thus formation of voids in the dielectric is rendered impossible. Such cables are known as *single-core conductor channel cables.* Its disadvantage is due to the fact that the channel is at the middle of the cable and is at full voltage with respect to earth, so requiring a complicated system of jointing. It is advantageous from the point of view of potential gradient due to larger diameter of conductor and its hollow construction.

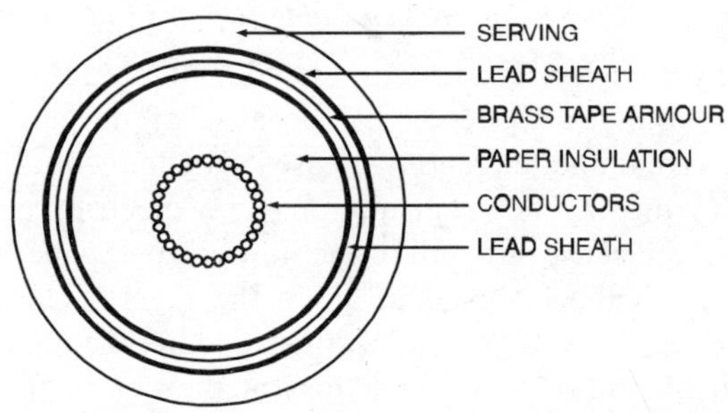

Fig. 8.6. ***Single-Core Conductor Channel Oil-Filled Cable***

The other type of single-core oil-filled cable is *sheath channel cable* (Fig. 8.7). In this type of cable, the conductor is made solid similar to that of solid cables and is paper insulated. In this type of cables the oil channels are produced either by grooving the sheath or by arranging spacers between dielectric and lead sheath. The latter arrangement is more sound mechanically in comparison to former one but due to high resistance to oil flow (6 to 8 times that of former type) requires large number of feeding points. In this case since the channels are at earth potential, system of joints and installation are simpler.

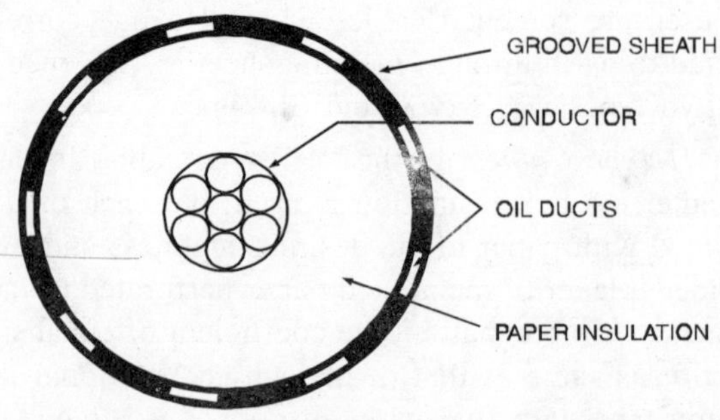

Fig. 8.7. ***Single-Core Sheath Channel Oil-Filled Cable***

In 3-core oil-filled cables (Fig. 8.8) the oil ducts are accommodated in the hollow filler spaces. These channels are made of perforated metal-ribbon tubing and are at earth potential. While jointing such a cable great care is required to be taken.

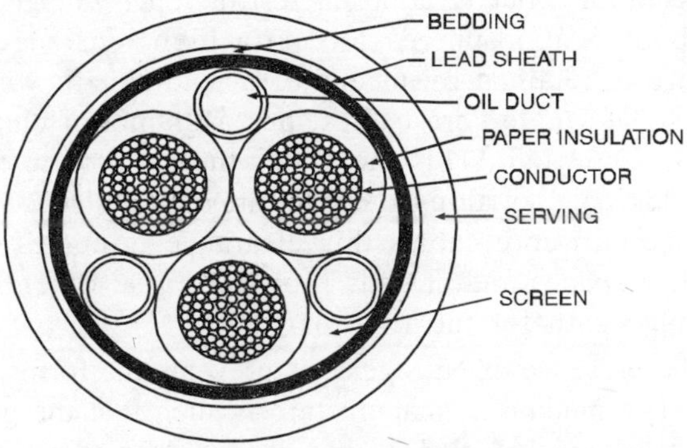

Fig. 8.8. ***Three-Core Filler Space Channel Oil-Filled Cable***

Another design of 3-core oil-filled cable is flat type, as illustrated in Fig. 8.9. The flat sides are reinforced with metallic tapes and binding wires so that during increase in oil pressure, due to heating, the flat side is deformed and the cable section becomes slightly elliptical.

In another construction of 3-core oil-filled cables use of 3-core paper insulated cable without a lead sheath is made. The cable is pulled into a steel pipe which is then filled with oil. Pumps are then used to maintain a specified oil pressure and permit it to expand and contract with the loading cycle.

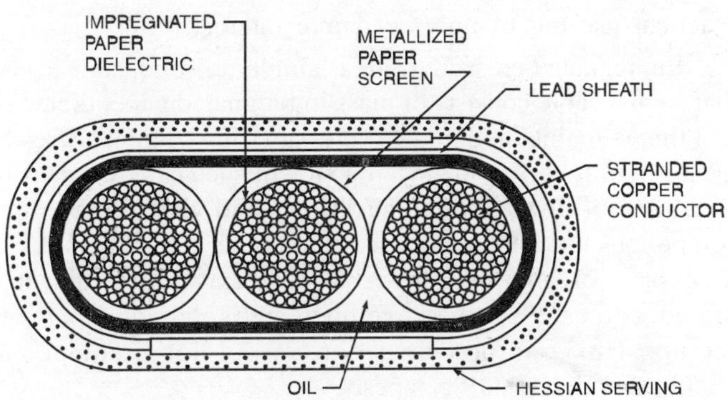

Fig. 8.9. ***Flat Oil-Filled Pressure Cable***

Leakage of oil in oil-filled cables is a very serious problem. Automatic signalling is provided to indicate the decrease in oil pressure in any of the phases. The pressure within the cable must not go below atmospheric and a lower limit of about 21,700 N/m^2 gauge is set. The transient pressure, developed due to sudden increase of load, should not increase beyond about 8.6×10^6 N/m^2 gauge. The joints in the cable are to be properly made so that there is no restriction to the oil flow.

The breakdown inception stress for oil-filled cables is 30–40 kV/mm which is much higher than that for a solid type cable. So the normal operating stresses are raised considerably. Maximum stresses up to 13 kV/mm are acceptable for 275 kV and 400 kV oil-filled cables. The current carrying capacity of oil-filled cables is determined by allowing a normal maximum conductor temperature of 85ºC.

The advantages and disadvantages of oil-filled cables are given below:

Advantages. (*i*) Smaller overall size and smaller weight for given voltage and kVA rating due to reduction in the thickness of dielectric required.

(*ii*) No ionisation, oxidation and formation of voids.

(*iii*) More perfect impregnation.

(*iv*) Possibility of increased temperature range in service.

(*v*) Smaller thermal resistance due to decrease in dielectric thickness, so higher current rating.

(*vi*) Impregnation possible after sheathing.

(*vii*) More maximum permissible stresses.

(*viii*) Detection of fault easy, as it will come to notice immediately as oil will start leaking.

Disadvantages. (*i*) Greater cost.

(*ii*) Complicated laying of cables and maintenance.

(*b*) *Gas-Pressure Cables.* Gas pressure cables are of two types (*i*) external pressure cables and (*ii*) gas-filled cables.

(*i*) *External Pressure Cables.* The external pressure cable, is a serious rival to the oil-filled type, and is being developed for the highest voltages. In such a cable the pressure, is applied externally and raised to such an extent that no ionisation can take place. At the same time the radial compression due to this increased pressure tend to close any voids. Also the working power factor of such a cable is improved.

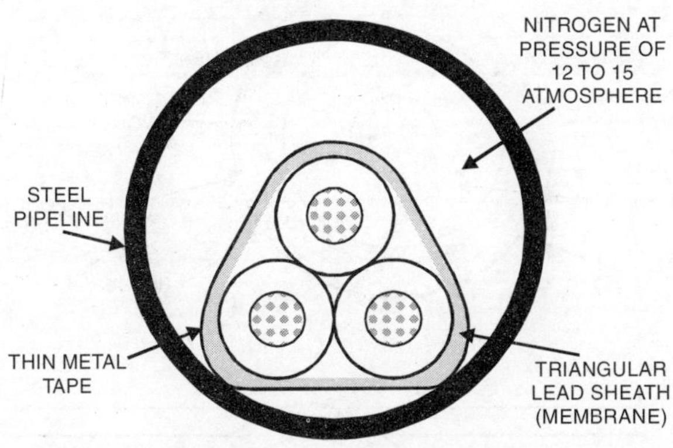

Fig. 8.10. ***Pipeline Type Cable***

The external pressure cable originally designed by Hochstadter, Vogel and Bowden is illustrated is Fig. 8.10. This cable is similar in construction to that of ordinary solid type except that it is triangular instead of circular in section and thickness of lead sheath is 75% of that of solid type cable. The triangular section reduces the weight, gives a low thermal resistance and at the same time lead sheath acts as a pressure membrane. Also there is no bedding and serving so thermal resistance is reduced. The cable is armoured with a thin metal tape so that the formation of any abnormal ties over its surface is avoided. The core may have circular or oval section but modern practice is to employ an oval section, which gives an improved copper space factor. The cable is laid in gas-tight metal pipes of somewhat larger section. The pipe is filled with nitrogen at a pressure of 12 to 15 atmospheres which continually compresses the cable radially from outside so that radial breathing of the cable occurs and any voids etc. are closed. The steel pipes used are coated with a special paint in order to avoid corrosion and it is further protected with an impregnated felt.

In comparison with a normal cable, such cables can carry 1.5 times the load current, double the operating voltage and thus transmit 3 times power. The maximum potential gradient is 10 kV/mm and the dielectric power factor at 15°C is 0.6 per cent. The steel pipes provide mechanical protection to the cables. The nitrogen in steel pipes help in quenching any flame. Moreover, maintenance cost of such cables is small. The disadvantage of such cables is that cost is very high.

The above cable is pipeline type. The second type is self contained type, in which an additional reinforced lead sheath is employed, the principle being the same as that of former type. Dried nitrogen is fed into the small annular space between dielectric and sheath from gas cylinders, and finding its way along the cable maintains the whole of the dielectric at a pressure of about 14 atmospheres.

(*ii*) *Internal Pressure Cables.* In internal pressure cables, the gas is in contact with the dielectric. Such cables are of three types :

1. High pressure gas-filled cables.
2. Gas cushion cables.
3. Impregnated pressurized cables.

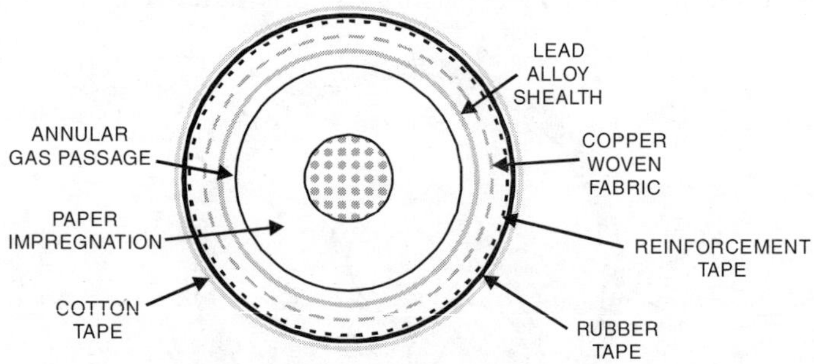

(a) High Pressure Gas-Filled Cable

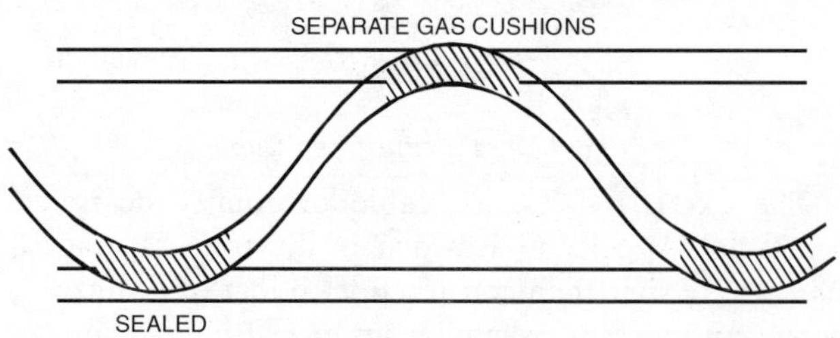

(b) Gas Cushion Type Cables

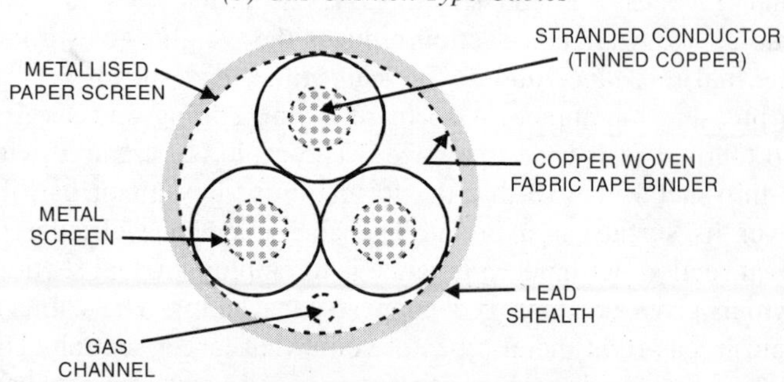

(c) Impregnated Pressurized Cables

Fig. 8.11

In high pressure gas-filled cables spaces are provided in the dielectric itself for the gas which is inert gas like nitrogen at pressure of about 12 atmospheres for super-voltage cables and about 6 atmospheres for extra high tension cables. Pressure is retained by means of a lead sheath which in the case of single-core cables, has a diametral clearance of about 0.63 mm. This facilitates the axial flow of gas, which also passes along the unimpregnated strand. In the case of multi-core cables this clearance is not essential; the filler spaces and strands providing a sufficiently low resistance path for the flow of gas.

The presence of Nitrogen gas prevents the formation of voids. The method of manufacture is such that the gas can move freely inside pockets but cannot diffuse outside the insulation.

In gas cushion cable, screened space is provided all along the length of cable, in between the lead sheath and the dielectric. The screened space is subdivided by means of barriers and thus inert gas is stored at various points along the run of a cable. This facilitates the jointing of cable without losing gas from the entire cable, because while cutting the cable, loss of gas will be local only. The special feature of the cable is thus that no arrangement is required for the transmission of pressure to the cables from outside, so that the cable is a complete unit with its own armouring, requiring no external pipe protection whatever. This type of cable is not of much practical use but of only academic interest.

Impregnated pressure cable is similar to solid cable except that such a cable consists of mass impregnated paper dielectric and this is maintained under a pressure of 14 atmospheres by means of nitrogen. In the manufacture of such cables, provision is made for longitudinal gas flow, and the impregrating compounds used are suitable for the higher dielectric stresses necessary for high voltage cables. Special reinforcement is provided to cater for the large hoop and longitudinal stresses set up. This reinforcement consists of longitudinal and circumferential metallic tapes.

The advantages of the internal pressure cables are :

(*i*) External accessories are eliminated.

(*ii*) With suitable designs the cable can be used for vertical run without any fear of drainage.

(*iii*) There is marked improvement in the power factor of the cable dielectric with the increased pressure.

Gas SF_6, because of its good thermal characteristics and high dielectric strength, is also used for insulating the cables. Cables insulated with SF_6 gas can be matched to overhead lines and can be operated corresponding to their surge impedance loading. Such cables can be employed for transmission of thousands of MVA even at ultra high voltage while the use of conventional cables is limited to 1,000 MVA and 500 kV.

8.7. INSULATION RESISTANCE OF A SINGLE-CORE CABLE

The cable conductor is provided with an insulation of suitable thickness so as to avoid leakage of current. The path for leakage current is radial, as shown in Fig. 8.12 (*a*), through the insulating material. The opposition offered by the insulation to the leakage current is called the *insulation resistance* of the cable.

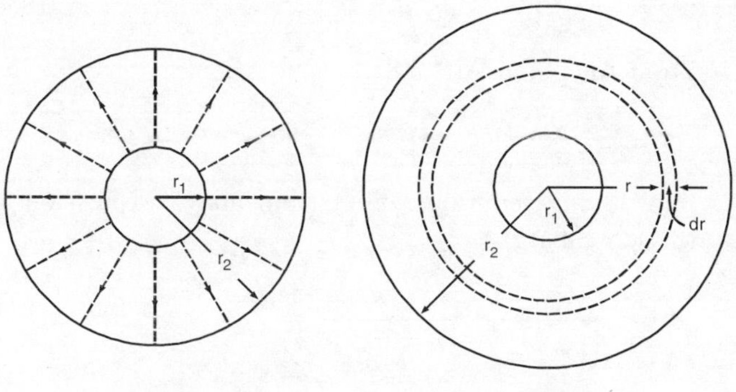

(*a*) (*b*)

Fig. 8.12

Consider a single core cable of conductor radius r_1, internal sheath radius r_2, length l and insulation material resistivity ρ.

Let an elementary cylindrical section of the insulation of radius r and thickness dr be considered. Now the length through which the leaking current will flow is dr and area of crosssection is $2\pi rl$.

The insulation resistance offered to the leakage current by elementary cylindrical section of the insulation under consideration is $\frac{\rho dr}{2\pi rl}$.

Insulation resistance of the cable,

$$R_{INS} = \int_{r_1}^{r_2} \frac{\rho dr}{2\pi rl} = \frac{\rho}{2\pi l}\left[\log_e r\right]_{r_1}^{r_2} = \frac{\rho}{2\pi l}\log_e \frac{r_2}{r_1} \quad ...(8.1)$$

i.e., the insulation resistance of the cable varies inversely as the length of the cable $\left(R_{INS} \propto \frac{1}{l}\right)$.

If ρ is in Ω-m, r_2, r_1, and l are in metres then insulation resistance of the cable is in ohms.

Average value of ρ for impregnated paper is about $5 \times 10^{12} - 8 \times 10^{12}$ Ω-m at 15ºC and decreases exponentially with temperature so that

$$\rho_t = \rho_0 e^{-\alpha t} \quad ...(8.2)$$ α is about 0.04 or 0.05.

Example 8.1. Calculate insulation resistance of 5 km length of single-core cable whose insulation resistance is 5 × 10^{14} Ω-cm, insulation thickness is 1 cm and radius of conductor is 1.25 cm.

[*U.P. Technical Univ. Elements of Power System 2009-10*]

Solution : Specific resistance of insulation,

$$\rho = 5 \times 10^{14} \text{ Ω-cm} = 5 \times 10^{12} \text{ Ω-m}$$

Length of cable, l = 5 km = 5,000 m

Core radius, r_1 = 1.25 cm

Internal sheath radius, r_2 = r_1 + insulation thickness = 1.25 + 1 = 2.25 cm

$$\log_e \frac{r_2}{r_1} = \log_e \frac{2.25}{1.25} = 0.5878$$

Insulation resistance of cable

$$= \frac{\rho}{2\pi l}\log_e \frac{r_2}{r_1} = \frac{5 \times 10^{12}}{2\pi \times 5{,}000} \times 0.5878 = 93.55 \text{ MΩ} \quad \textbf{Ans.}$$

Example 8.2. Calculate the resistivity of the dielectric in a cable, the core diameter of which is 1.5 cm and sheath diameter is 5 cm. For a length of 3,000 metres of cable, the insulation resistance is 1,820 MΩ.

[*Rajasthan Univ. Transmission and Distribution of Electrical Power 2007*]

Solution : Core radius, $r_1 = \frac{1.5}{2} = 0.75$ cm

Sheath radius, $r_2 = \frac{5}{2} = 2.5$ cm

$$\log_e \frac{r_2}{r_1} = \log_e \frac{2.5}{0.75} = 1.2$$

Installation resistance, R_{INS} = 1,820 MΩ = 1,820 × 10^6 Ω

Resistivity,

$$\rho = \frac{R_{INS} \times 2\pi l}{\log_e \frac{r_2}{r_1}} = \frac{1{,}820 \times 10^6 \times 2\pi \times 3{,}000}{1.2}$$

$$= 28.6 \times 10^{12} \text{ Ω-m} \quad \textbf{Ans.}$$

8.8. CAPACITANCE OF A SINGLE-CORE CABLE

The cable can be considered to be two coaxial cylinders of inner diameter d and outer diameter D. In actual cable, d represents the diameter of core and D represents the inner diameter of lead sheath which is at earth potential. Let the relative permittivity of dielectric in between the core and sheath be ϵ_r.

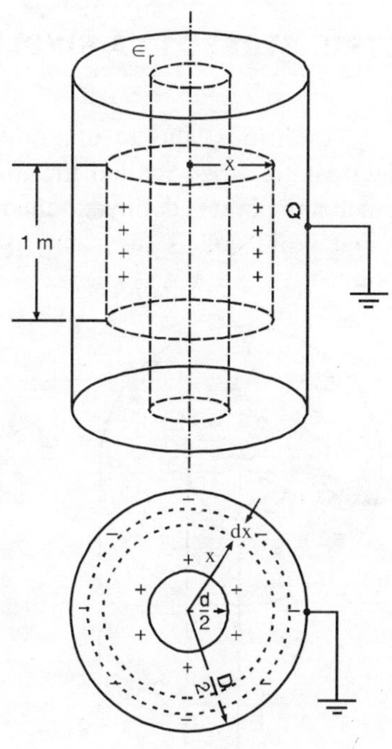

Fig. 8.13

Let the charge per metre length of the cable on the outer surface of the core be +Q coulombs and on the inner surface of the lead sheath be –Q coulombs. For all practical purposes, the charge of +Q coulombs/m on the surface of the core can be assumed to be located along its axis. The metal sheath is earthed.

Surface area of the coaxial cylinder of radius x metres and length one metre is $2\pi x$ metres^2.

Therefore, electric field intensity at a point x metres from the centre of the inner cylinder,

$$E_x = \frac{Q}{2\pi \epsilon_0 \epsilon_r x} = \frac{Q}{2\pi \epsilon_0 \epsilon_r} \cdot \frac{1}{x} \text{ V/m}$$

Potential difference between the capacitor plates, (between core and sheath),

$$V = \int_{d/2}^{D/2} E_x\, dx = \int_{d/2}^{D/2} \frac{Q}{2\pi \epsilon_0 \epsilon_r} \cdot \frac{1}{x}\, dx = \frac{Q}{2\pi \epsilon_0 \epsilon_r} \log_e \frac{D}{d} \text{ volts}$$

Capacitance of cable,

$$C = \frac{Q}{V} = \frac{Q}{\frac{Q}{2\pi \epsilon_0 \epsilon_r} \log_e \frac{D}{d}} \text{ F/m}$$

$$= \frac{2\pi \epsilon_0 \epsilon_r}{\log_e \frac{D}{d}}$$

$$= \frac{2\pi \times 8.854 \times 10^{-12}\, \epsilon_r}{2.303 \log_{10} \frac{D}{d}}$$

$$= \frac{\epsilon_r \times 10^{-9}}{41.4 \log_{10} \frac{D}{d}} \text{ F/m or } \frac{0.024\varepsilon_r}{\log_{10} \frac{D}{d}} \text{ μ F/km} \quad ...(8.3)$$

8.9. DIELECTRIC STRESS IN A SINGLE-CORE CABLE

Under operating conditions, the insulation of a cable is subjected to electrostatic stress, called the *dielectric stress.* Potential gradient at any point is defined as the rate of increase of potential at that point and is the same as the dielectric stress at that point.

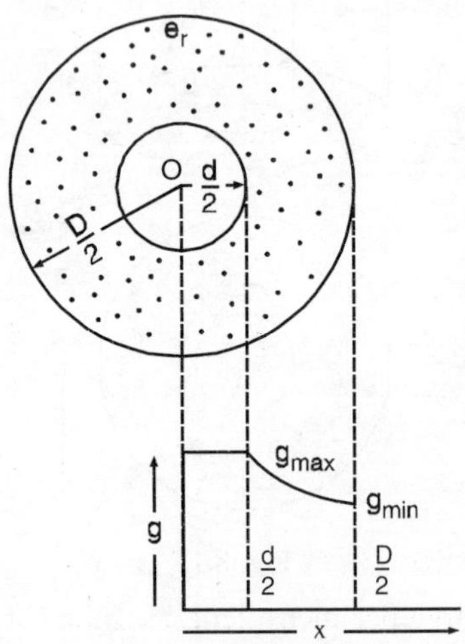

Fig. 8.14

Since cable is a form of cylindrical condenser, therefore, electric intensity at a distance x from the centre O of the cable is given by the expression

$$E_x = \frac{Q}{2\pi \epsilon_0 \epsilon_r} \cdot \frac{1}{x} \text{ V/m}$$

Since potential gradient = Electric intensity

$$\therefore g = \frac{Q}{2\pi \epsilon_0 \epsilon_r} \cdot \frac{1}{x} \text{ V/m} \quad ...(8.4)$$

From foregoing Art. 8.8

$$V = \frac{Q}{2\pi \epsilon_0 \epsilon_r} \log_e \frac{D}{d}$$

Substituting the value of Q from above expression in Eq. (8.4), we have

$$g = \frac{2\pi \epsilon_0 \epsilon_r V}{\log_e \frac{D}{d}} \cdot \frac{1}{2\pi \epsilon_0 \epsilon_r} \cdot \frac{1}{x}$$

$$= \frac{V}{x \log_e \frac{D}{d}} \text{ volts/metre} \quad ...(8.5)$$

Since potential gradient g varies inversely as x (as obvious from the above expression), therefore, potential gradient will be maximum when x is minimum *i.e.*, $x = \frac{d}{2}$ and potential gradient will be minimum when x is maximum *i.e.*, $x = \frac{D}{2}$.

Maximum and minimum values of potential gradient are given by

$$g_{max} = \frac{2V}{d \log_e \frac{D}{d}} \text{ volts/metre} \quad ...(8.6)$$

$$\text{and } g_{max} = \frac{2V}{D \log_e \frac{D}{d}} \text{ volts/metre} \quad ...(8.7)$$

Hence the ratio between the maximum potential gradient and the minimum potential gradient *i.e.*,

$$\frac{g_{max}}{g_{min}} = \frac{2V/d \log_e \frac{D}{d}}{2V/D \log_e \frac{D}{d}} = \frac{D}{d} \quad ...(8.8)$$

It is to be noted here that the Eq. (8.6) for the dielectric stress at the conductor surface is obtained on the assumption of a smooth cylindrical conductor, and thus only provides a nominal value for this stress. With the ordinary stranded conductors in common use, the dielectric stresses near the conductor are increased by about 20 per cent due to the greater curvature of the surface of the individual wires. This, however, can be avoided in practice by covering the conductor with a thin smooth metal sheath.

In high-voltage cables the value of g is taken to the highest possible limit in order to keep down the overall diameter, but since the dielectric losses increase rapidly with potential gradient, the maximum permissible value of g is only about one-fifth of the breakdown value viz 4-5 kV/mm.

Example 8.3. A single-core cable for use on 11 kV, 50 Hz system has conductor area of 0.645 cm² and the internal diameter of sheath is 2.18 cm. The permittivity of the dielectric used in the cable is 3.5. Find the maximum and minimum electrostatic stresses in the cable. Show also the curve of stress distribution in the cable v/s distance from the centre of the conductor.

Find the capacitance of the above cable per km length. Find the charging current.

[*Pb. Univ. Transmission & Distribution of Electrical Energy Dec. 1982; Rajasthan Technical Univ. Transmission & Distribution of Electrical Power 2011*]

Solution : Area of cross section of conductor, a = 0.645 cm²

Diameter of conductor,

$$d = \sqrt{\frac{4a}{\pi}} = \sqrt{\frac{4 \times 0.645}{\pi}} = 0.906 \text{ cm or } 0.00906 \text{ m}$$

Internal diameter of sheath,

D = 2.18 cm or 0.0218 m

System voltage,

V = 11,000 V assuming 1-phase system

Maximum electrostatic stress,

$$g_{max} = \frac{2\text{V}}{d \log_e \frac{\text{D}}{d}} = \frac{2 \times 11{,}000}{0.00906 \log_e \frac{0.0218}{0.00906}}$$

= 2,765,539 V/m or 27.6554 kV/cm (rms) **Ans.**

Minimum electrostatic stress,

$$g_{min} = \frac{2\text{V}}{\text{D} \log_e \frac{\text{D}}{d}} = \frac{2 \times 11{,}000}{0.0218 \log_e \frac{0.0218}{0.00906}}$$

= 1,149,350 V/m or 11.4935 kV/cm (rms) **Ans.**

Stress distribution curve is shown in Fig. 8.14.

Capacitance of the cable per km length,

$$\text{C} = \frac{0.024 \in_r}{\log_{10} \frac{\text{D}}{d}} \ \mu\text{F}$$

$$= \frac{0.024 \times 3.5}{\log_{10} \frac{0.0218}{0.00906}} \times 10^{-6} \text{ F} = 0.22 \times 10^{-6} \text{ F } \textbf{Ans.}$$

Charging current per phase per km length of cable,

$$I_C = 2\pi f\text{CV} = 2\pi \times 50 \times 0.22 \times 10^{-6} \times 11{,}000 = 0.762 \text{ A } \textbf{Ans.}$$

Example 8.4. A 33 kV, 3-phase, 50 Hz underground line, 3.4 km long, uses three single-core cables. Each cable has a core diameter of 2.5 cm and the radial thickness of insulation is 0.6 cm. The relative permittivity of the dielectric is 3.1. Find (*i*) maximum stress and (*ii*) total charging kVAR.

[*A.M.I.E. Sec. B. Power Systems Winter 1996*]

Solution : System peak voltage per phase,

$$\text{V} = \frac{33{,}000}{\sqrt{3}} \times \sqrt{2} = 26{,}944 \text{ V}$$

Inner diameter of sheath,

D = d + 2 × radial thickness of insulation

= 2.5 + 2 × 0.6 = 3.7 cm or 0.037 m

Maximum electrostatic stress,

$$g_{max} = \frac{2\text{V}}{d \log_e \frac{\text{D}}{d}} = \frac{2 \times 26{,}944}{0.025 \log_e \frac{0.037}{0.025}}$$

= 5.5 × 10⁶ V/m (peak) **Ans.**

Capacitance of cable,

$$\text{C} = \frac{0.024 \in_r}{\log_{10} \frac{\text{D}}{d}} \times l \times 10^{-6} = \frac{0.024 \times 3.1}{\log_{10} \frac{0.037}{0.025}} \times 3.4 \times 10^{-6} \text{ F}$$

$$= 1.4857 \times 10^{-6} \text{ F}$$

Charging current of cable,

$$I_C = 2\pi f\text{CV}_P = 2\pi \times 50 \times 1.4857 \times 10^{-6} \times 19{,}052 = 8.893 \text{ A}$$

Total charging kVA

$$= \sqrt{3}\, V_L I_C \times 10^{-3} = \sqrt{3} \times 33 \times 10^3 \times 8.893 \times 10^{-3}$$

= 508.23 kVAR **Ans.**

8.10. THE MOST ECONOMICAL DIAMETER OF CONDUCTOR

For a given voltage and overall diameter of a single-core cable there is a certain value of the conductor radius (or core diameter) that gives a minimum potential gradient at the surface. A small conductor radius allows a greater insulation thickness, but on the other hand, the smaller radius of curvature tends to increase the stress; while the effect of too large a diameter of core is to cause an increase of the stress through reduction of the total thickness of the insulation.

For fixed values of the voltage V and overall diameter D, the g_{max} will be minimum when $d \log_e \frac{\text{D}}{d}$ is maximum and $d \log_e \frac{\text{D}}{d}$ will be maximum if

$$\frac{d}{dd}\left[d \log_e \frac{\text{D}}{d}\right] = 0$$

or $$\log_e \frac{\text{D}}{d} = 1$$

or when $$\frac{\text{D}}{d} = e = 2.71828$$

or when $$d = \frac{\text{D}}{2.71828} \qquad ...(8.9)$$

and $$g_{max} = \frac{2\text{ V}}{d} \text{ volts/metre} \qquad ...(8.10)$$

For the low and medium voltage cables, the diameter of the conductor determined from the above consideration *i.e.*, $d = \frac{2\text{ V}}{g_{max}}$ is somewhat less than that determined from the consideration of the safe current density, therefore, the main criterion for determining conductor diameter for such cables is the current carrying capacity. On the other hand for high voltage cables the value of core diameter d determined from the above expression would, in general, give a conductor of cross section much larger than required from the consideration of current carrying capacity. The desired overall diameter of the conductor without increasing the cross-sectional area can be had in two ways, by making the conductor hollow or by building it with a lead sheath. The latter method has the advantage of eliminating at the same time the increase of stress due to stranding which may be as high as 20%.

Example 8.5. Show that for a given voltage V and maximum stress E_{max} in the dielectric of a single-core cable, the sheath diameter D is a minimum when D: *d* = *e* where *d* is conductor diameter and *e* is the base of natural logarithms. Determine D and *d* for V = 10 kV and E_{max} = 23 kV/cm.

[*Pb. Univ. Transmission & Distribution of Electrical Energy April 1982*]

Solution : Maximum stress,

$$g_{max} = E_{max} = 23 \text{ kV/cm} = 23 \times 10^5 \text{ V/m (assume rms)}$$

Applied voltage, V = 10,000 V

∴ Diameter of conductor,

$$d = \frac{2V}{g_{max}} = \frac{2 \times 10{,}000}{23 \times 10^5} = 0.0087 \text{ m or } 8.7 \text{ mm } \textbf{Ans.}$$

Internal diameter of sheath,

$$D = e \times d = 2.71828 \times 8.7 = 23.64 \text{ mm } \textbf{Ans.}$$

Example 8.6. Find the most economical value of the diameter of a single core-cable to be used on a 132 kV, 3-phase system. Find also the overall diameter of the insulation if the peak permissible stress is not to exceed 60 kV/cm.

[*A.M.I.E. Sec. B. Power Systems Winter 1998; Pb. Technical Univ. Power System-I (Transmission & Distribution) Dec. 2008*]

Solution : Maximum stress,

$$g_{max} = 60 \text{ kV/cm (peak)}$$

$$= \frac{60}{\sqrt{2}} \text{ kV/cm } (rms) \text{ or } 42.426 \times 10^5 \text{ V/m}$$

RMS value of phase voltage,

$$V = \frac{132}{\sqrt{3}} \times 1{,}000 = 76{,}210 \text{ V}$$

Diameter of conductor,

$$d = \frac{2V}{g_{max}} = \frac{2 \times 76{,}210}{42.426 \times 10^5} = 0.0359 \text{ m or } 35.9 \text{ mm } \textbf{Ans.}$$

Internal diameter of sheath,

$$D = ed = 2.71828 \times 35.9 = 97.65 \text{ mm } \textbf{Ans.}$$

8.11. GRADING OF CABLES

It has been observed that the electrostatic stress in a single core cable is maximum at the conductor surface and decreases towards the sheath. In a cable having homogeneous insulation, the maximum voltage which can be safely applied to a cable is limited by the electrostatic stress at the conductor surface. If a dielectric of high strength is used in a cable, then no doubt it is useful just over the conductor where the stress is maximum, but as we go away from the conductor surface, the value of dielectric stress decreases, so the dielectric will be unnecessarily strong and expensive. If, however, by some means the stress can be so distributed that its value in the outer layers of the dielectric is increased without increasing the stress near the conductor, insulation of less thickness will be required for any given working voltage, and an important economy will be a affected. *The process of achieving uniformity in dielectric stress is known as grading of cables.*

There are two methods of grading *viz.* (*i*) capacitance grading and (*ii*) intersheath grading.

(*i*) ***Capacitance Grading.*** In this method of cable grading, the uniformity in dielectric stress is achieved by using various layers of different dielectrics in such a manner that the permittivity, ϵ_r of any layer is inversely proportional to its radius of distance from the centre *i.e.*,

$$\epsilon_r \propto \frac{1}{x} \text{ or } \epsilon_r x = \text{constant or } g \propto \frac{Q}{2\pi \epsilon_0} \text{ which in constant}$$

Thus, we see that if such a condition is attained, the value of dielectric stress at any point is constant and is independent of distance from the centre, and the grading will be ideal one. But it is not possible to use infinite number of dielectrics for a single cable and in practice two or three dielectrics are used in such a manner that the permittivity of the dielectric near the core has got highest value and then it decreases and finally it has got minimum value in the outer most layer.

Now consider a cable in which 3 layers of dielectrics of outer diameters d_1, d_2 and D and of relative permittivity ϵ_{r1}, ϵ_{r2} and ϵ_{r3} have been used. If the permittivities are such that $\epsilon_{r1} > \epsilon_{r2} > \epsilon_{r3}$ and if the maximum stress to which each dielectric is subjected is constant then

$$\epsilon_{r1} d = \epsilon_{r2} d_1 = \epsilon_{r3} d_2$$

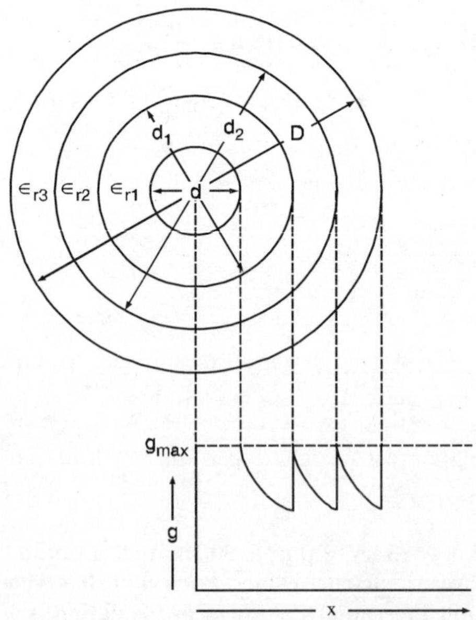

Fig. 8.15. ***Capacitance Grading of Cables***

For the *pd* across the inner layer, we have

$$V_1 = \int_{d/2}^{d_1/2} g\, dx = \int_{d/2}^{d_1/2} \frac{Q}{2\pi \epsilon_0 \epsilon_r} dx$$

$$= \frac{Q}{2\pi \epsilon_0 \epsilon_r} \log_e \frac{d_1}{d} = g_{max} \frac{d}{2} \log_e \frac{d_1}{d}$$

Similarly $V_2 = g_{max} \frac{d_1}{2} \log_e \frac{d_2}{d_1}$

and $V_3 = g_{max} \frac{d_2}{2} \log_e \frac{D}{d_2}$

and the *pd* between core and earthed sheath,

$$V = V_1 + V_2 + V_3$$

$$= \frac{g_{max}}{2}\left(d \log_e \frac{d_1}{d} + d_1 \log_e \frac{d_2}{d_1} + d_2 \log_e \frac{D}{d_2}\right) \quad ...(8.11)$$

The advantage of grading is that if the overall diameter is same for a non-graded and graded cable, then the permissible safe potential of the cable is increased *i.e., for* the same permissible safe potential, the size of the graded cable will be smaller than that of the non-graded cable.

(*ii*) ***Intersheath Grading.*** In this method of cable grading a homogeneous dielectric is used, which is divided into various layers, by suitably placing the metallic intersheaths. These metallic intersheaths are held at certain potentials which are in between the core potential and the earth potential. There is thus a definite potential difference between the inner and outer layers of each sheath, so that each sheath can be treated like a homogeneous single-core cable and we have

$$g_{1max} = \frac{V_1}{\frac{d}{2}\log_e \frac{d_1}{d}}, \quad g_{2max} = \frac{V_2}{\frac{d_1}{2}\log_e \frac{d_2}{d_1}}, \quad g_{3max} = \frac{V_3}{\frac{d_2}{2}\log_e \frac{D}{d_2}} \quad ...(8.12)$$

But for a homogeneous dielectric, we have

$$g_{1max} = g_{2max} = g_{3max}$$

$$\therefore \quad \frac{V_1}{\frac{d}{2}\log_e \frac{d_1}{d}} = \frac{V_2}{\frac{d_1}{2}\log_e \frac{d_2}{d_1}} = \frac{V_3}{\frac{d_2}{2}\log_e \frac{D}{d_2}} = g_{max} \text{ (say)} \quad ...(8.13)$$

Again, since all these potential differences are in phase, since the cable is, in effect, a series of capacitors in series, we also have

$$V = V_1 + V_2 + V_3 \quad ...(8.14)$$

Grading in Practice. Grading is only useful for very high-voltage cables for which the ratio D/*d* is large. If D/*d* is 2, grading will increase the permissible safe working voltage by only 30% but if D/*d* = 5, grading will increase the safe working voltage by 100%. With capacitance grading the chief handicap is due to the fact that there are such few high grade insulating materials of reasonable cost, whose permittivities can be made to vary over the required range. The use of rubber ($\epsilon_r = 4 - 6$) and impregnated paper ($\epsilon_r = 3 - 4$) have been suggested. However, rubber is very expensive and the cost of dielectric becomes prohibitive. A second difficulty is the possibility of change of permittivity with time that may completely alter the distribution of dielectric stress and lead to the breakdown of insulation even at the normal operating voltage.

With intersheath grading, the main objection is the possibility of damage to the intersheath during transportation and installation. Due to charging currents there are losses in the intersheaths. The modern trend is to avoid grading as far as possible and employ oil-filled or gas-pressure cables.

Note : While calculating the safe potential, the peak values of voltages are to be taken and not the rms values.

Example 8.7. A conductor of 1 cm diameter passes centrally through a porcelain cylinder of internal diameter 2 cm and external diameter 7 cm. The cylinder is surrounded by a tightly fitting metal sheath. The permittivity of porcelain is 5 and the peak voltage gradient in air must not exceed 34 kV/cm. Determine the maximum safe working voltage.

[U.P. Technical Univ. Elements of Power System 2007-08]

Solution : The configuration is shown below in Fig. 8.16.

Let q be charge per unit length of the conductor.

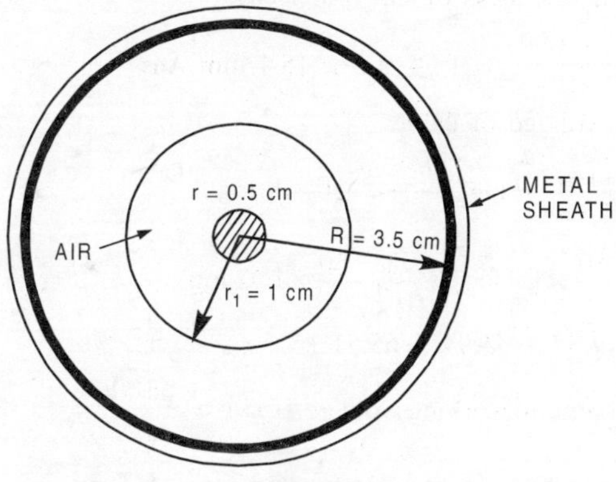

Fig. 8.16

Maximum value of potential gradient (at the surface of conductor),

$$g_{1\,max} = \frac{q}{2\pi\varepsilon_0 r}$$

Maximum value of potential gradient (at the surface of porcelain cylinder)

$$g_{2\,max} = \frac{q}{2\pi\varepsilon_0\varepsilon_r r_1}$$

Thus, $\quad r g_{1max} = \varepsilon_r r_1 g_{2max}$

or $\quad g_{2\,max} = \frac{r}{r_1 \varepsilon_r} \times g_{1max} = \frac{0.5}{1 \times 5} \times 34 = 3.4 \text{ kV/cm}$

Maximum safe working voltage (peak value),

$$V_{peak} = r g_{1max} \log_e \frac{r_1}{r} + r_1 g_{2max} \log_e \frac{R}{r_1}$$

$$= 0.5 \times 34 \log_e \frac{1}{0.5} + 1 \times 0.34 \log_e \frac{3.5}{1}$$

$$= 11.7835 + 4.2594 = 16.043 \text{ kV}$$

RMS value of maximum safe working voltage,

$$V_{rms} = \frac{16.043}{\sqrt{2}} = 11.44 \text{ kV} \quad \textbf{Ans.}$$

Example 8.8. A single-core lead sheathed cable has a conductor of 1 cm diameter and a sheath of 5 cm inside diameter. Two dielectric materials having permittivities of 4 and 2.5 and permissible potential gradients of 60 kV/cm and 50 kV/cm respectively are used. Determine the thickness of the dielectric materials and the maximum working voltage.

Solution : Maximum permissible stress, g_{1max} = 60 kV/cm

Maximum permissible stress, $g_{2\,max}$ = 50 kV/cm

Let the diameter over the insulation of inner dielectric be d_1 cm

$$\text{Now } \frac{g_{1max}}{g_{2max}} = \frac{Q/2\pi \epsilon_0 \epsilon_{r1}\, d}{Q/2\pi \epsilon_0 \epsilon_{r2}\, d_1} = \frac{\epsilon_{r2}\, d_1}{\epsilon_{r1}\, d} = \frac{2.5 \times d_1}{4 \times 1} = 0.625\, d_1$$

$$\text{or } \frac{60}{50} = 0.625\, d_1$$

$$\text{or } d_1 = \frac{1.2}{0.625} = 1.92 \text{ cm}$$

Radial thickness of inner dielectric

$$= \frac{1.92 - 1}{2} = 0.46 \text{ cm or } 4.6 \text{ mm } \textbf{Ans}$$

Radial thickness of outer dielectric

$$= \frac{5 - 1.92}{2} = 1.54 \text{ cm or } 15.4 \text{ mm } \textbf{Ans.}$$

Peak voltage of cable

$$= \frac{g_{1max}}{2} d \log_e \frac{d_1}{d} + \frac{g_{2max}}{2} d_1 \log_e \frac{D}{d_1}$$

$$= \frac{60}{2} \times 1 \times \log_e \frac{1.92}{1} + \frac{50}{2} \times 1.92 \log_e \frac{5}{1.92}$$

$$= 19.57 + 45.94 = 65.51 \text{ kV}$$

$$\text{Maximum working voltage } (rms) = \frac{65.51}{\sqrt{2}} = 46.32 \text{ kV } \textbf{Ans.}$$

Example 8.9. A single-core cable employing three layers of insulation with dielectric constants ϵ_{r1} = 5, ϵ_{r2} = 4 and ϵ_{r3} = 3 respectively has a conductor of radius 1 cm. Assuming that all the three insulating materials are worked at the same maximum potential gradient, workout the potential difference in kV between the core and earthed sheath. The inner radius of the sheath is 2.5 cm and the maximum potential gradient is 40 kV/cm.

Solution : Diameter of conductor, d = 2 × 1 = 2.0 cm

The diameter over the insulation, D = 2 × 2.5 = 5.0 cm

Maximum potential gradient, g_{max} = 40 kV/cm

Let the diameters over the insulations of relative permittivities 5 and 4 be d_1 and d_2 respectively.

As the maximum stress in the three dielectrics is the same

$$\therefore \epsilon_{r1} d = \epsilon_{r2}\, d_1 = \epsilon_{r3}\, d_2$$

$$\text{or } d_1 = \frac{\epsilon_{r1}}{\epsilon_{r2}} d = \frac{5}{4} \times 2.0 = 2.5 \text{ cm}$$

$$\text{and } d_2 = \frac{\epsilon_{r1}}{\epsilon_{r3}} \times d = \frac{5}{3} \times 2.0 = 3.33 \text{ cm}$$

Permissible peak voltage for the cable

$$= \frac{g_{max}}{2}\left(d \log_e \frac{d_1}{d} + d_1 \log_e \frac{d_2}{d_1} + d_2 \log_e \frac{D}{d_2} \right)$$

$$= \frac{40}{2}\left(2 \log_e \frac{2.5}{2} + 2.5 \log_e \frac{3.33}{2.5} + 3.33 \log_e \frac{5}{3.33} \right)$$

$$= 20(0.4463 + 0.7192 + 1.3516) = 50.34 \text{ kV}$$

$$\text{So working voltage } (rms) \text{ for the cable} = \frac{50.34}{\sqrt{2}} = 35.6 \text{ kV } \textbf{Ans.}$$

Example 8.10. A 66 kV single-core lead sheathed cable is graded by using two dielectrics of relative permittivities 5 and 3 respectively thickness of each being 1 cm. The core diameter is 2 cm. Determine the maximum stresses in the two dielectrics.

[*M.D. Univ. Power System-I, Dec. 2009*]

Solution : Conductor diameter, d = 2 cm

Diameter over the insulation of relative permittivity 5,

d_1 = 2 + 2 × 1 = 4 cm

Diameter over the insulation of relative permittivity 3,

D = 4 + 2 × 1 = 6 cm

Let the potential gradient at the conductor surface be $g_{1\,max}$ kV/cm
Then the maximum stress in outer dielectric is given by

$$g_{2\,max} = g_{1max} \times \frac{d\varepsilon_{r_1}}{d_1 \varepsilon_{r_2}} = g_{1max} \times \frac{2 \times 5}{4 \times 3} = \frac{5}{6} g_{1max}$$

System peak voltage per phase,

$$V_{max} = \frac{66}{\sqrt{3}} \times \sqrt{2} = 53.89 \text{ kV}$$

Since, system peak voltage, V_{max} is given by the expression

$$V_{max} = g_{1max} \times \frac{d}{2} \times \log_e \frac{d_1}{d} + g_{2max} \times \frac{d_1}{2} \log_e \frac{D}{d_1}$$

$$\text{so } 53.89 = g_{1max} \times \frac{2}{2} \times \log_e \frac{4}{2} + \frac{5}{6} g_{1max} \times \frac{4}{2} \log_e \frac{6}{4}$$

$$= g_{1max}(0.6931 + 0.6758) = 1.36888\, g_{1max}$$

$$\text{or } g_{1\,max} = \frac{53.89}{1.36888} \text{ kV/cm} = 39.37 \text{ kV/cm } \textbf{Ans.}$$

Maximum stress in the outer dielectric,

$$g_{2\,max} = \frac{5}{6} g_{1max} = \frac{5}{6} \times 39.37 = 32.8 \text{ kV/cm } \textbf{Ans.}$$

Example 8.11. A single-core lead covered cable is to be designed for 66 kV to earth. Its conductor radius is 0.5 cm and its three insulating materials A, B and C have relative permittivities of 4, 2.5 and 4 with maximum permissible stresses of 50, 30 and 40 kV/cm respectively. Determine the minimum internal diameter of lead sheath.

Solution : Peak value of cable voltage, $V_{max} = 66\sqrt{2}$ kV

In order to have minimum internal diameter of the lead sheath and consistency in the dielectric stresses, it is necessary that the material having maximum of the product of dielectric stress and the permittivity (*i.e.,* in this case the material A having this product of 50 × 4 = 200 as the maximum) is placed near the surface of the core. Material C having the product of dielectric stress and the permittivity of 40 × 4 = 160 (second maximum) be placed next and then the material B

Thus, $g_{1\,max}$ = 50 kV/cm; $g_{2\,max}$ = 40 kV/cm and $g_{3\,max}$ = 30 kV/cm

ϵ_{r1} = 4; ϵ_{r2} = 4 and ϵ_{r3} = 2.5.

For a given cable we have

$$g_{1\,max} = \frac{Q}{2\pi \epsilon_0 \epsilon_{r1}\, r}\,;\; g_{2\,max} = \frac{Q}{2\pi \epsilon_0 \epsilon_{r2}\, r_1} \text{ and } g_{3\,max} = \frac{Q}{2\pi \epsilon_0 \epsilon_{r3}\, r_2}$$

From the above equations we have

$$Q = 2\pi\epsilon_0 \epsilon_{r1} r g_{1\,max} = 2\pi\epsilon_0 \epsilon_{r2} r_1 g_{2\,max} = 2\pi\epsilon_0 \epsilon_{r3} r_2 g_{3\,max}$$

$$\text{or } \epsilon_{r1} r g_{1\,max} = \epsilon_{r2} r_1 g_{2\,max} = \epsilon_{r3} r_2 g_{3\,max}$$

$$\text{or } r_1 = \frac{\epsilon_{r1}\, r g_{1max}}{\epsilon_{r2}\, g_{2max}} = \frac{4 \times 0.5 \times 50}{4 \times 40} = 0.625 \text{ cm}$$

$$\text{and } r_2 = \frac{\epsilon_{r_2}\, r_1 g_{2\,max}}{\epsilon_{r3}\, g_{3\,max}} = \frac{4 \times 0.625 \times 40}{2.5 \times 30} = 1.333 \text{ cm}$$

Since $V_{max} = g_{1\,max} \times r \log_e \frac{r_1}{r} + g_{2\,max} \times r_1 \log_e \frac{r_2}{r_1}$

$$+ g_{3\,max}\, r_2 \log_e \frac{R}{r_2}$$

$$\text{or } 66\sqrt{2} = 50 \times 0.5 \log_e \frac{0.625}{0.5} + 40 \times 0.625 \log_e \frac{1.333}{0.625}$$

$$+ 30 \times 1.333 \log_e \frac{R}{r_2}$$

$$93.338 = 5.5786 + 18.942 + 40 \log_e \frac{R}{r_2}$$

$$\text{or } \log_e \frac{R}{r_2} = \frac{1}{40}[93.338 - 5.5786 - 18.942] = 1.7204$$

$$\text{or } R = r_2 \times 5.58676 = 1.333 \times 5.58676 = 7.45 \text{ cm}$$

and inner diameter of lead sheath, D = 2R = 2 × 7.45 = 14.9 cm **Ans.**

Example 8.12. A 66 kV, single-core metal sheathed cable is to be graded by metallic intersheath. Calculate diameter of intersheath and voltage to obtain minimum overall cable diameter. Maximum voltage gradient 60 kV/cm. [*M.D. Univ. Power System-I, May 2008*]

Solution : RMS value of cable voltage, V_{max} = 66 kV

Peak value of cable voltage, $V_{max} = 66\sqrt{2}$ kV

Maximum permissible potential gradient of dielectric,

g_{max} = 60 kV/cm (peak value)

Let d be the conductor diameter, d_1 be the outer diameter of the inner insulating layer and D be the outer diameter of the outer insulating layer.

If V_1 = potential difference between core and sheath

V_2 = potential difference between intersheath and the outer sheath

$$\text{then } g_{max} = \frac{V_1}{\frac{d}{2}\log_e \frac{d_1}{d}} = \frac{V_2}{\frac{d_1}{2}\log_e \frac{D}{d_1}}$$

For minimum overall diameter of cable,

$$\text{Conductor diameter, } d = \frac{2V_{max}}{e \times g_{max}} = \frac{2 \times 66\sqrt{2}}{2.71828 \times 60} = 1.144 \text{ cm}$$

Diameter of intersheath,

$d_1 = e \times d$ = 2.71828 × 1.144 = 3.1 cm **Ans.**

Voltage between conductor and intersheath,

$$V_1 = \frac{V}{e} = \frac{66}{2.71828} = 24.28 \text{ kV}$$

Voltage between the intersheath and sheath overall

= V – V_1 = 66.0 – 24.28 = 41.72 kV **Ans.**

Example 8.13. A single-core cable has a conductor diameter of 2.5 cm and a sheath of inside diameter 6 cm. Calculate the maximum stress. It is desired to reduce the maximum stress by using two intersheaths. Determine their best positions, the maximum stress and the voltage on each. System voltage is 66 kV, 3-phase.

Solution : System peak voltage per phase,

$$V_{max} = \frac{66}{\sqrt{3}} \times \sqrt{2} = 53.8888 \text{ kV}$$

Without intersheath the maximum stress,

$$g_{max} = \frac{2V}{d\log_e \frac{D}{d}} = \frac{2 \times 53.8888}{2.5 \log_e \frac{6}{2.5}} = 49.24 \text{ kV/cm } \textbf{Ans.}$$

For same maximum and minimum stresses,

$$\frac{d_1}{d} = \frac{d_2}{d_1} = \frac{D}{d_2} = \alpha \text{ (say)}$$

$$\text{So } \frac{d_1}{d} \times \frac{d_2}{d_1} \times \frac{D}{d_2} = \frac{D}{d} \text{ or } \alpha^3 = \frac{6}{2.5} \text{ or } \alpha = 1.339$$

The diameter of intersheaths,

$d_1 = \alpha d$ = 1.339 × 2.5 = 3.347 cm

and d_2 = 1.339 × 3.347 = 4.481 cm

From Eqs. (8.14) and (8.13), we have

$$V_1 + V_2 + V_3 = V$$

$$g_{max} \times \left[\frac{d}{2}\log_e \frac{d_1}{d} + \frac{d_1}{2}\log_e \frac{d_2}{d_1} + \frac{d_2}{2}\log_e \frac{D}{d_2}\right] = V$$

$$\text{or } g_{max} \times \left[\frac{2.5}{2}\log_e \alpha + \frac{3.347}{2}\log_e \alpha + \frac{4.481}{2}\log_e \alpha\right] = 53.8888 \text{ kV}$$

$$\text{or } g_{max} \times \frac{53.8888}{5.164 \log_e 1.339} = 35.75 \text{ kV/cm}$$

$$\text{Peak voltage } V_{1max} = g_{max}\frac{d}{2}\log_e \frac{d_1}{d} = 35.75 \times \frac{2.5}{2}\log_e 1.339$$

$$= 13 \text{ kV}$$

$$\text{Peak voltage, } V_{2max} = g_{max}\frac{d_1}{2}\log_e \frac{d_2}{d_1}$$

$$= 3.5.75 \times \frac{3.347}{2} \times \log_e 1.339 = 17.46 \text{ kV}$$

Peak voltage on conductor surface, V_{max} = 53.8888 kV

Peak voltage on first intersheath = $V_{max} - V_{1max}$

= 53.8888 – 13 = 40.8888 kV **Ans.**

Peak voltage on second intersheath

= Peak voltage on first intersheath – V_{2max}

= 40.8888 – 17.46 = 23.4288 kV **Ans.**

8.12. CAPACITANCE OF THREE-CORE BELTED TYPE CABLES

The capacitance of a cable system is of much greater importance than that of an overhead line of the same length firstly due to the narrow spacing between the conductors themselves and between the conductors and earthed sheath and secondly due to their separation by a dielectric medium of higher permittivity as compared to air.

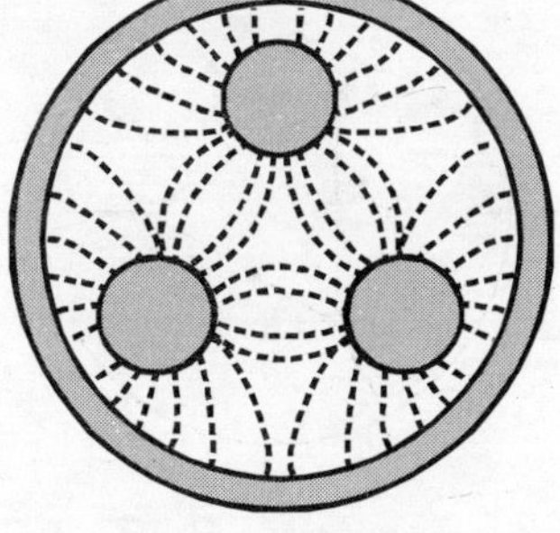

Fig. 8.17. *Electrostatic Field of a 3-core Belted Cable*

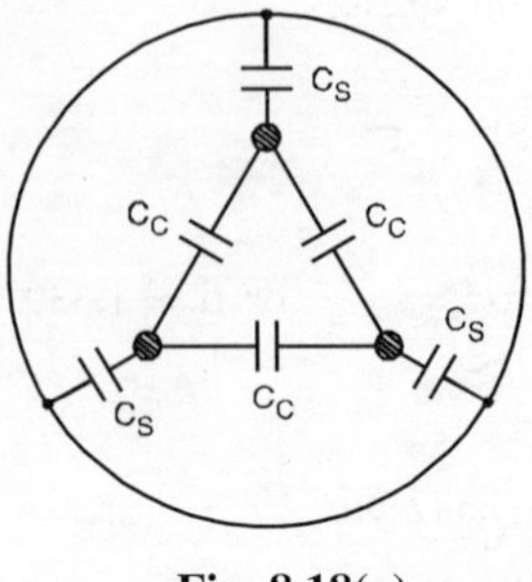

Fig. 8.18(*a*)

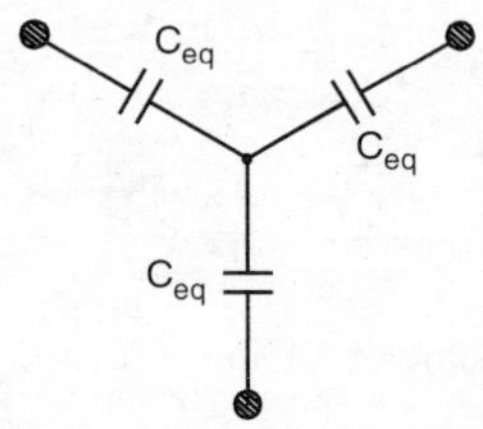

Fig. 8.18(*b*)

Since there is a potential difference between pairs of conductors and between each conductor and sheath there will be a system of electrostatic fields in the cross section of the cable somewhat as shown in Fig. 8.17. Consequently the three-core cable has capacitance between the cores and each core has capacitance with sheath, as shown in Fig. 8.18 (*a*).

Let the capacitance between cores be C_C and capacitance between cores and sheath be C_S.

The delta-connected capacitances can be converted into equivalent star-connected capacitances, as shown in Fig. 8.18 (*b*).

$$\text{The } C_{eq} = 3C_C \qquad \text{...(8.15)}$$

The star point may be assumed to be at zero potential and if the sheath is also at zero potential, the capacitance of each core to neutral

$$C_N = C_{eq} + C_S = 3C_C + C_S \qquad \text{...(8.16)}$$

If V_P is the phase voltage, then the charging current is given

$$I_C = 2\pi f V_P C_N \text{ amperes} \qquad \text{...(8.17)}$$

The values of capacitances C_C and C_S may be determined by measurement as follows:

1. Capacitance is measured between the three cores bunched together and the earthed sheath. (Fig. 8.19 (*a*)). This gives $3C_S$ because three capacitances C_C are eliminated leaving the three capacitances C_S in parallel

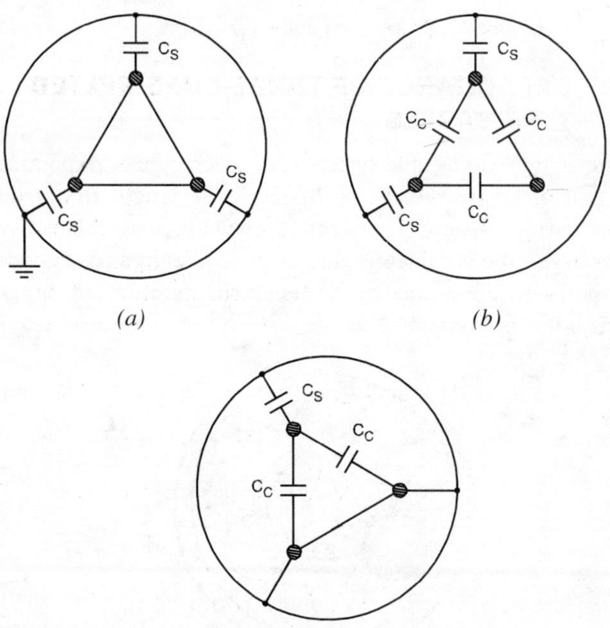

(*a*) (*b*)

(*c*)

Fig. 8.19

i.e., $C_1 = 3C_S$

$$\text{or } C_S = \frac{C_1}{3} \qquad \text{...(8.18)}$$

2. Capacitance between the two cores or lines is measured with the third core being either insulated or connected to sheath (Fig. 8.19 (*b*)). This eliminates one of the capacitors C_S and measured value C_2 gives $\frac{C_N}{2}$ directly.

$$C_2 = C_C + \frac{C_C}{2} + \frac{C_S}{2}$$

$$\text{or } C_2 = \frac{1}{2}(3C_C + C_S) = \frac{1}{2}C_N \qquad \text{...(8.19)}$$

3. The capacitance is measured between the two cores shorted with the sheath and the third core (Fig. 8.19 (*c*)). This gives $2C_C + C_S$

i.e., $C_3 = 2C_C + C_S$ ...(8.20)

From the above equations values of C_C and C_S can be determined.

It is very difficult to determine the capacitance C_N of the cable from the geometry of the figure. However, the empirical formulae given below may be used for determining the value of C_N.

$$C_N = \frac{0.03\epsilon_r}{\log_e\left[1+\frac{T+t}{d}\left(3.84-1.7\frac{t}{T}+0.52\frac{t^2}{T^2}\right)\right]} \ \mu\text{F/km} \qquad \text{...(8.21)}$$

where d = diameter of conductor t = thickness of belt insulation and T = thickness of conductor insulation.

The value of C_N is usually determined by experiments, as described above, not by the emperical formula.

Example 8.14. An 11 kV, 2.5 km, 3-phase, 3-wire belted cable gave the following results in 'capacitance' testing.

(1) Capacitance between two conductors joined to sheath and the third conductor = 1.5 μF.

(2) Capacitance between all three conductors joined and sheath = 1.8 μF. Find (*i*) capacitance of each core to neutral (*ii*) capacitance between any two core and (*iii*) charging current.

Solution : Capacitance between all three conductors joined and sheath [Fig. 8.19 (*a*)] gives $3C_S$

So $3C_S = 1.8\ \mu F$

$$\text{or } C_S = \frac{1.8}{3} = 0.6\ \mu F$$

Capacitance between two conductors joined to sheath and the third conductor [Fig. 8.19 (*c*)] gives $2C_C + C_S$

So $2C_C + C_S = 1.5\ \mu F$

$$\text{or } C_C = \frac{1.5 - C_S}{2} = \frac{1.5 - 0.6}{2} = 0.45\ \mu F$$

(*i*) Capacitance of core to neutral,

$$C_N = 3C_C + C_S = 3 \times 0.45 + 0.6 = 1.95\ \mu F \textbf{ Ans.}$$

(*ii*) Capacitance between any two core

$$= \frac{1}{2}C_N = \frac{1.95}{2} = 0.975\ \mu F \textbf{ Ans.}$$

(*iii*) Charging current per phase,

$$I_C = 2\pi f V_P C_N$$

$$= 2\pi \times 50 \times \frac{11{,}000}{\sqrt{3}} \times 1.95 \times 10^{-6} = 3.89 \text{ A } \textbf{Ans.}$$

Example 8.15. A 3-core cable gives on test a capacitance of 3.7 μF between two cores. Find the line charging current of the cable when it is connected to 11 kV, 50 Hz system.

[Pb. Technical Univ. Power System-I (Transmission & Distribution) May 2008]

Solution : Capacitance between any two core, C_2 = 3.7 μF

Capacitance of each core to neutral, $C_N = 2C_2 = 2 \times 3.7 = 7.4$ μF

Supply frequency, f = 50 Hz

$$\text{Phase voltage, } V_P = \frac{11{,}000}{\sqrt{3}} = 6{,}350.85 \text{ V}$$

Line charging current of cable,

$$I_C = 2\pi f V_P C_N = 2\pi \times 50 \times 6{,}350.85 \times 7.4 \times 10^{-6}$$

$$= 14.76 \text{ A } \textbf{Ans.}$$

Example 8.16. Calculate the kVA taken by a 10 km long, 3-phase, 3-core, cable, if the capacitance measured between any two cores is 0.3 μF/km when it is connected to 10 kV, 50 Hz bus-bar.

Solution : Capacitance between any two core for 10 km length,

$$C_2 = \frac{1}{2} C_N = 0.3 \times 10 = 3 \text{ μF}$$

$$\text{or } C_N = 2 \times 3 \times 10^{-6} = 6 \times 10^{-6} \text{ F}$$

kVA taken by the cable

$$= \frac{3 V_P I_C}{1{,}000}$$

$$= \frac{3 V_P \times 2\pi f V_P C_N}{1{,}000}$$

$$= \frac{\frac{3 \times 10{,}000}{\sqrt{3}} \times 2\pi \times 50 \times \frac{10{,}000}{\sqrt{3}} \times 6 \times 10^{-6}}{1{,}000}$$

$$= 188.5 \text{ kVAR } \textbf{Ans.}$$

Example 8.17. A 3-core, 3-phase metal-sheathed cable had (*i*) capacitance of 1 μF between shorted conductors and sheath and (*ii*) capacitance between two conductors shorted with the sheath and the third conductor 0.6 μF. Find the capacitance (*i*) between any two conductors and (*ii*) between any two shorted conductors and the third conductor.

Solution : (*i*) Capacitance between shorted conductor and sheath [Fig. 8.19 (*a*)] gives $3C_S$

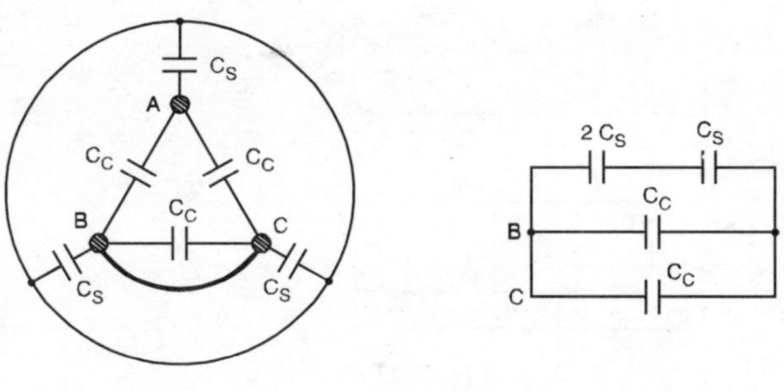

(*a*) (*b*)

Fig. 8.20

So $3C_S$ = 1 μF

$$\text{or } C_S = \frac{1}{3} = 0.333 \text{ μF}$$

(*ii*) Capacitance between two conductors shorted with sheath and the third conductor [Fig. 8.19 (*c*)] gives $2C_C + C_S$

So $2C_C + C_S$ = 0.6 μF

$$\text{or } C_C = \frac{0.6 - C_S}{2} = \frac{0.6 - 0.333}{2} = 0.1333 \text{ μF}$$

Capacitance between any two cores

$$= \frac{1}{2}[3C_C + C_S] = \frac{1}{2}[3 \times 0.1333 + 0.333] = 0.3666 \text{ μF } \textbf{Ans.}$$

Capacitance between any two shorted conductors and the third conductor (Fig. 8.20) gives

$$2C_C + \frac{2}{3} C_S = 2 \times 0.1333 + \frac{2}{3} \times 0.333 = 0.488 \text{ μF } \textbf{Ans.}$$

Example 8.18. A 33 kV, 3-phase, 50 Hz underground line, 3.4 km long, uses three single core cables. Each cable has a core diameter of 2.5 cm and the radial thickness of insulation is 0.6 cm. The relative permittivity of the dielectric is 3.1. Find (*i*) maximum stress and (*ii*) total charging kVAR.

[U.P. Technical Univ. Elements of Power System 2005-06]

Solution : Diameter of conductor, d = 2.5 cm

Internal diameter of sheath, D = 2.5 + 2 × 0.6 = 3.7 cm

Capacitance of cable,

$$C = \frac{0.024\varepsilon_r}{\log_{10}\frac{D}{d}} \times \text{length of cable μF}$$

$$= \frac{0.024 \times 3.1}{\log_{10}\frac{3.7}{2.5}} \times 3.4 \times 1{,}000 = 1{,}486 \times 10^{-6} \text{ F/phase}$$

Charging current,

$$I_C = 2\pi f C V$$

$$= 2\pi \times 50 \times 1{,}486 \times 10^{-6} \times \frac{33 \times 10^3}{\sqrt{3}} \text{ A/phase} = 8.895 \text{ A/phase}$$

Total charging kVAR

$$= 3 \times V_P \times I_C \times 10^{-3} = 3 \times \frac{33{,}000}{\sqrt{3}} \times 8.895 \times 10^{-3}$$

$$= 508.4 \text{ kVAR } \textbf{Ans.}$$

Maximum stress in the cable,

$$E_{max} = \frac{V}{\frac{d}{2}\log_e \frac{D}{d}} = \frac{33{,}000/\sqrt{3}}{\frac{2.5}{2}\log\frac{3.7}{2.5}} = 38.88 \text{ kV/cm (rms) } \textbf{Ans.}$$

8.13. MEASUREMENT OF INSULATION RESISTANCE OF CABLES

There are two methods of measuring insulation resistance of a cable namely direct deflection method and megger method. These two methods will be described in Art. 11.9.

8.14. METHODS OF LAYING OF UNDERGROUND CABLES

There are three main methods of laying underground cables *viz* direct laying, draw-in-system (or duct/pipe laying) and solid system (or trough laying).

1. *Direct Laying.* The cables with efficient steel tape or wire armouring are laid direct as they afford excellent protection from mechanical damage. The paper insulated cables for direct laying are usually served-finish with one or two layers of cotton, jute or hessian treated with bitumen compound. The PVC armoured cables are also laid direct.

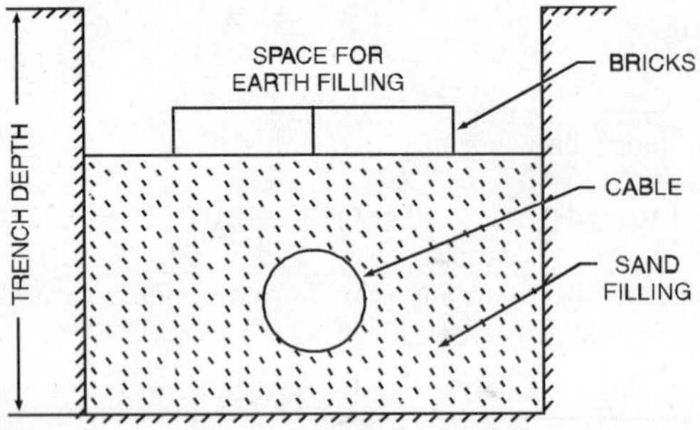

Fig. 8.21. ***Direct Laying of Underground Cable***

This method of laying underground cable is simple and cheap. In this method of laying underground cables a trench about 1.2 m deep and 0.5 m wide is dug, a layer of (of about 10 cm thickness) fine sand is spread and the cable is laid over this sand bed. The sand does not allow the moisture from ground and thus protects the cable from decay. After the cable has been laid in the trench, it is covered with another layer of sand of about 10 cm thickness and the cable is provided with a wall of bricks on either side and over the top. Then the remaining dug out soil is filled again in the trench. This arrangement is shown in Fig. 8.21. In cases, when more than one cable is to be laid in the same trench the minimum spacing between any two adjacent cables should be about 30-40 cm (both in horizontal or vertical formation of cable laying) filled with sand or riddled soil between the cables. This is done to avoid mutual heating and to ensure that fault occurring in one cable does not affect the other cables laid in the same trench. Cables are generally laid in a straight line, however loops are formed at both ends, and at least one at the centre so that in case of fault the length of cable required can be gained from the nearest loop and thus avoid jointing which it would have otherwise.

As recommended in ISI specifications the minimum depth of trench from the ground level or from the upper surface of the street pavement are as given below:

For cables below 1.1 kV rating – 0.45 m plus radius of complete cable.

For cables between 3.3 and 11 kV rating – 0.75 m plus radius of complete cable.

For cables between 22 and 33 kV rating – 1.0 m plus radius of complete cable.

Though this method of laying underground cables is simple, clean, safe and cheap and provides better conditions for heat dissipation resulting in higher current carrying capacity of cables, but suffers from the following drawbacks:

(*i*) Extension of load necessitates completely new excavation costing as much as the original work.

(*ii*) Localization of fault is difficult.

(*iii*) There is always a risk of the cable being damaged at a subsequent excavation work carried out for expansion or repair work.

(*iv*) High maintenance cost.

(*v*) The cable sheath may sometimes undergo chemical changes and get damaged due to impurities present in the soil.

(*vi*) It cannot be employed in congested areas where excavation is expensive and inconvenient. It is necessary to preserve a detailed dawing showing location, type and size of cable, details of junctions and joints for future references.

This method of cable laying is used in open areas where excavation can be done conveniently and at low cost

2. *Draw-in-System.* In congested areas where excavation is expensive and inconvenient this system of laying of underground cables is often adopted. In this method of laying of underground cables a line of conduits, ducts or tubes made of either iron, glazed stoneware, clay or cement concrete are laid in trenches side by side. Separate pipes and ducts are provided for each cable laid in the same duct. The pipes or ducts may also be built up together. The cables are then pulled into position from manholes, provided at suitable positions along the cable route, by means of a strong rope or a flexible steel wire rope attached to a chain of cane rods (IS. 1255–1967). The diameter of pipe selected for ducting a cable should always be larger than the outer diameter of the cable to be pulled-in which should be at least 2–3 cm larger than the diameter of the cable for obstruction free pulling. The ducts are made to run straight between manholes but sometimes they are to be curved to dodge obstacles. For curved ducts, the diameter of the pipe is kept larger to facilitate drawing cable into the ducts. The depth of the trench should be such that the top pipe remains at least 0.6 m below the ground level. The width of the trench depends upon the number of pipes to be accommodated. Normally a spacing of 0.25–0.75 m is maintained between two adjacent pipes. The bottom of the trench is dressed to smooth graded surface which forms the bedding of the pipes.

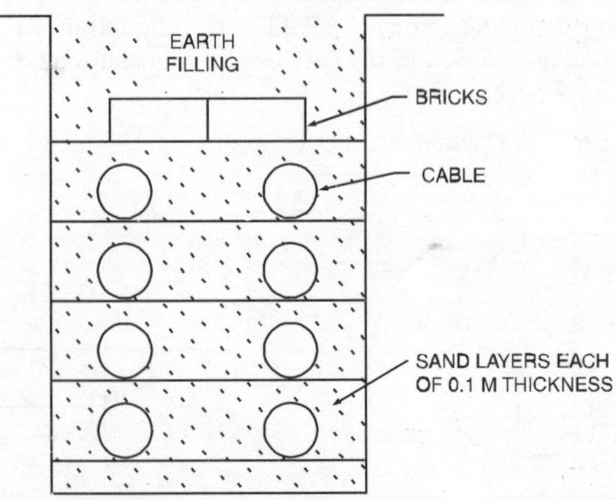

Fig. 8.22

Manhole is the underground chamber where the pipes or ducts terminate and which is employed for pulling-in cable through the ducts and jointing chamber for cables coming-in and going-out from it. Generally, the manholes are located at street crossings where from the ducts can radiate in all directions. Manhole should be large enough in size to permit a man to enter into without difficulty and its size also depends on the number of ducts terminating from different directions.

The cables to be laid in this way need not be provided with armouring but must be provided with serving of hessian tape and jute in order to protect them when drawn in. As the cables are not armoured, therefore, joints become simpler and maintenance cost is reduced considerably. In this method of laying of underground cables repairs, alterations or additions to the cable network are possible without going through the process of re-excavation of the trench. There are fewer chances of occurrence of fault as the duct affords excellent protection against mechanical damage. This system of laying cables offers long life to the cables as these do not come in direct contact with the soil, that may contain chemicals. This system of laying of underground cables is costlier in first cost and also the current carrying capacity of the cables is reduced on account of close grouping of cables and unfavourable conditions for dissipation of heat. Separate pipes and ducts (or multiple-way ducts) are provided for each cable laid in the same duct. This system of cable laying is unsuitable for secondary distribution as service mains are to be tapped frequently from them for electricity supply to consumers.

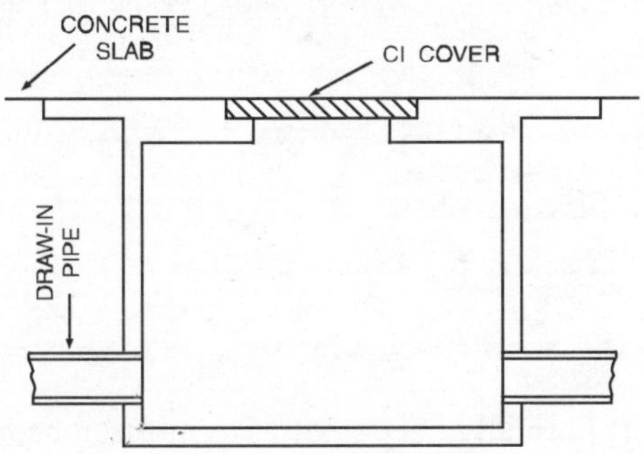

Fig. 8.23. ***Manhole***

This method of cable laying is most suitable for short length cable routes such as in workshops, railway bridge-crossing, road-crossing where frequent digging is costlier or impossible.

3. ***Solid System.*** In solid system of laying underground cables, the cable is laid in troughing, in an open trench/channel dug out in earth along the cable route. After the cable is placed in position, the troughing is filled with bituminous or asphaltic compound and covered over. The cover can be of the same material as the trough. Brick tile covers, placed in position while the compound is still liquid, may also be used. The troughing is of cast iron, stone asphalt or treated wood. Cables laid in troughs are shown in Fig. 8.24. The cable is placed on spacers, which in the case of stoneware troughs may be cast integral with the trough. The spacers ensure that the cable is completely surrounded by the compound. Cables laid in this method are usually plain lead covered as the troughing affords good mechanical protection. Also it is effective in preventing chemicals from attacking the cable and the electrolytic corrosion of the cable.

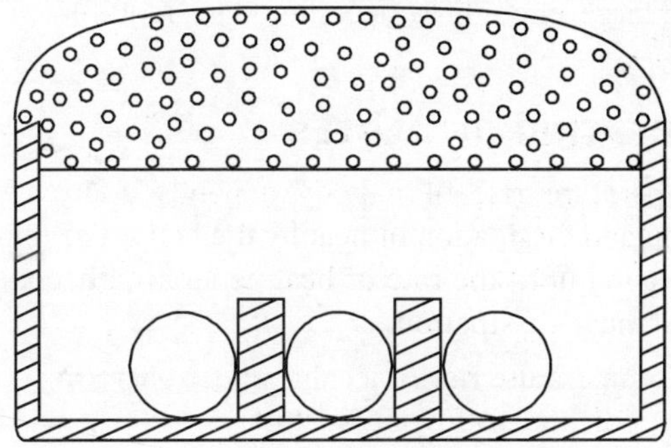

Fig. 8.24

This method of laying underground cables is rarely used because of its high cost. Moreover its success depends upon skilled supervision and favourable weather conditions while the job is being carried out. Current carrying capacity of the cable is also reduced on account of poor heat dissipation facilities. The other disadvantages are that (*i*) laying and repair require more time, and (*ii*) laying and repair cannot be carried out in rainy season.

8.14.1. Laying of Cables in Special Locations

1. ***Road Crossing.*** For laying underground cables under the roads, cast iron, galvanized steel or reinforced concrete pipes are employed for carrying them. Steel pipes are preferred as they can be easily installed and offer high resistance to mechanical damage. The pipe should be laid inclined in order to reduce the angle of bend as the cable enters and leaves the road crossing.

2. ***Railway Crossing.*** The laying of underground cables is to be carried out in accordance with the rules and specifications of the railway authorities and for that approval of railway authorities is essential. Normally, the cables are laid in steel, cast iron or spun–reinforced pipes. The pipes should be laid at a depth of 1 m below formation level and should extend to a minimum distance of 3 m from the centre of the outermost track on either side.

3. ***Water Main Crossing.*** While crossing water supply mains by a power cable, a minimum clearance of 0.4 m is to be provided between the two.

4. ***Communication Line Crossing.*** For laying power cables near communication lines it is desirable to provide a minimum clearance of 0.6 m between the two horizontally and vertically, so that disturbance to communication circuits are avoided. It is also desirable that both of the lines cross at right angles.

5. ***In Tunnels.*** In tunnels, the cables are normally cleated on walls or laid on earthed racks or brackets. The cables are run straight so as to avoid sag.

6. ***On Bridges.*** In concrete bridges, covered ducts in the form of pipes or piped holes for accommodating cables are provided. Inspection holes with concrete covers are provided at regular intervals for facilitating the laying of cables.

In steel bridges, steel pipes are laid on them and the cables are drawn-in through them. The pipes should be of diameter large enough to accommodate required number of cables.

8.15. HEATING OF CABLES

The temperature rise of a body depends upon the rate of generation and dissipation of heat by the body, The temperature goes on rising until the rate of heat generation becomes equal to that of heat dissipation.

The temperature rise of a cable under operating conditions depends upon the following factors:

(*a*) Total heat produced within the cable *i.e.,* up to its periphery.

(*b*) The conveyance of heat up to the periphery of the cable.

(*c*) The conveyance of heat through the medium, and thus away from the cable.

(*d*) The current carrying capacity of the cable, or cables, singly and collectively, where installed under the varying conditions encountered in practice.

(*e*) The nature of the load—continuous, intermittent or possibility of short circuits etc.

(*a*) *Heat Generation in Cables.* Within the cables the sources of heat generation are (*i*) I^2R losses in the conductor.

(*ii*) Dielectric losses in the cable insulation,

(*iii*) Losses in the metallic sheath and armourings.

(*i*) *Copper losses in the conductors.* Resistance of conductor at an operating temperture of 65°C (assumed) is determined from the resistance given in standard tables (usually at 15.6°C) from the following relation.

$$R_h = R_a\,[1 + \alpha_a(65 - 15.6)]$$

where R_h, R_a and α_a are the hot resistance (resistance at 65°C), resistance at ambient temperature of 15.6°C and the temperature coefficient of resistance at ambient temperature of 15.6°C.

To allow for stranding, the resistance of a single-core cable is multiplied by the factor 1.02. In case of multi-core cables the resistance is further multiplied by the factor 1.02 in order to allow for the lay of the whole conductor.

(*ii*) Dielectric Losses. The energy losses occurring in the dielectric of cables are due to leakage and so called *dielectric hysteresis.* The former loss, is due to passing of current by conduction through the resistance of dielectric, and is independent of supply frequency and, therefore, it occurs both with dc and ac. The leakage current is proportional to the applied voltage and, therefore, the loss is proportional to the square of the applied voltage.

With alternating voltages there is a further loss of energy, usually known as *hysteresis loss* and this loss under usually operating conditions is very much higher than the leakage loss. Apparently some energy is consumed in reversing the stresses in solid dielectrics, and this appears as heat, causing increase in temperature of the dielectric. At normal operating stresses the dielectric hysteresis loss is proportional to the square of the electrostatic stresses, that is to the square of the applied voltage.

In order to take into account the above losses, the charging current of a cable I is assumed to have two components :

One being true capacitance current which is equal to ωCV and leads the applied voltage by 90° and the other being the energy component which is in phase with the applied voltage and represents the dielectric loss component of current.

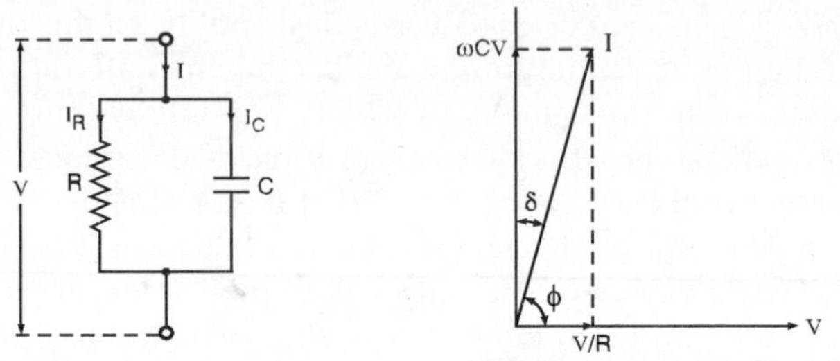

(*a*) *Equivalent Circuit of a Cable* (*b*) *Phasor Diagram of* (*a*)

Fig. 8.25

The equivalent circuit for this system is represented by a parallel combination of leakage resistance R, representing dielectric power loss, and a capacitance C, as shown in Fig. 8.25 (*a*). The phasor diagram is shown in Fig. 8.25 (*b*).

If V is the applied voltage C is the capacitance of cable, ϕ is the phase angle between voltage and current called the *power factor angle of the cable* and δ is the loss angle of the dielectric,

$$\text{Charging current, } I_C = \frac{V}{X_C} = \frac{V}{1/\omega C} = \omega CV$$

Dielectric loss per phase

$$= VI\cos\phi = VI\sin\delta \qquad \because \phi = 90^\circ - \delta$$

$$= V\frac{I_C}{\cos\delta}\sin\delta = V\omega CV\tan\delta = \omega CV^2\tan\delta$$

Sin δ is normally very small, so tan δ can be taken equal to δ (in radians) and dielectric loss per phase is given by the expression

$$P = V^2\omega C\delta \text{ watts} \qquad ...(8.22)$$

This shows that the dielectric losses are proportional to the square of the voltage, with the result that, although these are of little importance in low-voltage cables, these are of very great importance in high-voltage (say over 22,000 V) cables.

It has been found that, over the working temperature range, the dielectric losses in most cables fall to a minimum at about 40°C and then rise again. The variation when plotted gives rise to what is known as the 'V' curve of the dielectric losses, and the minimum temperature point corresponds very closely to the temperature of liquefaction of the resinous impregnating compound. A typical curve for the paper-insulated cable is shown in Fig. 8.26.

Since the variation of capacitance current with temperature is small, the dielectric power factor–temperature curve is of the same general shape as the dielectric loss–temperature curve. In this connection, it seems preferable to exhibit dielectric –loss data in terms of dielectric power factor rather than in power loss per km of cable in order to facilitate comparison between different types and sizes of cables.

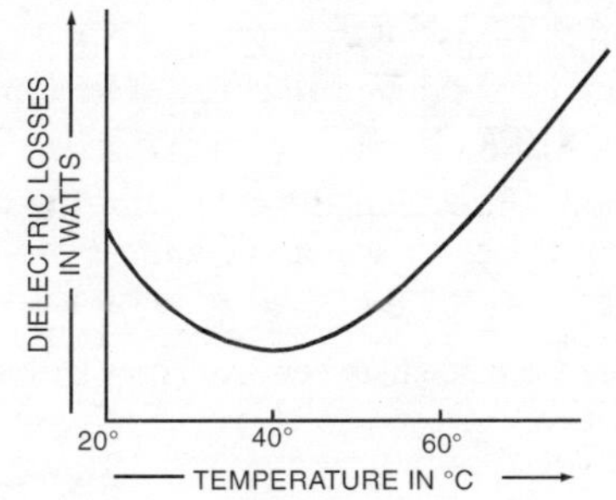

Fig. 8.26

The actual shape of 'V' curve depends on several factors and may be deliberately changed by using different papers, impregnating compounds, and methods of manufacture.

The operating temperature of a paper–insulated cable is about 65°C which is to the right of the lowest point on the 'V' curve. So around the operating temperature, if the temperature further increases, due to overload or any other reason, dielectric losses will increase causing further increase in heat generation and so rise in temperature. The increase in temperature will also cause increase in temperature gradient between the cable and the atmosphere that will result in greater heat dissipation. In case the rate of heat dissipation is less in comparison to that of heat generation, the temperature will go on increasing until the dielectric overheats and fails electrically. This is referred to as *thermal instability.*

Fortunately, action taken to reduce dielectric losses, makes the dielectric loss-temperature curve flat and reduces the tendency towards thermal instability.

(*iii*) *Sheath and Armouring Losses.* In case of power transmission by single-core cables, the presence of lead sheath around each conductor influences the electrical characteristics of the circuit. In 3-core cables the effect is negligible but for single-core cables the effect is of great importance. The electromagnetic fields produced by the currents flowing through the conductors induce emfs in the sheaths, and under certain conditions heavy currents are set up therein. The actual current flowing along the sheath depends, amongst other things, on the magnitude and frequency of the current in the conductor, the arrangement and spacing of the cables, the sheath resistance and upon the conditions that the sheath is bounded (two different cables having sheath electrically connected at both ends) or unbounded (cables having one end or no end electrically short-circuited).

The induced sheath currents are of two types:

(*i*) The currents, which have both outward and inward directions, called the *sheath eddies.*

(*ii*) The currents, which have outward and inward current paths in separate sheaths, as shown in Fig. 8.27, called the *sheath-circuit eddies.*

Thus, for unbounded cables sheath eddies will be there, but for bounded cables both sheath eddies and sheath circuit eddies will be there.

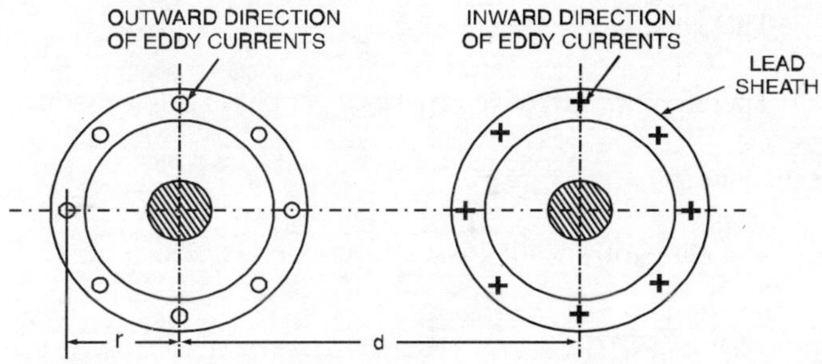

Fig. 8.27

The approximate formulae for eddy loss for unbounded cables given by Arnold is as follows:

$$\text{Sheath loss} = I^2\left[\frac{78\omega^2}{R_s}\left(\frac{r}{d}\right)^2 \times 10^{-9}\right] \text{ watts/phase} \quad ...(8.23)$$

where I is the current per conductor, r is the mean radius of the sheath, d is the inter-axial spacing of conductors, and R_s is the sheath resistance in ohms.

If f = 50 Hz, then $78\omega^2 \times 10^{-9} = 78\,(2\pi \times 50)^2 \times 10^{-9}$
$= 0.0077$

$$\text{and sheath loss} = \frac{0.0077\, I^2}{R_s}\left(\frac{r}{d}\right)^2$$

Conductor loss $= I^2R$ where R is the resistance of conductor

$$\therefore \quad \frac{\text{Sheath loss}}{\text{Conductor loss}} = \frac{78\omega^2}{RR_s}\left(\frac{r}{d}\right)^2 \times 10^{-9}$$

The sheath loss in open-circuited sheath is about 2% of the total loss and can be neglected.

The voltage induced in the sheath = I × ωM = IX_m where M is the mutual inductance between conductor and sheath.

The value of M between the conductor and sheath can be determined in a similar manner to that of self-inductance of the two conductors. For two sheaths with respect to the core conductors (Fig. 8.27) the mutual inductance is given as

$$M = 4 \log_e \frac{d}{r} \times 10^{-7} \text{ H/m}$$

Large voltages are induced in the sheaths if they are open circuited, and it is very probable that arcing will occur between them. It is, therefore, standard practice to bond the sheaths at both ends so that high voltages are avoided. The impedance of the sheath current path is due to the sheath resistance R_s and the sheath self-inductance, which is equal to M. Thus if the sheaths are bonded the sheath current is given by the expression

$$I_{sh} = \frac{E_{sh}}{\sqrt{(R_s^2 + X_m^2)}} = I\,\frac{X_m}{\sqrt{(R_s^2 + X_m^2)}}$$

The magnitude of the sheath current is independent of the distance between the bonds, since X_m and R_s, both, are proportional to the length.

The sheath losses/phase $= \dfrac{I^2X_m^2R_s}{R_s^2+X_m^2}$

Therefore the effective resistance per phase is the conductor resistance plus $\dfrac{X_m^2\,R_s}{R_s^2+X_m^2}$

and ratio of sheath losses to copper losses

$$= \frac{\dfrac{I^2\,X_m^2\,R_s}{R_s^2+X_m^2}}{I^2\,R} = \frac{X_m^2\,R_s}{R\left(R_s^2+X_m^2\right)}$$

where R is the resistance of the conductors.

With large conductors and close spacing X_m is very small in comparison with R_s and therefore, the ratio of sheath losses to copper losses

$$\simeq \frac{X_m^2\,R_s}{R\,R_s^2} \simeq \frac{X_m^2}{R\,R_s} \qquad \text{...(8.24)}$$

Example 8.19. A single-core cable 1.0 km long has a core diameter of 0.5 cm and diameter undersheath 2.0 cm. The relative permittivity of the insulating material is 3.5. The power factor on open circuit is 0.05 and the supply voltage is 11 kV, 50 Hz. Determine (*i*) the capacitance of the cable (*ii*) the charging current per conductor (*iii*) the dielectric loss and (*iv*) the equivalent insulation resistance.

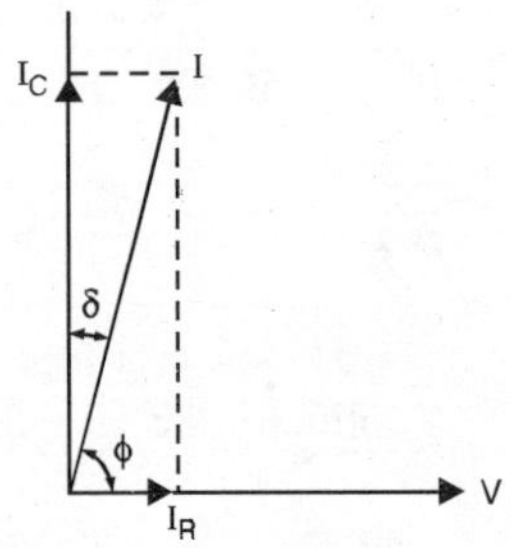

Fig. 8.28

Solution : (*i*) Capacitance of the cable,

$$C = \frac{0.024\,\epsilon_r}{\log_{10}\dfrac{D}{d}} \times 10^{-6}\text{ F/km} = \frac{0.024\times 3.5}{\log_{10}\dfrac{0.02}{0.005}} \times 10^{-6}$$

$$= 0.1395\ \mu\text{F } \textbf{Ans.}$$

(*ii*) Charging current,

$$I_C = \omega CV = 2\pi \times 50 \times 0.1395 \times 10^{-6} \times \frac{11{,}000}{\sqrt{3}}$$

$$= 0.278\text{ A } \textbf{Ans.}$$

Power factor on open circuit,

$\cos\phi = 0.05$

$\therefore\ \phi = \text{Cos}^{-1}\ 0.05 = 87.134$

Dielectric loss angle,

$\delta = 90° - 87.134 = 2.866°$

(*iii*) Dielectric loss

$= \omega CV^2 \tan\delta$

$$= 2\pi \times 50 \times 0.1395 \times 10^{-6} \times \left(\frac{11{,}000}{\sqrt{3}}\right)^2 \tan 2.866°$$

$= 88.5$ watts **Ans.**

(*iv*) The equivalent insulation resistance of the cable

$$= \frac{V^2}{\text{Dielectric loss}} = \frac{(11{,}000/\sqrt{3})^2}{88.5} = 0.456\text{ M}\Omega \textbf{ Ans.}$$

Example 8.20. For a certain length of a 3-core lead sheathed belted type cable following test results were obtained.

(*i*) With one core shorted to sheath capacitance between the other two core 2.5 μF and

(*ii*) With one core shorted to sheath resistance between the other two core : 810 k Ω.

Estimate the loss angle of the cable and the no-load current drawn by it when the cable is connected to a 3-phase, 11 kV 50-Hz supply. [*A.M.I.E. Sec. B. Power Systems Summer 1999*]

Solution : The capacitance between two cores with one core shorted to sheath gives $\frac{1}{2}C_N$

So $\frac{1}{2}C_N = 2.5\ \mu\text{F}$ or $C_N = 5\ \mu\text{F}$

Charging current,

$I_C = 2Vf\,V_PC_N$

$$= 2\pi \times 50 \times \frac{11\times 1{,}000}{\sqrt{3}} \times 5 \times 10^{-6} = 9.976\text{ A}$$

The equivalent resistance circuit for the cable with one core shorted to sheath is shown in Fig. 8.29. Measured resistance between two cores gives twice the insulation resistance of each core, *i.e.*,

$$2R_{INS} = 810\text{ k}\Omega \text{ or } R_{INS} = 405\text{ k}\Omega$$

$$\text{Loss angle } \delta = \text{Tan}^{-1}\frac{V/R_{INS}}{\omega CV} \text{ or } \text{Tan}^{-1}\frac{1}{R_{INS}\omega C}$$

$$i.e.,\ \delta = \text{Tan}^{-1}\frac{1}{405\times 10^3\times 2\pi\times 50\times 5\times 10^{-6}}$$

$$= \text{Tan}^{-1}\ 0.00157 = 0.09° \textbf{ Ans.}$$

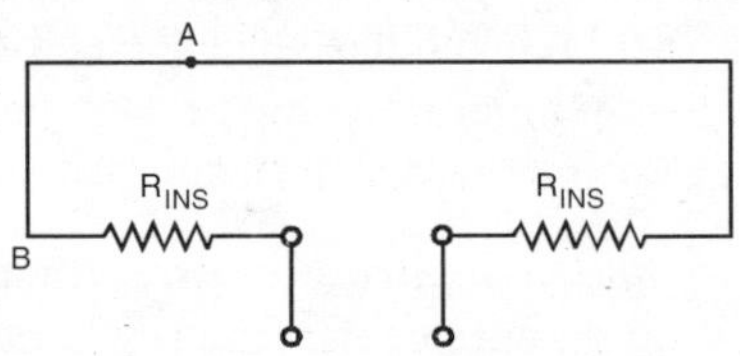

Fig. 8.29

Energy component of current,

$$I_e = \frac{V_P}{R_{INS}} = \frac{11\times 1{,}000}{\sqrt{3}\times 405\times 10^3} = 0.0157\text{ A}$$

No-load current drawn by cable,

$$I = \sqrt{I_C^2+I_e^2} = \sqrt{(9.976)^2+(0.0157)^2} = 9.976\text{ A } \textbf{Ans.}$$

8.16. THERMAL RESISTANCE OF CABLES

It is seen that heat flow in a cable is similar to leakage current flow (both of them flow radially out from the cable core through the insulating material, the sheath, bedding, and serving and finally to the ground)—the heat flow (watts) corresponds to the electric current while temperature difference (degree centrigrade) corresponds to electrical potential difference. The thermal form of Ohm's law gives the amount of heat transferred across a layer whose opposite faces have a temperature difference and is given as

$$\text{Heat flow in watts, H} = \frac{\text{Temperature difference in °C}}{\text{Thermal resistance, } R_\theta \text{ in thermal ohms}} \qquad \text{...(8.25)}$$

Thermal ohm is defined as the thermal resistance which requires a temperature difference of 1°C to cause a heat flow of one watt (*i.e.*, one joule per second). The unit of thermal resistance is therefore, the degree Celsius per watt (°C/W).

Like an electrical resistance, thermal resistance is directly proportional to length, 'l', in the direction of transmission and inversely proportional to cross-sectional area, 'a', at right angles to that direction and is given as

$$R_\theta = \rho_\theta \frac{l}{a} \qquad ...(8.26)$$

where ρ_θ is the constant of proportionality and is known as *thermal resistivity* of the dielectric.

The thermal resistivity is defined as the thermal resistance of a piece of material having unit length and unit cross section. Its unit is thermal ohms per metre cube or thermal ohm-metre.

The thermal resistance of a single-core cable having core diameter d and inner diameter of sheath D is given as

$$R_{\theta D} = \frac{\rho_{\theta D}}{2\pi} \log_e \frac{D}{d} \text{ thermal ohms per metre length} \qquad ...(8.27)$$

where $\rho_{\theta D}$ is the thermal resistivity of dielectric in thermal ohm-metre. The value of $\rho_{\theta D}$ is 7.5 thermal ohm-metres for solid cables up to and including 2.2 kV and 5.5 for solid cables above 2.2 kV. For oil-filled cables and impregnated pressure cables the usual values of $\rho_{\theta D}$ are 4.5 and 5.5 thermal ohm-m respectively.

Thermal Resistance of a 3-Core Cable. Because of the non-uniformity of the dielectric material it is quite difficult to derive an expression for the thermal resistance of a 3-core cable. However, an empirical formula, for determination of thermal resistance of a three-core belted type cable, suggested by Simon is given below:

$$R_{\theta D} = \frac{\rho_{\theta D}}{6\pi}\left(0.85 + \frac{0.2t}{T}\right) \log_e \left[1 + \left(\frac{T+t}{d}\right)\left(4.15 - \frac{1.1t}{T}\right)\right]$$

thermal ohms per metre length ...(8.28)

where d is the diameter of core, t is the thickness of belt, T is the thickness of conductor insulation and $\rho_{\theta D}$ is the thermal resistivity of dielectric.

*Thermal Resistance of Ground***.** The thermal resistance of the ground in which cable is buried depends on the nature of soil and also the depth to which the cable is buried. Logarithmic formula has been developed to ascertain its value and is given as

$$R_{\theta G} = \frac{\rho_{\theta G}}{2\pi} \log_e \frac{2h}{R}$$

where $\rho_{\theta G}$ is the thermal resistivity of the soil, h is the depth of the cable below the soil and R is the overall radius of the cable.

It is found that in practice, the thermal resistivity $\rho_{\theta G}$, as measured in the laboratory, is too large for use in the above expression and it must be applied by a correction factor of 2/3 in order to correspond with the results of experiments on buried cables. The amended expression becomes

$$G = \frac{\rho_{\theta G}}{3\pi} \log_e \frac{2h}{R} \qquad ...(8.29)$$

In practice $\rho_{\theta G}$ varies from 1.2 to 8 or 10 depending on the soil and its moistness.

In armoured cables the lead sheath is protected from the armour by a bedding of hessian or jute, and a relationship for the associated thermal resistance may be deduced from a knowledge of the dimensions and thermal resistivity of the protective covering.

Since the protective covering (bedding and serving) is in the form of a cylinder, the expression for its thermal resistance is similar to that given by Eq. (8.27) and is given as

$$R_{\theta p} = \frac{\rho_{\theta p}}{2\pi} \log_e \frac{R_2}{R_1} \text{ thermal ohms per metre length} \qquad ...(8.30)$$

where $\rho_{\theta p}$ is the thermal resistivity of bedding and serving, (neglecting the thickness of armour usually being very small, about 5°C-m/watt), R_1 is the radius over the sheath and R_2 is the external radius of cable.

8.17. DETERMINATION OF MAXIMUM CURRENT CARRYING CAPACITY OF CABLES

Cables should not be operated too hot for the following reasons:

1. Unequal expansion may create voids in the insulation which will lead to ionization.
2. Expansion of the oil may cause the sheath to burst.
3. The oil may lose its viscosity and it may start draining away from higher levels.
4. The thermal instability may rise due to the rapid increase of dielectric losses with temperature.

The current rating of cables, like that of any other electrical equipment, is governed by the maximum permissible rise in temperature which the insulation employed can bear. This obviously depends upon two factors (*i*) amount of heat produced (*ii*) the surrounding temperature of the soil. The current rating of the cables apart from the above factors also depends on a large number of other factors such as method of cable laying employed, spacing between the cables, number of cores, and the thermal conductivity of the soil. Cable manufacturers employ multiplying factors to take care of each of them. An approximate indication of the current carrying capacity for Indian conditions is given below which is based on a maximum conductor temperature of 80°C in the case of low voltage cables and 65°C for 11 kV cables and above. The ground ambient temperature of 30°C applies to most places in India. The approximate continuous current rating of a single-, 3- or 4-core cables laid directly in ground will be as indicated below for cables up to 1,100 V, based on a ground temperature of 15°C.

Area in mm^2	Current carrying capacity in amperes
4.52	51
9.35	77
14.52	100
25.80	135
38.71	170
64.52	225
96.77	280
161.29	375
322.58	530

The following multiplying factors are used:

1. For a ground temperature of 30°C = 0.88
2. For 3,300 and 6,600 V cables = 0.95
3. For 11,000 V cables = 0.9
4. For 22,000 V cables = 0.8

	15 cm	30 cm	61 cm
5. For 2 cables	0.84	0.88	0.92
6. For 3 cables	0.75	0.80	0.85

The above figures may be taken as the maximum permissible currents for cables already in service. In the planning of new installations, it will be better to allow at least 50 per cent margin for the future extension.

8.17.1. Calculations of Current Rating of Cables

When the thermal resistance of a cable is known, it is possible to calculate the value of current I which will raise the conductor temperature to the maximum permissible value. This is the cable rating under the specified conditions and under continuous loading, or loading sufficiently prolonged to enable steady thermal conditions to be attained.

Neglecting the dielectric losses, the insulation of the cable will be passing heat through due to the conductor losses, while bedding and serving will be passing heat due to the sheath losses in addition.

The power losses in the n cores of cable, each core having a resistance of R ohms per metre length are given as nI^2R watts per metre length where I is the current per core in amperes.

If θ_m is the maximum permissible temperature of the conductor of the cable and θ_s is the sheath temperature, then according to thermal ohm's law

$$nI^2R = \frac{\theta_m - \theta_s}{R_{\theta D}} \qquad ...(8.31)$$

where $R_{\theta D}$ is the thermal resistance of the dielectric. Assuming sheath losses λ times the conductor power losses and neglecting armour losses, the total losses become $(1 + \lambda)nI^2R$ and for an ambient temperature of θ_a, we have

$$(1 + \lambda)nI^2R = \frac{\theta_s - \theta_a}{R_{\theta P} + R_{\theta G}} \qquad ...(8.32)$$

where $R_{\theta P} + R_{\theta G}$ are the thermal resistances of protective covering and ground (or soil).

Since normally θ_s is not known, eliminating θ_s from the above two Eqs. (8.31) and (8.32), we have

$$nI^2RR_{\theta D} + (1 + \lambda)\, nI^2R(R_{\theta P} + R_{\theta G}) = \theta_m - \theta_a$$

$$\text{or } I = \sqrt{\frac{\theta_m - \theta_a}{nR[R_{\theta D} + (1+\lambda)(R_{\theta P} + R_{\theta G})]}} \text{ amperes} \qquad ...(8.33)$$

Effect of Dielectric Losses. It becomes necessary to consider dielectric losses for cables of voltages 66 kV and above. These losses are independent of load current except in so far as the current causes an increase in temperature of the dielectric. The method is to compute the dielectric loss per metre length at the maximum permissible temperature, assuming that it is all concentrated at the core, and then calculate the temperature rise θ_d it would produce if it were the only source of heat. This is subtracted from the maximum permissible temperature rise θ_m, and the difference $(\theta_m - \theta_d)$ in used in place θ_m in the above Eq. (8.33). Thus we have

$$I = \sqrt{\frac{\theta_m - \theta_d - \theta_a}{nR[R_{\theta D} + (1+\lambda)(R_{\theta P} + R_{\theta G})]}} \text{ amperes} \qquad ...(8.34)$$

Example 8.21. A single-core cable having core diameter of 25 mm and the dielectric thickness of 35 mm is buried in the ground. The thermal resistivity of the dielectric is 5 thermal ohm-metres, and the thermal resistance between the sheath and ground surface is 0.5 thermal ohm per metre length of cable. The electrical resistance of the cable is 100 μΩ per metre length. Estimate the maximum current loading if the temperature difference between the core and ground surface is not exceed 60°C. Dielectric and sheath losses may be neglected.

Solution : The thermal resistance of dielectric,

$$R_{\theta D} = \frac{\rho_{\theta D}}{2\pi} \log_e \frac{D}{d} = \frac{5}{2\pi} \log_e \frac{25 + 2 \times 35}{25}$$

= 1.06 thermal ohms per metre length

Soil thermal resistance, $R_{\theta G}$ = 0.5 thermal ohms per metre length

Total thermal resistance, $R_{\theta T} = R_{\theta D} + R_{G\theta} = 1.06 + 0.5 = 1.56$ thermal ohms per metre length

Electrical resistance of the cable, $R = 100 \times 10^{-6}\ \Omega$ per metre length

Neglecting the losses in the sheath and dielectric,

$$I^2R = t/R_{\theta T}$$

So maximum current loading,

$$I = \sqrt{\frac{t}{R_{\theta T} \times R}} = \sqrt{\frac{60}{1.56 \times 100 \times 10^{-6}}} = 620 \text{ A } \textbf{Ans.}$$

8.18. SELECTION OF CABLES

For selection of cable of correct size and type for a particular application, the factors to be considered are as follows:

1. System Voltage. The type of operating system such as dc (2-wire or 3-wire) or ac (single-phase, three-phase), earthed or unearthed and operating voltage such as 415/240 V, 11, 33, 66, 132 kV etc.

2. Current Carrying Capacity. The current rating is the most important factor, already discussed in Art. 8.17.

3. Permissible Voltage Drop. The cable should be selected of such a size that the voltage drop in the cable is within permissible limits.

4. Short-circuit Rating. For selection of *lt* cables this factor is not important but for *hv* cables, this factor is the most important. It is because in high-voltage systems the fault current (phase-to-phase or phase-to-earth) is very high. The cable should be so selected that it can withstand the stresses and the resulting increase in temperature caused by the maximum short-circuit current produced by the phase-to-phase short circuit up to a period of 1 second (IS : 1255 – 1967).

The last, but not the least, factor is the *economy consideration.* The cable selected satisfying the above requirements-continuous current rating, voltage grade, permissible voltage drop and short-circuit current rating (in case of *ht* cables), should also give minimum operating cost (power losses, the interest on capital cost and depreciation).

8.19. FAILURE OF UNDERGROUND CABLES

8.19.1. Causes of Failure of Underground Cables

1. The most common point of failure is at the cable sealing box mostly due to poor workmanship of the cable jointer when the end was sealed. Damage caused to the insulation on account of improper handling at the time of laying will also ultimately result in a failure.

2. Another very common cause is the mechanical puncturing of the lead sheathing of a cable, such as by a crowbar, especially in industrial installations where excavation and building operations are carried on in areas having several underground cables.

3. Although the lead sheathing of underground cables has a very high degree of resistance against corrosion, but inspite of it certain soils are chemically active and may cause severe pitting and corrosion.

4. The cable may also get damaged due to vibration fatigue or overheating.

5. Where a cable is laid vertically up posts, or walls, the impregnating oil is likely to leak from the cable box at the lower level due to the hydrostatic pressure.

8.19.2. Breakdown Voltage

The electric stress which can cause the breakdown of a certain dielectric depends upon many factors such as time duration of application, temperature, pressure, the presence of moisture or gaseous spaces, shape of electrodes etc.

The dependence of stress on time duration is very important, and tests are conducted to determine the breakdown curve giving relation between electric stress (voltage) and time duration of application. Figure 8.30 shows the electric stress versus time duration curves for two dielectrics. The curve I is for a very well impregnated paper of 1 mm thickness while curve II is for a poorly impregnated paper which contains air spaces. In case of very well impregnated paper, the short-time breakdown voltage is about 48 kV giving electric stress of 45 kV/mm, but the breakdown voltage attains a steady value of 35 kV in about 6 hours. If a voltage of 34 kV is applied, the dielectric never breaks down. It is of the greatest advantage if the asymptotic value (*i.e.,* final value) is attained in a short time, for then decisive tests may be short; presumably also the cable is stable and likely to give long service. In case of poorly impregnated paper containing air spaces, the short-time breakdown voltage is not much less than for the good dielectric, but the asymptotic value is much lower and is not reached even in 50 hours. With slow deterioration of this kind it is difficult to say what voltage the dielectric can maintain definitely. Moisture has the same effect as gaseous voids on the breakdown curve of dielectric.

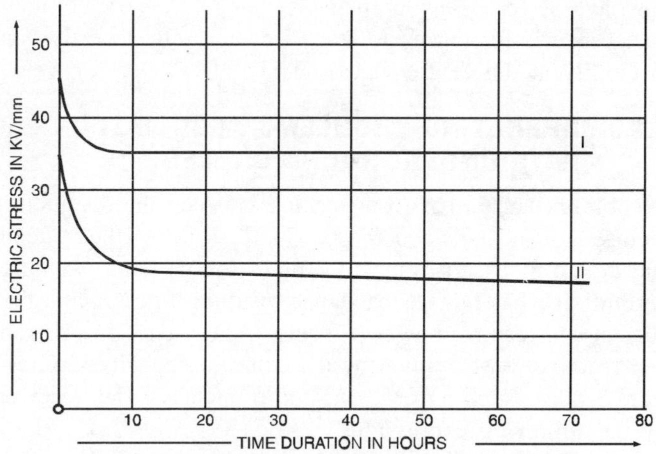

Fig. 8.30. ***Breakdown Curves For Dielectrics***

A well impregnated cable will initially contain no voids and so it will have a breakdown curve, as represented by curve I. When, however, this cable is subjected to fluctuating loads, the heating causes the oil to expand and sheath to stretch; when the cable cools, the sheath does not recover and small voids are formed by cavitation. After a number of fluctuations the voids may be such that the breakdown curve is like curve II, and eventual failure will occur if the applied voltage is greater than the new asymptotic value. In order to ascertain the stability of the curve, the cable is subjected to the operating voltage or a higher voltage whilst the cable is alternately heated and allowed to cool. Such a test is called a *stability test* and is applied to all types of *ht* cables.

The formation of voids is accompanied by ionization by collision and a rise of power factor of the cable. The voids are eliminated in the oil-filled and pressure cables, whilst the gas cable prevents ionization by the application of hydrostatic pressure. All such cables have good breakdown curves.

Useful life of a cable depends on the electric stress and degree of homogeneity of the dielectric. Lower the electric stress greater is the life. Greater the degree of homogeneity, greater is the life. The temperature variations displace the breakdown curve downwards, the degree of displacement depending on the maximum temperature and number of heat cycles per unit time.

8.19.3. Mechanism of Breakdown

There are two ways in which breakdown of cables usually occurs. One way is by a progressive coring and tracking, which always starts from the core or sheath, and ultimately bridges the electrodes. Another way is by *thermal instability* which occurs due to rapid increase in power factor with the rise in temperature (a small rise of temperature causes the dielectric losses to increase by a greater amount than can be dissipated out). A marked difference between the methods of breakdown is that coring, once it occurs, will continue until the cable breaks down, although the time duration may be considerable for complete action. In thermal instability, however, no damage occurs until just before breakdown, so that if the load is reduced before breakdown the cable will not have suffered any permanent change. A very common occurrence is for coring to start and then introduce thermal instability at the centre of coring.

8.20. OPERATING PROBLEMS ASSOCIATED WITH UNDERGROUND CABLES

A cable, because of narrow spacing between the conductors themselves and between the conductors and earthed sheath and due to their separation by a dielectric medium of higher permittivity, has large capacitance resulting in high charging current and reactive power (capacitive). The charging current and reactive power drawn by the cable of capacity C farads from 3-phase supply of V volts and frequency f Hz are given by the following expressions

$$I_C = 2\pi f C V_P = \frac{2\pi f C V_L}{\sqrt{3}} \text{ amperes} \quad ...(8.35)$$

$$\text{Reactive power} = \sqrt{3}\, V_L I_C$$

$$= \sqrt{3}\, V_L \times \frac{2\pi f C V_L}{\sqrt{3}}$$

$$= 2\pi f C V_L^2 \text{ VARs (capacitive)} \quad ...(8.36)$$

Flow of charging current causes heating of cable. This reduces the load current capability of the cable. Sometimes current carrying capability of the cable is also reduced due to dielectric losses resulting in increase in temperature rise. The use of cable, in practice, is reduced to a length of less than 50 km because of these factors.

Because of capacitive current/capacitive reactive power drawn from the system, an increase in receiving-end voltage may occur under no-load conditions (Ferranti effect). Thus the receiving-end voltage may vary considerably from full-load to no-load conditions.

Generation of capacitive reactive power may hamper the stability conditions of large generators as large size generators have high synchronous reactance (about 1.5 pu) and their leading MVAR capacity is low.

The interruption of cable capacitive currents may give rise to overvoltage,

8.21. HIGH VOLTAGE DIRECT CURRENT (HVDC) CABLES

Main operational features of hvdc transmission system with cables are no charging current; very low dielectric loss; the stress distribution in dc cables by dielectric resistance, not by dielectric capacitance as in case of ac cables.

The resistance of dielectric decreases appreciably with the rise in temperature and so the electric stress distribution in the dielectric is considerably influenced by thermal considerations.

In dc cables the electric stress increases from conductor to sheath and becomes maximum at the sheath while in ac cables it is reverse of it. This phenomenon is known as *stress inversion* and may lead to troubles at terminations and joints where longitudinal stresses are developed.

Because of above features of underground *hv* dc transmission system, there is practically no limitation regarding distance of transmission. The dc cables have lesser thickness of insulation and are more compact. These cables are always single core and always have stranded conductor. A cotton tape covers the plastic oversheath and a galvanized steel armour is put over the cotton tape for mechanical protection. Impregnated fibrous wrapping or serving is provided over the armour as an anti-corrosion measure. DC cables may have plastic or impregnated paper insulation.

8.22. TESTING OF CABLES

The tests to be carried on paper insulated cables with lead sheaths are:

1. Acceptance tests at works
   (*a*) Conductor resistance test
   (*b*) Voltage test
   (*c*) Dielectric power factor/voltage test
2. Sample tests at works
3. Performance tests
4. Tests on oil-filled and gas-filled cables
   (*a*) Load cycle tests
   (*b*) Thermal stability test for 132 kV cables only
   (*c*) Impulse test
   (*d*) Dielectric thermal resistance test
   (*e*) Cold power factor/voltage test
   (*f*) Mechanical test
   (*g*) Binding test

1. Acceptance Tests at Works

(*a*) Conductor Resistance Test. The dc resistance of the cable core is measured at room temperature and corrected to the standard reference temperature of 20°C.

(*b*) Voltage Test. In this test, a sinusoidal voltage of frequency 40–60 Hz is applied to the cable and is increased gradually to full value and maintained continuously for 15 minutes between conductors and between each conductor and sheath. For an 11 kV system, for belted type cables, the voltage applied between the conductors varies from 24 kV to 36 kV while the value of voltage applied between the conductor and sheath varies between 14 and 21 kV. For 'SL' type and 'H' type cables, the voltage applied between the conductor and sheath varies between 15 and 22 kV. It is to be noted that the lower range for applied voltage refers to the cable as manufactured and the higher range meant for cables after a bending test. In bending test the cable has to be bent around a cylinder of specified diameter to make one complete turn, it is then, unwound and the process is repeated in the opposite direction. This sequence of operation is normally carried out two to three times.

(*c*) **Dielectric Power Factor/Voltage Test.** This test is performed on 33-kV cables only. After each cable drum is complete, its core is tested for dielectric power factor at room temperature on 50 Hz single-phase ac supply of voltage 10, 20, 30, 39 kV. The measured value of power factor at normal operating voltage should not exceed the value declared by the manufacturer and the maximum tolerance shall be only 0.01.

Also the difference in pf between half normal operating voltage and twice normal operating voltage is not to exceed the value as claimed by the manufacturer. The tolerances for above for a 3-core screened cable and for a single-core and 'SL' type cables are 0.0006 and 0.001 respectively. The manufacturer can also be asked to produce evidence to show that the pf at normal operating voltage does not exceed 0.01 at a series of temperatures ranging from 15°C to 65°C.

2. Sample Tests at Works. On the finished cables after bending, voltage tests are conducted on certain samples, in order to ascertain the declared values. For cables to be installed vertically, drainage tests are conducted in order to ensure that the paper does not become dry at the higher points, otherwise, the dried-out paper will be a weak spot in the cable and the cable is likely to get damaged at that spot. It can be said that this test is performed to ascertain the viscosity of the compound.

3. Performance Tests. After installation of cable, voltage tests are conducted in the same manner but with somewhat reduced voltages. For belted type cables the values of reduced voltages are normally 25 kV, 15 kV and 11 kV. For single-core cables the voltage value is 15 kV and for 'SL' type and screened type cables, the values of voltage are 20 kV and 12 kV respectively.

4. Tests on Oil-Filled and Gas-Filled Cables. On the oil-filled and gas-filled cables *i.e.*, pressure cables, instead of applying 3-4 times the normal operating voltage for a certain duration, impulse tests are conducted. This is done so that no unnecessary limitations on the dielectric stresses are imposed. Even the type approval tests are different from the tests already discussed.

The tests, which are normally conducted on the pressure-cables are:

(*a*) **Load Cycle Tests.** A test loop formed out of the cable and each type of accessory, is subjected to 20-load cycles to a minimum conductor temperature 5°C in excess of the design value. The temperature limit is increased in certain cables. The test voltage is 1.5 times the normal operating voltage. The cable is to be tested at a stipulated minimum internal pressure.

(*b*) **Thermal Stability Test.** This test is necessary for 132-kV cables only. After performing load cycle tests the cable is energized at 1.5 times the normal operating voltage and the load current is adjusted to give a maximum temperature 5°C in excess of the design value. The load current is maintained at this value for a period of 5–7 hours with the provision that all the other conditions remain unchanged. If the cable can withstand the above, it is thermally stable. For 275 kV cables 1.33 times the normal operating voltage is applied. This test is conducted in order to ascertain the thermal stability of the cable under operating conditions.

(*c*) **Impulse Test.** In this test, a loop is formed out of the cable and each type of accessory is subjected to 10 positive and 10 negative impulses at the voltage values given below

Operating Voltage in kV	Impulse Test Peak Operating Voltage Ratio	Impuls Test Voltage in kV
33	194	7.2
66	342	6.4
132	640	6.0
275	1,050	4.7

(*d*) **Dielectric Thermal Resistance Test.** In this test the thermal resistance of the cable is measured.

(*e*) **Cold Power Factor/Voltage Test.** The power factor of a 100 m length of cable is measured at 0.5, 1.0, 1.5 and 2 times normal operating voltage with the cable at the stipulated minimum internal pressure. The measured values should not exceed the guaranteed values.

(*f*) **Mechanical Test.** This test is performed for measuring mechanical strength of the metallic reinforcement. In this test, a cable sample is subjected to twice the maximum specified internal pressure for a period of one week.

(*g*) **Binding Test.** In this test the cable is subjected to three binding cycles around a drum of diameter 24 times the diameter of the pressure retaining sheath. The cable is then made to withstand the routine voltage test carried out on all production lengths of cable.

Exercises

1. Describe general construction of the cable. What are the necessary requirements of a cable ?
[Rajasthan Technical Univ. Transmission & Distribution of Electrical Power 2009]
2. Discuss the qualities the dielectrics used for cable insulation must have. Also name three dielectrics with their properties used in the manufacture of underground cables.
[Jodhpur Univ. Power Systems–I 1978]
3. Explain various insulating materials used for cables.
[M.D. Univ. Power System-I, Dec. 2010]
4. What should be the desirable characteristics of insulating materials used in cables ? Discuss the classification of cables according to voltage levels.
[G.B. Technical Univ. Elements of Power System 2011-12]
5. State the classification of cables (according to voltage) and discuss their general construction.
[Pb. Technical Univ. Power System (T&D), Dec. 2007]
6. Discuss in brief about the types of cables.
[U.P. Technical Univ. Elements of Power System 2009-10]
7. What are types of *hv* and *lv* cables ? Explain.
[M.D. Univ. Power System-I, May 2008]
8. Explain the construction of H-type cable. What are limitations of solid type cables ?
[M.D. Univ. Power System-I, May 2009]

9. Mention criterion of the cable classification alongwith justification. Draw the extra high tension cable schematic diagram and explain the various terminology.
[U.P. Technical Univ. Elements of Power System 2005-06]

10. Describe with a neat sketch the construction of a 3-core belted type cable. Discuss the limitations of such a cable.
[U.P. Technical Univ. Elements of Power System 2007-08]

11. Explain different types of insulating materials for cables. Explain oil-filled cables. **[M.D. Univ. Power System-I, Dec. 2008]**

12. Derive an expression for the insulation resistance of a single-core metal sheathed cable.
[A.M.I.E. Sec. B. Power Systems Summer 1994]

13. Show that insulation resistance of a cable is inversely proportional to its length.
[Pb. Technical Univ. Power System-I (T&D), May 2010]

14. Find expression for the capacitance of a cable per km length.
[Pb. Univ. Dec. 1982; Rajasthan Technical Univ. Transmission & Distribution of Electrical Power 2009]

15. Derive a formula for the electric stress in a single core cable. Where is maximum stress ? Where is it minimum ?
[Transmission & Distribution of Electrical Power, Rajasthan Univ. 2005; Rajasthan Technical Univ. 2008, 2009]

16. Show that the most economical size of conductor in a single core cable is obtained when radius of cable sheath (R) equals *er* where *e* is the base of natural log and *r* is the radius of conductor. Explain dielectric loss and heating of a cable.
[U.P. Technical Univ. Elements of Power System 2006-07]

17. Derive relation between conductor radius '*r*' and inner radius of sheath R for stable and satisfactory working of a cable. Discuss the effect of *r*/R on the stability of the cable.
[U.P. Technical Univ. Elements of Power System 2006-07]

18. What is meant by grading of cables ? Explain why and how the grading of cable is done.
[Rajasthan Univ. T&D of Electrical Power 1999, 2004, 2007]

19. Explain capacitance grading method used in cables.
[Pb. Technical Univ. Power System-I (T&D), Dec. 2008]

20. What is meant by capacitance grading of a cable ? Derive expressions for capacitance of and maximum potential gradients in two dielectrics of a graded cable in terms of dielectric constants and radius of core and overall radius.
[M.D. Univ. Transmission & Distribution, May 2011]

21. Why is capacitance grading required in cables ? How is it done ?
[U.P. Technical Univ. Elements of Power System 2009-10]

22. With appropriate derivations compare the performance of cables using (*i*) capacitance grading (*ii*) intersheath grading.
[U.P. Technical Univ. Elements of Power System 2005-06]

23. Explain intersheath grading method used in cables.
[Pb. Technical Univ. Power System-I (T&D), May 2009]

24. Prove that the ratio of gradient with and without intersheath will be $\frac{2}{1+\alpha}$, when there is only one larger, where $\alpha = \frac{r_1}{r} = \frac{R_1}{r_1}$. The core radius is *r* and outer sheath radius is R. Intersheath is at r_1 distance from the centre.
[U.P. Technical Univ. Elements of Power System 2007-08]

25. What is the purpose of using intersheaths in a cable ? Show that in a cable using two intersheaths the maximum stress in the (homogeneous) dielectric is reduced to $\frac{3}{1+\alpha+\alpha^2}$ of maximum stress without intersheaths where $\alpha = \frac{d_1}{d} = \frac{d_2}{d_1} = \frac{D}{d_2}$ (D and *d* are the external sheath diameter and core diameter and d_1 and d_2 are the internal diameters of intersheath 1 and 2, respectively.
[U.P. Technical Univ. Elements of Power System 2008-09]

26. Explain, how you can find the capacitance of a three core cable. **[U.P. Technical Univ. Elements of Power System 2002-03]**

27. Describe an experiment to determine the capacitance of a belted cable.

28. What are the various methods of laying underground cables ? Explain.
[Pb. Technical Univ. Power System-I (T&D), Dec. 2009]

29. Explain direct laying method of laying of underground cables.
[Pb. Technical Univ. Power System-I (T&D), May 2009]

30. Explain the dielectric power factor.
[U.P. Technical Univ Elements of Power System 2004-05]

31. Derive the expression for the thermal resistance of a single-core cable and a 3-core cable.
[Rajasthan Univ. T&D of Electrical Power, 2003]

32. Write short notes on the following :
(*i*) General construction of underground cable.
(*ii*) Classification of underground cables. **[M.D. Univ. Dec. 2007]**
(*iii*) PILCDSTA cable.
(*iv*) Oil-filled cables. **[M.D. Univ. Dec. 2009; Dec. 2010]**
(*v*) Gas pressure filled cables. **[Rajasthan Univ. 2003]**
(*vi*) Grading of cables. **[Rajasthan Univ. 2005; G.B. Technical Univ. 2011-12]**
(*vii*) Laying of underground cables.
(*viii*) Thermal rating of cables. **[M.D. Univ. Dec. 2007]**

Short Answer Type Questions With Answers

Q. 1. For what purpose metallic sheath is provided in underground cables ?
Ans. The metallic sheath is provided around the insulation to protect it against ingress of moisture, gas or other damaging liquids (acid or alkalies) from the soil and atmosphere.

Q. 2. Why is a layer of bedding provided over the metallic sheath in underground cables ?
Ans. A layer of bedding is provided over the metallic sheath for the protection of metallic sheath against corrosion and from mechanical injury.

Q. 3. Why are copper conductors not used in power cables ?
Ans. Copper conductors are not used in power cables because of scarcity of copper.

Q. 4. What are the main requirements of insulating materials used for underground cables ?
Ans. The main requirements of the insulating materials used for underground cables are high insulation resistivity, high dielectric strength, good mechanical properties, non-hygroscopic, non-inflammable, immune to attacks by acids and alkalies.

Q. 5. Why VIR cables preferred over paper cables for smaller installations ? **[Pb. Technical Univ. Dec. 2008]**
Ans. VIR cables are preferred over paper cables for smaller installations as they are smaller in size, stronger, durable and economical.

Q. 6. Why single-core cables are usually not provided with armouring ?
Ans. Single-core cables are usually not provided with armouring in order to avoid excessive loss in the armour.

Q. 7. Give classification of underground cables.
[Pb. Technical Univ. Dec. 2009]

Ans. According to voltage, the cables may be classified as:

(*i*) Low voltage (LT) cables for operating voltage up to 1,000 V.

(*ii*) High voltage (HT) cables for operating voltage up to 11,000 V.

(*iii*) Super-tension (ST) cables for operating voltage up to 33,000 V.

(*iv*) Extra high tension (EHT) cables for operating voltage up to 66,000 V.

(*v*) Extra super-voltage cables for operating voltage beyond 1,32,000 V.

A cable may be a single-core or multi-core depending upon the type of service for which it is intended to be used. A multi-core cable may be two-core, three-core or four-core. For 3-phase, 3-wire systems either three single-core cables or three-core cables can be used depending upon the operating voltage and load demand. For 3-phase, 4-wire systems 4-core cables may be used.

Q. 8. List three groups of underground cables used. **[Pb. Technical Univ. Dec. 2006]**

Ans. The underground cables may be grouped into three groups as (*i*) *lt* cables, (*ii*) *ht* cables and (*iii*) EHT cables.

Q. 9. What are voids ? **[Pb. Technical Univ. Dec. 2007]**

Ans. Empty spaces in cables are called voids. The formation of voids in cables causes unequal voltage stresses (which may exceed the safe limits) and also temperature rise due to leakance current.

Q. 10. How void formation can be eliminated in cables ? **[Pb. Technical Univ. Dec. 2007]**

Ans. In EHT and super-voltage cables useful for 132 kV and above, the formation of voids are eliminated by increasing the pressure of compound in it.

Q. 11. Give the limitations of 3-core belted type cable. **[Pb. Technical Univ. Dec. 2007]**

Ans. The belted type construction is not suitable for the cables used for voltages above 22 kV because of development of both the radial and tangential stresses. The tangential stresses act along the layers of insulation. The electric resistance, therefore, dielectric strength of the impregnated paper is much higher across the layers than along the layers. The leakage current on account of tangential stresses along the impregnated paper insulation causes power loss at the centre filling and local heating resulting in breakdown at any moment. Further owing to non-homogenity of dielectric in belted construction, when cables are loaded and unloaded some portions of the dielectric are less stressed whereas some portions are overstressed resulting in formation of vacuous spaces and voids. These vacuous spaces are ionised when voltage is applied and ultimately deteriorate the cable insulation.

Q. 12. Draw the cross-sectional view of SL type cables. **[Pb. Technical Univ. Dec. 2006]**

Ans. Refer to Fig. 8.5.

Q. 13. On what factors does the insulation resistance of a cable depend ?

Ans. The insulation resistance of a cable depends upon core radius, insulation thickness, length of cable and resistivity of insulating material.

Q. 14. How does the insulation resistance of a cable vary with its length ?

Ans. Insulation resistance of a cable varies inversely as the length of the cable.

Q. 15. Write the expression of insulation resistance of a single-core cable. **[Pb. Technical Univ. Dec. 2006]**

Ans. Insulation resistance of a single-core cable is given by expression

$$R_{ins} = \frac{\rho}{2\pi l} \log_e \frac{r_2}{r_1}$$

where ρ is the resistivity of the insulating material, l is the length of cables and r_1 and r_2 are radii of cable conductor and internal sheath respectively.

Q. 16. On what factors does the capacitance of a single core depend ?

Ans. The capacitance of a single core depends upon its length, relative permittivity of its dielectric and the ratio of core diameter to sheath inner diameter.

Q. 17. How does electric stress vary in the dielectric of a cable ?

Ans. The electric stress in the dielectric of a cable varies inversely as the distance from the centre of the cable and hence minimum at the sheath and maximum at the conductor surface.

Q. 18. For what value of R/*r* the minimum value of stress is obtained in a cable ?

Ans. For value of R/*r* = *e i.e.*, 2.71828 the minimum value of stress is obtained in a cable.

Q. 19. What do you understand by the term grading of cables ? **[Pb. Technical Univ. May 2009]**

Ans. The process of achieving uniformity in dielectric stress is known as grading of cables.

Q. 20. What is meant by capacitance grading of cables ? **[Pb. Technical Univ. May 2010]**

Ans. In the capacitance grading of cables, the uniformity in dielectric stress is achieved by using various layers of different dielectrics in such a manner that the permittivity, $\in_r$ of any layer is inversely proportional to its radius of distance from the centre *i.e.*,

$$\in_r \alpha \frac{1}{x} \text{ or } \in_r x = \text{constant or } g \,\alpha\, \frac{Q}{2\pi \in_0} \text{ which in constant.}$$

Q. 21. What is the modern practice adopted to avoid grading of cables ?

Ans. The modern practice adopted to avoid grading of cables is to employ oil-filled or gas-pressure cables.

Q. 22. Name the sources of heat generation in cables. **[Pb. Technical Univ. May 2008]**

Ans. Within the cables the sources of heat generation are

(*i*) I^2R losses in the conductor.

(*ii*) Dielectric losses in the cable insulation,

(*iii*) Losses in the metallic sheath and armourings.

Q. 23. Why loss angle of a cable should be very small ?

Ans. As the dielectric losses occurring in a cable are very small, the value of loss angle δ is small.

Q. 24. How does dielectric loss vary with the change in voltage, frequency of supply and capacitance of cable ?

Ans. The dielectric loss in a cable varies in proportion with the capacitance of a cable and frequency of supply and varies as the square of the voltage.

Q. 25. Mention important factors to be considered while selecting a cable. **[Pb. Technical Univ. Dec. 2008]**

Ans. The factors to be considered in selection of a cable are continuous current rating, voltage grade, permissible voltage drop and short-circuit current rating (in case of *ht* cables only) and operating cost.

Q. 26. What is the cause of underground cable failure ?

[Pb. Technical Univ. December 2009]

Ans. 1. The most common point of failure is at the cable sealing box mostly due to poor workmanship of the cable jointer when the end was sealed. Damage caused to the insulation on account of improper handling at the time of laying will also ultimately result in a failure.

2. Another very common cause is the mechanical puncturing of the lead sheathing of a cable, such as by a crowbar, especially in industrial installations where excavation and building operations are carried on in areas having several underground cables.

3. Although the lead sheathing of underground cables has a very high degree of resistance against corrosion, but inspite of it certain soils are chemically active and may cause severe pitting and corrosion.

4. The cable may also get damaged due to vibration fatigue or overheating.

5. Where a cable is laid vertically up posts, or walls, the impregnating oil is likely to leak from the cable box at the lower level due to the hydrostatic pressure.

Problems

1. A single-core cable has the core diameter of 2.5 cm, thickness of insulation as 1.25 cm. Calculate the insulation resistance per km assuming the resistivity of insulation as 1.5×10^{14} Ω-cm. Derive any formula used.

[Jodhpur Univ. Power System-I 1978] [**Ans.** 495 MΩ]

2. Calculate the insulation resistance for a 5 km length of a 1-core cable. Specific resistance of insualtion (impregnated paper) is 5×10^{14} Ω-cm, insulation thickness is 1 cm and radius of conductor is 2.5 cm.

[U.P. Technical Univ. Elements of Power System 2004-05]

[**Ans.** 53.55 MΩ]

3. Three single-core cables are used for a 132 kV three-phase underground line. The core diameter of the cable is 2.5 cm and the insulation thickness is 4 cm. Calculate the maximum and minimum electrical stress in the dielectric. Derive formula used if any. **[A.M.I.E. Sec. B. Power Systems Summer 1999]**

[**Ans.** 4.248×10^6 V/m; 1.01×10^6 V/m]

4. A single core lead-sheathed cable has core diameter of 2.5 cm and the sheath diameter is 5.3 cm. The oil impregnated paper having a relative permittivity of 3.2 has been used as insulation between the core and sheath. Determine the maximum and minimum voltage gradients in the dielectric when the voltage applied between the conductor and the sheath is 75 kV (rms).

[Allahabad Univ. Power System-I, 1976 Supp.]

[**Ans.** 83.3 kV/cm; 39.2 kV/cm]

5. Calculate the charging current taken by an oil-filled cable of 7 km length having a core diameter 3.5 cm and inside sheath diameter of 7.8 cm. The cable is used as one phase of a 3-phase, 66 kV, 50 Hz system. Permittivity of the insulating material is 2.8. [**Ans.** 17.4 A]

6. Calculate the maximum field strength of a concentric single-core cable, operating at 33 kV, having an insulation of 1.0 cm and a conductor radius of 0.4 cm. For the same maximum potential gradient, find the most economical size of the conductor.

[U.P.S.C. I.E.S. Generation, Transmission and Distribution of Power 1972]

[**Ans.** 66 kV; 1 cm diameter]

7. Find the overall diameter of single-core cable and its most economical core diameter when working on 220 kV, 3-phase system. The maximum permissible stress in the dielectric is not to exceed 250 kV cm.

[Agra Univ. Transmission & Distribution of Elec. Power 1979 Supp; M.D. Univ. Power System-I, Dec. 2007]

[**Ans.** 3.9 cm 1.44 cm]

8. Determine the 'economic' overall diameter of a-1-core cable metal sheathed for an operating voltage of 85 kV, assuming the dielectric strength of insulation to be 65 kV/cm. Derive any formula used. [**Ans.** 7.109 cm]

9. A 110 kV single-core metal sheathed cable is to be graded by means of a metallic intersheath. Find the diameter of the intersheath and the voltage at which it must be maintained in order to obtain the minimum overall diameter D. The insulating material can be worked up to 55 kV/cm.

[Allahabad Univ. Power System-I, 1976]

[**Ans.** 5.66 cm; 69.5 kV]

10. A single-core lead sheathed cable has a conductor of 1 cm diameter and two layers of different insulating materials each 1 cm thick. The relative permittivities are 3.0 (inner) and 2.5 (outer). Calculate the potential gradient at the surface of the conductor and at other points. System voltage is 66 kV.

[Pb. Univ. Transmission and Distribution of Electrical Energy April 1981]

[**Ans.** 63 kV/cm; 25.2 kV/cm; 15.12 kV/cm]

11. A single-core lead covered cable is designed for 66 kV to earth. Its conductor radius is 1.0 cm and its 3 insulating materials A, B and C have permittivities of 5, 4 and 3 respectively with corresponding maximum safe working stress of 38 kV/cm (rms), 26 kV/cm and 20 kV/cm respectively. Find the minimum diameter of the lead sheathed cable. **[Agra Univ. Transmission and Distribution of Electrical Power 1976]**

[**Ans.** 8.3 cm]

12. Estimate the charging current drawn by a cable with 3 cores and protected by a metal sheath when switched on to 11 kV, 3-phase, 50 Hz supply. The capacitance measured between two cores with the third connected to the sheath is 3.7 μF.

[Pb. Univ. Elec. Power II, Nov., 1974]

[**Ans.** 14.78 A]

13. A kilometre of a 3-core, 3-phase, metal-sheathed cable gave the following results on a test for capacitance (*i*) Capacitance between bunched conductors and sheath is 1.0 μF, (*ii*) Capacitance between two conductors bunched with the sheath and third conductor 0.6 μF. With the sheath insulated find the capacitance (*a*) between any two cores (*b*) between any two bunched conductors and the third conductor.

[A.P.S. Univ. Elec. Power Generation, Transmission and Distribution 1977]

[**Ans.** (*a*) 0.367 μF (*b*) 0.489 μF]

14. The capacitance of a 3-core belted type cable are measured and found to be as follows:

(*i*) Between cores bunched together and the sheath = 8 μF

(*ii*) Between conductor and the other two connected together to the sheath = 5 μF.

Calculate the capacitance to neutral and the total charging kVA when the cable is connected to an 11 kV, 50 Hz, 3-phase system. [**Ans.** $\frac{37}{6}$ μF; 234.4 KVAR]

❑❑❑

CHAPTER 9

DC Distribution

INSIDE THIS CHAPTER

9.1. INTRODUCTION

The conductor system, by means of which electrical energy is conveyed from bulk power source or sources (generating stations or major substations supplied over transmission lines) to the consumers is known as *distribution system.* The distribution system may be divided into two systems known as *high voltage (or primary) distribution* and *low volage (or secondary) distribution.*

Primary Distribution System. From generating stations the electrical power is usually transmitted to various substations through extra high tension transmission lines at voltage from 33 to 765 kV. At these substations this voltage is stepped down to 11, 6.6 or 3.3 kV (usually to 11 kV) and power at this voltage is conveyed to different substations for distribution and to the bulk supply consumers. Such a system is known as *high voltage or primary distribution system.* The voltage employed for primary distribution system depends upon the amount of power to be conveyed and the distances of the substations required to be fed*.

Secondary Distribution System. At distribution substations the voltage is stepped down to 415 volts. From these substations various low voltage (415 volts between phases and 240 volts between phase and neutral) distributors radiate out and feed the consumers. This system of distribution is known as *low voltage or secondary distribution system.*

The distribution system may be subdivided into feeders; distributors and service mains, as already explained in Art. 1.2.

9.1.1. Design Considerations

The distribution system should be designed so as to be economical but conforming to the Indian Electricity Rules. Thus it is essential to design each distributor with the minimum volume of the conductor material consistent with the voltage regulation needs.

Both voltage drop and current rating, in addition to overall economy, are the important considerations in the design of distribution systems. The cross-sectional area of a feeder** is determined on the basis of current to be carried and for overall economy (the minimum cost of conductor material and power losses). The consideration of voltage drop is not important in the design of feeders, because no consumer is tapped off from it and the receiving-end voltage can be raised to the desired value and kept within permissible limits.

On contrary, the voltage regulation is very important in the design of distributor. It is because of the statutory regulation giving the voltage variations permissible at the consumer's terminals (±6% of declared or rated value). The size of the distributor should be such that voltage at the consumer's terminals remains within the permissible limits.

According to IE rules the declared*** consumer voltage is 415/240 V (*i.e.*, 415 V between phases and 240 V between phase and neutral). Thus a consumer near a distribution substation may have a voltage as high as 440/254 V during light load hours while a consumer at the farthest end from the distribution substation may have a voltage as low as 400/230 V.

* The primary distribution voltage is usually 11 kV but in case the load densities are very high (say of the order of 5 MVA/km^2), as in cities like Mumbai, Kolkata, Chennai, Delhi, Kanpur, Ghaziabad, Noida, Faridabad, Ludhiana the primary distribution voltages may be higher (33 or 66 kV) even.

** A line carrying power from secondary substation to the distribution substation or a primary distribution line.

*** Declared voltage means the voltage the supplier guarantees to maintain at the consumer's terminals.

9.1.2. Determination of Size of Conductor For Feeders and Distributors

In practice, ACSR conductors are universally employed for distribution systems (feeders as well as distributors). However, in case of distributors all aluminium conductors (AAC) can also be used provided the spans are small.

The conductor size for a *feeder* is mainly governed by the current carrying capacity and overall economy. The current carrying capacity of a conductor depends on the conductor losses and surroundings. The current carrying capacity is usually determined for a maximum operating temperature of 75°C (current carrying capacity for ACSR conductors of different sizes are available in IS : 398–1976). After determining the size of the conductor on the basis of current carrying capacity it should be checked that voltage drop in the feeder is not out of the range of the regulating equipment. If the voltage drop comes out to be too high, the conductor size of the feeder is increased to the next higher standard value and so on till the voltage drop in the feeder comes down to reasonable value. This value of conductor size so determined now be checked for overall economy, as discussed in Art. 1.14.

The main consideration for the determination of size of conductor for a distributor, as already mentioned in Art. 9.1.1, is the permissible voltage drop in the line. Other considerations to be made are current carrying capacity, tensile strength of the conductor and overall economy. An initial guess of conductor size is made from the considerations of current carrying capacity. The next step in the design of a distributor is the determination of voltage drop*. The actual value of voltage drop is checked with the permissible value. If the calculated voltage drop is found more than permissible, the conductor size of the distributor is then increased to the next higher standard value and voltage drop calculation is made again. The process is continued till the voltage drop in the distributor is within limits. Sometimes change in layout or increase in the number of distributors may be required so as to bring voltage drop within permissible limits. After the conductor size has been selected on the basis of current carrying capacity and permissible voltage drop, it is ensured that the tensile strength of the selected conductor size is sufficient to withstand mechanical forces acting upon it owing to its own weight, wind and other local atmospheric conditions. In case the conductor size determined on the basis of current carrying capacity and permissible voltage drop is found mechanically weak then either the span is reduced or conductor size of the distributor is increased to the next higher standard value, whichever is practical and economical.

9.2. CLASSIFICATION OF DISTRIBUTION SYSTEMS

Distribution systems may be classified in various ways :

According to current the distribution systems may be classified as (*i*) ac distribution and (*ii*) dc distribution. As already discussed in chapter 1, ac system is universally employed for distribution of electrical power.

According to the character of service, the distribution systems may be classified as (*i*) general light and power (*ii*) industrial power (*iii*) railway (*iv*) street lighting etc.

According to the type of construction the distribution systems may be classified as (*i*) overhead distribution system and (*ii*) underground distribution system. As already discussed in chapter 1, due to low cost overhead type system is usually employed. Underground distribution system is used where overhead system becomes impracticable.

According to number of wires the distribution systems may be classified as (*i*) two-wire (*ii*) three-wire and (*iii*) four-wire distribution systems. In case of dc supply 3-wire distribution system is usually employed owing to its advantages over two-wire distribution system given in Art. 9.10. In case of ac supply 3-phase 3-wire system is employed for balanced loads such as power loads, 3-phase 4-wire system is employed for unbalanced loads such as light and power loads combined and single phase two-wire system is used for lighting and small power appliances.

According to the scheme of connections the distribution systems may be classified as (*i*) radial (*ii*) parallel (*iii*) ring and (*iv*) interconnected distribution systems, as illustrated in Fig. 9.1.

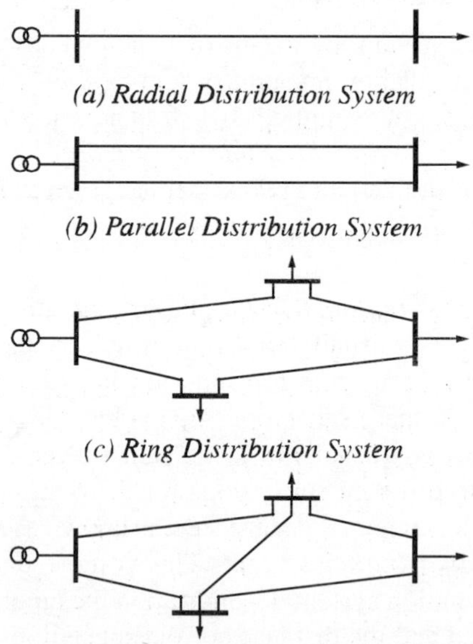

(*a*) *Radial Distribution System*

(*b*) *Parallel Distribution System*

(*c*) *Ring Distribution System*

(*d*) *Interconnected Distribution System*

Fig. 9.1. ***Arrangements of Feeders and Distributors***

The *radial system* is the simplest and lowest in first cost but has poorest service reliability. The *parallel system* provides improved reliability of supply, particularly if the circuits follow different routes. The *ring system* provides alternative supplies to a number of scattered substations. The interconnected network is a common development of the simple ring system.

9.3. TWO-WIRE DC DISTRIBUTION SYSTEM

In dc system the electrical energy may be fed and distributed either by two-wire or by three-wire system.

* In case of ac supply, it is necessary to assume suitable values of power factors (as unity for domestic and commercial loads and 0.8 lagging for industrial and irrigation loads) in calculation of voltage drops along the distributor.

A 2-wire dc distribution system from generating station or substation to the consumer's terminals is shown in Fig. 9.2 (*a*). Untapped feeders run to bus-bars in suitable feeding points in the distribution area. Distributor cables are connected to the bus-bars through fuses or links and run through the streets past the premises requiring supply. Each separate consumer is fed from the distributor by a service cable tapped on to the distributor at the nearest convenient point.

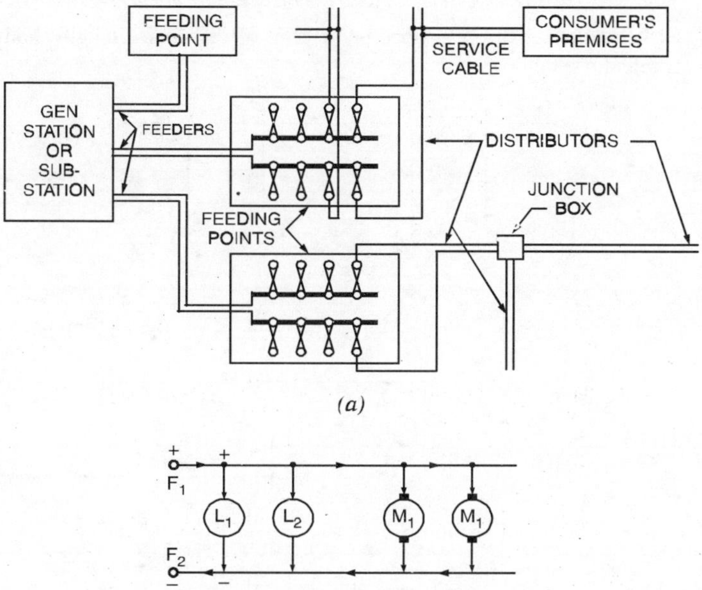

Fig. 9.2. ***Two-Wire DC Distribution System***

It consists of two wires, one outgoing, also known as the *positive wire,* and another returning one, known as *negative wire.* In this system lamps, motors and other electrical appliances are connected in parallel between the two wires, as shown in Fig. 9.2(*b*).

This system is never employed for transmission purpose because of poor efficiency.

9.4. TYPES OF DISTRIBUTORS

The distributors are of the following types :

(*i*) Distributors fed at one end

(*ii*) Distributors fed at both ends.

(*iii*) Distributors fed at the centre.

(*iv*) Ring main distributor.

1. ***Distributors Fed at One End.*** In this type of feeding, the distributor is connected to supply mains at one end and loads are tapped at different points along the length of the distributor, as shown in Fig. 9.4. In this type of distributor current in the sections away from the feeding point and voltage across the loads away from the feeding point goes on decreasing. The minimum voltage occurs on the farthest load point.

If fault occurs in any section of distributor, the whole distributor is required to be disconnected from the supply mains and thus supply continuity is disturbed.

2. ***Distributors Fed From Both Ends.*** In this type of feeding, the distributor is connected to supply mains at both ends, as shown in Fig. 9.11. The voltage at both feeding points may be different or equal. In this type of distributor, load voltage first goes on decreasing, reaches the minimum value, then starts increasing and reaches the maximum value when we reach the other feeding point while going from one load point to another load point. The point of minimum voltage is never fixed. It always shifts with the variation of load on the different sections of the distributor.

Advantages. (*i*) In case of fault in any one feeder feeding the distributor, the continuity of supply is maintained by feeding it from other end.

(*ii*) If any section of the distributor is isolated in case of fault, the continuity of supply is maintained to the remaining sections.

(*iii*) Since cross section required for doubly-fed distributors is much less as compared to singly fed one, hence it is economical.

3. ***Distributors Fed at The Centre.*** In this type of feeding, the centre of the distributor is connected to the supply mains, as shown in Fig. 9.3. In fact this type of distributor is equivalent to two singly-fed distributors, each distributor being of one half of its total length and having common feeding point.

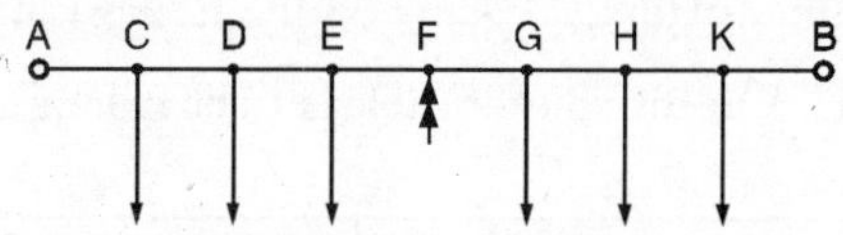

Fig. 9.3. ***Distributor Fed At The Centre***

4. ***Ring Main Distributor.*** When the two ends of a distributor fed at equal voltages brought together, then such distributor is known as *ring main* (Fig. 9.20). It has got all the advantages of doubly-fed distributor, while the feeder required is only one. Ring main may be fed at one or more points.

9.5. TYPES OF LOADING

There are various types of loading such as :

(*i*) Concentrated loading

(*ii*) Uniform loading, and

(*iii*) Uniform loading in addition to concentrated loading.

9.6. DC DISTRIBUTOR FED AT ONE END

9.6.1. Concentrated Loading

Consider a distributor AB with concentrated loads of I_1, I_2, I_3, I_4 and I_5 at points C, D, E, F, and G respectively and fed at one point (say A), as shown in Fig. 9.4.

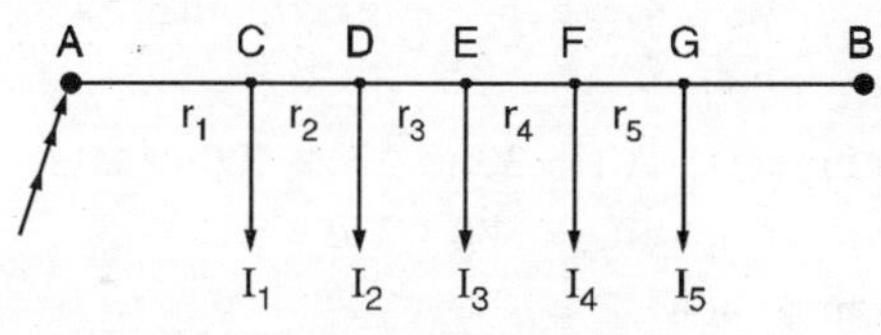

Fig. 9.4. ***Distributor Fed At One End With Concentrated Loading***

In this figure only for simplicity one wire is shown instead of two but resistances marked on diagram are for both wires for respective sections.

Total current fed from feeding point $A = I_1 + I_2 + I_3 + I_4 + I_5$

The currents flowing through different sections are given below :

Sections	Current flowing
AC	$I_1 + I_2 + I_3 + I_4 + I_5$
CD	$I_2 + I_3 + I_4 + I_5$
DE	$I_3 + I_4 + I_5$
EF	$I_4 + I_5$
FG	I_5

...(9.1)

$$
\begin{aligned}
&\text{Voltage drop in section AC} = (I_1 + I_2 + I_3 + I_4 + I_5)r_1\\
&\text{Voltage drop in section CD} = (I_2 + I_3 + I_4 + I_5)r_2\\
&\text{Voltage drop in section DE} = (I_3 + I_4 + I_5)r_3\\
&\text{Voltage drop in section EF} = (I_4 + I_5)r_4\\
&\text{Voltage drop in section FG} = I_5 r_5
\end{aligned}
\quad ...(9.2)
$$

Total voltage drop in the distributor

$$= (I_1 + I_2 + I_3 + I_4 + I_5)r_1 + (I_2 + I_3 + I_4 + I_5)r_2 + (I_3 + I_4 + I_5)\, r_3 + (I_4 + I_5)\, r_4 + I_5 r_5$$

From Eqs. (9.1) and (9.2) it can be seen that current in the sections away from the feeding point and voltage across the load points away from the feeding point goes on decreasing and the minimum voltage occurs on the farthest load point G.

Example 9.1. A dc distributor cable is 1,000 m long and is loaded as under :

Distance from the feeding point (metres) "A"	250	750	1,000
Load in amperes	100	200	300

The resistance of each conductor is 0.025 Ω per 1,000 m (km). Find the voltage at each load point if the voltage at the feeding point A is maintained at 250 V.

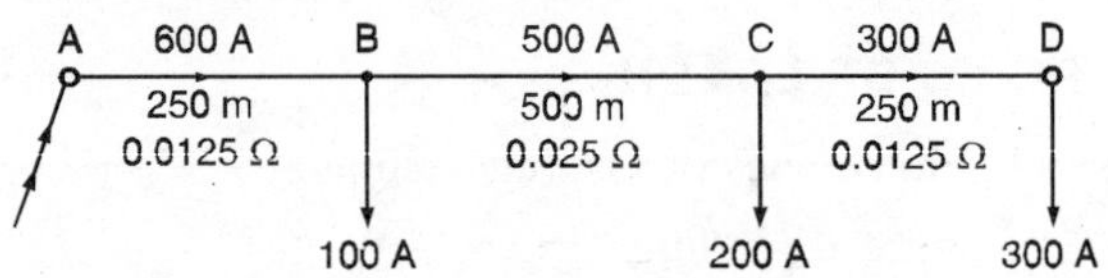

Fig. 9.5

Solution : Resistance of section AB (both wires),

$$R_{AB} = \frac{0.025}{1,000} \times 250 \times 2 = 0.0125\ \Omega$$

Resistance of section BC (both wires),

$$R_{BC} = \frac{0.025}{1,000} \times 500 \times 2 = 0.025\ \Omega$$

Resistance of section CD (both wires),

$$R_{CD} = \frac{0.025}{1,000} \times 250 \times 2 = 0.0125\ \Omega$$

Current in section AB = 300 + 200 + 100 = 600 A

Current in section BC = 300 + 200 = 500 A

Current in section CD = 300 A

The current and resistance distribution is shown on Fig. 9.5.

Voltage at load point B,

V_B = Voltage at feeding point A – voltage drop in section AB

$= V_A - I_{AB}R_{AB} = 250 - 600 \times 0.0125 = 242.5$ V **Ans.**

Voltage at load point C,

$V_C = V_B - I_{BC}R_{BC} = 242.5 - 500 \times 0.025 = 230$ V **Ans.**

Voltage at load point D,

$V_D = V_C - I_{CD}\,R_{CD} = 230 - 300 \times 0.0125 = 226.25$ V **Ans.**

Example 9.2. Find the cross-sectional area of the distributor shown in Fig. 9.6. The distances are given in metres. Take ρ = 1.78 × 10^{-8} Ω-m. The maximum voltage drop is not to exceed 10 V. The conductor is fed from point A.

[Pb. Univ. Transmission and Distribution of Electrical Energy April 1982]

Solution : Let the resistance per 100 m of distributor length (both conductors) be of *r* ohms.

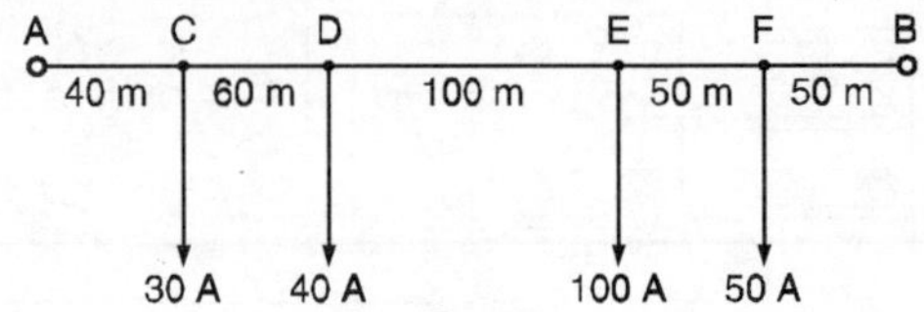

Fig. 9.6

$$\text{Resistance of section AC, } R_{AC} = \frac{40}{100} \times r = 0.4\, r\ \Omega$$

$$\text{Resistance of section CD, } R_{CD} = \frac{60}{100} \times r = 0.6\, r\ \Omega$$

$$\text{Resistance of section DE, } R_{DE} = \frac{100}{100} \times r = r\ \Omega$$

$$\text{Resistance of section EF, } R_{EF} = \frac{50}{100} \times r = 0.5\, r\ \Omega$$

$$\text{Resistance of section FB, } R_{FB} = \frac{50}{100} \times r = 0.5\, r\ \Omega$$

Current distribution in the various sections of the distributor is shown in Fig. 9.7.

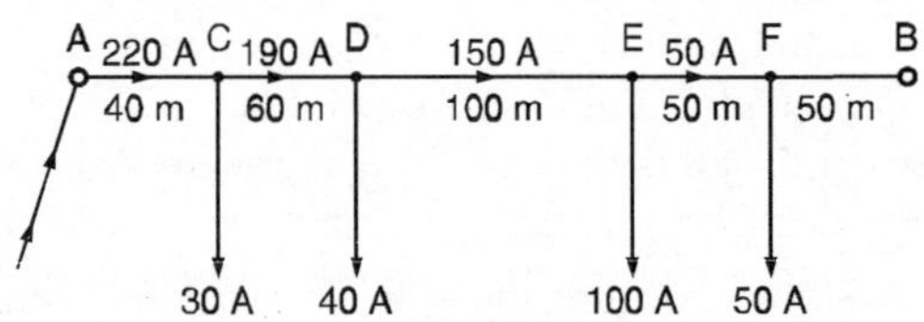

Fig. 9.7

Total voltage drop in the distributor,

$$v = I_{AC}\,R_{AC} + I_{CD}\,R_{CD} + I_{DE}\,R_{DE} + I_{EF}\,R_{EF}$$

$$= 220 \times 0.4r + 190 \times 0.6r + 150 \times r + 50 \times 0.5r = 377r$$

Since maximum permissible voltage drop is 10 V

$\therefore\ 377r = 10$

or $r = \dfrac{10}{377}\ \Omega$

Area of cross section of the distributor,

$$a = \frac{\rho l}{r} = \frac{1.78 \times 10^{-8} \times 2 \times 100}{10/377}$$

$= 0.13421 \times 10^{-3}\ m^2$ or 134.21 mm^2 **Ans.**

Example 9.3. A 2-wire distributor AG, 1200 metres long is fed at 250 V. Loads of 80, 70, 100, 90 and 60 A are tapped from points B, C, D, E and F whose distances from A are 100, 250, 600, 800 and 1,100 metres respectively. If the voltage at G is 210 V, find the resistance per metre of distributor.

[M.D. Univ. Power Systems-I, May 2008]

Solution : Let the resistance per 100 metre of distributor length (both conductors) be of *r* Ω

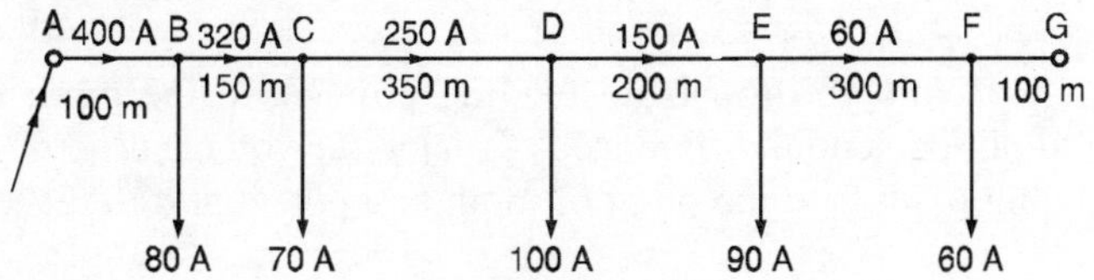

Fig. 9.8

$$\text{Resistance of section AB, } R_{AB} = \frac{100}{100} \times r = r\ \Omega$$

$$\text{Resistance of section BC, } R_{BC} = \frac{150}{100} \times r = 1.5r\ \Omega$$

$$\text{Resistance of section CD, } R_{CD} = \frac{350}{100} \times r = 3.5r\ \Omega$$

$$\text{Resistance of section DE, } R_{DE} = \frac{200}{100} \times r = 2r\ \Omega$$

$$\text{Resistance of section EF, } R_{EF} = \frac{300}{100} \times r = 3r\ \Omega$$

Current distribution in the various sections of the distributor is shown in Fig. 9.7.

Total voltage drop in distributor,

$$v = I_{AB}R_{AB} + I_{BC}R_{BC} + I_{CD}R_{CD} + I_{DE}R_{DE} + I_{EF}R_{EF} + I_{FG}R_{FG}$$
$$= 400r + 320 \times 1.5r + 250 \times 3.5r + 150 \times 2r + 60 \times 3r + 0 \times r$$
$$= 400r + 480r + 875r + 300r + 180r + 0 = 2{,}235r$$

Since voltage drop in feeder, $v = V_A - V_G = 250\ V - 210\ V = 40\ V$

$\therefore\ 2235r = 40$

or $r = \dfrac{40}{2235} = 0.0179$ per 100 m length of distributor

Resistance per metre of distributor

$= \dfrac{0.0179}{100} = 0.000179\ \Omega/m$ or $0.179 \times 10^{-3}\ \Omega/m$ **Ans.**

9.6.2. Uniformly Loaded Distributor

Let the distributor AB of length l, resistance r ohms per unit run of both of the conductors and loaded with i amperes per unit run be fed from feeding point A at V_A volts, as shown in Fig. 9.9(a).

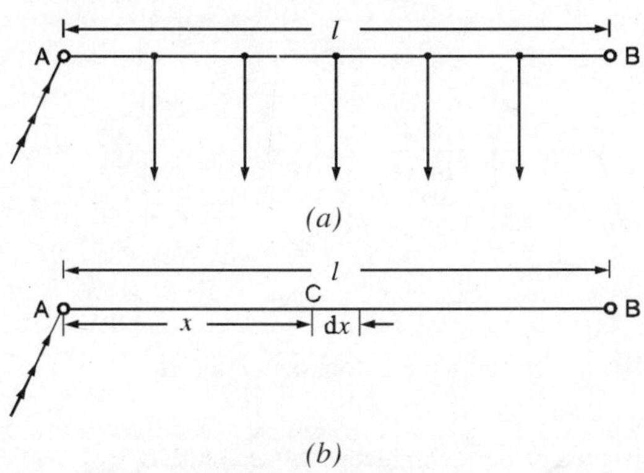

Fig. 9.9. ***Distributor Fed At One End With Uniformly Distributed Loading***

Let us consider point C, x units from the feeding point A and find the voltage drop in section AC.

The current at point C = $il - ix = i(l - x)$

Consider a section of very small length dx near point C. Its resistance will be rdx and therefore voltage drop in this small length dx will be given by

$$dv = i(l - x)rdx = ir(l - x)\,dx$$

Total voltage drop up to point C is given by

$$v = \int_0^x ir(l-x)\,dx = ir\left(lx - \frac{x^2}{2}\right) = irlx - \frac{1}{2}irx^2 \quad ...(9.3)$$

which is the equation of a parabola.

The voltage drop up to point B can be determined by substituting $x = l$ in the above equation

$\therefore$ Voltage drop up to point B

$$= ir\left(l^2 - \frac{l^2}{2}\right) = \frac{irl^2}{2} = \frac{(il)(lr)}{2} = \frac{1}{2}\text{IR} \quad ...(9.4)$$

where $i \times l = I$, total current fed from feeding point A;

$r \times l = R$, total loop resistance of the distributor AB.

From the expression of voltage drop up to point B it is clear that in a uniformly loaded distributor total voltage drop is equal to that caused by the whole of the load assumed concentrated at the middle point.

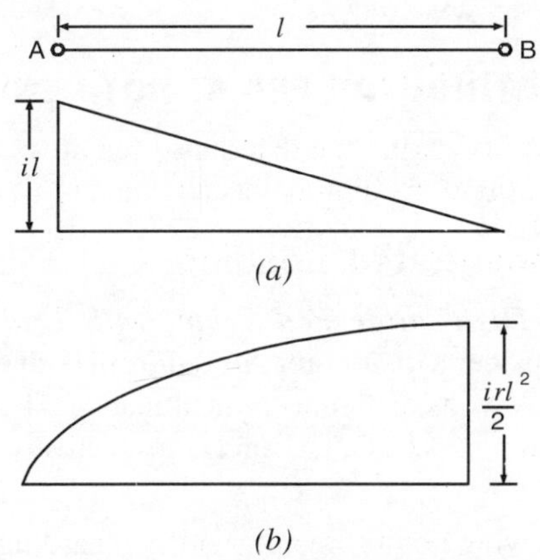

Fig. 9.10

The current loading diagram and voltage drop diagram are shown in Figs. 9.10 (a) and 9.10 (b) respectively.

Example 9.4. A uniformly distributed load on a distributor of length 500 metres is rated at 1 A per metre length. The distributor is fed from one end at 220 V. Determine the voltage drop at a distance of 400 metres from the feeding point. Assume a loop resistance of 2 × 10⁻⁵ Ω/m.

Solution : Current loading, $i = 1$ A/m

Resistance per metre run, $r = 2 \times 10^{-5}\ \Omega$

Length of the distributor, $l = 500$ metres

Distance of the point, at which voltage drop is to be determined, from feeding point A, $x = 400$ metres

Voltage drop at a distance of 400 metres from the feeding point

$$= ir\left(lx - \frac{x^2}{2}\right) = 1 \times 2 \times 10^{-5}\left(500 \times 400 - \frac{400 \times 400}{2}\right)$$

$= 2.4$ V **Ans.**

Example 9.5. A 250 m, 3-wire distributor fed from one end is loaded uniformly at the rate of 1.6 A/metre. The resistance of each conductor is 0.0002 Ω/m. Find the voltage necessary at feed point to maintain 250 V, (*i*) at far end, (*ii*) at the mid point of distributor.

[*M.D. Univ. Power System-I, December 2009*]

Solution : Current load, $i = 1.6$ A/m

Resistance per metre run for both conductors of distributor

$$r = 0.0002 \times 2 = 0.0004\ \Omega/\text{m}$$

Length of feeder, l = 250 m

(*i*) Necessary voltage at feeding point

= Voltage at the far end + voltage drop in feeder up to far end

$$= 250 + \frac{irl^2}{2}$$

$$= 250 + \frac{1.6 \times 0.0004}{2} \times (250)^2$$

= 250 + 20 = 270 V **Ans.**

(*ii*) Necessary voltage at the mid point of distributor (*i.e.*, x = 125 m)

= Voltage at the mid point + voltage drop in feeder up to mid point

$$= 250 + ir\left(lx - \frac{x^2}{2}\right)$$

$$= 250 + 1.6 \times 0.0004\left(250 \times 125 - \frac{125^2}{2}\right)$$

= 250 + 15 = 265 V **Ans.**

9.7. DC DISTRIBUTOR FED AT BOTH ENDS

As mentioned in Art. 9.4 the distributor fed at two points may have voltage at both feeding points (*i*) equal or (*ii*) different.

9.7.1. Concentrated Loading

(a) DC Distributor Fed From Both Ends at Equal Voltages. Consider a distributor AB fed from both ends at equal voltages, say V volts and concentrated loads I_1, I_2, I_3, I_4 and I_5 are supplied at points C, D, E, F and G respectively, as shown in Fig. 9.11.

The following points are noteworthy regarding this type of distributor.

(*i*) As we move away from feeding point A, load voltage goes on decreasing, reaches the minimum value at some load point, say E, and then starts increasing and becomes V volts as we reach the other feeding point B. The point of minimum voltage is never fixed. It always shifts with the variation of load on the different sections of the distributor.

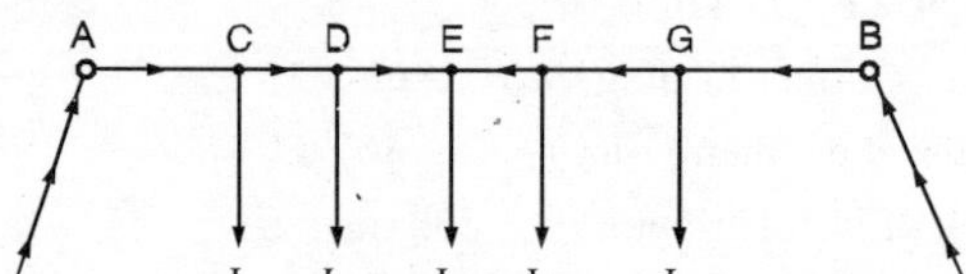

Fig. 9.11. ***Distributor Fed at Both Ends***

(*ii*) All the currents tapped off between feeding point A and point of minimum potential difference are supplied from feeding point A where as the currents tapped off between feeding point B and point of minimum potential are supplied from feeding point B.

(*iii*) The current tapped at the point of minimum potential difference is supplied partly by feeding point A and partly by feeding point B.

(*iv*) The sum of currents fed from feeding points A and B is equal to the total current tapped off from the distributor.

i.e., $$I_A + I_B = I_1 + I_2 + I_3 + I_4 + I_5 \qquad ...(9.5)$$

(*v*) Voltage drop from feeding point A to the point of minimum potential difference is equal to the voltage drop from feeding point B to the point of minimum potential difference.

(b) DC Distributor Fed From Both Ends With Unequal Voltages. Consider a distributor AB fed at unequal voltages, say end A at voltage of V_A volts and end B at voltage of V_B volts.

In order to determine the point of minimum potential, assume that current of I_A amperes is supplied from feeding point A.

Currents in sections AC, CD, DE, EF, FG and GB will be I_A, $(I_A - I_1)$, $(I_A - I_1 - I_2)$, $(I_A - I_1 - I_2 - I_3)$, $(I_A - I_1 - I_2 - I_3 - I_4)$, $(I_A - I_1 - I_2 - I_3 - I_4 - I_5)$ respectively and voltage drop from feeding point A up to feeding point B will be equal to

$$I_AR_{AC} + (I_A - I_1)R_{CD} + (I_A - I_1 - I_2)R_{DE} + (I_A - I_1 - I_2 - I_3)R_{EF} + (I_A - I_1 - I_2 - I_3 - I_4)R_{FG} + (I_A - I_1 - I_2 - I_3 - I_4 - I_5)R_{GB}$$

i.e., $V_A - V_B$

$$= I_AR_{AC} + (I_A - I_1)R_{CD} + (I_A - I_1 - I_2)R_{DE} + (I_A - I_1 - I_2 - I_3)R_{EF} + (I_A - I_1 - I_2 - I_3 - I_4)R_{FG} + (I_A - I_1 - I_2 - I_3 - I_4 - I_5)R_{GB}$$

From the above equation the value of unknown current I_A can be determined, the other quantities being known. In case of both the voltages being same $V_A - V_B$ be put equal to zero.

The load point at which currents are coming from both sides of the distributor will be the point of minimum potential.

Example 9.6. A two-conductor street main AB, 500 metres in length is fed from both ends at 250 volts. Loads of 50 A, 60 A, 40 A and 30 A are tapped at distances of 100 m, 250 m, 350 m and 400 m from A, respectively. If the cross section of the conductors be 1 cm² and specific resistance of the material of the conductors is 1.7 × 10⁻⁸ Ω-m, determine the minimum consumer voltage.

Solution : Resistance per 100 metres of street main both conductors,

$$r = \rho\frac{l}{a} = \frac{1.7 \times 10^{-8} \times 2 \times 100}{1.0 \times 10^{-4}} = 0.034\ \Omega$$

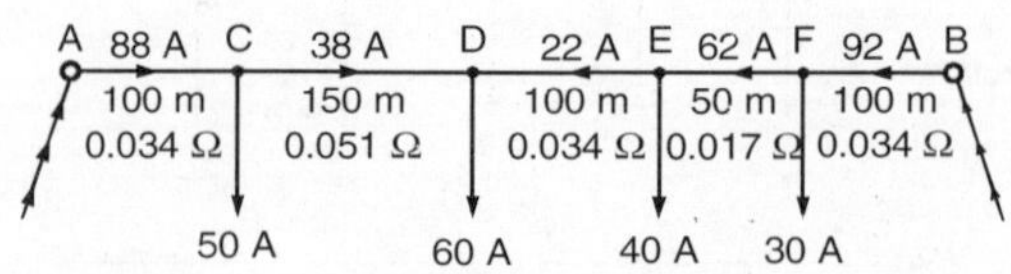

Fig. 9.12

Resistance of both conductors of section AC,

R_{AC} = 0.034 Ω

Resistance of both conductors of section CD,

$$R_{CD} = 0.034 \times \frac{150}{100} = 0.051\ \Omega$$

Resistance of both conductors of section DE,

R_{DE} = 0.034 Ω

Resistance of both conductors of section EF,

$$R_{EF} = 0.034 \times \frac{150}{100} = 0.017\ \Omega$$

Resistance of both conductors of section FB,

$R_{FB} = 0.034\ \Omega$

Let current supplied from feeding point A be I_A amperes then currents in various sections will be as follows :

Current in section AC, $I_{AC} = I_A$

Current in section CD, $I_{CD} = I_A - 50$

Current in section DE, $I_{DE} = I_A - 50 - 60 = I_A - 110$

Current in section EF, $I_{EF} = I_A - 50 - 60 - 40 = I_A - 150$

Current in section FB, $I_{FB} = I_A - 50 - 60 - 40 - 30$

$= I_A - 180$

Since voltage at other feeding end B

= Voltage at feeding end A – voltage drops in sections AC, CD, DE, EF and FB respectively

$$\text{or } V_B = V_A - (I_{AC}R_{AC} + I_{CD}R_{CD} + I_{DE}R_{DE} + I_{EF}R_{EF} + I_{FB}R_{FB})$$

$$\text{or } 250 = 250 - [0.034\, I_A + 0.051\,(I_A - 50) + 0.034\,(I_A - 110) + 0.017\,(I_A - 150) + 0.034\,(I_A - 180)]$$

or $0.17 I_A = 14.96$

or $I_A = 88$ A

Now current in section AC, $I_{AC} = I_A = 88$ A from A to C

Current in section CD, $I_{CD} = I_A - 50 = 38$ A from C to D

Current in section DE, $I_{DE} = I_A - 110$

$= -22$ A from D to E

or 22 A from E to D

Current in section EF, $I_{EF} = I_A - 150 = -62$ A from E to F

or 62 A from F to E.

Current in section FB, $I_{FB} = I_A - 180 = -92$ A from F to B

$= 92$ A from B to F

Since towards load point D the currents are incoming from both sides hence minimum voltage will occur at point D, at a distance of 250 metres from end A.

Minimum consumer voltage,

$V_D = V_A - I_{AC}R_{AC} - I_{CD}R_{CD}$

$= 250 - 88 \times 0.034 - 38 \times 0.051 = 245.07$ volts **Ans.**

Example 9.7. A distributor AB is fed from both ends. At feeding point A the voltage is maintained at 235 V and at B 236 V. The total length of feeder is 200 metres and loads are tapped as under.

20 A at 50 metres from A; 40 A at 75 metres from A ; 25 A at 100 metres from A and 30 amperes at 150 metres from A.

The resistance per 1,000 metres of one conductor is 0.4 Ω. Calculate the current in various sections of the feeder, the minimum voltage and the point at which it occurs in the system.

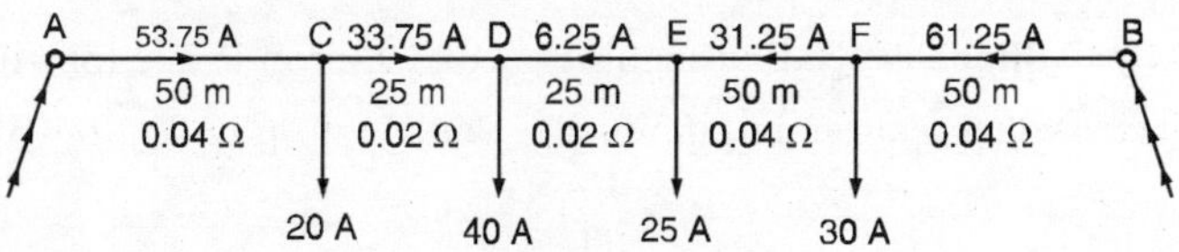

Fig. 9.13

Solution : Let current supplied from feeding point A be I_A then current in various sections will be given as.

Current in section AC, $I_{AC} = I_A$

Current in section CD, $I_{CD} = I_A - 20$

Current in section DE, $I_{DE} = I_A - (20 + 40) = I_A - 60$

Current in section EF, $I_{EF} = I_A - (20 + 40 + 25) = I_A - 85$

Current in section FB, $I_{FB} = I_A - (20 + 40 + 25 + 30) = I_A - 115$

Resistance of both the conductors of section AC,

$$R_{AC} = 2 \times 50 \times \frac{0.4}{1{,}000} = 0.04\ \Omega$$

Similarly $R_{CD} = 0.02\ \Omega$; $R_{DE} = 0.02\ \Omega$; $R_{EF} = 0.04\ \Omega$ and $R_{FB} = 0.04\ \Omega$

Voltage at other end B,

V_B = Voltage at feeding end A – voltage drops in sections AC, CD, DE, EF and FB respectively.

$$\text{or } V_B = V_A - I_{AC}R_{AC} - I_{CD}R_{CD} - I_{DE}R_{DE} - I_{EF}R_{EF} - I_{FB}R_{FB}$$

$$\text{or } 236 = 235 - I_A \times 0.04 - (I_A - 20) \times 0.02 - (I_A - 60) \times 0.02 - (I_A - 85) \times 0.04 - (I_A - 115) \times 0.04$$

$$\text{or } 236 = 235 - 0.16 I_A + 9.6$$

$$\text{or } I_A = \frac{8.6}{0.16} = 53.75 \text{ A}$$

Now Current in section AC, $I_{AC} = I_A = 53.75$ A

Current in section CD, $I_{CD} = I_A - 20 = 33.75$ A

Current in section DE,

$I_{DE} = I_A - 60 = 53.75 - 60$

$= -6.25$ A from D to E $= 6.25$ A from E to D.

Current in section EF,

$I_{EF} = I_A - 85 = 53.75 - 85$

$= -31.25$ A from E to F $= 31.25$ A from F to E

and current in section FB,

$I_{FB} = I_A - 115 = 53.75 - 115$

$= -61.25$ A from F to B $= 61.25$ A from B to F

Current distribution in various sections is shown in Fig. 9.13.

Since towards loading point D the currents are incoming from both sides, *i.e.* from point C, as well as from point E so minimum voltage will occur on loading point D at which a load of 40 amperes is tapped off.

Minimum voltage

= Voltage at point load D

$= V_A - I_{AC}R_{AC} - I_{CD}R_{CD} = 235 - 53.75 \times 0.04 - 33.75 \times 0.02$

$= 232.175$ V **Ans.**

Example 9.8. A dc two line distributor AB 600 metre long is fed at 440 V from substation A and at 430 V from substation B, the loads are :

100 A at C, 150 metres from A

200 A at D, 150 metres from C

250 A at E, 50 metres from D

300 A at F, 100 metres from E

If each line conductor has a resistance of 0.01 Ω per 100 metres, calculate the current supplied from substations A and B and the voltage across each load.

Solution : Let current supplied from feeding end A be I_A then current in sections AC, CD, DE, EF and FB will be I_A, $(I_A - 100)$, $(I_A - 300)$, $(I_A - 550)$ and $(I_A - 850)$ amperes respectively.

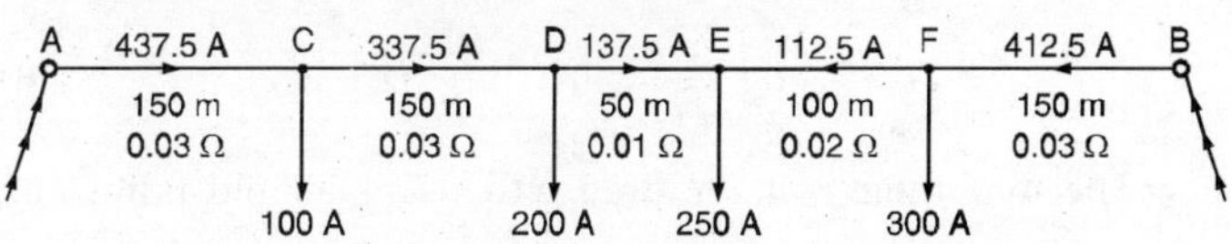

Fig. 9.14

Resistances of the various sections of the distributor (both conductors) are shown in Fig. 9.14.

$$\text{Now} \quad V_B = V_A - I_{AC} R_{AC} - I_{CD} R_{CD} - I_{DE} R_{DE} - I_{EF} R_{EF} - I_{FB} R_{FB}$$

$$\text{or } 430 = 440 - 0.03\, I_A - 0.03\,(I_A - 100) - 0.01\,(I_A - 300) - 0.02\,(I_A - 550) - 0.03\,(I_A - 850)$$

$$\text{or } 0.12 I_A = 52.5$$

$$\text{or } I_A = 437.5 \text{ A}$$

Now currents in sections AC, CD, DE, EF and FB will be 437.5 A from A to C, 337.5 A from C to D, 137.5 A from D to E, 112.5 A from F to E and 412.5 A from B to F respectively.

∴ Current supplied from end A, I_A = 437.5 A **Ans.**

Current supplied from end B, I_B = 412.5 A **Ans.**

Voltage across load point C,

$$V_C = V_A - I_{AC} R_{AC} = 440 - 437.5 \times 0.03 = 426.875 \text{ V} \textbf{ Ans.}$$

Voltage across load point D,

$$V_D = V_C - I_{CD} R_{CD} = 426.875 - 337.5 \times 0.03 = 416.75 \text{ V} \textbf{ Ans.}$$

Voltage across load point E,

$$V_E = V_D - I_{DE} R_{DE} = 416.75 - 137.5 \times 0.01 = 415.375 \text{ V} \textbf{ Ans.}$$

Voltage across load point F,

$$V_F = V_E - I_{EF} R_{EF} = 415.375 - (-112.5) \times 0.02 = 417.625 \text{ V} \textbf{ Ans.}$$

9.7.2. Uniform Loading

(*a*) DC Distributor Fed From Both Ends at Equal Voltages. Let the distributor AB of length *l*, resistance *r* ohms per unit run of both of the conductors and loaded with *i* amperes per unit run be fed from feeding points A and B at equal voltages, say V volts, as shown in Fig. 9.15.

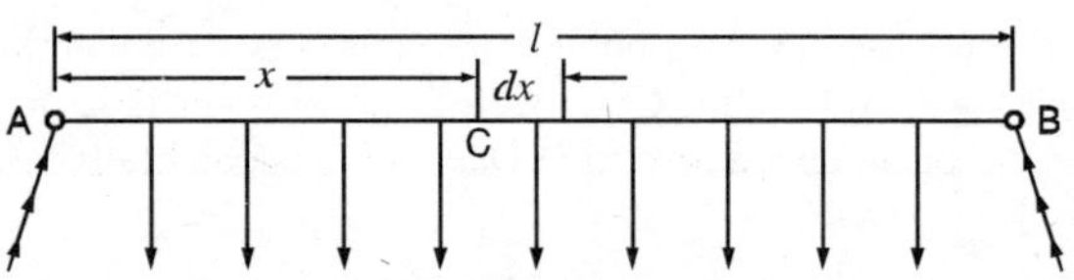

Fig. 9.15. ***Distributor Fed at Both Ends at Equal Voltages With Uniform Distributed Loading***

Let us find the voltage drop in section AC, C being a point at a distance of *x* units from feeding point A.

Total current on conductor AB $= il$

Current supplied from each end $= \dfrac{il}{2}$

(Since feeding points are at the same voltage)

The current at point C $= \dfrac{il}{2} - ix = i\left(\dfrac{l}{2} - x\right)$

Consider a section of very small length *dx* near point C. Its resistance is *rdx*; Hence voltage drop in length *dx* is given by

$$dv = i\left(\frac{l}{2} - x\right) rdx = ir\left(\frac{l}{2} - x\right) dx$$

Voltage drop in section AC

$$= \int dv = \int_0^x ir\left(\frac{l}{2} - x\right) dx = \frac{ir}{2}(lx - x^2) \qquad \text{...(9.6)}$$

The maximum voltage drop will occur at mid point *i.e.*, when $x = \dfrac{l}{2}$

So maximum voltage drop, V_{max}

$$= \frac{ir}{2}\left[l \times \frac{l}{2} - \left(\frac{l}{2}\right)^2\right] = \frac{irl^2}{8} = \frac{il \times lr}{8} = \frac{IR}{8} \qquad \text{...(9.7)}$$

where il = I, total current fed to the distributor and lr = R, total loop resistance of the distributor.

Minimum voltage occurring at mid point

$$= V - \frac{irl^2}{8} = V - \frac{IR}{8} \qquad \text{...(9.8)}$$

Example 9.9. A uniform 2-wire distributor is 50 m long. It is loaded with 4 A/m. Resistance of single wire is 1 Ω/km. Calculate the maximum voltage drop when the distributor is fed from both ends at equal voltages.

Solution : Length of distributor, l = 50 m

Current loading, i = 4 A/m

Resistance per metre run of both of the conductors,

$$r = \frac{2 \times 1}{1{,}000} = 0.002 \ \Omega/\text{m}$$

Maximum voltage drop will occur at mid point and is given by the Eq. (9.7)

Maximum voltage drop

$$= \frac{irl^2}{8} = \frac{4 \times 0.002 \times (50)^2}{8} = 2.5 \text{ V} \textbf{ Ans.}$$

(*b*) DC Distributor Fed From Both Ends at Different Voltages. Let the distributor AB of length *l*, resistance *r* ohms per unit run of both of the conductors and loaded with *i* amperes per unit run be fed from feeding points A and B at voltages V_A and V_B respectively (Fig. 9.15).

Let point, C at a distance of *x* units fron feeding point A be at the minimum voltage, so there will be no current at point C and the current fed from feeding point A will be the current in section AC *i.e.*, *ix*.

Hence voltage drop in section AC $= \dfrac{irx^2}{2}$ volts

Point C is also at a distance of $(l - x)$ units from B, so current fed by feeding point B is $i(l - x)$ and voltage drop in section BC is $ir(l - x)^2/2$ volts

Now voltage at point C

= Voltage at feeding point A – voltage drop in section AC

also = Voltage at feeding point B – voltage drop in section BC

$$\therefore \quad V_A - \frac{irx^2}{2} = V_B - \frac{ir(l - x)^2}{2}$$

$$\text{or } x = \frac{V_A - V_B}{irl} + \frac{l}{2} \qquad \text{...(9.9)}$$

Example 9.10. An 800 metre distributor fed from both ends A and B is loaded uniformly at the rate of 1.2 A per metre run, the resistance of each conductor being 0.05 Ω per km. Determine the minimum voltage and the point where it occurs if feeding points A

and B are maintained at 255 V and 250 V respectively. Find also the current supplied from feeding points A and B.

Solution : Voltage at feeding end A, V_A = 255 V

Voltage at feeding end B, V_B = 250 V

Length of distributor, l = 800 metres

Current loading, i = 1.2 A/metre run

Resistance per metre run of both of the conductors,

$$r = \frac{2 \times 0.05}{1,000} = 0.0001\ \Omega/\text{m}$$

Distance of point of minimum potential difference from feeding point A,

$$x = \frac{V_A - V_B}{irl} + \frac{l}{2} = \frac{255 - 250}{1.2 \times 0.0001 \times 800} + \frac{800}{2}$$

$$= 452.08 \text{ m}$$

Minimum voltage

$$= V_A - \frac{irx^2}{2} = 255 - \frac{1.2 \times 0.0001 \times (452.08)^2}{2}$$

$$= 242.74 \text{ V} \quad \textbf{Ans.}$$

Current fed from feeding point A

$= ix = 1.2 \times 452.08 = 542.5$ A **Ans.**

Current fed from feeding point B

$= i(l - x) = 1.2 \times 347.92 = 417.5$ A **Ans.**

Example 9.11. A 2-wire distributor, 800 metre long is fed from both ends. It is uniformly loaded at the rate of 1.1 A/m run. Calculate the voltage at feeding points A and B if the minimum potential of 230 V occurs at point C at a distance of 500 metres from the end A. Resistance of the distributor is 0.1 Ω/km (go and return). [*West Bengal Univ. of Technology, Power System-I, 2009-10*]

Solution : Length of distributor, l = 800 m

Current loading, i = 1.1 A/m

Resistance of distributor per metre run of both conductors

$$= \frac{0.1}{1,000} = 1 \times 10^{-4}\ \Omega/\text{m}$$

Distance of point of minimum pd from feeding point A,

x = 500 m

Distance of point of minimum potential difference from feeding point B,

$l - x$ = 800 – 500 = 300 m

Voltage at feeding point A,

$V_A = V_C$ + voltage drop in section AC

$$= 230 + \frac{irx^2}{2}$$

$$= 230 + \frac{1.1 \times 1 \times 10^{-4} \times 500^2}{2} = 243.75 \text{ V} \quad \textbf{Ans.}$$

Voltage at feeding point B,

$V_B = V_C$ + Voltage drop in section BC

$$= 230 + \frac{ir(l - x)^2}{2}$$

$$= 230 + \frac{1.1 \times 1 \times 10^{-4} \times 300^2}{2} = 234.95 \text{ V} \quad \textbf{Ans.}$$

Example 9.12. A two-wire dc distributor 500 m long has a uniformly distributed load of 1 A/m. There are concentrated loads of 200 A, 150 A, 50 A and 100 A at 100, 200, 300 and 400 m respectively from the end A. The distributor resistance is 0.1 Ω/km (go and return). Determine (*a*) the value of the minimum potential and (*b*) the position of the minimum potential on the distributor if the distributor is fed at both ends at 250 V.

[*A.M.I.E. Sec. B. Power Systems Winter 1993*]

Solution : Let the current supplied from end A be I_A amperes.

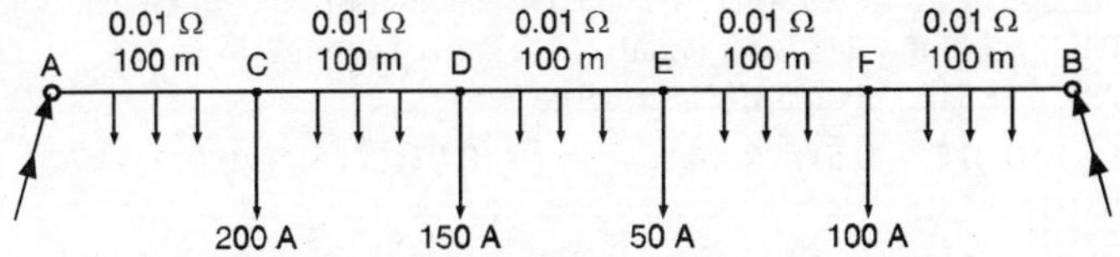

Fig. 9.16

Resistance per 100 m of (go and return) distributor length

$$= \frac{0.1}{1,000} \times 100 = 0.01\ \Omega$$

Resistances of the various sections of the distributor (both conductors) are shown in Fig. 9.16.

Since at every metre current of 1 A is supplied, therefore, as we go towards load point C current is reduced by 1 A for every metre run.

Hence current just before load point C

$= I_A - 100 \times 1 = (I_A - 100)$ amperes

and current just after load point C

$= I_A - 100 - 200 = (I_A - 300)$ amperes

Similarly current just before load point D

$= I_A - 300 - 100 \times 1 = (I_A - 400)$ A

Current just after load point D

$= I_A - 400 - 150 = (I_A - 550)$ A

Current just before load point E

$= I_A - 550 - 100 \times 1 = (I_A - 650)$ A

Current just after load point E

$= I_A - 650 - 50 = (I_A - 700)$ A

Current just before load point F

$= I_A - 700 - 100 \times 1 = (I_A - 800)$ A

Current just after load point F

$= I_A - 800 - 100 = (I_A - 900)$ A

Current just before feeding point B

$= I_A - 900 - 100 \times 1 = (I_A - 1,000)$ A

Voltage drop in section AC

= Voltage drop due to concentrated load

+ voltage drop due to uniform loading of the section

$$= (I_A - 100) \times 0.01 + \frac{1}{2} \times 1 \times 100 \times 0.01$$

$$= (0.01\ I_A - 0.5) \text{ V}$$

Similarly voltage drop in section CD

$$= (I_A - 400) \times 0.01 + \frac{1}{2} \times 1 \times 100 \times 0.01 = (0.01 I_A - 3.5) \text{ V}$$

Voltage drop in section DE

$$= (I_A - 650) \times 0.01 + \frac{1}{2} \times 1 \times 100 \times 0.01 = (0.01 I_A - 6) \text{ V}$$

Voltage drop in section EF

$$= (I_A - 800) \times 0.01 + \frac{1}{2} \times 1 \times 100 \times 0.01 = (0.01I_A - 7.5)\ V$$

Voltage drop in section FB

$$= (I_A - 1{,}000) \times 0.01 + \frac{1}{2} \times 1 \times 100 \times 0.01 = (0.01I_A - 9.5)\ V$$

Since voltage at feeding point B = Voltage at feeding point A less voltage drops in sections AC, CD, DE, EF and FB and voltage at both feeding point A and B are equal, so the sum of voltage drops in sections AC, CD, DE, EF and FB is equal to zero

i.e., $(0.01I_A - 0.5) + (0.01I_A - 3.5) + (0.01I_A - 6) + (0.01I_A - 7.5) + (0.01I_A - 9.5) = 0$

or $0.05I_A = 27$

or $I_A = \frac{27}{0.05} = 540$ A

Since current fed from point A is 540 A, current will be zero at load point D because current just after load point D is 540 – 550 *i.e.*, 10 A flowing from load point E to load point D. Hence D is the point of minimum potential **Ans.**

Minimum potential

= Potential of point D

$= V_A$ – voltage drop in section AC – voltage drop in section CD

$= V_A - (0.01I_A - 0.5) - (0.01I_A - 3.5)$ volts

$= 250 - (0.01 \times 540 - 0.5) - (0.01 \times 540 - 3.5) = 243.2$ V **Ans.**

Example 9.13. A two-wire dc distributor is 300 m long. The end is fed at 205 V and end B at 200 V. The distributor is uniformly loaded at 0.15 A/metre and has concentrated loads of 50 A, 60 A and 40 A at points distant 75, 125 and 225 metres respectively from end A. The resistance of each conductor is 0.15 Ω/km. Calculate (*i*) the point of minimum potential, (*ii*) currents fed at A and B. [*M.D. Univ. Power Systems-I, Dec. 2008*]

Solution : Let the current supplied from end A be I_A amperes

Resistance per 100 m of (go and return) distributor length

$$= \frac{0.15}{1{,}000} \times 2 \times 100 = 0.03\ \Omega$$

Resistances of various sections of the distributor (both conductors) are shown in Fig. 9.17.

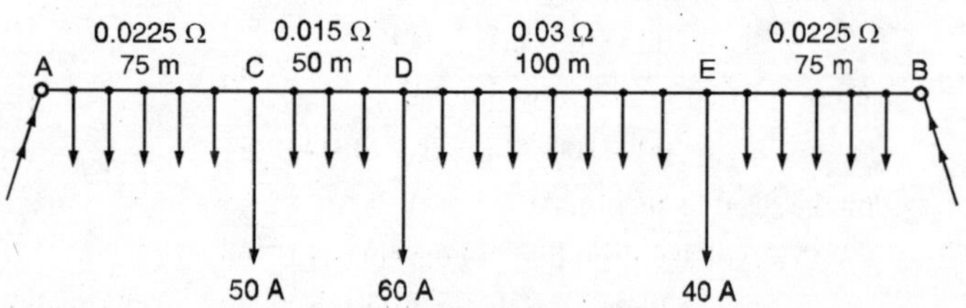

Fig. 9.17

Since at every metre current of 0.15 A is supplied, therefore, as we go towards load point C, current is reduced by 0.15 A for every metre run.

Hence current just before load point C

$= I_A - 75 \times 0.15 = (I_A - 11.25)$ amperes

and current just after load point C

$= I_A - 11.25 - 50 = (I_A - 61.25)$ amperes

Similarly, current just before point D

$= I_A - 61.25 - 50 \times 0.15 = (I_A - 68.75)$ amperes

Current just after load point D $= I_A - 68.75 - 60$
$= (I_A - 128.75)$ amperes

Current just before load point E $= I_A - 128.75 - 100 \times 0.15$
$= (I_A - 143.75)$ amperes

Current just after load point, E $= I_A - 143.75 - 40$
$= I_A - 183.75$ A

Current just before feeding point B $= I_A - 183.75 - 75 \times 0.15$
$= (I_A - 195)$ amperes

Voltage drop in section AC

$$= (I_A - 11.25) \times 0.0225 + \frac{1}{2} \times 0.15 \times 75 \times 0.0225$$
$$= (0.0225\ I_A - 0.253125 + 0.1265625)\ V$$
$$= (0.0225\ I_A - 0.1265625)\ V$$

Voltage drop in section CD

$$= (I_A - 68.75) \times 0.015 + \frac{1}{2} \times 0.15 \times 50 \times 0.015$$
$$= (0.015I_A - 1.03125 + 0.05625)\ V$$
$$= (0.015I_A - 0.975)\ V$$

Voltage drop in section DE

$$= (I_A - 143.75) \times 0.03 + \frac{1}{2} \times 0.15 \times 100 \times 0.03$$
$$= (0.03I_A - 4.3125 + 0.225)\ V$$
$$= (0.03I_A - 4.0875)\ V$$

Voltage drop in section EB

$$= (I_A - 195) \times 0.0225 + \frac{1}{2} \times 0.15 \times 75 \times 0.0225$$
$$= (0.0225I_A - 4.3875 + 0.1265625)\ V$$
$$= (0.0225I_A - 4.2609375)\ V$$

Since voltage at feeding point A is 205 V and that at point B = 200 V, sum of voltage drops in sections AC, CD, DE and EB will be equal to $V_A - V_B$ *i.e.*,

$$V_A - V_B = (0.0225I_A - 0.1265625) + (0.015I_A - 0.975) + (0.03I_A - 4.0875)\ V + (0.0225I_A - 4.2609375)\ V$$

or $205 - 200 = 0.09I_A - 9.45$

or $I_A = \frac{5 + 9.45}{0.09} = 160.555$ A **Ans.**

Since current fed from point A is 160.555 A, current will be zero at point E because current just after load point E is 160.555 – 183.75 *i.e.*, 23.1944 amperes will be from feeding point B to load point E. Hence E is the point of minimum potential. **Ans.**

Minimum potential

= Potential of point E

$= V_A$ – Voltage drops in sections AC, CD and DE

$= 205 - (0.0225 \times 160.555 - 0.1265625)$
$- (0.015 \times 160.555 - 0.975) - (0.03 \times 160.555 - 4.0875)$

$= 205 - 3.6125 + 0.1265625 - 2.408333 + 0.975 - 4.816665 + 4.0875$

$= 205 - 10.8375 + 5.1890625 = 199.35$ V **Ans.**

Current supplied from feeding point B,

$I_B = -(I_A - 195) = 195 - I_A = 195 - 160.555 = 34.445$ A **Ans.**

Example 9.14. A two-wire dc distributor AB, 500 m long, fed at both ends at equal potential of 250 V, is loaded with a uniformly distributed load of 0.5 A/m along with concentrated loads of 100 A, 200 A, 300 A and 400 A at distances of respectively 100 m, 200 m, 300 m and 400 m from end A. Calculate the distance from A where the potential of the line would be minimum. Assuming the resistivity of the conductor material to be 1.7 × 10^{-8} Ω-m, calculate the minimum cross section of the conductors so that the maximum drop may not exceed 20 V.

[A.M.I.E. Sec. B. Power Systems Summer 1999]

Solution : Let the resistance per 100 m of distributor length (both conductors) be of r ohms and current supplied from end A be I_A amperes.

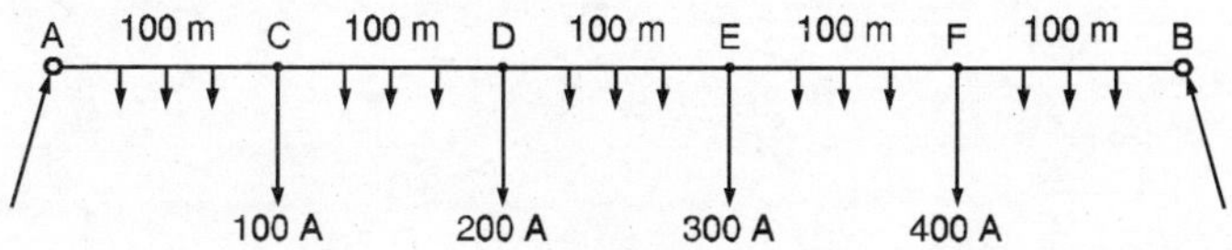

Fig. 9.18

Since at every metre current of 0.5 A is supplied, therefore, as we go towards load point C current is reduced by 0.5 A for every metre run.

Thus current just before load point C $= I_A - 100 \times 0.5 = (I_A - 50)$ A

and current just after load point C $= (I_A - 50) - 100 = (I_A - 150)$ A

Similarly current just before and just after load point D are $(I_A - 200)$ and $(I_A - 400)$ A respectively; current just before and just after load point E are $(I_A - 450)$ A and $(I_A - 750)$ A respectively; current just before and after load point F are $(I_A - 800)$ A and $(I_A - 1{,}200)$ A respectively and current just before feeding point B is $(I_A - 1{,}250)$ amperes.

Voltage drop in section AC

= Voltage drop due to concentrated load + voltage drop due to uniform loading of the section

$$= (I_A - 50) \times r + \frac{1}{2} \times 0.5 \times 100 \times r = r(I_A - 25) \text{ volts}$$

Similarly voltage drop in section CD

$$= (I_A - 200) \times r + \frac{1}{2} \times 0.5 \times 100 \times r = r\,(I_A - 175) \text{ volts}$$

Voltage drop in section DE

$$= (I_A - 450) \times r + \frac{1}{2} \times 0.5 \times 100 \times r = r\,(I_A - 425) \text{ volts}$$

Voltage drop in section EF

$$= (I_A - 800) \times r + \frac{1}{2} \times 0.5 \times 100 \times r = r\,(I_A - 775) \text{ volts}$$

Voltage drop in section FB

$$= (I_A - 1{,}250) \times r + \frac{1}{2} \times 0.5 \times 100 \times r = r\,(I_A - 1{,}225) \text{ volts}$$

Since voltage at feeding point B equals voltage at feeding point A less voltage drops in sections AC, CD, DE, EF and FB and voltages of both feeding points are equal, the sum of voltage drops in sections AC, CD, DE, EF and FB is equal to zero *i.e.*,

$$r(I_A - 25) + r(I_A - 175) + r(I_A - 425) + r(I_A - 775) + r(I_A - 1{,}225) = 0$$

$$\text{or } I_A = \frac{1{,}225 + 775 + 425 + 175 + 25}{5} = 525 \text{ A}$$

Since current fed from point A is 525 A, current will be zero at load point E because current just after load point E is 525 – 750 *i.e.*, 225 A flowing from load point F to load point E. Hence E is the point of minimum potential *i.e.* point of minimum potential is at a distance of 300 metres from point A **Ans.**

Maximum potential drop

= Sum of voltage drops in sections AC, CD and DE

= Sum of voltage drops in sections EF and FB.

$$= -[r(I_A - 775) + r(I_A - 1{,}225)]$$

$$\text{or } 20 \text{ V} = -[r\,(525 - 775) + r\,(525 - 1{,}225)] = 950\,r$$

$$\text{or } r = \frac{20}{950}\ \Omega$$

Area of cross section of conductor,

$$a = \frac{\rho l}{r} = \frac{1.7 \times 10^{-8} \times 2 \times 100}{20/950}$$

$$= 1.615 \times 10^{-4} \text{ m}^2 \text{ or } 1.615 \text{ cm}^2 \textbf{ Ans.}$$

9.8. STEPPED OR TAPERED DISTRIBUTOR

As already mentioned in Art. 9.1.1, it is necessary to design each distributor with the minimum volume of conductor material consistent with the voltage regulation needs.

When a series of currents is tapped off from any conductor, the current, obviously, varies along the length of the conductor. For example, a uniformly loaded distributor supplied at one end carries a current varying progressively from a maximum at the feeding point to zero at the far end. If the conductor of uniform cross section, from the point of view of current carrying capacity, is used throughout, there is a great deal too much conductor material in use, and an appreciable saving can be affected if the cross section is determined in accordance with the current to be carried *i.e.* if the conductor is stepped or tapered. The minimum volume of conductor material is required when the conductor is tapered so that the cross section at any point is proportional to the square root of the distance from the far end of the distributor. This is not possible in practice. The practical optimum is a stepped distributor (Fig. 9.19).

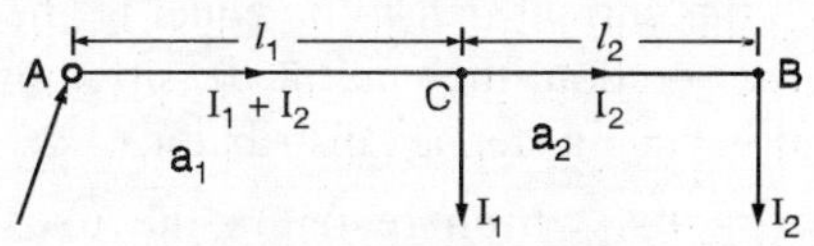

Fig. 9.19. ***Stepped Distributor***

Consider a distributor AB with loads I_1 and I_2 tapped off at points C and B respectively (Fig. 9.19). Let the lengths and cross sections of sections AC and CB be l_1, a_1 and l_2, a_2 respectively.

$$\text{Resistance of section AC (both wires), } R_{AC} = \frac{2\rho l_1}{a_1}$$

$$\text{Resistance of section CB (both wires), } R_{CB} = \frac{2\rho l_2}{a_2}$$

$$\text{Voltage drop in section AC, } v_1 = (I_1 + I_2)R_{AC}$$

$$= \frac{2\rho l_1}{a_1}(I_1 + I_2) \quad ...(9.10)$$

$$\text{or } a_1 = \frac{2\rho l_1}{v_1}(I_1 + I_2)$$

Voltage drop in section CB,

$$v_2 = \text{Total permissible voltage drop along the distributor} - \text{voltage drop in section AC}$$

So $$\frac{2\rho l_2 I_2}{a_2} = v - v_1$$

or $$a_2 = \frac{2\rho l_2 I_2}{v - v_1} \quad ...(9.11)$$

Volume of conductor material in the distributor, reckoning both sides,

$$\text{Volume} = 2l_1a_1 + 2l_2a_2 = \frac{4\rho l_1^2}{v_1}(I_1 + I_2) + \frac{4\rho l_2^2 I_2}{v - v_1}$$

Differentiating both sides with respect to v_1

$$\frac{d\,\text{Volume}}{dv_1} = -\frac{4\rho l_1^2}{(v_1)^2}(I_1 + I_2) + \frac{4\rho l_2^2 I_2}{(v - v_1)^2}$$

$$= 4\rho\left[\frac{l_2^2 I_2}{(v - v_1)^2} - \frac{l_1^2(I_1 + I_2)}{v_1^2}\right]$$

Substituting the values of v_1 and $(v - v_1)$ from Eqs. (9.10) and (9.11) respectively in above equation we have for the condition for the minimum volume of conductor material

$$0 = 4\rho\left[\frac{l_2^2 I_2 a_2^2}{4\rho^2 l_2^2 I_2^2} - \frac{l_1^2(I_1 + I_2)a_1^2}{4\rho^2 l_1^2 (I_1 + I_2)^2}\right]$$

or $$\frac{a_1}{a_2} = \sqrt{\frac{I_1 + I_2}{I_2}} \quad ...(9.12)$$

From above equation it is obvious that for economy the cross sections of the various sections of a stepped conductor should be proportional to the square roots of the appropriate currents. Since the current density is equal to the quotient of current and cross section, the current density, obviously, will not be uniform in such a stepped distributor.

The drawbacks, which restricts the use of stepped distributor, are given below :

1. Availability of few standard cross sections makes selection of required sections impossible without departing considerably from the economy aimed at.
2. The scheme entails more jointing than a scheme with uniform cross section while minimum number of joints are desirable.
3. The last but the most important drawback of using stepped distributor is that future additions to any system may entirely alter the current distribution in the various sections of any given distributor and may therefore render a stepped conductor useless.

The use of a stepped conductor, however, can be justified for a long distributor fed from one end and terminating at the other end at the boundary of the supply undertaking.

9.9. RING MAIN DISTRIBUTOR

A distributor arranged to form a closed circuit and fed at one or more than one points is called the *ring main distributor.* The advantages of such an arrangement are greater reliability and better flexibility. In the event of fault on any section, it can be isolated from both sides and the continuity of supply is maintained for the remaining sections. The result is that only a faulty section suffers discontinuity of supply. The ring distributor fed at one point, for the purpose of calculations, can be considered equivalent to a straight distributor fed from both ends at equal voltages. A ring main distributor is shown in Fig. 9.20.

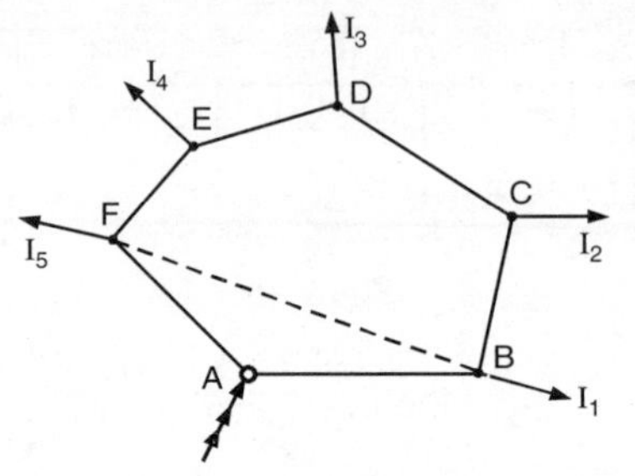

Fig. 9.20. ***Ring Main Distributor Fed at One End***

Sometimes two or more ends of a ring main are connected through interconnector(s), shown dotted in Fig. 9.20, so as to form a complete network. The addition of an interconnector to a ring main reduces the voltage drop and power loss in the system.

The current distribution and voltage drop in such an arrangement can be determined by applying Kirchhoff's laws directly.

Example 9.15. A 2-wire dc ring distributor is 300 m long and is fed at 240 V at point A. At point B, 150 m from A, a load of 120 A is taken and at C, 100 m in opposite direction, a load of 80 A is taken. If resistance per 100 m of a single conductor is 0.03 Ω, find : (*i*) Current in each section of distributor, (*ii*) Voltage at points B and C. [*M.D. Univ. Power Systems-I May 2009*]

Solution : Resistance of both conductors of section AB of ring main,

$$R_{AB} = \frac{2 \times 0.03}{100} \times 150 = 0.09\ \Omega$$

Similarly, $$R_{BC} = \frac{2 \times 0.03}{100} \times (300 - 150 - 100) = 0.03\ \Omega$$

and $$R_{CA} = \frac{2 \times 0.03}{100} \times 100 = 0.06\ \Omega$$

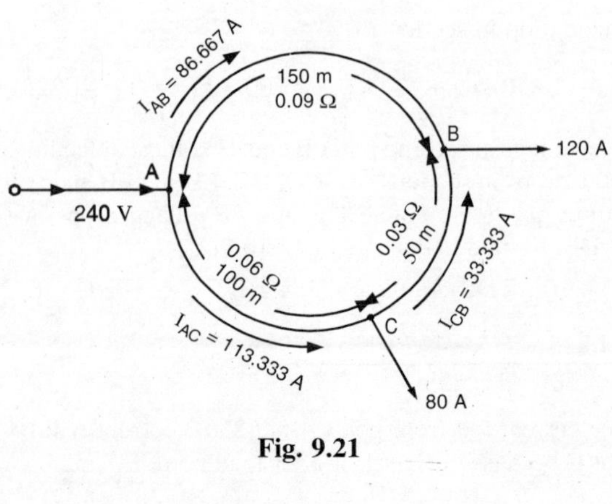

Fig. 9.21

Let current I_A be flowing in section AB, then current in section BC will be $(I_A - 120)$ A and in section CA will be $(I_A - 120 - 80)$ *i.e.*, $(I_A - 200)$ A.

According to Kirchhoff's second law, voltage drop in this closed loop will be equal to zero.

$$\therefore \quad I_{AB} R_{AB} + I_{BC}R_{BC} + I_{CA}R_{CA} = 0$$

or $0.09I_A + 0.03(I_A - 120) + 0.06(I_A - 200) = 0$

$$\text{or } 0.18I_A = 15.6$$

$$\text{or } I_A = \frac{15.6}{0.18} = 86.667 \text{ A}$$

Current in section AB,

$I_{AB} = I_A = 86.667$ A from feeding point A to load point B **Ans.**

Current in section BC,

$I_{BC} = I_A - 120 = 86.667 - 120$

$= -33.333$ A *i.e.*, 33.333 A from load point C to load point B **Ans.**

Current in section CA,

$I_{CA} = I_A - 200 = 86.669 - 200$

$= -113.333$ A *i.e.*, 113.333 A from feeding point to load point C **Ans.**

Voltage at load point B,

$V_B = V_A - I_{AB}R_{AB} = 240 - 86.667 \times 0.09 = 232.2$ V **Ans.**

Voltage at load point C,

$V_C = V_B - I_{BC}R_{BC} = 232.2 - [(-33.333) \times 0.03] = 233.2$ V **Ans.**

Example 9.16. A ring distributor is supplied through a feeder AB and is loaded as shown below: Calculate (*i*) the cross section of AB and (*ii*) that of the ring main for minimum volume of copper. Assume that the maximum voltage drop from A to the point of minimum potential is 15 volts. Take ρ = 1.73 μΩ per cm cube.

[*Agra Univ. Transmission and Distribution of Electrical Power 1979*]

Solution : Let the resistance per 100 m of ring distributor length (both conductors) be r ohms.

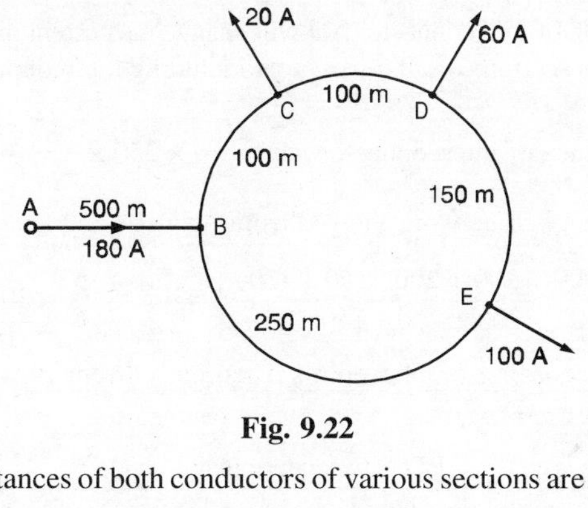

Fig. 9.22

Resistances of both conductors of various sections are determined as below :

$$R_{BC} = \frac{r}{100} \times 100 = r\,\Omega$$

$$\text{Similarly} \quad R_{CD} = \frac{r}{100} \times 100 = r\,\Omega$$

$$R_{DE} = \frac{r}{100} \times 150 = 1.5r\,\Omega$$

$$\text{and } R_{EB} = \frac{r}{100} \times 250 = 2.5r\,\Omega$$

Let current I_x be flowing in section BC then currents in sections CD, DE and EB will be $(I_x - 20)$, $(I_x - 80)$ and $(I_x - 180)$ amperes respectively.

According to Kirchhoff's second law voltage drop in this closed loop will be equal to zero.

$$\therefore \quad I_{BC}R_{BC} + I_{CD}R_{CD} + I_{DE}R_{DE} + I_{EB}R_{EB} = 0$$

or $I_x r + (I_x - 20)r + (I_x - 80) \times 1.5r + (I_x - 180) \times 2.5r = 0$

$$\text{or } 6I_x r = 590r$$

$$\text{or } I_x = 98.333 \text{ A}$$

Hence currents in sections BC, CD, DE and EB will be equal to 98.333 A, 78.333 A, 18.333 A and –81.667 A respectively.

Hence E is the point of minimum potential.

Let a_F and a_D be the cross-sectional areas of the feeder and distributor respectively resistance of section BE for both go and return

$$= \frac{1.73 \times 10^{-6} \times 250 \times 2 \times 100}{a_D} = \frac{0.0865}{a_D}\,\Omega$$

Resistance of feeder AB for both go and return

$$= \frac{1.73 \times 10^{-6} \times 500 \times 2 \times 100}{a_F} = \frac{0.173}{a_F}\,\Omega$$

Maximum voltage drop

= Voltage drop in feeder + voltage drop in distributor

$= v_F + v_D = 15$ volts (given)

$$\text{and } v_F = \frac{180 \times 0.173}{a_F} \quad ...(i)$$

$$\text{and } v_D = \frac{81.667 \times 0.0865}{a_D} \quad ...(ii)$$

Volume of copper used in feeder

$= 500 \times 100 \times 2 \times a_F$ cm^3

$$= 500 \times 100 \times 2 \times \frac{180 \times 0.173}{v_F} \text{ cm}^3$$

$$= \frac{3{,}114{,}000}{v_F} = \frac{3{,}114{,}000}{15 - v_D} \text{ cm}^3 \quad ...(iii)$$

Volume of copper used in distributor

$= 600 \times 100 \times 2 \times a_D$

$$= \frac{120{,}000 \times 81.667 \times 0.0865}{v_D} = \frac{847{,}700}{v_D} \text{ cm}^3 \quad ...(iv)$$

$$\text{Total volume of copper required, Vol} = \frac{3{,}114{,}000}{15 - v_D} + \frac{847{,}700}{v_D}$$

For volume of copper to be minimum $\frac{d(\text{vol})}{dv_D}$ should be zero

$$\text{i.e.,} \quad \frac{3{,}114{,}000}{(15 - v_D)^2} - \frac{847{,}700}{v_D^2} = 0$$

or $2.2663v_D^2 + 25.431v_D - 190.7325 = 0$

$$\text{or } v_D = \frac{-25.431 \pm \sqrt{(25.431)^2 + 4 \times 2.2663 \times 190.7325}}{2 \times 2.2663} = 5.143 \text{ V}$$

$v_F = 15 - 5.143 = 9.857$ V Neglecting –ve sign giving higher value

Substituting the values of v_D and v_F in Eqs. (*i*) and (*ii*), we get

$$a_F = \frac{180 \times 0.173}{v_F} = \frac{180 \times 0.173}{9.857} = 3.16 \text{ cm}^2 \text{ **Ans.**}$$

and $a_D = \frac{81.667 \times 0.0865}{v_D} = \frac{81.667 \times 0.0865}{5.143} = 1.3736 \text{ cm}^2$ **Ans.**

Example 9.17. Three loads A, B, C are connected to 240 volt supply point through separate cables having resistances of 0.06, 0.15 and 0.12 ohms respectively. A is joined with B through 0.08 Ω connector and B is joined with C through a connector of resistance 0.05 Ω. Determine the voltage at points A, B and C if the loads are 40, 60 and 25 amperes at load points A, B and C respectively.

Solution : Total supply current, I = 40 + 60 + 25 = 125 amperes

Let current in feeders OA, OB and OC be I_1, I_2 and I_3 respectively. Applying Kirchhoff's first law at supply point O, we get

$$I = I_1 + I_2 + I_3$$

$$\text{or } I_3 = I - I_1 - I_2$$

$$\text{or } I_3 = 125 - I_1 - I_2 \quad ...(i)$$

Applying Kirchhoff's second law to mesh OABO, we get

$$0.06I_1 + 0.08(I_1 - 40) - 0.15I_2 = 0$$

$$\text{or } 0.14I_1 - 0.15I_2 - 3.2 = 0 \quad ...(ii)$$

Applying Kirchhoff's second law to mesh OBCO, we get

$$0.15I_2 + 0.05(I_1 + I_2 - 100) - 0.12(125 - I_1 - I_2) = 0$$

$$\text{or } 0.17I_1 + 0.32I_2 - 20 = 0 \quad ...(iii)$$

Multiplying Eq. (*ii*) by 32 and Eq. (*iii*) by 15, we get

$$4.48I_1 - 4.8I_2 - 102.4 = 0 \quad ...(iv)$$

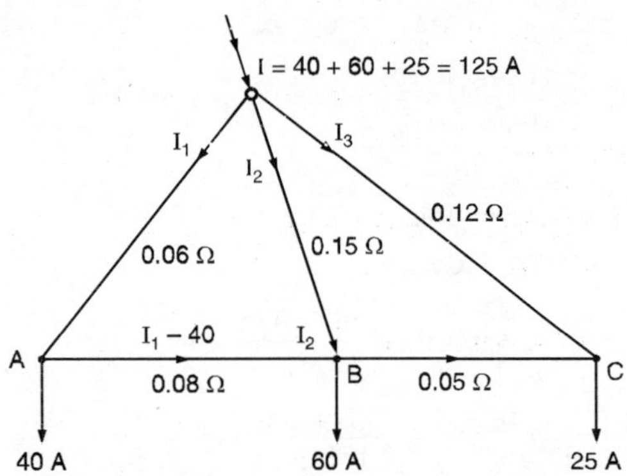

Fig. 9.23

$$\text{and } 2.55I_1 + 4.8I_2 - 300 = 0 \quad ...(v)$$

Adding Eqs. (*iv*) and (*v*), we get

$$7.03I_1 - 402.4 = 0$$

$$\text{or } I_1 = \frac{402.4}{7.03} = 57.24 \text{ A}$$

Substituting $I_1 = \frac{402.4}{7.03}$ in Eq. (*ii*), we get

$$0.14 \times \frac{402.4}{7.03} - 0.15I_2 - 3.2 = 0$$

$$\text{or } I_2 = 32.09 \text{ A}$$

$$I_3 = 125 - 57.24 - 32.09 = 35.67 \text{ A}$$

Voltage at point A, $V_A = 240 - 0.06I_1$

$= 240 - 0.06 \times 57.24 = 236.6$ V **Ans.**

Voltage at point B, $V_B = 240 - 0.15 \times 32.09 = 235.2$ V **Ans.**

Voltage at point C, $V_C = 240 - 0.12 \times 35.67 = 335.72$ V **Ans.**

Example 9.18. For the same voltage drop between the supply point A and the load point D in the network shown below, compare the weights of copper required with and without an interconnector between A and D. Conductors have the same cross sections in each scheme.

Solution : First consider the case without interconnector

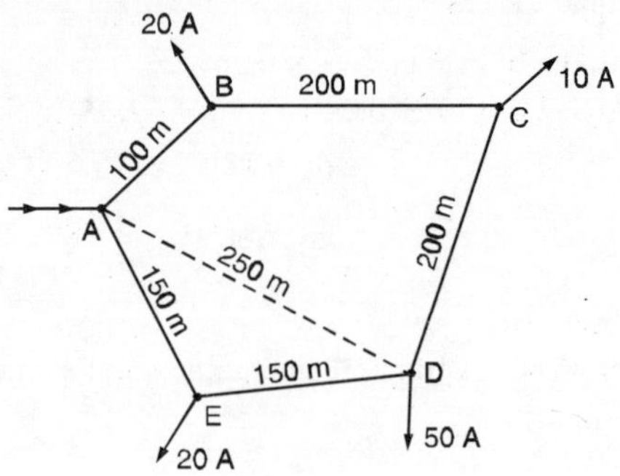

Fig. 9.24

Let the current flowing in section AB = x; then currents in various sections will be as follows :

Current in section AB = x

Current in section BC = $x - 20$

Current in section CD = $x - 30$

Current in section DE = $x - 80$

Current in section EA = $x - 100$

Applying Kirchhoff's second law round the closed loop ABCDEA, we get

$$\frac{2\rho}{a_1}[100x + 200(x - 20) + 200(x - 30) + 150(x - 80) + 150(x - 100)] = 0$$

$$\text{or } 800x = 37{,}000$$

$$\text{or } x = 46.25 \text{ amperes}$$

Voltage drop from A to D

$$= \frac{2\rho}{a_1}[46.25 \times 100 + 26.25 \times 200 + 16.25 \times 200] = 26{,}250\frac{\rho}{a_1} \quad ...(i)$$

where a_1 is the cross section of various sections.

Addition of interconnector AD will change the current distribution. Let the cross sections of all conductors including interconnectors be a

$$\text{Resistance of interconnector AD} = \frac{\rho}{a} \times 250 \times 2 = \frac{500\rho}{a}$$

Resistance of network viewed from points A and D

$$\frac{2\rho}{a}\left[\frac{(100 + 200 + 200)(150 + 150)}{100 + 200 + 200 + 150 + 150}\right] = \frac{2\rho}{a} \times \frac{5}{8} \times 300 = \frac{375\rho}{a}$$

Now according to Thevenin's Theorem current flowing in the interconnector AD

$$= \frac{\text{PD between terminals A and D}}{\text{Resistance of distribution network + Resistance of interconnector}}$$

$$= \frac{26{,}250\rho / a}{(375 + 500)\rho / a} = \frac{26{,}250}{875} = 30 \text{ A}$$

In the second case *i.e.*, with interconnector let the cross section of the conductors including interconnector be a_2. Then voltage drop between A and D

$$= 30 \times \frac{\rho}{a_2} \times 250 \times 2 = 15{,}000\frac{\rho}{a_2} \quad ...(ii)$$

For same voltage drop between terminals A and D, equating Eqs. (*i*) and (*ii*), we get

$$26{,}250 \frac{\rho}{a_1} = 15{,}000 \frac{\rho}{a_2}$$

$$\text{or} \quad \frac{a_2}{a_1} = \frac{15{,}000}{26{,}250} = \frac{4}{7}$$

i.e., the conductor cross section with interconnector is $\frac{4}{7}$ times the corresponding cross section without interconnector.

However, total lengths of conductor with and without interconnector are 1,050 and 800 metres respectively so that ratio of volumes (or weights) in the two cases

$$= \frac{4}{7} \times \frac{1{,}050}{800} = 0.75 \textbf{ Ans.}$$

9.10. THREE-WIRE DC DISTRIBUTION SYSTEM

Since high voltage increases the efficiency of transmission and reduces the cost of conductor material in feeder cables, but in dc system it is not possible to change the voltage between feeders and distributors without using rotating machinery which is undesirable, so the feeders and distributors are operated at same voltage *i.e.*, consumer voltage is the same as the voltage at feeding points.

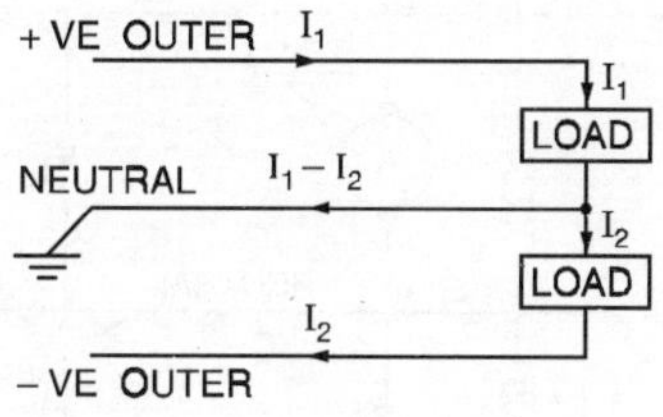

(*a*)

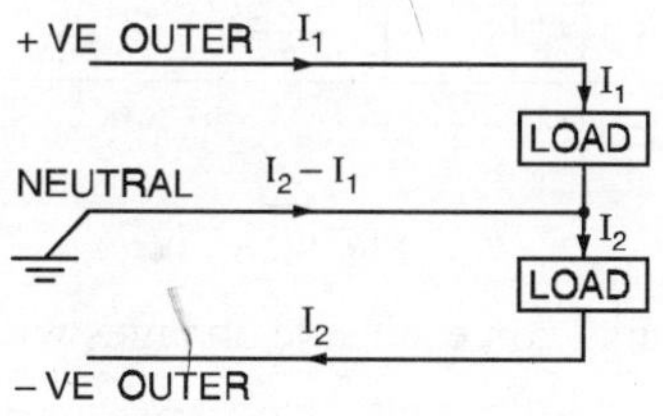

(*b*)

Fig. 9.25. *3-Wire DC System*

The above difficulty is largely overcome by employing 3-wire system which consists of two outers and an earthed* middle wire known as *neutral wire*. In this system consumer apparatus or lamps are connected between the neutral and one of the outers. The voltage between the outers being twice that of consumer's terminals increases the transmission efficiency and reduces the cost of conductor material. If the load on both sides is equal, no current flows through the neutral and such a load is known as balanced load. But if the load on two sides is different then difference of load currents on two sides, known as *out of balance* current, flows in the neutral. Since only out of balance current flows through the neutral wire, generally neutral wire is taken of half cross section of either outer. If currents flowing through +ve and –ve outers are I_1 and I_2 respectively then current in the middle wire will be difference of currents I_1 and I_2 flowing in the outers.

When the load on +ve side is greater than that on –ve side, current equal to difference of outer currents will flow in the neutral wire from load end to supply end, as shown in Fig. 9.25 (*a*).

When the load on –ve side is greater than that on +ve side, current equal to difference of outer currents will flow in the neutral wire from supply end to load end, as shown in Fig. 9.25 (*b*).

Typical layout of 3-wire dc distribution system is shown in Fig. 9.26.

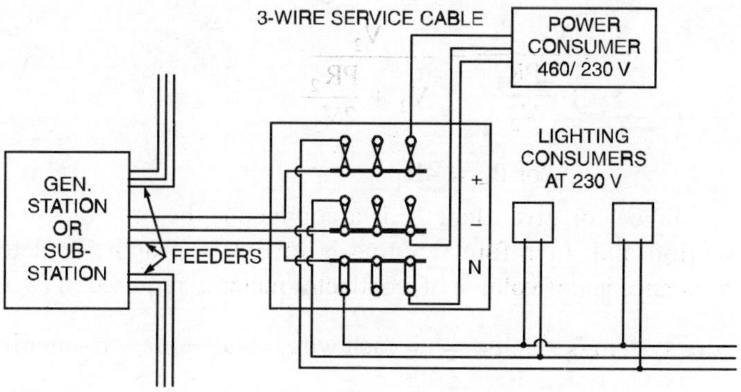

Fig. 9.26. *Typical Layout of 3-Wire DC Distribution System*

Advantages of 3-*Wire DC Distribution.* (*i*) Two different voltages are available–high voltage for power loads and low voltage for domestic and commercial loads.

(*ii*) With the same volume of conductor material, 3-wire dc distribution is able to deliver more power *i.e.,* twice of that with two-wire distribution without affecting the efficiency.

(*iii*) If both of the 3-wire and 2-wire feeders transmit the same amount of power over the same distance with the same efficiency, the three-wire feeder, if load is balanced and the cross section of middle wire is half of the cross section of either outers, requires only 0.3125 times conductor material as required by two wire feeder, which can be proved as follows.

Let power delivered to consumers be P watts

Voltage at power station and feeding point be V_1 and V_2 volts respectively.

$$\text{Resistance of each conductor} = R_1 \text{ ohms}$$

$$\text{Then in two-wire feeder current, } I = \frac{P}{V_2}$$

$$\text{Voltage drop in both of the conductors} = 2IR_1 = \frac{2P}{V_2} R_1$$

$$\text{Voltage at power station, } V_1 = V_2 + 2IR_1 = V_2 + \frac{2P}{V_2} R_1$$

$$\text{Transmission efficiency, } \eta_1 = \frac{V_2}{V_2 + \frac{2P}{V_2} R_1}$$

* Earthed at the substation

In a 3-wire system the total voltage is double *i.e.*, $2V_2$, and, therefore, each outer will carry current of $\frac{P}{2V_2}$ so long as the load is balanced.

$$\text{Voltage drop in both of the conductors} = 2 \times \frac{P}{2V_2} R_2 = \frac{PR_2}{V_2}$$

$$\text{Voltage at power station} = 2V_2 + \frac{PR_2}{V_2}$$

$$\text{Transmission efficiency, } \eta_2 = \frac{2V_2}{2V_2 + \frac{PR_2}{V_2}} = \frac{V_2}{V_2 + \frac{PR_2}{2V_2}}$$

Since $\eta_1 = \eta_2$

$$\therefore \quad \frac{V_2}{V_2 + \frac{2PR_1}{V_2}} = \frac{V_2}{V_2 + \frac{PR_2}{2V_2}}$$

or $R_2 = 4R_1$

Since for given length of a conductor, the area of cross section and, therefore, volume is inversely proportional to resistance and if volume of conductor material required in two-wire system is v units, $\frac{v}{2}$ in each wire, volume of each outer in the 3-wire system $= \frac{v/2}{4} = \frac{v}{8}$ and volume of middle wire $= \frac{v}{16}$

Total volume of copper required in 3-wire system

$$= \frac{v}{8} + \frac{v}{8} + \frac{v}{16} = \frac{5}{16} v$$

Hence in a 3-wire system to transmit the same amount of power with same efficiency over same distance with same consumer voltage we require 5/16th or 0.3125 times conductor material as required in a two wire system.

Disadvantages. Requirement of additional wire (third wire) partially reduces safety and necessity of a balancer set causing an increase in cost. In an overhead system, the cross-arms of greater length are required.

9.11. METHODS OF OBTAINING THREE-WIRE SYSTEM

Three-wire system can be obtained by using *two-generator method* or *balancer set method.*

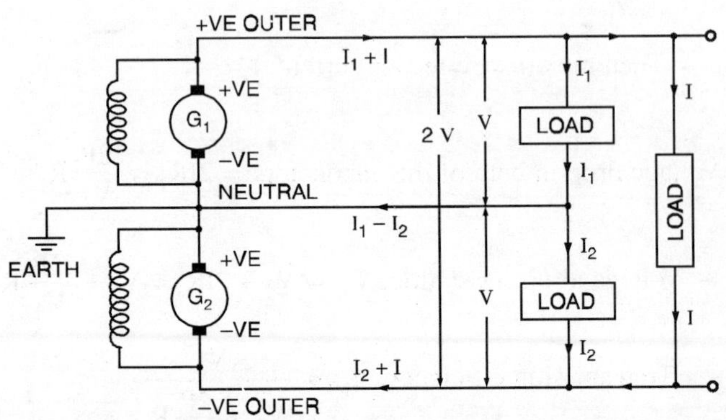

Fig. 9.27. *Two Generator Method of Obtaining 3-Wire System*

9.11.1. Two Generator Method

In this method two shunt wound dc generators G_1 and G_2 are connected in series by connecting +ve terminal of G_2 with –ve terminal of G_1, as shown in Fig. 9.27. Both the generators may be driven by the same prime mover. In this type of combination each generator supplies the load on its own side. The main disadvantage of this method is that two separate generators are required.

9.11.2. Balancer Set

The generator supplying a 3-wire system is connected across the outers. If load is balanced, the voltage across the loads on both sides *i.e.*, between +ve outer and middle wire and between –ve outer and middle wire will be equal, but if the load is unbalanced then load voltages on the two sides will become unequal. In order to equalise the voltages on the two sides (+ve and –ve sides) use of balancer is made as discussed below.

The most general form of a balancer set consists of two identical dc shunt wound machines A and B mechanically coupled having their armatures and fields connected in series across the outers, as shown in Fig. 9.28. The middle wire is taken from the junction of the armatures.

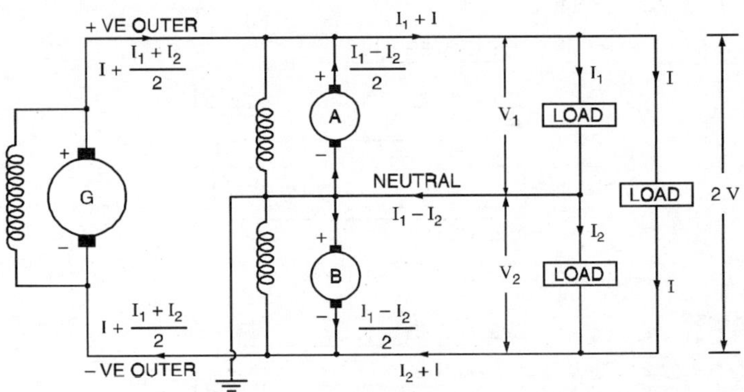

Fig. 9.28

The load voltage on each side changes with the variation in load.

When the load on two sides is same *i.e.*, the system is balanced, both machines run as motor on no load and their back emfs, are the same due to equal excitation and equal speed.

When the load is unbalanced, say $I_1 + I$ amperes on + ve side and $I_2 + I$ on –ve side then out of balance current $I_1 - I_2$ will flow through the neutral wire. If +ve is heavily loaded *i.e.*, I_1 is greater than I_2; potential difference on + ve side will tend to fall hence machine A will start working as a generator and potential difference on other side *i.e.*, on –ve side will tend to increase hence machine B will start working as a motor. Thus energy is transferred from lower-load side to higher-load side by means of a balancer.

If the internal losses of the machines are negligible, then motor input will be equal to generator output and armature current of both machines will be equal to $\frac{I_1 - I_2}{2}$ amperes.

Since both the machines have same excitation and run at same speed, therefore induced emfs of both machines will be the same. Let it be E volts.

If R_a is the armature resistance of each machine then terminal voltage across machine A, working as a generator,

$$V_1 = E - \left(\frac{I_1 - I_2}{2}\right) R_a$$

and terminal voltage across machine B, working as a motor,

$$V_2 = E + \left(\frac{I_1 - I_2}{2}\right) R_a$$

Difference of terminal voltage on two sides

$$= V_2 - V_1 = E + \left(\frac{I_1 - I_2}{2}\right) R_a - \left[E - \left(\frac{I_1 - I_2}{2}\right) R_a\right]$$

$$= (I_1 - I_2) R_a$$

Hence the difference of terminal voltages on the two sides of the system is proportional to the (*i*) out of balance current and (*ii*) armature resistance.

That is why for keeping the voltage on the two sides equal, armature resistance of machines is kept small and efforts are made to equalise the loads on the two sides in order to make out of balance current as small as possible.

The difference of terminal voltages, $(V_1 - V_2)$ may further be reduced by cross connecting the field windings of the machines. A better way for improving the performance of the balancer is to use a series field with each machine. The series field shall be so connected that the machine acting as a generator is cummulatively compounded and that as a motor differentially compounded.

Example 9.19. A dc 3-wire system with 500 V between outers supplies 300 kW on the +ve side and 500 kW on the –ve side. Neglecting losses in the balancer machines, calculate (*a*) the current in each balancer armature (*b*) current in middle wire and (*c*) total current supplied by the generator.

Solution : Connections are shown in Fig. 9.29.

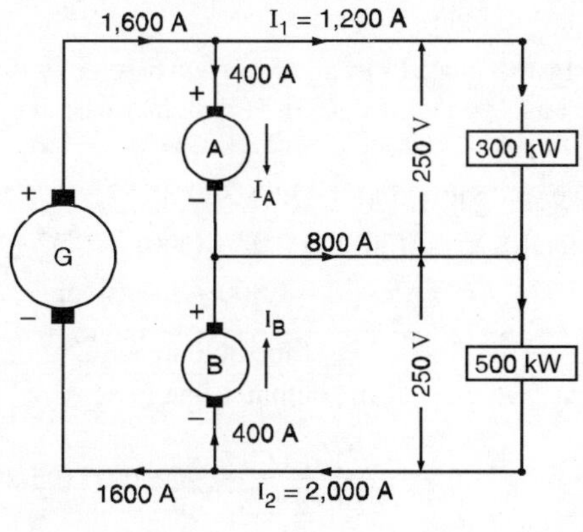

Fig. 9.29

$$\text{Load current on + ve side, } I_1 = \frac{300 \times 1{,}000}{250} = 1{,}200 \text{ A}$$

$$\text{Load current on – ve side, } I_2 = \frac{500 \times 1{,}000}{250} = 2{,}000 \text{ A}$$

$$\text{Current in neutral wire, } I_N = I_2 - I_1 = 2{,}000 - 1{,}200 = 800 \text{ A } \textbf{Ans.}$$

So load on + ve side is less and on – ve side is more, therefore, machine A will run as a motor and machine B will run as a generator.

Total load on the main generator = 300 + 500 = 800 kW

$$\text{Main generator current, } I_g = \frac{800 \times 1{,}000}{500} = 1{,}600 \text{ A } \textbf{Ans.}$$

Current in machine A, $I_A = I_g - I_1 = 1{,}600 - 1{,}200 = 400$ A **Ans.**

Current in machine B, $I_B = I_2 - I_g = 2{,}000 - 1{,}600 = 400$ A **Ans.**

Example 9.20. A load on 3-wire dc system with 500 volts between outers consists of a lighting current of 1,500 A on positive and 1,300 A on negative side while motors connected across outers absorb 500 kW. Assuming each balancer machine having loss of 3.75 kW, calculate load on main generator and each balancer machine.

Solution : Connections are shown in Fig. 9.30.

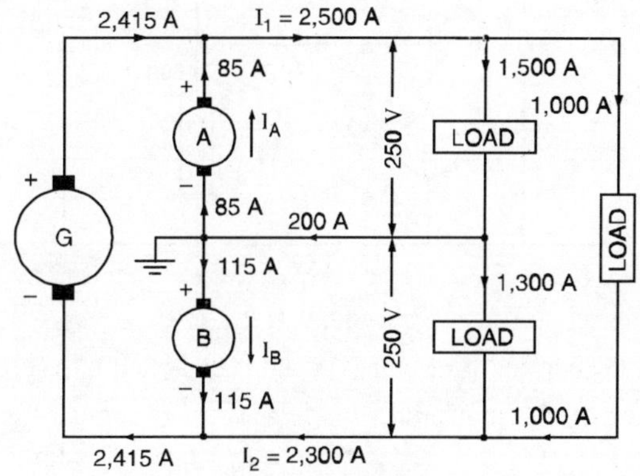

Fig. 9.30

$$\text{Load on + ve side, } P_1 = \frac{1{,}500 \times 250}{1{,}000} = 375 \text{ kW}$$

$$\text{Load on – ve side, } P_2 = \frac{1{,}300 \times 250}{1{,}000} = 325 \text{ kW}$$

Load on outers, $P_3 = 500$ kW

Losses in balancer machines = 2 × 3.75 = 7.5 kW

Total load on main generator,

$$P = P_1 + P_2 + P_3 + \text{losses}$$

$$= 375 + 325 + 500 + 7.5 = 1{,}207.5 \text{ kW } \textbf{Ans.}$$

$$\text{Current supplied by main generator, } I_g = \frac{1{,}207.5 \times 1{,}000}{500} = 2{,}415 \text{ A}$$

Out of balance current = $I_1 - I_2 = 2{,}500 - 2{,}300 = 200$ A

Current through machine A, $I_A = I_1 - I_g = 2{,}500 - 2{,}415 = 85$ A

Current through machine B, $I_B = I_g - I_2 = 2{,}415 - 2{,}300 = 115$ A

$$\text{Load on machine A} = \frac{85 \times 250}{1{,}000} = 21.25 \text{ kW } \textbf{Ans.}$$

$$\text{Load on machine B} = \frac{115 \times 250}{1{,}000} = 28.75 \text{ kW } \textbf{Ans.}$$

Example 9.21. A dc 3-wire system with 500 volts across outers supplies 1,000 A on + ve side, 800 A on – ve side and 1,000 A across outers. The rotary balancer has each an armature resistance of 0.1 Ω and takes 10 A on no load. Find (*a*) the voltage across each balancer (*b*) the load on the main generator and (*c*) current loading of each balancer unit.

Solution : Total current on + ve outer,

$$I_1 = 1{,}000 + 1{,}000 = 2{,}000 \text{ A}$$

Total current on – ve outer,

$$I_2 = 1{,}000 + 800 = 1{,}800 \text{ A}$$

Current in middle wire,

$$I_N = 2{,}000 - 1{,}800 = 200 \text{ A}$$

Since load on + ve side is more, therefore, machine A will run as a generator and machine B will run as a motor.

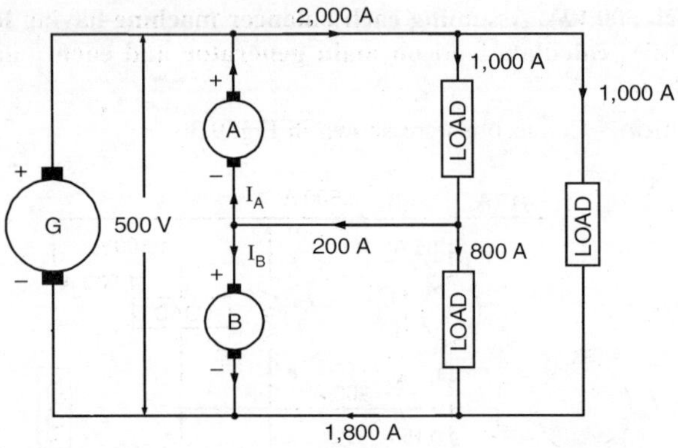

Fig. 9.31

Let the current of machines A and B be I_A and I_B respectively.

Since current in neutral wire, $I_N = I_A + I_B$

$$\therefore \quad I_A = I_N - I_B = 200 - I_B$$

Let the terminal voltage across machines A and B be V_A and V_B respectively. Since machine B is driving machine A.

∴ Output of motor = Output of generator + losses in the set

or
$$V_B I_B = V_A(200 - I_B) + 500 \times 10 + 0.1 I_B^2 + 0.1\,(200 - I_B)^2 \quad ...(i)$$

Since field current and speed of both machines are same, their back emfs will be the same. Let it be E volts.

Back emf on no load, $E = 250 - 0.1 \times 10 = 249$ V

Terminal voltage across machine A,

$$V_A = E - I_A R_a = 249 - (200 - I_B) \times 0.1 = 229 + 0.1 I_B$$

Terminal voltage across machine B,

$$V_B = E + I_B R_a = 249 + 0.1 I_B$$

Substituting $V_A = 229 + 0.1 I_B$ and $V_B = 249 + 0.1 I_B$ in Eq. (*i*), we have

$$(249 + 0.1\, I_B) I_B = (229 + 0.1 I_B)(200 - I_B) + 5{,}000 + 0.1 I_B^2 + 0.1(40{,}000 - 400 I_B + I_B^2)$$

or
$$249 I_B + 0.1 I_B^2 = 45{,}800 - 0.1 I_B^2 - 209 I_B + 5{,}000 + 0.1 I_B^2 + 4{,}000 - 40\, I_B + 0.1 I_B^2$$

or $498 I_B = 54{,}800$

or $I_B = \dfrac{54{,}800}{498} = 110$ A **Ans.**

and $I_A = 200 - 110 = 90$ A **Ans.**

(*a*) Voltage across machine A,

$$V_A = 229 + 0.1\, I_B = 229 + 0.1 \times 110 = 240 \text{ V} \textbf{ Ans.}$$

Voltage across machine B,

$$V_B = 249 + 0.1\, I_B = 249 + 0.1 \times 110 = 260 \text{ V} \textbf{ Ans.}$$

(*b*) Load on main generator,

$$I_g = I_1 - I_A = 2{,}000 - 90 = 1{,}910 \text{ A} \textbf{ Ans.}$$

Example 9.22. The input pd to a 3-wire distributor is 250 V between each outer and the mid point. At the far end of the cable a lighting load of 500 A is taken between the positive outer end and the middle wire, and one of 300 A between the mid wire and the negative outer. The resistance of each outer wire is 0.02 Ω and of the mid wire is 0.04 Ω. Calculate the pd across each of the loads and the transmission efficiency. If the mid wire becomes disconnected at the input end, what would be the pd across each load assuming that their resistances remain unchanged.

Solution : Resistances of outers, $R_{AA'} = R_{CC'} = 0.02\ \Omega$

Resistance of mid wire, $R_{BB'} = 0.04\ \Omega$

Current in +ve outer, $I_{AA'} = 500$ A

Current in –ve outer, $I_{CC'} = 300$ A

Current in mid wire, $I_{BB'} = 500 - 300 = 200$ A

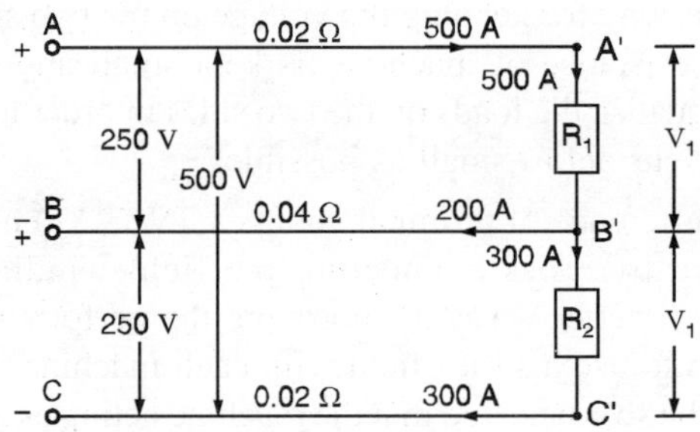

Fig. 9.32

Voltage on load end on +ve side,

$$V_{A'B'} = V_{AB} - (I_{AA'} R_{AA'} + I_{BB'} R_{BB'})$$
$$= 250 - (500 \times 0.02 + 200 \times 0.04) = 232 \text{ V} \textbf{ Ans.}$$

Voltage on load end on negative side,

$$V_{B'C'} = V_{BC} - (I_{BB'} R_{BB'} + I_{CC'} R_{CC'})$$
$$= 250 - [(-200) \times 0.04 + 300 \times 0.02) = 252 \text{ V} \textbf{ Ans.}$$

(Since in closed circuit BB′ C′C current in middle wire is in anticlockwise direction)

Note : In the closed circuit if current is clockwise the voltage drop be taken as +ve and if current is anticlockwise the voltage drop be taken as –ve.

$$\text{Power output} = 500 \times 232 + 300 \times 252 = 191{,}600 \text{ W}$$

$$\text{Line losses} = (500)^2 \times 0.02 + (300)^2 \times 0.02 + (200)^2 \times 0.04$$
$$= (5{,}000 + 1{,}800 + 1{,}600) \text{ or } 8{,}400 \text{ W}$$

$$\text{Line efficiency, } \eta = \frac{\text{Line output}}{\text{Line output + line losses}} \times 100$$

$$= \frac{191{,}600}{191{,}600 + 8{,}400} \times 100 = 95.8\% \textbf{ Ans.}$$

$$\text{Load resistance, } R_1 = \frac{232}{500} = 0.464\ \Omega$$

$$\text{Load resistance, } R_2 = \frac{252}{300} = 0.84\ \Omega$$

If the mid-wire is disconnected, the circuit becomes as shown in Fig. 9.33.

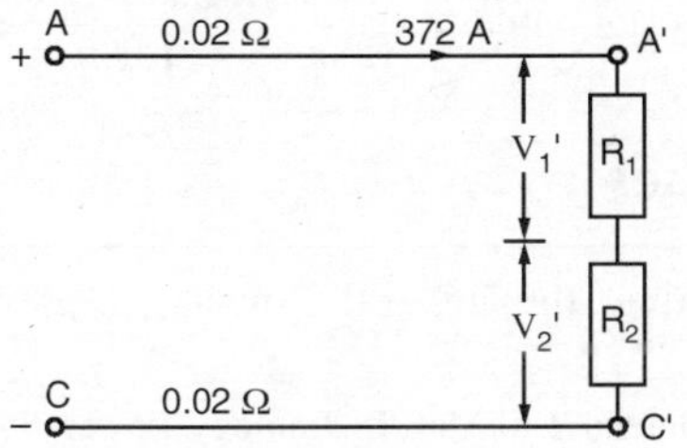

Fig. 9.33

Total resistance in the circuit,

R_T = Resistance of outers + $R_1 + R_2$

$= 2 \times 0.02 + 0.464 + 0.84 = 1.344\ \Omega$

Current supplied from 500 V supply, $I = \frac{V}{R_T} = \frac{500}{1.344} = 372$ A

PD across load resistor R_1, $V_{1'} = IR_1 = 372 \times 0.464 = 172.61$ V **Ans.**

PD across load resistor R_2, $V_{2'} = IR_2 = 372 \times 0.84 = 312.48$ V **Ans.**

Example 9.23. The three-wire dc system supplies a load of 4 Ω resistance across + ve wire and the neutral wire and load of 6 Ω resistance across – ve outer and neutral at the far end of distributor. The resistance of each conductor is 0.15 Ω and voltage across each outer is 240 volts at load end. Calculate the feeding end voltage.

Also determine the load current and load voltages when there is a break in the (*a*) neutral (*b*) positive outer and (*c*) negative outer. Assume the load resistance and feeding-end voltage remaining the same.

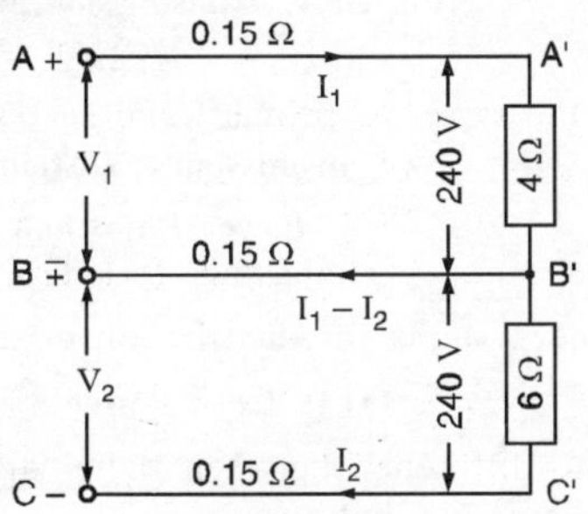

Fig. 9.34

Solution : Current in +ve outer, $I_1 = \frac{240}{4} = 60$ A

Current in – ve outer, $I_2 = \frac{240}{6} = 40$ A

Current in neutral wire = 60 – 40 = 20 A

Voltage at feeding end on + ve side $= V_L + I_{AA'}R_{AA'} + I_{B'B}R_{BB'}$

$= 240 + 60 \times 0.15 + 20 \times 0.15 = 252$ volts **Ans.**

Voltage across feeding end on – ve side

$= V_L + I_{BB'}R_{BB'} + I_{C'C}R_{CC'}$

$= 240 + (-20) \times 0.15 + 40 \times 0.15 = 243$ volts **Ans.**

(*a*) *When there is a break on neutral wire*

When there is a break on neutral wire, the system is just equivalent to a two-wire system having a load of (4 + 6) Ω connected across it at load end and pd (252 + 243) *i.e.*, 495 volts at feeding end.

Let the current in load resistance be of I amperes

Now $0.15I + 4I + 6I + 0.15I = 495$

or $I = \frac{495}{10.3} = 48.06$ A

Voltage across 4 Ω resistance = $4 \times 48.06 = 192.24$ V **Ans.**

Voltage across 6 Ω resistance = $6 \times 48.06 = 288.36$ V **Ans.**

(*b*) *When break is on +ve wire.* When there is break on +ve outer, no current will flow through the load of 4 Ω connected across +ve outer and neutral wire.

The system will be just equivalent to a two-wire system having a load of 6 Ω connected across it at the load end and a pd of 243 volts at feeding end.

Load current $= \frac{243}{0.15 + 6 + 0.15} = 38.57$ A **Ans.**

Voltage across the load = $6 \times 38.57 = 231.43$ volts **Ans.**

(*c*) *When break is on –ve outer.* When there is a break on –ve outer, no current will flow through the load of 6 Ω connected across –ve outer and neutral. The system will be just equivalent to a two-wire system having a load of 4 Ω connected across it at the load end and pd of 252 volts at feeding end.

Load current $= \frac{252}{0.15 + 4 + 0.15} = 58.6$ A **Ans.**

Load voltage = $58.6 \times 4 = 234.4$ volts **Ans.**

Example 9.24. A 3-wire dc distributor, with the neutral of half the cross section of either outers, 300 metres long is fed at one end, as shown in Fig. 9.35 (*a*). The resistance of each outer conductor is 0.005 Ω per metre. Calculate the potential difference at the farthest load point of 15 A.

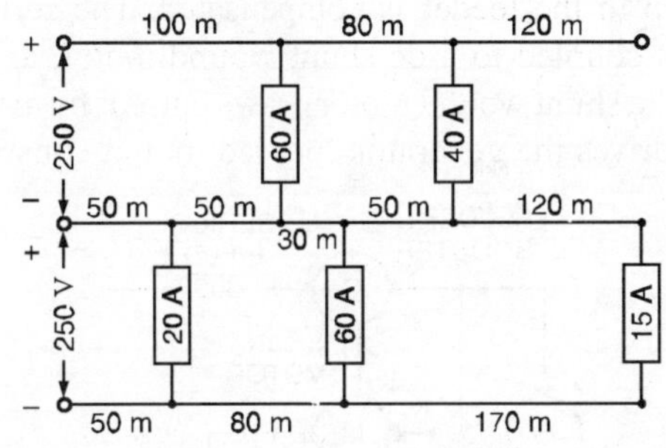

(*a*)

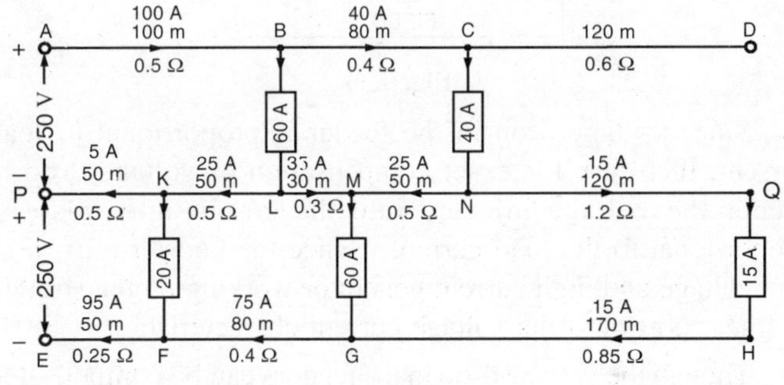

(*b*)

Fig. 9.35

Solution : Resistance of outers = 0.005 Ω per metre

Resistance of middle wire = $2 \times 0.005\ \Omega = 0.01\ \Omega$ per metre

Resistance of section AB of +ve outer, $R_{AB} = 0.005 \times 100 = 0.5\ \Omega$

Similarly resistances of other sections BC and CD of positive outer, EF, FG and GH of negative outer and PK, KL, LM, MN and NQ of middle wire are calculated. The resistances of various sections are indicated in Fig. 9.35 (*b*).

Current in section AB of +ve outer, $I_{AB} = 40 + 60 = 100$ A from A to B

Current in section BC of +ve outer, I_{BC} = 40 A from B to C

Current in section HG of –ve outer, I_{HG} = 15 A from H to G

Current in section GF of –ve outer, I_{GF} = 15 + 60 = 75 A from G to F

Current in section FE of –ve outer, I_{FE} = 75 + 20 = 95 A from F to E

Current in section QN of middle wire, I_{QN} = 15 A from N to Q

Current in section NM of middle wire, I_{NM} = 40 – 15 = 25 A from N to M

Current in section ML of middle wire, I_{ML} = 60 – 25 = 35 A from L to M

Current in section LK of middle wire, I_{LK} = 60 – 35 = 25 A from L to K

Current in section KP of middle wire, I_{KP} = 25 – 20 = 5 A from K to P

PD at the farthest load point of 15 A,

V_{QH} = Supply voltage – voltage drops in sections PK, KL, LM, MN, NQ, HG, GF and FE

$$= V_s - I_{PK}R_{PK} - I_{KL}R_{KL} - I_{LM}R_{LM} - I_{MN}R_{MN} - I_{NQ}R_{NQ} - I_{HG}R_{HG} - I_{GF}R_{GF} - I_{FE}R_{FE}$$

$$= 250 - (-5) \times 0.5 - (-25) \times 0.5 - 35 \times 0.3 - (-25) \times 0.5 - 15 \times 1.2 - 15 \times 0.85 - 75 \times 0.4 - 95 \times 0.25 = 250 - 67.5 = 182.5 \text{ V}$$ **Ans.**

Current in sections, PK, KL and MN being in counter-clockwise direction have been taken negative.

9.12. BOOSTERS

A booster is a series wound dc generator, which is inserted into a circuit to add or inject a certain voltage so that the excessive voltage drop in the feeder is compensated. The series wound generator is coupled to a dc shunt wound motor, as shown in Fig. 9.36. The shunt wound motor draws current from the supply mains and drives the generator coupled to it at constant speed.

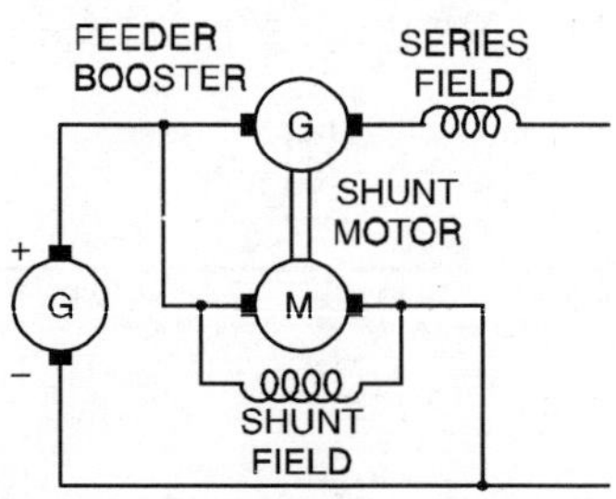

Fig. 9.36

Since voltage drop in the feeder is proportional to load current, therefore for exact compensation of voltage drop in feeder the voltage injected into the circuit must also be proportional to the load current. Hence the booster must be a low voltage and high current generator working on the straight or linear portion of its voltage-current characteristic.

Though the voltage drop in the feeders can be compensated by using compound wound generators but in case of feeders of different lengths the use of booster is necessary and convenient.

Example 9.25. A 2-wire system has the voltage at the supply end maintained at 500 V. The line is 5 km long. If the load current is 100 A, what must be the booster voltage and output so that voltage at the far end be maintained at 500 V. The resistance of the cable is 0.4 Ω/km.

Solution : Total resistance of the line = 5 × 0.4 = 2.0 Ω

Voltage drop in the line at full load = 100 × 2 = 200 V

Hence the booster voltage, V = 200 V **Ans.**

$$\text{Booster output} = \frac{V \times I}{1{,}000} = \frac{200 \times 100}{1{,}000} = 20 \text{ kW}$$ **Ans.**

Exercises

1. Write down the general aspects of power distribution system.
   [West Bengal Univ. of Technology Power System-I, 2009-10]
2. Name the different methods of distribution and the principle of distributor design.
   [A.M.I.E. Sec. B. Power Systems Summer 1993]
3. What are the basic considerations which influence the design of distribution system ?
   [Rajasthan Technical Univ. Transmission and Distribution of Electrical Power 2008]
4. Discuss the design considerations in distribution system.
   [M.D. Univ. Power System-I, Dec. 2010]
5. What is a distribution system ? Describe briefly the design considerations in distribution system.
   [M.D. Univ. Power System-I May 2009]
6. Explain different types of distribution systems with the help of neat sketches.
7. Compare (*i*) radial distribution system (*ii*) ring or loop distribution system and (*iii*) network or grid distribution system. State applications.
   [Pb. Univ. Transmission and Distribution of Electrical Energy April 1982]
8. What are the types of primary and secondary distribution systems ? **[Transmission & Distribution of Electrical Power Rajasthan Univ. 2004, 2006; Rajasthan Technical Univ. 2008, 2009]**
9. Classify the systems for distribution of dc power.
   [M.D. Univ. Power System-I, Dec. 2006]
10. Explain the difference between radial distribution system and ring main distribution system.
    [M.D. Univ. Power System-I, May 2008]
11. Explain about mesh distribution network.
    [Pb. Technical Univ. Power System-I, Dec. 2009]
12. Discuss the radial and mesh distribution networks.
    [Pb. Technical Univ. Power System-I, Dec. 2007]
13. Explain 2-wire dc system of distribution of electrical power.
14. What are the advantages of a doubly fed distributor over singly fed distributor ?
15. Prove that the voltage drop diagram for a uniformly loaded distributor fed at one end is parabola.
    [A.P.S. Univ. Elec. Power Generation, Transmission and Distribution 1977, Agra Univ. Transmission and Distribution of Electrical Power 1979 Suppl.]
16. Derive an expression for the voltage drop for a uniformly loaded distributor fed at one end.
    [M.D. Univ. Power System-I, Dec. 2010]
17. Compare the 2-wire and 3-wire systems of dc distribution in respect of weights of copper conductors, all other conditions being the same.

18. State the advantages of a 3-wire dc distribution system over a 2-wire dc distribution. What is balancer ? State its function, with suitable examples. **[U.P.S.C I.E.S. Elec. Engg. 1971]**

19. How balancers are used to maintain equal voltages on two sides of neutral in a 3-wire dc distribution system ? Explain.

20. Explain the location, action and use of a booster with suitable diagrams.

21. Write short notes on the following :

(*i*) Systems of dc distribution. **[U.P.S.C I.E.S. Elec. Engg. 1973]**

(*ii*) Radial and ring-main distributor. **[B.T.E. U.P. 1998]**

(*iii*) Radial and mesh distribution networks. **[P.T.U. May 2008]**

(*iv*) 3-wire dc distribution system.

(*v*) Balancer set

(*vi*) Boosters

(*vii*) Design of distributors.

Short Answer Type Questions With Answers

Q. 1. What is distribution system ? **[M.D. Univ. May 2009]**

Ans. The part of power system which distributes electric power for local use is known as distribution system.

In general, the conductor system, by means of which electrical power is conveyed from bulk power source or sources (generating stations or major substations supplied over transmission lines) to the consumers is known as distribution system. It generally consists of feeders distributors and service mains, as shown in Fig. 1.

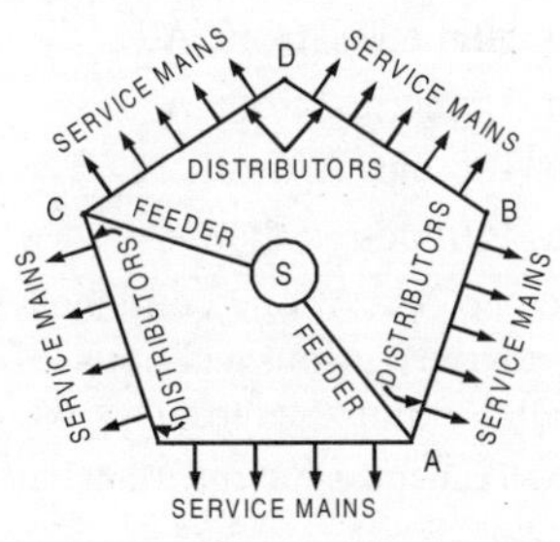

Fig. 1

Q. 2. On what factors does the primary distribution voltage depend ?

Ans. The voltage employed for primary distribution system depends upon the amount of power to be supplied and distances of the substations required to be supplied.

Q. 3. What is/are the primary distribution voltage/voltages adopted in India ?

Ans. The usual value of primary distribution voltage in India is 11 kV but in case of high load density area it may be higher (33 or 66 kV) even.

Q. 4. What is the declared consumer voltage as per IE rules ?

Ans. The declared consumer voltage, according to IER is 415/240 V (*i.e.*, 415 V between phases and 240 V between phase and neutral) with a permissible voltage variation of $\pm 6\%$.

Q. 5. Why is it necessary to keep voltage variation at the consumer's terminals as low as possible ?

Ans. All the appliances and motors are designed to be operated at a particular or declared value of voltage. Large voltage variation means either a too high or a too low voltage. Low voltage will cause not only in burning of motors/appliances/equipment and insufficient lighting but also loss of revenue to the supplier. On the other hand high voltage will cause the lamps to burn out permanently and also may cause failure of other appliances. So the primary requirement of a distribution system is that voltage variations at the consumer's terminals should be the lowest possible (not more than $\pm 6\%$ of the rated value).

Q. 6. What is a feeder ?

Ans. Feeder is a line which supplies a point of distribution network without being tapped at any intermediate point. In simple terms, the feeder may be defined as the line carrying current from the secondary substation to the distribution substations or as a primary distribution line.

Q. 7. What is a distributor ?

Ans. Distributor is a line from which tappings are taken along its length for providing supply to the consumers.

Q. 8. Why the voltage drop is of primary importance in the design of the distributors ?

Ans. The voltage drop is of primary importance in the design of distributors because of the statutory regulation providing the voltage variations permissible at the consumers terminals (±6% of declared or rated value). The size of the distributor should be such that the voltage at the consumers terminals remains within the permissible limits.

Q. 9. What is meant by primary and secondary distribution systems ?

Ans. ***Primary Distribution System.*** From generating stations the electrical power is usually transmitted to various substations through extra high tension transmission lines at voltage from 33 to 220 kV. At these substations this voltage is stepped down to 11 or 6.6 or 3.3 kV and power at this voltage is conveyed to different substations for distribution and to the bulk supply consumers. Such a system is known as *high voltage or primary distribution system.*

Secondary Distribution System. At distribution substations the voltage is stepped down to 415 volts. From these substations various low voltage (415 volts between phases and 240 volts between phase and neutral) distributors radiate out and feed the consumers. This system of distribution is known as *low voltage or secondary distribution system.*

Q. 10. What is meant by radial and ring main (mesh) systems of distribution ?

Ans. In radial distribution system, the feeder radiates from the secondary substation and branches into subfeeders and lateral which extends into all parts of the area served. Radial feeder is the simplest, most economical and most commonly used one. It has the drawbacks of poor reliability and greater voltage drop feeder. In the ring main (mesh) scheme, the distributor makes a closed loop or ring, as shown in Fig. 9.22 and fed at or more than one points. Such an arrangement has the advantages of greater reliability and better flexibility.

Q. 11. Why dc 2-wire system is never employed for transmission purpose ?

Ans. Because of poor efficiency.

Q. 12. Why is radial system of distribution is not generally used ?

Ans. Because of poorest service reliability.

Q. 13. What are the advantages of ring main distributor ?

Ans. Better flexibility, greater reliability, better voltage regulation and less power losses.

Q. 14. Why is neutral wire taken of half cross section of either outer in a 3-wire dc distribution system ?

Ans. Since only out of balance current flows through the middle or neutral wire, so generally neutral wire is taken of half cross section of either outer.

Q. 15. What is a booster ?

Ans. Booster is a low-voltage, high-current series wound dc generator inserted into a dc circuit to add or inject a certain voltage proportional to circuit current.

Q. 16. For what purpose a booster is employed ?

Ans. A booster (low-voltage, high-current series wound dc generator) is used for compensating voltage drop in the feeder.

Problems

1. A 2-wire dc distributor cable AB is 2 km long and supplies loads of 100 A, 150 A, 200 A and 50 A situated 500 m, 1,000 m, 1,600 m and 2,000 m from the feeding point A. Each conductor has a resistance of 0.01 Ω per 1,000 m. Calculate the voltage at each load point if a voltage of 300 V is maintained at point A.

[**Ans.** 295 V, 291 V, 288 V and 284 V]

2. A 2-wire distribution AG, 1,200 metres long is fed at A at 250 V. Loads of 80, 70, 100, 90, and 60 A are tapped from points B, C, D, E and F whose distances from A are 250, 400, 600, 800 and 1100 metres respectively. If the voltage at G is 210 V, find the resistance per metre of the distributor.

[**Ans.** 1.625×10^{-4} Ω/m]

3. A 2-wire dc distributor AB is 300 metres long. It is fed at point A. The various loads and their positions are given below :

At Point	*Distance from A in metres*	*Concentrated loads in amperes*
C	40	30
D	100	40
E	150	100
F	250	50

If the maximum permissible voltage drop is not to exceed 10 V, find the cross-sectional area of the distributor. Take $\rho = 1.78 \times 10^{-8}$ Ω-m.

[Bombay Univ. Elec. Power-I, April 1971] [**Ans.** 1.163 cm^2]

4. Figure 1 shows a tramway distribution system with two

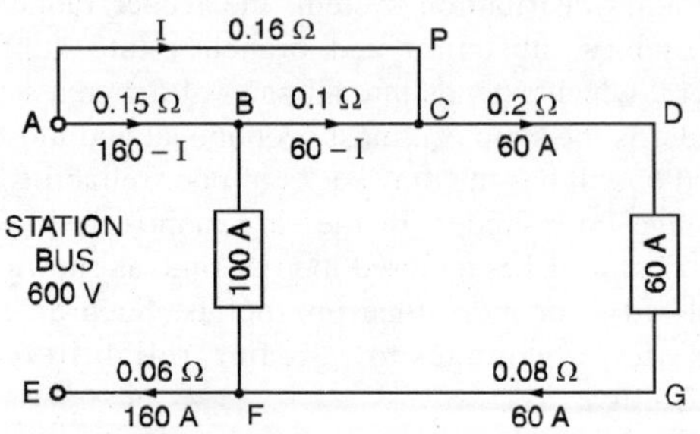

Fig. 1

tram cars B and D taking 100 A and 60 A respectively. The voltage at station bus is 600 V. The resistance of overhead trolley and track system is given below. Calculate the voltage at each car. R_{AB} = 0.15 Ω, R_{BC} = 0.10 Ω, R_{CD} = 0.20 Ω, R_{APC} = 0.16 Ω, R_{EF} = 0.06 Ω and R_{FG} = 0.08 Ω.

[**Ans.** 577.38 V; 548.87 V]

5. A 200 m long distributor is fed from both ends A and B at the same voltage of 250 V. The concentrated loads of 50, 40, 30 and 25 A are coming on the distributor at distances of 50, 75, 100 and 150 m respectively from end A. Determine the minimum potential and locate its position. Also determine the current in each section of the distributor. It is given that the resistance of the distributor conductor is 0.08 Ω per 100 metres for go and return.

[Gujarat Univ. Elec. Power System October, 1970]

[**Ans.** 245.98 V at point D; 83.75 A, 33.75 A, 6.25 A, 36.25 A, 61.25 A]

6. A two-wire distributor is loaded as shown in the Fig. 2, the two ends A and F being fed at 248 and 250 volts respectively. If the resistance of the distributor per metre is 400 micro ohms for both lines, calculate the minimum voltage and show the position where it occurs.

Distances are shown in metres and the loads in amperes.

[U.P.S.C. I.E.S. Electrical Engineering 1970]

[**Ans.** 244.16 V at loading point D]

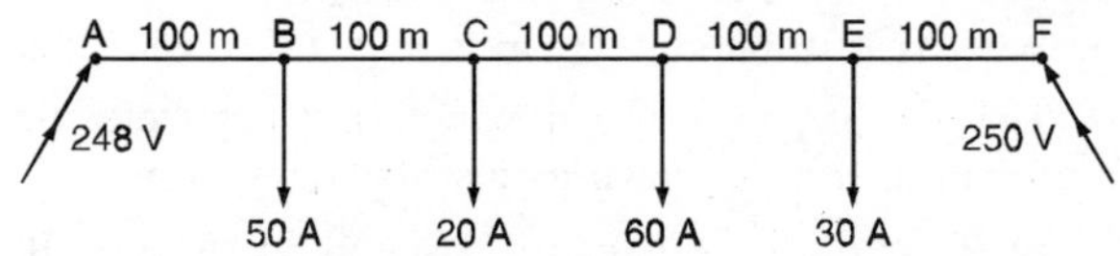

Fig. 2

7. A distributor AB is fed from both ends. At feeding point A, the voltage is maintained at 236 V and at B 237 V. The total length of the distributor is 200 metres and loads are tapped off as under :

(*i*) 20 A at 50 metres from A

(*ii*) 40 A at 75 metres from A

(*iii*) 25 A at 100 metres from A

(*iv*) 30 A at 150 metres from A

The resistance per km of one conductor is 0.04 Ω. Calculate the currents in the various sections of the distributor, the minimum voltage and the point at which it occurs.

[Madras Univ. Transmission and Distribution September 1969]

[**Ans.** 53.75 A, 33.75 A, 6.25 A, 31.25 A, 61.25 A, 233.18 V at point D]

8. A distributor of 500 metre length fed at both ends at 230 V is loaded with loads of 100, 60, 40 and 100 A at distances of 50, 150, 250 and 400 metres respectively. If the maximum voltage drop of 5.5 volts is to be allowed, find the cross-sectional area of each conductor and point of minimum voltage. Specific resistance of conductor material may be taken as 1.73×10^{-6} ohms per cm cube.

[**Ans.** 1.06 cm^2, Minimum voltage point is E]

9. Calculate the voltage at a point 300 metres from the feeding point maintained at 250 volts when the cable is 1,000 metres long, has a resistance of 0.05 Ω per km for each conductor and supplies a uniform load of 0.5 A/metre.

Find also the voltage at this point when the cable is supplied at 250 volts at both ends. **[Pb. Univ. April 1965]**

[**Ans.** 237.25 V, 244.75 V]

10. A 2-core distributor cable 400 metres long supplies a uniformly distributed lighting load of 1 A per metre. There are concentrated loads of 120, 72, 48 and 120 A at 40, 120, 200 and 320 metres respectively from the end A. The cable has a resistance of 0.15 Ω/km run. Determine the position of the lowest-run lamp and its voltage when the cable is fed at 250 V from both ends A and B.

[A.P.S Univ. Elec. Power Generation, Transmission and Distribution 1977]

[**Ans.** 48 A load point ; 239.1 volts]

11. In a dc ring main a voltage of 400 V is maintained at A. At B, 500 metres away from A, a load of 150 A is taken; at C 300 metres from B, a load of 200 A is taken. The distance between A and C is 700 metres. The resistance of each conductor of the mains is 0.03 Ω per 1,000 metres. Find the voltage at B and C and also find the current in section BC. **[Kerala Univ. 1969]**

[**Ans.** V_B = 394.2 V; V_C = 393.42 V; I_{BC} = 43.3 A flowing from B and C]

12. A ring main is fed at the point A at a constant voltage of 240 volts. The loads are as shown. Find the potential difference at each load point. The resistances given are those of each conductor. [**Ans.** 234 V, 228.65 V, 227.13 V, 230.63 V]

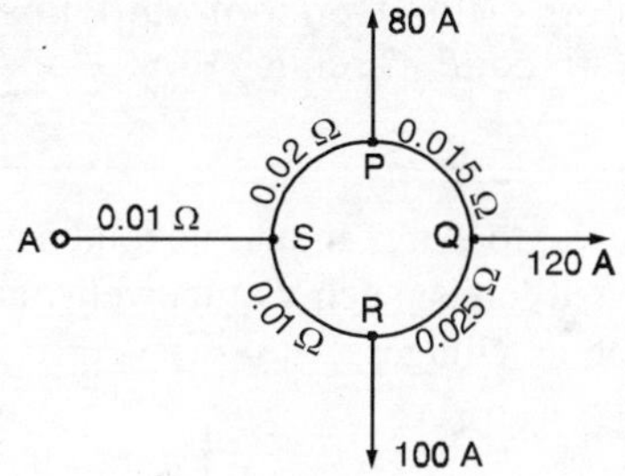

Fig. 3

13. On a 3-wire dc generating station with 500 V between outers, there is a load of 1,500 kW on the +ve side and 2,000 kW on the –ve side. Calculate the current in the neutral and in each of the balancer armatures and the total current supplied by the generator. Neglect losses.

[Madras Univ. Sept. 1968]

[**Ans.** 2,000 A; 1,000 A and; 7,000 A]

14. The load on a dc 3-wire system employing a rotary balancer with 500 V between outers consists of lighting loads of 300 A on the positive side and 500 A on the negative side. Power loads taking 600 A are connected across the outers. Calculate the load in kW on the main generator and on each of the balancer machines. Assume a loss of 2.0 kW in each balancer machine.

[U.P.S.C I.E.S. Generation, Transmission and Distribution of Power 1972]

[**Ans.** 504 kW; 27 kW; 23 kW]

15. A 3-wire feeder carries a current of 120 A on the +ve side and 100 A and on the –ve side. The supply voltage is maintained at 220 volts between each outer and middle wire at the sending end. Determine the voltage at the receiving end if the resistance of each outer is 0.11 Ω. The cross section of middle wire is half of that of either outer.

[**Ans.** 202.4 V; 213.4 V]

16. In a 3-wire system of dc distribution the supply voltage is 250 V on each side. The load on one side is 3 Ω and on the other side is 4 Ω. The resistance of each of the three conductors is 0.05 Ω. Find the load voltage.

[**Ans.** 244.9 V; 247.9 V]

17. A 3-wire direct current system is loaded as shown in Fig. 4 Determine the voltage between the middle wire and the outer at each load point.

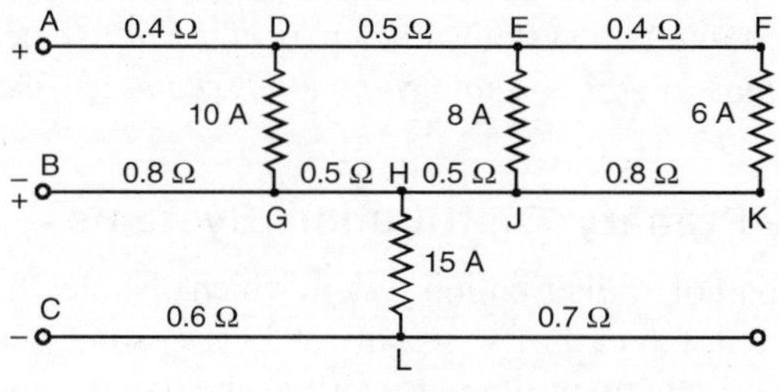

Fig. 4

[U.P.S.C. I.E.S. Electrical Engineering 1971]

[**Ans.** V – 16.8; V – 30.3; V – 37.5, V – 2.3 V]

❑❑❑

CHAPTER 10

AC Distribution

INSIDE THIS CHAPTER

10.1. INTRODUCTION

As already mentioned in Art. 9.1, the distribution systems may be divided into two systems known as *high voltage* or *primary distribution systems* and *low voltage or secondary distribution systems*.

10.1.1. Primary Distribution Systems

It is that part of ac distribution system which operates at voltages (such as 3.3, 6.6 or 11 kV; usually at 11 kV) somewhat higher than that of general utility (415/240 V) and handle large blocks of power. As already mentioned in chapter 1, the primary distribution is universally carried out by 3-phase, 3-wire system.

Electric power from the generating station is transmitted, through extra high tension transmission lines at voltages from 33 to 765 kV, to various substations located in or near the city. At these substations, called the *secondary substations*, the voltage is stepped down to 11, 6.6 or 3.3 kV (usually to 11 kV) with the help of power transformers for *primary distribution.*

The 33/11 kV secondary substations are usually located in the area having load requirement of the order of 5 MVA and normally a primary distribution line or a feeder is designed to carry a load of 1 – 2 MVA, so the number of feeders emanating from a secondary substation of 33/11 kV is 3-4. For the load exceeding 8 MVA the secondary transmission is carried out at 66 kV so as to reduce the line losses and, therefore, secondary substations are 66/11 kV. The number of feeders emanating from a 66/11 kV secondary substation is 6 – 8. The feeders may be radial, parallel, ring or *interconnected*.

1. Radial Feeders. It derives its name from the fact that the feeder radiates from the secondary substation and branches into subfeeders and laterals which extend into all parts of the area served. The distribution transformers (11 kV/415 V) are connected to the primary feeders, subfeeders and laterals, usually through fuse cut-outs. Radial feeder is the simplest, most economical and most commonly used one. It is advantageous for supplying power to heavy industrial load near the secondary substation, isolated loads, such as tubewells, and areas of low load density such as villages.

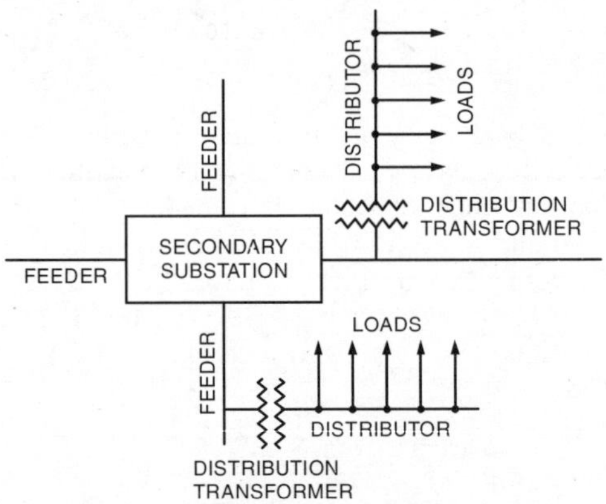

Fig. 10.1. ***Radial Feeder System***

However, it suffers from the following drawbacks:

1. *Poor reliability.* When a fault occurs at any point on the feeder, supply to all consumers beyond the fault point towards the tail end gets interrupted.
2. In case of increase in load demand, the length of the feeder has to be extended and it may result in a greater voltage drop. It may cause the voltage towards the tail end to reach a value below the permissible value. This can be avoided by bifurcating the feeder but it will not prove economical.

2. Parallel Feeders. In a parallel feeder system, two radial feeders originating from the same or different secondary substations are run in parallel. Each feeder, though capable of supplying the entire load, shares the total load equally in normal conditions. Though this system is expensive but reliability is increased as in case of fault on one feeder, the total load can be supplied by the healthy feeder. Interruption of supply is only for the time duration, that is taken in transferring the load from the faulty feeder to healthy one either by manual or automatic switches. Such a system is employed wherever the continuity of supply is of greater importance.

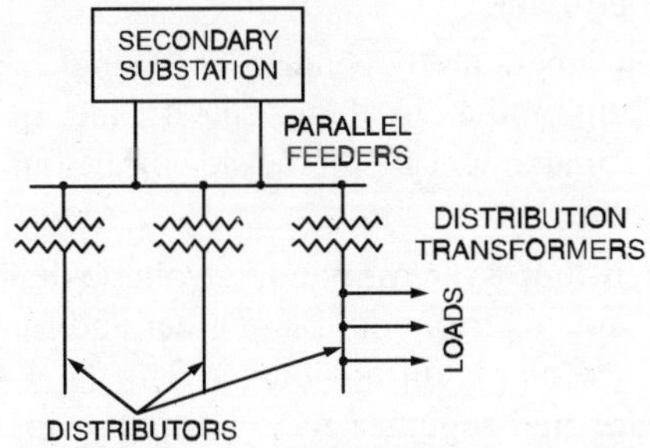

Fig. 10.2. *Parallel Feeder System*

3. Loop Feeders. In a loop feeder system, two or more radial feeders originating from the same or different secondary substations are laid on different routes of load areas. The arrangement having the ends of two feeders tied together through normally open switching devices is known as *open-loop system* while the arrangement having the ends of two feeders tied together through normally closed switching devices is called the *ring main feeder* or the *ring loop*. This system is by far the most reliable for continuity of supply and gives better voltage regulation and less power losses.

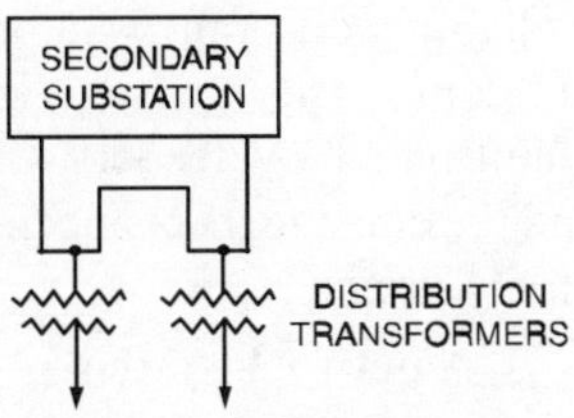

Fig. 10.3. *Loop Feeder System*

The use of this system is prohibitive on account of very large investment involved. It is used where continuity of supply is the priority factor and thus large investment may be warranted.

Feeders and loop components must have sufficient reserve capacity to handle the load that may be transferred under abnormal conditions.

4. Interconnected Network System. When the feeder ring main is energized from two or more than two generating stations or substations it is called the *interconnected network system.* Because of interconnected feeders, power can be supplied to all the distribution transformers even though a part of network may be out of service. Such a system provides better reliability and flexibility and is employed in large metropolitan cities where continuity of supply is the most important.

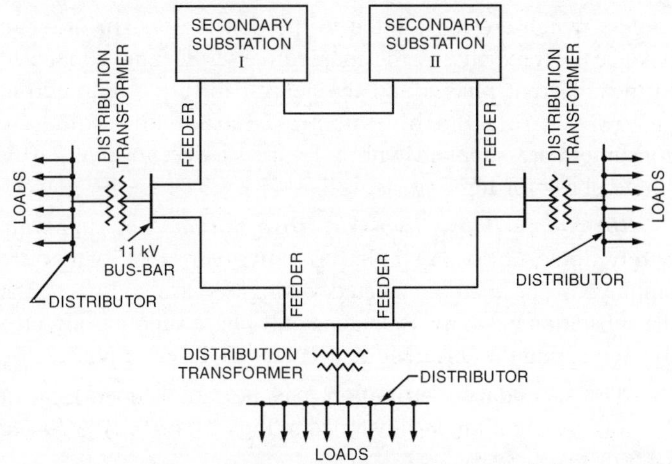

Fig. 10.4. *Interconnected Network System*

10.1.2. Secondary Distribution Systems

As already said in chapter 9 that in case of ac supply 3-phase 3-wire system is employed for balanced loads such as power loads, 3-phase 4-wire system is employed for unbalanced loads such as light and power loads combined and single-phase two-wire system is used for lighting and small power appliances.

(a) Three-Phase 4-Wire Distribution. Three-phase, 4-wire distribution layout is shown in Fig. 10.5. Transformer substations are built at or near the load centres of the area where the load is to be supplied. The substation contains high voltage switchgear

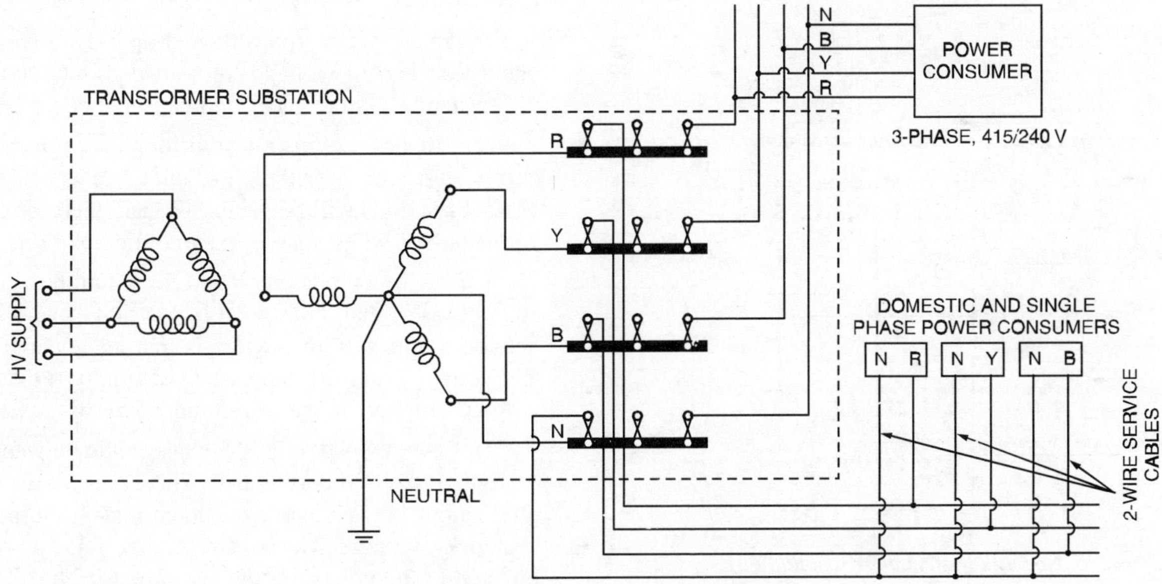

Fig. 10.5. *Three-Phase, Four-Wire Distribution*

and bus-bars, and low voltage fuses or links. The supply to the primary is by high voltage or extra high voltage feeder cable from the generating station. The low voltage bus-bars are distinguished by colour marking : red, yellow, and blue, for the phases, and black for the neutral. The local 4-wire distributors are connected in parallel to the bus-bars through the fuses or links, and the separate cores are also distinguished by similar colour marking. Domestic and other low voltage consumer's premises are connected to the distributors by 2-wire service cables, which are tee jointed to the distributors. In order to balance the three phases, the consecutive services are connected to the different phases and the neutral in turn : red-neutral, yellow-neutral, and blue-neutral. Larger non-domestic consumers are supplied with a 4-wire service, to give 240 V single phase for lighting, as well as 415 V 3-phase for power.

(b) Single-Phase Two-Wire Distribution. For supplying purely domestic loads single phase two-wire distributors are employed. These are connected to single phase and neutral at the substation bus-bars. Balancing of 3-phase supply is affected by connecting the different distributors to different phases.

The secondary distribution may be radial, open-loop or network distribution, as illustrated in Figs. 10.6(*a*), 10.6(*b*) and 10.6(*c*) respectively.

In **radial system,** the low voltage distributors radiate out from the distribution transformer and run through the areas to be supplied by them. This system is the simplest and the least expensive system. This system generally not used because of the following drawbacks.

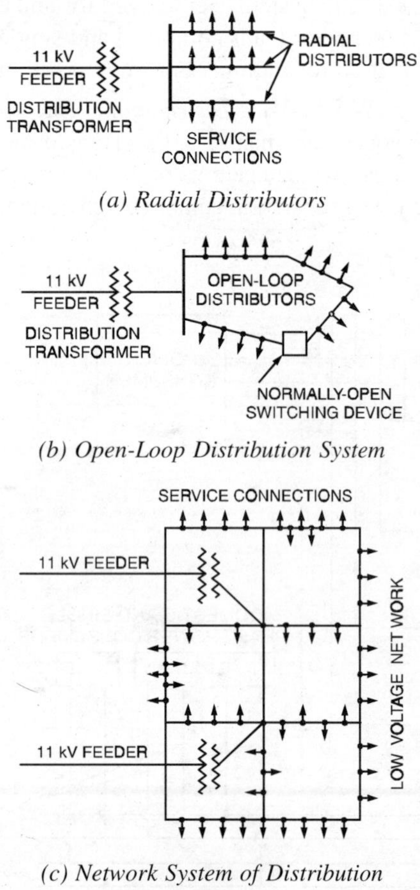

(a) Radial Distributors

(b) Open-Loop Distribution System

(c) Network System of Distribution

Fig. 10.6

(*i*) The end of the distributor nearest to the supply end would be heavily loaded.

(*ii*) The consumers at the farthest end of the distributor would be subjected to serious voltage fluctuations with the variation in load.

(*iii*) The consumers are dependent on a single distributor so that a fault on the distributor cuts off the supply from all the consumers who are on the side of the fault away from the station.

In an **open-loop system,** two low voltage distributors take off from the same distribution transformer and run in different directions and so supply different areas. The far ends of the two distributors are tied together by a normally-open switching device. In the case of a fault on one line, the load can be supplied, though partially, from the other line. Thus in this system, the reliability of supply is improved and cost is increased marginally in comparison to those in case of radial system. However, a fault on 11 kV feeder will cause the interruption of supply to all the consumers connected to this feeder through a distribution transformer.

A **network system of distribution** consisting of a number of interconnected distributing lines is supplied from two or more distribution transformers operating in parallel, as illustrated in Fig. 10.6 (*c*). This arrangement has the following advantages:

1. The reliability of supply is improved because in case one transformer fails, load can be supplied by the other one.
2. The voltage drop in this system is reduced resulting in a better voltage regulation.
3. A better load distribution on transformers is achieved, by diversifying the load on a bank of transformers.

Such a system is quite suitable for serving high load density metropolitan areas.

10.2. AC DISTRIBUTION CALCULATIONS

AC distribution differs from the dc distribution in the following respects:

(*i*) In dc system, the voltage drop is due to resistance alone but in ac system the voltage drops are due to the combined effects of resistances, inductances and capacitances.

(*ii*) In dc system, all additions and subtractions of the currents are simply arithmetical but in a ac system all additions and subtractions will be vectorially and therefore, currents and voltages will be expressed in symbolic notations.

(*iii*) All methods employed for solution of dc distributors and networks shall hold good in solution of ac distributors and networks, the only difference is that all impedances, currents and voltages shall be expressed vectorially (*i.e.*, symbolically) with current or voltage as reference phasor.

(*iv*) At the load point the phase angle between voltage and current plays an important part. There are two ways of referring phase angle between voltage and current: (*a*) it may be referred to supply voltage which is reference phasor, (*b*) it may be referred to the voltage at the load point itself.

10.3. AC DISTRIBUTORS WITH CONCENTRATED LOADS

As mentioned above there are two ways of referring power factor namely: (*i*) with reference to receiving-end voltage; and (*ii*) with reference to respective load voltages.

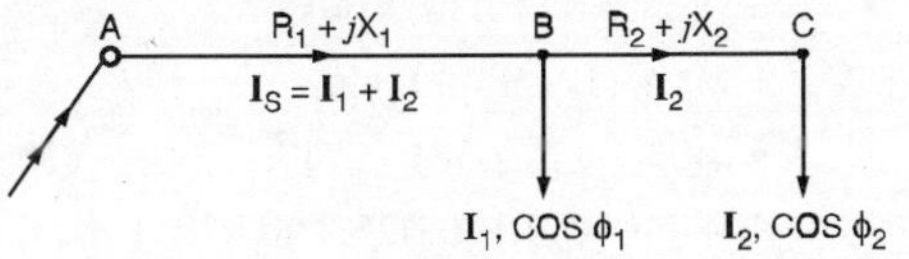

Fig. 10.7

1. *Power Factors Referred to Receiving-End Voltage.* Consider an ac distributor AC with concentrated loads I_1 and I_2 at points B and C respectively, as shown in Fig. 10.7. Taking voltage at receiving end (V_C) as reference phasor, let the power factors of the loads at points B and C be cos ϕ_1, and cos ϕ_2 lagging respectively. Let R_1, X_1 and R_2, X_2 be the resistances and reactances of sections AB and BC of the distributor. Phasor diagram is shown in Fig. 10.8, which does not require any further explanation.

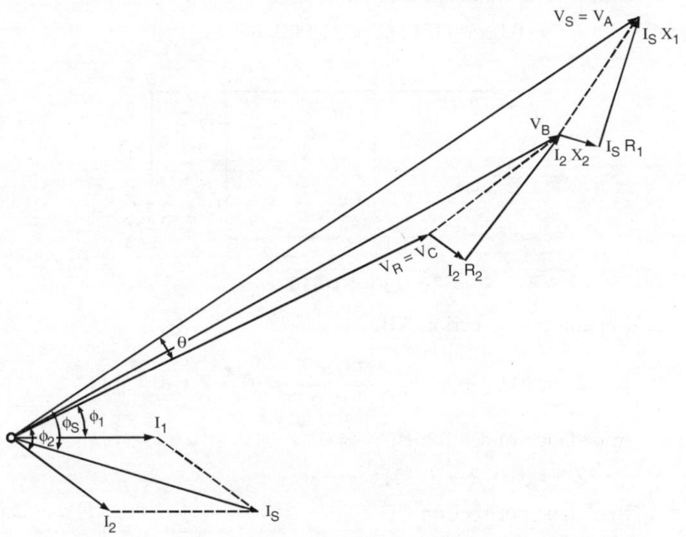

Fig. 10.8. *Phasor Diagram*

Load current at load point B, $\mathbf{I_1} = I_1(\cos \phi_1 - j \sin \phi_1)$

Load current at load point C, $\mathbf{I_2} = I_2(\cos \phi_2 - j \sin \phi_2)$

Impedance of section AB, $\mathbf{Z_1} = (R_1 + jX_1)$

Impedance of section BC, $\mathbf{Z_2} = (R_2 + jX_2)$

Current in section BC $= \mathbf{I_2} = I_2 (\cos \phi_2 - j \sin \phi_2)$

Current in section AB,

$$\mathbf{I_S} = \mathbf{I_1} + \mathbf{I_2} = I_1 (\cos \phi_1 - j \sin \phi_1) + I_2(\cos \phi_2 - j \sin \phi_2)$$

Voltage drop in section BC

$$= \mathbf{I_2 Z_2} = I_2 (\cos \phi_2 - j \sin \phi_2)(R_2 + jX_2)$$

Voltage drop in section AB

$$= (\mathbf{I_1} + \mathbf{I_2})\, \mathbf{Z_1}$$
$$= [I_1 (\cos \phi_1 - j \sin \phi_1) + I_2(\cos \phi_2 - j \sin \phi_2)](R_1 + jX_1)$$

Sending-end current, $\mathbf{I_S}$

$$= \mathbf{I_1} + \mathbf{I_2} = I_1 (\cos \phi_1 - j \sin \phi_1) + I_2(\cos \phi_2 - j \sin \phi_2)$$

Sending-end voltage, $\mathbf{V_S}$

$$= \mathbf{V_C} + \mathbf{I_2 Z_2} + (\mathbf{I_1} + \mathbf{I_2})\mathbf{Z_1}$$
$$= V_C + I_2 (\cos \phi_2 - j \sin \phi_2)(R_2 + jX_2) + [I_1 (\cos \phi_1 - j \sin \phi_1) + I_2(\cos \phi_2 - j \sin \phi_2)](R_1 + jX_1)$$

Sending-end power factor = Cos ϕ_S

2. *Power Factors Referred to Respective Load Voltages.* Let the power factors of the loads at point B and C be cos ϕ_1 and cos ϕ_2, lagging behind their respective load voltages V_B and V_C respectively. The phasor diagram is shown in Fig. 10.9. Voltage drops in sections BC and AB and sending-end voltage, current and power factor can be determined as in case 1.

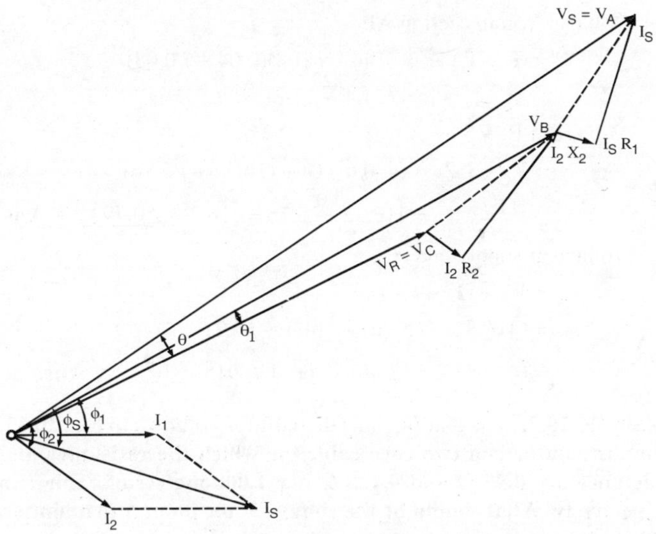

Fig. 10.9. *Phasor Diagram*

10.4. AC INTERCONNECTED SYSTEMS

The usual methods employed for solution of network problems in interconnected systems, in addition to a method of trial and error are: (1) direct application of Kirchhoff's laws; (2) Superposition of currents; (3) Circulating current method; and (4) Thevenin's theorem. The trial and error method is useful only when a single operating condition is required. In certain cases star-delta transformation makes the calculations simplified.

Example 10.1. A 2-wire feeder ABC has a load of 120 A at C and of 60 A at B both at pf 0.8 lagging. The impedance AB is (0.04 + *j* 0.08) Ω and that of BC is (0.08 + *j*0.12) Ω. If the voltage at the far end C is to be maintained at 400 V, determine the voltage at A and B. [*M.D. Univ. Power System-I, Dec. 2007*]

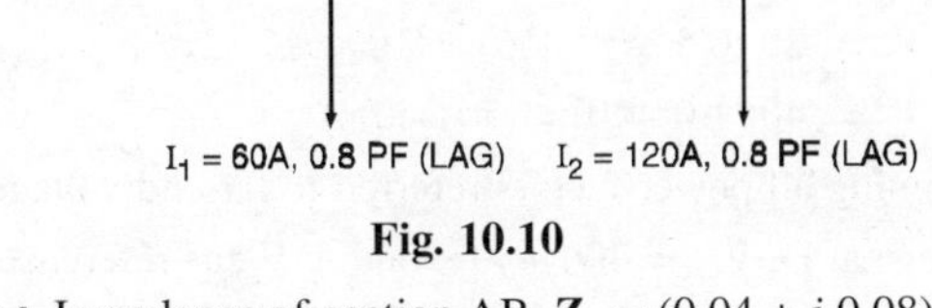

Fig. 10.10

Solution : Impedance of section AB, $\mathbf{Z_1} = (0.04 + j\,0.08)\ \Omega$

Impedance of section BC, $\mathbf{Z_2} = (0.08 + j\,0.12)\ \Omega$

Taking voltage at far end C, V_C as reference phasor and assuming that power factors of currents at load points B and C are given referred to the receiving-end voltage at C, V_C

$$\mathbf{V}_C = 400 + j0$$

Current through section BC

$$= \mathbf{I}_2 = 120\,(0.8 - j\,0.6) = (96 - j\,72)\text{ A}$$

Current through section AB

$$= \mathbf{I}_1 + \mathbf{I}_2 = 60\,(0.8 - j\,0.6) + 120\,(0.8 - j\,0.6) = (144 - j\,108)\text{ A}$$

Voltage drop in section BC

$$= \mathbf{I}_2\mathbf{Z}_2 = (96 - j\,72)(0.08 + j\,0.12) = (16.32 + j\,5.76)\text{ volts}$$

Voltage drop in section AB

$$= (\mathbf{I}_1 + \mathbf{I}_2)\,\mathbf{Z}_1 = (144 - j\,108)(0.04 + j\,0.08) = (14.4 + j\,7.2)\text{ volts}$$

Voltage at B, $\mathbf{V}_B$

$$= \mathbf{V}_C + \mathbf{I}_2\mathbf{Z}_2 = (400 + j\,0) + (16.32 + j\,5.76)$$

$$= 416.32 + j\,5.76 = 416.36\ \angle 0.793°\text{ V}$$ **Ans.**

Voltage at supply end A,

$$\mathbf{V}_A = \mathbf{V}_B + (\mathbf{I}_1 + \mathbf{I}_2)\,\mathbf{Z}_1 = (416.32 + j\,5.76) + (14.4 + j\,7.2)$$

$$= (430.72 + j\,12.96)\text{ V or } 430.915\ \angle 1.72°\text{ V}$$ **Ans.**

Example 10.2. The loading on a distributor is shown in Fig. 10.11. The distributor is a two-core cable for which the resistance and reactance are 0.25 Ω and 0.125 Ω per 1,000 metres of cable run respectively. What should be the voltage at the point A to maintain 400 V at point D ?

Solution : The impedance of cable

$$= (0.25 + j0.125)\ \Omega \text{ per 1,000 metres}$$

Impedance of section AB,

$$\mathbf{Z}_1 = (0.25 + j0.125) \times \frac{100}{1{,}000} = (0.025 + j0.0125)\ \Omega$$

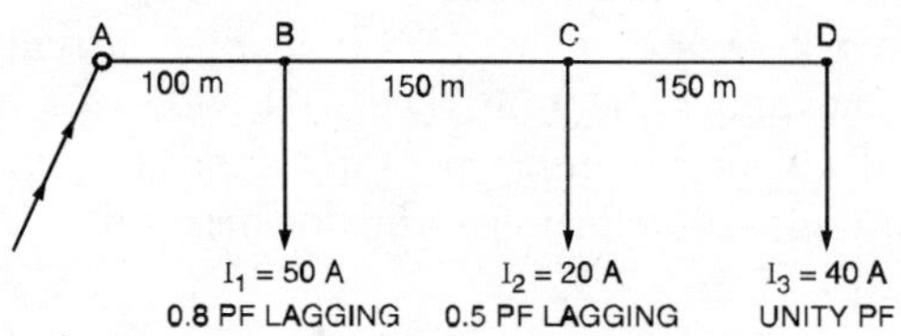

Fig. 10.11

Impedance of section BC,

$$\mathbf{Z}_2 = (0.25 + j0.125) \times \frac{150}{1{,}000} = (0.0375 + j0.01875)\ \Omega$$

Impedance of section CD,

$$\mathbf{Z}_3 = (0.0375 + j0.01875)\ \Omega$$

$$\mathbf{I}_1 = 50\,(0.8 - j0.6) = (40 - j30)\text{ A}$$

$$\mathbf{I}_2 = 20\,(0.5 - j0.866) = (10 - j17.32)\text{ A}$$

$$\mathbf{I}_3 = 40\,(1.0 - j0) = (40 - j0)\text{ A}$$

Assuming all power factors referred to far-end voltage V_D

Voltage at D, $\mathbf{V}_D = 400 + j0$ Taking V_D as reference phasor

Voltage at C, $\mathbf{V}_C = \mathbf{V}_D + \mathbf{I}_3\mathbf{Z}_3 = (400 + j0) + (40 - j0)\,(0.0375 + j0.01875) = (401.5 + j0.75)\text{ V}$

Voltage at B, $\mathbf{V}_B = \mathbf{V}_C + (\mathbf{I}_2 + \mathbf{I}_3)\,\mathbf{Z}_2$

$$= (401.5 + j0.75) + [(10 - j17.32) + (40 - j0)]\,(0.0375 + j0.01875)$$

$$= (401.5 + j0.75) + (50 - j17.32)\,(0.0375 + j0.01875)$$

$$= (403.7 + j1.038)\text{ volts}$$

Voltage at A, $\mathbf{V}_A = \mathbf{V}_B + (\mathbf{I}_1 + \mathbf{I}_2 + \mathbf{I}_3)\,\mathbf{Z}_1$

$$= (403.7 + j1.038) + [(40 - j30) + (10 - j17.32) + (40 - j0)]\,(0.025 + j0.0125)$$

$$= (403.7 + j1.038) + (90 - j47.32)\,(0.025 + j0.0125)$$

$$= (406.542 + j0.98) = 406.542\ \angle 0.14°\text{ volts}$$ **Ans.**

Example 10.3. A single phase two-wire feeder, 1500 m long, supplies a load of 60 A at 0.8 pf; 40 A at 0.85 pf; and 50 A at 0.88 pf lagging at distances of 600, 1,200 and 1,500 metres respectively from the feeding point. The resistance and reactance of the feeder per km length are 0.06 and 0.1 ohms, respectively. If the voltage at the far end is to be maintained at 220 V, calculate the voltage at the sending end, and its phase angle with respect to the receiving-end voltage. [*Gorakhpur Univ. 1975*]

Solution : The impedance of cable

$$= (0.06 + j\,0.1)\ \Omega \text{ per 1,000 metres.}$$

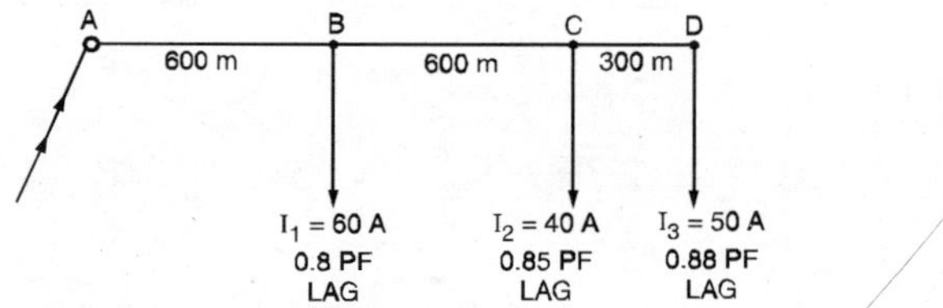

Fig. 10.12

Impedance of section AB,

$$\mathbf{Z}_1 = (0.06 + j0.1) \times \frac{600 \times 2}{1{,}000} = (0.072 + j0.12)\ \Omega$$

Impedance of section BC,

$$\mathbf{Z}_2 = (0.072 + j0.12)\ \Omega$$

Impedance of section CD,

$$\mathbf{Z}_3 = (0.06 + j0.1) \times \frac{300 \times 2}{1{,}000} = (0.036 + j0.06)\ \Omega$$

Taking voltage at far end D, $\mathbf{V}_D$ as reference phasor, we have

$$\mathbf{V}_D = (220 + j0)\text{ volts}$$

Current in section CD,

$$\mathbf{I}_3 = 50(0.88 - j0.475) = (44 - j23.75)\text{ A}$$

Voltage at load point C,

$$\mathbf{V}_C = \mathbf{V}_D + \mathbf{I}_3\mathbf{Z}_3 = (220 + j0) + (44 - j23.75)\,(0.036 + j0.06) = (223.009 + j1.795)\text{ volts}$$

Phase angle between V_C and V_D,

$$\theta_1 = \text{Tan}^{-1}\frac{1.795}{223.009} = 0°26'$$

Phase angle between load current I_2 and load voltage,

$$V_C = \text{Cos}^{-1}\,0.85 = -31°48'$$

Phase angle between load current I_2 and voltage at far end

$$V_D = -31°48' + 0°26' = -31°22'$$

Load current at point C,

$\mathbf{I}_2 = 40\,(\cos 31°22' - j\sin 31°22')$ or $(34.21 - j20.73)$ A

Load current in section BC

$= \mathbf{I}_3 + \mathbf{I}_2$

$= (44 - j23.75) + (34.21 - j20.73) = (78.21 - j44.48)$ A

Voltage at load point B,

$\mathbf{V}_B = \mathbf{V}_C + (\mathbf{I}_2 + \mathbf{I}_3)\,\mathbf{Z}_2$

$= (233.009 + j1.795) + (78.21 - j44.48)(0.072 + j0.12)$

$= (233.969 + j6.183)$ volts

Phase angle between V_B and V_D,

$$\theta_2 = \text{Tan}^{-1}\,\frac{6.183}{223.969} = 1°\,31'$$

Phase angle between load current I_1 and load voltage V_B

$= \text{Cos}^{-1}\,0.8 = -36°\,52'$

Phase angle between load current I_1 and voltage at far end

$= -36°52' + 1°31' = -35°21'$

Load current at point B,

$I_1 = 60(\cos 35°21' - j\sin 35°21') = (48.94 - j34.71)$ A

Load current in section AB

$= \mathbf{I}_1 + \mathbf{I}_2 + \mathbf{I}_3 = (48.94 - j34.71) + (78.21 - j44.48)$

$= (127.15 - j79.19)$ A

Voltage at sending-end,

$\mathbf{V}_S = \mathbf{V}_B + (\mathbf{I}_1 + \mathbf{I}_2 + \mathbf{I}_3)\,\mathbf{Z}_1$

$= (233.969 + j6.183) + (127.15 - j79.19)\,(0.072 + j0.12)$

$= (252.63 + j15.74)$ or $253.12\,\angle 3.56°$ V

i.e., Voltage at sending-end, $V_S = 253.12$ V **Ans.**

Phase angle of sending-end voltage with respect to the receiving-end voltage,

$\theta = 3.56°$ **Ans.**

Example 10.4. A single phase ac distributor 500 m long has a total impedance of (0.02 + *j*0.04) Ω and is fed from one end at 250 V. It is loaded as under :

(*i*) 50 A at unity power factor 200 m feeding point.

(*ii*) 100 A at 0.8 pf lagging 300 m from feeding point.

(*iii*) 50 A at 0.6 pf lagging at the far end.

Calculate the voltage drop and voltage at the far end.

Solution :

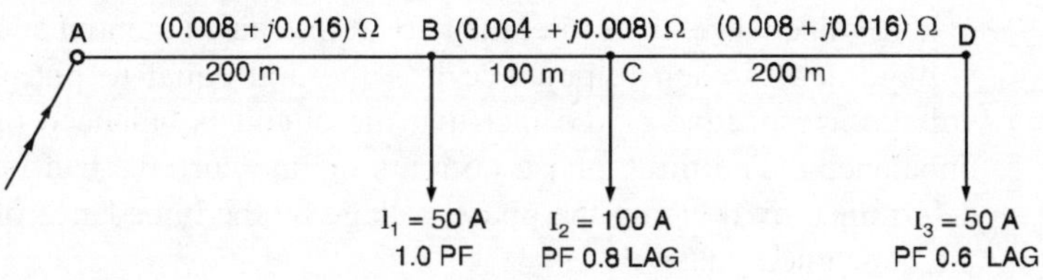

Fig. 10.13

Impedance of section AB;

$$\mathbf{Z}_1 = (0.02 + j0.04) \times \frac{200}{500} = (0.008 + j0.016)\ \Omega$$

Impedance of section BC,

$$\mathbf{Z}_2 = (0.02 + j0.04) \times \frac{100}{500} = (0.004 + j0.008)\ \Omega$$

Impedance of section CD,

$$\mathbf{Z}_3 = (0.02 + j0.04) \times \frac{200}{500} = (0.008 + j\,0.016)\ \Omega$$

Assuming all power factors referred to far-end voltage V_D

$\mathbf{I}_1 = 50\,(1 - j0) = (50 - j0)$ A

$\mathbf{I}_2 = 100\,(0.8 - j0.6) = (80 - j60)$ A

$\mathbf{I}_3 = 50\,(0.6 - j0.8) = (30 - j40)$ A

Let voltage at load point D, $\mathbf{V}_D = (V + j0)$

Total voltage drop in distributor

$= (\mathbf{I}_1 + \mathbf{I}_2 + \mathbf{I}_3)\,\mathbf{Z}_1 + (\mathbf{I}_2 + \mathbf{I}_3)\,\mathbf{Z}_2 + \mathbf{I}_3\,\mathbf{Z}_3$

$= (160 - j100)(0.008 + j0.016) + (110 - j100)\,(0.004 + j0.008)$
$+ (30 - j40)\,(0.008 + j0.016)$

$= (4.9 + j2.4) = 5.4562\,\angle 26.1°$ volts **Ans.**

Voltage at sending end,

$\mathbf{V}_A$ = Voltage at far-end + voltage drop in distributor

$= [(V_D + 4.9) + j\,2.4]$ volts

or $250 = \sqrt{(V_D + 4.9)^2 + (2.4)^2}$

or $V_D = 245.1$ V **Ans.**

Example 10.5. A single phase ring distributor ABC is fed at A. The loads at B and C are 20 A at 0.8 pf lagging and 15 A at 0.6 pf lagging respectively, both expressed with reference to voltage at A. The total impedance of the sections AB, BC and CA are (1 + *j*1), (1 + *j*2) and (1 + *j*3) ohm respectively. Find the current in each section and total current fed at A.

[*Pb. Technical Univ. Power System-I (Transmission and Distribution), December 2006*]

Solution : The given ring distributor is shown in Fig. 10.14. Assume that feeder BC is removed as shown in Fig. 10.15 (*a*).

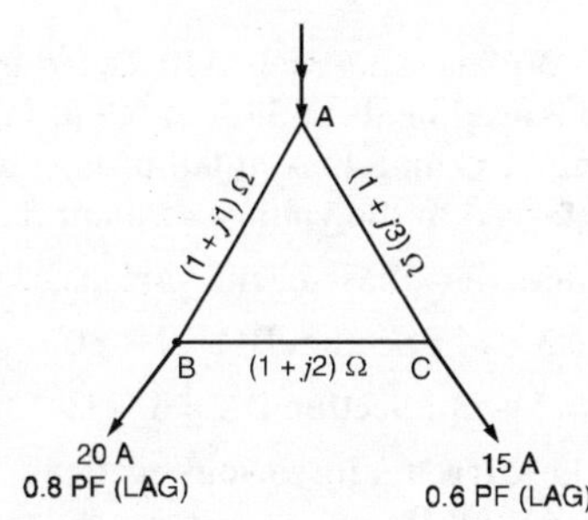

Fig. 10.14

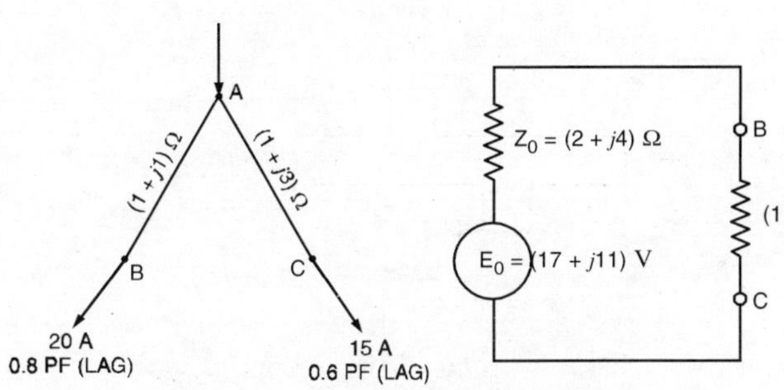

Fig. 10.15

Current in section AB, $\mathbf{I}_{AB} = \mathbf{I}_B = 20(0.8 - j0.6)$ A $= (16 - j12)$ A

Current in section AC, $\mathbf{I}_{AC} = \mathbf{I}_C = 15(0.6 - j0.8)$ A $= (9 - j12)$ A

Voltage drop in section AB, $\mathbf{V}_{AB} = \mathbf{I}_{AB} \times \mathbf{Z}_{AB} = (16 - j12)(1 + j1) = (28 + j4)$ V

Voltage drop in section AC, $\mathbf{V}_{AC} = \mathbf{I}_{AC} \times \mathbf{Z}_{AC} = (9 - j12)(1 + j3) = (45 + j15)$

PD between terminals B and C, B being at higher potential than C

$$\mathbf{V}_{BC} = (45 + j15) - (28 + j4) = (17 + j11) \text{ V}$$

This is Thevenin's equivalent emf, E_0

i.e., $E_0 = (17 + j11)$ V

Impedance of the network looking back from terminals B and C *i.e.*, Thevenin's equivalent impedance,

$$Z_0 = (1 + j1) + (1 + j3) = (2 + j4)\ \Omega$$

The equivalent Thevenin's circuit is shown in Fig. 10.15 (*b*).

$$\text{Current in section BC, } \mathbf{I}_{BC} = \frac{\mathbf{E}_0}{\mathbf{Z}_0 + \mathbf{Z}_{BC}} = \frac{(17 + j11)}{(2 + j4) + (1 + j2)} = (2.6 - j1.53) \text{ A}$$

Current in section AB, $\mathbf{I}_{AB} = I_B + I_{BC}$
$= (16 - j12) + (2.6 - j1.53)$
$= (18.6 - j13.53)$
$= 23 \angle -36.03°$ A **Ans.**

Current in section AC, $\mathbf{I}_{AC} = \mathbf{I}_C - \mathbf{I}_{BC}$
$= (9 - j12) - (2.6 + j1.53)$
$= (6.4 - j10.47)$ A
$= 12.27 \angle -58.56°$ A **Ans.**

Total current fed at A $= \mathbf{I}_{AB} + \mathbf{I}_{AC}$
$= (18.6 - j13.53) + (6.4 - j10.47)$
$= (25 - j24) = 34.65 \angle -43.83°$ A **Ans.**

Example 10.6. A 3-phase ring main ABCD, fed from the end A at 11 kV supplies balanced loads of 30 A at 0.9 pf lagging at B, 50 A at 0.707 pf lagging at C and 40 A at 0.8 pf lagging at D, the load currents being referred to the voltage at point A.

The impedances per phase of the various sections are :

Section AB = 3 + *j*4 ; Section BC = 1 + *j*1

Section CD = 2 + *j*3 ; Section DA = 1 + *j*2

Determine the currents in various sections and station bus-bar voltages at B, C and D.

Solution : One phase of the main ring is shown in Fig. 10.16.

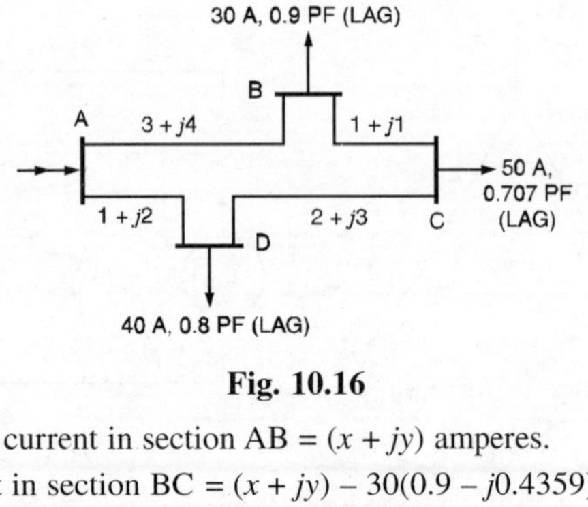

Fig. 10.16

Let the current in section AB = $(x + jy)$ amperes.

Current in section BC $= (x + jy) - 30(0.9 - j0.4359)$
$= (x - 27) + j(y + 13.1)$

Current in section CD $= (x - 27) + j(y + 13.1) - (35.35 - j35.35)$
$= (x - 62.35) + j(y + 48.45)$

Current in section DA $= (x - 62.35) + j(y + 48.45) - 40(0.8 - j0.6)$
$= (x - 94.35) + j(y + 72.45)$

Applying Kirchhoff's voltage law to the closed loop ABCDA, we have

$$(x + jy)(3 + j4) + [(x - 27) + j(y + 13.1)](1 + j1) + [(x - 62.35) + j(y + 48.45)](2 + j3) + [(x - 94.35) + j(y + 72.45)](1 + j2) = 0$$

or $(7x - 10y - 549.4) + j(10x + 7y - 220.3) = 0$

Since the real (or active) and imaginary (or reactive) parts have to be separately zero,

$\therefore\ 7x - 10y - 549.4 = 0$

and $10x + 7y - 220.3 = 0$

Solving for x and y, we get

$x = 40.55$ and $y = -26.45$

$\therefore$ Current in section AB = $(40.55 - j\,26.45)$ amperes **Ans.**

Current in section BC
$= (x - 27) + j(y + 13.1)$
$= (40.55 - 27) + j(-26.45 + 13.1) = (13.55 - j13.35)$ amperes **Ans.**

Current in section CD
$= (40.55 - 62.35) + j(-26.45 + 48.45) = (-21.8 + j22)$ amperes **Ans.**

Current in section DA
$= (40.55 - 94.35) + j(-26.45 + 72.45) = (-53.8 + j46)$ amperes **Ans.**

$$\text{Supply voltage at A} = \frac{11{,}000}{\sqrt{3}} = 6{,}351 \text{ volts/phase}$$

Voltage at station B
$= (6{,}351 + j0) - (40.55 - j26.45)(3 + j4)$
$= (6{,}130.6 - j82.85)$ volts/phase or $6{,}131 \angle -0.77°$ V **Ans.**

Voltage at station C
$= (6{,}130.6 - j82.85) - (13.55 - j13.35)(1 + j1)$
$= (6{,}103.7 - j83.05)$ volts/phase or $6{,}104.3 \angle -0.78°$ V **Ans.**

Voltage at station D
$= (6{,}103.7 - j83.05) - (-21.8 + j22)(2 + j3)$
$= (6{,}213.3 - j61.65)$ volts/phase or $6{,}213.6 \angle -0.57°$ V **Ans.**

10.5. 3-PHASE, 4-WIRE, STAR-CONNECTED UNBALANCED LOAD CIRCUITS

In 3-phase, 4-wire, star-connected load circuits the star points of load and the generator are tied together through neutral wire of zero impedance, therefore, they are at the same potential and voltage across each impedance is same and equal to phase voltage irrespective of the fact that the circuit is balanced or unbalanced. The three phase currents or line currents can be determined by dividing the phase voltage by the impedance of the concerned phase

$$i.e.,\ \mathbf{I}_R = \frac{\mathbf{V}_R}{\mathbf{Z}_R}\ ;\ \mathbf{I}_Y = \frac{\mathbf{V}_Y}{\mathbf{Z}_Y} \text{ and } \mathbf{I}_B = \frac{\mathbf{V}_B}{\mathbf{Z}_B}$$

The current in neutral wire can be determined by applying Kirchhoff's first law at star point N according to which

$$\mathbf{I}_R + \mathbf{I}_Y + \mathbf{I}_B + \mathbf{I}_N = 0$$

or current in neutral wire, $\mathbf{I}_N = -(\mathbf{I}_R + \mathbf{I}_Y + \mathbf{I}_B)$

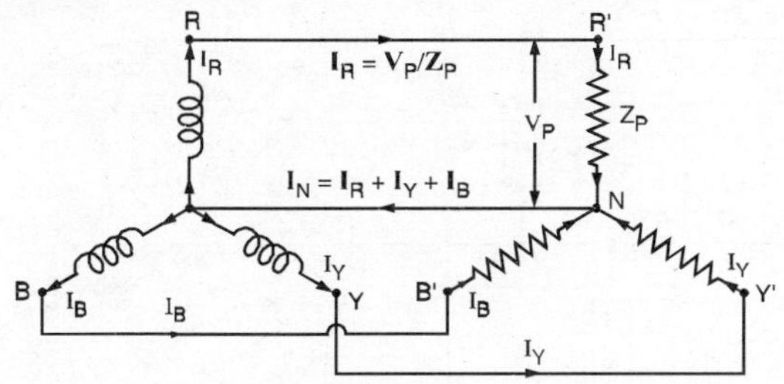

Fig. 10.17

In balanced load circuits the line currents are equal and have the same phase angles with their respective phase voltages, so their phasor sum is zero and current in neutral wire is zero. Since in balanced star-connected load circuits, neutral wire carries no current and therefore it is usually omitted.

Example 10.7. In a 3-phase, 4-wire system, the line voltage is 400 volts and non-inductive loads of 10, 8 and 5 kWs are connected between the three line conductors and the neutral, as shown in Fig. 10.18. Calculate (*a*) the current in each line and (*b*) the current in the neutral conductor.

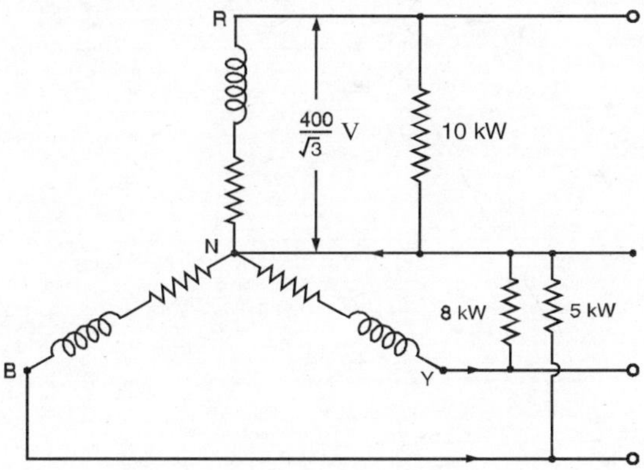

Fig. 10.18

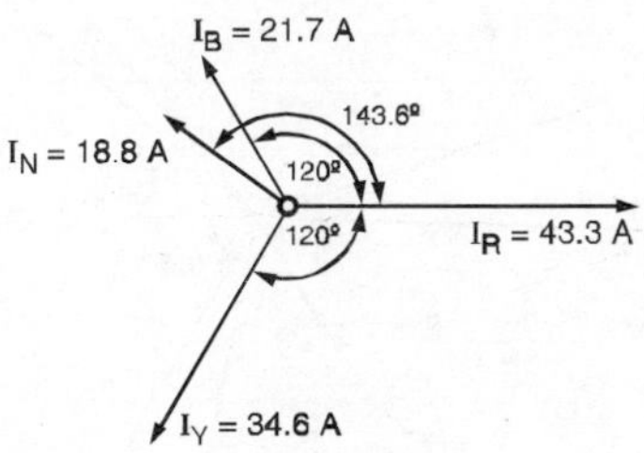

Fig. 10.19

Solution : Phase voltage across each load,

$$V_P = \frac{V_L}{\sqrt{3}} = \frac{400}{\sqrt{3}} = 231 \text{ V}$$

$$\text{Current in red phase, } I_R = \frac{\text{kW} \times 1{,}000}{V_P \cos\phi} = \frac{10 \times 1{,}000}{231 \times 1}$$

$$= 43.3 \text{ A } \textbf{Ans.}$$

Similarly current in yellow phase,

$$I_Y = \frac{8 \times 1{,}000}{231 \times 1} = 34.6 \text{ A } \textbf{Ans.}$$

$$\text{Current in blue phase, } I_B = \frac{5 \times 1{,}000}{231 \times 1} = 21.7 \text{ A } \textbf{Ans.}$$

These currents are mutually 120º out of phase as the circuit is purely resistive. Taking I_R as reference phasor, the line currents may be given as

$$\mathbf{I}_R = 43.3\,(1 + j\,0)$$

$$\mathbf{I}_Y = 34.6\,(-0.5 - j0.866)$$

$$\mathbf{I}_B = 21.7\,(-0.5 + j0.866)$$

(*b*) Current in neutral wire,

$$\mathbf{I}_N = -(\mathbf{I}_R + \mathbf{I}_Y + \mathbf{I}_B)$$

$$= -[43.3(1 + j0) + 34.6(-0.5 - j0.866) + 21.7(-0.5 + j0.866)]$$

$$= -15.15 + j11.17$$

$$I_N = \sqrt{(-15.15)^2 + (11.17)^2} = 18.8 \text{ A } \textbf{Ans.}$$

$$\phi = \text{Tan}^{-1}\frac{11.17}{-15.15} = 143.6^\circ \textbf{ Ans.}$$

Example 10.8. A 3-phase star-connected system with 230 volts between each phase and neutral has resistances of 4, 5 and 6 Ω respectively in the three phases. Estimate the current flowing in each phase and the neutral current. Find the total power absorbed.

Solution : PD across each resistance, V = 230 volts

$$\text{Current in 4 } \Omega \text{ resistance, } I_R = \frac{230}{4} = 57.5 \text{ A } \textbf{Ans.}$$

$$\text{Current in 5 } \Omega \text{ resistance, } I_Y = \frac{230}{5} = 46 \text{ A } \textbf{Ans.}$$

$$\text{Current in 6 } \Omega \text{ resistance, } I_B = \frac{230}{6} = 38.33 \text{ A } \textbf{Ans.}$$

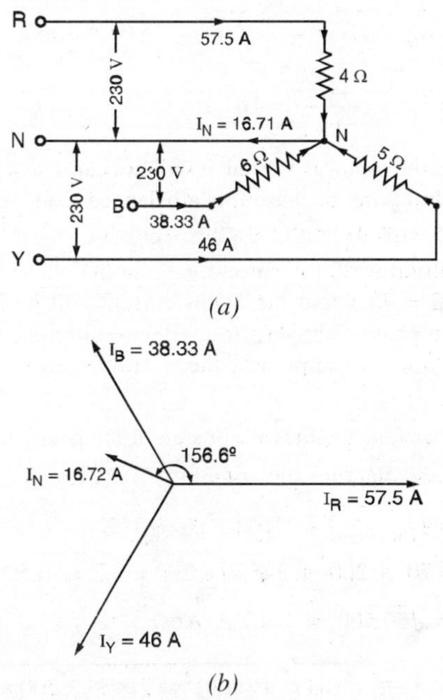

Fig. 10.20

Since circuit is purely resistive, so currents will be in phase with their respective voltages and will be mutually 120º apart. So taking $\mathbf{I}_R$, the current in 4 Ω resistance along X-axis, the line currents are given as

$$\mathbf{I}_R = 57.5\,(1 + j0) \text{ A}$$

$$\mathbf{I}_Y = 46\,(-0.5 - j0.866) \text{ A}$$

$$\mathbf{I}_B = 38.33\,(-0.5 + j0.866) \text{ A}$$

Central in neutral wire

$$\mathbf{I}_N = -[57.5(1 + j0) + 46(-0.5 - j0.866) + 38.33(-0.5 + j0.866)]$$

$$= -15.34 + j6.64 \text{ A}$$

$$I_N = \sqrt{(-15.34)^2 + (6.64)^2} = 16.72 \text{ A } \textbf{Ans.}$$

$$\phi = \text{Tan}^{-1} \frac{6.64}{-15.34} = 156.6^\circ$$

Total power absorbed, P

$$= V(I_1 + I_2 + I_3) = 230\ (57.5 + 46 + 38.33)$$

$$= 32{,}621 \text{ watts or } 32.621 \text{ kW } \textbf{Ans.}$$

Example 10.9. A 3-phase 4-wire system supplies power at 400 volts and lighting at 230 volts. If the lamps in use require 70, 84 and 33 amperes in each of the three lines, what will be the current in the neutral wire ? If the motor now starts taking 200 A from the lines at a pf of 0.2 lagging, what will be the total current in each line and in the neutral wire ? Find also the total power supplied to the lamps and motor.

Solution : (*a*) *When motor is not started*

Taking current in red phase along axis of reals, the current in different phases may be expressed as

$$\mathbf{I}_{LR} = 70(1 + j0)$$

$$\mathbf{I}_{LY} = 84(-0.5 - j0.866)$$

$$\mathbf{I}_{LB} = 33(-0.5 + j0.866)$$

Current in neutral wire will be phasor sum of these line currents but with its phase changed by 180°

$$\therefore \quad \mathbf{I}_N = -(\mathbf{I}_{LR} + \mathbf{I}_{LY} + \mathbf{I}_{LB})$$

$$= -[70\ (1 + j0) + 84(-0.5 - j0.866) + 33(-0.5 + j0.866)]$$

$$= (-11.5 + j44.2) \text{ A}$$

$$I_N = \sqrt{(-11.5)^2 + (44.2)^2} = 45.67 \text{ A } \textbf{Ans.}$$

$$\phi = \text{Tan}^{-1} \frac{44.2}{-11.5} = 104.58^\circ$$

(*b*) When the motor is started, it will not cause any change in the current of neutral wire, because it is a balanced load, so neutral-wire current will be same as before starting of motor *i.e.,* 45.67 A **Ans.**

Motor will draw 200 A current lagging the phase voltage by an angle $\cos^{-1} 0.2 = 78.5°$ but the lamps current will be in phase with their respective phase voltages. So the phase difference between the line currents due to lamps and the motor is 78.5° as shown in Fig. 10.21 (*c*).

The line currents are the phasor sum of lamp and motor currents.

From phasor diagram shown in Fig. 10.21 (*c*)

$$I_R^2 = I_{LR}^2 + I_M^2 + 2I_{LR}I_M \cos\phi$$

$$= 70^2 + 200^2 + 2 \times 70 \times 200 \times 0.2 = 50{,}500$$

$$\text{or } I_R = \sqrt{50{,}500} = 224.7 \text{ A } \textbf{Ans.}$$

$$\text{Similarly } I_Y = \sqrt{(84)^2 + (200)^2 + 2 \times 84 \times 200 \times 0.2}$$

$$= 231.9 \text{ A } \textbf{Ans.}$$

$$I_B = \sqrt{(33)^2 + (200)^2 + 2 \times 33 \times 200 \times 0.2}$$

$$= 209.1 \text{ A } \textbf{Ans.}$$

Power taken by lamps

$$= V\ (I_{LR} + I_{LY} + I_{LB}) = 230\ (70 + 84 + 33) = 43{,}010 \text{ watts } \textbf{Ans.}$$

Power taken by motor

$$= \sqrt{3}\ V_L I_L \cos\phi = \sqrt{3} \times 400 \times 200 \times 0.2 = 27{,}713 \text{ watts } \textbf{Ans.}$$

Total power supplied,

$$P = 43{,}010 + 27{,}713 = 70{,}723 \text{ W or } 70.723 \text{ kW } \textbf{Ans.}$$

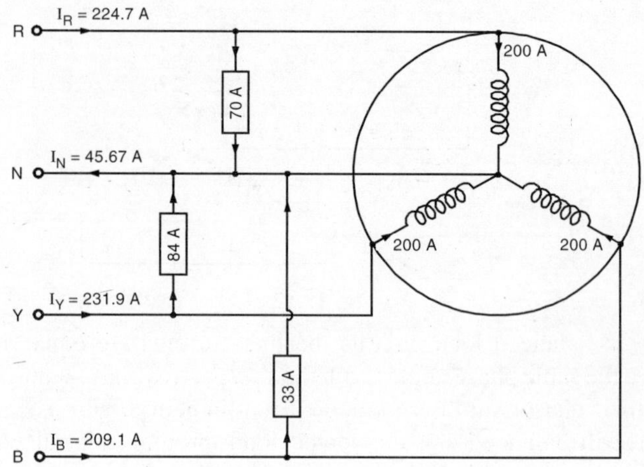

(*a*)

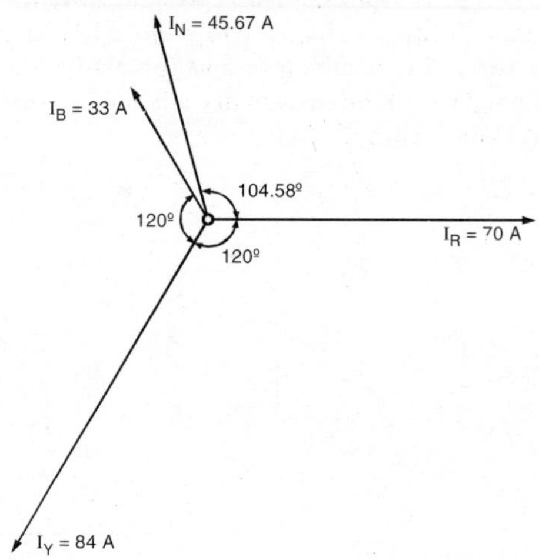

(*b*)

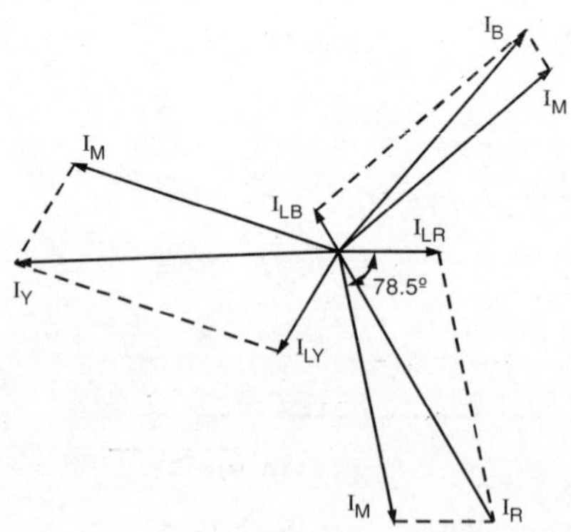

(*c*)

Fig. 10.21

Example 10.10. A 440/254 volt, 3-phase, 4 core supplies an unbalanced load represented by the following impedances in ohms connected between R, Y and B phases respectively and the neutral 16 + *j*12; 14 – *j*21 and 25. The phase sequence is RYB. Calculate the current in each conductor of the cable.

Solution : The potential differences between the three lines and neutral are given by:

$$\mathbf{V}_{LR} = 254\ (1 + j0)$$

$$\mathbf{V}_{LY} = 254\ (-0.5 - j0.866)$$

$$\mathbf{V}_{LB} = 254\ (-0.5 + j0.866) \text{ taking } V_R \text{ as the reference phasor}$$

$$\mathbf{I}_R = \frac{\mathbf{V}_R}{\mathbf{Z}_R} = \frac{254 + j0}{16 + j12} = (10.16 - j7.62) \text{ amperes}$$

Similarly $\mathbf{I}_Y = \dfrac{254\,(-0.5 - j0.866)}{14 - j21} = (4.46024 - j\,9.01954)$ amperes

$$\mathbf{I}_B = \frac{254\,(-0.5 + j0.866)}{25} = (-5.08 + j\,8.79856) \text{ amperes}$$

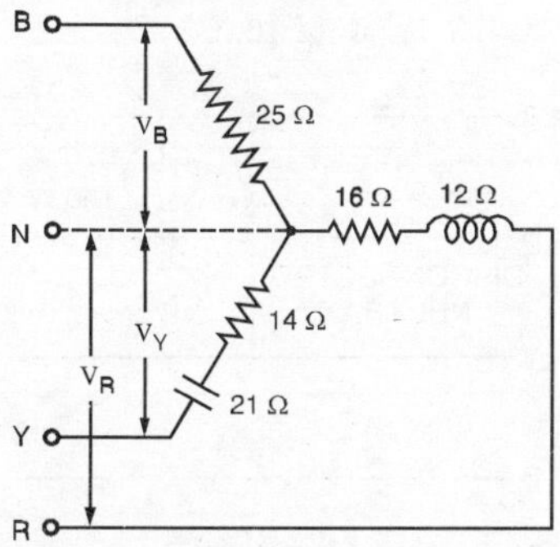

(a) Circuit Diagram

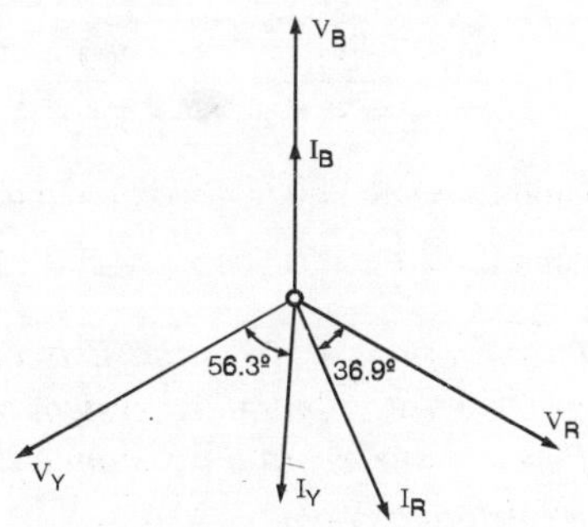

(b) Phasor Diagram

Fig. 10.22

Now current in neutral wire,

$$\mathbf{I}_N = -(\mathbf{I}_R + \mathbf{I}_Y + \mathbf{I}_B)$$
$$= -[(10.16 - j7.62) + (4.46024 - j9.01954) + (-5.08 + j8.79856)]$$
$$= (-9.54024 + j\,7.84098) \text{ amperes}$$

Evaluating these currents, we get

$$I_R = \sqrt{(10.16)^2 + (-7.62)^2} = 12.7 \text{ A } \textbf{Ans.}$$

$$I_Y = \sqrt{(4.46024)^2 + (-9.01954)^2} = 10.06 \text{ A } \textbf{Ans.}$$

$$I_B = \sqrt{(-5.08)^2 + (8.79856)^2} = 10.16 \text{ A } \textbf{Ans.}$$

$$I_N = \sqrt{(-9.54024)^2 + (7.84098)^2} = 12.35 \text{ A } \textbf{Ans.}$$

Example 10.11. A 3-phase, 4-wire system with 400 V line voltage supplies the following loads :

(*a*) A 400 V, 3-φ, 75-HP motor at 0.8 pf lagging and with an efficiency of 88%.

(*b*) A 15 kW unity power factor load connected between phase A and neutral

(*c*) A 12.5 kW 0.866 leading power factor load connected between phase B and neutral. Calculate (*i*) the neutral current (*ii*) the three line currents (*iii*) total power delivered by the supply. Phase sequence being ABC. [*Pb. Univ. Nov. 1976*]

Solution : The potential differences between the three lines and neutral are given as:

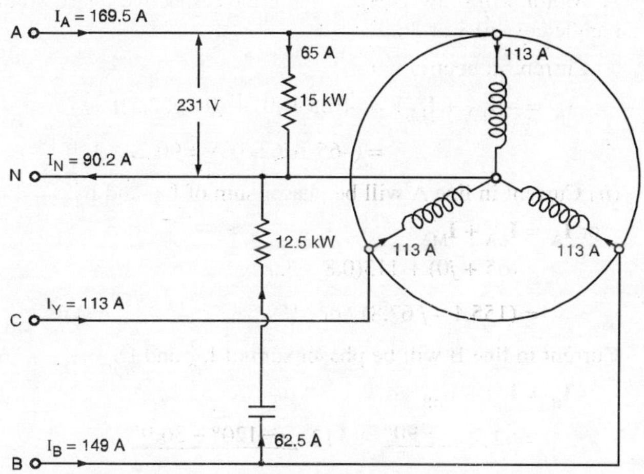

Fig. 10.23

$$\mathbf{V}_{AN} = \frac{400}{\sqrt{3}}(1 + j0) \text{ or } \frac{400}{\sqrt{3}}\angle 0° \text{ volts}$$

$$\mathbf{V}_{BN} = \frac{400}{\sqrt{3}}(-0.5 - j0.866) \text{ or } \frac{400}{\sqrt{3}}\angle -120° \text{ volts}$$

and $\mathbf{V}_{CN} = \dfrac{400}{\sqrt{3}}(-0.5 + j0.866)$ or $\dfrac{400}{\sqrt{3}}\angle 120°$ volts

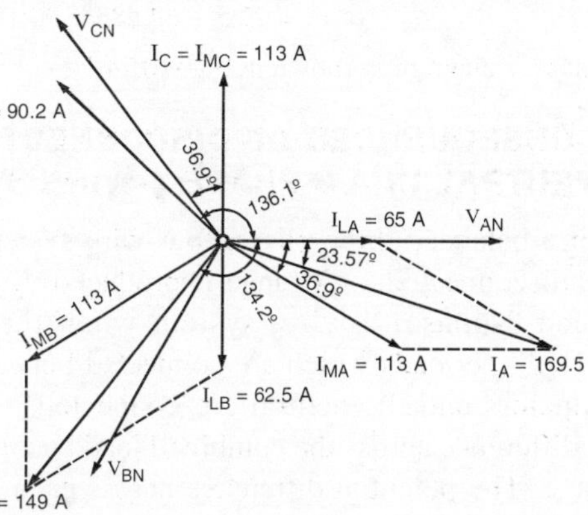

Fig. 10.24

Current drawn by 15 kW load at unity power factor,

$$\mathbf{I}_{LA} = \frac{\text{Load in watts}}{V_P \times \cos\phi} = \frac{15{,}000}{\frac{400}{\sqrt{3}} \times 1} = 65 \text{ A in phase with } V_{AN}$$

$$\therefore\ \mathbf{I}_{LA} = 65\angle 0° \text{ or } (65 + j0) \text{ A}$$

Current drawn by 12.5 kW load at 0.866 power factor leading,

$$\mathbf{I}_{LB} = \frac{\text{Load in watts}}{V_P \times \cos\phi} = \frac{12{,}500}{\frac{400}{\sqrt{3}} \times 0.866}$$

$$= 62.5 \text{ A leading } V_{BN} \text{ by an angle } \cos^{-1} 0.866 \text{ or } 30°$$

$$\therefore\ \mathbf{I}_{LB} = 62.5\angle -120° + 30° = 62.5\angle -90° \text{ or } (0 - j62.5) \text{ A}$$

Current drawn by the motor per phase

= Line current drawn by motor

$$\therefore\ I_L = I_P = \frac{\text{Motor output in watts}}{\sqrt{3} \times V_L \times \cos\phi \times \eta}$$

$$= \frac{75 \times 735.5}{\sqrt{3} \times 400 \times 0.8 \times 0.88} = 113 \text{ A}$$

∴ Motor will draw 113 A lagging the respective phase voltage by an angle $\cos^{-1} 0.8$ or 36.9°.

(*i*) Current in neutral wire,

$$\mathbf{I}_N = -(\mathbf{I}_{LA} + \mathbf{I}_{LB}) = -[(65 + j0) + (0 - j62.5)]$$

$$= (-65 + j62.5) \text{ A} = 90.2 \angle 136.1° \text{ A}$$

(*ii*) Current in line A will be phasor sum of I_{LA} and I_{MA}

$$\text{or } \mathbf{I}_A = \mathbf{I}_{LA} + \mathbf{I}_{MA}$$

$$= (65 + j0) + 113(0.8 - j0.6)$$

$$= (155.4 - j\,67.8) \text{ or } 169.5 \angle -23.57° \text{ A } \textbf{Ans.}$$

Current in line B will be phasor sum of I_{LB} and I_{MB}

$$\text{or } \mathbf{I}_B = \mathbf{I}_{LB} + \mathbf{I}_{MB}$$

$$= 62.5 \angle -90° + 113 \angle -120° - 36.9°$$

$$= 62.5 \angle -90° + 113 \angle -156.9°$$

$$= (0 - j62.5) + 113(-0.92 - j0.393)$$

$$= (0 - j62.5) + (-104 - j44.39)$$

$$= (-104 - j106.89) = 149 \angle -134.2° \text{ A } \textbf{Ans.}$$

Current in line C will be due to motor only and will be equal to 113 A lagging V_{CN} by 36.9° or lead V_{AN} by 83.1°.

$$(iii) \text{ Total power delivered} = 15 + 12.5 + \frac{75 \times 0.7355}{0.88}$$

$$= 27.5 + 62.685 = 90.185 \text{ kW } \textbf{Ans.}$$

The phasor diagram is shown in Fig. 10.24.

10.6. CONSEQUENCES OF DISCONNECTING NEUTRAL IN A 3-PHASE 4-WIRE SYSTEM

In case of a balanced three-phase, four-wire system when the neutral is disconnected, no change is produced. But in case of unbalanced 3-phase, 4-wire, system when the neutral is disconnected, the loads which are connected between any two line conductors and the neutral are connected in series and potential difference across the combined load becomes equal to line voltage. The potential difference across each load is thus changed as per rating of the load.

The effect of disconnecting the neutral in a 3-phase, 4-wire unbalanced system will be clear from the following example.

Example 10.12. In a 3-phase, 4-wire, 400/230 V system, a lamp of 100 watts is connected to one phase and neutral and a lamp of 200 watts is connected to the second phase and the neutral. Calculate the voltage across each lamp when the neutral wire is disconnected.

Solution : Rating of lamp L_1 = 100 watts

Rating of lamp L_2 = 200 watts

$$\text{Resistance of lamp } L_1, R_1 = \frac{(V_P)^2}{\text{Wattage of lamp } L_1} = \frac{(230)^2}{100}$$

$$= 529\ \Omega$$

$$\text{Resistance of lamp } L_2, R_2 = \frac{(230)^2}{200} = 264.5\ \Omega$$

When the neutral wire is disconnected, as shown in Fig. 10.26 the lamp L_1 (a resistance of 529 Ω) and lamp L_2 (a resistance of 264.5 Ω) are connected in series and the potential difference across them would be equal to line voltage *i.e.*, 400 V.

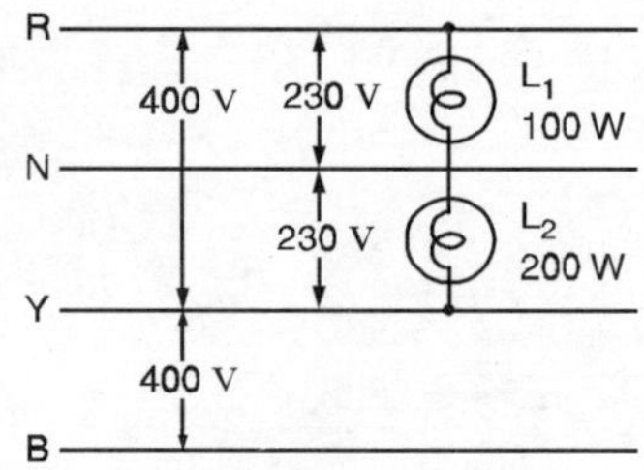

Fig. 10.25

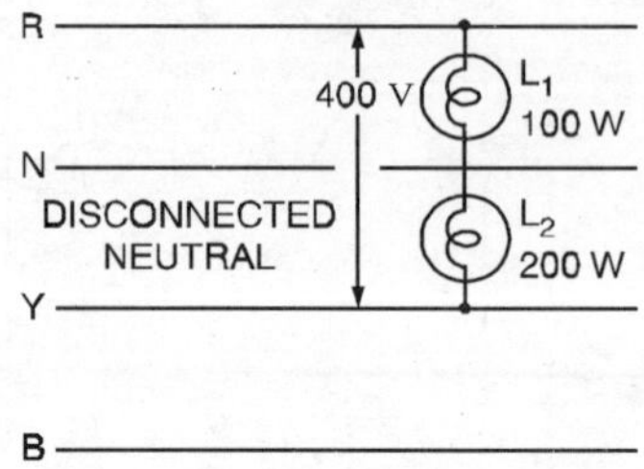

Fig. 10.26

Current flowing through lamps,

$$I = \frac{V_L}{R_1 + R_2} = \frac{400}{529 + 264.5} = 0.504 \text{ A}$$

Voltage across lamp L_1 = IR_1 = 0.504 × 529 = 266.67 V **Ans.**

Voltage across lamp L_2 = IR_2 = 0.504 × 264.5 = 133.33 V **Ans.**

Note : *From the above illustration it is clear that if the neutral in 3-phase, 4-wire unbalanced system is disconnected, the potential difference across high resistance is increased and that across low resistance is decreased.*

Exercises

1. Explain the following systems of distribution :
   (*i*) Radial system
   (*ii*) Ring main system
   (*iii*) Interconnected system
   [M.D. Univ. Power System-I, Dec. 2009]
2. Classify the systems for distribution of ac power. **[Pb. Technical Univ. Power System-I, (Transmission and Distribution) Dec. 2008]**
3. List the various systems available for distribution of ac power. **[Pb. Technical Univ. Power System-I (Transmission and Distribution); Dec. 2006]**
4. Discuss in brief various methods of ac distribution. **[Agra Univ. Power Systems II, 1981]**
5. Explain 3-phase 4-wire system of distribution of electrical power.
6. Write short notes on the following :
   (*i*) Radial, ring main and interconnected systems. **[Rajasthan Univ. 1999, 2000, 2002; R.T.U. 2009]**
   (*ii*) AC interconnected systems. **[M.D. Univ. Dec. 2007]**
   (*iii*) Three-phase 4-wire ac distribution system.
   (*iv*) Single-phase 2-wire ac distribution system.
   (*v*) Methods of solving ac distributors.

Short Answer Type Questions With Answers

Q. 1. What is primary distribution system and what is normal operation voltage of primary distribution ?

Ans. Primary distribution system is that part of the distribution system which operates at voltages such as 3.3, 6.6 or 11 kV. The normal operation voltage of primary distribution is 11 kV.

Q. 2. What is usable voltage for secondary distribution ?

Ans. The secondary distribution usable voltage is 415/240 V (415 V for 3-phase loads and 240 V for single-phase loads).

Q. 3. Differentiate between open-loop system and ring-loop system.

Ans. In a loop feeder system, two or more radial feeders originating from the same or different secondary substations are laid on different routes of load areas. The arrangement having the ends of two feeders tied together through normally open switching devices is known as *open-loop system* while the arrangement having the ends of two feeders tied together through normally closed switching devices is called the *ring main feeder* or the *ring-loop*.

Q. 4. What is interconnected system and where is it employed and why ?

Ans. When the feeder ring main is energized from two or more than two generating stations or substations it is called the *interconnected network system.* Because of interconnected feeders, power can be supplied to all the distribution transformers even though a part of network may be out of service. Such a system provides better reliability and flexibility and is employed in large metropolitan cities where continuity of supply is the most important.

Q. 5. What is radial system of distribution ? What are its advantage and drawbacks ?

Ans. In **radial system,** the low voltage distributors radiate out from the distribution transformer and run through the areas to be supplied by them. This system is the simplest and the least expensive system. This system generally not used because of the following drawbacks.

(*i*) The end of the distributor nearest to the supply end would be heavily loaded.

(*ii*) The consumers at the farthest end of the distributor would be subjected to serious voltage fluctuations with the variation in load.

(*iii*) The consumers are dependent on a single distributor so that a fault on the distributor cuts off the supply from all the consumers who are on the side of the fault away from the station.

Problems

1. A single-phase distributor has a resistance of 0.4 Ω and a reactance of 0.6 Ω. At far end 2 the voltage V_2 is 240 V and the current is 50 A at power factor of 0.8 lagging. At mid point the current is 50 A at 0.6 lagging power factor with respect to voltage V_1. Find the supply voltage V_s and phase angle between V_s and V_2.

[Rajasthan Univ. Transmission and Distribution of Electrical Power 2003]

[**Ans.** 344.97 V; 4.58°]

2. A 2-wire feeder, 1000 metre long supplies a load of 100 amperes at 0.8 pf at its far end and a load of 60 amperes at 0.85 pf at its mid-point. Both power factors are lagging and refer to voltage at the respective load points. The resistance and reactance of the feeder per km (go and return) are 0.08 Ω and 0.12 Ω respectively. If the voltage at the far end is to be maintained at 250 V, calculate the voltage at the supply end and the phase angle between the voltages at two ends. [**Ans.** 267 V ; 1°26′ (lead)]

3. A single-phase distributor has a resistance of 0.2 Ω and reactance of 0.3 Ω. At the far end the voltage V_B = 240 V, the current is 100 A and the power factor is 0.8. At the mid-point A, a current of 100 A is supplied at a power factor 0.6 with reference to voltage V_A at A. Find the

(*i*) supply voltage V_S,

(*ii*) supply-end current I_S,

(*iii*) power factor at supply end and

(*iv*) phase angle between V_S and V_B.

[**Ans.** (*i*) 292 volts (*ii*) 198.25 A (*iii*) 0.595 (*iv*) 2.6°]

4. A single-phase ring distributor ABC is fed at A. The loads at B and C are 20 A at 0.8 pf lagging and 15 A at 0.6 pf lagging respectively, both expressed with reference to voltage at A. The impedances of the sections AB, BC and CA are (1 + *j*1), (1 + *j*2) and (1 + *j*3) ohms respectively. Find the total current fed at A and the current in each section.

[Madras Univ. Transmission and Distribution II April, 1970]

[**Ans.** $23.1 \angle -32^\circ$, $3.01 \angle 30.1^\circ$, $13.1 \angle -60.8^\circ$, $34.6 \angle -34.8^\circ$ amperes]

5. A 3-phase ring main ABCD, fed from the end A supplies balanced loads of 50 A at 0.8 pf lagging at B, 120 A at unity pf at C and 70 A at 0.866 pf lagging at D, the load currents being referred to the voltage at point A.

The impedances per phase of the various line sections are :

Section AB = (1 + *j*0.6) Ω

Section BC = (1.2 + *j*0.9) Ω

Section CD = (0.8 + *j*0.5) Ω

Section DA = (3 + *j*2)

Determine the currents in the various sections.

[Bombay Univ. Elec. Power-I March 1971]

[**Ans.** (139.7 – *j*42.8); (99.7 – *j*12.8); (–20.3 – *j*12.8), (–80.9 + *j* 22.2) amperes]

6. In a 4-wire, 3-phase distribution system with 240 V between lines and neutral, there is a balanced motor load of 500 kW at pf 0.8 lagging. Lamp loads connected between the respective lines and neutral absorbs 50 kW, 150 kW and 200 kW. Calculate the current in each line and in the neutral wire of the feeder cable supplying this load.

[Agra Univ. Transmission & Distribution of Electrical Power 1982]

[**Ans.** $1{,}042.2 \angle -30^\circ$ A; $1{,}128.64 \angle -70.2^\circ$ A; $342.76 \angle 35.9^\circ$ A; $550.4 \angle -19.1^\circ$ A]

❑❑❑

CHAPTER 11

Insulation Resistance of a System

INSIDE THIS CHAPTER

11.1. Introduction **11.2.** Measurement of Insulation Resistance of Live Mains **11.3.** Types of Cable Faults **11.4.** Causes of Cable Faults **11.5.** Location of Cable Faults **11.6.** Loop Tests **11.7.** Fall of Potential Test **11.8.** Test For Open-Circuit Faults **11.9.** Measurement of Insulation Resistance of a Cable.

11.1. INTRODUCTION

The leakage of electric current taking place under operating conditions in a distribution system is mainly due to surface leakage across the apparatus and fittings though there is also leakage across the insulation of the cables. These leakages cause currents to flow even when the load circuits are open. Imagine the system to be perfect and resistances connected between the various bus-bars and earth, the values of these resistances being adjusted until the currents flowing through them are equal, both in magnitude and direction, to the actual values of the earth currents due to leakages. The values of these resistances are called the *leakage or insulation resistances* of the system. The various methods for determining the leakage or insulation resistances of the power system during operation *i.e.,* without interrupting the supply are given below :

11.2. MEASUREMENT OF INSULATION RESISTANCE OF LIVE MAINS

1. ***Two-Wire System.*** For measurement of insulation resistance to earth of a two-wire distribution system, while the power is on, the voltages between the two mains (+ve and –ve mains), between positive main and earth and between –ve main and earth are measured with a high resistance voltmeter whose resistance should be comparable with the insulation resistance under test.

Let the voltage between +ve and –ve mains be V (From measurements)

The voltage between +ve main and earth be V_1 (From measurements)

The voltage between –ve main and earth be V_2 (From measurements)

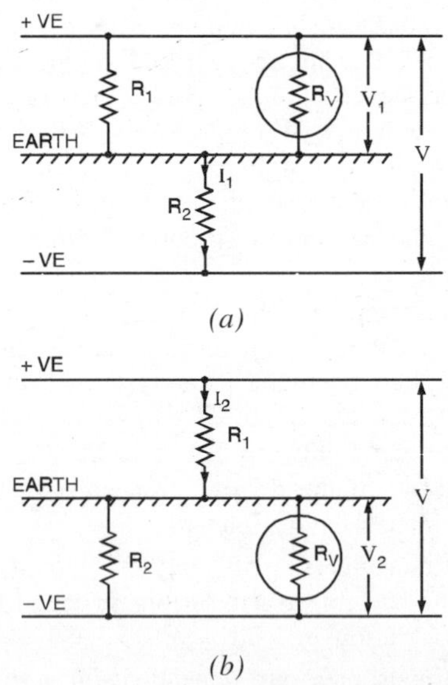

Fig. 11.1. ***Measurement of Insulation Resistance of 2-Wire Live Mains***

The resistance of the voltmeter be R_V (known)

The insulation resistance between +ve outer and earth be R_1 (To be determined)

The insulation resistance between –ve outer and earth be R_2 (To be determined)

If the current flowing from +ve main to the –ve main in the first case is I_1 then

$$V_1 = I_1 \times \frac{1}{\frac{1}{R_1} + \frac{1}{R_V}} = I_1 \frac{R_1 R_V}{R_1 + R_V} \qquad ...(11.1)$$

Since R_1 and R_V are in parallel across +ve main and earth and $V - V_1 = I_1 R_2$ ...(11.2)

Dividing Eq. (11.2) by Eq. (11.1), we have

$$\frac{V - V_1}{V_1} = \frac{R_2}{\frac{R_1 R_V}{R_1 + R_V}} = \frac{R_2 (R_1 + R_V)}{R_1 R_V}$$

$$\text{or} \quad \frac{V}{V_1} = \frac{R_1 R_V + R_2 (R_1 + R_V)}{R_1 R_V}$$

$$= \frac{R_1 R_2 + R_V (R_1 + R_2)}{R_1 R_V} \quad ...(11.3)$$

Similarly if in second case the current flowing from +ve main to –ve main is I_2, then we have

$$V_2 = I_2 \frac{R_2 R_V}{R_2 + R_V} \quad ...(11.4)$$

and $V - V_2 = I_2 R_1$ ...(11.5)

Dividing Eq. (11.5) by Eq. (11.4), we have

$$\frac{V - V_2}{V_2} = \frac{R_1 (R_2 + R_V)}{R_2 R_V}$$

$$\text{or} \quad \frac{V}{V_2} = \frac{R_2 R_V + R_1 (R_2 + R_V)}{R_2 R_V}$$

$$= \frac{R_1 R_2 + R_V (R_1 + R_2)}{R_2 R_V} \quad ...(11.6)$$

Dividing Eq. (11.3) by Eq. (11.6), we have

$$\frac{V_2}{V_1} = \frac{R_2}{R_1} \quad \text{or} \quad R_2 = R_1 \frac{V_2}{V_1}$$

Substituting $R_2 = R_1 \frac{V_2}{V_1}$ in Eq. (11.3), we have

$$\frac{V}{V_1} = \frac{R_1 \times R_1 \frac{V_2}{V_1} + R_V \left(R_1 + R_1 \frac{V_2}{V_1} \right)}{R_1 R_V}$$

$$= \frac{R_1 \frac{V_2}{V_1} (R_1 + R_V) + R_1 R_V}{R_1 R_V}$$

$$= \left(\frac{R_1 + R_V}{R_V} \right) \frac{V_2}{V_1} + 1$$

or $R_V V = R_1 V_2 + R_V V_2 + R_V V_1$

or $R_1 V_2 = R_V (V - V_1 - V_2)$

$$\text{or} \quad R_1 = R_V \left[\frac{V - V_1 - V_2}{V_2} \right] \quad ...(11.7)$$

$$\text{Similarly} \quad R_2 = R_V \left[\frac{V - V_1 - V_2}{V_1} \right] \quad ...(11.8)$$

The combined insulation resistance of the system can be determined by assuming that the two mains are bunched together so that the resultant value is the combined resistance of R_1 and R_2 in parallel.

$$\therefore \text{ Insulation resistance of the system} = \frac{R_1 R_2}{R_1 + R_2} \quad ...(11.9)$$

2. *Three-Wire System (Russell's Method).* For measurement of insulation resistance of 3-wire live mains use of an electrostatic voltmeter is made and connection of middle wire to earth is temporarily removed. First the voltmeter is connected between the middle wire and earth, as shown in Fig. 11.2(*a*) and reading of voltmeter is noted: let it be V_1. Next a known resistance *r* of a fairly high value (about 2,000 Ω) is connected between the middle wire and earth and the new reading of the voltmeter is noted; let it be V_2 [Fig. 11.2(*b*)] Since the resistance of an electrostatic voltmeter can be taken as infinity, so there will not be any current flowing through it.

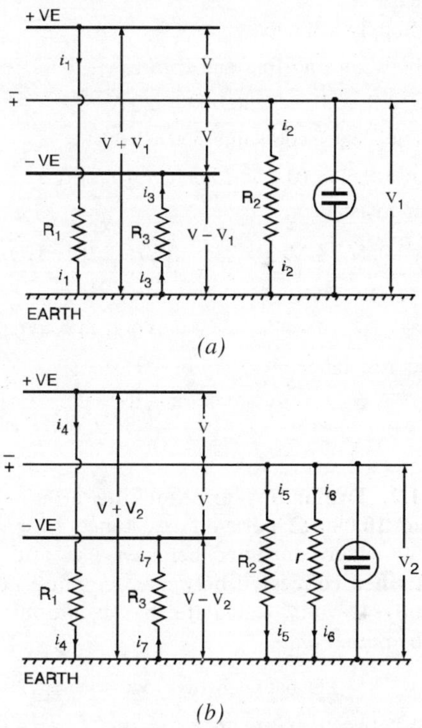

Fig. 11.2. ***Measurement of Insulation Resistance of 3-Wire Live Mains***

According to Kirchhoff's laws

For the first test

$$i_1 + i_2 = i_3$$

$$\text{or} \quad \frac{V + V_1}{R_1} + \frac{V_1}{R_2} = \frac{V - V_1}{R_3} \quad ...(11.10)$$

and for the second test

$$i_4 + i_5 + i_6 = i_7$$

$$\text{or} \quad \frac{V + V_2}{R_1} + \frac{V_2}{R_2} + \frac{V_2}{r} = \frac{V - V_2}{R_3} \quad ...(11.11)$$

Subtracting Eq. (11.11) from Eq. (11.10), we have

$$(V_1 - V_2) \left(\frac{1}{R_1} + \frac{1}{R_2} + \frac{1}{R_3} \right) = \frac{V_2}{r} \quad ...(11.12)$$

Since the combined insulation resistance of the system is given by the resultant of the three resistances R_1, R_2 and R_3 connected in parallel

$$\text{i.e.,} \quad R = \frac{1}{\frac{1}{R_1} + \frac{1}{R_2} + \frac{1}{R_3}}$$

Hence Eq. (11.12) becomes

$$\frac{V_1 - V_2}{R} = \frac{V_2}{r}$$

$$\text{or } R = r\left(\frac{V_1}{V_2} - 1\right) \qquad ...(11.13)$$

Example 11.1. The insulation resistances R_1 and R_2 of a live 2-wire network are measured by earthing each main in turn through a miliammeter in series with a high resistance r. Evaluate R_1 and R_2 for r = 20,000 ohms, i_1 = 1 mA and i_2 = 0.5 mA when the supply voltage is 500 volts.

Solution : Supply voltage, V = 500 V

Voltage between one line and earth,

$$V_1 = i_1 r = 1 \times 10^{-3} \times 20{,}000 = 20 \text{ V}$$

Voltage between second line and earth,

$$V_2 = i_2 r = 0.5 \times 10^{-3} \times 20{,}000 = 10 \text{ V}$$

Insulation resistance

$$R_1 = \frac{V - (V_1 + V_2)}{V_2} \times r = \frac{500 - (20 + 10)}{10} \times 20{,}000$$

$$= 940{,}000\ \Omega = 0.94\ \text{M}\Omega \ \textbf{Ans.}$$

Insulation resistance

$$R_2 = \frac{V - (V_1 + V_2)}{V_1} \times r = \frac{470}{20} \times 20{,}000 = 0.47\ \text{M}\Omega \ \textbf{Ans.}$$

Example 11.2. Two mains are working at a pd of 220 V. A voltmeter of 10,000 Ω per volt resistance with a maximum scale of 250 V when connected between + ve and earth reads 149 volts but when connected between –ve main and earth the reading is only 42 volts. Calculate the insulation resistance to earth of each main.

Solution : PD between two mains, V = 220 V

Voltage between +ve main and earth, V_1 = 149 V

Voltage between –ve main and earth, V_2 = 42 V

Total resistance of voltmeter,

$$R_V = 250 \times 10{,}000 = 2.5 \times 10^6\ \Omega$$

Insulation resistance,

$$R_1 = \frac{V - (V_1 + V_2)}{V_2} \times R_V = \frac{220 - (149 + 42)}{42} \times 2.5 \times 10^6$$

$$= 1.726\ \text{M}\Omega \ \textbf{Ans.}$$

Insulation resistance,

$$R_2 = \frac{V - (V_1 + V_2)}{V_1} \times R_V = \frac{29 \times 2.5 \times 10^6}{149} = 0.487\ \text{M}\Omega \ \textbf{Ans.}$$

Example 11.3. The voltage between the neutral point and earth of a 2 × 230 volt dc 3-wire insulated system was measured by an electrostatic voltmeter and found to be 40 volts. When an ammeter in series with a resistance was connected between the neutral point and earth the current was 0.05 A and the voltage 8 volts. Develop an expression for the fault resistance in the above test and use it to calculate the values with given figure.

Solution : Voltage between outer and middle wire, V ≐ 230 V

Voltage between middle wire and earth before connecting ammeter in series with resistance between middle wire and earth,

$$V_1 = 40 \text{ volts}$$

Voltage between middle wire and earth after connecting ammeter in series with resistance between middle wire and earth,

$$V_2 = 8 \text{ volts}$$

Current through ammeter,

$$i_6 = 0.05 \text{ A}$$

Resistance in series with ammeter,

$$r = \frac{V_2}{i_6} = \frac{8}{0.05} = 160\ \Omega$$

Fault of insulation resistance,

$$R = r\left(\frac{V_1}{V_2} - 1\right) = 160\left(\frac{40}{8} - 1\right) = 640\ \Omega \ \textbf{Ans.}$$

11.3. TYPES OF CABLE FAULTS

The faults which are most likely to occur in the cables are :

1. ***Ground or Earth Faults.*** When the insulation of the cable gets damaged, the current starts flowing from core to earth or to cable sheath. Such faults are known as *ground or earth faults.*

2. ***Cross or Short-Circuit Faults.*** When the insulation between two cables or between two cores of a multi-core cable gets damaged, the current starts flowing from one cable to another cable or from one core to another core of a multi-core cable directly (without passing through load). Such faults are known as *short-circuit faults.*

3. ***Open-Circuit Faults.*** When the conductor of a cable is broken or joint is pulled out there is no current in the cable. Such faults are known as *open-circuit faults.*

First the nature of fault is determined and then the point of fault is located. For determination of nature of faults, the insulation resistance of each core to ground and between cores is measured with the help of a megger. The low value of insulation resistance between any core and earth indicates the ground fault whereas the low value of insulation resistance between two cores (with far ends of the cable isolated from load) indicates short-circuit fault.

For determination of open-circuit fault, the far end of the cable, in case of single cable, is earthed (or the far ends of the cables are interconnected), an ammeter is inserted in series with the cable and low voltage supply is connected across the cable and earth or cables. Zero deflection in ammeter indicates open-circuit fault.

11.4. CAUSES OF CABLE FAULTS

Most faults are caused by dampness in the paper insulation of the cable due to porous or otherwise damaged lead sheath. The causes of damage to the lead sheath are : crystallization of the lead through vibration; chemical action on the lead when buried in the earth and insufficiently protected; and mechanical damage.

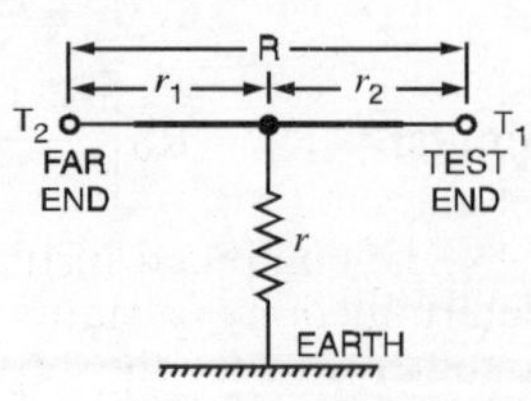

Fig. 11.3. ***Ground Fault of a Single Cable***

11.5. LOCATION OF CABLE FAULTS

There are various methods for locating the point of faults.

Ground Fault of a Single Cable. Blavier's test is used to locate the ground fault of a single cable.

This test is performed with the aid of a low-voltage supply and either an ammeter or voltmeter or a bridge network. In this test resistance between one end of the cable T_1 and earth is measured first with the far end T_2 isolated from earth and then with the far end T_2 earthed. Let the two readings be R_1 and R_2 respectively.

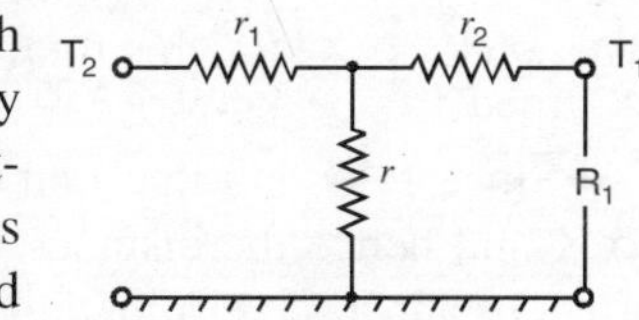

Fig. 11.4

If r_1 and r_2 are the conductor resistances of the lengths of cable "far end" to fault and "test end" to fault respectively and r is the resistance of fault to earth then

$$R_1 = r_2 + r \qquad ...(11.14)$$

$$\text{and } R_2 = \frac{1}{\frac{1}{r} + \frac{1}{r_1}} + r_2 = \frac{r\,r_1}{r + r_1} + r_2 \qquad ...(11.15)$$

The total resistance of conductor,

$$R = r_1 + r_2 \qquad ...(11.16)$$

By solving the above three equations, we can get the value of r, r_1 and r_2.

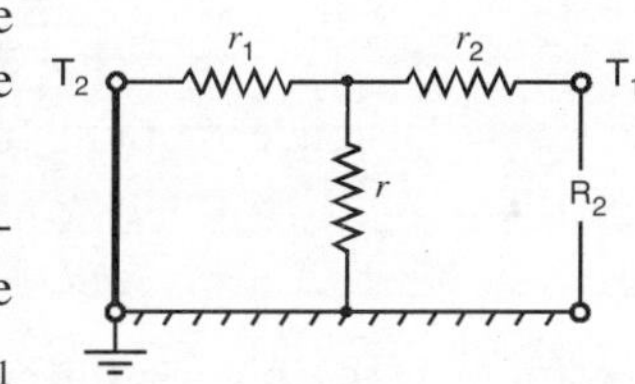

Fig. 11.5

If the total length of the cable is L metres, the length of cable between far end and fault is L_1 metres, length of cable between test end and fault is L_2 metres and cross section of conductor is uniform then

$$\frac{L_1}{L_2} = \frac{r_1}{r_2}$$

$$\text{and} \quad \frac{L_1 + L_2}{L_2} = \frac{r_1 + r_2}{r_2}$$

$$\text{or } L_2 = L \times \frac{r_2}{r_1 + r_2} \qquad ...(11.17)$$

Thus the distance of fault from test end can be determined.

Example 11.4. A feeder cable 500 metres long has a fault to earth. The fault is localised by the following resistance measurements between earth and one of the cable.

(*a*) Distant end insulated, 7.0 Ω

(*b*) Distant end earthed 1.7 Ω

The cable has total resistance of 1.8 Ω. Find the resistance of the fault and its distance from the testing end.

Solution : In this problem $R_1 = 7.0\ \Omega$, $R_2 = 1.7\ \Omega$, $R = 1.8\ \Omega$

If r_1 and r_2 are the conductor resistances of the length of cable far end to fault and test end to fault respectively and r is the resistance of fault to earth then

$$r_1 + r_2 = R = 1.8\ \Omega \qquad ...(i)$$

$$r + r_2 = R_1 = 7.0\ \Omega \qquad ...(ii)$$

$$\frac{rr_1}{r + r_1} + r_2 = R_2 = 1.7\ \Omega \qquad ...(iii)$$

Substituting $r_2 = 1.8 - r_1$ from Eq. (*i*) and $r = 7.0 - r_2 = 7.0 - (1.8 - r_1) = 5.2 + r_1$ from Eqs. (*i*) and (*ii*) in Eq. (*iii*), we have

$$\frac{(5.2 + r_1)r_1}{5.2 + r_1 + r_1} + 1.8 - r_1 = 1.7$$

$$\text{or } 5.2r_1 + r_1^2 = (r_1 - 0.1)(2r_1 + 5.2)$$

$$\text{or } 5.2r_1 + r_1^2 = 2r_1^2 + 5r_1 - 0.52$$

$$\text{or } r_1^2 - 0.2r_1 - 0.52 = 0$$

$$\text{or } r_1 = \frac{0.2 \pm \sqrt{(0.2)^2 - 4 \times (-0.52)}}{2}$$

$$= \frac{0.2 \pm 1.455}{2}$$

$= 0.8275\ \Omega$ as $-0.6275\ \Omega$ is impracticable figure

$$\therefore r_1 = 0.8275\ \Omega$$

$$r_2 = 1.8 - r_1 = 1.8 - 0.8275 = 0.9725\ \Omega$$

$$r = 5.2 + r_1 = 5.2 + 0.8275 = 6.0275\ \Omega$$

∴ Resistance of fault, $r = 6.0275\ \Omega$ **Ans.**

$$\text{Resistance per metre length} = \frac{1.8}{500} = 0.0036\ \Omega$$

∴ Distance from testing end

$$= \frac{r_2}{\text{Resistance per metre length}} = \frac{0.9725}{0.0036} = 270.14 \text{ metres } \textbf{Ans.}$$

11.6. LOOP TESTS

These tests are performed for the location of either an earth fault or a short-circuit fault in underground cables provided that a sound cable runs along with the grounded cable or a short-circuited cable. In these tests the resistance of fault does not affect the results obtained except when the resistance of fault is very high. There are two loop tests usually used and are known as *Murray loop and Varley loop tests.* These tests employ the principle of Wheatstone bridge.

11.6.1. Murray Loop Test

The connection diagrams to locate earth fault and short-circuit fault by Murray loop test method are shown in Figs. 11.6 and 11.7 respectively. As already said Wheatstone bridge

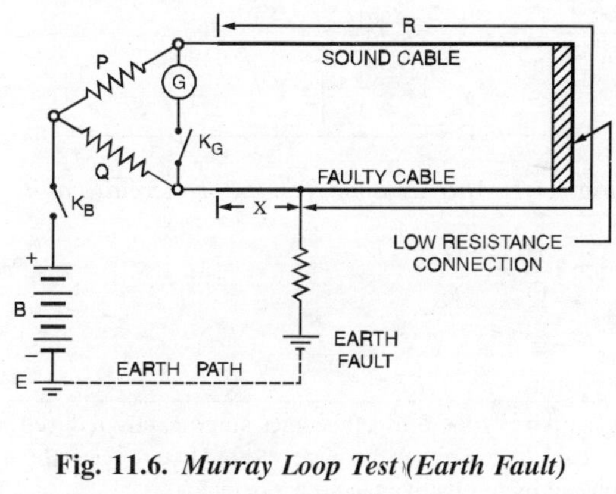

Fig. 11.6. ***Murray Loop Test (Earth Fault)***

principle is used in these tests. P and Q are the two ratio arms consisting of step resistors or slide wire, G is the galvanometer, B is the battery, K_B is a battery key and K_G is a galvanometer key. The faulty cable is connected to a second cable through low resistance link at the far end. Bridge is balanced by adjusting the resistances of ratio arms P and Q until the galvanometer indicates zero.

In balanced condition of bridge

$$\frac{P}{Q} = \frac{R}{X}$$

$$\text{or } \frac{P}{Q} + 1 = \frac{R}{X} + 1$$

$$\text{or } \frac{P+Q}{Q} = \frac{R+X}{X} = \frac{2r}{X}$$

$$\text{or } X = \frac{Q}{P+Q} \times 2r$$

where r is resistance of one of the cables when free from fault.

Since resistance per metre length of cable = $\frac{r}{l}$ where l is the length of cable in metres.

∴ Distance of fault from testing end

$$= \frac{X}{r/l} = \frac{lX}{r} = \frac{Q}{P+Q} \times 2r\frac{l}{r}$$

$$= \frac{Q}{P+Q} \times 2l \text{ metres} \qquad \text{...(11.18)}$$

In the above test since the fault resistance is in the battery circuit (not in bridge circuit) so balancing is not affected by it unless it is very high and reduces the sensitivity.*

The connections for short-circuit test are practically the same as in earth fault test except, that a portion of one of the short-circuiting cables is used instead of earth as return path in the battery circuit. Balance is obtained as before.

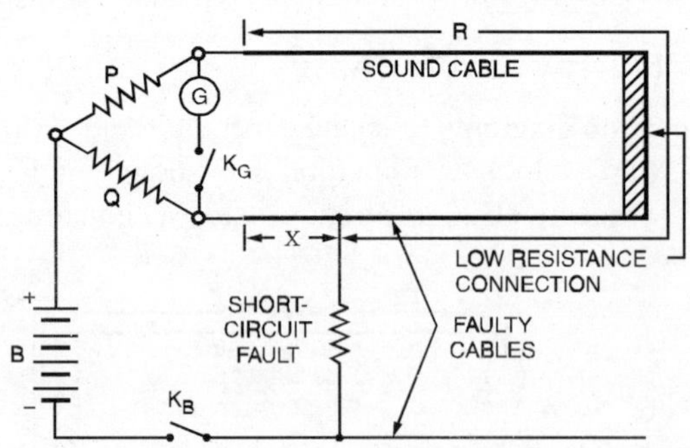

Fig. 11.7. ***Murray Loop Test (Short-Circuit Fault)***

$$\text{At balance } \frac{P}{Q} = \frac{R}{X}$$

$$\text{or } \frac{Q}{P+Q} = \frac{X}{R+X}$$

$$\text{or } X = \frac{Q}{P+Q}(R+X) \qquad \text{...(11.19)}$$

where (R + X) is the total resistance of the loop and is assumed to be known.

Since P, Q and the total resistance of the loop is known so X and hence the distance of fault from lower end of ratio arm Q can be determined.

Example 11.5. In a test by Murray loop for ground fault on 1,000 metres of cable having a resistance of 1.6 Ω per km, the faulty cable is looped with a sound cable of the same length and cross section. If the ratio of the other two arms of testing network at balance are in the ratio of 3:1, calculate the distance of the fault from the testing end of the cable. [*Gorakhpur Univ.* 1977]

Solution : The ratio of two arms,

$$\frac{P}{Q} = 3$$

$$\text{or } \frac{P+Q}{Q} = 3 + 1 = 4$$

Total resistance of loop, $(R + X) = \frac{1.6}{1,000} \times 2 \times 1,000 = 3.2\ \Omega$

$$X = \frac{Q}{P+Q}(R+X) = \frac{1}{4} \times 3.2 = 0.8\ \Omega$$

Distance of fault from testing end

$$= \frac{X}{r/l} = \frac{0.8}{1.6/1,000} = 500 \text{ metres } \textbf{Ans.}$$

Example 11.6. In a test for an earth fault on a 500 metre cable having a resistance of 1 Ω per 1,000 metres the faulty cable is looped with a sound cable of the same length but having a resistance of 2.25 Ω per 1,000 metres. The resistance of the other two arms of the testing network, at balance are in the ratio of 2.75:1. Calculate the distance of the fault from the testing end of the cable.

Solution : The ratio of two arms,

$$\frac{P}{Q} = 2.75$$

$$\text{or } \frac{P+Q}{Q} = 2.75 + 1 = 3.75$$

Total resistance of loop, (R + X)

$$= \frac{1}{1,000} \times 500 + \frac{2.25}{1,000} \times 500 = 1.625\ \Omega$$

$$X = \frac{Q}{P+Q}(R+X) = \frac{1.625}{3.75} = 0.4333\ \Omega$$

Distance of fault from testing end

$$= \frac{X}{r/l} = \frac{0.4333}{1/1,000} = 433.3 \text{ metres } \textbf{Ans.}$$

* The sensitivity of the bridge becomes singificantly reduced when the fault resistance is so high as to reduce the battery circuit current to 10 mA or less. In such a case either the battery should be replaced by a high-voltage dc supply obtained through a rectifier circuit or the galvanoneter should be replaced by a sensitive electronic detector.

11.6.2. Varley Loop Test

In this test also sound cable is required in addition to the existing cable. The circuit diagrams to determine the location of ground fault and short-circuit fault by method of Varley loop tests are shown in Figs. 11.8 and 11.9 respectively.

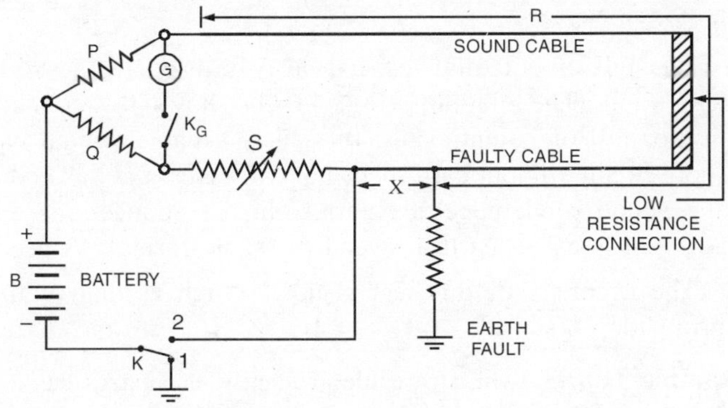

Fig. 11.8. *Varley Loop Test (Earth Fault)*

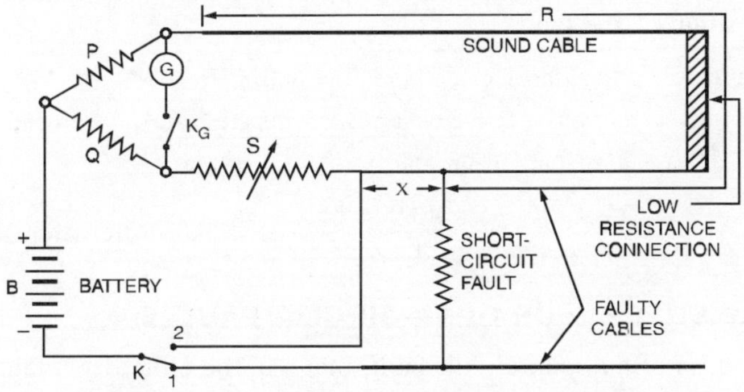

Fig. 11.9. *Varley Loop Test (Short-Circuit Fault)*

The difference between Murray loop and Varley loop tests is that in Varley loop test provision is made for the measurement of total loop resistance instead of obtaining it from the relation $R = \rho \frac{1}{a}$.

In this test the ratio arms P and Q are fixed and balance is obtained by varying the known variable resistance S. Balance is obtained once with key K in the battery circuit on stud 1 and then on stud 2.

At balance for earth fault or short-circuit fault with key K on stud 1.

$$\frac{P}{Q} = \frac{R}{X + S_1}$$

where S_1 is the setting of variable resistance S

$$\text{or} \quad \frac{P+Q}{Q} = \frac{R + X + S_1}{X + S_1}$$

$$\text{or } X = \frac{Q(R+X) - PS_1}{P+Q} \qquad \text{...(11.20)}$$

At balance for earth fault or short circuit fault when key K on stud 2.

$$\frac{P}{Q} = \frac{R+X}{S_2}$$

where S_2 is the new setting of variable resistance S.

$$\text{or } (R + X)Q = PS_2 \qquad \text{...(11.21)}$$

From Eqs. (11.20) and (11.21), we have

$$X = \frac{P(S_2 - S_1)}{P+Q} \qquad \text{...(11.22)}$$

Thus X can be determined from the known values of P, Q, S_2 and S_1.

Example 11.7. In locating a ground fault by Varley loop test (Fig. 11.8) the following observations were recorded.

(*i*) With key K at stud 1, a balance was obtained when P = 10, Q = 80 and S = 2,400 Ω.

(*ii*) With key K at stud 2, a balance was obtained when P = 10, Q = 80 and S = 3,400 Ω.

Determine the distance of the fault from the testing end. Each conductor is 5 km long.

Solution : $P = 10\ \Omega$; $Q = 80\ \Omega$; $S_2 = 3{,}400\ \Omega$; $S_1 = 2{,}400\ \Omega$

$$X = \frac{P(S_2 - S_1)}{P+Q} = \frac{10(3{,}400 - 2{,}400)}{10+80}$$

$$= 111.11\ \Omega$$

$$\text{Loop resistance, } (R + X) = \frac{P}{Q} S_2 = \frac{10}{80} \times 3{,}400 = 425\ \Omega$$

$$\text{Resistance per km length of cable} = \frac{425}{10} = 42.5\ \Omega$$

$$\text{Distance of fault from testing end} = \frac{111.11}{42.5} = 2.623 \text{ km } \textbf{Ans.}$$

11.7. FALL OF POTENTIAL TEST

Such a test can be made when a second cable, free from fault, run along with the faulty cable. The sound cable is used as a potential lead to the voltmeter, as shown in Figs. 11.10 and 11.11. The test circuit for testing twin core cable with an earth fault is shown in Fig. 11.10. The two cores are joined together at the distant end, as shown, by the low resistance connection. The battery circuit serves to pass current through the length of cable between end A and the fault.

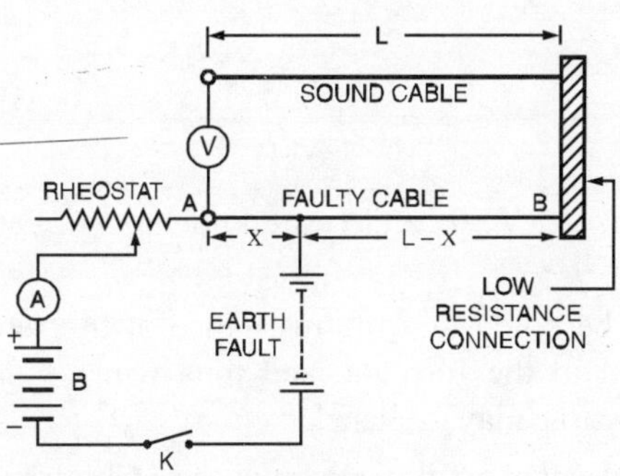

Fig. 11.10. *Fall of Potential Test (Earth Fault)*

The current is measured by the ammeter A and the voltage drop across the ends of this part of the cable is measured by voltmeter V. The current through the fault is adjusted by means of an adjustable resistance to suit the range of the instruments. The resistance of the fault itself plays no part in the test, except to tend to limit the test current. Let the readings of ammeter and voltmeter be I_1 and V_1, respectively. The apparatus is then moved to the end B and the whole test is repeated with the end A now short-circuited, let these readings be V_2 and I_2.

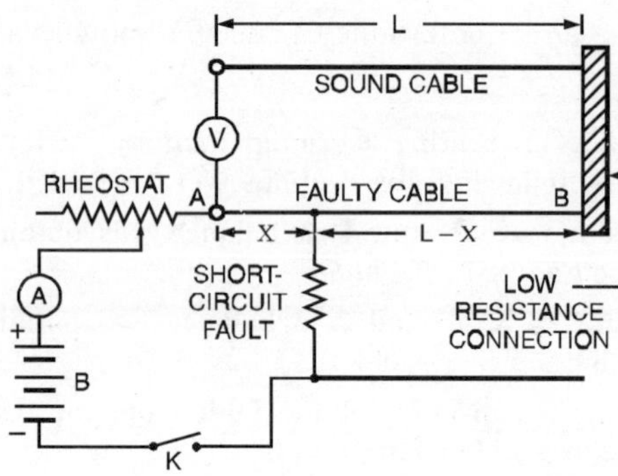

Fig. 11.11. *Fall of Potential Test (Short-Circuit Fault)*

Let the total length of the faulty core is L and length of core between fault and end A be X.

$$\text{Now } \frac{V_1}{I_1} = R_1 = \frac{\rho X}{a}$$

$$\text{or } X = \frac{a}{\rho} \cdot \frac{V_1}{I_1} \quad ...(11.23)$$

$$\text{Also } \frac{V_2}{I_2} = R_2 = \frac{\rho(L - X)}{a}$$

$$\text{or } L - X = \frac{a}{\rho} . \frac{V_2}{I_2} \quad ...(11.24)$$

Dividing Eq. (11.24) by Eq. (11.23), we have

$$\frac{L - X}{X} = \frac{V_2}{I_2} \times \frac{I_1}{V_1}$$

$$\text{or } \frac{L}{X} = \frac{V_1 I_2 + V_2 I_1}{V_1 I_2}$$

$$\text{or } X = \frac{L V_1 I_2}{V_1 I_2 + V_2 I_1} \quad ...(11.25)$$

Since L, V_1, V_2, I_1 and I_2 are known so X can be calculated.

It may be noted that the resistance of the return wire is not included in the formula, and thus wire of any suitable size and section may be used.

The voltmeter employed should be of high resistance as compared with the resistance of cable under test, if corrections for the voltmeter current are to be avoided. If electrostatic voltmeter is used, no corrections are, of course, necessary. Care must be taken that the current passed through the cable is not sufficient to produce appreciable heating as the resistance per unit length will then be different from that used in the calculation of length. If the cross section is uniform throughout, however, the proportionality renders the precaution less necessary, although it is not advisable to pass such a current that will produce appreciable heating, even in this case.

The fall of potential test can also be used for a short-circuit fault. The diagram (Fig. 11.11) shows the connections for fall of potential test on a 3-core cable, which has a short circuit without earth between two cores, the third core being sound (undamaged). Apart from the connections the whole test is as for earth test and the same formula is used.

The accuracy of this test is not as high as that of the Murray loop test.

Example 11.8. A twin-core cable of length 800 metres has developed a low resistance earth on one core. A fall of potential test is made from both ends, A and B. The readings taken at A are 1.5 V and 5 A and those at B are 3 V and 6 A. Find the position of the fault.

Solution : Length of cable core L = 800 metres

V_1 = 1.5 volts; V_2 = 3 volts; I_1 = 5 amperes and I_2 = 6 amperes

Distance of fault from end A

$$= \frac{L V_1 I_2}{V_1 I_2 + V_2 I_1} = \frac{800 \times 1.5 \times 6}{1.5 \times 6 + 3 \times 5} = 300 \text{ metres } \textbf{Ans.}$$

11.8. TEST FOR OPEN-CIRCUIT FAULTS

In case of an open-circuit fault, that may be either by a fault burning clear without causing an earth fault or on account of the cable being pulled out at a joint, it can be located by a capacity test.

In this test the capacitance of the faulty cable is measured from both the ends either by means of a ballistic galvanometer or by an ac bridge method.

Since the capacitance of the cable to the ground or to another parallel conductor is proportional to the length of the cable, so the ratio of the capacitances measured from two ends is equal to the ratio of the two lengths of the cable.

$$\textit{i.e.,} \quad \frac{C_1}{C_2} = \frac{L_1}{L_2}$$

$$\text{or } \frac{C_1 + C_2}{C_2} = \frac{L_1 + L_2}{L_2}$$

$$\text{or } L_2 = \frac{L_1 + L_2}{C_1 + C_2} C_2 = \frac{C_2 L}{C} \quad ...(11.26)$$

where L = $L_1 + L_2$, the total length cable and C is the capacitance of total length of cable.

Thus the distance of fault from either end can be determined. In case the total capacitance C, which is equal to $C_1 + C_2$ is known, measurement of capacitance from one end is sufficient.

The connections for testing a twin-core cable of total length L, of which one core is broken and other core sound are shown in Fig. 11.12. A charge and discharge key is connected to one end of faulty cable and the faulty cable is charged for a definite period of time, say 10 seconds, by making contact to the battery. The cable is then discharged through a ballistic galvanometer and the deflection is noted. Let it be D_1. The sound core is then charged for the same time period and discharged through the ballistic galvanometer and the deflection is noted. Let it be D_2. Then assuming the capacitance to earth per metre length to be the same for the two cores.

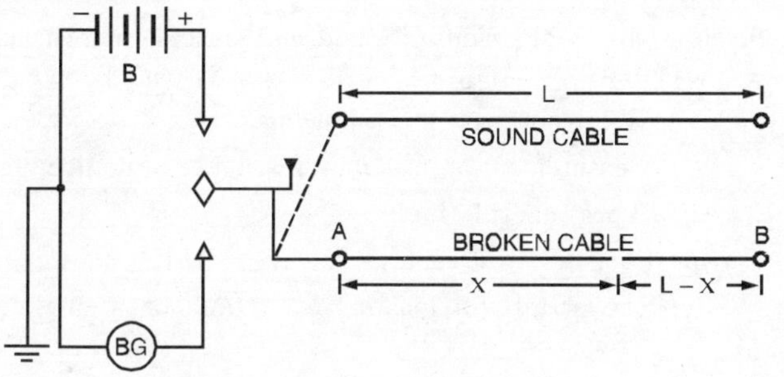

Fig. 11.12. *Test For Open Circuit Fault*

$$\text{Distance of fault from test end, } X = L \times \frac{D_1}{D_2} \quad \text{...(11.27)}$$

The test may be repeated from the other end of the faulty cable as a check.

If all cores are broken, tests are made from both ends of one core. Let the galvanometer readings be D_1 and D_2 then

$$X = \frac{D_1 L}{D_1 + D_2} \quad \text{...(11.28)}$$

The essential requirement for these capacity tests is that the core to be tested has a good insulation resistance.

11.9. MEASUREMENT OF INSULATION RESISTANCE OF A CABLE

1. ***Direct Deflection Method.*** In this method of measuring insulation resistance a very sensitive and high-resistance (1,000 Ω or more) moving coil galvanometer is connected in series with the resistance to be measured, and to a battery supply. The deflection of the galvanometer gives a measure of the insulation resistance. This is a very simple but rough method of measuring insulation resistance.

The direct deflection method for measuring insulation resistance of a cable is shown in Fig. 11.13. The galvanometer G measures the current between core and metal sheath; leakage currents over the surface of the insulating material are carried by the guard wire wound on the insulation and do not flow through the galvanometer. The ratio of the voltage applied between the core and metal sheath and current flowing between them gives the insulation resistance of the cable.

Cables without metal sheath can be tested in a similar way if the cable, except for the end or ends to which connections are made, is immersed in water in a tank. The water and tank then form the return path for the current.

Before testing, the cable is kept immersed in water for at least 24 hours and the temperature of water is kept constant throughout this period and during the experiment. The water enters the pores if any, and soaks through the defects and at the same time the cable attains the temperature of the water. The galvanometer should be well shunted in the earlier stage of the measurement, and it is normally desirable to include a series resistance of high value (several mega ohms). The true value of the insulation resistance of the cable is determined by subtracting the value of series resistance from the observed resistance value, Further, in tests on cables the galvanometer should actually be short-circuited before applying the voltage. After applying voltage, the short-circuiting connection is removed only after sufficient time has elapsed for the charging and absorption currents to have ceased to flow. The galvanometer is thus protected from the sudden initial rush of current which charges the cable the latter acting of course, as a capacitor.

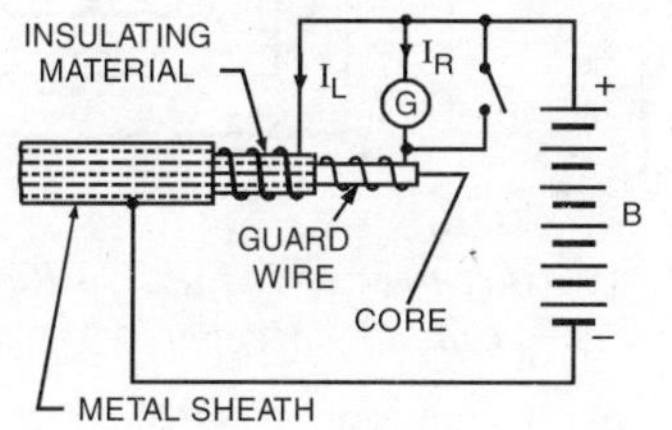

(a)

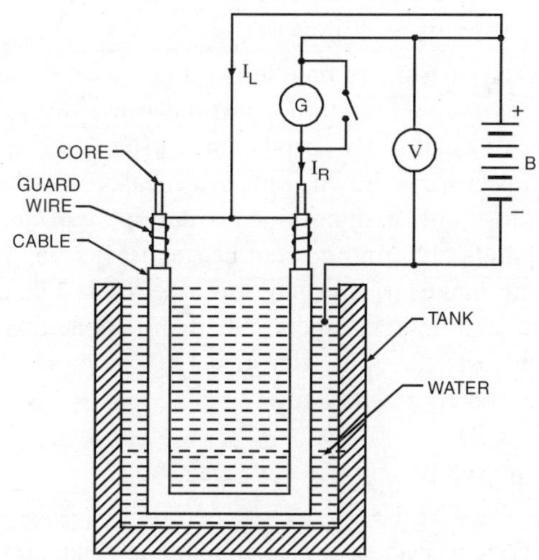

(b)

Fig. 11.13. *Direct Deflection Method of Measurement of Insulation Resistance of a Cable*

The galvanometer employed, which is of the D'Arsonval type, should be very sensitive (at least 1,000 mm per microampere at a scale distance of 1 metre), should be of high resistance, and, also have uniform scale. The galvanometer and its circuit, together with the keys, must be well insulated to prevent leakage currents.

The battery should be of about 500 volts and its emf should remain constant.

2. ***Megger Method.*** The insulation resistance of a cable may also be measured by a self-contained insulation tester, known as *megger**. Such an arrangement is shown in Fig. 11.14. Megger is provided with the three terminals known as line, earth and guard terminals. The 'line' terminal is connected to the core of the cable, the 'earth' terminal is to a plate immersed in water, or the tank side if suitable and the 'guard terminal' to the guard wire wound tightly round the insulation, as shown. Rotating of handle at steady speed then indicates the value of insulation resistance of the cable by the final position taken up by the instrument pointer on its scale.

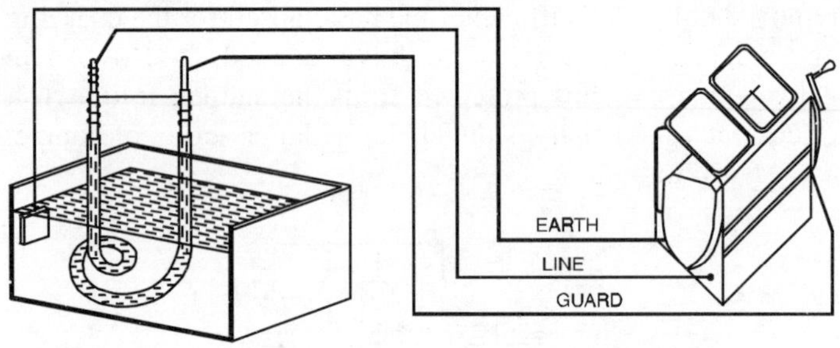

Fig. 11.14. ***Measurement of Insulation Resistance of Cable by Megger***

During the test it may be possible that the instrument pointer fall at first, as the handle is turned and, after several seconds rise to the correct position. This happens due to electrostatic capacitance of the cable. This phenomenon is described in detail as follows:

When the instrument handle is rotated with steady speed, voltage is generated and the current flows to charge the condenser; (the cable has capacitance effect) the magnitude of the initial current drawn is thus a measure of the capacitance of the cable and not its insulation resistance. After several seconds charging current ceases to flow and conduction current makes the pointer to rise to indicate correct value of insulation resistance. When the capacitance of the cable under test is of the order of 0.1 μF or more, the above initial effect is more marked.

Exercises

1. Describe a method of measuring the insulation resistance of a dc service main. **[U.P.S.C. I.E.S. Generation, Transmission & Distribution of Power 1971]**
2. Enumerate the faults which are most likely to occur in underground cables. Describe Blavier's test for locating ground fault on a single core cable.
3. What are the faults that usually occur in cables ? How do you localize them ? Describe using diagrams. **[Pb. Univ. April 1979]**
4. Discuss the procedure to locate fault in an underground three-phase cable:
   (*i*) With one conductor broken.
   (*ii*) With one conductor faulted to ground.
5. Explain the Murray loop method to find the faults in underground cables.
6. Describe the Murray's loop for localizing of earth and short-circuit faults in cables. **[Pb. Univ. June 1992]**
7. What is essential difference between Murray and Varley loop tests ? Describe a suitable test for location of earth and short circuit faults in a cable.
8. What is Varley loop test ? Explain the procedure of locating a fault on a cable by this test. **[Pb. Univ. June 1993]**
9. How are 'cable faults' located and 'insulation resistance' measured ? **[A.M.I.E. Sec. B. Power Systems Winter 1997]**
10. Write short notes on the following ;
    (*i*) Measurement of insulation resistance of dc live mains.
    (*ii*) Types of cable faults.
    (*iii*) Loop tests on cables.
    (*iv*) Blavier test for locating earth fault on a single-core cable.
    (*v*) Murray loop cable.
    (*vi*) Localization of cable faults by Varley loop tests.
    (*vii*) Testing of cables.
    (*viii*) Open circuit fault test.

Short Answer Type Questions With Answers

Q. 1. What are the faults that are likely to occur in cables ?

Ans. The faults which are likely to occur in cables are open circuit, short circuit and ground faults.

Q. 2. What is meant by 'ground fault' in cables ?

Ans. When the current starts flowing from core to earth or to cable sheath because of damage to the cable insulation, the fault is known as ground or earth fault.

Q. 3. What is meant by 'short-circuit' fault in cables ?

Ans. When the current starts flowing from one cable to another cable or from one core to another core of a multi-core cable due to damage to the cable insulation, the fault is called the cross or short-circuit fault.

Q. 4. What is meant by 'open-circuit fault' in cables ?

Ans. When the conductor of a cable is broken or joint is pulled out, the fault is called the open-circuit fault.

Q. 5. What principle do Murray loop and Varley loop tests employ ?

Ans. Murray loop and Varley loop tests employ the principle of Wheatstone bridge.

Q. 6. What is the essential difference between Murray and Varley loop tests ? **[Pb. Univ. Nov. 1977]**

Ans. The difference between Murray and Varley loop tests is that in Varley loop test provision is made for the measurement of total loop resistance instead of obtaining it from the relation $R = \rho \frac{l}{a}$.

* For description of the instrument refer to Art. 10.7 of *A Course in Electrical and Electronic Measurements and Instrumentation* by J B Gupta.

Q. 7. In what way Varley loop test is advantageous over Murray loop test?

Ans. In Murray loop test sound cable resistance must be known whereas in Varley loop test it is not so.

Q. 8. What is the essential requirement for performing capacity tests on cables ?

Ans. The essential requirement for performing capacity tests on cables is that the core to be tested has a good insulation resistance.

Problems

1. A 60,000 Ω voltmeter when connected between the +ve main and earth on a 250 V system reads 25 volts. When connected between the –ve main and earth, the reading is 50 volts. Calculate the insulation resistance of each main to earth. [**Ans.** 0.21 MΩ; 0.42 MΩ]

2. What is the insulation resistance of each main when the respective voltages to earth measured on the voltmeter having a resistance of 80,000 Ω are 80 V and 30 V. Potential difference between mains is 250 V.

[City and Guilds, London] [**Ans.** 0.373 MΩ, 0.14 MΩ]

3. A feeder cable 300 metres long has a fault to earth. The fault is localised by the following resistance measurements between earth and one of the cable.

(*i*) Distant end insulated, 6.0 Ω.

(*ii*) Distant end earthed, 1.7 Ω.

The cable has a total resistance of 1.8 Ω. Find the resistance of the fault and its distance from the testing end.

[**Ans.** 4.3725 Ω ; 162 metres from testing end]

4. In a test by Murray loop for ground fault on 500 metres of cable having a resistance of 1.6 Ω/km the faulty cable is looped with a sound cable of the same length and cross section. If the ratio of the other two arms of the testing network at balance are in the ratio of 3 : 1, find the distance of fault from the testing end of the cables. [**Ans.** 250 metres]

5. In a test for a fault to earth on a 520 metre length of cable having a resistance of 1.1 Ω per 1,000 metres, the faulty cable is looped with a sound cable of the same length but having a resistance of 2.29 Ω per 1,000 metres. The resistance of the other two arms of the testing network, at balance, are in the ratio 2.7 : 1. Calculate the distance to the fault from the testing end of the cable.

[Allahabad Univ. 1976 Supp.]

[**Ans.** 433 metres]

❑❑❑

CHAPTER 12

Interference of Power Lines With Neighbouring Communication Lines

INSIDE THIS CHAPTER

12.1. Introduction **12.2.** Electromagnetic Effect **12.3**. Effect of Earth Fault Currents **12.4.** Electrostatic Effect **12.5.** Transposition of Lines.

12.1. INTRODUCTION

It is usual practice to run telephone lines along the same route as the power lines. The transmission lines transmit bulk power at relatively higher voltages and, therefore, these lines give rise to electromagnetic and electrostatic fields of sufficient magnitude which induce currents and voltages in the neighbouring telephone lines. The currents so induced are superimposed on the true speech currents in the neighbouring telephone wires and set up distortion while the voltages so induced raise the potential of the communication circuit as a whole. In extreme cases, the effects of these fields may make it impossible to transmit any message faithfully and may raise the potential of the telephone receiver above the ground to such an extent to render the handling of the telephone receiver extremely dangerous and in such cases elaborate precautions are required to be observed to avoid this danger.

12.2. ELECTROMAGNETIC EFFECT

12.2.1. Single-Phase Circuit Line and Telephone Line

Consider two power conductors A and B of a single-phase single circuit line and two telephone conductors C and D of a telephone line running on the same pole as the power conductors as shown in Fig. 12.1.

Let the distances between the power conductors and telephone conductors be d_{AC}, d_{AD}, d_{BC} and d_{BD} respectively.

First of all consider the loop formed by the conductors A and C. The self-induction of this loop is given by one-half of the expression of the self-induction of a pair of parallel conductors, because for this loop only conductor A is contributing to the emf induced in conductor C. Denoting this self-induction by L_{AC}, we have

$$L_{AC} = 0.2 \log_e \frac{d_{AC}}{r} \text{ mH/km}$$

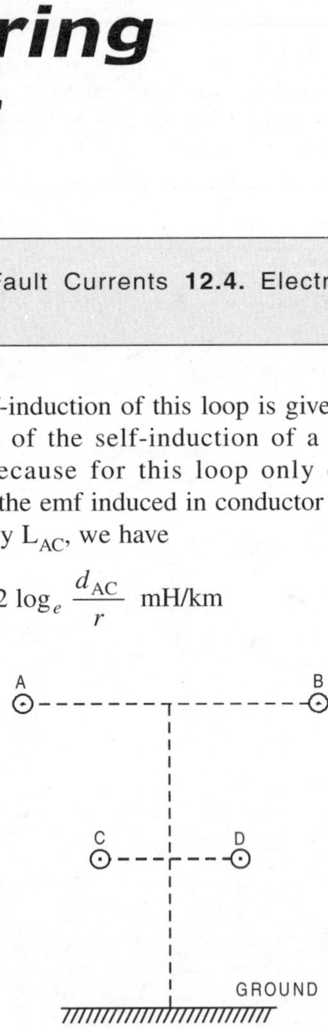

Fig. 12.1

Similarly the self-induction of the loop formed by conductors A and D,

$$L_{AD} = 0.2 \log_e \frac{d_{AD}}{r} \text{ mH/km}$$

Mutual inductance between conductor A and and the loop CD,

$$M_A = L_{AD} - L_{AC}$$

$$= \left(0.2 \log_e \frac{d_{AD}}{r} - 0.2 \log_e \frac{d_{AC}}{r}\right) \text{ mH/km}$$

$$= 0.2 \log_e \frac{d_{AD}}{d_{AC}} \text{ mH/km}$$

Similarly the mutual inductance between conductor B and the loop CD, M_B is given as

$$M_B = 0.2 \log_e \frac{d_{BD}}{d_{BC}} \text{ mH/km}$$

The net mutual inductance, which is phasor sum of two inductances is given as

$$\mathbf{M} = \mathbf{M}_A + \mathbf{M}_B = 0.2 \log_e \frac{d_{AD}}{d_{AC}} - 0.2 \log_e \frac{d_{BD}}{d_{BC}}$$

$\because$ M_A and M_B are in phase opposition

$$= 0.2 \log_e \frac{d_{AD} d_{BC}}{d_{AC} d_{BD}} \text{ mH/km}$$

$$= 0.4 \log_e \sqrt{\frac{d_{AD}\, d_{BC}}{d_{AC}\, d_{BD}}} \text{ mH/km}$$

The voltage induced in the loop CD is given by

$$V_m = 2\pi f \text{ MI volts per km}$$

where I is the current flowing through the power conductor and f is the supply frequency.

12.2.2. Three-Phase Circuit Line and Telephone Line

Consider three power conductors A, B, and C of a 3-ϕ single circuit line and two telephone conductors P and Q of a telephone line running on the same transmission tower as the power conductors, as shown in Fig. 12.2.

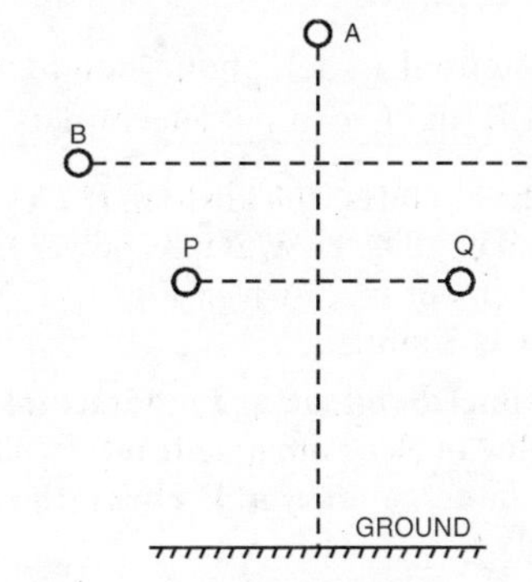

Fig. 12.2

Let the distances between the power conductors and telephone conductors be d_{AP}, d_{AQ}, d_{BP}, d_{BQ}, d_{CP} and d_{CQ} respectively.

First of all consider the loop formed by the conductors A and P. The self-induction of this loop is given by one-half of the expression for the self-induction of a pair of parallel conductors, because for this loop only conductor A is contributing to the emf induced in conductor P. Denoting this self-induction by L_{AP}, we have

$$L_{AP} = 0.2 \log_e \frac{d_{AP}}{r} \text{ mH/km}$$

Similarly the self-induction of the loop formed by the conductors A and Q.

$$L_{AQ} = 0.2 \log_e \frac{d_{AQ}}{r} \text{ mH/km}$$

$\therefore$ Mutual inductances between conductor A and the loop PQ,

$$M_A = L_{AQ} - L_{AP}$$

$$= \left(0.2 \log_e \frac{d_{AQ}}{r} - 0.2 \log_e \frac{d_{AP}}{r}\right) \text{ mH/km}$$

$$= 0.2 \log_e \frac{d_{AQ}}{d_{AP}} \text{ m H/km}$$

Similarly the mutual inductance between conductor B and the loop PQ, M_B and the mutual inductance between conductor C and loop PQ, M_C are given as

$$M_B = 0.2 \log_e \frac{d_{BQ}}{d_{BP}} \text{ mH/km}$$

and $$M_C = 0.2 \log_e \frac{d_{CQ}}{d_{CP}} \text{ mH/km}$$

These three mutual inductances are the result of fluxes which have a phase displacement of 120º. Therefore, the net effect of the magnetic field will be

$$M = M_A + M_B + M_C$$

where M is the net mutual inductance which is the phasor sum of the three inductances.

The voltage induced in the loop PQ is given by

$$V_m = 2\pi f \text{MI volts per km} \quad \text{...(12.1)}$$

where I is the current flowing through the power conductor and f is the supply frequency.

It is to be noted that since the net mutual inductance M is the phasor sum of mutual inductances M_A, M_B, and M_C, therefore, there is a partial cancellation of the induced voltages due to fundamental of the power current and as the distance between the power conductors and the telephone conductors is increased, the mutual inductances, M_A, M_B and M_C progressively become equal in magnitude and thus M diminishes. But the presence of third harmonics and multiples of third harmonics will not cancel as they are in phase in all the power conductors and, therefore, are dangerous for communication circuits. These harmonics are troublesome in two ways: firstly because the induced voltage is proportional to the frequency, and secondly, because the higher frequencies come within the audible range.

Example 12.1. A single-phase, 50 Hz power line has horizontal configuration with 1.2 m spacing. A telephone line is run horizontally on the same support and is located symmetrically 0.85 m below the power line. Determine the voltage induced per km in the telephone line if the current in the power line is 120 A. Take the spacing between telephone conductors as 0.4 m.

[U.P. Technical Univ. Elements of Power System 2003-04]

Solution : Refer to Fig. 12.3

$$d_{AD} = \sqrt{0.85^2 + (0.6+0.2)^2} = 1.1673 \text{ m}$$

$$d_{AC} = \sqrt{0.85^2 + (0.6-0.2)^2} = 0.9394 \text{ m}$$

$$d_{BD} = d_{AC} = 0.9394 \text{ m}$$

$$d_{BC} = d_{AD} = 1.1673 \text{ m}$$

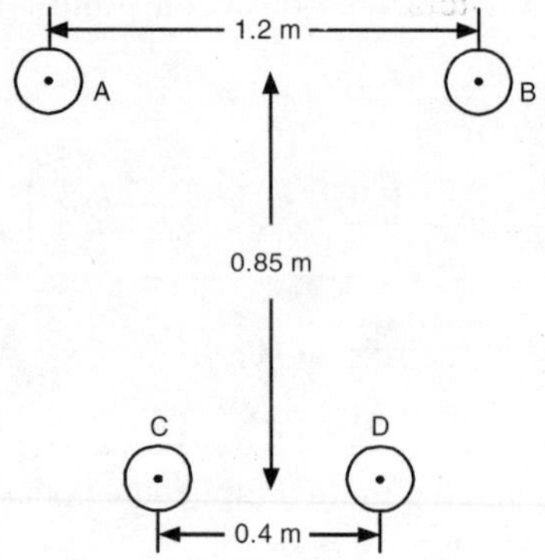

Fig. 12.3

Mutual inductance,

$$M = 0.4 \log_e \sqrt{\frac{d_{AD} d_{BC}}{d_{AC} d_{BD}}} \text{ mH/km}$$

$$= 0.4 \log_e \sqrt{\frac{1.1673 \times 1.1673}{0.9394 \times 0.9394}} \text{ mH/km}$$

$$= 0.086883 \text{ mH/km}$$

The voltage induced per km in the telephone line,

$$V_m = 2\pi f M I = 2\pi \times 50 \times 0.086883 \times 10^{-3} \times 120$$

$$= 3.275 \text{ V/km} \quad \textbf{Ans.}$$

12.3. EFFECT OF EARTH FAULT CURRENTS

There may be a large voltage drop in the earth itself, due to high soil resistivity when an earth fault occurs from a line to earth and there is a large return fault current. In case the telephone-circuit earth is within the resistance area of the power system earth, the potential of the telephone-circuit earth may be raised well above true earth potential and so a serious hazard may be present.

12.4. ELECTROSTATIC EFFECT

We know that when a conductor, such as A (Fig. 12.4), runs parallel to an infinite plane (earth for example), the potential

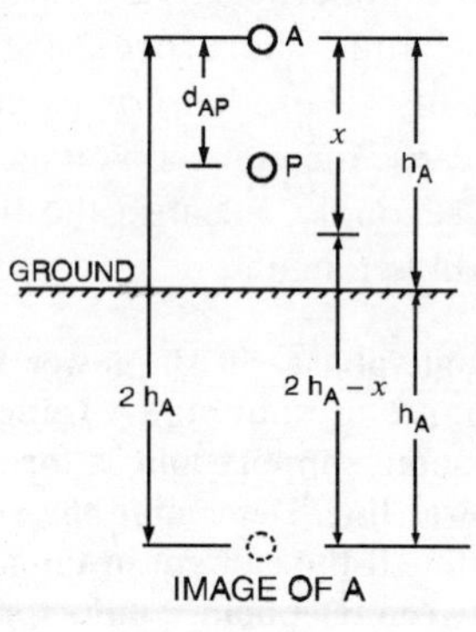

Fig. 12.4

distribution between this conductor and the plane is exactly similar to that between its image and the plane. Hence if q is the charge per unit length of the power conductor A and h_A is the height of the conductor A from the ground, the potential of conductor A with respect to earth is given by

$$V_A = \frac{1}{2\pi \epsilon_0} \int_r^{h_A} \left(\frac{q}{x} + \frac{q}{2h_A - x} \right) dx$$

where r is the radius of conductor A

$$\text{or } V_A = \frac{q}{2\pi \epsilon_0} \log_e \frac{2h_A - r}{r} \quad \text{...(12.2)}$$

Similarly potential of conductor P due to the conductor A

$$V_{PA} = \frac{1}{2\pi \epsilon_0} \int_{d_{AP}}^{h_A} \left(\frac{q}{x} + \frac{q}{2h_A - x} \right) dx$$

$$= \frac{q}{2\pi \epsilon_0} \log_e \frac{2h_A - d_{AP}}{d_{AP}}$$

Substituting $q = \dfrac{2\pi \epsilon_0 . V_A}{\log_e \dfrac{2h_A - r}{r}}$ from Eq. (12.2) in the above equation, we have

$$V_{PA} = \frac{2\pi \epsilon_0 . V_A}{\log_e \dfrac{2h_A - r}{r}} \times \frac{1}{2\pi \epsilon_0} \log_e \frac{2h_A - d_{AP}}{d_{AP}}$$

$$= V_A \frac{\log_e \dfrac{2h_A - d_{AP}}{d_{AP}}}{\log_e \dfrac{2h_A - r}{r}} \quad \text{...(12.3)}$$

Similarly, we can determine the potential of telephone conductor P due to power conductors B and C and hence the potential of telephone conductor P due to power conductors A, B and C will be

$$\mathbf{V_P = V_{PA} + V_{PB} + V_{PC}}$$

Similarly the potential of telephone conductor Q due to power conductors A, B and C can be determined.

Example 12.2. A 3-phase, 50 Hz, 200 km long 132 kV transmission line delivers a total load of 28,000 kW at 0.85 power factor lagging. The conductor arrangement is shown in Fig. 12.5. The radius of each power conductor is 5 mm.

Determine the induced voltage at fundamental frequency in the telephone circuit due to electromagnetic effect. Also determine the potential of telephone conductor P above the earth due to electrostatic effect only.

Solution : From Fig. 12.5

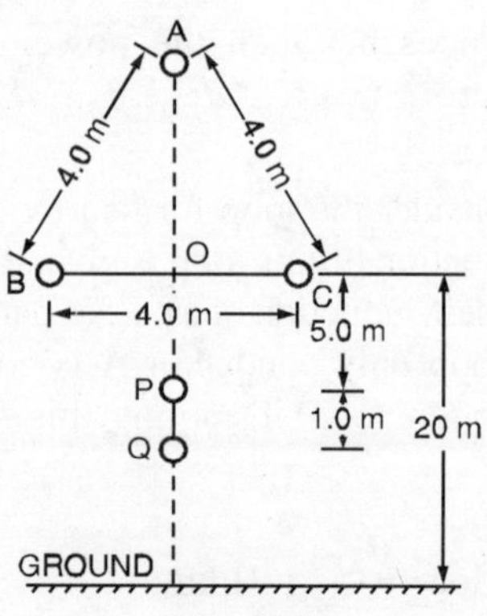

Fig. 12.5

$$AO = \frac{4\sqrt{3}}{2} = 3.46 \text{ m}$$

$$d_{AP} = 3.46 + 5 = 8.46 \text{ m}$$

$$d_{AQ} = 3.46 + 5 + 1.0 = 9.46 \text{ m}$$

$$d_{BP} = \sqrt{5^2 + 2^2} = 5.39 \text{ m}$$

$$d_{BQ} = \sqrt{6^2 + 2^2} = 6.32 \text{ m}$$

∴ Mutual inductance between conductor A and the loop PQ,

$$M_A = 0.2 \log_e \frac{d_{AQ}}{d_{AP}} \text{ mH/km} = 0.2 \log_e \frac{9.46}{8.46} = 0.022 \text{ mH/km}$$

$$M_B = M_C = 0.2 \log_e \frac{d_{BQ}}{d_{BP}} = 0.2 \log_e \frac{6.32}{5.39} = 0.032 \text{ mH/km}$$

Since M_A, M_B and M_C are mutually displaced by 120°

∴ $M = (0.032 - 0.022)$ mH/km = 0.01 mH/km

Current flowing through the power conductors,

$$I = \frac{\text{Load in kW} \times 1{,}000}{\sqrt{3} \times \text{Line voltage} \times pf} = \frac{28{,}000 \times 1{,}000}{\sqrt{3} \times 132{,}000 \times 0.85} = 144 \text{ A}$$

Supply frequency, $f = 50$ Hz

∴ Induced voltage at fundamental frequency in the telephone circuit due to electromagnetic effect

$$V_m = 2\pi f MI = 2\pi \times 50 \times 0.01 \times 10^{-3} \times 144 = 0.452 \text{ V/km}$$

Thus, for the whole route of 200 km the induced voltage at fundamental frequency in the telephone circuit due to electromagnetic effect

= 200 × 0.452 = 90.4 volts **Ans.**

For the electrostatic effect we have for conductor P,

$$V_{PA} = \frac{V_A \log_e \frac{2h_A - d_{AP}}{d_{AP}}}{\log_e \frac{2h_A - r}{r}}$$

$$= \frac{132{,}000}{\sqrt{3}} \frac{\log_e \frac{2(20 + 3.46) - 8.46}{8.46}}{\log_e \frac{2(20 + 3.46) - 0.005}{0.005}} = 12{,}617 \text{ volts}$$

and $$V_{PB} = V_{PC} = V_B \frac{\log_e \frac{2h_A - d_{BP}}{d_{BP}}}{\log_e \frac{2h_A - r}{r}}$$

$$= \frac{132{,}000}{\sqrt{3}} \frac{\log_e \frac{46.92 - 5.39}{5.39}}{\log_e \frac{46.92 - 0.005}{0.005}} = 17{,}013 \text{ volts}$$

The potential of telephone conductor P with respect to earth, V_P being the phasor sum of above three components mutually displaced by 120°

V_P = 17,013 – 12,617 = 4,396 volts **Ans.**

12.5. TRANSPOSITION OF LINES

As already said, in extreme cases the effects of the electromagnetic and electrostatic fields set up by power lines become so strong to make it impossible to transmit any message faithfully and render the handling of the telephone receiver extremely dangerous. In order to reduce the induced voltages due to electromagnetic and electrostatic induction, the conductors of both the power line and the telephone line are regularly transposed, as shown in Fig. 12.6. By transposition of power lines the capacitances of the lines are balanced and, therefore, electrostatically induced voltages balance out in the length of a complete set of transpositions, known as *barrel.* Also electromagnetically induced voltages on the conductors of communication lines get diminished as the fluxes due to the positive and negative phase sequences cancel out along the barrel. This, however, does not apply to zero-sequence currents and that is why the telephone line is also transposed along with the transposition of power line, as shown in Fig. 12.6. Each transposition of a telephone line consists of a complete cross-over of the two wires, while each transposition of a 3-ϕ line consists of a twist, in a plane at right angles to the run of the line, of one-third of a revolution. Thus three transpositions are then necessary to bring the phases back to their original positions.

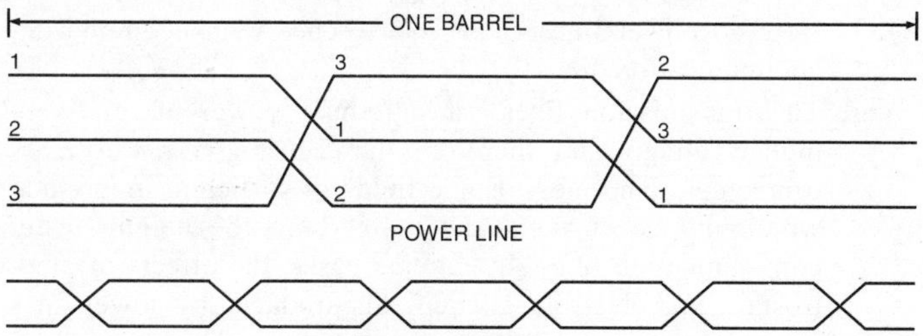

Fig. 12.6

Electrostatic charging of the telephone line also results in the flow of currents which tends to interfere with the clarity of the speech. This effect cannot be eliminated entirely by transposition of lines.

Although, transposition of conductors is helpful under normal operating conditions but does not provide complete solution of the problem under abnormal conditions caused due to such faults as a short circuit to earth, In such a case high emf is induced due to electromagnetic induction since the fault results in large zero-sequence currents flowing along the wires in parallel and through the earth return.

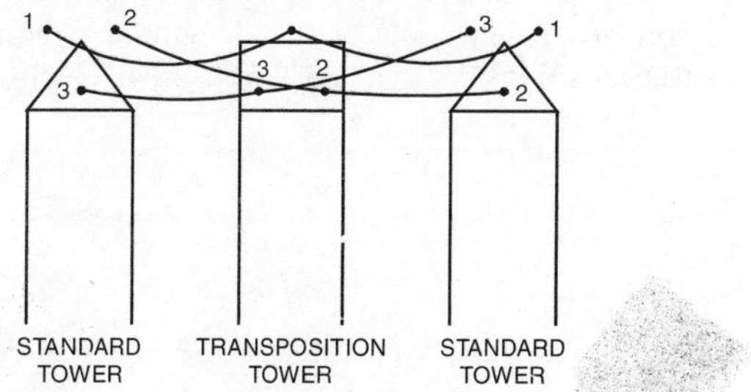

Fig. 12.7

With communication lines perfectly transposed and balanced no voltage between two conductors may exist and yet a considerable voltage will be induced between each conductor and earth. So it is necessary that the communication line insulation is so arranged that insulators can withstand such a voltage.

In extreme cases of electrostatic charging it may become necessary to isolate the telephone apparatus from the telephone line completely by means of highly insulated transformers, and also, to ensure the dissipation of the induced charges by means of such devices as earthed 'drainage' coils and lightning arrestors.

Exercises

1. Explain how does the power circuit interfere with the normal operation of the nearby communication circuit and what steps are taken to avoid this interference ?
2. Write short notes on the following :
   (*i*) Interference of power lines with communication circuits. **[Rajasthan Univ. 2003; 2007]**
   (*ii*) Radio interference. **[M.D. Univ. Dec. 2008; Dec. 2010]**
   (*iii*) Reduction of radio interference.
   (*iv*) Transposition of lines.

Short Answer Type Questions With Answers

Q. 1. How do power lines cause interference with neighbouring communication lines ?

Ans. The transmission lines transmit bulk power at relatively higher voltages and, therefore, these lines give rise to electromagnetic and electrostatic fields of sufficient magnitude which induce currents and voltages in the neighbouring communication lines. In extreme cases, the effects of electrostatic and electromagnetic fields produced by power lines may make it impossible to transmit any message faithfully and may raise the potential of telephone receiver above the ground to such an extent to render the handling of the telephone receiver extremely dangerous.

Q. 2. How the radio interference caused by power lines in the neighbouring communication lines be reduced ?

Ans. Radio interference caused by power lines in the neighbouring communication lines can be reduced by regular transposition of both the power lines and communication lines. By transposition of both power lines and communication lines voltages induced due to electrostatic and electromagnetic effects are reduced.

Q. 3. How do regular transposition of power lines and communication lines reduce the radio interference ?

Ans. By transposition of power lines the capacitances of the lines are balanced and, therefore, electrostatically induced voltages balance out in the length of a complete set of transpositions known as barrel. Also electromagnetically induced voltages on the communication line conductors get diminished as the fluxes due to positive and negative phase sequence cancel out along the barrel. This, however, does not apply to zero-sequence currents and communication line is also transposed regularly along with the transposition of power line so as reduce the effects of zero-sequence currents.

Q. 4. Why is it necessary to arrange the communication line insulation so that insulators can withstand high voltages.

Ans. With communication lines perfectly transposed and balanced no voltage between two conductors may exist and yet a considerable voltage will be induced between each conductor and earth. So it is necessary that the communication line insulation is so arranged that insulators can withstand such a voltage.

Q. 5. How can the radio interference be reduced in extreme cases of electrostatic charging ?

Ans. In extreme cases of electrostatic charging it may become necessary to isolate the telephone apparatus from the telephone line completely by means of highly insulated transformers, and also, to ensure the dissipation of the induced charges by means of such devices as earthed 'drainage' coils and lightning arrestors.

Problems

1. A 3-phase, 50 Hz, 100 km long 132 kV transmission line delivers a total load of 30 MW at 0.9 power factor lagging. The conductor arrangement is shown in the Fig. 1. The radius of each conductor is 6 mm.
   Determine the induced voltage at fundamental frequency in the telephone circuit due to electromagnetic effect. Also determine the potential of telephone conductor P above the earth due to electrostatic effect. [**Ans.** 28.67 V; 3,655 V]

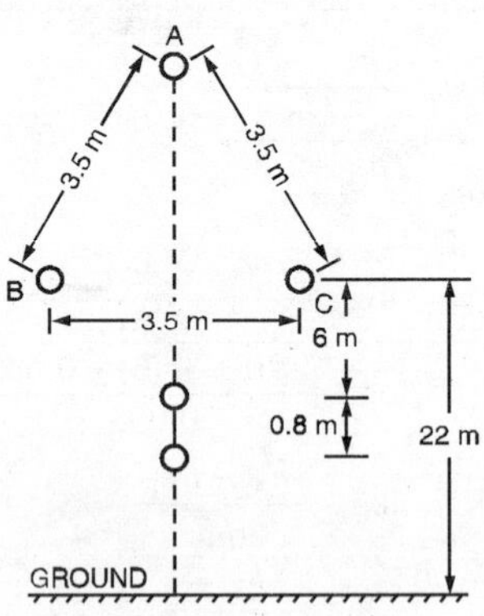

Fig. 1

□□□

CHAPTER 13

Extra High Voltage (EHV) AC Transmission

INSIDE THIS CHAPTER

13.1. INTRODUCTION

The basic function of a transmission system is to transfer electrical power from one place to another place or from one network to another network. A transmission system includes terminal substations, transmission lines and intermediate substations.

Transmission systems are required for (*i*) transfer of bulk power from large group of generating stations up to the main transmission network, (*ii*) the main transmission network, (*iii*) system interconnection and (*iv*) transfer of power from the main transmission network to the secondary substations.

A transmission system is employed either for transfer of power from sending end to the receiving end or for system interconnection for exchange of electrical power between independently controlled networks.

The increasing need of transmitting greater and greater amounts of power over long distances has led to a continuous increase in the transmission voltages. As a rule, larger the amount of power to be transferred, higher is the requirement of transmission voltage. Similarly longer the lines, higher is the transmission voltage.

Up to 1970's, the choice was exclusively in favour of ac high voltage (up to 220 kV) and extra high voltage (above 220 kV and up to 750 kV). The present trend is to use transmission voltage from high voltages (up to 300 kV), extra high voltages (300-765 kV) to ultra high voltages (above 765 kV), depending upon the technical and economical considerations, for the transmission of large amounts of power over long distances.

In India, the transmission voltages in use are 33, 66, 110, 132, 220 and 400 kV. The next higher voltage of 765/800 kV has been selected for transmission of huge blocks of power from generating stations to the load centres, as already mentioned in introductory Art. 1.1.

13.2. NECESSITY OF EHV TRANSMISSION

As already mentioned above in Art. 13.1, modern trend is to use extra high voltage (EHV) and ultra high voltage (UHV) for transmission of huge blocks of power over long distances. The reasons for adopting of EHV/UHV range for transmission purposes are given below :

1. ***Reduction of electrical losses, increase in transmission efficiency, improvement of voltage regulation and reduction in conductor material requirement***—For transmission of given amount of power over a given distance through the conductors of a given material and at a given power factor as the transmission voltage increases,

(*a*) line losses are reduced since line losses are inversely proportional to the transmission voltage,

(*b*) transmission efficiency increases because of reduction in line losses,

(*c*) voltage regulation is improved because of reduction of percentage line drop, and

(*d*) lesser conductor material is required being inversely proportional to the square of transmission voltage. (Refer to Art. 1.13).

2. Economic considerations have led to the construction of power stations of large capacity and so need of transfer of bulk power over long distances arose. Transmission of bulk power from generating stations to the load centres is technically and economically feasible only at voltages in the EHV/UHV range.

3. Generating stations (Steam, hydro and nuclear power stations) are located in remote areas (far away from load centres) because of the reasons of economy, feasibility and from the point of view of safety and environmental conditions. EHV transmission is, therefore, inevitable for transmission of huge blocks of power over long distances from these power plants to load centres.

4. ***Flexibility for future system growth***—There is flexibility of future system growth.

5. ***Increase in transmission capacity of the line***—Power transferred is expressed as

$$P = \frac{V_S \,.\, V_R}{X} \sin \delta \qquad ...(13.1)$$

where V_S and V_R are the two terminal voltages, δ is the load angle and X is the line reactance.

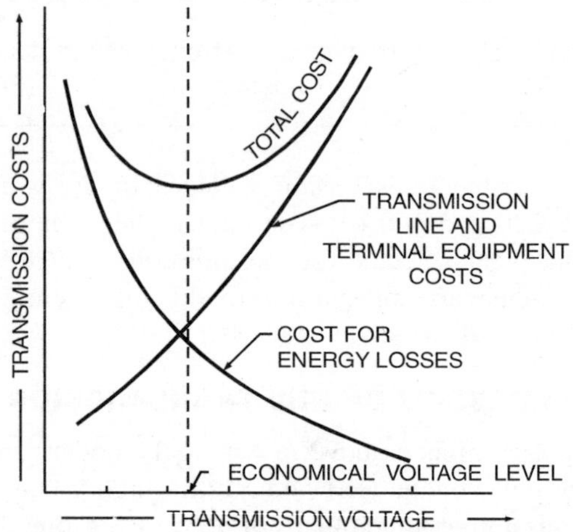

Fig. 13.1

Thus the power transmission capacity of a transmission line increases with the increase in transmission voltage. No doubt the cost of transmission line and terminal equipment also increases with the increase in the transmission voltage but in general these costs are proportional to the transmission voltage rather than the square of the transmission voltage. Moreover there is also a saving in cost due to reduction in energy losses occurring in transmission lines. As a consequence the total cost of transmission decreases with the increase in transmission voltage, as depicted in Fig. 13.1.

6. ***Possibility of interconnections of power systems***—It is practically not possible to have interconnections of two or more power systems, which is necessary to achieve sharing of installed reserves and for development of integrated systems and grids, without EHV transmission.

7. ***Increase of surge impedance loading***—Load carrying capability of a line is usually expressed in terms of "*surge impedance loading*" (SIL). Surge impedance loading (SIL) is the power that a line carries when each phase is terminated by a load equal to the surge impedance of the line. For a transmission line, the surge impedance is given as $Z_C = \sqrt{\frac{L}{C}}$ where L and C are respectively the series inductance and shunt capacitance per unit length. The surge impedance loading (SIL), for a transmission line is given as $\frac{3V^2}{Z_C}$ where V is the line-to-neutral voltage.

It is evident that SIL varies as the square of the operating voltage, and, therefore, with the increase in voltage level, SIL itself increases. Thus power transfer capability of the line increases with the increase in voltage level.

The surge impedance of a line can be determined from its conductor configuration. The approximate values of surge impedances for lines with single, double, triple and quadruple conductors are 400, 300, 280 and 260 ohms respectively.

8. ***Reduction in right of way***—In some countries 'rights of way' are paid for at a rate proportional to the total width of the transmission lines. Even in countries where right of way is not directly paid, there are usually strong pressures from the public towards fewer and fewer transmission lines. With the passage of time right of way becomes either more costlier or difficult to obtain and therefore it is becoming necessary to have fewer transmission lines operating at EHV/UHV. The noteworthy point here is that with the increase in operating voltage, number of circuits and requirement of land is reduced considerably.

13.3. PROBLEMS ASSOCIATED WITH EHV TRANSMISSION

The major problems associated with EHV transmission are discussed below :

1. Corona Loss and Radio Interference. As we know that corona is not only a source of power loss but it is also a source of interference with radio and television. The corona loss depends on various factors such as system frequency, system voltage, air conductivity, air density, conductor radius, conductor surface, load conditions, atmospheric conditions etc. as already explained in Art. 4.9. The problem is more acute in case of EHV transmission. When the electric field at the surface of an energized conductor exceeds 2-3 kV/mm, audible and sometimes visible corona discharge takes place, causing a loss of power and radiation of electrical noise. Corona loss varies through the year depending upon weather conditions. The corona loss under bad weather conditions may be as high as 100 times the fair weather condition loss.

Corona inception voltage gradient is an important parameter for conductor design. For limiting the corona loss, audible noise and radio interference it is necessary to limit the electric stress at the surface of the conductor to 1.8 kV/mm (rms), preferably to 1.5 kV/mm (rms). The corona loss can be reduced either by increasing the spacing between the conductors or the diameters of conductors but the spacing between the conductors cannot be increased to a large extent as already explained in Art. 4.10. Large diameter conductors (hollow conductors or ACSR conductors) have been used to bring down the corona loss and radio interference but the cost of manufacture of such conductors is high and their handling is both difficult and expensive. Another difficulty with the use of such conductors is large wind and snow loadings. However, use of ACSR conductors is quite economical for line voltage up to 400 kV. Corona and radio interference are also kept within permissible limits for such lines with this type of conductors. Bundled conductors, explained in Art. 5.8, are invariably employed for EHV (400 kV or more) lines. The use of bundled conductors reduces the voltage gradient in the vicinity of line and thus, reduces the possibilities of corona discharge. Although the bundled conductors are used on EHV transmission lines primarily to reduce corona loss and radio interference, but they have several other advantages, as explained in Art. 5.8. *The choice of configuration and cross section of ac conductors with reference to corona considerations is generally above the economic cross section based on thermal considerations or transient stability considerations i.e., the design of ac conductors based on corona limitations gives a cross section much larger than that with respect to economical power transfer limit imposed by stability limit.*

The corona loss is quite large in EHV transmission lines. For example, in a twin-Moose bundled conductor, $2 \times 235 \text{ mm}^2$ (copper equivalent area) at 1.618 kV/mm of the 400 kV line, designed by UPSEB, the annual corona loss per km of line is about 3,700 kWh. If the bundled conductors were not have been employed in EHV lines, then corona loss would have been comparatively very high. It is quite difficult to estimate corona loss with reasonable accuracy, as such loss depends on weather conditions. Attempts are made towards designing the transmission lines such that the fair weather corona loss is kept low and bad weather conditions do not seriously affect the line performance.

When bundled conductors are employed, spacers have to be installed at intervals of 80-100 m in each span to maintain the spacing between sub-conductors. This is necessary to prevent the conductors from getting stuck up due to wind forces or electromagnetic induction introduced by heavy currents.

The increase in SIL with the increase in the spacing between the sub-conductors is small and, therefore, the intra bundle spacings are not very critical and 0.45 m is the most commonly used value.

Transmission lines with bundled conductors need somewhat stronger towers. It is due to the fact that bundled conductors are subjected to large wind and snow loadings.

2. Heavy Supporting Structures and Erection Difficulties. EHV transmission lines have large mechanical loading on towers because of use of bundled conductors, as already mentioned above, large air and ground clearances, considerable dynamic forces due to broken conductors etc. Transmission line towers with fabricated steel members are usually employed in EHV transmission. A typical standard suspension tower for a 400 kV line is shown in Fig. 13.2.

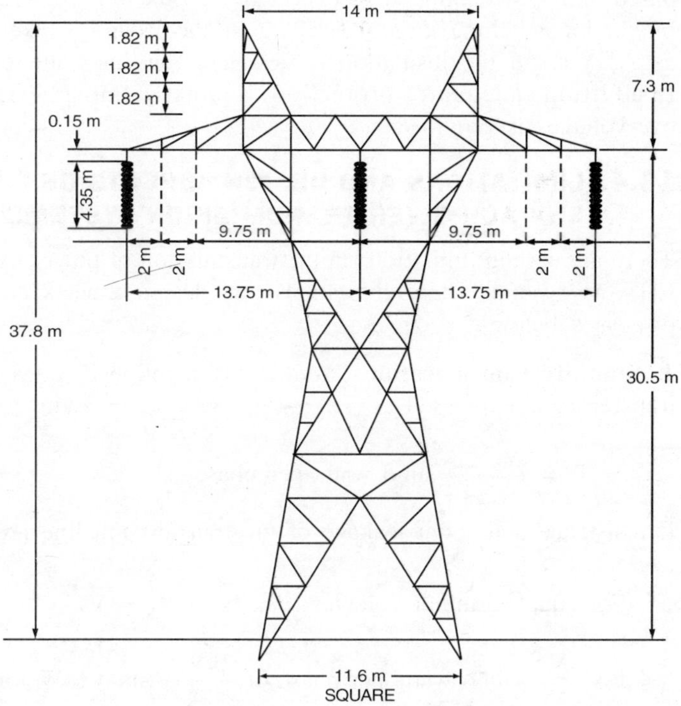

Fig. 13.2. ***Standard Suspensation Tower For 400 kV Line Showing Various Arrangements For Locations of Conductors and Ground Wires***

The transmission lines are made more wind-resistant as they are to bear out the wind pressures during storms and cyclones. Recent designs for line supports include any of the following concepts:

(*i*) Composite design employing standard structural steel for lightly loaded members and high tensile steel for heavily loaded members.

(*ii*) H-frame structures supported by steel guy cables.

(*iii*) Guyed aluminium towers in place of steel towers.

(*iv*) Suspension towers using reinforced concrete tubes.

Problems of transportation and erection arise as the supporting structures are to be transported over long distances and high standard workmanship is required for erection of EHV transmission lines.

3. Insulation Requirements. The level of insulation required depends upon the magnitude of likely voltage surges due to internal causes (switching operations) or due to external causes (lightning etc.) The lines are usually protected against lightning etc. by use of ground wire and rapid autoreclosing circuit breakers. The ground wire, the line insulation and the tower footing are properly coordinated for adequate lightning protection. Switching surges are, however more dangerous as they may cause overvoltages of 2-3 times the normal operating voltage. With the developments in the design of relay-breaker systems, however it is possible to control and minimise switching overvoltages.

Line insulation is designed to take care of switching overvoltages, temporary overvoltages and atmospheric overvoltages. The insulation level of a transmission line is based on the switching surge expectancy on the system. The maximum switching surge overvoltage to the ground is taken as 2.5 pu and the insulation is designed for this voltage. In addition adequate protection against atmospheric overvoltages is provided.

13.4. LIMITATIONS AND DESIGN ASPECTS OF EHV-AC POWER TRANSMISSION SYSTEMS

There are certain limitations in the transmission of power by EHV ac system. These limitations and design aspects are discussed below :

1. Stability Considerations. As already mentioned, power transferred is expressed as

$$P = \frac{V_S V_R}{X} \sin \delta \text{ watts per phase}$$

if resistance and shunt leakage of the transmission line are neglected.

For equal terminal voltages (*i.e.*, $V_S = V_R = V$)

$$P = \frac{V^2}{X} \sin \delta \text{ watts per phase or } \frac{(kV)^2}{X} \sin \delta \text{ MW for}$$

3 phases where kV is the line voltage in kilovolts.

Maximum steady state transfer of power takes place theoretically for $\delta = 90°$ but in actual practice δ is kept within limits of 20° to 30° at full load. Taking $X = 0.3475\ \Omega$/km and $\delta = 30°$, we have

$$P = \frac{kV^2}{0.3475l} \sin 30° = 1.43 \frac{(kV)^2}{l} \text{ MW where } l \text{ is the}$$

length of line in km.

If the line operates at natural load given by the expression

$$P = VI = V \times \frac{V}{Z_C} \text{ watts or } \frac{(kV)^2}{Z_C} \text{ MW}$$

where Z_C is surge impedance and is equal to $\sqrt{\frac{L}{C}}$ ohms.

then $$\frac{1.43 \times (kV)^2}{l} = \frac{(kV)^2}{Z_c}$$

or $l = 1.43\ Z_C$ kilometres

Thus, we see that the theoretical length of a transmission line in km, that can be operated on its natural load and load angle of 30° without loss of stability only is 1.43 times surge impedance (which comes out to be about 500 km). The transmission distance in EHV ac systems is, therefore, limited. This limitation is overcome by use of special equipment such as series capacitors or shunt reactors.

2. Current Carrying Capacity. The loading of overhead line conductors does not depend on the thermal considerations. However, for overhead transmission lines operating at voltage up to 220 kV, the conductor size is determined on the basis of its continuous and short-term current carrying capacity (*i.e.*, on the basis of thermal considerations). In case of underground cables, however, both thermal current limit and charging current are very important and that is why the underground cable is never operated beyond its natural load. Also the cost of cable is very high in comparison to that of an overhead line, therefore use of underground cables is limited in high voltage systems.

3. Ferranti Effect. We know that with the capacitive load on the line the receiving-end voltage is higher than sending-end voltage. This increase is of the order of 1.5 per cent for 160 km, 13 per cent for 500 km and 100 per cent for 960 km. Also there is a rise in sending-end voltage whenever the load on the generator is thrown off suddenly. Care is also to be taken from these aspects in case of ac transmission system.

Voltage rise, due to Ferranti effect, is controlled by using shunt reactors at the load end.

4. Surge Impedance Loading. Surge impedance loading (SIL) of a transmission line is defined as the load at the receiving end which is equivalent to $\sqrt{\frac{L}{C}}$. The surge impedance loading for a transmission line is given as $\frac{V^2}{Z_0}$ watts per phase. SIL is also called the natural load. When the line carries natural load, the voltage along the entire length of line is the same.

However, in practice a line is not always operated at its natural load. In such a case there will be a voltage drop along the line for inductive load and a voltage rise for capacitive load. Although the reactive voltage drop/rise and the natural load do not impose any restriction on the distance over which power may be transmitted but fix the voltage that may be employed for the transmission of a certain amount of power. Typical values of surge impedance and natural load for overhead transmission line are given below in tabular form.

Voltage (Line) in kV	132	220	275	380	400	500	765
Surge impedance, Z_0 in ohms	400	375	370	335	320	312	345
Natural Load in MW	44	130	205	430	500	800	1,700
Current in Amperes	190	340	430	655	720	925	1,280

5. Mechanical Vibrations and Oscillations. With the increase in number of sub-conductors in bundled conductor transmission lines, there may be a considerable effect on vibrations of conductors. Hence the mechanical design of the system (tower dimensions, phase spacing, sub-conductor spacing, conductor height etc.) needs modifications so as to counter the problem of conductor vibrations. These modifications affect the electrical design such as voltage gradient.

Other problems include Aeolian vibrations and galloping. The problem of galloping can, however, be controlled by applying weights at different locations of the line span.

6. Environment and Biological Aspects. The biological effect of electric field of EHV lines and EHV substations has been studied extensively during 1970s. The electrostatic and electromagnetic fields produced by EHV/UHV transmission lines induce currents and voltages in animals, birds and human beings. EHV/UHV lines are designed such that maximum electrostatic field gradient does not exceed 9 kV/m at mid span under the line near the ground level.

It has been found that very low frequency magnetic fields even of weak intensities can affect certain cellular processes. Induced current densities below 10 mA/m^2 have no significant biological effect. Currents 10-100 mA/m^2 cause minor biological effects. Current densities exceeding 100 mA/m^2 cause health hazards.

Safe line to ground clearance of 20 m at mid span is recommended for 400 kV lines and 24 m for 1,100 kV lines. This permits movement of vehicles safely.

7. Audible Noise. EHV transmission lines and substations also produce audible noise. The sources of noise generation are corona, humming of transformers, cooling systems and mechanical and electrical auxiliaries.

The audible noise (AN) generated by a transmission line is a function of the following :

(*i*) Voltage gradient on the surface of a conductor.

(*ii*) Number of sub-conductors in a bundle.

(*iii*) Conductor diameter.

(*iv*) Atmospheric conditions.

(*v*) Lateral distance between the line and the point of measurement of noise.

The audible noise is caused by vibrations produced in air due to change in the air pressure.

Below 500 kV, audible noise (AN) levels do not exceed the permissible noise but the lines of 500 kV and above must be designed for a satisfactory audible noise level.

The width of right of way for the line corridor has a reference to the decision about AN. Line geometry is based upon 50 dB at the edge of ROW.

8. Radio Interference and Television Interference. Operation of EHV transmission lines and substations can cause radio interference. Since radio noise is associated with corona, it mainly depends on the potential gradients at the conductors. It is also influenced by all those factors which influence corona, such as air density, humidity, wind contaminants, imperfections and precipitation. The radio noise in bad weather is about 10-25 dB higher than the radio noise during fair weather. The other causes for radio interference are partial discharges on insulators and spark across gaps.

The noise due to corona in EHV lines does not cause a serious problem for frequency modulated (FM) radio receivers or TV receivers. However, the TV reception in the immediate vicinity of a UHV line may get affected especially during bad weather.

Radio interference and television interference is caused by electromagnetic waves in the frequency range of broad cast frequencies.

Radio interference : 0.5–1.6 MHz

Television interference : 54–216 MHz

The line is designed such that radio noise within the width of the line corridor should be below permissible limits, say 40 dB at 1 MHz. Radio interference is more important factor in the design of line and in deciding right of way. The radio interference can be eliminated and/or minimised by appropriate design of line conductor and hardware. By using bundled conductors, surface voltage stress, corona and radio interference can be reduced.

9. Route Clearance. EHV/UHV transmission lines are generally constructed over different terrain including dense forests, hilly areas and agricultural land, and clearance of minimum ROW is very important. Generally the trees are not taller than 10 m. Only dry trees need to be cleared from the ROW of transmission lines because a live tree is found to be well grounded through its sap, has small power loss and little chance of burning. It has also been found that the dry tree burning can occur if an induced voltage exceeding 100 kV occurs and potential of 100 kV can exist within a radius of 6 m from 765 kV line. So a minimum clearance of 6 m from 765 kV line conductors is essential.

10. Conductor Material. ACSR conductors are universally employed for overhead transmission lines. Necessary data about ACSR bundle conductors used for ac transmission lines is given below in tabular form.

Normal voltage in kV	Highest Voltage in kV	Phase Spacing in Metres	Sub-conductor Diameter in mm	Number of Sub-conductors Per Phase
230	245	6	25	1
345	365	8.3		2
500	550	11		3
765	800	15.3		4
1,100	1,200	20		8

11. Span, Number of Circuits, Conductor Configuration. The considerations governing the choice of span has already been discussed in Art. 2.9. For a high voltage line the economical value of span is 200-300 m, whereas for 400 kV line it is 350-400 m.

If two three-phase lines are on the same tower and are identical in construction and electrically in parallel then they have the same reactance X. The inductive reactance of the single equivalent circuit becomes $\frac{X}{2}$ and hence the power transfer capability is doubled. Double circuit lines are in common use. They have additional advantage that continuity of supply is maintained over one line, though with reduced capability, while the other line is out of service for maintenance or repair.

Each circuit of a double circuit line is usually designed for 75% of the line capacity. In India, both single and double circuit lines exist in the high voltage and extra high voltage class (66, 132, 220 and 400 kV). The number of circuits is determined from the SIL.

Vertical configuration has been found the most economical for double circuit lines. For single circuit lines, horizontal or L type configurations are most suitable.

12. Insulation Coordination. Transmission lines and substations are subjected to overvoltages due to switching, lightning, faults, resonance and other causes. The surge arrestors are provided at strategic locations to protect the line insulation and substation equipment from transient and temporary overvoltages.

In general there are three types of overvoltages viz. switching overvoltages, temporary overvoltages and overvoltages due to lightning. Switching overvoltages are caused due to energization of lines, reclosing, fault interruption, load throw, out of phase switching and sudden switch-off of a line. The temporary overvoltages at power frequency are due to sudden loss of load, disconnection of inductive loads, connections of capacitive loads etc.

For EHV and UHV transmission lines the switching overvoltages become the limiting factor in the design of insulation levels and arrester characteristics. Once the line is designed adequately against switching overvoltages, it is capable of encountering temporary overvoltages. It has also been found that overvoltages due to lightning are not very important for EHV/UHV transmission lines.

The waveshapes of switching surges are practically infinite. Since the waveshape has a definite influence on the flash-over strength, a standard waveshape has been selected for analysing the line performance. This standard switching surge has a *crest time* (*i.e.*, the time to attain the peak value) of 175 μs, and a *tail time* (*i.e.*, the time to attain half the peak value on the tail) of 3,200 μs. It has been found desirable and possible to reduce the switching surges for EHV systems to about 2.5 pu and insulation of EHV system is usually designed for this strength. The temporary overvoltages can be limited to 1.5 pu if shunt reactors are employed.

The insulation characteristics of various equipment is defined in terms of rated voltage, standard switching impulse, standard lightning impulse, power frequency withstand voltage etc. These are co-related with characteristic of surge arresters.

The insulation levels of various substation apparatus and the transmission line should be coordinated with various surge arresters such that the overvoltages are discharged to earth without causing damage to the equipment insulation.

13. Series Compensation. Series compensation is an important method of improving the performance of EHV transmission lines. It consists of capacitors connected in series with the line at suitable locations and thus opposes directly the effect of series inductive reactance of the line. It increases the power handling capacity and reduces the voltage regulation as explained below :

The power transfer capability of a transmission line is given as

$$P = \frac{V_S V_R}{X_L} \sin \delta \qquad \text{...Refer to Eq. (13.1)}$$

With the insertion of capacitors having capacitive reactance X_C in the line, the net reactance of the line becomes $(X_L - X_C)$ and the power transfer capability of the line is given as

$$P' = \frac{V_S V_R}{X_L - X_C} \sin \delta \qquad \text{...(13.2)}$$

From above Eqs. (13.1) and (13.2) it is obvious that for the same magnitudes of V_S, V_R and δ, P′ is much higher than P and the increase in power transfer capability is given as

$$\frac{P'}{P} = \frac{X_L}{X_L - X_C} = \frac{X_L}{X_L\left(1 - \frac{X_C}{X_L}\right)} = \frac{1}{1-K} \qquad \text{...(13.3)}$$

where K is the *degree of compensation* and equals $\frac{X_C}{X_L}$.

For example, if the line is compensated to an extent of 50% *i.e.*, K = 0.5, the power transfer capability of the line becomes $\frac{1}{1-0.5}$ *i.e.*, double of that without series compensation.

Economic Considerations. The growing cost of ROW, aesthetic considerations calling fewer transmission lines, increase in the lengths of transmission lines and in transmission voltages, decrease in series capacitor costs, have all contributed to an increase in the rate of use of series capacitors in India and other countries.

In view of the line cost, the effect of lines operating in parallel and problems of system protection, the economic degree of compensation is found to be in the range of 0.4 to 0.7.

Improvement of System Stability. From the power flow equation it is evident that for the transfer of same amount of power and the same values of sending-end and receiving-end voltages V_S and V_R, the power angle δ is lower for compensated line as compared to that in case of an uncompensated line. Lower values of power angle means better system stability.

Various measures have been adopted for improvement of power system stability. These include reducing the reactance of synchronous generators and transformers, increasing the number of parallel lines, use of bundled conductors and series compensation. Series capacitors usually offer the most economic solution with respect to both the steady state and transient stability.

Location of Series Capacitors. Series capacitors can be located either along the lines, in intermediate substations or switching stations. Line location has many advantages such as better voltage profile along the line, reduced short-circuit current through the capacitor during a fault. The small short-circuit current contributes to a much simpler protection of the capacitor. The capacitor stations are generally unattended type.

Installation of series capacitors in intermediate attended type substations or switching stations may appear advantageous, since staff for maintenance and service, auxiliary power etc. are available. However, these advantages must be weighted against the higher cost for protection and control and sometimes also for the space required as well as the increase of transmission losses.

Capacitor Bank. A capacitor bank is built up of a number of capacitor units connected in series and in parallel. The units for series capacitors are designed, manufactured and tested with due regard for the specific service conditions, such as high overvoltages and capacitor discharge currents.

A capacitor unit consists of a number of capacitor elements in a common container. The units are equipped with fuses, which may be either external ones for each unit or internal individual element fuses. The capacitor units are mounted in simple frames, called the racks, placed on supporting insulators and stacked on top of each other.

Problems Associated With Series Compensation. There are some major problems associated with series compensation. These problems are summarized below.

(i) Subsynchronous Resonance. The series capacitor introduces a subsynchronous frequency, proportional to the square root of the compensation in the system. In some cases this frequency may interact with turbo-generator shafts and develop high torsional stresses. The risk of subsynchronous resonance in hydrogenerators is quite small as the torsional frequencies are about 10 Hz or even less.

(ii) Ferroresonance. When a lightly loaded transformer is energized through a series compensated line, ferroresonance may occur. This can be suppressed by using shunt resistors across the capacitors or by short circuiting the capacitor temporarily through an isolator or bypass breaker.

(iii) Maloperation of Line Protection. Series compensation may lead to maloperation of the protection system leading to unnecessary trippings.

(iv) High Recovery Voltage. High recovery voltage may develop across the circuit breaker contacts due to series compensation.

14. Shunt Compensation. In EHV transmission systems, shunt compensation is invariably essential. Shunt compensation with capacitive VARs is used to inject reactive power and control the receiving-end voltage whereas shunt reactor compensation is used to neutralize the Ferranti effect. For shunt compensation, static compensation employing capacitors and reactors or synchronous compensation using synchronous phase modifier may be employed. Static compensation is being preferred to the synchronous compensation because of its inherent advantages of higher speed of response, absence of fault infeed to the system, lower maintenance, lower cost, greater reliability and simpler erection. In addition to shunt capacitors, shunt reactors and phase modifiers, a thyristor controlled static shunt compensator to meet reactive power generation and absorption demand has appeared in recent years.

One primary reason for using shunt reactors or reactive control devices on EHV lines is to control steady-state overvoltages when energizing the long EHV lines or when operating under light load conditions. If the shunt reactors are not employed, the reactive power generated by the capacitance can cause high voltages at the receiving end of the line (Refer to Ferranti effect Art. 6.7). To restrict insulation stresses caused by overvoltages following sudden load rejection a substantial part of the shunt reactive compensation is usually left permanently connected.

It should be noted that with shunt reactors, the power transfer capability of the line is reduced owing to increase of transfer reactance X. Thus at fault load when stability is most important, the shunt compensation reduces the maximum power limit of the long line. In case line shunt reactors are switched out under heavy load conditions, the maximum power transfer capability can be considerably increased, but voltage variations due to sudden load thrown off are likely to be unacceptably high. In actual practice some of the shunt reactors are kept connected permanently so as to avoid voltage increase due to sudden fall in load from heavy load conditions.

15. Series and Shunt Compensation. In EHV long transmission lines both series capacitors and shunt reactors

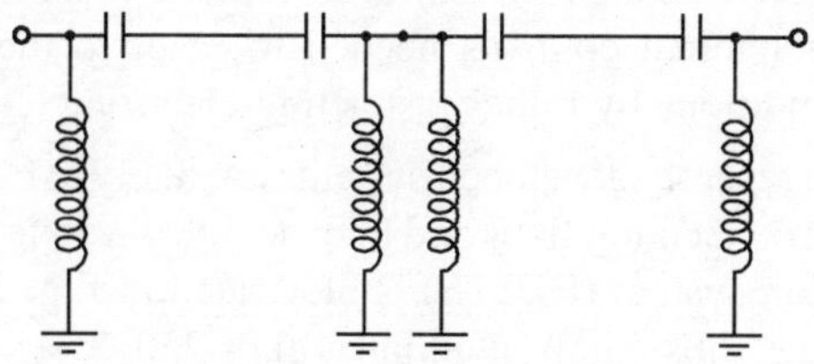

(a)

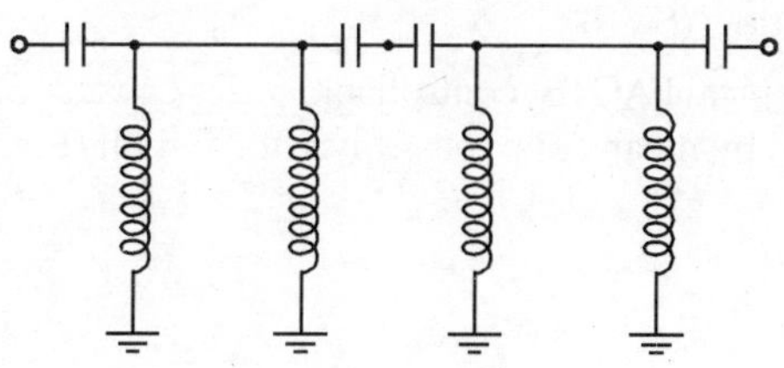

(b)

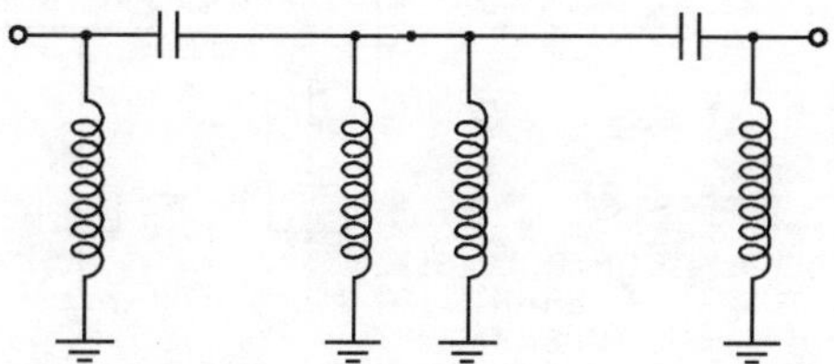

(c)

Fig. 13.3. ***Series and Shunt Compensation Schemes***

are provided. The purpose of providing series capacitors is to artificially reduce series reactance of the line, so as to improve stability and efficiency of transmission. The shunt reactors are provided to artificially reduce the susceptance of the line so as to improve voltage regulation under light load conditions.

Practical and economical reasons lead to concentration of the compensating elements at a few points along the line. For a given total MVARs of series and shunt compensation of a transmission system, maximum power transfer capability, voltage control conditions and efficiency of power transmission depend on the number, location and circuit schematic of the series capacitor and the shunt reactor stations. Series and shunt compensation schemes are illustrated in Figs. 13.3. There are also other possible schemes.

13.5. FLEXIBLE AC TRANSMISSION SYSTEM (FACTS)

With everincreasing demand of electric power, the existing transmission networks are found to be weak resulting in poor quality of unreliable supply. Also, it is realized that for expanding or enhancing the power transfer capability of the present transmission networks, huge sum of finances are required and sometimes even difficulties are encountered in finding right of way for the new transmission lines. Lot of research has been done for developing new technologies over the past few years to gain increased efficiency and reliability from the existing transmission network. As a result, new transmission system, known as *Flexible AC Transmission System* (abbreviated as FACTS) has been developed. The new technologies make use of high speed thyristors for switching in or out transmission line components such as capacitors, reactors or phase shifting transformer for some desirable performance of the system. The main objective of this system is to replace the existing slow acting mechanical controls needed to react to the changing system conditions by rather fast acting electronic controls.

These recently developed systems are likely to be employed for ac interconnecting lines and long ac lines. A long flexible ac transmission system (FACTS) is illustrated in Fig. 13.4. It has intermediate substations at an interval of 250–350 km. 1, 1 are controllable series capacitors and 1′, 1′ are controllable shunt compensation (SVS).

In a typical FACTS, controllable series capacitor installation is obtained by means of bypass thyristor switch (Fig. 13.5). The amount of series compensation is controlled by controlling the bypass current I_B through the thyristor controlled switch (TCS) and current I_C through series capacitor C_{se}.

Power flow is given by expression

$$P = \frac{V_S V_R}{X_L - X_{CF}} \sin \delta$$

where X_L is series reactance of the transmission line and X_{CF} is the controllable series reactance of series capacitor of FACTS.

So, by controlling X_{CF} by means of bypass thyristor switch, the power flow can be controlled. The control over power flow P and power angle δ is provided by the phase control of bypass thyristor switch. The basic limitation of ac transmission line and series compensation had been the lack of controllability, which has been overcome by controllable series capacitors of FACTS. In addition to controllable series capacitor; controllable shunt compensation is obtained by means of SVS in FACTS. Thus, voltage at both ends of transmission line is controlled dynamically. This provides voltage stability and improves dynamic stability of the transmission link.

In a typical FACTS, controllable series capacitors and controllable SVS combinations are installed at an interval of about 250 km along a long ac transmission line.

Single line schematic diagram of FACTS is shown in Fig. 13.4. Figure 13.5 shows a schematic diagram of the details of controllable series capacitors 1. All the equipment of controllable series capacitors are installed on insulating platform. The control unit senses the phase-to-ground voltage V at the location of controllable series capacitors and adjusts phase angle of thyristor control switch (TCS) and so bypass current I_B is controlled. By varying I_B, I_C is controlled because $I_L = I_C + I_B$. This results in variable series compensation which can be controlled from load control centre through compensation, the power flow P is controlled. By controlling static VAR source (SVS), the voltage at sending end, receiving end and at intermediate substations are controlled.

Series and shunt compensations are controllable and control is fast and over a wide range. Hence FACTS improves the dynamic stability of the ac transmission system.

By simultaneous control of 1, 1, ... and 1′, 1′ ... the power flow P, sending-end voltage V_S, receiving-end voltage V_R and power angle δ between V_S and V_R are controlled. The oscillations in P, V, δ are damped by controlling 1 and 1′.

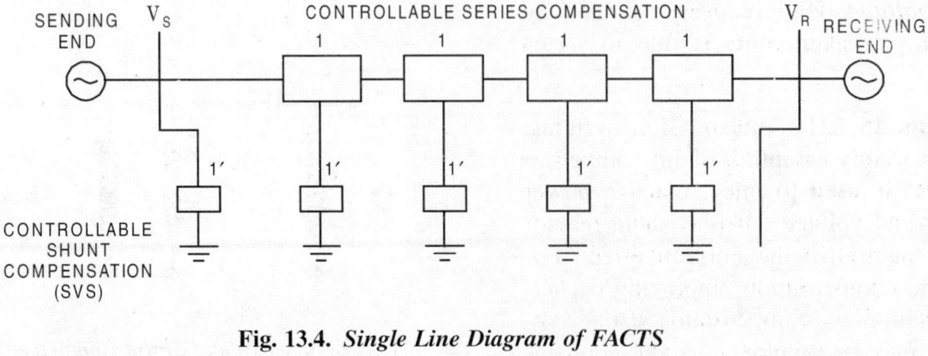

Fig. 13.4. ***Single Line Diagram of FACTS***

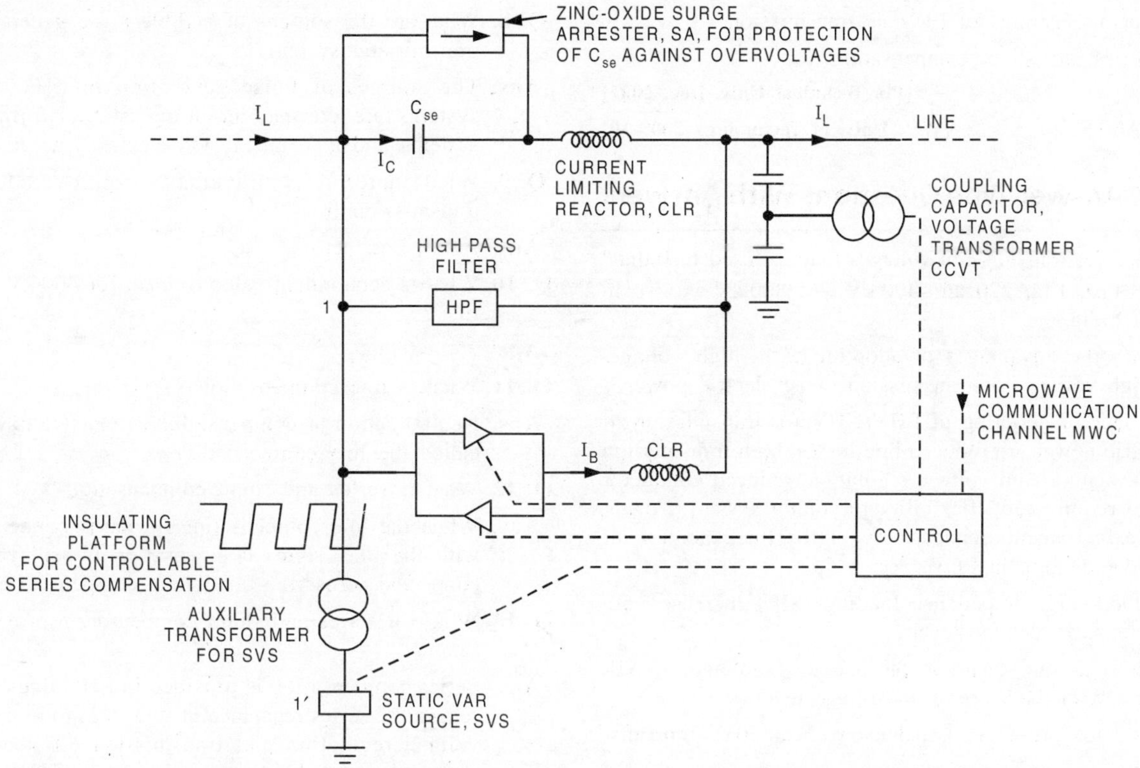

Fig. 13.5. ***Details of controllable Series Compensation in The FACT Installation (Only One Phase)***

Exercises

1. Give the advantages, drawbacks and limitations of EHV ac transmission.
2. Answer the following :
   Relative merits and demerits of EHV transmission system.
   [U.P. Technical Univ. Elements of Power System 2009-10]
3. What are the problems involved in transmission of power using EHV ac system ? Discuss in detail. Also state how these problems are being solved.
4. What do you understand by hollow, ACSR and bundled conductors ?
   Explain in detail the reasons for adoption of bundled conductors in EHV (extra high voltage) ac transmission lines. **[Pb. Univ. April 1980]**
5. What are disadvatnages of HVAC transmission.
   [U.P. Technical Univ. Elements of Power System 2007-08]
6. Discuss the various problems of EHV power transmission.
   [U.P. Technical Univ. Elements of Power System 2005-06]
7. Describe the design considerations of EHV transmission lines. **[U.P. Technical Univ. Elements of Power System 2007-08]**
8. Discuss the points which are considered while designing a transmission line.
   [U.P. Technical Univ. Elements of Power System 2008-09]
9. Describe the various conductor configurations and choice of number of circuits for EHV transmission lines.
   [U.P. Technical Univ. Elements of Power System 2009-10]
10. Discuss the advantages of using 'bundled conductors' in EHV overhead transmission lines.
    [U.P.S.C. I.E.S. Elec. Engineering II, 1998]
11. What is series compensation ? What for it is provided ? Give its advantages and problems associated with it.
12. Discuss the series and shunt compensation of a transmission line. **[Pb. Technical Univ. Power System-I (Transmission and Distribution), May 2008]**
13. What is passive compensation ? Compare series and shunt compensators.
    [W.B. Univ. of Technology Power System-III, 2009-10]
14. Identify the utility of FACTS devices in power network.
    [U.P. Technical Univ. Elements of Power System 2005-06]
15. Explain the principle and usefulness of flexible ac transmission system.
    [U.P. Technical Univ. Elements of Power System 2003-04]
16. Discuss the principle and advantages of using Flexible AC transmission (FACT) system.
    [U.P. Technical Univ. Elements of Power System 2002-03]
17. Explain in brief the role of flexible ac transmission systems.
    [U.P. Technical Univ. Elements of Power System 2006-07]
18. What are the various FACTS controllers ? Explain the operating principle of any one of them.
    [U.P. Technical Univ. Elements of Power System 2006-07]
19. How FACTS controllers make ac transmission flexible ? Describe working principle of TCSC and show how it enhances power transfer capability of lines.
    [U.P. Technical Univ. Elements of Power System 2004-05]
20. Write short notes on the following :
    (*a*) Necessity of EHV ac transmission.
    (*b*) Problems associated with EHV ac transmission.
    (*c*) Limitations of EHV ac transmission.

(*d*) Special features of EHV ac transmission.

(*e*) Series and shunt compensation.

[Pb. Technical Univ. Dec. 2007]

(*f*) FACTS. **[W.B. Univ. of Technology 2009-10]**

Short Answer Type Questions With Answers

Q. 1. What are the transmission voltages that are used in India ?

Ans. 33, 66, 110, 132, 220 and 400 kV. 765/800 kV is also in used in India.

Q. 2. What are the advantages of adopting extra high voltage/ultra high voltage for transmission of ac electric power ?

Ans. Advantages of adopting of EHV/UHV for transmission of ac electric power are reduced line losses, high transmission efficiency, improved voltage regulation, reduced conductor material requirement, flexibility for future system growth, increase in transmission capacity of the line, increase of SIL and reduction in ROW.

Q. 3. Why does surge impedance loading (SIL) increase with the increase in voltage level ?

Ans. SIL varies as the square of the operating voltage, so SIL increases with the increase in voltage level.

Q. 4. What are the problems associated with an EHV transmission ?

Ans. The problems associated with EHV transmission are corona loss and radio interference, requirements of heavy supporting structures and erection difficulties and insulation requirements.

Q. 5. Why does the use of bundle conductors reduce corona loss in an EHV line ? **[U.P.S.C. I.E.S. E.E.-II, 2002]**

Ans. A bundled conductor is a conductor made of two or more sub-conductors and is used as one phase conductor. The inductive reactance of the bundle conductors is reduced because the self GMD of the conductor is increased.

$$L = 2 \times 10^{-7} \log_e \left[\frac{D_m}{D_{SA}} \right] \text{ H/m}$$

By bundling the conductors the self GMD of the conductors is increased thereby, the critical disruptive voltage is increased and hence corona loss is reduced.

Q. 6. On what basis the conductor size is determined for transmission lines up to 220 kV ?

Ans. For overhead transmission lines operating at voltages up to 220 kV, the conductor size is determined on the basis of its current carrying capacity.

Q. 7. On what basis the conductor size of EHV line is determined ?

Ans. The conductor size of an EHV line is decided on the basis of corona.

Q. 8. What are the sources of audible noise generation in EHV transmission systems ?

Ans. The sources of noise generation in EHV transmission systems are corona, humming of transformers, cooling systems and mechanical and electrical auxiliaries.

Q. 9. What material is universally employed for overhead transmission lines ?

Ans. ACSR conductors.

Q. 10. What is economical value of span for 400 kV transmission line ?

Ans. 350–400 m.

Q. 11. What is line compensation ?

Ans. Neutralization or compensation of reactive power of line is called the line compensation.

Q. 12. What is series and shunt compensation ?

Ans. When the line compensating device is connected in series with the line, series compensation is achieved and when connected across the line, shunt compensation is achieved.

Q. 13. What for series and shunt compensation provided in EHV lines ?

Ans. Series compensation is provided in EHV lines to artificially reduce the series reactance of line so as to improve stability, voltage regulation and transmission efficiency where as shunt compensation is provided to artificially reduce the line susceptance so as to improve voltage regulation under light load conditions.

Q. 14. Why shunt compensation is preferred over series compensators ?

Ans. There are some major problems associated with series compensation as summarized below :

1. Series compensation introduces a subsynchronous frequency, proportional to the square root of the compensation in the system. In some cases, this frequency may interact with turbo-generator shafts and develop high torsional stresses.
2. Switchings-in of an unloaded transformer through a series compensated line may produce ferroresonance.
3. Series compensation may lead to maloperation of the protection system leading to unnecessary tripping.
4. High recovery voltage may develop across the circuit breaker contacts due to series compensation.

Q. 15. Give advantages of series compensation.

[Pb. Technical Univ. May 2008]

Ans. Series compensation is provided in EHV transmission lines to artificially reduce the series reactance of the line, that results in improvement of stability, voltage regulation and transmission efficiency.

❑❑❑

CHAPTER 14

High Voltage Direct Current (HVDC) Transmission

INSIDE THIS CHAPTER

14.1. INTRODUCTION

In the initial stages dc used to be employed for generation, transmission and distribution of electric power. The first 110 V dc Central Electric Station was installed by Edison in New York in 1882. With the introduction of transformers and the 3-ϕ ac system the situation changed in favour of ac, especially where the electric energy used to be harnessed from water power which is usually available far from the load centres. For many reasons, such as facility of transformation of voltages from one value to another, better performance of ac motors and superiority of ac generators in comparison to dc generators, the power was generated, transmitted, distributed and utilized in the form of ac. The supporters of dc, however, did not forget the advantages of dc transmission and extensive research has been carried out in this field and as a result it has staged a sort of comeback to the field of electric power transmission at high and extra high dc voltages.

The earlier dc schemes used mercury-arc convertors. Extensive research has been carried out especially in Sweden for the development of high voltage converters. With the successful development of high power (50 kV, 100 A) thyristors in early 1970's the HDVC transmission has become technically and commercially viable alternative to EHV/UHV AC transmission particularly for distance bulk power transmission, cable transmission and system interconnection. For these applications HVDC transmission systems have a distinct superiority over EHV AC transmission systems and are being increasingly preferred. However, HVDC transmission cannot be substitute of ac transmission because for backbone ac network, generation, transmission and distribution is definitely superior. The choice of transmission systems and operating voltages for a transmission line is made from HV-AC (up to 220 kV), EHV-AC (between 400 kV and 765 kV), UHV-AC (exceeding 765 kV) and HVDC (up to 1,600 kV) on the basis of technical and economical studies for each particular line and associated ac systems.

The first dc link was set up in 1954 between Mainland of Sweden and the island of Gotland. This was a monopolar, 100 kV, 20 MW, cable system making use of sea return. The England-France cross-channel dc link was commissioned in 1961. This was a bipolar, ±100 kV, 160 MW, cable system over a distance of about 65 km. Since then more and more HVDC systems have been set up in the world.

HVDC transmission has also been introduced in India. ±500 kV has been selected as the voltage for HVDC transmission. A ± 500 kV, 1500 MW, 810 km bipolar HDVC line has already been set up between Rihand/Singrauli complex and Delhi. It is designed to operate in the bipolar, monopolar-ground return, and monopolar metallic return modes. Another HVDC system set up in India is Vindhyachal HVDC back to back system. This back to back link is for exchange of power between Northern and Western regions. Each block of 250 MW is capable of operating independently in either direction and can transfer power in the range of 25-250 MW depending on the conditions of the system. In addition, Chandarpur back to back project (2 × 250 MW), Chandarpur-Padghe bipolar system (± 500 kV, 1,500 MW, 736 km), Jeypore back to back project (2 × 250 MW) and Mau back to back project (2 × 250 MW) are also proposed.

At present the world has over 50 HVDC schemes in operation for a total capacity of more than 50,000 MW and the capacity is increasing by about 2,000 MW every year.

14.2. PRINCIPLE OF HVDC SYSTEM OPERATION

When using direct current to provide an asynchronous link between two ac systems, it is necessary to have two convertor stations one at each end, connected by a dc transmission line. The main equipments in a convertor station are transformers and thyristor valves. Chokes and filters are provided at each end to ensure smooth direct current and suppress harmonics. At the sending end the thyristor valves act as rectifiers to convert ac into dc which is transmitted over the line. At the receiving end the thyristor valves act as inverters to convert dc into ac which is utilized at the receiving end.

Single line diagram of a HVDC transmission system is shown in Fig. 14.1, where A and B are the two converter stations. Converter station A is supplied from the generating station G. In converter station at the sending end the voltage is stepped up to appropriate value by step-up transformer and then converted into direct current by the thyristor valves. Thus at the start of transmission line, we have high voltage direct current. This rectified current flows along the transmission line to the receiving-end converting station B, where it is converted into 3-phase ac current by the thyristor valves and then stepped down by the step-down transformer to low voltage for further distribution.

The power despatched from the generating station P_S less the power received at the receiving end P_R *i.e.*, $(P_S - P_R)$ represents the power losses due to conversion and transmission. The convertor at the sending end acts as a rectifier and is suitable for power frequency (*i.e.*, frequency of generator) on its ac side while the converter at the receiving end acts as an inverter and its frequency is determined by the frequency of the load system. This frequency is independent of the sending-end frequency provided the two ends A and B are not additionally connected by the 3-phase lines.

The dc output voltage magnitude is controlled by varying the firing angle of the thyristor valves in the converter. In rectifier the firing angle is between 0° and 90° while in inverter it is between 90° and 180°. As the dc output voltage is a function of cosine of the firing angle hence the converter voltage becomes negative when firing angle α exceeds 90°. This makes the converter to operate as an inverter. The two converters at sending end and receiving end are identical and whether they have to work as rectifier or inverter is determined by the direction of power flow.

In practical HVDC converter stations three-phase bridge converters are employed at both ends (sending as well as receiving ends). Reversible operation of converters as well as bidirectional power flow in HVDC link is possible simply by the control of firing angle.

14.2.1. Power Control in HVDC System

In dc systems the power transferred from one station to another station is governed only by the magnitudes of terminal dc voltages at the two ends while in ac transmission systems the power transfer is governed by phasor difference (magnitude as well as phase) of the voltages at the two ends. Thus the controllability of HVDC power is fast and stable. The current flows from higher voltage to lower voltage which are set by adjustment of firing/extinction angles of the two converters (rectifier and inverter).

If V_S is the voltage at the sending and V_R is the voltage at the receiving end then the line current is given as

$$I_{dc} = \frac{V_S - V_R}{R} \qquad ...(14.1)$$

where R is the resistance of the entire transmission link

The sending-end voltage is given as

$$V_S = \left[\left(\frac{3\sqrt{2}}{\pi} V_{ac\,S}\right)\cos\alpha - \frac{3X_{CS}}{\pi} I_{dc}\right] \qquad ...(14.2)$$

and the receiving-end voltage is given as

$$V_R = \left[\left(\frac{3\sqrt{2}}{\pi} V_{ac\,R}\right)\cos\beta - \frac{3X_{CR}}{\pi} I_{dc}\right] \qquad ...(14.3)$$

where α is the firing angle of rectifier, β is the extinction angle of inverter, $V_{ac\,S}$ is the ac side line-to-line rms voltage at the sending end, $V_{ac\,R}$ is the ac side line-to-line rms voltage at the receiving end, X_{CS} is the commutation reactance at the sending end and X_{CR} is the commutation reactance at the receiving end.

Thus, the power transferred is given as

$$P = V_S I_{dc} = \frac{V_S - V_R}{R} V_S \text{ watts} \qquad ...(14.4)$$

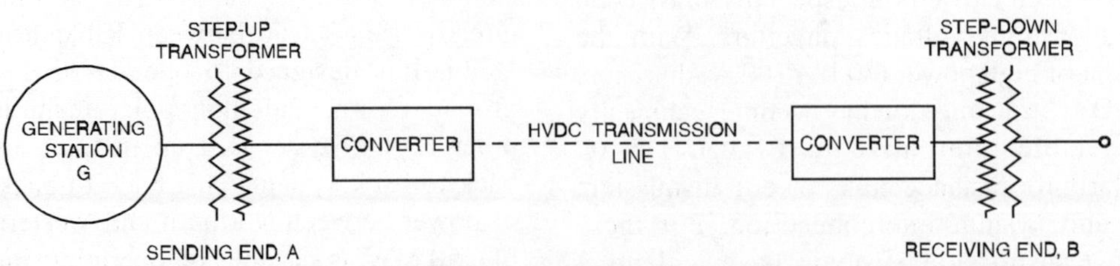

Fig. 14.1. *Line Diagram of Typical HVDC Transmision System*

Tap-changers on the ac side take care of voltage variations on ac side and dc power is controlled by controlling the sending-end and receiving-end voltages V_S and V_R which is possible by control of firing and extinction angles α and β respectively.

14.3. ADVANTAGES OF HVDC TRANSMISSION

DC transmission possesses many technical and economic advantages over ac transmission. Some of these are given below:

1. Cheaper in Cost. Bipolar HVDC transmission lines require two-pole conductors while ac system requires 3 conductors to carry power.

HVDC transmission can utilize earth return and, therefore, does not require a double circuit while EHV-AC transmission always require a double circuit.

The potential stress on the insulation in case of dc system is $\frac{1}{\sqrt{2}}$ times of that in ac system for the same operating voltage. Hence for the same operating voltage less insulation is required.

The phase-to-phase clearances, phase-to-ground clearances and tower size are smaller in case of dc transmission as compared to those for ac transmission. Because of lesser load on the supporting structures tower designs are simpler and cheaper. Requirement of 'right of way' for a dc line is about 20-40 per cent lesser than that for an ac line of the same power transmission capability.

EHV-AC transmission needs intermediate substations at an interval of 300 km for compensation but HVDC transmission system does not require any such intermediate substation for compensation.

However, the dc line terminal equipment because of high cost of converters is considerably more expensive than ac line terminal equipment. HVDC transmission becomes economical over ac transmission above the break-even distance when the saving in overall costs of conductors, towers etc. compensates the additional cost of the terminal equipment (rectifiers and inverters). The break-even distance is generally different for different projects. For overhead lines it may be about 600-800 km while for submarine cables and underground cables it may lie in the range of 20 to 50 km and 50 to 100 km respectively. Sometimes it becomes necessary to adopt dc transmission, even for distances smaller than the break-even distances, because of technical advantages.

2. An HVDC line can be built in stages. The dc line can be built as a monopolar line with ground return in the initial stage and may be converted into a bipolar line on a later date when the load requirement increases.

3. No Skin Effect. There is no skin effect in dc, so there is a uniform distribution of current over the section of the conductor. Thus there is full utilization of line conductor in case of dc transmission while it is not so in case of ac transmission.

4. Lower Transmission Losses. HVDC transmission system needs only two conductors and, therefore, the power losses in a dc line are lesser than the losses in ac line of the same power transfer capability.

5. Voltage Regulation. There is no inductance, hence the voltage drop due to inductive reactance does not exist in dc transmission line. Thus voltage regulation is better in case of dc transmission.

6. Line Loading. The permissible loading on an EHV-AC line is limited by transient stability limit and line reactance to almost one-third of thermal rating of conductors. No such limit exists in case of HVDC lines.

7. Surge Impedance Loading. Long EHV-AC lines are loaded to less than 80 per cent of natural load. No such condition is applicable to HVDC lines.

8. Greater Reliability. A two-conductor bipolar dc line is more reliable than a 3-wire 3-phase ac line because the dc line may be operated in a monopolar mode with ground return when the other line develops a fault.

9. Rapid Change of Energy Flow. The control of converter (thyristor) valves permit rapid changes in magnitude and direction of power flow when the two ac systems are interconnected by a dc line. Thus a more flexible coordination of system control at the two ends, a more economical use of cheap power generation in either of the two ac systems, and reduction of system reserve and standby capacity are available. Transient stability is also increased.

10. Independent Control. The ac systems interconnected by a dc line can be controlled independently. They can be completely independent as regards frequency, system control, short-circuit rating, future extension etc.

Independence of frequencies of sending-end and receiving-end networks, makes it possible to generate the power at one frequency (say 50 Hz) and to utilise it at some other frequency (say 60 Hz).

11. Lesser Dielectric Power Loss and Higher Current Carrying Capacity. The cables have lesser dielectric power loss with dc in comparison with ac and, therefore, have higher current carrying capacity.

12. Negligible Sheath Losses. In case of dc, only leakage current flows in the cable sheath where as in case of ac charging, circulating and eddy currents flow through the sheath of cable. Thus in case of dc transmission with cables, the sheath losses in the cables are negligible.

13. Higher Natural Dielectric Strength and Longer Life of Cable Insulation. The natural dielectric strength of cable insulation with dc is substantially greater than that with ac. Phenomenon of dielectric fatigue is also absent, therefore the cable insulation has longer life.

14. Absence of Charging Current and Limitations of Cable Lengths. Because of large charging currents, the use of EHV-AC for underground transmission over long distances is prohibited but because of absence of charging current in dc systems there is no limit on the length of the cable.

In ac systems there is limit on length of cable depending upon rated voltage—about 60 km in case of 145 kV, 40 km in case of 245 kV and 25 km in case of 400 kV.

15. Low Short-Circuit Currents. In ac transmission, addition of parallel lines results in larger short-circuit currents in the system. This is due to reduced equivalent reactance. When an existing ac system is interconnected with another ac system by ac transmission line, the fault level of both the systems increases. Sometimes this necessitates the replacement of the existing circuit breakers by the ones of higher rating. However, when the two ac systems are interconnected by a dc line, the contribution of a dc line to the short circuit current is only up to the rated current of the dc line.

During fault in dc line, grid control of converter can drastically reduce the fault current.

16. Lesser Corona Loss and Radio Interference. The corona loss is directly proportional to $(f + 25)$, where f is the supply frequency. So corona loss in dc system are lesser than those in ac systems for the same conductor diameter and operating voltage. This is because in dc system f is equal to zero. Corona loss and radio interference are directly related and, therefore, radio interference in case of dc is less as compared to ac. Also corona and radio interference slightly decrease by foul weather conditions (snow, rain or fog) in case of dc where as they increase appreciably in case of ac system.

17. Higher Operating Voltages. Modern high voltage transmission lines are designed on the basis of expected switching surges rather than lightning surges because the former are more severe in comparison to the latter. The level of switching surges due to dc is lower than that due to ac and, therefore, the same size of conductors and string insulators can be employed for higher voltages in case of dc as compared to ac.

18. Reactive Power Compensation. Long distance ac power transmission is feasible only with the use of series and shunt compensation, provided at intervals along the line [Art. 13.4(15)]. For such lines shunt compensation is required to absorb the line charging VARs during light load conditions and series compensation is provided for stability reasons. HDVC lines do not require any reactive power compensation like EHV-AC lines. It is because of absence of charging currents and unity power factor operation.

19. No Stability Limit. There is no stability limit in dc systems and, therefore, transmission line may be of any length whereas in ac systems the theoretical length of a transmission line which can be operated without loss of stability is only 1.43 times its natural impedance (mostly it is about 500 km). However, with series compensation the length of ac line can be somewhat longer than this value.

20. Use as a Asynchronous Link Between Two AC Systems. Parallel operation of ac with dc which increases the stability limit of the system or interconnection of two large ac systems by a dc transmission line. Here the dc line is a asynchronous link between two rigid (frequency constant) systems where otherwise slight difference in frequency of the two large systems would produce serious problems of power transfer controi in the small capacity link.

14.4. LIMITATIONS OF HVDYC TRANSMISSION

Although dc transmission has so many advantages over ac transmission, yet it will never completely overtake ac transmission, at the most it will work in conjunction with it. The problems associated with and limitations of dc transmission are given below:

1. Costly Terminal Equipment. The converters required at both ends of the line have proved to be reliable but they are much more expensive than the conventional ac equipments.

The converters have very little overload capacity and absorb considerable reactive power—the reactive power can be as high as 25% of true power.

It is obvious that no reactive power can be transmitted by HVDC transmission system and, therefore, equipment such as static or synchronous capacitors are installed at the receiving end to generate reactive power (MVAR) according to the demand of the load and converting apparatus.

The converters produce lot of harmonics both on dc and ac sides which may cause interference with the audio-frequency communication lines. Filtering and smoothing equipments are provided at the convertor stations in order to remove ripples from the dc output. It is also possible to provide filters on the ac side to absorb the harmonic currents. Thus there is considerable increase in cost of the converter substation.

HVDC converters need complex cooling systems.

HVDC circuit-breaker system comprises several components including the main circuit-breaker capacitors, reactors etc., and the total cost is likely to be several times that of an ac circuit breaker of equivalent rating.

2. More Maintenance of Line Insulators. Maintenance of insulators in HVDC transmission line is more.

3. Circuit breaking in multi-terminal dc systems is difficult and costlier.

4. Voltage Transformation. Voltage transformation is not easier in case of dc and hence it has to be accomplished on the ac side of the system. DC system cannot be employed for distribution, sub-transmission and backbone transmission.

14.5. COMPARISON BETWEEN EHV-AC AND HVDC TRANSMISSION

Comparison between EHV-AC and HVDC for long distance high power transmission is given in tabular form in Table 14.1.

Table 14.1. Comparison of Long Distance Overhead EHV-AC and HDVC Transmission

S. No.	Characteristics	EHV-AC System	HVDC System	Remarks
1.	**Capital Cost** Line cost Substation Cost Number of circuits Intermediate Substations	 Higher Lower More Required	 Lower Higher One Not required	 HVDC system becomes economical above 800 MW. Choice based on economics.
2.	Skin Effect	Present	Absent	
3.	Ground Return	Not possible	Possible	Only single bipolar HVDC line adequate in most cases.
4.	Line Losses	High	Low	HVDC substation has power losses.
5.	Power Transfer	Limits imposed by power angle and line inductive reactance.	No limit	Single HVDC link adequate up to 3,000 MW
6.	Voltage Control	Difficult for long lines due to series inductance and shunt capacitance	Easier	—
7.	Compensation Requirements	Series and shunt compensations are necessary	No compensation is required	
8.	Corona and Radio Interference	More	Less	
9.	Reliability	Lesser	More	
10.	Short-circuit Currents	Large	Small	
11.	Control System	Simpler, cheaper and limitations of control	Difficult, costlier but fast and accurate control	For interconnections, HVDC system is advantageous due to superior control
12.	Stability of AC Networks	Very low due to line inductive reactance	Much higher	AC line is loaded up to 50% of its thermal rating while dc line is loaded up to 90% of its thermal rating

The dc transmission links so far used mostly are based on the factors like long river crossings, frequency conversions and asynchronous ties between large *ac* systems. To give an approximate idea of economic distance for HVDC transmission is that for a distance of 400 km the power of atleast 100 MW is required to be transmitted. With the development of converting equipment, the break-even distances will be reduced.

14.6. TYPES OF DC LINKS

The dc links may be classified into (*i*) monopolar link (*ii*) bipolar link and (*iii*) homopolar link.

1. Monopolar DC Link. As the name suggests, monopolar link has only one conductor and return path is provided by permanent earth or sea. The line usually operates with negative polarity with respect to ground so as to reduce corona loss and radio interference. The monopolar *dc* link is shown in Fig. 14.2.

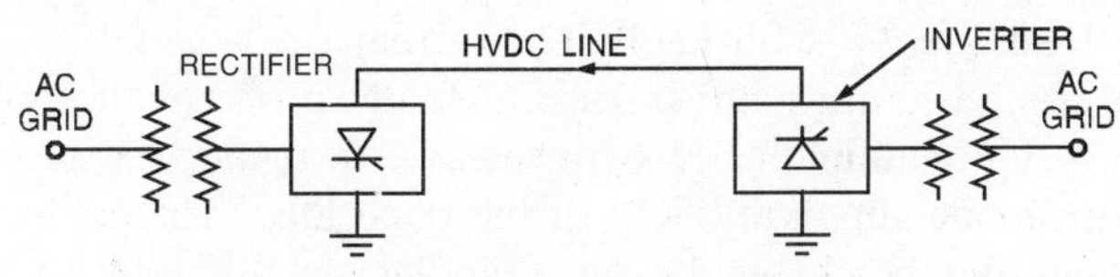

Fig. 14.2. *Monopolar DC Link*

The earth electrodes are designed for continuous rated current operation and for any overload capacity required in the specific case. The sea or ground return is permanent and of continuous rating. The ground return path has a low resistance and, therefore, low power loss in comparison with a metallic line conductor of economical size and equal length provided the ground electrodes are of proper design. Monopolar line is more economical than a bipolar line because the ground return saves the cost of the one metallic conductor and losses in it.

Monopolar HVDC links were used only for low power rating and mainly for cable transmission. In some cases the monopolar lines installed earlier are converted into bipolar systems by adding additional substation pole and transmission pole.

Monopolar HVDC line has only the rating equal to half of corresponding bipolar line rating and is, therefore, not economically competitive with EHV—ac schemes for submarine cables longer than 25 km and of power rating of about 250 MW. For such cable transmission high voltage ac scheme is not technically feasible due to large charging currents with ac cables beyond thermal limit. Bipolar cable is not justified for ratings up to about 500 MW. Recent HVDC cable schemes are bipolar.

2. Bipolar DC Link. This is most widely used *dc* link for overhead long distance HVDC transmission systems and also for back-to-back HVDC system. This link has two conductors—one operating with positive polarity and the other with negative polarity with respect to the earthed tower structure. There are two converters of equal voltage rating and connected in series at each end of the *dc* line, as shown in Fig. 14.3. The neutral points *i.e.*, the junction between converters may be grounded at one end or at both the ends. If it is grounded at both ends each pole can operate independently. The rated voltage of a bipolar link is expressed as ± 500 V. Power rating of one pole is about half of bipole power rating. The earth carries only a small out-of-balance current during the normal operation. When the currents in the two conductors are equal, the ground current is zero. During fault or trouble on one of the lines, the other line along with ground return can supply half of the rated load. Thus continuity of supply is maintained. After taking corrective measures, the system is switched over to normal bipolar operation. Thus the reliability of a bipolar line is equal to that of a double circuit 3-phase line although it has only two conductors instead of 6 for 3-phase line. Example of bipolar HVDC link is Ranchi-Delhi single bipolar overhead line of length 810 km, capacity 1,500 MW and operating at ± 500 kV for transmission of bulk power. This line is designed to operate in bipolar mode under normal conditions, both the poles sharing the load equally with negligible current (less than 10 A) in the ground return path. In case of fault on one of the poles, the system automatically switches over to monopolar ground return mode supplying 50% of the rated load. Thereafter the system may be changed over to bipolar metallic return mode when the other conductor is used as return conductor.

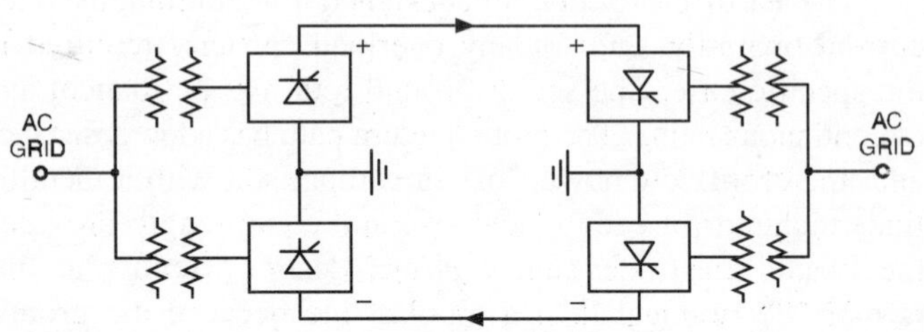

Fig. 14.3. ***Bipolar DC Link***

3. Homopolar DC Link. A homopolar link has two or more conductors having the same polarity, usually negative, and always operates with ground as the return conductor. In case of a fault on any one of the conductors, the converter equipment can be reconnected so that the healthy conductor can supply power. Such a scheme is very complicated and is preferred to a bipolar link provided continuous ground return does not pose additional problems. Such a scheme may be used for the following :

Two homopolar overhead lines supplying to a common monopolar cable termination.

One overhead transmission tower carrying insulator strings supporting two homolar transmission line conductors.

Thus homopolar dc link has limited applications.

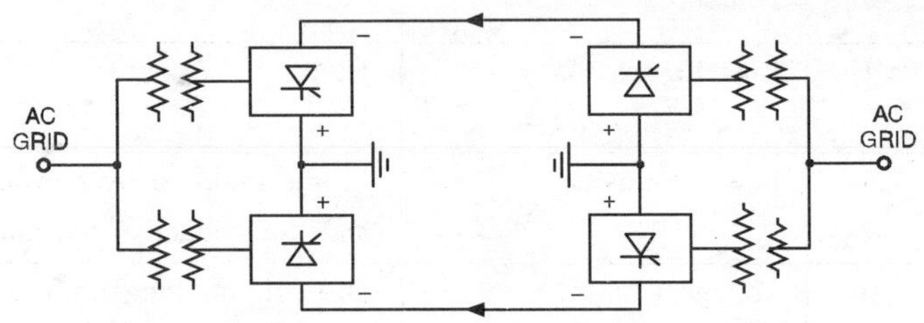

Fig. 14.4. ***Homopolar DC Link***

14.7. GROUND RETURN

A ground return means the ground or sea water or both as the return conductor. Most *dc* transmission lines use ground return path for reasons of economy and reliability. The monopolar link and homopolar link use the ground return continuously for carrying the return current where as the bipolar link uses ground return for short duration of emergency (when one of the lines is under repair or maintenance). For the same length of transmission the resistance offered by the ground in case of *dc* is much less than that in case of ac transmission. This is because the direct current in the earth, unlike alternating current, does not follow closely the route of the line but spreads over a very large cross-sectional area in both depth and width. The resistance of ground return path in case of *dc* is independent of the length of lines (for long lines) and is essentially the sum of grounding resistances of the earthing electrodes at the two ends of the line.

Besides the advantage of economy, as already mentioned in Art. 14.6(1), there are two more definite advantages of using ground return. The first advantage is that a dc line can be built in two stages—the line can be built as a monopolar line with ground return in the initial stages and may be converted into a bipolar line on a later date when the load requirement increases. Thus a considerable part of the total investment can be deferred until the second stage. The second advantage is the reliability of the system *i.e.*, in the event of an outage of one conductor of the bipolar line, it can be operated temporarily at almost half of its rated power, by using the healthy line and the ground, as already mentioned in Art. 14.6(2).

Ground return lines have some problems also. The main problems are designing of ground electrodes for low resistance and low cost of installation and maintenance, location and screening of electrodes so that ground currents cause negligible electrolytic corrosion of buried or immersed metallic structures, prevention of interference of the ground currents with the operation of other services such as ac power transmission, railway signals, ship's compasses etc. and control of surface gradient near the electrodes for the safety of persons, live stock and fish.

14.8. EARTH ELECTRODE AND STATION EARTH

The mid point of converters, called the *neutral point*, in each station is grounded with a suitable switching arrangement. The earthing is independent of the station earthing. This electrode earthing is through electrode installed at a safe distance (about 5 to 25 km) from the terminal station, major pipelines, substations and populated areas. The earth electrode is installed away from the substations and major pipelines so as to avoid the galvanic corrosion of the substation earthing system, underground pipes, buried cables and structures.

Earth electrode is an array of conducting elements placed in the earth or sea which provide a low resistance path between the dc circuit and the earth and which is capable of carrying continuous current for some extended period. The earth electrode at one of the stations acts as an anode* while at the other station it acts as a cathode*. The terminal station is connected to the earth electrode through an insulated cable, called the *earth electrode line*.

Station earth is an array of conducting elements placed in earth at the substation location and which provides connection between the earthed parts of the station equipment and neutral of converter transformers and the earth. Station earth is provided for the protection of equipment from overvoltages and safety of personnel. There is no direct metallic connection between earth electrode and station earth.

Graphite electrodes buried in a pit filled with crushed coke are used. The electrode transfers current to coke which distributes it to earth. This arrangement reduces the loss of material due to corrosion to a negligible value. Iron and graphite cannot be employed as material for earth electrode because iron has high rate of corrosion and graphite when buried in earth causes significant loss of material due to corrosion.

The design aspects of earth electrode are low earth resistance, low current density (not exceeding 1.5 A/m^2) at electrode surface, temperature rise of electrode and surroundings not exceeding 60°C and step voltage on the ground surface above the electrode not exceeding the safe limits.

An earth electrode may be a straight electrode, a ring electrode or a radial star electrode buried horizontally in earth at a depth of about one metre. The radial star arrangement is commonly employed as it uses land area more effectively.

14.9. HVDC TRANSMISSION SYSTEMS

No doubt, 3-phase ac systems have a definite superiority over HVDC systems for generation, transmission, distribution and utilisation of electrical power and are, therefore, universally employed.

However, HVDC transmission is a strong alternative to EHV-AC transmission in some of the applications such as long distance high power transmission by overhead lines, underwater transmission, transmission by underground cables, system interconnection by means of overhead lines or underground cables or back-to-back HVDC coupling stations, frequency conversion (50/60 Hz) and hence preferred.

On the basis of applications, HVDC transmission systems can be classified as follows:

1. Long Distance High Power Transmission. DC overhead lines are cheaper than three-phase transmission lines of the given transmission capacity. With large distances, the saving in cost becomes greater than the additional expenditure on the terminal equipment. HVDC becomes favourable above 800 km, 1,000 MW when the cost of EHV-AC line/substation exceeds that of equivalent HVDC line/substation.

A two-terminal dc link employing overhead lines is used for transmission of large blocks of relatively cheap power from remote sources (say hydroelectric) to load centres situated far away from the generating stations.

2. Underwater Transmission. In case of 3-ϕ transmission, the maximum length of the underwater cable is 25-40 km. This is due to the problem of charging current, and in case of oil-filled cables it is difficult to ensure oil flow over long distances. For longer distances HVDC is the only solution. In the absence of dielectric losses, the dc cables can be of the impregnated type for voltages of at least up to 400 kV, and practically no limit is set to its length.

3. Underground Transmission. Sometimes, the power transmission in congested areas has to be done through underground cables. Wherever the distances involved exceed 40 km, dc transmission is the only feasible solution. DC transmission in such cases offers definite advantages, since the equipment for compensating charging kVA is not required and also there is no reduction in the transmission capacity of cables. Moreover, there is no dielectric heating of the insulation or inductive heating of armouring. Hence the heat dissipation problem in dc cables is far less as compared to the 3-ϕ ac cables. The extra cost of the converters is compensated by saving in cable costs at distances from 50 km–100 km depending on the conditions in each individual case.

4. DC Link in Parallel With AC Link. For interconnection between two ac systems having their own load frequency control, HVDC links have several advantages over ac links. HVDC links form an asynchronous-tie *i.e.*, the two ac systems interconnected by HVDC tie-line need not in synchronism with each other.

HVDC interconnection is superior to EHV-AC interconnection in many respects and is preferred due to its technical superiority. With HVDC interconnection, power flow can be controlled, the frequency disturbances are not transferred, short-circuit levels remain unaffected at both ends, transient stability of an existing ac link can be considerably improved.

* At anode, direct current enters earth from the terminal station while at cathode direct current returns from earth to HVDC circuit.

A dc link has an extremely rapid speed of response (of the order of 50 ms) and, therefore, it can change its transmitted power much more rapidly than, say a turbine generator can adjust its output. Thus the system stability can be significantly improved.

HVDC interconnection can provide a weak tie (of lesser capacity) between strong and a weak ac network. This is difficult with ac interconnection.

HVDC system control can be modified to dampen oscillations in load or power angle δ and thereby improve the stability of an existing ac link.

Most important task of interconnector is transfer of required amount of power in required direction and to assist the interconnected ac network to maintain transient stability. AC interconnectors have severe limitations where as dc interconnectors do not.

5. HVDC Back-To-Back System. Such a system provides an asynchronous interconnection between two adjacent ac systems through converter station and zero (or negligible) length dc line. Such schemes are employed for coupling ac systems of different frequencies (*e.g.*, 50/60 Hz) or of different system controls. The back to back coupling stations can be located at any suitable location, where two networks meet geographically and exchange of required amount of electrical power is desired.

Vindhyachal HVDC back to back link is for exchange of power between Northern and Western regions. Each block of 250 MW is capable of operating independently in either direction and can transfer power in the range of 25 MW to 250 MW depending on the system conditions.

6. Multi-Terminal HVDC Interconnection or Links. This is the new HVDC possibility. Three or more ac networks can be interconnected asynchronously by means of a multi-terminal HVDC link. Power flow from each connected ac network can be controlled suitably. Large power can be transferred.

Almost all dc links set up so far are two-terminal links except *Kingsnorth-London scheme* in which each half of the sending-end power is independently controlled and delivered to two different receiving-end stations. One of the essential requirements in a multi-channel dc scheme is the use of large dc circuit breakers.

14.10. CONVERTER STATION EQUIPMENT

In an HDVC transmission system or HVDC back-to-back coupling station, ac is converted into dc by means of a combination of converter transformers and converter valves. The conversion of ac into dc is called *rectification.* At the receiving end of the HVDC transmission line, dc is converted back to ac by means of converter valves and converter transformers. The conversion of dc into ac is called the *inversion.*

An HVDC link comprises an ac substation and conversion substation at each end of an HVDC transmission line but in case of back-to-back HVDC link, there is only a conversion substation between two ac substations and there is no dc transmission line.

Thyristor valves are the most important converter station equipment and the other main equipments include converter transformers, dc reactor, harmonic filtering equipment, control equipment and reactive power compensation equipment.

14.10.1. Thyristor Valves

Since the early 1970's, the thyristor valves are universally employed in all new HVDC schemes while earlier mercury-arc valves were used for conversion/inversion purposes. A thyristor valve has a better operating life and higher consistency of performance than those of a mercury-arc valve. The use of thyristors has resulted in a considerable simplification in the design of conversion substation. The ratings of thyristors have increased remarkably during the last one decade and the current ratings of the thyristors available now are of the magnitudes of line current required for transmission. However, the voltage capability of individual thyristor is quite small in comparison to the line voltage. Thus for obtaining the proper voltage rating it becomes imperative that a large number (typically 5 to 200) of thyristors are connected in series. A thyristor valve is always provided with a number of additional or redundant thyristors in series so that even if some of the thyristors fail, the valve operation remains unaffected. The valves used are of indoor design, air-insulated and air/water cooled. Thyristor assemblies are protected from $\frac{dv}{dt}$ and $\frac{dI}{dt}$ protection. Six-pulse or twelve-pulse converters are used in modern HVDC schemes—12-pulse is more preferred as it contains minimum ripple. There are two 3-phase transformers connecting each 12-pulse converter to the ac bus-bars. One of the transformers is star/star connected and the other is star/delta connected so as to give a phase shift of 30° between the lower and upper 6-pulse bridge of the converter.

The structural design of the converters is very simple. Each thyristor module consists of the thyristors and the converter pole is composed of series connection of several thyristor modules. In addition to the thyristor elements, each module contains damping and overvoltage protection circuitry, voltage dividers and firing control circuits. The modules are mounted in layer form in a box-like structure to form the unit.

Basically, the converters are of 12-pulse operation. However, the use of a single 12-pulse converter per pole as well as multiple valves per pole are available in practice. The former design, though, is very simple in construction and operation but it needs a heavy transformer causing transportation and erection problems. Use of single-phase transformers has been attempted in the West to overcome this problem.

Use of four valve units per phase of a 12-pulse converter configuration is common. A typical *Quadruple valve* comprises four valves placed vertically one above the other to form one

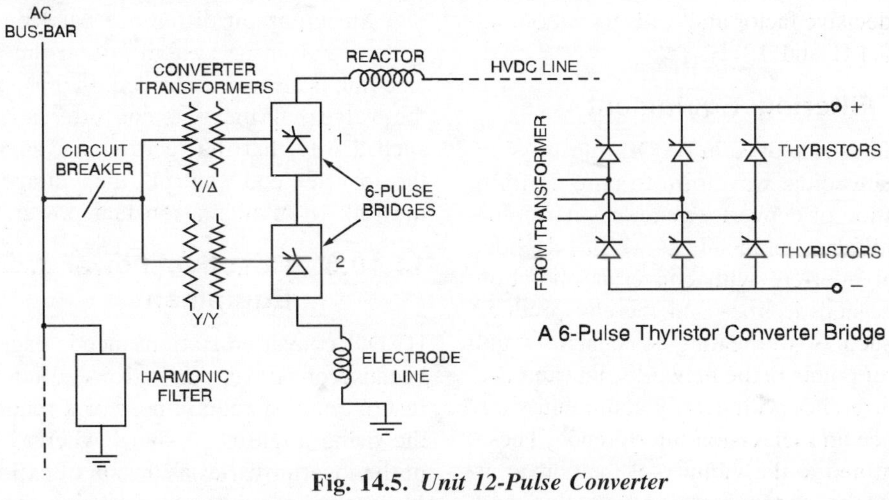

Fig. 14.5. ***Unit 12-Pulse Converter***

limb of the converter. Such an arrangement provides the most compact and economical layout of the valves and the valve housing. Three quadruple valves constitute a 12-pulse converter. Valve surge arresters may be directly mounted on the valve structure.

Thyristors produce heat of 30-40 W/cm^2 and, therefore, need effective cooling system. The valves are cooled by air, SF_6 gas, oil or a combination. The cooling system and insulation are independent. The temperature of silicon wafer joint should be held below critical value (90°–125°C) to prevent change in characteristics and damage to thyristors.

The schematic unit of a 12-pulse bridge is shown in Fig. 14.5.

In practice two 12-pulse converters per pole (connected in series or in parallel) are used so as to enhance the reliability. In parallel connection, a smoothing reactor is required for each 12-pulse converter and current balancing controls are required to be used. A parallel configuration gives less line loss during the period when full-line currents are not yet established.

The thyristors are required to be triggered in a desired sequence and desired instants. Though conventional firing circuitry is still in use but the most modern method of triggering of thyristors is called the "*optronics*" which employs light guide system. In this system the electrical pulses are converted into light signal pulses by fibre-optic techniques. The light pulses are transmitted through glass-fibre optical cables up to the individual thyristor. Each thyristor is provided with a small separate thyristor which is triggered by the light pulse. Thereby the main thyristor is triggered.

For the protection of thyristors the protections provided are *dv*/*dt* protection, over-temperature protection, forward voltage firing, forward recovery protection and current sharing protection between two parallel thyristors.

14.10.2. Converter Transformers

Converter transformers are connected between converter valves and the ac bus-bars and their main function is to transform the ac voltage to a suitable value for feeding the converter. The other functions served by the converter transformers is to supply the reactive power to the converter through tap changing; control of fault level by suitable reactance offered by converter transformers and help in harmonic suppression. Twelve-pulse operation is feasible due to transformer connections—by suitable star-star and star-delta connections the required phase shift of 30° for 12-pulse operation is achieved.

Converter transformers are either single-phase units, or three-phase units with either two-winding type or three-winding type. These are specially designed so as to withstand direct voltage stresses. Moreover, in a converter transformer the currents flowing through the winding have high harmonic contents. Thus the eddy current losses are high. Noise due to magnetostriction is higher and may require special design tank to suppress the same. On-load tap changing is normally used to reduce the demand of reactive power in steady-state operation. The number of tap-changer operations is higher in HVDC systems and, therefore, proper quality control of components of tap-changers is required. Converter transformers employed in large HVDC schemes are of high MVA ratings (about 300 MVA or even more).

14.10.3. DC Reactor

In HVDC transmission systems the dc reactor, connected in series with each pole of the converter plays an important role. It prevents commutation failures in inverter by limiting the rate of increase of direct current during commutation in one bridge when the direct voltage of another bridge collapses. It reduces the incidence of commutation failures in the inverter during dips in the alternating voltage. It reduces harmonic voltages and currents in the dc line, smoothens the ripple in the direct current and limits the current in the rectifier following a short circuit in the line.

The dc reactor is usually of air-core and oil-cooled type and has the nonlinear magnetic characteristics.

The value of inductance of the reactor is very important. With the increase in inductance the current waveform on the dc side improves but the control response slows down and the resonance frequency falls making the stabilization of current control more difficult. The limitation of dc line short-circuit currents is normally the decisive factor and leads to smoothing reactances of between 0.4 H and 1.0 H.

14.10.4. Harmonic Filtering Equipment

The 3-phase bridge converter employed in HVDC transmission should convert pure ac sinusoidal waveform to pure dc form but in practice the operation of converter generates harmonic currents and harmonic voltages on ac side as well as dc side. These harmonics do not interfere with converter operation but they flow through ac and dc lines and thereby produce several harmful effects such as overheating of capacitors and generators, overvoltages at points in the networks, interference with protective gear, interference with nearby communication systems, radio interference and television interference. These disturbances are not confined to the vicinity of the converter station but spread over the ac network and dc line and surrounding residential areas.

An n-pulse converter generates harmonics predominantly of the order of $nx \pm 1$ on ac side and nx on dc side where x is an integer and n is usually 6 or 12. Thus in case of a 12-pulse valve group operation, the orders of main harmonics are 11th, 13th, 23rd and 25th on the ac side and 12th and 24th on the dc side. The amplitudes of harmonics decrease with increasing order. It is necessary to provide filters on both the ac and dc sides.

Harmonic filters provided on the ac side serve the following purposes :

1. Harmonic voltages and currents in the ac power network are reduced to acceptable levels.

2. All or a part of the reactive power required by the converter is provided, the additional reactive power being supplied by the shunt capacitor banks or by the ac power system.

The ac filters employed in HVDC systems are tuned filter which is sharply tuned to a harmonic frequency, damped filter which, if shunt connected, offers a low impedance over a broad band of frequencies (say 11th and harmonics of higher orders). The adoption of 12-pulse valve group operation has resulted in reduction of harmonic current generation and considerable simplification in ac filters. In a 12-pulse valve group operation only two damped filters are employed while 6-pulse valve group operation requires four tuned filters and one damped filter. Damped filters are simpler to design and have a lower risk of the resonant overvoltages and currents in comparison to tuned filters.

In bipolar operation, under ideal conditions, the induced voltages would be negligible and filters are not required. However, in practice, assuming certain unbalance (in transformer reactances etc.), the induced voltage level would increase. So use of high pass dc filter tuned to 12th harmonic is usually made for a 12-pulse converter circuit.

14.10.5. Control Equipment

The control of firing angle is very important in HVDC systems. This is accomplished by optic fibre based hardware circuitry.

An important difference between the operations of a dc system and an ac system is that the power transmitted over a dc link is always controlled. In most practical HVDC systems the valves of the two converter stations are controlled in such a way that the rectifier end controls the current while the inverter end controls the voltage. These controls allow the link to maintain constant power.

14.10.6. Reactive Power Compensation Equipment

HVDC converter stations need reactive power, particularly because of converter control and commutation process. The reactive power requirement of a rectifier varies as the sin of the firing angle (*i.e.*, sin α) whereas the power requirement of the inverter varies as the sin of extinction angle (*i.e.*, sin β). Moreover, converter transformers consume reactive power. The reactive power requirement is usually in the range of 50 to 60 per cent of the transmitted power. As no reactive power is transmitted over the dc line and maintenance of reactive power balance at the two ends is essential to keep ac voltage within specified limits, installation of reactive power compensation equipment in the HVDC terminal stations becomes imperative. This equipment may consist of ac filters, static shunt capacitors, synchronous condensers, static VAR compensators etc. A combination of these equipments is also used.

14.11. TRANSIENT OVERVOLTAGES AND INSULATION COORDINATION

Convertor overvoltages are due to transient in converter itself, disturbance on dc side, disturbance on ac side, and transient due to external agents.

Transient overvoltages may occur in the converter itself. This occurs due to convertor's own problems like commutation failure or repetitive misfire. Sudden blocking of inverter firing may also cause overvoltage across the converter.

Disturbance on dc side that causes overvoltages may be due to short circuit of one pole in a bipolar link and transient overvoltage induced in the healthy pole. Re-energization of thyristor bridges can cause large transient overvoltages.

Severe overvoltages may be caused, on the ac side, due to resonance between the filter and the system for a low order harmonic. The other factors that may cause overvoltages on the ac side, are transformer switching, load rejection, ac system faults etc. Transformer saturation due to harmonic current may also result in overvoltage.

Lightning and switching surges are external agents that may cause transient overvoltages in the HVDC systems. Also the switching surges and lightning surges in the ac system can be transferred to the dc side through the converter transformers.

Zinc oxide surge arresters (ZnO arresters)* are connected in dc yard, valve hall, ac yard filter circuits etc. HVDC system require a large number of arresters. These are coordinated with surge arresters in ac yard.

* Zinc oxide surge aresters have superior characteristics and high discharge capability.

The rated voltage switching impulse withstand characteristics, lightning impulse withstand characteristics, test voltages, test procedures, connection surge arresters etc. form a part of insulation coordination studies.

14.12. CORONA IN HVDC LINES

In dc corona, the charges released from one conductor are carried to the ground or the other conductor because of the opposite polarity. Thus the movement of charged particles is characterised by electric field distribution rather than the surface gradient. In case of ac voltage, the peak value of voltage wave is $\sqrt{2}$ times the rms value but in case of dc there is no such factor and, therefore, in HVDC lines the corona loss is smaller. In case of HVDC lines, corona over insulators and metal surface are negligible. Corona loss in HVDC lines during bad weather conditions does not increase abruptly as in case of ac line. Corona loss in case of dc lines does not increase with the increase in operating voltage as rapidly as in case of ac lines.

In case of ac lines, the ions going away from the conductor surface due to like polarity of charging during a half cycle get attracted towards the conductor during the next half cycle of ac wave. But in case of dc lines, the ions with the same polarity as that of the conductor get a time to go away from the conductor. Thus the corona phenomenon acts differently with ac and dc lines.

However, the corona loss in HVDC system is an important factor and cannot be neglected in the design of the line. The corona behaviour of a monopolar HVDC line is different from that of a bipolar line. This is due to the difference in release of the charge from the vicinity of the conductor surface. Corona loss in the monopolar HVDC line is minimum.

The critical line voltages V_C for monopolar and bipolar HVDC line are given by Eqs. (14.5) and (14.6) respectively.

For monopolar HVDC line,

$$V_C = E_C rm \log_e \frac{2h}{r} \text{ kV (dc)} \qquad \text{...(14.5)}$$

and for bipolar HVDC line,

$$V_C = 2E_C rm \log_e \frac{D}{r} \frac{1}{\sqrt{1+\left(\frac{D}{2h}\right)^2}} \text{ kV} \qquad \text{...(14.6)}$$

where E_C is critical stress in kV/cm (14 kV/cm in case of dc), m is the surface irregularity factor (0.8–0.92), h is the height of the conductor above earth in cm, r is conductor radius in cm and D is the pole to pole spacing in cm.

The corona loss in a bipolar HVDC line is given as

$$P_{dc} = P_0 + K(E - E_C)^2 \text{ kW/km} \qquad \text{...(14.7)}$$

where P_0 is corona loss at critical gradient (0.3–0.5 kW/km), E is actual surface stress in kV/cm and K is a constant.

14.13. PROTECTION ASPECT

The malfunctioning of a converter may occur due to the following reasons:

1. Misfire and back fire in converter valves.
2. Commutation failure, the disturbance that usually occur in inversion operation.
3. Short circuit in converter station.

Many of the valve faults are of temporary nature and can be eliminated by interrupting conduction temporarily. The faults can be eliminated by the control action used in controlling firing angle.

The faults (short circuit or earth faults) occurring on ac side of the HVDC system cause dip in the voltage at the terminals of the converter and are required to be cleared in the shortest possible time duration. It is necessary for keeping dc power transfer normal. Inverse relays and differential protections are employed for clearing the faults that are generated in the filter banks.

14.14. HVDC CIRCUIT BREAKING

The process of arc extinction in ac circuit breakers is easier because of natural zero value of alternating currents. But direct current being a steady unidirectional current does not have a natural zero and, therefore, it is difficult to interrupt large direct currents at high voltages. The lack of dc circuit breakers has limited the network of dc lines.

Almost all dc transmission projects till this date are two terminal projects and it is not difficult to interrupt the fault currents. The transient faults can be cleared using valve control methodology while for clearing permanent faults combination of valve control, fault locators and isolating switches be employed.

A vacuum interrupter is employed for breaking direct currents in HVDC systems. An auxiliary device is attached to the vacuum interrupter such that the arc current is opposed by the auxiliary device current (oscillatory discharge of capacitor). However, in practice, the fault current is shaped before being interrupted. Commutation of load current from a saturable reactor to an R-C circuit is made such that after commutation the current is opposed by the high inductance of the reactor. Researches on HVDC circuit breakers are still in progress and use of these circuit breakers is limited to multi-terminal dc systems only.

Exercises

1. What is the principle of HVDC system operation ? Discuss the technical and economic advantages of dc systems over ac systems.
2. Briefly discuss the advantages and the operational problems of HVDC transmission. **[U.P.S.C I.E.S. Elec. Engineering 1999]**
3. Explain the advantages of HVDC transmission. **[M.D. Univ. Power System-I, May 2009]**
4. Explain the advantages and limitations of HVDC transmission system. **[M.D. Univ. Power System-I, Dec. 2009]**

5. What are the problems in EHVAC transmission ? How are they overcome in HVDC transmission ?

[U.P. Technical Univ. Elements of Power System 2002-03]

6. Write merits and demerits of high voltage dc transmission systems.

[U.P. Technical Univ. Elements of Power Sysetm 2006-07]

7. Discuss merits and demerits of HVDC transmission over EHV-AC transmission.

[U.P. Technical Univ. Elements of Power System 2007-08]

8. Compare EHV AC and HVDC options for an integrated power network.

[U.P.S.C. I.E.S. Electrical Engineering II, 2002]

9. Discuss types of HVDC links. **[U.P. Technical Univ. Elements of Power System 2004-05; 2007-08]**

10. Discuss layouts, relative merits and demerits of different types of HVDC links.

[U.P. Technical Univ. Elements of Power System 2008-09]

11. Give a comparisons of monopolar and bipolar HVDC links.

[U.P. Technical Univ. Elements of Power System 2006-07]

12. Explain with necessary diagrams kinds of HVDC links used in HVDC systems. Also explain the relative advantages and disadvantages of using these links. What are the precautions needed for series and parallel connections of thyristors to construct a HVDC system ?

[U.P.S.C. I.E.S. Electrical Engineering II, 2005]

13. What is ground return and why is it employed ? Give the problems associated with the use of ground as the return conductor.

14. What are the types of HVDC transmission system applications ? **[U.P.S.C. I.E.S. Electrical Engineering-II, 2010]**

15. Give a neat sketch of a 12-pulse bipolar HVDC station indicating the position of various equipment. Give the uses and limitations of each equipment.

[U.P.S.C. I.E.S. Electrical Engineering-II, 2003]

16. Write short notes on the following :

(*i*) HVDC transmission. **[M.D. Univ. Dec. 2008; Dec. 2010]**

(*ii*) Special features of HVDC transmission.

(*iii*) Techno-economic aspects of HVDC transmission.

[A.M.I.E. Sec. B. Summer 1996]

(*iv*) EHV DC transmission versus EHV AC transmission.

(*v*) Terminal components of HVDC transmission system.

(*vi*) HVDC links and its types. **[M.D. Univ. May 2008]**

Short Answer Type Questions With Answers

Q. 1. Why is voltage regulation better in case of dc transmission ?

Ans. Because of absence of inductance in dc systems.

Q. 2. Why there is no limit on the length of the cable for dc system while it is in case of ac system ?

Ans. Because of absence of charging currents in dc system.

Q. 3. Why HVDC line do not require any reactive power compensation ?

Ans. Because of absence of charging currents and unity power factor operation.

Q. 4. Why dc systems invariably use ground or sea return ?

Ans. Because of economy and reliability.

Q. 5. Why is dc used for high voltage transmission of electric power ? **[U.P.S.E. I.E.S. E.E.-II, 2004]**

Ans. There is a limit to the distance that the ac can be transmitted by overhead lines because of the surge impedance limitation producing high voltages at the receiving end. That is why dc is used for high voltage transmission of electric power as an alternative.

Q. 6. Why iron and graphite cannot be employed as material for earth electrode ?

Ans. Because iron has high rate of corrosion and graphite when buried in earth causes significant loss of material due to corrosion.

Q. 7. Why dc transmission is economical and preferable over ac transmission for large distances only ?

Ans. Because with larger distances, the saving in cost of dc overhead lines becomes greater than the additional expenditure on terminal equipment.

Q. 8. What is the most modern method of triggering of thyristors ?

Ans. Optronics employing light guide system.

Q. 9. What are the precautions needed for series and parallel connections of thyristors to construct a HVDC system ?

[U.P.S.C. I.E.S. E.E.-II, 2005]

Ans. In case of hvdc transmission, the voltage and current levels are so high that a single thyristor cannot meet these requirements. Under such circumstances, it is essential to connect more than one thyristors in series to achieve higher voltages and in parallel to meet higher current needs.

Series Connection. When the required voltage rating exceeds the SCR voltage rating, a number of SCRs are required to be connected in series to share the forward and reverse voltage. As it is not possible to have SCRs of completely identical characteristics, deviation in characteristics lead to the following two major problems during series connections of the SCRs :

(*i*) Unequal distribution of voltage across SCRs.

(*ii*) Difference in recovery characteristics.

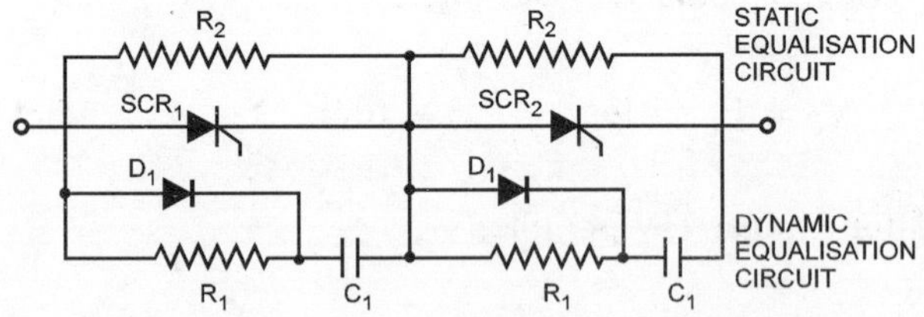

Fig. 1. ***Equalisation For Series Connection***

Care must be taken to share the voltage equally. For steady-state conditions, voltage sharing is achieved by using a resistance or a zener diode in parallel with each SCR. For transient voltage sharing a low non-inductive resistor and capacitor in series are placed across each SCR, as shown in Fig. 1. Diodes D_1 connected in parallel with resistor R_1, helps in dynamic stabilisation. This circuit reduces differences between blocking voltages of the two devices within permissible limits. Additionally the R-C circuit can also serve the function of 'snubber circuit'. Values of R_1 and C_1 can primarily be calculated for snubber circuit and a check can be made for equalization. If ΔQ is the difference in recovery charge of two devices arising out of different recovery current for different time and ΔV is the permissible difference in blocking voltage then $C_1 = \frac{\Delta Q}{\Delta V}$.

The value of resistance R_1 should be sufficient to overdamp the circuit.

Since the capacitor C_1 can discharge through the SCR during turn-on, there can be excessive power dissipation, but the switching current from C_1 is limited by the resistor R_1. This resistance also serves the purpose of damping out 'ringing' which is oscillation of C_1 with the circuit inductance during commutation. All the SCRs connected in series should be turned *on* at the same time when signals are applied to their gates simultaneously.

Parallel Connection. When the load current exceeds the SCR current rating, SCRs are connected in parallel to share the load current. But when SCRs are operated in parallel, the current sharing between them may not be proper. The device having lower dynamic resistance will tend to share more current. This will raise the temperature of that particular device in comparison to other, thereby reducing further its dynamic resistance and increasing current through it. This process is cumulative and continues till the device gets punctured.

Some other factors which directly or indirectly add to this problem are difference in turn-on time, delay time, finger voltage* and loop inductance. Arrangement of SCRs in the cubicle also plays vital role.

When the SCRs are connected in parallel, it must be ensured that the latching current level of all the SCRs is such that when gate pulse is applied, all of them turn *on* and remain on when the gate pulse is removed. Further the holding currents of the devices should not be so much different that at reduced load current one of the device gets turned *off* because of fall of current through it below its holding current value. This is particularly important because on increase in load current, the device which has stopped conducting cannot start in the absence of gate pulse.

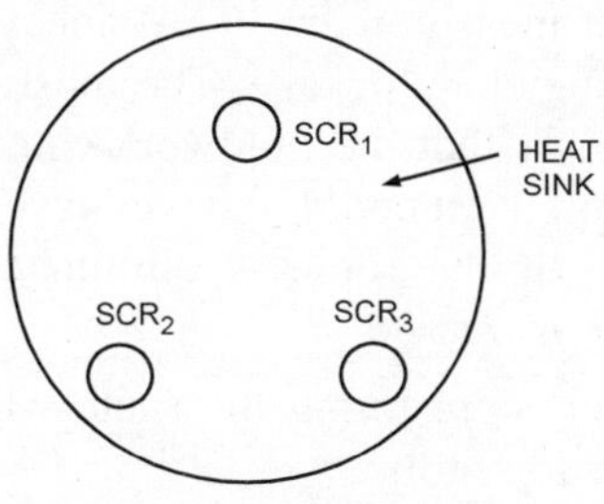

Fig. 2. ***Common Heat Sink***

Another point to be considered is the on-state voltage across the device. For equal sharing of currents by the devices voltage drop across the parallel paths must be equal.

For operation of all the SCRs connected in parallel at the same temperature, it becomes necessary to use a common heat sink for their mounting, as illustrated in Fig. 2.

Resistance compensation used for dc circuits is shown in Fig. 3. In this circuit the resistors R_1 and R_2 are chosen so as to cause equal voltage drop in both arms.

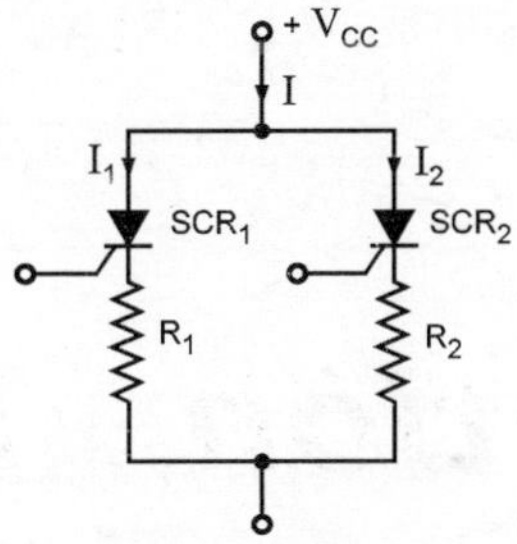

Fig. 3

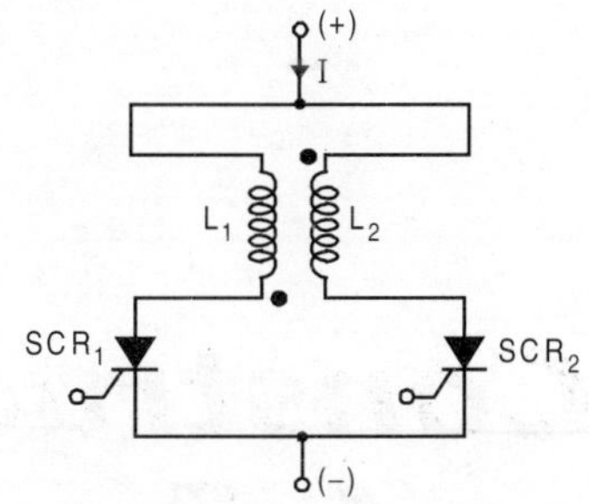

Fig. 4

Inductive compensation used for ac circuits is shown in Fig. 4. The difference in characteristics due to different turn-on time, delay time, finger voltage, latching current, holding current can be minimized by using inductive compensation.

Firing circuits giving high rate of rise can be used to reduce mismatch of gate characteristics and delay time if currents I_1 and I_2 are equal.

Current sharing circuits must be designed so as to distribute current equally at maximum temperature and maximum anode current. This is done to ensure that the devices share current equally under worst operating conditions.

Mechanical arrangement of SCRs also plays an important role in reducing mismatching. Cylindrical construction is perhaps the best from this point of view.

Q. 10. What is normal value of inductance of the dc reactor ?

Ans. The value of inductance of the dc reactor should be such that a resonance of the dc circuit does not occur at power frequency and is in the range of 0.4–1.0 H.

Q. 11. What are the equipments that supply reactive power in HVDC converter stations ?

Ans. AC filters, static shunt capacitors, synchronous condensers, static VAR compensators etc.

Q. 12. Can corona loss be ignored in the design of HVDC transmisson lines ?

Ans. No.

❑❑❑

* Finger voltage is the minimum voltage required between anode to cathode of an SCR before it can be triggered into conduction mode. Finger voltage is slightly greater than normal on-state voltage.

CHAPTER

15

Construction, Testing and Commissioning of Overhead Lines

INSIDE THIS CHAPTER

15.1. INTRODUCTION

We have already discussed the necessity of transmission and distribution system for bulk power transportation over long distances (from generating stations to load centres). We have also discussed the main components of an overhead line, various types of supports (poles/towers), conductor materials, line insulators etc. used in overhead lines. Here, we will discuss survey of lines, planning of construction work, erection of supports, fixing of insulators, erection of conductors, earthing, guarding, testing and commissioning of overhead lines and maintenance of overhead lines.

15.2. SURVEY OF OVERHEAD LINES

In order to determine the route along which the overhead line is to be laid it is necessary that the area is surveyed. The survey of the line must be carried out accurately and expeditiously. A mistake in the field or subsequent office work may pass undetected to an advanced stage in construction, thus causing unnecessary expenditure and inconvenience. The proposed route of the line should be the shortest practicable distance. The first thing to be done is to select the route and plot it on a map to a large scale. Second step is to conduct a walk-over survey to determine the topography of the area and any obstacles likely to come across. This is to be followed by a detailed survey so as to determine the quantity of material required for the construction of the line. The detailed survey can be conducted by theodolite, dumpy level, levelling staff and chain or tacheometry, using the instrument and staff only. The method adopted depends upon the topography of the area and the number of lines to be surveyed etc. Tacheometry is the indirect measurement of distances by optical methods instead of using chain or tape. The most important advantage of this method is that the fieldwork can be carried out much faster than with the theodolite, level, levelling staff and chain. The chaining of the route is eliminated and the levelling speed greatly increased.

The factors concerning the route which could affect the cost of the works are:

1. Excessive deviations from a straight line; the cost of the deviation depends on the size of the conductor and the type of line construction employed.
2. Excessive rock blasting.
3. A route through a swamp or peat bog where expensive special foundations have to be employed and access is difficult for the greater part of the year.
4. A route involving awkward contours where there is the expense of long spans, long poles or section points in excess of about one for every 10 spans.

In the selection of route for the line, the areas to be avoided as far as possible are wells, bridges, steep slopes, low-lying areas, waterlogged areas, gardens, wooded areas, urban development areas, buildings containing explosives, slaughter houses, aerodromes and graveyards. The number of crossings of grid lines, telephone lines, telegraph wires, railways or rivers should be kept minimum possible.

The route of the line should be selected preferably along an available road in the area as it will facilitate in delivery of material as well as the construction and maintenance of the line. Local authorities should be consulted before deciding the route and permission must be obtained from the railway authorities wherever the line is likely to cross over a railway line. Permission must also be had from the power telecommunication coordination committee if the power lines being constructed are to be laid parallel or across the telegraph/ telephone lines.

Surveying of power lines for low and medium, high and extra high voltages are discussed below:

15.2.1. Low and Medium Voltage (240-415 V) Lines

These lines are confined within the limits of towns and cities and within small distances (say, 3–5 km) in rural areas because consumer services are directly tapped from such lines.

The first step in survey work is to mark the position of the distribution lines along the streets in case of towns and cities; and along the lanes in case of villages and to locate the position of consumers and mark the route of the distribution line in case of agricultural load.

After marking the route of the line on the map, the lines and poles* are plotted to scale on a map by carrying out walk-over survey of the route.

Lines in Urban Areas. The power lines are plotted on the map of town/city after consulting the municipal authorities. The points to be considered are road crossings, railway track crossings, telecommunication line crossings, private lands and crossings over buildings.

Lines in Rural Areas. For conducting a survey of distribution lines in a village, a scaled map, called the *khasra map,* of the village is taken and following markings are done on the map.

(*i*) The nearest *ht* line from which the tapping is to be taken for the distribution transformer.

(*ii*) The location of distribution transformer keeping in view the distribution of load.

(*iii*) The layout of *lt* line network with location of angle poles and road crossings.

Agricultural Field Areas. In case of agricultural loads, the transformer should be located at the load centre so as to keep the length of the line and voltage drop minimum possible.

15.2.2. HT (11-33 kV) Lines

The survey of overhead *ht* lines can be divided broadly into two heads : Preliminary walk-over survey and detailed survey.

The steps to be followed for the preliminary survey of *ht* line routes is the same as detailed for *lt* distribution lines. Detailed survey of *ht* lines will be discussed along with for EHV lines.

The points to be considered in deciding the route of the *ht* lines are given below:

(*i*) An *ht* line should run along a road as far as possible so as to facilitate in construction and maintenance.

(*ii*) The number of angle poles should be the minimum.

(*iii*) Crossing of highway and railway tracks should be avoided as far as possible.

(*iv*) Wherever unavoidable, *ht* line should intersect the road, railway track, river or telephone line at right angles.

(*v*) At crossings, adequate clearance as per IE rules should be maintained and guards should be provided for safety of personnels and materials.

15.2.3. EHT (Exceeding 33 kV) Lines

As in case of *ht* lines, the survey of extra high tension lines can be broadly divided into two heads: Preliminary walk-over survey and detailed survey,

The *preliminary survey* is carried out after provisionally fixing the line route on the survey map. After ensuring the availability of right of way (*i.e.*, the land owner consent for laying of line and erection of towers) the position of angle towers and the various crossings is marked on the map.

After marking the position of the line on the map on the basis of preliminary survey a detailed survey is conducted for marking the exact position of poles/towers. A wooden peg is driven into the ground at the starting point of the line. A standard level of the area called *bench mark* is located as a reference point. The level of ground where the peg is fixed is determined with reference to the bench mark. Thereafter the line route is surveyed with the help of level. The position of the supports (poles or towers) with a straight run or at angle locations is determined, as illustrated in Fig. 15.1.

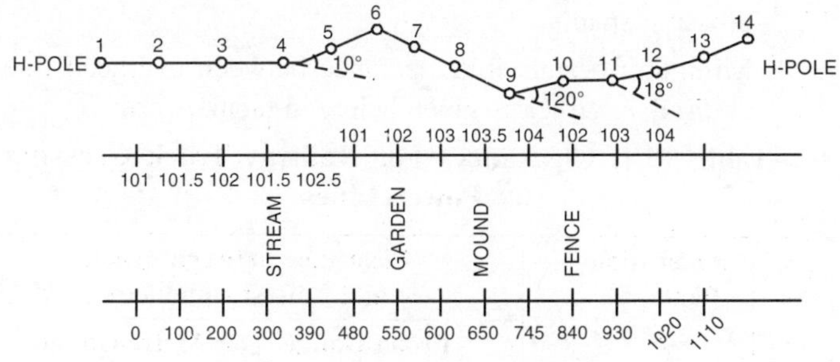

Fig. 15.1. ***Route Survey of Line***

At some places along the route of the line, it may not be possible to erect an overhead line due to one or other difficulty/

* The survey party actually walks along the route of line marked on the map and finalizes the position of poles by making physical measurements with the help of chains/tapes.

obstactle. In such a case the line may be terminated there and an underground cable is laid till a suitable position is again found for resuming the construction of overhead line.

There are some problems which must be considered when the route of the line crosses a river, railway line, national highway, telecommunication lines or power lines or passes in the vicinity of an aerodrome or passes through a jungle.

1. River Crossings. A data of high flood levels of last 20-25 years should be obtained from the revenue authorities and the structures should be so erected that they remain approachable even during the periods of floods. In case of navigable rivers the structures should be designed so as to provide sufficient clearance between the lowest conductor and the highest flood level. In case of non-navigable rivers, the transmission line structures are designed to give a clearance of 3 metres between the lowest conductor and the highest flood level. This is necessary as the flood water may carry trees and their branches. In case of navigable rivers consultation with navigation authorities is also essential.

2. Railway Crossings. The regulations for the passage of overhead as well as underground power lines across a railway track have been framed by the railway authorities. The points to be kept in mind in this regard are:

1. Normally an overhead line should cross the railway track at right angles. However, a deviation to the extent of 30° may be permitted under unavoidable circumstances.
2. The minimum distance between the supporting structures for a crossing span is determined on the basis of the number of tracks to be crossed. According to rules the distance between one of the supporting structures and the nearest track should be equal to the height of the structure above ground level plus 3 metres. The crossing span should not exceed 80 per cent of normal span of the line.
3. All guards wires should be of galvanized steel and of size not less than 6 SWG for carriers and 10 SWG for cross lacings.

   All guy wires should be of galvanized steel and of size not less than 7/11 SWG. No jointing is permitted in the crossing span.
4. Minimum permissible clearance between overhead line and railway track is given below in tabular form.

Table 15.1. Clearances For Railway Track Crossing by Power Lines

S. No.	Operating Voltage	Clearance between Track and Lowest Conductor			
		Broad Gauge		Metre Gauge	
		Inside Station	Outside Station	Inside Station	Outside Station
1	Up to 650 V	7.2 m	7.2 m	6.1 m	6.1 m
2	Above 650 V and Up to 33 kV	10 m	7.6 m	8.6 m	6.4 m

3. Road Crossings. The guards are provided below the line conductors of the overhead lines when they cross roads. This is done to make the snapped line conductors dead before falling down to ground. In this arrangement two GI wires of size 3.15 mm are stretched under tension between the line supports on either sides of the road and fixed to the cross arms. The supports of adequate strength and height have to be erected if the existing supports are found unsuitable. The line should not cross the road at an angle less than 60°. The recommendations in this regard are given below in Table 15.2.

Table 15.2

S. No	Type of Road	Width of Road	Minimum Horizontal Clearance Between The Centre of Road and The Line Support
1	National Highway	11.6 m	7.6 m
2	State Highway	9.8 m	6.7 m
3	Major District Roads	7.3 m	5.5 m

4. Telecommunication Line Crossings. The communication lines under the influence of electric and magnetic fields set up by the nearby high-voltage power lines experience greater induction which may prove dangerous to the communication equipments as well as to the workers engaged in the maintenance work of telecommunication lines.

The Power Telecommunication Coordination Committee (PTCC) has set the code of practice for crossing between the telecommunication and power lines. The essential points to be kept in view in this regard are:

1. The telecommunication line is always kept below the power line as it needs constant attention for maintenance.
2. The most desired angle of crossing between the overhead power line and telecommunication line is 90°. In no case it should be less than 60°.
3. A safe clearance between the two lines (power and telecommunication lines) has to be maintained.
4. The guard between the telecommunication line and the power line may be fixed either to the supports of the power line or the supports of telecommunication line.

Minimum clearances to be a maintained between the two lines are:

Above 33 kV and up to 66 kV ... ... 2.4 m

Above 66 kV and up to 132 kV ... ... 2.7 m

Above 132 kV and up to 220 kV ... ... 3.0 m

5. Crossing Over Other Power Lines. This is governed by IER 78 which is reproduced here. No conductor of an overhead line crossing a tramway or trolley bus route using trolley wires shall have less than the following clearances above any trolley wire.

(*a*) low and medium voltage lines...1.22 m (4 ft)

Provided that where an insulated conductor suspended from a bare wire crosses over a trolley wire the minimum clearance for such insulated conductor shall be 0.6 m (2 ft)

(*b*) high-voltage lines up to and including 11,000 volts... 1.83 m (6 ft)

(*c*) high-voltage lines above 11,000 volts ... 2.44 m (8 ft)

(*d*) extra high-voltage lines ... 3.05 m (10 ft)

6. Forest Area. While passing through forest area permission is required from the forest department for carrying out the survey and for falling of trees so that minimum tree clearance of 1.8 m is had from conductor.

15.3. PLANNING OF CONSTRUCTION WORK

1. Preparation of Estimate. After completion of survey work and working out of length of line, the estimate of material required for construction work is to be prepared. The detail of material is prepared on the basis of the detailed survey. Then the entire material is procured. Site offices are built keeping in mind the availability of water and transportation.

2. Transportation. In case the route of the line is along the road, the poles should be unloaded at different points along the line route (not at one place). In case the line passes through the field, the poles may be transported by a trailer to the nearest road point and then distributed through handcarts or bullock carts. PCC and RCC poles need special care in loading and unloading otherwise they may get damaged.

The conductor drums should also be unloaded at such distances that the conductor length in a drum is equal to the length of the line. Other materials such as insulators, cross arms, clamps, channels, earthing pipes, barbed wires, MS flats, nuts and bolts etc are stored at the site store.

3. Staff. The number of field staff such as line superintendents/line inspectors/JEs, foremen, draftsmen, surveyors, linemen and unskilled labour should be arranged matching the construction schedule. Efforts should also be made to provide adequate facilities of housing, foods, medical treatment near the camp.

4. Tools and Plants. Adequate tools and plants (T&P), which are most essentials for carrying out the job with safety and efficiency, should be procured well in advance.

5. Marking of Trees For Falling. A good method of marking of trees is to put a ring around the tree to be fallen at the level where falling is required. If lopping (branches of trees) only is required, a cross should be put on the site to be lopped.

15.4. POINTS TO BE CONSIDERED AT THE TIME OF ERECTION OF OVERHEAD LINES

1. Continuous ground wire earthed at the substation should be provided. Every 4th span of the pole and first and last pole should be properly earthed and wire should be joined properly to the continuous earth wire.
2. The clearance of conductor from the ground and adjoining should be according to IER 77, 78, 79 and 80.
3. All metallic parts, line components (except the main line conductor) should be earthed properly according to IER 90.
4. Permissible spacing between the conductors should be maintained throughout the lines.
5. Anti-climbing device and danger plate (or caution notice) should be fixed on each pole.
6. One-sixth of the pole length should be properly buried and rammed in the soil.
7. The joints of the conductor should be mechanically strong and in proper sleeves.
8. The conductor should be tightened on the insulator properly with binding wires.
9. Guard wires should be provided where essential.
10. Pole steps should be provided in the towers for the lineman to climb the pole.
11. Proper jumper should be used wherever required.

The points to be kept in view in the design of an overhead line are :

1. The line should be able to meet the desired load demands efficiently.
2. The line should be able to withstand adverse atmospheric conditions.
3. The voltage drop along the line should be within prescribed limits (IER 54).
4. The cost of the overhead line should be tolerable.

15.5. ERECTION OF SUPPORTS

Once the collection and transportation of material work is completed, the next step is to commence the pole foundation excavation work and then erect the poles.

Excavations for single poles in ordinary earth present no serious problem. Excavation is generally done by pickaxes, crowbars, shovels etc. In case of rocky areas, blasting is done with gunpowder to make the pole foundation. The holes are excavated in the direction of line, as this, in addition to giving greater lateral stability, facilitates pole erection.

In excavating for "H" poles a single hole large enough to accommodate the poles and bottom brace blocks is used, or, if foundation baulks have to be fitted near the surface, separate holes are excavated.

As already mentioned, the depth of the pole to be buried in the ground is usually taken as one-sixth of the pole length.

Wooden Poles. The portion of the wooden pole *i.e.*, 2 metres from the bottom which is to be buried in the ground should be painted with bitumen to avoid deterioration. After the erection of the pole, the pit should be partially filled with small pieces of bricks, stones and rammed well with a crowbar. After that earth should be filled-in and again ramming is done.

Steel Tubular Poles, Braced Channel Poles and Rails. A suitable pad of cement concrete, stone, or steel is provided at the bottom of the pit. This is to be done before erection of pole. A cement-concrete muff extending 0.2 m above the ground with a tapered top to shed rain water is provided so that corrosion is avoided and better grouting is achieved.

Prestressed and RCC Poles. RCC poles are usually of larger cross sections than PCC poles and, therefore, there is no need of providing base plates or muffs for such poles. However, in case of PCC poles, a base plate (40 cm × 40 cm × 7 cm) concrete block is provided.

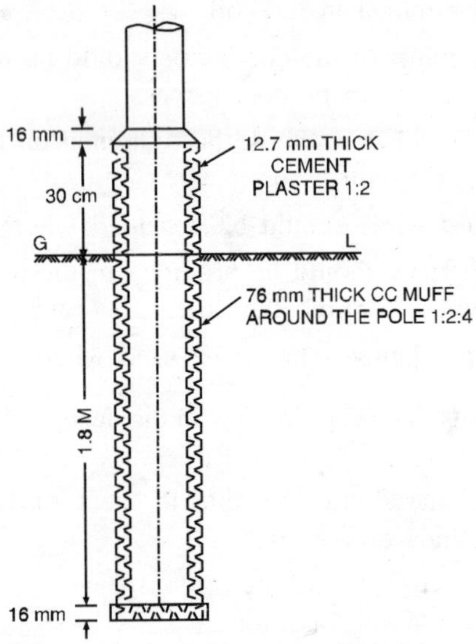

Fig. 15.2. *Steel Pole With Muff*

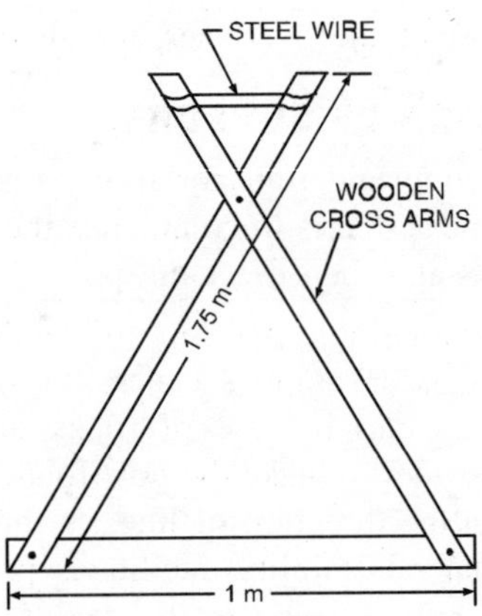

Fig. 15.3. *Dead-Man Structure*

Erection of Supports. For carrying out the pole erection work with maximum of safety and as expeditiously as possible there should be a special erection squad consisting of 1 lineman, 1 assistant lineman and 8 T-mates and fully equipped with necessary T&P [8 ladders or pole pikes (2 – 2.5 m; 2 – 3.5 m; 2 – 5 m; 2 – 6.5 m) fitted with studs to prevent slipping; 6 – 25 mm diameter ropes 12 m long; 6 – 10 cm × 7.5 cm angle stubs (pointed); 1 dead-man structure*, 2 – 5 kg hammers; 1 – 1 kg hammer; 1 – 25 mm × 22 mm spanner; 1 – 22 mm × 14 mm spanner; 1 – 22 mm and 1 – 14 mm box spanner; 1 hand saw; 6 round shovels; 2 earth rammers; 1 bucket and short rope; 1 cant hook for twisting poles; 1 pair pliers; 1 plumb bob; 2 skid boards.]

Generally two methods, called the *derrick-pole method* and *dead-man method,* are used for raising and setting of poles. Derrick-pole method is usually used for erecting heavier poles whereas dead-man method is employed for erection of lighter poles

1. Dead-Man Method. The pole is laid down along the line route. A wooden board is placed vertically at the back of the foundation pit and the pole is moved forward until it rests against the board. The pole is then raised manually and is supported on the dead-man. Guy ropes of diameter 25 mm are attached to the upper part** of the pole in three different directions and a winch rope of 32 mm diameter is attached in the direction as shown in Fig. 15.4. The pole is raised and the dead-man structure is moved forward until the 2.5 m ladder or pole pike can take its place. The 2.5 m ladder is employed to take the first lift and the dead-man is moved forward. Then the 2.5 m ladder is moved forward and the 3.5 m ladder or pike is introduced. The dead-man now will be dispensed with and the side guys are tightened to prevent pole from swinging. The 5 metre ladder is introduced and lifts are taken alternately with each ladder until the pole reaches an angle of approximately 70° with the ground. The ladder is then removed and the side guys are tightened and back guy is loosened till the pole is pushed up to the vertical position. The pole is then carefully plumbed by the guy ropes and the butt lined with the poles already erected and the next to be erected. If the butt requires moving over into the line, the simplest method is to insert a short wooden baulk between the pole and the side of the foundation. The pole is allowed to go off the plumb in a direction away from the baulk and then pulled over by the guy rope so that the butt slides into the required position. The pole is then plumbed and tied off with the guys.

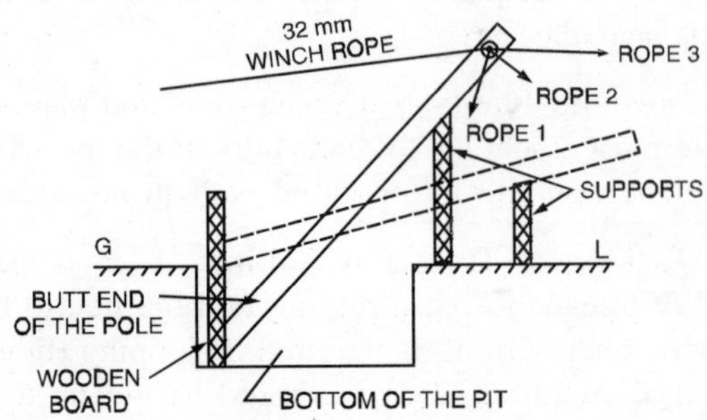

Fig. 15.4. *Dead-Man Method of Erection of Pole*

About 0.5 m of earth is put in round the butt and rammed and the pole is then checked for twist *i.e.*, to see that the cross arms are at right angles to the line, or, in the case of angle poles, bisecting the angle. If necessary, it is twisted by

* The dead-man structure is a temporary x-shaped structure made of two wooden cross arms.

** Five-eights of the length of pole from the butt end.

a cant hook and replumbed. Two of the guys and two of the stubs are then removed.

Half of the squad moves forward to the next pole with the necessary gear and prepares for erection. The rest of the squad complete the back filling and ramming and then remove the two guy ropes and stakes and move forward.

Erection of Concrete Poles. The erection of concrete poles is not as simple as wooden, tubular or rail poles. It is due to greater weight involved and liability to crack or fracture. Concrete poles must not be mishandled in any way during erection, and a wrongly fitted guy rope or winch rope can do a lot of damage. The remarkable point is that the pole can be subjected to much greater stresses in erection than are ever likely to be experienced in service. An important point is that concrete poles are much stronger on the across-line axis than they are on the in-line axis, and advantage should be taken of this fact during handling and erection. There are various methods used for concrete pole erection but derrick-pole method has been very successful. A skid board is placed vertically at the back of the foundation pit and a piece of wood is placed in the bottom of the pit to facilitate the turning of the pole. The pole is pushed against the skid board and should be laying with its stronger axis in the line of route. It is then raised manually and supported on the wooden blocks. A derrick pole* supported by rope guys is erected vertically so that its leg is near the bottom of the pole. The pole is lifted with the help of a winch, as illustrated in Fig. 15.5. When the pole comes in the vertical position, it is plumbed and adjusted if necessary by means of various ropes so that the pole comes in complete alignment and is in the plumb. The pit is filled with earth, layer by layer, taking care to ram the earth one layer at a time.

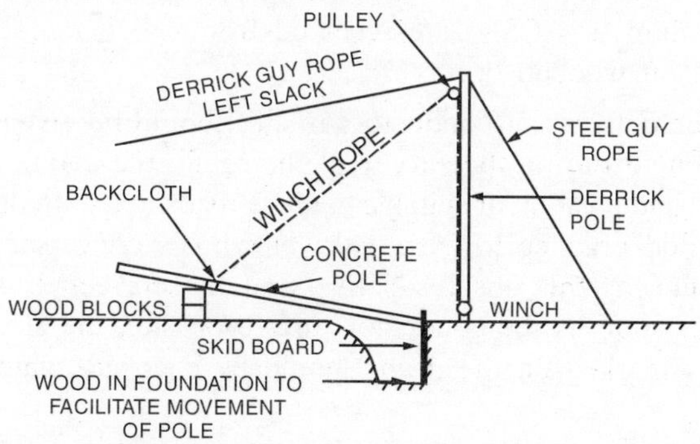

Fig. 15.5. ***Derrick-Pole Method of Erection of Pole***

H-Pole Structures. Normally, single poles are used in the straight runs of an overhead power line. H poles are employed in the line only at places where the line takes a turn at an angle, or where the line terminates, or where a branch line is teed off. Even in overhead lines running in a straight line, one H pole is provided after every one km. This makes the line stable.

Erection procedure for H-pole structure is similar to that of single pole except that in this case two pits are dug in the ground. The pits so dug should be along the bisection of the angle of deviation in the case of H poles used at a turn. A cross-bracing and top channel are fixed, as shown in Fig. 15.6 (*a*) so that the structure remains in position.

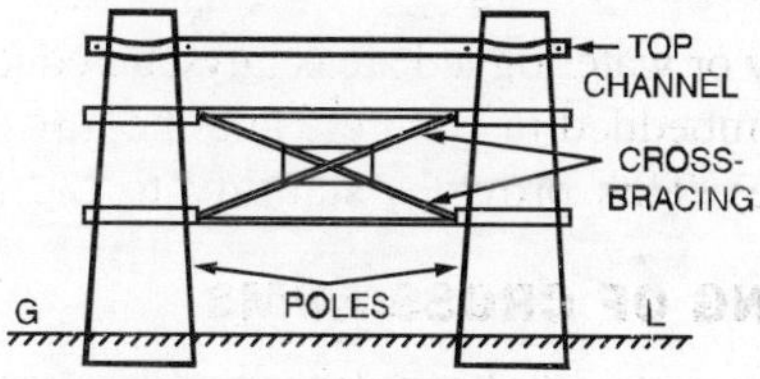

(a) H-Pole Structure

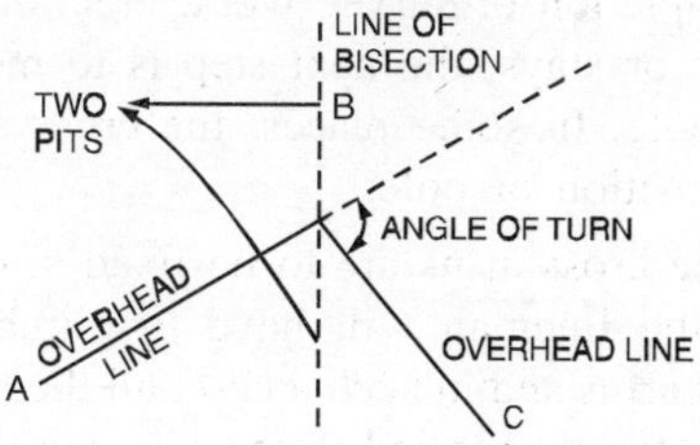

(b) Pits for H Poles of an Overhead Line Taking a Turn

Fig. 15.6.

15.6. SETTING OF STAYS

Guys and stays have already been described in chapter 2. (refer to Art. 2.7).

The marking of stay holes for excavation, the excavation of the stay holes and the setting of the stays must be carefully executed. Before marking of hole for excavation, the inclination of the stay** relative to the ground is roughly determined. This enables the position of stay hole to be fixed so that when the stay is set the stay rod will have the correct inclination and will come out of the ground at the correct distance from the pole.

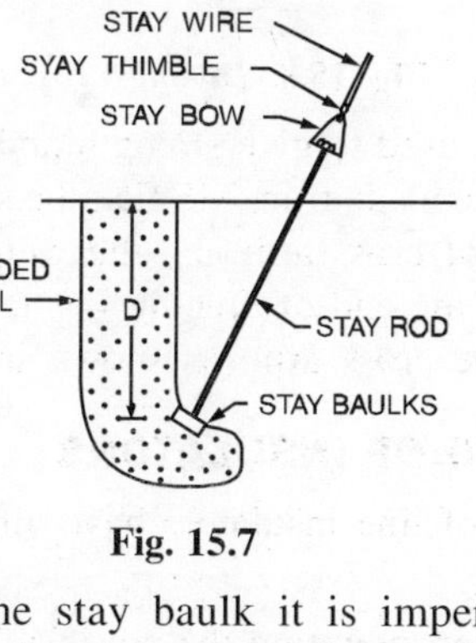

Fig. 15.7

When setting the stay baulk it is imperative that the bank is undercut, so that the stay baulk is pulling against the virgin ground. The stay rod and the baulk should be placed in position and the slot to accommodate the rod adjusted, so that the rod is in the line of sight with the stay wire attachment on the pole. Back filling and ramming must be well done. In setting stays the whole screwed portion of the rod plus 13 cm should protrude through the ground. When the stays

* The derrick pole (a steel tubular pole), being lighter is erected first. Then the concrete pole is erected.

** The angle of guy may be either 30° or 45° and where it is not possible to have either of these two angles, any other suitable angle within these two limits (30° – 45°) be used.

are finally tightened the pole should lean slightly towards the stays (about half the width of the pole at the top) and when the load comes on, the pole will pull plumb due to the straightening out of the conductors. A stay rod and baulk in position are shown in Fig. 15.7.

In marshy or waterlogged areas guys are concreted (*i.e.*, the guy plate is embedded in concrete and the guy rod is covered with concrete). This provides stability to the line.

15.7. FIXING OF CROSS ARMS

Cross arms have already been described in chapter 2 (refer to Art. 2.5).

After completion of survey work, erection of poles and fixing of guys or stays, the next step is to mount the cross arms on the pole. In some places, the cross arms are fixed prior to the erection of pole.

In case the cross arms are to be fixed after the erection of the pole, the lineman will have to climb the pole for fixing of cross arms and in such a case, all the tools necessary for the job are to be supplied to the lineman through a hand line (13 mm rope running around a pulley). The ground man helps the lineman in lowering and raising the material.

The cross arm is fixed to the pole by clamps and bolts and nuts as shown in Fig. 2.14. Half clamp is used for fixing the cross arm to the intermediate poles in the straight route while at terminal and angle poles, a double cross arm fixed with long bolts and nuts to the pole is used (Fig. 15.8).

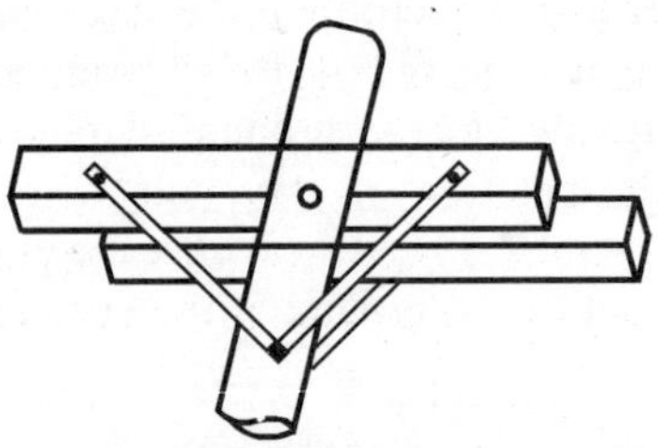

Fig. 15.8. ***Double Cross arm***

Braces are used to give strength and rigidity to the cross arms. Braces are made from MS flat, the size of which depends upon the size of cross arm and the weight of the conductor to be carried. One end of bracing is attached to the pole and the other to the cross arm, as shown in Fig. 2.6 (*a*).

15.8. FIXING OF INSULATORS

Various types of line insulators have already been discussed in chapter 3.

The pins for mounting the pin insulators are fixed in the holes provided on the cross arm and on the bracket at the top of the pole. The insulators are then mounted in their places and tightened. Figures 2.6 (*a*) and 2.6 (*b*) show pin type insulators fitted on 11 kV wooden poles.

On *lt* poles, insulators are attached directly to the poles with the help of D-clamps. Figures 3.12 (*a*) and 3.12 (*b*) show shackle type insulators fitted on the pole.

In the case of disc insulators, used on poles at terminals and at angles the straps used for fixing the insulators to the pole should be so connected that they can move freely in horizontal direction. The arrangement is shown in Fig. 15.9.

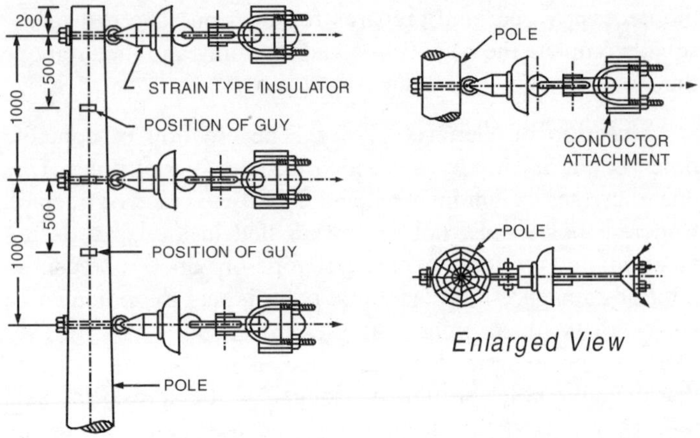

Fig. 15.9. ***Fixing of Insulators in Vertical Configuration For HV Lines (With a Bend of 30° to 60° Angle)***

15.9. CONDUCTOR ERECTION

The erection of conductors is a very important phase in construction. The appearance of the line and the future security depend to a great extent on the proper erection of the conductors. This phase of work should never be undertaken without the proper tool and plants (T&P), otherwise the safety of the erection personnel may be endangered.

The erection of conductors can be subdivided into four separate activities *viz* (*i*) delivery of conductors (*ii*) running out and stringing of conductors (*iii*) tensioning and sagging of conductors and (*iv*) jointing of conductors in bows.

1. Delivery of Conductors. The line conductors (All Aluminium or ACSR) are received from the manufacturers wound on wooden drums.

The delivery of conductors to site should be given some forethought during the survey of the route; the nature of the ground and position of suitable haulage roads noted. In delivery it is good practice to have a driver who is conversant with the route, as this ensures delivery of the correct drums on site. As soon as the conductors arrive on site, the sizes and lengths marked on the drum should be checked against the order.

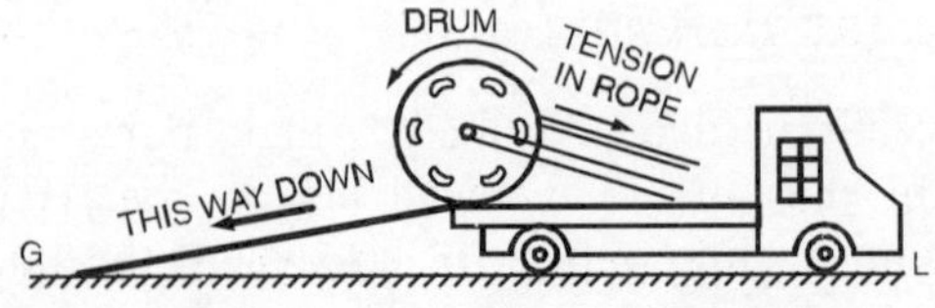

Fig. 15.10. ***Unloading of Conductor Drum From Truck***

It is important to handle the line conductor drum in such a manner as to prevent nicking, cutting, smashing or otherwise damaging the conductor especially when loading, transporting and unloading from trucks. Unloading of conductor drum from a truck is shown in Fig. 15.10. The truck is secured by check blocks and the drum is allowed to roll down on two strong boards, the movement being controlled by means of a rope as shown in the figure. The conductor drums should never be dropped. They may be rolled down as illustrated in figure. Even when lying in the store, the conductor drums should be kept off the ground and protected from injury.

(*ii*) Running Out (or Paying Out) and Stringing of Conductors. After removing the covering, often called the 'lagging', from the drum, it should be checked if any nails or sharp metallic parts are on the inside of the reel heads so as to prevent damage to the conductor during running out or unreeling operation.

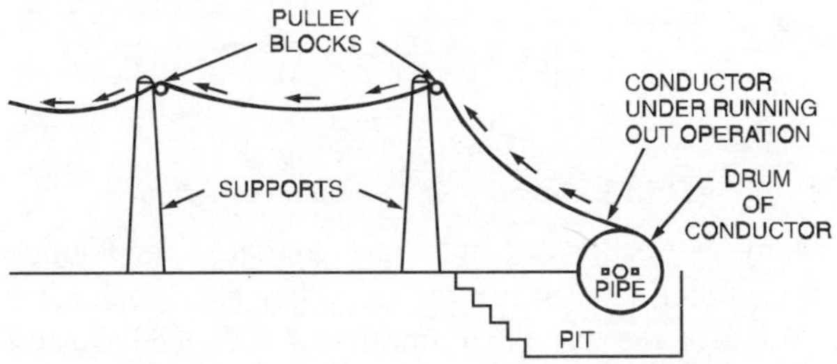

Fig. 15.11. ***Unreeling or Running Out of Conductor From Drum and Stringing of Conductor***

Various methods are adopted to support the conductor drums during running-out operation. Some of these are heavy jacks, special durm carriages, and ground excavation method; the latter is by far the most popular. In this method, holes are excavated in the ground, the length of the hole being 23 cm longer than the diameter of the drum, the breadth 8 cm greater than the drum diameter, and the depth 15 cm greater than the drum radius. Two 1.25 m planks, approximately 10 cm thick and 20 cm wide are placed, one on each side of the excavation, to support the drum spindles. The drum, with spindle in position, is lowered to the planks and two pieces of wood 4 cm thick are nailed on each side of the spindle so as to prevent it from travelling along the plank. A wooden plank is used for braking the drums during running out operations. A good practice is to run the conductors from the top of the drums, as it prevents any damage due to chafing, and this is very important when running out aluminium or ACSR conductors.

The reel is rotated for running out the conductor, so as to prevent kinks and other damages which would otherwise be caused if the drum was kept stationary and the conductor was removed from it turn by turn. As the conductor rolls out, it is passed through gloved hands and examined for defects and damage by feel. If and when a defect is noticed, rolling is discontinued and the faulty section is either removed or repaired. It should be ensured that the conductor does not rub against the ground while being pulled.

Stringing of conductors means their mounting on poles. When the conductor is running out of drum, it is simultaneously raised to the level of the cross arm of the pole and then passed through the grooves of aluminium rolling blocks (Fig. 15.12) installed one for each phase, on each pole.

Derricks are employed where roads, rivers, communication circuits, canals, railways and power circuits have to be crossed during stringing operations. This prevents, as far as practicable, interference with normal services. Ladders can be suitably rigged up, complete with stays. Care should be taken to ensure that the running blocks are properly fixed to prevent any of the conductors coming adrift.

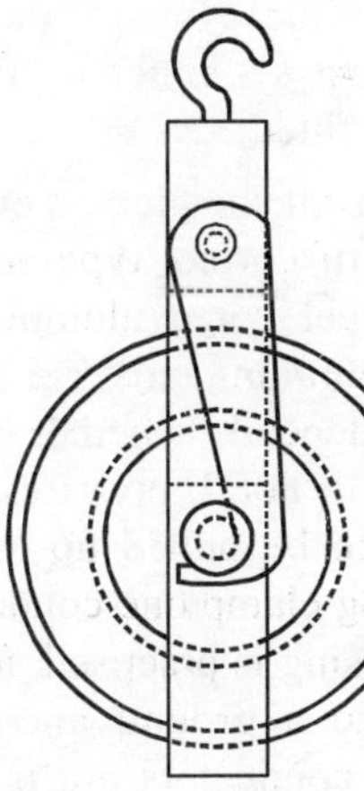

Fig. 15.12. ***Aluminium Roller Block Employed For Stringing of Conductor***

At important crossings of roads, canals, navigable rivers, railways etc, flagmen should be in attendance to ensure that normal services are not unduly interrupted. These crossings should only be carried out in conjunction with, and with the approval of, the proper authorities concerned.

In the case of tower lines, where the aerial earth wire is usually located at the top of the tower, the running block should be so placed that the conductor will not rub against the tower steelwork. Precautions must be taken at all barbed-wire fences, walls, etc to prevent damage during running-out operations.

During stringing operations there must be an efficient code of signals for the stringing party and these are usually given by whistles. A horse or tractor is generally employed for pulling out the conductors. The former has advantages, particularly on arable land. The conductors should be pulled out evenly in preference to pulling out a complete section of one wire, and so on.

(*iii*) Tensioning and Sagging of Conductors. On the completion of stringing, tensioning operations commence. The tension insulator sets, complete with clamped conductors, are erected at the end remote from the tensioning points. At the tensioning end the running blocks are fitted on the cross arms. Underneath the slings, fixing the blocks to the cross arm, sacking is wrapped to prevent the slings being cut. During the preparations, all the midspan tension joints are made, and care should be taken to keep all joints at least 3 m from a support. The tensioning gear depends to a greater extent upon the sizes of conductors being handled. Tensioning is often done with block and tackle but more often with a winch, which, in many instances, is incorporated in the lorry fittings and operated from a separate gearbox. Sufficient tension is applied to the conductors to clear fields, etc.; and allow the phasing to be checked. All derricks, etc, are removed and running blocks adjusted at angle points. It is important that this should be done at positions where the formation changes and at tee-off or other special points. The foreman/junior engineer should be provided with a diagram showing the proposed arrangements. It is necessary that the conductors are straightened out before any tensioning takes place, otherwise kinked conductors are unavoidable. Where necessary, additional stays, either on the structures or the cross arms, should be fitted.

It is important that all the clamps employed for pulling up the conductors are of correct type and dimensions. This applies specially to steel cored aluminium clamps, which must grip both the aluminium and steel properly. The actual tensioning procedure depends whether pre-stressing of the conductors is practised or not. If pre-stressing is not practised then the conductors can be pulled up to sag, By means of winch and a come-along clamp one conductor is pulled up to sag, or, where pre-stressing is practised, to the predetermined tension and secured to a ground anchor. The procedure continues until all the conductors are pulled up.

A span should be chosen for sagging approximately in the middle of the section. This should be as near the length of the basic span as practicable and, where suitable, supports of equal height should be chosen. The sag is obtained from the sag chart for the corresponding temperature and span length, and the figure so obtained is the sag to which the conductors are erected. In the case of pin insulator lines the distance corresponding to the sag is measured from the insulator groove down the support and marked; in the case of suspension insulator lines, the distance measured down the support must be the sag value plus the length of the insulator assembly. In the case of tower lines the sagging level is usually marked on the tower legs, but in wooden and concrete structures wooden sighting battens are erected.

There are two methods of checking of sag. One is to sight over the batten on one support to the batten on the next, as illustrated in Fig. 15.13, and judge when the bottom of the conductor sag is level with the line of sight. The second method, generally used for long span work, as in the case of broad-based lattice tower lines, is the use of a dumpy level. In this method, the level is clamped to the tower leg so that the collimation line is at the sag mark on the tower leg. The level is adjusted until the cross-hairs are focused on the sag mark on the next tower and the instrument can be rotated in this plane. The conductors are adjusted until the bottom of the sag is in the line of sight. With double-circuit lines this greatly simplifies checking.

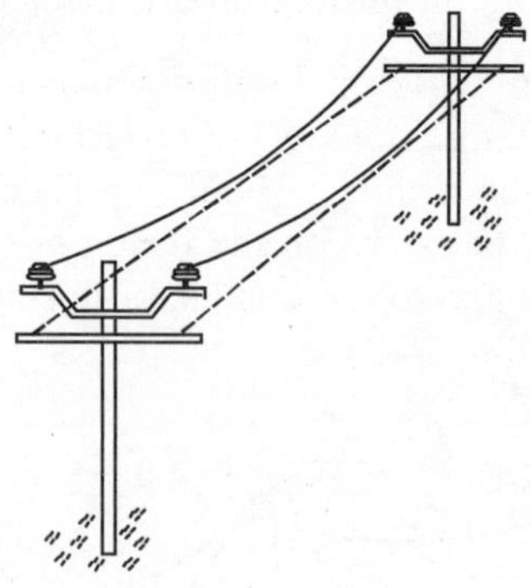

Fig. 15.13. *Sagging of Conductors by Battens*

Many engineers sag only one conductor and judge the rest by paralleling. But in practice it has been found that the time to check the sag is so small that it is well worthwhile to check all the sags. Excellent results are often obtained by paralleling even, but it is difficult to locate a position to view the conductors in anything like a favourable plane.

After completion of work of sagging of conductors the tension insulator clamps are to be fitted, and the method used for it depends to a great extent on the sizes of the conductors under erection and the type of clamps being employed. In case of high-tension conductors it is preferable to lower the conductors to the ground and fit the clamps. A mark is put on the conductor at a distance from the cross arm equal to the length of the complete insulator assembly. The clamp is fitted either on the conductor under tension or on the ground on the slacked-out conductor. It is important that all the clamps should be fitted strictly in accordance with the manufacturer's instructions. Before the insulator sets are raised to position all the nuts should be free, all phosphor bronze springs in correct position and all split pins opened. If the clamps are fitted under tension, a pull lift or similar device is used in the operation of connecting the clamp to the insulator set. A come-along clamp is placed on the conductor beyond the conductor clamp and to this is attached the pull lift. This is ancohored on the support and the conductors are pulled-in sufficiently to allow the insulator assembly to be fitted to the clamp. If the clamps have been fitted on the

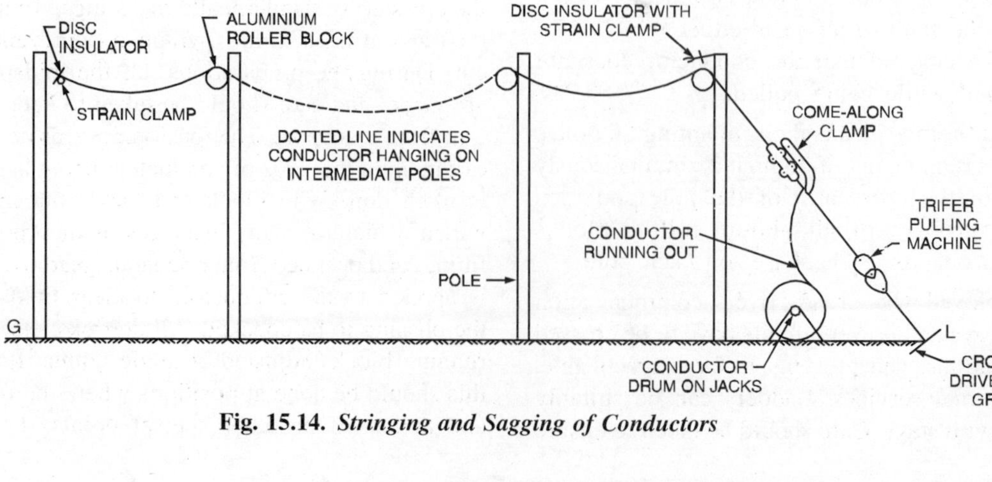

Fig. 15.14. *Stringing and Sagging of Conductors*

ground then the tensioning gear will be used to pull up the conductors again, and can be pulled-in to allow the insulator set to be attached to the clamps. There is one point that needs special mention, and this deals with long insulator sets. When these are being raised to the horizontal they should be supported at a central position in addition to the end support, otherwise the pins get bent and the insulators may get damaged. The clamping of the conductors is then carried out at intermediate supports on suspension insulator lines. On pin insulator lines the conductors are bound on the insulators.

***(iv)* Jointing of Conductors in Bows.** When tensioning operations are completed, the non-tension joints in the bows at section points are made. It is important that sufficient clearance is given between the conductor jumper and the ironwork. It may be necessary to weight the jumpers at angle towers on *ht* lines to reduce swinging. On stayed supports the jumpers may need to be carried on pilot insulators.

15.10. REPAIRING AND JOINTING OF CONDUCTORS

15.10.1. Repairing of Conductors

Repair sleeves are employed for repairing of ACSR or all aluminium conductors which have sustained surface damage causing a few of aluminium strands to be damaged or broken. When the damage is more severe, it is not advisable to use repair sleeves because it may be necessary to cut the conductor completely and make a joint with the help of jointing sleeve.

The most popular type of repair sleeve is the *swage type* which is designed for use with small size conductors. Such a sleeve is made of aluminium and applied on the damaged part of the conductor with a swage-and-die made of steel.

For large size conductors, bigger sleeves are employed and a hydraulic compressor (instead of swage-and-die) is used to the press the sleeve on the damaged conductor, working first towards one end and then towards the other end.

A repair sleeve is shown in Fig. 15.15.

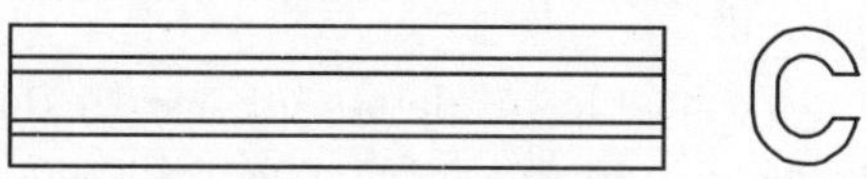

Fig. 15.15. ***Repair Sleeve For Repair of ACSR Conductors***

A jointing sleeve consists of an oval, thin walled aluminium sleeve large enough to slip over the overlapped ends of conductors to be jointed. Jointing sleeves are employed for AA and ACSR conductors of area of cross section of up to 30 mm^2.

15.10.2. Jointing of Conductors

The use of mid-span tension joints is a subject on which there are different opinions. Some engineers do not use mid-span joints except for repairs, while others do use joints on new construction.

Too much stress is put on the mechanical properties of joints and too little on the electrical requirements. Some joint failures have occurred under fault conditions when the conductor was under normal tension. Joints which behave perfectly under ordinary conditions of current loading often show signs of distress under heavy fault currents, and for this reason new designs are subjected to electrical as well mechanical tests.

Essential Features of Good Joints. 1. Conductivity of joint should not be less than an equivalent length of the conductor.

2. Joints should be capable of carrying maximum fault current without failure or deterioration for the time required for the protective system to operate.
3. Mechanical strength should not be less than 95% of the breaking load of the conductor.
4. Life of joint should be equal to that of the conductors without deterioration either electrically or mechanically.
5. Simple in design and assembly, since joints have often to be made under difficult working conditions in poor light and bad weather.

Joints are of splice, Britannia, married, sleeve and compression types.

1. Splice Joint. This joint is generally used with single-strand conductors. The ends of the conductors to be jointed are scrapped and cleaned and placed together overlapping up to 200–300 mm. The conductors are held firmly together and then the overlapped conductors are twisted. The number of twists that are usually given is four.

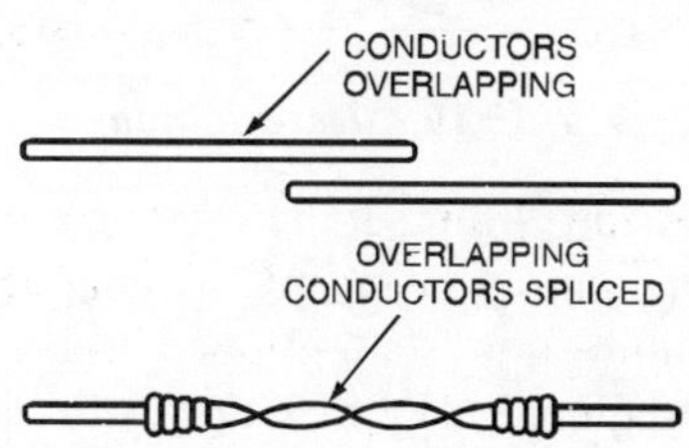

Fig. 15.16. ***Splice Joint***

2. Britannia Joint. Such a joint can be used for both solid and stranded conductors. In this case the two ends of the conductors are scrapped and cleaned and then they are made to overlap one above the other. The length of overlapping portion depends on the size of conductors and it may vary from 150 mm to 350 mm. Temporary binding of few turns are applied over the overlapping portion. The binding wire used for this purpose is soft copper wire for copper conductor and aluminium wire for aluminium conductor. The free ends of the line conductors are also bound and then soldered.

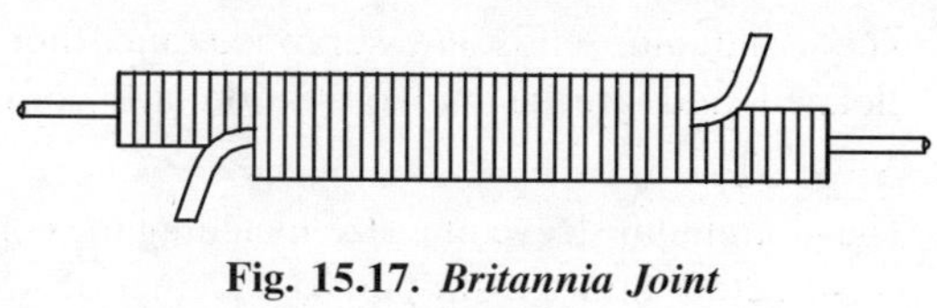

Fig. 15.17. ***Britannia Joint***

3. Married Joint. Such a joint is employed only for stranded conductors. The strands of the conductors to be jointed are first opened one by one and then straightened up, then the strands of one conductor are fitted into the strands of the other conductor. The strands are pressed generally by hand to make them lie close to the conductor. Each strand is served with few turns over the conductor one after the other. If this method is used for jointing stranded conductors then joint should also be soldered. In case of aluminium conductor there is no need of soldering.

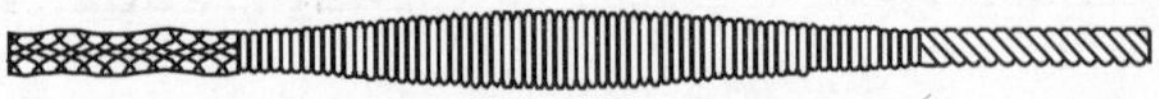

Fig. 15.18. *Married Joint*

4. Sleeve Joint. This joint is preferred for jointing all aluminium and ACSR conductors because it provides a good electrical contact and sufficient mechanical strength. If in this case the ends of the conductors are scrapped and cleaned then the inside part of an aluminium sleeve is cleaned thoroughly to make it free from dirt or grease. The sleeve is an overshaped aluminium tube available in different lengths for jointing different types of conductors. Only one sleeve is used for jointing AA conductors but two sleeves are required in case of ACSR conductors. The cleaned conductors are placed inside the aluminium sleeve from opposite ends and are drawn beyond the end of the sleeve. Two twisting wrenches with dies of desired size are fitted on the sleeve about 12 mm from its ends. The sleeve is then given two or three and half turns by twisting the wrenches in opposite directions. The conductor ends are turned up and cut close to the sleeve. When two sleeves are used then each sleeve is given four and half turns.

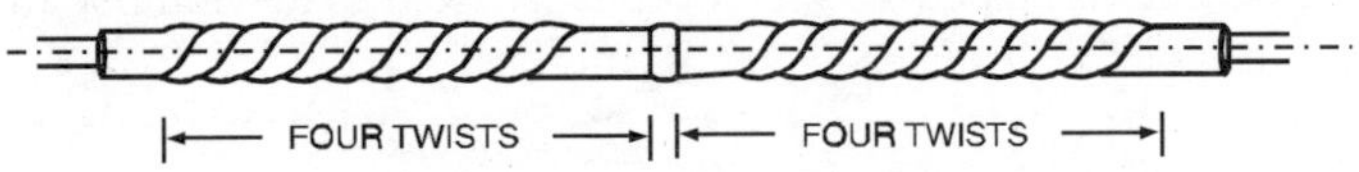

Fig. 15.19. *Sleeve Jointing*

5. Compression Joint. Joints of the compression type are excellent for jointing heavy ACSR conductors. The steel core and aluminium strands are gripped separately and filling of joints is not required. The making of this joint requires a portable compressor with special dies. Care should be taken that the joint is rotated a little before each fresh grip, otherwise

Fig. 15.20

the tendency is for the joint to be banana shaped. In such a joint 95% of the ultimate strength of the conductor is retained. The procedure for making compression joint is given below :

1. The steel and aluminium sleeves are thoroughly cleaned.
2. The conductor ends are scrapped and thoroughly cleaned and grease and oxidation are completely removed.
3. The aluminium sleeve of a size matching the conductor is selected and is slipped over one of the two conductors.
4. The length equal to half length of the steel compression sleeve plus 12 mm is marked off and the steel reinforcement is exposed by cutting aluminium strands from each conductor by a hacksaw. Take care that hacksaw does not damage the steel core. Moreover, before cutting the aluminium strands the conductor is bound with aluminium binding wire just before the cut.
5. The exposed reinforcement of the conductors is inserted into the steel sleeve in such a way that the two ends meet at the centre of the sleeve. The sleeve is then compressed by a compression tool at approximately 70 tonne pressure starting from the middle to the other end in such a way that the two sections of the die meet completely and a very thin paper cannot be inserted between them.
6. Now the binding from the conductors to be jointed is removed.
7. The aluminium sleeve, that was inserted right at the beginning on one conductor, is slipped back over the compressed steel sleeve till it covers the aluminium strands of both conductors. It should be ensured that the centre of the aluminium sleeve is coincident with the midpoint between the steel cores of the two conductors.
8. Filler paste is injected through the two filler holes provided on the aluminium sleeve until the holes are completely filled. Plugs are inserted in the filler holes and they are hammered firmly.
9. Finally the aluminium sleeve is compressed with the compression machine in the same way as was done in case of steel sleve.

15.11. DEAD-END CLAMPS

In the entire network of transmission and distribution system the conductors have to be dead-ended many times, especially at such places where the line terminates.

There are several methods of dead ending a conductor. The dead-end assembly should be mechanically strong enough to take all pull of the conductors and should be of correct size and curvature.

For *lt* lines, shackle insulators are used extensively for dead ending because in this arrangement the conductors are not subjected to undue stress.

In *ht* lines, different types of strain clamps are used for dead ending.

1. Single-Bolt Strain Clamp. The single-bolt clamps are extensively used for dead ending conductors of medium or light duty transmission lines. The salient features of this type of clamps are that they must have adequate holding power to prevent slipping of the line conductors at normal line tension. Such clamps are normally used for dead ending ACSR conductors.

2. Two-Bolt Strain Clamp. These clamps are used for dead ending line conductors of main distribution lines and light transmission lines. The gripping power of such clamps is more than that of single-bolt strain clamps.

3. Multi-Bolt Strain Clamp. These clamps are used for dead ending heavy duty distribution and transmission lines. Such clamps provide sufficient gripping power of holding heavy duty line conductors under tension.

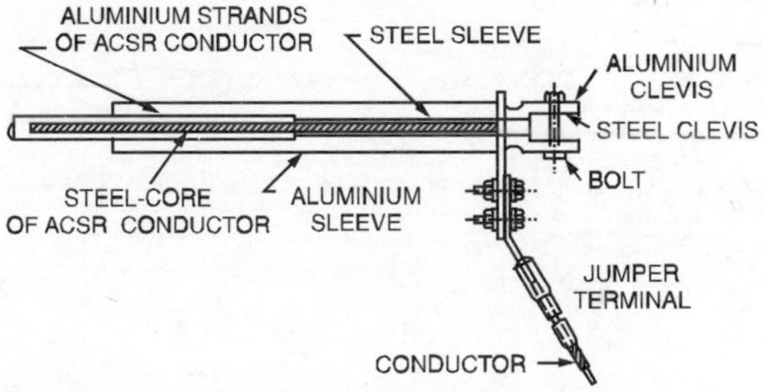

Fig. 15.21. ***Compression Type Dead-End Clamp For ACSR Conductor Line***

4. Compression Type Dead-End Clamp. Such clamps are used for dead ending ACSR conductors of transmission lines of voltages exceeding 66 kV. It has been noticed that there is fault at jumpers in case of multi-bolt dead-end assembly in case of lines operating at voltages exceeding 66 kV.

Compression type dead-end clamp for use in ACSR conductor overhead line consists of the following two parts:

1. A steel clevis made integral with a steel sleeve; and
2. an aluminium clevis made integral with an aluminium sleeve.

A single steel bolt is passed through the eyes of the steel and aluminium clevises for attachment to the insulator.

When employed in AA conductor lines, steel sleeve and steel clevis are not required. Such a clamp is shown in Fig. 15.22.

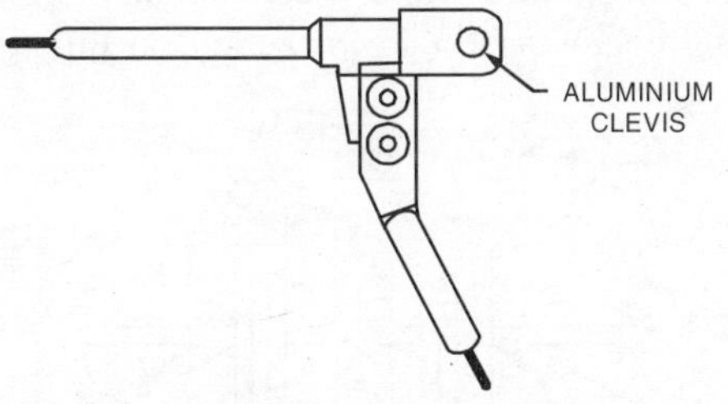

Fig. 15.22. ***Compression Type Dead-End Clamp For AA Conductor Line***

15.12. POSITIONING OF CONDUCTOR AND ATTACHMENT TO INSULATORS

15.12.1. Positioning of Conductor

The next step is to remove the conductor from the aluminium roller blocks and to place it in the groove on the top of the pin insulator in case of straight runs [Fig. 15.23 (*a*)] while at an angle pole, the conductor is to be placed in the side groove of the insulator [Fig. 15.24 (*a*)].

15.12.2. Attachment of Conductors to Insulators

There are several methods of attaching the conductors to the insulators. With copper, cadmium copper, steel-cored copper and steel conductors either a soft wire binding or a stirrup type binding is used. With ordinary earthed lines a soft copper wire binding has the advantages, particularly for solid conductors, as it can be of the slip type, so that when a conductor fracture occurs the conductor will slip through the binder and thereby relieve the cross arm from excessive pressure.

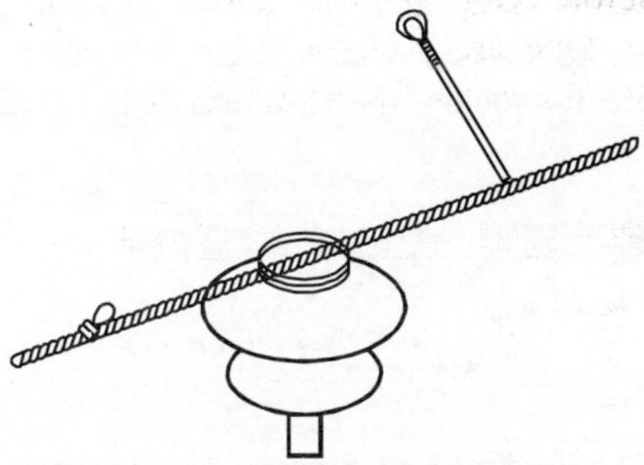

(a) Conductor in Top Groove of Pin Insulator (Straight Run)

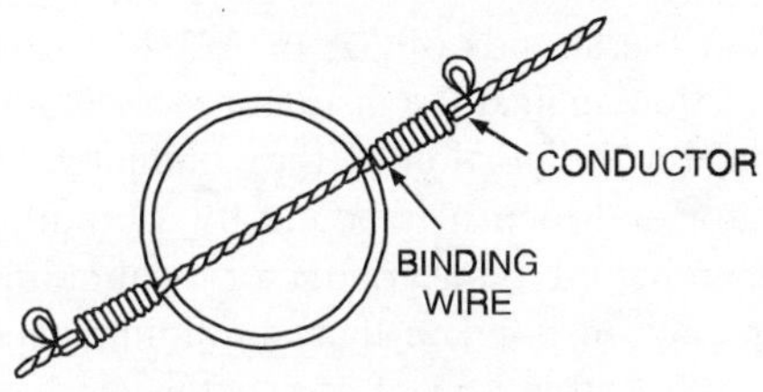

(b) Conductor and Binding Wire

Fig. 15.23

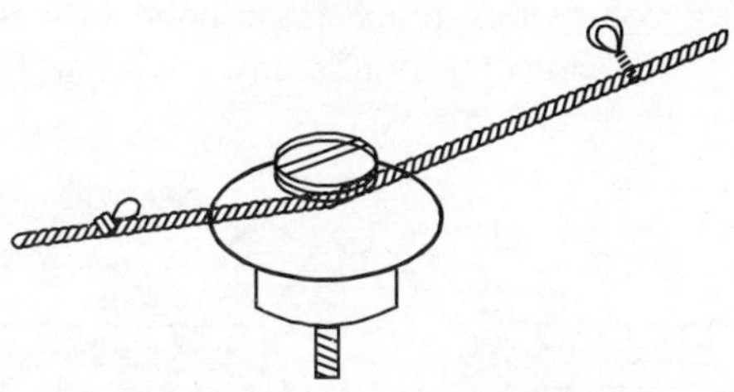

(a) Conductor in side Groove of Pin Insulator (Angle Pole)

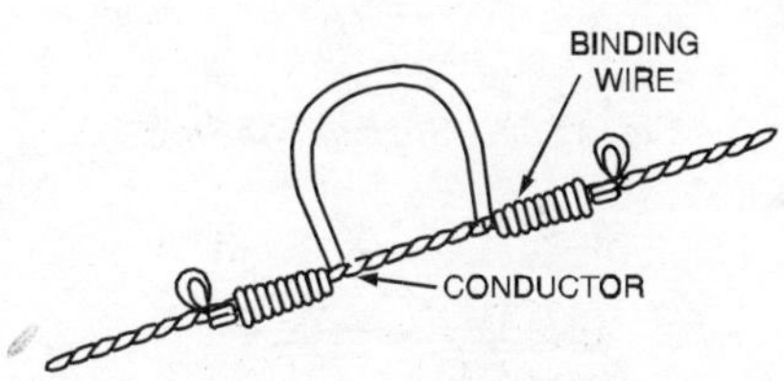

(b) Conductor and Binding Wire

Fig. 15.24

1. Stirrup Binding Method. Stirrup binders are used on all types of conductors as it has certain advantages such as, in case of breaking of a conductor this binder will hold together the conductor and prevent it from coming into contact with the cross arm. The ACSR conductors are wrapped with a soft aluminium tape before fitting the stirrups. On straight-line positions, the conductor is bound in the top groove of the insulator; at angle positions binding is done in the side groove so that the lateral pull is taken by the insulator and not the binder. Soft copper binding wire is used for binding conductors on *lt* lines. At intermediate and angle positions the conductor is bound on the side of the insulator nearest the pole, so that if it gets away from the pole it will remain near the insulator and not fall on the ground.

The conductors are attached to the insulators by means of an 'aluminium' binding wire in case of AA or ACSR conductor and with copper binding wire in case of copper conductor. Before tying the conductor to insulators two layers of aluminium tape are wrapped over the conductor in the portion where it touches the insulator (Fig. 15.25).

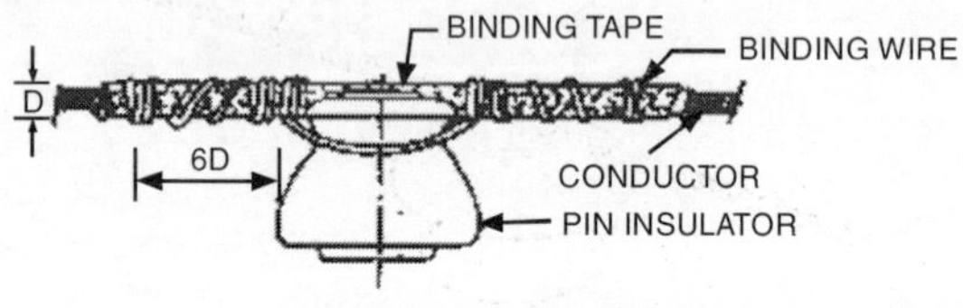

Fig. 15.25. ***Binding Tape Wrapped on Conductor Before Tying To Insulator***

To prevent the strands of AA or ACSR conductors from damage by friction against the insulator groove, an aluminium tape is wound on the portion of the conductor to be placed on the insulator and then it is bound by a binding wire. The insulator is then bound to the conductor by aluminium binding wire. The precaution is taken that aluminium tape is wound in the direction of stranding of conductor, so that the stands do not get unwcund.

The top conductor groove contour should be such that the conductor completely touches the conductor throughout, This prevent load concentration at any point and the slipping of conductor from the groove.

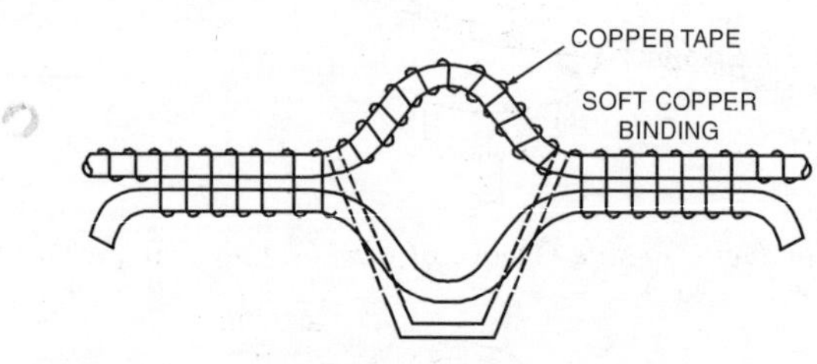

Fig. 15.26. ***Stirrup Binding***

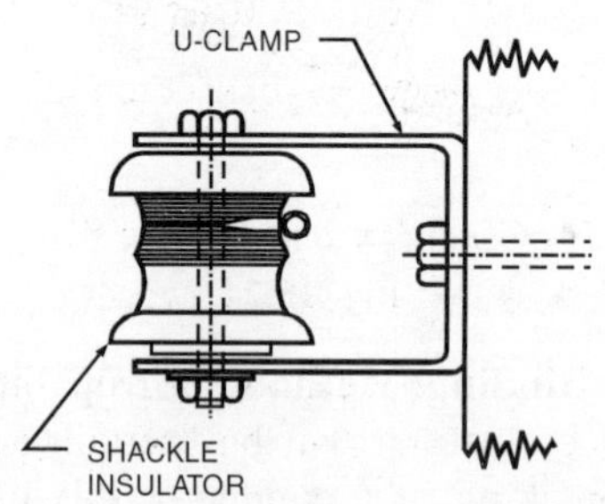

Fig. 15.27. ***Positioning and Typing Conductor With LT Insulator***

On *lt* lines the conductor is tied with the insulator between the U-clamp and the shackle insulator, as shown in Fig. 15.27. The conductor is tied around the central portion of the insulator if it is used at dead end and tied with separate wires of the same metal as conductor where distribution line changes its angle. The insulator will be under compression in that case.

2. Western Union Method. The simplest and most widely used method of binding of line conductors with pin insulators is illustrated in Fig. 15.28. This method is known as *Western union method.*

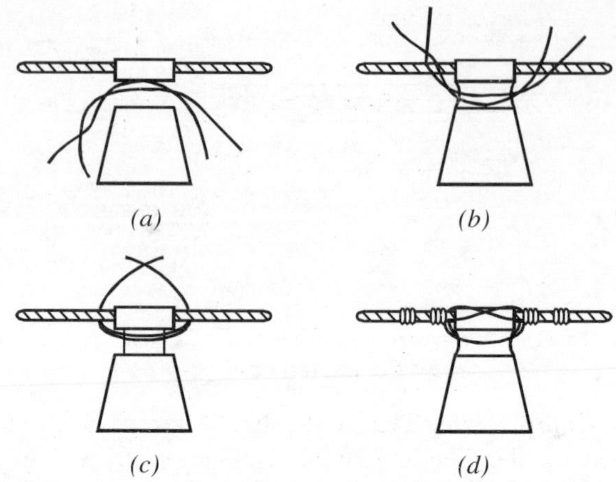

Fig. 15.28. ***Western Union Method of Binding of Line Conductors With Pin Insulators***

15.13. JUMPERS

In a straight run, one terminal pole is provided after every one kilometer so as to facilitate sagging. The short length of the conductor used to connect the line conductor on one side of the terminal pole to the line conductor on the other side of the terminal pole is known as the *jumper*. A jumper is made of the same material and has the same current carrying capacity as that of the line conductor. Jumpers are fixed to the line conductors with a suitable clamp. For *hv* lines the jumpers are arranged in such a way that under maximum deflection condition there is a minimum clearance of 0.3 m between the line jumpers and the other metallic parts. PVC sleeves or porcelain beads are provided on jumpers to avoid birdage.

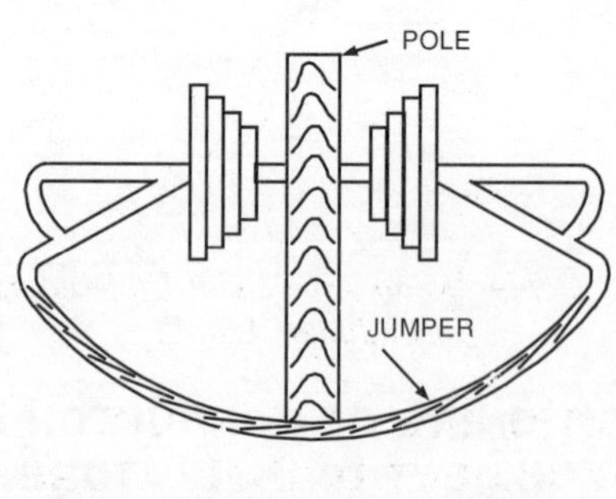

Fig. 15.29. ***Jumper Used on Terminal Pole***

15.14. TEE-OFFS

The tee-off from a line should be taken only from a pole and not in the middle of the span. While taking branch line from an existing line for supplying villages/farms etc, it is essential that the correct phasing, phase-to-phase clearance and phase-to-earth clearance are observed. Dissimilar materials should not be directly connected. Where metals such as copper and aluminium are to be joined a special connector with an interaction fibre washer should be used. Parallel groove clamps are used for connecting similar conductors at branch positions.

1. Single Pole Tee-Off Arrangement. A single cross-arm is fitted to take the two lower conductors and the third conductor is taken from an eye-bolt. A second eye-bolt may be required for taking the earth wire. The pole should be of sufficient height to enable various attachments to be fitted and still maintain the required mid-span ground clearances from the conductors and the earth wire.

2. H-Pole Tee-Off Arrangement. If the tee-off line is approximately at right angles to the main line then the tee-off cross arm is fitted on the pole away from the branch line with suitable bracing angle to tie both legs of the H-pole together. The third phase is taken from an eye-bolt. If the branch line is at an acute angle to the main line then all tee-off fittings are placed on the limb of the support next to the branch line. Figure 15.30 shows H-pole tee-off arrangement.

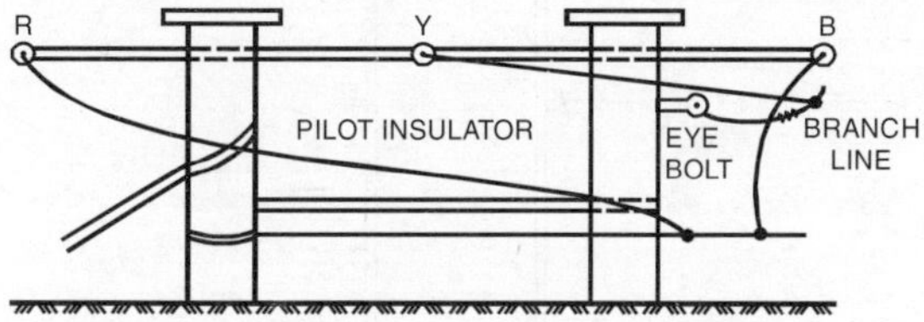

Fig. 15.30. *H-Pole Tee-Off Arrangement*

3. A-Pole Tee-Off Arrangement. The arrangement of A-pole tee-off is shown is Fig. 15.31. Where the A-pole is of normal height the cross arm is fitted on one limb only and the eye-bolt for the centre phase is made to go from between the poles. On very high A poles the cross arms can be fitted between the legs of the support. An 11 kV tee-off arrangement is shown in Fig. 15.32.

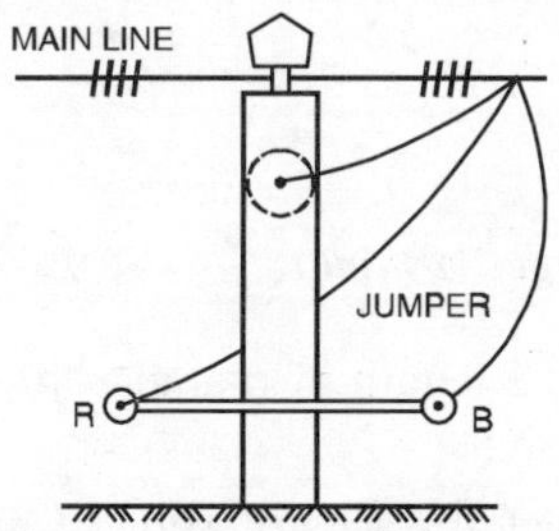

Fig. 15.31. *A-Pole Tee-Off Arrangement*

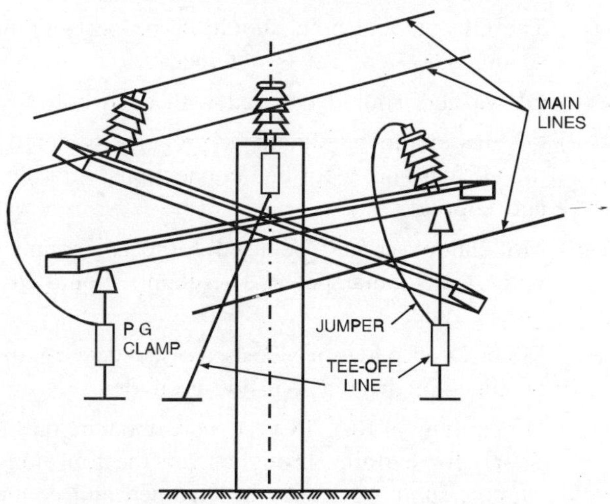

Fig. 15.32. *11 kV Line Tee-off Arrangement*

4. Parallel Groove (PG) Clamps. These clamps are used at locations in the overhead system where the jointed conductors are not under mechanical stress. These clamps should not be used for connecting the bridle and the main conductor as such joints are subjected to tension. The clamps are made of specially treated non-corrosion aluminium alloy in two pieces and provided with galvanized steel nuts, bolts and washers. The clamps retain their efficiency for a long time if they are installed with care.

PG clamps are extensively used for jumpers and tee-off connections.

15.15. RIGHT OF WAY

The land that is required for installation of an overhead line is known as right of way (ROW) which is a stretch of land over which the electrical supply authorities have absolute rights.

In most of the cases relating to low, medium or *ht* lines a continuous stretch of land is not required but only small pieces of land for installation of supports (poles and towers) are required. The table below gives a rough idea of the width of way for different categories of lines.

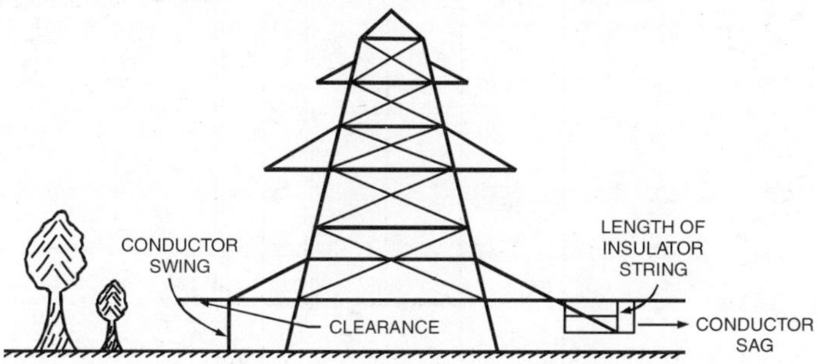

Fig. 15.33. *Clearing of the "Right of Way"*

Table 15.3. Width of Right of Way

Type of Line	Type of Circuit	Width
1-Pole Line	Single Circuit	7.5 – 15 m
	Double Circuit	9 – 22.5 m
2-Pole Line		13.5 – 30 m
Tower Line	Single Circuit	9 – 18 m
	Double Circuit	12 – 22.5 m

15.16. EARTHING OF TRANSMISSION LINE

The earthing is done (*i*) to ensure that no current carrying conductor rises to a greater potential with respect to general mass of earth than its designed insulation and (*ii*) to ensure that the potential of non-current carrying metalwork does not rise to a value such that a person may get shock on touching it. The supply system neutral point is earthed at power station or substation to serve the former purpose whereas the latter purpose is served by the regulations which specify that all non-current carrying metalwork of high voltage lines shall be permanently and efficiently connected to earth. For this

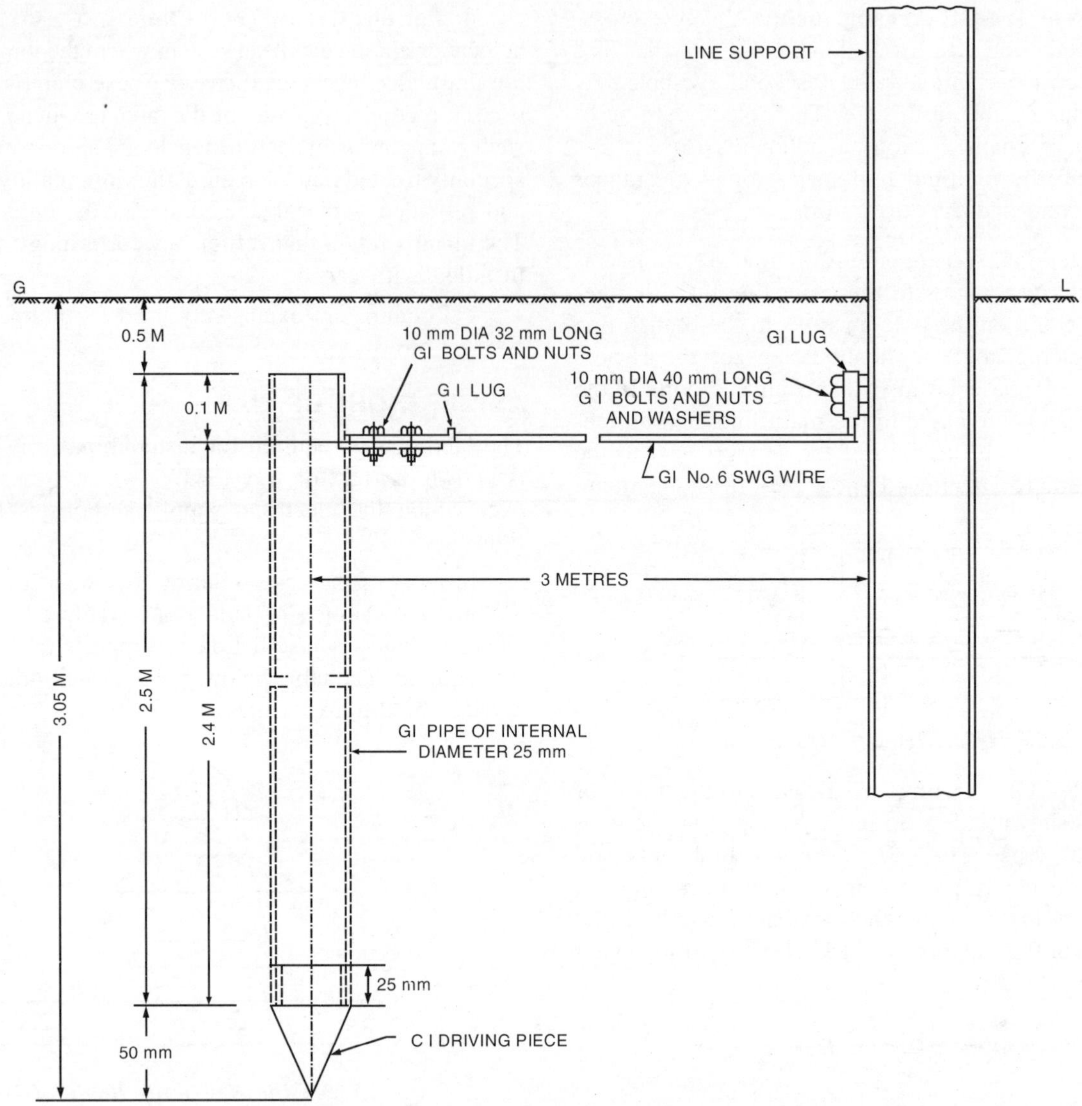

Fig. 15.34. ***Earthing of Transmission Line Supports (Pipe Earthing)***

purpose a continuous earth wire shall be run and connected to earth at four points in every 1.6 km, the spacing between the points being as nearly equal as possible, or alternatively, the metalwork shall be connected to an effective earthing device at each individual support.

The design and construction of earthing system should be such that when contact is made between line conductor and metal connected to earth, the resulting leakage current shall not be less than twice the leakage current required to operate the devices installed to make the line dead.

The arrangement of earthing of transmission line supports (pipe earthing) is shown in Fig. 15.34.

(*i*) The earthing shall be done with 2.5 metre long, 25 mm internal diameter GI pipe, by driving it vertically inside the ground for its full length with top at least 0.6 metre below the ground level.

(*ii*) In case where moist earth is not met with 2.5 metre long pipe or the soil resistivity is higher or the earth resistance after earthing of support is not less than 5 ohms, the length or diameter of pipe may suitably be increased.

(*iii*) The contact surfaces of support and pipe should be thoroughly cleaned of paint, rust etc.

(*iv*) The GI nuts and bolts should be properly tightened so that good contact is obtained.

(*v*) GI washers should be used with each bolt.

(*vi*) GI lugs employed should be properly soldered with the GS wire no. 6 for end connections with the pipe and support.

(*vii*) MS clamps should be used for earth connections with steel tubular poles. The clamp should fit tight on the pole.

(*viii*) MS or wooden lugs may be used to facilitate hammering on pipe for driving it inside ground.

(*ix*) For earthing of RCC/PCC supports the wire (GS No. 6 SWG) for earthing brought from the top along the support should properly be clamped and connected to earth electrode directly.

15.17. GUARDING OF OVERHEAD LINES

A gaurding is provided for the safety of the life, installations, and of communication circuits. The guarding for 11 kV lines is provided at road crossings, canal crossings, railway crossings, crossings over *lt* lines or communication lines. Regarding guarding of line crossings or approaching each other IER 87 provides all important guide lines, IER 88 provides that every guard wire shall be connected with earth at each point at which electrical continuity is broken. Every guard wire should have sufficient current carrying capacity to ensure the circuit rendering dead, without risk of fusing of the guard wire or wires till the contact of any line wire has been removed. There are various other rules relating to guarding where lines cross trolley wires as mentioned in subrule (5) of IER 88.

Various guarding arrangements are shown in Fig. 15.35.

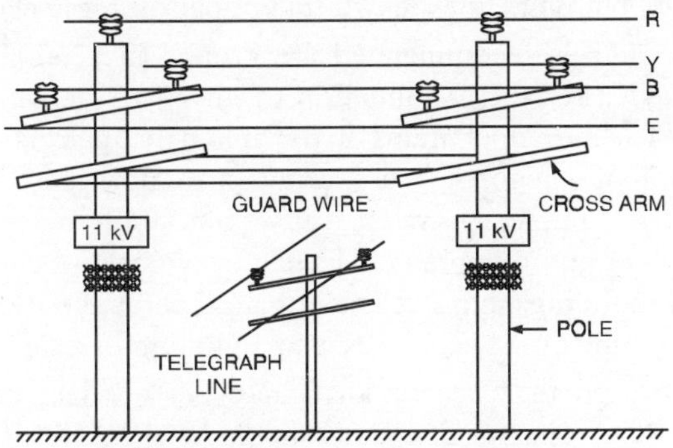

(a) Gaurding For Telegraph Line

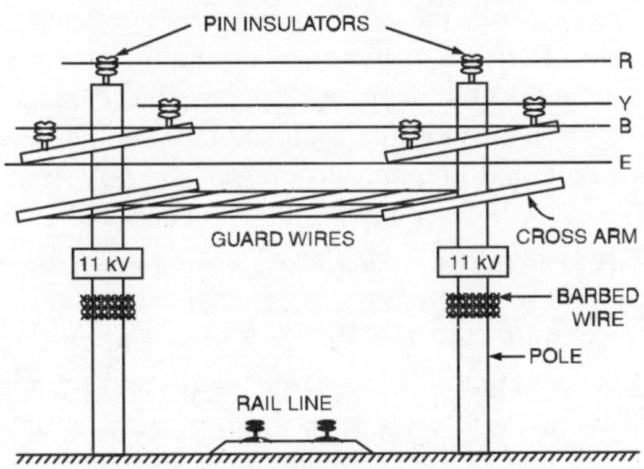

(b) Gaurding For Railway Line

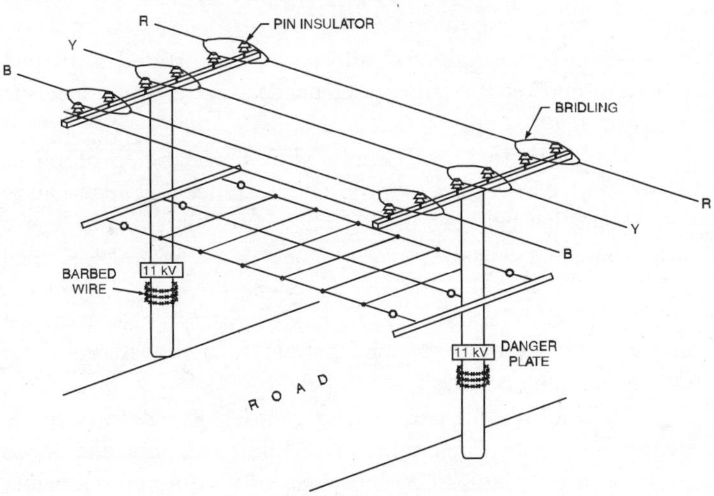

(c) Guarding at Road Crossing

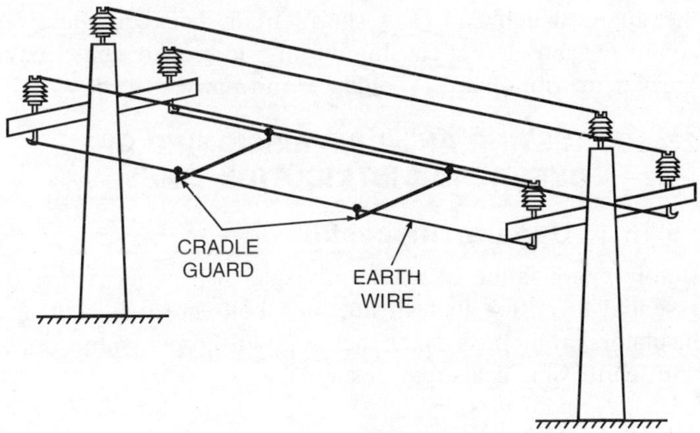

(d) Cradle Guards Provided on an 11 kV Line

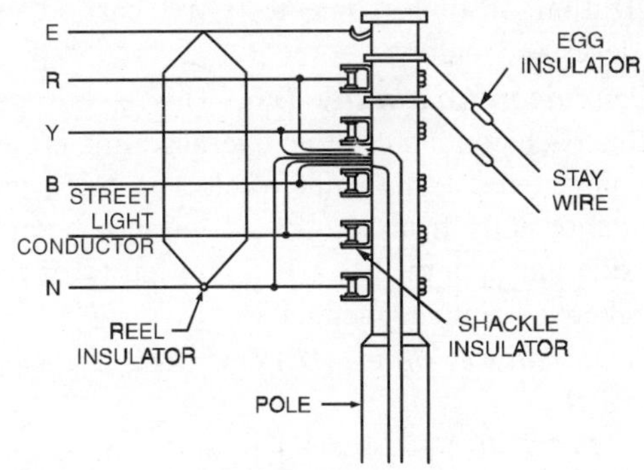

(e) Cage Guard System

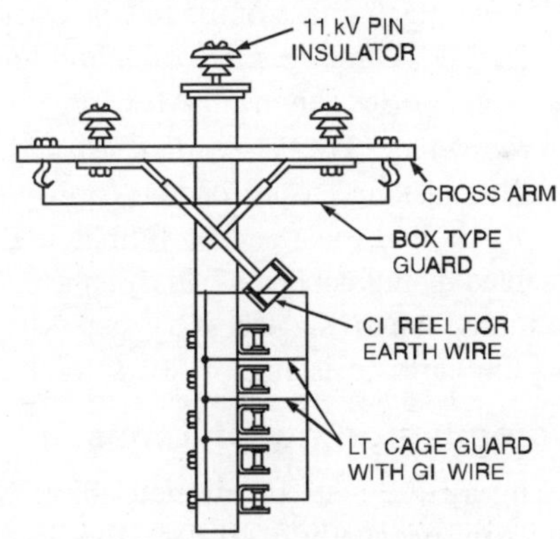

(f) Box type Guard System

Fig. 15.35

Cradle guarding is provided when the conductors are in horizontal or delta formation as in the case of 11 kV lines. The cradle guard is directly connected to earth wire. If a line conductor breaks, it will fall on the guard, thereby blowing the fuse or tripping the coil of a circuit breaker.

When an *lt* line crosses a road or a railway line it is in horizontal formation, and therefore, at such places cradle guard is provided for *lt* lines also.

Cage guarding is provided on *lt* lines with vertical formation. The guard should be made of the same material as used for earth wire. In *lt* lines guard is connected with the neutral and insulated from the earth by means of porcelain reel. In case a live conductor breaks it comes in contact with the guard which causes a short circuit between the live conductor and neutral causing the fuse to blow or the circuit breaker to trip. Guards should be uniformly spaced.

15.18. TESTING AND COMMISSIONING OF OVERHEAD DISTRIBUTION LINES

15.18.1. General Inspection

Before energization of a line, a visual inspection of line is essential to ensure that all nuts and bolts are tight and the insulators are in position at each support. Proper connections with earth should also be ensured.

15.18.2. Testing

After completion of erection work of 11 kV/lt lines but before the energization of lines some tests are carried out. Such tests are described below:

(*a*) *Conductor Continuity Test.* This test is performed to verify that each conductor of an overhead line is electrically properly connected *i.e.*, the value of electrical resistance does not differ abnormally from that of a continuous conductor of the same size and length. Electrical continuity of the line is normally checked with a megger.

(*b*) *Insulation Resistance (IR) Test.* This test is preformed with a megger.

(*i*) *LT (415/240 V) Lines.* Each phase and neutral of the line is tested for leakage with a 500 V megger. IR of each conductor should not be less than 2 MΩ.

(*ii*) *11 kV Line.* Each phase of the line is tested for leakage with a 1,000 V megger. IR of individual conductor should not be less than 50 MΩ.

(*c*) Earth *Resistance Test.* The earth resistance test is carried out with an earth-resistance tester on the earth wire, on three poles per km of the line, on metallic fittings and on stays at locations frequented by human beings. In case the earth resistance is found to be more than 10 Ω, necessary steps should be taken to bring down the earth resistance to 10 Ω or less.

15.18.3. Commissioning of Lines

After conducting all the tests mentioned above if the results are found within prescribed limits, the line is fit for commissioning. However, before doing so, the chief electrical inspector of the state or his representative is informed about the readiness of the line for energization. After getting his approval in writing, 11 kV lines and above are first energized at low voltage. After a few hours of satisfactory operation, the line may be energized at the rated voltage after informing the residents of the area.

In case of *lt* lines, these may be energized at the rated voltage after performing the tests mentioned above in Art. 15.18.2. provided the test results are within the prescribed limits.

15.19. MAINTENANCE OF OVERHEAD LINES

For maintaining continuity of supply through overhead lines, regular maintenance is essential. Maintenance work pertaining to the overhead lines consists of the following:

1. A pre-planned routine inspection followed by necessary preventive maintenance.
2. A special inspection, usually before the difficult operating periods of the system (such as before monsoons), followed by the preventive maintenance.
3. An emergency inspection, after each interruption of power supply, to locate the fault, followed by necessary repairs.

15.19.1. Routine Inspection and Maintenance of Overhead Lines

All overhead lines should be patrolled regularly at ground level while the line is live. There are varied opinions as to the number of line patrols which should be carried out each year. Many engineers are of the opinion that *ht* lines should be patrolled every month while others 3 or 4 times a year. On *ht* lines of major importance frequent patrolling may be desirable, but on *lt* lines less frequent patrolling is necessary.

The patrols recommended are ground inspections using field glasses for the examinaticn of any parts above normal height. The principal items to be noted on patrols are the condition of supports. cross arms, insulators and fittings, conductors and earth wires, guys and their accessories, foundations, anti-climbing devices, infringement of clearances by trees, buildings, haystacks etc; gang-operated (GO) switch and fuse unit on 11 kV lines and lightning arresters.

1. Supports. The supports may be wooden, metallic, PCC or RCC. The *wooden poles* should be checked for correct alignment and also the underground portion of the pole should be checked to verify that it is healthy and not eaten by white ants. In case of *metallic supports* the condition of the concrete foundation should be examined for possible damage. If any soil erosion is noticed at the foundation, it should be set right by earth filling and ramming to prevent the pole from tilting/falling. The PCC/RCC poles should be checked for cracks

2. Cross arms. Wooden cross arms should be checked for tilting and /or decaying owing to rot while the metallic cross arms should be checked for tilting and rusting.

3. Insulators and Fittings. Insulators and fittings should be checked for the defects such as (*i*) accumulation of dust, coal etc on insulators, (*ii*) burnt and fused spots on the insulator glaze (*iii*) broken and chipped porcelain (*iv*) tilted insulators and (*v*) rusting of fittings.

Damaged insulators should be replaced, tilted insulators set right and dirty insulators cleaned but after disconnecting the line from supply.

4. Conductors and Earth Wires. Clearance of phase wires and earth wire to ground, clearance between phase wires and that between the various phase wires and the earth wire should be checked to verify that they are within prescribed limits under IE rules. In case if any discrepancy is noticed, the sag should be adjusted and any foreign objects found on the conductors or earth wire should be removed but after disconnecting the line from supply.

5. Conductor Fittings and Joints. The defects to be noted are (*i*) broken strands (*ii*) burnt jumpers and loose fittings of PG clamps (*iii*) looseness of binding on insulators and at joints and (*iv*) the slipping of conductor from the insulator seat.

Broken strands should be repaired by providing aluminium conductor repair sleeves, burnt jumpers should be replaced and loose fittings of clamps should be tightened; slipped conductors should be placed on the insulators and bound again but after disconnecting of line from supply.

6. Guys and Their Accessories. These should be checked for the defects (*i*) loose guys (*ii*) broken or chipped guy insulators (*iii*) the corrosion of guy rods (*iv*) the condition of the guy anchoring and (*v*) earthing arrangement of guys, broken or missing.

Loose guys should be tightened; guy anchoring should be filled with earth, the broken guy insulators should be replaced and the earthing arrangement should be restored.

7. GO Switch and Fuse Unit. Checking should be made for the defects (*i*) missing or broken earth wire (*ii*) defective switch (*iii*) cracked or chipped insulators (*iv*) damaged arcing contacts and (*v*) burning or overheating of contacts.

Adjustments should be made for contact alignment and adequate contact pressure. Damaged contacts should be repaired/ replaced; earth connection, if broken, should be restored; bearings should be lubricated for correct and smooth operation of the moving parts.

8. Lightning Arresters. It should be ensured that the line lead of the lightning arrester is firmly connected with the phase wires, the earth connection is tight and earth resistance of earth wire is below 10 Ω,

9. Tree Clearance. Trees or branches coming close to the lines should be trimmed/removed – a minimum clearance of 1.8 m should be maintained on each side of the line but after disconnecting the line from the supply.

10. Bird Nests. Bird nests coming on the overhead lines should be removed.

15.19.2. Special Inspection

Such inspections of lines are required to be carried out before and after the monsoon/heavy rains because interruptions and breakdowns on overhead lines are usually more frequent during the rainy season owing to rains and thunderstorms. Any defect(s) noticed during inspection should be removed/set right.

15.19.3. Emergency Inspection

Whenever any overhead line trips and the fault sustains, such inspection is required to be carried out to locate the fault and determine the nature of fault. Such faults may be loose sag, snapping of conductors, conductors falling on cross arms, tree branches coming over the lines, birdage etc. The necessary repairs should be carried out and remedial measures be taken so that re-occurrence of such faults may be avoided.

15.19.4. Concept of Hot Line Technique

This is a technique with which the faults are rectified without interrupting the power supply. The work is carried out wearing rubber slippers, insulated tools and without coming in contact with walls etc. The applications are (*i*) removal and replacement of cross arms on energised towers and lines (*ii*) connecting temporary jumpers to line and also in substations (*iii*) replacement of damaged and flashed-over discs on the live line (*iv*) testing of power transformers, bushings and other equipment for leakages etc.

15.20. MEASUREMENT AND CHECKING OF SAG

Most common methods used for checking of sag are sight method and stop-watch or return-wave method.

1. Sight Method. In this case wooden strips are used as targets, the length of each of such target is equal to the minimum clearance required from the ground. These wooden strips are placed along the poles vertically at each end of the span so that the distance of the upper end of the targets and the conductors at the poles is equal to the required sag. The conductor is said to be correctly sagged when its lowest point in the span is in line with top of the targets when viewed from one end of the span to the other.

2. Stop-Watch or Return-Wave Method. In this method a sharp blow is given to the conductor. This will cause a wave to travel to the other pole, from where the wave is again reflected back. Thus wave will travel back and forth until it finally dies out by absorption of energy by the inductor. There is a definite relation between the time in seconds taken by the wave to return to the first support and the amount of sag. The return wave can be felt by a person on the first pole by placing his finger lightly on the conductor.

Exercises

1. Explain the factors concerning the route which could affect the cost of the line.
2. What planning is done before the commencement of erection of an overhead line ?
3. Give the points to be considered at the time of erection of overhead lines.
4. Discuss the elementary ideas about transmission line construction and erection.
   [Pb. Technical Univ. Power System-I (Transmission and Distribution, May 2008]
5. Explain, with the help of diagrams, the methods of repairing and jointing of damaged conductors.
6. Explain the procedure for earthing 11 kV line supports.
7. What is hot line technique concept and its application in the transmission lines ?
8. State the routine inspections and maintenance required for supports, cross arms, insulators, conductors, guys, earth wire, GO switch etc used in the overhead lines.
9. Discuss the various methods employed for measuring and checking sag during erection of a transmission line.
10. Write short notes on the following:
    (*i*) Salient features of survey conducted for determination of route for *ht*/*lt* lines.
    (*ii*) Stringing of conductors.
    (*iii*) Earthing of transmission line.
    (*iv*) Maintenance of overhead lines.
    (*v*) Checking of sag.
    (*vi*) Testing and commissioning of overhead lines.

❑❑❑

MULTIPLE CHOICE TYPE QUESTIONS

Note: Each of the following questions comprises multiple responses (two or more than two answers). Select the correct answers. In case you feel that there is more than one correct answer, select the answer which you consider the best. In any case choose ONLY ONE answer or response for each question.

1. Transmission line connects
(a) generating station to a switching station/step-down transformer station.
(b) step-down transformer station to service transformer banks.
(c) distribution transformer to consumer premises.
(d) service points to consumer premises.

2. A 3-phase, 4-wire system is commonly used for
(a) primary distribution. (b) secondary distribution.
(c) primary transmission. (d) secondary transmission.

3. The rated voltage of a 3-phase power system is given as
(a) rms phase voltage. (b) peak phase voltage.
(c) rms line to line voltage. (d) peak line to line voltage.
[GATE E.E. 2004]

4. Which of the following is usually not the generating voltage ?
(a) 6.6 kV. (b) 9.9 kV. (c) 11 kV. (d) 13.2 kV.
[A.M.I.E. Sec B. Power Systems Winter 1999]

5. In a transmission system the feeder supplies power to
(a) transformer substations (step-up).
(b) service mains.
(c) distributors. (d) all of the above.

6. Feeder is designed mainly from the point of view of
(a) its current carrying capacity.
(b) voltage drop in it.
(c) operating voltage. (d) operating frequency.

7. Distributors are designed from the point of view of
(a) its current carrying capacity.
(b) operating voltage.
(c) voltage drop in it. (d) operating frequency.

8. Transmission and distribution of electric power by underground system is superior to overhead system in respect of
(a) appearance and public safety.
(b) maintenance cost.
(c) frequency of faults, power failure and accidents.
(d) all of the above.

9. The main drawback(s) of underground system over overhead system is/are
(a) exposure to lightning. (b) heavy initial cost.
(c) exposure to atmospheric hazards such as smoke, ice, wind etc.
(d) inductive interference between power and communication circuits.

10. The main drawback(s) of overhead system over underground system is/are
(a) underground system is more flexible than overhead system.
(b) higher charging current.
(c) surge problem. (d) high initial cost.

11. By increasing the transmission voltage double of its original value, the same power can be despatched keeping the line loss
(a) equal to its original value.
(b) half of original value.
(c) double the original value.
(d) one-fourth of original value.
[A.M.I.E. Sec B. Summer 1993; W.B. Univ. of Technology Power System-I, 2009-10]

12. If a fixed amount of power is to be transmitted over certain length with fixed power loss, it can be said that volume of conductor is
(a) inversely proportional to magnitude of the voltage and that of power factor of the load.
(b) inversely proportional to square of the voltage and square of power factor of the load.
(c) proportional to square of voltage and that of power factor of the load.
(d) proportional to magnitude of the voltage only.
[U.P.S.C. I.E.S. E.E.-II, 2010]

13. For the same voltage drop, increasing the voltage of a distributor n-times
(a) reduces the x-section of the conductor by n times.
(b) increases the x-section of the conductor by n times.
(c) reduces the x-section of the conductor by n^2 times.
(d) increases the x-section of the conductor by n^2 times.
[A.M.I.E. Sec B. Summer 1994]

14. In a transmission system, the weight of copper used is proportional to
(a) E^2. (b) E. (c) $1/E^2$. (d) 1/E.
(e) none of the above. **[A.M.I.E. Sec B. Winter 1994]**

15. The volume of copper required for an ac transmission line is inversely proportional to
(a) current. (b) voltage.
(c) pf. (d) both (b) and (c).

16. Improving pf
(a) reduces current for a given output.
(b) increases losses in line.
(c) increases the cost of station equipment.
[A.M.I.E. Sec B. Power Systems Summer 1997]

17. For a given amount of power to be transmitted over a certain distance with fixed power loss, the volume of copper required is
(a) directly proportional to voltage.
(b) inversely proportional to voltage.
(c) inversely proportional to the square of voltage and pf of the load.
(d) directly proportional to the square of the voltage and pf of the load.

18. For the same conductor length, same amount of power, same losses and same maximum voltage to earth, which system requires minimum conductor area?
(a) single phase ac (b) 3 phase ac
(c) 2 wire ac (d) 3 wire dc
[A.M.I.E Sec B. Design of Electrical Systems Winter 2006]

19. Which of the following distribution systems is preferred for good efficiency and high economy?
(a) Single-phase, 2-wire system. (b) 2-phase, 3-wire system.
(c) 3-phase, 3-wire system. (d) 3-phase, 4-wire system.

Answers: **1.** (a), **2.** (b), **3.** (c), **4.** (b), **5.** (c), **6.** (a), **7.** (c), **8.** (d), **9.** (b), **10.** (c), **11.** (d), **12.** (b), **13.** (a), **14.** (c), **15.** (d), **16.** (a), **17.** (c), **18.** (d), **19.** (d),

20. The approximate cost ratio of a 220 kV, underground cable transmission and 220 kV O/H transmission is
(a) 50 (b) 25 (c) 13 (d) 5
[A.M.I.E. Sec B. Power Systems Summer 2000]

21. With the same maximum voltage to earth, which of the following ac systems with 0.8 pf will need more copper in comparison to dc 2-wire system?
(a) Single-phase, 2-wire (midpoint earthed).
(b) Single-phase, 3-wire (neutral half of outer).
(c) Three-phase, 3-wire.
(d) Three-phase, 4-wire (neutral = outer).

22. The main reason for using high voltage for long distance power transmission is
(a) reduction in transmission losses.
(b) reduction in time of transmission.
(c) increase in system reliablility.
(d) none of the above.

23. 66 kV is suitable for transmission of power over
(a) 30 km (b) 60 km (c) 120 km (d) 200 km
[A.M.I.E. Sec B. Power Systems Winter 1993]

24. If 3 MW power is to be transmitted over a distance of 30 km, the desirable transmission voltage will be
(a) 11 kV (b) 33 kV (c) 66 kV (d) 3.3 kV
[A.M.I.E. Sec B. Power Systems Winter 1999]

25. The highest transmission voltage used in India is
(a) 400 kV (b) 220 kV
(c) 132 kV (d) 765 kV
[U.P.T.U. Odd Semester 2008-09; W.B. Univ. of Technology Power System-III, 23009-10]

26. If variable part of annual cost on account of interest and depreciation on the capital out lay is equal to the annual cost of electrical energy wasted in the conductors, the total annual cost will be minimum and the corresponding size of conductor will be the most economical. This statement is known as
(a) Lenz's law. (b) Kelvin's law.
(b) Faraday's law. (d) Ohm's law.

27. Which of the curves shown in the figure represents the relation between capital cost and the transmission voltage ?
(a) Curve I.
(b) Curve II.
(c) Curve III.
(d) Curve IV.

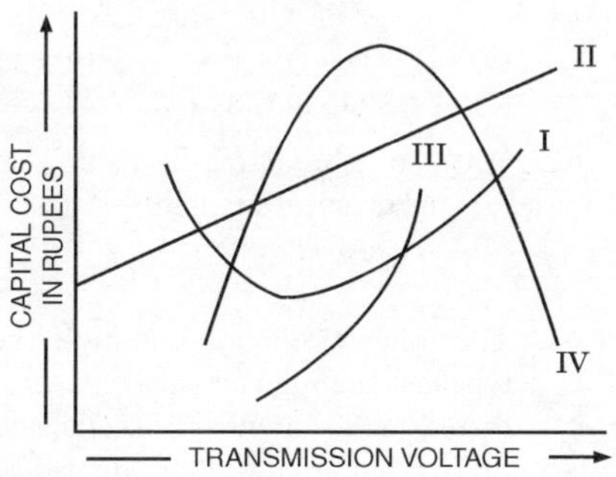

28. For high-voltage transmission lines, conductors are suspended from towers so as to
(a) increase clearance from ground.
(b) reduce clearance from ground.
(c) take care of increase in length.
(d) reduce wind and snow effects.

29. The supports used for transmission lines should have the characteristic(s) of
(a) high mechanical strength and longer life.
(b) good looking, light in weight and easily accessible for painting and erection of line conductors.
(c) cheap in initial as well in maintenance cost.
(d) all of the above.

30. The wooden poles well impregnated with creosite oil or any preservative compound have life of
(a) 25-30 years. (b) 20-25 years.
(c) 10-15 years. (d) 5-10 years.

31. Steel poles for transmission lines need protection against
(a) borer. (b) termites.
(c) corrosion. (d) all of these.

32. RCC poles usually have the spans of
(a) 250-400 m. (b) 80-150 m.
(c) 50-80 m. (d) 25-50 m.

33. In India for distribution of electric power we usually use
(a) wooden poles. (b) steel poles.
(c) RCC poles. (d) both (b) and (c).

34. Conductors used in *ht* transmission lines are stranded because of
(a) increased tensile strength. (b) ease in handling.
(c) cheaper in cost. (d) reduced resistivity.

35. Consider the following materials for line conductors :
1. Hard drawn copper. 2. Cadmium copper.
3. Aluminium. 4. Galvanised steel.
The correct sequence of the descending order of their electrical conductivities is
(a) 1, 2, 4, 3 (b) 2, 1, 4, 3 (c) 2, 1, 3, 4 (d) 1, 2, 3, 4
[U.P.S.C. I.E.S. E.E.-II, 1995]

36. Which of the following properties has got higher value for aluminium in comparison to that of copper ?
(a) Electrical resistivity. (b) Melting point.
(c) Thermal conductivity. (d) Specific gravity.

37. In a 7/30 ACSR conductor why is grease put between steel and aluminium conductors?
(a) To reduce corrosion by electrolytic action between zinc (galvanising agent on steel) and aluminium.
(b) To reduce friction between the strands.
(c) To reduce leakage of current from aluminium strands to steel strands.
(d) To eliminate air pockets. **[U.P.S.C. I.E.S. E.E.-II, 2006]**

38. ACSR is used in place of copper in overhead lines because of
(a) higher current carrying capacity.
(b) being lighter in weight.
(c) economy. (d) higher tensile strength.

39. ACSR conductors have
(a) all conductors made of aluminium.
(b) outer conductors made of aluminium.
(c) inner conductors made of aluminium.
(d) no conductors made of aluminium. **[A.M.I.E. Sec B. Summer 1993]**

40. "Expanded ACSR" are conductors composed of
(a) larger diameter individual strands for a given cross section of the aluminium strands.
(b) larger diameter of the central steel strands for a given overall diameter of the conductor.
(c) larger diameter of the aluminium strands only for a given overall diameter of conductor.
(d) A filter between the inner steel and the outer aluminium strands to increase the overall diameter of the conductor.
[U.P.S.C. I.E.S. E.E.-II, 1998]

41. The function of steel wire in an ACSR conductor is to
(a) compensate for skin effect.
(b) take care of surges.
(c) provide additional mechanical strength.
(d) reduce inductance.

Answers: **20.** (c), **21.** (d), **22.** (a), **23.** (b), **24.** (c), **25.** (d), **26.** (b), **27.** (a), **28.** (a), **29.** (d), **30.** (a), **31.** (c), **32.** (b), **33.** (d), **34.** (b), **35.** (d), **36.** (a), **37.** (a), **38.** (c), **39.** (b), **40.** (b), **41.** (c),

42. The material used for the manufacture of ground wire is
(a) aluminium. (b) galvanised steel.
(c) cast iron. (d) stainless steel.

43. Guy wire is employed for
(a) providing protection against surges.
(b) providing emergency earth route.
(c) supporting the pole. (d) all of the above.

44. The sag of a transmission line is least affected owing to
(a) weight of the conductor.
(b) current through the conductor.
(c) atmospheric temperature.
(d) ice deposition on the conductor.

45. Effect of temperature rise in overhead lines is to
(a) increase the sag and decrease the tension.
(b) decrease the sag and increase the tension.
(c) increase both.
(d) decrease both. **[A.M.I.E. Sec B. Power Systems Winter 1996]**

46. The sag of a transmission line conductor in summer is
(a) less than that in winter. (b) more than that in winter.
(c) same as in winter. **[A.M.I.E. Sec B. Power Systems Summer 1999]**

47. In a transmission line, sag depends upon
(a) span length. (b) tension in conductors.
(c) weight of the conductor per unit length.
(d) all of the above.

48. Which of the following statements is correct ?
(a) Ice on conductors increases skin effect.
(b) Wind pressure reduces corona effect.
(c) Wind pressure is taken to act at perpendicular to that for ice.
(d) Ice on conductors reduces sag.

49. Wind loading in coastal regions is in the range of
(a) 40-50 kg/m^2 (b) 150 kg/m^2
(c) 96 kg/m^2 **[A.M.I.E. Sec B. Power Systems Summer 1997]**

50. The maximum tension in a section of overhead line conductor between two supports of unequal height occurs at
(a) the higher support. (b) the lower point.
(c) the midpoint of the conductor.
(d) None of the above. **[A.M.I.E. Sec B. Power Systems Winter 1996]**

51. Stringing chart is useful
(a) for finding the sag in the conductor.
(b) in the design of tower.
(c) in the design of insulator string.
(d) finding the distance between towers.
[A.M.I.E. Sec. B Summer 1995]

52. Hot template curves are plots of
(a) temperature and humidity.
(b) conductor sag and span lengths.
(c) conductor weight and sag.
(d) none of the above.

53. The effect of wind pressure is more predominant on
(a) insulators. (b) transmission lines.
(c) supporting towers. (d) none of the above.

54. Galloping in transmission line conductors arises due to
(a) asymmetrical layers of ice formation.
(b) vortex phenomenon in light winds.
(c) heavy weight of the line conductors.
(d) adoption of horizontal conductor configuration.
[U.P.S.C. I.E.S. E.E.-II, 1997]

55. Which one of the following is reduced by using stock bridge dampers on power overhead transmission lines?
(a) Sag. (b) Conductor vibration.
(c) Line losses. (d) Mechanical tension.
[U.P S.C. I.E.S. E.E.-II, 2009]

56. The sag of the conductors of a transmission line is 2.5 m when the span is 250 m. Now if the height of supporting tower is increased by 25%, the sag will
(a) reduce by 25% (b) increase by 25%
(c) reduce by 12.5% (d) remain unchanged.

57. In an overhead transmission line with the conductors supported between two supports, the variation in tension T with atmospheric temperature t is represented by

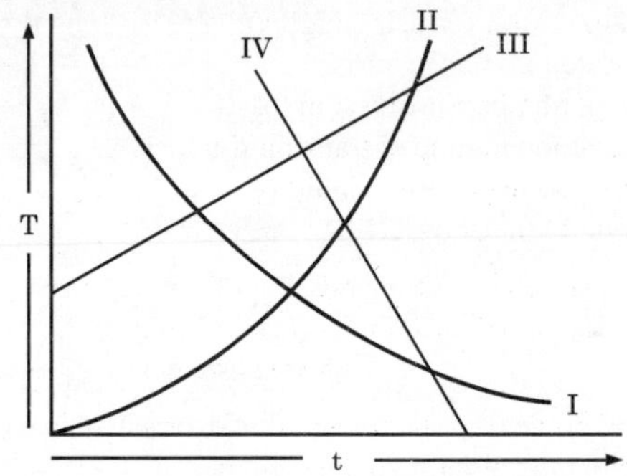

(a) curve I. (b) curve II. (c) curve III. (d) curve IV.

58. For a 400 kV line, the spacing between phase conductors is around
(a) 8 m. (b) 11 m. (c) 14 m. (d) 17 m.
[A.M.I.E. Sec B. Power Systems Summer 1996, 1998]

59. Stranded conductors usually have a central wire around which there are successive layers of 6,12,18,24 wires. For n-layers, the total number of individual wires is
(a) $3n(n+1)$ (b) $2n(n+1)$
(c) $3n(n+1)+1$ (d) $2n(n+1)+1$

60. The diameter of each strand is d then the diameter of n-layer stranded conductor will be
(a) $(2n+1)d$ (b) $3(n+1)d$
(c) $(2n-1)d$ (d) $3(n-1)d$

61. Strain type insulators are used
(a) at dead ends. (b) at intermediate anchor towers.
(c) on straight runs. (d) any of (a) or (b).

62. Wavy structure of pin insulator increases its
(a) mechanical strength. (b) puncture strength.
(c) flash-over voltage. (d) thermal strength.
[A.M.I.E. Sec B. Power Systems Summer 1993]

63. The voltage rating of a multiple shell (petticoat or rainshed) pin type insulator unit cannot be increased beyond a limiting value by increasing the number of shells, because
(a) the internal voltage distribution between shells becomes unequal.
(b) the leakage path resistance starts diminishing.
(c) the disruptive critical voltage for the material of the insulator is reached.
(d) the puncture voltage of the material of the insulator is reached.
[U.P.S.C. I.E.S. E.E.-II, 1993]

64. Which type of insulators are used on 132 kV transmission lines ?
(a) Pin type. (b) Disc type.
(c) Shackle type. (d) Pin and Shackle type.
[A.M.I.E. Sec B. Power Systems Winter 1999]

65. Whenever the conductors are dead-ended or there is a change in the direction of transmission line, the insulators used are of the

Answers: **42.** (b), **43.** (c), **44.** (b), **45.** (a), **46.** (b), **47.** (d), **48.** (c), **49.** (b), **50.** (a), **51.** (a), **52.** (b), **53.** (c), **54.** (a), **55.** (b), **56.** (d), **57.** (a), **58.** (a), **59.** (c), **60.** (a), **61.** (d), **62.** (c), **63.** (a), **64.** (b), **65.** (c),

(a) Pin type. (b) Suspension type.
(c) Strain type. (d) Shackle type.
[U.P.S.C. I.E.S. E.E.-II, 1998]

66. Post type insulators are generally used in lines operating
(a) above 100 kV. (b) below 33 kV.
(c) at any voltage level, *hv* or *ehv*.
[A.M.I.E. Sec B. Power Systems Summer 1999]

67. The number of discs in a string of insulators for 400 kV ac overhead transmission line lies in the range of
(a) 32 to 33 (b) 22 to 23 (c) 15 to 16 (d) 9 to 10
[U.P.S.C. I.E.S. E.E.-II, 1999]

68. The non-uniform distribution of voltage across the units in a string of suspension type insulators is due to
(a) unequal self-capacitance of the units.
(b) non-uniform distance of separation of the units from the tower body.
(c) the existence of stray capacitance between the metallic junctions of the units and the tower body.
(d) non-uniform distance between the cross-arms and the units.
[U.P.S.C. I.E.S. E.E.-II, 1998]

69. The voltages across the various discs of a string of suspension insulators having identical discs is different due to
(a) surface leakage currents. (b) series capacitance.
(c) shunt capacitance to ground.
(d) series and shunt capacitances. [U.P.S.C. I.E.S. E.E.-II, 1995]

70. The string efficiency of a string of suspension insulators is dependent on
(a) size of the insulators. (b) number of discs in the string.
(c) size of tower. [A.M.I.E. Sec B. Winter 1994]

71. 100 per cent string efficiency means
(a) one of the insulator discs shorted.
(b) zero potential across each disc.
(c) equal potential across each insulator disc.
(d) none of the above.

72. In a suspension type insulator the potential drop is
(a) maximum across the lowest disc.
(b) maximum across the topmost disc.
(c) uniformly distributed over the discs.
[A.M.I.E. Sec B. Power Systems Summer 1999]

73. If the frequency of a transmission system is changed from 50 Hz to 100 Hz, the string efficiency
(a) will increase. (b) will decrease.
(c) remain unchanged.
(d) may increase or decrease depending on the line parameters.
[A.M.I.E. Sec B. Winter 1994]

74. The string efficiency of a high-voltage line is around
(a) 100% (b) 80% (c) 40% (d) 10%
[A.M.I.E. Sec B. Power Systems Summer 1996]

75. In three-unit insulator string, voltage across the lowest unit is 17.5 kV and string efficiency is 84.28%. The total voltage across the string will be equal to
(a) 8.285 kV (b) 44.25 kV (c) 88.25 kV (d) 442.5 kV
[U.P.S.C. I.E.S. E.E.-II, 2001]

76. Two-insulator discs of identical capacitance value C makes up a string for a 22 kV, 50 Hz, single-phase overhead line insulation system. If the pin to earth capacitance is also C, then the string efficiency is
(a) 50% (b) 75% (c) 90% (d) 86%
[U.P.S.C. I.E.S. E.E.-II, 2003]

77. The equivalent capacitor arrangement of a two-string insulator is shown in the figure. The maximum voltage that each unit can withstand should not exceed 17.5 kV. The line voltage of the complete string is
(a) 17.5 kV (b) 33 kV
(c) 35 kV (d) 37.3 kV
[U.P.S.C. I.E.S. E.E.-II, 1999]

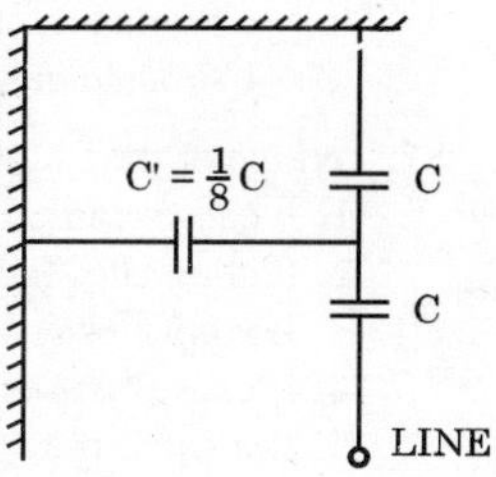

78. Consider a three-phase, 50 Hz, 11 kV distribution system. Each of the conductors is suspended by an insulator string having two identical porcelain insulators. The self-capacitance of the insulator is 5 times the shunt capacitance between the link and the ground, as shown in the figure. The voltage across the two insulators are

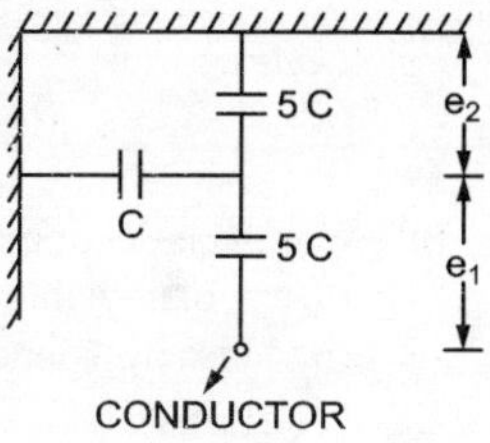

(a) $e_1 = 3.74$ kV, $e_2 = 2.61$ kV
(b) $e_1 = 3.46$ kV, $e_2 = 2.89$ kV
(c) $e_1 = 6.0$ kV, $e_2 = 4.23$ kV
(d) $e_1 = 5.5$ kV, $e_2 = 5.5$ kV [GATE, E.E. 2010]

79. In a cable the sheath radius is R and conductor radius is *r*. As r changes from 0.5 R to 0.25 R the maximum voltage gradient in the dielectric
(a) decreases by about 6%. (b) increases by about 6%.
(c) increases by about 15%. (d) decreases by about 15%
[A.M.I.E. Sec B. Power Systems Summer 2000]

80. Consider the following statements :
In the case of suspension type insulators, the string efficiency can be improved by
1. using a longer cross arm. 2. using a guard ring.
3. grading the insulator discs. 4. reducing the cross-arm length.
Of these statements
(a) 1, 2, and 3 are correct. (b) 2, 3 and 4 are correct.
(c) 2 and 4 are correct. (d) 1 and 3 are correct.
[U.P.S.C. I.E.S. E.E.-II, 1998]

81. The ratio of puncture voltage to the flash-over voltage of a line insulator is
(a) equal to 1 (b) lower than 1
(c) much greater than 1 [A.M.I.E. Sec B. Winter 1993]

82. The insulators may fail due to
(a) flash over. (b) short-circuits.
(c) deposition of dust. (d) any of the above.

83. The purpose of guard ring in transmission lines is to
(a) reduce the earth capacitance of the lowest unit.
(b) increase the earth capacitance of the lowest unit.
(c) reduce the transmission line losses.
(d) none of the above.

84. The use of a guard ring
(a) equalises the voltage division between insulator discs.
(b) is unnecessary complication.
(c) decreases string efficiency.
[A.M.I.E. Sec B. Power Systems Summer 1997]

85. Corona is
(a) partial breakdown of air. (b) complete breakdown of air.
(c) sparking between lines. [A.M.I.E. Sec B. Power Systems Summer 1997]

86. Which of the following statements is true regarding corona ?
(a) Corona takes place at a voltage lower than breakdown voltage.
(b) Corona takes place at a voltage higher than breakdown voltage.

Answers: **66.** (c), **67.** (b), **68.** (c), **69.** (c), **70.** (b), **71.** (c), **72.** (a), **73.** (c), **74.** (b), **75.** (b), **76.** (b), **77.** (b), **78.** (b), **79.** (b), **80.** (a), **81.** (c), **82.** (d), **83.** (a), **84.** (a), **85.** (b), **86.** (a).

(c) Corona is a current phenomenon.
(d) Corona increases the transmission line efficiency.

87. Which of the following statements regarding corona are true ?
1. It causes radio interference.
2. It attenuates lightning surges.
3. It amplifies switching surges.
4. It causes power loss.
5. It is more prevalent in the middle conductor of a transmission line employing a flat conductor configuration.

Select the correct answer using the codes given below.
Codes:
(a) 1, 3, 5 (b) 2, 3, 4 (c) 1, 2, 4, 5 (d) 2, 3, 4, 5
[U.P.S.C. I.E.S. E.E.-II, 1993]

88. Corona is accompanied by
(a) violet visible discharge in darkness
(b) hissing sound. (c) vibration.
(d) power loss. (e) radio-interference.
(f) ozone. (g) all of the above.
[A.M.I.E. Sec B. Power Systems Winter 1993]

89. Ozone effect can be detected by
(a) presence of ozone detected by odour.
(b) hissing sound.
(c) faint luminous glow of bluish colour.
(d) all of the above.

90. Corona loss increases with
(a) decrease in conductor size and increase in supply frequency.
(b) increase in both conductor size and supply frequency.
(c) decrease in both conductor size and supply frequency.
(d) increase in conductor size and decrease in supply frequency.
[U.P.S.C. I.E.S. E.E.-II, 2004]

91. The good effect of corona on overhead lines is to
(a) increase the line carrying capacity due to conducting ionised air envelope around the conductor.
(b) increase the power factor due to corona loss.
(c) reduce the radio interference from the conductor.
(d) reduce the steepness of surge fronts. **[U.P.S.C. I.E.S. E.E.-II, 2001]**

92. Compared with a solid conductor of the same radius, corona appears on a stranded conductor at a lower voltage, because stranding
(a) assists ionisation.
(b) makes the current flow spirally about the axis of the conductor.
(c) produces oblique sections to a plane perpendicular to a axis of the conductor.
(d) produces surfaces of smaller radius. **[U.P.S.C. I.E.S. E.E.-II, 1997]**

93. Corona loss can be reduced by the use of hollow conductors because
(a) the current density is reduced.
(b) the eddy current in the conductor is eliminated.
(c) for a given cross section, the radius of the conductor is increased.
(d) of better ventilation in the conductor. **[U.P.S.C. I.E.S. E.E.-II, 1994]**

94. Bundled conductors reduce
(a) surface electric stress of conductor.
(b) increases the line reactance.
(c) decreases the line capacitance.
[A.M.I.E. Sec B. Power Systems Summer 1997]

95. Corona loss in a transmission line is dependent on
(a) diameter of the conductor.
(b) material of the conductor.
(c) height of the conductor. **[A.M.I.E. Sec B. Winter 1994]**

96. Corona occurs between two transmission conductors when they
(a) have high potential difference. (b) are closely spaced.
(c) carry dc power. (d) both (a) and (b).

97. In humid weather, the corona occurs at a voltage
(a) much less than that needed in fair weather.
(b) much higher than that needed in fair weather.
(c) equal to that needed in fair weather.
(d) none of the above.

98. The effect of corona is
(a) increased energy loss. (b) increased reactance.
(c) increased inductance. (d) all of the above.
[A.M.I.E. Sec B. Power Systems Winter 1999]

99. The charging current in a transmission line increases due to corona effect because corona increases
(a) line current. (b) effective line voltage.
(c) power loss in lines. (d) the effective conductor diameter.
[U.P.S.C. I.E.S. E.E.-II, 1996]

100. The chances of occurrence of corona are maximum during
(a) humid weather. (b) dry weather.
(c) winter. (d) hot summer.

101. Corona is likely to occur maximum in case of
(a) distribution lines. (b) transmission lines.
(c) domestic wiring. (d) service mains.

102. Corona is affected by
(a) size of conductor.
(b) shape and surface condition of the conductor.
(c) operating voltage. (d) all of the above.

103. The only advantage of corona is that it
(a) produces a pleasing luminous glow.
(b) makes line current sinusoidal.
(c) works as a safety valve for surges.
(d) ozone gas is produced.

104. Presence of ozone owing to corona
(a) improves the pf. (b) reduces the pf.
(c) corrodes the material. (d) improves regulation.

105. What is the approximate breakdown strength of atmospheric air at N.T.P.?
(a) 0.3 kV/cm. (b) 1.0 kV/cm.
(c) 3 kV/cm. (d) 30 kV/cm.
[U.P.S.C. I.E.S. E.E.-II, 2007]

106. The dielectric strength of air under normal condition is about
(a) 100 kV_p/cm (b) 21.1 kV_p/cm
(c) 30 kV_p/cm (d) 200 kV_p/cm
[W.B. Univ. of Technology Power System-I, 2009-10]

107. Disruptive corona begins in smooth cylindrical conductors in air at NTP if the electric field intensity at the conductor surface goes up to
(a) 21.1 kV (rms)/cm. (b) 21.1 kV (peak)/cm.
(c) 21.1 kV (average)/cm (d) 21.1 kV (rms)/m.
[U.P.S.C. I.E.S. E.E.-II, 2000]

108. The dielectric strength of air is
(a) proportional to barometric pressure.
(b) proportional to absolute temperature.
(c) inversely proportional to barometric pressure.
(d) none of the above.

109. Visual critical voltage is
(a) lower than disruptive critical voltage.
(b) higher than disruptive critical voltage.
(c) equal to critical voltage. (d) none of the above.

Answers: **87.** (c), **88.** (g), **89.** (d), **90.** (a), **91.** (d), **92.** (d), **93.** (c), **94.** (a), **95.** (a), **96.** (d), **97.** (a), **98.** (a), **99.** (d), **100.** (a), **101.** (b), **102.** (d), **103.** (c), **104.** (c), **105.** (d), **106.** (c), **107.** (a), **108.** (a), **109.** (b),

110. Critical voltage limit of a transmission line is increased by
(a) increasing the radius of the conductors.
(b) increasing the spacing between conductors.
(c) reducing the spacing between conductors.
(d) reducing the radius of the conductors.

111. The maximum permissible value of fair weather corona loss for an HV line is
(a) 0.6 kW/3-phase km. (b) 1.2 kW/3-phase km.
(c) 0.3 kW/3-phase km. (d) 2.4 kW/3-phase km.
[A.M.I.E. Sec B. Power Systems Summer 1998]

112. Corona losses are minimised when
(a) conductor size is reduced. (b) smooth conductor is used.
(c) sharp points are provided in the line hardware.
(d) current density in conductors is reduced. **[GATE E.E. 1999]**

113. The corona loss on a particular system at 50 Hz is 1 kW/km per phase. What is the corona loss at 60 Hz in kW/km per phase?
(a) 0.83 (b) 1.0 (c) 1.13 (d) 1.2
[GATE E.E., 2000; U.P.S.C. I.E.S. E.E.-II, 2005]

114. Corona loss can be reduced by using
(a) solid conductor of diameter 'd'.
(b) hollow conductor of diameter '$d + \delta d$'.
(c) bundle conductor. (d) both (a) and (b).
(e) both (b) and (c).
[W.B. Univ. of Technology Power System-I, 2009-10]

115. In a transmission line the distributed constants are
(a) resistance and shunt conductance only.
(b) resistance and inductance only.
(c) resistance, inductance and capacitance only.
(d) resistance, inductance, capacitance and shunt conductance.

116. Skin effect depends upon
(a) x-section of conductor. (b) supply frequency.
(c) permeability of conductor material.
(d) all of the above.

117. Skin effect in transmission line is due to
(a) supply frequency. (b) self inductance of conductor.
(c) high sensitivity of material in the centre.
(d) both (a) and (b).

118. The effective resistance of a conductor will be the same as ohmic resistance when
(a) voltage is low. (b) current is true sinusoidal.
(c) current is uniformly distributed in the x-section of the conductor.

119. When an alternating current flows through a conductor
(a) entire current passes through the core of the conductor.
(b) portion of conductor near the surface carries more current in comparison to the core.
(c) current remains uniformly distributed over the whole x-section of the conductor.
(d) portion of conductor near the surface carries less current in comparison to the core.

120. The conductor carries more current on the surface in comparison to its core. This phenomenon is called the
(a) skin effect. (b) corona.
(c) Ferranti effect. (d) Lenz's effect.

121. Increasing the frequency of transmission line will
(a) increase shunt reactance. (b) decrease line resistance.
(c) increase line resistance. (d) decrease series reactance.
[A.M.I.E. Sec B. Power Systems Summer 1994]

122. Skin effect exists in
(a) cable carrying dc current. (b) dc transmission line only.
(c) ac transmission line only.
(d) dc as well as a c transmission lines.

123. Skin effect in a conductor becomes more pronounced
(a) at higher frequency. (b) at lower frequency.
(c) at dc. **[A.M.I.E. Sec B. Winter 1996]**

124. The skin effect in conductor results in
(a) increases in its dc resistance. (b) decrease in its ac resistance.
(c) increase in its ac resistance.
[A.M.I.E. Sec B. Power Systems Summer 1999]

125. Skin effect
(a) increases the effective resistance and effective internal reactance.
(b) reduces the effective resistance and effective internal reactance.
(c) increases the effective resistance but reduces the effective internal reactance.
(d) reduces the effective resistance but increases the effective internal reactance.

126. The skin effect of a conductor reduces with the increase in
(a) supply frequency.
(b) resistivity of the conductor material.
(c) x-section of conductor.
(d) permeability of conductor material.

127. Skin effect in conductor is proportional to
(a) (diameter of conductor)$^{1/2}$. (b) diameter of conductor.
(c) (diameter of conductor)^2. (d) (diameter of conductor)^4.

128. In order to reduce the skin effect at UHF
(a) copper tubes with silver plating are used.
(b) copper rods with silver plating are used.
(c) anodised conductors are used.
(d) painted conductors are used.

129. A 100 km long transmission line is loaded at 110 kV. If the loss of line is 15 MW and the load is 150 MVA, the resistance of the line is
(a) 8.06 ohms per phase. (b) 0.806 ohms per phase.
(c) 0.0806 ohms per phase. (d) 80.6 ohms per phase.
[U.P.S.C. I.E.S. E.E.-II, 2003]

130. The component inductance due to the internal flux-linkage of a non-magnetic straight solid circular conductor per metre length, has a constant value, and is independent of the conductor diameter, because
(a) All the internal flux due to a current remains concentrated on the peripheral region of the conductor.
(b) The internal magnetic flux density along the radial distance from the centre of the conductor increases proportionately to the current enclosed.
(c) The entire current is assumed to flow along the conductor axis and the internal flux is distributed uniformly and concentrically.
(d) The current in the conductor is assumed to be uniformly distributed throughout the conductor cross section.
[U.P.S.C. I.E.S. E.E.-II, 2003]

131. Consider a long, two-wire line composed of solid round conductors. The radius of both conductors is 0.25 cm and the distance between their centres is 1 m. If this distance is doubled, then the inductance per unit length.
(a) doubles. (b) halves.
(c) increases but does not double.
(d) decreases but does not halve. **[GATE E.E. 2002]**

Answers: **110.** (a), **111.** (d), **112.** (b), **113.** (c), **114.** (e), **115.** (d), **116.** (d), **117.** (d), **118.** (c), **119.** (b), **120.** (a), **121.** (c), **122.** (c), **123.** (a), **124.** (c), **125.** (c), **126.** (b), **127.** (c), **128.** (a), **129.** (a), **130.** (d), **131.** (c),

132. The inductance of single-phase two-wire power transmission line per km gets doubled when the
(a) distance between the wires is doubled.
(b) distance between the wires is increased four fold.
(c) distance between the wires is increased as square of the original distance.
(d) radius of the wire is doubled. **[U.P.S.C. I.E.S E.E.-II, 1994]**

133. Inductance of an overhead line, in comparison to that of a cable of same capacity is
(a) larger. (b) smaller. (c) of the same order.
[A.M.I.E. Sec B. Power Systems Summer 1999]

134. The inductance of a single-phase two-wire line is given by (D is the distance between conductors and r is the radius of conductor)
(a) $0.4 \log_e \frac{D}{r}$ mH/km (b) $0.55 \log_e \frac{D}{r}$ mH/km
(c) $0.4 \log_e \frac{r}{D}$ mH/km (d) $0.55 \log_e \frac{r}{D}$ mH/km
[A.M.I.E. Sec B. Power Systems Winter 1998]

135. For 11 kV line, the inductance per km per phase will be of the order of
(a) 1 H (b) 0.1 H (c) 1 mH (d) 0.1 mH
[A.M.I.E. Sec B. Power Systems Winter 1999]

136. For a single-phase overhead line having solid copper conductors of diameter 1 cm spaced 60 cm between centres, the inductance in mH/km is
(a) 0.05 + 0.2 log 60 (b) 0.2 log 60
(c) $0.05 + 0.2 \log \frac{60}{0.5}$ (d) $0.2 \log \frac{60}{0.5}$ **[GATE E.E. 1999]**

137. A 3-phase transmission line has its conductors at the corners of an equilateral triangle with side 3 m. The diameter of each conductor is 1.63 cm. The inductance of the line per phase per km is
(a) 1.232 mH (b) 1.182 mH (c) 1.093 mH (d) 1.043 mH
[U.P.S.C. I.E.S. E.E.-II, 2002]

138. The conductors of a 10 km long, single-phase, two-wire line are separated by a distance of 1.5 m. The diameter of each conductor is 1 cm. If the conductors are of copper, the inductance of the circuit is
(a) 50.0 mH (b) 45.3 mH (c) 23.8 mH (d) 19.6 mH
[GATE E.E.-II, 2004]

139. The inductance of a power transmission line increases with
(a) decrease in line length.
(b) increase in diameter of conductor.
(c) increase in spacing between the phase conductors.
(d) increase in load current carried by the conductors.
[GATE E.E. 1992]

140. The inductance of a transmission line is minimum when
(a) GMD is high. (b) GMR is high.
(c) both GMD and GMR are high.
(d) GMD is low and GMR is high. **[A.M.I.E. Sec B. Summer 2000]**

141. For a given conductor the value of GMR is
(a) larger for capacitance calculation.
(b) larger for inductance calculation.
(c) same for both capacitance and inductance calculation.
[A.M.I.E. Sec B. Power Systems Summer 1999]

142. The self GMD of a conductor with three strands each of radius r and touching each other is
(a) $r\,(0.7788 \times 2 \times 2)^{1/3}$ (b) $r\,(0.7788 \times 2 \times 2 \times 2)$
(c) $r\,(0.7788 \times 2 \times 2 \times 2)^3$ (d) $r\,(0.7788 \times 2 \times 2)^3$
[U.P.S.C. I.E.S. E.E.-II, 2000]

143. Hollow conductors are used in transmission lines to
(a) reduce weight of copper. (b) improve stability.
(c) reduce corona.
(d) increase power transmission capacity.
[U.P.S.C. I.E.S. E.E.-II, 1997]

144. Bundled conductors are employed to improve
(a) appearance of the transmission line.
(b) mechanical stability of the line.
(c) current carrying capacity of the line.
(d) corona performance of the line. **[GATE E.E. 1997]**

145. In hv transmission, the spacing between subconductors of a bundle is approximately
(a) 20 cm (b) 40 cm (c) 80 cm (d) 3.5 cm
[A.M.I.E. Sec B. Power Systems Summer 2000]

146. For a stranded conductor, the ratio of GMR to actual radius is
(a) equal to 1. (b) more than 1.
(c) equal to 0.7788. (d) less than 0.7788
[A.M.I.E. Sec B. Power Systems Summer 2000]

147. Bundled conductors in EHV transmission system provide
(a) reduced capacitance. (b) increased capacitance.
(c) increased inductance. (d) increased voltage gradient.
[U.P.S.C. I.E.S. E.E.-II, 1996]

148. Which one of the following statements is not correct for the use of bundled conductors in transmission lines?
(a) Control of voltage gradient.
(b) Reduction in corona loss.
(c) Reduction in radio interference.
(d) Increase in interference with communication lines.
[U.P.S.C. I.E.S. E.E. 2003]

149. Bundled conductors are mainly used in high voltage overhead transmission lines to
(a) reduce transmission line losses.
(b) increase mechanical strength of the line.
(c) reduce corona. (d) reduce sag. **[GATE E.E. 2003]**

150. D_s is the GMR of each subconductor of a four subconductor bundle conductor and d is the bundle spacing. What is the GMR of the equivalent single conductor?
(a) $1.09\sqrt{D_s \times d^3}$ (b) $1.09\sqrt{D_s^3 \times d^3}$
(c) $1.09\sqrt[4]{D_s^3 \times d^3}$ (d) $1.09\sqrt[4]{D_s \times d^3}$
[GATE E.E. 2004]

151. Consider a bundled conductor of an overhead line, consisting of three identical subconductors placed at the corners of an equilateral triangle as shown in the figure. If we neglect the charges on the other phase conductors and ground, and assume that spacing between subconductors is much larger than their radius, the maximum electric field intensity is experienced at

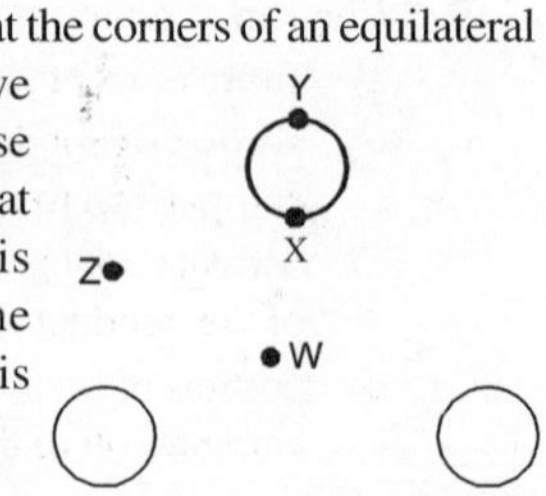

(a) Point X (b) Point Y
(c) Point Z (d) Point W
[GATE E.E. 2007]

152. Proximity effect
(a) is more pronounced for large conductors, high frequencies and close proximity.
(b) increases the resistance of the conductors and reduces the self reactance.
(c) is substantially eliminated with stranded conductors.
(d) all of the above.

Answers: **132.** (c), **133.** (a), **134.** (a), **135.** (c), **136.** (c), **137.** (a), **138.** (c), **139.** (c), **140.** (d), **141.** (a), **142.** (a), **143.** (c), **144.** (d), **145.** (b), **146.** (d), **147.** (b), **148.** (d), **149.** (c), **150.** (d), **151.** (b), **152.** (d),

153. Following effects are associated with transmission lines
1. Skin effect. 2. Corona effect. 3. Proximity effect.
The effective resistance of a conductor is increased by
(a) 1 only. (b) 2 and 3 only.
(c) 1 and 3 only. (d) 1, 2 and 3.

154. Capacitance in equivalent circuit of a transmission line is due to
(a) current in the line. (b) difference in potential of line.
(c) leakage of current. (d) presence of magnetic flux.

155. If the separation between the three phases of a transmission line is increased then
(a) the inductance will increase and capacitance will remain unchanged.
(b) both the inductance and capacitance will increase.
(c) the inductance will increase and the capacitance will decrease.
(d) the inductance will decrease and the capacitance will increase.
[A.M.I.E. Sec B. Power Systems Winter 1994]

156. The capacitance of an overhead transmission line increases with
1. increases in mutual geometrical mean distance.
2. increase in height of conductors above ground.
Select the correct answer from the following:
(a) Both 1 and 2 are true. (b) Both 1 and 2 are false.
(c) Only 1 is true. (d) Only 2 is true.
[U.P.S.C. I.E.S. E.E.-II, 2002]

157. The charging reactance of 50 km length of the line is 1500 Ω. What is the charging reactance for 100 km length of the line?
(a) 1500 Ω (b) 3000 Ω (c) 750 Ω (d) 600 Ω
[U.P.S.C. I.E.S. E.E.-II, 2005]

158. Bundled conductors are used to
(a) reduce inductance of the line.
(b) reduce both inductance and capacitance.
(c) reduce corona loss.
(d) reduce corona loss and the line inductance.
[A.M.I.E. Sec B. Power Systems Winter 1994]

159. A 3-phase overhead transmission line has its conductors horizontally spaced with spacing between adjacent conductors equal to '*d*.' If now the conductors of the line are rearranged to form an equilateral triangle of sides equal to '*d*' then
(a) average capacitance and inductance will increase.
(b) average capacitance will decrease and inductance will increase.
(c) average capacitance will increase and inductance will decrease.
(d) surge impedance loading of the line increases.
[GATE E.E. 1993]

160. Proximity of a line to the earth surface
(a) does not affect its capacitance to neutral.
(b) increases the capacitance to neutral.
(c) decreases the capacitance to neutral.
[A.M.I.E. Sec B. Power Systems Summer 1999]

161. If the effect of earth is taken into account, then the capacitance of line to ground
(a) decreases. (b) increases.
(c) remains unaltered. (d) becomes infinite.
[U.P.S.C. I.E.S. E.E.-II, 1993]

162. h = half centre to centre spacing, r = conductor radius and $\in$ = permittivity of the medium. Which one of the following is equal to the capacitance per unit length of a two-wire transmission line?

(a) $\log_e \dfrac{\pi\in}{\left\{\left(\dfrac{h}{r}\right)+\left(\sqrt{\dfrac{h^2}{r^2}-1}\right)\right\}}$ (b) $\log_e \dfrac{2\pi\in}{\left\{\left(\dfrac{h}{r}\right)+\left(\sqrt{\dfrac{h^2}{r^2}-1}\right)\right\}}$

(c) $\log_e \dfrac{3\pi\in}{\left\{\left(\dfrac{h}{r}\right)+\left(\sqrt{\dfrac{h^2}{r^2}-1}\right)\right\}}$ (d) $\log_e \dfrac{4\pi\in}{\left\{\left(\dfrac{h}{r}\right)+\left(\sqrt{\dfrac{h^2}{r^2}-1}\right)\right\}}$
[U.P.S.C. I.E.S. E.E.-I, 2007]

163. For equilateral spacing of conductors of an untransposed 3-phase line, we have
(a) balanced receiving-end voltage and no communication interference.
(b) unbalanced receiving-end voltage and no communication interference.
(c) balanced receiving-end voltage and communication interference.
(d) unbalanced receiving-end voltage and communication interference. **[GATE E.E. 1996]**

164. Transmission lines are transposed to
(a) reduce corona loss. (b) reduce skin effect.
(c) prevent interference with neighbouring telephone lines.
(d) prevent short-circuit between any two lines.
[U.P.S.C. I.E.S. E.E.-II, 1996]

165. Transposition of transmission line is done to
(a) reduce line loss. (b) reduce skin effect.
(c) balance line voltage drop.
(d) reduce corona. **[W.B. Univ. of Technology Power System-I, 2009-10]**

166. High voltage transmission lines are transposed because then
(a) corona losses can be minimized.
(b) computation of inductance becomes easier.
(c) voltage drop in the lines can be minimized.
(d) phase voltage imbalances can be minimized.
[A.M.I.E. Sec B. Winter 1994]

167. The concept of an electrically short, medium and long line is primarily based on the
(a) nominal voltage of the line. (b) physical length of the line.
(c) wavelength of the line.
(d) power transmitted over the line. **[GATE E.E. 2006]**

168. Equivalent π model is quite suitable for analyzing the performance of transmission line of
(a) 50 km length. (b) 150 km length.
(c) 250 km length. (d) All of the above lengths.
[U.P.S.C. I.E.S. E.E.-II, 2005]

169. Which of the following is neglected while analysing a short transmission line ?
(a) Shunt admittances. (b) Power losses.
(c) Series impedance. (d) None of the above.

170. In modelling the equivalent circuit of a short length overhead transmission line, the line resistance and inductance are only considered because line capacitance to ground is
(a) equal to zero. (b) finite but very small.
(c) finite but very large. (d) infinite.
[U.P.S.C. I.E.S. E.E.-II, 2010]

171. For some given transmission line expression for voltage regulation is given $\dfrac{|V_S|-|V_R|}{|V_R|}\times 100\%$. Hence,
(a) this must be a 'short line'.
(b) this may either be a 'medium line' or 'short line'.
(c) this expression is true for any line.
(d) this may either be a 'medium line' or 'long line'.
[U.P.S.C. I.E.S. E.E.-II, 2002]

Answers: **153.** (d), **154.** (b), **155.** (c), **156.** (b), **157.** (c), **158.** (d), **159.** (c), **160.** (b), **161.** (b), **162.** (a), **163.** (d), **164.** (c), **165.** (c), **166.** (d), **167.** (b), **168.** (c), **169.** (a), **170.** (b), **171.** (a),

172. As compared to sending-end voltage, the receiving-end voltage of a short line under no-load condition is
(a) higher. (b) lower. (c) remains the same.
[U.P.S.C. I.E.S. E.E.-II, 2000]

173. If a short transmission line is delivering to lagging pf load, the sending-end pf would be (notations having their usual meaning)

(a) $\dfrac{V_R \cos\phi + IR\sin\phi}{V_S}$ (b) $\dfrac{V_R \cos\phi + IR}{V_S}$

(c) $\dfrac{V_R \sin\phi + IR}{V_S}$ (d) $\dfrac{V_R \sin\phi + IR\cos\phi}{V_S}$

[U.P.S.C. I.E.S. E.E.-II, 2000]

174. Which of the following voltage regulation is considered to be the best ?
(a) 2% (b) 30% (c) 70% (d) 98%
[A.M.I.E. Sec B. Power Systems Winter 1999]

175. The regulation of a line at full load 0.8 pf lagging is 12%. The regulation at full-load 0.8 pf leading can be
(a) 24% (b) 18% (c) 12% (d) 4%
[A.M.I.E. Sec B. Power Systems Summer 1996]

176. If in a short transmission line, resistance and inductive reactance are found to be equal and regulation appears to be zero, then the load will
(a) have unity power factor. (b) have zero power factor.
(c) be 0.707 leading. (d) be 0.707 lagging.
[U.P.S.C. I.E.S. E.E.-II, 1997]

177. A short transmission line, having its line impedance angle as θ, is delivering a given power at the receiving end at a lagging power factor angle of ϕ. Which one of the following is a set of conditions for which this line will have maximum and zero regulation ?

	Maximum Regulation	Zero Regulation
(a)	$\phi = \theta$	$\phi - \theta = \dfrac{\pi}{2}$
(b)	$\phi - \theta = \dfrac{\pi}{2}$	$\phi = \theta$
(c)	$\phi = \theta$	$\phi + \theta = \dfrac{\pi}{2}$
(d)	$\phi + \theta = \dfrac{\pi}{2}$	$\phi = \theta$

[U.P.S.C. I.E.S. E.E.-II, 1996]

178. In a short transmission line, voltage regulation is zero what the power factor angle of the load at the receiving end side is equal to

(a) $\tan^{-1}\left(\dfrac{X}{R}\right)$ (b) $\tan^{-1}\left(\dfrac{R}{X}\right)$

(c) $\tan^{-1}\left(\dfrac{X}{Z}\right)$ (d) $\tan^{-1}\left(\dfrac{R}{Z}\right)$

[U.P.S.C. I.E.S. E.E.-II, 2001]

179. If X is the inductive reactance/phase and R is the resistance/phase of a short transmission line, what is the power factor angle of the load for maximum voltage regulation?
(a) $\cos^{-1}$ X/R (b) $\tan^{-1}$ X/R
(c) $\cos^{-1}$ R/X (d) $\tan^{-1}$ R/X
[U.P.S.C. I.E.S. E.E.-II, 2006]

180. For a short line if the receiving-end voltage is equal to sending-end voltage under loaded conditions
(a) the sending-end power factor is unity.
(b) the receiving-end power factor is unity.
(c) the sending-end power factor is leading.
(d) the receiving-end power factor is leading.
[A.M.I.E. Sec B. Summer 1994]

181. A single phase transmission line of impedance j 0.8 ohm supplies a resistive load of 500 A at 300 V. The sending-end power factor is
(a) unity. (b) 0.8 lagging.
(c) 0.8 leading (d) 0.6 lagging.
[U.P.S.C. I.E.S. E.E.-II, 1996]

182. For an ac transmission line of length not exceeding 80 km, it is usual to lump the line capacitance at
(a) the sending end. (b) the receiving end.
(c) the midpoint. (d) any convenient point.

183. The active and the reactive power delivered at the receiving end of a short transmission line of impedance $Z\angle\Psi$ are respectively given by

$$P_R = \frac{V_S V_R}{Z}\cos(\Psi - \delta) - \frac{V_R^2}{Z}\cos\Psi, \text{ and}$$

$$Q_R = \frac{V_S V_R}{Z}\sin(\Psi - \delta) - \frac{V_R^2}{Z}\sin\Psi, \text{ with}$$

V_S and V_R being the magnitude of voltages at the sending and receiving ends, δ the power-angle. At the power-limit condition i.e. for maximum P_R
(a) leading VARs (Q_R) goes to the load for any value of V_S and V_R.
(b) leading VARs (Q_R) goes to the load ONLY for $V_S = V_R$.
(c) lagging VARs (Q_R) goes to the load for any value of V_S and V_R.
(d) lagging VARs (Q_R) goes to the load ONLY for $V_S = V_R$.
[U.P.S.C. I.E.S. E.-II, 2001]

184. If L,C and Y are the inductance, capacitance and shunt admittance of a line per unit length, then for length l.
(a) the shunt admittance is Y.l. (b) the inductance is L/l.
(c) the capacitance is C/l. (d) the shunt admittance is Y/l.
[A.M.I.E. Sec B. Summer 1994]

185. Transmission efficiency of a transmission line increases with the
(a) decrease in power factor and voltage.
(b) increase in power factor and voltage.
(c) increase in power factor but decrease in voltage.
(d) increase in voltage but decrease in power factor.

186. Consider two parallel short transmission lines of impedances Z_A and Z_B respectively. Currents I_A and I_B are both lagging and the sending-end voltage is V_S. If the reactance to resistance ratio of both the impedances Z_A and Z_B are equal, then the total current I will

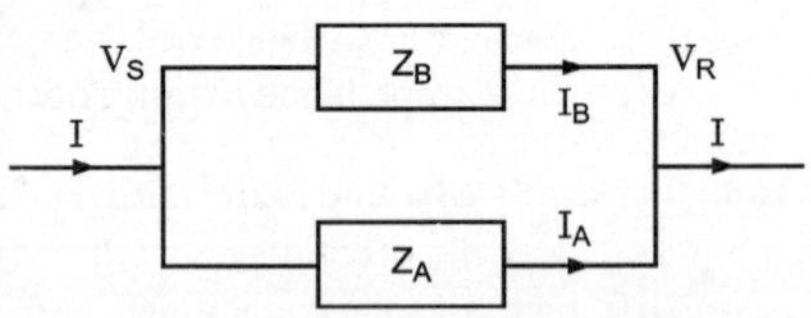

(a) lag both I_A and I_B. (b) lead both I_A and I_B.
(c) lag one of I_A and I_B but lead the other.
(d) be in phase with both I_A and I_B. [U.P.S.C. I.E.S. E.E.-II, 1997]

187. Under no-load conditions, the current in a transmission line is because of
(a) capacitance effect. (b) corona effect.
(c) proximity effect. (d) back flow from earth.

188. Which of the following statements are correct ?
(a) Flow of unduly heavy current is Ferranti effect.
(b) Ferranti effect occurs under unloaded condition of line.
(c) The rise in receiving-end voltage is Ferranti effect.
(d) Both (b) and (c) combined is Ferranti effect.

Answers: **172.** (c), **173.** (b), **174.** (a), **175.** (d), **176.** (c), **177.** (c), **178.** (b), **179.** (b), **180.** (d), **181.** (d), **182.** (b), **183.** (a), **184.** (a), **185.** (b), **186.** (d), **187.** (a), **188.** (d),

189. In a long transmission line under no-load condition
(a) the receiving-end voltage is less than the sending-end voltage.
(b) the sending-end voltage is less than the receiving-end voltage.
(c) the sending-end voltage is equal to the receiving-end voltage.
(d) none of these. **[A.M.I.E. Sec B. Winter 1993]**

190. When is the Ferranti effect on long overhead lines experienced?
(a) The line is lightly loaded. (b) The line is heavily loaded.
(c) The line is fully loaded. (d) The power factor is unity.
[U.P.S.C. I.E.S. E.E.-II, 2008]

191. Ferranti effect happens in transmission line when the line is
(a) short and loaded. (b) long and loaded.
(c) long and unloaded. (d) none of these.
[W.B. Univ. of Technology Power Systems-II, 2009-10]

192. The ABCD constants of a 3 phase transposed transmission line with linear and passive elements
(a) are always equal. (b) never equal.
(c) A and D are equal. (d) B and C are equal.
[A.M.I.E. Sec B. Winter 1993]

193. The values of A, B, C and D constants for a short transmission line are respectively
(a) Z, 0, 1 and 1. (b) 0, 1, 1 and Z.
(c) 1, Z, 0 and 1. (d) 1, 1, Z and 0.
[U.P.S.C. I.E.S. E.E.-II, 2000]

194. The angle of A, constant of the transmission line normally lies between
(a) 90º-70º (b) 70º-40º (c) 40º-10º (d) 10º-0º
[A.M.I.E. Sec B. Power Systems Summer 1993]

195. For a transmission line with resistance R, reactance X and negligible capacitance, the generalised constant A is
(a) 0 (b) 1 (c) $R + jX$ (d) $R + X$
[U.P.S.C. I.E.S. E.E.-II, 1996]

196. The value of the 'A' parameter of a transmission line
(a) increases with the increase in length of the line.
(b) decreases with the increase in line length.
(c) is independent of line length.
[A.M.I.E. Sec B. Power Systems Winter 1996]

197. For a medium length transmission line, A is
(a) equal to B. (b) equal to C. (c) equal to D.
(d) not equal to any of the above. **[A.M.I.E. Sec B. Summer 1994]**

198. The generalised constant A for a medium line has magnitude
(a) close to but less than 1. (b) close to but greater than 1.
(c) nearly equal to the series impedance of the line.
[A.M.I.E. Sec B. Power Systems Summer 1999]

199. For transmission line which one of the following relations is true?
(a) AD – BC = 1 (b) –AD – BC = 1
(c) AD – BC = –1 (d) AD – BC = 0
[U.P.S.C. I.E.S. E.E.-II, 2002]

200. Which one of the following equations is correct?
(a) –AB + CD = – 1 (b) AD + CD = 1
(c) AB — CD = – 1 (d) – AD + BC = – 1
where A, B, C and D are generalized circuit constants.
[U.P.S.C. I.E.S. E.E.-II, 2004]

201. A 10 km long lossless transmission line has a reactance of 0.3 Ω/km and negligible shunt capacitance. The value of $\begin{bmatrix} A & B \\ C & D \end{bmatrix}$ is

(a) $\begin{bmatrix} 1 & 0 \\ j3 & 1 \end{bmatrix}$ (b) $\begin{bmatrix} 1 & 0 \\ 0.3 & 1 \end{bmatrix}$ (c) $\begin{bmatrix} 1 & j3 \\ 0 & 1 \end{bmatrix}$ (d) $\begin{bmatrix} j3 & 0 \\ 1 & 1 \end{bmatrix}$

[U.P.S.C. I.E.S. E.E.-II, 2010]

202. The ABCD constants of a 3-phase transmission line are
$A = D = 0.8 \angle 1°$ $B = 170 \angle 85°\ \Omega$ $C = 0.002 \angle 90.4°$ mho
The sending-end voltage is 400 kV. The receiving-end voltage under no-load condition is
(a) 400 kV (b) 500 kV (c) 320 kV (d) 417 kV
[U.P.S.C. I.E.S. E.E.-II, 2001]

203. The A, B, C, D constants of a 220 kV line are:
$A = D = 0.94 \angle 10°$, $B = 130 \angle 73°$, $C = 0.001 \angle 90°$. If the sending-end voltage of the line for a given load delivered at a nominal voltage is 240 kV, the % voltage regulation of the line is
(a) 5 (b) 9 (c) 16 (d) 21
[U.P.S.C. I.E.S. E.E.-II, 2006]

204. The ABCD parameters of a 3-phase overhead transmission line are $A = D = 0.9 \angle 0°$. $B = 200 \angle 90°\ \Omega$ and $C = 0.95 \times 10^{-3} \angle 90°$ S. At no-load condition a shunt inductive reactor is connected at the receiving end of the line to limit the receiving-end voltage to be equal to the sending-end voltage. The ohmic value of the reactor is
(a) ∞ Ω (b) 2000 Ω (c) 105.26 Ω (d) 1052.6 Ω
[GATE E.E. 2003]

205. The generalized circuit constants of a 3-phase, 220 kV rated voltage medium length transmission line are

$$A = D = 0.936 + j\,0.016 = 0.936 \angle 0.98°$$
$$B = 33.5 + j\,138 = 142.0 \angle 76.4°\ \Omega$$
$$C = (-5.18 + j\,914) \times 10^{-6}\ ℧$$

If the load at the receiving end is 50 MW at 220 kV with a power factor of 0.9 lagging, then magnitude of line to line sending-end voltage should be
(a) 133.23 kV (b) 220.00 kV
(c) 230.78 kV (d) 246.30 kV **[GATE E.E. 2004]**

206. A 220 kV, 20 km long 3-phase transmission line has the following A, B, C, D constants.
$A = D = 0.96 \angle 3°$, $B = 55 \angle 65°$ Ω/phase,
$C = 0.0005 \angle 80°$ S/phase. Its charging current per phase is
(a) $\frac{11}{\sqrt{3}}$ A (b) 11 A (c) 220 A (d) $\frac{220}{\sqrt{3}}$ A
[GATE E.E. 1999]

207. The velocity of propagation of electromagnetic waves on overhead lines is
(a) 3×10^8 m/s (b) 3×10^8 km/s
(c) 3×10^{10} m/s (d) 3×10^8 km/hour
[A.M.I.E. Sec B. Power Systems Summer 1996]

208. A cable has the following characteristics :
L = 0.20/μH/m and C = 196.2 pF/m.
The velocity of wave propagation through the cable is
(a) 32 m/s (b) 159.24 m/μs
(c) 0.0312 m/s (d) 159.24 m/s

209. The characteristic impedance of a transmission line with series impedance Z ohms per unit length and shunt admittance Y mhos per unit length, is given by
(a) $(Z + Y)^2$ (b) $\sqrt{(Z.Y)}$ (c) $\sqrt{\left(\frac{Z}{Y}\right)}$ (d) $\sqrt{\left(\frac{Z}{Y}\right)^2}$
[U.P.S.C. I.E.S. E.E.-II, 1999]

Answers: **189.** (b), **190.** (a), **191.** (c), **192.** (c), **193.** (c), **194.** (d), **195.** (b), **196.** (a), **197.** (c), **198.** (a), **199.** (a), **200.** (d), **201.** (c), **202.** (b), **203.** (c), **204.** (b), **205.** (c), **206.** (a), **207.** (a), **208.** (b), **209.** (c),

210. If the open-and short-circuit input impedance of a lossless line are respectively Z_1 and Z_2, its characteristic impedance will be

(a) $Z_1 + Z_2$ (b) $\sqrt{Z_1^2 + Z_2^2}$ (c) $\dfrac{Z_1^2 + Z_2^2}{Z_1 + Z_2}$ (d) $\sqrt{Z_1 Z_2}$

[U.P.S.C. I.E.S. E.E.-I, 2000]

211. A transmission line section shows an input impedance of 36 Ω and 64 Ω respectively, when short-circuited and open-circuited. What is the characteristic impedance of the transmission line?

(a) 100 Ω (b) 50 Ω
(c) 45 Ω (d) 48 Ω **[U.P.S.C. I.E.S. E.E.-I, 2009]**

212. The open-circuit and short-circuit impedances of a line are 100 Ω each. What is the characteristic impedance of the line?

(a) $100\sqrt{2}$ Ω (b) 100 Ω (c) $\dfrac{100}{\sqrt{2}\ \Omega}$ (d) 50 Ω

[U.P.S.C. I.E.S. E.E.-I, 2005]

213. A lossless transmission line of characteristic impedance 300 Ω and length λ/2 is shorted at one end and is terminated in its characteristic impedance at the other. The input impedance measured at the mid section of the line is

(a) 0 Ω (b) 100 Ω
(c) 300 Ω (d) 150 Ω **[U.P.S.C. I.E.S. E.E.-I, 2011]**

214. A cable has inductance of 0.22 mH per km and capacitance of 0.202 μF per km. The surge impedance of the cable is

(a) 28 Ω (b) 33 Ω (c) 42 Ω (d) 50 Ω

[U.P.S.C. I.E.S. E.E.-II, 2002]

215. A 50 Ω distortionless transmission line has a capacitance of 10^{-10} F/m. What its inductance per metre?

(a) 0.25 μH (b) 500 μH (c) 5000 pH (d) 50 μH

[U.P.S.C. I.E.S. E.E.-I, 2005]

216. Consider the following statements:
The characteristic impedance of a transmission line can increase with the increase in

1. resistance per unit length.
2. conductance per unit length.
3. capacitance per unit length.
4. inductance per unit length.

Which of the statements are correct?

(a) 1 and 2 (b) 2 and 3 (c) 1 and 4 (d) 3 and 4

[U.P.S.C. I.E.S. E.E.-I, 2001]

217. Which of the following statements is correct ?

(a) Surge impedance is the impedance at the time of breakdown of voltage.
(b) Surge impedance and characteristic impedance for a transmission line are the same.
(c) Surge impedance is the impedance of transmission line when corona takes place.
(d) None of the above.

218. The characteristic impedance of a transmission line depends upon

(a) shape of the conductor.
(b) conductivity of the conductor material.
(c) geometrical configuration of the conductors.
(d) none of the above.

219. For a lossy transmission line, the characteristic impedance does *not* depend on

(a) the operating frequency of the line.
(b) the conductivity of the conductors.
(c) conductivity of the dielectric separating the conductors.
(d) length of the line. **[U.P.S.C. I.E.S. E.E.-I, 2011]**

220. In case the characteristic impedance of a transmission line is equal to the load impedance

(a) the system will resonate badly.
(b) all the energy sent will be absorbed by the load.
(c) all the energy sent will pass to the earth.
(d) all the energy will be lost in transmission line as transmission losses.

221. In a transmission line having negligible resistance, the surge impedance is

(a) $\sqrt{L+C}$ (b) $\sqrt{C/L}$
(c) $\sqrt{\dfrac{1}{LC}}$ (d) $\sqrt{L/C}$

[W.B. Univ. of Technology Power System-I, 2009-10]

222. Characteristics of an overhead line is usually in the range of

(a) 100 ohms–200 ohms. (b) 200 ohms–300 ohms.
(c) 400 ohms–500 ohms. (d) 0 ohm–100 ohms.

[W.B. Univ. of Technology Power System-I, 2009-10]

223. Surge impedance of a transformer is in the range of

(a) 80-100 Ω. (b) 400-500 Ω.
(c) 1,000-2,000 Ω. **[A.M.I.E. Sec B. Summer 1997]**

224. The surge impedance of a 400 kV, 100 km transmission line is 300 ohms. For a 200 km length the line surge impedance will be

(a) 300 ohms. (b) 150 ohms.
(c) 600 ohms. (d) 400 ohms.

[U.P.S.C. I.E.S. E.E.-II, 2005]

225. The characteristic impedance of a 250 km long overhead line is 400 Ω. For a line of 200 km length of the same line, the characteristics impedance will be

(a) 50 Ω (b) 400 Ω
(c) 200 Ω (d) 800 Ω

[W.B. Univ. of Technology Power Systems-III, 2009-10]

226. Surge impedance values for cables are

(a) higher than those of overhead lines.
(b) lower than those of overhead lines.
(c) of the same order as those of overhead lines.

[A.M.I.E. Sec B. Winter 1996]

227. Multiple conductor lines have

(a) lower surge impedance than single conductor lines.
(b) higher surge impedance than single line conductors.
(c) higher surge impedance than single conductor lines for short lengths and lower surge impedance for long lines.

[A.M.I.E. Sec B. Winter 1996]

228. A transmission line is distortionless if

(a) RG = LC (b) RC = GL
(c) $\dfrac{R}{C} = \dfrac{G}{L}$ (d) R = G

[U.P.S.C. I.E.S. E.E.-I, 2011]

229. The propagation constant of a transmission line is given as

(a) $j\omega\sqrt{LC}$ (b) $j\sqrt{LC}$
(c) $j\sqrt{L/C}$ (d) $j\sqrt{C/L}$

230. The transmission capacity of a line at 50 Hz frequency as compared to that at 60 Hz frequency is

(a) lower. (b) higher.
(c) the same. **[A.M.I.E. Sec B. Power Systems Summer 1998]**

231. As the line inductance increases the transmission capacity

(a) increases. (b) decreases.
(c) remains the same. **[U.P.S.C. I.E.S. E.E. 2000]**

Answers: **210.** (d), **211.** (d), **212.** (b), **213.** (c), **214.** (b), **215.** (a), **216.** (c), **217.** (b), **218.** (c), **219.** (d), **220.** (b), **221.** (d), **222.** (c), **223.** (c), **224.** (a), **225.** (b), **226.** (b), **227.** (a), **228.** (b), **229.** (a), **230.** (b), **231.** (b),

232. What is the characteristic impedance Z_0 of a line having resistance R, inductance L, capacitance C and conductance G?

(a) $\sqrt{\dfrac{R + j\omega L}{G + j\omega C}}$ (b) $\sqrt{\dfrac{G + j\omega C}{R + j\omega L}}$

(c) $R + j\omega L - \dfrac{G}{\omega C}$ (d) $\sqrt{R^2 + \left(\omega L - \dfrac{1}{\omega C}\right)^2}$

[U.P.S.C. I.E.S. E.T.E.-I, 2006]

233. A transmission line has R, L, G, C distributed parameters per unit length of line. If γ is the propagation constant of the line, which one of the following expressions represents the characteristic impedance of the line?

(a) $\dfrac{\gamma}{R + j\omega L}$ (b) $\dfrac{R + j\omega L}{\gamma}$

(c) $\dfrac{G + j\omega C}{\gamma}$ (d) $\sqrt{\dfrac{G + j\omega C}{R + j\omega L}}$

[U.P.S.C. I.E.S. E.E.-I, 2006]

234. When the load on a transmission line is equal to the surge impedance loading
(a) the receiving-end voltage is less than the sending-end voltage.
(b) the sending-end voltage is less than the receiving-end voltage.
(c) the receiving-end voltage is equal to the sending-end voltage.
(d) none of these. [A.M.I.E. Sec B. Winter 1993]

235. The recieving-end voltage of a transmission line will be greater than the sending-end voltage if the load is
(a) greater than SIL (surge impedance loading).
(b) less than SIL. (c) equal to SIL.
[A.M.I.E. Sec B. Winter 1996]

236. The SIL of a single circuit 220 kV line is around
(a) 120 MW (b) 90 MW (c) 220 MW (d) 400 MW
[A.M.I.E. Sec B. Summer 1998]

237. What is the surge impedance loading of a lossless 400 kV, 3-phase, 50 Hz overhead line of average of surge impedance of 400 ohms?

(a) 400 MW (b) 400 $\sqrt{3}$ MW (c) $\dfrac{400}{\sqrt{3}}$ MW (d) 400 kW

[U.P.S.C. I.E.S. E.E.-II, 2008]

238. Consider the following statements:
Surge impedance loading of a transmission line can be increased by
1. increasing its voltage level.
2. addition of lumped inductance in parallel.
3. addition of lumped capacitance in series.
4. reducing the length of the line.
Of these statements
(a) 1 and 3 are correct. (b) 1 and 4 are correct.
(c) 2 and 4 are correct. (d) 3 and 4 are correct.
[U.P.S.C. I.E.S. E.E.-II, 2001]

239. The L/C ratio for 132 kV and 400 kV lines are typically 160×10^3 and 62.5×10^3 respectively. The natural 3-phase loading for the two lines will be respectively
(a) 108.9 and 2,560 MW (b) 44 and 2,560 MW
(c) 44 and 640 MW (d) 640 and 44 MW
[U.P.S.C. I.E.S. E.E.-II, 1996]

240. Which one of the following is the correct expression for the propagation constant in a transmission line?

(a) $(R - j\omega L)(G - j\omega C)$ (b) $\sqrt{\dfrac{(R - j\omega L)}{(G - j\omega C)}}$

(b) $\sqrt{\dfrac{(R + j\omega L)}{(G + j\omega C)}}$ (d) $\sqrt{(R + j\omega L)(G + j\omega C)}$

[U.P.S.C. I.E.S. E.E.-I, 2004]

241. A lossless radial transmission line with surge impedance loading
(a) takes negative VAR at sending end and zero VAR at receiving end.
(b) takes positive VAR at sending end and zero VAR at receiving end.
(c) has flat voltage profile and unity power factor at all points along it.
(d) has sending-end voltage higher than receiving-end voltage and unity power factor at sending end.
[U.P.S.C. I.E.S. E.E.-I, 2001]

242. What does the standing wave ratio (SWR) of unity imply ?
(a) Transmission line is open-circuited.
(b) Tran....ssion line is short-circuited.
(c) Transmission line's characteristic impedance is equal to load impedance.
(d) Transmission line's characteristic impedance is not equal to load impedance. [U.P.S.C. I.E.S. E.E.-I, 2007]

243. Consider the following statements:
1. Equivalent-T circuit of a long line is preferred to equivalent-π circuit.
2. The nature of reactive power compensation is different for peak load and off-peak load conditions.
3. Ferranti effect is significant only on medium and long lines.
Which of the statements given above are correct?
(a) 1 and 2 only (b) 1 and 3 only
(c) 2 and 3 only (d) 1, 2 and 3
[U.P.S.C. I.E.S. E.E.-II, 2008]

244. The propagation constant of a transmission line is $0.15 \times 10^{-3} + j\,1.5 \times 10^{-3}$.
The wavelength of the travelling wave is

(a) $\dfrac{1.5\times10^{-3}}{2\pi}$ (b) $\dfrac{2\pi}{1.5\times10^{-3}}$

(c) $\dfrac{1.5\times10^{-3}}{\pi}$ (d) $\dfrac{\pi}{1.5\times10^{-3}}$

[U.P.S.C. I.E.S. E.E.-II, 1997]

245. For a good voltage profile under no-load condition, a long line needs
(a) shunt capacitors at receiving end.
(b) shunt reactors at the receiving end.
(c) shunt resistance at receiving end. [A.M.I.E. Sec B. Summer 1996]

246. No-load compensation of a high voltage line involves
(a) shunt capacitors. (b) shunt reactors.
(c) series capacitors. [A.M.I.E. Sec B. Summer 1997]

247. Full-load compensation in a line requires
(a) shunt capacitors. (b) series capacitors.
(c) transformers. (d) shunt reactors.
[A.M.I.E. Sec B. Summer 1997]

248. Use of additional shunt capacitor can be made for increasing the capability of line as it
(a) reduces surge impedance Z_0.
(b) increases phase shift β.
(c) increase in α. (d) all of the above.

249. To increase the transmission capability of a high voltage long line
(a) the resistance can be increased.
(b) the resistance can be decreased.

Answers: **232.** (a), **233.** (b), **234.** (c), **235.** (b), **236.** (a), **237.** (a), **238.** (a), **239.** (c), **240.** (d), **241.** (c), **242.** (c), **243.** (c), **244.** (b), **245.** (b), **246.** (b), **247.** (a), **248.** (a), **249.** (c),

(c) the series reactance can be reduced.
(d) the shunt admittance can be reduced.
[A.M.I.E. Sec B. Summer 1994]

250. Power despatch through a line can be increased by
(a) installing series capacitors. (b) installing shunt capacitors.
(c) installing series reactor. (d) installing shunt reactor.
[A.M.I.E. Sec B. Summer 1993]

251. The power transmitted will be maximum when
(a) line reactance is high. (b) corona losses are minimum.
(c) sending-end voltage is more.
(d) receiving-end voltage is more.

252. For constant voltage transmission, the voltage drop along the line is maintained constant by installing
(a) capacitors. (b) inductors.
(c) resistors.
(d) synchronous phase modifiers at the receiving end.

253. Constant voltage transmission have the drawback(s) of
(a) increase of short-circuit current of the system.
(b) lower reserve of lines in case of line trouble.
(c) increased risk of interruption of supply due to falling of synchronous motors out of synchronism.
(d) all of the above.

254. Constant voltage transmission have the advantage(s) of
(a) availability of steady voltage at all loads at the receiving end.
(b) possibility of better protection for the line due to possible use of higher terminal reactances.
(c) possibility of carrying increased power for a given conductor size in case of long distance heavy power transmission.
(d) all of the above.

255. Constant voltage transmission have the advantage(s) of
(a) increase of short-circuit current of the system.
(b) large reserve of lines in case of line trouble.
(c) improvement of power factor at the times of moderate and heavy loads.
(d) all of the above.

256. A synchronous condensor is a
(a) synchronous generator. (b) paper condensor.
(c) synchronous motor. (d) none of these.
[W.B. Univ. of Technology Power System-III, 2009-10]

257. A synchronous compensator absorbs inductive reactive power. It is
(a) overexcited. (b) normally excited.
(c) underexcited. (d) none of these.
[W.B. Univ. of Technology Power Systems-I, 2009-10]

258. A synchronous phase modifier supplies
(a) both active and reactive powers.
(b) both lagging and leading reactive power.
(c) inductive reactive power only. **[A.M.I.E. Sec B. Summer 1999]**

259. A synchronous phase modifier as compared to synchronous motor of the same rating has
(a) larger shaft diameter and higher speed.
(b) smaller shaft diameter and higher speed.
(c) larger shaft diameter and smaller speed.
(d) smaller shaft diameter and smaller speed.
[A.M.I.E Sec B. Winter 1993]

260. Phase modifier is normally installed in case of
(a) short transmission lines. (b) medium length lines.
(c) long lines. (d) for any length of lines.
[A.M.I.E Sec B. Winter 1994; Summer 1995]

261. Which of the following statements is not true ?
(a) Synchronous phase modifiers are installed at sending end.
(b) Synchronous phase modifiers are installed at load end.
(c) Synchronous phase modifiers are nothing but synchronous motors (specially designed).
(d) Synchronous phase modifiers do not carry load.

262. Capacitors are used in power system to
(a) improve supply power factor.
(b) improve voltage regulation.
(c) change the load characteristics. (d) all of the above.

263. Series capacitors on transmission lines are of little use when
(a) the load VAR requirement is small.
(b) the load VAR requirement is large.
(c) the load VAR requirement is fluctuating.
(d) series capacitors are never used in transmission lines.
[A.M.I.E. Sec B. Winter 1998]

264. The reactive power transfer over a line mainly depends on
(a) power angle δ. (b) $|V_S| - |V_R|$
(c) V_S. (d) V_R. **[A.M.I.E. Sec B. Summer 1998]**

265. For a fixed value of complex power flow in a transmission line having a sending end voltage V, the real power loss will be proportional to
(a) V (b) V^2 (c) $\frac{1}{V^2}$ (d) $\frac{1}{V}$
[GATE E.E. 2009]

266. Consider the model shown in figure of a transmission line with a series capacitor at its mid-point. The maximum voltage on the line is at the location

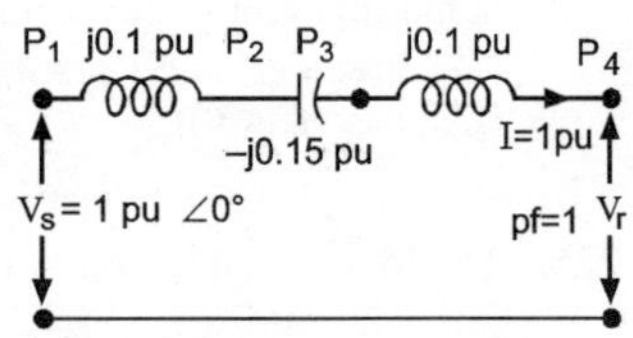

(a) P_1 (b) P_2 (c) P_3 (d) P_4 **[GATE E.E. 2001]**

267. A transmission line has equal voltages at the two ends, maintained constant by two sources. A third source is to be provided to maintain constant voltage (equal to end voltages) at either the midpoint of the line or at 755 of the distance from the sending end. Then the maximum power transfer capabilities of the line in the original case and the other two cases respectively will be in the following ratios.
(a) 1 : 1 : 1. (b) 1 : 2 : 1/0.75.
(c) 1 : 2 : 4. (d) 1 : 4 : 16. **[GATE E.E. 2000]**

268. For a fixed receiving-end and sending-end voltages in a transmission system, what is the locus of the constant power?
(a) A straight line. (b) An ellipse.
(c) A parabola. (d) A circle.
[U.P.S.C. I.E.S. E.E.-II, 2004]

269. Two networks are connected in cascade as shown in the figure. With the usual notations the equivalent A, B, C and D constants are obtained. Given that, C = 0.025 ∠45°, the value of Z_2 is

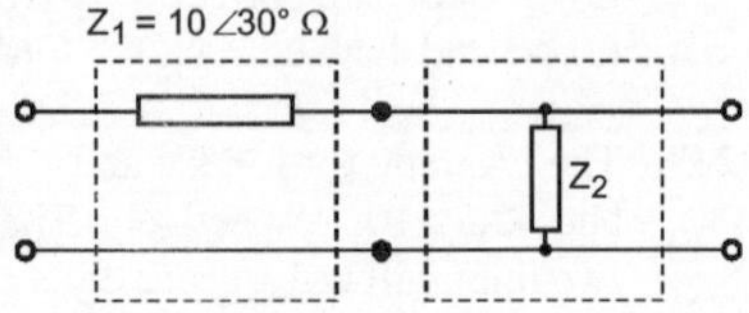

(a) 10 ∠30° Ω (b) 40 ∠–45° Ω
(c) 1 Ω (d) 0 Ω **[GATE E.E.-II, 2005]**

270. Transmission of power by ac cables is impossible beyond
(a) 35-45 km (b) 500 km
(c) 300 km **[A.M.I.E. Sec B. Power Systems Winter 1998]**

Answers: 250. (a), **251.** (d), **252.** (d), **253.** (d), **254.** (d), **255.** (c), **256.** (c), **257.** (c), **258.** (b), **259.** (b), **260.** (c), **261.** (a), **262.** (b), **263.** (a), **264.** (b), **265.** (b), **266.** (c), **267.** (a), **268.** (d), **269.** (b), **270.** (a),

271. Conduit pipe is generally employed for protection of
(a) unsheathed cables. (b) armoured cables.
(c) PVC sheathed cables. (d) all of above.

272. Sheaths are used in cables to
(a) provide proper insulation. (b) provide mechanical strength.
(c) prevent ingress of moisture. (d) none of the above.
[A.M.I.E Sec B. Winter 1993; Summer 1995]

273. The bedding on a cable consists of
(a) jute strands. (d) hessian tape.
(c) paper tape compounded with a fiberous material.
(d) any of the above.

274. The material(s) used for armouring of an underground cable is/are
(a) galvanised steel wire. (b) steel tape.
(c) aluminium. (d) either (a) or (b).

275. Metallic shielding is provided on underground cables to
(a) reduce thermal resistance.
(b) reduce corona effect.
(c) control the electrostatic voltage stress.
(d) all of the above.

276. Metallic shielding provided on cables is usually of thickness.
(a) 0.1–0.8 mm. (b) 3–5 mm.
(c) 10–15 mm. (c) 15–25 mm.

277. The effect of bonding the cable is
(a) to increase the effective resistance and inductance.
(b) to increase the effective resistance but reduce inductance.
(c) to reduce the effective resistance and inductance.
(d) to reduce the effective resistance but increase the inductance.
[A.M.I.E Sec B. Winter 1994]

278. The thickness of insulation layer provided on the conductor, in cables, depends upon
(a) operating voltage. (b) current to be carried.
(c) power factor. (d) both (a) and (b).

279. The insulating material used for cables should have all of the following except
(a) high dielectric strength, high mechanical strength, high tensile strength and plasticity, high resistivity and high viscosity at impregnation temperature.
(b) low thermal coefficient, low permittivity, acid proof, non-inflammable and non-hygroscopic.
(c) high water absorption.
(d) capability of withstanding high rupturing voltages.

280. Empire tape is
(a) varnished cambric. (b) impregnated paper.
(c) vulcanised rubber. (d) enamel insulation.

281. Paper as an insulating material has the main drawback that it
(a) is hygroscopic. (b) has poor dielectric strength.
(c) has low insulation resistivity.
(d) has high capacitance.

282. In paper insulated cables, the conductor x-section is usually limited to
(a) 50 mm^2 (b) 250 mm^2 (c) 600 mm^2 (d) 1200 mm^2

283. The dielectric strength of impregnated paper is about
(a) 30 kV/mm (b) 20kV/mm
(c) 15 kV/mm (d) 5 kV/mm

284. Paper used as an insulating material is usually treated with oily compound because it
(a) is hygroscopic.
(b) gets electrostatically charged at high voltage.
(c) is porous. (d) all of the above.

285. Single-core cables are usually not provided with armouring in order to
(a) avoid excessive loss in the armour.
(b) make the cable more flexible.
(c) make the cable non-hygroscopic.
(d) none of the above.

286. Single-core cable should have armour made of
(a) magnetic material.
(b) non-magnetic and non-conducting material.
(c) non-magnetic but conducting material.
[A.M.I.E. Sec B. Summer 1993]

287. Multi-core cables generally use
(a) oval shaped conductors. (b) sector shaped conductors.
(c) square conductors. (d) either (a) or (b).

288. In a 3-phase, 4-wire cable, the x-sectional area of neutral conductor is
(a) half of the area of phase conductor.
(b) equal to the area of phase conductor.
(c) double the area of phase conductor.
(d) 1.5 times the area of phase conductor.

289. The belted type construction is not suitable for cables used for voltages exceeding 22 kV because of
(a) development of both radial and tangential stresses.
(b) formation of vacuous spaces and voids on loading and unloading owing to non-homogenity of dielectric in belted construction.
(c) local heating caused by power loss at the centre filling owing to leakage current produced by tangential stresses along the impregnated paper insulation resulting in breakdown at any time.
(d) all of the above.

290. SL type cables, over H-type cables, have the advantage(s) of
(a) possibility of bending of cables owing to no overall lead sheath.
(b) less tendency for oil drainage on hilly routes owing to elimination of filler spaces containing compound.
(c) easy manufacturing.
(d) both (a) and (b).

291. Screened type cables, over belted cables, have the advantage(s) of
(a) reduced possibility of core to core faults.
(b) uniform radial electric stresses in all sections of the dielectric.
(c) no possibility of formation of voids within the dielectric.
(d) increased current carrying capacity.
(e) all of the above.

292. Solid type cables are not considered suitable for operating voltages exceeding 66 kV because
(a) skin effect dominates on the conductor.
(b) there is a danger of breakdown of insulation because of formation of voids in the layers of dielectric.
(c) there is a corona loss between conductor and sheath material.
(d) insulation may melt due to heating.

293. In a 3-core extra high voltage cable, a metallic screen around each core insulation is provided to
(a) facilitate heat dissipation. (b) give mechanical strength.
(c) obtain radial electric stress.
(d) obtain longitudinal electric stress. **[U.P.S.C. I.E.S. E.E.-II, 1997]**

Answers: **271.** (a), **272.** (c), **273.** (d), **274.** (d), **275.** (d), **276.** (b), **277.** (b), **278.** (a), **279.** (c), **280.** (a), **281.** (a), **282.** (c), **283.** (a), **284.** (b), **285.** (a), **286.** (c), **287.** (d), **288.** (b), **289.** (d), **290.** (d), **291.** (e), **292.** (b), **293.** (c),

294. As the operating voltage and consequently the electric stress on the dielectric of solid type cable is increased from a low value, the dielectric power factor cos ϕ remains almost unchanged up to a certain value of the stress beyond which cos ϕ increases very rapidly. This is due to increase in
(a) resistivity of dielectric material.
(b) ionisation in the voids present in the dielectric.
(c) core-to-core capacitance of the cable.
(d) core-to-earth capacitance of the cable. **[U.P.S.C. I.E.S. E.E.-II, 1993]**

295. At higher voltages, belted cables are more suitable than screened cables
(a) yes, because
(b) no, because at high voltages a breakdown may occur due to non-radial dielectric stresses. **[A.M.I.E. Sec B. Winter 1996]**

296. Oil-filled cables have the advantage(s) of
(a) no ionisation, oxidation and formation of voids.
(b) possibility of increased temperature range in service.
(c) more maximum permissible stresses .
(d) all of the above.

297. Oil-filled cables have the advantage(s) of
(a) smaller overall size. (b) most perfect impregnation.
(c) easy detection of fault. (d) all of the above.

298. Oil-filled cables have the drawback(s) of
(a) no possibility of impregnation after sheathing.
(b) greater cost and complicated laying of cables and maintenance.
(c) high thermal resistance.
(d) all of the above.

299. Cables used for 220 kV lines are invariably
(a) compressed oil or compressed gas insulated.
(b) paper insulated.
(c) mica insulated. (d) none of the above.

300. The insulation used in a cable designed for use on 1,000 kV is usually
(a) impregnated paper. (b) compressed SF_6 gas.
(c) PVC. (d) any of the above.

301. Sulphur hexafluoride cable is insulated by
(a) impregnated paper. (b) polyvinyl chloride.
(c) high pressure oil. (d) compressed gas.
[A.M.I.E. Sec B. Winter 1993]

302. In compressed gas insulated cable SF_6 has the gas pressure in the range of
(a) 10–20 mm Hg (b) 80–100 mm Hg
(c) 3–5 kg/cm^2 (d) 40–50 kg/cm^2

303. At bridge crossings and near the railway track ternary lead cables are used because they
(a) are of high tensile strength.
(b) are of low coefficient of thermal expansion.
(c) are of low specific gravity.
(d) can withstand shocks and vibrations.

304. Internal pressure cables have the advantage(s) of
(a) elimination of external accessories.
(b) suitability for vertical run without any fear of drainage with suitable designs.
(c) marked improvement in the pf of the cable dielectric with the increased pressure.
(d) all of the above.

305. With the rise in temperature, the insulation resistivity
(a) remains unchanged. (b) decreases linearly.
(c) increases linearly. (d) reduces exponentially.

306. The insulation resistance of a single-core cable is 200 MΩ/km. The insulation resistance for 5 km length is
(a) 40 MΩ (b) 1000 MΩ (c) 200 MΩ (d) 8 MΩ
[U.P.S.C. I.E.S. E.E.-II, 1999]

307. The power factor of an open-ended cable can be improved by
(a) increasing the capacitance.
(b) decreasing the capacitance.
(c) increasing the conductor resistance.
(d) increasing the insulation resistance. **[A.M.I.E. Sec B. Summer 1994]**

308. The capacitance of a cable increases
(a) linearly with the increase in cable length.
(b) linearly with the decrease in cable length.
(c) exponentially with the increase in cable length.
(d) none of the above.

309. The capacitance of a cable depends upon the
(a) length of the cable.
(b) relative permittivity of dielectric used in cable.
(c) ratio of sheath diameter and core diameter.
(d) all of the above.

310. The charging current drawn by the cable
(a) lags behind the voltage by 90°. (b) leads the voltage by 90°.
(c) leads the voltage by 180°. (d) none of the above.

311. In underground cables, the electrostatic stress is
(a) maximum at conductor surface and minimum at the sheath.
(b) minimum at conductor surface and maximum at the sheath.
(c) same at the conductor and sheath.
(d) zero at the conductor as well as on the sheath.

312. In a cable of conductor diameter 'd' and overall diameter with dielectric material 'D', the maximum dielectric stress.
(a) occurs at the conductor surface and is proportional to d.
(b) occurs at the conductor surface and is proportional to 1/d.
(c) occurs at the middle of the dielectric and is proportional to 1/D.
(d) occurs at the outer surface of the dielectric and is proportional to D. **[A.M.I.E Sec B. Winter 1994]**

313. To obtain the minimum value of stress in cables, the ratio (R/*r*) should be
(a) 2.13 (b) 2.718 (c) 1.96 (d) 1.5
[W.B. Univ. of Technology Power System-I, 2009-10]

314. The surge impedance of a 50 miles long underground cable is 50 Ω. For a 25 miles length it will be
(a) 25 Ω (b) 50 Ω (c) 100 Ω (d) none of these.
[A.M.I.E. Sec B. Summer 1995]

315. The breakdown of insulation of a cable can be avoided economically by using
(a) insulation layers of different dielectrics.
(b) intersheath.
(c) either (a) or (b). (d) none of the above.

316. Capacitance grading of cable means
(a) use of dielectrics in different concentrations.
(b) introduction of capacitances at various lengths of cable to counter the effect of inductance.
(c) use of dielectrics of different permittivities.
(d) grading according to capacitance per km length of the cable.

317. Grading of cables
(a) reduces insulation cost and increases current rating.
(b) reduces insulation cost but decreases current rating.
(c) increases both. (d) none of the obove.
[A.M.I.E. Sec B. Winter 1996]

Answers: 294. (b), **295.** (b), **296.** (d), **297.** (d), **298.** (b), **299.** (a), **300.** (b), **301.** (d), **302.** (c), **303.** (c), **304.** (d), **305.** (d), **306.** (a), **307.** (d), **308.** (a), **309.** (d), **310.** (b), **311.** (a), **312.** (b), **313.** (b), **314.** (b), **315.** (c), **316.** (c), **317.** (a),

318. The intersheaths in cables are used to
(a) provide proper stress distribution.
(b) minimize the stress. (c) use inferior insulation.
(d) provide protection against moisture and voltage surges.
(e) provide protection against current and voltage surges.

319. The desired overall diameter of the conductor without increasing its x-sectional area can be had by
(a) using aluminium core instead of copper.
(b) stranding the copper conductors around a hemp centre.
(c) stranding the copper conductors over a lead tube.
(d) any of the above.

320. In a 3-core cable, the capacitance between two conductors (with sheath earthed) is 3 μF. The capacitance per phase will be
(a) 1.5 μF (b) 3 μF (c) 6 μF (d) 12 μF
[A.M.I.E. Sec B. Summer 2000]

321. Consider a three-core, three-phase, 50 Hz, 11 kV cable whose conductors are denoted as R, Y and B in the figure. The inter-phase capacitance (C1) between each pair of conductors is 0.2 μF and the capacitance between each line conductor and the sheath is 0.4 μF. The per-phase charging current is

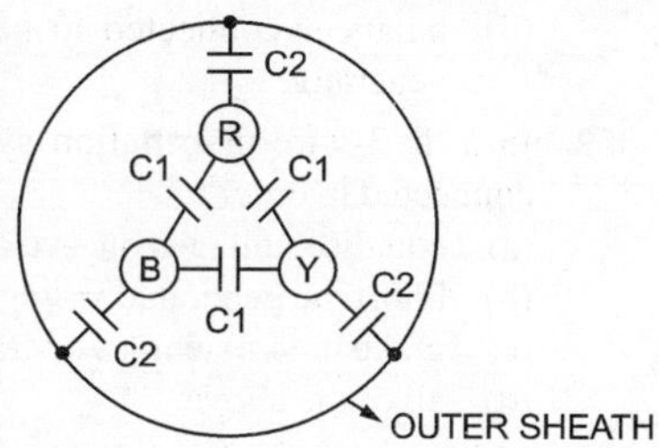

(a) 2.0 A (b) 2.4 A (c) 2.7 A (d) 3.5 A
[GATE E.E. 2010]

322. For the circuit shown in figure, the capacitance measured between terminals *B* and *Y* will be

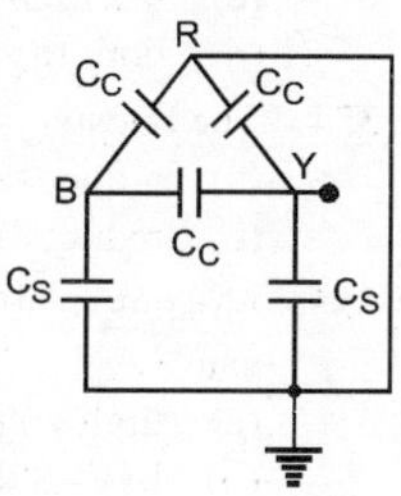

(a) $C_c + (C_s/2)$
(b) $C_s + (C_c/2)$
(c) $(C_s + 3C_c)/2$
(d) $3C_c + 2C_s$ **[GATE E.E. 1999]**

323. Underground cables are laid at sufficient depth so as to
(a) minimise temperature stresses.
(b) minimise effects of shocks and vibrations owing to passing vehicles etc.
(c) avoid being unearthed easily owing to removal of soil.
(d) both (a) and (c).

324. In case the communication cables are to be laid parallel to power cables the distance between the two should be at least so that there is no interference.
(a) 0.5 m (b) 2.0 m (c) 4.0 m (d) 0.05 m

325. While crossing the road the cable should be
(a) buried in trenches.
(b) surrounded by sawdust to absorb vibrations.
(c) laid in conduits or pipes. (d) none of the above.

326. Consider the following statements :
1. The insulation resistance of cable will increase if the length of cable is increased.
2. For the same overall diameter of cable, the grading of cable will increase the safe working voltage.
3. The normal operating temperature of PVC cable is 70°C.
4. The thermal resistance of soil increases as the moisture content of soil increases.

Of these statements
(a) 1 and 2 are correct. (b) 2 and 3 are correct.
(c) 3 and 4 are correct. (d) 1 and 4 are correct.
[U.P.S.C. I.E.S. E.E.-II, 1995]

327. The source(s) of heat generation in cables is/are
(a) copper loss in conductor.
(b) dielectric losses in cable insulation.
(c) losses in metallic sheathings and armourings.
(d) all of the above.

328. A cable carrying ac has
(a) leakage losses only. (b) hysteresis losses only.
(c) hysteresis and leakage losses only.
(d) hysteresis, leakage and friction losses.

329. Dielectric hysteresis loss in a cable varies as
(a) impressed voltage. (b) $(\text{impressed voltage})^2$.
(c) $(\text{impressed voltage})^{1/2}$. (d) $(\text{impressed voltage})^{3/2}$.

330. A 110 kV, single-core coaxial, XLPE insulated power cable delivering power at 50 Hz, has a capacitance of 125 nF/km. If the dielectric loss tangent of XLPE is 2×10^{-4}, then dielectric power loss in this cable in W/km is
(a) 5.0 (b) 31.7 (c) 37.8 (d) 95.0
[GATE E.E. 2004]

331. The current carrying capacity of cables in dc is more than that in ac. It is mainly due to
(a) smaller hysteresis losses. (b) absence of harmonics.
(c) absence of ripples. (d) none of the above.

332. The fault(s) which are likely to occur in cables is/are
(a) breakdown of cable insulation.
(b) cross- or short-circuit fault.
(c) open-circuit fault. (d) all of the above.

333. The lead sheath of the cable may get damaged due to
(a) mechanical injury.
(b) crystallisation of lead through vibrations.
(c) chemical action with impurities present in the soil when buried in earth.
(d) any of the above.

334. The cables should not be operated too hot otherwise
(a) expansion of oil may cause sheath to burst.
(b) the oil may loose its viscosity and it may start drawing off from higher levels.
(c) unequal expansion may create voids in the insulation leading to ionization.
(d) rapid increase in dielectric losses with temperature may cause thermal instability.
(e) all of the above.

335. The breakdown voltage of a cable depends upon
(a) presence of moisture. (b) operating temperature.
(c) time of application of the voltage.
(d) all of the above.

336. Breakdown of cable insulation may occur due to
(a) thermal instability. (b) puncture.
(c) tracking. (d) any of the above.

337. Fibre-optic cables are used in power system applications mainly for
(a) SCADA.
(b) communication between power station and substation.
(c) communication between power station and load control centre.
(d) all of these. **[U.P.S.C. I.E.S. E.E.-II, 2010]**

Answers: 318. (a), **319.** (d), **320.** (c), **321.** (a), **322.** (c), **323.** (b), **324.** (a), **325.** (c), **326.** (b), **327.** (d), **328.** (c), **329.** (b), **330.** (d), **331.** (a), **332.** (d), **333.** (d), **334.** (e), **335.** (d), **336.** (d), **337.** (d),

338. The main criterion for selection of the size of a distributor for a radial distribution system is
(a) voltage drop. (b) corona loss.
(c) temperature rise. (d) capital cost. [GATE E.E. 1995]

339. The distribution systems in India are mostly
(a) radial. (b) parallel.
(c) network. (d) either (b) or (c).
[A.M.I.E Sec B. Design of Electrical Systems Winter 2006]

340. Why is a ring main distribution system preferred to a radial system?
1. Voltage drop in the feeder is less.
2. Power factor is higher.
3. Supply is more reliable.

Select the correct answer using the code given below:
(a) 1 and 2 only. (b) 2 and 3 only.
(c) 1 and 3 only. (d) 1, 2 and 3.
[U.P.S.C. I.E.S. E.E.-II, 2007]

341. As per electricity regulations in India the maximum permissible voltage change at consumer premises should be
(a) ±5% of declared voltage. (b) ±6% of declared voltage.
(c) ±10% of declared voltage.
(d) ±12% of declared voltage.
[A.M.I.E Sec B. Design of Electrical Systems Winter 2006]

342. The cost of material used in a distribution circuit per kVA of distributed power varies as
(a) square of linear dimensions of supply area
(b) directly according to (supply area)^2
(c) cube of linear dimensions of supply area
(d) none of the above
[A.M.I.E Sec B. Design of Electrical Systems Summer 2005]

343. An approximate formula for loss factor in a distribution system is
(a) Loss factor = 0.3 × load factor + 0.7 (load factor)^2
(b) Loss factor = 0.5 (load factor)
(c) Loss factor = load factor
(d) Loss factor = 0.7 × load factor + 0.3 (load factor)^2
[A.M.I.E Sec B. Design of Electrical System Summer 2005]

344. In a distribution system, which of the following items shares the major cost?
(a) Conductors. (b) Earthing system.
(c) Distribution transformer. (d) Insulators.

345. The distributors in residential areas are
(a) single phase, two-wire. (b) three phase, three-wire.
(c) three phase, four-wire. (d) two phase, four-wire.

346. The loads on 3-phase, 4-wire distributors are usually
(a) balanced. (b) unbalanced.
(c) either of above. (d) none of above.

347. A uniformly loaded dc distributor is fed at both ends with equal voltages. In comparison to a similar distributor fed at one end only, the maximum voltage drop will be
(a) one-fourth. (b) one-half. (c) one-third. (d) one-sixth.

348. The null point of a uniformly loaded distributor fed at equal voltage at both ends lies at
(a) midpoint. (b) either end.
(c) two-third distance from one end.
(d) one-fourth distance from one end. [A.M.I.E. Sec B. Summer 1993]

349. A uniformly loaded dc distributor is fed at both ends with equal voltages. In comparison to a similar distributor fed at one end only, the drop at the mid point will be
(a) one-fourth. (b) one-half.
(c) one-third. (d) one-sixth.

350. A dc distribution system is shown in figure with load currents as marked. The two ends of the feeder are fed by voltage sources such that $V_P V_Q = 3$ V. The value of the voltage V_P for a minimum voltage of 220 V at any point along the feeder is

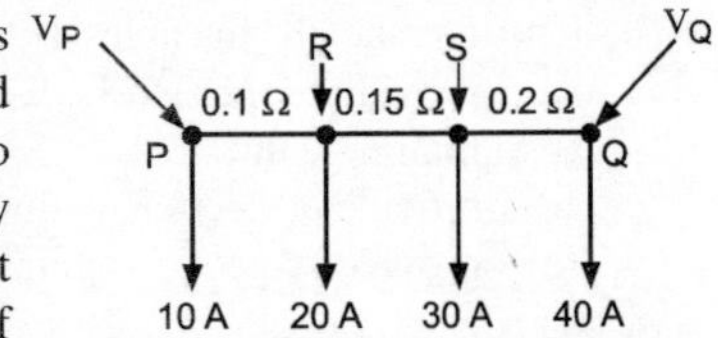

(a) 225.89 V (b) 222.89 V (c) 220.0 V (d) 228.58 V
[GATE E.E. 2003]

351. In a dc 3-wire distributor using balancers and having unequal loads on the two sides
(a) both balancers operate as generators.
(b) both balancers operate as motors.
(c) balancer connected to lightly loaded side operates as a generator.
(d) balancer connected to heavily loaded side operates as a generator.

352. In a dc 3-wire distribution system, balancer fields are cross-connected in order to
(a) equalise voltages on +ve and –ve outers.
(b) boost the generated voltage.
(c) make both machines operated as unloaded motors.
(d) all of the above.

353. The booster
(a) is a series wound dc generator driven by dc shunt motor.
(b) is a low voltage and high current generator operating on straight or linear portion of its voltage-current characteristic.
(c) is a high voltage and low circuit machine.
(d) both (a) and (b).

354. The booster is inserted in the circuit to
(a) reduce current. (b) increase current.
(c) reduce voltage drop. (d) compensate for voltage drop.

355. Floating neutral in a 3-phase supply is considered undesirable as it causes
(a) high voltage across the load.
(b) low voltage across the load.
(c) unequal line voltages across the load.
(d) none of the above.

356. A single-phase ac distributor supplies two single-phase loads as shown in the figure. The voltage drop from A to C is

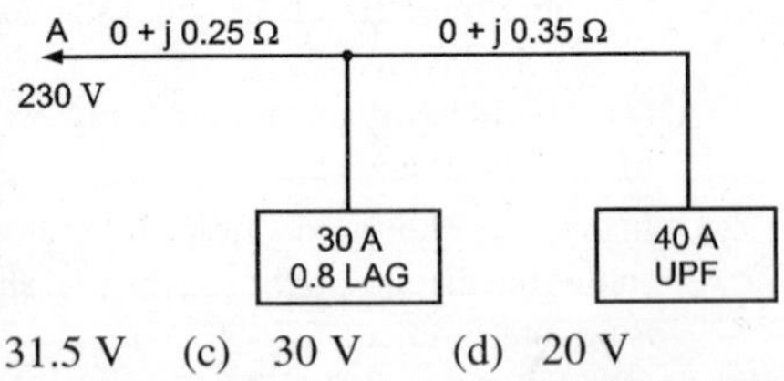

(a) 4.5 V (b) 31.5 V (c) 30 V (d) 20 V
[GATE E.E. 1999]

357. Single line diagram of a 4-bus single source distribution system is shown in figure. Branches e_1, e_2, e_3 and e_4 have equal impedances. The load current values indicated in the figure are in per unit. Distribution company's policy requires radial system operation with minimum loss. This can be achieved by opening of the branch

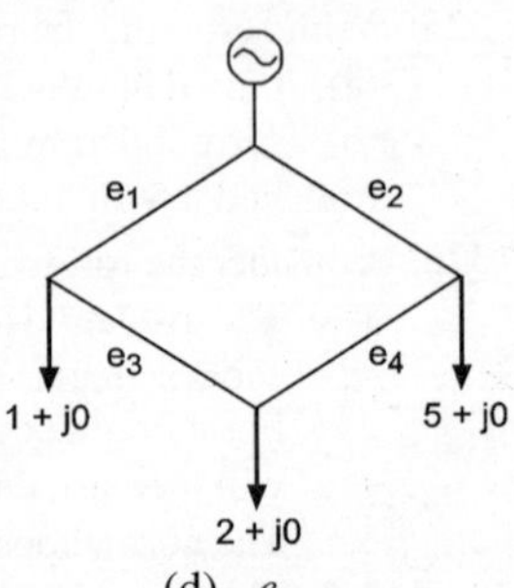

(a) e_1 (b) e_2 (c) e_3 (d) e_4
[GATE E.E. 2008]

358. A 230 V (Phase), 50 Hz, three-phase, 4-wire, system has a sequence ABC. A unity power-factor load of 4 kW is connected

Answers: **338.** (a), **339.** (a), **340.** (c), **341.** (b), **342.** (b), **343.** (a), **344.** (c), **345.** (c), **346.** (b), **347.** (a), **348.** (a), **349.** (c), **350.** (a), **351.** (d), **352.** (a), **353.** (d), **354.** (d), **355.** (c), **356.** (c), **357.** (d), **358.** (b),

between phase A and neutral N. It is desired to achieve zero neutral current through the use of a pure inductor and impure capacitor in the other two phases. The value of inductor and capacitor is
(a) 72.95 mH in phase C and 139.02 μF in Phase B.
(b) 72.95 mH in phase B and 139.02 μF in Phase C.
(c) 42.12 mH in phase C and 240.79 μF in Phase B.
(d) 42.12 mH in phase B and 240.79 μF in Phase C.
[GATE E.E. 2007]

359. In order to reduce the inductive interference to parallel communication circuits to the minimum.
(a) the conductors are transposed at regular intervals.
(b) bundle conductors are used.
(c) thick conductors are used.
(d) either (a) or (b).

360. When there is interference in an overhead communication line running parallel and in close proximity to an overhead power line, the voltage induced in the communication line in the longitudinal and lateral directions by the power line are due to
(a) magnetic induction and electric induction respectively.
(b) electric induction and magnetic induction respectively.
(c) both magnetic induction and electric induction.
(d) magnetic induction only. **[U.P.S.C. I.E.S. E.E.-2000]**

361. A single-phase transmission line and a telephone line are both symmetrically strung one below the other, in horizontal configurations, on a common tower. The shortest and logest distances between the phase and telephone conductors are 2.5 m and 3 m respectively. The voltage (volt/km) induced in the telephone circuit, due to 50 Hz current of 100 amps in the power circuit is
(a) 4.81 (b) 3.56 (c) 2.29 (d) 1.27
[GATE E.E. 2006]

362. A long wire composed of a smooth round conductor runs above and parallel to the ground (assumed to be a large conducting plane). A high voltage exists between the conductor and the ground. The maximum electric stress occurs at
(a) the upper surface of the conductor.
(b) the lower surface of the conductor.
(c) the ground surface
(d) midway between the conductor and ground. **[GATE E.E. 2002]**

363. Extra high voltage transmission means
(a) voltage less than 400 kV.
(b) voltage in the range of 400-750 kV.
(c) voltage greater than 750 kV. **[A.M.I.E. Sec B. Summer 1997]**

364. Bundled conductors are used for EHV transmission lines primarily for reducing the
(a) corona loss. (b) surge impedance of the line.
(c) voltage drop across the line.
(d) I^2R losses. **[U.P.S.C. I.E.S. E.E.-II, 2001]**

365. For an extra high voltage overhead transmission line, four conductors are used per phase (in a bundle) at the corners of a square of side s metre. The GMR (Geometric Mean Radius) of each conductor is r'_m metre.
What is the GMR of the bundle conductor?
(a) $\left(r'_m \times s^2 \times \sqrt{2s}\right)^{1/4}$ (b) $\left(r'_m \times s^3\right)^{1/4}$
(c) $\left(r'_m \times 3s^3\right)^{1/4}$ (d) $[r'_m \times (\sqrt{2}\, s^3]^{1/4}$
[U.P.S.C. I.E.S. E.E.-II, 2008]

366. Use of bundled conductors reduces
(a) power loss due to corona.
(b) surge impedance.
(c) radio interference. (d) line inductance.
(e) all of these.

367. With bundled conductors
(a) the corona inception voltage increases.
(b) the corona inception voltage decreases.
(c) the corona inception voltage remains unaffected.
[A.M.I.E. Sec B. Winter 1993]

368. Consider the following statements :
1. By using bundle conductors in an overhead line, the corona loss is reduced.
2. By using bundle conductors, the inductance of transmission line increases and capacitance reduces.
3. Corona loss causes interference in adjoining communication lines.
Which of the these statements are correct ?
(a) 1 and 2. (b) 2 and 3. (c) 1 and 3. (d) 1, 2 and 3.
[U.P.S.C. I.E.S. E.E.-II, 1999]

369. The conductor of an EHV line is selected on the basis of
(a) current carrying capacity.
(b) corona and RI performance.
(c) line voltage. (d) none of these.
[A.M.I.E. Sec B. Summer 1998]

370. Frequency of resonant vibrations lies in the range of
(a) 10-20 Hz (b) 100-200 Hz
(c) 50-100 Hz (d) 200-300 Hz
[A.M.I.E. Sec B. Summer 1996]

371. On what basis is the insulation level of a 400 kV, EHV overhead transmission line decided?
(a) Lightning overvoltage. (b) Corona inception voltage.
(c) Switching overvoltage. (d) Radio and TV interference.
[U.P.S.C. I.E.S. E.E.-II, 2005]

372. The total reactance and total susceptance of a lossless overhead EHV line, operating at 50 Hz, are given by 0.045 pu and 1.2 pu respectively. If the velocity of wave propagation is 3×10^5 km/s, then the approximate length of the line is
(a) 122 km. (b) 172 km. (c) 222 km. (d) 272 km.
[GATE E.E. 2007]

373. An extra high voltage transmission line of length 300 km can be approximated by a lossless line having propagation constant β = 0.00127 radians per km. Then the percentage ratio of line length to wavelength will be given by
(a) 24.24% (b) 12.12% (c) 19.05% (d) 6.06%
[GATE E.E. 2008]

374. A lossless transmission line of length 50 cm with L = 10 μH/m, C = 40 pF/m is operated at 30 MHz. What is the electric length (βl)?
(a) 20 λ (b) 0.2 λ (c) 108° (d) 40 π
[U.P.S.C. I.E.S. E.E.-I, 2004]

375. A surge of 20 kV magnitude travels along a lossless cable towards its junction with two identical lossless overhead transmission lines. The inductance and the capacitance of the cable are 0.4 mH and 0.5 μF per km. The inductance and capacitance of the overhead transmission lines are 1.5 mH and 0.015 μF per km. The magnitude of the voltage at the junction due to surge is
(a) 36.72 kV (b) 18.36 kV (c) 6.07 kV (d) 33.93 kV
[GATE E.E. 2003]

Answers: **359.** (d), **360.** (c), **361.** (d), **362.** (b), **363.** (b), **364.** (a), **365.** (a), **366.** (e), **367.** (a), **368.** (c), **369.** (b), **370.** (c), **371.** (c), **372.** (b), **373.** (d), **374.** (c), **375.** (a),

376. Series capacitor is used in a transmission line to
(a) compensate the voltage drop. (b) reduce line loss.
(c) limit short-circuit current.
(d) improve load power factor.

377. Series capacitive compensation in EHV transmission lines is used to
(a) reduce the line loading.
(b) improve the stability of the system.
(c) reduce the voltage profile.
(d) improve the protection of the line. **[GATE E.E. 1998]**

378. Series capacitors
(a) increase the power transmission capability of an EHV long line.
(b) provide voltage control.
(c) increase the shunt admittance of the line.
[A.M.I.E. Sec B. Summer 1997]

379. Series capacitors can be used in distribution lines
(a) to provide reactive power compensation.
(b) to reduce the receiving-end voltage under light load conditions.
(c) to reduce the voltage drop.
(d) to reduce line losses. **[A.M.I.E. Sec B. Summer 1994]**

380. Shunt compensation in an EHV line is used to improve
(a) stability and fault level.
(b) fault level and voltage profile.
(c) voltage profile and stability.
(d) stability, fault level and voltage profile.
[U.P.S.C. I.E.S. E.E.-II, 2000]

381. The most accurate and versatile method of achieving reactive power compensation is by using
(a) switched capacitors.
(b) fixed capacitor with controlled reactor.
(c) saturable reactor with capacitor bank.
(d) switched capacitor with controlled reactors.
[U.P.S.C. I.E.S. E.E.-II, 2000]

382. Installation of capacitors at suitable locations and of optimum size in a distribution system results in
1. improved voltage regulation.
2. reduction in distribution power losses.
3. reduction of kVA rating of distribution transformers.

Select the correct answer from the codes given below :
(a) 1 alone (b) 1 and 2 (c) 1, 2 and 3 (d) 3 alone
[U.P.S.C. I.E.S. E.E.-II, 2001]

383. Insulation coordination for UHV lines (above 500 kV) is done based on
(a) lightning surges.
(b) lightning surges and switching surges.
(c) switching surges. (d) none of the above.
[U.P.S.C. I.E.S. E.E.-II, 1999]

384. At an industrial substation with a 4 MW load, a capacitor of 2 MVAR is installed to maintain the load power factor at 0.97 lagging. If the capacitor goes out of service, the load power factor becomes
(a) 0.85 (b) 1.00 (c) 0.80 lag (d) 0.90 lag
[GATE E.E. 2005]

385. Consider two buses connected by an impedance of $(0 + j5)\ \Omega$. The bus 1 voltage is $100\angle 30°$ V, and bus 2 voltage is $100\angle 0°$ V. The real and reactive power supplied by bus 1, respectively, are
(a) 1000 W, 268 VAr. (b) –1000 W, –134 VAr.
(c) 276.9 W, –56.7 VAr. (d) –276.9 W, 56.7 VAr.
[GATE E.E. 2010]

386. For enhancing the power transmission in a long EHV transmission line, the most preferred method is to connect a
(a) series inductive compensator in the line.
(b) shunt inductive compensator at the receiving end.
(c) series capacitive compensator in the line.
(d) shunt capacitive compensator at the sending end.
[GATE E.E. 2011]

387. In a transmission line, the midpoint voltage is maintained to V by a compensating device as shown in the circuit above.

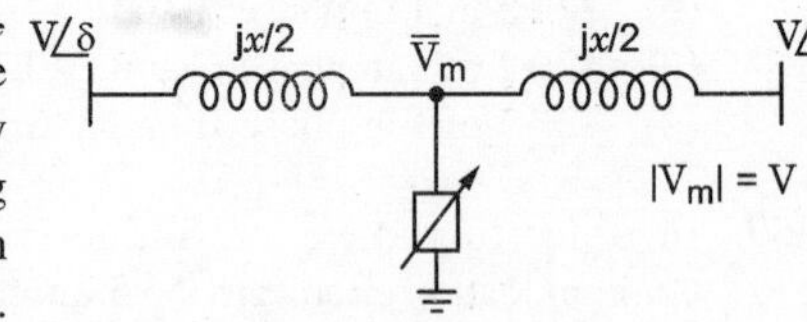

What is the real power flow through the line?
(a) $\frac{V^2}{X}\sin\frac{\delta}{2}$ (b) $\frac{2V^2}{X}\sin\delta$
(c) $\frac{V^2}{X}\sin\delta$ (d) $\frac{2V^2}{X}\sin\frac{\delta}{2}$
[U.P.S.C. I.E.S. E.E.-II, 2009]

388. A short transmission line having zero resistance and total series reactance of 0.4 pu is provided with reactive power compensation at the midpoint of the line such that the midpoint voltage is held at 0.96 pu when the voltage at both ends are 1.0 pu. What is the steady state power transmission limit of such a system?
(a) 4.8 pu (b) 0.0 pu
(c) 2.4 pu (d) 9.6 pu
[U.P.S.C. I.E.S. E.E.-II, 2009]

389. A nuclear power station of 500 MW capacity is located at 300 km away from a load center. Select the most suitable power evacuation transmission configuration among the following options

(a) 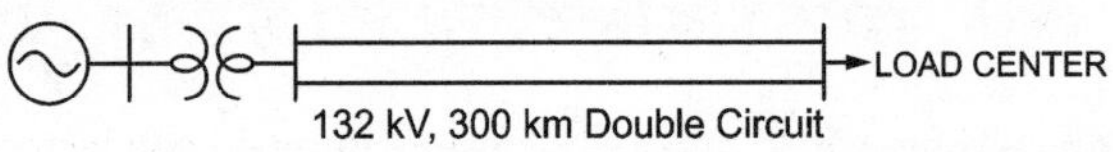

(b) 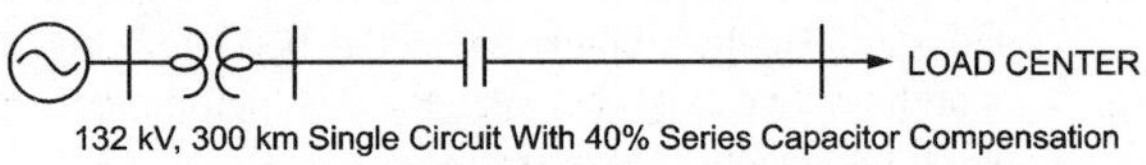

(c) 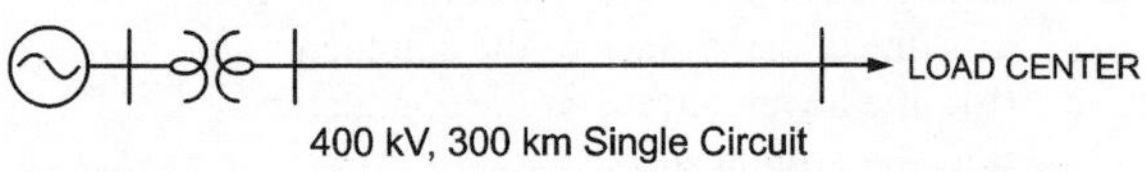

(d) 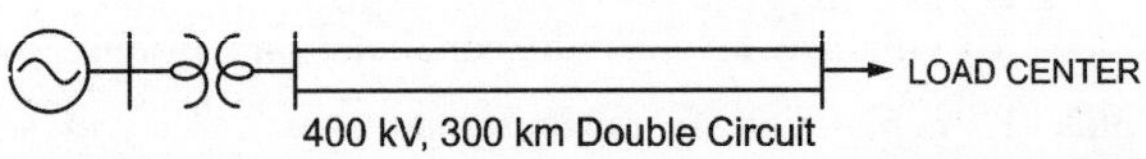

[GATE E.E. 2011]

390. A lossless transmission line having Surge Impedance Loading (SIL) of 2,280 MW is provided with a uniformly distributed series capacitve compensation of 30%. Then, SIL of the compensated transmission line will be
(a) 1,835 MW (b) 2,280 MW
(c) 2,725 MW (d) 3,257 MW **[GATE E.E. 2008]**

391. As compared to ac transmission dc transmission is free from
(a) inductance. (b) capacitance.'
(c) phase displacement. (d) all of the above.
[A.M.I.E. Sec B. Winter 1998]

392. Out of considerations (i) to (iv) listed below :
(i) no distance limitation related to steady-state stability.
(ii) no reactive power requirement from the system at the two terminals.
(iii) no substantial effect on fault level of the two systems at the terminals inspite of the interconnection.
(iv) no corona problems.

The considerations which constitute advantages of HVDC transmission are

Answers: 376. (a), **377.** (b), **378.** (a), **379.** (a), **380.** (c), **381.** (d), **382.** (c), **383.** (c), **384.** (c), **385.** (a), **386.** (c), **387.** (c), **388.** (c), **389.** (d), **390.** (b), **391.** (d), **392.** (a),

(a) all of the above. (b) (i) and (iii).
(c) (iii) and (iv). (d) (ii) and (iv). [GATE E.E. 2000]

393. In an HVDC system:
(a) both generation and distribution are dc.
(b) generation is ac and distribution is dc.
(c) generation is dc and distribution is ac.
(d) both generation and distribution are ac.
[U.P.S.C. I.E.S. E.E.-II, 2011]

394. In a dc transmission line
(a) it is necessary for the sending end and receiving end to be operated in synchronism.
(b) the effects of inductive and capacitive reactances are greater than in an ac transmission line of the same rating.
(c) there are no effects due to inductive and capacitive reactances.
(d) power transfer capability is limited by stability considerations.
[GATE E.E. 1999]

395. HVDC transmission is preferred to EHV AC because
(a) HVDC terminal equipment are inexpensive.
(b) VAR compensation is not required in HVDC system.
(c) system stability can be improved.
(d) harmonics problem is avoided. [GATE E.E. 1994]

396. High Voltage DC (HVDC) transmission is mainly used for
(a) bulk power transmission over very long distances.
(b) interconnecting two systems with the same nominal frequency.
(c) eliminating reactive power requirement in the operation.
(d) minimizing harmonics at the converter stations.
[GATE E.E. 2005]

397. Consider the following statements regarding HVDC power transmission:
1. The modern HVDC systems use 12-pulse converters.
2. DC systems never use ground or sea return.
3. Most of present-day dc schemes are two-terminal links.
Which of the statements given above is/are correct?
(a) 1, 2 and 3. (b) 1 only.
(c) 2 and 3 only. (d) 1 and 3 only.
[U.P.S.C. I.E.S. E.E.-II, 2008]

398. Choose the appropriate auxiliary components of HVDC transmission system from the following
1. DC line inductor. 2. AC line inductor.
3. Reactive power sources. 4. Distance relays on dc line.
5. Series capacitance on AC line.
(a) 1 and 2 (b) 1 and 3 (c) 2 and 4 (d) 4 and 5
[GATE E.E. 2003]

399. Bulk power transmission over long HVDC lines are preferred an account of
(a) low cost of HVDC terminals.
(b) no harmonic problems. (c) minimum line power losses.
(d) simple protection. [GATE E.E. 1998]

400. Consider the following statements:
HVDC transmission is superior to HVAC transmission due to
1. Lack of reliable dc circuit breakers.
2. Lesser number of conductors for same power carrying capacity.
3. Non-synchronous link between two different systems simplifying the problem of voltage stability and frequency control.
4. No costly terminal equipments such as converters and inverters are required.
Which of the above statements is/are correct?
(a) 4 only. (b) 4 and 3.
(c) 1 and 2. (d) 2 and 3.
[U.P.S.C. I.E.S. E.E.-II, 2009]

401. For bulk power transmission over long distance HVDC transmission is
(a) is economically and technically superior to ac transmission.
(b) not good alternative to ac transmission.
(c) not economically viable.
[A.M.I.E. Sec B. Power Systems Winter 1998]

402. Corona loss in a dc line is
(a) less than that in ac line operating at same rms voltage.
(b) more than that in ac line at the same rms voltage.
(c) same as in ac linear equal rms voltage.
[A.M.I.E. Sec B. Power Systems Summer 1999]

403. In the case of an HVDC system, there is
(a) charging current but no skin effect.
(b) no charging current but skin effect.
(c) neither charging current nor skin effect.
(d) both charging current and skin effect.
[U.P.S.C. I.E.S. E.E.-II, 2000]

404. Which one of the following statements in respect of HVDC transmission line is not correct ?
(a) The power transmission capability of bipolar line is almost the same as that of single circuit ac line.
(b) HVDC link can operate between ac systems whose frequencies need not be equal.
(c) There is no distance limitation for HVDC transmission by UG cable.
(d) Corona loss is much higher in HVDC transmission line.
[U.P.S.C. I.E.S. E.E.-II, 1999]

405. A dc line carries
(a) less power than an equivalent ac line.
(b) more power than the ac line.
(c) same power as an ac line.
[A.M.I.E. Sec B. Power Systems Summer 1997]

406. The pu impedance of a line to 50 MVA, 132 kV base is 0.4, the pu impedance to a 100 MVA, 132 kV base will be
(a) 0.2 (b) 0.4 (c) 0.8 (d) 1.6
[A.M.I.E. Sec B. Power Systems Summer 2000]

407. Back to back HVDC is used to
(a) increase the transmission capability.
(b) decrease line losses.
(c) provide stable interconnection.
(d) reduce voltage drop. [A.M.I.E. Sec B. Summer 1994]

408. HVDC transmission needs
(a) pulse converters. (b) dc generators.
(c) ac filters. (d) dc filters.
(e) transformers. [A.M.I.E. Sec B. Power Systems Summer 1997]

409. In an HVDC link, power flows
(a) from the rectifier end to the inverter end.
(b) from the inverter end to the rectifier end.
[A.M.I.E. Sec B. Power Systems Winter 1996]

410. In HVDC system a converter acts as a rectifier when the firing angle α has a value

Answers: **393.** (d), **394.** (c), **395.** (c), **396.** (a), **397.** (d), **398.** (b), **399.** (c), **400.** (d), **401.** (a), **402.** (a), **403.** (c), **404.** (d), **405.** (b), **406.** (c), **407.** (c), **408.** (a), **409.** (a), **410.** (b),

(a) $0 < \alpha < 120^o$ (b) $0 < \alpha < 90^o$
(c) $90^o < \alpha < 180^o$ **[A.M.I.E. Sec B. Power Systems Winter 1996]**

411. In a 3-phase converter used in HVDC transmission, the three anodes conduct sequentially. Due to overlap caused by circuit inductance, two anodes conduct simultaneously during overlap period. The output voltage waveform during this period is the
(a) voltage of the 1st anode, because the second anode has not completely taken over.
(b) mean of the two anode voltages, as they conduct together.
(c) voltage of second anode, because the voltage of this anode is greater than that of the 1st.
(d) sum of 1st and the second anode voltages because both the anodes are conducting. **[U.P.S.C. I.E.S. E.E.-II, 1999]**

412. The capital cost on a HVDC converter is
(a) less than ac substation.
(b) more than ac substation.
(c) same as that of ac substation.
[A.M.I.E. Sec B. Power Systems Summer 1997]

413. Consider the following statements regarding the suitable choice of HVDC converter configuration:
1. Pulse number should be high.
2. Ratio of peak inverse voltage to no-load dc output voltage should be as high as possible
3. Transformer utilization factor should be nearly unity.

Which of the statements given above are correct?
(a) 1 and 2 (b) 1, 2 and 3 (c) 2 and 3 (d) 1 and 3
[U.P.S.C. I.E.S. E.E.-II, 2004]

414. Consider the following statements:
The suitable configuration of HVDC converter is decided on the requirements of
1. high pulse number.
2. converter transformer utilization factor should be near to unity.

Which of the above statements is/are correct?
(a) 1 only. (b) Both 1 and 2.
(c) Neither 1 nor 2. (d) 2 only.
[U.P.S.C. I.E.S. E.E.-II, 2010]

415. In HVDC transmission there are predominant
(a) voltage harmonics on dc side and current harmonics on ac side of converters.
(b) current harmonics on dc side and voltage harmonics on ac side of converters.
(c) current harmonics only on the dc side of converters.
(d) voltage harmonics only on the ac side of converters.
[U.P.S.C. I.E.S. E.E.-II, 2003]

416. High voltage dc transmission requires filters for
(a) current harmonics on ac side and voltage harmonics on dc side.
(b) voltage harmonics on ac side and current harmonics on dc side.
(c) voltage harmonics on both ac and dc sides.
(d) current harmonics on both ac and dc sides.
[U.P.S.C. I.E.S. E.E.-II, 2006]

417. An HVDC link consists of rectifier, inverter, transmission line and other equipments. Which one of the following is true for this link?
(a) The transmission line produces/supplies reactive power.
(b) The rectifier consumes reactive power and the inverter supplies reactive power from/to the respective connected ac systems.
(c) Rectifier supplies reactive power and the inverter consumes reactive power to/from the respective connected ac systems.
(d) Both the converters (rectifier and inverter) consume reactive power from the respective connected ac systems.
[GATE E.E. 2006]

418. In the HVDC system, the ac harmonics which gets effectively eliminated with 12 pulse bridge converters, are
(a) triplen harmonics.
(b) triplen and 5th harmonics.
(c) triplen, 5th and 7th harmonics.
(d) 5th and 7th harmonics. **[U.P.S.C. I.E.S. E.E.-II, 2001]**

419. Power is transferred from system A to system B by an HVDC link as shown in the figure. If the voltages V_{AB} and V_{CD} are as indicated in the figure, and $I > 0$, then

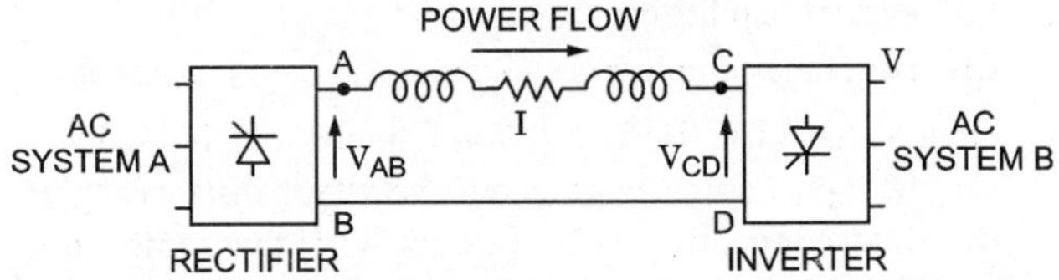

(a) $V_{AB} < 0, V_{CD} < 0, V_{AB} > V_{CD}$
(b) $V_{AB} > 0, V_{CD} > 0, V_{AB} < V_{CD}$
(c) $V_{AB} > 0, V_{CD} > 0, V_{AB} > V_{CD}$
(d) $V_{AB} > 0, V_{CD} < 0$ **[GATE E.E. 2010]**

420. Two regional systems, each having several synchronous generators and loads are interconnected by an ac line and a HVDC link as shown in the figure. Which of the following statements is true in the steady state:

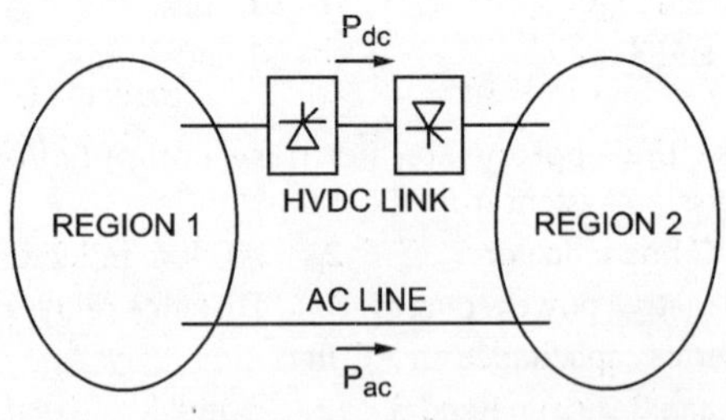

(a) Both regions need not have the same frequency.
(b) The total power flow between the regions ($P_{ac} + P_{dc}$) can be changed by controlling the HVDC converters alone.
(c) The power sharing between the ac line and the HVDC link can be changed by controlling the HVDC converters alone.
(d) The direction of power flow in the HVDC link (P_{dc}) cannot be reversed. **[GATE E.E. 2007]**

Answers: 411. (b), **412.** (b), **413.** (d), **414.** (d), **415.** (b), **416.** (b), **417.** (b), **418.** (d), **419.** (c), **420.** (c).

ASSERTION (A)/REASON (R) TYPE QUESTIONS

Note : Each of the following questions consists of two statements, one labelled the 'Assertion (A)' and the other labelled the 'Reason (R)'. Examine the two statements carefully and decide if the Assertion (A) and Reason (R) are individually true and if so, whether the reason (R) is correct explanation of the Assertion (A). Select your answers to these questions using the codes given below:

Codes: (a) Both A and R are true and R is a correct explanation of A.
(b) Both A and B are true and R is not a correct explanation of A.
(c) A is true but R is false. (d) A is false but R is true.

1. (A): Feeders are designed mainly from the point of view of the voltage drop in them.
(R): Current loading of the feeder remains the same along its length.

2. (A): Transmission of electrical power by dc is economical than that with ac.
(R): In case of dc transmission, there is no charging current which contributes to continuous load, even on no load.

3. (A): The maximum operating temperature of overhead line conductors made of aluminium or copper is restricted to 75°C.
(R): Conductor temperature beyond 75°C may shatter procelain insulators. [U.P.S.C. I.E.S. E.E.-II, 2000]

4. (A): Glazing is done on ceramic insulators to make the surface smooth and non-absorbent.
(R): Moisture from the atmosphere can collect on the surface discontinuities on a ceramic and result in electrical breakdown. [U.P.S.C. I.E.S. E.E.-I, 2003]

5. (A): A glazed ceramic insulator has a lower breakdown strength than an unglazed one.
(R): Glazing covers pores and cracks and makes the surface non-absorbent. [U.P.S.C. I.E.S. E.E.-I, 1998]

6. (A): The string efficiency can be improved by using long cross arms.
(R): Use of long cross arms reduces the ratio of capacity to earth to capacity per insulator thereby increasing the string efficiency.

7. (A): Capacitance grading method of improving string efficiency is not used above 220 kV.
(R): In capacitance grading method the units used are of different capacitances and so are not interchangeable.

8. (A): Corona reduces the effects of transients produced by lightning and other causes.
(R): Charges induced on the line by lightning or other causes are partially dissipated as a corona loss.

9. (A): Corona causes interference with neighbouring communication circuits.
(R): Owing to distortion of waveform, the harmonics predominately third harmonics are introduced into the transmission line.

10. (A): Normally it is not advisable to operate a line above its fair weather disruptive critical voltage.
(R): It is usually more economical to pay for corona loss for a fraction of the year (storms are experienced at intervals) than if corona was to be entirely avoided by using heavy conductors.

11. (A): The electric field intensity E is proportional to x, the distance from the axis of the conductor of radius r, for $x \geq r$.
(R): Bundled conductors increase the line capacitance from that of a single conductor of the same cross-sectional area. [U.P.S.C. I.E.S. E.E.-II, 1993]

12. (A): The concept of self GMD is applicable for the calculation of inductance of transmission lines.
(R): The radius of a conductor of transmission lines, in the expression for inductance is not the self GMD of the conductor. [U.P.S.C. I.E.S. E.E.-II, 1998]

13. (A): Transposition of conductors in a transmission line is necessary.
(R): Corona losses are reduced by transposition of conductors.

14. (A): With a constant applied voltage at the sending end of a long transmission line, the voltage at the receiving end rises to a high value at light load at the other end.
(R): The large charging capacitive current flowing through the line causes the voltage rise. [U.P.S.C. I.E.S. E.E.-II, 1993]

15. (A): In a good power supply, the percentage of voltage regulation should be close to zero.
(R): Zero percentage regulation means that there will be no change in output voltage if the load resistance varies between the limit 0 to ∞. [U.P.S.C. I.E.S. E.E.-II, 2006]

16. (A): In the modelling of medium and long transmission lines the nominal-π and T circuits are not equivalent to each other.
(R): A star-delta transformation can be used to derive the one circuit from the other. [U.P.S.C. I.E.S. E.E.-II, 2002]

17. (A): One-phase supply can be converted to a 3-phase supply with the help of a suitable transformer.
(R): 3-phase supply can be converted to a 1-phase supply by a suitable transformer. [U.P.S.C. I.E.S. E.E.-II, 1998]

18. (A): The bulk supply into the large urban areas if often carried out by underground cables operating at 132 kV and above. This results in large generation of VARs by the cable and consequent rise in voltage at the receiving end.
(R): These VARs may be neutralised by using series capacitance in the line at appropriate intervals. [U.P.S.C. I.E.S. E.E.-II, 1998]

19. (A): High and extra high voltages are adopted for transmission of bulk power over long distances.
(R): The maximum true power transferred over a given line increases with the increase in V_S and V_R.

20. (A): Control of reactive power is essential for maintaining a desired voltage profile.
(R): The reactive power transferred over a transmission line depends on load or power angle.

Answers: **1.** (d), **2.** (b), **3.** (c), **4.** (a), **5.** (a), **6.** (a), **7.** (d), **8.** (a), **9.** (b), **10.** (a), **11.** (d), **12.** (b), **13.** (c), **14.** (a), **15.** (a), **16.** (c), **17.** (a), **18.** (c), **19.** (a), **20.** (c),

21. (A): The conductor used in cables should be stranded one.
(R): The skin effect is much smaller with stranded conductor than with solid conductors.

22. (A): In a high-voltage cable, concentric conducting surfaces are provided in an insulating medium throughout the cross-section of the cable to reduce the ratio of maximum to minimum electric stress.
(B): The potential differences between the conducting surfaces redistribute the charge distribution. [U.P.S.C. I.E.S. E.E.-II, 1993]

23. (A): Screened type cables have lesser possibilities of core to core fault in comparison to belted type cables.
(R): The metal sheathed core cable has a greater core to core thickness for a given overall diameter in comparison to a plain cable.

24. (A): The modern trend is to avoid grading of cables as far as possible and use gas pressure or oil-filled cables.
(R): With intersheath grading, the main objection is the possibility of damage to the intersheath during transportation and installation and with capacitance grading the chief handicap is non-availabilities of high grade insulating materials whose permittivities can be made to vary over the required range.

25. (A): 3-wire dc distribution system is preferred over 2-wire dc distribution system.
(R): The 3-wire dc system of distribution is more safe.

26. (A): EHV transmission lines make use of bundled conductors.
(R): Bundled conductors reduce the line inductance per phase and increase the capacitance.

27. (A): The insulation of modern EHV lines is designed based on the lightning voltage.
(R): With the developments in the design of relay breaker systems, it is possible to control and minimise switching over-voltages.

28. (A): It is not necessarily desirable to compensate by a series capacitance for all the line reactance.
(R): Power can be transformed only if reactance is present. [U.P.S.C. I.E.S. E.E.-II, 2011]

29. (A): The determination of cross section of conductor for EHV lines is based on corona, not on thermal considerations.
(R): The choice of cross section of conductor with reference to corona considerations is generally above the economic cross-section based on thermal considerations.

30. (A): In lossless transmission line the voltage and current distributions along the line are always constant.
(R): The voltage and current distributions in an open line are such that at a distance $\lambda/4$ from the load end, the line looks like a series resonant circuit. [U.P.S.C. I.E.S. E.E.-II, 2004]

31. (A): In HVDC systems, the dc voltage should be as high as possible and the rectifier control angle α should be as low as possible.
(R): Control of dc voltage is exercised by the rectifier and inverter control angles α and γ respectively. [U.P.S.C. I.E.S. E.E.-II, 1997]

Answers: **21.** (b), **22.** (a), **23.** (a), **24.** (a), **25.** (c), **26.** (b), **27.** (a), **28.** (c), **29.** (a), **30.** (d), **31.** (c).

MATCHING TYPE QUESTIONS

Note: Each of the following questions consists of two lists. Match the items given in list 1 and those in list 2. Select your answers to these questions using the codes given in each of the questions.

1.

List 1	List 2
A. Skin effect	1. Increase in resistance but decrease in self inductance.
B. Proximity effect	2. Increase in ac resistance.
C. Ferranti effect	3. Owing to voltage drop across line inductance due to flow of a charging current.
D. Surge impedance	4. Square root of ratio of line impedance and shunt admittance.

Codes:	A	B	C	D
(a)	2	1	3	4
(b)	1	2	4	3
(c)	3	4	2	1
(d)	4	3	1	2

2.

List 1 (Design parameters)	List 2 (Factor(s) on which they depend)
A. Number of suspension insulator discs.	1. Voltage and tower footing resistance.
B. Permissible sag of transmission line conductor for a given tower.	2. Voltage.
C. Corona discharge.	3. Voltage and conductor configuration.
D. Inductance of transmission line conductor.	4. Conductor configuration and tower configuration.

Codes:	A	B	C	D
(a)	1	2	3	4
(b)	4	3	2	1
(c)	1	3	2	4
(d)	4	2	3	1

[U.P.S.C. I.E.S. E.E.-II, 2002]

3.

List 1	List 2
A. Per cent power lost in transmission.	1. Decreases with system voltage.
B. For a given current density, the conductor size.	2. Reduces with line length.
C. Power handling capacity of a line at a given voltage.	3. Remains independent of line length.
D. Surge impedance of a transmission line.	4. Increases with line length.

Answers: **1.** (a), **2.** (a), **3.** (d),

Codes:	A	B	C	D
(a)	1	2	4	3
(b)	3	4	2	1
(c)	3	2	4	1
(d)	1	4	2	3

[U.P.S.C. I.E.S. E.E.-II, 1998]

4.

	List 1		**List 2**
A.	improve power factor.	1.	shunt reactor.
B.	reduce the current ripples.	2.	shunt capacitor.
C.	increase the power flow in line.	3.	series capacitor.
D.	reduce the Ferranti effect.	4.	series reactor.

(a) $a \to 2, b \to 3, c \to 4, d \to 1$
(b) $a \to 2, b \to 4, c \to 3, d \to 1$
(c) $a \to 4, b \to 3, c \to 1, d \to 2$
(d) $a \to 4, b \to 1, c \to 3, d \to 2$ **[GATE E.E. 2009]**

5.

	List 1		**List 2**
A.	Greatz Bridge Converter.	1.	EHV/UHV AC Transmission.
B.	Series Compensation.	2.	HVDC Transmission.
C.	Sag Templates.	3.	Insulators.
D.	Grading Ring.	4.	Tower Location.

Codes:	A	B	C	D
(a)	2	1	3	4
(b)	1	2	4	3
(c)	2	1	4	3
(d)	1	2	3	4

[U.P.S.C. I.E.S. E.E.-II, 2003]

6. **List 1 (Different Configurations for Asynchronous Interconnection)** — **List 2 (Corresponding Figures)**

A. Two-terminal DC Link — 1.

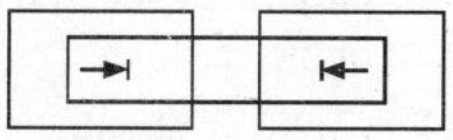

B. Back to back HVDC Link along with AC feeder — 2.

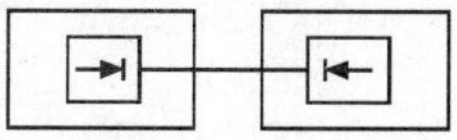

C. Back to back HVDC Link at border — 3.

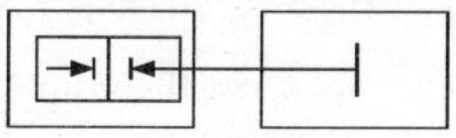

Codes:	A	B	C
(a)	3	2	1
(b)	2	3	1
(c)	3	1	2
(d)	2	1	3

[U.P.S.C. I.E.S. E.E.-II, 2003]

Answers: **4.** (b), **5.** (c), **6.** (d).

TRUE(T)/FALSE(F) TYPE QUESTIONS

Note : Identify whether the following statements are TRUE (T) or FALSE (F).

1. Single phase ac power pulsates around the average value at the supply frequency.
2. Normally distribution of electrical energy is through 3-phase 4-wire system.
3. In India the primary transmission voltages are 110, 132, 220, 400 or 765 kV depending upon the distance, the amount of power to be transmitted and the system stability.
4. The voltages for primary distribution are 11, 6.6 or 3.3 kV.
5. For secondary distribution usuable voltage is 230 V.
6. In a distribution system, the feeder supplies power to distributors.
7. The electric power can be transmitted by either overhead system or underground system.
8. The best method is to employ dc system for generation and distribution and ac system for transmission.
9. Overhead system is more safer than under ground system.
10. Underground system is more flexible than overhead system.
11. The underground system cannot be operated above 66 kV.
12. Underground system is less expensive in comparison to overhead system.
13. The voltage drop in underground system is very low as compared to that in overhead system.
14. Maintenance cost of underground system is quite high in comparison with that of overhead system.
15. Mostly the high voltage transmission is carried out by overhead system due to low cost.
16. Underground system is used for distribution only either in conjested areas or modern cities.
17. 3-phase 4-wire system is usually adopted for transmission of electrical power.
18. With the increase in voltage of transmission, the insulation needed between the conductors and earthed tower increases.
19. With the increase in voltage of transmission the line support cost reduces.
20. With the increase in voltage of transmission, longer cross arms and higher towers are required.
21. Economical cross section of conductor can be determined by Ferranti effect.
22. Wooden poles are used as line supports for voltages up to 22 kV.
23. The conductors used in *ht* overhead lines are stranded ones.
24. ACSR conductor implies aluminium conductor steel reinforced.
25. In ACSR conductors, the resistance of the conductor due to the steel core decreases.
26. The skin effect is very predominating in ACSR conductor and so the resistance of the composite conductor is taken equal to that of the aluminium alone.
27. The ACSR conductor gets deteriorated in service to the atmospheric corrosion due to bimetallic action of zinc and aluminium, electrochemical in nature.
28. ACSR conductor being of high tensile strength and lighter in weight produces small sag and so longer spans can be used.

Answers: **1.** (F), **2.** (T), **3.** (T), **4.** (T), **5.** (F), **6.** (T), **7.** (T), **8.** (F), **9.** (F), **10.** (F), **11.** (T), **12.** (F), **13.** (T), **14.** (F), **15.** (T), **16.** (T), **17.** (F), **18.** (T), **19.** (F), **20.** (T), **21.** (F), **22.** (T), **23.** (T), **24.** (T), **25.** (F), **26.** (T), **27.** (T), **28.** (T),

29. The ACSR conductor has largest diameter than any other type of conductor of same resistance and so corona losses are reduced but stronger supports are required for a given span.

30. Composite conductors are principally used for shorter spans, not exceeding 100 m.

31. Galvanised iron and steel conductors have been used for comparatively shorter spans.

32. Guy wires are generally galvanised.

33. Guy wires generally make an angle of 45° to 60° with the surface of the earth.

34. Guy wire should have high electrical conductivity.

35. Addition of 1 or 2% of cadmium in copper increases the conductivity but reduces the tensile strength.

36. Addition of about 1/2% each of silicon and magnesium to aluminium increases its tensile strength but reduces its resistance to corrosion.

37. Increase in temperature in overhead transmission line causes decreases in stress but increase in length.

38. Copper clad steel conductors are quite suitable for shorter spans.

39. The choice of economic span length of transmission line depends upon working voltage, size, material and number of conductors.

40. Supports for conductors may have unequal heights in hilly areas.

41. Heavier the conductors smaller will be the sag.

42. Sag is directly proportional to the square of the span.

43. Sag reduces and tension in the conductors increases with the rise in temperature.

44. For transmission line conductors, maximum sag occurs when temperature is maximum and there is no wind pressure.

45. For transmission line conductors the lowest tension and maximum sag occurs when temperature is minimum and wind maximum.

46. The top most conductor in HV transmission line is earth conductor.

47. V-strings are suitable for single circuit lines.

48. Dancing of conductors can be seen in high temperature regions.

49. Eddies produced behind the conductors by the action of low velocity winds result in resonant vibrations.

50. Armour rods and stock bridge dampers are used to minimise vibrations in a conductor.

51. Overhead lines should be earthed at every 1.6 km.

52. Allowable variation in declared voltage is 12½ % in case of high or extra high voltage.

53. For 400 kV transmission lines the clearance from ground should be nearly 8.5 m.

54. The material used for insulator materials should be nonporous, free from internal impurities and cracks and impervious to the fluids and gases present in the atmosphere.

55. As an insulator material, toughened glass insulator have greater puncture strength.

56. Glass is cheaper and have longer life in comparison to porcelain.

57. Pin type insulators give more flexibility to the line and the mechanical stresses are reduced.

58. Suspension type insulators are used on EHT transmission lines.

59. The insulators used in guy cables are egg or stay insulators.

60. The corona increases with the increase in diameter of the conductor.

61. The corona effect decreases with the decrease in spacing between the conductors.

62. A solid conductor gives rise to more corona than a stranded conductor.

63. At low voltage, there is no corona effect.

64. Line voltage largely affects the corona.

65. If the spacing between the conductors is made very large in comparison to their diameters, there may not be any corona effect.

66. Corona increases the surge effects.

67. Corona discharges emit radiations inducing noise signals in the communication channels.

68. Due to corona, the probability of flashover is increased.

69. Corona increases the effects of transients produced due to lightning and other causes.

70. Corona does not involve any power loss.

71. The value of disruptive critical voltage is considerably increased during bad atmospheric conditions such as fog, sleet, rain and snow storms.

72. Corona occurs on short transmission lines.

73. Corona losses are minimized when current density in conductors is reduced.

74. At high frequencies the skin effect is very small.

75. Capacitance of a transmission line increases with the increase in its length.

76. The proximity effect results in uniform distribution of current in cross section of a conductor.

77. The charging current of a 400 kV transmission line is more than that of a 220 kV line of the same length. **[GATE E.E. 1994]**

78. The presence of earth in case of overhead lines increases the capacitance.

79. Transposition of transmission line is done to reduce corona.

80. In short transmission lines, the shunt capacitance is totally neglected and the series impedance is a lumped parameter.

81. The phenomenon of rise in voltage at the receiving end of the open-circuited or lightly loaded line is known as the proximity effect.

82. In short transmission lines

$$\mathbf{A = D = 1;\ B = Z \text{ and } C = 0}$$

83. In nominal T line

$$\mathbf{A = D = 1 + \frac{ZY}{2};\ B = Z + \frac{Z^2Y}{4} \text{ and } C = Y}$$

84. In nominal π-line

$$\mathbf{A = D = 1 + \frac{1}{2}YZ;\ B = Z \text{ and } C = 1 + \frac{1}{4}YZ}$$

Answers: **29.** (T), **30.** (F), **31.** (F), **32.** (T), **33.** (T), **34.** (F), **35.** (F), **36.** (F), **37.** (T), **38.** (F), **39.** (T), **40.** (T), **41.** (F), **42.** (T), **43.** (F), **44.** (T), **45.** (F), **46.** (T), **47.** (T), **48.** (F), **49.** (T), **50.** (T), **51.** (F), **52.** (T), **53.** (T), **54.** (T), **55.** (T), **56.** (T), **57.** (F), **58.** (T), **59.** (T), **60.** (F), **61.** (F), **62.** (F), **63.** (T), **64.** (T), **65.** (T), **66.** (F), **67.** (T), **68.** (F), **69.** (F), **70.** (F), **71.** (F), **72.** (F), **73.** (F), **74.** (F), **75.** (T), **76.** (F), **77.** (T), **78.** (T), **79.** (F), **80.** (T), **81.** (F), **82.** (T), **83.** (T), **84.** (F),

85. Power loss in each of the line conductor of open-circuited line due to charging current is equal to 1/3 I_C^2 R wattswhere $I_C/\sqrt{3}$ is the rms value of charging current and R is the resistance of conductor.
86. On inductive loads, as usual in practice, the effect of capacitance is to reduce voltage drop and also the copper losses of the line.
87. For a lossless line the characteristic impedance is called the surge impedance.
88. Surge impedance of a transmission line is independent of its length.
89. The reactive power transfer over a line mainly depends on $|V_S| - |V_R|$.
90. Insulation resistance of the cable increase with the increase its length.
91. A metallic sheath of lead or lead alloy or of aluminium is provided around the insulation in an underground cable for reducing the capacitance.
92. The thickness of metallic sheath on the cables is usually of the order of 0.5 mm.
93. A layer of bedding consisting of paper tape compounded with a fibrous material is provided over the metallic sheaths in underground cables in order to protect the metallic sheath from mechanical injury from the armouring.
94. The size of power cable conductor depends upon the operating voltage and type of insulation used.
95. The operating voltage for low tension cables is up to 400 V.
96. In single core cable copper used is tinned stranded.
97. Multi-core cables are used up to 11 kV.
98. For voltages up to 6,600 V, electrostatic stresses developed in cable are very small.
99. The belted type construction is not suitable for the cables for use on operating voltage exceeding 22 kV.
100. The high tension cables are so designed that the ionization effects are maximum.
101. In the belted type cables the dielectric does not remain homogeneous and may result into formation of voids.
102. H-type cables are used up to 66 kV.
103. The disadvantage of SL type cable is that the manufacturing is difficult because of thinner lead sheaths.
104. Maintenance of oil-filled cables is difficult.
105. The operating voltage for extra supertension power cables is beyond 66 kV.
106. The sheath loss in open-circuited sheath is about 2 % of the total loss.
107. Capacitance is important in cables because of narrow spacing between the conductors.
108. The disadvantages of oil-filled cables are higher cost and complicated laying and maintenance.
109. The insulation resistance of a single-core sheathed cable increases with the increase in its length.
110. The breakdown strength of the cable depends upon the maximum stress it can bear.
111. The electrostatic stress in a single core cable is minimum at the conductor surface.
112. In capacitance grading of cables a homogeneous dielectric is used.
113. Grading of cables is only useful for very high voltage cables where the ratio of external diameter to internal diameter is very small.
114. When a cable is to cross a road, it should be laid in conduits, ducts or tubes made up of iron, glazed stone wares, clay or cement concrete.
115. Direct laying of underground cables is rarely used because of its high maintenance cost.
116. When two or more generating stations are connected together, the system is known as the ring main system.
117. A distributor arranged to form a closed circuit and fed at one or more than one point is called the ring distributor.
118. For transmission of same amount of power with same efficiency over a given distance with same consumer voltage, three-wire dc distribution system takes much more copper than that required by two-wire dc distribution system.
119. The cross-sectional area of the neutral wire in a 3-wire dc distribution system is usually one-half of that of outer.
120. The effect of corona can be reduced by making use of stranded conductors.
121. Bundle conductors are mainly used for reducing corona losses and radio interference.
122. Surge impedance of a transmission line is independent of its length.
123. The size of conductor for modern EHV lines is determined on the basis of current density.
124. The insulation for EHV lines is designed on the basis of corona.
125. The insulation for EHV lines is designed on the basis of switching surges.
126. Series compensation is provided in EHV transmission lines to overcome the Ferranti effect.
127. Synchronous compensation is being preferred over static compensation in EHV lines.
128. The dc output voltage magnitude is controlled by varying the firing angle of the thyristor valves in the converter.
129. Corona loss and radio interference are much more in dc system in comparison to those in ac system.
130. The dc line terminal equipment is considerably less costly than ac line terminal equipment.
131. In dc system, skin effect is absent and line losses are less than that ac system.
132. The ac systems interconnected by a dc line can be controlled independently.
133. Maintenance of insulators in HVDC transmission line is less.
134. In case of agricultural loads, the distribution transformer should be located at the load centre.
135. The overhead lines should not cross the road at an angle less than 60°.

Answers: **85.** (T), **86.** (T), **87.** (T), **88.** (T), **89.** (T), **90.** (F), **91.** (F), **92.** (F), **93.** (T), **94.** (F), **95.** (F), **96.** (T), **97.** (T), **98.** (T), **99.** (T), **100.** (F), **101.** (T), **102.** (T), **103.** (T), **104.** (T), **105.** (F), **106.** (T), **107.** (T), **108.** (T), **109.** (F), **110.** (T), **111.** (F), **112.** (F), **113.** (F), **114.** (T), **115.** (F), **116.** (F), **117.** (T), **118.** (F), **119.** (T), **120.** (F), **121.** (T), **122.** (T), **123.** (F), **124.** (F), **125.** (T), **126.** (F), **127.** (F), **128.** (T), **129.** (F), **130.** (F), **131.** (T), **132.** (T), **133.** (F), **134.** (T), **135.** (T).

FILL IN THE BLANKS TYPE QUESTIONS

Note : Fill in the blanks with appropriate word or words.

1. Generation voltages are 3.3 kV, 6.6 kV, 11 kV or 33 kV, most usual adopted voltage for generation in practice is ________ kV.
2. The maximum generation voltage in advance countries is ________ kV while that in India is 11 kV.
3. Secondary transmission is usually carried at ________ or 66 kV.
4. The highest transmission voltage in India is ________ kV. (220/132/400/765)
5. Large industrial consumers are supplied power at ________ kV. (11/66/132/400)
6. 750 kV falls under the category of ________ high voltage.
7. The conductor connecting consumer's terminals to the distributor is called the ________. (feeder/distributor/service mains)
8. The permissible variation of frequency in power system is ________.
9. The distribution by ac system is undoubtedly ________ to that by dc system.
10. ________ poles are generally used for distribution purposes in rural areas, steel tubular poles and ________ poles are usually used for distribution purposes in urban areas to give good appearance and steel rails, or narrow base lattice steel towers are used for transmission at ________ kV or 33 kV and broad base lattice steel towers are used for transmission at ________ kV and above.
11. Transmitted power remaining the same, if the supply voltage of a two-wire distributor is doubled saving in copper will be ________ %. (25, 50, 75, 100)
12. In a 2-wire dc feeder, voltage drop per wire is 2.5%, transmission efficiency of the feeder will be ________ %. (93.75, 95, 97.5, 98.75)
13. With the same maximum voltage between conductors, the ratio of copper volume required is 3-phase, 3-wire system and 1-phase, two-wire system is ________. $\left(\frac{4}{3}, \frac{3}{4}, \frac{5}{3}, \frac{3}{5}\right)$
14. The most economical voltage for transmitting given amount of power over a known distance by overhead transmission lines is about ________ kV/km.
15. Maximum permissible span with wooden poles is ________ m. (20, 30, 50, 100)
16. Steel poles need galvanizing for longer life. Galvanizing is the process of applying a layer of ________. (zinc/tarcoal/paint/varnish)
17. RCC poles are ________ in weight and need ________ maintenance.
18. In an overhead system, the number of cross arms carried by a pole depends on the number of ________ it has to carry.
19. For installations of switchgear and power transformers we usually use ________ type structure. (A, L, H, V)
20. The conductor material used for transmission and distribution of electrical power must be of ________ conductivity, ________ tensile strength, ________ specific gravity and ________ cost.
21. Stranded conductors are preferred over solid conductors for overhead lines owing to ________.
22. ________ conductor is the best conductor owing to its high electrical conductivity and greater tensile strength for all types of transmission.
23. The conductivity of aluminium is ________ times of that of copper.
24. The tensile strength of aluminium is ________ times of that of copper.
25. ________ conductor consists of a core of galvanised steel strand surrounded by a number of aluminium strands.
26. In ACSR conductor aluminium carries the bulk of ________ and steel core takes a greater percentage of ________.
27. In overhead lines we generally use ________ conductors.
28. ACSR conductors having 70 aluminium conductors and 6 steel conductors will be specified as ________. (6/70, 6/76, 70/6, 70/76)
29. The difference in level between the points of supports and the lowest point is called the ________.
30. Between two supports, due to sag the conductor takes the form of ________. (semicircle, catenary, parabola)
31. The minimum clearance of hv lines from ground across streets is ________ m. (3, 5, 6, 8)
32. Minimum horizontal clearance of a lv line from residential building is ________ m. (0.6, 0.9, 1.2, 1.6)
33. If a 132 kV line passes over a residential building, the minimum vertical clearance from the roof of the building shall be ________ m. (3, 4.57, 6, 9)
34. The minimum clearance between 132 kV transmission line and ground is about ________ m. (3.2, 6.4, 7.5, 10.5)
35. The horizontal spacing between phase conductors of a 132 kV transmission line is about ________ m. (8, 11, 14, 17)
36. The colour of the neutral wire of 3-core fluxible cable is ________. (black, blue, red, brown)
37. ________ chart is helpful in knowing the sag and tension at any temperature.
38. The insulators employed in overhead lines must be mechanically ________ and of ________ relative permittivity and ________ insulation resistance.
39. ________ is extensively used material for insulators and is produced by firing at a controlled temperature a mixture of kaolin, feldspar and quartz.
40. Glass insulators can be used up to ——— kV under ordinary atmospheric conditions and well up to ________ kV in dry atmosphere.
41. Pin type insulators are generally not used for voltages exceeding ________ kV. (11, 25, 33, 66)
42. Shackle insulators are used on ________ lines.
43. ________ or ________ insulators are used in guy cables.
44. Wet flash-over voltage is ________ than dry flash-over voltage of an insulator. **[A.M.I.E. Sec B. Summer 1993]**
45. Lowest disc in a string of suspension insulators has ________ pd across it. **[A.M.I.E. Sec B. Summer 1994]**
46. The whole phenomenon, the hissing noise, the violet glow and production of ozone gas is known as ________.
47. In case of ac system, the current due to corona is ________.

48. In stormy weather corona occurs at ________ voltage as compared with fair weather.
49. The effect of dirt on the surface of the conductor is to ________ irregularity and thereby ________ the breakdown voltage.
50. Corona ________ the probability of flashover and ________ the system performance.
51. In case of overhead lines spacing between the conductors is to be increased in order to provide adequate insulation and avoid ________ loss.
52. An electric transmission line has four parameters viz resistance, inductance, capacitance and ________.
53. The increase in resistance due to non-uniform distribution of current in a conductor is known as ________ effect. [GATE E.E. 1994]
54. Any two conductors separated by an insulating medium constitute a ________.
55. The transmission lines are basically electrical circuits having ________ constants.
56. The ratio of line-to-line capacitance and line-to-neutral capacitance is ________. $\left(\frac{1}{4}, \frac{1}{2}, \frac{1}{3}, 2\right)$
57. In short transmission lines the effect of ________ is negligible.
58. 120 km long transmission line is considered to be ________ line.
59. A 25 km, 33 kV transmission line is considered to be a ________ transmission line.
60. A 160 km, 220 kV transmission line falls under the category of ________ transmission lines.
61. The effect of capacitance can be neglected when the length of overhead transmission line does not exceed ________ km. (20, 60, 120, 250)
62. For a 500 Hz frequency excitation, a 50 km long power line will be modelled as ________ line. (short, medium, long)
63. For 11 kV transmission line the capacitance per km will be about ________ μF. (0.01, 0.1, 1, 10)
64. The phenomenon of rise in receiving-end voltage of the open-circuited or lightly loaded line is called the ________ effect.
65. The regulation of a transmission line, for any given load power factor can be graphically determined by ________ regulation diagram.
66. The velocity of propagation of electromagnetic waves is ________ on overhead lines and on cables it depends on ________. [A.M.I.E. Sec B. Winter 1997]
67. The velocity of propagation of electromagnetic waves on underground cables with relative permittivity of 3 will be ________ m/s. [GATE E.E. 1997]
68. The square root of the ratio of line impedance and shunt admittance is called the ________ impedance.
69. The load (of unity pf) that can be delivered by the line of negligible resistance is called the surge impedance loading or ________.
70. The rated load of an underground cable is always ________ its natural load. [GATE E.E. 1995]
71. A line which has same voltage all along its length is known as ________ line. [A.M.I.E. Sec B. Summer 1994]
72. For constant voltage transmission specially designed synchronous motors called the synchronous ________ are installed at the receiving end which maintains the voltage drop along the line constant.
73. Overexcited synchronous phase modifier draws ________ current.
74. The real power transfer over a line mainly depends on ________ angle.
75. The SIL of a line is proportional to ________. (V^2, V, $\sqrt{V}$)
76. The most economical load on an overhead line is ________ than the natural load.
77. The most economical load on an underground cable is ________ than the surge loading.
78. In underground cables, over and above armouring a layer of fibrous material is again provided, which is similar to that of bedding in order to protect the armouring. Such a layer of fibrous material is called the ________.
79. The breakdown strength of air at NTP is ________. [A.M.I.E. Sec B. Winter 1997]
80. Insulating materials are classified for different temperature rise such as ________.
81. The insulating materials most commonly used in a cable are ________. [A.M.I.E. Sec B. Winter 1997]
82. The multi-core cables for use up to 11,000 V are of ________ type.
83. Single core oil-filled cables can be employed up to ________ kV.
84. ________ gas is often used in external pressure cables.
85. The energy losses occuring in the dielectric of cables are due to leakage and so called ________.
86. In internal pressure cables the pressure of nitrogen is about ________ atmosphere for EHT cables and about ________ atmosphere for super voltage cables.
87. The process of achieving uniformity in dielectric stress is known as ________ of cables.
88. There are two methods of grading viz intersheath grading and ________ grading.
89. The material commonly used for sheaths of underground cables is ________. (copper/lead/steal/rubber)
90. ________ protects the underground cables against mechanical injury. (Armouring/Sheath/Bedding)
91. Cables generally used beyond 66 kV is ________. (oil-filled, belted, SL type)
92. The cables used for 132 kV are ________.
93. ________ system of laying underground cables is rarely used because of its higher cost.
94. While laying 11 kV underground cables the minimum bending radius must be ________ D. (3, 6, 12, 18)
95. The depth of trench required for laying underground cables by direct laying method is about ________ m. (0.2, 0.5, 0.8, 1.2)
96. While laying 33 kV underground cable the minimum bending radius should be ________ D. (12, 30, 100)
97. A 250 V cable after installation should be checked for about ________ minutes at a voltage not less than ________ V.
98. An 11 kV cable after installation should be checked for short duration at a voltage of ________ kV. (11, 20, 33, 110)
99. A booster is connected in ________ with the feeder.
100. Balancers are used for ________. [A.M.I.E. Sec B. Winter 1997]
101. The low value of insulation resistance between any core of cable and earth indicates the ________ fault.
102. The highest transmission voltage in the world is ________ kV.
103. One of the advantage of high voltage dc transmission is that there is no ________ effect.
104. The main source of radio and television interference near EHV system is ________.
105. The size of conductor for EHV transmission line is based on ________.

106. SIL means _______.

107. By using bundled conductors, the critical voltage for formation of corona will ________. (decrease/increase)

108. Bundle conductors in transmission lines ________ radio interference. (reduce/increase)

109. The dc transmission is mainly at ________ kV. (50, 66, 132, 500)

110. For a 400 kV line, the spacing between phase conductors is about ________ m. (8, 11, 14, 20)

111. _______ conductors are invariably employed for overhead transmission lines.

112. For a 400 kV the economical value of span is _______ metres.

113. _______ configuration has been found the most economical for double circuit lines.

114. _______ and _______ voltage lines are confined within the limits of towns and cities and within small distance in rural areas.

115. Normally an overhead line should cross the railway track at _______ angle. However, a deviation to the extent of _______ can be permitted under unavoidable circumstance.

ANSWERS

1. 11 **2.** 33 **3.** 33 **4.** 765
5. 11 **6.** ultra **7.** service mains **8.** ± 3%
9. superior **10.** Wooden; RCC. 11; 66 **11.** 50 **12.** 95
13. 3/4 **14.** 0.6 **15.** 50 **16.** zinc
17. heavy; little **18.** wires **19.** H **20.** high; high; low; low
21. flexibility **22.** Hard drawn copper **23.** 0.6 **24.** 0.45
25. ACSR **26.** current; mechanical stresses **27.** ACSR **28.** 70/6
29. sag **30.** catenary **31.** 6 **32.** 1.2
33. 4.57 **34.** 6.4 **35.** 8 **36.** black
37. Stringing **38.** very strong; high; high **39.** Porcelain **40.** 25; 50
41. 33 **42.** low voltage **43.** Egg; stay **44.** lower
45. maximum **46.** corona **47.** nonsinusoidal **48.** lower
49. increase; reduce **50.** reduces; improves **51.** corona **52.** shunt conductance
53. skin **54.** capacitor **55.** distributed **56.** 1/2
57. capacitance **58.** medium **59.** short **60.** long
61. 60 **62.** long **63.** 0.01 **64.** Ferranti
65. Kapp's **66.** 3×10^8 m/s; $\in_r$ of dielectric **67.** 1.732×10^8 **68.** surge
69. natural load **70.** higher than **71.** constant voltage **72.** phase modifiers
73. leading **74.** power **75.** V^2 **76.** greater
77. lesser **78.** serving **79.** 3 kV/mm **80.** Y, A, E, B, F, H, C
81. VIR; PVC; impregnated paper **82.** belt **83.** 132
84. Nitrogen **85.** dielectric hysteresis **86.** 6, 12 **87.** grading
88. capacitance **89.** lead **90.** Armouring **91.** oil-filled
92. extra super voltage **93.** Solid **94.** 12 **95.** 1.2
96. 30 **97.** 15, 500 **98.** 20 **99.** series
100. balancing the load **101.** ground **102.** 1,500 **103.** skin
104. corona **105.** corona **106.** surge impedance loading **107.** increase
108. reduce **109.** 500 **110.** 11 **111.** ACSR
112. 350 – 400 **113.** Vertical **114.** Low; medium **115.** right; 30°

❑❑❑

PART III

SWITCHGEAR AND PROTECTION

CHAPTER 1

Introductory

INSIDE THIS CHAPTER

1.1. GENERAL CHARACTERISTICS OF ELECTRICAL EQUIPMENT

All electrical machines, apparatus and other forms of electrical equipment must satisfy two main requirements—they must be able to operate continuously under normal service conditions and must be able to withstand short-time overcurrents and overvoltages such as may arise during emergent conditions.

For operation of electrical machines and apparatus with full reliability in normal operating conditions, the following two requirements must be met.

$$\text{Rated current} \geq \text{Actual load current} \quad ...(1.1)$$

and $$\text{Rated voltage} \geq \text{Working voltage of the piece of equipment} \quad ...(1.2)$$

Nominal or *rated current* is that maximum current with which a given piece of electrical apparatus or machine will be able to operate continuously throughout its service life (provided the equipment is rated for continuous operation).

When a piece of apparatus or machine operates with a load current that does not exceed its current rating, the temperature rise of any part will never exceed the permissible limits, thus providing continuous and reliable operation.

The *rated voltage* of a piece of electrical apparatus or a machine is the value of the line voltage given on its name plate, this value being equal to the nominal line voltage of the circuit it must operate in.

The rated voltage must be selected so that the maximum working voltage during operation does not exceed the rated voltage by more than 10–15 per cent.

Owing to the fact that conditions under which the insulation of equipment intended for operation in the open air are much more severe than that of the equipment which is installed within enclosed premises, due to atmospheric precipitations, corrosive chemical deposits, fly ash and dust. Electrical apparatus and machines are designed for either indoor or outdoor service.

All electrical installations and circuits are subject to the danger of a fault leading to short-circuiting and flow of current in the equipment greatly exceeding the rated value.

In normal operating conditions all the circuit elements of an electrical system carry currents, whose magnitude depends upon the value of the generator voltage and the effective impedances of all the power transmission/distribution system elements, including the impedances of the loads usually relatively large. However, when, due to a fault (known as a *short circuit*) the conductors of different phases come into contact with each other in a power line, power transformer, or any other circuit element, part of the impedance of the system will be shunted out of the circuit. This reduces the effective impedance of the system while the current rises. The lesser the effective impedance in a faulty circuit, the greater the short-circuit current.

Short-circuit currents are harmful for two reasons–the first is that even a short-time flow of heavy current will overheat the equipment, the second is that the flow of short-circuit currents through the current carrying parts produces forces of electrodynamic interaction which may destroy or damage the equipment. This is why all the elements of any electrical installation are designed and selected for a thermal and a dynamic stability sufficient to withstand the largest possible flow of short-circuit current that may occur in the given installation.

The severest conditions in service are those of the switching devices during a short circuit. These devices are circuit breakers and the fuses, whose function is to interrupt the fault current within a short period or time (0.05 – 0.3 second) and thereby switch out the faulty section.

The ability of a switching device to interrupt a short-circuit current is expressed by its *rated breaking current*, the maximum value of the current it can safely interrupt at the normal service voltage of the installation in which it is designed to operate.

For satisfactory operation of a circuit breaking device the following condition should be fulfilled:

Rated breaking current ≥ Maximum current flowing through the circuit during the short circuit ...(1.3)

The ability of a switching device to interrupt short circuits is expressed by their rated MVA breaking capacity.

Rated breaking MVA $= \sqrt{3} \times$ rated breaking current in kA × rated voltage of the given device in kV ....(1.4)

Another condition to be fulfilled in order that a switching device will operate reliably is that

Rated breaking MVA ≥ Maximum MVA developed in the circuit during the short circuit (also called the fault power) ...(1.5)

The fault power is given as

$S_{max\ sc} = \sqrt{3} \times$ maximum value of short-circuit current in kA × rated voltage of the given installation in kV ...(1.6)

1.2. FAULTS IN POWER SYSTEM

A fault in an electrical equipment/apparatus is defined as a defect in the electrical circuit due to which current is diverted from the intended path.

The nature of a fault simply implies any abnormal condition which causes a reduction in the basic insulation strength between phase conductors or between phase conductors and earth or any earthed screen surrounding the conductors. Actually the reduction of insulation strength is not considered as a fault until it creates some effect on the system *i.e.,* until it results either in excessive current or in the reduction of the impedance between conductors or between conductors and earth to a value below that of the lowest load impedance normal to the circuit.

In an electrical power system comprising of generators, switchgear, transformers, power receivers and transmission and distribution circuits, it is inevitable that sooner or later some failure will occur somewhere in the system. The probability of the failure or occurrence of abnormal condition is more on power lines—about one-half of the faults occur on the power lines. This can be explained by the fact that the power lines are widely branched, have greater length, operate under variable weather conditions and are subject to the action of atmospheric disturbances of electrical nature.

According to the causes of incidence, causes of failures may be classified, as mentioned below.

(*i*) Breakdown may occur at normal voltage due to (*a*) the deterioration or ageing of insulation and (*b*) the damages caused by unpredictable happenings such as blowing of heavy winds, tree falling across lines, vehicles colliding with towers or poles, birds shorting out lines, aircraft colliding with lines, line breaks etc.

(*ii*) Breakdown may occur due to abnormal voltages caused by switching surges or lightning strokes which may be either direct or induced.

The current practice is of providing a high insulation level of the order of 3 to 5 times the nominal value of the voltage but still the insulation strength is reduced because of pollution on an insulator string, commonly caused by deposited soot or cement dust in industrial areas and by windborne sea-spray in coastal areas. Initially the insulation resistance is lowered and small leakage currents are diverted and thus the deterioration is hastened. Even in enclosed installations such as sheathed and armoured cables and metal-clad switchgear, insulation gets deteriorated because of ageing. Void formation in the insulating compound of underground cables due to unequal expansions and contractions caused by the increase and decrease in temperatures is another cause of insulation failure.

The line and apparatus insulation may be subjected to transient overvoltages because of the switching operations. The voltage which rises at a rapid rate may achieve a peak value approaching three times phase-to-neutral voltage. This is the reason that a higher insulation level is provided initially. In case the insulation levels have been correctly chosen and they have not been impaired in a way described under (*i*) above, the system will withstand these routine overvoltages. But if the insulation gets deteriorated due to one or other reason, it is at the time of switching that failure may occur.

Lightning produces a very high voltage surge in the power system of the order of million of volts and thus it is not feasible to provide insulation which can withstand this abnormality. These surges travel with the velocity of light in the power circuits, the limiting factors are the surge impedance and the line reactance.

The faults can be minimised by improving the system design, quality of the equipment and maintenance. However, the faults cannot be eliminated completely.

1.2.1. Fault Statistics

It is useful to have some ideas of percentages of various causes of faults and frequency of fault occurrence in different links of a power system.

Table 1.1 gives approximately percentagewise the various causes of faults.

Table 1.1. Approximately Percentage of Various Causes of Faults

Causes	Percentage of Total
1. Lightning	12
2. Sleet, wind, mechanical (jumping conductors)	20
3. Apparatus failure	20
4. Switching to a fault	20
5. Miscellaneous (tree falling on line, birdage, sabotage, accidents etc.)	28

Distribution of faults in the various sections of a power system are shown in Table 1.2.

Table 1.2. Frequency of Fault Occurrence in Different Links of a Power System

Equipment	Percentage of Total
Overhead lines	50
Cables	10
Transformers	10
Switchgear	15
Control equipment	3
Instrument transformers (CTs and PTs)	2
Miscellaneous	10

1.2.2. Main Kinds of Fault

The most common and dangerous fault, that occurs in a power system, is the *short circuit* or *shunt fault*. They occur as a result of breakdowns in the insulation of current carrying phase conductors relative to earth or in the insulation between phases.

In 3-phase ac power circuits, the short-circuit faults can be classed as follows:

1. Single phase to ground (L-G) — due to breakdown of insulation between one of the phases and earth.
2. Phase to phase (L-L) — due to breakdown of insulation between either of the two phases.
3. Two phases to ground (L-L-G) — due to breakdown of insulation between two phases and earth.
4. Phase to phase and third phase to ground — a combination of faults given at no. 1 and 2 *i.e.,* due to breakdown of insulation between two phases and simultaneous breakdown of insulation between the third phase and earth.
5. All the three phases to ground (L-L-L-G) — due to breakdown of insulation between all the phases as well as to earth.
6. All the 3-phases short-circuited (L-L-L) — due to breakdown of insulation between all the three phases.

The relative probability of occurrence of various kinds of short circuits in overhead and cable power transmission circuits is given in Table 1.3.

The first four types of faults are of an unsymmetrical nature and give rise to unsymmetrical currents *i.e.,* different currents in the three phases. The latter two types of faults are of symmetrical nature and give rise to symmetrical current *i.e.,* equal fault currents in the three phases with 120º displacement.

The line to ground (L-G) fault occurs most commonly in overhead line practice as indicated in Table 1.3. Fortunately a large number of such faults are transitory in nature and may vanish within a few cycles as would be the case when a twig falls across a line and a cross arm and burns itself out or just falls down. The balanced three-phase fault is very rare in occurrence, accounting for only about 5 per cent of the total, but is the severest of all types of faults and thus imposes the most severe duty on circuit breakers and is used in the determination of the circuit breaker rating. The symmetrical three-phase faults, generally occur due to carelessness of operating personal.

Table 1.3. Relative Probability of Occurrence of Most Frequent Kinds of Short-Circuit Faults in Overhead and Cable Power Transmission Circuits.

S. No.	Type of Short-Circuit Fault	Representation	Percentage Occurrence
1.	Single phase to ground (L-G)	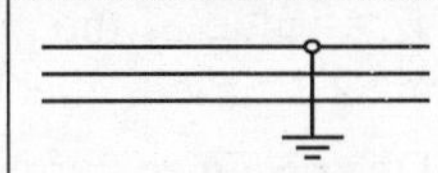	70
2.	Phase to phase (L-L)	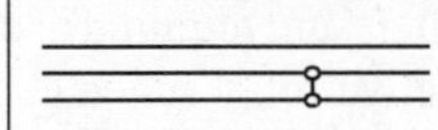	15
3.	Two phases to ground (L-L-G)	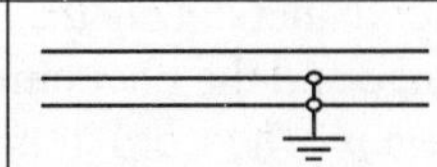	10
4.	Phase to phase and third phase to ground	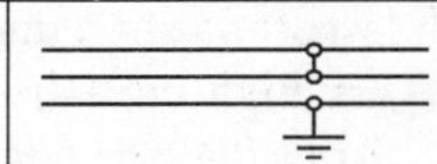	2 or 3
5.	All the three phases to ground (L-L-L-G)	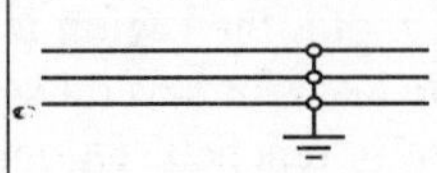	2 or 3
6.	All the three phases shorted	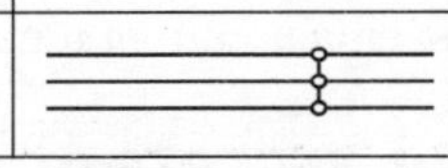	2 or 3

1.3. SHORT-CIRCUIT CURRENTS

Whenever a short circuit occurs in a power network, it causes heavy current to flow in the network.

Figure 1.1 shows how a short-circuit current varies on occurrence of a short-circuit fault. During the first moment of the fault the current attains its maximum value and then decays to some steady-state value. For checking the thermal and dynamic stability of the equipment in electrical installations with respect to the fault currents which may arise in the power networks, it is necessary to know the following values of the short-circuit currents.

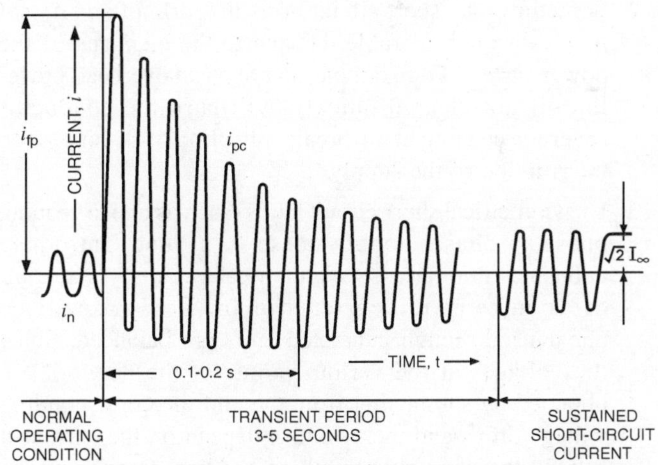

Fig. 1.1. ***Process of Change in a Short-Circuit Current***

Sustained Current, I_∞. Sustained current is the fault current that will flow in the circuit if it is allowed to persist after the transient process ends. This is the value used in designing of electrical equipment for thermal stability.

First Peak Short-Circuit Current, i_{fp}. This is the maximum instantaneous value of the current that will reach during the first moment of a short circuit. The value of this current is used in designing electrical equipment for dynamic stability. It is determined from the following equation.

First peak short-circuit current, $i_{fp} = 1.8\,(\sqrt{2}\,I_{pc})$ ...(1.7)

where I_{pc} is the rms value of the periodic component of the short-circuit current 2-3 periods after the short circuit is initiated.

The values of the sustained current I_∞ and the initial periodic component current I_{pc} are required in carrying out the calculations for a given protective relay scheme.

Thus we observe that during the fault, the current undergoes a continuous change and the phenomenon observed is called the *transient phenomenon*. The word 'transient' refers to a 'temporary happening' which lasts for a short duration of time. The zone in which current is very high but falls very rapidly is called the *subtransient state*. After the first few cycles when the decrease in current is less rapid, the region is called the *transient state*. The transient state lasts for several cycles and after the transient state, *steady state* is reached. During the steady state the rms value of the short-circuit current remains almost constant.

1.4. HARMFUL EFFECTS OF SHORT-CIRCUIT FAULTS ON POWER SYSTEM

Depending upon power circuit voltage and configuration, method of neutral connection, presence of regulating devices and the speed of disconnection of the faulted circuit section, various types of short circuits may differently influence the power system as a whole. It is well known that the flow of heavy short-circuit currents incident to the occurrence of interphase short circuits near the generating units frequently results in substantial disturbance to normal operation of power system. The short circuit may have any of the following consequences:

1. The heavy currents due to short circuit cause excessive heating which may result in fire or explosion.
2. Sometimes the short circuit takes the form of an arc that may cause considerable damage to the elements of the power system. For example, an arc on an overhead power line, if not cleared quickly, will burn the conductor severely causing it to break, resulting in a long time interruption of the supply.
3. Unsymmetrical short-circuit faults (such as line to ground, phase to phase, double phase to ground) introduce unbalance in otherwise symmetrical circuits. As a result of asymmetry the currents of the network will be unbalanced though generated emfs are balanced. Since the voltages at the various points of the network are obtained by subtracting from the emf the corresponding voltage drops and these in turn depend on the current it follows that the voltages (phase and line) throughout the system will be unbalanced.
4. Stability of the power system may be adversely affected and even the complete shutdown of the power system may occur.
5. Damage to other apparatus in the system may be caused due to overheating and due to abnormal mechanical forces set up.
6. A marked reduction in the voltage which may sometimes be so large that relays having pressure coils tend to fail.
7. There may be an interruption in power supply to consumers when power circuits or generating units are switched out with resultant outage of connected consumer equipment and in a number of cases formidable breakdowns.
8. There may be a considerable reduction in voltage on healthy feeders connected to the system having fault. This may cause either an abnormally high current being drawn by the motors or the operation of no-voltage coils of motors. In the latter case considerable loss of industrial production may result as the motors will have to be restarted.
9. Sometimes in an interconnected system, when a fault develops, it is accompanied by a fall in voltage or frequency, with the result that other loads such as motors, which normally take power from the supply, starts feeding the faults and from the fault point of view, they may be termed as sources of power.

1.5. SOURCES OF FAULT POWER

The sources of fault power originate in all the generating and other connected plants which under normal operating conditions draw power from the system. Mainly there are three sources which can feed into the short circuit; they are:

1. Generators.
2. Synchronous motors and synchronous condensers.
3. Induction motors.

1. ***Generators as a Source of Fault Power.*** The generators are driven by prime movers and when the short circuit occurs, the generators continue to run at normal speed and the excitation is also maintained. Thus the normal voltage is continued to be generated, and due to this generated voltage an enormous current is fed to the fault point which is limited by the impedance of the generator and that of the circuit up to the fault point.

2. ***Synchronous Motor as a Source of Fault Power.*** Synchronous motors are similar to synchronous generators in construction. They have dc excitation and draw ac power from the power system and convert it into mechanical power. On occurrence of a short circuit, the system voltage is considerably reduced (almost to zero) and so supply of electrical power to the motor is stopped. As a result, the motor starts slowing down, but the flywheel and inertia of the motor prevents it from slowing down instead drives it as a prime mover. As the dc excitation is maintained, the synchronous motor starts operating as a synchronous generator. Mechanical energy stored in the flywheel and the rotor is thus converted into electrical energy and fed to the fault point. The magnitude of short-circuit current which it can supply to the fault point depends on the output kW, reactance and voltage rating of the motor.

3. ***Induction Motor as a Source of Fault Power.*** Like synchronous motor the induction motor also behaves like a generator at the time of short circuit. The difference between them is that the induction motor does not have dc excitation, but during normal operation the flux from the stator appears to be as if is being produced from the dc winding.

On occurrence of a short circuit on a system, the terminal voltage across the motor terminals drops to almost zero suddenly, the flux cutting the rotor does not change instantaneously, and the inertia of the motor continues to drive it, thus it also behaves as a generator and starts feeding the fault point until the stator flux decays.

Although, synchronous motors and induction motors act as generators only for a short intervals; but they definitely produce sufficient power to cause momentary duty on the circuit breakers.

1.6. NECESSITY FOR PROTECTION SYSTEM

Modern power systems are growing fast with more generators, transformers and large network in the systems. For system operation a high degree of reliability is required. In order to protect the system (lines and equipment) from damage due to undue currents and/or abnormal voltages caused by faults (such as short circuits) the need for reliable protective devices, such as relays and circuit breakers arises. As already mentioned, the most common electrical hazard against which protection is required is the short circuit. However, there are many other abnormal conditions—*e.g.,* overloads, undervoltage and overvoltage, open phase, unbalanced phase currents, reversal of power, underfrequency and overfrequency, overtemperature, power swings, instability—for which some protection is desired.

As a rule, on the occurrence of short circuits which may lead to heavy disturbances in normal operation (damage to equipment, impermissible drop in voltage etc.), the protective scheme is designed to disconnect or isolate the faulty section from the system without any delay. The protective scheme is designed to *energise an alarm* or *signal* whenever the overloads and short circuits do not present a direct danger to the faulted circuit element and the entire installation, for example, on occurrence of a single phase fault to earth in overhead circuits operated with an insulated neutral. In such cases it is possible for the operating personnel to take the necessary measures for removal of the abnormality and prevent any interruption in power supply to consumers.

The main functions of protective relaying are to detect the presence of faults, their locations and initiate the action for quick removal from service of any element of a power system when it suffers a short circuit or when it starts to operate in any abnormal manner that might cause damage or otherwise interfere with effective operation of the rest of the system. The relaying equipment is added in this task by the circuit breakers that are capable of disconnecting the faulty element when they are called upon to do so by the relaying equipment.

Circuit breakers are generally located so that each generator, transformer, bus, transmission line etc., can be completely disconnected from the rest of the system. The circuit breakers must have sufficient capacity so that they can carry momentarily the maximum short-circuit current that can flow through them and then interrupt this current; they must also withstand closing in on such a short circuit and then interrupting it according to certain prescribed norms.

Fuses are employed where protective relays and circuit breakers are not economically justifiable.

1.7. FAULT CLEARING PROCESS

The protective relays are connected in the secondary circuits of current transformers (CTs) and/or potential transformers (PTs). Under normal operating conditions, the voltage induced in the secondary of the CT is small and therefore current flowing in the relay operating coil is insufficient in magnitude to close the relay contacts. This keeps the trip coil of the circuit breaker de-energised. Consequently the breaker contacts remain closed and it carries the normal load current. On occurrence of fault, a large current flows through the primary of the CT. This increases the voltage induced in the secondary and hence the current flowing through the relay operating coil. The relay contacts are closed and the trip coil of the breaker gets energised to open the breaker contacts. The circuit breaker opens its contacts. An arc is drawn between the contacts as they separate. The arc is extinguished at a natural current zero of the ac wave by suitable medium and technique. After final arc extinction and final current zero, a high voltage wave appears across the circuit breaker contacts tending to re-establish this arc. This transient voltage wave is called transient recovery voltage (TRV). The transient recovery voltage comprises a high frequency transient component superimposed on a power frequency recovery voltage. These phenomena have a profound influence on the behaviour of the circuit breakers and the associated equipment (Refer to chapters 7 and 8).

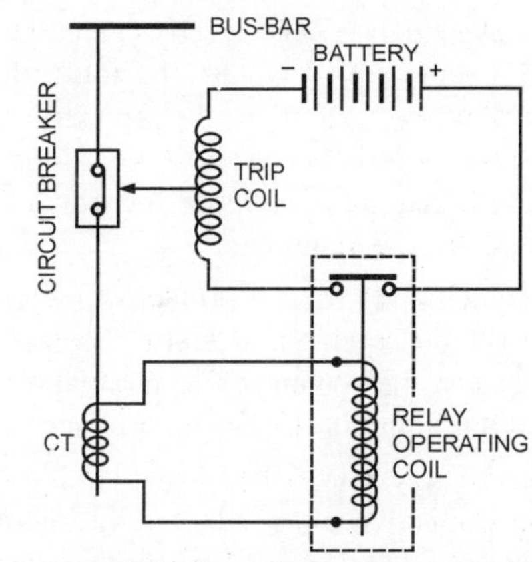

Fig. 1.2. ***Typical Relay Circuit***

1.8. BASIC REQUIREMENTS OF PROTECTION SYSTEM

Protection system is an extremely important part of the power system as it is provided to operate under abnormal conditions to prevent failure or isolate trouble and limit its effect. Every protection system which isolates a faulty element is required to satisfy four basic requirements: (*i*) reliability; (*ii*) selectively; (*iii*) sensitivity; and (*iv*) speed of operation. Without reliability and selectivity the protection would be rendered largely ineffective and could even become liability.

1. *Reliability and Simplicity.* Reliability is the basic requirement of the protection system as it is provided to improve the reliability of the whole system. Reliability means that the protection system must be ready to function, correctly at all times and under all conditions of any fault and abnormal conditions of the power system for which it has been designed.

Reliability is a qualitative term. Reliability of protective systems is assessed from statistical data. It cannot be easily specified in terms of a mathematical expression with certainty. Statistical survey and records give idea about reliability of protective systems. With the growing size of systems, use of EHV lines, interconnections and use of large generators and transformers, the importance of reliability of protective systems has increased.

Inherent reliability is a matter of design based on long experience. This can be achieved partly by (*i*) simplicity and robustness in construction; (*ii*) high contact pressure; (*iii*) dust free enclosures; (*iv*) good contact material; (*v*) good workmanship, and (*vi*) careful maintenance.

Simplicity is closely associated with reliability. Simplicity of construction and good quality of relay, correctness of design and installation and good maintenance and supervision etc., are the main basic factors which influence the protective reliability. As a rule, the simple the protective scheme and lesser the number of relays, circuits and contacts it contains, the greater will be its reliability.

Reliability is rather closely associated with simplicity. Therefore one of the basic requirements of the protection technique is simplicity.

2. *Selectivity and Discrimination.* Selectively is the ability of the protective system to select correctly that part of the system in trouble and disconnect or isolate the faulty part without disturbing the rest of the system *i.e.,* the protective system must be able to *discriminate* between those conditions for which prompt operation is required and those for which no operation or time-delay operation is required.

A well designed and efficient protective system should be selective *i.e.,* it should be able to detect the point at which the fault occurs and cause the opening of the circuit breakers nearest to the fault with minimum or no damage to the system.

If the protective system is not selective and operates for faults beyond its protective zones, a larger portion of the system will get disconnected unnecessarily, causing embarrassment to supplier and consumers.

Selectivity is *absolute* if the protection responds only to the faults within its own zone, and *relative* if it is obtained by grading the settings of the protective relays of several zones all of which may respond to a given fault. Systems of protection which in principle are *absolutely selective* are called the *unit systems.* Systems in which selectivity is relative are *non-unit* systems.. Differential protection and frame leakage protection fall in the former category while the current time graded protection and distance protection fall in the latter category.

3. *Sensitivity.* Any protective system must be sufficiently *sensitive* so that it will operate reliably when required under the actual condition that produces the least operating tendency.

Sensitivity of any protective scheme refers to the smallest value of actuating quantity at which it starts operating in relation with the minimum value of fault current in the protected zone.

Sensitivity of a protective relay is a function of volt-amperes input to the relay coil necessary to cause its operation. The smaller the volt-ampere input required to cause relay operation, the more sensitive is the relay.

A protection with high degree of sensitivity is more complex and uses more equipment and circuitry and is, therefore, more expensive. Such protection is employed only in that case where simpler arrangement cannot be applied because of low degree of sensitivity.

4. *Speed of Operation.* The protective system should disconnect the faulty section as quick as possible. This is desirable for many reasons such as (*i*) reduction in the amount of damage incurred; (*ii*) improvement in power system stability; (*iii*) less annoyance to electric power consumers and decrease in total outage time for power consumers; (*iv*) decrease in likelihood of development of one type of fault into other more severe type and; (*v*) permit use of rapid reclosure of circuit breakers to restore service to consumers.

However, too fast operation may result in undesired operation during transient faults.

1.8.1. Economic Considerations

A modern power system cannot operate justifiably without protective scheme but its cost has also to be considered. It is futile to achieve all important general requirements together, so compromises become necessary. Too much protection is as bad as too little and the protection engineer must make a sensible compromise with due regard to practical situation considered. As a rule, the protective gear should not cost more than 5% of the total cost of the plant. However, when the apparatus to be protected is of vital importance (such as generators, main transmission lines) economic considerations are subordinated to reliability.

1.9. PROTECTION ZONES

In order to provide selectivity to the system it is usual practice to divide the entire system into several protection zones. When a fault occurs in a given zone, then only the circuit breakers within that zone will be opened and thus only the faulty element will be isolated without disturbing the rest of the system.

Depending upon the rating of the machine, its location, relative importance, probability of faults and abnormal conditions etc., each power system component (generator, transformer, transmission lines, bus-bar, cables, capacitors, individual loads, etc.) is covered by a protective zone. A part of the system protected by a certain protective scheme is called *protective zone* or zone of protection. The entire power system is covered by several protective zones and no part of the system is left unprotected. The protection zone usually contains one or at the most two elements of the power system.

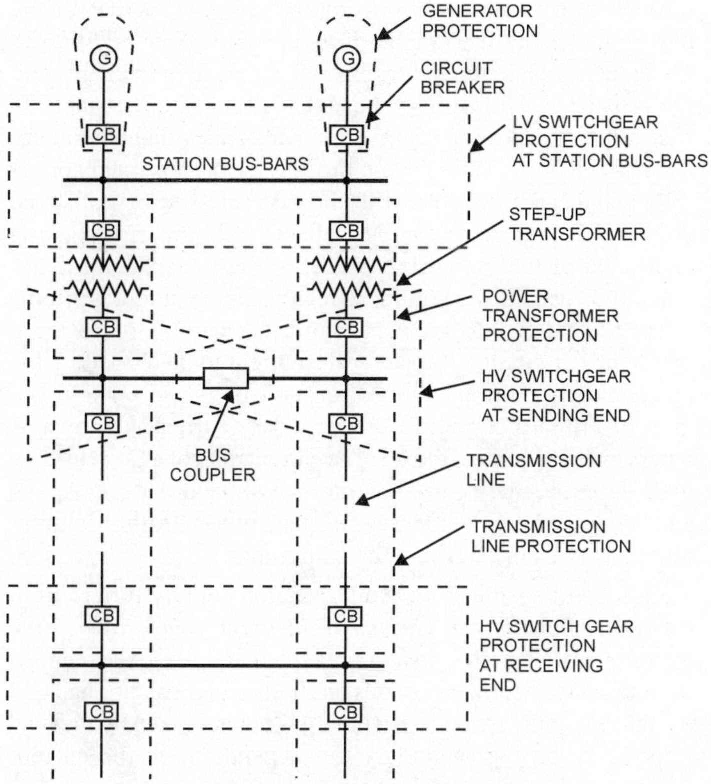

Fig. 1.3. ***Overlapping Zones of Protection***

It may be seen from Fig. 1.3 that there is a certain amount of overlapping between the adjacent protection zones. For failures within the region where two adjacent protective zones overlap, more breakers will be tripped than the minimum necessary to disconnect the faulty portion. But, if there were no overlap, a failure in a region between zones would not lie in either zone and therefore no breaker would be tripped. The overlap is the lesser evil. The extent of overlap is relatively small, and the probability of failure in this region is low; consequently, the tripping of too many breakers will be quite infrequent.

Protective scheme which permits sensitivity to faults only within a certain unit part of the system allows a higher quality of system protection by virtue of its high degree of sensitivity and its adaptability to fast speed of operation. Figure 1.3 indicates how protection may be set up to operate for faults only in the protected area with overlapping of the areas around the connecting circuit breakers.

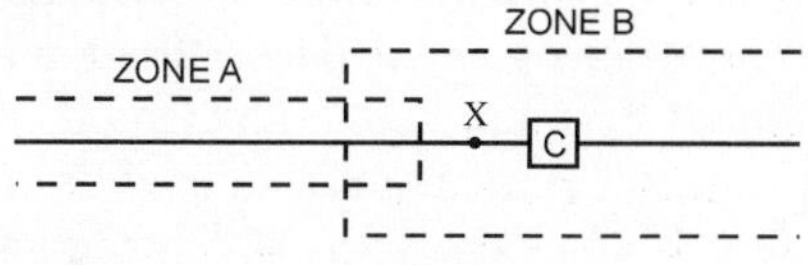

Fig. 1.4

When it becomes desirable for economic or space saving reasons to overlap on one side of a breaker, the relaying equipment of the zone that overlaps the breaker must be arranged to trip not only the breakers within its zone but also one or more breakers of the adjacent zone, in order to completely disconnect such faults. This is illustrated in Fig. 1.4, where it can be seen that for a fault at X the circuit breakers of zone B, including breaker C, will be tripped; however, this does not interrupt the flow of fault current from zone A, the relaying equipment of zone B must also trip certain breakers in zone A. This is not a disadvantage for a fault at X, but the same breakers in zone A will be tripped unnecessarily for other faults in zone B to the right of breaker C. How for this unnecessary tripping can be tolerated will depend on the particular application.

1.10. TYPES OF PROTECTION

Very frequently, for attaining higher reliability, speeded-up action and improvements in operating flexibility of the protection schemes, the separate elements of a power system, in addition to *main* or *primary protection,* are provided with a *backup* and *auxiliary* protection.

The *main protection* is the first line of defence and ensures quick acting and selective clearing of faults within the boundary of the circuit section or element it protects. Main protection, as a rule, is provided for each section of an electrical installation.

Backup protection is the name given to a protection which backs up the main protection whenever the latter fails in operation or is cut out for repairs etc.

Backup protection is important to the proper functioning of a good system of electrical protection since cent per cent reliability not only of the protective scheme but also of the associated CTs, PTs and circuit breakers cannot be guaranteed. It is the second line of defence which functions to isolate a faulty section of the system in case the main protection fails to function properly. The failure of the main protection may be due to any of the following reasons :

(*i*) The dc supply to the tripping circuit fails.

(*ii*) The current or voltage supply to the relay fails.

(*iii*) The tripping mechanism of circuit breaker fails.

(*iv*) The circuit breaker fails to operate.

(*v*) The main protective relay fails.

Backup protection may be provided either on the same circuit breakers which would be normally opened by the main protection or, still better, if the second line of protection makes use of different circuit breakers. It is generally satisfactory to isolate more than the faulty section alone in case "backup" protection is called upon to function. Very often the main protection of a given circuit also serves as a backup for the protection of an adjacent circuit. Backup protection is only provided where the main protection of the adjacent circuit is unable to back up the main protection of the given circuit. In a number of cases, for simplification, the backup protection can have a lower sensitivity factor and be operative over a limited backup zone *i.e.,* be operative for only part of the protected circuit.

An example of remote backup protection is provided by simple time graded relays, as shown in Fig. 1.5. For any fault at F normally relay R_4 will operate the circuit breaker at D and isolate the faulty section. Now if for some reasons circuit breaker at D fails to operate, the faulty section would be isolated by the operation of the relay R_3 at C.

Experience has shown that backup protection for other than short circuits is not economically justifiable. The extent to which backup protection is provided, depends upon economic and technical considerations. The cost of backup protection is justified on the basis of probability of failure of individual component in protection system, cost of protected equipment, importance of protected equipment, location of protected equipment etc.

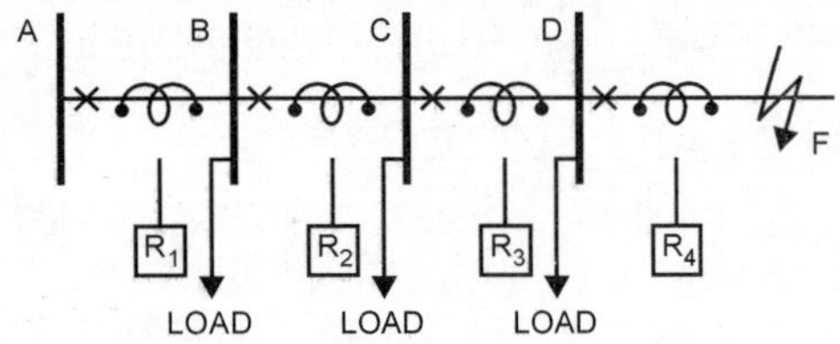

Fig. 1.5

The backup protection is normally of a form different from the main protection and should preferably be of the *non-unit* type, *e.g.*, overcurrent or distance protection. It is usually for economic reasons not so fast or as discriminative as the main protection.

Auxiliary protection is employed in separate cases to accelerate main protection operation during faults within the protected zone and to also protect what is termed the *dead zone* of the main protection scheme. The auxiliary protection devices usually back up the main protection scheme in the events of its failure.

1.11. INSTRUMENT TRANSFORMERS

AC type protective relays are actuated by current and voltage supplied by current and potential (or voltage) transformers, known as *instrument transformers*. The main function of instrument transformers are:

(*i*) To provide insulation against the high voltage of the power circuit and to protect the apparatus and the operating personnels from contact with the high voltages of the power circuits.

(*ii*) To supply protective relays with current and voltage of magnitude proportional to those of the power circuit but sufficiently reduced in magnitude so that the relays can be made relatively small and inexpensive.

(*iii*) Possibility of different types of secondary connections to obtain the required currents and voltages.

For safety purposes, the secondaries of current and potential transformers (CTs and PTs) are grounded.

For the proper applications of CTs and PTs, required considerations are:

Mechanical construction, type of insulation (dry or liquid), ratio in terms of primary and secondary currents or voltages, continuous thermal rating, short-time thermal and mechanical ratings, insulation class, impulse level, service conditions, accuracy and connections.

1.11.1. Current Transformers (CTs)

Current transformers are connected in ac power circuits to feed the current coils of indicating and metering instruments (ammeters, wattmeters, watt-hour meters) and protective relays. Thus the CTs broaden the limits of measurements and maintain a watch over the currents flowing in the circuits and over the power loads. In high voltage installations CTs in addition to above, also isolate the indicating and metering instruments and protective relays from high voltage. The current transformer (CT) basically consists of an iron core on which are wound a primary and one or two secondary windings. The primary winding of the CT is connected in series with the load and carries the actual power system current (normal or fault) while the secondary is connected to the measuring circuit or the relay. The primary winding is usually single turn winding and the number of turns on the secondary winding depend upon the current to be carried by the power circuit. The larger the current to be carried by the power circuit, more the number of turns on the secondary. The ratio of primary current to the secondary current is known as *transformation ratio* of the CT. The current ratio of a CT is usually high. The secondary current ratings are of the order of 5 A, 1 A and 0.1 A the latter being used for static relays. Primary current ratings vary from 10 A to 3,000 A or more.

The current transformer in operation slightly differs from the power transformer. In case of a current transformer as the load impedance or "*burden*" on the secondary is very small, so the current transformer operates on short-circuit conditions. Also the current in the secondary winding is not governed by the load impedance on the secondary but depends upon the current flowing in the primary winding (*i.e.*, power circuit current).

The amount of power which the CT handles is small. The product of voltage and current on the secondary side when the CT is supplying the instrument or relay with its maximum rated value of current, is known as *rated burden* and is measured in volt-amperes (VA). The volt-ampere rating of CTs is low (5-150 VA) as compared with that of power transformers (a few kVA to several MVA).

All types of CTs are employed for protective relaying purposes. The bushing CT is most common for hv circuits as it is less expensive than the other types. However, it is not employed for circuits below 5,000 V or in metal-clad equipment. The bushing type CT consists only of an angular shaped core with a secondary winding. Such CT is built into equipment such as circuit breakers, power transformers, generators or switchgear, the core being arranged to encircle an insulating bushing through which a power conductor passes.

There is only one primary turn and the internal diameter of the bushing CT core is large. Hence the mean length of turn of the magnetic path is greater than those in other CTs. The cross section of the core is made larger. The saturation in the core is less and a bushing CT is more accurate than other CTs at high multiples of the primary current rating. At low currents, a bushing CT is usually less accurate due to its large exciting current.

Ratio and Phase Angle Errors of CTs. CTs introduces two errors—ratio error and a phase angle error.

Current ratio error is mainly due to energy component of excitation current and is given as

$$\text{Ratio error} = \frac{K_T I_S - I_P}{I_P} \qquad \text{...(1.8)}$$

where I_P is the primary current. K_T is the turn ratio and I_S is the secondary current.

Table 1.4. Characteristics of CTs According to Their Accuracy Class.

(A): ***For Measuring CTs***

Accuracy class	Permissible current ratio error in percentage at % rated current					Permissible phase angle error in min. at % rated current			
	5	20	50	100	120	5	20	100	120
0.1	± 0.4	± 0.2		± 0.1	± 0.1	± 15	± 8	± 5	± 5
0.2	± 0.75	± 0.35		± 0.2	± 0.2	± 30	± 15	± 10	± 10
0.5	± 1.5	± 0.75		± 0.5	± 0.5	± 90	± 45	± 30	± 30
1.0	± 3.0	± 1.5		± 1.0	± 1.0	± 180	± 90	± 60	± 60
3.0			± 3.0		± 3.0	Not standardised			
5.0			± 5.0		± 5.0				

(B): ***For Protective CTs***

Accuracy Class	Permissible current ratio error in % age at rated primary current	Permissible phase displacement in min., at rated primary current	Permissible composite error at rated accuracy limit primary current in percentage
5 *p*	± 1	± 60	5
10 *p*	± 3	—	10
15 *p*	± 5	—	15

Phase angle error is introduced on account of magnetising component of excitation current and a phase angle error δ is given in minutes of angle of departure of the secondary current from exact opposition to the primary current. For relaying purposes, the phase angle error has negligible effect on the accuracy of the CT. The load on the secondary of the CT is usually highly inductive so that the secondary current is practically in phase with the exciting current.

The CTs are rated for rated voltage of the installation, the rated currents of the primary and secondary windings and the accuracy class. The accuracy class indicates the limit of the error in percentage of the rated turn ratio of the given CT.

Table 1.4 gives the classification of CTs according to their accuracy class and the corresponding limits of errors at different percentages of rated current.

1.11.2. Potential Transformers (PTs)

The potential transformers are employed for voltages above 380 V to feed the potential coils of indicating and metering instruments (voltmeters, wattmeters, watt-hour meters) and relays. These transformers make the ordinary low voltage instruments suitable for measurement of high voltage and isolate them for high voltage.

The primary winding of the potential transformer is connected directly to the power circuits either between two phases or between a phase and ground depending on the rating of the transformer and its application. To the secondary windings, various indicating and metering instruments and relays are connected. The primary winding has a large number of turns and the secondary winding, which has a much smaller number of turns, is coupled magnetically through the magnetic circuit to the primary winding. The primaries of PT are rated from 400 V to several thousand volts and secondaries always for 110 V (sometimes for $110\sqrt{3}$ V). The ratio of the rated primary voltage to the rated secondary voltage is known as *turn* or *transformation ratio*.

The ends of the windings of potential transformers are marked according to the same rules as used for power transformers to ensure correct connection of the windings to each other, to the high-voltage power circuit and to various indicating and metering instruments and relays they serve to supply

The theory of operation of a potential transformer (PT) is essentially the same as that of the power transformer. The main difference between a PT and a power transformer is due to the fact that the load current of the former is of the order of its exciting current and secondary impedance of the PT is mostly resistive being made up of the potential circuit of the instrument. The PT is rated in terms of the maximum burden (volt-ampere output) it delivers without exceeding specified limits of error, whereas the power transformer is rated by the secondary output it delivers without exceeding a specified temperature rise. The output of PTs is usually limited to a few hundred volt-amperes while the output of a power transformer may be of the order of several MVA.

Ratio and Phase Angle Errors of PTs. Ideally a PT should give secondary voltage exactly proportional to the primary voltage and exactly in phase opposition. This cannot obviously be achieved in practice due to primary and secondary voltage drops. Thus both the ratio and phase angle errors are introduced by a PT.

The *voltage ratio error* is generally expressed in terms of the measured voltage and is given as

$$\text{Percentage ratio error} = \frac{K_N V_S - V_P}{V_P} \times 100 \qquad ...(1.9)$$

where K_N is the nominal ratio — the ratio of rated primary voltage and rated secondary voltage.

The phase angle error δ is usually given in minutes and indicates by what small angle δ the secondary voltage V_S departs from exact opposition to the applied primary voltage V_P.

With the increase in the number of instruments and relays connected to the secondary of a PT, *i.e.,* with the increase in burden of the secondary circuit, both of the errors will increase.

The 'burden' is the total external volt-ampere load on the secondary at rated secondary voltage. Where several loads (instruments and relays) are connected in parallel, it is usually sufficient to add their individual volt-amperes arithmetically to determine the total volt-ampere burden.

The ratio and phase angle errors are also influenced by the primary voltage. Operation in excess of 10% overvoltage may cause increased errors and excessive heating. However, if a PT has acceptable accuracy at its rated voltage, it is suitable over the range from zero to 110% of rated voltage.

The ratio and phase angle errors of any standard ASA PT are so small that they may be neglected for the protective relaying purpose if the burden is within the 'thermal' volt-ampere rating of the PT. This thermal VA rating corresponds to the full-load rating of the power transformer.

The *rated burden* of a PT is the VA burden which must not be exceeded if the transformer is to operate with its rated accuracy (this burden is indicated on the name plate and in the data sheet).

The *limiting* or *maximum burden* is the greatest VA load at which the PT will operate continuously without overheating its windings beyond the permissible limits. This burden is several times greater than the rated burden.

In practice, use is made of single-phase and three-phase potential transformers, the latter being of the two-winding or three-winding type.

Table 1.5 gives the classification of PTs according to their accuracy class and the corresponding limits of errors.

Table 1.5. Characteristics of Potential Transformers According to Their Accuracy Class

Accuracy class	Permissible voltage ratio error in percentage	Permissible phase angle error in minutes
0.2	± 0.2	± 10
0.5	± 0.5	± 20
1.0	± 1.0	± 40
3.0	± 3.0	± 120
5.0	± 5.0	± 300

There are two types of potential transformers; the conventional wound type (or electromagnetic type) and the capacitor voltage (potential) transformer. For voltages exceeding 100 kV (phase) the conventional type of potential transformer becomes extremely expensive owing to the insulation requirements. For voltages above 100 kV a *capacitor potential transformer,* a combination of a capacitor potential divider and a magnetic potential transformer (known as the intermediate transformer) of relatively small ratio is used. The circuit diagram is shown in Fig. 1.6. A stack of high-voltage capacitors forms the potential divider, the capacitors of the two sections being C_1 and C_2 respectively and Z is the burden. The voltage applied to the primary of the intermediate transformer T is usually about 10 kV. Both the potential divider and the intermediate transformer will have ratios and insulation requirements which are suitable for economical construction. The intermediate transformer must be of very small ratio error and phase angle to give satisfactory performance of the complete unit.

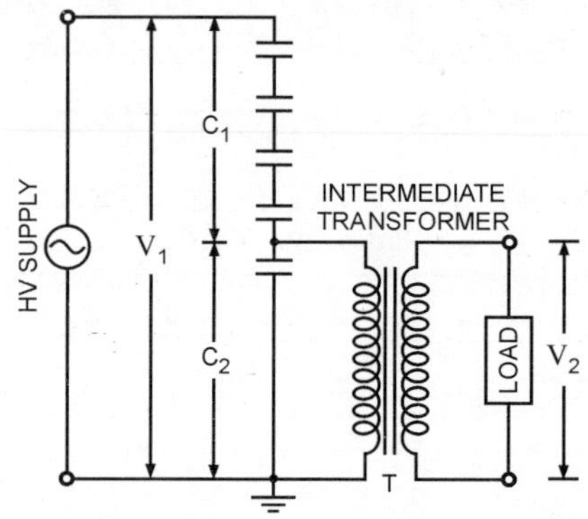

Fig. 1.6. ***Capacitor Potential Transformer Circuit Diagram***

If the current flowing in the output or secondary circuit is negligible then

$$V_2 = V_1 \times \frac{C_1}{C_1 + C_2} \quad ...(1.10)$$

Exercises

1. Explain why fault currents are more than load currents ? What are the harmful effects of faults ? What is meant by fault level or fault MVA ? Explain.
2. Classify the different faults which occur in power systems. Which of these are more frequent ?
3. Differentiate between symmetrical and unsymmetrical faults. Explain different types of faults and discuss which is the most common amongst them.
4. Which types of faults are taken into account while determining the rating of the circuit breaker ?

   What is symmetrical fault current and unsymmetrical fault current ? Which type of fault causes them ?
5. What are the principal types of faults in a power system ? In what way a fault may be harmful to the power system ?

   [G.B. Technical Univ. Switchgear and Protection 2011-12]
6. Explain clearly the sources of short-circuit power in a power system and the nature of such power.
7. Explain briefly the role of protection in a power system. Describe the essential features of a protective relay with reference to reliability, selectivity, speed of operation and discrimination.

8. "Relay is the brain of protection." Justify this statement using trip circuit.

[U.P. Technical Univ. Switchgear and Protection 2009-10]

9. What is a protective relay ? Explain the functional characteristics of a protective relay ?

[U.P. Technical Univ. Switchgear and Protection 2006-07]

10. Explain the process of fault clearing with the help of a neat sketch.

[U.P. Technical Univ. Switchgear and Protection 2006-07]

11. What is meant by short circuit ? What are the possible causes of short circuit in a power system ? Explain the harmful effects of short circuit on a power system and state the steps to be taken to protect the equipment from such fault.
12. What are the causes of faults in power systems ? Explain why the fault studies are carried out ?
13. What is a zone of protection ? Discuss various zones of protection of a power system with the help of line diagram.

[G.B. Technical Univ. Switchgear and Protection 2010-11; 2011-12]

14. What are primary and backup protections ? What are various kinds of backup protections ?

[G.B. Technical Univ. Switchgear and Protection 2011-12]

15. Describe about primary and backup protection.

[U.P. Technical Univ. Switchgear and Protection 2009-10]

16. Differentiate the following with examples :

Primary and backup protections.

[U.P. Technical Univ. Switchgear and Protection 2004-05; 2007-08]

17. Sometimes faults are not cleared by primary relay system because of trouble within the relay, under such conditions, what protection is required ? Explain it.
18. Write an essay about the nature, causes and consequences of faults in power system. Explain the basic essential qualities of a protective scheme.
19. What are the desirable features of a power system protection system ?
20. What are basic requirements of protective relaying schemes ?

[G.B. Technical Univ. Switchgear and Protection 2010-11]

21. What are the basic characteristics of a protective relay ? Explain clearly. **[U.P. Technical Univ. Switchgear and Protection 2004-05; 2007-08]**
22. Write short note on faults on power system.

[Rajasthan Univ. 2005]

Short Answer Type Questions With Answers

Q. 1. What is meant by short circuit ?

Ans. When conductors of different phases come into contact with each other in a power line, power transformer, or any other circuit element, part of the impedance of the system is shunted out of the circuit, a large current flows in one or more phases, a short circuit is said to have occurred.

Q. 2. What is difference between a short circuit and an overload ?

Ans. On occurrence of a short circuit, the voltage at the fault point is reduced to zero and current of abnormally high magnitude flows through the network to the fault point. On the other hand, an overload means that load greater than the designed value have been imposed on the system. In case of an overload, the voltage at the overload point may become low, but cannot be zero. The undervoltage conditions may extend to some distance beyond the overload point in the remainder of the system. In case of an overload, the currents are high but are substantially lower than in the case of a short circuit.

Q. 3. The probability of the failure or occurrence of abnormal conditions is more on the power line. Why ?

Ans. The probability of the failure or occurrence of abnormal conditions is more on the power line. This can be explained by the fact that the power lines are widely branched, have greater length, operate under variable weather conditions and are subject to the action of atmospheric disturbances of electrical nature.

Q. 4. Name the different faults likely to occur in an overhead transmission line.

Ans. The different faults likely to occur in an overhead transmission line are (1) Single phase to ground, (2) Phase to phase, (3) Two phases to ground, (4) Phase to phase and third phase to ground, (5) All the three phases shorted and (6) All the three phases to ground.

Q. 5. Which type of fault occurs more frequently in overhead lines ?

Ans. Single phase to ground (L-G) fault occurs more frequently in overhead lines.

Q. 6. Which type of faults are taken into account while determining the rating of the circuit breaker ?

Ans. The balanced three-phase fault, though its occurrence is quite rare (only about 5 per cent of the total), imposes most severe duty on circuit breakers and is, therefore, used in determination of the circuit breaker rating.

Q. 7. What are the sources of short-circuit power in a power system ?

Ans. Mainly there are three sources viz synchronous generators, synchronous motors and synchronous condensers and induction motors, which can feed into the short circuit.

Q. 8. What is a protective relay ?

Ans. Protective relay is an electrical device designed to initiate isolation of a part of an electrical installation, or to operate an alarm signal, in the event of an abnormal condition or a fault.

Q. 9. What are the functions of a protective relay ?

Ans. The main functions of protective relays are to detect the presence of faults, their locations and initiate the action for quick removal from service of the faulty section of the system. Relays operate to trip the power circuit breakers to disconnect the faulty section.

Q. 10. What are the essential elements of a protective relay ?

Ans. The essential elements of a protective relay are (*i*) sensing element (*ii*) comparison element, and (*iii*) control element.

Q. 11. What is a zone of protection ?

Ans. A part of the system protected by a certain protective scheme is called *protective zone* or zone of protection. The entire power system is covered by several protective zones and no part of the system is left unprotected. The protection zone usually contains one or at the most two elements of the power system.

Q. 12. What is meant by primary protection and backup protection ? **[G.B. Technical Univ. 2011-12]**

Ans. The *main protection* is the first line of defence and ensures quick acting and selective clearing of faults within the boundary of the circuit section or element it protects. Main protection as a rule, is provided for each section of an electrical installation.

Backup protection is the name given to a protection which backs up the main protection whenever the latter fails in operation or is cut out for repairs etc.

Backup protection is important to the proper functioning of a good system of electrical protection since cent per cent reliability not only of the protective scheme but also of the associated CTs, PTs and circuit breakers cannot be guaranteed. It is the second line of defence which functions to isolate a faulty section of the system in case the main protection fails to function properly.

Q. 13. What are the basic requirements of a protection system ?

Ans. The basic requirements of a protection system are essentially reliability and simplicity, selectivity and discrimination, sensitivity, speed of operation and economy.

Q. 14. Why current transformers are required in a protective relaying scheme ?

Ans. The primary circuit currents which are of high magnitudes are to be reduced to values suitable for relay operation and current transformers are used for this purpose. In addition to this, CTs provide insulation against the high voltage of the power circuit and protect the apparatus and the operating personnels from contact with the high voltages of the primary circuits.

Q. 15. Why potential transformers are required in a protective relaying scheme ?

Ans. It is not possible to connect the potential coils of the protective devices directly into the system in case of high voltage systems. It is, therefore, necessary to step down the voltage and also to insulate the protective equipment from the primary circuit (high voltage power circuit). This is achieved by employing potential (or voltage) transformers (PTs).

❑❑❑

CHAPTER 2

Representation of Power System Components

INSIDE THIS CHAPTER

2.1. INTRODUCTION

Electric power systems may be of great complexity and spread over large geographical area. An electric power system consists of generators, transformers, transmission lines and consumer equipment (loads). The system must be protected against flow of heavy short-circuit currents (which can cause permanent damage to major equipment) by disconnecting the faulty section of system by means of circuit breakers and protective relaying.

It is necessary to know the maximum short-circuit currents that can occur at the different points of a system in order that circuit breakers may be selected that are adequate to withstand the currents and operate successfully to cut out the faulty section, and also in order that the protective relays may be selected for correct operation. The design of machines, bus-bars, isolators, circuit breakers etc., is based on the considerations of normal and short-circuit currents.

It is also necessary to be able to calculate, approximately at least, the size of the protective reactors which must be inserted in the system to limit the short-circuit currents to a value which is not beyond that capable of being withstood by the circuit breakers.

The short-circuit currents in an ac system are determined mainly by the reactance of the alternators, transformers and lines up to the point of the fault in the case of phase-to-phase faults. When the fault is between phase and earth, the resistance of the earth path play an important role in limiting the currents.

In case of circuit breakers their rupturing capacities are based on the symmetrical short-circuit current which is the most simple calculation among all types of short circuits. However, for determination of settings of relays it is absolutely necessary to know fault current due to unsymmetrical fault condition too for which knowledge of symmetrical components etc., is required. Only symmetrical short-circuit currents will be dealt here. Discussion to other types of faults involving an unsymmetrical condition will be made in chapter 5.

2.2. POWER SYSTEM NETWORK

Power System can be subdivided into three major components namely :

(*i*) Generation System,

(*ii*) Transmission System, and

(*iii*) Distribution system.

Nowadays, mostly, power is generated by three-phase synchronous generators, transmitted by three-phase lines and then distributed through three-phase distribution networks. Three-phase synchronous generators are always designed to produce three-phase balanced excitation *i.e.,* to generate three-phase voltages which are equal in magnitude and displaced from each other by an equal angle of 120 electrical degrees. The three-phase transmission lines and three-phase distribution network are either arranged symmetrically or if arrangement is unsymmetric, they are transposed at regular intervals to balance their electrical characteristics resulting in balanced three-phase network comprising transmission and distribution networks. Thus three-phase generation or supply as well as the three-phase transmission and networks are balanced and, therefore, so far as the calculations are concerned, they can be treated as a single-phase system for the analysis.

However, if there is a fault on the system, the excitation becomes unbalanced *i.e.,* the generated three phase voltages may not be equal in magnitude or may not be displaced by an equal angle of 120 electrical degrees but the three-phase transmission and distribution networks are balanced. Thus, it becomes a case of unbalanced supply connected to the three phase balanced network. Such situations are usually handled by transforming the actual phasor quantities to component quantities by linear time dependent power invariant transformations because after transformation, these phasor quantities, which were coupled earlier, become uncoupled and the differential equations, describing the dynamic behaviour of the system, which have time dependent *i.e.,* variable coefficients, after transformation, become differential equations with constant coefficients. In practice, a large number of such transformations are available but all of them have the same common features.

In case, in addition to the unbalanced supply or excitation, the three-phase network is also unbalanced, their transformation into component quantities will not serve the purpose because even after transformation, the phasor quantities remain coupled. But, such situations are not very common.

2.3. REPRESENTATION OF POWER SYSTEM COMPONENTS

For the purpose of the power system studies such as transient stability studies, load flow studies or short-circuit studies, there is a need for the development of mathematical model of the power system network. For the development of fairly accurate models of the power system network, it is necessary that model reflects correctly the terminal behaviour of each component of the network for the purpose of study for which the model has been developed. This is because the terminal behaviour of the power system components differ from the normal *i.e.,* steady-state conditions to abnormal *i.e.,* transient conditions. The different components of power system are : synchronous machines, transmission network, transformers, distribution network including static loads, composite loads and dynamic loads such as induction machines etc.

Now, the representation of different components (synchronous machines, transformers and transmission lines) of the power system will be discussed in brief here.

2.3.1. Representation of Synchronous Machines

The synchronous machine is the most important component of a power system. It converts mechanical power into electrical form and supplies it into the power system network or, in the case of a motor, it draws electrical power from the network and converts it into the mechanical form.

Most simplified model *i.e.,* representation of a synchronous generator for the purpose of transient stability studies is a constant voltage source behind proper reactance. The voltage source may be subtransient, transient or steady-state voltages and the reactance may be the corresponding reactances. However, for understanding this model, let us consider a synchronous generator operating at no load before occurrence of a 3-phase short circuit at its terminals. The current flowing in the synchronous generator just after occurrence of the three-phase short circuit at its terminals is similar to the current that flows in an R-L circuit upon which sudden an ac voltage is applied (Refer to Art. 4.2). Hence the current will have both ac (*i.e.,* steady-state) component as well as the dc (*i.e.,* transient) component which decays exponentially with time constant τ equal to L/R. However, if the dc component is neglected the oscillograph of the ac component of the current that flows in the synchronous generator just after the fault occurs, will have the waveshape as shown in Fig. 2.1. For analysis of short circuit oscillogram, refer to Art. 4.3.3.

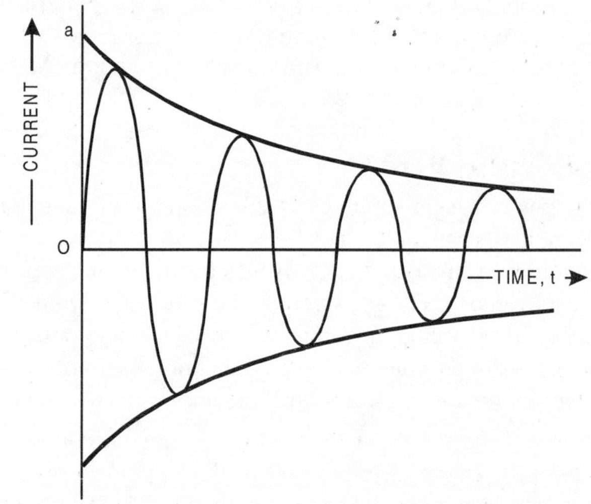

Fig. 2.1. ***Oscillogram of The Current That Flows in The Unloaded Synchronous Generator Just After a 3-Phase Fault at Its Terminals Excluding DC Component***

2.3.2. Representation of Transformer

As we know the transformer is a static machine which is used to transform electric power from one circuit to another without change in frequency. It may raise or lower the voltage in a circuit

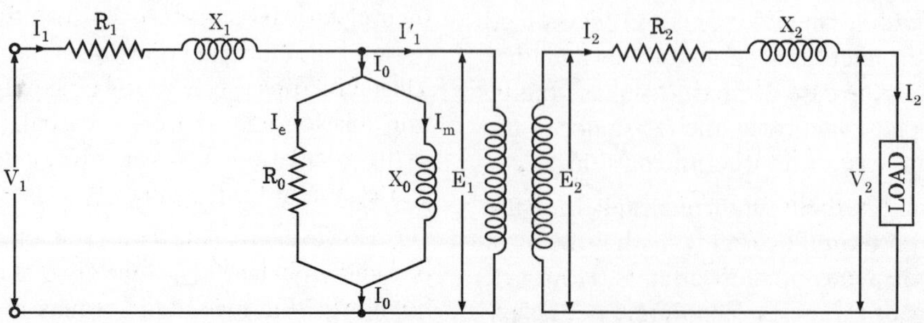

Fig. 2.2. ***Equivalent Circuit of a Transformer***

but with a corresponding decrease or increase in current. Thus transformer plays a vital role in power systems. Practically every transformer is provided with taps for ratio control, and thus it is also used for controlling secondary voltage level. Most of the transformers are provided with *off-load tap-changers* while some of the transformers are provided with *on-load tap-changers*.

A power transformer with turn ratio $K = \frac{E_2}{E_1} = \frac{N_2}{N_1}$ can be represented as illustrated in Fig. 2.2.

The induced emf in primary winding E_1 is primary applied voltage V_1 less primary voltage drop. This voltage causes iron loss current I_e and magnetising current I_m and we can, therefore, represent these two components of no-load current by the current drawn by a non-inductive resistance R_0 and pure reactance X_0 having the voltage E_1 or (V_1—primary voltage drop) applied across them, as shown in Fig. 2.2.

$$\text{Secondary current, } I_2 = \frac{I_1'}{K} = \frac{I_1 - I_0}{K}$$

Terminal voltage V_2 across load is induced emf E_2 in secondary winding less voltage drop in secondary winding.

The equivalent circuit can be simplified by transferring the voltage, current and impedance to the primary side. After transferring the secondary voltage, current and impedance to primary side equivalent circuit is reduced to that shown in Fig. 2.3.

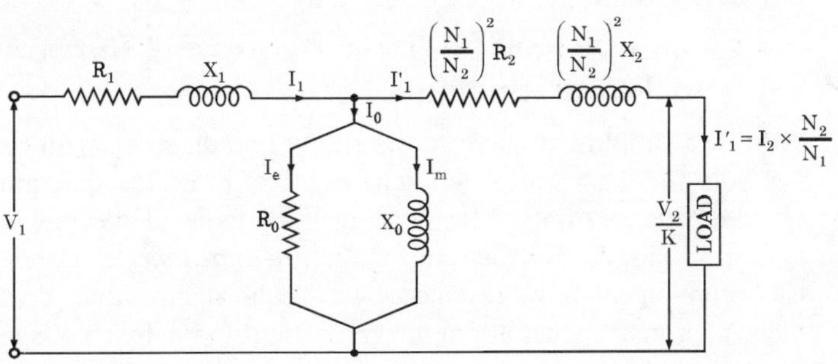

Fig. 2.3. ***Equivalent Circuit of a Transformer With All Secondary Impedances Transferred To Primary Side***

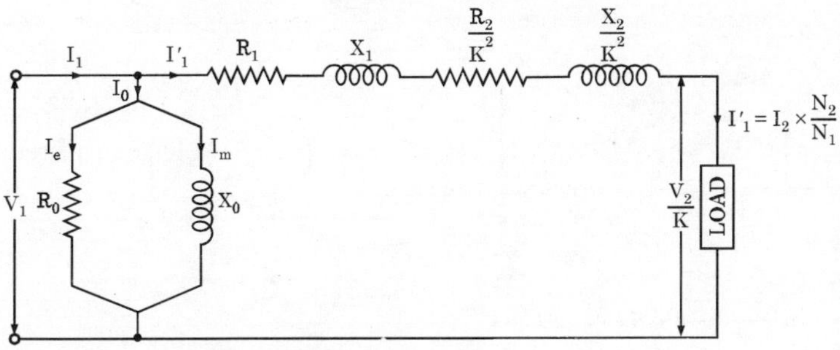

Fig. 2.4. ***Approximate Equivalent Circuit of a Transformer***

The equivalent circuit diagram can further be simplified by transferring the resistance R_0 and reactance X_0 towards left end, as shown in Fig. 2.4. The error introduced by doing so is very small and can be neglected.

No-load current I_0 is hardly 3 to 5 per cent of the full-load rated current, the parallel branch consisting of resistance R_0 and reactance X_0 can be omitted without introducing any appreciable error in the behaviour of the transformer under loaded condition. The equivalent circuit referred to primary side (neglecting no-load current I_0) is shown in Fig. 2.5.

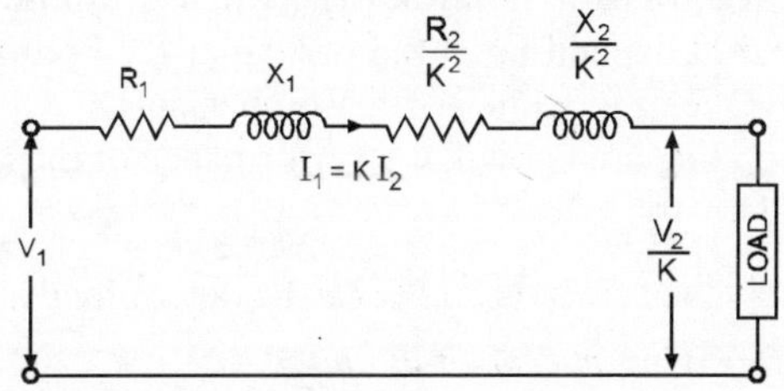

Fig. 2.5. ***Equivalent Circuit of a Transformer Referred to Primary***

2.3.3. Representation of Transmission Lines

The three-phase transmission system is either symmetrical (*i.e.,* balanced) in arrangement or if it is unsymmetrical, the lines are transposed in order to balance the electrical characteristics of different phases so as to have the same parameters for all the three phases. With such an arrangement the 3-phase network consisting of transmission system and also the distribution system are assumed to be symmetrical or balanced.

The performance of a transmission line is governed by its four parameters—series resistance R, and inductance L, shunt capacitance C and the shunt conductance G. The resistance R is due to the fact that every conductor offers opposition to the flow of current. The inductance L is due to the fact that the current carrying conductor is surrounded by the magnetic lines of force. The capacitance of the line is due to the fact that the conductor carrying current forms a capacitor with the earth which is always at lower potential than the conductor and the air between them forms a dielectric medium. The shunt conductance G is mainly due to flow of leakage currents over the surface of the insulators especially during bad weather. However, the conductance is normally neglected in the case of transmission line calculations since leakage at normal frequency are negligible.

Generalised Circuit Constants. In any passive, bilateral and linear network with two input and two output terminals, the input voltage and the input current can be expressed in terms of output voltage and output current. Incidently, a transmission line is a 4-terminal network; two input terminals where power enters the network, and two output terminals where the power leaves the network. Such a circuit is passive as it does not contain any source of emf, linear as its impedance is independent of current flowing and bilateral as its impedance is independent of direction of current flowing.

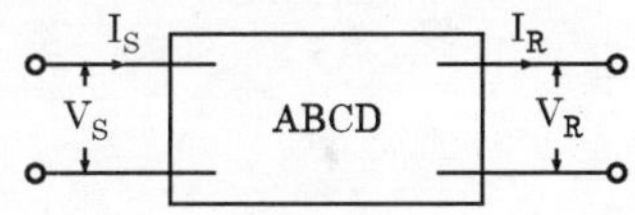

Fig. 2.6

The input voltage per phase (voltage at sending end) and the input current (current at sending end) of a transmission line can be expressed as

$$V_S = AV_R + BI_R \qquad ...(2.1)$$

$$I_S = CV_R + DI_R \qquad ...(2.2)$$

where $\mathbf{V_S}$ is the sending-end voltage per phase, $\mathbf{I_S}$ is the sending-end current, $\mathbf{V_R}$ is the receiving-end voltage per phase, $\mathbf{I_R}$ is the receiving-end current, and **A**, **B**, **C** and **D** (generally complex numbers) are the constants, called the *generalised circuit constants* of a transmission line. The values of these constants A, B, C and D depend upon the particular method of solution employed. Once the values of these constants for a particular transmission line are known, the performance of the line can be determined easily.

The values of A, B, C and D can be determined as follows :

With receiving end open circuited

Receiving-end current $\mathbf{I_R}$ is zero, therefore, from Eqs. (2.1) and (2.2)

$$\mathbf{A} = \mathbf{V_S}/\mathbf{V_R} \qquad \text{...(2.3)}$$

$$\text{and } \mathbf{C} = \mathbf{I_S}/\mathbf{V_R} \qquad \text{...(2.4)}$$

With receiving end short circuited

Receiving-end voltage $\mathbf{V_R}$ is zero, therefore, from Eqs. (2.1) and (2.2)

$$\mathbf{B} = \mathbf{V_S}/\mathbf{I_R} \qquad \text{...(2.5)}$$

$$\text{and } \mathbf{D} = \mathbf{I_S}/\mathbf{I_R} \qquad \text{...(2.6)}$$

From the above expressions **A**, **B**, **C** and **D** can be defined as follows :

A can be defined as the ratio of sending-end voltage $\mathbf{V_S}$ to receiving-end voltage $\mathbf{V_R}$ when the line is open circuited on receiving end. **A** is dimensionless and usually a complex number.

B is defined as the ratio of sending-end voltage $\mathbf{V_S}$ to receiving-end current $\mathbf{I_R}$ when the line is short circuited on receiving end. **B** is also the complex number and has the unit of ohm.

C is defined as the ratio of sending-end current $\mathbf{I_S}$ to receiving-end voltage $\mathbf{V_R}$ when the line is open circuited on the receiving end. **C** is also the complex number and has the unit of *siemen.*

D is defined as the ratio of sending-end current $\mathbf{I_S}$ to receiving-end current $\mathbf{I_R}$ when the line is short circuited on the receiving end. **D** is dimensionless and usually a complex number

For passive network $\mathbf{AD} - \mathbf{BC} = 1$ ...(2.7)

and in symmetrical network $\mathbf{A} = \mathbf{D}$ ...(2.8)

2.4. REPRESENTATION OF POWER SYSTEMS

A complete diagram of power system representing all the three phases becomes too complicated and cumbersome for a system of practical size, so much so that it may no longer convey the information it is intended to convey. It is much more practical to represent a power system by means of simple symbols for each component resulting in what is called a single line diagram.*

2.4.1. Single Line Diagram

The single line diagram of a power system network shows the *main connections* and *arrangements* of the system components along with their data (such as output rating, voltage, resistance and reactance etc.). In case of transmission lines sometimes the conductor size and spacings are given. It is not necessary to show all the components of the system on a single line diagram, *e.g.,* circuit breakers need not be shown in a load flow study but are must for a protection study. In a *single line diagram,* the system components are usually drawn in the form of their symbols. Generators and transformer connections—star, delta and neutral earthing are indicated by symbols drawn by the side of the representation of these elements. Circuit breakers are represented by rectangular blocks. Figure 2.7 represents the single line diagram of a typical power system. The ratings of generators, motors and transformers are given below the diagram.

2.4.2. Impedance Diagram Representation of a Power System

A further simplification from the single line diagram, with its symbols for the various components, is to draw the diagram with impedances only. The impedance diagram of the power system of Fig. 2.7 is shown in Fig. 2.8. In *impedance diagram,* each component is represented by its equivalent circuit, *e.g.,* the synchronous generator at the generating station by a voltage source in series with a resistance and reactance, the transformer by its equivalent circuit and the transmission line by nominal π-equivalent circuit. Loads are assumed to be passive

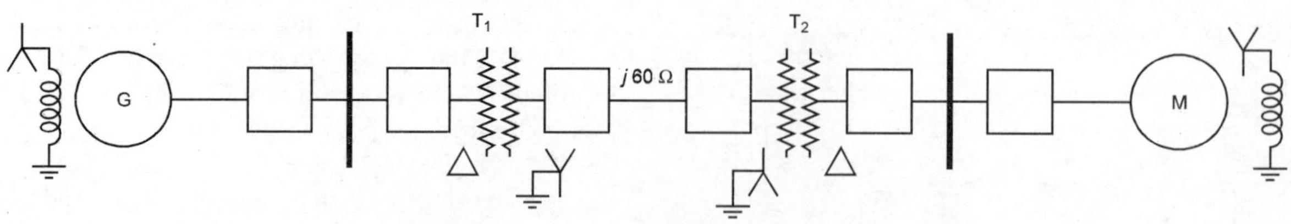

Generator	:	40 MVA, 11 kV, X" = 20 %
Motor	:	30 MVA, 11 kV, X" = 30 %
Transformer T_1	:	40 MVA, 11/220 kV, X = 15 %
Transformer T_2	:	40 MVA, 220/11 kV, X = 15 %

Fig. 2.7. ***Single Line Representation of a Typical Power System***

* Any 3-phase network consisting of generators, transformers, transmission lines etc., can be solved as a single-phase network composed of one of the three phases and a neutral return, if it is balanced under normal operation. Many times, the components of the system are shown in a single line diagram omitting the neutral also.

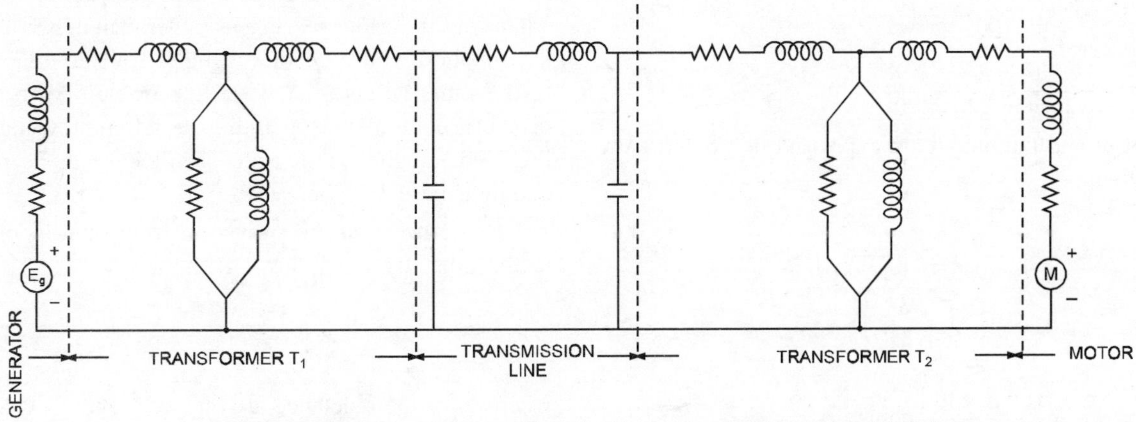

Fig. 2.8. *Impedance Diagram For The Power System Shown in Fig. 2.7*

(not involving rotating machines) and are represented by resistance and inductive reactance in series. Neutral earthing impedances do not appear in the diagram as balanced conditions are assumed.

The impedance diagram shown in Fig. 2.8 is known as *positive sequence diagram* since it is drawn for a balanced 3-phase system. Still another possibility would be to use three separate diagrams to represent the positive, negative and zero sequence networks separately. These three separate impedance diagrams are used in short-circuit studies of unsymmetrical faults.

2.4.3. Reactance Diagram Representation of a Power System

The impedance diagram can further be simplified by making certain assumptions and reduced to simplified reactance diagram. Reactance diagram is drawn by neglecting effective resistance of generator armature, transformer winding resistance, transmission line resistance, line charging and the magnetising circuit of transformers. Reactance diagram of power system of Fig. 2.7 is shown in Fig. 2.9.

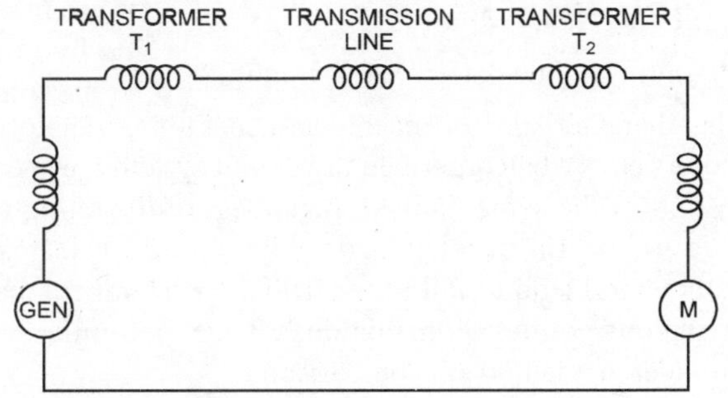

Fig. 2.9. *Reactance Diagram For Power System Shown in Fig. 2.7*

Assumptions made for drawing reactance diagram give results quite accurate for many power system studies, such as short-circuit studies etc., as winding resistances including the line resistances are quite small in comparison with leakage reactances and shunt path which includes line charging and transformer magnetizing circuit provide a very high parallel impedance with fault. In general it is to be noted that if the resistance is less than one-third of the reactance, and resistance is ignored, the error introduced will not be more than 5%. If, however, the resistance is of the order of one-half that of reactance and resistance is ignored errors up to 12 % may be introduced. By "errors" it is meant that calculations will result in values higher than is actually the case being obtained and in some cases, lead to the purchase of protective gear with a higher rating than required.

2.5. PERCENTAGE RESISTANCE AND REACTANCE AND BASE KVA AND KV

Resistance and reactance (and consequently impedance) may be expressed in percentage or ohmic terms. In the case of rotating machines and transformers, manufacturer values are always in percentage terms while tables for cables and overhead lines are always in ohmic terms. Calculations can be made using either but not a mixture. For short-circuit calculations, generally the percentage values are employed and, therefore, their understanding is essential.

Percentage Resistance. It is the voltage drop across the given resistance expressed as percentage of normal voltage when carrying full-load current pertaining to normal rating, *i.e.*,

$$\%R = \frac{IR}{V} \times 100 \qquad \text{...(2.9)}$$

where R is the resistance in ohms, I is the full-load current and V is the rated voltage.

Percentage Reactance. Percentage reactance can be defined in the same way as percentage resistance. It is the voltage drop across the given reactance expressed as percentage of normal voltage when carrying full-load current pertaining to normal rating, *i.e.*,

$$\%X = \frac{IX}{V} \times 100 \qquad \text{...(2.10)}$$

where X is the reactance in ohms, I is the full-load current and V is the rated voltage.

From Eq. (2.10) we have

$$X = \frac{(\%X) \times V}{I \times 100} \text{ ohms}$$

Multiplying and dividing the right hand expression by V, we have

$$X = \frac{(\%X) \times V \times V}{V \times I \times 100}$$

$$= \frac{(\%X)V^2}{(\text{Output in VA}) \times 100} \text{ ohms} \qquad ...(2.11)$$

when the voltage and the output are expressed in kV and kVA respectively, then

$$X = \frac{(\%X) \times \left(\frac{V}{1,000}\right) \times \left(\frac{V}{1,000}\right) \times 1,000}{100 \times \left(\frac{V}{1,000}\right) \times I}$$

$$= \frac{(\%X)\,(kV)^2 \times 10}{kVA} \text{ ohms} \qquad ...(2.12)$$

$$\text{or } \%X = \frac{(X)\,(kVA)}{10(kV)^2} \qquad ...(2.13)$$

Thus if actual reactance in ohms is given, percentage reactance can be determined and vice versa.

Base KVA. If number of equipments such as generators, transformers, transmission lines etc. are connected in parallel and their percentage resistances and reactances also refer to their respective kVA ratings, it is difficult to compare these percentage resistances and reactances and their combined effect until and unless they are all referred to a common kVA. This common kVA, which is taken as an arbitrary one, is known as the *base* kVA of the system.

A base kVA may be chosen in the following manner :

(*i*) Equal to the kVA rating of the largest unit connected in the network.

(*ii*) Equal to the sum of the kVA ratings of all the units connected in the network.

(*iii*) Any arbitrary value.

It must, however, be clearly understood that the value of the base kVA, has no bearing whatsoever on the results; since in the ultimate formula for the calculation of short-circuit current base MVA is to be taken into consideration.

The conversion of percentage reactance at rated kVA to the percentage reactance at base kVA can be made by using the following expression:

Percentage reactance at base kVA

$$= \frac{\text{Base kVA}}{\text{Rated kVA}} \times \text{percentage reactance at rated kVA} \qquad ...(2.14)$$

Thus if a transformer is rated for 10,000 kVA and has percentage reactance of 7.5%, then the percentage reactance at the base kVA of 25,000 shall be

$$= \frac{25,000}{10,000} \times 7.5\ \% = 18.75\%$$

Base kV. In some cases it is convenient to work in ohmic values of the various reactances rather than in percentage values. The method would become simple if all the reactances relate to the same voltage but if step-up or step-down transformers or other equipments operating at different voltages are also included, all the ohmic values will have to be reduced to a common base voltage.

Reactances can be converted from one operating voltage to the other by the following relation

$$X_2 = \left(\frac{E_2}{E_1}\right)^2 \times X_1 \qquad ...(2.15)$$

where X_1 is the reactance at voltage E_1 and X_2 is the reactance at voltage E_2.

It must, however, be remembered clearly that all the values discussed above *i.e.,* resistances, reactances and voltages refer to the phase values and not to the line values.

2.6. PER UNIT (PU) METHOD OF REPRESENTING QUANTITIES

The quantities involved in power system are kVA, voltage, current and impedance of the equivalent circuits of the various system components. The equivalent circuits are at different voltages and are connected together in the system by means of transformers and interconnections. Each apparatus is rated in kVA and its impedance in actual ohms or in percentage value* referred to its rated kVA and rated voltage, as already explained in Art. 2.5. In power system analysis, it is usual to express voltage, current, kVA and impedance in per unit of base or reference values of these quantities. Such a method simplifies the calculations.

The per unit value of any quantity is defined as

$$\frac{\text{the actual value of the quantity in any unit}}{\text{the base or reference value in the same unit}}$$

2.6.1. Advantages and Drawbacks of Per Unit (PU) System

Advantages. 1. Calculations are simplified.

2. The characteristics of machines (generators, transformers, motors etc.) when described in per unit system are specified by almost the same number, regardless of the rating of the machines. In other words, the characteristics (or parameters) tend to fall in relatively narrow range, making erroneous values conspicuous. Thus per unit system provides a method of comparison.

3. For circuits connected by transformers, per unit system is particularly suitable. By choosing suitable base kV's for the circuits the per unit reactance remains the same, referred to either sides of the transformer. Therefore, the various circuits can be connected in the reactance diagram.

4. This method is useful to eliminate ideal transformers as circuit components since the typical power system contains

* Per cent value = Per unit value × 100

Per cent value is not convenient for use as the factor of 100 has to be carried in computations.

hundreds, if not thousands of transformers, and this is a non-trivial savings.

Drawbacks. 1. Some equations that hold in the unscaled cases are modified when scaled into per unit. Factors such as $\sqrt{3}$ and 3 are removed or added by this method.

2. Equivalent circuits of the components are modified, making them somewhat more abstract. Sometimes phase shifts that are clearly present in the unscaled circuit vanish in the per unit circuit.

2.6.2. Selection of Bases

For a common representation, base kVA and base voltage are to be chosen. Then the base current and base impedance can be expressed as follows :

$$\text{Base current, } I_B = \frac{\text{Base kVA (kVA}_B)}{\text{Base kV (kV}_B)} \text{ amperes} \quad ...(2.16)$$

$$\text{Base impedance, } Z_B = \frac{\text{Base voltage in volts, } V_B}{\text{Base current in amperes, } I_B} \text{ ohms}$$

$$= \frac{V_B \times V_B}{I_B \times V_B}$$

$$= \frac{V_B^2}{V_B\, I_B}$$

$$= \frac{\left(\frac{V_B}{1,000}\right) \times \left(\frac{V_B}{1,000}\right) \times 1,000}{\frac{V_B\, I_B}{1,000}}$$

$$= \frac{(\text{kV}_B)^2 \times 1,000}{\text{kVA}_B}$$

$$= \frac{(\text{kV}_B)^2}{\text{MVA}_B} \text{ ohms} \quad ...(2.17)$$

For a single phase, phase-to-neutral voltage, kVA per phase are taken as bases while in a three-phase system, three-phase line-to-line voltage and three-phase kVA are used as bases. This simplifies the calculations.

The base values in a system are so selected that the per unit voltages and currents in the system are approximately unity. Sometimes the base kVA is chosen equal to the sum of the ratings of the various equipments in the system or equal to the kVA rating of the largest unit connected in the system.

If in the network there is no transformer present, the same base voltage is used throughout, but if the transformers are present, the rule is to change the base voltage in proportion to the transformation ratio of the transformer when transformer is reached. Thus, all impedances in the network before the transformer is reached, including the transformer primary leakage impedance, are expressed in per unit, to the voltage base chosen for the primary side; all impedances beyond the transformer, including the transformer secondary leakage impedance, are expressed in per unit to a new base voltage which is the primary side base voltage multiplied by the transformer transformation ratio. This is very important.

2.6.3. Determination of Base Impedance

Voltage, current, volt-amperes and impedances are so related that selection of base values for any two of them determines the base values of the remaining two. For example for a **single-phase circuit** we have the following relations

$$S_B = V_B I_B \quad ...(2.18)$$

$$Z_B = \frac{V_B}{I_B} \quad ...(2.19)$$

where S_B, V_B, I_B and Z_B represent the base power, base voltage, base current and base impedance expressed in volt-amperes, volts, amperes and ohms respectively. Thus from above two equations, if any two of the four quantities are specified, the remaining two may be determined without any problem. As already mentioned in Art. 2.6.2, base kVA and base kV are usually specified. Let kVA_B and kV_B be the base kVA and base kV, then

$$\text{Per unit kV} = \frac{\text{Actual kV}}{\text{Base kV}} = \frac{\text{kV}_{\text{actual}}}{\text{kV}_B} \quad ...(2.20)$$

$$\text{Base current, } I_B = \frac{\text{Base kVA}}{\text{Base kV}} = \frac{\text{kVA}_B}{\text{kV}_B} \text{ amperes} \quad ...(2.21)$$

$$\text{Per unit current, } I_{pu} = \frac{\text{Actual current}}{\text{Base current}}$$

$$= \frac{\text{Actual current}}{\text{kVA}_B} \times \text{kV}_B \quad ...(2.22)$$

$$\therefore \text{ Base impedance, } Z_B = \frac{\text{Base kV} \times 1,000}{\text{Base current}}$$

$$= \text{kV}_B \times \frac{\text{kV}_B}{\text{kVA}_B} \times 1,000$$

$$= \frac{(\text{kV}_B)^2 \times 1,000}{\text{kVA}_B} \text{ ohms}$$

...already worked out [Eq. (2.17)]

$$\text{Base power, } S_B = \text{Base kVA} \quad ...(2.23)$$

Per unit impedance,

$$Z_{pu} = \frac{\text{Actual impedance}}{\text{Base impedance}}$$

$$= \text{Actual impedance} \times \frac{\text{kVA}_B}{(\text{kV}_B)^2 \times 1,000} \quad ...(2.24)$$

In a **3-phase system** rather than obtaining the per unit values using per phase quantities, the per unit values can be obtained directly by using three-phase base quantities (total kVA, line-to-line voltage and line currents). Let kVA_B be the base kVA (total output of three phases) and kV_B be the base line-to-line voltage in kV.

Assuming star connection (equivalent star can always be found)

$$\text{Base current, } I_B = \frac{\text{kVA}_B}{\sqrt{3}\, \text{kV}_B} \text{ amperes}$$

Base impedance,

$$Z_B = \frac{kV_B \times 1{,}000}{\sqrt{3} \times I_B}$$

$$= \frac{kV_B}{\sqrt{3}} \times \frac{\sqrt{3}\, kV_B}{kVA_B} \times 1{,}000$$

$$= \frac{(kV_B)^2 \times 1{,}000}{kVA_B} \text{ ohms} \quad \text{...(2.25)}$$

Per unit impedance,

$$Z_{pu} = \frac{\text{Actual impedance}}{\text{Base impedance}}$$

$$= \frac{\text{Actual impedance} \times kVA_B}{(kV_B)^2 \times 1{,}000} \quad \text{...(2.26)}$$

It should be noted that the same expression is obtained for a single-phase and three-phase systems. In power system problems the data are given in terms of 3-phase kVA, line-to-line kV, the machine kVA and machine phase-to-phase (or line-to-line) kV rating. Further the direct-axis synchronous reactance is also known as positive sequence reactance of the machine.

2.6.4. Change of Base

Normally the per unit impedance of various components corresponding to its own rating voltage and kVA are given and since we choose one common base kVA and base kV for the whole system, therefore, it becomes imperative to determine the per unit impedance of the various equipments corresponding to the common base kV and base kVA. If the individual quantities are $Z_{pu\ old}$, kVA_{old}, kV_{old} and the common base quantities are $Z_{pu\ new}$, kVA_{new} and kV_{new}, then making use of Eq. (2.24), according to which per unit impedance is directly proportional to the base kVA and inversely proportional to the square of the base kV, we have

Per unit impedance to new base,

$$Z_{pu\ new} = Z_{pu\ old} \times \frac{kVA_{new}}{kVA_{old}} \times \frac{(kV_{old})^2}{(kV_{new})^2} \quad \text{...(2.27)}$$

This is very important relation used in power system analysis.

2.6.5. Per Unit Impedance of a Two Winding Transformer

The approximate equivalent circuit of a two winding transformer with all impedances referred to primary (low-voltage) side is illustrated in Fig. 2.10.

According to Eq. (2.24)

$$Z_{01\ pu} = \frac{Z_{01} \times kVA_B}{(kV_{B1})^2 \times 1{,}000} \quad \text{...(2.28)}$$

where Z_{01} is the total impedance of the transformer referred to primary side in ohms, kVA_B is the rated kVA of the transformer and kV_{B1} is the rated voltage of the transformer on the primary side in kV.

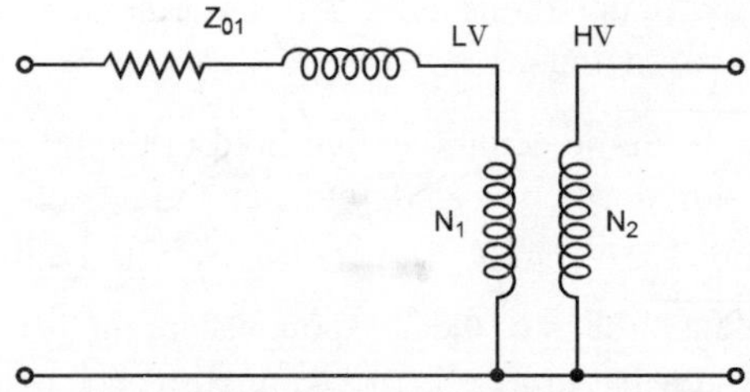

Fig. 2.10. ***Simplified Equivalent Circuit of a Single-Phase Two Winding Transformer***

Now let us consider the transformer with all its impedances referred to secondary side. The total impedance of the transformer referred to secondary (high-voltage side).

$$Z_{02} = Z_{01} \times \left(\frac{N_2}{N_1}\right)^2 = Z_{01} \frac{(kV_{B2})^2}{(kV_{B1})^2} \quad \text{...(2.29)}$$

Thus the impedance of the transformer in per unit viewed from the secondary (high-voltage) side,

$$Z_{02\ pu} = \frac{Z_{02} \times kVA_B}{(kV_{B2})^2 \times 1{,}000} \quad \text{...(2.30)}$$

where Z_{02} is the total impedance of the transformer referred to secondary side in ohms and kV_{B2} is the rated voltage of the transformer on the secondary side in kV.

Substituting the value of Z_{02} from Eq. (2.29) in Eq. (2.30) we have

$$Z_{02\ pu} = Z_{01} \times \frac{(kV_{B2})^2}{(kV_{B1})^2} \times \frac{kVA_B}{(kV_{B2})^2 \times 1{,}000}$$

$$= \frac{Z_{01} \times kVA_B}{(kV_{B1})^2 \times 1{,}000} \quad \text{...(2.31)}$$

Comparing Eqs. (2.28) and (2.31) we conclude that the per unit impedance is the same regardless of the side from which it is viewed. Thus the simple equivalent circuit for the two-winding transformer in which $Z_{01\ pu} = Z_{02\ pu} = Z_{pu}$ is depicted in Fig. 2.11.

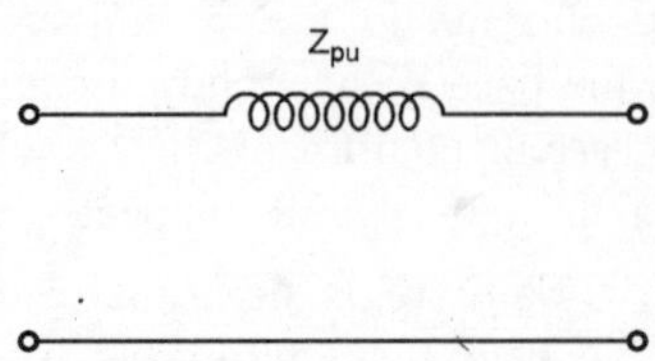

Fig. 2.11. ***Simplified Equivalent Circuit of a Two Winding Transformer With Impedance Expressed in Per Unit***

2.6.6. Mutual Impedance in Per Unit Between Lines of Different Voltage Levels

Let us consider two three-phase lines of different voltage levels running together, with mutual inductance X_m present, as depicted in Fig. 2.12.

There is one value $X_{m\ pu}$ (mutual reactance in per unit), which will serve both lines.

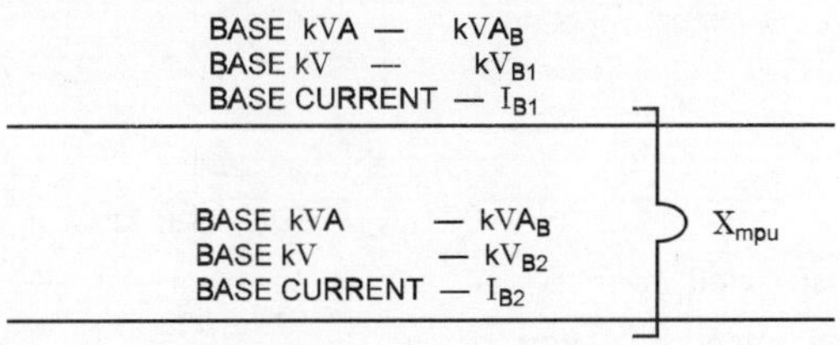

Fig. 2.12. ***Two Parallel Lines of Same MVA But of Different Voltage Levels***

Thus in terms of line 2, we have

$$X_{m\,pu} = \frac{X_m}{X_{m\,B}} = \frac{X_m}{1{,}000 \times kV_{B_2} / I_{B_1}}$$

$$= \frac{X_m}{kV_{B_2} \times 1{,}000} \times I_{B_1}$$

$$= \frac{X_m}{kV_{B_2} \times 1{,}000} \times \frac{I_{B_1} \times kV_{B_1}}{kV_{B_1}}$$

$$\text{or } X_{m\,pu} = \frac{X_m \cdot kVA_B}{kV_{B_1} \times kV_{B_2} \times 1{,}000} = \frac{X_m\ MVA_B}{kV_{B_1} \times kV_{B_2}} \quad ...(2.32)$$

2.7. PER UNIT IMPEDANCE DIAGRAM OF A POWER SYSTEM

Per unit impedance diagram can be drawn directly from the single line diagram of a power system by following the procedure given below :

Choose an appropriate common kVA (or MVA) base for the whole system. Consider the system to be divided into a number of sections by the transformers; choose an appropriate kV base in one of the sections and determine kV bases of other sections in the ratio of transformation. Now calculate per unit values of voltages and impedances for each section and connect them as per the topology of the single line diagram. The result is the single-phase per unit impedance diagram. The method is illustrated by Examples 2.11 and 2.12.

Example 2.1. Base voltage = 1,100 volts. Base kVA = 10^6. What is base impedance ? [*Madurai Kamaraj Univ. Power System Analysis Nov. 1996*]

Solution : Base kVA $= 10^6$

$$\text{Base kV} = \frac{1{,}100}{1{,}000} = 1.1 \text{ kV}$$

$$\text{Base current, } I_B = \frac{kVA_B}{kV_B} = \frac{10^6}{1.1}$$

$$\text{Base impedance, } Z_B = \frac{kV_B \times 1{,}000}{I_B} = \frac{1.1 \times 1{,}000 \times 1.1}{10^6}$$

$$= 0.00121\ \Omega \quad \textbf{Ans.}$$

Base impedance can also be determined directly by substituting the values of base kVA and base kV in Eq. (2.17) and we get

$$Z_B = \frac{(kV_B)^2 \times 1{,}000}{kVA_B} = \frac{(1.1)^2 \times 1{,}000}{10^6}$$

$$= 0.00121\ \Omega \quad \textbf{Ans.}$$

Same as above.

Example 2.2. If the resistance in ohms is 5 Ω, find the per unit value. Given base kVA = 10 and base kV = 11.

[*Madras Univ. Power System Analysis October 1996*]

Solution : Resistance, R = 5 Ω

$$\text{Base resistance, } R_B = \frac{(kV_B)^2 \times 1{,}000}{kVA_B} = \frac{(11)^2 \times 1{,}000}{10}$$

$$= 12{,}100\ \Omega$$

$$\text{Per unit resistance, } R_{pu} = \frac{R}{R_B} = \frac{5}{12{,}100} = 0.000413 \text{ pu} \quad \textbf{Ans.}$$

Example 2.3. A single-phase transformer is rated as 2.5 kVA, 11/0.4 kV. If the leakage reactance is 0.96 Ω when referred to low-voltage side, then determine its leakage reactance in per unit.

Solution : Base voltage, kV_B = 0.4 kV

Base kVA, kVA_B = 2.5 kVA

Low-voltage side base impedance,

$$Z_{BLV} = \frac{(kV_B)^2 \times 1{,}000}{kVA_B} = \frac{(0.4)^2 \times 1{,}000}{2.5} = 64\ \Omega$$

Leakage reactance in per unit,

$$Z_{pu} = \frac{\text{Actual reactance}}{\text{Base impedance}} = \frac{0.96}{64} = 0.015 \text{ pu} \quad \textbf{Ans.}$$

Example 2.4. For a 110/440 V, 25 kVA, single-phase transformer, primary and secondary leakage reactances are 0.04 Ω and 0.10 Ω respectively. Show that net pu leakage reactance of the transformer referred to *lv* side is same as that referred to *hv* side.

[*U.P. Technical Univ. Power System Analysis 2004-05*]

Solution : Base kVA, kVA_B = 25 kVA

$$\text{Transformation ratio, } K = \frac{E_2}{E_1} = \frac{440}{110} = 4$$

For low-voltage side

$$\text{Base kV, } kV_{BLV} = \frac{110}{1{,}000} \text{ kV} = 0.11 \text{ kV}$$

$$\text{Base current, } I_{BLV} = \frac{kVA_B}{kV_{BLV}} = \frac{25}{0.11} \text{ A}$$

$$\text{Base impedance, } Z_{BLV} = \frac{kV_{BLV} \times 1{,}000}{I_{BLV}}$$

$$= \frac{0.11 \times 1{,}000}{\frac{25}{0.11}} = 0.484\ \Omega$$

Actual impedance referred to low-voltage side,

$$Z_{LV} = 0.04 + \frac{0.1}{4^2} = 0.04625\ \Omega$$

Per unit equivalent impedance,

$$Z_{pu\,LV} = \frac{Z_{LV}}{Z_{BLV}} = \frac{0.04625}{0.484} = 0.095558 \text{ pu} \quad ...(i)$$

For high-voltage side

Base kV, kV_{BHV} = 0.44 kV

$$\text{Base current, } I_{BHV} = \frac{kVA_B}{kV_{BHV}} = \frac{25}{0.44} \text{ A}$$

$$\text{Base impedance, } Z_{BHV} = \frac{kV_{BHV} \times 1{,}000}{I_{BHV}}$$

$$= \frac{0.44 \times 1{,}000}{\frac{25}{0.44}} = 7.744\ \Omega$$

Actual impedance referred to high-voltage side,

$$Z_{HV} = 0.04 \times 4^2 + 0.1 = 0.74\ \Omega$$

Per unit equivalent impedance,

$$Z_{pu\ HV} = \frac{Z_{HV}}{Z_{BHV}} = \frac{0.74}{7.744} = 0.095558 \text{ pu} \quad ...(ii)$$

From Eqs. (*i*) and (*ii*), it is seen that per unit values referred to either side (HV or LV) of the transformer are the same.

Example 2.5. An 11/0.4 kV, 200 kVA transformer has an equivalent impedance of (2.4 + *j* 12.4) ohms referred to the hv side. Determine the base values for the pu system, the per unit equivalent impedance and the equivalent impedance drop at one-half rated current.

Solution : Base kVA, kVA_B = 200 kVA **Ans.**

For high-voltage side,

Base kV, kV_{BHV} = 11 kV **Ans.**

$$\text{Base current, } I_{BHV} = \frac{kVA_B}{kV_{BHV}} = \frac{200}{11} = 18.18 \text{ A } \textbf{Ans.}$$

$$\text{Base impedance, } Z_{BHV} = \frac{kV_{BHV} \times 1{,}000}{I_{BHV}}$$

$$= \frac{11 \times 1{,}000}{18.18} = 605\ \Omega \textbf{ Ans.}$$

Per unit equivalent impedance,

$$Z_{pu\ HV} = \frac{Z_{HV}}{Z_{BHV}}$$

$$= \frac{2.4 + j12.4}{605}$$

$$= (0.004 + j\,0.0205) \text{ pu } \textbf{Ans.}$$

At one-half rated current,

Equivalent impedance drop, $(IZ)_{pu} = 0.5\ (0.004 + j\,0.0205) = 0.0104$ pu

Thus impedance drop is 0.0104 times the rated voltage *i.e.,* 0.104 × 11,000 = 115 V **Ans.**

For low-voltage side,

Base kV, kV_{BLV} = 0.4 kV **Ans.**

$$\text{Base current, } I_{BLV} = \frac{kVA_B}{kV_{BLV}} = \frac{200}{0.4} = 500 \text{ A } \textbf{Ans.}$$

$$\text{Base impedance, } Z_{BLV} = \frac{kV_{BLV} \times 1{,}000}{I_{BLV}} = \frac{0.4 \times 1{,}000}{500}$$

$$= 0.8\ \Omega \textbf{ Ans.}$$

Actual impedance referred to low-voltage side,

$$Z_{LV} = \frac{2.4 + j12.4}{(11/0.4)^2}$$

$$= (0.00317 + j\,0.0164) \text{ ohm}$$

Per unit equivalent impedance,

$$Z_{pu\ LV} = \frac{Z_{LV}}{Z_{BLV}} = \frac{0.00317 + j0.0164}{0.8}$$

$$= (0.004 + j\,0.0205) \text{ pu } \textbf{Ans.}$$

At one-half rated current.

Equivalent impedance drop,

$$IZ_{pu} = 0.5\ (0.004 + j\,0.0205) = 0.0104 \text{ pu}$$

This impedance drop is 0.0104 times the rated voltage, or 0.0104 × 400 = 4.16 V **Ans.**

Remarks. From the above it is seen that per unit values referred to either side (hv or lv) of the transformer are the same. This will hold true as long as the voltage bases on the two sides are chosen in the ratio of transformation.

Example 2.6. Determine the per unit impedance of a transmission line having an impedance of (30 + *j* 110) Ω on 100 MVA and 132 kV base voltage.

Solution : Impedance of line, Z = (30 + *j* 110) Ω

Base kVA, kVA_B = 100 MVA or 100,000 kVA

Base kV, kV_B = 132 kV

$$\text{Per unit impedance, } Z_{pu} = \frac{Z \times kVA_B}{(kV_B)^2 \times 1{,}000}$$

$$= \frac{(30 + j110) \times 100{,}000}{(132)^2 \times 1{,}000}$$

$$= (0.172 + j\,0.631) \text{ pu } \textbf{Ans.}$$

Example 2.7. A 30 MVA, 11 kV generator has a reactance of 0.2 pu referred to its ratings as bases. Determine the per unit reactance when referred to base kVA of 50,000 kVA and base kV of 33 kV.

Solution : Old base kVA = 30,000 kVA

Old base kV = 11 kV

Old per unit impedance,

$Z_{pu\ old}$ = 0.2 pu

New base kVA = 50,000 kVA

New base kV = 33 kV

New per unit impedance,

$$Z_{pu\ new} = Z_{pu\ old} \times \frac{\text{New base kVA}}{\text{Old base kVA}} \times \left(\frac{\text{Old base kV}}{\text{New base kV}}\right)^2$$

$$= 0.2 \times \frac{50{,}000}{30{,}000} \times \left(\frac{11}{33}\right)^2 = 0.037 \text{ pu } \textbf{Ans.}$$

Example 2.8. The one line diagram of a simple power system is given below. The transformation ratio of the step-up transformer is 11/220 kV. The base MVA = 100 and base kV = 11 on the generator side. What is the base kV on the transmission side ?

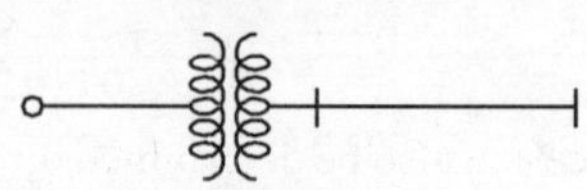

Fig. 2.13

Solution : Base kV on the transmission side

= Base kV on the generator side × transformation ratio

$$= 11 \times \frac{220}{11} = 220 \text{ kV } \textbf{Ans.}$$

Example 2.9. By choosing 3-ϕ power and line voltage as base values for a 3-ϕ system, demonstrate how the per unit method performs power system calculation on single phase basis.

[M.D. Univ. Power System-I, 2010-11]

Solution : Base impedance for a single-phase system is

$$Z_{B1\phi} = \frac{(kV_{1\phi})^2}{MVA_{1\phi}}$$

Base impedance for a three-phase system is

$$Z_{B3\phi} = \frac{(kV_{L3\phi})^2}{MVA_{3\phi}}$$

Assuming system to be star-connected

$$kV_{line} = \sqrt{3} \times kV \text{ phase} = \sqrt{3}\, kV_{1\phi}$$

$$MVA_{3\phi} = 3 \times MVA_{1\phi}$$

$$\text{So, } Z_{B3\phi} = \frac{(\sqrt{3}\, kV_{1\phi})^2}{3 \times MVA_{1\phi}} = \frac{(kV_{1\phi})^2}{MVA_{1\phi}} = Z_{B1\phi}$$

$$\text{i.e., } Z_{B3\phi} = Z_{B1\phi}$$

Example 2.10. Three generators are rated as follows :

Generator 1 : 100 MVA, 33 kV, reactance = 10 %

Generator 2 : 150 MVA, 32 kV, reactance = 8 %

Generator 3 : 110 MVA, 30 kV, reactance = 12 %

Choosing 200 MVA and 35 kV as base quantities, compute per unit reactances of the three generators referred to these base quantities. Draw reactance diagram and mark per unit reactances. The three generators are connected to common bus-bars.

[Bharathiyar Univ. Power System Analysis April 1997]

Solution : Using Eq. (2.27) we have

Per unit reactance to new base,

$$X_{pu\ new} = X_{pu\ old} \times \frac{\text{New base kVA}}{\text{Old base kVA}} \times \left(\frac{\text{Old base kV}}{\text{New base kV}}\right)^2$$

So the reactances on 200 MVA and 35 kV base are :

Per unit reactance of generator 1,

$$X_{pu\ new\ 1} = \frac{10}{100} \times \frac{200 \times 1{,}000}{100 \times 1{,}000} \times \left(\frac{33}{35}\right)^2 = 0.1778 \text{ pu } \textbf{Ans.}$$

Per unit reactance of generator 2,

$$X_{pu\ new\ 2} = \frac{8}{100} \times \frac{200 \times 1{,}000}{150 \times 1{,}000} \times \left(\frac{32}{35}\right)^2 = 0.0892 \text{ pu } \textbf{Ans.}$$

Per unit reactance of generator 3,

$$X_{pu\ new\ 3} = \frac{12}{100} \times \frac{200 \times 1{,}000}{110 \times 1{,}000} \times \left(\frac{30}{35}\right)^2 = 0.1603 \text{ pu } \textbf{Ans.}$$

Reactance diagram is shown in Fig. 2.14. The values indicate reactances in per unit and E_1, E_2 and E_3 are the generated emfs of generators G_1, G_2 and G_3 respectively.

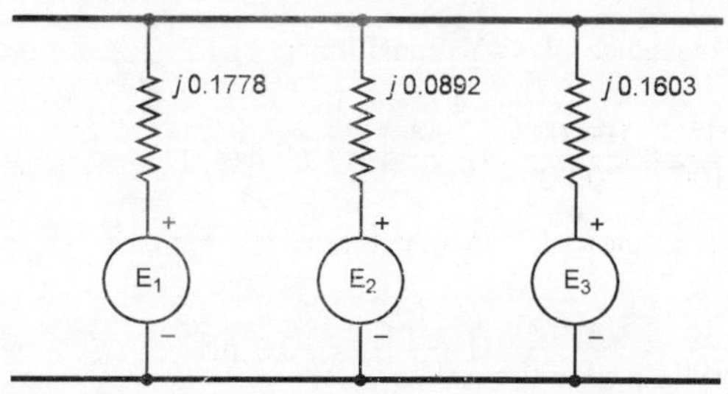

Fig. 2.14. ***Reactance Diagram For Example 2.10***

Example 2.11. Draw the pu impedance diagram for the power system shown in Fig. 2.15. Neglect resistance and use a base of 100 MVA, 220 kV in 50 Ω line. The rating of the generator, motor and transformers are :

Generator	:	**40 MVA, 25 kV, X" = 20 %**
Motor	:	**50 MVA, 11 kV, X" = 30 %**
Y-Y Transformer	:	**40 MVA, 33Y-220 Y kV, X = 15 %**
Y-Δ Transformer	:	**30 MVA, 11 Δ-220 Y kV, X = 15 %**

[Madras Univ. Power System Analysis October 1996]

Solution : Base kVA for the complete circuit = 100,000 kVA

Base kV in 50 Ω line = 220 kV

$$\text{Per unit impedance of line} = \frac{X_L \times kVA_B}{(kV_B)^2 \times 1{,}000}$$

$$= \frac{j50 \times 100{,}000}{(220)^2 \times 1{,}000} = j\,0.1033 \text{ pu}$$

Base kV in generator circuit

$$= 220 \times \frac{1}{\text{Transformation ratio of Y-Y transformer}}$$

$$= 220 \times \frac{33}{220} = 33 \text{ kV}$$

Per unit reactance of the generator

$$= \frac{20}{100} \times \frac{100{,}000}{40{,}000}\left(\frac{25}{33}\right)^2 = j\,0.287 \text{ pu}$$

Base kV in motor circuit

= 220 × transformation ratio of Y–Δ transformer

$$= 220 \times \frac{11}{220} = 11 \text{ kV}$$

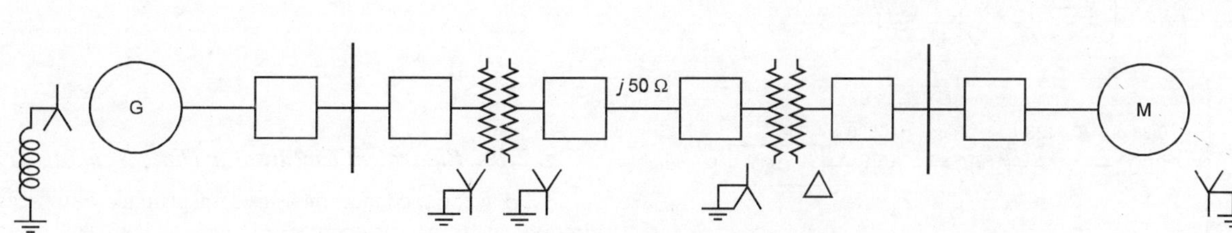

Fig. 2.15

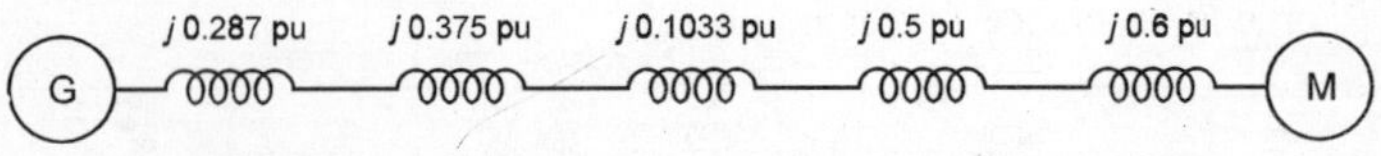

Fig. 2.16. ***Reactance Diagram For Example 2.11***

Per unit reactance of the motor

$$= \frac{30}{100} \times \frac{100{,}000}{50{,}000} \times \left(\frac{11}{11}\right)^2 = j\,0.6 \text{ pu}$$

Per unit reactance of Y–Y transformer

$$= \frac{15}{100} \times \frac{100{,}000}{40{,}000} \times \left(\frac{33}{33}\right)^2 = j\,0.375 \text{ pu}$$

Per unit reactance of Y–Δ transformer

$$= \frac{15}{100} \times \frac{100{,}000}{30{,}000} \times \left(\frac{11}{11}\right)^2 = j\,0.5 \text{ pu}$$

The reactance diagram for the system is shown in Fig. 2.16.

Example 2.12. A simple power system is shown in Fig. 2.17. Redraw this system where the per unit impedances are represented on a common 5,000 VA base and common system base voltage of 250 V. *[Pb. Univ. Power System Analysis December 1994]*

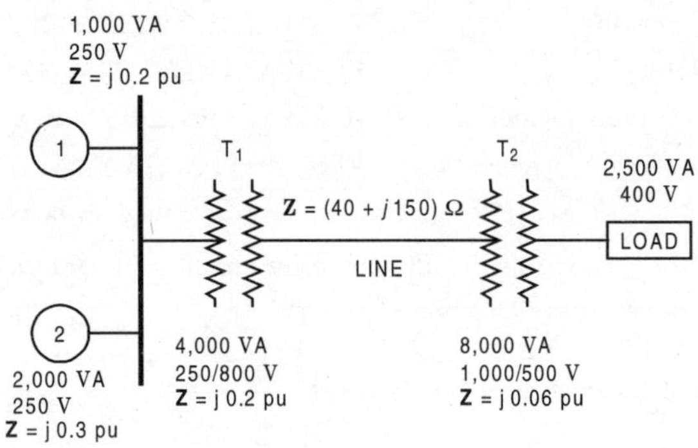

Fig. 2.17

Solution : Base kVA for the complete circuit $= \frac{5{,}000}{1{,}000} = 5$ kVA

Base voltage in generator circuits

$$= \frac{250}{1{,}000} = 0.25 \text{ kV}$$

Per unit reactance of machine 1,

$$X_{\text{pu}\,M_1} = j0.2 \times \left(\frac{0.25}{0.25}\right)^2 \times \left(\frac{5.0}{1.0}\right) = j\,1.0 \text{ pu}$$

Per unit reactance of machine 2,

$$X_{\text{pu}\,M_2} = j0.3 \times \left(\frac{0.25}{0.25}\right)^2 \times \frac{5}{2} = j\,0.75 \text{ pu}$$

Per unit reactance of transformer T_1,

$$X_{\text{pu}\,T_1} = j0.2 \times \left(\frac{0.25}{0.25}\right)^2 \times \left(\frac{5}{4}\right) = j\,0.25 \text{ pu}$$

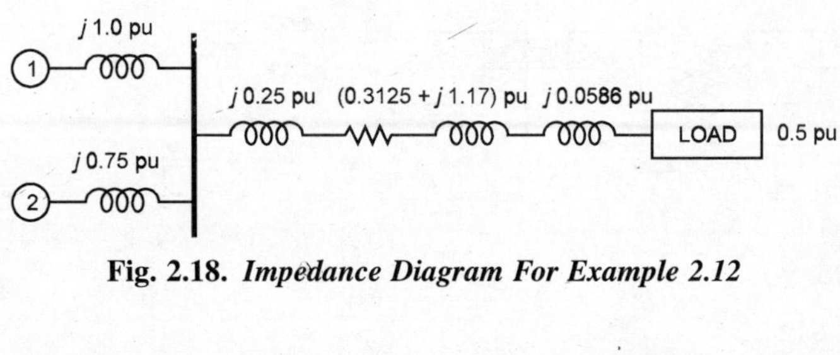

Fig. 2.18. ***Impedance Diagram For Example 2.12***

Base voltage in line circuit

= Base voltage in generator circuit × transformation ratio of transformer T_1

$$= 0.25 \times \frac{800}{250} = 0.8 \text{ kV}$$

Per unit impedance of line,

$$X_{\text{pu L}} = (40 + j\,150) \times \frac{\text{kVA}_B}{(\text{kV}_B)^2 \times 1{,}000} \qquad \text{...Refer to Eq. (2.24)}$$

$$= \frac{(40 + j150) \times 5}{(0.8)^2 \times 1{,}000} = (0.3125 + j\,1.17) \text{ pu}$$

Per unit impedance of transformer 2,

$$X_{\text{pu}\,T_2} = j0.06 \times \left(\frac{1}{0.8}\right)^2 \times \frac{5}{8} = 0.0586 \text{ pu}$$

Per unit load,

$$\text{kVA}_{\text{pu}} = \frac{2.5}{5} = 0.5 \text{ pu}$$

The entire system is now represented by the impedance diagram shown in Fig. 2.18, in which all components are represented on a common 5,000 VA (or 5 kVA) base and base voltage of 250 V (or 0.25 kV).

Example 2.13. The three-phase ratings of a three-winding transformer are:

Primary — Y-connected, 66 kV, 15 MVA

Secondary — Y-connected, 13.2 kV, 10 MVA

Tertiary — Δ-connected, 2.3 kV, 5 MVA

Neglecting resistances, the leakage impedances are :

Z_{PS} **= 0.08 per unit on 15 MVA, 66 kV base**

Z_{PT} **= 0.10 per unit on 15 MVA, 66 kV base**

Z_{ST} **= 0.09 per unit on 10 MVA, 13.2 kV base**

Find the per unit impedances of the star-connected equivalent circuit for a base of 15 MVA, 66 kV base in the primary circuit.

[Bharathidasan Univ. Power System Analysis November 1996]

Solution : The equivalent circuit of a 3-winding transformer is shown in Fig. 2.19. The impedances Z_P, Z_S and Z_T are the impedances of primary, secondary and tertiary windings respectively. These may be given in ohms or in per unit based on the ratings of respective windings. The kVA and kV ratings of the three windings may be different. However, the per unit impedances in the impedance diagram are expressed on the same base kVA.

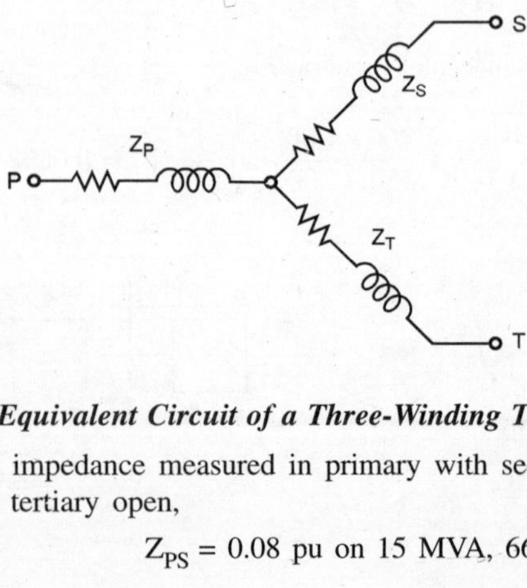

Fig. 2.19. ***Equivalent Circuit of a Three-Winding Transformer***

Leakage impedance measured in primary with secondary short circuited and tertiary open,

$$Z_{PS} = 0.08 \text{ pu on 15 MVA, 66 kV base.}$$

Leakage impedance measured in primary with tertiary short circuited and secondary open,

$$Z_{PT} = 0.1 \text{ pu on 15 MVA, 66 kV base}$$

Leakage impedance measured in secondary with tertiary short circuited and primary open,

$$Z_{ST} = 0.09 \text{ pu on 10 MVA, 13.2 kV base}$$

$$\text{Base for primary} = 15 \text{ MVA, 66 kV}$$

$$\text{Base for secondary} = 15 \text{ MVA, 13.2 kV}$$

$$\text{Base for tertiary} = 15 \text{ MVA, 2.3 kV}$$

The first two results *i.e.*, Z_{PS} and Z_{PT} are on the above base

$$\text{and } Z_{ST} = 0.09 \times \frac{15{,}000}{10{,}000} = 0.135 \text{ pu}$$

It is seen from the equivalent circuit shown in Fig. 2.19

$$Z_P + Z_S = Z_{PS} = 0.08 \text{ pu} \quad ...(i)$$

$$Z_P + Z_T = Z_{PT} = 0.1 \text{ pu} \quad ...(ii)$$

$$\text{and } Z_S + Z_T = Z_{ST} = 0.135 \text{ pu} \quad ...(iii)$$

Solving Eqs. (*i*), (*ii*) and (*iii*), we have

$$Z_P = \frac{1}{2}[Z_{PS} + Z_{PT} - Z_{ST}]$$

$$= \frac{1}{2}[0.08 + 0.1 - 0.135] = 0.0225 \text{ pu } \textbf{Ans.}$$

$$Z_S = \frac{1}{2}[Z_{PS} + Z_{ST} - Z_{PT}]$$

$$= \frac{1}{2}[0.08 + 0.135 - 0.1] = 0.0575 \text{ pu } \textbf{Ans.}$$

$$\text{and } Z_T = \frac{1}{2}[Z_{PT} + Z_{ST} - Z_{PS}]$$

$$= \frac{1}{2}[0.1 + 0.135 - 0.08] = 0.0775 \text{ pu } \textbf{Ans.}$$

Example 2.14. A 100 MVA, 33 kV, 3-phase generator has a subtransient reactance of 15%. The generator is connected to the motors through a transmission line and transformer as shown in Fig. 2.20. The motors have rated inputs of 30 MVA, 20 MVA and 50 MVA at 30 kV with 20% subtransient reactance. The 3-φ transformers are rated at 110 MVA 32 kV Δ/110 kV Y with leakage reactance 8%. The line has a reactance of 50 Ω. Selecting the generator rating as the base quantities in the generator circuit, determine the base quantities in other parts of the system and evaluate the corresponding pu values.

[*Rajasthan Technical Univ. Power System Analysis 2011*]

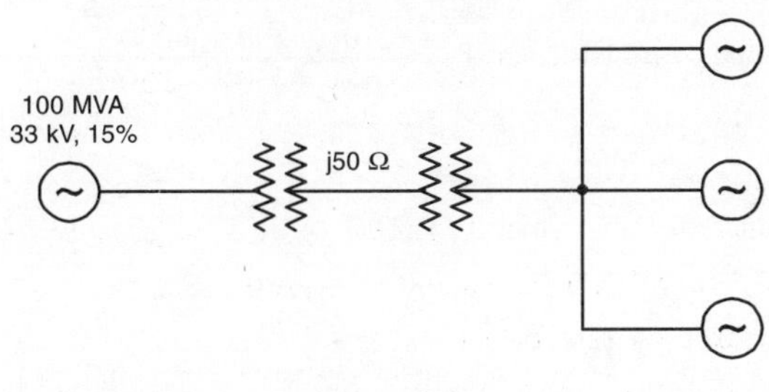

Fig. 2.20

Solution : Taking generator's rating as values, Base MVA for the whole circuit,

$$MVA_B = 100 \text{ MVA}$$

Base kV in generator circuit, $kV_B = 33$ kV

PU reactance of the generator = *j*0.15 pu

The base value of the line,

$$kV_{BL} = 33 \times \frac{110}{32} = 113.4375 \text{ kV } \textbf{Ans.}$$

The base value in the motor circuit,

$$kV_{BM} = 113.4375 \times \frac{32}{110} = 33 \text{ kV } \textbf{Ans.}$$

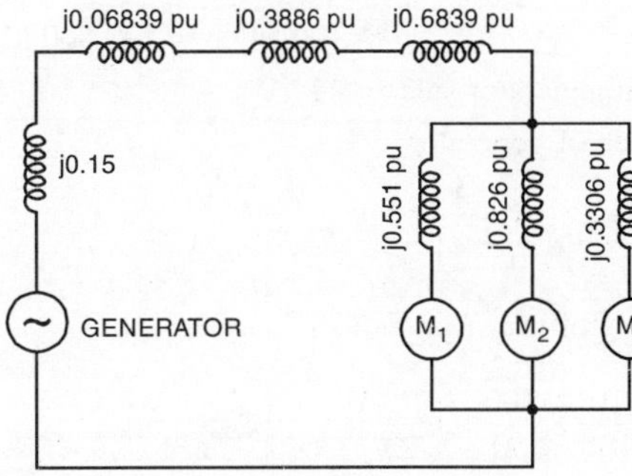

Fig. 2.21. ***PU Reactance Diagram of The System***

PU reactance of transformers

$$= 0.08 \times \frac{100}{110} \times \left(\frac{32}{33}\right)^2 \quad \text{Refer to Eq. (2.27)}$$

$$= 0.06839 \text{ pu } \textbf{Ans.}$$

$$\text{PU impedance of line} = j50 \times \frac{100}{(113.4375)^2} \quad \text{Refer to Eq. (2.24)}$$

$$= 0.3886 \text{ pu } \textbf{Ans.}$$

$$\text{PU reactance of motor 1} = 0.2 \times \frac{100}{30} \times \left(\frac{30}{33}\right)^2 = 0.551 \text{ pu } \textbf{Ans.}$$

$$\text{PU reactance of motor 2} = 0.2 \times \frac{100}{20} \times \left(\frac{30}{33}\right)^2 = 0.826 \text{ pu } \textbf{Ans.}$$

$$\text{PU reactance of motor 3} = 0.2 \times \frac{100}{50} \times \left(\frac{30}{33}\right)^2 = 0.3306 \text{ pu } \textbf{Ans.}$$

The pu reactance diagram for the system is shown in Fig. 2.21.

Example 2.15. The one line diagram of a 3-phase power system is shown in Fig. 2.22. Select a common base of 100 MVA and 22 kV on the generator side.

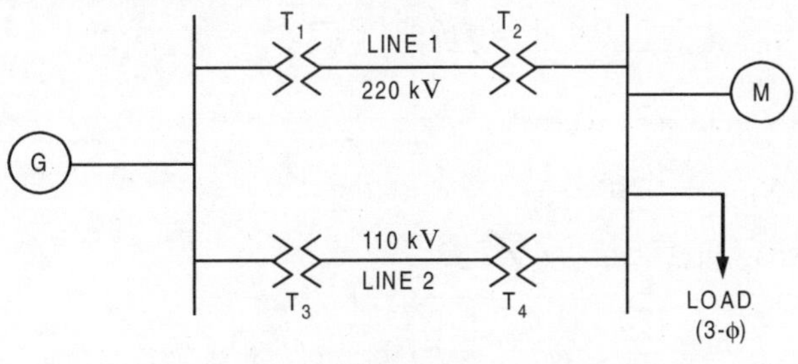

Fig. 2.22

Draw its impedance diagram with all impedances including the load impedance marked in per unit.

The manufacturer's data for each device are given below :

G : 90 MVA, 22 kV, X = 18%

T_1 : 50 MVA, 22/220 kV, X = 10%

T_2 : 40 MVA, 220/11 kV, X = 6%

T_3 : 40 MVA, 22/110 kV, X = 6.4%

T_4 : 40 MVA, 110/11 kV, X = 8%

M : 66.5 MVA, 10.45 kV, X = 18.5%

The 3-phase load absorbs 57 MVA, 0.6 pf lagging at 10.45 kV. Line 1 and Line 2 have reactances of 48.4 and 65.43 ohms respectively. *[U.P. Technical Univ. Power System Analysis 2007-08]*

Solution : Base kVA for complete circuit

$$= 100 \text{ MVA} = 100{,}000 \text{ kVA}$$

Base kV on generator side = 22 kV

Per unit reactance of generator

$$= \frac{18}{100} \times \frac{100}{90} \times \left(\frac{22}{22}\right)^2 = j0.2 \text{ pu}$$

Per unit reactance of transformer T_1

$$= \frac{10}{100} \times \frac{100}{50} \times \left(\frac{22}{22}\right)^2 = j0.2 \text{ pu}$$

Per unit reactance of transformer T_3

$$= \frac{6.4}{100} \times \frac{100}{40} \times \left(\frac{22}{22}\right)^2 = j0.16 \text{ pu}$$

Base kV in line 1

$$= 22 \times \frac{220}{22} = 220 \text{ kV}$$

Per unit impedance of line 1

$$= \frac{X_{L1} \times kVA_B}{(kV_{B1})^2 \times 1{,}000} = \frac{48.4 \times 100{,}000}{(220)^2 \times 1{,}000} = j0.1 \text{ pu}$$

Base kV in line 2

$$= 22 \times \frac{110}{22} = 110 \text{ kV}$$

Per unit impedance of line 2

$$= \frac{X_{L2} \times kVA_B}{(kV_{B2})^2 \times 1{,}000} = \frac{65.43 \times 100{,}000}{(110)^2 \times 1{,}000} = j0.54 \text{ pu}$$

Per unit reactance of transformer T_2

$$= \frac{6}{100} \times \frac{100}{40} \times \left(\frac{220}{220}\right)^2 = j0.15 \text{ pu}$$

Per unit reactance of transformer T_4

$$= \frac{8}{100} \times \frac{100}{40} \times \left(\frac{110}{110}\right)^2 = j0.2 \text{ pu}$$

Base kV in motor circuit,

$$kV_{BM} = \left(220 \times \frac{11}{220}\right) \text{ or } \left(110 \times \frac{11}{110}\right) = 11 \text{ kV}$$

Per unit reactance of motor

$$= \frac{18.5}{100} \times \frac{100}{66.5} \times \left(\frac{10.45}{11}\right)^2 = j0.251 \text{ pu}$$

Base voltage for load

= Base voltage for motor = 11 kV

$$\text{Load} = \left(\frac{10.45}{11}\right)^2 \times \frac{100}{57} \angle -\cos^{-1} 0.6$$

$$= 1.583 \angle -53.13° \text{ pu}$$

Reactance diagram is shown in Fig. 2.23.

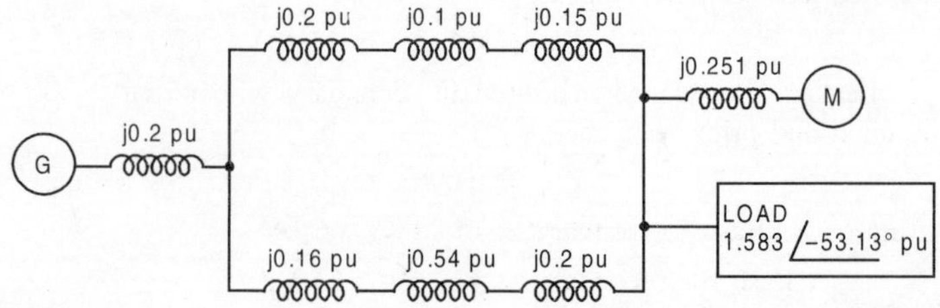

Fig. 2.23

Example 2.16. A 300 MVA, 20 kV, 3-phase generator has a subtransient reactance of 20%. The generator supplies a number of synchronous motors over a 64 km transmission line having transformers at both ends as shown in one line diagram of Fig. 2.24. The motors all rated 13.2 kV, are represented by just two equivalent motors. The neutral of one motor M_1 is grounded through reactance. The neutral of the second motor M_2 is not grounded. Rated inputs to the motors are 200 MVA and 100 MVA for M_1 and M_2 respectively. For both motors X″ = 20%. The 3-phase transformers T_1 is rated 350 MVA, 230/20 kV with leakage reactance of 10%. Transformer T_2 is composed of three single-phase transformers each rated 127/13.2 kV, 100 MVA with leakage reactance of 10%. Series reactance of transmission line is 0.5 Ω/km. Draw the reactance diagram with all reactances marked in per unit. Select the generator rating as base in generator circuit.

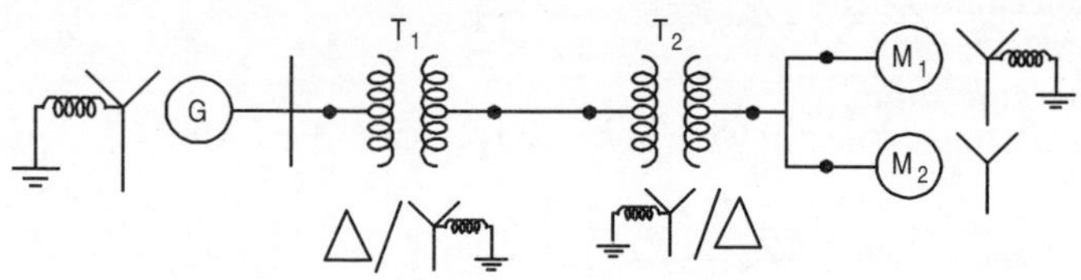

Fig. 2.24

[U.P. Technical Univ. Power System Analysis 2005-06]

Solution : Base kVA, kVA_B = 300 MVA

Base voltage for generator = 20 kV

$$\text{Base voltage for line} = 20 \times \frac{230}{20} = 230 \text{ kV}$$

Line to line voltage ratio of transformer T_2

$$= \sqrt{3} \times \frac{127}{13.2} = 16.664$$

$$\text{Base voltage for motors} = \frac{230}{16.664} = 13.8 \text{ kV}$$

PU reactance of generator = 20% = $j0.2$ pu

$$\text{PU reactance of transformer } T_1 = \frac{10}{100} \times \frac{300}{350} \times \left(\frac{230}{230}\right)^2$$

$$= j0.0857 \text{ pu}$$

The reactance of transformer T_2 is 10% on $\sqrt{3} \times 127$: 13.2 kV and 3 × 100 *i.e.,* 300 MVA base. For 300 MVA, 230 : 13.8

So pu reactance of transformer T_2

$$= \frac{10}{100} \times \left(\frac{13.2}{13.8}\right)^2 \times \frac{300}{300} = j0.0915 \text{ pu}$$

Reactance of line = 64 × 0.5 = 32 Ω

$$\text{Base reactance of line} = \frac{(\text{Base kV})^2}{\text{Base MVA}} = \frac{(230)^2}{300} = 176.333\ \Omega$$

$$\text{Per unit reactance of line} = \frac{32}{176.333} = j0.1815 \text{ pu}$$

$$\text{Per unit reactance of motor } M_1 = \frac{20}{100} \times \frac{300}{200} \times \left(\frac{13.2}{13.8}\right)^2 = j0.2745 \text{ pu}$$

$$\text{Per unit reactance of motor } M_2 = \frac{20}{100} \times \frac{300}{100} \times \left(\frac{13.2}{13.8}\right)^2 = j0.549 \text{ pu}$$

Reactance diagram is shown in Fig. 2.25. E_g is the emf of generator and E_1 and E_2 are the emfs of motors.

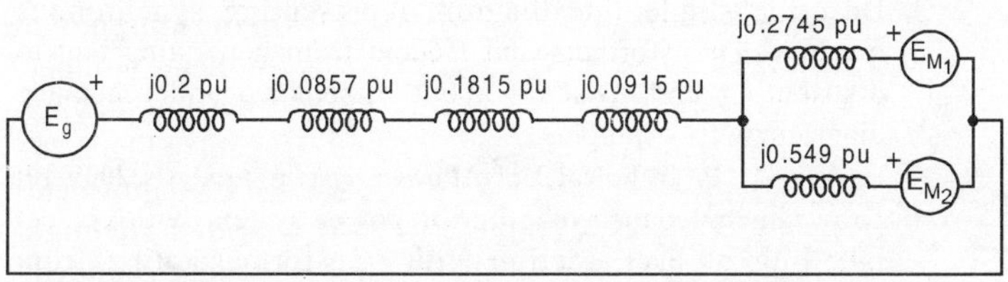

Fig. 2.25. Reactance Diagram for Example 2.16

Example 2.17. A 90 MVA, 11 kV, 3-phase generator has a reactance of 25%. The generator supplies two motors through transformers and transmission line as shown in Fig. 2.26. The transformer T_1 is a 3-phase transformer, 100 MVA, 10/132 kV, 6% reactance. The transformer T_2 is composed of 3 single-phase units each rated at 30 MVA, 66/10 kV with 5% reactance. The connections of T_1 and T_2 are as shown. The motors are rated 50 MVA and 40 MVA both 10 kV and 20% reactance. Taking the generator rating as base, draw reactance diagram and indicate the reactances in pu. The reactance of the line is 100 Ω.

[*Power System Analysis; Rajasthan Univ. 2000, 2002; Rajasthan Technical Univ. 2010*]

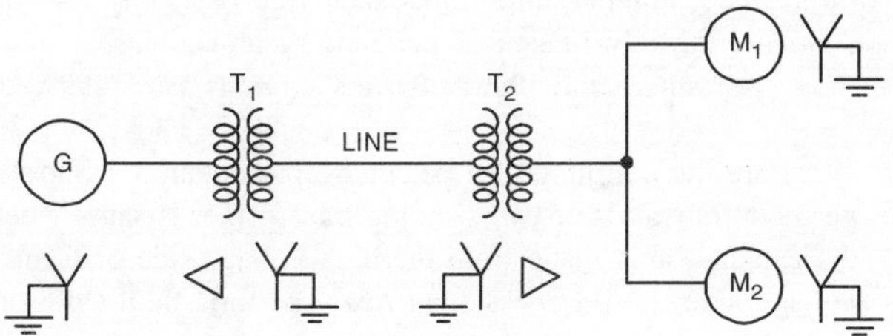

Fig. 2.26

Solution : Taking generator rating as base values,

Base MVA for the whole circuit, MVA_B = 90 MVA

Base kV for the generator circuit = 11 kV

PU reactance of the generator = $j0.25$ pu

$$\text{Base voltage for line} = 11 \times \frac{132}{10} = 145.2 \text{ kV}$$

$$\text{Line-to-line voltage ratio of transformer } T_2 = \sqrt{3} \times 66 : 10 = 114.32 : 10$$

$$\text{Base voltage for motors} = 10 \times \frac{145.2}{114.32} = 12.7 \text{ kV}$$

$$\text{PU reactance of the line} = j100 \times \frac{90}{(145.2)^2} = 0.427 \text{ pu}$$

$$\text{PU reactance of transformer } T_1 = 0.06 \times \frac{90}{100} \times \left(\frac{10}{11}\right)^2 = 0.0446 \text{ pu.}$$

The reactance of transformer T_2 is 5% (*i.e.*, 0.05 pu) on $\sqrt{3} \times 66$: 10 kV

i.e., 114.31 : 10 kV and 3 × 30 × 90 MVA base. So

PU reactance of transformer T_2

$$= 0.05 \times \frac{90}{90} \times \left(\frac{10}{12.7}\right)^2 = 0.031 \text{ pu}$$

$$\text{PU reactance of motor } M_1 = 0.2 \times \frac{90}{50} \times \left(\frac{10}{12.7}\right)^2 = 0.223 \text{ pu}$$

$$\text{PU reactance of motor } M_2 = 0.2 \times \frac{90}{40} \times \left(\frac{10}{12.7}\right)^2 = 0.279 \text{ pu}$$

PU reactance diagram for the system is shown in Fig. 2.27.

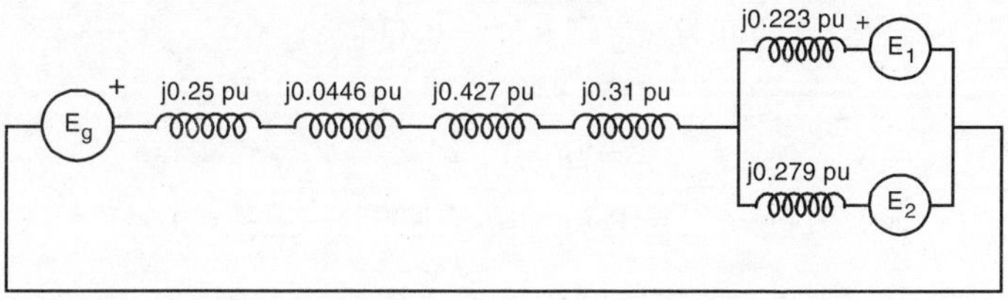

Fig. 2.27. ***PU Reactance Diagram For Example 2.17***

Example 2.18. Obtain the per unit reactance diagram of the power system shown below (Fig. 2.28).

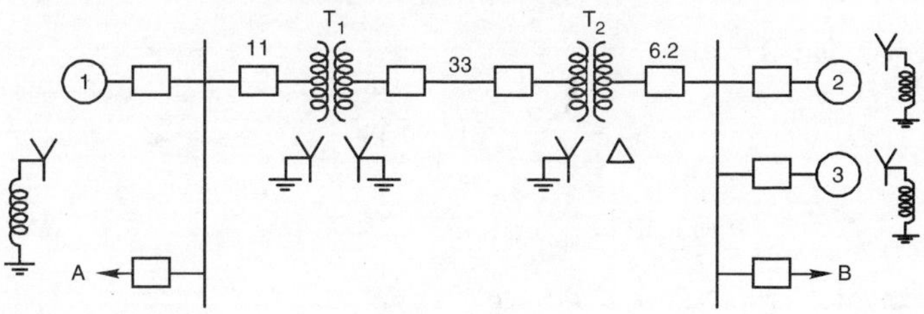

Fig. 2.28

Specifications of the system components are :

Generator 1 : 30 MVA, 10.5 kV, X″ = 1.6 Ω

Generator 2 : 15 MVA, 6.6 kV, X″ = 1.2 Ω

Generator 3 : 25 MVA, 6.6 kV, X″ = 0.56 Ω

Transformer T_1 : 15 MVA, 33/11 kV; X = 15.2 Ω/phase on ht side

Transformer T_2 : 15 MVA, 33/6.2 kV, X = 16 Ω/phase on ht side

Transmission line : 20.5 Ω/phase

Load A : 40 MW, 11 kV, 0.9 pf lag

Load B : 40 MW, 6.6 kV, 0.5 pf lag.

[*U.P.S.C. I.E.S. Electrical Engineering-II, 2004*]

Solution : Let the base MVA be 30 MVA

and base MVA for the line be 33 kV

So base kV in the circuit of generator 1 is 11 kV and base kV in the circuit of generator 2 is 6.2 kV.

PU ıeactance for the transmission line

$$= 20.5 \times \frac{30}{33^2} = 0.565 \text{ pu} \qquad \text{Refer to Eq. (2.24)}$$

PU reactance for the transformer T_1

$$= 15.2 \times \frac{30}{33^2} = 0.419 \text{ pu}$$

PU reactance for the transformer $T_2 = 16 \times \frac{30}{33^2} = 0.44$ pu

PU reactance for the generator $G_1 = 1.6 \times \frac{30}{11^2} = 0.397$ pu

PU reactance for the generator $G_2 = 1.2 \times \frac{30}{6.2^2} = 0.937$ pu

PU reactance for the generator $G_3 = 0.56 \times \frac{30}{6.2^2} = 0.437$ pu

For Load A

$$\text{Voltage} = \frac{11}{11} = 1 \text{ pu}$$

$$\text{PF, } \cos\phi = 0.9 \text{ lag}$$

$$\text{Sin } \phi = \sqrt{1-0.9^2} = 0.436$$

$$\text{Load, P} = \frac{40}{30} = 1.333$$

$$\text{Reactive power, Q} = \text{P} \times \tan\phi = 1.333 \times \frac{0.436}{0.9} = 0.6456 \text{ pu}$$

$$\text{Per unit resistance, } R_{pu} = \frac{V_{pu}^2}{P_{pu}} = \frac{1^2}{1.333} = 0.75 \text{ pu}$$

$$\text{Per unit reactance, } X_{pu} = \frac{V_{pu}^2}{Q_{pu}} = \frac{1^2}{0.6456} = 1.549 \text{ pu}$$

For Load B

$$\text{Voltage} = \frac{6.6}{6.2} = 1.065 \text{ pu}$$

$$\text{Power, P} = \frac{40}{30} = 1.333$$

$$\text{Power factor, } \cos\phi = 0.5 \text{ lag}$$

$$\sin\phi = \sqrt{1-0.5^2} = 0.866$$

$$\text{Reactive power, Q} = \text{P} \times \tan\phi = 1.333 \times \frac{0.866}{0.5} = 2.31 \text{ pu}$$

$$\text{Per unit resistance, } R_{pu} = \frac{V_{pu}^2}{P_{pu}} = \frac{1.065^2}{1.333} = 0.85 \text{ pu}$$

$$\text{Per unit reactance, } X_{pu} = \frac{V_{pu}^2}{Q_{pu}} = \frac{1.065^2}{2.31} = 0.49 \text{ pu}$$

Per unit reactance diagram of the power system given in Fig. 2.28 is given in Fig. 2.29.

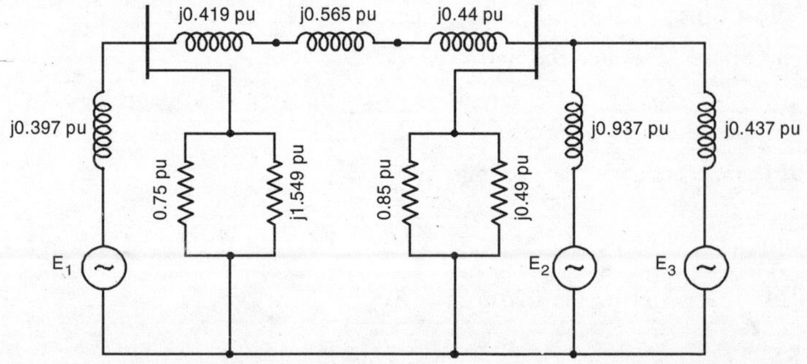

Fig. 2.29. ***Per Unit Reactance Diagram For Example 2.18***

Exercises

1. Draw steady state model for the following power system components: (*i*) Synchronous machine, (*ii*) Transformer, (*iii*) Transmission line, (*iv*) Line reactor.
[U.P. Technical Univ. Power System Analysis 2004-05]
2. What do you understand by one line diagram ? What is the difference between one line diagram and impedance diagram ? Explain with the help of examples.
[U.P. Technical Univ. Power System Analysis 2008-09]
3. Describe single line diagram representing synchronous machines, transformers and feeders from generating end to distributing end. Discuss about impedance and reactance diagram with examples.
[U.P. Technical Univ. Power System Analysis 2009-10]
4. Draw an impedance diagram of power system with correct labelling of each section with transformer rating from generation to load point.
[G.B. Technical Univ. Power System Analysis 2010-11]
5. What do you understand by per unit representation of a power system ? Discuss.
[U.P. Technical Univ. Power System Analysis 2005-06]
6. What is meant by percentage reactance ? Why is it preferable to express the reactances of various elements in percentage values for short-circuit calculations ?
7. Explain the terms base kVA, base kV and percentage reactance. State the relationship between them.
8. Define (*i*) Percentage reactance, (*ii*) Percentage reactance at base kVA and (*iii*) Short-circuit kVA.
[W.B. Univ. of Technology Power System-II, 2010]
9. Define per unit impedance and then give expression for base impedance and per unit impedance referred to a new base. Also list the advantages of per unit system.
[Rajasthan Univ. Power System Analysis 1995, 1999, 2001, 2002, 2003, 2005, 2007]
10. What are the advantages of per unit representation ? Show that per unit equivalent of two winding transformer is same whether the calculation is made from the high voltage side or from low voltage side. **[Power System Analysis; Rajasthan Univ. 2008; Rajasthan Technical Univ. 2010]**
11. What are the advantages of per unit representation.
[Anna Univ. Chennai Power System Analysis May/June 2011]
12. Why the per unit system is used in power system analysis ? Which of the electrical quantities are chosen as base values ?
[Pb. Univ. Power System Analysis December 1994]
13. Explain the advantages of per unit system and derive the fundamental relations between the actual values and per unit values on chosen base values for a three-phase system.
[Pb. Univ. Power System Analysis May 1995]
14. Discuss per unit system. How are the base values chosen in representation of a power system ?
[U.P. Technical Univ. Power System Analysis 2009-10]
15. Discuss per unit system. Explain the per unit representation of a transformer.
[U.P. Technical Univ. Power System Analysis 2008-09]
16. Discuss the representation of a power system network by reactance diagram.
Show that the per unit impedance of a transformer computed from primary or secondary side is same if the voltage bases on two sides are in the ratio of transformation.
[U.P. Technical Univ. Power System Analysis 2007-08]

17. Write short notes on:

(*i*) Single line diagram.

(*ii*) Change of base.

(*iii*) Reactance of synchronous machines.

[Anna Univ. Chennai Power System Analysis May/June 2011]

Short Answer Type Questions With Answers

Q. 1. What is meant by percentage reactance ?

Ans. Percentage reactance can be defined as the voltage drop across the given reactance expressed as percentage of normal voltage when carrying full-load current pertaining to normal rating.

Q. 2. Which of the electrical quantities are chosen as base values ?

Ans. For common representation, base kVA (or MVA) and base voltage (kV) are chosen as base values.

Q. 3. What is meant by per unit value of any quantity ?

Ans. Per unit value of any quantity is defined as the ratio of the actual value of the quantity in any unit to the base or reference value in the same unit.

Q. 4. What is the relationship between base kVA, base kV and percentage reactance ?

Ans. Percentage reactance

$$= \text{Actual reactance} \times \frac{\text{base kVA}}{(\text{base kV})^2 \times 1{,}000} \times 100$$

Q. 5. Write expression for per unit impedance (Z_{pu}) in a 3-phase system. **[M.D. Univ. 2010-11]**

Ans.
$$Z_{pu} = \frac{Z \text{ in ohms} \times MVA_B}{(kV_B)^2}$$

Q. 6. Why the per unit method is considered superior to per cent method for short-circuit calculations ?

Ans. Per cent value is not convenient for use as the factor of 100 has to be carried in computations.

Q. 7. Define base impedance of a power system element and write an expression for it in terms of base MVA and kV.

[Bharathidasan Univ. November 1996]

Ans. Base impedance is defined as the ratio of base voltage in volts and base current in amperes and it is expressed in terms of base MVA and kV as follows :

$$\text{Base impedance in ohms} = \frac{(\text{Base kV})^2}{\text{Base MVA}}$$

Q. 8. What is meant by base quantities ? **[Bharathiyar Univ. April 1997]**

Ans. For common representation, the quantities (voltage, current, power or impedance) choosen as common or reference quantities are called the base quantities.

Q. 9. Mention two advantages of pu system.

[Bharathiyar Univ. Nov. 1996, April 1997]

Ans. Refer to Art. 2.6.1.

Q. 10. Define per unit power. **[Bharathiyar Univ. April 1997]**

Ans. Per unit power is defined as the ratio of actual value of power in any unit (in kVA, MVA, kW or MW) to the base or reference value of power in the same unit.

Q. 11. Define pu impedance. **[Bharathiyar Univ. Nov. 1996]**

Ans. Per unit impedance is defined as the ratio of voltage drop across the given impedance to the normal voltage when carrying full-load current pertaining to normal rating.

Q. 12. What is relation between percentage impedance and pu impedance ? **[Bharathiyar Univ. November 1996]**

Ans. Percentage impedance = PU impedance × 100

Q. 13. List advantages of using per unit values in power system calculation. **[M.D. Univ. 2010-11; U.P.S.C. I.E.S. 2001]**

Ans. Refer to Art. 2.6.1.

Q. 14. Draw the equivalent circuit of a 3-winding transformer.

[Bharathidasan Univ. November 1996; Madras Univ. April 1997]

Ans. Refer to Fig. 2.19.

Problems

1. A single-phase transformer is rated 100 kVA, 11/0.4 kV. Its leakage reactance is 0.15 Ω when referred to low-voltage side. Determine its leakage reactance in per cent and per unit. **[Ans.** 9.375 % ; 0.09375 pu**]**

2. Convert an impedance of (4 + *j* 24) into per unit impedance with base voltage of 11 kV and base kVA of 500.

[Ans. (0.0165 + *j* 0.0992) pu**]**

3. The per unit impedance of 100 MVA, 11/132 kV step-up transformer is 0.3 Ω on its own base. Determine (*i*) actual impedances of the transformer referred to hv as well as lv side and (*ii*) base impedance referred to both sides.

[Ans. (*i*) 52.272 Ω ; 0.363 Ω (*ii*) 174.24 Ω ; 1.21 Ω**]**

4. The primary and secondary windings of a single-phase 1,000 kVA, 2,000/1,000 V transformer have leakage reactances each of 2 Ω. Find the pu reactance of the transformer.

[Rajasthan Univ. Power System Analysis 2004] **[Ans.** 2.5 pu**]**

5. Three parts of a single-phase electric system are designated A, B and C and are connected to each other through transformers, as shown in Fig. 1. The transformers are rated as follows :

A-B 10,000 kVA, 138/13.8 kV, leakage reactance 10%

B-C 10,000 kVA, 138/69 kV, leakage reactance 8%

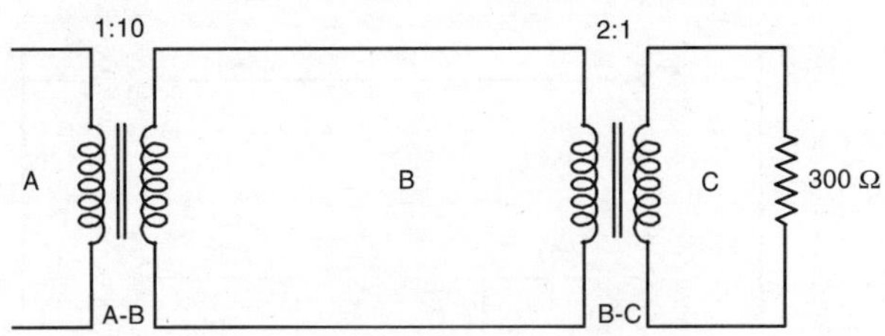

Fig. 1

If the base in the circuit B is chosen as 10,000 kVA, 138 kV, find the per unit impedance of the 300 Ω resistive load in circuit C referred to circuits C, B and A. Draw the impedance diagram neglecting magnetizing current, transformer resistances and line impedances.

[Rajasthan Univ. Power System Analysis 2000]

[Ans. (0.63 + *j*0) pu; Fig. 2**]**

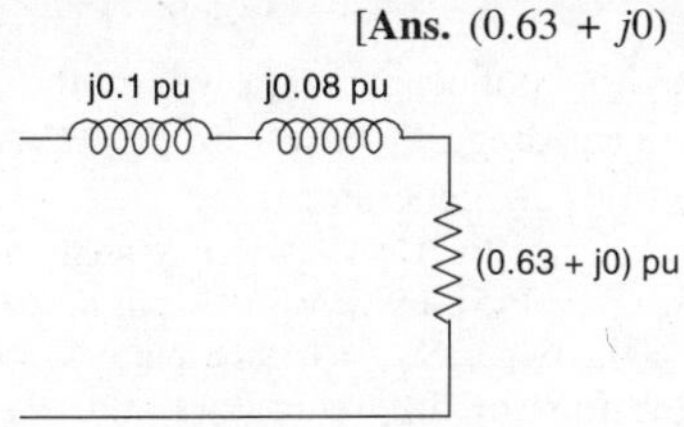

Fig. 2

6. Two generators rated 10 MVA, 13.2 kV and 15 MVA, are connected to common bus-bars. Two synchronous motors are connected to these bus-bars. These two motors have 8 and 12 MVA capacity respectively. Both motors are rated 12.5 kV. Choosing 50 MVA and 13.8 kV as base quantities, compute reactances of all the machines referred to these base quantities. The generators have 15 % reactance each and motors have 20 % reactance each on the basis of the respective ratings of the machines.

[Bharathiyar Univ. Power System Analysis April 1997]

[Ans. 68.62 % ; 45.75 % ; 102.56 % ; 68.37 %**]**

7. Redraw the power system shown in Fig. 3 showing the pu values of each component on a common base of 100 MVA, 11 kV at generator bus. All impedances are given in their own base.

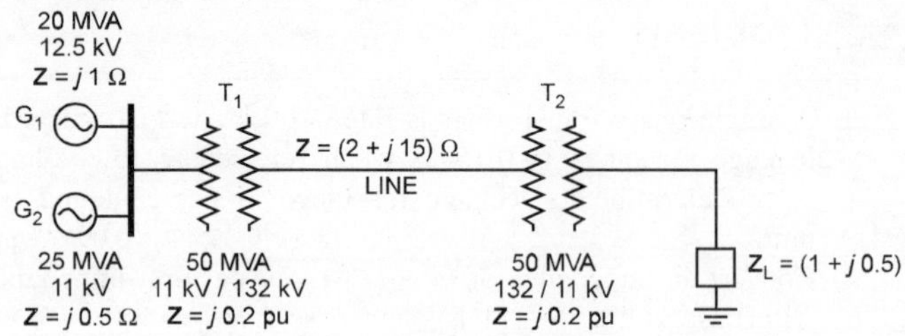

Fig. 3

[Pb. Univ. Power System Analysis April 1993]

[**Ans.** $X_{G1} = j\,0.826$ pu ; $X_{G2} = j\,0.413$ pu ; $X_{T1} = j\,0.4$ pu ; $X_{T2} = j\,0.4$ pu ; $\mathbf{Z} = (j\,0.0115 + j\,0.086)$ pu; $\mathbf{Z_L} = (0.826 + j\,0.413)$ pu]

8. A 3-bus system is given in the figure (Fig. 4). The ratings of the various components are listed below :

Generator 1 : 50 MVA, 13.8 kV, X″ = 0.15 pu
Generator 2 : 40 MVA, 13.2 kV, X″ = 0.20 pu
Generator 3 : 30 MVA, 11 kV, X″ = 0.25 pu
Transformer 1 : 45 MVA, 11 kV Y/110 kV Y, X = 0.1 pu
Transformer 2 : 25 MVA, 12.5 kV Y/115 kV Y, X = 0.15 pu
Transformer 3 : 40 MVA, 12.5 kV Δ/115 kV Y, X = 0.1 pu

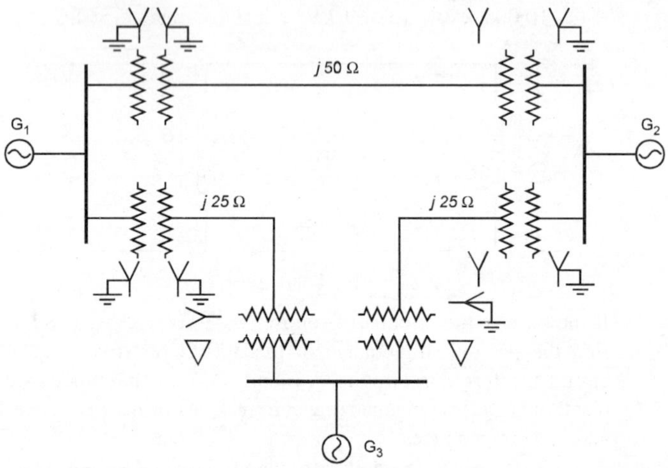

Fig. 4

Line impedances are shown in the figure. Determine the reactance diagram based on 50 MVA and 13.8 kV as base quantities in generator 1.

[Pb. Univ. Power System Analysis December 1993]

[**Ans.** $X_{G1} = 0.15$ pu; $X_{G2} = 0.1936$ pu; $X_{G3} = 0.224$ pu; $X_{T1} = 0.0706$ pu; $X_{T2} = 0.208$ pu; $X_{T3} = 0.0868$ pu; $X_L = j\,0.1312$ pu]

9. The ratings of the generators and transformers of the power system shown in Fig. 5 are given below ;

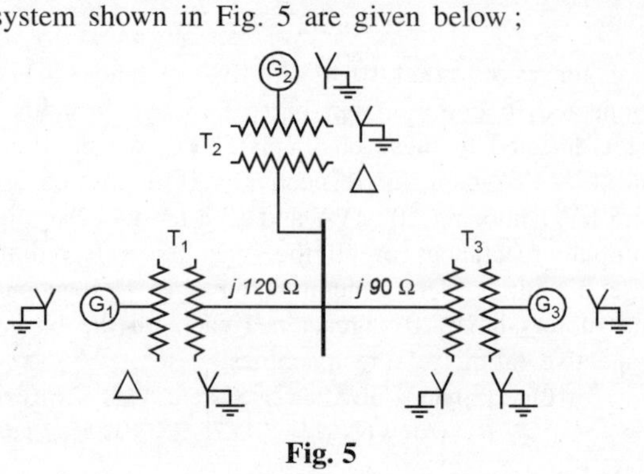

Fig. 5

Generator G_1 : 25 MVA, 6.6 kV, $j\,0.20$ pu
Generator G_2 : 15 MVA, 6.6 kV, $j\,0.15$ pu
Generator G_3 : 30 MVA, 13.2 kV, $j\,0.15$ pu
Transformer T_1 : 30 MVA, 6.6 Δ – 115 Y kV, $j\,0.1$ pu
Transformer T_2 : 15 MVA, 6.6 Δ – 115 Y kV, $j\,0.1$ pu
Transformer T_3 : Single phase units each rated 10 MVA, 6.9/69 kV, $j\,0.1$ pu

Draw an impedance diagram and all values in pu choosing a base of 30 MVA, 6.6 kV in the circuit of generator 1.

[Pb. Univ. Power System Analysis December 1995]

[**Ans.** $X_{G1} = j\,0.24$ pu; $X_{T1} = j\,0.1$ pu; $X_{L1} = j\,0.2722$ pu; $X_{T2} = j\,0.2$ pu; $X_{G2} = j\,0.3$ pu; $X_{L2} = j\,0.204$ pu; $X_{T3} = j\,0.108$ pu; $X_{G3} = j\,0.1976$ pu]

10. A 3-phase 3-winding Y-Y-Δ transformer is rated as under :

Primary 132 kV, 30 MVA
Secondary 11 kV, 20 MVA
Tertiary 6.6 kV, 10 MVA

The results of the short-circuit tests are as under :

Secondary short circuited, tertiary open circuited, primary excited: X = 7% on 132 kV, 30 MVA base

Secondary open circuited, tertiary short circuited, primary excited : X = 9% on 132 kV, 30 MVA base

Tertiary short circuited, primary open circuited, secondary excited : X = 4% on 11 kV, 20 MVA base

Find X_1, X_2, X_3 in pu. Neglect resistances.

[Rajasthan Univ. Power System Analysis 2000]

[**Ans.** 0.05 pu, 0.02 pu and 0.04 pu]

11. A 100 MVA, 33 kV, 3-phase generator has a subtransient reactance of 15%. The generator connected to the motors through transmission line and transformers are shown in Fig. 6. The motors have rated inputs of 30 MVA, 20 MVA and 50 MVA at 30 kV with 18% subtransient reactance. The three-phase transformers are rated at 100 MVA, 33 kVΔ/110 kV⅄ with leakage reactance of 9%. The line has a reactance of 50 Ω. Selecting the generator rating as base quantitites, obtain pu reactance diagram of the system.

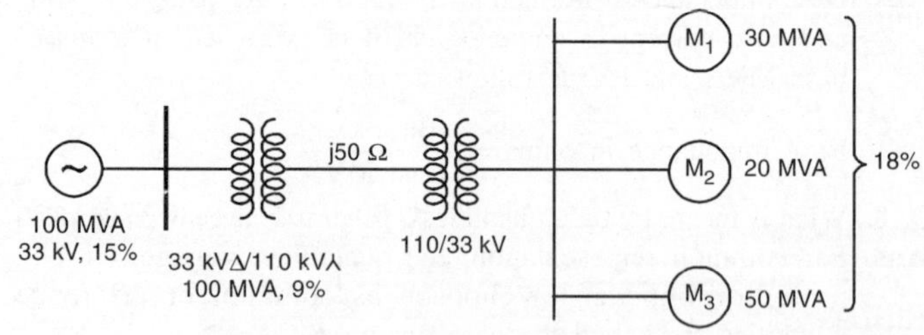

Fig. 6

[U.P.S.C. I.E.S. Electrical Engineering II, 2001]

[**Ans.** Fig. 7]

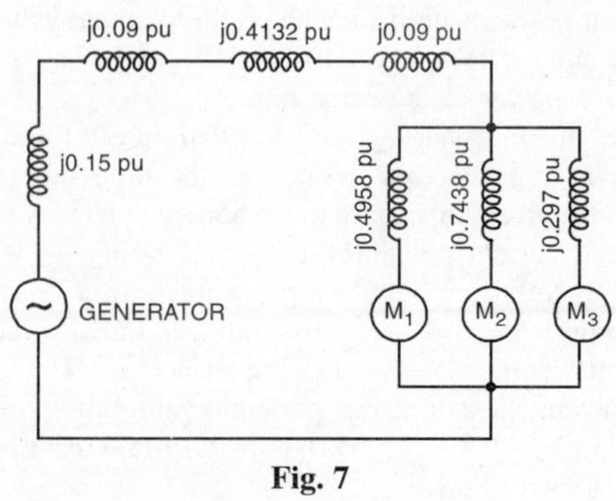

Fig. 7

❑❑❑

CHAPTER 3

Symmetrical Components

INSIDE THIS CHAPTER

3.1. INTRODUCTION

The method of symmetrical components is very powerful approach and has simplified the procedure for solving problems on unbalanced polyphase systems. The method of symmetrical components was proposed by CL Fortesque in the year 1918. Although this method is valid for any number of phases, the three-phase system is of main interest.

According to Fortesque theorem, any unbalanced three-phase system of currents, voltages or other sinusoidal quantities can be resolved into three balanced systems of phasors which are called *symmetrical components* of the original unbalanced system. Such three balanced systems constitute three sequence networks which are solved separately on a single phase basis. Once the problem is solved in terms of symmetrical components, it can be transferred back to the actual circuit condition by superposition or phasor additions of these quantities (currents or voltages) easily.

3.1.1. Symmetrical Components of 3-Phase Systems

The symmetrical components differ in phase sequence, that is, the order in which the phase quantities go through a maximum. There may be a positive phase sequence, a negative phase sequence and a zero or uniphase sequence. Thus the balanced sets of components can be given as positive sequence component, negative sequence component and zero or uniphase sequence component.

The *positive phase sequence system* is that system in which the phase or line currents or voltages attain a maximum in the same cyclic order as those in a normal supply, *e.g.*, assuming the conventional counter-clockwise rotation, then the positive phase sequence phasors are those as shown in Fig. 3.1 (*a*). A balanced system corresponding to normal conditions contains a positive phase sequence only. It is also the condition for a 3-phase fault.

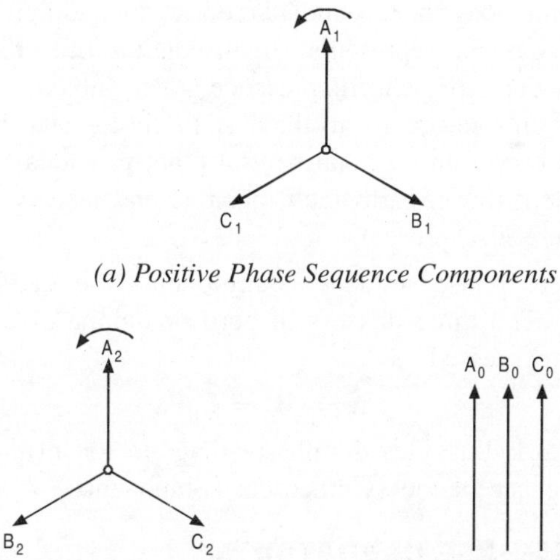

(*a*) *Positive Phase Sequence Components*

(*b*) *Negative Phase Sequence Components*

(*c*) *Zero Phase Sequence Components*

Fig. 3.1. ***Symmetrical Components of an Unbalanced Three Phase System***

The positive phase sequence components are marked by subscript 1. The three phasors of the positive phase sequence system are of equal magnitude, spaced 120° apart and may be represented in complex form as follows:

$$\mathbf{A_1};\ \mathbf{B_1} = \mathrm{A}_1 \angle -120^\circ \text{ or } \mathbf{A_1} e^{-j2\pi/3} \text{ or } a^2 \mathbf{A_1}$$

$$\text{and } \mathbf{C_1} = \mathrm{A}_1 \angle 120^\circ \text{ or } \mathbf{A_1} e^{j2\pi/3} \text{ or } a\,\mathbf{A_1}$$

The complex number $e^{j2\pi/3}$ in above expression is called the *operator* of the three-phase system and is denoted by letter '*a*'. Multiplication of a phasor by a or $e^{j2\pi/3}$ turns it through 120° in counter-clockwise direction. Multiplying a phasor by a^2 or $e^{-j2\pi/3}$ turns it through 240° in the counter-clockwise direction and the position of the phasor remains unchanged if multiplied by a^3.

The *negative phase sequence system* is that system in which the phasors still rotate counter-clockwise but attains a maximum in reverse order *i.e.,* A – C – B as shown in Fig. 3.1 (*b*). This sequence only arises under conditions of unbalance as when an unsymmetrical fault occurs. Such faults contain also the positive sequence system as mentioned in the preceding para and, in the case of faults to earth, a zero phase sequence, as explained in the following para.

The negative phase sequence components are marked by subscript 2. The three phasors of the negative phase sequence, alike positive phase sequence are of equal magnitude and spaced 120° apart. They are represented in complex form as follows:

$$\mathbf{A}_2;\ \mathbf{B}_2 = \mathbf{A}_2 \angle 120^\circ \text{ or } \mathbf{A}_2\, e^{j2\pi/3} \text{ or } a\,\mathbf{A}_2$$

and $$\mathbf{C}_2 = \mathbf{A}_2 \angle -120^\circ \text{ or } \mathbf{A}_2\, e^{-j2\pi/3} \text{ or } a^2\,\mathbf{A}_2$$

The *zero phase sequence system* is a single phase phasor system combining three equal phasors in phase, as illustrated in Fig. 3.1 (*c*), and represents the residual current or voltage present under fault conditions on a 3-phase system with a fourth wire or earth return present. Clearly the zero phase sequence embraces the ground, therefore, in addition to the three line wires and represents a fault condition to ground or to a fourth wire if present. Its presence arises only where fault to earth currents can return to the system via the star point of that system or via an artificial neutral point provided to earth a delta system. In an earth fault, positive and negative phase sequences are also present.

The zero phase sequence components are marked by subscript 0. The three phasors of zero or uniphase sequence system are expressed as

$$\mathbf{A}_0 = \mathbf{B}_0 = \mathbf{C}_0$$

Figure 3.1 illustrates the phasor diagram for all the three phase sequences of the symmetrical components.

3.2. THE PHASE OPERATOR "a"

For convenience in notation and manipulation a phasor operator is introduced. Through usage it has come to be known as the phasor a and is defined as $a = -0.5 + j\,0.866 = e^{j2\pi/3}$. It indicates that the phasor a has unit length and is oriented $120^\circ \left(\text{or } \dfrac{2\pi}{3} \text{ radians}\right)$ in a positive (counter-clockwise) direction from the reference axis. A phasor operated upon by a is not changed in magnitude but is simply rotated in position 120° in the forward direction. For example $\mathbf{V}' = a\,\mathbf{V}$ is a phasor having the same length as the phasor **V**, but rotated 120° forward from the phasor **V**. This relationship is illustrated in Fig. 3.2. The square of the phasor a is another unit phasor oriented 120° in a negative (clockwise) direction from the reference axis, or oriented 240° forward in positive direction.

$$a^2 = \left(e^{\frac{j2\pi}{3}}\right)^2 = e^{\frac{j4\pi}{3}} = e^{\frac{-j2\pi}{3}} = -0.5 - j\,0.866$$

As illustrated in Fig. 3.2, the resultant a^2 operating on a phasor **V** is the phasor **V″** having the same length as **V**, but located 120° in a clockwise direction from **V**.

The three phasors $(1 + j\,0)$, a^2 and a (taken in this order) form a balanced, symmetrical set of phasors of positive phase sequence rotation, since the phasors are of equal length, displaced by equal angles of 120° from each other, and cross the reference line in the order 1, a^2 and a (following the usual convention of counter-clockwise rotation for the phasor diagram). The phasor 1, a and a^2 (taken in this order) form a balanced, symmetrical, set of phasors of negative phase sequence, since the phasors do not cross the reference line in the order named, keeping the same convention of counter-clockwise rotation, but third name following the first etc.

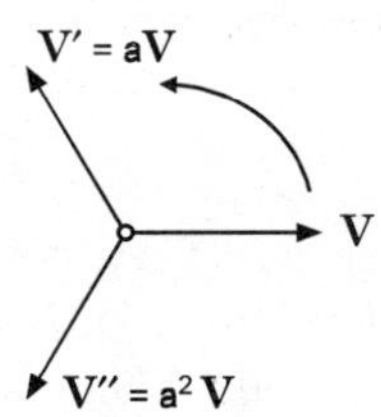

Fig. 3.2

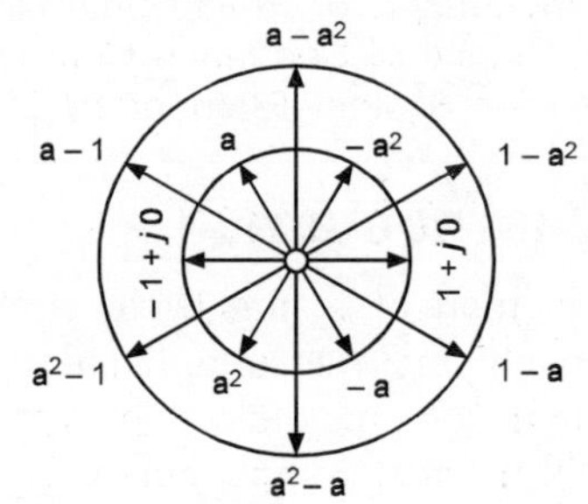

Fig. 3.3

Fundamental properties of the phasors are given below and are shown on phasor diagram (Fig. 3.3)

$$a = -0.5 + j\,0.866 = e^{j2\pi/3}$$
$$a^2 = -0.5 - j\,0.866 = e^{-j2\pi/3}$$
$$a^3 = 1 + j\,0 = e^{j\,0}$$
$$a^4 = e^{4(j2\pi/3)} = a$$
$$a^5 = e^{5(j2\pi/3)} = a^2 \text{ and so on}$$

Thus a is numerically equivalent to the cube root of unity.

$$a + a^2 + 1 = (-0.5 + j\,0.866) + (-0.5 - j\,0.866) + 1 = 0$$
$$a + a^2 = -1 + j\,0 = e^{j\pi}$$
$$a^2 - a = (-0.5 - j\,0.866) - (-0.5 + j\,0.866)$$
$$= 0 - j\,1.732 = \sqrt{3}\, e^{j\,3\pi/2}$$
$$a - a^2 = 0 + j\,1.732 = \sqrt{3}\, e^{j\,\pi/2}$$
$$a - 1 = \sqrt{3}\, e^{j\,5\pi/6}$$
$$1 - a = \sqrt{3}\, e^{j\,11\pi/6}$$
$$a^2 - 1 = \sqrt{3}\, e^{j\,7\pi/6}$$
$$1 - a^2 = \sqrt{3}\, e^{j\,\pi/6}$$

3.3. EVALUATION OF THE COMPONENTS

Let us express the phasors of an unbalanced three-phase system in terms of their symmetrical components

$$\mathbf{A} = \mathbf{A}_1 + \mathbf{A}_2 + \mathbf{A}_0 \quad ...(i)$$
$$\mathbf{B} = \mathbf{B}_1 + \mathbf{B}_2 + \mathbf{B}_0 \quad ...(ii)$$
$$\mathbf{C} = \mathbf{C}_1 + \mathbf{C}_2 + \mathbf{C}_0 \quad ...(iii)$$

Expressing all the phasors of the symmetrical system in the above equations in terms of $\mathbf{A}_1$, $\mathbf{A}_2$ and $\mathbf{A}_0$ we get

$$\mathbf{A} = \mathbf{A}_1 + \mathbf{A}_2 + \mathbf{A}_0 \quad ...(iv)$$
$$\mathbf{B} = a^2\mathbf{A}_1 + a\mathbf{A}_2 + \mathbf{A}_0 \quad ...(v)$$
$$\text{since } \mathbf{B}_1 = a^2\mathbf{A}_1 \text{ and } \mathbf{B}_2 = a\mathbf{A}_2$$
$$\text{and } \mathbf{C} = a\mathbf{A}_1 + a^2\mathbf{A}_2 + \mathbf{A}_0 \quad ...(vi)$$
$$\text{since } \mathbf{C}_1 = a\mathbf{A}_1 \text{ and } \mathbf{C}_2 = a^2\mathbf{A}_2$$

Adding Eqs. (*iv*), (*v*) and (*vi*), we get

$$\mathbf{A} + \mathbf{B} + \mathbf{C} = \mathbf{A}_1(1 + a^2 + a) + \mathbf{A}_2(1 + a + a^2) + 3\mathbf{A}_0$$

$$\text{or } \mathbf{A}_0 = \frac{\mathbf{A}+\mathbf{B}+\mathbf{C}}{3} \quad ...(vii) \text{ since } 1 + a + a^2 = 0$$

Multiplying Eq. (*v*) by a and Eq. (*vi*) by a^2, we get

$$a\mathbf{B} = a^3\mathbf{A}_1 + a^2\mathbf{A}_2 + a\mathbf{A}_0 \quad ...(viii)$$
$$a^2\mathbf{C} = a^3\mathbf{A}_1 + a^4\mathbf{A}_2 + a^2\mathbf{A}_0 \quad ...(ix)$$

Adding Eqs. (*iv*), (*viii*) and (*ix*), we get

$$\mathbf{A} + a\mathbf{B} + a^2\mathbf{C} = \mathbf{A}_1 (1 + a^3 + a^3) + \mathbf{A}_2 (1 + a^2 + a^4) + \mathbf{A}_0 (1 + a + a^2)$$
$$= 3\mathbf{A}_1 + \mathbf{A}_2 (1 + a^2 + a) + \mathbf{A}_0 (1 + a + a^2)$$
$$\text{since } a^4 = a \text{ and } a^3 = 1$$

$$\text{or } \mathbf{A}_1 = \frac{\mathbf{A}+a\mathbf{B}+a^2\mathbf{C}}{3} \quad ...(x) \text{ since } 1 + a + a^2 = 0$$

Multiplying Eq. (*v*) by a^2 and Eq. (*vi*) by a, we get

$$a^2\mathrm{B} = a^4\mathbf{A}_1 + a^3\mathbf{A}_2 + a^2\mathbf{A}_0 \quad ...(xi)$$
$$a\mathbf{C} = a^2\mathbf{A}_1 + a^3\mathbf{A}_2 + a\mathbf{A}_0 \quad ...(xii)$$

Adding Eqs. (*iv*), (*xi*) and (*xii*), we get

$$\mathbf{A} + a^2\mathbf{B} + a\mathbf{C} = \mathbf{A}_1(1 + a^4 + a^2) + \mathbf{A}_2(1 + a^3 + a^3) + \mathbf{A}_0(1 + a^2 + a)$$
$$= \mathbf{A}_1(1 + a + a^2) + 3\mathbf{A}_2 + \mathbf{A}_0(1 + a + a^2)$$
$$\text{since } a^4 = a \text{ and } a^3 = 1$$

$$\text{or } \mathbf{A}_2 = \frac{\mathbf{A}+a^2\mathbf{B}+a\mathbf{C}}{3} \quad ...(xiii) \text{ since } 1 + a + a^2 = 0$$

In matrix form we can express as

$$\begin{bmatrix} \mathbf{A} \\ \mathbf{B} \\ \mathbf{C} \end{bmatrix} = \begin{bmatrix} 1 & 1 & 1 \\ 1 & a^2 & a \\ 1 & a & a^2 \end{bmatrix} \begin{bmatrix} \mathbf{A}_0 \\ \mathbf{A}_1 \\ \mathbf{A}_2 \end{bmatrix} \quad ...(3.1)$$

$$\text{and } \begin{bmatrix} \mathbf{A}_0 \\ \mathbf{A}_1 \\ \mathbf{A}_2 \end{bmatrix} = \frac{1}{3}\begin{bmatrix} 1 & 1 & 1 \\ 1 & a & a^2 \\ 1 & a^2 & a \end{bmatrix} \begin{bmatrix} \mathbf{A} \\ \mathbf{B} \\ \mathbf{C} \end{bmatrix} \quad ...(3.2)$$

Resolution of unbalanced phasor system into a symmetrical components is essentially a mathematical technique used for solving three-phase circuit problems under unbalanced conditions which may be due to unbalanced loads, unequal phase voltages, single- and two-phase short circuits, open line wire etc.

3.4. SOME PROPERTIES OF 3-PHASE CIRCUITS WITH REGARD TO SYMMETRICAL COMPONENTS OF CURRENTS AND VOLTAGES

In a 3-phase, 4-wire system, current in the neutral wire is equal to the sum of the line currents and is consequently equal to three times the zero phase sequence component of currents.

The sum of the line currents in a 3-phase, 3-wire delta-or star-connected system is known to be zero. It follows from Eq. (*vii*) that line currents of such a system do not carry any zero phase component.

The above applies equally to any 3-phase system in which the sum of line voltages is zero. It means that a set of unbalanced 3-phase line-to-line voltages may be represented by a positive system and a negative system of balanced voltages.

The degree of unbalance is given by dissymmetry coefficient in terms of the ratio of the negative phase sequence component to the positive phase sequence component *i.e.*,

$$\sigma = \frac{V_{L_2}}{V_{L_1}} \times 100 \text{ per cent}$$

The system of line voltages is considered to be balanced when this coefficient is less than 5 per cent.

The absence of current in one or two phases under unbalanced condition means the sum of the three symmetrical components of the currents in these phases is equal to zero.

The above statement will be better understood from examination of the following examples. Let in a system the phases B and C be open (Fig. 3.4). Then

$$\mathbf{I}_B = \mathbf{I}_C = 0$$

In accordance with Eqs. (*x*), (*xiii*) and (*vii*) the symmetrical components of the system are as follows:

(*a*) The component of the positive phase sequence current,

$$\mathbf{I}_{A1} = \frac{\mathbf{I}_A + a \times 0 + a^2 \times 0}{3} = \frac{\mathbf{I}_A}{3}$$

(*b*) The component of the negative phase sequence current,

$$\mathbf{I}_{A2} = \frac{\mathbf{I}_A + a^2 \times 0 + a \times 0}{3} = \frac{\mathbf{I}_A}{3}$$

(*c*) The component of the zero phase sequence current,

$$\mathbf{I}_{A0} = \frac{\mathbf{I}_A + 0 + 0}{3} = \frac{\mathbf{I}_A}{3}$$

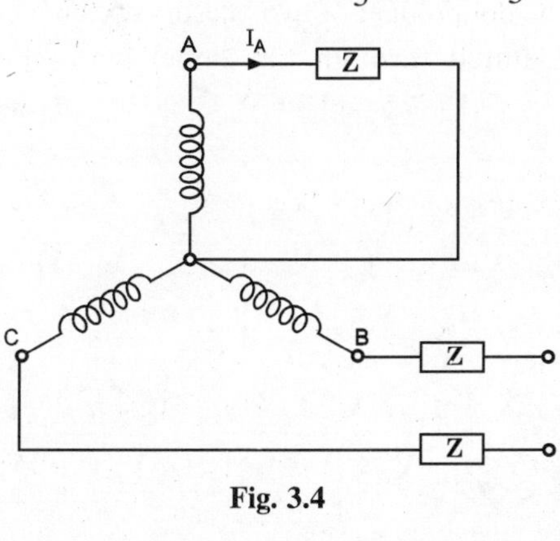

Fig. 3.4

The geometric addition of symmetrical components of positive, negative and zero phase gives

$$\mathbf{I}_{A1} + \mathbf{I}_{A2} + \mathbf{I}_{A0} = \frac{\mathbf{I}_A}{3} + \frac{\mathbf{I}_A}{3} + \frac{\mathbf{I}_A}{3} = \mathbf{I}_A$$

$$\mathbf{I}_{B1} + \mathbf{I}_{B2} + \mathbf{I}_{B0} = a^2\frac{\mathbf{I}_A}{3} + a\frac{\mathbf{I}_A}{3} + \frac{\mathbf{I}_A}{3}$$

$$= \frac{\mathbf{I}_A}{3}(a^2 + a + 1) = 0$$

and $$\mathbf{I}_{C1} + \mathbf{I}_{C2} + \mathbf{I}_{C0} = a\frac{\mathbf{I}_A}{3} + a^2\frac{\mathbf{I}_A}{3} + \frac{\mathbf{I}_A}{3}$$

$$= \frac{\mathbf{I}_A}{3}(a + a^2 + 1) = 0$$

3.5. THREE-PHASE POWER IN TERMS OF SYMMETRICAL COMPONENTS

The power consumed in a 3-phase circuit can be determined directly from the symmetrical components provided the currents and voltages are given. Total complex power in a 3-phase circuit is given as

$$\mathbf{S} = \mathrm{P} + j\mathrm{Q} = \mathbf{V}_p^{\mathrm{T}}\mathbf{I}_p^* = \mathbf{V}_a\mathbf{I}_a^* + \mathbf{V}_b\mathbf{I}_b^* + \mathbf{V}_c\mathbf{I}_c^* \quad ...(3.3)$$

where $\mathbf{V}_a$, $\mathbf{V}_b$ and $\mathbf{V}_c$ are the voltages to neutral and $\mathbf{I}_a$, $\mathbf{I}_b$ and $\mathbf{I}_c$ are the currents flowing into the circuit in the three lines, the 3-phase circuit being assumed to be star-connected. The superscript T stands for transposition.

$$\text{or } \mathbf{S} = (\mathbf{V}_{a1}+\mathbf{V}_{a2}+\mathbf{V}_{a0})(\mathbf{I}_{a1}+\mathbf{I}_{a2}+\mathbf{I}_{a0})$$

$$+(a^2\mathbf{V}_{a1}+a\mathbf{V}_{a2}+\mathbf{V}_{a0})(a^2\mathbf{I}_{a1}+a\mathbf{I}_{a2}+\mathbf{I}_{a0})$$

$$+(a\mathbf{V}_{a1}+a^2\mathbf{V}_{a2}+\mathbf{V}_{a0})(a\mathbf{I}_{a1}+a^2\mathbf{I}_{a2}+\mathbf{I}_{a0})$$

$$= \mathbf{V}_{a1}\cdot\mathbf{I}_{a1}+\mathbf{V}_{a1}\cdot\mathbf{I}_{a2}+\mathbf{V}_{a1}\cdot\mathbf{I}_{a0}+\mathbf{V}_{a2}\cdot\mathbf{I}_{a1}+\mathbf{V}_{a2}\cdot\mathbf{I}_{a2}+\mathbf{V}_{a2}\cdot\mathbf{I}_{a0}$$

$$+\mathbf{V}_{a0}\cdot\mathbf{I}_{a1}+\mathbf{V}_{a0}\cdot\mathbf{I}_{a2}+\mathbf{V}_{a0}\cdot\mathbf{I}_{a0}$$

$$+a^2\mathbf{V}_{a1}\cdot a^2\mathbf{I}_{a1}+a^2\mathbf{V}_{a1}\cdot a\mathbf{I}_{a2}+a^2\mathbf{V}_{a1}\cdot\mathbf{I}_{a0}$$

$$+a\mathbf{V}_{a2}\cdot a^2\mathbf{I}_{a1}+a\mathbf{V}_{a2}\cdot a\mathbf{I}_{a2}+a\mathbf{V}_{a2}\cdot\mathbf{I}_{a0}+\mathbf{V}_{a0}\cdot a^2\mathbf{I}_{a1}$$

$$+\mathbf{V}_{a0}\cdot a\mathbf{I}_{a2}+\mathbf{V}_{a0}\cdot\mathbf{I}_{a0}$$

$$+a\mathbf{V}_{a1}\cdot a\mathbf{I}_{a1}+a\mathbf{V}_{a1}\cdot a^2\mathbf{I}_{a2}+a\mathbf{V}_{a1}\cdot\mathbf{I}_{a0}+a^2\mathbf{V}_{a2}\cdot a\mathbf{I}_{a1}$$

$$+a^2\mathbf{V}_{a2}\cdot a^2\mathbf{I}_{a2}+a^2\mathbf{V}_{a2}\cdot\mathbf{I}_{a0}$$

$$+\mathbf{V}_{a0}\cdot a\mathbf{I}_{a1}+\mathbf{V}_{a0}\cdot a^2\mathbf{I}_{a2}+\mathbf{V}_{a0}\cdot\mathbf{I}_{a0}$$

Now the dot product of two phasors does not change when both are rotated through the same angle. For example $a^2\mathbf{V}_{a1}\cdot a^2\mathbf{I}_{a1} = \mathbf{V}_{a1}\mathbf{I}_{a1}$; $a\mathbf{V}_{a1}.a\mathbf{I}_{a1} = \mathbf{V}_{a1}\mathbf{I}_{a1}$; $a^2\mathbf{V}_{a1}\cdot a\mathbf{I}_{a2} = a\mathbf{V}_{a1}.\mathbf{I}_{a2}$ and so on.

Thus the expression for **S** becomes

$$\mathbf{S} = \mathbf{V}_{a1}\cdot\mathbf{I}_{a1}+\mathbf{V}_{a1}\cdot\mathbf{I}_{a2}+\mathbf{V}_{a1}\cdot\mathbf{I}_{a0}+\mathbf{V}_{a2}\cdot\mathbf{I}_{a1}+\mathbf{V}_{a2}\cdot\mathbf{I}_{a2}+\mathbf{V}_{a2}\cdot\mathbf{I}_{a0}$$

$$+\mathbf{V}_{a0}\cdot\mathbf{I}_{a1}+\mathbf{V}_{a0}\cdot\mathbf{I}_{a2}+\mathbf{V}_{a0}\cdot\mathbf{I}_{a0}+\mathbf{V}_{a1}\cdot\mathbf{I}_{a1}+a\mathbf{V}_{a1}\cdot\mathbf{I}_{a2}$$

$$+a^2\mathbf{V}_{a1}\cdot\mathbf{I}_{a0}+a^2\mathbf{V}_{a2}\cdot\mathbf{I}_{a1}+\mathbf{V}_{a2}\cdot\mathbf{I}_{a2}+a\mathbf{V}_{a2}\cdot\mathbf{I}_{a0}+a\mathbf{V}_{a0}\cdot\mathbf{I}_{a1}$$

$$+a^2\mathbf{V}_{a0}\cdot\mathbf{I}_{a2}+\mathbf{V}_{a0}\mathbf{I}_{a0}+\mathbf{V}_{a1}\cdot\mathbf{I}_{a1}+a^2\mathbf{V}_{a1}\mathbf{I}_{a2}+a\mathbf{V}_{a1}\mathbf{I}_{a0}$$

$$+a\mathbf{V}_{a2}\mathbf{I}_{a1}+\mathbf{V}_{a2}\mathbf{I}_{a2}+a^2\mathbf{V}_{a2}\mathbf{I}_{a0}+a^2\mathbf{V}_{a0}\mathbf{I}_{a1}+a\mathbf{V}_{a0}\mathbf{I}_{a2}+\mathbf{V}_{a0}\mathbf{I}_{a0}$$

$$= 3\mathbf{V}_{a1}\cdot\mathbf{I}_1+3\mathbf{V}_{a2}\cdot\mathbf{I}_{a2}+3\mathbf{V}_{a0}\cdot\mathbf{I}_{a0}+\mathbf{V}_{a1}\cdot\mathbf{I}_{a2}(1+a+a^2)$$

$$+\mathbf{V}_{a1}\cdot\mathbf{I}_{a0}(1+a+a^2)+\mathbf{V}_{a2}\cdot\mathbf{I}_{a1}(1+a+a^2)$$

$$+\mathbf{V}_{a2}\cdot\mathbf{I}_{a0}(1+a+a^2)+\mathbf{V}_{a0}\cdot\mathbf{I}_{a1}(1+a+a^2)$$

$$+\mathbf{V}_{a0}\cdot\mathbf{I}_{a2}(1+a+a^2)$$

$$= 3[\mathbf{V}_{a1}\cdot\mathbf{I}_{a1}+\mathbf{V}_{a2}\cdot\mathbf{I}_{a2}+\mathbf{V}_{a0}\cdot\mathbf{I}_{a0}]$$

$$\text{or P} = 3[\mathrm{V}_{a1}\mathrm{I}_{a1}\cos\theta_1 + \mathrm{V}_{a2}\mathrm{I}_{a2}\cos\theta_2 + \mathrm{V}_{a0}\mathrm{I}_{a0}\cos\theta_0] \quad ...[3.4(a)]$$

The same power expression can be derived very easily using matrix manipulations.

$$\mathbf{S} = \mathrm{P} + j\mathrm{Q} = \mathbf{V}_a\mathbf{I}_a^* + \mathbf{V}_b\mathbf{I}_b^* + \mathbf{V}_c\mathbf{I}_c^*$$

$$= [\mathbf{V}_a \ \mathbf{V}_b \ \mathbf{V}_c]\begin{bmatrix}\mathbf{I}_a\\ \mathbf{I}_b\\ \mathbf{I}_c\end{bmatrix}^* = \begin{bmatrix}\mathbf{V}_a\\ \mathbf{V}_b\\ \mathbf{V}_c\end{bmatrix}^{\mathrm{T}}\begin{bmatrix}\mathbf{I}_a\\ \mathbf{I}_b\\ \mathbf{I}_c\end{bmatrix}^*$$

Now from Eq. (3.1)

$$\begin{bmatrix}\mathbf{V}_a\\ \mathbf{V}_b\\ \mathbf{V}_c\end{bmatrix} = \begin{bmatrix}1 & 1 & 1\\ 1 & a^2 & a\\ 1 & a & a^2\end{bmatrix}\begin{bmatrix}\mathbf{V}_{a0}\\ \mathbf{V}_{a1}\\ \mathbf{V}_{a2}\end{bmatrix} = \mathrm{AV}$$

and $$\begin{bmatrix}\mathbf{V}_a\\ \mathbf{V}_b\\ \mathbf{V}_c\end{bmatrix}^{\mathrm{T}} = (\mathrm{AV})^{\mathrm{T}} = \mathbf{V}^{\mathrm{T}}\mathrm{A}^{\mathrm{T}}$$

$$\text{So P} + j\mathrm{Q} = \begin{bmatrix}\mathbf{V}_a\\ \mathbf{V}_b\\ \mathbf{V}_c\end{bmatrix}^{\mathrm{T}}\begin{bmatrix}\mathbf{I}_a\\ \mathbf{I}_b\\ \mathbf{I}_c\end{bmatrix}^*$$

$$= [\mathbf{V}_{a0} \ \mathbf{V}_{a1} \ \mathbf{V}_{a2}]\begin{bmatrix}1 & 1 & 1\\ 1 & a^2 & a\\ 1 & a & a^2\end{bmatrix}\begin{bmatrix}\mathbf{I}_a\\ \mathbf{V}_b\\ \mathbf{V}_c\end{bmatrix}^*$$

Now substituting for phase currents the corresponding symmetrical components

$$\begin{bmatrix}\mathbf{I}_a\\ \mathbf{I}_b\\ \mathbf{I}_c\end{bmatrix}^* = \begin{bmatrix}1 & 1 & 1\\ 1 & a^2 & a\\ 1 & a & a^2\end{bmatrix}^*\begin{bmatrix}\mathbf{I}_{a0}\\ \mathbf{I}_{a1}\\ \mathbf{I}_{a2}\end{bmatrix}^*$$

$$= \begin{bmatrix}1 & 1 & 1\\ 1 & a & a^2\\ 1 & a^2 & a\end{bmatrix}\begin{bmatrix}\mathbf{I}_{a0}\\ \mathbf{I}_{a1}\\ \mathbf{I}_{a2}\end{bmatrix}^*$$

$\because$ a and a^2 are conjugate

Thus, P + jQ

$$= [\mathbf{V}_{a0} \ \mathbf{V}_{a1} \ \mathbf{V}_{a2}]\begin{bmatrix}1 & 1 & 1\\ 1 & a^2 & a\\ 1 & a & a^2\end{bmatrix}\begin{bmatrix}1 & 1 & 1\\ 1 & a & a^2\\ 1 & a^2 & a\end{bmatrix}\begin{bmatrix}\mathbf{I}_{a0}\\ \mathbf{I}_{a1}\\ \mathbf{I}_{a2}\end{bmatrix}^*$$

$$= [\mathbf{V}_{a0} \ \mathbf{V}_{a1} \ \mathbf{V}_{a2}]\begin{bmatrix}3 & 0 & 0\\ 0 & 3 & 0\\ 0 & 0 & 3\end{bmatrix}\begin{bmatrix}\mathbf{I}_{a0}\\ \mathbf{I}_{a1}\\ \mathbf{I}_{a2}\end{bmatrix}^*$$

$$= 3[\mathbf{V}_{a0} \quad \mathbf{V}_{a1} \quad \mathbf{V}_{a2}]\begin{bmatrix}\mathbf{I}_{a0}\\ \mathbf{I}_{a1}\\ \mathbf{I}_{a2}\end{bmatrix}^*$$

$$= 3[\mathbf{V}_{a0}\mathbf{I}_{a0}^* \quad \mathbf{V}_{a1}\mathbf{I}_{a1}^* \quad \mathbf{V}_{a2}\mathbf{I}_{a2}^*]$$

and $P = 3[V_{a0}I_{a0}\cos\theta_0 + V_{a1}I_{a1}\cos\theta_1 + V_{a2}I_{a2}\cos\theta_2]$...[3.4(*b*)]

The above expression for power does not have any cross terms (such as $\mathbf{V}_{a1}\cdot\mathbf{I}_{a2}$, $\mathbf{V}_{a1}\cdot\mathbf{I}_{a0}$ etc.) because there is no mutual coupling between the three sequence networks. The factor of 3 arises due to the total number of 9 components (3 for each phase). It is seen that three-phase power is equal to the sum of powers of the three symmetrical components.

Example 3.1. A generator supplying an unbalanced load measures the following phase-to-ground voltages :

$$\mathbf{V}_a = 18.0\angle 0^\circ \text{ kV}$$

$$\mathbf{V}_b = 13.3\angle -132^\circ \text{ kV}$$

$$\mathbf{V}_c = 12.0\angle +110^\circ \text{ kV}$$

Find the symmetrical components of the set of phasor voltages.

[*U.P. Technical Univ. Power System Analysis 2008-09*]

Solution : $\mathbf{V}_a = 18.0\angle 0^\circ \text{ kV} = (18 + j0) \text{ kV}$

$$\mathbf{V}_b = 13.3\angle -132^\circ \text{ kV} = (-8.9 - j9.884) \text{ kV}$$

$$\mathbf{V}_c = 12.0\angle +110^\circ \text{ kV} = (-4.104 + j11.276) \text{ kV}$$

Assuming phase sequence *abc*,

Positive sequence components

$$\mathbf{V}_{a_1} = \frac{1}{3}\left[\mathbf{V}_a + a\mathbf{V}_b + a^2\mathbf{V}_c\right]$$

$$= \frac{1}{3}\left[(18 + j0) + 13.3\angle -132^\circ + 120^\circ + 12.0\angle 110^\circ + 240^\circ\right] \text{kV}$$

$$= \frac{1}{3}\left[(18 + j0) + (13.009 - j2.765) + (11.818 - j2.084)\right] \text{kV}$$

$$= (14.276 - j1.616) \text{ kV} = 14.37\angle -6.46^\circ \text{ kV} \quad \textbf{Ans.}$$

$$\mathbf{V}_{b_1} = a^2\mathbf{V}_{a_1} = 14.37\angle -6.46^\circ + 240^\circ \text{ kV}$$

$$= 14.37\angle 233.54^\circ \text{ kV} \quad \textbf{Ans.}$$

$$\mathbf{V}_{c_1} = a\mathbf{V}_{a_1} = 14.37\angle -6.46^\circ + 120^\circ$$

$$= 14.37\angle 113.54^\circ \text{ kV} \quad \textbf{Ans.}$$

Negative sequence components

$$\mathbf{V}_{a_2} = \frac{1}{3}\left[\mathbf{V}_a + a^2\mathbf{V}_b + a\mathbf{V}_c\right]$$

$$= \frac{1}{3}\left[18 + j0 + 13.3\angle -132^\circ + 240^\circ + 12\angle 110^\circ + 120^\circ\right] \text{kV}$$

$$= \frac{1}{3}\left[(18 + j0) + (-4.11 + j12.65) + (-7.713 - j9.193)\right] \text{kV}$$

$$= (2.059 + j1.152) \text{ kV} = 2.359\angle 29.227^\circ \text{ kV} \quad \textbf{Ans.}$$

$$\mathbf{V}_{b_2} = a\mathbf{V}_{a_2} = 2.359\angle 29.227^\circ + 120^\circ \text{ kV}$$

$$= 2.359\angle 149.227^\circ \text{ kV} \quad \textbf{Ans.}$$

$$\mathbf{V}_{c_2} = a^2\mathbf{V}_{a_2} = 2.359\angle 29.227^\circ + 240^\circ \text{ kV}$$

$$= 2.359\angle 269.227^\circ \text{ kV} \quad \textbf{Ans.}$$

Zero sequence components

$$\mathbf{V}_0 = \frac{1}{3}(\mathbf{V}_a + \mathbf{V}_b + \mathbf{V}_c)$$

$$= \frac{1}{3}\left[(18 + j0) + (-8.9 - j9.884) + (-4.104 + j11.276)\right] \text{kV}$$

$$= (1.665 + j0.464) \text{ kV} = 1.728\angle 15.572^\circ \text{ kV} \quad \textbf{Ans.}$$

Example 3.2. Compute symmetrical components of the three phase voltages

$$\mathbf{V}_A = 100\angle 0^\circ;\ \mathbf{V}_B = 110\angle -100^\circ \text{ and } \mathbf{V}_C = 115\angle +110^\circ$$

[*Madurai Kamaraj Univ. Power System Analysis Nov. 1996*]

Solution : Positive sequence components,

$$\mathbf{V}_1 = \frac{1}{3}[\mathbf{V}_A + a\mathbf{V}_B + a^2\mathbf{V}_C]$$

$$= \frac{1}{3}\left[100\angle 0^\circ + 110\angle -100^\circ + 120^\circ + 115\angle 110^\circ + 240^\circ\right]$$

$$= \frac{1}{3}[(100 + j0) + (103.37 + j37.62) + (113.25 - j19.97)]$$

$$= (105.54 + j5.88) \text{ or } 105.7\angle 3.19^\circ \text{ V } \textbf{Ans.}$$

Negative sequence components,

$$\mathbf{V}_2 = \frac{1}{3}[\mathbf{V}_A + a^2\mathbf{V}_B + a\mathbf{V}_C]$$

$$= \frac{1}{3}\left[100\angle 0^\circ + 110\angle -100^\circ + 240^\circ + 115\angle 110^\circ + 120^\circ\right]$$

$$= \frac{1}{3}[(100 + j0) + (-84.26 + j70.71) + (-73.92 - j88.1)]$$

$$= (-19.39 - j5.8) \text{ or } 20.24\angle -163.35^\circ \text{ V } \textbf{Ans.}$$

Zero sequence components,

$$\mathbf{V}_0 = \frac{1}{3}[\mathbf{V}_A + \mathbf{V}_B + \mathbf{V}_C]$$

$$= \frac{1}{3}\left[100\angle 0^\circ + 110\angle -100^\circ + 115\angle 110^\circ\right]$$

$$= \frac{1}{3}[(100 + j0) + (-19.10 - j108.33) + (-39.33 + j108.06)]$$

$$= (13.86 - j0.09) \text{ or } 13.86\angle -0.37^\circ \text{ V } \textbf{Ans.}$$

Example 3.3. In a three phase, four wire system, the currents in R, Y and B lines under abnormal conditions of loading are as under :

$$\mathbf{I_R} = 200 \angle 30° \text{ A}$$

$$\mathbf{I_Y} = 100 \angle 300° \text{ A}$$

$$\mathbf{I_B} = 60 \angle 180° \text{ A}$$

Calculate the zero, positive and negative sequence currents in R line and return current in the neutral wire.

[U.P. Technical Univ. Power System Analysis 2009-10]

Solution : $\mathbf{I}_R = 200 \angle 30° \text{ A} = (173.205 + j100) \text{ A}$

$$\mathbf{I}_Y = 100 \angle 300° \text{ A} = (50 - j86.6) \text{ A}$$

$$\mathbf{I}_B = 60 \angle 180° \text{ A} = (-60 + j0) \text{ A}$$

Assuming phase sequence RYB,

Zero sequence current in line R

$$I_0 = \frac{1}{3}(\mathbf{I}_R + \mathbf{I}_Y + \mathbf{I}_B)$$

$$= \frac{1}{3}\left[(173.205 + j100) + (50 - j86.6) + (-60 + j0)\right] \text{A}$$

$$= (54.4 + j4.4667) \text{ A} = 54.583 \angle 4.694° \text{ A}$$

Positive sequence current in line R

$$\mathbf{I}_{R_1} = \frac{1}{3}\left[\mathbf{I}_R + a\mathbf{I}_Y + a^2\mathbf{I}_B\right]$$

$$= \frac{1}{3}\left[200 \angle 30° + 100 \angle 300° + 120° + 60 \angle 180° + 240°\right] \text{A}$$

$$= \frac{1}{3}\left[(173.205 + j100) + (50 + j86.6) + (30 + j51.96)\right] \text{A}$$

$$= (84.4 + j79.52) \text{ A} = 115.96 \angle 43.29° \text{ A} \quad \textbf{Ans.}$$

Negative sequence current in line R

$$\mathbf{I}_{R_2} = \frac{1}{3}\left[\mathbf{I}_R + a^2\mathbf{I}_Y + a\mathbf{I}_B\right]$$

$$= \frac{1}{3}\left[200 \angle 30° + 100 \angle 300° - 120° + 60 \angle 180° + 120°\right] \text{A}$$

$$= \frac{1}{3}\left[(173.205 + j100) + (-300 + j0) + (30 - j51.96)\right] \text{A}$$

$$= (-32.265 + j16.0133) \text{ A} = 36.02 \angle 153.6° \text{ A} \quad \textbf{Ans.}$$

Current in neutral wire,

$$\mathbf{I}_N = \mathbf{I}_R + \mathbf{I}_Y + \mathbf{I}_B$$

$$= (173.205 + j100) + (50 - j86.6) + (-60 + j0) \text{ A}$$

$$= (163.2 + j13.4) \text{ A} = 163.75 \angle 4.694° \text{ A} \quad \textbf{Ans.}$$

Example 3.4. Determine the positive sequence components of the three phase currents $\mathbf{I}_a = 10 \angle 0°$ A, $\mathbf{I}_b = 10 \angle 230°$ A and $\mathbf{I}_c = 10 \angle 130°$ A.

[U.P.S.C. I.E.S. Electrical Engineering-II, 2010]

Solution : Assuming phase sequence RYB

Positive sequence components,

$$\mathbf{I}_{a_1} = \frac{1}{3}[\mathbf{I}_a + a\mathbf{I}_b + a^2\mathbf{I}_c] \text{A}$$

$$= \frac{1}{3}\left[10 \angle 0° + 10 \angle 230° \times a + 10 \angle 130° \times a^2\right] \text{A}$$

$$= \frac{1}{3}\left[10 \angle 0° + 10 \angle 230° + 120° + 10 \angle 130° + 240°\right] \text{A}$$

$$= \frac{1}{3}\left[10 \angle 0° + 10 \angle -10° + 10 \angle 10°\right] \text{A}$$

$$= \frac{1}{3}[(10 + j0) + 10(0.9848 - j0.1736) + 10(0.9848 + j0.1736)]$$

$$= 9.899 \angle 0° \text{ A} \quad \textbf{Ans.}$$

$$\mathbf{I}_{b_1} = a^2\mathbf{I}_{a_1} = 9.899 \angle 0° + 240° = 9.899 \angle 240° \text{ A} \quad \textbf{Ans.}$$

$$\mathbf{I}_{c_1} = a\mathbf{I}_{a_1} = 9.899 \angle 0° + 120° = 9.899 \angle 120° \text{ A} \quad \textbf{Ans.}$$

Example 3.5. Three resistors of 50 Ω, 100 Ω and 200 Ω are connected in delta across the three phases of a balanced 100 volts supply. What are the sequence components in the resistors and in supply lines ?

[W.B. Univ. of Technology Power System-II, 2010]

Solution : Let the voltages be expressed as (Fig. 3.5)

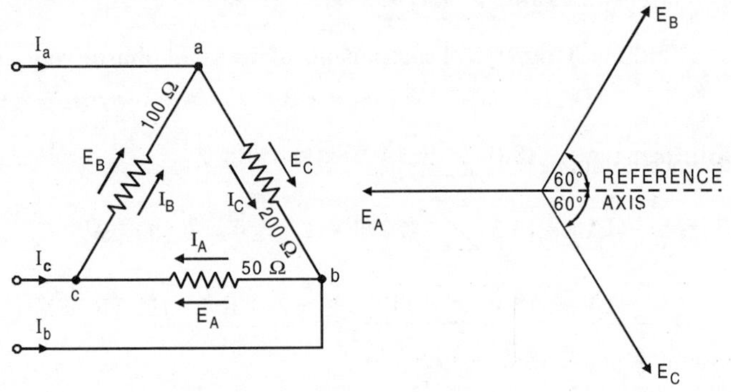

Fig. 3.5

$$\mathbf{E}_A = 100 \angle 180° \text{ V}$$

$$\mathbf{E}_B = 100 \angle 60° \text{ V}$$

$$\mathbf{E}_C = 100 \angle -60° \text{ V}$$

Current through resistors

$$\mathbf{I}_A = \frac{\mathbf{E}_A}{50} = \frac{100 \angle 180°}{50} = 2 \angle 180° \text{ A}$$

$$\mathbf{I}_B = \frac{\mathbf{E}_B}{100} = \frac{100 \angle 60°}{100} = 1 \angle 60° \text{ A}$$

$$\mathbf{I}_C = \frac{\mathbf{E}_C}{200} = \frac{100 \angle -60°}{200} = 0.5 \angle -60° \text{ A}$$

Sequence component currents in resistors

$$\mathbf{I}_{A_1} = \frac{1}{3}[\mathbf{I}_A + a\mathbf{I}_B + a^2\mathbf{I}_C]$$

$$= \frac{1}{3}\left[2 \angle 180° + 1 \angle 60° + 120° + 0.5 \angle -60° + 240°\right] \text{A}$$

$$= \frac{1}{3}\left[2 \angle 180° + 1 \angle 180° + 0.5 \angle 180°\right] \text{A}$$

$$= \frac{3.5}{3} \angle 180° = 1.1667 \angle 180° = (-1.1667 + j0) \text{ A} \quad \textbf{Ans.}$$

$$\mathbf{I}_{A_2} = \frac{1}{3}[\mathbf{I}_A + a^2\mathbf{I}_B + a\mathbf{I}_C]$$

$$= \frac{1}{3}\left[2\angle 180° + 1\angle 60° + 240° + 0.5\angle -60° + 120°\right] \text{A}$$

$$= \frac{1}{3}\left[2\angle 180° + 1\angle -60° + 0.5\angle 60°\right] \text{A}$$

$$= \frac{1}{3}[(-2 + j0) + (0.5 - j0.866) + (0.25 + j0.433)] \text{ A}$$

$$= -\frac{1.25}{3} - j\frac{0.433}{3} = (-0.4167 - j0.144) \text{ A} \quad \textbf{Ans.}$$

$$\mathbf{I}_{A_0} = \frac{1}{3}(\mathbf{I}_A + \mathbf{I}_B + \mathbf{I}_C)$$

$$= \frac{1}{3}\left[2\angle 180° + 1\angle 60° - 0.5\angle 60°\right] \text{A}$$

$$= \frac{1}{3}[(-2 + j0) + (0.5 + j0.866) + (0.25 - j0.433)] \text{ A}$$

$$= \frac{-1.25}{3} + j\frac{0.433}{3} = (-0.4167 + j0.144) \text{ A} \quad \textbf{Ans.}$$

Line currents

$$\mathbf{I}_a = \mathbf{I}_C - \mathbf{I}_B$$
$$\mathbf{I}_b = \mathbf{I}_A - \mathbf{I}_C$$
$$\mathbf{I}_c = \mathbf{I}_B - \mathbf{I}_A$$

Sequence components of line currents

$$\mathbf{I}_{a_1} = \frac{1}{3}(\mathbf{I}_a + a\mathbf{I}_b + a^2\mathbf{I}_c)$$

$$= \frac{1}{3}(\mathbf{I}_C - \mathbf{I}_B + a\mathbf{I}_A - a\mathbf{I}_C + a^2\mathbf{I}_B - a^2\mathbf{I}_A)$$

$$= \frac{1}{3}[-(a^2\mathbf{I}_A + \mathbf{I}_B + a\mathbf{I}_C) + (a\mathbf{I}_A + a^2\mathbf{I}_B + \mathbf{I}_C)]$$

$$= \frac{1}{3}[-a^2(\mathbf{I}_A + a\mathbf{I}_B + a^2\mathbf{I}_C) + a(\mathbf{I}_A + a\mathbf{I}_B + a^2\mathbf{I}_C)]$$

$$= \frac{1}{3}[(-a^2 + a)(3\mathbf{I}_{A_1})] = j\sqrt{3}\,\mathbf{I}_{A_1} = j\sqrt{3} \times 1.1667\angle 180°$$

$$\text{or } \mathbf{I}_{a_1} = \sqrt{3} \times 1.1667\angle 180° + 90° \text{ A} = 2.02\angle 270° \text{ A} = j2.02 \text{ A} \quad \textbf{Ans.}$$

$$\mathbf{I}_{a_2} = \frac{1}{3}(\mathbf{I}_a + a^2\mathbf{I}_b + a\mathbf{I}_c)$$

$$= \frac{1}{3}[\mathbf{I}_C - \mathbf{I}_B + a^2\mathbf{I}_A - a^2\mathbf{I}_C + a\mathbf{I}_B - a\mathbf{I}_A]$$

$$= \frac{1}{3}[(a^2\mathbf{I}_A + a\mathbf{I}_B + \mathbf{I}_C) - (a\mathbf{I}_A + \mathbf{I}_B + a^2\mathbf{I}_C)]$$

$$= \frac{1}{3}(a^2 - a) \times 3\mathbf{I}_{A_2} = (a^2 - a)\mathbf{I}_{A_2} = -j\sqrt{3}\mathbf{I}_{A_2}$$

$$\text{or } \mathbf{I}_{a_2} = -j\sqrt{3} \times (-0.4167 - j0.144) = (-0.249 + j0.722) \text{ A}$$

$$= 0.7637\angle 109° \text{ A} \quad \textbf{Ans.}$$

$$\mathbf{I}_{a_0} = \mathbf{I}_a + \mathbf{I}_b + \mathbf{I}_c = (\mathbf{I}_C - \mathbf{I}_B) + (\mathbf{I}_A - \mathbf{I}_C) + (\mathbf{I}_B - \mathbf{I}_A) = 0 \quad \textbf{Ans.}$$

The above example reveals that the star-delta sequence currents are related as follows :

Positive sequence currents, $\mathbf{I}_{a_1} = j\sqrt{3}\,\mathbf{I}_{A_1}$

Negative sequence currents, $\mathbf{I}_{a_2} = -j\sqrt{3}\mathbf{I}_{A_2}$

Zero sequence currents, $\mathbf{I}_{a_0} = 0$

where $\mathbf{I}_{a_1}, \mathbf{I}_{a_2}, \mathbf{I}_{a_0}$ are the star currents and $\mathbf{I}_{A_1}, \mathbf{I}_{A_2}$ are the delta currents.

Example 3.6. A single-phase load of 100 kVA connected across lines *bc* of a 3-phase supply of 3.3 kV. Determine symmetrical components of line current.

[U.P.S.C. I.E.S. Electrical Engineering-II, 1998]

Solution : The system is shown in Fig. 3.6.

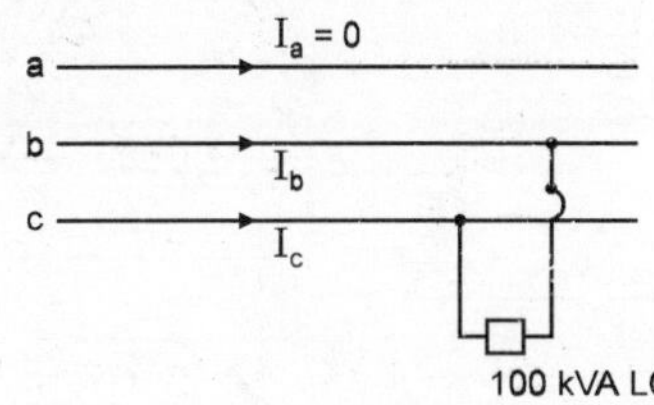

Fig. 3.6

Now $\mathbf{I}_a = 0$

$$\mathbf{I}_b = \frac{100 \times 1{,}000}{3.3 \times 1{,}000} = 30.3 \text{ A}$$

$$\mathbf{I}_c = -\mathbf{I}_b = -30.3 \text{ A}$$

Positive sequence component,

$$\mathbf{I}_{a1} = \frac{1}{3}(\mathbf{I}_a + a\mathbf{I}_b + a^2\mathbf{I}_c)$$

$$= \frac{1}{3}\mathbf{I}_b(a - a^2)$$

$$= \frac{1}{3}\mathbf{I}_b \times 1.732 = \frac{30.3}{\sqrt{3}} = 17.5 \text{ A} \quad \textbf{Ans.}$$

Negative sequence component,

$$\mathbf{I}_{a2} = \frac{1}{3}(\mathbf{I}_a + a^2\mathbf{I}_b + a\mathbf{I}_c)$$

$$= \frac{1}{3}(0 + a^2\mathbf{I}_b - a\mathbf{I}_b)$$

$$= \frac{1}{3}(a^2 - a) = \frac{30.3}{3} \times (-1.732) = -17.5 \text{ A} \quad \textbf{Ans.}$$

Zero sequence component,

$$\mathbf{I}_{a0} = \frac{1}{3}(\mathbf{I}_a + \mathbf{I}_b + \mathbf{I}_c) = \frac{1}{3}(0 + \mathbf{I}_b - \mathbf{I}_b) = 0 \quad \textbf{Ans.}$$

3.6. PHASE SHIFT IN STAR-DELTA TRANSFORMERS

The angular difference between phasors representing the voltages induced between high voltage and low voltage terminals having same marking letters and the corresponding neutral points (real or imaginary), expressed with reference to high voltage side is known as *phase displacement* (*or shift*) of the transformer. In a star-star or delta-delta 3-phase transformer there is no phase shift between the corresponding voltages of any phase to neutral on either side. However, a delta-connected winding is desirable in many power transformers for reasons of harmonic elimination. As such most of the power transformers are either star-delta or delta-star connected. In such transformers, even under normal operating condition, the phase-to-phase voltages and phase-to-neutral voltages of *hv* side are displaced from the corresponding voltages of *lv* side. The phase shift of 30°

comes in group 4 and that of –30° comes in group 3 (IS: 2026, 1962 reprint 1972). Similarly the currents on two sides are displaced. Generally in short-circuit calculations the phase shift needs not to be considered.

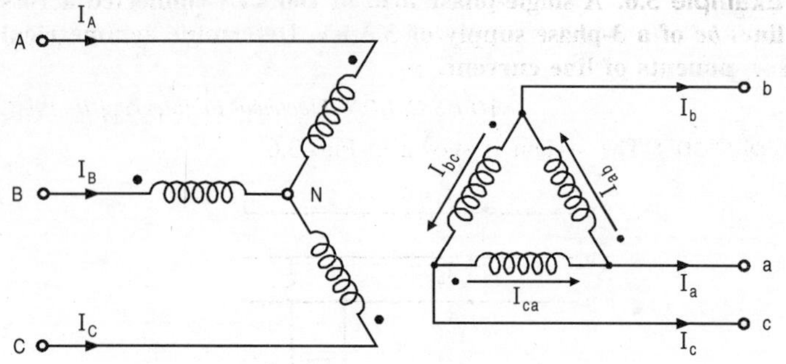

Fig. 3.7. ***Labelling of a Star-Delta Transformer***

Now let us consider 3-phase star-delta transformer with primary side Y-connected and secondary side delta-connected, as shown in Fig. 3.7. Windings shown parallel to each other, being wound on the same core, are magnetically coupled. The polarity markings are indicated on each phase. The dots at the windings indicate the terminals which are positive at the same time with respect to the undotted terminals. With phase marked as ABC on the star side, there are number of ways of labelling the phases *a b c* on the delta side. The labelling indicated on the diagram corresponds to + 90° connection in which the positive sequence phase a to neutral voltage (delta side) leads phase A to neutral voltage (star side) by 90° and so the line currents in *a* and A. This labelling is computationally convenient: The alternative way is to label delta as $b \to a$, $c \to b$ and $a \to c$, and thus we get standard Yd_1, – 30° connection. If the polarities on the delta side are also reversed, we have standard Yd_{11}, 30° connection.

Double suffixes are used for line-to-line voltages and delta currents and single suffix are used for line currents and phase (line-to-neutral) voltages. Line-to-line transformation ratio is being taken to be unity.

The positive and negative sequence voltages on primary (star) and secondary (delta) sides of the transformer are shown in Fig. 3.8 while the positive and negative sequence currents on the two sides of the transformer are shown in Fig. 3.9.

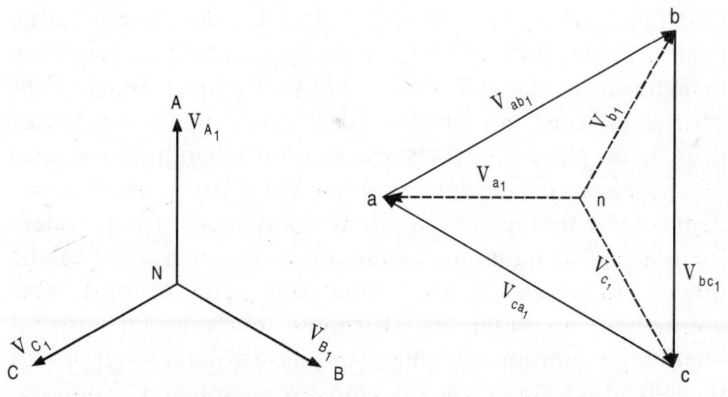

Fig. 3.8(*a*). ***Positve Sequence Voltages in a Y-Δ Transformer***

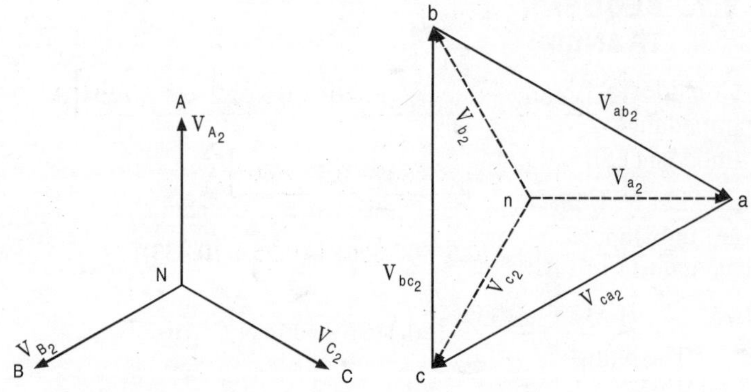

Fig. 3.8(*b*). ***Negative Sequence Voltages in a Y-Δ Transformer***

From these figures it is observed that

$$\mathbf{V}_{a_1} = j\mathbf{V}_{A_1};\ \mathbf{I}_{a_1} = j\mathbf{I}_{A_1};\ \mathbf{V}_{a_2} = -\,j\mathbf{V}_{A_2} \text{ and } \mathbf{I}_{a_2} = -j\mathbf{I}_{A_2} \quad ...(3.5)$$

In case of reversal of power flow *i.e.,* when delta acts as primary and star as secondary, the voltage phasors do not change but all the current phasors reverse. The phasor relationship between star and delta voltages and currents therefore remain the same.

Thus, we see that magnitude of phase shift is same for positive sequence components and negative sequence components. However, the direction of phase shift in case of negative phase sequence components is reverse of that applicable to the positive sequence components (due to reverse phase sequence). The magnitude and direction of phase shift depends on the transformer group and allocation of phase references. Phase shift of zero sequence quantities need not be considered in star-delta transformers because the zero sequence currents do not flow in lines on delta-connected side.

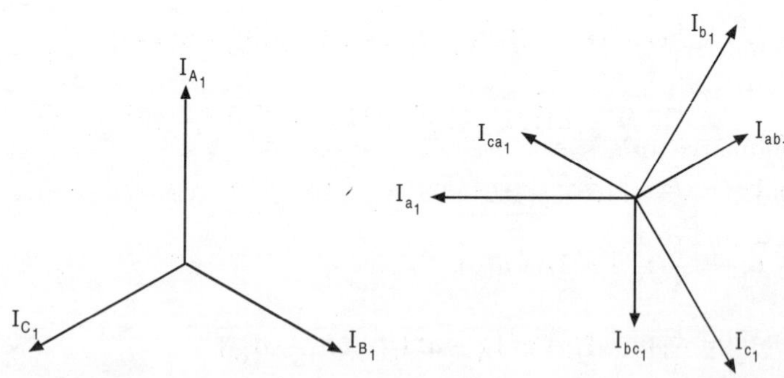

Fig. 3.9(*a*). ***Positive Sequence Currents in a Y-Δ Transformer***

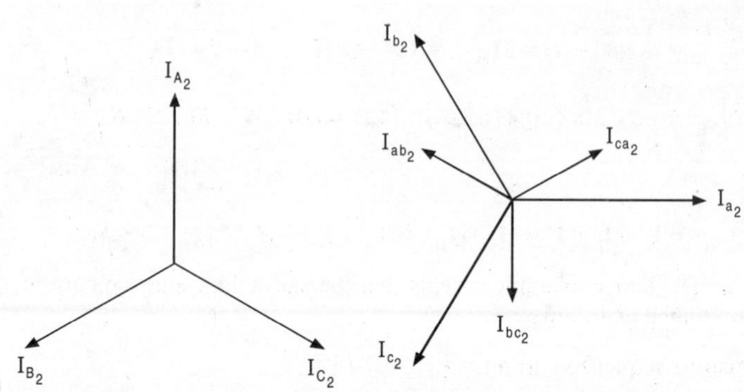

Fig. 3.9(*b*). ***Negative Sequence Currents in a Y-Δ Transformer***

3.7. SEQUENCE IMPEDANCES OF TRANSMISSION LINES

Consider a fully transposed 3-φ line (Fig. 3.10) with self impedance Z_s per phase and mutual impedance Z_m (mutual impedance of any line pair).

Let I_a, I_b and I_c be the line currents. If the system voltages are unbalanced, we have a neutral current, I_n flowing through the neutral (ground) having impedance Z_n. Thus,

$$\mathbf{I}_n = \mathbf{I}_a + \mathbf{I}_b + \mathbf{I}_c$$

The phase voltages at the two ends of the line are V_a, V_b, V_c and V'_a, V'_b, V'_c respectively. The voltage drop in line a is caused by the current I_a due to self reactance Z_s and by currents I_b and I_c due to mutual reactances Z_m between lines b and a and between lines c and a.

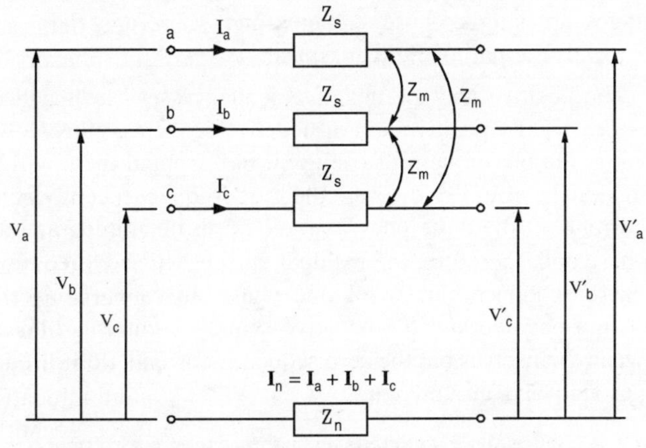

Fig. 3.10. ***Unsymmetrically Loaded Transmission Line***

According to Kirchhoff's voltage law, we have

$$\left.\begin{aligned} &\mathbf{V}_a - \mathbf{V}'_a = \mathbf{Z}_s\mathbf{I}_a + \mathbf{Z}_m\mathbf{I}_b + \mathbf{Z}_m\mathbf{I}_c + \mathbf{Z}_n(\mathbf{I}_a + \mathbf{I}_b + \mathbf{I}_c) \\ \text{Similarly, } &\mathbf{V}_b - \mathbf{V}'_b = \mathbf{Z}_m\mathbf{I}_a + \mathbf{Z}_s\mathbf{I}_b + \mathbf{Z}_m\mathbf{I}_c + \mathbf{Z}_n(\mathbf{I}_a + \mathbf{I}_b + \mathbf{I}_c) \\ \text{and } &\mathbf{V}_c - \mathbf{V}'_c = \mathbf{Z}_m\mathbf{I}_a + \mathbf{Z}_m\mathbf{I}_b + \mathbf{Z}_s\mathbf{I}_c + \mathbf{Z}_n(\mathbf{I}_a + \mathbf{I}_b + \mathbf{I}_c) \end{aligned}\right\} \quad ...(3.6)$$

In matrix form, we may write

$$\begin{bmatrix} \mathbf{V}_a \\ \mathbf{V}_b \\ \mathbf{V}_c \end{bmatrix} - \begin{bmatrix} \mathbf{V}'_a \\ \mathbf{V}'_b \\ \mathbf{V}'_c \end{bmatrix} = \begin{bmatrix} \mathbf{Z}_s + \mathbf{Z}_n & \mathbf{Z}_m + \mathbf{Z}_n & \mathbf{Z}_m + \mathbf{Z}_n \\ \mathbf{Z}_m + \mathbf{Z}_n & \mathbf{Z}_s + \mathbf{Z}_n & \mathbf{Z}_m + \mathbf{Z}_n \\ \mathbf{Z}_m + \mathbf{Z}_n & \mathbf{Z}_m + \mathbf{Z}_n & \mathbf{Z}_s + \mathbf{Z}_n \end{bmatrix} \begin{bmatrix} \mathbf{I}_a \\ \mathbf{I}_b \\ \mathbf{I}_c \end{bmatrix} \quad ...(3.7)$$

In more compact form, we may write

$$\mathbf{V}_{abc} - \mathbf{V}'_{abc} = \mathbf{Z}_{abc}\mathbf{I}_{abc} \quad ...(3.8)$$

The above equations reveal that the currents of one phase causes voltage drops in other phases also. These impedances are said to be coupled so far as phase voltages and phase currents are concerned. Applying symmetrical transformation to Eq. (3.8), we have

$$\mathbf{A}\mathbf{V}_{012} - \mathbf{A}\mathbf{V}'_{012} = \mathbf{Z}_{abc}\,\mathbf{A}\mathbf{I}_{012}$$

Multiplying both sides by $\mathbf{A}^{-1}$, we have

$$\mathbf{V}_{012} - \mathbf{V}'_{012} = \mathbf{A}^{-1}\mathbf{Z}_{abc}\mathbf{A}\mathbf{I}_{012} \quad ...(3.9)$$

We may define

$$\mathbf{Z}_{012} = \mathbf{A}^{-1}\mathbf{Z}_{abc}\,\mathbf{A} \quad ...(3.10)$$

where $\mathbf{Z}_{012}$ is known as the symmetrical components impedance matrix.

Equation (3.10) is employed for conversion of transmission line matrix. Z_{abc} into a form that is consistent with the use of zero, positive and negative sequence voltages (V_{abc}) and currents (I_{abc}).

Combining Eqs. (3.9) and (3.10), we have

$$\mathbf{V}_{012} - \mathbf{V}'_{012} = \mathbf{Z}_{012}\,\mathbf{I}_{012} \quad ...(3.11)$$

The symmetrical components impedances matrix $\mathbf{Z}_{012}$ for the transmission line may be determined as follows :

$$\mathbf{Z}_{012} = \mathbf{A}^{-1}\mathbf{Z}_{abc}\mathbf{A}$$

$$= \frac{1}{3}\begin{bmatrix} 1 & 1 & 1 \\ 1 & a & a^2 \\ 1 & a^2 & a \end{bmatrix} \begin{bmatrix} \mathbf{Z}_s + \mathbf{Z}_n & \mathbf{Z}_m + \mathbf{Z}_n & \mathbf{Z}_m + \mathbf{Z}_n \\ \mathbf{Z}_m + \mathbf{Z}_n & \mathbf{Z}_s + \mathbf{Z}_n & \mathbf{Z}_m + \mathbf{Z}_n \\ \mathbf{Z}_m + \mathbf{Z}_n & \mathbf{Z}_m + \mathbf{Z}_n & \mathbf{Z}_s + \mathbf{Z}_n \end{bmatrix} \begin{bmatrix} 1 & 1 & 1 \\ 1 & a^2 & a \\ 1 & a & a^2 \end{bmatrix}$$

For convenience sake, let us assume that

$$\mathbf{Z}_s + \mathbf{Z}_n = x \text{ and } \mathbf{Z}_m + \mathbf{Z}_n = y$$

$$\mathbf{Z}_{abc}\,\mathbf{A} = \begin{bmatrix} x & y & y \\ y & x & y \\ y & y & x \end{bmatrix} \begin{bmatrix} 1 & 1 & 1 \\ 1 & a^2 & a \\ 1 & a & a^2 \end{bmatrix}$$

$$= \begin{bmatrix} x + y + y & x + y(a + a^2) & x + y(a + a^2) \\ y + x + y & y + a^2x + ay & y + ax + a^2y \\ y + y + x & y + a^2y + ax & y + ay + a^2x \end{bmatrix}$$

$$= \begin{bmatrix} x + 2y & x - y & x - y \\ x + 2y & a^2(x - y) & a(x - y) \\ x + 2y & a(x - y) & a^2(x - y) \end{bmatrix}$$

$$= \begin{bmatrix} \mathbf{Z}_s + 2\mathbf{Z}_m + 3\mathbf{Z}_n & \mathbf{Z}_s - \mathbf{Z}_m & \mathbf{Z}_s - \mathbf{Z}_m \\ \mathbf{Z}_s + 2\mathbf{Z}_m + 3\mathbf{Z}_n & a^2(\mathbf{Z}_s - \mathbf{Z}_m) & a(\mathbf{Z}_s - \mathbf{Z}_m) \\ \mathbf{Z}_s + 2\mathbf{Z}_m + 3\mathbf{Z}_n & a(\mathbf{Z}_s - \mathbf{Z}_m) & a^2(\mathbf{Z}_s - \mathbf{Z}_m) \end{bmatrix}$$

Let us substitute $p = \mathbf{Z}_s + 2\mathbf{Z}_m + 3\mathbf{Z}_n$ and $q = \mathbf{Z}_s - \mathbf{Z}_m$

$$\mathbf{Z}_{abc}\mathbf{A} = \begin{bmatrix} p & q & q \\ p & a^2q & aq \\ p & aq & a^2q \end{bmatrix}$$

$$\mathbf{A}^{-1}\mathbf{Z}_{abc}\,\mathbf{A} = \frac{1}{3}\begin{bmatrix} 1 & 1 & 1 \\ 1 & a & a^2 \\ 1 & a^2 & a \end{bmatrix} \begin{bmatrix} p & q & q \\ p & a^2q & aq \\ p & aq & a^2q \end{bmatrix}$$

$$= \begin{bmatrix} p + p + p & q(1 + a^2 + a) & q(1 + a + a^2) \\ p(1 + a + a^2) & q(1 + a^3 + a^3) & q(1 + a^2 + a^4) \\ p(1 + a + a^2) & q(1 + a^4 + a^2) & q(1 + a^3 + a^3) \end{bmatrix}$$

or $\quad \mathbf{Z}_{012} = \mathbf{A}^{-1}\mathbf{Z}_{abc}\,\mathbf{A}$

$$= \frac{1}{3}\begin{bmatrix} 3p & 0 & 0 \\ 0 & 3q & 0 \\ 0 & 0 & 3q \end{bmatrix} = \begin{bmatrix} p & 0 & 0 \\ 0 & q & 0 \\ 0 & 0 & q \end{bmatrix} \quad \because\ 1 + a + a^2 = 0 \text{ and } a^4 = a$$

or $\mathbf{Z}_{012} = \mathbf{A}^{-1}\,\mathbf{Z}_{abc}\,\mathbf{A}$

$$= \begin{bmatrix} \mathbf{Z}_s + 2\mathbf{Z}_m + 3\mathbf{Z}_n & 0 & 0 \\ 0 & \mathbf{Z}_s - \mathbf{Z}_m & 0 \\ 0 & 0 & \mathbf{Z}_s - \mathbf{Z}_m \end{bmatrix}$$

$$\text{or} \quad \mathbf{Z}_{012} = \begin{bmatrix} \mathbf{Z}_0 & 0 & 0 \\ 0 & \mathbf{Z}_1 & 0 \\ 0 & 0 & \mathbf{Z}_2 \end{bmatrix} \qquad ...(3.12)$$

where in

$$\mathbf{Z}_1 = \mathbf{Z}_s - \mathbf{Z}_m = \text{Positive sequence impedance}$$
$$\mathbf{Z}_2 = \mathbf{Z}_s - \mathbf{Z}_m = \text{Negative sequence impedance}$$
$$\mathbf{Z}_0 = \mathbf{Z}_s + 2\mathbf{Z}_m + 3\mathbf{Z}_n = \text{Zero sequence impedance}$$

$$\text{and} \quad \begin{bmatrix} \mathbf{V}_{a_0} \\ \mathbf{V}_{a_1} \\ \mathbf{V}_{a_2} \end{bmatrix} - \begin{bmatrix} \mathbf{V}'_{a_0} \\ \mathbf{V}'_{a_1} \\ \mathbf{V}'_{a_2} \end{bmatrix} = \begin{bmatrix} \mathbf{Z}_0 & 0 & 0 \\ 0 & \mathbf{Z}_1 & 0 \\ 0 & 0 & \mathbf{Z}_2 \end{bmatrix} \begin{bmatrix} \mathbf{I}_{a_0} \\ \mathbf{I}_{a_1} \\ \mathbf{I}_{a_2} \end{bmatrix} \qquad ...(3.13)$$

In expanded form, we may write

$$\left.\begin{aligned} \mathbf{V}_{a_0} - \mathbf{V}'_{a_0} &= \mathbf{Z}_0 \mathbf{I}_{a_0} \\ \mathbf{V}_{a_1} - \mathbf{V}'_{a_1} &= \mathbf{Z}_1 \mathbf{I}_{a_1} \\ \mathbf{V}_{a_2} - \mathbf{V}'_{a_2} &= \mathbf{Z}_2 \mathbf{I}_{a_2} \end{aligned}\right\} \qquad ...(3.14)$$

The above relations reveal that for a fully transposed line

1. Positive and negative sequence impedances are equal.
2. Zero sequence impedance is much larger than the positive or negative sequence impedance (it is usually 2 to 3.5 times)

It is further revealed that the sequence circuit. Equations (3.13) are in decoupled form, *i.e.*, there are no mutual sequence impedances. Equations (3.13) can be represented in network form as in Fig. 3.11.

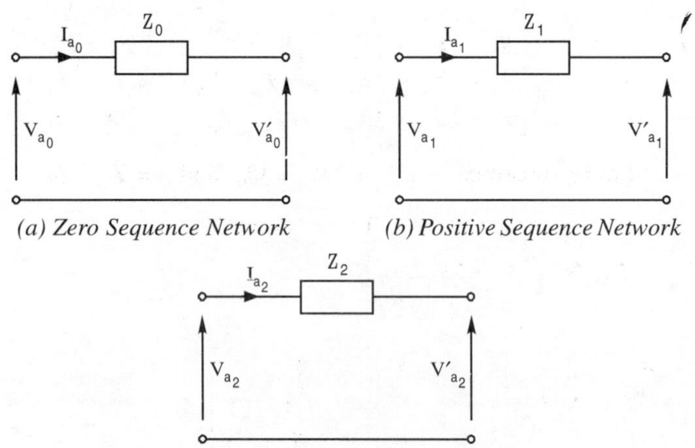

(a) Zero Sequence Network *(b) Positive Sequence Network*

(c) Negative Sequence Network

Fig. 3.11. ***Sequence Networks of a Transmission Line***

The decoupling between sequence networks of a fully transposed transmission line holds also in 3-ϕ synchronous machines and three-phase transformers. This fact leads to considerable simplifications in the use of symmetrical components method in unsymmetrical fault analysis.

In case of three static unbalanced impedances, coupling appears between sequence networks and the method is no more helpful than a straight forward three-phase analysis.

3.8. PHYSICAL SIGNIFICANCE OF SEQUENCE COMPONENTS

A physical significance of sequence voltages can be well understood by considering the fields developed on application of such sequence voltages to the stator of a 3-phase ac machine (a synchronous machine or an induction machine). If a set of positive sequence voltages is applied to the stator winding of a 3-phase induction motor, a magnetic field revolving in a certain direction will be developed. Now if two of the three phase supply leads to the stator winding are interchanged, the direction of revolution of the field developed would be reversed. It is seen that for this condition the relative phase positions of the voltages applied to the stator of the motor are the same as for the negative sequence set. Thus the field developed by the set of negative sequence voltages will rotate in a direction opposite to that of the field developed by the set of positive sequence voltages. Phase sequence should not be confused with the rotation of phasors. For both sets of positive and negative sequence voltages, the standard convention of counter-clockwise rotation is followed. The zero sequence voltages are single phase voltages and, therefore, they give rise to an alternating field in space.

The positive and the negative sequence sets are balanced ones *i.e.,* if only positive and/or negative sequence currents are flowing, the phasor sum of each will be zero and there will be no residual current. However, the zero sequence components of currents in the three phases are equal in magnitude and are in phase and, therefore, the residual current will be three times the zero sequence current of one phase. In case of an earth fault, the positive and the negative sequence currents balance between themselves but the zero sequence currents flow through the ground and ground wires.

3.9. SEQUENCE IMPEDANCES AND SEQUENCE NETWORKS

So far we have discussed the symmetrical components for the currents, voltages and power. Now we will discuss something about the sequence impedances of the power system. The sequence impedances of a network or a component of power system are the positive, negative, and zero sequence impedances. They are defined as below :

The impedance of the network offered to the flow of positive sequence currents is called the *positive sequence impedance.* Similarly if only negative sequence currents flow the impedance of the network offered to these currents is called the *negative sequence* impedance. Also the impedance offered to zero sequence currents is called the *zero sequence impedance* of the network.

If $\mathbf{Z}_a$, $\mathbf{Z}_b$ and $\mathbf{Z}_c$ are the impedances of the load between phases *a*, *b* and *c* to neutral *n*, then sequence impedances are given as

$$\left.\begin{aligned} &\text{Positive sequence impedance,} \\ &\quad \mathbf{Z}_1 = \frac{1}{3}(\mathbf{Z}_a + a\,\mathbf{Z}_b + a^2\mathbf{Z}_c) \\ &\text{Negative sequence impedance,} \\ &\quad \mathbf{Z}_2 = \frac{1}{3}(\mathbf{Z}_a + a^2\mathbf{Z}_b + a\mathbf{Z}_c) \\ &\text{Zero sequence impedance,} \\ &\quad \mathbf{Z}_0 = \frac{1}{3}(\mathbf{Z}_a + \mathbf{Z}_b + \mathbf{Z}_c) \end{aligned}\right\} \qquad ...(3.15)$$

For a 3-phase, symmetrical static circuit without internal voltages like transformers and transmission lines, the impedances offered to the currents of any sequence are the same in the three phases; also the currents of a particular sequence will cause voltage drop of the same sequence or a voltage of a particular sequence will give rise to current of the same sequence only, which means that there is no mutual coupling between the sequence networks. Since in case of a static device, the sequence has no significance, the positive and negative sequence impedances are equal; the zero sequence impedance which includes impedance of the return path through the ground is usually different from the positive and negative sequence impedances. The impedances offered by rotating machine to positive sequence components of currents, differ from those offered to negative sequence components of currents.

The single-phase equivalent circuit composed of the impedances to current of any one sequence only is called the *sequence network* for that particular sequence. Hence corresponding to positive, negative and zero sequence currents we have positive, negative and zero sequence networks. Thus, for every power system three sequence networks can be formed and these sequence networks, carrying currents I_{a1}, I_{a2} and I_{a0} are then interconnected in different ways to represent different unbalanced fault conditions. The sequence currents and voltages during the fault are then calculated from which actual fault currents and voltages can be determined.

The positive sequence network is considered in the analysis of symmetrical faults. Positive sequence network is the same as impedance or reactance diagram. The negative sequence network is, in general, quite similar to the positive sequence network but differs in the following aspects:

(*i*) Normally there are no negative sequence emf sources.

(*ii*) Negative sequence impedances of rotating machines are generally different from their positive sequence impedances.

The phase displacement of transformer banks for negative sequence is of opposite sign to that of positive sequence.

The *zero sequence network* likewise will be free of internal voltages, the flow of current being caused by the voltage at the fault point. The impedances to zero sequence currents are very frequently different from the positive or negative sequence currents. Zero sequence reactance of transmission lines is higher than that for positive sequence. The impedances of transformers and generators will depend upon the type of connections [delta or star (grounded or isolated)].

3.10. SEQUENCE IMPEDANCES OF POWER SYSTEM ELEMENTS

For determination of the behaviour of the power system under unbalanced conditions (unbalanced loads or unsymmetrical faults), it is necessary to know the impedances offered by the different elements of the power system to the flow of the different phase sequence components of current. We shall now discuss in detail about the sequence impedances of some important elements (such as synchronous and induction machines, transmission lines, transformers, and static loads) encountered in power systems.

3.10.1. Sequence Impedances and Networks of Synchronous Machines

An unloaded synchronous machine (generator or motor) grounded through a reactor of impedance Z_n is shown in Fig. 3.12. E_a, E_b and E_c are the induced emfs in the three phases. When an unsymmetrical fault occurs on the machine terminals, unbalanced currents I_a, I_b and I_c flow in the lines. If the fault involves ground, a current I_n (equal to phasor sum of line currents I_a, I_b and I_c) flows to neutral from ground *via* reactor Z_n. Depending on the type of fault one or more of the line currents may be zero. Unbalanced line currents can be resolved into their symmetrical components.

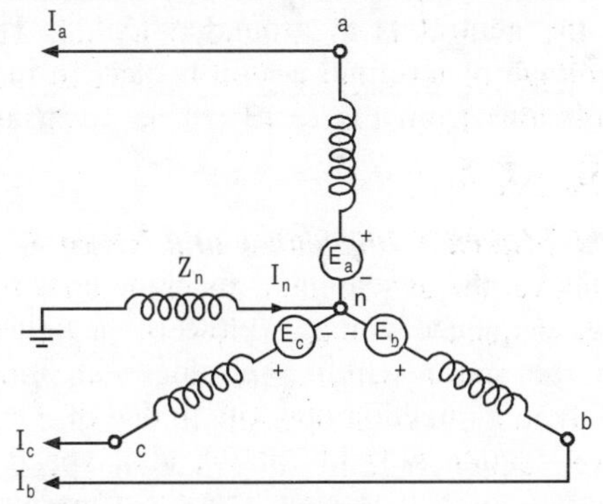

Fig. 3.12. ***Three-Phase Unloaded Synchronous Generator With Grounded Neutral Through Impedance Z_n***

Positive Sequence Impedance and Network. Since a synchronous machine is designed with symmetrical windings, it has induced emfs of positive sequence only, *i.e.*, no negative or zero sequence voltages are induced in it. The armature reaction field set up by positive sequence currents rotates at synchronous speed in the same direction as the rotor, *i.e.*, it is stationary with respect to field excitation. The machine equivalently offers a direct axis reactance whose value reduces from subtransient reactance X''_d to transient reactance X'_d and finally to steady state (synchronous) reactance X_d as the short circuit transient progresses in time. Subtransient reactance is used in a circuit where a sudden value of current under switching of fault condition is to be obtained. Where current after a few cycles (3 or 4) is desired, transient reactances are used, and for steady-state condition, the steady state or synchronous reactances are used. Positive sequence reactances together with the negligible resistance make up the positive sequence subtransient, transient or steady state positive sequence impedances, respectively.

The positive sequence network for a synchronous machine can be represented by the source emf on no load and the positive sequence impedance Z_1 in series with it, as shown in Fig. 3.13 (*a*). The neutral impedance Z_n does not appear in the circuit because the phasor sum of I_{a1}, I_{b1} and I_{c1} is zero and no positive sequence current can flow through Z_n. Since it is a balanced network, it can be drawn on single phase basis, as shown in Fig. 3.13 (*b*), for purpose of analysis.

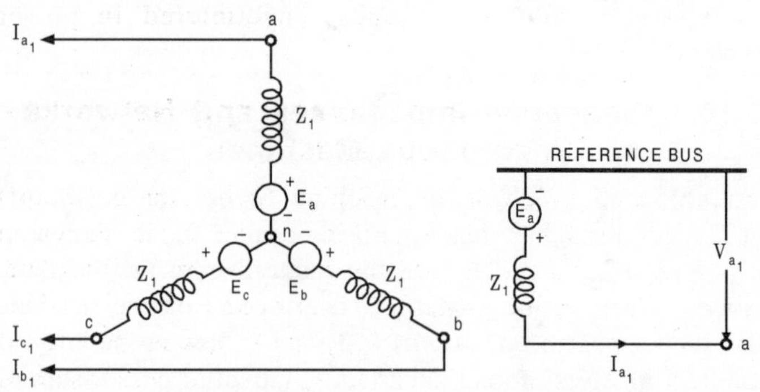

(a) Three-Phase Model (b) Single-Phase Model

Fig. 3.13. ***Positive Sequence Network of a Synchronous Machine***

The reference bus for a positive sequence network is at neutral potential. Further, since no current flows from ground to neutral, the neutral is at ground potential. The positive sequence voltage of terminal a with respect to the reference bus, as is obvious from Fig. 3.13 (*b*), is given as

$$\mathbf{V}_{a_1} = \mathbf{E}_a - \mathbf{I}_{a_1}\mathbf{Z}_1 \qquad ...(3.16)$$

Negative Sequence Impedance and Network. As already mentioned above, the synchronous machine does not generate any negative sequence voltage. Flow of negative sequence currents in the stator winding produces an mmf rotating synchronously in a direction opposite to that of the rotor. Thus the negative sequence field rotates at a speed twice the synchronous speed with respect to rotor. Currents at double the stator frequency are, therefore, induced in rotor field and damper winding. In sweeping over the rotor surface, the negative sequence mmf is alternately presented with reluctances of direct and quadrature axes. Thus the negative sequence reactance X_2 is found to oscillate between X''_d and X''_q and the value taken is usually the average. Thus, the negative sequence reactance

$$X_2 = \frac{X''_d + X''_q}{2} \qquad ...(3.17)$$

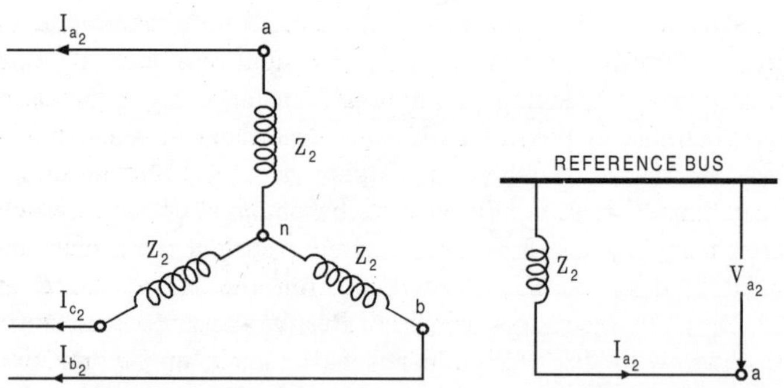

(a) Three-Phase Model (b) Single-Phase Model

Fig. 3.14. ***Negative Sequence Network of a Synchronous Machine***

The negative sequence networks of a synchronous machine, on a three-phase and single-phase basis are shown in Figs. 3.14 (*a*) and 3.14 (*b*) respectively. The reference bus for the negative sequence network is also the neutral of the machine. The negative sequence voltage of terminal a with respect to reference bus is given as

$$\mathbf{V}_{a_2} = -\,\mathbf{I}_{a_2}\mathbf{Z}_2 \qquad ...(3.18)$$

Zero Sequence Impedance and Network. No zero sequence voltage is induced in synchronous machine. The flow of zero sequence currents in the stator windings produces three mmfs which are in time phase but are distributed in space by 120°. The resultant air-gap field produced by zero sequence currents is therefore zero. Hence, the rotor windings offer leakage reactance only to the flow of zero sequence currents.

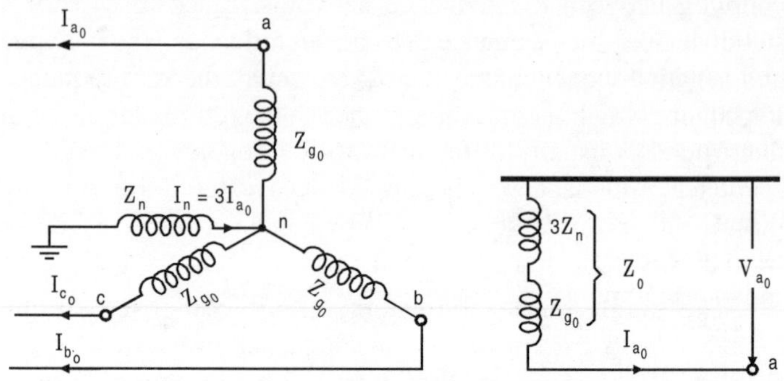

(a) Three-Phase Model (b) Single-Phase Model

Fig. 3.15. ***Zero Sequence Network of a Synchronous Machine***

As the current flowing in the reactor impedance Z_n is the sum of the zero sequence currents in all the three phases, hence voltage drop caused by it will be $3\,I_{a0}Z_n$. Now voltage drop of zero sequence from terminal a of the synchronous machine to ground is $3I_{a0}Z_n + I_{a0}Z_{g0}$ where Z_{g0} is the zero sequence impedance of the generator. Hence total zero sequence impedance, through which zero sequence current I_{a0} flows is

$$\mathbf{Z}_0 = 3\mathbf{Z}_n + \mathbf{Z}_{g0} \qquad ...(3.19)$$

The zero sequence voltage of terminal *a* of the synchronous machine with respect to ground is, therefore, given as

$$\mathbf{V}_{a0} = -\,\mathbf{I}_{a0}\,\mathbf{Z}_0 = -\mathbf{I}_{a0}\,(\mathbf{Z}_{g0} + 3\mathbf{Z}_n) \qquad ...(3.20)$$

3.10.2. Sequence Impedances of Induction Machines

In rotating machines, the impedances of the three phase sequences will generally be different. For example in case of induction machines, the positive sequence impedance is represented by the normal equivalent circuit of the machine. Negative sequence stator currents produce a magnetic field rotating at the same speed as under normal conditions but in opposite direction (against the rotor). As a result the field rotates relative to rotor at nearly twice its speed relative to the stator and many times its speed relative to the rotor under normal conditions. The currents induced in the rotor also go up considerably. By Lenz's law, they tend to reduce the magnetic field in a greater measure that under normal conditions, thereby bringing about a reduction in the emfs induced by the magnetic field in the stator windings. Thus the stator currents increase and as a consequence, the motor offers a low impedance to the negative sequence than to positive sequence currents (*i.e.*, $Z_2 < Z_1$) for the same applied positive and negative sequence voltages and the same speed and direction of rotation.

Because the zero sequence stator currents do not produce a rotating field the equivalent circuit may be considered of magnetizing and leakage impedances only. However, in practice, the presence of third harmonic currents and fluxes is an important consideration in the zero sequence analysis of induction machines.

3.10.3. Sequence Impedances of Transmission Lines

A fully transposed three-phase line is completely symmetrical and, therefore, the positive and negative sequence impedances of a transmission line are independent of phase sequence and are equal. The expression for inductive reactance, derived in Chapter 5 of Part II "Transmission and Distribution of Electrical Power" is valid for both positive and negative sequences. When only zero sequence currents flow in a transmission line, the currents in each phase are identical in both magnitude and phase. Such currents return partly through ground and the rest through overhead ground wires. The magnetic field due to flow of zero sequence currents through the transmission lines, ground wires and ground is very different from the magnetic field set up by the flow of positive or negative sequence currents. The zero sequence impedance (particularly the reactance) is about 2 to 4 times the positive sequence impedance.

3.10.4. Sequence Impedances and Networks of Transformers

The positive sequence series impedance of a transformer is equal to its leakage reactance (the resistance of the winding is usually small in comparison to the leakage reactance). Transformer being a static device, the positive and negative sequence impedances are identical, because the impedance is independent of phase order, provided the applied voltages are balanced. Thus, for a transformer

$$\mathbf{Z}_1 = \mathbf{Z}_2 = \mathbf{Z}_{\text{leakage}} \qquad ...(3.21)$$

The situation with 3-phase transformer is more complex with regard to zero sequence impedance because of the possibility of variety of connections. Assuming such transformer connections that zero sequence currents can flow on both sides, a transformer offers a zero sequence impedance slightly different from positive sequence impedance but the difference is so small that zero sequence impedance can be assumed equal to positive or negative sequence impedance. However, the zero sequence currents can flow through the winding connected in star only if the star point is grounded. Moreover, the zero sequence currents cannot flow in the windings if the star point is isolated. No zero sequence currents can flow in the lines connected to a delta-connected winding as no return path is available for these zero sequence currents. Zero sequence currents can, however, flow through the delta-connected windings themselves if any zero sequence voltages are induced in delta. This fact is illustrated in Fig. 3.16. These various conditions can be taken into account by the use of general circuit shown in Fig. 3.17. Z_0 is the zero sequence impedance of windings of the transformer. There are two series and two shunt switches—one series and one shunt switch for each side. The series switch of a particular side is closed if it is star grounded and the shunt switch is closed if that side is delta connected, otherwise they remain left open.

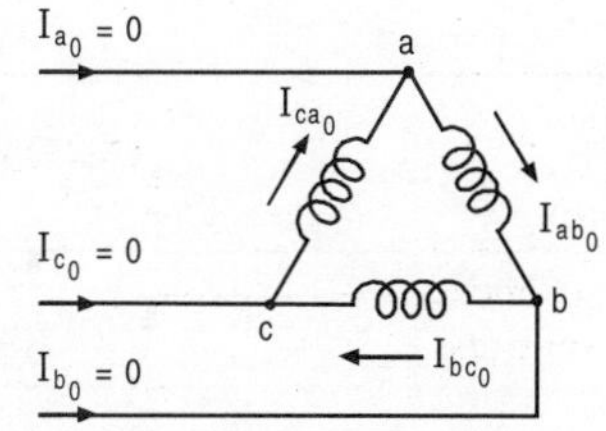

Fig. 3.16. ***Flow of Zero Sequence Current in Delta-Connected Windings***

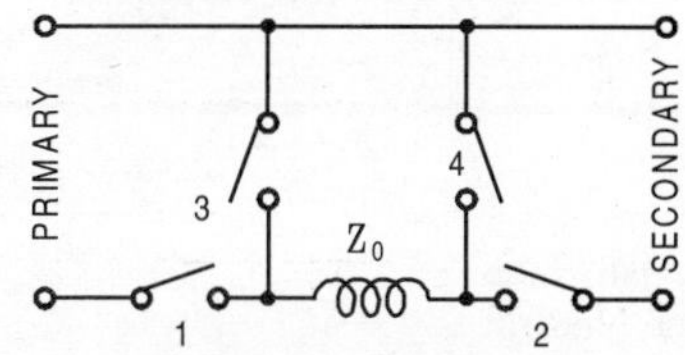

Fig. 3.17. ***General Circuit For Determination of Zero Sequence Network of a Transformer***

The switches need not be shown in the final network. The zero sequence representations of transformers for various winding arrangements are given in Table 3.1.

3.10.5. Sequence Impedance and Networks of Load

For a star-connected load with isolated neutral, there is no path for the flow of zero sequence currents and the star point of the load will not be connected to the reference bus, as shown in Fig. 3.18. Thus the zero sequence impedance is infinite beyond the neutral point (*i.e.*, $Z_n = \infty$). This fact is indicated by an open circuit in the zero sequence network between the neutral of the star-connected load and the reference bus in Fig. 3.18 (*b*).

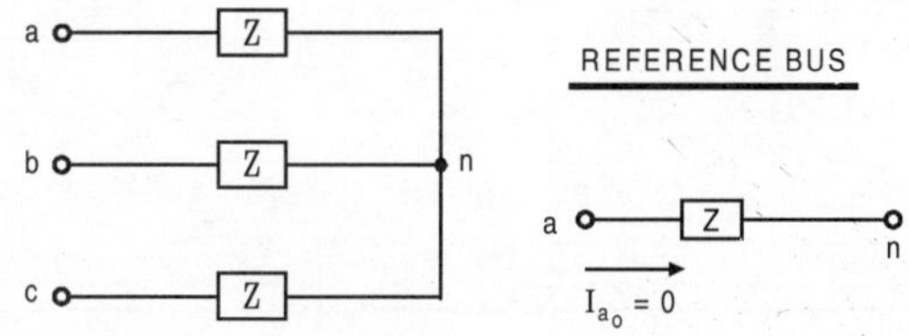

(a) Star-Connected Load With Isolated Neutral *(b) Zero Sequence Network*

Fig. 3.18

When the star point of the load is grounded, there will be continuity of the load circuit to ground and the star point of the load, *n* will be connected to the reference bus. The current flowing through the neutral under unbalanced condition of system is zero sequence current. If the neutral point of load is grounded through a reactor of impedance Z_n, the zero sequence voltage drop caused by flow of current $3I_{a0}$ through Z_n will be the same as if current I_{a_0} flows through $3Z_n$. Thus impedance $3Z_n$ is introduced between neutral point *n* and the reference bus, when representing on the zero sequence network.

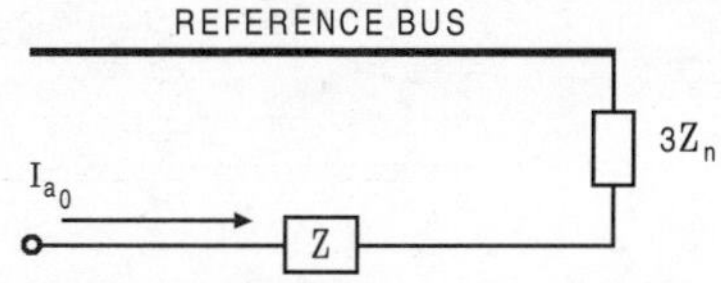

Fig. 3.19. ***Zero Sequence Network For Star-Connected Load With Neutral Grounded Through Impedance Z_n***

Table 3.1

Transformer Winding Arrangements	Connection Diagrams	Zero Sequence Equivalent Circuit	Remarks
1. Star-Star Transformer with Isolated Neutrals	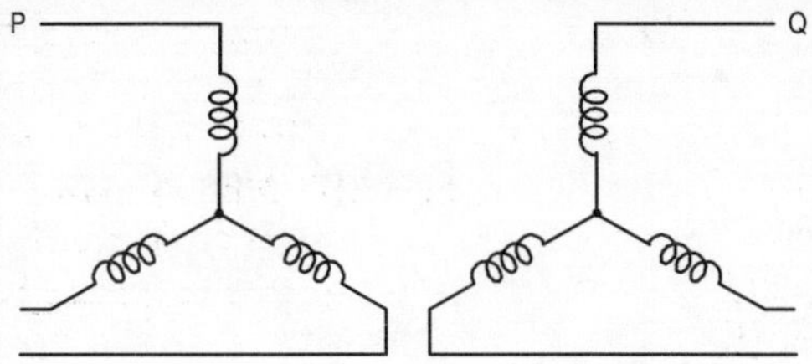	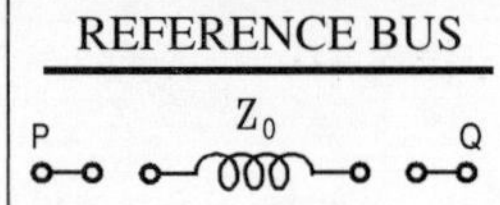	Since neutrals of both primary and secondary windings are isolated, zero sequence currents cannot flow. All the four switches are kept open.
2. Star-Star Transformer with Any One Neutral Grounded	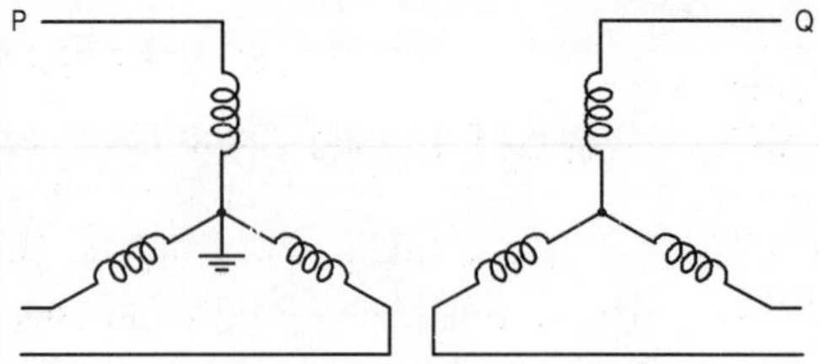	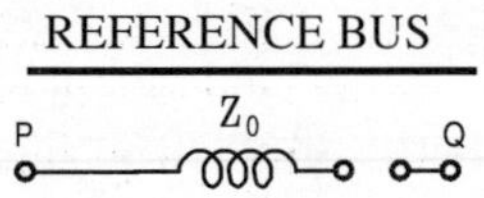	If any one of the two neutrals of a star-star transformer is ungrounded, zero sequence currents cannot flow in the ungrounded (or isolated) star and consequently, these cannot flow in the grounded star. Hence an open circuit exists in the zero sequence network between P and Q. Series switch 1 is closed and switches 2,3 and 4 are left open.
3. Star-Star Transformer with both Neutral Grounded	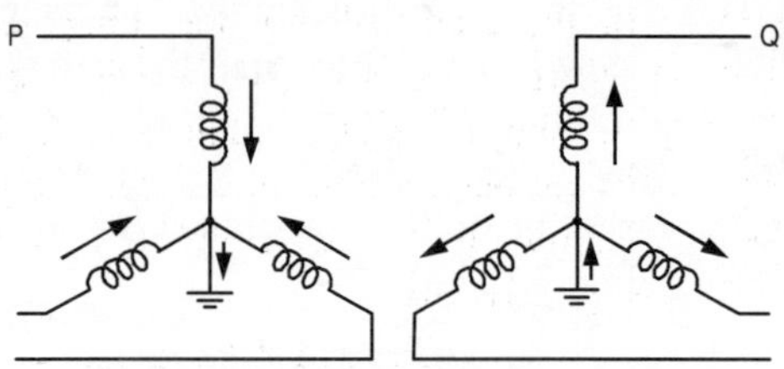	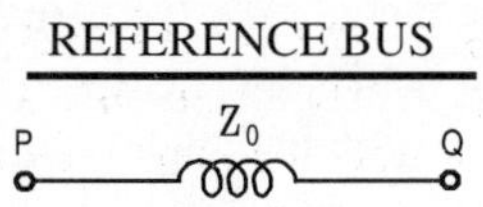	Since both of the neutrals of a star-star transformer are grounded, a path through the transformer exists for zero sequence currents in both windings via two grounded neutrals. Hence both series switches 1 and 2 are closed and shunt switches 3 and 4 are kept open *i.e.,* in the zero sequence network P and Q are connected by the zero sequence impedance (Z_0) of the transformer.
4. Star-Delta Transformer with Grounded Y-Neutral	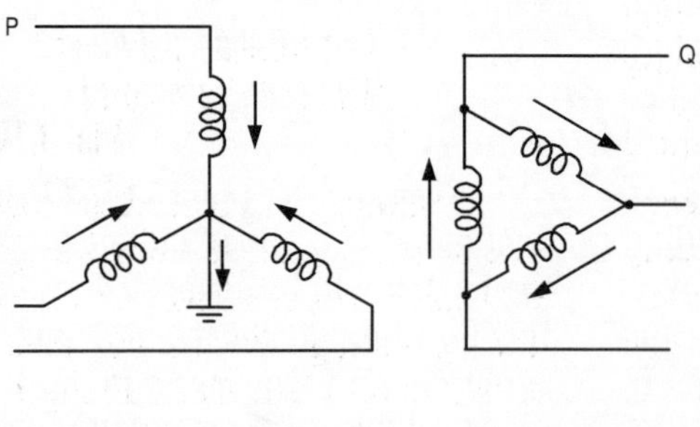 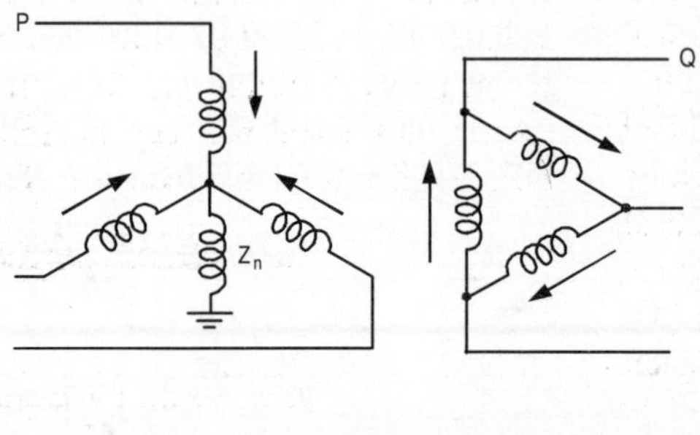	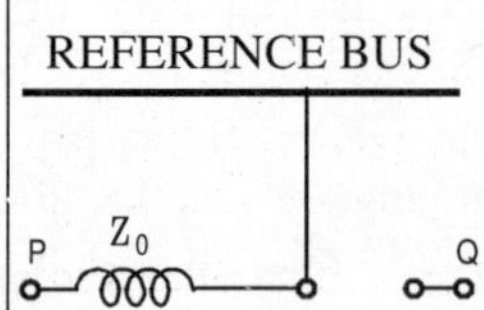 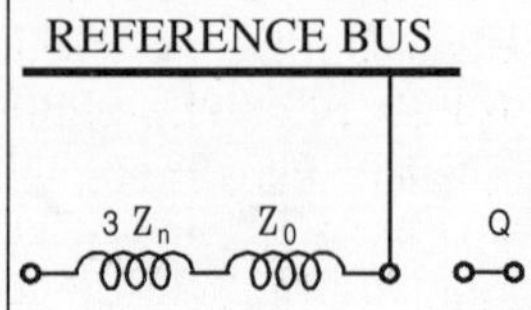	In this arrangement, the neutral of star side is grounded so zero sequence currents can flow in star and the balancing zero sequence currents can flow in delta. Of course no zero sequence currents can flow in the lines on delta side. Series switch 1 and shunt switch 4 are closed while series switch 2 and shunt switch 3 are kept open *i.e.*, the zero sequence network has a path from the line P on the star side through the zero sequence impedance of the transformer to the reference bus, while an open circuit exists on the line Q of delta side. If the star neutral is grounded through reactor of impedance Z_n, an impedance $3Z_n$ appears in series with Z_0 in the sequence network.

Transformer Winding Arrangements	Connection Diagrams	Zero Sequence Equivalent Circuit	Remarks
5. Star-Delta Transformer with Isolated Star	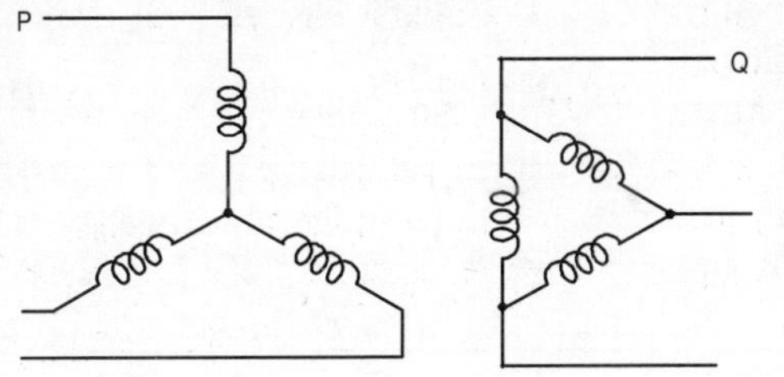	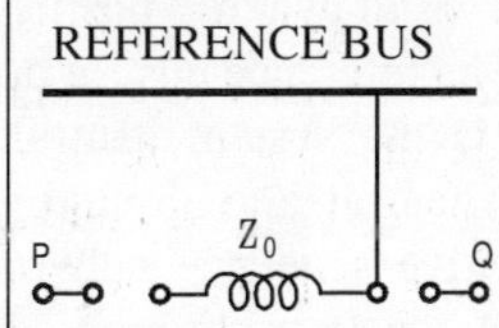	Since primary is star connected with neutral isolated, switches 1 and 3 are kept left open and secondary side is delta connected, the shunt switch 4 of secondary side is closed and series switch 2 is left open. In this arrangement no zero sequence current can flow in the transformer windings.
6. Delta-Delta Transformer	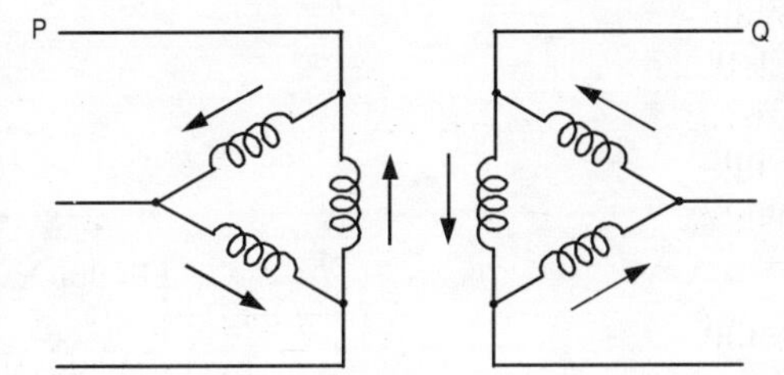	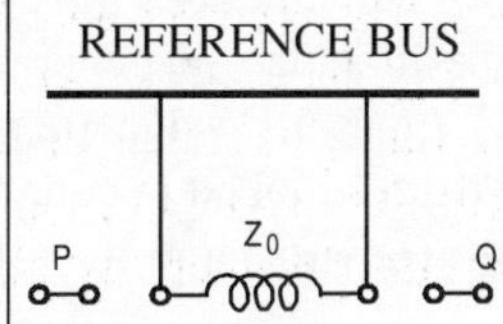	Since both the sides are delta connected, both series switches 1 and 2 are kept open and both shunt switches 3 and 4 are closed. Delta circuit does not provide return path, zero sequence currents cannot flow in or out of delta-delta transformer. However, zero sequence currents can circulate in the delta windings. Zero sequence impedance Z_0 is connected to the reference bus on both sides to account for any circulating current in the two deltas.
7. Star-Star with Any one Neutral Grounded and Tertiary	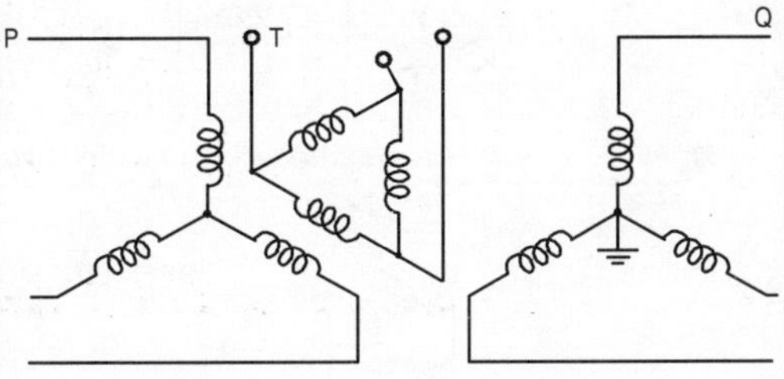	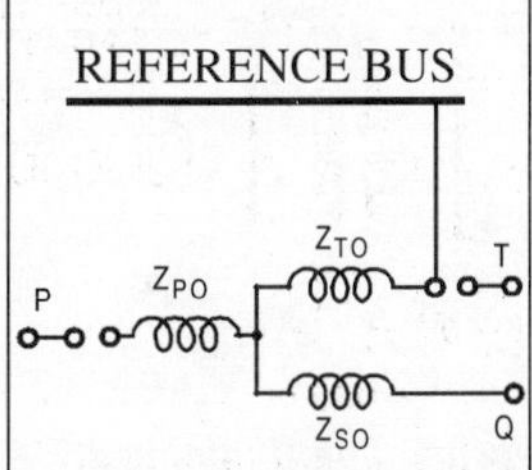	In this arrangement tertiary winding provides path for zero sequence currents.

The zero sequence impedance of the load is equal to its positive or negative sequence impedance if the load is balanced.

The positive sequence network is composed of positive sequence impedances only and since positive sequence currents do not flow to ground, therefore, neutral point *n* and the ground are at the same potential. Thus impedance inserted between neutral and ground has no effect on positive sequence currents.

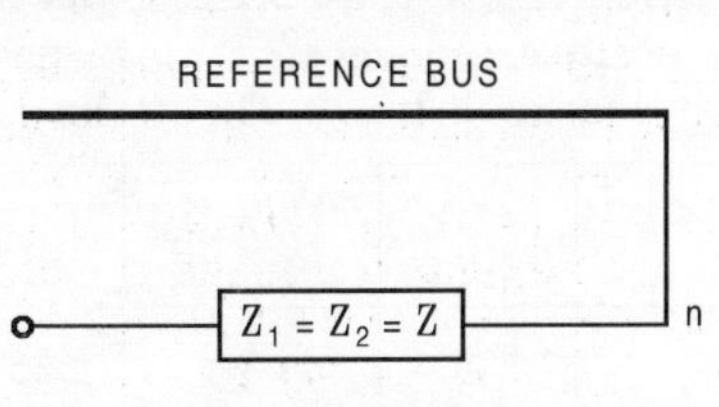

Fig. 3.20. ***Positive/Negative Sequence Network For Star-Connected Load***

The negative sequence impedance of a static load is the same as positive sequence impedance and so the negative sequence network is the same as the positive sequence network (Fig. 3.20).

3.11. FORMATION OF SEQUENCE NETWORKS OF POWER SYSTEMS

Sequence networks of power systems are very useful for computing unsymmetrical faults at different points of a power system network. The knowledge of positive sequence network is necessary for load studies on power systems. If the stability studies involve unsymmetrical faults, then negative and zero sequence networks are required also.

A power system network consists of synchronous machines, transformers and lines. The sequence networks of all these components have already been discussed in the preceding sections. Using these, complete sequence networks of a power system can be easily drawn. The positive sequence network is drawn by examining one line diagram of the power system. In fact, the single line reactance diagram, as employed for calculation of symmetrical fault current, is the positive sequence diagram of the power system.

The negative sequence network is quite similar to positive sequence network—only generators or rotating machines may have different sequence impedances and the negative sequence network does not contain any voltage source. The negative sequence impedances for transmission lines and transformers are the same as the positive sequence impedances. In many cases only one sequence network is drawn for positive and negative sequence representation. The reference bus for positive and negative sequence networks is the system neutral. Any impedance connected between a neutral and ground is not included in the positive and negative sequence networks as neither of these sequence currents can flow through such an impedance.

Zero sequence subnetworks for different elements of a power system can be easily combined to form complete zero sequence network. The zero sequence network does not contain any voltage source. Any impedance included in generator or transformer neutral becomes three times its value in a zero sequence network. Special care needs to be taken in connecting the zero sequence impedance of transformer, as already discussed in Art. 3.10.4.

All the sequence impedances are expressed in per unit values and referred to the same base MVA and base kV.

The procedure for drawing sequence networks is illustrated through the following Example 3.7.

Example 3.7. Draw the sequence networks for the power system shown in Fig. 3.21.

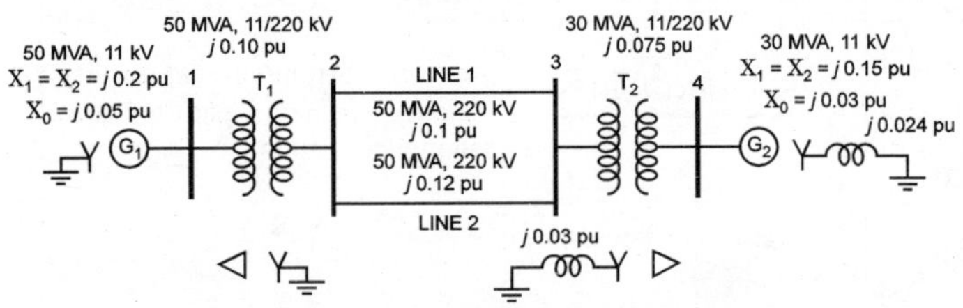

Fig. 3.21. *Power System Network For Example 3.7*

Solution : Let base MVA for complete system be 50 MVA, base kV for generator side 11 kV and transmission side 220 kV.

Per unit reactances of generator G_1

$$X_1 = X_2 = j\,0.2 \text{ pu}$$

$$\text{and } X_0 = j\,0.05 \text{ pu} \qquad \text{...(Given)}$$

Per unit reactances of generator G_2

$$X_1 = X_2 = j\,0.15 \times \frac{50}{30} = j\,0.25 \text{ pu}$$

$$\text{and } X_0 = j\,0.03 \times \frac{50}{30} = j\,0.05 \text{ pu}$$

Per unit reactances of transformer T_1

$$X_1 = X_2 = X_0 = j\,0.10 \text{ pu}$$

Per unit reactances of transformer T_2

$$X_1 = X_2 = X_0 = j\,0.075 \times \frac{50}{30} = j\,0.125 \text{ pu}$$

Per unit reactances of transmission line 1

$$X_1 = X_2 = 0.1 \text{ pu} \qquad \text{...(Given)}$$

Assuming zero sequence reactance of transmission line 3.5 times of its positive sequence reactance

$$X_0 = 0.35 \text{ pu}$$

Per unit reactances of transmission line 2

$$X_1 = X_2 = j\,0.12 \text{ pu} \qquad \text{...(Given)}$$

$$X_0 = j\,0.12 \times 3.5 = j\,0.42 \text{ pu}$$

The neutral reactances of $j\,0.024$ and $j\,0.03$ comes out to be $j\,0.024 \times \frac{50}{30}$ *i.e.*, 0.04 pu and $j\,0.03 \times \frac{50}{30}$ *i.e.*, j 0.05 pu. These neutral reactances appear as $3 \times j\,0.04$ *i.e.*, $j\,0.12$ pu and $3 \times j\,0.05$ *i.e.*, j 0.15 pu in the zero sequence networks.

The positive, negative and zero sequence networks are shown in Figs. 3.22 (*a*), 3.22 (*b*) and 3.22 (*c*) respectively.

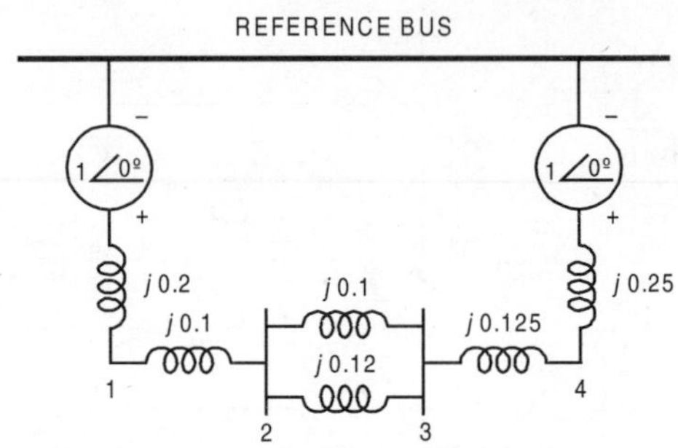

(*a*) *Positive Sequence Network For Example 3.7*

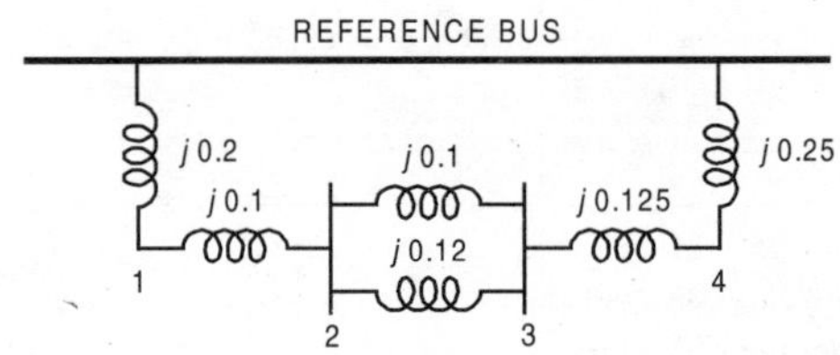

(*b*) *Negative Sequence Network For Example 3.7*

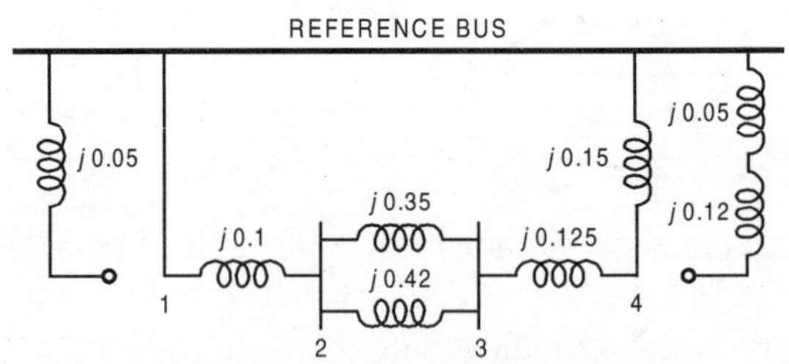

(*c*) *Zero Sequence Network For Example 3.7*

Fig. 3.22

Example 3.8. Given figure (Fig. 3.23) shows a power system network. Draw positive sequence network, negative sequence network and zero sequence network. The system data are as under:

Equipment	MVA Rating	Voltage Rating	X_1 pu	X_2 pu	X_0 pu
Generator-1(G_1)	100	11 kV	0.25	0.25	0.05
G_2	100	11 kV	0.2	0.2	0.05
T_1	100	11/220 kV	0.06	0.06	0.06
T_2	100	11/220 kV	0.07	0.07	0.07
Transmission Line-1	100	220 kV	0.1	0.1	0.3
Transmission Line-2	100	220 kV	0.1	0.1	0.3

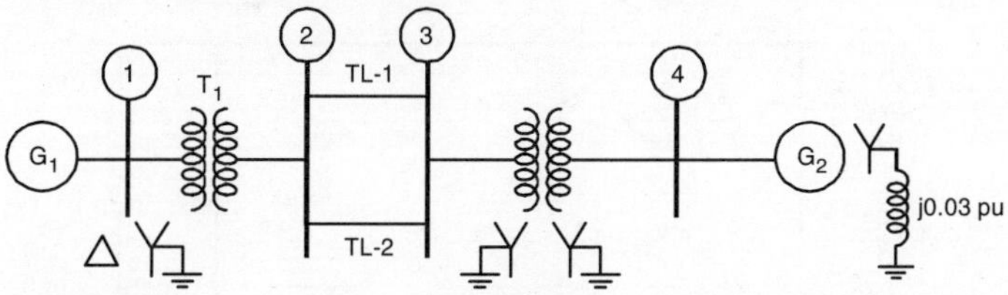

Fig. 3.23

Solution : Let Base MVA be 100 MVA, Base voltage 11 kV for LT side of transformer and 220 kV for HT side of transformers. The positive, negative sequence networks are shown in Fig. 3.24.

Since positive and negative sequence impedances of all the elements are equal, the positive and negative sequence networks are similar except that the voltage sources are present only in the positive sequence network. The neutral reactance of $j0.03$ pu appears as three times *i.e.*, $j0.09$ pu in the zero sequence network.

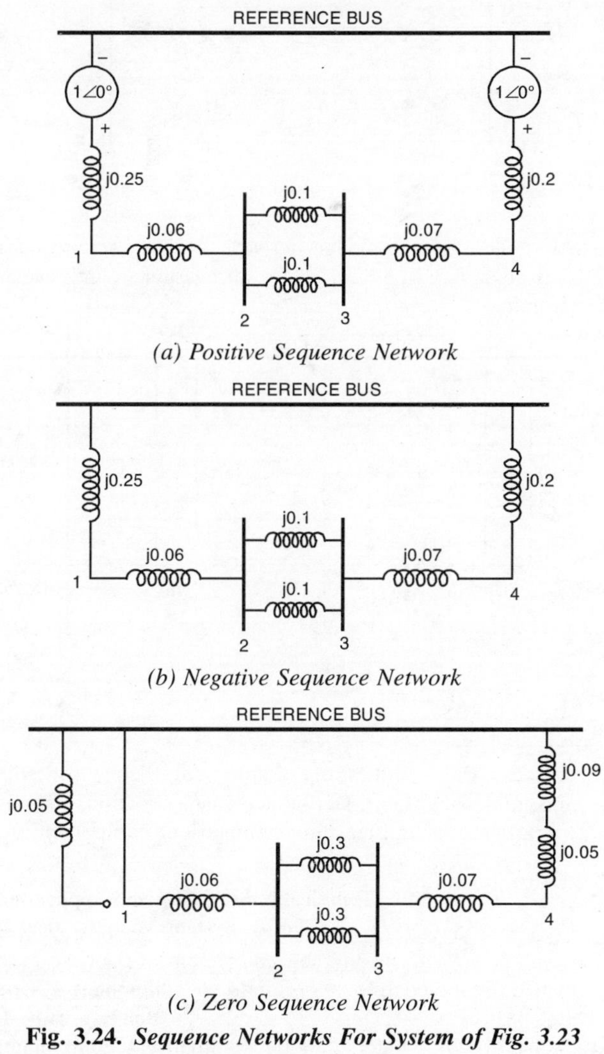

Fig. 3.24. ***Sequence Networks For System of Fig. 3.23***

Example 3.9. A 25 MVA, 11 kV, 3-phase generator has a subtransient reactance of 20%. The generator supplies two motors over a transmission line with transformers at both ends. The motors have rated inputs of 15 and 7.5 MVA, both 10 kV with 25% subtransient reactance. The 3-phase transformers are both rated 30 MVA, 10.8/121 kV connected in Δ-Y with leakage reactance of 10% each. The series reactance of the line is 100 ohms. Draw the positive, and negative, sequence networks of the system with reactances marked in per unit.

[U.P. Technical Univ. Power System Analysis 2007-08]

Solution : Let the base MVA for the complete system be 25 MVA and base kV for generator side 11 kV.

Per unit reactance of generator $X_1 = X_2 = 20\% = 0.2$ pu

Per unit reactance of transformers $X_1 = X_2 = \frac{10}{100} \times \frac{25}{30} \times \left(\frac{10.8}{11}\right)^2$

$= 0.0803$ pu

Transmission line voltage base $= 11 \times \frac{121}{10.8} = 123.24$ kV

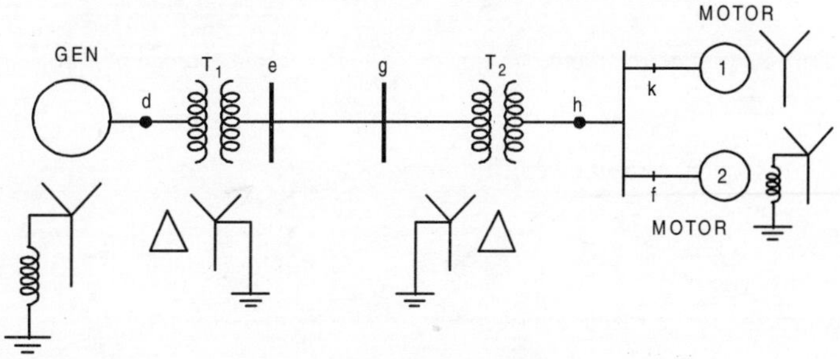

Fig. 3.25. ***Generator***

$$\text{Line reactance} = 100 \times \frac{25}{(123.2)^2} = 0.165 \text{ pu}$$

$$\text{Motor base voltage} = 123.24 \times \frac{10.8}{121} = 11 \text{ kV}$$

$$\text{Per unit reactance of motor 1} = \frac{25}{100} \times \frac{25}{15} \times \left(\frac{10}{11}\right)^2 = 0.3444 \text{ pu}$$

$$\text{Per unit reactance of motor 2} = \frac{25}{100} \times \frac{25}{7.5} \times \left(\frac{10}{11}\right)^2 = 0.689 \text{ pu}$$

The required positive sequence network is shown in Fig. 3.26.

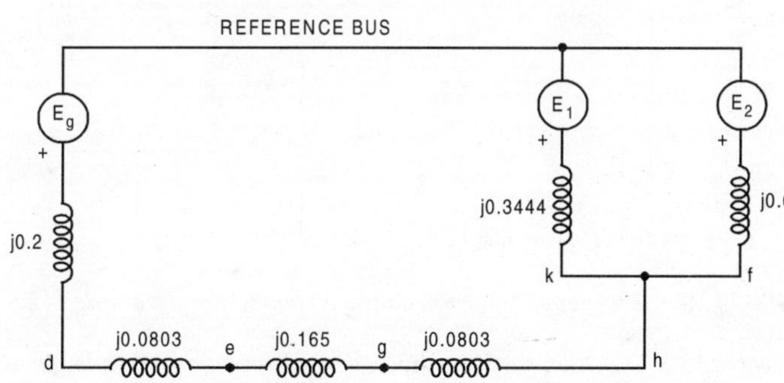

Fig. 3.26. ***Positive-Sequence Network For Example 3.9***

Since all the negative sequence reactances of the system are the same as the positive sequence reactances, the negative sequence network is identical to the positive sequence network but for the omission of voltage sources. The negative sequence network is shown in Fig. 3.27.

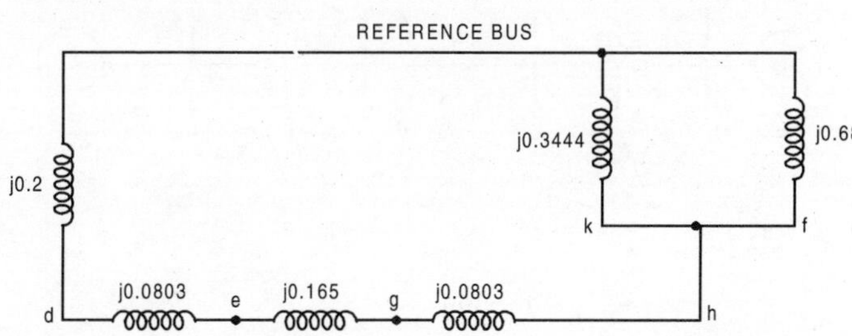

Fig. 3.27. ***Negative-Sequence Network For Example 3.9***

Example 3.10. Draw the per unit zero sequence impedance network for the power system represented in Example 2.16. The zero sequence reactance of each machine is 8% based on its own rating. Assume that the zero sequence reactance of transmission line is 300% of its positive sequence reactance. Neutrals of machines G_1 and M_1 are grounded through 5% reactance of their own ratings. [*U.P. Technical Univ. Power System Analysis 2005-06*]

Solution : Zero sequence reactance of generator $= j8\% = j0.08$ pu

Zero sequence reactance of transformer, $T_1 = j0.0857$ pu

Zero sequence reactance of transformer, $T_2 = j0.0915$ pu

$$\text{Zero sequence reactance of line} = \frac{300}{100} \times \text{ pu} \times \text{reactance of line}$$

$$= 3 \times j0.1815 \text{ pu} = j0.5445 \text{ pu}$$

Zero sequence reactance of motor,

$$M_1 = \frac{8}{100} \times \frac{300}{200}\left(\frac{13.2}{13.8}\right)^2 = j0.11 \text{ pu}$$

Zero sequence reactance of motor,

$$M_2 = \frac{8}{100} \times \frac{300}{100} \times \left(\frac{13.2}{13.8}\right)^2 = j0.22 \text{ pu}$$

The neutral reactance for generator = 5% = 0.05 pu

The neutral reactance for motor,

$$M_1 = \frac{5}{100} \times \frac{300}{200} \times \left(\frac{13.2}{13.8}\right)^2 = j0.0686 \text{ pu}$$

The per unit zero sequence impedance network for the given power system is shown in Fig. 3.28.

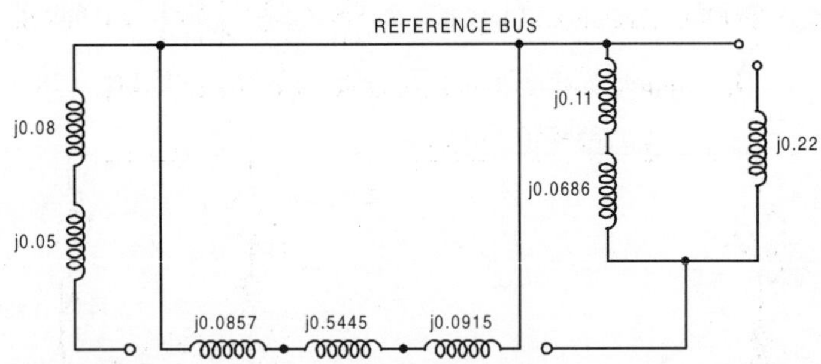

Fig. 3.28. ***Zero-Sequence Impedance Network For Example 3.10***

Example 3.11. One line diagram of a small power system is given below (Fig. 3.29). Draw the zero sequence network. The zero reactances of various components are denoted as follows :

Generator 1 = X_{10}; Generator 2 = X_{20}; Transformer $T_1 = X_{T10}$; Transformer $T_2 = X_{T20}$.

[*U.P.S.C. I.E.S. Electrical Engineering-II, 2009*]

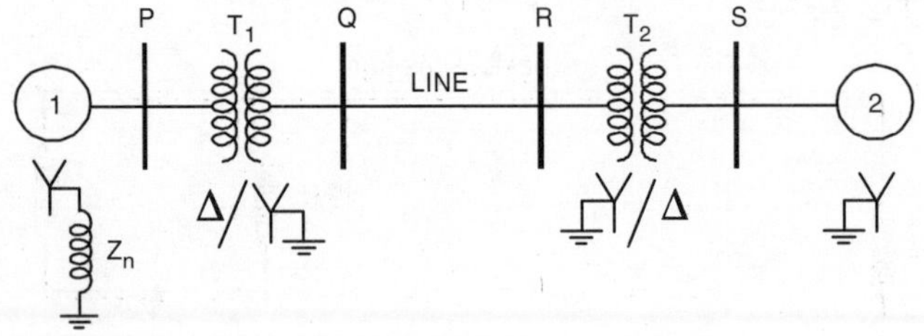

Fig. 3.29

Solution : The zero sequence network for the system shown in Fig. 3.29 is given in Fig. 3.30.

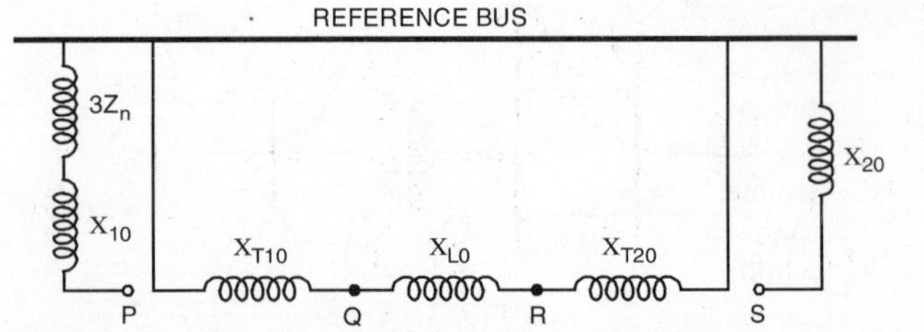

Fig. 3.30. ***Zero Sequence Network For System of Fig. 3.29***

Example 3.12. Draw the zero sequence network for the system shown in Fig. 3.31. [*Rajasthan Univ. Power System Analysis 2000*]

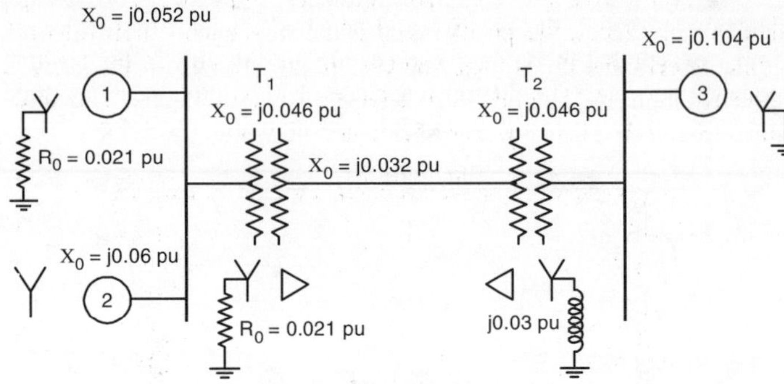

Fig. 3.31

Solution : The zero sequence network for the system given in Fig. 3.31 is given in Fig. 3.32 in which all resistances and reactances are in per unit.

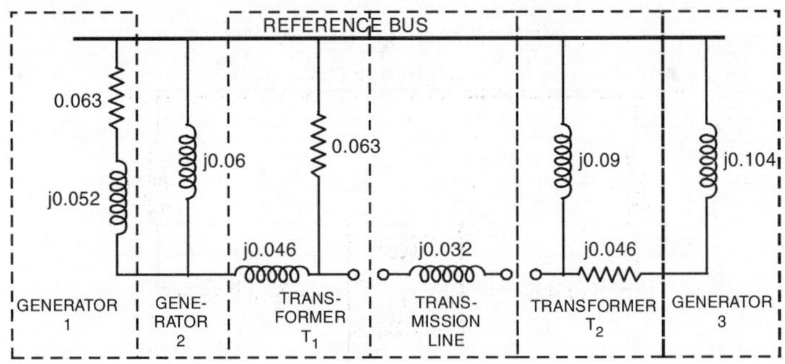

Fig. 3.32

Exercises

1. Discuss the symmetrical component theory for solving unsymmetrical fault. Derive necessary equations to convert:
   (*i*) Phase quantities into symmetrical components.
   (*ii*) Symmetrical components into phase quantities.
   [Pb. Technical Univ. Computer Applications to Power System Analysis, Dec. 2010]
2. Explain how an unsymmetrical system of 3-phase currents can be resolved into 3 symmetrical component systems.
   [Bombay Univ. 1986]
3. What do you understand by symmetrical components of unbalanced phasors ? Deduce the expressions for symmetrical components.
   [U.P. Technical Univ. Power System Analysis 2006-07]
4. How 3-phase unbalanced phasors are transformed into their symmetrical components ? Discuss.
   [U.P. Technical Univ. Power System Analysis 2005-06]
5. Explain what is the difference between symmetrical components of positive, negative and zero phase sequence.
6. How are the zero phase sequence components of the line currents related to the current in the neutral wire of a 3-phase, 4-wire circuit ?

7. Discuss the principle of symmetrical components. Derive equations to convert phase quantities into symmetrical components.
[Rajasthan Technical Univ. Power System Analysis 2011]

8. Explain why does the line current of a 3-phase, 3-wire circuit has no zero phase sequence component.

9. Explain the symmetrical component theory. Derive the phase currents in terms of the sequence currents and vice versa. Show that in the delta-connected system, the zero phase sequence voltage is absent in the line voltages.
[Agra Univ. 1970]

10. Show that the current through the neutral wire is equal three times zero sequence current.
[U.P. Technical Univ. Power System Analysis 2004-05]

11. Derive an expression for total complex power in a 3-phase system in terms of symmetrical components of voltages and currents. **[U.P.S.C. I.E.S. Electrical Engineering-II, 2001]**

12. Derive expression for 3-ϕ system in terms of symmetrical components. **[M.D. Univ. Power System-I, 2010-11]**

13. Find sequence impedances of transmission line's of a power system carrying unbalanced currents.
[Power System Analysis Rajasthan Univ. 2003, 2008; Rajasthan Technical Univ. 2010]

14. Prove that for the transmission line's zero sequence impedance is much larger than the positive sequence impedance. Also discuss the effect of mutual inductances.
[G.B. Technical Univ. Power System Analysis 2011-12]

15. Find sequence impedances and draw sequence networks of a synchronous machine.
[Rajasthan Univ. Power System Analysis 2003]

16. Show that the sequence impedances of a star-connected with neutral grounded through an impedance Z_n is given as :

$$Z^{0,\,1,\,2} = \begin{bmatrix} Z_s + 3Z_n & 0 & 0 \\ 0 & Z_s & 0 \\ 0 & 0 & Z_s \end{bmatrix}$$

Where Z_s is the impedance of the load per phase.
[U.P. Technical Univ. Power System Analysis 2007-08]

17. Show that the zero sequence impedance of a generator with neutral grounded through an impedance of Z_n is $(Z_s + 3Z_n)$, where Z_s is the synchronous impedance of the generator.
[U.P. Technical Univ. Power System Analysis 2006-07]

18. Draw a general circuit, which can be used to determine the zero sequence network of a two winding transformer. Draw the zero sequence network of : (*i*) Star-Star transformer with star point grounded, (*ii*) Delta-Delta transformer.
[Rajasthan Univ. Power System Analysis 2007, 2008]

19. Discuss sequence impedances of transformers. Also draw the transformer connection and their equivalent sequence networks. **[Rajasthan Univ. Power System Analysis 2005, 2006]**

20. Draw the zero sequence networks of transformer banks.
[Anna Univ. Chennai Power System Analysis, May/June 2011]

21. What do you understand by sequence networks ? What is their importance in unsymmetrical fault calculations ?
[U.P. Technical Univ. Power System Analysis 2009-10]

22. For the power system whose line diagram is given below, draw the zero sequence network.

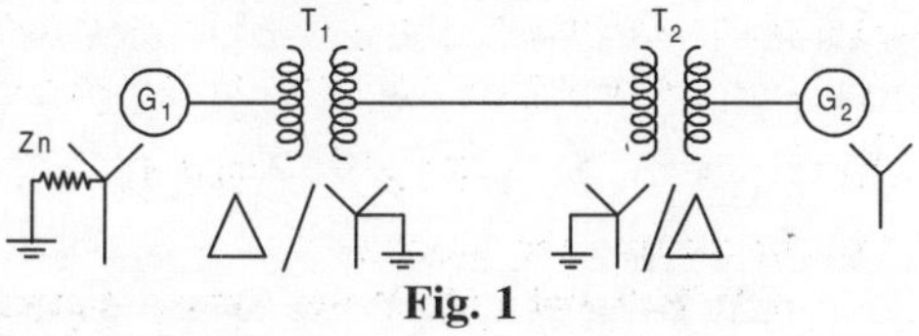

Fig. 1

[U.P. Technical Univ. Power System Analysis 2007-08]

Short Answer Type Questions With Answers

Q. 1. What do you mean by Fortescue theorem.
[Rajasthan Univ. 2004, 2007, 2008; Rajasthan Technical Univ. 2011]

Ans. According to Fortesque theorem, any unbalanced three-phase system of currents, voltages or other sinusoidal quantities can be resolved into three balanced systems of phasors which are called *symmetrical components* of the original unbalanced system. Such three balanced systems constitute three sequence networks which are solved separately on a single phase basis.

Q. 2. What are symmetrical components ?
[Madurai Kamaraj Univ. November 1996]

Ans. According to the theory of symmetrical components, any unbalanced system of phasors (voltages, currents or impedances etc.) can be resolved into balanced systems of phasors equal in number to the number of phases. These balanced systems of phasors are called the symmetrical components of the original unbalanced system.

Q. 3. What are the uses of symmetrical components in power systems ? **[Bharathidasan Univ. November 1996]**

Ans. Symmetrical component method provides a useful analytical tool for the protection engineer and is useful in system fault calculation and system stability problems etc.

Q. 4. Differentiate between the positive sequence system and zero sequence system.

Ans. The positive sequence system is a set of three symmetrical voltages (*i.e.,* all numerically equal and all displaced from each other by 120°) having the same phase sequence as the original set of three unbalanced voltages *i.e.,* A-B-C whereas the zero sequence system is a set of three voltages, all equal in magnitude and in phase with each other.

Q. 5. Evaluate $a^2 + a + 1$

Ans. $a^2 + a + 1 = (-0.5 - j\,0.866) + (-0.5 + j\,0.866) + 1 = 0$

Q. 6. Define negative sequence impedance. **[Madras Univ. April 1997]**

Ans. The impedance of the network offered to the flow of negative sequence currents is called the negative sequence impedance of the network.

Q. 7. What are sequence networks ?
[Madurai Kamaraj Univ. November 1996]

Ans. The single phase equivalent circuit composed of impedances to current of any one sequence only is called the sequence network for that particular sequence. Hence corresponding to positive, negative and zero sequence currents we have positive, negative and zero sequence networks.

Q. 8. Draw the zero sequence network for a star-connected generator earthed through a resistance R. **[Madras Univ. April 1997]**

Ans.

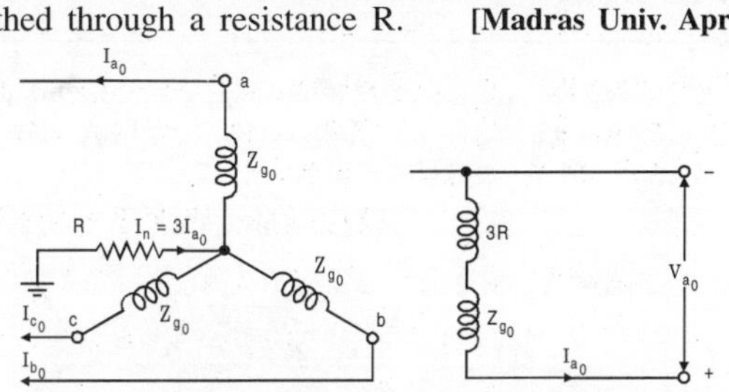

(*a*) *Three-Phase Model* (*b*) *Single-Phase Model*

Fig. 2

Q. 9. The neutral grounding impedance Z_n appears as 3 Z_n in the zero sequence equivalent circuit why ?
[Bharathidasan Univ. November 1996]

Ans. As the current flowing in the neutral grounding impedance Z_n is the sum of the zero sequence currents in all the three phases, hence voltage drop caused by it will be $3I_{a0}\,Z_n$. That is why neutral grounding impedance Z_n appears as 3 Z_n in the zero sequence equivalent circuit as shown in Fig. 3.15 (*b*).

Q. 10. Draw the zero sequence networks for

(*a*) 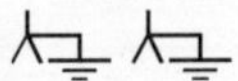 (*b*) Δ – Δ connected transformers

[Madras Univ. April 1997]

Ans. (*a*) 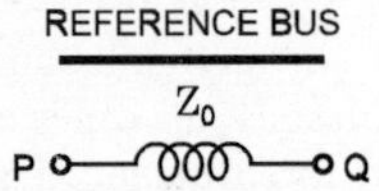

(*b*) 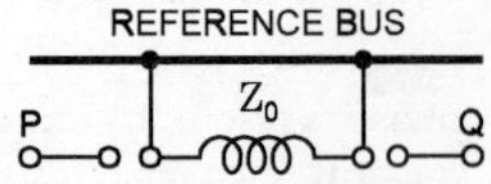

Q. 11. Draw zero sequence impedance networks for the below three-phase transformer connections.

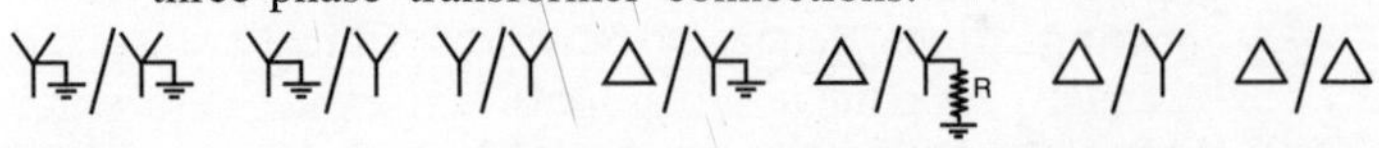

[U.P.S.C. I.E.S. E.E.-II, 2007]

Ans. Zero sequence equivalent circuits are given below :

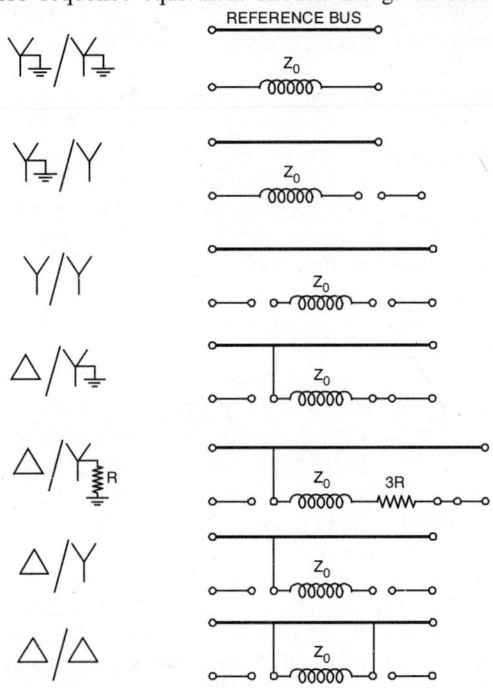

Q. 12. Draw the zero sequence networks of the transformer connection shown below:

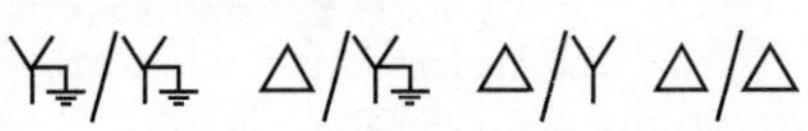

[U.P.S.C. I.E.S. E.E.-II, 2011]

Ans. Refer to Answer of Q. 11.

Problems

1. Determine the values of symmetrical components of a system of currents $\mathbf{I}_R = 0 + j\,120$; $\mathbf{I}_Y = 50 - j\,100$ A and $\mathbf{I}_B = -100 - j\,50$ A. Phase sequence is RYB.

[Agra Univ. 1976; Bombay Univ. 1986]

[**Ans.** $\mathbf{I}_{R1} = 110.7 \angle 78.2°$ A; $\mathbf{I}_{Y1} = 110.7 \angle -41.8°$ A and $\mathbf{I}_{B1} = 110.7 \angle 198.2°$ A; $\mathbf{I}_{R2} = 22.54 \angle 105.7°$ A; $\mathbf{I}_{Y2} = 22.54 \angle 225.7°$ A and $\mathbf{I}_{B2} = 22.54 \angle -14.3°$ A $\mathbf{I}_0 = 19.44 \angle -149°$ A]

2. The currents in a 3-phase system are $\mathbf{I}_R = 12 + j6$; $\mathbf{I}_Y = 12 - j12$ and $\mathbf{I}_B = -15 + j10$. Find the symbolic expression for the three symmetrical components. **[Allahabad Univ. 1974]**

[**Ans.** $\mathbf{I}_{R1} = 10.85 + j10.13$; $\mathbf{I}_{Y1} = 3.66 + j14.44$;
$\mathbf{I}_{B1} = -14.2 + j4.31$ amperes.
$\mathbf{I}_{R2} = -1.85 - j5.47$; $\mathbf{I}_{Y2} = 5.7 + 1.13$;
$\mathbf{I}_{B2} = -3.83 + j4.36$ amperes.
$\mathbf{I}_{R0} = \mathbf{I}_{Y0} = \mathbf{I}_{B0} = (3 + j1.33)$ amperes]

3. Determine the positive, negative and zero phase sequence components of the following set of three phase unbalanced voltage phasors:

$\mathbf{V}_A = 10 \angle 30°$; $\mathbf{V}_B = 30 \angle -60°$ and $\mathbf{V}_C = 15 \angle 145°$;

[**Ans.** $\mathbf{V}_1 = 17.6 \angle 45°$; $\mathbf{V}_2 = 8.24 \angle -156.2°$ and $\mathbf{V}_0 = 5.6 \angle -47.4°$]

4. The following currents were in R, Y and B lines of a 3-phase system under abnormal conditions:

$\mathbf{I}_R = 450 \angle 300°$ A; $\mathbf{I}_Y = 750 \angle 240°$ A and $\mathbf{I}_B = 1{,}500 \angle 60°$ A

Determine positive, negative and zero phase sequence components.

[**Ans.** $804 \angle -44.3°$ A; $558 \angle 171°$ A; $217.5 \angle 23.3°$ A]

5. For the circuit shown in Fig. 1, find symmetrical components of current in line *a*.

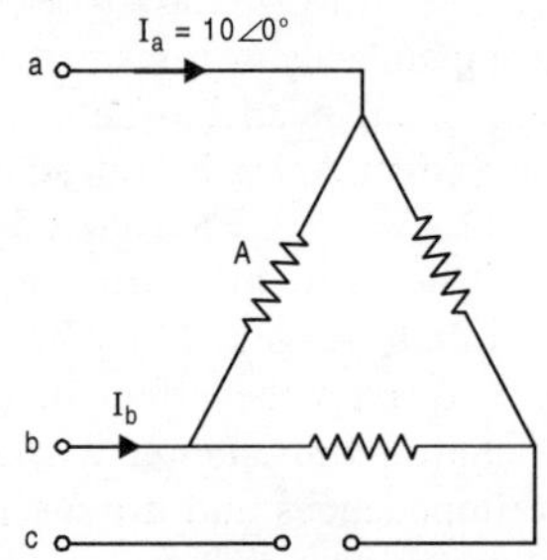

Fig. 1

[U.P. Technical Univ. Power System Analysis 2004-05]

$$\left[\textbf{Ans. } \frac{10}{\sqrt{3}} \angle 150°\text{ A}, \frac{-10}{\sqrt{3}} \angle 30°\text{ A}, 0\right]$$

6. Draw the sequence networks for the power system shown in Fig. 2.

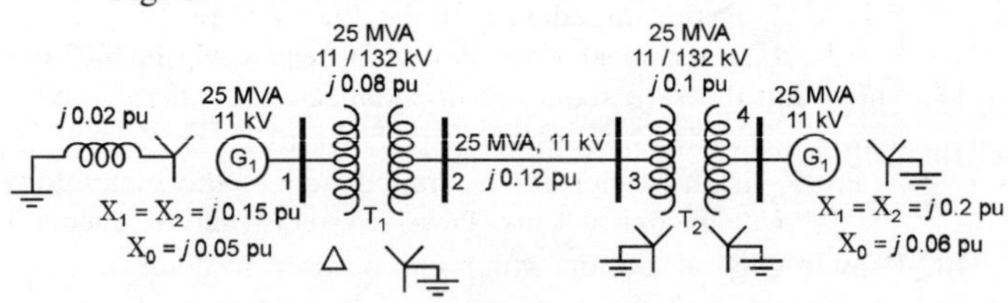

Fig. 2

7. Draw the positive sequence, negative sequence and zero sequence reactance diagrams for the system shown below (Fig. 3).

[Madurai Kamaraj Univ. Power System Analysis Nov. 1996]

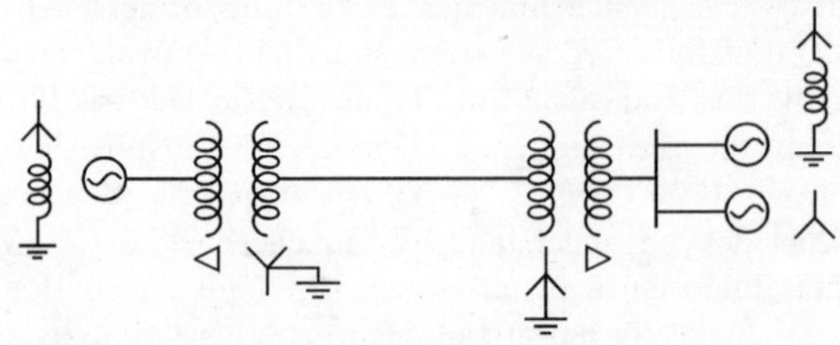

Fig. 3

8. When a generator has terminal 'A' open and the other two terminals are short circuited to ground, typical value for the symmetrical components of current in phase 'A' are:

$I_{a_1} = 600 \angle -90°$ A, $I_{a_2} = 250 \angle 90°$ A and $I_{a_0} = 350 \angle 90°$ A.

Find the current into the ground.

[U.P. Technical Univ. Power System Analysis 2005-06]

[**Ans.** $1{,}050 \angle 90°$ A]

❑❑❑

CHAPTER 4

Symmetrical Fault Analysis

INSIDE THIS CHAPTER

4.1. INTRODUCTION

The steady state operation of a 3-phase ac power system is balanced one. However, sudden external or internal changes in the system, disrupt this condition. Failure of insulation of the system at one or more points or coming of a live point in contact with a conducting object resuits in a short circuit or occurrence of fault. The causes of faults are numerous as already detailed in chapter 1. As already mentioned in chapter 1, the faults may be symmetrical or unsymmetrical. A fault involving all the three phases is known as *symmetrical (balanced) fault* while one involving only one or two phases is known as *unsymmetrical fault*. Majority of the faults are unsymmetrical. The system needs protection against flow of heavy short-circuit currents (which may cause permanent damage to the major equipment) by isolation of the faulty portion of the system by means of circuit breakers operated by protective relaying. For proper choice of circuit breakers and protective relaying, it is necessary to perform short circuit calculations. Such calculations enable the fault MVA due to a symmetrical 3-phase fault at a point of interest to be determined.

4.2. SWITCHING OPERATION IN AN R-L SERIES CIRCUIT

As already mentioned, the selection of the circuit breaker does not depend only on the current carrying capacity under normal conditions ; but also depends upon the momentary maximum current it may have to carry under abnormal or short-circuit conditions. Thus it becomes necessary to determine the initial maximum value of current at the time of short circuit. To approach the problem of determining the initial value when a short circuit occurs, there is a need of studying the transient in an R-L series circuit which is the most fundamental type of ac circuit.

Consider an R-L series circuit, which is connected, at the instant $t = 0$, to a source of alternating voltage $v = V_{max} \sin(\omega t + \alpha)$ where α is the phase displacement between the voltage v and the reference wave which passes through zero at the time $t = 0$.

The equation relating the applied voltage and the current in the given circuit is given as

$$v = iR + L\frac{di}{dt}$$

or $$L\frac{di}{dt} + iR = V_{max} \sin(\omega t + \alpha) \qquad ...(4.1)$$

The complete solution of the above equation consists of two parts which are called the particular integral and complementary function. The particular integral is the solution corresponding to the steady-state conditions, namely

$$i = \frac{V_{max}}{\sqrt{R^2 + \omega^2 L^2}} \sin(\omega t + \alpha - \phi)$$

$$= \frac{V_{max}}{Z} \sin(\omega t + \alpha - \phi)$$

where $\sqrt{R^2 + (\omega L)^2} = Z$, the circuit impedance, ϕ is the phase angle between the current and voltage determined by $\phi = \text{Tan}^{-1}\frac{\omega L}{R}$.

The above solution is obtained without any constant of integration and cannot, therefore, represent initial conditions. Complementary function is, therefore, necessary to represent the initial conditions. This is the solution of the equation with zero substituted for $V_{max} \sin(\omega t + \alpha)$, namely the solution of

$L\frac{di}{dt} + Ri = 0$ which is $i = Ae^{\frac{-R}{L}t}$ and complete solution of Eq. (4.1) is given as

$$i = \frac{V_{max}}{Z}\sin(\omega t + \alpha - \phi) + Ae^{\frac{-R}{L}t} \qquad ...(4.2)$$

Now when $t = 0$; $i = 0$ hence substituting these values in above Eq. (4.2), we have

$$0 = \frac{V_{max}}{Z}\sin(\alpha - \phi) + A$$

or $$A = -\frac{V_{max}}{Z}\sin(\alpha - \phi)$$

Hence Eq. (4.2) may be written as

$$i = \frac{V_{max}}{Z}\sin(\omega t + \alpha - \phi) - \frac{V_{max}}{Z}\sin(\alpha - \phi)\, e^{\frac{-R}{L}t}$$

or $$i = I_{max}\sin(\omega t + \alpha - \phi) - I_{max}\sin(\alpha - \phi)\, e^{-t/\lambda}. \qquad ...(4.3)$$

The first term in the above Eq. (4.3) for i is current corresponding to the steady-state condition and the second term is a transient which vanishes theoretically after infinite time. But practically, it vanishes very quickly after two or three cycles.

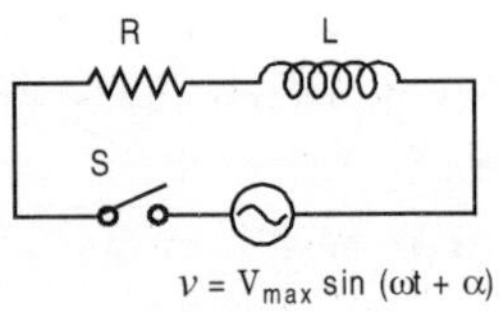

(*a*) *R-L AC Series Circuit*

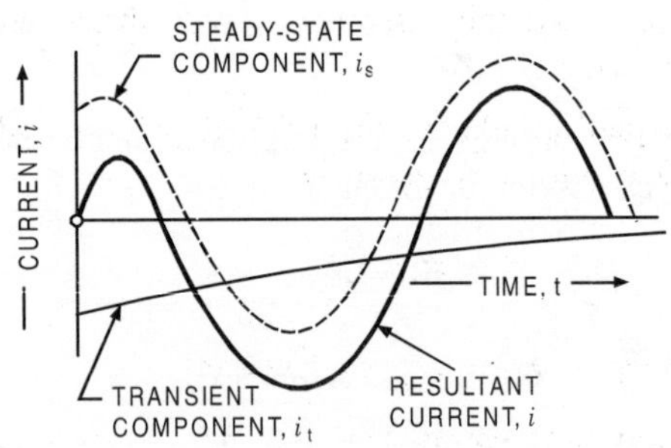

(*b*)

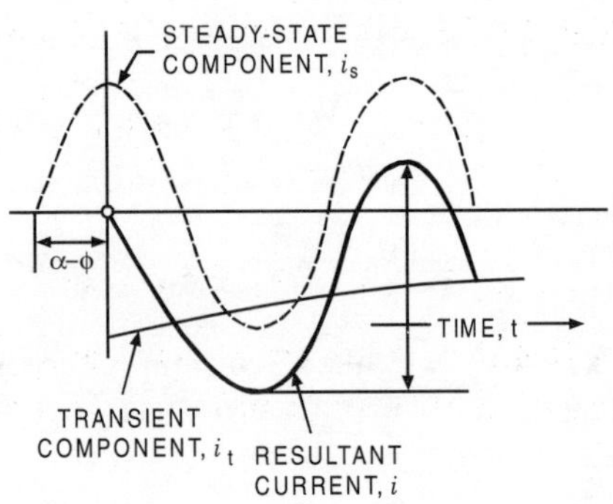

(*c*)

Fig. 4.1

The total current (i) curve is shown in Fig. 4.1 (*b*). As obvious from the figure with decreasing transient current i_t, the total current i tends towards the forced* current value. However, in the time interval between T/4 and 3 T/4 after switching, depending upon the phase angle α, the current value may exceed the peak value of the forced current.

The total (or resultant) current can take the maximum possible value when at the closure of the switch the forced current is at its peak and the time constant is large (*i.e.*, when R is approximately zero, the time constant tends to infinity, and the phase angle ϕ is about 90°), or when the transient decay very slowly. Under such conditions α is about equal to 180°, and the applied voltage should pass through zero at switch closure. The current waveform for $\alpha - \phi = 90°$ and for sufficiently large value of time constant is shown in Fig. 4.1 (*c*). In about half a cycle after switch closure, the resultant current is twice the peak value of forced current. This is known as *doubling effect.*

Thus when a sinusoidal voltage is applied to an R-L network, the resultant current cannot exceed *twice the peak value* of the forced current under any circumstances.

The initial value of the free current is equal and opposite to that of the forced current. Therefore, if at the closure of the switch the forced current passes through zero, the initial value of free component is also zero. Thus there will be no transient current flowing and the circuit will attain a forced state at once. According to Eq. (4.3) this occurs at

$$\alpha - \phi = 0 \quad \text{or} \quad \alpha - \phi = \pi.$$

Thus if the switch happens to be closed when v is maximum, a current wave which is symmetrical and has a maximum instantaneous value equal to I_{max} results. It may also be noted that closing of switch at other instants will give *asymmetrical current* waves with small transients.

4.2.1. Transient on a Transmission Line

If some simplifying assumptions are made, then a transmission line can be represented by the circuit model shown in Fig. 4.2 (*a*), which is equivalent to an R-L series circuit shown in Fig. 4.1 (*a*).

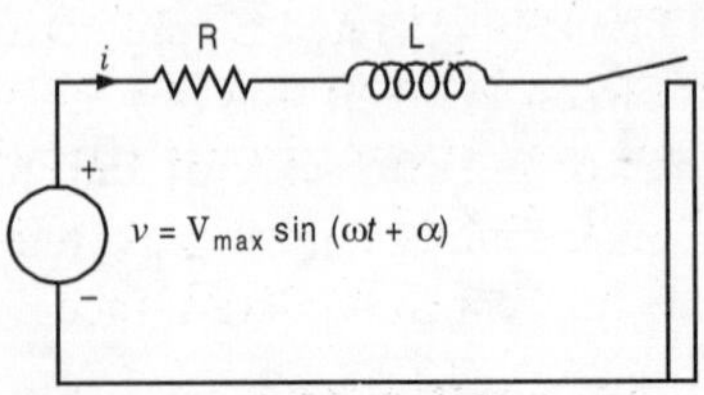

Fig. 4.2(*a*)

Simplifying assumptions made are :

1. The line is supplied from a constant voltage source.
2. Short circuit takes place when the line is unloaded.
3. Line capacitance is negligible and the total resistance and inductance of the line are assumed to be lumped at one place.

* The particular integral represents *forced* or *steady-state component* of the response (which may be current or voltage) and the complementary solution is a *force-free* or simply *free*, or *transient component* of the response.

Short-circuit current is given by Eq. (4.3)

$$i = I_{max} \sin(\omega t + \alpha - \phi) - I_{max} \sin(\alpha - \phi)\, e^{-t/\lambda}$$

$$i = I_{max} \sin(\omega t + \alpha - \phi) + I_{max} \sin(\phi - \alpha)\, e^{-t/\lambda}$$

where $I_{max} = \dfrac{V_{max}}{Z} = \dfrac{V_{max}}{\sqrt{R^2 + (\omega L)^2}}$;

$$\Phi = \text{Tan}^{-1} \frac{\omega L}{R} \text{ and } \lambda = \frac{L}{R}$$

In power system terminology, the first term of the above equation, *steady-state current* is called the *symmetrical short-circuit current* and the second term, *unidirectional transient component* is called the *dc offset current*, which causes the total short circuit to be unsymmetrical till the transient decays.

The maximum momentary short-circuit current I_{mm} corresponds to the first peak, as obvious from Fig. 4.2 (*b*). Neglecting the decay of transient current in this short period we have

$$I_{mm} = I_{max} + I_{max} \sin(\phi - \alpha)$$

Since transmission line resistance is small as compared to its reactance, $\phi \simeq 90°$.

$$\therefore\ I_{mm} = I_{max} + I_{max} \cos \alpha$$

This has the maximum possible value for $\alpha = 0$ *i.e.*, short circuit occurring when the voltage wave is going through zero. Thus

I_{mm} (maximum possible)

$$= 2I_{max} = \frac{2V_{max}}{Z} = \frac{2V_{max}}{\sqrt{R^2 + (\omega L)^2}}$$

= Twice the maximum of symmetrical short-circuit current

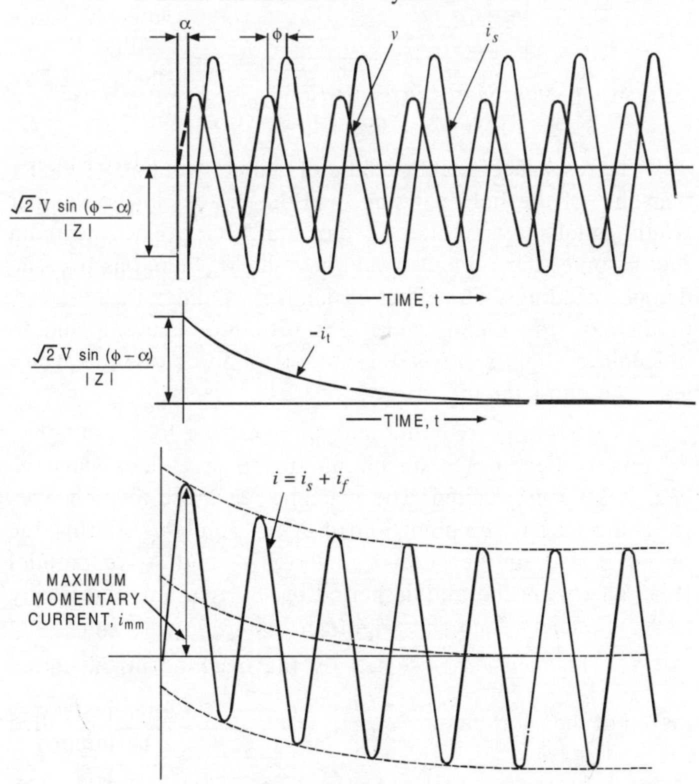

Fig. 4.2 (*b*). *Waveform of a Short-Circuit Current on a Transmission Line*

For a selection of circuit breakers momentary short-circuit current is taken corresponding to its maximum possible value (a safe choice).

Example 4.1. A transmission line of inductance 0.1 H and resistance 5 Ω is suddenly short-circuited at the far end, as shown in the Fig. 4.3. Write the expression for the short-circuit current *i*(*t*). Find approximately the value of the maximum momentary short-circuit current.

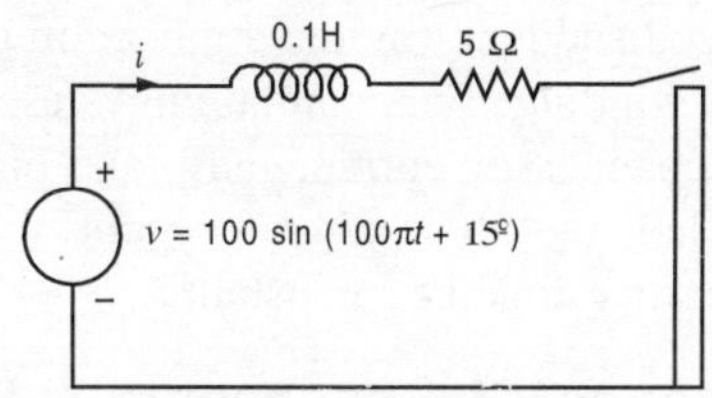

Fig. 4.3

Solution : Applied voltage,

$$v = 100 \sin(100\pi t + 15°).$$

So V_{max} = Coefficient of the sine of the time angle = 100 V

ω = Coefficient of time, $t = 100\,\pi$

$$Z = \sqrt{R^2 + (\omega L)^2} = \sqrt{5^2 + (100\pi \times 0.1)^2} = 31.81\ \Omega$$

$$\phi = \text{Tan}^{-1} \frac{\omega L}{R} = \text{Tan}^{-1} \frac{10\pi}{5} = \text{Tan}^{-1} 2\pi = 80.96°$$

$$I_{max} = \frac{V_{max}}{Z} = \frac{100}{31.81} = 3.1435 \text{ A}$$

Expression for short-circuit current is given as

$$i = I_{max} \sin(\omega t + \alpha - \phi) + I_{max} \sin(\phi - \alpha)\, e^{-t/\lambda}$$

$$= 3.1435 \sin(100\,\pi\, t + 15° - 80.96°) + 3.1435 \sin(80.96° - 15°)\, e^{\frac{-5t}{0.1}}$$

$$= 3.1435 \sin(100\pi t - 65.96°) + 2.871\, e^{-50t} \textbf{ Ans.}$$

Assuming that the first current maximum occurs at the same time as the first current maximum of the symmetrical short-circuit current,

$$I_{mm} = 3.1435 + 2.871 = 6 \text{ A } \textbf{Ans.}$$

4.3. 3-Phase Sudden Short Circuit of an Alternator (on No Load)

In abnormal operation, the alternator may be subjected to transient conditions which cannot be explained from steady-state theory. These transients may occur from (*i*) switching (*ii*) sudden change of load, (*iii*) *sudden short circuit* (either line-to-line or line-to-neutral or from a symmetrical short circuit). These short circuits may develop severe mechanical stresses on the armature coils and their end turns or large torques, which may damage the alternator or its prime mover. The analysis of a synchronous machine under conditions of such transients is useful in predicting the possible conditions that may result from abnormal operation resulting from short circuits. However, here the discussions will be limited to symmetrical short circuit (short circuit on all the output terminals of a 3-phase alternator).

Effects of short-circuit currents can be conveniently studied by using the concept of constant flux linkage, which may be stated as below.

It is impossible to change the flux linkages in a closed electrical circuit that contains no resistance or capacitance (i.e., in a pure inductive circuit).

In case of a synchronous machine, the armature and field windings can be assumed almost pure inductive because they do not contain any capacitance and their resistances are almost negligible in comparison to their inductive reactances. So the flux linkages in the armature circuit and the field circuit cannot be changed suddenly by the application of short circuit to the armature winding. For maintaining these flux linkages constant, large changes of current may take place in both of the windings when the short circuit occurs in order to keep their respective flux linkages constant.

4.3.1. Reactance of Synchronous Machines

The current flowing in the armature of a synchronous generator when its terminals are short circuited is similar to that flowing when a sinusoidal voltage is suddenly applied to an R-L series circuit. However, there is one important difference, that is, in case of an R-L series circuit, reactance X (ωL) is a constant quantity where as in case of the synchronous generator the reactance is not a constant one but is a function of time. In practice three discrete values are assigned and thus we have three reactances—direct-axis subtransient reactance symbolised as X''_d, direct-axis transient reactance symbolised as X'_d and direct-axis synchronous reactance symbolized as X_d.

In short-circuit studies only the above direct-axis reactances are involved. This can be justified as follows:

Short-circuit currents, in general, feed through lines or transformers where the inductive reactance is quite large in comparison to resistance, making the power factor approach zero as currents go lagging and when the armature current is in quadrature (lagging) with the excitation voltage (or no-load voltage) the entire armature mmf acts directly upon the magnetic paths through the salient poles and all of the armature mmf is directly opposing the mmf of the salient pole field windings. Now these direct-axis reactances of the synchronous machine will be briefly described.

Under normal no-load condition of operation, there is no mmf due to armature reaction. When a sudden three-phase short circuit occurs at the terminals of a synchronous generator, the current in the armature circuit increases suddenly to a large value (the symmetrical short-circuit current is limited only by the leakage reactance X_l of the machine) and since the resistance of the circuit is negligible as compared to its reactance, the current is highly lagging and the power factor is approximately zero. A sudden increase in armature current is accompanied by armature reaction. Since the air-gap flux (main flux) cannot change instantaneously (concept of constant flux linkage), to counter the demagnetization of the armature short-circuit current (armature reaction acts on the direct axis in a direction to oppose the main excitation) currents appear in the field winding as well as in damper windings in a direction to help the main flux. This happens due to the transformer action* at the instant of short circuit. Thus during the initial part of the short circuit, the equivalent circuit appears as shown in Fig. 4.4 (*a*)—field winding reactance X_f, damper winding reactance X_{dw} appear in parallel with X_a.

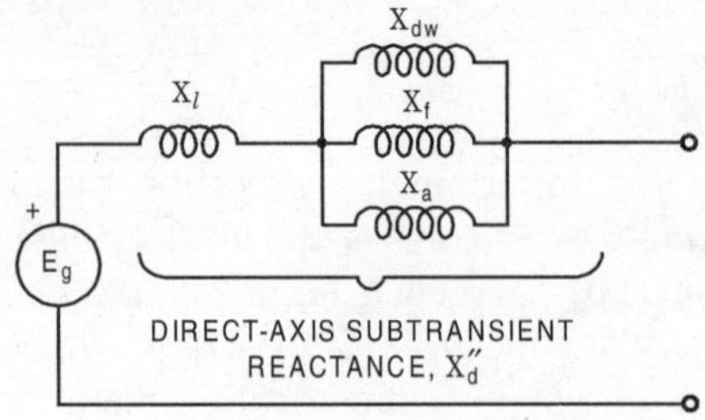

(a) Approximate Circuit Model During Subtransient Period of Short Circuit

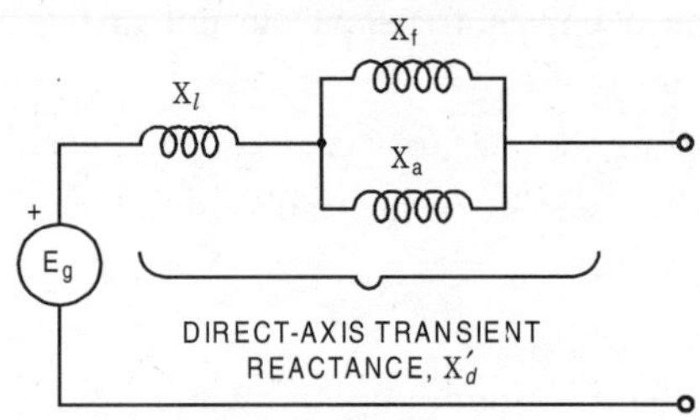

(b) Approximate Circuit Model During Transient Period of Short Circuit

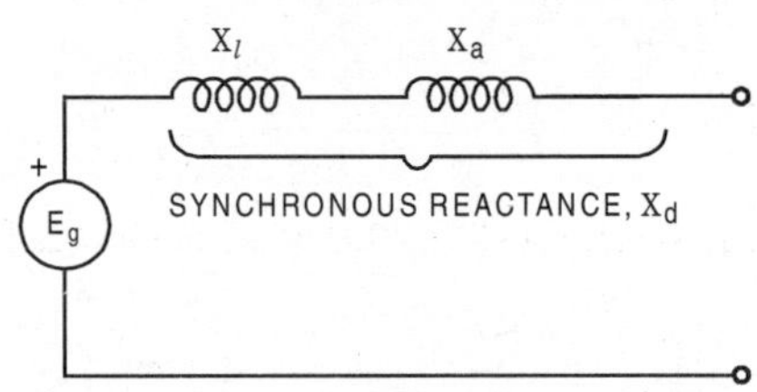

(c) Steady-State Short-Circuit Model of a Synchronous Machine

Fig. 4.4

The resistance/reactance ratio of damper windings is higher than that of the field winding, and the current in the damper windings falls to zero after three or four cycles, when the main flux assumes the value it would have had if there has been no damper windings. The effect of damper windings is, therefore, to increase the initial value of short-circuit current, and to maintain a slightly increased current for three or four cycles after the short circuit.

As the damper winding currents are first to die out, X_{dw} effectively becomes open circuited and at a later stage X_f becomes open circuited. The machine reactance thus changes from the parallel combination of X_a, X_f and X_{dw} during the initial period of the short circuit to X_a and X_f in parallel [Fig. 4.4 (*b*)] in the middle period of short circuit, and finally to X_a in steady state [Fig. 4.4 (*c*)].

The reactance represented by the machine in the initial period of the short circuit *i.e.*, $\left[X_l + \dfrac{1}{\dfrac{1}{X_a} + \dfrac{1}{X_f} + \dfrac{1}{X_{dw}}}\right]$ is called

* In case of machines provided with damper windings on the rotor poles, the whole system acts as a three-winding transformer in which stator is the primary and the rotor field winding and damper windings form the secondaries of the transformer. It is to be noted that the transformer action is there with respect to the dc component of current only.

the *subtransient reactance* (X''_d) of machine ; while the reactance effective after the damper winding currents have died out, *i.e.*, $\left[X_l + \dfrac{1}{\dfrac{1}{X_a} + \dfrac{1}{X_f}}\right]$ is called the *transient reactance* X'_d of the machine. Of course, the reactance under steady-state short-circuit conditions is the *synchronous reactance* of the machine. Obviously $X''_d < X'_d < X_d$. The machine thus offers a time-varying reactance which changes from X''_d to X'_d and finally to X_d.

4.3.2. Analysis of Short-Circuit Oscillogram

As already explained in Art. 4.3.1, when a short circuit occurs across the terminals of a synchronous generator the initial short-circuit current is limited by the subtransient reactance for a few cycles, later on it is controlled by the transient reactance. Finally the short-circuit current settles down to the steady-state short-circuit value limited by the synchronous reactance of the machine.

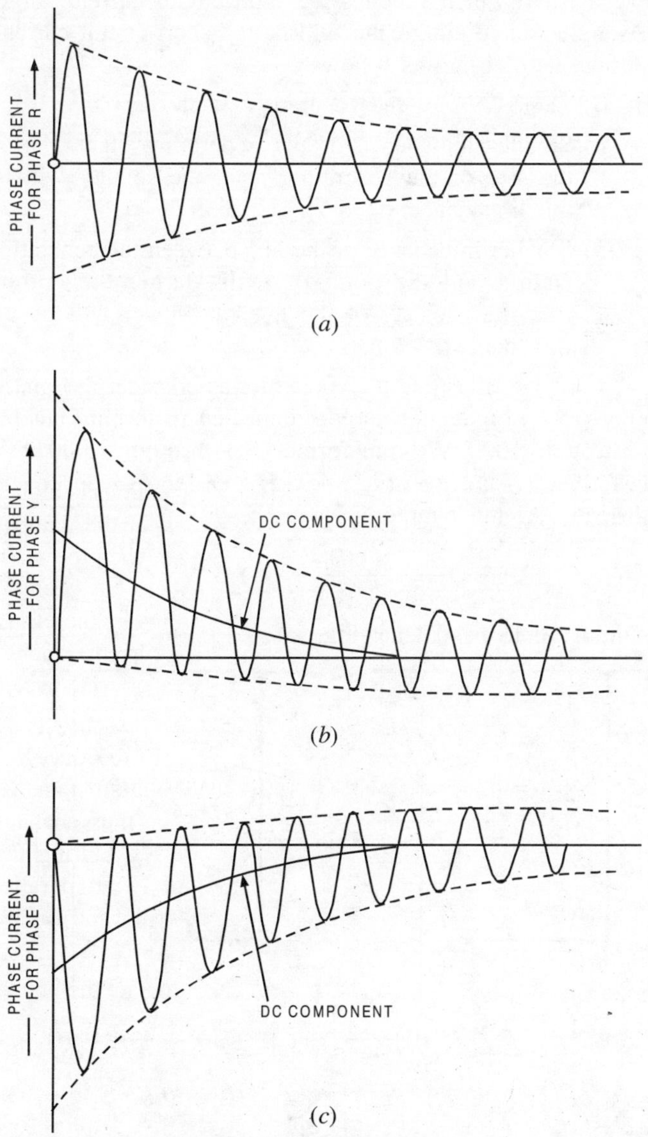

Fig. 4.5. ***Short-Circuit Current Waveforms in Three Phases of a Synchronous Generator***

Besides, as explained in Art. 4.2, at least two phases will have asymmetry in the short-circuit waves, the extent to which asymmetry occurs depending upon the value of the phase voltage at the instant of short circuit. Those phases in which the voltage is near its maximum value at the instant of short circuit will have least asymmetry while those where the voltage is at or near its zero will have a greater or lesser degree of asymmetry, the maximum asymmetry being equal to the otherwise short-circuit current.

The current can be split up into unidirectional or asymmetrical component, and an alternating component of supply frequency. The dc component represents the displacement of the ac wave from the zero axis and is therefore, sometimes known as *dc offset current.* If dc component is subtracted from the short-circuit current wave, we will be left with the ac symmetrical component as shown in Fig. 4.6.

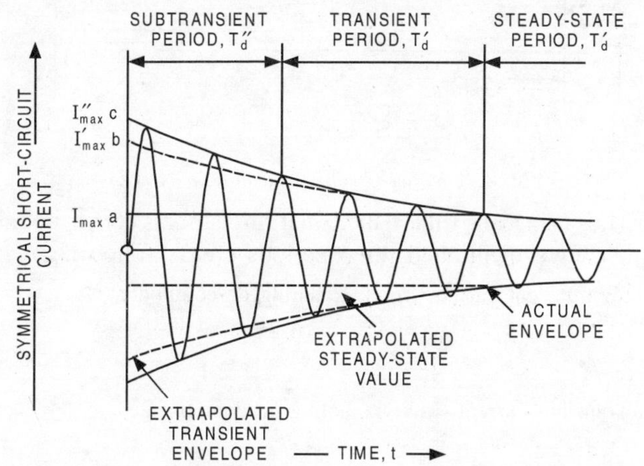

Fig. 4.6. ***Oscillogram of Symmetrical Short-Circuit Current***

The characteristics of short-circuit currents of a 3-phase synchronous generator are shown in Fig. 4.5. These curves indicate that the short circuit occurred at the instant when the voltage of phase R was maximum (assuming that the machine was not delivering any current to the load prior to fault). The transient reactances can be obtained from any one of the three oscillograms by graphical methods. In case of using Figs. 4.5 (*b*) or 4.5 (*c*), it is necessary to eliminate the dc components and plot the envelopes of ac components only.

Figure 4.6 illustrates the short-circuit current for one phase with dc component eliminated.

The short-circuit period can be divided into three periods—initial *subtransient period*, lasting only for the first few cycles, during which the current decrement is very rapid, the middle *transient period*, covering a relatively longer time, during which the current decrement is moderate; and finally the steady-state period. In accordance with the foregoing theory (Art. 4.3.1) of changing reactance, we have

Subtransient current (maximum) excluding dc component,

$$I''_{max} = Oc = \frac{E_{g\,max}}{X''_d} \qquad ...(4.4)$$

Transient current (maximum) excluding dc component,

$$I'_{max} = Ob = \frac{E_{g\,max}}{X'_d} \qquad ...(4.5)$$

Steady-state current (maximum),

$$I_{max} = Oa = \frac{E_{g\,max}}{X_d} \quad ...(4.6)$$

where $E_{g\,max}$ is the maximum voltage from one terminal to neutral on no load.

Note that over any one cycle the current is approximately sinusoidal. So for using rms or effective values, it is quite reasonable to divide the maximum values of Fig. 4.6 by $\sqrt{2}$. The new curve so obtained is shown in Fig. 4.7 and is known as a *decrement curve*.

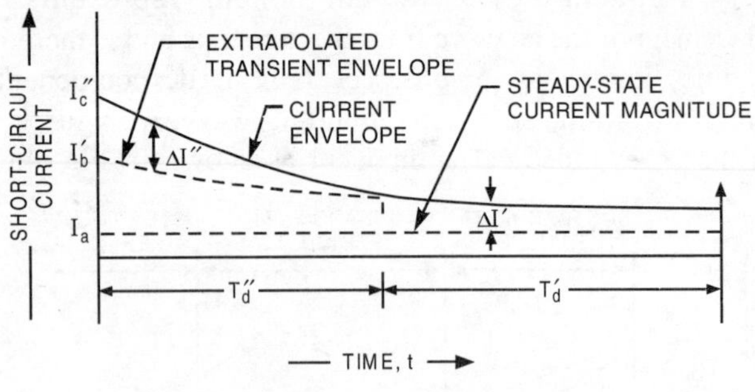

Fig. 4.7. ***Short-Circuit Decrement Curve***

Thus using rms values, the initial rms currents corresponding to the above mentioned three periods are as follows:

Subtransient current (rms) excluding dc component,

$$I'' = \frac{E_g}{X''_d} \quad ...(4.7)$$

Transient current (rms) excluding dc component,

$$I' = \frac{E_g}{X'_d} \quad ...(4.8)$$

and steady-state current (rms),

$$I = \frac{E_g}{X_d} \quad ...(4.9)$$

where $E_g = \frac{E_{g\,max}}{\sqrt{2}}$, the rms voltage from one terminal to neutral.

The subtransient and transient decay periods have time constants T″ and T′ respectively. By determining these time constants it is possible to express the instantaneous current by

$$i = (I'' - I')\, e^{-t/T''} + (I' - I)\, e^{-t/T'} + I \quad ...(4.10)$$

$$\text{or} \quad i = \left(\frac{E_g}{X''_d} - \frac{E_g}{X'_d}\right) e^{-t/T''} + \left(\frac{E_g}{X'_d} - \frac{E_g}{X_d}\right) e^{-t/T'} + \frac{E_g}{X'_d} \quad ...(4.11)$$

If dc component is to be included, then we find the effective value of armature short-circuit current, which is given as

$$i_{rms} = \sqrt{i_{dc}^2 + i_{ac}^2} \quad ...(4.12)$$

where maximum dc current,

$$i_{dc\,max} = \frac{\sqrt{2}\,E_g}{X''_d} \quad ...(4.13)$$

$$\text{and} \quad i_{ac\,max} = \frac{E_g}{X''_d} \quad ...(4.14)$$

$$\text{Thus,} \quad i_{rms\,max} = \sqrt{\left(\frac{\sqrt{2}E_g}{X''_d}\right)^2 + \left(\frac{E_g}{X''_d}\right)^2} = \sqrt{3}\,\frac{E_g}{X''_d} \quad ...(4.15)$$

The momentary short-circuit which a breaker should be capable of carrying is given by Eq. (4.15) but according to AIEE Switchgear Committee, the multiplying factor should be taken as 1.6 instead of $\sqrt{3}$. This multiplying factor can be further reduced to 1.5 for 5 kV breaker or less.

4.4. LIMITATIONS TO FAULT CURRENT

On occurrence of a short circuit at any point in a system, the short-circuit current is limited by the impedance of the system up to the point of fault. Thus referring to Fig. 4.8, if a fault occurs on the feeder at point F, then the short-circuit current from the generating station will have a value limited by the impedance of the generator and transformer and the impedance of the line between the generating station and the point of fault.

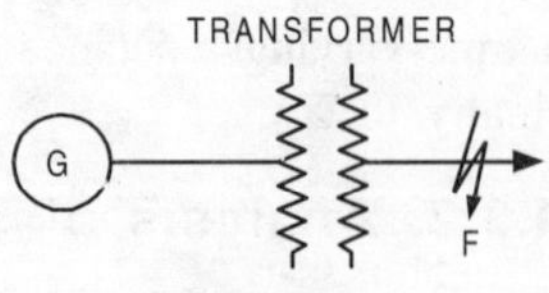

Fig. 4.8

Also, the short-circuit current depends on the size of the power system which supplies the short-circuit current, larger the supply source the greater will be the short-circuit current. The reasons of it are as follows :

(*i*) Large kVA of the generating plant.

(*ii*) Smaller impedance of the generating plant—larger the size of the generating plant the smaller will be the impedance.

(*iii*) Smaller impedance of the line between the generating station and the point of fault—larger the current carrying capacity of the line, the lower will be the impedance.

In order to illustrate the above, let us consider a 3-phase, 400 V, 75 kW induction motor connected to infinite bus-bar (*i*) through 100 kVA transformer (*ii*) through 4,000 kVA transformer as shown in Figs. 4.9 (*a*) and 4.9 (*b*) respectively.

Full-load current of motor,

$$I_L = \frac{75 \times 1{,}000}{\sqrt{3} \times 400 \times 0.8} = 135 \text{ A}$$

assuming pf to be 0.8 lagging.

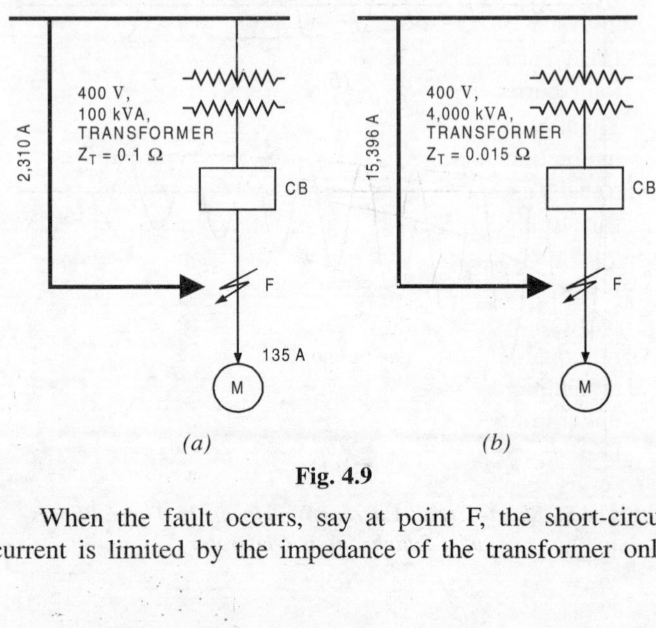

Fig. 4.9

When the fault occurs, say at point F, the short-circuit current is limited by the impedance of the transformer only.

The short-circuit current, which can be supplied by the source in each case, is worked out as follows :

In first case, *i.e.*, with 100 kVA transformer,

$$I_{sc_1} = \frac{\text{Phase voltage}}{\text{Impedance up to point F}}$$

$$= \frac{400/\sqrt{3}}{0.1} = 2{,}310 \text{ A}$$

In second case, *i.e.*, with 4,000 kVA transformer

$$I_{sc_2} = \frac{400/\sqrt{3}}{0.015} = 15{,}396 \text{ A}$$

Thus, it is observed that as the system grows, the capability of supply of short-circuit will be increased along with it.

4.5. ANALYSIS OF SYMMETRICAL FAULTS IN POWER SYSTEM NETWORKS

Symmetrical (L-L-L) fault occurs infrequently, as for example, when a line, which has been made safe for maintenance and/or repairs by clamping all the three phases to earth, is accidentally made alive or when, due to slow fault clearance, an earth fault spreads across to the other two phases or when a mechanical excavator cuts quickly through a whole cable. It is an important type of fault in that it results in an easy calculation and generally, a pessimistic answer.

The analysis of symmetrical (L-L-L) faults includes the determination of the voltage at any point (or bus) in the power system network, the current in any branch and value of reactance necessary to limit the fault current to any desired value. Such calculations provide the necessary data for selection of circuit breakers and design of protective scheme. The circuit breaker MVA breaking capacity is based on 3-phase fault MVA. Since the circuit breakers are manufactured in preferred standard sizes, *e.g.*, 250, 500, 750 MVA, high precision is not required in calculations of 3-phase fault level at a point in a power system. Moreover, the system impedances are also never known accurately. To make complete and accurate calculations is virtually impossible, and certainly too tedious for operational work. It is customary to perform the short-circuit analysis under the following simplifying assumptions:

1. Load currents are considered negligible as compared to fault currents.
2. Shunt elements in the transformer model that account for magnetizing current and core loss are neglected. The transformer is represented by a reactance in series, as transformer resistance is quite low in comparison with its reactance.
3. Shunt capacitances of the transmission lines are neglected.
4. System resistance is neglected and only inductive reactance of the system is taken into account. This assumption cannot be applied in case overhead lines or underground cables of considerable length are included in the network. A transmission line is represented by series reactance (and resistance).
5. The emfs of all the generators are assumed to be equal to $1\angle 0°$ per unit. This means that the system voltage is at its nominal value and the system is operating on no load at the time of occurrence of fault. The selection of zero phase for one source is arbitrary and convenient. Assuming that all sources are in phase and of the same magnitude is equivalent to neglecting prefault load current. When desirable, the load current can be taken into account, at a later stage by superposition.
6. The effect of dc component is accounted for by using correction factors. The correction or multiplying factor for determination of breaking capacity of a circuit breaker depends on the speed of the circuit breaker. For example, a two-cycle circuit breaker might require a factor 1.4 whereas with an eight-cycle breaker a factor 1.0 would be sufficient.

Generator reactances are normally taken as their subtransient values in order to depict the most pessimistic condition. However, if transient current is to be determined, then transient reactances should be used.

For simple systems, calculations can be made by *network reduction technique*, which will be discussed here. However, for modern complex systems, ac network analyzers or digital computers are used for fault calculations.

4.5.1. Network Reduction Technique

Because of the balanced nature of fault and the system, any condition which apply to one phase apply equally to the remaining two phases. Thus the problem is reduced itself to a single phase problem involving a single supply source acting through the equivalent network impedance up to the fault. The equivalent network impedance up to the fault can be obtained by network reduction, that involves series-parallel combinations and star/delta or delta/star conversion of reactances.

Various steps involved in the short-circuit calculations are given below :

1. Make out a single line diagram of the complete network indicating on each component, its rating, voltage, resistance and reactance.
2. Choose a common base kVA (or MVA) and convert all the resistances and reactances in per unit values as referred to common base kVA (or MVA).
3. From the single line diagram draw a single line reactance (or impedance) diagram showing one phase and neutral. In this diagram write down the reactances (or impedances) of the elements in per unit values, determined under step 2.
4. Reduce the reactance (or impedance) diagram, by network reduction technique keeping the identity of the fault point intact. Find the reactance of the system as seen from the fault point (Thevenin reactance).
5. Determine the fault current and fault MVA in per unit, as detailed in Art. 4.5.2. Convert these per unit values to actual values.
6. Retrace the steps of calculations to work out the current and voltage distribution through out the network.

4.5.2. Fault MVA and Fault Current (Steady State)

Per unit fault (or short-circuit current),

$$I_{sc\ pu} = \frac{\text{PU voltage at fault point}}{\text{PU } X_{equivalent}} \quad ...(4.16)$$

Per unit fault level (MVA)

$$= \sqrt{3} \text{ per unit fault current} \times \text{per unit source voltage}$$

$$\text{or Fault MVA} = \frac{\text{Base MVA}}{\text{PU } X_{equivalent}} \text{ MVA (lagging)} \quad ...(4.17)$$

$$\text{Fault current } I_{sc} = \frac{\text{Fault MVA} \times 10^3}{\sqrt{3} \times \text{base kV}} \text{ amperes} \quad ...(4.18)$$

Example 4.2. A 3-phase, 10,000 kVA, 11 kV alternator has a subtransient reactance of 8%. A 3-phase short circuit occurs at its terminals. Determine the fault current and fault MVA.

Solution : Alternator percentage reactance is based on its own voltage and kVA ratings. Let us choose 10,000 kVA as base kVA and 11 kV as base kV

$$\text{Base current, } I_B = \frac{\text{Base kVA}}{\sqrt{3} \times \text{base kV}} = \frac{10{,}000}{\sqrt{3} \times 11} = 525 \text{ A}$$

$$\text{Per unit fault current, } I_{sc\ pu} = \frac{\text{PU voltage}}{\text{PU reactance}} = \frac{1}{0.08}$$

$$= 12.5 \angle -90° \text{ A}$$

$$\text{Fault current, } I_{sc} = I_{sc\ pu} \times I_B = 12.5 \times 525 = 6{,}562 \text{ A } \textbf{Ans.}$$

$$\text{Fault power} = \sqrt{3} \times I_{sc} \times \text{source voltage} \times 10^{-6} \text{ MVA}$$

$$= \sqrt{3} \times 6{,}562 \times 11{,}000 \times 10^{-6}$$

$$= 125 \text{ MVA } \textbf{Ans.}$$

Alternative Method :

$$\text{Fault MVA} = \frac{\text{Base MVA}}{X_{pu}} = \frac{10{,}000 \times 10^{-3}}{8/100}$$

$$= 125 \text{ MVA } \textbf{Ans.}$$

$$\text{Fault current, } I_{sc} = \frac{\text{Fault MVA} \times 10^3}{\sqrt{3} \times \text{base kV}}$$

$$= \frac{125 \times 1{,}000}{\sqrt{3} \times 11} = 6{,}562 \angle -90° \text{ A } \textbf{Ans.}$$

Example 4.3. A symmetrical 3-phase short circuit occurs on the 22 kV bus-bars of the circuit shown as one line diagram in Fig. 4.10. Calculate the fault current and the fault apparent power.

[U.P.S.C. I.E.S. Electrical Engineering-II, 2005]

Solution : Let base kVA for the complete system be 100 MVA.

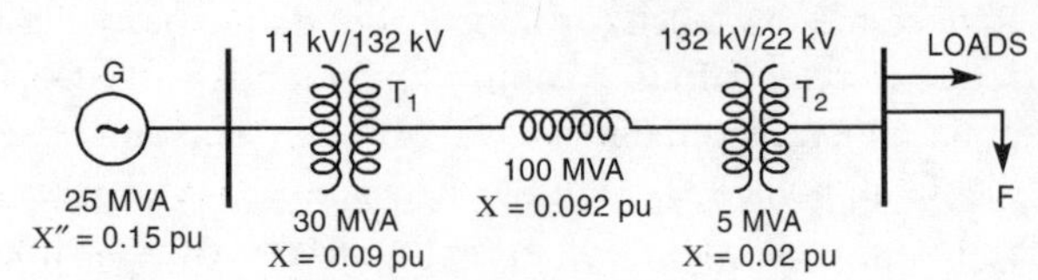

Fig. 4.10

Base kV on generator side = 11 kV

Base kV on feeder side = 132 kV

Base kV for load = 22 kV

$$\text{Per unit reactance of generator} = 0.15 \times \frac{100}{25} = 0.6 \text{ pu}$$

$$\text{Per unit reactance of transformer, } T_1 = 0.09 \times \frac{100}{30} = 0.3 \text{ pu}$$

$$\text{Per unit reactance of transformer, } T_2 = 0.02 \times \frac{100}{5} = 0.4 \text{ pu}$$

Per unit reactance of transmission line = 0.092 pu

$$\text{Total per unit reactance up to fault} = j(0.6 + 0.3 + 0.4 + 0.092)$$

$$= j1.392 \text{ pu}$$

Per unit reactance diagram is shown in Fig. 4.11.

Assuming prefault voltage of the generator to be $1 \angle 0°$ pu

Per unit fault current,

$$I_{sc\ pu} = \frac{\text{Per unit prefault voltage}}{\text{Per unit reactance (total)}}$$

$$= \frac{1 \angle 0°}{1.392 \angle 90°}$$

$$= 0.7184 \angle -90° \text{ pu}$$

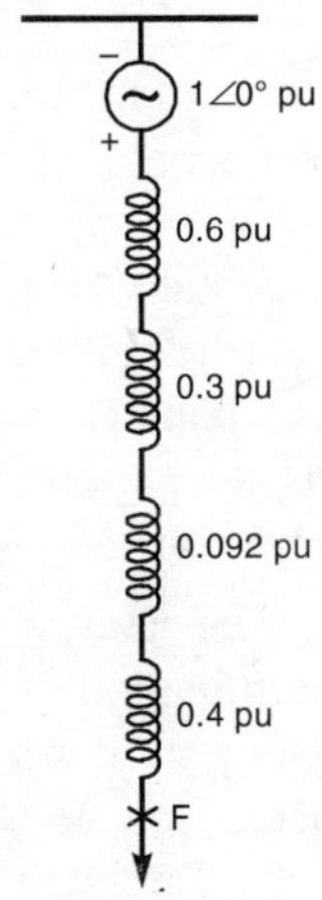

Fig. 4.11

$$\text{Base current, } I_B = \frac{\text{Base MVA} \times 1{,}000}{\sqrt{3} \times \text{base kV}}$$

$$= \frac{100 \times 1{,}000}{\sqrt{3} \times 22} = 2{,}624 \text{ A}$$

$$\text{Fault current, } I_{sc} = I_{sc\ pu} \times I_B$$

$$= 0.7184 \times 2{,}624$$

$$= 1{,}885 \text{ A } \textbf{Ans.}$$

Fault apparent power

$$= I_{sc\ pu} \times \text{per unit prefault voltage}$$

$$= 0.7184 \times 1$$

$$= 0.7184 \text{ pu} = 0.7184 \times 100 = 71.84 \text{ MVA } \textbf{Ans.}$$

Example 4.4. A 3-phase transmission line operating at 10 kV and having a resistance of 1 Ω and reactance 4 Ω is connected to the generating station bus-bars through 5 MVA step-up transformer having a reactance of 5%. The bus-bars are supplied by 10 MVA alternator having 10% reactance. Calculate the short-circuit kVA fed to symmetrical fault between phases if it occurs : (*i*) at the load end of transmission line, (*ii*) at the high voltage terminals of the transformer.

[Power System Analysis Rajasthan Univ. 2003; Rajasthan Technical Univ. 2010]

Solution : Single line diagram for the network is given in Fig. 4.12.

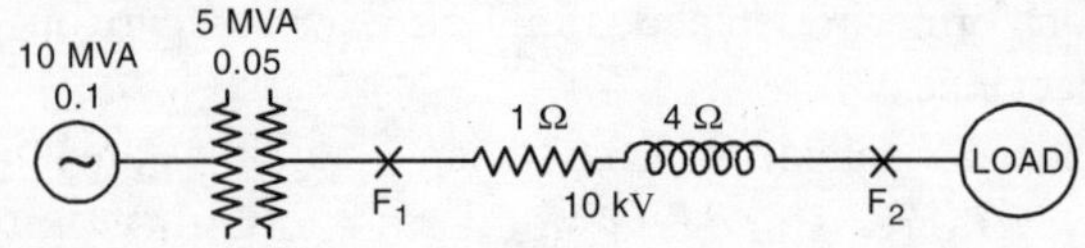

Fig. 4.12

Let the base MVA for the complete system be 10 MVA

$$\text{Per unit reactance of alternator} = \frac{10}{100} = 0.1 \text{ pu}$$

$$\text{Per unit reactance of transformer} = \frac{5}{100} \times \frac{10}{5} = 0.1 \text{ pu}$$

Actual impedance of line $= (1 + j4)\ \Omega$

$= 4.1231\ \angle 75.96°\ \Omega$

$$\text{PU impedance of line} = \text{Actual impedance} \times \frac{\text{Base MVA}}{(\text{kV}_B)^2}$$

$$= 4.1231\ \angle 75.96° \times \frac{10}{10^2}$$

$$= 0.41231\ \angle 75.96°\ \text{pu}$$

$$= (0.1 + j0.4)\ \text{pu}$$

The pu reactance diagram is shown in Fig. 4.13.

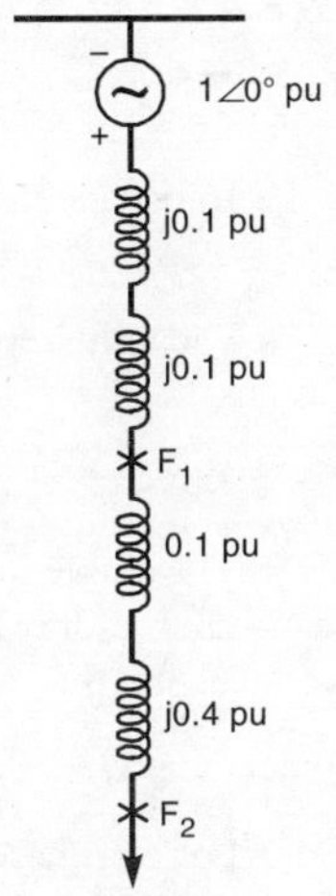

Fig. 4.13

(i) For a fault at the load end of the transmission line

Total per unit impedance

$$= j0.1 + j0.1 + 0.1 + j0.4$$

$$= (0.1 + j0.6)\ \text{pu}$$

$$= 0.6083\ \angle 80.54°\ \text{pu}$$

Short-circuit kVA

$$= \frac{\text{Base MVA} \times 1{,}000}{\text{Total per unit impedance}}$$

$$= \frac{10 \times 1{,}000}{0.6083}$$

$= 16{,}440$ kVA **Ans.**

(ii) For a fault at high voltage terminals of transformer

Total per unit impedance $= j0.1 + j0.1 = j0.2$ pu

$$\text{Short circuit kVA} = \frac{10 \times 1{,}000}{0.2} = 50{,}000\ \text{kVA}$$ **Ans.**

Example 4.5. A 3-ϕ fault occurs at middle point F on the transmission line as shown in Fig. 4.14. Determine transfer reactance between generator and infinite bus.

[M.D. Univ. Power System-I, 2010-11]

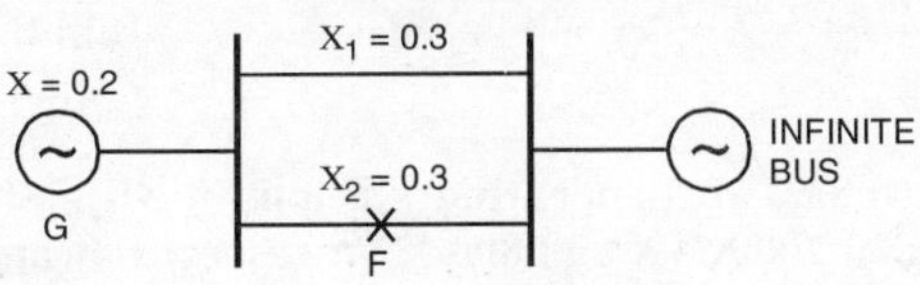

Fig. 4.14

Solution : Figure 4.15 depicts the per unit reactance diagram that appears between generator and infinite bus.

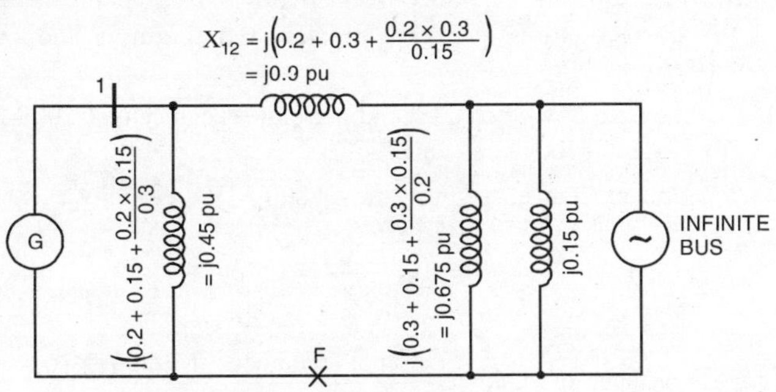

Fig. 4.15

Per unit transfer reactance,

$$X_{12} = j0.2 + j0.3 + \frac{j0.2 \times j0.3}{j0.15} = j0.9\ \text{pu}$$ **Ans.**

Example 4.6. Two generators are connected in parallel to the low voltage side of 3-phase Δ-Y transformer as shown in Fig. 4.16.

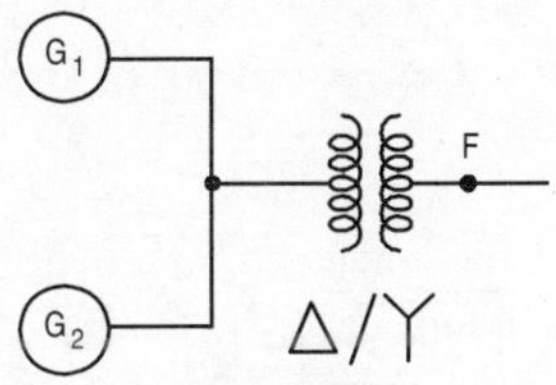

Fig. 4.16

Ratings of the generators are :

for $G_1 \rightarrow$ 50 MVA, 13.8 KV, X″ = 25%

for $G_2 \rightarrow$ 25 MVA, 13.8 KV, X″ = 25%

Transformers ratings are 75 MVA, 13.8Δ/69Y kV with the leakage reactance of 10%.

Before fault occurs, the voltage on high tension side of the transformer is 66 kV at F.

The transformer is unloaded and there is no circulating current between the generators. Find the subtransient current in each generator when a 3-phase short circuit occurs on the high tension side of the transformers.

[U.P. Technical Univ. Power System Analysis 2005-06]

Solution : Single line diagram of the network is shown in Fig. 4.17 *(a)*.

Let base MVA be 50 MVA, Base kV on generator side be 13.8 kV and on feeder side 69 kV.

Per unit reactance of generator $G_1 = 0.25$ pu

$$\text{Per unit reactance of generator } G_2 = \frac{50}{25} \times \left(\frac{13.8}{13.8}\right)^2 \times 0.25 = 0.5\ \text{pu}$$

$$\text{Per unit reactance of transformer} = \frac{50}{75} \times \frac{10}{100} \times \left(\frac{13.8}{13.8}\right)^2 = 0.0667\ \text{pu}$$

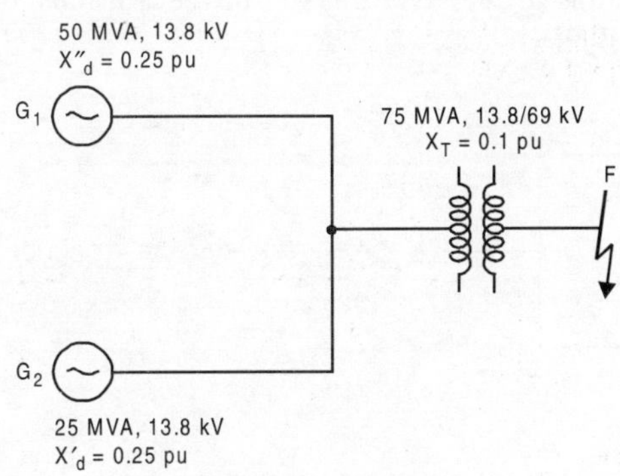

Fig. 4.17 *(a)*. *Single Line Diagram For Example 4.6*

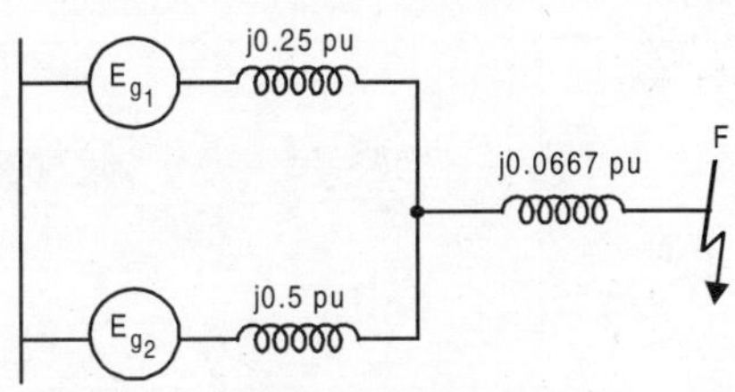

Fig. 4.17 *(b)*

The equivalent circuit during subtransient fault state is shown in Fig. 4.17 *(b)*.

The net subtransient reactance of the circuit (shown in Fig. 4.17) up to fault point F

$$= (0.25 \parallel 0.5) + 0.0667 \text{ pu} = \frac{0.25 \times 0.5}{0.25 + 0.5} + 0.0667 \text{ pu}$$

$$= 0.1666 + 0.0667 = 0.2333 \text{ pu}$$

Per unit fault current,

$$I_{sc\,pu} = \frac{\text{Per unit voltage}}{\text{Per unit impedance}} = \frac{\frac{66}{69}}{0.2333} = 4.1 \text{ pu}$$

Base current,

$$I_B = \frac{\text{Base MVA} \times 1{,}000}{\sqrt{3} \times \text{base kV}} = \frac{50 \times 1{,}000}{\sqrt{3} \times 69} = 418.4 \text{ A}$$

Fault current $= I_{sc\,pu} \times I_B = 4.1 \times 418.4 = 1{,}715$ A

Subtransient current in generator, G_1

$$= \text{Fault current} \times \frac{0.5}{0.5 + 0.25} \times \text{transformation ratio}$$

$$= 1{,}715 \times \frac{0.5}{0.75} \times \frac{69}{13.8} = 5{,}716 \text{ A}$$

Subtransient current in generator G_2

$$= 1{,}715 \times \frac{0.25}{0.5 + 0.25} \times \text{transformation ratio}$$

$$= 1{,}715 \times \frac{0.25}{0.75} \times \frac{69}{13.8} = 2{,}858 \text{ A} \quad \textbf{Ans.}$$

Example 4.7. Two 3-phase 11 kV generators of capacities 8 MVA and 4 MVA and subtransient reactances of 8% and 4% respectively operate in parallel. The generating station is connected to a transmission line of 200 km length, through a step-up transformer of capacity 4 MVA, and having percentage reactance of 3.5%. The resistance and reactance of the transmission line per km of its length are 0.025 Ω and 0.1 Ω respectively and it operates at 66 kV. Calculate the short-circuit MVA for a phase-to-phase faults at the receiving end of the transmission line and at the sending end.

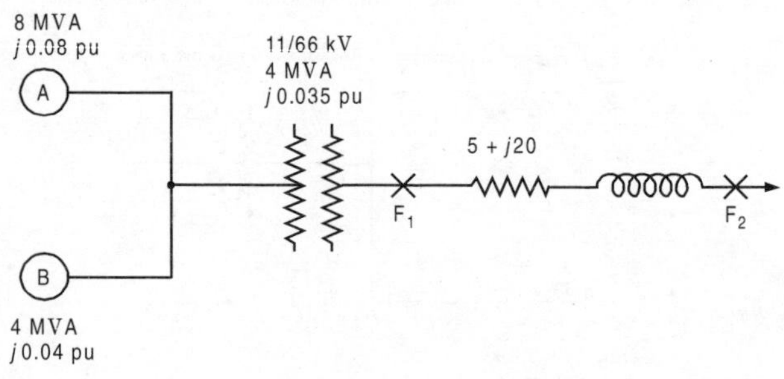

(a) Single Line Diagram of System

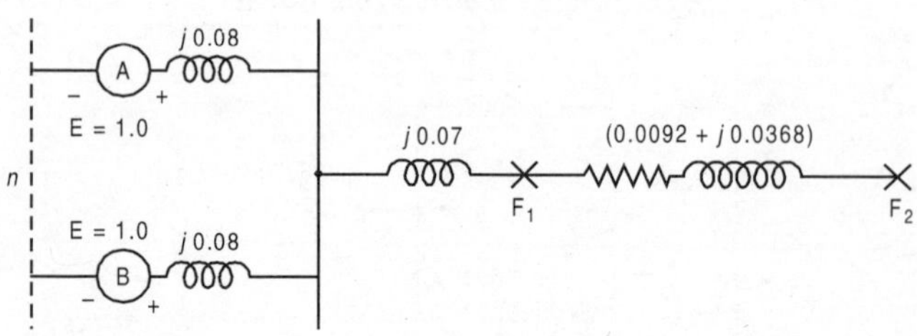

(b) Reactance Diagram

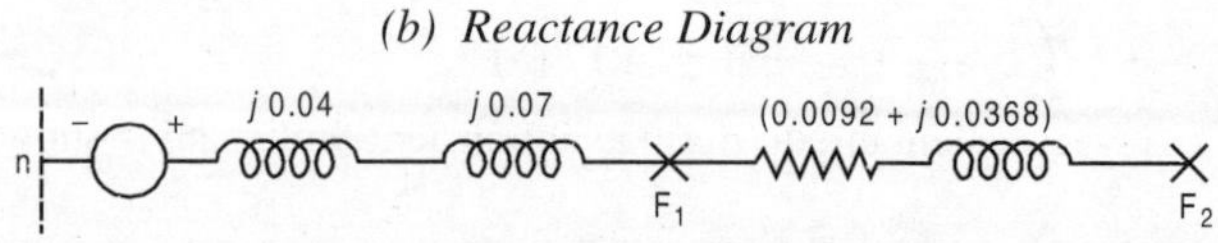

(c) Reduced Reactance Diagram

Fig. 4.18

Solution : Single line diagram of the network is shown in Fig. 4.18 (*a*). Let base kVA for complete system be 8,000 kVA, base kV for generator side be 11 kV and transmission line side 66 kV.

Per unit reactance of generator A,

$$= \frac{8}{100} = 0.08$$

Per unit reactance of generator B,

$$= \frac{4}{100} \times \frac{8{,}000}{4{,}000} = 0.08$$

Per unit reactance of transformer

$$= \frac{3.5}{100} \times \frac{8{,}000}{4{,}000} = 0.07$$

Impedance of transmission line

$$= 200\,(0.025 + j\,0.1) = (5 + j\,20)\ \Omega$$

Per unit impedance of transmission line

$$= \frac{(5 + j20) \times 8{,}000}{(66)^2 \times 1{,}000} = (0.0092 + j\,0.0368) \text{ pu}$$

The single line impedance diagram corresponding to Fig. 4.18 (*a*) can now be drawn as in Fig. 4.18 (*b*), which when further simplified reduces to that shown in Fig. 4.18 (*c*) because equivalent of two reactances (j 0.08 each) in parallel comes out to be $\frac{j\,0.08}{2} = j\,0.04$ pu.

For a 3-phase fault at the sending end (F_1) side of the transmission line, the total impedance up to the fault point $= j\,0.04 + j\,0.07 = j\,0.11$.

$$\text{Short-circuit MVA} = \frac{\text{Base MVA}}{\text{Per unit reactance}} = \frac{8{,}000 \times 10^{-3}}{0.11}$$

$$= 72.73 \quad \textbf{Ans.}$$

For a 3-phase fault at the receiving end (F_2) side of the transmission line, the total impedance up to the fault point

$$= j\,0.04 + j\,0.07 + (0.0092 + j\,0.0368)$$

$$= (0.0092 + j\,0.1468) \text{ pu} = 0.1471 \angle 86.4^\circ$$

$$\text{Short-circuit MVA} = \frac{\text{Base MVA}}{\text{Per unit impedance}} = \frac{8{,}000 \times 10^{-3}}{0.1471}$$

$$= 54.4 \quad \textbf{Ans.}$$

Example 4.8. Two generating stations having short-circuit capacities of 1,200 MVA and 800 MVA respectively and operating at 11 kV are linked by an interconnected cable have a reactance of 0.5 Ω per phase. Determine the short-circuit capacity of each station. [*U.P.S.C. I.E.S. Electrical Engineering-II, 2006*]

Solution : Single line diagram of the network is shown in Fig. 4.19 (*a*) and the corresponding per unit reactance diagram is shown in Fig. 4.19 (*b*).

Let the base MVA be 1,200 and per unit reactance of 1,200 MVA capacity generator be 1 per unit.

Per unit reactance of generator of 800 MVA capacity

$$= 1 \times \frac{1{,}200}{800} = 1.5 \text{ pu}$$

$$\text{Per unit reactance of cable} = \frac{\text{Actual reactance} \times \text{MVA}_B}{(\text{kV}_B)^2}$$

$$= \frac{0.5 \times 1{,}200}{11^2} = 4.9587 \text{ pu}$$

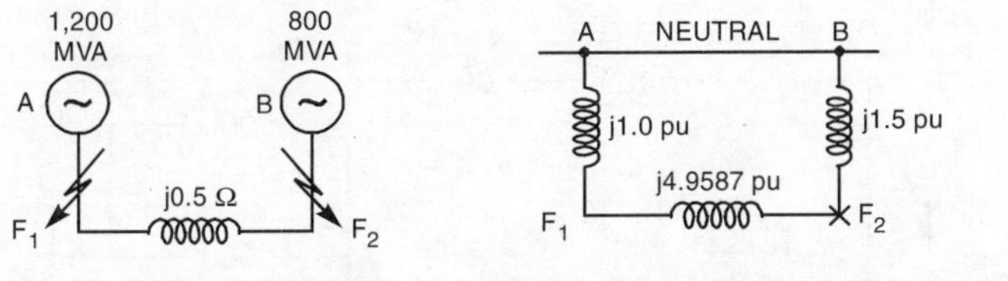

(a) Single Line Diagram *(b) Per Unit Reactance Diagram*

Fig. 4.19

(*a*) *Fault on Generating Station A.* When a fault occurs on generating station A (point F_1), the total per unit reactance from generator neutral to fault point F_1 is the parallel combination of $j(1.5 + 4.9587)$ pu and $j1.0$ pu *i.e.*,

Total per unit reactance up to fault point F_1

$$= j\frac{(1.5 + 4.9587) \times 1}{1.5 + 4.9587 + 1} = j0.8666 \text{ pu}$$

$$\text{So short-circuit MVA} = \frac{\text{Base MVA}}{\text{Per unit reactance}}$$

$$= \frac{1{,}200}{0.8666} = 1{,}385 \text{ MVA} \quad \textbf{Ans.}$$

(*b*) *Fault on Generating Station B.* When a fault occurs on generating station B (point F_2), the total per unit reactance from generator neutral to fault point F_2 is the parallel combination of $j(1 + 4.9587)$ pu and $j1.5$ pu *i.e.*,

Total per unit reactance up to fault point F_2

$$= \frac{j(1 + 4.9587) \times 1.5}{1 + 4.9587 + 1.5} \text{ pu} = j1.1983 \text{ pu}$$

$$\text{So short-circuit MVA} = \frac{1{,}200}{1.1983} = 1{,}001 \text{ MVA} \quad \textbf{Ans.}$$

Example 4.9. Generators G_1 and G_2 are identical and rated 11 kV, 20 MVA and have a transient reactance of 0.25 pu at own MVA base. The transformers T_1 and T_2 are also identical and are rated 11/66, 5 MVA and have a reactance of 0.06 pu to their own MVA base. The tie line is 50 km long, each conductor has a reactance of 0.848 Ω/km. The three-phase fault is assumed at F_1 = 20 km from generator G_1 as in Fig. 4.20. Find the short-circuit current. [*Rajasthan Univ. Power System Analysis 2004*]

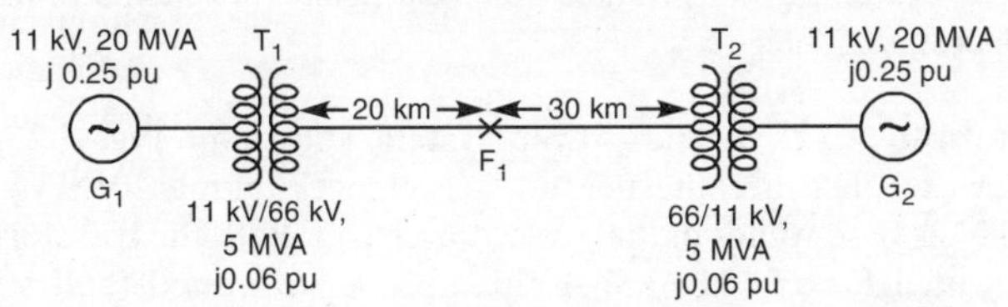

Fig. 4.20

Solution : Let the base MVA of the system be 20 MVA, base kV of generator side 11 kV and for transmission circuit side 66 kV. Per unit reactances of different elements corresponding to this common base MVA are :

For generator A and B, $X_{pu\,g} = j0.25$ pu (as given)

$$\text{For transformers } T_1 \text{ and } T_2, \quad X_{pu\,t} = j0.06 \times \frac{20}{5} = j0.24 \text{ pu}$$

$$\text{Per unit reactance of 20 km line} = \frac{20 \times 0.848 \times 20}{66^2} = 0.0779 \text{ pu}$$

$$\text{Per unit reactance of 30 km line} = \frac{30 \times 0.848 \times 30}{66^2} = 0.1168 \text{ pu}$$

Per unit reactance diagram is shown in Fig. 4.21 (*a*), which on simplification is reduced to those shown in Figs. 4.21 (*b*) and 4.21 (*c*).

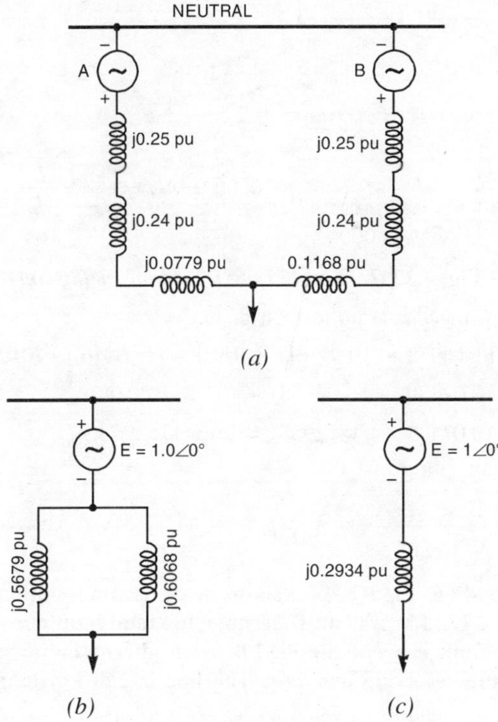

Fig. 4.21

From equivalent circuit [Fig. 4.21 (*c*)], per unit equivalent reactance is given as

$$X_{eq\,pu} = j(0.25 + 0.24 + 0.0779) \parallel j(0.25 + 0.24 + 0.1168) \text{ pu}$$

$$= j(0.5679) \parallel j(0.6068) \text{ pu}$$

$$= j\frac{0.5679 \times 0.6068}{0.5679 + 0.6068} = j0.2934 \text{ pu}$$

$$\text{Fault MVA} = \frac{\text{Base MVA}}{X_{eq\,pu}} = \frac{20}{0.2934} = 68.17 \text{ MVA}$$

$$\text{Fault current, } I_{sc} = \frac{\text{Fault MVA} \times 1{,}000}{\sqrt{3} \times \text{base kV}} = \frac{68.17 \times 1{,}000}{\sqrt{3} \times 66}$$

$$= 596.3 \text{ A} \quad \textbf{Ans.}$$

Example 4.10. Determine the required MVA rating of the circuit breaker CB for the system shown in Fig. 4.22 (*a*). Consider the grid as infinite bus. Choose 6 MVA as base.

Transformer : 3-phase, 33/11 kV, 6 MVA, 0.01 + *j* 0.08 pu impedance.

Load : 3-phase, 11 kV, 5,800 kVA, 0.8 lag, *j* 0.2 pu impedance.

Impedance of each feeder : 9 + *j* 18. [*GATE 1999*]

Solution : Base MVA = 6

Per unit impedance of transformer = $0.01 + j\,0.08$

Per unit impedance of feeder

$$= \frac{(9 + j18) \times 6}{33^2} = (0.04958 + j\,0.09916) \text{ pu}$$

Per unit impedance of load

$$= \frac{j\,0.2 \times 6}{5.8} = j\,0.207$$

Single line diagram of the system is shown in Fig. 4.22 (*a*) and the corresponding single line impedance diagram is shown in Fig. 4.22 (*b*).

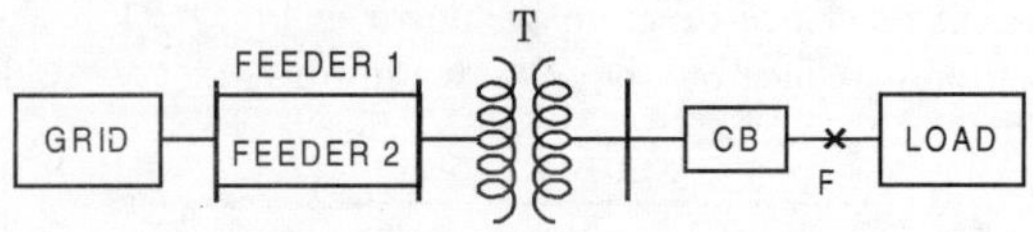

Fig. 4.22 (*a*)

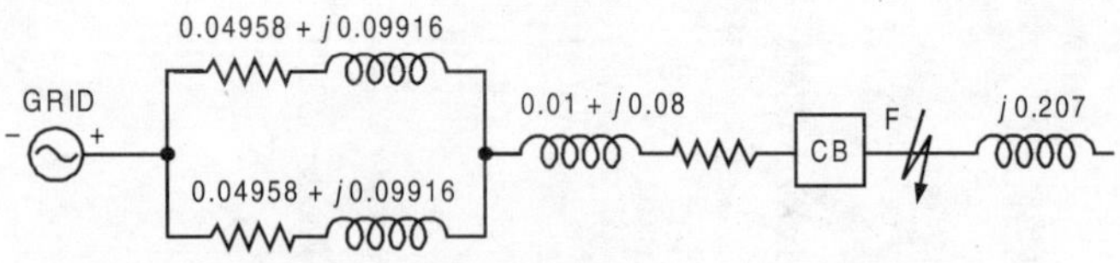

Fig. 4.22 (*b*) *Single Line Impedance Diagram*

Equivalent impedance up to fault point,

$$\mathbf{Z}_{eq\ pu} = [(0.04958 + j\,0.09916) \parallel (0.04958 + j\,0.09916)] + (0.01 + j\,0.08)$$

$$= (0.02479 + j\,0.04958) + (0.01 + j\,0.08)$$

$$= (0.03479 + j\,0.12958) = 0.13417 \angle 74.97^\circ \text{ pu}$$

Short-circuit rating of CB

$$= \frac{\text{Base MVA}}{Z_{eq\ pu}} = \frac{6}{0.13417} = 44.72 \text{ MVA } \textbf{Ans.}$$

Example 4.11. Figure 4.23 shows a generating station feeding power to a 132 kV system. Determine the total fault current, fault level and fault current supplied by each alternator for a 3-phase fault at the receiving end bus. The line is 200 km long.

[*Bharathidasan Univ. Power System Analysis Nov. 1996*]

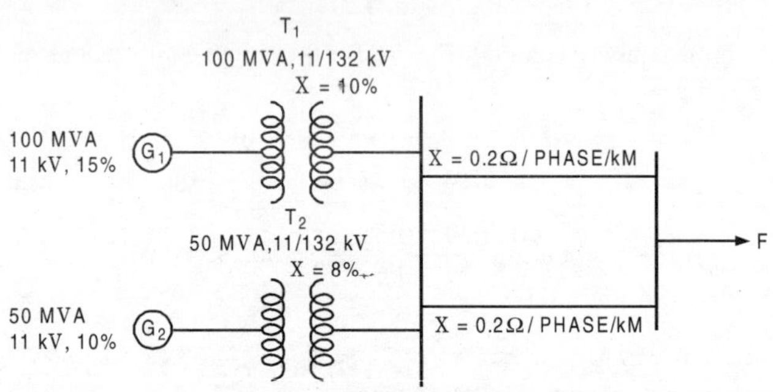

Fig. 4.23. *System For Example 4.11*

Solution : Let the base MVA for the complete system be 100 MVA, base kV be 11 kV for generator side and 132 kV for the feeder side.

Per unit reactance of generator G_1 = $j\,0.15$ pu (given)

Per unit reactance of generator $G_2 = j\,0.1 \times \dfrac{100}{50} = j\,0.2$ pu

Per unit reactance of transformer T_1 = $j\,0.1$ pu (given)

Per unit reactance of transformer $T_2 = j\,0.08 \times \dfrac{100}{50} = j\,0.16$ pu

Per unit reactance of each line $= j\,0.2 \times 200 \times \dfrac{\text{MVA}}{(\text{kV})^2}$

$$= j\,40 \times \frac{100}{(132)^2} = j\,0.23 \text{ pu}$$

Single line reactance diagram is shown in Fig. 4.24 (*a*), which when further simplified is reduced to those shown in Fig. 4.24 (*b*) and 4.24 (*c*) respectively.

Thevenin equivalent reactance of the system when viewed from the fault point F comes out to be $(j\,0.25 \parallel j\,0.36) + j\,0.115$ *i.e.,*

$$\frac{0.25 \times 0.36}{0.25 + 0.36} + j\,0.115 \text{ or } j\,0.26254$$

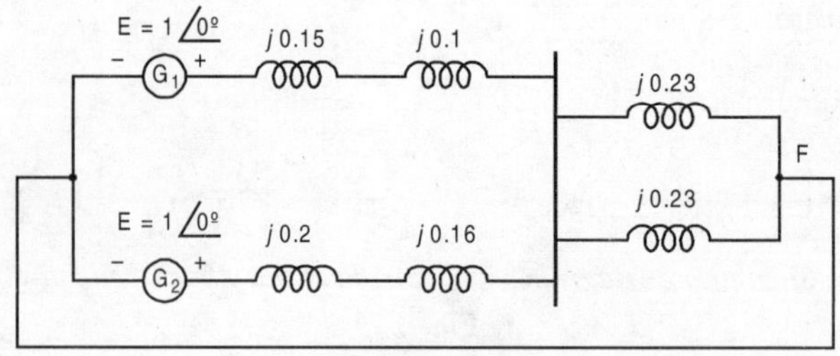

(*a*) *Reactance Diagram*

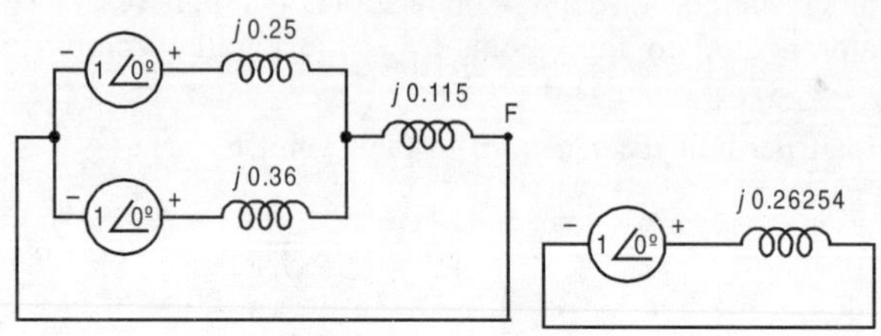

(*b*) (*c*)

Fig. 4.24

$$\text{Fault level} = \frac{\text{Base MVA}}{X_{eq\ pu}} = \frac{100}{0.26254} = 380.9 \text{ MVA } \textbf{Ans.}$$

$$\text{Total fault current} = \frac{\text{Fault level in MVA} \times 1{,}000}{\sqrt{3} \times \text{base kV of feeder side}}$$

$$= \frac{380.9 \times 1{,}000}{\sqrt{3} \times 132} = 1{,}666 \text{ A } \angle -90^\circ \text{ A}$$

Total fault current supplied by two generators

= Total fault current on feeder side × transformation ratio

$$= 1{,}666 \angle -90^\circ \times \frac{132}{11} = 19{,}992 \angle -90^\circ \text{ A } \textbf{Ans.}$$

Fault current supplied by generator G_1

$$= \frac{19{,}992 \angle -90^\circ \times j0.36}{j(0.36 + 0.25)} = 11{,}799 \angle -90^\circ \text{ A } \textbf{Ans.}$$

Fault current supplied by generator G_2

$$= \frac{19{,}992 \angle -90^\circ \times j0.25}{j(0.36 + 0.25)} = 8{,}193 \angle -90^\circ \text{ A } \textbf{Ans.}$$

Note: Currents get distributed amongst parallel generators in inverse proportion of their reactances.

Example 4.12. In the power system shown in Fig. 4.25, the values marked are the per unit reactances taking 20 MVA and 11 kV as base values in the generator circuit. Both the transformers are rated for 11 / 110 kV. A three-phase to ground fault with a fault impedance of *j* 0.088 pu occurs at bus 2. Determine the actual values of fault current and the currents supplied by the generators. [*U.P.S.C. I.E.S. Electrical Engineering II, 1999*]

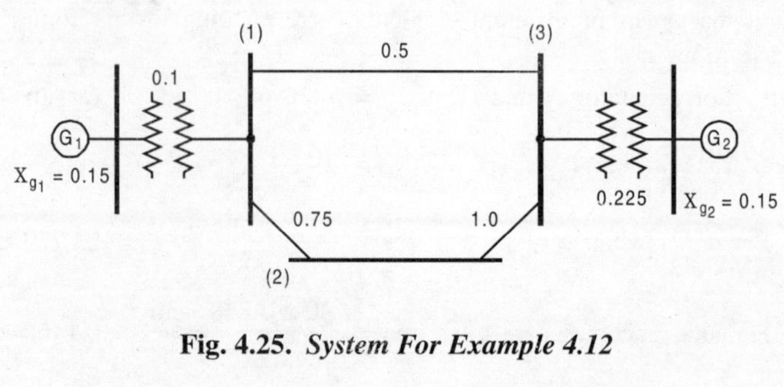

Fig. 4.25. *System For Example 4.12*

Solution : Single line reactance diagram is shown in Fig. 4.26 (*a*). Using a delta-star transformation this diagram reduces to Fig. 4.26 (*b*). By series-parallel combinations the diagram is further simplified to that shown in Fig. 4.26 (*c*).

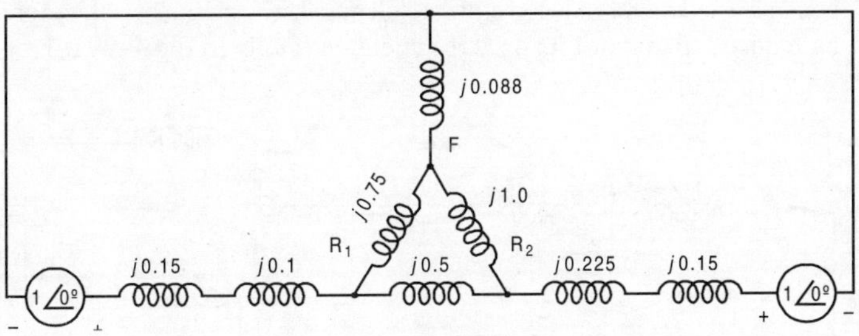

(*a*) *Reactance Diagram*

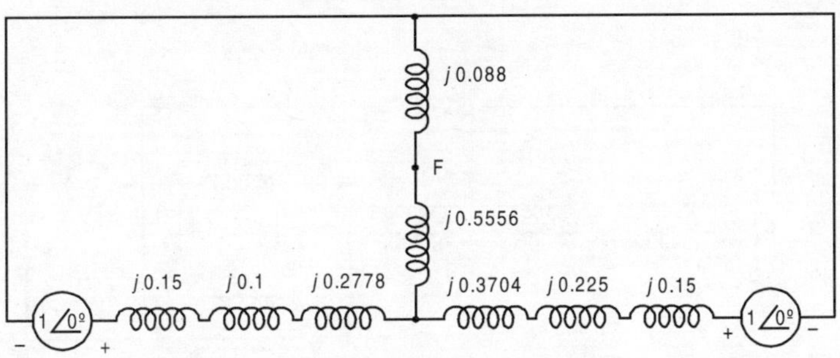

(*b*)

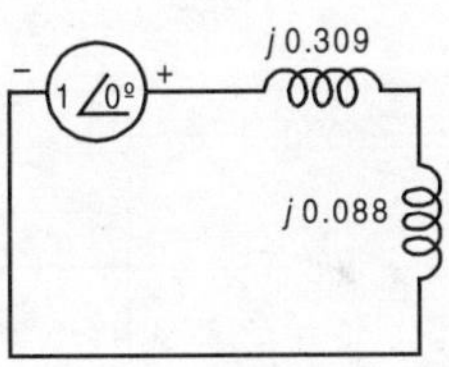

(*c*)

Fig. 4.26

$$\text{Fault current} = \frac{1}{j0.309 + j0.088} \times \frac{20 \times 1{,}000}{\sqrt{3} \times 110} = 264.4 \angle -90^\circ \text{ A}$$

Total fault current supplied by two generators

$$= 264.4 \angle -90^\circ \times \frac{110}{11} = 2{,}644 \angle -90^\circ \text{ A } \textbf{Ans.}$$

Fault current supplied by generator G_1

$$= \frac{2{,}644 \angle -90^\circ \times j\,0.7454}{j\,0.5278 + j\,0.7454} = 1{,}548 \angle -90^\circ \text{ A } \textbf{Ans.}$$

Fault current supplied by generator G_2

$$= \frac{2{,}644 \angle -90^\circ \times j\,0.5278}{j\,0.5278 + j\,0.7454} = 1{,}096 \angle -90^\circ \text{ A } \textbf{Ans.}$$

Example 4.13. The one line diagram of a three-phase power system is shown in Fig. 4.27. Using per unit method of analysis calculate the 3-phase short circuit MVA at fault point F.

[*U.P. Technical Univ. Power System Analysis 2006-07*]

Solution : Assume that all the generators are solidly grounded.

Let the base MVA be 50 MVA

Per unit reactance of different elements as referred to common base MVA of 50 MVA, we have

$$X_{g1\,pu} = j0.05 \times \frac{50}{20} = j0.125 \text{ pu}$$

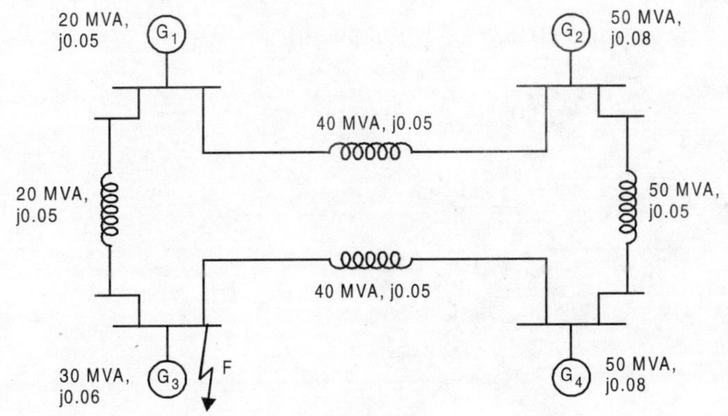

Fig. 4.27

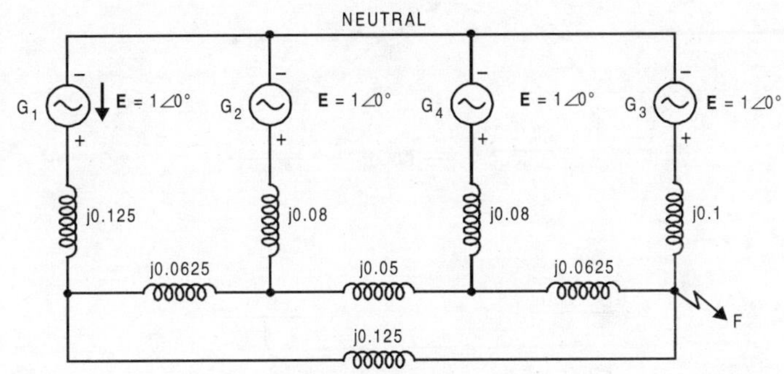

(*a*) *Reactance Diagram*

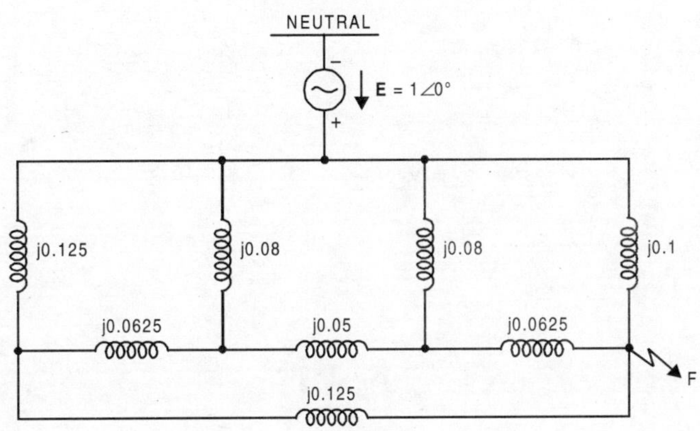

(*b*)

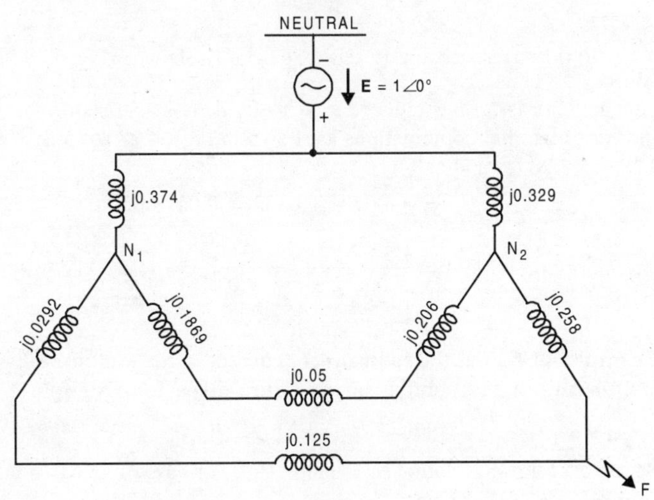

(*c*)

Fig. 4.28

$$X_{g2\,pu} = j0.08 \times \frac{50}{50} = j0.08 \text{ pu}$$

$$X_{g3\,pu} = j0.06 \times \frac{50}{30} = j0.1 \text{ pu}$$

$$X_{g4\,pu} = j0.08 \times \frac{50}{50} = j0.08 \text{ pu}$$

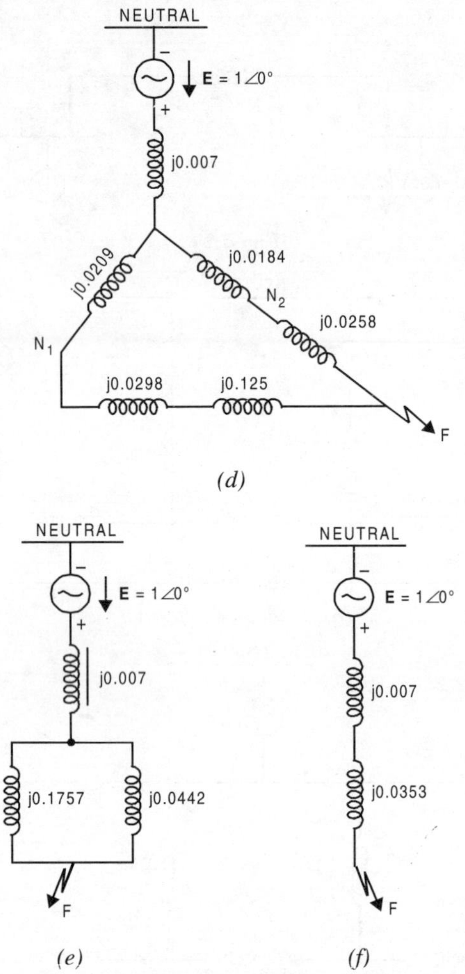

Fig. 4.28

Interconnector reactances are $\frac{50}{40} \times j0.05, j0.05, \frac{50}{40} \times 0.05$ and $\frac{50}{20} \times j0.05$ pu respectively. Reactance diagram is shown in Fig. 4.28 (*a*). This network is further reduced in steps by delta-star transformation and series-parallel combinations as shown in Figs. 4.28 (*b*), 4.28 (*c*), 4.28 (*d*), 4.28 (*e*) and 4.28 (*f*).

So $X_{eq\,pu} = j0.007 + j0.0353 = j0.0423$ pu

$$\text{Short circuit MVA} = \frac{\text{Base MVA}}{X_{eq\,pu}} = \frac{50}{0.0423} = 1{,}182 \text{ MVA} \quad \textbf{Ans.}$$

Example 4.14. Find the subtransient current in the fault if 3-phase fault occurs at the point F in the figure given for Example 3.9.

[*U.P. Technical Univ. Power System Analysis 2007-08*]

Solution : From Solution of Example 3.9, For base of 25 MVA and base kV of 11 kV, we have

Reactance of generator = 0.2 pu

Reactance of transformers = 0.0803 pu

Line reactance = 0.165 pu

Reactance of motor 1 = 0.3444 pu

Reactance of motor 2 = 0.689 pu

The circuit model of the system for fault calculations is shown in Fig. 4.29 (*a*). The system being initially on no load (assumed), the generator and motor induced emfs are identical. The circuit can, therefore, be reduced to that of Fig. 4.29 (*b*) and then finally to Fig. 4.29 (*c*).

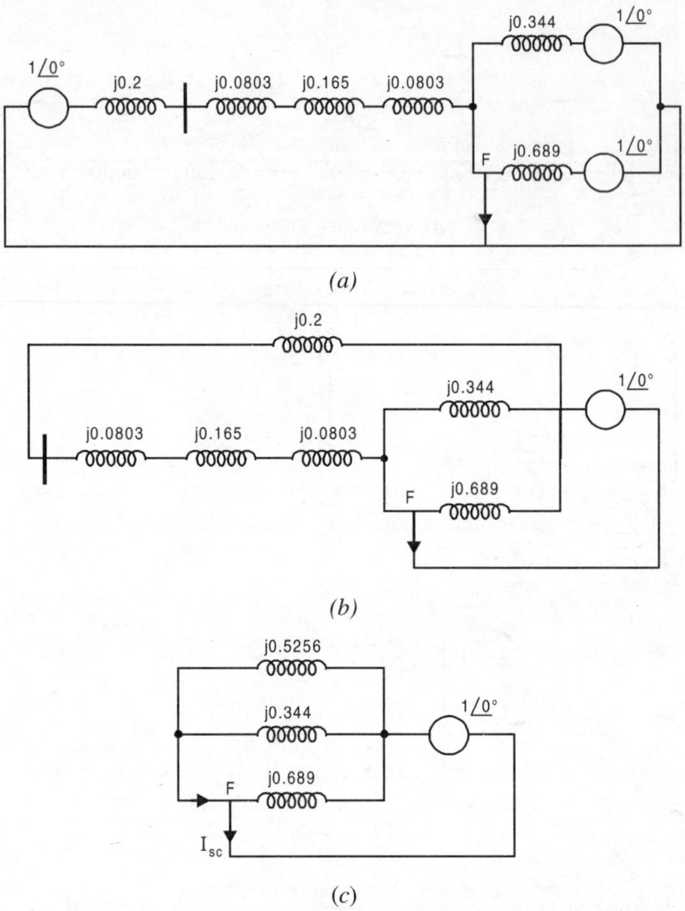

Fig. 4.29

From circuit shown in Fig. 4.29

$$\text{Short-circuit current, } I_{sc} = \frac{1\angle 0°}{j0.5256} + \frac{1\angle 0°}{j0.344} + \frac{1\angle 0°}{j0.689}$$

$$= (-j1.9026 - j2.907 - j1.4513) \text{ pu} = -j6.261 \text{ pu}$$

$$\text{Base current, } I_B = \frac{\text{Base kVA}}{\sqrt{3} \times kV_B} = \frac{25 \times 1{,}000}{\sqrt{3} \times 11} = 1{,}312 \text{ A}$$

Subtransient current = 6.261 × 1,312 = 8,148 A **Ans.**

Example 4.15. A 3-phase, 50 Hz generator is rated at 500 MVA, 20 kV with X''_d = 0.2 pu. It supplies pure resistive load of 400 MW at 20 kV. The load is connected directly across the terminals of the generator. If all the three phases of the load are short circuited simultaneously, find the initial symmetrical rms current in the generator in per unit on a base of 500 MVA, 20 kV.

[*U.P.S.C. I.E.S. Electrical Engineering-II, 2009*]

Solution : Let the base kVA be 500 MVA and the base voltage be 20 kV

Load = 400 MW at unity power factor ∵ Load is pure resistive

$$= \frac{400}{500} = 0.8\angle 0° \text{ pu}$$

Terminal voltage, V_0 = 20 kV

$$\text{or } V_0 = \frac{20}{20} = 1 \text{ pu}$$

$$\text{Prefault current, } I_0 = \frac{\text{Load}}{V_0} = \frac{0.8 \angle 0°}{1 \angle 0°} = 0.8 \angle 0° \text{ pu}$$

Voltage behind subtransient reactance for the generator,

$$E''_g = V_0 + jI_0X''_d = (1 + j0) + j0.8 \angle 0° \times 0.2$$

$$= 1 + j0.16 = 1.01272 \angle 9.09° \text{ pu}$$

Initial symmetrical rms current in the generator on a base of 500 MVA, 20 kV,

$$I'' = \frac{E''_g}{X''_d} = \frac{1.01272 \angle 9.09°}{0.2 \angle 90°}$$

$$= 0.50636 \angle -80.9° \text{ pu } \textbf{Ans.}$$

Example 4.16. A synchronous generator rated 500 kVA, 400 V, 0.1 pu subtransient reactance is supplying a passive load of 400 kW at 0.8 lagging power factor. Calculate the initial symmetrical rms current for a three-phase fault at generator terminals.

[U.P. Technical Univ. Power System Analysis 2008-09]

Solution : Let base kVA be the 500 kVA and base voltage 400 V

Load = 400 kW at 0.8 lagging pf

$$= \frac{\frac{400}{0.8}}{500} = 1.0 \text{ pu, 0.8 lagging pf}$$

$$\text{Prefault current, } \mathbf{I}_0 = \frac{\text{Load}}{V_0} \angle \cos^{-1} \text{pf}$$

$$= \frac{1}{1.0} \angle \cos^{-1} 0.8 = 1.0 \angle -36.87° \text{ pu}$$

Prefault voltage, $\mathbf{V}_0$ = 400 V or 1 $\angle 0°$ pu (reference)

Voltage behind subtransient reactance for the generator,

$$\mathbf{E}''_g = \mathbf{V}_0 + j\mathbf{I}_0X_d = (1 + j0) \times j0.1 \times (0.8 - j0.6)$$

$$= (1.06 + j0.08) \text{ pu}$$

$$\text{Initial fault current, } \mathbf{I}'' = \frac{\mathbf{E}''_g}{X''_d} = \frac{1.06 + j0.08}{j0.1}$$

$$= (0.8 - j10.6) \text{ pu}$$

$$= 10.63 \angle -85.68° \text{ pu}$$

$$\text{Base current, } I_B = \frac{500 \times 1{,}000}{\sqrt{3} \times 400} = 721.7 \text{ A}$$

Initial symmetrical fault current

$$= 721.7 \times 10.63 \angle -85.68° \text{ A}$$

$$= 7{,}672 \angle -85.68° \text{ A } \textbf{Ans.}$$

4.6. CONSIDERATION OF PREFAULT LOAD CURRENT

In the discussions made so far, the effects of prefault system loading have been neglected. Usually this is justified because fault currents are much larger than load currents. However, in some situations it may become necessary to consider the effect of load current. The straight forward approach would be to connect load impedances and simply deal with a more elaborate network, but usually it is not done for two reasons :

1. All loads are not correctly modelled as simple impedances.
2. If effects for several different loads are to be considered, load impedances are required to be recalculated and a new fault study is to be made for each case.

The prefault line loadings are typically available as output data from a load flow study. The voltages at different points in the power network depart from unity and zero in magnitude and phase value due to line loadings. It is possible to calculate proper values for a given load condition and use these values in a fault study. However, the same result can be obtained by Thevenin's theorem, by which we determine the changes in currents in the branches and changes in the voltages at different buses of the network. By superposition of the prefault branch currents and changes in branch currents, post-fault branch currents are determined. Similarly the post-fault bus voltages can be determined.

If the load happens to be a synchronous motor, it continues to run due to its stored energy and supplies power to the fault. The motor circuit, under fault conditions, will have current equal to that of phasor sum of motor current during prefault condition and the fault current contribution of the motor. The noteworthy points are that in case of a generator the directions of load current and fault current are same (both the currents flow out of generator terminals) while in case of a motor the directions of the two currents are opposite to each other because the load current flows into the motor terminals and the fault current flows out of the motor terminals.

The above facts will be illustrated by the following example.

Example 4.17. A synchronous generator and motor are rated as 30,000 kVA, 13.2 kV, and both have subtransient reactance of 20%. The line connecting them has a reactance of 10% on the base of the machine rating. The motor is drawing 20,000 kW at 0.8 pf leading and terminal voltage of 12.8 kV when a symmetrical 3-phase fault occurs at the motor terminals. Find the subtransient current in the generator, motor and fault by using the internal voltage of the machine.

[U.P.S.C. I.E.S. Electrical Engineering-II, 2004]

Solution : Single line diagram for the given system, prefault equivalent circuit and equivalent circuit during fault condition are shown in Figs. 4.30 (*a*), (*b*) and (*c*) respectively. All the reactances are given on a common base of 30 MVA.

$$\text{Prefault voltage, } V_0 = \frac{12.8}{13.2} = 0.9697 \angle 0° \text{ pu}$$

Load = 20 MW at 0.8 pf leading

$$= \frac{20}{30} = 0.6667 \text{ pu, 0.8 pf leading}$$

$$\text{Prefault current, } I_0 = \frac{\text{Load}}{V_0 \times \text{pf}} \angle \cos^{-1} \text{pf}$$

$$= \frac{0.6667}{0.9697 \times 0.8} \angle \cos^{-1} 0.8 = 0.8594 \angle 36.87° \text{ pu}$$

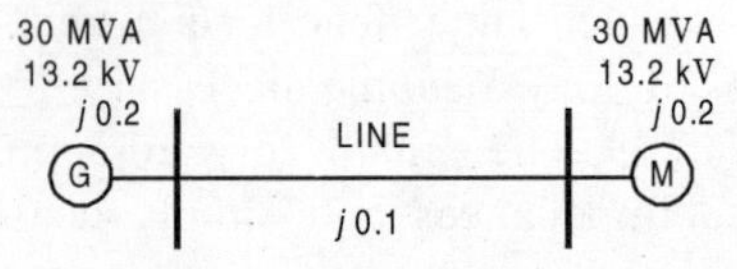

(a) Single Line Diagram For The System

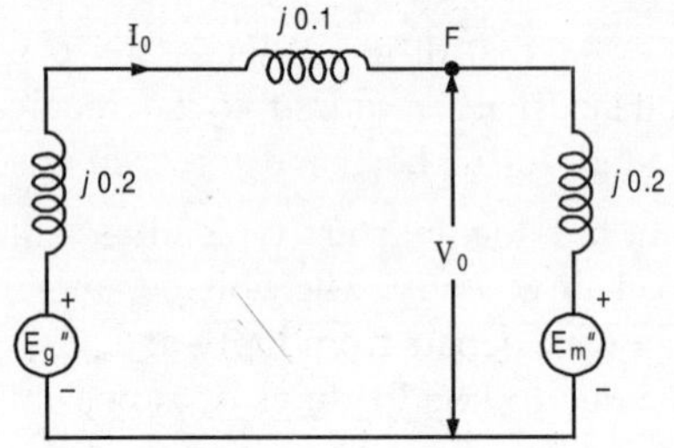

(b) Prefault Equivalent Circuit

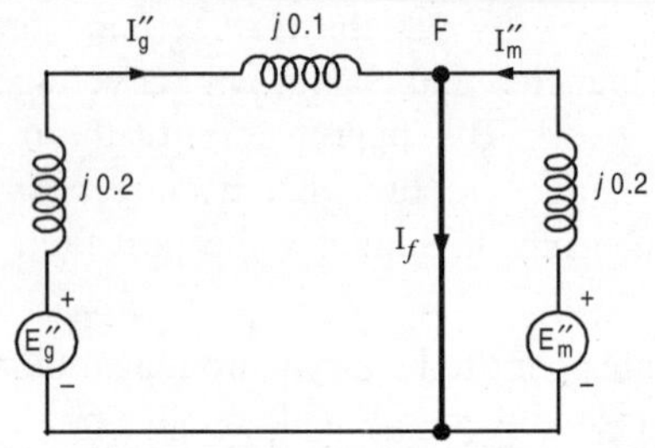

(c) Equivalent Circuit During Fault Condition

Fig. 4.30

(a) *By Direct Calculations.* Voltage behind subtransient reactance for the generator,

$$E''_g = V_0 + jI_0X''_d = 0.9697 \angle 0^\circ + j0.3 \times 0.8594 \angle 36.87^\circ$$
$$= 0.9697 + j0.2063 - 0.1547$$
$$= 0.815 + j0.2063 \text{ pu}$$

Voltage behind subtransient reactance for the motor,

$$E''_m = V_0 - jI_0X''_d = 0.9697 \angle 0^\circ - j0.2 \times 0.8594 \angle 36.87^\circ$$
$$= 0.9697 + 0.1031 - j0.1375$$
$$= (1.0728 - j0.1375) \text{ pu}$$

Under fault conditions [Fig. 4.30 (*c*)]

$$I''_g = \frac{E''_g}{j0.2 + j\,0.1} = \frac{0.815 + j0.2063}{j0.3} = (0.6876 - j2.717) \text{ pu}$$

$$I''_m = \frac{E''_m}{j\,0.2} = \frac{1.0728 - j0.1375}{j0.2} = (-0.6876 - j5.364) \text{ pu}$$

Total fault current,

$$I_f = I''_g + I''_m = (0.6876 - j2.717) + (-0.6876 - j5.364)$$
$$= -j8.081 \text{ pu}$$

Base current,

$$I_B = \frac{30 \times 10^3}{\sqrt{3} \times 13.2} = 1{,}312.16 \text{ A}$$

So $I''_g = 1{,}312.16 \times (0.6876 - j2.717) = (902.3 - j\,3{,}565)$ A

$= 3{,}677 \angle -75.8^\circ$ A **Ans.**

$I''_m = 1{,}312.16 \times (-0.6876 - j5.364) = (-902.3 - j\,7{,}038)$ A

$= 7{,}096 \angle -97.3^\circ$ A **Ans.**

$I_f = 1{,}312.16 \times (-j\,8.081)$

$= -j10{,}603$ A or $10{,}603 \angle -90^\circ$ A **Ans.**

(b) *By Thevenin's and Superposition Theorems.* The circuit model for the system for computation of post fault condition is shown in Fig. 4.31. Thevenin's reactance when viewed from fault terminals FG

$$= (j0.2 + j0.1) \,||\, j0.2 = \frac{j0.3 \times j0.2}{j0.3 + j0.2} = j0.12 \text{ pu}$$

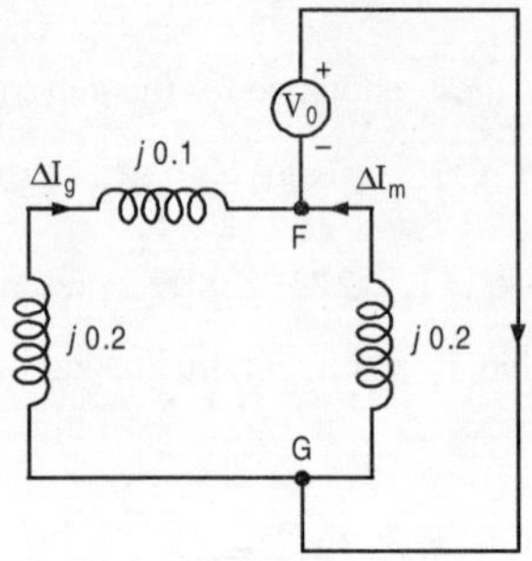

Fig. 4.31

Total fault current,

$$I_f = \frac{V_0}{\text{Thevenin's reactance}} = \frac{0.9697}{j0.12} = -j8.081 \text{ pu}$$

Change in generator current due to fault,

$$\Delta I_g = I_f \times \frac{j0.2}{j0.5} = -j\,8.081 \times 0.4 = -j\,3.2324 \text{ pu}$$

Change in motor current due to fault,

$$\Delta I_m = -j8.081 \times \frac{j0.3}{j0.5} = -j\,4.8486 \text{ pu}$$

Generator subtransient current,

$$I''_g = I_0 + \Delta I_g = 0.8594 \angle 36.87^\circ - j3.2324$$
$$= (0.6876 - j2.717) \text{ pu}$$

Motor subtransient current,

$$I''_m = -I_0 + \Delta I_m = -0.8594 \angle 36.87^\circ - j4.8486$$
$$= (-0.6876 - j5.364) \text{pu}$$

Which are the same, calculated already. This is as expected.

This method, indeed, is a powerful method for solution of large networks.

Example 4.18. A 625 kVA, 2.4 kV generator with transient reactance of 8% is connected to a bus through a CB, as shown in Fig. 4.32. The synchronous motors each rated 250 hp, 2.4 kV, upf, 90% efficiency are connected to the bus. The subtransient reactance of each of the motors being 20%. Find the subtransient fault current in pu in the fault for a 3-phase load short circuit at bus at E. Determine the fault current rating of the generator breaker (A) and each motor CB (B) discarding prefault currents.

[*Rajasthan Univ. Power System Analysis 2006*]

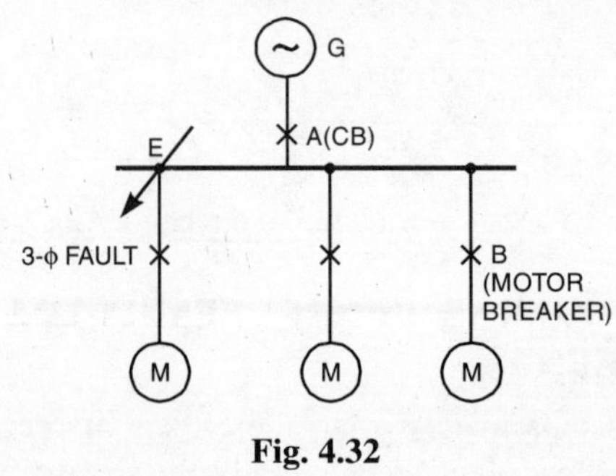

Fig. 4.32

Solution : Let the base kVA be 625 kVA and base voltage be 2.4 kV

Per unit subtransient reactance of generator,

$$X''_g = \frac{8}{100} = 0.08 \text{ pu}$$

$$\text{Input to each motor} = \frac{250 \times 0.7355}{0.9} = 204.3 \text{ kVA}$$

Per unit subtransient reactance of motor

$$= j0.2 \times \frac{625}{204.3} = 0.612 \text{ pu}$$

Per unit reactance diagram is shown in Fig. 4.33.

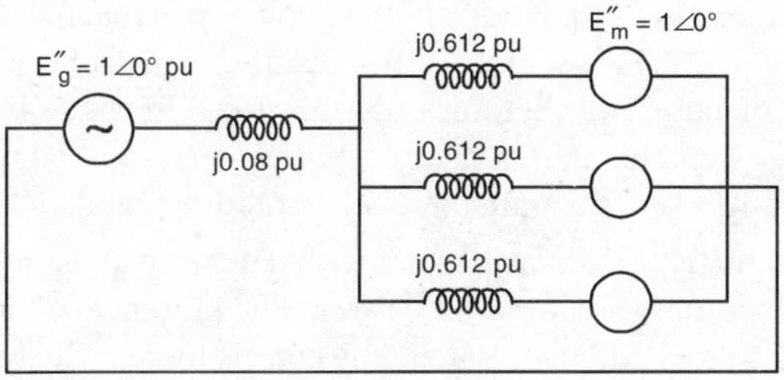

(a) Per Unit Reactance Diagram

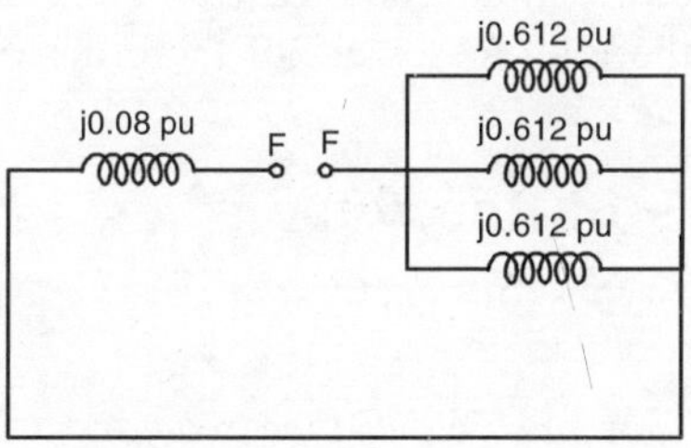

(b) Thevenin Equivalent Circuit

Fig. 4.33

Thevenin reactance when viewed from fault terminals FF

$$X_{Th} = \frac{j\dfrac{0.612}{3} \times j0.08}{j\dfrac{0.612}{3} + j0.08} = j0.0575 \text{ pu}$$

Initial symmetrical current at fault F,

$$I'' = \frac{1}{j0.0575} = -j17.4 \text{ pu} \quad \textbf{Ans.}$$

The generator contribution in I″,

$$I''_g = -j17.4 \times \frac{j\dfrac{0.612}{3}}{j\dfrac{0.612}{3} + j0.08} = -j12.5 \text{ pu} \quad \textbf{Ans.}$$

Contribution of motors in I″

$$3I''_m = I'' - I''_g = -j17.4 - (-j12.5) = -j4.9 \text{ pu}$$

$$\text{Contribution of each motor, } I''_m = \frac{-j4.9}{3} = -j1.633 \text{ pu}$$

Thus fault current rating of generator = 12.5 pu **Ans.**

Fault current rating of each motor = 1.633 pu **Ans.**

4.7. CURRENT LIMITING REACTORS

A current limiting reactor, also sometimes called a *series reactor*, is an inductive coil having a large inductive reactance in comparison to its resistance and is used for limiting short-circuit currents during fault conditions. These are installed in feeders and ties, in generator leads, and between bus sections to reduce the magnitude of short-circuit currents and the effect of the resulting voltage disturbances. Their cost is often more than offset by the saving in circuit breaker cost as a result of the lower short-circuit ratings that can be used, their use is, therefore, restricted to the interconnection of large power systems. The reactors allow free interchange of power under normal conditions but under short-circuit conditions the disturbance is confined to the faulty section. As the resistance of reactors in comparison to their reactance is very small, the efficiency of the system is not affected appreciably.

4.7.1. Principle of Operation of Current Limiting Reactors

It is clear that the short-circuit currents are reduced by an increase of the percentage reactance in the system. The short-circuit currents depend upon the generating capacity, voltage at the fault point and the total reactance between the generators and the fault point. The circuit breakers should have enough breaking current capacity—more than the expected fault currents. If the fault currents are beyond the capacity of the circuit breaker, the fault currents may not be interrupted. In large interconnected systems the total rating of the generators is very high and, therefore, the fault currents are very high. There is a possibility that the circuit breakers of suitable breaking capacity may not be available. So it becomes necessary to limit the fault currents by some means so that available or existing circuit breakers can be used safely for handling them. Further, when the system is extended by the addition of more generating units, more generating stations or by a supply from the grid, the fault currents to be interrupted by the same circuit breaker will become greater than before. It is uneconomical to scrap the existing switchgear, which was adequate for the former output but is not of sufficient capacity to meet the extensions. In such a case there is a need of providing a protective reactor between the old system and the extensions for limiting the short-circuit currents to a permissible value. By including a reactor or a few reactors at strategic locations, the short-circuit currents at different points can be reduced. Hence current limiting reactors are of considerable importance in limiting the currents so that the various circuit breakers are not called upon to break currents above their rated value.

4.7.2. Primary Functions of Current Limiting Reactors

The current limiting reactors are used to perform the following functions :

1. Protective reactors are used to reduce the flow of current into a short circuit so as to protect the apparatus from excessive mechanical stresses and from overheating and thus protect the system as a whole.
2. Protective reactors are used to reduce the magnitude of voltage disturbances caused by short circuits.

3. They also localise the fault by limiting the current that flows into the fault from other healthy feeders or parts of the system, thereby avoiding the fault from spreading. This increases the chances of continuity of supply.
4. They reduce the duty imposed on switching equipment during short circuits to be within economical ratings. So they are used (*i*) in systems where extensions have been made and the circuit breaker rupturing capacities have become inadequate and (*ii*) in large systems, so as to limit the short-circuit MVA to match with the rupturing capacity of the circuit breakers.

In general, reactors should be located at points in the network where they can be most effective. Very few occasions arise where it is necessary or desirable to introduce reactance in the generator circuit as modern alternators have sufficient inherent reactance to enable them to withstand the forces of short circuit. However, when modern machines are to be operated in parallel with older ones, a case may arise where added reactance in the circuits of the older machines will provide protection and give them roughly the same characteristics as the new machines.

Reactors installed in individual feeder circuits are not an economical proposition as often a considerable number of feeders are involved. A good case, however, can be that of a group feeder where the insertion of additional reactance is necessary to protect a group of circuit breakers of low rupturing capacity. Similarly, an interconnector between new and old sections of an installation may profitably include a reactor and thus eliminate the need of replacing old circuit breakers.

4.7.3. Drawbacks of Current Limiting Reactors

Current limiting reactors have some drawbacks too as with the introduction of the reactors, total percentage reactance of the circuit increases, thereby causing increase in reactive volt drop and decrease in power factor due to increased angle of lag. Thus regulation becomes poorer.

4.7.4. Reactor Ratings

The rating of a reactor is usually expressed in terms of percentage and on a three-phase system operating at 11 kV, a 20 per cent is one which will have a voltage drop of 1,270 volts across it with full-load current flowing through it.

Other ratings are given below:

Continuous Rated Current. It is the rms value of current which the reactor can carry continuously, with temperature rise of current carrying parts and other parts, within specified limits [*e.g.,* 1,000 A].

Rated Short-Time Current. It is the symmetrical rms value of fault current which the reactor can carry for specified short-time duration (*e.g.,* 60 kA for 1 second).

Rated Voltage. It is the line-to-line service voltage for which the reactor is designed.

Short-Circuit Rating. The reactors should be capable of withstanding the mechanical and thermal stresses during short circuit at its terminals for a specified period of time.

Rated Overcurrent Factor. It is the ratio of rated short-time current to continuous current.

Rated Through-Put kVA. It equals $\sqrt{3}$ times the product of rated voltage and rated current in case of 3-phase reactors.

4.7.5. Design Features of Current Limiting Reactors

The essential requirement of a current limiting reactor is that its reactance should not decrease on account of saturation under short-circuit conditions (when a large current flows through its windings) if short-circuit current has to be limited to a predetermined value. If fault current exceeds about three times rated full-load current an iron core reactor having essentially constant permeability would require large cross section of core and therefore, the reactor will turn out to be very costly and heavy. Because of this reason, air-cored reactors are usually used to limit the short-circuit or fault currents.

Alternatively an iron core with an air gap is used for current limiting reactors. Moreover, the presence of an iron core introduces eddy current and hysteresis losses which means that power consumed by an iron-cored reactor is more than the power consumed in an air-cored reactor. Normally in an iron-cored reactor, the total losses are of the order of 5 % of kVA rating of the reactor.

4.7.6. Types of Reactors

There are two types of reactors in use viz, bare type and shielded type.

In the *bare* or *unshielded type reactors,* circular coils or bars of stranded copper are embedded in a number of specially designed concrete slabs. Such an arrangement affords a very rigid mechanical support against the mechanical forces developed due to the flow of short-circuit currents through the reactor windings. The mechanical forces act in the form of a couple that tries to compress the winding axially and gives rise to radial expansion and for this reason only circular coils are used as in the case of transformers. The individual turns of the winding are inclined with respect to the horizontal plane. The necessary insulation to earth is provided by a concrete base and porcelain post insulators which also serve the purpose of supports. Such reactors are also known as *cast concrete type* or *dry type reactors.*

Dry type reactors are usually cooled by natural ventilation and are sometimes designed with forced air and heat exchanger auxiliaries. Dry type reactors are very simple from the constructional point of view and are robust but have the following disadvantages :

(*i*) Large space is required because the magnetic field on account of load current is practically unrestricted.

(*ii*) Difficulty is experienced in cooling of large coils by fans.

(*iii*) These are unsuitable for outdoor services.

(*iv*) Their use is limited to the voltages of 33 kV.

The *shielded* or *oil-immersed type reactors* employ insulation and cooling arrangement similar to those of the ordinary transformers. In an air-cored construction, there must be laminated iron shields or copper shields around the outside of the conductors in order to prevent the magnetic flux entering the tank walls and causing excessive losses and heating. In iron-cored construction air gaps are introduced in the core to prevent saturation and to give a magnetising current of the desired value. Such current limiting reactors can be used up to any voltage level, for outdoor or indoor constructions. Other advantages of oil-immersed current-limiting reactors are :

1. They have size smaller than the dry type air-cored reactors because of ease of cooling due to oil immersion.
2. They have higher factor of safety against flash-over.
3. They have higher thermal capacity.
4. There is no magnetic field outside the tank to cause heating or magnetic forces in adjacent reactors or metal structures during short circuit.

The air-cored type, having no iron, has a constant reactance at all currents, but the reactance of the iron-core or iron-shielded types may drop by about 10 per cent due to saturation during short circuits.

4.7.7. Selection of Reactors

The aspects to be considered in the selection of a current limiting reactors are : (*i*) Reactance in ohms or in percentage. (*ii*) Normal current rating. (*iii*) Short-time current rating (*iv*) Rated voltage (*v*) Rated through-put kVA (*vi*) Type (dry or oil-immersed, air-cored or iron-cored) (*vii*) Number of phases (single or three) (*viii*) Indoor or outdoor (*ix*) Circuit characteristics (frequency, voltage etc).

4.8. LOCATION OF REACTORS

Reactors in a power system can be located : (*i*) in series with generators; (*ii*) in series with feeders or (*iii*) in busbars (in ring system or in tie-bar system), as explained below :

1. Generator Reactors. When the reactors are inserted between the generator and the generator bus, as shown in Fig. 4.34 (*a*), the reactors are known as *generator reactors*. Such reactors protect the machines individually.

Since modern machines have transient reactance sufficient to protect themselves against dead 3-phase short circuit at its terminals, hence separate reactors are not required in the modern installations. However, when new machines are installed in an old power station, generator reactors may be added for the older generators. The magnitude of such reactors is very approximately about 5 per cent or 0.05 per unit. The main disadvantage of this method is that if a short circuit occurs on one feeder, the voltage at the generator bus may drop to such a low value that the synchronous machines connected to this common bus-bar may fall out of step. Thus the whole system will be adversely affected. Moreover, a full-load current is always flowing through these reactors during normal operation resulting in a constant voltage drop and constant power loss.

2. Feeder Reactors. When the reactors are connected in series with the feeders, as shown in Fig. 4.34 (*b*), the reactors are known as *feeder reactors*. In the event of fault on any one feeder, in this case, the main voltage drop is in its reactor only and the bus-bar voltage is not affected much, hence other machines continue supplying load. The other advantage is that the fault on a feeder will not affect other feeders and thus the effects of fault are localised.

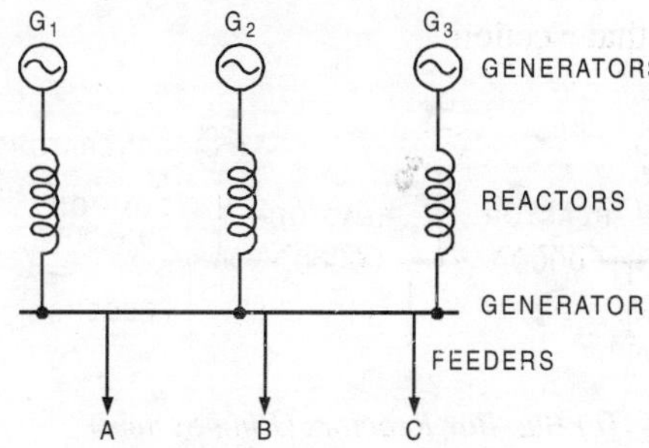

(a) Generator Reactors

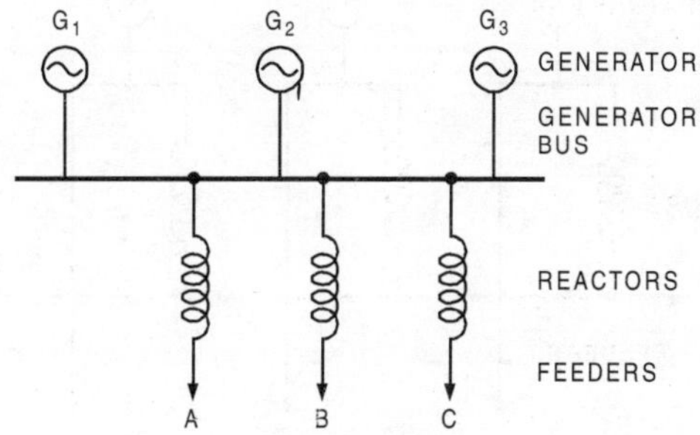

(b) Feeder Reactors

Fig. 4.34

The disadvantage of locating reactors in this position is that these do not provide any protection to the generators against short-circuit faults across the bus-bars. This is, however, of no importance, because such faults are rare and also modern alternators have large transient reactance for their protection against short circuits.

The other drawback is of constant voltage drop and constant power loss in reactors even during normal operation.

As already discussed individual feeder reactors are very expensive and must be resorted to only in the case of group feeders. The usual value of reactance for feeder circuits is of the order of 3 to 5 per cent based on the normal current loading of the circuit. Here it is to be noted that with highly inductive loads, serious interference with regulation can arise but at or near unity power factor the effect on regulation is negligible.

3. Bus-Bar Reactors. The constant voltage drop and constant power loss in reactors may be avoided by inserting the reactors in the bus-bars, as shown in Figs. 4.34 (*c*) and 4.34 (*d*). The former is the *ring system*, and the latter is the tie-bar *system*.

(a) Bus-Bar Reactors (Ring System). Bus-bar reactors are used to tie together separate bus sections. In this system sections are made of generators and feeders and these sections are connected to each other at the common bus-bar, as shown in Fig. 4.34 (*c*). In this system normally one feeder is fed from one generator. Under normal operating conditions small amount of power flows through the reactors, therefore, voltage drop and power loss in the reactors is low. The reactors can, therefore, be made with a fairly high ohmic resistance and there is not much voltage drop across it. In case of fault on any one feeder, only one generator feeds the fault while the current from other generators is limited because of presence of the bus-bar reactors. Thus heavy currents and voltage disturbances caused by a short circuit on a bus section are reduced and confined primarily to that section.

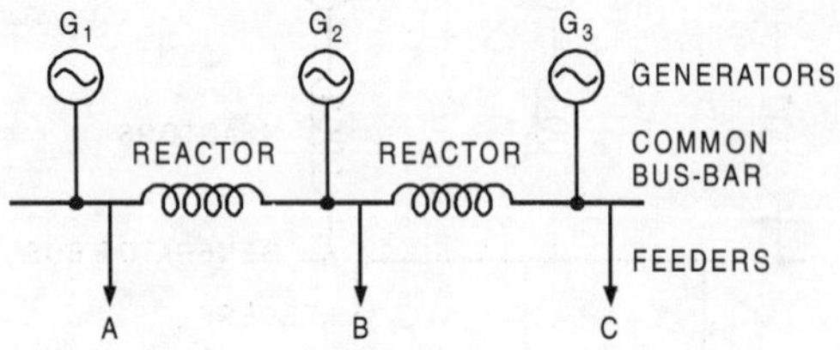

(c) Bus-Bar Reactors (Ring System)

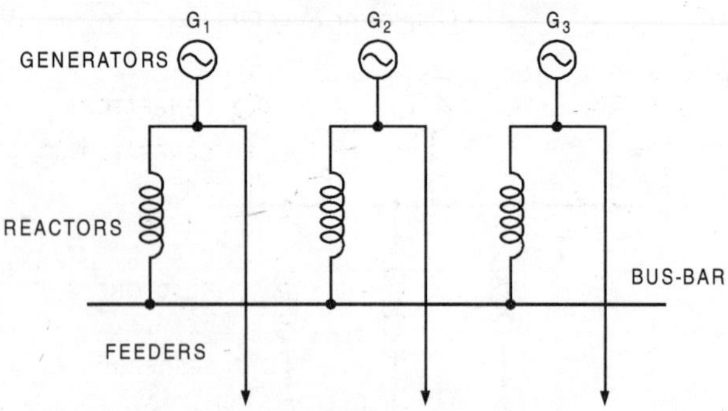

(d) Bus-Bar Reactors (tie-Bar System)

Fig. 4.34

Bus-bar reactors, however, do not protect the generators connected to the faulty sections. They facilitate the parallel operation of large systems and are extensively employed.

(b) Bus-Bar Reactors (Tie-Bar System). This is a modification of the above system. In general, with this system the voltage regulation between feeder sections is better than the above system. This system is ideally suited to the generating systems where frequently new generators are being added. In this system the generators are connected to the common bus-bar through the reactors but the feeders are fed from the generator side of the reactors. The operation of this system is similar to ring system. It has got an additional advantage. If the number of sections is increased, the fault current will not exceed a certain value, which is fixed by the size of individual reactors. Thus the switchgear designed to operate successfully on this limiting current will continue operating successfully for any number of extensions of the sections and require no modification.

Example 4.19. Find the value in ohms of the reactance per phase external to a 20 MVA, 10 kV, 50 Hz, 3-phase generator such that the steady-state current on short circuit shall not exceed 8 times the full-load current. The internal reactance of the generator is 5%.

Solution : Reactance required to limit the short-circuit current to 8 times of full-load current

$$= \frac{\text{Full-load current}}{\text{Short-circuit current}} = \frac{1}{8} = 0.125 \text{ pu}$$

Internal synchronous reactance per phase

$$= 5\ \% = 0.05 \text{ pu}$$

External reactance required per phase,

$$X_{pu} = 0.125 - 0.05 = 0.075 \text{ pu}$$

External reactance required per phase in ohms

$$= X_{pu} \times \frac{(kV_B)^2}{MVA_B} = \frac{0.075 \times (10)^2}{20}$$

$$= 0.375\ \Omega \textbf{ Ans.}$$

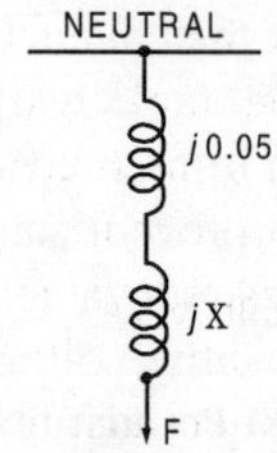

Fig. 4.35

Reactance diagram is shown in Fig. 4.35.

Example 4.20. A 100 MVA generator with 10 % reactance and a 200 MVA generator with 8 % reactance (reactances on their own bases) are connected as shown in Fig. 4.36 (*a*). The fault level on bus 1 is to be restricted to 1,500 MVA. Calculate, on 100 MVA base (*i*) the reactance of bus-bar reactor X (*ii*) fault level of bus 2 and (*iii*) MVA rating of circuit breaker C. [*GATE 1991*]

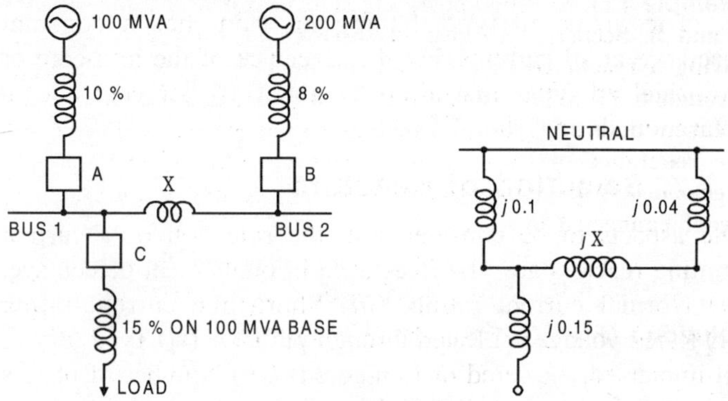

(a) *(b) Reactance Diagram*

Fig. 4.36

Solution : Single line reactance diagram is shown in Fig. 4.36 (*b*). Let the base MVA be 100.

$$\text{Per unit reactance of 100 MVA generator} = \frac{10}{100} = 0.1 \text{ (given)}$$

$$\text{Per unit reactance of 200 MVA generator} = \frac{8}{100} \times \frac{100}{200} = 0.04$$

$$\text{Per unit reactance of feeder} = \frac{15}{100} = 0.15 \quad \text{(given)}$$

(*i*) When fault occurs on bus 1, reactance of 200 MVA generator in series with bus-bar reactor acts in parallel with the reactance of 100 MVA generator.

So per unit reactance between fault point and the neutral bus

$$= j\,\frac{0.1 \times (0.04 + X)}{0.1 + (0.04 + X)} = j\,\frac{0.1X + 0.004}{0.14 + X}$$

Since short-circuit MVA

$$= \frac{\text{Base MVA}}{\text{Per unit reactance}}$$

$$\therefore\ 1{,}500 = \frac{100 \times (0.14 + X)}{0.1X + 0.004}$$

$$\text{or} \quad X = \frac{100 \times 0.14 - (1{,}500 \times 0.004)}{1{,}500 \times 0.1 - 100} = 0.16 \text{ pu } \textbf{Ans.}$$

(*ii*) When fault occurs on bus 2, reactance of 100 MVA generator in series with bus-bar reactor acts in parallel with the reactance of 200 MVA generator. So per unit reactance between fault point and the neutral bus

$$= j\frac{(0.1+X)\times 0.04}{(0.1+X)+0.04} = j\frac{0.04\,(0.1+0.16)}{0.1+0.16+0.04}$$

$$= \frac{j0.04\times 0.26}{0.3} = j0.0346 \text{ pu}$$

$$\text{Fault MVA} = \frac{\text{Base MVA}}{\text{Per unit reactance}} = \frac{100}{0.0346} = 2{,}885 \text{ MVA } \textbf{Ans.}$$

(*iii*) Per unit reactance from the neutral bus to fault point 0,

$$X_{eq\ pu} = j\left[0.15+\frac{0.1\times(0.04+0.16)}{0.1+0.04+0.16}\right] = j0.15+\frac{0.02}{0.3}$$

$$= 0.2166 \text{ pu}$$

MVA rating of circuit breaker C

$$= \frac{\text{Base MVA}}{\text{Per unit reactance, } X_{eq\ pu}} = \frac{100}{0.2166}$$

$$= 461.6 \text{ MVA } \textbf{Ans.}$$

Example 4.21. A station operating at 33 kV is divided into sections A and B. Section A consists of three generators 15 MVA each having a reactance of 15% and section B is fed from the grid through a 75 MVA transformer of 8 % reactance. The circuit breakers have each a rupturing capacity of 750 MVA. Determine the reactance of the reactor to prevent the breakers being overloaded if a symmetrical short circuit occurs on an outgoing feeder connected to A.

[*Pb. Univ. Power System Analysis December 1994; A.M.I.E. Sec. B. Advanced Power Systems Winter 1993*]

Solution : Given system is shown in Fig. 4.37.

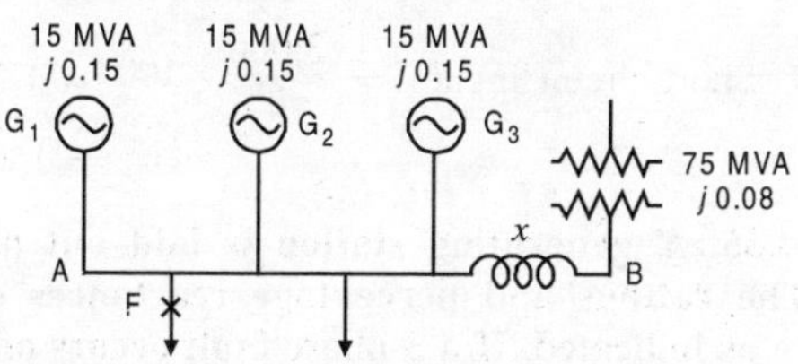

Fig. 4.37

Let the base MVA be 75 MVA.

$$\text{Per unit reactance of each generator} = j\frac{15}{100}\times\frac{75}{15} = j\,0.75.$$

Per unit reactance of transformer $= j\,0.08$ (given)

Let x per unit be the reactance of the reactor for base MVA of 75 and base kV of 33.

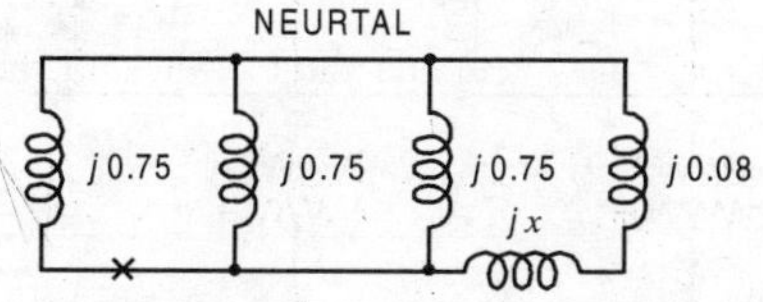

Fig. 4.38. *Reactance Diagram*

Single line reactance diagram is shown in Fig. 4.38. The reactances of generators G_1, G_2 and G_3 are in parallel so their combined reactance is $\frac{j0.75}{3}$ *i.e.*, $j\,0.25$ pu. This reactance is in parallel with reactance of transformer in series with reactance of the reactor *i.e.*, $j(0.08 + x)$. Thus the per unit reactance between the fault point and the neutral bus

$$= j\,0.25 \parallel (j\,0.08+x) = j\frac{0.25\,(x+0.08)}{0.25+x+0.08} = j\frac{0.25x+0.02}{x+0.33}$$

$$\text{Since short-circuit MVA} = \frac{\text{Base MVA}}{\text{Per unit reactance}}$$

$$\therefore\ 750 = \frac{75\,(x+0.33)}{0.25x+0.02}$$

$$\text{or } x = \frac{75\times 0.33-(750\times 0.02)}{750\times 0.25-75}$$

$$= \frac{24.75-15}{187.5-75} = \frac{9.75}{112.5} = 0.08666 \text{ pu}$$

Actual value of reactance of reactor

$$= \frac{0.08666\times 33^2}{75} = 1.2584\ \Omega \textbf{ Ans.}$$

Example 4.22. Figure 4.39 (*a*) shows a system having four alternators each rated at 11 kV, 50 MVA and each having a subtransient reactance of 15 %. Find (*a*) fault level for a fault on one of the feeders (near the bus) with zero value of X (*b*) the reactance of the current limiting reactor X to limit the fault level to 800 MVA for a fault on one of the feeders (near the bus).

[*Rajasthan Univ. Power System Analysis 2005*]

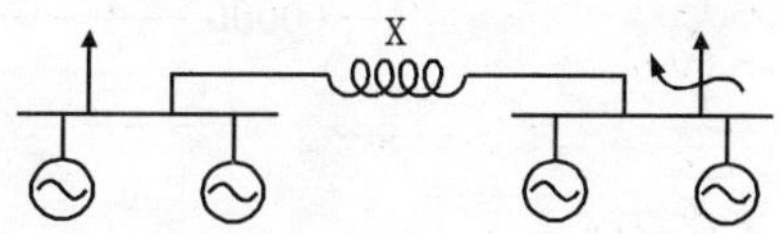

(*a*)

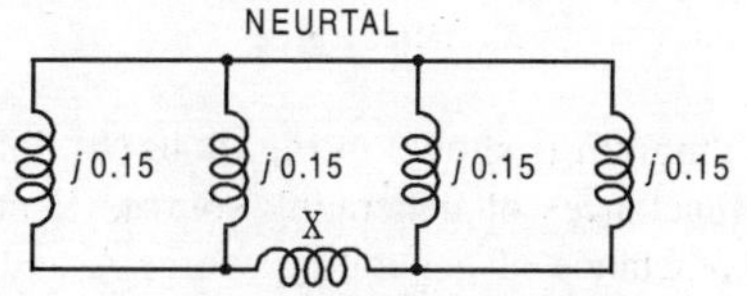

(*b*) *Reactance Diagram*

Fig. 4.39

Solution : Let the base MVA be 50 and base kV be 11.

Single line reactance diagram is shown in Fig. 4.39 (*b*).

Per unit reactance between the fault point and the neutral bus with zero value of X.

$$X_{pu} = \frac{j0.15}{4} = j\,0.0375 \text{ pu}$$

$$\text{Fault level} = \frac{\text{Base MVA}}{X_{pu}} = \frac{50}{0.0375} = 1{,}333.3 \text{ MVA } \textbf{Ans.}$$

Per unit reactance between the fault point and the neutral bus with current limiting reactor X,

$$X'_{pu} = j[(0.15 \parallel 0.15)] \parallel j[(0.15 \parallel 0.15) + X]$$

$$= j\,0.075 \parallel (j\,0.075 + X)$$

$$= j\frac{0.075\times(0.075+X)}{0.075+(0.075+X)} = j\frac{0.075\times(0.075+X)}{0.15+X}$$

$$\text{Fault level} = \frac{\text{Base MVA}}{X'_{pu}} = \frac{50}{\frac{0.075(0.075+X)}{0.15+X}} = 800 \quad \text{...(given)}$$

or $X = 0.3$ pu **Ans.**

$$\text{Base reactance} = \frac{(\text{Base line voltage in kV})^2}{\text{Base MVA}} = \frac{11^2}{50} = 2.42\ \Omega$$

So X in ohms $= X'_{pu} \times$ base reactance $= 0.3 \times 2.42 = 0.726\ \Omega$ **Ans.**

Example 4.23. In the power system circuit diagram shown in Fig. 4.40 (*a*), the current limiting reactor X is to be chosen such that the feeder breaker rating does not exceed 425 MVA. The system data is as follows:

Feeder transformer reactance : 10 % on 50 MVA base. The generating sources A, B, C have individual fault levels of 1,000 MVA with respective generator breakers open. Ignore prefault currents and assume 1.0 pu voltages through out before fault. Assume common base of 100 MVA. [*GATE 1996*]

Solution : Common base MVA = 100 ...(given)

$$\text{Per unit reactance of transformer} = \frac{100}{50} \times \frac{10}{100} = 0.2 \text{ pu}$$

Per unit reactance of each generator

$$= \frac{\text{Base MVA}}{\text{Short-circuit MVA}} = \frac{100}{1{,}000} = 0.1 \text{ pu}$$

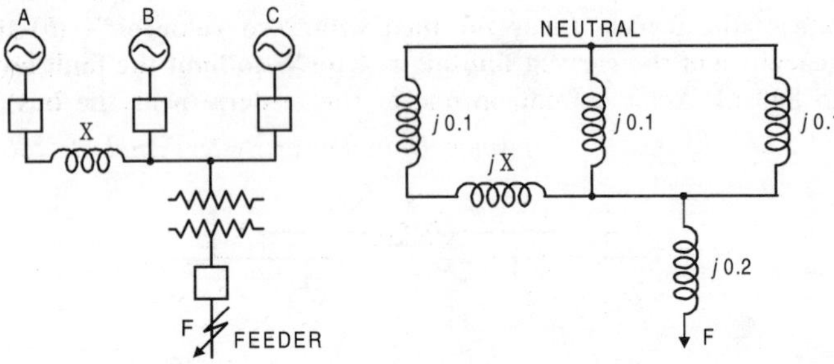

(*a*) (*b*) *Reactance Diagram*

Fig. 4.40

Reactance diagram is shown in Fig. 4.40 (*b*). When fault occurs at point F, the reactances of generating sources B and C and series combination of reactance of generating source A and bus-bar reactor act in parallel and therefore, their combined reactance

$$= j[(0.1 \parallel 0.1) \parallel (0.1 + X)]$$

$$= j\left[\frac{0.1}{2} \parallel (0.1 + X)\right]$$

$$= j\frac{0.05 \times (0.1 + X)}{0.05 + 0.1 + X} = j\frac{0.05X + 0.005}{X + 0.15}$$

This reactance is in series with transformer reactance of $j0.2$, so the per unit reactance between fault point and neutral bus,

$$X_{eq\ pu} = j\frac{0.05X + 0.005}{X + 0.15} + j0.2 = j\frac{0.25X + 0.035}{X + 0.15}$$

Since short-circuit MVA

$$= \frac{\text{Base MVA}}{X_{eq\ pu}}$$

$$\therefore \quad 425 = \frac{100\,(X + 0.15)}{0.25X + 0.035}$$

$$\text{or} \quad X = \frac{100 \times 0.15 - (425 \times 0.035)}{425 \times 0.25 - 100} = \frac{15 - 14.875}{106.25 - 100}$$

$$= 0.02 \text{ or } 2\ \% \textbf{ Ans.}$$

Example 4.24. A generating station has five section bus-bars connected with a tie-bar through 7.5% reactors rated at 3,000 kVA. Generators aggregating 3,000 kVA with 10% reactance are connected to each section bus-bar. Find the total steady input to a dead short circuit between the lines on one of the section bus-bars (*i*) with and (*ii*) without reactors.

Solution : Let the fault occur on section 5, as shown in Fig. 4.41.

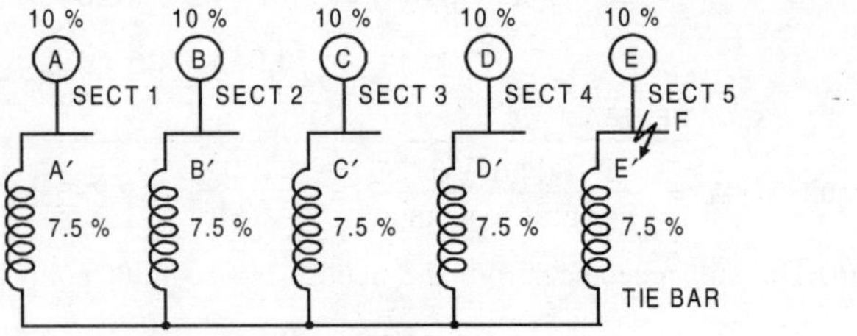

Fig. 4.41

(*a*) The reactances of generators A, B, C, D and their associated reactors are in parallel so that their combined reactance is 1/4 (10 + 7.5) *i.e.*, 4.375 %. This is in series with reactance E. Therefore combined reactance up to fault due to generator A, B, C, D and reactors A, B, C, D and E,

$$= 4.375 + 7.5 = 11.875\ \%.$$

This reactance is in parallel with that of generator E (10%) and total reactance up to fault point

$$= \frac{1}{\frac{1}{10} + \frac{1}{11.875}} = 5.428\ \%$$

$$\therefore \text{ Short-circuit input} = \frac{3{,}000 \times 10^{-3}}{5.428} \times 100 = 55.3 \text{ MVA } \textbf{Ans.}$$

(*b*) When there is no reactor,

$$\text{Total reactance up to fault point} = \frac{10}{5} = 2\%$$

$$\therefore \quad \text{Short-circuit input} = \frac{3{,}000}{2} \times 100 = 1{,}50{,}000 \text{ kVA}$$

$$= 150 \text{ MVA } \textbf{Ans.}$$

Example 4.25. A generating station is laid out as shown in Fig. 4.42. The ratings and percentage reactances of different elements are as indicated. If a 3-phase fault occurs on any feeder near transformer secondary let at F, then find the short-circuit MVA fed to the fault. Choose 5 MVA as the base value.

[*Rajasthan Technical Univ. Power System Analysis, 2011*]

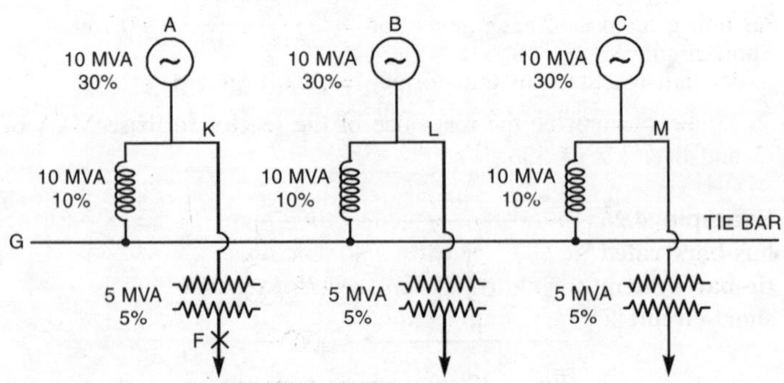

Fig. 4.42

Solution : Base MVA = 5 MVA (Given)

Per unit reactance of each generator on the base MVA

$$= \frac{30}{100} \times \frac{5}{10} = 0.15 \text{ pu}$$

Per unit reactance of each reactor on the base MVA

$$= \frac{10}{100} \times \frac{5}{10} = 0.05 \text{ pu}$$

Per unit reactance of each transformer on the base MVA

$$= \frac{5}{100} \times \frac{5}{5} = 0.05 \text{ pu}$$

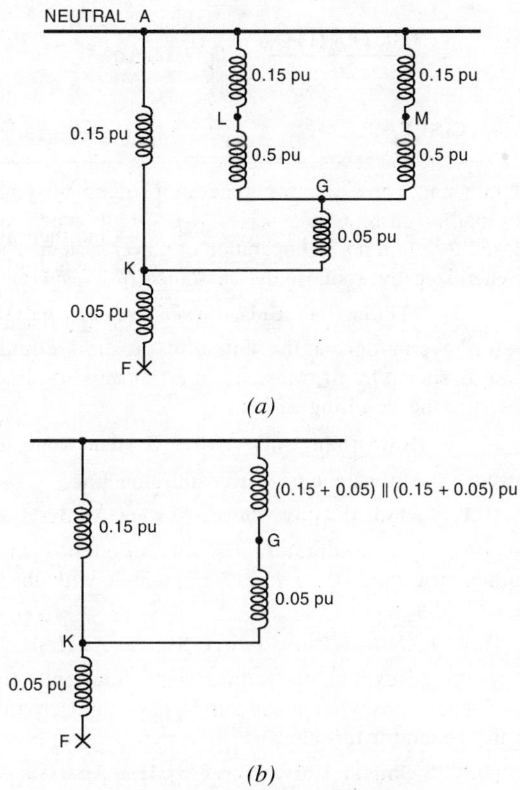

Fig. 4.43. ***Reactance Diagram***

When a three-phase short circuit occurs at point F of the feeder near the secondary of the transformer, the reactance diagram will be as shown in Fig. 4.43 (*a*). This circuit can be further simplified as shown in Fig. 4.43 (*b*).

Total per unit reactance from generator neutral up to fault point F

$$= (0.1 + 0.05) \parallel 0.15 + 0.05 = \frac{0.15 \times 0.15}{0.15 + 0.15} + 0.05 = 0.125 \text{ pu}$$

Short-circuit MVA

$$= \frac{\text{Base MVA}}{\text{Per Unit fault reactance}} = \frac{5}{0.125} = 40 \text{ MVA } \textbf{Ans.}$$

Example 4.26. Show that a generating plant having N section bus-bars rated at Q kVA with *x*% reactance, connected on the tie-bar system through bus-bar reactances of *b*% has a total short-circuit kVA on one section of

$$\left[\frac{Q}{x} + \frac{Q(N-1)}{bN + x}\right] \times 100$$

If the section rating is 50,000 kVA; *x* = 20 % and *b* = 10 %, find the short-circuit kVA with (*i*) 3 sections (*ii*) 9 sections and (*iii*) show that with a very large number of sections the short-circuit kVA would not exceed 7,50,000 kVA.

[*Rajasthan Univ. Power System Analysis 1999*]

Solution : The equivalent circuit diagram is shown in Fig. 4.44 (*a*).

If Nth section is short-circuited, there will be (N – 1) paths in parallel between the neutral and tie-bar, each path having a total reactance of (*x* + *b*) per cent.

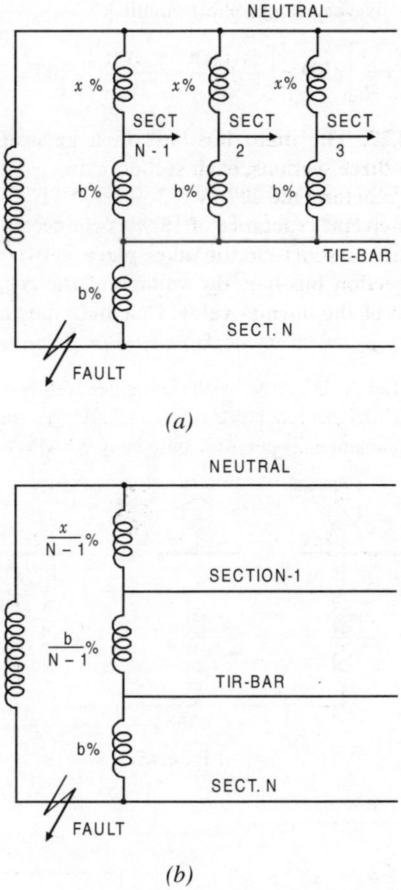

Fig. 4.44

Total reactance between neutral and tie-bar $= \dfrac{x+b}{N-1}$

Total reactance between the neutral and fault via tie-bar

$$= \frac{x+b}{N-1} + b = \frac{x+bN}{N-1}$$

This reactance is in parallel with the reactance *x* of the short-circuited machine.

Total reactance between the neutral and fault

$$= \frac{1}{\frac{1}{x} + \frac{1}{\frac{x+bN}{N-1}}} = \frac{1}{\frac{1}{x} + \frac{N-1}{x+bN}} = \frac{x(x+bN)}{(x+bN)+x(N-1)} \text{ per cent}$$

Total short-circuit kVA

$$= \frac{Q \times 100}{\frac{x(x+bN)}{(x+bN)+x(N-1)}} = \left[\frac{Q(x+bN)}{x(x+bN)} + \frac{Qx(N-1)}{x(x+bN)}\right] \times 100$$

$$= \left[\frac{Q}{x} + \frac{Q(N-1)}{(bN+x)}\right] \times 100 \textbf{ Proved}$$

(*i*) Short-circuit kVA with 3 sections [Q = 50,000 kVA ; *x* = 20%, *b* = 10 %]

$$= \left[\frac{50{,}000}{20} + \frac{50{,}000 \times (3-1)}{10 \times 3 + 20}\right] \times 100 = 450{,}000 \text{ kVA } \textbf{Ans.}$$

(*ii*) Short-circuit kVA with 9 sections

$$= \left[\frac{50{,}000}{20} + \frac{50{,}000 \times (9-1)}{10 \times 9 + 20}\right] \times 100 = 613{,}600 \text{ kVA } \textbf{Ans.}$$

(*iii*) When N is very large, short-circuit kVA

$$= \left[\frac{Q}{x}+\frac{Q}{b}\right]\times 100 = \left[\frac{50{,}000}{20}+\frac{50{,}000}{10}\right]\times 100 = 750{,}000\,\text{kVA}\ \textbf{Ans.}$$

Example 4.27. The main bus-bars in a generating station are divided into three sections, each section being connected to tie bar by a similar reactor. One 20 MVA, 3-phase, 50 Hz, 11 kV generator with a short-circuit reactance of 15 % is connected to each section bus-bar. When a short-circuit takes place between the phases of one of the section bus-bar, the voltage of the remaining sections falls to 55 % of the normal value. Calculate the reactance of each reactor. [*Pb. Univ. Switchgear and Protection May 1983*]

Solution : Let A, B and C be the three generators, each of 20 MVA and of 15 % (0.15 pu) reactance (Fig. 4.45). A′, B′, and C′ are the three reactors of reactance X per unit based on 20 MVA.

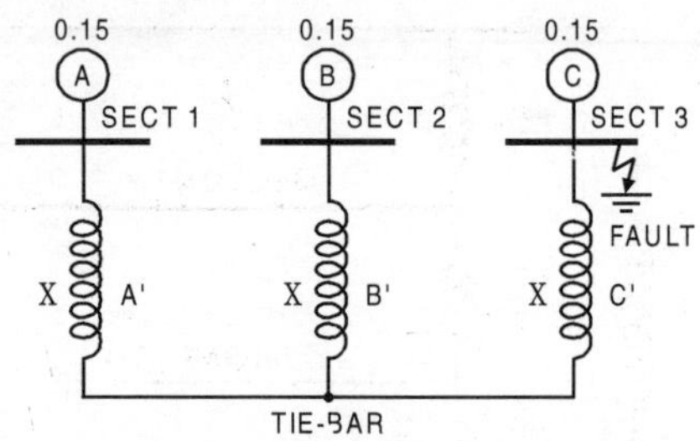

Fig. 4.45

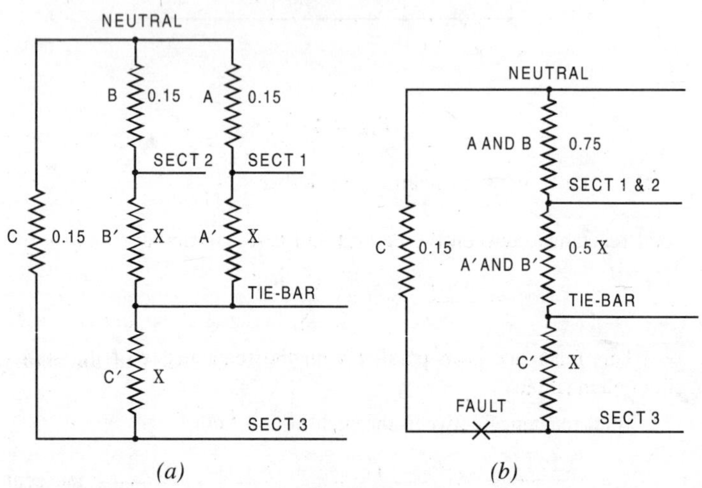

Fig. 4.46

The circuit diagram is shown in Fig. 4.46 assuming that the fault occurs in section 3. A and A′ in series are in parallel with B and B′ in series.

∴ Reactance of A, B, A′ and B′ between the tie-bar and neutral

$$= \frac{0.15 + X}{2} \quad \textit{i.e.,}\ (0.5X + 0.075)\ \text{pu}$$

Reactance from neutral to fault via tie-bar

$$= [(0.5X + 0.075) + X] = (1.5X + 0.075)\ \text{pu}$$

The voltage of sections 1 and 2 drops to 55% *i.e.,* 0.55 times of the normal value, therefore, 0.45 times of the normal value is dropped in the reactance of A and B *i.e.,* reactance of A and B $\left(\textit{i.e.,}\ \frac{0.15}{2}\right)$ is 0.45 times the total reactance from neutral to fault

$$\textit{i.e.,}\quad \frac{0.15}{2} = 0.45\,(1.5\,X + 0.075)$$

$$\text{or } X = 0.0611\ \text{pu.}$$

Hence reactance of each reactor

$$= \frac{\text{Per unit reactance} \times (\text{kV})^2}{\text{Base MVA}}$$

$$= \frac{0.0611 \times (11)^2}{20} = 0.37\ \Omega\ \textbf{Ans.}$$

Exercises

1. Obtain an expression for generation of an unsymmetrical alternating current in a series R-L circuit when switch is closed to retain normal operation of power system. The circuit is energized by a sinusoidal ac source of $V_m \sin(\omega t + \alpha)$.
[Rajasthan Univ. Power System Analysis 2005]
2. Describe and derive the equation of short circuit when there is short circuit transient on a transmission line ? Also describe the doubling effect.
[Rajasthan Univ. Power System Analysis 2002]
3. Discuss the transient on a transmission line.
[U.P. Technical Univ. Power System Analysis 2008-09]
4. What do you understand by instantaneous maximum momentary current for line ? Explain it with the help of suitable diagram.
[U.P. Technical Univ. Power System Analysis 2009-10]
5. Discuss and explain the subtransient, transient and steady state reactances when a sudden 3-phase short circuit occurs at the generator terminals.
[U.P. Technical Univ. Power System Analysis 2005-06]
6. Write short note on the reactances of an alternator when sudden 3-phase short circuit occurs on its terminals.
[U.P. Technical Univ. Power System Analysis 2006-07]
7. Draw equivalent circuits of synchronous machine under subtransient, transient and steady-state conditions.
[Rajasthan Technical Univ. Power System Analysis 2010]
8. A 3-ϕ short circuit occurs at the terminals of a synchronous motor. Explain with justification whether the machine will continue to act as motor ?
[U.P. Technical Univ. Power System Analysis 2004-05]
9. A three-phase alternator running unloaded with resistance and transient inductance per phase R and L respectively is suddenly short circuited across all its terminals at $t = 0$. If the alternator is considered as an emf $\sqrt{2}\,V \sin(\omega t + \alpha)$ per phase in series with R and L, derive an expression for the short circuit current per phase. Hence explain the following terms—symmetric short circuit current, dc offset current and doubling effect.
10. Draw a figure showing the variation of current in any one phase of an unloaded 3-phase alternator with respect to time for a sudden three-phase short circuit at the terminals of the alternator. Explain why does the current follow the shape shown in figure.
11. What are the different steps for the calculation of short-circuit kVA and short-circuit fault current.
[Rajasthan Technical Univ. Power System Analysis 2011]
12. Explain the steps involved in the symmetrical fault calculations. How can the load current be taken into account in fault calculations ?
[Anna Univ. Chennai. Power System Analysis May/June 2011]

13. List out the methods of reducing short-circuit current.

[Anna Univ. Chennai Power System Analysis, May/June 2011]

14. Why we do use reactors in power system ? Discuss their advantages. **[Rajasthan Univ. Power System Analysis 2002]**

15. What are limiting reactors ? Discuss their applications and location aspects.

[U.P. Technical Univ. Power System Analysis 2006-07]

16. Discuss the various strategic locational aspects of reactors for limiting the fault current and their advantages.

[U.P. Technical Univ. Power System Analysis 2007-08]

17. Discuss the use and location of current limiting reactors in a power system.

[U.P. Technical Univ. Power System Analysis 2005-06]

18. Describe various applications, locations and design features of current limiting reactors.

[Pb. Univ. Switchgear and Protection Dec. 1992]

19. Compare the performance of ring bus-bar reactors and tie-bar reactors in power systems for limiting short-circuit currents. **[Pb. Univ. Switchgear and Protection May 1983]**

20. Write short notes on the following :
    (*i*) Short-circuit capacity. **[Rajasthan Univ. 2008]**
    (*ii*) Transient on a transmission line.
    (*iii*) Analysis of symmetrical faults in power system networks.
    (*iv*) Bus-bar reactors and their locations.

Short Answer Type Questions With Answers

Q. 1. What is meant by fault level ?

[Bharathidasan Univ. November 1996]

Ans. When fault occurs at a point in a power system, the corresponding MVA is referred to as the fault level at that point.

Q. 2. What are the approximities in short circuit analysis ?

[Madurai Kamaraj Univ. November 1996]

Ans. See Art. 4.5.

Q. 3. What are current limiting reactors ?

Ans. Current limiting reactors are inductive coils having large inductive reactances in comparison to their resistances and are commonly installed on large power systems to limit the magnitude of short-circuit currents and reduce the resulting voltage disturbances on the rest of the system.

Q. 4. Why it is not necessary to insert current limiting reactor between the generator and the generator bus ?

Ans. Since modern generators have sufficient inherent transient reactance to protect themselves against dead 3-phase short circuit at their terminals, insertion of current limiting reactors between the generators and their bus is not necessary.

Problems

1. A 3-phase, 5 MVA, 6.6 kV alternator has a subtransient reactance of 12.5 %. A 3-phase short circuit occurs at its terminals. Determine fault MVA and fault current.

[Ans. 40 MVA, 3,500 $\angle -90°$ A]

2. A 3-phase, 6 MVA, 6.6 kV generator has a reactance of 0.1 pu. It is connected through a 6 MVA, 6.6/33 kV, 0.09 pu reactance transformer to a transmission line having a resistance of 0.09 Ω and a reactance of 0.36 Ω per km. A 3-phase dead short circuit occurs at a distance of 10 km from the transformer. The generator voltage at the time of fault is 7.2 kV. Find the fault current in the generator.

[Anna Univ. Chennai Power System Analysis May/June 2011]

[Ans. 2,728 $\angle -88.65°$ A]

3. For the radial network shown in Fig. 1, three-phase fault occurs at F. Determine the fault current and the line voltage at 11 kV base under fault conditions.

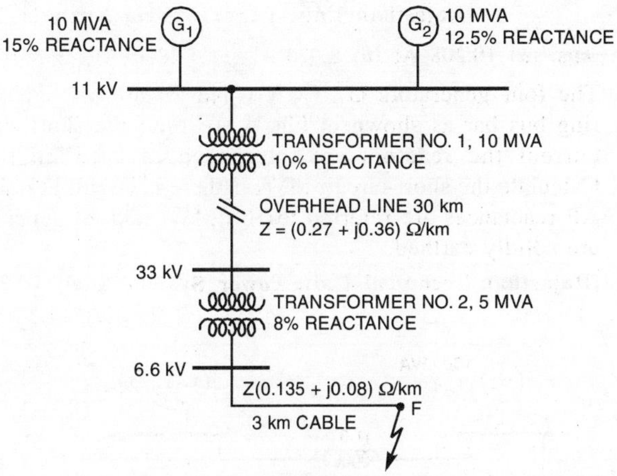

Fig. 1

[Rajasthan Univ. Power System Analysis 2003]

[Ans. 1,715 A; 9.68 kV]

4. Two 3-phase, 6.6 kV, 5 MVA generators having subtransient reactances of 12.5 % operate in parallel. The generators supply power to a transmission line through a 10 MVA transformer of ratio 6.6/33 kV and having a leakage reactance of 4%. Calculate fault current and fault MVA for 3-phase fault on (*i*) *lt* side (*b*) *ht* side of the transformer.

[Ans. (*i*) 7,000 $\angle -90°$ A; 80 MVA

(*ii*) 1,060 A $\angle -90°$ A; 60.61 MVA]

5. The estimated short circuit MVA at the bus-bars of a generating station is 1,500 and of another station 1,000. The generated voltage of each station is 13.2 kV. Calculate the possible short circuit MVA at each station when they are linked by an interconnector cable having a reactance of 0.5 Ω.

[Pb. Univ. Power System Analysis May 1994]

[Ans. 1,760 MVA; 1,283 MVA]

6. Two generating stations each having short-circuit capacities of 1,300 MVA and 900 MVA respectively are operating at 11 kV and are linked by an interconnected cable having a reactance of 0.55 Ω/phase. Determine the short-circuit capacity of each station.

[Anna Univ. Chennai Power System Analysis May/June 2011]

[Ans. 1,477 MVA; 1,088 MVA]

7. A 10 MVA, 132/6.6 kV transformer having a reactance of 0.15 pu is fed from an infinite bus. The transformer feeds 2 motors each 5 MVA, 6.6 kV. Each motor has a transient reactance of 0.3 pu and a subtransient reactance of 0.2 pu based on its own rating. A 3-phase fault occurs at the terminals of one motor as shown in Fig. 2. Find (*a*) subtransient fault

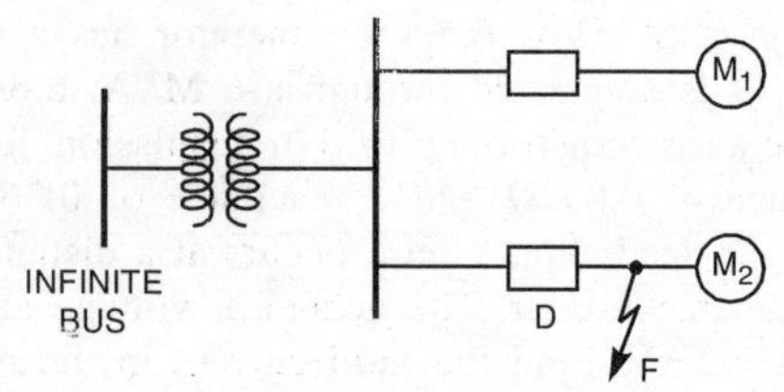

Fig. 2

current, (*b*) subtransient current in breaker D, (*c*) momentary current rating of breaker, (*d*) current to be interrupted by breaker which has a breaking time of 5 cycles. Assume suitable values of factors to account for dc offset.

[Rajasthan Univ. Power System Analysis 2005]

[**Ans.** (*a*) 10,208 A; (*b*) 8,020 A, (*c*) 12,830 A, (*d*) 8,015 A]

8. The four generators G_1, G_2, G_3 and G_4 are arranged in a ring bus bar as shown in Fig. 3. To limit the short-circuit current the reactors are interposed in between them. Calculate the short-circuit MVA if there is a fault at point F. All reactances are referred to 100 MVA and all generators are solidly earthed.

[Rajasthan Technical Univ. Power System Analysis 2010]

[**Ans.** 714.3 MVA]

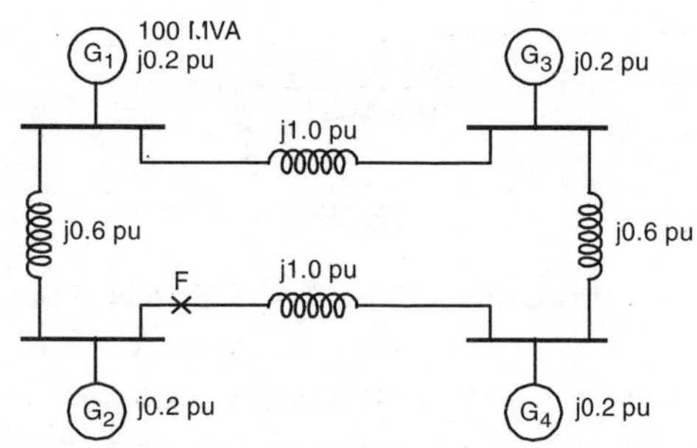

Fig. 3

9. The single line diagram shown in Fig. 4 represents a 33 kV interconnected 3-phase system. The relevant ratings and per unit reactances of the system are as follows :

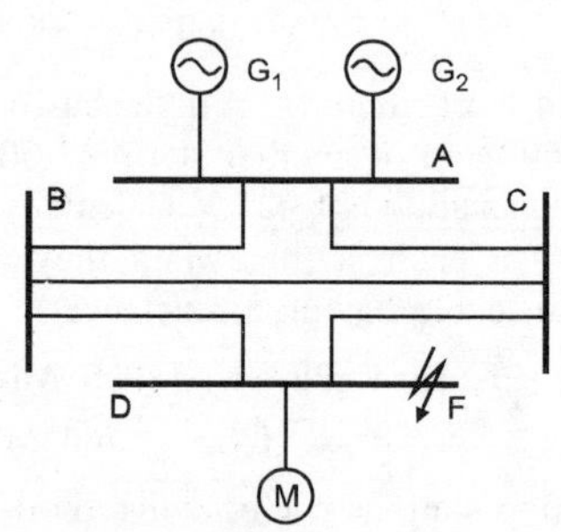

Fig. 4

Generators G_1 and G_2, both 25 MVA, 0.1 pu. Synchronous motor M, 20 MVA, 0.1 pu. Transmission lines, AB, BC, CD, AC, BD 10- Ω each. There is symmetrical 3-phase short circuit fault at bus bar D. Ignoring losses throughout, determine the fault level at busbar D. Assume that the excitation of each machine is such that it would be producing normal rated voltage at no load, and that each machine is operating at synchronous speed during the short circuit period.

[**Ans.** 290 MVA]

10. A 600 kVA, 450 V alternator supplies a purely resistive load of 480 kW at 450 V. The subtransient reactance of the generator is 8%. Assuming the load is directly connected across the generator terminals, find the initial symmetrical rms current in pu at the generator terminals for a 3-phase dead short at its terminals.

[U.P. Technical Univ. Power System Analysis 2009-10]

[**Ans.** 9,642 $\angle -86.34°$ A]

11. A synchronous generator and a synchronous motor each rated 25 MVA, 11 kV having 15% subtransient reactances are connected through transformer and a line as shown in Fig. 5. The transformers are rated 25 MVA, 11/66 and 66/11 kV with leakage reactance of 10% each. The line has a reactance of 10% on a base of 25 MVA, 66 kV.

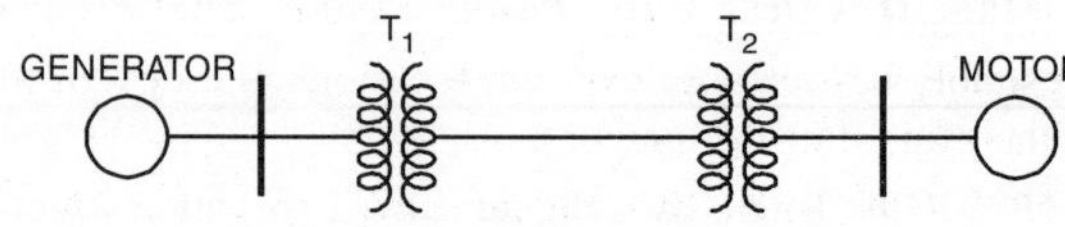

Fig. 5

The motor is drawing 15 MW, at 0.8 pf leading and a terminal voltage of 10.6 kV when a symmetrical 3-ϕ fault occurs at the motors terminals. Find the subtransient current in the generator, motor and fault.

[Rajasthan Technical Univ. Power System Analysis, 2011]

[**Ans.** 2,624 A; 7,860 A, 1922.8 A]

12. An 11 kV, 100 MVA alternator having a subtransient reactance of 0.25 pu is supplying a 50 MVA motor having a subtransient reactance of 0.2 pu through a transmission line. The line reactance is 0.05 pu on a base of 100 MVA. The motor is drawing 40 MW at 0.8 pf leading with a terminal voltage of 10.95 kV when a 3-phase fault occurs at the generator terminals. Calculate the total current in the generator and motor under fault conditions.

[Rajasthan Univ. Power System Analysis 2006]

[**Ans.** 19,106 $\angle -82.4°$ A, 13,148 $\angle 261.8°$ A]

13. A synchronous generator rated 625 kVA, 480 V, 0.1 per unit subtransient reactance is supplying a passive load of 500 kW at 0.8 lagging power factor. Calculate the initial symmetrical rms current for a 3-phase fault at the generator terminals. [**Ans.** 7,991 $\angle -85.68°$ A]

14. A synchronous generator rated 500 kVA, 440 V, 0.1 per unit, subtransient reactance is supplying a passive load of 400 kW at 0.8 lagging power factor. Calculate the initial symmetrical rms current for a 3-phase fault at the generator terminals.

[Madras Univ. Power System Analysis October 1996]

[**Ans.** 6,974 $\angle -85.68°$]

15. A 3-phase, 30 MVA, 33 kV alternator has 4 % reactance (internal) and negligible resistance. Find the (external) reactance per phase to be connected in series with the alternator so that the steady-state current on short circuit does not exceed 10 times the full-load current. [**Ans.** 2.178 Ω]

16. A 30 MVA generator with 20 % reactance is connected to a bus-bar. A 25 MVA transformer with 15 % reactance is also connected through a 10 % bus-bar reactor to the same bus-bar. Both of these reactances are based on 25 MVA rating.

If a feeder taken out from the bus-bars through a circuit breaker develops a line to ground fault, what should be the approximate rating of a circuit breaker ? [**Ans.** 250 MVA]

17. The plant capacity of a 3-phase power station consists of two 8 MVA generators, each of 14% reactance voltage and one 4 MVA generator of 10% reactance voltage. These are connected to common bus-bars from which load is taken through a number of 3 MVA step-up transformers, each having a reactance voltage of 4%. Determine the maximum MVA which the circuit breakers on (*i*) low voltage side and (*ii*) the high voltage side, may have to deal with.

[**Ans.** (*i*) 154.3 MVA (*ii*) 50.5 MVA]

18. A power station has two alternators of 2,500 kVA and 5,000 kVA capacities and the reactances are 6% and 8% respectively. The circuit breakers are rated 150 MVA. The system is to be extended by taking supply from a grid via transformer of 5,000 kVA rating having a reactance of 7.5 %. If the same circuit breakers are to be used, find the necessary reactance to be placed between the bus bar section of the station and the supply from the grid to protect the switchgear. The voltage of the system at the bus-bars is 3-phase, 3.3 kV. The layout is shown in Fig. 6.

[Agra Univ. Generation & Switchgear Practice 1979; Jodhpur Univ. Power System I, 1978; Allahabad Univ. Power System II, 1975]

[**Ans.** 0.074 Ω]

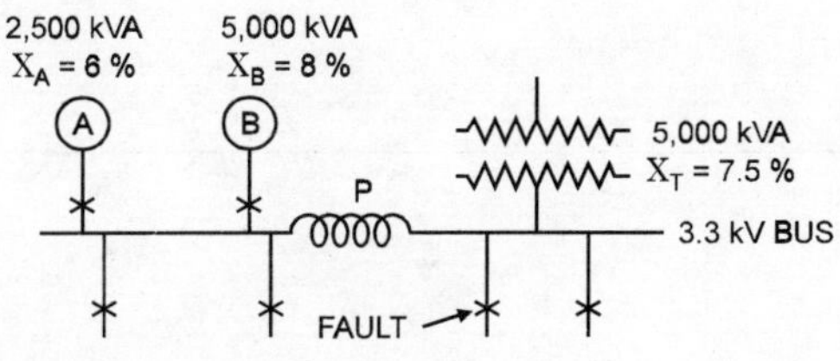

Fig. 6

19. Three similar alternators of 50 MVA, 50% reactance are connected to separate bus sections. Two reactors each of 10% reactance (on 50 MVA base) separate the bus sections. find fault MVA for a fault at middle bus.

[**Ans.** 266.67 MVA]

[**Hint** : Per unit reactance between the fault point and the neutral bus $= j\,[(0.5 + 0.1) \parallel (0.5 + 0.1) \parallel 0.5] = j\,0.1875$ pu]

20. A generating station has three section bus-bars connected with a tie-bar through 6% reactors rated at 5,000 kVA. Each generator is of 5,000 kVA with 12% reactance and is connected to one section of bus-bars. Find the total steady input to a dead short circuit between the lines on one of the sections of bus-bar, (*i*) with and (*ii*) without reactors.

[Rajasthan Univ. Power System Analysis 2001, 2007]

[**Ans.** (*i*) 74,960 MVA, (*ii*) 125 MVA]

❑❑❑

CHAPTER 5

Unsymmetrical Fault Analysis

INSIDE THIS CHAPTER

5.1. Introduction **5.2.** Fault Analysis of Unloaded Synchronous Generator **5.3.** Faults in Power Systems **5.4.** Series or Open Conductor Faults **5.5.** Computer Method For Short-Circuit Calculations.

5.1. INTRODUCTION

It has been shown in the preceding chapter how calculations are made for the condition of a symmetrical fault, this being the accepted method for the selection of suitably rated circuit breakers or fuses and for the stresses developed in bus-bars and connection structures under short-circuit conditions. Such calculations are also adequate for determining the overcurrent factors for current transformers and protective gear stability on through faults.

Fault conditions however, are not confined to the three-phase condition and indeed the majority of faults that occur in power systems are *unsymmetrical faults*, *e.g.,* single line-to-ground faults, line-to-line faults, double line-to-ground faults, breaking of one or two conductors etc. It becomes necessary, for engineers to calculate the fault currents in such conditions. The information obtained is necessary for the determination of protective gear relay settings and in studies of transient stability in an interconnected power system. Not only it is essential to know the current at the point of fault but also how it is distributed (and in what magnitude) throughout the network behind the fault. A full study of this subject also involves the unsymmetrical voltages which arise, a knowledge of which is essential particularly in protective systems of the directional type where the voltage element is important. Here we shall concern ourselves only with the current aspect.

For unsymmetrical faults, simple single phase representation is not valid and analysis has to be carried out on a three phase basis. The analysis of unsymmetrical fault conditions may alternatively be carried out using the method of *symmetrical components*, which is a means by which an unbalanced 3-phase system of impedances may be resolved into three equivalent single-phase systems having independent impedance parameters. The three independent impedance systems may be suitably connected and constrained to represent all types of fault conditions.

It is possible to solve unsymmetrical fault problems by direct evaluation using phase coordinates. Such methods have advantage in certain untransposed overhead line calculations because mutual impedance effect may be more accurately represented, and they have been used in computer studies. For manual solution of unsymmetrical faults the symmetrical component method is simpler and enables more generality to be given to fault performance studies, the results of which are sufficiently accurate for most purposes. Symmetrical component method provides a useful analytical tool for the protection engineer, particularly in connection with the tracing out of fault current paths. Majority of the faults in power systems involve earth, and it will be shown that, by tracing out the zero sequence paths in associated apparatus (such as transformers) a rapid assessment of the behaviour of the system under such fault conditions may be made.

5.2. FAULT ANALYSIS OF UNLOADED SYNCHRONOUS GENERATOR

The synchronous generators may be subjected to different types of faults at its terminals and they are as follows:

(*i*) Single line-to-ground (L-G) fault

(*ii*) Line-to-line (L-L) fault

(*iii*) Double line-to-ground (L-L-G) fault

(*iv*) Symmetrical three–phase (L-L-L or L-L-L-G) fault

These are listed in increasing order of severity, and the symmetrical fault follows these as the most severe. If the fault is symmetrical, it is unimportant whether there is earth connection at the fault or not.

5.2.1. Three-Phase Fault

The simplest case of a three-phase symmetrical fault is considered here. An unloaded star-connected 3-phase alternator is shorted by a 3-phase fault. The generator neutral is earthed through an impedance. The connections are shown in Fig. 5.1. The generator terminals being shorted V_a, V_b and V_c are equal to zero and the sum of the currents I_a, I_b and I_c is zero.

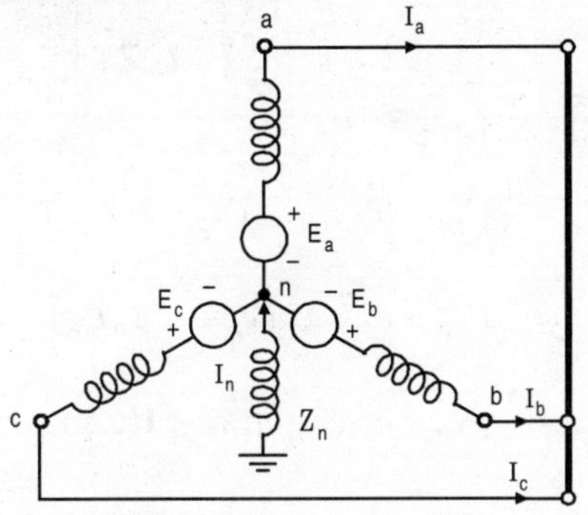

Fig. 5.1. *3-Phase Fault*

As the generator develops only positive sequence voltage hence with reference to Art. 3.10.

$$\mathbf{V}_{a1} = \mathbf{E}_a - \mathbf{I}_{a1}\,\mathbf{Z}_1$$

$$\mathbf{V}_{a2} = -\,\mathbf{I}_{a2}\,\mathbf{Z}_2$$

$$\mathbf{V}_{a0} = -\,\mathbf{I}_{a0}\,\mathbf{Z}_0 = -\,\mathbf{I}_{a0}(\mathbf{Z}_{g0} + 3\mathbf{Z}_n)$$

In matrix form

$$\begin{bmatrix}\mathbf{V}_{a1}\\ \mathbf{V}_{a2}\\ \mathbf{V}_{a0}\end{bmatrix} = \begin{bmatrix}\mathbf{E}_a\\ 0\\ 0\end{bmatrix} - \begin{bmatrix}\mathbf{Z}_1 & 0 & 0\\ 0 & \mathbf{Z}_2 & 0\\ 0 & 0 & \mathbf{Z}_0\end{bmatrix}\begin{bmatrix}\mathbf{I}_{a1}\\ \mathbf{I}_{a2}\\ \mathbf{I}_{a0}\end{bmatrix} \qquad ...(5.1)$$

Solving above equations we have, $\mathbf{V}_{a0} = 0$ as $\mathbf{Z}_0$ is finite; $\mathbf{I}_{a2} = 0$ and

$$\mathbf{E}_a - \mathbf{I}_{a1}\mathbf{Z}_1 = 0 \text{ or } \mathbf{I}_{a1} = \frac{\mathbf{E}_a}{\mathbf{Z}_1} \qquad ...(5.2)$$

This means only positive sequence network is present for the solution of three-phase fault current. The sequence network is shown in Fig. 3.13 (*b*).

5.2.2. Single Line-To-Ground (L-G) Fault

An unloaded generator with its phase *a* grounded is shown in Fig. 5.2. The generator neutral is assumed to be grounded through an impedance Z_n. Let the fault impedance be Z_f.

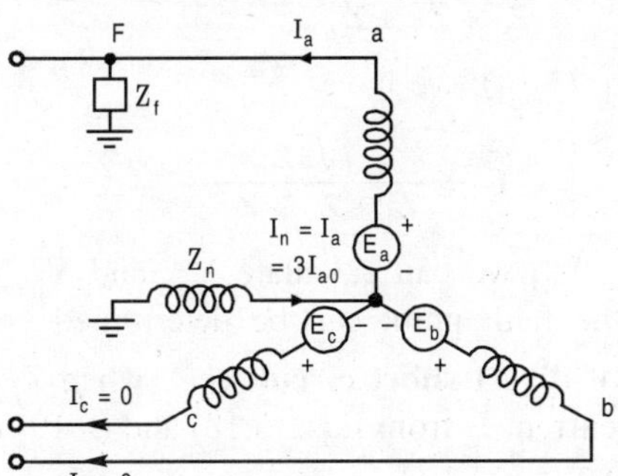

Fig. 5.2. *Line-To-Ground (L-G) Fault On an Unloaded Generator*

Under fault condition, the currents and voltages are given as

$$\mathbf{I}_b = 0;\ \mathbf{I}_c = 0 \text{ and } \mathbf{V}_a = \mathbf{I}_a\mathbf{Z}_f \qquad ...(5.3)$$

The symmetrical components of the fault current are given by substituting above values of currents in Eq. (3.2)

$$\begin{bmatrix}\mathbf{I}_{a0}\\ \mathbf{I}_{a1}\\ \mathbf{I}_{a2}\end{bmatrix} = \frac{1}{3}\begin{bmatrix}1 & 1 & 1\\ 1 & a & a^2\\ 1 & a^2 & a\end{bmatrix}\begin{bmatrix}\mathbf{I}_a\\ 0\\ 0\end{bmatrix} \quad \because \mathbf{I}_b = \mathbf{I}_c = 0 \qquad ...(5.4)$$

From above equation we have

$$\mathbf{I}_{a0} = \mathbf{I}_{a1} = \mathbf{I}_{a2} = \frac{1}{3}\mathbf{I}_a \qquad ...(5.5)$$

From Eqs. (5.3) and (5.5), we have

$$\mathbf{V}_a = \mathbf{V}_{a0} + \mathbf{V}_{a1} + \mathbf{V}_{a2} = \mathbf{I}_a\mathbf{Z}_f = 3\mathbf{I}_{a1}\mathbf{Z}_f \qquad ...(5.6)$$

Substituting above values in Eqs. (3.20), (3.16) and (3.18), we have

$$\mathbf{V}_{a0} = -\mathbf{I}_{a1}\mathbf{Z}_0$$

$$\mathbf{V}_{a1} = \mathbf{E}_a - \mathbf{I}_{a1}\mathbf{Z}_1$$

$$\text{and} \quad \mathbf{V}_{a2} = -\mathbf{I}_{a1}\mathbf{Z}_2 \qquad ...(5.7)$$

From Eqs. (5.6) and (5.7), we have

$$-\,\mathbf{I}_{a1}\mathbf{Z}_0 + \mathbf{E}_a - \mathbf{I}_{a1}\mathbf{Z}_1 - \mathbf{I}_{a1}\mathbf{Z}_2 = \mathbf{I}_a\mathbf{Z}_f$$

$$\text{or } \mathbf{E}_a - \mathbf{I}_{a1}(\mathbf{Z}_1 + \mathbf{Z}_2 + \mathbf{Z}_0) = 3\mathbf{I}_{a1}\mathbf{Z}_f \qquad ...(5.8)$$

From above Eq. (5.8), we have

$$\mathbf{I}_{a1} = \frac{\mathbf{E}_a}{\mathbf{Z}_1 + \mathbf{Z}_2 + \mathbf{Z}_0 + 3\mathbf{Z}_f} \qquad ...(5.9)$$

Therefore, fault current,

$$\mathbf{I}_f = \mathbf{I}_a = 3\mathbf{I}_{a1} = \frac{3\mathbf{E}_a}{\mathbf{Z}_1 + \mathbf{Z}_2 + \mathbf{Z}_0 + 3\mathbf{Z}_f} \qquad ...(5.10)$$

It may be recalled that

$$\mathbf{Z}_0 = \mathbf{Z}_{g0} + 3\mathbf{Z}_n$$

The Eqs. (5.5) and (5.8) show that the positive, negative and zero sequence networks are to be connected in series for the solution of currents under fault conditions. Sequence network representing the L-G fault on phase *a* is given in Fig. 5.3.

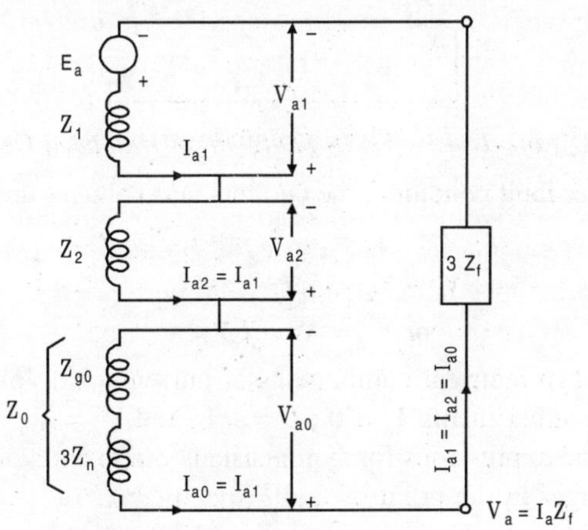

Fig. 5.3. *Sequence Network Representing L-G Fault on Phase a of an Unloaded Generator*

In case of **direct short circuit** (*i.e.,* when $Z_f = 0$) we get fault current from Eq. (5.10) for $Z_f = 0$. Thus in this case fault current is given as

$$\mathbf{I}_f = \mathbf{I}_a = 3\mathbf{I}_{a1} = \frac{3\mathbf{E}_a}{\mathbf{Z}_1 + \mathbf{Z}_2 + \mathbf{Z}_0} \quad ...(5.11)$$

In Fig. 5.3, the fault impedance Z_f equals zero.

In case of *generator with isolated neutral,* the zero sequence impedance Z_0 becomes infinite and, therefore, from Eq. (5.9)

$$I_{a1} = \frac{\mathbf{E}_a}{\mathbf{Z}_1 + \mathbf{Z}_2 + \infty + 3\mathbf{Z}_f} = 0 \quad ...(5.12)$$

The same result can be envisaged by looking at the system when the neutral is isolated: there is no return path for the current and, therefore, $I_{a1} = I_{a2} = I_{a0} = 0$. This means that for such a system the fault current is zero.

If the generator was supplying balanced load during prefault condition, the load current was of positive sequence only. A generator does not generate negative or zero sequence voltage and, therefore during prefault condition, there was no negative or zero sequence component of current to the load. Under the fault condition, the positive sequence current to the fault is to be added to the prefault current. However, only the positive sequence current flows into the fault provided the prefault current into the fault is zero.

5.2.3. Line-To-Line (L-L) Fault

An unloaded generator with line-to-line fault on its phases *b* and *c* is shown in Fig. 5.4. The generator neutral is assumed to be grounded through an impedance Z_n. The phase *a* is open. Let the fault impedance be Z_f.

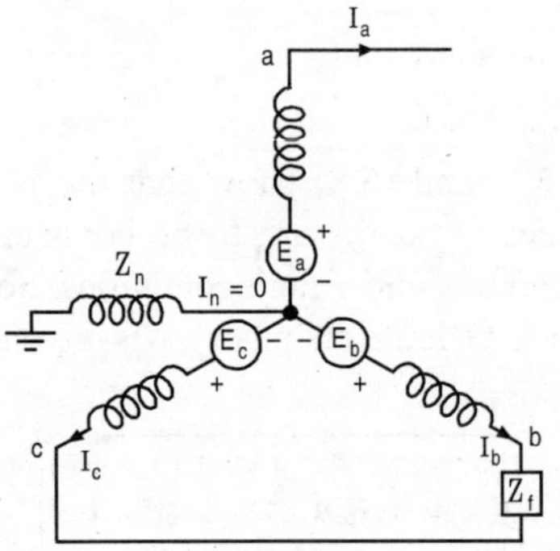

Fig. 5.4. ***Line-To-Line (L-L) Fault On an Unloaded Generator***

Under fault condition, the currents and voltages are given as

$$\mathbf{I}_a = 0;\ \mathbf{I}_b = -\mathbf{I}_c$$

$$\text{and } \mathbf{V}_b - \mathbf{V}_c = \mathbf{I}_b\mathbf{Z}_f$$

$$\text{or } \mathbf{V}_c = \mathbf{V}_b - \mathbf{I}_b\mathbf{Z}_f$$

The symmetrical components of currents and voltages are given by substituting $\mathbf{I}_a = 0$; $\mathbf{I}_c = -\mathbf{I}_b$ and $\mathbf{V}_c = \mathbf{V}_b - \mathbf{I}_b \mathbf{Z}_f$ in the general expressions for symmetrical components of currents and voltages at the point of fault.

$$\begin{bmatrix} \mathbf{I}_{a0} \\ \mathbf{I}_{a1} \\ \mathbf{I}_{a2} \end{bmatrix} = \frac{1}{3}\begin{bmatrix} 1 & 1 & 1 \\ 1 & a & a^2 \\ 1 & a^2 & a \end{bmatrix}\begin{bmatrix} 0 \\ \mathbf{I}_b \\ -\mathbf{I}_b \end{bmatrix}$$

from which we have

$$\mathbf{I}_{a1} = \frac{1}{3}[0 + a + a^2]\,\mathbf{I}_b$$

$$\text{and } \mathbf{I}_{a2} = \frac{1}{3}(0 + a^2 + a) \times (-\mathbf{I}_b)$$

$$\text{or } \mathbf{I}_{a1} = -\mathbf{I}_{a2} \text{ and } \mathbf{I}_{a0} = 0 \quad ...(5.13)$$

$$\begin{bmatrix} \mathbf{V}_{a0} \\ \mathbf{V}_{a1} \\ \mathbf{V}_{a2} \end{bmatrix} = \frac{1}{3}\begin{bmatrix} 1 & 1 & 1 \\ 1 & a & a^2 \\ 1 & a^2 & a \end{bmatrix}\begin{bmatrix} \mathbf{V}_a \\ \mathbf{V}_b \\ \mathbf{V}_b - \mathbf{I}_b\mathbf{Z}_f \end{bmatrix} \quad ...(5.14)$$

from which we have

$$\mathbf{V}_{a1} = \frac{1}{3}[\mathbf{V}_a + a\mathbf{V}_b + a^2(\mathbf{V}_b - \mathbf{I}_b\mathbf{Z}_f)]$$

$$= \frac{1}{3}[\mathbf{V}_a + (a + a^2)\mathbf{V}_b - a^2\mathbf{I}_b\mathbf{Z}_f]$$

$$\text{and } \mathbf{V}_{a2} = \frac{1}{3}[\mathbf{V}_a + a^2\mathbf{V}_b + a(\mathbf{V}_b - \mathbf{I}_b\mathbf{Z}_f)]$$

$$= \frac{1}{3}[\mathbf{V}_a + (a + a^2)\mathbf{V}_b - a\mathbf{I}_b\mathbf{Z}_f]$$

Solving above two equations we have

$$3(\mathbf{V}_{a1} - \mathbf{V}_{a2}) = (a - a^2)\,\mathbf{I}_b\,\mathbf{Z}_f = j\sqrt{3}\,\mathbf{I}_b\mathbf{Z}_f \quad ...(5.15)$$

$$\text{Now } \mathbf{I}_b = (a^2 - a)\,\mathbf{I}_{a1} = -j\sqrt{3}\,\mathbf{I}_{a1} \quad ...(5.16)$$

$$\because\ \mathbf{I}_{a2} = -\mathbf{I}_{a1} \text{ and } \mathbf{I}_a = 0$$

Substituting $\mathbf{I}_b$ from above Eq. (5.16) in Eq. (5.15), we have

$$\mathbf{V}_{a1} - \mathbf{V}_{a2} = j\frac{\sqrt{3}}{3} \times (-j\sqrt{3}\,\mathbf{I}_{a1}\,\mathbf{Z}_f) = \mathbf{I}_{a1}\mathbf{Z}_f \quad ...(5.17)$$

Equations (5.13) and (5.17) suggest parallel connection of positive and negative sequence networks through a series impedance Z_f, as shown in Fig. 5.5. Since $\mathbf{I}_{a0} = 0$, the zero sequence network is unconnected.

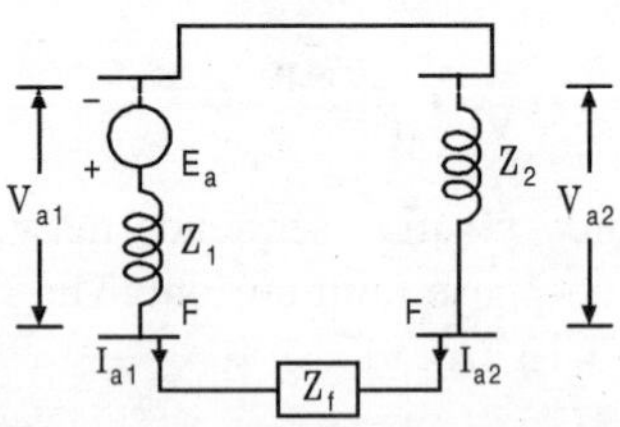

Fig. 5.5. ***Sequence Network Representing L-L Fault On Phases b and c of an Unloaded Generator***

In terms of the Thevenin equivalents, from Fig. 5.5, we have

$$\mathbf{I}_{a1} = \frac{\mathbf{E}_a}{\mathbf{Z}_1 + \mathbf{Z}_2 + \mathbf{Z}_f} \quad ...(5.18)$$

From Eq. (5.16), we have

$$\mathbf{I}_b = -\mathbf{I}_c = \frac{-j\sqrt{3}\mathbf{E}_a}{\mathbf{Z}_1 + \mathbf{Z}_2 + \mathbf{Z}_f} \quad ...(5.19)$$

Knowing $\mathbf{I}_{a1}$, we can calculate $\mathbf{V}_{a1}$ and $\mathbf{V}_{a2}$ from which voltages at the fault point can be determined.

In case of **direct short circuit** (*i.e.,* when $Z_f = 0$) we get $\mathbf{I}_{a1}$ and fault current $\mathbf{I}_b$ from Eqs. (5.18) and (5.19) respectively. Thus, we have

$$\mathbf{I}_{a1} = \frac{\mathbf{E}_a}{\mathbf{Z}_1 + \mathbf{Z}_2} \quad ...(5.20)$$

and fault current

$$\mathbf{I}_f = \mathbf{I}_b = \frac{-j\sqrt{3}\,\mathbf{E}_a}{\mathbf{Z}_1 + \mathbf{Z}_2} \quad ...(5.21)$$

5.2.4. Double Line-To-Ground (L-L-G) Fault

The circuit diagram for double line-to-ground (L-L-G) fault through an impedance Z_f on an unloaded synchronous generator is depicted in Fig. 5.6. The generator neutral is assumed to be grounded through an impedance Z_n.

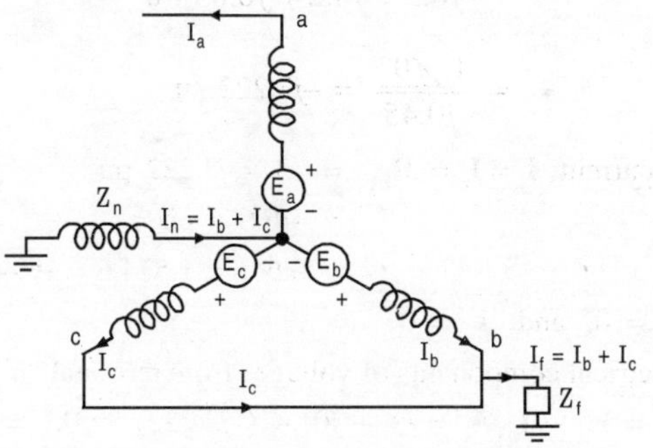

Fig. 5.6. ***Double Line-To-Ground (L-L-G) Fault On an Unloaded Generator***

Under fault conditions, currents and voltages are given as

$$\mathbf{I}_a = 0;\ \mathbf{I}_f = \mathbf{I}_b + \mathbf{I}_c \quad ...(5.22)$$

$$\mathbf{V}_b = \mathbf{V}_c = \mathbf{I}_f\mathbf{Z}_f = (\mathbf{I}_b + \mathbf{I}_c)\,\mathbf{Z}_f = 3\,\mathbf{I}_{a0}\,\mathbf{Z}_f \quad ...(5.23)$$

The symmetrical components of voltages are given as

$$\begin{bmatrix}\mathbf{V}_{a0}\\ \mathbf{V}_{a1}\\ \mathbf{V}_{a2}\end{bmatrix} = \frac{1}{3}\begin{bmatrix}1 & 1 & 1\\ 1 & a & a^2\\ 1 & a^2 & a\end{bmatrix}\begin{bmatrix}\mathbf{V}_a\\ \mathbf{V}_b\\ \mathbf{V}_c\end{bmatrix}$$

$$\text{or } \mathbf{V}_{a1} = \mathbf{V}_{a2} = \frac{1}{3}\left[\mathbf{V}_a + (a + a^2)\,\mathbf{V}_b\right]$$

$$= \frac{\mathbf{V}_a - \mathbf{V}_b}{3} \quad ...(5.24) \qquad \because a + a^2 = -1$$

$$\text{and } \mathbf{V}_{a0} = \frac{\mathbf{V}_a + 2\mathbf{V}_b}{3} \quad ...(5.25)$$

Subtracting Eq. (5.24) from Eq. (5.25), we have

$$\mathbf{V}_{a0} - \mathbf{V}_{a1} = \frac{1}{3}[2\mathbf{V}_b + \mathbf{V}_b] = \mathbf{V}_b = 3\mathbf{I}_{a0}\,\mathbf{Z}_f$$

$$\text{or } \mathbf{V}_{a0} = \mathbf{V}_{a1} + 3\mathbf{I}_{a0}\,\mathbf{Z}_f \quad ...(5.26)$$

Equations (5.22), (5.24) and (5.26) suggest the sequence network, as shown in Fig. 5.7.

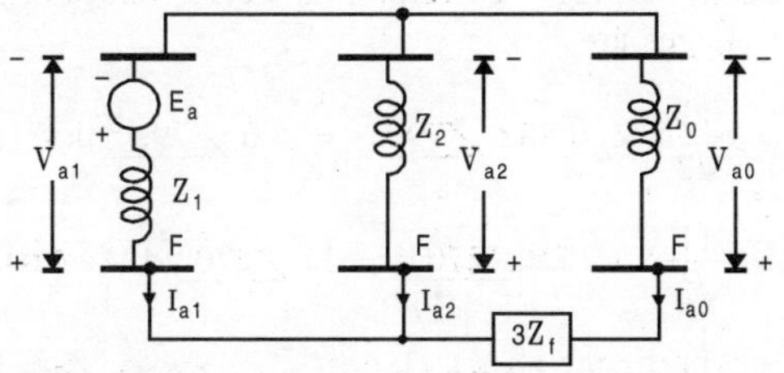

Fig. 5.7. ***Sequence Network Representing L-L-G Fault On Phases b and c of an Unloaded Generator***

In terms of the Thevenin equivalents from Fig. 5.7, we have

$$\mathbf{I}_{a1} = \frac{\mathbf{E}_a}{\mathbf{Z}_1 + \mathbf{Z}_2 \parallel (\mathbf{Z}_0 + 3\mathbf{Z}_f)}$$

$$= \frac{\mathbf{E}_a}{\mathbf{Z}_1 + \mathbf{Z}_2\dfrac{\mathbf{Z}_0 + 3\mathbf{Z}_f}{\mathbf{Z}_2 + \mathbf{Z}_0 + 3\mathbf{Z}_f}} \quad ...(5.27)$$

In case of direct short circuit (*i.e.*, when $Z_f = 0$), we have

$$\mathbf{I}_{a1} = \frac{\mathbf{E}_a}{\mathbf{Z}_1 + \mathbf{Z}_2 \parallel \mathbf{Z}_0} = \frac{\mathbf{E}_a}{\mathbf{Z}_1 + \dfrac{\mathbf{Z}_2\,\mathbf{Z}_0}{\mathbf{Z}_2 + \mathbf{Z}_0}} \quad ...(5.28)$$

For an ungrounded generator, $Z_n = \infty$, therefore $Z_0 = \infty$ and we have

$$\mathbf{I}_{a1} = \frac{\mathbf{E}_a}{\mathbf{Z}_1 + \mathbf{Z}_2} \quad ...(5.29)$$

Example 5.1. A 3-phase, 15 MVA, 11 kV, 50 Hz generator with solidly earthed neutral has subtransient reactance X''_d of 20%; direct-axis transient reactance X'_d of 25% and synchronous reactance X_d of 60%. Negative sequence reactance X_2 = 20% and zero sequence reactance X_0 = 8%. The generator is operated on open circuit when fault occurs. Take E_a = 1.0.

If the fault is a three-phase short without an impedance, compute the initial symmetrical subtransient, transient and sustained pu values of the line current under faulty condition.

Solution : From Eq. (5.2)

Initial symmetrical transient current,

$$\mathbf{I}''_{a1} = \frac{\mathbf{E}_a}{\mathbf{Z}''_1} = \frac{1.00}{j\,0.20} = -j\,5.0 \text{ per unit } \textbf{Ans.}$$

Transient symmetrical rms value of current,

$$\mathbf{I}'_{a1} = \frac{\mathbf{E}_a}{\mathbf{Z}'_1} = \frac{1.0}{j\,0.25} = -j\,4.0 \text{ per unit } \textbf{Ans.}$$

Sustained value of the symmetrical rms current,

$$\mathbf{I}_{a1} = \frac{\mathbf{E}_a}{\mathbf{Z}_1} = \frac{1.0}{j\,0.6} = -j\,1.667 \text{ per unit } \textbf{Ans.}$$

$\mathbf{I}_b = \mathbf{I}_a \angle 240°$ and $\mathbf{I}_c = \mathbf{I}_a \angle 120°$; equal in numerical value

As there is complete dead short

$$\mathbf{V}_a = \mathbf{V}_b = \mathbf{V}_c = 0.$$

Example 5.2. A short circuit to earth occurs near the terminals of phase A of a 3-ϕ alternator, star-connected with neutral point earthed, the current to earth being 1,000 A.

If the alternator is not supplying any normal current, calculate positive, negative and zero sequence components of currents of all phases. [*Rajasthan Univ. Power System Analysis 2007*]

Solution : Taking current in phase '*a*' as reference.

Current in phase *a*, $\mathbf{I}_a = 1{,}000 \angle 0°$ A

$$\mathbf{I}_b = 0$$

$$\mathbf{I}_c = 0$$

Positive sequence components are

$$\mathbf{I}_{a1} = \frac{1}{3}[\mathbf{I}_a + a\mathbf{I}_b + a^2\mathbf{I}_c]$$

$$= \frac{1}{3}(1{,}000 \angle 0° + 0 + 0) \qquad \because \mathbf{I}_b = \mathbf{I}_c = 0$$

$$= 333.3 \angle 0° \text{ A}$$

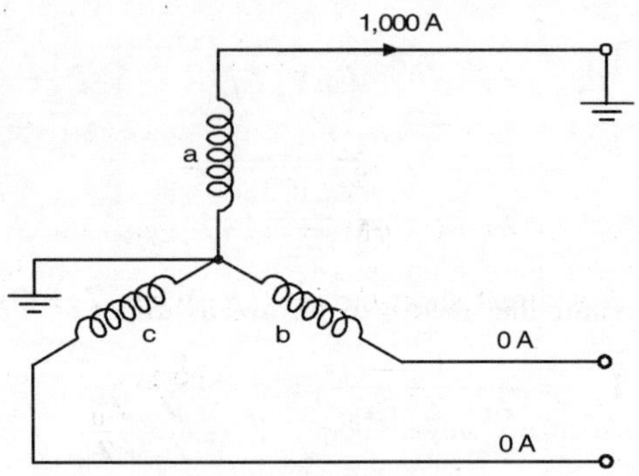

Fig. 5.8

$\mathbf{I}_{b1} = a^2\mathbf{I}_{a1} = 333.3 \angle 240°$ A **Ans.**

$\mathbf{I}_{c1} = a\mathbf{I}_{a1} = 333 \angle 120°$ A

Negative sequence components are :

$\mathbf{I}_{a2} = \mathbf{I}_{a1} = 333.3 \angle 0°$ A **Ans.**

$\mathbf{I}_{b2} = a\mathbf{I}_{a2} = 333.3 \angle 120°$ A **Ans.**

$\mathbf{I}_{c2} = a^2\mathbf{I}_{a2} = 333 \angle 240°$ A **Ans.**

Zero sequence components

$$\mathbf{I}_{a0} = \mathbf{I}_{b0} = \mathbf{I}_{c0} = \frac{1}{3}[\mathbf{I}_a + \mathbf{I}_b + \mathbf{I}_c]$$

$$= \frac{1}{3}(1{,}000 \angle 0°) = 333.3 \angle 0° \text{ A} \quad \textbf{Ans.}$$

Example 5.3. A single line-to-ground fault of 0.05 Ω resistance occur in a 3-phase system supplied by a synchronous generator with a generated emf of 11 kV between lines. The positive, negative and zero sequence reactances of the generators and network up to the fault are 0.5 Ω, 0.2 Ω and 0.1 Ω respectively. Find the fault current.

[*Anna Univ. Chennai Power System Analysis May/June 2011*]

Solution : From Eqs. (5.5) and (5.9), we have

$$\mathbf{I}_{a0} = \mathbf{I}_{a1} = \mathbf{I}_{a2} = \frac{\mathbf{E}_a}{\mathbf{Z}_1 + \mathbf{Z}_2 + \mathbf{Z}_0 + 3\mathbf{Z}_f}$$

Assuming the fault to be on phase *a*

$$= \frac{\frac{11}{\sqrt{3}} \times 1{,}000}{j0.5 + j0.2 + j0.1 + 3 \times 0.05}$$

$$\text{or } \mathbf{I}_{a0} = \mathbf{I}_{a1} = \mathbf{I}_{a2} = \frac{6{,}350.853}{\sqrt{0.8^2 + 0.15^2}} = 7{,}802.6 \text{ A}$$

Fault current, $\mathbf{I}_f = \mathbf{I}_a = 3\mathbf{I}_{a1} = 3 \times 7{,}802.6 = 23{,}407.8$ A **Ans.**

Example 5.4. A 30 MVA, 11 kV generator has $Z_1 = Z_2 = 0.2$, $Z_0 = j0.05$ pu. A line to ground fault occurs on the generator terminals (*a*) Find the fault current and line to line voltage during faulty conditions. Assume that the generator neutral is solidly grounded and that the generator is operating at no load and at rated voltage at the occurrence of fault, (*b*) Find the line current for a 3-phase fault. [*Rajasthan Univ. Power System Analysis 2004*]

Solution : Let base MVA be 30 MVA and base voltage be 11 kV.

$$\text{Base current} = \frac{30 \times 10^6}{\sqrt{3} \times 11 \times 10^3} = 1{,}574.6 \text{ A}$$

Let *a*, *b*, *c* be the three terminals of the alternator. Let fault occur between terminal "*a*" and ground, as shown in Fig. 5.2.

Fault impedance, $Z_f = 0$

Let the induced voltage of phase "*a*" (line-to-neutral) be 1 per unit and be taken as reference phasor. So

$$\mathbf{E}_a = 1 \angle 0° \text{ pu}$$

From Eqs. (5.5) and (5.9), we have

$$\mathbf{I}_{a0} = \mathbf{I}_{a1} = \mathbf{I}_{a2} = \frac{\mathbf{E}_a}{\mathbf{Z}_1 + \mathbf{Z}_2 + \mathbf{Z}_0 + 3\mathbf{Z}_f}$$

$$= \frac{1 \angle 0°}{j0.2 + j0.2 + j0.05 + 0} \quad \because Z_f = 0$$

$$= \frac{1 \angle 0°}{j0.45} = -j2.222 \text{ pu}$$

(*a*) Fault current, $\mathbf{I}_f = \mathbf{I}_a = 3\mathbf{I}_{a1} = -3 \times j2.222$ pu

$= -j6.666$ pu

$= 6.666 \times 1{,}574.6 = 10{,}497$ A **Ans.**

$\mathbf{I}_b = 0$ and $\mathbf{I}_c = 0$

(*b*) Symmetrical components of voltages from terminal "*a*" to ground.

$\mathbf{V}_{a1} = \mathbf{E}_a - \mathbf{I}_{a1}\mathbf{Z}_1 = (1 + j0) - (-j2.222) \times j0.2 = 0.5556$ pu

$\mathbf{V}_{a2} = -\mathbf{I}_{a2}\mathbf{Z}_2 = -(-j2.222) \times j0.2 = -0.4444$ pu

$\mathbf{V}_{a0} = -\mathbf{I}_{a0}\mathbf{Z}_0 = -(-j2.222) \times j0.05 = -0.1111$ pu

Line-to-ground voltages,

$\mathbf{V}_a = \mathbf{V}_{a1} + \mathbf{V}_{a2} + \mathbf{V}_{a0} = 0.5556 - 0.4444 - 0.1111 = 0$

$\mathbf{V}_b = a^2\mathbf{V}_{a1} + a\mathbf{V}_{a2} + \mathbf{V}_{a0}$

$= 0.5556(-0.5 - j0.866) - 0.4444(-0.5 + j0.866) - 0.1111$

$= [(-0.2778 + 0.2222 - 0.1111) - j(0.4812 + 0.3848)]$

$= (-0.1667 - j0.866)$ pu

$\mathbf{V}_c = a\mathbf{V}_{a1} + a^2\mathbf{V}_{a2} + \mathbf{V}_{a0}$

$= 0.5556(-0.5 + j0.866) - 0.4444(-0.5 - j0.866) - 0.1111$

$= [(-0.2778 + 0.2222 - 0.1111) + j(0.4812 + 0.3848)]$

$= (-0.1667 + j0.866)$ pu.

Line-to-line voltages at fault points

$\mathbf{V}_{ab} = \mathbf{V}_a - \mathbf{V}_b = 0 - (-0.1667 - j0.866) = 0.1667 + j0.866$

$= 0.882 \angle 79.1°$ pu

$\mathbf{V}_{bc} = \mathbf{V}_b - \mathbf{V}_c = (-0.1667 - j0.866) - (-0.1667 + j0.866)$

$= -j1.732 = 1.732 \angle 270°$ pu

$\mathbf{V}_{ca} = \mathbf{V}_c - \mathbf{V}_a = -(0.1667 + j0.866) - 0$

$= (-0.1667 + j0.866) = 0.882 \angle 100.9°$ pu

Since the generated line to neutral voltage has been taken as 1 pu in the calculation of sequence currents, the above voltages are in pu of line-to-neutral voltage. Therefore, the actual values of line-to-line post fault voltages are

$$\mathbf{V}_{ab} = \frac{11}{\sqrt{3}} \times 0.882 \angle 79.1° = 5.6 \angle 79.1° \text{ kV} \quad \textbf{Ans.}$$

$$\mathbf{V}_{bc} = \frac{11}{\sqrt{3}} \times 1.732 \angle 270° = 11 \angle 270° \text{ kV} \quad \textbf{Ans.}$$

$$\mathbf{V}_{ca} = \frac{11}{\sqrt{3}} \times 0.882 \angle 100.9° = 5.6 \angle 100.9° \text{ kV} \quad \textbf{Ans.}$$

Before the fault each line voltage was equal to 11 kV. It is seen that a fault causes the post fault voltages to be different from prefault voltages.

(*b*) For a 3-phase fault,

$$\text{Line current} = \frac{1\angle 0^\circ}{j0.2} = 5\angle{-90^\circ} \text{ pu}$$

Actual value of line current

$$= 5 \times 1574.6 \angle{-90^\circ} = 7{,}873 \angle{-90^\circ} \text{ A} \quad \textbf{Ans.}$$

Example 5.5. An 11 kV, 25 MVA alternator has positive, negative and zero sequence reactances of 0.12, 0.12 and 0.08 per unit respectively. The generator neutral is grounded through a reactance of 0.03 per unit. Determine the fault current when a single line to ground fault occurs at generator terminal. Also calculate the line to line fault voltage. Assume generator was unloaded before the fault.

[*U.P. Technical Univ. Power System Analysis 2006-07*]

Solution : Let *a*, *b*, *c* be the three terminals of the alternator and fault occur between terminal *a* and ground as illustrated in Fig. 5.2. Let the induced voltage of phase a line to neutral $\mathbf{E}_a$ be 1 per unit *i.e.*,

$$\mathbf{E}_a = (1 + j0) \text{ pu}$$

From Eqs. (5.5) and (5.9), we have

$$\mathbf{I}_{a0} = \mathbf{I}_{a1} = \mathbf{I}_{a2}$$

$$= \frac{\mathbf{E}_a}{\mathbf{Z}_1 + \mathbf{Z}_2 + \mathbf{Z}_0 + 3Z_f}$$

$$= \frac{1 + j0}{j0.12 + j0.12 + j0.08 + 3 \times j0.03}$$

$$\because \ Z_0 = Z_{g0} + 3Z_n \text{ and } Z_f = 0 \text{ (assumed)}$$

$$= \frac{1 + j0}{j0.41} = -j2.439 \text{ pu or } 243.9\%$$

Fault current,

$$\mathbf{I}_f = \mathbf{I}_a = 3 \times \mathbf{I}_{a1}$$

$$= 3 \times (-j2.439 \text{ pu}) = -j7.317 \text{ pu or } -j731.7\% \quad \textbf{Ans.}$$

Currents $\mathbf{I}_b = 0$ and $\mathbf{I}_c = 0$

Symmetrical components of voltages from terminal *a* to ground

$$\mathbf{V}_{a1} = \mathbf{E}_a - \mathbf{I}_{a1}\mathbf{Z}_1 = (1.0 + j0) - (-j2.439) \times (j0.12) = 0.7073 \text{ pu}$$

$$\mathbf{V}_{a2} = -\mathbf{I}_{a2}\mathbf{Z}_2 = -(-j2.439) \times (j0.12) = -0.2927 \text{ pu}$$

$$\mathbf{V}_{a0} = -\mathbf{I}_{a0}\mathbf{Z}_0 = -(-j2.439) \times (j0.17) = -0.4146 \text{ pu}$$

Line-to-ground voltages

$$\mathbf{V}_a = \mathbf{V}_{a1} + \mathbf{V}_{a2} + \mathbf{V}_{a3} = 0.7073 - 0.2927 - 0.4146 = 0$$

$$\mathbf{V}_b = a^2\mathbf{V}_{a1} + a\mathbf{V}_{a2} + \mathbf{V}_{a0}$$

$$= 0.7073\,(-0.5 - j0.866) - 0.2927(-0.5 + j0.866) - 0.4146$$

$$= (-0.6219 - j0.866) \text{ pu}$$

$$\mathbf{V}_c = a\mathbf{V}_{a1} + a^2\mathbf{V}_2 + \mathbf{V}_{a0}$$

$$= 0.7073(-0.5 + j0.866) - 0.2927(-0.5 - j0.866) - 0.4146$$

$$= (-0.6219 + j0.866) \text{ pu}$$

Line-to-line voltages

$$\mathbf{V}_{ab} = \mathbf{V}_a - \mathbf{V}_b = 0 - (-6219 - j0.866)$$

$$= 0.6219 + j0.866 = 1.066 \angle 54.32^\circ \text{ pu} \quad \textbf{Ans.}$$

$$\mathbf{V}_{bc} = \mathbf{V}_b - \mathbf{V}_c = (-0.6219 - j0.866) - (-0.6219 + j0.866)$$

$$= -j1.732 \text{ pu} = 1.732 \angle{-90^\circ} \text{ pu} \quad \textbf{Ans.}$$

$$\mathbf{V}_{ca} = \mathbf{V}_c - \mathbf{V}_a = (-0.6219 + j0.866) - 0$$

$$= -0.6219 + j0.866$$

$$= 1.066 \angle 125.68^\circ \text{ pu} \quad \textbf{Ans.}$$

Example 5.6. A 3-phase 10 MVA, 11 kV generator with solidly earthed neutral point supplies a feeder. The relevant impedances of the generator and feeder in pu are as under :

	Generator	**Feeder**
Positive sequence impedance	***j*1.2**	***j*1.0**
Negative sequence impedance	***j*0.9**	***j*1.0**
Zero sequence impedance	***j*0.4**	***j*3.0**

If a fault from one phase to earth occurs on the far end of the feeder, calculate

(*i*) the magnitude of fault current

(*ii*) Line to neutral voltage at the generator terminals.

[*Rajasthan Technical Univ. Power System Analysis 2011*]

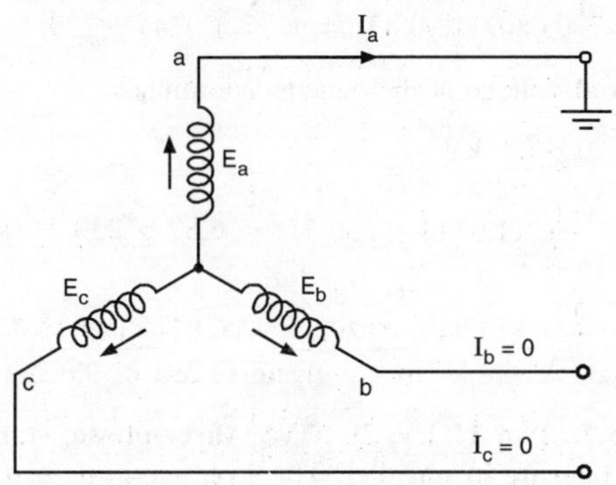

Fig. 5.9

Solution : Sequence impedances,

$$\mathbf{Z}_1 = j1.2 + j1.0 = j2.2$$

$$\mathbf{Z}_2 = j0.9 + j1.0 = j1.9$$

$$\mathbf{Z}_3 = j0.4 + j3.0 = j3.4$$

Fault impedance, $\mathbf{Z}_f = 0$

Let base MVA be 10 MVA and base voltage be 11 kV

$$\text{Base current} = \frac{10 \times 10^6}{\sqrt{3} \times 11 \times 10^3} = 524.86 \text{ A}$$

Let the induced voltage in phase '*a*' (line-to-neutral) be 1 per unit and be taken as reference phasor. So

$$\mathbf{E}_a = 1\angle 0^\circ \text{ pu}$$

From Eqs. (5.5) and (5.9), we have

$$\mathbf{I}_{a0} = \mathbf{I}_{a1} = \mathbf{I}_{a2} = \frac{\mathbf{E}_a}{\mathbf{Z}_1 + \mathbf{Z}_2 + \mathbf{Z}_0 + 3\mathbf{Z}_f}$$

$$= \frac{1\angle 0^\circ}{j2.2 + j1.9 + j3.4 + 0} \qquad \because \ Z_f = 0$$

$$= \frac{1\angle 0^\circ}{j7.5} = -j0.1333 \text{ pu}$$

(*i*) Fault current, $\mathbf{I}_f = \mathbf{I}_a = 3\mathbf{I}_{a1} = 3 \times 0.1333$

$$= 0.4 \text{ pu} = 0.4 \times 524.86$$

$$= 210 \text{ A} \quad \textbf{Ans.}$$

(*ii*) $I_b = 0$, $I_c = 0$

Symmetrical components of voltages from terminal "*a*" to ground

$$V_{a1} = E_a - I_{a1}Z_1 = (1 + j0) - (-j0.1333) \times j2.2 = 0.707 \text{ pu}$$

$$V_{a2} = -I_{a2}Z_2 = -(-j0.1333) \times j1.9 = -0.2533 \text{ pu}$$

$$V_{a0} = -I_{a0}Z_0 = -(-j0.1333) \times j3.4 = -0.4533 \text{ pu}$$

Line to ground voltages

$$V_a = 0$$

$$V_b = a^2V_{a1} + aV_{a2} + V_{a0}$$
$$= 0.707(-0.5 - j0.866) - 0.2533(-0.5 + j0.866) - 0.4533$$
$$= [(-0.3535 + 0.1266 - 0.4533) - j(0.6123 + 0.2194)]$$
$$= (-0.6802 - j0.8317) \text{ pu} = 1.0744 \angle 230.7° \text{ pu}$$

$$V_c = aV_{a1} + a^2V_{a2} + V_{a0}$$
$$= 0.707(-0.5 + 0.866) - 0.2533(-0.5 - j0.866) - 0.4533$$
$$= [(-0.3535 + 0.1266 - 0.4533) + (0.6123 + 0.2194)$$
$$= (-0.6802 + j0.8317) \text{ pu} = 1.0744 \angle 129.3° \text{ pu}$$

Line to neutral voltage at the generator terminals

$$V_a = 0 \angle 0° \text{ kV}$$

$$V_b = \frac{11}{\sqrt{3}} \times 1.0744 \angle 230.7° = 6.82 \angle 230.7° \text{ kV} \quad \textbf{Ans.}$$

$$V_c = \frac{11}{\sqrt{3}} \times 1.0744 \angle 129.3° = 6.82 \angle 129.3° \text{ kV} \quad \textbf{Ans.}$$

Example 5.7. Two 11 kV, 20 MVA, three-phase, star-connected generators operate in parallel. The +ve, –ve and zero reactances of each being respectively *j*0.18, *j*0.15, *j*0.10 pu. The star point of one of the generator is isolated and that of the other is earthed through a 2 Ω resistor. A single line-to-ground fault occurs at the terminals of one of the generators. Estimate (*a*) the fault current, (*b*) current in the grounding resistor and (*c*) voltage across the grounding resistor. [*U.P.S.C. I.E.S. Electrical Engineering-II, 2004; Rajasthan Univ. 1997, 2004*]

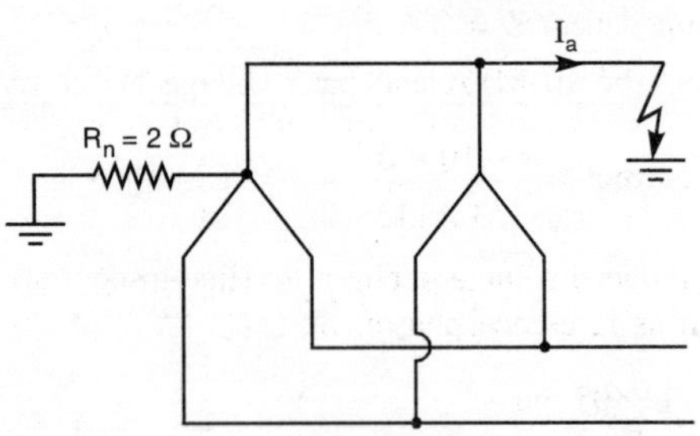

Fig. 5.10

Solution : Let the base MVA be 20 and base kV be 11 kV

Induced voltage, $E_a = (1 + j0) \text{ pu} = \frac{11}{\sqrt{3}}$ kV

Full-load current of each alternator $= \frac{20 \times 10^6}{\sqrt{3} \times 11 \times 1{,}000} = 1.05$ kA

Since the two identical alternators are operating in parallel

$$Z_1 = j\frac{0.18}{2} = j0.09 \text{ pu}$$

$$Z_2 = j\frac{0.15}{2} = j0.075 \text{ pu}$$

Since the star point of the second generator is isolated, its zero sequence reactance does not come into picture. Therefore,

$$Z_0 = j0.10 + 3R_n = j0.1 + 3 \times \frac{2 \times 20}{11^2} = j0.1 + 0.992$$

For an L-G fault

Fault current,

$$I_f = 3I_{a1} = \frac{3E_a}{Z_1 + Z_2 + Z_0}$$
$$= \frac{3 \times 1 \angle 0°}{j0.09 + j0.075 - j0.1 + 0.992}$$
$$= \frac{3 \angle 0°}{0.992 + j0.265} = \frac{3 \angle 0°}{1.02678 \angle 14.96°}$$
$$= 2.92 \angle -14.96° \text{ pu}$$

(*a*) Fault current, $I_f = 2.92 \times 1.05 = 3.066$ **Ans.**

(*b*) Current in grounding resistor, $I_f = 3.066$ kA **Ans.**

(*c*) Voltage drop across grounding resistor,

$$= I_f \times R_n = 3.066 \times 2 = 6.132 \text{ kV} \quad \textbf{Ans.}$$

Example 5.8. Three 6.6 kV, 3-phase 10 MVA alternators are connected to a common bus. Each alternator has a positive sequence reactance of 0.15 pu. The negative and zero sequence reactances are 75% and 30% of positive sequence reactance. A single line to ground fault occurs on the bus. Find the fault current if (*a*) all the alternator neutrals are solidly ground, (*b*) one neutral is solidly grounded and the other two neutrals are isolated, (*c*) one alternator neutral is grounded through 0.3 Ω resistance and the other two neutrals are isolated.

[*Rajasthan Univ. Power System Analysis 2000*]

Solution : Let the base MVA be 10 and base kV be 6.6 kV

Induced voltage, $E_a = (1 + j0)$ or $1 \angle 0°$ pu $= \frac{6.6}{\sqrt{3}}$ kV

Full-load current of each alternator

$$= \frac{10 \times 10^6}{\sqrt{3} \times 6.6 \times 1{,}000} = 875 \text{ A}$$

(*a*) *When all the three alternators are solidly grounded as shown in Fig. 5.11.*

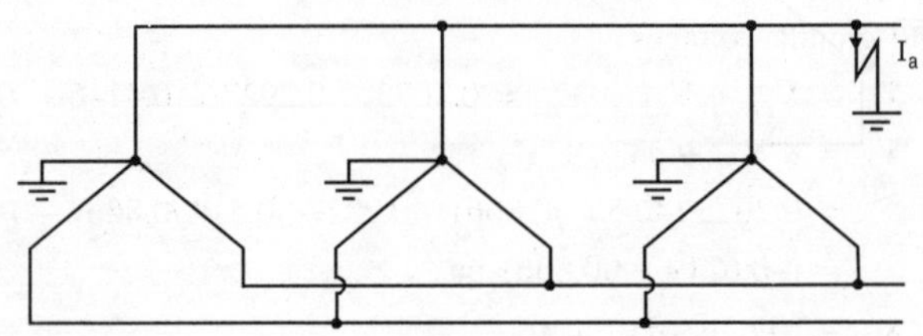

Fig. 5.11

Since all the alternators are operating in parallel, the resultant reactances will be one-third *i.e.*,

$$Z_1 = j\frac{0.15}{3} = j0.05 \text{ pu}$$

$$Z_2 = j0.15 \times \frac{75}{100} \times \frac{1}{3} = j0.0375 \text{ pu}$$

$$Z_0 = j0.15 \times \frac{30}{100} \times \frac{1}{3} = j0.015 \text{ pu}$$

$$\text{Fault current, } \mathbf{I}_f = \mathbf{I}_a = \frac{3 \times \mathbf{E}_a}{\mathbf{Z}_1 + \mathbf{Z}_2 + \mathbf{Z}_0}$$

$$= \frac{3 \times 1 \angle 0^\circ}{j0.05 + j0.0375 + j0.015} = -j29.2683 \text{ pu}$$

or $-j29.2683 \times 875 = -j25{,}610$ A **Ans.**

(b) When only one of the alternators is solidly grounded and the others are isolated as shown in Fig. 5.12.

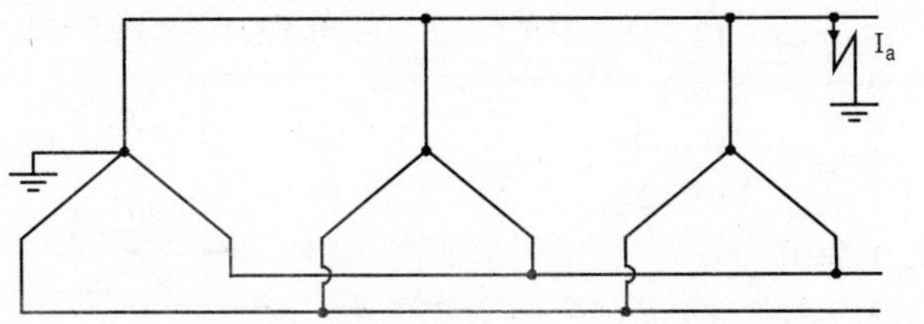

Fig. 5.12

$$\mathbf{Z}_1 = j\frac{0.15}{3} = j0.05 \text{ pu}$$

$$\mathbf{Z}_2 = j0.15 \times \frac{75}{100} \times \frac{1}{3} = j0.0375 \text{ pu}$$

$$\mathbf{Z}_0 = j0.15 \times \frac{30}{100} = j0.045 \text{ pu}$$

$$\text{Fault current, } \mathbf{I}_f = \mathbf{I}_a = \frac{3 \times \mathbf{E}_a}{\mathbf{Z}_1 + \mathbf{Z}_2 + \mathbf{Z}_0}$$

$$= \frac{3 \times 1 \angle 0^\circ}{j0.05 + j0.0375 + j0.045}$$

$$= \frac{3 \angle 0^\circ}{j0.1325} = -j22.64 \text{ pu}$$

or $-j22.64 \times 875 = -j19{,}811$ A **Ans.**

(c) When one of the alternators is grounded through an inductor of 0.3 Ω, as shown in Fig. 5.13.

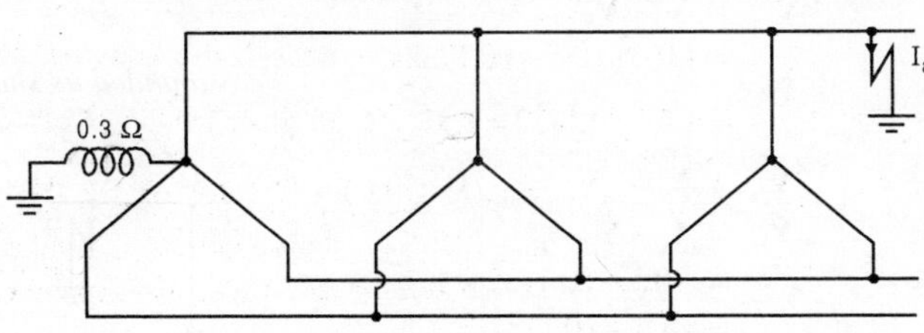

Fig. 5.13

$$\mathbf{Z}_1 = 0.05 \text{ pu}$$

$$\mathbf{Z}_2 = 0.0375 \text{ pu}$$

$$\mathbf{Z}_0 = j0.045 + j3 \times 0.3 \times \frac{10}{(6.6)^2} = j0.2516 \text{ pu}$$

$$\text{Fault current, } \mathbf{I}_f = \mathbf{I}_a = \frac{3\mathbf{E}_a}{\mathbf{Z}_1 + \mathbf{Z}_2 + \mathbf{Z}_0} = \frac{3 \times 1 \angle 0^\circ}{0.05 + 0.0375 + 0.2516}$$

$$= \frac{3 \angle 0^\circ}{j0.339} = -j8.8466 \text{ pu}$$

or $-j8.8466 \times 875 = 7{,}741$ A **Ans.**

Example 5.9. A 10 MVA, 13.8 kV alternator has positive, negative and zero sequence reactances of 30%, 40% and 5% respectively.

(*i*) What reactance must be put in the generator neutral so that the fault current for a line-to-ground fault of zero fault impedance will not exceed the rated line current.

(*ii*) What value of resistance in the neutral will serve the same purpose ?

Express both resistance and reactance in pu and in ohms.

[U.P.S.C. I.E.S. Electrical Engineering II 2000; A.M.I.E. Sec. B. Advanced Power Systems Summer 1994]

Solution : Let base MVA be 10, base kV = 13.8; $E_a = (1 + j0)$ pu

$$\text{Base impedance} = \frac{(\text{Base kV})^2}{\text{Base MVA}} = \frac{(13.8)^2}{10} = 19.044\ \Omega$$

$$(i) \text{ Fault current, } I_f = \frac{3\mathbf{E}_a}{\mathbf{Z}_1 + \mathbf{Z}_2 + \mathbf{Z}_{go} + 3\mathbf{Z}_n + 3\mathbf{Z}_f}$$

$$= \frac{3 \times 1}{j(0.3 + 0.4 + 0.05 + 3X_n + 0)} = -j1$$

$$\text{or } X_n = \frac{3 - (0.3 + 0.4 + 0.05)}{3} = 0.75 \text{ pu } \textbf{Ans.}$$

X_n = Base impedance × per unit reactance

= 19.044 × 0.75 = 14.283 Ω **Ans.**

$$(ii) \text{ Fault current, } \mathbf{I}_f = \frac{3\mathbf{E}_a}{\mathbf{Z}_1 + \mathbf{Z}_2 + \mathbf{Z}_{go} + 3\mathbf{Z}_n + 3\mathbf{Z}_f}$$

$$= \frac{3 \times 1}{j(0.3 + 0.4 + 0.05) + 3R_n} = -j1$$

$$\text{or } R_n = \frac{1}{3}\sqrt{3^2 - (0.3 + 0.4 + 0.05)^2}$$

= 0.968 pu **Ans.**

= 0.968 × 19.044 = 18.44 Ω **Ans.**

Example 5.10. A 3-phase generator is rated 25 MVA, 13.2 kV and X''_d = 0.2 pu; X_2 = 0.030 pu and X_0 = 0.10 pu. The neutral is solidly grounded. A line-to-line fault occurs on phases B and C of the generator. Calculate :

(*a*) I_b in amperes, (*b*) I_c in amperes, (*c*) V_{AB} in volts, (*d*) V_{BC} in volts and (*e*) V_{CA} in volts.

[U.P.S.C. I.E.S. Electrical Engineering-II, 2007]

Solution : Assume that the generator is operating on no load and at rated voltage at the occurrence of fault.

Let the induced voltage of phase A (line-to-neutral) voltage be $1 \angle 0^\circ$ pu. Sequence network is shown in Fig. 5.5.

$$\text{So, } \mathbf{E}_a = (1 + j0) \text{ or } 1 \angle 0^\circ \text{ pu} = \frac{13.2}{\sqrt{3}} \text{ kV}_{\text{actual}}$$

$$\text{Current, } \mathbf{I}_{a_1} = \frac{\mathbf{E}_a}{\mathbf{Z}_1 + \mathbf{Z}_2 + \mathbf{Z}_f} \qquad \text{Refer to Eq. (5.18)}$$

$$= \frac{1 \angle 0^\circ}{j0.2 + j0.3} = -j2 \text{ pu} \qquad \because \mathbf{Z}_f = 0$$

$$\mathbf{I}_{a_2} = -\mathbf{I}_{a_1} = j2 \text{ pu}$$

$$\mathbf{I}_{a_0} = 0$$

$$\mathbf{I}_a = \mathbf{I}_{a_1} + \mathbf{I}_{a_2} + \mathbf{I}_{a_0} = -j2 + j2 + 0 = 0 \quad \textbf{Ans.}$$

$$\mathbf{I}_b = \frac{-j\sqrt{3}\mathbf{E}_a}{\mathbf{Z}_1 + \mathbf{Z}_2 + \mathbf{Z}_f} \quad \text{Refer to Eq. (5.19)}$$

$$= \frac{-j\sqrt{3} \times 1 \angle 0^\circ}{j0.2 + j0.3} = -j2\sqrt{3} \text{ pu} \quad \textbf{Ans.}$$

$$\mathbf{I}_c = -\mathbf{I}_b = j2\sqrt{3} \text{ pu}$$

$$\text{Base current} = \frac{25 \times 1{,}000}{\sqrt{3} \times 13.2} = 1{,}093.5 \text{ A}$$

So, $\mathbf{I}_a = 0$

(*a*) $|\mathbf{I}_b| = 2\sqrt{3} \times 1{,}093.5 = 3{,}788$ A **Ans.**

(*b*) $|\mathbf{I}_c| = |\mathbf{I}_b| = 3{,}788$ A **Ans.**

Symmetrical components of the voltages from terminal A to ground,

$$\mathbf{V}_{a1} = \mathbf{E}_a - \mathbf{I}_{a1}\mathbf{Z}_1 = (1 + j0) - (-j2) \times j0.2 = 0.6 \text{ pu}$$

$$\mathbf{V}_{a2} = -\mathbf{I}_{a2}\mathbf{Z}_2 = -j2 \times j0.3 = 0.6 \text{ pu}$$

$$\mathbf{V}_{a0} = 0 \quad \because \text{ generator is grounded}$$

$$\mathbf{V}_a = \mathbf{V}_{a1} + \mathbf{V}_{a2} + \mathbf{V}_{a0} = 0.6 + 0.6 + 0 = 1.2 \text{ pu}$$

$$\mathbf{V}_b = a^2\mathbf{V}_{a1} + a\mathbf{V}_{a2} + \mathbf{V}_{a0}$$

$$= 0.6(-0.5 - 0.866) + 0.6(-0.5 + j0.866) = -0.6 \text{ pu}$$

$$\mathbf{V}_c = a\mathbf{V}_{a1} + a^2\mathbf{V}_{a2} + \mathbf{V}_{a0}$$

$$= 0.6\,(-0.5 + 0.866) + 0.6(-0.5 - 0.866) = -0.6 \text{ pu}$$

Line-to-line voltages

$$\mathbf{V}_{ab} = \mathbf{V}_a - \mathbf{V}_b = 1.2 - (-0.6) = 1.8 \text{ pu}$$

$$\mathbf{V}_{bc} = \mathbf{V}_b - \mathbf{V}_c = (-0.6) - (-0.6) = 0$$

$$\mathbf{V}_{ca} = \mathbf{V}_c - \mathbf{V}_a = (-0.6) - 1.2 = -1.8 \text{ pu}$$

So, (*c*) $\mathbf{V}_{ab} = 1.8 \times \dfrac{13.2}{\sqrt{3}} = 13.718$ kV **Ans.**

(*d*) $\mathbf{V}_{bc} = 0$ kV **Ans.**

(*e*) $\mathbf{V}_{ca} = -1.8 \times \dfrac{13.2}{\sqrt{2}} = -13.718$ kV **Ans.**

Example 5.11. A 15 MVA, 13.2 kV alternator has X″ = X_2 = 20% and X_0 = 8%. The neutral of this machine is grounded through a reactance of 0.5 Ω. Compute initial symmetrical rms current in all the three phases of the machine when L-L fault occurs at its terminals under unloaded condition. The fault impedance is *j* 0.1 pu. Compute voltages at the terminals of the generator under faulted condition and express these voltages in the form of percentages of rated voltage.

[*Bharathiyar Univ. Power System Analysis Nov; 1996*]

Solution : Let *a*, *b*, *c* be the three terminals of the alternator and let line-to-line fault occur between terminals *b* and *c*, as shown in Fig. 5.4. Let induced voltage of phase *a* line-to-neutral voltage be 1 per unit.

So $\quad \mathbf{E}_a = (1 + j0) \text{ pu} = \dfrac{13.2}{\sqrt{3}} \text{ kV}_{\text{actual}}$

Now $\mathbf{V}_b - \mathbf{V}_c = \mathbf{I}_b\mathbf{Z}_f$; $\mathbf{I}_a = 0$; $\mathbf{I}_b = -\mathbf{I}_c$; $\mathbf{Z}_1 = j\,0.20$ pu; $\mathbf{Z}_2 = j\,0.2$ pu;

$\mathbf{Z}_f = j\,0.1$ pu and $\mathbf{Z}_{go} = j\,0.08$ pu

$$\mathbf{Z}_0 = \mathbf{Z}_{go} + j3 \times 0.5 \times \frac{15}{(13.2)^2}$$

$$= (j\,0.08 + j\,0.129) \text{ pu} = j\,0.209 \text{ pu}$$

$$\mathbf{I}_{a1} = \frac{\mathbf{E}_a}{\mathbf{Z}_1 + \mathbf{Z}_2 + \mathbf{Z}_f} = \frac{1 + j0}{j0.20 + j0.20 + j0.1} = -j\,2.0 \text{ pu}$$

$$\mathbf{I}_{a2} = -\mathbf{I}_{a1} = j\,2.0 \text{ pu}$$

$$\mathbf{I}_{a0} = 0$$

$$\mathbf{I}_a = \mathbf{I}_{a1} + \mathbf{I}_{a2} + \mathbf{I}_{a0} = -j\,2.0 + j\,2.0 + 0 = 0$$

$$\mathbf{I}_b = a^2\mathbf{I}_{a1} + a\mathbf{I}a_2 + \mathbf{I}_{a0}$$

$$= -j\,2.0\,(-0.5 - j\,0.866) + j\,2.0\,(-0.5 + j\,0.866) + 0 = -3.464 \text{ pu}$$

$$\mathbf{I}_c = a\mathbf{I}_{a1} + a^2\mathbf{I}_{a2} + \mathbf{I}_{a0}$$

$$= -j\,2.0\,(-0.5 + j\,0.866) + j\,2.0\,(-0.5 - j\,0.866) + 0 = 3.464 \text{ pu}$$

Base current

$$= \frac{15 \times 1{,}000}{\sqrt{3} \times 13.2} = 656 \text{ A}$$

$\therefore \mathbf{I}_a = 0$

$\mathbf{I}_b = -3.464 \times 656 = -2{,}272$ A **Ans.**

$\mathbf{I}_c = 3.464 \times 656 = 2{,}272$ A **Ans.**

Symmetrical components of the voltages from terminal *a* to ground,

$$\mathbf{V}_{a1} = \mathbf{E}_a - \mathbf{I}_{a1}\mathbf{Z}_1 = (1.0 + j\,0) - (-j\,2.0) \times j\,0.20 = 0.6 \text{ pu}$$

$$\mathbf{V}_{a2} = -\mathbf{I}_{a2}\mathbf{Z}_2 = -j\,2.0 \times j\,0.20 = 0.4 \text{ pu}$$

$$\mathbf{V}_{a1} - \mathbf{V}_{a2} = \mathbf{I}_{a1}\mathbf{Z}_f \quad \text{(for check)}$$

$$(0.6 - 0.4) \text{ pu} = -j\,2.0 \times j\,0.1 = 0.2 \text{ pu}$$

$$\mathbf{V}_{a0} = 0 \quad \because \text{ generator is grounded}$$

Line-to-ground voltages,

$$\mathbf{V}_a = \mathbf{V}_{a1} + \mathbf{V}_{a2} + \mathbf{V}_{a0} = 0.6 + 0.4 + 0 = 1.0 \text{ pu}$$

$$\mathbf{V}_b = a^2\mathbf{V}_{a1} + a\mathbf{V}_{a2} + \mathbf{V}_{a0}$$

$$= 0.6\,(-0.5 - j\,0.866) + 0.4\,(-0.5 + j\,0.866) + 0$$

$$= -0.5 - j\,0.1732$$

$$\mathbf{V}_c = a\mathbf{V}_{a1} + a^2\mathbf{V}_{a2} + \mathbf{V}_{a0}$$

$$= 0.6\,(-0.5 + j\,0.866) + 0.4\,(-0.5 - j\,0.866) + 0$$

$$= -0.5 + j\,0.1732$$

$$\mathbf{V}_b - \mathbf{V}_c = \mathbf{I}_b\mathbf{Z}_f = -3.464 \times j\,0.1 = -j\,0.3464$$

Line-to-line voltages,

$$\mathbf{V}_{ab} = \mathbf{V}_a - \mathbf{V}_b$$

$$= 1.0 - [-0.5 - j\,0.1732]$$

$$= 1.5 + j\,0.1732 = 1.51 \angle 6.6^\circ \text{ pu or } 151 \angle 6.6^\circ \,\% \quad \textbf{Ans.}$$

$$\mathbf{V}_{bc} = \mathbf{V}_b - \mathbf{V}_c = -j\,0.3464 \text{ pu or } -j\,34.64\,\% \quad \textbf{Ans.}$$

$$\mathbf{V}_{ca} = \mathbf{V}_c - \mathbf{V}_a$$

$$= -0.5 + j\,0.1732 - 1$$

$$= -1.5 + j\,0.1732$$

$$= 1.51 \angle 173.4^\circ \text{ pu or } 151 \angle 173.4^\circ \,\% \quad \textbf{Ans.}$$

Example 5.12. A synchronous generator is rated 20 MVA, 13.8 kV. It has X_1 = 0.25 pu, X_2 = 0.35 pu and X_0 = 0.10 pu. The neutral is solidly grounded. The machine is on no load and is operating at rated voltage when a D-L-G fault occurs at the terminals. Find the subtransient currents in all the phases and the fault (all in amps) and the line-to-line voltages in all the phases (in kV).

[*U.P.S.C. I.E.S. Electrical Engineering-II, 2011*]

Solution : Let *a*, *b*, *c* be the three terminals of the alternators and let L-L-G fault occur between terminals *b* and *c* to ground, as shown in Fig. 5.6.

Let induced voltage of phase *a* line-to-neutral voltage be 1 per unit.

So $\mathbf{E}_a = (1 + j0)$ pu $= \dfrac{13.8}{\sqrt{3}}$ kV$_{\text{actual}}$

$$\text{Base current, } \mathbf{I}_B = \frac{\text{MVA} \times 1{,}000}{\sqrt{3}\ \text{kV}} = \frac{20{,}000}{\sqrt{3} \times 13.8} = 836.74 \text{ A}$$

$$\text{From Eq. (5.28), } \mathbf{I}_{a1} = \frac{\mathbf{E}_a}{\mathbf{Z}_1 + \dfrac{\mathbf{Z}_2\mathbf{Z}_0}{\mathbf{Z}_2 + \mathbf{Z}_0}} = \frac{1 + j0}{j0.25 + \dfrac{j0.35 \times j0.10}{j0.35 + j0.10}}$$

$$= -j\,3.051 \text{ pu}$$

Now for L-L-G fault, from sequence diagram shown in Fig. 5.7, with $Z_f = 0$,

$$\mathbf{V}_{a1} = \mathbf{V}_{a2} = \mathbf{V}_{a0}$$

$$\text{Also } \mathbf{V}_{a1} = \mathbf{E}_a - \mathbf{I}_{a1}\mathbf{Z}_1$$

For determining $\mathbf{I}_{a2}$ and $\mathbf{I}_{a0}$ we should first determine $\mathbf{V}_{a1}$ and $\mathbf{V}_{a2}$

$$\mathbf{V}_{a0} = \mathbf{V}_{a2} = \mathbf{V}_{a1} = \mathbf{E}_a - \mathbf{I}_{a1}\mathbf{Z}_1 = (1 + j0) - (-j\,3.051) \times j\,0.25$$
$$= 0.2373 \text{ pu}$$

$$\mathbf{I}_{a2} = -\frac{\mathbf{V}_{a2}}{\mathbf{Z}_2} = -\frac{0.2373}{j0.35} = j\,0.678 \text{ pu}$$

$$\text{and } \mathbf{I}_{a0} = -\frac{\mathbf{V}_{a0}}{\mathbf{Z}_0} = -\frac{0.2373}{j0.1} = j\,2.373 \text{ pu}$$

$$\mathbf{I}_a = \mathbf{I}_{a0} + \mathbf{I}_{a1} + \mathbf{I}_{a2} = j\,2.373 - j\,3.05 + j\,0.677 = 0$$

$$\mathbf{I}_b = a^2\mathbf{I}_{a1} + a\mathbf{I}_{a2} + \mathbf{I}_{a0}$$
$$= -j\,3.05\,(-0.5 - j\,0.866) + j\,0.678\,(-0.5 + j\,0.866) + j\,2.373$$
$$= -3.228 + j\,3.56 = 4.8\,\angle 132.2° \text{ pu}$$

$$\mathbf{I}_c = a\mathbf{I}_{a1} + a^2\mathbf{I}_{a2} + \mathbf{I}_{a0}$$
$$= -j\,3.051\,(-0.5 + j\,0.866) + j\,0.678\,(-0.5 - j\,0.866) + j\,2.373$$
$$= 3.228 + j\,3.56 = 4.8\,\angle 47.8° \text{ pu}$$

Subtransient currents,

$$\mathbf{I}_a = 0$$
$$\mathbf{I}_b = 4.8\,\angle 132.2° \times 836.74 = 4{,}016\,\angle 132.2° \text{ A } \textbf{Ans.}$$
$$\mathbf{I}_c = 4.8\,\angle 47.8° \times 836.74 = 4{,}016\,\angle 47.8° \text{ A } \textbf{Ans.}$$

Fault current,

$$\mathbf{I}_f = \mathbf{I}_b + \mathbf{I}_c = j\,7.12 \text{ pu} = 7.12\,\angle 90° \times 4{,}016$$
$$= 5{,}958\,\angle 90° \text{ A } \textbf{Ans.}$$

$$\mathbf{V}_{a0} = \mathbf{V}_{a1} = \mathbf{V}_{a2} = 0.2373 \text{ pu, as determined above}$$

From Fig. 5.6, with $Z_f = 0$

$$\mathbf{V}_b = \mathbf{V}_c = 0$$
$$\mathbf{V}_a = \mathbf{V}_{a0} + \mathbf{V}_{a1} + \mathbf{V}_{a2} = 3 \times 0.2373 = 0.7119 \text{ pu}$$

Line-to-line voltages in pu,

$$\mathbf{V}_{ab} = \mathbf{V}_a - \mathbf{V}_b = 0.7119 - 0 = 0.7119\,\angle 0° \text{ pu}$$
$$\mathbf{V}_{bc} = \mathbf{V}_b - \mathbf{V}_c = 0$$
$$\mathbf{V}_{ca} = \mathbf{V}_c - \mathbf{V}_a = 0 - 0.7119 = 0.7119\,\angle 180° \text{ pu}$$

Line-to-line voltage in kV,

$$\mathbf{V}_{ab} = 0.7119\,\angle 0° \times \frac{13.8}{\sqrt{3}} \text{ kV} = 5.672\,\angle 0° \text{ kV } \textbf{Ans.}$$
$$\mathbf{V}_{bc} = 0$$
$$\mathbf{V}_{ca} = 0.7119\,\angle 180° \times \frac{13.8}{\sqrt{3}} \text{ kV} = 5.672\,\angle 180° \text{ kV } \textbf{Ans.}$$

Example 5.13. A 10 MVA, 13.8 kV turbo-generator having $X''_d = X_2 = 15\%$ and $X_0 = 5\%$ is about to be connected to a power system. The generator has current limiting reactor of 0.7 Ω in the neutral. Before the generator is connected to the system, its voltage is adjusted to 13.2 kV when a double line-to-ground fault develops at terminals *b* and *c*. Find the initial symmetrical rms currents in the ground and in line *b*.

[*Bharathiyar Univ. Power System Analysis April 1997*]

Solution : Let the base MVA be 10 and base kV be 13.8

$$\text{Base current, } \mathbf{I}_B = \frac{\text{MVA} \times 1{,}000}{\sqrt{3} \times 13.8} = \frac{10{,}000}{\sqrt{3} \times 13.8} = 418 \text{ A}$$

$$\mathbf{Z}_0 = \mathbf{Z}_{go} + j3 \times 0.7 \times \frac{10}{(13.8)^2}$$
$$= j\,0.05 + j\,0.11 = j\,0.16 \text{ pu}$$

$$\mathbf{E}_a = \frac{13.2}{13.8} + j0 = (0.9565 + j\,0) \text{ pu}$$

From Eq. (5.28)

$$\mathbf{I}_{a1} = \frac{\mathbf{E}_a}{\mathbf{Z}_1 + \dfrac{\mathbf{Z}_2\mathbf{Z}_0}{\mathbf{Z}_2\mathbf{Z}_0}} = \frac{0.9565}{j0.15 + \dfrac{j0.15 \times j0.16}{j0.15 + j0.16}} = -j\,4.206 \text{ pu}$$

$$\text{Now } \mathbf{V}_{a0} = \mathbf{V}_{a2} = \mathbf{V}_{a1} = \mathbf{E}_a - \mathbf{I}_{a1}\mathbf{Z}_1$$
$$= 0.9565 - (-j\,4.206) \times j\,0.15 = 0.3256 \text{ pu}$$

$$\mathbf{I}_{a2} = -\frac{\mathbf{V}_{a2}}{\mathbf{Z}_2} = -\frac{0.3256}{j0.15} = j\,2.171 \text{ pu}$$

$$\mathbf{I}_{a0} = -\frac{\mathbf{V}_{a0}}{\mathbf{Z}_0} = -\frac{0.3256}{j0.16} = j\,2.035 \text{ pu}$$

Initial symmetrical rms current in line *b*,

$$\mathbf{I}_b = a^2\mathbf{I}_{a1} + a\mathbf{I}_{a2} + \mathbf{I}_{a0}$$
$$= -j4.206\,(-0.5 - j0.866) + j2.171\,(-0.5 + j0.866) + j2.035$$
$$= -5.52 + j\,3.0525$$
$$= 6.3\,\angle 151° \text{ pu} = 6.3\,\angle 151° \times 418 = 2{,}633\,\angle 151° \text{ A } \textbf{Ans.}$$

Initial symmetrical rms current in ground wire

$$= 3\mathbf{I}_{a0} = 3 \times j\,2.035 \times 418 = 2{,}552\,\angle 90° \text{ A } \textbf{Ans.}$$

Example 5.14. Two identical 11 kV, 50 MVA, 3-phase alternators are connected in parallel and supply a substation by a feeder having an impedance of (0.4 + *j*0.7) Ω to positive and negative sequence currents and (0.7 + *j*3.0) Ω to zero sequence current. The positive, negative and zero sequence reactances of each of the generators are 0.6 Ω, 0.4 Ω and 0.2 Ω respectively. Both machines have their neutrals earthed through resistances of 0.2 Ω. Calculate the potential of the alternator neutral with respect to earth, if an earth fault occurs simultaneously on blue and yellow phases of the substation. Neglect prefault power and generator resistances.

[*U.P.S.C. I.E.S. Electrical Engineering-II, 2002*]

Solution : Let the base MVA be 50 MVA and base kV be 11 kV.

For synchronous generators

Per unit positive sequence reactance,

$$X_1 = j\frac{0.6 \times 50}{11^2} = j0.248 \text{ pu}$$

Per unit negative synchronous reactance

$$X_2 = j\frac{0.4 \times 50}{11^2} = j0.165 \text{ pu}$$

Per unit zero synchronous reactance

$$X_0 = j\frac{0.2 \times 50}{11^2} = j0.083 \text{ pu}$$

For feeder

$$X_1 = X_2 = \frac{(0.4 + j0.7) \times 50}{11^2} = (0.165 + j0.289) \text{ pu}$$

$$X_0 = \frac{(0.7 + j3.0) \times 50}{11^2} = (0.289 + j1.24) \text{ pu}$$

Grounding resistance,

$$R_G = \frac{0.2 \times 50}{11^2} = 0.0826 \text{ pu}$$

Two identical generators operate in parallel, therefore

$$X_{1\text{ eff}} = j\frac{0.248}{2} + 0.165 + j0.289 = (0.165 + j0.413) \text{ pu}$$

$$X_{2\text{ eff}} = j\frac{0.165}{2} + (0.165 + j0.289) = (0.165 + j0.3715) \text{ pu}$$

$$X_{0\text{ eff}} = \frac{j0.083 + 3 \times 0.0826}{2} + (0.289 + j1.24)$$

$$= (0.4129 + j1.2815) \text{ pu}$$

For L-L-G fault on the generator terminals

Assuming prefault voltage $\mathbf{E}_a$ be $1 \angle 0°$ pu

$$\mathbf{I}_{a1} = \frac{\mathbf{E}_a}{X_{1\text{eff}} + \dfrac{X_{2\text{eff}} \times X_{0\text{eff}}}{X_{2\text{eff}} + X_{0\text{eff}}}} \text{ pu}$$

$$= \frac{1 \angle 0°}{0.165 + j0.413 + \dfrac{(0.165 + j0.3715)(0.4129 + j1.2815)}{(0.165 + j0.3715) + (0.4129 + j1.2815)}} \text{ pu}$$

$$= \frac{1 \angle 0°}{0.165 + j0.413 + 0.120 + j0.2894} \text{ pu}$$

$$= \frac{0.285 - j0.7024}{0.285^2 + 0.7024^2} \text{ pu} = \frac{0.285 - j0.7024}{0.57459}$$

$$= (0.496 - j1.2224) \text{ pu}$$

$$\mathbf{I}_{a0} = \frac{\mathbf{E}_a - \mathbf{I}_{a_1} X_{1\text{eff}}}{X_{0\text{eff}}}$$

$$= \frac{1 - (0.496 - j1.2224) \times (0.165 + j0.413)}{(0.4129 + j1.2815) \text{ pu}}$$

$$= \frac{0.4127 \angle 0°}{1.3464 \angle 72.14°} = 0.3065 \angle -72.14° = (0.094 - j0.2917) \text{ pu}$$

$\mathbf{I}_{a0}$ in each generator

$$= \frac{0.094 - j0.2917}{2} = (0.047 + j0.146) \text{ pu}$$

Potential of alternator neutral with respect to ground

$= 3 \times \mathbf{I}_{a0}$ per generator $\times R_G$

$= 3(0.047 - j0.146) \times 0.0826$

$= (0.01165 - j0.0362)$ pu $= 0.03802 \angle -72.2°$ pu

Potential of alternator neutral with respect to ground in volts

$$= 0.03802 \angle 72.2° \times \frac{11{,}000}{\sqrt{3}} = 241.5 \angle -72.2° \text{ V} \quad \textbf{Ans.}$$

Example 5.15. Compare the fault currents of a synchronous generator with three-phase fault current.

Solution : Let *a*, *b*, *c* be the three terminals of the synchronous generator and $\mathbf{E}_a$ be 1 pu. Assume that the impedances $\mathbf{Z}_1$, $\mathbf{Z}_2$ and $\mathbf{Z}_0$ are pure reactances and $Z_f = 0$. The fault currents are

For 3-phase fault,

$$I_f = \frac{E_a}{X_1} = \frac{1}{X_1} \text{ pu} \qquad \text{...(i) Refer to Eq. (5.2)}$$

For fault on terminal *a* to ground,

$$I_f = \frac{3E_a}{X_1 + X_2 + X_0} = \frac{3}{X_1 + X_2 + X_0} \text{ pu} \qquad \text{...(ii) Refer to Eq. (5.11)}$$

For line-to-line fault between terminals *b* and *c*,

$$I_f = \frac{\sqrt{3}E_a}{X_1 + X_2} = \frac{1.732}{X_1 + X_2} \text{ pu} \qquad \text{...(iii) Refer to Eq. (5.21)}$$

Ratio of fault currents $\dfrac{\text{L-G fault}}{\text{3-phase fault}} = \dfrac{\text{Expression (ii)}}{\text{Expression (i)}}$

$$= \frac{3X_1}{X_1 + X_2 + X_0} \quad \textbf{Ans.}$$

$$\frac{\text{L-L fault}}{\text{3-phase fault}} = \frac{\text{Expression (iii)}}{\text{Expression (i)}}$$

$$= \frac{1.732 X_1}{X_1 + X_2} \quad \textbf{Ans.}$$

Example 5.16. A synchronous generator is rated 25 MVA, 11 kV. It is star connected with the neutral point solidly grounded. The generator is operating at no load rated voltage. Its reactances are X″ = X$_2$ = 0.20 pu and X$_0$ = 0.08 pu. Calculate the symmetrical subtransient line currents for

(*i*) single line-to-ground fault

(*ii*) double line-to-ground fault

(*iii*) double line fault

(*iv*) symmetrical three phase fault.

Compare these currents and comment.

[U.P. Technical Univ. Power System Analysis 2008-09]

Solution : Let the induced voltage of phase a line to neutral $\mathbf{E}_a$ be 1 per unit *i.e.*

$$\mathbf{E}_a = (1 + j0) \text{ pu}$$

Let Base MVA = 25 MVA

Base kV = 11 kV (line-to-line)

$$\text{Base current, } I_B = \frac{\text{Base MVA} \times 1{,}000}{\sqrt{3} \times \text{base kV}} = \frac{25 \times 1{,}000}{\sqrt{3} \times 11} = 1{,}312 \text{ A}$$

(i) Single Line-To-Ground Fault

Let *a*, *b*, *c* be three terminals of the generator. Let fault occur between terminal and ground as illustrated in Fig. 5.2.

From Eqs. (5.5) and (5.9)

$$\mathbf{I}_{a0} = \mathbf{I}_{a1} = \mathbf{I}_{a2} = \frac{\mathbf{E}_a}{\mathbf{Z}_1 + \mathbf{Z}_2 + \mathbf{Z}_0} = \frac{1 + j0}{j0.20 + j0.20 + j0.08}$$

$$= -j2.083 \text{ pu}$$

Fault current,

$$\mathbf{I}_f = \mathbf{I}_a = 3 \times \mathbf{I}_{a1} = 3 \times (-j2.083)$$

$$= -j6.25 \text{ pu} = 1{,}312 \times (-j6.25)$$

$$= 8{,}200 \angle -90° \text{ A} \quad \textbf{Ans.}$$

(ii) Double Line-To-Ground Fault

$$\text{Base voltage (line-to-neutral)} = \frac{11}{\sqrt{3}} = 6.35 \text{ kV}$$

$$\text{Let } \mathbf{E}_a = (1 + j0) \text{ pu} = 6.35 \text{ kV}$$

$$\text{From Eq. (5.28), } \mathbf{I}_{a1} = \frac{\mathbf{E}_a}{\mathbf{Z}_1 + \dfrac{\mathbf{Z}_2 \mathbf{Z}_0}{\mathbf{Z}_2 + \mathbf{Z}_0}}$$

$$= \frac{(1 + j0)}{j0.20 + \dfrac{j0.20 \times j0.08}{j0.20 + j0.08}} = -j3.889 \text{ pu}$$

Now for L-L-G fault, from sequence diagram shown in Fig. 5.7 with $Z_f = 0$

$$\mathbf{V}_{a1} = \mathbf{V}_{a2} = \mathbf{V}_{a0} = \mathbf{E}_a - \mathbf{I}_{a1}\mathbf{Z}_1 = (1 + j0) - (-j3.889) \times (j0.20) = -0.222 \text{ pu}$$

$$\mathbf{I}_{a2} = -\frac{\mathbf{V}_{a2}}{\mathbf{Z}_2} = \frac{-0.222}{j0.20} = j1.111 \text{ pu}$$

$$\mathbf{I}_{a0} = -\frac{\mathbf{V}_{a0}}{\mathbf{Z}_0} = \frac{-0.222}{j0.08} = j2.778 \text{ pu}$$

$$\mathbf{I}_a = \mathbf{I}_{a_1} + \mathbf{I}_{a_2} + \mathbf{I}_{a_0} = -j3.889 + j1.111 + j2.778 = 0$$

$$\mathbf{I}_b = a^2\mathbf{I}_{a_1} + a\mathbf{I}_{a_2} + \mathbf{I}_{a_0}$$
$$= -j3.889(-0.5 - j0.866) + j1.111(-0.5 + j0.866) + j2.778$$
$$= (-4.3 + j4.167) \text{ pu}$$

$$\mathbf{I}_c = a\mathbf{I}_{a1} + a^2\mathbf{I}_{a2} + \mathbf{I}_{a0}$$
$$= -j3.889(-0.5 + j0.866) + j1.111(-0.5 - j0.866) + j2.778$$
$$= (4.3 + j4.167) \text{ pu}$$

Fault current,

$$\mathbf{I}_f = \mathbf{I}_b + \mathbf{I}_c$$
$$= (-4.3 + j4.167) + (4.3 + j4.167)$$
$$= j8.334 \text{ pu} = j8.334 \times I_B = j8.334 \times 1{,}312 = j10{,}934 \text{ A}$$
$$= 10{,}934 \angle -90° \text{ A} \quad \textbf{Ans.}$$

(iii) Line-to-line fault current between terminals b and c (Fig. 5.4)

$$\text{Fault current, } I_f = \frac{\sqrt{3}E_a}{X_1 + X_2} \qquad \text{Refer to Eq. (5.21)}$$

$$= \frac{\sqrt{3} \times 1.0}{0.2 + 0.2} = 4.33 \text{ pu} = 4.33 \times 1{,}312 = 5{,}681 \text{ A} \quad \textbf{Ans.}$$

(iv) For symmetrical three-phase fault

$$I_f = \frac{E_a}{X_1} = \frac{1}{0.2} = 5 \text{ pu} = 5 \times 1{,}312 = 6{,}560 \text{ A} \quad \textbf{Ans.}$$

Example 5.17. A generator of negligible resistance having 1.0 per unit voltage behind transient reactance is subjected to different types of faults.

Type of Fault	Resulting fault current in pu
3-phase	3.33
L-L	2.23
L-G	3.01

Calculate the per unit values of 3 sequence-reactances.

[*Pb. Technical Univ. Computer Applications To Power System Analysis May 2010; Madras Univ. Power System Analysis October 1996*]

Solution : Let *a*, *b*, *c* be the three terminals of the synchronous generator and let fault impedance Z_f be zero.

From Eq. (5.2) for three-phase fault

$$\text{Fault current, } I_f = \frac{E_a}{X_1} = \frac{1}{X_1} = 3.33$$

$$\text{or positive sequence reactance, } X_1 = \frac{1}{3.33} = 0.3 \text{ pu} \quad \textbf{Ans.}$$

From Eq. (5.21) for L-L fault

$$\text{Fault current, } I_f = \frac{\sqrt{3}\,E_a}{X_1 + X_2} = \frac{1.732}{X_1 + X_2} = 2.23$$

$$\text{or } X_1 + X_2 = \frac{1.732}{2.23} = 0.7767$$

Substituting the value of $X_1 = 0.3$ pu in the above equation we have Negative sequence reactance, $X_2 = 0.7767 - 0.3 = 0.4767$ pu **Ans.**

From Eq. (5.11) for L-G fault

$$\text{Fault current, } I_f = \frac{3E_a}{X_1 + X_2 + X_0} = \frac{3}{X_1 + X_2 + X_0} = 3.01$$

$$\text{or } X_1 + X_2 + X_0 = \frac{3}{3.01} = 0.9967$$

Substituting the value of $X_1 + X_2 = 0.7767$ in the above equation we have,

Zero sequence reactance, $X_0 = 0.9967 - 0.7767 = 0.22$ pu **Ans.**

Normally the positive sequence impedance is greater than the negative sequence one but since the given positive sequence impedance corresponds to the transient state, it may be less than the negative sequence impedance.

Example 5.18. Three faults simulated on a power system give the following results :

$I_{f3\phi} = 1{,}700$ A, $I_{f\text{L-L}} = 1{,}500$ A and $I_{f\text{SLG}} = 2{,}000$ A

The base value at fault are : 1,000 MVA, 132 kV. Find pu values of all sequence impedances of Thevenin Equivalent network.

[*M.D. Univ. Power System-I, 2010-11*]

Solution : Let *a*, *b*, *c* be the terminals of synchronous generator and fault impedance Z_f be zero.

$$\text{Base impedance, } Z_B = \frac{(\text{kV})_B^2}{(\text{MVA})_B} = \frac{132^2}{1{,}000} = 17.424\ \Omega$$

$$\text{For 3-phase fault, } I_{f3\phi} = \frac{\text{Line-to-neutral voltage } i.e., E/\sqrt{3}}{Z_1}$$

$$= \frac{132{,}000/\sqrt{3}}{Z_1}$$

$$\text{or } Z_1 = \frac{132{,}000}{\sqrt{3} \times 1{,}700} = 44.83\ \Omega$$

For line-to-line fault

$$\text{Fault current, } I_{f\text{L-L}} = \frac{\sqrt{3} \times \dfrac{132{,}000}{\sqrt{3}}}{Z_1 + Z_2}$$

$$\text{or } Z_1 + Z_2 = \frac{132{,}000}{1{,}500} = 88\ \Omega$$

$$Z_2 = 88 - Z_1 = 88 - 44.83 = 43.17\ \Omega$$

For single line-to-ground fault,

$$\text{Fault current, } I_{f\text{SLG}} = \frac{\sqrt{3} \times 132{,}000}{Z_1 + Z_2 + Z_0}$$

$$\text{or } Z_1 + Z_2 + Z_0 = \frac{132{,}000 \times \sqrt{3}}{2{,}000} = 114.32\ \Omega$$

$$\text{or } Z_0 = 114.32 - (Z_1 + Z_2) = 114.32 - 88 = 26.32\ \Omega$$

In terms of per unit value

$$Z_{0\,\text{pu}} = \frac{26.32}{17.424} = 1.51 \text{ pu} \quad \textbf{Ans.}$$

$$Z_{1\,\text{pu}} = \frac{44.83}{17.424} = 2.573 \text{ pu} \quad \textbf{Ans.}$$

$$Z_{2\,\text{pu}} = \frac{43.17}{17.424} = 2.478 \text{ pu} \quad \textbf{Ans.}$$

5.3. FAULTS IN POWER SYSTEMS

So far we have considered the faults at the terminals of an unloaded synchronous generator but the results obtained can also be used in the analysis of faults in power systems.

Consider a short section of a transmission line in which phase *a* is grounded. Only for theoretical purpose (for convenience of representation of fault currents), think of the fault as occurring on a short stub line, built out from the line at the fault point, as illustrated in Fig. 5.14. Let I_a, I_b and I_c be currents in three phases of the stub line.

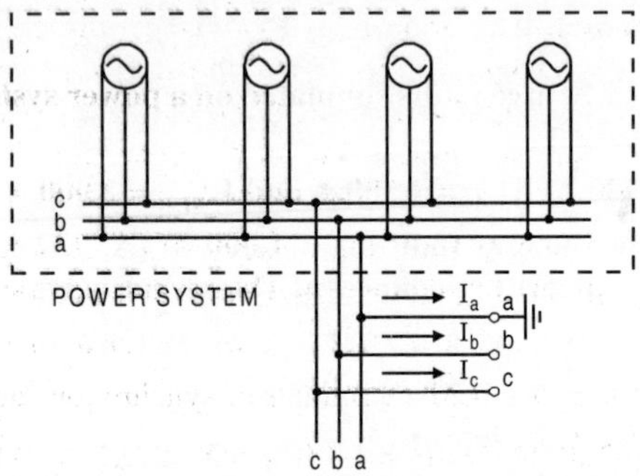

Fig. 5.14. ***Representation of Power System Under Loaded Condition***

Assuming that the power system consists of linear elements, the whole system can be reduced to a single generator (Fig. 5.15) with an equivalent internal impedance by using Thevenin's theorem. The emfs E_{a1}, E_{a2}, E_{a0} are the symmetrical components of the open-circuit voltage to ground before the fault occurred. In the usual case the prefault voltages are balanced and therefore, $E_{a1} = E_a$; $E_{a2} = 0$ and $E_{a0} = 0$. The impedances Z_1, Z_2 and Z_0 are the positive, negative and zero sequence impedances, viewed from the fault point.

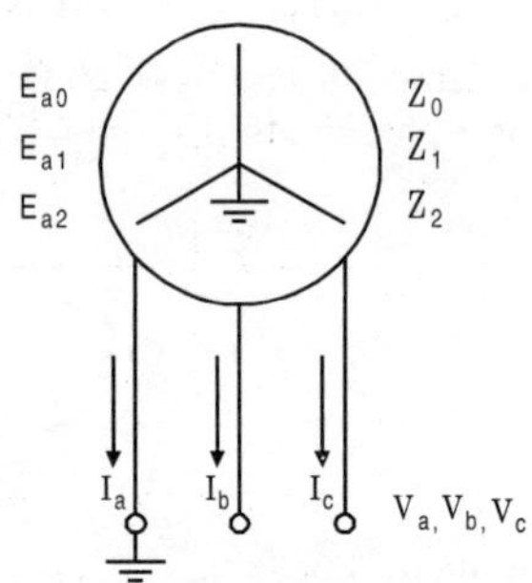

Fig. 5.15. ***Equivalent Generator***

For illustration of the application of Thevenin's theorem for determining the equivalent sequence networks, consider a simple power system shown in Fig. 5.16.

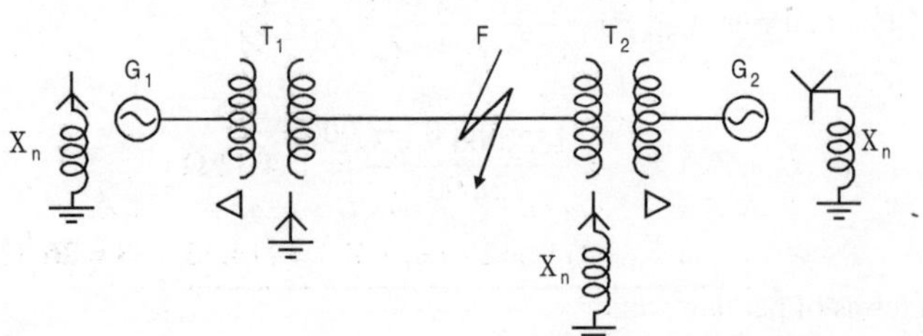

Fig. 5.16. ***Single Line Diagram of a Balanced 3-Phase System***

On occurrence of fault, say at point F in Fig. 5.16, unbalanced conditions will be introduced. For analysis following assumptions are made.

1. The impedances are all constant and independent of currents.
2. The synchronous generator is a salient pole type which may generate under fault condition, negative and zero sequence emfs which are small and are negligible. Thus for all purposes the machine is assumed to generate only positive phase sequence emfs.

The network under such conditions can be represented by three independent single-phase sequence networks as follows:

Sequence networks of the system are shown in Fig. 5.17. The point where a fault is assumed to occur is marked F on the single line diagram and on the sequence networks. Sequence networks are drawn, as already discussed in Arts. 3.10 and 3.11.

Each of the networks can be replaced by its Thevenin's equivalent between the two terminals composed of its reference bus and the point of occurrence of the fault. The Thevenin's equivalent circuit of each sequence network is given below the corresponding network in Fig. 5.17.

Thevenin's equivalent of positive sequence network is shown in Fig. 5.17 (*d*). The internal voltage V_f of the single generator of the equivalent circuit for the positive sequence network is the prefault voltage to neutral at the fault point, Z_1 is the impedance measured between fault point F and the reference bus of the positive sequence network with all internal emfs short-circuited.

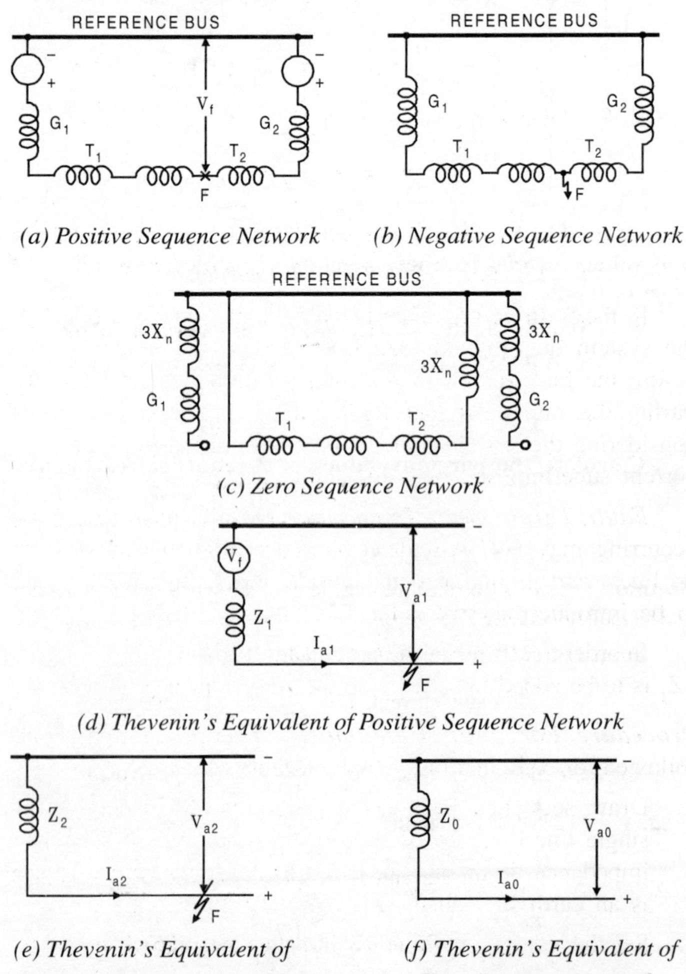

(*a*) *Positive Sequence Network* (*b*) *Negative Sequence Network*

(*c*) *Zero Sequence Network*

(*d*) *Thevenin's Equivalent of Positive Sequence Network*

(*e*) *Thevenin's Equivalent of Negative Sequence Network* (*f*) *Thevenin's Equivalent of Zero Sequence Network*

Fig. 5.17. ***Sequence Networks For Power System Shown in Fig. 5.16***

Similarly, the Thevenin's equivalent negative and zero sequence networks are obtained from the negative and zero sequence networks respectively. Since the system is balanced, no negative or zero sequence currents are flowing before the occurrence of the fault. So no emfs appear in the Thevenin's equivalent circuit. The impedances Z_2 and Z_0 are measured between the fault point F and the reference bus in their respective networks.

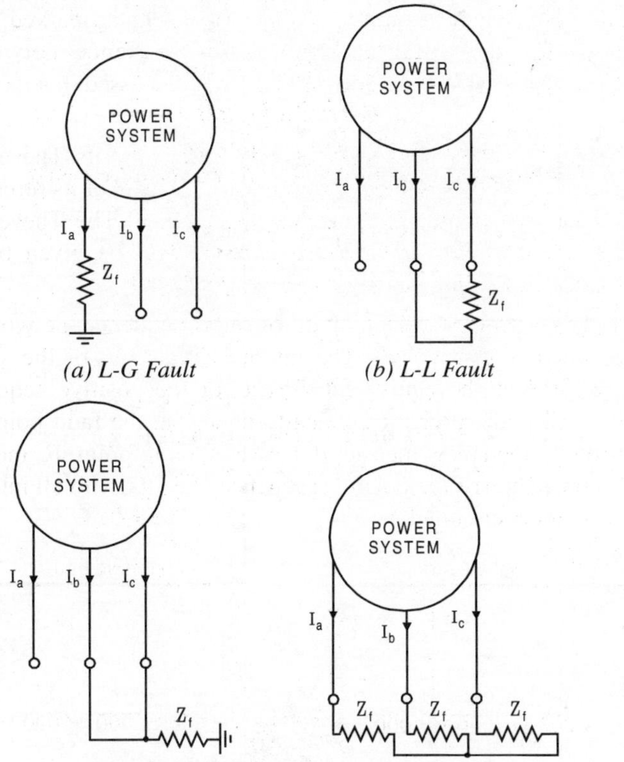

(a) L-G Fault *(b) L-L Fault*

(c) L-L-G Fault *(d) L-L-L Fault*

Fig. 5.18. ***Fault Through Impedance***

In the positive sequence network, the currents throughout the system due to fault can be added to the load currents before the fault to obtain the total positive sequence current during the fault. The net fault current is the fault current considering the system under no-load condition plus the load current superimposed over the fault currents.

Fault Through an Impedance. Sometimes the fault occurring may not be a dead short circuit between the line (or lines) and ground but may result in a fault impedance Z_f to be introduced between them, as illustrated in Fig. 5.18.

In order to afford for the ground fault through impedance, $3Z_f$ is to be added in series with the zero sequence of network.

Procedure For Fault Calculation. The procedure to be followed for computation of fault currents etc., is given below:

1. Draw sequence impedance diagrams by inspection of single line diagram of the power system. Zero sequence impedance diagram is required to be drawn only if there is an earth fault also.
2. Set the values of sequence impedances in each network.
3. Reduce the networks into their equivalent impedances Z_1, Z_2 and Z_0.
4. Connect the equivalent networks, as given in Table 5.1.
5. Obtain the sequence components of currents and voltages using equations given in Table 5.1.
6. Obtain the currents in different phases from sequence-components of currents using equations given in Table 5.1.
7. Voltage, if required can be obtained from knowledge of impedances and sequence currents.

Example 5.19. A single line diagram of power system is shown in Fig. 5.19, where the sequence reactance of generator (G), synchronous motor (M) and transformers T_1 and T_2 are given in per unit. The neutrals of the generator and transformers are solidly grounded. The motor neutral is grounded through a reactance X_n = 0.05 per unit. Draw the positive, negative and zero sequence networks with reactance values in per unit on a 100 MVA, 13.8 kV base in the zone of the generator. The prefault voltage is 1.05 per unit.

Calculate the per unit fault current for a three-phase to ground fault at bus '*d*'. The system data are as follows:

G – 100 MVA; 13.8 kV;

X_1 = 0.15; X_2 = 0.17; X_0 = 0.05

T_1, T_2 – 120 MVA, 13.8 kV/138 kV Δ/Y; X = 0.12

M – 100 MVA; 13.8 kV; X_1 = 0.2 ;

X_2 = 0.21 ; X_0 = 0.1; X_n = 0.05

Line—X_1 = X_2 = 20 Ω; X_0 = 60 Ω. *[GATE 1992]*

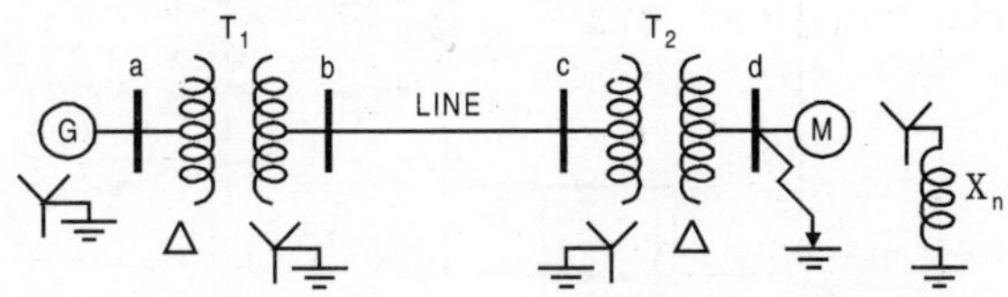

Fig. 5.19. ***Power System Network For Example 5.19***

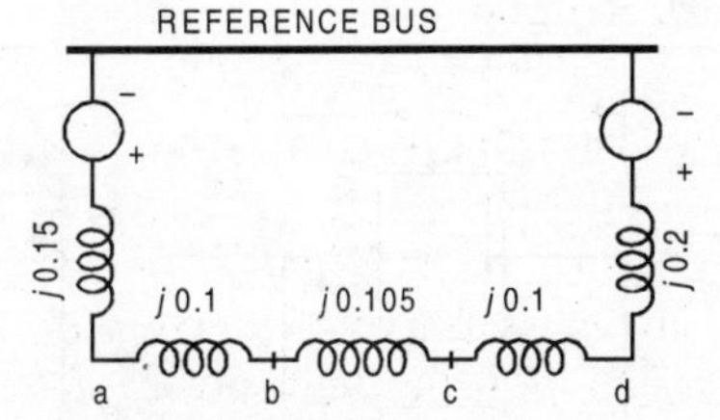

(a) Positive Sequence Network for Example 5.19

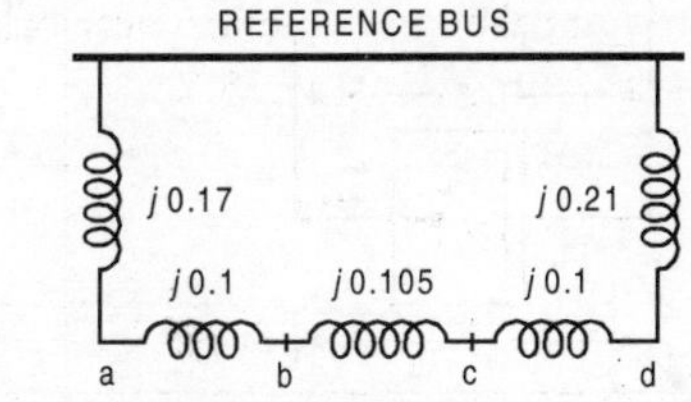

(b) Negative Sequence Network For Example 5.19

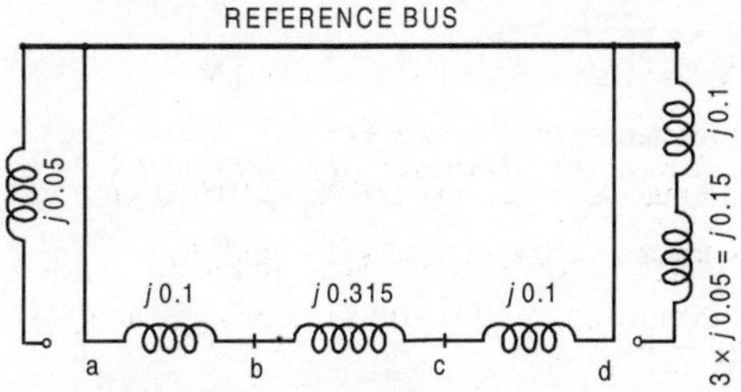

(c) Zero Sequence Network For Example 5.19

Fig. 5.20

Table 5.1

Type of Fault	Network Interconnection	General Equivalent Equations	Sequence Components of Currents at Fault Point	Sequence Components of Voltages Across Fault
L-G	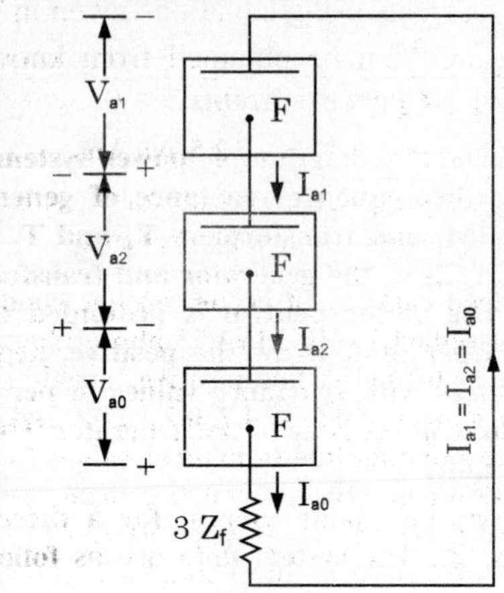	$\mathbf{I}_{a1} = \mathbf{I}_{a2} = \mathbf{I}_{a0}$ $\mathbf{V}_{a1} + \mathbf{V}_{a2} + \mathbf{V}_{a0} = \mathbf{V}_a = 3\,\mathbf{I}_{a1}\,\mathbf{Z}_f$ $\mathbf{I}_a = 3\,\mathbf{I}_{a1}$ $\mathbf{I}_b = 0$ $\mathbf{I}_c = 0$	$\mathbf{I}_{a1} = \dfrac{\mathbf{V}_f}{\mathbf{Z}_1 + \mathbf{Z}_2 + \mathbf{Z}_0 + 3\,\mathbf{Z}_f}$ $\mathbf{I}_{a2} = \mathbf{I}_{a1}$ $\mathbf{I}_{a0} = \mathbf{I}_{a1}$	$\mathbf{V}_{a1} = \mathbf{V}_f - \mathbf{I}_{a1}\,\mathbf{Z}_1$ $\mathbf{V}_{a2} = -\mathbf{I}_{a2}\,\mathbf{Z}_2$ $\mathbf{V}_{a0} = -\mathbf{I}_{a0}\,\mathbf{Z}_0$
L-L	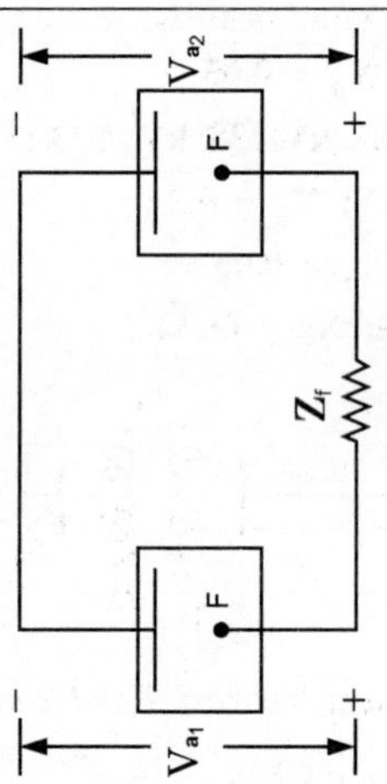	$\mathbf{I}_{a1} = -\,\mathbf{I}_{a2}$ $\mathbf{I}_{a0} = 0$ $\mathbf{I}_a = 0$ $\mathbf{I}_b = -\mathbf{I}_c = \mathbf{I}_f$ $\mathbf{V}_b - \mathbf{V}_c = \mathbf{I}_b\,\mathbf{Z}_f$	$\mathbf{I}_{a1} = \dfrac{\mathbf{V}_f}{\mathbf{Z}_1 + \mathbf{Z}_2 + \mathbf{Z}_f}$ $\mathbf{I}_{a2} = -\,\mathbf{I}_{a1}$ $\mathbf{I}_{a0} = 0$	$\mathbf{V}_{a1} = \mathbf{V}_f - \mathbf{I}_{a1}\,\mathbf{Z}_1$ $\mathbf{V}_{a2} = -\,\mathbf{I}_{a2}\,\mathbf{Z}_2$ $\mathbf{V}_{a0} = 0$ $\mathbf{V}_{a1} - \mathbf{V}_{a2} = \mathbf{I}_{a1}\,\mathbf{Z}_f$
L-L-G	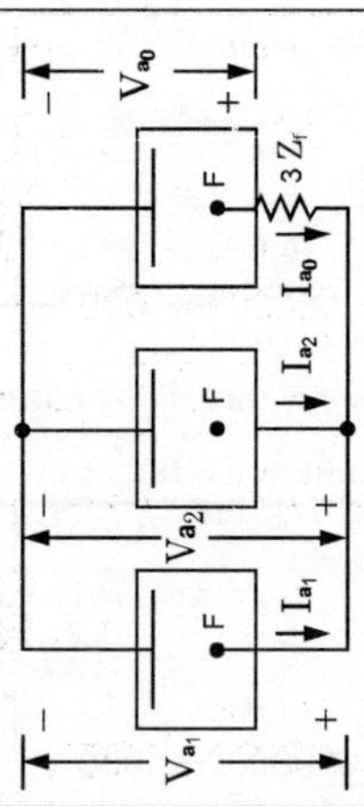	$\mathbf{I}_{a0} = 0$ $\mathbf{I}_f = \mathbf{I}_b + \mathbf{I}_c$ $\mathbf{V}_b = \mathbf{V}_c = \mathbf{I}_f\,\mathbf{Z}_f$ $\mathbf{I}_{a1} + \mathbf{I}_{a2} + \mathbf{I}_{a0} = 0$	$\mathbf{I}_{a1} = \dfrac{\mathbf{V}_f}{\mathbf{Z}_1 + \dfrac{\mathbf{Z}_2(\mathbf{Z}_0 + 3\mathbf{Z}_f)}{\mathbf{Z}_2 + \mathbf{Z}_0 + 3\mathbf{Z}_f}}$	$\mathbf{V}_{a1} = \mathbf{V}_f - \mathbf{I}_{a1}\,\mathbf{Z}_1$ $\mathbf{V}_{a2} = -\,\mathbf{I}_{a2}\,\mathbf{Z}_2 = \mathbf{V}_{a1}$ $\mathbf{V}_{a0} = \mathbf{V}_{a1} + 3\mathbf{I}_{a0}\,\mathbf{Z}_f$

Solution : Base MVA = 100 and Base kV = 13.8 in the zone of generator

Base kV for transmission side = 138 kV

Per unit reactances of generator G :

$X_1 = 0.15$ pu; $X_2 = 0.17$ pu; and $X_0 = 0.05$ pu since base is the same

Per unit reactances of transformers T_1 and T_2 :

$$X_1 = X_2 = X_0 = 0.12 \times \frac{100}{120} = 0.1 \text{ pu}$$

Per unit reactances of motor M :

$X_1 = 0.2$ pu; $X_2 = 0.21$ pu; $X_0 = 0.1$ pu and $X_n = 0.05$ pu

Per unit reactances of transmission line :

$$X_1 = X_2 = 20 \times \frac{\text{Base MVA}}{(\text{Base kV})^2} = \frac{20 \times 100}{(138)^2} = 0.105 \text{ pu}$$

$$X_0 = \frac{60 \times 100}{(138)^2} = 0.315 \text{ pu}$$

The positive, negative and zero sequence networks are given in Figs. 5.20 (*a*), 5.20 (*b*) and 5.20 (*c*) respectively,

When a fault occurs at bus *d*, the positive sequence diagram during the fault condition becomes, as given in Fig. 5.21.

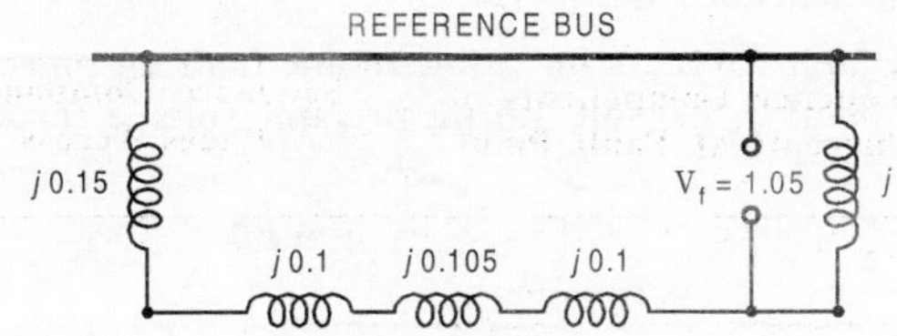

Fig. 5.21

Equivalent positive sequence impedance,

$$\mathbf{Z}_1 = j(0.15 + 0.1 + 0.105 + 0.1) \parallel j0.2$$

$$= \frac{j0.455 \times 0.2}{0.455 + 0.2} = j0.139$$

$$\text{Fault current, } \mathbf{I}_f = \frac{\mathbf{V}_f}{\mathbf{Z}_1} = \frac{1.05}{j0.139} = -j7.558 \text{ pu } \textbf{Ans.}$$

$\because$ V_f = 1.05 pu (given)

Example 5.20. A 3-phase star-connected generator supplies a star-connected inductive load through a transmission line. The star point of the load is grounded and the generator neutral is ungrounded. The load reactance is j 0.5 pu per phase and the line reactance is j 0.1 pu per phase. The positive, negative and zero sequence reactances of the generator are j 0.5, j 0.5 and j 0.05 pu respectively.

A single line-to-ground fault takes place in phase 'A' halfway down the line. Prior to fault the network is balanced and the voltage at the fault location is 1 $\angle 0°$ pu. Calculate the current through the fault path. *[GATE 1993]*

Solution : Sequence networks are given in Fig. 5.22.

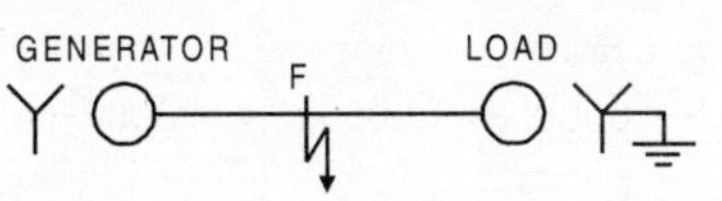

Fig. 5.22. ***Single Line Diagram of a Balanced 3-Phase System***

Equivalent positive sequence impedance,

$$\mathbf{Z}_1 = j(0.5 + 0.05) \parallel j(0.5 + 0.05)$$

$$= \frac{j(0.5 + 0.05) \times (0.5 + 0.05)}{(0.5 + 0.05) + (0.5 + 0.05)} = j0.275 \text{ pu}$$

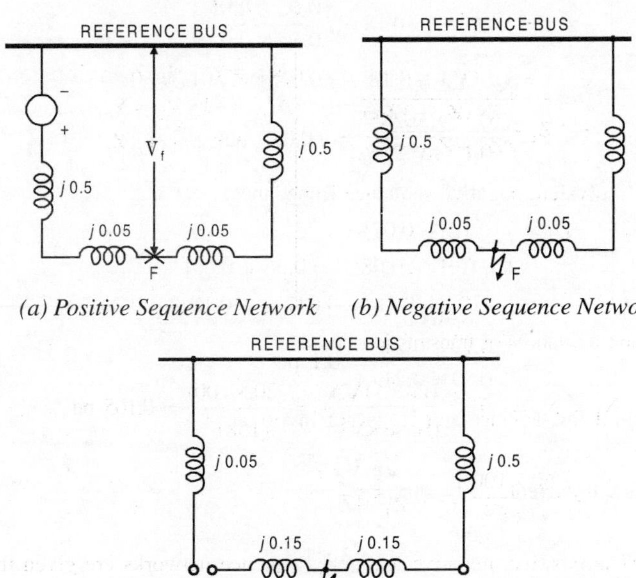

(a) Positive Sequence Network *(b) Negative Sequence Network*

(c) Zero Sequence Network

Fig. 5.23

Equivalent negative sequence impedance,

$$\mathbf{Z}_2 = \mathbf{Z}_1 = j0.275 \text{ pu}$$

Equivalent zero sequence impedance,

$$\mathbf{Z}_0 = j3 \times 0.05 + j0.5 = j0.65$$

$$\text{Fault current, } \mathbf{I}_f = \frac{3\mathbf{V}_f}{\mathbf{Z}_1 + \mathbf{Z}_2 + \mathbf{Z}_0} = \frac{3 \times (1 + j0)}{j(0.275 + 0.275 + 0.65)}$$

$$= -j2.5 \text{ pu } \textbf{Ans.}$$

Note: Normally the zero sequence impedance of transmission line is about 2 to 4 times the positive sequence impedance.

Example 5.21. A dead earth fault occurs on one conductor of a 3-conductor cable supplied by 10 MVA, 3-phase alternator with earthed neutral. The alternator has positive, negative and zero phase sequence impedances of 0.5 +j 4.7, 0.2 +j 0.6 and j 0.48 ohm per phase. The corresponding line-to-neutral values for the cable up to the fault position are (0.36 + j 0.25), (0.36 + j 0.25) and (2.9 + j 0.95) Ω. Find *(i)* the fault current *(ii)* the sequence component in each line, and *(iii)* the voltages of the sound lines to earth at the fault. The generator is excited to give 6,600 V between lines on open circuit.

Solution : Equivalent positive sequence impedance:

$$\mathbf{Z}_1 = \mathbf{Z}_{g1} + \mathbf{Z}_{l1}$$

$$= (0.5 + j4.7) + (0.36 + j0.25)$$

$$= (0.86 + j4.95)\ \Omega = 5.024 \angle 80.14°\ \Omega$$

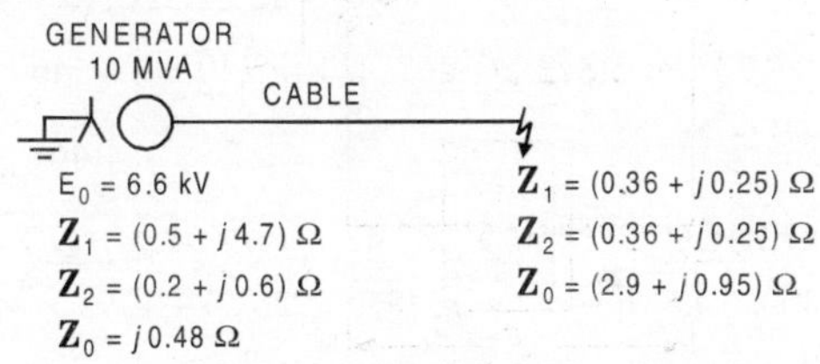

Fig. 5.24. ***Power System For Example 5.21***

Equivalent negative sequence impedance,

$$\mathbf{Z}_2 = \mathbf{Z}_{g2} + \mathbf{Z}_{l2}$$

$$= (0.2 + j0.6) + (0.36 + j0.25)$$

$$= (0.56 + j0.85)\ \Omega = 1.018 \angle 56.62°\ \Omega$$

$$\mathbf{Z}_0 = \mathbf{Z}_{g0} + \mathbf{Z}_{l0} = j0.48 + (2.9 + j0.95)$$

$$= (2.9 + j1.43)\ \Omega = 3.23 \angle 26.25°\ \Omega$$

(i) Fault current,

$$\mathbf{I}_f = 3\mathbf{I}_{a1} = \frac{3 \times \mathbf{V}_f}{\mathbf{Z}_1 + \mathbf{Z}_2 + \mathbf{Z}_0}$$

$$= \frac{3 \times 6{,}600/\sqrt{3}}{(0.86 + j4.95) + (0.56 + j0.85) + (2.9 + j1.43)}$$

$$= \frac{6{,}600\sqrt{3}}{4.32 + j7.23} = \frac{6{,}600\sqrt{3}}{8.422 \angle 59.14°}$$

$$= 1{,}357 \angle -59.14° \text{ A } \textbf{Ans.}$$

(ii) $\mathbf{I}_{a0} = \mathbf{I}_{a1} = \mathbf{I}_{a2}$

$$= \frac{\mathbf{I}_a}{3} = \frac{\mathbf{I}_f}{3} = \frac{1{,}357 \angle -59.14°}{3}$$

$$= 452.33 \angle -59.14° \text{ A } \textbf{Ans.}$$

$\mathbf{I}_a = 1{,}357 \angle -59.14°$ **Ans.** $\mathbf{I}_b = 0,\ \mathbf{I}_c = 0$

(*iii*) $\mathbf{V}_{a1} = \dfrac{6{,}600}{\sqrt{3}} - \mathbf{I}_{a1}\mathbf{Z}_1$

$= 3{,}810.5 - 452.33 \angle -59.14° \times 5.024 \angle 80.14°$

$= 3{,}810 - 2{,}272 \angle 21° \;=\; 1{,}686.35 - j813.6$ V

$= 1{,}872 \angle -26°$ V

$\mathbf{V}_{a2} = -\mathbf{I}_{a2}\,\mathbf{Z}_2 = -452.33 \angle -59.14° \times 1.018 \angle 56.62°$

$= -460 \angle -2.52° \;=\; (-459.55 + j\,20.23)$ V

$\mathbf{V}_{a0} = -\mathbf{I}_{a0}\mathbf{Z}_0 = -452.33 \angle -59.14° \times 3.23 \angle 26.25°$

$= -1{,}461 \angle -32.89° \;=\; (-1{,}226.80 + j793.37)$ V

$\mathbf{V}_a = \mathbf{V}_{a1} + \mathbf{V}_{a2} + \mathbf{V}_{a0}$

$= (1{,}686.35 - j813.6) + (-459.55 + j20.23)$

$+ (-1{,}226.80 + j793.37) = 0$ **Ans.**

$\mathbf{V}_b = a^2\mathbf{V}_{a1} + a\mathbf{V}_{a2} + \mathbf{V}_{a0}$

$= 1 \angle 240° \times 1{,}872 \angle -26° + 1 \angle 120° \times -460 \angle -2.52°$

$+ -1{,}461 \angle -32.89°$

$= 1{,}872 \angle 214° - 460 \angle 117.48° - 1{,}461 \angle -32.89°$

$= -1{,}552 - j1{,}047 + 212 - j\,408 - 1{,}226.80 + j\,793$

$= (-2{,}566.80 - j\,662)$ V or $2{,}650 \angle -165.5°$ V **Ans.**

$\mathbf{V}_c = 1 \angle 120° \times 1{,}872 \angle -26° + 1 \angle 240° \times -460 \angle -2.52°$

$- 1{,}461 \angle -32.89°$

$= 1{,}872 \angle 94° - 460 \angle 237.48° - 1{,}461 \angle -32.89°$

$= -130 + j\,1{,}867 + 247 + j\,388 - 1{,}226.80 + j\,793$

$= -1{,}109.8 + j\,3{,}048 = 3{,}243 \angle 110°$ V **Ans.**

Example 5.22. A single line diagram of a power network is shown in Fig. 5.25.

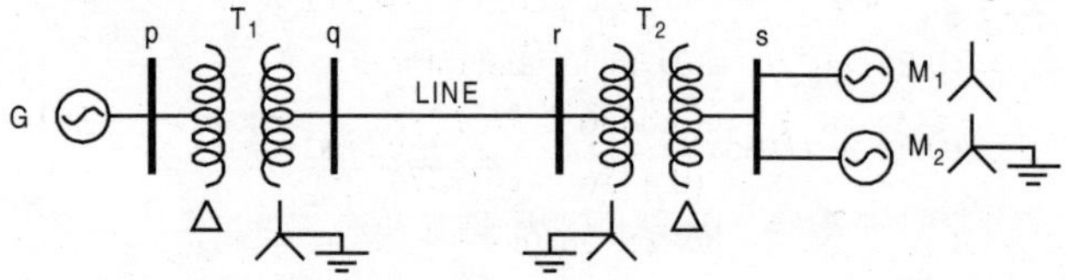

Fig. 5.25

The system data is given in the table below:

Element	Positive Sequence Reactance (PU)	Negative Sequence Reactance (PU)	Zero Sequence Reactance (PU)
Generator G	0.1	0.12	0.05
Motor M_1	0.05	0.06	0.025
Motor M_2	0.05	0.06	0.025
Transformer T_1	0.07	0.07	0.07
Transformer T_2	0.08	0.08	0.08
Line	0.10	0.10	0.10

Generator grounding reactance is 0.5 pu.

(*a*) **Draw sequence networks.**

(*b*) **Find fault currents for a line-to-line fault on phase *b* and *c* at point *q*. Assume 1.0 pu prefault voltage throughout.**

[*GATE 1996*]

Solution : Sequence networks are given in Fig. 5.26.

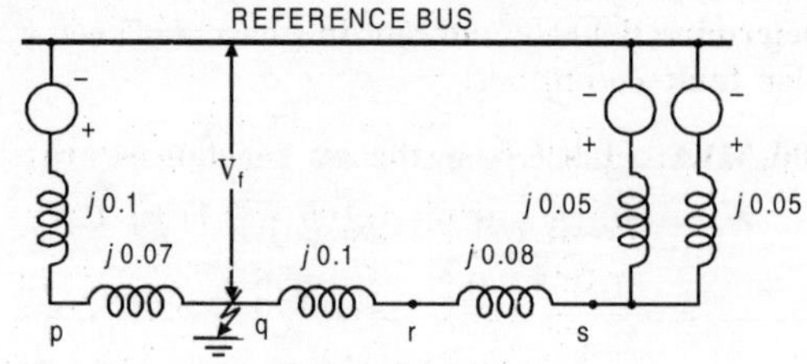

(a) Positive Sequence Network

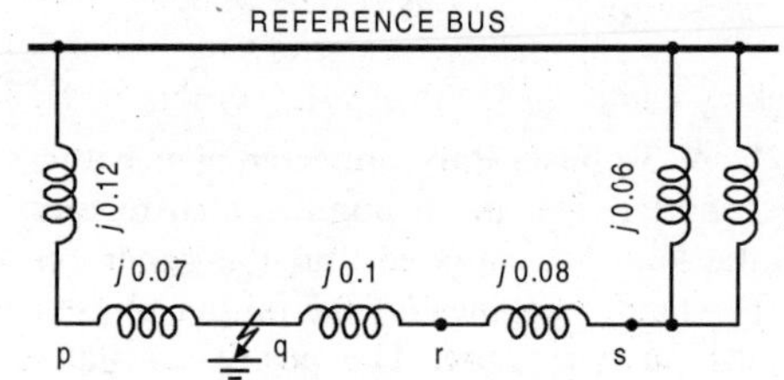

(b) Negative Sequence Network

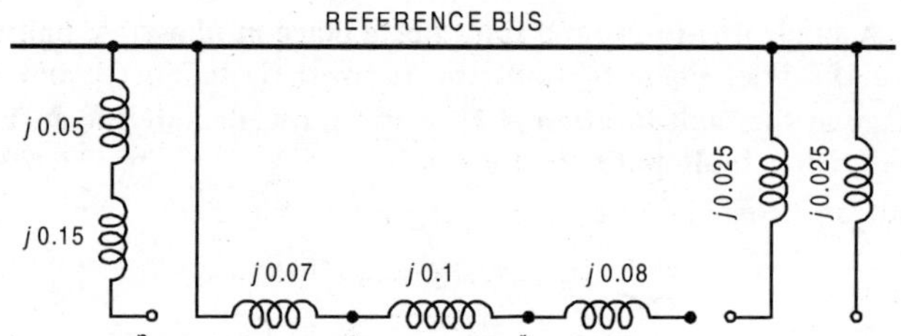

(c) Zero Sequence Network

Fig. 5.26

Equivalent positive sequence impedance,

$\mathbf{Z}_1 = j\,(0.1 + 0.07)\text{a}$

$\parallel [j\,(0.1 + 0.08) + j\,0.05 \parallel j\,0.05]$

$= j\,0.17 \parallel \left(j0.18 + j\dfrac{0.05 \times 0.05}{0.05 \times 0.05} \right)$

$= j\,0.17 \parallel (j\,0.18 + j\,0.025) = j\,017 \parallel j\,0.205$

$= j\,\dfrac{0.17 \times 0.205}{0.17 + 0.205} = j\,0.093$ pu

Equivalent negative sequence Impedance,

$\mathbf{Z}_2 = j\,(0.12 + 0.07)$

$\parallel [j\,(0.1 + 0.08) + j\,0.06 \parallel 0.06]$

$= j\,0.19 \parallel j\,(0.18 + 0.03) = \simeq 0.19 \parallel j\,0.21$

$= j\,\dfrac{0.19 \times 0.21}{0.19 + 0.21} = 0.1$ pu

In a line-to-line fault Z_0 is not involved.

Fault current, $\quad I_f = \dfrac{\sqrt{3}\,\mathbf{V}_f}{\mathbf{Z}_1 + \mathbf{Z}_2}$

$= \dfrac{\sqrt{3} \times 1}{j0.093 + j0.1}$

$= -j\,8.974$ pu **Ans.**

Example 5.23. A 3-phase, 100 MVA, 11 kV star-connected synchronous generator with solidly grounded neutral is supplying a 3-phase, 50 MVA, 11 kV synchronous motor with solidly grounded neutral through a short line. The motor is drawing 40 MW at 0.8 pf leading with a terminal voltage of 10.95 kV when a single line-to-ground fault occurs at the generator terminals of phase a. Determine the total current in phase *a* of generator and motor under fault conditions.

On 100 MVA, 11 kV base the pu reactances are

	Positive sequence	Negative sequence	Zero sequence
Generator	0.2	0.2	0.05
Motor	0.4	0.4	0.1
Line	0.05	0.05	0.15

Solution : Taking the motor terminal voltage as the reference phasor we have

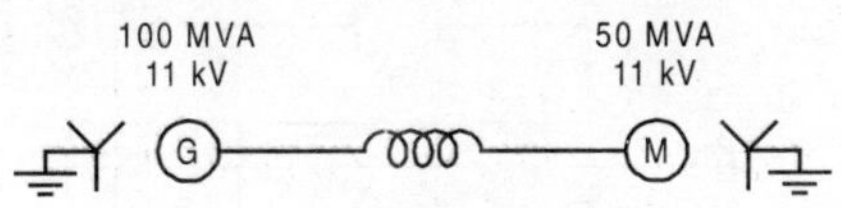

Fig. 5.27

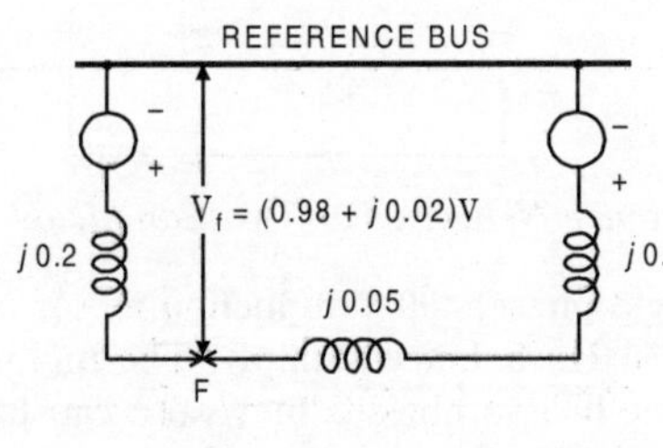

(a) Positive Sequence Network

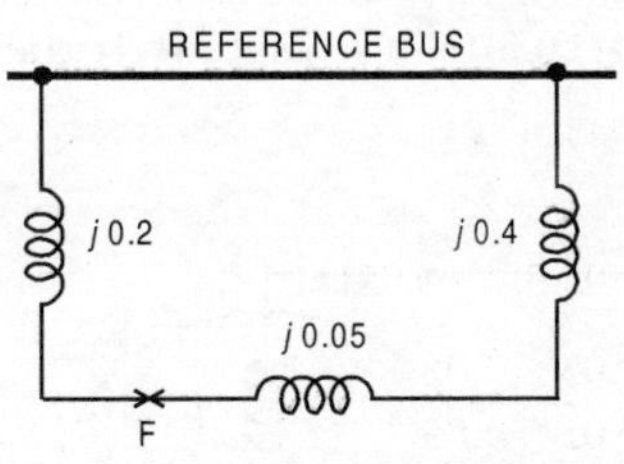

(b) Negative Sequence Network

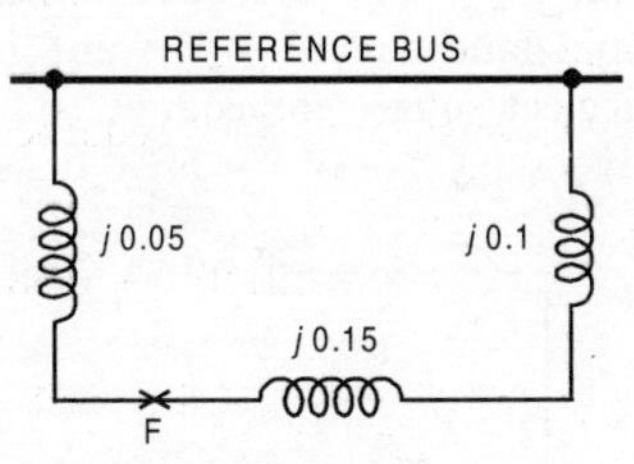

(c) Zero Sequence Network

Fig. 5.28

$$\text{Motor terminal voltage} = \frac{10.95}{11} = 0.9955 \angle 0^\circ \text{ pu}$$

$$\text{Base current, } \mathbf{I_B} = \frac{\text{Base MVA} \times 1{,}000}{\sqrt{3} \times \text{base kV}}$$

$$= \frac{100 \times 1{,}000}{\sqrt{3} \times 11} = 5{,}248 \text{ A}$$

$$\text{Load current, } \mathbf{I_L} = \frac{40 \times 10^3}{\sqrt{3} \times 10.95 \times 0.8} = 2{,}636 \text{ A}$$

$$= \frac{2{,}636}{5{,}248} \angle \cos^{-1} 0.8 = 0.5023 \angle 36.87^\circ$$

$$= (0.402 + j\,0.301) \text{ pu}$$

Prefault voltage at the terminals of the synchronous generator

$$\mathbf{V}_f = (0.9955 + j\,0) + \mathbf{I_L} \times \mathbf{Z_L}$$

$$= (0.9955 + j\,0) + 0.5023 \angle 36.87^\circ \times j\,0.05$$

$$= 0.98 + j\,0.02 \text{ pu}$$

Sequence networks are given in Fig. 5.28.

Equivalent positive sequence impedance,

$$\mathbf{Z}_1 = j\,0.2 \parallel (j\,0.05 + j\,0.4) = j\frac{0.2 \times 0.45}{0.2 + 0.45} = j\,0.1385 \text{ pu}$$

Equivalent negative sequence impedance,

$$\mathbf{Z}_2 = j\,0.2 \parallel (j\,0.05 + j\,0.4) = j\,0.1385 \text{ pu}$$

Equivalent zero sequence impedance,

$$\mathbf{Z}_0 = j\,0.05 \parallel (j\,0.15 + j\,0.1) = j\frac{0.05 \times 0.25}{0.05 + 0.25} = j\,0.0417 \text{ pu}$$

$$\mathbf{I}_{a0} = \mathbf{I}_{a1} = \mathbf{I}_{a2} = \frac{\mathbf{V}_f}{\mathbf{Z}_1 + \mathbf{Z}_2 + \mathbf{Z}_0}$$

$$= \frac{0.98 + j0.02}{j0.1385 + j0.1385 + j0.0417}$$

$$= \frac{0.98 + j0.02}{j0.3187} = (0.063 - j\,3.075) \text{ pu}$$

$$\mathbf{I}_{ag1} = \mathbf{I}_{a1} \times \frac{j0.05 + j0.4}{j0.2 + j0.05 + j0.4}$$

$$= (0.063 - j\,3.075) + \frac{j0.45}{j0.65} = (0.0436 - j\,2.1288) \text{ pu}$$

$$\mathbf{I}_{am1} = \mathbf{I}_{a1} \times \frac{j0.20}{j\,0.65} = (0.063 - j\,3.075) \times \frac{j0.20}{j\,0.65}$$

$$= (0.0194 - j\,0.9462) \text{ pu}$$

$$\mathbf{I}_{ag2} = \mathbf{I}_{ag1} = (0.0436 - j\,2.1288) \text{ pu}$$

$$\mathbf{I}_{am2} = \mathbf{I}_{am1} = (0.0194 - j\,0.9462) \text{ pu}$$

$$\mathbf{I}_{ag0} = \mathbf{I}_{a0} \times \frac{j0.15 + j0.1}{j\,0.05 + j0.15 + j0.1}$$

$$= (0.063 - j\,3.075) \times \frac{j0.25}{j0.30} = (0.0525 - 2.5625) \text{ pu}$$

$$\mathbf{I}_{am0} = \mathbf{I}_{a0} \times \frac{j0.05}{j0.3} = (0.063 - j\,3.075) \times \frac{j0.05}{j0.3}$$

$$= (0.0105 - j\,0.5125) \text{ pu}$$

Total positive sequence current in generator

$$= \mathbf{I}_{ag1} + \mathbf{I_L} = (0.0436 - j\,2.1288) + (0.402 + j\,0.301)$$

$$= (0.4456 - j\,1.8278) \text{ pu}$$

Total positive sequence current in motor

$$= \mathbf{I}_{am1} - \mathbf{I_L} = (0.0194 - j\,0.9462) - (0.402 + j\,0.301)$$

$$= (-0.3826 - j\,1.2472) \text{ pu}$$

Current in phase *a* of generator

$$= \mathbf{I}_{ag1} + \mathbf{I_L} + \mathbf{I}_{ag2} + \mathbf{I}_{ag0}$$

$$= (0.4456 - j\,1.8278) + (0.0436 - j\,2.1288) + (0.0525 - j\,2.5625)$$

$$= (0.5417 - j\,6.5191) \text{ pu}$$

$$\text{or } 6.542 \angle -85.25^\circ \text{ pu or } 6.542 \angle -85.25^\circ \times 5{,}248$$

$$= 34{,}332 \angle -85.25^\circ \text{ A}$$ **Ans.**

Current in phase *a* of motor

$$= \mathbf{I}_{am1} - \mathbf{I_L} + \mathbf{I}_{am2} + \mathbf{I}_{am0}$$

$$= (-0.3826 - j\,1.2472) + (0.0194 - j\,0.9462) + (0.0105 - j\,0.5125)$$

$$= (-0.3527 - j\,2.7059) \text{ pu or } 2.73 \angle 277.43^\circ \text{ pu}$$ **Ans.**

$$= 2.73 \angle 277.43^\circ \times 5{,}248 = 14{,}327 \angle 277.43^\circ \text{ A}$$ **Ans.**

5.4. SERIES OR OPEN CONDUCTOR FAULTS

The faults, which we have discussed so far, involve power conductor or conductors-to-ground or short circuit between conductors and are termed as *shunt faults.* Such faults are characterized by increase in current and fall in voltage and frequency. On the other hand, a plain break in one or two conductors in a 3-phase circuit leads to an *open-conductor* or *series fault.* When fuses are used in 3-phase circuits, the fuses in one or two phases may blow off due to overload, the fault incurred is known as open–conductor fault. Series faults are characterized by increase in voltage and frequency and fall in current in the faulted phases.

The series faults are classified as (*i*) one conductor open and (*ii*) two conductor open faults. Such faults also disturb the symmetry in one or two phases and are, therefore, unbalanced faults. The unbalanced conditions so developed can be analysed by the method of symmetrical components.

(i) One Conductor Open Fault. A 3-phase line with conductor *a* broken leading to one conductor open condition is shown in Fig. 5.29. The terminal conditions at the fault point give following equations

$$\mathbf{I}_a = 0 \qquad \text{...(5.30)}$$

$$\mathbf{V}_{bb}' = \mathbf{V}_{cc}' = 0 \qquad \text{...(5.31)}$$

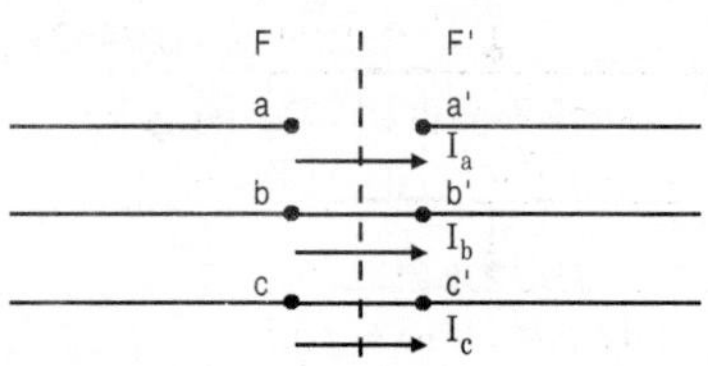

Fig. 5.29. ***One Conductor Open Fault***

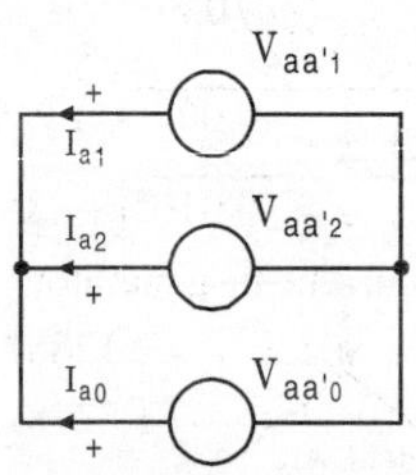

Fig. 5.30. ***Sequence Network For One Conductor Open Fault***

In terms of symmetrical components these conditions can be represented as

$$\mathbf{I}_{a1} + \mathbf{I}_{a2} + \mathbf{I}_{a0} = \mathbf{I}_a = 0 \qquad \text{...(5.32)}$$

$$\text{or } \mathbf{I}_{a1} = -(\mathbf{I}_{a2} + \mathbf{I}_{a0})$$

$$\text{and } \mathbf{V}_{aa1}' = \mathbf{V}_{aa2}' = \mathbf{V}_{aa0} = \frac{1}{3}\mathbf{V}_{aa}' \qquad \text{...(5.33)}$$

The above Eqs. (5.32) and (5.33) suggest a parallel connection of sequence networks, as shown in Fig. 5.30.

(ii) Two Conductors Open Fault. A 3-phase line with conductors *b* and *c* broken leading to two conductors open condition is shown in Fig. 5.31. The terminal conditions at the fault point give the following equations.

$$\mathbf{V}'_{aa} = 0 \qquad \text{...(5.34)}$$

$$\mathbf{I}_b = \mathbf{I}_c = 0 \qquad \text{...(5.35)}$$

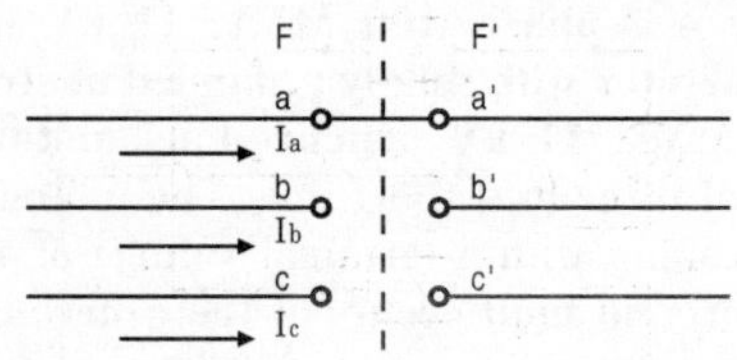

Fig. 5.31. ***Two Conductors Open Fault***

In terms of symmetrical components these conditions can be represented as

$$\mathbf{I}'_{aa1} = \mathbf{I}'_{aa2} = \mathbf{I}'_{aa0} = \frac{\mathbf{I}_a}{3} \qquad \text{...(5.36)}$$

$$\mathbf{V}'_{aa} + \mathbf{V}'_{aa2} + \mathbf{V}'_{aa0} = 0 \qquad \text{...(5.37)}$$

The above Eqs. (5.36) and (5.37) suggest a series connection of sequence networks, as shown in Fig. 5.32.

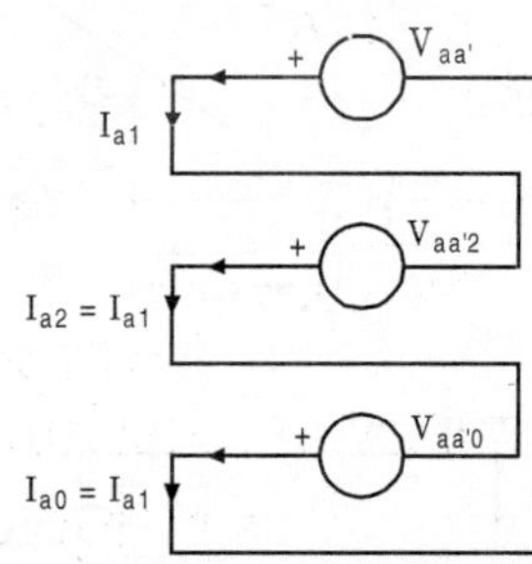

Fig. 5.32. ***Sequence Network For Two Conductors Open Fault***

Example 5.24. A 3-phase, 400 V induction motor is operating on 3-phase, 400-V, 50 Hz balanced supply. The motor continues to run even when the fuse in phase *a* blows out causing open circuit in it. Determine the line currents and voltages $\mathbf{V}_{a'a'}$, $\mathbf{V}_{an}$, $\mathbf{V}_{bn}$, $\mathbf{V}_{cn}$ and $\mathbf{V}_{Nn}$ (Fig. 5.33). Assume positive and negative sequence impedances as (3 + *j* 1) and (0.1 + *j* 0.5) ohms respectively and neglect source impedance.

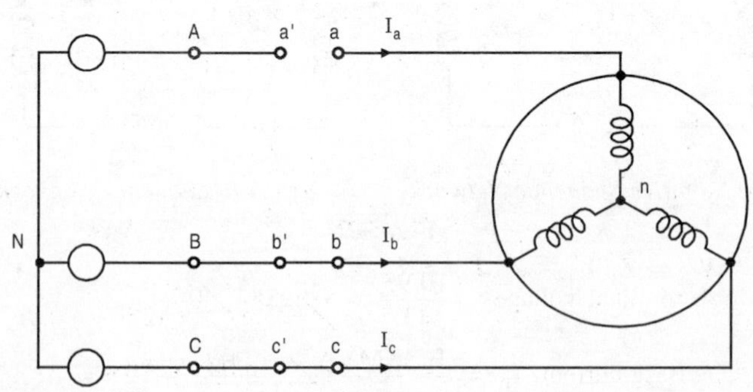

Fig. 5.33

Solution : Sequence network connections are shown in Fig. 5.34. The zero sequence network is not connected because of absence of neutral wire.

From Eqs. (5.32) and (5.33) we have

$$\mathbf{I}_{a2} = -\mathbf{I}_{a1} \qquad \text{...}(i)$$

$$\text{and } \mathbf{V}_{a'a1} = \mathbf{V}_{a'a2} = \mathbf{V}_{a'a0} \qquad \text{...}(ii)$$

Applying Kirchhoff's second law to positive and negative sequence networks, we have

$$\mathbf{V}_{a'a1} = \mathbf{E} - \mathbf{I}_{a1}\,\mathbf{Z}_1 \qquad \text{...}(iii)$$

$$\text{and } \mathbf{V}_{a'a2} = -\mathbf{I}_{a2}\,\mathbf{Z}_2 \qquad \text{...}(iv)$$

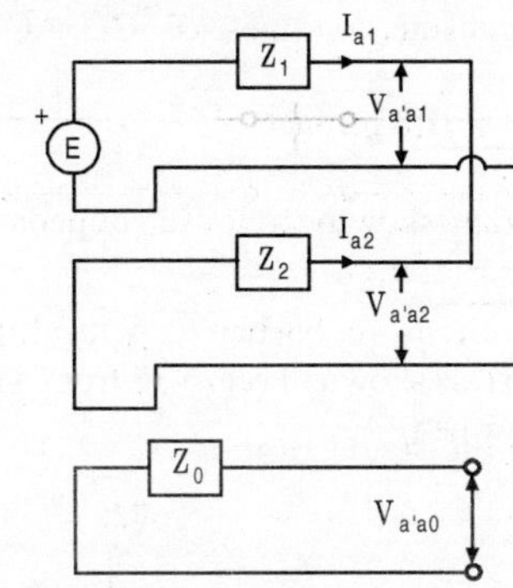

Fig. 5.34. ***Sequence Networks***

Substituting $\mathbf{I}_{a2} = -\mathbf{I}_{a1}$ from Eq. (*i*) and $\mathbf{V}_{a'a1} = \mathbf{V}_{a'a2}$ from Eq. (*ii*) in Eq. (*iv*), we have

$$\mathbf{V}_{a'a1} = \mathbf{I}_{a1}\,\mathbf{Z}_2 \qquad ...(v)$$

Substituting $\mathbf{V}_{a'a1} = \mathbf{I}_{a1}\,\mathbf{Z}_2$ in Eq. (*iii*), we have

$$\mathbf{I}_{a1}\mathbf{Z}_2 = \mathbf{E} - \mathbf{I}_{a1}\mathbf{Z}_1$$

or $\mathbf{I}_{a1}(\mathbf{Z}_1 + \mathbf{Z}_2) = \mathbf{E}$

$$\text{So } \mathbf{I}_{a1} = \frac{\mathbf{E}}{Z_1 + Z_2} = \frac{\frac{400}{\sqrt{3}}\angle 0^\circ}{(3 + j1) + (0.1 + j0.5)}$$

$$= \frac{230.94\angle 0^\circ}{3.1 + j1.5}$$

$$= 67.06\angle -25.82^\circ \text{ A}$$

$$\mathbf{I}_{a2} = -\mathbf{I}_{a1} = -67.06\angle -25.82^\circ \text{ A}$$

$$\mathbf{I}_a = \mathbf{I}_{a1} + \mathbf{I}_{a2} = 67.6\angle -25.82^\circ - 67.06\angle -25.82^\circ$$

$$= 0 \textbf{ Ans.}$$

$$\mathbf{I}_b = a^2\mathbf{I}_{a1} + a\mathbf{I}_{a2} + \mathbf{I}_{a0} = a^2\mathbf{I}_{a1} - a\,\mathbf{I}_{a1} + 0$$

$$= \mathbf{I}_{a1}(a^2 - a) = 67.06\angle -25.82^\circ \times \sqrt{3}\angle -90^\circ$$

$$= 116.15\angle -115.82^\circ \text{ A } \textbf{Ans.}$$

$$\mathbf{I}_c = -\mathbf{I}_b = -116.15\angle -115.82^\circ = 116.15\angle 64.18^\circ \text{ A}$$

$$\mathbf{V}_{a'a_2} = -\mathbf{I}_{a2}\,\mathbf{Z}_2 = 67.06\angle -25.82^\circ \times (0.1 + j\,0.5)$$

$$= 67.06\angle -25.82^\circ \times 0.51\angle 78.69^\circ \text{ A}$$

$$= 34.2\angle 52.87^\circ \textbf{ Ans.}$$

$$\mathbf{V}_{a'a} = \mathbf{V}_{a'a_1} + \mathbf{V}_{a'a_2} + \mathbf{V}_{a'a_0} = 3\mathbf{V}_{a'a_2}$$

$$= 3 \times 34.2\angle 52.87^\circ = 102.6\angle 52.87^\circ \text{ V } \textbf{Ans.}$$

$$\mathbf{V}_{an} = \mathbf{Z}_1\,\mathbf{I}_{a1} + \mathbf{Z}_2\,\mathbf{I}_{a2} = \mathbf{I}_{a1}(\mathbf{Z}_1 - \mathbf{Z}_2)$$

$$= 67.06\angle -25.87^\circ \times (2.9 + j\,0.5)$$

$$= 197.34\angle -16.04^\circ \text{ V } \textbf{Ans.}$$

$$\mathbf{V}_{bn} = a^2\mathbf{Z}_1\,\mathbf{I}_{a1} + a\,\mathbf{Z}_2\,\mathbf{I}_{a2} = (a^2\mathbf{Z}_1 - a\,\mathbf{Z}_2)\,\mathbf{I}_{a1}$$

$$= [(3 + j\,1)(-0.5 - 0.866) - (0.1 + j\,0.5)(-0.5 + j\,0.866)]\,\mathbf{I}_{a1}$$

$$= 67.06\angle -25.82^\circ\,[3.1623\angle 18.435^\circ \times 1\angle 240^\circ - 0.51\angle 78.69^\circ \times 1\angle 120^\circ]$$

$$= 67.06\angle -25.82^\circ \times 3.447\angle -108.9^\circ$$

$$= 231.2\angle -133.02^\circ \text{ V } \textbf{Ans.}$$

$$\mathbf{V}_{cn} = a\mathbf{Z}_1\mathbf{I}_{a1} + a^2\mathbf{Z}_2\mathbf{I}a_2 = \mathbf{I}_{a1}(a\mathbf{Z}_1 - a^2\mathbf{Z}_2)$$

$$= 67.06\angle -25.82^\circ \times 3.44\angle 145.91^\circ$$

$$= 230.7\angle 125.09^\circ \text{ V } \textbf{Ans.}$$

$$\mathbf{V}_{Nn} = \mathbf{V}_{a'a0} = \mathbf{V}_{a'a2} = 34.2\angle 52.87^\circ \text{ V } \textbf{Ans.}$$

5.5. COMPUTER METHOD FOR SHORT-CIRCUIT CALCULATIONS

Digital computer may be defined as a device that can compute by arithmetic processes. The basic processes used are addition and subtraction. These processes are also used in successive approximation or interaction to obtain function values, integration, solution of algebraic equations and linear and nonlinear differential equations. The input data describing the system is specified using the names of power stations, names of substations, data of system components, voltage levels, reactances etc.

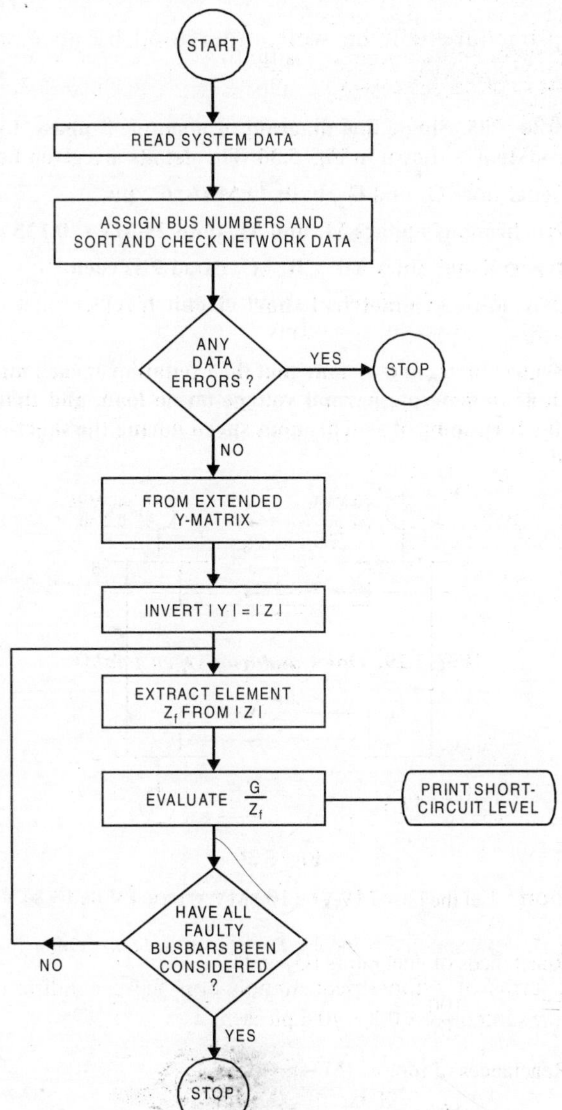

Fig. 5.35. ***Flow Chart Diagram For Digital Computer Short-Circuit Calculations***

Since the short-circuit level is normally required at all busbars it is imperative to use the system neutral as reference. Source in-feeds are represented with their equivalent impedances connected between the source bus-bar and the system neutral. An admittance matrix, extended to include the source admittances, is formed, and the computation process reduces the extended network to a single impedance connected between the neutral (zero node) and the point of fault. During this phase sequential bus numbers are assigned, network data is rearranged to facilitate formation of admittance matrices and excessive data checks are performed. Flow chart diagram for digital computer short-circuit calculations is shown in Fig. 5.35. If the faulty bus-bar voltage is assumed to be 1 pu, then as previously discussed the pu short-circuit MVA is given by $\frac{1}{Z_f}$ where Z_f has been appropriately extracted from the inverse of | Z | of the extended admittance matrix | Y |. The actual MVA is given by $\frac{\text{Base MVA}}{Z_{f\,pu}}$.

The procedure will be well understood by an example (Example 5.25).

Example 5.25. Single line diagram of a simple 3-phase 13.8 kV power system is shown in Fig. 5.36. The details are given below :

Generators G_1 and G_2, both 40 MVA, 0.2 pu

Synchronous motors M_1 and M_2 both 25 MVA, 0.125 pu

Transmission lines AB, CD, AC, BD, 1.9 Ω each.

Determine symmetrical short-circuit level at each of the busbars.

Neglect losses and assume that the excitation of each machine is such as to generate normal voltage on no load, and that each machine is running at synchronous speed during the short-circuit period.

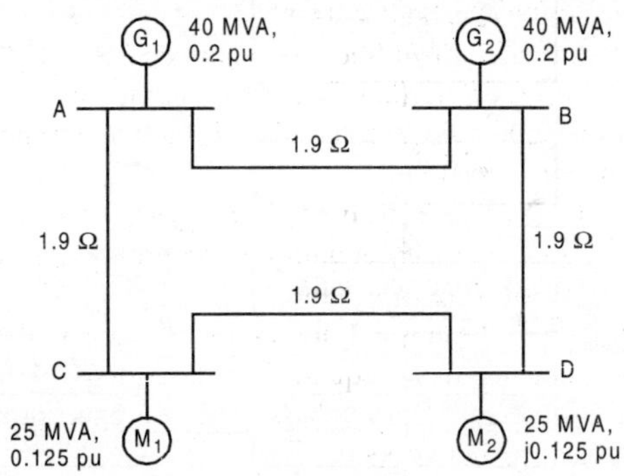

Fig. 5.36

Solution : Let the base MVA be 100 MVA, base kV be 13.8 kV (line-to-line)

Reactances of generators (G_1 and G_2)

$$= \frac{100}{40} \times 0.2 = j0.5 \text{ pu each}$$

Reactances of motors (M_1 and M_2)

$$= \frac{100}{25} \times 0.125 = j0.5 \text{ pu each}$$

Impedances of transmission lines AB, CD, AC and BD

$$= j\frac{1.9 \times 100}{13.8^2} = j1.0 \text{ pu each}$$

Admittances of generators (G_1 and G_2)

$$= \frac{1}{j0.5} = -j2.0 \text{ pu each}$$

Admittances of motors (M_1 and M_2)

$$= \frac{1}{j0.5} = -j2.0 \text{ pu each}$$

Admittances of transmission lines AB, CD, AC and BD

$$= \frac{1}{j1.0} = -j1.0 \text{ pu each}$$

Single line diagram showing relevant impedances is shown in Fig. 5.37.

Consider a short circuit on busbar A. A resulting network to be solved for nodal analysis is shown in Fig. 5.38 from which the extended admittance matrix is formed.

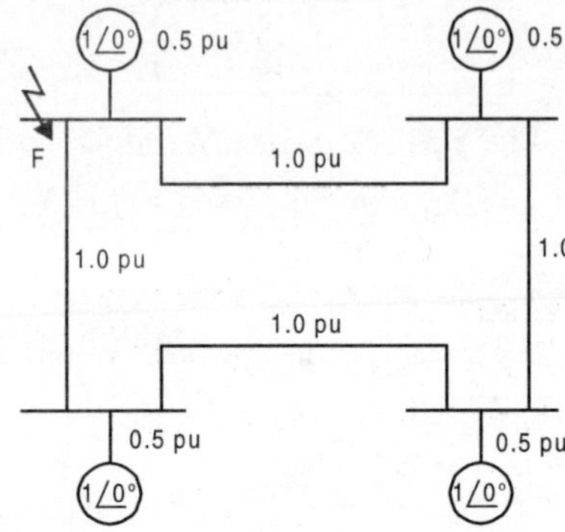

Fig. 5.37. ***Single Line Diagram Indicating Relevant Impedances***

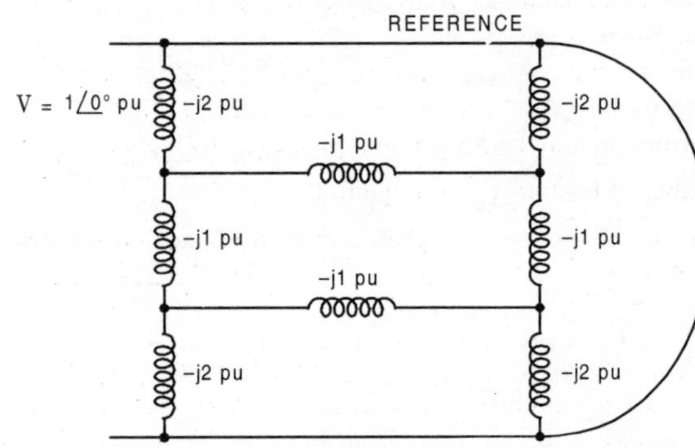

Fig. 5.38. ***Fault Circuit Indicating Admittances***

The standard matrix representation for nodal analysis becomes of the form

$$\begin{bmatrix} I_A \\ I_B \\ I_C \\ I_D \end{bmatrix} = \begin{bmatrix} y_{AA} & -y_{BA} & -y_{CA} & -y_{DA} \\ -y_{AB} & y_{BB} & -y_{CB} & -y_{DB} \\ -y_{AC} & -y_{BC} & y_{CC} & -y_{DC} \\ -y_{AD} & -y_{BD} & -y_{CD} & y_{DD} \end{bmatrix} \begin{bmatrix} V_A \\ V_B \\ V_C \\ V_D \end{bmatrix} \quad \text{...(5.38)}$$

When applying such analysis to a particular network, unconnected members will be zero inside the admittance matrix :

$$\begin{bmatrix} I_A \\ I_B \\ I_C \\ I_D \end{bmatrix} = \begin{bmatrix} -j(2+1+1) & j1 & j1 & 0 \\ j1 & -j(2+1+1) & 0 & j1 \\ j1 & 0 & -j(2+1+1) & j1 \\ 0 & j1 & j1 & -j(2+1+1) \end{bmatrix} \begin{bmatrix} V_A \\ V_B \\ V_C \\ V_D \end{bmatrix} \quad \text{...(5.39)}$$

Now, at all nodes except the faulty node, the injected current is zero.

Inversion of the above admittance matrix provides

$$\begin{bmatrix} V_A \\ V_B \\ V_C \\ V_D \end{bmatrix} = \begin{bmatrix} j0.291 & j0.082 & j0.082 & j0.042 \\ j0.082 & j0.291 & j0.042 & j0.082 \\ j0.082 & j0.042 & j0.291 & j0.082 \\ j0.042 & j0.082 & j0.082 & j0.291 \end{bmatrix} \begin{bmatrix} I \\ 0 \\ 0 \\ 0 \end{bmatrix} \quad \text{...(5.40)}$$

Extracting V_A from above equation, we have

$$V_A = 0.291I$$

or $I = \frac{1}{0.291}$ if $V_A = 1\angle 0°$ pu

i.e. the system has effectively been reduced to a single impedance between the zero node and the point of fault as shown in Fig. 5.39.

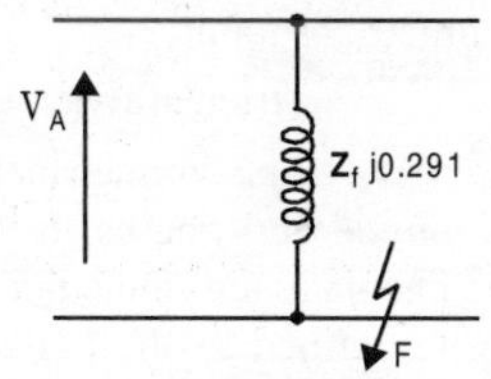

Fig. 5.39

In this example, the symmetry of the values was intentionally chosen so as to make the short-circuit level at each busbar the same.

$$\text{Short circuit MVA} = \frac{1}{0.291} = 3.436 \text{ pu}$$

or $3.436 \times 100 = 343.6$ MVA **Ans.**

Provision of Data : It will be appreciated from Fig. 5.35 that data must be compiled in tabular form in the order in which the information is needed. Busbars should be numbered as 1, 2, 3 etc. with a zero designation for the neutral. Lines may be referred to in terms of the connecting busbars A-B where A-B may be read as 1-2, 2-3, 4-5 etc. Data information may be drawn up typically as follows :

Number of busbars (except neutral), N = ...

Total number of connecting line admittances and generator infeeds, M = ...

Based MVA, S = ...

A programme may be written which reads in the data provided and makes use of a standard subroutine for matrix inversion. Print instructions may be given to print out the short-circuit level at all the busbars. If, for any reason, a short-circuit level were required part way along a line this could be arranged by introducing a fictitious node at that point.

Exercises

1. Obtain positive, negative and zero sequence networks of an unloaded generator having its neutral grounded through impedances Z_n.

   [U.P. Technical Univ. Power System Analysis 2004-05]

2. A 3-phase synchronous generator with its neutral solidly grounded and operating at no load develops an L-G fault in one of the phases having fault impedance Z_f. Derive expressions for the fault currents and the line-to-ground voltages at the location of the fault at all the phases.

3. Derive an expression for fault current for single line-to-ground fault by symmetrical component method.

   [U.P. Technical Univ. Power System Analysis 2009-10]

4. Develop the connection diagram of sequence networks for a single line-to-ground (L-G) fault. Also find the expression for fault current in terms of sequence currents in the fault conditions.

   [G.B. Technical Univ. Power System Analysis 2010-11]

5. Derive the sequence network of an unloaded generator for a single line-to-ground fault through fault impedance Z_f.

   [Rajasthan Univ. Power System Analysis 2000, 2004, 2007]

6. Derive the necessary equation to determine the fault current for L-L fault. Draw a diagram showing the interconnection of sequence networks.

   [Pb. Technical Univ. Computer Applications To Power System Analysis, May 2010]

7. What is relationship between V_{a1} and V_{a2} for solid short circuit between lines *b* and *c* ? Does this relationship remain valid if fault is through impedance Z_f. If yes, justify. If not, determine new relationship.

   [U.P. Technical Univ. Power System Analysis 2004-05]

8. Derive the expression for the symmetrical components of fault current of a power system for L-L fault through impedance. How symmetrical networks are connected to have equivalent circuit ?

   [U.P. Technical Univ. Power System Analysis 2005-06]

9. Show that the three symmetrical component sequence-networks are connected in :

   (*i*) series for a single line-to-ground fault. (*ii*) parallel for a bold double line-to-ground fault.

   [U.P.S.C. I.E.S. Elec. Engineering II 1998]

10. A double line-to-ground fault occurs on phase *b* and *c* of an unloaded generator. Derive a sequence network representation of this condition and determine the fault current.

11. A double line fault occurs between the two phases of a transmission line with an impedance Z_f between them. Show how the sequence-networks may be connected at the fault point to represent this type of fault. Derive the formula for determining the fault current.

12. Draw connection of sequence networks for L-L-G fault at the terminals of an unloaded generator. Show that this network is identical to that for L-L fault in absence of neutral grounded.

    [U.P. Technical Univ. Power System Analysis 2004-05]

13. Derive the necessary equation to determine the fault current for a double line-to-ground fault.

    [Rajasthan Technical Univ. Power System Analysis 2010-11]

14. Derive the necessary equation to determine the fault current for a L-L-G fault. Draw a diagram showing the interconnection of sequence networks.

    [Pb. Technical Univ. Computer Applications To Power System Analysis, December 2010]

15. Why is 3-ϕ symmetrical fault more severe than a 3-ϕ unsymmetrical fault ?

    [Rajasthan Univ. Power System Analysis 2001, 2003, 2008]

16. Give reasons : For a fault at alternator terminals, a single line-to-ground fault is more severe than a 3-ϕ fault.

    [Rajasthan Univ. Power System Analysis 2001, 2002]

17. Discuss the assumptions made for short-circuit analysis of a power system. Deduce the expression for system impedance matrix in bus frame of reference (Z_{BUS}) using singular transformation.

    Discuss the representation of a 3-phase power system for (taking the assumptions into account) short circuit studies.

    [U.P. Technical Univ. Power System Analysis 2006-07]

18. Discuss the different kinds of faults occurring under unloaded conditions of synchronous generator. Explain the interpretation of the Z_{bus} matrix. Discuss step-by-step Z_{bus} formulations.

    [U.P. Technical Univ. Power System Analysis 2008-09]

19. Describe in detail with flow chart diagram, the computational method for short circuit calculations.

    [U.P. Technical Univ. Power System Analysis 2009-10]

20. Write short notes on :

(*i*) Analysis of single line-to-ground fault.

(*ii*) Analysis of double line-to-ground fault.

(*iii*) Formation of Z_{bus} using singular transformation.

(*iv*) Computer method for SC calculations.

Short Answer Type Questions With Answers

Q. 1. Draw the sequence network for single line-to-ground fault without fault impedance. **[Madras Univ. October 1996]**

Ans.

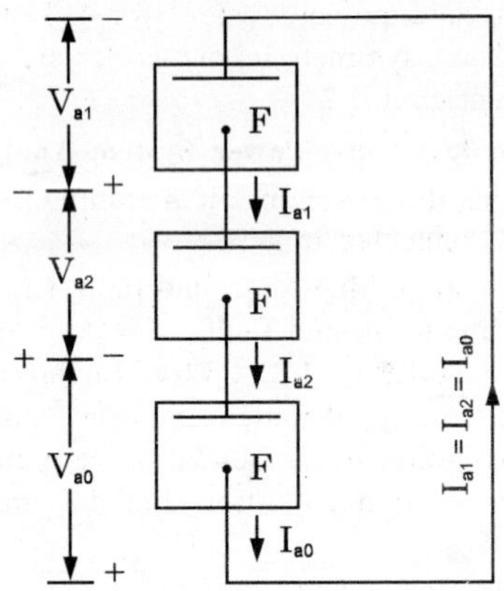

Fig. 1

Q. 2. How are sequence networks connected in case of L-G fault ? **[Bharathiyar Univ. November 1996]**

Ans. Three symmetrical component sequence-networks are connected in series for L-G fault as shown in Fig. for Q. 1.

Q. 3. Draw the sequence network connections corresponding to L-L fault at a bus. **[Bharathiyar Univ. November 1996]**

Ans.

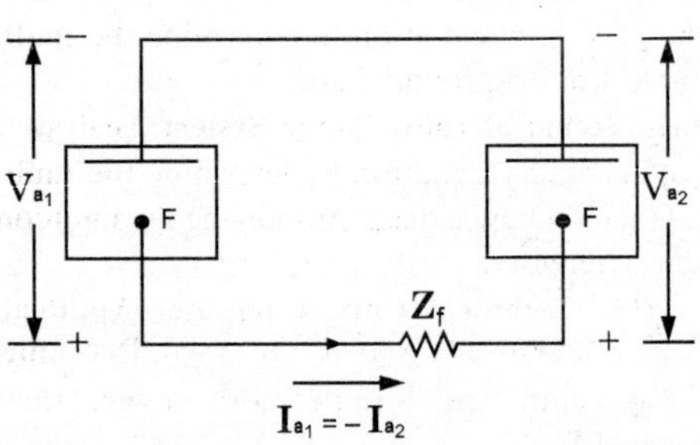

Fig. 2

Q. 4. How are sequence networks interconnected in the case of L-L-G fault ? **[Bharathiyar Univ. April 1997]**

Ans. Three symmetrical component sequence networks are connected in parallel for a bold L-L-G fault, as shown in the fig. 3

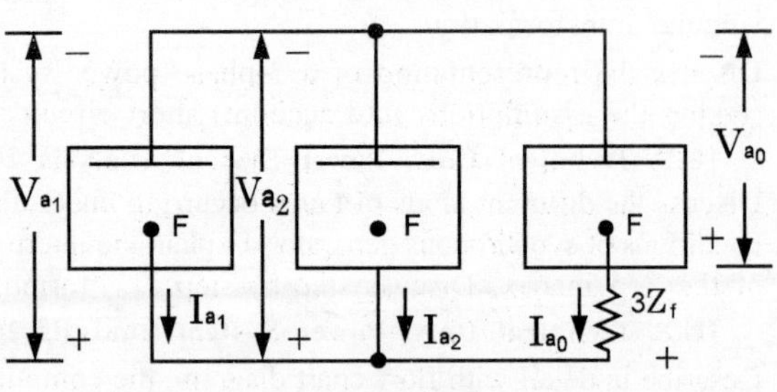

Fig. 3

Q. 5. What are the different types of faults which occur in a power system ? **[Madurai Kamaraj Univ. November 1996]**

Ans. The most common and dangerous faults that occur in a power system are the short circuit or shunt faults such as L-G, L-L, L-L-G, L-L-L and L-L-L-G faults. The other type of faults that occur in a power system are series type such as one conductor open fault and two conductor open fault.

Q. 6. What are unsymmetrical faults ? **[Bharathiyar Univ. November 1996]**

Ans. The faults giving rise to unsymmetrical currents *i.e.*, currents differing in magnitude and phases in the three phases of the power system are known as unsymmetrical faults. Unsymmetrical faults may also be defined as the faults involving one or two phases (not involving 3 phases) such as L-G, L-L, L-L-G faults.

Q. 7. Distinguish between symmetrical and unsymmetrical faults. **[Rajasthan Technical Univ. 2011]**

Ans. A fault involving all the three phases is known as symmetrical (balanced) fault while one involving only one or two phases is known as unsymmetrical fault. Single line-to-ground (L-G), line-to-line (L-L), double line-to-ground (L-L-G) faults are unsymmetrical (unbalanced) faults. Majority of the faults are unsymmetrical ones.

Q. 8. For a fault at a given location rank the various faults in the order of severity. **[Madras Univ. October 1996]**

Ans. The faults in increasing order of severity are L-G, L-L, L-L-G and symmetrical three-phase faults (L-L-L or L-L-L-G).

Q. 9. When a synchronous machine is suddenly short circuited, what is the maximum possible instantaneous current, in terms of symmetrical short circuit current ? **[Madras Univ. October 1996]**

Ans. $i_{fp} = 1.8\sqrt{2}\ I_{pc}$ Refer to Eq. (1.7).

Q. 10. Name the faults which do not have zero sequence currents flowing. **[Madras Univ. October 1996]**

Ans. In line-to-line faults and symmetrical faults (L-L-L or L-L-L-G faults) no zero sequence current flows.

Q. 11. Name the fault in which all the three sequence currents are present and are equal to each other. **[Madras Univ. October 1996]**

Ans. In L-G fault all the three sequence currents are present and are equal to each other.

Q. 12. Which of the faults would cause a greater electromagnetic interference between the power line and a nearby communication line ? **[Bharathidasan Univ. November 1996]**

Ans. Single phase-to-ground (L-G) fault, due to heavy unbalancing of line currents, causes greater electromagnetic interference between the power line and a nearby communication line.

Problems

1. A 15 MVA, 11 kV, 3-phase synchronous generator with solidly grounded neutral has a subtransient reactance of 25%. The negative and zero sequence reactances are 35% and 10% respectively. Determine the subtransient current in the generator and line-to-line voltages when a single line-to-ground fault occurs at the terminals of an unloaded generator operating at rated voltage. Resistance may be neglected. **[Ans.** 3,377 $\angle 270°$ A; 6.41 $\angle 77.7°$ kV; 12.57 $\angle 270°$ kV; 6.41 $\angle 102.3°$ kV**]**

2. For a single line to ground fault at the terminals of an unloaded generator, positive sequence current was found to be 50 A. Determine sequence currents in phases b and c.

[U.P. Technical Univ. Power System Analysis 2004-05]

[Ans. 50 $\angle 240°$ A, 50 $\angle 120°$ A, 50 $\angle 0°$ A and 50 $\angle 120°$ A, 50 $\angle 240°$ A, 50 $\angle 0°$ A]

3. A 25 MVA, 13.2 kV alternator with solidly grounded neutral has a subtransient reactance of 0.25 pu. The negative and zero sequence reactances are 0.35 pu and 0.1 pu respectively. A single phase to ground fault occurs at the terminals of this unloaded generator. The fault impedance is j 0.1 pu. Compute fault current, terminal voltages under faulted condition. Express all currents and voltages under faulted condition as a percentage of rated values. Express sequence components of currents in the form of percentage of full-load current. **[Bharathiyar Univ. Power System Analysis Nov. 1996]**

[Ans. $I_{a_0} = I_{a_1} = I_{a_2}$ = 100%; I_f = 300%, $I_b = I_c$ = 0, V_{a_1} = 75%, V_{a_2} = –35%, V_{a_0} = –10%; V_a = 30%, V_{ab} = 112.6%, V_{bc} = 190.52%, V_{ca} = 112.6%]

4. A 25 MVA, 13.2 kV alternator with solidly grounded neutral has a subtransient reactance of 0.25 pu. The negative and zero sequence reactances are 0.35 pu and 0.1 pu respectively. A single line-to-ground fault occurs at the terminals of an unloaded alternator. Determine the fault current and the line-to-line voltages. Neglect resistance.

[Rajasthan Univ. Power System Analysis 2005]

[Ans. 4,686 $\angle -90°$ A; 7.7 $\angle 77.8°$ kV; 15.1 $\angle 270°$ kV; 7.7 $\angle 102.2°$ kV]

5. Two 3-phase, 25 MVA, 11 kV, 50 Hz alternators operate in parallel, as shown in Fig. 1. The positive, negative and zero sequence reactances of each being respectively j 0.20, j 0.15 and j 0.10 pu. The star point of one of the alternators is isolated and that of the other is grounded through an inductor of 2.0 Ω. A single line-to-ground fault occurs at the terminals of one of the alternators. Compute (i) the fault current (ii) currents in the grounding inductor and (iii) voltage across grounding inductor.

[Ans. (i) 2,598 $\angle 270°$ A (ii) 2,598 $\angle 270°$ A (iii) 5.196 kV]

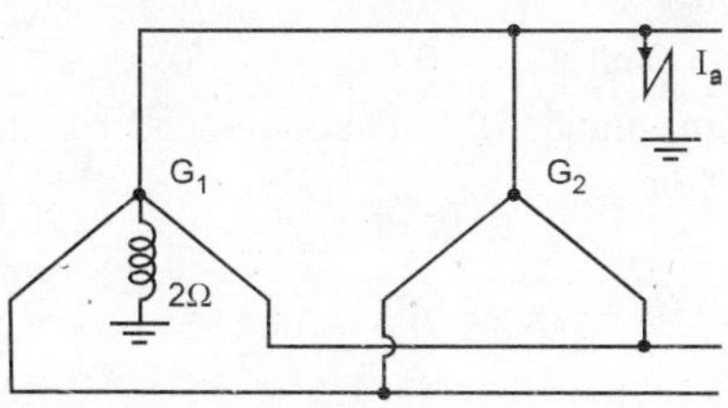

Fig. 1

[Hint : $X_{1eq} = \dfrac{j0.20}{2} = j\,0.1$; $X_{2eq} = \dfrac{j0.15}{2} = j\,0.075$;

$X_{0eq} = j\,0.1 + j\,3 \times \dfrac{2 \times 25}{(11)^2} = j\,1.34$ pu

$I_f = \dfrac{3E_a}{X_{1eq} + X_{2eq} + X_{0eq}}$ and Voltage across $X_g = I_f \times X_g$]

6. A 3-phase generator rated 15 MVA, 13.2 kV has a solidly grounded neutral. It has positive, negative and zero sequence reactances of 30%, 40% and 5% respectively.

(i) What value of reactance must be placed in the generator neutral so that the fault current for line-to-ground fault of zero fault impedance shall not exceed the rated line current ?

(ii) What value of resistance in neutral will serve the same purpose ?

[Pb. Univ. Power System Analysis May 1994]

[Ans. (i) 8.712 Ω (ii) 11.244 Ω]

7. A 30 MVA, 11 kV generator has $Z_1 = Z_2 = j0.2$ pu, $Z_0 = j0.05$ pu. A line-to-line fault occurs on the terminals of the generator. Find the line currents and line-to-neutral voltages under fault conditions. Assume that the generator neutral is solidly grounded and that the generator is operating at no load and rated voltage at the occurrence of fault.

[Rajasthan Univ. Power System Analysis 2007]

[Ans. 0; 6,818 $\angle 180°$ A; 6,818 $\angle 0°$ A; 6.35 kV; 3.175 kV, – 3.175 kV]

8. The positive, negative and zero sequence reactances of a 20 MVA 13.2 kV synchronous generator are 0.3, 0.2 and 0.1 pu respectively. The generator is solidly grounded and is not loaded. A double line-to-ground fault occurs on generator terminals b and c. Calculate the fault current and line voltages.

[Ans. 4,772 $\angle 90°$ A; 4.157 kV; 0; – 4.157 kV]

9. A 50 Hz, 13.2 kV, 15 MVA alternator has $X_1 = X_2$ = 20% and X_0 = 8% and the neutral is grounded through a reactor of 0.5 Ω. Determine the initial symmetrical rms current in the ground reactor when a double line-to-ground fault occurs at the generator terminals at the time when the generator voltage was 12 kV.

[Madras Univ. Power System Analysis April 1997]

[Ans. 2,910 $\angle 90°$ A]

10. A 15 MVA, 6.9 kV generator, star-connected has +ve, –ve and zero sequence reactances of 25%, 25% and 8% respectively. A reactor with 6% reactance based on the rating of the generator is placed in line from neutral to ground. A line-to-line fault occurs at the terminals of the generator when it is operating at rated voltage and disconnected from supply. Find the initial symmetrical rms line and ground wire currents and line-to-line neutral voltages (i) if the fault does not involve ground (ii) if the fault is solidly grounded at the instant of its occurrence.

[Ans. (i) 0; – 4,320 A; 4,320 A and zero; 3,900 $\angle 0°$ V; –1,995 $\angle 180°$ V; 1,995 $\angle 180°$ V (ii) 0; 4,977 $\angle 150.6°$ A; 4,977 $\angle 29.4°$ A; 4,890 $\angle 90°$ A; 4,077 V; 0; 0]

11. The per unit values of positive, negative and zero sequence reactance of a network at fault are 0.16, 0.14 and 0.2. Determine the fault current if fault is double line to ground.

[U.P. Technical Univ. Power System Analysis 2009-10]

[Ans. 5.097 $\angle 90°$ pu]

12. A 3-phase star-connected alternator is rated 30 MVA, 13.8 kV and has the following sequence reactance values: X_1 = 0.25 pu; X_2 = 0.35 pu and X_0 = 0.10 pu. The neutral of the alternator is solidly grounded. Determine the alternator line currents when a double line-to-ground fault occurs on its terminals. Assume that the alternator is unloaded and is operating at rated voltage when the fault occurs. **[GATE 1995]**

[Ans. 0; 6,024 $\angle 132.2°$ A; 6,024 $\angle 47.8°$ A; 8,936 $\angle 90°$ A]

13. A 25 MVA, 13.2 kV alternator with solidly grounded neutral has a subtransient reactance of 0.25 pu. The negative and zero sequence reactances are 0.35 and 0.1 pu. respectively. Determine the fault current and line to line voltages at the fault when a double line to ground fault occurs at the terminals of the alternator. Deduce also the expression used for calculating fault current.

[U.P. Technical Univ. Power System Analysis 2007-08]

[**Ans.** 7,785.5 $\angle 90°$ A; 5,428 kV; 0; 5,428 $\angle 180°$ kV]

14. A 25 MVA, 11 kV, 3-phase alternator was subjected to the following faults : 3-phase fault = 1,000 A; line-to-line fault = 1,400 A; line to ground fault = 2,200 A. The generator neutral is solidly grounded. Ignoring resistances calculate the values of three reactances of alternator.

[U.P. Technical Univ. Power System Analysis 2009-10]

[**Ans.** 6.35 Ω; 1.507 Ω, 0.803 Ω]

15. A 50 MVA, 11 kV, 3-ϕ alternator was subjected to different types of faults. The fault currents were

3-phase fault = 1,870 A; Line-to-line fault = 2,590 A; single line-to-ground fault = 4,130 A.

The alternator is solidly grounded. Find pu values of three sequence reactances of the alternator.

[U.P.S.C. I.E.S. Electrical Engineering-II, 2001]

[**Ans.** 0.151 pu; 1.403 pu; 0.352 pu]

16. A 20 MVA, 33 kV, 3-phase alternator is subjected to different types of short circuits and the following are the values of fault current.

3-phase short circuit = 319 A

Single line-to-ground fault = 659 A

Line-to-line fault = 435 A

Determine the positive, negative and zero sequence reactances of the generator in per unit and ohms. Neglect resistive part. **[U.P.S.C. I.E.S. Electrical Engineering-II, 2003]**

[**Ans.** 54.45 Ω; 0.1995 pu; 59.725 Ω; 1.0968 pu; 16.137 Ω; 0.2963 pu]

17. A 50 MVA, 11 kV three-phase alternator was subjected to different types of faults. The fault currents are as under :

3-phase fault = 2,000 A, Line-to-line fault = 2,600 A

Line-to-ground fault = 4,200 A

The generator is solidly grounded, Find the values of the three sequence reactances of the alternator. Ignore resistances.

[Rajasthan Technical Univ. Power System Analysis 2011]

[**Ans.** 1.1536 pu; 1.312 pu; 1.7157 pu]

18. A 50 MVA, 13.2 kV star-connected alternator is connected to a transmission line as shown below (Fig. 2):

The positive, negative and zero sequence reactances of alternator are 0.1, 0.1 and 0.05 pu respectively. The transformer ratings are 50 MVA, 13.2 kV/120 kV with star point solidly grounded. The sequence reactance of transformer are $X_1 = X_2 = X_0 = 0.1$ pu.

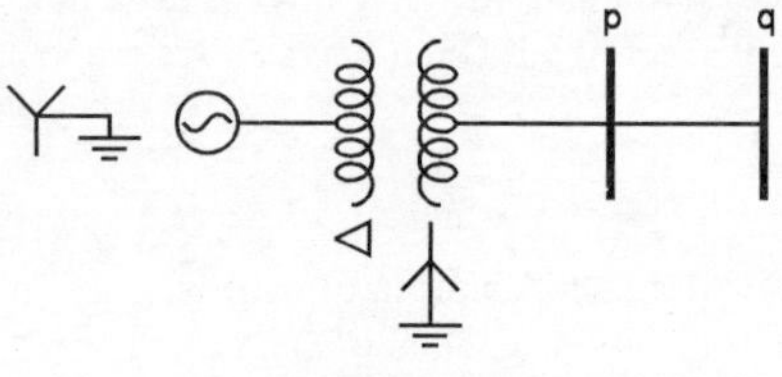

Fig. 2

The transmission line sequence–reactances are X_1 = 0.03, X_2 = 0.03 and X_0 = 0.09 pu.

A single phase-to-ground fault-occurs at bus *q*. Compute fault current and express this current as a percentage of full-load current of alternator.

[Bharthiyar Univ. Power System Analysis April 1997]

[**Ans.** 10,094 A; 461.54 %]

19. A generator supplies a motor through a ⅄ – D transformer. A fault occurs between motor terminals and transformer terminals. The symmetrical components of subtransient current flowing from motor towards fault are $\mathbf{I}_{a_1}$ = (– 0.8 – *j* 2.6) pu; $\mathbf{I}_{a_2}$ = – *j* 2.0 pu and $\mathbf{I}_{a_0}$ = *j* 3.0 pu. The symmetrical components of currents flowing from transformer towards the fault are $\mathbf{I}_{a_1}$ = 0.8 – *j* 0.4 pu; $\mathbf{I}_{a_2}$ = –*j*1.0 pu and $\mathbf{I}_a$ = 0. Assume X_1 = X_2 for both motor and generator. Describe the type of fault and compute the subtransient current in each phase of generator.

[Madurai Kamaraj Univ. Power System Analysis Nov. 1996]

[**Ans.** L-G fault; – *j* 9.0 pu; 0,0]

[**Hint** Symmetrical component of fault current [(– 0.8 – *j* 2.6) + (0.8 – *j* 0.4)], (– *j* 2.0– *j* 1.0) and – *j* 3.0 per unit *i.e.*, – *j* 3.0, – *j* 3.0 and – *j* 3.0 per unit]

20. A single line-to-earth fault occurs on red phase at the load end of a 132 kV transmission line. The line side of transformer is in star with star point earthed and the generator side is in delta with a 90° connection. On 50 MVA, 132 kV base, the pu reactances are

	Positive	Negative	Zero sequences
Generator	0.1	0.1	0.1
Transformer	0.1	0.1	0.1
Line up to fault	0.4	0.4	0.1

Find the magnitude of line currents under fault in amperes in the generator. [**Ans.** 2.143 pu; zero; zero]

❑❑❑

CHAPTER 6

Fuses

INSIDE THIS CHAPTER

6.1. INTRODUCTION

Fuse is perhaps the simplest and cheapest device used for interrupting an electrical circuit under short circuit, or excessive overload, current magnitudes. As such, it is used for overload and/or short-circuit protection in high voltage (up to 66 kV) and low voltage (up to 400V) installations/circuits. In high voltage circuits their use is confined to those applications where their performance characteristics are particularly suitable for current interruption.

The action of a fuse is based upon the heating effect of the electric current. In normal operating conditions, when the current flowing through the circuit is within safe limits, the heat developed in the fuse element carrying this current is readily dissipated into the surrounding air, and therefore, fuse element remains at a temperature below its melting point. However, when some fault, such as short circuit, occurs or when load connected in a circuit exceeds its capacity, the current exceeds the limiting value, the heat generated due to this excessive current cannot be dissipated fast enough and the fusible element gets heated, melts and breaks the circuit. It thus protects a machine or apparatus or an installation from damage due to excessive current.

The time for blowing out of fuse depends upon the magnitude of the excessive current. Larger the current, the more rapidly the fuse will blow *i.e.,* the fuse has inverse time-current characteristic, as shown in Fig. 6.1. Such a characteristic is desirable for protective gear.

Essentially, a fuse consists of a fusible element in the form of a metal conductor of specially selected small cross-sectional area, a case or cartridge to hold the fusible element, and in some cases, provided with a means to aid arc extinction. The part which actually melts and opens the circuit is known as the *fuse element.* It forms a series part of the circuit to be protected against short circuit or excessive overloads.

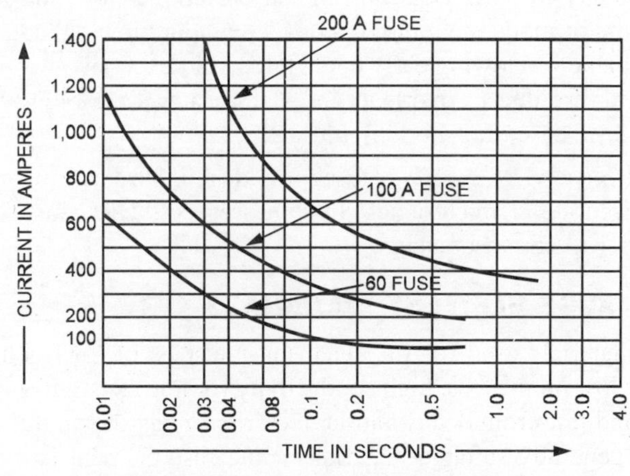

Fig. 6.1. ***Time-Current Characteristic***

Fuses has got following advantages and disadvantages.

Advantages. (*i*) It is the cheapest form of protection available.

(*ii*) It needs no maintenance.

(*iii*) Its operation is inherently completely automatic unlike a circuit breaker which requires an elaborate equipment for automatic action.

(*iv*) It interrupts enormous short-circuit currents without noise, flame, gas or smoke.

(*v*) The minimum time of operation can be made much smaller than that with the circuit breakers.

(*vi*) The smaller sizes of fuse element imposes a current limiting effect under short-circuit conditions.

(*vii*) Its inverse time-current characteristic enables its use for overload protection.

Disadvantages. (*i*) Considerable time is lost in rewiring or replacing a fuse after operation.

(*ii*) On heavy short circuits, discrimination between fuses in series cannot be obtained unless there is considerable differences in the relative sizes of the fuses concerned.

(*iii*) The current-time characteristic of a fuse cannot always be correlated with that of the protected device.

The function of fuse wire is (*i*) to carry the normal working current safely without heating and (*ii*) to break the circuit when the current exceeds the limiting current.

Distribution circuits are protected from ground and short-circuit currents by fuses or circuit breakers so arranged as to disconnect the faulted equipment promptly from its source of supply. Fuses are used almost exclusively for the protection of cables in low-voltage light and power circuits and for transformers of rating not exceeding 200 kVA, in primary distribution systems. Circuit breakers are employed for larger amounts of power and in cases where the operation of the overload device is so frequent as to make the use of fuses impractical.

Nowadays several types of fuses are available which find extensive use in low and moderate voltage applications where frequent operations are not expected or where the use of circuit breakers is uneconomical. Most important among these is the *motor protective fuse* which may be chosen to have a suitable time delay to permit starting inrush current to flow and also carrying of moderate motor overloads without blowing, and yet be capable of blowing early enough to avoid excessive winding temperature due to overload and at the same time blow very quickly in case of a short circuit.

Modern HRC cartridge fuses provide reliable discrimination and accurate characteristics. In some respects HRC fuses are superior to circuit breakers.

6.2. FUSE ELEMENT MATERIALS

The materials used for fuse elements must be of low melting point, low ohmic loss, high conductivity (or low resistivity), low cost and free from deterioration. Experience has shown that the most generally suitable material for the fuse element is a low melting point material such as tin, lead or zinc. A low melting point is, however, available with a high specific resistance metal (refer to Table 6.1).

Table 6.1

Metal	Melting Point in °C	Specific Resistance in μΩ - mm	Value of Fuse Constant k For d in m
Silver	980	16	—
Tin	240	112	12.8
Zinc	419	60	—
Lead	328	210	10.8
Copper	1,090	17	80
Aluminium	665	28	59

A high specific resistance metal entails, for a given rating of fuse element, a greater mass metal than would be required if the elements were made from high melting point metal of low specific resistance. Thus the use of low melting point metals introduces the problems of handling large mass of vaporised metal set free on fusion.

Oxidation of the fuse element tends to generate an outer layer which may be strong enough to support the metallic core of the fuse element when in a molten state. The effect is more pronounced with certain high melting point metals such as copper and iron than it is with low melting point metals such as tin or lead.

The above discussion reveals that no material possesses all the desirable characteristics. For instance lead has low melting point but it has high specific resistance and it is liable to oxidation. Similarly, copper has high conductivity and low cost but oxidises rapidly. Thus a compromise is required to be made in the selection of material for a fuse element.

The materials commonly used for fuse elements are tin, lead, silver, copper, zinc, aluminium and alloy of lead and tin. An alloy of lead and tin (lead 37% and tin 63%) is used for small current rating fuses (say not beyond 15 A). For current exceeding 15 A this alloy is not used as the diameter of the wire will be larger and after fusing the metal released will be excessive. Beyond 15 A rating circuits copper wire fuses are employed. Copper wire suffers from the drawback that it operates at a rather high temperature if a reasonably low fusing factor is desired. There is thus a tendency for the wire to overheat with the result that its cross-sectional area and fusing currents are gradually reduced and premature melting of the wire may occur. It is usual practice to tin the copper for its protection against oxidation. Zinc (in strip form only) is good if a fuse with considerable time delay is required *i.e.,* one which does not melt very quickly with a small overload.

The present trend is to use silver as fuse element material despite its higher cost owing to its following advantages.

(*i*) It does not get oxidised and its oxide is unstable.

(*ii*) It does not deteriorate when used in dry air and it remains bright but when the air is moist and contains hydrogen sulphide, the silver surface is attacked and a layer of silver sulphide is formed at the top which shields the metal from further attack.

(*iii*) Owing to its very small coefficient of expansion no critical fatigue occurs and thus continuous full normal current capacity rating is assured for a long time.

(*iv*) The conductivity of silver does not deteriorate with oxidation.

(*v*) Owing to its high conductivity the mass of molten metal to be handled is minimum and thus operating speed is fast.

(*vi*) Vaporization temperature of silver is very much below the one at which the vapour will readily ionize. This means that on the formation of an arc through the vaporized portion of the element, the arc path has comparatively high resistance and the short-circuit current is quickly reduced.

(*vii*) Owing to its low specific heat silver fusible elements can be raised from normal operating temperature to vaporization very much quicker than other fusible elements. Also as the temperature of the silver reaches the melting point its resistance increases abruptly and thus the transition from melting to vaporization becomes nearly instantaneous. As a result, operation becomes very much faster at higher currents.

Either copper or lead-tin alloy is mostly used as an ordinary fuse wire.

6.3. DEFINITIONS

1. *Fuse.* Fuse is a current interrupting device which breaks or opens the circuit (in which it is inserted) by fusing the element when the current in the circuit exceeds a certain value.

2. *Fuse Element or Fuse Wire.* It is that part of the fuse which actually melts when an excessive current flows in the circuit and thus isolates the faulty device from the supply circuit.

3. *Fuse Link.* It is that part of the fuse which needs replacement when the fuse blows out.

4. *Current Rating.* This rating is specified by the manufacturer. It is defined as the rms value of the current which the fuse wire can carry continuously without deterioration, and with temperature rise within specified limits. It depends upon the permissible temperature rise of the contacts of the fuse holder, fuse material and the surroundings of the fuse.

5. *Fusing Current.* It is defined as the minimum value of current at which the fuse element or fuse wire melts. Obviously, its value will be more than the current rating of the fuse element. It is the asymptotic value of current from the characteristic of total operating time.

For a round wire the approximate value of fusing current is given by

$$I = kd^{3/2} \qquad ...(6.1)$$

where k is a constant, called the *fuse constant*, depending upon the metal of the fuse element. Sir W.H. Preece has given the value of the constant k, as indicated in Table 6.1.

The fusing current depends upon various factors such as:

(*i*) Type of material used.

(*ii*) The x-sectional area *i.e.*, whether round or rectangular section.

(*iii*) Length—the shorter the fuse the greater the current since with the fuse of smaller length heat is easily conducted away.

(*iv*) Size and location of the terminals (a bigger size will dissipate more heat).

(*v*) Diameter of wire.

(*vi*) Type of enclosure employed.

(*vii*) Type of surface—the fusing current for a standard fuse will be less than the product of the fusing current for one strand and the number of strands, as shown in Table 6.2.

Table 6.2

Number of Wires	1	2	3	4	7
Fusing Current	1	1.667	2.25	2.75	4

6. *Fusing Factor.* The ratio of minimum fusing current and the current rating of fuse element is known as fusing factor and it is always greater than unity.

$$\text{Fusing factor} = \frac{\text{Minimum fusing current}}{\text{Current rating of fusing element}} \qquad ...(6.2)$$

The smaller the fusing factor, the greater is the difficulty in avoiding deterioration due to overheating and oxidation at rated carrying current. For a semi-enclosed or rewirable fuse which employs copper wire as the fuse element, fusing factor is equal to 1.9 to 2.0. For the standard duty cartridge fuses fusing factor is equal to 1.45.

7. *Prospective Current.* Figure 6.2 shows how an ac current is cut off by a fuse. The fault current would normally have a very large first loop, but since it generates sufficient energy, the fuse element melts well before the peak of the loop is reached. The rms value of the first loop of the fault current is called the *prospective current* which may, therefore be defined as below:

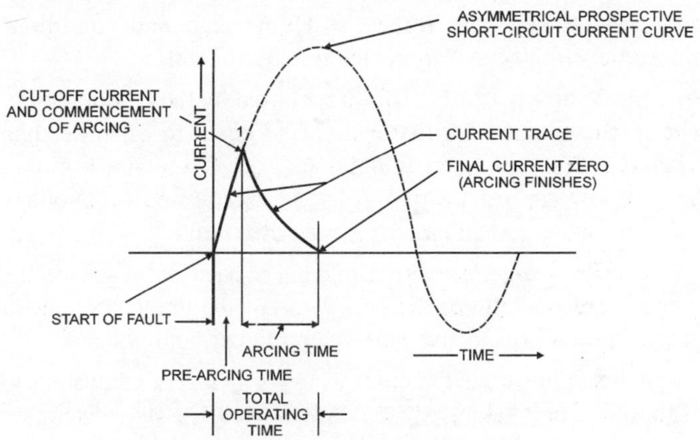

Fig. 6.2. *Cut-off Characteristic of a Fuse*

Prospective current may be defined as the current that would flow in the circuit under fault conditions when the fuse were replaced by a link of negligible impedance.

8. *Cut-off Current.* The melting of fuse element before the current reaches the prospective peak is called the *cut off*. The value of current at which the cut off occurs is called the cut-off value. Cut-off current is of instantaneous value.

On the occurrence of a fault, the fault current has a very large first loop due to a fair degree of asymmetry. The heat generated is sufficient to melt the fuse element well before the peak of first loop is reached (point 1 in Fig. 6.2). The current corresponding to point 1 is called the cut-off current. The cut-off value depends upon (*i*) current rating of fuse (*ii*) the value of prospective current and (*iii*) asymmetry of short-circuit current.

The action of breaking the circuit before the fault current attains its first peak is the unique feature of the fuse. This gives the fuse a great advantage over a circuit breaker since the most severe thermal and electromagnetic effects of short-circuit currents (which occur at the peak value of prospective current) are not experienced with fuses. So the circuits protected by fuses can be designed to withstand maximum current equal to cut-off value. This consideration together with the relative cheapness of fuses allows much saving in cost.

9. *Fuse Link Operation.* It is a process of pre-arcing and arcing resulting in 'blowing' of fuse link.

10. *Pre-arcing or Melting Time.* On occurrence of a fault, the fault current rises rapidly and heat is generated in the fuse element. As the fault current attains it cut-off value, the fuse melts and an arc is initiated. Pre-arcing time is the time between the commencement of the current which tends to cause a break in a fuse element and the instant that the arc is initiated.

11. *Arcing Time.* This is the time accounted from the instant of arc initiation (end of the pre-arcing time) to the instant of arc being extinguished or the arc current becomes zero.

12. *Total Operating Time.* This is the sum of pre-arcing time and arcing time. The noteworthy point is that the operating time of a fuse is usually quite low (say 0.002 second) as compared to a circuit breaker (say 0.05 second or so). This is an additional advantage of a fuse over a circuit breaker. A fuse in a series with a circuit breaker of low breaking capacity is useful and economical arrangement for providing adequate short-circuit protection. It is because the fuse will blow under fault conditions before the circuit breaker has the time to operate.

13. *Voltage Rating.* This is specified by the manufacturer. The rated voltage of a fuse should be equal to or more than (*i*) voltage of the circuit in a single-phase ac or two-wire dc circuit (*ii*) line voltage in case of 3-phase ac circuit and (*iii*) voltage between two outers in case of 3-wire dc circuit.

14. *Breaking Capacity.* Breaking capacity of a fuse is the rating corresponding to the rms value of the ac component of the maximum prospective current and the system voltage.

A fuse almost never actually passes a current equivalent to its rupturing or breaking capacity. We, however, assign this higher value to a fuse because it prevents the development of this current by its cut-off action.

6.4. TYPES OF FUSES

A fuse unit essentially consists of a metal fuse element or link, a set of contacts between which it is fixed and a body to support and isolate them. Many types of fuses also have some means for extinguishing the arc which appears when the fuse element melts.

In general, there are two categories of fuses viz, (*i*) low voltage fuses and (*ii*) high voltage fuses.

Usually isolating switches are provided in series with fuses where it is necessary to permit fuses to be replaced or rewired with safety. In absence of such isolation means, the fuses must be so shielded as to protect the user against accidental contact with the live metal when the fuse carrier is being inserted or removed.

6.5. LOW VOLTAGE FUSES

Low voltage fuses can be further divided into two classes namely semi-enclosed or rewirable type and totally enclosed or cartridge type.

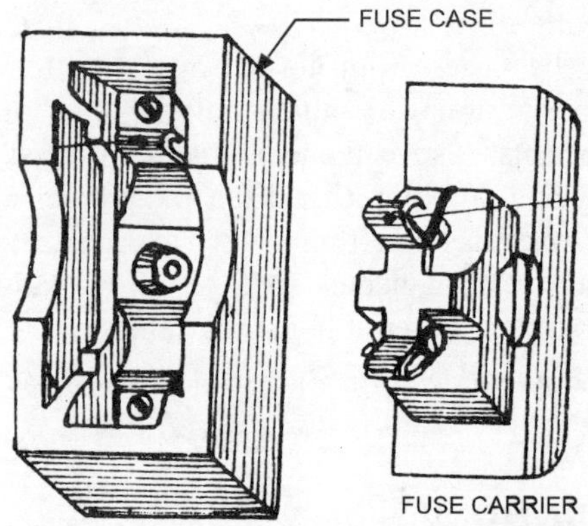

Fig. 6.3. ***Rewirable or Kit-Kat Type Fuse***

1. *Rewirable Fuses.* The most commonly used fuse in 'house wiring' and small current circuits is the semi-enclosed or rewirable fuse (also sometimes known as kit-kat type fuse). It consists of a *porcelain base* carrying the fixed contacts to which the incoming and outgoing live or phase wires are connected and a *porcelain fuse carrier* holding the fuse element, consisting of one or more strands of fuse wire, stretched between its terminals. The fuse carrier is a separate part and can be taken out or inserted in the base without risk, even without opening the main switch. If fuse holder or carrier gets damaged during use, it may be replaced without replacing the complete unit.

The fuse wire may be of lead, tinned copper, aluminium or an alloy of tin-lead. The actual fusing current will be about twice the rated current. When two or more fuse wires are used, the wires should be kept apart and a derating factor of 0.7 to 0.8 should be employed to arrive at the total fuse rating.

The specifications for rewirable fuses are covered by IS : 2086-1963. Standard ratings are 6, 16, 32, 63 and 100 A. A fuse wire of any rating not exceeding the rating of the fuse may be used in it *i.e.,* a 80 A fuse wire can be used in a 100 A fuse, but not in the 63 A fuse.

On occurrence of fault, the fuse element blows off and the circuit is interrupted. The fuse carrier is pulled out, the blown out fuse element is replaced by new one and the supply is restored by re-inserting the fuse carrier in the base.

Though such fuses have the advantages of easy removal or replacement without any danger of coming into contact with a live part and negligible replacement cost but suffers from the following disadvantages:

(*a*) *Unreliable Operation.* The operation of the rewirable fuses is unreliable because of the following factors :

(*i*) There is a possibility of renewal by the fuse wire of wrong size.

(*ii*) The fuse wire deteriorates, over a period, due to oxidation through the continuous heating up of the element. After a relatively short period the current causes the metal to deteriorate and the fuse operates at a lower current than originally rated.

(*iii*) A fuse with normal rated capacity of say 60 A would fuse at nearly 120 amperes. Should that amount of current damage the apparatus in the circuit, obviously the size of the fuse wire must be smaller but, as its rated capacity will also be lower, the same dependence could not be placed upon it.

(*iv*) The protective capacity is uncertain *i.e.,* no dependence can be placed upon the wire to interrupt the circuit when a given current flows. For instance although theoretically a No. 17 SWG wire should interrupt a circuit when a current of 135 A is flowing, the circuit may be interrupted when a lower current flows or in other circumstances when only a much higher current flows.

(*v*) Accurate calibration of the fuse wire is impossible, as a longer fuse operates earlier than one of shorter length.

Single phasing of three-phase induction motors is a common occurrence where rewirable type fuses are employed in motor circuits.

(b) Lack of Discrimination. Due to unreliable operation, discrimination cannot be always ensured. However, it will be possible to achieve some measure of discrimination by using a fuse which has twice the rating of the next fuse ahead of it in series. For example, a fuse of 80 A will discriminate with a fuse of 40 A, but not with a fuse of 55 A.

(c) Small Time Lag. Due to the small time lag, such fuses can blow with large transient currents which are encountered during the starting of motors and switching-on operation of transformers, capacitors, fluorescent lights etc., unless fuses of sufficiently high rating are used.

(d) Low Rupturing Capacity. Rewirable fuses have limited breaking or rupturing capacity. For example, according to IS : 2086-1963 the rewirable fuse of 16 A normal current have a breaking current of 2 kA and those up to 200 A normal current have a breaking current of 4 k A.

(e) No Current-Limiting Feature.

(f) Slow Speed of Operation. No special means are employed to extinct the arc that blows after the fuse melts. Thus arcing time is more in such fuses.

(g) Risk of Flame and Fire.

Because of simplicity and cheapness rewirable type fuses are extensively used as protective device against overload and short circuit on low and medium voltage circuits.

2. *Totally Enclosed or Cartridge Type Fuses.* The fuse element is enclosed in a totally enclosed container and is provided with metal contacts on both sides. These fuses are further classified as (*i*) D-type and (*ii*) link type.

Link type cartridge fuses are again of two types viz., knife blade or bolted type.

(a) D-Type Cartridge Fuses. It is a non-interchangeable fuse comprising a fuse base, adapter ring, cartridge and a fuse cap. The cartridge is pushed in the fuse cap and the cap is screwed on the fuse base. On complete screwing the cartridge tip touches the conductor and circuit between two terminals is completed through the fuse link. The standard ratings are 6, 16, 32 and 63 amperes. The breaking or rupturing capacity is of the order of 4 kA for 2 and 4 ampere fuses and 16 kA for 6 to 63 A fuses.

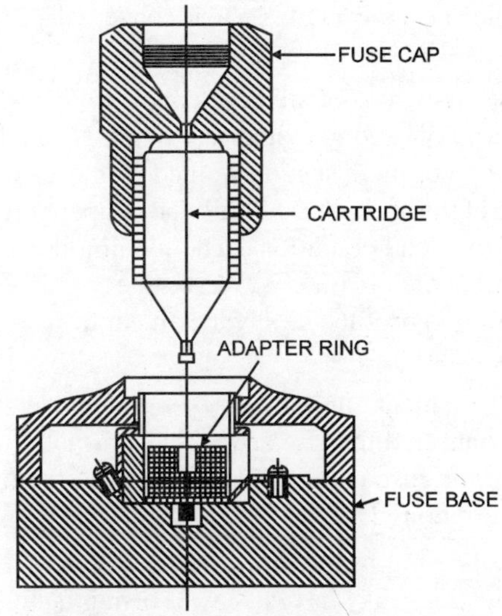

Fig. 6.4. ***D-Type Cartridge Fuse***

D-type cartridge fuses have none of the drawbacks of the rewirable fuses. Their operation is reliable. Coordination and discrimination to a reasonable extent are achieved with them.

(b) Link Type Cartridge or High Rupturing Capacity (HRC) Fuses. Where large concentrations of power are concerned, as in the modern distribution systems, it is essential that fuses should have a definite known breaking capacity and also this breaking capacity should have a high value. High rupturing capacity cartridge fuse, commonly called HRC cartridge fuses, have been designed and developed after intensive research by manufacturers and supply engineers in this direction.

An HRC fuse is shown in Fig. 6.5. The outer body is usually of steatite, a ceramic material having good mechanical strength. Epoxy resins have been recently introduced and are replacing the ceramic material. The fuse elements are connected to two end caps of brass or copper inside the body. The metal end caps are screwed to the ceramic body by means of special forced screws to withstand the pressure developed under short-circuit condition. End contacts are welded to the metal end caps. An indicating device is provided to distinguish the blown fuse. The space between the fuse element and the cartridge body is filled with powdered pure quartz which acts as an arc extinguishing agent.

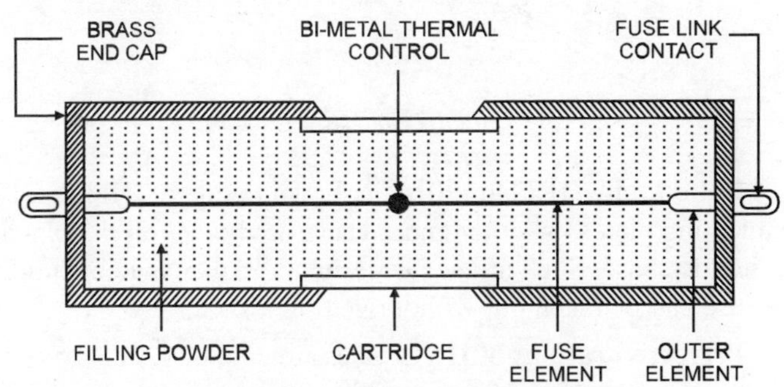

Fig. 6.5. ***HRC Cartridge Fuse***

Here quartz powder avoids oxidation of silver element. Quartz sand is used as a filler, because it can absorb heat at a very high rate and it does not involve appreciable amount of gas.

After absorbing heat the quartz sand is converted into glass beads which obstruct the path of restriking voltage.

The fuse element is of silver or copper with a special shape (not in the form of a long cylindrical wire). Normally, the fuse element has two or more sections joined by means of a tin joint. The purpose of the tin joint is to limit the temperature of the fuse under small overload conditions. The melting point of silver is 980ºC while that of pure tin is 240ºC. As the circuit is overloaded the melting of tin prevents the silver element from attaining high temperature.

The fuse element consists of several identical strips which are so interconnected that the arc spreads instantly to all strips. The shape of the fuse element is governed by the time-current characteristic required.

The usual fusing factor for link fuses is 1.45. The fuses for special applications may have as low a fusing factor as 1.2.

The specifications for medium voltage HRC link fuses are covered under IS : 2208-1962. There are two types of such fuses (*i*) knife blade type and (*ii*) bolted type.

Knife blade type HRC link fuse is shown in Fig. 6.6. It can be replaced on a live circuit at no load with the help of a special insulated fuse puller, as illustrated in the figure.

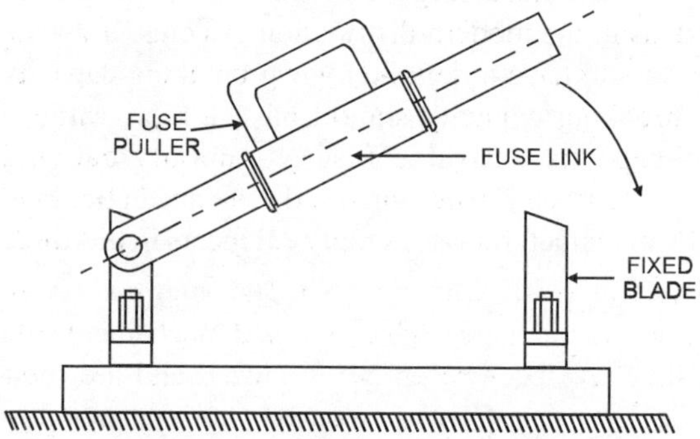

Fig. 6.6. ***Knife Blade Type HRC Link Fuse***

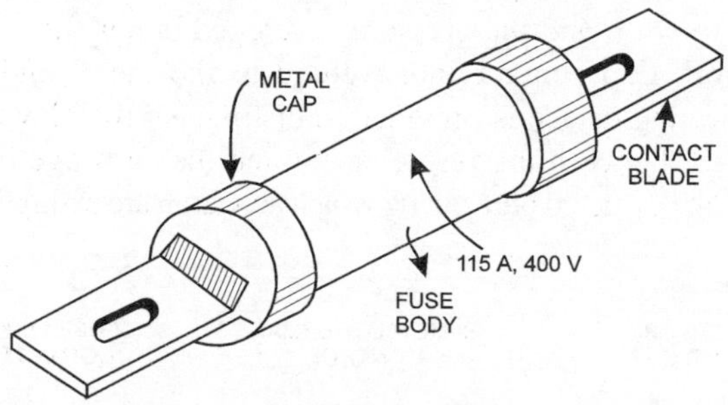

Fig. 6.7. ***Bolted Type HRC Link Fuse***

Bolted type HRC link fuse is shown in Fig. 6.7. It has two conducting plates on either ends. These are bolted on the plates of the fuse base. Such a fuse needs an additional switch so that the fuse can be taken out without getting a shock.

Preferred ratings of HRC fuses are 2, 4, 6, 10, 16, 25, 30, 50, 63, 80, 100, 125, 160, 200, 250, 320, 400, 500, 630, 800, 1000, and 1,250 amperes.

The *process of fusing* comprises the following steps.

(*i*) Pre-arcing operation *i.e.,* melting of silver elements (*ii*) Arcing operation *i.e.,* vaporisation of the elements (*iii*) Fusion of silver vapour and filling powder and (*iv*) Extinction of arc under fusion process.

The electrical phenomena associated with the operation of the fuse are :

(*i*) Establishment of high resistance due to chemical reaction between silver vapour and the filling powder.

(*ii*) Change over of the high resistance into an insulator as the current is cut off.

(*iii*) Creation of a transient voltage within the fuse at the instant of fault current interruption on account of sudden release of energy.

The physical phenomena associated with the process of fusing includes a sudden rise of temperature and the generation of a high internal pressure.

HRC Fuse Action. Normally the fuse elements are in parts which are connected in middle by tin bridge. The melting point of the tin bridge is precise and about 230° C. The bridge does not melt at temperatures below the melting point. Since the melting point is higher than ambient temperature, the melting is not affected at the ambient temperature.

With a certain current, the temperature rises and the bridge melts producing a break in the circuit. Thereby an arc is produced. This arc immediately spreads over the neighbouring elements and they too melt. The metal vapour diffuses with quartz powder and the chemical reaction between the two (silver vapour and quartz powder) produces a substance of high resistance. This high resistance substance acts as an insulator because the fault current decreases along with the high pressure created within the fuse by the fault (excessive) current. There after a transient voltage is created at the instant of interruption of fault current on account of sudden release of energy.

Rated values are rated voltage, rated frequency, rated current, rated breaking or rupturing current (symmetrical or asymmetrical), rated let-through current (the largest momentary value of the current during the breaking time), rated impulse current, rated natural frequency and rated amplitude factor.

Duty Category of HRC Fuses. HRC fuses have been sub-divided into four classes according to their rupturing capacity, as indicated below in Table 6.3.

Table 6.3

Category	Prospective Fault Current in kA	Power Factor not Exceeding
Class 1 AC and DC	10	0.3
Class 2 AC and DC	33	0.3
Class 3 AC only	46	0.15
Class 4 AC only	80	0.15

HRC link fuses may be *quick acting* (*i.e.,* having quick response), *slow action* (*i.e.,* having delayed action) or a combination of the two. Quick response fuses are suitable for steady loads, and delayed action fuses for motor loads. Link fuses having a combination of the properties of quick response and delayed action are suitable for either type of load.

Characteristics of HRC Fuses. The most important characteristics of HRC fuses are given below.

(i) I^2t Characteristics. Curves drawn for total ampere squared-second and pre-arcing ampere squared-seconds against fusing current are shown in Fig. 6.8. Such a characteristic helps in determination of the maximum amount of energy the fuse will pass to the appliance under protection. Also such a characteristic is useful for expressing cut-off characteristics.

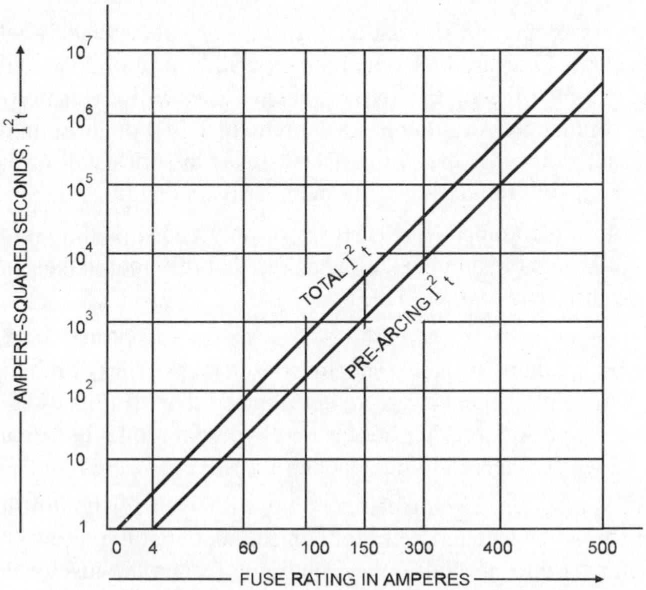

Fig. 6.8. *I^2t Characteristics*

(ii) Cut-off Current Characteristics. When an HRC fuse interrupts a heavy fault current it exhibits an ability to limit the short-circuit current. This ability is known as a 'cut-off', as illustrated in Fig. 6.9. Cut-off property has a great advantage that the short-circuit current does not attain the prospective current. Hence the circuit is not subjected to electrodynamic stresses corresponding to peak prospective current. Cut-off is in fact one of the main reasons why HRC fuse is so successful as a protective device and it is at times preferred over circuit breakers of low ratings. Because of this property of HRC fuse the operating time is as low as one-fourth of a cycle. The breaking or rupturing capacity of a fuse is generally given as the product of its service voltage and rms value of prospective current it can handle while the cut-off current is the maximum current to which the fault current reaches before the fuse melts.

The cut-off action is explained below :

On occurrence of fault, heavy fault current starts flowing through the fuse element, and the silver fuse element starts melting at one or more points depending upon the design of the fuse. On the blowing out of fuse an arc is formed between the two ends and a transient current is superimposed on the prospective current. When the sum of transient current and prospective current is zero the arc is quenched. Figure 6.9 represents the cut-off and occurrence of restriking voltage in case of an HRC fuse. The arc voltage at the start may be several times the supply voltage depending upon the length and cross section of the fuse.

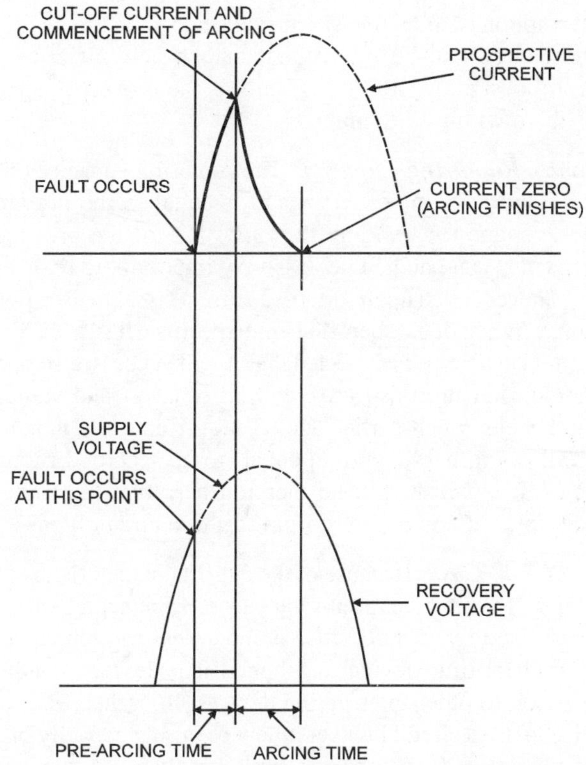

Fig. 6.9. *Current Cut-off Characteristic and Recovery Voltage in an HRC Fuse*

(iii) Fast Operation. The HRC fuse interrupts the short-circuit current long before it attains the maximum value which is not possible in case of circuit breakers. This property of HRC fuse reduces both thermal and magnetic stresses on the equipment under protection and the short-circuit current is interrupted well within the first quarter of a cycle.

(iv) Time-Current Characteristic. HRC fuse has inverse time-current characteristic, as shown in Fig. 6.1. It makes it useful for overload protection.

(v) Non-Deteriorating Characteristics. As the HRC fuses are called upon to function only once in while, the property of non-deterioration (*i.e.,* property of preservation of their characteristics throughout their useful life) is most essential. This also implies that they should not operate inadvertently when carrying normal load currents as happens in case of rewirable fuses due to oxidation and reduction of cross-reactional area. This property of HRC fuses is very important and is achieved by the hermetic sealing of the silver element within the fuse body with the help of special cementing and the soldering of the end caps. It has been found that HRC fuses maintain non-deterioration property unimpaircd even after approximately 20 years of their manufacture.

(vi) Low Temperature Operation. This property is necessary so as to avoid the deterioration of the fuses and prevent overheating of associated contacts. This is achieved in HRC fuses by using fabricated elements of pure silver, specially designed to give a low temperature rise while carrying their normal load current.

(vii) Arc Voltage. The magnitude of voltages induced due to interruption of inductive circuits depends upon the magnitude of short circuit to be interrupted and the circuit constants. A careful design of HRC fuses controls these overvoltages and keeps them within safe limits.

(viii) Rupturing Capacity. The rupturing capacity of fuse is expressed in terms of MVA and is equal to the product of service voltage (in kV) and the *rms* value of the prospective current (in kA) it can handle. It is very important to realize that a fuse is never called upon to carry a current equal to its rupturing capacity. When it is specified that the fuse has a particular rupturing capacity it is expected that it will successfully operate in a circuit having prospective current equal to said value, but the fuse melts much earlier due to cut-off action. Thus a fuse never allows to pass a current equal to its specified rupturing capacity. It is worthwhile to mention here that the rupturing capacity of a fuse is different from that of a circuit breaker.

(ix) Low Cost. Because of the cut-off characteristics of the HRC fuse, for the given rupturing capacity the actual current to be interrupted by an HRC fuse is much less in comparison to other interrupting device and hence it is less expensive in comparison to other interrupting devices. It is, therefore, usual practice to use a circuit breaker of low rupturing capacity backed up by an HRC fuse where circuit breakers are necessary otherwise. In such a scheme circuit breaker trips in case of overload while HRC fuse operates in case of short circuits.

(x) Discrimination. Discrimination means the breaking of the faulty section of the line to an extent restricted to minimum; healthy sections of the circuits remaining in operation.

When more than one protective device (two or more fuses, a fuse and a circuit breaker etc.) is used in series, it is necessary to have discrimination between these protective devices on occurrence of a fault. Discrimination is concerned with correct operation of correct device. It means the coordination between the fuse and the other protective device or devices should be such that only the necessary device operates, the other remaining unaffected.

In case of an HRC fuse it is found that the time of operation is inversely proportional to the prospective short-circuit current over a much wider range of fault condition and , within practical limits, while the values of prospective short-circuit current increase, the time of operation will continue to decrease without reaching a definite minimum. This means that a fuse of low current rating will blow before a fuse of a higher rating, no matter how heavy the fault is.

Proper coordination of fuses is necessary where fuses are used in series. The maximum number of line-sectionalizing fuses permissibly used in series varies from one to six. When fuses are properly coordinated, a fault on any part of the radial circuit will cause the fuse nearest the fault on the supply side to blow, thereby isolating the faulty section. No other fuse nearer the supply end should blow or begin to melt if fuses are chosen correctly.

In determination of sizes of fuses that will coordinate properly it is necessary to calculate maximum short-circuit current at the several points on the circuit where fuses are to be provided. Short-circuit currents can be obtained by use of charts of approximated by addition of impedances of various sections of the circuit arithmetically and assuming supply voltage maintained constant at the supply bus. Values obtained by this approximate method are 10 to 20% greater than actual values. Having determined maximum short-circuit currents at the various points where fuses are to be installed, the next step is to determine the correct size of fuses that will coordinate with each other when a fault occurs.

For proper discrimination, the time-current curves of all the fuses must be parallel to each other. A difference of 25% in the pre-arcing time of the two adjacent fuses will ensure close discrimination. As a thumb rule, a ratio of 1.5 : 1 or above in the current ratings of the two adjacent fuses in series will ensure discrimination under all fault and overload conditions.

It is recommended that fuses of only one make should be used throughout an installation, as fuses of different makes may have different characteristics.

HRC fuses are sometimes used as backup protection for circuit breakers. In such a scheme, the characteristics of the fuse and the circuit breaker are so coordinated that all faults within the range of the breaker are cleared by it and the faults beyond its range are cleared by the backup fuses.

Very often, circuit breakers are employed for controlling incoming feeders and fuses for controlling out going feeders. In such a case for a fault on any outgoing feeder, the fuse of that feeder should operate first. Circuit breaker should be as a backup. The characteristics of the fuse and circuit breaker should be matched such that the fuse takes less time for operation than circuit breaker (Fig. 6.10). A time lag of 50-150 ms will be generally sufficient for the purpose.

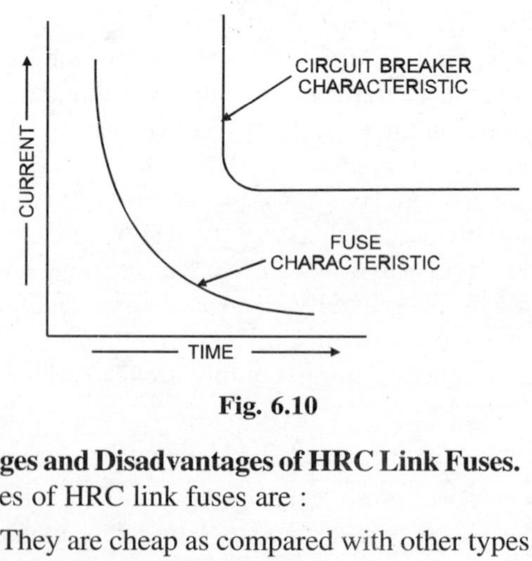

Fig. 6.10

Advantages and Disadvantages of HRC Link Fuses. The main advantages of HRC link fuses are :

(*i*) They are cheap as compared with other types of circuit interrupters of same breaking capacity.

(*ii*) Simple and easy installation.

(*iii*) No maintenance is required.

(*iv*) High breaking capacity.

(*v*) They are consistent in performance.

(*vi*) The operation is quick and sure.

(*vii*) They do not deteriorate with time.

(*viii*) They have inverse time-current characteristic, which makes them suitable for overload protection.

(*ix*) They are capable of clearing high as well as low currents.

(*x*) They are quite reliable and can be selected for proper discrimination.

(*xi*) Closer control on sustained overloads due to low fusing factor and the time lag feature.

(*xii*) Current limitation by cut-off property. During severe faults, an HRC fuse breaks the circuit before the peak of the prospective fault current is reached, thus protecting the equipment against maximum thermal and dynamic stresses.

HRC fuses suffer from the following disadvantages:

(*i*) Time is lost in replacing fuse after each operation.

(*ii*) Interlocking is not possible.

(*iii*) It causes overheating of adjacent contacts.

High Capacity HRC Cartridge Fuse. In order to increase the breaking capacity of a cartridge fuse having two or more widely separated silver fuse elements in parallel has been developed by G.E.C. The sizes of various elements are so adjusted that first of all only one element will melt and then other elements will follow it in succession. The body is made of ceramic material cylindrical in shape and closed by metallic end caps which carry the copper terminals. The fuse elements are surrounded by silica as the arc quenching medium. In order to control the restriking voltage a resistance wire is used. Also an indicator is provided and consists of a fine resistance wire, connected in parallel with the fuse elements and led through a small quantity of a mild explosive held in a pocket in the side of the fuse and covered by a label. The fine wire is automatically fused when the fuse operates and the resulting combustion of the explosive material chars with the label.

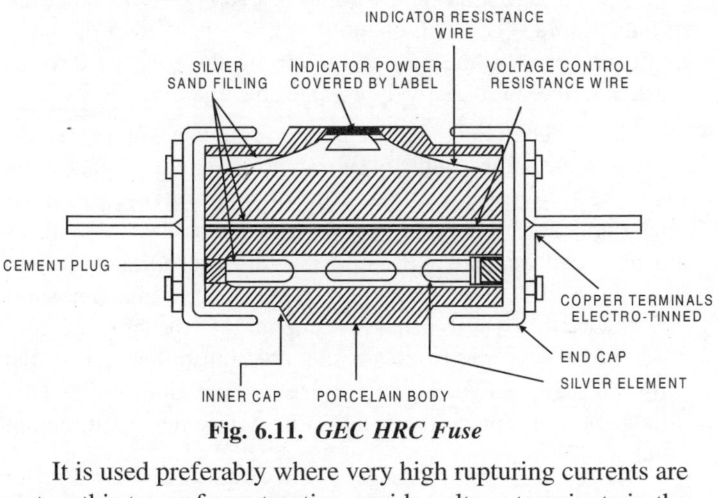

Fig. 6.11. ***GEC HRC Fuse***

It is used preferably where very high rupturing currents are met as this type of construction avoids voltage transients in the circuit. This is due to the fact that in such a fuse whole of the current is not shed immediately.

HRC Fuse With Tripping Device. HRC cartridge fuse incorporating tripping device is shown in Fig. 6.12. The body of the fuse is made of ceramic material with a cap rigidly fixed at either end. These are connected by a number of silver fuse elements. At one end is a plunger which acts as a tripping device for the associated circuit breaker. The plunger is electrically connected through a weak fusible link, chemical charge and tungsten wire to the other end of the cap as shown in figure. When a fault occurs, the fuse elements are the first to be blown off which is followed by fusing of weak link in series with tungsten wire and the chemical charge is detonated and the plunger is pushed out with a force to operate the circuit breaker. A limit is set to its travel to prevent it from being ejected from the fuse body, and at the same time to trap it so that it cannot be reset. HRC fuse with a tripping device has the following advantages over an HRC fuse without the tripping device.

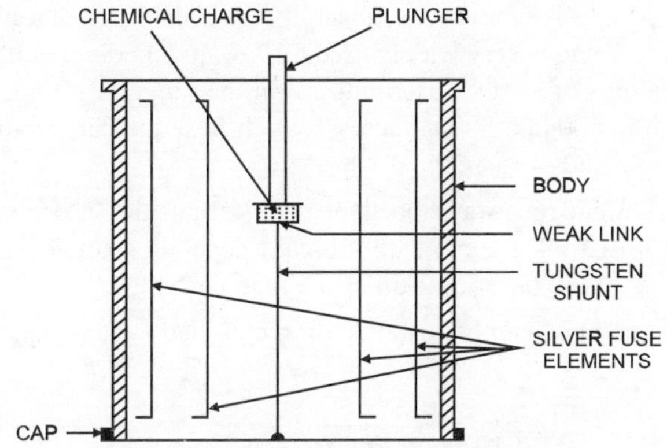

Fig. 6.12. ***HRC Fuse With Tripping Device***

1. In case of 3-phase system having a motor etc., if, a single phase earth fault occurs, its fuse will be blown off and the motor will continue to run on two phases resulting into overheating or damaging of the motor and the wiring. So when one fuse operates it must be ensured that the other two phases are also disconnected. This is achieved by employing a fuse incorporated with the tripping device.
2. The fuse-tripped circuit breaker is generally capable of dealing with fairly small fault currents itself. This avoids the necessity for replacing the fuse except after highest fault currents for which it is intended.
3. The effects of full short-circuit current need not be considered in the choice of circuit breaker. This allows the use of a relatively inexpensive circuit breaker.

Time Delay Fuses. These are also known as *slow acting fuses i.e.,* they operate at quite slow rate even for large short-circuit currents; but for smaller short-circuit currents their operation time may be same as those of ordinary fuses which is achieved by increasing the thermal capacity of the fuse. Such fuses are recommended as a backup protection for motors having long starting operations.

Low voltage HRC fuses may be built with a rupturing capacity of 16,000 A to 33,000 A at 400 V. They are extensively used on low voltage distribution systems against overload and short-circuit conditions.

6.6. DROP-OUT FUSE

The melting of fuse causes the fuse element to drop under gravity about its lower support—an additional isolation. Such fuses are employed for protection of outdoor transformers. On blowing off fuse the complete tube can be lifted from the hinge by means of a special insulator rod and brought down for replacement of fuse element. After replacing the element the tube is placed in the hinge and the device is closed in a way similar to closing of isolators. Drop-out fuse-isolator combination is usually pole mounted.

6.7. STRIKER FUSE

It is a mechanical device having enough force and displacement which can be used for closing tripping/indicator circuits.

6.8. SWITCH FUSE

It is a compact combination, generally metal enclosed of a switch and a fuse and is very widely used for low and medium voltages. The ratings of switch fuse units are in the range of 30,60, 100, 200, 400, 600 and 800 amperes. Switch fuse units are available as 3-pole and 4-pole units.

They have been developed for making capacities (prospective) up to 46 kA. They can safely break, depending upon ratings, currents of the order of 3 times the load current.

Such units can be installed on metal-clad switchgear.

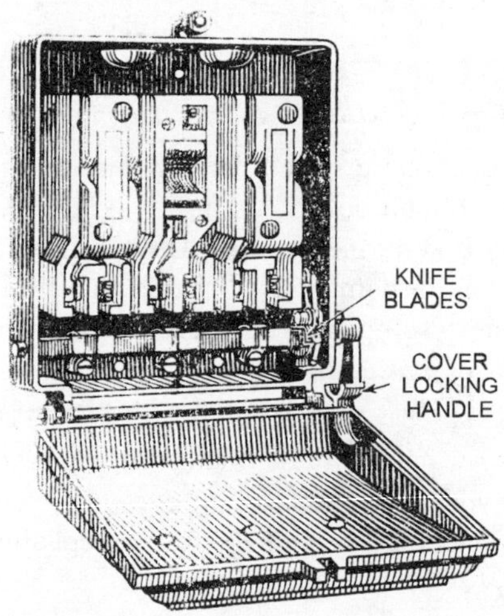

Fig. 6.13. ***Modern Three-Phase Knife Switch and Fuses***

6.9. HIGH VOLTAGE HRC FUSES

Low voltage fuses, discussed so far, have low normal current rating and rupturing capacity. They cannot be therefore, employed successfully on modern high voltage circuits.

The main problem that starts at high voltage is that of corona. Therefore high voltage fuses have got special designs. These are of two types viz., cartridge type and liquid type.

1. Cartridge Type HV HRC Fuse. This is similar in general construction to the low voltage type except that some special design features are incorporated . In some designs the element is wound in the shape of a helix so as to avoid corona effect at higher voltages. In some designs two fuse elements in parallel are employed—one of low resistance carries the normal current.; the other is of high resistance (tungsten wire) and, after the low resistance has blown off, reduces the short-circuit current and finally breaks the circuit.

High voltage cartridge fuses are used up to 33 kV, with rupturing capacity of about 8,760 A at that voltage (500 MVA three phase). Ratings of the order of 200 A at 6.6 kV and 11 kV, and 50 A at 33 kV are also available.

2. Liquid Type HV HRC Fuse. Such fuses are filled with carbon tetra chloride and have the widest range of applications in high voltage circuits. They may be employed for voltage transformer protection or for circuits up to about 400 A rated current on systems up to 132 kV or higher and may have breaking capacity of 6,100 A at 33 kV (350 MVA, 3-phase).

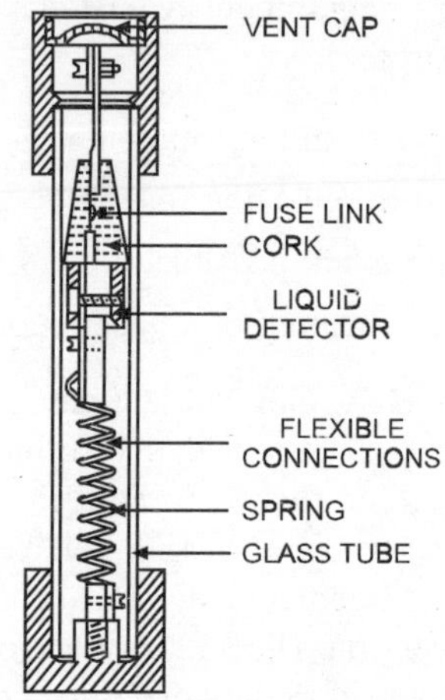

Fig. 6.14. ***Liquid type HV HRC Fuse***

Liquid fuse is shown in Fig. 6.14. It consists of a glass tube filled with carbon tetra chloride and sealed at both ends with brass caps. The fuse wire is sealed at one end of the tube and the other end of the element is held by a strong phosphor bronze spiral spring fixed at the other end of the glass tube. On occurrence of a fault or on overload the current exceeds the permissible limit, the fuse element is blown out, the spring retracts part of it through a baffle (or liquid detector) and draws it well into the liquid. The small quantity of gas generated at the point of fusion forces some part of liquid into the passage through baffle and there it effectively extinguishes the arc.

High voltage HRC fuses are used for the backup protection to circuit breakers, whose short-circuit capacity has been increased owing to extension of generating plant or interconnection to a value beyond their rated MVA. The fuses are provided to deal with short circuits exceeding a certain magnitude while the circuit breakers continue to interrupt short circuits within their breaking capacity. The time-current characteristics of the fuses are suitably adjusted to that of the circuit breaker in order to ensure correct operation. This arrangement eliminates the necessity of replacement of the circuit breaker.

6.10. EXPULSION TYPE HV FUSE

High voltage expulsion type fuses are widely used for protection of feeders and transformers because of their low cost. Such fuses are developed for 11 kV, 250 MVA rupturing capacity. The fuses are operated by a special operating rod from ground below. A blown fuse can be readily spotted from a distance as the fuse-carrying barrel drops when the fuse blows.

Such a fuse comprises of hollow open ended tube made of synthetic resin-bonded paper. The fuse element is placed in the tube and the ends of the element are connected to suitable fittings at each end. The length of the tube is generally longer than conventional enclosed fuses. The arc produced by blowing of fuse causes decomposition of inner coating of the tube and the gases thus formed assist arc extinction.

6.11. SELECTION OF FUSE

In order to ensure that the fuses will correctly and reliably protect any given section or element of an electrical circuit, their characteristics and ratings must be selected in accordance with the following requirements.

Rated Voltage : Fuses and their fusible elements must be selected so that their voltage rating is the same as the service voltage of the circuit they are to be inserted in.

Maximum Current Rupturing Capacity : To extinguish the arc resulting from blowing of the element in a fuse safely, the latter and its element must have a current interrupting capacity at least equal to or greater than the largest short-circuit current which can flow in the circuit to be protected.

Rated Current : The fuse element to be used in a fuse must be selected for a rated current large enough to ensure that the element will not fuse when the circuit is carrying its sustained maximum load current and when the circuit is carrying a short-time overload current.

To satisfy the first of these requirements the following condition must be observed to continue normal operation when practically a continuous maximum load current is being carried.

Rated current = K times sustained maximum load current

where K is a factor of safety with respect to the load current and depends upon the nature of the load. For practically constant loads such as lighting load K = 1.1 – 1.2. If the load current varies considerably as in the case of motors which, during starting periods draw large currents greatly in excess of the normal operating current, the current rating of the fuse element must be selected so as to satisfy the second requirement. This consists in selecting the current rating so that the element will not blow during short-time overload (of 2–10 second duration). It is, therefore, necessary to select an element with such a time-current characteristic that the duration of the overload current will be shorter than the period in which the element will blow at the same value of current. This means that it will carry the short-time overload current without blowing, although its continuous current rating may be less than short-time overload current.

To select the fuse element as mentioned above, its time-current characteristic must be available. As this is not convenient in practice, the element current rating is determined from the following simple expression.

Continuous current rating = Short-time overload current/K

where K is the factor of safety with respect to the overload current.

Suitable values for K have been determined on the basis of extensive operating experience and are valid on condition that the overload, and the resultant greater heating of the element due to it, does not lead to oxidation and premature aging of the element.

Depending upon the duty conditions, for overload duration 2-3 seconds (moderate duty) K = 2.5; for overload duration of 10 seconds (severe duty) K = 1.5 – 2. When motors are started frequently the element current rating should be selected by taking the factor for severe duty conditions.

It is noteworthy that a fuse with an element selected keeping in view the short-time overload current will not protect the equipment supplied through it from overload and only serves as a protection against short circuits. Protection against overloads should be provided by separate devices like overload trips.

In houses, blocks of flats, personnel amenity rooms and public buildings, *i.e.,* wherever the electrical circuits are not constantly supervised by trained personnel the fuse elements must satisfy the condition.

Continuous current rating of fuse element = 0.8 × permissible continuous current carrying capacity of the conductors forming the circuits to be protected.

In case of protection for *transformers and fluorescent lighting*, the fuse selected should generally have a normal rating 25 to 50 per cent above the normal full-load current of the apparatus to be protected so as to withstand the transient current surge on switching in.

In case of protection of capacitors it is desirable to use a cartridge fuse link having rated current about 25-50% greater than the rated current of capacitors. Protection of capacitors is a difficult task for any protective device due to presence of the higher harmonics in the voltage wave and also due to the fact that the fuses protecting the capacitors are subject to transient currents during switching operation.

Fuses will provide protection against *earth faults* only when the earth resistance is sufficiently low to enable their operation. In case the earth resistance is high, the fuses will not be able to provide the desired protection and safety.

Exercises

1. What is meant by a fuse? What are the advantages of fuses? Name some of the commonly used materials as fuse element. Why silver is preferred as fuse element inspite of its higher cost ?
2. Explain the difference between fusing current and current carrying capacity of a fuse.

   Name some important materials for fuse elements. Describe the essential requirements of a fuse material.
3. Discuss the different types of fuses in brief as used in electrical circuits.
4. Distinguish between fuse and circuit breaker. Discuss the advantages and disadvantages of a fuse. What are the desirable characteristics of fuse elements ? Why do we prefer silver as a fuse element ?
5. What are the different types of fuses? Discuss with a neat sketch the construction and working principle of a HRC fuse. State the advantages and disadvantages of it over other types of fuses.
6. Silver, inspite of its high cost is used as a fuse element in a HRC fuse, explain why ? What are the purposes of using quartz sand as filler material in the construction of HRC fuse ?

7. What is a HRC fuse ? Explain its construction and operation. Also mention its applications.
8. Describe the construction and operation of a HRC cartridge fuse. What are its advantages and disadvantages ?
   Explain the following terms in connection with a fuse:
   (*i*) rated current. (*ii*) fusing factor.
   What do you mean by discrimination ? Discuss discrimination between a fuse and an overcurrent relay.
9. With a neat sketch explain the principle of operation of a HRC fuse. Name at least two application areas of such a fuse. Is there any better type of fuse as compared to HRC fuses available in the market ? What is drop-out fuse ?
10. Explain in brief the principle of working of low voltage HRC cartridge fuse. How can this fuse ease the duty on circuit breakers ?
11. Describe the construction and operation of the HRC cartridge fuse. What are its advantages and disadvantages ?
    What are considerations in selecting a fuse for
    (*i*) transformer protection (*ii*) motor protection (*iii*) capacitor protection (*iv*) heaters (*v*) lighting loads ?
    What do you mean by discrimination ? Discuss discrimination between (*i*) two fuses; (*ii*) a fuse and an overcurrent relay.
    [A.M.I.E. Sec. B. Advanced Power Systems Summer 2000]
12. In relation to a fuse, explain what do you mean by (*i*) prospective current (*ii*) cut-off current and (*iii*) time-delay fuse. Give the advantages of HRC fuses.
    [Pb. Univ. Switchgear & Protection June 1991]
13. Describe the HRC fuse with a tripping device with a neat sketch. State its advantages over an HRC fuse without tripping device.
14. Write short notes on the following :
    (*i*) Time-current characteristic of a fuse.
    (*ii*) Fuse element materials.
    (*iii*) HRC fuses. **[Pb. Univ. Dec., 1991]**
    (*iv*) HRC fuse with tripping device.
    (*v*) High voltage fuse.
    (*vi*) Short-circuit protection using HRC fuses.

Short Answer Type Questions With Answers

Q. 1. What is meant by a fuse ?

Ans. Fuse is a device used in an electrical circuit for protecting electrical equipments against overloads and/or short circuits.

Q. 2. On what principle does the fuse operate ?

Ans. The action of a fuse is based upon the heating effect of the electric current. During overloads or short circuits the fuse gets heated up to melting point due to excessive current and breaks the circuit.

Q. 3. What kind of protection is offered by a fuse ?

Ans. The fuse provides protection against short circuits and overloads.

Q. 4. What function does the fuse perform ?

Ans. The function of a fuse wire is (*i*) to carry the normal working current safely without heating and (*ii*) to break the circuit when current exceeds the limiting current.

Q. 5. Give the names of five materials with their melting points which are used for fuse element.

Ans. The materials commonly used for fuse elements are silver, tin, lead, copper and aluminium. Their melting points are 980° C, 240°C, 328°C, 1,090° C and 665° C respectively.

Q. 6. What material is mostly used for an ordinary fuse wire ?

Ans. Copper or lead-tin alloy.

Q. 7. What is fuse link ?

Ans. Fuse link is that part of the fuse which needs replacement when the fuse blows out.

Q. 8. On what factors does the current carrying capacity of a fuse depend ?

Ans. The current carrying capacity of a fuse depends upon the permissible temperature rise of the contacts of the fuse holder, size of fuse wire, fuse material and the surrounding of the fuse.

Q. 9. On what factors does the fusing current depend ?

Ans. The fusing current depends upon (*i*) diameter of wire (*ii*) length of wire (*iii*) fuse material (*iv*) x-sectional area (round or rectangular) (*v*) size and location of terminals (*vi*) type of enclosure used (*vii*) type of surface etc.

Q. 10. What are the purposes of using quartz sand as filler material in the construction of HRC fuse ?

Ans. The purposes of using quartz sand as filler material in the construction of HRC fuse are to absorb heat at a very high rate, to produce a substance of high resistance and avoid the oxidation of silver element.

Q. 11. What is meant by discrimination ?

Ans. Discrimination means the breaking of the faulty section of the line to an extent restricted to minimum; healthy sections of the circuit remaining in operation.

Q. 12. Can the HRC fuse be used for both ac and dc circuits. If so under what condition ?

Ans. In general an HRC fuse meant for ac circuit may not be used for dc circuits; but however, a fuse satisfying the requirement of arc voltage can be used on both ac and dc circuits.

❑❑❑

CHAPTER 7

Circuit Breaking

INSIDE THIS CHAPTER

7.1. INTRODUCTION

Everyone is familiar with low voltage switches and rewirable fuses. A switch is used for opening and closing of an electric circuit while a fuse is used for overcurrent protection. Every electric circuit needs a switching device and protective device. Switching and protective devices have been developed in different forms. *Switchgear* is a general term covering a wide range of equipment concerned with switching and protection.

In a power system switchgear serves two basic purposes :

(*i*) Switching during normal operating conditions for the purpose of operation and maintenance.

(*ii*) Switching during abnormal conditions such as short circuits and interrupting the fault currents.

The first of above could be served by relatively simple switches because it is relatively simple as it involves normal currents which are easy to interrupt. The second function is, however, complex. With the advancement of electrical power system the lines and other equipment operate at very high voltage and carry large currents. Whenever a short circuit occurs, a heavy current flows through the equipment causing considerable damage to the equipment and interruption of service. In order to avoid such a damage every part of the power system is provided with a protective relaying system and an associated switching device. The function of protective relaying system is to cause the prompt removal from service of any element of a power system when it suffers a short circuit, or when it starts to operate in any abnormal manner that might cause damage or otherwise interfere with the effective operation of the rest of the system. The relaying equipment is aided in this task by *circuit breakers* that are capable of disconnecting the faulty element when they are called upon to do so by the relaying equipment. In addition to circuit breakers and protective relays, the associated equipment for controlling, regulating and measuring can also be considered as switchgear devices. Thus the apparatus including its associated auxiliaries employed for controlling, regulating or switching *on* or *off* the electrical circuits in the electrical power system is known as *switchgear.* It includes switches, fuses, circuit breakers, isolators, relays, current and potential transformers, indicating instruments, lightning arresters and control panels.

A circuit breaker is a device that interrupts the abnormal or fault currents and in addition performs the function of a switch. Circuit breakers and fuses both possess a time element of operation, whereby they operate practically instantaneously on short circuits but with definite time lag on overloads. Circuit breakers can be reset in less time and with less trouble that are required to replace blown fuses and spare parts are seldom required. For low voltage circuits fuses may be used to isolate the faulty circuits; but for higher voltages, say from 3.3 kV upwards, isolation is achieved by circuit breaker. Circuit breakers may also be preferred where continuity of service is an important consideration or where frequent fuse replacement may be expected. However, for low capacities a fuse with circuit breaking arrangement is quite useful and economical.

As the circuit breakers have to be accessible for inspection and maintenance, the Indian Standard regulations require the provision for separation of the breaker from the live parts by means of isolating switches. *Isolating switches* are employed for separating the disconnected element from the live portion of the system. They should be employed only after the corresponding circuit breakers have been opened. Isolating switches are employed for isolating a piece of equipment from adjacent live elements for purposes of safety.

7.2. CIRCUIT BREAKERS

Circuit breakers are mechanical devices designed to close or open contact members, thus closing or opening of an electrical circuit under normal or abnormal conditions.

Automatic circuit breakers, which are usually employed for the protection of electrical circuits, are equipped with a trip coil connected to a relay or other means, designed to open the breaker automatically under abnormal conditions, such as overcurrent. The automatic circuit breakers perform the following duties :

(*i*) It carries the full-load current continuously without overheating or damage,

(*ii*) It opens and closes the circuit on no load,

(*iii*) It makes and breaks the normal operating current and

(*iv*) It makes and breaks the short-circuit currents of magnitude up to which it designed for.

The circuit breaker performs first three duties satisfactorily but in performing fourth duty *i.e.,* when it is to make or break short-circuit currents, it is subjected to mechanical and thermal stresses. The circuit breakers are rated in terms of maximum voltage, number of poles, frequency, maximum continuous current carrying capacity, maximum interrupting capacity and maximum momentary and 4-s current carrying capacity.

The interrupting or rupturing capacity of a circuit breaker is the maximum value of current which can be interrupted by it without any damage. The circuit breakers are also rated in MVA which is the product of interrupting current, rated voltage and 10^{-6}.

7.2.1. Operating Principle

A circuit breaker is a switching and current interrupting device. It consists, essentially, of fixed and moving contacts, which are touching each other and carry the current under normal conditions *i.e.,* when circuit breaker is closed. When the circuit breaker is closed, the current carrying contacts, called the electrodes, engage each other under the pressure of a spring.

During the normal operating condition the circuit breaker can be opened or closed by a station operator for the purpose of switching and maintenance. To open the circuit breaker, only a small pressure is required to be applied on a trigger. Whenever a fault occurs on any part of the power system, the trip coils of the breaker get energized and the moving contacts are pulled apart by some mechanism, thus opening the circuit. The separation of current carrying contacts produces an arc. The current is thus able to continue until the discharge ceases. The production of arc not only delays the current interruption process but it also generates enormous heat which may cause damage to the system or to the breaker itself. Therefore, the main problem in a circuit breaker is to extinguish the arc within the shortest possible time so that heat generated by it may not reach a dangerous value.

The basic construction of a circuit breaker requires the separation of contacts in an insulating fluid which serves two functions :

1. Extinguishes the arc drawn between the contacts when the circuit breaker opens.
2. Provides insulation between the contacts and from each contact to earth.

The insulating fluids commonly used for this purpose are as follows :

1. Air at atmospheric pressure.
2. Compressed air.
3. Oil producing hydrogen for arc extinction.
4. Ultra high vacuum.
5. Sulphur hexafluoride (SF_6).

The fluids used in circuit breakers should have the properties of high dielectric strength, noninflammability, high thermal stability, arc extinguishing ability, chemical stability, and commercial availability at moderate cost.

Of the simple gases air is the cheapest and most widely used for circuit breaking. Hydrogen has better arc extinguishing property but it has lower dielectric strength as compared to air. Also if hydrogen is contaminated with air, it forms an explosive mixture. Nitrogen has similar properties as air. CO_2 has almost the same dielectric strength as air but is a better arc extinguishing medium at moderate currents. Oxygen is a good extinguishing medium but is chemically active. SF_6 has outstanding arc-quenching properties and good dielectric strength. Of all these gases SF_6 and air are used in commercial gas blast circuit breakers.

7.3. ARC PHENOMENA

The arc consists of a column of ionized gas having molecules which have lost one or more electrons. The electrons being negatively charged are attracted towards the positive contact (*i.e.,* anode) with a high velocity and on the way they detach more electrons by impact. The positive ions are attracted towards the negative contact (*i.e.,* cathode), but as they comprise almost the entire weight of the atom, they move towards it relatively slowly. Thus current flow is caused due to movement of electrons.

7.3.1. Initiation of an Arc

For the initiation of an arc it is necessary that the electrons are emitted from the cathode as soon as the contacts begin to separate on occurrence of fault. Initiating electrons are thought of produced by the following two processes :

(*i*) By high voltage gradient at the cathode resulting into *field emission.*

As the moving contact is withdrawn, the contact area and the pressure between the separating contacts decreases and due to decrease in contact area the resistance increases (but it is still much less than a ohm). Although contact resistance is quite small but due to large magnitude of fault current a sufficiently high potential drop, of the order of 10^6 V/cm, is caused between the separating contacts so as to dislodge the electrons from the cathode surface.

(*ii*) By increase of temperature resulting into *thermionic emission.*

As the contacts apart, the decrease in contact area causes increase in current density to very high values, of the order of

10^6 A/cm^2. These very high current densities raise the temperature of the contact (cathode) surface resulting into *thermal emission*.

In case of circuit breakers the contacts used are usually of copper, the thermionic emission from such a metal is quite low and so for initiation of arc, the field emission is mostly responsible.

7.3.2. Maintenance of Arc

The electrons so emitted from the cathode make many collisions with the atoms and molecules of gases and vapours existing between the two contacts during their journey towards the anode. Such collisions cause ionization of atoms and the molecules, thus dislodging more electrons. The ionization is further facilitated by :

(*i*) high temperature of the medium around the contacts caused by high current densities, with high temperature the kinetic energy gained by the moving electrons is increased.

(*ii*) the field strength or voltage gradient which increases the kinetic energy of moving electrons and increases the chances of detaching electrons from neutral molecules.

(*iii*) an increase of mean free path—the distance through which the electron moves freely. As the contacts apart, the mean path increases and the number of neutral molecules increases, also, the increase in mean path decreases the density of gas which further increases the free path movement of the electrons.

All the above three processes (thermal emission, ionization and field emission) may start either one after the other or almost simultaneously and enable the arc to be initiated and maintained and finally if the arc current is high, the arc may attain a temperature high enough for thermal ionization to become the main source of electrical conductivity.

7.3.3. Arc Voltage

As the contacts of the circuit breaker apart, an arc is formed. The voltage that appears across the contacts of the circuit breaker is called the *arc voltage*.

For moderate values of current and voltage, the arc characteristic can be expressed by Ayrton's equation

$$e_a = A + \frac{B}{i_a} \qquad ...(7.1)$$

Thus with the increase in arc current, the voltage drops as a hyperbola. The constants A and B vary linearly with the arc length l

$$A = \alpha + \gamma l$$

$$\text{and } B = \beta + \delta l \qquad ...(7.2)$$

Average values of α, γ, β and δ for arcs in air between copper electrodes are as follows :

$$\alpha = 30 \text{ V}; \gamma = 10 \text{ V/cm}; \beta = 10 \text{ VA}; \delta = 30 \text{ VA/cm}$$

From above Eq. (7.1) it is obvious that the volt-ampere characteristic of an arc voltage is negative *i.e.*, arc voltage is high when the arc current is low and vice versa. This is, of course, a well known property of arcs.

Figure 7.1 illustrates the characteristics of ac current and voltage with respect to time. It is seen from Fig. 7.1 that the arc voltage is almost constant during the time when the current is near its peak values. At current zero, the arc voltage rises rapidly to peak value and this peak value tends to maintain the current flow in the form of arc.

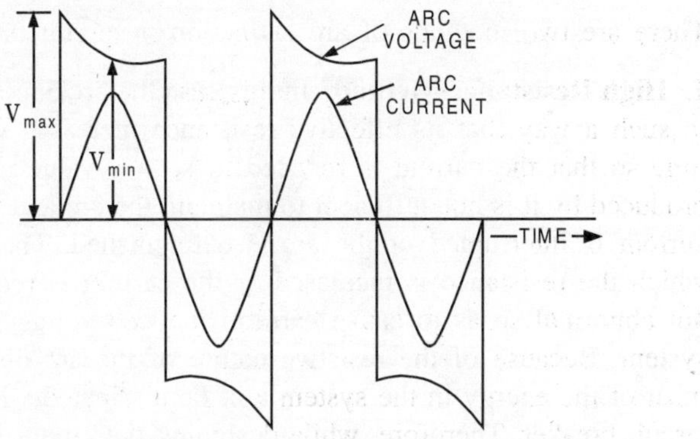

Fig. 7.1. ***Voltage and Current Waveforms of an AC Arc***

The voltage across the arc is in phase with arc current as the arc current is predominantly resistive. The magnitude of arc voltage increases in each successive current loop. This is because the circuit breaker contacts are assumed to be separating thereby increasing the arc length and therefore the arc voltage.

7.4. ARC EXTINCTION

When current carrying contacts of a circuit breaker are parted, an arc is formed, which persists during the brief period after separation of contacts. The arc provides a gradual transition from the current carrying to the voltage isolating states of the contacts, but it is dangerous on account of the energy generated in it in the form of heat which may result in explosive force. The circuit breaker should be capable of extinguishing the arc without causing any damage to the equipment or danger to personnel. The arc plays a vital role in the behaviour of the circuit breaker. The interruption of dc arcs is relatively more difficult than ac arcs. In ac arcs, as the current becomes zero during the regular wave, the arc vanishes and it is prevented from restriking.

Before discussing the methods of arc extinction, it is necessary to examine the factors responsible for the maintenance of arc between the contacts. These are :

(*i*) potential difference between the contacts, and

(*ii*) ionised particles between the contacts.

The potential drop between the separating contacts is just sufficient to maintain the arc and is quite small. One way to extinguish the arc is to separate the contacts to such a distance that the pd becomes inadequate to maintain the arc. However, this method is impracticable in high voltage systems where a separation of many metres would be required for this purpose.

The conductance of the arc is proportional to the number of electrons per cubic centimetre produced by ionization, the square of the diameter of the arc, and the reciprocal of the length. As already stated above, we cannot do much by increasing the length of arc to any reasonable value. What can be done is to reduce the density of free electrons, *i.e.,* reduce the ionization, and decrease the diameter of the arc. The arc extinction can, therefore, be facilitated by deionizing the arc path. This may be achieved by cooling the arc or by bodily removing the ionised particles from the space between the circuit breaker contacts.

7.4.1. Methods of Arc Extinction

There are two methods of arc extinction in circuit breakers.

1. High Resistance Method. In this case the arc is controlled in such a way that its effective resistance increases with the time so that the current is reduced to such a value that heat produced by it is not sufficient to maintain the arc and thus the current is interrupted or the arc is extinguished. The rate at which the resistance is increased or the current is reduced is not abnormal so as to cause harmful induced voltages in the system. Because of the resistive nature of the arc discharge, most of the energy in the system will be dissipated within the circuit breaker. Therefore, while designing the circuit breaker, provision of mechanical strength to withstand such sudden release of large quantities of energy must be made. This is the main drawback of this method of arc extinction and so its use is restricted to dc circuit breakers and air break type ac circuit breakers of relatively low capacities of the order of a few hundred MVA.

The resistance of the arc can be increased by

(*i*) ***Cooling of arc.*** Cooling of arc brings about recombination of ionized particles. This increases the arc resistance. Cooling removes the heat from the arc. Cooling is brought about by bringing the arc in contact with cool air.

(*ii*) ***Increasing the length of arc.*** The length of the arc can be increased by increasing the gap length between the contacts but it is not practicable to draw the arc out to such a length (may be in metres for ht system) that the voltage available becomes insufficient to maintain the arc.

(*iii*) ***Reducing the cross section of arc.*** The cross section of an arc can be reduced by having a small area of contacts or by letting the arc pass through a narrow opening. By reducing the area of cross section of the arc, the voltage necessary to maintain the arc is increased.

(*iv*) ***Splitting of arc.*** The resistance of the arc can be increased by splitting the arc into a number of small arcs in series. Each one of these arcs experiences the effect of lengthening and cooling. The arc can be splitted up by introducing some conducting plates between the contacts. In the other method of splitting of arc the arc is forced into an arrangement of splitters by which the arc is lengthened and the cooling is improved because of contacts with the splitters.

2. Low Resistance or Current Zero Interruption. This method is applicable only in ac circuit interruption because there is natural zero of current 100 times in a second for 50 Hz 3-phase supply system. This property of ac circuit is exploited for interruption purposes and the current is not allowed to rise again after a zero occurs. Also it is neither necessary nor desirable to cut off the current at any other point on the ac wave because this will induce high voltages in the system.

In this method the arc resistance is kept low until the current is zero where the arc extinguishes naturally and is prevented from restriking after it has gone out at a current zero. This method of arc extinction is employed in all modern high power ac circuit breakers.

The phenomenon of arc extinction is explained by two theories as follows:

(*i*) ***Energy Balance or Cassie Theory.*** This theory states that if the rate of heat dissipation between the contacts is greater than the rate at which heat is generated, the arc will be extinguished, otherwise it will restrike. The heat generated varies from time to time depending upon the separation of breaker contacts. Initially when the contacts are about to open, the restriking voltage is zero and, therefore, the heat generated is zero. Again when the contacts are fully open, the resistance between the contacts is also infinite and hence the heat generated is zero. Between these two limits the heat generation reaches a maximum. Now if the heat so generated could be removed by cooling, lengthening and splitting the arc at a rate higher than that of generation, the arc is extinguished.

(*ii*) ***Recovery Rate or Slepian's Theory.*** This theory states that if the rate at which the ions and electrons combine to form or are replaced by neutral molecules *i.e.,* the rate at which the gap recovers its dielectric strength is faster than the rate at which voltage stress rises, the arc will be extinguished ; if otherwise the arc may be interrupted for a brief period but it again restrikes. This theory assumes that the restriking voltage and build-up of dielectric strength are independent quantities. This assumption is not quite true because the dielectric strength calculations do not agree with the observed value.

In an ac system current drops to zero after every half cycle. At every current zero, the arc extinguishes for a brief period. Now the medium between the breaker contacts contains ions and electrons so that it has small dielectric strength and can be easily broken down by the rising contact voltage called the *restriking voltage*. If such a breakdown does occur, the arc will persist for another half cycle when the process will be repeated. If immediately after the current zero, the dielectric strength of the medium between breaker contacts is built up more rapidly than the voltage across the contacts, the arc fails to restrike and the current will be interrupted. The rapid increase of dielectric strength of the medium near current zero can be achieved by either causing the ionised particles in the space between contacts to recombine into neutral molecules or sweeping the ionized particles away and replacing them by un-ionised particles. The problem is, therefore, to remove the ions and electrons either by causing them to recombine into neutral molecules or by sweeping them away, as soon as the current becomes zero, so that rising contact voltage or restriking voltage cannot breakdown the space between the contacts. This can be achieved by the following methods :

(*a*) ***Lengthening of the gap.*** The dielectric strength or post-zero resistance is proportional to the length of the gap between the breaker contacts. So lengthening by rapid opening of the breaker contacts is an obvious process. The permissible arc length is limited, however, by other considerations, *e.g.,* arc energy and possibility of transient voltages due to current chopping.

(*b*) ***Increasing the pressure in the vicinity of the arc.*** By increasing the pressure the density of particles constituting the discharge also increases. The increased density of particles causes higher rate of deionization and thus the dielectric strength of the medium between the contacts is increased.

(*c*) ***Cooling.*** If the particles are allowed to cool the natural combination of ionized particles will take place more rapidly resulting increase in dielectric strength of the medium. Cooling by conduction to adjacent parts, *e.g.,* baffles or by the use of gas such as hydrogen that has a high diffusion and heat absorption rate is, therefore, effective.

(*d*) ***Blast Effect.*** By blowing a stream of air through the arc ionized particles between the contacts are swept away and replaced by un-ionised particles. These un-ionised particles increase the dielectric strength of the medium considerably.

Extinction of dc arc is much difficult than that of an ac arc because in an ac circuit the current wave passes through the zero point twice during each cycle *i.e.,* 100 times in a second but in dc circuit the full current has to be broken. When a direct current passing through a highly inductive circuit is broken, the arc tends to persist, this tendency is more marked if the voltage is high. Once an arc is formed between two points, the air molecules in the path become extremely hot and get ionized, *i.e.,* the normal insulation property of the air is destroyed and the hot air molecules become conductors of electricity. Therefore, the arc is maintained even if the contacts are further drawn apart, resulting ultimately in a flash-over. The heat generated by an arc is intense. To overcome this difficulty, several methods have been devised. When small currents are involved, the contacts are broken quickly before the air can get ionised. When heavier currents are to be broken, the arc is blown off by creating a magnetic field having the property of deflecting the arc. The magnetic field is created by *blow-out coils* through which the current to be broken flows. The arc path is deflected until it is blown upwards in special arc chutes.

7.5. IMPORTANT TERMS ASSOCIATED WITH CIRCUIT BREAKING

7.5.1. Restriking Voltage and Recovery Voltage

In ac circuit breakers, the current interruption takes place invariably at the natural zero of current wave.

At current zero, a high frequency transient voltage appears across the breaker contacts and is caused by the rapid distribution of energy between the magnetic and electric fields associated with the plant and transmission lines of the power system. This transient voltage is known as *restriking voltage. This* voltage appearing across the breaker contacts at the moment of final current zero has a profound influence on the arc extinction process. Under the influence of this voltage the arc tries to restrike and hence it is named as the *restriking voltage.* After current zero, the arc gets extinguished if the rate of rise of restriking voltage between the contacts is less than the rate at which dielectric strength of the medium between the contacts gains.

Thus *restriking voltage may be defined as the resultant transient voltage which appears across the breaker contacts at the instant of arc extinction.*

In 3-phase circuits the restriking voltage refers to the voltage across the first pole to clear because this voltage is generally higher than that which appears across each of the other two.

Immediately after final current interruption, the voltage that appears across the breaker contacts is a transient voltage of higher natural frequency (restriking voltage) superimposed on the power frequency system voltage (recovery voltage). However, the transient component vanishes rapidly (after a short time of the order of less than 0.1 ms) due to damping effect of system resistance and the normal frequency system voltage is established. This voltage across the breaker contacts is called the *recovery voltage.* Figure 7.2 illustrates the waveforms of the recovery voltage and the restriking voltage. After current zero the voltage appearing across the breaker contacts is composed of transient restriking voltage and power frequency recovery voltage.

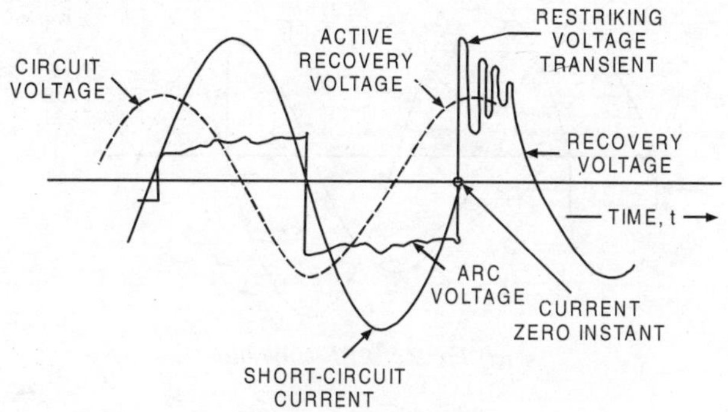

Fig. 7.2. *Arc Extinction*

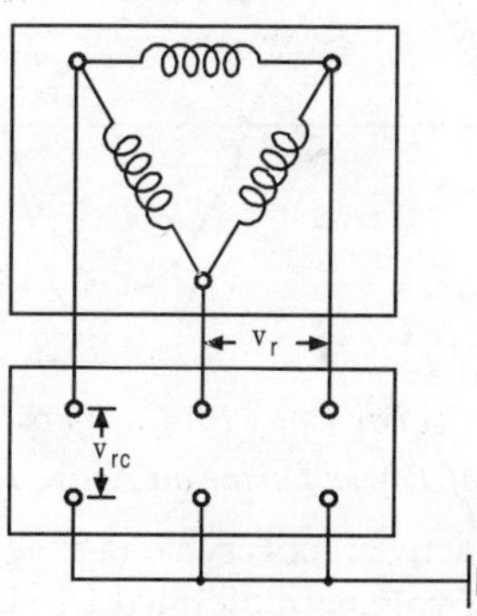

Fig. 7.3

Thus recovery voltage may be defined as the power frequency rms voltage which appears across the breaker contacts after the transient oscillations die out and final extinction of arc has resulted in all the poles.

In case of multipole circuit breakers it is expressed as the voltage between the phases of the circuit. In such cases another term, *recovery voltage component* (V_{rc}) is used to express the voltage across the breaks of each pole.

In addition to above there is another term which is called as *active recovery voltage* which is defined as the instantaneous recovery voltage at the instant of arc extinction.

7.5.2. Conditions of Severity

The active recovery voltage (*i.e.,* the instantaneous value of recovery voltage at the instant of final current zero) depends upon the following factors:

1. Effect of Power Factor of Circuit on Recovery Voltage. The value of active recovery voltage depends on the power factor of the circuit, as illustrated in Fig. 7.4. For unity power factor loads the voltage and current are in phase and both are zero at the same instant. For zero power factor currents, the peak of the voltage is impressed on the circuit breaker pole at the instant of current zero. Such sudden application of voltage results in severe transient and has a high rate of transient recovery voltage (or restriking voltage). It is, therefore, obvious that a reactive short circuit is much more difficult to interrupt than a resistive circuit of the same current and voltage, as the active recovery voltage in the former case is high (maximum voltage) as compared to latter (zero voltage). Unfortunately all serious short circuits are almost purely reactive.

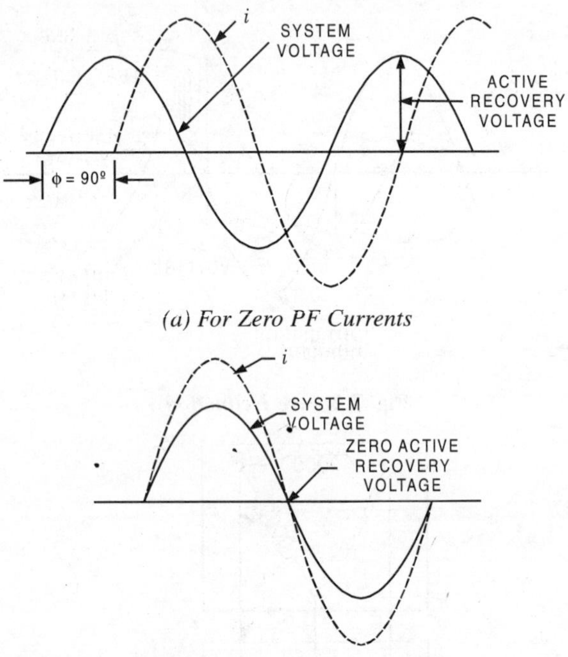

(a) For Zero PF Currents

(b) For Unity PF Currents

Fig. 7.4. ***Effect of Power Factor on Active Recovery Voltage***

In general, the active recovery voltage equals the maximum value of the system voltage multiplied by sin φ where φ is the power factor angle of the circuit.

$$i.e., \quad \text{Recovery voltage} = V_{max} \sin \phi \qquad ...(7.3)$$

2. Effect of Armature Reaction on Recovery Voltage. The short-circuit currents flowing in the generator windings, being of lagging power factor, have a demagnetising armature reaction. As a result the induced emf of an alternator is reduced during short-circuit currents. The emf needs sometime to regain its value. Hence the power frequency component of recovery voltage (*i.e.,* active recovery voltage) is slightly less than the normal value of system voltage.

3. Effect of Current Asymmetry on Recovery Voltage. The existence of a large dc component in a short-circuit current may cause the recovery voltage wave to start near the zero point instead of at its peak. The instantaneous value of recovery voltage is thus considerably reduced.

Thus by introducing asymmetry in the current wave to be interrupted it is possible to vary the value of active recovery voltage.

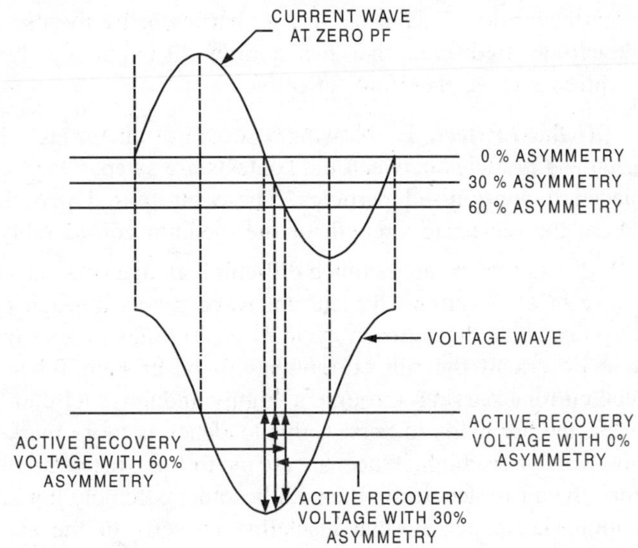

Fig. 7.5. ***Effect of Current Asymmetry on Active Recovery Voltage***

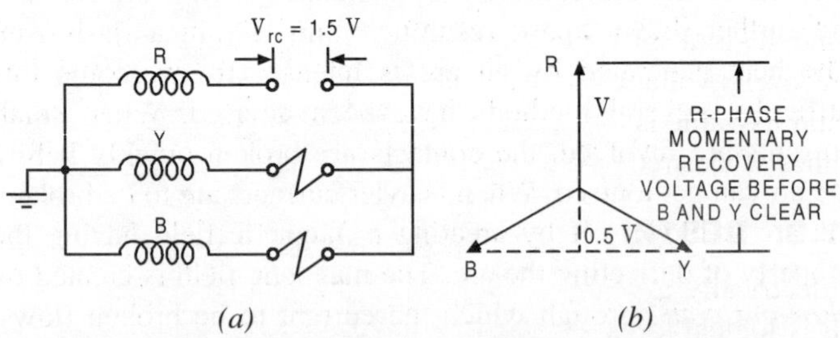

(a) *(b)*

Fig. 7.6. ***Effect of Fault on Recovery Voltage***

4. Effect of Circuit Conditions on Recovery Voltage. Another factor that influences the active recovery voltage is the circuit conditions *e.g.,* the type of fault and the condition of the neutral point (whether grounded or insulated). Three-phase faults that are insulated from earth, either at neutral or at fault, produce recovery voltages in the first phase to clear which are normally more severe than those produced by single-phase or three-phase faults on systems with grounded neutrals. This is explained below:

Consider an ungrounded 3-phase fault on a 3-phase system with neutral grounded, being cleared by a circuit breaker.

On opening of circuit breaker, an arc is drawn out in each phase. Let the arc in R phase be cleared first as shown in Fig. 7.6 (*a*). At the instant of extinction of this arc, Y and B phases are still arcing and have the same instantaneous phase voltage, – 0.5 V where V is the phase-to-neutral voltage of the system voltage. Now the arc resistances in the Y and B phases at this instant are negligible and therefore, the fault itself is momentarily at the potential of −0.5 V. Since the fault is common to all the three phases, the momentary value of the recovery voltage component in the first phase R must be V + 0.5 V *i.e.,* 1.5 V, as shown in Fig. 7.6 (*b*). This means that the recovery voltage component in the first phase to clear on a 3-phase ungrounded fault is 1.5 times that on ground fault assuming grounded neutral.

7.5.3. Mathematical Expression For Active Recovery Voltage in a 3-Phase Circuit

Let phase-to-neutral (rms) value of the system voltage be V volts. Hence maximum value of phase-to-neutral voltage V_{max} is given as

$$V_{max} = \sqrt{2}\ V \qquad ...(7.4)$$

In view of Eq. (7.3), the active recovery voltage between phase and neutral is given as

$$V_{max} \sin \phi \qquad ...(7.5)$$

In general the active recovery voltage can be given as

$$V_r = K_1 K_2 K_3 V_{max} \sin \phi \qquad ...(7.6)$$

where V_r is the phase-to-phase value of active recovery voltage;

K_1 is the demagnetizing factor due to which the recovery voltage will be less than the system voltage

$$\text{or } K_1 = \frac{\text{Recovery voltage}}{\text{System voltage}} \qquad ...(7.7)$$

K_2 is a condition or phase factor *i.e.,* it depends on the condition whether the symmetrical fault is grounded or not *i.e.,* its value is unity or 1.5, as shown in Art 7.5.2(4).

and K_3 is a third constant whose value is unity if the active recovery voltage between phase and neutral is to be obtained and its value is $\sqrt{3}$ if the active recovery voltage between two lines is desired.

7.6. RESTRIKING VOLTAGE TRANSIENT

Electrically a power system is an oscillatory network so that it is logical to expect that the interruption of fault current will give rise to a transient whose frequency depends on the constants of the circuit.

Let us consider a simple circuit, having a circuit breaker CB, as illustrated in Fig. 7.7 (*a*) and that a short circuit occurs on the feeder close to the bus-bars. The equivalent circuit is shown in Fig. 7.7 (*b*). Let L be the inductance per phase of the system up to the fault point, R be the resistance per phase of the system up to the fault point and C be the capacitance to earth of circuit breaker porcelain bushing.

Consider the opening of a circuit breaker under fault conditions shown in simplified form in Fig. 7.7 (*b*). Before current interruption, the capacitance C is short-circuited by the fault and the short-circuit current through the breaker is limited by resistance R and inductance L of the system. If R is negligible compared to L, the short-circuit current *i* will lag behind the system voltage *v* by 90°, as illustrated in Fig. 7.7 (*c*).

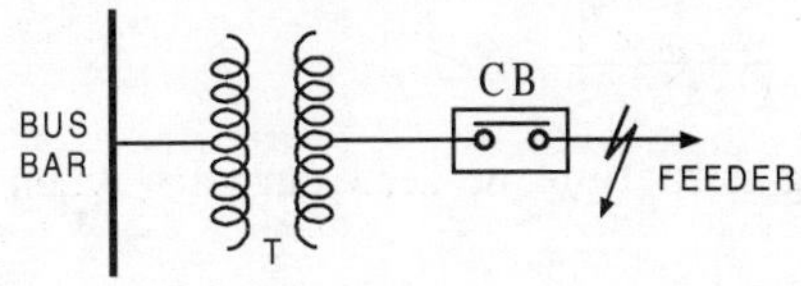

(*a*)

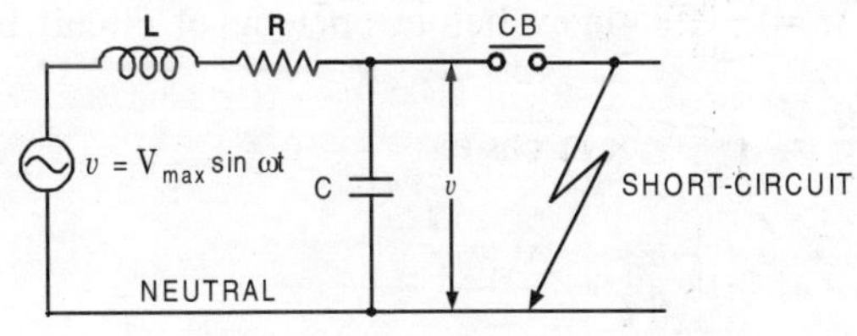

(*b*) *Fault and its Equivalent Circuit*

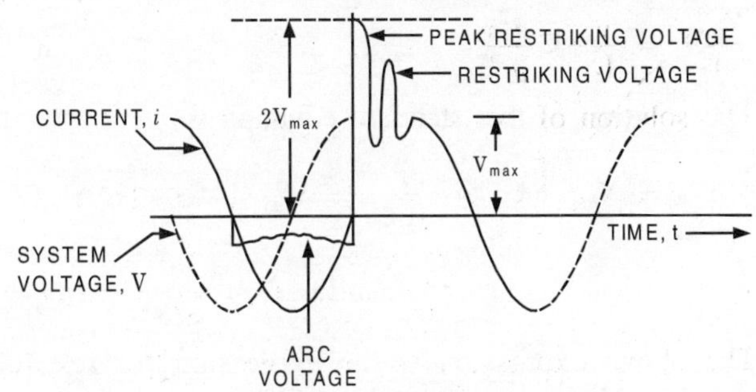

(*c*) *Operation of a Circuit Breaker*

Fig. 7.7

With the contacts opened and the arc broken, current *i* is diverted through capacitance C so that the voltage *v*, which has so far been effective only across the inductance L, is suddenly applied to the inductance L and capacitance C in series which form an oscillatory circuit, having a natural frequency.

$$f_n = \frac{1}{2\pi}\sqrt{\frac{1}{LC}}\ \text{Hz} \qquad ...(7.8)$$

The initial charging current surge tends to carry the voltage across the capacitor, and therefore across the circuit breaker contacts to double its equilibrium value *i.e.,* $2V_{max}$; this is the *restriking voltage transient* which tends to re-establish the arc in the circuit breaker.

These frequencies are of the order of 10 to 10 kHz depending upon the values of L and C. The actual power system is composed of distributed capacitances and inductances. The circuit configuration is also complex. The restriking voltage transient for such circuits can have several component frequencies ranging from a few Hz to several kHz.

7.6.1. Expression For Restriking Voltage Transient

When the breaker contacts are opened and the arc finally extinguishes at some current zero, a voltage *v* is suddenly applied across capacitor and therefore, across the circuit breaker contacts. The current *i* which would flow to the fault is not injected in the capacitor and inductor. Thus

$$i = i_L + i_c$$

$$\text{or } i = \frac{1}{L}\int v\, dt + C\frac{dv}{dt}$$

$$\therefore \quad \frac{di}{dt} = \frac{v}{L} + C\frac{d^2v}{dt^2} \qquad ...(7.9)$$

Assuming zero time at zero currents when $t = 0$, and further

$$v = V_{max} \cos \omega t$$

$$i = \frac{V_{max}}{\omega L} \sin \omega t \text{ before opening of circuit breaker}$$

$$\frac{di}{dt} = \frac{V_{max}}{\omega L} \times \omega \cos \omega t$$

$$t = 0; \quad \left|\frac{di}{dt}\right| = \frac{V_{max}}{L}$$

Substituting in Eq. (7.9) we have

$$\frac{V_{max}}{L} = \frac{v}{L} + C\frac{d^2v}{dt^2} \qquad ...(7.10)$$

The solution of this standard equation is

$$v = V_{max}\left[1 - \cos\frac{t}{\sqrt{LC}}\right] = V_{max}(1 - \cos 2\pi f_n t) \quad ...(7.11)$$

$$\because \text{ from Eq. (7.8) } \frac{1}{\sqrt{LC}} = 2\pi f_n = \omega_n$$

The above expression is the expression for restriking voltage where V_{max} is the peak value of recovery voltage (phase-to-neutral), t is time in seconds, L is inductance in henrys, C is the capacitance in farads and v is the restriking voltage in volts.

The maximum value of restriking voltage is $2V_{max}$ and occurs at $t = \frac{\pi}{\omega}$ or $t = \pi\sqrt{LC}$.

7.6.2. Classification of Restriking Transients

Restriking voltage transients, and consequently their respective circuits can generally be placed under two main categories.

(*i*) ***Single Frequency Oscillatory Transients.*** The single frequency restriking voltage transient is produced in the circuit illustrated in Fig. 7.7(*b*). The voltage waveform is shown in Fig. 7.7 (*c*).

(*ii*) ***Double Frequency Transients.*** It is quite possible that the circuit breaker may have L and C on its both sides, as illustrated in basic circuit given in Fig. 7.8 (*a*). Before clearing the fault, both terminals 1 and 2 are at the same potential. After the fault is cleared, *i.e.*, the arc has been extinguished, both the circuits oscillate at their own natural frequencies and a composite double frequency transient appears across the circuit breaker pole. This is illustrated in Fig. 7.8. (*b*).

7.6.3. Characteristics of Restriking Voltage

The important characteristics of restriking voltage which affect the performance of the circuit breaker are: (*i*) amplitude factor and (*ii*) rate of rise of restriking voltage, abbreviated as RRRV.

(*i*) ***Amplitude factor.*** The amplitude factor is defined as the ratio of the peak of transient voltage to the peak system frequency voltage.

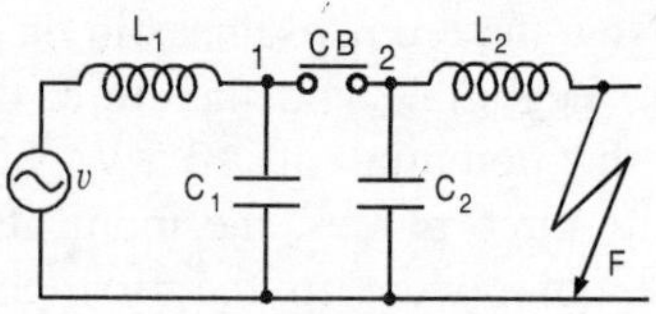

(a) Basic Circuit

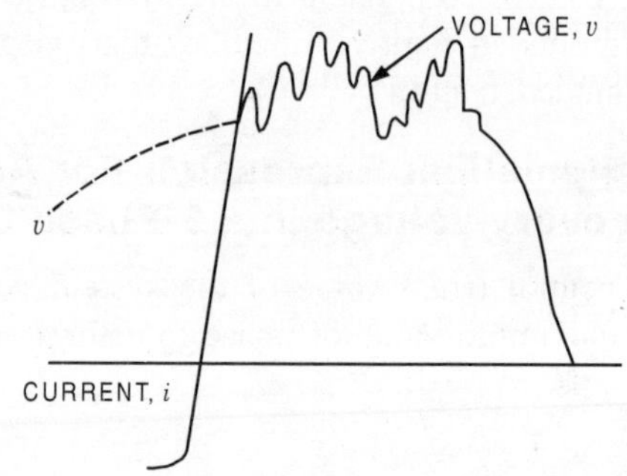

(b) Current and Voltage Waveforms Double Frequency Restriking Transient

Fig. 7.8

(*ii*) ***Rate of Rise of Restriking Voltage (RRRV).*** It is the rate of rise of restriking voltage and is expressed in kV/μs. It may be defined as the slope of the steepest tangent to the restriking voltage curve. For a restriking voltage having a single frequency transient component the RRRV is obtained by dividing the maximum amplitude of the oscillation by the duration of the first half wave. Higher values of natural frequencies can be related with higher rates of rise of restriking voltage. It is clear that other things being equal, the duty of circuit breaker is much more severe when employed in a network of higher natural frequency than on a network of low natural frequency. This is because the average RRRV is much greater in the former case. In the latter case the voltage across the circuit breaker contacts rises slowly thereby giving longer time for building up of the dielectric strength.

7.6.4. Expression For RRRV

The expression for the restriking voltage has already been derived in Art. 7.6.1 and is given as

$$v = V_{max}\left(1 - \cos\frac{t}{\sqrt{LC}}\right)$$

$$\text{and RRRV} = \frac{dv}{dt} = \frac{V_{max}}{\sqrt{LC}} \sin\frac{t}{\sqrt{LC}} \qquad ...(7.12)$$

RRRV will be maximum when

$$\sin\frac{t}{\sqrt{LC}} = 1 \text{ or } t = \frac{\pi}{2}\sqrt{LC} \qquad ...(7.13)$$

Hence maximum value of RRRV,

$$RRRV_{max} = V_{max}/\sqrt{LC} \qquad ...(7.14)$$

Further, the peak restriking voltage occurs when v is maximum *i.e.*, when

$$\frac{dv}{dt} = 0 \text{ or } \sin\frac{t}{\sqrt{LC}} = 0 \text{ or } t = \pi\sqrt{LC} \qquad ...(7.15)$$

7.6.5. Factors Affecting Restriking Voltage Characteristics

After current zero, the initial rate of rise and peak value of the restriking voltage stressing the contact gap depend upon the configuration of the network, its natural frequency and on the relative position of the resistances (in series or in parallel with the circuit main capacitance), as shown in Fig. 7.9. Because of presence of resistance dampening of the rate of rise of restriking voltage is quite logical. The true nature of attenuation is quite complicated because the losses depend on several factors such as conductor resistance, iron loss, dielectric loss, corona, etc. These factors depend on frequency and voltage in different ways.

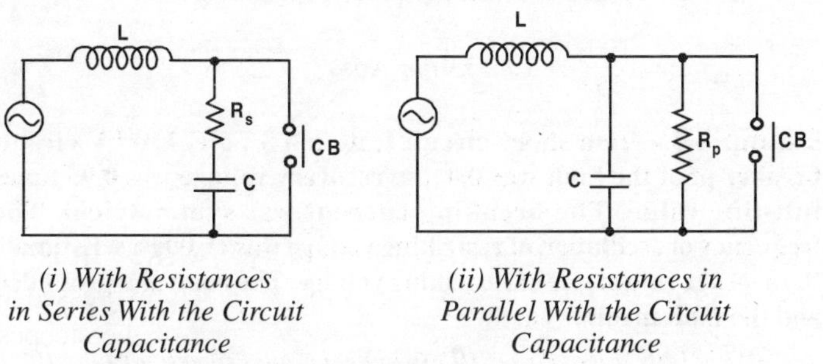

(i) With Resistances in Series With the Circuit Capacitance *(ii) With Resistances in Parallel With the Circuit Capacitance*

(a) Typical Switching Circuits

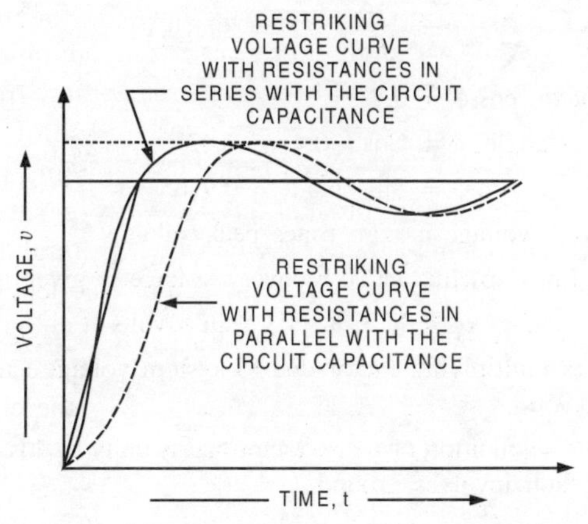

(b)

Fig. 7.9

In a network consisting of generators, transformers, reactors and transmission line, each of them exerts its own damping. Usually the attenuation due to them is too small to be relied upon for improvement in the breaker performance. Where high RRRV are expected circuit breakers with shunt resistances are employed. Now for ensuring exponential build up of voltage across the breaker to the 50 Hz recovery voltage, without overshoot, instead of exhibiting the oscillatory doubling effect associated with an undamped circuit, the value of resistance R_P required to achieve critical damping is $\frac{1}{2}\sqrt{L/C}$. Figure 7.10 illustrates the striking voltage critically damped and other related waveshapes.

It can thus be inferred that the shunt resistance across the breaker modifies the oscillatory restriking voltage into an periodic wave. This entails the arc to be extinguished even though the dielectric strength of the gap increases only relatively slowly as a result of severe short circuit. Inclusion of the shunt resistor thus increase the rupturing capacity of the breaker.

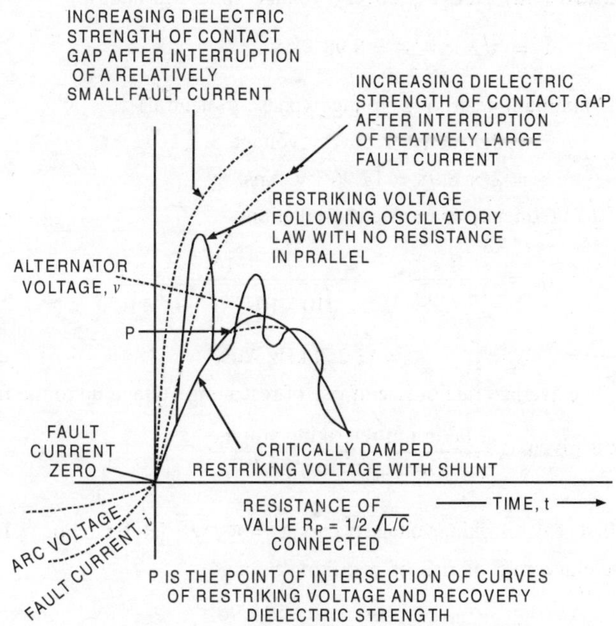

Fig. 7.10. ***Critically Damped Restriking Voltage and Other Related Waveshapes***

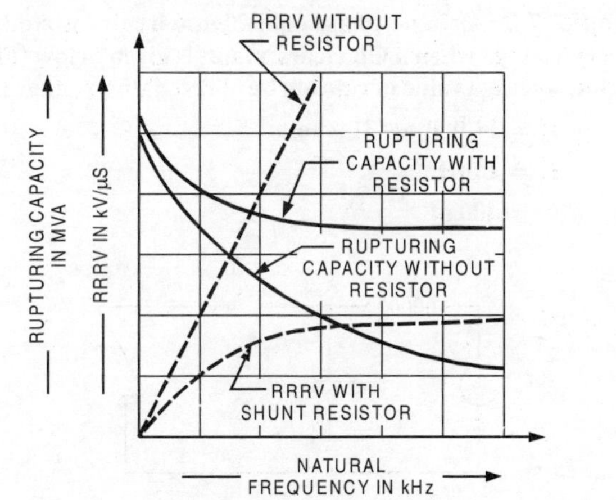

Fig. 7.11. ***RRRV and Rupturing Capacity of an ABCB Expressed as a Function of Natural Frequency***

Figure 7.11 shows the relation of RRRV and rupturing capacity of an air blast circuit breaker with and without shunt resistors as a function of natural frequency. Without shunt resistance RRRV is directly proportional to the natural frequency of the circuit, and the rupturing capacity of the breaker therefore drops rapidly with the increase in frequencies. In the case of breaker with shunt resistance the RRRV cannot exceed a certain value determined by the resistor and hence the rupturing capacity does not drop to that extent. For higher values of natural frequencies the advantage gained in the rupturing capacity is more.

Example 7.1. A 50 Hz, 11 kV generator is connected to a power system. The system inductance and capacitance per phase are 10 mH and 0.02 μF respectively. Calculate *(a)* the maximum voltage across the contacts of the circuit breaker at an instant when it passes

through zero (*b*) frequency of transient oscillation (*c*) average rate of rise of voltage up to the first peak of oscillation. Neglect resistance.

[*Pb. Univ. Elec.Power III (Switchgear and Protection) Dec. 1991*]

Solution : (*a*) Active recovery voltage (phase-to-neutral)

$$= \sqrt{2} \times \frac{11}{\sqrt{3}} = 8.98 \text{ kV}$$

Maximum restriking voltage (phase-to-neutral)

$= 2 \times$ active recovery voltage

$= 2 \times 8.98 = 17.96$ kV **Ans.**

(*b*) Frequency of transient oscillation

$$f = \frac{1}{2\pi\sqrt{LC}} = \frac{1}{2\pi\sqrt{10 \times 10^{-3} \times 0.02 \times 10^{-6}}}$$

$= 11.254$ kHz **Ans.**

The average rate of rise of rise of restriking voltage up to the first peak is given as $\frac{\text{Maximum restriking voltage}}{\text{Time up to first peak}}$

and first peak restriking voltage occurs at $t = \pi\sqrt{LC}$ [Refer to Eq. (7.15)]

∴ Average rate of rise of restriking voltage

$$= \frac{2V_{max}}{\pi\sqrt{LC}} = \frac{17.96}{\pi\sqrt{10 \times 10^{-3} \times 0.02 \times 10^{-6}}}$$

$= 0.404$ kV/μs **Ans.**

Example 7.2. A single-phase equivalent circuit for studying the recovery voltage when a CB clears a fault is given below (Fig. 7.12). Compute average value of rate of rise of restriking voltage (RRRV).

$v(t) = 11{,}000 \cos 314t$ **volts**

L = 1 mH

C = 400 pF

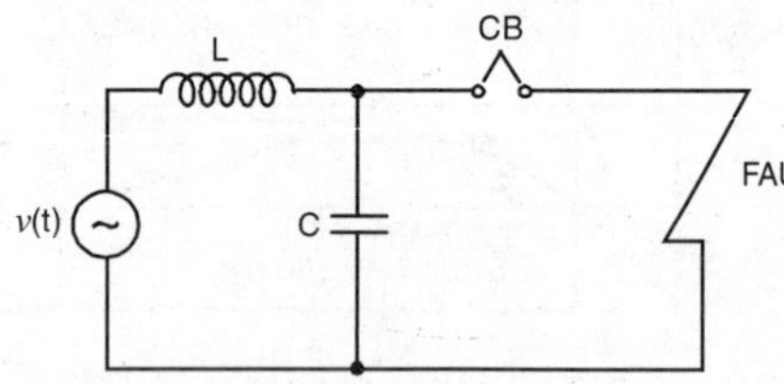

Fig. 7.12

[*U.P.S.C. I.E.S. Electrical Engineering-II, 2009*]

Solution : Maximum restriking voltage (phase-to-neutral)

$= 2 \times 11{,}000 = 22{,}000$ V $\quad \because V_{max} = 11{,}000$ V

$L = 1$ mH $= 0.001$ H

$C = 400$ pF $= 400 \times 10^{-12}$ F

Average rate of rise of restriking voltage

$$= \frac{\text{Maximum restriking voltage}}{\pi\sqrt{LC}}$$

$$= \frac{22{,}000}{\pi\sqrt{0.001 \times 400 \times 10^{-12}}}$$

$= 11{,}072$ V/μs or 11.072 kV/μs **Ans.**

Example 7.3. In a short-circuit test on a circuit breaker, the following data was obtained on a frequency transient.

(*i*) Time to reach the peak restriking voltage 55 μs.

(*ii*) The peak restriking voltage 100 kV.

Determine the (*a*) natural frequency of the circuit and (*b*) average rate of rise of restriking voltage.

Solution : From Eq. (7.15) time to reach the first peak restriking voltage is given as

$$t = \pi\sqrt{LC} = 55 \times 10^{-6} \text{ second}$$

(*a*) Natural frequency of the circuit

$$= \frac{1}{2\pi\sqrt{LC}} = \frac{1}{2 \times 55 \times 10^{-6}}$$

$= 9{,}091$ Hz or 9.091 kHz **Ans.**

(*b*) Average rate of rise of restriking voltage

$$= \frac{\text{Peak restriking voltage}}{\text{Time to reach peak restriking voltage}}$$

$$= \frac{100}{55} = 1.82 \text{ kV/μs}$$ **Ans.**

Example 7.4. In a short-circuit test on a 3-pole, 110 kV circuit breaker pf of the fault was 0.4, the recovery voltage was 0.95 times full-line value. The breaking current was symmetrical. The frequency of oscillation of restriking voltage was 15,000 c/s. Estimate the average rate of rise of restriking voltage. The neutral is grounded and the fault involves earth.

[*Pb. Univ.·Elec. Power III (Switchgear and Protection) June 1993*]

Solution : The peak value of line-to-neutral voltage,

$$V_{max} = \frac{110}{\sqrt{3}} \times \sqrt{2} = 89.8 \text{ kV}$$

Power factor, $\cos\phi = 0.4$

Power factor angle, $\phi = \text{Cos}^{-1}\, 0.4 = 66.42°$

$\text{Sin}\,\phi = \text{Sin}\, 66.42° = 0.9165$

Recovery voltage is 0.95 times peak value.

From Eq. (7.6) the active recovery voltage is given as

$$V_r = K_1K_2K_3\, V_{max} \sin\phi \text{ volts}$$

where K_1 is multiplying factor due to system voltage and is equal to 0.95

K_2 is a condition or phase factor and is unity in this case since the fault involves ground.

K_3 is unity for recovery voltage between phase and neutral.

So recovery voltage (from line-to-neutral),

$V_r = 0.95 \times 1 \times 1 \times 89.8 \times 0.9165$

$= 78.2$ kV (instantaneous).

Time to reach the peak restriking voltage,

$$t = \pi\sqrt{LC} = \frac{2\pi\sqrt{LC}}{2}$$

$$= \frac{1}{2f_n} = \frac{1}{2 \times 15{,}000}$$

$= 33.33$ μs

$$\text{Average RRRV} = \frac{2 \times \text{recovery voltage}}{t} = \frac{2 \times 78.2}{33.33}$$

$= 4.692$ kV/μs **Ans.**

Example 7.5. In a short-circuit test on a 3-pole, 132 kV CB, the following observations are made :

Pf of fault 0.4, the recovery voltage 0.90 times full-line value, the breaking current symmetrical, the frequency of oscillation of

restriking voltage 16 kHz. Assume that the neutral is grounded and the fault does not involve ground, determine the average rate of rise of restriking voltage.

[*Pb. Univ. Elec. Power III* (*Switchgear & Protection July 1992*]

Solution : Peak value of line-to-neutral voltage,

$$V_{max} = \frac{132}{\sqrt{3}} \times \sqrt{2} = 107.78 \text{ kV.}$$

Multiplying factor due to system voltage,

$$K_1 = 0.9$$

Condition or phase factor,

$$K_2 = 1.5 \text{ as the fault does not involve ground}$$

$$K_3 = 1$$

Power factor, $\text{Cos } \phi = 0.4, \phi = \text{Cos}^{-1} 0.4 = 66.42°$

$$\text{Sin } \phi = \text{Sin } 66.42° = 0.9165$$

Active recovery voltage,

$$V_r = K_1, K_2, K_3 V_{max} \sin \phi$$

$$= 0.9 \times 1.5 \times 1 \times 107.78 \times 0.9165 = 133.35 \text{ kV}$$

Time to reach the peak restriking voltage,

$$t = \frac{1}{2 f_n} = \frac{1}{2 \times 16{,}000} = 31.25 \text{ μs}$$

$$\text{Average RRRV} = \frac{2 V_r}{t} = \frac{2 \times 133.35}{31.25} = 8.535 \text{ kV/μs } \textbf{Ans.}$$

Example 7.6. A 50 Hz 3-phase alternator with grounded neutral has an inductance of 1.6 mH per phase and is connected to the bus-bars through a circuit breaker. The capacitance to earth of the circuit between the alternator and the circuit breaker is 0.0032 μF per phase. Due to a short on the bus-bars the breaker opens when the rms value of the current is 8,000 A. Determine the following :

(*i*) Frequency of oscillations (*ii*) Active recovery voltage (*iii*) Time for maximum RRRV and (*iv*) Maximum RRRV.

Solution : (*i*) Frequency of oscillation,

$$f_n = \frac{1}{2\pi \sqrt{LC}} = \frac{1}{2\pi \sqrt{1.6 \times 10^{-3} \times 0.0032 \times 10^{-6}}}$$

$$= 70.34 \text{ kHz } \textbf{Ans.}$$

Short-circuit current = 8,000 A

$$\text{Reactance} = 2\pi f L = 2\pi \times 50 \times 1.6 \times 10^{-3} = 0.5026 \ \Omega$$

$$\text{Recovery voltage} = \text{Short-circuit current} \times \text{reactance}$$

$$= 8{,}000 \times 0.5026 = 4{,}021 \text{ V rms}$$

(*ii*) Active recovery voltage,

$$V_{max} = \sqrt{2} \times 4{,}021 = 5{,}687 \text{ V (line-to-neutral) } \textbf{Ans.}$$

(*iii*) Time to attain maximum RRRV, according to Eq. (7.13),

$$t = \frac{\pi}{2} \sqrt{LC}$$

$$= \frac{\pi}{2} \sqrt{1.6 \times 10^{-3} \times 0.0032 \times 10^{-6}}$$

$$= 3.55 \text{ μs } \textbf{Ans.}$$

(*iv*) Maximum RRRV, according to Eq. (7.14),

$$\text{RRRV}_{max} = \frac{V_{max}}{\sqrt{LC}} = \frac{5{,}687}{\sqrt{1.6 \times 10^{-3} \times 0.0032 \times 10^{-6}}}$$

$$= 2.513 \text{ kV/μs } \textbf{Ans.}$$

7.7. DUTIES OF CIRCUIT BREAKERS

The circuit breakers may be subjected to widely varying stresses under different operating conditions. The current may vary from a few amperes due to no-load magnetizing current of a transformer to the heaviest short-circuit currents amounting to a hundred kilo amperes. The circuit impedance may also change within the first 10^{-4} to 10^{-3} second because open lines or cables connected to the busbars at the breaker behave initially like resistors, of values of their surge impedances and later like capacitors. Whereas load currents are more or less ohmic, short-circuit currents are purely inductive and the currents of unloaded or open-circuited lines are mainly capacitive.

Circuit breakers not only must interrupt but also must close the circuit. This may prove sometimes troublesome, especially if the breaker closes on a short circuit, because in that case the voltage breakdown that bridges the contact gap before the contacts touch produces a high current arc, which melts the contacts before closure. Such a situation is not desirable as the breaker must be able to open the contacts again. Automatic reclosing is often required, because the faults are usually of temporary nature. In case after reclosing short circuits persist (may be in about 20% cases) the breaker again has to interrupt the short-circuit current. This is very severe duty, especially if there are extremely large currents requiring heavy contacts to be accelerated and decelerated in both directions within hundredths of a second without bouncing, which would cause contact welding and wear.

The main duties which a circuit breaker has to perform in addition to satisfying the rated breaking capacities and rated making and breaking times are :

1. Interruption of small inductive currents.
2. Switching of unloaded transmission lines and unloaded cables.
3. Switching of capacitor banks.
4. Interruption of terminal faults.
5. Interruption of short-line faults (kilometric fault)
6. Asynchronous switching.

7.7.1. Interruption of Small Inductive Currents, Current Chopping

The *deionization* is an important requirement only if it occurs immediately after current zero. The deionization at any other stage of the arc current cycle is disadvantageous and unnecessary. If a deionizing force is applied in the middle of the current loop, more voltage will be required to maintain the arc current and it will cause release of energy which is required to be safely disposed of.

The other drawback of deionization is that it causes **current chopping**, *the phenomenon of current interruption before the natural current zero is reached.* In order to understand current chopping phenomenon consider that the same deionizing force is applied by the circuit breaker for all values of the short-circuit current within the capacity of the circuit breaker as in case of air blast circuit breakers which retain the same extinguishing force irrespective of the magnitude of the current to be interrupted. The deionizing force to be applied will naturally be high so as to be sufficient to interrupt highest value of short-circuit current. Now if the current to be interrupted is quite low then the available deionizing force will be sufficient to force the arc from its high value straight to zero before the arc current actual reaches its natural zero, this phenomenon is termed as *current chopping*.

The current chopping is a serious drawback as it produces high voltage transient across the contacts of the breaker.

The necessity of interrupting small inductive current arises while disconnecting transformers on no load. No-load currents of transformer, *i.e.,* magnetizing currents are almost at zero power factor lag. The current is smaller than normal current rating of the breaker but breaking of such a low current presents a severe duty on the circuit breaker because of current chopping. For example if a 220 kV, 50 MVA transformer is disconnected on no load, it may produce voltage transient of the order of as high as 10,000 kV.

The current chopping phenomenon is explained below.

Referring to Fig. 7.13 (*a*), the arc current is seen to approach zero in a normal way initially with low arc voltage so that there is virtually no capacitance current. At a certain arc current, due to a large deionizing force instability occurs and the arc current immediately collapses to zero or there occurs the first chop. The current in the arc was flowing from the source through the inductance and the circuit breaker contacts. The energy contained in the electromagnetic field cannot become zero instantaneously. It changes into some other form of energy. The only possibility is the conversion from electromagnetic to electrostatic form of energy *i.e.,* the energy stored in the inductor will discharge into the stray capacitance C charging the latter to a prospective voltage v. If i is the instantaneous value of arc current where the chop takes place, the prospective value of voltage to which the capacitor C will be charged, is given as

$$v = i\sqrt{L/C} \qquad ...(7.16) \qquad \because \frac{1}{2}Li^2 = \frac{1}{2}Cv^2$$

This prospective voltage may be extremely high as compared to the normal system voltage.

For example consider a 220 kV circuit breaker interrupting a transformer having magnetising current of 10 A (rms value). Let the current be chopped at an instantaneous value of 5 A. If inductance and capacitance values are 100 H and 0.004 μF, the prospective voltage developed would be $5\sqrt{\dfrac{100}{0.004 \times 10^{-6}}}$ *i.e.,* 790 kV.

The prospective voltage v is very high, as seen above, as compared to the dielectric strength gain by the breaker gap so that the breaker restrikes. As the deionizing force is still in action, therefore, chop occurs again but the arc current this time is smaller than the previous value. This induces a lower prospective voltage to re-ignite the arc. In fact several chops may occur until a low enough current is interrupted that produces insufficient induced voltage to restrike across the breaker gap. Consequently the current is finally interrupted as shown in the figure. No further restrike takes place since the gap gets deionized completely by this time.

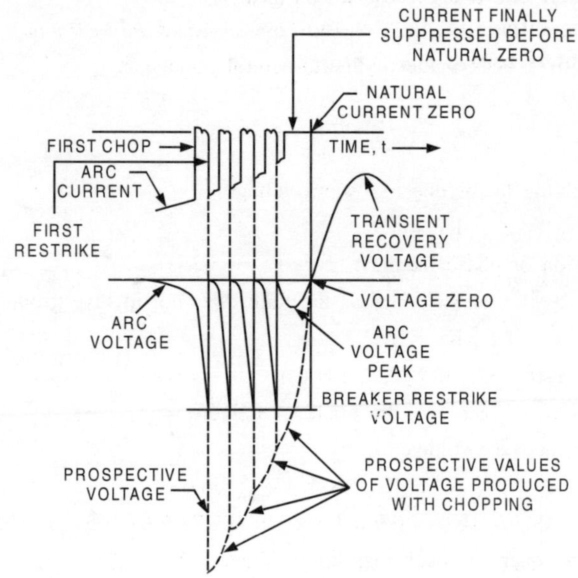

(*a*) *Current Chopping*

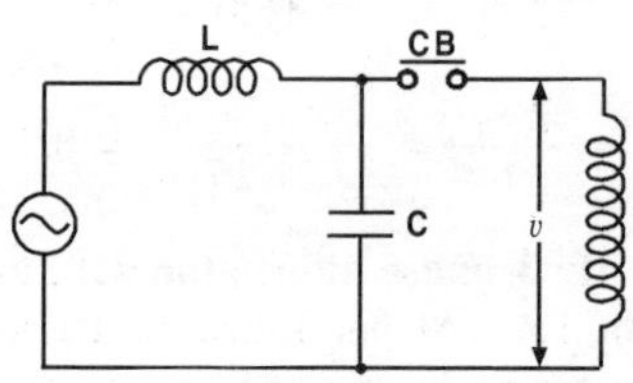

(*b*)

Fig. 7.13

Resistance switching is used to overcome the effect of overvoltages due to current chopping. The value of resistance used for resistance switching may of the order of thousands of ohms. Resistance switching will be explained in Art. 7.8.

Current chopping is not common in oil circuit breakers because in most of the oil circuit breakers, arc control is proportional to the fault current to be interrupted. In other words, the extinguishing power in such breakers is proportional to the current to be interrupted.

7.7.2. Interruption of Capacitive Currents (Switching of Unloaded Long Transmission Lines, Unloaded Underground Cables and Capacitor Banks)

Another cause of excessive voltage transients across the circuit breaker contacts is the interruption of capacitive currents. Examples of such instances are opening of unloaded long transmission lines and unloaded underground cables and disconnecting of capacitor banks employed in the network to provide reactive power at leading power factor. The magnitude of capacitive currents encountered in practice are:

Unloaded lines	:	Charging currents up to 10 A
Underground cables	:	Charging currents up to 100 A
Capacitor banks	:	Currents up to 1,400 A

Consider the simple equivalent circuit of an unloaded transmission line shown in Fig. 7.14 (*a*). Such a line, although unloaded in the normal sense will actually carry a capacitive current i on account of appreciable amount of capacitance C between line and earth. Let us assume that the line is opened by the circuit breaker at the instant t_1 when line capacitive current is zero [Fig. 7.14 (*b*)].

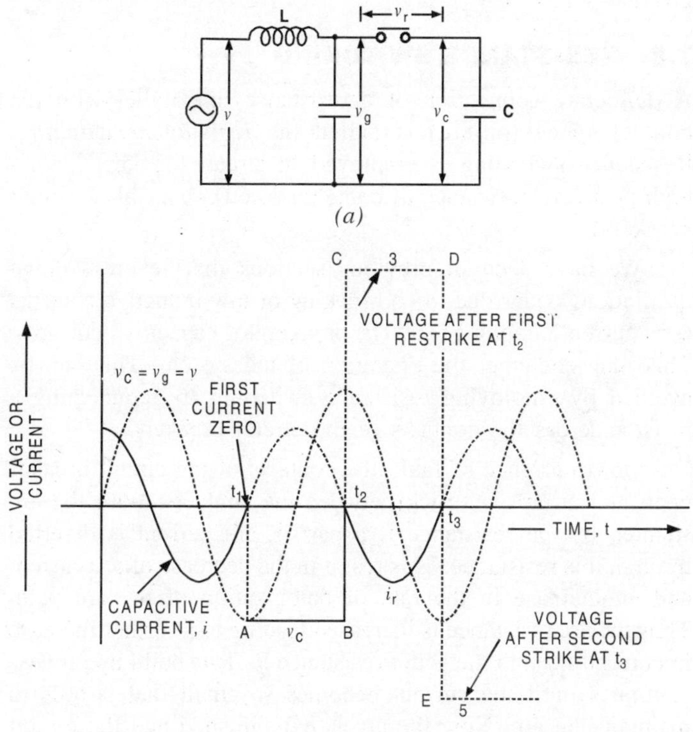

(*a*)

(*b*)

Fig. 7.14. ***Capacitive Current Interruption***

It can be seen from Fig. 7.14 (*b*) that at instant t_1 when capacitive current i is zero, the transmission line is at negative maximum voltage so that at the instant of current interruption (at t_1) the line is left in a fully charged condition to this negative maximum value of the generated voltage V_g. After instant t_1 the breaker gap is subjected to the difference of voltage V_g and V_c. After a time interval of half cycle from A, *i.e.*, at instant t_2 the voltage across the breaker is 2 times maximum value of V_g *i.e.*, $2V_{gm}$. Within such a short interval of half cycle the breaker has been subjected to a severe condition, with the result that the breaker might restrike. If such a condition occurs the voltage across the breaker falls almost instantaneously from two times maximum value of V_g to zero. In doing so high frequency oscillations are set up which build up the voltage to 3 times maximum value of V_g. The restrike current i_r reaches zero value which provides an opportunity to interrupt. The line is charged to a voltage of 3 times maximum value of V_g to earth after interruption of restrike current i_r. At this stage immediately after C the voltage across the breaker contact is twice the maximum value of V_g since generated voltage V_g itself is positive maximum. The voltage across the breaker contacts now continues to increase and at D this reaches a value 4 times maximum value of V_g. If the breaker restrikes again at this point the events of B will be repeated on an even more formidable scale as the voltage swing will now be 8 times maximum V_g and the line may then be left isolated at a potential of 5 times maximum V_g to earth. Theoretically this phenomena may proceed indefinitely increasing the voltage by successive increments of 2 times maximum V_g. However, due to leakage and corona loss, the maximum voltage on the line in such cases is limited to 5 V_{gm}.

Although these extreme conditions caused due to capacitive current breaking are improbable and rare, they do sometimes occur causing serious damage. The sole cause of setting up this type of voltage transients is the inability of the circuit breaker to provide adequate dielectric strength in the gap after current interruption.

The circuit breaker employed for a particular application should be capable of performing opening and closing operations without getting damaged and with overvoltages within specified limits.

Vacuum, SF_6 and air blast circuit breakers are suitable for capacitor current interruption duty.

7.7.3. Interruption of Terminal Faults

A fault occurring very near to the terminal of the circuit breaker (Fig. 7.15) is known as the terminal fault. Under this condition the fault or short-circuit current depends upon the source voltage V and source impedance X (ωL) as the impedance between the breaker and the fault is negligible. After the arc extinguishes at natural zero of the 50 Hz waveform, the circuit recovers and a restriking transient voltage or transient recovery voltage (TRV) appears across the breaker pole. The magnitude and shape of the restriking transient voltage waveform is very important to the circuit breaker and has already been discussed in previous sections.

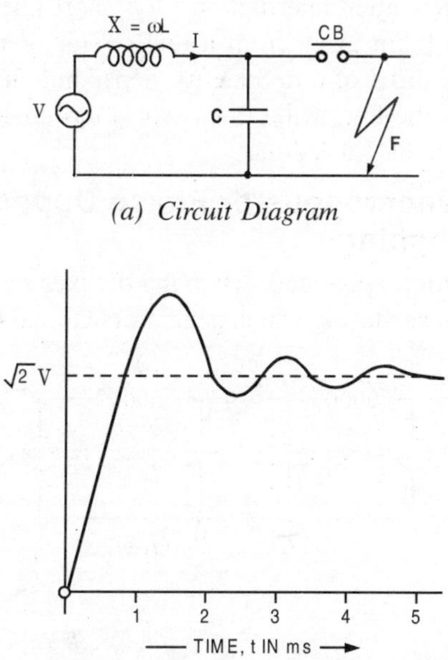

(*a*) *Circuit Diagram*

(*b*) *Simplified TRV Waveform*

Fig. 7.15. ***Interruption of Terminal Faults***

7.7.4. Interruption of Short-Line Faults (or Kilometric Fault)

The faults occurring between a distance of a few km to a few tens km from the circuit breaker are called the *short-line* or *kilometric faults*. Such faults are characterised by high frequency of restriking voltage of the order of 10 to 100 kHz depending upon the line length and fault location.

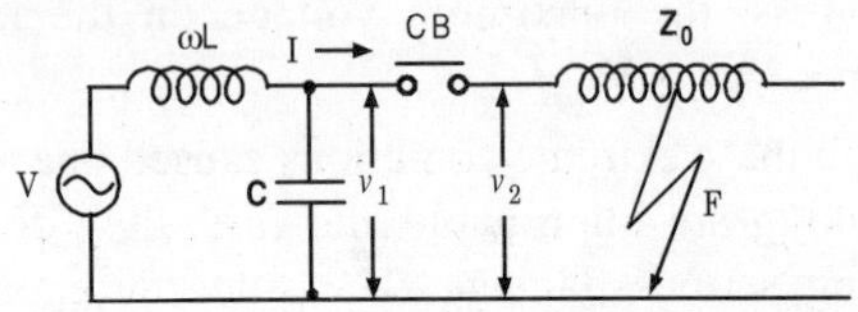

Fig. 7.16. ***Interruption of Kilometric Fault***

The interruption of fault current due to kilometric faults on overhead lines imposes a serious duty on the circuit breaker. This is because the transient recovery voltage (TRV) across the breaker terminals is accompanied by a high-frequency line-side component, whereas the reduction of fault current due to the inductance of the short-circuited line is only slightly less than that of a terminal fault.

The transient voltage of a short-circuited line is proportional to the magnitude of the fault current, and the frequency is inversely proportional to the length of the short-circuited line. After the fault current interruption the voltage drop along the line is left behind in the form of a line charge. This charge decays in the form of a travelling wave oscillating at its natural frequency. The rate of rise of these oscillations is quite high owing to the effective surge impedance of the short-circuited line. The RRRV is given as

$$\text{RRRV} = \sqrt{2}\,\omega I Z_0 \qquad \text{...(7.17)}$$

where ω is the supply or source angular frequency, I is the short-circuit current and Z_0 is the effective surge impedance of the short-circuited line.

The RRRV after interruption of a short-line fault (short circuit about 1 km away from a line circuit breaker) current may be of the order of 6 to 8 kV/μs, depending upon the surge impedance of the line while in case of a terminal fault RRRV is of the order of 1.5 kV/μs.

7.7.5. Asynchronous or Phase Opposition Switching

Phase opposition may occur when the breaker recloses after a fairly long pause, during which generators G_1 and G_2 fall out of synchronism (Fig. 7.17). When the switch opens the peak value of the TRV is determined by the sum of V_1 and V_2 and approaches two times that of the short-circuit or terminal fault interruption. The recovery voltage has a maximum value when the two voltages are in phase opposition (*i.e.*, 180° out of phase) and is given as $\sqrt{2}\,(V_1 + V_2)$.

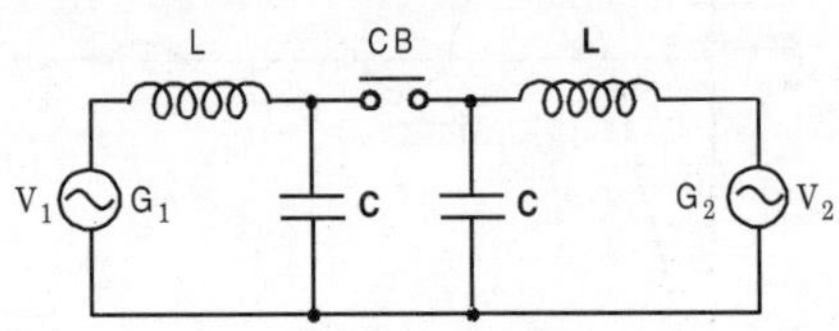

(a) Circuit Diagram

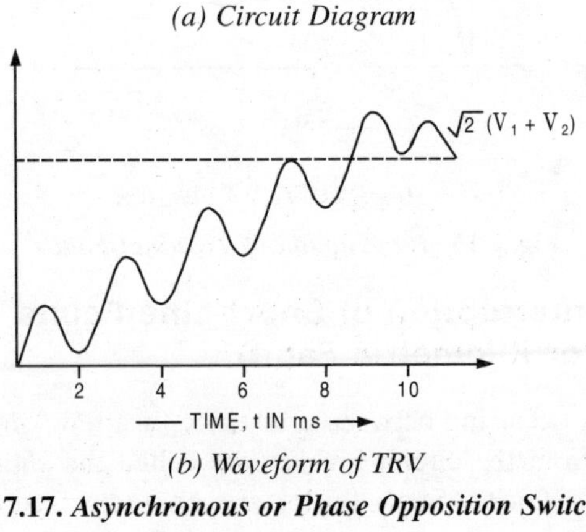

(b) Waveform of TRV

Fig. 7.17. ***Asynchronous or Phase Opposition Switching***

7.8. RESISTANCE SWITCHING

A deliberate connection of a resistance in parallel with the contact space (or arc) is called the *resistance switching*. Resistance switching is employed in circuit breakers having high post zero resistance of contact space (*i.e.*, air blast circuit breakers).

We have seen in previous sections that severe voltage oscillations occur due to (*i*) breaking of low inductive currents (*i.e.*, current chopping) and (*ii*) breaking of capacitive currents. This may endanger the operation of the system. This can be avoided by employing resistance switching (by connecting a resistor across the contacts of the circuit breaker).

On occurrence of fault, the contacts of the circuit breaker open and an arc is struck between the contacts. With the arc shunted by the resistance R a part of arc current is diverted through this resistance. This result in the decrease of arc current and an increase in the rate of deionization of the arc path. Thus the arc resistance is increased leading to a further increase in current through the shunt resistance R. This build up process continues until the current becomes so small that it fails to maintain the arc. Now the arc is extinguished and the circuit current gets interrupted.

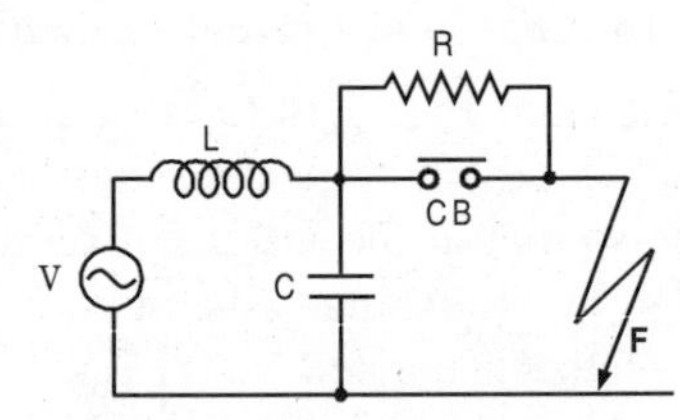

(a) Circuit Diagram

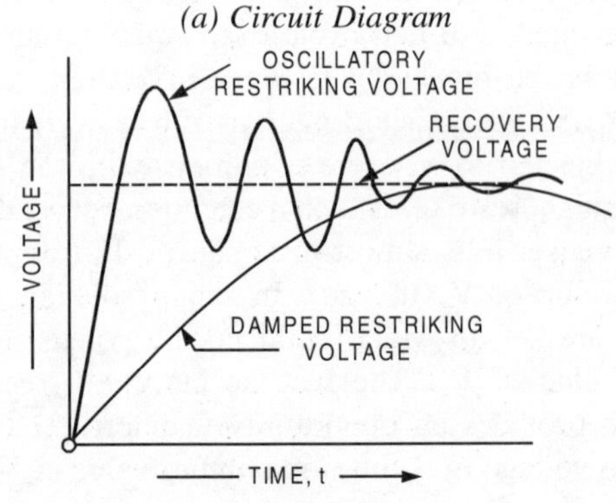

(b)

Fig. 7.18. ***Resistance Switching***

Alternatively, the resistance may be automatically switched in by transference of the arc from the main contacts to the probe contact as in the case of an axial blast circuit breaker, the time required for this action is very small (usually less than one half-cycle of the current wave). Having the arc path substituted by a metallic path, the current flowing through the resistance is limited and then easily broken.

Typical resistor connections are shown in Fig. 7.19. In Fig. 7.19 (*a*) a second break is provided to break the resistor current. In Fig. 7.19 (*b*) the gaps are so arranged that the moving contact finally breaks the resistor elements. In Fig. 7.19 (*c*) the arc first appears across fixed and moving contacts F and M which is then transferred across fixed and probe contacts F and P and then broken there.

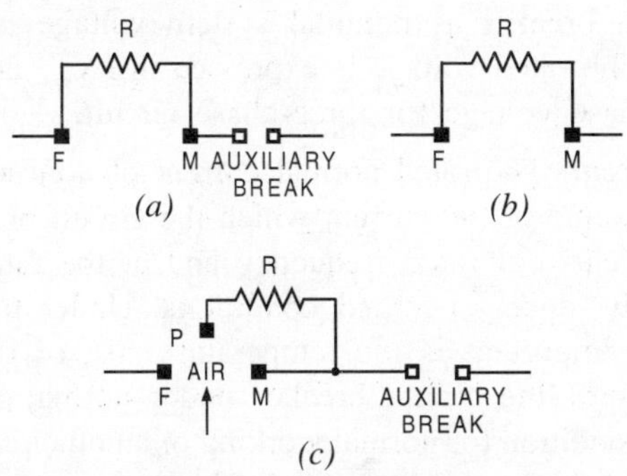

Fig. 7.19. ***Typical Resistor Connections For Arc Interruption***

The shunt resistor also helps in limiting the oscillatory growth of restriking voltage transients. It can be proved mathematically that the natural frequency of oscillations of the circuit shown in Fig. 7.18 (*a*) is given as

$$f_n = \frac{1}{2\pi}\sqrt{\frac{1}{LC} - \frac{1}{4C^2R^2}} \qquad \text{...(7.18)}$$

The effect of shunt resistor R is to prevent the oscillatory growth of restriking voltage and cause it to grow exponentially up to recovery voltage. This is being most effective when the value of R is so chosen that the circuit is critically damped. The value of R required for critical damping is $0.5\sqrt{L/C}$. Figure 7.18 (*b*) shows the oscillatory growth and exponential growth when the circuit is critically damped.

To sum up, resistors across breaker contacts may be used to perform any one or more of the following functions.

1. It reduces the RRRV and thus reduces the burden on the circuit breaker.
2. It ensures the damping of the high frequency restriking voltage transients during switching out inductive or capacitive loads.
3. In a multi-break circuit breaker it helps in distributing the transient recovery voltage more uniformly across all the contact gaps.

The resistors employed may be either nonlinear or wire wound. Nonlinear resistors are suitable both from space and reliability considerations for small shunt currents where wire-wound resistors tend to be less satisfactory from mechanical considerations. Where heavy currents are involved there may be difficulty in accommodating the relatively large volume of required resistor material.

Nonlinear resistors are not suitable for modification of the RRRV and of the voltage peak as are linear resistors, but they are especially suited to voltage equalization and overvoltage suppression applications in which relatively small currents of the order of 1–10 A at normal peak voltage are adequate.

In the plain break oil circuit breakers (tank type) the post-zero resistance of the contact space is low. Hence resistance switching is not necessarily required. The performance at low currents can, however, be improved by employing resistance switching and it is sometimes employed; when interrupting a small current, the value of reactance in the circuit will tend to be so high that the inductance L in the expression for the critical resistance will be larger, resulting in resistors of the order of thousands of ohms.

The post-zero resistance of air blast circuit breaker is high. This may result in severe voltage transients due to current chopping. Hence the resistance switching is employed. The auxiliary contacts here are replaced by isolating contacts, which are parts of air circuit breakers.

Example 7.7. In a system of 132 kV, the circuit phase to ground capacitance is 0.02 μF and the circuit inductance is 5 H. The circuit breaker interrupts a magnetising current of 5 A (peak). Find :

(*i*) The voltage across the circuit breaker contacts after the current interruption.

(*ii*) The value of resistance to be used across the contacts to suppress restriking voltage.

Solution : (*i*) Voltage across the circuit breaker contacts after the interruption of 5 A current

$$v = i\sqrt{\frac{L}{C}} = 5\sqrt{\frac{5}{0.02 \times 10^{-6}}} = 79{,}000 \text{ V or } 79 \text{ kV } \textbf{Ans.}$$

...Refer to Eq. (7.16)

(*ii*) Value of resistance to be used across the breaker contacts to suppress restriking voltage

$$R = 0.5\sqrt{\frac{L}{C}} = 0.5\sqrt{\frac{5}{0.02 \times 10^{-6}}} = 7.9 \text{ k}\Omega \ \textbf{Ans.}$$

Example 7.8. In a 220 kV system, the reactance and capacitance up to the location of circuit breaker are 8 Ω and 0.025 μF respectively. A resistance of 600 Ω is connected across the contacts of the circuit breaker. Determine the following :

(*i*) Natural frequency of oscillation ;

(*ii*) Damped frequency of oscillation ;

(*iii*) Critical value of resistance which will give no transient oscillation ;

(*iv*) The value of resistance which will give damped frequency of oscillation, one fourth of the natural frequency of oscillation.

[*A.M.I.E. Sec. B. Advanced Power Systems Summer 2000*]

Solution : System capacitance up to CB, $C = 0.025 \times 10^{-6}$ F

System inductance up to CB, $L = \frac{X_L}{2\pi f} = \frac{8}{2\pi \times 50} = 0.0255$ H

Resistance connected across breaker contacts, R = 600 Ω

(*i*) Natural frequency of oscillation,

$$f_n = \frac{1}{2\pi\sqrt{LC}} = \frac{1}{2\pi\sqrt{0.025 \times 10^{-6} \times 0.0255}}$$

$$= 6.3 \text{ kHz } \textbf{Ans.}$$

(*ii*) Damped frequency of oscillation,

$$f'_n = \frac{1}{2\pi}\sqrt{\frac{1}{LC} - \frac{1}{4C^2R^2}}$$

$$= \frac{1}{2\pi}\sqrt{\frac{1}{0.0255 \times 0.025 \times 10^{-6}} - \frac{1}{4 \times (0.025 \times 10^{-6})^2 \times 600^2}}$$

= 3.41 kHz **Ans.**

(*iii*) Critical value of resistance to eliminate transient oscillation,

$$R = 0.5\sqrt{L/C} = 0.5\sqrt{\frac{0.0255}{0.025 \times 10^{-6}}} = 504.6\ \Omega \textbf{ Ans.}$$

(*iv*) Let the resistance of R ohms be connected across the breaker contacts so that damped frequency of oscillation becomes, one-fourth of the natural frequency of oscillation *i.e.,*

$$\frac{6.3}{4} = 1.575 \text{ kHz or } 1{,}575 \text{ Hz}$$

Substituting L = 0.0255 H, C = 0.025 × 10^{-6} and F_n = 1,575 Hz in Eq. (7.18), we have

$$1{,}575 = \frac{1}{2\pi}\sqrt{\frac{1}{0.0255 \times 0.025 \times 10^{-6}} - \frac{1}{4 \times (0.025 \times 10^{-6})^2 R^2}}$$

or R = 521.5 Ω **Ans.**

7.9. CIRCUIT BREAKER RATINGS

The rating of a circuit breaker is given according to the duties that are performed by it. For complete specifications, standard ratings and various tests of switches and circuit breakers IS 375/1951 may be consulted. Apart from the normal working of circuit breakers the circuit breaker is required to perform following 3 major duties under short-circuit conditions.

1. A circuit breaker must be capable of breaking the circuit and isolating the faulty section in case of a fault. This is described as *breaking capacity* of a circuit breaker.
2. Since in practice a circuit breaker is put on 2-3 times in order to ensure the permanency of the fault *i.e.,* it must be capable of making circuit in the greatest asymmetrical peak in current wave. This refers to *making capacity* of a circuit breaker.
3. When a circuit breaker operates in conjunction with the other circuit breakers and in case of a fault on any one section the breakers in the sound sections should not trip *i.e.,* a circuit must be capable of carrying fault currents safely for a short time while another circuit breaker (in series) is clearing the fault. This refers to *short-time capacity* of a circuit breaker.

In addition to the above ratings, a circuit breaker should be specified in terms of (*i*) the number of poles (*ii*) rated voltage (*iii*) rated current (*iv*) rated frequency and (*v*) operating duty.

The *number of poles* per phase of a breaker is a function of the operating voltage.

Rated Voltage. During normal operating conditions the voltage at any point of the power system is not constant. Due to this the manufacturer guarantees perfect operation of the circuit breaker at rated maximum voltage, which as a rule is higher than rated nominal voltage.

The rated maximum voltage of a circuit breaker is the highest rms voltage, above nominal system voltage, for which the circuit breaker is designed and is the upper limit for operation. The earlier practice of specifying the rated voltage of a circuit breaker as nominal system voltage is no more followed. The rated voltage is expressed in kV_{rms} and refer to phase-to-phase voltage for three-phase circuit.

Rated Current. The rated normal current of a circuit breaker is the rms value of the current which the circuit breaker shall be able to carry at rated frequency and at the rated voltage continuously, under specified conditions. Under the specific conditions, important is the temperature rise of the various components of the circuit breaker under normal loads. The important condition for normal working of an oil circuit breaker is that the temperature rise of oil should not be more than 40°C and that of contacts should not exceed 35°C.

Rated Frequency. The rated frequency of a circuit breaker is the frequency at which it is designed to operate. Standard frequency is 50 Hz. Applications at other frequencies need special considerations.

Operating Duty. The operating duty of a circuit breaker consists of the prescribed number of unit operations at stated intervals.

The *operating sequence* denotes the sequence of opening and closing operations which the circuit breaker can perform under specified conditions.

Breaking Capacity. This term expresses the highest rms value of short-circuit current that the circuit breaker is capable of breaking under specified conditions of transient recovery voltage and power frequency voltage. It is expressed in kA rms at contact separation.

It can be seen from the short-circuit current wave shown in Fig. 7.20 that the rms value of the current varies with time on account of the presence of dc component of the current which decays with time.

It is known that in a particular phase the current is maximum at the instant of fault, after which the current decays. In addition, owing to relaying time the circuit breaker starts to open its arcing contacts only some time later, after the initiation of short circuit. Hence the actual current interrupted by the circuit breaker is less than initial value of short-circuit current I_1.

Let at the instant of separation of contacts.

AC component of short-circuit current, $I_{ac} = x$

DC component of short-circuit current, $I_{dc} = y$

Now symmetrical breaking current

= RMS value of ac component of short-circuit current at the instant of separation of contacts

$$= \frac{x}{\sqrt{2}} \qquad \text{...(7.19)}$$

Asymmetrical breaking current

= RMS value of the combined sums of ac and dc components

$$= \sqrt{\left(\frac{x}{\sqrt{2}}\right)^2 + y^2} \quad ..(7.20)$$

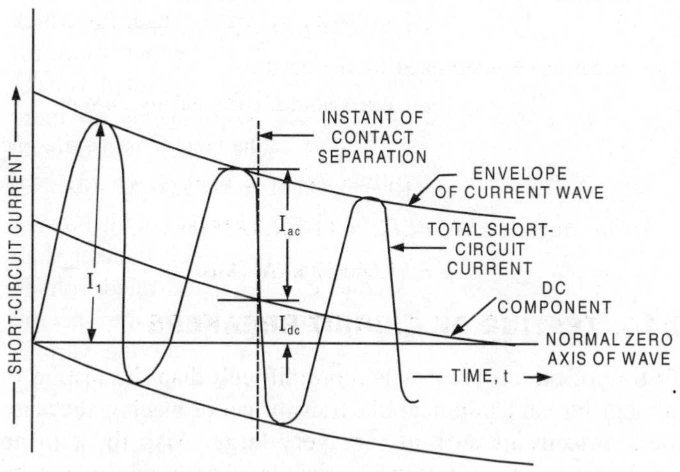

Fig. 7.20. ***Short-Time Current Rating***

Now according to these two values of breaking currents there are two corresponding values of breaking capacities. Conventionally the breaking capacity of a circuit breaker in MVA is given as $\sqrt{3}$ × rated voltage in kV × rated breaking current in kA.

This practice of specifying breaking capacity in terms of MVA is convenient while determining the fault level. However, as per revised standards the breaking capacity is expressed in kA for specified conditions of TRV, and this method takes into account both breaking current and TRV.

The two breaking capacities can now be defined as follows.

(*i*) The *symmetrical breaking capacity* of a circuit breaker is the value of symmetrical breaking current which the circuit breaker is capable of breaking at a stated recovery voltage and a stated reference restriking voltage under prescribed conditions.

(*ii*) The asymmetrical breaking capacity of a circuit breaker is the value of the asymmetrical breaking current which the circuit breaker is capable of breaking at a stated recovery voltage and a stated reference restriking voltage under prescribed conditions.

Making Capacity. There is always a possibility that the circuit breaker is closed under short-circuit condition. The making capacity of the circuit breaker depends upon its ability to withstand the effects of electromagnetic forces which are proportional to the square of the peak value of the making current. The making current of a circuit breaker when closed on a short circuit is the peak value of the maximum current wave (including dc component) in the first cycle of the current after the circuit is closed by the circuit breaker.

For determination of making current of a circuit breaker, we must multiply symmetrical breaking current by $\sqrt{2}$ to convert the rms value to peak value, and then by 1.8 to take into account the "doubling effect" of maximum asymmetry.

Thus rated making current

$= 1.8 \times \sqrt{2}$ rated short-circuit breaking current

$= 2.55$ rated short-circuit breaking current

or Making capacity = 2.55 × symmetrical breaking capacity ...(7.21)

Short-Time Current Rating. As already pointed out, a circuit breaker is sometimes required to carry short-circuit current for short intervals without tripping. This happens in case of momentary faults like birdage on the transmission lines and the fault is automatically cleared and persists only for 1 or 2 seconds. For this reason the circuit breakers are short-time rated and they trip only when the fault persists for a duration longer than the specified time limit.

The short-time current of a circuit breaker is the rms value of current that a circuit breaker can carry in a fully closed position without damage, for the specified time interval under prescribed conditions. It is normally expressed in terms of kA for a period of 1 second or 4 seconds, known as one-second rating and four-second rating respectively. These ratings are based on thermal limitations.

Low-voltage breakers do not have any such short-time rating because these are normally equipped with direct acting series overload trips.

Example 7.9. A 3-phase circuit breaker is rated at 2,000 MVA, 33 kV. Find its making current. [*M.D. Univ. Power Systems-I, 2010-11*]

Solution : Rated symmetrical breaking current

$$= \frac{\text{Rupturing capacity in MVA}}{\sqrt{3} \times \text{operating voltage in kV}} \text{ kA}$$

$$= \frac{2{,}000}{\sqrt{3} \times 33} = 35 \text{ kA} \quad \textbf{Ans.}$$

Making current = 2.55 × rated symmetrical breaking current

= 2.55 × 35 = 89.25 kA **Ans.**

Example 7.10. A 3-phase circuit is rated at 1,250 A, 2,000 MVA, 33 kV, 4 seconds. Find the rated symmetrical breaking current, making current and short-time rating.

[*G.B. Technical Univ. Switchgear and Protection 2011-12*]

Solution : Rated normal current = 1,250 A

Rupturing capacity = 2,000 MVA

Operating voltage = 33 kV

Time = 4 seconds

(*i*) Rated symmetrical breaking current

$$= \frac{\text{Rupturing capacity in MVA}}{\sqrt{3} \times \text{operating voltage in kV}} \text{ kA}$$

$$= \frac{2{,}000}{\sqrt{3} \times 33} = 35 \text{ kA}$$

(*ii*) Rated making current = 2.55 × rated symmetrical breaking current

= 2.55 × 35 = 89.25 kA **Ans.**

(*iii*) Short-time rating = 35 kA for 4 seconds **Ans.**

Example 7.11. An 11 kV, 500 MVA circuit breaker suddenly closes on to a fault. Determine (*i*) the symmetrical breaking current ; (*ii*) the asymmetrical breaking current assuming 50 %

dc component; (*iii*) the peak making current; (*iv*) the short-time current rating. [*W.B. Univ. of Technology Power System-II, 2010*]

Solution : Rupturing capacity = 500 MVA

Operating voltage = 11 kV

(*i*) Symmetrical breaking current $= \dfrac{\text{Rupturing capacity in MVA}}{\sqrt{3} \times \text{operating voltage in kV}}$ kA

$$= \frac{500}{\sqrt{3} \times 11} = 26.24 \text{ kA (rms)} \textbf{ Ans.}$$

AC component of short-circuit current,

$$x = \sqrt{2} \times \text{symmetrical breaking current}$$
$$= \sqrt{2} \times 26.24 = 37.1 \text{ A}$$

DC component of short-circuit current,

$$y = 50\% \text{ of } x = \frac{1}{2} \times 37.1 = 18.55 \text{ A}$$

(*ii*) Asymmetrical breaking current

$$= \sqrt{\left(\frac{x}{\sqrt{2}}\right)^2 + y^2} = \sqrt{(26.24)^2 + (18.55)^2}$$

$$= 32.13 \text{ kA (rms)} \textbf{ Ans.}$$

(*iii*) Peak making current = 2.55 × symmetrical breaking current

= 2.55 × 26.24 = 66.9 kA **Ans.**

(*iv*) Short-time current rating = 32.13 kA **Ans.**

Example 7.12. A 3-phase generator is rated 7,500 kVA and 13.8 kV. It has X''_d = 9%, X'_d = 15%, X_d = 100%. This generator is connected to a 3-phase transformer (step up) through a 3-phase circuit breaker which has speed of 5 cycles. The generator is operating on no load and at rated voltage. Under this condition, a 3-phase fault occurs between the breaker and the transformer. Calculate : (*a*) Sustained short-circuit current in the CB, (*b*) Initial symmetrical rms current in the breaker, (*c*) Maximum possible dc component of short-circuit in the CB, (*d*) Momentary current rating of CB, (*e*) Current to be interrupted by the breaker, (*f*) Interrupting kVA. [*U.P.S.C. I.E.S. Electrical Engineering-II, 2007*]

Solution : Steady-state reactance, X_d = 100% or 1.0 pu

So, the steady-state short-circuit kVA = 7,500 kVA

$$\because \text{ SC MVA} = \frac{\text{Rated MVA}}{\text{PU reactance}}$$

(*a*) Steady-state short-circuit current

$$= \frac{\text{Short-circuit kVA}}{\sqrt{3} \times \text{Rated voltage in kV}}$$

$$= \frac{7{,}500}{\sqrt{3} \times 13.8} = 313.78 \text{ A}$$

(*b*) The initial symmetrical rms current is the current to subtransient state where the reactance is 9% or 0.09 pu

Initial symmetrical rms current

$$= \frac{\text{Rated kVA}}{\text{PU } X''_d \times \sqrt{3} \times \text{rated voltage in kV}}$$

$$= \frac{7{,}500}{0.09 \times \sqrt{3} \times 13.8} = 3{,}486.41 \text{ A} \textbf{ Ans.}$$

(*c*) Maximum possible dc component of SC current in CB

= Peak value of subtransient current

$= \sqrt{2} \times 3{,}486.41 = 4{,}930.53$ A **Ans.**

(*d*) Momentary current rating = 1.6 × subtransient current

= 1.6 × 3,486.41 = 5,578.26 A **Ans.**

(*e*) Current to be interrupted by the breaker

= 1.1 × symmetrical breaking current

∵ the breaker is 5-cycle one

= 1.1 × 3,486.41 = 3,835.06 A **Ans.**

(*f*) Interrupting kVA $= \sqrt{3} \times 13.8 \times 3{,}835.06$

= 91,666.67 kVA **Ans.**

7.10. TESTING OF CIRCUIT BREAKERS

Testing of circuit breaker is more difficult than the testing of other electrical equipment like transformer or machine because the short-circuit currents are very large. Also there is no satisfactory method of testing circuit breaker at reduced power.

Testing of circuit breakers can be classified into two main groups, viz., (*i*) type tests and (*ii*) routine tests.

7.10.1. Types Tests

These tests are conducted on first few prototype circuit breakers of each type for the purpose of proving the capabilities and confirming the rated characteristics of the circuit breaker of that design. Such tests are conducted in specially built testing laboratories. Type tests are performed as per recommendations of standards (IEC) or (IS).

Type tests can be broadly classified as (*i*) mechanical performance tests (*ii*) thermal tests (*iii*) dielectric or insulation tests and (*iv*) short-circuit tests in order to check making capacity, breaking capacity, short-time rating current and operating duty.

Mechanical Tests. These are mechanical endurance type tests involving repeated opening and closing of the breaker. A circuit breaker must open and close at the correct speed and perform its designated duty and operation without mechanical failure.

Thermal Tests. Thermal tests are carried out to check the thermal behaviour of the breakers. The breaker under test is subjected to steady-state temperature rise due to flow of its rated current through its poles in closed condition. The temperature rise for rated current should not exceed 40° C for current less than 800 A normal current and 50° C for normal value of current 800 A and above.

In such tests the contact drops or the contact resistances are also measured as these contact surfaces are responsible for generation of heat and subsequent temperature rise.

Dielectric Tests. These tests are performed to check power frequency and impulse voltage withstand capacity.

Power frequency tests are conducted on a clean new circuit breaker, the test voltage varies with circuit breaker rated voltage. The test voltage with a frequency between 15–100 Hz is applied as follows:

(*i*) between poles with circuit breaker closed, (*ii*) between poles and earth with circuit breaker open, and (*iii*) across terminals with circuit breaker open.

The voltage is gradually increased and maintained at test value for one minute.

In *impulse tests* impulse voltage of specified shape and magnitude is applied to the breaker. For outdoor circuit dry and wet tests are conducted.

Short-Circuit Tests. Circuit breakers are subjected to sudden short-circuits in short-circuit test laboratories and oscillograms are taken to know the behaviour of the circuit breakers at the time of switching- in, during contact breaking, and after arc extinction. The oscillograms are studied with particular reference to the making and breaking currents, both symmetrical- and asymmetrical-restriking voltages, and switchgears are tested a number of times at rated conditions.

7.10.2. Routine Tests

Once type tests are conducted and a particular design is found to be satisfactory the product becomes prototype and a large number of circuit breakers of similar design are manufactured. However, each and every circuit breaker is still subjected to a few more tests before commissioning. These tests are called routine tests.

Routine tests are also performed as per recommendations of the standards (IEC/IS). These tests are performed in the manufacturer's premises. Routine tests confirm the proper functioning of the circuit breaker.

The routine tests include (*i*) power frequency voltage test (*ii*) millivolt drop test and (*iii*) operational tests.

Power frequency voltage test being the same as mentioned under the heading of type tests, millivolt drop test is performed in order to determine the voltage drops within the current path of the breaker mechanism. Operational test is performed on the breaker by simulating its tripping by artificially closing the contacts of the relays.

7.11. TESTING STATION AND EQUIPMENTS

There are three types of testing stations viz. field type testing station, laboratory type testing station; composite testing station.

In the *field type of testing*, the power required for testing is directly drawn from a large power system, the breaker being tested is connected in the system. Though this type of testing provides the most convincing method of testing hv circuit breakers but it suffers from the drawback of limited available flexibility of the system. It is difficult to set the system for the specified RRRV for hv breakers.

In the *laboratory type testing station,* special generators, called the short-circuit generators provide the power for testing. In this type of testing station, it is possible to vary the test conditions at will. Establishment of a short-circuit testing plant particularly of the laboratory type is an exceedingly costly project, and it is not possible for all switchgear manufacturers to have such a facility of their own.

A *composite testing station* is a combination of the field type testing station and laboratory type testing station.

Description of a Simple Test Plant. In such stations, the short-circuit power is supplied by specially built short-circuit generators. There are normally two or more generators though only one is shown in Fig. 7.21. The short-circuit generators are driven by three-phase induction motor and special type of excitation, known as impulse excitation, is provided. Series resistors and reactors are provided for adjustment of magnitude of short-circuit current and pf. The master CB has capacity higher than that of CB being tested and pf. In case, the circuit breaker being tested fails, the master CB opens and protects the circuit. Making switch is a specially designed circuit closing device which can close at the desired moment and can withstand the making currents. Transformers are provided to give test voltages other than the generator voltage. In addition to above equipment there is equipment for

(*i*) Measurement, record, control;

(*ii*) Sequence switch to provide sequential operation;

(*iii*) Auxiliaries etc.

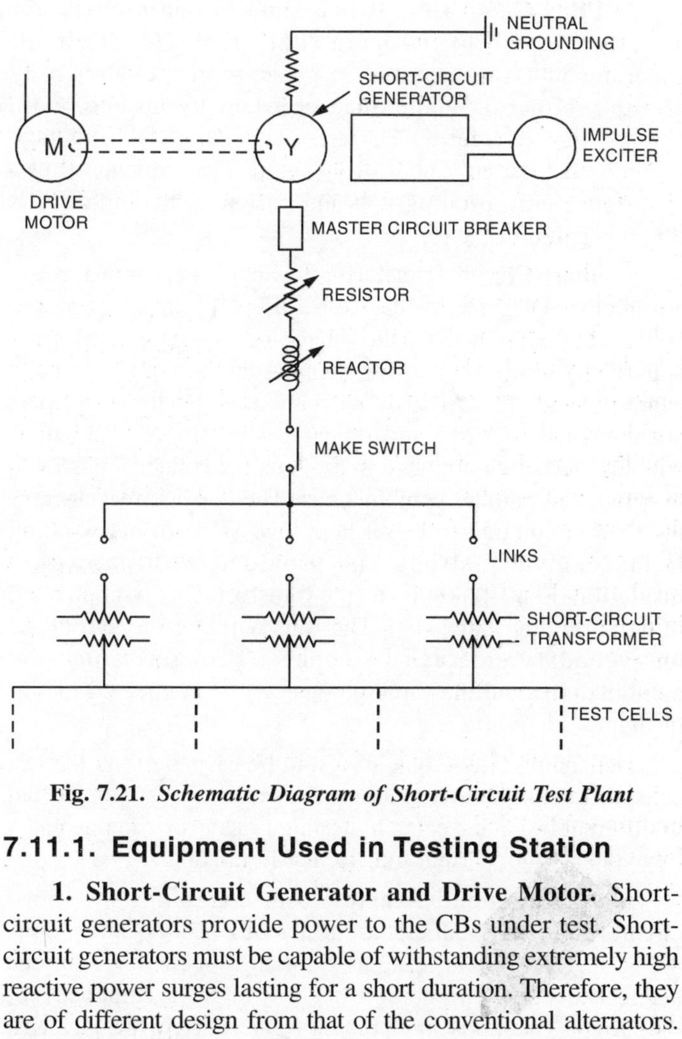

Fig. 7.21. *Schematic Diagram of Short-Circuit Test Plant*

7.11.1. Equipment Used in Testing Station

1. Short-Circuit Generator and Drive Motor. Short-circuit generators provide power to the CBs under test. Short-circuit generators must be capable of withstanding extremely high reactive power surges lasting for a short duration. Therefore, they are of different design from that of the conventional alternators. Short-circuit generators must have low reactance in order to provide maximum short-circuit output. Short-circuit generator is a 3-phase alternator. Each winding is made in two or more parts which can be connected in series, parallel compensations of star or delta to provide different terminal voltages.

The generator is driven by a 3-phase induction motor connected through a resilient shaft. The generator is equipped with a built-in flywheel which provides the kinetic energy during short circuit and speed regulation of the set.

2. Impulse Excitation. A separate dc converter set with a high peak output provides the "impulse excitation." Impulse excitation or super excitation is provided to counteract the demagnetizing effect of armature reaction. The SC currents which are at lagging pf have demagnetizing effect. This results in a reduction of total field, hence in the reduced emf. As a result recovery voltage is less than the voltage before the short circuit. The effect is reduced by boosting the generator field current by means of *impulse excitation.* The converter set employed for providing impulse excitation is fitted with a large flywheel. The motor is disconnected from the supply before the application of excitation. The field current is increased shortly to about 10 times its normal value at the instant of short circuit. This takes care of the demagnetising effect of short circuit and provides the desired recovery voltage.

3. Pilot Generator. It is a small 3-phase alternator directly coupled to the main shaft of the short-circuit generator and synchronised in phase with the latter. Any present voltage is maintained constant by an automatic voltage regulator during SC tests. This dependable voltage is required to supply control power to the sequence timer, electromagnetic oscillograph and various actuating circuits for conduction of the test.

4. Short-Circuit Transformers. Short-circuit transformers are employed when test voltages are different from the generator voltage. For stepping down the voltage a three-phase transformer is normally used, whereas for stepping up the voltage, usually banks of single-phase transformers are used. Such transformers are designed to withstand repeated short circuit and their windings are often arranged in sections for voltage adjustment in series and parallel combinations. The leakage reactance of the short-circuit transformer is kept low. Transformer winding is mechanically strong and provided with extra-turn insulation. For 3-phase tests, the transformers are connected in delta on alternator side. The four winding of each phase on secondary side can be connected in series/parallel combinations and the three phases can be connected in star or delta.

Depending on the type of test to be carried out in the test cells, different types and ratings of transformers are connected in different test cells, e.g. a hv testing transformer in one and a low voltage high current transformer in the other.

5. Resistors and Reactors. For controlling the short-circuit test current, three-phase banks of resistors and reactors are used. The reactors being used for adjustment of magnitude and resistors for control of rate of decay of the dc component of the current. The short-circuit pf is also controlled by these means. There are a number of coils per phase and so many variations can be had by connecting them in series/parallel combinations.

6. Master Circuit Breaker. This is normally an air-blast type of capacity more than that of the breakers under test. It is provided mainly as a backup circuit breaker. In case of failure of the test circuit breakers, the master circuit breaker opens. In addition after every test, it isolates the specimen under test from the supply source and must be capable of handling the full SC power of the test circuit.

7. Make Switch. The making switch or making device, as the name indicates, is employed to ensure that the short-circuit currents are applied correctly at the desired moment. The equipment is characterised by close making time and high making capacity. However, the breaking capacity is negligible as the making device is not used for circuit interruption. At the moment of making, the backup CB and the test CB are already closed. As such on closure if there is even a slight contact rebound the contacts might get burnt or even welded together. In order to ensure, that the contact separation is minimum on closure, a high air pressure is maintained in the chamber. The closing speed is also so high that the contacts are fully closed before the short-circuit current attains its peak value.

8. Capacitors. Capacitor banks serve two purposes :

1. Provide leading current for testing the performance of CB in interruption of charging currents.
2. Control the rate of rise of restriking voltage given by

$$f_n = \frac{1}{2\pi\sqrt{LC}}$$

In synthetic testing and other indirect tests, capacitors are vital items in the test circuit. These are single-phase banks and can be connected in series or parallel as desired, both individually and in any combination of the three.

9. Test Cells (or Cubicals). These are constructions of RCC or strong brickwork. There is a provision for observation. Supply of compressed air and oil purifying system is given to the test cells for facilitating testing of air blast circuit breaker and oil circuit breakers respectively. Separate test cubicles are provided for testing LV; HV; and EHV circuit breakers. Each test cell has got an end box where control and measuring cables are terminated.

10. Test Control Room. It is situated nearly at a distance of 20 m from test cubicles and is provided with a control desk for remote control operation of the excitation of the SC generators, etc. There are also an oscillograph, distribution panel, an electromagnetic oscillograph; a cathode ray oscilloscope (CRO), an electric sequence switch and an electronic sequence switch. The control room is designed to provide facilities for the observation of tests by both the station staff as well as visitors.

11. Safety and Signalling System. A foolproof interlocking system is provided for the safety of the working personnel, as enormous electrodynamic forces are brought into play during short-circuit testing. This system makes the excitation of the machines impossible, if any of door leading to machine hall basement reactor room, etc. remains open. A system of audio visual alarms are also provided as a warning before SC tests.

7.12. TESTING PROCEDURE

A preliminary checkup of various equipments is made before the actual commencement of tests. The correct values of resistance and reactance coils are inserted as per the magnitude of the short-circuit current. Transformers are properly set and the measuring circuits are connected and oscillograph loops are calibrated. The sequential switching of equipment measurement and control circuits is accomplished by a sequence switch. Sequence switch is a drum switch with several pairs of contacts. The drum is rotated by a motor. Once the drum is rotated, it closes and opens several control circuits according to a certain sequence. For example, the sequences for breaking capacity test in one test were as follows :

1. Drive motor of SC generator and the exciting machines is switched off.
2. Impulse excitation is switched on.
3. Master circuit breaker is closed.
4. Oscillograph circuit is connected.
5. Make switch is closed.
6. Circuit breaker under test is opened.
7. Master circuit breaker is switched off.
8. Exciter is switched off and its field is suppressed.

The above operations take a very short time of the order of 0.2 second. After the completion of short circuit, the starters of the drive motors of the SC generator and the exciting machines are automatically reset to a lower degree. The motors are then switched on and are allowed to run at full speed, so that in a short time a further short circuit can be initiated. A schematic representation of test circuit is shown in Fig. 7.22.

7.13. DIRECT TESTING

The direct testing of CBs in a plant enables us to test it under conditions largely representing those in actual network as well as tests of greater serverity. The circuit breaker under test is subjected to the value of transient restriking voltage to which it is expected to be put in practice rather than testing it under most severe conditions.

The circuit for direct test is shown in Fig. 7.23.

The preliminary preparation for testing of a CB include connecting the equipment adjusting the magnitude of reactors, connecting transformers to give the desired test voltages etc. The contacts on sequence switch are adjusted to have desired timings. The oscillographs are adjusted and calibrated. The operations of test follow automatically by means of sequence switch as mentioned earlier.

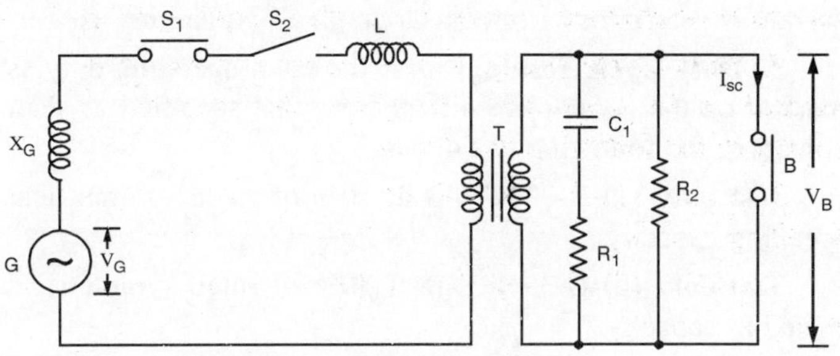

G = Short-circuit generator
V_G = Generator voltage
V_B = Voltage across test circuit breaker
I_{SC} = Short-circuit test current
X_G = Generator reactance
S_1 = Backup (master) circuit breaker
S_2 = Making switch
L = Current control reactor
T = Transformer
B = Breaker under test
C_1, R_1, R_2 = Capacitance, resistance for adjustment of transient recovery voltage

Fig. 7.23. ***Short-Circuit Test Arrangement (Single Line Representation)***

The various tests carried out for circuit breaker testing are described below :

1. Making Capacity Test : Making capacity test is necessary type test. All CBs are tested for their ability to make

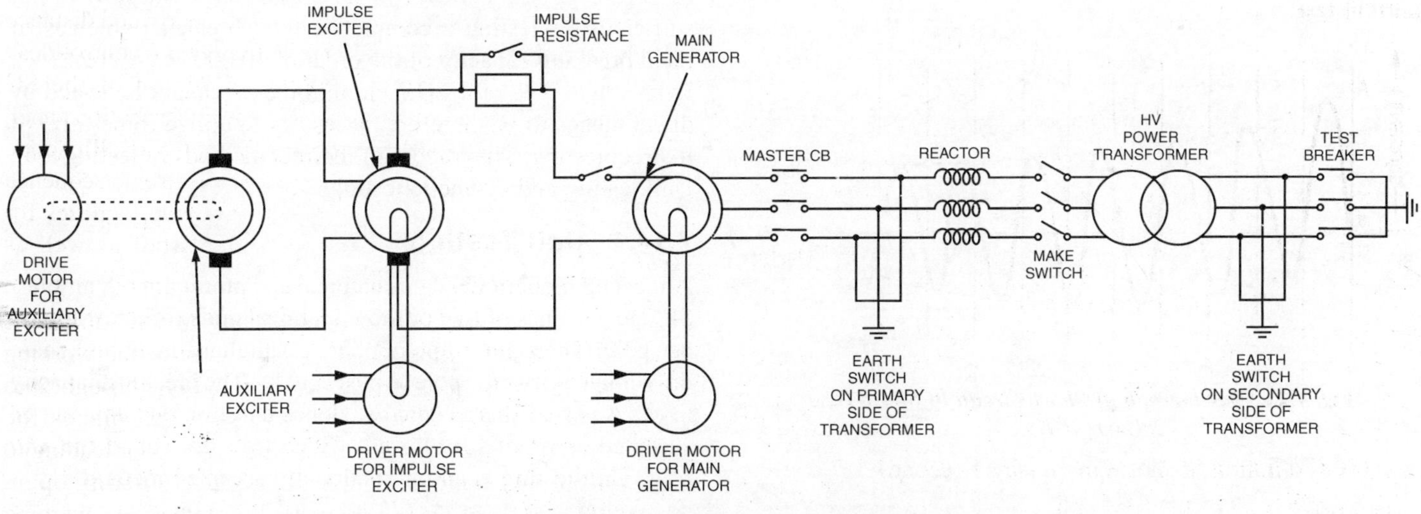

Fig. 7.22. ***Schematic Representation of the Test Circuit***

on to a short circuit. The master circuit breaker and the make switch are closed first, then the breaker under test is closed on a 3-phase short-circuit test. The making current is determined as explained in Art. 7.9.

2. Braking Capacity Test : For testing breaking capacity, Master CB and circuit breaker under test are closed first and then the short circuit is applied by closing the main switch. The breaker under test is opened at desired moment and the breaking current is determined from oscillograph as explain in Art. 7.9.

3. Duty Cycle Tests : Unless the rated operating duty as marked on the name plate differs from that specified, it shall consist of the following test duties.

Test duty (1) B-3′-B-3′-B at 10% of rated symmetrical breaking capacity.

Test duty (2) B-3′-B-3′-B at 30% of rated symmetrical breaking capacity.

Test duty (3) B-3′-B-3′-B at 60% of rated symmetrical breaking capacity

Test duty (4) B-3′-MB-3′-MB at not less than 100% of rated symmetrical breaking capacity and not less than 100% of rated making capacity.

Test duty (4) may be carried out as two separate duties as follows :

Test duty (4a) M-3′-M (make test)

Test duty (4b) B-3′-B-3′-B (break test)

Test duty (5) B-3′-B-3′-B at not less than 100% rated asymmetrical breaking capacity.

B and M in the above duty cycle indicate break and make operations respectively. MB means a making operation followed by a breaking operation without any intentional time lag. 3′ denotes the time duration in minutes between successive operations of an operating duty.

4. Short Time Current Tests : The CB is tested for this by means of a separate high current and low voltage transformer. The current is passed through the breaker for a short time (1 second or 3 seconds) and the oscillograph is taken. Figure 7.24 shows an example of oscillograph taken during the short time current test.

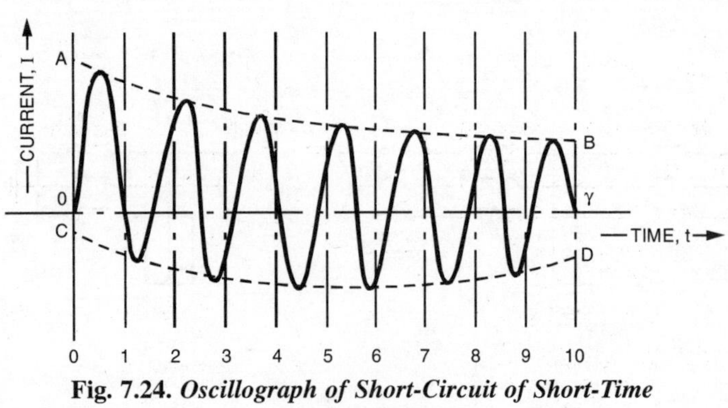

Fig. 7.24. ***Oscillograph of Short-Circuit of Short-Time Current Test***

OT is duration of short-circuit (say 1 second)

AB, CD — Envelope of short-circuit current wave

$I_0, I_1, I_2, ..., I_{10}$ rms value of asymmetrical current at each instant. The rms value of current during the time interval 0 to T of such a wave is given by the expression

$$I_{rms} = \sqrt{\frac{1}{T}\int_0^T i^2\, dt}$$

where i is the variable current, t is the variable time in seconds, and T is duration of current in seconds.

Equivalent rms value of short-time current is determined as follows :

The time interval 0T is divided into 10 equal parts marked by 0, 1, 2, ..., 10. The rms values at these instants are $I_0, I_1, I_2, ..., I_{10}$.

$$I_{rms} \text{ at this instant} = \sqrt{(I_{sym})^2 + (I_{dc})^2}$$

where I_{sym} is rms value of ac component at this instant and I_{dc} is dc component at this instant.

It is rms value of current at this instant. Thus I_0, I_1, I_2, ... etc up to I_{10} are computed. From these values, the rms value of short-time current is determined by Simson formula *i.e.*,

$$I = \sqrt{\frac{1}{3}\left[I_0^2 + 4(I_1^2 + I_3^2 + I_5^2 + I_7^2 + I_9^2) + 2(I_2^2 + I_4^2 + I_6^2 + I_8^2 + I_{10}^2)\right]}$$

Temperature rise limits are not specified for short-time current tests. It is very difficult to record the transient temperature during 1 second duration.

Good contact is required to be maintained in spite of forces developed by the short-circuit current so as to avoid welding. The heat produced by the short-circuit current over the period of test (1 second or 3 seconds) should not cause damage to any insulation in contact with the current carrying parts. When the breaker has cooled down to the ambient temperature, it should be capable of carrying full-load current continuously without excessive temperature rise.

7.14. INDIRECT TESTING

The short-circuit power available in earlier testing stations (of the order of 4,000 MVA in laboratory type station) is not sufficient for testing a complete circuit breaker (which is of rated breaking capacity of the order of 10,000 MVA at 245 V). Even single pole of a EHV circuit breaker cannot be tested by direct means. It is, therefore, necessary to utilize some form of indirect testing. The important indirect methods of testing are : Unit testing and synthetic testing.

7.14.1. Unit Testing

Almost all modern EHV circuit breakers, minimum oil, air blast SF_6 etc. consists of two or more identical units (or interrupters) per pole. These interrupters operate simultaneously and share the voltage across the pole almost equally. The breaking capacity in MVA is also shared equally. Hence a test on one unit can be accepted as proof for all units. Such tests are called the *unit tests.* Unit testing is an internationally accepted method.

While applying unit test the voltage is to be reduced by a factor k and all the impedances should be reduced by the same factor k in order to have test voltage across the unit same as that following expressions :

$$k = \frac{1}{n} \text{ when one unit is tested together}$$

$$= \frac{m}{n} \text{ when m units are tested together}$$

where n is the number of units per pole

For instance, consider a 3-pole 220 kV circuit breaker with three units per pole. Test is to be performed at normal voltage of 230 kV. Voltage across each pole is $\frac{230}{\sqrt{3}}$ *i.e.,* 132.8 kV.

$$k = \frac{1}{n} = \frac{1}{3} \qquad \because n = 3$$

Voltage required for testing one unit

$$= k \times 132.8 \text{ kV} = \frac{1}{3} \times 132.8 \text{ kV} = 44.3 \text{ kV}$$

Also, L and C test circuit should be reduced to have the same natural frequency as that direct testing *i.e.,*

$$f_n = \frac{1}{2\pi\sqrt{LC}} \text{ in direct testing}$$

$$f_n = \frac{1}{2\pi\sqrt{kL \times \frac{C}{k}}} = \frac{1}{2\pi\sqrt{LC}}$$

The natural frequency of the transient restriking voltage remains the same. Time scale also remains the same.

With circuit breakers in which the voltage distribution across the pole is not evenly distributed amongst the units, some units will be stressed more and the others less. The test should be performed so as to test the highest stress coming over the unit. Statically, unit test has been established as a reliable testing method.

7.14.2. Synthetic Testing

Synthetic testing is a practical and economical solution for testing of CBs of high rupturing capacities, without actually using the corresponding short-circuit capacity of the testing station. The synthetic circuit is designed to simulate as accurately as possible the electrical stresses impressed on the circuit breaker during the interruption of fault current under system conditions.

Figure 7.25 depicts the principle of synthetic testing. Synthetic testing makes use of two sources viz. (1) current source (of relatively low voltage) and (2) voltage source (of relatively low current). The current source is to provide the short-circuit current while the voltage source is to provide restriking voltage in addition to recovery voltage. Others r, L, C etc are used for having the desired conditions. The switch S_1 is closed for supplying short-circuit current i_G. At near final current zero switch S_2, usually a spark gap, is closed and V_s is applied to the CB under test at an appropriate moment. The voltage will have transient because of L and C of the circuit.

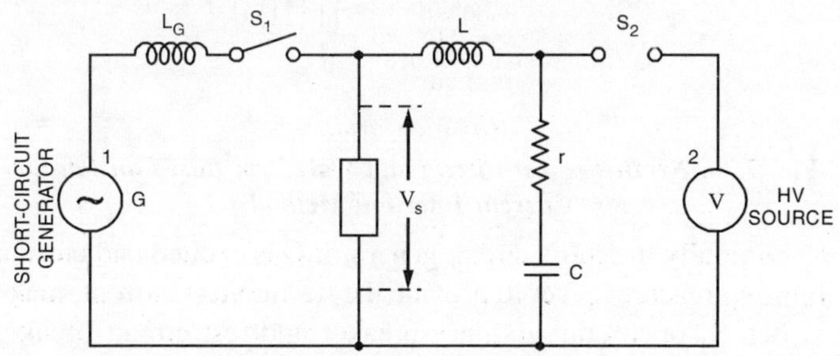

Fig. 7.25. *Synthetic Testing Test Circuit (Simplified)*

This testing method has the following advantages :

1. This method is simple and can be applied to unit testing also.
2. This method makes possible testing of CB of capacity (MVA) of five times that of the capacity of the test plant.
3. The CB can be tested for desired TRV and RRRV.
4. The short-circuit generator has to supply currents at a relatively lower voltage (as compared to direct testing).
5. Both test current and test voltage can be independently varied. This provides flexibility to the test.

Synthetic test circuits are of two types viz. parallel current injection method and series current injection method.

Parallel current injection method is widely employed for testing CBs because it can provide high frequency transient voltages as needed by standards. In parallel current injection method [Fig. 7.26 (*a*)], the voltage circuit (2) is effectively connected in parallel with current circuit (1) and the test circuit

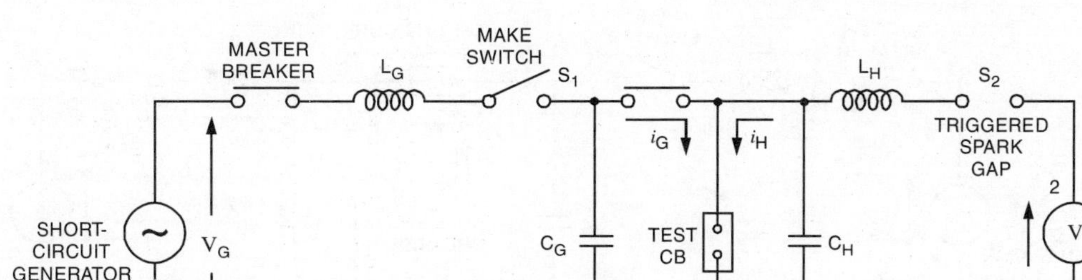

Fig. 7.26 *(a) Parallel Current Injection*

breaker before the main current i_G in test breaker current is properly simulated.

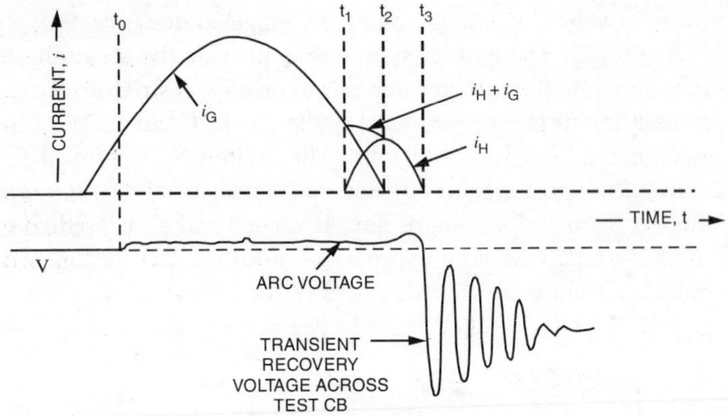

(b) Waveforms

Fig. 7.26. ***Synthetic Test Circuit and Waveform Based on Parallel Current Injection Method***

Initially the short-circuit generator G is excited and current limiting reactor L_G set to provide the desired test current, make switch S_1 open and auxiliary breaker and test circuit breaker are closed.

In the voltage circuit C_H is charged to provide the required voltage and the spark gap S_2 is set ready to be fired. L_H is so chosen that when i_H flows consequent to the triggering of the gap, the rate of change of current through zero is the same, as that of test current. C_H is selected to provide the required transient restriking voltage frequency.

The make switch S_1 is closed and the normal frequency 50 Hz current i_G flows in the circuit breaker. At time t_0, the auxiliary breaker and test circuit breaker begin to open so that they are fully open at the following current zero (t_2) in Fig. 7.26 (*b*). At time t_1, the spark gap S_2 is fired and the current i_H flows in the test circuit breaker. The frequency of this current is determined by L_H and C_H and timed so that its peak approximately coincides with zero of current i_G. At time t_2 when the current i_G becomes zero it is interrupted by the auxiliary breaker and the test circuit breaker carries current i_H from the voltage circuit. When this current becomes zero at t_3 the restriking voltage transient appears across the circuit breaker. The magnitude and frequency of this restriking voltage transient depend on voltage to which the capacitor C_H is charged and the parameters C_H and i_H. This enables the supply of current and voltage at the moment of zero current from one and the same source.

Thus breaking capacities up to ten or more times the short-circuit capacity of the short-circuit generator can be achieved and the test circuit breaker can be subjected to exactly the same stresses as in an actual system.

Series Current Injection Method. Figure 7.27 (*a*) represents series current injection method.

This circuit differs from the parallel current injection circuit in a way in which the small loop of current from the voltage source is introduced. The voltage circuit is connected across auxiliary breaker, instead of test circuit breaker and the source capacitance C_H is charged to the opposite polarity, with the result that the injected current flows in the direction opposite to that of 50 Hz current, and thus subtracts from it (Fig. 7.27). As the voltage and current circuits are series connected, it is rather difficult to choose circuit parameters to suit both the current and voltage circuits which would at the same time provide required restriking voltage transient. This arrangement is particularly well suited for low-frequency circuits.

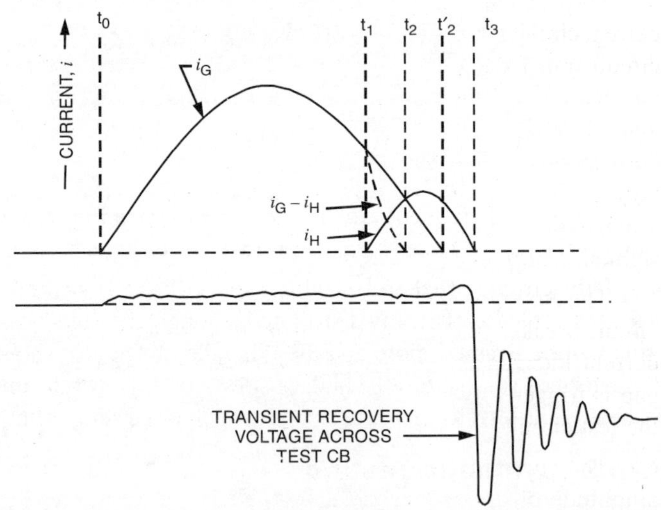

(b) Waveforms

Fig. 7.27. ***Synthetic Test Circuit and Waveforms Based on Series Current Injection Method***

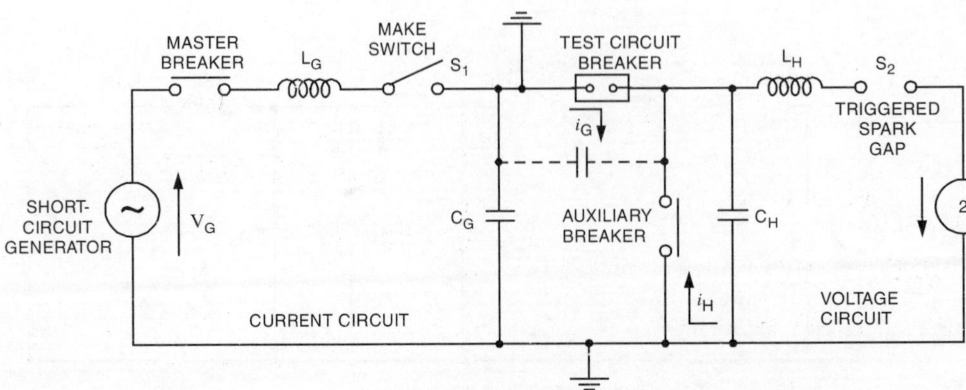

Fig. 7.27 *(a) Series Current Injection*

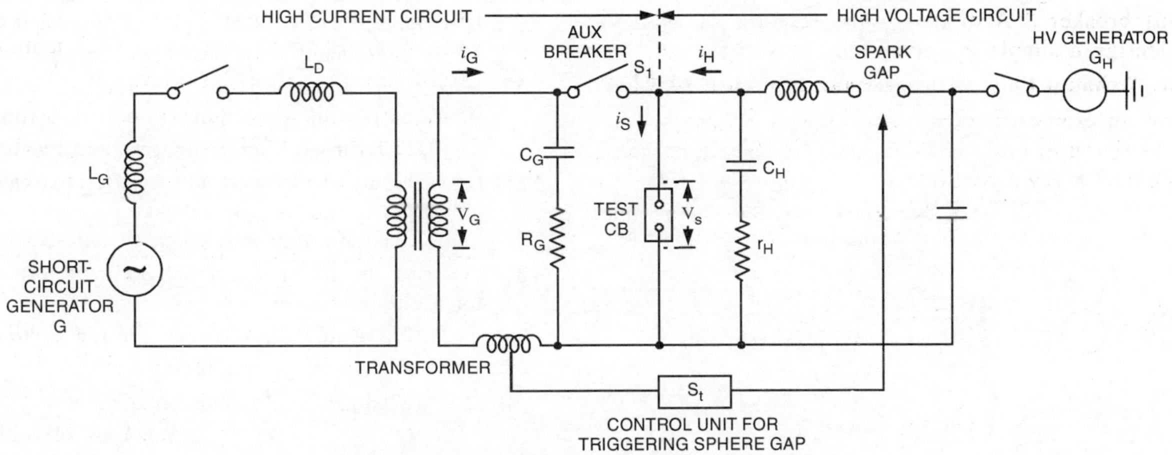

Fig. 7.28. ***Brown Boveris Synthetic Testing Circuit***

Brown Baveries Synthetic Testing Circuit. Synthetic test circuit shown in Fig. 7.28 is used by Brown Boveri, Switzerland. In this circuit, the short-circuit current is supplied from a circuit at a relatively low voltage while the restriking and recovery voltage is provided by a separate HV circuit.

The circuit on the left side of the test circuit breaker is high current circuit which consists of short-circuit generator G, short-circuit transformer and also capacitor C_G and resistor R_G. C_G and R_G control the natural frequency of high current circuit. The short-circuit power is supplied at voltage V_G which corresponds to about 30 kV, this voltage is smaller than recovery voltage V_s required for testing the specimen. The recovery voltage V_s is supplied from a separate voltage circuit on the right side of the test circuit breaker.

The auxiliary breaker S_1 is opened simultaneously with the circuit breaker under test and a few microseconds before the current interruption (i_G) in the test circuit breaker. The spark gap is triggered by control S_1 and the voltage V_s is applied to the test circuit breaker.

The current i_H has a natural frequency of 500 Hz and an amplitude of one-tenth of that of current i_G. The currents are superimposed in current zero zone in such a way that during last 100 μs only current i_H is flowing through test circuit breaker. The auxiliary circuit breaker S_1 interrupts high current circuit from HV circuit before current $i_S = i_G + i_H$ has to interrupt only current i_H. The restriking voltage across test circuit breaker is, therefore, given by that of HV circuit.

Exercises

1. What is a circuit breaker ? Explain its functions.
2. What are the major duties that a circuit breaker is required to perform ? Explain them clearly.
3. Explain the phenomena of arc formation.
4. Discuss the factors which influence the arc voltage.
5. Discuss different methods of interrupting the arc in circuit breakers. Explain two main theories of current zero interruption.

   [G.B. Technical University Switchgear and Protection 2010-11]
6. Explain different methods of arc extinction in a circuit breaker. **[W.B. Univ. of Technology Power System-II, 2010]**
7. Explain the following terms :

   (*i*) Restriking voltage, (*ii*) Recovery voltage, (*iii*) *RRRV.*

   [W.B. Univ. of Technology Power System-II, 2010]
8. Discuss recovery rate theory and energy balance theory of arc interruption in a CB.

   [U.P. Technical Univ. Switchgear and Protection 2005-06]
9. Define the term "active recovery voltage". Discuss the factors which influence the active recovery voltage of a circuit breaker.
10. Explain the terms recovery voltage, restriking voltage and RRRV. Derive an expression for restriking voltage in terms of system capacitance and inductance.

    Explain the Slepian's theory of arc interruption and discuss its limitations.
11. Explain the terms (*i*) restriking voltage (*ii*) recovery voltage (*iii*) RRRV.

    Derive an expression for the restriking voltage in terms of system voltage, inductance and capacitance across a circuit breaker contact when a 3-phase fault takes place. Assume the neutral of the system to be solidly grounded.
12. Define recovery voltage and active recovery voltage.

    Discuss the effect of power factor and the amount of asymmetry in the short-circuit wave upon the active recovery voltage.

    What is a restriking voltage ? Explain why severe voltage oscillations may occur while a circuit breaker is de-energizing a long transmission line which is open circuited at the receiving end.
13. Derive the expression for restriking voltage and RRRV. What measures are taken to reduce it ?

    [U.P. Technical Univ. Switchgear and Protection 2005-06]
14. Define the terms recovery voltage and rate of rise of restriking voltage. Derive an expression for the latter.
    [U.P. Technical Univ. Switchgear and Protection 2006-07]
15. Discuss about RRRV. Draw a neat diagram to show the voltages and current during relay and circuit breaker operations. Discuss about recovery voltage, relay operating time, circuit breaking time and circuit making time.

    [U.P. Technical Univ. Switchgear and Protection 2009-10]

16. What is the significance of RRRV in operation of a circuit breaker ? Also derive the relation for RRRV considering a simple power system.

[U.P. Technical Univ. Switchgear and Protection 2008-09]

17. Derive an expression for transient recovery voltage due to fault as shown in Fig. 1. Assume stray capacitance between circuit breaker and earth. Circuit breaker opens at $t = 0$.

[U.P. Technical Univ. Switchgear and Protection 2004-05, 2007-08]

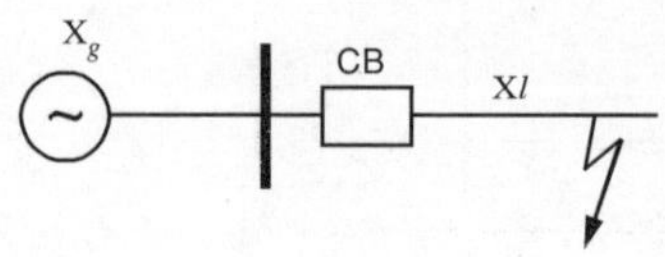

Fig. 1

18. Write short notes with reference to current interruption on : (*i*) Kilometric fault, (*ii*) Current chopping phenomena, (*iii*) Effect of power factor.

19. Explain, 'successive restrikes' and current chopping as applied to the interruption of capacitive and low inductive currents.

How does resistance switching help in such conditions ?

20. Explain the phenomena of current chopping associated with axial blast circuit breaker. How this dangerous phenomena is rectified by resistance switching ?

21. Explain the phenomenon of current chopping in a circuit breaker.

[G.B. Technical Univ. Switchgear and Protection 2010-11]

22. Discuss the problems associated with the interruption of (*i*) capacitive current (*ii*) fault current, if fault is very near to the substation.

[G.B. Technical Univ. Switchgear and Protection 2010-11]

23. What are the problems associated with short-line interruption ? How is the problem eliminated ?

[U.P. Technical Univ. Switchgear and Protection 2009-10]

24. Show how the restriking transients can be damped by connecting a resistance across the contacts of a circuit breaker.

[U.P. Technical Univ. Switchgear and Protection 2006-07]

25. Describe resistance switching. What do you understand by pre-insertion resistance and pre-insertion time ?

[U.P. Technical Univ. Switchgear and Protection 2004-05; 2007-08]

26. Explain various ratings of a circuit breaker.

[U.P. Technical Univ. Switchgear and Protection 2006-07]

27. Explain the following terms referred to a circuit breaker :
(*a*) Symmetrical braking current, (*b*) Asymmetrical breaking current, (*c*) RRRV, (*d*) Making current, (*e*) Breaking capacity.

[U.P.S.C. I.E.S. Electrical Engineering-II, 2002]

28. Explain (*i*) Making capacity (*ii*) Breaking capacity (*iii*) Restriking voltage (*iv*) Short-line fault (*v*) Rated operating duty.

[U.P. Technical Univ. Switchgear and Protection 2005-06]

29. Discuss in detail the different types of rating of a circuit breaker bringing out clearly their physical significance.

30. What are the usual ratings of a circuit breaker ? Why a circuit breaker has different types of voltage, current and power ratings ?

31. What are the different tests carried out on a circuit breaker ? Differentiate and explain type and routine tests.

[U.P. Technical Univ. Switchgear and Protection 2008-09]

32. What is the difference between direct and indirect testing ? Also describe the procedure of indirect testing.

[U.P. Technical Univ. Switchgear and Protection 2008-09]

33. How are circuit breakers rated ? Describe the direct and indirect testing of circuit breakers.

[U.P. Technical Univ. Switchgear & Protection 2004-05; 2007-08]

34. Describe about the direct and indirect testing procedure for a circuit breaker.

[U.P. Technical Univ. Switchgear and Protection 2009-10]

35. Write short notes on the following :

(*a*) Switchgear and circuit breaker.

[W.B. Univ. of Technology 2010]

(*b*) 'High resistance' and 'zero point' interruption of arc in a circuit breaker. **[W.B. Univ. of Technology 2010]**

(*c*) Duties of circuit breaker.

(*d*) Resistance switching.

(*e*) Circuit breaker ratings.

Short Answer Type Questions With Answers

Q. 1. What is a circuit breaker ?

Ans. Circuit breaker is a mechanical device designed to close or open contact members, thus closing or opening an electrical circuit under normal or abnormal conditions.

Q. 2. How does a circuit breaker differ from a switch ?

Ans. A circuit breaker is a device that interrupts the abnormal or fault currents and in addition performs the function of a switch.

Q. 3. Which of the gases are employed in commercial gas-blast circuit breakers ?

Ans. SF_6 and air are used in commercial gas-blast circuit beakers.

Q. 4. What is meant by current chopping ?

Ans. Current chopping is a phenomenon of current interruption before the natural current zero is reached.

Q. 5. Why is current chopping considered a serious drawback in a circuit breaker ?

Ans. The current chopping is considered a serious drawback because it sets up high voltage transient across the breaker contacts.

Q. 6. How is the effect of overvoltage developed due to current chopping overcome ?

Ans. By using resistance switching *i.e.*, by connecting a resistance across the circuit breaker contacts.

Q. 7. Why current chopping is not common in oil circuit breakers ?

Ans. Current chopping is not common in oil circuit breakers because in most of the oil circuit breakers the arc extinguishing power is proportional to the magnitude of current to be interrupted.

Q. 8. What is meant by kilometric fault ?

Ans. The fault occurring between a distance of a few km to few tens km from the circuit breaker arc is called kilometric fault.

Q. 9. What is meant by breaking or rupturing capacity of a circuit breaker ?

Ans. Breaking capacity of a circuit breaker is the highest rms value of current that it is capable of breaking under specified conditions of transient recovery voltage and power frequency voltage. It is expressed in kA_{rms} at contact separation.

Q. 10. Why asymmetrical breaking current is higher than symmetrical breaking current in a circuit breaker ?

Ans. Asymmetrical breaking current is the rms value of the combined sum of ac and dc components of short-circuit current at the instant of separation of breaker contacts while symmetrical breaking current is the rms value of the ac component alone. Thus asymmetrical breaking current is higher than symmetrical breaking current.

Q. 11. What is meant by making capacity of a circuit breaker ?

Ans. The making capacity of a circuit breaker, when closed on a short circuit, is the peak value of the maximum current wave (including dc component) in the first cycle of the current after the circuit is closed by the circuit breaker.

Rated making capacity

$$= 2.55 \times \text{symmetrical rated breaking capacity}$$

Q. 12. What is meant by rated voltage of a circuit breaker ?

Ans. The rated maximum voltage of a circuit breaker is the highest rms voltage, above nominal system voltage, for which the circuit breaker is designed and is the upper limit for operation.

Q. 13. Why testing of a circuit breaker is more difficult than the testing of other electrical equipment like transformer or machine ?

Ans. Testing of circuit breaker is more difficult than the testing of other electrical equipment like transformer or machine because the short-circuit currents are very large. Also there is no satisfactory method of testing circuit breaker at reduced power.

Q. 14. What are type tests which are conducted on circuit breakers ?

Ans. Type tests are conducted on first few prototype circuit breakers of each type for the purpose of proving the capabilities and confirming the rated characteristics of the circuit breaker of that design. Such tests are conducted in specially built testing laboratories. Type tests are performed as per recommendations of standards (IEC) or (IS).

Type tests can be broadly classified as (*i*) mechanical performance tests (*ii*) thermal tests (*iii*) dielectric or insulation tests and (*iv*) short-circuit tests in order to check making capacity, breaking capacity, short-time rating current and operating duty.

Q. 15. What are routine tests which are conducted on circuit breakers ?

Ans. Once type tests are conducted and a particular design is found to be satisfactory the product becomes prototype and a large number of circuit breakers of similar design are manufactured. However, each and every circuit breaker is still subjected to a few more tests before commissioning. These tests are called routine tests.

Routine tests are also performed as per recommendations of the standards (IEC/IS). These tests are performed in the manufacturer's premises. Routine tests confirm the proper functioning of the circuit breaker.

The routine tests include (*i*) power frequency voltage test (*ii*) millivolt drop test and (*iii*) operational tests.

Q. 16. What are the types of testing stations for CBs ?

Ans. There are three types of testing stations viz. field type testing station, laboratory type testing station; composite testing station.

Q. 17. List the main equipment required in the testing stations for circuit breakers.

Ans. Equipment used in testing stations are : (*i*) Short-circuit generator and drive motor (*ii*) Impulse exciter (*iii*) Pilot generator, (*iv*) Short-circuit transformer (*v*) Master circuit breaker, (*vi*) Resistors and reactors (*vii*) Make switch (*viii*) Capacitors (*ix*) Test cubicles (*x*) Test control room (*xi*) Safety and signalling system.

Q. 18. What is direct testing in case of circuit breakers ?

Ans. The direct testing of CBs in a plant enables us to test it under conditions largely representing those in actual network as well as tests of greater serverity. The circuit breaker under test is subjected to the value of transient restriking voltage to which it is expected to be put in practice rather than testing it under most severe conditions.

Q. 19. Why does indirect testing of CBs become necessary ?

Ans. The short-circuit power available in earlier testing stations (of the order of 4,000 MVA in laboratory type station) is not sufficient for testing a complete circuit breaker (which is of rated breaking capacity of the order of 10,000 MVA at 245 V). Even single pole of a EHV circuit breaker cannot be tested by direct means. It is, therefore, necessary to utilize some form of indirect testing. The important indirect methods of testing are : Unit testing and synthetic testing.

Q. 20. What is unit testing and synthetic testing ?

Ans. Almost all modern EHV circuit breakers, minimum oil, air blast, SF_6 etc. consists of two or more identical units (or interrupters) per pole. These interrupters operate simultaneously and share the voltage across the pole almost equally. The breaking capacity in MVA is also shared equally. Hence a test on one unit can be accepted as proof for all units. Such tests are called the *unit tests.* Unit testing is an internationally accepted method.

Synthetic testing is a practical and economical solution for testing of CBs of high rupturing capacities, without actually using the corresponding short-circuit capacity of the testing station. The synthetic circuit is designed to simultate as accurately as possible the electrical stresses impressed on the circuit breaker during the interruption of fault current under system conditions.

Problems

1. In a 132 kV circuit breaker the bushing to ground capacitance is 0.01 μF and the transformer inductance succeeding the circuit breaker is 5 H. Calculate the voltage appearing across the poles of the circuit breaker if it breaks a magnetising current of 10 A flowing in the transformer. [**Ans.** 223.6 kV]
2. In 220 kV system, reactance and capacitance up to location of CB is 8 Ω and 0.025 μF. A resistance of 600 Ω is connected across the contacts of CB. Determine (*i*) natural frequency of oscillation (*ii*) damped frequency of oscillation, (*iii*) critical value of resistance.

   [U.P. Technical Univ. Switchgear and Protection 2005-06]

   [**Ans.** (*i*) 6.3 kHz, (*ii*) 3.41 kHz, (*iii*) 504.6 Ω]
3. In a power system the rms voltage is 38.1 kV, L is 10 mH and C = 0.02 μF. Determine

(*a*) peak restriking voltage across the circuit breaker (*b*) frequency of restriking voltage transient (*c*) average rate of restriking voltage up to peak restriking voltage, and (*d*) maximum RRRV.

[**Ans.** (*a*) 107.8 kV (*b*) 11.254 kHz (*c*) 2.43 kV/μs (*d*) 3.81 kV/μs]

4. From the following data of a 50 Hz generator–emf to neutral 7.5 kV (rms), reactance of generator and connected system 4 Ω, distributed capacitance to neutral 0.01 μF, resistance negligible: find (*i*) the maximum voltage across the contacts of the circuit breaker when it breaks a short-circuit current at zero current; (*ii*) the frequency of transient oscillation; and (*iii*) average rate of rise of voltage up to the first peak of oscillation.

[**Ans.** (*i*) 21.21 kV (*ii*) 14.1 kHz (*iii*) 0.6 kV/μs]

5. A 50 Hz, 11 kV, three-phase alternator with earthed neutral has a reactance of 5.0 Ω per phase and is connected to bus--bar through a circuit breaker. The capacitance to earth between the alternator and the circuit breaker is 0.02 μF be per phase. Assuming the resistance of the generator to be negligible, calculate the following:

(*i*) Maximum voltage across the contacts of the circuit beaker. (*ii*) Frequency of oscillation. (*iii*) The average rate of rise of restriking voltage. [**Ans.** (*i*) 17.963 kV (*ii*) 8.92 kHz (*iii*) 0.32 kV/μs]

6. Calculate the RRRV of a 132 kV circuit breaker with earthed neutral. The short-circuit test data obtained are as follows:

The current broken is symmetrical and the restriking voltage has an oscillatory frequency of 20 kHz. The power factor of the fault is 0.3. Assume the short-circuit to be a grounded fault.

[**Ans.** 8.225 kV/μs]

7. In a short-circuit test, with earthed neutral on a 132 kV, 3-phase circuit breaker, the power factor of the fault was 0.3, the recovery voltage was 0.95 of full-line value, the breaking current was symmetrical and the restriking transient had a natural frequency of 16,000 Hz. Estimate the rate of rise of restriking voltage.

[**Ans.** 6.25 kV/μs]

8. In a short-circuit test on a 132 kV, 3-phase system, the breaker gave the following results: pf of the fault 0.4, recovery voltage 0.95 of full-line value; the breaking current is symmetrical and restriking transient had a natural frequency of 16 kHz. Determine the rate of rise of restriking voltage. Assume that the fault is grounded. **[Pb. Univ. Electrical Power III (Switchgear and Protection) Dec. 1991]**

[**Ans.** 6 kV/μs]

9. In a 132 kV, 3-phase, 50 Hz power system the line to ground capacitance is 0.04 μF and the inductance is 4 H. Calculate (*i*) voltage appearing across the breaker pole when a magnetising current of 6 A is interrupted (*ii*) resistance to be connected across the breaker contacts to eliminate the restriking voltage.

[**Ans.** (*i*) 60 kV (*ii*) 5 kΩ]

10. A circuit breaker is rated 2,500 A, 1,500 MVA, 33 kV, 3 s, 3-phase oil-circuit breaker. Determine the rated normal current, breaking current, making current and short-time rating current.

[Pb. Univ. Switchgear and Protection April 1988]

[**Ans.** 2.5 kA, 26.24 kA (rms); 66.9 kA; 26.24 kA for 3 seconds]

❑❑❑

CHAPTER 8

Circuit Breakers

INSIDE THIS CHAPTER

8.1. INTRODUCTION

As already explained in Chapter 7, a circuit breaker is a device that interrupts the abnormal or fault currents and in addition performs the function of a switch. The duties, ratings and testing of circuit breakers have already been dealt with in the preceeding chapter, classification of circuit breakers and circuit breakers of different types will be dealt with here.

8.2. CLASSIFICATION OF CIRCUIT BREAKERS

There are several methods of classification of circuit breakers. On the basis of type of current they may be classified as *ac circuit breakers* and *dc circuit breakers.*

AC circuit breakers may be classified on the basis of rated voltages. Circuit breakers below rated voltage of 1,000 V are known as the *low-voltage circuit breakers* and above 1,000 V are called the *high-voltage circuit breakers*. However, the most general way of classification of the circuit breakers is on the basis of medium of arc extinction such as air break circuit breakers/miniature circuit breakers, oil circuit breakers (tank type or bulk oil), minimum oil circuit breakers, air blast circuit breakers, sulphur hexafluoride circuit breakers and vacuum circuit breakers. All high-voltage circuit breakers may be classified under two main categories viz., oil circuit breakers and oil-less circuit breakers.

Oil circuit breakers utilize dielectric oil (transformer oil) for arc extinction. Oil circuit breakers can further be subdivided into two classes viz., *bulk oil circuit breakers* and *low oil content circuit breakers* or *minimum oil circuit breakers*. In *bulk oil circuit breakers* transformer oil, with which they are filled, is used to extinguish the arc during opening of the breaker contacts. The oil also serves as an insulator for the current conducting parts from one another and from the earthed tank. The ratings range from 25 MVA at 2.5 kV to 5,000 MVA at 230 kV. Various types of arrangements are available for indoor as well as outdoor working at different voltage levels. In *low oil content* or *minimum oil circuit breakers*, oil is used for arc extinction by blast action, and is used chiefly for this function and not so much for insulating live parts from earth. In such breakers, the oil-filled arc enclosing chamber is arranged within the hollow of the porcelain insulator which insulates the live parts of the breaker from earth. These circuit breakers can be employed for a voltage range of 33 kV to 220 kV and breaking capacities of 1,500 MVA to 7,500 MVA. Another type of oil circuit breaker is the *oil impulse breaker*. In this breaker, an arc extinguishing oil jet is produced by a piston pump which is externally energised by means of a spring or compressed air. The jet of oil is aimed at the gap formed between the separating contacts of the breaker to extinguish the arc. Oil-impulse breaker has many features and characteristics similar to those of the air blast circuit breakers.

The main types of *oil-less circuit breakers* are :

1. *Air circuit breakers* in which the arc is initiated and extinguished in substantially static air in which the arc moves. Such breakers are used for low voltages, generally up to voltages of 15 kV and rupturing capacities of 500 MVA.
2. *Air blast circuit breakers* in which a blast of air is used to blow out the arc. In modern air blast circuit breakers, compressed air is stored in a tank and released through a nozzle to produce a high velocity jet, and this is used to extinguish the arc. Air blast circuit breakers are used for indoor services in the medium high voltage field and for medium rupturing capacity—generally up to voltages of 15 kV and capacities of 2,500 MVA. The air blast circuit

breakers are now employed in high voltage circuits in outdoor switchyards for 220 kV lines.

3. *Sulphur hexafluoride circuit breakers* in which SF_6 under pressure is used to extinguish the arc. SF_6 gas have excellent dielectric, arc quenching, chemical and other physical properties and has proved its superiority over other arc quenching mediums such as oil or air.
4. *Vacuum circuit breakers* in which the fixed and moving contacts are housed inside a permanently sealed vacuum interrupter. The arc is quenched as the contacts are separated in high vacuum. Vacuum circuit breakers are more efficient, less bulky, cheaper in cost, negligible maintenance and longer life.

Each type of circuit breaker has its own merits, demerits and applications. Now each type of circuit breakers will be discussed in details.

8.3. OIL CIRCUIT BREAKERS

These are the oldest type of circuit breakers. The separating contacts of the breakers are made to separate within an insulating oil, which has better insulating properties than air. On occurrence of fault as the breaker contacts open under oil, an arc is struck between them and the heat of the arc evaporates the surrounding oil and dissociates it into a substantial volume of gaseous hydrogen (hydrogen gas along with a small percentage of methane, ethylene and acetylene) at high pressure. The pressure built-up and flow of gases is influenced by the design of arc control device, speed of contact travel, the energy liberated by the arc etc. The oil is pushed away from the arc and an expanding hydrogen gas bubble surrounds the arc region and adjacent portion of the contacts, as shown in Fig. 8.1.

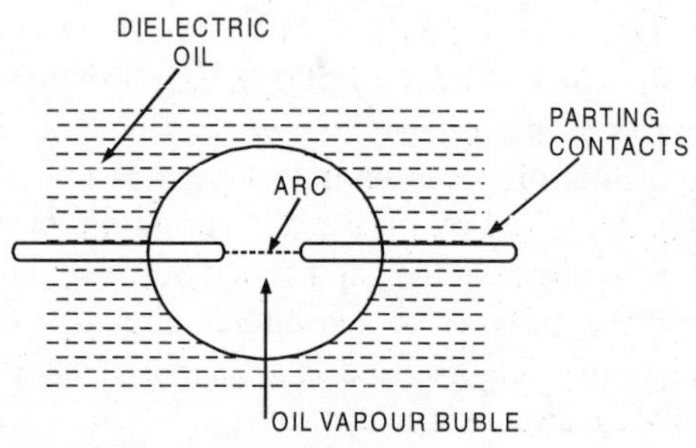

Fig. 8.1

The arc extinction is facilitated chiefly by two processes. Firstly, the hydrogen gas has high heat conductivity and cools the arc. Secondly, the gas sets up turbulence in oil and forces it into the space between contacts after the final arc interruption at a current zero and thus arcing products from the arc path are eliminated. The result is that arc is extinguished and the circuit current is interrupted.

Oil circuit breakers have the virtues of reliability, simplicity and relative cheapness. The oil circuit breakers, as already mentioned in the previous Art, can be divided into.

1. ***Bulk oil circuit breakers*** using a large quantity of oil, also called the *dead tank type,* because the tank is held at ground potential. Such breakers are available in all classifications of voltage and interrupting rating for indoor and outdoor applications.
2. ***Low oil circuit breakers*** which operate with a minimum amount of oil, so that is why sometimes called *minimum oil circuit breakers* or small oil circuit breakers. These circuit breakers are also sometimes called the *live tank* circuit breakers because the oil tank is insulated from the ground.

The oil can be moved into arc zone after the current reaches zero by the following actions :

(*i*) by the pressure caused by the natural head of the oil,
(*ii*) by the pressure generated by the action of the arc itself,
(*iii*) by the pressure caused by external means.

Thus, the oil circuit breakers may be classified as :

(*i*) Plain break oil circuit breakers.
(*ii*) Self blast or self-generated oil circuit breakers
(*iii*) Externally generated pressure oil circuit breakers or forced blast oil circuit breakers or impulse oil circuit breakers.

Oil, as an arc quenching medium, has the following advantages and disadvantages :

Advantages. (*i*) Arc energy is absorbed in decomposing of oil.

(*ii*) The gas formed, which is mainly hydrogen, has a high diffusion rate and high heat absorption in changing from the diatomic to monotomic state and thus provides good cooling properties.
(*iii*) The oil has a high dielectric strength and provides insulation between the contacts after the arc has been finally extinguished and there has been time for the oil to flow into the gap between contacts.
(*iv*) Cooling oil presents the cooling surface in close proximity to the arc.
(*v*) The oil used (such as transformer oil) is a very good insulator and allows smaller clearance between line conductors and earth components.

Disadvantages. (*i*) Oil is inflammable and may cause fire hazards, if a defective oil circuit breaker should fail under pressure and cause an explosion.

(*ii*) There is a risk of formation of explosive mixture with air.
(*iii*) Due to decomposition of oil in the arc, the oil becomes polluted by carbon particles, which reduce its dielectric strength. Hence periodical maintenance and replacement are required.

8.3.1. Plain Break Oil Circuit Breakers

The plain oil circuit breaker is very simple in construction. It consists of current carrying contacts enclosed in a strong (in order to withstand the large gaseous pressure developed owing to dissipation of large energies), weather tight (to keep moisture out), earthed metal tank and immersed in oil, called the *transformer oil.* The oil acts both as an arc extinguishing medium and as an insulator between the live parts and earth.

At the top of the oil air is filled in the seal vessel to serve as cushion to accommodate the displaced oil on formation of gas around the arc and also to absorb the mechanical shock of the upward movement of oil. The breaker tank is securely bolted to an adequate foundation to bear out the vibrations caused on interrupting very heavy currents.

An ample head of oil above the arcing contacts is necessary to provide substantial oil pressure at the arc; and to prevent occurrence of the chimney effect. A chimney of gas from the arc to the oil surface is produced which comes in contact with the earthed tank. If this gas is partially ionized and is of low dielectric strength, an arc will strike between the contact and the earthed tank with serious consequences. Hence an appreciable amount of oil depending upon the operating voltage should always exist between the contact and the tank.

A gas outlet from the tank is essential and some form of vent is fitted in the tank cover. The position of the vent is carefully chosen so that the partially ionized gases which come out of the vent do not harm the personnel and also do not cause flash-over to the neighbouring equipments. Important feature of such breakers is that no special device is used for controlling the arc caused by the moving contact.

Figure 8.2 shows a double break plain oil circuit breaker which is perhaps the most familiar of all oil circuit breakers. It is called a *double break* breaker because it provides two breaks in series. The two breaks in series provide rapid arc lengthening without the need for a specially fast moving contact speed, and the total gap distance at the end of stroke can conveniently be made ample. The vertical break principle also permits the use of a cylindrical oil tank requiring relatively small floor area.

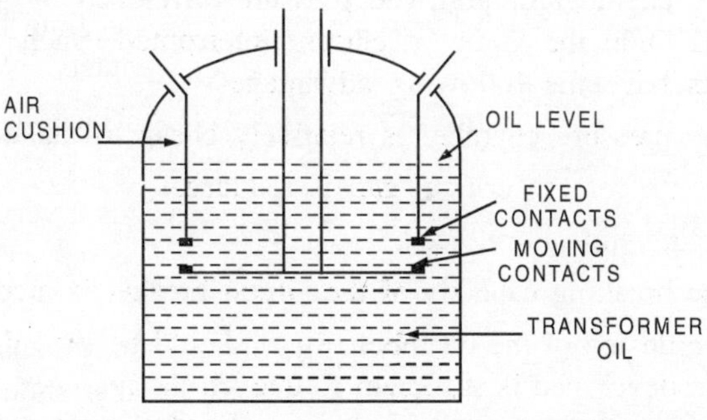

Fig. 8.2. ***Schematic Diagram of a Plain-Break OCB***

The *plain break principle* involves the simple process of separating the current carrying contacts under oil with no special control over the resulting arc other than the increase in length caused by the moving contact. For successful interruption, a comparatively long arc length is essential so that the turbulence in the oil caused by the pressures generated by the arc, may assist in quenching it.

Under normal operating conditions, the fixed and moving contacts remain closed and the breaker carries the normal circuit current. On occurrence of a fault the moving contacts are pulled down by the protective system and an arc is struck between the contacts and a large amount of heat is liberated and a temperature of about 5,000° absolute is reached which vaporizes the surrounding oil into gas. The gas thus liberated surrounds the arc and its explosive growth around it displaces the oil violently. As the moving contacts separate, the length of the arc increases, the rate of gas formation decreases due to fall in temperature. The arc is extinguished when the distance between the fixed and moving contacts reaches a certain critical value, depending upon the arc current and recovery voltage. Based on energy balance principle, final extinction of arc takes place at a current zero when the power input to the arc is less than that dissipated between the contacts. Such breakers thus suffers from the defect of permitting rather long and inconsistent arcing time (up to 100–150 ms).

From practical point of view the speed of the break should be as high as possible because a certain break distance has to be reached before interruption is likely to occur and the sooner it is achieved the smaller the energy released in the breaker and the less mechanically strong a breaker will have to be designed.

The main drawback of a double break plain oil circuit breaker is that a problem of unequal voltage distribution across the breaker is introduced and thus uneven sharing of the total interrupting duty is caused. One break may share 70 to 80 per cent of the interrupting duty.

In order to explain the above statement consider Fig. 8.3 (*a*) which represents the capacitance C_1 between the fixed and moving contacts and capacitance C_2 between the moving contact and earth; while Fig. 8.3 (*b*) represents its equivalent circuit.

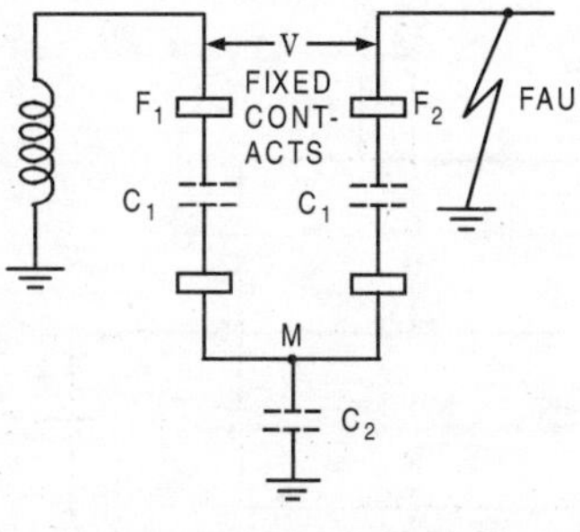

(a) Illustration of Capacitances Between Contact and Earthed Steel Tank

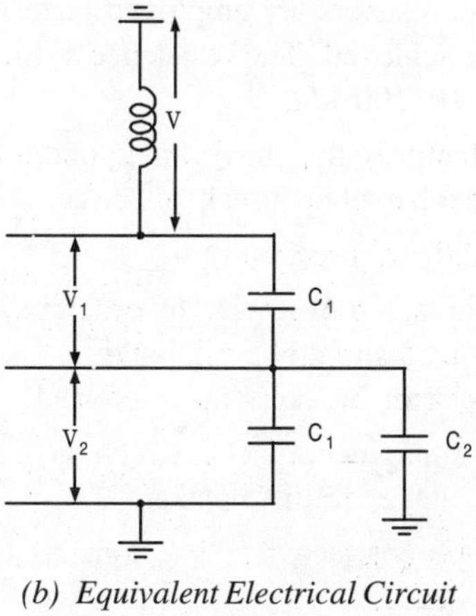

(b) Equivalent Electrical Circuit

Fig. 8.3

If I is the fault current, then in Fig. 8.3 (*b*),

$$V_1 = \frac{I}{\omega C_1} \text{ volts}$$

$$\text{and } V_2 = \frac{1}{\omega(C_1 + C_2)} \text{ volts} \qquad ...(8.1)$$

$$\text{or } \frac{V_1}{V_2} = \frac{C_1 + C_2}{C_1}$$

If we take C_2 four times C_1 we get $V_1 = 5V_2$. Thus, the voltage across one gap is about 83.4% of the system voltage. In order to equalise this voltage across the gaps, high resistances or capacitors are connected across them, as shown in Fig. 8.4. If $C_1 = 10\ \mu\mu F$, $C_2 = 40\ \mu\mu F$ and $C_3 = 50\ \mu\mu F$ are connected as shown in Fig. 8.4, Now

$$\frac{V_1}{V_2} = \frac{C_1 + C_2 + C_3}{C_1 + C_3}$$

$$= \frac{10 + 40 + 50}{10 + 50}$$

$$= \frac{100}{60} = 1.66$$

Thus V_1 becomes only 1.66 times that of V_2. However, if C_3 be taken of higher value the difference between V_1 and V_2 can be reduced further.

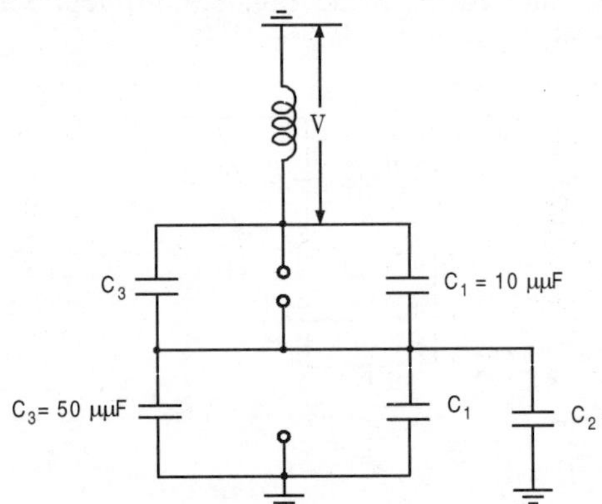

Fig. 8.4. ***Capacitance Voltage Grading***

If high value resistors are employed instead of capacitors, damping will be achieved. The resistance values are generally of the order of 10–100 kΩ.

The main features that have an important bearing upon the performance of a plain break oil circuit breaker are:

1. Critical length of break.
2. Speed of contact movement. In order to quench the arc earlier the speed should be high and this is the reason that two break circuit breakers are preferred.
3. Head of oil above contacts, which is governed by the operating voltage of the breaker.
4. The clearance between the live contacts and the earthed pressurized tank.

Among the factors mentioned above that contribute to increased rupturing capacity are a considerable head of oil and large clearances; for this reason large rupturing capacity tends to require large circuit breakers. The plain break oil circuit breakers are widely employed on low voltage dc and ac circuits. Such circuit breakers are not considered suitable for short-circuit rating exceeding 250 MVA at 11 kV. Also such breakers are not considered suitable for high speed interruption, therefore, these cannot be employed for autoreclosing.

8.3.2. Arc Control Oil Circuit Breakers

In case of a plain break oil circuit breaker, discussed above, there is no control over the arc other than increase in length caused by the moving contact. However, it is necessary and desirable that final arc extinction should occur while the contact gap is still short. For this purpose, some arc control is to be provided and the breakers provided with arc control are called the *arc control circuit breakers. Arc control breakers* are of two types, namely.

1. Self blast or self-generated pressure oil circuit breakers in which arc control is provided by internal means *i.e.,* the arc itself is used for its own extinction efficiently.
2. Externally generated pressure or forced blast or impulse oil circuit breakers in which arc control is provided by mechanical means external to the circuit breaker.

1. *Self Blast or Self-generated Pressure Oil Circuit Breakers.* In such breakers use of pressure generated by the arc itself is made in speeding up the movement of the oil into contact space immediately after the current zero. The high pressure generated by the arc causes an immediate displacement of oil into the space between the contacts after the arc current goes to zero. This is achieved by surrounding the contacts by a pressure chamber or pot. The pressure developed by the oil depends upon the value of current interrupted. Such circuit breakers have the following advantages:

1. The pressure chamber is relatively cheap to make.
2. Length for the critical gap is reduced.
3. Arcing time is reduced.
4. The breaking capacity of the circuit breaker is increased.

The design of the chamber or pot should be such that the pressure developed is sufficient to quench the arc even at low values of current but not so much as to break the pot on heavy current. This has led to the manufacture of a variety of pots, some of which are discussed below.

(a) Plain Explosion Pot. It is a rigid cylinder of insulating material enclosing the fixed and moving contacts, closed at the top but with a restricted opening, called throat, at the bottom. The moving contact is a cylindrical rod passing through the throat. On occurrence of fault, the contacts separate, an arc is struck between them and oil is decomposed, by the heat of the arc, into a gas at very high pressure in the pot. This high pressure forces the oil and gas through and round the arc to extinguish it. If the arc is not extinguished while the moving contact is still within the pot, it happens just after the moving contact leaves the pot due to axial high velocity blast, released through the slot. Since the plain explosion pot performs axial extinction of the arc, it is sometimes called the *axial extinction pot.*

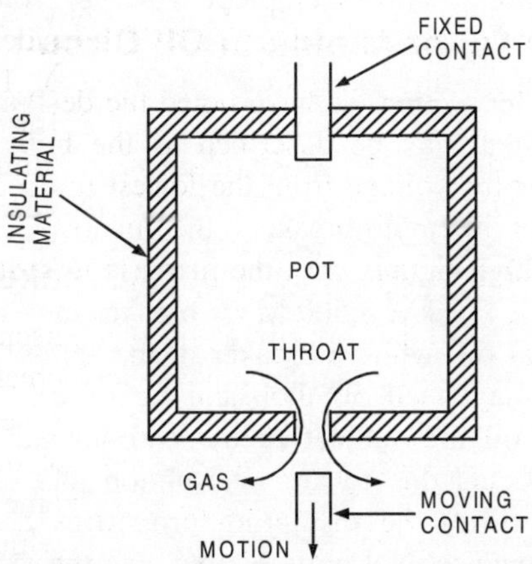

Fig. 8.5. ***Plain Explosion or Axial Extinction Pot***

This device has the effect of controlling the nature of the turbulence to a certain extent, thereby giving a certain degree of consistency.

This type of pot cannot be used either for very low or for very large currents. With low fault currents, the pressure developed is small, thereby increasing the arcing time. On the other hand, with large fault currents, the gas is produced so rapidly that explosion pot is liable to burst due to high pressure. Thus the breaking capacity of such a circuit breaker is limited.

(*b*) Cross-Jet Explosion Pot. This type of pot is just a modification of plain explosion pot and is illustrated in Fig. 8.6.

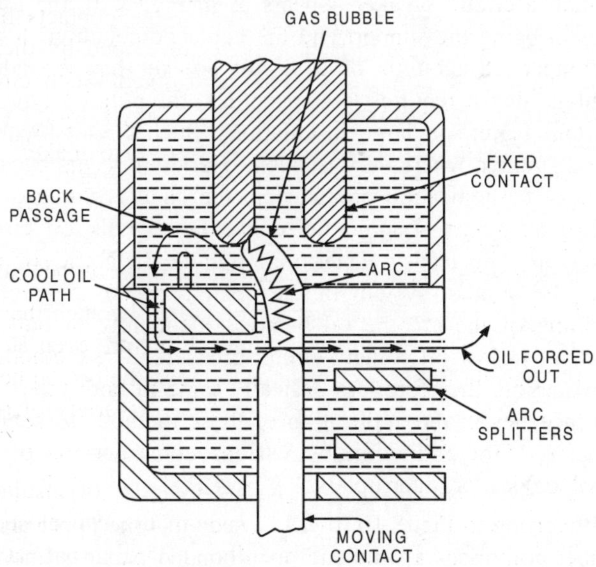

Fig. 8.6. ***Cross-Jet Explosion Pot***

It is made of insulating material and has channels on one side which act as arc splitters. The function of the arc splitters is to provide an increased arc length corresponding to a given amount of contact travel, and also to provide cutting edges across which the arc is attenuated, weekend and finally interrupted. Oil movement is checked by the pressure of the arc itself until a current zero (the pressure then is very low) when the contacts of the pot are ejected across the arc path. The pot is called the cross-jet explosion pot because it performs cross or transverse extinction of the arc.

On occurrence of a fault, the breaker moving contact begins to separate and with the separation of contacts, the arc is initially struck in the top of the pot. The gas generated by the arc exerts pressure on the oil in the backpassage. When the moving contact uncovers the arc splitter ducts, fresh dielectric oil is forced across the arc path. The arc is, thus, driven sideways into the arc splitters which increase the arc length, causing arc extinction.

The performance of cross–jet explosion pot is very good when heavy fault currents are to be interrupted. However, for low fault currents, the gas pressure is small and consequently the pot does not give a satisfactory operation. This type of oil circuit breaker is suitable for interrupting heavy currents at high voltage (66,000V).

(c) Self Compensated Explosion Pot. It is combination of the cross-jet explosion pot and the plain explosion pot and thus operates quite satisfactorily both at heavy currents as well as low currents. Such a pot is shown in Fig. 8.7. It consists of two chambers—the upper being a cross-jet explosion pot with two lateral orifices while the lower is a plain explosion pot. When the short-circuit current is low, the rate of generation of the gas is also quite low and as the pressure builds up the tip of the moving contact has time to reach the bottom chamber and there is only a small pressure leakage through the lateral jets because the movement of the oil to them is obstructed by the right angle bends and also by the presence of the arc. The arc is quenched by the plain explosion pot action. When the short-circuit current is heavy the rate of generation of gas is very high and the device operates as a cross-jet explosion pot, arc extinguishes when the first or second lateral orifice is uncovered by the moving contact. Hence the arc is being self compensated for low as well as high values of short-circuit currents.

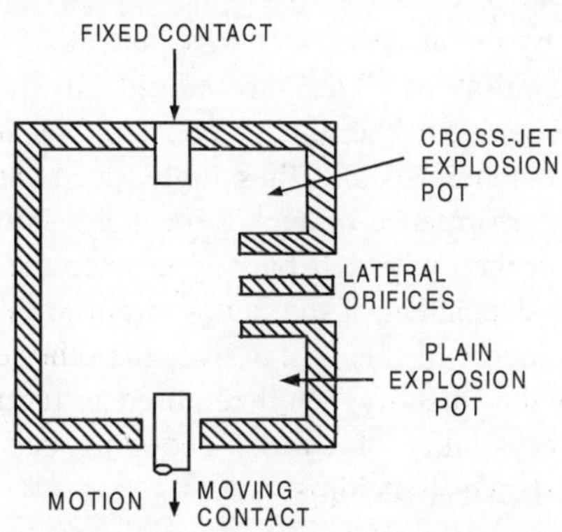

Fig. 8.7. ***Self Compensated Explosion Pot***

(d) Oil Blast Explosion Pot. This is an example of axial extinction and double break. Such a pot consists of mainly three compartments—the upper fixed contact, intermediate contact and low hollow moving contact, which are in contact under pressure. It has two chambers—the upper and the lower and they are connected through holes, when the circuit breaker is closed. On occurrence of fault instead of one, the two lower contacts move downwards together and an arc is struck between the top fixed and intermediate moving contact. Because of this arc a high pressure is developed in the upper tank and there is no relief from this pressure until the intermediate contact reaches its top after its maximum travel, now, the lower moving contact detaches from the intermediate contact and an other arc is struck between the intermediate and lower moving contacts. The pressure developed by the arc is subsided by the movement of the oil through the lower moving contact which is hollow. When the arc current goes to zero, the oil is forced through the arc and it is extinguished. Long arcing time is the only drawback of this type of oil circuit breaker.

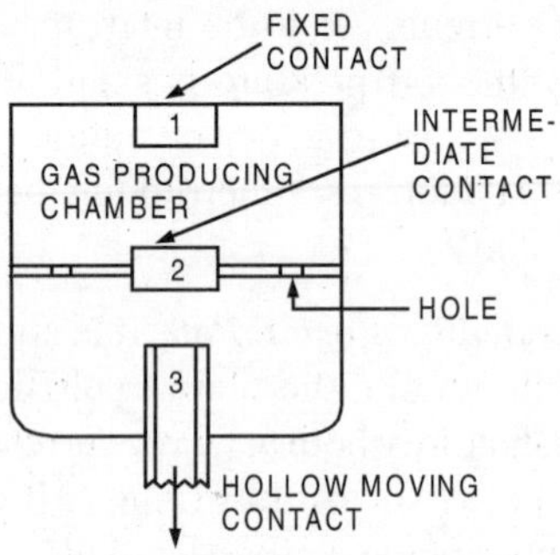

Fig. 8.8. ***Oil Blast Explosion Pot***

2. Externally Generated Pressure or Forced Blast or Impulse Type Oil Circuit Breakers. In the self blast oil circuit breakers discussed above, the arc itself generates the pressure required to force the oil across the arc path. The major drawback of such breakers is that arcing times tend to be long and inconsistent with fault currents considerably less than rated currents. It is due to generation of reduced gas at low values of fault currents. This problem is overcome in forced blast oil circuit breakers in which the necessary pressure is developed by external mechanical means independent of the magnitude of the fault currents to be interrupted.

In this type of circuit breaker, the oil pressure instead of being created by the arc is created mechanically by the piston-cylinder arrangement. The movement of the piston is mechanically coupled to the moving contact, thus automatically oil pressure is generated and thus high speed interruption is attained. The performance of such a breaker at low currents is more consistent than with self blast type since the oil pressure developed is independent of the fault current to be broken. An important advantage of this design over the other conventional design is that the quantity of oil required is reduced to one-quarter. The possibility of current chopping can be avoided by using resistance switching.

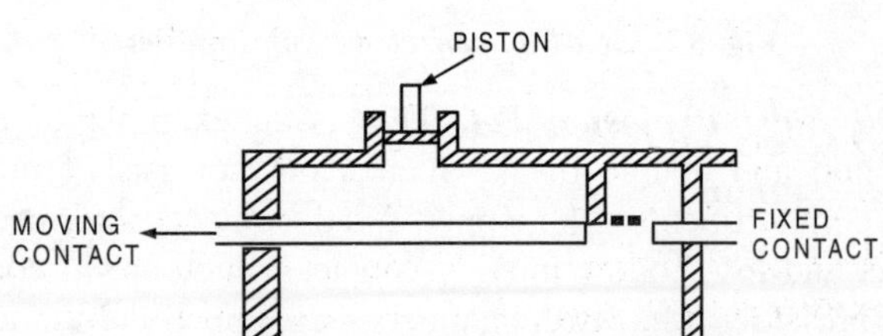

Fig. 8.9. ***Externally Generated Pressure Oil Circuit Breaker***

Circuit breakers have also been built up which combine the self blast and forced blast.

8.3.3. Low Oil or Minimum Oil Circuit Breakers

In the preceding section we have noted the design and special features of what may be described as the bulk oil type of circuit breaker for voltage from the lowest up to 380 kV. But in case of bulk oil circuit breakers the quantity of oil required reaches very high figures with the increase in system voltage. For example, a 110 kV, 3,500 MVA breaker may need 8 to 12 thousand kg of oil, while a breaker of the same rating output for 220 kV may need 50 thousand kg of oil. These large quantities of oil are subject to the carbonisation, sludging, etc., which occurs due to arc interruption and other causes, reducing (in time) the insulating properties and requiring regular maintenance. Not only this but also the expenses, tank size and weight of the breaker is increased. In case of a floor mounted large bulk oil circuit breaker, access to the contact system is not easy and is obtained only through access parts in the tank side after the oil has been pumped away to storage.

We have noted that in bulk oil circuit breakers the oil performs two functions. Firstly, it acts as an arc extinguishing medium and secondly it insulates the live parts from earth. Fortunately the quantity of oil actually required for arc extinction is only about one–tenth of the total, the rest being used for insulation. This fact led to the development of *low oil content* or live tank circuit breaker in which the arc chamber contains a minimum quantity of oil and is mounted on a porcelain insulator to insulate it from the earth, the arc chamber itself is, in fact of bakelised paper enclosed in porcelain, so that such a circuit breaker appears as in Fig. 8.10, the lower porcelain being the support and the upper porcelain enclosing the contacts. Such type of a circuit breaker has the added advantage that it requires less space than the bulk oil type, an important feature in large installations. Such circuit breakers are less suitable where very frequent operation occurs because degree of carbonisation produced in the small volume of oil is far more severe than in the conventional bulk oil circuit breakers and this leads to deterioration in the dielectric strength across the contact system in the open position. As regards quenching of the arc the oil behaves identically in bulk oil as well as in low-oil content circuit breakers. The methods of arc extinction, therefore, are basically same in both types, the difference lies in the application of these methods to achieve the desired arc extinction by various constructions of arc control device or interrupters.

Reverting to Fig. 8.10 it will be seen that the upper arcing chamber comprises a synthetic resin bonded paper cylindrical enclosure within a porcelain insulator. An annular space between these is filled with oil as an insulating medium but, again, this is physically separated from that in the arcing chamber and thus cannot be contaminated.

The circuit breaker is of the single break type in which a moving contact tube moves in a vertical line to make or break contact at the upper fixed contacts mounted within the arc control device. A lower ring of fixed contacts are in permanent contact with the moving arm to provide the other terminal of the phase unit. Within the moving contact tube is a fixed piston, which, as the tube moves downwards on opening, forces the column of oil inside the tube into the arc control device. This has two effects, firstly, a partial pressure balance is ensured, so that the pressure generated inside the arc control device has little effect on the acceleration of the moving contact and, secondly, the amount of cavitation caused by the removal of the moving contact is controlled and the efficiency of arc extinction is increased.

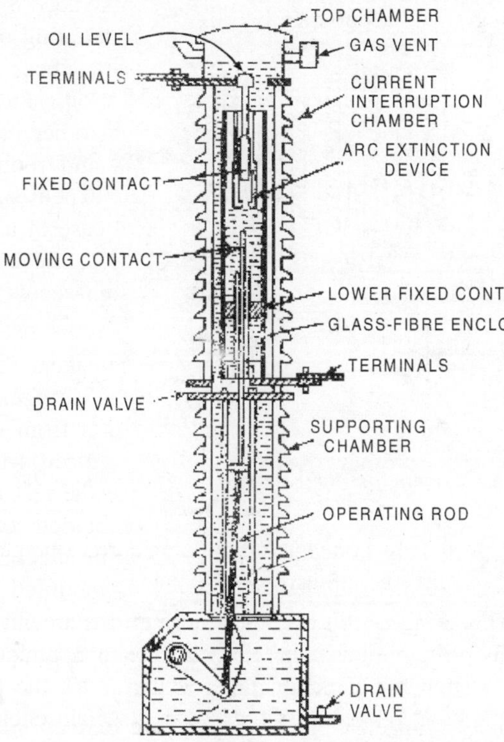

Fig. 8.10. ***Low Oil Content Circuit Breaker***

The "turbulator" arc control device is built up of oil–impregnated vulcanised fibre plates held under compression by tension members, the plates being arranged to form a series of vents on one side of the arc, and a series of oil pockets on the other. The fixed arcing tip is so arranged that when the circuit breaker opens, the arc is drawn in front of the vents where it can be most easily extinguished. To ensure that oil vaporised by the arc is replaced quickly when the arc is extinguished a filling valve is fitted in the top casting of the arc-control device. This valve, lightly spring-loaded, closes when pressure is set up within the device, but as this pressure dies away when the arc is extinguished, the valve opens to allow an inrush of oil.

When a circuit breaker closes on to a fault there is always a danger that pre-arcing will occur as the moving contact approaches the fixed contacts. In this design this possibility is minimised by fitting inserts of anti-tracking material on the inside edges of the "turbulator" plates.

The upper arcing chamber contains a separator which eliminates, by centrifugal action, loss of oil when the breaker operates under fault conditions. It has a spring-loaded vent valve, a breather to prevent moisture from entering the circuit breaker, and a safety diaphragm, under the domed cover, designed to lift and protect the circuit breaker from damage if excessive pressure should arise in the circuit breaking compartment.

Circuit breakers of this type can be used either out-of-doors or indoors, and be mounted either on pedestals or at ground level. At certain voltages and interrupting capacities, two breaks per phase are used.

Current transformers, which are incorporated in the bushings of the dead tank type, usually, have to be mounted separately in the live tank type. These are usually of post type with low oil content, the transformer being housed within the porcelain housing and filled with oil under vacuum.

Low oil circuit breakers are now available for all voltages and for the highest breaking capacities.

The low oil circuit breakers have the advantages of requirement of lesser quantity of oil, smaller space requirement, smaller tank size, smaller weight, low cost, reduced risk of fire and reduced maintenance problems. However, it suffers from the following drawbacks when compared with a bulk oil circuit breaker:

(*i*) Increased degree of carbonisation due to smaller quantity of oil.

(*ii*) Rapid deterioration of dielectric strength of oil due to high degree of carbonisation.

(*iii*) Difficulty in removal of gases from the contact space in time.

8.3.4. Maintenance of Oil Circuit Breakers

After a circuit breaker has interrupted short-circuit currents a few times or load currents several times, the contacts may get burnt due to arcing. In addition the dielectric oil gets carbonized in the vicinity of the contacts, thereby losing some of its dielectric strength. This results in the reduced breaking capacity of the breaker. The maintenance work of oil circuit breaker, therefore requires checking and replacement of contacts and oil. It is good practice to inspect the circuit breaker at regular intervals of 3 or 6 months. According to ISS 335–1963 oil in good condition should withstand 40 kV for one minute in a standard oil testing cup with 4 mm gap between the spherical electrodes.

During inspection of the oil circuit breaker it is recommended to check the following:

1. All current carrying parts be checked and arcing contacts be attended if necessary.
2. Dielectric strength, condition and level of oil should be checked. In case the dielectric strength is low or the oil is badly discoloured, replace it.
3. Inspect the insulation for possible damage. Clean the surface and remove deposits of carbon. Never use loose cotton waste for this purpose.
4. Check closing, tripping and interlock mechanism.
5. Check indicating devices and lamps.
6. Ensure, before closing the tank, that no tool have been left behind, that the tank linings and barriers are in position and secure, and that the tank gasket is in good condition.

8.4. AIR BREAK CIRCUIT BREAKERS

The arc interruption in oil is due to generation of hydrogen gas owing to oil decomposition. This fact led to study the arc interruption in air. No doubt, arc interruption properties of hydrogen are much superior to air, but air has several advantages, over the oil, as quenching medium. These are:

1. Elimination of fire risk and maintenance associated with the use of oil.
2. Absence of mechanical stresses that are set up by gas pressure and oil movement.
3. Elimination of cost of regular oil replacement that arises due to deterioration of oil with successive breaking operation.

Relatively inferior arc extinguishing properties of air may be offset by using various principles of arc control and operating air at high pressures.

In the *air break circuit breaker* the contact separation and arc extinction takes place in air at atmospheric pressure. High resistance principle is employed in such circuit breakers. The arc is rapidly lengthened by means of the arc runners and arc chutes and arc resistance is increased by cooling*, lengthening** and splitting*** the arc. The arc resistance is increased to such an extent that the voltage drop across the arc becomes more than the system voltage and the arc gets extinguished at current zero of ac wave.

Air break circuit breakers are employed in dc circuits and ac circuits up to 12,000 volts. Such breakers are usually of indoor type and installed on vertical panels or indoor draw-out switchgear. AC circuit breakers are widely employed in indoor medium voltage and low voltage switchgear.

1. *Plain break type air break circuit breaker* shown in Fig. 8.11 is the simplest one in which contacts are made in the shape of two horns. The air initially strikes across the shortest distance between

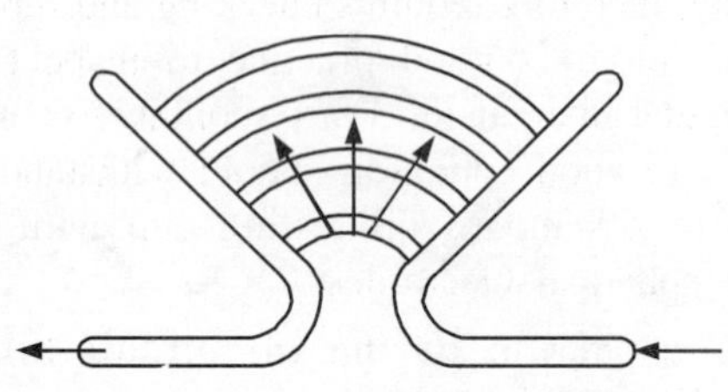

Fig. 8.11. ***A Simple Air Break Circuit Breaker***

the horns and is then driven steadily upwards by the convection currents caused by heating of air during arcing and the interaction of magnetic and electric fields. The arc extends from one tip to the other when the horns are fully separated resulting in lengthening and cooling of arc. The relative slowness of the process and the possibility of arc spreading to adjacent metal work limits the application to about 500 V and to low power circuits.

2. *Magnetic Blow-Out Type Air Break Circuit Breaker.* In a number of air circuit breakers employed in circuits up to 11 kV the arc extinction is accomplished by means of magnetic field provided by current in blow-out coils connected in series with the circuit being interrupted. Such coils are called *blow-out coils*, because they assist in the arc being magnetically *blown out*. The magnetic field itself does not extinguish the arc. It simply moves the arc into arc chutes where the arc is lengthened, cooled and extinguished. The arc shields prevent arc spreading to adjacent metalwork.

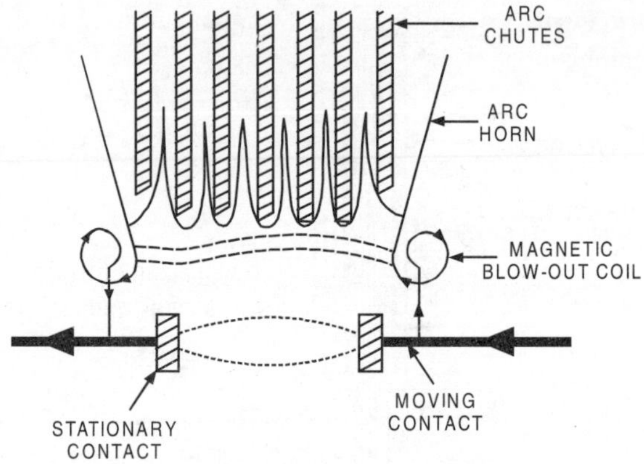

Fig. 8.12. ***Principal Scheme of a Magnetic Blow-Out Air Break Circuit Breaker***

It is important to connect the coils at correct polarity so that the arc is directed upwards.

As the breaking action becomes more effective with large currents, this principle has resulted in increasing the rupturing capacities of such breakers to higher values.

Arc chute is an efficient device for arc extinction in air and performs the following three interrelated functions:

1. It confines the arc within a restricted space.
2. It provides magnetic control over the arc movement so as to make arc extinction within the device.
3. It provides for the rapid cooling of arc gases to ensure arc extinction by deionzation.

3. *Arc-chute Air Break Circuit Breaker.* The normal arrangement of an arc-chute air break circuit breaker employed for low and medium voltage circuits is illustrated in Fig. 8.13. There are two sets of contacts called the *main contacts* and *arcing* or *auxiliary contacts.* Main contacts are usually of copper and conduct the current in closed position of the breaker. They have low contact resistance and are silver plated. The arcing contacts are hard, heat resistant and usually of copper alloy. Arcing contacts are used to relieve the main contacts from damage due to arcing. The arcing contacts are easily renewable when required. The auxiliary and arcing contacts close before and open after the main contacts during operation. Here the blow-outs consist of steel inserts in the arcing chutes. These are so arranged that the magnetic field

* The voltage required to maintain ionization increases with a fall in temperature so that cooling effectively increases the resistance.

** Resistance increases approximately in proportion to the arc length.

*** An appreciable voltage is absorbed at the two contact surfaces so that on splitting of an arc into a number of small arcs in series the voltage available for the actual arc column in reduced.

induced in them by the current in the arc moves it upwards still faster. The steel plates divide the arc into a number of short arcs in series. The distribution of voltage along the arc length is not linear but is accompanied by a rather large anode and cathode drops. In case the total sum of anode and cathode drops of all short arcs in series exceeds the system voltage, conditions for quick extinction of the arc are automatically established. When the arc comes into contact with the relatively cool surfaces of the steel plates, it gets rapidly and effectively cooled. The movement of the arc may be natural or aided by a magnetic blow-out (being employed for heavy duties up to 500 MVA at 16 kV). Thus the arc is extinguished by lengthening and increasing the voltage gradient *i.e.,* power loss of the arc. In this breaker relatively high arc resistance is obtained near current zeros. This effect plays an important role in having high rupturing capacity by modifying the circuit power factor near current zero such that voltage available to restrike the arc is considerably less than the peak value.

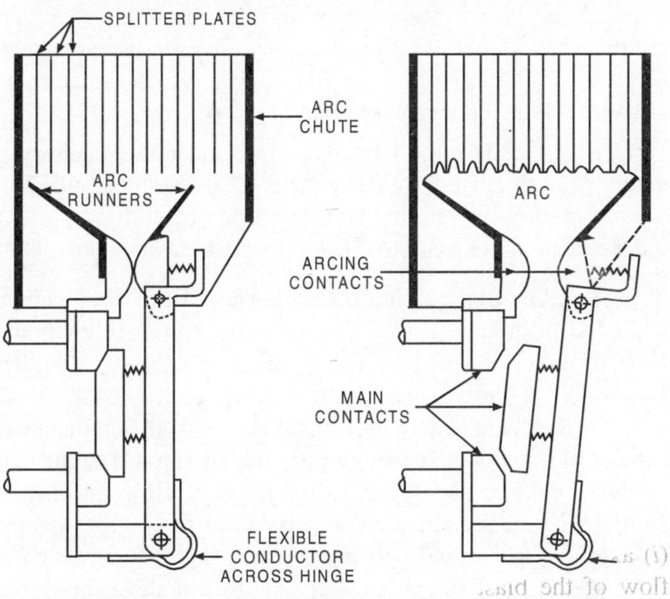

Fig. 8.13. ***Elementary Arrangement of Air Break Circuit Breaker***

A drawback of arc chute principle is the inefficiency at low currents where the electromagnetic fields are weak. The chute itself is not necessarily less efficient in its lengthening and deionizing action than at high currents, but the arc movement into the chute tends to become slower and high speed interruption is not necessarily obtained.

With the increase in operating voltage and rupturing capacity, such a breaker becomes bulky, the arc chutes become more complex and the initial cost increases. Further the time of operation is not small enough to make such a breaker suitable for modern power system.

Action. On occurrence of fault, the main contacts dislodge first and the current is shifted to the arcing contacts. Now the arcing contacts dislodge and the arc is drawn between them. This arc is forced upwards by the electromagnetic forces and thermal action. The arc ends travel along the arc runner (arcing horns). The arc moves upwards and is split by the arc splitter plates. The arc is extinguished by lengthening, cooling, splitting etc.

Applications. Air break circuit breakers are generally suitable for the control of power station auxiliaries and industrial plants. They do not require any associated equipment such as compressors, etc. As they have no oil, they are recommended where there are possibilities of fire or explosion hazards.

Air break principle of lengthening of arc, arc runners, magnetic blow-up is employed for dc circuit breakers up to 15 kV.

8.5. AIR BLAST CIRCUIT BREAKERS

The drawbacks of the oil circuit breakers are the fire risk due the inflammable oil, the deterioration of the oil, necessitating periodic replacement, and the difficulty of reaching the contacts for maintenance purposes. This led to the development of circuit breakers using compressed air or gas as the interrupting medium. Though gases such as nitrogen, carbon dioxide, hydrogen or freon can be used as the arc interrupting medium but compressed air is the accepted circuit breaking medium for gas blast circuit breakers. The reasons are given below:

Nitrogen has circuit breaking properties similar to compressed air and there is no added advantage of using it. Carbon dioxide has the drawback of its being difficult to control owing to freezing at valves and other restricted passages. No doubt hydrogen has increased breaking capacity but it is costlier and needs ancillary apparatus. Freon has high dielectric strength and good arc extinguishing properties, but it is expensive and it is decomposed by the arc into acid forming elements.

The desirable features to be found in air blast circuit breakers are:

1. ***High Speed Operation,*** which is very necessary on large interconnected networks in order that system stability can be maintained and in the air blast circuit breaker this is achieved because the time interval between the receipt of a tripping impulse (derived from the protective gear applied) and contact separation is very short. Once the contacts part and an arc is drawn it should, ideally, be interrupted in the shortest possible time and this time duration should be reasonably consistent at all values of current which the circuit breaker may be called upon to break *i.e.,* from small line charging or transformer magnetising current up to the highest value of fault current. This the air blast circuit breaker does and arc durations throughout the current range are of the order of one-half to one cycle. In the oil circuit breakers this consistency rarely exists and because the deionisation processes are largely dependent on the current value, it is usual to find longer arcing times at low current values than at the higher values, and that the whole range of arcing times is higher in the air blast design.

2. ***Suitability For Frequent Operation.*** Repeated switching by an air blast circuit breaker is possible simply because of absence of oil, which rapidly carbonises with frequent operation, and because there is an insignificant amount of wear and tear at the current-carrying contact surfaces. But it must be remembered that if frequent switching is anticipated, the maintenance of an adequate air supply is essential.

3. ***Facility of High-speed Reclosure.*** High-speed reclosure by automatic means is an advantage on hv interconnected networks to assist and maintain system stability during the clearance of transient faults, a type of fault which is perhaps in majority on overhead lines. Provided that the time interval between fault interruption and reclosure is chosen to permit insulation recovery, then a system can often be restored to normal by breaker reclosure, the cause of the interruption (insulator or line flash-over) having disappeared. The low inertia of the moving contacts in air blast circuit breakers and the relative ease with which compressed air mechanisms can be reversed, all help in very short restoration times being achieved.

4. ***Negligible Maintenance.*** The ability of the air blast circuit breaker to cope with repeated switching also means that negligible maintenance is required. For example, the relatively large quantities of dielectric oil essential in the oil circuit breaker needs an installation of oil filtering plant and very regular treatment of oil. No such requirement arises with an air blast circuit breaker.

5. ***Elimination of Fire Hazard.*** Because of absence of oil the risk of fire is eliminated.

6. ***Reduced Size.*** The growth of dielectric strength is so rapid in air blast circuit breakers that final gap required for arc extinction is very small. This reduces the size of the device.

Drawbacks of Air Blast Circuit Breakers. Against the formidable array of advantages mentioned above, however, it is worthwhile to note that compressed air at the correct pressure, clean and dry, must be available at all times, involving in the largest installation of a plant with two or more compressors and an excessive air supply network in the form of a ring main or duplicate bus system. The maintenance and upkeep of this plant and the problem of air leakages at the pipe fittings are factors which operate against air blast circuit breakers in many (smaller) installations and it is a costly adjunct for low-voltage systems as compared with the use of oil or air break circuit breakers.

Because of fixed air pressure, it is obviously available regardless of the magnitude of current to be interrupted by the breaker. It must naturally, be sufficient to deal with the highest value of anticipated fault current, but this means that it can be very drastic in its effect on small currents and the problem of current chopping arises leading to serious overvoltages, as explained in Art. 7.7.1.

Sensitivity of the air blast circuit breaker to circuit severity, i.e., the rate of rise of restriking voltage. Many of the air blast circuit breakers (and some very high voltage circuit breakers) overcome this problem by resort to resistance switching wherein external resistors are automatically connected in shunt with the contact gap and thereby damp out very high restriking voltage transients caused by current chopping.

Not all air blast circuit breakers incorporate resistance switching. It is usual in the design known as axial blast and is less frequently employed in the design known as cross blast.

During the period 1950–1970, air blast circuit breakers were preferred for 220 kV and above. However today SF_6 circuit breakers, which are maintenance free and of superior switching performance, are preferred for this range. For 11 kV and 33 kV applications vacuum circuit breakers are preferred. Thus air blast circuit breakers have become almost obsolete.

Principle of Arc Excitation in Air Blast Circuit Breakers. The air blast circuit breaker requires an auxiliary compressed air system which supplies air to the breaker air receiver. When opening is required, compressed air is admitted to the arc extinction chamber. It pushes away the moving contacts. In doing so the contacts are separated and the air blast takes away the ionized gases along with it and assist arc extinction. Air blast extinguish the arc within one or two cycles and the arc chamber is filled with high pressure air, which prevents restrike. In some low capacity circuit breakers the isolator is an integral part of the circuit breaker. The circuit breaker opens and immediately after that isolator opens to provide additional gap. In EHV switchyards, isolators are generally independently mounted.

The air blast circuit breakers fall under the category of external extinguishing energy type. The energy supplied for arc extinction is obtained from high pressure air and is independent of current to be interrupted.

Types of Air Blast Circuit Breakers. All air blast circuit breakers follow the principle of separating their contacts in a flow of air established by the opening of a blast valve. The arc which is drawn is usually rapidly positioned centrally through a nozzle where it is kept to a fixed length and is subject to maximum scavenging by the air flow. The air blast circuit breakers, according to the type of flow of blast of compressed air around the contacts are of three types namely (*i*) axial (*ii*) radial and (*iii*) cross blast. The different ways of flow of the blast of compressed air around the contacts are shown in Fig. 8.14.

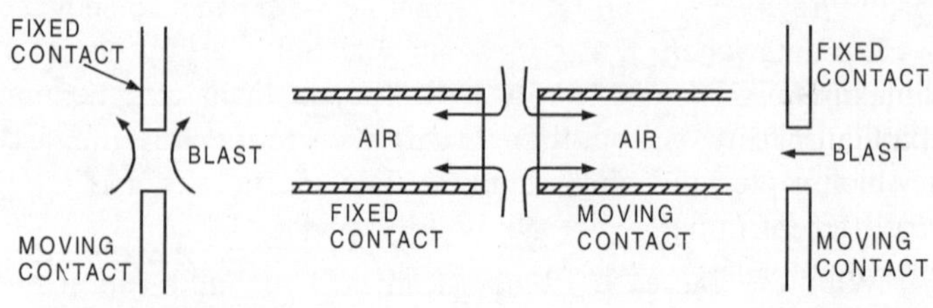

(*a*) *Axial Blast* (*b*) *Radial Blast* (*c*) *Cross Blast*

Fig. 8.14. ***Methods of Flow of Blast Compressed Air Around The Contacts***

Axial or radial blast seems to be favoured for the higher voltages although cross blast breakers particularly for voltages of about 15 kV and heavy current (up to 100,000 A) have proved satisfactory and need less air than would an axial blast circuit breaker at such high currents.

1. ***Axial Blast Air Circuit Breaker.*** In axial blast type circuit breaker, the flow of air is longitudinal along the arc

[Fig. 8.14 (*a*)]. Axial blast circuit breakers may be *single blast* or *double blast* [Fig. 8.14 (*b*)]. Breakers employing double blast arrangement are sometimes called *radial blast circuit breakers* as the air blast flows radially into the nozzle or space between the contacts.

The essential components of a typical axial blast circuit breaker are shown in Fig. 8.15. The fixed and moving contacts are held in closed position by spring pressure under normal operating conditions. The air reservoir tank is connected to the arc chamber through an air valve, which is opened by a tripping impulse.

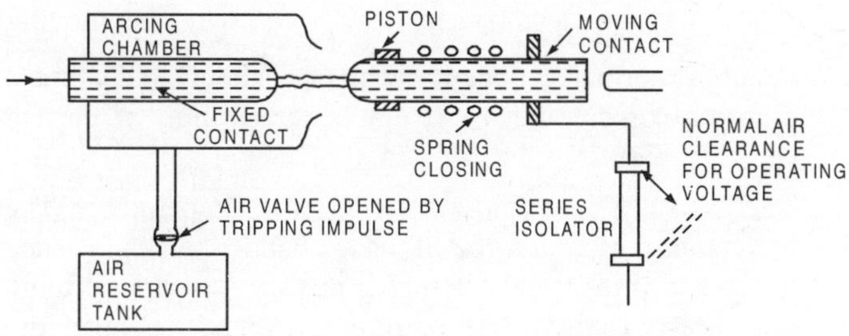

Fig. 8.15. ***Axial Blast Air Circuit Breaker***

On occurrence of a fault, the tripping impulse causes opening of the air valve connecting the reservoir to the arcing chamber. The air entering the arc chamber exerts pressure on the moving contacts which moves when the air pressure exceeds the spring force. The moving contact is separated and an arc is struck. The air flowing at a high speed axially along the arc causes removal of heat from the periphery of the arc and the diameter of the arc reduces to a low value at current zero. At this instant the arc is interrupted and the contact space is flushed with fresh air following through the nozzle.

The flow of fresh air through the contact space ensures removal of hot gases and rapid building up of dielectric strength. After the brief duration of air flow, the interrupter is filled with high pressure air. The dielectric strength of air increases with pressure. Thus the fresh high pressure air in the contact space is capable of withstanding the transient recovery voltage.

It is noteworthy here that the air pressure form the reservoir is maximum initially and falls thereafter. It is known that for a particular reservoir pressure there is a certain optimum contact at which rupturing capacity is maximum. This gap is usually small (ten of mm) and may reach very quickly if the moving parts are of small inertia. The shorter the gap, relatively smaller amounts of energy are released in the arcing chamber. The arc is kept in the high velocity blast of air converging into the nozzle throat. The falling reservoir pressure and short optimum gap result in three important features of the axial blast principle.

1. The interruption must take place at the first current zero after the optimum gap has reached otherwise restrikes may take place at subsequent zeros of current due to decreasing pressure. In oil circuit breakers it is otherwise, *i.e.,* chances of interruption increase if arcing exists beyond the first current zero.
2. The axial blast circuit breaker gives high speed clearance because of the small gap required for interruption. This is desirable for improving transient stability on hv transmission and interconnection networks.
3. The small contact gap after interruption may constitute inadequate clearance for the normal system voltage. Therefore, an isolating switch is incorporated as a part of this circuit breaker. This switch opens immediately after fault interruption to give the necessary clearance for insulation.

For low voltages the isolating switch is not needed, an adequate travel is provided instead for the moving contact.

The arcing time of arc controlled circuit breaker varies considerably depending upon the breaking current. The higher the breaking current (within the rating of the breaker), the smaller the arcing time. The arcing time in case of air blast circuit breaker is independent of the breaking current due to fixed air pressure and the optimum small contact gap. The short gap along with an isolating switch provides a total break time of 2 to 5 cycles.

The operation of the air blast circuit breaker is very much affected by the circuit natural frequency. When the current is passing through zero value the residual column has relatively high resistance which reduces the chances of the restriking voltage transient being damped. Now the effect of RRRV during this zero current condition is more serious especially where the chance of extinction decreases after the optimum gap has reached. The effect of natural frequency on the performance of the air blast circuit breaker is overcome by shunting the arc with resistors of suitable values.

2. ***Cross Blast Air Circuit Breaker.*** In such a breaker, an air blast is directed at right angles to the arc.

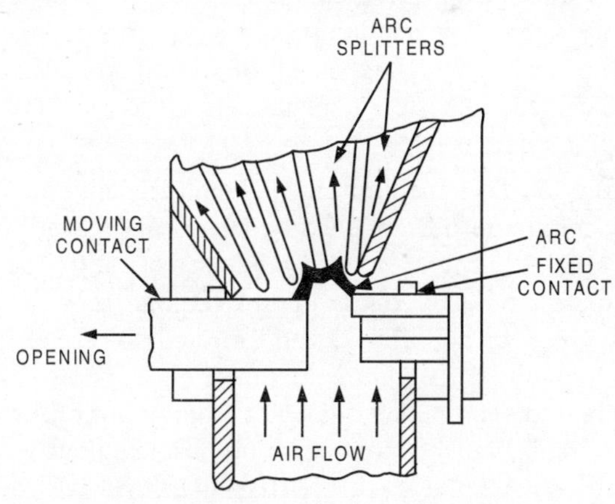

Fig. 8.16. ***Principle of Cross Blast Air Circuit Breaker***

The principle used in the cross blast type air circuit breakers is fundamentally different from the axial blast one. The schematic representation of the principle of cross blast air circuit breaker is given in Fig. 8.16. A moving contact arm operates in close proximity to an arc chute to draw an arc which is forced by a transverse blast of air into the splitter plates within the arc chute, thereby lengthening it to the point when it cannot restrike after current zero. The consistent high speed operation of the axial blast type is not reproduced in this type, but as the air blast is constant regardless of current magnitude, it is quite efficient in switching small currents.

Because the moving arm is not restricted (relatively) in its travel, full isolation is obtained without the need for a series isolator as in other types.

Resistance switching is not normally required as the lengthening of arc automatically introduces some resistance to control the restriking voltage transient but if extra resistance is thought desirable, it is possible to introduce it by connecting it in sections across the arc splitters.

8.6. SULPHUR HEXAFLUORIDE (SF_6) CIRCUIT BREAKERS

In circuit breakers discussed so far (oil circuit breakers, air break circuit breakers and air blast circuit breakers) the extinguishing force builds up relatively slowly after the moment of contact separation, and hence the arc is usually extinguished after a few half cycles of current have passed zero. The prevention arc re-ignition needs a high dielectric strength of the arc path and its fast recovery after current zero. In case of hv circuit breakers, these properties are particularly required to have quick arc extinction and have less time for quick recovery voltage build up. Vacuum circuit breakers and SF_6 circuit breakers have better properties in this regard compared to conventional bulk oil, minimum oil as well as air blast circuit breakers. Hence modern trend is to employ vacuum circuit breakers and SF_6 circuit breakers in hv systems.

Oil, an obviously inflammable substance for extinguishing the hot arc, is a well proven medium because it releases hydrogen which by virtue of its low mass and high velocity is an excellent cooling medium. But modern circuit breakers employ heavy gas SF_6 as the medium for quenching the arc. SF_6 gas, because of its excellent dielectric, arc quenching, chemical and other physical properties, has proved its superiority over other mediums such as oil, or air for use in circuit breakers. Several types of SF_6 circuit breakers have been developed by different manufacturers during last three decades for rated voltages 3.6 to 760 kV.

Before 1970's in medium and high voltage range, air-break, bulk oil, minimum oil, air blast circuit breakers were in use. During 1970's vacuum circuit breakers were introduced for applications up to rated voltage of 36 kV. Single pressure puffer type SF_6 breakers were introduced for rated voltages from 3.3 to 760 kV. Fault levels and rated voltages in the power system have increased. The bulk oil breakers, minimum oil breakers, air blast circuit breakers are becoming obsolete now.

8.6.1. Properties of Sulphur Hexafluoride Gas

Sulphur hexafluoride (SF_6) gas is prepared by burning coarsely crushed sulphur in fluorine gas, in a gas-tight steel box, provided with staggered horizontal shelves, each bearing about 4 kg of sulphur. The gas thus obtained contains other fluorides such as S_2F_{10}, SF_4 and must be purified further. SF_6 gas is generally supplied by chemical firms. The gas costs less if manufactured on a large scale.

The gas is transported in liquid form in cylinders. Before filling the gas, the circuit breaker is evacuated to the pressure of about 4 mm of mercury so as to remove the moisture and air. The gas is then filled in the circuit breaker. The gas can be reclaimed by the gas handling unit.

The sulphur hexafluoride gas possesses very good insulating properties and outstanding arc-quenching characteristics which make it an ideal medium for circuit interruption.

The physical, chemical and dielectric properties and arc-quenching characteristics of SF_6 gas are given below:

1. It is colourless, odourless, non-toxic and noninflammable gas.
2. This gas is extremely stable and inert, and its density is five times that of air.
3. It has a thermal conductivity higher (about 2–2.5 times) than that of air and assists in better cooling of current carrying parts.
4. The gas starts liquefying at certain low temperatures. The temperature at which the SF_6 gas changes to liquid state depends on pressure. With higher pressure, this temperature increases. To avoid the liquefaction of SF_6 gas the temperature of SF_6 should be maintained above certain level. For 15 atm pressure SF_6 gas starts liquefying at a temperature of about 10°C. Thermostatically controlled heaters are, therefore, provided so that the gas temperature above about 16°C could be maintained in case of high pressure system.
5. Heat content property at temperatures below 6,000 K is much higher than that of nitrogen. This assists in cooling of arc space after current zero, due to continuous removal of heat from the contact space by the surrounding gas.
6. The thermal time constant of SF_6 is low and as a result the pressure at which it should be stored and used are relatively smaller as compared to air.
7. Due to the electronegativity* of SF_6 gas the arc time constant (the time between current zero and the instant the conductance of contact space reaches zero value) of SF_6 gas is very low ($\leq$ 1 μs) and rate of rise of dielectric strength is high. Hence SF_6 circuit breakers can withstand severe RRRV and thus are more suitable for short line faults without switching resistors and can interrupt capacitive currents without restriking.
8. It is chemically stable up to 500°C and does not react with structural material up to 500°C.
9. The gas is strongly electronegative, which means that free electrons are readily removed from a discharge by the formation of negative ions through processes by which a free electron is attached to a neutral gas molecule. The attachment of the electron with the neutral gas molecule may occur in two ways:

   (*i*) As direct attachment $SF_6 + e^- \rightarrow SF_6^-$

   (*ii*) As dissociative attachment $SF_6 + e^- \rightarrow SF_5 + F^-$

   The negative ions formed are relatively heavier and

* The ability of an atom to attract and hold electrons has been designated as its electronegativity.

immobile as compared to free electrons and are therefore, ineffective as current carriers so that ionized SF_6 has as high an electric strength as unionized gases such as nitrogen at equal density.

10. The SF_6 gas not only possesses a good dielectric strength but it has the unique property of fast recombination after the source energizing the spark is removed. This property of SF_6 makes it very effective in quenching arc. SF_6 is approximately 100 times as effective as air quenching arc.
11. It has exceptionally low reactivity and does not attack metals, plastics etc. The chemical inertness of SF_6 gas is advantageous in switchgear. The life of metallic part, contacts is longer in SF_6 gas. The components do not get oxidised or deteriorated. Hence the maintenance requirements are reduced—the breaker can be operated for several years without opening for the purpose of overhauling.
12. Moisture is very harmful to the properties of SF_6 gas. In the presence of moisture, hydrogen fluoride is formed during arcing which can attack the metallic and insulating parts of the circuit breaker.
13. During arc extinction process SF_6 is decomposed, into atoms, electrons and ions. These atomic components do not recombine completely to form the original SF_6 gas on cooling. They form low molecular gaseous sulphur fluorides and compounds with the contact metals, *e.g.*, copper fluorides. The percentage of gaseous decomposition products have been found extremely small. These products and any other secondary gaseous reaction products are removed from the gas circuit by filters containing activated alumina (Al_2O_3) when the gas is pumped back into the high pressure tank.

    The metallic fluorides are good dielectric materials and are, therefore, safe for electric equipment. However, they must be removed during periodic maintenance as they absorb moisture and loose their dielectric property. This happens if the breaker is dismantled during rainy season.
14. Dielectric strength of SF_6 gas at atmospheric pressure is 2.35 times that of air and 30% less than that of dielectric oil. At higher pressure the dielectric strength of the gas increases. At pressure about three times the atmospheric pressure, it is more than that of oil. This property permits smaller clearances and small size of equipments for the same voltage level.

    Actually speaking the dielectric strength of the SF_6 gas depends upon the nature of field existing between the electrodes, which in turn depends on the shape and configuration of electrodes, and the gap between the electrodes. The dielectric strength may actually increase to about 5 times depending upon the non–homogeneity of the field.
15. SF_6 gas maintains high dielectric strength even when diluted by air (nitrogen). 30% SF_6 and 70% air by volume has a dielectric strength twice that of air at a given pressure. Below 30% by volume the dielectric strength falls sharply.

8.6.2. Arc Extinction in SF_6 Circuit Breakers

The arc extinction process in SF_6 circuit breakers is different from that in air blast circuit breakers.

During the arcing period, SF_6 gas is blown axially along the arc. The heat is removed from the arc by axial convection and radial dissipation. Consequently, the arc diameter reduces during the decreasing node of the current wave. The diameter becomes small during current zero. Turbulent flow is introduced around current zero for extinguishing the arc.

As already mentioned in the preceding section, SF_6 gas, due to its electronegativity and low arc time constant, regains its dielectric strength rapidly after the final current zero, the rate of rise of dielectric strength is very high and the time constant is very small. SF_6 is a remarkable medium for arc extinction. The arc extinguishing properties are improved by moderate rates of forced gas flow through the arc space. The SF_6 gas at atmospheric pressure can interrupt currents of the order of 100 times the value of those can be interrupted in air with a plain breaker interrupter.

The basic requirement in arc extinction is not primarily the dielectric strength, but high rate of recovery of dielectric strength. In SF_6 gas, the dielectric strength is quickly regained, as already explained. Problems connected with current chopping are, therefore minimum.

In SF_6 circuit breakers, the gas is made to flow from a high pressure zone to a low pressure zone through a convergent–divergent nozzle. The nozzle is located such that the gas flows axially over the arc length. The flow of gas attains almost supersonic speed in the divergent portion of the nozzle, thereby the gas takes away the heat from the arc periphery and arc diameter is reduced. Finally the arc diameter becomes almost zero at current zero and the arc is extinguished. The contact space is filled with fresh SF_6 gas and its dielectric strength is rapidly regained because of electronegativity of the gas and turbulent flow of fresh gas in the contact space.

8.6.3. Construction

Such a circuit breaker essentially consists of two parts, namely (*i*) the interrupter unit and (*ii*) the gas system.

The Interrupter Unit. This unit consists of moving contact and fixed contacts comprising a set of current carrying fingers and an arcing probe. In the close position of the breaker, the fingers make contact round the circumference of the moving contact having the arcing probe enclosed within its hollow end. The contacts are surrounded by interrupting nozzles and blast shield to control the arc displacement and the movement of the hot gas. The moving contact is in the form of hollow nozzle sliding in a second set of spring loaded fingers. Side vents in the moving contact permits the high pressure gas into the main tank. When the moving contact is withdrawn from the fixed finger contacts an arc is struck between the moving nozzle and the arcing probe. With the increasing separation between the contacts the arc is extended and attenuated and it is finally extinguished by the flow of gas from the high pressure to low pressure systems.

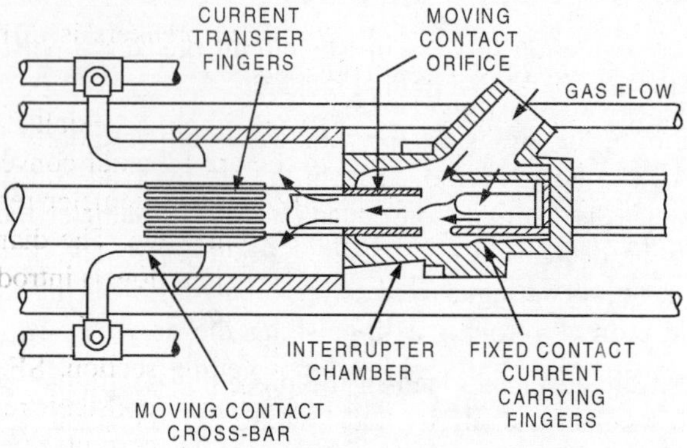

(a) Interrupter Head

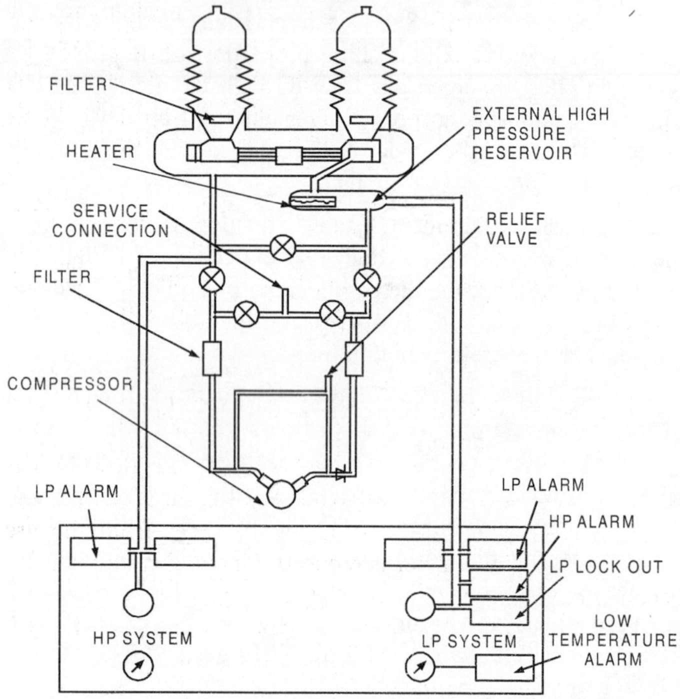

(b) Gas System SF_6 Circuit Breaker

Fig. 8.17

The Gas System. The closed circuit gas system [Fig. 8.17 (*b*)] is employed in SF_6 circuit breakers. Since the gas is costly, it is reconditioned and reclaimed after each operation of the breaker. Necessary auxiliary system is provided for such purpose. The low and high pressure systems are provided with low pressure alarms and a set of lockout switches which give a warning the moment the gas pressure drops below a certain value, failing which dielectric strength will be reduced and arc quenching ability of the circuit breaker will be endangered. On reaching danger limit the safety devices immobilise the breaker. The gas is stored in the high pressure chamber at 16 atmospheres while the gas pressure on the low pressure side is 3 atmospheres. Lot of care is required to prevent gas leakages at joints by providing perfect sealing. The temperature is kept 20°C. A heater backed with a thermostat at 16°C is provided in the high pressure chamber to prevent liquefaction of the gas in the high pressure chamber at low temperature.

SF_6 Circuit Breaker With Puffer Piston Principle. SF_6 Circuit breaker employing puffer piston principle is shown in Fig. 8.18. With the puffer–piston principle the SF_6 arc quenching pressure is produced during an opening operation by means of a piston attached to the moving contacts. The arc extinction takes place in the insulating nozzle. A generously dimensioned current carrying path concentrically arranged around the arcing contacts permits high load currents to be accommodated.

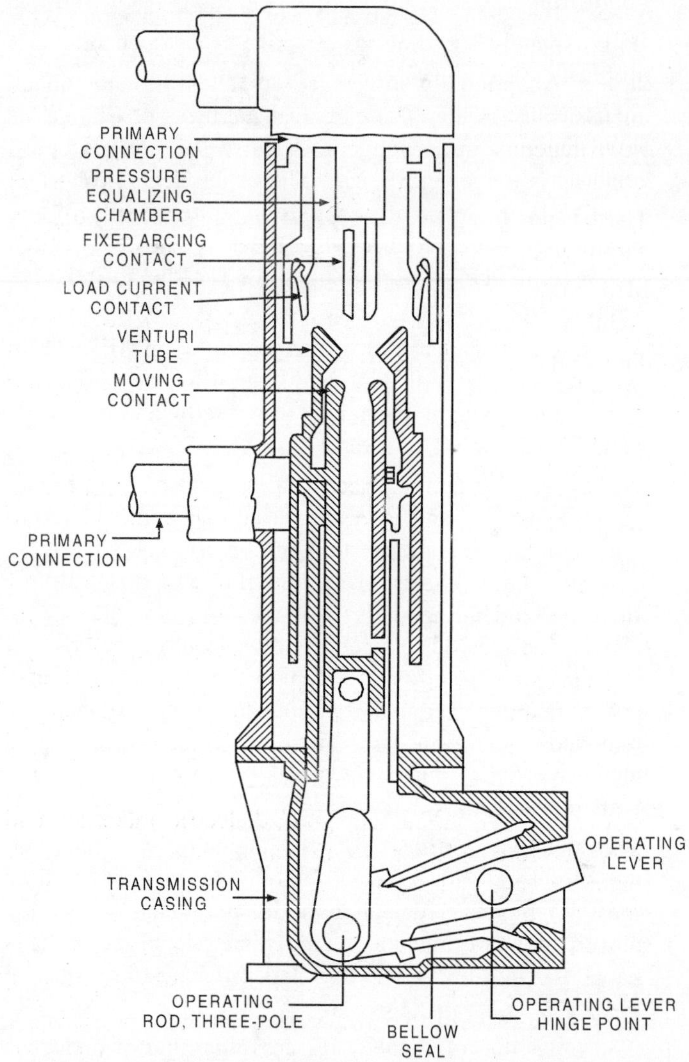

Fig. 8.18. ***SF_6 Circuit Breaker With Puffer-Piston Principle***

SF_6 breakers are made from 72–550 kV with rated interrupting current of 20–63 kA and rated current of 1,200–12,000 A.

8.6.4. Advantages of SF_6 Circuit Breakers

SF_6 circuit breakers have the following advantages over conventional breakers.

1. Excellent insulating, arc-extinguishing, physical and chemical properties of SF_6 gas is the greater advantage of SF_6 circuit breakers.
2. The gas is noninflammable and chemically stable. The decomposition products are non-explosive *i.e.,* there is no risk of fire or explosion.
3. Electrical clearances are very much reduced because of high dielectric strength of SF_6.

4. Outdoor EHV SF_6 circuit breaker has less number of interrupters per pole in comparison to air blast circuit breaker and minimum oil circuit breaker. Outdoor SF_6 circuit breaker is simple, comparatively cheaper in cost, maintenance free and compact.
5. Its performance is not affected due to variation in atmospheric conditions.
6. It gives noiseless operation—it does not make sound like air blast circuit breaker during operation.
7. No frequent contact replacement–arcing time is small owing to outstanding arc quenching properties of SF_6 and, therefore contact erosion is less. Hence contacts do not suffer oxidation.
8. There is no reduction in dielectric strength of SF_6 since no carbon particle is formed during arcing.
9. Minimum maintenance. The breaker may require maintenance once in four to ten years.
10. The sealed construction avoids the contamination by moisture, dust, sand etc. No costly compressed air system is required as in case of air blast circuit breaker.
11. Same gas is recirculated in the circuit thereby reducing the requirement of SF_6 gas. One triple pole, 145 kV SF_6 circuit breaker needs only about 30 kg of SF_6 gas for first filling and no replacement is required for about 5 years.
12. No overvoltage problem. The arc is extinguished at natural current zero without current chopping and associated overvoltage originating in circuit breakers.
13. The SF_6 gas circuit breaker can perform various duties like clearing short line faults, opening unloaded transmission lines, capacitor switching, transformer reactor switching etc., without any problem.
14. Ample overload margin. For the same size of conductors, the current carrying capability of SF_6 circuit breakers is about 1.5 times that of air blast circuit breakers because of superior heat transferability of SF_6 gas.

8.6.5. Problems Associated With SF_6 Circuit Breakers

Problems associated with SF_6 circuit breakers are given below:

1. Sealing problems. Imperfect joints lead to leakage of SF_6 gas.
2. SF_6 gas is suffocating to some extent. In case of leakage in the breaker tank, SF_6 gas being heavier than air settles in the surroundings and may lead to suffocation of the operating personnel. However, it is non-poisonous.
3. Arced SF_6 gas is poisonous and should not be inhaled or let-out.
4. Influx of moisture in the breaker is very harmful to SF_6 gas circuit breakers. Several failures are reported because of it.
5. Mechanism of higher energy level is required for puffer type SF_6 circuit breakers. Lower speeds due to friction, misalignment can cause failure of the breaker.
6. The internal parts need thorough cleaning during periodic maintenance under clean and dry environment. Dust of Teflon and sulphides should be removed.
7. Special facilities are required for transportation of gas, transfer of gas and maintenance of quality of gas. The deterioration of quality of gas affects the performance and hence reliability of the SF_6 circuit breaker.

8.7. VACUUM CIRCUIT BREAKERS

High vacuum has two outstanding properties:

1. Highest insulating strength: In comparison to various other insulating media in use in circuit breakers vacuum is a superior dielectric medium. It is better than all other media except air and SF_6 gas which are generally employed at high pressure.
2. When an ac circuit is opened by separating the contacts in a vacuum, interruption occurs at the first current zero with the dielectric strength across the contacts building up at a rate thousands of times higher than that obtained with other circuit breakers. This is because with the increase in gap due to separation of contacts and movement of breaker contacts, the breakdown kV peak increases. The breakdown voltage of vacuum compared with air for one pair of 9.4 mm diameter tungsten contacts are given in Table 8.1.

Table 8.1

Gap in mm	kV peak in air	kV peak in vacuum
1.25	2.50	100
2.50	7.50	140
4.50	15.00	200

The above two properties obviously make the vacuum circuit breakers more efficient, less bulky and cheaper in cost. The service life also is much greater than that of conventional circuit breakers and almost no maintenance is required. Vacuum breakers are ideally suited for most duties encountered in typical electric utility and industrial applications. Their voltage and interrupting rating is such that with little modifications, they can be made to perform specific switching duties on hv ac systems.

8.7.1. Vacuum Medium

We have seen earlier that during contact separation of circuit breaker arc is formed due to ionization of particles in the medium between the contacts. Idea behind the vacuum circuit breaker is to eliminate the medium between the contacts *i.e.,* vacuum. Every medium that has a pressure below atmospheric, which is 760 mm of Hg, is known as *vacuum*. Low pressures are measured in torr, where 1 torr = 1 mm of Hg. The pressures below about 10^{-5} torr are considered to be high vacuum.

The breakdown voltage of certain contact gap varies with the absolute pressure in the vacuum interrupters. When vacuum is used to serve the purpose of insulation, it implies that pressure of gas at which breakdown voltage becomes independent of pressure. Such a condition is believed to be realized at values of pressures from 10^{-4} to 10^{-6} torr.

8.7.2. Vacuum Arc

If an arc is to be formed in vacuum it is essential that the pressure should be low because only then it approaches the ideal arc. The ideal arc should have a low pressure, when it is initiated during its burning period and after its extinction.

Arc formed in vacuum is of different type from those formed in other mediums as it is formed by the neutral atoms, ions and electrons emitted from the electrodes themselves. The cathode surface normally, is not perfectly smooth but have many micro-projections. During the separation of contacts the current will be concentrated in these micro-projections as they are last points of contact. Due to their small area of cross section the projections will suffer explosive evaporation by resistive heating and supply sufficient quantity of vapour for the arc formation. In vacuum arc the electron emission occurs only at the cathode spots and not from the entire surface of the cathode. For this reason the vacuum arc is also known as the *cold cathode arc*. In cold cathode the emission of electrons could be due to any of the combinations of the following mechanisms:

(*i*) Field emission; (*ii*) Thermionic emission; (*iii*) Field and thermionic emission; (*iv*) Secondary emission by positive ion bombardment; (*v*) Secondary emission by photons, and (*vi*) Pinch effect emission.

At high currents the ionized metal vapour spreads through quite a large volume surrounding the electrodes. At low currents the quantity of vapour produced is greatly reduced. Due to rapid expansion of vapour evolved from the cathode spots in vacuum, the probability of maintaining a charge carrier density sufficient to retain adequate conductivity in the column and sustain the emission process becomes increasing small. At current zero the cathode spots extinguish in times of the order of 10^{-8} second.

8.7.3. Vacuum Arc Stability

In a 50 Hz ac circuit the current passes through zero value 100 times in a second (*i.e.,* once every 10 ms). It is desirable that the current is interrupted when it is passing through zero value, otherwise overvoltage will be induced due to current chopping. Therefore, for successful arc interruption it is necessary that the arc be stable for a half-cycle duration and particularly that it continues to exist at currents approaching zero. The arc stability is found to depend upon (*i*) the contact material and its vapour pressure and (*ii*) circuit parameters such as voltage, current, capacitance and inductance. In low current circuits most of the evaporation takes place at discrete points, called the cathode spots, while at higher currents the evaporation takes place from cathode and anode spots. In addition to these sources gas is added to the enclosure of the contacts when it is stripped from other parts of the enclosure because of high temperature and impinging metal vapour.

Vapour pressure and arc stability are found to be interrelated. The higher the vapour pressure at lower temperature, the better is the stability of the arc *i.e.,* the longer the lifetime of the arc). There are certain metals like Zn, Bi, which provide such characteristics and are better electrode materials for vacuum circuit breakers. Besides the vapour pressure, the thermal conductivity of the metal also affects the current chopping level. A good heat conducting metal will cool very rapidly and its contact surface temperature will fall. This will reduce the evaporation rate and the arc will be chopped because of insufficient vapour. On the other hand, a bad heat conductor will maintain its temperature and vaporization for a longer time and the arc will be more stable.

Shunting the contacts with different values of capacitance reduces the average lifetime of the arc. The higher the value of the capacitance, the lower is the average lifetime of the arc. An inductance in series, on the other hand increases the lifetime of the arc. Similarly, higher the system operating voltage, longer is the duration of the arc. This is because more restoring voltage is available to keep the arc alive.

8.7.4. Vacuum Breakdown

The major insulating media are atmosphere air, oil, paper and porcelain. These can withstand appreciable voltage but they are very small as compared to the withstand voltage of a gap in vacuum.

In the vacuum arc the neutral atoms, ions and electrons do not come from the medium in which the arc is drawn but they are obtained from the electrodes themselves by evaporating its surface material. Because of the large mean free path for the electrons, the dielectric strength of the vacuum is a thousand times more than when the gas is employed for arc interruption. In the vacuum of range from 10^{-4} to 10^{-6} torr, the breakdown strength is independent of the gas density, but varies with gap length only. The actual value of breakdown voltage for a given gap depends upon the condition of electrode surfaces. Highly polished and thoroughly degassed electrodes have shown particularly high breakdown voltages. Contacts get roughened after use and thus the dielectric strength decreases which can be improved by applying successive high voltage impulse sparks. This does not alter the arc roughened surfaces but removes the loosely adhering metal particles from the electrodes deposited there by the vapour blast during arcing. It has been found that for a vacuum of 10^{-6} torr some of the metals like silver, bismuth, copper etc. attain their maximum breakdown strength when the gap is slightly less than 3 mm. This property permits the use of short gaps in vacuum switches which results in simplifying the operating mechanism and increasing the speed of operation over conventional switches.

8.7.5. Arc Extinction in Vacuum Interrupters

The arc interruption process in vacuum interrupters is quite different from that in other types of circuit breakers. The vacuum as such is a dielectric medium and arc cannot persist in ideal vacuum. However, separation of current carrying contacts causes the vapour to be released from the contacts giving rise to plasma.* Thus on separation of current carrying

* Plasma is conducting medium for electric current. The plasma consists of charged particles. The conductivity of plasma depends upon number of ions per unit volume.

contacts, the contact space is filled with vapour of positive ions liberated from the contact material; the vapour density depending on the current in the arcing. During the decreasing mode of current wave the rate of release of the vapour falls and after the current zero, the medium regains its dielectric strength provided vapour density around the contacts has substantially reduced.

While interrupting a current of the order of a few hundred amperes by separating the flat contacts under high vacuum, the arc usually has several parallel paths and each arc path originates and sinks in a hot spot of current. Thus the total current is divided in several parallel arcs. The parallel arcs repel each other so that the arc tends to spread over the contact surface. Such an arc is called the *diffused arc*. The diffused arc can get interrupted easily. At higher values of currents of the order of a few thousand amperes, the arc gets concentrated on a small region and becomes self sustained arc. The concentrated arc around a small area causes rapid vaporisation of the contact surface. The transition from diffused arc to the concentrated arc depends upon the material and shape of contact, the value of current and the condition of electrodes. The interruption of arc is possible when vapour density varies in phase with the current and the arc remains in the diffused state. The arc does not restrike again if the metal vapour is quickly removed from the contact zone.

Thus the arc extinction process in a vacuum circuit breaker is related to a great extent to the material and shape of the contacts and the technique used in condensing the metal vapour. The contact geometry is so designed that the arc root keeps on moving so that the temperature at one point on the contact does not attain a very high value.

The rapid building up of dielectric strength after final arc extinction is a unique advantage of vacuum circuit breaker. There are limitations of oil circuit breakers in switching capacitive currents. In oil circuit breakers after an interruption takes place, the dielectric strength builds up so slowly that delayed restrikes take place often as much as half a cycle later. The vacuum switch is free from this trouble since after half a cycle, corresponding to 2.5 mm contact separation, a dielectric strength of the order of 100 kV (rms) is available. This type of switch has the advantage of restrike-free performance and is ideally suited to capacitor switching. Its operation is also practically independent of contact parting time.

8.7.6. Current Chopping in Vacuum Circuit Breakers

We have seen that current chopping in air and oil circuit breakers occurs because of instability in the arc column whereas in case of vacuum circuit breakers it depends upon vapour pressure and the electron emission properties of the contact material. The chopping level is also influenced by the thermal conductivity—lower the thermal conductivity, lower is the chopping level. It is possible to reduce the current level at which chopping occurs by selecting a contact material which gives out sufficient metal vapour to allow the current to come to a very low value or zero value, but this is rarely done as it affects the dielectric strength adversely.

8.7.7. Vacuum Arc Recovery Phenomenon

It has already been mentioned that high vacuum possesses extremely high dielectric strength. At current zero the cathode spot extinguishes within 10^{-8} second and after this the original dielectric strength is established very soon. This quick return of high dielectric strength is, of course, due to the fact that the vaporized metal which is localized between the contacts, diffuses rapidly due to absence of gas molecules. The metal molecules are blown at high speeds to the glass walls and condense there. After arc interruption the recovery strength during the first few microseconds is 1 kV/μs for an arc current of 100 A, as compared with 50 V/μs in case of air gap.

Because of the above mentioned attributes of vacuum interrupters, vacuum circuit breakers can be employed without reservations for fault clearing in any location on a system. They are capable of handling the severe recovery transients associated with short line faults or faults close to a transformer without any difficulty.

All forms of load switching can be performed with the same ease.

8.7.8. Contact Materials

The contact material for vacuum circuit breakers should have the following properties:

1. High electrical conductivity so as to pass normal load currents without overheating.
2. Low contact resistance.
3. High thermal conductivity so as to dissipate rapidly the large heat generated during arcing.
4. High cold and hot hardness to prevent wear and tear during normal opening and closing operations.
5. High density.
6. Sufficiently low vapour pressure so as to reduce the amount of inseparable metal vapour in the chamber.
7. High thermionic function to enable early arc extinction.
8. Low tendency to weld.
9. High boiling point to reduce arc erosion.
10. Low gas content (1 part in 10^7 or lesser) to ensure longer service life.
11. High arc withstand ability.
12. Low current chopping level.
13. High arc withstand ability.
14. Sufficient mechanical strength to retain structural integrity under high voltage gradients at the surface.
15. The material should preferably not have a surface film. If the film cannot be avoided then it should be conducting.
16. The heat of vapourization should be such that (*i*) arc energy generates enough metal vapour to sustain arcing till the first natural current zero, and (*ii*) there is insufficient vapour present to cause re-ignition after first current zero.
17. Easy to manufacture and economical.

From the above it becomes clear that a number of conflicting requirements have to be satisfied by a material to be suitable for electrodes. No single metal gives all the desirable properties. A high vapour pressure and low conductivity metal is more desirable to limit the current chopping whereas low vapour pressure metals are more desirable from the arc extinction point of view. Materials having high boiling and melting points have low vapour pressure at high temperatures but are poor conductors whereas metals having low boiling and melting points have high vapour pressure at high temperatures, low electron functions and have good thermal and electrical conductivities. Therefore, to combine these contradictory properties in one single material composites of two or more metals or a metal and a non-metal have to be made. Copper-bismuth, copper-lead, copper-tellurium, copper-thallium, silver-bismuth, silver-lead, and silver-tellurium are some of the alloys used as contact materials.

8.7.9. Construction of Vacuum Circuit Breaker

As far as construction is concerned vacuum circuit breaker is very simple in comparison to an air or oil circuit breaker. A schematic diagram of such a circuit breaker is shown is Fig. 8.19. The outer envelope is normally made of glass due to the ease of jointing it to the metallic end caps and also because the glass envelope facilitates the examination of the

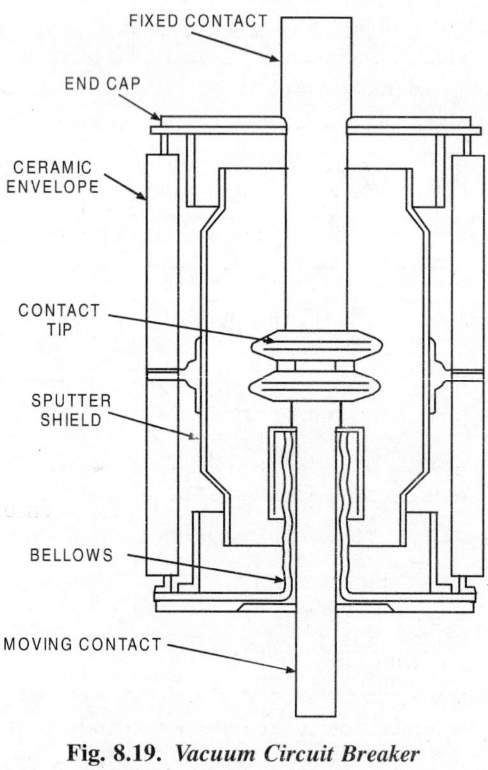

Fig. 8.19. *Vacuum Circuit Breaker*

breaker from outside after the operation. If it is becomes milky white from its original finish of silvery mirror then it indicates that the baffle is losing its vacuum. A sputter shield, usually, made of stainless steel, is placed between the contacts and the envelope in order to prevent the metal vapour reaching the envelope as it reduces the breakdown strength between the contacts. Inside the sputter shield the breaker has two contacts—one fixed and the other moving contact. The moving contact moves through a short distance of 5 to 10 mm depending upon the operating voltage. The metallic bellows made of stainless steel is used to move the lower contact. The design of the bellows is very important because the life of the vacuum breaker depends upon the ability of this component to perform repeated operations satisfactorily. The periphery of the end cap is sealed to the envelope and the fixed contact stem is an integral part of one end cap. One end of the fixed as well as moving contact is brought out of the chamber for external connections.

Operating Mechanism. The lower end of the breaker is fixed to a spring-operated or solenoid operated mechanism so that the metallic bellows inside the chamber are moved upward and downward during closing and opening operations respectively. The contact movement should be such as to avoid bounce. It is noteworthy that the operating mechanism should provide sufficient pressure for a good connection between the contacts.

The pressure in a vacuum interrupter at the time of sealing off is kept about 10^{-6} torr. The interrupting rating is between 250 and 1,000 MVA. The normal current carrying capacity for a single interrupter is 800–3,000 A; 4.2 kV–7.6 kV.

8.7.10. Advantages of Vacuum Circuit Breakers

1. Vacuum circuit breaker is a self contained and does not require filling of oil or gas. They do not need auxiliary air system, oil handling system etc. No need of periodic refilling.
2. Rapid recovery of very high dielectric strength on current interruption so that only half cycle or less arcing occurs after proper contact separation.
3. Current interruption occurs at the first current zero after contact separation with no restriking, making it exceptionally good for capacitor and cable switching and long line dropping.
4. Very high power frequency and impulse withstand voltages with small contact spacing, allowing ease of actuation and timing.
5. No emission of gases—pollution free.
6. Non-explosive and silent operation.
7. Breaker unit is compact and self contained. It can be installed at any required orientation.
8. Larger number of operations on load; or short circuit. Suitable for repeated operating duty. Long life.
9. Usable on any voltage up to 230 kV and higher where exceptionally long life and maintenance free operation is desired.
10. There are no gas decomposition products in vacuum and hermetically sealed vacuum interrupter keeps out all environmental effects.
11. *Constant contact resistance.* In vacuum the contacts cannot oxidise, therefore, their very small resistance is maintained throughout their life.
12. *High total current switched.* Because of small erosion of contact piece, rated normal current can be interrupted up

to 30,000 times and rated short-circuit rupturing current on average a hundred times.

Because of the above reasons together with the economic advantages offered, vacuum circuit breakers have high acceptance.

8.7.11. Disadvantages of Vacuum Circuit Breakers

1. Requirement of high technology for production of vacuum interrupters.
2. The vacuum interrupter is costlier than the interrupting devices in other types of circuit breakers and its cost is affected by production volume. It is uneconomical to manufacture vacuum interrupters in small quantities.
3. Rated voltage of single interrupter is limited to about $\frac{36}{\sqrt{3}}$ *i.e.,* 20 kV. Above 36 kV, two interrupters are required to be connected in series. Thus the vacuum circuit breaker becomes uneconomical for rated voltages exceeding 36 kV.
4. Loss of vacuum, due to transit damage or failure, makes the entire interrupter useless and it cannot be repaired at site.
5. It needs additional surge suppressors in parallel with each phase for interruption of low magnetising currents in a certain range.

8.7.12. Applications of Vacuum Circuit Breakers

1. Because of the short gap and excellent recovery characteristics of vacuum breakers, they are very useful as very high speed making switches in many industrial applications.
2. There are many applications where a simple load–break switch is not enough and at the same time the devices employed should not be costly. They include shunt reactor switching, transformer switching, line dropping, capacitor bank switching. These applications give a fast RRRV and vacuum circuit breakers are the best solutions.
3. Where voltages are high and current to be interrupted is low, these breakers have definite superiority over the oil or air circuit breakers.
4. For low fault interrupting capacities the cost is low in comparison to other interrupting devices.
5. The vacuum switches can be employed for capacitor switching which is a very difficult task for oil circuit breakers.
6. These can be used along with static overcurrent relays and given an overall clearance time of less than 40 ms on phase-to-phase faults.
7. Because of least requirements of maintenance these breakers are very suitable in a country like India, with 11 to 33 kV network extending into vast rural complex. Even with limited MVA rating of say 60 to 100 MVA, it should be suitable for a majority of applications in rural areas.

8.8. COMPARISON BETWEEN FUSE AND CIRCUIT BREAKER

S. No.	Particulars	Fuse	Circuit Breaker
1.	Functions	It performs both of functions—detection as well as interruption.	It performs the function of interruption only. Fault is detected by relay system provided for this purpose.
2.	Operation	Inherently complete automatic.	Elaborate equipment, such as relays, are required for automatic action.
3.	Operating Time	Very small (say 0.002 second or so)	Comparatively large (0.02–0.05 second).
4.	Breaking Capacity	Small.	Very large.
5.	Replacement	Needs replacement after every operation.	No replacement after operation.

8.9. HVDC CIRCUIT BREAKERS

Light duty dc circuit breakers have been in use since long. However, there is lack of suitable circuit breakers for HVDC systems. At present most of the HVDC systems are with two terminals and in a two terminal HVDC system, HVDC circuit breakers are not required because fault current can be controlled or eliminated by controlling the firing angle of the converters. In multi-terminal HVDC systems, the need of HVDC circuit breakers will arise.

Problems of Direct Current Interruption. AC circuit breaker easily interrupts the arc at natural current zero in the ac wave. At current zero, the energy $\left(\frac{1}{2}Li^2\right)$ to be interrupted is also zero. The contact gap has to cool and recover the dielectric strength to withstand natural transient recovery voltage. With dc circuit breakers, the problem is more complex as the dc waveform does not have natural current zeros. Forced arc interruption would produce high transient recovery voltage and restrike without arc interruption and ultimate destruction of the breaker contacts.

In designing of HVDC circuit breakers, there are three main problems to be overcome. These are (*i*) creation of artificial current zero (*ii*) prevention of restrikes and (*iii*) dissipation of stored energy.

The *artificial current zero principle* is used in HVDC circuit breakers for arc quenching. By introducing a parallel L-C circuit, the arc currents are subjected to oscillations. These oscillations are severe and have several artificial current zeros. The breaker extinguishes the arc at one of the artificial current zeros. The crest currents of the oscillation must be greater than the direct current to be interrupted. Figure 8.20 shows the schematic diagram of such a scheme.

A series resonant circuit with L and C is connected across the main contact M of a conventional ac circuit breaker through an auxiliary contact S_1 and resistor R is connected through contact S_2. Under normal operating conditions, main contact M and charging contact S_2 remain closed and the capacitor C is charged to line voltage through the high resistance R. Contact S_1 is open and has line voltage across it. For interrupting main circuit current I_d, the operating mechanism opens contact S_2 and closes contact S_1. This initiates discharge of capacitor C through inductance L, main contact M and auxiliary contact S_1 setting up an oscillatory current shown in Fig. 8.20 (*b*). Thus artificial current zeros are created and the circuit breaker main contact M is opened at a current zero Z. Thereafter, contact S_1 is opened and contact S_2 is closed.

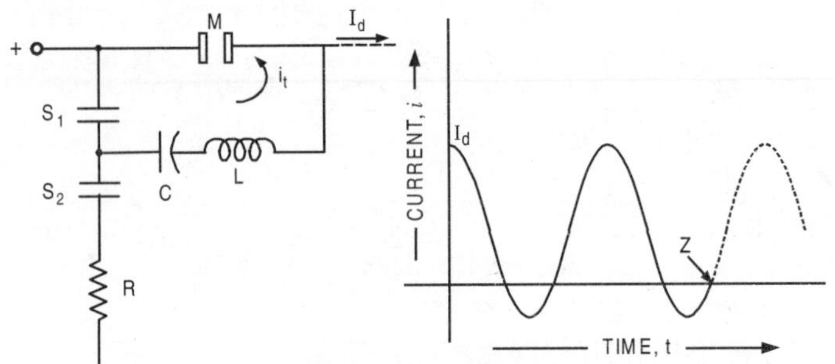

(*a*) *Schematic Circuit* (*b*) *Waveform of Current Through Main Contact M*

Fig. 8.20. ***HVDC Circuit Breaking***

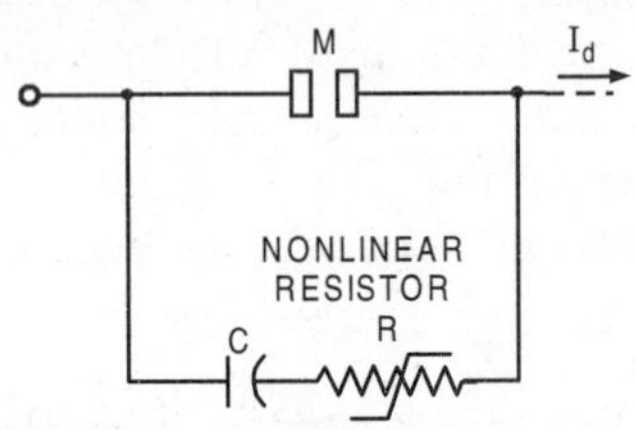

Fig. 8.21. ***HVDC Breaker With Nonlinear Resistor***

Another way of interrupting main circuit direct current is by its diversion to the capacitor so that the magnitude of current to be interrupted by the circuit breaker becomes smaller. This is illustrated in Fig. 8.21. The capacitor C is initially uncharged. When the main contact M opens, the main circuit current is diverted to the capacitor C. Thus the current to be interrupted by the main contact M becomes smaller. The rate of rise of the recovery voltage across M is $\frac{dV_c}{dt} = \frac{I_d}{C}$. The nonlinear resistor R absorbs energy without greatly adding to the voltage across the main contact M.

The problem of *prevention of restrikes* is more acute in oscillating current dc circuit breakers, where the time in which the current is chopped is very small (of the order 100 μs). Thus a steep surge of restriking voltage across the breaker terminals is produced and the circuit breaker must be capable of withstanding this voltage.

For producing a good deionizing arc, the space between two walls of arc chute can be narrowed for restricting the arc and simultaneously it may be splitted into a number of smaller arcs by inserting a grating of vertical metal plates.

A large amount of energy stored in the circuit inductance at the start of the interruption and that supplied by the rectifier during the interruption duration has to be dissipated, otherwise it will be transferred to the system capacitance and set up overvoltages. A protective spark gap may be connected across the circuit breaker so as to reduce the commutating capacitor size. It will also keep the abnormal voltage caused at the switching instant at desired level. By means of high frequency currents the spark gap acts as an energy dissipating device. Alternatively a Zno arrester may be connected across the breaker which will limit the transient recovery voltage and absorb associated energy.

8.10. AUTORECLOSURE

It has been found that most of the line faults on overhead transmission system are transient in nature. Statistical evidence shows that about 90 % of the faults are caused by lightning, birds, vines, tree branches etc. These conditions result in such arcing faults that if the fault energy infeed is interrupted for a short period, the arc extinguishes and the line can be re-energised. This fact is employed as a basis for autoreclosure schemes. In such schemes, after the relays at both ends of the line have picked up, the circuit breakers are tripped as far as possible at the same time and reclosed after time has been allowed for deionization. The fault disappears if it is transient, and line is fully restored to service after the reclosure. If the fault is not cleared after the first reclosure, a double or triple attempt of isolation and reclosure can be made. If the fault still persists, the breaker may permanently open till it is reset manually.

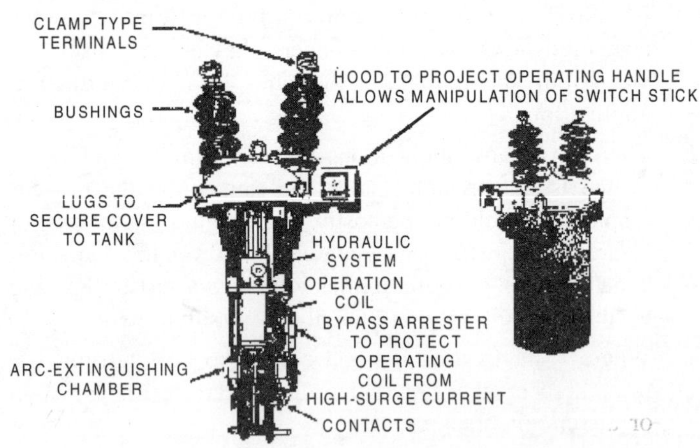

Fig. 8.22. ***Oil Circuit Reclosure (Single-Phase)***

An autoreclosure consists essentially of an oil switch or breaker actuated by relays which make it to open when predetermined current values flow through it. Reclosures are usually connected to protect portions of primary circuits and may take the place of line fuses. The switch or breaker is arranged to reclose after a short interval of time and reopen again should the fault or overload responsible for excess current flow persist. The reclosure can be set for 3 or 4 operations before it locks itself open for manual operation.

Oil circuit reclosures are increasingly employed in unattended substations and rural distribution schemes, where the circuit breakers are installed in remote areas. They obviate the need for an operator to proceed to the point to close the breaker manually every time it trips. Outages are thereby greatly reduced. In case of persisting fault and getting the reclosure locked, necessitating manual resetting, the technician after investigating and clearing the fault closes the reclosure.

Autoreclosing may be single- or three-phase type. Mostly single-phase autoreclosing breakers are preferred as most of the transmission faults are single phase to ground faults. Autoreclosing in single phase also improves stability as the power remains transmitted through the two healthy phases when one phase is interrupted.

Like any other circuit breaker, the rupturing capacity of reclosure should be properly chosen. In large distribution network, reclosures may be provided to look after each separate zone and fuse protection provided for the subsidiary branch lines in each zone.

The breakers may be rapid autoreclosing type (about 20 cycles or 0.4 second), or delayed autoreclosing (5 to 30 s) type. It is not necessary to check synchronism with high speed reclosures while with delayed autoreclosing breakers, it is necessary to check synchronism before reclosing. For this purpose, synchronising relays are employed.

8.11. BREAKER OPERATING MECHANISMS

Each circuit breaker is provided with an operating mechanism. The functions of this mechanism are (*i*) to close the breaker contacts rapidly without hesitation at all currents from zero to rated making current capacity (*ii*) to hold the breaker contacts together against the action of powerful opening springs and (*iii*) to open the breaker contacts without delay immediately on receiving tripping signal.

The small low-voltage circuit breakers are manually operated while medium and large circuit breakers are power operated either electrically or pneumatically.

The force for opening a circuit breaker must be instantaneously available on receipt of the tripping impulse from the protective gear, and the universal practice is to employ a previously charged spring for oil circuit breakers and compressed air for air blast circuit breakers. The operating time between instant of receiving trip signal and final contact separation is of the order of 1.5 cycles or 0.03 s in the modern EHV circuit breakers and in slow circuit breakers employed in distribution systems it may be about 3 cycles or 0.06 second.

With spring opening, the circuit breaker closing mechanism compresses the spring and latches the moving contacts in the closed position. The tripping impulse then energises a small solenoid that releases the latch and allows the spring to exert its full force to open the contacts; suitable buffers must be provided to absorb the kinetic energy of the moving parts at the end of the stroke.

With an air blast circuit breaker the contacts are normally held closed by a spring and forced open by the admission of compressed air, the compressed air thus performing the double function of opening the contacts and extinguishing the arc. The air supply must therefore be maintained until after the auxiliary isolating switch (also operated by compressed air) has opened, after which it is cut off and the main contacts return to the closed position under the action of their spring.

The force for closing a circuit breaker may be applied by hand , by a solenoid, by a spring or by compressed air. With spring-opened circuit breakers, the closing force has to compress the spring and must therefore be considerable. For a large circuit breaker the power required for closing may be of the order of 25 kW, but the time occupied is only a fraction of a second so that the energy is relatively small *e.g.,* 6.25 kW-s or 1.75 Wh. Manual operation is thus impracticable for circuit breakers having short-circuit ratings above about 150 MVA.

Solenoid closing is the most widely used method, the solenoid being fed from a battery, usually of 110 or 220 V. For large circuit breakers, the current may be as high as 600 A at 110 V dc.

For medium-sized circuit breakers, spring closing may be used, the spring being precharged by hand or by a small motor. In such cases the provision of a battery may not prove economical.

The use of compressed air in connection with the air blast circuit breaker has been referred to above, closing of circuit breakers being affected by the isolating switch. With oil circuit breakers in extensive substations, where the heavy cabling and voltage drop resulting from the solenoid currents may cause difficulties, compressed air closing is sometimes employed.

In all cases the closing mechanism must be trip-free *i.e.,* if a tripping impulse comes during the closing operation the circuit breaker must immediately trip even though the closing force be still maintained; with spring-opened circuit breakers this is affected by a mechanical toggle mechanism, while with air blast circuit breakers it is inherent in their operation.

8.12. TYPES OF CIRCUIT BREAKER MOUNTINGS AND ENCLOSURES

Circuit breakers are usually connected to a common busbar (as shown in Fig. 8.23 by single line diagram). On both sides

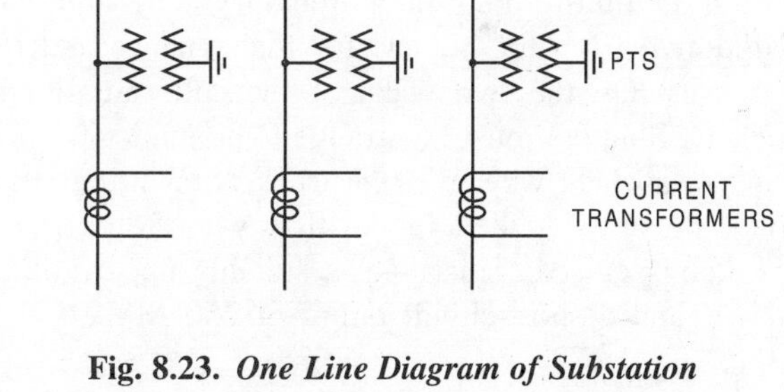

Fig. 8.23. ***One Line Diagram of Substation***

Table 8.2.

Type of Circuit Breaker	Merits	Demerits
1. Bulk Oil Circuit Breakers	Simplicity in construction, high rupturing capacity, suitability for automatic as well as manual operation; possibility of locating CTs in bushings.	Larger size and greater weight, fire hazard, explosion hazard, unsuitability for indoor installation, requirement of large volume of high grade mineral oil (scarcely available in our country), necessity for periodic inspection of the quantity and quality of oil in the tank, greater wear and tear of the contacts resulting in their frequent replacement, unsuitability for repeated cycles of operation in connection with automatic reclosing. Becoming obsolete now.
2. Minimum Oil Circuit Breakers	Requirement of lesser quantity of oil, smaller space requirement, smaller tank size, less weight, low cost, reduced risk of fire and reduced maintenance problems, easier access to the contacts, suitability for both automatic and manual operation.	Increased degree of carbonisation due to smaller quantity of oil, rapid deterioration of oil dielectric strength due to high degree of carbonisation, greater contact damage, comparatively low rupturing capacity, difficulties in removal of gases from the contact space in time and locating CTs, necessity of more frequent inspection and replacement of oil and total unsuitability for repeated cycles of operation. Now superseded by SF_6 circuit breakers.
3. Air Blast Circuit Breakers	High speed operation, suitability for frequent operation, facility for high speed reclosure, negligible maintenance, elimination of fire hazard, reduced size, comparatively less weight, less contact damage, very easy access to the contacts, very high rupturing capacity.	Requirement of a complete compressed air system consisting of air compressor, electric motor, air-tight tubes etc., more complicated construction, higher cost, requirement of highly trained personnel for maintenance, problem of current chopping, sensitivity to circuit severity *i.e.,* RRRV. Becoming obsolete now.

of circuit breakers isolators are provided to make dead or disconnect the circuit breakers from live parts whenever required for the purpose of maintenance and repairs. Some electrical or mechanical interlocks must be provided so that circuit breaker is open before opening of its isolators and isolators are open before the opening or removal of any cover or barrier.

Current transformers (CTs) and potential transformers (PTs) are usually connected, as shown in Fig. 8.23.

There are four types of circuit breaker mountings, which are described below:

1. ***Cubicle Type or Cellular Type.*** In this type of mountings the circuit breakers and their ancillary equipment are mounted in individual chambers or cells made of slate marble or concrete with doors of sheet steel or in a completely sheet steel cubicles. In this arrangement complete isolation of the cell or cubicle from bus-bars may be obtained. Every part in the cell is accessible and any visible fault can be traced out easily but is not very safe.

2. ***Truck Type.*** In this type of mounting the circuit breaker and its auxiliaries are mounted on the movable truck, which is maintained in a masonry cell and can be withdrawn from the cubicle for maintenance and repairs purposes. On the two walls of the cells are fitted two sockets. Shutters must be arranged automatically to cover the live sockets whenever the truck is taken out, and the truck must be locked in position whenever the circuit breaker is closed. This type is suitable up to voltage of 11 kV and a short-circuit rating of 250 MVA.

3. ***Metal-Clad Type.*** In this case each and every live part of the switchgear is completely enclosed in metallic enclosures which are earthed. The enclosures are air, oil or compound filled. It provides proper safety, protects the switchgear from the ingress of dirt and dust from outside and reduces the space occupied to minimum. The circuit breaker can be moved away from the bus-bars either horizontally or between the guides provided by this purpose after disconnecting from live parts. This type is suitable up to voltage of 66 kV and a short-circuit rating of 1,000 MVA.

4. ***Open Type.*** For the largest ratings, it is economical to install the circuit breakers in open to avoid building costs. Bulk oil, low oil content and air blast circuit breakers are mainly used for outdoor installations. In some cases it is found economical to install open type switchgear in a building, the switchgear in this case being of low oil content type.

8.13. COMPARATIVE MERITS AND DEMERITS OF DIFFERENT TYPES OF CONVENTIONAL CIRCUIT BREAKERS

Relative merits and demerits of different types of conventional circuit breakers (bulk oil circuit breakers, minimum oil circuit breakers and air blast circuit breakers) are given above in tabular form in Table 8.2.

8.14. COMPARISON OF DIFFERENT TYPES OF CIRCUIT BREAKERS

Arc quenching medium, voltage, breaking capacity, design features, applications of various types of circuit breakers are given on next page in tabular form in Table 8.3.

Table 8.3

Type of Circuit Breaker	Arc Quenching Medium	Voltage, Breaking Capacity	Design Features	Applications	Remarks
1. Air break Circuit Breaker	Air at atmospheric pressure	430-600 V, 5–15–35 MVA 3.6-12 kV, 500 MVA	Arc runners, arc splitters, magnetic coils.	For medium low voltages ac/dc. Industrial circuit breakers	Have current limiting features.
Miniature Circuit Breaker	—do—	430-600 V	Small size, current limiting feature.	For low and medium voltages	—
2. Oil Circuit Breakers					
(*i*) Bulk Oil	Dielectric oil (Transformer oil)	12 kV, 3.6 kV, 500 MVA	One tank up to 36 kV, three tanks above 36 kV, fitted with arc control devices.	Used up to 12 kV, 500 MVA	Getting obsolete now.
(*ii*) Minimum Oil	—do—	3.6 kV to 145 kV preferably	Small size, arc control device used.	For metal enclosed switchgear up to 36 kV, outdoor type between 36 and 245 kV	Now superseded by SF_6 circuit breakers.
3. Air Blast Circuit Breakers	Compressed air (20-30 kg/cm^2)	245 kV, 35,000 MVA up to 1,100 kV, 50,000 MVA	Unit type construction, several units per pole.	Suitable for all EHV applications, Also for arc furnace duty.	High speed operation, Requirement of auxiliary compressed air system. Now superseded by SF_6 circuit breakers for 145 kV and above.
4. SF_6 Circuit Breakers (Single Pressure Puffer Type)	SF_6 gas (5 kg/cm^2)	12 kV, 1,000 MVA 36 kV, 2,000 MVA 145 kV, 7,500 MVA 245 kV, 10,000 MVA 420 kV, 40 kA	One interrupter pole up to 245 kV	Most popular over wide range of voltages from 3.6 kV to 760 kV	Maintenance free.
5. Vacuum Circuit Breakers	Vacuum	Up to 36 kV, 750 MVA	Variety of designs	Suitable for variety of applications from 3.6 kV to 36 kV. Preferred for indoor switchgear up to 36 kV, 750 MVA	Long life, modest maintenance.
6. HVDC Circuit Breaker	Oil or Air blast	33 kV, 2 kA	Artificial current zero by switching in capacitors.	For HV direct current interruption.	—

Exercises

1. What are the different types of circuit breakers when the arc quenching medium is the criterion ? Mention the voltage range for which a particular type of circuit breaker is recommended.

2. Explain the principle of circuit interruption. State the various types of circuit breakers used for HV and EHV systems.
[Pb. Univ. Switchgear & Protection (Elec. Power III) July 1992]

3. Classify circuit breakers. Explain the basic difference between bulk oil circuit breaker and minimum oil circuit breaker.

4. Explain the construction, principle and working of an oil circuit breaker with a neat sketch.

5. Discuss the principle of arc extinction in an oil circuit breaker with reference to restriking and recovery voltages.
[G.B. Technical Univ. Switchgear and Protection 2011-12]

6. Explain how an arc is initiated and sustained in an oil circuit breaker (CB) when the CB contacts separate. Also explain the terms : Restriking voltage, Recovery voltage, Rate of rise of restriking voltage.
[U.P.S.C. I.E.S. Electrical Engineering-II, 2008]

7. Explain with the help of a neat sketch the function of a plain explosion pot. What are its advantages over the plain break circuit breaker ? What are its limitations ?

8. Discuss the process of arc extinction in an oil circuit breaker. Draw a neat sketch of an oil circuit breaker and explain its working.

9. Show under what distribution of capacitance in a double break type plain-break oil circuit breaker, will about 83.4% of the total voltage appear across the break nearer to the source of supply.

10. Describe with a neat sketch the working of cross-jet explosion pot in bulk oil circuit breakers.
Enumerate the advantages of minimum oil circuit breaker over the bulk oil circuit breaker.

11. Describe with a neat sketch the principle of operation of a minimum oil circuit breaker. Why is it called so ?

12. Describe the construction and operation of a minimum oil circuit breaker.
[U.P. Technical Univ. Switchgear and Protection 2009-10]

13. Explain the construction, principle of operation and application of a minimum oil circuit breaker.
[Pb. Univ. Switchgear & Protection September 1990]

14. Explain the features of an air break circuit breaker by means of simple sketches.
[Pb. Univ. Switchgear & Protection September 1990]

15. What are the advantages and disadvantages associated when air is used as an arc quenching media instead of oil ? Discuss with necessary circuit diagram the principle of operation of an air-break circuit breaker with magnetic blow-out coil and "arc chute" used for 11 kV system.
16. With a neat sketch explain the principle of operation of an air-break circuit breaker. Why an arc chute and a blow-out coil are used in a circuit breaker ? Explain briefly why is it easier to break current in an ac circuit breaker than in a dc circuit breaker ? What technique is adopted for dc circuit breaker?
17. Discuss the principle of arc interruption in (*i*) an oil circuit breaker (*ii*) air blast circuit breaker.
[U.P. Technical Univ. Switchgear and Protection 2006-07]
18. Explain with necessary diagram the phenomenon of current chopping in an air blast circuit breaker. How this problem can be rectified by using resistance switching ? Why mostly air blast circuit breakers are susceptible to current chopping ?
[U.P.S.C. I.E.S. Electrical Engineering-II, 2005]
19. What are the usual methods of arc quenching in a circuit breaker ? Why arcs are formed during breaking of large currents ?
What is the difference between cross-blast and axial-blast technology in air blast circuit breaker? Show with a neat sketch the air blast mechanism used there.
20. Discuss with a neat sketch the principle of operation of an axial blast circuit breaker and hence explain why an air blast circuit breaker is very much susceptible to current chopping ?
21. Explain the working of an air blast circuit breaker with the help of suitable diagram.
Explain current chopping phenomenon associated with ABCB.
22. Discuss the following circuit breakers :
(*i*) Air blast circuit breaker (*ii*) SF_6 circuit breaker.
[U.P. Technical Univ. Switchgear and Protection 2007-08]
23. Draw appropriate diagrams to show the complete operation of a SF_6 circuit breaker. Give the merits of SF_6 circuit breakers.
[G.B. Technical Univ. Switchgear and Protection 2011-12]
24. Describe the construction, principle of operation and applications of SF_6 circuit breaker. How is it different from air blast circuit breaker ?
[U.P. Technical Univ. Switchgear and Protection 2006-07]
25. Explain the SF_6 circuit breaker. What are its advantages and disadvantages over air blast circuit breaker ?
[U.P. Technical Univ. Switchgear and Protection 2004-05]
26. Discuss briefly, the construction, principle of operation and applications of SF_6 CB.
[U.P. Technical Univ. Switchgear and Protection 2005-06]
27. Explain the construction and working of a puffer type SF_6 circuit breaker. Also enumerate the properties of SF_6 gas which make at a good dielectric and arc quenching medium.
[U.P. Technical Univ. Switchgear and Protection 2008-09]
28. Draw appropriate diagrams to show the complete operation of a SF_6 circuit breaker. Give the merits of SF_6 circuit breakers.
[U.P. Technical Univ. Switchgear and Protection 2009-10]
29. Describe the working principle of a vacuum circuit breaker. How contact movement and electrical connections with the contacts are made in a vacuum circuit breaker ? What are the shortcomings of a vacuum circuit breaker ?
30. Discuss the behaviour of electric arc in vacuum with the help of a neat diagram. Explain the construction of a vacuum circuit breaker.
[U.P. Technical Univ. Switchgear and Protection 2008-09; 2011-12]
31. Discuss the merits and demerits of a vacuum circuit breaker for low voltage applications.
[U.P.S.C. I.E.S. Electrical Engineering-II, 2003]
32. Discuss the problems encountered in HVDC circuit breaking. Suggest remedies for them.
33. Explain with a suitable example the phenomenon of autoreclosing.
[G.B. Technical Univ. Switchgear and Protection, 2011-12]
34. What are the special features of an autoreclosing CB ? What are its advantages ?
[A.M.I.E. Sec. B. Advanced Power Systems Winter 1998]
35. Name different types of circuit breakers with their relative merits.
36. Compare the performances and characteristics of (*i*) minimum-oil breakers and air blast circuit breakers (*ii*) air blast circuit breaker and bulk oil circuit breaker.
[Pb. Univ. Elec. Power III (Switchgear & Protection) Dec. 1991]
37. Compare the performance and characteristics of minimum oil breakers and air blast circuit breakers.
[G.B. Technical Univ. Switchgear and Protection 2011-12]
38. What are the merits and drawbacks of air blast circuit breakers over bulk oil circuit breakers.
[G.B. Technical Univ. Switchgear and Protection 2011-12]
39. Compare vacuum and SF_6 circuit breakers.
[A.M.I.E. Sec. B. Power System Protection & Communication Winter (1998)
40. Write short notes on the following :
(*i*) Oil circuit breakers. **[Pb. Univ. Sept. 1991]**
(*ii*) Minimum oil circuit breaker. **[Pb. Univ. May 1987]**
(*iii*) Air circuit breaker.
(*iv*) Air blast circuit breaker. **[Pb. Univ. October 1986]**
(*v*) SF_6 circuit breakers. **[Pb. Univ. June 1991]**
(*vi*) Constructional details and principle of operation of SF_6 circuit breaker.
(*vii*) Vacuum circuit breakers.
[Pb. Univ. April 1986; Sept. 1997; May 1989]
(*viii*) Oil circuit reclosers.

Short Answer Type Questions With Answers

Q. 1. What do you understand by switchgear ?
Ans. The apparatus employed for switching, controlling and protecting the electrical circuits and equipment is known as switchgear.
Q. 2. Which of the gases are employed in commercial gas-blast circuit breakers ?
Ans. SF_6 and air are used in commercial gas-blast circuit beakers.
Q. 3. Name the materials used for contacts of vacuum circuit breakers ?
Ans. Copper-bismuth, copper-lead, copper-tellurium, copper-thallium, silver-bismuth, silver-lead and silver-tellurium are some of alloys employed as contact materials in vacuum circuit breakers.
Q. 4. Name the three problems encountered in HVDC circuit breakers ?
Ans. The three problems encountered in HVDC circuit breakers are:
1. How to create artificial current zero ?
2. How to prevent restriking of arc ?
3. How to dissipate the stored energy ?

❑❑❑

CHAPTER 9

Protective Relays

INSIDE THIS CHAPTER

9.1. INTRODUCTION

In modern power system to have a normal operation of the system without electrical failure and damage to the equipment two alternatives are available with the designer, one is to design the system so that faults cannot occur and the other is to accept the possibility of faults and take steps to guard against the ill effects of such faults. Although it is possible to eliminate faults to a large extent by careful system design, careful insulation coordination, proper operation and maintenance, it is obviously not possible to ensure cent per cent reliability and therefore the possibility of faults must be accepted and the necessity of protection scheme must be realized.

Protection scheme required for the protection of power system components against abnormal conditions such as faults etc. essentially consists of protective relaying and circuit breaker. Protective relay functions as a sensing device, it senses the fault, then determines its location and finally, it sends tripping command to the circuit breaker. The circuit breaker after getting the command from the protective relay, disconnects the faulted element. From this, it can be concluded, that, the protective relay which is the brain behind the scheme, plays a vital role. Therefore, proper care should be taken in designing and selecting an appropriate protective relay which is reliable, efficient and fast in operation. As a matter of fact by clearing the fault fast with the help of fast acting protective relay and associated circuit breaker, the damage to the apparatus is reduced, subsequent hazards like fire, risk to the life are reduced, by removing the particular faulted section, the continuity of supply is maintained through remaining healthy section, by clearing the fault fast, fault arising time is reduced and therefore, the system can be restored to the normal state sooner, transient state stability limit of the system is greatly improved, permanent damage to the equipment is avoided and possibility of developing most simple fault such as single phase-to-ground into more severe fault such as double phase-to-ground fault is reduced.

All the above objectives can be achieved only if the protective relay is reliable, maintainable and sensitive enough to distinguish between normal and abnormal (or faulty) conditions. The relay must come into action whenever there is a fault and must not operate if there is no fault. A number of relays are employed for the protection of the power system. Some of them are primary relays meaning that they are the first line of defence. Such relays sense the fault and send signal to the proper circuit breaker to trip and clear the fault. The fault may not be cleared if the circuit breaker fails to open or relay maloperates. The circuit breaker failure is because of two reasons, first, failure of supply to its trip coil and second, struck mechanical lever. The relay failure is because of three reasons such as wrong setting, bad contacts and open circuit in the relay coil. In such cases, a second line of defence is provided by the backup relays. These backup relays have longer operating time even though they sense the fault along with the primary relays. Backup relays are arranged in two ways viz., they are located at the same place as the primary relay (local backup) and operate the same breaker or they are located at different stations and operate different breakers. The latter scheme has higher reliability.

In order to attain the desired reliability, the power system network is divided into different protection zones. The overall system protection is divided into (*i*) generator protection (*ii*) transformer protection (*iii*) bus protection (*iv*) transmission line protection and (*v*) feeder protection.

The relays employed for protection of apparatus and transmission lines are (*i*) overcurrent relays (*ii*) undervoltage relays (*iii*) under-frequency relays (*iv*) directional relays (*v*) thermal relays (*vi*) phase sequence relays such as negative sequence or zero sequence relays (*vii*) differential relays including percentage-differential relays (*viii*) distance (or impedance) relays such as plane impedance relays, angle impedance *i.e.,* ohm (or reactance) relays, angle admittance *i.e.,* Mho's relays, offset or restricted relays (*ix*) pilot relays such as wire pilot, carrier channel pilot or microwave pilot relays. All the existing relaying schemes employ either one or more of these relaying schemes with slight modifications.

The protective relays do not eliminate the possibility of fault occurrence on the power system rather their action starts only after the fault has occurred on the system. It would be ideal if protection could anticipate and prevent fault occurrence but this is obviously impossible except where the original cause of a fault creates some effect which can operate a protective relay. So far only type of relay falls within this class, this is the gas detector (Buchholz relay) employed for protection of transformers. Such a relay operates when the oil level in the conservator pipe of a transformer is lowered due to accumulation of gas caused by a poor connection or by an incipient breakdown of insulation (slowly developing fault).

The function of the protective relays, as already discussed, is to sense the fault and then to energize the circuit breaker trip coil. Circuit breakers normally employed for protection of transmission lines and equipment are (*i*) air circuit breakers (*ii*) oil circuit breakers (*iii*) minimum oil circuit breakers (*iv*) air blast circuit breakers (*v*) SF_6 circuit breakers and (*vi*) vacuum switches. All of these breakers have been described in chapter 8.

The first of the above is employed for the apparatus protection, the second is used for the feeder protection and the third and the fourth are employed for the EHV/UHV transmission lines. The air blast circuit breakers on EHV/UHV lines have become standard practice. In place of air, SF_6 is used and these SF_6 circuit breakers have several advantages, as described in Art. 8.6.4. For important lines breakers are also provided with a reclosing feature.

The main features of a good protective relaying are (*i*) reliability (*ii*) selectivity (*iii*) sensitivity (*iv*) simplicity (*v*) speed and (*vi*) economy, all have been described in Art. 1.8.

9.2. BASIC TERMINOLOGY

It is desirable to define and explain some important terms concerned with protective relaying for the sake of familiarity.

Protective Relay. It is an electrical device designed to initiate isolation of a part of an electrical installation, or to operate an alarm signal, in the event of an abnormal condition or a fault.

Energizing Quantity. It is an electrical quantity, *i.e.,* current or voltage either alone or in combination with other electrical quantities, required for the operation of the relay.

Trip Circuit. It is a circuit that controls the circuit breaker for opening operation and comprises of trip coil, relay contacts, auxiliary switch, seal-in-coil, battery supply etc.

Characteristic Quantity. This is a quantity to which the relay is designed to respond, *e.g.,* current in an overcurrent relay, impedance in an impedance relay, phase angle in a directional relay etc. Some relays have a calibrated response to one or more quantities, called the **characteristic quantities.**

Operating Force or Torque. It is a force (or torque) which tends to close the relay contacts.

Restraining Force or Torque. It is a force (or torque) which opposes the operating force (or torque) and tends to prevent the closure of relay contacts.

Setting. It is an actual value of the energizing or characteristic quantity at which the relay is designed to operate under given conditions.

Power Consumption (or Burden) of a Relay. It is the value of power consumed by the relay circuits at the rated current or voltage and expressed in VA for ac and watts for dc.

Pick Up. A relay is said to pick up when it moves from the *off* position to the *on* position or the operation of the relay is called the *relay pick-up.*

Operating or Pick-up Level. It is the value of actuating quantity (current or voltage) which is on threshold (border) above which the relay operates and closes its contacts. So long as the current in the relay is less than the pick-up value, the relay does not operate and the breaker controlled by it remains in the closed position. However, when the relay coil current is equal or exceeds the pick-up value, the relay operates to energise the trip coil which opens the circuit breaker.

Drop Out or Reset Level. This is the value of current or voltage etc. below which a relay opens its contacts and comes back to its original position. The ratio of the drop-out or reset value to the pick-up or operating value is called the drop-out or reset ratio.

Flag or Target. It is a visual device, usually spring- or gravity-operated, for indicating the operation of a relay.

Operating Time. It is given by the time which elapses between the instant when the actuating quantity exceeds the pick-up value to the instant when the relay contacts close.

Reset Time. It is given by the time which elapses between the instant when the actuating quantity becomes less than the reset value to the instant when the relay returns to its original position.

Seal-in-Coil. It is a coil that is provided not to allow the relay contacts to open when the current is flowing through them.

Maximum Torque Angle (MTA) or Characteristic Angle of Relay. This is the design angle of the relay that will yield maximum torque.

Overshoot Time. This is the time during which stored operating energy is dissipated after the characteristic quantity has been suddenly restored from a specified value to the value which it had at the initial position of the relay.

Fault Clearing Time. Time elapsed between the instant of occurrence of fault and instant of final arc extinction in circuit breaker is called the *fault clearing time.*

Breaker Time. Time interval between closure of trip circuit and final arc extinction in circuit breaker is called the *breaker time.*

Relay Time. The interval between occurrence of fault and closure of relay contacts is called the *relay time.*

Relay time plus breaker time is equal to *fault clearing time.*

Consistency. It is accuracy with which the relay can repeat its electrical or time characteristics.

Reach. A distance relay operates whenever the impedance seen by the relay is less than a pre-specified value. This impedance or the corresponding distance is known as the *reach* of the relay.

Reach of a relay may be defined as the limiting distance covered by the protection, the faults beyond which are not within the reach of the protection and should be covered by the other relay.

Overreach. The tendency of the relay to operate at impedances larger than its setting is known as overreach. This can occur with some high speed relays if the current applied is not symmetrical, as is frequently the case during the first few cycles following a fault.

Underreach. The extra impedance introduced by an arc into the fault loop affects the distance measured by the relay and causes it to underreach. Underreach is just reverse of overreach and may be defined as the failure of distance relay to operate within the set protected distance (say 90%).

9.3. OPERATING PRINCIPLE OF PROTECTIVE RELAYS

All the relays employed for protection against short circuits operate by virtue of the current and/or voltage supplied to them by CTs or PTs. Failures in the system are indicated by the individual or relative changes in the currents or voltages supplied to the protective relaying equipment. For every type and location failure, there is some distinctive difference in these quantities and there are various types of protective relays available, each of which is designed to recognise a particular difference and to operate in response to it. The difference may be in terms of the magnitude, frequency, phase angle, rate of change, harmonics or waveshape, duration of the conditions etc.

The main principle employed in the operation of the relay is either electromagnetic attraction or electromagnetic induction. In an electromagnetic attraction relay, a plunger is drawn into a solenoid or an armature is attracted to the poles of an electromagnet. Such relays can be operated either by dc or ac. In the case of electromagnetic induction relays, the principle of induction motor is used and the torque is developed by electromagnetic induction. Such relays are operated by the ac quantities only.

9.4. CLASSIFICATION OF RELAYS

The protective relay may be defined as an electrical device interposed between the main circuit and the circuit breaker in such a manner that any abnormality in the circuit acts on the relay, which in turn, if the abnormality is of a dangerous character, causes the breaker to open and so to isolate the faulty element. The relay ensures the safety of the circuit equipment from any damage which might be otherwise caused by the fault.

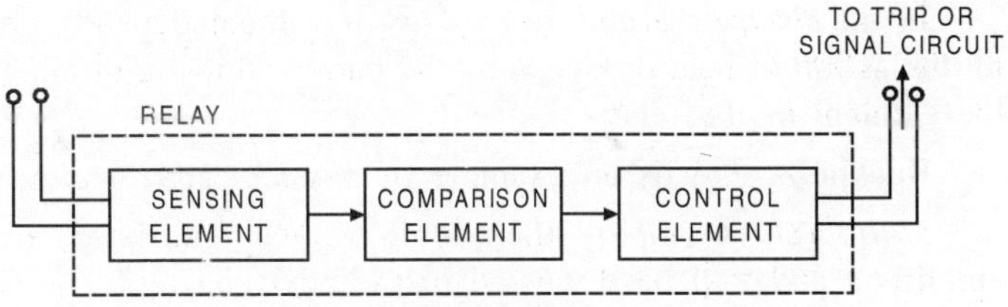

Fig. 9.1. ***Basic Elements of a Relay***

All the relays have three essential fundamental elements, as illustrated in Fig. 9.1.

1. ***Sensing element,*** sometimes also called the measuring element, responds to the change in the actuating quantity, the current in a protected system in case of overcurrent relay.

2. ***Comparing element*** serves to compare the action of the actuating quantity on the relay with a preselected relay setting.

3. ***Control element*** on a pick up of the relay, accomplishes a sudden change in the control quantity such as closing of the operative current circuit.

Relays may be classified as to which kind of *physical quantity* the *sensing element* respond to and may be classed as *electrical* (actuated by some electrical quantity such as current, voltage, power etc.); *mechanical* (actuated by pressure, velocity of outflow of a liquid or gas etc.), *thermal* (actuated by heating effect), *optical*, *acoustical*, and other types of relays.

In power system protective relaying, the most widely applied are electrical type relays. Generally speaking the electrical protective relays can be broadly, classified into two categories: (*i*) electromagnetic relays and (*ii*) static relays. In this chapter electromagnetic relays will be discussed. The study of static relays will be taken up in chapter 10.

According to principle of operation and construction the relays may be classified such as electromagnetic attraction type, electromagnetic induction type, electrodynamic type, moving coil type, electro-thermal type, physicoelectric type and static relays.

Electromagnetic attraction relays are of two types namely attracted armature type and solenoid type. In an *attracted armature type relay* operation depends on the movement of an armature under the influence of attractive force due to magnetic field set up by the current flowing through the relay winding while in a *solenoid type relay* operation depends on the movement of an iron plunger core along the axis of solenoid. Moving iron, moving plunger, attracted armature hinged, and balanced beam types of relays fall under this category. Such relays are actuated by dc or ac quantities.

In an *electromagnetic induction type relay* operation depends on the movement of a metallic disc or cylinder free to rotate by the interaction of induced currents and the alternating magnetic fields producing them.

In an *electrodynamic type relay* moving member consists of a coil free to rotate in an electromagnetic field as in case of a moving coil instrument.

In a *moving coil type relay* moving member consists of a coil free to rotate in the air gap of a permanent magnet.

In an *electro-thermal type relay* movement depends up on the action of heat produced by the current flowing through the element of the relay.

Buchholz relay is an example of *physicoelectric relay.*

Static relays employ thermionic valves, transistors or amplifiers and will be discussed in the next chapter.

According to their functions in the relay protective scheme, relays may be divided into main, auxiliary (or supplementary) and signal relays. The *main relays* are the protective elements which respond to any change in the actuating quantity. *e.g.,* current, voltage, power etc. The *auxiliary* or *supplementary relays* are those which are controlled by other relays to perform some auxiliary or supplementary function such as introduction of a time delay, multiplying the number of contacts, increasing the making or breaking capacity of the contacts of another relay, passing a command pulse from one relay to another relay, acting upon a circuit breaker closing or opening coil, energizing a signal or an alarm, etc. *Signal relays* function to register (by flag or target) the operation of some relay or relay protection, and control warning (visible) and alarm (audible) signal devices. The choice of signal relay is governed by the importance of the associated switchgear, the method of control and the number of alarm indications to be displayed.

According to actuating quantity, electrical relays are classified into current, voltage, power, reactance, impedance, frequency relays, and, according to the direction of the change they respond to, as minimum (or under) and maximum (over) relays. A *maximum (or over) relay* is one which responds to the actuating quantity when it exceeds a predetermined value as in the case of an overcurrent relay which operates when the current exceeds a predetermined value. A *minimum (or under) relay* is one which operates when the value of actuating quantity drops below a predetermined value as in case of an undervoltage relay which operates when the voltage falls below a particular level.

According to the connection of sensing element, relays are classed as primary and secondary relays. *Primary relays* are those whose sensing (or measuring) elements are directly connected in the circuit or element they protect, while the *secondary relays* are those whose measuring elements are connected to the circuit they protect through instrument transformers (CTs or PTs). Normally secondary relays are used in power system protection because of the high values of line voltages and currents.

As to how the relays act upon the circuit breaker, they are subdivided into *direct-acting relay* whose control element acts mechanically to operate a circuit breaker and *indirect-acting relay* whose control element operates an auxiliary (operative) power source circuit to control the circuit breaker.

Primary relays, despite the substantial simplification of the protective schemes afforded by them are employed only in the simplest schemes for low-voltage circuits. The application of primary relays to high-voltage circuits is rendered difficult by complications arising from their conditions of operation and servicing (adjustment, checking, supervision at the high voltage). Protective schemes using indirect-acting primary relays need, furthermore, the use of an operative current. Such arrangements are bulky and costly. Direct-acting secondary relays do not require an operative current, their armatures act directly upon the circuit breaker tripping device. Protection schemes using direct-acting secondary relays, notwithstanding difficulty in checking and adjustment, have found application because of their simple design.

The relay most frequently used is the indirect-acting secondary relay, notwithstanding the greater complexity of the protection scheme and the need for operative power. Such relays allow changes to be made in their settings and can be removed, or cut out of the circuit for repair or checking during operation of the circuit element they protect. Such relays, furthermore, possess a relatively high sensitivity and, due to the fact that the connections to the protected circuit elements are made through instrument transformers, are electrically isolated from the high primary circuit voltage. They are also to a much lesser degree than primary relays subject to the action of short-circuit currents.

As to time of action relays can be distinguished as relays without time delay and relays with time delay. Among the relays which fall under the category of non-time delay are *non-inertia types* with operating times of several thousandths of a second (milliseconds)—usually electronic relays; *quick-acting relays* with operating times ranging from 5 to 40 ms, and *ordinary relays* with operating times of 0.04 to 0.2 second. Among the time delay or time lag relays are relays with *non-regulable* time delay, obtained mainly by the use of short-circuited windings, eddy current sleeves etc., and also relays with *regulable* time delay. The latter type is termed time lag relays. Protective relay schemes use time lag relays having time delays ranging from about 0.2 s up to 5 to 20 s.

As to type of contacts, relays can be differentiated as relays with normally-open (n-o) and normally-closed (n–c) contacts. The relay contacts are called normally-open (n-o) if they are open (separated) when the relay winding is not energised, and normally-closed (n-c) if they are closed (in contact) when the relay winding is not energised. Use is also made of relays provided with two-way or double-throw contacts and with sliding contacts (which during relay operation close for a brief interval of time). When relays are operated, their contacts may close or open instantaneously or with time delay. Relay contacts, depending upon the type and function of the relay, as well as type and material of the contacts themselves, are capable of satisfactorily withstanding multitudes of operations in operative circuits of the corresponding voltage (usually 12 to 220 V) and switch circuits handling power flows ranging from several to hundreds of volt-amperes (VA).

According to applications the relays may be classified as :

(*i*) ***Overvoltage, Overcurrent and Overpower Relay.*** The relay operates when the voltage, current or power rises above a specified value.

(*ii*) ***Undervoltage, Undercurrent and Underpower Relay.*** The relay operates when the voltage, current or power falls below a specified value.

(*iii*) ***Directional or Reverse Current Relay.*** The relay operates when the applied current assumes a specified phase displacement with respect to the applied voltage and the relay is compensated for fall in voltage.

(*iv*) ***Directional or Reverse Power Relay.*** The relay operates when the applied current and voltage assume specified phase displacement and no compensation is allowed for fall in voltage.

(*v*) ***Differential Relay.*** The relay operates when some specified phase or magnitude difference between two or more electrical quantities occurs.

(*vi*) ***Distance Relay.*** In this relay the operation depends upon the ratio of the voltage to the current.

According to timing characteristics the relays can be divided into following classes :

(*i*) ***Instantaneous Relays.*** In these relays complete operation takes place after a very short (negligible) time duration from the incidence of the current or other quantity resulting in operation. The time of operation of such relays is lesser than 0.2 second.

(*ii*) ***Definite Time Lag Relays.*** In these relays the time of operation is sensibly independent of the magnitude of the current or of other quantity causing operation.

(*iii*) ***Inverse Time Lag Relays.*** In these relays the time of operation is approximately inversely proportional to the magnitude of the current or other quantity causing operation.

(*iv*) ***Inverse Definite Minimum Time (IDMT) Lag Relays.*** In these relays the time of operation is approximately inversely proportional to the smaller values of current or other quantity causing operation and tends to be a definite minimum time as the value increases without limit.

In induction type of relays the time lag may be achieved by employing a permanent magnet which is so arranged that the relay rotor cuts the flux between the poles of this magnet. Such a magnet is called the "drag magnet".

In other types of relays this may be achieved in two ways. Firstly, when a series connected overload trip coil is used, an oil dash pot can be attached in which is a piston connected to the lower end of the solenoid plunger of the relay. In case of fault when the plunger is pulled, the piston, immersed in the oil, retards plunger motion, thus provides the necessary drag or delay in the relay operation, as shown in Fig. 9.2 (*a*).

When the trip coil is transformer operated, time delay is achieved by connecting a time-limit fuse across the trip coil *i.e.*, in parallel. This fuse should have an inverse time-current characteristic and will normally carry the CT secondary current to bypass the trip coil. As and when sufficient current occurs to "blow" this fuse, the whole current is then transferred to the trip coil which operates to trip the circuit breaker. This is illustrated in Fig. 9.2 (*b*).

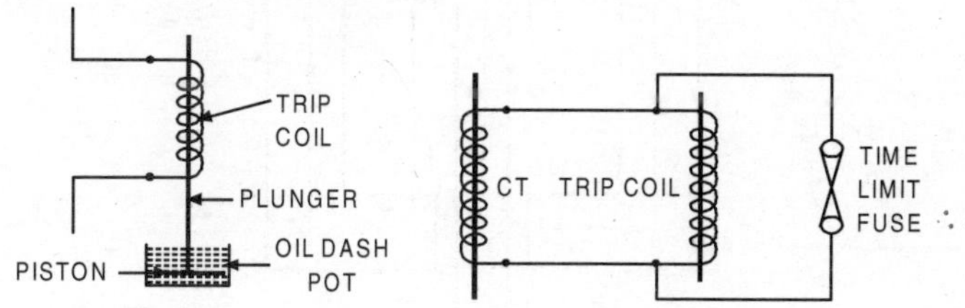

(*a*) (*b*)

Fig. 9.2

9.5. ELECTROMAGNETIC ATTRACTION RELAYS

These are the simplest type of relays and include plunger (or solenoid), hinged armature, rotating armature (or balanced beam) and moving iron polarized relays—all such relays are shown in Fig. 9.3. All of these relays operate on the same principle *i.e.*, in such relays the operation is obtained by virtue of an armature being attracted to the poles of an electromagnet or a plunger being drawn into a solenoid. The electromagnetic force exerted on the moving element is proportional to the square of flux in the air gap or the square of the current flowing through the coil. It is basically a single actuating quantity relay. Such relays respond to both ac and dc because operating torque is proportional to the square of current flowing through the coil. With dc the torque developed is constant and if it exceeds the restraining torque or force caused by the controlling spring, the relay operates reliably.

In case of ac quantity the electromagnetic force developed is given as

$$F_e = KI^2 = K(I_{max} \sin \omega t)^2$$

$$= \frac{1}{2} K[I^2_{max} - I^2_{max} \cos 2\omega t] \quad ...(9.1)$$

It shows that the electromagnetic force consists of two components, one constant independent of time ($\frac{1}{2} KI^2_{max}$) and another dependent upon time and pulsating at double the supply frequency ($\frac{1}{2} KI^2_{max} \cos 2\omega t$). The total electromagnetic force, therefore, pulsates at double the supply frequency. Since the restraining force produced by the spring is constant and the electromagnetic force developed is pulsating, the relay armature vibrates at double the power supply frequency. This causes the relay to hum and produce noise and also is a source of damage to the relay contacts. This leads to sparking and unreliable operation of the relay operative circuit contacts due to make and break of the circuit.

To overcome this difficulty in ac electromagnetic attraction relays, the flux developing the electromagnetic force is splitted into two fluxes acting simultaneously but differing in time phase, so that the resultant deflecting force is always positive and constant. This can be easily achieved either by

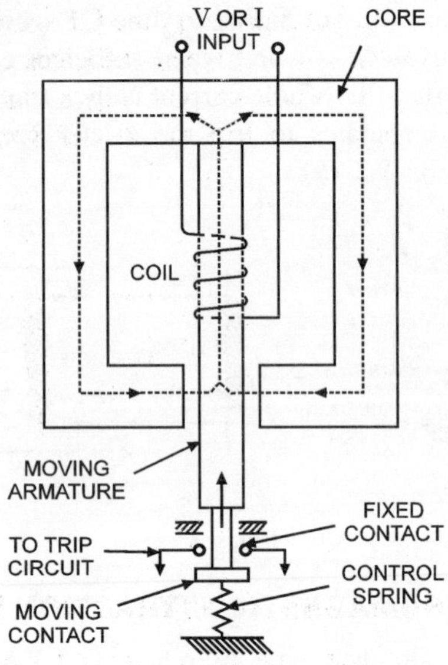

(a) Plunger Type

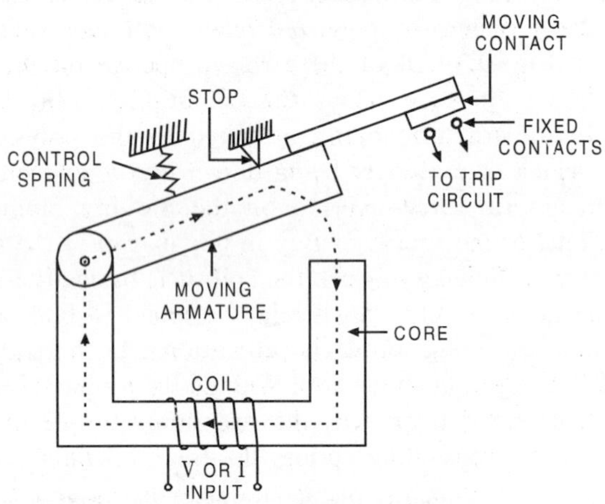

(b) Hinged Armature Type

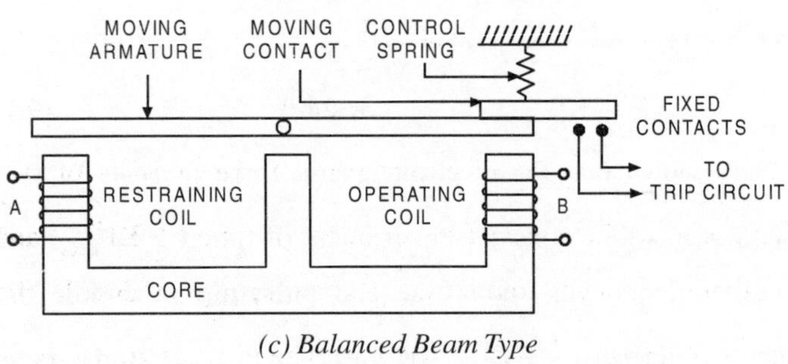

(c) Balanced Beam Type

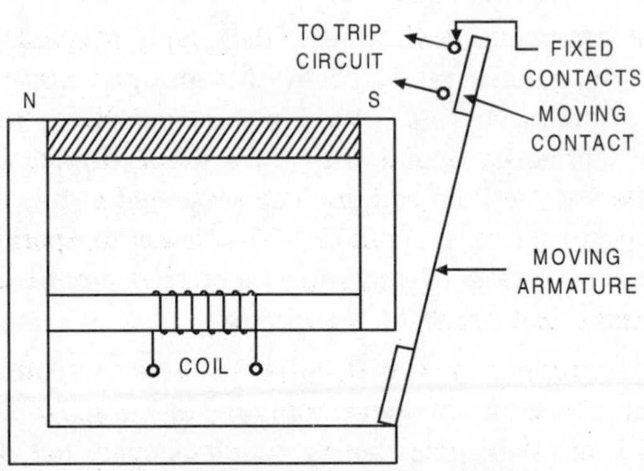

(d) Polarized Moving Iron Type

Fig. 9.3. *Electromagnetic Attraction Armature Relays*

providing two windings on the electromagnet having a phase shifting network, as shown in Fig. 9.4 or by putting shading rings on the poles of the electromagnet, as shown in Fig. 9.5. The latter method is more simple and is widely used.

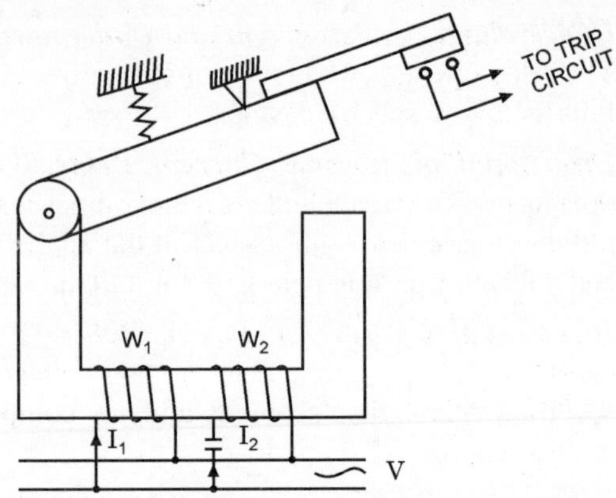

Fig. 9.4. *Hinged Armature Electromagnetic Relay With Two Windings*

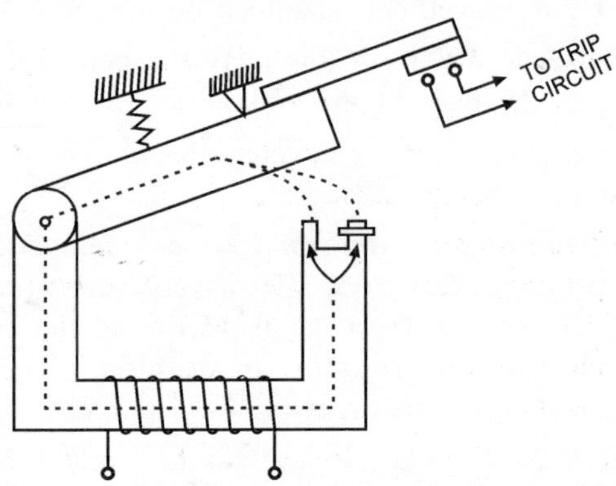

Fig. 9.5. *Hinged Armature Electromagnetic Relay With a Shading Coil on the Pole*

Hinged armature relays are mainly employed as auxiliary relays, *e.g.*, tripping relays, ac and dc voltage and current relays.

In the case of a *balanced beam type relay* shown in Fig. 9.3 (*c*) two quantities $|A|^2$ and $|B|^2$ are compared because electromagnetic forces developed vary as the square of the ampere-turns. Ratio of reset/operating current for such relays is low. If set for fast operation, it will have a tendency to overreach on transient conditions.

The sensitivity of the hinged armature relays can be increased for dc operation by the addition of a permanent magnet, as illustrated in Fig. 9.3 (*d*). This is known as a *polarized moving iron relay.* It is more robust in construction; most of these employ leaf-spring supported armatures.

Modern attraction armature relays are compact, robust, reliable and fast. VA burden depends on construction, setting etc. For a typical relay it is of the order of 0.2 to 0.6 VA for current range of 0.1 to 0.4 A. Such relays are usually instantaneous type but can be made a definite time lag or inverse-time lag by using an oil dash pot, an air-escapement chamber, a clockwork mechanism or by placing a fuse in parallel with the relay. These relays do not have directional feature unless they are provided with additional polarized coil.

The attraction armature relays have fast operation and fast reset because of small length of travel and light moving parts.

As these relays are fast and operate on dc as well as on ac they are affected by transients. The transients contain dc component in addition to ac wave. Therefore, though the steady-state value may be less than relays pick-up value, the relay may pick up during transient state.

Attraction armature relays have many applications in protection of ac and dc equipment. They are, however, sensitive to starting currents, load fluctuations and current surges.

Attraction armature relays can be designed to respond to over and under current, over/under voltage, for both dc and ac operations. They are employed as measuring or auxiliary relays.

Balanced beam relay is difficult to be designed over a wide range current because the force is proportional to square of current. Such relays are fast and instantaneous but can be made a definite time lag or inverse time lag by using an oil dash pot, an air-escapement chamber, a clockwork mechanism or by placing a fuse in parallel with the relay. Such relays can have operating time of the order of 0.02 second. High ratio of resetting quantity can be had. VA burden depends on application. In current balance type the VA burden is of the order of 0.2, 0.4, 0.6 VA for 0.1 to 0.6 A current range. This relay is being largely superseded by permanent magnet moving coil relay having better accuracy and lower VA burden.

9.6. ELECTROMAGNETIC INDUCTION RELAYS

Electromagnetic induction relays are the most widely used relays for protective relaying purposes involving only ac quantities. These relays operate on the simple principle of split-phase induction motor. Actuating force is developed on a moving element, that may be a disc or other form of rotor of non-magnetic current conducting material (such as aluminium), by the interaction of electromagnetic fluxes with eddy currents that are induced in the rotor by these fluxes.

Figure 9.6 illustrates how force is developed in a section of a rotor that is pierced by two adjacent ac fluxes. Various quantities are shown at an instant when both fluxes are directed downward and are increasing in magnitude. Each flux induce voltage around itself in the rotor, and currents flow in the rotor under the influence of the two voltages. The current produced by one flux reacts with the other flux, and vice versa, to produce forces that act on the rotor.

The quantities shown in Fig. 9.6 may be expressed as

$$\phi_1 = \phi_{1max} \sin \omega t$$

$$\phi_2 = \phi_{2\,max} \sin (\omega t + \theta)$$

where θ is the angle by which ϕ_2 leads ϕ_1

It may be assumed with negligible error that the paths in which the rotor currents flow have negligible self-inductance, and therefore rotor currents are in phase with their respective induced voltages. The phasor diagram is shown in Fig. 9.7. Now

$$i_1 \propto e_1 \propto \frac{d\phi_1}{dt} \propto \phi_{1\,max} \cos \omega t$$

$$i_2 \propto e_2 \propto \frac{d\phi_2}{dt} \propto \phi_{2\,max} \cos (\omega t + \theta)$$

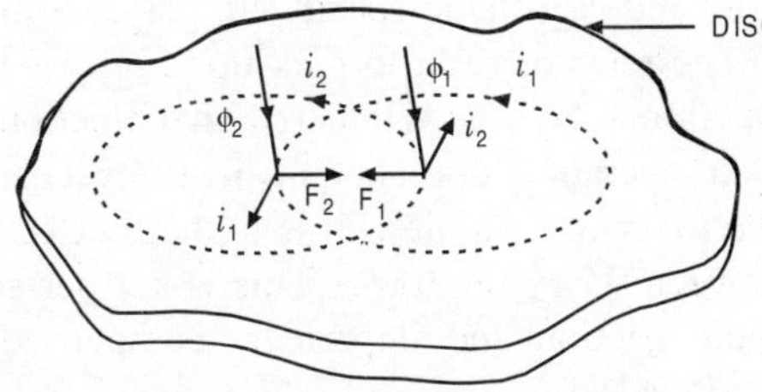

Fig. 9.6. *Production of Torque in an Induction Relay*

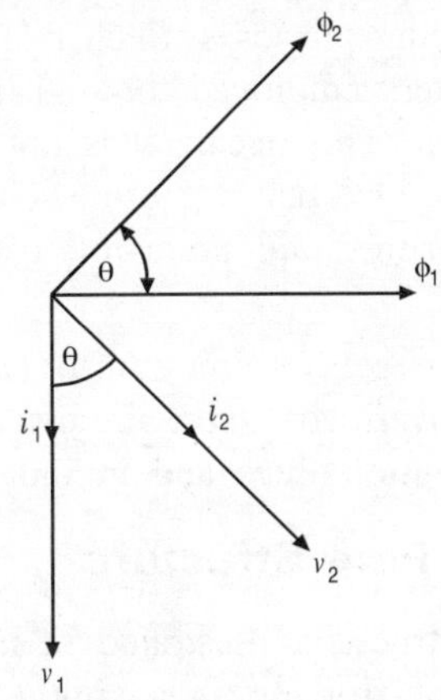

Fig. 9.7. *Phasor Diagram For an Induction Relay*

Since the two forces (F_1 and F_2) developed are in opposition, as illustrated in Fig. 9.6, therefore net force acting on the movable element is given as

$$F = (F_2 - F_1)$$
$$\propto \phi_2 i_1 - \phi_1 i_2$$
$$\propto \phi_{1max}\, \phi_{2max}\left[\cos \omega t \sin(\omega t + \theta) - \sin \omega t \cos(\omega t + \theta)\right]$$

$$\text{or } F \propto \phi_{1\,max}\, \phi_{2\,max} \sin \theta \quad \text{...(9.2)}$$

The following points may be noted from Eq. (9.2)

(*i*) The net force is same at every instant as the term ωt is not involved in the expression for the force developed. It is most significant. This fact does not depend upon the assumptions made in arriving at Eq. (9.2). The action of a relay under the influence of such a force is positive and free from vibration.

(*ii*) The greater the phase angle θ between the two fluxes, the greater is the net force applicable to the movable element. Obviously the force developed will be maximum when phase angle θ is 90°.

(*iii*) Also the force developed will be more when the resistance R of the annular ring is low because $i \propto v/R$ *i.e.*, the movable element must be of low resistance material such as copper or aluminium. From torque-weight ratio point of view the movable element should be of aluminium alone.

(*iv*) The direction of net force and hence the direction of motion of movable element depends upon which flux is leading.

The different types of structures that have been used for obtaining the phase difference in the fluxes and hence the operating torque in induction relays are (*i*) shaded pole structure (disc type) (*ii*) watt-hour meter or double winding structure (disc type) and (*iii*) induction cup structure.

High, low and adjustable speeds are possible in induction type relays. The ratio of reset to pick-up is inherently high in case of induction relays as compared to attraction armature relays as their operation does not involve any change in the air gap of the magnetic circuit as it is in the case of latter. The ratio lies between 95% and 100%. This is not perfectly 100% because of the friction and imperfect compensation of the control spring torque.

The accuracy of an induction type relay recommends it for protective relaying purposes. Such relays are comparable in accuracy to meters employed for registration of electrical energy consumption. This accuracy is not a consequence of the induction principle, but because such relays invariably employ jewel bearings and precision parts that minimize friction.

Induction type relays are employed extensively for protection against overloads, short circuits, and earth faults on transmission/distribution lines and in industrial plants.

9.6.1. Shaded Pole Structure

The shaded pole structure, illustrated in Fig. 9.8, is usually actuated by current flowing in a single coil wound on a magnetic structure containing an air gap. The air-gap flux produced by the actuating current is split into two fluxes displaced in time and space by a so-called *shading ring*, generally of copper, that encircles part of the pole face of each pole, as shown in Fig. 9.8. The disc is normally made of aluminium so as to have low inertia and, therefore, needs less deflecting torque for its movement. The two rings have currents induced in them by the alternating flux of the electromagnet and the magnetic fields developed by these induced currents cause the flux, in the portions of the iron surrounded by the rings to lag in phase by 40° to 50° behind the flux in the unshaded portions of the pole. As already proved, the driving torque is given as

$$T \propto \phi_s \phi_u \sin \theta$$

Assuming that the flux in the shaded portion of the pole, ϕ_s and the flux in the unshaded portion of the pole, ϕ_u to be proportional to the actuating current in the coil, I.

$$T \propto I^2 \sin \theta$$

i.e., the driving torque varies as the square of the current flowing through the relay coil.

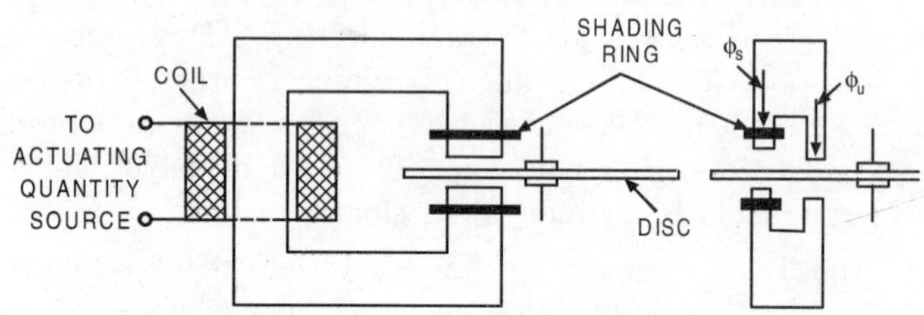

Fig. 9.8. ***Shaded Pole Construction of Induction Disc Relay***

The shading rings may be replaced by coils if control of the operation of a shaded pole relays is desired. If the shading coils are short-circuited by a contact of some other relay, torque will be developed; but, if the coils are open-circuited, no torque will be developed because there will be no phase-splitting of the flux. Such torque control is used where directional feature is required, which will be described later.

The control torque is provided with the help of a control spring attached to the disc spindle. With the movement of the disc towards closing of the contacts spring torque increases slightly with the winding of the spring. The relay disc is so shaped that as it turns towards the pick-up position (closing of contacts), there is increase in the area of the disc between the poles of the actuating structure which causes increase in eddy currents and, therefore, increase in electrical torque that just balances the increase in the control spring torque. The shape of the disc usually is that of a spiral. Since the shape of disc is not perfectly circular, suitable balance weight is provided on the part of the disc which has smaller area. A permanent magnet of high retentivity steel is provided to give eddy current braking to the disc. This is necessary to reduce to a minimum the over-run of the disc in case the current or voltage providing the driving torque stops before the operation has been completed. A modern induction disc relay will have an over-run of not more than 2 cycles on interruption of 20 times the setting quantity.

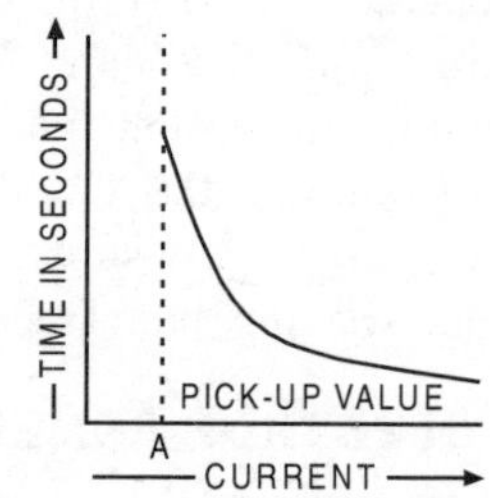

Fig. 9.9. ***Inverse Characteristic***

Modern induction disc relays are robust and reliable. VA burden depends upon rating. It is of the order of 2.5 VA. The time-current characteristics of induction relays are inverse characteristics, as shown in Fig. 9.9, *i.e.*, the time reduces as current increases. The current setting can be changed by taking the suitable number of turns. Higher current setting will require smaller number of turns so as to give fixed ampere-turns required for developing the minimum torque for the movement of the disc. The time setting can be changed by changing the relative position of contacts by adjusting the length of travel of moving contacts. This is known as *time multiplier setting*. The higher the time multiplier setting the greater is the operating time. The effect of dc offset may be neglected with inverse time single quantity induction relays, because they are generally slow. The dc offset may affect fast relays. Ratio of reset to pick-up is high (above 95%) because operation does not involve any change in air gap.

9.6.2. Watt-hour Meter Structure

Watt-hour meter type induction disc relay gives the same results as given by a shaded pole type. The construction of this relay is similar to the watt-hour meter commonly used everywhere. It consist of an E-shaped electromagnet and a U-shaped electromagnet with a disc free to rotate in between. A phase displacement between the fluxes produced by the two magnets is obtained either by having different resistance and inductance for the two circuits or by energizing them from two different sources whose outputs are relatively displaced in phase. Many variations in the design and construction are possible to suit the required conditions.

The E-shaped electromagnet carries two windings; the primary and the secondary. The primary winding carries relay current I_1 while the secondary winding is connected to the windings of U-shaped electromagnet (Fig. 9.10). The primary current induces emf in the secondary and so circulates a current I_2 in it. The flux ϕ_2 induced in the U-shaped or lower magnet by the current in the secondary winding of the E-shaped or upper magnet will lag behind flux ϕ_1 by an angle θ. The two fluxes ϕ_1 and ϕ_2 induced in upper and lower magnets respectively differing in phase by angle θ will develop a driving torque on the disc proportional to $\phi_1\phi_2 \sin\theta$.

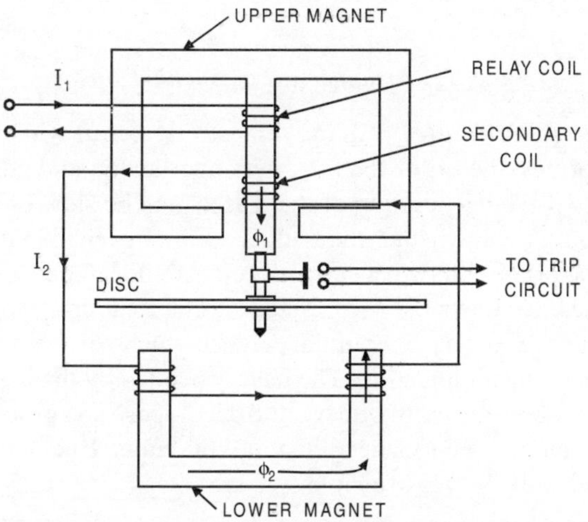

Fig. 9.10. ***Watt-hour Meter Type Induction Disc Relay***

Most of the modern induction relays are of this type. The advantage of this type of construction is that it can provide larger phase angle between ϕ_1 and ϕ_2 and thus higher torque, being proportional to $\sin\theta$. An important feature of relay of this type is that its operation can be controlled by opening or closing the secondary winding circuit. If this circuit is opened, no torque will be developed and thus the relay can be made inoperative.

9.6.3. Induction Cup (Electromagnetic) Relay

These relays operate on the same principle as the induction motor. The relay has two, four or more electromagnets energized by the relay coils. A stationary iron core is placed between these electromagnets (four in number) as illustrated in Fig. 9.11. The rotor is a hollow metallic cylindrical cup which is free to rotate in the gap between the electromagnets and the stationary iron core. The rotating field is produced by two pairs of coils wound on four poles as shown. The rotating field induces currents in the cup causing it to rotate in the same direction. The rotation depends on the direction of rotation of the field and the magnitude of the applied voltage and/or currents and the phase angle between them. A control spring, and the back stop or closing of the contacts carried on an arm are attached to the spindle of the cup so as to prevent the continuous rotation.

The operating time characteristic depends on the type of structure. These relays have inverse time characteristics.

Induction cup structures are more efficient torque producers than either the shaded pole or the watt-hour meter structures. Therefore, such relays are very fast in operation and may have an operating time of less that 0.01 second.

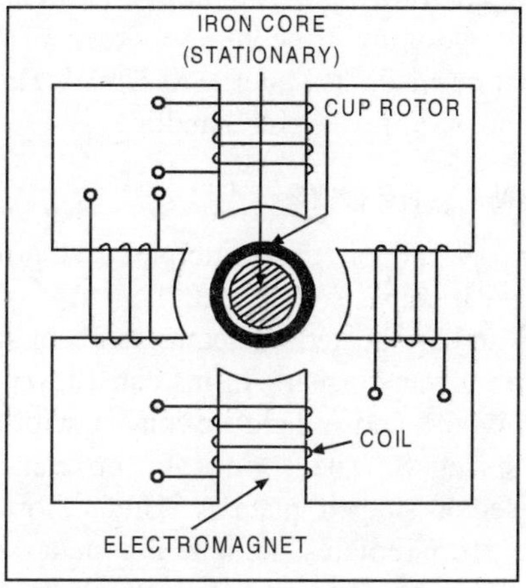

Fig. 9.11. ***Induction Cup Relay***

9.7. PERMANENT MAGNET MOVING COIL RELAY

The arrangement is shown in Fig. 9.12. In this relay, the coil is free to rotate in the magnetic field of a permanent magnet. The torque is developed by the interaction between the field of the permanent magnet and the coil field developed due to flow of actuating current in the coil. The time-current characteristic of such relays is an inverse-time characteristic. The relay responds to dc only but it can be used with ac in conjunction with a rectifier. The characteristic can be varied by adjusting the control spring. The time setting is obtained by adjusting the position of the contact.

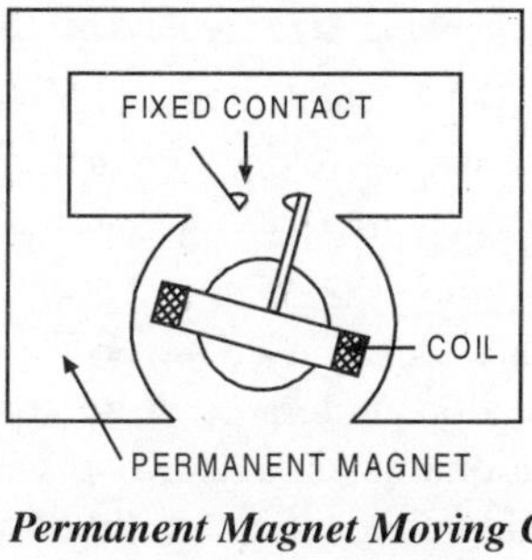

Fig. 9.12. ***Permanent Magnet Moving Coil Relay***

The operating torque is proportional to the actuating current. Such a relay has uniform torque for different positions of the coil and therefore, can be accurately set. Theoretically the reset value is equal to the operating value.

9.7.1. Moving Coil Relay With Axial Moving Coil

Such a relay is shown in Fig. 9.13. The coil is supported

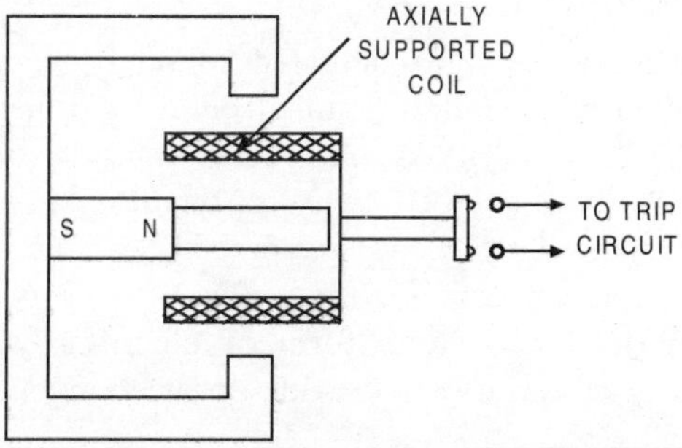

Fig. 9.13. ***Moving Coil Relay With Axial Moving Coil***

axially and moves horizontally when energized by the actuating current. Such relay is faster than the rotating coil type owing to small travel and lighter parts. Relay of this type may have an operating time of the order of 0.03 second. Sensitivity can be made as small as 0.1 mW. These relays are delicate and so require careful handling.

9.8. THERMAL RELAYS

These relays operate on the principle of thermal effect of electric current. It consists of bimetallic strips which are used in small sizes and are heated by heating coils or strips supplied through a current transformer. An insulated lever arm carrying a contact is pivoted and is held in contact with this trip with the help of spring S. The spring tension can be varied by rotating the sector shaped plate A. Under normal operating condition the strip remains straight but under the action of fault current the strip is heated and bent and the tension of the spring is released. Thus the relay contacts are closed which energises the trip circuit.

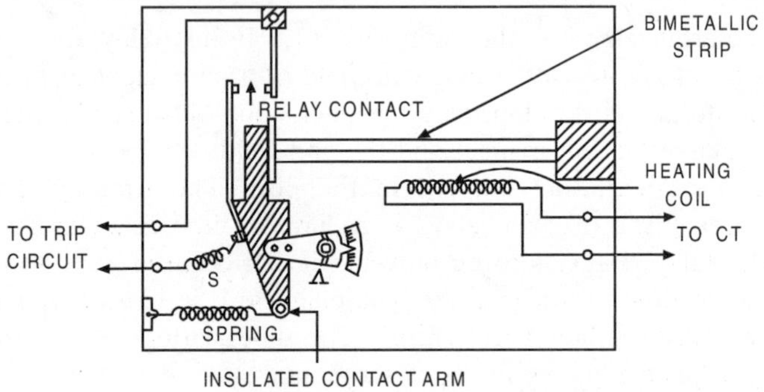

Fig. 9.14. ***Schematic Arrangement of Thermal Relay***

The construction of the bimetal element consists of two nickel alloyed steel strips, welded together. These strips have a high heat resistivity and are free from thermal secondary effects and ageing.

These relays are mostly used for protection of low-voltage squirrel cage induction motors or dc motors of lower output ratings. The limitation of such a relay is the short-time overload withstand. Its heating element is usually designed to withstand short-time overload say up to 6 or 7 times full-load current. The thermal relay is not suitable for operation on short circuit as it will burn the element sufficiently before the strip may deflect so as to close the contacts. This type of relay is used in conjunction with instantaneous short-circuit relays of high setting or suitably graded time limit fuses.

9.9. BUCHHOLZ RELAY

Buchholz relay is a gas actuated relay. It is practically universally used on all oil immersed transformers having rating more than 500 kVA*. Such relay can only be fitted to the transformers equipped with conservator tanks as it is installed in between the conservator tank and the main tank *i.e.*, in the pipe connecting the two (Fig. 9.15). It is employed in conjunction with some form of electrically operated protective gear because it provides protection only against transformer internal faults and does not respond to external bushing or cable connection faults.

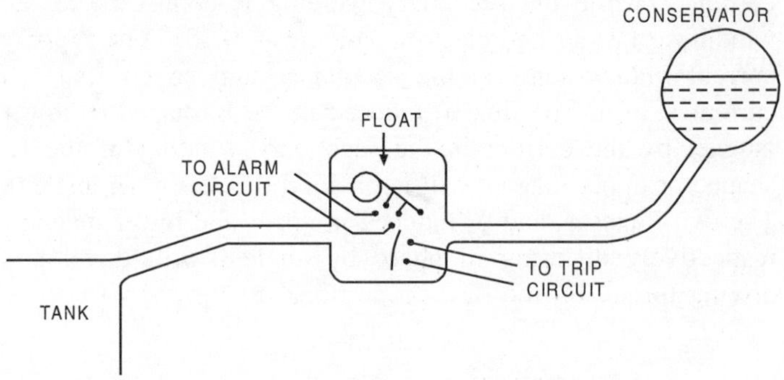

Fig. 9.15. ***Arrangement of Buchholz Relay***

Working Principle. Whenever a fault occurs inside the transformer, the oil of the tank gets overheated and gases are generated. The generation of the gases may be slow or violent depending upon whether the fault is a minor or incipient one or heavy short circuit. Most short circuits are developed either by impulse breakdown between adjacent turns at the end turns of the winding or as a very poor initial contact which will immediately heat to arcing temperature. The heat generated by the high local current causes the transformer oil to decompose and produce gas which can be used to detect the winding faults. Buchholz relay operates only on this principle.

Construction. The construction of a Buchholz relay is shown in Fig. 9.16. It consists of two hinged floats in a metallic chamber located in the pipe connection between the conservator and the transformer tank. One of the floats is near the top of the chamber and actuates the mercury switch connected to the external alarm circuit. The other float is opposite the orifice of the pipe to the transformer and actuates the mercury switch connected to the tripping circuit.

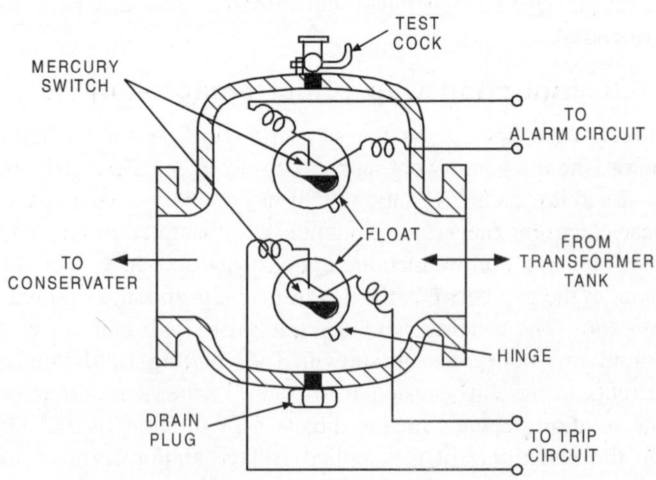

Fig. 9.16. ***Buchholz Relay***

Operation. When a minor fault occurs, heat is produced due to current leakage, some of the oil in the transformer tank evaporates and some vapours collect in the top of the chamber while passing to the conservator tank. When a predetermined amount of vapours accumulate in the top of the chamber, the

* Buchholz relays are not provided for transformers below 500 kVA from the point of view of economic considerations.

oil level falls, the mercury type switch attached to the float is tilted and so closes the alarm circuit and rings the bell. Thereby the operator knows that there is some incipient fault in the transformer. The transformer is disconnected at the earliest possible and the gas sample is tested. The testing of gas provides clue regarding the type of insulation failure. Buchholz relay gives an alarm so that the transformer can be disconnected before the incipient fault grows into a severe one. A release cock is provided at the top of the chamber so that after operation the pressure in the chamber can be released and the gas emitted to allow the chamber to refill with oil.

When a severe fault occurs large volume of gas is evolved so that the lower float containing a mercury switch mounted on a hinged type flat is tilted and the trip coil is energized. There after the transformer is removed from the service.

The main advantage of Buchholz relays is that they indicate incipient faults, for example, between turns faults or core heating and so may enable a transformer to be taken out of service before serious damage occurs.

However, Buchholz relays have some limitations also, as given below.

Only faults below oil levels are detected. For faults above the oil level, this relay is ignorant. These relays do not protect the connecting cables which must therefore have a separate protection.

Setting of the mercury switch cannot be very sensitive, otherwise there can be a false operation by vibrations, earthquakes, mechanical shocks to the pipe, sitting of birds etc.

The relay is slow, minimum operating time is about 0.1 second, average time 0.2 second. This is not desirable.

9.10. OVERCURRENT RELAYS

Depending upon the time of operation, already defined in Art. 9.2, overcurrent relays may be categorized as (*i*) instantaneous overcurrent relays (*ii*) inverse time overcurrent relay (*iii*) definite time overcurrent relay (*iv*) inverse definite minimum time (IDMT) overcurrent relay (*v*) very inverse overcurrent relay and (*vi*) extremely inverse overcurrent relay.

(*i*) ***Instantaneous Overcurrent Relay.*** An instantaneous overcurrent relay is one in which no intentional time delay is provided for operation. In such a relay, the relay contacts close immediately after the current in the relay coil exceeds that for which it is set. Although there will be a short time interval between the instant of pick-up and the closing of the relay contacts, no intentional time delay is provided. This characteristic can be achieved with the help of hinged armature relays. Such relay has a unique advantage of reducing the time of operation to a minimum for faults very close to the source where the fault current is the greatest. The instantaneous relay is effective only where the impedance between the relay and source is small compared with the impedance of the section to be protected.

One of the most important considerations in overcurrent and overvoltage protection is the speed of operation. With hinged armature relays, the time of operation of 0.01 second at three times the setting can be obtained. Such relays are employed for restricted earth fault and other types of circulating current protection. With so fast an operation it is likely that the relay may operate on transients beyond the normal range of setting.

(*ii*) ***Inverse-Time Overcurrent Relay.*** An inverse time relay is one in which the operating time is approximately inversely proportional to the magnitude of the actuating quantity. Figure 9.9 illustrates the time-current characteristics of an inverse current relay. At values of current less than pick-up value, the relay never operates. At higher values, the operating time of the relay decreases steadily with the increase of current. The more pronounced the effect is the more inverse the characteristic is said to be. In fact, all time-current curves are inverse to a greater or lesser degree. They are normally more inverse near the pick-up value of the actuating quantity and become less inverse as it is increased.

The operating time of all overcurrent relays tends to become asymptotic to a definite minimum value with increase in the value of actuating quantity. This is inherent in electromagnetic relays due to saturation of the magnetic circuit. So by varying the point of saturation different characteristics are obtained. These are (*i*) definite time (*ii*) inverse definite minimum time (*iii*) very inverse and (*iv*) extremely inverse, as shown in Fig. 9.17.

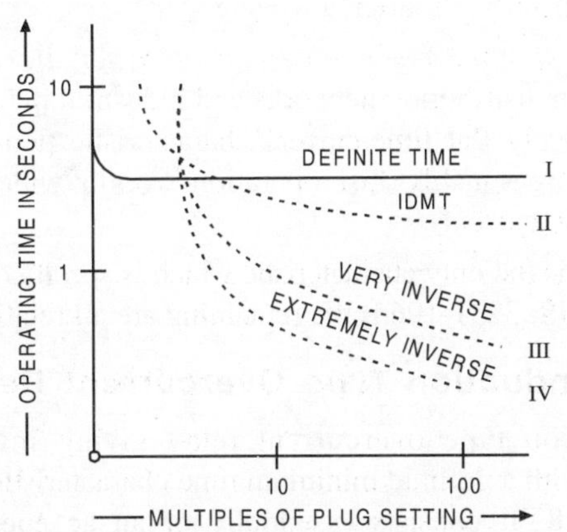

Fig. 9.17. ***Characteristics of Various Overcurrent Relays***

These characteristics can be obtained by induction disc and induction cup relays, already explained in Art. 9.6.

(*iii*) If the core is made to saturate at a very early stage, the time of operation remains same over the working range. This characteristic is shown by curve I in Fig. 9.17 and is known as **definite time characteristic.** Such a relay operates after a specified time irrespective of the magnitude of the fault current.

The definite time relays are used in (*i*) radial or loop circuits having a few sections (*ii*) as backup protection for other types of protection and (*iii*) on systems with wide variations of fault current due to source impedance.

Selectivity amongst such relays is obtained if there is difference of 0.5 s in the time settings of the two successive relays.

(*iv*) ***Inverse Definite Minimum Time (IDMT) Relays.*** Such a relay is one in which operating time is approximately inversely proportional to fault current near pick-up value and becomes substantially constant slightly above the pick-up value of the relay, as illustrated by curve II in Fig. 9.17. This is achieved by using a core of the electromagnet which gets saturated for currents slightly greater than the pick-up current.

(*v*) ***Very Inverse Relay.*** In such a relay the saturation of the core occurs at a still later stage, as illustrated by curve III in Fig. 9.17. This curve is known as *very inverse characteristic* curve. The time-current characteristic is inverse over a greater range and after saturation tends to definite time. Relays with very inverse time-current characteristics are employed on feeders and long sub-transmission lines.

(*vi*) ***Extremely Inverse Relay.*** The curve IV in Fig. 9.17 illustrate extremely inverse characteristic *i.e.,* core saturation occurs at a very late stage. The equation describing the curve IV in the figure is approximately of the form $I^2t = K$ where I is the operating current and t is the operating time. Such relays are quite suitable for the protection of transformers, cables etc, as it is possible to achieve accurate discrimination with fuses and autoreclosures in their case, which can seldom be made selective with standard IDMT relays. This is because of their ability to ride through starting currents and surges, providing at the same time fast operation under fault conditions. They are, thus more suitable for installations with large inrush currents after an outage.

Relays with *inverse time-current characteristics* are widely employed in distribution networks and industrial plant systems. Their relatively flat time-current characteristic permits them to achieve reasonably fast operation over a wide range of short-circuit currents.

IDMT is the only characteristic which is specified by Indian Standards (IS: 3231-1965) the remaining are all relative curves.

9.10.1. Induction Type Overcurrent Relay

An induction type overcurrent relay giving inverse-time operation with a definite minimum time characteristic is shown in Fig. 9.18. It consists essentially of an ac energy meter mechanism with slight modification to give required characteristics. The relay has two electromagnets. The upper electromagnet has two windings, one of these is primary and is connected to the secondary of a CT in the line to be protected and is tapped at intervals. The tappings are connected to a plug setting bridge by which the number of turns in use can be adjusted, thereby giving the desired current setting. The plug bridge is usually arranged to give seven sections of tappings to give overcurrent range from 50% to 200% in steps of 25%. If the relay is required to response for earth fault the steps are arranged to give a range from 10 % to 70 % or 20 to 80% in steps of 10%. The values assigned to each tap are expressed in terms of percentage of full-load rating of CT with which the relay is associated and represents and value above which the disc commences to rotate and finally closes the trip circuit. Thus pick-up current equals the rated secondary current of CT multiplied by current setting. For example suppose that an overcurrent relay having a current setting of 150% is connected to a supply circuit through a CT of 500/5 A. The rated secondary current of CT is 5 A and therefore, the pick-up value will be 1.5 × 5 *i.e.,* 7.5 A. It means that with above current setting, the relay will actually operate for a relay current equal to or greater than 7.5 A. Similarly for current settings of 50,100 and 200% the relay will operate for relay currents of 2.5 A, 5 A and 10 A respectively. Adjustment of current setting is made by inserting a pin between the spring loaded jaw of the bridge socket at the tap value required. When the pin is withdrawn for the purpose of changing the setting value while the relay in service, the relay automatically adopts higher setting, thus the CT's secondary is not open-circuited.

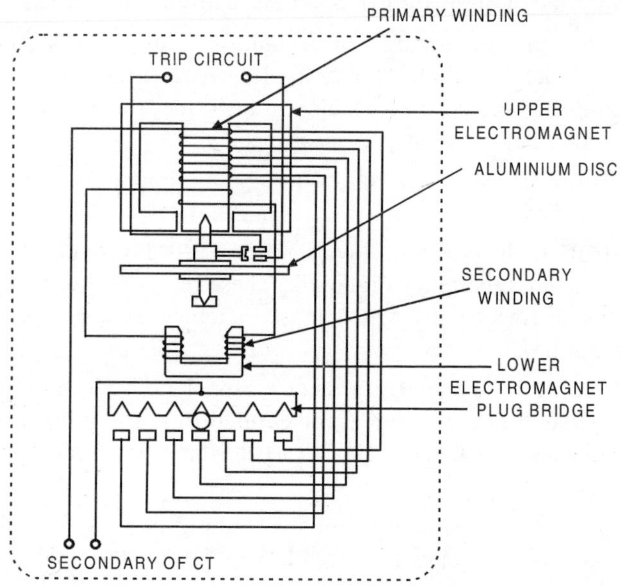

Fig. 9.18. ***Induction Type Non-Directional Overcurrent Relay***

The second winding is energized by induction from the primary, and is connected in series with the winding on the lower magnet. By this arrangement, leakage fluxes of upper and lower electromagnets are sufficiently displaced in space and phase to set up a rotational torque on the aluminium disc suspended between the two magnets, as in the shaded pole induction disc motor. This torque is controlled by the spiral spring and also sometimes by a permanent magnet brake on the disc. The torque is given by the expression

$$T = K_1 I^2_{rms} - K_2$$

where I_{rms} is the current through the coil and K_2 is the restraining torque of the spring. The disc spindle carries a moving contact which bridges two fixed contacts (trip circuit contacts) when the disc has rotated through a preset angle. The angle can be set to any value between 0° and 360° and thereby giving desired time setting. This adjustment is known as *time setting multiplier.* Time multiplier setting is generally in the form of an adjustable backstop which decides the arc length through which the disc travels, by reducing the length of travel, the operating time is reduced. The time setting multiplier is calibrated from 0 to 1 in steps of 0.05. These figures do not represent the actual operating times but are multipliers to be used to convert the time known from the relay name plate curve (time-PSM curve) into the actual operating time. Thus if time setting is 0.2 and the operating time obtained from the time-PSM curve of the relay is 5 seconds, then actual operating time of the relay will be equal to 0.2 × 5 *i.e.,* 1 second.

Since the time required to rotate the disc through a preset angle depends upon the torque which varies as the current in the primary circuit, therefore, more the torque lesser will be the time required. So the relay has inverse time characteristic.

In more recent designs the definite minimum time characteristic is obtained by saturating iron in the upper electromagnet so that there is practically no increase in flux after the current has reached a certain value and any further increase in current will not affect the relay operation.

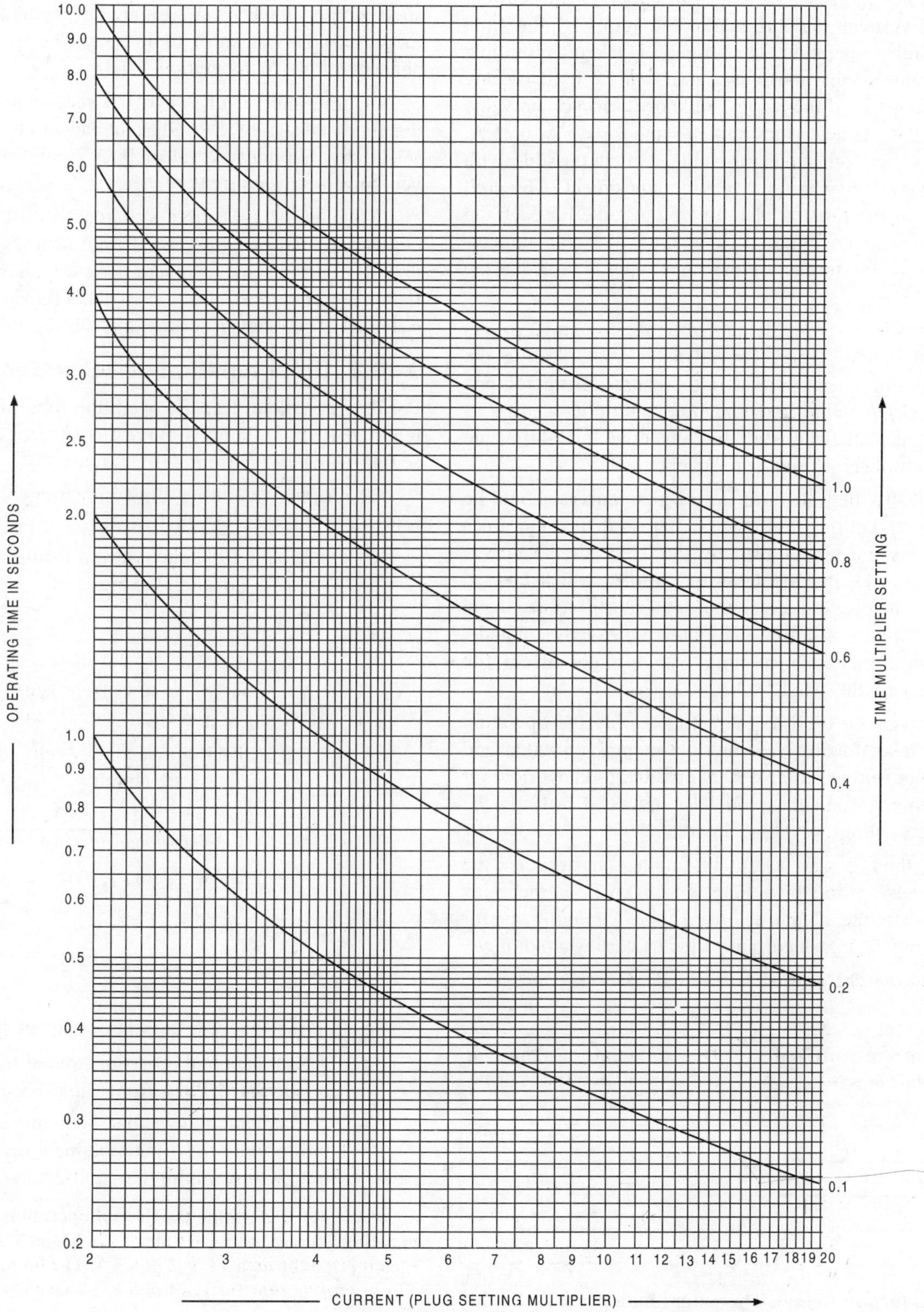

Fig. 9.19. ***Current-Time Characteristics of an Overcurrent Induction Disc Relay***

The ratio of reset to pick-up is inherently high in induction relays because their operation does not involve any change in the air gap. It lies between 95% and 100%.

Current-Time Characteristics of an Overcurrent Induction Disc Relay. A set of typical time-current characteristics of the above relay is given in Fig. 9.19. The horizontal scale is marked in terms of plug setting multiplier and represents the number of times the relay current is in excess of the current setting. The vertical scale is marked in terms of the time required for relay operation. The abscissa is taken as multiple of pick-up value so that the same curves can be used for any value of pick-up *i.e.,* if the curves are known for pick-up value of 5 A, then the characteristics remain same for 2.5 A, 6.25 A, 7.5 A, 10 A or any other pick-up value. This is possible with induction type relays where the pick-up adjustment is by coil, because the ampere-turns at pick-up are the same for each tap and hence at a given multiple of pick-up, the coil ampere-turns, and hence the torque are the same regardless of the tap used.

These curves are normally plotted on log-log graph papers as illustrated in the figure. The advantages of plotting the curves on log-log sheets is that if the characteristic for one particular pick-up value and one time multiplier setting is known, then the characteristics for any other pick-up value and time multiplier settings can be obtained.

Theoretically the time ordinates of these curves should be in proportion to the time multiplier setting so that, if the times for a given current are divided by the TMS all the curves should coincide. Owing to inertia of the disc which takes a little time for the disc to accelerate from standstill to its steady speed at low values of current they may not exactly coincide. It introduces some error in time which might affect the discrimination of the whole scheme.

The curves are used to estimate not only the operating time of the relay for a given multiple of pick-up value and time multiplier setting but also it is possible to know how far the relay moving contact would have travelled towards the fixed contacts within any time interval.

This method is also useful in finding out whether the relay will pick up and how long it will take for the relay operation when the actuating quantity is changing as for example during the inrush current period of starting a motor etc.

Current-time characteristics of an IDMT relay are given below:

Operating current expressed as a multiple of setting	**Operating time in seconds at the maximum time setting *(TMS = 1.0)***
20	2.2
10	3.0
5	4.0
2	10.0

Setting of Induction Relays. The setting for a phase fault relay is usually of the order of 150-200% of the full-load current. An induction relay will not operate at a current equal to or less than its setting and minimum operating current of the relay must not exceed 130% of the setting.

The setting of the earth fault relays should be in the range of 20 to 80%. An earth fault relay is subject to maloperation if its setting is too low. This maloperation can be due to unbalance in load currents, unbalance in output of CTs, switching surges or saturation of CTs during phase faults. As such, settings lower than 20% are not recommended. Also, the more sensitive an earth fault relay, the greater will be the chances of its maloperation.

A time interval of 0.4 to 0.5 second may be allowed in the time settings of two adjacent induction relays for proper selectivity. Shorter interval of 0.35 second may be used with very inverse overcurrent relays.

Whenever a protective system comprises of a series of overcurrent relays with graded time settings, it is imperative that the current-time characteristics of all such relays must flatten out in the same fashion so that a flawless relay operation is ensured howsoever severe may be the nature of the fault.

9.10.2. Determination of Relay Operating Time

For determination of actual operating time of a relay, the data required is (*i*) Time-PSM curve (*ii*) Current setting (*iii*) Time setting (*iv*) Fault current (*v*) CT ratio.

The actual operating time of a relay is determined by following the steps given below:

(*i*) Determination of relay current from fault current I_f and CT ratio $x : y$ from the expression

$$\text{Relay current, } I_R = \frac{I_f \times y}{x} \qquad \text{...(9.3)}$$

(*ii*) Determination of relay current setting multiplier (*i.e.,* PSM) which is given as

$$\text{PSM} = \frac{\text{Fault current in relay coil}}{\text{Pick-up value}}$$

$$= \frac{I_R}{\text{Current setting} \times \text{rated secondary current of CT}}$$

$$= \frac{I_R}{\frac{I_p}{100} \times y} = \frac{I_R \times 100}{I_p \times y}$$

$$= \frac{I_f \times 100 \times y}{x \times I_p \times y} = \frac{I_f \times 100}{x \times I_p} \qquad \text{...(9.4)}$$

where I_p is the per cent current setting of the relay

(*iii*) Determination of operating time of relay corresponding to calculated PSM from Time-PSM curve.

(*iv*) Determination of actual operating time of relay by multiplying the time obtained in step (*iii*) by the time setting multiplier in use.

Example 9.1. Determine the time of operation of an IDMT relay of rating 5 A and having setting of 125% and TMS = 0.5. The relay is connected through a CT of 400/5 A. The fault current is 4,000 A. The operating time for PSM of 8 is 3.2 seconds.

[*U.P. Technical Univ. Switchgear and Protection 2007-08*]

Solution : Fault current, I_f = 4,000 A

CT ratio = 400:5

$$\text{Relay current, } I_R = \frac{I_f}{\text{CT ratio}} = \frac{4{,}000}{400/5} = 50 \text{ A}$$

Pick-up value of relay

= Current setting × Rated secondary current of CT

$$= \frac{125}{100} \times 5 = 6.25 \text{ A}$$

Plug setting of multiplier of the relay, PSM

$$= \frac{I_R}{\text{Pick-up value of relay}} = \frac{50}{6.25} = 8$$

Time corresponding to the PSM of 8 from the given data is 3.2 seconds.

So actual operating time

= 3.2 × Time setting multiplier

= 3.2 × 0.5 = 1.6 seconds **Ans.**

Example 9.2. An IDMT type overcurrent relay is used to protect a feeder through 500/1 A CT. The relay has a PS of 125% and TMS = 0.3. Find the time of operation of the said relay if a fault current of 5,000 A flows through the feeder. Make use of the following characteristic

PSM	**2**	**3**	**5**	**8**	**10**	**15**
Time for unity TMS	**10**	**6**	**4.5**	**3.2**	**3**	**2.5**

(100% current = 1 A)

Solution : Fault current, I_f = 5,000 A

CT ratio = 500 : 1

$$\text{Relay current, } I_R = \frac{I_f}{\text{CT ratio}} = \frac{5{,}000}{500} = 10 \text{ A}$$

Pick-up value of relay

= Current setting × rated secondary current of CT

$$= \frac{125}{100} \times 1 = 1.25 \text{ A}$$

Plug setting multiplier of the relay,

$$\text{PSM} = \frac{\text{Fault current in relay coil, } I_R}{\text{Pick-up value of relay}} = \frac{10}{1.25} = 8$$

Time corresponding to the PSM of 8 from the given data is 3.2 seconds

So actual operating time = 3.2 × Time setting multiplier

= 3.2 × 0.3 = 0.96 second **Ans.**

Example 9.3. Reference to Fig. 9.20, given that

Fault current = 2,000 A; Relay 1 set on 100%; CT ratio = 200/1; Relay 2 set on 125%. For discrimination the time gradient margin between the relays in 0.5 second.

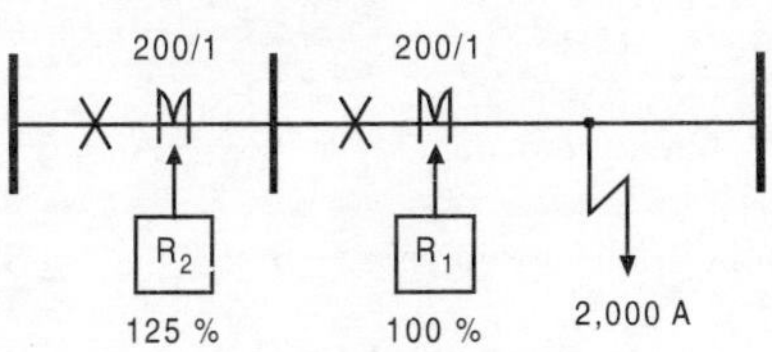

Fig. 9.20

Determine the time of operation of the two relays assuming that both the relays have the characteristic as shown in the following table and the relay no. 1 has a time multiplier setting = 0.2.

Plug setting multiplier	**2**	**3.6**	**5**	**8**	**10**	**15**	**20**
Time in seconds for a time multiplier of 1	**10**	**6**	**3.9**	**3.15**	**2.8**	**2.2**	**2.1**

Also determine the time setting multiplier of relay no 2.

[A.M.I.E. Sec. B. Power System Protection and Communication Summer 2000]

Solution : Relay 1 Relay current, $I_{R1} = \frac{2{,}000}{200} = 10$ A

Pick-up current = 1 A

PSM = 10/1 = 10

Time of operation of relay 1 corresponding to PSM of 10 from the given data = 2.8 seconds

Actual operating time of relay 1 with TSM of 0.2

= 2.8 × 0.2 = 0.56 second **Ans.**

Relay 2 Relay current, $I_{R2} = \frac{2{,}000}{200} = 10$ A

Pick-up current = 1 × 1.25 = 1.25 A

$$\text{PSM} = \frac{10}{1.25} = 8$$

Time of operation of relay 2 corresponding to PSM of 8 from the given data is 3.15 seconds

But actual time of operation of relay 2

= Time of operation of relay 1 + time grading margin

= 0.56 + 0.5 = 1.06 second **Ans.**

Time-setting multiplier

$$= \frac{\text{Actual time of operation}}{\text{Time of operation corresponding to PSM}}$$

$$= \frac{1.06}{3.15} = 0.3365 \text{ **Ans.**}$$

Example 9.4. It is given that fault current level at 33 kV side is 2,700 A; CT ratio at 33 kV side is 200 : 1 and 132 kV side is 100 : 1 (Refer to Fig. 9.21). If both the relays R_1 and R_2 are set for 100% plug setting, determine the operating time for both the relays when time grading margin of 0.6 second is given and TMS for relay R_1 is 0.15.

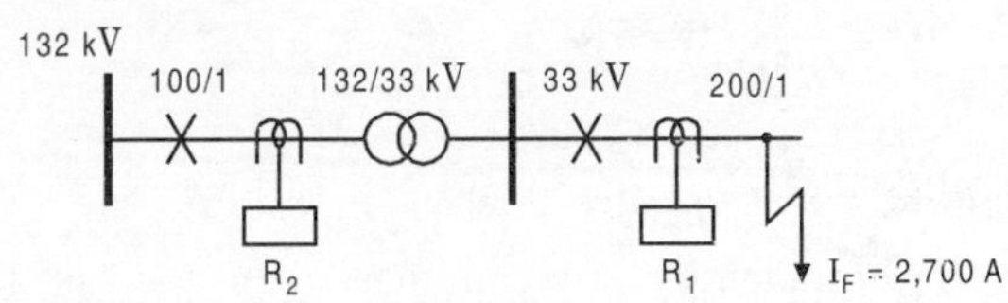

Fig. 9.21

Solution : Relay 1 Relay current, $I_{R1} = \frac{2{,}700}{200} = 13.5$ A

Pick-up current $= 1 \times \frac{100}{100} = 1$ A ∵ Relay R_1 is set for 100%

$$\text{PSM} = \frac{13.5}{1} = 13.5$$

From IDMT curve given in Fig. 9.19, the operating time corresponding to PSM of 13.5 is 2.6 seconds

∴ Actual operating time = 2.6 × TMS = 2.6 × 0.15

= 0.39 second **Ans.**

Relays 2 Fault current on 132 kV side

$$= \frac{2{,}700 \times 33}{132} = 675 \text{ A}$$

$$\text{Relay current, } I_{R_2} = \frac{675}{100} = 6.75 \text{ A}$$

$$\text{Pick-up current} = 1 \times \frac{100}{100} = 1 \text{ A}$$

∵ Relay R_2 is set for 100%

$$PSM = \frac{6.75}{1} = 6.75$$

From IDMT curve given in Fig. 9.19, the operating time corresponding to PSM of 6.75 is 3.6 seconds.

But this relay should operate 0.6 second after the operation of relay R_1 *i.e.*,

Total operating time required = 0.39 + 0.6 = 0.99 second **Ans.**

Hence TMS for relay $R_2 = \frac{0.99}{3.6} = 0.275$ **Ans.**

Example 9.5. Provide time-current grading for the following system :

Relay Point	CT Ratio	Fault current
a	400/5	6,000 A
b	200/5	5,000 A
c	200/5	4,000 A

Use standard 2.2 second IDMT characteristic.

Time in seconds	30	10	7	5	3	2.2
PSM	1.0	2	3	5	10	20

[*U.P. Technical Univ. Switchgear and Protection 2006-07*]

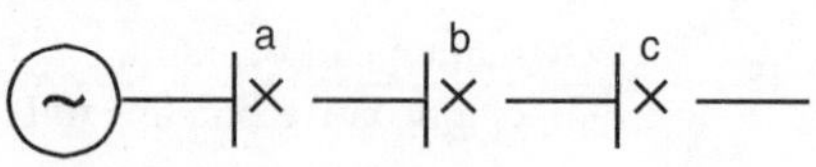

Fig. 9.22

Solution : Relay current with maximum fault current at *c*

$$I_R = \frac{4000}{200/5} = 100 \text{ A}$$

For 100% setting, pick-up value of relay

= Current setting × rated secondary current of CT

$$= \frac{100}{100} \times 5 = 5 \text{ A}$$

$$PSM = \frac{I_R}{\text{Pick-up value of relay}} = \frac{100}{5} = 20$$

Time of operation of relay corresponding to PSM of 20 from the given data is 2.2 seconds.

Actual operating time of relay with TSM of 0.1 (assumed)

= 2.2 × 0.1 = 0.22 s **Ans.**

To achieve discrimination between relays at *b* and at *c* when the fault occurs just before *c* or just after *c* when there is no change in fault current. Let the discriminating time between relays be 0.5 second which includes the time for operation of relay at *c*, the operation of CB at *c* and over travel of relay at *b*.

Thus the operating time of relay at *b* when fault occurs near *c* will be 0.22 + 0.5 = 0.72 second.

The relay current at location *b* when fault occurs near *c* will be

$$\frac{4,000}{40} = 100 \text{ A}$$

Pick-up current = 1.25 × 5 = 6.25 assuming relay setting 125%

$$PSM = \frac{100}{6.25} = 16$$

The operating time from the given datas is 2.5 seconds approximately. Actual operating time of relay at *b* when graded w.r.t. relay at *c* is 0.72 s.

$$\text{TSM of relay at } b = \frac{0.72}{2.5} = 0.29$$

When fault occurs near *b*, the PSM with operating current 6.25 A is

$$PSM = \frac{5,000}{\frac{200}{5} \times 6.25} = 20$$

The operating time corresponding to PSM of 20 is 2.2 seconds. Actual operating time of relay at *b* = 2.2 × 0.29 = 0.638 second **Ans.**

Since CT ratio at *a* is $\frac{400}{5}$ which is high in comparison to relay at *b*, the current discrimination is inherent. Let the relay setting be 125%.

$$\text{The PSM of relay at } a \text{ when fault occurs near } b = \frac{5,000}{\frac{400}{5} \times 6.25} = 10$$

The operating time corresponding to PSM of 10 is 3 seconds whereas the operating time of relay at *a* w.r.t. that at *b* will be 0.638 + 0.5 = 1.138 seconds

$$\text{So, TMS} = \frac{1.138}{3} = 0.379$$

When fault occurs near *a*, the PSM will be

$$\frac{6,000}{\frac{400}{5} \times 6.25} = 12$$

The operating time corresponding to PSM of 12 is 2.6 seconds

So, the actual operating time of relay at *a* will be

2.6 × 0.379 = 0.985 second **Ans.**

Example 9.6. A 20 MVA transformer which is used to operate at 30% overload feeds an 11 kV bus-bar through a circuit breaker. The transformer circuit breaker is equipped with a 1,000/5 current transformer and the feeder circuit breaker with 400/5 current transformer and both the current transformers feed IDMTL relays having the following characteristics.

Plug Setting Multiplier	2	3	5	10	15	20
Time (Seconds)	10	6	4.1	3	2.5	2.2

The relay on the feeder circuit breaker has 125% plug setting and a 0.3 time multiplier setting. If a fault current of 5,000 A flows from the transformer to the feeder, determine

(*i*) operating time of feeder relay.

(*ii*) Suggest suitable plug setting and time multiplier setting of the transformer relay to ensure adequate discrimination of 0.5 s between the transformer relay and feeder relay

[*U.P.S.C. I.E.S Electrical Engineering II, 2000*]

Solution : Feeder relay current, $I_{R_F} = \frac{5,000}{400/5} = 62.5$ A

$$\text{Feeder relay pick-up current} = 5 \times \frac{125}{100} = 6.25 \text{ A}$$

$$\text{Feeder relay PSM} = \frac{62.5}{6.25} = 10$$

From given data, the operating time corresponding to PSM of 10 is 3 seconds

Actual operating time for feeder relay

= 3 × TMS = 3 × 0.3 = 0.9 s **Ans.**

Transformer overload current

$$= 1.3 \times \frac{20 \times 10^6}{\sqrt{3} \times 11,000} = 1,365 \text{ A}$$

Transformer relay current,

$$I_{RT} = \frac{1{,}365}{1{,}000/5} = 6.825 \text{ A}$$

Transformer relay pick-up current

= PS × 5 where PS means plug setting

Since the transformer relay must not operate to overload current,

$$\text{Plug setting (PS)} > \frac{6.825}{5} > 1.365 \text{ or } 136.5\%.$$

The plug settings are restricted to standard values in steps of 25%, so the nearest value but higher than 136.5% is 150%.

$$\text{So transformer relay pick-up current} = \frac{150}{100} \times 5 = 7.5 \text{ A}$$

Transformer relay current corresponding to fault current of 5,000 A

$$= \frac{5{,}000}{1{,}000/5} = 25 \text{ A}$$

$$\text{Transformer relay PSM} = \frac{25}{7.5} = 3.33 \textbf{ Ans.}$$

Time corresponding to PSM of 3.33 from PSM-time curve shown in Fig. 9.19 is 5.6 seconds

Time setting for transformer relay

= Actual operating time of feeder relay + time grading margin

= 0.9 + 0.5 = 1.4 seconds

Time setting multiplier

$$= \frac{1.4}{5.6} = 0.25 \textbf{ Ans.}$$

Example 9.7. A 100/1 A bar primary current transformer supplies an overcurrent relay setting at 50% pick-up value and has a burden of 2 VA. Determine the knee point voltage and cross section of the core if the CT has 50 turns on its secondary and the fault current is 15 times the relay setting. Assume the flux density as 1.4 Wb/m² and frequency = 50 Hz.

[*U.P.S.C. I.E.S. Electrical Engineering-II, 2006*]

Solution : Rated current of CT secondary = 1 A

Operating current of the relay

= Current setting × rated secondary current of CT

= 0.5 × 1 = 0.5 A

Rated burden = 2 VA

$$\text{Rated secondary voltage} = \frac{\text{Rated burden}}{\text{Relay operating current}} = \frac{2}{0.5} = 4 \text{ V}$$

The secondary voltage of the CT when the current is 15 times the relay setting = 4 × 15 = 60 V.

The knee voltage must be slightly greater than 60 V.

Now since, Induced emf,

$$E = 4.44\, \phi_{max} fN$$

$$= 4.44\, B_{max} \times A \times f \times N \quad \because \phi_{max} = B_{max} \times A$$

So cross section of the core,

$$A = \frac{E}{4.44\, B_{max} fN}$$

$$= \frac{60}{4.44 \times 1.4 \times 50 \times 50}$$

$$= 38.61 \times 10^{-4} \text{ m}^2 \text{ or } 38.61 \text{ cm}^2 \textbf{ Ans.}$$

9.11. INDUCTION TYPE DIRECTIONAL POWER RELAY

The non-directional relay discussed in Art. 9.10.1 can operate for fault flow in either direction. In order to achieve operation for the fault flowing in a specific direction, it is necessary to add a directional element (Fig. 9.23) to the non-directional element.

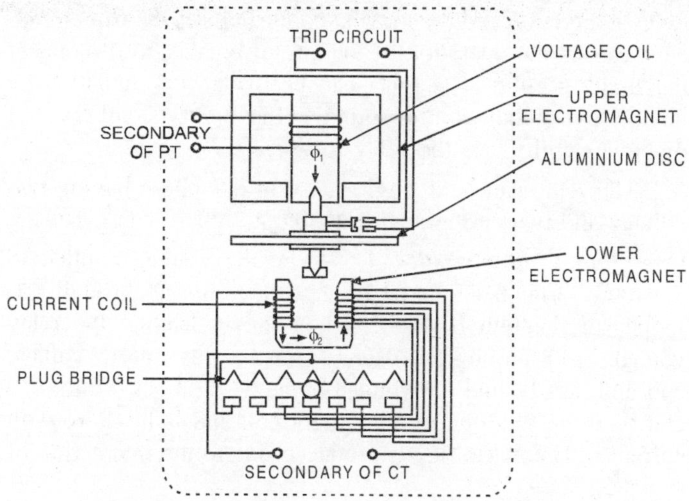

Fig. 9.23. ***Induction Type Directional Power Relay***

The directional (or reverse) power relay operates when the power through the relay will be reversing *i.e.,* generator supply to the network fails and the power from the other sources in the system try to feed the power to this unit in the reverse direction. In the case of motors such a relay is employed to prevent the motors reversing the direction of rotation.

The principle of operation of this relay is similar to that of an overcurrent (non-directional) induction relay shown in Fig. 9.18. The difference lies in the fact that in case of overcurrent relay the torque is developed due to interaction of magnetic fields obtained from the current in the circuit through CT, while in case of directional power relay the driving torque is derived from the interaction of the fields produced from both voltage and current sources of the circuit it protects. Since the relay has both voltage and current coils, the relay is essentially a wattmeter and the direction of the torque developed in the relay depends upon the direction of current in relation to the voltage with which it is associated *i.e.,* the relay recognizes the phase difference between voltage and current.

Constructional Details. The induction type directional power relay is shown in Fig. 9.23. It essentially consists of an aluminium disc which is free to rotate in between the poles of two electromagnets. The upper electromagnet has a winding, called the *voltage* or *potential coil,* on the middle limb connected to the circuit voltage source through a potential transformer (PT). The lower electromagnet has a separate winding, called the *current coil,* connected to the secondary of CT in the line to be protected. The current coil is provided with a number of tappings connected to the plug bridge, so as to give the desired current setting. The restraint torque is provided by a spiral spring.

Operation. The torque developed on the disc suspended between the two magnets is proportional to VI. When the power flows in the normal direction, the torque developed on the disc assisted by the spring tends to turn away the moving contact from the fixed trip circuit contacts. Thus the relay remains inoperative. A reversal of current in the circuit reverses the torque produced on the disc and when this is large enough to overcome the control spring torque, the disc rotates in the reverse direction and the moving contact closes the trip circuit. This causes the operation of the circuit breaker to disconnect the faulty section. The relay can be made very sensitive by having a very light control spring so that a very small reversal of power will cause the relay to operate.

The relay can be a single-phase or a 3-phase having two voltage and two current elements like a 3-phase energy meter.

Operating Characteristics. Let V be the voltage applied to the relay through PT and I be the relay current through CT. In phasor diagram (Fig. 9.24) I is shown leading the relay voltage V by an angle θ. Here ϕ_V is the flux due to voltage coil and lags behind the voltage by angle ϕ (about 60° to 70°) and ϕ_I is the flux due to the current coil and is in phase with current I. The net torque is produced due to the interaction of ϕ_I and ϕ_V.

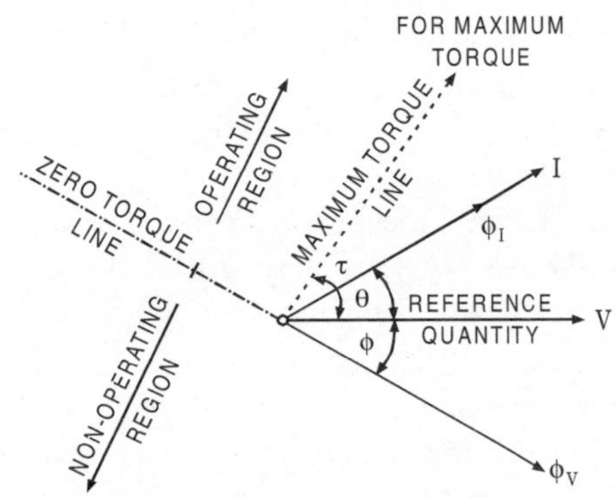

Fig. 9.24. ***Phasor Diagram For Directional Element***

Torque, therefore, is given as

$$T \propto \phi_V \phi_I \sin(\phi + \theta) \text{ where } \phi_V \propto V \text{ and } \phi_I \propto I$$

So the torque equation for the relay can be given as

$$T = KVI \sin(\phi + \theta) \qquad ...(9.5)$$

The torque is maximum when the two fluxes are displaced by 90° *i.e.,* when $(\phi + \theta) = 90°$. Here dotted line in the phasor diagram represents the desired position of ϕ_I for maximum torque. Since V is the reference quantity and ϕ_V has fixed position with respect to V for a particular design, the angle between the dotted line and the reference quantity V is known as the *maximum torque angle* and let it be denoted by τ. Zero torque will occur when $\sin(\phi + \theta) = 0$ *i.e.,* $(\phi + \theta) = 0°$ or 180°, this being satisfied when the relay current phasor lies along the chain dotted line which is at right angles to the maximum torque line. The directional element will, therefore, operate provided the current phasor lies with in ± 90° of the maximum torque line. If the current phasor is displaced by more than 90° the directional element will restrain. The operating and the non-operating regions are shown in the figure.

It may be seen that

$$\tau = 90° - \phi$$
$$\text{or } \phi = 90° - \tau \qquad ...(9.6)$$

and the torque equation becomes

$$T = KVI \sin(\theta + 90° - \tau)$$
$$= KVI \cos(\theta - \tau) \qquad ...(9.7)$$

When the relay is about to start, neglecting the spring constant,

$$VI \cos(\theta - \tau) = 0$$
$$\text{or } \theta - \tau = 90°$$
$$\text{or } \theta = \tau + 90° \qquad ...(9.8)$$

This is the equation describing the polar characteristic (Fig. 9.25) of the directional relay.

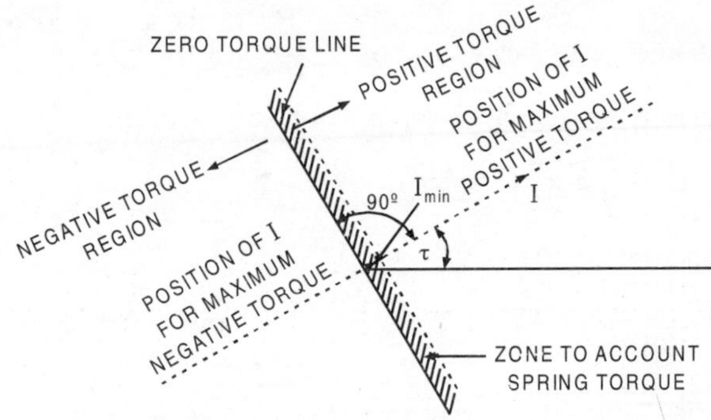

Fig. 9.25. ***Polar Characteristic of Directional Relay***

The zone between the dotted line and the line parallel to it corresponds to the spring torque. If the current phasor lies within these lines the torque developed is less than spring torque and hence the relay does not operate. If the current phasor crosses the dotted line the operating torque exceeds the spring torque and hence the relay operates. Relay will not pick up or it will reset for any current phasor lying in the negative torque region.

It may be noted that the system current usually lags behind the system voltage but the relay current is made to lead the relay voltage by inserting resistance or capacitance or a combination of the two in series with the voltage or potential coil.

Such relays are very suitable for protection of parallel feeders. The directional overcurrent relay suffers from the drawback that the feeder voltage falls to a much lower value when a fault occurs resulting into non-operation of the relay. This shortcoming may be overcome by compensating the relay secondary winding on the lower magnet. The compensating winding ampere-turns on the lower magnet opposes the ampere-turns produced by the current coil. Therefore turns of current coil will have to be appropriately increased. When the voltage falls due to the fault on the feeder, the resultant ampere-turns provided by the windings on the lower electromagnet jointly increase, compensating the reduced ampere-turns provided by the voltage coil.

9.12. INDUCTION TYPE DIRECTIONAL OVERCURRENT AND EARTH FAULT RELAY

The directional power relay, discussed in Art. 9.11, cannot be employed as a directional protective relay under short-circuit conditions because under short-circuit conditions the system voltage drops to a low value and therefore the torque developed in the relay may be insufficient to cause its operation. This difficulty is overcome in the directional overcurrent relay which is designed to be almost independent of system voltage and power factor.

Constructional Details. Constructional details of a typical induction type directional overcurrent and earth fault relay are shown in Fig. 9.26. It consists of two relay elements. viz., (*i*) directional element and (*ii*) non-directional element, mounted in a common case.

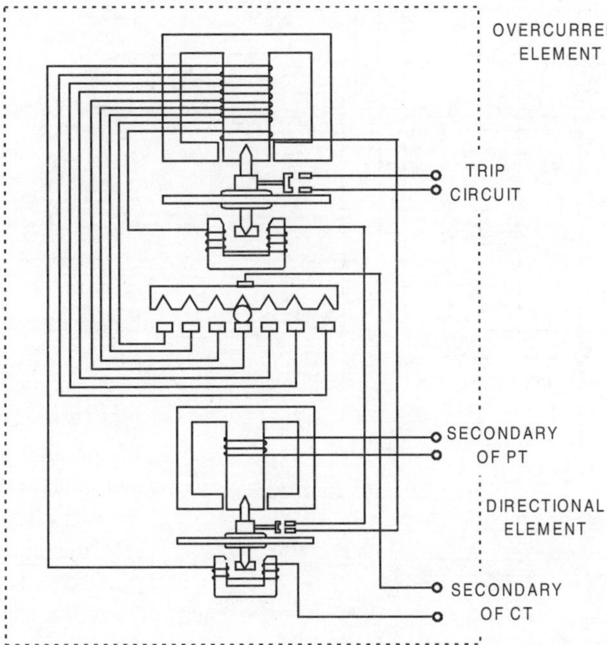

Fig. 9.26. ***Induction Type Directional Overcurrent and Earth Fault Relay***

Directional element is essentially a *directional power relay,* already discussed in Art. 9.11. The voltage coil of this element is connected to the circuit voltage through a PT while its current coil is energized through a CT by circuit current. This winding is carried over the upper magnet of the non-directional element. The trip contacts of the directional element are connected in series with the secondary circuit of the overcurrent element. Thus overcurrent element cannot start to operate until its secondary circuit is completed *i.e.,* the directional element must operate first in order to operate the overcurrent element.

Non-directional element is an overcurrent element similar in all respects to a non-directional overcurrent relay described in Art. 9.10.1. The spindle of the disc of this element carries a moving contact which closes the trip circuit contacts after the operation of directional element. The tappings are provided over the upper magnet of the overcurrent element and are connected to the bridge, thereby provide facility for current setting.

Under normal operating conditions, power flow is in the normal direction in the circuit protected by the relay. Thus the directional power relay (lower element) does not operate, thereby keeping the overcurrent element (upper element) unenergised. But as soon as there is a reversal of current or power the disc of the reverse power relay (lower element) starts, rotating and completes the circuit for overcurrent element. Due to overcurrent a torque is produced in the disc and the action closes the trip circuit, thereby enabling the circuit breaker to operate and isolate the faulty section.

The directional element is made as sensitive as possible to ensure positive operation—even 20% of the power in the reverse direction operates it. The relay operates only when (*i*) *the direction of current is in reverse direction* (*ii*) *current in the reverse direction exceeds the preset value and* (*iii*) *excessive current* (*greater than the preset value*) *persists for duration longer than its time setting.*

Directional relays must have the following features:

(*i*) high speed of operation (*ii*) high sensitivity (*iii*) adequate short-time thermal rating (*iv*) ability to operate with low values of voltage (*v*) burden must not be excessive and (*vi*) there should be no voltage and current creep *i.e.,* if either the voltage coil alone or the current coil alone is energized with the other one denergized there should be no movement.

Induction cup units satisfy the above requirements and are, therefore, very popular.

Such relays are employed when graded time overload protection is applied to ring mains and interconnected networks, since fault current can flow in either direction.

9.12.1. Directional Overcurrent Relay Connections

Relay connections must be made so that the currents and voltages applied to the relay during different fault conditions which may arise on the protected circuit section afford the relay a positive and sufficiently large operating torque. To achieve this for all types of faults the relays cannot be connected to operate on true watts since for some faults the voltage will be extremely small and also the power factor will be very small which will result in a negligible small torque. To overcome this difficulty, and thus ensure that sufficient torque is available, each relay is supplied with current and voltage as described below.

There are two types of relay connections in use. Directional element connections are conveniently and popularly described in terms of the angle by which unity power factor (UPF) balanced load current flowing in the tripping direction leads the applied voltage applied to the relay voltage coil with due consideration given to the polarity of the relay coils.

The two types of relay connections used are:

(*i*) The 30°-relay with a maximum torque angle of 0°.

(*ii*) The 90°-relay with a maximum torque angle of 45°.

The relay angle is defined as the angle between the voltage and current supplied to the relay under balanced three- phase unity power factor conditions.

In phasor diagram for 0° directional relay with zero maximum torque angle shown in Fig. 9.27 I_R, I_Y, I_B and V_R, V_Y, V_B represent the phase currents, and phase voltages of a 3-phase balanced system with unity power factor conditions. Phasor V_{RB} representing the system red phase to blue phase voltage lags behind phasors I_R and V_R by 30°. Let the relay element be supplied with current I_R and voltage V_{RB}, in phase relation as shown *i.e.*, I_R leads V_{RB} by 30°. The connections are, therefore, referred to as 30° relay.

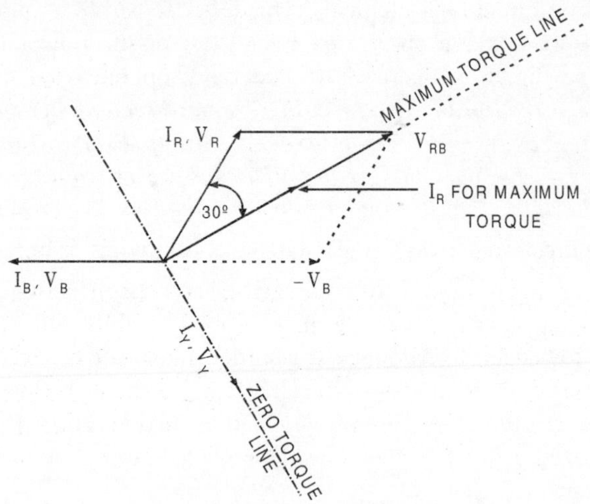

Fig. 9.27. ***Phasor Diagram For 0° Directional Relay With Zero Maximum Torque Angle***

The angle between the current and voltage supplied to the relay for maximum torque, T is zero so that the position of the relay current phasor for maximum torque will be along V_{RB}. Also since according to Eq. (9.7).

$$T = KVI \cos(\theta - \tau) = KVI \cos\theta \qquad \text{for } \tau = 0$$

Now for unity power factor condition *i.e.*, system current I_R in phase with system voltage, the relay current I_R leads the relay voltage V_{RB} by 30° *i.e.*, θ = 30° so that torque developed

$$T = KVI \cos 30° = 0.866\ KVI \qquad ...(9.9)$$

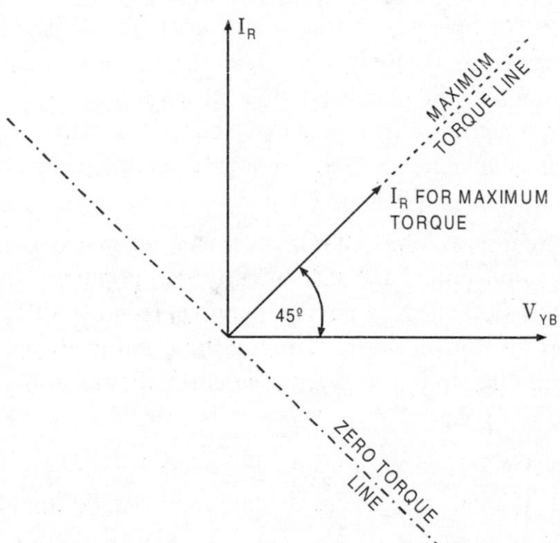

Fig. 9.28. ***Phasor Diagram For 90° Directional Relay With 45° Maximum Torque Angle***

While if system current I_R lags behind the system voltage V_R by 30°, the relay current I_R will be in phase with relay voltage V_{RB} so that θ = 0° and the torque developed will be maximum and is given as

$$T = KVI \qquad ...(9.10)$$

On occurrence of fault, I_R may lag V_R, say by 90°, and in such case θ will be 60° and the torque developed will be 0.5 T_{max}.

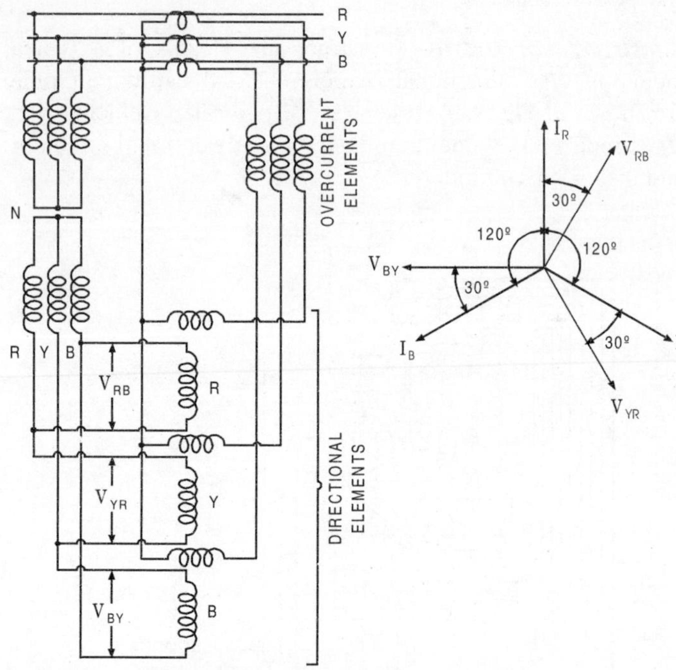

Fig. 9.29. *30° Connection Diagram*

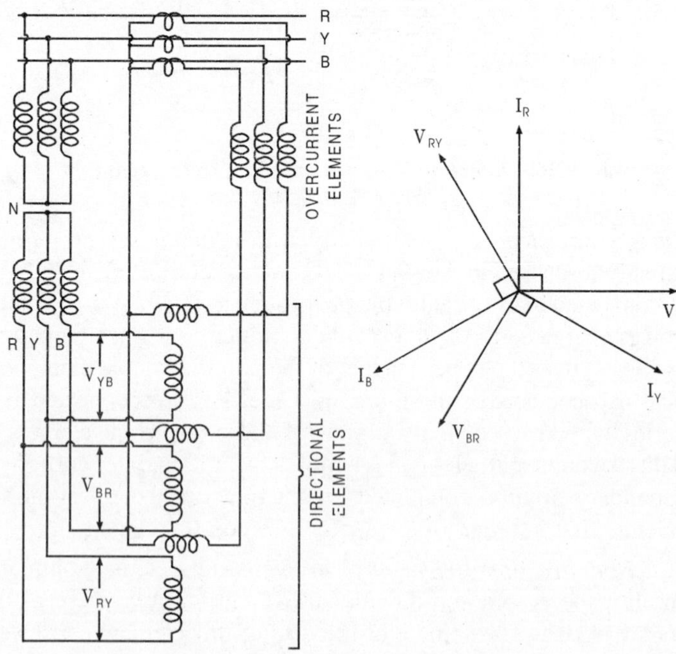

Fig. 9.30. *90° Connection Diagram*

We thus see that the initial lead of 30° makes the relay more sensitive at low power factors. Such relays are usually satisfactory for plain feeders.

The phasor diagram for the *90°-relay with a maximum torque angle of 45°* is shown in Fig. 9.28.

Since the angle between the current and voltage supplied to relay for maximum torque, τ is 45°, the torque developed will be maximum for θ = 45° and is given as

$$T_{max} = KVI \cos(\theta - \tau) = KVI \cos(45° - 45°) = KVI$$

and torque for 90° connection will be

$$T = KVI\cos(90° - 45°) = 0.707\ KVI = 0.707\ T_{max}$$

The connections for 30° and 90° relays are shown in Figs. 9.2) and 9.30 respectively.

9.12.2. Voltage Source Connection For Providing Residual Voltage

Directional relay controlling earth faults on all the three phases of a 3-ϕ circuit is energized by residual current and residual voltage. The residual current is obtained by summing the currents in the three phases, employing line CTs or a core balance current transformer. The former is more common. For three line currents I_R, I_Y, I_B

$$I_R + I_Y + I_B = 0 \quad \text{for healthy conditions}$$

$$I_R + I_Y + I_B = I_{RES} \quad \text{for earth fault on one phase}$$

A suitable residual voltage source can be a broken delta secondary winding on three single-phase transformers [Fig. 9.31(*a*)] or an equivalent three-phase five-limb voltage transformer [Fig. 9.31 (*b*)]. The application of a broken delta to provide a residual voltage is shown in Fig. 9.32. Under normal operating (or healthy) conditions the secondary voltage phasors form a closed triangle and therefore no voltage appears across the relay potential coil, as illustrated in Fig. 9.32(*a*). On occurrence of earth fault on any line say at point F on line B in Fig. 9.32(*b*) the voltage phasors are as shown, leaving a residual voltage V_{RES} across the open delta secondary which is the voltage appearing at the voltage coil of the directional element of the relay. The magnitude and phase of this residual voltage will depend on the magnitude of the earth fault current and the impedance of the fault path.

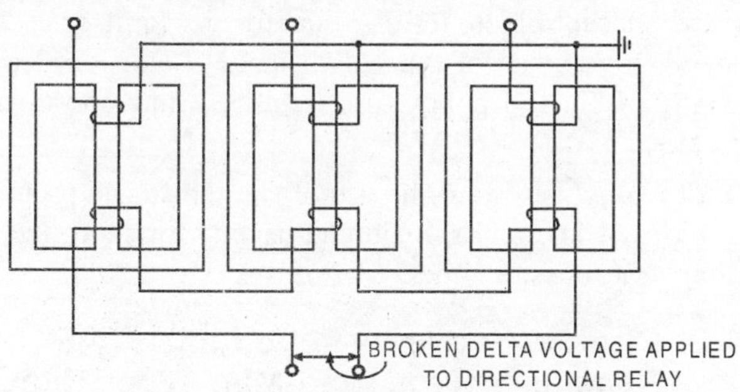

(a) Three Single-Phase Transformer Bank

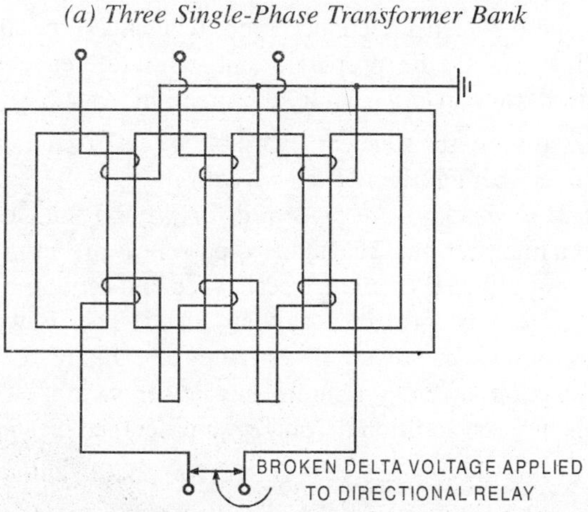

(b) Five-Limbed Core Three-Phase Transformer

Fig. 9.31. ***Voltage Source Connection For Providing Residual Voltage***

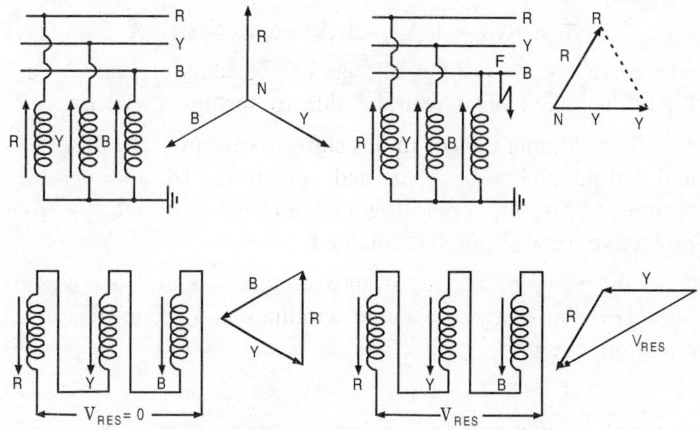

(a) Under Normal Operating Conditions *(b) Under Faulty Conditions*

Fig. 9.32. ***Broken Delta Connection For Providing Residual Voltage***

The connections for directional earth fault relay are shown in Fig. 9.33. The torque developed on the relay will be proportional to $V_{RES}\ I_{RES}\cos(\theta - \tau)$ where τ is the maximum torque angle of the relay and θ is the angle between V_{RES} and I_{RES}.

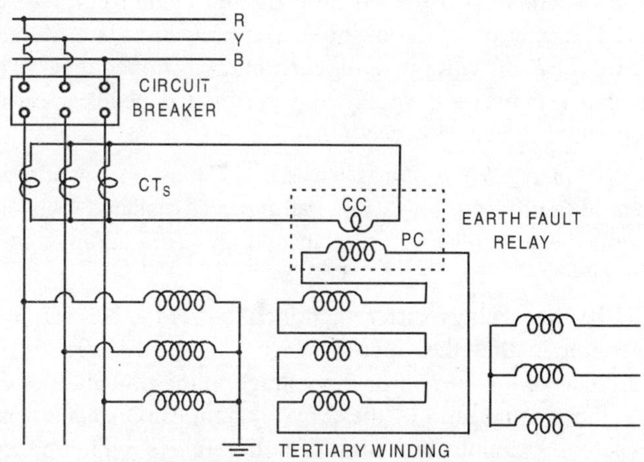

Fig. 9.33. ***Connections For Directional Earth Fault Relay Residual Voltage From a Tertiary Delta Winding***

9.13. UNIVERSAL RELAY TORQUE EQUATION

Most of the protection relays consist of some arrangement of electromagnets. These electromagnets have either current windings or voltage windings and in some cases both the windings. Currents through the windings produce magnetic fluxes and the torque is produced by interaction between the fluxes of the same winding or between the fluxes of both the windings.

$$\text{Torque developed by current winding} = K_1 I^2$$

$$\text{Torque developed by voltage winding} = K_2 V^2$$

If both the current and voltage windings are used, the torque developed by the interaction between the fluxes

$$= K_3 VI\cos(\theta - \tau) \quad \text{...Refer to Eq. (9.7)}$$

where θ is the angle between V and I and τ is the relay maximum torque angle.

If a relay consists of all the three elements, the torque will be developed due to all the three causes and therefore, its torque in general will be given as

$$T = K_1I^2 + K_2V^2 + K_3VI \cos (\theta - \tau) + K_4 \quad ...(9.11)$$

where K_1, K_2, K_3 are tap settings or constants of I and V and K_4 is the mechanical restraint due to spring or gravity.

By assigning plus or minus signs to certain of the constants and letting others be zero, and sometimes by adding other similar terms, the operating characteristics of all types of protective relays can be obtained.

For example, in case of an overcurrent relay $K_2 = K_3 = 0$ because of absence of voltage windings and therefore torque equation becomes

$$T = K_1I^2 - K_4 \quad ...(9.12)$$

–ve sign is assigned to K_4 as the torque produced by spring is restraining one.

Similarly, for directional relay $K_1 = K_2 = 0$ and the torque developed will be given as

$$T = K_3VI \cos (\theta - \tau) - K_4 \quad ...(9.13)$$

9.14. DISTANCE PROTECTION

Distance protection is the name given to the protection, whose action depends upon the distance of the feeding point to the fault. The time of operation of such a protection is a function of the ratio of voltage and current *i.e.,* impedance. This impedance between the relay and the fault depends upon the electrical distance between them.

Distance relay group is perhaps the most interesting and versatile family of relays. Principal types of distance relays are (*i*) impedance relays (*ii*) reactance relays (*iii*) admittance or mho relays.

Distance relays differ in principle from other forms of protection in that their performance is not governed by the magnitude of the current or the voltage in the protected circuit but rather on the ratio of these two quantities. Distance relays are actually double actuating quantity relays with one coil energized by voltage and the other coil by current. The current element produces a positive or pick-up torque while the voltage element produces a negative or reset torque. The relay operates only when the V/I ratio falls below a predetermined value (or set value). During a fault on a transmission line the fault current increases and the voltage at the fault point decreases. The V/I ratio is measured at the location of CTs and PTs. The voltage at PT location depends on the distance between PT and the fault. If the fault is nearer, measured voltage is lesser and if the fault is farther, measured voltage is more. Hence assuming constant fault impedance each value of V/I measured from relay location corresponds to distance between relaying point and the fault along the line. Hence such protection is called the *distance protection* or *impedance protection.*

Distance protection is non-unit type protection, the protection zone is not exact. The distance protection is high speed protection and is simply to apply. It can be employed as a primary as well as backup protection. It can be employed in carrier aided distance schemes and in autoreclosing schemes. Distance protection is very commonly used in protection of transmission lines.

Distance relays are used where overcurrent relaying is too slow and is not so selective. Distance relays are used for both phase fault and ground fault protection and they provide higher speeds for clearing faults than overcurrent relays. Distance relays are also independent of changes in magnitude of the short-circuit currents and hence they are not much affected by changes in the generation capacity and the system configuration. Thus they eliminate long clearing times for faults near the power sources required by overcurrent relays if used for the purpose.

9.14.1. Application of Distance Protection

Distance protection schemes are widely employed for the protection of high-voltage ac transmission lines and distribution lines. They have replaced the overcurrent protection of the transmission lines. The reasons are faster protection, simpler coordination; simpler application, permanent setting without need for readjustments; less effect of amount of generation and fault levels, fault current magnitude; permits the high line loading.

Static distance relays have superior and versatile characteristics and enlist several additional merits of static distance protection schemes.

Distance protection schemes are commonly employed for providing the primary or main protection and backup protection for ac transmission and distribution lines against 3-phase faults, phase-to-phase faults and phase-to-ground faults.

In some schemes for short lines, the phase-to-ground fault protection sensing may be by distance relay and measurement by overcurrent relays because distance protection for shorter lines are susceptible to errors due to arc fault resistance. In general, the choice of type of distance protection depends on length of line, tripping time required and coordination requirements.

Today trend is toward the use of static distance protection for all types of line faults, main and backup for short, medium and long transmission lines.

9.14.2. Choice Between Impedance Type, Reactance Type, and Mho Type Relays

The choice between impedance type, reactance type and mho type relays should be properly done with reference to their application to transmission line protection.

The ground resistance is variable. The ground distance relays must be unaffected by variation in fault resistance. Reactance type relays are preferred for ground fault relaying. For short line sections, reactance type relays are preferred as they are not affected by the arc resistance and more percentage of the line can be protected at high speed. The drawback of reactance relays, however, is that they are likely to operate undesirably on severe synchronising power swings at certain locations unless additional relay equipment is provided.

Mho type relays are best suited for phase fault relaying for long lines and particularly where a severe synchronising power surge may occur. They are most sensitive of all the distance relays. The mho relay is reliable because it combines both the directional and distance measuring functions in one unit.

The impedance type relay is suitable for phase fault relaying for lines of moderate length. The arc affects the impedance relay more that the reactance relay but less than the mho relay. The impedance relay along with the directional unit can be made to change its characteristics and the modified impedance relay may resemble the reactance or mho unit. There is considerable overlap in the application of these relays.

9.15. IMPEDANCE TYPE DISTANCE RELAY

It is worthwhile to note here that in electrical engineering, the term 'impedance' can be applied to resistance alone or reactance alone, or a combination of the two. In protective relaying terminology, however, an impedance relay has a characteristic that is different from that of a relay responding to any component of impedance and therefore, the "impedance relay" is very specific.

An impedance relay is a voltage restrained overcurrent relay. The relay measures impedance up to the point of fault and gives tripping command if this impedance is less than the relay setting Z. Relay setting Z is known as *replica impedance* and it is proportional to the set impedance *i.e.,* impedance up to the reach of the relay. The relay monitors continuously the line current I through CT and the bus voltage V through PT and operates when the V/I ratio falls below the set value.

Principle of Operation. The principle of operation of an impedance relay is illustrated in Fig. 9.34. The voltage element of the relay is excited through a potential transformer (PT) from the line under protection and current element of the relay is excited from a current transformer (CT) in series with the line. The portion AB of the line is the protected zone. Under normal operating conditions, the impedance of the protected line is Z. The relay is so designed it closes its contacts whenever impedance of the protected section falls below the set value *i.e.,* Z in this case.

Now assume that a fault occurs at point F_1 in the protected zone. The impedance, the ratio of the bus voltage and fault current (V/I), between the point where the relay is located and the point of fault will become less than Z and hence the relay operates. Should the fault occur beyond the protected zone (say at point F_2) the impedance will be more than Z and the relay contacts do not close.

Operating Characteristic of an Impedance Relay. As already mentioned, the impedance relay is a double actuating quantity relay and essentially consists of two elements—current-operated element and voltage-operated element. The current element produces a positive or pick-up torque while the voltage element develops a negative or reset torque. Taking spring control effect as $-K_3$, the torque equation of the relay is

$$T = K_1 I^2 - K_2 V^2 - K_3 \qquad ...(9.14)$$

where V and I are the rms values of voltage and current respectively. At balance point, when the relay is on the verge of operating, the net torque is zero, and

$$K_2 V^2 = K_1 I^2 - K_3$$

$$\text{or} \quad \frac{V}{I} = Z = \sqrt{\frac{K_1}{K_2} - \frac{K_3}{K_2 I^2}}$$

It is customary to neglect the effect of control spring, since its effect is noticeable only at current magnitudes well below those normally encountered. Hence taking $K_3 = 0$, the relay torque equation becomes

$$Z = \sqrt{\frac{K_1}{K_2}} = \text{Constant} \qquad ...(9.15)$$

The operating characteristic in terms of voltage V and current I is shown in Fig. 9.35, where the effect of control spring is shown as causing a noticeable bend in the characteristic only at the low-current end. For all practical purposes, the dash line, which represents a constant value of Z, may be considered the operating characteristic. The relay will pick up for any combination of V and I represented by a point above the characteristic in the positive-torque region, or, in other words, for any value of impedance less than constant value represented by the operating characteristic. By adjustment, the slope of the operating characteristic can be changed so that the relay will respond to all values of impedance less than any desired upper limit.

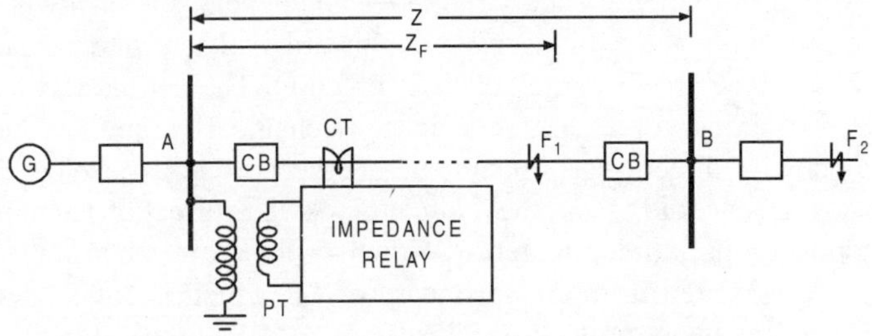

Fig. 9.34. ***Principle of Operation of an Impedance Relay***

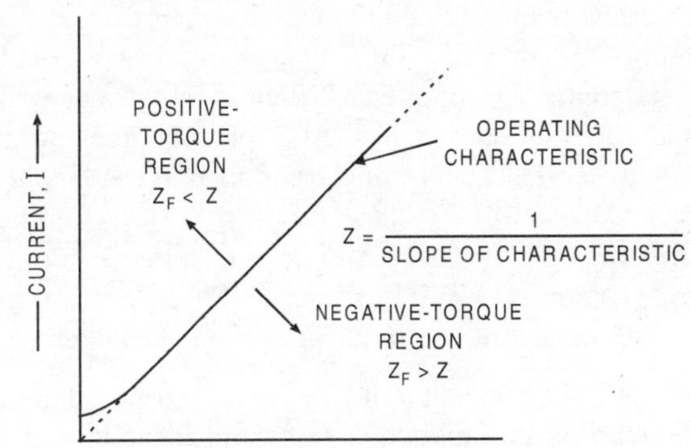

Fig. 9.35. ***Operating Characteristic of an Impedance Relay***

The more convenient way of describing the operating characteristic of a distance relay is by means of 'impedance diagram' or R-X diagram, as illustrated in Fig. 9.36. The numerical value of the ratio of V to I is shown as the length of the radius vector, such as Z, and the phase angle θ between V and I determines the position of vector, as shown. Since the operation of the impedance relay is practically or actually independent of the phase angle θ, the operating characteristic is a circle with its centre at the origin. Any value of impedance less than the radius of the circle will result in positive torque and any value of impedance greater that this radius will result in negative torque.

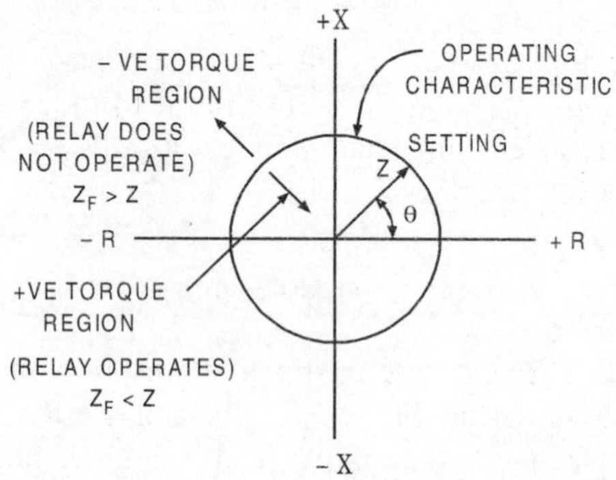

Fig. 9.36. ***Operating Characteristic of an Impedance Relay on an R-X Diagram***

The impedance relays normally used are high speed relays. These relays may be of electromagnetic or induction type described below.

1. ***Electromagnetic Type Impedance Relay.*** Such a relay is shown in Fig. 9.37 (*a*). In this relay two torques are produced by electromagnetic action of the voltage and current and these two torques are compared. The solenoid B is voltage excited from the secondary of the PT and develops a clockwise torque pulling the plunger P_2 downward and tends to rotate the balance arm in the clockwise direction. The spring acts as a restraining force and sets up a mechanical torque in the clockwise direction as illustrated. Another torque is developed by the solenoid A in the counterclockwise direction which tends to pull the plunger P_1 downwards. This solenoid is current excited from the secondary of CT which is connected to the line under protection. This torque is called the deflecting or pick-up torque.

In normal operating conditions *i.e.,* when there is no fault and equilibrium prevails, the balance arm is horizontal and the relay contacts are open. When a fault occurs in the protected zone the current in the CT primary goes up and so in the relay coil. Thus torque developed by solenoid A is increased. Also due to voltage drop with the occurrence of fault the magnitude of restraining torque developed by solenoid B decreases. Thus the balance arm rotates in counterclockwise direction and the relay contacts are closed.

The pull of the solenoid A (*i.e.,* current element) is proportional to I^2 and that due to solenoid B (voltage element) to V^2. Consequently the relay will operate when

$$K_1 I^2 > K_2 V^2$$

$$\text{or} \quad \frac{V}{I} < \sqrt{\frac{K_1}{K_2}} \quad \text{or} \quad Z < \sqrt{\frac{K_1}{K_2}}$$

The values of the constants K_1 and K_2 depend upon the ampere-turns of the two solenoids, and the ratios of the instrument transformers. By providing tappings on the coil, the setting of the relay can be changed.

The time-impedance characteristic of the relay is shown in Fig. 9.37 (*b*). The ordinate represents the time of operation for the relay and the abscissa represents impedance which is proportional to distance. As may be seen the operating time for almost the entire length (for which the relay is designed to operate) is constant. Towards the end of the predetermined length, the curve rises gradually because the pulls of the voltage and current elements become nearly equal, and after the point A has been passed the operating time rapidly becomes infinite.

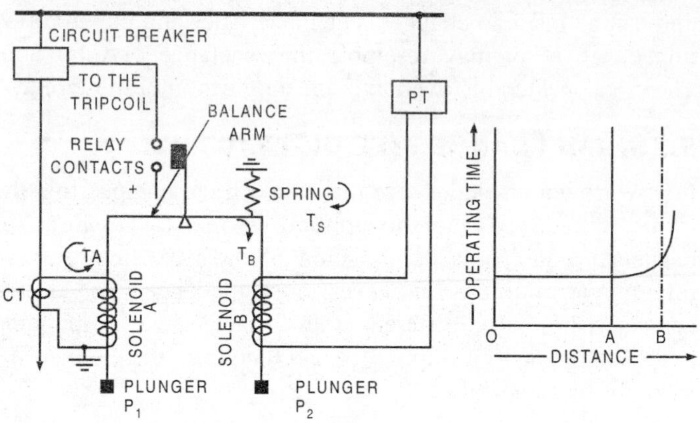

(*a*) *Electromagnetic Type Impedance Relay* (*b*)

Fig. 9.37

Such a relay is classified as *definite distance type impedance relay,* because it operates instantaneously for faults up to a predetermined distance from the relay.

The quality of protection provided by the distance impedance system is superior to that provided by overcurrent (time graded) relays. Since the operating time is same for all relays, the number of feeders in series that can be protected is unlimited. It lacks, however, one important feature of the latter in that the backup protection is not available except for short length AB.

2. ***Induction Type Impedance Relay.*** Induction type impedance relay consists of a combination of an overcurrent element of the type illustrated in Fig. 9.18 with a voltage-restraint element, the circuit being illustrated in Fig. 9.38(*a*).

The induction type impedance relay consists of a metallic disc usually made of aluminium or copper which is capable of rotating between two electromagnets—upper magnet and lower magnet. The upper electromagnet has two separate windings similar to that of overcurrent relay. The primary winding is connected to the secondary of the CT connected in the line to be protected. The winding has a number of tappings so as to vary the current settings, the tappings are connected to a plug bridge. The secondary winding on the upper electromagnet is connected in series with the windings on the lower electromagnet. The currents in the secondary are due to electromagnetic induction. By this arrangement leakage fluxes of upper and lower electromagnets are sufficiently displaced in space and phase to set up a rotational torque on the induction disc, as in the shaded pole induction disc motor. The torque is controlled by a permanent magnet (not shown in the figure). The controlling or braking torque caused by the permanent magnet varies directly as the driving torque. The spindle which carries the induction disc is shown connected by means of spiral spring coupling to the second spindle which carries the bridging piece or the moving contact of the relay trip constacts. The bridge is normally held in the open position by an armature held against the pole face of an electromagnet energized by the voltage of the line. It is to be pointed out that in actual practice the spindle carrying the induction disc is attached to the inner end of the spiral spring through a geared counter shaft in order to obtain the required characteristic for the relay operating time.

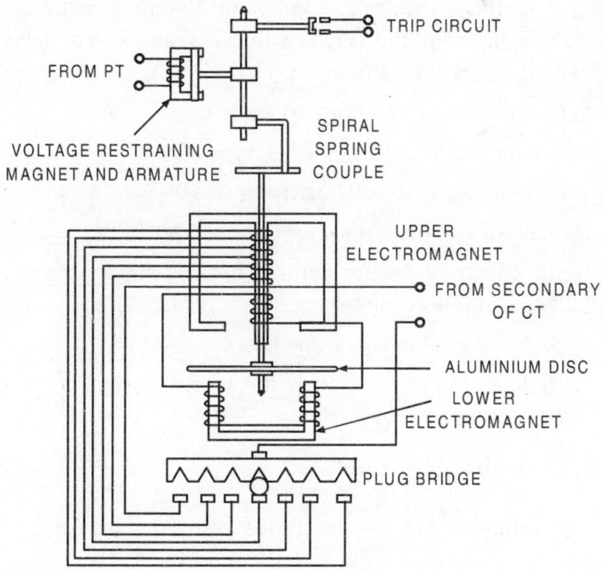

(a) Connection Diagram

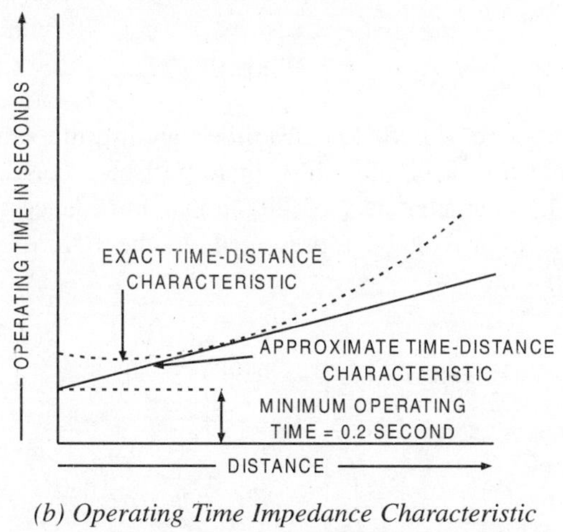

(b) Operating Time Impedance Characteristic

Fig. 9.38. ***Induction Type Impedance Relay***

Under normal operating conditions the pull exerted by the armature is more than that of the induction element and thus the trip circuit contacts remain open. When the fault occurs, the induction disc starts to rotate with a speed approximately proportional to the operating current, neglecting the effect of the control spring. Hence the time taken by the disc to turn through a given angle varies inversely as the current. Also as the disc rotates the spring is wound. The disc continues rotating till the tension of the spring is sufficient to overcome pull of the voltage restraint electromagnet over its armature and as soon as this armature is released the trip contacts are closed. Thus the angle through which the induction disc is to rotate for the operation of the relay depend upon the value of the pull on the restraint armature. The greater this pull, the greater would be the travel of the disc. However, this pull is also approximately proportional to the voltage, therefore, the angle through which the disc is to rotate for the operation of the relay is directly proportional to voltage V. Thus in this type relay the time required is directly proportional to line voltage V and inversely proportional to current I *i.e.,* the time of operation is proportional to V/I or impedance of the line or section.

The simple impedance relay described earlier [Fig. 9.37(*a*)], is instantaneous type *i.e.,* it operates as soon as the impedance of the line or section falls below a certain value. But the time of operation in case of induction type impedance relay described above [Fig. 9.38(*a*)] is proportional to the distance of fault from the relay point. A fault nearer to the relay will operate it earlier than a fault farther away from the relay. Because of such characteristic, this relay is called the time-distance relay.

The approximate time-distance characteristic of time distance relay is shown in Fig. 9.38(*b*). The exact time-distance characteristic is curved, as illustrated dotted in the figure. This is explained as below.

The torque on the disc tends to be proportional to the square of the circuit current *i.e.,* I^2 while the pull of the restraint magnet tends to be proportional to the square of the voltage *i.e.,* V^2. Thus the time of operation tends to be proportional to the square of the impedance *i.e.,* Z^2 or proportional to the square of the distance. Thus we get the time-distance characteristic as a curve.

Drawbacks of Plain Impedance Relay. The plain impedance relay, though very simple in theory and in construction, has several drawbacks such as given below:

1. The plain impedance relay can operate in either direction. It responds to the faults on both sides of CT, PT location. So it cannot discriminate between internal and external faults.
2. The plain impedance relay is sensitive to power swing as a large area is covered by the circle on each side on R-X plane. During power swing which is caused by severe faults, the relay sees a fictitious impedance and if this impedance is less than the relay setting, the relay may operate.
3. The plain impedance relay is affected by arc resistance of line fault and results in underreach.

Time-Characteristics of High Speed Type Impedance Relay. Although impedance relays with inherent time delay are encountered occasionally, only high-speed type impedance relays will be considered. The operating-time characteristic of a high-speed impedance relay is shown in Fig. 9.39. The curve shown in the figure is for a particular value of current magnitude. For other current values similar characteristics are obtained. Curves for higher currents will lie below this curve, and curves for lower currents will lie above it. It is observed that for impedance values above 100% pick-up impedance, the relay does not operate The curve I represents the actual characteristic while curve II is simplified representation of the same (right angle instead of curve).

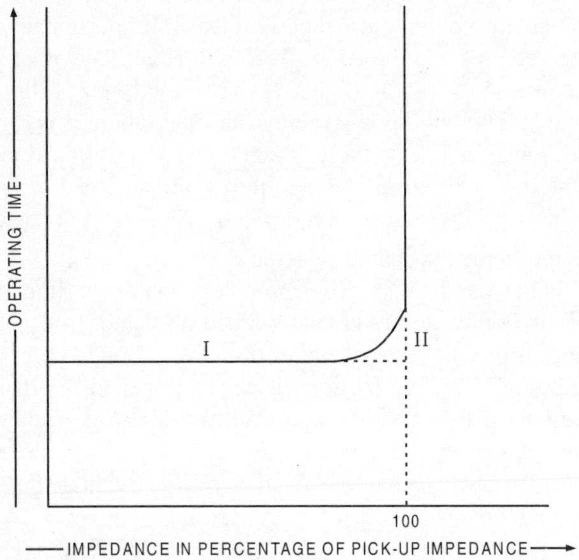

Fig. 9.39. ***Operating-Time Vs Impedance Characteristic of a High Speed Impedance Relay For One Value of Current***

9.15.1. Directional Impedance Relays

The directional feature to the impedance relay can be provided by employing the impedance relay along with a directional unit as is done in the case of a simple overcurrent relay to operate as a directional overcurrent relay. This means the impedance unit will operate only when the directional unit has operated.

Directional feature senses the direction in which fault power flows with respect to the location of CT and PT. Directional impedance relays operate for the following conditions:

1. Impedance between the fault point and relay location is less than the relay setting Z.
2. The fault power flows in a particular direction from relay.

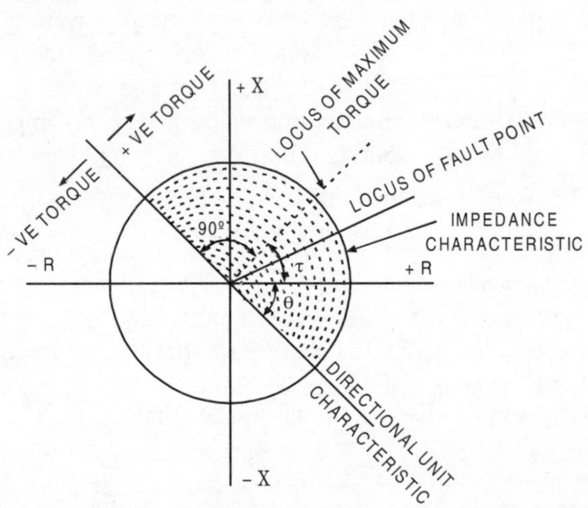

Fig. 9.40. ***Operating Characteristic of an Impedance Relay With Directional Unit***

The operating characteristic of an impedance relay with directional unit will be as shown in Fig. 9.40. Since the directional unit permits tripping only in the positive torque region the active portions of the impedance unit characteristic is shown shaded. The net result is that the relay will operate only for fault points that are both within the circle and above directional unit characteristic.

Torque Equation For Directional Impedance Relay. Torque equation for directional element is given as

$$T = K_1 VI \cos(\theta - \tau) \qquad \text{Refer to Eq. (9.7)}$$

where θ is the phase angle between V and I and τ is the impedance angle of the relay. Values of angles are taken as positive in clockwise direction.

When the relay is about to pick up,

the torque $T = 0$; $\cos(\theta - \tau) = 0$

or $(\theta - \tau) = \pm 90°$ or $\theta = \tau \pm 90°$

Hence for positive torque θ should be $\pm 90°$.

If the effect of spring control is taken into account, the torque equation becomes

$$T = K_1 VI \cos(\theta - \tau) - K_2$$

At balance point *i.e.*, when the relay is about to pick up

$$T = 0;$$

$$K_1 VI \cos(\theta - \tau) = K_2 \qquad ...(9.16)$$

Substituting $I = \frac{V}{Z}$ in above equation, we have

$$K_1 \times V \times \frac{V}{Z} \cdot \cos(\theta - \tau) = K_2$$

$$\text{or } Z = \frac{K_1}{K_2} V^2 \cos(\theta - \tau) \qquad ...(9.17)$$

The above Eq. (9.17) describes an infinite number of circles and for each value of V there will be a corresponding circle. The characteristics of directional impedance relay for one value of V can be represented on the R-X diagram as shown in Fig. 9.41.

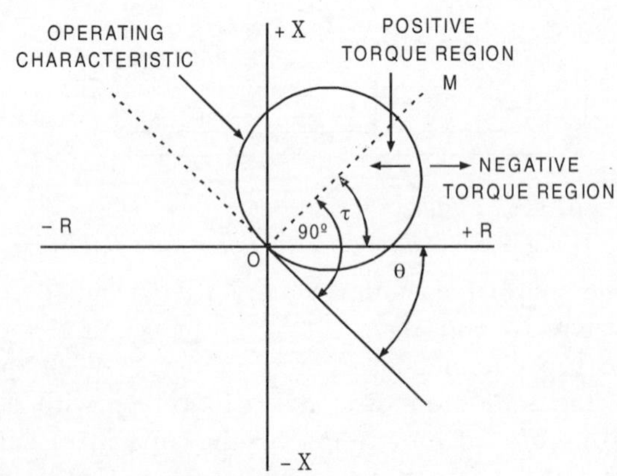

Fig. 9.41. ***Characteristics of a Directional Impedance Relay for One Value of Voltage***

The fact that for some values of θ, impedance Z will be negative which should be ignored. Negative Z has no significance and cannot be shown on R-X diagram.

When $\cos(\theta - \tau) = 1$ or $\theta = \tau$, impedance Z will have maximum value and will lie on the line of maximum torque OM. When $\cos(\theta - \tau) = 0$, $Z = 0$ and $\theta = 90° \pm \tau$. The characteristic will therefore be a circle with Z as diameter and the centre of the circle will lie on the line OM.

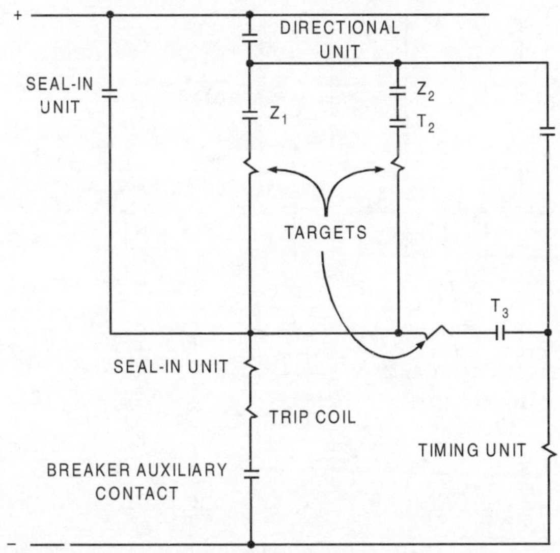

Fig. 9.42. ***Schematic Contact of Circuit Connections of an Impedance Type Distance Relay***

The diameter of the circle will be proportional to the square of the voltage. At normal voltage, and even at considerably reduced voltages, the diameter will be so large that for all practical purposes we may assume the straight line characteristic of Fig. 9.43.

9.15.2. Single-Phase Impedance Type Distance Relay For Transmission Line Protection

A single-phase impedance type distance relay for protection of transmission line consists of a single-phase directional unit, three high speed impedance relay units, and a time unit, together with the usual targets, seal-in-unit, and other auxiliaries. The contact circuits of the principal units are shown very schematically in Fig. 9.42. The three impedance units are labelled Z_1, Z_2 and Z_3 respectively. These can be adjusted for any desired value independently.

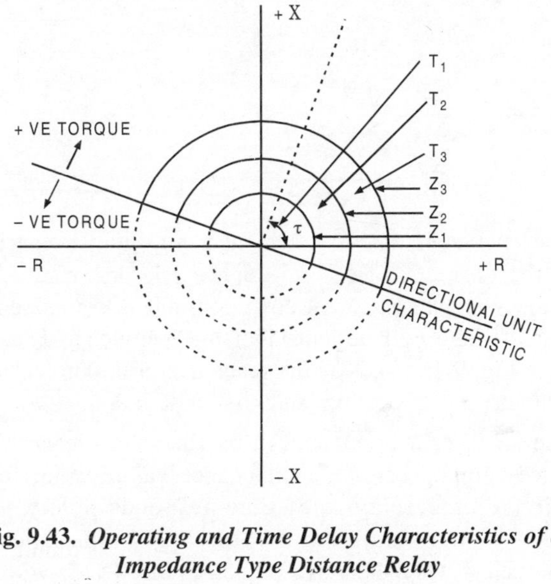

Fig. 9.43. ***Operating and Time Delay Characteristics of an Impedance Type Distance Relay***

On R-X diagram shown in Fig. 9.43 the circle for Z_1 is the smallest, the circle for Z_3 is the largest, and the circle for Z_2 is an intermediate. Thus we have divided the transmission line under protection in three zones. It can be seen from Fig. 9.43 that any value of impedance within the Z_1 circle will cause all the three impedance units to operate. The operation of Z_1 and the directional unit will trip a breaker directly in time T_1 which is very short. Whenever Z_3 and the directional unit operate, the timing unit is energised. After a definite delay, the timing unit will first close its T_2 contact, and later its T_3 contact, both time delays being independently adjustable. Therefore, it can be seen that a fault within zone Z_2 but outside zone Z_1 (*i.e.,* between the circles of Z_1 and Z_2) will result in tripping in T_2 time. And finally, a value of Z outside the Z_1 and Z_2 circles, but within the Z_3 circle, will result in tripping in T_3 time.

It will be noted that, if tripping is somehow blocked, the relay will make as many attempts to trip as there are characteristic circles around a given impedance point. However, use may not be made of this possible feature.

Figure 9.43 shows also the relation of the directional-unit operating characteristic to the impedance-unit characteristics on the same R-X diagram. Since the directional unit permits tripping only in its positive-torque region, the inactive portions of the impedance unit characteristics are shown dotted. Thus tripping will occur only for points that are both within the circles and above the directional unit characteristic.

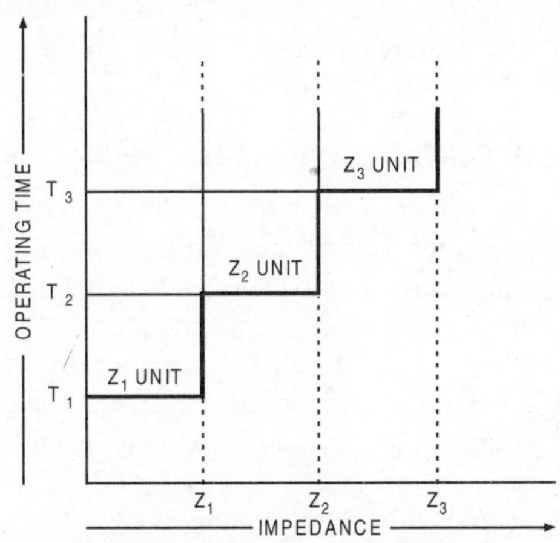

Fig. 9.44. ***Operating Time Vs Impedance Characteristic For an Impedance Type Distance Relay***

Operating time-impedance characteristic for an impedance type distance relay is shown in Fig. 9.44. This characteristic is usually called a "stepped" time-impedance characteristic. It is noteworthy that the Z_1 and Z_2 units provide the primary protection for a given transmission line section while Z_2 and Z_3 provide backup protection for adjoining busses and line sections.

9.15.3. Modified Impedance Type Distance Relay

The modified impedance type distance relay is like the impedance type except that the operating characteristics are shifted as illustrated in Fig. 9.45. It is shifted by introducing a current bias' which merely consists of introducing into the voltage supply an additional voltage proportional to the square of the current. Hence the modified torque equation is given as

$$T = K_1I^2 - K_2 (V + CI)^2 \qquad ...(9.18)$$

The term (V + CI) is the rms magnitude of the vector addition of V and CI involving angle θ between V and I as well as a constant angle in the constant C term. By such biasing, the characteristic circle can be shifted in any direction from the origin, and by any desired amount, even to the extent that the origin is outside the circle. Slight variations may occur in the biasing, due to circuit elements saturation. That is why, it is not the practice to make the circles pass through the origin, and therefore a separate directional unit is required as indicated in Fig. 9.45.

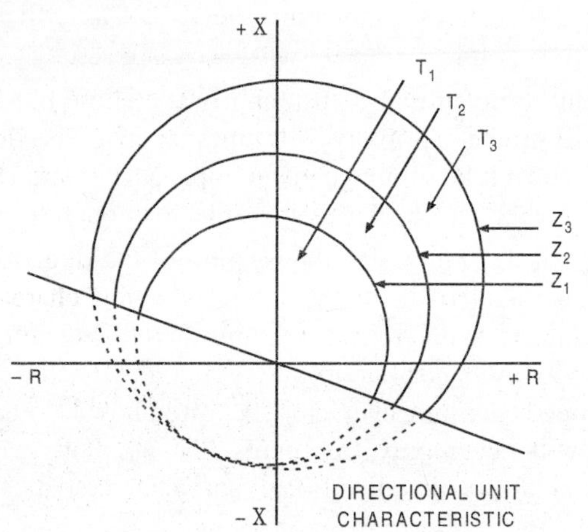

Fig. 9.45. ***Operating Characteristic of a Modified Impedance Type Distance Relay***

9.16. REACTANCE TYPE DISTANCE RELAY

Reactance type relay is a high speed relay. It consists of two units—an overcurrent element developing positive torque and a current-voltage directional element which either aids or opposes the overcurrent element depending on the phase angle between the current and voltage. This means, a reactance relay is an overcurrent relay with directional restraint. The directional element is arranged to develop maximum negative torque when its current lags behind its voltage by 90°. The induction cup or double induction loop structures are best suited for actuating reactance type distance relays.

A typical reactance relay using induction cup structure is shown in Fig. 9.46 (*a*). It has a 4-pole structure carrying operating, polarizing, and restraining coils, as shown in the figure. The operating torque is developed by the interaction of fluxes due to current carrying coils *i.e.,* interaction of fluxes of poles 2, 3 and 4 and the restraining torque is produced by the interaction of fluxes due to the poles 1, 2 and 4. Thus operating torque will be proportional to I^2 while the restraining torque will be proportional to VI cos (θ – 90°). The desired maximum torque angle is obtained with the help of resistance-capacitance circuits, as illustrated in the figure.

If the spring control effect is indicated by – K_3, the torque equation becomes

$$T = K_1I^2 - K_2VI \cos (\theta - 90^\circ) - K_3 \qquad ...(9.19)$$

where θ is defined as positive when I lags behind V

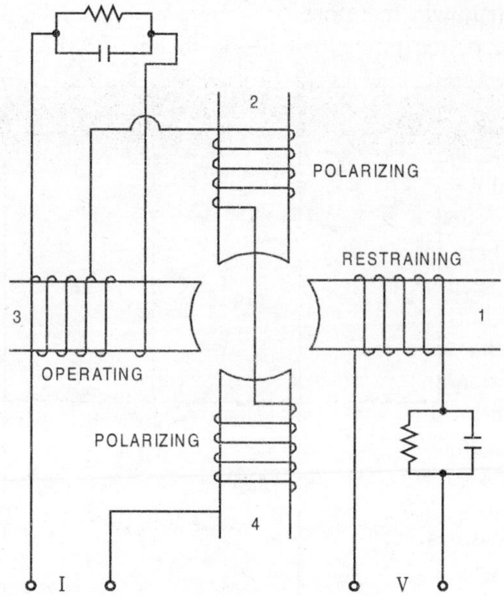

(a) Schematic Diagram

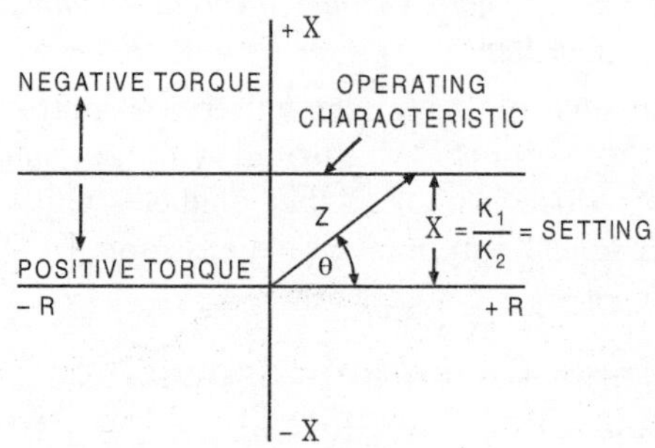

(b) Operating Characteristic

Fig. 9.46. ***Reactance Type Distance Relay***

$$\text{or } T = K_1I^2 - K_2VI \sin \theta - K_3$$

At balance point, the net torque is zero, and hence

$$K_1I^2 - K_2VI \sin \theta - K_3 = 0$$

$$\text{or } K_1I^2 = K_2VI \sin \theta + K_3$$

$$\text{or } K_1 = K_2 \frac{V}{I} \sin\theta + \frac{K_3}{I^2}$$

$$\text{or } \frac{V}{I} \sin \theta = \frac{K_1}{K_2} - \frac{K_3}{K_2I^2}$$

$$\text{or } Z \sin \theta = X$$

$$= \frac{K_1}{K_2} \text{ if the spring control effect is neglected.} \qquad ...(9.20)$$

The characteristic of a reactance relay will be such that for all the positions of the impedance phasors, whose heads lie on this characteristic will have constant X component *i.e.,* the characteristic will be a straight line parallel to R-axis, as shown in Fig. 9.46 (*b*). X is the reactance of the protected line between the relay location and the fault point.

The significant point about this characteristic is that the resistance component of the impedance has no effect on the operation of the relay, the relay responds solely to the reactance component. Any point below the operating characteristic—whether above or below the R axis-will lie in the positive torque region. If the value of τ, in the general torque equation $T = K_1I^2 - K_2V\,I\cos(\theta - \tau) - K_3$, is made any other than 90°, a straight line operating characteristic will still be obtained, but it will not be parallel to R-axis. Such a relay is called an "angle impedance relay".

This relay, as can be seen from its characteristic, is a non-directional relay. It is not capable of discriminating when used on transmission lines, whether the fault has taken place in the section where the relay is located or it has taken place in the adjoining section. The directional unit used with the reactance relay will not be same as used with the impedance type relay because the reactance relay will trip even under normal operating conditions at or near unity power factor. This is so because the restraining reactive volt-ampere in that case will be nearly equal to zero. Therefore the reactance type distance relay needs a directional unit that is inoperative under normal load conditions. The type of the unit employed for this purpose has voltage-restraining element opposing the directional element, and it is called an "admittance" or "mho" unit or relay. This will be dealt in the forthcoming section.

Reactance type relay is very suitable as a ground relay for ground fault because its reach is not affected by fault impedance (*i.e.,* arcing resistance).

9.17. MHO TYPE DISTANCE RELAY

Mho relay is a high speed relay and is also known as *admittance relay.* It is also sometimes called an *angle impedance* relay. In this relay operating torque is obtained by the volt-ampere element and the restraining torque is developed due to the voltage element. It means a mho relay is a voltage restrained directional relay.

A typical mho relay using induction cup structure is shown in Fig. 9.47 (*a*). The operating torque is developed by the interaction of fluxes due to poles 2, 3 and 4 and the restraining torque due to poles 1, 2 and 4.

If the spring control effect is indicated by $-K_3$, the torque equation becomes

$$T = K_1VI\cos(\theta - \tau) - K_2V^2 - K_3 \quad \text{...(9.21)}$$

where θ and τ are defined as positive when I lags behind V

At balance point, the net torque is zero, and hence

$$K_1VI\cos(\theta - \tau) - K_2V^2 - K_3 = 0$$

$$\text{or} \quad \frac{K_1}{K_2}\cos(\theta-\tau) - \frac{K_3}{K_2VI} = \frac{V}{I} = Z$$

$$\text{or } Z = \frac{K_1}{K_2}\cos(\theta - \tau) \quad \text{...(9.22)}$$

if the spring control effect is neglected.

The above equation is similar to that of a directional relay when the control spring effect is taken into account but the difference is that there is no voltage term in the expression. Hence the relay has but one circular characteristic. The operating characteristic described by Eq. (9.22) is given in Fig. 9.47(*b*). The diameter of this circle is practically independent of V or I, except at very low magnitudes of current or voltage when the control spring effect is considered, which causes the diameter to decrease. The diameter of the circle is given as

$$\frac{K_1}{K_2} = Z_R = \text{ohmic setting of the relay.}$$

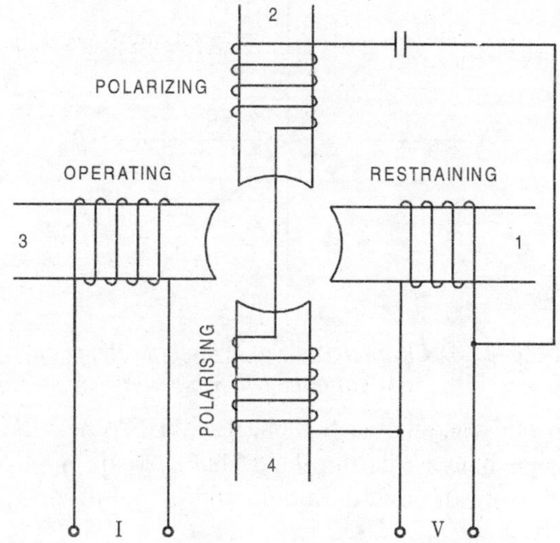

(*a*) *Schematic Diagram*

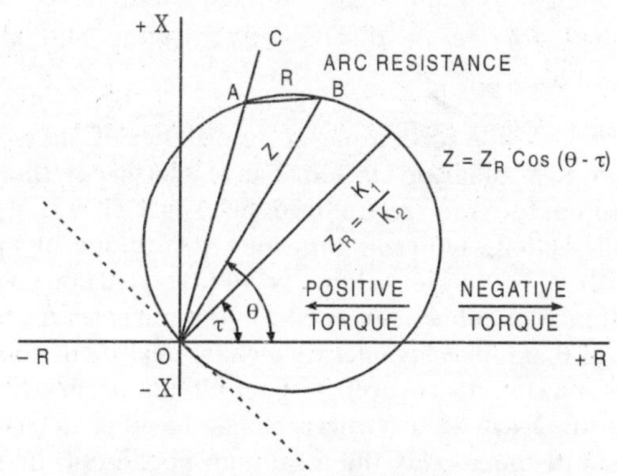

(*b*) *Operating Characteristic*

Fig. 9.47. ***Mho Type Distance Relay***

The relay operates when the impedance seen by the relay falls within the circle. It is the fact that the circle passes through the origin which makes the relay inherently directional. The relay because of its inherently directional characteristic needs only one pair of contacts which makes it fast tripping for fault clearance and reduces the VA burden on the CTs. The impedance angle of the protected line is normally 60 to 70° which is shown by line OC in Fig. 9.47 (*b*). The arc resistance R, is represented by length AB which is horizontal to OC from the extremity of the chord Z. By making τ equal to, or little less lagging than θ, the circle is made to fit very closely around the faulty area so that the relay is insensitive to power swings and therefore particular applicable to the protection of long or heavily loaded lines.

The Eq. (9.22) can also be written as

$$\frac{1}{Z}\cos(\theta-\tau) = \frac{K_2}{K_1}$$

$$\text{or} \quad Y\cos(\theta-\tau) = \frac{K_2}{K_1} = \text{Constant} \quad \text{...(9.23)}$$

where Y is the admittance in mho

For a given relay τ is constant, and the heads of the admittance phasor Y will lie on a straight line. The characteristic of mho relay on admittance diagram is, therefore, a straight line and is shown in Fig. 9.48.

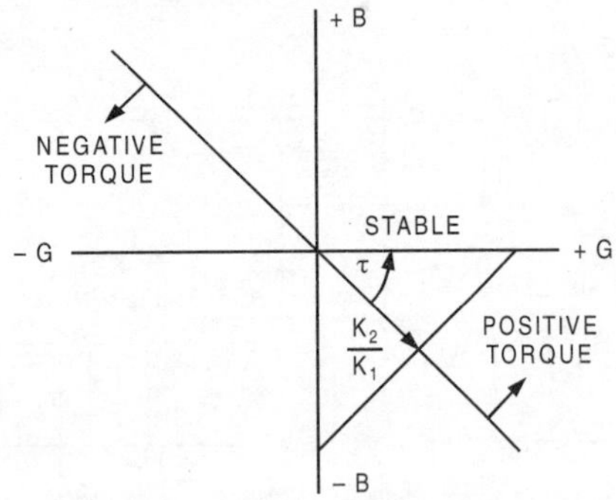

Fig. 9.48. ***Characteristic of Mho Type Relay on Admittance Diagram***

Mho relay is suitable for long EHV/UHV heavily loaded transmission lines as its threshold characteristic in Z-plane is a circle passing through the origin and its diameter is Z_R *i.e.,* $\frac{K_2}{K_1}$. Because of this, the threshold characteristic is quite compact enclosing fault area compactly and hence, there is lesser chance to operate during power swing and also it is directional.

Example 9.8. A line section has an impedance of 3.6 + j6 ohms. Show it on R-X diagram as impedance phasor. If the relay is adjusted to operate for zero impedance short circuit at the end of the line section, show on the same R-X diagram operating characteristics of (*i*) an impedance relay (*ii*) a reactance relay and (*iii*) admittance or mho relay, used for the purpose. Assume that the centre of the mho relay operating characteristic lies on the line impedance phasor. If an arcing short-circuit occurs having an impedance of $2 + j0\ \Omega$ anywhere along the line, determine for each type of distance relay the maximum portion of the line that can be protected.

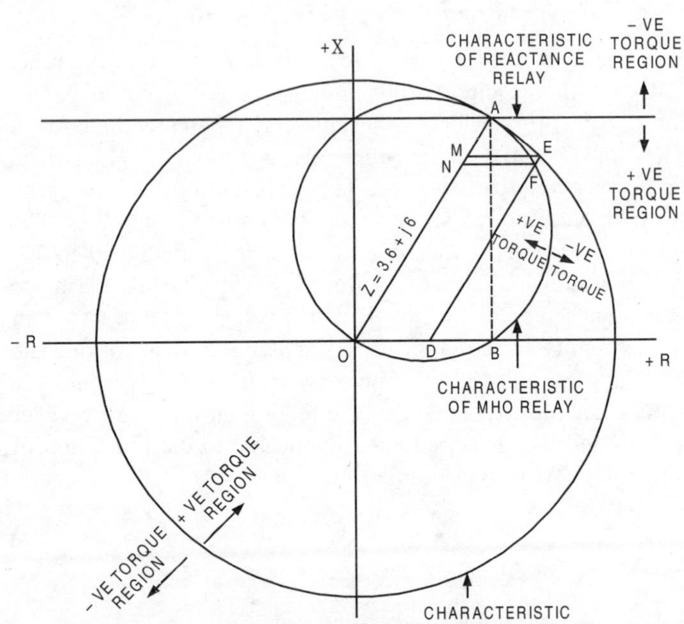

Fig. 9.49

Solution : The characteristics of the relays are shown in Fig. 9.49. OA is the impedance phasor of (3.6 + *j*6) ohms; OB = 3.6 units and BA = 6 units. The circle with O as centre and OA as radius represents the characteristic of impedance relay. A straight line drawn parallel to the R-axis at a distance of 6 units corresponding to the reactance of the line represents the characteristic of the reactance relay. The circle drawn on OA as diameter represents the characteristic of the admittance or mho relay for the protection of the given line.

If the impedance of short circuit due to arcing is (2 + *j* 0) ohms, this is represented by taking OD = 2 units and drawing a line parallel to the impedance phasor OA cutting the mho relay characteristic at F and impedance relay characteristic at E. Then draw EM and FN parallel to R-axis to cut impedance phasor OA at M and N respectively.

Then $\frac{OM}{OA} \times 100$ gives the percentage of line protected by the impedance relay and $\frac{ON}{OA} \times 100$ gives the percentage of line protected by the mho relay.

On measurement the impedance relay is found to protect 82% of the line while the mho relay is found to protect 77% of the line.

The reactance relay is unaffected by the presence of the arc resistance and hence even with arc present, the percentage of the line that is protected by the reactance relay is 100%.

9.18. CLASSIFICATION OF DISTANCE RELAYS

Distance relays used for the protection of power circuits may be divided into two groups viz., (*i*) definite distance relays and (*ii*) time distance relays.

Definite distance relays operate instantaneously when the impedance (reactance or admittance) falls below a specified value. These relays may be of impedance, reactance or mho type. A definite distance type impedance relay is shown in Fig. 9.37(*a*).

Time distance relays have the time of operation dependent upon the value of impedance (reactance or admittance) *i.e.,* upon the distance of the fault from the relay point. A fault nearer to the relay will operate it earlier than a fault farther away from the relay. These relays may be of impedance, reactance or mho type. Induction type time distance impedance relay is shown in Fig. 9.38(*a*).

9.19. DIFFERENTIAL PROTECTION

Differential protective relaying is the most positive in selectivity and in action. It operates on the principle of comparison between the phase angle and magnitudes of two or more similar electrical quantities. Comparing two electrical quantities in a circuit by means of differential relays is simple in application and positive in action. For example, in comparison of the current entering a line and the current leaving it, if more current enters the protected line than leaves it, the extra current must flow in the fault. The difference between the two electrical quantities can operate a relay to isolate the circuit.

A *differential relay* is defined as the relay that operates when the phasor difference of two or more similar electrical quantities exceeds a predetermined amount. This means that for a differential relay, it should have (*i*) two or more similar electrical quantities and (*ii*) these quantities should have phase displacement (normally approximately 180°) for the operation of the relay. Almost any type of relay, when connected in a certain way, can be made to operate as a differential relay. In other words, it is not so much the relay construction as the way the relay is connected in a circuit that makes it a differential relay. Most of the differential relays are of the "current differential" type in which phasor difference between the current entering the winding and current leaving the winding is used for sensing and relay operation.

Differential protection is generally unit protection. The protected zone is exactly determined by the location of CTs and PTs. The phasor difference is achieved by suitable connections of secondaries of CTs or PTs.

Differential protection principle is employed for the protection of generators, generator-transformer units, transformers, feeders (transmission lines), large motors and bus-bars.

9.19.1. Current Differential Relays

An arrangement of an overcurrent relay connected to operate as differential really is shown in Fig. 9.50. The dotted line represents the system element that is to be protected by the differential relay. This system element might be a length of circuit, a portion of the bus or a winding of a generator or that of a transformer. A pair of CTs are fitted on either end of the section to be protected. The secondaries of CTs are connected in series with the help of pilot wires in such a way that they carry the induced currents in the same direction. The operating coil of an overcurrent relay is connected across the CT secondary circuit, as shown in Fig. 9.50.

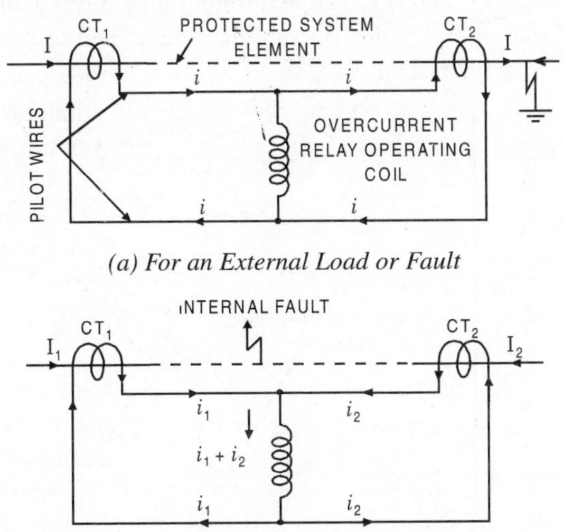

(a) For an External Load or Fault

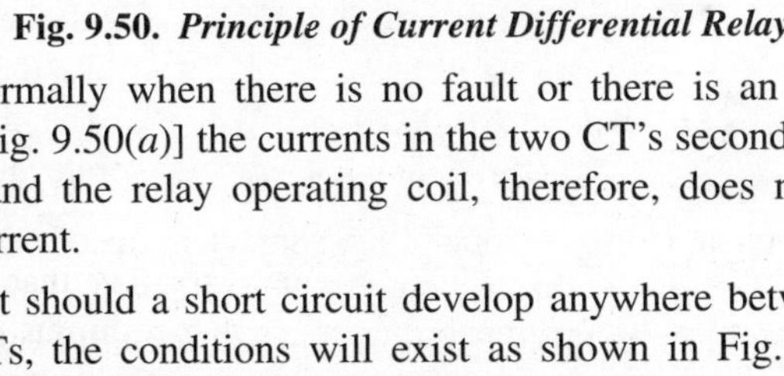

(b) For an Internal Fault

Fig. 9.50. ***Principle of Current Differential Relay***

Normally when there is no fault or there is an external fault [Fig. 9.50(*a*)] the currents in the two CT's secondaries are equal and the relay operating coil, therefore, does not carry any current.

But should a short circuit develop anywhere between the two CTs, the conditions will exist as shown in Fig. 9.50(*b*). If the current flows to the fault from both sides as shown, the sum of the CT secondary currents will flow through the differential relay, It is not necessary that fault current flows to the fault from both sides to cause secondary current flow to the differential really. A flow on one side only, or even some current flowing out of one side while a larger current entering the other side will cause a differential current to flow through the relay operating coil. In the other words, the differential relay current will be proportional to the phasor difference between the currents entering and leaving the protected element; and, if the differential current exceeds the relay's pick-up value, the relay will operate.

Differential Protection For 3-Phase Circuits. The above principle can be extended to a system element having several connections. Figure 9.51 shows the case for a three-phase circuit. It is only necessary, as before, that all the CTs have the same ratio, and that they be connected so that the differential relay carries no current when the total current entering the circuit is vectorially equal to that current leaving the circuit.

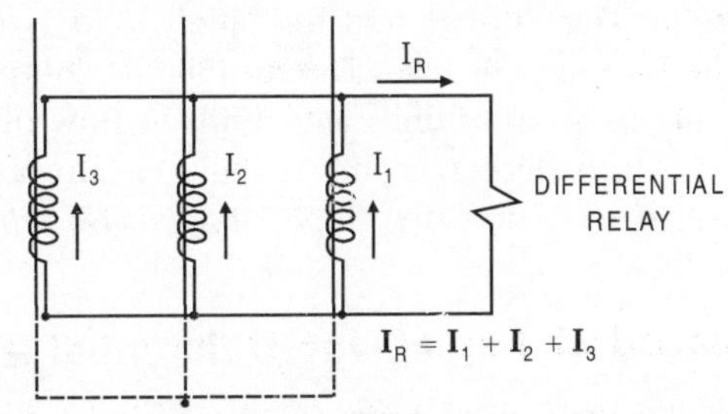

Fig. 9.51. ***Differential Protection For 3-Phase Circuit***

During normal operating conditions the three secondary currents of CTs are balanced and no current flows through the relay coil. But during fault in the protected zone, the balance is disturbed and differential current flows through the relay operating coil and when the differential current exceeds the relay's pick-up value, the relay operates.

The principle can still be applied for the protection of a 3-phase power transformers, but in this case, the ratios and connections of the CTs on the opposite sides of the power transformer must be such as to compensate for the magnitude and phase angle change between the power transformer currents on either side. This subject will be treated in detail in chapter 12.

Difficulties Associated With Differential Protection

1. *Difference in Length of Pilot Wires.* The power system element under protection and CTs are located at different places and normally it is not possible to connect the relay operating coil to the equipotential points. However, this difficulty can be overcome by connecting adjustable resistors in series with the pilot wires.

2. *CT Ratio Errors During Short Circuits.* The CTs used may have almost equal ratio at normal currents, but during short-circuit conditions, the primary currents are unduly large and the ratio errors of CTs on either side differ. This is due to (*i*) inherent difference in characteristic of CTs arising out of difference in magnetic circuit, saturation conditions etc. (*ii*) unequal dc components in the short-circuit currents.

Saturation of Magnetic Circuits of CTs Under Short-Circuit Condition. The differential relay of the type explained above is likely to operate inaccurately with heavy through (*i.e.,* external) faults. The relay may loose its stability for through faults. This drawback is overcome by using *percentage differential relay* or *biased differential relay* which will be explained in the forthcoming section.

Magnetizing Current Inrush at the Switching Instant. When the power transformer is connected to the supply, a large current (about 6 to 10 times full-load current) inrush takes place. The differential relay operates due to such inrush current, though the transformer has no fault. However, this difficulty is overcome by providing *harmonic restraint* for the differential relay. This relay filters the harmonic component from the inrush current and supplies it to the restraining coil. The harmonic content of the magnetizing current is used to obtain restraining torque during switching-in of transformer.

Tap Changing. Transformer transformation ratio is changed whenever the taps are changed. Due to this CT ratios do not match with the new tap settings and result in flow of current in pilot wires even during healthy condition. However, this problem is also overcome by employing *biased differential relay*.

9.19.2. Biased or Percentage Differential Relay

The most extensively used form of differential relay is the percentage differential or beam biased relay. This is essentially the same as the overcurrent type of current-balance relay that was described in Art. 9.19.1, but it is connected in a differential circuit, as illustrated in Fig. 9.52 (*a*). Schematic arrangement is shown in Fig. 9.52 (*a*) while equivalent circuit is shown in Fig. 9.52 (*b*). This system consists of an additional restraining coil connected in the pilot wires, as shown in the figure and current induced in both CTs flows through it. The operating coil is connected to the midpoint of the restraining coil. The reason for using this modification in circulating current differential relay is to overcome the trouble arising out of differences in CT ratios for high values of external short-circuit currents.

The torque due to restraining coil prevents the closing of trip circuit contacts, while the torque due to operating coil tends to close the trip circuit contacts. Under normal operating conditions and through load conditions, the torque developed by restraining coil is greater than the operating coil torque. Thus the relay remains inoperative. When an internal fault occurs, the operating torque exceeds the restraining torque. Consequently the trip circuit contacts are closed to open the circuit breaker. The restraining torque may be adjusted by varying the number of turns of the restraining coil.

The differential current required to operate this relay is a variable quantity, owing to the effect of the restraining coil. The differential current in the operating coil is proportional to $(I_1 - I_2)$ and the equivalent current in the restraining coil is proportional to $\frac{(I_1 + I_2)}{2}$ as the operating coil is connected to the midpoint of the restraining coil.

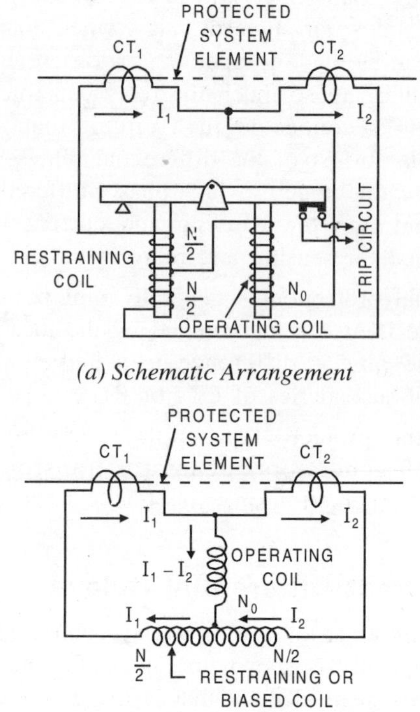

(*a*) *Schematic Arrangement*

(*b*) *Equivalent Circuit Diagram*

Fig. 9.52. ***Biased or Percentage Differential Relay***

The torque developed by the operating coil is proportional to the ampere-turns *i.e.,* $T_0 \propto (I_1 - I_2) N_0$ where N_0 is the number of turns on the operating coil. The torque due to restraining coil $T \propto \frac{(I_1 + I_2)}{2} N$ where N is the number of turns on the restraining coil. For external faults both I_1 and I_2 increase and thereby the restraining torque increases which prevents the maloperation.

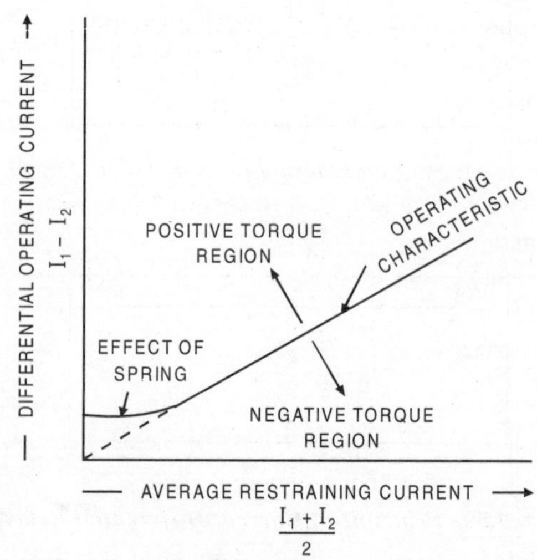

Fig. 9.53. ***Operating Characteristic of a Percentage Differential Relay***

It is clear from the operating characteristic of the relay shown in Fig. 9.53, that except for the effect of the control spring at low currents, the ratio of the differential operating current to the average restraining current is a fixed percentage. This is why it is known as *percentage differential relay.* The relay is also called the *biased differential relay* because the restraining coil is also called a biased coil as it provides additional flux.

The percentage or biased differential relay has a rising pick-up characteristic. So with the increase of magnitude of through current, the relay is restrained against maloperation.

Figure 9.54 shows the comparison of a simple overcurrent relay with a percentage differential relay under such conditions. An overcurrent relay having the same minimum pick-up as a percentage differential relay would operate undesirably when the differential current merely exceeded the value X, while there would be no tendency for the percentage differential relay to operate.

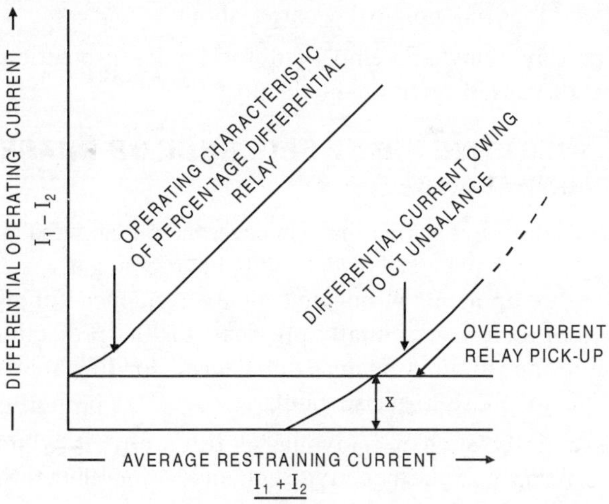

Fig. 9.54

Three-Terminal System Application of a Percentage Differential Relay. Percentage differential relay protection can be applied to the system elements having more than two terminals, as in the three-terminal application shown in Fig. 9.55. Each of the three restraining coils has the same number of turns, and each coil develops a torque independently of others. Their torques are added arithmetically. The per cent slope characteristic for such a relay will vary with the distribution of currents between the three restraining coils. Percentage differential relays are generally instantaneous or high speed. Time delay is not required for selectivity because the percentage differential characteristic makes these relays immune to the effects of transients when the relays are properly applied.

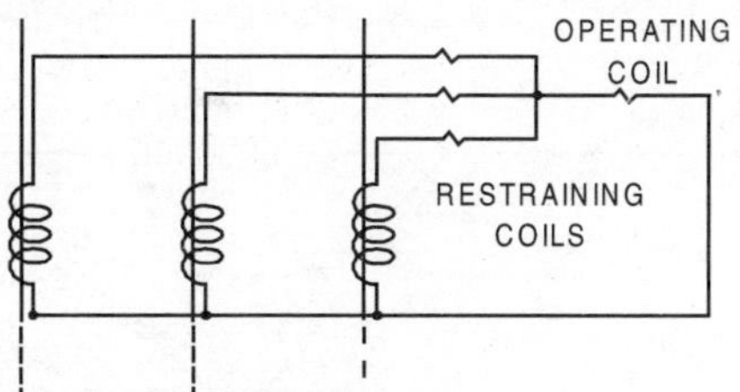

Fig. 9.55. ***Three-Terminal System—Application of Percentage Differential Relay***

Induction Type Biased Differential Relay. In the simplest electromagnetic form the relay is shown in Fig. 9.56. Such a relay consists of a pivoted disc free to rotate in the air gaps of two electromagnets, a portion of each pole of which is fitted with a copper shading ring. This ring can be moved further into, or out of the pole. The disc experiences two torques—one due to the operating element and the other due to restraining element. If the shading rings were in the same position on each element, the resulting torque experienced by the disc would be zero. But if the shading rings of restraining element were moved further into the iron core, the torque exerted by the restraint element will exceed than that of the operating element.

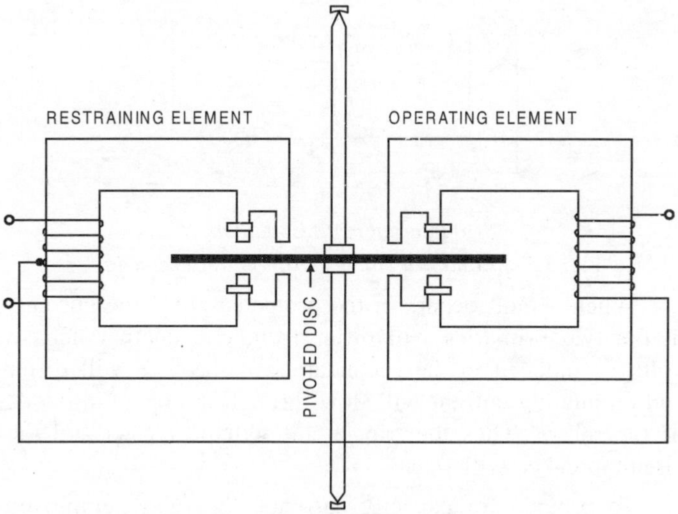

Fig. 9.56. ***Induction Type Biased Differential Relay***

The advantages of such a relay over the beam type are (*i*) the induction element is not susceptible to operation due to transients and (*ii*) a slight time delay can be obtained and the biasing feature can be finally adjusted merely by changing the position of the shading rings on either or both elements.

9.19.3. Voltage Balance Differential Relay

The differential relays described above are known as current balance relays. Such relays are convenient where both ends of the protected element are close together *e.g.,* with generator or transformer protection but do not suit for the protection of feeders. If such relays are used for feeder protection of several km length, the secondary emfs of the CTs would be required to circulate about 1 or 5 A at full load or several times the current during external fault conditions, round a pilot loop of fairly high impedance. Such a burden is impracticable for any economic design of CT. Another class of relays are the voltage balance differential relays, which are preferred for the feeder protection.

In this arrangement, two similar current transformers are connected at either end of the system element under protection (such as a feeder) by means of pilot wires. The relays are connected in series with the pilot wires, one at each end. The relative polarity of the CTs is such that there is no current through the relay under normal operating conditions and through fault conditions [Fig. 9.57 (*a*)]. The CTs used in such protection should be such that they should induce voltages in the secondary linearly with respect to the current. Since the magnitude of fault current is very large, in order that the voltage should be a linear function of such large currents the CTs should be air-cored.

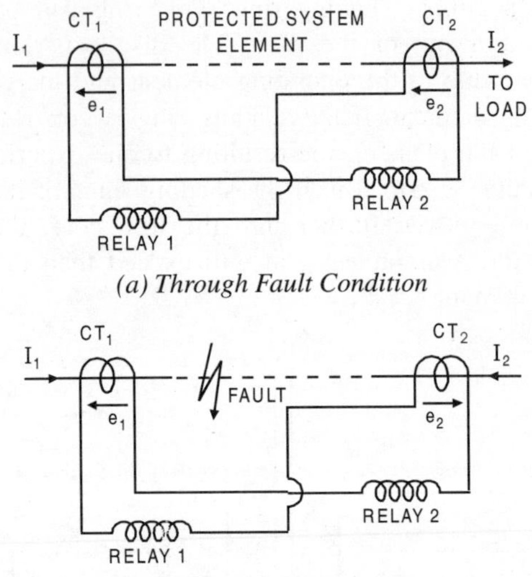

(a) Through Fault Condition

(b) Internal Fault Condition

Fig. 9.57. ***Balanced Voltage Differential Protection***

When a fault occurs in the protected zone, the currents in the two primaries will differ from one another and so voltages induced in the secondaries of the CTs will differ and circulating current will flow through the operating coils of the relays. Thus the trip circuit will be closed and the circuit breaker will open.

To provide for capacity currents, the relays employed may be overcurrent type which should operate only when the difference of the currents on both sides exceeds certain value.

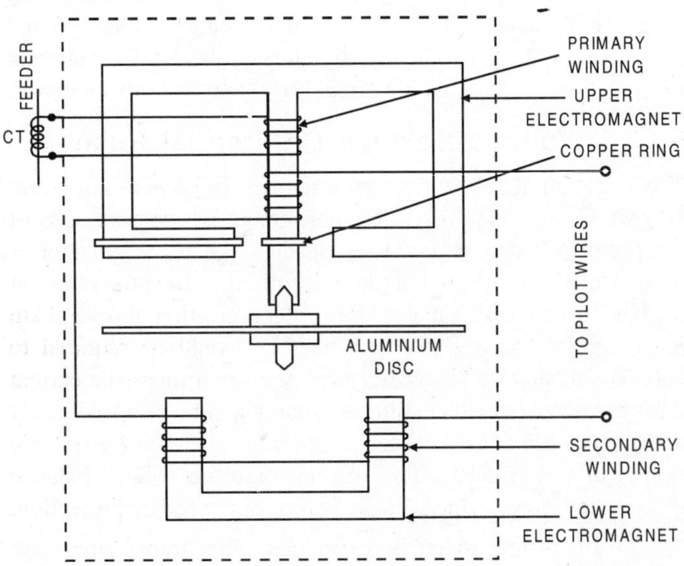

Fig. 9.58. ***Transfer Relay***

In this system no restraining coil or balancing resistance or overload coil is required.

Though this method is more reliable that current balance or circulating current system but has great disadvantage that CTs do not carry current so acts as open circuited and inserts a high impedance in the circuit. This method may be employed for protection of feeders, alternators and transformers. For use on transformers, the turn-ratio of power transformers must be kept in view.

9.20. TRANSLAY RELAY

The Translay relay is quite similar in construction to an overcurrent induction type relay with the following exceptions:

(*i*) The secondary winding distributed over the central limb of the upper electromagnet and the lower electromagnet does not close upon itself but is open to be connected to pilot wires.

(*ii*) The central limb and the left hand side limb of the upper electromagnet are shaded with short-circuited thick copper rings. These rings serve as compensating devices neutralizing the effects of capacitance currents in pilot wires and thus any unbalance between the CTs need not to be cared about.

Translay relay are employed for feeder protection, as will be discussed later on in details.

9.21. NEGATIVE PHASE SEQUENCE OR PHASE UNBALANCE RELAY

It has been discussed earlier that whenever there is an unbalanced circuit, the unbalanced currents will have a negative phase sequence component. A negative phase sequence (or phase unbalance) relay is essentially provided for the protection of generators and motors against unbalanced loading that may arise due to phase-to-phase faults.

Essentially such a relay has a filter circuit which is responsive only to the negative sequence components. Since small magnitude overcurrent can cause dangerous conditions, it becomes necessary to have low setting of such relays. No doubt, an earth relay can also provide the desired protection but only in case when there is a fault between any phase and earth. For phase-to-phase faults an earth relay cannot provide necessary protection and hence negative phase sequence relay is required.

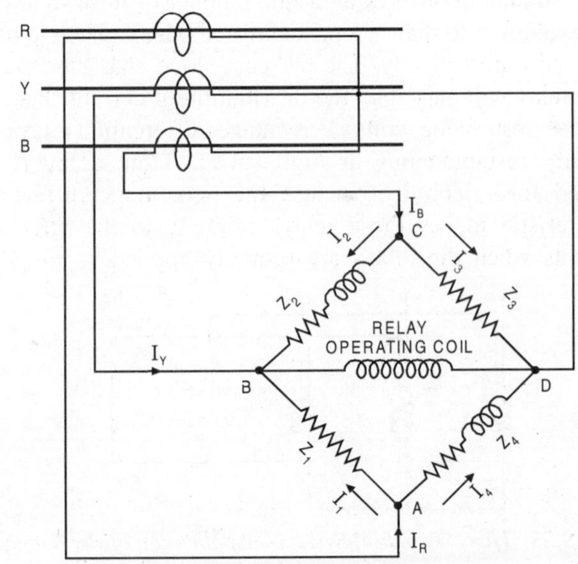

Fig. 9.59. ***Negative Phase Sequence Relay***

Figure 9.59 illustrates the scheme used for negative phase sequence relay. A network consisting of four impedances Z_1, Z_2, Z_3 and Z_4 of equal magnitude connected in a bridge formation is energized from three CTs. A single pole relay having an inverse-time characteristic is connected across the circuit, as illustrated in the figure. Z_1 and Z_3 are non-inductive resistors while Z_2 and Z_4 are composed of both resistance and inductance. The values of Z_2 and Z_4 are so adjusted that currents flowing through them lag behind those in impedances Z_3 and Z_1 by 60°. The relay is assumed to have negligible impedance. The current from phase R at junction A is equally divided into two branches as I_1 and I_4 but I_4 will lag behind I_1 by 60°.

From Fig. 9.60

$$I_1 = I_4 = \frac{I_R}{\sqrt{3}} \quad \because \; I_R^2 = I_1^2 + I_4^2 + 2 I_1 I_4 \cos 60°$$

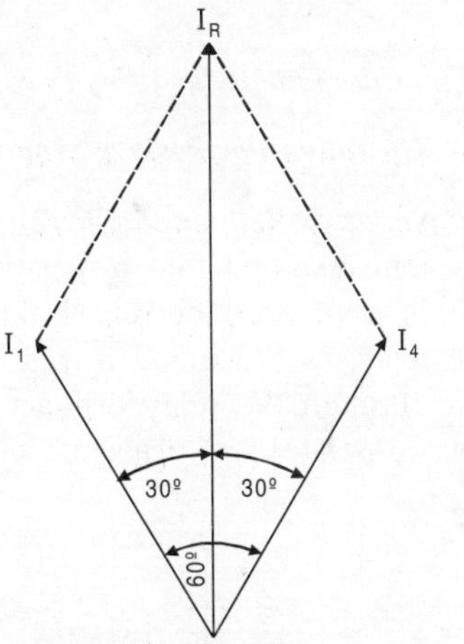

(a) Division of Current in Phase R

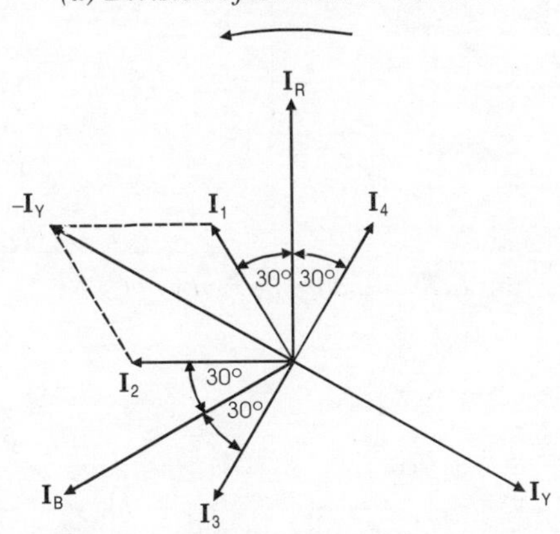

(b) Positive Sequence Currents in Normal Operating Condition

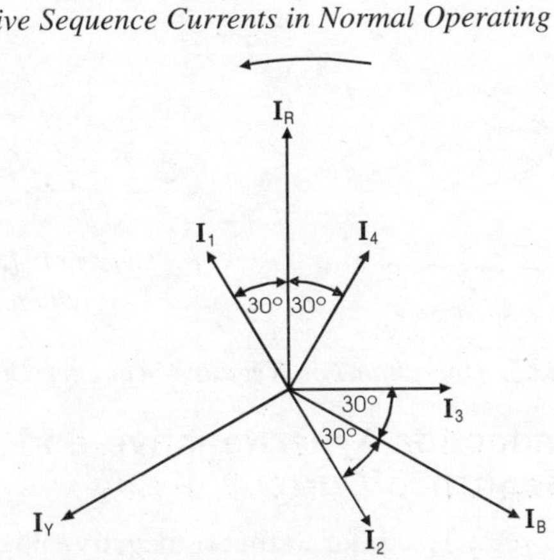

(c) Negative Sequence Currents

Fig. 9.60

Similarly the current from phase B divide at junction C into two equal components I_3 and I_2; I_2 lagging behind I_3 by 60°.

$$I_2 = I_3 = \frac{I_B}{\sqrt{3}}$$

Note that I_1 leads I_R by 30° while I_4 lags behind I_R by 30°. Similarly I_2 lags behind I_B by 30°, whereas I_3 leads I_B by 30°.

The current through relay operating coil at junction B will be equal to phasor sum of I_1, I_2 and I_Y.

i.e., $\quad \mathbf{I}_{RELAY} = \mathbf{I}_1 + \mathbf{I}_2 + \mathbf{I}_Y$

$$= \frac{I_R}{\sqrt{3}} \text{ leading } I_R \text{ by } 30° + \frac{I_B}{\sqrt{3}} \text{ lagging behind } I_B \text{ by } 30° + I_Y \qquad ...(9.24)$$

Flow of + ve Sequence Currents. Figure 9.60 (*b*) represents the phasor diagram when the load is balanced or when there is no negative sequence current. Since the current through the relay is $\mathbf{I}_1 + \mathbf{I}_2 + \mathbf{I}_Y = 0$ because $\mathbf{I}_1 + \mathbf{I}_2 = -\mathbf{I}_Y$

So the relay remains in operative for a balanced system.

Flow of – ve Sequence Currents. Figure 9.60 (*c*) represents the phasor diagram for negative sequence currents. It is noted that at junction B current I_1 and current I_2 are equal but opposite to each other, so they cancel each other and current I_Y flows through the relay operating coil. Thus the relay operates due to flow of current I_Y through it. A low setting value well below the normal full-load rating of the machine is provided since comparatively small values of unbalance currents produce a great danger.

Flow of Zero Sequence Currents. The current at junction B

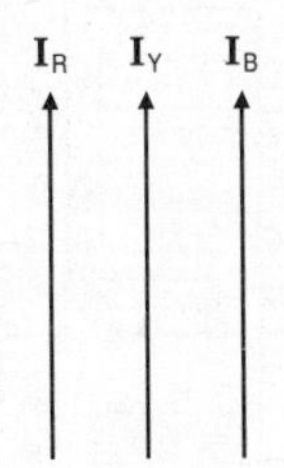

(a) Zero Phase Sequence Components of the Secondaries of CTs

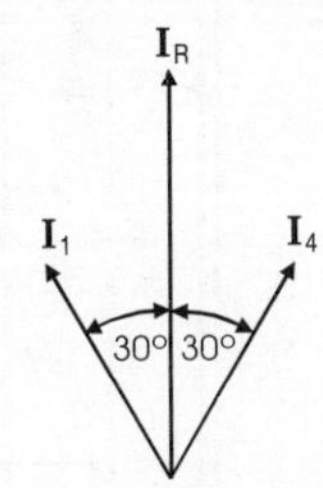

(b) Phasor Representation of Currents I_1 and I_4

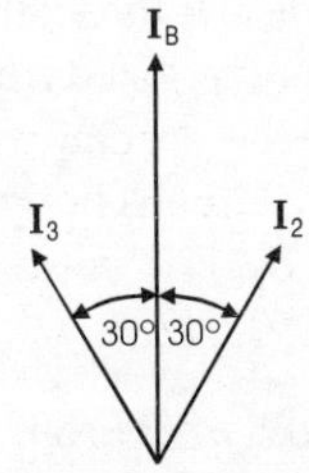

(c) Phasor Representation of Currents I_2 and I_3

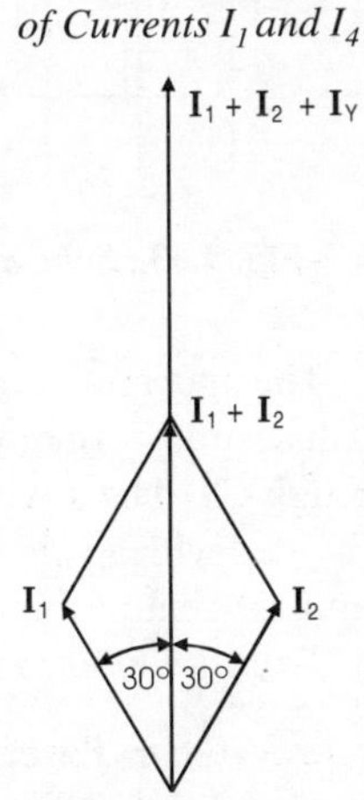

(d) Current Through Relay

Fig. 9.61

of the relay is represented in phasor diagram [Fig. 9.61 (*d*)] from which it is observed that the currents I_1 and I_2 are displaced from each other by 60°, so that the resultant of these currents is in phase with the current in phase Y. Thus a total current of twice the zero sequence current would flow through the relay and would therefore cause its operation.

To make the relay inoperative under the influence of zero sequence current, the CTs are connected in delta as shown in Fig. 9.62 because then no zero sequence current can flow in the network circuit.

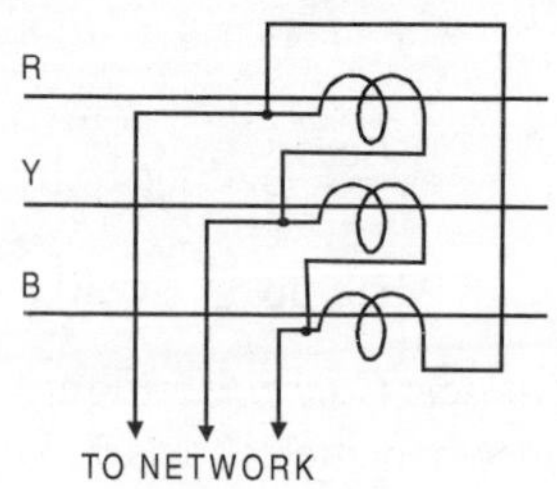

Fig. 9.62

9.21.1. Induction Type Negative Phase Sequence Relay

This type of relay has similar construction as that of an induction type overcurrent relay. The difference lies only in the primary winding provided on the central limb of the upper electromagnet which is provided with a central tap resulting into three terminals 1, 2 and 3 of this winding, as shown in Fig. 9.63.

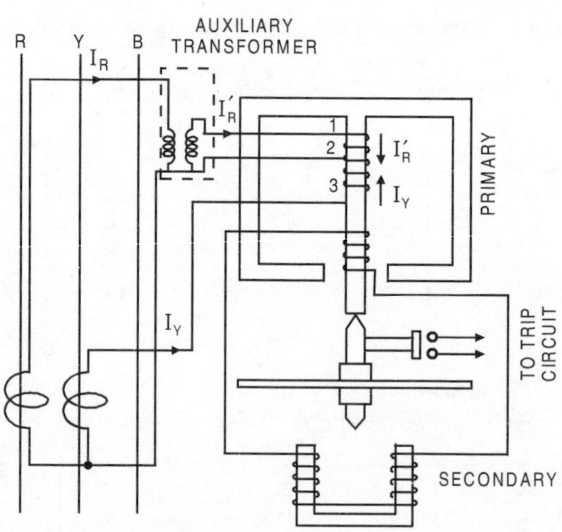

Fig. 9.63. ***Schematic of Induction Type Negative Phase Sequence Relay***

The upper half is energized from phase R through CT and an auxiliary transformer while lower half is energized from phase Y through CT. The auxiliary transformer has special construction (provided with an air gap in its magnetic circuit) in order that the output current of this transformer lags by 120° instead of usual 180° from the input, as illustrated in Fig. 9.64 (*a*).

Operation For Positive Sequence Currents. From Fig. 9.64, it is obvious that the currents I'_R and I_Y flowing through the primary winding of the relay are in opposition, the auxiliary transformer is so arranged that I'_R and I_Y are of equal magnitude. Thus the relay remains inoperative for a balanced system.

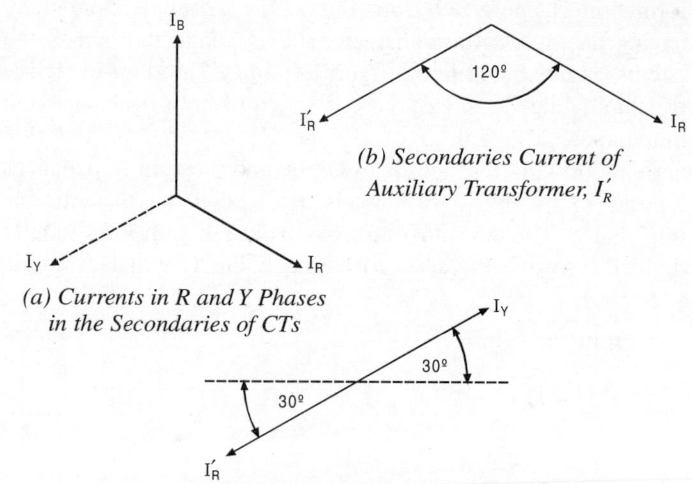

(*a*) *Currents in R and Y Phases in the Secondaries of CTs*

(*b*) *Secondaries Current of Auxiliary Transformer, I'_R*

(*c*) *Current in Relay Primary Winding*

Fig. 9.64. ***Operation For Positive Sequence Currents***

Operation For Negative Sequence Currents. When there is a fault on the system resulting into a negative phase sequence currents, there is a flow of current I through the primary winding of the relay, as illustrated in Fig. 9.65(*c*). When the current flowing through the relay primary exceeds the relay setting, the relay operates and trips the circuit.

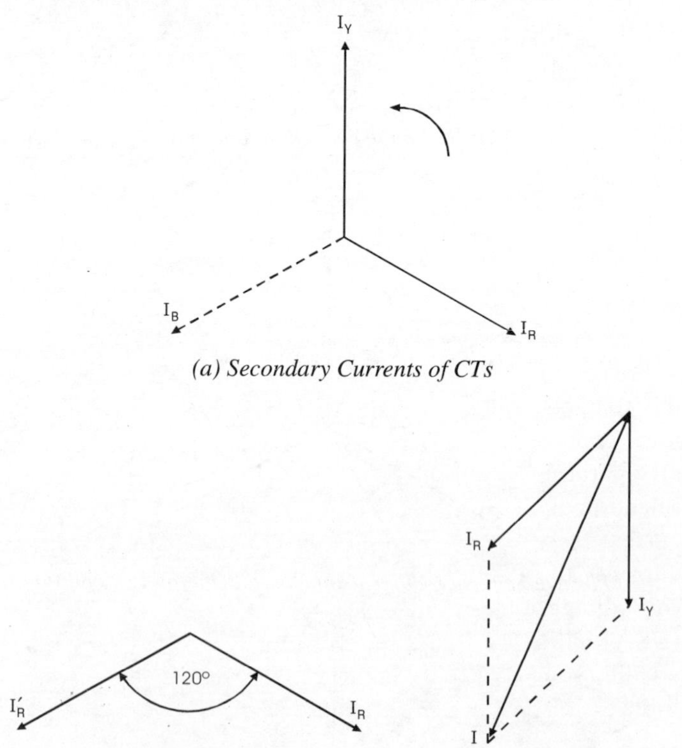

(*a*) *Secondary Currents of CTs*

(*b*) *Secondary Current of Auxiliary Transformer, I'_R*

(*c*) *Current in Relay Primary Winding, I*

Fig. 9.65. ***Operation For Negative Sequence Currents***

9.21.2. Induction Type Negative and Zero Sequence Current Relay

This type of relay is similar to that of negative phase sequence relay discussed above. The relay is made to respond to the flow of zero sequence currents also by providing another winding on the central limb of the upper electromagnet, connected in the residual circuit of three line CTs, as illustrated in Fig. 9.66.

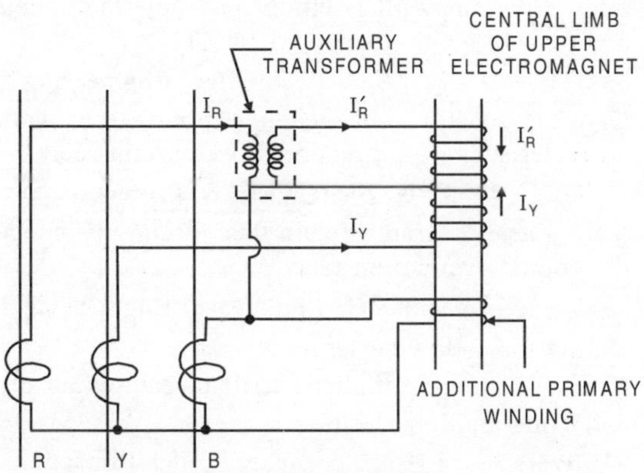

Fig. 9.66. ***Arrangement For Response To Negative and Zero Sequence Currents (Only the Central Limb of the Upper Magnet has been Illustrated)***

9.22. RELAY DESIGN CONSIDERATION

The most significant considerations in the design for construction are reliability, simplicity of construction and circuitry. The relay construction is divided into the following :

(*i*) Contacts (*ii*) Bearings (*iii*) Electromechanical design and (*iv*) Terminations and housing.

1. Contacts. Contact performance is probably the most significant part that affects the relay reliability. Corrosion or deposition of dust may result in relay non-operation. Thus, the material and shape of contacts are of considerable importance. A good contact system design gives limited contact resistance and reduced contact wear. Gold, gold alloys, platinum, silver and palledium may be used as contact material. The selection of the contact material depends upon several factors such as voltage per contact, break, the current to break and the type of atmospheric pollution under which the contacts are to be operated. However, there is no one ideal contact material which can be used universally under all operating conditions. The factors to be considered in selection of a suitable contact material are : (*i*) Nature of the current to be interrupted (ac or dc); (*ii*) Magnitude of the currents to be interrupted; (*iii*) Frequency of operation; (*iv*) Voltage at break and make operations; (*v*) Contact shape; (*vi*) Actual speed of contact at make or break including the arc duration and contact bounce and (*vii*) the contact closing force.

2. Bearings. The bearings used may be single ball bearings; Multi-ball bearings; Pivot and jewel bearings; or knife edge bearings. *Single ball bearing* is used for high sensitivity and low friction. A single ball bearing between two cup-shaped sapphire jewels is in use. *Multi-ball bearings,* available in less than 1.6 mm diameter, provide low friction and greater resistance to shock and combine side-thrust and end thrust in a single bearing. *Pivot and jewel bearings* are the most common for precision relays such as induction relays. Modern relays have spring-mounted jewels in order to take shocks on a shoulder and not on the jewel. *Knife edge bearings* are used in case of hinged armature relays which normally operate many contacts.

3. Electromechanical Design. Electromechanical design includes design of magnetic circuit and the mechanical fixtures of core, yoke and armature.

The reluctance of the magnetic path is kept to a minimum by enlarging the pole face that makes the magnetic circuit more efficient. DC electromagnets are usually economical and more efficient than ac electromagnets. However, small ac electromagnets made from soft iron, low carbon steel core having a slot for mounting a shaded ring are quite common. The relay coil current is usually restricted to 5 A and the coil voltage to 220 V. However, still the insulation for the relay coil is designed to withstand at least 4,000 V. The relay coil is designed to carry about 15 times the normal current for one second. From the mechanical strength considerations, it is desirable that the conductor diameter must not be less than 0.5 mm even if the above considerations are met.

4. Terminations and Housing. The assembly of armature with magnet and the base is made with the help of a spring. The spring is insulated from the armature by moulded blocks which provide dimensional stability and better appearance economically. Stainless steel, phosphorous bronze, nickel silver or berillium copper may be used as material for springs and nylon is used as material for moulded blocks. The fixed contacts are usually rivetted or spot welded on the terminal link.

Exercises

1. What is a protective relay ? Explain its function in an electrical system.
2. What are the fundamental requirements of protective relaying ? Depending upon their (*i*) construction and principle of operation and (*ii*) time of operation, how relays are classified ?
3. Explain briefly the role of protection in a power system.

   Describe the essential features of a protective relay with reference to reliability, selectivity, speed of operation and discrimination. With reference to protective relays define the terms :

   Pick-up, operating time, seal-in-relay and burden.
4. What are the shortcomings in electromagnetic relays ? Discuss briefly. Discuss also how these shortcomings are overcome in induction cup (induction cylinder) relays.

   Discuss briefly any five of the following terms :

   (*i*) Pick-up (*ii*) Threshold operation (*iii*) Balance point *i.e.,* Reach of a relay. (*iv*) Burden of a relay (*v*) Blinder (*vi*) Flag indicator (*vii*) Power swing.
5. Derive an expression for torque produced in an induction relay.

   [G.B. Technical Univ. Switchgear and Protection 2010-11]
6. Draw a neat sketch of an induction disc or cup type relay and explain its principle of operation.

   [U.P. Technical Univ. Switchgear and Protection 2008-09]

7. Explain the working principle of electromagnetic induction type relays. What is use of shading ring ?
[U.P. Technical Univ. Switchgear and Protection 2007-08]

8. Discuss about wattmetric type induction disc relay in detail.
[U.P. Technical Univ. Switchgear and Protection 2009-10]

9. Discuss with necessary circuit diagram, the principle of operation of an induction disc relay. What are the advantages of induction cup relays over induction disc relays ?

10. For what type of protection will you recommend (*i*) induction disc type (*ii*) induction cup type construction ?
What measures are taken to minimise the overrun of the disc ?
Derive an expression for torque produced by an induction relay.

11. What are the different types of electromagnetic relays ? What are their field of applications ?
[G.B. Technical Univ. Switchgear and Protection 2011-12]

12. Derive the torque equation of an induction disc relay ?
[W.B. Univ. of Technology Power System II, 2010]

13. What do you mean by symmetrical and unsymmetrical faults in a 3-ϕ system ? Deduce the mathematical expression for the force developed in an induction cup relay.
[W.B. Univ. of Technology Power System II, 2010]

14. Explain with a sketch the operation of a thermal relay and its field of application.

15. What is Buchholz relay ? Which equipment is protected by it ? For what tyne of fault is it employed ? With neat diagram discuss its working principle.

16. Draw a neat diagram of gas actuated relay. What is its location and use in the device ?
[U.P. Technical Univ. Switchgear and Protection 2009-10]

17. Explain the functions of Buchholz relay.
[U.P. Technical Univ. Switchgear and Protection 2004-05]

18. Describe the construction, principle of operation and application of Buchholz relay.
[U.P. Technical Univ. Switchgear and Protection 2005-06]

19. Give the various types of overcurrent relays and give their approximate characteristics.
[Pb. Univ. Elec. Power-III (Switchgear & Protection) 1991]

20. Explain the functions of a protective relay in a power system. Discuss the principal features of a good protective relay.
Draw and explain the principle of operation of an induction type overcurrent relay.

21. Describe the different types of overcurrent relays.
[U.P. Technical Univ. Switchgear and Protection 2004-05]

22. Write a short note on time-current characteristics of an overcurrent relay. Classify various types of overcurrent relays and give their applications.
[U.P. Technical Univ. Switchgear and Protection 2006-07]

23. Classify various types of overcurrent relays and give their applications alongwith their approximate characteristics and general time-current equations.
[U.P. Technical Univ. Switchgear and Protection 2008-09]

24. Describe with a neat sketch the operating principle of an inverse type overcurrent relay. How plug settings control the pick-up current of such a relay ? Define PSM and explain its significance.

25. Describe the difference between definite characteristic and inverse characteristic of relays.
[Pb. Univ. Elec. Power III (Switchgear & Protection) June 1993]

26. Define the terms 'Plug Setting Multiplier' and "Time Multiplier Setting" as used in the context of an IDMTL relay. How many plug settings and time multiplier settings are available with a standard IDMTL relay ?
[U.P.S.C. I.E.S. Elec. Engineering II, 2000]

27. Give the constructional features, principle of working and characteristics of a directional overcurrent relay.
[Pb. Univ. Switchgear & Protection Sept. 1991]

28. With a neat diagram explain the working of induction type directional overcurrent relay.
[U.P.S.C. I.E.S. Elec. Engineering II, 1993]

29. Define the following terms :
(*i*) Plug setting multiplier, (*ii*) Current setting of a relay
(*iii*) Time-multiplier setting
[U.P. Technical Univ. Switchgear and Protection 2004-05]

30. What do you mean by time multiplier setting (TMS) and plug multiplier setting of an overcurrent relay ? Explain with the help of relay characteristics.
[G.B. Technical Univ. Switchgear and Protection 2010-11]

31. Give the constructional features, principle of working and characteristics of a directional overcurrent relay.
What do you understand by IDMT relay. Draw the typical characteristic of an IDMT relay.
[U.P. Technical Univ. Switchgear and Protection 2007-08]

32. Describe the principle of a directional overcurrent relay. How does it help in discrimination in protection of parallel feeders, ring mains ?
[Pb. Univ. Switchgear & Protection December 1992]

33. Explain V-I and polar characteristics of a directional relay.

34. Define 'reset' and 'pick-up' value of an overcurrent relay. Why reset to pick-up ratio is low in dc attracted armature type relay as compared to ac relay ?
Explain with reasons why an unpleasant noise (relay chattering) sometimes occurs in an ac attracted armature type relay. Discuss the remedial measures adopted to minimise this unpleasant noise.
Discuss with necessary circuit diagram a typical directional overcurrent relay. Draw with necessary phasor diagrams the (*i*) 30° connection, (*ii*) 90° connection as applied to directional relay.

35. Explain the distance relay protection scheme.
[U.P.S.C. I.E.S. Electrical Engineering II, 1996]

36. Describe the principle of impedance type distance relays.
[Pb. Univ. Switchgear & Protection December 1992]

37. What is distance relay ? Draw its characteristics. How is directional features added with overcurrent relays ? Why is it required ? **[G.B. Technical Univ. Switchgear and Protection 2009-10, 2011-12]**

38. Describe different types of distance relays used for protection of transmission line of power system. Which one is the best and why ?
[G.B. Technical Univ. Switchgear and Protection 2009-10, 2011-12]

39. What are the different types of distance relays ? Compare their merits and demerits. Discuss their fields of application.
[U.P. Technical Univ. Switchgear and Protection 2005-06]

40. Explain the principle of distance relays stating clearly the difference between impedance relay, reactance relay, and mho relay.
[U.P. Technical Univ. Switchgear and Protection 2006-07]

41. Explain the applications of impedance, reactance and mho relays for line protection.

[U.P. Technical Univ. Switchgear and Protection 2004-05]

42. What is an angle impedance relay ? Discuss how the characteristic is realized using phase comparison scheme.

[U.P. Technical Univ. Switchgear and Protection 2005-06]

43. Draw and explain the characteristics of a Mho relay. Why a Mho relay is preferred for protection of long lines against phase faults ?

[U.P. Technical Univ. Switchgear and Protection 2008-09]

44. Derive mathematically impedance relay characteristic. Draw the characteristic and show zones of operation and no operation.

[U.P. Technical Univ. Switchgear and Protection 2005-06]

45. Discuss the differential reactance scheme for distance protection.

[G.B. Technical Univ. Switchgear and Protection 2011-12]

46. Describe with a schematic diagram the construction of an impedance relay.

Obtain on the R-X diagram the null characteristic of the impedance relay and hence describe its principle of operation. Mark on this diagram the impedance line representing the impedance as seen by the relay during metallic short-circuit and also the fault area where the measured impedance may lie during arcing fault.

Explain the disadvantages of the above relay for the protection of a short and a long transmission line.

47. Explain why 1st-zone cover only 80% of the protected line section. Why is the line impedance locus set not to pass through the diameter of the mho circle ?

[A.M.I.E. Sec. B. Power System Protection & Communication Summer 1999]

48. Using universal torque equation of relays draw the characteristic of a mho relay.

[U.P. Technical Univ. Switchgear and Protection 2004-05]

49. What is Universal torque equation ? Using this equation derive the following characteristics:

(*i*) Impedance relay (*ii*) reactance relay (*iii*) mho relay.

Draw the characteristics and indicate clearly the zones of operation and no-operation.

[Pb. Univ. Switchgear & Protection (Electrical Power III) Dec. 1991]

50. Explain how you provide directional feature to (*i*) impedance and (*ii*) reactance relay. Explain why directional feature provided for impedance relay cannot be used for reactance relay.

51. Name different types of distance relaying schemes you are familiar with. Why an impedance (*i.e.*, ohm) relay is preferred as a ground relay for ground fault ? Write its signals both as a phase comparator as well as an amplitude comparator in the complex B - (*i.e.*, Z) plane and draw the threshold characteristic of this relay. What are the main shortcomings of ohm (*i.e.*, reactance) relay and what additional precautions are taken to overcome these shortcomings ?

52. What do you understand by phase distance relays and ground distance relays ? State why and for what type of faults these relays are used ? How many phase relays and ground relays covering all three zones of protection of transmission lines are used at any bus ?

Discuss about switched distance relays and explain why these are preferred over conventional phase and ground relays for transmission line.

53. How many single-phase units are required to protect a 3-phase system by distance relays ? Why the input for distance relays are changed when phase faults and ground faults are catered respectively ? Deduce the input required for phase fault protection.

What is a reactance relay ? Discuss its efficiency in the protection of transmission lines. What is its main snag ? How is it overcome ?

54. Explain what is meant by distance protection. What arrangement is made to make the relay measure positive sequence impedance only for L-L and 3-phase fault.

55. What is the principle of differential relays ? What are their limitations ? Explain their characteristics.

[G.B. Technical Univ. Switchgear and Protection 2011-12]

56. Distinguish between time overcurrent relays, directional relays and differential relays.

[G.B. Technical Univ. Switchgear and Protection 2011-12]

57. Why a biased differential relay is preferred over a simple differential relay ? Make a list of applications of a biased differential relay.

58. Why biasing is needed in differential relays ? What is meant by percentage bias ? Show the complex current plane characteristic of such a relay.

59. What is meant by per cent bias ? How is this achieved in practice in a differential relay ? Under what circumstances is a percentage-differential relay preferred over the differential one ?

[Pb. Univ. Switchgear & Protection April 1986]

60. Write short notes on the following :

(*i*) Classification of relays.

(*ii*) Electromagnetic attraction type relays.

(*iii*) Thermal relays. **[Pb. Univ. Sept. 1988, April 1986, Nov. 1985, April 1982]**

(*iv*) Buchholz relay. **[Pb. Univ. April 1988]**

(*v*) Induction relays.

(*vi*) Reverse power relay.

[Pb. Univ. October 1986, Nov. 1985]

(*vii*) Directional overcurrent relay. **[Pb. Univ. April 1986]**

(*viii*) Reactance relay. **[Pb. Univ. Nov. 1985]**

(*ix*) Mho relay.

(*x*) Translay relay. **[Agra Univ. 1981]**

Short Answer Type Questions With Answers

Q. 1. What is a protective relay ?

Ans. Protective relay is an electrical device designed to initiate isolation of a part of an electrical installation, or to operate an alarm signal, in the event of an abnormal condition or a fault.

Q. 2. What are the functions of a protective relay ?

Ans. The main functions of protective relays are to detect the presence of faults, their locations and initiate the action for quick removal from service of the faulty section of the system. Relays operate to trip the power circuit breakers to disconnect the faulty section.

Q. 3. What are the essential elements of a protective relay ?

Ans. The essential elements of a protective relay are (*i*) sensing element (*ii*) comparison element, and (*iii*) control element.

Q. 4. What are the fundamental requirements of protective relaying ?

Ans. The fundamental requirements of protective relaying are speed, selectivity, sensitivity, reliability, simplicity, and economy.

Q. 5. Define the terms 'sensitivity' and 'selectivity' of a relay.

Ans. Sensitivity of a relay means its capacity to operate reliably under the actual conditions that produce the least operating tendency while the selectivity means the ability to determine the point at which the fault occurs and select the nearest circuit breaker in the system to trip so that the fault is cleared with minimum or no damage to the system.

Q. 6. What is meant by "relay setting" ?

Ans. Relay setting means actual value of the energizing or characteristic quantity at which the relay is designed to operate under given conditions.

Q. 7. What is meant by operating time of a protective relay ?

Ans. The operating time of a protective relay means the time which elapses between the instant when the actuating quantity exceeds the pick-up value to the instant when the relay contacts close.

Q. 8. What is meant by 'relay time' ?

Ans. Time interval between occurrence of fault and closure of relay contacts is called the relay time.

Q. 9. How are relay time, breaker time and fault clearing time related ?

Ans. Relay time plus breaker time equals the fault clearing time.

Q. 10. What are the shortcomings in electromagnetic relays ?

Ans. The vibration of relay armature at double the supply frequency, due to pulsating nature of electromagnetic force, causes the relay to hum and produce noise and also is a source of damage to the relay contacts. This leads to sparking and unreliable operation of the relay operative circuit contacts due to make and break of the circuit.

Q. 11. How definite time lag is achieved in attraction armature relays ?

Ans. The instantaneous type attraction armature relays can be made a definite time lag or inverse-time lag by using an oil dash pot, an air-escapment chamber a clockwork mechanism or by placing a fuse in parallel with it.

Q. 12. How induction cup type construction is superior to induction disc type ?

Ans. Induction cup type construction develops more efficient torque than either shaded pole or watt-hour meter constructions. Such relays are very fast in operation and may have an operating time of less than 0.01 second.

Q. 13. What is Buchholz relay ? Which equipment is protected by it ?

Ans. Buchholz relay is a gas actuated relay and is used for the protection of power transformers of ratings exceeding 500 kVA.

Q. 14. For what type of fault Buchholz relay is employed ?

Ans. Buchholz relay provides protection only against transformer internal faults.

Q. 15. Define PSM ?

Ans. PSM stands for plug setting multiplier and is defined as the ratio of fault current in relay coil to the pick-up value.

Q. 16. What are the main features of directional relays ?

Ans. Directional relays must have the following features: (*i*) High speed operation, (*ii*) high sensitivity, (*iii*) adequate short-time thermal rating, (*iv*) ability to operate with low values of voltage (*v*) burden must not be excessive and (*v*) there should be no voltage and current creep.

Q. 17. Where is directional relay used?

Ans. Directional relays are used when graded time overload protection is applied to ring mains and interconnected networks.

Q. 18. What is distance protection ?

Ans. Distance protection is the name given to the protection, whose action depends upon the distance of the feeding point to the fault point. It is non-unit type protection and the protection zone is not exact. It can be used as primary as well as backup protection.

Q. 19. In what way a distance relay is superior to overcurrent protection for the protection of transmission lines ?

Ans. Distance relay is superior to overcurrent protection for the protection of transmission lines. The reasons are faster protection, simpler coordination, simpler application, permanent setting without need for readjustments, less effect of amount of generation and fault levels, fault current magnitude, permits the high line loading.

Q. 20. Why directional feature provided for impedance relay cannot be used for reactance relay ?

Ans. The directional feature provided for impedance relay cannot be used for reactance relay because the reactance relay will trip even under normal operating conditions at or near unity power factor. The reactance type distance relay needs a directional unit that is inoperative under normal load conditions.

Q. 21. Why reactance type relay is very suitable as a ground relay for ground fault ?

Ans. Reactance type relay is very suitable as a ground relay for ground fault because its reach is not affected by fault impedance (*i.e.,* arcing resistance).

Q. 22. For what type of protection will you recommend (*i*) impedance relay (*ii*) reactance relay and (*iii*) mho relay ?

Ans. The impedance type relay is suitable for phase fault relaying for lines of moderate length; reactance type relays are preferred for ground fault relaying while mho type relays are best suited for long lines and particularly where a severe synchronising power surge may occur.

Q. 23. What is meant by a differential relay ?

Ans. A differential relay is defined as the relay that operates when phasor difference of two or more similar electrical quantities exceeds a predetermined amount.

Q. 24. Where differential protection principle is employed ?

Ans. Differential protection principle is employed for the protection of generators, generator-transformer units, transformers, feeders, large motors and bus-bars.

Q. 25. Why is percentage differential relay named so ?

Ans. Percentage differential relay is known by this name because the operating current required to trip can be expressed as a percentage of load current.

Q. 26. Why a biased differential relay is preferred over a simple differential relay ?

Ans. Biased differential relay is preferred over a simple differential relay because its operation is not affected by the trouble arising out of difference in CTs ratios for high values of external short-circuit currents.

Q. 27. Where is negative phase sequence relay employed ?

Ans. Negative phase sequence relay is essentially employed for the protection of generators and motors against unbalanced loading that may arise due to phase-to-phase faults.

Q. 28. How a negative phase sequence relay is made to respond to the flow of zero-sequence currents also ?

Ans. The negative phase sequence relay is made to respond to the flow of zero-sequence currents also by providing additional winding on the central limb of the upper electromagnet of the relay connected in the residual circuit of three line CTs, as illustrated in Fig. 9.66.

Problems

1. The calculated short-circuit current through a feeder is 1,200 A. An overcurrent relay of rating 5 A is connected for the protection of the feeder through a 1,000/5 A CT. Calculate the operating time of the relay when it has a plug setting of 50% and time setting of 0.8. The characteristic of the relay is as follows :

PSM	1.3	2	4	6	10	20
Time (in seconds)	30	10	6.5	3.5	3	2.2

[**Ans.** 6.8 seconds]

2. An IDMT relay operates in 5 seconds and 3 seconds for PSM's of 4 and 10 respectively. The relay is used to protect a feeder through a 1,000/5 A CT. Calculate the time of operation of the relay when the feeder current is 1,500 A. The relay has a plug setting of 75% and time setting 0.4. The nominal current rating of the relay is 5 A.

[**Ans.** 4 seconds]

3. Figure shows a radial distribution system having identical IDMTL overcurrent relay at A, B and C. For a time delay step of 0.5 second, calculate the time multiplier settings at A and B.

IDMTL relay characteristic is given below :

PSM	2	3	5	10	20
Time (in seconds)	10	6	4.5	3	2

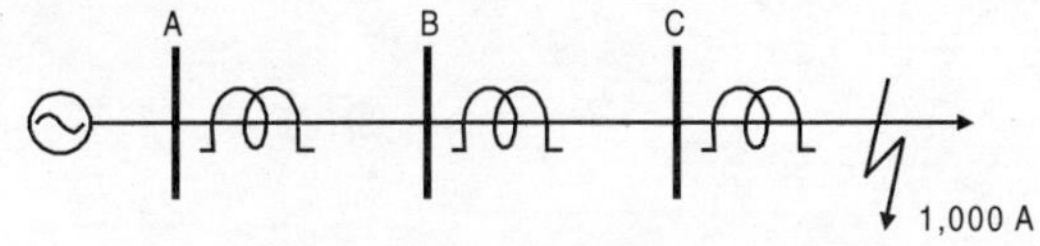

Fig. 1

CT Ratios	200/5	200/5	100/5
Plug setting current	5 A	2.5 A	2.5 A
Time multiplier setting	—	—	0.1

[**Ans.** 0.2333; 0.2666]

4. A current-voltage directional relay has spring constant of 30 units and the constant in the torque equation is 0.5 unit per volt-ampere. Sketch the polar characteristic of the relay if the polarising voltage is 110 V and the angle of the maximum torque is 30° lag.

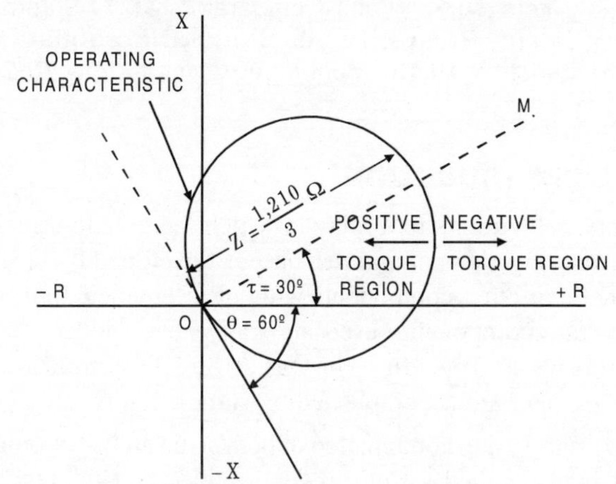

Fig. 2

□□□

CHAPTER 10

Static Relays

INSIDE THIS CHAPTER

10.1. INTRODUCTION

A static relay refers to a relay in which there is no armature or other moving element and response is developed by electronic, magnetic or other components without mechanical motion. The solid state components used are transistors, diodes, resistors, capacitors and so on. The function of comparison and measurement are accomplished by static circuits.

A relay using combination of both static and electromagnetic units is also called a *static relay* provided that the response is accomplished by static units.

In static relays, the measurement is performed by electronic, magnetic, optical or other components without mechanical motion. However, additional electromechanical relay units may be employed in output stage as auxiliary relays. A protective system is formed by static relays and electromechanical auxiliary relays.

10.1.1. Needs For Static Relays

With the rapid growth of electrical transmission and distribution systems during last fifty years and with the advent of much larger power stations and interconnected systems, the duty imposed upon protective gear became more and more severe. Thus relay's functions became more and more complicated and many types of relays tend to become very complex mechanically and hence costly and difficult to test and maintain.

The basis of the so called static relaying is the use of the circuits and components to obtain a variety of functions and operating characteristics which for protection purposes have traditionally been obtained using electromechanical devices. The need of fast and reliable protective schemes was realized because short-circuit levels, circuit ratings and complexity of interconnection have increased. Shorter operating times have become more essential for preserving dynamic stability of the system as the character and loading approach design limits. The satisfaction of these requirements has left little scope for further improvements in conventional electromechanical relays. Experience shows that such requirements can readily be met by using static relays, which are capable of performing electronic circuit control functions in a manner similar to that of an electromagnetic relay without using moving parts or elements.

The transistors have made it possible to achieve greater sensitivity and simultaneously excellent mechanical stability which would have never been possible with electromechanical relays. The noteworthy point is that it is usually not economical to replace existing electromechanical relays with their static counterparts just to reduce maintenance.

It is interesting to note that the static relays have first been commercially manufactured for the distance and differential protective schemes while the much simpler overcurrent relays have not been developed. This is because the distance and differential schemes are more amenable to mathematical analysis whereas the overcurrent characteristics are more of empirical nature. With the use of static relays it has been possible to achieve many varied and complex distance protection characteristics which are impossible with the conventional electromechanical relays.

10.1.2. Essential Components of Static Relays

The essential components of a static relay are shown in Fig. 10.1. Here the relaying quantity *i.e.,* the output of a CT or PT or a transducer is rectified by a rectifier. The rectified output is supplied to a measuring unit comprising of comparators, level detectors, filters, logic circuits. The output is actuated when the dynamic input (*i.e.,* the relaying quantity) attains the threshold value. This output of the measuring unit is amplified by amplifier and fed to the output unit device, which is usually an electromagnetic one. The output unit energizes the trip coil only when relay operates.

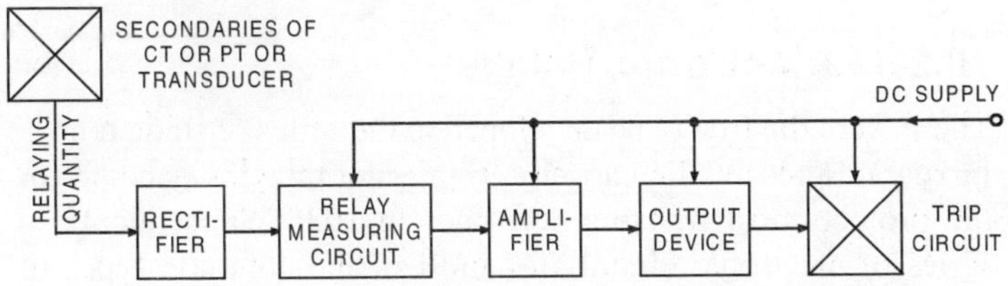

Fig. 10.1. ***Simplified Block Diagram of a Static Relay***

In a static relay the measurement is carried out by static circuits consisting of comparators, level detectors, filters etc., while in a conventional electromagnetic relay it is done by comparing operating torque (or force) with restraining torque (or force). In individual relays there is a wide variation. The relaying quantity such as voltage/current is rectified and measured. When the quantity under measurement attains certain well defined value, the output device is triggered and thereby the circuit breaker trip circuit is energized.

Static relays can be arranged to respond electrical inputs. The other types of inputs such as heat, light, magnetic field, travelling waves etc, can be suitably converted into equivalent analogue or digital signals and then supplied to the static relay. A multi-input static relay can accept several inputs. The logic circuit in the multi-input digital static relay can determine the conditions for relay response and sequence of events in the response.

10.1.3. Comparison of Static and Electromagnetic Relays

The conventional electromagnetic relays are robust and quite reliable, but are required to operate under different forces under fault conditions. This leads to delicate setting small contact gaps, special bearing systems, special clutch assemblies and several measuring problems. These require instrument transformers (CTs and PTs) with high burden and are bulky in size also.

The static relays in comparison to the corresponding electromagnetic relays have many advantages and a few limitations.

Advantages of Static Relays

1. The power consumption in case of static relays is usually much lower than that in case of their electromechanical equivalents. Hence burden on the instrument transformers (CTs and PTs) is reduced and their accuracy is increased, possibility of use of air-gaped CTs is there, problems arising out of CT saturation are eliminated, and there is an overall reduction in the cost of CTs and PTs.
2. Quick response, long life, shockproof, fewer problems of maintenance, high reliability and a high degree of accuracy.
3. Absence of moving contacts and associated problems of arcing, contact bounce, erosion, replacement of contacts etc.
4. Quick reset action—a high reset value and absence of overshoot can be easily achieved because of the absence of mechanical inertia and thermal storage.
5. There is no effect of gravity on operation of static relays and therefore they can be installed in vessels, aircrafts etc.
6. Ease of providing amplification enables greater sensitivity to be obtained.
7. Use of printed (or integrated) circuits avoids wiring errors and facilitates rationalization of batch production.
8. The basic building blocks of semiconductor circuitry permit a greater degree of sophistication in the shaping of operating characteristics, enabling the practical realization of relays with threshold characteristics more closely approaching the ideal requirements.
9. By combining various functional circuits, several conventional relays can be substituted by a single static relay. For example a single static relay can provide overcurrent, undervoltage, single phasing, short-circuit protection in an ac motor by incorporating respective functional blocks.
10. Static relays are very compact. A single static relay can perform several functions. A single microprocessor based system can substitute several independent protection and control relay units. The space required for installation of protective relays and control relays etc., is reduced.
11. The characteristics of static relays are accurate and superior. They can be altered within certain range as per protection needs.
12. Static relays assisted by power line carrier can be employed for remote backup and network monitoring.
13. Static relays can be designed for repeated operations. This is possible because of absence of moving parts in the measuring circuits.
14. The risk of unwanted tripping is less with static relays.
15. Static relays are quite suitable for earth quake prone areas, ships, vehicles, locomotives, aeroplanes etc. This is because of high resistance to shock and vibration.
16. The static relays are provided with integrated features for self-monitoring, easy testing and servicing. Defective module can be replaced easily.
17. A static protection control and monitoring system can perform several functions such as protection, monitoring, data acquisition, measurement, memory, indication, data-communication etc.

Limitations of Static Relays:

1. Auxiliary dc supply is required. However, this drawback is not very important as auxiliary dc supply can be obtained from station battery supply and conveniently changed as per local needs.
2. Semiconductor components are sensitive to electrostatic discharges. Some components are more sensitive than others. Even small discharges can damage the components and therefore precautions are necessary in the manufacturing

of static relays to avoid component failures due to electrostatic discharges.

3. Static relays are sensitive to voltage spikes or voltage transients. Special measures (such as use of filter circuits in relays, screening of cables connected to the relays etc.,) are taken to avoid such problem.
4. The characteristics of static relays are influenced by ambient temperature and ageing. However, temperature compensation can be provided by using thermistor circuits and digital measuring techniques etc., while ageing may be minimised by pre-soaking for several hours at a relatively high temperature.
5. The reliability of the system depends upon a large number of small components and their electrical connections.
6. The static relays have low short-time overload capacity compared with electromagnetic relays.

   Both of the above factors [factors (5) and (6)] are the design features of the circuit and careful design of static relay circuitry can compensate for these limitations.
7. Static relays are costlier, for simple and single function, than their equivalent electromechanical counterparts. But for multi-function protection, static relays prove economical. The production technology of plug-in-type static relays on the panel permits the manufacturing of standard relays in mass and the customer's needs can be met quickly by incorporating required relay units on the panel. Static relays with ICs are cheaper than those with discrete components.
8. Static relay characteristic is likely to be affected by the operation of the output device but this is not so in case of electromagnetic relay because its operation is based on the comparison between operating torques/forces.
9. Highly trained personnel are required for their servicing.
10. Static relays are not very robust in construction and easily affected by surrounding interference.

For simple protective schemes, conventional electromechanical relays provide economic and satisfactory choice. Electromechanical units are also used as components in total predominantly static relays such as for auxiliary relay functions, output functions.

For complex protective functions requiring accurate characteristics for various protection systems and for protection of costly large equipment or machines static relays are preferred technically and economically. These may be hardwired or programmable. As static relays perform protective and monitoring functions, the additional cost is justified on the basis of improved system stability and reliability.

For integrated protection and monitoring systems programmable microprocessor controlled static relays are preferred.

10.2. CLASSIFICATION OF STATIC RELAYS

Static relays are classified according to the type of the measuring unit or comparator as follows:

(*a*) Electronic relays
(*b*) Transductor or magnetic amplifier relays
(*c*) Rectifier bridge relays
(*d*) Transistor relays
(*e*) Hall effect relays and
(*f*) Gauss effect relays.

10.2.1. Electronic Relays

These were the first to be developed in the series of static relays. Fitzgerald presented a carrier current pilot relaying scheme for the protection of transmission lines in 1928. Subsequently a series of electronic circuits for most of the common types of protective gear relays were developed. The components used were electronic valves for measuring unit.

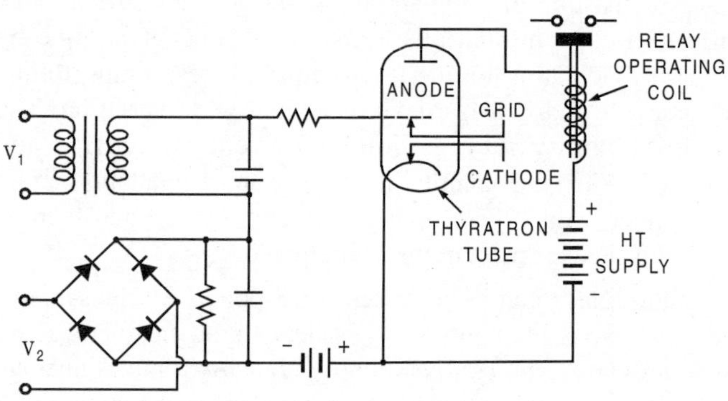

(*a*) *Electronic Amplitude Comparator*

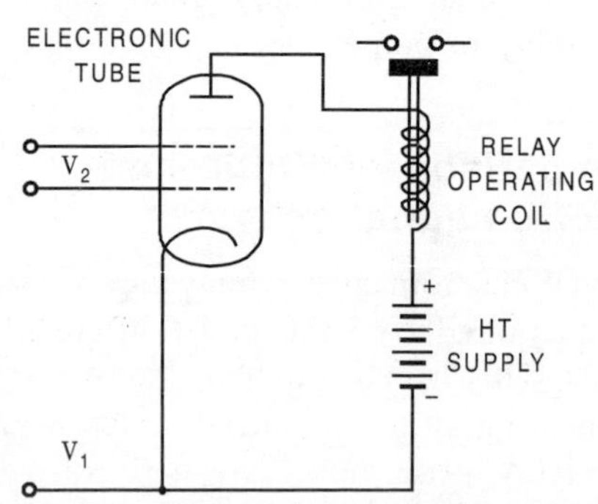

(*b*) *Electronic Phase Comparator*

Fig. 10.2. ***Basic Arrangements of Electronic Relays***

Two basic arrangements are shown in Fig. 10.2. One uses an amplitude comparator [Fig. 10.2 (*a*)] and the other makes use of a phase comparator [Fig. 10.2 (*b*)]. In the former arrangement two ac quantities under comparison are rectified and applied in opposition in the control grid circuit of an electronic tube and the operation occurs when one quantity exceeds the other by an amount depending on the bias. In the latter arrangement one ac quantity is connected to the control grid and the other to the screen grid of the electronic tube. Operation occurs when the two ac quantities under comparison are in phase.

Inspite of the advantages of fast operation, low maintenance, low burden on CTs and PTs, absence of mechanical inertia and bouncing contacts, they suffered inherently from the requirements of *ht* supply, short life, large power consumption, *lt* supply for the heater elements. These relays could not meet practical requirements and hence never reached the commercial stages.

10.2.2. Transductor (Magnetic Amplifier Relays)

A transductor comprises essentially a magnetic core carrying two groups of windings usually known as operating windings and control windings. Each group may comprise only one winding but if there is more than one winding in a group all those windings are magnetically linked. On the other hand the windings of the different groups are not magnetically linked. The control windings are energized with dc and the operating windings with ac. A transductor operates so as to present a variable impedance to currents flowing through the operating windings the magnitude of this impedance being varied by the current flowing through the control winding. A single core transductor is illustrated in Fig. 10.3.

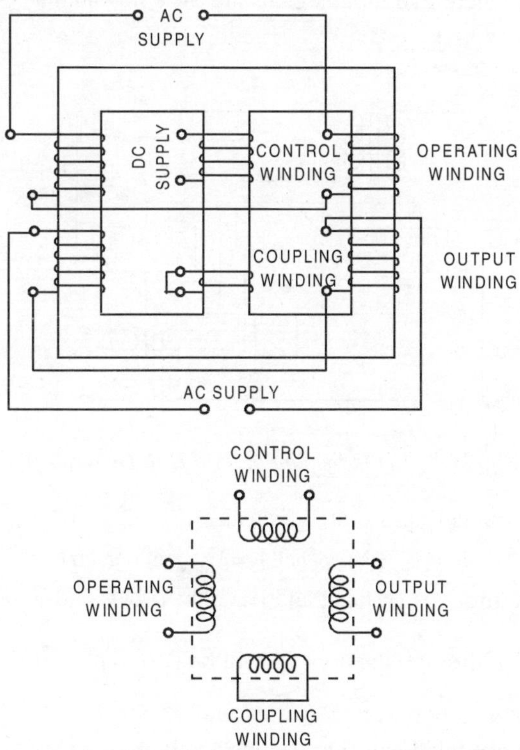

Fig. 10.3. ***Single Core Transductor***

If transductor is employed as an amplitude comparator its sensitivity is limited due to the sensitivity of the slave relay in its output circuit. On the other hand if it is used as a phase comparator to achieve higher sensitivity it will depend on an external ac supply which is sometimes difficult to arrange.

A transductor relay is shown in Fig. 10.4. The restraining voltage is applied to the control winding through an impedance Z_R and a rectifier. The restraining current so obtained is rectified and partially smoothed by the short-circuit effect of the rectifier on the control winding and also due to the short-circuited coupling winding. The effect of this is to drive both the limbs on which the operating winding is wound into saturation. The operating winding is wound in such a way that the ampere-turns so produced opposes the restraining ampere-turns in one limb and strengthen them in the other. Assuming equality of restraining and operating turns, and neglecting the finiteness of permeability, the operating current would drive out of saturation, the limb in which operating current opposes the restraining ampere-turns, when the peak value of the operating current exceeds the magnitude of restraining current. When this happens a voltage is induced in the output winding and the relay gets energised. The same operation is repeated on the other limb during next half cycle. The slave relay current is very low until the above mentioned condition exists. Then it suddenly increases and becomes very large for a small value of operating current. That is why a heavy duty telephone type relay is employed with large setting.

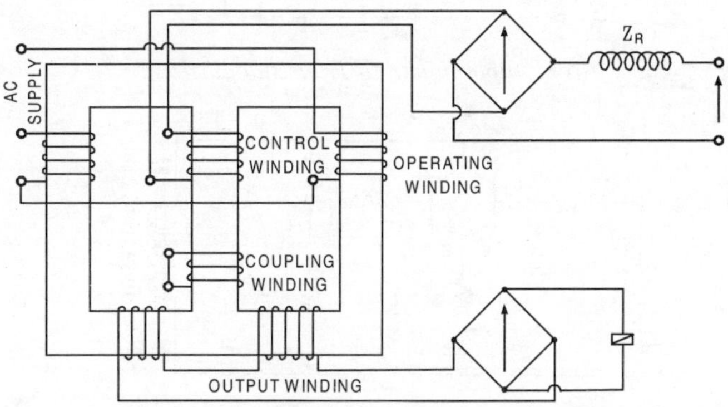

Fig. 10.4. ***Transductor Relay***

The transductor relays are mechanically very simple and are quite reliable. Due to smoothing and rectifying a signal, a delay is introduced because of the time constant of the smoothing circuit and hence the relays are slow in operation and, therefore, are discarded for protection applications.

10.2.3. Rectifier Bridge Relays

Such relays became popular because of development of semiconductor diodes. This relay consists of two rectifier bridges and a moving coil or polarized moving iron relay. The most common are relay comparators based on rectifier bridges, which can be arranged as either amplitude or phase comparators. These will be discussed later on in this chapter.

10.2.4. Transistor Relays

Transistor relays are the most widely used static relays. In fact when we talk of static relays we generally mean transistor relays. Transistor which acts like a triode can overcome most of the limitations posed by the electronic valves and thus has made possible to develop the electronic relays more commonly known as static relays. The fact that a transistor can be employed both as an amplifying device and a switching device, makes it suitable for achieving any functional characteristic. The transistor circuits cannot only perform the essential functions of a relays (such as comparison of inputs, summation and integrating them) but they also provide necessary flexibility to suit the various relay requirements.

Two basic arrangements of relays based on transistor comparators are shown in Fig. 10.5.

In either of the circuits shown in Fig. 10.5 (*a*) and 10.5 (*b*) current of constant magnitude flows in the collector circuit only when the input ac quantities are simultaneously positive; a relay in the collector circuit will pick up when the overlap angle exceeds a certain value, *i.e.,* when the mean dc level in the collector circuit exceeds the relay pick up as a consequence of phase coincidence.

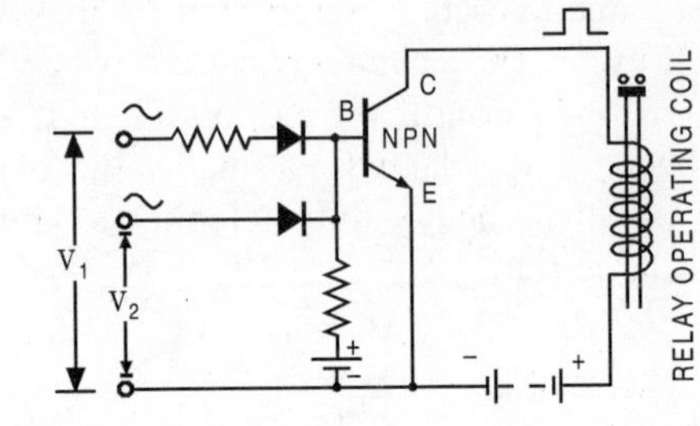

(a) Multiple Inputs To Transistor Base

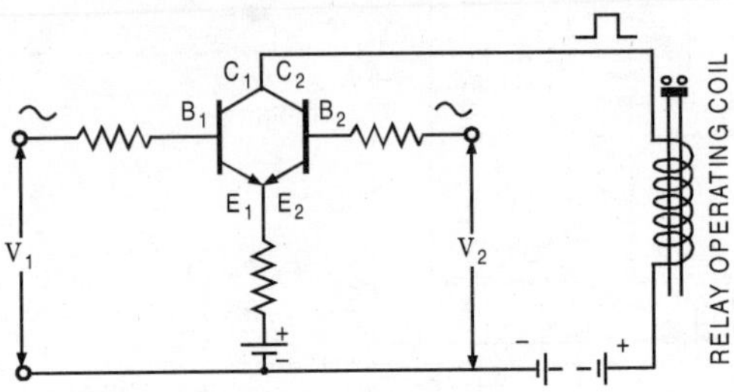

(b) Two Back-To-Back Transistors

Fig. 10.5. ***Static Comparators***

10.2.5. Hall Effect Relays

When a conductor is kept perpendicular to the magnetic field and a direct current is passed through it, it results in an electric field perpendicular to the directions of both the magnetic field and current with a magnitude proportional to the product of the magnetic field strength and current. The voltage so developed is very small and it is difficult to detect it. But in some semiconductors such as indium arsenide, indium antimonide, indium phosphate, germanium etc., this voltage is enough for measurement with a sensitive moving coil instrument. This phenomenon is called the *Hall effect.*

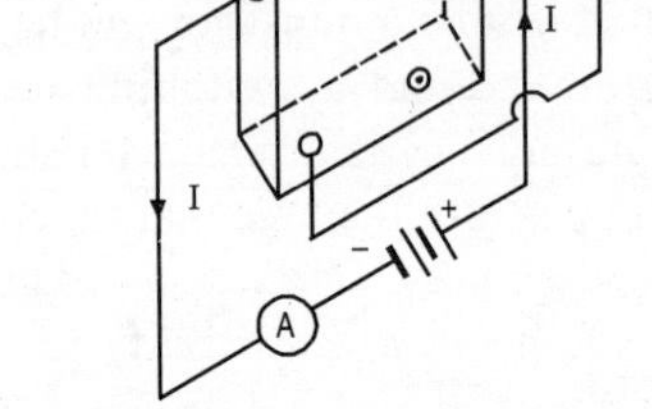

Fig. 10.6. ***Basic Hall Generator***

Figure 10.6 illustrates the basic Hall generator where the Hall crystal in the form of a slab is connected to a battery so that a current I flows through the slab in the manner shown in the figure and the magnetic field is applied so that it is perpendicular to the slab of the Hall crystal. A potential difference V_H, called the *Hall voltage*, is developed between the top and bottom of the slab. The magnitude of this voltage is proportional to the product of strength of the magnetic field and current.

$$\text{Let } \phi = \phi_{max} \sin \omega t$$

$$\text{and } i = I_{max} \sin (\omega t - \alpha)$$

then Hall voltage V_H is given as

$$V_H \;\alpha\; \phi i \;\alpha\; \phi_{max} I_{max} \sin \omega t \sin (\omega t - \alpha)$$

$$= K_H \phi_{max} I_{max} [\cos \alpha - \cos (2\omega t - \alpha)] \qquad ...(10.1)$$

In the above expression the first term on the right hand side is a dc voltage proportional to the vectorial product of the two inputs—flux and current, the second term is an ac voltage of double frequency. The double frequency term can be eliminated by cross-connecting two Hall generators, as illustrated in Fig. 10.7 where two input signals are the sinusoidal ac currents I_1 and I_2.

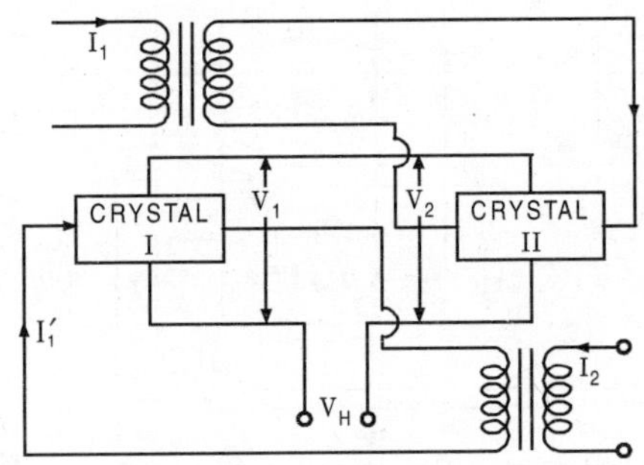

Fig. 10.7. ***Cross Section of Two Hall Generators***

If currents I_1 and I_2 are given as

$$I_1 = I_{1max} \sin \omega t \text{ and } I_2 = I_{2max} \sin (\omega t - \alpha) \qquad ...(10.2)$$

Flux through crystals I and II will be $\phi_1 \alpha\, I_1$ and $\phi_2 \,\alpha\, I_2$

and current through the crystals will be $I_1' \;\alpha\; \frac{dI_2}{dt}$ and $I_2' \;\alpha\; \frac{dI_1}{dt}$ ...(10.3)

Two crystals I and II are connected in such a way that their output voltages are in opposition. So the output voltage will be

$$V_H \;\alpha\; (V_1 - V_2) \;\alpha\; \left(I_1 \frac{dI_2}{dt} - I_2 \frac{dI_1}{dt} \right)$$

$$\alpha\; I_{1max} \sin \omega t \times I_{2max}\, \omega \cos (\omega t - \alpha)$$

$$- I_{2\,max} \sin (\omega t - \alpha).\, \omega\, I_{1max} \cos \omega t$$

$$\text{or } V_H \;\alpha\; I_{1max} I_{2max} \sin [\omega t - (\omega t - \alpha)]$$

$$\alpha\; I_{1max} I_{2max} \sin \alpha \qquad ...(10.4)$$

The device thus acts as a pure phase comparator. A telephone type relay can be directly operated from the Hall voltage. Hall device is much simpler in construction and does not require auxiliary dc supply while the transistorized circuits require about 10V dc supply.

Because of the high cost of Hall element, low output and large temperature error, this comparator is normally not used.

10.2.6. Gauss Effect Relays

The resistivity of some metals and semiconductors at low temperature changes when exposed to the magnetic field. This phenomenon is called the *magnetoresistivity or Gauss effect.* This effect depends upon the ratio of depth to width and increases with the increase in this ratio. In some metals, this effect is noticeable at room temperature, as in case of bismuth. Magnetoresistors are sensitive to the magnetic field strength and not to the rate of variation. Indium antimonide and indium asrenide are more sensitive in comparison to bismuth or mumetal.

If one ac voltage V_1 develops a magnetic field through the crystal in the form of disc having large diameter and another ac voltage V_2 sends a current radially through the disc, then the current will be proportional to $V_1V_2 \cos\theta$ where θ is the angle between the two voltages V_1 and V_2 *i.e.,* current is maximum when the two voltages are in phase and zero when they are in quadrature. Such a relay is considered better than Hall effect relay because of simpler construction and circuitry [Fig. 10.8]. Polarizing current is not required and output is relatively higher. But its use in static relays is limited because of high cost of crystal.

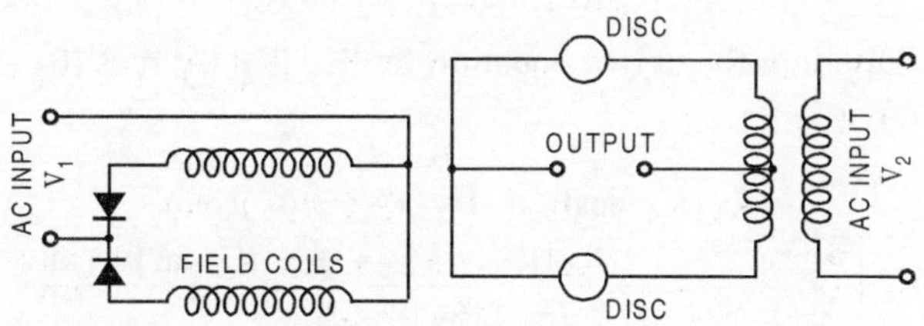

Fig. 10.8. ***Magnetoresistivity Phase Comparator***

10.3. OUTPUT DEVICES

Amongst the static units the thyristor is an excellent means for tripping the circuit breaker from a low power signal of short duration. It operates with signals of μs duration but once triggered stays in conduction until broken by a switch. A typical thyristor (SCR) tripping circuit is shown in Fig. 10.9. The measuring circuit of the relay sends a pulse to the SCR gate when threshold condition is reached. The SCR triggers and the current from the battery flows through the trip coil of the circuit breaker. The circuit breaker opens and the auxiliary switch also opens as it is interlocked with the circuit breaker and thereby the SCR is turned off.

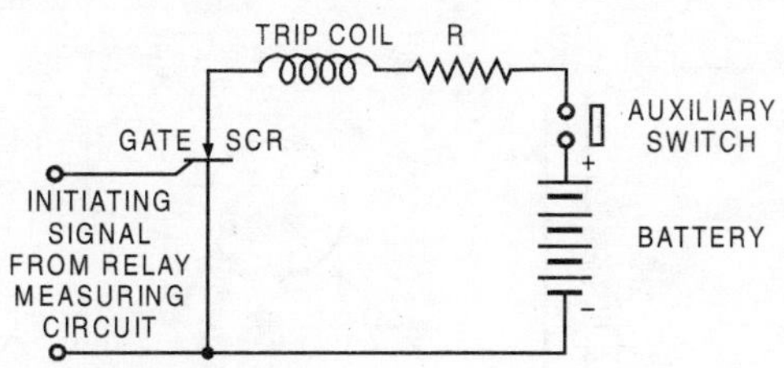

Fig. 10.9. ***Thyristor Tripping Circuit***

10.4. COMPARATORS

The function of a protective relay is to sense any abnormal condition in the system and send a signal to the breaker which in turn isolates the faulty section of the feeder from the healthy one. The relay does all this by comparing two quantities either in amplitude or in phase. The amplitude or phase relation depends on the conditions of the system and for a predetermined value of this relation, indicative of a particular type and location of fault, the relay operates. Except in relays, such as overcurrent relays, where only one electrical quantity overcomes a mechanical quantity such as the restraint from a spring, usually two electrical quantities are compared. The device that makes such comparison is called the *comparator* and forms the heart of a protective relay. The operating characteristics of any relay are governed by the comparator.

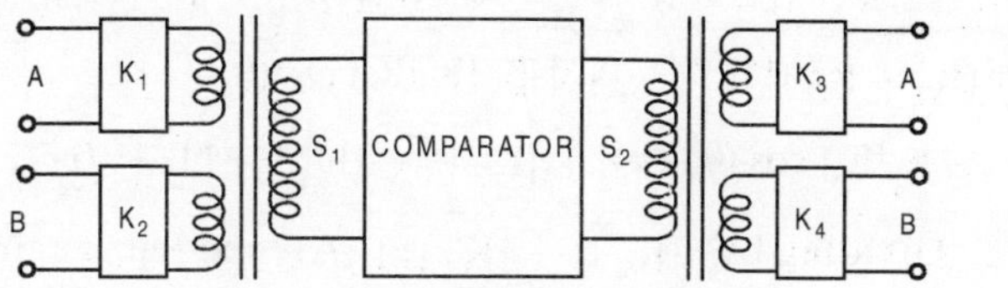

(a) Comparison of Mixed Signals

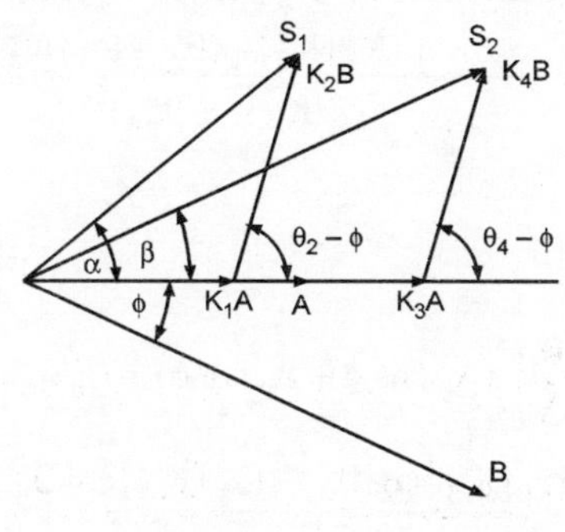

(b) Phasor Diagram For a Comparator

Fig. 10.10

10.4.1. General Equation For Comparators

Let us first derive the general threshold equation assuming that there are two input signals S_1 and S_2 such that when the phase relationship or magnitude relationship fulfils predetermined threshold conditions, tripping is initiated. The input signals are derived from the system through instrument transformers (CTs and/or PTs). In case the two quantities to be compared are different (*i.e.,* voltage and current), some form of mixing device, such as *current voltage transactor,* is required.

Let the two input signals be represented as

$$\text{and} \quad \left.\begin{aligned} \mathbf{S_1} &= \mathrm{K_1}\mathbf{A} + \mathbf{K_2B} \\ \mathbf{S_2} &= \mathrm{K_3}\mathbf{A} + \mathbf{K_4B} \end{aligned}\right\} \quad \ldots(10.5)$$

where A and B are the primary system quantities, K_1 and K_3 are the scalar numbers and K_2 and K_4 are the complex numbers with angles θ_2 and θ_4 respectively. Taking A as the reference phasor and phasor B to lag A by an angle ϕ. Then the above equation can be rewritten as

$$S_1 = K_1\,|A| + |K_2|\,|B|\,\{\cos(\theta_2 - \phi) + j\sin(\theta_2 - \phi)\}$$

$$\text{and} \quad S_2 = K_3\,|A| + |K_4|\,|B|\,\{\cos(\theta_4 - \phi) + j\sin(\theta_4 - \phi)\} \quad \ldots(10.6)$$

10.4.2. Analysis For Amplitude Comparator

If the operating criterion is given by $|S_1| \geq |S_2|$ then at the threshold of operation

$$|S_1| = |S_2|$$

or $\{K_1|A| + |K_2||B|\cos(\theta_2-\phi)\}^2 + \{|K_2||B|\sin(\theta_2-\phi)\}^2$

$$= \{K_3|A| + |K_4||B|\cos(\theta_4-\phi)\}^2 + \{|K_4|^2|B|\sin(\theta_4-\phi)\}^2$$

Rearranging the terms we have

$$(K_1^2-K_3^2)|A|^2 + 2|A||B|\{K_1|K_2|\cos(\theta_2-\phi) - K_3|K_4|\cos(\theta_4-\phi)\} + |K_2|^2|B|^2\cos^2(\theta_2-\phi) - |K_4|^2|B^2|\cos^2(\theta_4-\phi) + |B|^2\{|K_2|^2\sin^2(\theta_2-\phi) - |K_4|^2\sin^2(\theta_4-\phi)\} = 0$$

or $(K_1^2-K_3^2)|A|^2 + 2|A||B|\{K_1|K_2|\cos(\theta_2-\phi) - K_3|K_4|\cos(\theta_4-\phi)\} + (|K_2|^2 - |K_4|^2)|B^2| = 0$ ...(10.7)

Dividing by $(|K_2|^2 - |K_4|^2)|A|^2$ and rearranging the terms we have

$$\left|\frac{B}{A}\right|^2 + 2\left|\frac{B}{A}\right| \frac{(K_1|K_2|\cos\theta_2 - K_3|K_4|\cos\theta_4)\cos\phi + (K_1|K_2|\sin\theta_2 - K_3|K_4|\sin\theta_4)\sin\phi}{|K_2|^2-|K_4|^2} + \frac{K_1^2-K_3^2}{|K_2|^2-|K_4|^2} = 0$$

or $$\left|\frac{B}{A}\right|^2 + 2\left|\frac{B}{A}\right|(A_0\cos\phi + B_0\sin\phi) + C_0 = 0 \qquad ...(10.8)$$

where $A_0 = \dfrac{K_1|K_2|\cos\theta_2 - K_3|K_4|\cos\theta_4}{|K_2|^2-|K_4|^2}$;

$$B_0 = \frac{K_1|K_2|\sin\theta_2 - K_3|K_4|\sin\theta_4}{|K_2|^2-|K_4|^2}$$

and $C_0 = \dfrac{K_1^2-K_3^2}{|K_2|^2-|K_4|^2}$

Equation (10.8) represents the equation of a circle on the β-plane having $\left|\frac{B}{A}\right|\cos\phi$ and $j\left|\frac{B}{A}\right|\sin\phi$ as coordinates. This circle has radius

$$r = \frac{\sqrt{K_1^2|K_4|^2 + |K_2|^2K_3^2 - 2K_1|K_2|K_3|K_4|\cos(\theta_2-\theta_4)}}{|K_2|^2-|K_4|^2}$$

and coordinates of the centre are (c, ϕ) where

$$c = \frac{\sqrt{K_1^2|K_2|^2 + K_3^2|K_4|^2 - 2K_1|K_2|K_3|K_4|\cos(\theta_2-\theta_4)}}{|K_2|^2-|K_4|^2}$$

The characteristic is shown in Fig. 10.11.

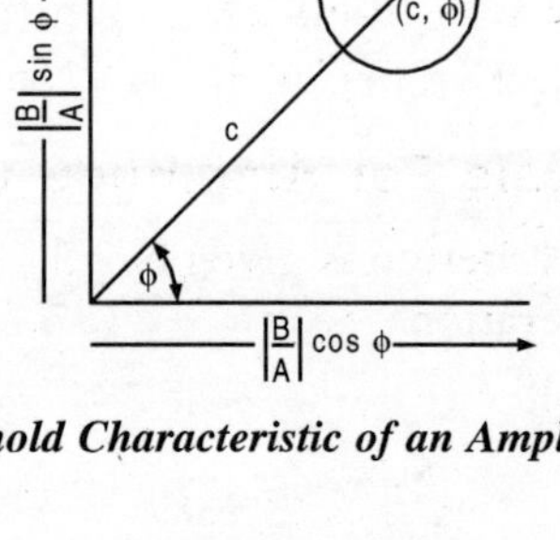

Fig. 10.11. ***Threshold Characteristic of an Amplitude Comparator***

10.4.3. Analysis For Phase Comparator

The two quantities to be compared are S_1 and S_2. Let the phase angles of S_1 and S_2 with respect to a reference axis be α and β respectively. The relay operates when the product of S_1 and S_2 is positive. The product of S_1 and S_2 is maximum when they are in phase and the threshold condition *i.e.*, positive torque will be obtained when $\alpha - \beta = \pm 90°$ or $\pm\pi/2$ radians.

Under this condition $\tan(\alpha-\beta) = \pm\infty$

or $$\frac{\tan\alpha - \tan\beta}{1+\tan\alpha\tan\beta} = \pm\infty$$

or $1 + \tan\alpha\tan\beta = 0$

Substituting for tan α and tan β from phasor diagram shown in Fig. 10.10

$$1 + \frac{|K_2||B|\sin(\theta_2-\phi)}{K_1|A| + |K_2||B|\cos(\theta_2-\phi)} \times \frac{|K_4||B|\sin(\theta_4-\phi)}{K_3|A| + |K_4||B|\cos(\theta_4-\phi)} = 0$$

or $K_1K_3|A|^2 + K_1|K_4||A||B|\cos(\theta_4-\phi) + K_3|K_2||A||B|\cos(\theta_2-\phi) + |B|^2|K_2||K_4|\cos(\theta_2-\phi)\cos(\theta_4-\phi) + |K_2||K_4||B^2|\sin(\theta_2-\phi)\sin(\theta_4-\phi) = 0$ ...(10.9)

Dividing the above equation by $|K_2||K_4||A|^2\cos(\theta_2-\theta_4)$, we have

$$\left|\frac{B}{A}\right|^2 + \left|\frac{B}{A}\right| \frac{(K_1|K_4|\cos\theta_4 + |K_2|K_3\cos\theta_2)\cos\phi + (K_1|K_4|\sin\theta_4 + |K_2|K_3\sin\theta_2)\sin\phi}{|K_2||K_4|\cos(\theta_2-\theta_4)} + \frac{K_1K_3}{|K_2||K_4|\cos(\theta_2-\theta_4)} = 0$$

or $$\left|\frac{B}{A}\right|^2 + \left|\frac{B}{A}\right|\{A_0'\cos\phi + B_0'\sin\phi\} + C_0' = 0 \qquad ...(10.10)$$

where $A_0' = \dfrac{K_1|K_4|\cos\theta_4 + |K_2|K_3\cos\theta_2}{|K_2||K_4|\cos(\theta_2-\theta_4)}$

$$B_0' = \frac{K_1|K_4|\sin\theta_4 + |K_2|K_3\sin\theta_2}{|K_2||K_4|\cos(\theta_2-\theta_4)}$$

and $C_0' = \dfrac{K_1K_3}{|K_2||K_4|\cos(\theta_2-\theta_4)}$

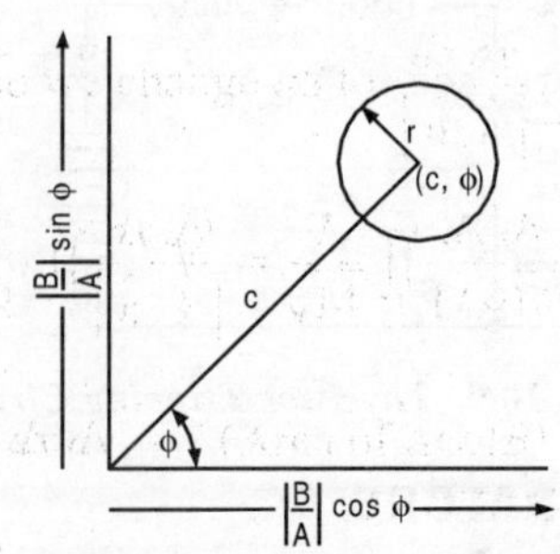

Fig. 10.12. ***Threshold Characteristic of a Phase Comparator***

The above Eq. (10.10) represents an equation to a circle with radius

$$r = \frac{\sqrt{K_1^2 |K_4|^2 + |K_2|^2 K_3^2 - 2K_1 |K_2| K_3 |K_4| \cos(\theta_2 - \theta_4)}}{2|K_2||K_4| \cos(\theta_2 - \theta_4)}$$

and the coordinates of the centre are (c, ϕ) where

$$c = \frac{\sqrt{K_1^2 |K_4|^2 + |K_2|^2 K_3^2 + 2K_1 |K_2| K_3 |K_4| \cos(\theta_2 - \theta_4)}}{2|K_2||K_4| \cos(\theta_2 - \theta_4)}$$

as given in Fig. 10.12

In most of the relays, at least one of the constants K_1, K_2, K_3 and K_4 is zero and two of them are often equal. Also the angle of the two phasor constants is usually the same. This makes the practical case relatively simple. If $\theta_2 = \theta_4$, the values of r and c in the two cases are tabulated below.

Quantity	Amplitude Comparator	Phase Comparator
Radius r	$\dfrac{K_1 \lvert K_4\rvert - \lvert K_2\rvert K_3}{\lvert K_2\rvert^2 - \lvert K_4\rvert^2}$	$\dfrac{K_1 \lvert K_4\rvert - \lvert K_2\rvert K_3}{2\lvert K_2\rvert \lvert K_4\rvert}$
c	$\dfrac{K_1 \lvert K_2\rvert - K_3 \lvert K_4\rvert}{\lvert K_2\rvert^2 - \lvert K_4\rvert^2}$	$\dfrac{K_1 \lvert K_4\rvert + \lvert K_2\rvert K_3}{2\lvert K_2\rvert \lvert K_4\rvert}$

10.4.4. Duality Between Amplitude and Phase Comparators

It can be shown with the help of phasor diagrams that an inherent amplitude comparator becomes a phase comparator and vice versa if the input quantities to the comparator are changed to the sum and difference of the original two input quantities.

Consider the operation of an amplitude comparator with input signals S_1 and S_2 such that it operates when $|S_1| > |S_2|$.

If the inputs are changed to $|S_1 + S_2|$ and $|S_1 - S_2|$ so that it operates when $|S_1 + S_2| > |S_1 - S_2|$.

If these quantities are fed to an amplitude comparator, the comparator essentially compares the phase relation between S_1 and S_2. This is illustrated in Fig. 10.13.

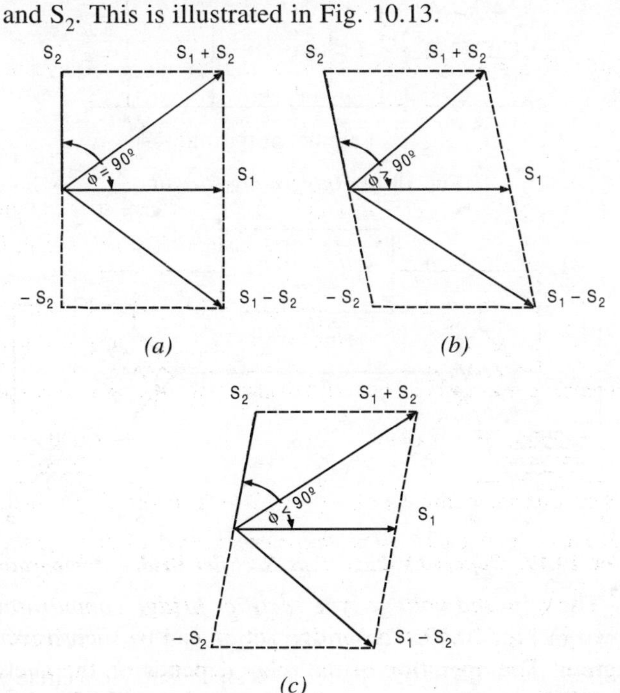

Fig. 10.13. *Phase Comparison Using an Amplitude Comparator*

It is observed that the requirement $|S_1 + S_2| < |S_1 - S_2|$ puts a condition on the phase relation between S_1 and S_2 *i.e.*, unless the phase difference between original phasors S_1 and S_2 exceeds 90°, $|S_1 + S_2|$ cannot be less than $|S_1 - S_2|$. So the original amplitude comparator with inputs now $|S_1 + S_2|$ and $|S_1 - S_2|$ is a phase comparator *i.e.*, a *converted phase comparator.*

It is to be noted that the phase comparator in case of static circuits is a cosine comparator as opposed to a sine comparator in case of electromechanical relays.

The amplitude comparison using a phase comparator is explained with the help of phasor diagram shown in Fig. 10.14.

From the phasor diagram it is clear that if the original inputs S_1 and S_2 to the phase comparator are with such a phase relation that they will operate the relay and if now the inputs are changed to $|S_1 + S_2|$ and $|S_1 - S_2|$ and supplied to the same comparator, the comparator essentially compares the amplitude relation between S_1 and S_2. Unless $|S_1| > |S_2|$, the phase relation between $|S_1 + S_2|$ and $|S_1 - S_2|$ will not be less than 90° and hence the phase comparator with inputs $|S_1 + S_2|$ and $|S_1 - S_2|$ will now be an amplitude comparator.

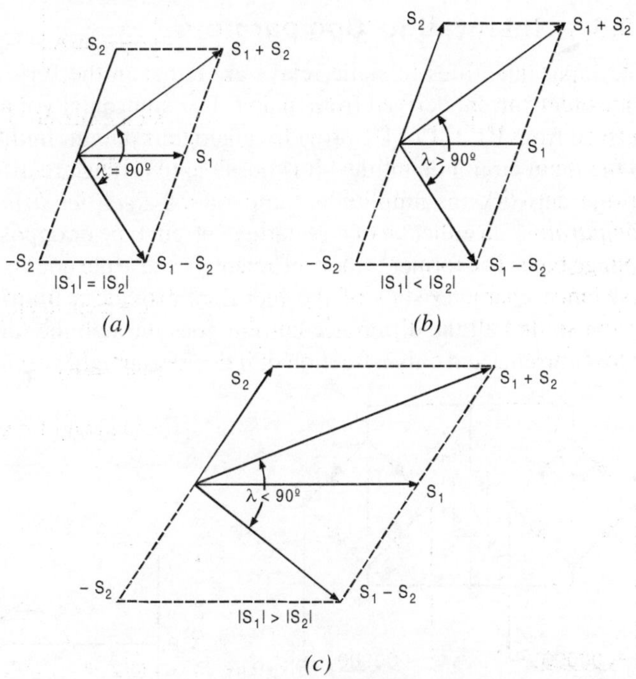

Fig. 10.14. *Phase Comparison Using a Phase Comparator*

Though a given relay characteristic can be obtained using either of the two comparators, consideration of the constants computed for required characteristics would indicate which type of comparator is preferable. In general an inherent comparator is better than the converted type because if one quantity is very large in comparison with the other, a small error in the larger quantity may cause an incorrect comparison when their sum and difference are fed as inputs to the relay.

10.5. AMPLITUDE COMPARATORS

Amplitude comparator compares the amplitudes of two (or more) input quantities. The phase angle between the quantities under comparison (inputs) is not recognized or noticed by the amplitude comparator.

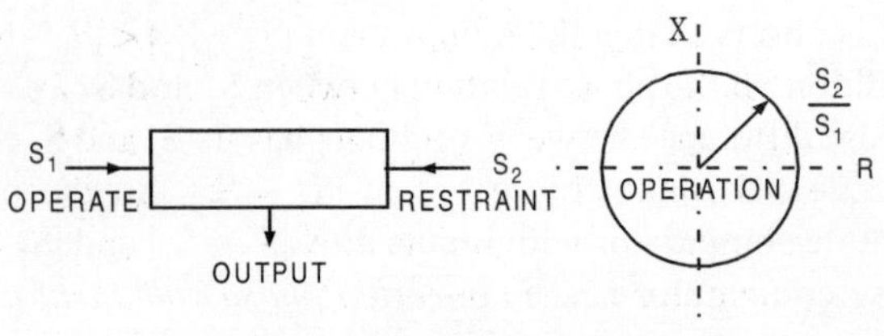

Fig. 10.15. *Amplitude Comparator*

If the two input signals are S_1 and S_2 (say S_1 the operating and S_2 restraining), the amplitude comparator gives positive output only if

$$S_2/S_1 \le K.$$

The function is represented by a circle in the complex plane with its centre at the origin. This defines the boundary of the marginal operation.

The main purpose of amplitude comparators is to provide direction and distance protection.

Mainly there are three types of amplitude comparators viz., integrating comparators, instantaneous comparators, and sampling comparators. These are discussed in brief as follows.

10.5.1. Integrating Comparators

The input quantities to static relays are either in the form of sinusoidal current derived from main CT or sinusoidal voltage derived from PT. CT or PT provides analogous output, faithful to the main circuit quantities. It is possible to arrange rectifier bridge networks as amplitude comparators. *Rectifier bridge comparator* can either be of circulating current type or opposed voltage type. The former is more efficient, because the nonlinear resistance characteristics of the rectifiers provide a limiting action so that all the difference current goes through the relay at low currents and only a fraction of it does so at high currents.

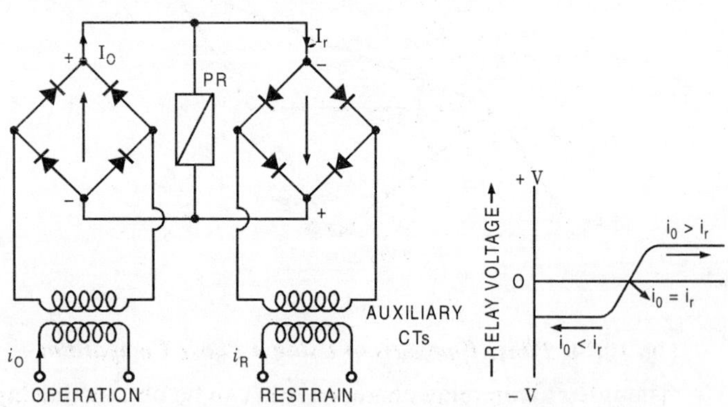

(a) Basic Circuit *(b) Output Voltage*

Fig. 10.16. *Circulating Current Type Rectifier Bridge Comparator*

The basic circuit for the *circulating current type rectifier bridge* comparator is shown in Fig. 10.16. Current input signals are used in this comparator. Two full-wave rectifier bridges are connected in opposition and the output relay is connected in parallel to the two rectifier bridges. The output of the rectifier bridge comparator is received by the output stage continuously. The output device should be an integrating device operating on the average value of the difference between i_0 and i_r. This could be a polarized relay or an integrating circuit followed by a level detector. The operation of the polarised relay (PR) occurs when $S_1 > S_2$ where $S_1 = Ki_0$ and $S_2 = Ki_r$. The output voltage characteristics across the relay are shown in Fig. 10.16 (*b*). The voltage across the relay will never exceed twice the forward voltage drop of one of the rectifiers and this will normally of the order of 1 V. The comparatively low voltage level required for the operation enables reasonable accuracy in impedance measurement to be achieved even for line or source impedance ratios of the order of 30.

A static integrating circuit consisting of an averaging circuit, and polarity detecting circuit is shown in Fig. 10.17. The two currents i_0 and i_r are rectified and their difference $(i_0 - i_r)$ is averaged. The output is obtained only if the averaged value is positive.

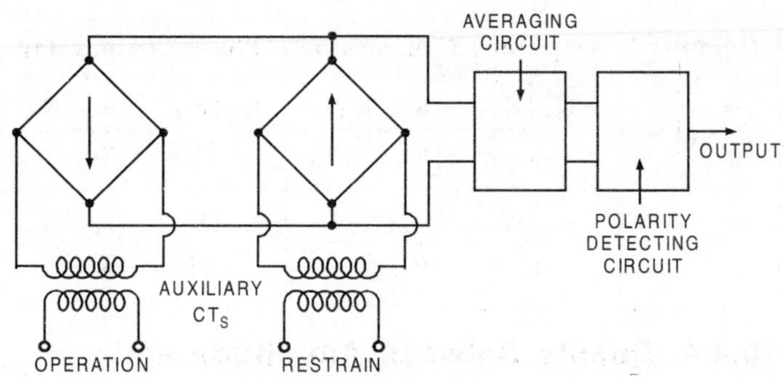

Fig. 10.17. *Rectifier Bridge Comparator With Static Output Device*

An integrator circuit is shown in Fig. 10.18. The tripping occurs when the capacitor voltage attains the setting value of the level detector and triggers a thyristor.

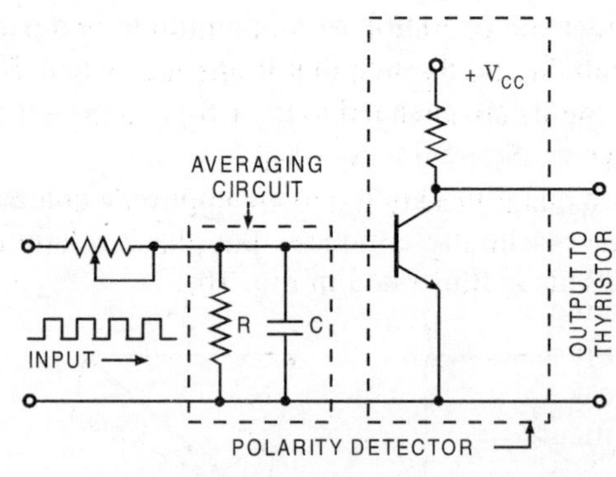

Fig. 10.18. *Integrating Circuit*

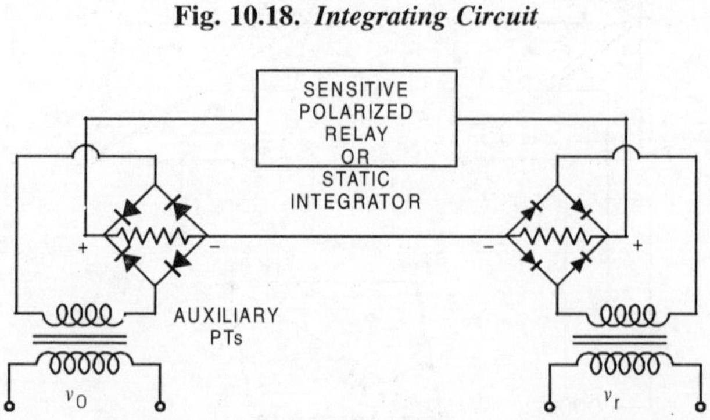

Fig. 10.19. *Opposed Voltage Type Rectifier Bridge Comparator*

The *opposed voltage type rectifier bridge comparator* is shown in Fig. 10.19. The bridge action can be seen from the diagram. The operation of the relay depends on the average of the difference between v_0 and v_r derived from PTs. The bridge is less sensitive at low voltage inputs and the comparator has no limiting action on both voltage and current in the output device.

10.5.2. Instantaneous Amplitude Comparators

Instantaneous amplitude comparators, also called the *direct amplitude comparators*, are of two types viz., averaging type and phase splitting type.

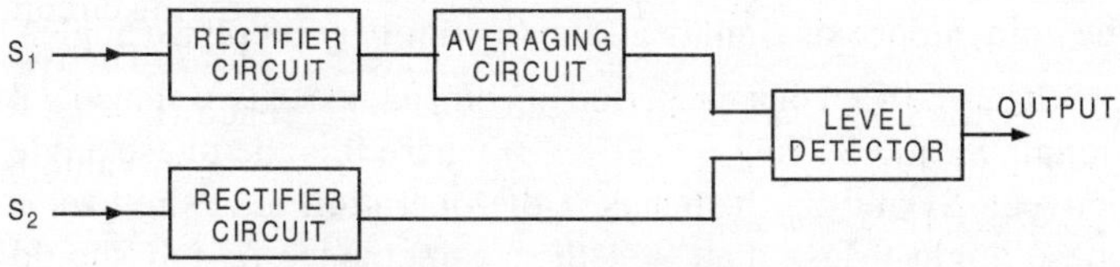

Fig. 10.20. ***Block Diagram of Averaging Type Instantaneous Amplitude Comparator***

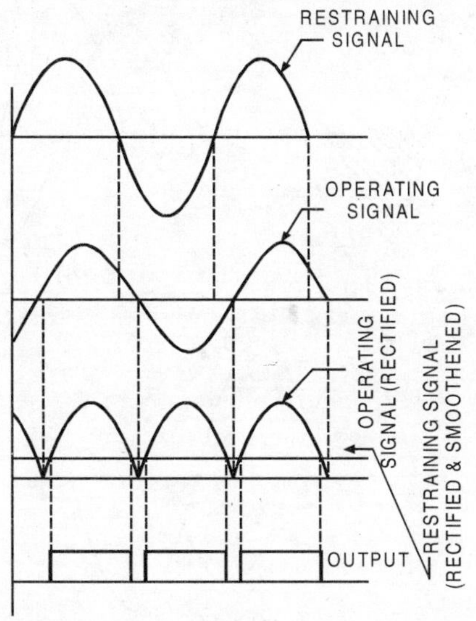

Fig. 10.21. ***Waveshapes of an Instantaneous Amplitude Comparator***

In the *averaging type instantaneous amplitude comparator* the restraining signal is rectified and smoothened completely in order to provide a level of restraint. This is then compared with the peak value of the operating signal, which may or may not be rectified, but is not smoothened. The peak of the operating input signal should exceed the level of restraint for the tripping output. The block diagram is shown in Fig. 10.20. Smoothing is done with the help of a capacitor; so there is a delay in operation. The waveshapes are shown in Fig. 10.21.

Phase-splitting type instantaneous amplitude comparator is faster in operation. Here phase splitting is done before rectification, as shown in Fig. 10.22 *i.e.,* the input is split into six components 60° apart, so that it is smoothened within 5%. In this both operating and restraining input signals are smoothened out before being compared so that a continuous output signal is obtained. The averaging circuit can be eliminated. The operating time is determined by the time constant of the slowest arm of the phase-splitting circuit and by the speed to the output device.

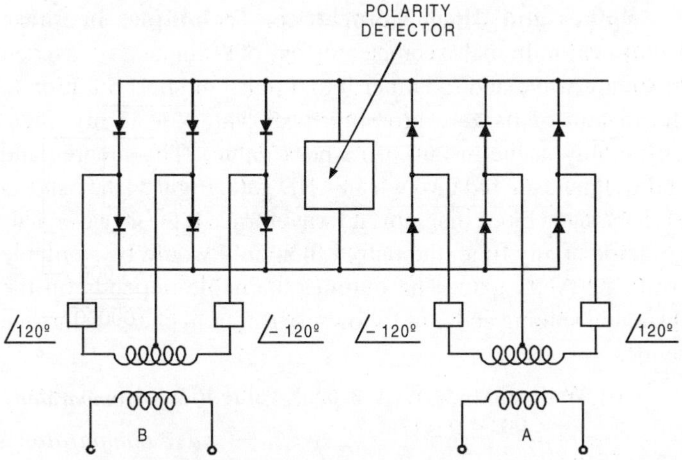

Fig. 10.22. ***Phase-Splitting of Inputs (Six Phase) To Amplitude Comparator***

10.5.3. Sampling Comparators

Sometimes it is convenient to obtain the required characteristics by comparing the amplitude of one input signal at a certain point on its wave with the rectified and smoothed value of the second signal. In case of a reactance relay the instantaneous value of the voltage at the instant of current zero is compared with the rectified and smoothed value of current. Let the phase angle of the circuit be ϕ lagging. (*i.e.,* current lagging behind voltage by ϕ). The value of voltage at current zero will be V sin ϕ. The reactance relay operates when X < K where X is the reactance seen by the relay and K is design reactance of the system.

$$\text{Since } X = Z \sin\phi$$

$$\text{so } Z \sin\phi < K$$

$$\text{or } \frac{V}{I} \sin\phi < K$$

$$\text{or } \frac{V_{max}}{\sqrt{2}} \sin\phi < KI_{av} \times 1.11$$

$$\text{or } V_{max} \sin\phi < K'I_{av}$$

The block diagram for the sampling comparator is given in Fig. 10.23.

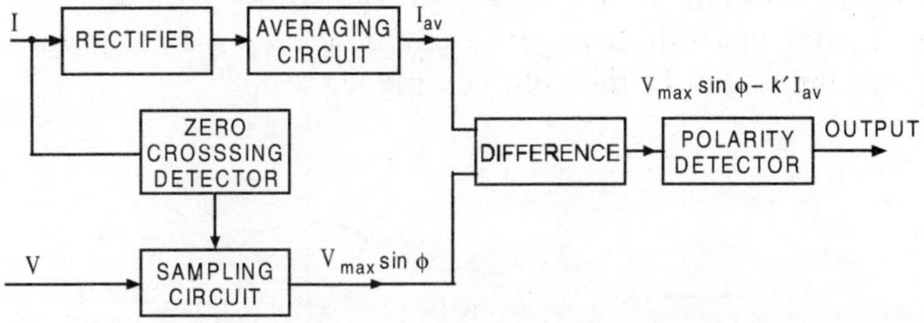

Fig. 10.23. ***Sampling Comparator For a Reactance Relay***

It is also possible to compare the instantaneous magnitude of one signal at a certain instant with the instantaneous magnitude of the second signal at that very instant or at certain other instant. It simply means that one or both signals can be sampled. Voltage is sampled again when the current is passing through zero value and current is sampled after a delay of say α.

For the same reactance relay, for operation we have

$$\frac{V}{I} \sin\phi < K$$

$$\text{or } V_{max} \sin\phi < K' I_{max} \sin\alpha$$

The amplitudes of these two signals are converted into proportional pulse widths and these pulses are compared in an AND gate. If the two samples are taken at different instants, the pulse width representing the one taken first in time sequence is delayed by the time difference between the two sampling instants, before feeding to the AND gate. The scheme is shown in Fig. 10.24.

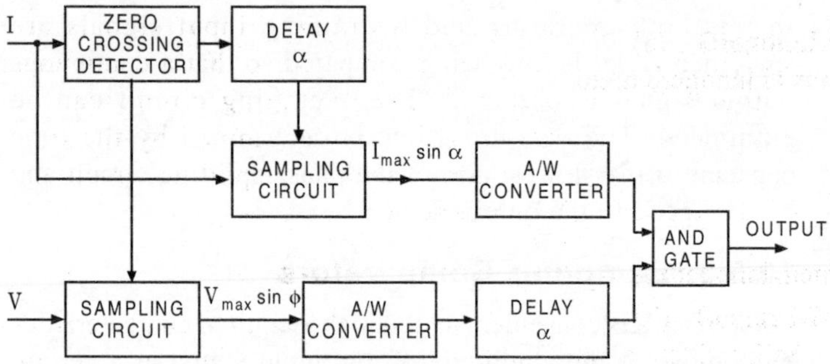

Fig. 10.24. ***Block Diagram When Both Signals are Sampled***

By using sampling techniques, the phase shifting and mixing circuits are eliminated resulting in saving in space and cost but the sampling techniques require a high degree of sophistication in the relay circuitry. The measuring circuits are isolated from the transformer secondaries, except during 50 μs sampling period in each half cycle, and the possibility of maloperation due to spikes on the input waveforms is remote.

10.6. PHASE COMPARATORS

Phase comparison technique is the most widely used technique for all practical directional, distance, differential and carrier relays.

In a phase comparator, the operation of the relay takes place when the phase relation between two inputs S_1 and S_2 varies within certain specified limits. Both inputs must exist for an output to occur; ideally, operation is independent of their amplitudes, and depends only on their phase relationship. Phase comparator, in its simple form, is illustrated in Fig. 10.25. The function, as defined by the boundary of marginal operation, is represented by two straight lines from the origin of the complex plane.

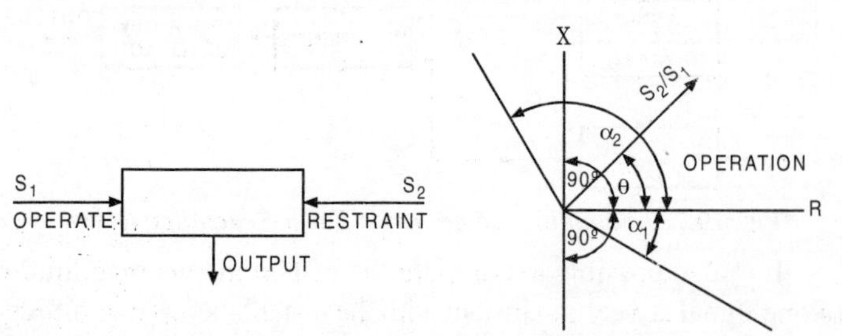

Fig. 10.25. ***Phase Comparator Output When Angle Between S_1 and S_2 is Within Limits α_1 and α_2***

Mathematically, the condition of operation is given as

$$-\alpha_1 \le \theta \le \alpha_2$$

where θ is the angle by which S_1 lags behind S_2. If $\alpha_1 = \alpha_2 = 90°$ the comparator is called the cosine comparator and if $\alpha_1 = 0$ and $\alpha_2 = 180°$, it is called the sine comparator.

Static phase comparators are of two types viz., coincidence type phase comparator and vector product phase comparator.

10.6.1. Coincidence Type Phase Comparators

The basic concept of phase comparison is simpler in that it is possible to deal with signals of equal strength whose coincidence or non-coincidence is readily measurable. Let us consider two sinusoidal signals S_1 and S_2. Their period of coincidence depends upon their phase difference. Figure 10.26 illustrates the coincidence of signals S_1 and S_2 for different phase angles.

It can be seen that the period of coincidence of two sinusoidal signals S_1 and S_2 is $\psi = (180° - \theta)$ where θ is the phase angle between S_1 and S_2. It means if the operation is desired for a phase angle θ less than 90°, then coincidence period should be greater than 90°. Thus, the criterions for operation becomes $-90° \le \theta \le 90°$.

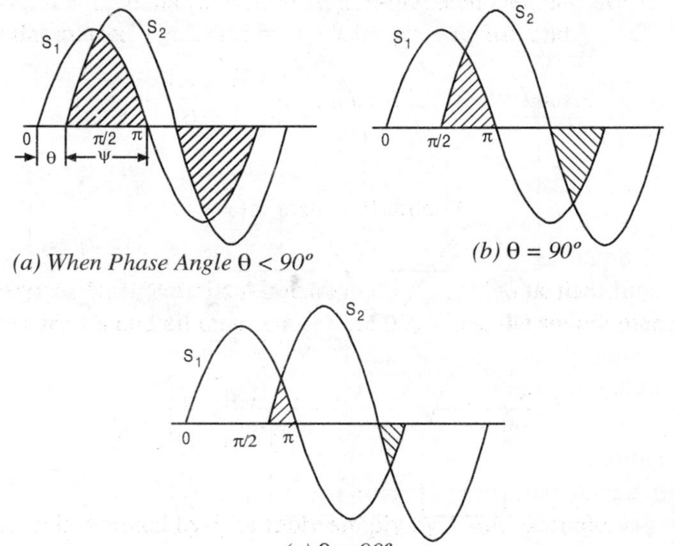

(a) When Phase Angle θ < 90° *(b) θ = 90°* *(c) θ > 90°*

Fig. 10.26. ***Coincidence of Signals S_1 and S_2***

By measuring the period of coincidence, it is possible to design the circuit to give an output a YES or a NO depending upon the phase relation of the input signals. Different techniques can be used to measure the period of coincidence, some of them are given below:

Spikes and Block Coincidence Techniques in Phase Comparator. In such a comparator, one of the inputs is converted to square wave and the other into a pulse of short duration at the instant of its zero crossing, peak value or at any angle (preferably at the instant of its peak value). The squared and spike signals are fed through an AND gate. Figure 10.27 shows the schematic block diagram and waveforms. If these two signals coincide at any time, the output then only would be available from the AND gate. The output obtainable depends on the instant of spiking and is as follows for different phase difference ranges.

(*i*) With spike derived at peak value, output is available for $-90° \le \theta \le 90°$

(*ii*) With spike derived at zero value, output is available for $0 \le \theta \le 180°$.

(*iii*) With spike derived at any other instant α, output is available for $0 \leq \theta \leq 180^\circ - \alpha$.

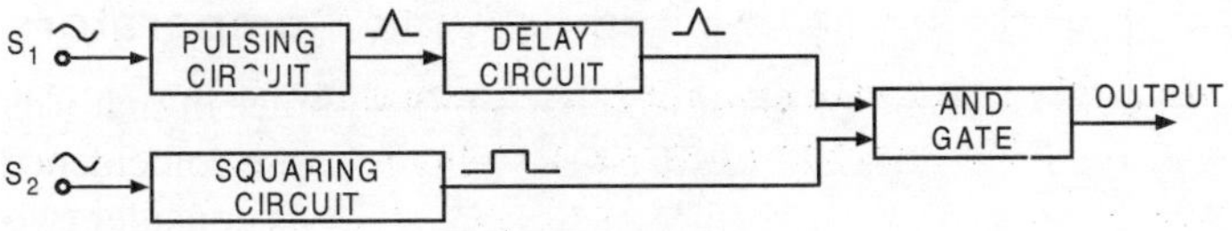

(a) Block Diagram

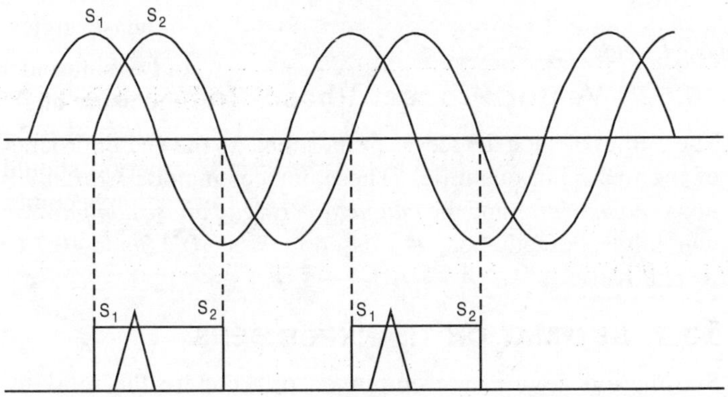

(b) Waveforms

Fig. 10.27. ***Coincidence Phase Comparator With Spike and Block Technique***

Squaring of one input signal is obtained by first limiting it and then amplifying it so that the rectangular block is in phase with the original sine wave. The spike is produced by a peaking transformer or by squaring the wave through the diodes and feeding it through a transformer. The spike is obtained at zero value of the sine wave in each case. This is shifted as per requirement through the delay circuit. A comparison is made after every half cycle. The main drawback of such a comparator is that in case of spurious spikes due to any switching or external interference, the relay may operate which is not desirable. Shielding of the circuit against electric and magnetic fields is, therefore, essential.

Phase Comparator With Phase Splitting Technique. Figure 10.28 illustrates this method in which both the input signals S_1 and S_2 are split into two components $S_1 \angle -45^\circ$, $S_1 \angle 45^\circ$, $S_2 \angle -45^\circ$ and $S_2 \angle 45^\circ$ with respect to the original signal. The four components are then fed into an AND gate which gives an output when all the four inputs are simultaneously positive at any time in the cycle. The coincidence of all the four signals is possible when the phase angle θ satisfies the condition $-90^\circ \leq \theta \leq 90^\circ$.

Integrating Phase Comparator. In this comparator the two input voltage signals are fed into an AND gate the output of which is integrated to measure the period of coincidence of the two signals. If the period of coincidence exceeds 90°, the output is obtained so that the condition is $-90^\circ \leq \theta \leq 90^\circ$ for operation. In earlier integrator phase comparators, transistor type AND gate used to be employed. But in recent types the periods of coincidence are integrated and then fed into a level detector, as illustrated in Fig. 10.29. The critical operating threshold occurs when the periods of concidence and non-coincidence are equal *i.e.*, $\theta = \pm 90^\circ$.

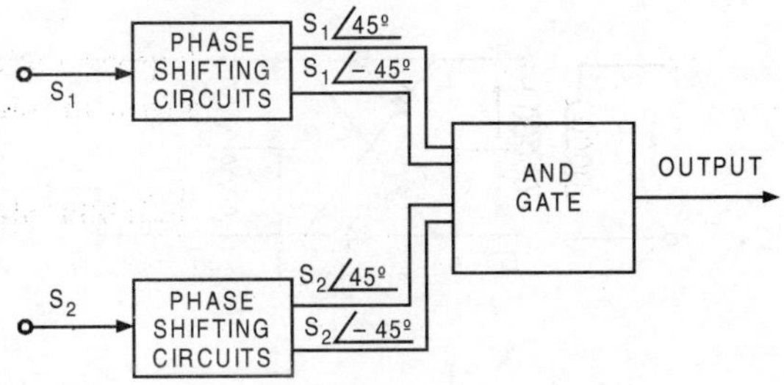

(a) Block Diagram

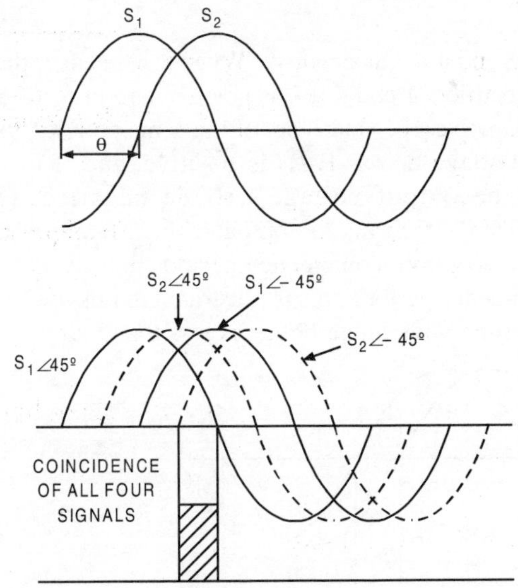

(b) Waveforms

Fig. 10.28. ***Coincidence Phase Comparator With Phase Splitting Technique***

Fig. 10.29. ***Block Diagram For Integration of Coincidence Blocks***

The input signals S_1 and S_2 are compared in a coincidence circuit producing standard output pulses, which are positive when both the signals are of the same polarity (either both positive or both negative) and negative when they are of opposite polarity. The output pulses are fed into an integrating circuit whose output increases linearly during the period when the pulse is positive and falls at the same rate when the polarity reverses. The level detector switches when the output from the integrator exceeds some preset value, and resets when the output falls below some second value.

Integrating Type Phase Comparator With Rectifier Bridge AND Gate. Such a phase comparator has the advantage of simplicity and economy. The circuit is illustrated in Fig. 10.30 (*a*). It consists of a rectifier phase comparator followed by a polarity detection circuit, R-C charging circuit and a level detector. It is a circulating current bridge whose output current is supplied to a centre tapped resistance R-R. The output current is equal to the smaller of the two input currents. The path of the current through the bridge is established by the larger of two input currents and depends upon their relative instantaneous polarity. If i_1 exceeds i_2, the current will flow in top and bottom rectifiers 1 and 2 if i_1 is positive, and in diagonal

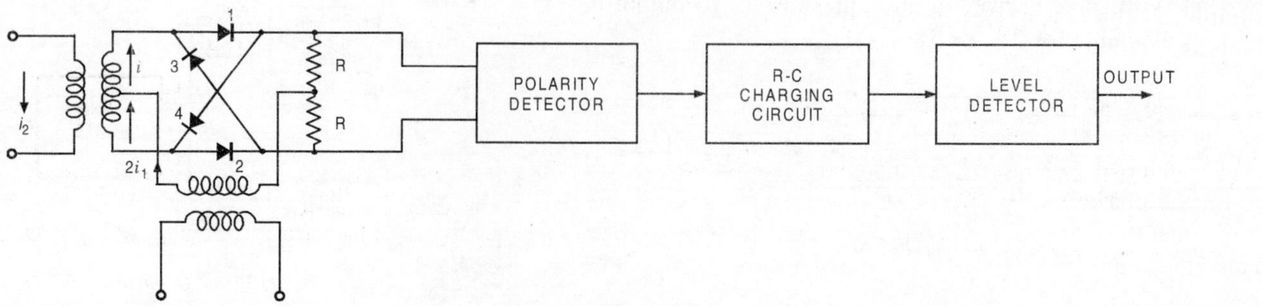

Fig. 10.30 ***(a) Basic Circuit***

rectifiers 3 and 4 if i_1 is negative. When i_2 exceeds i_1 the current flows in rectifiers 1 and 4 if i_2 is positive and in rectifiers 2 and 3 if i_2 is negative. If i_1 and i_2 are of the same polarity, the output voltage (voltage across R-R) is positive, and if of opposite polarity the output voltage will be negative. [refer to Fig. 10.30 (*b*)]. It means the output voltage is positive during positive or negative coincidence periods and negative during non-coincidence period. This is illustrated in Fig. 10.30 (*b*). The operating time with single bridge is less than half a cycle.

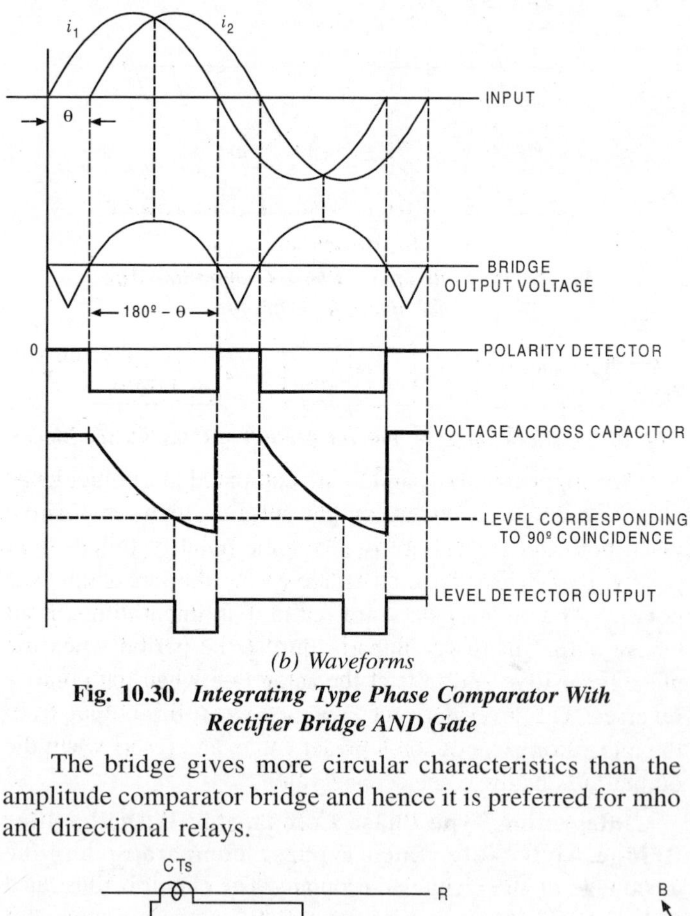

(b) Waveforms

Fig. 10.30. ***Integrating Type Phase Comparator With Rectifier Bridge AND Gate***

The bridge gives more circular characteristics than the amplitude comparator bridge and hence it is preferred for mho and directional relays.

10.6.2. Vector Product Phase Comparators

The output of such devices is proportional to the vector product of the two input quantities. These devices include *Hall effect phase comparator* and the *magnetoresistivity phase comparator* which have already been discussed in Arts. 10.2.5 and 10.2.6. respectively.

10.7. SUMMATION TRANSFORMERS

Summation devices are employed to compare the relaying quantities derived from the currents in the three phases of the primary circuit. The effectiveness of the comparison of the derived relaying quantities may be considered in terms of the effective ratio of the primary currents under comparison, such as at the two ends of a feeder. Summation devices are employed for converting 3-phase quantities into single-phase quantities.

The basic arrangement is shown in Fig. 10.31. The three line CTs are connected to the primary of an auxiliary CT (summation transformer), as illustrated in the figure. Each line CT energizes a different number of turns on the primary of the auxiliary CT and the resulting single-phase output appears across the secondary.

In case the device is to be employed for unbalanced conditions in the system and due to fault conditions to operate the relays properly, it is preferable to convert the unbalanced currents into positive, negative and zero sequence components. A simple arrangement of the phase sequence current segregating network is shown in Fig. 10.32. The summated output will be according to fault conditions. Summation devices have their limitations because of loss of discrimination due to condensation of information.

The *output ampere-turns* from any summation device may be expressed in terms of symmetrical components in the form $MI_0 + NI_1 + PI_2$ where M, N and P depend on the choice of summated quantities. The ratio of the compared quantities, for

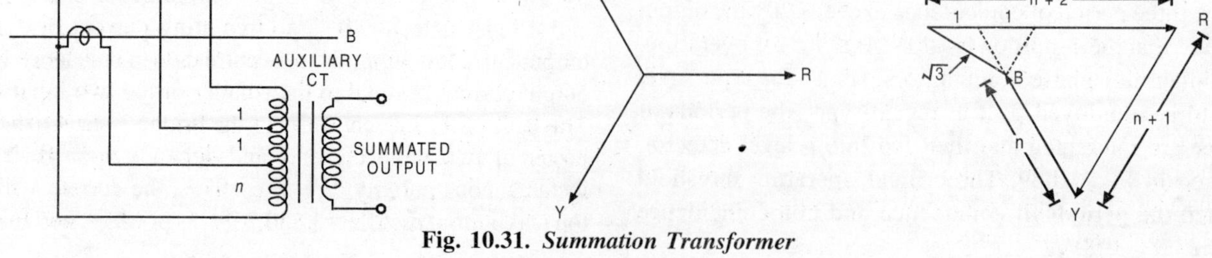

Fig. 10.31. ***Summation Transformer***

example, at the end of a protected feeder say A and B is given as

$$\frac{I_A}{I_B} \angle\phi = \frac{MI_{A0} + NI_{A1} + PI_{A2}}{MI_{B0} + NI_{B1} + PI_{B2}} \quad ...(10.11)$$

The values of M, N and P depend on the summation device and on the phase affected by the fault. The ideal conditions assume that the current transducers are linear in response and so also the summation devices.

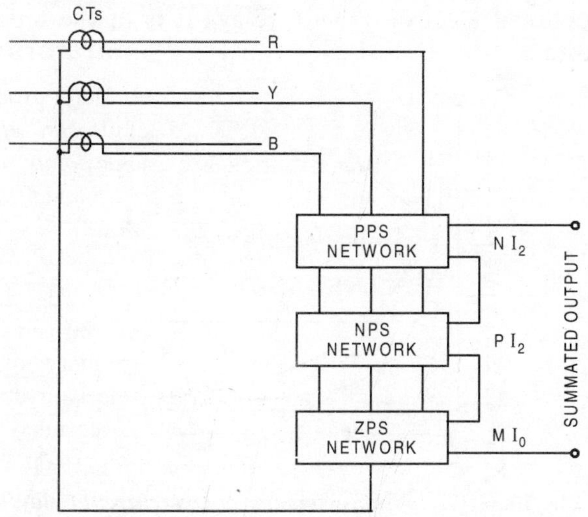

Fig. 10.32. *Phase Sequence Current Segregating Network*

The effective output from the secondary winding of the summation transformer having primary turn ratio 1 : 1 : *n* (Fig. 10.31) is given as

$$(n+2)I_R + (n+1)I_Y + nI_B$$

If this is to be expressed in symmetrical components, it becomes

$$3(n+1)I_0 + (2+a^2)I_1 + (2+a)I_2$$

This defines the constants M, N and P for the reference phase fault conditions *i.e.,* ground fault, double line to ground fault etc. For the faults on the phases other than the reference phase R, the N and P should be multiplied by a and a^2 operators respectively, to define the appropriate output.

The values of M, N and P for various types of faults referred to the phase R for summation transformer having turn ratio 1 :1 : *n* are given below in tabular form [Table 10.1].

Table 10.1

Type of Fault	*M*	*N*	*P*
Shunt R-E, Y-B or Y-B-E	$3(n+1)$	$(2+a^2)$	$(2+a)$
Shunt Y-E, B-R or B-R-E	$3(n+1)$	$(1+2a)$	$(1+2a^2)$
Shunt B-E, R-Y or R-Y-E	$3(n+1)$	$(2a^2+a)$	$(2a+a^2)$

In case of phase sequence network filters, the recent studies show that the preferred values of M : N : P are 0 : –1 : 6 for reference phase R. For faults on other phase, the value are given below in tabular form [Table 10.2]

Table 10.2

Type of Fault	*M*	*N*	*P*
Shunt R-E, Y-B or Y-B-E	0	– 1	6
Shunt Y-E, B-R or B-R-E	0	$-a$	$6a^2$
Shunt B-E, R-Y or R-Y-E	0	$-a^2$	$6a$

Summation transformer is a simpler summation device and has less burden while sequence network filters are more versatile.

10.8. OVERCURRENT RELAYS

The overcurrent relays, even though simplest of all types of electromechanical relays, are the most difficult static relays. This is because the induction disc characteristics of the overcurrent relays (inverse characteristics) are not amenable to simple mathematical analysis. The first static relays developed were the high speed differential relays and the distance relays.

Fault current level detectors are termed *overcurrent relays*. They are more complicated in static form as compared to their electromagnetic counterparts. However, static overcurrent relays offer several advantages over the electromagnetic form.

1. *Low CT burden.* The VA consumption of static overcurrent relays is quite low (7 mVA to 100 mVA) as compared with that of electromagnetic relays (1VA to 3VA) so that smaller CTs are required. The performance of CT under short-circuit condition is also improved.
2. *Compact in size.* The size of a single three-phase overcurrent relay may be about one-fourth of three electromagnetic relays. Hence less panel space is required in case of static overcurrent relays.
3. *Possibility of Instantaneous Reset.* In case of static overcurrent relays instantaneous reset is possible. This is due to the absence of moving parts, which facilitates the application of automatic reclosing of breakers.
4. No overreaching tendencies and more accurate time-current characteristic.
5. Less maintenance, long life and not affected by shock and vibration.

The technique of overcurrent relaying is still widely used as a means of fault detection on distribution system and on transmission lines supplied from one end. In the case of transmission lines supplied from both ends, it is employed with directional relays. Overcurrent relays are also employed in conjunction with distance relays to provide backup protection.

The protective relays are either *single actuating quantity relays* such as overcurrent, undervoltage, earth fault relay or *double actuating quantity relays* such as distance relays, differential relays.

Static overcurrent relays are of two types viz., (*i*) instantaneous overcurrent relays and (*ii*) time overcurrent relay.

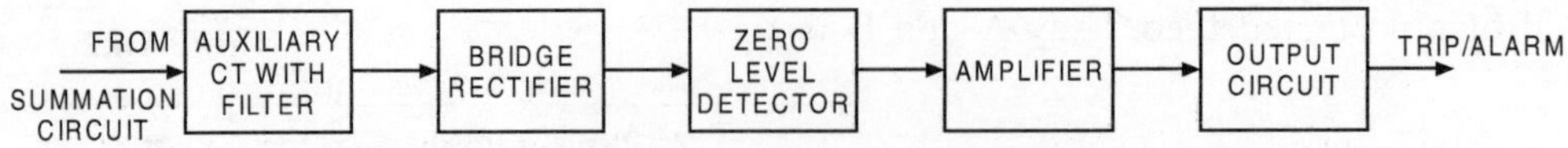

Fig. 10.33. ***Simplified Block Diagram of Static Overcurrent Relay***

10.8.1. Static Overcurrent Relays

The block diagram of an instantaneous overcurrent relay is shown in Fig. 10.33. The same construction may be used for undervoltage, overvoltage or earth fault relays too.

The secondaries of the line CTs are connected to a summation circuit (not shown in the figure). The output of this summation CT is fed to an auxiliary CT, whose output is rectified, smoothened and supplied to the measuring unit (level detector). The measuring unit determines whether the quantity has attained the threshold value (set value) or not. When the input to measuring unit is less than the threshold value, the output of the level detector is zero. For an overcurrent relay

$$\text{For } I_{input} < I_{threshold}\,;\ I_{output} = 0$$

$$\text{For } I_{input} \geq I_{threshold}\,;\ I_{output} = \text{Present}$$

In an actual relay, $I_{threshold}$ can be adjusted.

After operation of the measuring unit, the output is amplified by the amplifer. The amplified output is given to the output circuit to cause trip/alarm.

If time delay is desired, a timing circuit is introduced before the level detector.

Smoothing circuit and filters are introduced in the output of the bridge rectifier.

Static overcurrent relay is made in the form of a single unit in which diodes, transistors, resistors, capacitors etc., are arranged on printed board and are bolted with epoxy resin.

The general equation for time characteristic is given as

$$I^n t = K \qquad ...(10.12)$$

where I is the relay current, t is the time of operation and n is the characteristic index of relay and K is constant.

In conventional electromagnetic relays, n may vary between 2 and 8. The characteristic becomes a straight line parallel to current axis for $n = 0$. Such a characteristic is known as *definite time characteristic* of overcurrent relay.

With $n = 1$; $It = K$. The characteristic becomes *inverse characteristic.*

With higher value of n, the characteristic becomes more and more inverse. With $n = 7$ or 8, the characteristic becomes extremely inverse.

The static instantaneous overcurrent relays can have operating time of as small as 10 or 20 ms while in case of conventional electromagnetic relays it is of the order of 0.1 second.

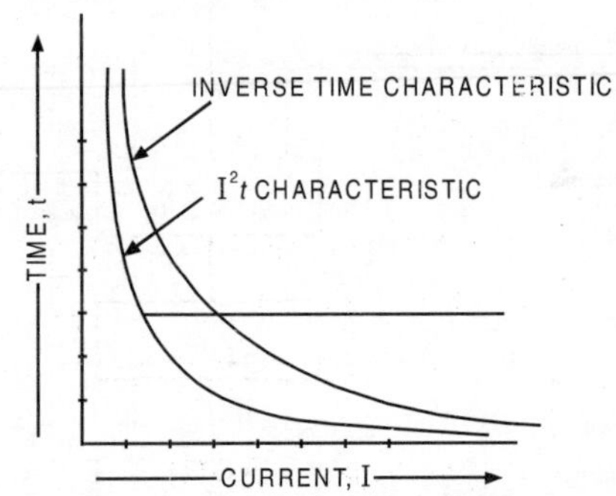

Fig. 10.34. ***Time Characteristics of Overcurrent Relays***

Definite time overcurrent relays are used for wide variations of systems conditions, as backup relays for differential and distance protection, and differential protection of transformers to avoid maloperation during magnetization inrush currents. *Inverse time relays* are used where the source impedance is much smaller than the line impedance. *Extremely inverse overcurrent relays* are used for the fuse coordination and thermal protection of transformers and induction motors.

The general expression for the operating time of a **static time current relays** may be given as

$$t = \frac{KM}{I^n - I_p^n} \qquad ...(10.13)$$

where M is time-multiple setting, I is multiple of tap current; I_p is the multiple of tap current at which pick-up occurs, n is characteristic index of relay, t is time of operation in seconds and K is design constant of the relay.

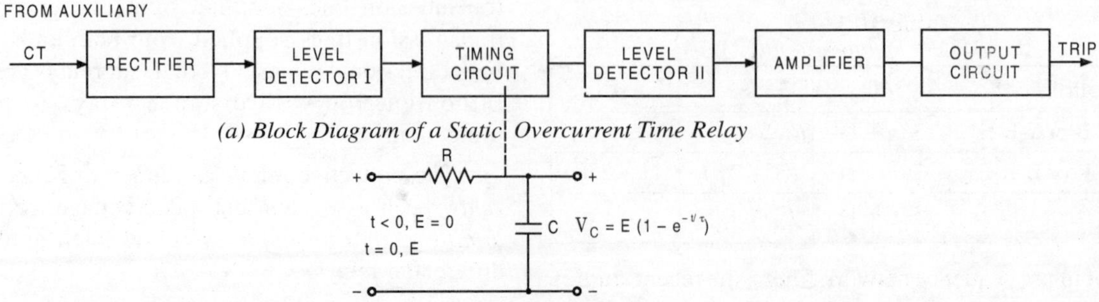

(a) Block Diagram of a Static Overcurrent Time Relay

(b) R-C Timing Circuit

Fig. 10.35

If the relay picks up at top value current *i.e.*, $I_p = 1$. then

$$t = \frac{KM}{I^n - 1}$$

The static overcurrent time relays can have the following typical characteristics

$$\text{IDMT standard inverse } t = \frac{0.14}{I^{0.02} - 1}$$

$$\text{Very inverse } t = \frac{13.5}{I - 1}$$

$$\text{Extremely inverse } t = \frac{80}{I^2 - 1}$$

In static relays it is advantageous to choose a circuit accommodating a wide range of alternative inverse time characteristics, precise minimum operating levels, definite minimum times and additional high set features if necessary. The block diagram of a static overcurrent time relay is shown in Fig. 10.35 (*a*).

The current from the line CT is reduced to 1/1,000th by an auxiliary CT, the auxiliary CT has taps on the primary for selecting the desired pick-up and current range and its rectified output is supplied to level detector I (overload level detector) and an R-C timing circuit. When the voltage on the timing capacitor V_C attains the threshold value of the level detector II, tripping occurs.

Time delay given by the timing circuit shown in Fig. 10.35 is given as

$$T_C = RC \log_e \frac{E}{E - V_T} \quad ...(10.14)$$

where V_T is the threshold value of the level detector II.

By varying values of R and C the time can be varied without difficulties. The basic R-C circuit can also be arranged in several series-parallel combinations to have different values of T_C.

Nonlinear resistors are used to have other time characteristics (Fig. 10.36).

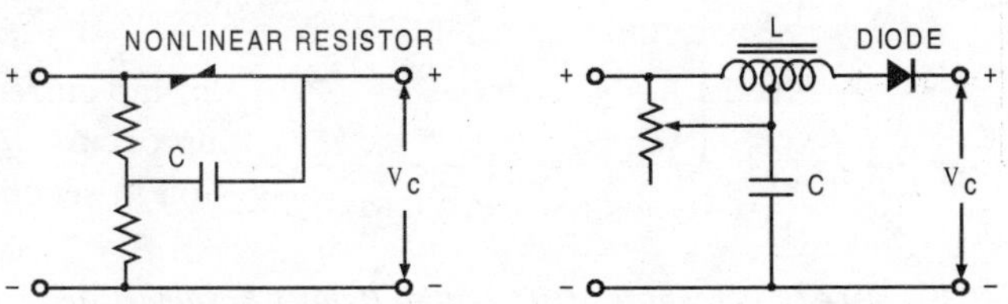

(*a*) *Timing Circuit with Nonlinear Resistor* (*b*) *Timing Circuit With Diode*

Fig. 10.36. ***Timing Circuit***

Time delay overcurrent relays are used in overcurrent protection of utility equipment, distribution circuits, protection of transformers, generators, motors etc.

In instantaneous overcurrent relays, the time delay circuit shown in Fig. 10.35 (*a*) is deleted. Such a circuit would need only one level detector. As there is no moving part, operating time of the order of 0.02 s (1 cycle) can be achieved. Instantaneous overcurrent relays are used for short-circuit protection of large equipment. Instantaneous overcurrent relays are also useful in other protective relay systems.

10.8.2. Overcurrent Ground Protection

Overcurrent ground fault protection can be obtained with only one overcurrent relay by connecting it across a zero phase sequence current filter, as shown in Fig. 10.37. On occurrence of ground fault, the zero sequence current will flow through the relay coil and relay will operate if the relay current exceeds the threshold value.

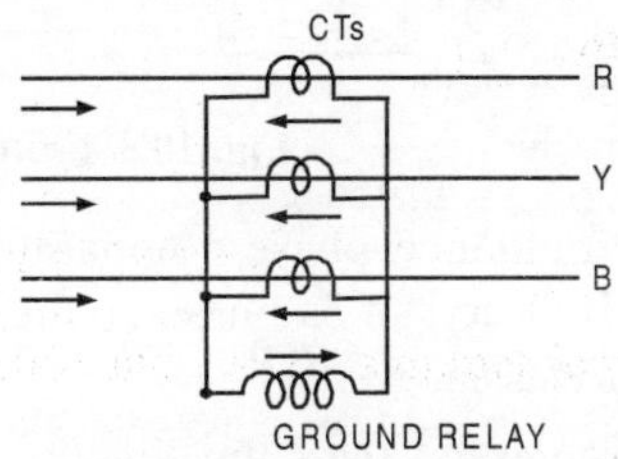

Fig. 10.37. ***Overcurrent Ground Fault Protection***

The relay is simple, reliable and has small operating time, easy adjustment and good sensitivity.

For complete protection against phase and ground faults, generally three CTs and two phase and one ground overcurrent relays are employed in the system. The connections are shown in Fig. 10.38.

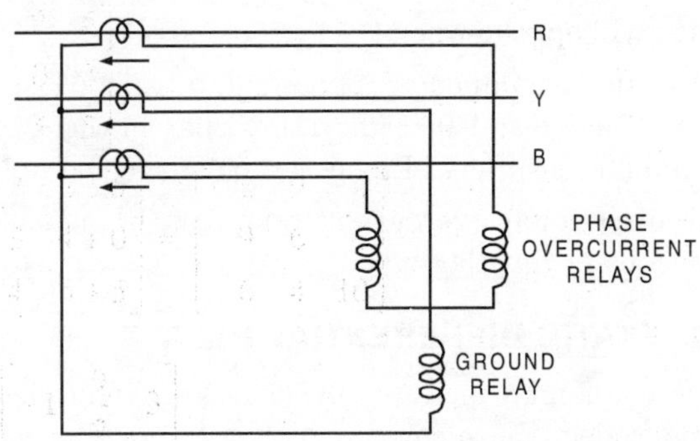

Fig. 10.38. ***Combined Phase and Ground Fcult Protection***

10.9. DIRECTIONAL STATIC OVERCURRENT RELAYS

For obvious reasons of obtaining selectivity overcurrent relays are made directional. Directional relay senses direction of power flow by means of phase angle between V and I. When the phase angle between V and I exceeds certain predetermined value, the directional relay operates with a condition that the current is above the pick-up value. Thus directional relay is a double actuating quantity relay with one input as current I from CT and the other input V from PT.

The induction cup type of electromagnetic relays, even though are very sensitive, have dead zone in their operation. In static directional relays this problem is less serious because static comparators are very sensitive and it is comparatively easier to make a static directional unit reliable down to 1% of system voltage which is well within the minimum fault voltage.

The choice of directional unit depends upon the type of comparator used. The comparator used in directional overcurrent relay may be Hall effect generator (described in Art. 10.2.5); Rectifier bridge comparator (described in Art. 10.5.1); Instantaneous concidence comparator (described in Art. 10.5.2)

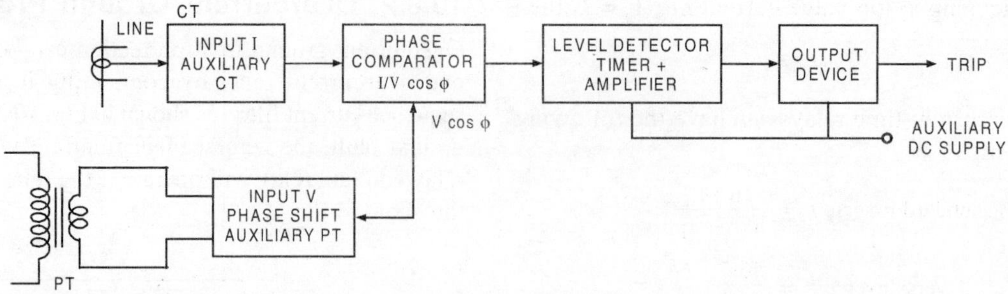

Fig. 10.39. *Block Diagram of Directional Static Overcurrent Relay*

or Integrating coincidence phase comparator (described in Art. 10.6.1). The Hall crystal as phase comparator has gained popularity only in Russia.

Figure 10.39 represents the static directional relay with two inputs (V and I). The inputs are supplied to the phase comparator. A phase shifter is included in the voltage input circuit, whose output is fed to the phase comparator; so that the output from phase comparator under phase faults/earth fault condition is maximum. The output of phase comparator is fed to the level detector. The output of the level detector is amplified and in case a timer is necessary, the output is applied to the output device through the timer.

Relay will operate when $I_s \le I \cos(\phi - \theta)$.

where I_s is the magnitude of set current, ϕ is the phase angle between V and I and θ is the relay characteristic angle. For maximum sensitivity of relay ϕ should be equal to θ.

The directional overcurrent relay can be made either instantaneous or time lag type.

10.10. STATIC DIFFERENTIAL RELAYS

The differential relay measures the phasor difference between two similar electrical quantities (voltage-voltage or current-current). The rectifier bridge amplitude comparator is the most common static form for applications as a differential relay element. The block diagram for such a relay is shown in Fig. 10.40.

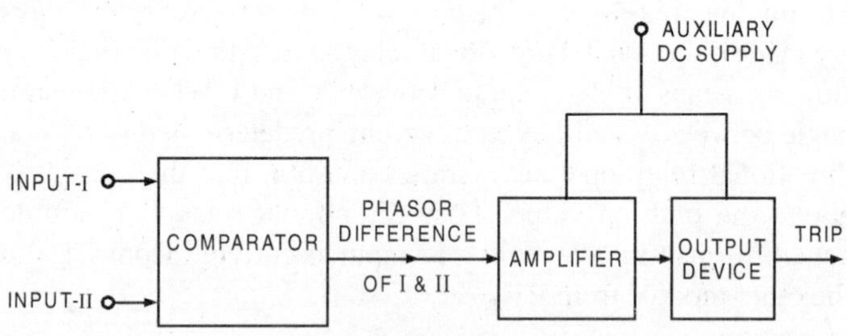

Fig. 10.40

Inputs I and II are supplied to the comparator. The output of the comparator (phasor difference of inputs I and II) is amplified and used to operate the relay.

The static differential relays are most commonly used for the protection of generators and transformers for any type of internal faults (two- and three-phase faults, earth faults with solidly grounded neutral or low resistance grounded neutral inter-turn faults).

These relays are advantageous over electromagnetic differential relays as they are very compact, highly sensitive for internal faults and have absolute stability for heavy through faults, extremely short tripping times (20-50 ms) regardless magnitude of auxiliary voltage, accurate and absolutely stable tripping characteristic even for asymmetrical faults as each phase can have its own relay, low VA burden, inrush current proof characteristic even during high starting currents, inrush currents. The selection of auxiliary voltage is also easy.

A permanent magnet moving coil relay is usually employed as tripping device.

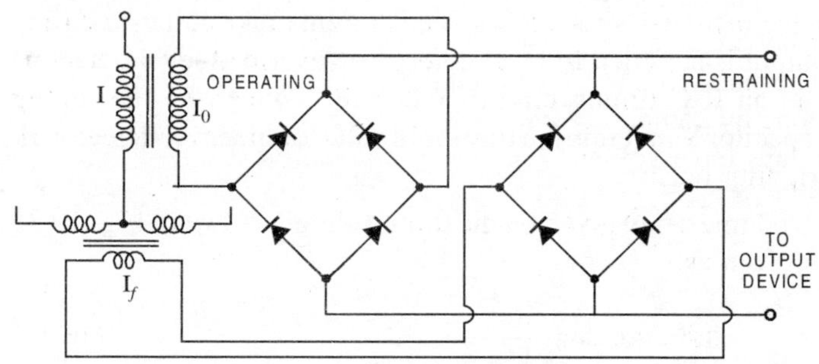

Fig. 10.41. *Static Type Percentage Differential Relay*

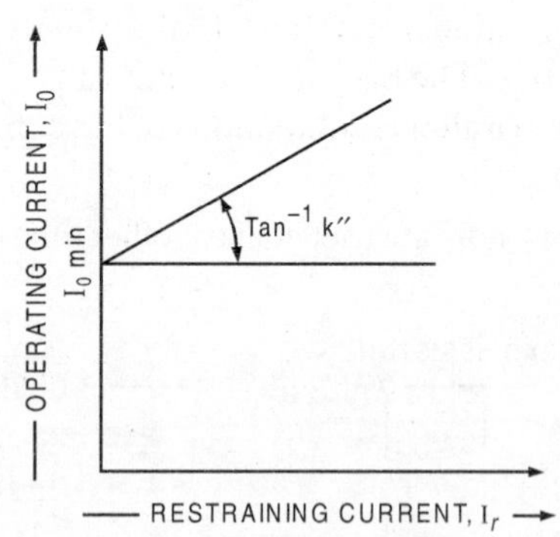

Fig. 10.42. *Percentage Differential Relay Characteristic*

Static type percentage differential relay is shown in Fig. 10.41. The difference of the currents in the operating coil and restraining coil is fed to the output element for the relay operation. They relay operates when

$$K_0 n_0 I_0 > K_r n_r I_r + K'$$

where n_0 and n_r are the number of turns on the operating and restraining coils respectively and K_0 and K_r the design constants and K' the spring control torque constant.

At the threshold of operation $K' = K_0 n_0 I_{0min}$

Now equation under threshold condition becomes

$$K_0 n_0 I_0 = K_r n_r I_r + K_0 n_0 I_{0min}$$

$$\text{or } I_0 = \frac{K_r n_r}{K_0 n_0} I_r + I_{0\min}$$

$$= K'' I_r + I_{0\min} \text{ where } K'' = \frac{K_r n_r}{K_0 n_0} \quad \text{...(10.15)}$$

The above equation is an equation of a straight line of the form $y = mx + c$, the intercept $c = I_{0\min}$ and the slop $m = K''$. The characteristic is shown in Fig. 10.42.

The differential current schemes do not react to the peak currents caused by overloads or swings, also due to dissimilarity in CTs, inrush magnetizing current in transformer protection.

10.11. STATIC DISTANCE RELAYS

Distance relays are characterised by having two input quantities proportional to the voltage and current at a particular point in the power system, referred to as the relaying point. Ideal static distance relays have characteristics independent of actual magnitudes of voltage and current but dependent only on their ratio and phase angle between them.

The versatile family of distance relays includes impedance relays, reactance relays and mho relays as discussed earlier. The measurement of impedance, reactance or admittance is done by comparing input current and voltage. Hence distance relays have voltage and current as input quantities.

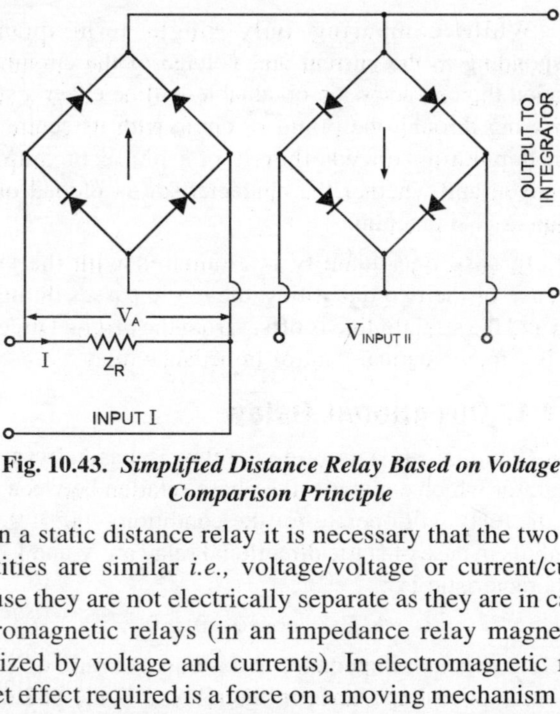

Fig. 10.43. ***Simplified Distance Relay Based on Voltage Comparison Principle***

In a static distance relay it is necessary that the two input quantities are similar *i.e.,* voltage/voltage or current/current because they are not electrically separate as they are in case of electromagnetic relays (in an impedance relay magnets are energized by voltage and currents). In electromagnetic relays the net effect required is a force on a moving mechanism and it can be equally obtained either by a voltage or a current which is not true in case of static relays. So the comparators employed in static distance relays can be of either voltage comparator or current comparator. In a voltage comparator line current is converted into equivalent voltage V_A by passing it through an impedance $Z_R \angle\theta$ and the voltage drop IZ_R is then compared with the line voltage V.$Z_R \angle\theta$ is the design impedance or a replica of the impedance of the line to be protected on a secondary basis. Distance relay based on voltage comparison principle is shown in Fig. 10.43.

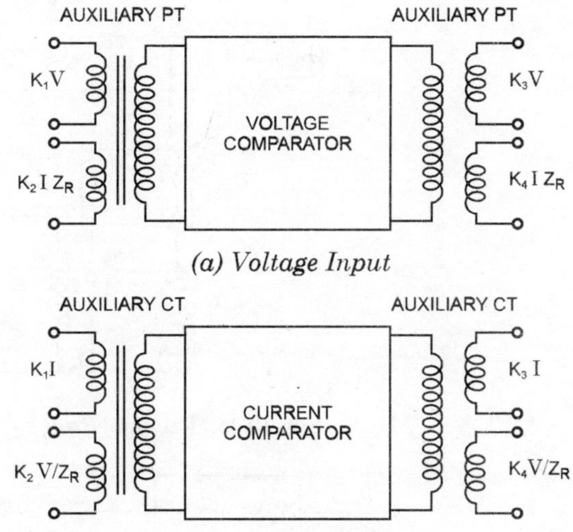

Fig. 10.44. ***Two Input Comparator***

Similarly in case of a current comparator, a current is derived from CT and the voltage from PT is converted into equivalent current V/Z_R by connecting replica impedance in series with secondary of PT. The current in secondary of PT corresponds to V/Z_R which is compared with I.

Sometimes it may be convenient to compare the two voltages V and IZ_R in a current comparator which is accomplished by connecting resistance in series with each voltage.

The arrangements of inputs for two input comparators (*i*) with voltage inputs and (*ii*) with current inputs are shown in Fig. 10.44.

It is to be noted that the transient dc component of current passing through line impedance produces a faithful voltage waveform which is derived from line PT and the secondary current of line PT $\left(\frac{V}{Z_R}\right)$ has faithful transient. The comparator compares $\frac{V}{Z_R}$ and I, both having identical transient (assuming faithful reproduction). Hence the effect of transients is cancelled out from impedance measurement. So use of replica impedance is not only convenient but permits fast tripping also as it eliminates error due to transients in the fault current.

A rectifier bridge current comparator, shown in Fig. 10.45, receives two current inputs (operating current I_0 and restraining current I_r). The output of comparator is applied to a permanent magnet coil relay or a static level detector.

In distance relays I_0 and I_r may be supplied either by the CT or by a PT through a series impedance or by both sources in a particular combination to have particular relay characteristic.

If the restraining current I_r is supplied by PT and operating current I_0 is supplied by CT, as shown in Fig. 10.45, the relay operates when the ratio V/I is less than a certain value and is therefore a *minimum impedance relay*.

A practical static distance protection scheme includes a starting, measuring and timing elements made up of solid state devices. The output unit is usually a moving coil relay. The starting element is usually an overcurrent relay. The output is given to the measuring element. Phase comparators are employed

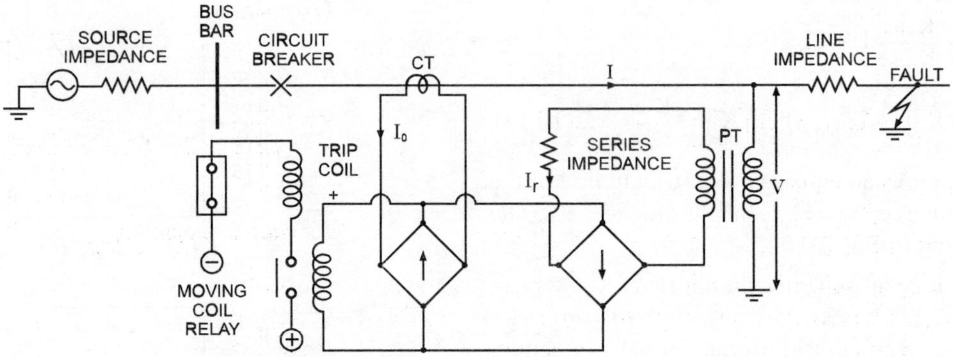

Fig. 10.45. *Rectifier Bridge Current Comparator Used in an Impedance Relay*

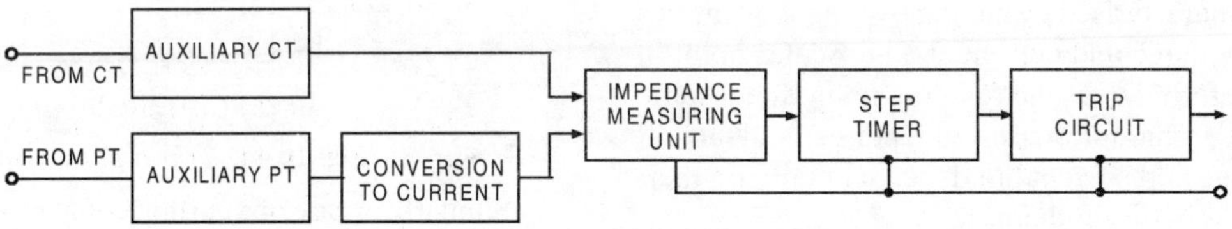

Fig. 10.46. *Distance Relay Based on Current Comparison Principle*

in the measuring devices. The measuring device determines whether the fault is within the protected zone or not. Corresponding a tripping signal is initiated in case the fault is within the protected zone. In case the fault is outside the protected zone, the timer unit starts. This initiates zone-wise protection.

A block diagram of a distance relay based on current comparison principle is given in Fig. 10.46. The line PT secondary is connected to auxiliary PT and the output of auxiliary PT is converted into current and this current is compared with the output of the auxiliary CT.

Static distance relays do not have any moving part so they operate much faster (operating time of the order of some milliseconds) and without risk of incorrect tripping as compared to electromagnetic relays. With semiconductor devices it is possible to obtain other distance characteristics than the conventional ones. Static distance relays are accurate over a wider range of fault currents and line lengths and require much lower burden as compared to their counterparts in electromagnetic relays. Static distance relays are compact in size and have better stability under power swing conditions.

The main problem with impedance relays is smoothening of one of the inputs so that the pick-up does not vary from zero to infinity during the cycle as first the voltage and then the current passes through zero. Normally, the voltage is smoothened as it is easier to do so as compared to the smoothening of current. This is done by a phase-splitting circuit.

Static distance relays are extensively used for protection of medium and long transmission lines, parallel feeders and unit backup protection as well as interconnected and T-connected lines.

10.12. DISTANCE RELAY CHARACTERISTICS

Distance relay characteristics are normally drawn on an R-X diagram but sometimes it becomes convenient to draw them on G-B diagram as well. While discussing and deriving the characteristics of distance relays (impedance, reactance and mho relays), including those of directional relay the following points are noteworthy.

1. While comparing only single term quantities corresponding to the current and voltage of the circuit under protection the characteristic obtainable will be either a straight line passing through the origin or circle with its centre at the origin depending on whether it is a phase or amplitude comparison and whether the characteristic is plotted on R-X diagram or G-B diagram.

2. In case one quantity is compared with the sum or difference of the two quantities, the circle passes through the origin and the straight line is offset from the origin. Directional relay is a mathematical dual of impedance relay.

10.12.1. Directional Relay

Phase Comparison. Directional relay is basically a phase comparator which compares the phase relation between V and I and the relay will operate for the condition $-90° \le \theta \le 90°$. The inputs in case of static directional relay are V and I Z_R and the characteristic is

$$Z Z_R \cos(\phi - \theta) \ge 0$$

where Z is the fault impedance and is equal to ratio of V and I, Z_R is the replica impedance in the relay, ϕ is the phase angle between V and I and θ is the relay characteristic angle.

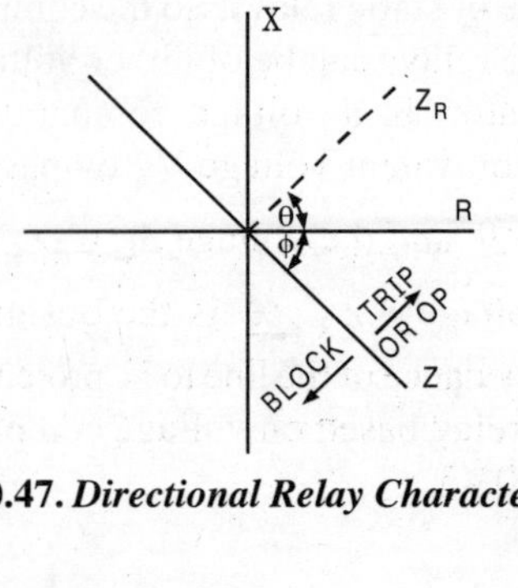

Fig. 10.47. *Directional Relay Characteristic*

Since Z and Z_R cannot be zero.

$$\therefore \qquad \cos(\phi - \theta) \geq 0 \text{ or } \phi - \theta = \pm\frac{\pi}{2}$$

Amplitude Comparison. The inputs for amplitude comparison will be

$$V + IZ_R \text{ and } V - IZ_R$$

For operation of relay

$$|V + I\,Z_R| > |V - I\,Z_R|$$
$$\text{or } |Z + Z_R| > |Z - Z_R|$$

and for no operation $|Z + Z_R| < |Z - Z_R|$

10.12.2. Impedance Relay

Amplitude Comparison. Impedance relay is inherently an amplitude comparator and the inputs are IZ_R and V.

For operation of relay

$$|IZ_R| > |V|$$
$$\text{or } |Z| < |\,Z_R|$$
$$\text{or } R + jX < Z_R$$

For threshold condition $R + jX = Z_R$

which is an equation of a circle on impedance (R-X) diagram.

The circle having radius equal to Z_R and centre at the origin, as illustrated in Fig. 10.48 (*a*), is the characteristic of the impedance relay.

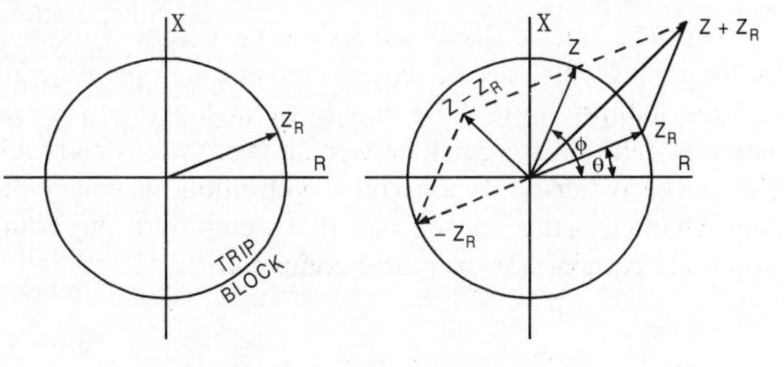

(*a*) (*b*)

Fig. 10.48. ***Impedance Relay Characteristic***

Phase Copmparison. The inputs are $(V + IZ_R)$ and $(V - IZ_R)$. The characteristic is given in Fig. 10.48 (*b*). It can be seen that as long as Z_R lies along the circumference of the circle with radius Z_R, the two quantities $(Z + Z_R)$ and $(Z - Z_R)$ make an angle of $\pm 90^\circ$. This gives the same characteristic as shown in Fig. 10.48 (*a*).

10.12.3. Angle Impedance Relay

Amplitude Comparison. The two input quantities are $(2IZ_R - V)$ and V and for operation of relay

$$|2IZ_R - V| > |V|$$

$$\text{or } |2Z_R - Z| > |Z|$$

The characteristic is shown in Fig. 10.49 (*a*).

Phase Comparison. The two input quantities are $(IZ_R - V)$ and IZ_R and for operation of relay the phase angle between $(Z_R - Z)$ and Z_R should be within $\pm 90°$. It can be seen that the characteristic is a straight line normal to Z_R. As long as Z lies below the line, the angle between $(Z_R - Z)$ and Z lies within the limits of $\pm 90^\circ$ [Fig. 10.49 (*b*)].

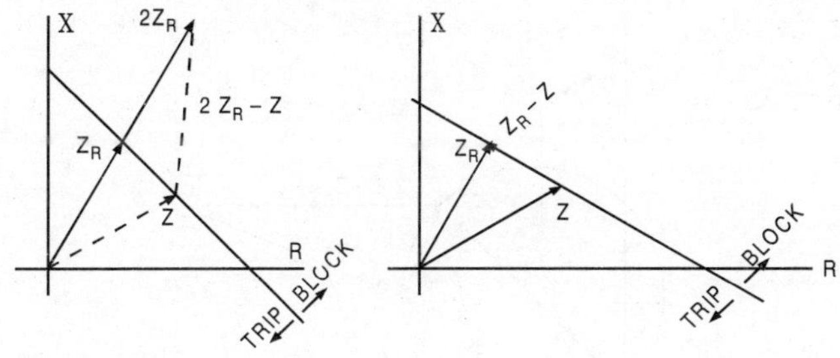

(*a*) (*b*)

Fig. 10.49. ***Angle Impedance Relay Characteristic***

10.12.4. Reactance Relay

Amplitude Comparision. This relay is a particular case of an angle impedance relay in which the reactance component of the impedance is measured and, therefore, for operatiton of relay

$$|2X_R - Z| > |Z|$$

The two inputs are V and $(2IZ_R - 2IR_R - V)$ where R_R is made equal to resistance of Z_R, thus leaving only its reactive component X_R.

Phase Comparison. The two inputs are IZ_R and $(IZ_R - V)$ as in case of an angle impedance relay. The relay will trip when Z is below the characteristic *i.e.*, when $(\psi + \theta) < 180^\circ$ [Fig. 10.50 (*b*)]. For Z to be purely reactive ψ would be 90° under threshold condition and the relay would trip when Z sin φ is less than X_R on the R-X diagram.

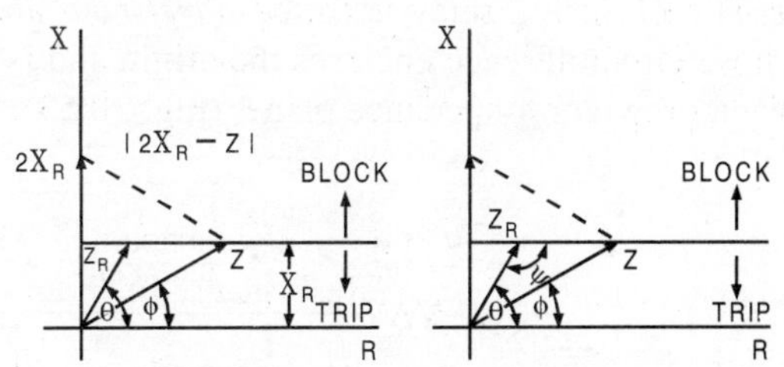

(*a*) (*b*)

Fig. 10.50. ***Reactance Relay Characteristic***

10.12.5. Mho or Angle Admittance Relay

This is the inverse of the angle impedance relay. The two relays are dual of each other. The equation of one type on an impedance diagram corresponds to the equation of other type on admittance diagram and vice versa. The characteristic for mho or admittance relay is a straight line offset from the origin on G-B diagram while on R-X diagram it will be a circle passing through the origin.

Amplitude Comparison. The two inputs are $|IZ_R|$ and $|2V - IZ_R|$. For relay operation.

$$|2V - IZ_R| < |IZ_R|$$
$$\text{or } |2Z - Z_R| < |Z_R|$$

The characteristic is illustrated in Fig. 10.51. The relay will operate as long as the fault impedance Z lies within the circle having diameter Z_R.

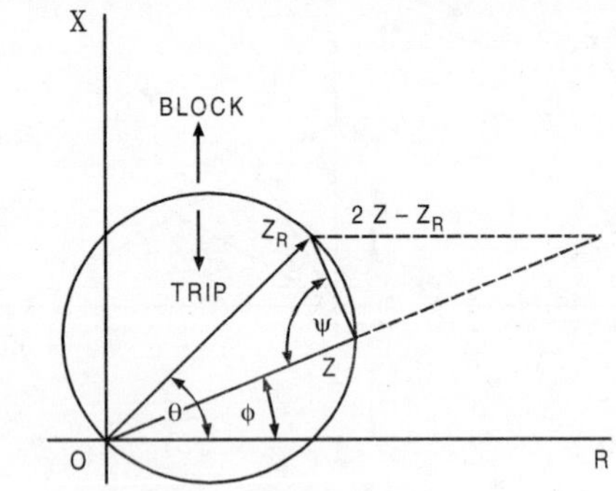

Fig. 10.51. ***Mho Relay Characteristic***

Phase Comparison. The two inputs are $|IZ_R - V|$ and V and the relay will operate when the phase angle between them is less than 90° *i.e.,* when 90° > ψ > –90°.

Since mho relay is inherently directional relay, phase comparator is the more convenient construction.

10.12.6. Offset Mho Relay

Use of a comparator as an offset mho relay is shown in Fig. 10.52. The inputs are fed to the comparator through a mixing transformer. The circuit is designed such that the relay operates when V and I have phase angle within certain limits and ratio $\frac{V}{I}$ is less than a certain value of Z_n, impedance setting of the relay. However, if the direction of power flow is reversed, the phase angle between I and V changes and then the relay operates when ratio $\frac{V}{I}$ is less than KZ_n which is less than unity. The characteristic of such a relay is called *offset mho* and it is a circle whose circumference encloses the origin and is slightly offset when drawn on impedance plane, (Fig. 10.53).

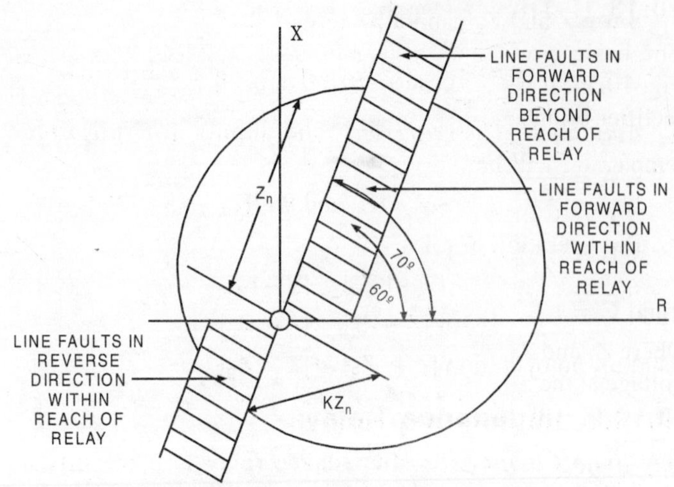

Fig. 10.53. ***Offset Mho Characteristic Relay***

10.13. MULTI-INPUT COMPARATORS

The comparators discussed so far are two-input comparators (phase or amplitude). Distance relays can be either single phase or polyphase using *multi-input comparators*. Such multi-input comparators may be instantaneous or integrating type and compare either phase or amplitude or both.

The characteristics of conventional two-input comparators are in the forms of straight lines, circles or sectors of circle on complex plane. Multi-input comparators may have elliptical, conical or quadrilateral characteristics on complex plane.

Distance relays, discussed so far, have some limitations for long lines which are heavily loaded and are operating at or near the stability limit, these may cause maloperation because they may fail to distinguish between power swings and faults. This can be avoided if we use a relay with elliptic characteristics. Such characteristics can be had if we employ a three-input amplitude comparator or hybrid comparator.

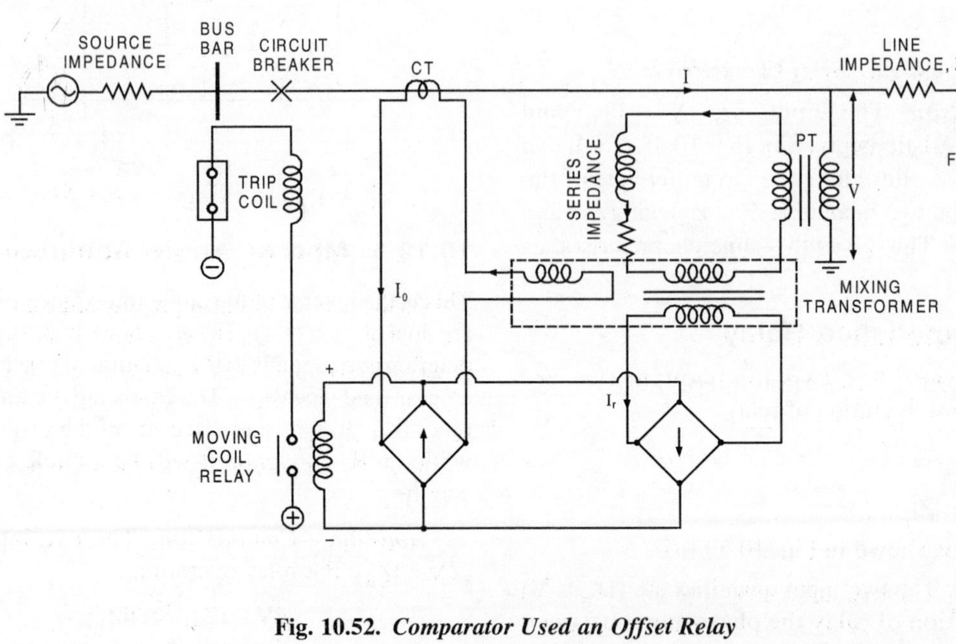

Fig. 10.52. ***Comparator Used an Offset Relay***

10.13.1. Three-Input Amplitude Comparator

The basic circuit for a three-input comparator is shown in Fig. 10.54. It is an amplitude comparator and comprises three rectifier bridges. Three inputs A, B and C are

$$A = \frac{V_r}{Z_1 + Z_2} - I\frac{Z_1}{Z_1 + Z_2}$$

$$B = \frac{V_r}{Z_1 + Z_2} - I\frac{Z_2}{Z_1 + Z_2}$$

and $C = I$

where Z_1 and Z_2 are the design or replica impedances, V_r is the voltage at the relay point during fault and I is the fault current.

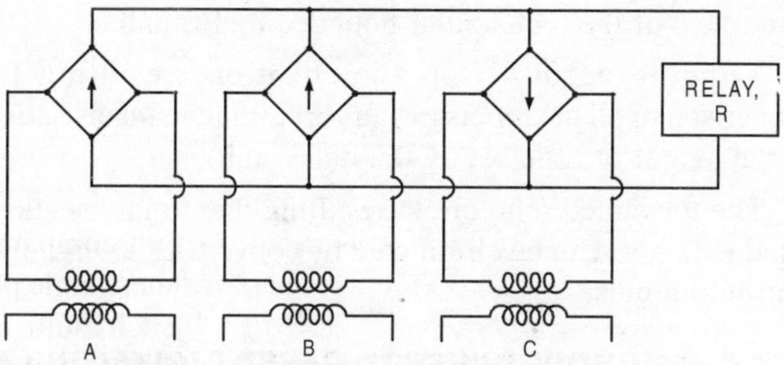

Fig. 10.54. *Three-Input Amplitude Comparator*

If Z_1 and Z_2 represent the phasors, the tips of which coincide with the foci of the ellipse drawn on complex plane and if $2a$ represents the major axis of ellipse then

$$|Z_L - Z_1| + |Z_L - Z_2| = 2a \quad ...(10.16)$$

where Z_L is the line impedance.

If the characteristic is passing through origin, as shown in Fig. 10.55, then

$$|Z_L - Z_1| + |Z_L - Z_2| = |Z_1 + Z_2|$$

Multiplying above equation by I we have

$$|IZ_L - IZ_1| + |IZ_L - IZ_2| = |I(Z_1 + Z_2)|$$

Now since $IZ_L = V_r$, the above equation can be written as

$$|V_r - IZ_1| + |V_r - IZ_2| = |IZ_1 + IZ_2|$$

$$\text{or} \left|\frac{V_r}{Z_1 + Z_2} - I\frac{Z_1}{Z_1 + Z_2}\right| + \left|\frac{V_r}{Z_1 + Z_2} - I\frac{Z_2}{Z_1 + Z_2}\right| = [I] \quad ...(10.17)$$

The above Eq. (10.17) represents the operating characteristic of an elliptic relay. The relay operating region is inside the ellipse.

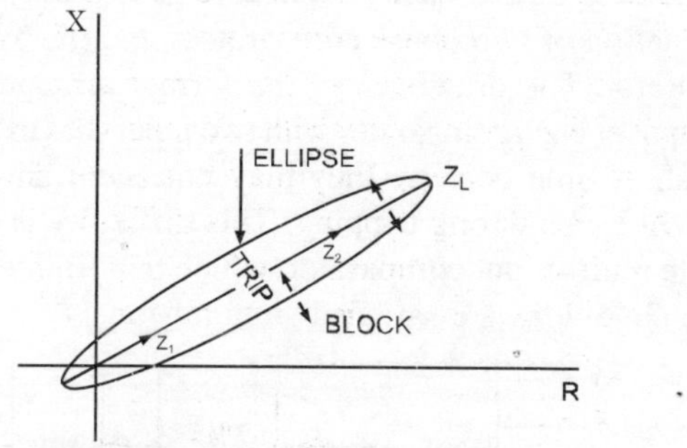

Fig. 10.55. *Elliptical Characteristic*

We will recall that when V/I measured by the relay is beyond the characteristic, the relay does not operate. During power swings, the elliptical characteristic with narrow coverage across R-axis is less liable for tripping than circular characteristic.

10.13.2. Hybrid Comparator

The hybrid comparator is a combination of amplitude comparator and phase comparator. The hybrid comparators are usually multi-input comparators. The three (or more) inputs are derived from output of CT and PT by means of replica impedances, mixing transformers, auxiliary CT's and PT's. The characteristic of hybrid comparator depends upon the three inputs.

In the case of a hybrid comparator the general equation for a conic characteristic is

$$Z = \frac{Z_R}{1 - K\cos(\phi - \theta)} \quad ...(10.18)$$

The above equation represents an ellipse if $K < 1$, a parabola if $K = 1$, a hyperbola if $K > 1$ and circle if $K = 0$.

The inputs required for having the elliptical characteristic represented by Eq. (10.18) are IZ_R, V and KV cos $(\phi - \theta)$. The first two inputs are supplied directly to the amplitude comparator but the third input is obtained from an auxiliary phase comparator which makes use of voltage IZ_R to polarize the voltage V. The scheme is illustrated by block diagram in Fig. 10.56.

10.13.3. Four-Input Phase Comparator With Quadrilateral Characteristics

The quadrangular (or quadrilateral) characteristic can be obtained by using four relays having straight line characteristics. The static arrangement for obtaining such a characteristic is

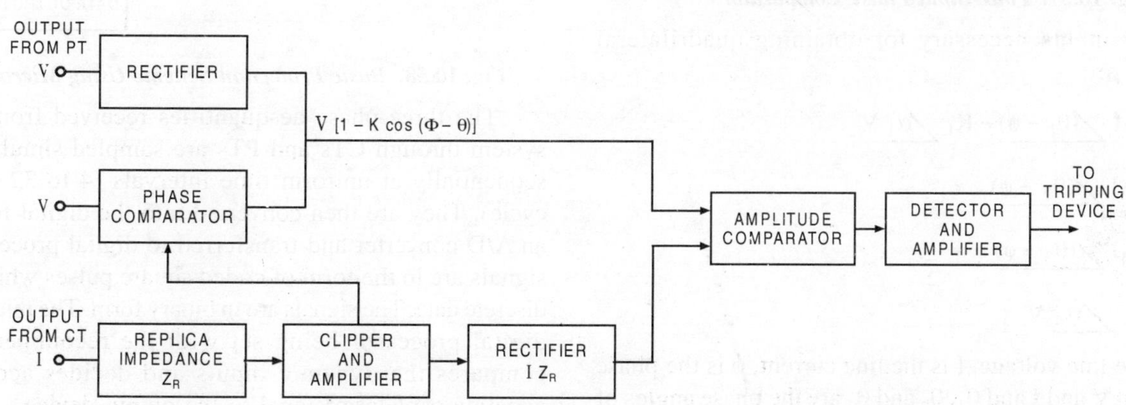

Fig. 10.56. *Block Schematic of Hybrid Comparator For Conic Characteristics*

very flexible and is easily realizable in two ways viz., (*i*) by combination of two phase comparators and (*ii*) by multi-input comparator. The drawback of the former arrangement is that the output of each comparator with two inputs has to be prolonged for a short time because they may not occur simultaneously. This will cause wrong tripping. This difficulty is overcome in a single multi-input comparator, which trips immediately when all the conditions are satisfied simultaneously.

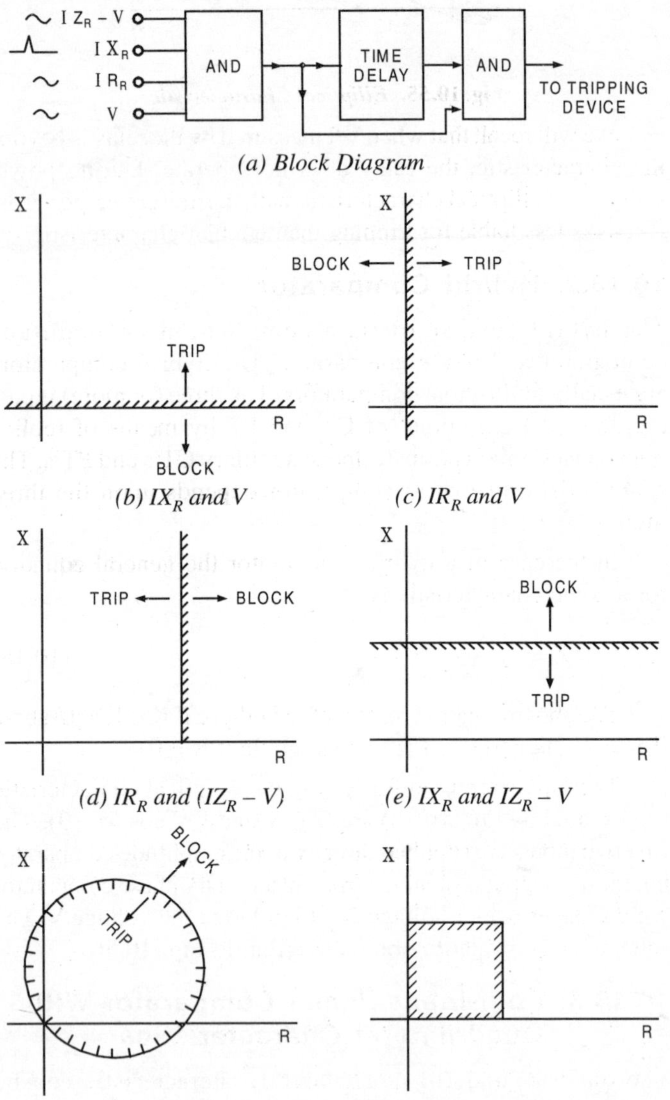

(a) Block Diagram

(b) IX_R and V *(c) IR_R and V*

(d) IR_R and $(IZ_R - V)$ *(e) IX_R and $IZ_R - V$*

(f) V and $(IZ_R - V)$ *(g) Composite Characteristic*

Fig. 10.57. ***Four-Input Phase Comparator***

The four inputs necessary for obtaining quadrilateral characteristic are

$$S_1 = Z_1 I \angle(\theta_1 - \phi) - K_1 \angle\alpha_1 V$$

$$S_2 = Z_2 I \angle(\theta_2 - \phi)$$

$$S_3 = Z_3 I \angle(\theta_3 - \phi)$$

$$S_4 = K_4 \angle\alpha_4 V$$

where V is the line voltage, I is the line current, ϕ is the phase angle between V and I and θ_1, θ_2 and θ_3 are the phase angles of the impedances Z_1, Z_2 and Z_3 respectively which are connected in the current circuit, α_1 and α_4 are the phase shifts of the voltage where required for locating the impedance characteristic.

To enclose the fault area, let

$$Z_2 = X_R, Z_3 = R_R \text{ and } Z_1 = R_R + jX_R = Z_R$$

and let $\alpha_1 = \alpha_4 = 0$ and $K_1 = K_4 = 1$.

The above inputs become

$$S_1 = IZ_R - V;\ S_2 = IX_R,\ S_3 = IR_R \text{ and } S_4 = V.$$

These are given to AND gate.

The mho circle caused by the intersection of S_1 and S_2 will not interfere with the rectangular tripping area if $Z_R = R_R + jX_R$, because the circle of diameter Z_R goes through the corners of the rectangular bounded by R_R and X_R.

Tripping occurs if all the equations resulting from comparison of all the inputs in pairs are simultaneously satisfied for the length of time set by the delay unit.

The unwanted mho circle resulting due to intersection of V and $(IZ_R - V)$ can be eliminated by converting at least one of them into a pulse.

10.14. MICROPROCESSOR BASED DIGITAL RELAYS

With the fast development in large scale integrated (LSI) technology, sophisticated and fast microprocessors are now available. With the rapid growth of modern complex large power system networks, fast, accurate and reliable protective schemes are essential. Microprocessor based schemes are becoming more and more popular for power system protection as they offer attractive compactness and flexibility. They reduce the number of types of relay units. An interface employing op-amps, analog multiplexer analog-digital (A/D) converter, voltage comparators and passive elements have been developed to provide the characteristics of various types of relays such as definite time relays, inverse-time overcurrent relays, phase comparators and reverse power relays etc.

A block diagram of a microprocessor based digital programmable static relay is shown in Fig. 10.58.

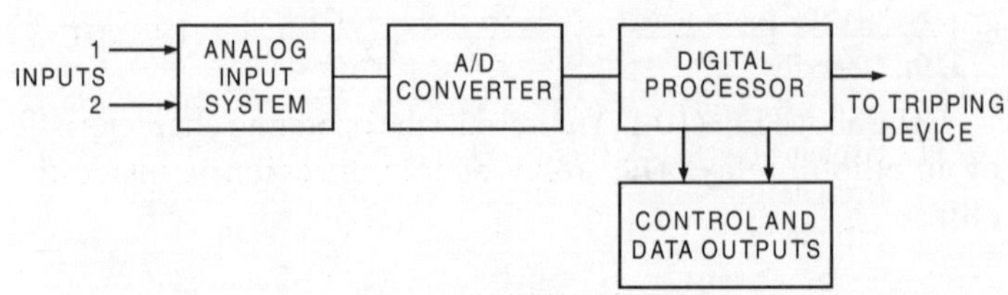

Fig. 10.58. ***Basic Protection Scheme Using Microcomputer***

The three-phase ac quantities received from the power system through CTs and PTs are sampled simultaneously or sequentially at uniform time intervals (4 to 32 samples per cycle). They are then converted into the digital form through an A/D converter and transferred to digital processor. Digital signals are in the form of coded square pulses which represent discrete data. The signals are in binary form. The microprocessor/digital processor being set with the recommended values compares the dynamic inputs and decides accordingly to generate trip/alarm signal to the output device.

Microprocessor based relays have numerous advantages. They have a very small burden on the CTs and PTs. Saturation can be avoided by using air-gap CT having a limited output. They can process and display the signals very efficiently, accurately and in a fastest possible manner. Due to their programmable characteristics, they can be applied extensively in the protection of electrical power systems. Moreover, one microprocessor unit may be able to perform relaying function of several systems. Microprocessor relays are more reliable and secure to relay engineers because they can alert the user to a malfunction before a false trip or failure to trip occurs. However, the microprocessors should be properly shielded from external influences and the system earthing must be very good from which they receive their control voltage.

Exercises

1. Compare a static relay with electromagnetic relays.
[Switchgear and Protection; U.P. Technical Univ. 2009-10, G.B. Technical Univ. 2011-12]
2. What are the advantages and limitations of static relays over electromagnetic relays. Also discuss the main components of a static relay.
[U.P. Technical Univ. Switchgear and Protection 2008-09]
3. Compare electronic relays with static relays.
[U.P. Technical Univ. Switchgear and Protection 2007-08]
4. Explain the concept of duality in static comparators.
[G.B. Technical Univ. Switchgear and Protection 2010-11]
5. Discuss about phase and amplitude comparators in detail.
[U.P. Technical Univ. Switchgear and Protection 2009-10]
6. Describe various types of phase comparators used in static relay.
[G.B. Technical Univ. Switchgear and Protection 2011-12]
7. What do you understand by amplitude comparator and phase comparator? Prove the duality between them with the help of phasor diagrams.
8. Describe various types of phase comparators used in static relays.
[U.P. Technical Univ. Switchgear and Protection 2008-09]
9. What are the types of phase comparators ? Explain integrating type phase comparators.
[U.P. Technical Univ. Switchgear and Protection 2005-06]
10. Describe the coincidence principle used in phase comparators.
[G.B. Technical Univ. Switchgear and Protection 2010-11]
11. Explain how inverse time overcurrent relay is different in operation from definite time overcurrent relay. With neat circuit diagram explain the working of inverse-time overcurrent static relay.
[U.P.S.C. I.E.S. Elec. Engineering II, 1992]
12. Explain clearly how a phase comparator is used in the protective relay systems.
13. Realize impedance relay using static components.
[U.P. Technical Univ. Switchgear and Protection 2004-05]
14. Realize mho relay characteristic using static components.
[U.P. Technical Univ. Switchgear and Protection 2007-08]
15. By taking an example, explain the dual input comparator relay threshold characteristic.
[U.P. Technical Univ. Switchgear and Protection 2007-08]
16. By taking an example, explain the threshold characteristics of a dual input comparator relay.
[U.P. Technical Univ. Switchgear and Protection 2004-05]
17. What is comparator ? Explain with phasor diagrams characteristic of multi-input comparators.
[U.P. Technical Univ. Switchgear and Protection 2005-06]
18. Prove that two input comparators are duals of each other. How would your realise a reactance relay characteristic using static comparators.
19. With the help of neat diagram explain the principle of static differential relay.
20. Show that the polar curve of a biased differential relay using a static comparator is a circle. Why does the area outside the circle represent the operating zone ?
21. Discuss the advantages of digital relays. Describe the basic functional blocks of a digital relay.
22. Write short notes on the following :
    (*i*) Essential components of a static relay.
    (*ii*) Merits of static relays over electromagnetic relays.
    (*iii*) Hall-effect relays.
    (*iv*) Magnetoresistivity phase comparator.
    (v) Level detector.
    (*vi*) Averaging type instantaneous amplitude comparator.
    (*vii*) Integrating type phase comparator with rectifier type AND gate.
    (*viii*) Static relays. **[A.M.I.E. Sec. B. Summer 2000]**

Short Answer Type Questions With Answers

Q. 1. What is static relay ?

Ans. A static relay refers to a relay in which there is no armature or other moving element and response is developed by electronic, magnetic or other components without mechanical motion.

A relay using combination of both static and electromagnetic units is also called a static relay provided that the response is accomplished by static units.

Q. 2. How are static relays classified ?

Ans. Static relays are classified according to the type of the measuring unit or comparator as follows:

(*a*) Electronic relays
(*b*) Transductor or magnetic amplifier relays
(*c*) Rectifier bridge relays
(*d*) Transistor relays
(*e*) Hall effect relays and
(*f*) Gauss effect relays.

Q. 3. Why electronic relays could not be adopted commercially ?

Ans. Inspite of the advantages of fast operation, low maintenance, low burden on CTs and PTs, absence of mechanical inertia and bouncing contacts, they suffered inherently from the requirements of *ht* supply, short life, large power consumption, *lt* supply for the heater elements. These relays could not meet practical requirements and hence never reached the commercial stages.

Q. 4. Why are transconductor relays not employed for protection applications ?

Ans. The transductor relays are mechanically very simple and are quite reliable. Due to smoothing and rectifying a signal, a delay is introduced because of the time constant of the smoothing circuit and hence the relays are slow in operation and, therefore, are discarded for protection applications.

Q. 5. What is comparator ?

Ans. The function of a protective relay is to sense any abnormal condition in the system and send a signal to the breaker which in turn isolates the faulty section of the feeder from the healthy one. The relay does all this by comparing two quantities either in amplitude or in phase. The amplitude or phase relation depends on the conditions of the system and for a predetermined value of this relation, indicative of a particular type and location of fault, the relay operates. Except in relays, such as overcurrent relays, where only one electrical quantity overcomes a mechanical quantity such as the restraint from a spring, usually two electrical quantities are compared. The device that makes such comparison is called the *comparator* and forms the heart of a protective relay. The operating characteristics of any relay are governed by the comparator.

Q. 6. Differentiate between amplitude comparators and phase comparators.

Ans. Amplitude comparator compares the amplitudes of two (or more) input quantities. The phase angle between the quantities under comparison (inputs) is not recognized or noticed by the amplitude comparator.

If the two input signals are S_1 and S_2 (say S_1 the operating and S_2 restraining), the amplitude comparator gives positive output only if

$$S_2/S_1 \leq K.$$

The function is represented by a circle in the complex plane with its centre at the origin. This defines the boundary of the marginal operation.

The main purpose of amplitude comparators is to provide direction and distance protection.

Phase comparison technique is the most widely used technique for all practical directional, distance, differential and carrier relays.

In a phase comparator, the operation of the relay takes place when the phase relation between two inputs S_1 and S_2 varies within certain specified limits. Both inputs must exist for an output to occur; ideally, operation is independent of their amplitudes, and depends only on their phase relationship. Phase comparator, in its simple form, is illustrated in Fig. 10.25. The function, as defined by the boundary of marginal operation, is represented by two straight lines from the origin of the complex plane.

Mathematically, the condition of operation is given as

$$-\alpha_1 \leq \theta \leq \alpha_2$$

where θ is the angle by which S_1 lags behind S_2. If $\alpha_1 = \alpha_2 = 90°$ the comparator is called the cosine comparator and if $\alpha_1 = 0$ and $\alpha_2 = 180°$, it is called the sine comparator.

Q. 7. What is hybrid comparator ?

Ans. The hybrid comparator is a combination of amplitude comparator and phase comparator. The hybrid comparators are usually multi-input comparators. The three (or more) inputs are derived from output of CT and PT by means of replica impedances, mixing transformers, auxiliary CT's and PT's. The characteristic of hybrid comparator depends upon the three inputs.

Q. 8. What are advantages of microprocessor base relays ?

Ans. Microprocessor based relays have numerous advantages. They have a very small burden on the CTs and PTs. Saturation can be avoided by using air-gap CT having a limited output. They can process and display the signals very efficiently, accurately and in a fastest possible manner. Due to their programmable characteristics, they can be applied extensively in the protection of electrical power systems. Moreover, one microprocessor unit may be able to perform relaying function of several systems. Microprocessor relays are more reliable and secure to relay engineers because they can alert the user to a malfunction before a false trip or failure to trip occurs. However, the microprocessors should be properly shielded from external influences and the system earthing must be very good from which they receive their control voltage.

❑❑❑

CHAPTER 11

Protection of AC Generators and Motors

INSIDE THIS CHAPTER

11.1. INTRODUCTION

The modern electric power system consists of several elements such as generators, transformers, station bus-bars, transmission lines and other associated equipment. It is imperative to protect these elements from different types of faults, which are likely to occur sooner or later. The protective relays discussed in chapters 9 and 10 can be advantageously used to detect the improper behaviour of any element of the power system and initiate corrective measures.

Protection of generators is the most complex and elaborate because of following reasons:

1. Generator is a costly equipment and one of the major links in a power system.
2. Generator is not a single equipment but is associated with the unit-transformers, auxiliary transformer, station bus-bars, excitation system, prime mover, voltage regulating equipment, cooling system etc. The protection of generator, is therefore, to be coordinated with the associated equipment.
3. The generator capacity has sharply risen in recent years from 30 MW to 500 MW with the result that loss of even a single machine may cause overloading of the associated machines in the system and eventual system instability.

The basic function of protection applied to generators is, therefore, to reduce the outage period to a minimum by rapid discriminative clearance of faults. Unlike other apparatus, opening a breaker to isolate the faulty generator is not sufficient to prevent further damage, since the generator will continue to supply power to a stator winding fault until its field excitation is suppressed. It is, therefore, necessary that the field is opened, fuel supply to the prime mover is stopped and sometimes braking application also becomes imperative.

While selecting the scheme for generator protection, the protection of the complete unit and the stability of the system due to disturbance in the generator should be considered in addition to the protection of the generator itself.

The type of protection used for a particular motor depends on the switchgear employed for its control (starting, stopping, speed variation etc.). In general two basic protections are provided for every motor which are (*i*) thermal overload protection and (*ii*) short-circuit protection.

11.2. GENERATOR FAULTS

Generator faults can be considered under the following heads :

1. Stator Winding Faults. Such faults occur mainly due to the insulation failure of the stator coils. The main types of stator winding faults are (*i*) phase-to-earth faults (*ii*) phase-to-phase faults and (*iii*) inter-turn faults involving turns of the same phase winding.

The stator winding faults are the most dangerous and are likely to cause considerable damage to the expensive machinery. So automatic protection is absolutely necessary to clear such faults in the shortest possible time in order to minimise the extent of damage.

Phase-to-phase faults and phase inter-turn faults are less common, these usually develop into an earth fault. Inter-turn faults are more difficult to be detected.

The effect of earth fault in the stator is two fold:

(*i*) Arcing to core, which welds laminations together, causing eddy current hot spots on subsequent use.

Repairs to this condition involve expenditure of considerable money and time.

(*ii*) Severe heating in the conductors damaging them and the insulation with possible fire breaks.

2. Field Winding or Rotor Circuit Faults. Faults in the rotor circuit may be either earth faults (conductor-to-earth faults) or inter-turn faults, which are caused by severe mechanical and thermal stresses.

The field system is normally not grounded (*i.e.*, remains isolated from the earth) and, therefore, a single fault between field winding and rotor body due to insulation breakdown does not give rise to any fault current. However, a second earth fault will short-circuit some part of the rotor winding and may thereby develop an unsymmetrical field system, giving unbalanced force on the rotor. This can cause severe vibration of the rotor with possible damage to the bearings. Thus a single earth fault can be tolerated for a while but it should not be allowed to continue. Rotor earth fault protection is provided in case of large generators.

Owing to a fault, there may be an unbalance in the three-phase stator currents. According to the theory of symmetrical components, unbalance three-phase currents have a negative sequence component, which rotates at synchronous speed in a direction opposite to the direction of rotation of rotor. So double frequency currents are induced in the rotor. This causes overheating of rotor and possible damage to the rotor. Unbalanced currents may also cause severe vibration, but the overheating problem is more acute. *Rotor temperature indicators* are used with large generators for detecting rotor overheating due to unbalanced loading of generator.

Rotor open-circuit faults, though rare, can cause arcing and thus result on serious conditions.

Reduced excitation may occur due to open circuit or short circuit in the field or exciter circuits or a fault in automatic voltage regulator. When a generator loses its field excitation it speeds up slightly and continues to run as an induction generator deriving excitation from the system and supplying power at a leading power factor. A fall in voltage will also occur due to loss to excitation which may result in loss of synchronism and system stability. There is also the possibility of overheating of the rotor due to induced currents in the rotor and damper windings. This can be avoided by using a tripping scheme which is so arranged that opening of field circuit breaker causes the tripping of generator unit breaker.

3. Abnormal Operating Conditions. The abnormal operating conditions that are likely to occur in a generator are: (*i*) failure of prime mover (turbine) resulting in operation of the generator as a synchronous motor (*ii*) failure of field (*iii*) unbalanced loading and subsequent heating of generator (*iv*) overloading (*v*) overvoltage at generator terminals (*vi*) overspeed (*vii*) ventilation failure and (*viii*) current leakage in the body of the generator.

11.3. STATOR PROTECTION

Different types of stator faults, which are likely to occur, have already been discussed in the previous Art. The earth fault current is usually limited by the resistance of neutral earthing resistor. When the fault current is less than 20 A, negligible burning of the iron core will occur if the machine is tripped within some seconds. The repair work then amounts to replacement of the damaged coil without restacking of core laminations. If, however, the earthing resistor is selected to pass a much larger earth fault current (greater than 200 A) severe burning of the stator core will take place, necessitating restacking of laminations. Even with a high speed earth fault differential protection severe damage may be caused owing to the large time constant of the field circuit and the relatively long time required to completely suppress the field flux. In the case of high earth fault currents it is therefore normal practice to install a circuit breaker in the neutral of the generator in order to reduce the total fault clearance time.

Circulating current biased differential protection provides the earth fault protection but the sensitivity of such a protection for earth fault depends upon the resistance in neutral to earth connection and the position of earth fault in the winding.

A separate and sensitive earth fault protection is generally necessary for generators with resistance earthing.

The occurrence of short-circuiting between the stator windings is quite rare because the insulation in a slot between coils of different phases is at least twice as thick as the insulation between one coil and the iron core. However, a phase-to-earth fault may result in a phase-to-phase fault within the slots. If a phase-to-phase fault should occur, this is most likely to occur at the end connections of the stator winding *i.e.*, in the overhanging parts outside the slots. Such a fault causes severe arcing with high temperatures, melting of copper and risk of fire if the insulation is not made of fire-resistant, noninflammable material. In this case there will be no damage to the core laminations, and the repair work will, therefore, be limited to replacement of damaged coils and mechanical parts of the end structure.

Circulating current biased differential protection provides adequate and speedy protection against phase-to-phase faults in the generator zone.

Short circuits between the turns of one coil may occur in case the stator winding makes use of multiturn coils. Such faults may develop due to incoming current surges with a steep wave-front that may causes a high voltage $\left(L\frac{di}{dt}\right)$ across the turns at the entrance of the stator winding. Differential protection and overcurrent protection does not sense the inter-turn faults. Stator inter-turn fault protection is separately provided.

11.3.1. Differential Protection For Generators

The most common system employed for the protection of stator windings against earth faults and phase-to-phase faults makes use of circulating current principle. In this scheme of protection, currents at the two ends of the protected section are compared. Under normal operating conditions, these currents are equal but may differ on the occurrence of a fault in the protected section. The difference of the currents under faulty conditions is made to flow through the relay operating coil. The relay then closes its contacts and makes the circuit breaker to trip and thus isolate the protected section from the system. Such a protection is called a *Merz-Price circulating current system*. Such a protective scheme is very effective for earth faults and faults between phases.

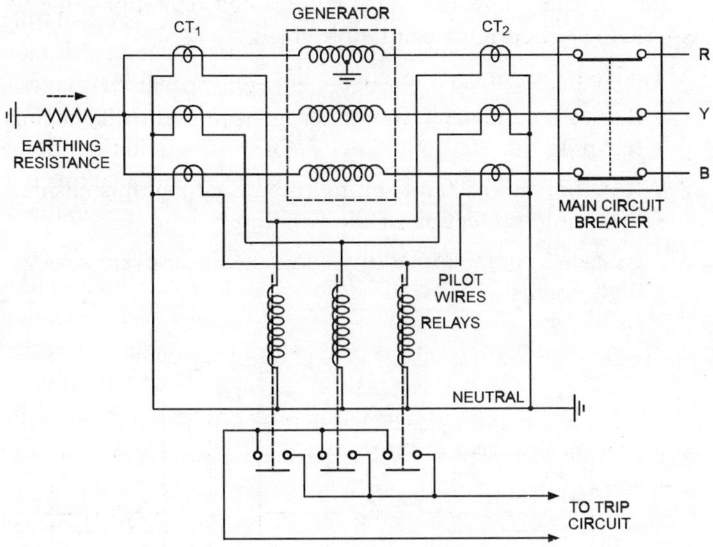

Fig. 11.1. ***Merz-Price Protection With Relay Being Connected in the Midpoints of Two Sets of CTs***

The schematic arrangement of differential protection scheme for a 3-phase generator is illustrated in Fig. 11.1. There are two sets of identical CTs, each set is mounted on either side of the stator phase windings. The secondaries of these current transformer sets are connected in star and their ends are connected through pilot wires. The relay coils are connected in star, the neutral point being connected to the current transformer common neutral and the outer ends one to each of the three pilot wires. The relays are connected across equipotential points of the three pilot wires so that the burden on each CT is the same. The equipotential points of the pilot wires would naturally be located at the centre of these wires and so the relays are located midway as illustrated in Fig. 11.1.

However, it is always convenient to locate the relay coils adjacent to CTs near the main circuit breaker. This can be accomplished by inserting balancing resistances in series with the pilot wires to make equipotential points located near the main circuit breaker, as illustrated in Fig. 11.2. These resistances are usually adjustable in order to obtain the exact balance.

The relays employed in this protection scheme are generally of electromagnetic type and are arranged for instantaneous operation as faults are expected to be cleared as quickly as possible.

Under healthy conditions, the currents at both ends of each winding will be equal, emfs induced in secondaries of CTs will be equal and so no current will flow through the operating coils of relays. When an earth fault or phase-to-phase fault occurs, this condition no longer holds good and the differential current flowing through the relay operating coil makes the circuit breaker to trip.

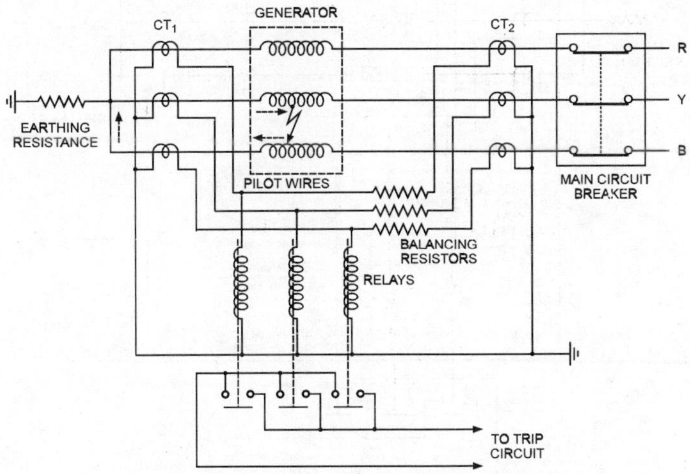

Fig. 11.2. ***Merz-Price Protection With Relay Being Connected Adjacent To One Set of CTs***

Now assume that there is an earth fault on phase R due to breakdown of insulation to earth, as illustrated in Fig. 11.1. The currents in the secondaries of the two CTs in phase R will become unequal and the difference of the two currents will flow through the corresponding relay coil and the circuit breaker will get tripped.

If there is a short circuit between the two phases Y and B, as illustrated in Fig. 11.2, it will cause fault current to flow through two phases, as illustrated in the figure. The currents in the secondaries of the two CT's in each affected phase will become unequal and the differential current will flow through the operating coils of the relays connected in these phases. The relay then closes its contacts to trip the circuit breaker.

11.3.2. Modified Differential Protection For Generators

It is a general practice to use neutral resistance earthing in order to avoid the adverse effects of earth fault currents. In such a case it will not be possible to protect whole of the stator windings of a star-connected generator against earth faults with the protection schemes discussed above. When an earth fault occurs near the neutral point, it will result in only a small fault current because of the small emf of short-circuited portion of the winding. This current, which is further reduced by the resistance of the neutral earthing, may not cause the relay to operate. The magnitude of the unprotected zone depends upon the value of resistance employed in neutral earthing and the relay setting.

Makers of protective gear speak of "protecting 80 or 85 per cent of the winding" which means that fault in the 20 or 15 per cent of the winding near the neutral point cannot cause tripping *i.e.,* this portion of the winding remains unprotected. If the relays with very low settings are used for the protection of adequate portion of the generator stator winding, it will not be desirable for reliable stability on heavy through phase faults.

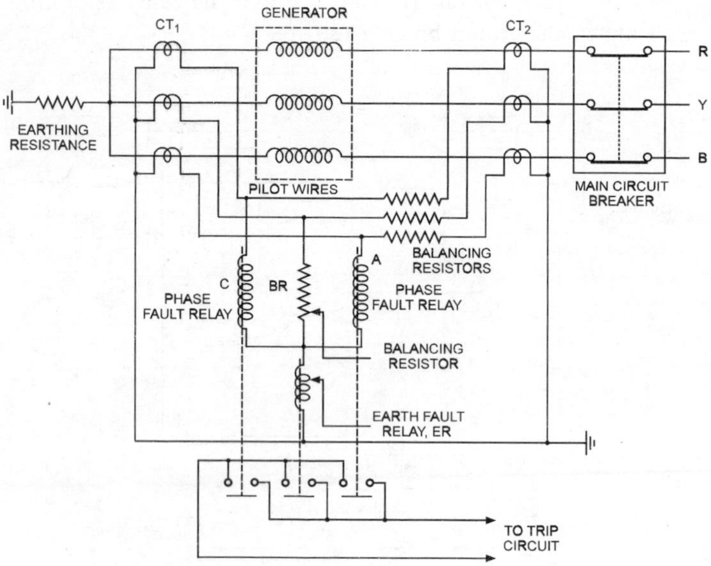

Fig. 11.3. ***Modified Scheme of Differential Protection For Generators Grounded Through High Resistance***

In order to overcome above difficulty, the modified scheme shown in Fig. 11.3 has been developed. In this modified form of differential protection the setting of earth faults is reduced without impairing stability. In this scheme two elements are arranged for phase-fault protection and the third for earth-fault protection only. The two phase elements (A and C) together with a balancing resistor (BR) are connected in star and the earth fault relay (ER) is connected between this star point and the neutral pilot wire.

The star-connected circuit is symmetrical as regards impedance and any symmetrical spill current from the circulating current circuit, due to high through phase-fault conditions, will cancel at relay star point and will not flow through the earth fault relay. It is, therefore, possible with this scheme of protection to operate with a sensitive earth fault relay and still maintain a high degree of stability. This sensitive operation of earth fault relay enables low setting for the element which provides protection to a greater percentage of stator windings.

11.3.3. Biased Circulating Current Protection (Percentage Differential Relay Protection)

This system, also called the *Merz-Price protection system,* is the most common type of protection used for stator windings against phase-to-phase or phase-to-ground faults. It is the standard practice of manufacturers to recommend differential protection for generators rated 1 MVA or higher, and most of such generators are protected by differential relays. Above 10 MVA, it is almost universally the practice to use differential relays. Generally, percentage differential relaying is used, protecting about 80% of the stator winding.

When differential relaying is used for protection, the CTs at both ends of a generator winding must be of equal accuracy; otherwise if the error is excessive it will cause a maloperation of the relay. To safeguard against such an operation, biased circulating current protection is used. Such a protection provides a biasing feature which automatically increases the relay setting in proportion to the load or through fault current, *i.e.,* the relay is set to operate not at a definite current, but at certain percentage of the through current.

By suitably proportioning the ratio of the restraining coil turns to the operating coil turns, any amount of biasing can be achieved and compensation for unwanted operations due to spurious spill currents can be provided.

The advantages of such a scheme are enumerated below.

1. It does not require CTs with air gaps or special balancing features.
2. It permits a low fault setting to be used and this ensures maximum protection of the windings.
3. It ensures complete stability under the most severe through fault conditions.

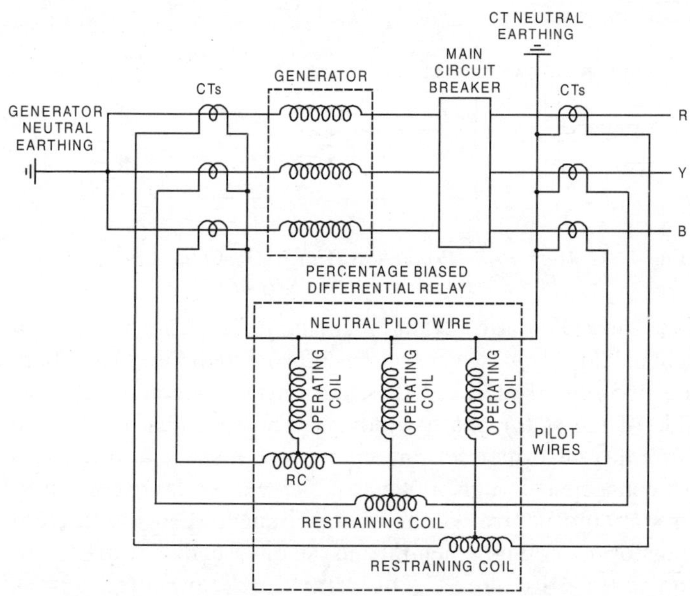

Fig. 11.4. ***Schematic Diagram For Percentage Differential Relay Protection of a Star-Connected Generator***

The protection of a three-phase star-connected generator by means of percentage differential relay is illustrated in Fig. 11.4. CTs connected in star are provided on both the out going sides and the machine winding connection to earth. The restraining coils are energised by the secondary connections of CTs in each phase, and the operating coils are connected to the central tapping on the restraining coils and the neutral pilot wire as illustrated in the figure.

For the greatest sensitivity of differential relaying the primary current rating of the CTs must be equal to the rated full-load current of the generator. In practice it is about 25% higher.

Under normal operating conditions the secondary outputs of the line CTs are equal, at any instant, to the outputs of the CTs at the neutral end. Thus there is balanced circulating current in the phase pilot wires and the relay restraining windings and so no current flows through the operating coils or in the common return pilot wire. If there is a fault due to a short circuit in the protected zone of the winding it produces a difference between the currents in the primary windings of CTs on both sides of the generator winding of the same phase. This results in a difference between the secondary currents of the two CTs. Thus, under fault conditions, a current flows through the operating coils of those phase elements corresponding to primary phases on which the fault has occurred. If this current attains the preset magnitude, the relay operates and causes the circuit breaker to trip. The tripping can be arranged for a certain setting, and the restraint adjusted to a certain percentage of the winding to be protected. Generally 80% of the winding is protected.

Percentage differential relaying for a 3-phase delta-connected generator is illustrated in Fig. 11.5.

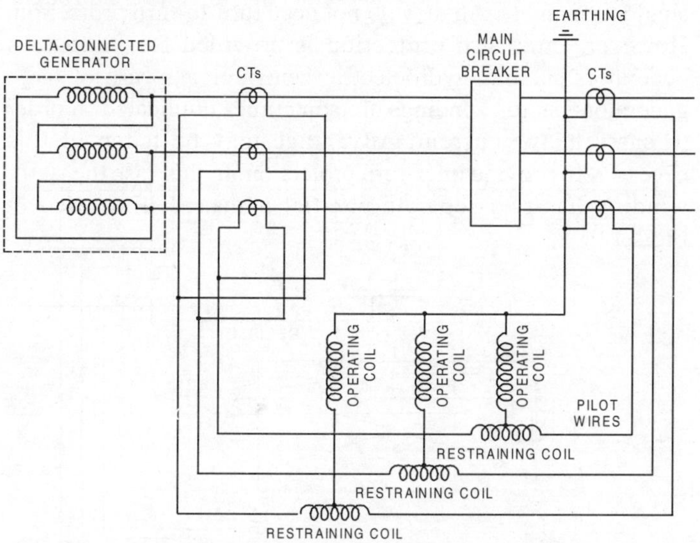

Fig. 11.5. ***Schematic Diagram For Percentage Differential Relay Protection of a Delta-Connected Generator***

11.3.4. Self Balance Protection System

It necessarily consists of two cables connected to the two ends of each phase and the cables are passed through the circular aperture of the ring type CTs, as shown in Fig. 11.6. Under normal operating conditions the current flowing through two cables in the central aperture of the CT will be in opposite directions and so there will be no magnetization of the CT. In case there is an earth fault on any phase, the fault current passes only once through the circular ring of the CT of that particular phase. This will set up magnetic flux and so induce an emf in the relay circuit which will trip the circuit breaker. This system is also effective for phase-to-phase faults but not useful for protection against fault between the turns of the same phase for ordinary generator winding. In case the winding of each phase is designed in two parallel paths such a protection will be useful. This is very sensitive type of earth fault protection scheme but it has got some drawbacks, as enumerated below:

1. Under heavy short circuits large electromagnetic forces are developed in the CTs ring.

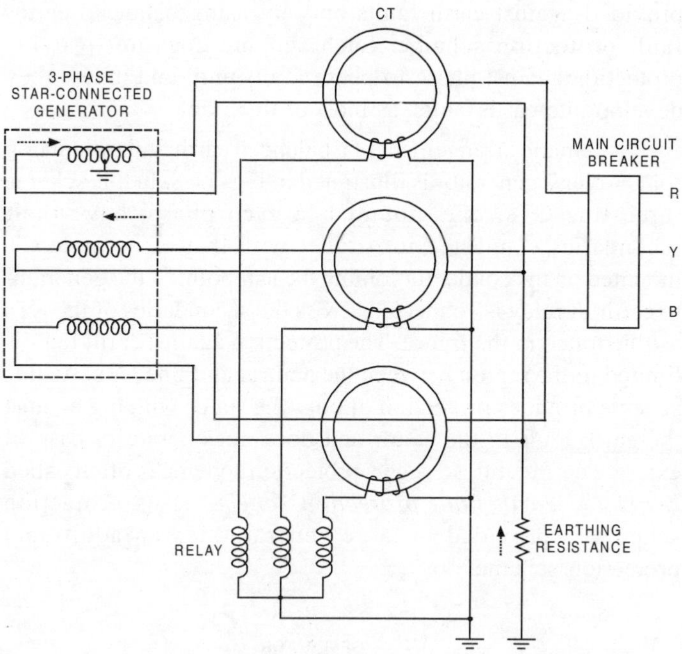

Fig. 11.6. ***Schematic Diagram For Self Balance Protection Scheme of Generator***

2. A different layout of the cable leads is required in this scheme.
3. For protection of the lead cable from the circuit breaker to stator terminal the ring CTs are to be placed near the circuit breaker which is again quite cumbersome. However, this difficulty can be overcome by providing a metallic sheath around the cable (but insulated from it) and earthing it in a manner illustrated in Fig. 11.7. In case the lead punctures anywhere, there will be flow of current through the sheet, as illustrated in the figure and relay will be energized resulting in tripping of the circuit breaker. In Fig. 11.7 connections for only one phase are shown.

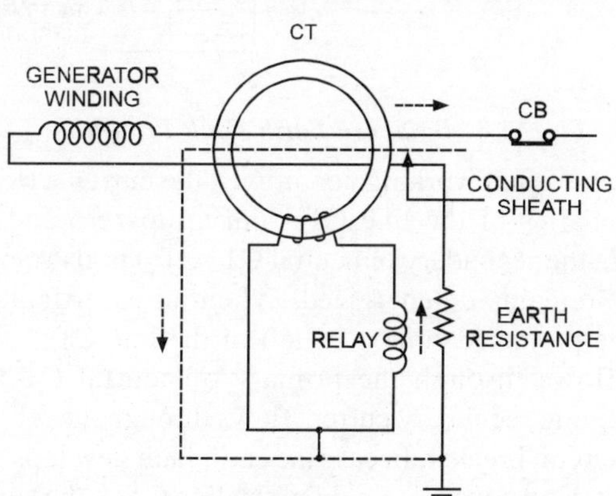

Fig. 11.7

11.3.5. Balanced Earth Fault Protection

In small size generators, the neutral ends of the three phase windings are sometimes connected internally to a single terminal. Under such conditions, it is not possible to use the circulating current protection described in previous Arts because neutral end is not accessible and protection is therefore, provided against earth faults only by using balanced earth-fault protection scheme. Such scheme does not provide protection against phase-to-phase faults until and unless they develop into earth faults, as most of them do.

Schematic arrangement of a balanced earth fault protection for a 3-phase generator is illustrated in Fig. 11.8. In this scheme three line CTs, one mounted in each phase, have their secondaries connected in parallel with that of a single CT mounted on the conductor joining the star point of the generator to earth. A relay is connected across the secondaries of the CTs as illustrated in the figure. The protection against earth fault is limited to the region between the neutral and line CTs. As this scheme provides protection of only the stator winding against the earth fault in the stator and does not operate in case of external earth fault, so such a protection scheme is often called *restricted earth fault protection scheme.* This protection scheme is provided in large generators as an additional protection scheme.

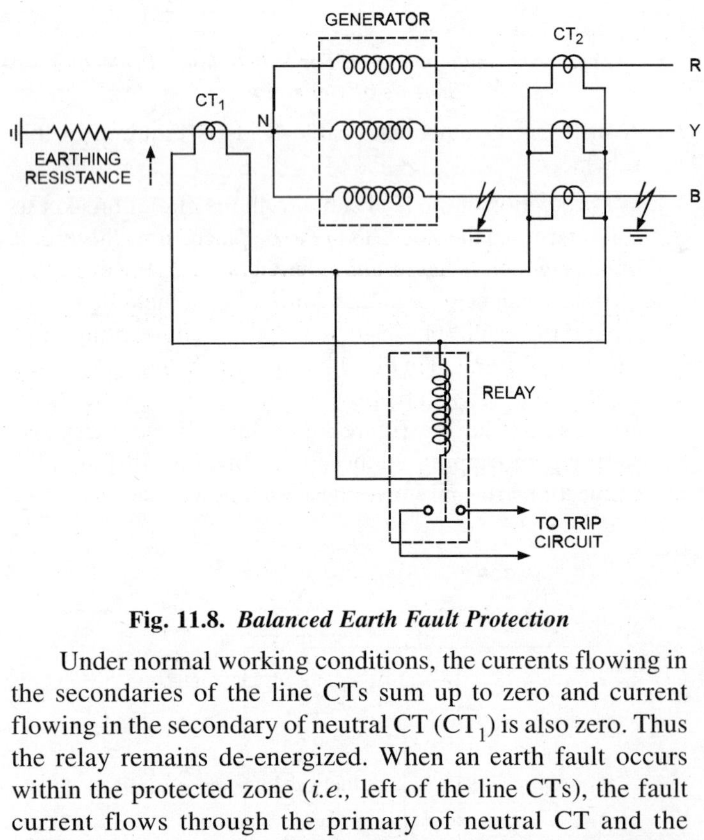

Fig. 11.8. ***Balanced Earth Fault Protection***

Under normal working conditions, the currents flowing in the secondaries of the line CTs sum up to zero and current flowing in the secondary of neutral CT (CT_1) is also zero. Thus the relay remains de-energized. When an earth fault occurs within the protected zone (*i.e.,* left of the line CTs), the fault current flows through the primary of neutral CT and the corresponding secondary current flows through the relay which trips the circuit breaker. In case the earth fault develops external to the protected zone (*i.e.,* right of the line CTs), the sum of the currents at the terminals of the generator is exactly equal to the current in the neutral connection and as such no current flows through the relay operating coil.

However, such a scheme has a drawback. In case the earth fault occurs nearer the neutral terminal or when grounding of neutral is through a resistance or through a distributing transformer, the fault current may be so low that the secondary current of the CT becomes lower than the pick-up current of the relay. The relay thus remains inoperative and the fault continues to persist in the generator winding which is highly undesirable.

11.3.6. Stator Inter-Turn Protection

Merz-Price circulating current protection system does not provide protection against turn-to-turn faults (short circuits between the turns) on the same phase winding of the stator because the current produced by such a fault flows in a local circuit between the turns involved and does not create the difference between the currents entering and leaving the winding at its two ends where the CTs are mounted. However, it is usually considered unnecessary to provide such a protection since short circuits between the turns on the same phase invariably develop into earth fault.

The coils of modern large turbogenerators are usually single turn and hence they do not need turn-to-turn protection. However, inter-turn protection is provided for multi-turn generators such as hydroelectric generators. In case of large generators, stator windings are sometimes duplicated in order to carry heavy current. Advantage may be taken of this necessity to provide inter-turn protection. In this case the stator winding has two separate parallel paths, as illustrated in Fig. 11.9.

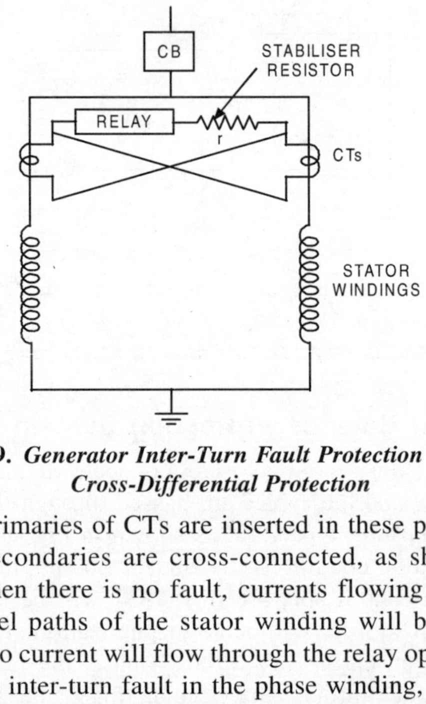

Fig. 11.9. ***Generator Inter-Turn Fault Protection Based on Cross-Differential Protection***

The primaries of CTs are inserted in these parallel paths and the secondaries are cross-connected, as shown in the figure. When there is no fault, currents flowing through the two parallel paths of the stator winding will be equal and therefore no current will flow through the relay operating coil. But during inter-turn fault in the phase winding, the currents flowing through the two parallel paths will be different and a current proportional to the difference of two currents will flow through the relay operating coil which will close the trip circuit and isolate the machine from the power system. Such a protection can be extremely sensitive.

Generators having single winding per phase or those generators whose parallel windings are not accessible can be protected by using zero sequence component of voltage caused by the reduction of the emf induced in the faulty phase. One such circuit is given in Fig. 11.10. In this arrangement a voltage transformer is connected between each phase terminal and the neutral of the winding. The secondary leads are connected in an open delta. The zero sequence voltage (residual voltage of the generator terminals) appears across the tertiary winding of the voltage transformer which is connected to the operating winding of a three element directional relay. This winding in quadrature to this operating winding of the relay is energized by the secondary of the voltage transformer. During normal operation, the residual voltage is zero *i.e.,*

$$\mathbf{V}_{RES} = \mathbf{V}_{RN} + \mathbf{V}_{YN} + \mathbf{V}_{BN} = 0$$

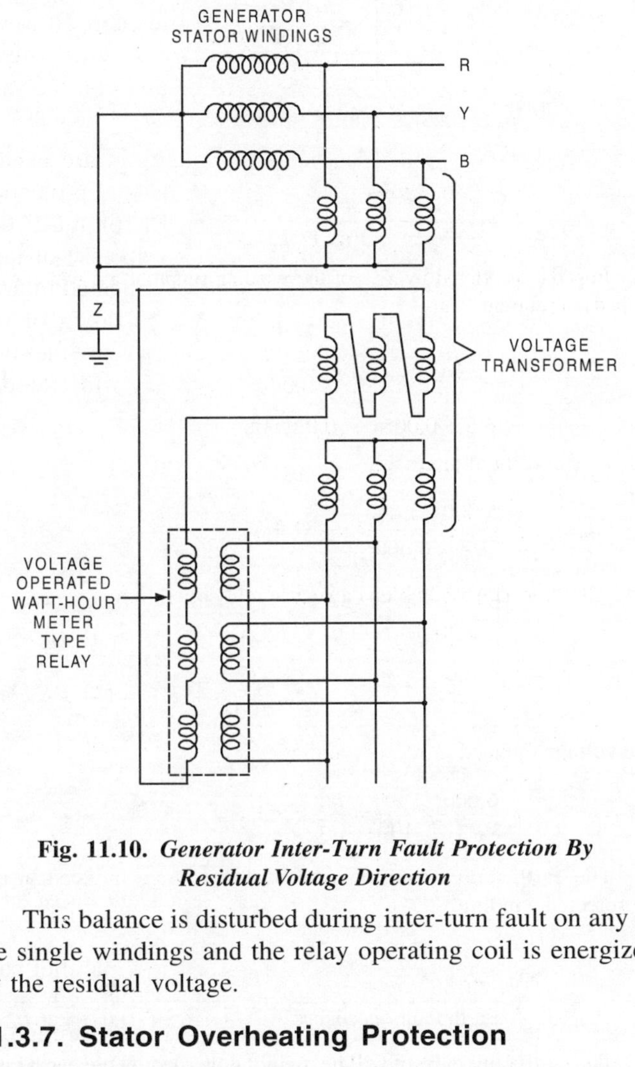

Fig. 11.10. ***Generator Inter-Turn Fault Protection By Residual Voltage Direction***

This balance is disturbed during inter-turn fault on any of the single windings and the relay operating coil is energized by the residual voltage.

11.3.7. Stator Overheating Protection

Generally stator overheating is caused by sustained overloads or by cooling system failure. Overheating because of short-circuited laminations is very localized, and it is just a matter of chance whether it can be detected before serious damage is caused. It is not practicable to provide overload protection by backup stator fault overcurrent protection as backup overcurrent protection is usually set for sensing fault currents and should not trip for overloads. Electrical overcurrent relays cannot sense the winding temperature accurately because temperature rise depends on I^2Rt and also on cooling. Electrical protection cannot detect failure of the cooling system.

The practice is to embed resistance temperature-detector coils or thermocouples in the slots below the stator coils of the generators of large capacity (above 1 MVA) for protection against overheating. Enough of such detectors are located at different places in the windings so that an indication can be obtained of the temperature conditions throughout the stator. Several of the detectors that provide the highest temperature indication are selected for use with temperature indicator or recorder usually having alarm contacts, or the detector providing the highest indication may be arranged to operate a temperature relay to sound alarm. Supplementary temperature devices may be provided for monitoring the cooling system; such devices would give the earliest alarm in the event of failure of cooling system. But it is usually realized that the stator temperature detectors and alarm devices are sufficient

Detector operated relaying system using a Wheatstone bridge circuit and a directional relay is shown in Fig. 11.11.

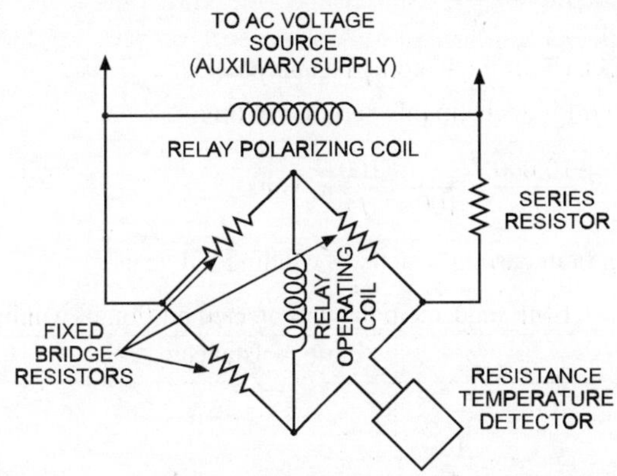

Fig. 11.11. ***Stator Overheating Protection With Resistance Temperature Detectors***

Example 11.1. Calculate the required value of neutral resistance for a 3-phase 11 kV alternator, so as to protect 70% of the winding against earth fault by a relay with pick-up current of 1 A. The neutral CT has a ratio of 250/5.

[*G.B. Technical Univ. Switchgear and Protection 2011-12*]

Solution : The minimum current which will operate the relay during fault conditions

= Relay pick-up current × CT ratio

$$= 1 \times \frac{250}{5} = 250 \text{ A} \qquad \ldots(i)$$

Let the earthing resistance required to protect the 70% of the winding *i.e.,*

$$x = (100 - 70) \text{ i.e., } 30\%$$

The emf induced in 30% of the winding

$$= \frac{11,000}{\sqrt{3}} \times \frac{30}{100} = 1,100\sqrt{3} \text{ V}$$

Earth fault current which 30% of the winding will cause

$$= \frac{1,100\sqrt{3}}{R} \qquad \ldots(ii)$$

Equating expressions (*i*) and (*ii*), we have

$$\frac{1,100\sqrt{3}}{R} = 250$$

$$\text{or } R = \frac{1,100\sqrt{3}}{250} = 7.621 \ \Omega \ \textbf{Ans.}$$

Example 11.2. An alternator rated at 10 kV protected by the balanced circulating current system has its neutral grounded through a resistance of 10 ohms. The protective relay is set to operate when there is an out-of-balance current of 1.8 amp in the pilot wires, which are connected to the secondary windings of 1,000/5 ratio current transformers.

Determine

(*i*) the percentage winding which remains unprotected.

(*ii*) the minimum value of earthing resistance required to protect 80% of the winding.

[*U.P. Technical Univ. Switchgear and Protection 2006-07*]

Solution : The minimum current which will operate the relay during fault condition

$$= 1.8 \times \frac{1,000}{5} = 360 \text{ A} \qquad ...(i)$$

(*i*) Let *x*% of the winding be unprotected.

The emf induced in this part of winding

$$= \frac{10,000}{\sqrt{3}} \times \frac{x}{100} = \frac{100}{\sqrt{3}} x \text{ volts}$$

Earth fault current which *x*% winding will cause

$$= \frac{\text{EMF induced in the unprotected portion of winding}}{\text{Earth resistance}}$$

$$= \frac{\frac{100}{\sqrt{3}} x}{10} = \frac{10}{\sqrt{3}} x \qquad ...(ii)$$

Equating expressions (*i*) and (*ii*), we have

$$\frac{10}{\sqrt{3}} x = 360$$

$$\text{or } x = \frac{360\sqrt{3}}{10} = 62.35\%$$

i.e., Percentage of winding which remains unprotected = 62.35% **Ans.**

(*ii*) Let the earthing resistance required to protect the 80% of the winding (*i.e.*, unprotected portion is 20%) be R ohms

EMF induced in 20% of the winding

$$= \frac{10,000}{\sqrt{3}} \times \frac{20}{100} = \frac{2,000}{\sqrt{3}} \text{ V}$$

Earth fault current which 20% of the winding will cause

$$= \frac{2,000 / \sqrt{3}}{\text{R}} \qquad ...(iii)$$

Equating expressions (*i*) and (*iii*), we have

$$\frac{2,000}{\sqrt{3}\text{R}} = 360$$

$$\text{R} = \frac{2,000}{360\sqrt{3}} = 3.2 \ \Omega \ \textbf{Ans.}$$

Example 11.3. A 5,000 kVA, 6,600 V, star-connected alternator has a synchronous reactance of 2 ohms per phase and 0.5 ohm resistance. It is protected by a Merz-Price balanced current system which operates when out-of-balance current exceeds 30% of load current. Determine what proportion of alternator winding is unprotected if the star point is earthed through a resistance of 6.5 ohms.

Solution : Let *x* % of the winding be unprotected

The equivalent impedance of *x* % of the winding

$$= (0.5 + j2) \times \frac{x}{100} \ \Omega$$

Earthing resistance = 6.5 Ω

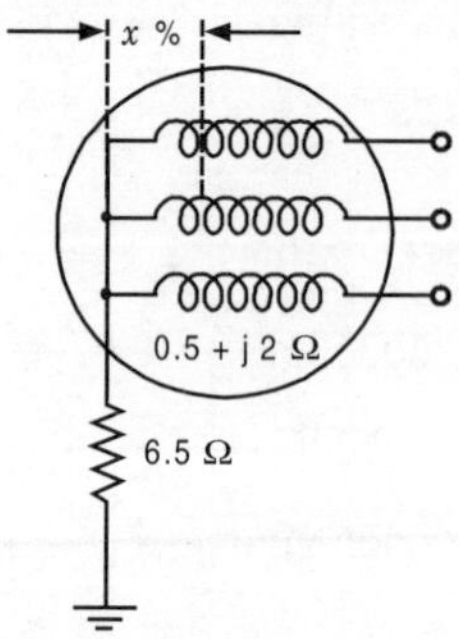

Fig. 11.12

Impedance offered by *x* % of the winding (unprotected winding) to earthed fault current

$$= 6.5 + (0.5 + j2) \times \frac{x}{100}$$

$$= (6.5 + 0.005x + j0.02x) \text{ ohms}$$

Full-load current of alternator

$$= \frac{5,000 \times 1,000}{\sqrt{3} \times 6,600} = 437.4 \text{ A}$$

Out-of-balance current which can cause relay operation

= 30 per cent of 437.4 A

$$i.e., \ \frac{30}{100} \times 437.4 = 131.2 \text{ A} \qquad ...(i)$$

The voltage induced in *x* % of winding

$$= \frac{6,600}{\sqrt{3}} \times \frac{x}{100} = \frac{66}{\sqrt{3}} x \text{ volts}$$

The earth fault current caused by the voltage induced in the unprotected winding

$$= \frac{\frac{66x}{\sqrt{3}}}{\text{Earth fault impedance}} = \frac{66x}{\sqrt{3} \times (6.5 + 0.005x + j0.02x)}$$

The earth fault current will be mainly governed by the earth fault resistance, so neglecting the generator reactance for the unprotected portion of winding,

$$\text{Fault current} = \frac{66x}{\sqrt{3} \times (6.5 + 0.005x)} \qquad ...(ii)$$

Equating expressions (*i*) and (*ii*), we have

$$\frac{66x}{\sqrt{3} \times (6.5 + 0.005x)} = 131.2$$

or *x* = 22.77 % **Ans.**

Example 11.4. Find the portion of winding protected against earth fault on a 11 kV, 50 MVA star-connected alternator having a reactance of 1.5 ohm per phase and negligible resistance. The generator neutral is grounded with 10 ohm. Merz-Price protection

scheme is used which operates when the out-of-balance current exceeds 5% of the full-load current.

[*U.P. Technical Univ. Switchgear and Protection 2007-08*]

Solution : Let x% of the winding be unprotected.

The equivalent impedance of x% of the winding

$$= (0 + j1.5) \times \frac{x}{100}\ \Omega$$

Earthing resistance = 10 Ω

Impedance offered by x% of the winding (unprotected winding) to earth fault current

$$= 10 + \frac{j1.5x}{100} = 10 + j0.015x$$

$$\text{Full-load current of alternator} = \frac{50 \times 10^6}{\sqrt{3} \times 11{,}000} = 2{,}624 \text{ A}$$

Out-of-balance current which can cause operation of relay

$$= 5\% \text{ of } 2{,}624 \text{ A } i.e., \frac{5}{100} \times 2{,}624 = 131.2 \text{ A}$$

The voltage induced in x% of winding

$$= \frac{11{,}000}{\sqrt{3}} \times \frac{x}{100} = \frac{110}{\sqrt{3}} x \text{ volts}$$

The earth fault current caused by the voltage induced in the unprotected winding

$$= \frac{110x/\sqrt{3}}{\text{Earth fault impedance}} = \frac{110x}{\sqrt{3}(10 + j0.015x)}$$

The earth fault current will be mainly governed by the earth fault resistance, so neglecting the generator reactance for the unprotected portion of the winding,

$$\text{Fault current} = \frac{110x}{\sqrt{3} \times 10} = 131.2$$

$$\text{or } x = \frac{131.2 \times \sqrt{3}}{11x} = 20.66\%$$

Portion of winding protected

$$= \left(1 - \frac{x}{100}\right) \times 100 = \left(1 - \frac{20.66}{100}\right) \times 100$$

$$= 79.34\% \quad \textbf{Ans.}$$

Example 11.5. An 11 kV, 100 MVA alternator is provided with differential protection. The percentage of winding to be protected against phase-to-ground fault is 85%. The relay is set to operate when there is 20% out-of-balance current. Determine the value of resistance in the neutral to ground connection.

[*U.P. Technical Univ. Switchgear and Protection 2005-06*]

Solution : Let the resistance in the neutral to ground connection be of R ohms

Unprotected winding, x = (100 – 85)% *i.e.* 15 per cent

$$\text{Full-load current of alternator} = \frac{100 \times 10^6}{\sqrt{3} \times 11{,}000} = 5{,}249 \text{ A}$$

Out-of-balance current which can cause relay operation

$$= 20 \text{ per cent of } 5{,}249 \text{ A } i.e., \frac{20}{100} \times 5{,}249 = 1{,}050 \text{ A} \quad ...(i)$$

The voltage induced in 15 per cent of winding

$$= \frac{11{,}000}{\sqrt{3}} \times \frac{15}{100} = \frac{1{,}650}{\sqrt{3}} \text{ V}$$

The earth fault current caused by the voltage induced in the unprotected winding

$$= \frac{1{,}650/\sqrt{3}}{\text{R}} \quad ...(ii)$$

Equating expressions (*i*) and (*ii*), we have

$$\frac{1{,}650}{\sqrt{3}\text{R}} = 1{,}050$$

$$\text{or R} = \frac{1{,}650}{1{,}050 \times \sqrt{3}} = 0.907\ \Omega \quad \textbf{Ans.}$$

Example 11.6. A 6.6 kV, 10 MVA star-connected alternator has a reactance of 2 ohms per phase and negligible resistance. Merz-Price protection is used for protection of winding. The neutral grounding resistance is 5 ohms. If only 10% of the winding is to remain unprotected, determine the setting of the relay.

[*GATE 1996*]

Solution : Voltage per phase, $V_p = \frac{6.6 \times 1{,}000}{\sqrt{3}} = 3{,}810.5$ V

Voltage induced in the unprotected portion (10%) of the winding

$$= 3,810.5 \times \frac{10}{100} = 381.05 \text{ V}$$

Impedance offered by 10 % of winding (unprotected winding) to earthed fault current

$$= 5 + j2 \times \frac{10}{100} = (5 + j\,0.2)\ \Omega$$

The earth fault current caused by voltage induced in the unprotected winding

$$= \frac{381.05}{5 + j0.2} = \frac{381.05}{5.00\,4} = 76.15 \text{ A}$$

$$\text{Full-load current} = \frac{\text{MVA rating} \times 10^6}{\sqrt{3} \text{ kV rating} \times 1{,}000}$$

$$= \frac{10 \times 10^6}{\sqrt{3} \times 6.6 \times 1{,}000} = 874.77 \text{ A}$$

$$\text{Relay setting} = \frac{\text{Earth fault current}}{\text{Full-load current}} \times 100 = \frac{76.15}{874.77} \times 100$$

$$= 8.7\ \% \quad \textbf{Ans.}$$

Example 11.7. A 13.8 kV, 125 MVA, star-connected alternator has a synchronous reactance of 1.4 per unit per phase and a negligible resistance. It is protected by a Merz-Price balanced current system which operates when out-of-balance current exceeds 10% of the full-load current. If the neutral point is earthed through a resistance of 2 Ω, determine what proportion of winding is protected against earth fault.

[*Switchgear and Protection; Pb. Univ. April 1988; G.B. Technical Univ. 2010-11*]

Solution : Let x% of the winding be unprotected. The arrangement is shown in simplified diagram (Fig. 11.13).

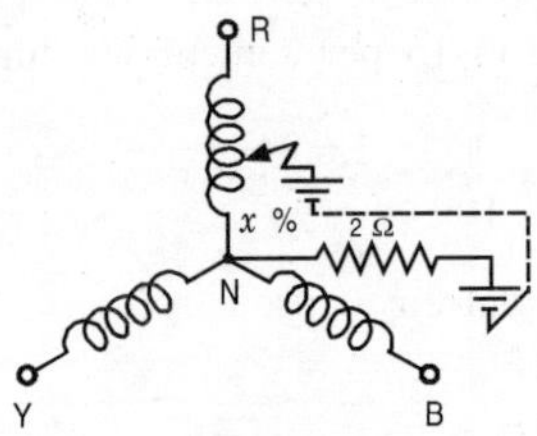

Fig. 11.13

$$\text{Voltage per phase, } V_p = \frac{13.8 \times 1{,}000}{\sqrt{3}} = 7{,}967 \text{ V}$$

$$\text{Full-load current, I} = \frac{\text{MVA} \times 10^6}{\sqrt{3}\ V_L} = \frac{125 \times 10^6}{\sqrt{3} \times 13{,}800} = 5{,}230 \text{ A}$$

$$\text{Reactance per phase, X} = \frac{\text{Reactance per unit} \times V_L}{\sqrt{3} \times I}$$

$$= \frac{1.4 \times 13{,}800}{\sqrt{3} \times 5{,}230} = 2.133\ \Omega$$

Voltage induced in unprotected winding

$$= 7{,}967 \times \frac{x}{100} = 79.67\,x \text{ volts}$$

Earth fault current caused by unprotected winding

$$= I \times \frac{10}{100} = 5{,}230 \times \frac{10}{100} = 523 \text{ A}$$

$$\text{Earth impedance} = \frac{\text{Voltage induced}}{\text{Earth fault current}} = \frac{79.67x}{523}\ \Omega$$

$$= 0.1523x\ \Omega$$

Reactance of unprotected winding

$$= \frac{x \times X}{100} = \frac{x}{100} \times 2.133 = 0.02133\,x\ \Omega$$

Earth resistance, R = 2 Ω

Since (earthing impedance)^2

= (earthing resistance)^2 + (earthing reactance)^2

$$\therefore (0.1523\,x)^2 = (2)^2 + (0.02133\,x)^2$$

or $x = 13.26$

∴ Protected winding = 100 – 13.26 = 86.74 % **Ans.**

Example 11.8. A 3-phase, star-connected 20 MVA, 11 kV alternator is protected by circulating current system, the star point being grounded through a resistor R. Plot a curve to a base of current up to 180% of full-load current through R indicating the percentage of winding protected for an earth fault setting of 30%. Determine the value of grounding resistance if 80% of winding is to be protected.

Solution : An earth fault setting of 30% means that if a terminal fault current is limited to the full-load current, then 30% of the winding from the neutral will be unprotected. Now for different values of currents through resistor R under terminal fault conditions by simple proportion it is possible to determine percentages of winding unprotected and the protected winding. This is illustrated in Table 11.1 for a 30% setting. The curve drawn from this data is illustrated in Fig. 11.14.

Table 11.1

% of full-load current through R for a terminal fault	% of winding unprotected	% of winding protected
30	100	0
60	50	50
90	33.33	66.67
120	25	75
150	20	80
180	16.66	83.34

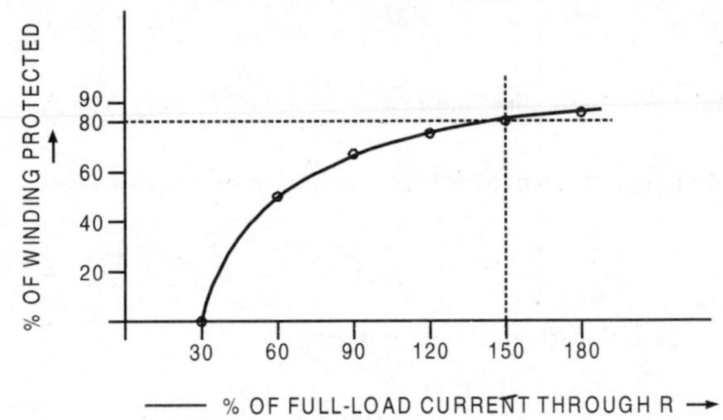

Fig. 11.14. ***Setting Characteristics For Earth Fault Relay***

For protecting 80% of the winding the current through R for a terminal fault comes out to be 150%.

Full-load current for the given alternator

$$= \frac{20 \times 10^6}{\sqrt{3} \times 11 \times 10^3} = 1{,}050 \text{ A}$$

The minimum current which will operate the relay during fault conditions

$$= \frac{150}{100} \times 1{,}050 = 1{,}575 \text{ A}$$

The emf induced in the unprotected (*i.e.,* 100 – 80 = 20 %) of winding

$$= \frac{20}{100} \times \frac{11{,}000}{\sqrt{3}} = 1270.2 \text{ V}$$

Earth-fault current which 20% of winding will cause

$$= \frac{1270.2}{R} \text{ A and it should be equal to 1,575 A}$$

$$\therefore \frac{1270.2}{R} = 1{,}575$$

$$\text{or} \quad R = \frac{1{,}575}{1{,}270.2} = 1.24\ \Omega \textbf{ Ans.}$$

Example 11.9. An alternator stator winding protected by a percentage differential relay is illustrated in Fig. 11.15. The relay has a 15% slope of characteristic $I_1 - I_2$ versus $\left(\frac{I_1 + I_2}{2}\right)$. A high resistance earth fault has occurred nearer the grounded neutral end of the generator stator winding while the generator is carrying load. The currents flowing at each end of the generator winding under this fault condition are shown in the figure. Assuming CT ratio of 500/5 A, will the relay trip the generator circuit breaker under this fault condition.

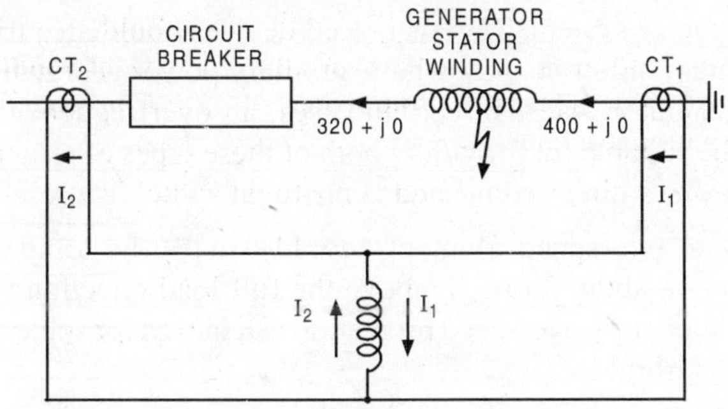

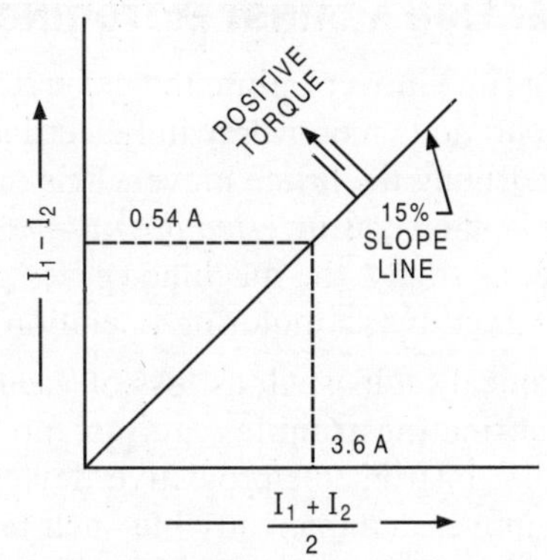

Fig. 11.15

Solution : CT ratio = 500 : 5

$$\text{Secondary current of } CT_1, I_1 = 400 \times \frac{5}{500} = 4 \text{ A}$$

$$\text{Secondary current of } CT_2, I_2 = 320 \times \frac{5}{500} = 3.2 \text{ A}$$

Direction of current flowing through secondaries of CTs is shown in figure.

Out-of-balance current (current flowing through the relay operating coil)

$$= I_1 - I_2 = 4 - 3.2 = 0.8 \text{ A}$$

$$\frac{I_1 + I_2}{2} = \frac{4 + 3.2}{2} = 3.6 \text{ A}$$

Corresponding point on $(I_1 - I_2)$ versus $\left(\frac{I_1 + I_2}{2}\right)$ characteristic is 3.6 × 0.15 A *i.e.,* 0.54 A [from the known slope]. The relay will operate if out-of-balance current is above the characteristic.

Since the point would be above 15% slope characteristic *i.e.,* it would be on the positive torque region and, therefore, the relay would operate.

Example 11.10. (*a*) The scheme shown in Fig. 11.16 is used for protection of the generator winding by an ordinary differential relay, having a minimum pick-up current of 100 mA. The CTs on the grounded neutral and the live ends actual ratios of 79 and 81 respectively though their nominal ratios are same. Show that the differential relay will trip even when the generator is delivering, full-load current of (400 + *j*0.0) A, to the bus and there is no fault in the differential zone.

(*b*) To overcome the defect in the question (*a*) a percentage differential relay having the same minimum pick-up current of 100 mA and 20% slope was used. Drawing the slope characteristic of the relay, show that the undesired tripping has been avoided.

[*U.P.S.C. I.E.S. Electrical Engineering-II, 2003*]

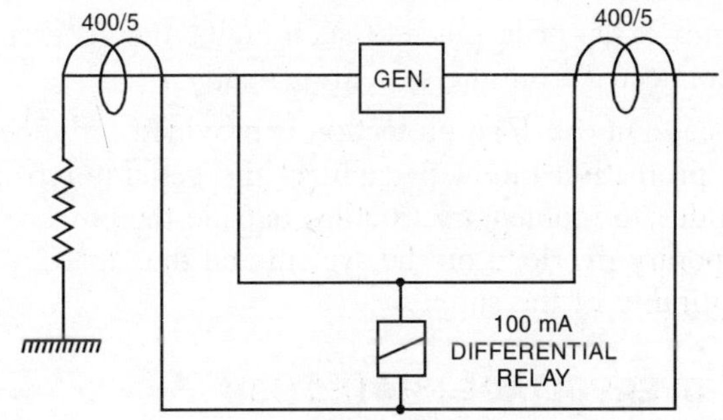

Fig. 11.16

Solution : (*a*) Full-load current = (400 + *j*0.0) A

$$\text{CT ratios are } \frac{1}{79} \text{ and } \frac{1}{81} \text{ respectively}$$

$$\text{Secondary current of } CT_1, I_1 = \frac{1}{79} \times 400 = 5.0633 \text{ A}$$

$$\text{Secondary current of } CT_2, I_2 = \frac{1}{81} \times 400 = 4.9383 \text{ A}$$

Out-of-balance current flowing through the relay operating coil

$$= I_1 - I_2 = 5.0633 - 4.9383 = 0.125 \text{ or } 125 \text{ mA}$$

As out-of-balance current exceeds 100 mA, the relay will operate even when no fault in differential zone.

(*b*) The current in operating coil = 125 mA

The averaging current through the restraining coil

$$= \frac{I_1 + I_2}{2} = \left(\frac{400}{79} + \frac{400}{81}\right) = 5 \text{ A}$$

Operating current with 20% slope

$$= 0.2 \times 5 + I_{min\ (pick\text{-}up)}$$

$$= 1.0 + 100 \times 10^{-3} = 1.1 \text{ A}$$

Since the current in operating coil is 125 mA, the relay will not operate and undesired tripping is avoided.

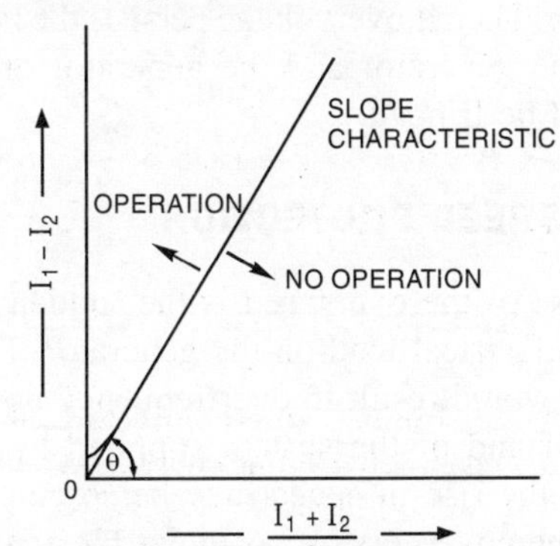

Fig. 11.17

11.4. OVERLOAD OR OVERCURRENT PROTECTION

Overloading of the generator may be caused either due to partial breakdown of winding insulation or due to excessive load on the power supply system. Overcurrent protection for alternators is not considered necessary, since modern alternators are capable of withstanding complete short circuit at their terminals for sufficient time without much overheating and damage. On occurrence of such faults, the generator can be disconnected from the system manually.

In case an overload protection is provided for generators such a protection might disconnect the generator from the system due to momentary troubles outside the power station or temporary overload on the system and thus interfere with the continuity of the supply.

11.5. OVERVOLTAGE PROTECTION

The field excitation system of modern generators is so designed that under normal operating conditions overvoltage conditions cannot occur. However, overvoltage in a generator occurs when speed of the prime mover increases due to sudden loss of the load on the generator. If there is an overvoltage in a generator, and if it persists, the generator main circuit breaker and the generator or exciter field breaker should be tripped, as it is not safe for the generator to continue to operate under such condition.

Generated overvoltage of significant duration or magnitude does not generally occur in turbogenerators because the control governors are very sensitive to speed variations. So overvoltage protection is not generally required with turbogenerators. Overvoltage protection is recommended for all hydroelectric or gas turbine generators that are subject to overspeed and consequent overvoltage on loss of load.

The overvoltage protection is provided with an overvoltage relay which has two units—one instantaneous relay set for pick up at about 130-150 % of the rated voltage and another IDMT relay set for pick up at about 110% of the rated voltage. Both relay units should be compensated against the effect of variations in frequency. The relay should be energized from a PT (potential transformer) other than the one used for the automatic voltage regulator. The operation of the relay should first insert additional resistance in the field circuit of the generator. Then if overvoltage persists the main generator breaker and the generator and the generator or exciter field breaker should be tripped.

11.6. OVERSPEED PROTECTION

The main cause of the overspeed is the sudden loss of all or major part of electrical load on the generator. The overspeed of a generator would result in overfrequency operation of the generator itself and of the system supplied by it. In certain circumstances the rise in speed may be so considerable that the action of the governors or closing of the emergency valve will not prevent rise in speed. Hence overspeed protection is recommended for all prime mover driven generators.

The overspeed element may be furnished as part of the prime mover, or of its speed governor, or of the generator, it should operate the speed governor, or whatever other shut-down means is provided, to shutdown the prime mover. It should also trip the generator circuit breaker; this is to prevent overfrequency operation of equipment and machinery connected to the system supplied by the generator, and also to prevent possible overfrequency operation of the generator itself from the ac system. The overspeed device should also trip the auxiliary circuit breaker where auxiliary power is drawn from the generator leads. In certain cases, an overfrequency relay may be suitable for providing both of these types of protection. However, a direct-connected centrifugal switch is preferred.

The overspeed element should usually be adjusted to operate at about 3 to 5 % above the full-load rejection speed. Supplementary overspeed protection is required for some forms of gas turbines.

11.7. PROTECTION AGAINST MOTORING

In the event of prime mover failure the generator continues to rotate as a synchronous motor drawing electrical power from the system and driving the prime mover. This motoring action of the machine is known as *inverted operation.* This operation is not desirable, as firstly, the machine ceases generation and no purpose is served by its motoring operation.

The mechanical faults such as loss of vacuum, failure of bearings, or lubricating troubles etc. are protected by the special indicating instruments/protective devices which may stop the prime mover automatically but such faults cannot be detected on the electrical side. However, the inverted operation of the machine can be protected by employing reverse power relay.

Protection against motoring is beneficial to the prime movers and to the system as a whole, but not to the generator.

During the motoring action of the generator, the power flows from the bus-bars to a machine and the conditions in the three-phases are balanced. Hence a single element directional power relay (reverse power relay) sensing the direction of power flow in any one phase in sufficient. The CTs for reverse power protection may be either at the neutral end or the bus-bar end of the generator stator windings. The setting depends on the type of the prime mover. Intentional time lag is provided in the reverse power protection to prevent operation by synchronising surges and power oscillations following system disturbances.

A *steam turbine* needs protection against overheating when its steam supply is cut off and the generator coupled to it acts as a motor. Such overheating occurs because there is not enough steam in the turbine to carry away the heat generated owing to windage losses. Modern condensing turbines will even overheat at outputs of less than approximately 10% of rated load. A condensing turbine that operates normally at high vacuum will withstand motoring much longer than a topping turbine that operates normally at high back pressure. A sensitive power directional equipment is generally used, and it should operate at less than 3% of the power rating of the generator.

In case of *hydrogenerator sets* mechanical protection is provided on the waterwheel to disconnect the generator from the system when water flow drops to a value likely to causes cavitation. Hence electrical protection is not required. However, motoring protection may occasionally be desirable for unattended hydraulic turbines for their protection against cavitation of the turbine blades. Cavitation of blade occurs on lean water flow that might occur, for example, from blocking of the trash gates. Power directional relaying equipment operating on a motoring current of somewhat less than 2.5 % of the full-load rating of the generator can be provided for such protection.

The *diesel-engine driven generators* would draw about 15% of its rated power or more from the system when acting as motor, and the motoring action should be avoided by suitable protective relaying. This is because the power drawn by the generator during motoring action constitutes an undesirably high load on the system. Also, there may be danger of fire or explosion from unburned fuel.

The power required to drive a *gas turbine with the generator coupled to it* working as a motor varies from 10% to 50% of its full-load rating and such heavy load on the system must be avoided by providing protective relaying. There is usually no turbine requirement for protection against motoring.

11.8. ROTOR FAULT PROTECTION

The different protection schemes for rotor protection are described below.

11.8.1. Rotor Earth Fault Protection

As already mentioned in Art. 11.2, field circuits are operated unearthed and it is not necessary to trip such a circuit completely if there is only one earth fault but the relay should be provided to give an indication that the fault has occurred so that the generator may be taken out of service at leisure since the incidence of second fault would cause serious damage.

One method of detecting earth fault on rotor circuit is illustrated in Fig. 11.18 (*a*). In this arrangement a high resistance is connected across the rotor circuit and its midpoint is grounded through a sensitive relay. The relay detects the earth fault for most of the rotor circuit except the rotor centre point.

Other methods of rotor earth fault protection include dc injection method and ac injection method. The arrangement is illustrated in Fig. 11.18 (*b*). In this arrangement either dc or ac voltage is impressed between the field circuit and ground through a sensitive overvoltage relay and current limiting resistor or capacitor (in case of ac voltage). A single earth fault in the rotor circuit will complete the circuit comprising the voltage source (ac or dc), sensitive overvoltage relay and earth fault and thus earth fault will be sensed by the relay. DC injection method is simple and has no problems of leakage currents. If direct current is used, the overvoltage relay can be more sensitive than if alternating current is used, with ac, the relay must not pick up on the current that flows normally through the capacitance to ground, and care must be taken to avoid resonance between the capacitance and the relay inductance.

Detection of open-circuit fault in rotor circuit is the same as used for detection of loss of excitation, described below (Art. 11.8.2).

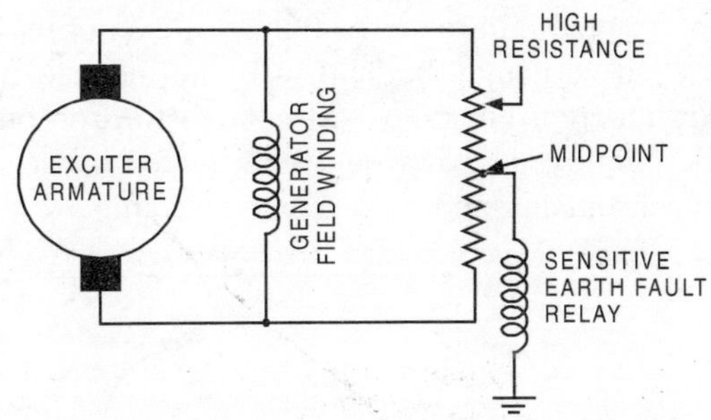

Fig. 11.18. (*a*) ***Rotor Earth Fault Protection***

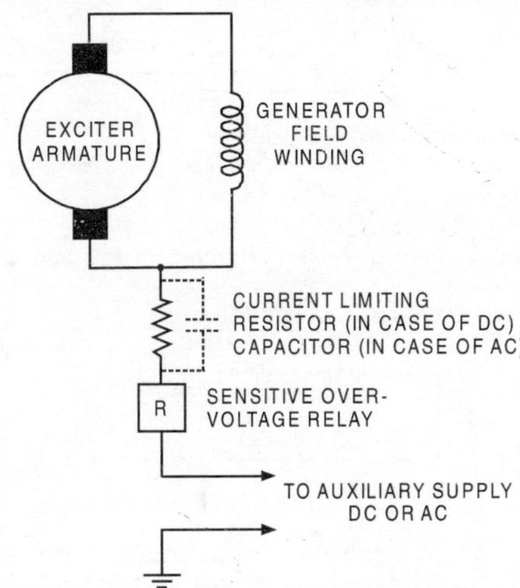

Fig. 11.18. (*b*) ***Rotor Earth Fault Protection***

11.8.2. Loss of Excitation (or Field Failure) Protection

Loss of excitation on the generator can endanger the generator or the connected system or both.

It is caused by accidental tripping of field breaker, short circuit in the field circuits, poor brush contact or operating errors. In absence of field current alternator runs as an induction generator and thus heavy currents are inducted in the rotor teeth and wedges. Most modern generators can only withstand these currents up to a period of 2 to 3 minutes. When a generator loses excitation, it draws reactive power from the system amounting as much as 2 to 4 times the generator's rated load. Before it lost excitation, the generator may have been delivering reactive power to the system. Thus, this large reactive load suddenly thrown on the system, together with the loss of generator's reactive power output may cause wide spread voltage reduction, which in turn, may cause extensive instability.

If the system is capable enough to tolerate the difference of reactive power then no automatic protection is required for the purpose as an operator has got 2 to 3 minutes to take care of it. But if there is chance of instability of the system, automatic protection is required.

Undercurrent moving coil relay connected across a shunt in series with the field winding for loss of excitation protection is illustrated in Fig. 11.19 (*a*).

In case of large generators which operate over a wide range of field excitation such a relay may be an embarrassment.

Furthermore, field failure due to failure of exciter may not be detected by it as it may be held in by ac induced from the stator. An undercurrent relay fast enough to drop out on ac cannot be employed as it would be affected by ac induced during synchronising and during external faults.

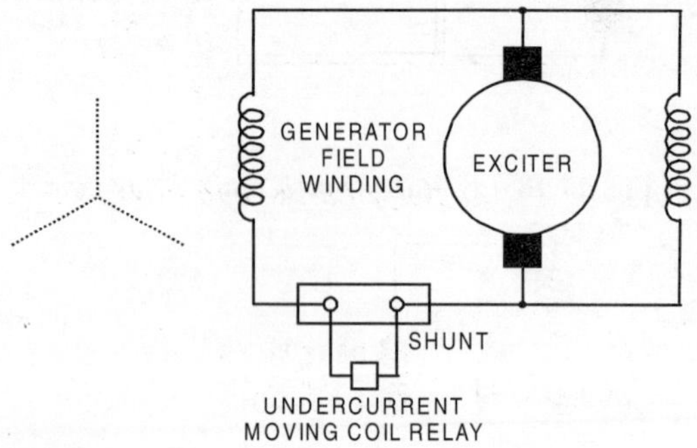

Fig. 11.19. ***(a) Loss of Excitation Protection***

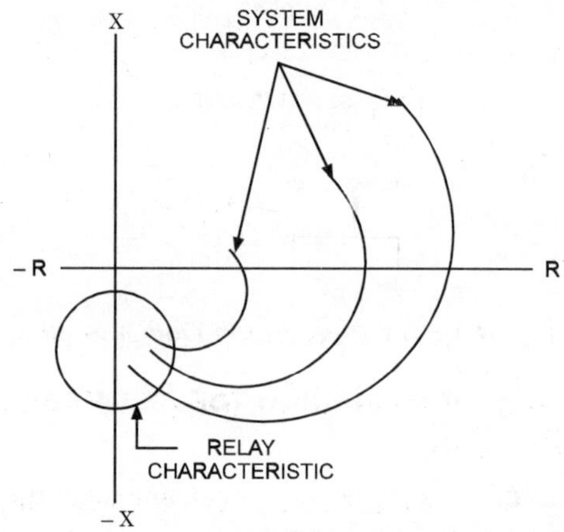

Fig. 11.19. ***(b) Loss of Excitation Relay and System Characteristic***

The most selective type of loss of excitation relay is a directional distance type relay operated from alternating current and voltage at the main generator terminals.

Several loss of excitation characteristics and the relay operating characteristic on an R-X diagram are illustrated in Fig. 11.19 (*b*). As soon on the excitation is lost, the equivalent generator impedance traces a curve from the first quadrant of R-X diagram into a region of the fourth quadrant. By enclosing this region within the relay operating characteristic, the relay will operate when the generator first starts to slip poles and will trip the field breaker as well as disconnect the generator from the system before either the generator or the system can be harmed. The generator may then be returned to service immediately when the cause of excitation failure is corrected.

11.8.3. Protection against Rotor Overheating Because of Overexcitation

In case of overcurrent due to overexcitation in the rotor circuit, a dc relay is usually provided. This relay senses and initiates alarm. However, application of such relay is very much limited as relaying of dc quantities is relatively uncommon and the rotor windings are designed to tolerate overcurrents due to overexcitation.

11.8.4. Rotor Temperature Alarm

Such a protection is provided only in case of large generators. It indicates the level of temperature and not the actual hot spot temperature. In the arrangement shown in Fig. 11.20, resistance is measured by comparing voltage and current by a double actuating quantity moving coil relay, the operating coil being used as the voltage coil and restraining coil as current coil. The relay measure the ratio of voltage and current *i.e.*, a resistance giving a measure of rotor temperature.

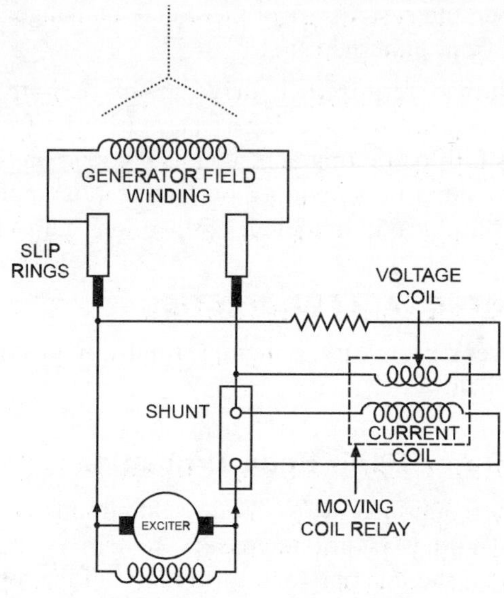

Fig. 11.20. ***Rotor Overheating Protection***

11.8.5. Automatic Field Suppression and Use of Neutral Circuit Breaker

In the event of a fault on a generator winding even though the generator circuit breaker is tripped, the fault continues to be fed as long as the excitation will exist because emf is induced in the generator itself. For the quick removal of the fault during emergency, it is necessary to disconnect the field simultaneously with the disconnection of the generator. Thus it is absolutely necessary to discharge its magnetic field in the shortest possible interval of time. Hence it is to be ensured that all the protection system not only trip the generator circuit breaker but also trip the automatic field discharge switch. The field discharge switch is an automatic control unit designed to remove the voltage from the generator after its isolation from the system.

The schematic diagram for field suppressing and opening of the neutral circuit breaker is illustrated in Fig. 11.21 (*a*). In the event of fault the circulating relay contact is closed and so the trip coils TC_1, TC_2, and TC_3 are energized. The trip coil TC_1 opens the main circuit breaker while the trip coil TC_3 opens the neutral circuit breaker. The trip coil TC_2 opens the upper contacts, shorts the lower contacts so as to short-circuit the field winding through resistor R. Thus the energy of the generator is dissipated in the resistor R.

An alternative arrangement of field suppression and dissipation of energy of the generator is illustrated in Fig. 11.21 (*b*).

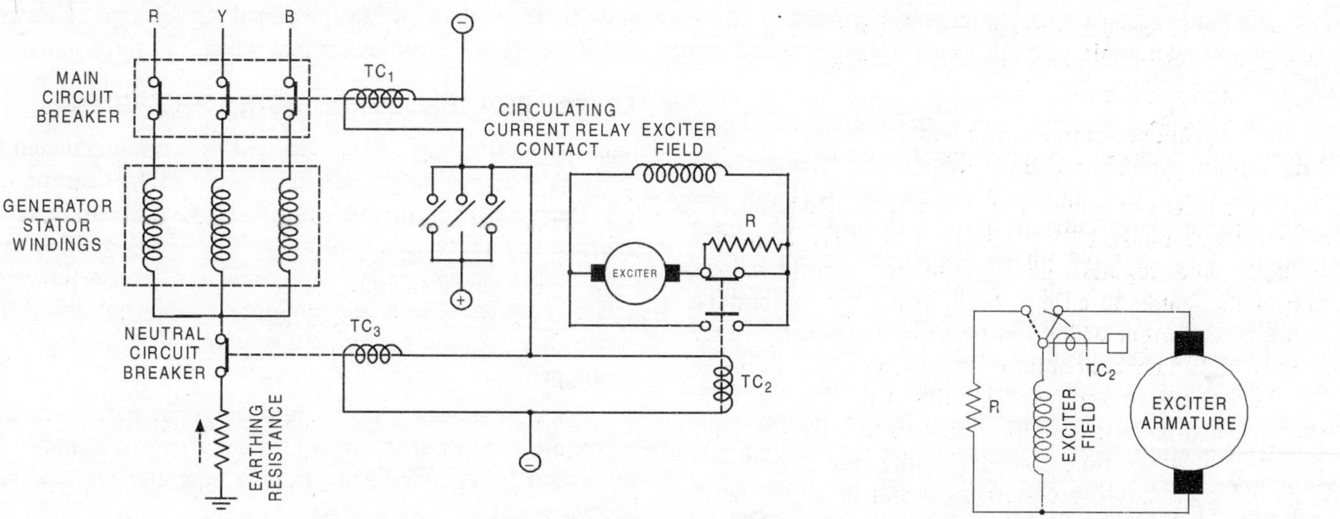

(a) Schematic Diagram For Field Suppressing and Opening of Neutral Circuit Breaker *(b) Alternative Arrangement of Field Discharge*

Fig. 11.21

Alternative Arrangement of Field Suppression. This arrangement is similar to that described above except that alternator field winding is also discharged through resistor R_2 by using trip coil TC_4, as illustrated in Fig. 11.22. This process of discharging consists of the isolation of the exciter from the generator rotor field winding and involves the dissipation of magnetic energy stored in the inductive reactance of the rotor and the main exciter windings. This is achieved by switching-y in the rotor and the dc exciter field windings across the field discharge resistors, as illustrated in Fig. 11.22. In order to avoid any overvoltages, the switching-over is performed immediately and without a break in the excitation system.

The quality of operation of an automatic field discharge system depends upon the rate of decrease of field, which in turn depends upon the time constant of the field winding. This term is also known as *the field discharge time constant* T_d. Normally the value of field discharge time constant lies between 3 and 8 seconds.

The discharge time T_d is defined as the time required for the drop in voltage at the generator stator terminals to a minimum value (usually 500 V).

The value of resistance R_2 in the rotor circuit is determined on the basis of a fall in generator voltage to 30% of normal voltage within 3 to 4 seconds. Resistance R_2 comes out to be roughly 4 to 5 times rotor winding hot resistance.

The resistance R_1 should be roughly equal to 10 times the exciter field winding resistance.

11.9. NEGATIVE SEQUENCE PROTECTION OF GENERATORS AGAINST UNBALANCED LOADS

If, owing to a fault, there is an imbalance in the three-phase stator currents, double frequency currents are induced in the rotor core. This causes overheating of rotor and possible damage to the rotor. Unbalanced stator currents also cause severe vibrations and heating of stator. Current balance relaying equipment would not be very effective for protection

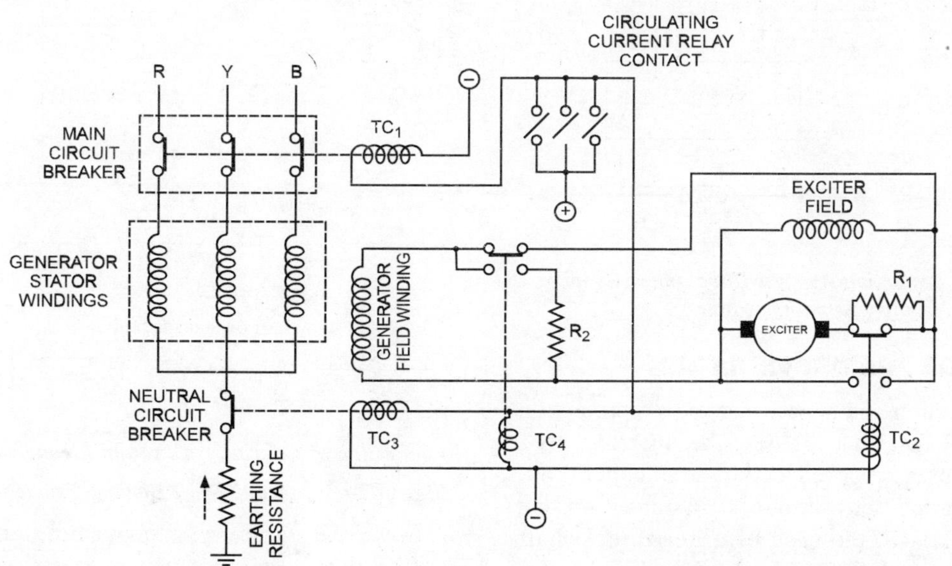

Fig. 11.22. ***Schematic Diagram For Field Suppressing and Opening of Neutral Circuit Breaker***

against such faults because such equipment energized by phase currents would operate quickly for small imbalances and too slowly for large imbalances.

Negative sequence current filter with overcurrent relay is used to provide protection against unbalanced loading.

From the theory of symmetrical components we know that unbalance three-phase currents have a negative sequence component. The negative phase sequence current causes heating of the stator. In case of high speed turbogenerators, the continuous current which can be carried is usually in the range of 10 and 15 per cent of the positive sequence continuous rating. The negative sequence heating follows a normal resistance law and so it is proportional to the square of the current. The heating time constant of the machine largely depends upon the cooling system used and is expressed as $I_2^2 t = K$ where I_2 is the negative sequence current expressed on the per unit basis of the *continuous maximum rating*, t is the *current duration* in seconds, and K is a constant which for turbogenerators usually lies between 3 and 20.

It is generally necessary to install negative sequence relays that match the above heating characteristic of the machine. The arrangement is shown in Fig. 11.23. Three CTs are connected in the three phases and the output from their secondaries is fed to the coil of the overcurrent relay through a *negative phase sequence filter*. Negative sequence filter circuit comprises resistors and inductors connected in such a way that negative sequence component flows through the relay coil. The overcurrent relay used is with inverse characteristics matching with $I_2^2 t$ rating curve of the machine. The relay can be set to operate at a particular value of the imbalanced current or the negative sequence component current.

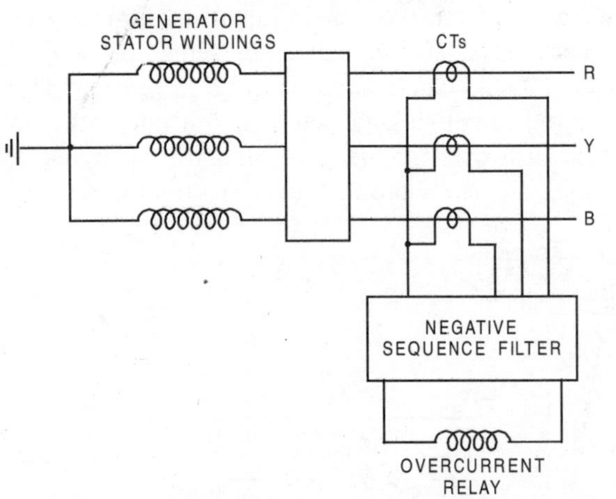

Fig. 11.23. ***Generator Protection Against Unbalanced Loading Using Negative Sequence Filter***

11.10. PROTECTION AGAINST VIBRATION

Rotor earth fault protection and negative sequence protection of generator against unbalanced loads, described in Arts. 11.8.1 and 11.9 respectively prevent or reduce vibration under those circumstances. A vibration detector may be mounted on one of the bearing pedestals in the case of a horizontal shaft generating set, or on the upper guide-bearing in case of a vertical shaft generating set. It may be set to trip the machine or initiate an alarm when the radial deflections of a certain duration exceed a perdetermined value.

11.11. BEARING OVERHEATING PROTECTION

Bearing overheating can be detected by a relay actuated by a thermometer type bulb inserted in a hole in the bearing, or by a resistance temperature detector relay, such as used for stator overheating protection, with the detector embedded in the bearing. In case lubricating oil is circulated through the bearing under pressure, the oil temperature may be monitored if the system has provision for giving an alarm if the oil circulation is stopped.

Such protection is provided for all unattended generators where the size or importance of the generator warrants it. Such protection for attended generators is generally limited only to sound an alarm.

11.12. EXTERNAL FAULT BACKUP PROTECTION

Overcurrent and earth fault protection is provided for backup protection of large sized (above 1 MVA) generators protected by differential protection against external phase-to-phase faults and earth faults. The schematic arrangement is illustrated in Fig. 11.24 (*a*). Induction type IDMT relays may be employed for this purpose.

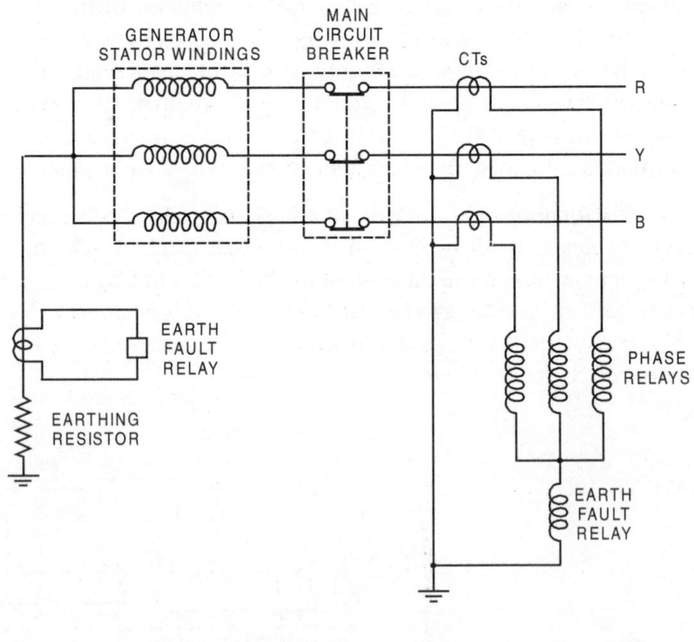

(a) Schematic Diagram

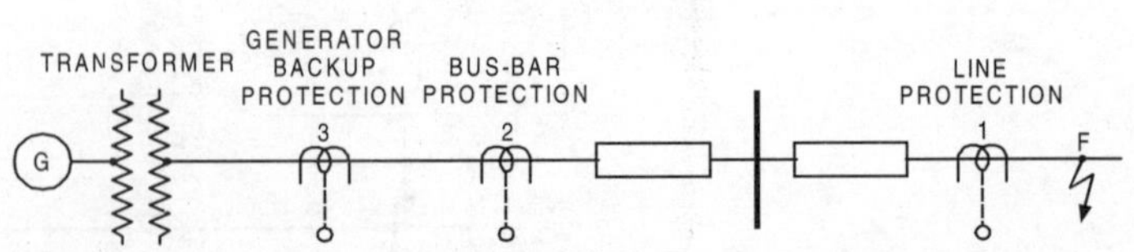

(b) Sequence of Operation Diagram (Sequence of Operation 1,2,3)

Fig. 11.24. ***External-Fault Backup Protection For Generators***

Since the faults in stator winding are fed by the stator winding itself, their influence on current in the outgoing terminals of generator is governed by the fault level of the main bus. Thus overcurrent and earth fault relays do not provide satisfactory protection against internal faults.

Sequence of operation is illustrated in Fig. 11.24 (*b*). The relays setting is selected that the generator overcurrent and earth fault protection does not normally respond to external faults such as at F.

However, overcurrent and earth fault protection of generator (3) may be set to operate with due time lag for larger values of external fault currents so that the fault F is not supplied by the generator in case it persists for a long time duration.

Hence high set, definite minimum time, induction type, inverse overcurrent, earth fault relays are required for backup protection of generators.

11.13. DIRECT CONNECTED GENERATOR PROTECTION

Though generators are usually connected to the grid through step-up power transformers, but if the generation and transmission/distribution voltages are the same (as in case of captive power plants), the generator may be connected directly to the load line. Direct connected generators are usually of small ratings (of the order of tens of MW).

A basic protection scheme for a direct connected 30 MVA generator is given in Fig. 11.25.

The scheme may consist of the following protective schemes:

1. Biased differential protection.
2. Negative phase sequence protection.
3. Standby earth fault protection.
4. Backup overcurrent and earth fault protection.

In addition to above protective schemes the following may also be provided.

1. Rotor earth fault protection.
2. Field failure protection.
3. Reverse power protection in case of back pressure turbine sets and engine driven sets.

11.14. TYPES OF PROTECTION FOR GENERATORS AND THEIR SETTINGS

Different types of protection used for different rating of generators are given in Table 11.2.

11.15. PROTECTION OF AC MOTORS

There is a wide range of ac motors such as three-phase induction motors, three-phase synchronous motors, single-phase motors, commutator motors etc. But here discussion will be limited to the protection of three-phase induction motors and synchronous motors.

11.15.1. Types of Faults

The types of faults, that occur in 3-phase induction and synchronous motors are summarised below.

1. Stator Faults. Stator faults include phase-to-phase faults, earth faults and inter-turn faults. Faults in the motor windings occur chiefly due to insulation failure that may occur due to excessive temperature rise. *Phase-to-phase* faults are rare because enough insulation is provided between phases. *Earth faults* are more likely to occur. *Inter-turn faults* invariably develop into earth fault, so it is considered unnecessary to provide separate protection against inter-turn faults.

2. Rotor Faults. Such faults are more likely to occur in wound rotor motors. Faults in wound rotors may be either earth faults (conductor-to-earth faults) or inter-turn faults, which are caused by severe mechanical and thermal stresses.

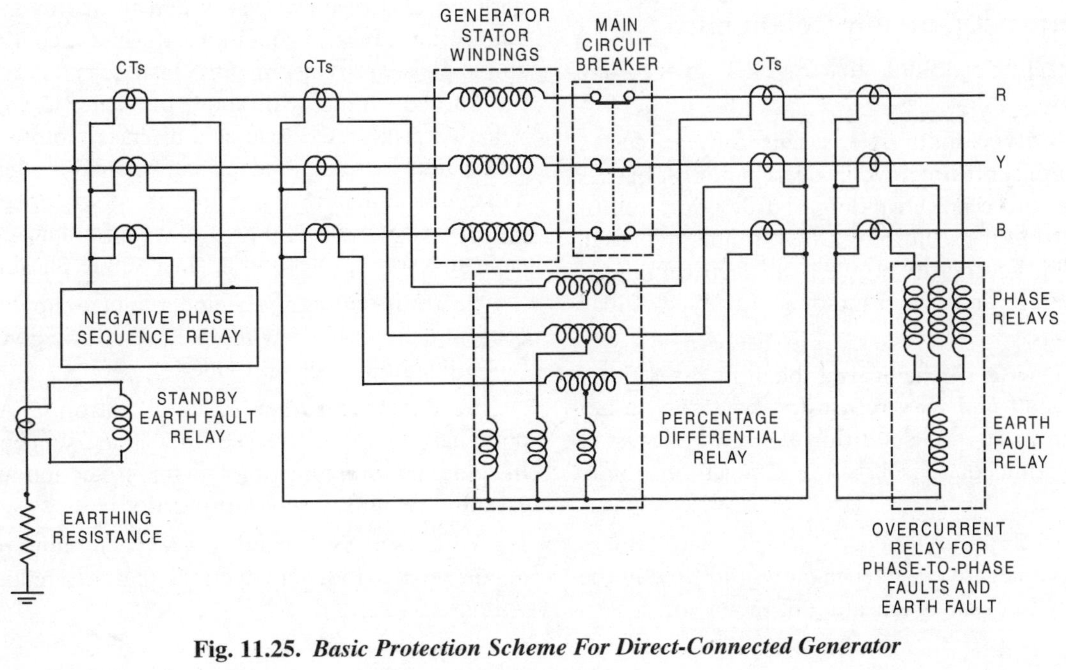

Fig. 11.25. *Basic Protection Scheme For Direct-Connected Generator*

Table 11.2. Types of Protection For Generators and Their Settings

Generator Rating	Type of Drive	Type of Protection Used	Recommended Fault Current Setting	Protection For Exciter Circuit and Neutral
Below 1 MVA	Steam Turbine, Steam Engine or Diesel Engine	1. Time graded earth fault. 2. Time graded overcurrent. 3. Reverse Power.	20 per cent full-load current. 125 per cent full-load current. 10-20 per cent of its normal power rating.	1. Neutral isolating switch. 2. Field switch in the exciter field circuit.
1 To 5 MVA	Steam Turbine	1. Balanced earth fault or Differential (Merz-Price). 2. Reverse power (only for back pressure turbines)	10-25 per cent full-load current. 10-25 per cent full-load current.	1. Neutral isolating switch. or Automatic neutral breaker.
	Hydraulic Turbine	1. Balanced earth fault or Differential (Merz-Price). 2. Overvoltage protection.	10-25 per cent full-load current. 10 per cent of normal voltage.	1. For unattended power stations, protection against field failure. Setting 20 to 30 per cent of normal excitation. 2. Neutral isolating switch or Automatic neutral breaker.
Above 5 MVA	Steam Turbine	1. Time graded earth fault. or Differential (Merz-Price). 2. Phase unbalance. 3. Reverse power (only for back pressure turbines).	20 per cent full-load current. 10 to 25 per cent negative phase sequence only. 10 to 25 per cent full-load current.	1. Field earth fault for protecting 90% of the field winding (only for generators above 20 MVA). 2. Automatic field suppression switch. 3. Automatic neutral breaker.
	Hydraulic Turbine	1. Time graded earth fault or Differential (Merz-Price). 2. Overvoltage.	20 per cent full-load current. 10 per cent of normal voltage.	1. For unattended power stations protection against field failure. Setting 20 to 30 per cent of normal excitation. 2. Field earth fault for protecting 90% of the field winding. 3. Automatic field suppression switch. 4. Automatic neutral breaker.

Other faults that are likely to occur include failure of bearings and faults in starters and associated circuits.

11.15.2. Abnormal Operating Conditions

The abnormal operating conditions that are likely to occur are summarised below.

1. Prolonged Overloading. It is caused by excessive mechanical loading, short-time cyclic overloading. It causes excessive temperature rise of winding and deterioration of insulation resulting in winding fault. Hence overload protection should be necessarily provided in electrical motors irrespective of the type and size of motor and the type of load which is being driven.

2. Stalling. Due to mechanical problem or overload at start motor may stall and may refuse to start. This is not desirable because in such a condition it draws excessive current. Hence the immediate isolation of motor from supply mains is essential.

3. Unbalanced Supply Voltages. The unbalanced three-phase supply causes negative sequence current to flow in the motor that is likely to cause overheating of the motor.

4. Single Phasing. This is caused when one of the supply lines get disconnected, may be due to any reason, such as blowing of a fuse or open circuit in one of the three supply connections. In such conditions, the motor continues to operate single phase, provided that the load does not exceed 57.7% of the normal rating, with about the same temperature rise as when carrying rated load as a 3-phase motor. Single phasing may cause extreme magnetic unbalance, reduction in torque and overheating due to negative phase sequence currents. This operation is not desired as it may cause damage to the motor and, therefore, protection against single phasing is provided.

5. Undervoltage. The undervoltage supply will generally cause the motor to draw more current for a given load, Hence this protection is also provided.

6. Reverse Phase. The direction of rotation of a polyphase motor is reversed if the phase sequence is changed. In some motor applications such a protection may become an essential feature of motor protection.

7. Loss of Synchronism. A synchronous motor may pull out of step due to severe overload or due to reduction in supply voltage.

11.15.3. Protection of Low-Voltage (Below 1,000 V) 3-Phase Induction Motors

These are the most widely used industrial motors. Circuit diagram for a magnetic contactor starter for a low-voltage 3-phase induction motor is shown in Fig. 11.26. Magnetic contactor starter essentially consists of a set of 'start' and 'stop' push buttons with associated contacts, overload and undervoltage protective devices. The start push button (usually green coloured) is a momentary contact switch that is held normally open by a spring. The stop push button (usually red coloured) is held normally closed by a spring. The operation is as follows.

When the start push button is pressed, the operating coil (or the main contactor) gets energized through the overload relay contacts OL (normally closed). This closes the three main contacts M that connect the motor to the supply. At the same time a set of auxiliary or maintaining contacts MC are closed. When the maintaining contacts MC are closed, a new circuit is established through stop push button, maintaining contacts MC and operating coil (or main contactor). Since operating coil circuit is now maintained by the auxiliary contacts MC, the starter button be released. Now the motor starts.

For stopping a motor, the stop push button is pressed, the operation coil gets de-energized, thereby opening all the main contacts and auxiliary contacts.

If the supply fails or line voltage drops below a certain value, the main contacts and the maintaining contacts are opened. Upon return of the supply, the contactors cannot

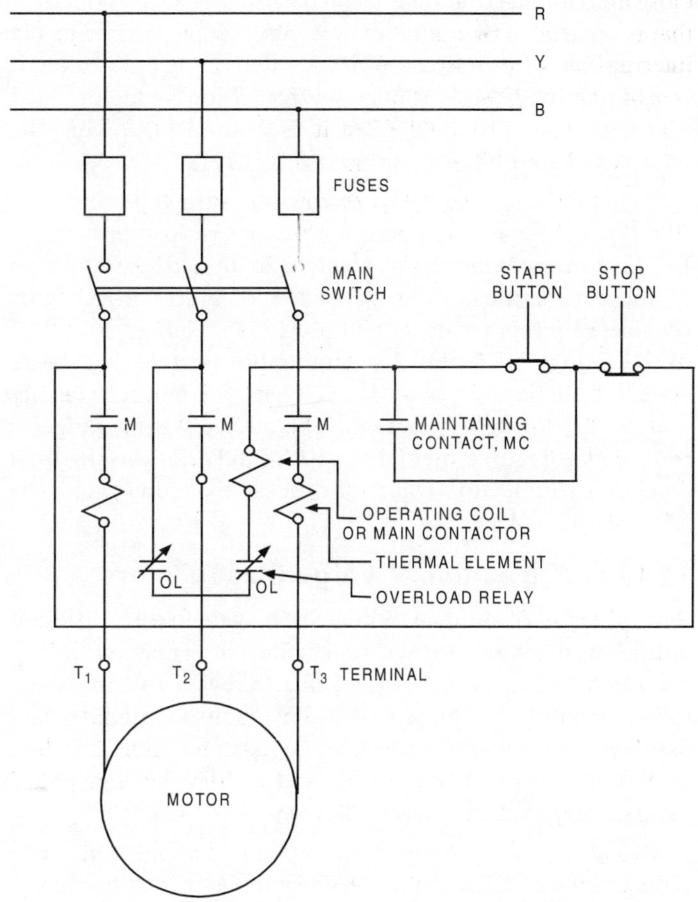

Fig. 11.26. *Magnetic Contactor Starter Circuit Diagram*

Table 11.3. Protection Chart For 3-Phase Induction Motors

Abnormal Operating Condition	Possible Forms of Protection	Remarks
Overloads	1. Thermal overload relays. 2. Inverse overcurrent relays. 3. Miniature circuit breaker with built-in trip coils. 4. Overload release.	1. Overload protection is provided almost for all types of motors. 2. Should not operate during starting periods.
Phase-to-phase faults and earth faults	1. HRC fuses. 2. High-set instantaneous overcurrent relays. 3. Differential protection.	Differential protection is feasible for large sized motors (*i.e.,* motors of rating exceeding 1,000 kW). Below this high set instantaneous protection preferred.
Undervoltage	1. Undervoltage relays. 2. Undervoltage release.	1. Undervoltage relay is employed in certain applications. 2. Undervoltage release is incorporated with every starter.
Unbalanced supply voltage	Negative phase sequence relays.	Only in certain applications.
Single phasing	1. Usually thermal overload relays. 2. Special single phase preventer.	1. Recently developed static single phasing devices. 2. Unbalance protection.
Reverse phase sequence	Phase reversal protection.	Generally at supply point.
Stalling	1. Thermal overload relays. 2. Instantaneous overcurrent relays.	Instantaneous Trip.

close until the start button is again pushed. Because a contactor that is controlled by a three-wire control circuit maintains the interruption of the circuit even after the supply is restored, it is said to provide *undervoltage protection* for the motor. Such a protection is provided when it is desired to prevent the unexpected starting of a motor.

Thermal overload relays (bimetallic strip type or solder film type) are commonly used for motor overload protection. Both act to open the motor control circuit and, therefore, to disconnect the motor from the source of supply. HRC fuses provide very rapid short-circuit protection. Current is cut off by HRC fuse even before it attains prospective peak. The selection of thermal relay is such that for normal starting conditions, the relay does not operate. A setting range is provided for adjustment for different variations in load conditions. Protection chart for 3-phase induction motors is given in Table 11.3.

11.15.4. Protection of Large Sized Motors

Several types of protective relays are available for different applications. Such relays sense the abnormal operating condition and make the motor circuit breaker to trip. Large motors are provided protection against faults in windings and associated circuits, excessive overloads and short circuits, undervoltage, phase unbalance and single phasing, phase reversal and switching overvoltages.

Characteristics of relays are such that the operating time reduces with the increase in the magnitude of fault current.

1. Stator Winding Protection

(*a*) Overcurrent Protection. This is the basic type of protection that is employed for short-circuit protection of stator windings. The devices for such protection range from fuses for motor voltages of 600 V and lower, through direct acting overcurrent tripping elements on circuit breakers, to separate overcurrent relays and circuit breakers for voltages of 2,200 V and higher. Where fuses or direct-acting devices are employed, there must be one protective element in each ungrounded conductor. Where relays and CTs are employed with ac tripping from the output of the CTs, a CT and relay are required for each ungrounded conductor. If battery or capacitor tripping is employed, 3 CTs with two phase relays and one ground relay will do for a 3-phase circuit whether or not the source neutral is grounded.

Motor may be grouped in two categories viz non-essential service motors and essential service motors.

For *non-essential service* (other than essential services) motors it is the practice to provide both inverse-time and instantaneous phase and ground overcurrent relays for automatic tripping. The inverse-time phase relays are adjusted to pick up at about 3.5 to 4 times rated motor current, but to have enough time delay so as not to operate during the motor starting period. The instantaneous phase relays are adjusted to pick up a little above the locked rotor current. The inverse-time ground relays are adjusted to pick up at about 20% of the rated motor current or about 10% of the maximum available ground fault current, whichever is smaller.

Percentage differential relaying is provided for large motors [2,200 V to 4,999 V inclusive 1,000 kW and above and 5,000 V and higher of capacity 400 kW and above]. The percentage differential relaying provides faster and more sensitive protection than overcurrent relaying but at the same time it does not operate on starting or other transient overcurrents.

For *essential service motors* the inverse-time phase overcurrent relays are usually not employed. The reason for this omission is to trip the motor breaker automatically only for short circuits and not to trip for any other reason. This is because the tripping of such a motor may force a partial or complete shutdown of a generator or any other device with which the motor is associated, any unnecessary tripping should be avoided.

(*b*) Stator Overheating of Motors. Overheating of motors is due to prolonged overloading, stalled rotors or unbalanced stator currents. For complete protection, three-phase motors should have an overload element in each phase as an open circuit in the supply to the power transformer feeding a motor will cause twice as much current to flow in one phase of the motor as in either of the remaining two phases. Inspite of the desirability of overload element in all the three phases, motors of rating of about 1,000 kW and lower are generally provided with overload elements in only two phases, on the assumption that the open phase condition will be detected and corrected before any motor can overheat.

For *non-essential service motors* it is the practice for motors of less than 1,000 kW rating to provide either replica type thermal overload relays or long time inverse-time overcurrent relays or direct acting tripping devices to disconnect a motor from its source of supply in the event of overload.

Other things being equal, the replica type relay will generally provide the best protection as its time-current characteristic nearly matches the heating characteristic of a motor over the full range of overcurrent. Also, it takes into account the heating effect of the load on the motor before the overload condition occurs. The inverse-time overcurrent relay will tend to overprotect at low currents and to underprotect at high currents, However, the overcurrent relay is very easy to adjust and test, and it is self-reset. The overload relays will also provide protection in the event of phase-to-phase faults. The tripping of relays may be adjusted as follows :

1. For continuous rated motors without service factor or short-time overload ratings—not more than about 115% of rated current of motor.
2. For motors with 115% service factor—about 125% of rated current of motor.
3. For motors with special short-time overload ratings, or with other service factors, the motor characteristic will determine the required tripping characteristic but the tripping current should not exceed about 140% of rated motor current.

Motors rated higher than 1,000 kW are generally provided with resistance temperature detectors embedded in the stator slots between the windings. In such cases, a single relay operating from these detectors is used instead of the replica type or inverse-time overcurrent relays. Also, current balance relays capable of operating at about 25% or less unbalance between the phase currents should be provided. If the resistance temperature detectors are not provided but current balance relays are provided, a single replica type thermal overload relay may be substituted for the resistance temperature detector relay.

Typical motor heating, replica relay and inverse-time relay characteristics are illustrated in Fig. 11.27.

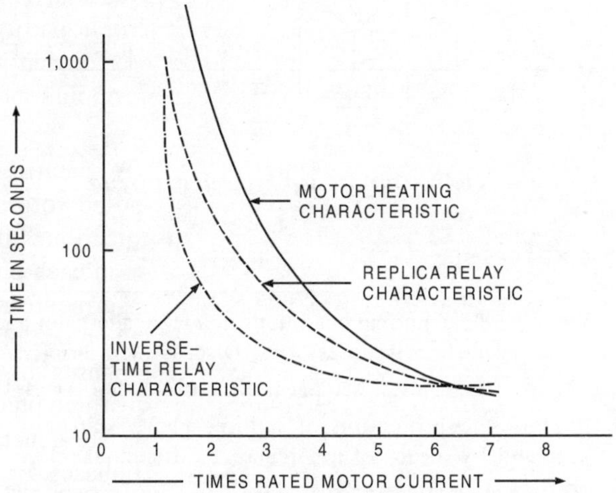

Fig. 11.27. ***Typical Motor Heating and Protective Relay Characteristics***

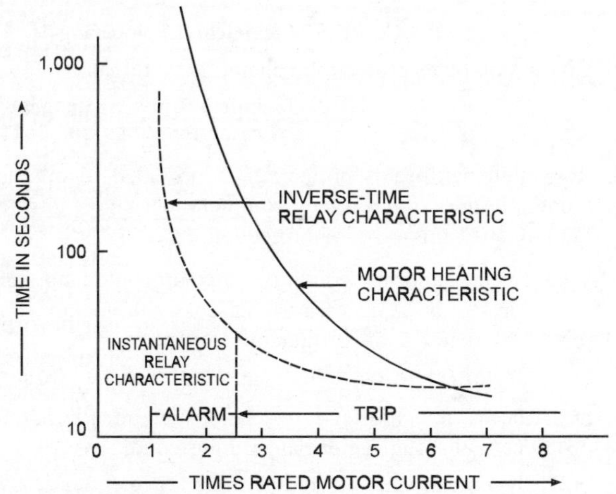

Fig. 11.28. ***Protection Characteristics For Essential Service Motors***

The protection recommended for some *essential service motors* is based on minimizing the possibility of unnecessarily tripping the motor, even though such practice may sometimes endanger the motor. So long time inverse-time overcurrent relays are used for motors of all ratings, but they merely control an alarm and leave tripping in the operator control. For motors that are likely to suffer locked rotor, supplementary instantaneous overcurrent relays are provided and their contacts are connected in series with those of the inverse-time overcurrent relays to trip the motor circuit breaker automatically. Supplementary instantaneous overcurrent relays are adjusted to pick up at about 2 to 3 times of the rated motor current. Protection characteristics for essential service motors are illustrated in Fig. 11.28.

For essential service motors for which automatic tripping is desired, in addition to alarm for overloads between about 1.15 times of rated current and the pick-up of the instantaneous overcurrent relays, thermal relays of either replica type or the temperature detector type should be employed depending on the size of the motor. Such relays permit operation for overloads as far as possible beyond the point where the alarm will be sounded but without damaging the motor to the extent that it must be repaired before it can be used again.

***(c)* Rotor Overheating Protection.** As already mentioned in Art. 11.15.1, rotor faults are more likely to occur in wound rotor motors. The increase in rotor current is reflected on stator current and the stator overcurrent protection thereby act. The setting of stator overcurrent relay is generally of the order of 1.6 times full-load current. This is enough to detect the rotor faults.

***(d)* Undervoltage Protection.** Motor draws excessive current when operated on undervoltage and so undervoltage protection can be provided by overload devices or temperature sensitive devices. However, a separate single element undervoltage relay energized with phase-earth or phase-to-phase voltage can be provided for protection against a three phase drop in voltage or an attempt to start with low voltage on all phases. A time delay feature is usually incorporated to avoid tripping by a transient voltage drop.

***(e)* Unbalance and Single Phasing Protection.** The unbalanced 3-phase supply causes negative sequence current to flow in the motor which may cause overheating of the stator and rotor windings of the motor. The unbalanced protection provided to a motor should be such as to avoid prolonged unbalanced condition, but should not disconnect the motor for permissible unbalance of short duration. The unbalance voltage protection can be based upon the following methods.

1. Bimettalic relays arranged to trip faster for unbalanced currents.
2. Single phase relays sensing overcurrent in heavily loaded phases.
3. Phase unbalance relays.

For small motors separate phase unbalance relay is not justified because of its cost. Additional phase failure relay (single phasing preventer) is provided where essential. For larger motors, additional unbalanced current relays are provided. The secondary currents of CTs are fed to a negative phase sequence filter. The output of the negative sequence filter is given to overcurrent unit or static level detector (Refer to Art. 11.9) which communicate tripping command to the starter or circuit breaker when the negative sequence current exceeds a pre-set limit.

***(f)* Reverse Phase Protection.** In some applications, phase reversal is dangerous such as in case of elevators, cranes, hoists, trams etc. In such applications reverse phase protection must be provided.

The phase reversal relay based on electromagnetic principle consists of a disc motor driven by magnetic system actuated by secondaries of two line CTs and PTs. For correct phase sequence the disc experiences a torque in positive direction and, therefore, keeps the auxiliary contacts closed. In the event of occurrence of phase reversal, the torque acting on the disc reverses and the disc starts rotating in opposite direction and so open the auxiliary contacts. Thus the magnetic coil of a starter can be de-energized or circuit breaker can be tripped.

The solid-state phase reversal relays and phase failure relay senses the phase reversal or phase failure. Under abnormal operating condition it sends tripping command to output stage, which is an auxiliary relay or static device.

11.15.5. Protection For Exclusively Synchronous Motors

Synchronous motor may be either a loaded-start motor or an unloaded-start motor. A *loaded-start motor* is any motor other than either a synchronous condenser, or a motor driving a generator, it includes any motor driving a mechanical load even though automatic unloading means may be used.

1. Amortisseur Overheating Protection. Such a protection during starting or loss of synchronism should be provided for all loaded-start motors. This protection is provided by a time-delay thermal overload relay connected in the field discharge circuit. This protection is not required in case of unloaded-start motors.

2. Field Winding Overheating Protection. Protection against field winding overheating because of prolonged overexcitation is provided for synchronous motors or condensers with automatic voltage regulators without automatic field current limiting features. A thermal overload relay with time delay or a relay that responds to an increase in field winding resistance with increasing temperature may be employed. In an attended station, the relay would merely control an alarm.

3. Loss of Synchronism Protection. A synchronous motor may pull out of step either due to excessive load or due to reduction in supply voltage. All loaded-start synchronous motors should be provided protection against loss of synchronism.

The condition of loss of synchronism can be detected by a relay that responds to the change in power factor that occurs when pole slipping occurs. One such typical circuit is illustrated in Fig. 11.29. In this arrangement the voltage between two phases is compared with the current in the third phase; an attracted armature relay energized from a full-wave bridge rectifier is differentially connected and is in the operated state so long as the motor is in synchronism. A non-linear resistor is provided for the protection of rectifiers and for extending the operating range of the relay.

4. Loss of Excitation Protection. All unloaded-start synchronous motors without automatic voltage regulators should be provided with loss of excitation protection. This is a low-set, time-delay-reset undercurrent relay whose coil is connected in series with the field winding.

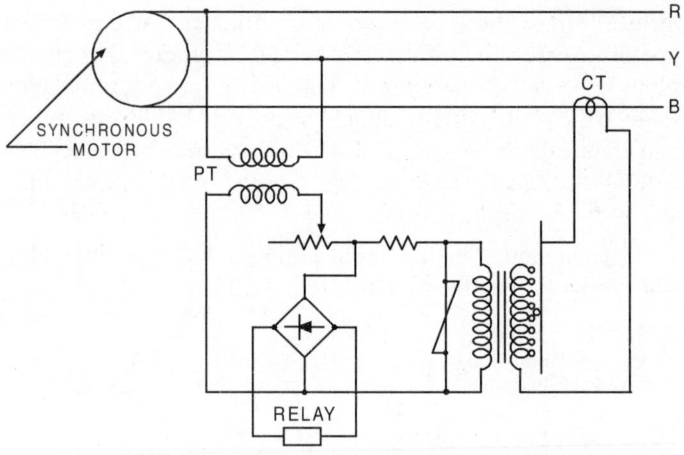

Fig. 11.29. *Out-of-Step Protection Circuit*

Exercises

1. What are the abnormal conditions in a large alternator against which protection is necessary ? Discuss them briefly.

   [Pb. Univ. Switchgear and Protection April 1986]

2. Explain the protection of a three-phase star-connected generator by means of a percentage differential relay.

   [U.P. Technical Univ. Switchgear and Protection 2008-09]

3. Describe the general principle of operation of generator differential protection scheme. Give the 3-phase connection of the biased scheme.

   [U.P.S.C. I.E.S. Electrical Engineering-II, 2003]

4. Give a complete protection scheme for alternator.

   [G.B. Technical Univ. Switchgear and Protection 2009-10, 2011-12]

5. Restricted earth-fault protection is provided to alternators though it does not protect complete wiring against earth fault. Justify this type of protection.

6. What are the abnormal conditions in a large alternator against which protection is necessary ? Discuss a protection scheme for an alternator against inter-turn faults.

   [U.P. Technical Univ. Switchgear and Protection 2005-06]

7. Describe various protection schemes of an alternator for different faults and abnormal running conditions.

   [U.P. Technical Univ. Switchgear and Protection 2009-10]

8. Describe with suitable diagrams how the stator windings of a 3-phase alternators are protected from

   (*a*) Earth faults on the stator winding.

   (*b*) Unbalanced current in the stator winding.

   (*c*) Earth fault on the rotor winding.

   (*d*) Overloading of alternator.

   [U.P.S.C. I.E.S. Electrical Engineering-II, 2007]

9. What types of protective device are used for the protection of an alternator against overheating of its (*i*) stator and (*ii*) rotor ? Discuss them in brief.

10. Discuss suitable protection schemes which are used for (*i*) rotor earth fault (*ii*) rotor open circuit of a synchronous generator.

11. A large turbogenerator is running in synchronism with an infinite bus when there is a sudden failure of its excitation system. What may be the consequence ? Describe a suitable protection scheme to prevent such consequence.

12. Explain : A loss of excitation relay used for alternator protection. **[U.P.S.C. I.E.S. Electrical Engineering-II, 2005]**

13. What are the types of faults that are likely to occur in a 3-phase induction motor ? If the motor is not fully loaded, is it necessary to provide protection against single phasing ? Explain why ?

[G.B. Technical Univ. Switchgear and Protection 2011-12]

14. What are the types of faults that are likely to occur in a three-phase induction motor ? Distinguish between short-circuit protection, overload protection and earth-fault protection of motor.

15. Explain the term "single phasing". In what form such protection is provided in case of a large 3-phase induction motor ? If the motor is not fully loaded, is it necessary to provide protection against single phasing ? Explain why.

16. Write short notes on the following :

(*i*) Generator faults.

(*ii*) Protection of alternator. **[Agra Univ 1982]**

(*iii*) Overload protection of alternator. **[Pb. Univ. May 1985]**

(*iv*) Self balance protection system.

(*v*) Inter-turn fault protection of alternators. **[Pb. Univ. Sept. 1987]**

(*vi*) Overspeed protection of alternators. **[Pb. Univ. April 1982]**

(*vii*) Loss of excitation protection of alternators. **[Pb. Univ. April 1988]**

(*viii*) Protection of small motors.

Short Answer Type Questions With Answers

Q. 1. What are the common types of generator faults ? **[A.M.I.E. Sec. B. Winter 1986]**

Ans. The common types of faults occurring in synchronous generators are enumerated below.

1. Stator winding faults—phase-to-phase faults, phase-to-earth faults and inter-turn faults.

2. Rotor winding faults—conductor-to-earth faults, open-circuit faults, inter-turn faults.

The other abnormal operating conditions are (*i*) failure of prime mover (*ii*) failure of field (*iii*) unbalanced loading (*iv*) overloading (*v*) overvoltage at generator terminals (*vi*) overspeed (*vii*) ventilation failure (*viii*) current leakage in the generator body.

Q. 2. The occurrence of short-circuiting between the stator windings is quite rare. Why so ?

Ans. The occurrence of short-circuiting between the stator windings is quite rare because the insulation in a slot between coils of different phases is at least twice as thick as the insulation between one coil and the iron core.

Q. 3. What is differential protection ?

Ans. The differential protection is that which responds to the phasor difference between two or more electrical quantities.

Q. 4. Which type of relays are used in Merz-Price protection system for alternator ?

Ans. The relays used in Merz-Price protection system of alternator are instantaneous electromagnetic type (attracted-armature type).

Q. 5. How does the inclusion of resistance in neutral grounding circuit of an alternator affect the performance of the differential protection of a 3-phase synchronous generator ?

Ans. With the inclusion of resistance in neutral grounding circuit of an alternator, it will not be possible to protect whole of the stator windings of a star-connected generator against earth faults. The percentage of winding protected depends on the value of neutral grounding resistor and earth fault relay setting.

Q. 6. Does the Merz-Price system of protection applied to alternators provide any protection for turn-to-turn faults on the same phase ? If not why ?

Ans. Merz-Price system of protection applied to alternators does not provide any protection against turn-to-turn faults on the same phase because the current produced by such a fault flows in a local circuit between the turns involved and does not create the difference between the currents entering and leaving the windings at its two ends where the high voltage CTs are mounted.

Q. 7. Why it is not necessary to provide protection for turn-to-turn fault in turbogenerators ?

Ans. The coils of modern large turbogenerators are usually single turn and, therefore, it is not necessary to provide protection for turn-to-turn faults.

Q. 8. What type of protective device is used for the protection of an alternator against overheating of its stator ?

Ans. Resistance temperature detector.

Q. 9. Why is overcurrent protection not necessary for modern alternators ?

Ans. Overcurrent protection is not considered necessary for modern alternators because these are capable of withstanding a complete short circuit at their terminals for sufficient time without much overheating and damage.

Q. 10. Why is it not necessary to provide overvoltage protection in turbogenerators ?

Ans. The overvoltage in a generator occurs when the prime mover speed increases due to sudden loss of the load on the generator and the speed control governors in case of turbogenerators are very sensitive to speed variations and, therefore, generated overvoltage of significant duration or magnitude does not generally occur. That is why it is not necessary to provide overvoltage protection in turbogenerators.

Q. 11. What types of protective device are used for the protection of an alternator against overheating of its rotor.

Ans. DC relay and resistance measuring circuit.

Q. 12. Why is it necessary to suppress the field immediately after disconnection of faulty alternator from the system?

Ans. In the event of a fault on a generator winding even though the generator circuit breaker is tripped, the fault continues to be fed as long as the excitation will exist because emf is induced in the generator itself. Hence it is necessary to suppress the field immediately after disconnection of faulty generator from the system.

Q. 13. What is field suppression ?

Ans. The opening of the field circuit breaker and discharge of stored energy in the field winding through a resistor is called the field suppression.

Q. 14. What type of relays are required for backup protection of alternators ?

Ans. High set, definite minimum time, induction type, inverse overcurrent, earth fault relays are required for backup protection of alternators ?

Q. 15. What are the types of faults that are likely to occur in operation of a 3-phase induction motor ?

Ans. The faults which are likely to occur in operation of a 3-phase induction motor are : Stator faults (phase-to-phase, phase-to-ground and inter-turn faults), rotor faults (earth faults and inter-turn faults), prolonged overloading, stalling, unbalanced supply voltages, single phasing, undervoltage, reverse phase.

Problems

1. A 3-phase, 20 MVA, 11 kV, star-connected generator is protected by the current balancing system of protection. If the ratio of CTs is 1,200/5, the minimum operating current of the relay is 0.75 A and the neutral earthing resistance is 6 Ω, calculate the percentage of each phase of stator winding which is unprotected against earth faults when the machine is operating at normal voltage. **[Pb. Univ. Switchgear and Protection Sept. 1991]**

[**Ans.** 17%]

2. A 3-phase, 10 MVA, 6.6 kV star-connected generator is protected by the current balancing system of protection. If the ratio of current transformers is 1,000/5, the minimum operating current of the relay is 0.5 A, and the neutral point earthing resistance is 5 Ω, determine the percentage of each phase of stator winding, which is protected against earth faults when the machine is operating at normal voltage.

[Pb. Univ. Switchgear and Protection December 1991]

[**Ans.** 86.88%]

3. A 11 kV, 100 MVA alternator is grounded through a resistance of 5 Ω. The CTs have a ratio of 1,000/5. The relay is set to operate when there in an out-of-balance current of 1 A. What percentage of the generator winding will be protected by the percentage differential scheme of protection. [**Ans.** 84.27%]

4. The neutral point of a 3-phase, 25 MVA, 11 kV alternator is earthed through a resistance of 5 Ω. The relay is set to operate when there is an out-of-balance current of 1.5 A. The CTs have a ratio of 1,000/5. What percentage of winding is protected against an earth fault and what should be the minimum value of earthing resistance to protect the 90 per cent of the winding ?

[**Ans.** 76.41%, 2.12 Ω]

❑❑❑

CHAPTER 12

Transformer Protection

INSIDE THIS CHAPTER

12.1. INTRODUCTION

The two major items of equipment in a power system are the generators and transformers. Chances of fault occurrence on them are very rare as compared to on lines, but the damaged caused by the faults usually takes much more time and money to repair than are required to repair the damage caused by the faults on the lines. Rapid reclosing of circuit breakers is feasible on lines and it helps in saving the amount of damage while a fault in a generator/transformer always requires some attention of the supervisory staff. Fast clearing of faults, however, assists in reducing the damage to the equipment and also reduces the interruption in power service caused by drop in voltage and from instability.

Power transformers are static devices, totally enclosed and usually oil immersed, and therefore, chances of fault occurrence on them are very rare. But the consequences of even a rare fault may be very serious unless the transformer is quickly disconnected from the system. Hence automatic protection of transformers against possible faults is essential and of utmost importance.

The faults occurring in power transformers are open-circuit faults (an open circuit in one phase of a 3-phase transformer), earth faults, phase-to-phase faults, inter-turn faults and overheating from overloading or from some internal cause such as core heating.

Interphase (phase-to-phase) short circuits are most frequent on leads of three-phase transformers, while the interphase short circuits within the windings are less frequent. Earth faults and inter-turn faults have the highest probability on the power transformers. Winding short circuits, also called the *internal faults,* generally result from failure of insulation due to temperature rise or deterioration of transformers oil.

An open circuit in one phase of a 3-phase transformer may cause undesirable heating but this condition is relatively harmless and so no relay protection is required against open circuits. On the occurrence of such a fault, the transformer can be disconnected manually from the system.

As already mentioned above, the most common types of internal winding faults are earth faults and inter-turn faults. Since the inter-turn faults develop rapidly into earth faults, only earth fault protection is essentially provided.

The choice of a protective gear for a transformer depends upon several factors such as (*i*) type of transformer *i.e.,* distribution transformer or power transformer (*ii*) size of the transformer (*iii*) type of cooling (*iv*) system where used *i.e.,* its electrical location in the network and (*v*) the importance of service for which it is required.

For distribution transformers employed in rural area, the normal practice is to use the fuses for its protection against external faults but for urban distribution networks, where discrimination is absolutely necessary, fuses will not serve the purpose.

For power transformers, the protection is to be provided usually against dangerous overloads and excessive temperature rise. Dangerous overloads may be due to external faults or the internal ones. External faults, however, are cleared by the relay system outside the transformer within the shortest possible time in order to avoid any danger to the transformer due to these faults. Hence the protection for internal faults is to be provided in such transformers. *Differential protection* is the most important type of protection used for protection against internal phase-to-phase and phase-to-earth faults. The other protection systems employed for protection of transformers against internal faults are Buchholz protection, core-balance leakage protection, combined leakage and overload protection, restricted earth fault protection. Different protection systems commonly used for protection of transformers will be discussed here.

12.2. BUCHHOLZ PROTECTION

On the occurrence of internal fault in an oil-immersed power transformer gas is usually generated, slowly for an incipient fault (such as sparking, small arcing, loose connections in conducting path etc.,) and violently for heavy faults. Most short circuits caused either by impulse breakdown between adjacent turns at the end turns of the winding or as a poor initial point contact which will immediately heat to arcing temperature. The heat generated by the large local currents causes the transformer oil to decompose and generate gases, which can be used in detection of winding faults. The relays based on this principle are (*i*) *pressure relays* and *pressure relief devices* which act on the measurement of the total accumulated pressure, (*ii*) *rate of pressure rise* relay, which acts on the measurement of the rate of formation of the gas and (*iii*) *gas accumulator relay,* most commonly known as *Buchholz relay*, actuated by the gas formed.

Buchholz protection employing Buchholz relay is the simplest form of protection and is most commonly used on all oil-immersed transformers provided with conservator. For details refer to Art. 9.9.

12.3. CORE-BALANCE LEAKAGE PROTECTION

An earth fault usually involves a partial breakdown of winding insulation to ground. The resulting leakage current is quite small as compared to short-circuit current. The earth fault may continue for a long time and cause considerable damage before it ultimately develops into a short circuit and removed from the system. Under such circumstances it is advisable to provide earth fault protection in order to ensure that the earth fault or leak is removed in the early stages. An earth fault relay used for it is essentially an overcurrent relay of low setting and operates as soon as earth fault or leak develops. One method of protection against earth faults in transformers is the *core-balance leakage protection* shown in Fig. 12.1.

This system consists of three primary conductors surrounded by the magnetic circuit of a current transformer. This has a single secondary winding which is connected to the relay operating coil. Under normal conditions *i.e.,* when there is no earth fault the instantaneous sum of the currents in the three phases is always zero, and there is no resultant flux in the core of the CT no matter how much the load is out of balance. Thus no current flows through the relay operating coil and trip circuit remains open. When an earth fault occurs, the sum of the three currents is no longer zero and a current is induced in the secondary of the CT causing the trip relay to operate and isolate the transformer from the bus-bars.

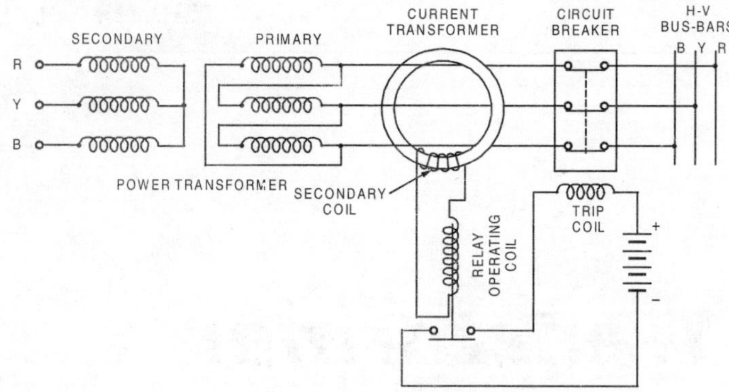

Fig. 12.1. ***Core-Balance Leakage Protection***

12.4. COMBINED LEAKAGE AND OVERLOAD PROTECTION

The core-balance protection described above suffers from the disadvantage that if the fault occurs between phases the relay does not operate. This shortcoming is overcome by employing three separate CTs. In this system of protection two overload relays and one earth leakage relay are connected, as shown in Fig. 12.2. The overload relays used are high current setting ones and are arranged to operate against phase-to-phase faults while the earth fault relay has low current setting and operates under earth or leakage faults only.

The two overload relays are sufficient to protect all the three phases while the earth-fault or leakage fault relay is energized by the resultant currents from all the three CTs in case of leakage fault. The trip contacts of the overload relays and earth-fault or leakage relay are connected in parallel, as illustrated in the figure. So the circuit breaker will trip in the event of energisation of either overload relay or leakage relay. Thus the protection against faults and short circuits either to earth or between phases is achieved.

12.5. DIFFERENTIAL PROTECTION FOR POWER TRANSFORMERS

The differential protection used for power transformers is based on Merz-Price circulating current principle. Such protection provides protection against internal phase-to-phase and phase-to-earth faults and is generally used for transformers of rating exceeding 2 MVA. This protection system as applied to power transformers is fundamentally the same as that for generators but with certain complicating features not encountered in the generator application.

Figure 12.3 shows the Merz-Price protection system as applied to a star/delta power transformer. The overload protection is provided by the use of overload fuses in the pilot wires between the three pairs of CTs. When the overload blows a fuse one of the pair of CTs is disconnected from the relay which receives the current from the remaining CTs. The balance is thus disturbed and the relay operates. Thus a backup protection is provided.

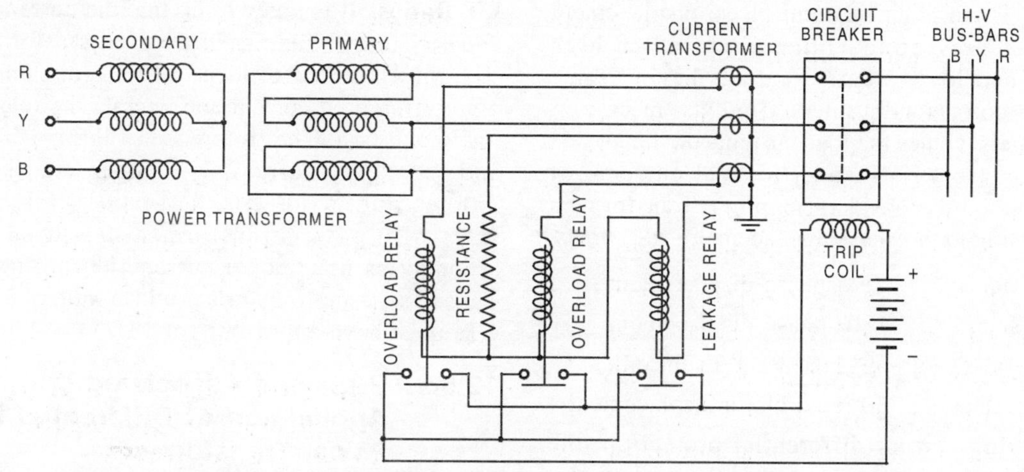

Fig. 12.2. ***Combined Leakage and Overload Protection***

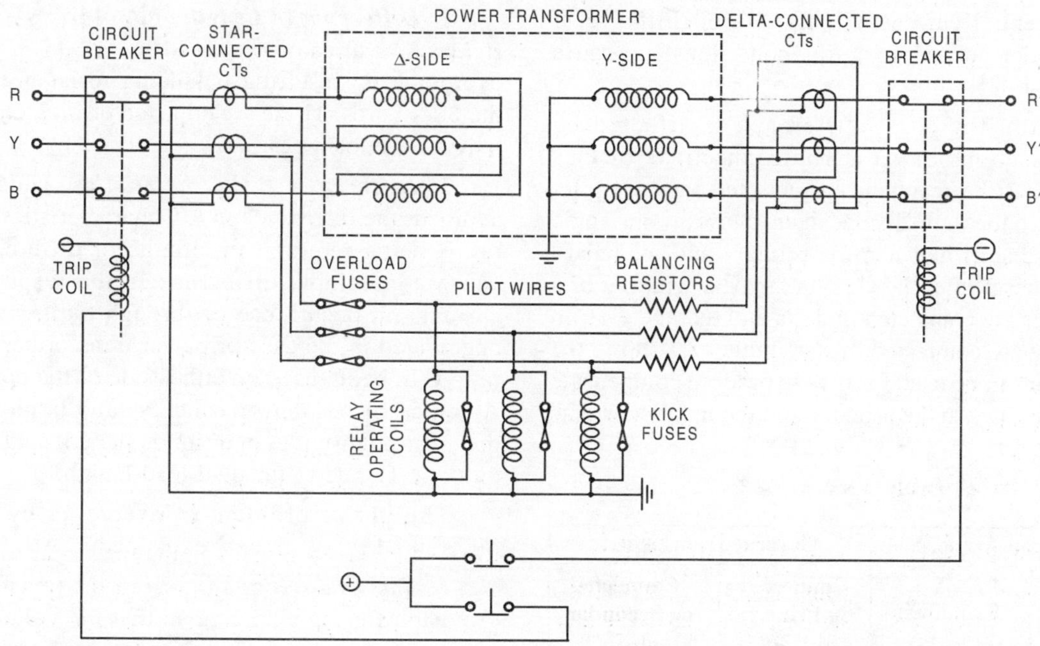

Fig. 12.3. ***Schematic Diagram of Differential Protection as Applied to a Star/Delta Power Transformer***

On energisation of transformer the transient inrush of magnetising current flowing into the transformer may be as large as 10 times full-load current and it decays relatively slowly. Since the large magnetising current flows only in primary windings, it causes difference in CTs output and makes the differential protection of transformer to operate falsely. In order to prevent operation of differential protection due to inrush of magnetising current, the "kick fuses" are provided across the relay coils, as shown in Fig. 12.3. These fuses are of the time-limit type with an inverse characteristic and do not operate in short-time duration of the switching-in surge. Under sustained fault conditions, however, the fuses operate (blow off) and the full current flows through the relay coils and operate the protection system. This scheme affects the relay setting and to avoid this, arrangements are sometimes made to have the "kick fuses" in circuit only during the period of switching-in. This problem can also be overcome by using a relay with an inverse and definite minimum time characteristic instead of an instantaneous type.

The circulating current protection system also provides protection against inter-turn faults. Short circuit between turns on the same phase alters the turn-ratio of the power transformer and so causes imbalance between CT pairs and makes the protection system to operate. However, the change in the turn-ratio of the transformer must be considerable so as to cause flow of sufficient current into the relay to operate. Such short circuits are dealt with more efficiently by Buchholz protection, already discussed in Art. 12.2.

There are two basic requirements which are to be satisfied by differential relay connections. The differential relay should not operate on overload or external faults, and it must operate on severe internal faults.

CT Connections. There is an inherent phase displacement between phasors representing voltage induced in high voltage windings and low voltage windings having same marking letter and corresponding neutral points, in case of star-delta transformers. Hence the load currents on the *hv* side are displaced in phase with respect to load currents of corresponding phase on the *lv* side. The power transformers are grouped, according to the phase displacement, as follows.

Group 1 (Star-star) : Zero phase displacement

Group 2 (Star-star) : 180° phase displacement

Group 3 (Delta-star) : 30° lag phase displacement

Group 4 (Delta-star) : 30° lead phase displacement

In the circulating current differential protection, the phase displacement in line currents on the two sides, introduces phase difference in the secondary currents of CTs on two sides. The result is that even if CTs of proper turn-ratio are used a differential current will flow through the relay operating coil under normal operating conditions and cause the relay operation.

The correction for phase difference is affected by appropriate connections of CTs. The connections of CTs should be such that the resultant currents fed into the pilot wires from either sides are displaced in phase by an angle equal to the phase shift between the primary and secondary currents. As a general rule, the CTs on any star winding of a power transformer are connected in delta and that the CTs on any delta winding are connected in star. Table 12.1 shows the type of connections to be used for CTs in order to compensate for the phase difference in the primary and secondary currents of power transformer.

Table 12.1

S. No.	Power Transformers		Current Transformers	
	Primary	Secondary	Connected on Primary side in	Connected on Secondary side in
1.	Star with Neutral Earthed	Delta	Delta	Star
2.	Delta	Star With Neutral Earthed	Star	Delta
3.	Star	Star With Neutral Earthed	Delta	Delta
4.	Delta	Delta	Star	Star
5.	Star With Neutral Earthed	Delta With Earthing Transformer on Secondary	Delta	Star
6.	Star	Star with a Tertiary Winding	Delta	Delta

CT Ratios. It is noteworthy that the currents in the primary and secondary windings of a power transformer are not equal and, therefore, use of identical CTs (of same turn-ratio) will give differential current and operate the relay even under no load condition. The difference in the magnitude of primary and secondary currents is compensated by using CTs of different turn-ratios. The turn-ratio of CTs used on the two sides of a power transformer should be such that their secondaries have equal currents during normal operation. Thus in star-star transformer with a ratio of 33 kV/132 kV, the CTs must have a turn-ratio of 132/33.

12.5.1. Problems Associated With The Application of Differential Protection to Power Transformers

Simple differential protection system is inadequate because of its following inherent drawbacks.

1. *Difference in Length of Pilot Wires on Either Sides of Relays.* As already explained in Art. 11.3, this difficulty is overcome by inserting balancing resistances in series with the pilot wires. These are adjusted on site to get equipotential points on pilot wires.

2. *Different CT Characteristics.* Unless saturation is avoided, the difference in CT characteristics due to different ratios being required in circuits of different voltages may cause appreciable difference in the respective secondary currents on occurrence of through faults.* This problem is aggravated in the case of power transformers due to unequal ratio CTs being used on either side of the power transformer. A source of ratio error which results in circulating currents under through-fault condition is the unequal burden imposed on the CTs due to unequal lead length.

The above problem is overcome by using a biased differential relay, already explained in Art. 9.19.2.

3. *Tap Changing.* In most of the transformers there are provision for tap changing so that the voltage can be varied as per requirement to keep the system voltage within limits. During tap changing voltage ratio differs from the original one and affects the operation of the differential relay. This problem is overcome by using *biasing coil* (*i.e.*, percentage differential relay protection).

4. *Magnetising Inrush Current.* Under normal operating conditions, the magnetising current is very small. However, when a transformer is energized after it has been taken out of service, the magnetising inrush current can be extremely large (as large as 10 times full-load current) for a short period (5 to 10 cycles, sometimes as large as 4 to 6 seconds). Since magnetising current flows only through primary winding, it causes difference in CTs output and makes the relay to operate, which is not desirable.

There are a number of ways of overcoming the above problem. Firstly, the relay may be provided with a setting larger than the maximum inrush current; secondly, the time setting may be made long enough for the magnetising current

* A through fault is one which is beyond the protected zone of the transformer but is fed through the transformer.

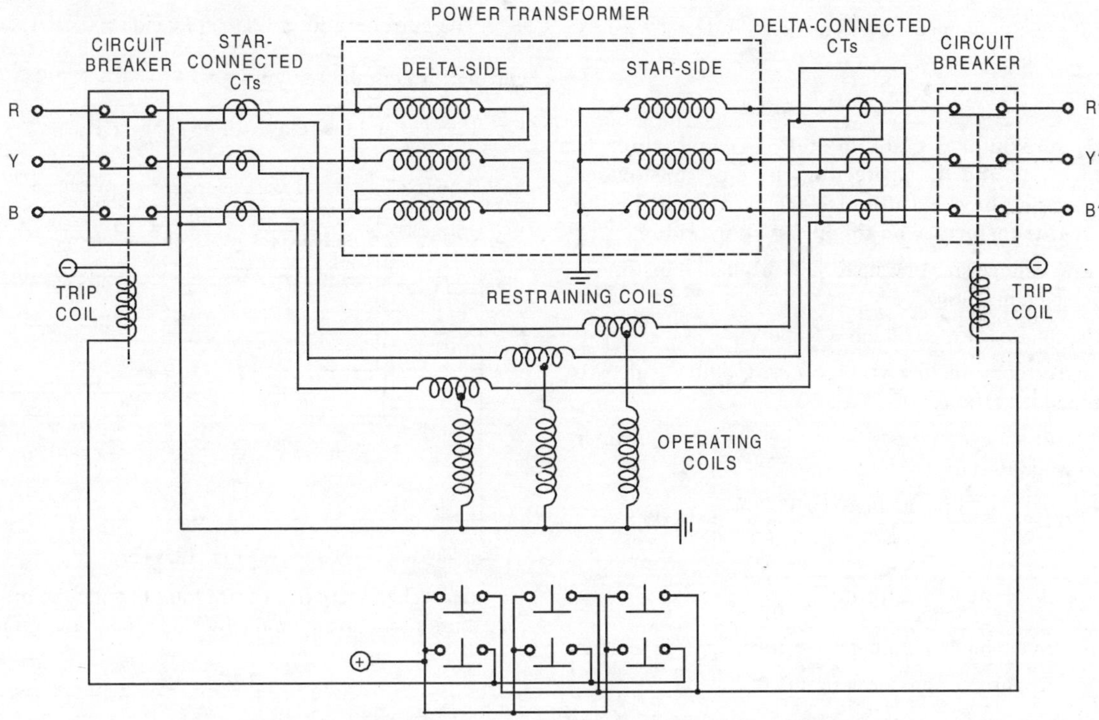

Fig. 12.4. *Biased Differential Protection For Power Transformers*

to subside (fall below the primary operating current) before the relay operates. But for EHV transformers these simple remedies are not incompatible. *The third alternative is the use of second harmonic restraint relay.**

12.6. BIASED DIFFERENTIAL PROTECTION

In order to avoid undesirable operation on heavy external faults due to CT's errors and ratio change as a result of tap changing use of biased or percentage differential relay (already described in Art. 9.19.2) is made, restraining winding being energized by the through current. Figure 12.4 shows the arrangement of percentage differential relaying for power transformers.

The power transformer is star connected on one side and delta connected on the other. The CTs on the star-connected side are delta-connected and those on delta-connected side are star-connected. The neutrals of CT star and power transformer star connections are grounded. The restraining coils are connected across the secondary windings of CTs. The operating coils are connected between the tapping points on the restraining coils and the star point of the CT secondary windings. The operating coils normally carry no current as they are balanced on both sides of the CTs. On the occurrence of internal fault in the power transformer windings, the balance is disturbed and the operating coils of the differential relays carry current corresponding to the difference of the current between the two sides of the power transformers and operate the relays to trip the main circuit breakers on both sides of the power transformer.

Example 12.1. A 3-phase, 66/11 kV star-delta connected transformer is protected by Merz-Price system. The CTs on lt side have a ratio of 420/5A. Find the ratio of the CTs on the ht side. [*Pb. Univ. Switchgear & Protection September 1988; September 1987; October 1986; U.P.S.C. I.E.S. Electrical Engineering-II, 2008*]

Solution : For star-delta power transformers, CTs will be connected in delta on star side of power transformer (*i.e.,* on 66 kV side) and in star on delta side of power transformer (*i.e.,* on 11 kV side), as shown in Fig. 12.5.

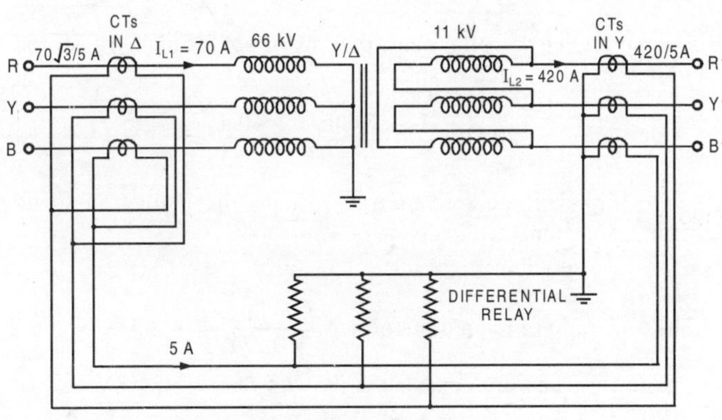

Fig. 12.5

Assume that line current on 11 kV side is 420 A

Then line current on 66 kV side $= \dfrac{\sqrt{3} \times 11{,}000 \times 420}{\sqrt{3} \times 66{,}000} = 70$ A

∴ Line current of star-connected CTs on 11 kV side

= Phase current of star-connected CTs on 11 kV side = 5 A

This current of 5 A will flow through the pilot wires. Obviously, this will be the current which flows through the secondary of CTs on the 66 kV side.

∴ Line current of delta-connected CTs on 66 kV = 5 A

* Refer to Art. 12.7.

Phase current of delta-connected CTs on 66 kV side = $5/\sqrt{3}$ A

Therefore turn-ratio of CTs on 66 kV side

$= 70 : 5/\sqrt{3} = 24.25 : 1$ **Ans.**

Example 12.2. A 3-phase 33,000/6,600 V transformer is connected in star/delta and the protecting current transformer on the low-voltage side have a ratio of 300/5. What will be the ratio of the current transformer on the high-voltage side ?

Solution : On low-voltage side of transformer the main transformer windings are connected in delta

So voltage per phase = Line voltage = 6,600 V

Current transformers on low voltage *i.e.,* 6,600 V side are connected in star and the ratio is 300/5 *i.e.,* 60

On high-voltage side of transformer, the main transformer windings are connected in star

$$\text{So phase voltage} = \frac{\text{Line voltage}}{\sqrt{3}} = \frac{33{,}000}{\sqrt{3}} = 19{,}050 \text{ V}$$

$$\text{The turn-ratio of the main transformer} = \frac{19{,}050}{6{,}600} = 2.887$$

Current transformers on high voltage side are connected in delta

High voltage side current transformer ratio

$$= 60 \times \frac{1}{2.887} = 20.8 : 1 \textbf{ Ans}$$

Example 12.3. A 3-phase, 200 MVA, 11/0.4 kV transformer is connected as delta/star. The protection transformer on the 0.4 kV side has turn ratio of 500/5. What will be CT ratio on the high voltage side? Also obtain the circulating current when the fault of 750 A of the following types occur on the low voltage side:

(*i*) Earth fault within the protection zone (*ii*) Earth fault outside the protection zone (*iii*) phase-to-phase fault within the protection zone (*iv*) phase-to-phase fault outside the protection zone.

Solution : On low-voltage side of transformer the main transformer windings are connected in star, So

$$\text{Phase voltage} = \frac{\text{Line voltage}}{\sqrt{3}} = \frac{0.4}{\sqrt{3}} = 0.231 \text{ kV}$$

On high-voltage side of transformer, the main transformer windings are connected in delta, so

Phase voltage = Line voltage = 11 kV

$$\text{The turn-ratio of main transformer} = \frac{11}{0.231} = 47.63$$

Current transformers on low-voltage *i.e.,* on 0.4 kV side are connected in delta and the turn-ratio = $\frac{500}{5} = 100$

High voltage side CTs turn-ratio

$$= \frac{\text{Turn-ratio of CTs on low-voltage side}}{\text{Turn-ratio of main transformer}}$$

$$= \frac{100}{47.63} = 2.1 \quad \textit{i.e.,} \quad 2.1 : 1 \quad \textbf{Ans.}$$

(*i*) The earth fault is within the protection zone and since the primary is isolated therefore, no zero sequence current will flow in the primary.

The line current on *hv* side (windings connected in delta) corresponding to 750 A

$$= 750 \times \frac{0.4}{11} = 27.27 \text{ A}$$

The current ratio of CTs on *hv* side = 2.1 : 1, so corresponding current through the relay will be $\frac{27.27}{2.1}$ = 12.99 A, as illustrated in Fig. 12.6 (*a*) and the relay will be energized.

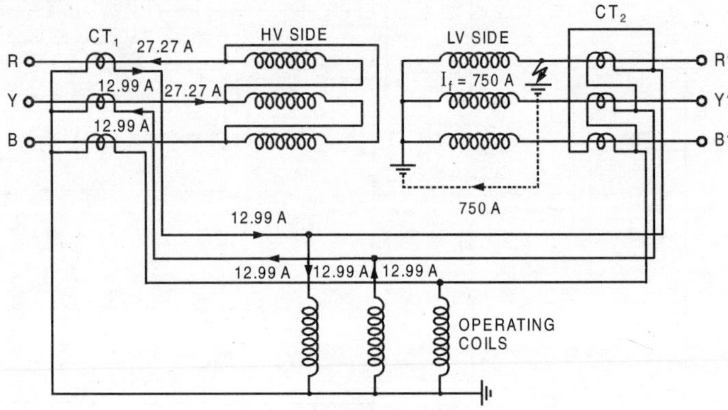

Fig. 12.6 (*a*)

(*ii*) When there is an earth fault outside the protection zone the distribution of current on the hv side remain the same as in case (*i*).

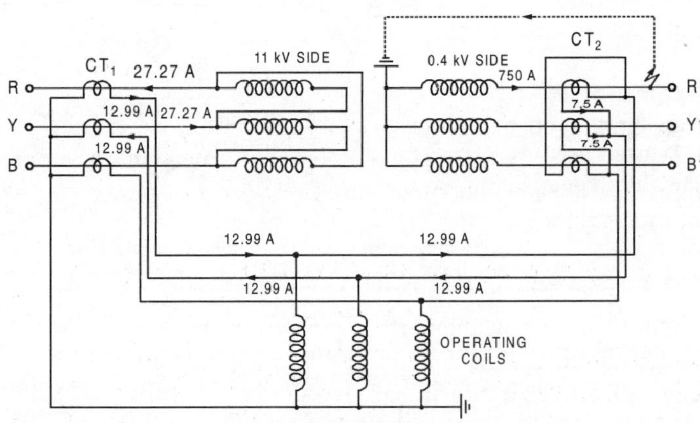

Fig. 12.6 (*b*)

The secondary current in CTs on lv side will 750×5/500 = 7.5 A, the direction of flow of current is shown in Fig. 12.6 (*b*).

The current in pilot wires = $7.5 \times \sqrt{3}$ = 12.99 A, CTs on *lv* side being connected in delta.

Thus the relay will carry no current and thus the circuit breaker for the through fault will not isolate the transformer.

(*iii*) When there is a phase-to-phase fault within protective zone, it causes flow of current in the two phases of the *lv* windings,

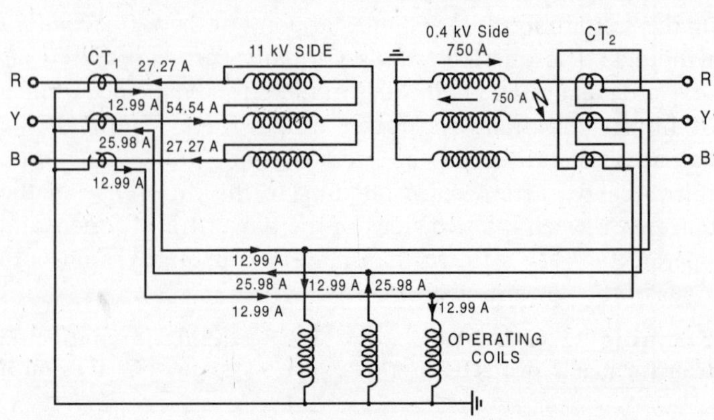

Fig. 12.6 (*c*)

correspondingly there will be flow of current in all the three phases of hv winding which will spill into the relay operating coils, as shown in Fig. 12.6 (*c*). Thus the relay operating coils will be energized, circuit breakers will get tripped off and the transformer will get isolated.

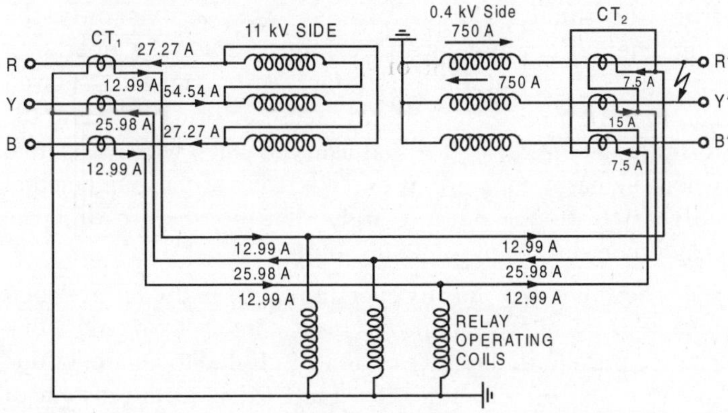

Fig. 12.6 (*d*)

(*iv*) When there is a phase-to-phase fault outside the protection zone, the current distribution becomes as illustrated in Fig. 12.6 (*d*). As obvious from the figure no current will flow through the operating coil of any relay and so the transformer will not be disconnected form the system.

Example 12.4. A 100 MVA Δ/Y-connected, 11/220 kV transformer is to be protected by percentage differential scheme. CTs used are of ratio 6,000/5 and 300/1 respectively. Draw the sketch of complete scheme. Relays are rated up to 1 A.

Solution : Rated current for star (220 kV) side

$$= \frac{\text{MVA rating} \times 10^6}{\sqrt{3}\,\text{V}} = \frac{100 \times 10^6}{220{,}000 \times \sqrt{3}} = 262 \text{ A}$$

Rated current for delta (11 kV) side

= Transformation ratio × Y-side current

$$= \frac{220}{11} \times 262 = 5{,}240 \text{ A}$$

$$\text{Current in } lv \text{ side CTs} = \frac{5{,}240 \times 5}{6{,}000} = 4.367 \text{ A}$$

$$\text{Current in } hv \text{ side CTs} = \frac{262}{300} \times 1 = 0.873 \text{ A}$$

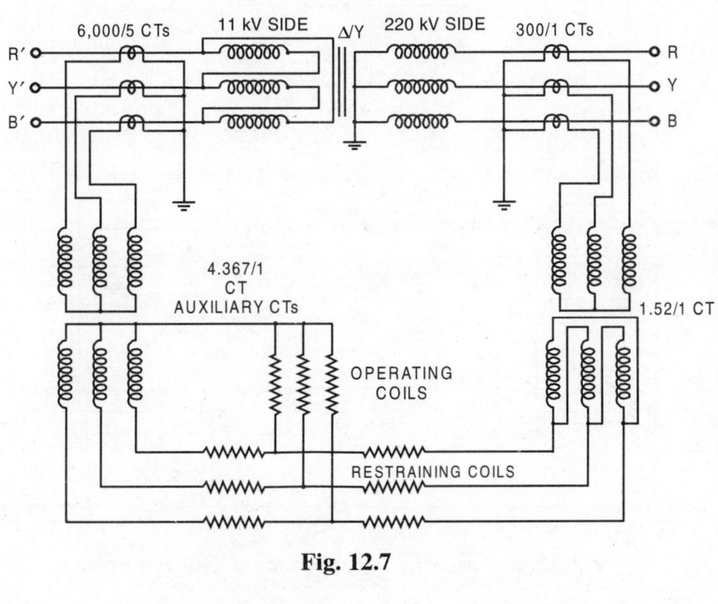

Fig. 12.7

Relays are rated up to 1 A hence all the currents exceeding 1 A will be reduced further by using auxiliary CTs.

For normal operation of transformer current output of CTs should be balanced (in phase as well as in magnitude) so that relay remains inoperative.

Take 1 A current for the rated current in transformer.

Auxiliary CTs used on *lv* side should be of ratio 4.367/1.

On hv side we compensate phase difference because of Y-Δ connection of transformer through using auxiliary CTs in reverse manner *i.e.*, Δ/Y.

If CTs used are of ratio *x*/1 then we will get $\sqrt{3}$ A output for *x* amperes in primary winding because of Δ-connected secondary winding of CTs. Hence to get 1 A output for 0.873 A input to CTs, ratio required is $\sqrt{3} \times 0.873 = 1.52$.

Example 12.5. A 50 MVA, 132/66 kV, Δ/Y, three-phase power transformer is protected by percentage differential relays. If the current transformers (CTs) located on delta and wye sides of the power transformer are 300/5 A and 1,200/5 A respectively, determine (*a*) the output current at full load (*b*) the relay current at full-load (*c*) the minimum relay current setting to permit 25 % overload. [*GATE 1991*]

Solution : Rated current for primary (Δ-connected side)

$$= \frac{\text{MVA rating} \times 10^6}{\sqrt{3} \times \text{V}} = \frac{50 \times 10^6}{\sqrt{3} \times 132{,}000} = 218.7 \text{ A } \textbf{Ans.}$$

(*a*) Rated current for secondary (Y-connected side)

$$= \frac{50 \times 10^6}{\sqrt{3} \times 66{,}000} = 437.4 \text{ A } \textbf{Ans.}$$

$$\text{Current in } hv \text{ side CTs} = 218.7 \times \frac{5}{300} = 3.645 \text{ A}$$

$$\text{Current in } lv \text{ side CTs} = 437.4 \times \frac{5}{1{,}200} = 1.822 \text{ A}$$

(*b*) Relay current at full load $= 3.645 - 1.822 \times \sqrt{3} = 0.489$ A **Ans.**

Assuming relay rating as 1 A

Percentage setting = 50 %

(*c*) **With 25% overload**

Line current in secondaries of CTs connected on *hv* side is 3.645 × 1.25 = 4.56 A and line current in secondaries of CTs connected in delta on *lv* side = $\sqrt{3} \times 1.822 \times 1.25 = 3.94$ A

∴ Relay current = 4.56 – 3.94 = 0.62 A

$$\text{So relay setting} = \frac{0.62}{1} \times 100 = 62\ \% \quad \textbf{Ans.}$$

Example 12.6. A 40 MVA, 3-phase 220/132 kV transformer is star/delta connected. Find the CT ratios on the two sides of the transformer for differential protection of the transformer. Draw the diagram for the same. Assume fault is more than 115% of full-load current and relay setting current is 5 Amp.

[*U.P. Technical Univ. Switchgear and Protection 2004-05*]

Solution : On low-voltage side of transformer the main transformer windings are connected in delta

So voltage per phase = Line voltage = 132 kV

On high-voltage side of transformer, the main transformer windings are connected in star.

$$\text{So voltage per phase} = \frac{\text{Line voltage}}{\sqrt{3}} = \frac{220}{\sqrt{3}} \text{ kV} = 127 \text{ kV}$$

$$\text{Turn-ratio of main transformer} = \frac{127}{132} = 0.962$$

Let the current transformation ratio of CTs on low-voltage side be k, then CT ratio on high voltage side will be $\frac{k}{0.962}$

Rated current for primary (star-connected side)

$$= \frac{40 \times 10^6}{\sqrt{3} \times 220 \times 1{,}000} = 104.97 \text{ A}$$

Rated current for secondary (delta-connected side)

$$= \frac{40 \times 10^6}{\sqrt{3} \times 132 \times 10^3} = 174.95 \text{ A}$$

$$\text{Current in hv side CTs} = \frac{104.97}{k/0.962} \text{ A} = \frac{109.12}{k} \text{ A}$$

$$\text{Current in lv side CTs} = \frac{174.95}{k} \text{ A}$$

$$\text{Relay current at full load} = \frac{174.95 - 109.12}{k} = \frac{65.83}{k} \text{ A}$$

$$\text{Relay current at 115\% of full load current} = \frac{65.83}{k} \times \frac{115}{100} = \frac{75.7}{k}$$

$$\text{So } \frac{75.7}{k} = 5$$

$$\text{or } k = \frac{75.7}{5} \text{ A}$$

i.e., Transformation ratio of CTs on lv side is $\frac{75.7}{5}$ A and that of CTs on hv side is $\frac{75.7/5}{0.962}$ A *i.e.*, $\frac{78.7}{5}$ A **Ans.**

12.7. HARMONIC RESTRAINT RELAY

The operation of the relays because of magnetising inrush current can be avoided by using kick fuses across the relay coils (Fig. 12.3) or using relays with inverse and definite minimum time (IDMT) characteristics, as mentioned in Art. 12.5.1 (4). However, for EHV transformers, the relay current and time ratings necessary to ensure stability on the magnetising inrush current caused by switching-in the transformer are not adequate for providing high speed protection.

A high speed biased differential relay incorporating a harmonic restraint feature is immune to the magnetising inrush current. The magnetising inrush currents have a high component of even and odd harmonics (about 63 % of second harmonics and 26.8 % of third harmonics) while harmonic component of short-circuit currents is negligible. The use of these facts is made for restraining the relay from operation during initial current inrush. The *harmonic restraint differential relay* is sensitive to fault currents but is immune to the magnetising currents. The operating coil of the relay carries only the fundamental component of current only while the restraining coil carries the sum of the fundamental and harmonic components.

Basic circuit of an harmonic restraint differential relay is illustrated in Fig. 12.8. The restraining coil is energized by a direct current proportional to bias winding current as well as the direct current due to harmonics. Harmonic restraint is had from the tuned circuit ($X_C - X_L$) that allows only the fundamental component of current to enter the operating circuit. The dc and higher harmonics (mostly second harmonics) are diverted into the rectifier bridge feeding the restraining coil. The relay is adjusted so that it will not operate when the harmonic current exceeds 15% of the fundamental current. Both the dc and higher harmonics are of large magnitude during magnetising inrush.

The relay may fail to operate due to harmonic restraint feature if an internal fault has considerable harmonics, that may be present in the fault current itself due to an arc, or due to saturation of CT. Also, if a fault exists at the instant of energization of transformer harmonics present in the magnetising current may prevent the operation of the relay. This problem can be overcome by providing instantaneous overcurrent relay in the differential circuit which is set above the maximum inrush current but will operate in less than one cycle on internal faults. Thus fast tripping is ensured for all internal faults.

The other method used is *harmonic blocking*. In this method the harmonic component of magnetising inrush current is used for blocking a separate relay, called the *blocking relay*, whose contacts are in series with the contacts of the differential relay. The blocking relay contains a 100 Hz blocking filter in operating coil and 50 Hz blocking filter in the restraining coil. During inrush currents the second harmonic component is predominant and the blocking relay is blocked. The blocking relay contacts remain open. During short circuits, fundamental component is predominant, so blocking relay operates and relay contact circuit is closed.

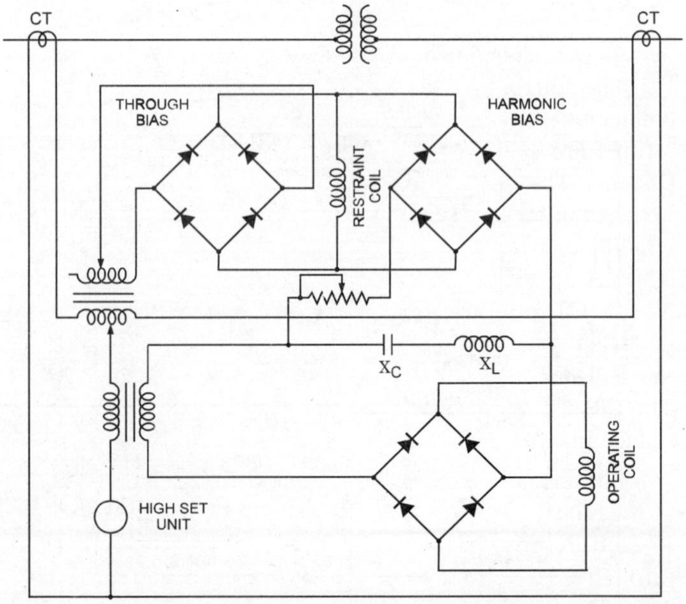

Fig. 12.8. ***Basic Circuit of Harmonic Restraint Relay***

12.8. SELF BALANCE PROTECTION SYSTEM

The self balance protection system discussed earlier (Art. 11.3.4) for the protection of alternators can also be employed to power transformers without any modification except that the same type of equipment has to be employed for both primary and secondary sides. The protective transformers (CTs) can be located in the oil of the transformer tank. This system of protection of power transformers is not much used because it cannot provide protection to transformer terminals and the connected cables up to switchgear.

12.9. DIFFERENTIAL MAGNETIC BALANCE PROTECTION SYSTEM

This system is necessarily a combination of circulating current and self balance protection systems. The main advantages of this protection system are increased stability and sensitivity and its application to power transformers provided with load tap changers. Figure 12.9 illustrates a scheme representing a power transformer having primary (low-voltage) side connected in delta and secondary (high-voltage) side connected in star.

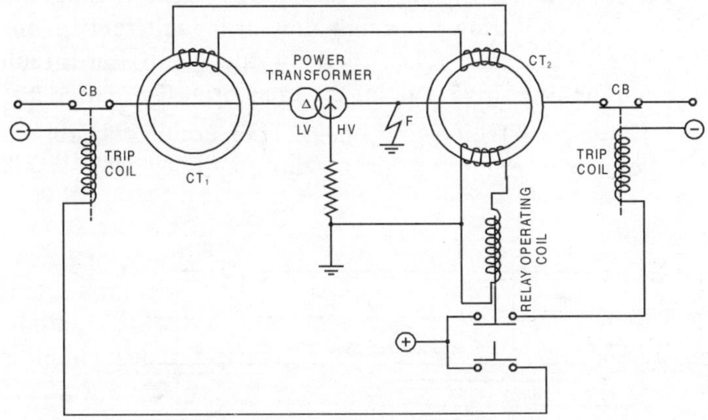

Fig. 12.9. ***Single Line Diagram of Differential Magnetic Balance Protection Scheme***

The current transformers used are bus type and are connected for circulating current protection. Their current ratings are so adjusted that they provide equal secondary currents. The relay operating coil instead of being connected to the equipotential point of the pilot wires is fed from another winding on hv side CT (CT_2). Since the core flux of *lv* side CT (CT_1) is more because of its more ampere-turns, the turns of hv side CT are increased so as to make core flux of CT_2 zero under all load conditions resulting in no current flowing in the relay operating coil under normal operating conditions. The CT_2 is made with high permeability alloy core so as to reduce magnetising current and provide an accurate balance.

For any through fault CT_2 continues to have no flux and so the relay operating coil remains inoperative. On occurrence of fault in protection zone, say at F, excessive current flows through CT_1 causing flow of current in the relay operating coil. Thus relay is energized and the circuit breaker gets tripped.

12.10. SELF STABILISING MAGNETIC BALANCE PROTECTION SYSTEM

For the protection of power transformer having tappings it is necessary that the protective CT connected on *hv* side (*i.e.*, CT_2) must also be capable of changing its current ratio whenever power transformer tappings are changed *i.e.*, CT_2 windings need some modification. This is explained below.

In Fig. 12.10 (*a*) plain magnetic balance protection system discussed in Art. 12.9 is illustrated, the relay connections are not shown for sake of simplicity. CTs used are bus-bar type. In Fig. 12.10 (*b*) self stabilizing circuit for magnetic balance protection system is shown. In this circuit the magnetic core of CT_2 is divided into two halves P_1 and P_2 and the secondary winding is so wound that the flux developed by the two halves P_1 and P_2 is equal and opposing each other. Thus in normal operating conditions no emf is induced in the secondary winding and the relay operating coil remains inoperative.

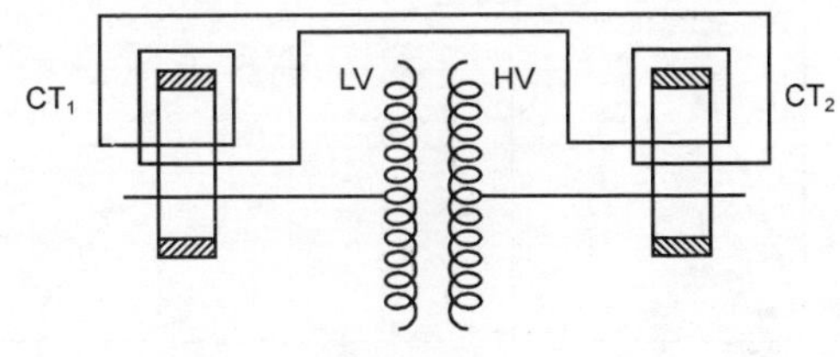

(a) Magnetic Balance Relay Without Stabilization Circuit

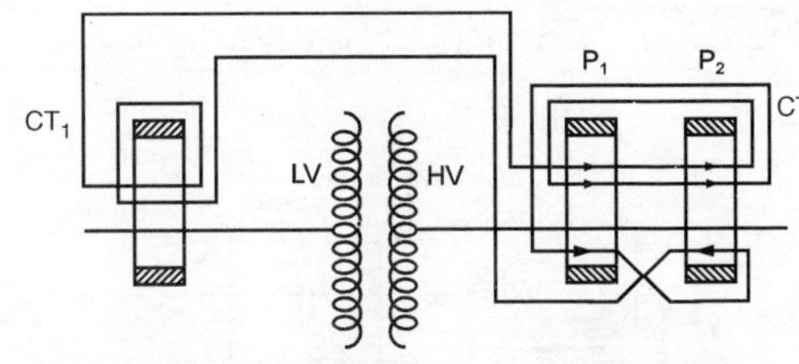

(b) Self Stabilizing Circuit For Magnetic Balance Protection System

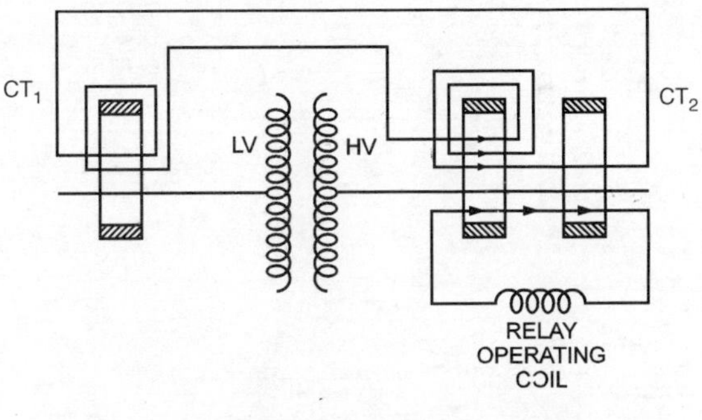

(c) Modified Circuit For Self Stabilizing Magnetic Balance Protection System

Fig. 12.10

When the transformer is operating under normal operating conditions and carrying full-load currents, the flux developed by the two halves is equal and relay winding coil is un-energized. Now when the tappings of the main transformer are changed, mmfs of the two halves are changed causing the flux developed by them to be different. So an emf, proportional to the difference of the two fluxes, will be induced in the relay coil, as shown in Fig. 12.11. If under this tap changed condition the load on the transformer is increased, mmf of the two halves will increase but the difference of fluxes developed will decrease. Thus with the increase in load on power transformer, the difference in fluxes developed by the two halves of core of CT_2 decreases, as shown in Fig. 12.11. Now if the relay is so designed that its minimum operating voltage is much more than the induced voltage under any desirable load condition but with no fault, as illustrated by OD in Fig. 12.11, then stability is ensured. In practice OD is made twice the minimum ordinate.

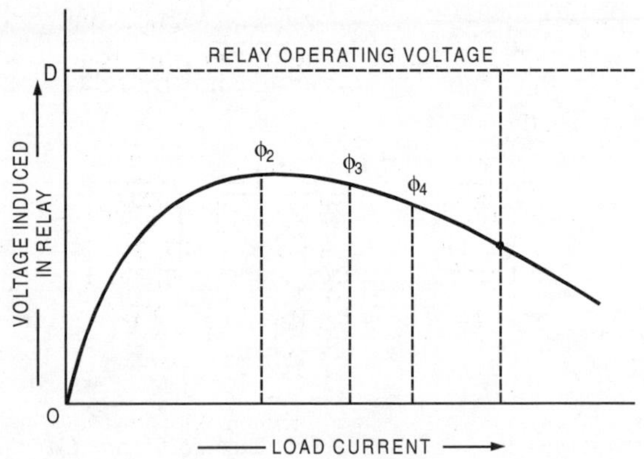

Fig. 12.11. ***Characteristic of Voltage Induced in the Relay***

12.11. RESTRICTED EARTH FAULT PROTECTION

Earth fault relays connected in residual circuit of line CTs provide protection against earth faults on the delta or unearthed star-connected windings of power transformers. The connections of restricted earth fault protection for star-connected and delta-connected windings are shown in Fig. 12.12. A CT is fitted in each connection to the protected and the secondaries of CTs are connected in parallel to a relay. Ideally, the output of the CTs is proportional to the sum of zero sequence currents in the line and the neutral earth connection if the latter is within the protected zone. For external faults zero sequence currents are either absent or sum to zero in the line and neutral earth connection. For internal faults, the sum of zero sequence currents is equal twice the total fault current.

In Fig. 12.13 the star-connected side is protected by *restricted* earth fault protection.

When there is an earth fault outside the protective zone, say at F_1, it causes the currents I_2 and I_1 in CT secondaries as illustrated in Fig. 12.13. So the resultant current in earth fault relay is negligible. For an earth fault within the protected zone, say at F_2, only current I_2 exists, I_1 being negligible. Thus current I_2 flows through the earth fault relay. Thus restricted earth fault relay does not operate for earth fault beyond the protective zone of the transformer.

For an earth fault near the neutral point of the transformer the voltage available for driving earth fault current is small.

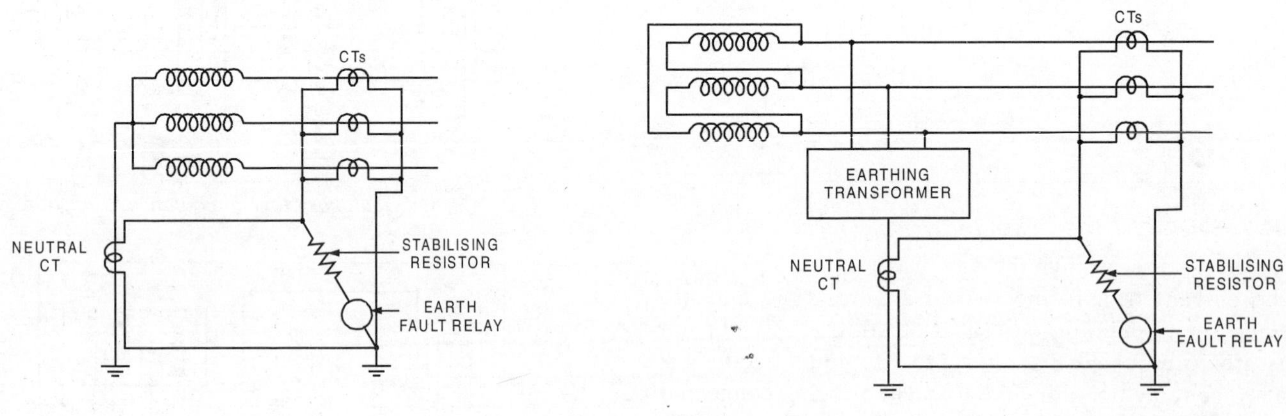

(a) Neutral Earthed Within the Protected Zone

(b) Neutral Not Earthed Within The Protected Zone

Fig 12.12. ***Restricted Earth Fault Protection***

For the relay to sense such fault, it has to be too sensitive and would, therefore, operate for spurious signals, external faults and switching surges. Hence the relay is set as per practice, so as to operate for earth fault current of the order of 15% of rated winding current. Such setting protects restricted portion of the winding, hence the name *restricted earth fault protection*.

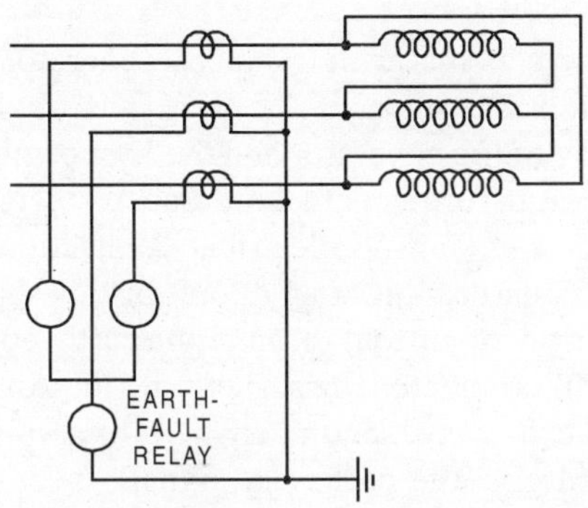

Earth Fault Protection by Residually Connected Earth Fault Relay

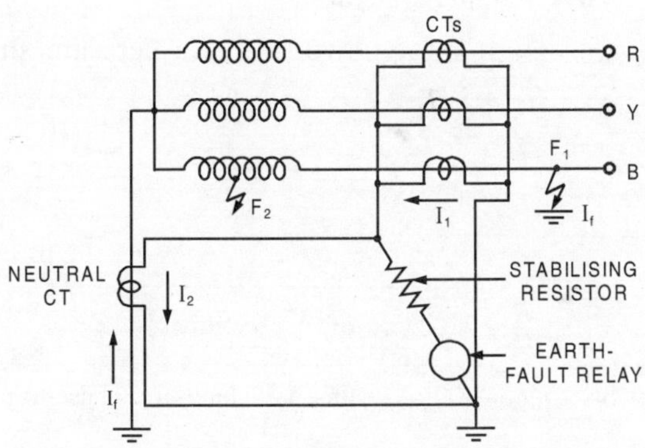

Restricted Earth Fault Protection of Star-Connected Neutral Grounded Side

Fig. 12.13

Stabilizing resistor is connected in series with the relay to avoid magnetising inrush current and also saturation of CT core.

12.12. FRAME LEAKAGE PROTECTION

The transformer is mounted insulated from the ground, as illustrated in Fig. 12.14. The transformer tank is connected to

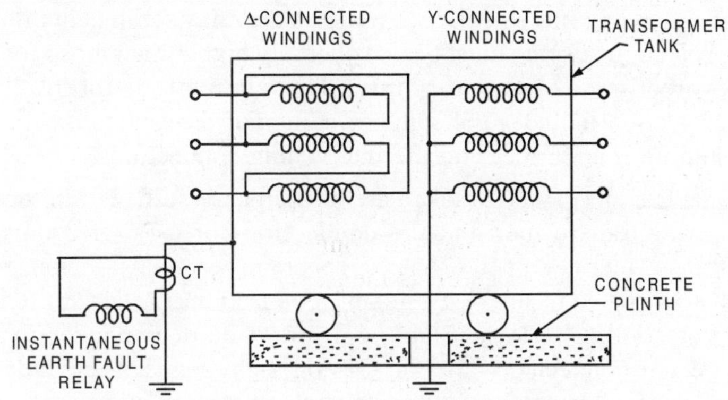

Fig. 12.14. ***Frame Leakage Protection***

earth through a CT to which an instantaneous earth fault relay is connected. In the case of an earth fault in the transformer (breakdown of insulation in any winding of the transformer), there is a flow of current to the earth over this connection causing the relay to operate. Such an arrangement is usually provided where banked transformers are provided with a single overall differential protection and it is difficult to find as to which transformer is faulty.

12.13. GENERATOR-TRANSFORMER UNIT PROTECTION

In hv transmission the bus-bars are operated at higher voltages than that of generation; it is common practice to connect the

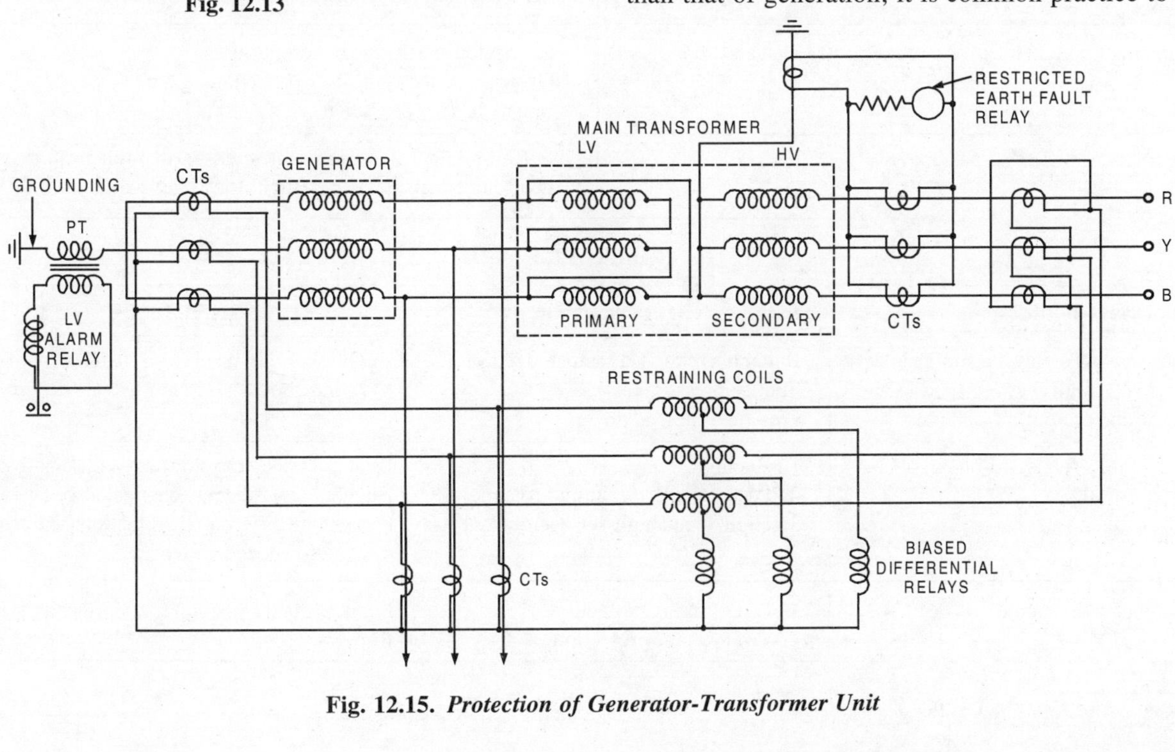

Fig. 12.15. ***Protection of Generator-Transformer Unit***

generators directly to step-up transformers. In this protection scheme no circuit breaker is interposed in between the generator and transformer. The main advantage of such protection is that it simplifies the protection, mainly the differential protection for both generator and transformer can be combined together by employing CTs on the neutral side of the generator and on the *hv* side of the power transformer, as illustrated in Fig. 12.15. Because of the occurrence of magnetising inrush current transients the relay settings in this protection scheme must be considerably higher than those for protecting a generator only. The zone of differential protection includes the stator windings of the generator, the step-up transformer and the intervening connections.

It is necessary to take care of the phase shift within the power transformer and the connections of CTs. If a unit transformer is tapped off at the generator terminals, this also has to be taken care of by suitable connections of the CTs for protection. CTs located on the neutral side of the generator are star-connected while the CTs on the secondary (*hv*) side of the main transformer are delta-connected so as to cancel the 30° displacement between line currents introduced by delta-connected primary of the main transformer. The unbalance caused between CT pairs due to load of unit transformer is avoided by providing another set of star-connected CTs in the primary leads of the latter. In healthy condition, the sum of secondary currents of these CTs and the secondary currents of the generator star-point CTs is equal to the currents in the pilot wires from the secondaries of the delta-connected CTs on the secondary side of the main transformer. On occurrence of fault differential relays are energized. The hv winding of the main transformer is protected against earth faults by the restricted earth fault protection scheme.

From the schematic diagram of generator-transformer unit protection shown in Fig. 12.15, it is obvious that the stator winding of the generator and the *lv* or primary windings of the main transformer and unit transformer* comprise a separate circuit having no electrical connection with the *hv* circuits. So an earth fault at any point of this separate circuit will cause a flow of current through the earth connection and through a PT connected in series with it. An alarm relay connected across the secondary winding of PT will get energized and give the necessary signal.

12.14. TYPES OF PROTECTION FOR POWER TRANSFORMERS

Types of protection for power transformers are shown in Tables 12.2 and 12.3.

Table 12.2

Type of Transformers	Type of Protection
Small distribution transformers	Only hv fuses for protection against earth faults and phase-to-phase faults. Overload protection generally not provided.
For more important transformers of about 500 kVA	Overcurrent relays. Instantaneous earth fault relays.
Transformers, in important locations, of rating 500 kVA and above	Restricted earth fault protection. Overcurrent and earth fault protection. Buchholz relay protection.
Transformers of rating of 2 MVA and above	Differential protection. Restricted earth fault protection. Overcurrent protection. Buchhloz relay protection

Table 12.3

Abnormal Condition	Type of Protection	Remarks
Incipient faults below oil level resulting in decomposition of oil (faults between phases and between phase and ground).	Buchholz relay (Gas actuated relay). Sudden pressure relay. Pressure relief valve.	Buchholz relay used for transformers of rating 500 kVA and above.
Large internal faults (phase-to-phase, phase-to-ground) below oil level. Faults in tap changer.	1. Buchholz relay. 2. Percentage differential protection. 3. High-speed high-set overcurrent relay.	Buchholz relay too slow and less sensitive. Buchholz relay for tap changer also. Percentage differential protection for transformers of rating 2 MVA and above.
Saturation of magnetic circuit.	1. Over fluxing protection. 2. Overvoltage protection.	For important generator transformer and feeder transformers.

* Unit transformer is connected to the generator terminals for supplying the power station auxiliaries.

Abnormal Condition	Type of Protection	Remarks
Earth faults	1. Earth fault relay. 2. Differential protection.	Instantaneous restricted earth fault relay. Time lag earth fault relay. For transformers of rating 2 MVA and above.
Overloads	1. Thermal overload relays. 2. Temperature relays sound alarm.	Generally temperature indicators are provided on the transformers.
Through faults	1. HRC fuses. 2. Graded time lag overcurrent relay.	Small distribution transformers of rating up to 500 kVA. Protection of distribution transformers.
High voltage surges due to lightning, switching etc.	1. Surge arresters 2. R-C surge suppressors	In addition to arresters for incoming lines.

Exercises

1. Discuss the different transformer faults. What are the various protection schemes available for transformers ?

2. Describe the construction, principle of operation and application of Buchholz relay. Why is this form of protection an ideal protection ?

[U.P. Technical Univ. Switchgear and Protection 2005-06]

3. What are the various protections usually recommended for power transformers ? In what respect 'core-balance leakage protection' is inferior to 'combined leakage and overload protection' for power transformers.?

4. Explain the protection of transformer by using Merz-Price system of protection with the help of neat diagram. How many faults develop in a power transformer ?

5. Describe 3-phase transformer differential protection. Show that Y-Δ transformer differential protection CTs are connected in Δ and Y respectively.

[U.P. Technical Univ. Switchgear and Protection 2007-08]

6. Draw a neat circuit diagram of Merz-Price per cent differential protection scheme for a 3-phase Δ-Y connected transformer.

[U.P.S.C. I.E.S. Electrical Engineering-II, 2012]

7. Explain with a neat circuit diagram the differential protection scheme used to protect Y-Δ transformers. Describe with a sketch the operation of Buchholz relay.

8. Describe the protection scheme for internal faults in a three-phase delta/star connected power transformer.

Draw a neat sketch and explain clearly why the CTs are to be connected in a particular fashion only.

[G.B. Technical Univ. Switchgear and Protection 2011-12]

9. For protecting a 132/33 kV, 50 MVA power transformer from internal faults, what are the different protection schemes normally used ? Discuss one of them, in brief, with sketch.

10. Discuss the special factors that are to be considered while designing the protection scheme for a large Y-Δ power transformer.

11. Describe the principle of differential system of protection applied to a power transformer. What are the shortcomings of this scheme and how are they overcome ?

12. Explain the protective scheme for the transformer that takes care of magnetizing inrush current without affecting the sensitivity.

13. Discuss the special problems of biased differential protection of a transformer. How are they overcome ? Develop a transformer differential protection scheme using the second harmonic restraint feature.

14. Why is a harmonic restrained differential relay required to be used for protecting a large size transformer? Describe the construction and working of such a relay.

15. Explain with a neat diagram the method of harmonic current restraint for protection of transformer.

[G.B. Technical Univ. Switchgear and Protection 2010-11]

16. Draw the schematic diagram of harmonic restraint differential protection scheme used for a Yd11 transformer. Explain the scheme.

[W.B. Univ. of Technology Power System-II, 2010]

17. What is the principle of harmonic restraint relay ? Explain its application.

18. Draw the connection diagram of a differential relay for the protection a star-delta transformer. How does the bias winding of a differential relay restricts malfunctioning of the relay against (*i*) CT mismatch (*ii*) on-load tap changing and (*iii*) magnetising current ? What is magnetising inrush current ? What is the principle used to make a differential relay insensitive to magnetising inrush current ?

19. Write short notes on the following :

(*i*) Buchholz relay. **[W.B. Univ. of Technology 2010]**

(*ii*) Core-balance leakage protection of power transformers.

(*iii*) Combined leakage and overload protection.

(*iv*) Biased differential protection.

(*v*) Harmonic restraint relay.

(*vi*) Restricted earth fault protection for power transformers.

20. Explain the following in brief.

(*i*) Why automatic protection of transformers against possible faults is essential and of utmost importance specially when chances of occurrence of faults in them are very rare ?

(*ii*) How does circulating current protection system also provide the protection against inter-turn faults in power transformers ?

(*iii*) How does the bias winding of a differential relay restrict the malfunctioning of the relay against (*i*) mismatching of CTs (*ii*) on load tap changing and (*iii*) magnetising inrush current ?

(*iv*) On what principle does the harmonic relay function ?

(*v*) Why is a harmonic restrained differential relay required to be used for protecting a large size transformer ?

(*vi*) Why is self balance protection system not much used for protection of power transformers ?

Short Answer Type Questions With Answers

Q. 1. What are the faults that are likely to occur in a power transformer ?

Ans. The faults likely to occur in a power transformer are open-circuit faults (an open circuit in one phase of a 3-phase transformer), earth faults, phase-to-phase faults, inter-turn faults and overheating.

Q. 2. What are incipient faults ?

Ans. Incipient faults are initially minor faults causing gradual damage. These faults include loose connections in conducting path, sparking, small arcing etc., and grow into major faults.

Q. 3. What is through fault ?

Ans. A through fault in one which is beyond the protected zone of the transformer but fed through it.

Q. 4. What are the various protective schemes available for power transformers ?

Ans. Different protective schemes available for the protection of power transformers are Buchholz protection, percentage differential protection, earth fault protection, overcurrent protection, combined leakage and overload protection, restricted earth fault protection, overfluxing protection etc.

Q. 5. On what factors does the choice of a protective gear for transformer depend ?

Ans. The choice of a protective gear for a transformer depends upon several factors such as (*i*) type of transformer (distribution transformer or power transformer), (*ii*) size of the transformer, (*iii*) type of cooling used (*iv*) systems where used *i.e.,* its electrical location in the network and (*v*) the importance of the service for which it is required.

Q. 6. In what respect core-balance leakage protection is inferior to combined leakage and overload protection ?

Ans. The core-balance leakage protection suffers from the drawback that it does not provide protection against phase-to-phase faults.

Q. 7. Does circulating current protection system provide protection against inter-turn fault in case of power transformers ?

Ans. Yes.

Q. 8. Name the two basic requirements that are to be fulfilled by differential relay.

Ans. The two basic requirements to be satisfied by differential relay on the protection of power transformers are that the differential relay should not operate on overload or external faults and it must operate on severe internal faults.

Q. 9. If T is the turn-ratio of power transformer, what will be the turn ratio of CTs on the *lv* side ?

Ans. If T is the turn-ratio of power transformer, then turn-ratio of CTs on the *lv* side will be T times that of CTs on the *hv* side ?

Q. 10. What are the shortcomings of differential protection scheme as applied to power transformer ?

Ans. Simple differential protection, as applied to a power transformer, has the following drawbacks.

(*i*) Unmatched characteristic of CTs (*ii*) Change in ratio due to tap changing and (*iii*) magnetising inrush current.

Q. 11. What is magnetising inrush current ?

Ans. When the transformer is energized initially there is no induced emf, the condition is similar to switching of an inductive circuit. The resistance being low, a transient rush of current takes place. This transient rush of current is known as magnetising inrush current and it may be equal to 6 to 8 times the rated current.

Q. 12. What is the reason behind providing a second harmonic restraint relay ?

Ans. Second harmonic restraint relay is provided to avoid maloperation of the relays due to magnetising inrush current.

Q. 13. What are the advantages of differential magnetic balance protection system, as applied to power transformers ?

Ans. The main advantages of differential magnetic balance protection system in its application to power transformers are increased stability and sensitivity and its application to transformers provided with load tap changers.

Q. 14. How is frame leakage protection provided in power transformers ?

Ans. The transformer tank is connected to earth through CT to which an instantaneous earth fault relay is connected.

Problems

1. A-3 phase transformer of 220/11,000 line volts is connected in star-delta. The protective transformers on 220 V side have a current ratio of 600/5. What should be CT ratio 11,000 V side? Show how shall they be connected. **[Ans** $12\sqrt{3}$ **: 5]**

2. A 3-phase transformer rated for 33/6.6 kV is star/delta connected and the protection current transformers on the low voltage side have a ratio of 400/5 A. Determine the ratios of CTs on the ht side.

[Pb. Univ. Switchgear & Protection May 1985]
[Ans. 27.7:1]

3. A 3-phase transformer having a line voltage ratio of 400/33,000 V is star-delta connected. The CTs on the 400 V side have a ratio of 800/5A. What must be the ratio of CTs on 33,000 V side ?

[Agra Univ. Generation & Switchgear Practice 1981]
[Ans. 5.6/5 A]

4. A three-phase delta-wye connected 30 MVA, 33/11 kV transformer is protected by a simple differential relaying scheme. The CT ratio on the primary side is 500 : 5 and that on the secondary side is 2,000 : 5. Sketch the CT connection diagram for the relaying scheme. Also calculate the relay current setting for fault drawing up to 200% of rated current.

[GATE 1995] [Ans. 33 %]

CHAPTER 13

Bus-Bar Protection

INSIDE THIS CHAPTER

13.1. INTRODUCTION

The bus-bars in electric power stations and substations are the one of the most vital elements. Although with the advent of modern design methods the probability of occurrence of faults has been reduced to negligible, but if a fault occurs on a bus-bar, considerable damage and distruption of supply will occur unless some form of quick-acting automatic protection is provided for isolation of the faulty bus-bar. The bus-bar area or zone, for the purpose of protection, includes not only the bus-bars themselves but also the isolating switches, circuit breakers, instrument transformers and bus-sectionalizing reactors, etc.

Bus-bar protection needs special attention because of the following reasons.

1. Fault level at bus-bars is very high.
2. The fault on the bus-bar would result in widespread supply interruption.
3. The system stability is adversely affected by fault in bus zone.

In the event of fault on any section of the bus-bar, all the circuit equipments connected to that section must be tripped out to give complete isolation in shortest possible time *e.g.,* (60 ms), so that damage is avoided to the installation due to heating of conductors. Internal bus faults are less frequent than line faults. A bus fault tends to be appreciably more severe, both with respect to the safety to personnel, system stability and the damage. A major system shutdown may be caused in the absence of adequate bus protection.

It is of utmost importance therefore the bus-zone protection should be quick acting, stable and most reliable. The desirable features of bus protection are given below:

1. High speed operation (less than 3 cycles).
2. Discrimination between fault in its protected section and fault elsewhere.
3. Stability for external faults.
4. Freedom from unwanted operation.
5. No operation due to CT saturation or power swings.
6. Separate control of trip circuit of each circuit breaker.
7. 'Main' and 'check' protection to ensure the isolation only when desirable.
8. Non-autoreclosure, no single pole tripping of circuit breakers for bus-fault.
9. Interlock overcurrent protection for tripping generator unit in the event of operation of bus-zone protection.

Some experts are of the opinion that the local bus protection should not be provided and the bus faults cleared by the backup relays at the neighbouring stations, as the provision of the local bus protection would certainly increase the risk of inadvertent tripping. In case local bus protection is provided care is taken by providing two independent protective circuits, both of which must be satisfied before tripping can occur.

Before discussing the nature and causes of faults occurrence in bus zones and different protection schemes we will like to discuss different bus-bar arrangements.

13.2. BUS-BAR ARRANGEMENTS

Bus-bar (or bus in short) term is used for a main bar or conductor carrying an electric current to which many connections may be made. The term bus is derived from the word *omnibus,* meaning collector of things. Thus electrical bus-bar is the collector of electrical energy at one location. Bus-bars are merely convenient means of connecting switches and other equipment into various arrangements. The usual arrangement of connections in most of the substations permits working on almost any piece of equipment without interruption to incoming or outgoing feeders.

The bus-bars used in substations are usually bare rectangular cross section bars but they can be of other shapes also, as round tubes, round solid bars, or square tubes. The bus-bars are usually of aluminium (less frequently of copper). The use of aluminium as an electric conductor is made due to its numerous advantages over copper such as higher conductivity on weight basis, lower cost for equal current carrying capacity, excellent corrosion resistance and ease of formability. For proper reliable electrical connections aluminium buses are coated with silver. The most common sizes of bus-bars are 40 × 4 mm (160 mm^2); 40 × 5 mm (200 mm^2); 50 × 5 mm (250 mm^2); 50 × 6 mm (300 mm^2); 60 × 8 mm (480 mm^2); 80 × 8 (640 mm^2) and 100 × 10 mm (1,000 mm^2) according to the maximum current to be carried. The bus-bars are of 5 or 6 metres in length.

There are several types of bus-bar arrangements. The choice of a particular arrangement depends upon different factors such as system voltage, position of substation in the system, reliability of supply, flexibility and cost. The other technical considerations governing the selection of any one particular arrangement are:

1. Simplicity which is a key note of a dependable system.
2. Maintenance possibility without interruption of supply or danger to the operating personnel.
3. Provision of extension with the load growth.
4. The installation should be as economical as possible keeping in view the needs and continuity of supply.
5. Availability of alternative arrangements in the event of an outage of any of the apparatus.
6. Importance of load and local conditions. Freedom from total shutdown and its period desired.

For a small and medium sized station where shutdown can be permitted at times, simple, single bus-bar arrangement can be favoured. But in case of major plants an elaborate bus-bar arrangement is essential in order that the duplicate supply is always available in case of fault, which might otherwise cause interruption of supply to a large area. Under these circumstances the heavy initial cost is justified. Economy is most important. The extra high-voltage equipment such as circuit breakers, isolators are usually costly. Hence unnecessary equipment should not be provided. Additional equipment to take care of remote contingencies should not be provided as it, in addition to increasing of installation cost will prove a source of weakness.

Different bus-bar arrangements will be discussed here. All the diagrams refer to 3-phase arrangement but are shown in single phase for simplicity.

1. *Single Bus-bar Arrangement.* This is the simplest arrangement consisting of a single set of bus-bars for the full length of the switchboard and to this set of bus-bars are connected all the generators, transformers and feeders, as illustrated by single line diagram in Fig. 13.1. Each generator and feeder is controlled by a circuit breaker. The isolators permit isolation of generators, feeders and circuit breakers from the bus-bars for maintenance. The chief advantages of such a bus-bar arrangement are low initial cost, less maintenance and simple operation.

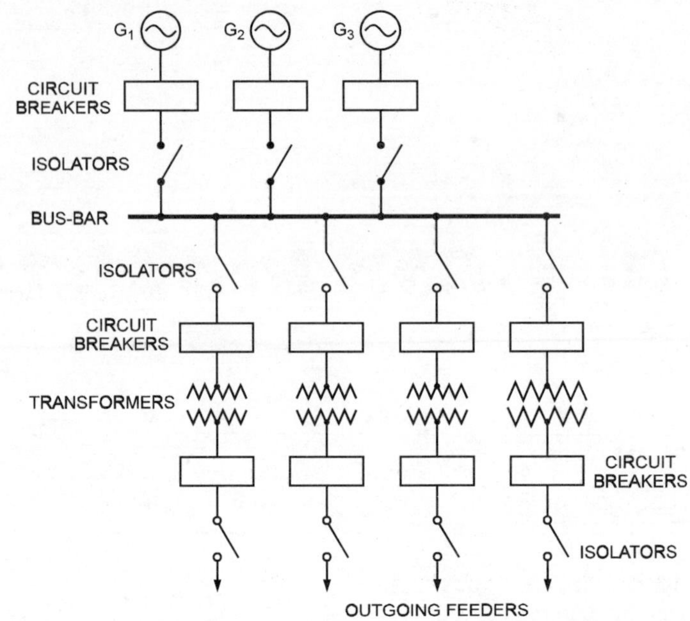

Fig. 13.1. ***Single Bus-Bar Arrangement***

The glaring drawback of this system is that in case of fault on the bus-bars, whole of the supply is affected and all the healthy feeders are disconnected. Moreover, when maintenance is to be carried out on any of the feeder sections or on a part of bus-bar the whole supply is to be disrupted. Thus such an arrangement provides least flexibility and immunity from total shutdown.

Such bus-bar arrangement is employed for switchboards, small and medium sized substations, small power stations and dc stations.

2. *Single Bus-Bar Arrangement With Bus Sectionalization.* The bus-bar may be sectionalized by a circuit breaker and isolating switches so that a fault on one part does not cause a complete shutdown. In large generating stations, where several units are installed, it is a common practice to sectionalize the bus as illustrated in Fig. 13.2. Normally the number of sections of a bus-bar are 2 to 3 in a substation, but actually it is limited by the short-circuit current to be handled. In a sectionalized bus-bar arrangement only one additional circuit breaker is required which does not cost much in comparison to the total cost of the bus-bar system. Such an arrangement provides three main advantages over simple single bus-bar arrangement. Firstly, in the event of occurrence of fault on any section of the bus-bar, the faulty section can be isolated without affecting the supply of other section or sections. Secondly, one section can be completely shutdown for maintenance and repairs without affecting the supply of the other section(s). Thirdly, by adding a current limiting reactor between the sections the fault level (MVA) can be reduced thereby circuit breakers of lower capacity can be used.

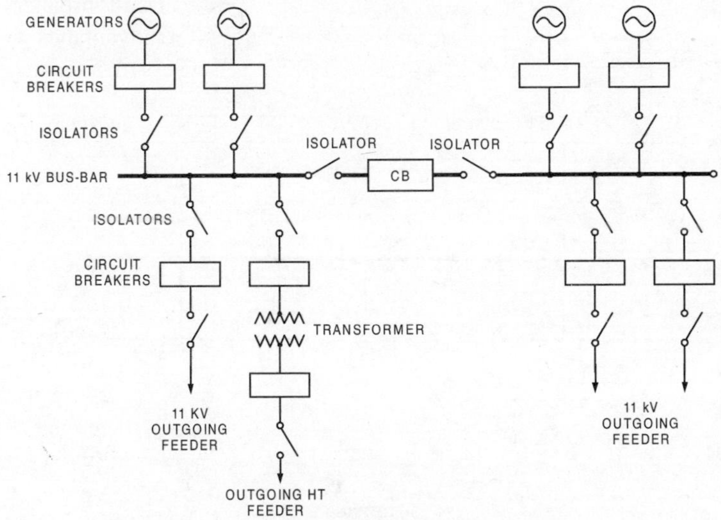

Fig. 13.2. ***Sectionalized Single Bus-Bar System***

At times air-break isolators were used in place of circuit breakers as bus-sectionalizer due to economy, but it must be remembered that any isolation affected by them must be affected under off-load conditions otherwise it may cause spark. It will be preferable to provide circuit breaker as a sectionalizing switch so that uncoupling of bus-bar may be carried out safely during load transfer. A double isolation is however necessary when the circuit breaker is employed as sectionalizing switch so that the maintenance work can be carried out on circuit breaker while the bus-bars are alive.

3. *Main and Transfer Bus Arrangement.* This arrangement has been quite frequently adopted where the loads and continuity of supply justify additional costs. This arrangement provides additional flexibility, continuity of supply and allows periodic maintenance without total shutdown. Such an arrangement is suitable for highly interconnected power network in which flexibility is very important.

Figure 13.3 illustrates the main and transfer bus arrangement in a generating station. Such an arrangement consists of two bus-bars, known as *main bus-bar* and *transfer bus-bar* used as an auxiliary bus-bar. Each generator and feeder may be connected to either bus-bar with the help of bus coupler which consists of a circuit breaker and isolating switches. In this arrangement a bus coupler is usually used so that change-over from one bus-bar to the other can be carried out under load conditions. While transferring the load to the reserve bus, the following steps may be taken.

1. Close the bus coupler (circuit breaker) so as to make the two buses at the same potential.
2. Close isolators on the reserve bus.
3. Open isolators on the main bus.

The load is now transferred to the reserve or auxiliary bus and main bus is disconnected.

The advantages and disadvantages of the arrangement are gives below.

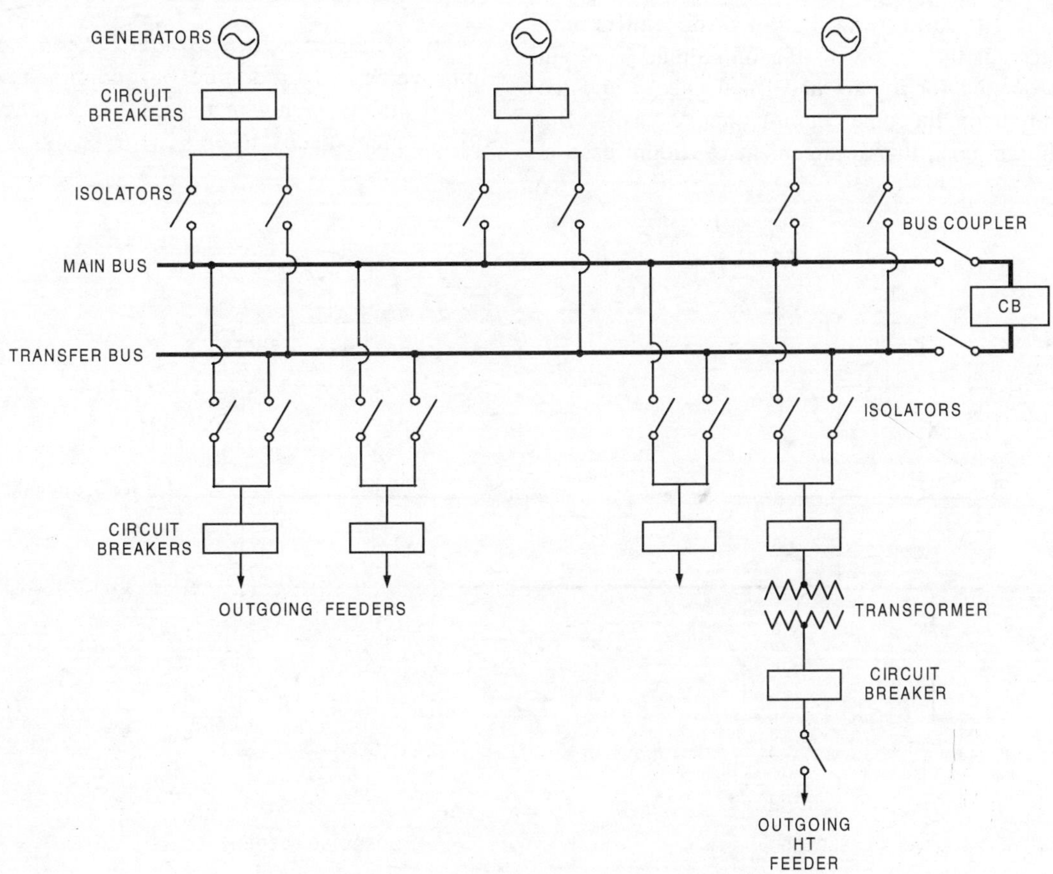

Fig. 13.3. ***Main and Transfer Bus Arrangement***

Advantages. 1. It ensures continuity of supply in case of bus fault. In the event of occurrence of fault on one of the bus, the entire load can be transferred to the other bus.

2. Repair and maintenance can be carried out on the main bus without interrupting the supply as the entire load can be transferred to the auxiliary bus.
3. Each load can be supplied from either bus.
4. The infeed and load circuit may be divided into two separate groups if required from operational considerations.
5. The testing and maintenance of feeder circuit breakers can be done by putting them on spare bus, thus keeping the main bus undisturbed.
6. The maintenance cost of substation is lowered.
7. The bus potential can be used for relays.

Disadvantages. 1. Additional costs.

2. The bus is maintained or expanded by transferring all of the circuits to the auxiliary bus depending upon the remote backup relays and breakers for removing faults of the circuits. During this condition a line fault on any of the circuits of the bus would shutdown the entire station.

4. *Double Bus Double Breaker Arrangement.* In very important power stations two circuit breakers are employed for each circuit, as illustrated in Fig. 13.4. Such a bus-bar arrangement does not require any bus coupler and permits switch-over from one bus to the other whenever desired, without interruption. This bus arrangement is very costly and its maintenance cost is also high. This arrangement provides maximum flexibility and reliability as the faults and maintenance interrupt the supply to the minimum. A circuit breaker can be opened for repairs and usual checks and the load can be shifted on the other circuit breaker easily. But because of its higher cost, this arrangement is seldom used at the substations.

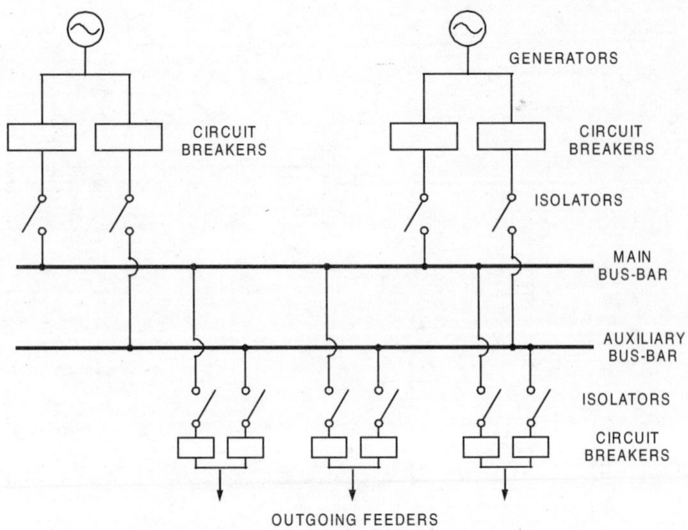

Fig. 13.4. ***Double Bus Double Breaker Arrangement***

For 400 kV switchyards two main buses plus one transfer bus scheme is preferred. The transfer bus is used for transferring power from main bus 1 to main bus 2 and vice versa.

5. *Sectionalized Double Bus Arrangement.* In this arrangement duplicate bus-bars are used with the main bus-bar in sections connected through a bus coupler, as illustrated in Fig. 13.5. In this arrangement, any section of bus-bar can be isolated for maintenance, while any section may be synchronised with any other through the auxiliary bus-bar. Sectionalization of auxiliary bus-bar is not required and would increase the cost if done.

6. *One-and-a-Half Breaker Arrangement.* This is an improvement over double bus double breaker arrangement and it affects saving in the number of circuit breakers. This arrangement needs three circuit breakers for two circuits.

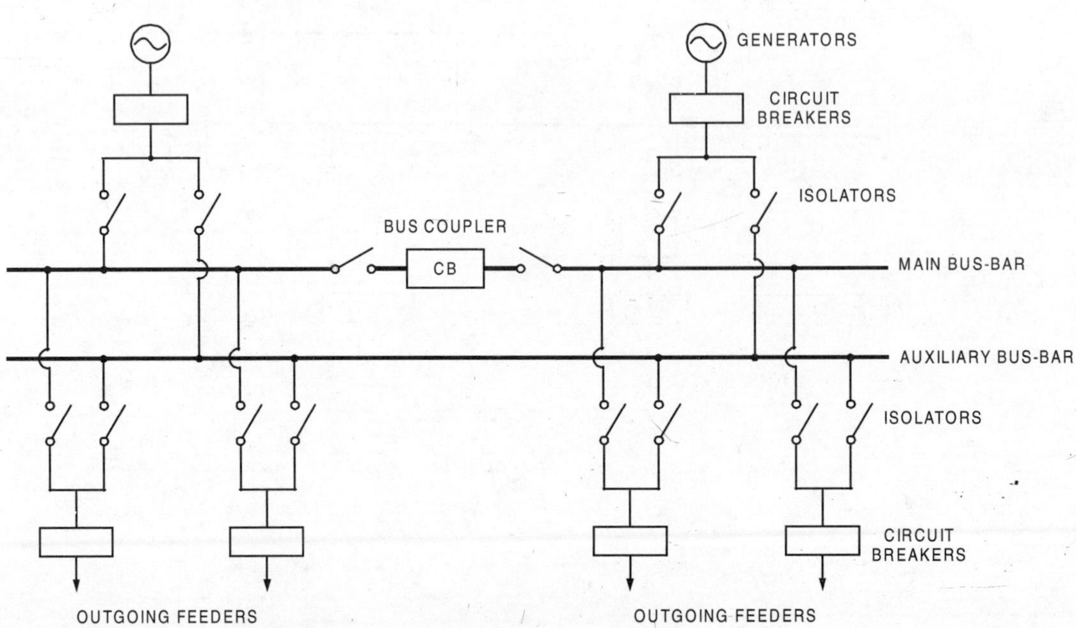

Fig. 13.5. ***Sectionalized Double Bus Arrangement***

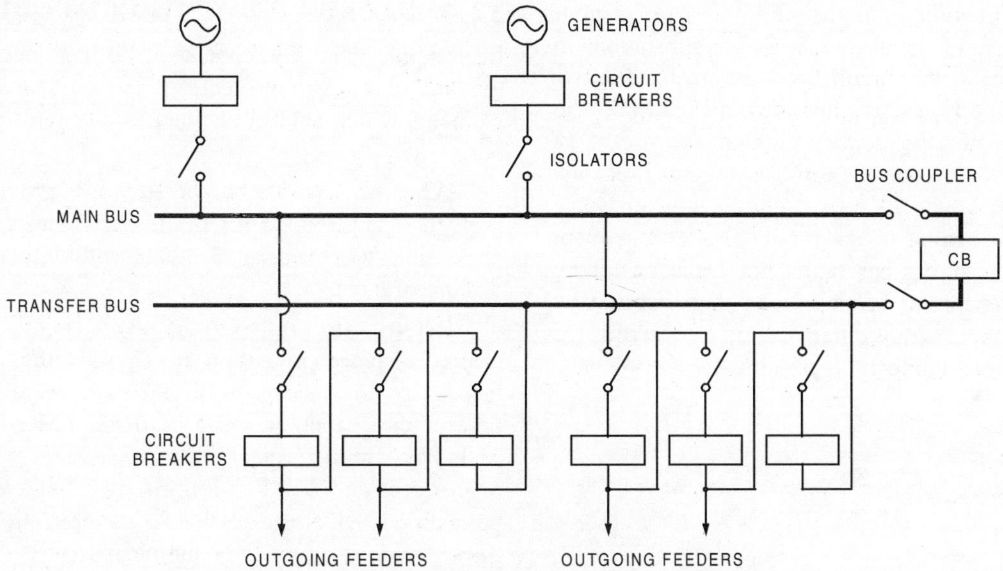

Fig. 13.6. ***One-and-a-Half Circuit Breaker Arrangement***

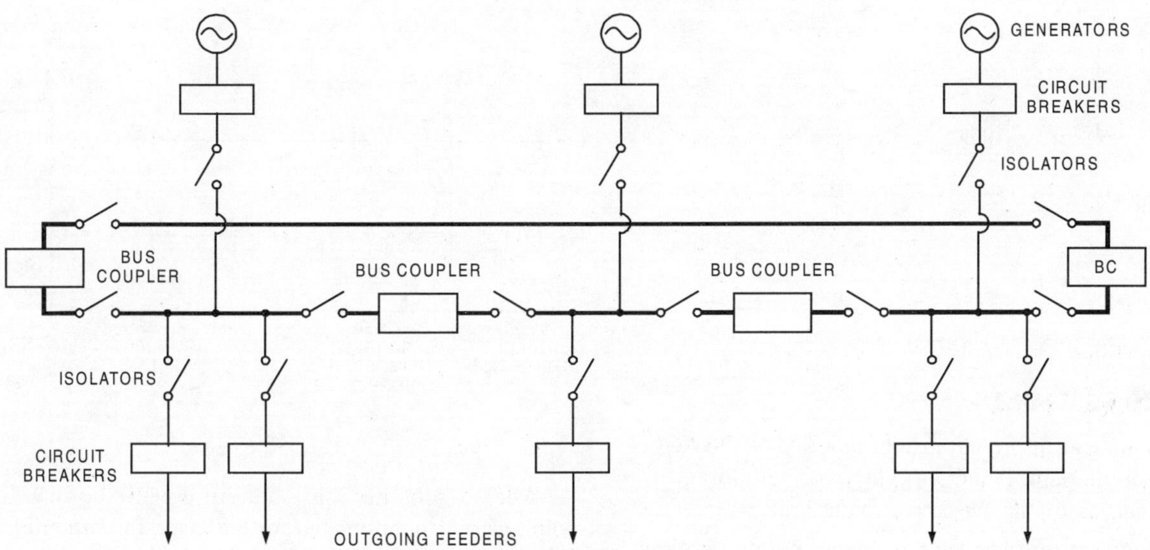

Fig. 13.7. ***Ring Main Arrangement***

The number of circuit breakers per circuit comes out to be 1½ hence the name. This arrangement is preferred in important large stations where power handled per circuit is large. This arrangement is shown in Fig. 13.6. This arrangement provides high security against loss of supply as a fault in a bus or in a breaker will not interrupt the supply. Possibility of addition of circuits to the system is another advantage. The bus potential can be used as supply to relays, however, at the time of bus fault such potential to the relay should be thrown off.

The main drawback of this arrangement is complications in relaying system because at the time of fault two breakers are to be opened. The other drawback is that for maintenance of circuit breakers, if the load shedding is not used, two breakers are to be opened in which case the other circuit in the line up will be operating with one breaker from one bus only. At the time of fault in that bus supply to the other circuit is also interrupted. The maintenance cost is higher.

The above arrangement has been used in important 400 kV and 750 kV substations.

7. *Ring Main Arrangement.* This is an extension of the sectionalized bus-bar arrangement where the ends of the bus-bars are returned upon themselves to form a ring, as illustrated in Fig. 13.7. This arrangement provides greater flexibility as each feeder is supplied by two paths, so that the failure of a section does not cause any interruption of the supply. The effect of fault in one section is localised to that section alone. The rest of the sections continue to operate. Circuit breakers can be maintained without interrupting the supply. The cost is also not much as the number of breakers used are nearly the same as that of a single bus-bar system. The drawbacks of the system are (*i*) difficulties in addition of any new circuit in the ring (*ii*) possibility of overloading of the circuits on opening of any section of the breaker and (*iii*) necessity of supplying potential to relays separately to each of the circuit.

8. ***Mesh Arrangement.*** This is another arrangement making economical use of circuit breakers in a substation. In this bus-bar arrangement, the circuit breakers are installed in the mesh formed by the buses, as illustrated in Fig. 13.8. The circuits are tapped from the node points of the mesh. In Fig. 13.8 eight circuits are controlled by four circuit breakers. When fault occurs on any section, two circuit breakers have to open, resulting in opening of the mesh. This arrangement provides security against bus-bar faults but lacks switching facility. It needs fewer circuit breakers than that required by one-and-a-half breaker arrangement. It is preferred for substations having large number of circuits.

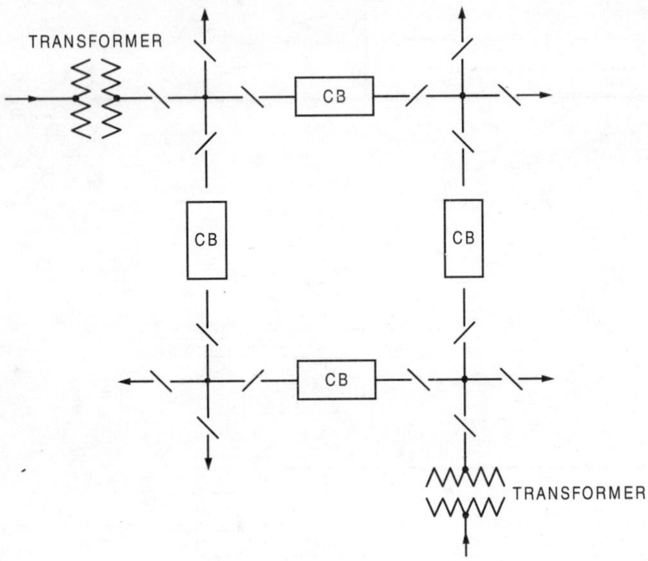

Fig. 13.8. *Mesh Arrangement*

13.3. BUS ZONE FAULTS

According to statistical information, the majority of the faults are single phase in nature and are, in part, temporary in character. The causes of the bus zone faults are:

1. Failure of support insulator, due to material deterioration, resulting in earth fault.
2. Flash-over across support insulator, caused by prolonged and excessive overvoltages.
3. Faulty operations performed by the attending personnel.
4. Heavily polluted insulator causing flash-over.
5. Foreign objects accidently falling across bus-bars.
6. Failure of circuit breakers to interrupt fault current or failure to clear under through fault conditions.

The clearing of a bus fault requires the opening of all the circuits branching from the faulty bus or bus section (leaving circuits having no backfeed). The most commonly used schemes for bus-zone protection are:

1. Backup protection, 2. Circulating current protection and 3. Frame leakage protection.

All the protection schemes must also be provided with a check system to ensure that the protection responds only to earth faults occurring within the bus zone and not to extraneous earth faults.

13.4. BACKUP PROTECTION FOR BUS-BARS

In principle it is the simplest of all to protect the buses with the aid of backup protections of the connected supplying elements which should respond to any fault appearing on the buses.

When no separate bus protection is provided but distance protection is provided for the feeders connected to the bus, it is possible to cover the bus-bars within zone 2 reach of the distance relays.

Referring to Fig. 13.9, the bus A is covered in the second step of distance protection B. Thus, in the event of fault on bus A, the distance protection B will operate. The operating time of the second step can be of the order of 0.4 second. In such a system the protection is slow and there can be unwanted disconnection of all incoming parallel circuits. Distance protection is widely employed for the protection of transmission lines, hence it is quite economical to use the same for bus protection.

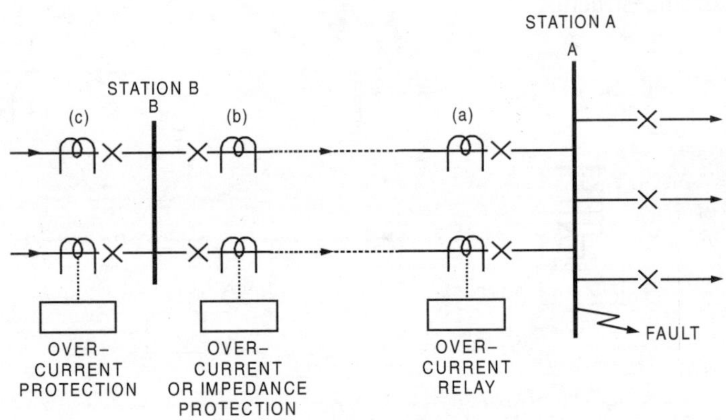

Fig. 13.9

Above scheme may be quite satisfactory for small switchgear installations but for large and important installations a separate bus-zone protection is provided.

Referring to Fig. 13.9

(*a*) The local overcurrent protection at station A provides the primary protection to bus-zone A.

(*b*) The remote overcurrent protection or impedance protection at station B provides a backup protection to bus-zone A. In case protection '*a*' fails protection '*b*' provides a backup.

(*c*) Local overcurrent protection of incoming lines or feeders at station B provides primary protection to bus B.

The drawbacks of such protection are: (*i*) delayed action (*ii*) disconnection of more circuits in case there are two or more incoming lines and (*iii*) exact discrimination not possible.

Bus backup protection may also mean that in case the breaker fails to operate for a fault on the outgoing feeder, then it must be regarded as a bus fault. It should then open all breakers on that bus. Such a backup protection can be provided with appropriate time delay through a timer.

13.5. FRAME LEAKAGE OR FAULT-BUS PROTECTION

It is one of the most simple form of protection and is applicable to small size metal-clad switchgear. This method consists of insulating the bus-supporting structure and its switchgear from ground, interconnecting all the framework, circuit breaker tanks etc. and providing a single ground connection through a CT that feeds an overcurrent relay as illustrated schematically in Fig. 13.10. The overcurrent relay controls a multi-contact auxiliary relay that trips the breakers of all circuits connected to the bus, as illustrated in Fig. 13.11. Sometimes an impedance is connected in the earth connection to limit the short-circuit current during line-to-earth fault. Also it is necessary to isolate the switchgear framework from lead cable sheaths, cable boxes and the conduit fittings so that when a leakage to the framework occurs, the only path for the leakage current is through the connection from the framework to earth. An external flash-over of an entrance bushing will also improperly trip all breakers unless the bushing support is insulated from the rest of the structure and independently grounded.

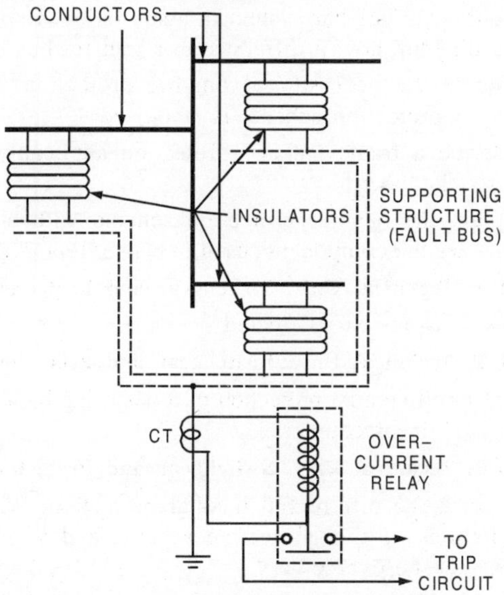

Fig. 13.10. ***Frame Leakage Protection***

In case of a sectionalized bus structure, the housing of each section must be insulated from adjoining sections, and separate fault bus-relaying is provided for each section. The fault-bus scheme does not provide overlapping of protective zones around circuit breakers; and consequently supplementary relaying is required to protect the regions between bus sections.

In the schematic arrangement of fault-bus protection illustrated in Fig. 13.10, the metal supporting structure or fault bus is grounded through a CT, the secondary of which is connected to an overcurrent relay. Under normal working conditions, the relay remains inoperative but fault involving a connection between a conductor and grounded supporting structure will result in current flow to ground through the fault bus, causing the relay to operate. The operation of the relay will trip all the breakers connecting equipment to the bus.

Figure 13.11 illustrates the scheme in which the station bus-bars are supplied from a power transformer having star-connected secondary. A CT is also connected in the earth connection of this star-connected secondary. The secondary of CT is connected to the operating coil of the check relay.

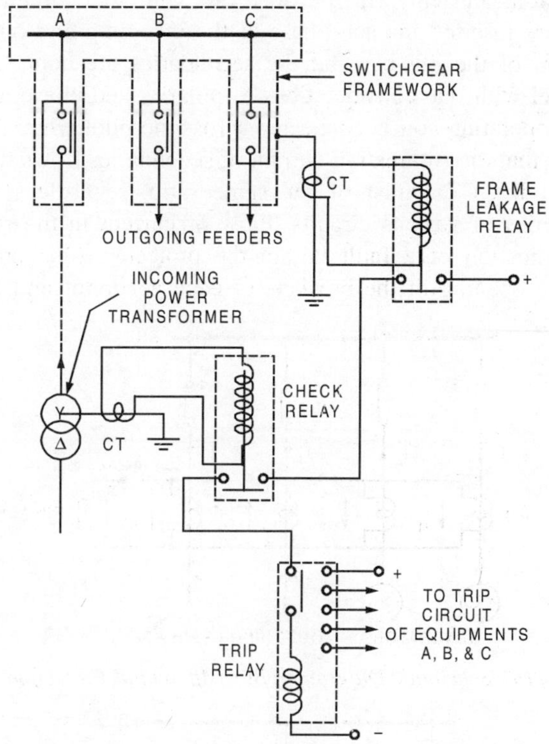

Fig. 13.11. ***Frame Leakage Protection Scheme***

From the scheme illustrated in the figure it is obvious that the multi-contact relay is energized only when both the leakage and check relay contacts are closed *i.e.,* there is an earth fault within the protected zone. However, if the earth fault occurs outside the protected zone, the earth leakage current will pass through only the earthed secondary of power transformer and the check relay contacts will close but the frame leakage relay contacts will not close and thus multi-contact trip relay will not operate.

Fault-bus protection is more favourable to indoor than to outdoor installations. Certain existing installations may not be adaptable to such protection, owing to the possibility of other paths for short-circuit current to flow to ground through concrete reinforcing rods or structural steel.

13.6. DIFFERENTIAL OVERCURRENT PROTECTION

For the main bus-bars in the power stations, due to their importance in the operating conditions, it is required that the disconnection be without any delay in the case of faults. Hence it is imperative to use a differential current protection without time delay.

The protection is based on simple circulating current principle that under normal operating conditions or under external fault conditions the sum of currents entering into a bus-bar will be equal to the sum of currents leaving the bus-bar. In case the sum of these currents (for a given conductor) is not zero, it must be due to a short circuit either a ground fault or phase-to-phase fault. Hence this protection scheme is applicable to both types of faults *i.e.,* phase-to-phase faults as well as ground faults.

Figure 13.12(*a*) shows the application of differential circulating current principle to a bus with four circuits. The CTs are inserted in each phase of the incoming and outgoing feeders of the bus-bar and the secondaries are connected in parallel with due considerations to polarity and phase and the relay operating coil is connected across the pilot wires in such a way that the summation current of secondaries flows through it. All the CTs must of the same ratio, regardless of the capacities of various circuits. Flow of current in the relay is an indication of a fault within the protected zone and will initiate opening of the breakers of each generator and feeder.

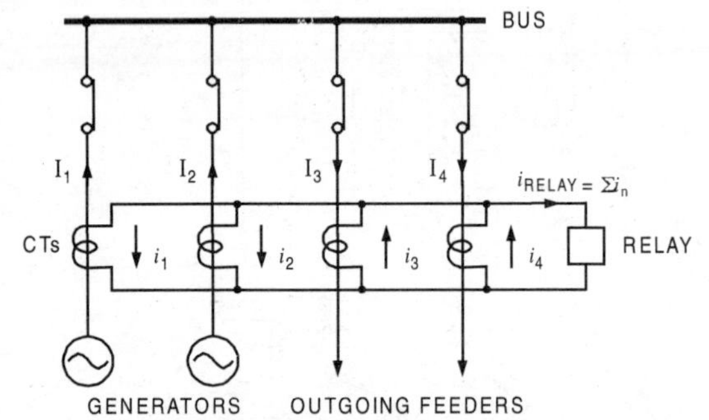

(a) Schematic Diagram of Bus Differential Protection

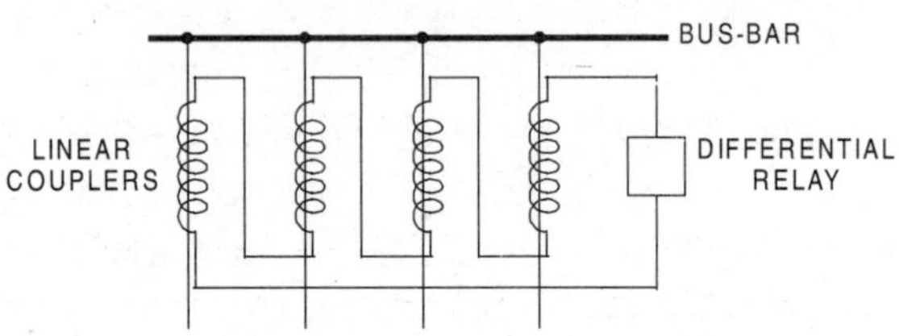

(b) Voltage Differential Relaying For Bus-bar

Fig. 13.12

The main drawback of differential overcurrent protection is the difference in the magnetic conditions of the iron-cored CTs which may cause false operation of the relay at the time of an external fault. Even with identical CTs having large iron cores to avoid the saturation with maximum fault currents the dc transient component creates problem due to its slow decay. Biasing of differential relays improves the stability considerably but does not solve the problem completely.

Better discrimination between internal and external faults can be had if high impedance bus differential relay is used in place of usual low impedance relay. *High impedance relay* is an overcurrent relay with a series resistance. Such a relay remains stable against spill currents due to external faults or CTs inaccuracies.

Another method of protecting bus-bar sections is by means of *voltage differential protection,* which overcomes the difficulties of iron-cored CTs. In this scheme CTs without iron cores, known as *linear couplers* are employed so that they have a much larger number of secondary turns than an iron-core CT. In this scheme secondary windings of CTs are connected together in series and the differential relay coil connected across them, as illustrated in Fig. 13.12 (*b*). Under normal operating conditions or under external fault conditions, the sum of voltages induced (proportional to the primary currents) in the secondary windings is zero but in the event of an internal fault on the bus-bar, the voltages of the CTs in all source circuits add to cause the flow of current through the secondary windings and the differential relay operating coil.

This scheme provides high speed protection for a relatively small net voltage in the differential circuit.

Exercises

1. What are the different types of bus-bar arrangements used in substations ? Illustrate your answer with suitable diagrams. **[M.D. Univ. Power System-I, May 2009]**
2. What are the different bus-bar arrangements possible in a substation ? Discuss them briefly with application areas.
3. Discuss and compare briefly various bus-bar arrangements in a power system.
4. Discuss why duplicate bus-bar system is used ? With a neat sketch develop the duplicate bus-bar system and explain why bus-coupler circuit breaker is used ?
5. Describe with neat line diagrams the principle of operation of duplicate bus-bar system in a substation. Explain with a neat diagram how it differs from a transfer bus-bar system.
6. What is the necessity of bus-bar protection ? How the bus-bar protection scheme is stabilised ?
7. Describe a frame leakage protection applicable to metal-clad switchboards.
8. Explain the bus-bar protection scheme with an example. What are the components used in carrier relaying schemes ?
[U.P. Technical Univ. Switchgear and Protection 2007-08]
9. How is a bus-bar protected ?
[U.P. Technical Univ. Switchgear and Protection 2009-10]
10. Explain differential protection of a bus using high impedance relay or linear couplers.
[G.B. Technical Univ. Switchgear and Protection 2010-11]
11. Describe the differential bus-bar protection. What are the limitations of this protective scheme and to what extent these can be overcome ?
12. Describe the principle of bus-bar protection based on voltage differential systems. How does it overcome the problems of saturation of CTs ?
13. Write short notes on the following :
    (*i*) Necessity of bus-bar protection.
    (*ii*) Bus-bar arrangements.
    (*iii*) One-and-a-half breaker arrangement.
    (*iv*) Ring main arrangement of bus-bars.
    (*v*) Frame leakage or fault-bus protection.
    (*vi*) Voltage differential protection of bus-bars.
14. Explain the followings:
    (*i*) Bus-bar protection need special attention. Why ?
    (*ii*) What is backup protection of bus-bars ?
    (*iii*) What is differential protection of bus-bars ?
    (*iv*) What is meant by high impedance relay and what for is it used in differential protection of bus-bars ?
    (*v*) How does linear coupler differ from ordinary CTs ?
    (*vi*) What is voltage differential protection of bus-bars and how is it superior to circulating current protection ?

Short Answer Type Questions With Answers

Q. 1. Why use of single bus-bar arrangement is restricted to small and medium sized substations and small power stations ?

Ans. The glaring drawback of this system is that in case of fault on the bus-bars, whole of the supply is affected and all the healthy feeders are disconnected. Moreover, when maintenance is to be carried out on any of the feeder sections or on a part of bus-bar the whole supply is to be disrupted. Thus such an arrangement provides least flexibility and immunity from total shutdown. That is why single bus-bar arrangement is restricted to small and medium sized substations and small power stations.

Q. 2. Why is circuit breaker preferred over an air-break switch or an isolator for sectionalizing the bus-bars ?

Ans. At times air-break isolators were used in place of circuit breakers as bus-sectionalizer due to economy, but it must be remembered that any isolation affected by them must be affected under off-load conditions otherwise it may cause spark. It will be preferable to provide circuit breaker as a sectionalizing switch so that uncoupling of bus-bar may be carried out safely during load transfer. A double isolation is however necessary when the circuit breaker is employed as sectionalizing switch so that the maintenance work can be carried out on circuit breaker while the bus-bars are alive.

Q. 3. For what a bus coupler is used in main and transfer bus arrangement ?

Ans. In main and transfer bus arrangement, a bus coupler is usually used so that the change-over from one bus-bar to the other can be carried out under load condition.

Q. 4. Why double bus double breaker arrangement is rarely used at substations ?

Ans. Because of its very heavy initial as well as maintenance costs, double bus double breaker arrangement is rarely used.

Q. 5. Name the most commonly used schemes for bus-bar protection.

Ans. The most commonly used schemes for bus-bar protection are : 1. Backup protection, 2. Frame leakage protection and 3. Differential overcurrent protection.

Q. 6. Why is it necessary to isolate the switchgear framework from lead cable sheaths, cable boxes and conduit fittings in frame leakage protection of bus-bars ?

Ans. It is necessary to isolate the switchgear framework from lead cable sheaths, cable boxes and conduit fittings so that when a leakage to the framework occurs, the only path for leakage current is through the connection from the framework to the earth.

Q. 7. What is the drawback of differential overcurrent protection for bus-bars and how is it overcome ?

Ans. The main drawback of differential overcurrent protection is the difference in the magnetic conditions of the iron-cored CTs which may cause false operation of the relay at the time of an external fault. Even with identical CTs having large iron cores to avoid the saturation with maximum fault currents the dc transient component creates problem due to its slow decay. Biasing of differential relays improves the stability considerably but does not solve the problem completely.

Better discrimination between internal and external faults can be had if high impedance bus differential relay is used in place of usual low impedance relay. *High impedance relay* is an overcurrent relay with a series resistance. Such a relay remains stable against spill currents due to external faults or CTs inaccuracies.

❑❑❑

CHAPTER 14

Protection of Feeders and Transmission Lines

INSIDE THIS CHAPTER

14.1. Introduction **14.2.** Overload Protection **14.3.** Overcurrent and Earth Fault Protection **14.4.** Time Graded Protection **14.5.** Current Graded Protection **14.6.** Differential Pilot Wire Protection **14.7.** Current Balance Differential Protection **14.8.** Merz-Price Voltage Balance System **14.9.** Translay Protection System **14.10.** Split-Conductor Protection of Feeders **14.11.** Carrier Current Protection **14.12.** Microwave Channel **14.13.** Distance or Impedance Protection **14.14.** Carrier Aided Distance Protection.

14.1. INTRODUCTION

The probability of faults occurrence on the overhead lines is much more due to their greater lengths and exposure to atmospheric conditions. The protective schemes employed for alternators and transformers, with slight modifications, may also be employed for protection of feeders and transmission lines. The modifications are required because of length of lines, capacitance currents and other reasons. Moreover, the necessity of discrimination in supply networks has called forward many protective schemes which have no application to the comparatively simple cases of transformers and generators.

The main requirements of line protection are:

(*i*) In the event of a short circuit, the circuit breaker nearest to the fault should open, all other circuit breakers remaining in a closed position.

(*ii*) If the breaker nearest to the fault fails to open, backup protection should be provided by the adjacent circuit breakers.

(*iii*) The relay operating time should be the smallest possible in order to preserve system stability but without unnecessary tripping of circuits.

The protection of lines presents a problem quite different from the protection of apparatus such as generators, transformers and busbars. While differential protection is an ideal protection for lines, but it is much more expensive to use. The two ends of a line may be many kilometres apart and for comparison of two currents, a costly pilot wire circuit is required. This expense may be justified but generally less costly protective schemes may be used.

There are several protective schemes for transmission lines and may be grouped into two groups viz, *non-unit type* and *unit type*.

The non-unit type of protection includes time graded overcurrent protection, current graded overcurrent protection, and distance protection, while the unit type protection includes pilot wire differential protection, carrier current protection based on phase comparison method etc.

Separate protection systems are necessary for ground faults because ground faults are more frequent on overhead transmission lines than phase fault, and ground fault current is different from phase fault current in magnitude.

The selection of a particular scheme of protection depends upon the following factors:

1. Economic justifiability of the scheme to ensure 100% continuity of supply.
2. Types of feeders—radial or ring mains.
3. Availability of pilot wires.
4. Number of switching stations in series between supply end and the far end of the system.
5. System earthing—whether the neutral is grounded or insulated.

Time graded overcurrent protection is normally employed for backup protection in large transmission systems or where a time lag can be permitted and instantaneous operation is not necessary *i.e.,* where time lag in fault clearance does not cause instability or damage to the lines. In distribution feeders, it plays a more important role and may be coordinated with fuses. Overcurrent protection is also employed for ground faults. Fuses are employed in distribution systems, where relays and circuit breakers are not necessary and fuses are preferred due to their low cost, current limiting features etc.

Distance protection is a high speed form of protection and is used when time lag cannot be permitted from considerations of stability. Today trend is toward the use of static distance protection for all types of line faults (phase-to-phase faults as well as phase-to-ground faults), main and backup for short, medium and long transmission lines. Distance protection is preferred to overcurrent protection because it is not nearly so much affected by changes in short-circuit current magnitude as overcurrent protection is, and therefore, is much less affected by changes in generation capacity and in system configuration. In distance protection there are three main types of measuring units viz, impedance, reactance and mho type distance relays, as already described in chapter 9. Each type has certain advantages and disadvantages and own field of applications. For very short lines reactance type is preferred, for medium length lines impedance relay suits and mho relays are suitable for phase faults of longer lines.

The pilot wire protection is a unit type of protection which would operate only for faults occurring within the protected zone providing no backup protection. It is used whenever high-speed protection is required for all types of short circuits and for any fault location. Pilot wire protection based on differential circulating current principle (Merz-Price) is employed for short lines where cost of pilot wires is not prohibitive. Carrier current protection is employed for long lines and interconnected lines.

Main transmission lines operating on 220 kV and above are provided with the fastest and the most reliable protection. Carrier pilot or high-speed distance protections are necessarily employed on main transmission circuits. 33 kV transmission lines are provided with directional time lag overcurrent relays or high-speed distance relays. Radial feeders operating on 11 kV and below are usually provided with time lag overcurrent relays supplemented by instantaneous relays. Sometimes only HRC fuses are used on these lines.

Autoreclosure schemes are incorporated in the protection of distribution and transmission lines. Autoreclosure of distribution lines is mainly for improving service continuity while autoreclosure of transmission lines is mainly for improving system stability.

14.2. OVERLOAD PROTECTION

This is the simplest way of protecting a line and is, therefore, widely used. Overload or overcurrent protection owes its application to the fact that the fault current in the event of a short circuit will rise to a value several times that of the maximum load current. The ratio of minimum fault current to maximum load current is chosen as criterion in order to prevent the possibility of maloperation under normal operating conditions. This form of protection can be applied only to simple systems. The overcurrent protection is provided at the supply end of the line.

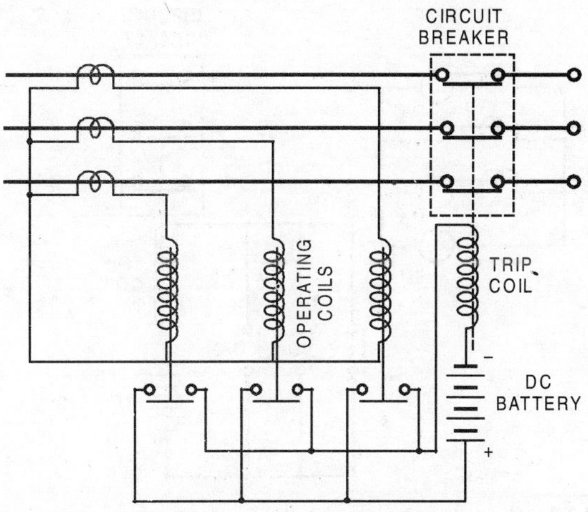

Fig. 14.1. *Overload Protection*

Figure 14.1 illustrates the protective scheme against overloading of feeders. Three CTs are mounted, one on each phase of the feeder and are connected across the three relay coils. In case of overloading the solenoid plunger system of the relays operate to close the trip coil circuit, which in turn opens the circuit breaker, thereby disconnecting the protected feeder. The relays need readjustment or even replacement whenever a change in the system is made. The operation times are generally large.

The alternative method for overload or overcurrent protection is the well known Z-connection which requires only two relays for the protection of a 3-phase circuit. This scheme is shown in Fig. 14.2.

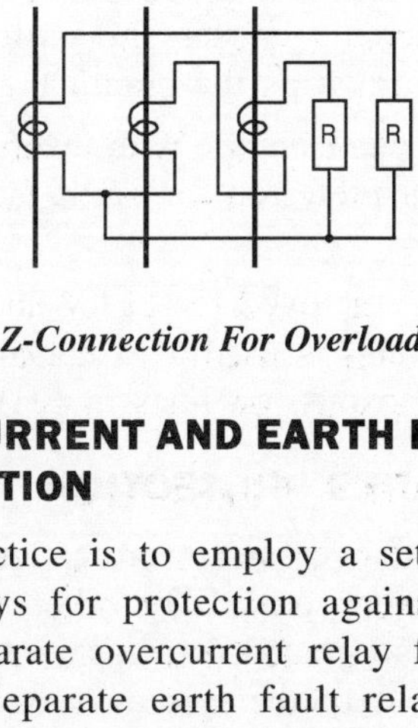

Fig. 14.2. *Z-Connection For Overload Protection*

14.3. OVERCURRENT AND EARTH FAULT PROTECTION

The general practice is to employ a set of two or three overcurrent relays for protection against phase-to-phase faults and a separate overcurrent relay for single line-to-ground faults. Separate earth fault relays are generally preferred because they can be adjusted to provide faster and more sensitive protection for single line-to-ground faults than that can be provided by the phase relays.

Earth fault current depends on the type of neutral earthing, *i.e.,* whether solidly earthed, insulated or earthed through some resistance or reactance. Where no neutral point is available, grounding transformer is employed.

Whatever the type of neutral earthing be employed, the earth fault current will be small as compared to phase fault currents in magnitude. The relay thus connected for earth fault protection is different from the ones provided for phase-to-phase faults.

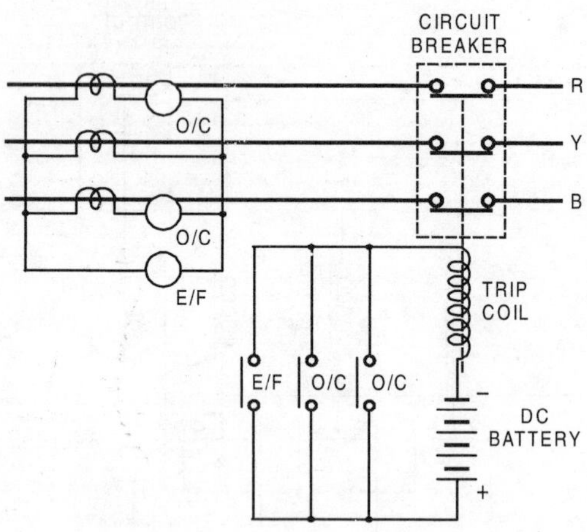

Fig. 14.3. ***Overcurrent and Earth Fault Protection Scheme***

In case of resistance earthed or solidly earthed systems, the overcurrent element connected in residual circuit of CTs is preferred. The setting of earth fault relays may be made less than rated full-load current of the line. The practice followed is to apply relays having a setting range of 10 to 40 %. A setting of 20 % on 300/ 5 A CT means the relay operates for primary earth fault current of $\frac{300}{5} \times \frac{20}{100} = 12$ A.

In the above protection scheme two IDMT type overcurrent relays are connected in two phases through CTs and one earth fault relay, as illustrated in Fig. 14.3. In case of phase-to-phase faults or overload the IDMT relays trip the circuit breaker. Under healthy conditions, the sum of all the three currents of CTs is zero and the earth fault relay remains inoperative. As soon as phase-to-ground fault occurs unbalancing in currents causes the earth fault relay to operate, which in turn trips the circuit breaker.

The earth fault elements are with inverse characteristics and time grading is preferred for earth fault protection of radial feeders.

This scheme is employed for 11 kV and 33 kV systems as main protection and is used as a backup protection for transformers and transmission lines in EHV systems.

14.4. TIME GRADED PROTECTION

This is a scheme of overcurrent protection in which time discrimination is incorporated. In other words, the time setting of relays is so graded that in the event of fault, the smallest possible part of the system is isolated.

The application of time graded protection to radial feeders, parallel feeders, and ring main system will be discussed here.

14.4.1. Protection of Radial Feeders

The main characteristic of a radial system is that power can flow in one direction only, from generator or supply end to the load end. It has the drawback that continuity of supply cannot be maintained at the load end in the event of fault.

In radial system when number of feeders are connected in series, as shown in Fig. 14.4 it is desired that the smallest possible part of the system should be off. This is conveniently achieved by employing time graded protection. In this system the time setting of overcurrent relays is adjusted in such a way that farther the relay from the generating station the lesser the time of operation. When the fault occurs on the line far side SS_4 the relay OC_5 should operate first and not any other, *i.e.,* the time required to operate the relay OC_5 should be less than that required for relay OC_4. Similarly the time required to operate the relay OC_4 must be less than that required for OC_3 and so on. This shows that the time setting required for these relays must be properly graded. The minimum interval of time which can be allowed for the two adjacent circuit breakers depends on its own clearance time, plus a small time for safety margin. With normal circuit breakers in use minimum discriminating time between adjacent breakers should be about 0.4 second. The time settings for relay OC_1, OC_2, OC_3, OC_4 and OC_5 will be 2.0 seconds, 1.5 seconds, 1.0 second, 0.5 second and instantaneous respectively.

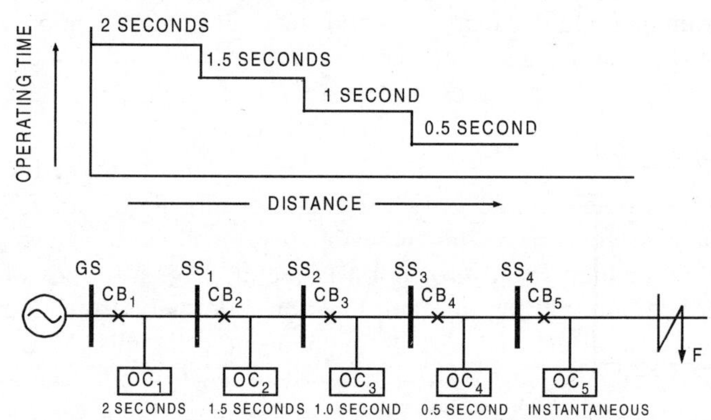

Fig. 14.4. ***Time Graded Protection For Radial Feeders***

In addition to this grading, it is also essential to have the time of operation dependent on the severity of fault. For severe fault the time of operation should automatically be less. This is achieved by using time limit fuses in parallel with the trip coils. Its additional advantage is that the relay will not operate under normal overload conditions of very short duration. The graded time lag relays when connected in series require that their time current characteristics are similar in shape. In no case they should cross each other at any point. From this point of view induction type inverse definite minimum time (IDMT) relays are most suitable. Their ratings provide complete discrimination under all fault conditions and they are most widely used.

Time graded overcurrent protection for phase faults is supplemented by time graded earth fault protection. The earth fault relay is residually connected. In general, two relays are used for phase faults and one for earth fault, as already discussed in Art. 14.3.

The drawbacks of graded time lag overcurrent protection are given below:

1. Time lag is to be provided which is not desirable on short circuits.
2. This scheme is suitable for radial feeders with supply at one end only; not for ring mains or interconnected lines.

3. It is difficult to coordinate and requires changes with the addition of loads.
4. It is not suitable for important long distance transmission lines where rapid fault clearance is necessary to ensure stability of the systems.

14.4.2. Protection of Parallel Feeders

For important installations continuity of supply is a matter of vital importance and at least two lines are used and are connected normally in parallel so as to share the load. These lines may or may not run on the same towers or the same right of way. Special means are available for protecting such parallel feeders. In the event of a fault occurring, protective device will select and isolate the defective feeder while the other instantly assumes the increased load.

The simplest method of obtaining such protection is by providing time graded overload relays with inverse time characteristic at the sending end and instantaneous reverse power or directional relays at the receiving end, as shown in Fig. 14.5.

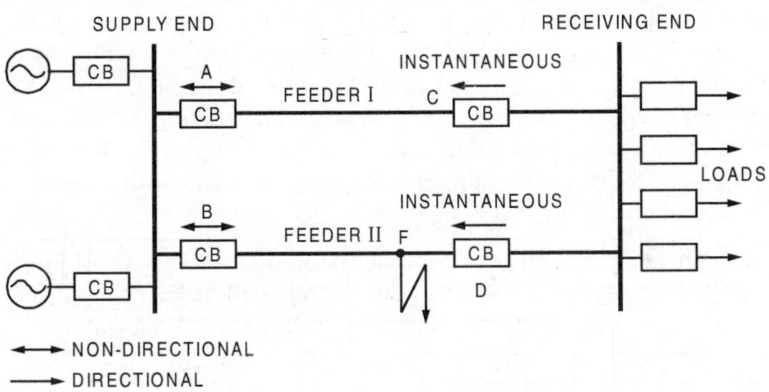

Fig. 14.5. ***Protection of Parallel Feeders***

When heavy fault or short circuit occurs on any one feeder, say at F on feeder No. II, the power is fed into the fault from the sending end as well as from the receiving end bus-bars. The direction of power flow will be reversed through the relay on D, which will open. The excess current is then confined to B until its overload relay operates and trips the circuit breaker, thus completely isolating the faulty feeder and supplying power through healthy feeder.

This method is not satisfactory unless the fault is heavy and reverses the power at D. Hence differential protection in addition to overload protection must be provided to both ends of the parallel feeders. A differential overcurrent relay is a centre zero induction type of instrument in which a pivoted spindle carries two discs. The torques on the two discs are caused by the currents flowing in the two feeders, and are in opposition. Under normal condition the currents in the two feeders will be equal and, therefore, torque acting on the relay will be zero. But in the event of fault, current in one of the feeders will increase and thus cause the relay to operate.

14.4.3. Protection of Ring Main System

The ring main is a system of interconnection between a series of power stations by an alternative route. The peculiarity of the ring main is that direction of power flow can be changed at will, particularly when an interconnector is used, consequently, the reverse power relay will not serve the purpose in this system. In case the reversal of power does not take place, reverse power relays can be employed.

An elementary diagram of such a system is shown in Fig. 14.6 where G is the generating station and A, B, C and D are substations. At the generating station, the power flows only in one direction *i.e.,* away from the bus-bars so non-directional time lag overload relays are used. The time graded directional relays are used at both ends of the substation and they are set so that they will only trip when an overload flows away from the substation which they protect. Going round the ring in the direction GABCD; the relays on the farther side of each station are set with decreasing time lags; for instance at generating station 2 seconds at stations A, B, C and D 1.5 seconds, 1.0 second, 0.5 second and instantaneous respectively. Similarly going round the ring in the opposite direction, the relays on the outgoing sides would be set as follows:

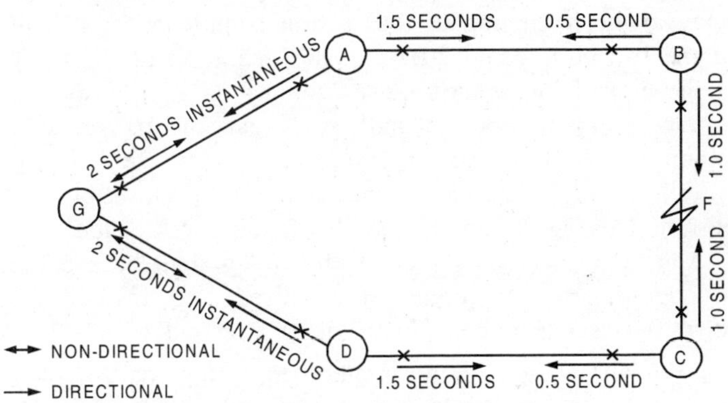

Fig. 14.6. ***Protection of Ring Main System***

Generating station G = 2 seconds; substations D = 1.5 seconds; C = 1.0 second; B = 0.5 second and A instantaneous. Now remembering that the reverse power relays are set so as to operate only when power flows away from the substation at which they are installed, a study of the diagram will make it clear that a short circuit occurring at any point on the system will cause only the two immediately adjacent circuit breakers to operate.

If a fault occurs at point F, the power is fed into the fault through two paths ABF and DCF. The relays to operate are that between substation B and fault point F and substation C and fault point F. Thus fault on any section will cause the relays on that section only to operate and the healthy sections will be operating uninterruptedly.

Limitations. In this system necessary stability of operation is achieved at the cost of slightly longer time intervals between adjacent relays. The time grading cannot be closer than 0.33 or 0.5 second and as the longest time that can be fed is 2 to 2.5 seconds, the maximum number of sections that can be protected in this way are six.

In order to overcome the above limitations, an *interlock system*, that can be employed for any number of sections with a very short time delay, has been designed. It may be

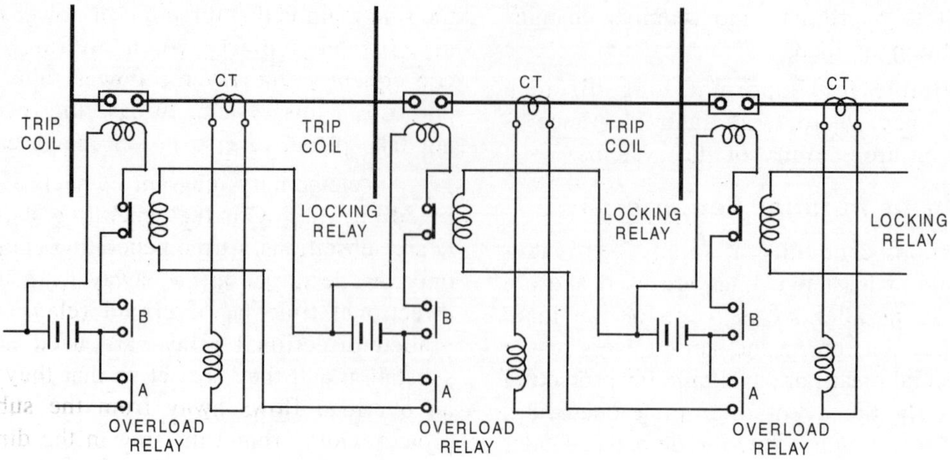

Fig. 14.7. ***Interlock System For Radial System***

employed for radial or ring main systems. Figure 14.7 illustrates the application of this method to a radial system.

If a section is healthy the same current flows at both ends and the overload relays operate. The operation of the relay at the beginning of the section completes the circuit of the trip coil (contact B), but the operation of the relay at the end of the section closes contact A, and causes the locking relay to operate and thus break the trip circuit. When the fault occurs within the section, the current entering is high, and causes the contact B to close, but the current leaving is small, and the locking relay does not operate. It is necessary that locking relay shall be capable of operating before contact B is closed, and as such the relay B is set to operate in 0.3 to 0.5 second.

For application of this method to a ring main or any interconnected system, it is necessary merely to have directional relays to close the circuit for the locking relays.

Example 14.1. A ring feeder with five sections and fed at one point is to be protected using directional overcurrent (DOC) relays and overcurrent (OC) relays with suitable time grading. Explain the working of the scheme. Show the location of DOCs and OCs and their time of operation. Assume a time grading of 5 ms between the relays. The fastest relay needs 5 ms for it to operate.

[U.P.S.C. I.E.S. Electrical Engineering-II, 2011]

Solution : A ring feeder with five sections and fed at one point is shown in Fig. 14.8 where G is the generating station and A, B, C and D are the substations. At the generating station, the power flows only in one direction *i.e.*, away from the bus-bars, so non-directional time lag overcurrent (OC) relays are used. The time graded directional relays are used at both sides of each substation and they are set so that they will only trip when an overload flows away from the substation which they protect. Going round the ring in the direction GABCD, the relays on the farther side of each section are set with decreasing time lags; for instance at generating station 25 ms and at substations A, B, C and D 20 ms, 15 ms, 10 ms and 5 ms respectively (providing time grading of 5 ms with the fastest relay with operating time of 5 ms at substation D). Similarly going round the ring in the opposite directions, the relays on the outgoing sides would be set as follows :

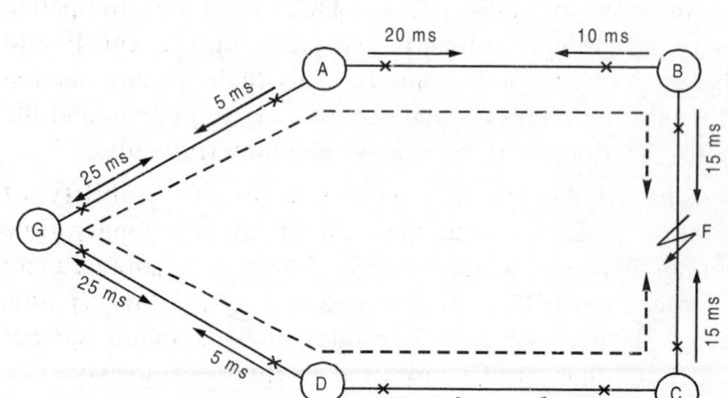

Fig. 14.8. ***Ring Feeder With Five Sections and Supplied at One Point***

Generating station G = 25 ms; substations D = 20 ms; C = 15 ms; B = 10 ms and A 5 ms A study of the diagram will make it clear that a short circuit occurring at any point on the ring feeder will cause only the two immediately adjacent circuit breakers to operate.

If a fault occurs at point F, the power is supplied into the fault through two paths ABF and DCF. The relays to operate are that between substation B and the fault point F and substation C and fault point F. Thus, fault on any section will cause the relays on that section only to operate and the healthy sections will be operating uninterruptedly.

14.5. CURRENT GRADED PROTECTION

An alternative to time grading or in addition to time grading, current grading protection can be applied when the impedance between two substations is sufficient. It is based on the fact that the short-circuit current along the length of the protected circuit decreases with the increase in distance between the supply end and the fault point. If the relays are set to operate at a progressively higher current towards the supply end then the drawback of long time delays occurring in graded time lag system can be partially overcome. This is known as *current grading*. Current graded systems normally employ *high-speed high-set overcurrent relays.*

A simple current graded protection scheme applied to a radial feeder is shown in Fig. 14.9. It consists of high-set overcurrent relays at A, B and C with settings such that relay at A would operate for faults between A and B, the relay at B for faults between B and C and the relay at C for faults beyond C. The current setting diminishes progressively from the supply end to the remote end of the line.

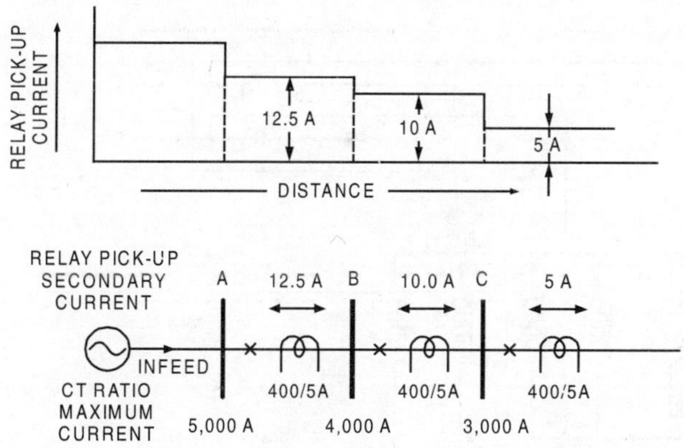

Fig. 14.9. ***Current Graded Protection Applied To a Radial Feeder***

In practice, however, this protection scheme poses some difficulties which are given below:

1. The relay cannot differentiate between faults which are very close to, but are on each side of B *i.e.,* if a fault is very near to station B in section BC, the relay at A may feel that it is in section AB because there may be very little difference in the fault currents and the relays do not discriminate between the fault in the next section and the end of first section. This is because:
   (*i*) the difference in the fault currents would be extremely small,
   (*ii*) the magnitude of fault currents cannot be accurately determined, and
   (*iii*) the accuracy of the relays under transient conditions is likely to be different.

   Hence for discrimination the relays are set to protect only part of the line, usually 80 %. For this reason current grading alone cannot be employed and this protection system should be supplemented by time graded inverse definite minimum time (IDMT) relay system.
2. The fault currents are different for different types of faults and so a certain difficulty is experienced in relay setting.
3. For ring mains, parallel feeders, interconnected systems, where power can flow to the fault from either direction, a system without directional control is not suited.

14.6. DIFFERENTIAL PILOT WIRE PROTECTION

The term "pilot" means that between the ends of a transmission line there is an interconnecting channel of some sort over which information can be conveyed. Three different types of such a channel are presently in use, and they are called "wire pilot", "carrier current pilot" and "microwave pilot". A wire pilot consists generally of a two-wire circuit of the telephone line type, either open wire or cable. A wire pilot is generally economical for distances up to 8 or 15 km, beyond which a carrier current pilot usually becomes more economical. Microwave pilots are employed when the number of services requiring pilot channels exceeds the technical or economic capabilities of carrier current.

The differential pilot wire protection is most satisfactory and is widely employed on account of the advantages such as simplicity, flexibility, a high stability ratio, rapid fault clearance (a time varying between 0.1 and 0.5 second according to the "break time" of the circuit breaker).

The differential pilot wire protection is based upon the principle that the currents compared at each end of the line or feeder by the use of pilot wires should be same under normal operating conditions and the equality is lost only when there is a fault in between the two ends. The system is quite similar to that employed for the protection of alternators and transformers and the difference lies only in the length of pilot wires. There are several differential protection schemes in use for lines but a few will be discussed here.

14.7. CURRENT BALANCE DIFFERENTIAL PROTECTION

Biased differential protection due to McColl is an example of current balance differential protection.

For understanding its principle let us first consider the protection of single phase feeders. The scheme is illustrated in Fig. 14.10. The current transformers CT_1, and CT_2 are mounted on the two ends of the protected line. The secondaries of the two CTs are connected through the restraining coils R_1 and R_2 and pilot wires P_1 and P_2. The operating coils of the relays are also connected across the secondaries of the CTs through diverting resistances DR_1 and DR_2. as illustrated in Fig. 14.10. Each of these two diverting resistances is equal to pilot wire resistance which will cause same current to flow through the operating coils O_1 and O_2. For equal number of turns on restraining coil and operating coil equal pull will be exerted on the sides of the beam relay. However, mechanical biasing can also be provided by

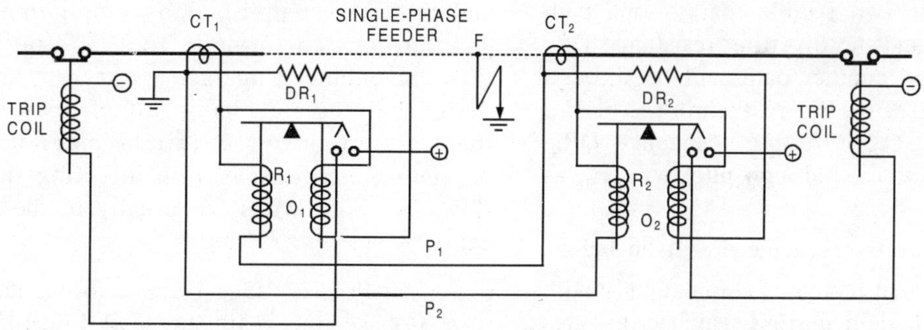

Fig. 14.10. ***McColl Circulating Current Protection For Single-Phase System***

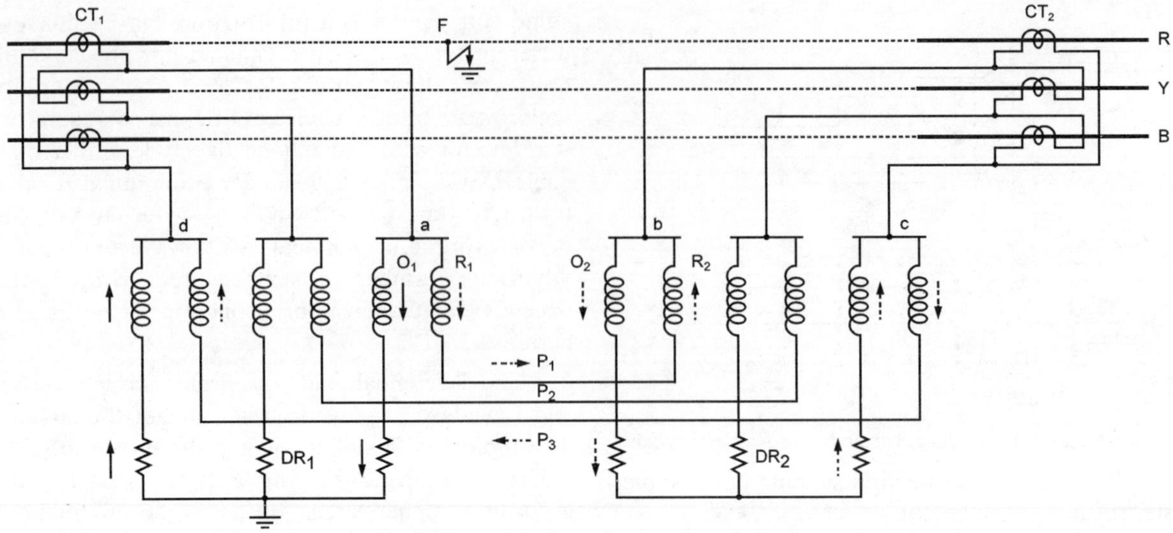

Fig. 14.11. *McColl Circulating Current Protection For Three-Phase System*

moving the fulcrum towards the operating coil and such an arrangement can allow 10 to 12 per cent greater current to flow in the operating coil before it can close contacts.

For an earth fault at point F, the current in CT_1 will exceed that in CT_2. The secondary current of CT_1 will be flowing in two parallel paths—one path will be through diverting resistance DR_1 and operating coil O_1 and the second path will be through restraining coil R_1, pilot wire P_1, diverting resistance DR_2, operating coil O_2, restraining coil R_2 and pilot wire P_2. Obviously the resistance of second path (consists of 3 coils, 2 pilot wires and one diverting resistance) is 3 times that of first path. Thus three-fourth of the total current will flow through the operation coil O_1 and one-fourth through the restraining coil R_1. If the current flowing through the operating coil O_1 exceeds the relay setting, the feeder will be disconnected from the supply end.

Figure 14.11 shows the application of biased differential protection to a 3-phase feeder. The CT sets CT_1 and CT_2 are connected in delta formation because the star-connection would require four pilot wires, the fourth pilot wire being connected between two star points. The diverting resistances DR_1 and DR_2 need only be one half of the resistance of one pilot wire, as the return pilot utilised for the fault current will also include a resistor equal to one half of this pilot resistance.

Let there be a fault in the R-phase at point F. The excess current in the secondary of CT_1 will flow to junction a where it will divide into two parallel paths—one path through operating coil O_1 and the diverting resistance DR_1 shown by solid arrows, and another path will be through restraining coil R_1, pilot wire P_1, restraining coil R_2, junction *b*, operating coil O_2, diverting resistance DR_2, junction *c* and then to junction *d* through pilot wire P_3, as shown by dotted arrows.

The advantages of this protective scheme are given below.

1. The relay operating current increases automatically with the increase of through fault current which eliminates the possibility of malfunctioning of the relay.
2. As the pilot capacitive current flows through the restraining coil instead of operating coil, this current adds to restraint.

14.8. MERZ PRICE VOLTAGE BALANCE SYSTEM

Probably the best known of the differential systems is the Merz-Price system, which, as applied to feeder protection, utilizes the principle of voltage balance. In 3-phase systems each conductor has its own pair of current transformers and relays. The secondaries of current transformers are connected in series by means of pilot wires. In normal conditions *i.e.,* when there is no fault on the feeder, equal currents flow at the two different ends, so induced voltages in the secondaries of current transformers are equal. As the secondaries are connected in opposition their secondary emfs are equalised resulting into no circulating current in the relays. But whenever fault occurs, currents differ at two ends, so induced emfs in the secondaries of current transformers will differ and circulating current will flow through the pilot wires and relays and the faulty feeder will be isolated.

It will be clear that the current transformers are critical feature of this system, since they have to be balanced exactly, not only initially but permanently. In order, that the induced voltage shall be proportional to line current, it is essential that the magnetic circuit shall not reach saturation, and this is accomplished by employing distributed air-gap current transformers. To secure initial matching, the CTs are balanced against a standard, and to ensure that there shall be no change of characteristics in service they are enclosed within a magnetic shield which prevents neighbouring iron affecting the distribution of flux. The pilot wires are usually in the form of a 3-core cable of size 7/0.73 mm.

Merz-Price voltage balance protection scheme applied to a 3-phase line is illustrated is Fig. 14.12 (*a*).

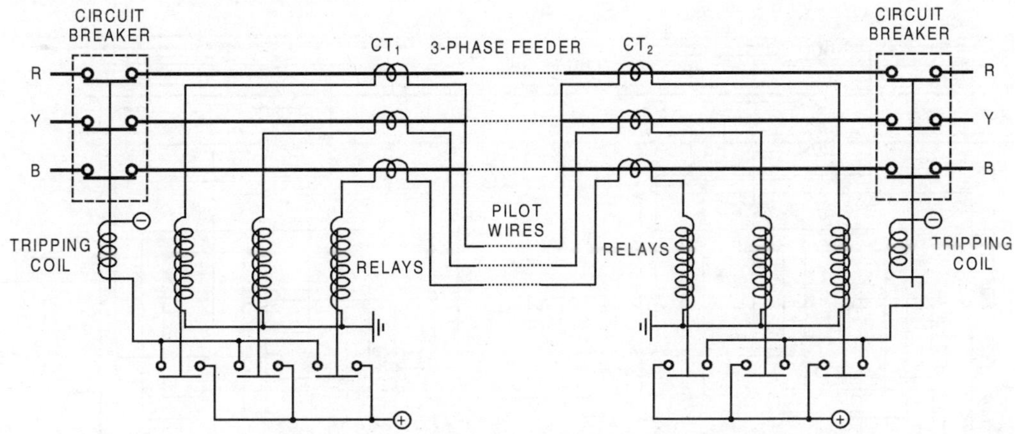

Fig. 14.12 (*a*) ***Merz-Price Voltage Protection Scheme Applied To 3-Phase Line***

This system has the following advantages and disadvantages:

Advantages. (*i*) This system is independent of operating voltage and fault power factor.

(*ii*) This system can be employed for protection of both, ring mains as well as parallel feeders.

(*iii*) This system provides instantaneous protection for ground faults, so the possibility of these faults involving other phases is reduced.

(*iv*) This system provides instantaneous relaying thereby reducing the amount of damage to overhead conductors resulting from arcing faults.

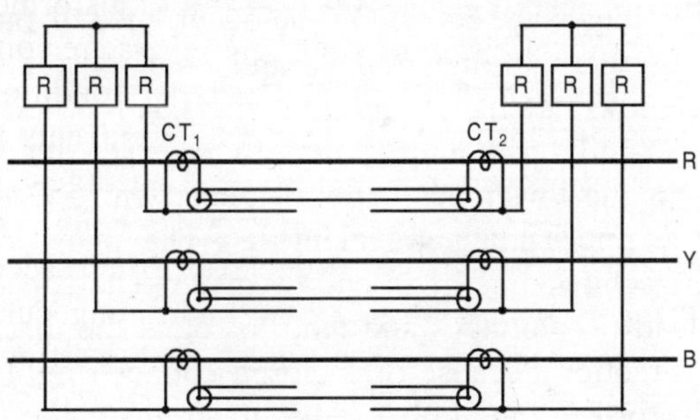

Fig. 14.12 (*b*) ***Beard-Hunter Sheathed Pilot System***

Disadvantages. (*i*) The trouble due to capacitance currents in the pilot circuit arises from the fact that, under through-fault conditions, voltages of the order of 1,000 volts or more are impressed on this circuit so that capacitance currents are comparatively heavy, and false operation may take place.

However this drawback can be overcome by the introduction of the *Beard-Hunter compensated pilot cable* in which means are provided for diverting the capacitance currents from the relays. This action is achieved by surrounding each pilot wire with an insulated metallic screen or sheath which is divided at the centre of its length so as to form two conductors of equal lengths. When a heavy overload comes on, a high voltage is induced in the current transformer secondaries, but the resulting capacitance current, instead of flowing in the relays, flows in the local circuit formed by the sheath, current transformer and pilot wire.

(*ii*) This system does not provide backup protection or overload protection.

(*iii*) Difficulties are experienced in balancing the secondaries of the two current transformers and that is why this system cannot be used beyond 33 kV.

(*iv*) The system will not operate in case a break in the pilot wire occurs.

(*v*) This system is very expensive owing to the greater length of pilot wires required.

(*vi*) There is no time delay.

14.9. TRANSLAY PROTECTION SYSTEM

The name "Translay" is evolved from the fact that the relay embodies a transformer feature. This system can be employed for protection of single-phase or 3-phase feeders, transformer feeders, feeders with a tee-off and parallel feeders against both earth and phase faults. This system is based on the established principle of the current entering at one end of the feeder being equal at any instant to that leaving at the other end. A simple form of Translay protection for a single-phase feeder is shown in Fig. 14.13. Under healthy conditions the line current transformers CT_1 and CT_2, at

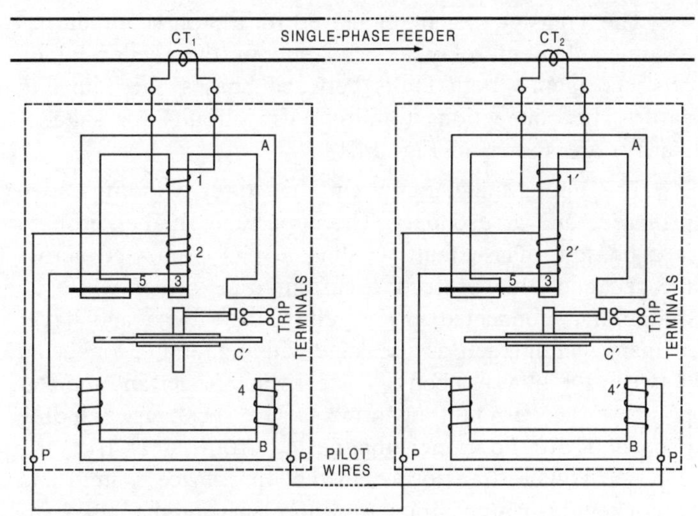

Fig. 14.13. ***Protection System For Single-Phase Feeders***

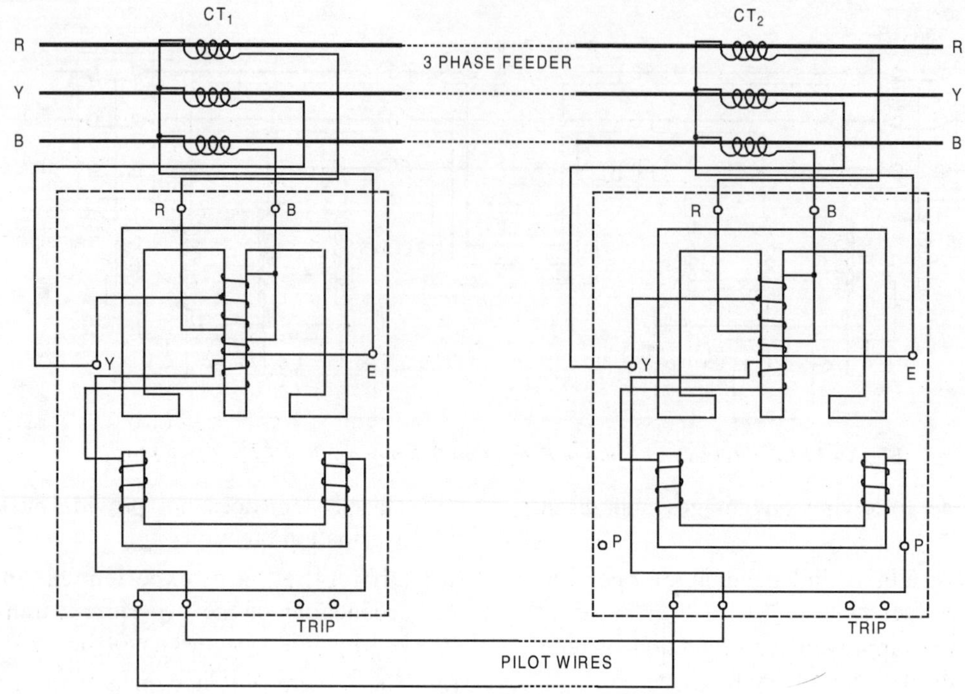

Fig. 14.14. ***Translay Protection System For 3-Phase Feeders***

opposite ends of the feeder carry equal currents and therefore the coils 1 and 1′ connected to them induce equal emfs in the windings 2 and 2′ respectively.

Since the windings 2 and 2′ are connected in opposition by means of pilot wires with the operating coils 4 and 4′ in series with them so no forward torque is exerted on the disc. When a fault occurs, the current through one CT is greater than that through the other so a small current circulates through the operating coils and pilot wires and when it attains the preset value the relay is caused to close the tripping circuit and thus disconnect the faulty feeder.

The Translay relay employed is quite similar in construction to an overcurrent induction type relay, as already described in Art. 9.20.

The Translay system, employed for a 3-phase circuit has a single element relay at each end of the feeder which protects against both faults between phases and faults to earth. The connections (omitting trip circuits for sake of clarity) are shown in Fig. 14.14.

The upper magnetic circuit has three windings, two primaries and a secondary. The upper and smaller primary is a phase-to-phase fault winding and is connected across the red and blue protective current transformer while its midpoint is connected to the yellow. The lower and larger primary winding acts as a leakage winding, and is connected between the blue protective current transformer and the star point of the current transformers. The secondary winding provided on the upper magnetic circuit acts like the opposed-voltage transformer in the Merz-Price system and is connected in opposition to a similar winding by means of two pilot wires, at the other end of the feeder. The secondary windings provided on the lower magnetic circuit are connected in series with the pilot wires. The rotating disc is composed of two sectors. Under normal operating conditions no current flows through the pilot wires as the opposing voltages are equal. On the occurrence of fault, the voltages in the windings differ and so a current flows through the lower elements and pilot wires. A forward torque is thus exerted on the disc due to interaction of flux produced in the lower magnetic elements with the leakage flux of the upper magnetic elements. The phase relation required is achieved as in an energy meter. The capacitance currents lead the voltages and tend to rotate the disc in the opposite direction because of a closed copper ring near the end of the projecting limb of the upper magnetic circuit. Thus the main disadvantages of Merz-Price system has been avoided.

The Translay relay can be biased by an unsymmetrical phase adjustment, which provides a backward torque when the flux in the upper element is large.

This system has got following advantages over Merz-Price system :

(*i*) The capacitive currents do not affect the operation of the relays.

(*ii*) Only two pilot wires are required.

(*iii*) The current transformers of normal designs *i.e.*, with air gap can be employed.

(*iv*) The pilot resistance does not affect the operation as major part of energy required to operate the relay is obtained from current transformer.

(*v*) The closed copper loop provided in the relay prevents the relay from operating for through-fault current.

14.10. SPLIT-CONDUCTOR PROTECTION OF FEEDERS

This system is another method of securing the benefits of a balanced method of protection without the necessity of using pilot wires. The principle of operation depends on the fact that two conductors of equal length and impedance, when connected in parallel will share the load equally, provided that the insulation of the system is sound. When a fault develops on any one conductor it will carry more current than the other, and this inequality of currents is arranged to operate a relay and thus isolate the faulty line.

In this system of protection each phase of the line is split into two sections having equal impedances. The two sections are lightly insulated from each other. In this system, as shown in Fig. 14.15, a single-turn current transformer is inserted at each end of the split conductor. The current transformers consist of laminated iron rings on which a secondary winding is wound all round the periphery.

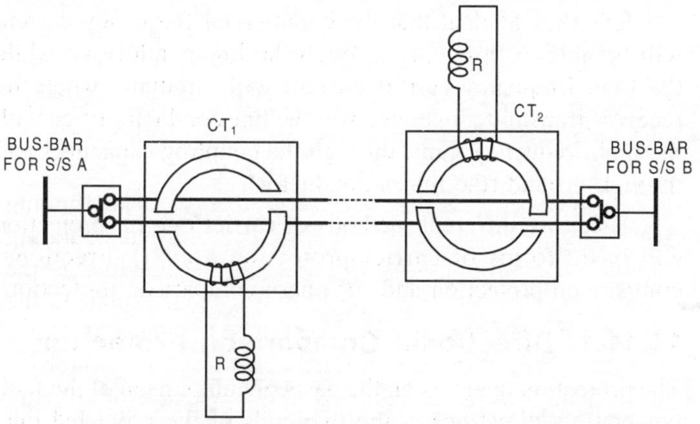

Fig. 14.15. *Merz-Hunter Split-Conductor Protection System*

Under healthy conditions the current flowing along the two splits are equal and since these are threaded through the current transformers CT_1 and CT_2 in the opposite directions hence the voltage across the terminals of the evenly spread secondary winding is zero. In fault conditions one of the split takes more current than the other, thereby giving rise to an unbalance of the primary side of the current transformers. Due to unbalancing of currents on the primary side of current transformers resultant flux will be set up in the core of the one of the current transformers and so the current will be induced in the evenly spread secondary and the relay coil R will be energised. The relay contacts will be closed and the trip coil will trip the circuit breaker and isolate the fault.

In the best arrangement the splits are carried into the circuit breakers on both sides of the feeder so that the splits are opened by the breakers. This is explained as follows :

Let the splits be not carried into the circuit breakers and a fault develop at the receiving end of a long line. Under these conditions, the impedance of the differential current transformer at this end may be insufficient to cause unbalance between the currents carried by each split conductor. Hence such a fault will not be cleared by the circuit breakers since the relay will not operate. But when the splits are carried into the circuit breakers the fault current is confined to the faulty split after the sending-end circuit breaker has tripped. In the former case, although the sending-end circuit breaker trips, the fault current is not confined to the faulty split but it would divide practically equal between the two splits being solidly connected, so the receiving-end circuit breaker will not trip. But in the latter case, the fault current is confined to the faulty split, the opening of the receiving-end circuit breaker takes place.

From above discussion it is obvious that the split circuit breakers impart a high sensitivity of operation to the protection and switchgear.

The disadvantage of this system is that we have to make use of a special type of cable with the lower limits for the voltages. In the case of overhead feeders, for each phase, a duplicate set of conductors, insulators, etc., have to be employed. The lines having step-up or step-down transformers or voltage regulators cannot be protected by this method.

14.11. CARRIER CURRENT PROTECTION

In modern high-power electrical systems it is necessary to have quick acting protections on long transmission lines. The requirements to be met by such protections are fully satisfied by the circulating current differential protection with its high sensitivity, quick action and independence upon the settings of the adjoining section protections. Not withstanding this, owing to the need for installing interconnecting conductors (cables) as already pointed out in previous Articles, circulating current differential protections are confined to lines up to 8 or 15 km long.

It is, however, possible to make use of the main line conductors as the interconnecting conductors of a circulating current differential protection. The need for special interconnecting conductors (cables) then disappears and it hence becomes possible to set up a circulating current differential protection on transmission lines of any length. This is the basis of what are called *carrier current protections*. The essential difference between carrier current protection and the voltage balance (Translay) pilot wire protection is that, in the former, only the phase angles of the currents at the two ends of a line are compared instead of actual currents as in the latter case and this phase angle decides whether the fault is internal or external.

To make possible the transmission of commercial frequency (50 Hz) load current, and at the same time use the main line conductors as the interconnecting conductors of the differential protection, it is necessary to use a current of higher frequency in order to be able to transmit current impulses from one end of the line to the other. High frequency signals in the range of 50 kHz to 400 kHz, commonly known as the *carrier*, are transmitted over the conductors of the protected line. To inject the carrier signal and to restrict it within the protected section of the line suitable coupling apparatus and line traps are used at both ends of the protected section. This obviously makes this protection scheme quite expensive and justifies its application only in transmission lines of 110 kV and above.

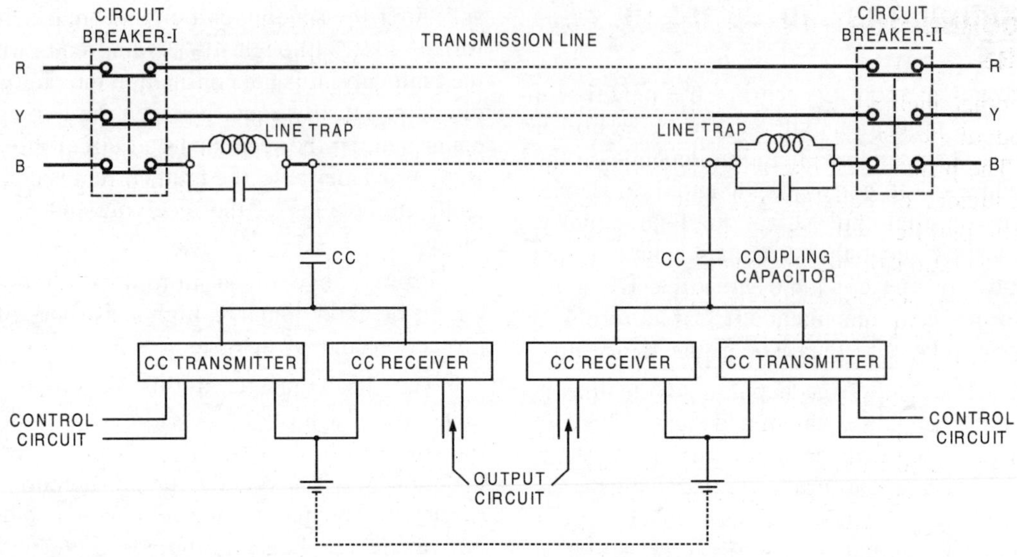

Fig. 14.16. *Schematic Circuit Diagram of Line Equipped For Carrier Current Transmission Line Protection (Line-Earth Scheme)*

The schematic diagram of carrier current protection is given is Fig. 14.16. The main elements of the carrier channel are (*i*) transmitter (*ii*) receiver (*iii*) coupling equipment and (*iv*) line trap.

Here we need not to go through the details of carrier current transmitters or receivers, all we need to know is that when a voltage of positive polarity is impressed on the control circuit of transmitter, it generates a high frequency output voltage. This output voltage is impressed between one phase conductor of the transmission line and the earth, as illustrated schematically in Fig. 14.16.

Each carrier current receiver receives carrier current from its local transmitter as well as from the transmitter at the distant end of the line. In effect, the receiver converts the received carrier current into a dc voltage that can be used in a relay or other circuit to perform any desired function. The voltage is zero when carrier current is not being received.

Line trap unit is inserted between the bus-bar and connection of coupling capacitor to the line. It is a parallel LC network tuned to resonance at the high frequency. It hence presents a high impedance to the high frequency carrier current, but a relatively low impedance (less than 0.1 Ω) to the power frequency (50 Hz) current. Traps are employed to confine the carrier currents to the protected section so as to avoid interference with or from other adjacent carrier current channels, and also to avoid loss of the carrier current signal in adjoining power circuits for any reason whatsoever, external short circuit being a principal reason. Consequently, carrier current can flow only along the line section between the traps.

The *coupling capacitor* (CC) connects the high frequency (carrier) equipment to one of the line conductors and simultaneously serves to isolate the carrier equipment from the high power line voltage. It presents a relatively low reactance to the high frequency currents (about 150 Ω at 500 kHz) and a high reactance to the power frequency (about 1.5 MΩ at 50 Hz). To reduce impedance further a low inductance is connected in series with the coupling capacitors to provide a resonance at carrier frequency.

It is thus evident that the commercial frequency current will be able to flow only through the line conductors, while the high frequency carrier current will circulate, when the receiver-transmitter operate, over the line conductor fitted with the high frequency traps, through the coupling capacitors and through ground (the return conductor).

There are different methods of carrier current protection and basic forms of carrier protection are: (*i*) directional comparison protection and (*ii*) phase comparison protection.

14.11.1. Directional Comparison Protection

The protection operates on the basis of comparison of the fault power flow directions at the two ends of the protected line. Operation takes place only when the flow of power at both ends of the line is in the bus-to-line direction, a condition which will evidently only arise in event of a fault on the protected section of the line. With directional comparison relaying, the carrier pilot informs the equipment at one end of the line how a directional relay at the other end responds to a short circuit.

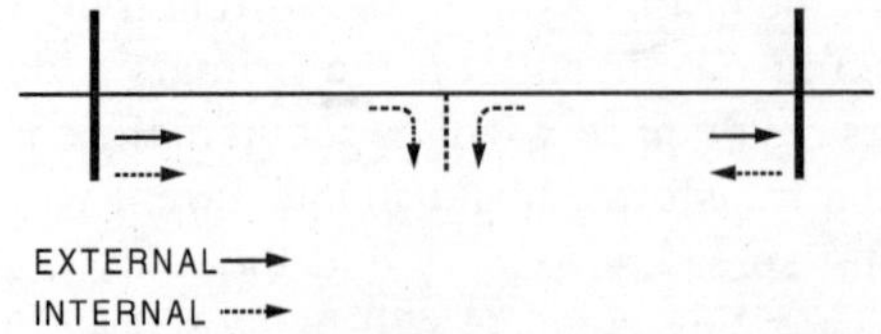

Fig. 14.17. *Direction of Current Flow For External and Internal Faults*

The conditions for internal and external faults are illustrated in Fig. 14.17. The relays at both ends of the protected section respond to fault power flowing away from the bus (tripping direction). For faults in the protected section, power flows in the tripping direction at both ends. For external faults power flow will be in opposite directions. A simple signal through carrier pilot is transmitted from one end to the other during faults.

The pilot scheme can be employed for transmitting either blocking, or permitting signals. Thus possible carrier protections are of two types viz, carrier blocking scheme and carrier permitting scheme.

In a *carrier blocking protection scheme,* the presence of carrier prevents or blocks operation of the protection. Carrier is, therefore, transmitted only upon the occurrence of a fault and is employed to prevent tripping in the event of an external fault. In carrier permitting scheme the presence of carrier permits operation of the protection. The carrier blocking scheme is more reliable than carrier permitting scheme. This is because a failure in the carrier permitting signal equipment will mean a failure in isolating the fault, where as a failure in carrier blocking signal equipment isolates the section on which no fault exists. However, such false operation is preferable to the failure to clear a faulted section.

In a *carrier blocking protection scheme,* normally no pilot signal is transmitted from any terminal. Should a short circuit occur in an immediately adjacent line section, a pilot signal is transmitted from any terminal where short-circuit current flows out of the line (*i.e.,* in the non-tripping direction). While any station is transmitting a pilot signal, tripping is blocked at all other stations. But should a short circuit occur on the protected section of the line, no pilot signal is transmitted and tripping occurs at any terminal where short-circuit current flows. Therefore, the pilot is blocking pilot, since the reception of a pilot signal is not required of permit tripping.

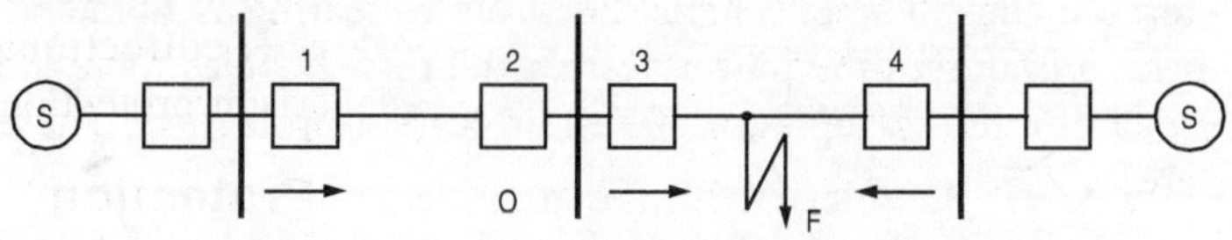

Fig. 14.18. ***Illustration of Directional Comparison Protection of a Power System***

Directional comparison protection scheme (carrier blocking type) is illustrated in Fig. 14.18. The operation of the directional element provided on each breaker is indicated by the arrow and the non-operation by the letter O. Occurrence of fault activates relays on each of the breakers near the fault. These relays unless blocked from operation, cause tripping of breakers. The blocking signal is controlled by the directional relays on each breaker, and is transmitted from one end of a protected section to the other by carrier. If a directional element determines that the fault is external to the protected section, a signal is transmitted blocking the operation of breakers at both ends of the section. In case the directional elements at both ends determine that fault is in the protected section, no blocking signal is transmitted from either end, and both breakers trip. The sequence of event for a fault at F is made clear by illustration in Fig. 14.18.

At breaker 1 the directional element shows that the fault may be in the section 1-2. This breaker trips if no blocking signal is received. No blocking signal is transmitted to breaker 2. At breaker 2 the directional element shows that the fault is not in section 1-2. A carrier signal is transmitted that blocks tripping of both the breakers 1 and 2.

At breakers 3 and 4 the directional elements show that the fault may be in section 3-4. No blocking signal is transmitted and after a very short time delay (1 to 3 cycles), both the breakers 3 and 4 trip.

14.11.2. Phase Comparison Carrier Protection

Phase comparison relaying equipment uses its pilot to compare the phase relation between current entering in the protected zone and the current leaving the protected zone. The current magnitudes are not compared. Phase comparison protection provides only main or primary protection, backup protection must be provided in addition.

Figure 14.19 shows schematically the principal elements of the equipment provided at both ends of a two-terminal transmission linc, employing a carrier current pilot. As in ac wire pilot protection, the transmission line CTs feed a network that transforms the CT output currents into a single-phase sinusoidal output voltage. This voltage is applied to a carrier current transmitter and to a 'comparer'. The output of a carrier current receiver is also applied to the comparer. The comparer controls the operation of an auxiliary relay for tripping the transmission line circuit breaker. These elements provide means for transmitting and receiving carrier current signals for comparing at each end of the relative phase relations of the transmission line current at both ends of the line.

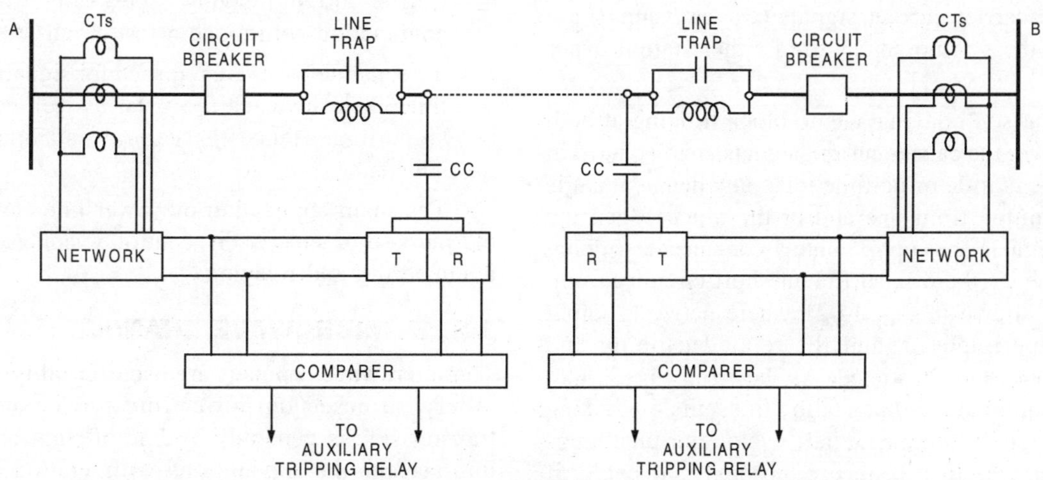

Fig. 14.19. ***Schematic Representation of Phase Comparison Carrier Protection***

For examining the relations between the network output voltages at both ends of the line and also the carrier current signals that are transmitted during external and internal fault conditions refer to Fig. 14.20. For an external fault at point D in Fig. 14.20, the network output voltages at stations A and B (waves *a* and *c*) are 180° out of phase. This is because the connections of CTs at the two stations are reversed. Since an ac voltage is used to control the transmitter, carrier current is transmitted only during the half cycles of the voltage wave when the polarity is positive. The carrier current signals transmitted from stations A and B (waves *b* and *d*) are displaced in time, so that there is always a carrier current signal being sent from one end or the other.

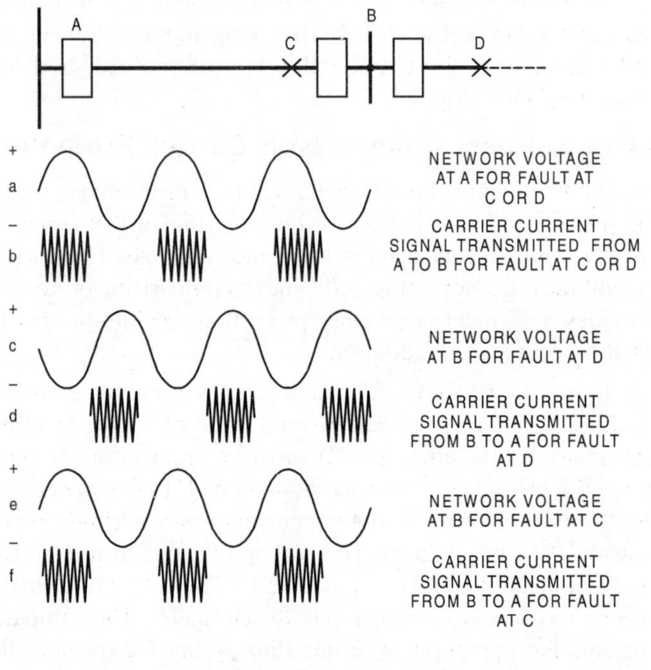

Fig. 14.20. ***Relations Between Network Output Voltages and Carrier Current Signals***

For internal fault at point C, the network output voltage at station B reverses because of reversal of power line currents at station B, the carrier current signals (waves *b* and *d*) are concurrent, and there is no signal from either station every other half cycle.

Phase comparison relaying acts to block tripping at both terminals whenever the carrier current signals are displaced in time so that there is little or no time interval when a signal is not being transmitted from one end or the other. When the carrier current signals are approximately concurrent, tripping will occur whenever there is sufficient short-circuit current flowing. This is shown in Fig. 14.21 where network output voltages are superimposed, and the related tripping and blocking tendencies are illustrated. As shown in Figs. 14.20 and 14.21, the equipment at one station transmits a blocking carrier current signal during one half cycle, and then stops transmitting and tries to trip during the next half cycle, if carrier current is not received from the other end of the line during this half cycle, the equipment operates to trip its breaker. But, if carrier current is received from the other end of the line during the interval when the local carrier current transmitter is idle, tripping does not occur.

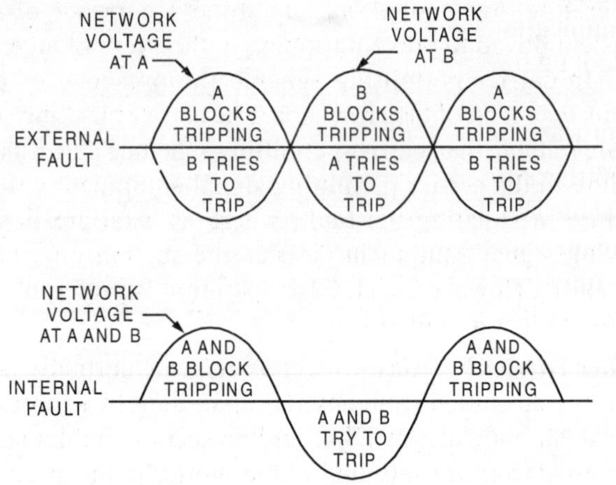

Fig. 14.21. ***Relation of Tripping and Blocking Tendencies To Network Output Voltages***

14.11.3. Advantages of Carrier Current Protection

Carrier current over the power line provides simultaneous tripping of circuit breakers at both ends of the line in one to three cycles. Thereby high speed fault clearing is obtained, which improves the power system stability. Besides there are several other advantages of carrier current protection. These are:

1. Fast, simultaneous operation of circuit breakers at both ends.
2. Autoreclosing simultaneous reclosing signal is sent thereby simultaneous (1 to 3 cycles) reclosing of circuit breaker is obtained.
3. Fast clearing prevents shocks to systems.
4. Tripping due to synchronising power surges does not occur, yet during internal fault clearing is obtained.
5. For simultaneous faults, carrier current protection provides easy discrimination.
6. Carrier current relaying is best suited for fast relaying in conjunction with modern fast circuit breakers.
7. No separate wires are required for signalling, as the power lines themselves carry power as well as communication signalling. Hence the capital and operating costs are smaller.

The main application of power line carrier has been for the purpose of supervisory control, telephone communication, telemetering and relaying.

14.12. MICROWAVE CHANNEL

The microwave channels are used for all types of protections otherwise based on power line carrier or pilot wire. The transmission is generally by line of sight and this must take into account the curvature of earth and topology of the route cover which the transmission takes place. This limits the maximum length of the simplest microwave channel to about 40 to 60 km.

It uses ultra high frequency (450 MHz to 10,000 MHz) transmitter-receiver system for connecting the relaying equipment located at the terminals of the protected line. The communication channel is space in this case; thus the line need not be fitted with additional equipment. The transmitters and receivers are controlled in the same way as the carrier current transmitter and receiver.

With radio links (microwave pilots) the signals are communicated by line of light antenna equipment. These are most expensive, but provide fast and reliable service. In USA radio links are used for communication, remote control and protection.

14.13. DISTANCE OR IMPEDANCE PROTECTION

The distance protection provides discrimination protection without making use of pilot wires. Distance protection is widely employed for protection of high voltage ac transmission lines because of its inherent advantages, as already mentioned in Art. 9.14.1. The principle and operation of distance relays have already been discussed in chapter 9.

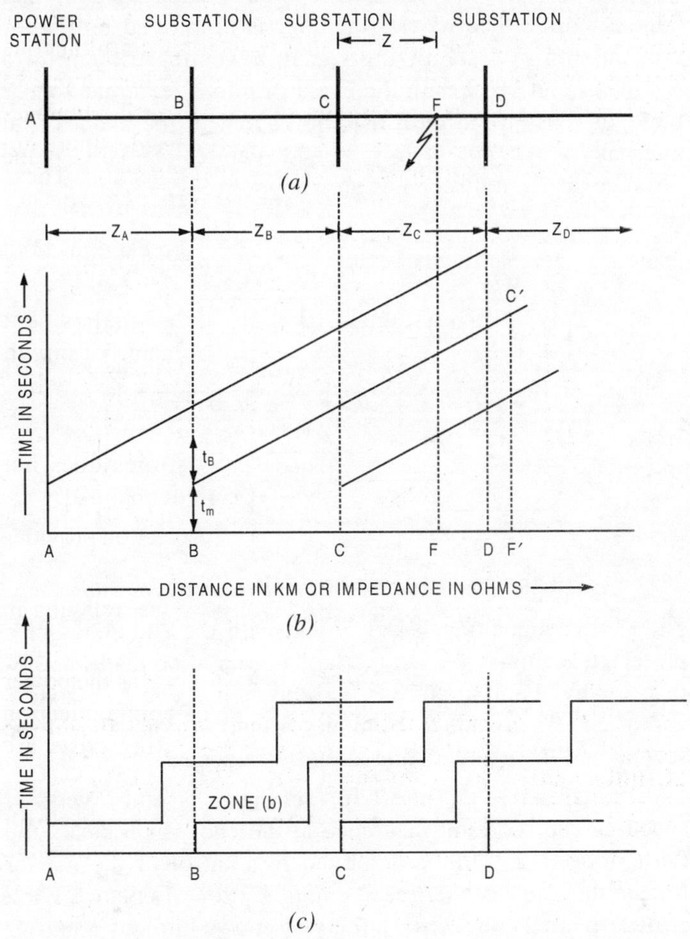

Fig. 14.22. ***Distance or Impedance Protection***

Figure 14.22 shows the simplest system consisting of feeders in series such that the power can flow only from left to right. The relays at A, B, C and D are set to operate with impedances less than Z_A, Z_B, Z_C and Z_D respectively. For a short-circuit fault at point F between substations C and D, the fault loop impedances at power station A and substations B and C are $(Z_A + Z_B + Z)$, $(Z_B + Z)$, and Z respectively. It is obvious that only relay at substation C will operate. Similarly for short-circuit faults between substations B and C, and power station and substation B only relays at substation B and power station A respectively will operate.

A system with instantaneous impedance relays, set to act on impedances less than or equal to the impedance of a section, as illustrated in Fig. 14.22 (*a*), would be difficult to adjust; a fault near the junction of two sections is likely to cause the operation of two relays. Furthermore, if a fault of finite resistance occurs near the end of a section, it is possible that total impedance is greater than that for relay operation. For these reasons it is advantageous to use impedance time relays, the characteristics of which are illustrated in Fig. 14.22 (*b*), for the power system illustrated in Fig. 14.22 (*a*).

If a fault occurs on right hand side of a substation B, say, relay at substation B operates in the minimum time t_m and the breaker at substation B operates t_B second later. If t_B is made less than the time difference between consecutive relays, only one relay will operate. Assume that the fault at F has a resistance causing total impedance at substation C represented by the point F′ (the fault resistance being FF′). Relay as substation C operates in time F′C′, where as in the previous system it would not operate at all.

An impedance time relay is a delicate mechanism and it is considered worthwhile to replace it by three simple impedance relays with a definite time of operation. The series combination can be arranged to provide a 3 step time characteristic, as illustrated in Fig. 14.22 (*c*), which does the same thing as the previous linear characteristic.

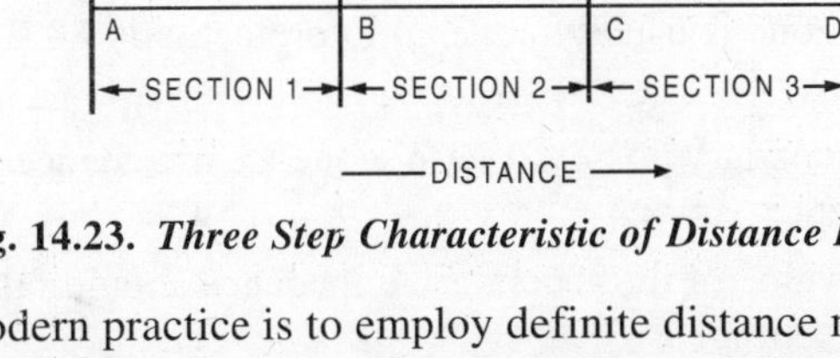

Fig. 14.23. ***Three Step Characteristic of Distance Relay***

Modern practice is to employ definite distance method of protection applied in 3 zones (steps). A number of distance relays are used in association with timing relays so that the power system is divided into a number of zones with varying tripping times associated with each zone. The first zone tripping which is instantaneous is normally set to 80 % of the protected section. The zone 2 protection with a time delay sufficient for circuit breaker operating time and discriminating time margins covers the remaining 20 % portion of the protected section plus 25 to 50 per cent of the next section. Zone 2 also provides backup protection for the relay in the next section for fault close to the bus. Zone 3 with still more time delay provides complete backup protection for all faults at all locations.

Thus the distance protection provided for line AB (section 1) serves as backup protection for sections 2 and 3, because in case of occurrence of faults in line BC (section 2) or line CD (section 3) it will clear those in their respective zone time from tripping the circuit breaker at end A.

14.13.1. Adjustment of Distance Relays

Distance relays are adjusted on the basis of the positive phase-sequence impedance between the relay location and the fault location beyond which operation of a given relay unit should stop. The impedance or the corresponding distance is called the "reach" of the relay or unit. For the purpose of rough approximation, it is customary to assume an average positive phase sequence reactance value of about 0.5 Ω per km for open transmission line construction and to neglect resistance.

All the relays are energised through CTs and PTs, the primary value of impedance must be converted to secondary values. For conversion of primary impedance to a secondary value for use in adjusting a phase or ground distance relay, the following relation is used.

Secondary impedance,

$$Z_{sec} = \frac{V_2}{I_2} = \frac{V_1}{\text{PT ratio}} \times \frac{\text{CT ratio}}{I_1}$$

$$= \frac{V_1}{I_1} \times \frac{\text{CT ratio}}{\text{PT ratio}} = Z_{prim} \times \frac{\text{CT ratio}}{\text{PT ratio}}$$

where the CT ratio is the ratio of high-voltage phase current to the relay phase current and the PT ratio is the ratio of the high-voltage phase-to-phase voltage and the relay phase-to-phase voltage—all under balanced 3-phase conditions.

Thus for a 100 km, 132 kV line with 600/5 Y-connected CTs and PT ratio $\frac{132,000}{110}$ the secondary positive phase-sequence impedance

$$= 100 \times 0.5 \times \frac{600}{5} \times \frac{110}{1,32,000} = 5\ \Omega$$

First zone is usually covered to protect only 80 % of first section.

So secondary positive phase sequence impedance to reach 80 % length = 0.8 × 5 = 4 Ω

This value of the secondary impedance should fall within the relay setting.

The relay reach may be severely limited if there is an appreciable infeed at the far end bus-bars as shown in Fig. 14.24. This is because an extra current due to the infeed while flowing through the line impedance between the far end bus-bars and the fault does not flow through the relay location whilst the extra voltage drop in the line due to this infeed current will increase the apparent ohmic impedance seen by the relay.

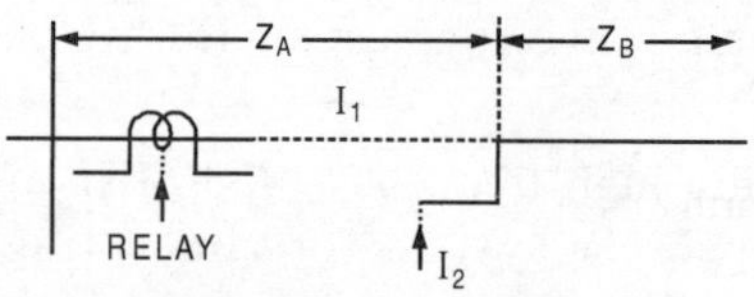

Fig. 14.24. ***Infeed in Line***

$$\text{Thus } V_R = (I_1 + I_2) Z_B + I_1 Z_A$$

$$I_R = I_1$$

$$\text{Therefore, } Z_R = \frac{V_R}{I_R} = \frac{(I_1 + I_2) Z_B + I_1 Z_A}{I_1}$$

$$= \left(1 + \frac{I_2}{I_1}\right) Z_B + Z_A$$

In case the infeed current I_2 equals the relay current I_1 (*i.e.*, $I_2 = I_1$) then the line impedance between the far end busbar and fault point will appear to the relay as twice its actual impedance value.

14.14. CARRIER AIDED DISTANCE PROTECTION

The directional comparison carrier pilot relay schemes presently employed are built around standard three-zone step type distance relays. This speeds up fault clearance for internal zone 2 faults. The carrier channel is employed either for transmission of a stabilizing signal preventing tripping of a remote circuit breaker in the event of a local external zone 2 fault, or for providing a tripping signal in the event of an internal zone 2 fault. The principal features of plain 3-zone distance protection schemes are shown in Fig. 14.25.

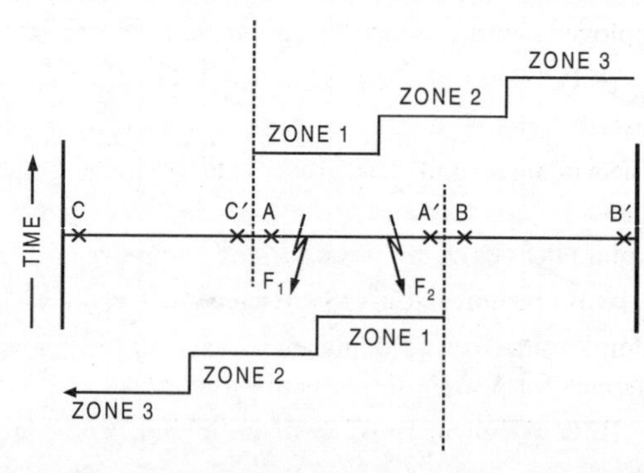

Fig. 14.25. ***Conventional 3-Zone Distance Protection***

Carrier signalling is concerned with the end zones of a protected section A A'. Let the faults occur at points F_1 and F_2 respectively. Fault at point F_1 will be seen at end A in zone 1 and at end A' in zone 2. Similarly a fault at point F_2 will be seen at end A′ in zone 1 and at end A in zone 2.

Transfer trip or intertrip technique is employed for speeding up the fault clearance at the end which clears the fault in zone 2. This is achieved by control of the carrier transmitter and a carrier receive relay by zone 1 contact. For a fault at point F_1 the zone 1 relay at end A initiates a carrier signal in addition to completing the zone 1 trip circuit of this end. Carrier signal on reaching end A' trips it immediately by shunting the zone 2 timer contacts with the help of a carrier receiver relay. A fault at point F_2 is also cleared in the same way.

Exercises

1. How earth fault protection is achieved in case of feeders ? Explain with sketches.
2. Explain overcurrent protection of feeders. How is the protection system graded with respect to the time of operation of relays for a radial feeder.
3. Describe in detail the protection of parallel feeder and ring mains. **[A.M.I.E. Sec. B. Power System Protection and Communication Summer 2000]**
4. Explain a scheme of protection for a ring main.
[Pb. Univ. Switchgear and Protection October 1989]
5. Describe a suitable protective system for a ring main formed by a power station and four substations, and explain clearly its working when a fault occurs. Explain the special features that a protective relay should possess in this case and also indicate the limitations of the system.
6. Describe the protection of ring mains feeder with an example.
[U.P. Technical Univ. Switchgear and Protection 2004-05]
7. Explain the time graded and current graded systems in overcurrent protection.
[U.P. Technical Univ. Switchgear and Protection 2004-05]
8. Explain the principle of time grading and current grading in a simple radial system equipped with IDMT overcurrent relays for protection.
9. Discuss the merits and demerits of various pilot-wire relaying schemes for protecting transmission lines.
[Pb. Univ. Switchgear and Protection April 1986]
10. Explain the term 'pilot' with reference to power line protection. What are the different types of pilot schemes which are presently employed in transmission line protection.
[U.P. Technical Univ. Switchgear and Protection 2008-09]
11. Describe pilot wire protection, its merits and demerits in detail.
[Switchgear and Protection; U.P. Technical Univ. 2009-10; G.B. Technical Univ. 2011-12]
12. Explain the operating principles of pilot wire protection.
[G.B. Technical Univ. Switchgear and Protection 2010-11]
13. Compare merits and demerits of various pilot wire relaying schemes for protecting transmission lines.
[U.P. Technical Univ. Switchgear and Protection 2006-07]
14. What is the Merz-Price voltage protection scheme ? How does it protect feeders ? Explain it with a neat diagram.
15. Discuss the problems of wire pilot differential relaying. Why are wire pilot relaying schemes used for short lines ? Discuss with necessary diagram the principle of operation of Translay feeder protection scheme.
16. Explain with the help of neat sketches the setup of carrier current relaying employed in transmission line protection.
Explain the utility of (*i*) line trap unit and (*ii*) coupling capacitor unit.
Discuss why carrier current protection is suitable for important interconnected lines.
17. Explain carrier current protection scheme with block diagram and neat sketches. Discuss how phase comparison scheme can be used to protect feeder from both ends.
[U.P. Technical Univ. Switchgear and Protection 2005-06]
18. Draw the schematic diagram of the carrier current protection scheme of lines. Also explain its working principle.
19. Explain the directional comparison method of carrier current protection.
20. Explain in detail carrier current protection scheme. Describe carrier phase comparison relay with neat sketches.
21. What are the advantages of distance protection over other types of protection of feeders ?
Describe any type of impedance relay and show how they provide discriminatory protection.
[Pb. Univ. Switchgear and Protection May 1990]
22. Discuss the directional reactance scheme for distance protection.
[G.B. Technical Univ. Switchgear and Protection 2011-12]
23. Explain the zonal protection scheme for feeder. Describe the reactance relay characteristic for 3-zone protection. Also draw the contact circuit for the same.
[U.P. Technical Univ. Switchgear and Protection 2007-08]
24. Explain with necessary diagram the principle of operation of a distance protection scheme. How a 3-zone stepped units help to achieve fast discrimination ? How many units are required for a complete protection scheme ? How the CTs and VTs associated with these relays are connected to a 3-phase transmission line ? Indicate the reasons of mode of connections of CTs and VTs in such a way.
[U.P.S.C. I.E.S. Electrical Engineering-II, 2005]
25. Describe the trip circuit diagram of 3-zone distance relay used for the protection of a transmission line, Why 3-zones are necessary ? How their reach setting and time setting is done ? **[A.M.I.E. Sec. B. Power System Protection and Communication Winter 1995]**
26. Discuss with a neat sketch the general principle of operation of a distance protection scheme. How many units are required for a complete distance protection scheme ?
Explain clearly the necessity of using 3-zone stepped distance protection scheme to achieve fast and discriminative protection of transmission line.
[A.M.I.E. Sec. B. Advanced Power Systems Summer 1998]
27. Explain stepped time-distance characteristics of three distance relaying units used for first, second and third zones of protection.
[G.B. Technical Univ. Switchgear and Protection 2010-11]
28. Discuss three zone protection scheme using Mho relay to protect trnasmission lines. Explain how static Mho relay is realized with phase comparator.
[U.P. Technical Univ. Switchgear and Protection 2005-06]
29. Give schemes of protection for a parallel feeder fed from (*i*) one end (*ii*) both the ends.
Using carrier system of protection, what is the basis of choice of frequency in power line carrier system.
[U.P. Technical Univ. Switchgear and Protection 2006-07]
30. Write short notes on the following:
(*i*) Time graded protection of feeders. **[Pb. Univ. June 1991]**
(*ii*) Protection of parallel feeders. **[Pb. Univ. Oct. 1986; Sept. 1988]**
(*iii*) Protection of ring mains. **[Pb. Univ. April 1986]**
(*iv*) Pilot wire protection of feeders using Merz-Price system.
(*v*) Translay protection to a 3-phase feeder.
(*vi*) Split-conductor protection.
(*vii*) Carrier current protection of transmission lines.
[Pb. Univ. Sept. 1998; Sept. 1990]

(*viii*) Carrier current protection.

[Agra Univ. 1981; Pb Univ. May 1990]

(*ix*) Distance protection of feeders. **[Pb. Univ. Sept. 1991]**

(*x*) Carrier aided distance protection.

Short Answer Type Questions With Answers

Q. 1. What is unit type protection ?

Ans. Unit system of protection is one in which the protection responds to faults in the protected zone alone and it does not respond to through faults (faults beyond the protected zone). Non-unit systems do not have exact zone boundary.

Q. 2. What are unit systems and non-unit systems of protection ?

[U.P. Technical Univ. Switchgear and Protection 2006-07]

Ans. Selectivity is *absolute* if the protection responds only to the faults within its own zone, and *relative* if it is obtained by grading the settings of the protective relays of several zones all of which may respond to a given fault. Systems of protection which in principle are *absolutely selective* are called the *unit systems*. Systems in which selectivity is relative are *non-unit* systems. Differential protection and frame leakage protection fall in the former category while the current time graded protection and distance protection fall in the latter category.

Q. 3. Give the examples of unit/and non-unit systems of protection.

Ans. Unit type protection includes pilot wire differential protection, carrier current protection based on phase comparison etc. while non-unit type protection includes time graded overcurrent protection, current graded overcurrent protection, and distance protection.

Q. 4. What is meant by time graded protection ?

Ans. Time graded protection is a scheme of overcurrent protection, in which discrimination is incorporated *i.e.,* the time setting of relays is so graded that in the event of fault, the smallest possible part of the system is disconnected.

Q. 5. What is the purpose of time grading of protection system and where is it employed ?

Ans. The purpose of time grading of protection system is to incorporate time discrimination. It is employed for the protection of radial feeders, parallel feeders and interconnected systems.

Q. 6. What type of relay is suitable for radial feeders ?

Ans. Induction type inverse definite minimum time (IDMT) relays are most suitable for radial feeders.

Q. 7. What is the minimum time interval which can be permitted for the two adjacent circuit breakers to operate in radial feeders ?

Ans. 0.4 second.

Q. 8. Why induction type IDMT relays are most suitable for protection of radial feeders ?

Ans. Induction type IDMT relays are most suitable for protection of radial feeders because their time-current characteristics are similar in shape and in no case they cross each other at any point.

Q. 9. On what fact is the current graded protection based ?

Ans. Current graded protection is based on the fact that the short-circuit current along the length of the protected circuit decreases with the increase in distance between the supply end and the fault point.

Q. 10. What type of relay is suitable for current graded protection ?

Ans. High-speed high-set overcurrent relays.

Q. 11. Can Merz-Price system be employed for protection of ring mains and duplicate feeders ?

Ans. Yes.

Q. 12. Does Merz-Price system provide backup protection or overload protection ?

Ans. No.

Q. 13. Why Merz-Price protection is not suitable beyond 33 kV ?

Ans. Difficulties are experienced in balancing the secondaries of the two CTs and that is why Merz-Price protection system cannot be used beyond 33 kV.

Q. 14. Is the operation of relays affected by capacitive currents in Translay system of protection of feeders?

Ans. No.

Q. 15. Why is use of wire pilot relaying scheme restricted to protection of short transmission lines ?

Ans. In case of medium and long transmission lines the pilot wire protection scheme becomes expensive because of cost of pilot wires.

Q. 16. What are the main elements of current carrier protection ?

Ans. The main elements of the current carrier protection scheme are: (*i*) transmitter (*ii*) receiver (*iii*) coupling equipment and (*iv*) line trap.

Q. 17. What purposes are served by line trap and coupling capacitors in carrier channels ?

Ans. Line trap is an LC network inserted between the bus-bar and connection of coupling capacitor to the line and tuned to resonance at the high frequency and are used to confine the carrier currents to the protected section so as to avoid interference with or from adjacent carrier current channels and also to avoid loss of the carrier current signal in adjoining power circuits for any reason whatsoever.

The coupling capacitor is used to connect the high frequency (carrier) equipment to one of the line conductors and simultaneously serves to isolate the carrier equipment from the high power line voltage.

❑❑❑

CHAPTER 15

Protection Against Overvoltages Due to Lightning

INSIDE THIS CHAPTER

15.1. INTRODUCTION

By *voltage surge* is meant any sudden excessive rise in voltage that may be dangerous to the electrical equipment of an installation. Among such surges are those whose value approaches the voltage at which the equipment in the installation has been tested. Overvoltages may occur both as a rise in voltages between phases and between a phase and ground.

All voltage surges in high-voltage installations are broadly classed under two headings; internal and external voltages.

Internal overvoltages originate in the system itself and may be transient, dynamic or stationary. Those of a *transient nature* will have a frequency unrelated to the normal system frequency and will persist a few cycles only. They can be caused by the operation of circuit breakers when switching inductive or capacitive loads, "current chopping" when interrupting very small currents or by the sudden grounding of one phase of a system operating with an insulated neutral.

Dynamic overvoltages occur at normal system frequency and persist only for a few seconds. They may be caused, for example, by the disconnection of a generator which overspeeds, or when a large portion of the load is suddenly thrown off.

Stationary overvoltages also occur at system frequency but they may persist for some time, perhaps hours. Such a condition can arise when an earth fault on one line is sustained, as indeed it may intentionally be when the neutral is grounded through an arc suppression coil, thereby leading to the overvoltages on the sound phases.

Overvoltages described above rarely exceed three to five times the normal phase-to-neutral peak voltage of the system and subject to apparatus having an adequate insulation level, should be relatively harmless.

External overvoltages are caused by atmospheric discharges such as static charges or lightning strokes and are therefore not related to the system. They are often of such magnitude as to cause considerable stress on the insulation and, in the case of lightning, will vary in intensity depending on how directly the line is struck, *i.e.,* directly by the main discharge, directly by a branch or streamer, or by induction due to a flash passing near to but not touching the line.

Switchgear specifications describe two classifications for an installation namely one which is "electrically exposed" resulting in the apparatus being subject to overvoltages of atmospheric origin, and another which is "electrically non-exposed" and therefore not subject to this type of overvoltage.

The first of these is generally recognized as an installation connected directly to overhead transmission lines, *e.g.,* outdoor switchgear or transformers but it can apply to indoor gear connected to overhead lines through outdoor to indoor through bushings or by a short length of cable. The second classification is one usually associated with underground cable networks and the switchgear will normally be of the indoor type.

Because of the stresses due to external voltages, it is now required that certain switchgear and transformers shall be impulse tested, and this is the case for circuit breakers of the outdoor type for service voltages of 22 kV and above. It is also intimated that suitable protective devices be installed to ensure that the amplitude of the surges imposed on the terminals of the outdoor apparatus is limited to a value not exceeding 80 per cent of the impulse test level.

A surge is the movement of a charge that is suddenly released in a conductor and which travels along that conductor in the form of a *wave*. The potential of a conductor being a measure of charge density, it follows that the potential wave must be accompanied by a current wave. The shape of the potential wave is affected by a number of factors, such as inductance and capacitance. Thus, on a line where the impedance is pure inductive, the front of the wave could be almost vertical, while if the impedance is pure capacitive then the wave front would slope steeply away from the vertical. From this it is clear that the waveshape is affected by the relation between L and C values, this relationship being expressed as the surge impedance,

$$Z_0 = \sqrt{L/C} \qquad ...(15.1)$$

where Z_0 is surge impedance in ohms, L is inductance in henries and C is capacitance in farads.

As the wave travels along the line its front is modified by the inductance of the line and the distributed capacitances to earth. It may be further modified by the capacitances of bushings, insulators, etc., which the wave encounters on its journey, thus reducing the steepness of the wave front.

A further point has also to be considered, namely that when the travelling wave reaches a point where there is a change in surge impedance of the line, reflection of the wave will occur, reflection being whole or partial depending on the amount of change in surge impedance. In the case, for example, of an open-ended feeder, the change in surge impedance is infinite and a travelling wave would be totally reflected, with the result that the pressure would be practically doubled at the point of reflection due to the front of the wave being reversed and adding itself to the remainder of the wave still travelling forward; the duration of doubled value will depend on the length of the wave tail. In other cases, such as one where an overhead line is joined to an underground cable, the change in surge impedance is not infinite and the reflection will depend on the relative values of the impedances of the overhead line and the cable.

It is obvious that surges when travelling along a line will, at some point reach terminal apparatus such as cable boxes, transformers, or switchgear, and, unless some means is provided to release the surge, there is a danger of a breakdown of the insulation of the terminal apparatus, unless the insulation level of the latter is high enough to be invulnerable. On smaller systems it could be an expensive solution and, therefore, it becomes necessary to install lightning arresters or surge diverters.

15.2. INTERNAL CAUSES OF OVERVOLTAGES

Internal overvoltages arise during sudden changes in the operating conditions of an installation, as, for example, a switch-in, switch-out or sudden change in load, the opening of circuits with large inductance or capacitance, the interruption of a short-circuit current, the appearance of an arc fault between one phase and earth, as well as other circuit disturbances.

In all the cases outlined above overvoltage appears as the result of the sudden release of a portion of the stored energy of the given installation or system.

The peak value of the overvoltage which may appear because of an internal circuit disturbance may exceed the operating voltage of the system by several times. It is expressed numerically by the *overvoltage ratio*, which is equal to the value of the overvoltage amplitude divided by the amplitude of the operating voltage.

Overvoltage characterized by a high ratio may become dangerous for both the line insulation and the insulation of the station and substation switchgear, bus-bar structures and apparatus.

Now the internal causes that give rise to overvoltages will be discussed in detail here.

1. ***Switching Operations on Unloaded Line.*** A switching operation produces a sudden change in the circuit conditions. When an open-ended line is connected to a source of voltage, travelling waves are set up which rapidly charge the line. On reaching the open end of the line, these waves are totally reflected without change of sign, thereby producing voltage—doubling at that end. These reflected waves travel back to the supply end, giving rise to further reflections. Actually, owing to losses, the waves gradually attenuate and the line ultimately attains the normal voltage. Hence the maximum potential attained by the line at the instant of switching-in cannot exceed twice the maximum supply voltage V_{max} or $2\sqrt{2}\ V_{rms}$ where V_{rms} is rms value of phase voltage. In practice, owing to the dissipation of energy, the potential attained by the line would be somewhat less than this theoretical maximum. If the surge impedance of the line is Z_0 then current will fluctuate between $\frac{+\sqrt{2}\ V_{rms}}{Z_0}$ and $\frac{-\sqrt{2}\ V_{rms}}{Z_0}$ sinusoidally.

Analogous phenomena is produced on switching out an open-ended line. Travelling waves are set up and the line momentarily attains the voltage of the magnitude not exceeding twice the supply voltage at the instant of disconnection.

2. *Sudden Opening of Loaded Line.* If a line carrying load is suddenly opened, a transient voltage of value given by $e = iZ_0$ is set up, where i is the instantaneous value of the current at the instant of opening of line, and Z_0 is the natural or surge impedance of the line.

If $i = 250$ A and $Z_0 = 500$ ohms, the transient rise of voltage, however, rapidly the current is interrupted cannot possibly exceed 250 × 500, *i.e.*, 125,000 volts, because this is the maximum value of the voltage wave necessary to store in the electrostatic field the whole of the energy stored in the magnetic field at the instant of break. If V_{max} is the peak value of the operating voltage, the maximum voltage to which the line may be subjected is V_{max} + 125,000 volts.

From the above discussion it is obvious that transient voltage rise by the sudden interruption of load is not a function of the line voltage and, therefore, low voltage transmission systems are liable to overvoltage of the same magnitude as high voltage systems. The higher safety factor of the insulation on low voltage systems is thus justified.

The worst case that could happen would be the interruption of the short-circuit current of the system at its peak value, when voltage rises of several hundred kilovolts might be caused. Fortunately, the use of oil circuit breakers has practically eliminated all danger on account of short circuits, as these breakers open the circuit when the current wave has its zero value.

3. *Insulation Failure.* The insulation failure in a power system may take place in various ways such as between the conductors of an overhead line or the cores of an insulated cable, or between one conductor, or core, and earth. The failure of insulation between the line and earth is very frequent. When there is a breakdown of insulation to earth, the potential at fault suddenly falls from maximum to zero and, therefore, a negative voltage wave of very steep front in the form of surge travels from the fault in both directions.

Let the line AB be earthed at point F. Let the instantaneous voltage of the line with respect to earth at the instant of failure of insulation be $2v$. The potential at the fault F suddenly falls to zero and therefore a negative voltage, of amplitude v, and of very steep front, travels from the fault in both directions. The reduction of voltage to one-half is due to the equal dimension of the energy into electromagnetic and electrostatic energy. Each voltage wave is accompanied by a current wave of amplitude v/Z_0 and consequently the current to earth, if there is an extremely low tower footing resistance—in the case of an overhead line is $-\dfrac{2v}{Z_0}$.

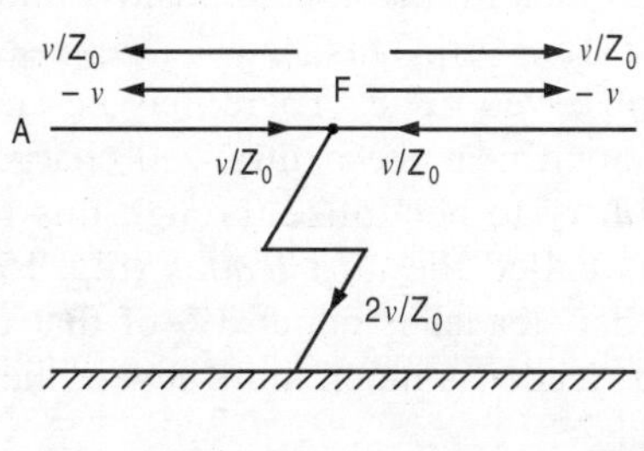

Fig. 15.1. ***Surge Due To Insulation Failure***

As these waves travel to the ends of the line they reduce the voltage to zero; and when they reach the open ends, reflected waves are set up which reduce the voltage to $v - v - v$, *i.e.*, $-v$ and the current is neutralized. When the reflected waves reach F, the portions of the line along which they have travelled will be charged to $-v$. The current at F can be reversed by a flash-over in the opposite direction, and the result is a periodic flash-over with reversals of potential on the line and currents at F until the stored energy is dissipated by damping.

4. *Arcing Grounds.* It is experienced in insulated neutral system. Consider an alternator, whose one phase has been connected to a long line which has got distributed inductance and capacitance to earth, as shown in Fig. 15.2. The alternator winding can be imagined to be connected to earth through its capacitance. Let a spark gap be connected between the line and earth. The alternating emf impressed on the circuit by the alternator will charge the line at each half cycle and if the line pd becomes high enough the gap will break down and the circuit will then resonate at its own natural frequency.

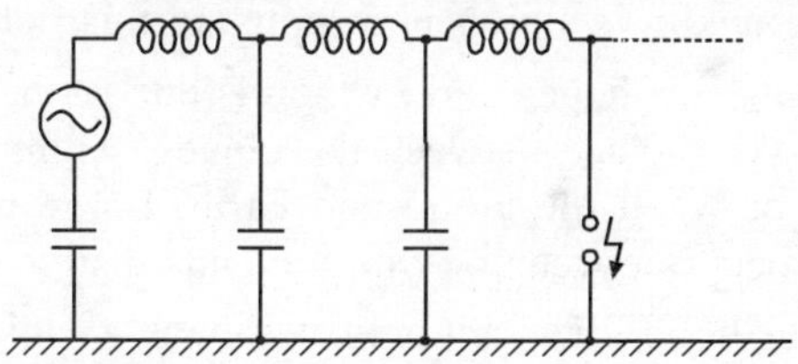

Fig. 15.2

Similar conditions arise when an earth fault develops on a transmission line connected to an all insulated system. On occurrence of flash-over, as by the puncture of an insulator on an overhead line, accidental contact with trees or by the breakdown of a weak spot in the insulation of an underground cable, the conductor is discharged through the agency of a transient high-frequency oscillation. The faulty conductor is thus brought down to earth potential, and the other two conductors of the three-phase system correspondingly rise in voltage from the phase to the line voltage. As soon as the conductor is discharged to ground, the spark gap to earth opens, and the conductor then charges again from the power supply of the system, and the cycle is repeated. In themselves, oscillations such as the above are not particularly dangerous, but, if continued, the resulting voltage rise may result in a breakdown of the insulation at some other point, and possibly on another conductor. In such a case double fault to earth will be formed, and it will be equivalent to a short circuit.

Surges and other transient phenomena due to switching operations are taking place in such a way that the initiating voltage is the system voltage, are rarely dangerous in themselves, partly because attenuation may limit serious voltage rise, and partly because the associated energy is small. The danger is when a flash-over may open a path for a power arc, much as the stepped leader of a lightning flash paves the way for the main discharge. An example is the double fault which may result from an arcing ground, as explained above. Because of the troubles due to arcing grounds power systems are rarely operated with insulated neutral. Instead the system neutral is grounded solidly (*i.e.*, without any impedance between the neutral and the ground) or through a resistance or reactance.

5. *Resonance.* Resonance in an electrical circuit implies that the impedance of the circuit is pure resistive and the power factor is unity. Thus at resonance the inductive reactances and capacitive reactances cancel out.

In usual transmission lines the capacitance is usually so small that resonance cannot occur at the fundamental supply frequency, but if the generator emf wave is distorted, trouble may be experienced due to resonance of one of the higher harmonics.

15.3. MECHANISM OF LIGHTNING

Benjamin Franklin performed his famous experiment of flying kite in thunder cloud in 1745. Before his discovery the lightning used to be considered to be "Act of God." He also proved that the lightning stroke is because of the discharge of electricity. He also invented lightning rods for fixation on tall buildings and grounding for their protection from lightning strokes. Thus Benjamin Franklin is a pioneer scientist in this field.

Lightning is a huge spark, which is due to the electrical discharge taking place between the clouds, within the same cloud and between the cloud and earth. Large number of discharges occur between or within clouds than to earth, and enough of them terminate on to the earth and result in serious hazards.

Many theories have been put forth to explain the phenomenon whereby a thundercloud becomes electrically charged. However, for our purpose it is quite sufficient to assume that as a result of certain atmospheric processes that take place during thunderstorms, charges are accumulated in clouds (or portions of clouds) and equal and opposite charges are induced in the earth beneath. According to another theory, the positive as well as negative ions in the air attach themselves to the small dust particles and small drops of water suspended in the air and because of polarisation by induction, they get charged to a certain potential under the storm conditions. Thus a cloud gets charged either positively or negatively. Whenever such a cloud passes over the earth, it induces opposite charges in the earth below.

Lightning discharge requires the puncture of air between the cloud and the earth. For breakdown of air at STP condition the electric field required is 30 kV/cm peak. But in a cloud where the moisture content in the air is large and also because of the high altitude (low pressure) it is reduced to 10 kV/cm. The mechanism of lightning discharge can be well explained with the help of Fig. 15.3. Whenever a local charge concentration near a cloud causes the potential gradient to exceed the critical breakdown value, the air surrounding gets ionized and an electric streamer starts from the clouds towards the earth. The current in the streamer is of the order of 100 A and the speed of the streamer is 0.15 m/μs. This streamer is known as *pilot streamer* because it leads to the lightning phenomenon. The process is shown in Fig. 15.3 (*a*). Depending upon the state of ionization of the air surrounding the streamer, it is branched in several paths and it is known as *stepped leader*. *This is illustrated* in Fig. 15.3 (*b*). The leader steps are of the order of 50 m in length and are accomplished in about a microsecond. The charge is brought from the cloud through the already ionized path to these pauses. The air surrounding these pauses is again ionized and the leader in this manner reaches the ground. This is illustrated in Fig. 15.3 (*c*).

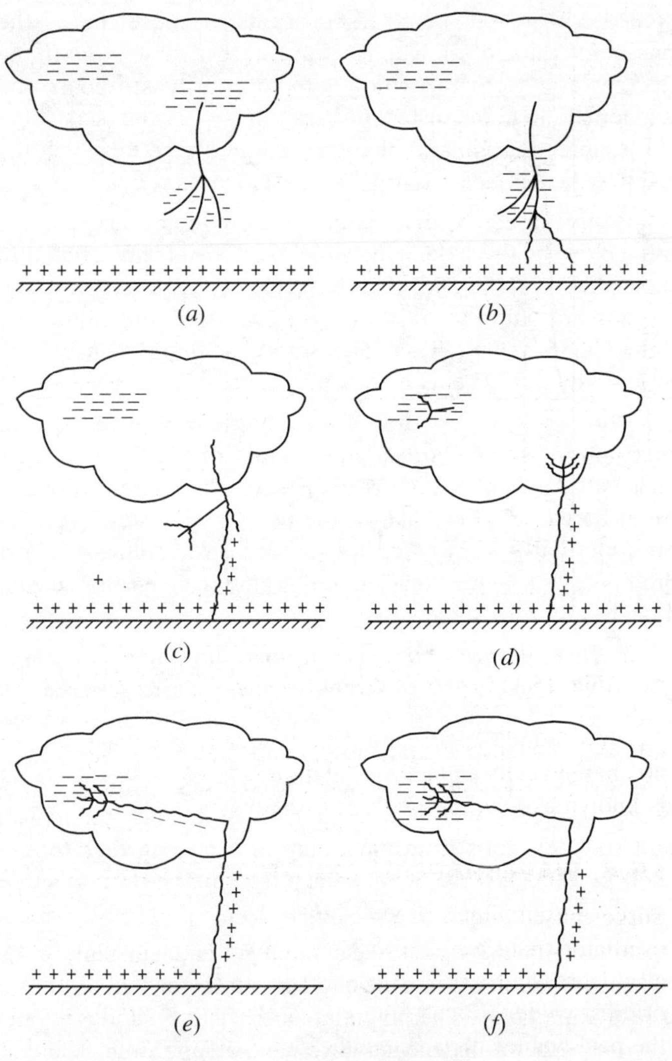

Fig. 15.3. ***Process of Lightning Discharge***

Once the stepped leader has made contact with the ground it is believed that a power return stroke [Fig. 15.3 (*c*)] moves very fast up towards the cloud through the already ionized path by the leader. This streamer is very intense where the current varies from 1 kA to 200 kA and the speed is about 10 per cent of that of light. It is here where the negative charge of the cloud is being neutralized by the positive induced charge on the earth [Fig. 15.3 (*d*)]. This instant gives rise to lightning flash which is observed by us. There may be another cell of charges in the cloud near the neutralized charged cell. This charged cell will try to neutralize through this ionized path. This streamer is called the *dart leader* [Fig. 15.3 (*e*)]. The velocity of the dart leader is about 3% of that of light. The effect of the dart leader is much more severe than that of the return stroke.

The discharge current in the return streamer is relatively very large but as it lasts only for a few microseconds the energy contained in it is small and, therefore, this streamer is called the *cold lightning stroke* while the dart leader is called the *hot lightning stroke* because even though current in this leader is relatively smaller but it lasts for some milliseconds and so the energy contained in it is relatively larger.

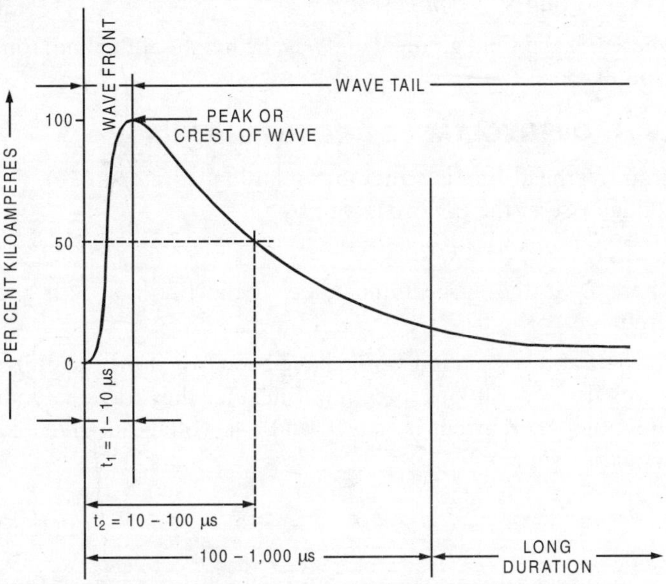

Fig. 15.4. ***Generalised Waveshape of Lightning Stroke***

It is found that each thundercloud may contain as many as 40 charged cells and a heavy lightning stroke may occur. This is known as *multiple* or *repetitive* stroke.

15.4. WAVESHAPE OF LIGHTNING STROKE

Surge caused due to lightning discharge are generally of non-oscillatory nature with a very steep wave front and have gradually decreasing tail. The wave is designated by (*i*) maximum or crest value (*ii*) the front time t_1 which is the time to rise to the peak value and (*iii*) the duration t_2 which is the time from 0 to the point where the voltage or current has decayed to one half its peak or crest magnitude. It has been observed that :

(*i*) t_1 varies from 1 µs to 10 µs

(*ii*) t_2 varies from 10 µs to 100 µs

(*iii*) Maximum value of peak current is 400 kA

(*a*) 2.5% of all the strokes are below 10 kA

(*b*) 84% of all the strokes are below 50 kA

(*c*) 11% of all the strokes are between 50 and 100 kA

(*d*) 2% of all the strokes are between 100 and 150 kA

(*e*) 0.5% of all the strokes are above 150 A.

15.5. LIGHTNING STROKES

There are two main ways in which lightning affects a line: by a direct stroke, and by electrostatic induction.

15.5.1. Direct Lightning Strokes

A direct stroke can take place in several ways. In one way a cloud attains a large amount of charge, it induces an opposite charge on the tall objects, such as temples, churches or mosques. In between these two charges an electrostatic field is set up and when the intensity of electrostatic field becomes sufficiently great to ionize the neighbouring air, the air breaks down and discharge takes place between the cloud and the object. Such a discharge is known as A *stroke*, and is characterised by the comparatively long time taken to produce it and the fact that it strikes the highest and most sharply pointed object in the neighbourhood. Another way results in a much more sudden stroke, which is produced in the manner shown in Fig. 15.6. Three clouds are involved. Clouds 1 and 3 are positively charged and cloud 2 is negatively charged. The potential of cloud 3 is reduced due to presence of the charged cloud 2. On flash-over from cloud 1 to cloud 2, both these clouds are discharged rapidly and cloud 3 assumes a much higher potential and flashes to earth very rapidly. This is the B *stroke*, and is characterized by its rapidity and the fact that it ignores tall objects and reaches ground in a random manner. Thus the stroke B is the result of stroke A between clouds 1 and 2. It is most dangerous stroke. No protection can be designed to avoid a stroke B where as protection can be provided against stroke A by channelising the charge through a lightning conductor placed at the highest point on the building and earthing properly at the other end. The stroke B is sometimes called the *induced stroke*.

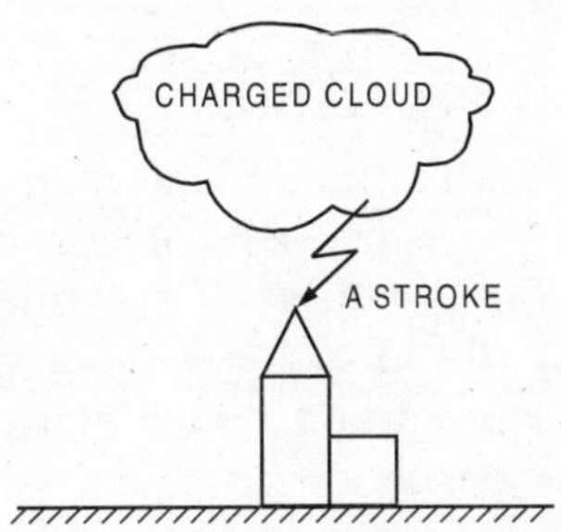

Fig. 15.5

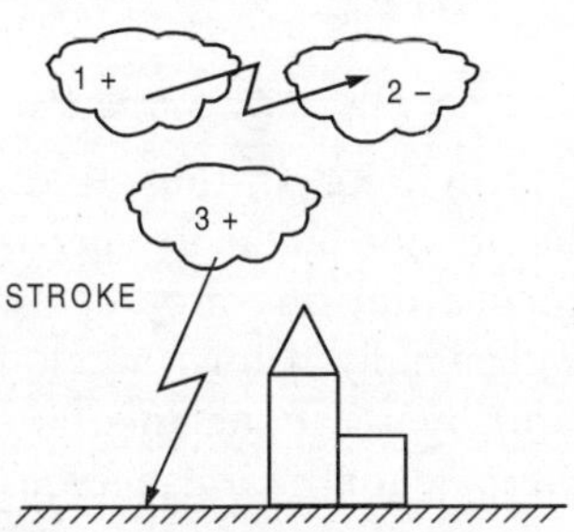

Fig. 15.6

For a transmission line, there is practically no protection against the direct lightning strokes. In case of a lightning stroke landing on a conductor, tremendous amount of overvoltage waves will travel on both the sides of the transmission line from the striking point and shall shatter the transmission line towers and insulators. In case of a stroke quite away from the generating station, the energy will be dissipated in the impedance of the line and the generating plant will not get damaged significantly. But when the stroke is quite near, the plant will get certainly damaged severely, if it is not protected against such overvoltages. Luckily direct strokes are very rare and if such strokes occur, it may be difficult to save the electrical equipment from such strokes completely. The most that can be done from protective devices is that they will limit the damage and prevent the resulting travelling waves affecting plant as far as possible.

A lightning discharge may have currents in the range of 10 kA to 90 kA and the duration of the flash may be of the order of one-millionth of a second. The real danger of the stroke consists in its short duration *i.e.,* the overvoltages will have travelling waves of very steep wave front which is actually responsible for the damage.

15.5.2. Surge Due To Electrostatic Induction

The most of the surges in a transmission system are due to lightning, and are caused by electrostatic induction in the manner illustrated in Fig. 15.7. A cloud electrostatically charged, and lying above a transmission line, will induce in the adjacent section of the line a corresponding charge of opposite sign known as a *bound charge*. This charge has the maximum value below the cloud and then gradually tails away. At the distant end of the line, charges of like sign to that in the cloud are located. This charge on the line will not flow since it is a bound charge. The positive charge on the far ends of the line will however leak to earth slowly through insulators, metallic parts etc., thus leaving only the negative charge on the line.

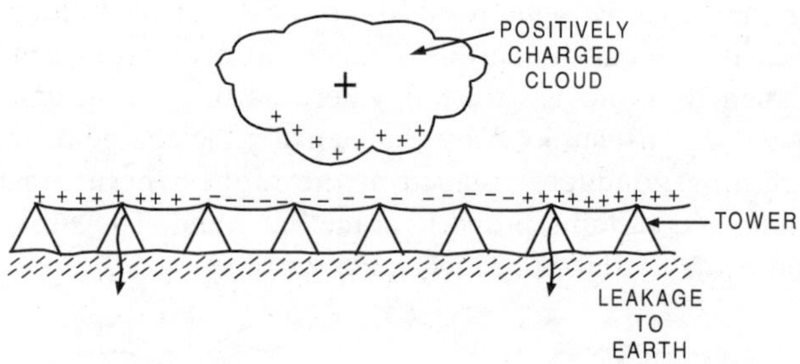

Fig. 15.7. ***Surge Due to Electrostatic Induction***

So long as the cloud remains stationary the static distribution on the line persists. In this static condition though no oscillations are set up but the potential of the line immediately below the cloud may rise to such a value that 'spillover' at the insulators is caused and thus a static overpotential is a dangerous phenomenon.

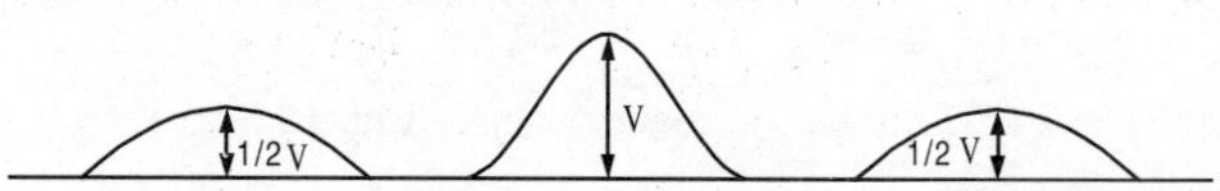

Fig. 15.8. ***Propagation of Voltage Distribution***

When the cloud discharges to earth or to another cloud, the negative charge on the line is isolated as it cannot passes quickly to earth over the insulators. The line thus acquires a high negative potential, which is maximum at the place nearest the cloud and drops slowly to a small value at a distance. The charge will flow from a higher to a lower potential and the result is travelling waves in both directions. The two waves will be equal and thus each wave will have half the potential of the charge at the time of discharge of the cloud; they will also have the space-voltage distribution of the original charge, as illustrated in Fig. 15.8. The waves travel in exactly the same way as the waves due to switching, so that current at any point of the line is the voltage divided by the surge impedance. In a line without resistance or leakage the waves travel without change in shape, but the effect of resistance and leakage is to attenuate the wave and to flatten the wave front.

The above travelling waves will be of quite large amplitude (10 to 15 kV) and shall have very steep wave fronts which can damage the unprotected equipment connected to the line and hence these must be passed to the earth.

The charge induced on the conductor is given as

$$Q = CV \qquad ...(15.2)$$

where C is the capacitance between conductor and ground

$$\text{and } V = Eh \qquad ...(15.3)$$

where E is the field gradient and h is the height of the line from ground.

15.6. OVERVOLTAGES DUE TO LIGHTNING

If an overhead line is struck by lightning (direct stroke), the voltage rise at the point is given by

$$V_d = I_d Z_0 \qquad ...(15.4)$$

where Z_0 is the surge impedance of the line and I_d is the discharge current.

The above equation would have been true if the travelling waves flow only in one direction. But since they travel in both directions, the current is halved and overvoltage is given by the expression

$$V_d = \frac{1}{2} I_d Z_0 \qquad ...(15.5)$$

A direct stroke may cause a potential of many million volts. For example if I_d is taken 30 kA and Z_0 = 500 Ω, then

$$V_d = \frac{1}{2} \times 30 \times 1{,}000 \times 500 = 7.5 \times 10^6 \text{ V.}$$

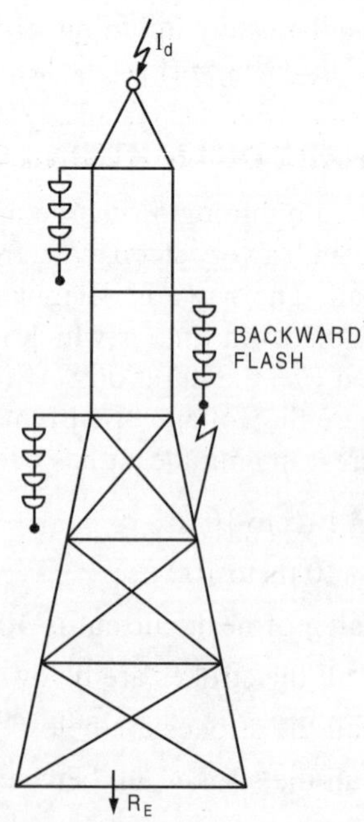

Fig. 15.9

Due to direct stroke on a line, the nearby lines are also subjected to overvoltages, but of less magnitude, through electromagnetic coupling. Experiences have shown that voltage rises induced by side strokes may attain a value of 2 million volts but that about 90% of such voltages are below 500 kV. These voltages may also spark-over on the insulators.

A direct stroke on the tower or earth conductor causes the voltage rise of the latter to a value given by the expression

$$V_d = I_d R_E + L\frac{di}{dt} \qquad ...(15.6)$$

where I_d is the discharge current, R_E is the tower footing resistance, L is the inductance of the tower and $\frac{di}{dt}$ is the slope of current wave.

For EHT towers of normal height, L is taken equal to 10 μH.

Taking $R_E = 5\ \Omega$, $I_d = 30$ kA and $\frac{di}{dt} = 10$ kA/μs, we have

$$V_d = 30 \times 10^3 \times 5 + 10 \times 10^{-6} \times \frac{10 \times 1{,}000}{1 \times 10^{-6}} = 250 \text{ kV}$$

If V_d exceeds the spark-over voltage of the insulators a backward flash-over from tower to conductors will give rise to travelling waves on the conductors in both directions which on reaching the terminal equipment may subject the same to dangerous voltage rises.

For all purposes only travelling waves through the lines are considered for protection.

15.7. LINE DESIGN BASED ON DIRECT LIGHTNING STROKES

The insulation level of a transmission line is governed by the surge voltages due to switching as well as lightning. Surge voltages are higher in magnitude in comparison to the power frequency voltages and in case the line is properly insulated to withstand surges it can easily withstand the highest expected power frequency voltages.

The induced strokes (or indirect strokes) due to lightning are important for 11 kV lines only. For high voltage transmission lines (up to 220 kV) the surges due to direct lightning stroke determine the line insulation design. A high voltage line designed to withstand the direct lightning strokes can easily withstand the switching surges and the induced lightning strokes. However, the severity of switching surges for lines in extra high voltage class (*i.e.,* 400 kV and above) is much more than that due to lightning.

For protection of a line against a direct lightning stroke, a shield is required to be provided to prevent lightning from striking the electrical conductors of the line. It also needs adequate drainage facilities and adequate insulation structures so that the lightning discharge can drain to the earth quickly without affecting the line conductors.

The basic principles in the design of a transmission line based on the direct lightning strokes theory are:

(*i*) Ground wires, of sufficient mechanical strength, must be provided to shield the line (or phase) conductors from direct lightning strokes.

(*ii*) Adequate clearance must be provided from the phase conductors to the tower or to the ground so as to achieve full effectiveness of the insulating structure.

(*iii*) Adequate clearances must also be provided from the ground wires to the phase conductors, especially at midspan, so as to avoid flashover between phase conductors and ground wire.

(*iv*) Tower footing resistance should be made as low as economically justifiable. A low tower footing resistance helps in maintaining the tower top voltage, under lightning discharge conditions, to a safe value.

The location of the ground wires on the towers should be such that the protective angle does not exceed 30°.

The tower footing resistance should be reduced to about 10 Ω as measured with ground wires removed. This is generally more economical than increasing the insulation. Adding insulators increases insulation level and provides increased protection.

Standard curves are available to fix the insulation level for a desired immunity against failures due to lightning.

15.8. PROTECTION AGAINST LIGHTNING

All the electrical equipment must be protected from severe damage due to the lightning strokes. The problem of protection of power systems against lightning can be studied under the following heads.

(*i*) Protection of power stations and substations from direct lightning strokes.

(*ii*) Protection of overhead transmission lines from direct lightning strokes.

(*iii*) Protection of electrical equipment from travelling waves.

15.9. PROTECTION OF POWER STATIONS AND SUBSTATIONS FROM DIRECT LIGHTNING STROKES

Power stations are usually indoor while substations may be indoor or outdoor. For protection of a structure from direct strokes there are three requirements which are to be fulfilled. These requirements are interception, conduction and dissipation. These requirements involve

1. an object in good electrical connection with the earth so that the leader stroke may get attracted,
2. a low impedance path joining this object to earth so that the discharge follows it in preference to any other path,
3. a low resistance connection with the earth body.

For 1, the upper portion of a metal structure may be employed. Alternatively a separate metallic system, often called the *shield*, either mounted on the structure or near to and above it may be provided. A particular shield configuration in the form of *masts* or *overhead ground* wires is considered to provide good shielding.

For 2, the requirements are:

(*i*) low resistance (*i.e.,* adequate conductivity and cross section, properly bounded joints, free from possible corrosion),

(*ii*) low reactance (*i.e.,* absence of sharp bends, or loops and short conductors),

(*iii*) and sufficient clearance from any other conducting object that might provide separate uncontrolled path to ground.

Outdoor substations have much of equipment carried on *metal gantries* and the interconnection of the upper portion of these will screen the apparatus. Usually, there is suitable grounding provided.

Shielding of the station and the incoming lines (about 0.8 km out from the station) to restrict the *severity of the waves that can enter the station through the lines* is a desirable supplement, particularly in the case of *hv* lines (66 kV and above) to the lightning arrester located in the station [Fig. 15.10 (*b*)].

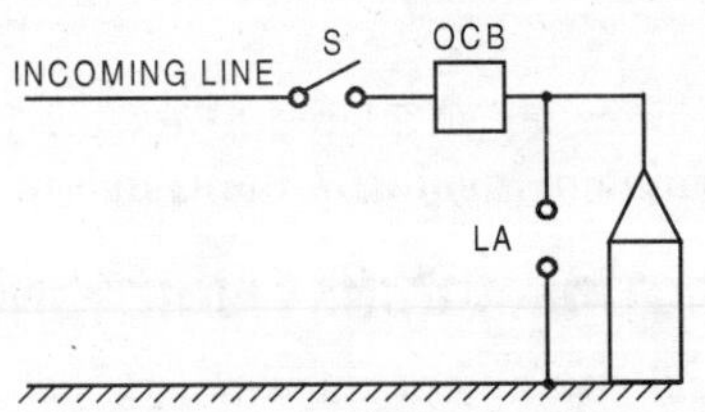

(a) Lightning Arrester at Station With No Direct Stroke Shielding

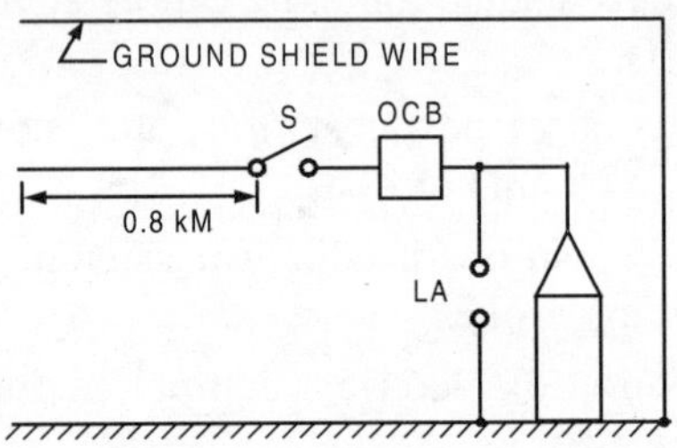

(b) Lightning Arrester At Station With Shielding Against Direct Lightning Strokes

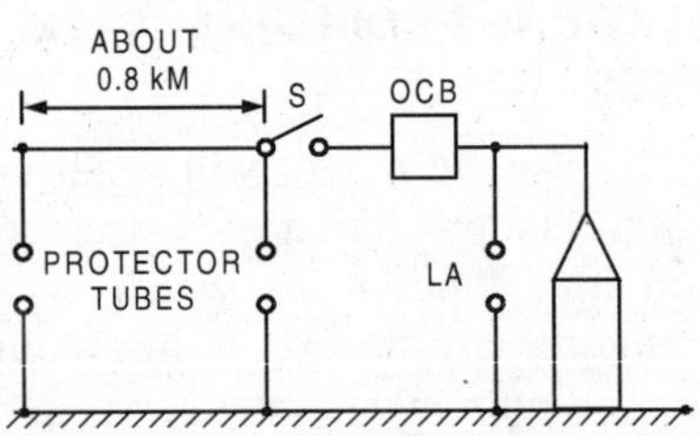

(c) Lightning Arrester At Station With Protector Tube Extending Out 0.8 kM Station Protection Schemes

Fig. 15.10

Where overhead ground wires cannot be provided on the incoming lines due to existing structure/construction, additional protection of the station equipment against direct lightning strokes can be provided by equipping each line with *protector tubes** at the entrance to the structure of the station and at each tower for a distance of about 0.8 km out from the station, as illustrated in Fig. 15.10 (*c*). However, shielding of the power station/substation is the only way of eliminating direct strokes to the station itself.

15.10. PROTECTION OF OVERHEAD TRANSMISSION LINES FROM DIRECT LIGHTNING STROKES

The two methods of protecting overhead transmission lines against lightning strokes are: (1) overhead ground wires and (2) expulsion protector tubes.

* Refer to Art. 15.10.2.

15.10.1. Protection of Overhead Transmission Lines From Direct Lightning Strokes By Ground Wires

A *ground wire* is a form of *lightning protection* employing a conductor or conductors, well grounded at regular intervals, preferably at each support (pole or tower), and attached from support to support above the transmission line conductors. No doubt the ground wire can be run below the line but running of ground wire above the line is considered better as it provides more effective shield. The ground wire shields the phase or line conductors by attracting itself the lightning strokes which, in the absence of ground wire, would strike the line conductors. The elevation at which the ground wire is strung is likewise determined by calculation and selected so that the line conductors come within the zone of protection of the respective ground wire. For reliable protection the protective angle α is taken equal to 20-30 degrees. This is the angle between the vertical line through the ground wire axis and the line passing from the ground wire axis to the outermost line or phase conductor.

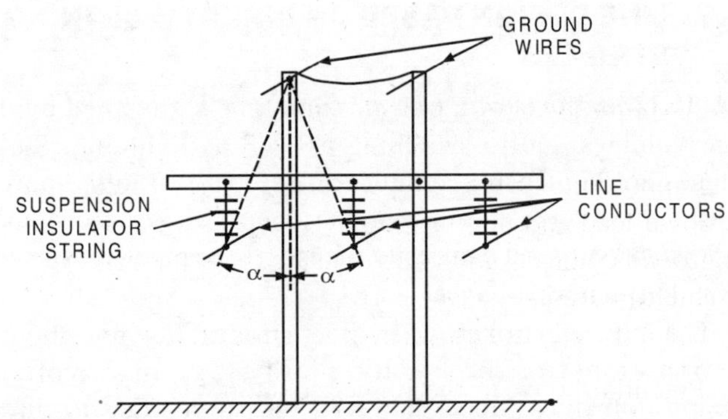

Fig. 15.11. ***Arrangement of Ground Wires on a Transmission Line For Direct Stroke Lightning Protection***

Coupling Factor. The entire ground wire or tower-top voltage does not appear across the line insulation as a voltage of the same polarity as the ground wire voltage is also induced in the line conductors. The ratio of this induced voltage on the line to the potential of ground wire is called the *coupling factor, i.e.,*

Coupling factor,

$$K = \frac{\text{Voltage induced in the line conductors}}{\text{Potential of ground wire}} = \frac{V_{Li}}{V_g} \quad ...(15.7)$$

The voltage across the line conductor and earth is the product of coupling factor and ground wire potential.

Consider an arrangement of line conductors and ground wire, as illustrated in Fig. 15.12. When the ground wire is struck by a lightning stroke at point A, currents flow as indicated in the figure. Currents I_1 and I_2 flow in opposite directions from point A. Again at point B, the current subdivides and let I_4 flow, through earth connecting wire, to the ground.

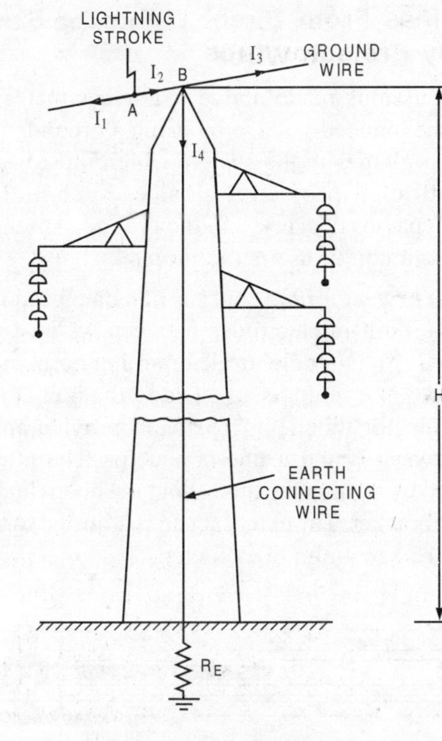

Fig. 15.12. *Arrangement of Conductors and Earth Representing Lightning Current*

Suppose the ground resistance R_E is high. The currents in the down lead and into the earth (I_4) causes a voltage drop in the ground connection, thereby raising the potential of ground (or shield) wire above true earth potential. Let this be V_g

$$\text{So, } V_g = I_4R_E \quad \text{...(15.8)}$$

Potential between the line conductor and earth,

$$V_1 = KV_g \quad \text{...(15.9)}$$

where K is the coupling factor and is given as

$$K = \frac{\log_e \frac{h}{h_1}}{\log_e \frac{2H}{r}} \quad \text{...(15.10)}$$

where h is the distance from phase conductor to ground wire, h_1 is the distance between conductor image and ground, H is the height of ground wire above ground and r is the radius of ground wire.

In the absence of corona the electrostatic and electromagnetic coupling factors are equal.

In case of corona, electrostatic coupling increases while the electromagnetic coupling remains unaffected.

The electromagnetic coupling factor is determined as above using the actual diameter of the ground wire while the electrostatic coupling factor is determined by using the increased diameter of the ground wire due to corona as follows.

$$K_{fs} = \frac{\log_e \frac{h}{h_1}}{\log_e \frac{2H}{r_1}} \quad \text{...(15.11)}$$

where r_1 is the increased radius of ground wire and K_{fs} is the electrostatic coupling factor

The resultant coupling factor,

$$K_{fr} = \sqrt{K_{fs}\, K_{fm}} \quad \text{...(15.12)}$$

where K_{fm} is the electromagnetic coupling factor.

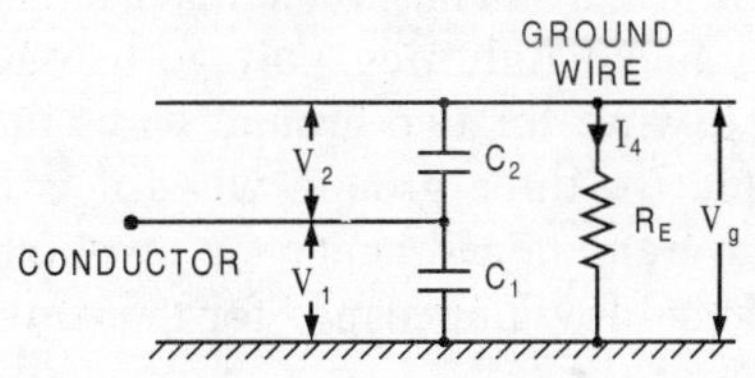

Fig. 15.13. *Equivalent Electrical Circuit*

Figure 15.13 represents the equivalent electrical circuit corresponding to arrangement shown in Fig. 15.12. C_1 and C_2 are the capacitances between conductor and earth and between ground wire and conductor respectively. Let V_1 and V_2 be the potential drops across these capacitances. Also from the equivalent electrical circuit diagram shown in Fig. 15.13, we have

$$V_2 = V_g - V_1 \quad \text{...(15.13)}$$

From Eqs. (15.13), (15.9) and (15.8), we have

$$V_2 = V_g - V_1 = V_g - KV_g = V_g(1 - K)$$
$$= I_4 R_E (1 - K) \quad \text{...(15.14)}$$

i.e., the potential difference between the ground wire and the line, V_2 may be high. If the clearance between the down lead and the line conductor is not sufficient or if the support is metallic and the insulator is small, a flash-over will be caused from the ground wire to the line conductor. This is just as bad as a direct hit on the line conductor with flash-over from it to ground because either may cause a short circuit. Sometimes, special measures such as counterpoise rods (horizontal rods buried in ground) may be necessary to obtain reasonable ground resistance (about 5–10 Ω) and sometimes it may be necessary to offset the down leads so that adequate clearances are obtained.

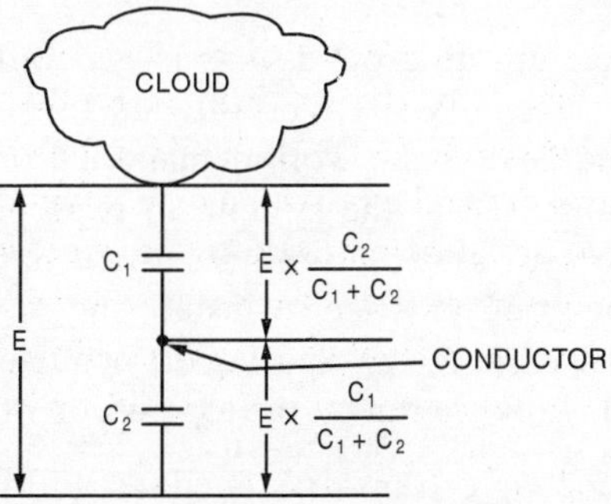

Fig. 15.14. *Reduction in Induced Voltage Due To Ground Wire*

Besides taking the burnt of a direct stroke, the ground wire reduces the voltage electrostatically or electromagnetically induced in the conductors by the discharge of a neighbouring cloud. For example, referring to Fig. 15.14, if C_1 is the capacitance of cloud to the line and C_2 the capacitance of the line to ground. The induced voltage on the line is $\frac{C_1}{C_1 + C_2}$

times the cloud voltage. The presence of the ground wire above the line causes a considerable increase in C_2 and therefore, a reduction of the induced voltage on the line.

The induced voltage could be very much reduced by an array of ground wires but this is too expensive to install in practice. Peak has made a laboratory study of the protective effect of ground wires and has found that a single ground wire reduces the induced lightning voltage to one-half of that without ground wire; for two ground wires the reduction is one-third, while for three ground wires it is to one-fourth. These results were obtained under the favourable conditions of good earths and low impedance for the earth connections.

The ground wire also affords an additional protective effect by causing an attenuation of any travelling waves that are set up in the lines by acting as a short-circuited secondary of the line conductors. For this reason its resistance should not be too large. Ground wire is usually made of steel, which has a high permeability and thus possesses a resistance which increases with frequency.

The objections to the ground wire are the additional cost, and the possibility of the wire breaking and falling across the line conductors, thus causing a direct short circuit. Failure due to the latter cause, however, is rare, as substantial galvanised stranded steel conductors are usually employed. Indeed, such a steel conductor joining the tops of the towers generally adds greatly to the mechanical strength and stability of the line, particularly if the line is of the flexible type.

Ground wires are extensively used for direct stroke protection of transmission lines for voltages of 110 kV and upward (up to 500 kV). Ground wires are usually strung on all vital transmission lines and on all sections of transmission lines running through regions subject to frequent lightning storms.

The selection of size, number and arrangement of ground wires is of great importance in line design. The selection of size of ground wire is based on the consideration of mechanical strength rather than electrical considerations.

In practice, one ground wire is generally used, which allows the intensity of overvoltages to be considerably reduced, and lessens the work demanded from the lightning arresters connected to the system. In exposed situations, or lines subject to severe lightning disturbances, it would probably be good practice to install one or two additional wires. In case of horizontal spacing of conductors on the tower, two ground wires are run on the top portion of the tower.

On wood pole transmission lines within the working voltage range of 33 to 110 kV, ground wires are provided only at the approaches to the power station and system substations. No special means of lightning protection is provided on the remaining sections of such lines as advantage can be taken of the insulation provided by the wood pole supports. By installing the ground wires on the substation approaches of the lines the latter will be guarded against the possibility of lightning strokes close to the substations.

15.10.2. Protection of Overhead Transmission Lines From Direct Lightning Strokes By Protector Tube

From the discussions made above it is clear that even after reduction in the induced voltage by using a ground wire, there still exist overvoltages in the system which must be removed by using additional protective devices such as *lightning* arrester that bypasses the surges to the ground. Another device that is quite common in use is the *protector tube*.

Expulsion protector tube consists of a bakelised fibre tube containing two built-in electrodes between which an internal gap is provided. An outer electrode or arcing horn, made from steel wire 5 or 6 mm in dia is attached to the bolt at the upper end of the protector tube (Fig. 15.15) to provide an external spark gap between it and a line conductor. The internal gap can be adjusted by turning the tube about the bolt which secures the tube to the bracket. The external gap is adjusted to templates made from wire 3 or 4 mm in diameter.

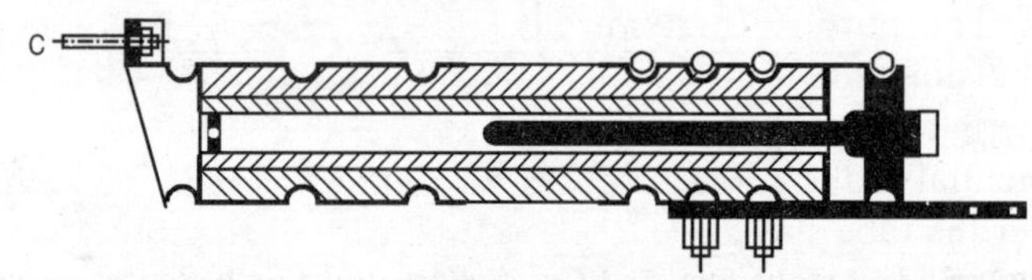

Fig. 15.15. ***Power Line Expulsion Protector Tube***

On a transmission tower one tube is often mounted below each line so that the upper electrode is connected to an arc-shaped horn located at the proper distance below the line, thus forming a series gap G_2 with it [Fig. 15.16 (*a*)]. The lower electrode is solidly grounded.

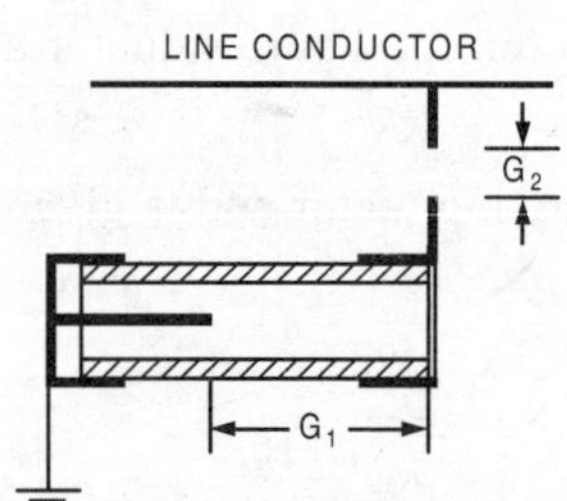

(a) Schematic Diagram of Protector Tube Connections

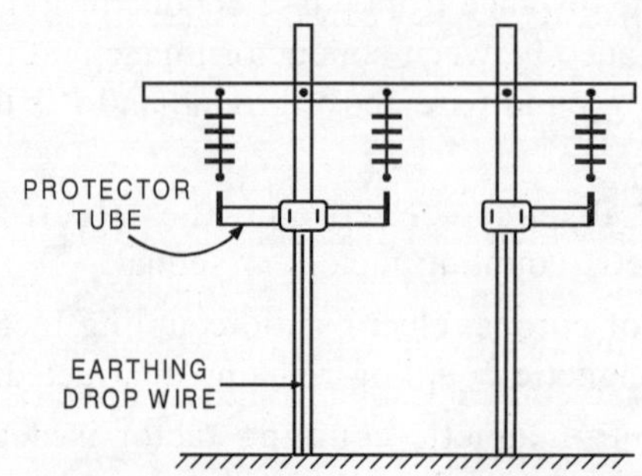

(b) Protector Tube Mounted on 110 kV Wooden Pole

Fig. 15.16. ***Expulsion Protector Tube***

When a surge of sufficient voltage travels along the phase conductor and reaches the point where the expulsion tube is mounted, both of the series air gaps (internal air gap G_1 and external air gap G_2) break down and drain the surge current to ground through the tube and its earthing conductor; thereby reducing the crest value of surge voltage. On breakdown of the tube gaps on two or three phases, or on one phase of solidly grounded neutral circuits, the operating voltage simultaneously initiates a flow of short circuit or power current. This current must flow through the tubes and set up arcs between their spark gaps. The high temperature of the arc across the gap in the tubes then produces a large amount of gases due to decomposition of some of the tube material. These gases flash out of the tube under pressures reaching from 100 to 500 atm and intensely deionize the arc. The latter is thus extinguished and the circuit insulation returned to its normal value with respect to earth.

Arc extinction duration will be only one or two half-periods. This interval is too short for the protective relays of the line to come into action, the circuit breaker remains closed and the line remains in operation. Immediately after the gases have been expelled and the arc suppressed, every tube is ready for a new operation.

The purpose of external air gap G_2 is to isolate the expulsion tube from the line conductor. Failure to provide the external gap would otherwise place the tube at the operating potential of the conductor and cause flow of leakage currents over the tube surface and to eventual carbonizing of the tube material and final destruction of the tube.

15.11. PROTECTION OF ELECTRICAL EQUIPMENT FROM TRAVELLING WAVES

As discussed in the foregoing Arts, the ground wire or earthing screen used for the protection of overhead lines and power stations and substations not only provides an adequate protection against lightning but also reduces the overvoltages induced electrostatically or electromagnetically, but such shielding is inadequate in providing protection against travelling waves which may reach the terminal equipment and cause damage to it. The damages that may be caused by travelling waves are:

1. The high peak or crest voltage of the surge may cause flash-over in the internal winding thereby spoil the winding insulation.
2. The steep wave front of the surge may cause internal flash-over between interturns of the transformer.
3. The high peak voltage of the surge may cause external flash-over, between the terminals of the electrical equipment which may result in damage to insulators.
4. The steep wave front resulting into resonance and high voltages may cause internal or external flashover of an unpredictable nature causing building up of the oscillation in the electrical apparatus.

Thus it is absolutely necessary to provide some protective device at the power stations or substations to prevent transformers and other equipment from being subjected to travelling surges reaching there. The most common devices used for protection of equipment at the substations against travelling waves are *lightning arresters* or *surge diverters*. A surge diverter is a device that is connected between line and earth, *i.e.*, in parallel with the equipment to be protected at the substation. The action of the surge diverter can be studied in reference to Fig. 15.17.

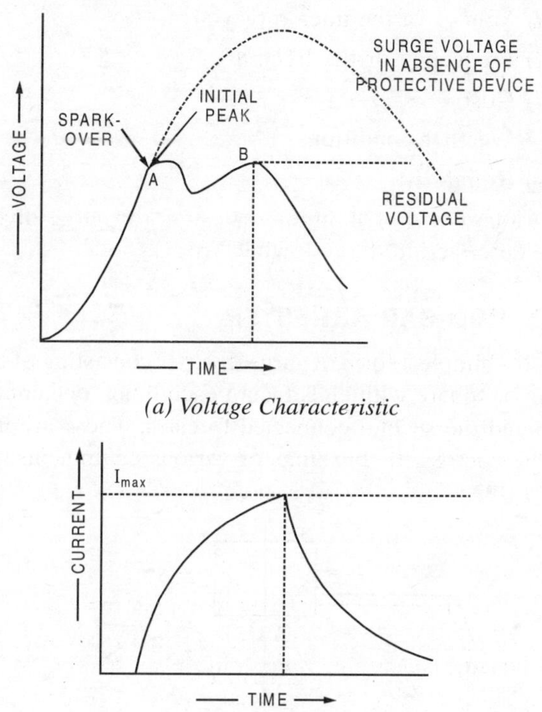

(a) Voltage Characteristic

(b) Current Characteristic

Fig. 15.17. ***Voltage and Current Characteristics of Surge Diverters***

When a travelling wave reaches the diverter, it sparks over at a certain prefixed voltage as illustrated by point A in the figure, and provides a conducting path of relatively low impedance between the line and ground. The surge impedance of the line restricts the amplitude of current flowing to ground. This is necessary in order to protect the insulation of the equipment. Figure 15.17 shows the shape of voltage and of current at the diverter terminals.

It should, however, be noted that the surge diverter should provide a path of low impedance only when the travelling surge reaches the surge diverter, neither before it nor after it.

An ideal surge diverter should have the following characteristics:

1. It should not draw any current during normal operating conditions, *i.e.*, its spark-over voltage must be above the normal or abnormal power frequency that may occur in the system.
2. Any abnormal transient voltage above the breakdown value must cause it to breakdown as quickly as possible so that it may provide a conducting path to ground.
3. When the breakdown have taken place, it should be capable of carrying the resulting discharge current without getting damaged itself and without the voltage across it exceeding the breakdown value.
4. The power frequency current following the breakdown must be interrupted as soon as the transient voltage has fallen below the breakdown value.

There are many types of surge diverters which are used to protect the power system. The choice of lightning arrester depends upon the following factors:

(*i*) Voltage of the line.

(*ii*) Frequency of the lightning.

(*iii*) Cost.

(*iv*) Weather conditions.

(*v*) Reliability.

Various types of lightning arresters (or surge diverters) will be described in the following Arts.

15.12. ROD GAP ARRESTER

This is the simplest form of surge diverter consisting of two 12 mm dia or square with ends facing each other, one connected to line and the second connected to earth. These are usually connected across the bushings of various equipments.

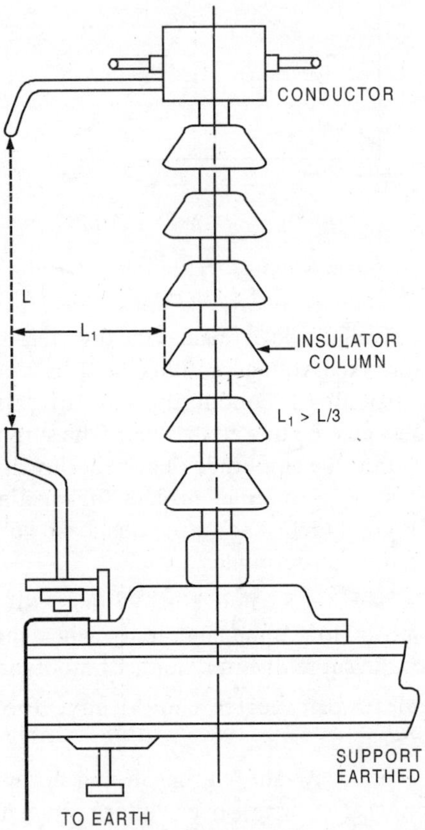

Fig. 15.18. ***Rod Gap***

To avoid cascading across the insulator surface of very steep fronted waves, the rod gap should be set to break down at about 20% below the impulse spark-over voltage of the insulation at the point where it is installed.

To protect the insulator from the arc, the distance between the rod gap and the insulator should be more than one-third of rod gap length.

The rod gap depends upon the operating voltage of the system. The typical figures are given below.

System Voltage	Rod Gap in mm
33	225 mm
66	350 mm
132	650 mm
275	1,230 mm

The rod gap has the advantage of low cost and easy adjustment on site.

The spark-over takes place at very high voltage due to lightning surges but it cannot flash-over at usual power supply voltage. The difficulty with the rod gap arrester is that once the spark having taken place may continue for sometime even at low supply voltage. To avoid it a current limiting resistance is used in series with the rod which limits the current to such an extent that it is sufficient to maintain the arc. Another difficulty is that the rod gaps are liable to be damaged due to high temperature of the arc which may cause the rods to melt. The performance of the rod gaps is badly affected due to climatic variations and also the polarity of the surge.

Thus, the rod gap is a very inferior type of surge diverter and cannot be relied upon as main protection in high voltage power systems where the continuity of supply and also the protection of equipment is given priority. It can, however, be employed as a second line of defence in view of its low cost.

15.13. SPHERE GAP ARRESTER

In such a device the air gap is provided by two similar spheres—one connected to the line and another grounded. The spacing between the two spheres is very small compared with their diameter and can be adjusted with the help of gauge. A choking coil is inserted between the phase winding of the transformer and sphere connected to the line. This is done to reflect off any overvoltage surges that might tend to enter the transformer winding. The minimum air gap is set so that a discharge does not take place at normal operating voltages but at predetermined excess voltages an arc is set up. This arc will travel up the spheres as the heated air near the arc tends to rise upwards. The arc will keep on travelling upwards and lengthening till it is interrupted automatically. When the source of voltages persists, a sound arc follows the first one and so on till the normal conditions of voltage are achieved.

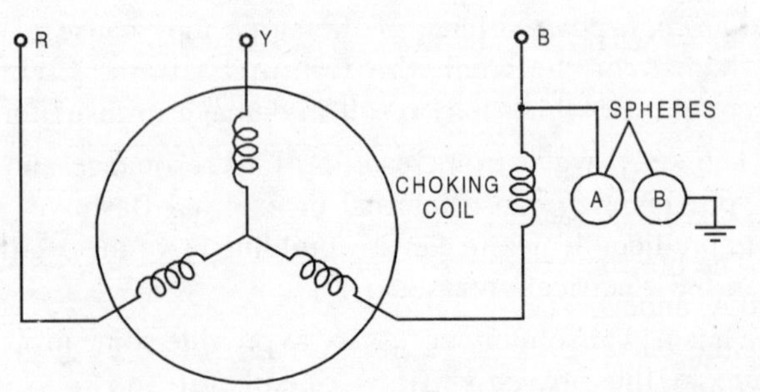

Fig. 15.19. ***Sphere Gap***

It has the advantage of impulse ratio of unity, *i.e.*, if the apparatus is protected against 50 cycle waves, it is protected against a wave of any duration. Unfortunately, when the sphere gap flashes over, the power current maintains the arc, which requires only a very low voltage to maintain it, and the arc is not self-extinguishing. The circuit breakers would have to intervene to interrupt the arc current and, therefore, the continuity of supply is interrupted. This is the reason that the sphere gap is not of use.

15.14. HORN GAP ARRESTER

This was one of the earliest types of surge diverters to be developed, and is still used to a certain extent on low-voltage lines on account of its great simplicity. It consists of two horn-shaped pieces of metal separated by a small air gap* and connected in shunt between each conductor and earth. The distance between the two electrodes is such that the normal voltage between the line and earth is insufficient to jump the gap, but abnormally high voltages (slightly less than twice normal operating voltage) will break down the gap and so find a path to earth. The arc thus formed by reason of heated air and electromagnetic action will rise up the horn and extinguish itself, thus preventing a follow on arc. The time taken for the complete operation is usually from 3 to 5 seconds. Usually a choking coil consisting of several turns of bare copper wire, is connected in the line between the arrester and the apparatus to be protected to reflect travelling waves back to the horns. The choke is without effect on low-frequency power wave because its reactance is negligible at normal power frequency. In practice 20 or 30 turns on a diameter of about 0.3 m are found to be quite effective for the purpose.

For the utmost protection to the terminal apparatus the arrester should be located as close to the end of transmission line as possible. Also the connections from choke coil to arrester, and arrester to earth, should be as short and direct as possible.

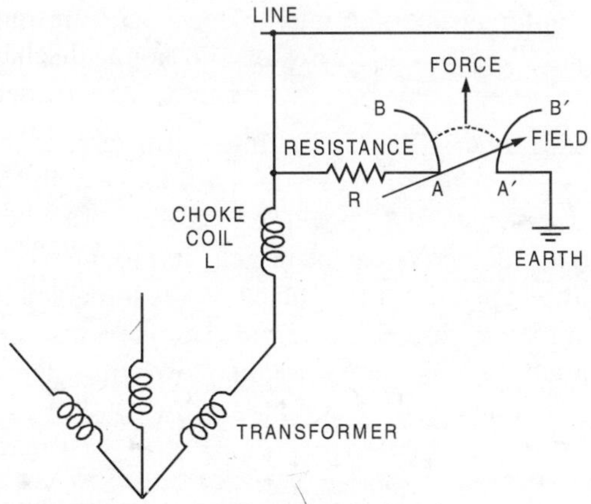

Fig. 15.20. ***Horn Gap With Choke Coil and Resistance***

The horn gap cannot rupture arc currents much in excess of 10 A, and as the arc is a dead short circuit it is necessary to limit the current to a small value. This is accomplished by inserting a non-inductive resistance, between the line and the horn on the line side. The efficiency of the horn gap is seriously reduced by the resistance. The selection of a value for the resistance is, therefore, a matter for compromise, the value generally used being such as to limit the dynamic current to between 1 and 5 A with normal voltage applied. The resistance is a water column, oil-immersed metal wire, carbon rod, or carborundum, and is made as non-inductive as possible.

Such an arrester has the following characteristics;

1. The voltage at which the breakdown occurs during transients, depends upon the impulse ratio** of the gap.
2. Breakdown voltage value is also affected by atmospheric conditions such as temperature and pressure of air.
3. The performance of the lightning arrester is also affected by any roughness of the horn gap and the frequent settings are required to be made for the gap.
4. At heights a longer air gap is required as it depends upon the air density and is inversely proportional to it.

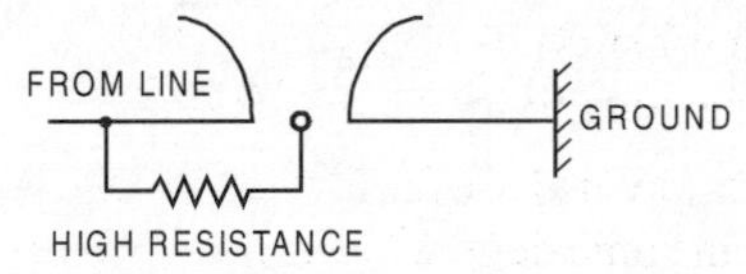

Fig. 15.21. ***Horn Gap With Auxiliary Electrode***

It is seen that on low-voltage installations the gap has to be set very closely, with consequent danger of accidental discharge by insects or other objects bridging the gap. Alteration of calibration also is liable to occur, due to corrosion or formation of small globules of melted metal. These difficulties are overcome in the arrester shown in Fig. 15.21. The main gap is set for a voltage well above that to be protected (1 cm for 2,000 V). The auxiliary gap, which is provided with an electrode having an adjustable platinum point is, however, set to discharge, say at 25% above the normal voltage. If a foreign body gets into the auxiliary gap the resulting discharge quickly dies out owing to the high resistance in series therewith. On the other hand, an arc formed by a voltage rise persists, and ionises the air in the main gap, causing the same to arc across and relieve the line.

Burke Arrester. This is another modification of the simple horn gap arrester. It embodies as a characteristic feature a triangular choke coil wound from copper strip, one side of the coil being used as one of the horns. The arrangement is illustrated in Fig. 15.22. It is seen that the line current passes round the triangular reactance coil and from the centre of the latter goes away to load. The main gap has a series resistance, and there is also an auxiliary gap which is connected direct to ground. Severe overvoltages flash across the main and auxiliary gap and go direct to ground while less severe voltages flash-over the main gap only, and the current is then limited by the resistance. The ohmic value of resistance is such that it limits the dynamic current with normal voltage to from 4 A at 3.3 kV to 2 A at 66 kV.

* The gap between the horns is less at the bottom and large at the top.

** Impulse ratio of any lightning arrester is the ratio of the breakdown voltage of a wave of special duration to the breakdown voltage of a 50 Hz wave. Thus, it is a function of time duration of the transient wave.

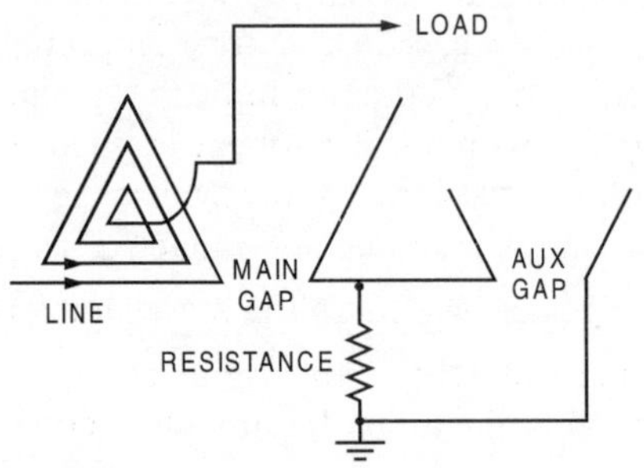

Fig. 15.22. ***Burke Horn Gap Arrester***

It is claimed that the choke coil in such an arrester has a magnetic blow-out effect, causing the arc to rise rapidly up the horns. Furthermore, a wave travelling along the line meets its first impedance at the first sharp upward bend of the triangle opposite to which the grounded horn is mounted. This piles up the voltage and permits a considerably wider gap to be used than with other types of horn arresters, without, however, diminishing the sensitivity of the device.

There are several other types of arresters which are modifications in some degree of the horn and resistance device. The graded-resistance horn arrester, for example, has a tapered or stepped gap. A moderate discharge takes place across the narrowest portion of the gap which has a high resistance in series therewith. If the current flow is not sufficient to relieve the line the arc jumps across to the next wider gap, and so on, the successive gaps having lower and lower resistances in series. Under the severest conditions the discharge will occur at the last gap where there is no series resistance.

15.15. MULTIPLE-GAP ARRESTER

Such an arrester consists of a series of small metal cylinders insulated from one another, and separated by air gaps of about 1 mm width. The first of the series is connected to the line, and the last to ground, and the number of gaps required depends upon the line voltage. To assist in the suppression of the arc and thereby prevent dynamic current from the line flowing through the arrester, the cylinders are made of an alloy of zinc known as the non-arcing metal. The vapour formed by an arc between these cylinders has a rectifying effect like that of mercury vapour, thereby suppressing the arc every other half cycle. It is common practice to knurl the surface of the cylinders so as to present as large a cooling surface as possible, this also helping to extinguish the dynamic current.

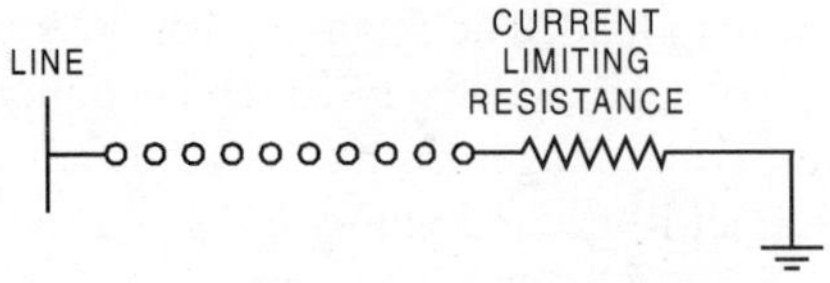

Fig. 15.23. ***Multiple-Gap Arrester***

There is a certain capacitance between consecutive cylinders, and also between each of these cylinders and ground. This leads to a non-uniform distribution of the potential between the various gaps—greatest at the line end of the arrester, and gradually diminishing towards the gaps at the grounded end. The result is, that when the voltage across the arrester attains a certain critical value, breakdown occurs between the first and second cylinders. The second cylinder is then connected to first one by an arc so that its potential rises accordingly until breakdown occurs between the second and third cylinders; and so on. The dynamic current then follows the discharge and in doing so produces a sensibly uniform fall of potential along the line of cylinders with the result that the maximum *pd* between cylinders is considerably less than that required for the initial breakdown. The dynamic current continues to flow until the line voltage passes through zero to the next half cycle, when the arc extinguishing quality of the metal electrodes come into action. Some of the zinc is vaporized by the heat of the discharge, and when the current in the gap passes through the zero point of the wave, the zinc vapour prevents its re-establishment in the opposite direction. Before the voltage again reverses, the arc vapour in the gap has cooled down to a non-conducting state, cutting off further flow of current.

As in the case of the horn arrester there is a current limit, although a higher one, beyond which the arc will be maintained despite the quenching action, so that limiting resistances must be used in most cases. Such resistances are usually of graphite mixture or metallic, and are connected in series with the group of gaps, as illustrated in Fig. 15.23. Their value is such as to limit he dynamic current to less than 20 A.

Such arresters are not satisfactory for use on systems with line voltages exceeding 33 kV, due to the fact that the necessary increase in the number of gaps to prevent arcing over by the normal voltage is out of all proportion to the increase in voltage. There is also much uncertainty as to the number of gaps required, this depending on the position of the arrester relative to surrounding grounded objects. In order to obtain a more equal distribution of potential over the arrester, and so allow of a reduction in the number of gaps necessary, a metal earth shield or antenna is sometimes placed near the gaps at the high potential end of the arrester and connected to the line conductor.

As with other types of lightning arresters, the multiple-gap arrester loses its efficiency as a discharge path for intense disturbances when it is provided with a series resistance. However, this difficulty can be largely surmounted by shunting some of the gaps with a resistance and using a smaller series resistance than with the unshunted arrangement. The arrester with shunted gaps is known as the 'low-equivalent arrester' and the principle of construction is shown in Fig. 15.24.

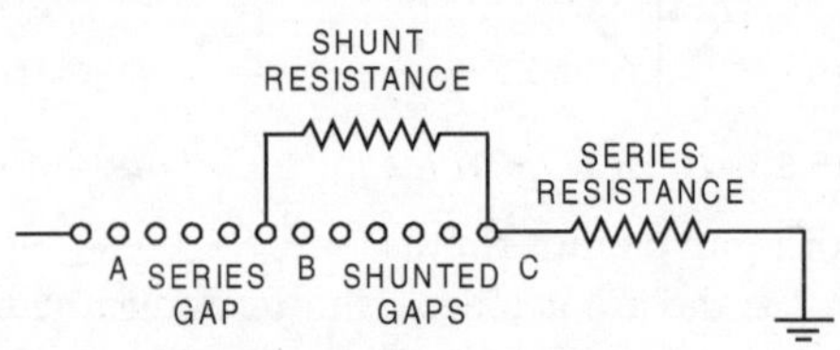

Fig. 15.24. ***Low-Equivalent Multiple-Gap Arrester***

From Fig. 15.24 it is obvious that the point B is at ground potential under normal conditions and therefore a discharge will take place when the voltage is sufficient to break down the series gaps between A and B. The impulsive rush of current following the breakdown, being of high frequency, will choose the straight through path to ground via the shunted gaps between B and C, instead of alternative path through the shunt resistance, finally passing to ground via the small series resistance. As soon as the impulsive rush of current is over, the arcs in the shunted gaps die out, with the result that any dynamic current attempting to follow will have to flow through both resistances, which are now in series. This current will be so reduced by the high resistance in its path that it will not be sufficient to maintain the arcs in the series gaps, which therefore go out as well.

The effect of shunting of some of the gaps is therefore to provide the arrester a certain amount of selective action, since it interposes a low resistance in the path of high-frequency disturbances, and a high resistance in the path of low-frequency discharges. In one form of construction the series resistance is dispensed with altogether, the arrester consisting simply of a large number of gaps some of which are shunted and some not.

In comparison to the horn arrester, the multiple-gap type arrester has the advantage of greater sensitivity to minor voltage rises. Furthermore, it has the property of extinguishing any dynamic current following the discharge at the zero value, instead of in the neighbourhood of the maximum value as with a horn gap. There is therefore no danger of the arrester becoming the seat of high-frequency oscillations after it has operated. On the other hand, the multiple-gap type arrester is more costly, especially for the higher line voltages.

15.16. IMPULSE PROTECTIVE GAP

It was pointed out that the sphere gap has an impulse ratio of unity, but suffers from the drawback that the arc between its electrodes is not self-extinguishing. The horn gap is self-extinguishing but it has a high impulse ratio of 2 or 3 unless the setting is small, as with low voltages.

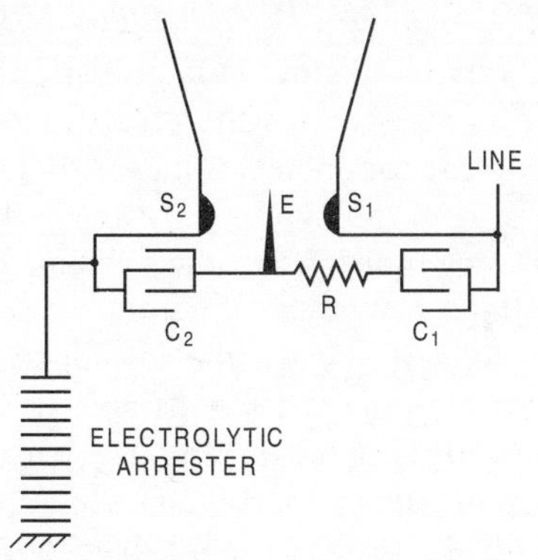

Fig. 15.25. ***Impulse Protective Gap With Electrolytic Arrester***

The impulse protective gap is designed to have a low impulse ratio, even less than unity and to extinguish the arc. In principle the impulse protective gap is very simple, as illustrated in Fig. 15.25. The gap proper consists of two sphere-horn electrodes S_1 and S_2 which are connected respectively to the line and to the arrester, the latter being usually of the electrolytic type. An auxiliary needle electrode E is placed midway between S_1 and S_2 and is connected to them via (R, C_1) and C_2. At normal line frequency the impedance of capacitance C_1 is quite large as compared to the impedance of resistance R; and if $C_1 = C_2$ the potential of the auxiliary electrode will be midway between those of S_1 and S_2 and the electrode has no effect on the flash-over between them.

In case of a transient occurring over the line, the impedance of capacitors C_1 and C_2 decreases and the impedance of the resistance R now becomes predominent. The result is that practically whole of the voltage is concentrated across the gap between E and S_1. The gap at once breaks down, the rest of the gap between E and S_2 immediately following. In effect, the length of the gap is halved the instant a high-frequency wave occurs. On the other hand, a slight time lag is introduced due to the fact that, in the initial breakdown, one of the electrodes is pointed. However, an apparent impulse ratio of 0.6 to 0.7 can be achieved with the commercial forms of the gap. The electrolytic arrester on the earth side extinguishes the arc.

15.17. ELECTROLYTIC ARRESTERS

Electrolytic arrester is the earliest type of arresters with a large discharge capacity. It operates on the fact that a thin film of aluminium hydroxide deposited on the aluminium plates immersed in electrolyte acts as a high resistance to a low voltage but a low resistance to a voltage above a critical value. Voltages more than 400 volts (critical breakdown voltage) causes a puncture and a free flow of current to earth. When the voltage regains its normal value of 400 volts, the arrester again offers high resistance in the path and leakage stops.

The insulating film of hydroxide is formed by applying a dc voltage up to the critical value to aluminium plates immersed in the electrolyte. During the formation of film, current passes fairly readily but on formation the current ceases.

To increase the total critical value of voltage, number of films are arranged one above the other. The total critical value of voltage of such an arrester is equal to the product of number of films and critical voltage of each film.

Such arresters are very delicate, need daily supervision, and the film is required to be reformed whenever destroyed. Serious difficulties may be encountered in the cold countries due to freezing of the electrolytes.

The electrolytic arrester is used in conjunction with an impulse gap.

15.18. EXPULSION TYPE LIGHTNING ARRESTER

Such an arrester is also known as a protector tube, already explained in Art. 15.10.2.

It consists of (*i*) a tube made of fibre which is very effective gas-evolving material (*ii*) an isolating spark gap (or external series gap) and (*iii*) an interrupting spark gap inside the fibre tube. During operation arc due to the impulse spark-over inside the fibrous tube causes some fibrous material of the tube volatized in the form of gas, which is expelled through a vent from the bottom of the tube, thus extinguishing the arc just like in circuit breakers. Since the gases generated have to be expelled, one of the electrode is hollow and the diverter is open at its lower end.

Expulsion type arrester is an improvement over the rod gap in that it seals the flow of power frequency follow current.

Its use is normally restricted to circuits in which the possible range of power current magnitude following a spark-over is not more than about three to one, since an excessive current may cause bursting of tube while a small current may not produce sufficient pressure to extinguish the arc.

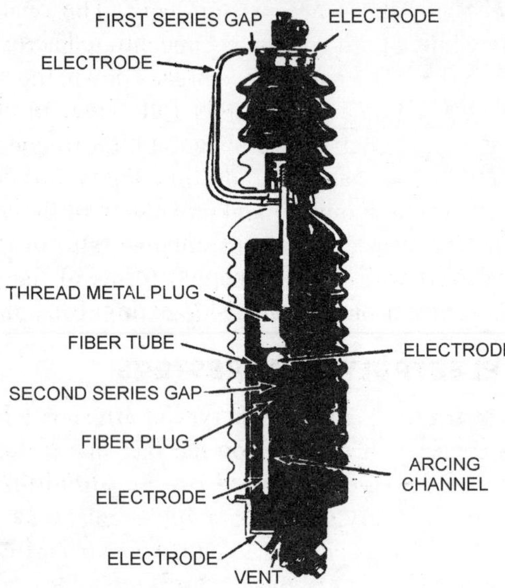

Fig. 15.26. ***Expulsion Type Lightning Arrester***

Though expulsion gaps offer a comparatively cheap form of surge protection but they posses the following disadvantages.

(*i*) They are totally unsuitable for protection of expensive station equipment on account of their poor volt-time characteristics.

(*ii*) There is a limit of the number of operations to which they can be safely subjected as some gas producing material is destroyed in each operation.

(*iii*) They are unsuitable for mounting within equipment enclosure or in close proximity to other electrical apparatus.

The expulsion type arresters can be considered as economical means of surge protection for small rural transformers, where the valve type arresters may prove expensive and the application of air gaps yield inadequate protection. These arresters can be also used on special transmission towers of extra height on river crossings where the possibility of lightning strokes are relatively high.

The expulsion type arrester can be considered very favourably for use on system operating at voltages not exceeding 33 kV.

15.19. VALVE TYPE LIGHTNING ARRESTERS

Valve type lightning arrester is also called a *nonlinear diverter*. Such an arrester consists essentially of a divided spark gap in series with a resistance element having nonlinear characteristics. The divided spark gap consists of a number of identical elements coupled in series, each of them consisting of two electrodes with pre-ionisation device; between each element a grading resistor of high ohmic value is connected in parallel and ensures even grading of the voltage between different elements. The system is similar to that of a number of capacitors connected in series and across each of these capacitors is a high value resistor.

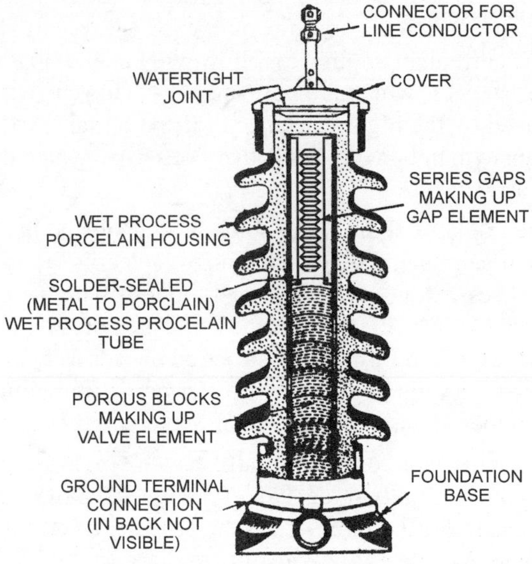

(a) Diagram

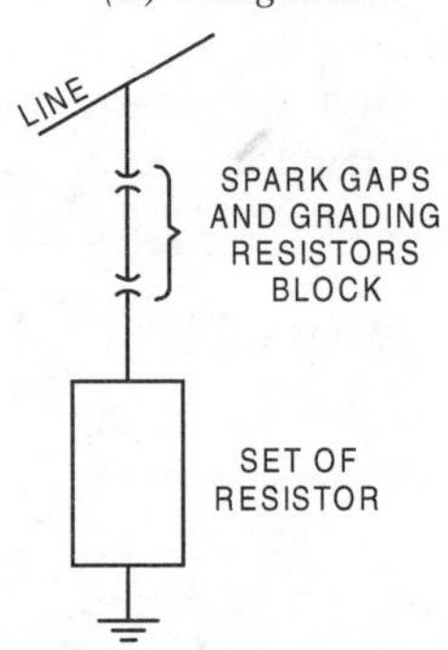

(b) Simplified Representation

Fig. 15.27. ***Valve Type Lightning Arrester***

In case of relatively slow voltage variations, there is no spark-over across the gaps as the effect of parallel resistors across these gaps predominates. It means that slow changes in applied voltage (internal voltages arising during coupling operations) are not injurious to the system. But when rapid changes in voltage occur, the potential is no longer evenly graded across the series gaps; the influence of unbalancing capacitances between the spark gaps and the ground prevails over the grading resistors. The impulse voltages are mainly concentrated on the upper spark-gaps which in sparking-over cause the complete arrester to spark-over too.

The voltage grading by means of resistors makes it possible to increase the interrupting capacity of the spark-gaps block. A leakage current which under normal operating conditions does not exceed 0.1 mA approximately permanently flows through these resistors. This is sufficient to maintain the enclosures at a temperature of 5° above that of the ambient air to offset the adverse effect by the ingress of moisture into the enclosure containing the arrester elements.

The ideal characteristic for the nonlinear resistance elements would be

$$RI = \text{constant} \qquad ...(15.15)$$

The materials used for construction of resistances named "thyrite" and "metrosil" are hard ceramic substances in the form of cylindrical blocks composed of small crystals of silicon carbide bound together by means of an inorganic binder, and the whole assembly subjected to heat treatment. The nonlinear characteristic is attributed to the properties of the electrical contacts between the grains of silicon carbide.

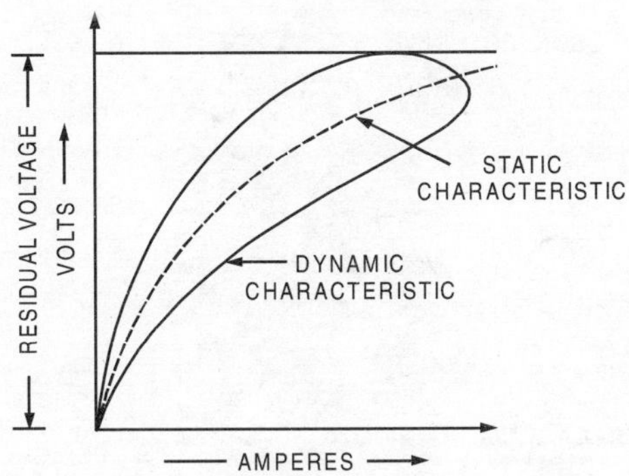

Fig. 15.28

Figure 15.28 illustrates the volt-ampere characteristics of a nonlinear resistor of the required type, the dotted curve being the static characteristic, and the closed curve the dynamic characteristic corresponding to the applications of a voltage surge. If a horizontal line, tangential to the dynamic characteristic is drawn, its intercept with the voltage axis gives the residual voltage. The residual voltage is defined as the peak value of the voltage appearing between the terminals of the surge diverter at the time of the discharge of a surge current wave. It varies from about 3 kV to 6 kV depending on the rate of discharge of current and whether the arrester is used in a substation or in a line and causes a breakdown of the divided gaps thus, the whole breaker starts sparking and pass current through the nonlinear resistor. When the gap flashes-over, the arrester cuts off the power flow current at voltage close to its rating. The voltage that causes flash-over depends on the rate of rise of surge current; if the incoming voltage is very steep, the flash-over may take place at about 4.5 times the arrester rating; however, if the wave is slow rising surge the flashover may take place at about 3 times the arrester rating. Fig. 15.29 (*a*) illustrates the incoming surge and the voltage across the nonlinear resistor, while Fig. 15.29 (*b*) illustrates V-I characteristic of the nonlinear resistor. The ratio of maximum residual voltage e_1 and the maximum power voltage e_2 is termed as *protective ratio*. Mathematically, protective ratio is given as

$$\text{Protective ratio} = \frac{e_1}{e_2} \qquad \text{...(15.16)}$$

The protective ratio may vary between 3 and 4.5 depending upon the magnitude of surge.

In order to illustrate the application of valve type lightning arrester, the successive stages of operation may be indicated, as illustrated in Fig. 15.30.

As the surge approaches the transformer, it meets the lightning arrester and in approximately 0.25 μs, the voltage attains the breakdown value of the series gap and the arrester discharges. But as the surge voltage increases, just as rapidly, the resistance of the nonlinear element drops, thus allowing further surge energy to discharge and so restricting the value of voltage transmitted to the terminal equipment, as illustrated in the figure. As the tail of the surge is passing the point, the voltage is decreasing and, in consequence, the current to ground decreases while the resistance increases, attaining a stage when the current flow is interrupted by the spark gaps and thus the lightning arrester is sealed again. The total time taken from the start of surge wave to sealing off the arrester is a few microseconds (approximately 25 to 30 μs). The maximum voltage developed across the arrester terminal and transmitted to the terminal equipment is known as the *discharge voltage* of the arrester.

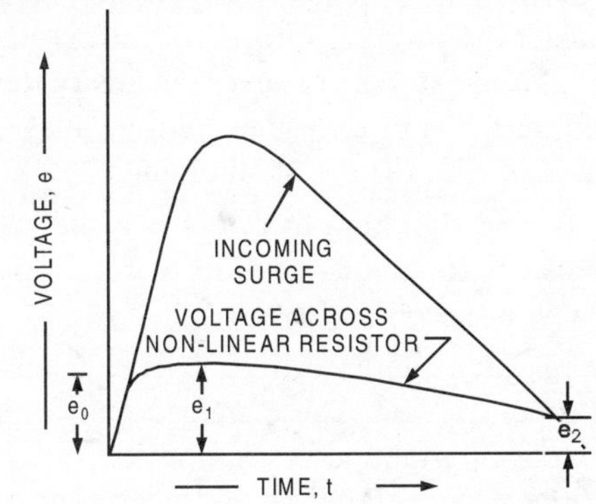

(a) Incoming Surge and Voltage Across Nonlinear Resistor

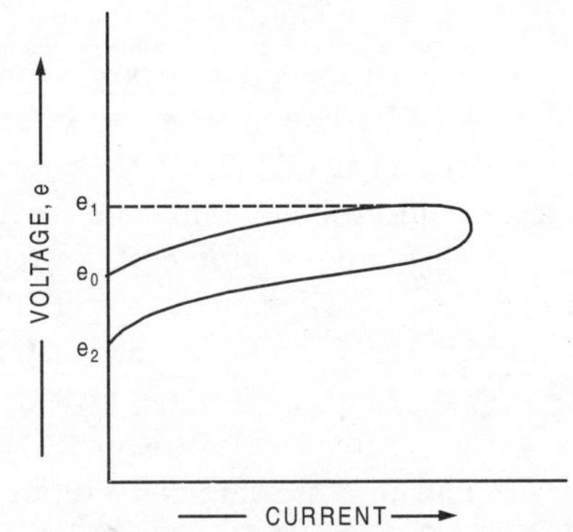

(b) V-I Characteristic of Nonlinear Resistor

Fig. 15.29

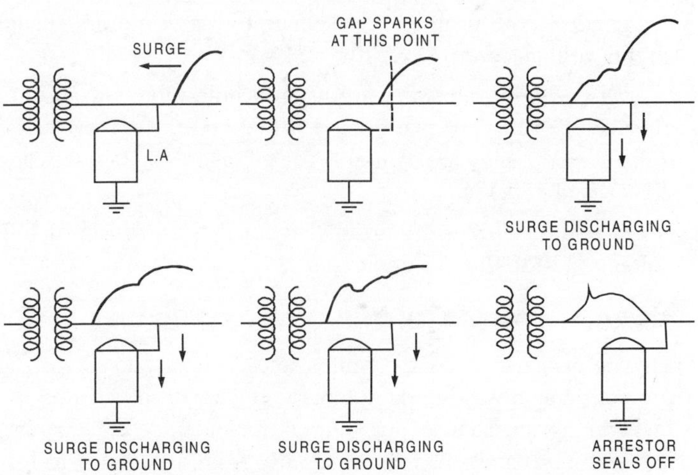

Fig. 15.30. ***Stage of Valve Type Lightning Arrester Operation***

The valve type lightning arresters have the following drawbacks:

1. The performance characteristic is adversely affected by ingress of moisture into the enclosure and so to overcome such problem the arrestor is hermetically sealed. In some cases, the enclosures are filled by an inert gas and contain some form of dehumidifier and in other cases a leakage current is permitted.
2. If the incoming surge is quite steep, the arrester may not be quick to respond to it. Fortunately, the lightning strokes on transmission systems get considerably attenuated as they travel along the line towards terminal stations. As such, normally the wave fronts of surges reaching the stations are within the reach of the protection afforded by such arresters. However, efforts are invariably made to prevent lightning strokes of severe intensity to the station equipment or the approach spans of transmission lines by means of an efficient type of shielding.

The valve type lightning arresters may be station type, line type, arresters for the protection of rotating machines, distribution type or secondary arresters.

Station type valve lightning arresters are the most efficient but the most expensive too. These are generally employed for the protection of important power equipment in the circuits of 2.2 kV to 400 kV and higher. They have high capacity of energy dissipation. The word "station" is an indication of their general application in large stations.

Line types are also used for the protection of power/substation equipment, like station type, but of relatively less importance where the cost of the latter is not considered justified. They are seldom employed on system of voltage exceeding 66 kV. These are constructed like station type but are smaller in cross section, lighter in weight and cheaper in cost. They permit higher surge voltages across their terminals in comparison to station types and have lower surge current capacity. They are generally employed for the protection of large transformers and intermediate substations; not for the protection of the lines as the name indicates. These are available in ratings up to 5 kA.

Arresters for the protection of rotating machines are designed for the protection of generators and motors (usual circuit voltages are from 2.2 to 22 kV).

Distribution arresters are usually pole-mounted and are employed for the protection of distribution circuits/transformers. They are available for voltages varying between 2.2 kV and 15 kV.

Secondary arrestors are meant for the protection of low voltage (120 to 750 V) apparatus.

15.20. THYRITE LIGHTNING ARRESTER

Thyrite arrester is most common and is mostly used for the protection against high dangerous voltages. It operates on the fact that thyrite, a dense inorganic compound of ceramic nature, has high resistance decreasing rapidly from high value to low value for currents of low value to those of high value. The current increases 12.6 times on doubling the voltage. It consists of discs of 15 cm diameter and 19 mm thickness. Both the sides are metal sprayed so as to give electrical contact between consecutive discs. These discs are assembled inside the glazed porcelain container. It is used in conjunction with porcelain container.

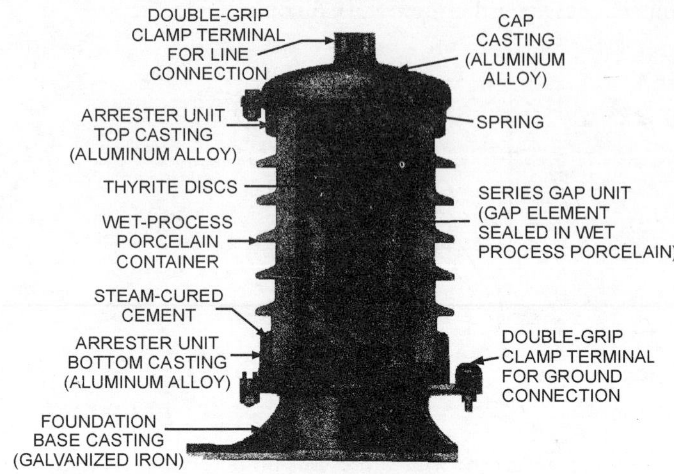

Fig. 15.31. ***11 kV Thyrite Arrester***

When lightning takes place, voltage is raised and breakdown of the gap occurs, the resistance falls to a very low value and wave is discharged to earth. After the surge has passed the thyrite again comes back to its original position, there being no chemical change occurring simultaneously.

The thyrite arrester discharges several thousand amperes without the slightest tendency to flash-over on the edges. Most important of all the advantages of this type of arrester is that there is absolutely no time lag in its performance.

The current-voltage relation for thyrite is given as

$$E = K I^n \qquad ...(15.17)$$

where E is the crest voltage, I is the crest current, K is a geometrical constant which depends on design, rating and dimensions of the element and the exponent n is typically in the range of 0.18 to 0.28

As per design of other lightning arresters, the discs are stacked one upon the other usually 11 in number and they are further placed in a porcelain container with suitable arrangements of gaps between them. These gaps serve the purpose of preventing any current flow during normal operating conditions. In the case of occurrence of transients, these gaps get punctured.

Example 15.1 A thyrite lightning arrester has a characteristic $RI^{0.72}$ = 72,000. Draw the volt-ampere characteristic of the arrester. What is the ratio of the voltage appearing at the end of line due to a 500 kV surge when the line is open-circuited to that when the line, is terminated by the arrester ? The surge impedance of the line is 500 Ω.

Solution :

$$RI^{0.72} = 72{,}000$$
$$\text{or } RI = 72{,}000 \times I^{0.28}$$
$$\text{or } E = 72{,}000\, I^{0.28}$$

The volt-ampere characteristic of the arrester is given in Fig. 15.32.

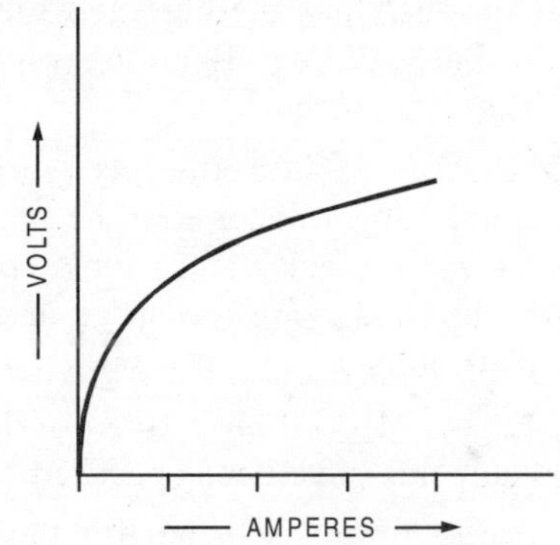

Fig. 15.32

When the line is open circuited

Voltage appearing at the end of line

$$= 2 \times 500 = 1{,}000 \text{ kV}$$

Ratio of voltages $= 2$

When the line is terminated by the arrester

$$\text{Surge current} = \frac{\text{Surge voltage}}{\text{Surge impedance of line}} = \frac{500 \times 1{,}000}{500}$$

$$= 1{,}000 \text{ A}$$

$$R = \frac{72{,}000}{I^{0.72}} = \frac{72{,}000}{(1{,}000)^{0.72}} = 498\ \Omega$$

The resistance of lightning arrester is almost equal to the surge impedance of the line and hence voltage appearing at the end of the line is almost the same as the surge.

So ratio of voltages = 1.

15.21. AUTOVALVE ARRESTER

This type of lightning arrester is very cheap, effective and robust and is, therefore, extensively used nowadays for high voltage systems. This consists of a number of flat discs of a porous material stacked one above the other and separated by thin mica rings. The porous material is made of specially prepared clay with a small admixture of powdered conducting substance. As the material of the discs is not homogeneous and conducting material has also been added, therefore, the glow discharge occurs between the discs at the time of overvoltage. The discharge occurs in the capillaries of the material and voltage drops to about 350 volts per unit. The discs are arranged in such a way that normal voltage may not cause the discharge to occur. The mica ring provides insulation during normal operation.

The valve type lightning arrester possesses the advantage over the electrolytic type is that the initial cost and maintenance charges are lower, and daily supervision is not required. It can be used for protection of isolated transformer stations, automatic substations, or similar plants where the installation of electrolytic arresters would not be justified from an economic point of view.

15.22. OXIDE FILM ARRESTER

It operates on the fact that certain dry chemicals have the property to change rapidly from a good conductor to almost perfect insulator when slightly heated. For example, lead peroxide, which has a specific resistance of 25 ohms per mm cube at normal temperature, becomes red lead at about 150ºC and has a specific resistance of the order of 600 mega ohms per mm cube.

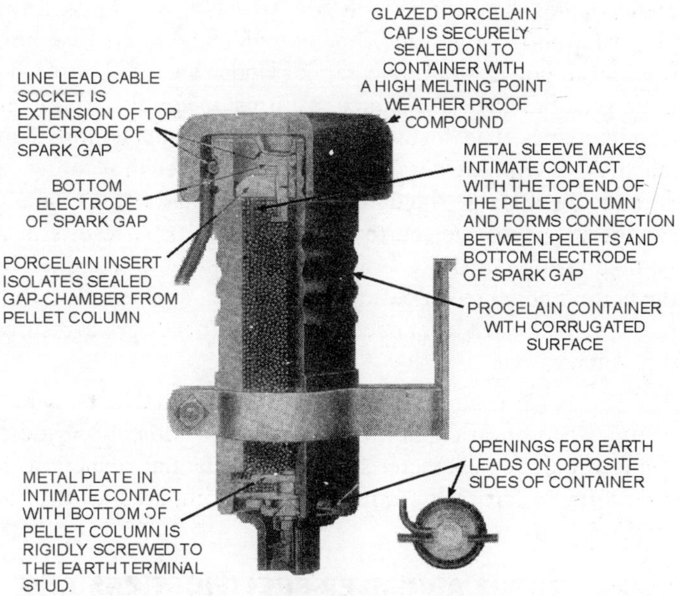

Fig. 15.33. *Oxide Film Pellet Lightning Arrester*

It consists of 2.4 mm diameter pellets of lead peroxide with a thin porous coating of litharge arranged in a column and enclosed in a tube of diameter of about 6 cm and of height of 5 cm per kV of rating. Out of the two leads of the arrester upper is connected to the line, while the lower is connected to earth. The tube contains a series spark gap. A single tube system is available for voltages up to 25 kV when the neutral is solidly grounded and 18 kV when the neutral is isolated or grounded through an inductive coil. For use on higher voltages several units in series are employed. The number of cells used in an arrester is such that the voltage per cell is about 300 volts.

When an overvoltage occurs an arc passes through the series spark gap and an additional voltage is applied to the pellet column and a discharge takes place. After the discharge, the resistance of the pellet column increases till only very small current can flow through it. This small current is finally interrupted by the series spark gap.

The great advantage of oxide film arrester is that it does not require daily charging, and it may thus be installed at points on transmission systems where daily attendance is difficult or expensive to provide.

The impulse ratio of such an arrester is more than unity.

15.23. METAL OXIDE LIGHTNING ARRESTERS

This type of an arrester belongs to the new generation of station metal oxide surge arresters developed by General Electric Co. in 1976. This is useful for the protection of systems rated 2.4 to 400 kV against overvoltages. Such an arrester consists of a stack of Zenox^{TM} disks mounted in a sealed porcelain housing. Each disk is wedged in place with silicone rubber which offers heat transfer and protection against physical damage. On the end faces of each disk, a conducting surface is applied to assure proper contact and uniform current distribution.

Zenox valve elements are made of a specially formulated compound of zinc oxide and small amounts of other selected metal oxides. The ingredients are mixed in powdered form pressed to form a disk, and fired at high temperature resetting in a dense polycrystalline ceramic. Under electric stress, the inter granular layers conduct resulting in highly nonlinear characteristics. The degree of non-linearity is much more than that in the thyrite arrester. The arrester is designed in accordance with the metal oxide standard IEEEC 62.11 and IEC TC-37. A change in the arrester current of 0.1 A to 10^4 A results in a voltage change of 54%. Under normal operating conditions, the current conducted does not exceed 1 mA. When the surge reaches the arrester, it conducts only current necessary to restrict the voltages.

Such an arrester is simple in construction and has superior energy absorption capability, better surge protection, more stable protective characteristics and substantial reduction in overvoltage across the equipment as compared to thyrite arrester.

15.24. SURGE ARRESTER SPECIFICATIONS AND TERMS

1. Surge Arrester. It is a device designed for the protection of electrical equipment from transient high voltages, and to limit the duration and amplitude of the follow current.

2. Nonlinear Resistor. It is that part of the arrester which offers a low resistance to the flow of discharge currents and thus limits the voltage across the arrester terminals and high resistance to power frequency voltage and thus limits the magnitude of follow current.

3. Rated Voltage. The rating of an arrester is defined as the rms voltage at which the arrester passes the duty cycle test as per standards.

Maximum permissible rms voltage between the line terminal and earth terminal of the arrester designated by the manufacturer is known as rated voltage of the arrester.

The noteworthy point here is that for the surge arrester voltage is in terms of phase to earth while for all other apparatus, the rated voltage is usually phase to phase.

The voltage rating of an arrester is given as

Voltage rating of an arrester

= Highest line-to-line rms voltage × coefficient of earthing.

The system highest voltage is generally assumed 1.1 times the corresponding nominal system voltage while the coefficient of coupling for effectively grounded system is taken as 0.8

Thus voltage rating of an arrester

= Line-to-line rms voltage × 1.1 × 0.8

= 0.88 × line-to-line rms voltage ...(15.18)

For non-effectively grounded systems and isolated or ungrounded neutral systems it is usual practice to apply arresters rated at 100% of system highest voltage (*i.e.*, coefficient of coupling is unity).

4. Follow Current. It is the current which flows from connected supply source through lightning arrester following the passage of discharge current.

5. Nominal Discharge Current. It is a surge current that flows through the lightning arrester after the spark-over and is expressed in crest value (peak value) for a specified wave. This term is used in classifying the surge arresters (station type, line type, distributor type).

In the B.S. 2914 : 1957 four ratings of 10,000, 5,000, 2,500 and 1,500 amperes have been specified.

10,000 A current rating is meant for the protection of major power stations and substations in areas in which the frequency of occurrence of lightning strokes is high on systems having a nominal system voltage exceeding 66 kV.

5,000 A current rating is meant for the protection of other power stations and large substations on systems having a nominal system voltage exceeding 66 kV.

2,500 A current rating is meant for the protection of small substations where diverters of higher rating are not economically justified. Their use is restricted to system having nominal system voltage not exceeding 22 kV.

1,500 A current rating is meant for the protection of rural distribution systems particularly where it is advisable to install diverters at frequent intervals for the protection of small transformers. Their use is restricted to systems having a nominal system voltage rating not exceeding 22 kV.

6. Discharge Current. It is the surge current that flows through the arrester after the spark-over.

7. Maximum Discharge Current. It is the peak value of the discharge current which the arrester can pass without damage or modifications of its characteristics. The rating is referred to a waveshape of 5/10 μs.

For lightning arresters of station type this current is 100 kA while for line and distribution types it is 65 kA.

Maximum discharge current can be determined as follows:

$$\text{Discharge current} = \frac{2.4 \times \dfrac{\text{critical flash-over}}{\text{voltage of the line}} - \dfrac{\text{discharge voltage}}{\text{of the arrester}}}{\text{Surge impedance of the line}}$$

8. Power Frequency Spark-Over Voltage. It is the rms value of power frequency (50 Hz) voltage applied between the line and ground terminals of the arrester and ground which causes spark-over of the series gap.

According to British standards it is 1.6 times the nominal or rated voltage of the lightning arrester.

9. Impulse Spark-Over Voltage. It is the highest value of voltage attained during an impulse of given polarity, of specified waveshape applied between the line terminal and ground terminal of an arrester before the flow of discharge current.

10. Residual or Discharge Voltage. The residual voltage is the peak value of the voltage appearing between the line terminals and ground, during the passage of discharge current.

Residual voltage depends for a given type and nominal voltage of lightning arrester on the wave front gradient, expressed in amperes per microsecond, and the amplitude of discharge current.

11. Rated Current. It is the maximum impulse current at which the peak discharge residual voltage is determined.

12. Coefficient of Earthing. It is defined as the ratio of highest rms voltage of healthy phase to earth to phase-to-phase normal rms voltage and multiplied by 100, *i.e.*,

Coefficient of earthing (CEF)

$$= \frac{\text{Highest rms voltage of healthy phase to earth}}{\text{Nominal phase-to-phase rms voltage}} \times 100$$

For effectively grounded systems, it is less than 80%. For non-effectively grounded systems the value of CEF is usually more than 80%. and therefore the voltage rating of the arrester should be at least equal to the line voltage of the system. For isolated neutral system, the arrester voltage should be further high (110% of the nominal voltage).

15.25. ARRESTER RATING AND LOCATION AND EFFECT OF CABLES

The selection of arrester rating for a particular application depends on the impulse level of equipment to be protected. Select the lowest rated surge arrester that will have a satisfactory service life on the power system while providing adequate protection for insulation. Both the maximum spark-over voltage and the residual voltage across the arrester should be lower than withstand voltage of the equipment under protection. The maximum residual voltage should be roughly 20% less than the impulse withstand voltage of the equipment.

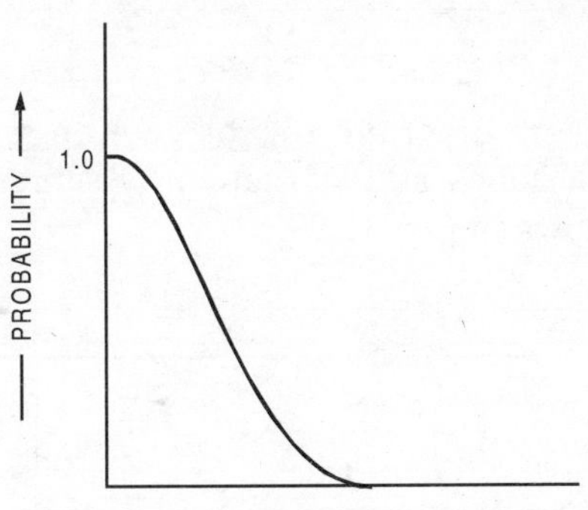

Fig. 15.34. *Typical Probability Curve of Peak Impulse Current*

The location of an arrester in a substation depends on the probability of flash-over on the line per 100 km per year, crest of surges, and arrester characteristics etc. A typical probability curve of peak impulse current is given in Fig. 15.34. It is not necessary to install one arrester adjacent to each piece of equipment in a substation because an arrester has the ability to protect up to suitable distances on both sides.

Lightning arrester requirement is reduced particularly where the equipments in an indoor substation are connected through a cable. Surges are attenuated in cables and, therefore, for indoor substations, particularly, in the medium voltage ranges, when fed through cables, the arrester duty reduces.

15.26. TESTS ON LIGHTNING ARRESTERS

The standard impulse test waves are shown in Figs. 15.35 (*a*) and 15.35 (*b*). Following tests are performed on a lightning arrester:

1. 1/50 impulse spark-over test.
2. Wave-front impulse spark-over test.
3. Impulse current withstand test.
4. Switching impulse voltage test.
5. Peak discharge residual voltage at rated current of the unit.

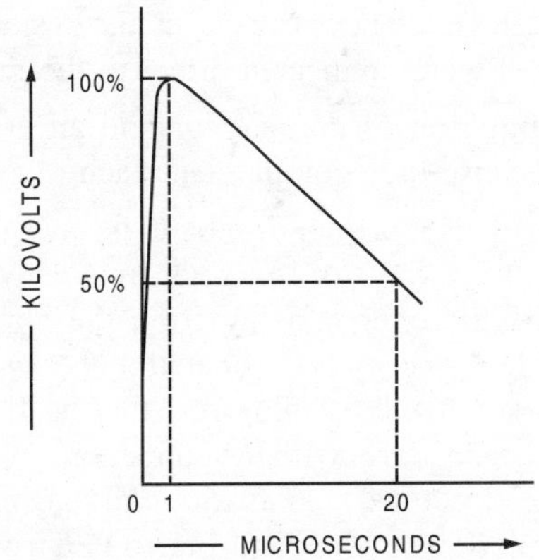

Fig. 15.35 *(a) 1/20 Impulse Current Wave*

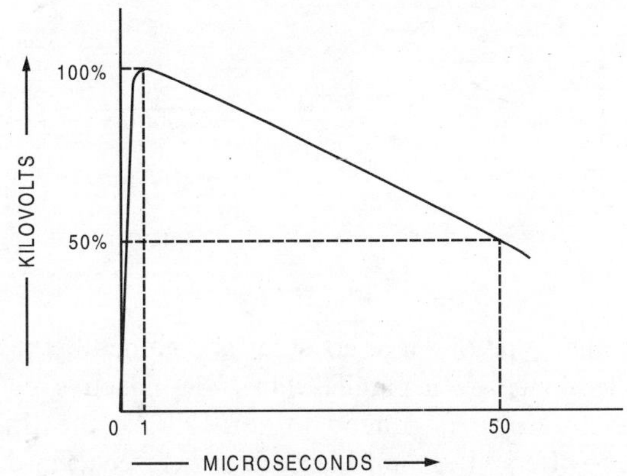

Fig. 15.35 *(b) 1/50 Impulse Voltage Wave*

6. Discharge capability test.
7. Transmission line discharge test.
8. Low current long-duration test.
9. Power duty cycle test
10. Pressure relief test.
11. Radio-interference voltage test at rated voltage.
12. Leakage current test.

15.27. SURGE CAPACITOR, SURGE REACTOR AND SURGE ABSORBER

Damage caused by overvoltages depends not only on the amplitude of an incoming wave but also on the steepness of its wave front. The device, which reduces the steepness of the wave front of a particular surge and thus minimises the danger due to overvoltages is known as *surge modifier* or *absorber*.

A condenser when placed between the line and earth reduces the steepness of the wave front to a considerable extent and hence protects the other apparatus from damage due to overvoltage.

The condenser also provides protection against comparatively low-voltage, high-frequency waves. Since the impedance of a condenser is inversely proportional to the frequency, it is low at high frequencies and large at low frequencies. The normal frequency voltage produces only a small current in the condenser, so that negligible loss is caused during normal operation. A condenser is very efficient in relieving a line from the effects of high frequency disturbances because for high frequency discharges the system behaves as though the lines were connected direct to the ground.

The combination of a capacitor and an arrester can be very useful because they complement each other.

A pure condenser cannot dissipate the energy in the wave front of a travelling wave or in a high frequency discharge. It merely reflects the energy away from the apparatus under protection, and the energy is dissipated in the resistance of the line conductors and the earthing resistances. It thus appears desirable to place in series with the condensers a non-inductive resistance [Fig. 15.36 (*a*)], so as to rapidly absorb the energy of a disturbance. Such a combination of resistance and capacitance is known as *surge absorber.*

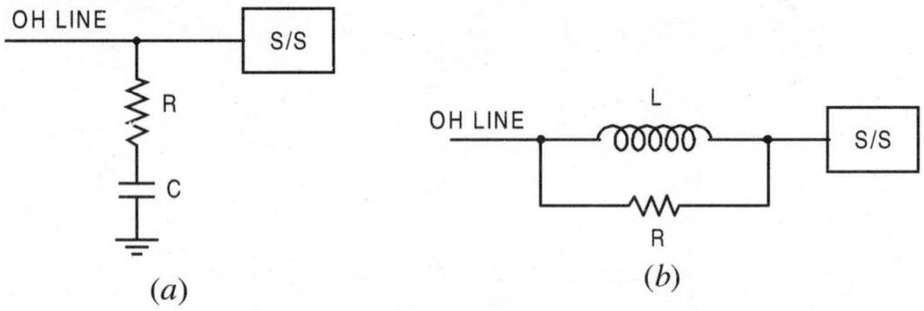

Fig. 15.36

Another type of surge absorber is obtained by having an inductance connected in parallel with a non-inductive resistance. The combination is placed in series with the line. The inductance offers a large impedance to higher frequency currents, which are forced to pass through the resistance, in which surge energy is dissipated. The normal frequency currents find the inductance a low impedance path and pass through it without much loss. The arrangement is depicted in Fig. 15.36 (*b*).

Another type of apparatus which has come into use is the *Ferranti surge absorber*, which consists of an air-cored inductor connected in series with the line and surrounded by, but insulated from, an earthed metallic sheet called a dissipator. For line voltages below 11 kV, it consists simply of a coil surrounded by a sheet-iron case; for voltages up to 33 kV there are two or more disc-shaped coils with dissipators in the form of metal sheets placed between the coils; for voltages exceeding 66 kV, the coil is disposed horizontally and is oil immersed, the dissipator being a metal cylinder inside the coil.

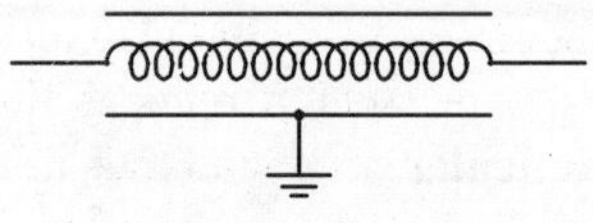

Fig. 15.37. ***Ferranti Surge Absorber***

The principle of all the three forms is the same *i.e.*, the coil can be considered as the primary of the transformer, and the dissipator as a short-circuited secondary of one turn. The energy of a surge is used up in the form of heat generated in the dissipator; firstly, due to the current set up in it by ordinary transformer action, and secondly, by eddy currents. The apparatus is used largely for the protection of transformers, and since there is distributed capacitance between coil and dissipator, just as there is distributed capacitance between transformer windings and grounded core, it is possible to regard the surge absorber as an extension of the transformer winding specially arranged to relieve the main winding from the initial very steep potential gradient. It is claimed that, by installing this apparatus, the voltage rise across the end turns of a transformer subjected to incoming high-frequency waves can be reduced to about 15% of the original value.

A more recent form of the surge absorber is the ERA (Electrical Research Association, UK) surge filter incorporating a gap G and an expulsion gap E. The arrangement is depicted in Fig. 15.38. When a high-frequency wave reaches the inductor L, a high voltage is induced across it and causes the gap G to breakdown and so the resistor R and the expulsion gap E are included in the circuit. An incoming wave is thus flattened by the inductor L and resistor R and its amplitude is reduced by the expulsion gap E.

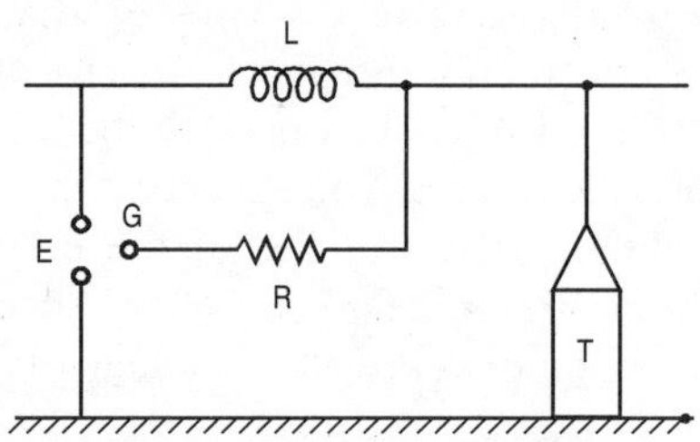

Fig. 15.38. ***ERA Surge Filter***

Connection diagram for protection of a 3-phase transformer installed at a substation by surge absorber and diverter is shown in Fig. 15.39.

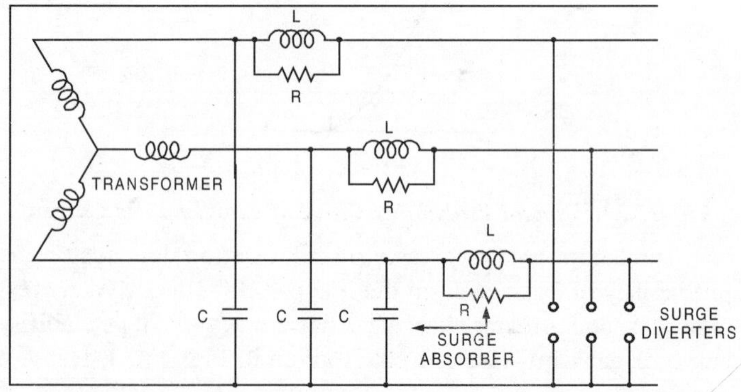

Fig. 15.39. ***Connection Diagram For Protection of 3-Phase Transformer From Lightning Surges***

Exercises

1. What are the causes of overvoltages in power systems ? Discuss. Bring out the functions of ground wire in transmission lines.

2. Discuss the phenomenon of lightning stroke. How can wave set up by such a stroke be represented ?
3. What is a voltage surge ? Draw a typical standard lightning voltage surge ?
4. How are travelling waves produced in a transmission line ? Explain earth wire protection scheme of transmission line.
5. What is the importance of a ground wire in protecting power conductors against direct lightning stroke? Discuss how these are achieved in practice.
6. What is a horn-gap arrester ? Explain how its works. What is the purpose of inserting a resistance between horn-gap arrester and the line ?
7. What are the causes of overvoltages and travelling waves in transmission lines ? Explain. Explain with neat diagrams different types of lightning arresters.
8. Describe the construction, principle of operation and application of (*i*) rod gaps (*ii*) expulsion gap, (*iii*) valve type lightning arresters.
9. What are the causes of dangerously high overvoltages on the transmission system ? Give their relative magnitudes.

   What are the means adopted to protect the power system against such overvoltages ?

   Describe the construction of a valve type lightning arrester and distinguish between station, line and distribution type lightning arresters.

   [A.M.I.E. Sec. B. Advanced Power Systems Winter 1996]
10. State the various causes of overvoltage in a power system ? Name the various devices used for protection against overvoltage due to lightning.

    Explain the working principle of thyrite arrester with the help of a neat sketch.
11. Draw neat labelled diagrams, of the following types of lightning arresters. (*i*) Rod gap (*ii*) Horn gap (*iii*) Multi-gap (*iv*) Expulsion type (*v*) Thyrite type.
12. Explain the working of different types of lightning arresters with particular reference to 'Spark-gap', 'Thyrite', and 'Horn-gap' types.

    What is impulse ratio and coupling factor and how are these utilised in the design of transmission line against lightning ?

    [A.M.I.E. Sec. B. Power Systems Winter 1997]
13. Describe the construction and explain the operation of thyrite type lightning arrester and explain for what part of the circuit it would be most satisfactory.

    [A.M.I.E. Sec. B. Advanced Power Systems Winter 1999]
14. Explain clearly why lightning arresters are used. Describe any type of oxide film arresters with a neat sketch, explaining underlying features. Name other types of lightning arresters used nowadays in protecting equipment and overhead line.
15. (*a*) Explain how a substation and the equipment in the substation are protected from lightning strokes.

    (*b*) What are surge absorbers and how do they differ from surge diverters ?

    (*c*) Explain how surge absorbers provide protection to equipments from destructive effects of travelling waves.
16. Give an idea about lightning voltages and currents that occur in power system. Hence sketch typical voltage wave and explain it.

    What is the function of a surge absorber ? In what way is it different from lightning arrester ?

    [A.M.I.E. Sec. B. Advanced Power Systems Winter 1995]
17. What are the requirements of a lightning arrester ?

    Differentiate between

    (*a*) a lightning arrester and a lightning conductor.

    (*b*) a surge diverter and a surge absorber.

    [Pb. Univ. Switchgear and Protection December 1992]
18. Write short notes on the following:

    (*i*) Causes of overvoltages in a power system.

    (*ii*) Switching surges. **[Pb. Univ. May 1987]**

    (*iii*) Lightning phenomenon and its effects on transmission lines.

    (*iv*) Protection against overvoltages.

    [Pb. Univ. May 1990, Oct 1990]

    (*v*) Lightning arresters. **[Pb. Univ. June 1991]**

    (*vi*) Lightning arresters and their uses.

    (*vii*) Thyrite arrester.

    (*viii*) Surge absorbers.

    [Panjabi Univ. 1978; Pb. Univ. Sept. 1990; Sept. 1991]

    (*ix*) Location, working principle and rating of lightning arrester.

    (*x*) ERA surge filter.

Short Answer Type Questions With Answers

Q. 1. What is meant by voltage surge ?

Ans. By voltage surge is meant any sudden excessive rise in voltage that may be dangerous to the electrical equipment of an installation.

Q. 2. What are the internal causes of overvoltages ?

Ans. Internal overvoltages arise during sudden changes in the operating conditions of an installation such as a switch-in, switch-out or sudden change in load, the opening of circuits with large inductance or capacitance, the interruption of a short-circuit current, the appearance of an arc fault between one phase and ground, as well as other circuit disturbances.

Q. 3. What is meant by lightning ?

Ans. Lightning is a huge spark, which is due to the electrical discharge taking place between the clouds, within the same cloud and between the clouds and ground.

Q. 4. What is difference between A stroke and B stroke ?

Ans. A stroke is characterised by the comparatively long time to produce discharge, while the B stroke is characterised by its rapidity. A stroke strikes the highest and most sharply pointed object in the neighbourhood while B stroke ignores the tall objects and reaches the ground in a random manner. B stroke is most dangerous.

Q. 5. What is earthing screen ? What type of protection is provided by it ? What is it application ?

Ans. Earthing screen consists of a network of copper conductors, earthed at least on two points, over all the electrical equipment in the substation. An electrostatic shield against external field is presented by such a screen and so protects the system. Earthing screen protects the system from direct lightning strokes, reduces the voltages induced electrostatically or electromagnetically but does not provide any protection against voltage waves which may still reach the terminal equipment.

Substations, interconnectors and powerhouses are protected from direct strokes by earthing screen.

Q. 6. What is an earth wire ?

Ans. An earth wire is a form of lightning protection using a conductor or conductors, well grounded at regular intervals, preferably at each support, and attached from support to support above the transmission line.

Q. 7. Why earth wire is provided in overhead transmission lines ?

Ans. Earth wire is provided above overhead transmission lines for protection from lightning strokes. The ground wire shields the phase or line conductors by attracting itself the lightning strokes which, in its absence would strike the phase conductors. Besides it, the ground wire reduces the voltage electrostatically or electromagnetically induced in the conductors by the discharge of a neighbouring cloud. It also provides additional protective effect by causing attenuation of travelling waves set in lines by acting as a short-circuited secondary of the line conductors.

Q. 8. What is meant by protective angle ? Give its value for reliable protection.

Ans. Protective angle is the angle between the vertical line through the ground wire axis and the line passing from the ground wire axis to the outermost phase conductor. Its value for reliable protection is taken equal to 20-30 degrees.

Q. 9. What is meant by coupling factor ?

Ans. The ratio of the induced voltage in the line conductors to the potential of ground wire is called the coupling factor.

Q. 10. What is meant by a lightning arrester or surge diverter ?

Ans. A surge diverter is a device that is connected between line and earth, *i.e.,* in parallel with the equipment under protection at the substation. It limits the duration and amplitude of the follow current.

Q. 11. What is meant by impulse ratio of any lightning arrester ?

Ans. Impulse ratio of any lightning arrester is the ratio of breakdown voltage of a wave of special duration to the breakdown voltage of a 50 Hz wave.

Q. 12. What is the application of an electrolytic arrester ?

Ans. Electrolytic arrester is very delicate, needs daily supervision, and the film is required to be reformed whenever destroyed. It is used in conjunction with an impulse gap.

Q. 13. What is the main advantage of oxide film arrester ?

Ans. The main advantage of an oxide film arrester is that it does not require daily replacement and it may thus be installed at points in transmission systems where daily attendance is difficult or expensive to provide.

Q. 14. How metal oxide lightning arrester is superior to thyrite arrester ?

Ans. Metal oxide lightning arrester is simple in construction and has superior energy dissipation capacity, better surge protection, more stable protective characteristics and substantial reduction in overvoltage across the equipment as compared to thyrite arrester.

Q. 15. What is meant by "follow current" ?

Ans. Follow current is the current that flows from connected supply source through lightning arrester following the passage of discharge current.

Q. 16. What is meant by surge absorber ?

Ans. The device, which reduces the steepness of the wave front of a particular surge and thus minimizes the danger due to overvoltage is called the surge absorber.

Q. 17. What is the difference between a lightning conductor and lightning arrester ?

Ans. Lightning conductor is installed at the top of tall structures and is connected to the general mass of earth through a system of connecting conductors and the earth electrodes so that the structure under protection is not subjected to the dangerously high potential of the cloud discharges. While lighting arrester is a device that is connected in parallel with the equipment, under protection from lightning strokes, at the substation.

Q. 18. What is the difference between a lightning arrester and surge absorber ?

Ans. Lightning arrester limits the duration and amplitude of the follow current while a surge absorber reduces the steepness of the wave front of a particular surge.

❑❑❑

CHAPTER

16

Travelling Waves

INSIDE THIS CHAPTER

16.1. INTRODUCTION

A transmission line is a distributed parameter circuit and a distinguishing feature of such a circuit is its ability to support travelling waves of voltage and current. A circuit with distributed parameters has a finite velocity of electromagnetic field propagation. In such a circuit the changes in voltage and current, owing to switching and lightning, do not occur simultaneously in all parts of the circuit but spread out in the form of *travelling waves* or surges. When a transmission line is suddenly connected to a voltage source by closing of a switch, the whole of the line is not energized all at once (the voltage does not appear instantaneously at the other end). This is due to presence of distributed constants (inductance and capacitance in a loss free line). When switch S is closed, the inductance L_1 acts as an open circuit and C_1 as short circuit instantaneously. The same instant next section cannot be charged because the voltage across capacitor C_1 is zero. So unless the capacitor C_1 is charged to some value whatsoever, charging of the capacitor C_2 through L_2 is not possible which, of course, will take some finite time. The same argument applies to the third section, fourth section and so on. So we see that the voltage at the successive sections builds up gradually. This gradual build-up of voltage over the transmission line conductors can be regarded as though a voltage wave is travelling from one end to the other end and the gradual charging of the capacitances is due to associated current wave. The current wave, which is accompanied by a voltage wave sets up a magnetic field in the surrounding space. At junctions and terminations these surges undergo reflections and refractions. In an extensive network with many lines and junctions, the number of travelling waves initiated by a single incident wave will mushroom at a considerable rate as the waves split and multiple reflections occur. It is true that the total energy of the resultant waves cannot exceed the energy of the incident wave. However, it is possible for the voltage to build up at certain junctions due to reinforcing action of several waves. For a complete study of the phenomenon the use of Bewley lattice diagram or digital computer is necessary. In this chapter we will study some of the elementary aspects of wave propagation, reflections and refractions.

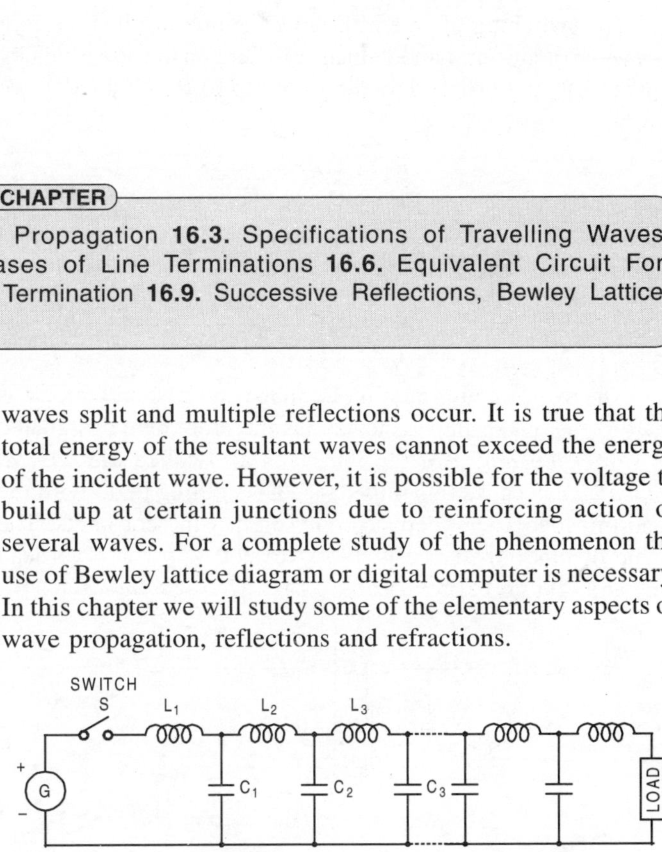

Fig. 16.1. ***Equivalent Π-Section of a Long Transmission Line***

16.2. SURGE IMPEDANCE AND VELOCITY OF PROPAGATION

As already mentioned in the preceding para, the gradual establishment of line voltage can be considered as due to voltage wave travelling from the supply source end towards the far end, and the progressive charging of the line capacitances will account for the associated current wave. Assume that in a very small time δt the conditions of a current I and a voltage E are established along a length δx of the line (Fig. 16.2). The emf E is balanced by the back emf generated by the magnetic flux which is produced by the current in this length of the line. The inductance of the length δx is $L\delta x$, (L is inductance of line per unit length) so that the flux built up is $IL\delta x$ and the back emf is the rate of build-up viz. $IL\frac{\delta x}{\delta t}$.

$$\text{So we have E} = \text{IL}\frac{\delta x}{\delta t} = \text{IL}v \quad ...(16.1)$$

where v is the velocity of propagation of wave.

The current I carries a charge Iδt in the time δt, and this charge remains on the line to charge it up to the potential of E. Since the capacitance of length δx of the line is C δx (C is the capacitance of the line per unit length), its charge is ECδx, so we have

$$\text{I}\delta t = \text{EC}\,\delta x$$

$$\text{or I} = \text{EC}\frac{\delta x}{\delta t} = \text{EC}v \quad ...(16.2)$$

The switching of an emf E on to the line results therefore in a wave of current I and velocity v where I and v are given by Eqs. (16.1) and (16.2). Dividing Eq. (16.1) by Eq. (16.2), we have

$$\frac{\text{E}}{\text{I}} = \frac{\text{IL}v}{\text{EC}v} = \frac{\text{I}}{\text{E}}\cdot\frac{\text{L}}{\text{C}}$$

$$\text{or } \frac{\text{E}^2}{\text{I}^2} = \frac{\text{L}}{\text{C}}$$

$$\text{or } \frac{\text{E}}{\text{I}} = \sqrt{\frac{\text{L}}{\text{C}}} = \text{Z}_n \text{ (say)} \quad ...(16.3)$$

The expression is a ratio of voltage and current which has the dimensions of impedance and is therefore here designated as *surge impedance* of the line. It is also called the *natural impedance* because this impedance has nothing to do with the load impedance, but depends only on the line constants. The value of this impedance is 400 Ω to 600 Ω for an overhead line and 40 to 60 Ω for a cable.

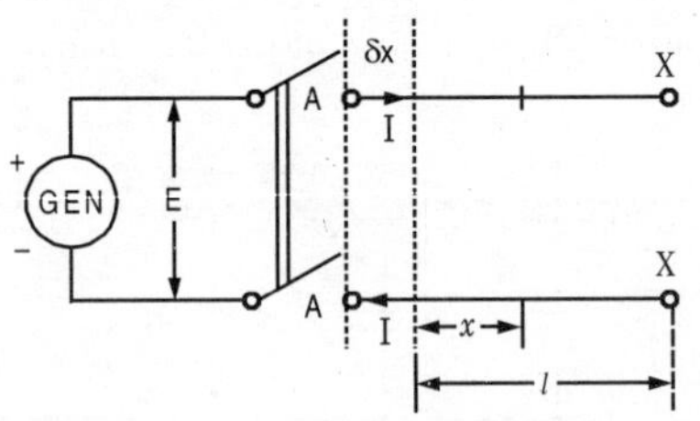

Fig. 16.2

Multiplying Eqs. (16.1) and (16.2), we have

$$\text{EI} = \text{IL}v \times \text{EC}v = \text{EILC}v^2$$

$$\text{or } v^2 = \frac{1}{\text{LC}}$$

$$\text{or } v = \sqrt{\frac{1}{\text{LC}}} \quad ...(16.4)$$

Substituting the values of L and C for overhead lines in above expression, we have

$$v = \frac{1}{2\times10^{-7}\log_e\frac{d}{r}\times\frac{2\pi\varepsilon_0\varepsilon_r}{\log_e\frac{d}{r}}}$$

$$\because \text{ For an overhead line L} = 2\times10^{-7}\log_e\frac{d}{r}\text{ H/m}$$

$$\text{and C} = \frac{2\pi\varepsilon_0\varepsilon_r}{\log_e\frac{d}{r}}\text{ F/m}$$

$$v = \frac{1}{\sqrt{4\pi\varepsilon_0\varepsilon_r\times10^{-7}}} = \frac{1}{\sqrt{4\pi\cdot\frac{1}{36\pi}\times10^{-9}\times10^{-7}}}$$

$$= 3\times10^8\text{ m/s} \quad ...(16.5)$$

$$\because \varepsilon_0 = \frac{1}{36\pi}\times10^{-9} \text{ and } \varepsilon_r = 1$$

Since the product of L and C is the same for all overhead lines, it follows that the velocity of propagation is also the same. This velocity is the same as the velocity of light, but as we have assumed a resistanceless line in the above analysis, the velocity in practice will be from 5 to 10 per cent less than this. Normally a velocity of approximately 285 m/μs is assumed.

The velocity of propagation over the cables will be smaller than that over the overhead lines because in case of overhead lines $\varepsilon_r = 1$ while for cables $\varepsilon_r > 1$, being the dielectric constant. The velocity of wave propagation in case of cables can be given as

$$v = \frac{3\times10^8}{\sqrt{\varepsilon_r}} \quad ...(16.6)$$

where ε_r varies from 2.5 to 4 in case of cables.

16.3. SPECIFICATIONS OF TRAVELLING WAVES

A travelling wave is characterized by four specifications, as illustrated in Fig. 16.3.

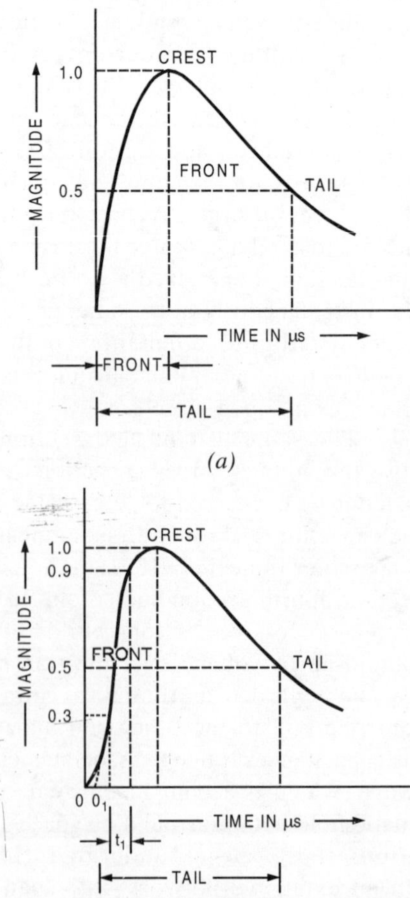

(b)

Fig. 16.3. *Specifications of a Travelling Wave*

1. Crest of the wave is the maximum amplitude of the wave and is usually expressed in kV or kA.

2. Front of the wave is the portion of the wave before crest and is expressed in time from beginning of the wave to the crest value in ms or μs.

However, for waves having a slow initial rate of rise, *i.e.,* a long toe, it is better to consider virtual front as determined by straight lines between 30% and 90% points [Fig. 16.3 (*b*)]. The virtual front is $1.667 \times t_1$. The extension of this straight line to X-axis gives O_1 *i.e.,* the virtual zero.

3. Tail of the wave is the portion beyond the crest. It is expressed in time (μs) from beginning of the wave to the point where the wave has reduced to 50% of its value at crest.

For waves having a slow initial rate of rise, *i.e.,* long toe, the tail time is measured from O_1 to 50% value on tail [Fig. 16.3 (*b*)].

4. Polarity is the polarity of crest voltage or current.

A positive wave of 500 kV crest, 1 μs front and 25 μs tail will be represented as +500/1.0/25.0.

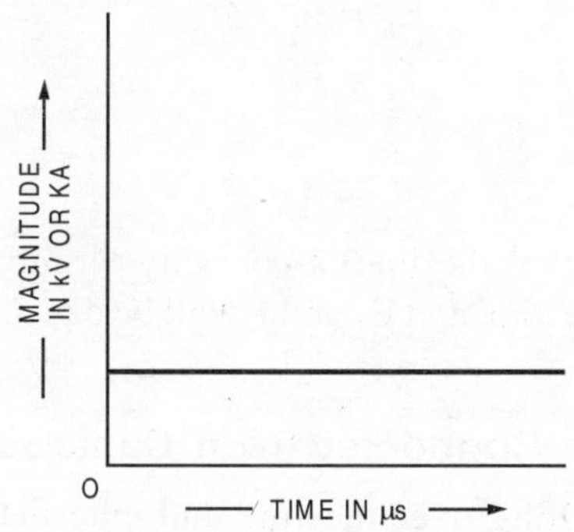

Fig. 16.4. *Step Wave*

A travelling wave can be represented mathematically in a number of ways. The simplest and most commonly used representation is the infinite rectangular or step wave illustrated in Fig. 16.4. Such a wave jumps suddenly from zero to full value and is maintained at that value thereafter. As this wave has front causing maximum gradients and sustained tail producing maximum oscillations in machine windings, it is most dangerous to apparatus/equipment. Hence, the analysis based on it is liable to err on the safer side.

16.4. REFLECTION AND REFRACTION OF WAVES

If a travelling wave arrives at a point where the impedance suddenly changes the wave is partly transmitted and partly reflected. Loading points, line-cable junctions and even faults constitute such discontinuities. Independent waves meeting along a line will combine in accordance with their polarity to provide different voltage and current levels at the meeting point. It is convenient to adopt a standard sign convention, and in what follows, forward waves of current and voltage are given the same polarity. If the wave is being reflected the corresponding current and voltage waves are given opposite polarity. This may be illustrated by considering waves of current and voltage being transmitted along a line of characteristic impedance Z_C terminated by an impedance Z (Fig. 16.5).

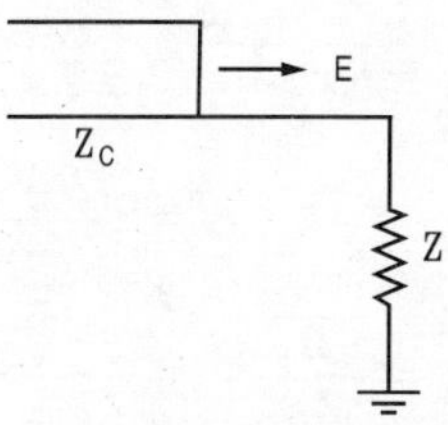

Fig. 16.5. *Line Terminated Through Impedance Z*

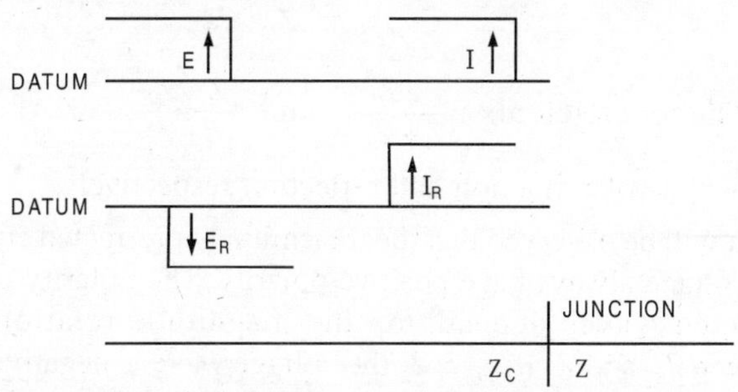

Fig. 16.6. *Transmission and Reflection at Discontinuities*

Let E and I represent the incident waves, E_T and I_T represent the transmitted (or refracted) waves and E_R and I_R the reflected waves. The state of affairs are illustrated in Fig. 16.6. The following relations hold good for incident, transmitted and reflected voltage and current waves.

$$E = IZ_C \qquad ...(16.7a)$$

$$E_T = I_T Z \qquad ...(16.7b)$$

$$E_R = -I_R Z_C \qquad ...(16.7c)$$

The negative sign in Eq. (16.7*c*) is because of the fact that E_R and I_R are travelling in the negative direction of *x* or backwards on the same line.

The transmitted voltage and current will be respectively the algebraic sum of incident and reflected voltage and current waves.

$$E_T = E + E_R \qquad ...(16.8\ a)$$

$$I_T = I + I_R \qquad ...(16.8\ b)$$

Substituting the values of I, I_R and I_T from Eqs. (16.7*a,b,c*) in Eq. (16.8 *b*), we have

$$\frac{E_T}{Z} = \frac{E}{Z_C} - \frac{E_R}{Z_C} \qquad ...(16.9)$$

From Eqs. (16.8 *a*) and (16.9), we have

$$\frac{E_T}{Z} = \frac{E}{Z_C} - \frac{E_T - E}{Z_C}$$

$$\text{or } E_T \times \frac{Z_C}{Z} + E_T = 2E$$

$$\text{or } E_T \left(1 + \frac{Z_C}{Z}\right) = 2E$$

$$\text{or } E_T = E\frac{2Z}{Z + Z_C} \qquad ...(16.10)$$

$$I_T = \frac{E_T}{Z} = \frac{2E}{Z + Z_C} = \frac{2Z_C I}{Z + Z_C} \qquad ...(16.11)$$

$$E_R = E_T - E = \frac{2ZE}{Z+Z_C} - E$$

$$= E\frac{Z-Z_C}{Z+Z_C} \quad ...(16.12)$$

$$\text{and } I_R = \frac{-E_R}{Z_C} = \frac{-E}{Z_C} \times \frac{Z-Z_C}{Z+Z_C}$$

$$= I\frac{Z_C - Z}{Z_C + Z} \quad ...(16.13)$$

The coefficients $\frac{2Z}{Z+Z_C}$ and $\frac{Z-Z_C}{Z+Z_C}$ are called coefficients of refraction and reflection respectively.

It will be observed that the transmitted or refracted current and voltage always have positive polarity. The polarity of the reflected waves depends on the magnitude relationship between Z_C and Z. If $Z_C > Z$, the voltage wave is negative and the current wave positive, but vice versa if $Z > Z_C$.

16.5. TYPICAL CASES OF LINE TERMINATIONS

16.5.1. Short-Circuited Line

If the line is short-circuited at the receiving end, *i.e.*, Z = 0, then the transmitted and reflected waves are given as

$$E_T = 0 \quad ...(16.14\ a)$$

$$I_T = 2I \quad ...(16.14\ b)$$

$$E_R = -E \quad ...(16.14\ c)$$

$$I_R = I \quad ...(16.14\ d)$$

The unique characteristic of the short circuit is that voltage across it is zero. When an incident voltage wave E arrives a short circuit, the reflected voltage wave must be – E to satisfy the condition that the voltage across the short circuit is zero. The waves are shown in Fig. 16.7.

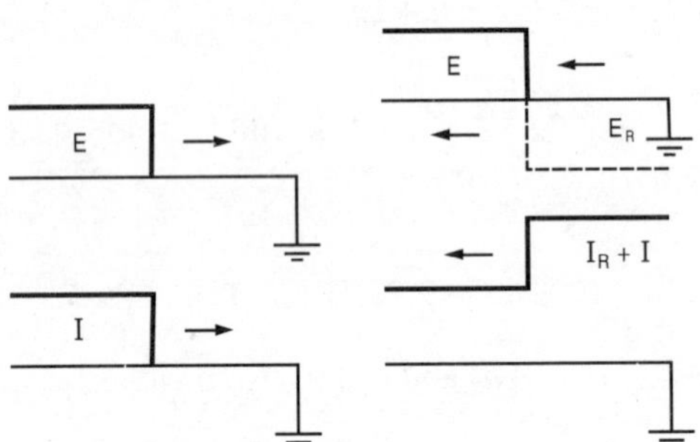

Fig. 16.7. ***Reflection of Waves at Short Circuit***

16.5.2. Open-Circuited Line

If the line is open-circuited at the receiving end, *i.e.*, Z is infinite, the transmitted and reflected waves are given as

$$E_T = 2E \text{ (doubling effect)} \quad ...(16.15\ a)$$

$$I_T = 0 \quad ...(16.15\ b)$$

$$E_R = E \quad ...(16.15\ c)$$

$$I_R = -I \quad ...(16.15\ d)$$

An open circuit at the end of a line demands that the current at that point is always zero. Thus when an incident current wave I arrives at the open circuit, a reflected wave equal to –I is at once initiated to satisfy the boundary condition. The waves are shown in Fig. 16.8.

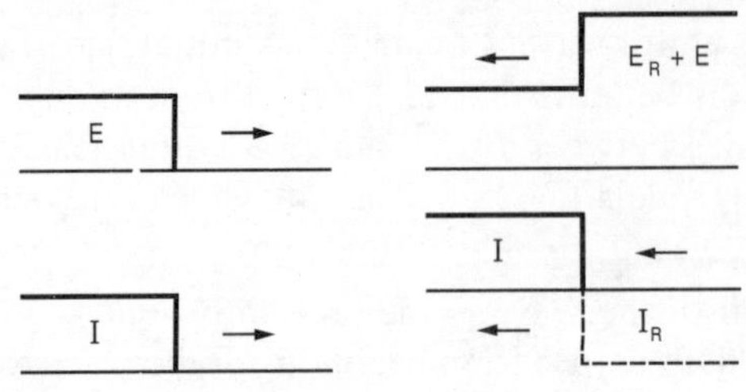

Fig. 16.8. ***Reflection of Waves at Open Circuit***

16.5.3. Line Terminated By an Impedance Equal To Surge Impedance

If the line is terminated with an impedance equal to surge impedance, *i.e.*, $Z = Z_C$, we have

$$E_T = E \quad ...(16.16\ a)$$

$$I_T = I \quad ...(16.16\ b)$$

$$E_R = 0 \quad ...(16.16\ c)$$

$$I_R = 0 \quad ...(16.16\ d)$$

It means that the line is correctly terminated and there will be no reflection and E_T and I_T will be equal respectively to E and I.

16.5.4. Line Connected To a Cable

A wave travelling over the line and entering the cable, as shown in Fig. 16.9, looks into a different impedance and, therefore, it suffers reflection and refraction at the junction. The refracted or transmitted voltage is given as

$$E_T = \frac{2Z_2E}{Z_1 + Z_2} \quad ...(16.17)$$

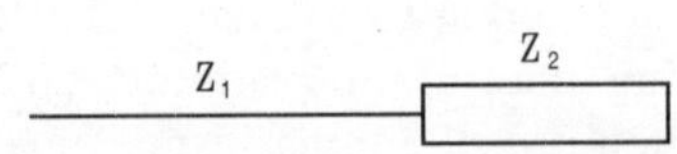

Fig. 16.9. ***Line Connected To a Cable***

The other waves can be had by employing the relations (16.11), (16.12) and (16.13). The surge impedances of the overhead line and cable are approximately 500 Ω and 50 Ω respectively. With these values it can be seen that the voltage entering the cable will be

$$E_T = E \times \frac{2 \times 50}{50 + 500} = \frac{2}{11} E$$

or roughly 20% of the incident voltage.

This is the reason that an overhead line is terminated near a station by connecting the station equipment to the overhead line through a short length of underground cable. Besides the reduction in magnitude of the voltage wave, it also reduces the steepness of the wave. It is because of capacitance of the cable and will be explained later on (Art. 16.8.3). The reduction in steepness is very important because it is one of the factors for reducing the voltage distribution along the equipment windings. In connecting the overhead line to a station equipment through a cable the important point to be remembered is that the length of the cable should not be shorter than the expected length of the wave otherwise successive reflections at the junction may cause piling up of voltage and the voltage at the junction may attain the value of incident voltage.

Example 16.1. A surge of 10 kV travels along the cable towards its junction with an overhead line. The surge impedances of the cable and the line are 50 Ω and 450 Ω respectively. Determine the surge voltage transmitted into the overhead line.

Solution : In this problem, the surge travels from the cable towards the overhead line and therefore, there will be positive voltage reflection at the junction.

Incident voltage, E = 10 kV

Surge impedance of cable, Z_1 = 50 Ω

Surge impedance of overhead line, Z_2 = 450 Ω

Surge voltage transmitted into the overhead line,

$$E_T = E \times \frac{2Z_2}{Z_1 + Z_2} = 10 \times \frac{2 \times 450}{50 + 450} = 18 \text{ kV} \quad \textbf{Ans.}$$

Example 16.2. A surge of 10 kV magnitude travels along a cable towards its junction with an overhead line. The inductance and capacitance of the cable and overhead line are respectively 0.3 mH, 0.4 μF and 1.5 mH, 0.012 μF per km. Find the voltage rise at the junction due to the surge.

[*U.P.S.C. I.E.S. Electrical Engineering-II, 2009*]

Solution : There will be positive voltage reflection at the junction of the cable and overhead line because the surge travels from the cable towards the overhead line.

Incident voltage, E = 10 kV

Natural impedance of the cable,

$$Z_1 = \sqrt{\frac{L}{C}} = \sqrt{\frac{0.3 \times 10^{-3}}{0.4 \times 10^{-6}}} = 27.386 \ \Omega$$

Natural impedance of the overhead line,

$$Z_2 = \sqrt{\frac{1.5 \times 10^{-3}}{0.012 \times 10^{-6}}} = 353.55 \ \Omega$$

The voltage rise at the junction is the voltage transmitted into the overhead line as the line voltage is zero before the arrival of the surge,

$$E_T = E \times \frac{2Z_2}{Z_1 + Z_2} = 10 \times \frac{2 \times 353.55}{27.386 + 353.55}$$

$$= 18.562 \text{ kV} \quad \textbf{Ans.}$$

Example 16.3. A surge of 100 kV is incident on a line having a surge impedance of 400 Ω. It meets a cable having a surge impedance of 40 Ω. Derive expression for the transmitted voltage and reflected voltage and compute their values.

Explain the practical importance of this situation.

[*U.P.S.C. I.E.S. Electrical Engineering-II, 2011*]

Solution : For derivation of expressions for transmitted voltage and reflected voltage refer to Art. 16.5.4. [(Eq. (16.17) and Eq. (16.12)].

Incident voltage, V = 100 kV

Surge impedance of overhead line, Z_1 = 400 Ω

Surge impedance of cable, Z_2 = 40 Ω

Transmitted voltage,

$$V_T = V \times \frac{2Z_2}{Z_1 + Z_2} = 100 \times \frac{2 \times 40}{400 + 40} = 18.18 \text{ kV} \quad \textbf{Ans.}$$

Reflected voltage,

$$V_R = V \times \frac{Z_2 - Z_1}{Z_2 + Z_1} = 100 \times \frac{40 - 400}{40 + 400} = -81.82 \text{ kV} \quad \textbf{Ans.}$$

With these values it can be seen that the voltage entering the cable will be 18.18% of the incident voltage. This is the reason that an overhead line is terminated near a station by connecting the station equipment to the overhead line through a short length of the underground cable. Besides, the reduction in the magnitude of the voltage wave, the steepness is also reduced because of the capacitance of the cable. The reduction in steepness of the voltage wave is very significant because it is one of the factors for reducing the voltage distribution along the windings of the equipment.

Example 16.4. An overhead line with surge impedance 400 ohms is connected to a terminal apparatus through a short length of cable of surge impedance 40 ohms. A travelling wave of constant magnitude of 100 kV and infinite duration originates in the overhead line and travels towards the junction with the cable. Calculate the energy transmitted into the cable during a period of 5 μsec after the arrival of the wave at the junction.

[*U.P. Technical Univ. Power System Analysis 2007-08*]

Solution : Incident voltage, E = 100 kV

Transmitted voltage,

$$E_T = \frac{2Z_2E}{Z_1 + Z_2} = \frac{2 \times 40 \times 100}{400 + 40} = 18.18 \text{ kV}$$

Transmitted current,

$$I_T = \frac{E_T}{Z_2} = \frac{18.18 \times 1{,}000}{40} = 454.55 \text{ A}$$

Energy transmitted into the cable during a period of 5 μs

$$= E_T I_T t = 18.18 \times 10^3 \times 454.55 \times 5 \times 10^{-6}$$

$$= 41.32 \text{ J} \quad \textbf{Ans.}$$

Example 16.5. Two long overhead transmission lines A and B having surge impedances of 400 Ω and 420 Ω respectively are connected by a short cable C of surge impedance 50 Ω. A rectangular surge of magnitude 100 kV and of infinite length travels along A towards the cable C. Find out the surge voltage which is transmitted to the cable at the junction A and C when the first reflected wave from the junction of C and B reaches the former junction. [*U.P.S.C. I.E.S. Electrical Engineering-II, 2002*]

Solution : Incident voltage, V = 100 kV

Refracted wave at junction of A and C,

$$V_T = V \times \frac{2Z_C}{Z_A + Z_C} = 100 \times \frac{2 \times 50}{400 + 50} = 22.22 \text{ kV}$$

At junction of B and C reflection coefficient,

$$r_{BC} = \frac{Z_B - Z_C}{Z_B + Z_C} = \frac{420 - 50}{420 + 50} = 0.787234$$

Magnitude of reflected wave at junction of C and B

$$V_{r_{BC}} = V_T \times r_{BC} = 22.22 \times 0.787234 = 17.494 \text{ kV}$$

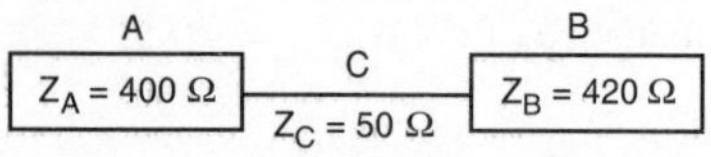

Fig. 16.10

Reflection coefficient at junction of A and C

$$r_{AC} = \frac{Z_A - Z_C}{Z_A + Z_C} = \frac{400 - 50}{400 + 50} = 0.7778$$

Magnitude of reflected wave at junction of A and C

$$V_{r_{AC}} = V_{r_{BC}} \times r_{AC} = 17.494 \times 0.7778 = 13.6 \text{ kV}$$

Total surge voltage transmitted into the cable

$$= V_T + V_{r_{AC}} = 22.22 + 13.6 = 35.82 \text{ kV} \quad \textbf{Ans.}$$

16.6. EQUIVALENT CIRCUIT FOR TRAVELLING WAVE STUDIES

An equivalent circuit for studies of travelling waves can be easily developed by making use of Thevenin's theorem that provides a mathematical technique for replacing a two terminal network by a voltage source V_T, and resistance R_T connected in series. The voltage source V_T, called the Thevenin's equivalent voltage, is the open-circuit voltage that appears across the load terminals when the load is removed or disconnected and resistance R_T, called the Thevenin's equivalent resistance, is equal to the resistance of the network looking back into the load terminals.

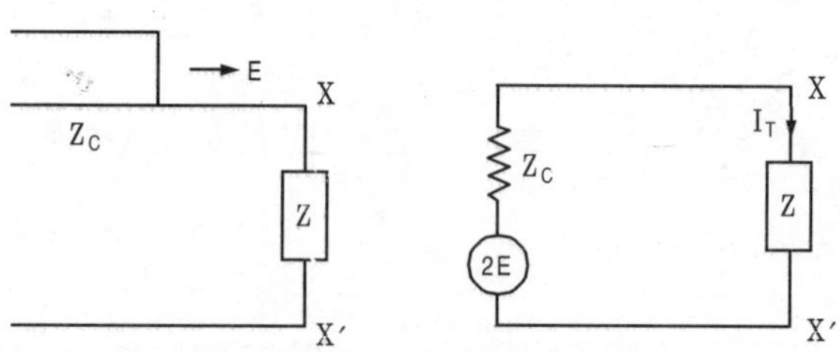

(a) Actual Circuit *(b) Equivalent Circuit*

Fig. 16.11. ***Equivalent Circuit For Travelling Waves***

A line with a surge impedance Z_C is shown in Fig. 16.11 (*a*). A surge wave E travels on the line towards the terminals XX′. The open-circuit voltage across the terminals XX′ is 2E (Art. 16.5.2) and the impedance seen from the terminals XX′, with the sending end of the line short-circuited is Z_C. Thus the circuit given in Fig. 16.11 (*a*) can be replaced by the equivalent circuit shown in Fig. 16.11 (*b*). From the equivalent circuit shown

$$\text{Current through Z, } I_T = \frac{2E}{Z + Z_C} \qquad \text{...(16.18)}$$

$$\text{and voltage across Z, } E_T = \frac{2EZ}{Z + Z_C} \qquad \text{...(16.19)}$$

The above results are the same as Eqs. (16.11) and (16.10). The values of E_R and I_R can be determined from Eqs. 16.8 (*a*) and (*b*). The equivalent circuit is very useful for the studies of travelling waves and can be employed for resistive as well as reactive terminations.

16.7. BIFURCATED LINE

Let a line of natural impedance Z_C bifurcate into two branches of natural impedances Z_1 and Z_2. As far as the voltage wave is concerned, the refracted (transmitted) portion will be the same for both branches as they are in parallel but the refracted currents will be different as $Z_1 \neq Z_2$.

Let the incident wave be (E, I) travelling to the right, the reflected wave (E_R, I_R) travelling to the left and the transmitted waves (E_T, I_{T_1}) and (E_T, I_{T_2}) travelling towards the right, as shown in Fig. 16.12 (*a*).

The analysis is exactly similar to that of a line terminated by impedance Z and the desired results can be had by using Eqs. (16.10), (16.11), (16.12), (16.13) and (16.8 *a*) and (16.8 *b*). Equivalent circuit is shown in Fig. 16.12 (*b*). From equivalent circuit shown in Fig. 16.12 (*b*), we have

$$Z = \frac{Z_1 Z_2}{Z_1 + Z_2} \qquad \text{...(16.20)}$$

$$E_T = \frac{2E}{\dfrac{Z_1 Z_2}{Z_1 + Z_2} + Z_C} \times \left(\frac{Z_1 Z_2}{Z_1 + Z_2}\right)$$

$$= \frac{2E / Z_C}{\dfrac{1}{Z_C} + \dfrac{1}{Z_1} + \dfrac{1}{Z_2}} \qquad \text{...(16.21)}$$

$$I_{T_1} = \frac{E_T}{Z_1} \qquad \text{...(16.22)}$$

$$\text{and } I_{T_2} = \frac{E_T}{Z_2} \qquad \text{...(16.23)}$$

$$E_R = E_T - E \qquad \text{...(16.24)}$$

$$I_R = I_{T_1} + I_{T_2} - I \qquad \text{...(16.25)}$$

The above procedure can be extended for the analysis of a line branching off into any number of lines.

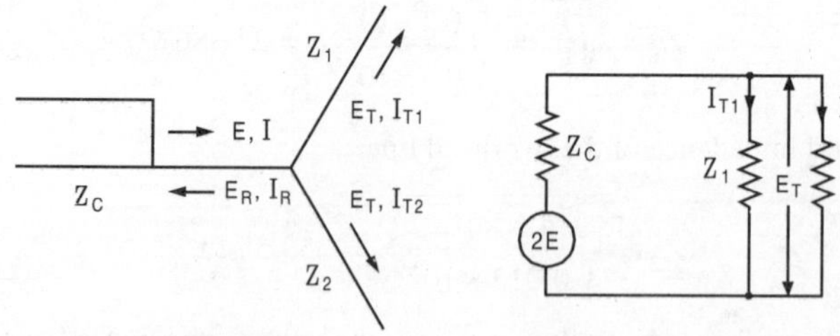

(a) Actual Circuit *(b) Equivalent Circuit*

Fig. 16.12. ***Bifurcated Line***

Example 16.6. A surge of 100 kV travelling in a line of natural impedance 600 ohms arrives at a junction of two lines of impedances 800 ohms and 200 ohms respectively. Find the surge voltages and currents transmitted into each of the branch lines.

[U.P.S.C. I.E.S. Electrical Engineering-II, 2007; W.B. Univ. of Technology Power Systems-III, 2009-10]

Solution : Incident voltage, E = 100 kV

Surge impedance of overhead line, $Z_C = 600\ \Omega$

Surge impedance of branch line, $Z_1 = 800\ \Omega$

Surge impedance of branch line, $Z_2 = 200\ \Omega$

The transmitted voltage,

$$E_T = \frac{2E/Z_C}{\frac{1}{Z_C}+\frac{1}{Z_1}+\frac{1}{Z_2}} = \frac{2\times 100/600}{\frac{1}{600}+\frac{1}{800}+\frac{1}{200}} = 42.1 \text{ kV } \textbf{Ans.}$$

$$\text{Transmitted current, } I_{T_1} = \frac{E_T}{Z_1} = \frac{42.1\times 1{,}000}{800} = 52.625 \text{ A } \textbf{Ans.}$$

$$\text{Transmitted current, } I_{T_2} = \frac{E_T}{Z_2} = \frac{42.1\times 1{,}000}{200} = 210.5 \text{ A } \textbf{Ans.}$$

Example 16.7. An overhead transmission line having a surge impedance of 500 Ω branches into two lines having a surge impedance of 40 Ω and 60 Ω respectively. If a travelling wave of vertical front and magnitude 100 kV travels along the overhead line, calculate the magnitude of voltage and current in the overhead line and in the two branches immediately after the travelling wave has reached the fork.

Solution : Incident voltage, E = 100 kV

Surge impedance of overhead line, Z_C = 500 Ω

Surge impedances of branched lines, Z_1 = 40 Ω

and Z_2 = 60 Ω

Transmitted voltage,

$$E_T = \frac{2E}{Z_C + \frac{Z_1\times Z_2}{Z_1+Z_2}} \times \frac{Z_1\times Z_2}{Z_1+Z_2}$$

$$= \frac{2\times 100}{500+\frac{40\times 60}{40+60}} \times \frac{40\times 60}{40+60} = \frac{200}{524}\times 24 = 9.16 \text{ kV } \textbf{Ans.}$$

Transmitted currents,

$$I_{T_1} = \frac{E_T}{Z_1} = \frac{9.16\times 1{,}000}{40} = 229 \text{ A } \textbf{Ans.}$$

$$I_{T_2} = \frac{E_T}{Z_2} = \frac{9.16\times 1{,}000}{60} = 152.67 \text{ A } \textbf{Ans.}$$

Reflected voltage, $E_R = E_T - E = 9.16 - 100 = -90.84$ kV

$$\text{Incident current, } I = \frac{E}{Z_C} = \frac{100\times 1{,}000}{500} = 200 \text{ A}$$

Reflected current, $I_R = I_{T_1} + I_{T_2} - I$ = 229 + 152.67 – 200

= 181.67 A **Ans.**

16.8. REACTIVE TERMINATION

16.8.1. Reflection From Terminal Inductance

Let the line be terminated with an inductance L, as illustrated in Fig. 16.13. The voltage across the inductance is given as

$$e_T = e_L = L\frac{di_L}{dt} \qquad ...(16.26)$$

Since from Eqs. (16.8 *b*), (16.7 *a*) and (16.7 *c*)

$$i_L = i_T = i + i_R = \left(\frac{e}{Z_C} - \frac{e_R}{Z_C}\right) \qquad ...(16.27)$$

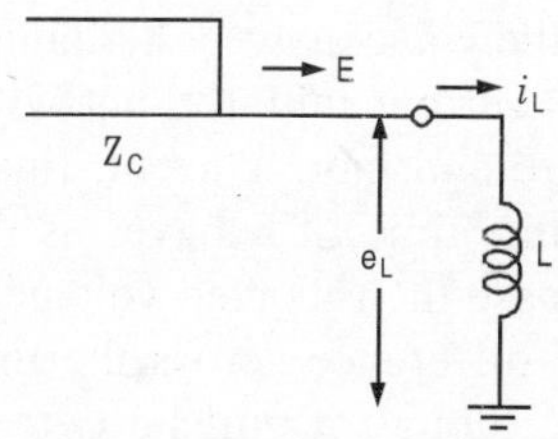

Fig. 16.13. *Transmission Line Terminated By Inductance L*

Substituting value of i_L from Eq. (16.27) in Eq. (16.26), we have

$$e_L = L\frac{d}{dt}\left(\frac{e}{Z_C} - \frac{e_R}{Z_C}\right) \qquad ...(16.28)$$

and since from Eq. (16.8 *a*)

$$e_L = e + e_R$$

$$\text{So } e + e_R = L\frac{d}{dt}\left(\frac{e}{Z_C} - \frac{e_R}{Z_C}\right)$$

$$\text{or } -e + \frac{L}{Z_C}\frac{de}{dt} = e_R + \frac{L}{Z_C}\frac{de_R}{dt} \qquad ...(16.29)$$

Let the incident wave *e* be of constant value E, so that its time derivative is zero. So, Eq. (16.29) becomes

$$\frac{L}{Z_C}\frac{de_R}{dt} = -(E + e_R)$$

$$\text{or } \frac{de_R}{(E+e_R)} = \frac{-Z_C}{L}dt \qquad ...(16.30)$$

Integrating both sides of above Eq. (16.30), we have

$$\log_e (E + e_R) = \frac{-Z_C}{L}t + A \qquad ...(16.31)$$

where A is a constant of integration which can be evaluated from initial condition *i.e.*, when $t = 0$, $i_L = 0$, because when the wave arrives at the inductance terminal the inductance L does not carry any current, the whole of it is reflected. It means that when $t = 0$, $e_R = E$ and therefore

$$\log_e (E + E) = 0 + A$$

$$\text{or } A = \log_e 2E$$

Substituing $A = \log_e 2E$ in Eq. (16.31), we have

$$\log_e (E + e_R) = \frac{-Z_C}{L}t + \log_e 2E$$

$$\text{or } \log_e \frac{E + e_R}{2E} = \frac{-Z_C}{L}t$$

$$\text{or } \frac{E+e_R}{2E} = e^{\frac{-Z_C}{L}t}$$

$$\text{or } e_R = E\left(2e^{\frac{-Z_C}{L}t} - 1\right) \qquad ...(16.32)$$

From the above Eq. (16.32) it is observed that the reflected voltage varies exponentially from + E to –E which is explained by the fact that as the wave arrives the inductance terminal it does not carry any current initially, but with the passage of time, it carries more and more current, finally (theoretically with elapse of infinite time) it behaves as if the line is short circuited in which case the reflected voltage will be – E. Also when the potential difference across the inductance is zero, the current passing though it will be twice that of incident current (refer to Eq. 16.14 *b*)

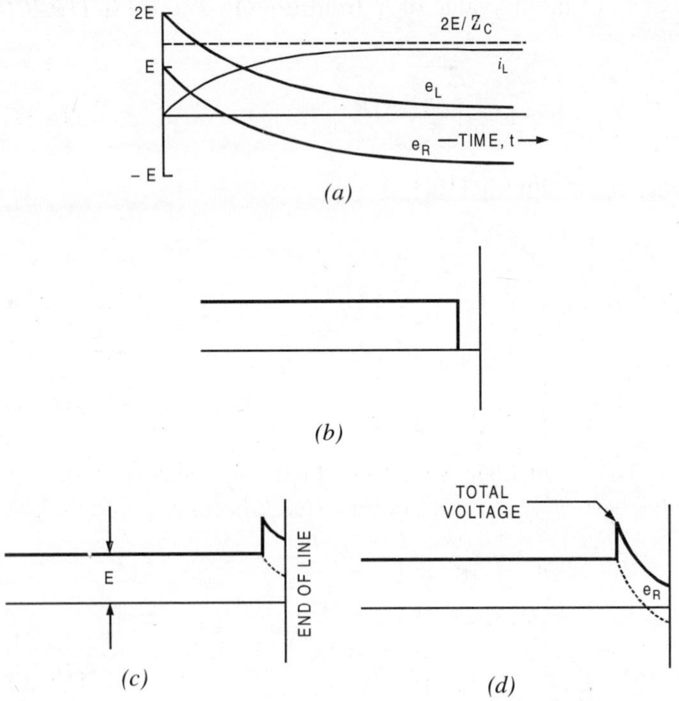

Fig. 16.14. ***Travelling Waves on a Line Terminated By Inductance L***

Again $e_L = e + e_R$ Refer to Eq. (16.8 *a*)

$$= E + E\left(2e^{\frac{-Z_C}{L}t} - 1\right) = 2Ee^{\frac{-Z_C}{L}t} \quad ...(16.33)$$

Current through inductance,

$$i_L = \frac{E}{Z_C} - \frac{e_R}{Z_C}$$

$$= \frac{E}{Z_C} - \frac{E}{Z_C}\left(2e^{\frac{-Z_C}{L}t} - 1\right)$$

$$= \frac{2E}{Z_C}\left(1 - e^{\frac{-Z_C}{L}t}\right) \quad ...(16.34)$$

From above Eq. (16.34) it is observed that the current through inductance also varies exponentially. e_L, i_L and e_R as functions of time are shown in Fig. 16.14 (*a*). The disposition of voltage waves at different instants is shown in Figs. 16.14 (*b, c* and *d*).

If the line is terminated with an inductance in parallel with resistance R, it can be shown that

$$e_R = \frac{E}{Z_C + R}\left(R - Z_C + 2Z_C e^{-\frac{R + Z_C}{L}t}\right) \quad ...(16.35)$$

16.8.2. Reflection From Terminal Capacitance

When the line is terminated with a capacitor, a charging current flows through it and at the moment of arrival of the incident wave (of value E volts) the capacitor acts as a short circuit. At this instant the reflected voltage wave is –E (because initially the charge across the capacitor is zero) and the current of the incident wave is momentarily doubled. As the capacitor gets charged the terminal voltage rises. Finally the capacitor becomes fully charged; it then behaves as an open circuit for the terminal current is zero and the voltage of the incident voltage is doubled.

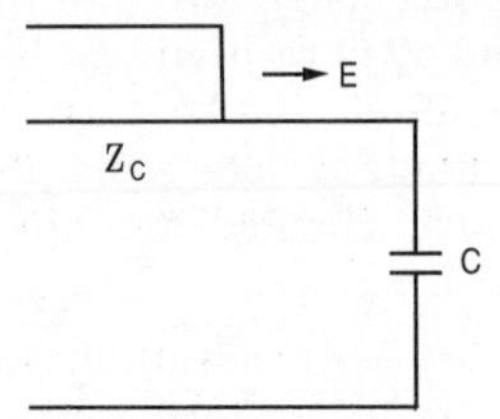

Fig. 16.15. ***Transmission Line Terminated By Capacitance C***

The reflected voltage varies exponentially and its mathematical expression can be arrived in the same way as in case of terminal inductance.

The expressions are given as

$$e_T = e_C = 2E\left(1 - e^{\frac{-t}{CZ_C}}\right) \quad ...(16.36)$$

$$e_R = e_T - E = E\left(1 - 2e^{\frac{-t}{CZ_C}}\right) \quad ...(16.37)$$

$$i_T = i_C = \frac{2E}{Z_C}e^{\frac{-t}{CZ_C}} \quad ...(16.38)$$

$$i_R = i_T - i = \frac{2E}{Z_C}e^{\frac{-t}{CZ_C}} - \frac{E}{Z_C}$$

$$= \frac{E}{Z_C}\left(2e^{\frac{-t}{CZ_C}} - 1\right) \quad ...(16.39)$$

e_T, i_T and e_R as functions of time are shown in Fig. 16.16 (*a*). The disposition of voltage wave at different instants is shown in Figs. 16.16 (*b, c, d*). In travelling wave studies, time *t* is usually expressed in μs, so L and C should be expressed in μH and μF respectively.

The *capacitance termination* is of more practical importance owing to the paradox that for lightning waves a transformer acts as a capacitance rather than as an inductance. The lightning surge arrives at a transformer as a travelling wave, and the front of the wave is so steep and rises so suddenly to a maximum, that there is practically no time for current to start to flow through the large inductance of the transformer winding. However, there is some small capacitance in the transformer (between turns and between winding and the core), and the transformer's reaction to the lightning surge is governed largely by this capacitance, rather than by inductance.

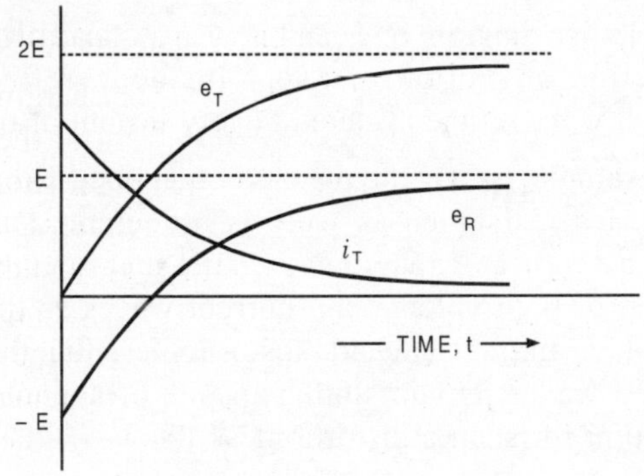

(a)

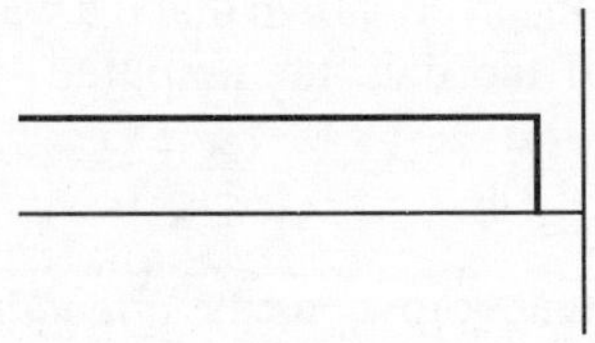

(b)

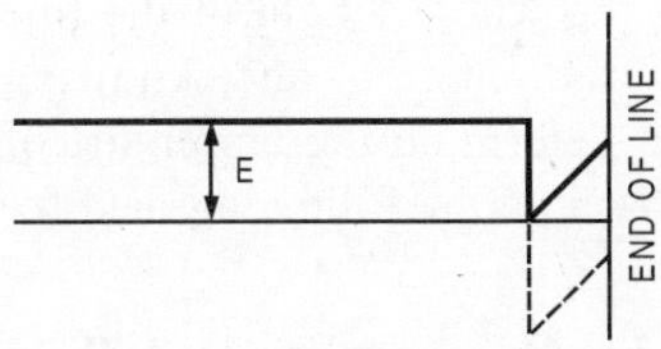

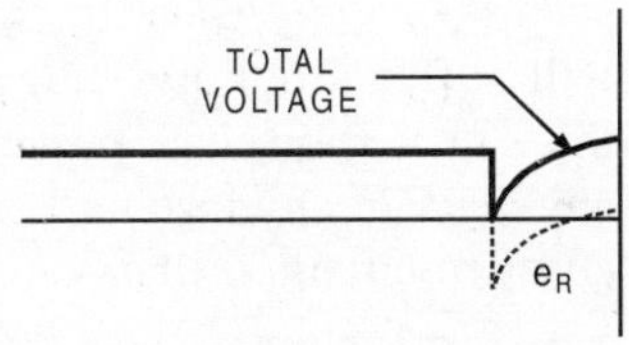

(c) (d)

Fig. 16.16. ***Travelling Waves on a Line Terminated by Capacitnace C***

16.8.3. Reflection From a Line Terminated With a Parallel Combination of Capacitance and Resistance

When a travelling wave meets a termination composed of parallel combination of a capacitance C and resistance R, as illustrated in Fig. 16.17 (*a*), the problem is the same as that illustrated in Fig. 16.16, except that Z in Eqs. (16.10) and (16.11) must be replaced by

$$\frac{\left(\frac{1}{pC}\right)R}{\frac{1}{pC}+R} = \frac{R}{1+pCR} \quad \text{where } p = \frac{d}{dt}$$

The voltage at the termination is thus

$$E_T = \frac{2RE}{R+Z_C(1+pCR)} \qquad ...(16.40)$$

It must be remembered that $p = \frac{d}{dt}$ and E is a voltage which is zero until $t = 0$ and E after $t = 0$, E_T may be determined as follows:

$$E = \frac{Z_C(1+pCR)+R}{2R} E_T$$

$$= \frac{1}{2}\left(pCZ_C + \frac{Z_C}{R} + 1\right) E_T$$

$$= \frac{1}{2} CZ_C\left(\frac{dE_T}{dt}\right) + \frac{1}{2}\left(\frac{Z_C}{R} + 1\right) E_T$$

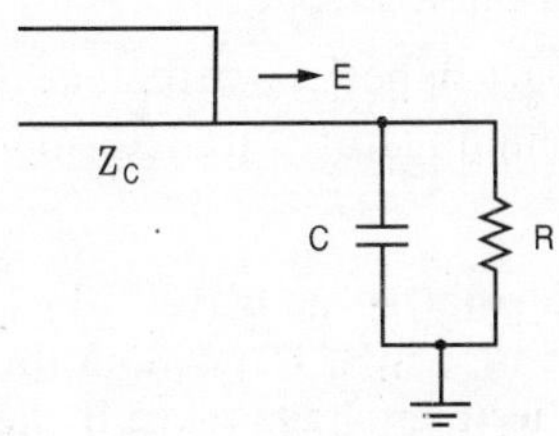

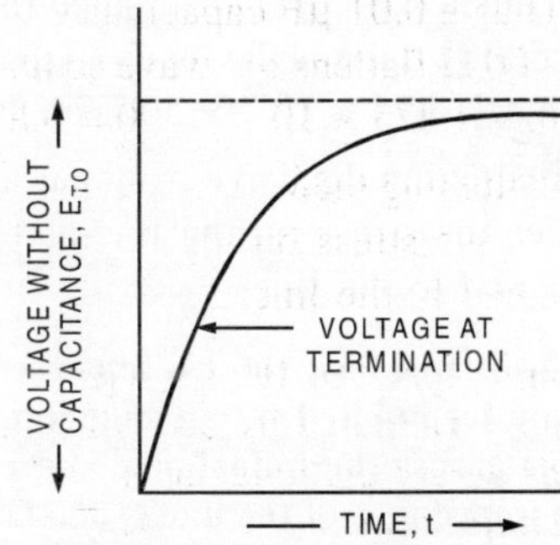

(a) Line Terminated With a Parallel Combination of R and C *(b) Effect of Shunt Capacitance At The Terminal*

Fig. 16.17

This is a linear differential equation for E_T of which the solution is

$$E_T = \frac{2E}{\frac{Z_C}{R}+1} + Ae^{-[(Z_C+R)/CZ_C R]t}$$

where A is an arbitrary constant and is determined by the fact that E_T can rise at a finite rate from its zero value. This provides

$$A = \frac{-2E}{\frac{Z_C}{R}+1}$$

$$\text{so that } E_T = \frac{2E}{\frac{Z_C}{R}+1}\left[1 - e^{-[(Z_C+R)/CZ_C R]t}\right]$$

$$= E_{T0}\left[1 - e^{-[(Z_C+R)/CZ_C R]t}\right] \qquad ...(16.41)$$

where E_{T0} is the voltage at the end when there is no capacitance. Figure 16.17 (*b*) shows the curve for voltage at the termination, E_T. The effect of capacitance is to cause the voltage at the end to rise to the full value gradually instead of abruptly, *i.e.,* it flattens the wave front. It is usual to specify the condition of the wave front by stating the time the wave takes to increase from 10 to 90 per cent of its value and multiplying by 1.25. If the wave attains x of its value in time t

$$1 - e^{-[(Z_C+R)/CZ_C R]t} = x$$

$$\text{so that } t = \frac{CZ_C R}{Z_C+R}\log_e\left(\frac{1}{1-x}\right)$$

The specifying time in this case is therefore

$$= 1.25\left[\frac{CZ_C R}{Z_C+R}\right]\left[\log_e 10 - \log_e 1.11\right] \text{ second}$$

$$= 2.75\,\frac{CZ_C R}{Z_C+R} \text{ second} \qquad ...(16.42)$$

In the case of a capacitance at a point of a line which stretches in both directions away from it, $Z_C = R$ and the time is 1.375 CZ_C second.

Thus a 0.01 μF capacitance in a line of surge impedance $Z_C = 500\ \Omega$ flattens the wave so that the time of the wave front becomes $1.375 \times 10^{-8} \times 500 = 6.875$ microsecond.

Flattening the wave front has a very beneficial effect, as it reduces the stress on the line-end windings of a transformer connected to the line.

Example 16.8. An 100 kV 2 μs rectangular surge travels along the line terminated by an inductance of 1 mH. Determine the voltage across the inductance and reflected voltage wave if the surge impedance of the line is 500 Ω.

Solution : Surge impedance, $Z_C = 500\ \Omega$

Inductance L = 1 mH = 1,000 μH

$t = 2$ μs

Incident voltage, E = 100 kV

Voltage across inductance,

$$e_L = e_T = 2E\, e^{\frac{-Z_C t}{L}} = 2 \times 100\, e^{\frac{-500 \times 2}{1{,}000}} \text{ kV}$$

$$= 200\, e^{-0.1} = 181 \text{ kV} \quad \textbf{Ans.}$$

Reflected voltage,

$$e_R = E\left(2e^{\frac{-Z_C t}{L}} - 1\right) = 100\left(2\, e^{\frac{-500 \times 2}{1{,}000}} - 1\right)$$

$$= 81 \text{ kV} \quad \textbf{Ans.}$$

Example 16.9. Explain surge impedance and velocity of propagation of travelling waves. A 500 kV, 2 μsec rectangular surge travels along the line terminated by a capacitor of 2,500 pF. Determine the voltage across the capacitance and reflected voltage wave if the surge impedance loading of line is 400 ohm.

[*U.P. Technical Univ. Power System Analysis 2009-10*]

Solution : Surge impedance, $Z_C = 400\ \Omega$

Capacitance, C = 2,500 pF = 0.0025 μF

$t = 2$ μs

Incident voltage, E = 500 kV

Voltage across the capacitance, $e_T = 2E\left(1 - e^{\frac{-t}{CZ_C}}\right)$

$$= 2 \times 500\left(1 - e^{\frac{-2}{400 \times 0.0025}}\right)$$

$$= 1{,}000\,(1 - e^{-2}) = 865 \text{ kV} \quad \textbf{Ans.}$$

Reflected voltage, $e_R = E\left(1 - 2e^{\frac{-t}{CZ_C}}\right)$

$$= 500\left(1 - 2e^{\frac{-2}{400 \times 0.0025}}\right)$$

$$= 500\,(1 - 2e^{-2}) = 365 \text{ kV} \quad \textbf{Ans.}$$

16.9. SUCCESSIVE REFLECTIONS, BEWLEY LATTICE OR ZIG-ZAG DIAGRAM

In an extensive network having many junctions and terminations, the number of transmitted and reflected waves, initiated by a single incident wave, increases as the wave meets different junctions. Generally it becomes difficult to keep track of the transmitted and reflected waves, but with the use of Bewley's lattice diagram one can know at a glance the position and direction of motion of every incident, reflected and transmitted wave on the system at every instant of time.

For example, let us take the case of an open-circuited line having series resistance as well as inductance, and shunt leakance as well as capacitance. With such a line there is attenuation of both voltage and current waves as they travel along the line. In the Heaviside distortionless line the voltage and current waves remain similar inspite of attenuation, and the condition for such a distortionless line is

$$RC = GL \quad \text{...(16.43)}$$

In such a line it can be shown that if a wave is of amplitude A_1 at any point of the line, the amplitude A_2 at some point distant x is

$$A_2 = A_1\, e^{-\beta x} \quad \text{...(16.44)}$$

where β is a constant, known as the *attenuation constant*. For the distortionless line the value of this constant is $\sqrt{RG}$. With such a line any wave travelling the whole length of the line will suffer a definite percentage attenuation, and, knowing this attenuation, it is possible to determine the amplitude of any wave at any instant. Let us assume the following data for an open-circuited line.

$R = 0.3\ \Omega$ per km, $G = 6.5 \times 10^{-7}$ S per km; $l = 640$ km

So attenuation constant,

$$\beta = \sqrt{RG} = \sqrt{0.3 \times 6.5 \times 10^{-7}} = 0.0004416$$

$$\text{and } e^{-\beta x} = e^{-0.0004416 \times 640} = e^{-0.2826} = 0.754$$

$$\because x = l = 640 \text{ km}$$

If this quantity is denoted by α and the initial value of the voltage at the generating end by unity, then following sequence of events, as far as the reflected wave is concerned, will occur. The time taken to make one tour of line,

$$t' = \frac{640 \times 1{,}000}{2.975 \times 10^8} = 0.00215 \text{ s.}$$

At zero time a wave of amplitude 1 start from generating or sending end G. At time t' a wave of amplitude $\alpha = 0.754$ strikes the open end and a reflected wave of $+\alpha = 0.754$ commences the return journey. At time $2t'$ this reflected wave is attenuated to $\alpha^2 = 0.5685$ and has reached G. Here it is reflected to $-\alpha^2 = -0.5685$ and after a time $3t'$ it reaches the open end attenuated to $-\alpha^3 = -0.4287$. It is then reflected without change of sign and reaches G after a time $4t'$ with an amplitude of $-\alpha^4 = -0.3232$. It is then reflected with a change of sign, thus starting off with an amplitude of $+\alpha^4 = +0.3232$, and so on. The Bewley lattice diagram is a space-time diagram with the space measured horizontally and time vertically, and the lattice for the above example is shown in Fig. 16.18.

Now, the increment of the voltage at the receiving end due to any reflection is twice the amplitude of the incident wave, because of the reflection without change of sign. Also, the final voltage at this end is the sum to infinity of all such increments. Thus, in the above example, it is

$$2(0.754 - 0.4287 + 0.2437 - \ldots).$$

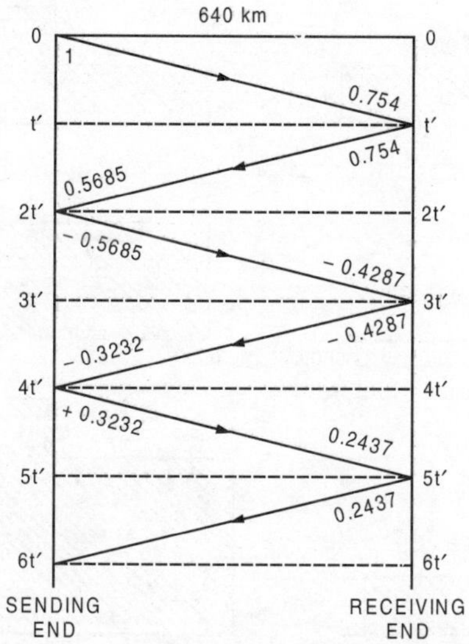

Fig. 16.18. ***Lattice Diagram For Line With Attenuation***

It is simpler to express the series generally in term of α, thus

$$2(\alpha - \alpha^3 + \alpha^5 - \alpha^7 + ...)$$
$$= 2[(\alpha + \alpha^5 + \alpha^9 + \alpha^{13} + ...) - (\alpha^3 + \alpha^7 + \alpha^{11} + \alpha^{15} + ...)]$$

$$= 2\left[\frac{\alpha}{1-\alpha^4} - \frac{\alpha^3}{1-\alpha^4}\right] = \frac{2\alpha}{1+\alpha^2} = \frac{2 \times 0.754}{1+0.5685} = 0.96.$$

Thus, even when open-circuited, such a line gives a far-end voltage less than the sending-end voltage, the reason being that the shunt leakance causes a drop along the series resistance.

As far as the sending end is concerned, the voltage at the moment of incidence of a wave sent from the far end is unity plus the amplitude of this wave. The lattice shows the amplitudes of these incident waves to be $+\alpha^2$, $-\alpha^4$, and so on, successive sending-end voltages thus being 1, $1 + \alpha^2$, 1, $1 - \alpha^4$, 1 and so on. Thus the voltage oscillates about unity, the amplitude of the oscillation rapidly diminishing so that eventually the voltage becomes the generator voltage of unity, as would be expected.

Now consider the case of the ideal resistanceless and leakanceless line joined at the far end by a resistance R. There is no attenuation as a wave travels along the line, but at the receiving end there is a reflection operator of $\frac{R - Z_C}{R + Z_C}$ which is less than unity. We can thus use this in place of α in the lattice diagram at the receiving end, and $\alpha = -1$ at the sending end. Suppose that $\frac{R - Z_C}{R + Z_C} = 0.4$, then the lattice diagram will be as illustrated in Fig. 16.19. At the receiving end the increment of voltage is the sum of the incident and reflected waves at each reflection, so that the ultimate voltage at this point is the sum to infinity of the series

$$(1 + \alpha) - (\alpha + \alpha^2) + (\alpha^2 + \alpha^3) - (\alpha^3 + \alpha^4) + ...$$

$$= (1 + \alpha)(1 - \alpha + \alpha^2 - \alpha^3 + ...) = (1 + \alpha)\left[\frac{1}{1-\alpha^2} - \frac{\alpha}{1-\alpha^2}\right] = 1$$

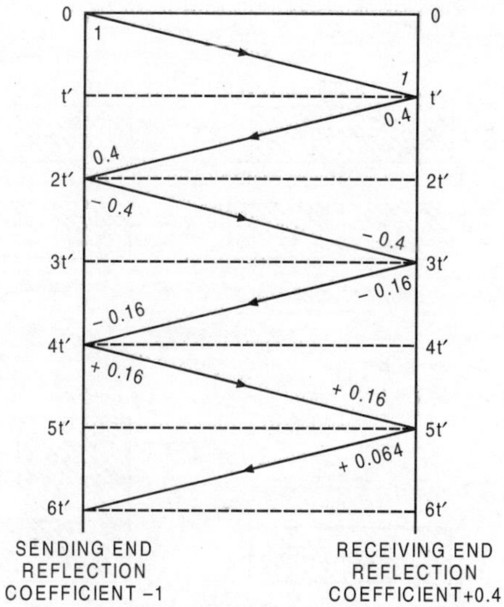

Fig. 16.19. ***Lattice Diagram For a Line Terminated Through a Resistance R***

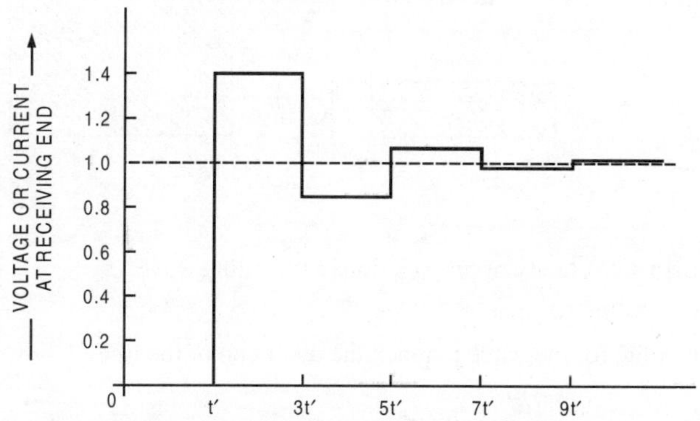

Fig. 16.20. ***Building-Up of Current and Voltage in a Line Terminated Through a Resistance***

Thus, the voltage at the receiving end finally settles down to that at the sending end, and consequently the current settles down to the simple Ohm's law value of E/R. The increments of current are obviously proportional to the increments of voltage at the receiving end, and therefore voltage-time and current-time curves for this end for $\alpha = 0.4$ are as illustrated in Fig. 16.20. The tabulated values are as shown, and it will be seen that both voltage and current oscillate about the value unity and finally settles down to this value.

Time	Increment Voltage or Current	Sum of Increments
0	0	0
t'	$1 + \alpha = 1.4$	1.4
$3t'$	$-(\alpha + \alpha^2) = -0.56$	0.84
$5t'$	$\alpha^2 + \alpha^3 = 0.224$	1.064
$7t'$	$-(\alpha^3 + \alpha^4) = -0.0896$	0.9744
$9t'$	$\alpha^4 + \alpha^5 = 0.03584$	1.01024

Example 16.10. An overhead transmission line 600 km having a surge impedance of 500 Ω is short-circuited at one end and a steady voltage of 5,000 V is applied at the other end. Neglecting the resistance, draw the diagrams of voltage and current waves. Calculate the sending-end current of the line 0.004 second after the voltage is applied.

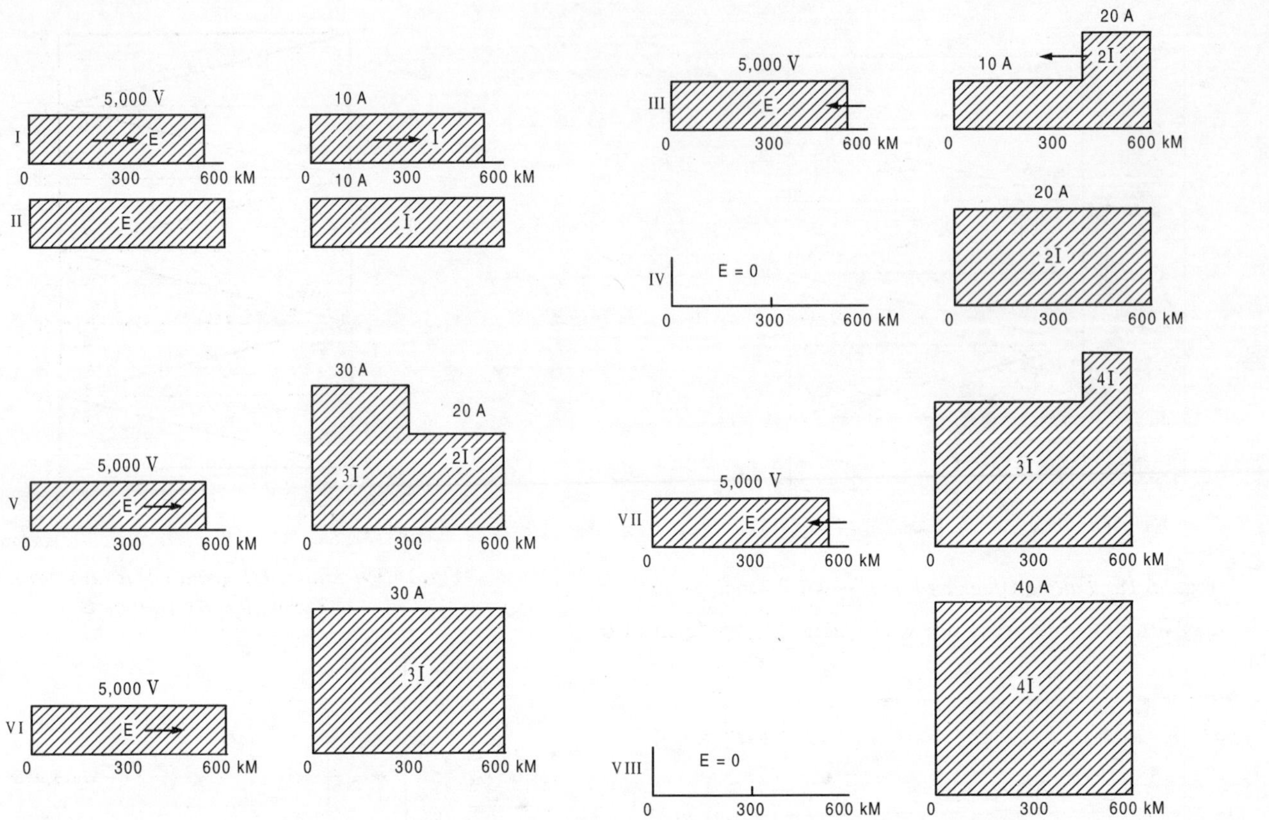

Solution : Velocity of propagation of travelling wave

$$= 3 \times 10^8 \text{ m/s}$$

So the time for the wave to reach the other end of the line

$$= \frac{600 \times 10^3}{3 \times 10^8} = 0.002 \text{ s}$$

$$E = 5{,}000 \text{ V}$$

$$\text{Current through the line} = \frac{E}{Z_C} = \frac{5{,}000}{500} = 10 \text{ A}$$

5,000 V; just before $t = 0.002$ s ...(I)

At $t = 0.002$ s, when reflection takes place ...(II)

Just after $t = 0.002$ s (Just after first reflection) ...(III)

When the line is short-circuited, the reflected voltage wave would be negative of the forward wave [refer to Eq. (16.14 *c*)], while the reflect current wave is of the same magnitude as of the incident wave. Thus the waves immediately after 0.002 s are given as :

At $t = 0.004$ s (the instant when the first reflection has reached the sending end) ...(IV)

The voltage wave is completely neutralized while the current wave becomes double the value *i.e.*, 20 A

At $t > 0.004$ s but < 0.006 s (after the second reflection has taken place) ...(V)

Voltage of E = 5,000 is being established along the line and the current value rises to

$$3I = (3 \times 10) = 30 \text{ A}$$

At $t = 0.006$ s (the instant when the second reflection has reached the far end.) ...(VI)

At $t > 0.006$ s < 0.008 s ...(VII)

At $t = 0.008$ s ...(VIII)

The waves are again reflected from the short-circuited end.

Example 16.11. A battery with an emf E and series resistance R are connected at $t = 0$ to the sending end of lossless transmission line which is short circuited at far end. Plot the sending-end current and voltage as the function of time for $R = \frac{R_0}{3}$. The time required for a wave to travel to full length for the line is T.

[U.P. Technical Univ. Power System Analysis 2005-06]

Solution : The circuit diagram is shown in Fig. 16.21.

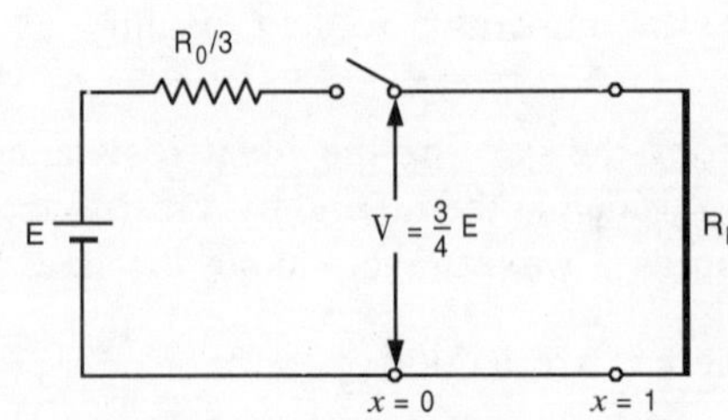

Fig. 16.21

$$\text{Sending-end voltage} = E - \frac{E}{R_0 + \frac{R_0}{3}} \times \frac{R_0}{3} = \frac{3}{4}E$$

$$\text{Sending-end current} = \frac{\frac{3}{4}E}{R_0} = 0.75\frac{E}{R}$$

Reflection coefficient for the load end is

$$K_L = \frac{R_L - R_0}{R_L + R_0} = -1 \qquad \because R_L = 0\ \Omega$$

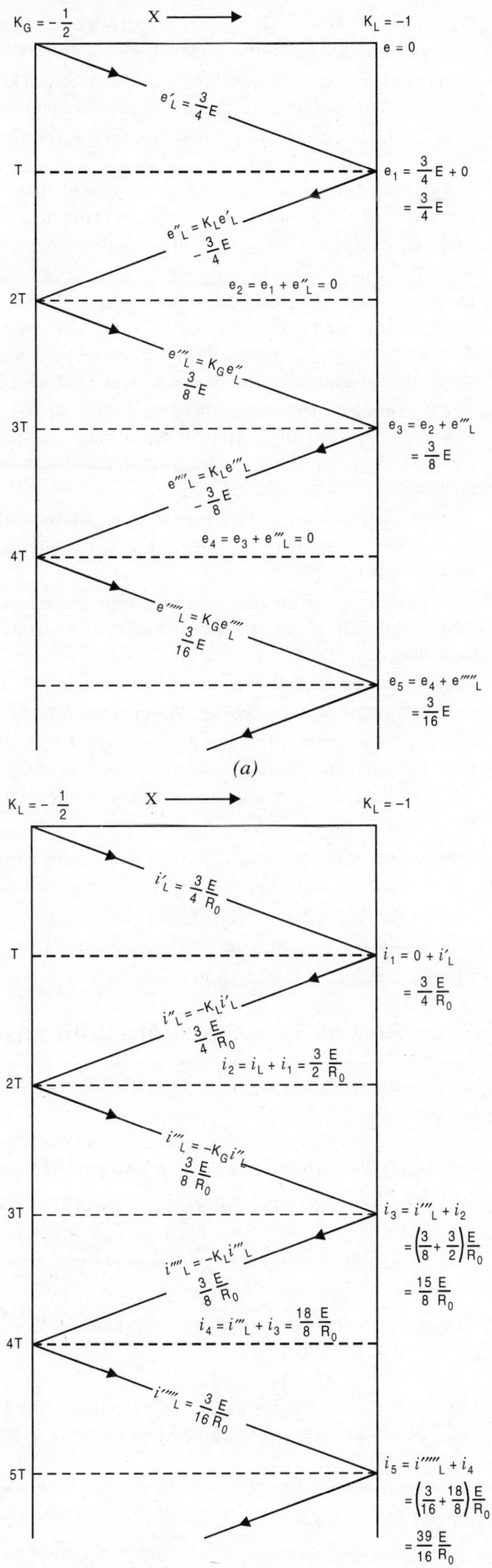

Fig. 16.22

Reflection coefficient for the generator end is

$$K_G = \frac{\dfrac{R_0}{3} - R_0}{\dfrac{R_0}{3} + R_0} = \frac{-2R_0}{4R_0} = -\frac{1}{2}$$

Zigzag diagrams for voltage and current are shown in Fig. 16.22. In a zigzag diagram distance is plotted horizontally and time is plotted downwards as shown in Fig. 16.22 (*a*) and (*b*).

The zigzag lines are wavefronts of the various reflections. The numbers assigned to the line indicate the magnitude of individual waves. The magnitude of each reflection is obtained by multiplying the magnitude of the preceding wave by reflection coefficient at the point where reflection occurs. The number in each intervening space is the sum of individual waves above the point, and represents the net voltage/current in that region of the diagram.

The voltage and current variations with time can now be easily plotted as shown in Fig. 16.23 (*a*) and 16.23 (*b*).

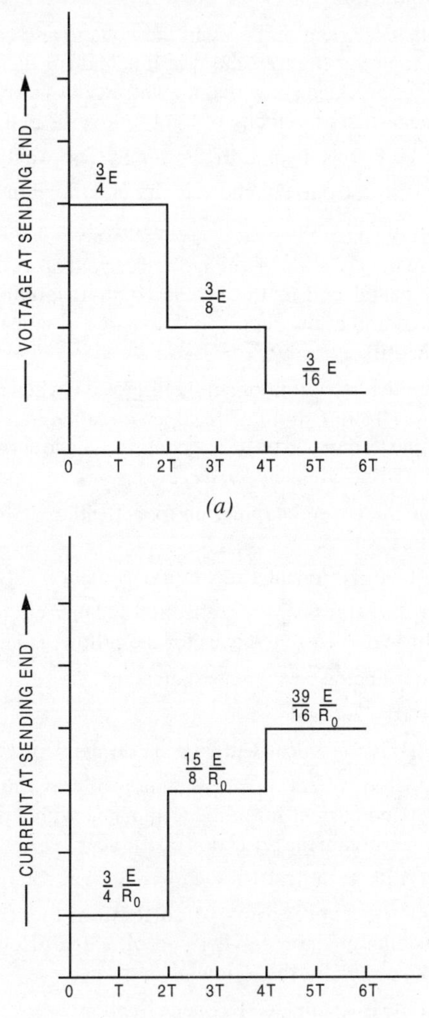

Fig. 16.23

Exercises

1. What is a travelling wave ? Explain the development of such a wave along an overhead transmission line.

2. Show that a travelling wave moves along an overhead line with a velocity of light and its speed is proportional to $\frac{1}{\sqrt{\varepsilon_r}}$ in case of a cable with dielectric material of relative permittivity ε_r.

3. Deduce expressions for surge impedance and velocity of propagation.

4. Show that the velocity of a travelling wave can be given by $v = \frac{1}{\sqrt{LC}}$.

[U.P. Technical Univ. Power System Analysis 2007-08]

5. Starting from first principles show that surges behave as travelling waves.

[U.P. Technical Univ. Power System Analysis 2006-07]

6. Develop wave equation for an uniform transmission line and find the velocity of its propagation.

[U.P. Technical Univ. Power System Analysis 2008-09]

7. Discuss the phenomenon of travelling wave over a transmission line conductors.

   Establish the relation between the voltage and current waves travelling over the transmission line and for their velocity of propagation. Also show that the velocity of wave propagation is same as that of velocity of light for overhead lines.

[U.P. Technical Univ. Power System Analysis 2005-06]

8. Define crest, front, tail and velocity of a travelling wave.

9. Discuss the phenomenon of wave reflection and refraction.

10. Explain the variation of voltage and current on an overhead line when the far end of the line is (*i*) short circuited (*ii*) open circuited and at the supply end a source of constant emf E is switched-in.

11. Discuss the behaviour of a travelling wave when it reaches the end of a (*i*) short circuited (*ii*) open circuited transmission line (*iii*) a line terminated by an impedance equal to surge impedance and (*iv*) a line connected to a cable.

12. Deduce the general expression for reflection and refraction of travelling waves.

[U.P. Technical Univ. Power System Analysis 2006-07]

13. Derive the expressions for reflection and refraction coefficients of voltage and current waves for the following cases :
    (*i*) terminated through resistance
    (*ii*) through a cable.

[U.P. Technical Univ. Power System Analysis 2008-09]

14. Discuss the reflection phenomenon of travelling waves of voltage and current for a line terminated with a resistive load and hence obtain the reflection coefficient.

[U.P. Technical Univ. Power System Analysis 2005-06]

15. Obtain the expressions of reflection and transmission coefficients at the receiving end of a line of characteristics impedance Z_{C_1} for the following cases :
    (*i*) Line is terminated by open circuit
    (*ii*) Line is terminated by short circuit
    (*iii*) Line is connected to two lines of characteristic impedance Z_{C_2} and Z_{C_3} respectively at the receiving end.

[U.P. Technical Univ. Power System Analysis 2004-05]

16. Discuss the behaviour of a travelling wave when it reaches (*i*) short circuited (*ii*) open circuited transmission lines and (*iii*) line terminated by an impedance equal to surge impedance (Z).

[U.P. Technical Univ. Power System Analysis 2006-07]

17. Develop an equivalent circuit for the analysis of the behaviour of travelling waves at transition points on a transmission line.

18. Derive an expression for travelling waves in overhead line, also mention interpretation from physical significance point of view.

[G.B. Technical Univ. Power System Analysis 2010-11]

19. Draw equivalent circuit for determining the transmitted voltage and current surges on a forked or bifurcated line. Derive expressions for the transmitted voltage and currents.

20. Make the analysis of a wave travelling along a line terminated with an inductance L.

    Derive an expression for the voltage across an inductance connected at the end of a transmission line of surge impedance Z_C when a step wave of magnitude E is travelling along it.

[U.P. Technical Univ. Power System Analysis 2009-10]

21. A unit-step voltage wave is travelling along a line of characteristic impedance Z_C. The line terminates in a shunt inductance L. Derive the expression for the voltage at the line and the inductance junction as a function of time.

[G.B. Technical Univ. Power System Analysis 2010-11]

22. What is the effect of shunt capacitance at the terminal of a transmission line ?

    Give an expression for the specifying time in case of a line terminated with a parallel combination of resistance and capacitance.

23. Explain Bewley's lattice diagram.

[W.B. Univ. of Technology Power System-III, 2009-10]

24. Explain the procedure for drawing Bewley's lattice diagram with the help of a suitable example.

[U.P. Technical Univ. Power System Analysis 2006-07]

25. Write short notes on the following :
    (*i*) Surge impedance and velocity of propagation.
    (*ii*) Specifications of travelling waves.
    (*iii*) Reflection and refraction of travelling waves.
    (*iv*) Equivalent circuit for travelling wave studies.
    (*v*) Bewley lattice or Zigzag diagram.

Short Answer Type Questions With Answers

Q. 1. What are the values of natural impedances for overhead lines and cables ?

Ans. The value of natural impedance for an overhead line is between 400 Ω and 600 Ω while for a cable it is between 40 Ω and 60 Ω.

Q. 2. Why velocity of propagation over all overhead lines is same ?

Ans. Velocity of propagation over all overhead lines is same because product of L and C is same for all overhead lines.

Q. 3. What is meant by crest of a wave ?

Ans. Crest of a wave is the maximum amplitude and is usually expressed in kV or kA.

Q. 4. What is meant by front of a wave ?

Ans. Front of a wave is the portion of the wave before crest and is expressed in time from the beginning of the wave to the crest value in ms or μs.

Q. 5. A transmission line of surge impedance Z_C is terminated through a resistance R. Give the coefficients of refraction and reflection.

Ans. Coefficients of refraction $= \frac{2R}{R + Z_C}$ and coefficient of reflection $= \frac{R - Z_C}{R + Z_C}$.

Q. 6. While connecting station equipment to an overhead line through a cable it is important that length of the cable should not be very small. Why ?

Ans. In connecting station equipment to an overhead line through a cable it is important that the length of cable should not be shorter than the expected length of the wave otherwise successive reflections at the junction may cause piling up of voltage and the voltage at the junction may attain the value of incident voltage.

Q. 7. What is the effect of shunt capacitance at the terminal of a transmission line ?

Ans. The effect of shunt capacitance at the terminal of a transmission line is to cause the voltage at the terminal to rise to full-value gradually instead of abruptly *i.e.,* to cause flattening of the wave front, which reduces the stress on the line-end windings of a transformer connected to the line.

Q. 8. What is the application of Bewley's lattice diagram ?

Ans. With the use of Bewley's lattice diagram one can know at a glance the position and direction of motion of every incident, reflected and transmitted wave on the system at every instant of time.

Problems

1. A transmission line has a capacitance of 0.0125 μF per km and an inductance of 1.5 mH per km. The overhead line is connected to an underground cable that has a capacitance of 0.3 μF per km and inductance of 0.25 mH per km.

   Determine the voltage rise that is developed at the junction of the line and cable by a wave with a crest value of 100 kV, travelling along the cable. [**Ans.** 184.6 kV]

2. A surge of 50 kV travels along an overhead line towards its junction with a cable. The surge impedances for the overhead lines and cable are 400 Ω and 40 Ω respectively. Determine the amplitude of the surge transmitted on the cable.

   [**Ans.** 9.1 kV]

3. An overhead transmission line having a surge impedance of 450 Ω runs between two substations A and B; at B it branches into two lines C and D, of surge impedances of 400 Ω and 50 Ω respectively. If a travelling wave of vertical front and magnitude 50 kV travels along the line AB, calculate the magnitude of voltage and current waves which enter the branches C and D.

   [**Ans.** 9 kV; 22.5A; 180 A]

4. An overhead line with surge impedance 500 Ω bifurcates into two lines of surge impedance 500 Ω and 50 Ω, respectively. If a surge voltage of 25 kV is incident on the overhead line, determine the magnitudes of voltage and current which enters the bifurcated lines.

   [G.B. Technical Univ. Power System Analysis 2010-11]

   [**Ans.** 4.167 kV; 8.334 A; 83.34 A]

5. An 100 kV 1 μs rectangular surge travels along the line terminated by an inductance of 4 mH. Determine the voltage across the inductance and reflected voltage wave if the surge impedance of the line is 400 Ω. [**Ans.** 181 kV, 81 kV]

6. A 200 kV 2 μs rectangular surge travelling along a line having a surge impedance of 400 Ω approaches a station at which the concentrated earth capacitance is 0.025 μF. Determine the maximum value of transmitted wave.

   [**Ans.** 72.5 kV]

❑❑❑

CHAPTER 17

Insulation Co-ordination

INSIDE THIS CHAPTER

17.1. INTRODUCTION

Each electrical equipment should have long service life (more than 25 or 30 years). The conductors are supported on insulators/embedded in insulation system. The insulation (internal and external) of each and every electrical equipment is exposed to continuous normal voltages and temporary abnormal overvoltages. The equipment insulation is designed such that it withstands the highest power frequency (PF) system voltage, occasional temporary power frequency overvoltages, occasional lightning/switching surges reaching the equipment (after the interception by lightning arresters). Each equipment has assigned *rated insulation level*, the capability being approved by type tests/routine tests for the equipment. The terms related with rated insulation levels are defined in the relevant IS (Indian Standards)/IEC (International Electrochemical Commission) standard specifications.

The **insulation requirements** are determined by considering the following:

1. Highest Power Frequency System Voltage (Continuous). AC network has different nominal power frequency voltage level (*e.g.,* 400 V, 3.3 kV, 6.6 kV, 11 kV, 22 kV, 33 kV, 66 kV, 110 kV, 132 kV, 220 kV, 400 kV rms continuous, at 50 Hz). During light load period, the power frequency voltage at the receiving end of transmission line rises. In a well-regulated system, the permissible maximum system voltage allowed is called the *highest system voltage*. Each nominal voltage level has certain corresponding highest system voltage (440 V, 3.6 kV, 7.2 kV, 12 kV, 24 kV, 36 kV, 72.5 kV, 123 kV, 145 kV, 245 kV and 420 kV rms continuous). Each equipment is designed and tested to withstand the corresponding highest power frequency system voltage of that voltage level continuously without internal/external insulation failure.

2. Temporary Power Frequency Overvoltages caused by load throw-off, faults, resonance etc. There is a difference between the characteristics of power frequency overvoltages and transit voltage surges and the corresponding stresses on equipment and surge arresters. The *temporary power frequency (PF) overvoltages* are of 50 Hz and of lesser peak, lesser rate of rise and of longer duration (seconds or even minutes). The protection against temporary PF overvoltages is provided by IDMT overvoltage relays (already explained in chapter 9) connected to secondary of bus potential transformer and circuit breakers. The relay and breakers action is within several tens of milliseconds to a few seconds. The circuit breakers open and the equipment (such as transformer or bus) is protected against the temporary PF overvoltage.

3. Transient Overvoltage Surges caused by lightning, switching, restrikes, travelling waves etc. Surges in the power system are of comparatively high peak, high rate of rise and last for a few tens/hundreds of microseconds and are therefore called the transients. Surges can cause spark-over and flash-over at sharp corners, flash-over between phase and earth at the weakest point, breakdown of gaseous/liquid/solid insulation, failure of transformers and rotating electrical machines.

The failure rate due to lightning and switching has been minimised by *proper insulation coordination* (explained later) and surge arrester protection (already explained in chapter 15).

Several protective devices are installed in the network to intercept lightning strokes and minimise the peak/rate of rise of surges reaching the equipment.

Figure 17.1 illustrates the range of the waveforms and durations of PF overvoltages, lightning surges and switching surges.

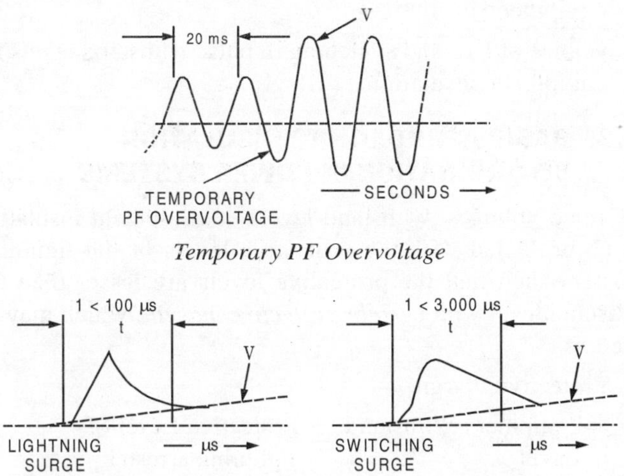

Temporary PF Overvoltage

Lightning Surges *Switching Surges*

Fig. 17.1. ***Representative Waveforms of PF Overvoltages and Surges Occuring in Network***

4. Withstand Levels of The Equipment. The BIL (basic impulse insulation level) is specified and other withstand levels are then selected from relevant tables provided in standard specifications.

Basic impulse insulation levels are reference levels expressed in impulse crest voltage with a standard wave not longer than 1.2/50 μs wave. Apparatus/equipment should be capable of withstanding test waves above BIL.

Table 17.1 gives the BIL for various reference class voltages (kV).

Table 17.1. Basic Impulse Insulation Levels

Reference Class kV	Standard Basic Impulse Levels kV	Reduced Insulation Level kV
23	150	
34.5	200	
46	250	
69	350	
92	450	
115	550	450
138	650	550
161	750	650
196	900	
230	1,050	900
287	1,300	1,050
345	1,550	1,300

17.2. CO-ORDINATION OF SYSTEM INSULATION

The co-ordination of the insulation of the different parts of a system is somewhat analogous to the grading of the fuses or other current limiting devices in the overall picture of protection against overcurrents. If there is an overcurrent in a subsidiary circuit, then it is essential that this circuit only shall be disconnected, and not the entire system. Similary with protection against overvoltages. It is obvious that the overinsulation of one part of the system may, in the event of severe overvoltage, result in the breakdown of the insulation of some vital and perhaps inaccessible part. Thus, it is preferable that the line insulators should flash-over rather than the transformer bushings breakdown, and likewise it would be preferable to have a transformer bushing breakdown rather than the damage to the winding themselves.

Insulation co-ordination implies the selection of the insulation of equipment with regard to its application in order to minimise undesired incidents due to voltage stresses (caused by transient overvoltages in the system) within the system. When an insulation system is co-ordinated, the solid insulation materials involved should have a voltage breakdown level higher than that of either the associated overvoltage protective devices or the clearance on which the co-ordination is based.

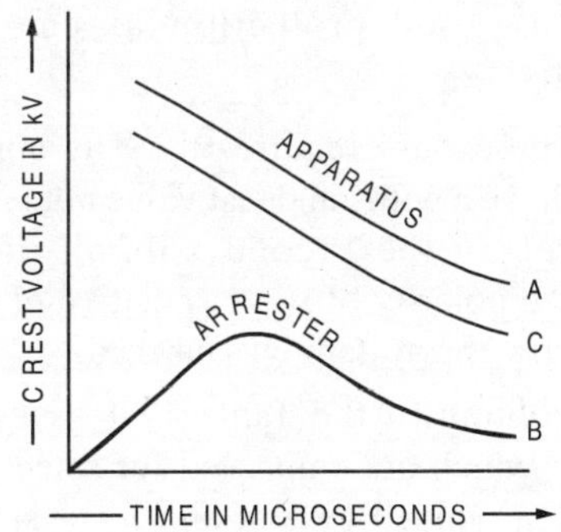

Fig. 17.2

Insulation co-ordination means the correlation of the insulation of the various components in a power system to the insulation of the protective devices used for the protection of those equipment against overvoltages. For instance in a substation, the insulation of the transformers, circuit breakers, bus-bar supports, etc. should have an insulation strength higher than the voltage levels that can be provided by the protective equipment such as lightning arresters, or surge absorbers. Thus if curve A represents the dielectric strength of the insulation of an electrical equipment/apparatus, the curve B for the lighting arresters protecting it must lie below the curve A. These curves have been plotted by the voltage and time. Any other electrical equipment having characteristics those of curve C will also be protected by the lightning arrester whose characteristics are given by the curve B. Hence we can generalise the above by stating that a protective device must have insulation characteristics which lie below the insulation characteristics of the equipment/apparatus under protection.

The problem of insulation co-ordination involves not only the protection of equipment but the protection of the protective devices too. To achieve this a lightning arrester must be applied and used on the system in such a way that it will discharge excessive voltage safely to the ground very quickly and then restore itself as an insulator and protect the equipment insulation.

For safe operation of the equipment, it should have insulation strength equal to or greater than the basic standard insulation level and the protective equipment for a station/substation should be chosen to give the insulation good protection corresponding to the working of these levels as economically as possible.

To assist in the process of insulation co-ordination, standard insulation levels have been recommended

The problem of insulation co-ordination can be studied under the following three steps:

1. Selection of a suitable insulation which is a function of reference class voltage (*i.e.*, 1.05 × working voltage of the system). Table 17.1 gives the basic impulse insulation levels (BIL) for various reference class voltages.
2. The design of the various equipments must be such that the breakdown or flash-over strength of all insulation in the station equals or exceeds the selected levels as in (1).
3. Selection of protective devices that will provide the apparatus as good protection as can be justified economically.

The above procedure requires that the apparatus under protection shall have a withstand test value not less than the kV magnitude given in the second column of Table 17.1, irrespective of the polarity of wave (positive or negative) and irrespective of how the system was earthed.

The third column of the Table 17.1 gives the reduced insulation levels which are employed for selecting insulation levels of solidly grounded system and for systems operating above 345 kV where switching surges are of more importance than the lightning surges.

17.2.1. Strategy of Insulation Co-ordination

The problem of overvoltages and insulation co-ordination can be solved by the following steps:

1. Each equipment/apparatus has specified power frequency *Withstand Level* and *Impulse Withstand Levels.*
2. The Withstand Levels of Equipment/apparatus/machines are co-ordinated with the *Protective Level* of the nearest *lightning* arrester. *Protective Levels* of lightning arrester at each voltage level shall be co-ordinated.
3. Every equipment is well protected and overall economy and reliability is achieved. In the event of occurrence of severe voltage-surge the damage is to least costly equipment.
4. Duplicate surge protection is provided in substations, one lightning arrester per phase at incoming bus and another lightning arrester at transformer terminals for each phase.
5. System neutral is grounded at every voltage level to reduce coefficient of grounding and to discharge the surges.

Insulation co-ordination covers the following aspects:

1. The causes and effects of transient overvoltages (surges) and the protection of electrical equipment insulation.
2. Standardisation of nominal voltage levels. Highest voltage levels in the network.
3. Choice of power frequency withstand values for equipment insulation.
4. Choice of BIL and switching impulse withstand levels for equipment insulation.

17.3. BASIC APPROACH TO INSULATION CO-ORDINATION IN POWER SYSTEMS

The rated voltages, withstand levels of equipment insulation are co-ordinated with the protective levels of the lightning arresters such that the protective levels are lesser than the withstand level with *certain protective margin*, which may be given as

$$\text{Protective margin} = \text{Equipment Withstand Level} - \text{Protective Level of lightning arrester.}$$

Protective margin may be expressed in terms of per cent of Equipment Withstand Level.

Protective levels are different for lightning impulse and switching impulse, *e.g.*,

$$\frac{\text{Lightning impulse voltage withstand level}}{\text{Switching impulse voltage withstand level}} = 1.2$$

The co-ordination between insulation levels of the equipment and protective levels of lightning arresters is further co-ordinated for different equipments at the same voltage level and further for equipment at different voltage levels. Such a grading of withstand values and the protective levels of lightning arresters at different voltage levels is known as the *insulation co-ordination*.

The protection of equipment against impulse voltage waves by means of lightning arresters is expressed in terms of protective ratio given as

$$\text{Protective ratio} = \frac{\text{Impulse Withstand Level of Equipment, kV}_P}{\text{Protection Level of Lightning Arrester, kV}_P}$$

Separate protective ratio is specified for (*i*) lightning impulse wave and (*ii*) switching impulse wave.

IEC and IS specifications on Insulation Co-ordination on HV Equipment; HV Testing specify the values of 1. Normal power frequency system voltages 2. Highest power-frequency system voltages 3. Required lightning impulse voltage levels 4. Required switching impulse voltage levels 5. Protective levels of lightning arresters and 6. Withstand levels of the equipment against 2, 3, 4.

System consultants/designers select the insulation levels at each voltage level from the Standards. These are co-ordinated for the entire network having different voltage levels. The equipment insulation and lightning arrester protection levels are graded. The equipment specifications are based on the co-ordinated values.

17.4. SELECTION OF ARRESTER RATING

The rating of an arrester is defined as the rms voltage at which the arrester passes the duty cycle test as per standards. In selecting the arrester for a given application, select the lowest rated surge arrester that will have a satisfactory service life on power system while providing adequate protection for insulation. Table 17.2 gives the typical rating of arresters for different system voltages (grounded neutral circuits).

Table 17.2. Typical Arrester Ratings For Different System Voltages (Grounded Neutral Circuits)

Nominal System Line-to-Line Voltage in kV	Arrrester Rating in kV
2.4	2.7
6.9	6.0
12.47	9.0
34.5	27
69	54
115	90
230	172
400	300

The recommended limit of the magnitude of voltage which may be continuously applied to the arrester is called the "maximum continuous operating voltage" of the arrester. The arrester rating must be selected such that the maximum continuous power system voltage applied to the arrester is less than the continuous voltage capacity of the arrester.

The lightning arrester selected should stand the magnitude of temporary overvoltages for one second. If temporary overvoltages cannot be cleared before the arrester capability is exceeded, faster time for clearance should be used or next higher rated lightning arrester should be used.

The magnitude of temporary overvoltages can be determined knowing the system conditions. The duration required before tripping can be found out by the response time of the backup system protection.

The ability of the arresters to dissipate overhead line surges is considered in terms of the maximum amount of energy that may be dissipated. Table 17.3 gives the energy capability of Tranquell XE arresters.

Table 17.3. Energy Capability of Tranquell XE Arresters

Arrester	Energy Capability	
	KJ/kV of Rating	kJ/kV of Maximum Continuous Operating Voltage
2.7 to 48	4.0	4.9
54 to 360	7.2	6.9

17.5. INSULATION CO-ORDINATION BETWEEN LINES AND SUBSTATIONS

Either the line or the substation is overinsulated with normal insulation for the other. Thus, there are two cases *viz.* Lines normally insulated, substation overinsulated and Lines overinsulated, substation normally insulated.

17.5.1. Lines Normally Insulated, Substation Overinsulated

This scheme is applicable to grounded overhead lines. It ensures flash-over occurrence at or near the point of strike. Precautions are taken to avoid damage to the line insulation. In case adequate shielding against direct strokes is provided at substation, the substation will be subjected only to chopped waves (received through the lines entering the substation). These chopped waves are limited to safer values.

17.5.2. Lines Overinsulated, Substation Normally Insulated

This scheme is employed in case of wooden pole overhead lines possibly with wooden cross arms and large conductor spacings. It aims at reducing to minimum, the risk of flash-over and damage to the overhead line. The substation in such a case is likely to be subjected to high surges and adequate protection at substations is necessary. This is provided through selection of proper rating and location of lightning arresters.

17.6. CHOICE OF INSULATION LEVELS FOR SUBSTATION EQUIPMENT

The insulation of substation equipment should withstand the overvoltages occurring owing to external and internal causes.

As already mentioned in Art. 17.1 the overvoltages may be either temporary power frequency overvoltages or impulse voltage surges due to lightning and switching.

Table 17.4. Insulation Levels of Substation/ Station Equipment

Normal Voltage, Line-to-Line in kV rms	Highest System Voltage, Line-to-Line in kV rms	Impulse Withstand Test Dry With Standard Full Wave Positive and Negative Polarities kV (Crest)	1 Minute Power Frequency Withstand Under Standard Condition kV rms
3.3	3.6	45	21
6.6	7.2	60	27
11	12	75	35
15	17.5	95	45
22	24	125	55
33	36	170	75
47	52	250	105
66	72.5	325	140
88	100	450	182
110	123	550	230
132	145	650	275
150	170	750	325
220	245	1,050	460
400	420	1,550	680

The performance of insulation is verified by power frequency tests and impulse tests.

The lightning arresters divert the transient overvoltages to ground and protect the substation insulation.

The condition for achieving the desired insulation levels in the substation are as follows:

1. Each equipment should have specified impulse withstand level.
2. Surge arresters should be of specified protective level.
3. Clearances should be according to the recommendations of standards. Such clearances are based on specified impulse withstand levels.
4. The protective ratio and protective margin should be correctly selected such that equipment design is economical and flash-over/damage does not cause major damage to costly and vital equipment.

Exercises

1. Why is insulation co-ordination required in a large power system ? What is meant by BIL of an equipment ?

2. What are the causes of dangerously high voltages on the transmission system? Give their relative magnitudes.

What are the means adopted to protect the power system against such overvoltages ? How is the insulation co-ordination carried out ?

[A.M.I.E. Sec. B. Advanced Power Systems Winter 1996]

3. What is the basic impulse level? Give its values for 33, 132, 220 and 400 kV effectively grounded systems.

[A.M.I.E. Sec. B. Advanced Power Systems Winter 1995]

4. What are BILs ? Explain their significance in power system studies.

5. State the difference between the "nominal system voltage" and "highest system voltage". Gives examples.

6. Explain the protective characteristics of a lightning arrester against the withstand characteristic of equipment on a voltage-time curve.

7. Define the terms :

(*a*) Insulation co-ordination.

(*b*) Rated voltage of surge arrester.

(*c*) Basic insulation level.

(*d*) Transient surges.

8. Discuss the insulation coordination and basic insulation levels adopted for EHV lines and equipment.

9. Write short notes on the following :

(*a*) Co-ordination of system insulation.

(*b*) Selection of arrester rating.

(*c*) Insulation co-ordination between lines and substations.

(*d*) Choice of insulation levels for substation equipment.

10. Explain the following :

(*i*) How insulation requirements are determined ?

(*ii*) What is difference between power frequency overvoltages and transient overvoltage surges ?

(*iii*) What is meant by basic impulse insulation level ?

(*iv*) What is meant by insulation coordination ?

(*v*) What is meant by "protective margins" and "protective ratio" ?

❑❑❑

CHAPTER 18

Power System Earthing

INSIDE THIS CHAPTER

18.1. INTRODUCTION

The term '*earthing*' means connecting of the non-current carrying parts of the electrical equipment or the neutral point of the supply system to the general mass of earth in such a manner that at all times an immediate discharge of electrical energy takes place without danger. Earthing is achieved by electrically connecting the respective parts in the installations to some system of electrical conductors or electrodes placed in intimate contact with the soil some distance below the ground level. This contacting assembly is called the *earthing*. The metallic conductors connecting the parts of the installations with the earthing are called *electrical connections*. When speaking of the earthing and the earthing connections as a whole, it is customary to use the term the *earthing system*.

The earthing is provided with the following objectives :

1. For the safety of the personnel from the electric shock—insuring that non-current carrying parts, such as equipment frames are always safely at ground potential even though insulation fails.
2. For the safety of the equipment and personnel against lightning and voltage surges providing the discharge path for surge arresters, gaps and other similar devices.
3. For providing the ground connections for grounded neutral systems.
4. For providing a means of positively discharging and de-energizing feeders or equipment before proceeding with maintenance on them.

Power system earthing is highly important. A *substantial and adequate ground* that will not burn off or permit dangerous rise in voltage under abnormal conditions is essential. Extensive damage and dangerous conditions have arisen when inadequate grounds have been provided. Multiple grounds and multiple connections to them are usually desirable to ensure ground protection, even though one ground or connection opens owing to burn-off or other condition such as high resistance. The station earthing system should have low resistance.

The earthing can be divided into *neutral* and *equipment* earthing. Neutral earthing (or grounding) deals with earthing of the system neutral to ensure system security and protection. Neutral earthing is also called the *system earthing*. Equipment earthing or safety earthing deals with earthing of non-current carrying parts of the equipment to ensure safety of personnel and protection against lightning.

Equipment earthing also helps in the earth fault protection. The earth fault current from the equipment flows through the earthing system to the earth and is sensed by protection system and circuit breakers are opened. The faulty equipment is then repaired and recommissioned. The earthed parts remain at approximately earth potential even during flow of fault current. In *neutral earthing* (or *neutral grounding*), the neutral points (or star points) of star-connected 3-phase windings of power transformers, generators, motors, earthing transformers are connected to low resistance earth. Neutral earthing has numerous advantages and will be discussed later on.

System earthing and equipment earthing have common earth mat* from which flat iron risers are taken out to connect all the non-current carrying metal parts of the equipment. At the time of fault the earth mat conductors rise to a voltage, which is equal to the resistance of the earth mat multiplied by ground fault current. This difference of potential results in voltage gradients**. The earth mat of the substation is so designed that the *step potential* (potential difference between two feet of a person at the time of fault) and *touch potential* (potential difference to which body is subjected to when a person comes in contact with an earthed body) are within safe limits. The average resistance of human body is 1,000 Ω and a current of 9 to 25 mA may be painful for the body for long duration. As fault is cleared very fast, *i.e.,* within 2 to 4 cycles (40 to 80 milliseconds) therefore 9 mA current is considered safe.

In electrical installations the following components must be earthed :

(*i*) The frames, tanks and enclosures of electric machines, transformers, and apparatus, lighting fittings, and other items of equipment;

(*ii*) the operating mechanism of the switchgear;

(*iii*) the framework of the switchboards, control boards, individual panel boards, cubicles;

(*iv*) the structural steelwork of indoor and outdoor substations, metal cable jointing boxes, the metal sheaths of the cable, the rigid metal conduit runs, and similar metalwork.

18.2. DEFINITIONS

1. Earth Electrode. Any wire, rod, pipe, plate or an array of conductors, embedded in ground horizontally or vertically is known as the *earth electrode*. In distribution systems the earth electrode may consist of a rod, about 1 metre in length and driven vertically into the ground. In generating stations and substations, grounding mats are used rather than individual rods.

2. Resistance-To-Earth or Earth Resistance. The resistance offered by the earth electrode to the flow of current into the ground is known as *resistance-to-earth*. The term resistance-to-earth (or earth resistance) does not imply the contact resistance between the earthing electrode and the soil (which is insignificant), but mainly the resistance of soil between the electrode and the point of zero potential. Numerically it is equal to the ratio of the potential of earth electrode with respect to remote point, to the current dissipated by it.

Mathematically, earth resistance is given as $\frac{V}{I}$ where V is the measured voltage between the electrode and the voltage spike and I is the injected current during the earth resistance measurement through the electrode.

As a general rule the lower the value of earth resistance better it is but even then the following values of earth resistance (maximum permissible) values will give satisfactory results.

Large power stations	— 0.5 Ω
Major power stations	— 1.0 Ω
Small substations	— 2.0 Ω
In all other cases	— 8.0 Ω

3. Earthing Lead. The wire which connects earth wire or any other apparatus to be earthed to the earth electrode is known as the earthing lead.

4. Earth Current. The current dissipated by the earth electrode into the ground is called the *earthing current.*

5. Step Potential. Step potential is the potential difference between the feet of a person standing on the floor of the substation, with 0.5 m spacing between two feet (one step), during the flow of earth fault current through the grounding system.

6. Touch Potential. Touch potential is a potential difference between the fingers of raised hand touching the faulted structure and the feet of the person standing on substation floor. The person should not get a shock even if the grounded structure is carrying fault current, *i.e.,* the touch potential should be very small.

7. Mesh Potential. It is the maximum touch potential within a mesh of the grid.

8. Transferred Potential. It is a special case of touch potential where a potential is transferred into or out of the substation.

18.3. EFFECTS OF ELECTRIC CURRENTS IN HUMAN BODY

Bruner (1967) states that the threshold of perception of electric shock is about 1 mA. At this level a tingling sensation is felt by the subject when there is a contact with an electrified object through intact skin. With the increase in magnitude of ac, the sensation of tingling gives way to contraction of muscles. The muscular contractions increase as the current is increased and finally a value of current is reached at which the subject cannot release his grip on the current carrying conductor. The maximum current at which the subject is still capable of releasing a conductor by using muscles directly stimulated by the current is called "*let go current*". The value of this current is significant because an individual can withstand, without serious after-effects, repeated exposures to his 'let go current' for at least the time required for him to release the conductor. Also, currents slightly in excess of 'let go current' would not permit the individual to release his grip from the conductor supplying current.

Based on the experiments conducted on males and females, it is generally accepted that the safe 'let go current'

* A number of rods when joined together through copper conductors constitute an earthing mat or earthing grid.

** The resistance of an earthing electrode is not concentrated at one point but is distributed over the soil around the electrode.

could be taken approximately 9 mA and 6 mA for men and women respectively.

At current levels higher than the 'let go current' the subject loses ability to control his own muscle actions and he is unable to release his grip on the electrical conductor. Such currents are very painful and hard to bear. This type of accident is called 'hold-on-type' accident, and is caused by currents in the range 20-100 mA. These currents may also cause physical injury due to powerful contraction of the skeletal muscles. However, the heart and respiratory function usually continue because of uniform spread of current through the trunk of the body.

If current contacts intact skin and passes through the trunk, at about 100 mA and above, there is a likelihood of pulling the heart into ventricular fibrillation. In this condition, the rhythmic action of the heart ceases, pumping action stops and the pulse disappears. Ventricular fibrillation is a serious cardiac emergency because once it starts, it practically never stops spontaneously. It proves *fatal* unless corrected within minutes, since the brain begins to die 2 to 4 minutes after it is robbed of its supply of oxygenated blood.

At very high currents of the order of 6 A and above, there is a danger of temporary respiratory paralysis and also of serious burns. However, if the shock duration is of only a very few seconds, there is a possibility of heart reverting to the normal rhythmic action.

The threshold of perception depends largely on the current density in the body tissues. It may vary widely depending upon the size of the current contact. At very small point contact, it is probable that even 0.3 mA current may be felt whereas a current in excess of perhaps 1 mA may not produce sensation if the contacts are somewhat larger. Similarly, depending on the size of contact, the threshold of pain may also be considerably above 1 mA, probably 10 mA if the contacts are large enough.

Besides the magnitude of current, the current duration and the relationship of current flow resistance are also important. Duration of less than 10 ms typically does not produce fibrillation whereas duration of 0.1 s or longer does. It has been found experimentally that the safe value of current in amperes (rms) which a human body can tolerate is given as

$$I = \frac{0.165}{\sqrt{t}} \quad \text{for } t < 3 \text{ s and } I = 9 \text{ mA for } t > 3 \text{ s.}$$

where t is the time duration in seconds of the flow of current.

The tolerable currents mentioned above are for power-frequency currents. It has been found that human body can tolerate about 5 times higher direct current. At high frequencies (3-10 kHz) still higher currents are tolerable.

18.4. SOIL RESISTIVITY

Before an attempt is made to discuss the design of the protective earthing, it is essential to understand the nature of the resistance of ground and the various factors which govern it. When an electrode is driven into the ground, the region around the electrode is known as *resistance area* or *potential gradient area* and the fault current tries to flow away from the electrode in all directions, as illustrated in Fig. 18.1. It follows that the rise of grounding potential or the flow of current in the ground depends upon the *resistivity of the soil* in which the earthing conductor is driven. The soil conductivity is purely electrolytic phenomenon and its resistivity is governed by nature and may vary from 1 to 10,000 Ω-m. Some typical values are given in Table 18.1.

Table 18.1

Type	Resistivity in Ω-m
Sea Water	2.5
Tap Water	20
Clay	50
Sand-clay Mixture	100
Sand	2,000
Wet Concrete	100
Dry Concrete	10,000
Rock	10,000
Chalk	250
Slates and Shales	300

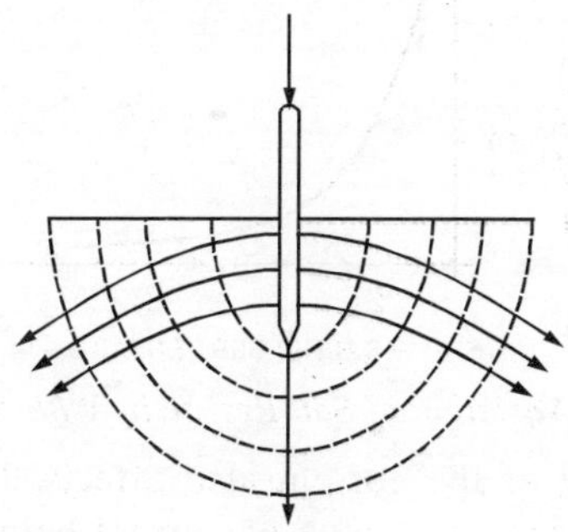

Fig. 18.1

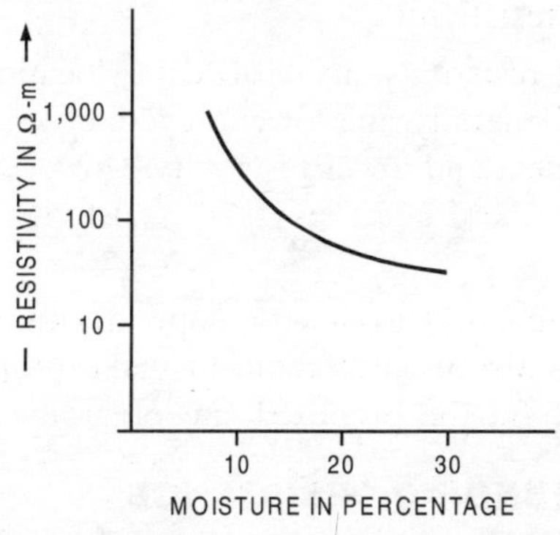

Fig. 18.2. ***Variation of Soil Resistivity With Moisture***

It may be a matter of interest to note that water has very poor conductivity. Hence it is useless to lay the grounding conductor in water, but its presence in the soil helps in dissolving the salts thereby improving soil conductivity. Soil conductivity increases with the increase in moisture content as depicted in Fig. 18.2. However, when moisture exceeds 20%, the change in resistivity is negligible.

The amount of water which a soil can absorb depends on the soil type, grain size and compactness.

The *temperature* of the soil also plays a vital role in varying its conductivity. The effect of the temperature on soil resistivity, as investigated by the US Bureau of Standards, is given in Table 18.2. When the temperature is more than 0°C, its effect on soil resistivity is negligible. At 0°C water in the soil starts freezing and resistivity increases.

Table 18.2

Temperature in °C	Resistivity in Ω-m
20	75
10	99
0 Water	138
0 Ice	300
–5	790
–15	3,300

The resistivity of the soil is also considerably affected by the composition and amount of soluble salts, as depicted in Fig. 18.3.

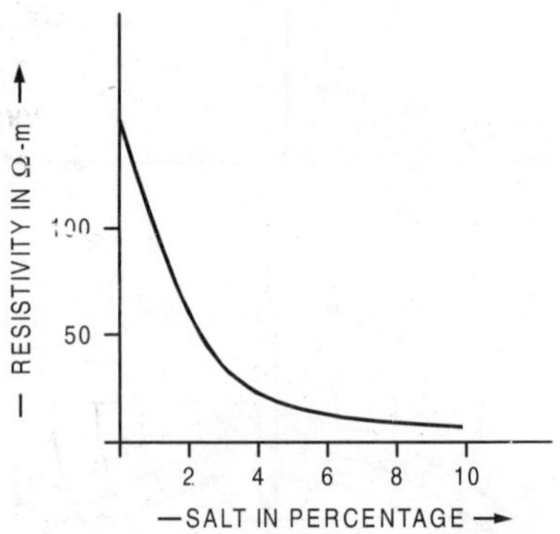

Fig. 18.3. ***Variation of Salt Resistivity With Salt Content***

Magnitude of the current also affects the resistivity of the soil as in case the value of current being dissipated by the soil is high, it may cause significant drying of soil and increase its resistivity.

The soil resistivity at a particular location also varies with depth. Generally, the lowers layers of soil have greater moisture content and lower resistivity. However, if the lower layer contains hard and rocky soil, resistivity may increase with depth.

Since the soil resistivity depends on a number of factors, it is always preferable to make proper resistivity measurements at the proposed site of grounding system.

18.5. GROUNDING RESISTANCE

18.5.1. Hemispherical Electrode

A hemisphere of radius b buried in ground is shown in Fig. 18.4. The total resistance is due to three constituents (*i*) resistance of conductor (*ii*) contact resistance between the surface of the electrode and main body of earth and (*iii*) resistance of body of earth surrounding the electrode. The first two constituents, *i.e.*, the resistance of the conductor and contact resistance are negligible. Thus the main component of the grounding resistance is that of body of earth surrounding the electrode.

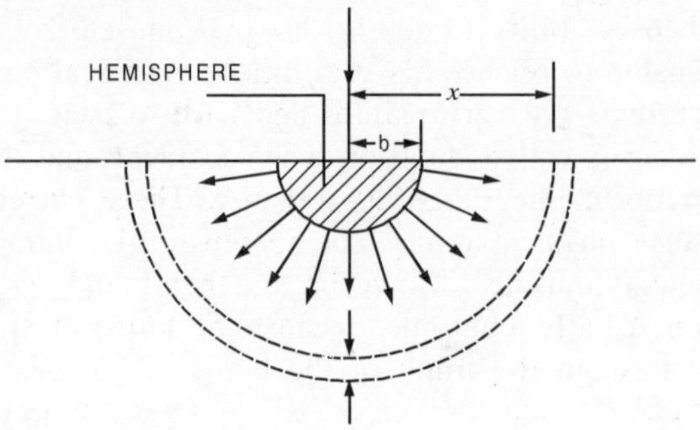

Fig. 18.4. ***Hemispherical Electrode***

Assume that the resistivity of the soil is uniform and currents are collected at an infinite distance. Therefore, when a current I flows to the sphere it distributes radially, as illustrated in Fig. 18.4 by arrows marked outwards

Current density at any radius

$$x = \frac{I}{2\pi x^2} \text{ as } 2\pi x^2 \text{ represents the area of the hemisphere}$$

$$\text{Voltage drop} = \frac{I}{2\pi x^2} \times \rho dx$$

where ρ is the resistivity of the soil

$$\text{Potential at } x = \int_x^{\infty} \frac{I\rho}{2\pi x^2} dx = \frac{I\rho}{2\pi x}$$

Potential at the surface of the electrode is obtained by substituting $x = b$ in above equation so the voltage between the hemispherical electrode and a point at infinity

$$E = \frac{I\rho}{2\pi b}$$

From the expression $\frac{I\rho}{2\pi x}$ it is obvious that the potential at any point is inversely proportional to the distance x.

Grounding resistance of the hemispherical electrode,

$$R_g = \frac{E}{I} = \frac{I\rho}{2\pi b I} = \frac{\rho}{2\pi b} \qquad ...(18.1)$$

18.5.2. General Equation

Let us consider a two electrode system and let V_1 and V_2 be the potentials of these two electrodes and V be the potential at any point in the soil of resistivity ρ.

The current flow normal to the surface at any point of the electrode is $\frac{1}{\rho}\frac{\delta V}{\delta n}$.

Total flow outwards from the electrode is

$$\frac{-1}{\rho}\iint \frac{\partial V}{\partial n} ds = -\frac{1}{\rho}\iint \frac{\partial \psi}{\partial n} ds$$

where ψ is the electrostatic potential in the analogous problem in electrostatics with air as medium, and ds is an element of the electrode surface.

If Q is the charge on this electrode in the analogous electrostatic case, then according to Gauss's theorem

$$-\iint \frac{\partial \psi}{\partial n} ds = 4\pi Q$$

So total flow of current, $I = \dfrac{4\pi Q}{\rho}$ ...(18.2)

If capacity between the electrodes in air in the analogous electrostatic case is C, then

$$\psi_1 - \psi_2 = V_1 - V_2 = \frac{Q}{C} \qquad ...(18.3)$$

The resistance between electrodes is given as

$$R = \frac{V_1 - V_2}{I} = \frac{Q}{C} \times \frac{\rho}{4\pi Q} = \frac{\rho}{4\pi C} \qquad ...(18.4)$$

If C is the capacitance of single electrode, the return electrode being at infinity, then R is the grounding resistance of single electrode.

Spherical Electrode. Consider a spherical electrode of radius b (Fig. 18.5). The capacity of the sphere, $C = b$, the radius of sphere, so ground resistance of a spherical electrode is given as

$$R_g = \frac{\rho}{4\pi b} \qquad ...(18.5)$$

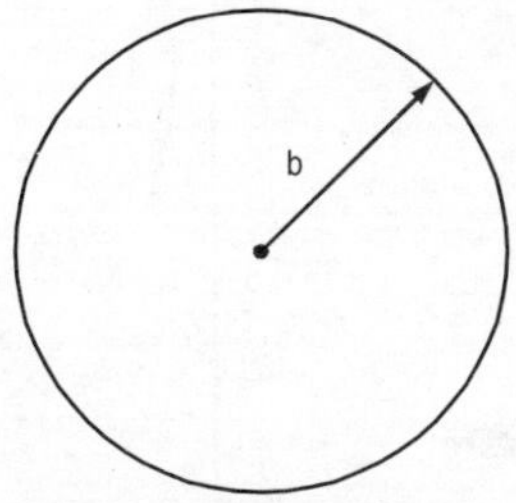

Fig. 18.5. ***Spherical Electrode***

Hemispherical Electrode. In case the electrode is a hemisphere buried with lower half in ground, the ground resistance would be doubled, *i.e.,* ground resistance of a hemispherical electrode is given as

$$R_g = \frac{\rho}{2\pi b} \qquad ...(18.6)$$

Thus general equation for any electrode can be written as

$$R_g = \frac{\rho}{2\pi C} \qquad ...(18.7)$$

where ρ is the resistivity of soil and C is the electrostatic capacity of the electrode along with its image above the ground surface, the combined electrode being considered as in air.

18.5.3. Rod Electrode

The resistance of a rod to ground is best determined by the method of images. Let l be the length of the rod and d be its diameter, as shown in Fig. 18.6. The dotted portion above the ground surface represents its image.

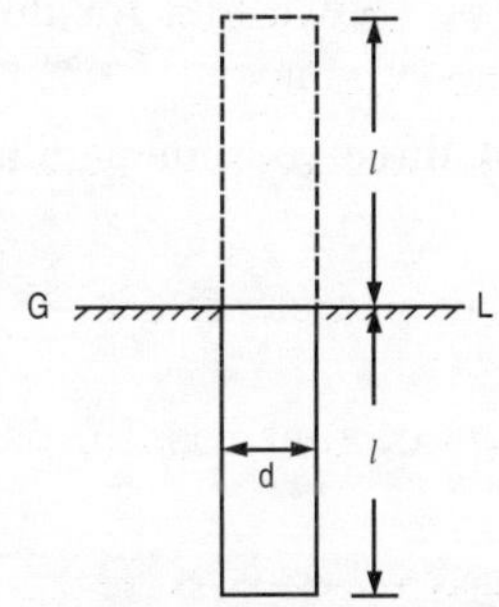

Fig. 18.6. ***Driven Rod***

The electrode can be considered equivalent to an ellipsoid of revolution in which major axis is very large in comparison with minor axis. On the basis of this equivalence, the capacity of rod and its image is given as

$$C = \frac{2l}{2\log_e \frac{4l}{d}} = \frac{l}{\log_e \frac{4l}{d}} \qquad ...(18.8)$$

Substituting the value of C from Eq. (18.8) in Eq. (18.7) we have ground resistance as

$$R_g = \frac{\rho}{2\pi l}\log_e \frac{4l}{d} \qquad ...(18.9)$$

18.5.4. Multiple Rod Electrodes

Let us consider two rods connected in parallel at a distance *S* from each other, as shown in Fig. 18.7. Each rod may be replaced by its equivalent hemisphere of radius b carrying a charge Q. The potential of either rod is given as

$$V = \frac{Q}{b} + \frac{Q}{S} = \frac{Q}{b}\left(1 + \frac{b}{S}\right) \qquad ...(18.10)$$

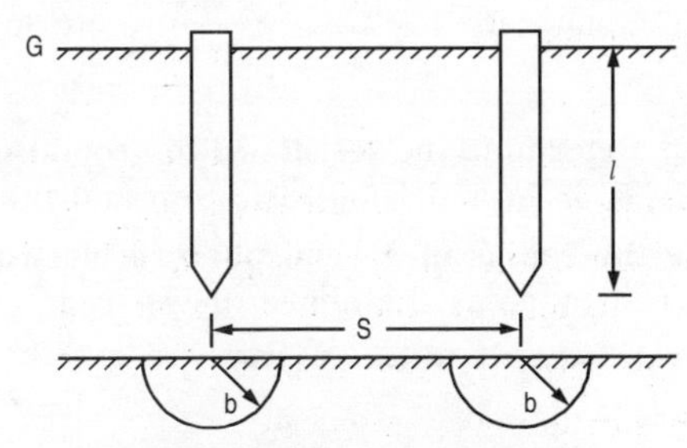

Fig. 18.7. ***Two Driven Rods***

Total charge is 2Q, so the capacity,

$$C = \frac{2Q}{V} = \frac{2b}{1 + \frac{b}{S}} \qquad ...(18.11)$$

Grounding resistance of two parallel rods is, therefore,

$$R_g = \frac{\rho}{2\pi C} = \frac{\rho}{4\pi b}\left(1 + \frac{b}{S}\right) \qquad ...(18.12)$$

Since the ground resistance of one rod is $\dfrac{\rho}{2\pi b}$, so we have

Ground resistance of two rods in parallel

$$\text{Ground resistance of one rod} \times \frac{\left(1 + \frac{b}{S}\right)}{2} \qquad ...(18.13)$$

Similarly it can be shown that for three rods, and four rods in a straight line, we have

Ground resistance of three rods in parallel

$$= \text{Ground resistance of one rod} \times \frac{2+\alpha-4\alpha^2}{6-7\alpha} \quad ...(18.14)$$

and Ground resistance of four rods in parallel

$$= \text{Ground resistance of one rod} \times \frac{12+6\alpha-21\alpha^2}{48-40\alpha} \quad ...(18.15)$$

$$\text{where } \alpha = \frac{b}{S}$$

Ground resistance for three rods placed at corners of an equilateral triangle

$$= \text{Ground resistance of one rod} \times \frac{\left(1+\frac{2b}{S}\right)}{3} \quad ...(18.16)$$

Ground resistance for four rods placed at the corners of a square

$$= \text{Ground resistance of one rod} \times \frac{1+\frac{2.707b}{S}}{4} \quad ...(18.17)$$

Example 18.1. Determine the resistance to ground for a hemispherical electrode on the surface of the ground. The diameter of the hemispherical electrode is 3 metres and the soil resistivity is 60 Ω-m.

Solution : Resistivity of soil, $\rho = 60$ Ω-m

Radius of hemispherical electrode, $b = \frac{3}{2} = 1.5$ m

So resistance-to-ground, $R_g = \frac{\rho}{2\pi b} = \frac{60}{2\pi \times 1.5} = 6.366\ \Omega$ **Ans.**

Example 18.2. Determine the resistance to ground of the rod whose diameter is 4 cm and length is 4.8 m and ρ = 80 Ω-m.

Determine the radius of a hemispherical electrode which has the same resistance as the above driven rod.

Solution : Soil resisvitiy, $\rho = 80$ Ω-m

Length of the rod, $l = 4.8$ m

Diameter of rod, $d = 4$ cm $= 0.04$ m

Ground resistance of rod,

$$R_g = \frac{\rho}{2\pi l}\log_e\frac{4l}{d} = \frac{80}{2\pi\times 4.8}\log_e\frac{4\times 4.8}{0.04} = 16.4\ \Omega \textbf{ Ans.}$$

Let the radius of a hemispherical electrode be b whose ground resistance is 16.4 Ω. Then

$$R_g = \frac{\rho}{2\pi b} = 16.4$$

$$\text{or } b = \frac{\rho}{2\pi\times 16.4} = \frac{80}{2\pi\times 16.4} = 0.8 \text{ m } \textbf{Ans.}$$

Example 18.3. Determine the ground resistance of the two driven rods each of length 4.8 m and diameter 4 cm. The spacing between rods is of 4 m and soil resistivity is 80 Ω-m.

Solution : Ground resistance of each rod = 16.4 Ω,

as determined in example 18.2

Ground resistance of two rods in parallel

$$= \text{Ground resistance of one rod} \times \frac{1+\frac{b}{S}}{2} = 16.4 \times \frac{1+\frac{0.8}{4}}{2}$$

$$= 9.84\ \Omega \textbf{ Ans.}$$

18.6. POTENTIAL GRADIENT

The resistance of an earthing electrode is not concentrated at one point but is distributed over the soil around the electrode. When fault current flows to ground, it results in a potential gradient around the electrode. This may be explained analytically as below.

Assuming that the base of structure through which fault current is flowing to ground is a hemisphere of radius 'b' as shown in Fig. 18.8. Then from Eq. (18.1)

$$e_x = \frac{\rho I_f}{2\pi \times b^2} \quad ...(18.18)$$

where e_x = Electric field strength at a distance x in V/m

ρ = Resistivity in Ω-m

I_f = Fault current in amperes

x = Distance from the surface of the hemisphere in metres.

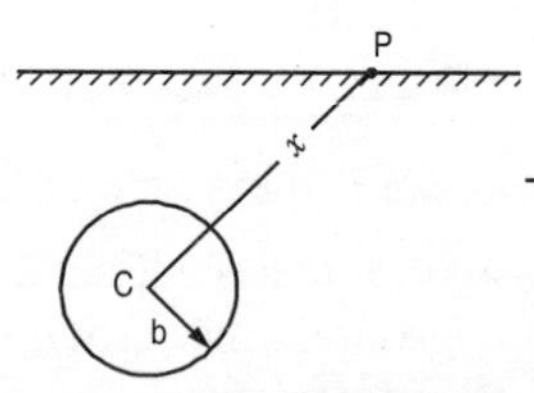

Fig. 18.8

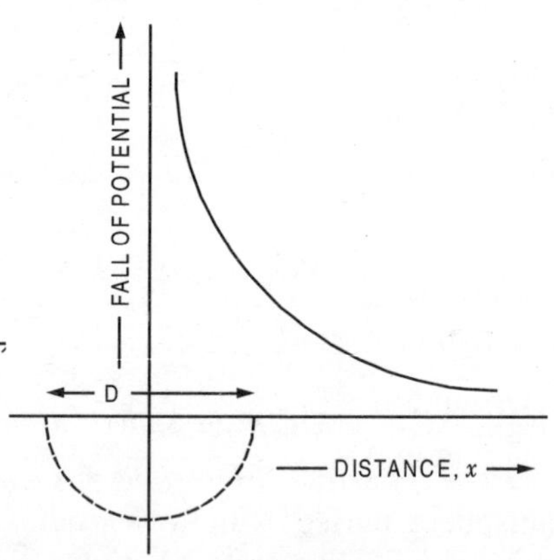

Fig. 18.9

If a curve between the fall of potential with distance from the base of structure is plotted, it is observed that the potential difference is quite pronounced near the electrode and the fault. Its magnitude which depends upon the resistivity and fault current may be of such order that it may prove dangerous. The fall of potential is shown in Fig. 18.9.

As such an earthing system should be provided such that, at the time the fault current flows to the ground, the potential difference thus produced may not prove to be dangerous to the person approaching that electrode or while touching the structure.

18.7. EARTHING MAT

A number of rods when joined together through copper conductors constitute an *earthing mat* (or earthing grid). It reduces overall grounding resistance. Such a mesh of conductors laid near the ground surface, incidentally helps, in limiting the potential gradient. In hard soils or in soils having low conductivity, driving of rods quite deep may prove expensive. If the rods are removed from the mesh the performance of the earthing system, *i.e.*, conductors laid in the ground in closed meshes as a ground mat, it has been observed, has not been affected by even 5 per cent. It can safely be summarized that in systems where large faults currents are to be experienced, it will be preferable to take care of the potential gradients by laying the earthing mat.

While designing a ground mat, the factors which have to be taken into account are :

1. Voltage, that may appear on the ground between the ground surface and the conductor buried in the ground in case of a fault;
2. This voltage should not be dangerous to a person who may at the time of occurrence of a fault happen to touch non-current carrying conducting surface of the plant;
3. The sustained fault current that may flow into the ground mat should be large enough to operate the protective relays, *i.e.*, grounding resistance should be low enough to permit the flow of fault current. The resistance of the mat should not be of such a magnitude as to permit the flow of fatal current in the live body;
4. The design of the grounding mat should be such that step voltage should be less than the permissible value which will depend upon the resistivity of the soil and the time required for isolating the faulty plant from live system.

Rational results, to fulfill the prerequisites of a protective earthing can be achieved by using the formula given below :

1. Grounding Current:

   Grounding mat resistance in ohms,

$$R_g = \frac{\rho_1}{4r} + \frac{\rho_1}{L} \quad ...(18.19)$$

2. Permissible step voltage,

$$E_{sp} = \frac{165 + \rho_2}{\sqrt{t}} \quad ...(18.20)$$

3. Permissible touch voltage,

$$E_{tp} = \frac{165 + 0.25\,\rho_2}{\sqrt{t}} \quad ...(18.21)$$

4. Mesh voltage,

$$E_m = \frac{K_m K_i^{\sqrt{t}} \rho_1 I_f}{L} \quad ...(18.22)$$

5. Step voltage,

$$E_s = \frac{K_s K_i \rho_1 I_f}{L} \quad ...(18.23)$$

6. Sustained fault current,

$$I_{sf} > (1{,}000 + 1.5\,\rho_2)\ \frac{9}{1{,}000} \times \frac{L}{K_m K_i \rho_1} \quad ...(18.24)$$

7. Mesh factor

$$K_m = \frac{1}{2\pi}\left[\log_e \frac{D^2}{16hd} + i \log\left(\frac{3}{4} \cdot \frac{5}{6} \cdot \frac{7}{8}\right)\right] \quad ...(18.25)$$

8. Step-factor,

$$K_s = \frac{1}{\pi}\left(\frac{1}{2h} + \frac{1}{D+h} + \frac{1}{2D} + \frac{1}{3D} + ...\right) \quad ...(18.26)$$

where L is the total length of buried conductors in metres, r is the radius of the circular plate in metres having an area equivalent to that covered by the mat, D is the grounding conductors spacing in metres, d is the grounding conductors spacing-dia, h is the depth under which conductors are buried, t is the total time taken by the circuit breaker and backup protection to isolate the section, K_i is the irregularity factor, I_f is fault current in amperes, ρ_1 is soil resistivity in Ω-m and ρ_2 is soil resistivity under the step in ohm-metres.

It will be observed from Eq. (18.19) that the low impedance of a protective grounding can be kept within limits, by covering sufficient area and by using adequate length of the conductors. The resistivity of the soil in view of its non-uniform texture varies throughout its mass, and to avoid complex mathematics and for a rational design, we presume the highest figure of resistivity which may come across during the investigation of the soil. From the formulae given above, it can be summarised that the potential gradient can be kept within limits by a judicious choice of the conductor length, size of the conductor, the depth at which it is to be buried, the size of the mesh and the use of high resistivity soil at the surface of the ground which will be trodden by the operating personnel.

In some cases, limitation of space does not permit extension of the grounding area, which may be required to limit the grounding resistance within permissible limits. In order to overcome such problem, grounding rods are also driven around the periphery of the *grounding grid*.

It is suggested, that few rods be driven near surge arresters and corners of the grid, to give a safety factor to the designing of the grounding mesh.

18.8. TOUCH AND STEP POTENTIALS

When a fault occurs, the flow of current to ground results in voltage gradient on the ground surface in the vicinity of grounding system. This voltage gradient may affect a person in two ways, viz., step or foot-to-foot contact and hand to both feet or touch contact.

Resistance of the human body is highly variable and the current that may pass through the body depends upon the resistance of the body and the position in which he is bridging the ground and the grounded structure. For calculation purposes, the resistance of the body may be taken as 1,000 Ω. The resistance of footwear, which is highly variable is negligible. It will be observed that the resistance of the ground under the foot affects the magnitude of the current which may flow through the body. Assuming the ground under one foot as a circular plate electrode with a diameter of 8 cm, the resistance of such an electrode can be determined as under :

$$R_f = \frac{\rho_2}{8b} \qquad ...(18.27)$$

where ρ_2 is the resistivity of the ground under the foot in Ω-m

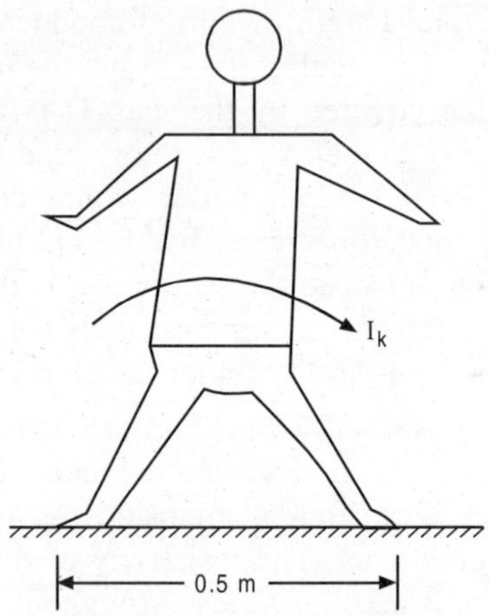

Fig. 18.10. ***Step Contact***

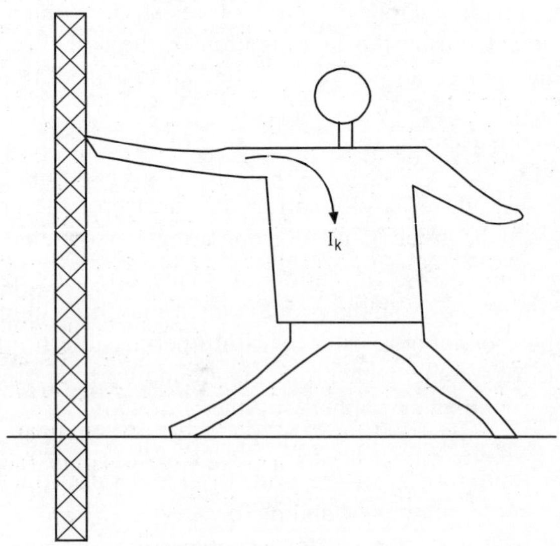

Fig. 18.11. ***Touch Contact***

$$\text{So } R_f = \frac{\rho_2}{8 \times 0.04}$$

$$= 3.125\ \rho_2 = 3\ \rho_2 \text{ (approximately)} \qquad ...(18.28)$$

Now referring to Fig. 18.12.

$$E_{TOUCH} = (R_k + R_f/2)I_k$$

where I_k is the portion of fault current flowing through the body and is given as

$$I_k = \frac{0.165}{\sqrt{t}}.$$

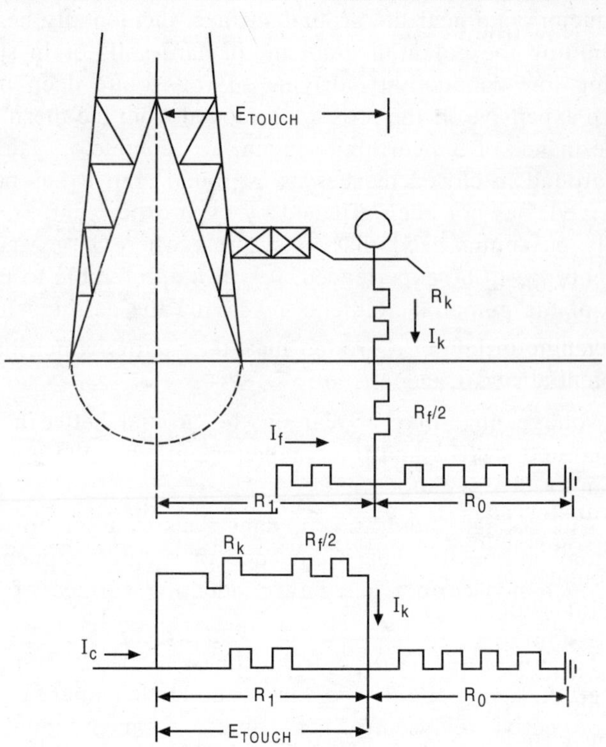

Fig. 18.12

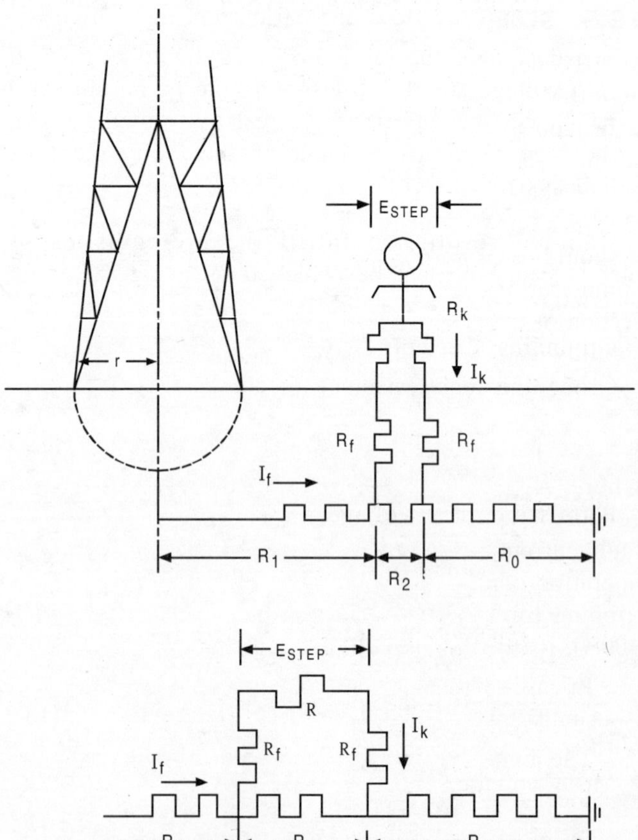

Fig. 18.13

$$\text{So } E_{TOUCH} = \frac{(1{,}000 + 1.5\rho_2) \times 0.165}{\sqrt{t}} \text{ volts}$$

$$= \frac{165 + 0.25\rho_2}{\sqrt{t}} \text{ volts} \qquad ...(18.29)$$

Now from Fig. 18.13,

$$E_{STEP} = (R_k + 2R_f)\, I_k$$

$$= \frac{(1{,}000 + 6\rho_2) \times 0.165}{\sqrt{t}}$$

$$= \frac{165 + \rho_2}{\sqrt{t}} \quad ...(18.30)$$

While calculating the length of the conductor, touch potential from a grounded structure to the centre of a rectangle of the grid mesh is used. It is known as the mesh potential and is expressed as under :

$$E_{MESH} = \frac{K_m K_i \rho_1 I_f}{L}$$

Transferred Potential. A special situation arise when a person standing within the station area touches a conductor grounded at a remote point or a person standing at a remote point touches a conductor connected to the station grounding grid. This special case of touch potential is known as *transferred potential* and can have a very large value. It is not possible to design an earthing grid for safe transferred potential. The only remedy is to isolate such conductors and to label them as *"live conductors."*

18.9. SIZE OF THE EARTHING CONDUCTOR

In EHV networks, the magnitude of earth fault current is quite large. The earthing conductor used should be able to withstand the heat which will be generated by the flow of earth fault current till the protective relays operate and isolate the faulty portion of the system. While temperature rise in case of steel conductor has been limited to 400°C, it should not exceed 300°C in case of copper. For aluminium conductor it may be limited to 200°C. Minimum cross section of the earthing conductor which should have thermal stability can be calculated from the following relation

$$A_{min} = \frac{I_F \sqrt{t}}{c}\ \text{mm}^2 \quad ...(18.31)$$

where A_{min} is the minimum required section of the conductor in mm^2 according to thermal stability, I_F is the fault current in amperes; t is the time in seconds for the operation of the protective relays including circuit breaker tripping time; and c is a constant; for steel conductors at 400°C, it is 70 and for copper conductors at 300°C is 166.

In case of aluminium conductor minimum cross section is usually taken as 1.54 times that of copper section.

The minimum size for mechanical strength is:

Steel—61 mm^2; Copper—107.2 mm^2 and Al—195 mm^2.

In most cases practical minimum size of the two conductors will be set by mechanical ruggedness when copper or aluminium is used by the magnitude of fault currents and their duration, *i.e.*, thermal stability consideration.

Size of steel grounding conductors used should be checked for corrosion. In moderately corrosive and severely corrosive soil steel strip should have minimum thickness of 6 mm and minimum cross section area of circular section steel conductor should be 200 mm^2.

18.10. MEASUREMENT OF EARTH RESISTANCE

The determination of resistance between the earthing plate and the surrounding ground in distribution systems is of utmost importance. This measurement is made by the potential fall method.

The resistance area of an earth electrode is the area of soil around the electrode within which a voltage gradient measurable with commercial instruments exists. In Fig. 18.14, E is the earth electrode under test, and A is an auxiliary earth electrode positioned so that two resistance areas do not overlap. B is a second auxiliary electrode placed half way between E and A.

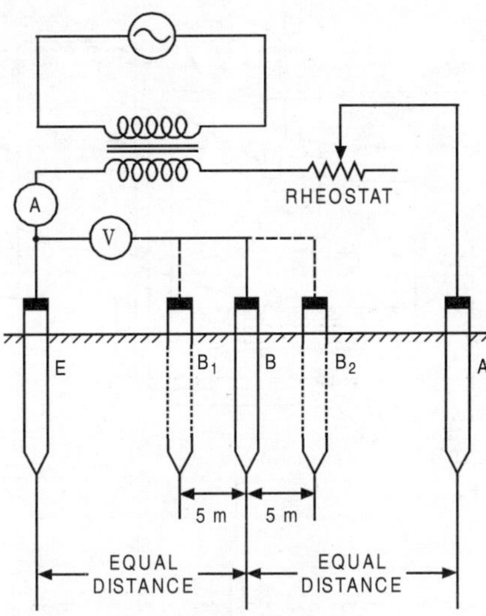

Fig. 18.14

An alternating current of steady value is passed through the earth path from E to A and the voltage drop between E and B is measured.

Then earth resistance,

$$R_e = \frac{\text{Voltage drop between E and B}}{\text{Current through earth path}} = \frac{V}{I} \quad ...(18.32)$$

To ensure that the resistance areas do not overlap, the auxiliary electrode B is moved to positions B_1 and B_2 respectively. If the resistance values determined are of approximately the same value in all three cases, the mean of the three readings can be taken as the earth resistance of earth electrode. Otherwise, the auxiliary earth electrode A must be driven in at a point further away from E and the above test repeated until a group of three readings obtained are in good agreement.

The use of alternating current source is necessary to eliminate electrolytic effect.

The test can be performed, with current at power frequency from a double wound transformer, by means of a voltmeter and an ammeter, as shown in Fig. 18.14 or by means of an *earth tester*.

The *earth tester* is a special type of megger which sends ac through earth and dc through the measuring instrument (Fig. 18.15). It has got four terminals, P_1, C_1, P_2 and C_2 outside. Two terminals P_1 and C_1 are shorted to form a common point which is connected to the earth electrode under test. The other two terminals C_2 and P_2 are connected to the auxiliary electrodes A and B respectively. The value of the earth resistance is indicated by the instrument directly when its handle is turned at uniform speed.

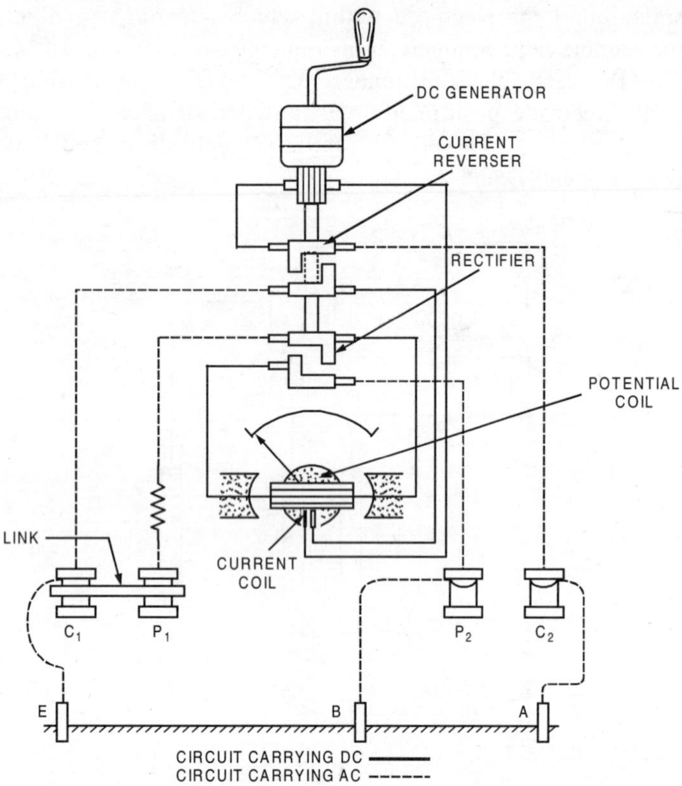

Fig. 18.15. ***Earth Resistance Measurement By Earth Tester***

18.11. MEASUREMENT OF SOIL RESISTIVITY

Detailed investigation of the soil resistivity is essential for the design of grounding system. Resistivity of the soil is usually measured by the four-spike method, in which four spikes arranged in a straight line are driven into the soil at equal distance, as illustrated in Fig. 18.16. A known current is passed between electrodes C_1 and C_2 and potential drop V is measured across electrodes P_1 and P_2. The current I develops an electrical field which is proportional to current density and soil resistivity. The voltage V is proportional to this field. Thus soil resistivity is proportional to the ratio of voltage V and current I (*i.e.*, V/I) and is given as

$$\rho = \frac{\dfrac{4\pi SV}{I}}{1+\dfrac{2S}{\sqrt{S^2+4b^2}}-\dfrac{2S}{\sqrt{4S^2+4b^2}}} \quad ...(18.33)$$

where ρ is soil resistivity in Ω-m; S is horizontal distance between two successive spikes (electrodes) in m and b is depth of burial in metre.

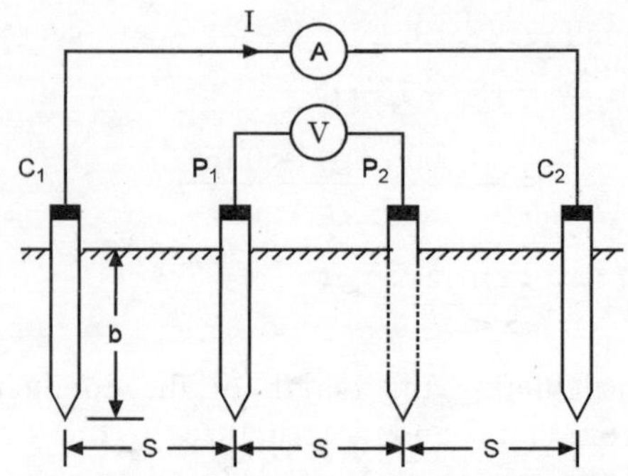

Fig. 18.16. ***Measurement of Soil Resistivity***

For $b << S$ ($b \simeq \frac{S}{20}$) the Eq. (18.33) becomes

$$\rho = 2\pi S\frac{V}{I} = 2\pi SR_g \quad ...(18.34)$$

For measurements in the field, earth testers based on above principle are available. They are calibrated to indicate R_g, *i.e.*, V/I directly. A number of readings with different values of spacing and the test set-up in different directions are taken. The average value of soil resistivity ρ so obtained should be used in the design.

If measurements are to be carried out using the mains supply, an isolating transformer should be connected between the mains supply and test set-up so that residual ground currents do not affect the results.

18.12. IMPULSE BEHAVIOUR OF GROUNDING SYSTEMS

The degree of protection provided by a grounding system under lightning discharge conditions depends upon its impulse impedance rather than on power frequency impedance. The impedance under impulse conditions, called the *impulse impedance*, is quite different from the power frequency impedance. This is because of effects of electrode inductance and soil ionization.

In case the electrode is of excessive length, its inductance may cause only a part of the grounding system to be effective in dissipating current and, therefore, increases impedance. With excessive impulse current, the soil in the vicinity of electrode may breakdown resulting in decrease in impedance.

The effects of these two factors depend on the type of grounding system. The effect of soil ionization is very predominant in driven rods and, therefore, its impulse impedance is always less than its power frequency impedance. The effect of electrode inductance is present in counterpoise and grids and, therefore, their impulse impedance is always larger than its power frequency impedance. With high impulse impedance, the protection against lightning strokes may not be satisfactory.

18.13. INTRODUCTION TO NEUTRAL EARTHING

There are two ways in which three-phase system can be operated and they are :

(*i*) With isolated or free or ungrounded neutral and (*ii*) With grounded neutral

The 3-phase 50 Hz ac power systems with *neutral earthed (or grounded) at every voltage level* are used for generation, transmission, distribution and utilization of electrical power. The neutral or star points of star-connected 3-phase windings of power transformers, generators, motors, grounding transformers are connected to low resistance ground (earth electrode/earth mat). Such a connection is called *neutral earthing* or *neutral grounding.*

Before 1950s the power systems were often without neutral grounding. Such systems are called *ungrounded systems* or *insulated systems*. The chief argument in favour of such a system is the possibility of maintaining continuity of supply in case one phase gets grounded until it is convenient to disconnect and repair the faulty line. This argument is particularly true for an overhead transmission system when the ground fault on one line is not likely to develop into a fault between two or more lines. But such systems experience repeated arcing grounds. The earth fault protection of ungrounded systems is difficult. Insulation failure may occur in several equipment and machines over entire voltage level during a single ground fault at remote location. The ungrounded systems must necessarily have equipment insulation withstand level corresponding to next higher system voltage so that cascade insulation failures can be avoided. Such a system needs a costlier insulation system of next higher voltage level (*e.g.*, 11 kV insulation for 6.6 kV bus-bars and motors, transformers, CTs, PTs etc.).

Ungrounded systems have advantage of negligible ground fault current but drawback of arcing grounds. Modern power systems are with grounded neutrals except some continuous process systems and essential protection/auxiliary supply systems where single phase to ground faults should not trip entire bus supply.

18.14. ISOLATED NEUTRAL OR UNGROUNDED SYSTEMS

The main feature of an ungrounded system is its ability, in some cases, to clear earth faults without interruption. The self-clearing feature disappears when the length of the line becomes excessive.

A simple three-phase system with isolated neutral is shown in Fig. 18.17 (*a*). The line conductors have capacitances between one another and to earth, the former being delta-connected, while the later are star-connected, as shown in the figure. The effect of line capacitances on the grounding characteristic of the system is little and, therefore, can be neglected.

First of all consider a 3-phase line (perfectly transposed) having same capacitances to ground. In such a line the charging currents for each line to earth capacitor lead the phase voltage by 90° and are equal. The magnitude of these currents is given by $\frac{V_P}{X_C}$ where V_P is the phase voltage and X_C is the reactance due to the capacitance of the line to ground.

The charging currents I_{CR}, I_{CY} and I_{CB} are balanced and their resultant is zero and no current flows to the earth.

Under balanced conditions, the potential of neutral is held at ground due to the presence of the shunt capacitances of the system, shown in Fig. 18.17 (*a*). Under balanced condition the phasor diagram is shown in Fig. 18.17 (*b*).

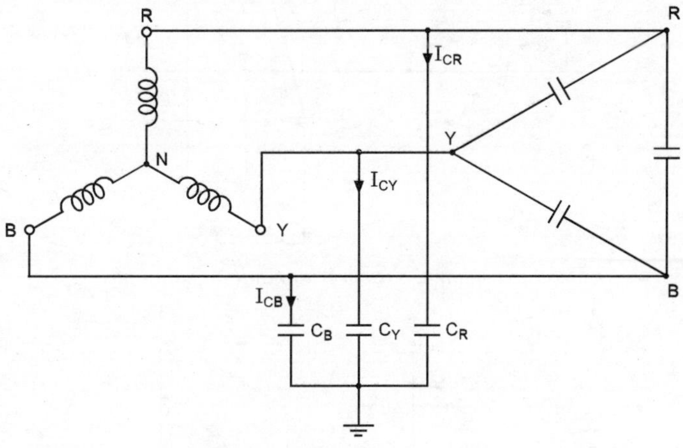

(a) Isolated Neutral System

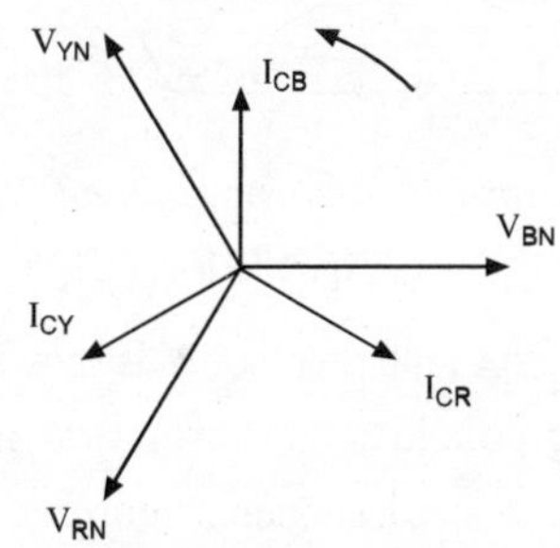

(b) Phasor Diagram For Isolated Neutral System

Fig. 18.17

Now consider a phase to earth fault in line Y say at point F. Under these circumstances, the faulty line takes up the earth potential while the potentials of the remaining two healthy lines R and B rise from phase values to line value. The capacitance currents become unbalanced and fault current I_F flows through the faulty line, into the fault and returns to the system via earth and through the earth capacitances C_R and C_B. Thus fault current I_F has two components I_{CR} and I_{CB} which flow through capacitances C_R and C_B respectively under the potential differences of V_{RY} and V_{BY} respectively. These currents lead their respective voltages by 90° and their phasor sum is equal to fault current I_F.

$$I_{CR} = \frac{V_{RY}}{X_{CR}} = \frac{\sqrt{3}V_P}{X_C} \qquad \text{...(18.35)}$$

$$\text{Similarly } I_{CB} = \frac{V_{BY}}{X_{CB}} = \frac{\sqrt{3}V_P}{X_C} \qquad \text{...(18.36)}$$

$\because$ phase voltages are equal, say, equal to V_P and capacitances from line to ground are also equal, say, equal to X_C.

Since the fault current, I_F is equal to phasor sum of I_{CR} and I_{CB}, therefore, fault current,

$$I_F = \sqrt{3}I_{CR} = \sqrt{3} \times \frac{\sqrt{3}V_P}{X_C} = \frac{3V_P}{X_C} \qquad \text{...(18.37)}$$

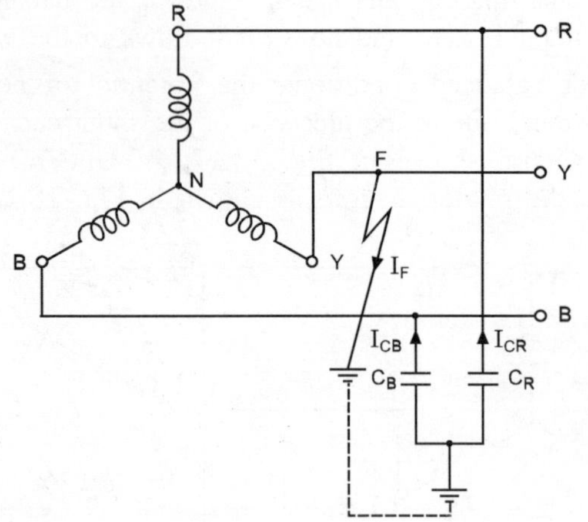

(a) Isolated Neutral System With Fault on One Phase

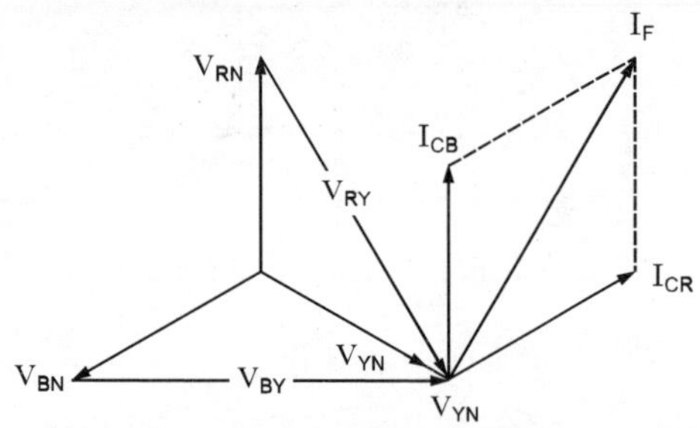

(b) Phasor Diagram For Fault on Phase Y

Fig. 18.18

Phasor diagram for fault on phase Y is shown in Fig. 18.18 (*b*).

The following conclusions can be drawn from the above.

(*i*) There is no zero sequence current.

(*ii*) There will be little interference with communication lines because of the absence of zero sequence currents.

(*iii*) In case of one phase becoming earthed the voltages of the remaining two healthy phases to earth rise from their normal phase-to-neural voltage to full line value ($\sqrt{3}$ times their normal value). This causes stress on the insulation of all the machines and equipment connected to the system. The voltage rise of phase above earth is sustained and thereby insulation failure is likely to occur on connected machines, though fault current in arcing ground may be negligible.

(*iv*) The capacitance currents in the two healthy phases increase to $\sqrt{3}$ times their normal values.

(*v*) The capacitance current in the faulty phase becomes 3 times the normal value.

(*vi*) For operating the protective devices, it is essential that the magnitude of the current supplied should be adequate to operate them. But in case of an earth fault on an isolated neutral system, the fault current may be too small to actuate the protective devices. Thus in ungrounded systems the earth fault relaying becomes complicated.

(*vii*) The overvoltages due to induced static charges are not discharged to ground in isolated neutral systems. The voltage due to lightning surges do not find path to earth.

(*viii*) The danger to equipment on the occurrence of line-to-line ground fault is appreciable and danger to life in the proximity of fault is often prolonged.

(*ix*) A capacitive fault current I_F flows into the earth. Such a current if exceeds 4–5 amperes, is sufficient to maintain an arc in the ionized path of the fault, even though the medium causing the fault has cleared itself. The persistency of the arc due to the flow of capacitance current gives rise to a condition known as "*arcing ground*" in which cyclic charging and discharging of the system capacity through the fault results in high frequency oscillations being superimposed on the whole system and a build-up of very high voltages can occur. This results in phase voltage to rise to 5 to 6 times of normal voltage. The build-up of high voltages may result in insulation breakdown.

(*x*) At the time of an earth fault in an isolated neutral system, the capacitive currents in each phase will become unbalanced as they are contributing towards the fault current. Due to unbalanced capacitive currents it is not possible to install discriminative type fault indicator. However, neutral displacement indicator can be connected across such system, but this will give only warning that there is a ground fault in one phase without indicating the actual location.

For the reasons mentioned above it is not desirable to operate a high voltage 3-phase system of considerable capacitance with an isolated neutral. Hence the above discussion brings us to a conception of earthed neutrals, which has got the following advantages :

1. Persistent arcing grounds can be eliminated by employing suitable protective gear. The arcing ground current flowing through the neutral to ground connections is made almost equal and opposite to the capacitive currents from healthy lines to ground. Thus sum of current flowing through neutral to ground connections and capacitive currents is zero and arcing grounds are eliminated. The system is thus not subjected to overvoltage surge due to arcing grounds.
2. In this system the neutral point is not shifted (*i.e.*, stable neutral point).
3. The voltages of healthy phases with respect to ground remain at normal value. They do not increase to $\sqrt{3}$ time the normal value as in the case of isolated neutral system.
4. Earth faults can be utilized to operate protective relays to isolate the fault in case of grounded neutral system.
5. The induced static charges do not cause any disturbance as they are conducted to ground immediately.
6. There is a possibility of installing discriminative protective gear on such systems.

7. By employing resistance or reactance in ground-connections, the ground fault current can be controlled.
8. Improved service reliability due to limitation of arcing grounds and prevention of unnecessary tripping of circuit breakers.
9. Such system provides greater safety to personnel and equipment. This is because of operation of fuses or relays on ground fault and limitation of voltages.
10. The life of insulation is long as voltage surges caused by arcing grounds are eliminated. Thereby maintenance, repairs and breakdowns are reduced and continuity of supply is improved.
11. Life of equipments, machines, installation is improved due to limitation of voltage. Thus overall economy is achieved.

18.15. GROUNDED NEUTRAL SYSTEMS

In grounded neutral systems, the neutral point is grounded directly or through resistance, reactance etc., depending on particular requirement.

Neutral grounding can broadly be classified in two categories *viz.* effective grounding (or solid grounding) and non-effective grounding. For non-effective grounding of neutral any of the following four methods can be used.

1. Resistance grounding.
2. Reactance grounding.
3. Peterson coil or arc suppression coil grounding.
4. Voltage transformer grounding.

18.16. EFFECTIVELY GROUNDED SYSTEM

The term 'effectively grounded' is now used in place of old term 'solidly grounded' for the reason of definition.

According to IEEE definition, a system or a portion of a system can be said to be effectively grounded when for all points on the system or specified portion thereof, the ratio of zero sequence reactance to positive sequence reactance is not greater than three (*i.e.*, $\frac{X_0}{X_1} \not> 3$) and the ratio of zero sequence resistance to positive sequence resistance is not greater than one (*i.e.*, $\frac{R_0}{R_1} \not> 1$) for any condition of operation and for any amount of generator capacity.

The effective grounded systems are less expensive than any other type of grounding for any operating voltage because for such a system in the event of a single line-to-ground fault, the maximum voltage of healthy phase does not exceed 80% of line-to-line voltage while for all other grounded systems the voltage of healthy phases rises to about 100% line-to-line voltage.

18.17. SOLID OR EFFECTIVE GROUNDING

In solid grounding, also called the *effective grounding*, a direct metallic connection is made from the system neutral to one or more earth electrodes [rods, pipes or plates buried or driven into the ground, as shown in Fig. 18.19 (*a*)].

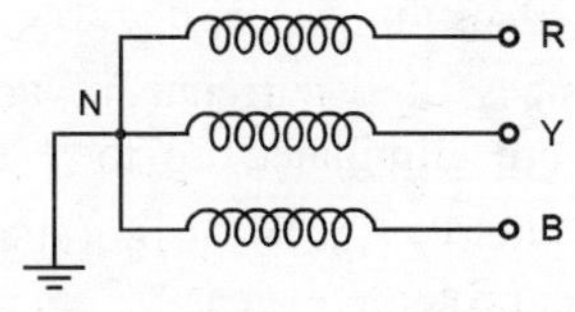

(a) Solid or Effective Grounding

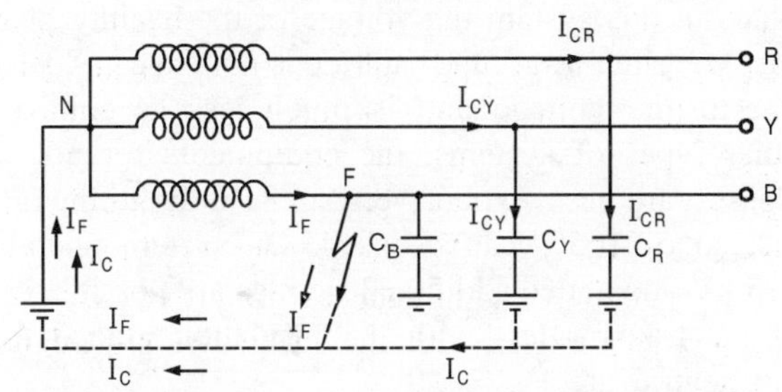

(b) Solidly Grounded System With Ground Fault on Line B

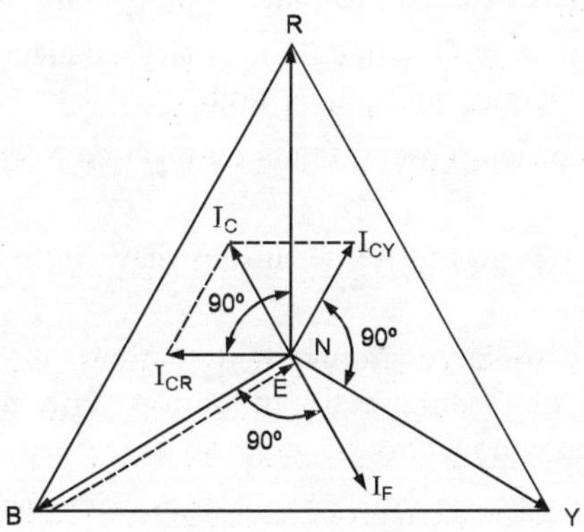

(c) Phasor Diagram For (b)

Fig. 18.19

When there is a ground fault over any phase, the phase to earth voltage of the grounded phase will become zero, but the voltage to earth of the remaining two healthy phases will be the normal phase voltages as in this case neutral point will not shift. Under an L-G fault on phase B, as shown in Fig. 18.19 (*b*), the neutral and terminal B are at earth potential. The phasor diagram for such a condition is shown in Fig. 18.19 (*c*). The reversed phasor is shown at V_B. Capacitive current I_{CR} leads V_{NR} by 90° and I_{CY} leads V_{NY} by 90°. The resultant capacitive current I_C will be phasor sum of I_{CR} and I_{CY}. It should be noted that in this system, in addition to capacitive currents the supply source also supplies the fault current I_F. This fault current will go to the fault point F through the faulty phase and then return back to supply source through the earth and neutral connection. The fault current I_F lags behind the faulty phase voltage by approximately 90° since the circuit is predominantly inductive (due to transformers, machines and line inductances). As illustrated in Fig. 18.19 (*c*), the fault current I_F will be in phase opposition to capacitive current I_C. Due to this effect, the capacitive current I_C will be fully neutralized by the large fault current.

The main features of effectively grounded neutral systems are given below :

1. Since fault current eliminates the effect of the capacitive currents, chances of occurrence of arcing grounds and overvoltages are eliminated up to a greater extent.
2. The flow of heavy fault current permits the use of discriminative protective gear.
3. Ground fault relaying is simple and satisfactory.
4. Since in this system the voltage of the healthy phases in case of a line-to-ground fault does not exceed 80% of the line-to-line voltage and is much less as compared to other types of systems, the equipments for all voltage classes are less expensive. An 84% lightning arrester instead of 105% can be used. On system operating at 115 kV and above additional savings are possible because of the transformers with the insulation graded towards the neutral are less costly.
5. The ground fault current is large. Its maximum values sometimes exceeds even the 3-phase short-circuit current.
6. Since the ground fault current is large, danger to personnel in the vicinity of fault is high.
7. Even transient ground faults may get converted into short circuit.
8. The heavy ground fault current may cause considerable damage.
9. Because of large ground fault current, the interference due to electromagnetic induction with neighbouring communication circuits may be high.
10. Because of large fault current, the current handling in the circuit breakers is a difficult proposition. This problem is, however, overcome by employing high rupturing and high speed circuit breakers and fast operating protective relays.

In order that the fault current remains within the limits, this system is used on the networks where normal impedance is quite large. Experience shows that combined impedance of the apparatus, circuit and ground return path in systems operating at voltages below 3.3 kV and those operating at voltages exceeding 33 kV is sufficiently large so as to limit the value of fault current to a safe value.

18.18. RESISTANCE GROUNDING

When it becomes necessary to limit the earth fault current, this type of grounding is used. In this method of neutral grounding a resistor is inserted between the neutral and earth to limit the fault current to a safe value, as illustrated in Fig. 18.20 (*a*). The resistor used for neutral grounding is usually a metallic resistance mounted on insulators in a metallic frame. For voltages below 6.6 kV liquid resistors are also used. Metallic resistors do not alter with time and need little or no maintenance. They are, however, slightly inductive and this is a drawback with overhead lines exposed to lightning, because travelling waves or impulses are subjected to positive reflection and this may cause undue stress to the insulation of the equipment and result in breakdown. Liquid resistors are free from this drawback and are normally of simple robust construction. They may be mounted outdoor. The value of resistance used for neutral grounding should be such that the ground fault current does not exceed 3-phase short-circuit current. This is necessary so that the power loss in the grounding resistance is not excessive. Neutral grounding resistors are normally designed to carry their rated current for a short period, usually 30 seconds.

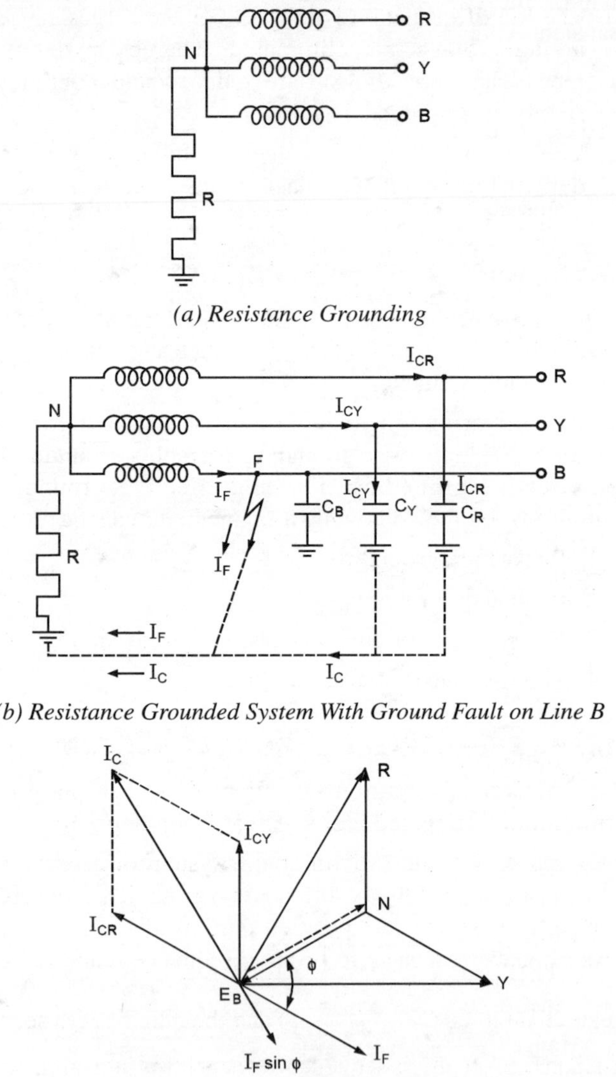

(a) Resistance Grounding

(b) Resistance Grounded System With Ground Fault on Line B

(c) Phasor Diagram For (b)

Fig. 18.20

When there is a ground fault over any phase, neutral is displaced and the maximum voltage across the healthy phases may become equal to line-to-line value. Figure 18.20 (*b*) shows a ground fault on phase B of a resistance grounded system. The phasor diagram for such a condition is illustrated in Fig. 18.20 (*c*). Capacitive currents I_{CR} and I_{CY} leads V_{BR} and V_{BY} respectively by 90°. Fault current I_F lags behind the phase voltage of the faulted phase by an angle ϕ, which depends upon the grounding resistance R and impedance of the system up to the fault point. The fault current I_F can be resolved into two components; one in phase with the faulty phase voltage and the other lagging the faulty phase voltage by 90° ($I_F \sin \phi$). The lagging component of fault current

($I_F \sin \phi$) is in phase opposition to the capacitive current I_C (phasor sum of I_{CR} and I_{CY}). By adjusting the value of grounding resistance R to a sufficiently low value it is possible to neutralize the effect of capacitive current I_C so as to avoid the occurrence of transient oscillations due to arcing ground. In case the grounding resistance R is made sufficiently large so that the lagging component of fault current becomes less than capacitive current I_C, then the system conditions approach that of the isolated neutral system with the risk of high transient voltage occurrence.

The main features of resistance grounded neutral system are given below :

1. The grounding of a system neutral through a low value resistance allows the use of discriminative protective gear and at the same time limits the ground fault current to a value that will cause no serious disturbance to the system and minimize the extent of damage at the fault location.
2. In general, a resistance grounded system will have ground fault current materially lower than that in effectively grounded system and therefore, the inductive interference to the neighbouring communication circuits is also lesser than that in effectively grounded system.
3. The power dissipation in the grounding resistance may improve system stability because the power dissipation reduces the accelerating power.
4. Transient ground faults are converted into controlled current faults.
5. There is an enormous energy loss in the neutral grounding resistance.
6. The system neutral will almost invariably be displaced in the case of an earth fault, thereby requiring the use of lightning arresters at an increased cost and sacrifice in protective performance.
7. The cost of equipment, particularly of transformers, increases as graded insulation towards neutral cannot be used.
8. The provision of ground resistor involves extra investment in the system.

Hence resistance grounding is usually employed for the systems operating on voltage exceeding 3.3 kV but not exceeding 33 kV. For circuits below 3.3 kV (*i.e.*, say 400 V distribution networks), the external resistance in the neutral circuit is unnecessary because the voltage available between phase and ground is only 230 V. The earth resistance of earthing electrode, earth connections etc. is of the order of 1.5 Ω. The earth current is limited to $\frac{230}{1.5}$, *i.e.*, 153 A even if the grounding resistance is not used. For networks above 33 kV the capacitive currents are large enough to neutralize the reactive fault currents, and therefore no external resistance is required in neutral connection to limit the fault current.

18.19. REACTANCE GROUNDING

Reactance grounding means grounding through impedance which is highly inductive.

For circuits between 3.3 kV and 33 kV the earth fault currents are likely to be excessive, if solid grounding is employed. Either resistance or reactance is connected in neutral to ground connection. There is no rule as regards which grounding should be used–resistance or reactance. If resistance is used fault current is limited and system reactance provides the necessary phase opposition between capacitive ground current and fault current. The reactance grounding provides additional reactance which provides a lagging current that nulifies the capacitive ground current.

As the value of reactance X connected in neutral to ground connection is increased, the ground fault current decreases and neutral displacement increases. If X is very small, the system behaves as an effectively grounded system. If X is very large, the system behaves as an isolated system. Reactance grounding lies between effective or solid grounding and resonant grounding discussed in Art. 18.20. The value of reactance required is to keep currents within safe limits. The transient voltages resulting from arcing increases as the reactance is increased. Similarly during switching operation higher values of reactance are expected to cause higher values of surge voltages.

A reactance grounded system ensures satisfactory relaying, partial grading of apparatus insulation, reduced interference to neighbouring communication circuits as compared with that in effectively ground systems. Other features of reactance grounding are :

1. Ground fault current is reduced but is much larger than capacitive ground fault current.
2. The voltages across healthy phases are between 80 to 100 per cent of line-to-line voltage.
3. Arcing grounds are avoided.
4. Transient ground faults are converted into controlled current faults.

Reactance grounding may be used for grounding the neutral of circuits where high charging currents are involved such as transmission lines, underground cables, synchronous motors, synchronous capacitors, etc. For networks where capacitance is relatively low, resistance grounding is preferred.

18.20. ARC SUPPRESSION COIL GROUNDING (OR RESONANT GROUNDING)

In the arc suppression coil method of grounding, the arcing-ground danger has been eliminated and the system is approximated to the isolated neutral system, in which one or two healthy phases continue to supply power and complete shutdown on the system is avoided till the fault was located and isolated.

It operates on the principle that if an inductance of appropriate value is connected in parallel with the capacitance, the fault current can be reduced considerably or even it can be neutralized. Thus the magnitude of inductance of the coil depends upon the capacitive currents flowing into the ground capacitances.

An arc *suppression coil*, also known as *Peterson coil* or *Ground fault neutralizer*, is an iron core tapped reactor connected in neutral to ground connection. The reactor is provided with tappings so that it can be tuned to the system capacitance. The function of arc suppression coil is to make the arcing ground faults self extinguishing and, in case of sustained faults, to reduce ground fault current to a comparatively low value so that the system can be kept in operation with one line grounded.

On occurrence of a ground fault (say on phase B), a lagging reactive current flows from the faulted phase to the ground and returns to the system through the inductive coil. Simultaneously, capacitive currents also flow from healthy phases to ground. The lagging fault current I_F and leading capacitive current I_C (phasor sum of I_{CR} and I_{CY}) are almost in phase opposition. By a proper selection of the value of inductance L of the arc suppression coil the two currents can be made almost equal so that there is no current through the ground fault and so there will be no arc. The combination of neutral reactance L and line capacitance C acts as a parallel resonant circuit. The circuit is shown in Fig. 18.21 (*a*). If V_P is the line-to-neutral voltage

$$I_{CR} = I_{CY} = \sqrt{3}\, V_P\, \omega C \quad ...(18.38)$$

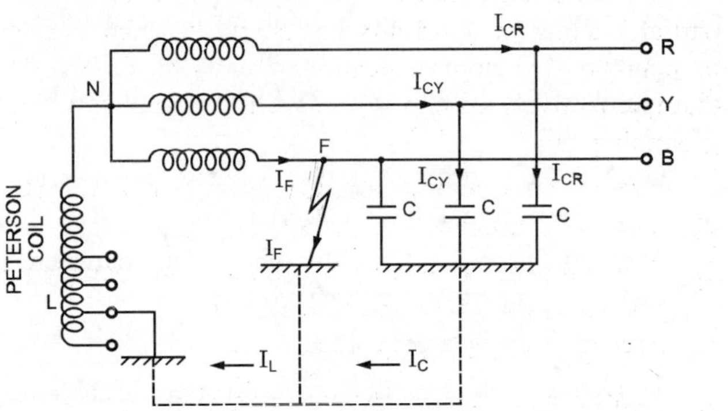

(a) Resonant Grounded System With Ground Fault on Phase B

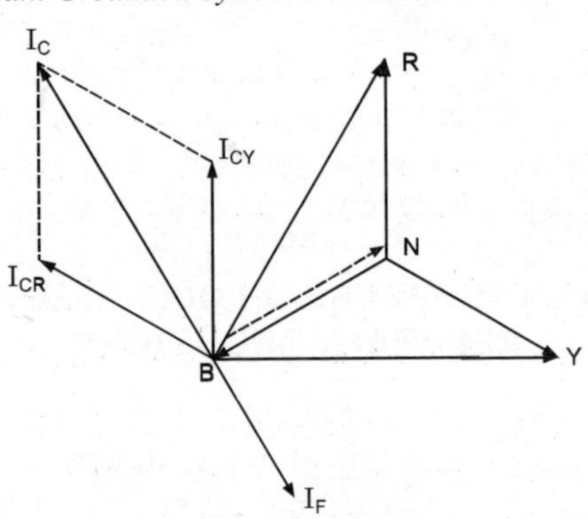

(b) Phasor Diagram For (a)

Fig. 18.21

Capacitive current,

$$\mathbf{I}_C = \mathbf{I}_{CR} + \mathbf{I}_{CY} = \sqrt{3} \times \sqrt{3}\ V_P\, \omega C = 3V_P\, \omega C \quad ...(18.39)$$

For balance condition,

$$I_L = I_C$$

$$\text{or} \quad \frac{V_P}{\omega L} = 3V_P\, \omega C$$

$$\text{or } L = \frac{1}{3\omega^2 C} \quad ...(18.40)$$

Phasor diagram is shown in Fig. 18.21 (*b*).

This method of neutral grounding is usually confined to medium voltage overhead transmission lines which are connected to the system generating source through power transformers. The reasons for this limitation are : (*i*) The higher insulation level is required on apparatus associated with arc suppression coil grounded system and it is comparatively difficult to provide on generators as compared on transformers. In addition, adequate winding protection cannot so readily be achieved for generators as for power transformers under arc suppression coil operating conditions. (*ii*) Overhead transmission lines are normally subject to transient ground faults due to lightning, birdage etc., of short duration.

It has been found that arc suppression coil grounding reduces the line outages from ground faults to 20 or 30 per cent of those obtainable with other types of grounding. As such this method of grounding is advantageously employed on radial lines as this avoids the construction of duplicate circuits for maintaining service continuity and thus there is overall economy.

Ground fault neutralizers should not be used where (*i*) fully graded insulation transformers are employed as the neutrals of such transformers are not sufficiently well insulated (*ii*) auto-transformers having a ratio greater than 0.95 to 1 are used.

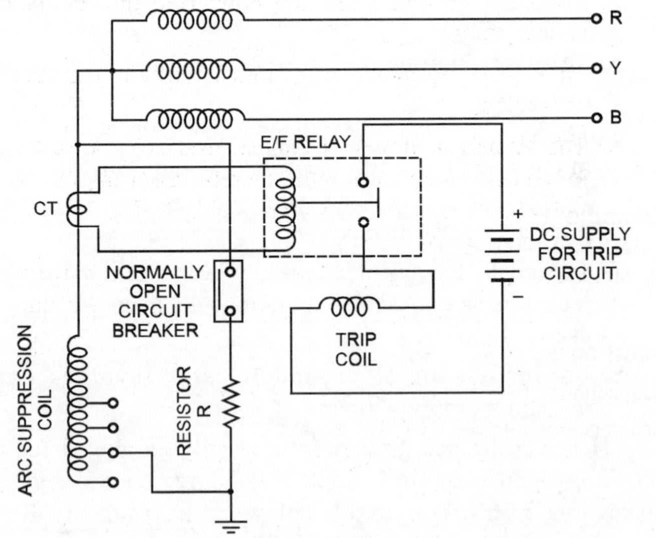

Fig. 18.22

The coils of the ground fault neutralizers are ten-minute time-rated on system where permanent ground faults can be located and removed promptly by ground relays or other suitable means. Otherwise, continuous time-rated neutralizers are used on all other systems. A short-time rated coil is equipped with an automatic circuit breaker, which by action of a relay shorts the arc-suppression coil after a certain time lag and connects the neutral directly to the ground, as shown in Fig. 18.22. Circuit breaker is normally open but is closed by the trip coil when the relay operates after a predetermined time. With this the fault current is bypassed through the resistor branch.

A short-time rated coil allows the clearing of intermittent faults, without any interruption of supply. Prolonged faults are removed by isolating the faulty section by the protective relay. A continuously rated coil allows a fault to remain in the system till it is located and removed.

An arc-suppression coil is also provided with an auxiliary winding for the energization of a relay to operate the short-circuiting device of a short-time rated coil.

Special features of arc-suppression coil grounding are given below :

1. During a single line-to-ground fault, the voltages across healthy phases rise to line-to-line value.
2. Continuity of supply can be maintained for long periods with one line grounded till the faulty section is isolated.
3. During sustained ground fault, the current is a small fraction of capacitive ground fault current. Burning and heating effects are reduced to minimum and damage to the equipment is limited.
4. Transient ground faults are suppressed. More than two-third of such faults are cleared without necessitating the operation of the circuit breaker.
5. Voltage gradient on ground surface in the vicinity of fault is greatly reduced and so the danger to the life in the proximity of fault.
6. Inductive interference to neighbouring communication circuits is lesser in magnitude but duration may be long.
7. Relaying needs special provision and is usually restricted to indication.
8. During normal operation, the arc suppression coil has little effect on balance of system voltages to ground and losses occurring in it are negligible.

Example 18.4. Calculate the reactance of a coil suitable for a 33 kV, 3-phase transmission system of which the capacitance to earth of each conductor is 4.5 μF.

[Pb. Univ. April 1988, October 1989]

Solution : Supply frequency = 50 Hz (assume)

Capacitance of each conductor to earth, $C = 4.5 \times 10^{-6}$ F

Reactance of the suitable coil,

$$\omega L = \frac{1}{3\omega C} = \frac{1}{3 \times (2\pi \times 50) \times 4.5 \times 10^{-6}} = \mathbf{235.8\ \Omega\ Ans.}$$

Example 18.5. A 132 kV, 3-phase, 50 Hz transmission line 192 km long consists of three conductors of effective diameter 20 mm arranged in a vertical plane with 4 m spacing and regularly transposed. Find the inductance and kVA rating of the arc suppressor coil. *[U.P. Technical Univ. Elements of Power System 2006-07]*

Solution : Conductor radius, $r = 10$ mm = 0.01 m

Spacings, $d_1 = 4$ m; $d_2 = 4$ m; $d_3 = 4 + 4 = 8$ m

Supply frequency, f = 50 Hz

Capacitance per phase is given by

$$C = \frac{2\pi\varepsilon_0}{\log_e \frac{\sqrt[3]{d_1 d_2 d_3}}{r}} \times \text{length of transmission line}$$

$$= \frac{2\pi \times 8.854 \times 10^{-12}}{\log_e \frac{\sqrt[3]{4 \times 4 \times 8}}{0.01}} \times 192 \times 10^3 = 1.717\ \mu\text{F}$$

Necessary inductance of Peterson coil

$$L = \frac{1}{3\omega^2 C} \qquad \text{Refer to Eq. (18.40)}$$

$$= \frac{1}{3 \times (2\pi \times 50)^2 \times 1.717 \times 10^{-6}} = 1.967\ \text{H}\ \mathbf{Ans.}$$

For ground fault, the current in the neutral is given by

$$I_L = \frac{V_P}{\omega L} = \frac{132{,}000/\sqrt{3}}{2\pi \times 50 \times 1.967} = 123.33\ \text{A}$$

Rating of suppressor coil

$$= \frac{V_P I_L}{1{,}000}\ \text{kVA} = \frac{132{,}000}{\sqrt{3}} \times 123.33 \times \frac{1}{1{,}000}$$

$$= 9{,}400\ \text{kVA} \quad \text{or} \quad 9.4\ \text{MVA per coil}\ \mathbf{Ans.}$$

18.21. VOLTAGE TRANSFORMER GROUNDING

In this system of neutral grounding the primary of a voltage transformer is connected between the neutral and the ground, as shown in Fig. 18.23 (*a*). A low resistor and a relay combination is connected across the secondary of the voltage transformer. An earth fault on the primary side will produce a voltage across the relay that will cause the protective device to operate.

This system of neutral earthing has got the following advantages :

1. The transient overvoltages owing to switching and arcing grounds are reduced because voltage transformer provides extremely high reactance to the earth path.
2. Arcing grounds are eliminated.
3. In this system of grounding, the neutral is grounded through a single-phase voltage transformer and so the system operates virtually like an isolated neutral system.

The major drawback of this system is that the grounded neutral acts as a reflection point for the travelling waves passing through the machine winding and in order to avoid high voltage build-up connection of a surge diverter between the machine neutral point and ground, as shown in Fig. 18.23 (*a*), becomes imperative.

The voltage transformer acts as a fault indicator, the extent of the fault can be determined by measuring the voltage on its secondary.

The use of this system of neutral grounding is normally confined to the generator equipments which are directly connected to step-up transformers. The generating networks of such equipments are physically isolated from the main distribution network. This is because interconnecting cables between the generator and transformer windings are usually of very short length, the electrostatic capacity of the circuit will be negligible.

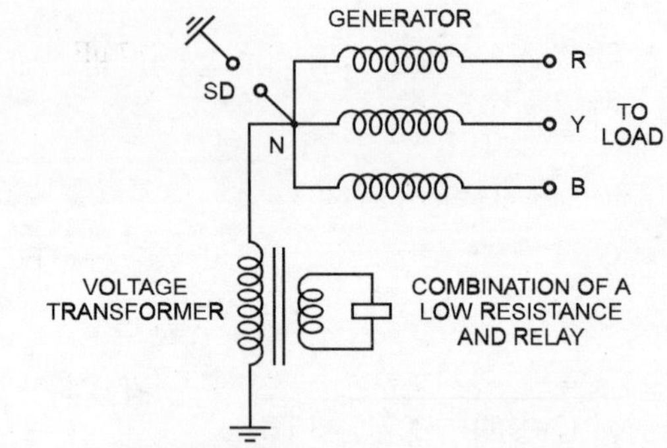

(a) Voltage Transformer Grounding

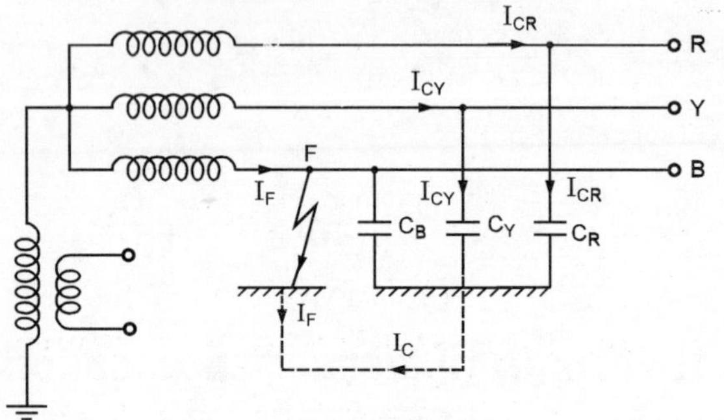

(b) Voltage Transformer Grounded System With Ground Fault on Phase B

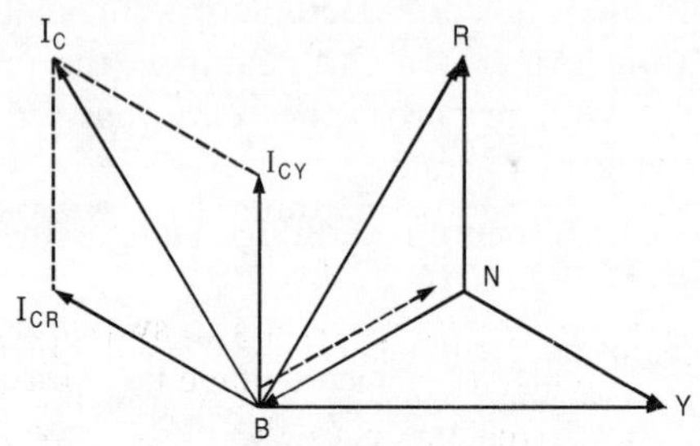

(c) Phasor Diagram For (b)

Fig. 18.23

18.22. GROUNDING TRANSFORMER

The neutral or star point is usually available at every voltage level from generator or transformer neutral. However, if no such point is available due to delta connections or if neutral point is desired on bus-bars, a zig-zag transformer is most commonly used. Such a transformer has no secondary. It is a core type transformer having three limbs built up in the same fashion, as illustrated in Fig. 18.24. The connections are also shown in Fig. 18.25. It is seen that the currents in the two halves of the winding on each limb are in opposite directions. As a result there will not be any undesirable harmonic prevailing in the circuit and so stresses on the insulation of the transformer are considerably reduced.

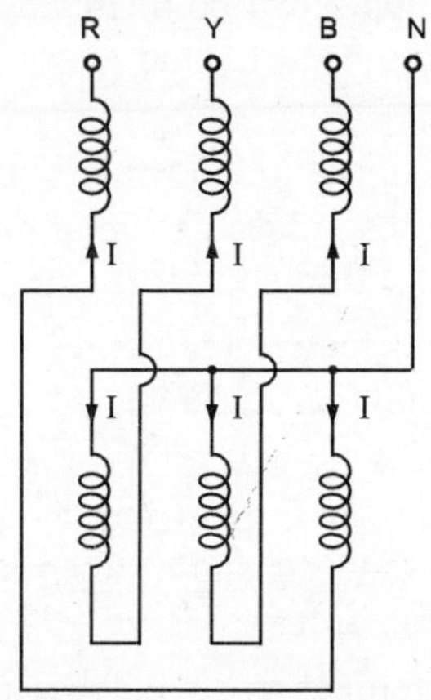

Fig. 18.24. ***Zig-Zag Transformer***

The fault current is quite high as the impedance of earthing transformer is quite low. The resistance or resistances are inserted either in the neutral circuit or in the windings of earthing transformer, as illustrated in Figs. 18.25 and 18.26 respectively in order to limit the magnitude of the fault current.

Under normal condition, only iron losses will be continuously occurring but at the time of fault copper losses will

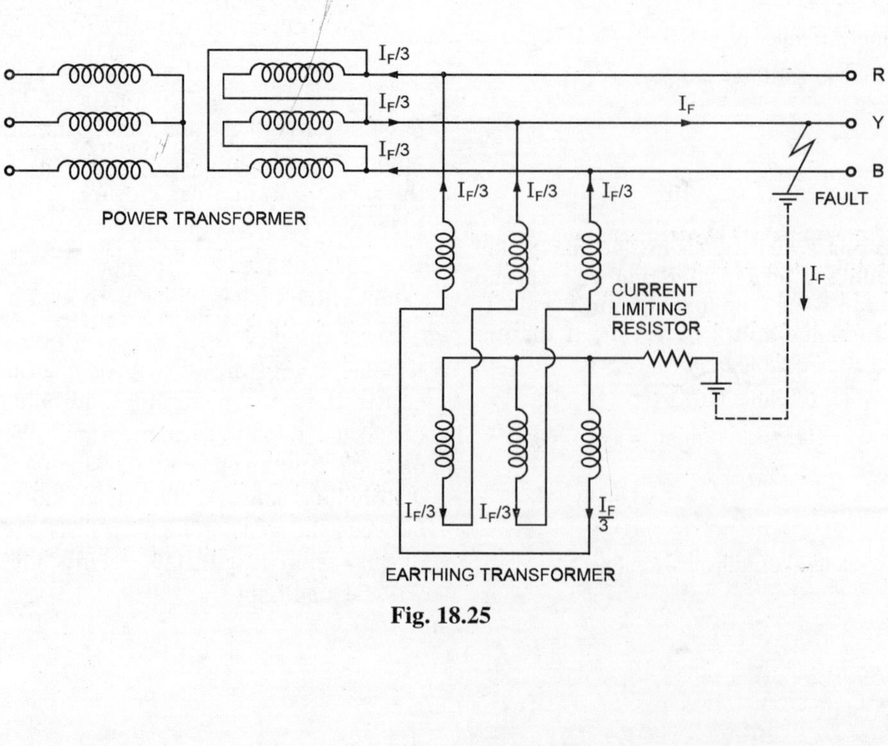

Fig. 18.25

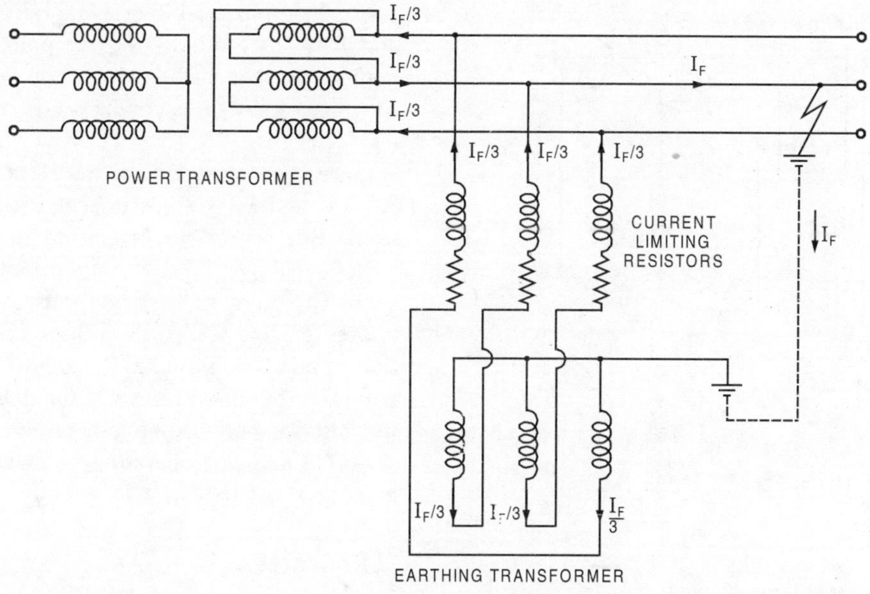

Fig. 18.26

also occur in addition, because of heavy fault current, in the earthing transformer. The duration of occurrence of copper losses is quite short, being the duration of fault, which is usually of 30–60 second duration.

As compared to the power transformer the rating of a zig-zag transformer is quite different. A grounding transformer is usually specified by the single-phase fault current that it handles. The two most common intervals specified for the fault current are 30 and 60 s. A 60 s grounding transformer will be more expensive as compared to a 30 one.

If a zig-zag transformer is not available, a star-delta transformer can be used without loading the delta side, as shown in Fig. 18.27. It is a step-down transformer. The star-connected primaries are connected to the bus-bars and its neutral is grounded. The secondaries are delta-connected and generally do not supply any load but provide a closed path for triple harmonic currents to circulate in them. Under normal operation the current in the grounding transformer is only its own exciting current. However, in the event of a single line-to-ground fault condition, large current may flow. So it should be of sufficient rating to handle the fault current.

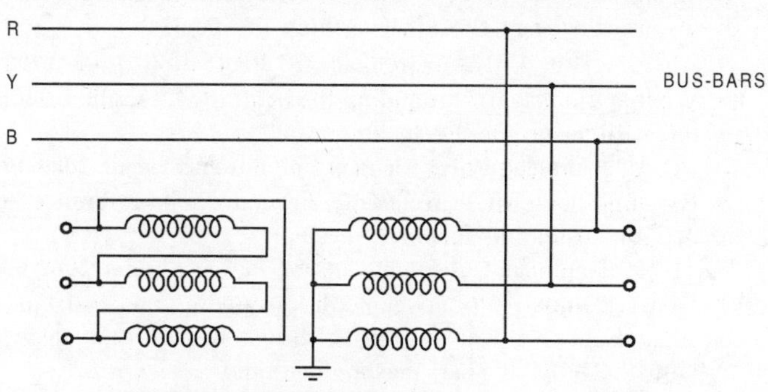

Fig. 18.27. ***Star-Delta Transformer Grounding***

18.23. HARMONIC SUPPRESSORS

Harmonic suppressor is an iron-cored reactor and is used in the neutral ground connection of a generator so that third harmonic currents* are reduced to safe value. If the third harmonics are not suppressed, they can cause dangerous overheating of grounding resistors of grounding transformers or erroneous operation of ground fault protective gear. The reactor is so designed that its core becomes saturated at very low system frequency fault currents. Thus it presents a high impedance to the flow of harmonic currents and low impedance path to the flow of fault currents. Since the harmonic suppressor acts as a reflection point for the travelling waves passing through the machine winding, it becomes necessary to connect a surge diverter between the machine neutral point and ground so as to avoid high voltage build-up. Grounding of generators through individual resistors and harmonic suppressors is illustrated in Fig. 18.28 (*a*).

18.24. NEUTRAL GROUNDING PRACTICE

1. Generally one neutral ground is provided at each voltage level. Between generation and distribution there are several voltage levels. One ground is provided at each voltage level [Fig. 18.28 (*b*)]
2. The grounding is provided at source and not at load end.
3. Each major bus section is grounded.
4. (*a*) The neutrals of generators are grounded through a resistance to limit the stator fault current. The value of resistor determines the percentage of generator winding left unprotected. Synchronous motors and synchronous capacitors are provided with reactance grounding.

* In case the generator output is nonsinusoidal, third harmonic currents are caused, by third harmonic voltage, to flow from the generator line terminals through the system and return via comparatively low zero sequence reactance path of the system capacitance or the windings of the transformer or other generators to its neutral terminal.

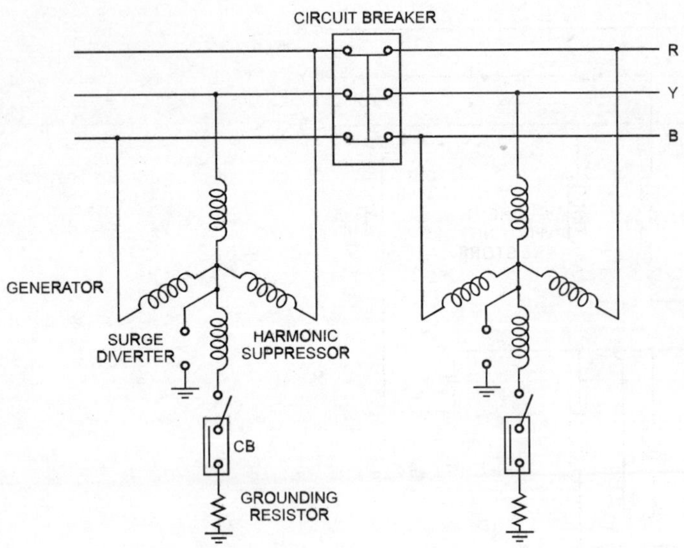

(*a*) *Grounding of Generators Through Individual Resistors and Harmonic Suppressors*

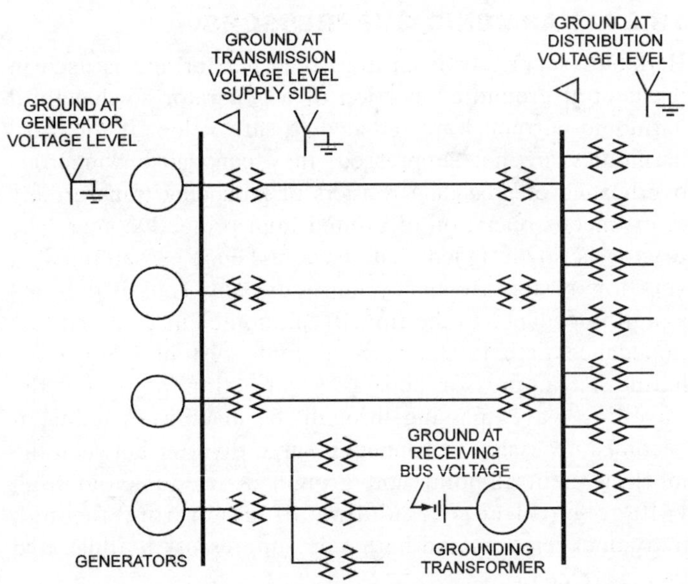

(*b*) *Neutral Ground at Every Voltage Level at Source END*

Fig. 18.28

(*b*) Where several generators are connected to a common neutral bus, the bus is connected to ground through a single grounding device. Disconnect switches can be used to ground the desired generators to the neutral bus.

(*c*) Where several generators are operating in parallel, only one generator neutral is earthed. This is done to avoid the interference of zero sequence currents.

(*d*) In generating station there is a provision to ground neutral of at least two generators, though only one is used at a time. The other generator neutral is grounded when the first generator is out of service.

(*e*) When there are one or two supply sources, no switching equipment is used in the grounding circuit.

5. For protection purpose, the neutral point of star side of power transformer is usually grounded.

6. The star-connected secondary sides of protective CTs and PTs are grounded at one point. This ensures stable neutral, proper measurement of voltages and currents, kWh and kVA on secondary side measuring instruments and controls.

7. As already mentioned in Art. 18.17, for circuits between 3.3 kV and 33 kV resistance or reactance grounding is used. But for low voltages up to 600 V and for high voltages above 33 kV solid or effective grounding is used. Effective grounding limits the voltages of healthy phases to line-to-neutral values in the event of ground faults and also eliminates the arcing grounds. The effective neutral grounding causes ground fault currents of high magnitudes but the modern protection systems are very sensitive and fast operating so that the faults are cleared in very short times.

Exercises

1. What is the purpose of earthing ? Distinguish between system earthing and equipment earthing. Define the terms earthing electrode and earth resistance.

2. Define the following terms :

(*i*) Earth electrode (*ii*) Earth resistance (*iii*) Earthing lead (*iv*) Step potential and (*v*) Touch Potential.

3. Describe the effects of electric currents on human body.

4. What is meant by soil resistivity ? On what factors does it depend ?

5. What is meant by "potential gradient" and "resistance area" ? How size of an electrode determined ?

6. How are the following measurements carried out ?

(*a*) Measurement of earth resistance. (*b*) Measurement of soil resistivity.

7. Discuss the advantages of (*i*) grounding the neutral (*ii*) keeping the neutral isolated of a power system.

[U.P. Technical Univ. Elements of Power System 2005-06]

8. What are the various advantages of neutral grounding ?

[U.P. Technical Univ. Elements of Power System 2007-08]

9. Explain the phenomenon of 'arcing ground' and suggest some method to minimise the effect of this phenomenon.

[Pb. Univ. Switchgear & Protection June 1996]

10. What is earthing of neutral ? Why is it done ? Briefly discuss different types of earthing the neutral.

[Pb. Univ. Switchgear & Protection June 1990]

11. What is the need of grounding the neutral ? Describe briefly the various grounding techniques.

[U.P. Technical Univ. Elements of Power System 2009-10]

12. Explain the term "Effectively grounded" in reference to neutral grounding in power system.

[U.P. Technical Univ. Elements of Power System 2002-03]

13. Give a comparison between solid grounding and resistance grounding of power systems neutrals. Justify your answer with the help of neat phasor diagrams.

[U.P. Technical Univ. Elements of Power System 2006-07]

14. Explain as to why the neutral point of a power system is earthed. Discuss the relative merits of earthing the neutral point through Peterson coil. Draw the vector diagram and derive the equation for the value of inductance.

[Agra Univ. Generation & Switchgear Practice 1981]

15. Explain the reasons leading to the general practice of earthing the neutral point of a power system and discuss the relative merits and demerits of earthing it (*i*) solidly (*ii*) through a resistance.

[Pb. Univ. Switchgear & Protection June 1991]

16. Discuss the following methods of neutral grounding :

(*i*) Peterson coil grounding

(*ii*) Resonant grounding

(*iii*) Solid grounding.

[U.P. Technical Univ. Elements of Power System 2008-09]

17. Classify the various methods of neutral earthing. Explain any one of them.

[G.B. Technical Univ. Elements of Power System 2010-11]

18. With the help of a neat phasor diagram and mathematical equations explain the method and advantages of resonant grounding.

[U.P. Technical Univ. Elements of Power System 2005-06]

19. Identify the criterion of grounding at 400 kV substation. If the total safety is to be ensured what will be the layout of grounding ? Explain alongwith diagram.

[U.P. Technical Univ. Elements of Power System 2005-06]

20. Explain different types of earthing the neutral point of a power system. Derive an expression for the reactance of the Peterson coil in terms of the capacitance of the protected line. **[Pb. Univ. Switchgear & Protection Sept. 1991]**

21. What is Peterson coil ? What protective function is performed by this device ?

[A.M.I.E. Sec. B. Advanced Power Systems Winter 1998]

22. Explain the role of earthing transformer in neutral grounding.

[U.P. Technical Univ. Elements of Power System 2006-07]

23. What is the difference between reactance and resonant grounding systems ? What are grounding practices ?

[U.P. Technical Univ. Elements of Power System 2004-05]

24. Why is it advantageous to ground the neutral points of generators and transformers ?

What are step and touch potentials ? How can they be reduced to safe values ?

How are transmission line towers grounded ?

Explain briefly the function of harmonic suppressor.

[A.M.I.E. Sec. B. Advanced Power Systems Summer 1999]

25. Discuss the following types of earthing :

(*i*) Voltage transformer earthing

(*ii*) Using interconnected star earthing transformer.

[Pb. Univ. Switchgear & Protection June 1991]

26. Write short notes on the following :

(*i*) Earthing and its objectives.

(*ii*) Effects of electric currents on human body.

(*iii*) Earthing mat.

(*iv*) Impulse behaviour of grounding systems.

(*v*) Methods of neutral grounding in a power system and their relative merits and demerits.

(*vi*) Peterson coil.

(*vii*) Earthing transformer.

(*viii*) Harmonic suppressors.

Short Answer Type Questions With Answers

Q. 1. What is difference between system earthing and equipment earthing ?

Ans. System earthing deals with earthing of the system neutral to ensure system security and protection while equipment earthing deals with earthing of non-current carrying parts of the equipment to ensure safety of personnel and protection against lightning.

Q. 2. What is average resistance of a human body ?

Ans. About 1,000 Ω.

Q. 3. What is the current that is considered safe for human body ?

Ans. About 9 mA.

Q. 4. What is meant by an earth electrode ?

Ans. Any wire, rod, pipe, plate and an array of conductors embedded in ground horizontally or vertically is called the earth electrode.

Q. 5. Give the maximum permissible values of ground resistance for (*a*) large power stations (*b*) small substations.

Ans. (*a*) 0.5 Ω; (*b*) 2.0 Ω.

Q. 6. What is meant by 'let go current' ?

Ans. The maximum current at which the subject is capable of releasing an electrical conductor by using muscles directly stimulated by the current is called 'let go current'.

Q. 7. On what factor does the resistivity of soil depend ?

Ans. Soil resistivity depends upon moisture content, soil temperature, composition and amount of soluble salts in the soil, magnitude of current flowing through the soil and depth of the layer.

Q. 8. What is meant by earthing mat ?

Ans. A number of rods when joined together through copper conductors constitute an earthing mat. This is also called the earthing grid.

Q. 9. What is earthing lead ?

Ans. The wire which connects earth wire or any other apparatus to be earthed to the earth electrode is known as the earthing lead.

Q. 10. What is meant by system grounding ?

Ans. The process of connecting some part of the electrical power system (*e.g.*, neutral point of a star-connected system, one conductor of the secondary of a transformer etc.) to ground (*i.e.*, soil) is known as the system grounding.

Q. 11. What is neutral grounding ?

Ans. The process of connecting neutral point of a 3-phase system to ground (*i.e.*, soil) either directly or through some circuit element (*e.g.*, resistance, reactance etc.) is called the neutral grounding.

Q. 12. What is the chief argument that favours use of isolated neutral system ?

Ans. The chief argument that goes in favour of isolated neutral system is the possibility of maintaining continuity of supply in case one phase gets grounded until it is convenient to disconnect and repair the faulty line.

Q. 13. What is arcing ground ? **[Pb. Technical Univ. May 2009]**

Ans. Arcing ground is a surge which is produced if the neutral is not connected to the earth. In a three-phase transmission line, each phase has a capacitance w.r.t. to the earth. When an earth fault occurs on any phase, a capacitive fault current I_F flows into the earth. Such a current if exceeds 4–5 amperes, is sufficient to maintain an arc in the ionized path of the fault, even though the medium causing the fault has cleared itself. The persistency of the arc due to the flow of capacitance current gives rise to a condition known as "*arcing ground*" in which cyclic charging and discharging of the system capacity through the fault results in high frequency oscillations being superimposed on the whole system and a build-up of very high voltages can occur. This results in phase voltage to rise to 5 to 6 times of normal voltage. The build-up of high voltages may result in insulation breakdown.

Q. 14. What are the various methods of neutral grounding ? **[U.P.T.U. 2006-07]**

Ans. The methods commonly used for grounding the neutral point of a 3-phase system are

(*i*) Solid or effective grounding
(*ii*) Resistance grounding
(*iii*) Reactance grounding
(*iv*) Peterson coil or arc suppression coil grounding
(*v*) Voltage transformer grounding.

Q. 15. What is effectively grounded system ?

Ans. When the neutral point of a 3-phase system (*e.g.*, 3-phase generator, 3-phase transformer etc.) is directly connected to ground (*i.e.,* soil) through a wire of negligible resistance and reactance, it is called the solid grounding or effective grounding.

According to IEEE definition, a system or a portion of a system can be said to be effectively grounded when for all points on the system or specified portion thereof, the ratio of zero sequence reactance to positive sequence reactance is not greater than three (*i.e.*, $\frac{X_0}{X_1} \ngtr 3$) and the ratio of zero sequence resistance to positive sequence resistance is not greater than one (*i.e.*, $\frac{R_0}{R_1} \ngtr 1$) for any condition of operation and for any amount of generator capacity.

Q. 16. What is arc suppression coil grounding ?

Ans. In the arc suppression coil method of grounding, the arcing-ground danger has been eliminated and the system is approximated to the isolated neutral system, in which one or two healthy phases continue to supply power and complete shutdown on the system is avoided till the fault was located and isolated.

An arc *suppression coil*, also known as *Peterson coil* or *Ground fault neutralizer*, is an iron core tapped reactor connected in neutral to ground connection. The reactor is provided with tappings so that it can be tuned to the system capacitance. The function of arc suppression coil is to make the arcing ground faults self extinguishing and, in case of sustained faults, to reduce ground fault current to a comparatively low value so that the system can be kept in operation with one line grounded.

This method of neutral grounding is usually confined to medium voltage overhead transmission lines which are connected to the system generating source through power transformer.

It has been found that arc suppression coil grounding reduces the line outages from ground faults to 20 or 30 per cent of those obtainable with other types of grounding. As such this method of grounding is advantageously employed on radial lines as this avoids the construction of duplicate circuits for maintaining service continuity and thus there is overall economy.

Q. 17. What is an earthing transformer ? **[U.P.T.U. 2006-07]**

or

What is the use of earthing transformer ? **[U.P.T.U. 2007-08]**

or

What is the role of earthing transformer in neutral grounding ?

Ans. In cases where the neutral points of a three-phase system are not accessible or where the transformers or generators are delta connected, an artificial neutral point is required to be created. This is accomplished by means of a zig-zag transformer, called the earthing transformer.

As compared to the power transformer the rating of a zig-zag transformer is quite different. A grounding transformer is usually specified by the single-phase fault current that it handles. The two most common intervals specified for the fault current are 30 and 60 s. A 60 s grounding transformer will be more expensive as compared to a 30 one.

Q. 18. What is harmonic suppressor ?

Ans. Harmonic suppressor is an iron-cored reactor and is used in the neutral ground connections of a generator so that third harmonic currents are reduced to safe value.

Problems

1. Determine the resistance to ground for a hemispherical electrode on the surface of the ground. The diameter of the hemispherical electrode is 2 m and the soil resistivity is 100 Ω-m. **[Ans.** 15.92 Ω]
2. Determine the resistance to ground of the rod whose diameter is 5 cm and length is 6 m and ρ = 100 Ω-m. Determine the radius of a hemispherical electrode which has the same resistance as the above driven rod. **[Ans.** 16.4 Ω; 0.97 m]
3. A 33 kV, 50 Hz network has a capacitance to neutral of 1.0 μF per phase. Calculate the reactance of an arc suppression coil suitable for the system to avoid adverse effect of arcing ground. **[Ans.** 1,061 Ω]
4. Calculate the reactance of a coil suitable for a 33 kV, 3-phase transmission system of which the capacitance to earth of each conductor is 4.5 μF. **[Ans.** 3.75 H; 308 kVA]
5. A 230 kV, 3-phase, 50 Hz, 200 km transmission line has a capacitance to earth of 0.01 μF/km per phase. Calculate the inductance and kVA rating of the Peterson coil used for earthing the above system. **[Ans.** 1.7 H; 33,238 kVA]
6. Derive an expression for the reactance of the Peterson coil in terms of the capacitance of the protected line. Calculate the reactance of a coil suitable for a 33 kV, 3-phase transmission system of which the capacitance to earth of each conductor is 5.0 μF. **[Pb. Univ. Switchgear & Protection May, 1990]** **[Ans.** 212.2 Ω]

❑❑❑

CHAPTER

19

Voltage Control

INSIDE THIS CHAPTER

19.1. INTRODUCTION

The electrical energy is normally generated at the power stations far away from the urban areas, where consumers are located, and delivered to the ultimate consumers through a network of transmission and distribution. For satisfactory operation of motors, lamps, and other loads it is desirable that the voltage at consumer's terminals is maintained substantially constant. Too wide variation of voltage may cause erratic operation or even malfunctioning of consumer's appliances. The main cause responsible for voltage variation is the variation in load on the supply system. With the increase in load on the supply system the voltage at the consumer premises falls due to increase in voltage drop in (*i*) alternator synchronous impedance (*ii*) transmission lines (*iii*) transformer impedance (*iv*) feeders and (*v*) distributors. The reverse would also happen should the load on the power system falls. Such voltage variations are undesirable and the supplier is required to maintain the voltage at the consumer's premises within prescribed limits (± 6% of declared supply voltage) so that the consumer's apparatus may operate satisfactorily. This is achieved by installing voltage control equipment at suitable places such as at (*i*) generating stations, (*ii*) transformer stations supplying to feeders and (*iii*) the feeder's end if the drop exceeds the permissible limit. The voltage regulating equipment is provided at more than one point in the power system because of extensive transmission and distribution network and different load characteristics of different load circuits in the power system.

In dc supply system voltage can be controlled by using overcompound generators in case of feeders of equal lengths fed from it, but in case of feeders of different lengths being fed by one generator the voltage at the end of each feeder is kept constant by means of feeder booster (series wound dc generator).

In ac supply system the voltage can be controlled by the following methods :

(*i*) Excitation control and voltage regulators in generating stations.

(*ii*) Use of tap changing transformers at sending end and receiving end of transmission lines, in factories, substations, distribution substations and transmission substations.

(*iii*) Use of booster transformers.

(*iv*) Use of induction regulators.

(*v*) Use of series capacitors in long EHV transmission lines.

(*vi*) Switching-in shunt reactors during light loads or while energizing long EHV lines.

(*vii*) Switching-in shunt capacitors during high load or low power factor load.

(*viii*) Use of static shunt compensation having shunt capacitors and thyristorized control for stepless control of reactive power and voltage.

(*ix*) Use of synchronous condensers.

Before dealing excitation control, we will like to discuss excitation systems.

19.2. EXCITATION SYSTEMS

The first step in the sophistication of the primitive excitation system was the introduction of an amplifier in the feedback path which amplified the error signal and made the system fast acting. With the increase in size of the unit and growth in the interconnection of the system, the excitation system have become more and more complex.

The excitation system is required to provide the necessary field current to the rotor winding of a synchronous machine. The availability of excitation at all times is of paramount importance. Loss of excitation of a unit on the bus results in a more serious disturbance than that resulting from outage of the generator unit from the bus, as the remaining units must not only pick up the load dropped but also supply the large reactive current drawn by the unexcited alternator. In view of this an excitation system with better reliability is preferable, even if the initial cost is more.

The main requirements of an excitation system are reliability under all conditions of service, simplicity of control, ease of maintenance, stability and fast transient response.

The amount of excitation required depends on the load current, load power factor and speed of the machine. Larger the load currents, lower the speeds and lagging power factors, more the excitation required.

An excitation system may be *individual* one, in which each alternator is provided with its own exciter in the form of small generator on an extension of the main shaft, or *centralised excitation system* having two or more exciters feeding a bus-bar to which field systems of all the alternators in the power plant are connected. Though centralised excitation system is a cheaper arrangement but a fault in this system adversely affects all the alternators in the power plant. As such individual excitation system is widely used.

The excitation systems can be broadly classified as: 1. DC excitation system, 2. AC excitation system, and 3. Static excitation system.

19.2.1. DC Excitation System

In dc *excitation system*, the system has two exciters—the main exciter (a separately excited dc generator providing the field current to the alternator) and a pilot exciter (a compound wound self excited dc generator providing the field current to the main exciter). The exciter output is regulated by an automatic voltage regulator (AVR) for controlling the output terminal voltage of the alternator. The current transformer input to the AVR ensures limiting of alternator current during fault. With the opening of field breaker, the field discharge resistor is connected across the field winding so as to dissipate the stored energy in the field winding which is highly inductive. The main and pilot exciters can be either driven by the main shaft (directly or through gearing) or separately driven by a motor. Direct driven exciters are usually preferred as these preserve the unit system of operation and the excitation is not affected by external disturbances. Exciters for small sized turbo-alternators are usually directly coupled to the shaft whereas for medium and large size turbo-alternators, the exciters are coupled to the main shaft through gear and are usually driven at 1,000 rpm. The voltage rating of the main exciter is about 400 V and its capacity is about 0.5% of the capacity of the alternator. Troubles in the exciters of turbo-alternators are quite frequent because of their high speeds and as such separate motor driven exciters are provided as standby exciters. In some cases the dc excitation system is equipped with amplidyne controller. The amplidyne is a high response cross-field dc generator, which has a very high amplification factor and requires very small control power. It is supplied from an automatic voltage regulator (AVR). The main shortcomings of dc excitation system are large time constant (around 3 seconds) and commutation difficulties. As such the dc excitation systems have been superseded by ac excitation systems and static systems.

19.2.2. AC Excitation System

This *system* consists of an alternator and thyristor rectifier bridge directly connected to the main alternator shaft. The main exciter may either be self excited or separately excited. A *rotating thyristor excitation system* employs self excited main exciter whereas the *brushless excitation system* employs a separately excited main exciter.

1. Rotating Thyristor Excitation System. Block diagram of a rotating thyristor excitation system is depicted in Fig. 19.2, the rotating portion being shown enclosed by dashed line rectangle. This system comprises of an ac exciter having a stationary field and a rotating armature. The exciter output, rectified by a full-wave thyristor bridge rectifier circuit, is supplied to the main alternator field winding. The alternator field winding is also supplied from the exciter output through

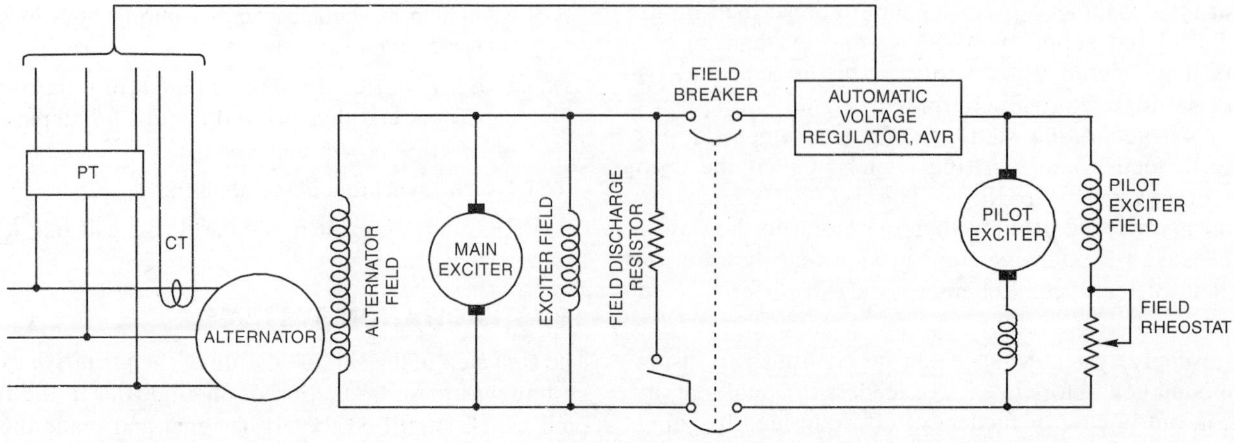

Fig. 19.1. *DC Excitation System*

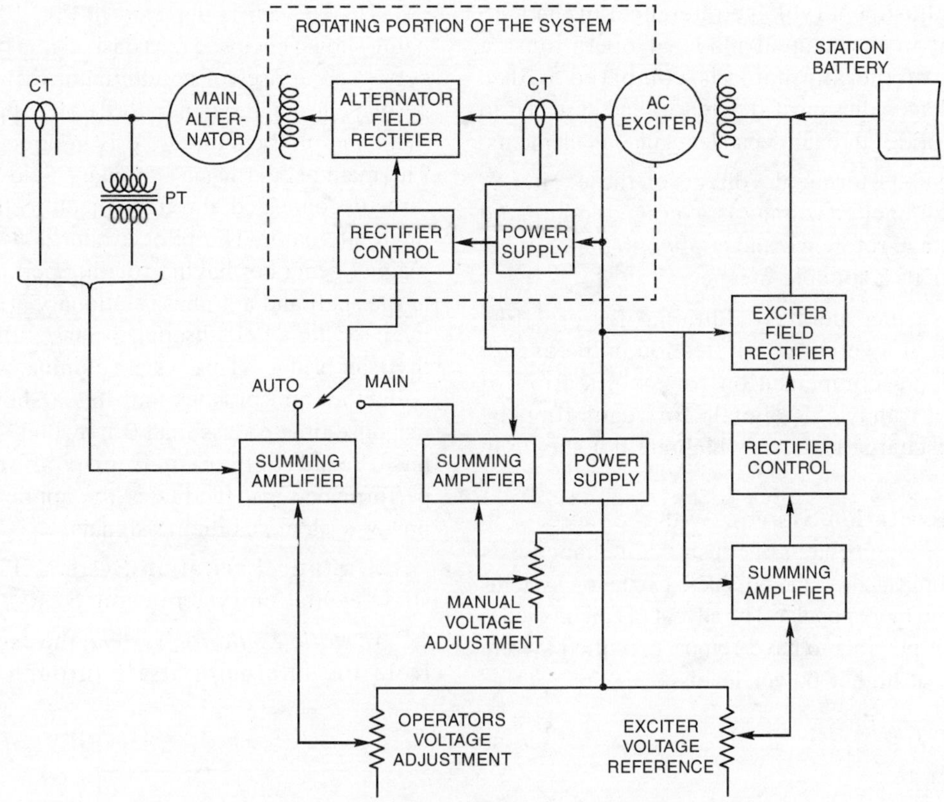

Fig. 19.2. ***Block Diagram For Rotating Thyristor Excitation System***

another rectifier circuit. Exciter voltage can be built up using its residual flux. For convenience and to shorten the build up time, stationary battery supply is also incorporated.

The power supply and rectifier control, incorporated as a part of rotating portion, generates the required phase controlled triggering signals in response to a dc control level fed by the voltage regulator.

An automatic voltage regulator (AVR) is also incorporated in the system. The alternator voltage signal (from the PT) is averaged and compared directly with the

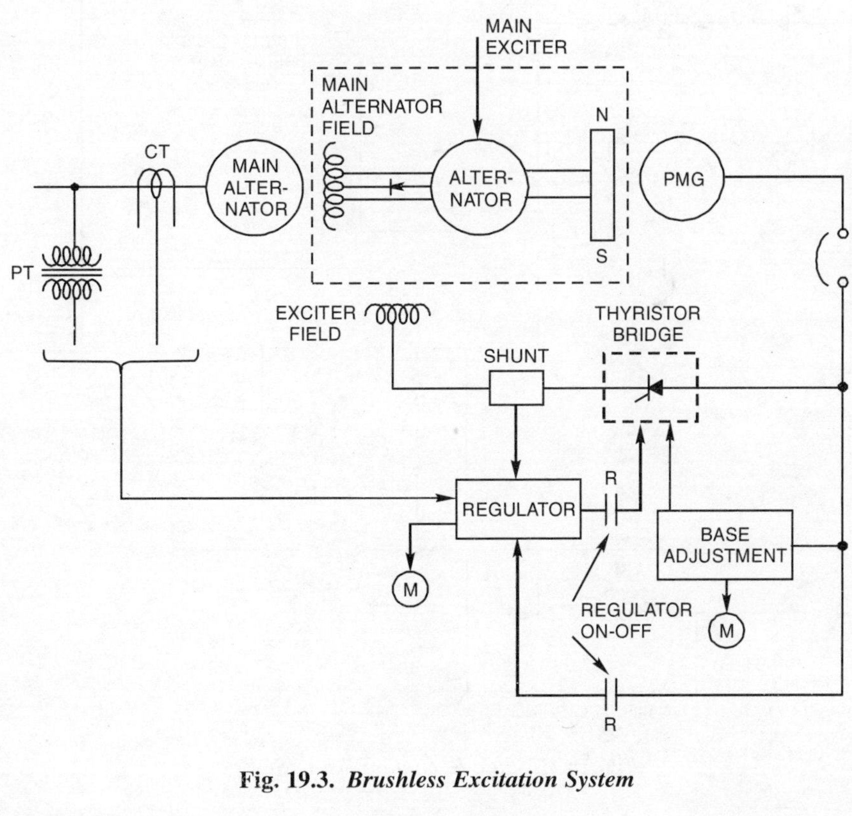

Fig. 19.3. ***Brushless Excitation System***

operators voltage adjustment (voltage reference) in the auto mode of operation. In the manual mode of operation the excitation current of the alternator is compared with a separate manual voltage adjustment (reference) and supplied to the rotating thyristor bridge through separate regulating elements.

For maintaining the terminal voltage of the exciter at the desired level, the exciter voltage is averaged, compared with the exciter voltage reference and applied to the rectifier control and rectifier in a suitable manner.

The system also incorporates additional features like overvoltage protection, overcurrent protection in the exciter field control, lead-lag compensation for stabilization of voltage control, watt and VAR signals for regulating the voltage and field discharge resistor, which are not shown in the block diagram.

2. Brushless Excitation System. With the increase in rating of alternators, the problem of brush maintenance gets more and more complicated, so the brushless excitation systems are becoming more and more popular. The advent of silicon diodes and thyristors made it possible to have compact rectified system converting ac to dc at higher power levels.

This system is depicted in Fig. 19.3, the rotating portion being shown enclosed by a dashed line rectangle. The excitation system comprises of an alternator rectifier main exciter and a permanent magnet generator (PMG) pilot exciter. Both of the main and pilot exciter are driven directly from the main shaft. The main exciter has a stationary field and a rotating armature directly connected, through silicon rectifiers, to the field of the main alternator. The pilot exciter is a shaft driven permanent magnet generator having rotating permanent magnets attached to the shaft and a 3-phase stationary armature, which feeds the main exciter's field through 3-phase full-wave phase controlled thyristor bridges. This system eliminates the use of commutator, collectors and brushes and has a short time constant and a response time of less than 0.1 second. The short time constant has the advantage in improved small signal dynamic performance and facilitates the application of supplementary power system stabilizing signals.

3. Static Excitation System. Figure 19.4 depicts the static excitation system using SCRs.

In *static excitation system*, the excitation supply is taken from the alternator itself through a 3-phase star/delta

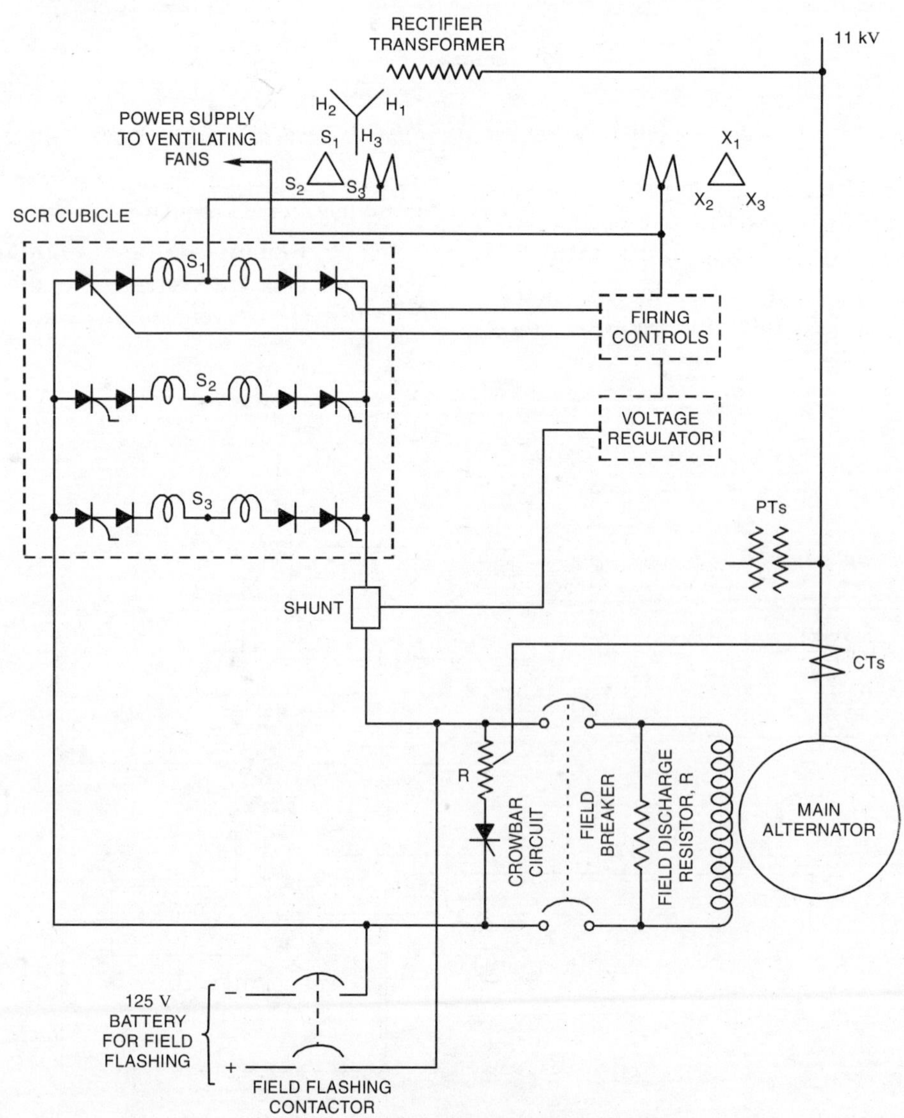

Fig. 19.4. ***Static Excitation Using SCRs***

connected, oil immersed, forced air cooled, indoor type step-down transformer and a rectifier system employing mercury-arc rectifiers or silicon controlled rectifiers. The star-connected primary is connected to the alternator bus, the delta-connected secondary supplies power to the rectifier system and the delta-connected tertiary feeds power to grid control circuits and other auxiliary equipment. The rectifiers are connected in parallel to give sufficient current carrying capacity. Each leg of rectifier is protected with series fuse, surge protection and fault indicating light. The rectifiers are forced air cooled. This system has a very small response time (about 20 milliseconds) and provides excellent dynamic performance. SCRs are ideally suited for a static excitation system because they have high speed of response, high power gain and can be easily paralleled. The advantages of the static excitation system are elimination of exciter windage loss and commutator wearing and winding maintenance resulting in reduced operating costs and *electronic speed response*. The fact that in static excitation the voltage is proportional to the speed, affords a major advantage in load rejection.

19.3. EXCITATION CONTROL

With the change in load on the supply system the terminal voltage of the alternator also changes due to variation in voltage drop in the synchronous impedance of the alternator. Since the alternators have to be run at a constant speed, the induced emfs, therefore, cannot be controlled by adjustment of speed. So to maintain constant voltage use of excitation regulation is made. The alternator excitation can be regulated by use of automatic or hand regulator acting in the field circuit of alternator exciter. Automatic method of excitation control is preferred in modern practice.

The excitation control system does not differ much for dc and ac excitation systems as in an automatic voltage regulators, the generator, AVR and excitation control circuit form a closed loop system. The AVR senses the terminal voltage and if required varies the excitation so as to keep the terminal voltage within the specified limits. A reactive volt ampere limiting device is built into the AVR. At a predetermined value of generator leading MVAR, the limiter overrides the basic AVR function and keeps the excitation current at a value to ensure that there is no further increase in rotor load angle and in the risk of instability.

A manual control system is invariably a part of the excitation control system and is employed whenever automatic control is not in action. When the AVR is in service, a circuitry is provided that makes the manual control to follow up the variations in excitation, so that in the event of failure of AVR, the generator excitation control is immediately transferred to manual control without any change in excitation current.

19.4. AUTOMATIC VOLTAGE REGULATORS

Synchronous generators have inherently large internal reactance and, therefore, high voltage regulation. So, in order to maintain terminal voltage of the generator under various operating conditions, provision of automatic voltage regulation equipment is essential.

An automatic voltage regulator operates on the principle of detection of error. The output voltage of an ac generator obtained through a PT is rectified, filtered and compared with a reference. The difference of the actual voltage and the reference voltage, known as *error voltage*, is amplified through an amplifier (rotary, magnetic or static) and supplied to the field circuit of the main exciter or pilot exciter. Thus, the amplified error signal controls the excitation of the main or pilot exciter through a buck or boost action. Exciter output control leads to the control of the main alternator terminal voltage.

The main functions of an AVR are as follows:

1. Control of system voltage within prescribed limits and have the operation of the machine nearer to the steady state stability limit.
2. Proper division of reactive load between the alternators operating in parallel.
3. Prevention of dangerous overvoltages on the occurrence of sudden loss of load on the system.
4. Increase of excitation under system fault conditions so that maximum synchronising power exists at the time of clearance of fault, to prevent loss of synchronism.

Modern alternators have high inherent reactance (more than 1 pu). The magnitudes of excitation for different load current and power factor conditions are very different. Hence automatic voltage regulators for modern alternators need stringent design characteristics.

On the occurrence of sudden change in load on the alternator, there should be a change in the excitation to provide the same terminal voltage under the new load conditions. The automatic voltage regulating equipment must operate on the exciter field which changes the exciter output voltage and then the alternator field current. The response of a voltage that controls the alternator terminal voltage is thus a function of the responses of the voltage regulator, the exciter, and the main alternator excitation current (or flux). The response time of the regulator is the time required for the regulator to take action following the change in voltage.

Due to high inductance of alternator either a variation of field circuit resistance or change in the exciter voltage will not be able to give quick response. To achieve quick response, which is required due to violent fluctuations in industrial load, quick acting voltage regulators based on the '*overshooting the mark*' principle are used. With the increase in load, the regulator of this type produces an increase in excitation more than that ultimately necessary. Before the voltage has the time to increase to the value corresponding to the increased excitation, the regulator reduces the excitation to the proper value.

For turbo-alternators a response time of 0.5 second is adequate, *i.e.*, a mean rate of rise of 110 V/second for nominal 220 V. The rate of rise of voltage per second required for hydroelectric generators is considerable larger, because of the overspeed to which these machines may be subjected.

19.4.1. Electromechanical Voltage Regulators

1. Tirril Automatic Voltage Regulator. It is a vibrating type voltage regulator, in which a fixed resistance is cut in and cut out of the exciter field circuit of the alternator. This is achieved by rapidly opening and closing a shunt circuit across the exciter rheostat. Its essential parts are shown in Fig. 19.5.

There are two levers at the top carrying main contacts at the facing ends. The lever on the left is controlled by the exciter magnet, which is excited by the current proportional to exciter voltage. The lever on the right is controlled by an ac magnet, known as *main control magnet* having both shunt and series excitations. The ac magnet is so adjusted that with normal load and voltage, the pulls of the two coils are equal and opposite, so the lever on the right hand side is held in horizontal position. Field circuit of the exciter is provided with a rheostat, the resistance of which can be adjusted to give the required excitation. The regulator is provided with a relay horse shoe magnet which consists of two identical windings, both of which are connected across the exciter armature. One on the left hand side permanently connected and while other on the right hand side has its circuit completed only when main contacts are closed. As long as the main contacts are open, the relay shoe magnet remains energized, lever carrying relay contact remains pulled down and therefore, relay contacts remain opened. When the main contacts are closed the flux in the relay magnet is destroyed because of two windings of relay connected differentially, it releases the lever carrying a relay contact, so the relay contacts provided across the rheostat are closed and short circuit the rheostat.

When the load on the alternator increases, the series excitation predominates and so pulls the ac control magnet. Thus the contacts of the two levers are closed, the relay shoe magnet is de-energised and so rheostat in the exciter field is short circuited. This increases the exciter voltage and hence the excitation of the alternator. The increase in excitation of the alternator causes an increase in alternator voltage rapidly. Now the excitation of the exciter control magnet increases due to increase in exciter voltage, pulling down the left hand lever, so opening the main contact energizing the relay and putting the rheostat again in the circuit of the exciter field before the alternator voltage has had time to increases too far.

The reverse action takes place when the load on the alternator decreases. Due to overshooting the mark principle the terminal voltage does not remain perfectly steady, but oscillates rapidly between its maximum and minimum values. The regulator is so quick acting that the variation in voltage is less than 1%. A capacitor is provided across the relay contacts to reduce the spark at the time of opening of the relay contacts.

This method can be employed to maintain the voltage at the consumer premises within permissible limits in case of short lines by designing the winding of the relay operated by the line voltage and current to allow for the regulation of a line end of the alternator.

This method is not suitable for long lines because to maintain voltage at the far end within permissible limits we will have to vary the terminal voltage of the alternator too much.

When loads are fed near the generator end and at the far end of the line, it is required to maintain the voltage at both ends of a transmission line constant. This cannot be achieved by this method. To achieve this condition other methods such as power factor control or tap changing transformers are employed.

2. Brown-Boveri Automatic Voltage Regulator. Brown-Boveri automatic voltage regulator also operates on the "*over-shooting the mark*", principle but it differs from the Tirril one. In this case the regulating resistance is gradually varied either continuously or in small steps, while in Tirril regulator the resistance is first completely inserted then completely cut out. Under steady conditions, all parts of the regulator are at rest and wear and tear is quite less as compared with that in Tirril regulator.

Brown-Boveri automatic voltage regulator is shown in Figs. 19.6 and 19.7.

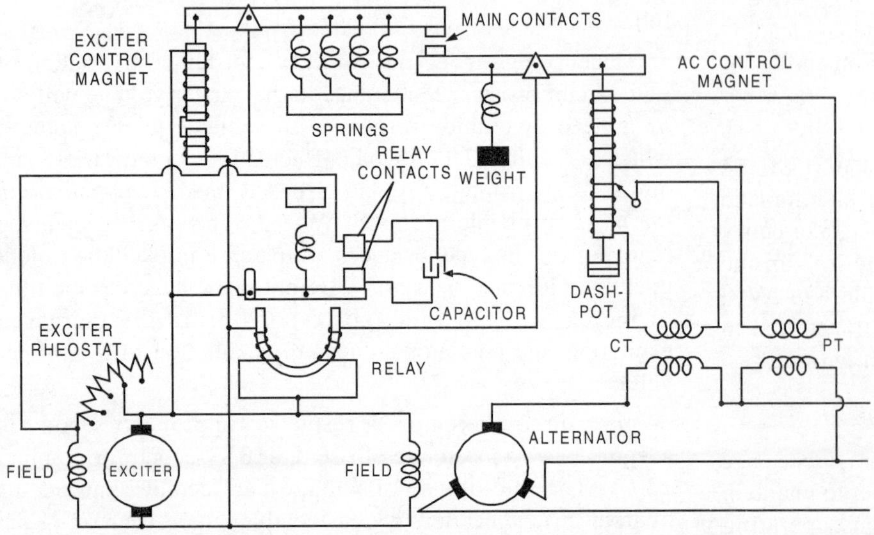

Fig. 19.5. ***Tirril Automatic Voltage Regulator***

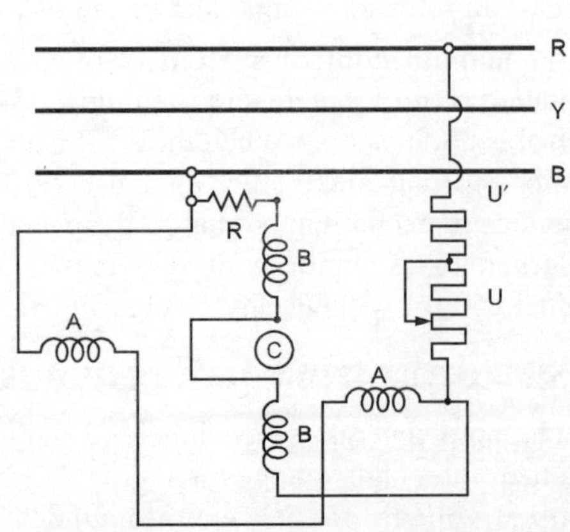

Fig. 19.6. ***Brown-Boveri Regulator***

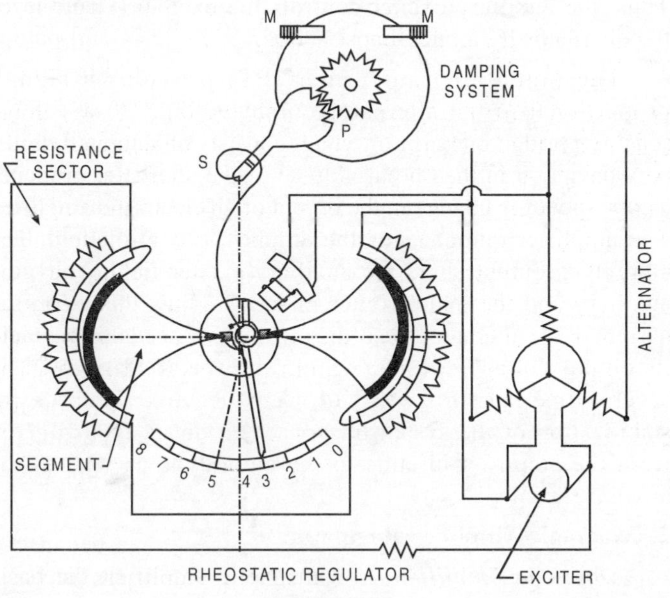

Fig. 19.7.

Brown-Boveri automatic voltage regulator consists of three important parts :

(a) Control System. It is essentially a split-phase induction motor, which consists of two windings A and B wound on an annular core of laminated sheet steel. The winding A is excited from two terminals of the generator (whose voltage is to be controlled) through coarse and fine resistances U and U′ and a resistance R is placed in the circuit of winding B. The ratio of resistance to reactance is made different in the two windings in order that phase angle difference may be in the currents flowing through them. Due to phase displacement between the currents of two windings, rotating magnetic field and an electromagnetic torque acting on an aluminium drum C, which is mounted on the spindle, are produced. The electromagnetic torque acting on the aluminium drum varies with the terminal voltage of alternator. If resistances U and U' in series with the windings are increased or decreased, the electromagnetic torque produced will decrease or increase respectively. Hence variable resistor U is a mean by which the automatic voltage regulator can be set to operate at different voltages.

(b) Mechanical Control Torque. Mechanical control torque is provided by the combination of two springs (a main spring and an auxiliary spring) which are so arranged that the mechanical torque provided by them is practically independent of the position of the rotor of control system. The mechanical torque under steady deflected state, is equal and in opposition to the electrical torque. The mechanical torque can be adjusted to a limited extent by a screw (which varies the torque due to an auxiliary spring).

(c) Operating System. It consists of a regulating resistance connected in series with the field circuit of the exciter. The regulating resistance consists of a pair of resistance elements, which are connected to the contact blocks on the inside surface of which roll contact segments. The good contact between the segments and resistance sectors is made by making use of springs which press the segments outside. The two segments roll on the inside surface of the resistance sectors because of torque acting upon the aluminium drum, which itself rotates clockwise or counter-clockwise when the terminal voltage of generator changes.

(d) Damping Torque. Damping system consisting of an aluminium disc rotating between two permanent magnets M is provided in order to prevent the oscillations of the moving system. The aluminium disc is geared to the rack P and is fastened to the aluminium drum by means of a flexible spring S acting as the *recall spring*. With the variations in the alternator voltage, the eddy currents induced in the disc produce the necessary damping torque, which resists the quick response of the moving system.

Operation. Suppose that the terminal voltage of the alternator is normal at position 3, for which the voltage regulator has been set by adjusting the resistances U and U′. At this position, the electromagnetic torque is exactly balanced by the mechanical torque provided by the springs. Hence moving system is in equilibrium.

Now let us consider that the terminal voltage of the alternator rises due to decrease in load. Due to increase in terminal voltage, the electromagnetic torque will increase and therefore, being greater than the mechanical torque due to springs, will cause rotation of moving system in a clockwise direction. When this happens, the segments move out of the resistance sectors and some resistance is inserted in exciter field rheostat, thereby decreasing the field current and hence less terminal voltage on the alternator.

When the generator voltage attains its normal value, the electromagnetic torque attains its original value and system becomes stable at this new position as the mechanical torque due to springs is independent of the position of the moving system.

The reverse operations take place when the terminal voltage of the alternator decreases due to increase in load.

The generator voltage will decrease or increase slowly because of inertia of the exciter and the generator field. The regulator is made quick responding by employing "overshooting the mark principle", *i.e.*, if the new position for equilibrium is 4, the moving system will first overshoot up to, say, position 5 and then return to position 4, which is a true position of equilibrium for new load, under the influence of mechanical torque provided by the tightening of recall spring.

Excitation control method used to be employed in early times. This method worked well in small isolated systems where there was no local load at the sending end. Also there are limits for the excitation as well. Excitation below a certain level may cause unstability of the system and excitation above certain level will cause overheating of the rotor. Thus the amount of regulation by this method is limited by the permissible voltage rise at the sending end and by the difficulty of designing efficient generating plant when the excitation range is too wide.

19.4.2. Static Voltage Regulators

Static voltage regulators are those which have no moving parts, the circuits comprising of static apparatus like transformers, capacitors, resistors, transistors, magnetic amplifiers etc.

Electromechanical voltage regulators described in foregoing section have the drawbacks of wear and friction and need fairly regular attention. Because of friction not only sensitivity is reduced, but also, a dead zone* of between 0.5 and 1% is caused. The contacts of the vibrating type AVR need regular attention because of erosion caused by the sparking. The rate of response of both types is limited by the mass of the moving parts. Even when providing full corrective action they cannot force the main exciter voltage to go negative—in order to limit alternator overvoltage on the occurrence of sudden loss of load on the system (an important consideration in case of hydroelectric generators)—as can an AVR using magnetic or rotating amplifiers. In brief, we can conclude that static voltage regulators are superior to electromechanical regulators in respect of accuracy of control, response, reliability and maintenance.

The important types of static voltage regulators are described below:

1. Servo Type Voltage Regulators. The main features of a servo type voltage regulator using an amplidyne is depicted in Fig. 19.8. The system comprises of a constant voltage main exciter driven from the alternator shaft and an auxiliary exciter whose field is controlled by amplidyne. Both the auxiliary exciter and amplidyne are driven by a dc motor coupled to both the machines. The dc motor is supplied from the main exciter. The main exciter has a saturated magnetic circuit and hence has a roughly constant output voltage. The armatures of main and auxiliary exciters are connected in series and this series combination excites the field winding of the alternator. Thus, the auxiliary exciter controls the excitation current of the alternator by a buck-boost action.

The potential transformer (PT) provides a signal, proportional to the alternator output voltage, to a voltage sensitive bridge one arm of which consists of saturated diode. Any deviation in the output voltage of the alternator produces a corresponding change in the output of an electronic amplifier. The amplifier output feeds the amplidyne control field. The amplidyne output alters the auxiliary exciter field. Thus, the auxiliary and the main exciter in series adjust the excitation current of the alternator, according to the load needs, to maintain its output voltage constant within the specified limits. The several time lags in the circuit needs negative feedback for stabilization of the system response. Feedback is also given from the amplifier circuits to the amplidyne as well as the auxiliary exciter.

2. Magnetic Amplifier Regulator.

Magnetic Amplifier. In a magnetic amplifier, the basic component is a steel-cored coil with an additional winding energized by direct current. It is intended for controlling alternating current of relatively large power by means of a low power dc.

The simplest practical magnetic amplifier has two identical steel cores with identical ac windings, called the *load winding,* on each. In addition they have a common dc winding called the *control winding.* The ac windings are connected together in series or in parallel, and in series with them there is a load. The series connection is used when a short-time response and a

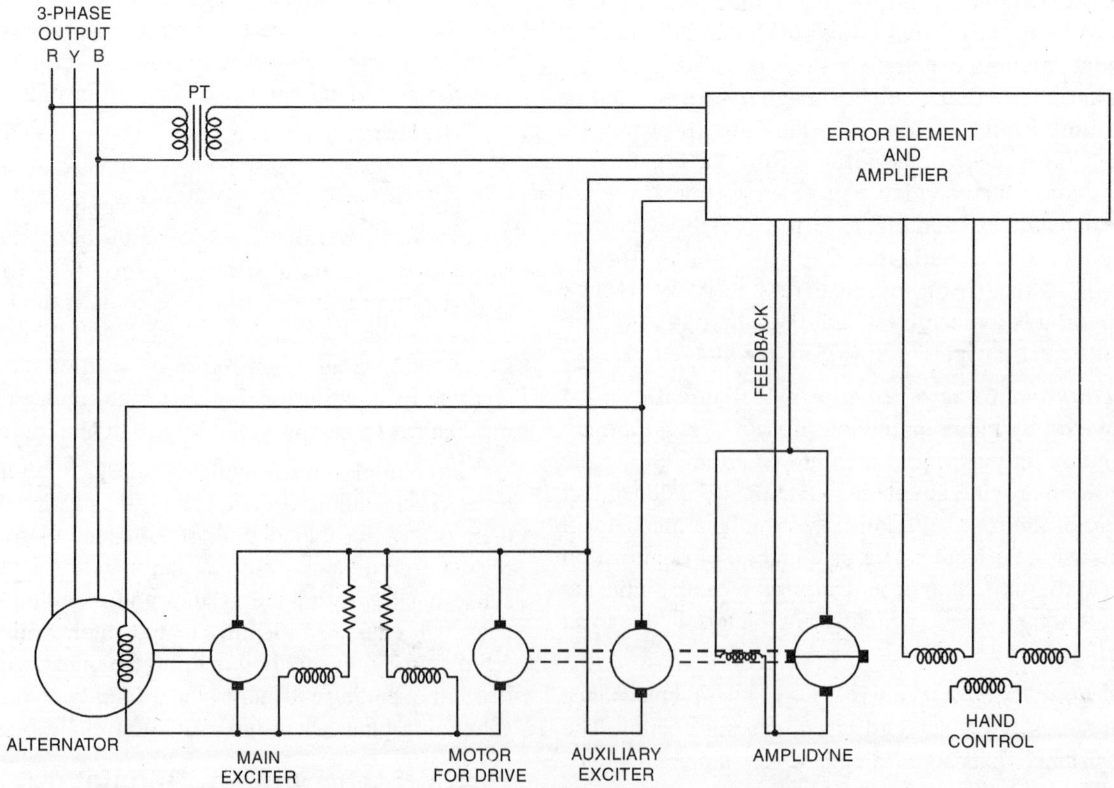

Fig. 19.8. ***Servo Type Voltage Regulator Using Amplidyne***

* In dead zone, no regulator response is expected.

high voltage are required, usually in controlling loads requiring low power. The parallel connection, on the other hand, has a slow response of approximate 1 to 3 seconds, but permits control of large current. The control winding is energized by dc.

With no current in control winding, the ac winding has a large inductance, and a high reactance is offered to an ac

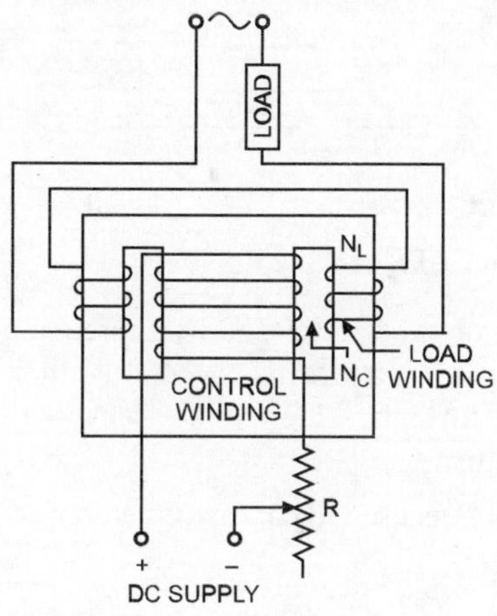

Fig. 19.9. *Simplest Magnetic Amplifier Circuit*

potential. As a result alternating current to the load is limited to a low value by the inductive reactance, and the load voltage is small. When a dc voltage is now applied to control winding the dc magnetic flux passes through the core and the core approaches magnetic saturation, even when a small value of dc voltage is applied. This causes the effective inductance of the ac winding to decrease, so that the ac winding impedance drops. Consequently the alternating current flowing through the ac winding (and the load voltage) increases as the flow of direct current through the control winding increases. Thus, varying the magnitude of control current over a small range causes the load voltage to vary over a wide range.

Magnetic Amplifier (or Transconductor) Voltage Regulator. Figure 19.10 depicts an excitation circuit using magnetic amplifier for regulating the alternator voltage. It is a simplified circuit that makes use of a magnetic amplifier (transconductor) for voltage control of an alternator. The field winding of the shunt wound exciter is connected in series with the resistor R and in parallel with this resistor is connected a single-phase full-wave rectifier I. Single-phase full wave rectifier I is supplied from the alternator through the transconductor ac winding. The two main dc windings of the transconductors arranged to provide opposite magnetic effects are supplied from the alternator through the rectifiers II and III. For a particular value of the output voltage of the alternator, the ampere-turns of the reference and feedback control windings are equal and opposite and, therefore, their resultant is zero. Stabilizing circuit transformer A connected across the exciter to feed an additional control winding on the transconductor is

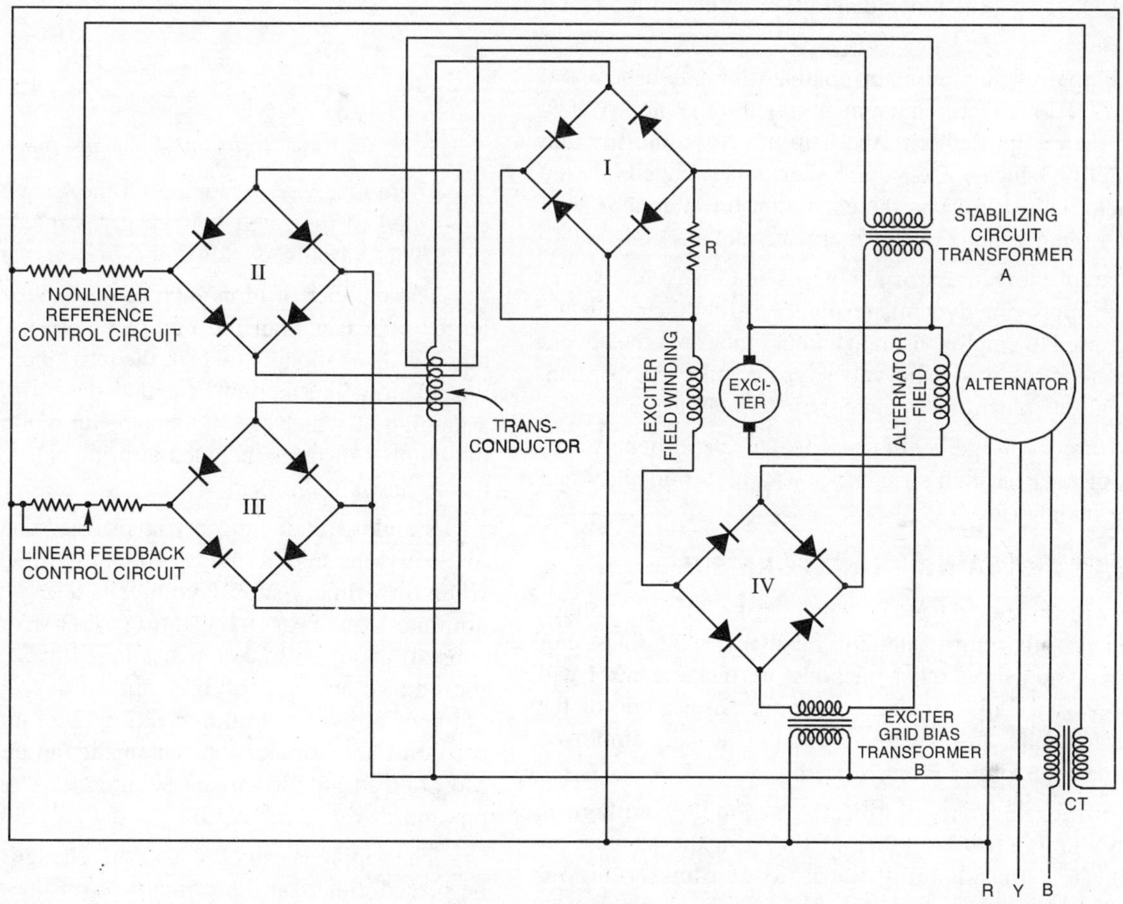

Fig. 19.10. *Magnetic Amplifier (Transconductor) Voltage Regulator*

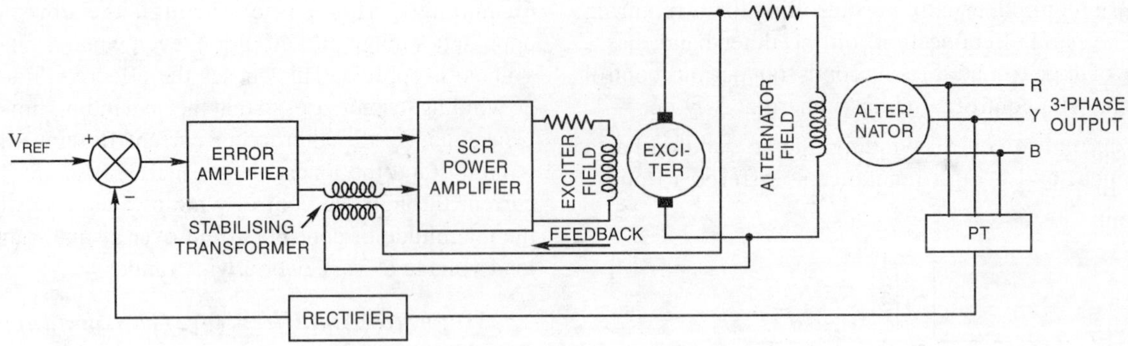

Fig. 19.11. ***Block Diagram of Solid State AVR***

provided so as to prevent hunting. Exciter reversal is avoided by means of a small transformer B supplied from the alternator output which feeds the exciter field to assist self excitation.

Assume that the terminal voltage of the alternator exceeds the preset value, may be due to decreased load. An increased voltage will appear across bridge rectifier I resulting in increased voltage drop across resistor R in opposition to that fed by the exciter armature. This reduces the field current of the exciter which in turn reduces the alternator excitation to reduce the output voltage of the alternator. Thus, the alternator output voltage is restored to the preset value.

3. Electronic or Solid State Automatic Voltage Regulator

Figure 19.11 depicts the block diagram of an electronic (solid state) AVR.

The output of the alternator obtained through a potential transformer (PT) is rectified, compared with reference voltage V_{REF} and then amplified with the help of error amplifier and power amplifier stages. The amplified error is supplied into the exciter field which provides the excitation for alternator such that the terminal voltage of the alternator remains constant.

The stabilising transformer gives feedback of exciter voltage for improving dynamic response. Earlier the electronic regulators used to employ vacuum and gas tubes but the advent of semiconductors made the valve type regulators obsolete. The present day electronic regulators invariably use ICs.

Electronic voltage regulators provides the benefits of accuracy of control, high speed response, high reliability and minimum maintenance.

19.5. TAP CHANGING TRANSFORMERS

As already mentioned, the excitation control method of voltage regulation is satisfactory only for relatively short lines. For long transmission lines, other methods of voltage control will have to be used. Tap changing transformer use is one of the important methods of voltage control and is usually employed where main transformer is necessary.

The principle of regulating the secondary voltage is based on changing the number of turns on the primary or secondary *i.e.,* on changing the ratio of transformation. Decrease in primary turns causes increase in emf per turn, and so in secondary output voltage. Secondary output voltage can also be increased by increasing secondary turns and keeping primary turns fixed. In other words, decrease in primary turns has the same effect as that of increase in secondary turns.

The tappings used in a transformer are shown in Fig. 19.12.

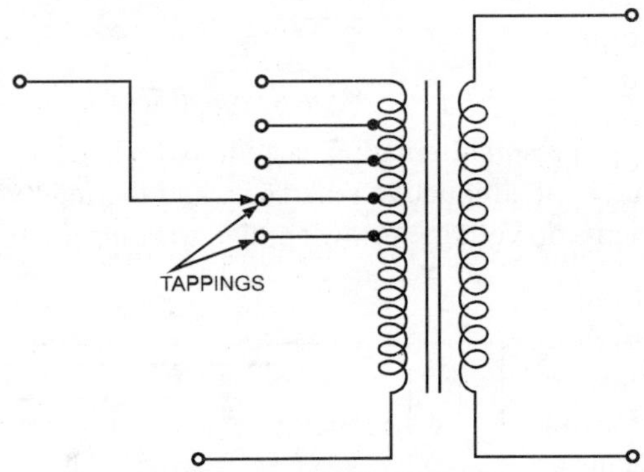

Fig. 19.12. ***Transformer Tappings***

There is always a tapping on the hv winding which when connected to the rated voltage gives rated voltage on the lv side. This tapping is called the *principal tapping.*

The number of turns included at other taps may be either larger or lesser than the number of turns used at principal tapping. The tapping at which the turns included are more than that of principal tapping is called the *positive tapping*, while a tapping at which the number of turns included is less than the number of turns included at principal tapping is called the *negative tapping*.

Standard distribution transformers have taps arranged in 2½% steps so that rated secondary voltage can be obtained when the primary supply voltage is 0, 2½, 5, 7½% below the nominal primary voltage rating. Transformers used in unit substations often have taps arranged in 2½% steps and may include some overvoltage taps to accommodate primary voltages above the nominal rating. Taps in 2¼% steps may be used on transformers with automatic tap changing equipment. Most of the transformers carry rated kVA on any overvoltage tap and/or rated current on a reduced voltage tap.

The tappings which cause the change in turn-ratio, may be placed either on the primary or on the secondary side. In transformers at the generating stations, tappings are provided on the secondary side while in a transformer installed at the end of a transmission line, the tappings are provided on the primary side.

Tappings are usually provided on the high voltage windings due to the following reasons :

1. The number of the turns on the hv winding is large, and therefore, a fine and smoother control of the output voltage is obtained.
2. It is not possible to tap other than an integral number of turns, and this may cause difficulty with *lv* windings. For example 260 V phase winding with 13 V per turn cannot be taped closer than 5%.
3. It is difficult to tap *lv* winding, since it is placed next to the core, due to insulation consideration. On the other hand the hv winding, which is placed outside the *lv* winding, is easily accessible and can, therefore, be tapped easily.
4. Tap changers on hv side will have to handle low currents though it will need more insulation.

   There is an additional advantage of providing tappings on hv winding in case of step-down transformers. On light load, for decreasing the voltage on the secondary side, the number of turns included on the hv side is increased resulting in reduced emf per turn and so in the flux density and the core loss.

The tappings may be changed with the transformer isolated from the supply on both sides. This is called *off-load tap changing*. This arrangement is used for occasional adjustments, as in distribution transformers which are provided with ± 5% and ± 2% taps. In some transformers special arrangements are made to avoid interruption of supply so that tap changing is affected with the transformer energized or on load. This is called the *on-load tap changing*. Daily and short-time adjustments are made with the help of *on-load tap changing* gear.

19.5.1. Off-Load Tap Changing

This is the cheapest method of changing the turn-ratio of a transformer. As the name indicates, it needs disconnection of transformer from the supply before changing the tap. This adjustment is made by tapping the respective windings as required and taking connections of tappings to some position near the top of the transformer. Tap changing is affected manually through hand holes provided in the core. Alternatively the reconnection can be made by carrying the tapping leads through the cover for changing either by hand or by manually operated switches. The commonly used switches are (*i*) vertical tapping switches and (*ii*) face-plate switches. *Vertical tapping switch* comprises an insulated bar, one for each phase, carrying knife contacts, to be raised or lowered by handwheel and gearing to make connections to the tapping contacts. In a *face-plate switch* the spindles of movable contact arms are geared to a shaft with a handwheel external to the tank. Such switches may be mounted on the upper yoke, or located nearer to the tapped positions on the windings.

A simple off-load tap clanger is illustrated in Fig. 19.13. It has six brass or copper studs. The winding is also tapped at six points (equal to number of studs). The tapping leads are connected to six corresponding marked stationary studs arranged in a circle. The face plate carrying the six studs can be mounted at a convenient place on the transformer such as upper yoke or located near the tapped position on the windings. The movable contact arm may be rotated by handwheel mounted externally on the tank.

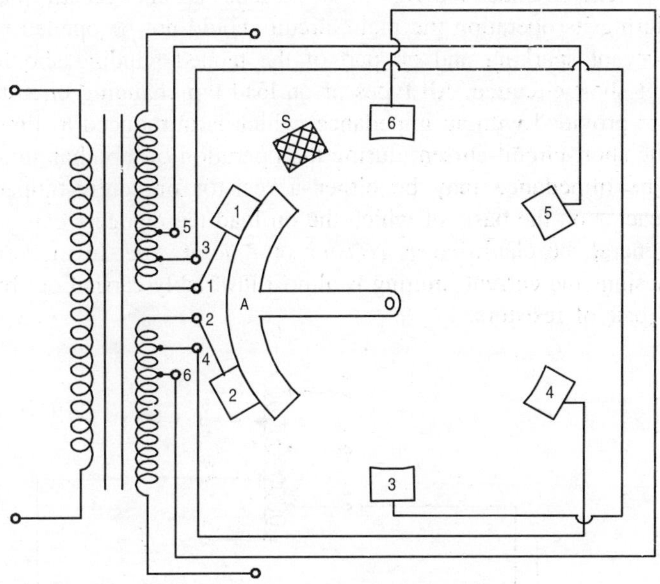

Fig. 19.13. ***Off-Load Tap Changer***

If the winding is tapped at 2½% interval, then the movement of the rotable arm A over studs 1,2; 2,3 ... 4,5; 5, 6 the winding in circuit reduces progressively from 100% at studs 1,2 to 90% at studs 5, 6. The stop S fixes the final position and prevents the arm A from being rotated beyond stud 6 in counter-clockwise direction or from stud 1 in clockwise direction so as to avoid bridging of studs 1 and 6 by the arm A. The tap changing is affected only after the transformer is disconnected from the supply. In Fig. 19.13 the arm A is bridging the studs 1 and 2 and the whole winding is in series. Now for bringing arm A at the studs 2 and 3, the transformer is first disconnected and then the arm A is rotated to bridge the studs 2 and 3. After this the transformer is switched on to the supply. Now 97.5% of the winding is in the circuit.

A mechanical lock is provided to prevent unauthorised operation of the off-load tap clanger. Further, for preventing inadvertent operation, an electromagnetic latching device or microswitch is provided to operate the circuit breaker so as to disconnect the transformer from the supply as soon as the movable arm is moved; well before the contact of the arm with the stud, with which it was in contact, opens.

19.5.2. On-Load Tap Changing

Such an arrangement is employed for changing the turn-ratio of the transformer to regulate the system voltage while the transformer is delivering load. Nowadays almost all the large power transformers are provided with on-load tap clanger. It is invariable practice to provide the tappings at the neutral end of hv windings of a generator transformer. The tappings on the windings are brought out through a terminal board to a separate oil-filled compartment in which the on-load tap changer switch is housed. The tap changer is in the form of a selector switch. The changer is operated by a motor operated driving mechanism by local or remote control and a handle is also fitted for manual operation in case of an emergency.

The essential features of an on-load tap changer are that during its operation the main circuit should not be opened to prevent sparking and no part of the tapped winding should get short-circuited. All types of on-load tap changing circuits are provided with an impedance, which is introduced to limit the short-circuit current during the operation of tap changing. The impedance may be either a resistor or centre-tapped reactor on the basis of which the on-load tap clangers can, in general, be classified as *resistor* or *reactor type*. In modern designs the current limiting is almost invariably carried out by a pair of resistors.

On-load tap changing gear with resistor transition, in which one winding tap is changed over for each operating position, is illustrated in Fig. 19.14. The sequence of operations during the transition from one tap to the next (adjoining) (in this case from tap 2 to tap 3) is shown. Backup main contactors are provided which short-circuit the resistors for normal operation.

In order to limit the loss of energy, it is essential that the resistors are in the circuit for as minimum of time durations as possible. The resistors are designed for short-time rating for economical consideration and therefore, it is desirable to reduce to minimum their duty time duration. So it becomes essential to incorporate some form of energy storage in the driving mechanism to ensure that a tap change, once initiated, shall be completed irrespective of failure of auxiliary control supply. All modern on-load tap changers use spring energy as storage elements. They reduce the time that a resistor is in the circuit

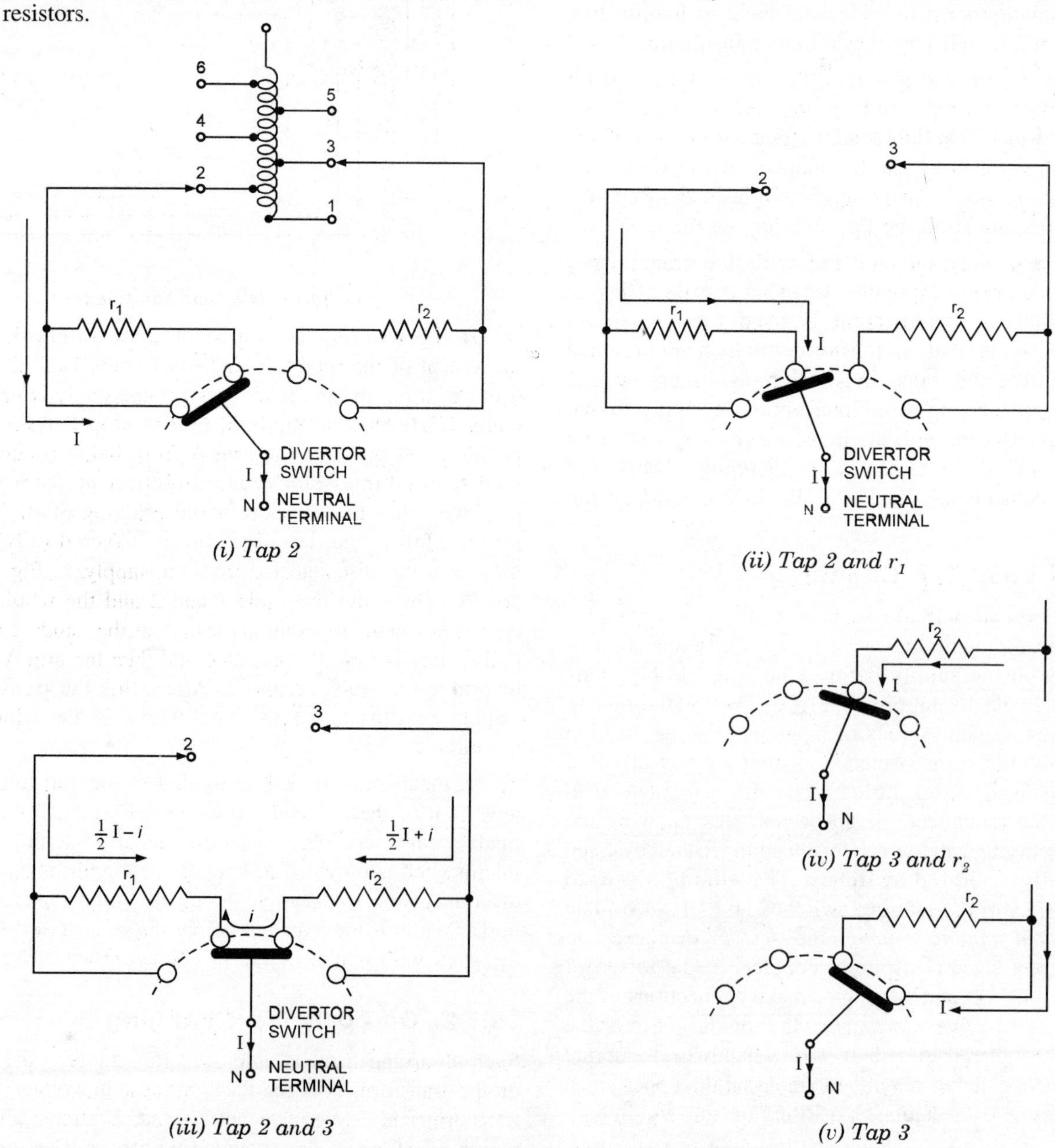

Fig. 19.14. ***Switching Sequence For On-Load Tap Changing***

to a few seconds. The current break is eased by the fact that the short-circuit resistor current has unity power factor.

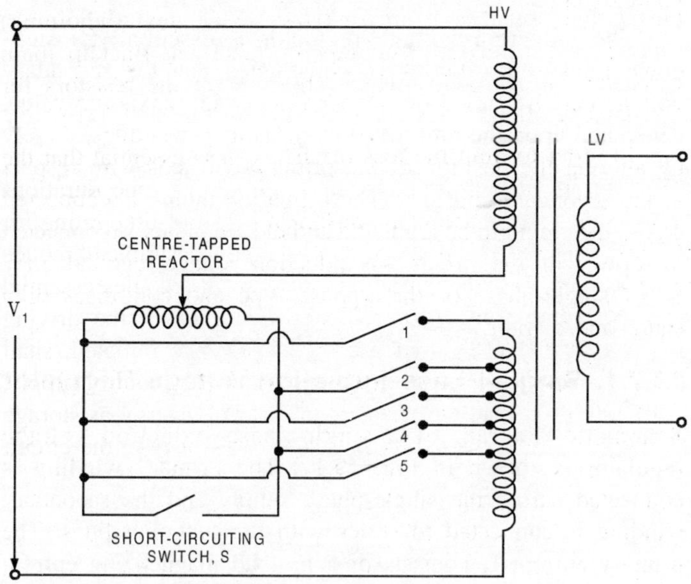

Fig. 19.15. ***On-Load Tap Changer***

On-load tap changer control gear may be from simple push-button initiation to complex automatic control of several transformers operating in parallel. The aim of control is to maintain a given voltage level within a specified tolerance, or to raise it with load to compensate for the voltage drop in the transmission line. The main components are an automatic voltage regulator, a time-delay relay, and compounding elements. The time-delay prevents initiation of a tap change due to minor transient voltage fluctuation. It may be set for a delay up to 60 seconds.

Another form of on-load tap changer, provided with a centre-tapped reactor is illustrated in Fig. 19.15. The function of the reactor is again to prevent the short-circuiting of the tapped windings. During normal operation the short-circuiting switch, S remains closed. The reactor used is suitably designed to prevent large values of current flowing in any section of the primary winding when two tapping switches are closed simultaneously. For understanding applications of this type of on-load tap changer assume that the tapping switch 1 is closed and the output voltage is minimum. For raising the output voltage, the short-circuiting switch S is opened, the second tapping switch is closed, the first tapping switch is opened and finally the short-circuiting switch S is closed. When the short-circuiting switch S is in open position and the two tapping switches are in closed position, a part of the transformer winding between the two tappings is shunted by the reactor, but large circulating current is not established on account of its high reactance. The line current remains unaffected by this reactance, as the current equally divides and flows in opposite directions in the two halves of the reactor. When only one tapping switch is closed the reactor carries the full primary current. Instead of switches sliding contacts may be used. Sliding contacts are so attached to the ends of reactor that one makes before the other breaks and in normal operating conditions both contacts touch the same tapping stud. If the reactor is designed to carry the full-load current continuously then a working position is possible with two consecutive tapping switches closed. This provides twice as many turn ratios as there are taps, minus one. Usually, the tappings are located midway between the end turns of the winding in order to prevent the surge voltages from getting into the load-ratio control elements.

19.6. BOOSTER TRANSFORMER

In some cases it is not convenient to have tap changing gear in the main transformer, and sometimes it may be required to

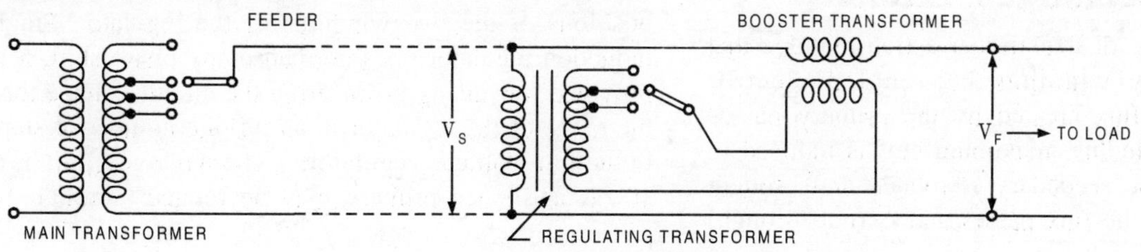

(a) Single-Phase Voltage Booster

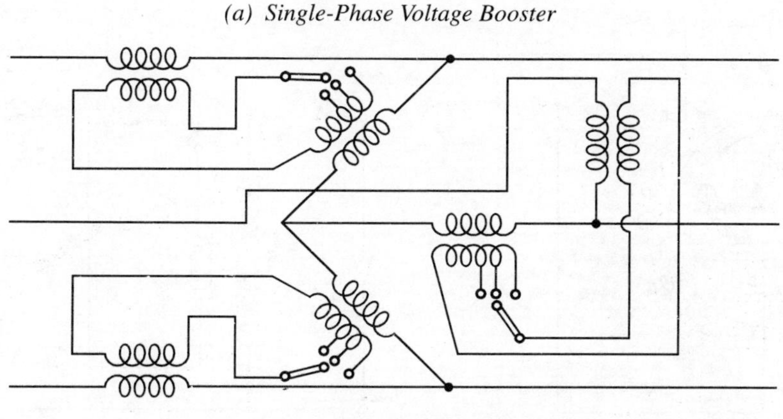

(b) Three-Phase Voltage Booster

Fig. 19.16

control the voltage of a feeder at a point far away from the main transformer. Booster transformer may be employed on such occasions. The secondary of the booster transformer is connected in series with the line and booster transformer primary is supplied from the secondary of the regulating transformer fitted with on load tap changing gear similar to that mentioned in Art. 19.5.2. The output (secondary) winding of the regulating transformer is so connected to the primary of the booster transformer that the voltage injected in the line, V_B is in phase with supply voltage V_S, as shown in the Fig. 19.16. By changing the tapping on the regulating transformer, magnitude of V_B can be changed and thus feeder voltage V_F can be regulated.

The rating of such regulating transformer is only the fraction (equal to a percentage boost) of that of the main transformer. It is given by the expressions

$$\frac{\text{Line current} \times \text{boost voltage}}{1{,}000} \text{ kVA for single-phase boosters}$$

and $$\frac{\sqrt{3} \times \text{Line current} \times \text{boost voltage}}{1{,}000} \text{ kVA for 3-phase voltage boosters.}$$

or kVA rating of booster

$$= \text{Load kVA} \times \text{percentage boost in voltage.}$$

The advantages of the above system are that the regulating equipment is independent of the main transformer so that a failure in the former will not throw the latter out of service for any length of time; and that it is much cheaper method when there is no main transformer at the point where regulation is desired. When the regulation is required at a point where a main transformer is to be placed, this system is costlier, requires more floor space, and increases the losses.

19.7. INDUCTION VOLTAGE REGULATORS

The fundamental principle of transformer action requires that primary and secondary windings be coupled electromagnetically. When the flux created by the primary passes through the secondary winding, maximum emf is induced in the latter. However, if the secondary is rotated and brought into such a position that the flux passes across (not through) it, no emf will be induced in it. The idea of rotating the primary with respect to secondary is applied in the induction voltage regulator.

The induction voltage regulator is essentially a step-down transformer, but is so constructed that the secondary voltage may be varied from zero to a certain maximum value, dependent upon the ratio of turns in the two windings, merely by altering the position of the primary-coil axis with respect to the secondary-coil axis. The primary winding is connected across the circuit to be regulated and the secondary is connected in series with the circuit. An induction voltage regulator may be of single-phase or three-phase type and comprises of a stator and a rotor.

19.7.1. Single-Phase Induction Voltage Regulator

Schematic diagram of a single-phase induction voltage regulator is shown in Fig. 19.17. The primary winding is connected across the single-phase supply and the secondary winding is connected in series with the outgoing lines. The primary current I_1 consists of a no-load magnetizing current together with a load component, the magnitude of which is determined by the main circuit current I_2 and the ratio of primary to secondary turns on the induction regulator. The phase of current I_2 with respect to output voltage V_2 depends, of course, on the nature of load connected on the output side.

The flux created is alternating and when the axis of the two windings coincide, the flux set up by the primary winding links with the whole of the secondary winding and maximum voltage is induced in the secondary. As the rotor is rotated through 90º, none of the flux set up by the primary links with the secondary and hence no emf is induced in secondary. And if rotor is further rotated, the direction of induced emf changes *i e.,* becomes negative. Thus the regulator either adds or subtracts from the circuit voltage depending upon the relative positions of the two windings of the regulator. Single-phase induction regulator does not cause any phase shift. Means are provided for raising or lowering the circuit voltage throughout the range of the regulator in an infinite number of steps in the induction voltage regulator. The voltage ratio is usually 10 : 1, and since primary may be rotated through 180º, the

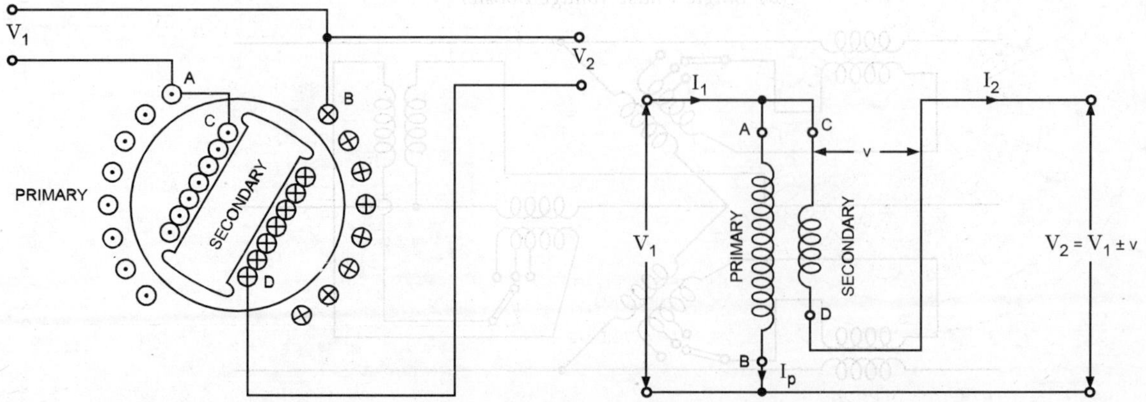

Fig. 19.17. ***Schematic Diagram of a Single-Phase Induction Regulator***

load voltage may be as much as 10% above oı below that across the primary. To provide a voltage regulation of 20% (or ± 10%) the induction regulator of rating 10% of the kVA rating of the circuit is required.

The actual single-phase induction regulator slightly differs from that shown in Fig. 19.17. The primary winding is placed in slots in the surface of the laminated cylindrical core since it has to carry smaller current and has small conductor area. No slip rings are required to feed the primary winding, since the rotor movement is restricted from 0° to 180°. Flexible leads are sufficient to feed the primary winding. The rotor also carries, in addition to the primary winding, a short-circuited compensating winding (also called the tertiary winding), whose magnetic axis is always 90° away from that of the primary winding, to neutralize the detrimental series reactance effect of the secondary. The rotor structure is mounted on a shaft so that it can be rotated on its axis and thereby change the relative position between the primary and secondary windings. Since the rotor is not required to rotate like that of a motor, the shaft is generally arranged vertically for convenience. The secondary winding connected in series with the outgoing lines is housed in the stator slots, due to larger conductor area. Both of the primary and secondary windings are usually designed to give 2-poles or 4-poles because the smaller the number of poles, the larger the mechanical angle corresponding to one pole-pitch and the easier the voltage adjustment. The spreadover of both windings is about 120°. Figure 19.18 shows diagrammatically the cross section of a bipolar single-phase induction regulator, and its connections.

If the regulator were not provided with the short-circuited compensating winding, its operation would not be satisfactory (voltage regulation would be very poor with adjustments for intermediate voltages).

19.7.2. Three-Phase Induction Voltage Regulator

The three-phase induction voltage regulators are somewhat similar to single-phase induction regulators in appearance and constructional features. They must, of course, have three primary and three secondary windings which must be spaced 120 degrees apart. Three primary windings are placed in slots of a laminated rotor core and connected across the 3-ϕ ac supply because of the few turns per phase as compared to those required in delta connection. Three secondary windings are placed in slots of a laminated stator core and are connected in series with the load as shown in Fig. 19.19. No compensating or tertiary winding is required in case of a 3-phase induction voltage regulator because, in this case, for every position of setting of the regulator each secondary winding is magnetically coupled to one or more of the primary windings. No slip rings are required and the primary windings are energized through flexible leads only as in case of single-phase induction voltage regulators. Since in this case rotating magnetic field of constant magnitude is produced, so induced voltages in the secondary are of constant magnitude but of varying phase, which changes with the change in position of the rotor with respect to stator. Rotor movement is obtained as in the case of a single-phase induction regulator.

The voltage phasor diagram for any particular phase is shown in Fig. 19.19 (*b*) where V_1 is the supply voltage, V_r is the induced voltage in the secondary and V_2 is the output voltage, each per phase. The output voltage V_2 is obtained as the phasor sum of V_1 and V_r for any rotor displacement angle θ. The locus of V_2 is consequently a circle drawn with centre on the extremity of V_1 and of radius V_r. The maximum output voltage $V_2 = V_1 + V_r$ is obtained when V_r is in phase with V_1 and minimum output voltage $V_1 = V_1 - V_r$ is obtained when V_r is in phase opposition to V_1. Thus output voltage varies from $(V_1 - V_r)$ to $(V_1 + V_r)$. For intermediate values of output voltage V_2, there is a phase displacement between V_1 and V_2. It may be seen from phasor diagram shown in Fig. 19.19 that

$$V_2 = \sqrt{(V_1 + V_r \cos\theta)^2 + (V_r \sin\theta)^2}$$

and the phase displacement angle θ is given as

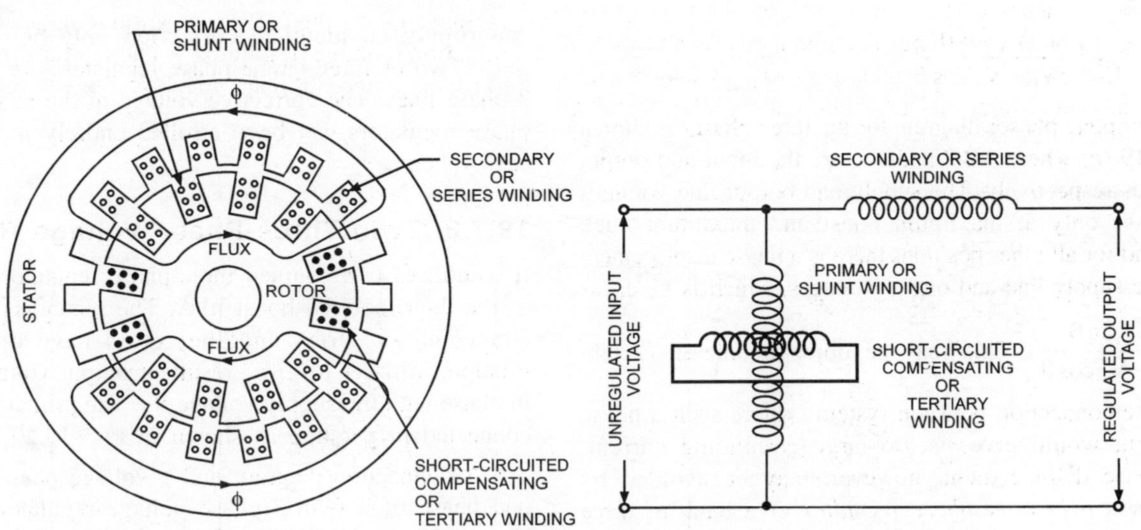

(*a*) *Cross Section* (*b*) *Connection Diagram*

Fig. 19.18. ***Single-Phase Induction Regulator***

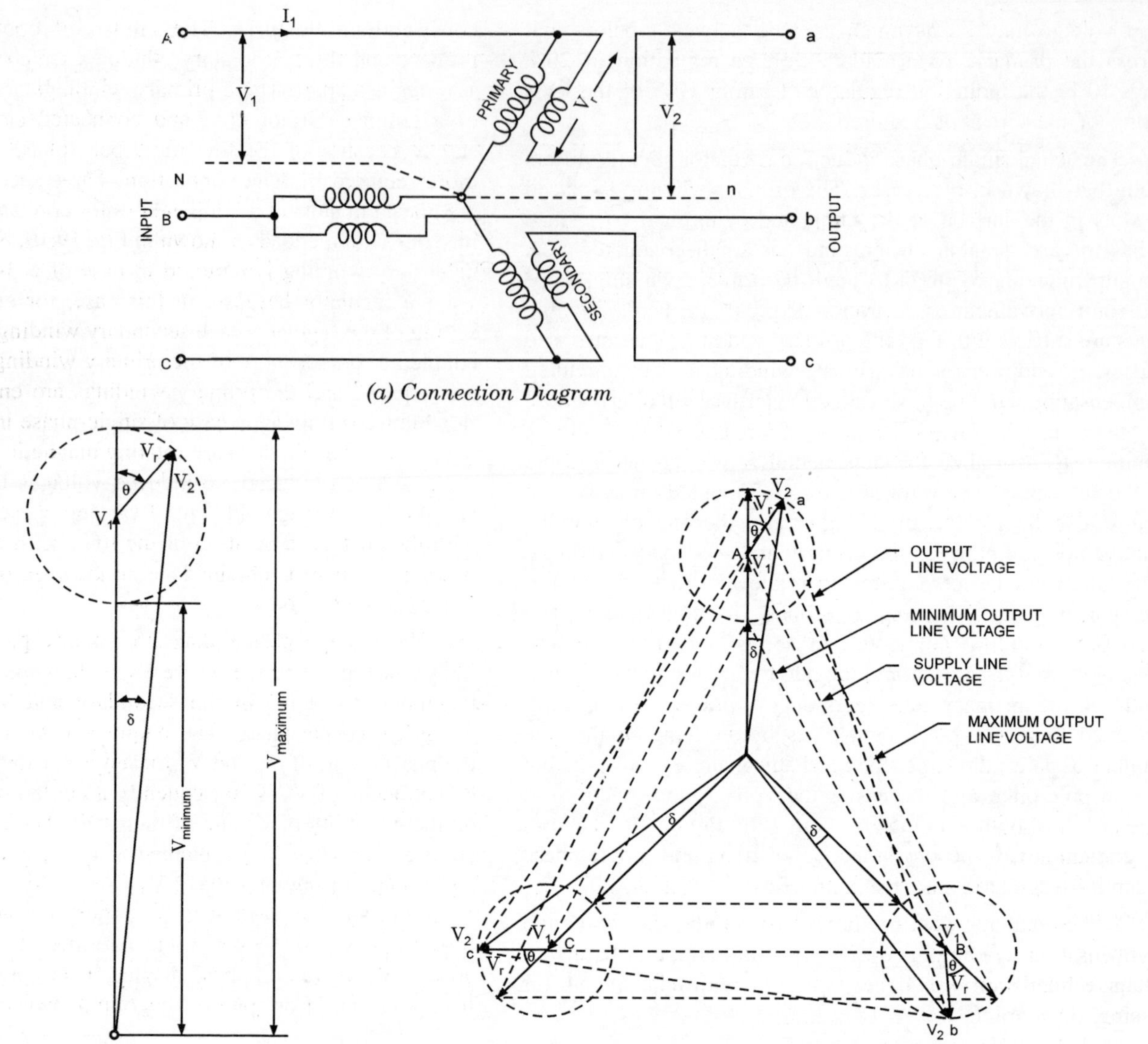

(a) Connection Diagram

(b) Voltage Phasor Diagram For One Phase

(c) Complete Phasor Diagram

Fig. 19.19. ***Three Phase Induction Voltage Regulator***

$$= \text{Tan}^{-1} \frac{V_r \sin \theta}{V_1 + V_r \cos \theta}$$

The complete phasor diagram for the three phases is shown in Fig. 19.19 (*c*) where ABC and abc are the input and output line voltages respectively. The supply and output line voltages are in phase only at maximum boost and maximum buck positions and for all other positions there is a phase displacement between the supply line and output voltages, which is given as $\tan^{-1} \frac{V_r \sin\theta}{V_1 + V_r \cos\theta}$. This may be objectionable in certain cases of interconnection between systems where such a phase displacement would give rise to large circulating current. Such a phase displacement, however, may be avoided by using either a *twin three-phase regulator* or a bank of three single-phase induction regulators.

The 3-phase induction regulator is unable to balance the line voltages when supplying unbalanced loads because it can supply an identical correction only between any two lines. Two or three single-phase regulators are often used on 3-phase lines. The corrective voltage of the respective single-phase regulators can be controlled entirely independently of each other.

19.7.3. Twin Three-Phase Voltage Regulator

It comprises two coupled three-phase regulators each of half of the total required boost kVA. The secondary windings are connected in series and the rotors have opposite phase rotation, in order to give resultant output voltage V_2 exactly in phase with the supply voltage V_1. The primary windings are connected in parallel, as shown in Fig. 19.20.

A connection diagram and a voltage phasor diagram for one phase of a twin 3-phase voltage regulator are given in Fig. 19.20. For the line voltages exceeding 11 kV, it is customary to use transformers to supply the regulator primary and to inject secondary induced voltage into the main circuit.

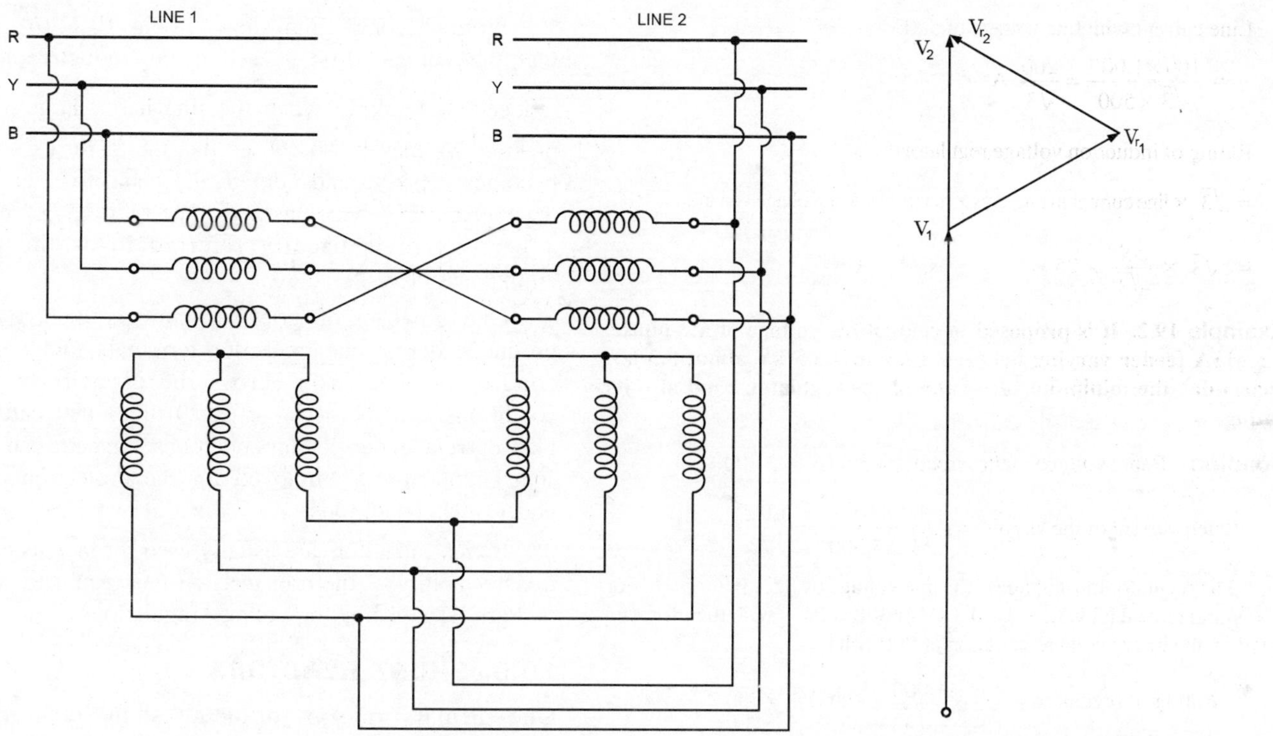

(a) Connection Diagram

(b) Voltage Phasor Diagram For One Phase

Fig. 19.20. ***Twin 3-Phase Voltage Regulator***

Induction voltage regulators like transformers are available in air-cooled or noninflammable liquid cooled designs for indoor use. Oil cooled regulators are ordinarily used outdoors.

Rating. An induction voltage regulator is rated by the voltage (either raise or lower) it provides, and the load current passing through the series winding. In other words, it is designed according to regulator capacity and not circuit capacity. In all cases, the induction regulator rating is much less than the circuit rating (10 or 15% of circuit rating).

Since an induction regulator inherently introduces into the circuit the same amount of voltage, whether boosting or bucking, select the higher of the two as the required range of regulation.

In single-phase circuits, the kVA rating of induction regulator is given by the expression

$$\frac{\text{Percentage raise or lower} \times \text{circuit rated voltage}}{100} \times \frac{\text{Maximum load in amps.}}{1{,}000}$$

In three-phase circuits, the kVA rating of induction regulator is given by the expression

$$\frac{\sqrt{3} \times \%\,\text{age raise or lower} \times \text{circuit rated voltage}}{100 \times 1{,}000} \times \text{Maximum load in amps}$$

A 3-phase induction regulator may be designed as one unit or may be formed by three single-phase induction regulators connected either in star or delta. When three single-phase induction regulators are connected in delta, the range of regulation is 50% greater than the regulation of the single unit. Single-phase regulators for ± 10% of voltage regulation will provide voltage regulation of ± 15% when connected in delta in a 3-phase circuit.

Advantages. (*i*) It is simple and rugged in construction.

(*ii*) It can withstand overloads well.

(*iii*) It is reliable in operation because of the absence of any part subject to wear.

(*iv*) Its operation is not affected by load and power factor variations.

(*v*) It provides stepless voltage variation without any arcing or short-circuiting of turns as in the case of transformer with tap-changing mechanism.

Disadvantages. The induction voltage regulator is more expensive than the transformer with tap-changing mechanism. The other advantages of tap-changing transformers over the induction voltage regulators are the small magnetizing currents, and the elimination of the shaft and bearings required by the regulator.

Applications. The induction voltage regulator finds its greatest application in distribution systems for raising or lowering the line voltage to maintain the load voltage at a constant value under all load conditions because it varies the voltage uniformly, not in steps.

Example 19.1. The voltage at the receiving end of a 3-phase feeder delivering 100 kVA varies between 475 V and 525 V. For what kVA capacity must an induction regulator be designed to maintain the voltage constant at 500 V ?

Solution : Total voltage variation

= (525 – 500) *i.e.,* 25 V or (475 – 500) *i.e.,* –25 V *i.e.,* ± 25 V

Line current with line voltage of 500 V

$= \frac{100 \times 1,000}{\sqrt{3} \times 500} = \frac{200}{\sqrt{3}}$ A

Rating of induction voltage regulator

$= \sqrt{3} \times$ line current in amperes × maximum line voltage variation $\times 10^{-3}$

$= \sqrt{3} \times \frac{200}{\sqrt{3}} \times 25 \times 10^{-3} = 5$ kVA **Ans.**

Example 19.2. It is proposed to control the voltage of a 3-phase, 3.3 MVA feeder varying between 3 kV and 3.5 kV about 3.3 kV. Determine the minimum turn-ratio of the regulator. Find also its rating.

Solution : Rated voltage of the circuit = 3.3 kV = 3,300 V

Rated current of the circuit $= \frac{3.3 \times 10^6}{\sqrt{3} \times 3,300} = \frac{1,000}{\sqrt{3}}$ A

The regulator must increase the line voltage by 3.5 – 3.3 = 0.2 kV or 200 V and reduce it by 3.3 – 3 = 0.3 kV or 300 volts. It will, therefore, be rated on the larger voltage variation of 300 volts.

$\therefore$ Rating of regulator $= \sqrt{3} \times \frac{1,000}{\sqrt{3}} \times 300$ VA = 300 kVA **Ans.**

Turn-ratio, $\frac{N_1}{N_2} = \frac{V_1}{V_r} = \frac{3,300}{300} = 11:1$ **Ans.**

19.8. SERIES CAPACITORS

In EHV and UHV transmission lines, series capacitors are connected in series with the line to reduce the effect of inductive reactance X_L between the sending end and the receiving end of the line.

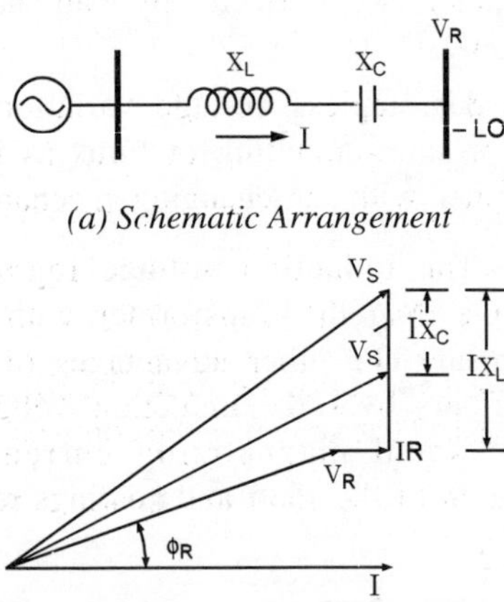

(a) Schematic Arrangement

(b) Phasor Diagram For the Line

Fig. 19.21. ***Effect of Series Capacitor in Series With the Line***

Voltage drop in the line is given as

$$\Delta V \simeq IR \cos \phi_R + IX_L \sin \phi_R \quad ...(i)$$

without series capacitor

$$\text{and } \Delta V = IR \cos \phi_R + I(X_L - X_C) \sin \phi_R \quad ...(ii)$$

with series capacitor

where I is the current flowing through the line.

Thus with series capacitors in the line, the voltage drop ΔV in the line is reduced and the receiving-end voltage V_R on load is improved.

From phasor diagram shown in Fig. 19.21 (*b*), it is obvious that the voltage drop caused by an inductive load can be reduced particularly when the line has a large $\frac{X}{R}$ ratio. In practice X_C may be so chosen that the factor $(X_L - X_C) \sin \phi_R$ becomes negative and numerically equal to R cos ϕ_R so that the voltage drop becomes zero. The ratio X_C/X_L expressed as a percentage is usually referred to as the percentage compensation.

One drawback of series capacitors is the high overvoltage produced across the capacitor terminals under short-circuit conditions. The drop across the capacitor under faulty conditions may be as large as 20 times that caused by full-load current under certain conditions. A spark gap with a high speed contactor is employed for the protection of capacitor under such conditions.

Series capacitors are usually used for increasing the power transfer ability of the transmission line [Refer to Art. 13.4(13) of Part II] and not for voltage regulation.

19.9. SHUNT REACTORS

One primary reason for using shunt reactors on EHV transmission lines is to control steady-state overvoltages when energizing the long EHV lines or when operating under light-load conditions. If the shunt reactors were not employed, the reactive power generated by the capacitance can cause high voltages at the receiving end of the line. The receiving-end voltage of a 400 kV, 1,000 km long transmission line may be as high as 800 kV. To restrict insulation stresses caused by overvoltages following sudden load rejection a substantial part of the shunt reactive compensation is usually left permanently connected.

It should be noted that with shunt reactors, the power transfer capability of the line is reduced owing to increase of transfer reactance X. Thus at fault when stability is most important, the shunt compensation reduces the maximum power limit of the line. In case line shunt reactors are switched out under heavy load conditions, the maximum power transfer capability can be considerably increased, but voltage variations due to sudden load thrown off are likely to be unacceptably high. In actual practice some of the shunt reactors are kept connected permanently so as to avoid voltage increase due to sudden fall in load from heavy load conditions.

Shunt reactors may be connected to the low-voltage tertiary winding of a transformer through a suitable circuit breaker. EHV shunt reactors may be connected to a transmission line without any EHV circuit breaker. Usually, oil-immersed magnetically shielded reactors with gaped core are employed. They appear like power transformers.

19.10. SHUNT CAPACITORS

Static shunt capacitors are installed near the load terminals, in factory substations, in the receiving substations, in switching substations etc. to provide leading volt-ampere-reactive (MVAR) and thus to reduce the line current and total kVA loading of the substation transformer. By using shunt capacitors line drop is reduced and the voltage regulation is improved.

Shunt capacitors are switched in when kVA demand on the distribution system rises and voltage of the bus drops.

19.11. SHUNT COMPENSATION—STATIC VAR SYSTEMS (SVS)

In EHV transmission practice, when the voltage at a bus falls below the reference value, capacitive reactive volt-amperes (VARs) are to be injected and when the bus voltage exceeds the reference value, inductive VARs are to be injected to lower the bus voltage. In conventional methods of shunt compensation, shunt capacitors are connected during heavy loads or low lagging power factor loads, as discussed in Art. 19.10, and shunt reactors are connected during light load periods, as discussed in Art. 19.9. Such switching operations are very slow because of the greater time (3-4 cycles) required for the operation of the circuit breakers. Moreover, circuit breakers are not suitable for frequent switching during variations of voltage. Such limitations can be overcome by using static var system (SVS). In a SVS, thyristors are employed as switching devices instead of circuit breakers. Thyristor switching is faster than mechanical switching and also it is possible to have transient-free operation by controlling the instant of switching. The advent of high-speed, high-current switching made possible by thyristors have introduced a new concept in providing reactive compensation for optimum EHV/UHV system performance.

A fast stepless variable compensation can be provided by thyristorised control of shunt reactors and capacitors. *Static VAR system* (SVS) acts within a few seconds and provides transient (fast) voltage control and improves voltage stability.

During peak-load period, the thyristors of capacitors control are made to conduct for a longer duration in each cycle while during light-load period, the thyristors in reactor circuit are made to conduct for longer duration in each cycle. Thus static compensation provides a stepless variation of shunt compensation.

Static VAR Compensators (SVCs) are those which have no rotating parts and are employed for surge impedance compensation and for compensation by sectionalizing a long transmission line. These are also employed for load compensation wherein these maintain constant voltage (*i*) under slowly varying loads, (*ii*) load rejection, outages of generator, and line (*iii*) under rapidly varying loads *e.g.*, arc furnaces, rolling mills etc. These improve system stability and system power factor. Since reactive power flow is controlled by SVCs, these are employed for minimization of transmission losses and control ac voltage near HVDC converter terminal.

A *static VAR compensator (SVC)* is basically a parallel combination of controlled reactor and fixed shunt capacitor, as shown in Fig. 19.22 and is capable of stepless adjustment of reactive power over an unlimited range (lagging and leading) without any time delay. Even though a fast response is usually desirable, but not essential for all operating conditions of the systems. For instance, fast compensation is not necessary for system stability when other factors limit the stability of the system. In actual practice, the reactive current is restricted in both lagging and leading regions because of the current carrying capacity of the compensator.

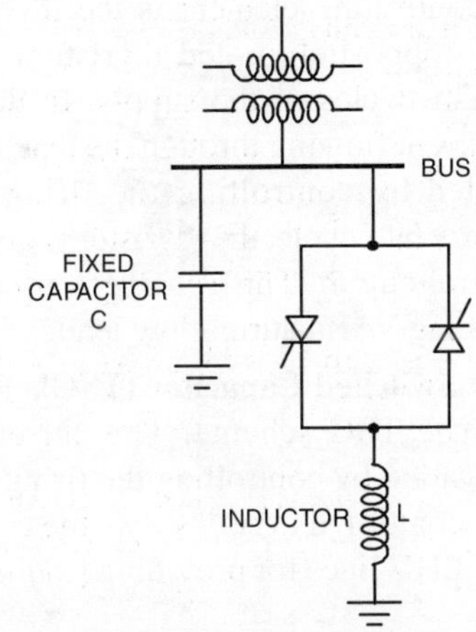

Fig. 19.22. ***Schematic of a Static VAR Compensator (SVC)***

The reactor control is accomplished by an anti-parallel thyristor switch assembly. The firing angle of the thyristors controls the voltage across the inductor and thus the current flowing through the inductor. In this way, the reactive power drawn by the inductor can be controlled. If Q_C is the reactive power generated by the fixed capacitor C and Q_L is the reactive power absorbed by the inductor, the net reactive power injection to the bus will be

$$Q = Q_C - Q_L$$

In SVR, Q_L is controllable and thus Q can be controlled. During light-load period, Q_L is made larger than Q_C while during heavy load condition, Q_L is made smaller than Q_C.

19.11.1. Static VAR Systems

Many SVS schemes are in operation. Some of commonly used schemes are as follows :

1. Thyristor controlled reactor (TCR).
2. Thyristor switched capacitor (TSC).
3. Self saturated reactor (SR).
4. Thyristor controlled reactor—Fixed capacitor (TCR-FC).
5. Thyristor switched capacitor—Thyristor controlled reactor (TSC-TCR).

1. Thyristor Controlled Reactor (TCR). Figure 19.23 (*a*) illustrates the basic TCR scheme. The controlled element is the

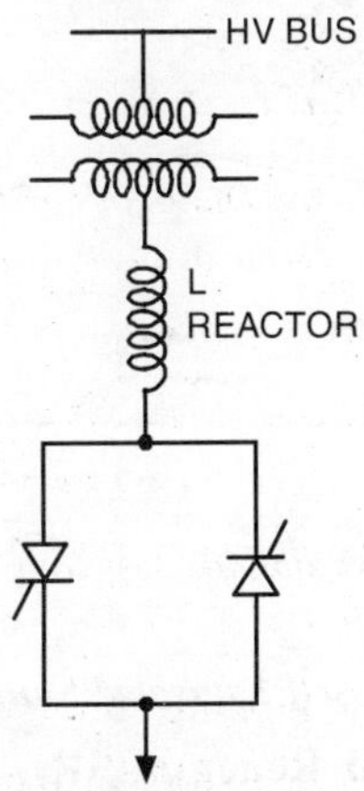

Fig. 19.23 (*a*) ***Single-Phase Thyristor Controlled Reactor (TCR)***

reactor and the controlling element is the thyristor controller consisting of two oppositely poled thyristors which conduct every alternate half cycles of the supply. In this scheme, the duration of the current flowing through the reactor during every cycle is controlled by controlling the firing angles of the thyristors. For every half cycle, the thyristor is given a triggering pulse by the control circuit. This scheme is used in EHV lines for providing lagging VARs during low loads or load rejection.

2. Thyristor Switched Capacitor (TSC). Figure 19.23 (*b*) illustrates the basic TSC scheme. The current through the capacitor can be varied by controlling the firing angles of back to back thyristors connected in series with the capacitor C. This scheme is used in EHV lines for providing leading VARs during heavy loads.

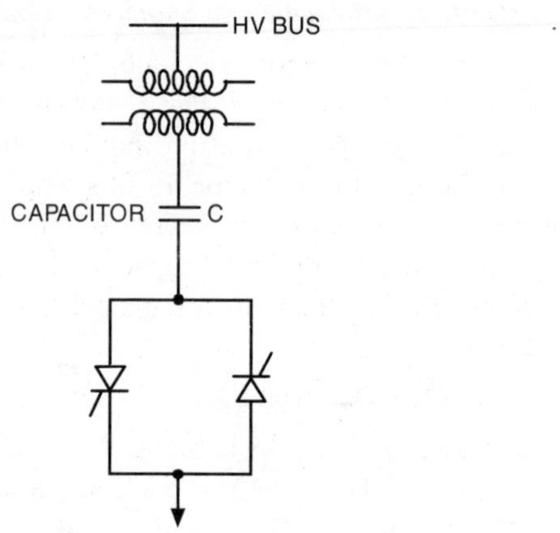

Fig. 19.23 (*b*). ***Single Phase Thyristor Switched Capacitor (TSC)***

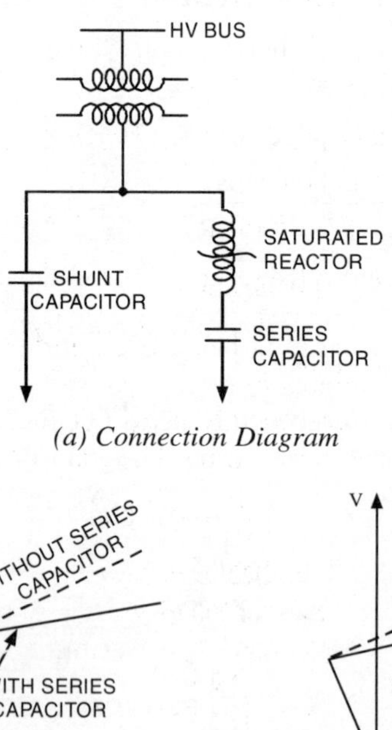

(a) Connection Diagram

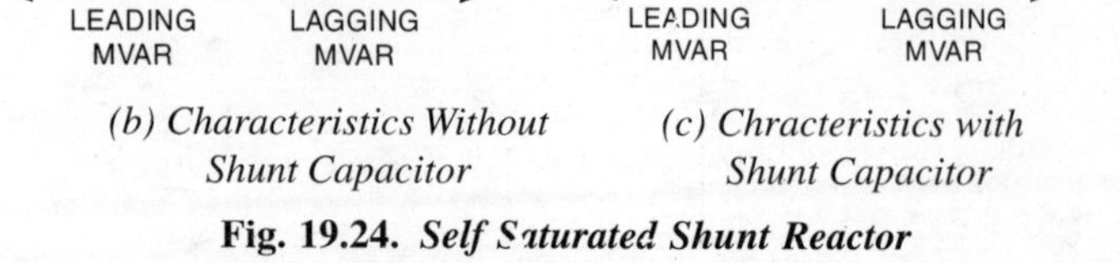

(b) Characteristics Without Shunt Capacitor *(c) Chracteristics with Shunt Capacitor*

Fig. 19.24. ***Self Saturated Shunt Reactor***

3. Self Saturated Reactor (SR). Self saturated shunt reactor is shown in Fig. 19.24 (*a*). The variable element is an iron-cored reactor operating in the saturated region and a capacitor bank is in parallel. In this scheme, loading of the reactor is automatically adjusted as per requirement of the system and no control system is required. For matching its characteristics to the needs of the system, connection of a slope correction capacitor in series may become necessary so that the slope reactance of the saturated reactor may be reduced as per needs.

The saturated reactor is self regulating device that responds only to the changes in its terminal voltage. The voltage regulation characteristics depend upon the natural saturation characteristics of the iron core reactor, the existence and rating of a capacitor in series for saturation slope correction and the existing and rating of shunt capacitor.

The saturated reactors for ratings up to 132 kV can be directly connected to bus but for higher voltages intermediate transformer is required.

The saturated reactor compensator (SR) even though is maintenance free, it has no control flexibility and it may need costly damping circuits to avoid any possibility of subharmonic instability. It has the overload capability which is useful in limiting overvoltages.

4. Thyristor Controlled Reactor-Fixed Capacitor (TCR-FC). The TCR-FC type VAR compensator is depicted in Fig. 19.25. This arrangement provides continuously controllable lagging to leading VARs through thyristor control of reactor current. Leading VARs are supplied by two or more capacitor banks. The TCR is usually rated larger than the total of fixed capacitance so that net lagging VARs can also be supplied. The current through the reactor can be varied by controlling the firing angles of back to back pairs of thyristors connected in series with reactor. Harmonics generated, because of switching operations, are injected into the system and results in several problems. So, the elimination of these harmonics becomes necessary. Since 6-pulse thyristors are to be used for 3-phase, fifth and seventh harmonics are generated in addition to third harmonics. Small reactors are usually connected in the fixed capacitor branches, so as to tune these branches as filters for the fifth and seventh harmonics. The TCR and coupling transformer secondary are connected in delta so as to cancel third harmonics. The steady state characteristics of TCR-FC system is shown in Fig. 19.26.

Here B_C = Susceptance of the capacitor

$B_L(\alpha)$ = Susceptance of the reactor at firing angle α

The steady-stage relationship between the output VARs (Q) and the applied voltage (V) can be linear over the control range AB. The slope of the characteristics in the regulation interval is determined by the control of the firing angle α. Outside the regulation interval, the characteristic is the same as that of an inductor (high voltage) or a capacitor (low voltage).

It is to be noted that the reactor current is controlled by the firing angle of the back to back connected thyristors. As the inductance is varied the susceptance varies over a wide range within limits $B_{L\,min}$ and $B_{L\,max}$ corresponding to $X_{L\,max}$ and $X_{L\,min}$. The voltage varies within the limits of $V_{REF} + \Delta V$ and $V_{REF} - \Delta V$, as shown in the figure.

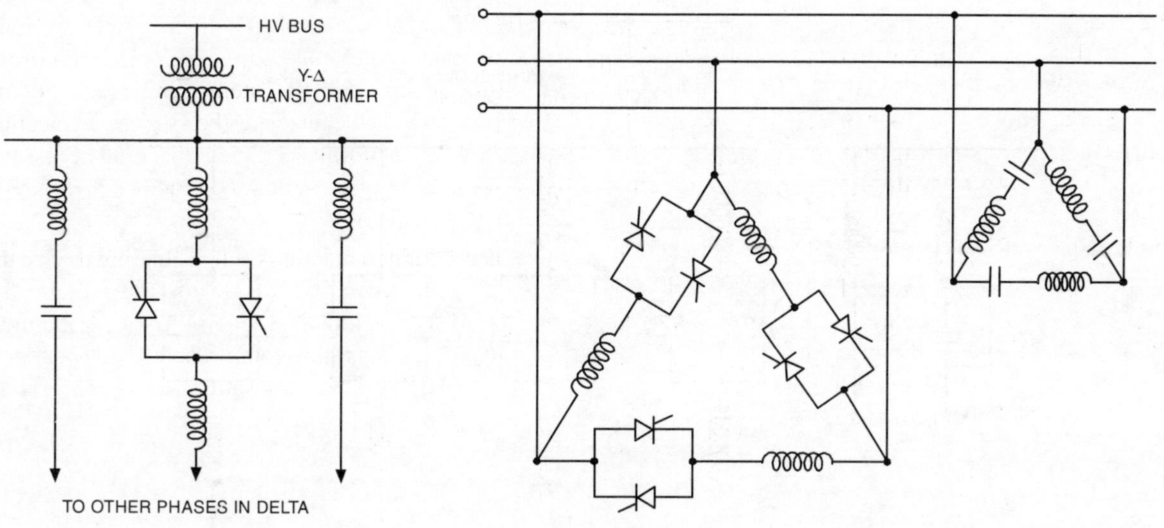

(a) TCR-FC Type VAR Compensator *(b) Delta Connection of TCR-FC Type VAR Compensator*

Fig. 19.25

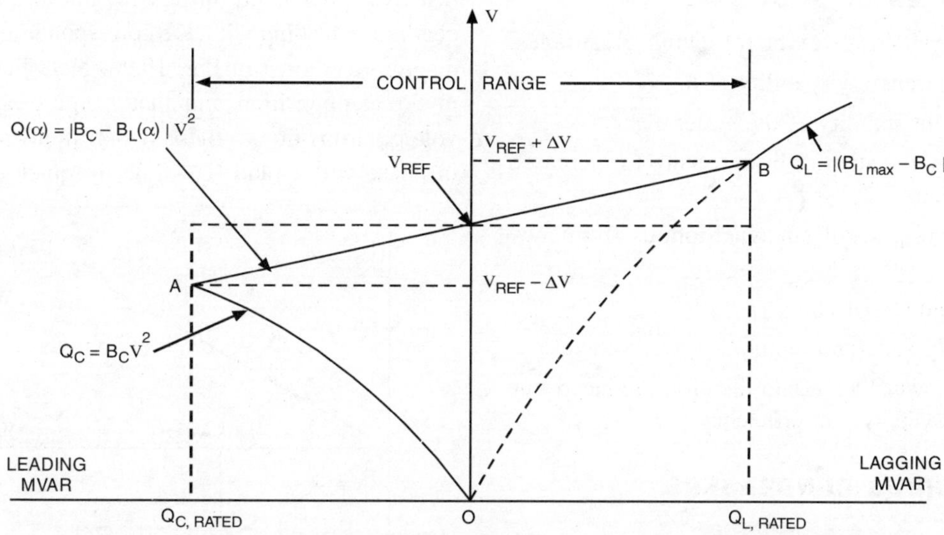

Fig. 19.26. ***Steady-State V-Q Characteristics of TCR-FC Type VAR Compensator***

5. Thyristor Switched Capacitor-Thyristor Controlled Reactor (TSC-TCR). The TSC-TCR VAR compensator is shown in Fig. 19.27. It consists of a thyristor controlled reactor bank(s) and a number of capacitor banks each in series with a thyristor switch. The number of capacitor banks is determined from practical considerations that include working voltage level, degree of compensation required, current rating of thyristors etc. The current through the reactors is varied by controlling the firing angles of back to back pairs of thyristors connected in series with each reactor. Similarly, the current through the capacitors is changed by controlling the firing angles of back to back pairs of thyristors connected in series with each capacitor. During heavy loads, the thyristors of TSC are made to conduct for longer duration in each cycle, so as to provide leading VARs. During low loads, the thyristors of TCR are made to conduct for longer duration in each cycle, so as to supply lagging VARs. Thus, TSC-TCR type VAR compensator is capable of providing the leading and lagging VARs both rapidly continuously, and independently. For continuous control of reactive power flow it is necessary that the TCR rating should be slightly higher than the individual capacitor bank. In addition to the switched capacitors, a small number of fixed capacitors, serving as tuned filters, are also connected in the circuit as shown in Fig. 19.27. It is usually necessary to connect inductors in series with switched capacitor bank to limit the current in the thyristor switches. The power transformer T is used to step down bus voltage (say 400 kV or higher) to a lower level such as 33 kV for economic design of SVC. The steady-state V-Q characteristic of TSC-TCR type VAR compensator is similar to that of TCR-FC type compensator (Fig. 19.26). The TSC-TCR VAR compensator is better than TCR-FC type compensator but is about 15-20% more costly as compared to TCR-FC type. The TSC-TCR scheme has the following advantages over TCR-FC scheme.

1. Improved performance during large system disturbances.
2. Lower power losses.

The TSC-TCR scheme of VAR compensation has replaced the earlier method of using synchronous phase modifiers for VAR compensation.

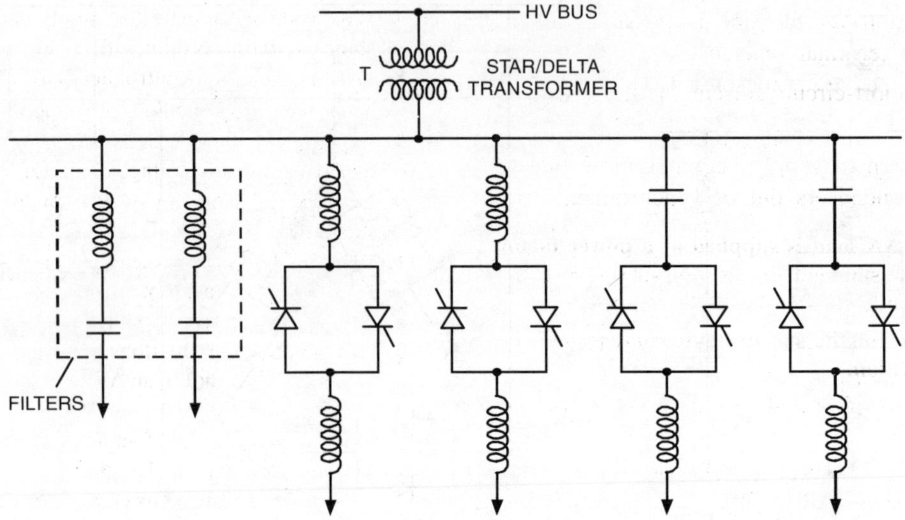

Fig. 19.27. ***TSC-TCR Type VAR Compensator***

19.11.2. Advantages of SVS

The static VAR system (SVS) offer the following advantages:

1. Increased power transfer capability of lines.
2. Improved transient stability of the system.
3. Improved dynamic system stability owing to increased damping provided.
4. Possibility of damping of subsynchronous frequency oscillations.
5. Very fast dynamic response.
6. Controlled steady state and temporary overvoltages.
7. Improved load power factor, and therefore, reduced line losses and improved system efficiency.

19.12. SYNCHRONOUS CONDENSERS

The voltage at the receiving end of a transmission line can be controlled by installing specially designed synchronous motors, known as *synchronous condensers** at the receiving end. The wattless leading kVA is automatically varied by the variation of its excitation, according to the load on the transmission line. The efficiencies of these machines are very high because values of the losses as a percentage of the kVA rating are very low (4 to 6 per cent). Hence current drawn by synchronous condensers can be assumed to lead the applied voltage by 90°.

For simplicity, consider transmission line having ohmic resistance and inductive reactance only.

Let load current be of I amperes at a pf of cos ϕ (lagging), current drawn by synchronous condenser be I_c amperes, the phase voltages at receiving and sending ends be V_R volts.

The three currents; I_p (in-phase component of load current), I_q (the quadrature component of load current) and I_c (the current drawn by synchronous condenser) cause resistive and reactive drops per phase of I_pR, I_pX; I_qR, I_qX; and I_cR, I_cX; the resistive drops being in phase with, and the reactive drops in quadrature leading with, the corresponding currents. The phasor diagram is shown in Fig. 19.28. Note that since I_q and I_c are in direct opposition and that I_c must be larger than I_q, the voltage drops due to these two currents simplify to $(I_c - I_q)R$ in phase with I_c and $(I_c - I_q)X$ in quadrature loading with I_c.

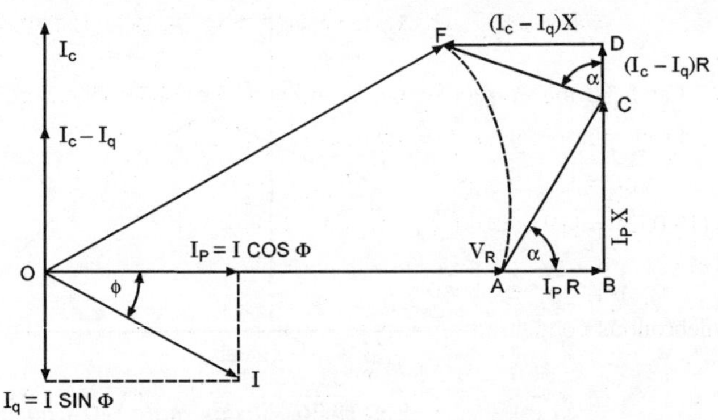

Fig. 19.28. ***Phasor Diagram***

From phasor diagram shown in Fig. 19.28.

$$OF^2 = (OA + AB - DF)^2 + (BC + CD)^2$$

$$\therefore \; V_R^2 = [V_R + I_PR - (I_c - I_q)X]^2 + [I_pX + (I_c - I_q)R]^2$$

In the above expression only I_c is unknown and, therefore, can be determined.

kVA capacity of condenser $= \dfrac{3V_PI_c}{1{,}000} = \dfrac{\sqrt{3}V_LI_c}{1{,}000}$ where V_L is line voltage.

The advantages and disadvantages of this method of voltage control are given below.

Advantages: (*i*) Same voltage can be maintained at both ends of a transmission line.

(*ii*) Power factor correction is affected at times of heavy loads.

* Also refer to Art. 7.13 page 158-159 of part II "Transmission and Distribution of Electrical Power".

(*iii*) A better protection for the line is possible due to possibility of using high terminal reactances.

Disadvantages: (*i*) Short-circuit current of the system is increased.

(*ii*) Risk of interruption of supply becomes more due to falling of synchronous condensers out of synchronism.

Example 19.3. A 12,500 kVA load is supplied at a power factor of 0.8 lagging by a 3-ϕ transmission line whose voltage is to be maintained at 33 kV at both ends.

Determine the capacity of the synchronous condenser to be installed for voltage regulation.

Given that the line resistance and reactance per phase are 4 Ω and 12 Ω respectively.

Solution : Load current, $I = \dfrac{\text{Load in kVA} \times 1{,}000}{\sqrt{3} \times \text{line voltage in volts}}$

$$= \frac{12{,}500 \times 1{,}000}{\sqrt{3} \times 33{,}000} = 218.7 \text{ A}$$

In-phase component of load current,

$$I_p = I \cos \phi = 218.7 \times 0.8 = 175 \text{ A}$$

Quadrature component of load current,

$$I_q = I \sin \phi = 218.7 \times 0.6 = 131 \text{ A}$$

Sending-end voltage per phase

$$= \text{Receiving end voltage per phase} = \frac{33{,}000}{\sqrt{3}} = 19{,}052 \text{ V}$$

Let I_c be the current drawn by the synchronous condenser.

From phasor diagram shown in Fig. 19.19.

$$V_R^2 = [V_R + I_pR - (I_c - I_q)X]^2 + [I_pX + (I_c - I_q)R]^2$$

or $(19{,}052)^2 = [19{,}052 + 175 \times 4 - (I_c - 131) \times 12]^2 + [175 \times 12 + (I_c - 131) \times 4]^2$

or $I_c = 201$ A

Synchronous condenser capacity

$$= \frac{3V_pI_c}{1{,}000} = \frac{3 \times 19{,}052 \times 201}{1{,}000} = 11{,}490 \text{ kVA} \quad \textbf{Ans.}$$

Exercises

1. What are the different methods used for voltage control of a power system ?
2. Enumerate the main requirements of an excitation system for the alternator.

   Enumerate the various systems of excitation and compare them on the basis of their merits and demerits.
3. Draw and describe a dc excitation system.

   [Rajasthan Univ. Power System Engineering 2000, 2007, 2008]
4. What are the type of excitation systems ? Draw the functional block diagrams of them and briefly describe the various elements of the block diagrams.

   Explain ac excitation system for synchronous machine in detail.

   [Rajasthan Technical Univ. Power System Engineering 2011]
5. Explain the rotating thyristor excitation scheme with its block diagram. **[Rajasthan Univ. Power System Engineering 2006]**
6. Draw the scheme of rotating thyristor excitation system and explain its operation. What type of excitation system is being used in new generating units in India ?

   [Rajasthan Univ. Power System Engineering 2005]
7. What is the necessity of dc supply in power system ? Explain static excitation system with its advantages.
8. What is excitation control and why is it required ?

   Name the various methods of voltage control in a power station.
9. Explain with the help of a neat diagram the working principle of Tirril automatic voltage regulator.
10. Explain the working of Brown-Boveri AVR giving neat sketches.
11. What are the drawbacks of electromechanical voltage regulators ? Name the important types of static voltage regulators. Describe with a neat diagram servo type AVR.
12. Explain the main functions of an automatic voltage regulator. What qualities should an AVR have ? Explain transconductor voltage regulator.
13. Give a neat diagram of an automatic voltage regulator using magnetic amplifier and explain its working.

    [Rajasthan Univ. Power System Engineering 2004, 2007]
14. Explain the working of solid state type automatic voltage regulator. **[Rajasthan Univ. Power System Engineering 2006]**
15. State whether the tap changer in transformer is provided on HV side or LV side.

    Explain the reason for its location and elaborate how lowering or raising of output voltage is actually done.
16. What are the different methods of voltage control in a power system ? Explain the working of an on-load tap changing transformer.

    [Rajasthan Technical Univ. Power System Engineering, 2011]
17. Why is it necessary to keep the receiving-end voltage constant within specified limits ?

    Name the different methods of controlling voltages at the receiving end.
18. Describe how series and shunt capacitors can minimise the voltage drop in the line.
19. Write the name of four methods of voltage control in a power system. Explain the method of voltage control using tap-changing transformer.

    [Rajasthan Univ. Power System Engineering 2004, 2007]
20. Explain the working of induction regulator.

    [Rajasthan Univ. 1999]
21. Describe various types of thyristorised static VAR compensators with their merits and demerits.

    [Rajasthan Univ. Power System Engineering 1999, 2000]
22. Explain a 3-phase static VAR compensator with necessary diagram. **[U.P.S.C. I.E.S. Electrical Engineering-II, 2003]**
23. Show with the aid of a phasor diagram, how the voltage at the receiving end of a transmission line can be maintained constant by the use of a synchronous phase modifier.
24. Write short notes on the following :

    (*i*) DC excitation system.

    (*ii*) AC excitation systems.

    (*iii*) Brushless excitation system.

    [Rajasthan Univ. 2002, 2003, 2004; Rajasthan Technical Univ. 2010]

    (*iv*) AC static excitation system.

    [Rajasthan Univ. 2003; 2008]

    (*v*) Servo type voltage regulator.

    (*vi*) Voltage regulator using magnetic amplifier.

    (*vii*) Location and function of transformer tap changer.

    (*viii*) Booster transformer.

    (*ix*) Induction regulator.

    (*x*) Static VAR systems.

Short Answer Type Questions With Answers

Q. 1. Why is it necessary to keep the receiving-end voltage constant within specified limits ?

Ans. For satisfactory operation of equipment/machinery supplied by the electrical power network it is necessary to keep the receiving-end voltage constant within specified limits. Too wide variation of voltage may cause erratic operation or even malfunctioning of consumers appliances. This is statuary requirement also.

Q. 2. What are the limits within which consumer's terminal voltage should be maintained ?

Ans. According to IE rules the declared consumer's voltage is 415/240 V (*i.e.*, 415 V between phases and 240 V between phase and neutral). The higher and lower limits are 440/254 during light load hours and 400/230 V during peak load hours.

Q. 3. How is voltage control achieved ?

Ans. Voltage control is achieved by installing voltage control equipment at suitable places such as at (*i*) generating stations (*ii*) transformer stations supplying to feeders and (*iii*) at load end.

Q. 4. Name the different methods of controlling voltages in ac supply systems.

Ans. The different methods used to control voltages in ac supply system are (*i*) excitation control at generating stations and by using (*ii*) tap-changing transformers (*iii*) induction regulators (*iii*) booster transformers (*iv*) series compensation (*v*) shunt compensation and (*vi*) synchronous phase modifiers.

Q. 5. Why the excitation control is not possible beyond certain limits ?

Ans. Excitation below a certain level may cause unstability of the system and excitation above a certain limit may cause overheating of the rotor. So excitation control is not possible beyond certain limits.

Q. 6. Why is excitation system required in a power station ?

Ans. The excitation system is required to provide the necessary field current to the rotor winding of a synchronous machine. The amount of excitation depends on the power factor, speed of the machine and load current. For large currents, lower speeds and lagging power factors, the excitation requirement is more.

Q. 7. What are the main requirements of excitation system ?

Ans. The main requirements of an excitation system are reliability under all conditions of service, simplicity of control, ease of maintenance, stability and high transient response.

The initial cost of excitation equipment should be a secondary consideration in the selection of excitation system. Loss of excitation of a unit on the bus results in a more serious disturbance than that resulting from outage of the generator unit from the bus as the remaining units must not only pick up the load dropped but also supply the large reactive current drawn by the unexcited generator. In view of this, an excitation system with better reliability is preferred, even if the initial cost is more. Simplicity of control and ease of maintenance are always preferred for any equipment and excitation system is no exception.

Q. 8. Distinguish between centralized and individual excitation systems.

Ans. An excitation system may be *individual* one, in which each alternator is provided with its own exciter in the form of small generator on an extension of the main shaft, or *centralised excitation system* having two or more exciters feeding a bus-bar to which field systems of all the alternators in the power plant are connected. Though centralised excitation system is a cheaper arrangement but a fault in this system adversely affects all the alternators in the power plant. As such individual excitation system is widely used.

Q. 9. Why have dc excitation systems been superseded by ac excitation systems ?

Ans. The main drawbacks of dc excitation system are large time constant (about 3 seconds) and commutation difficulties. As such the dc excitation systems have been superseded by ac excitation systems.

Q. 10. Why excitation circuit should not be opened suddenly ?

Ans. A sudden opening of the excitation circuit may cause a inducement of a high voltage in the inductive field winding because of sudden release of energy stored in it. As such the excitation circuit should not be opened suddenly.

Q. 11. In a rotating thyristor excitation system, the main alternator field rectifier usually consists of four parallel thyristors each having a fuse in its circuit. Why so ?

Ans. The main alternator field rectifier usually consists of four parallel thyristors each having a fuse in its circuit so that in case of failure of any thyristor, the fuse opens thereby allowing the system to continue operation. The thyristor assembly is so rated that with the loss of one thyristor unit, the alternator field needs are met with sufficient margin.

Q. 12. Why is brushless excitation system becoming more and more popular ?

Ans. With the increase in rating of alternators, the problem of brush maintenance gets more and more complicated, therefore, the brushless excitation system is becoming more and more popular.

Q. 13. What are the advantages of static excitation system ?

Ans. The advantages of the static excitation system are elimination of exciter windage loss and commutator wearing and winding maintenance resulting in reduced operating costs and *electronic speed response.* The fact that in static excitation the voltage is proportional to the speed, affords a major advantage in load rejection.

Q. 14. Why the excitation control is not possible beyond certain limits ?

Ans. Excitation below a certain level may cause unstability of the system and excitation above a certain limit may cause overheating of the rotor. So excitation control is not possible beyond certain limits.

Q. 15. Why is automatic voltage regulating equipment essential in operation of synchronous machines ?

Ans. Synchronous generators have inherently large internal reactance and, therefore, high voltage regulation. So, in order to maintain terminal voltage of the synchronous generators under various operating conditions, provision of automatic voltage regulation equipment is essential.

Q. 16. What is the operating principle of AVR ?

Ans. An automatic voltage regulator operates on the principle of detection of error. The output voltage of an ac generator obtained through a PT is rectified, filtered and compared with a reference. The difference of the actual voltage and the reference voltage, known as error voltage, is amplified through an amplifier (rotary, magnetic or static) and supplied to the field circuit of the main exciter or pilot exciter. Thus, the amplified error signals controls the excitation of the main or pilot exciter through a buck or boost action. Exciter output control leads to the control of the main alternator terminal voltage.

Q. 17. What are the main functions of AVRs ?

Ans. The main functions of an AVR are as follows:

1. Control of system voltage within prescribed limits and have the operation of the machine nearer to the steady state stability limit.
2. Proper division of reactive load between the alternators operating in parallel.
3. Prevention of dangerous overvoltages on the occurrence of sudden loss of load on the system.
4. Increase of excitation under system fault conditions so that maximum synchronising power exists at the time of clearance of fault, to prevent loss of synchronism.

Q. 18. What are the drawbacks of electromechanical AVRs ?

Ans. Electromechanical AVRs have the following drawbacks:

(*i*) The variations of resistances in the exciter circuit results in a variable time constant. Thus, the response time of the exciter will vary with different excitation conditions.

(*ii*) The response time of the regulator and main exciter field circuit is too slow. Dead band does not suit to generators when operating near steady-state limit.

(*iii*) With the increase in the rating of the machines, the excitation need is increased and it is difficult for regulator contacts to handle this power.

Electromechanical AVRs suffer from the wear and friction and need fairly regular attention. Friction reduces the sensitivity and causes a dead zone of between 0.5 and 1% within which no control action is available. The contacts of vibrating type AVR need regular attention because of erosion caused by the sparking. The rate of response is limited by the mass of the moving parts. Even when providing full corrective action they cannot force the main exciter to go negative in order to limit alternator voltage on occurrence of sudden loss of the load on the system.

Q. 19. What are the benefits of electronic voltage regulators ?

Ans. Electronic voltage regulators provide the benefits of accuracy of control, high speed response, high reliability and minimum maintenance.

Q. 20. What is a booster transformer ?

Ans. Booster transformer is one which is often used toward the end of a power line in order to raise the voltage to the desired value.

Q. 21. What special precautions must be taken when using a booster transformer ?

Ans. There must be no fuse on the *hv* side, or primary; since the booster transformer is similar to a current transformer, if the fuse should blow, an extremely high voltage could be built up on the secondary side.

Q. 22. State whether the tap changer in a transformer is provided on *hv* side or *lv* side ?

Ans. On *hv* side.

Q. 23. Why are both of the primary and secondary windings of a single-phase induction regulator usually designed to provide 2 poles or 4 poles ?

Ans. Both of the primary and secondary windings of a single-phase induction regulator are usually designed to provide 2 poles or 4 poles because the smaller the number of poles, the larger the mechanical angle corresponding to one pole pitch and the easier the voltage regulation.

Q. 24. Why is the single-phase induction regulator is provided with the short-circuited compensating winding ?

Ans. If the single-phase induction regulator were not provided with the short-circuited compensating winding, its operation would not be satisfactory (voltage regulation would be very poor with adjustments for intermediate voltages).

Q. 25. No compensating or tertiary winding is required in case of a 3-phase induction regulator. Why ?

Ans. No compensating or tertiary winding is required in case of a 3-phase induction regulator because, in this case, for every position of setting of the regulator each secondary winding is magnetically coupled to one or more of the primary windings.

Q. 26. What are the applications of induction regulators ?

Ans. The induction voltage regulator finds its greatest application in distribution systems for raising or lowering the line voltage to maintain the load voltage at a constant value under all load conditions because it varies the voltage uniformly, not in steps.

Q. 27. Name few commonly used static VAR schemes.

Ans. Many SVS schemes are in operation. Some of commonly used schemes are as follows :

1. Thyristor controlled reactor (TCR).
2. Thyristor switched capacitor (TSC).
3. Self saturated reactor (SR).
4. Thyristor controlled reactor-Fixed capacitor (TCR-FC).
5. Thyristor switched capacitor-Thyristor controlled reactor (TSC-TCR).

Q. 28. In case of TCR-FC scheme of VAR compensation, harmonics are generated because of switching operations. How these harmonics can be eliminated ?

Ans. Small reactors are usually connected in the fixed capacitor branches, so as to tune these branches as filters for the fifth and seventh harmonics. The TCR and coupling transformer secondary are connected in delta so as to cancel third harmonics.

Q. 29. What are the advantages of TSC-TCR scheme over TCR-FC scheme.

Ans. The TSC-TCR scheme has the following advantages over TCR-FC scheme.

1. Improved performance during large system disturbances.
2. Lower power losses.

Q. 30. Give the advantages of SVS.

Ans. The static VAR system (SVS) offer the advantages of (*i*) Increased power transfer capability of lines, (*ii*) Improved transient stability of the system, (*iii*) Improved dynamic system stability owing to increased damping provided, (*iv*) Possibility of damping of subsynchronous frequency oscillations, (*v*) Very fast dynamic response, (*vi*) Controlled steady state and temporary overvoltages, (*vii*) Improved load power factor, reduced line losses and improved system efficiency.

Q. 31. What are the advantages of synchronous compensation ?

Ans. Synchronous compensation has the following advantages: (*i*) Same voltage can be maintained at both ends of a transmission line. (*ii*) Power factor correction is affected at times of heavy loads. (*iii*) A better protection for the line is possible due to possibility of using high terminal reactances.

Q. 32. What are the shortcomings of synchronous compensation ?

Ans. Synchronous compensation has the following shortcomings : (*i*) Short-circuit current of the system is increased. (*ii*) Risk of interruption of supply becomes more due to falling of synchronous condensers out of synchronism.

□□□

CHAPTER 20

Substations

INSIDE THIS CHAPTER

20.1. INTRODUCTION

Substations serve as sources of energy supply for the local areas of distribution in which these are located. Their main functions are to receive energy transmitted at high voltage from the generating stations, reduce the voltage to a value appropriate for local distribution and provide facilities for switching. Some substations are simply switching stations where different connections between various transmission lines are made, others are converting substations which either convert ac into dc or vice versa or convert frequency from higher to lower or vice versa. Substations have some additional functions. They provide points where safety devices may be installed to disconnect equipment or circuit in the event of fault. Voltage on the outgoing distribution feeders can be regulated at a substation. A substation is convenient place for installing synchronous condensers at the end of the transmission line for the purpose of improving power factor and make measurements to check the operation of the various parts of the power system. Street lightning equipment as well as switching controls for street lights can be installed in a substation.

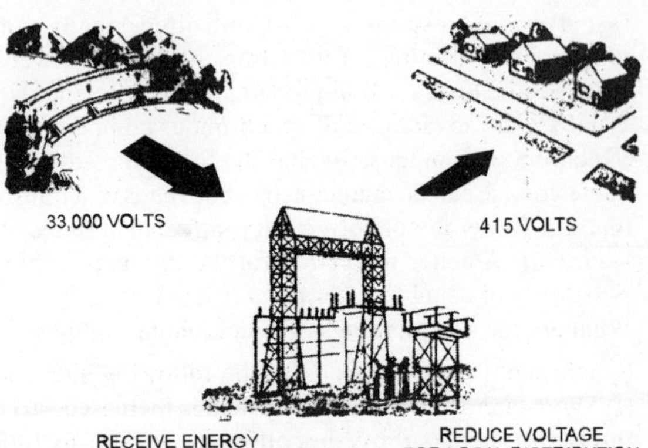

Fig. 20.1

20.2. CLASSIFICATION OF SUBSTATIONS

The substations may be classified in numerous ways such as on the basis of (*i*) nature of duties (*ii*) service rendered (*iii*) operating voltage (*iv*) importance and (*v*) design.

20.2.1. Classification of Substations on The Basis of Nature of Duties

The substations, on the basis of nature of duties, may be classified into the following three categories:

1. ***Step-Up or Primary Substations.*** Such substations are usually associated with generating stations. The generated voltage, which is usually low (3.3, 6.6, 11 or 33 kV), is stepped up to primary transmission voltage so that huge blocks of power can be transmitted over long distances to the load centres economically.

2. ***Primary Grid Substations.*** Such substations are located at suitable load centres along the primary transmission lines. In these substations, the primary transmission voltage is stepped down to different suitable secondary voltages. The secondary transmission lines are carried over to the secondary substations situated at the load centres where the voltage is further stepped down to sub-transmission or primary distribution voltages.

3. ***Step-Down or Distribution Substations.*** Such substations are located at the load centres, where the sub-transmission/primary distribution voltage is stepped down to secondary distribution voltage (415/240 V). These are the substations which feed the consumers through distribution network and service lines.

20.2.2. Classification of Substations on The Basis of Service Rendered

The substations, according to service rendered are:

1. ***Transformer Substations.*** Transformers are installed on such substations to transform the power from one voltage level to another level as per needs.

2. ***Switching Substations.*** Such substations are meant for switching operation of power lines without transforming the voltage. At such substations different connections are made between various transmission lines.

3. ***Converting Substations.*** Such substations are meant for either converting ac to dc or vice versa or converting frequency from higher to lower or vice versa.

20.2.3. Classification of Substations on The Basis of Operating Voltage

The substations, according to operating voltage, may be categorised as :

1. ***High Voltage Substations (HV Substations)*** involving voltages between 11 kV and 66 kV.

2. ***Extra High Voltage Substations (EHV Substations)*** involving voltages between 132 kV and 400 kV.

3. ***Ultra High Voltage Substations (UHV Substations)*** operating on voltage above 400 kV.

20.2.4. Classification of Substations on The Basis of Importance

1. ***Grid Substations.*** These are the substations from where bulk power is transmitted from one point to another point in the grid. These are important because any disturbance in these substations may cause the failure of the grid.

2. ***Town Substations.*** These substations step-down the voltages at 33/11 kV for further distribution in the towns and any failure in such substations results in the failure of supply for whole of the town.

20.2.5. Classification of Substations on The Basis of Design

1. ***Indoor Type Substations.*** In such substations the apparatus is installed within the substation building. Such substations are usually for a voltage up to 11 kV but can be erected for the 33 kV and 66 kV when the surrounding atmosphere is contaminated with impurities such as metal corroding gases and fumes, conductive dust etc.

2. ***Outdoor Substations.*** These substations are further subdivided into :

(*a*) *Pole Mounted Substations.* Such substations are erected for distribution of power in localities. Single stout pole or H-pole and 4-pole structures with suitable platforms are employed for transformers of capacity up to 25 kVA, 125 kVA and above 125 kVA (but up to 250 kVA) respectively.

(*b*) *Foundation Mounted Substations.* For transformers of capacity above 250 kVA the transformers are too heavy for pole mounting. Such substations are usually for voltages of 33,000 volts and above.

20.3. INDOOR SUBSTATIONS

Indoor Type Substations. In these substations the apparatus is installed within the substation building. Such substations are usually for a voltage up to 11,000 V but can be erected for 33,000 V and 66,000 volts when the surrounding atmosphere is contaminated with impurities such as metal corroding gases and fumes, conductive dust etc. The switchgear on supply or primary side will consist of oil circuit breakers only. The high voltage supply is given to the primary of the transformer through circuit breaker. From the bus-bar various feeders emerge out. The panel for each feeder consists of an isolator switch and circuit breaker. In addition to isolator and circuit breaker, the panel is provided with the measuring instruments. As regards protection of feeders usually reverse power relay is employed. For the protection of oil filled transformers with conservator tanks Buchholz's relay is most common. The auxiliaries of the indoor type of substations are (*i*) storage batteries (*ii*) firefighting equipment such as water buckets, fire extinguishers, etc. The batteries are used for the (*i*) operation of the protective gear and switch operating solenoids and (*ii*) emergency lighting in substations in case of failure of supply. The several compartments in which the indoor substation is divided are control compartment, indicating and metering instruments and protective device compartment, circuit breaker and operating mechanism compartment, main bus-bar compartment and current transformer and cable sealing box compartment, as shown in Fig. 20.2.

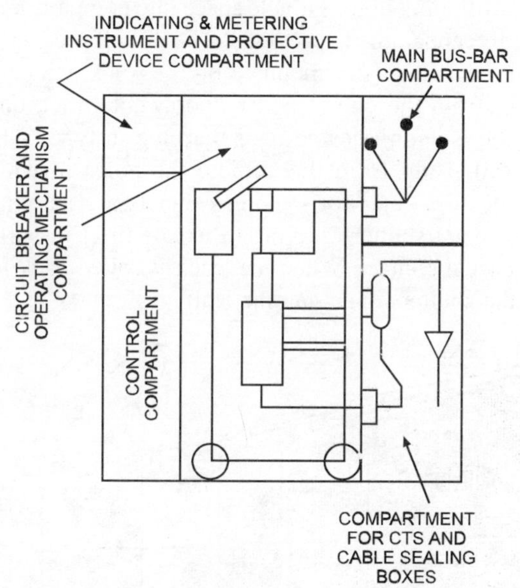

Fig. 20.2

Indoor distribution and transformer substations as well as high voltage switchboards consist of a series of open and enclosed chambers or compartments. The main equipment of the given installation is arranged in these compartments. The chamber space within which the equipment of any one main bus-bar connection is mounted, as a whole, is called as a *cell, cubicle or compartment*.

According to construction indoor distribution transformer substations and high voltage switchboards are further subdivided into the following categories.

1. ***Substations of The Integrally Built Type,*** in which the apparatus is installed on site. In such substations the cell structures are constructed of concrete or bricks.

2. ***Substations of The Composite Built-Up Type,*** in which the assemblies and parts are factory or workshop prefabricated, but are assembled on site within a substation switchgear room. The compartments of such substations take form of metal cabinets or enclosures, each of which contains the equipment of one main connection cell. Within such cabinets or enclosures an oil minimum circuit breaker, a load-interrupter switch and one or more voltage transformers may be mounted.

3. ***Unit Type Factory Fabricated Substations and Metal Clad Switchboards.*** These are built in electrical workshops and are shipped to the site of installations fully pre-assembled. After installations of substations and switchboards only connection to the incoming and outgoing power circuits are required to be made. Cubicles for unit type switchboards or substations take the form of fully enclosed metal-clad cabinets.

Metal-clad cubicles designed with withdrawable trucks and divided into several compartments are usually employed. The several compartments in which the cubicle is divided are control compartment, indicating and metering instrument and protective device compartment, circuit breaker and operating mechanism compartment, main bus-bar compartment and current transformers and cable sealing box compartment.

By partitioning of the cubicle space into compartments safe access to the apparatus is provided. The circuit breaker and its operating mechanism are mounted on the truck, which can be withdrawn from the cubicle. In withdrawable-truck unit-type cubicles the isolating device is of the plug-in type. When the truck is rolled out from the cubicle the holes in which the isolating devices enter for making contacts are automatically closed by metal shutters serving to isolate the live parts from possible casual contact. When the truck is rolled back into the cubicle, the shutters open automatically.

Fig. 20.3. ***General View of A Unit-Type Metal-Clad Switchboard***

To prevent any possible opening or closing of the disconnecting devices when the circuit breaker is closed, these cubicles are designed with interlocks which prevent the truck from being rolled in or withdrawn when the circuit breaker is closed.

A general view of a unit-type metal-clad switchboard assembled of several metal-clad cubicles is shown in Fig. 20.3.

20.4. OUTDOOR SUBSTATIONS

As already said, outdoor substations are of two types namely (*i*) pole-mounted substations and (*ii*) foundation-mounted substations.

1. ***Pole-Mounted Substations.*** Such substations are erected for mounting distribution transformers of capacity up to 250 kVA. Such substations are cheapest, simple and smallest of substations. All the equipment is of outdoor type and mounted on the supporting structure of *ht* distribution line. Triple Pole Mechanically Operated (TPMO) switch is used for switching "on" and "off" of *ht* transmission line. HT fuse unit is installed for protection of *ht* side. To control *lt* side iron clad low tension switch of suitable capacity with fuses is installed. Lightning arresters are installed over the *ht* line to protect the transformer from the surges. Substation is earthed at two or more places.

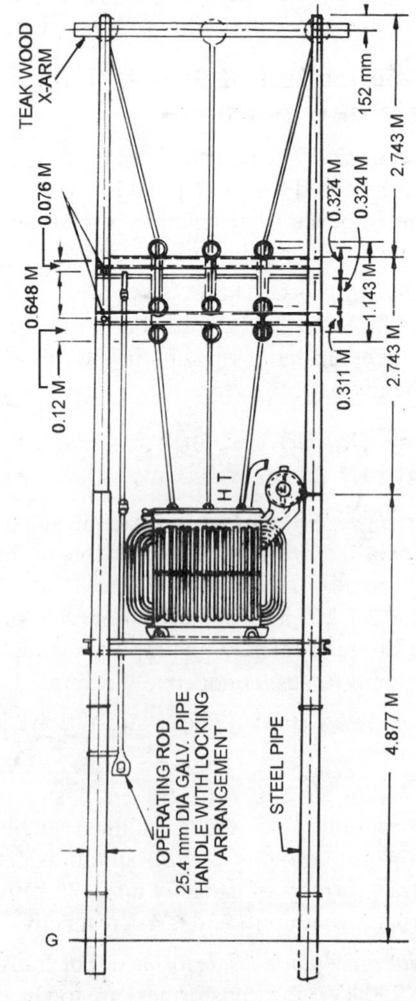

Fig. 20.4. ***200 kVA Pole-Mounted Substation***

Generally transformers of capacity up to 125 kVA are mounted on double pole structure and for transformers of capacity above 125 kVA but not exceeding 250 kVA 4-pole structure with suitable platform is used. This type of pole-mounted substation is erected in very thickly populated location.

The maintenance cost of such substations is low and by using a large number of such substations in a town it is possible to lay the distributors, at a lower cost. But owing to increase in number of transformers, total kVA is increased, no load losses increase and the cost per kVA is thus more. Economy is the main consideration when a choice is made for such substations. Typical 200 kVA pole-mounted substation is shown in Fig. 20.4.

2. *Foundation Mounted Substations*. These substations are built entirely in the open and in such substations all the equipment is assembled into one unit usually enclosed by a fence from the point of view of safety. Substations for primary and secondary transmission and for secondary distribution, (above 250 kVA) are foundation mounted outdoor type. Since equipment required for such substations is heavy, therefore, site selected for these substations must have a good access for heavy transport. Again, owing to exposed bus-bars and other associated equipment the clearances and spacings are not only to be governed by the operating voltage but also from the considerations of the encroachment from outside. Low level type substations are designed except when the space available is limited as these provide facility of easy inspection, cleaning and maintenance.

The switchgear consists of circuit breakers of suitable type on both the sides but with the increased reliability of the modern transformers, the practice is to dispense with the circuit breaker on the incoming side form the economic consideration. The isolating switches thus serve the purpose.

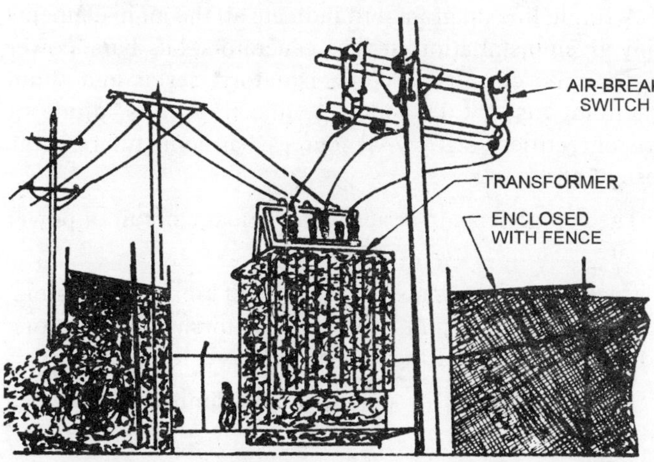

Fig. 20.5. ***Foundation-Mounted Outdoor Substation***

Foundation-mounted outdoor substation is shown in Fig. 20.5. 220 kV and 400 kV outdoor substations are shown in Figs. 20.6 (*a*) and 20.6 (*b*) respectively.

Advantages and Disadvantages of Outdoor Substations Over Indoor Substations

The outdoor substations have the following main advantages over indoor substations.

(*i*) All the equipment is within view and therefore fault location is easier.

(*ii*) The extension of the installation is easier, if required.

(*iii*) The time required in erection of such substations is lesser.

(*iv*) The smaller amount of building materials (steel-concrete) is required.

(*v*) The construction work required is comparatively smaller and cost of the switchgear installation is low.

(a) View of 220 kV Outdoor Substation

(b) View of 400 kV Outdoor Substation

Fig. 20.6

(*vi*) There is practically no danger of a fault which appears at one point being carried over to another point in the installation because the apparatus of the adjoining connections can be spaced liberally, without any appreciable increase in costs.

(*vii*) Repairing work is easy.

The disadvantages of outdoor installations in comparison of indoor installations are :

(*i*) The various switching operations with the isolators, as well as supervision and maintenance of the apparatus is to be performed in the open air during all kinds of weather.

(*ii*) More space is required for the substation.

(*iii*) Protection devices are required to be installed for protection against lightning surges.

(*iv*) The length of control cables required is more.

(*v*) The influence of rapid fluctuation in ambient temperature and dust and dirt deposits upon the outdoor substation equipment makes it necessary to install apparatus specially designed for outdoor service and, therefore, more costly.

Notwithstanding the disadvantages, outdoor substations are very widely used in power systems.

20.5. SELECTION AND LOCATION OF SITE FOR A SUBSTATION

The following factors are considered while making site selection for a substation.

1. *Type of Substation.* The category of substation is important for its location. For example a step-up substation, which is generally a point where power from various sources (generating machines or generating stations) is pooled and stepped up for long distance transmission, should be located as close to the generating stations as possible to minimize the transmission losses. Similarly a step-down substation should be located nearer to the load centre to reduce transmission losses, cost of distribution system and better reliability of supply.

2. *Availability of Suitable and Sufficient Land.* The land proposed for a substation should be normally level and open from all sides. It should not be water logged particularly in rainy season. The site selected for a substation should be such that approach of transmission lines and their take off can be easily possible without any obstruction.

According to the latest practice the land required for various types of substations is given below :

	Type of Substation	*Area Required*
(*a*)	400 kV substation	50 acres
(*b*)	220 kV substation	25 acres
(*c*)	132 kV substation	10 acres

The places nearer to aerodrome, shooting practice grounds etc. should be avoided.

3. *Communication Facility.* Suitable communication facility is desirable at a proposed substation both during and after its construction. It is better, therefore, to select the site along side on existing road to facilitate an easier and cheaper transportation.

4. *Atmospheric Pollution.* Atmosphere around factories, which, may produce metal corroding gases, air fumes, conductive dust etc., and nearer to sea coasts, where air may be more humid and may be salt loaden, is detrimental to the proper running of power system and therefore substations should not be located near factories or sea coast.

5. *Availability of Essential Amenities To The Staff.* The site should be such where staff can be provided essential amenities like school, hospital, drinking water, housing etc.

6. *Drainage Facility.* The site selected for the proposed substation should have proper drainage arrangement or possibility of making effective drainage arrangement to avoid pollution of air and growth of micro-organisms detrimental to equipment and health.

20.6. MAIN ELECTRICAL CONNECTIONS

The electric power in substations and switchgear installations is received and distributed by means of the main bus-bars to which the equipment is connected according to some given main circuit scheme.

It is necessary to distinguish two fundamental types of power station and substation circuit arrangements: (1) the main connections of primary (or power) circuits, and (2) the secondary (or control) circuit arrangements.

Primary circuit main connections are represented by either single line or three line schematic or elementary diagrams. The first type of diagram shows all three-phases of each circuit as a single line while the second type of diagram shows each phase of every circuit as a separate line, *i.e.,* shows each three-phase circuit as three lines.

The single line diagram represents the electrical circuit arrangement of any electrical power installation in its basic form. It is the diagram that is used by the designer in selecting the electrical equipment for any given installation, in drawing up the protective relay, control and signalling alarm system schemes and their elementary and wiring diagrams, and also in working up the constructional and mechanical design of the switchgear structures. In the everyday operation of electrical power installations single line diagrams, in the form of operative diagrams, play an important role in the performance of nearly all circuit switching operations.

A single line diagram will indicate all the main elements of any given installation such as generators, bus-bars, power transformers, circuit breakers, isolators, series and shunt capacitors, fuses, CTs and PTs, line trap units, diode or thyristor rectifiers, static VAR sources, harmonic filters, surge arresters etc.

The components in series with the main circuit of power flow are :

Bus-bars, power transformers, circuit breakers, isolators, fuses, CTs, line trap units, series capacitors, series reactors, diode or thyristor rectifiers etc.

The components in shunt circuits connected between phase and ground are :

Shunt capacitors, shunt reactors, static VAR sources, harmonic filters, PTs, lightning or surge arresters etc.

Special types of apparatus such as isolators (or disconnecting switches), circuit breakers, instrument transformers etc. are used for interconnecting high voltage power lines (overhead or cable), with the main bus-bars in the substations.

Connections may be divided as incoming, (power feeder connections) tie (lines interconnecting two substations or switchgear installations, each of which is fed through its own incoming feeder connection), outgoing (feeder connections for feeding other subsequent substations or switchgear installations), power transformers (connections made in a given substation), voltage transformers (connecting for control and metering).

The relative locations of circuit breakers, isolators and bus-bars follow the general practice and particular switching needs and maintenance and protection needs. Circuit breaker is connected between the bus-bar and each incoming and outgoing circuit. Isolator is provided on each side of the circuit breaker. CTs are provided for measurement and protection. CTs may be necessary on both sides of the circuit breaker so that protection zones overlap and cover the circuit breaker. PTs are generally connected to bus-bars and on incoming line side. Lightning or surge arresters are connected phase to ground at the incoming line as the first apparatus and also at the terminal of transformer, terminal of capacitor bank, terminal of shunt reactor, terminal of generator, terminal of large motor to divert switching/lightning surges to ground. Power transformers are connected between two voltage levels.

The main connection scheme is drawn keeping in view the following factors :

(*i*) General bus-bar arrangement (*ii*) operating voltage (*iii*) number of incoming and outgoing lines (*iv*) number of transformers, (*v*) safety to equipment (*vi*) safety to operating personnel (*vii*) future extension requirement.

The main connection diagram drawn for a substation shows the arrangements of all the circuits with its main bus-bars.

For simplicity and to facilitate reading, all the electrical connections of a substation can be represented by a single line diagram. It is understood that all the phases are connected identically.

20.7. GRAPHICAL SYMBOLS FOR VARIOUS TYPES OF APPARATUS AND CIRCUIT ELEMENTS ON SUBSTATION MAIN CONNECTION DIAGRAM

The main elements of the installations; circuit breakers, isolators, fuses, instrument transformers, power transformers etc. are represented by standard graphical symbols, as given below, on the single line diagrams.

Circuit Elements	Symbols
Earthing	
Circuit Breaker	
Current Transformer	
Bus Bar	
Potential Transformer	
Lightning Arrestor	
Fuse	
Transformer	
Isolator	
Tandom Isolator	
Auto-Transformer	

20.8. KEY DIAGRAMS OF TYPICAL SUBSTATIONS

The key diagrams of typical 11 kV, 33 kV, 66 kV and 220 kV substations are shown in Figs. 20.7, 20.8, 20.9 and 20.10 respectively.

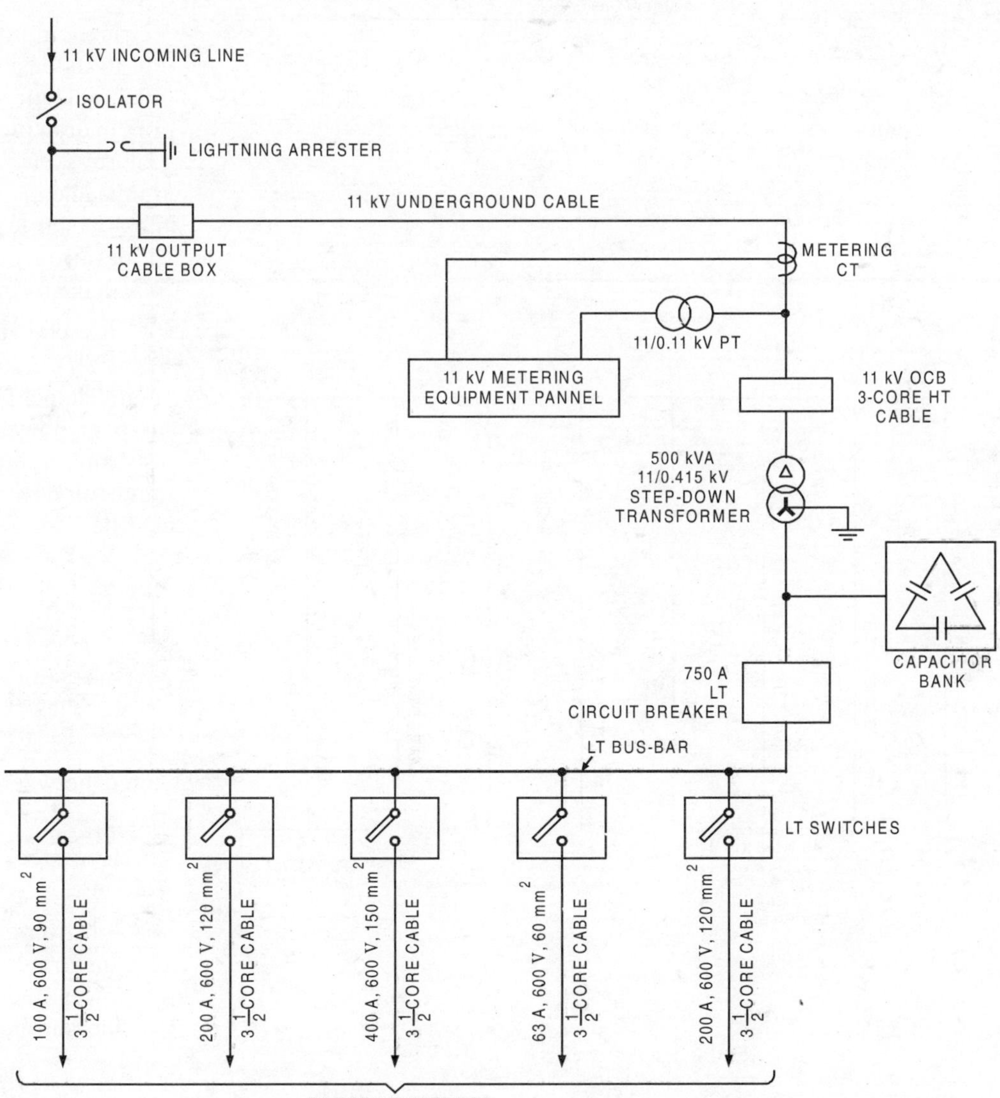

Fig. 20.7. ***Key Diagram For 11/0.4 kV Substation***

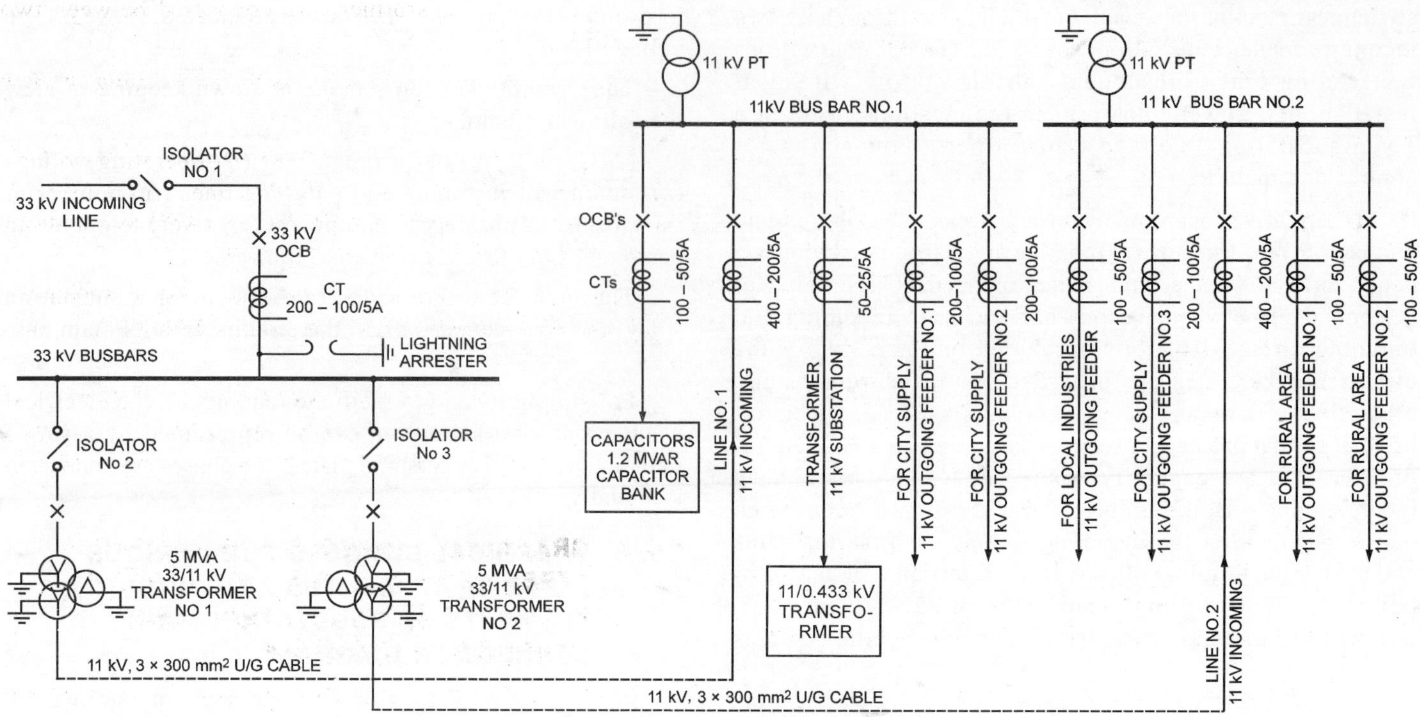

Fig. 20.8. *Key Diagram For 33 kV Substation*

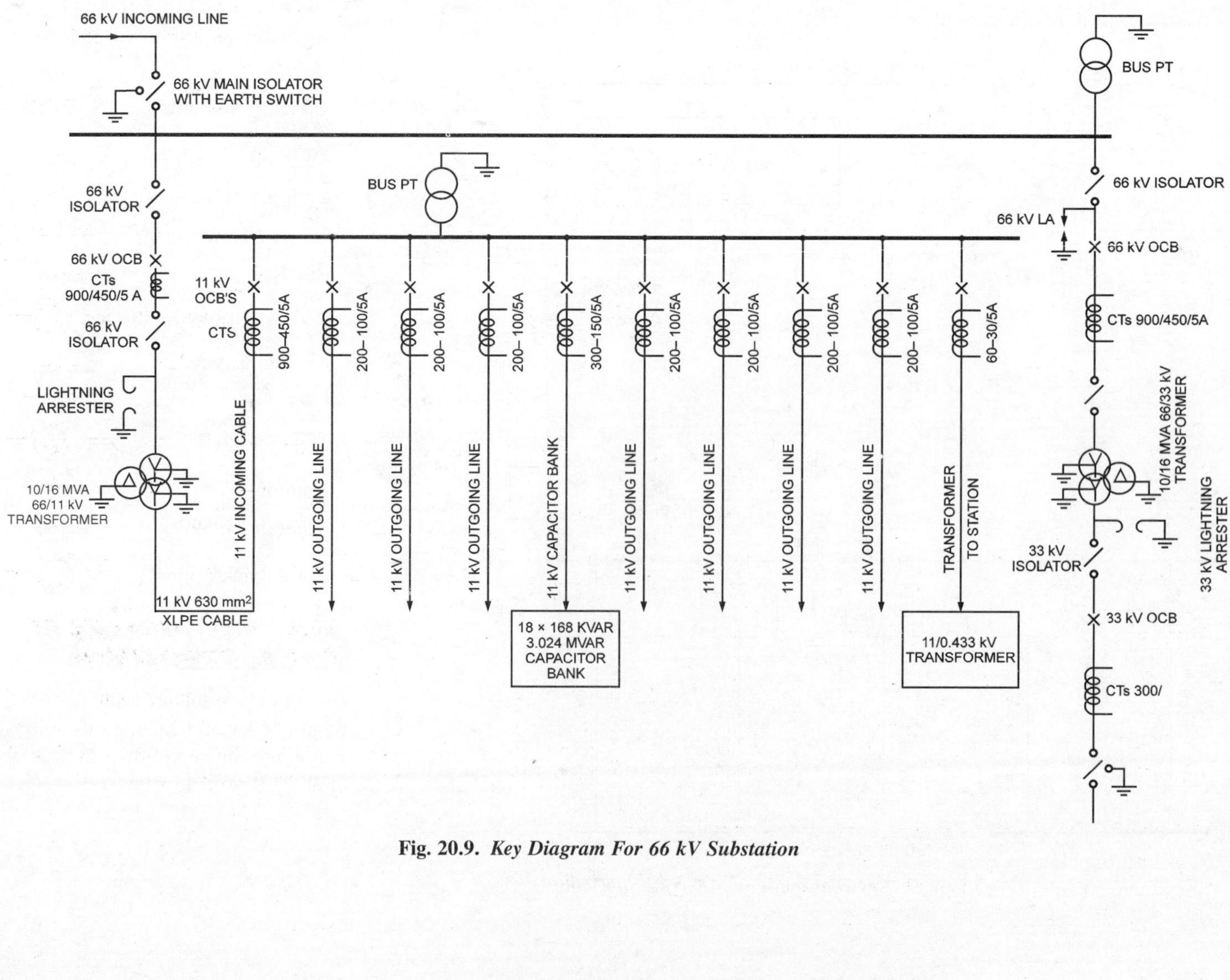

Fig. 20.9. *Key Diagram For 66 kV Substation*

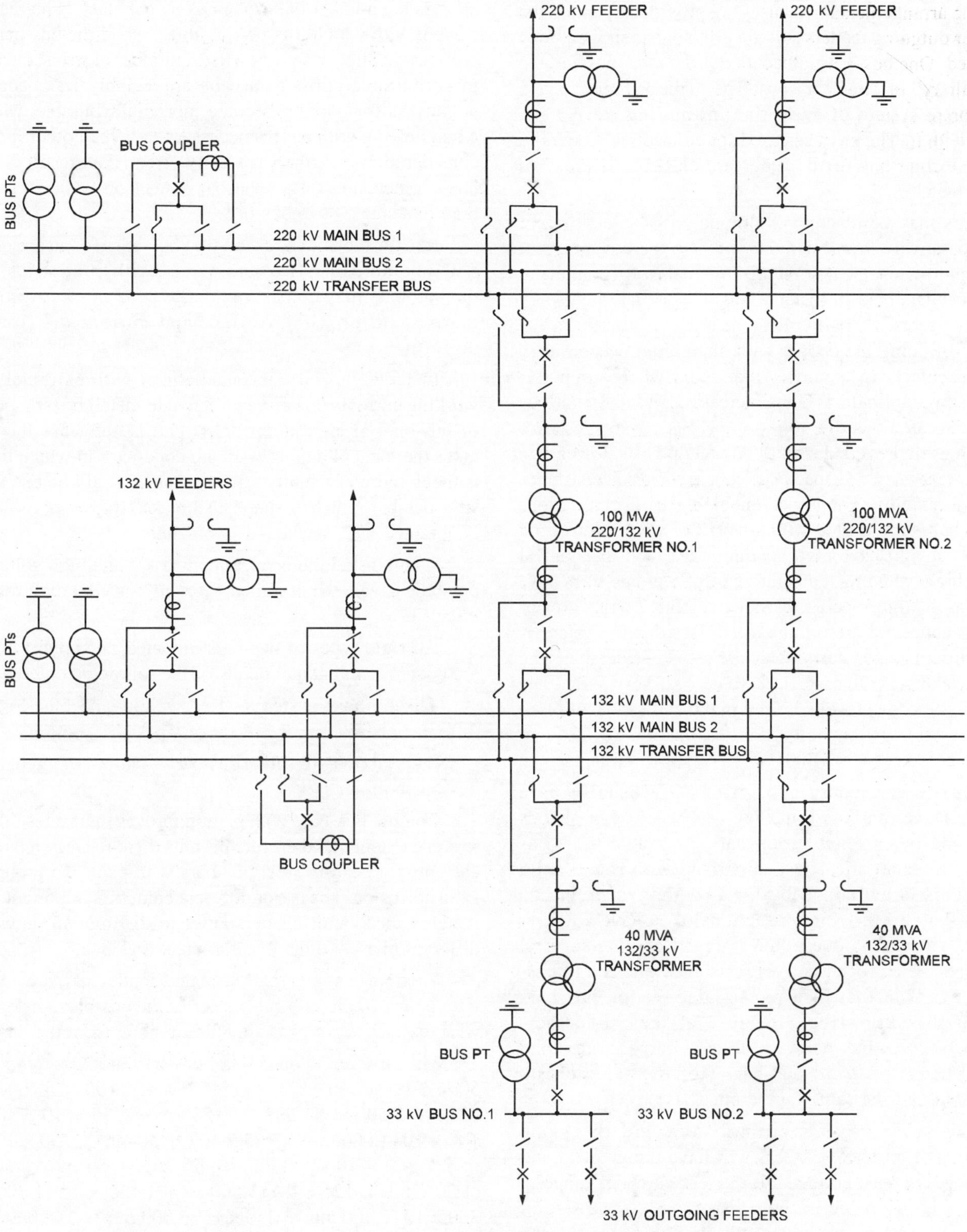

Fig. 20.10. ***Key Diagram For 220 kV Substation***

20.9. EQUIPMENT FOR SUBSTATIONS AND SWITCHGEAR INSTALLATIONS

The main equipment required for substations and switchgear installations is given below:

1. ***Main Bus-Bars.*** Bus-bar (or bus in short) term is used for a main bar or conductor carrying an electric current to which many connections may be made.

Bus-bars are merely convenient means of connecting switches and other equipment into various arrangements. The usual arrangement of connections in most of the substations permits working on almost any piece of equipment without interruption to incoming or outgoing feeders.

In some arrangements two buses are provided to which the incoming or outgoing feeders and the principal equipment may be connected. One bus is usually called the "*main*" bus and the other "auxiliary" or "transfer" bus. The main bus may have a more elaborate system of measuring instruments, relays etc. associated with it. The switches used for connecting feeders or equipment to one bus or the other are called "*selector*" or "transfer" switches.

Bus-bars may be of copper, aluminium or steel. *Copper* has a comparatively low resistivity and also the advantage of relatively high mechanical strength; this makes it economical to use copper bus-bars in installations of very large capacity where the currents are particularly heavy. During 1960's the need for substituting the copper with aluminium became very urgent, particularly in countries like India where copper is imported. Now aluminium is being increasingly used for various switchgear installations due to its numerous advantages over copper such as higher conductivity on weight basis, lower cost for equal current carrying capacity, excellent corrosion resistance and ease of formability. For proper reliable electrical connections aluminium buses are coated with silver. The **aluminium** used for bus-bars should have high conductivity, good mechanical properties, high softening temperature etc. **Steel bus-bars** have a comparatively high specific resistance (about 7 times greater than that of copper). Furthermore, the losses due to hysteresis and eddy currents when carrying ac are also considerable. The primary advantage is that steel bus-bars cost very little. The influence of the above factors is seen in the wide application of steel bus-bars in low-capacity installations where the load currents do not exceed 200-300 A.

The bus-bars used in substations are usually bare rectangular x-section bars (but they can be of other shapes such as round tubes, round solid bars, or square tubes) as they are more economical in comparison with round solid bus-bars. This is explained by the fact that *rectangular-section* bus-bars of the same x-sectional area have a higher rate of heat dissipation due to their greater cooling surface. Furthermore, the ac (or effective) resistance of a round bus-bar is greater than that of a rectangular-section bus-bars because of the skin effect. Because of these two facts rectangular-section bus-bars are able to carry larger load currents than round solid bus-bars (for the same cross-sectional area and the same temperature rise).

For the same cross-sectional area and the same temperature rise, copper bus-bars will have the greatest and steel bus-bars the least permissible current-carrying capacity because of the difference in their resistance. As the size of the bus-bars increases, their heat dissipation capacity falls off and the permissible current density must therefore be reduced.

If the load current to be carried exceeds the permissible current for a single-strip bus-bar of the greatest available size, each phase of the bus-bars shall have to be assembled of several strips arranged in a stack and clamped on post insulators. The air gaps left between the strips in a bus-bar stack are usually made equal to the thickness of the strips, this is necessary for adequate cooling.

As the number of strips in a bus-bar stack is increased, the current with which it is permissible to load the bus-bar cannot be raised in direct proportion to the number of strips; the increase in permissible current must be appreciably less because the conditions for cooling become more unfavourable. Moreover, when dealing with ac it is necessary to take another factor into consideration, *i.e.,* the **proximity effect**. The result is that the bus-bar metal in a stacked busbar is much less efficiently utilised than in a single-strip bus-bar.

It is for the above reason that bus-bars for ac are designed with not more than two, and, rarely, with three strips. For dc, the bus-bars may consist of a large number of strips because there is no proximity effect and current distribution is uniform.

In large-capacity ac installations with extremely heavy working currents it is more effective to install box-shaped types of bus-bars of aluminium or copper. In this case the bus-bar takes the form of a hollow square conductor in which the metal is much more efficiently utilized than it would be in a stacked-strip bus-bar. The box shape also makes for better cooling than is obtained with stacked-strip bus-bars.

When installations operate at 33 kV and high voltages, the bus-bars have to be designed with due consideration for corona effects.

All rigid types of bus-bars mounted on support insulators are coated with enamel paints of the following colours :

Three-phase systems—Red, yellow and blue to indicate different phases.

DC systems—Positive bus-bars—claret-coloured; negative bus-bars—blue.

Coating bus-bars with paint improves their rates of cooling to some extent and therefore permits them to carry a larger load current. By coating steel bus-bars with paint we protect them from corrosion. The use of different colours is important because it helps the operating personnel to distinguish between the different phases of the installation at a glance.

Flexible bus-bars (bare stranded conductors) are not coated with paint. To identify the phases of the bus-bars, disks painted with the respective phase colour are hung from the bus-bars.

The most common sizes of bus-bars are 25 × 6 (150 mm^2); 50 × 6 (300 mm^2), 75 × 6 (450 mm^2); 100 × 6 (600 mm^2); 125 × 6 (750 mm^2); 50 × 10 (500 mm^2); 75 × 10 (750 mm^2); 100 × 10 (1,000 mm^2); 125 × 10 (1,250 mm^2); 150 × 10 (1,500 mm^2); 200 × 10 (2,000 mm^2); 75 × 12 (900 mm^2); 100 × 12 (1,200 mm^2); 125 × 12 (1,500 mm^2); 150 × 12 (1,800 mm^2); 200 × 12 (2,400 mm^2); 250 × 12 (3,000 mm^2). The bus-bars are of 5 or 6 metres in length.

The early substations were generally with *flexible bus design*. A flexible bus consists of flexible ACSR or all-aluminium alloy stranded conductors supported by strain insulators from each end. The flexible bus is held at higher level above the different substation equipment. The connections between the flexible bus and the terminals of substation equipment are made by flexible conductors held in vertical or inclined plane.

Rigid bus-bars are easy to maintain. They are at lower height. Connections to substation equipment are easy. Aluminium tubes are preferred for rigid bus-bars.

A substation usually has a combination of rigid bus-bars and flexible bus-bars. ACSR conductors are preferred for flexible bus-bars.

The bus-bars are designed to carry certain normal current continuously. The cross section of conductors is designed on the basis of rated normal current and permissible temperature rise. The value of cross section so obtained is verified for temperature rise under short-time short-circuit current. The bus-bar conductors are supported on post insulators or strain insulators.

2. *Station Insulators.* Station insulators are used in generating stations and substations to fix and insulate the bus-bar systems. They may be subdivided into post and bushing (through) type.

A post insulator consists of porcelain body, cast iron cap and flanged cast iron base, as shown in Fig. 20.11 (*a*). The hole in the cap is threaded so that the bus-bars are either directly bolted to the cap or fixed by means of bus-bar clamp. Post insulators are available with round, oval, and square flanged bases for fixing respectively, with aid of one, two or four bolts. Each base, in addition, also has an earthing bolt.

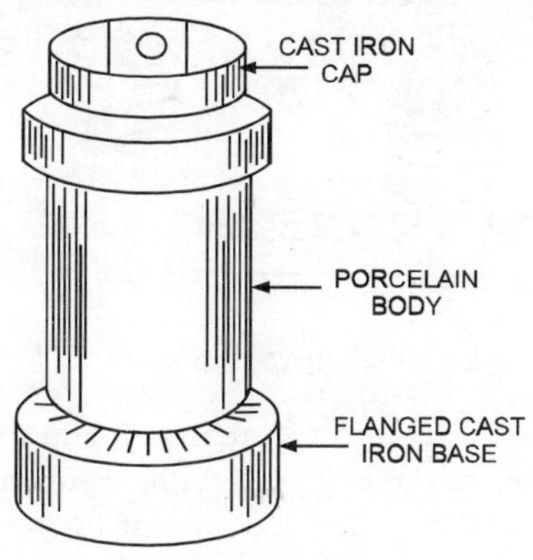

(a) Post Insulator

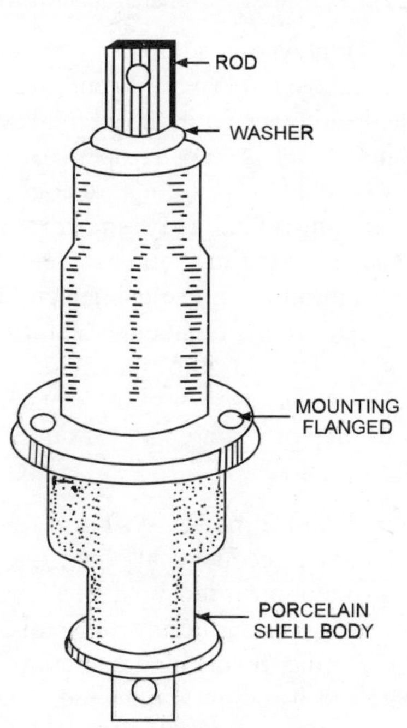

(b) Bushing

Fig. 20.11

A bushing or through insulator consists of porcelain-shell body, upper and lower locating washers used for fixing the position of bus-bar or rod in shell, and mounting flange with holes drilled for fixing bolts and supplied with an earthing bolt, as shown in Fig. 20.11 (*b*).

For current rating above 2,000 A, the bushings are designed to allow the main bus-bars to be passed directly through them.

Each phase of the bus-bars is coated with paint according to a fixed colour code—red, yellow and blue so that phase of the main bus-bars can be identified.

3. *Switchgear.* During the operation of the power system the generating plants, transmission lines, distributors and other electrical equipments are required to be switched on or off under both normal and abnormal operating conditions. The apparatus including its associated auxiliaries employed for controlling, regulating or switching on or off the electrical circuits in the electrical power system is known as *switchgear.* In home or office a simple tumbler switch with a fuse somewhere in the back ground serves to control and protect lights, domestic apparatus or other equipment and is in its own way a type of switchgear. With appliances of higher rating ordinary fuse may not give the desired result as such high rupturing capacity (HRC) fuses are used along with switches. But with the advancement of electrical power systems, the lines and other equipment operate at very high voltage and carry large current. Whenever a short circuit occurs, a heavy current flows through the equipment causing considerable damage to the equipment and interruption of service, so in order to protect the lines, generators, transformers and other electrical equipment from damage automatic *protective device or switchgear* is required. An automatic protective switchgear mainly consists of the relays and circuit breakers. When fault occurs on any section of the system, protective relay of that section comes in operation and close the trip circuit of the breaker, which disconnects the faulty section. The healthy section continues supplying loads as usual and thus there is no damage to equipment and no complete interruption of supply.

Broadly speaking switchgear is of two types viz. (*i*) outdoor type and (*ii*) indoor type. For voltage above 66 kV outdoor type is almost universally used because for such voltages building work will unnecessarily increase the installation cost owing to large spacing between conductors and large size of insulators. Below 66 kV there is no difficulty in providing building work for the switchgear at a reasonable cost. Moreover this type of switchgear is of metal-clad type and is compact. Owing to compactness, safety clearances for operation are also reduced thus reducing the area required.

Operation and functions of circuit breakers have been fully described in Chapters 7 and 8.

4. *Switches.* *Switches*, on the other hand, are used either for breaking a circuit where current is within the normal capacity or for breaking a circuit carrying no current. Where used for the latter purpose, switches are called *isolators*. Two forms of switchgear are in general use nowadays: air-break switches and oil-break switches.

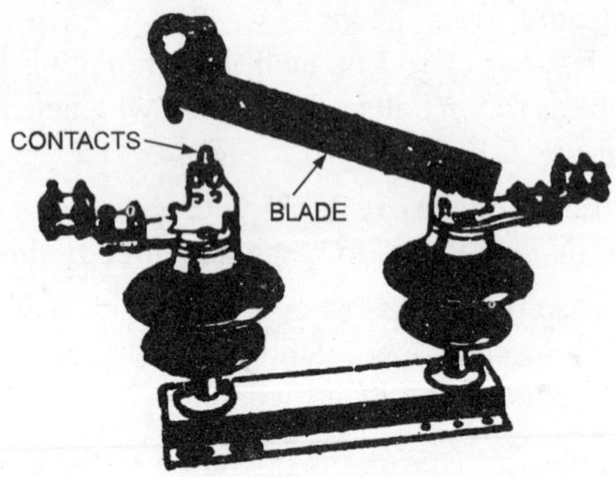

Fig. 20.12. ***Air Switch***

As their name imply, *air-switches* are those whose contacts are opened in air, while oil switches are those whose contacts are opened under oil. Oil switches are usually employed in very high voltage heavy current circuits.

Air switches are further classified as air-break switches and isolators (or disconnect switches).

(*i*) **Air-break Switches.** The air-break switch has both the blade and the contact equipped with arcing horns. Arcing horns are pieces of metal between which the arc resulting from opening a circuit carrying current is allowed to form. As the switch opens, these horns are spread farther and farther apart and the arc is lengthened until it finally breaks.

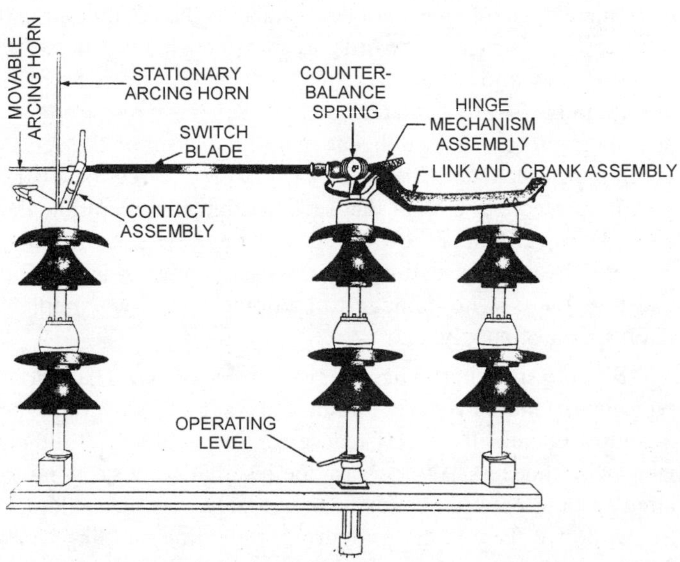

Fig. 20.13. ***Air-break Switch With Arcing Horns***

Air-break switches are of several designs. Some are operated from the ground by a hook on the end of a long insulated stick; some others through a system of linkages are opened by a crank at the foot of the pole. Where more than one conductor are opened, there may be several switches mounted on the same pole. These may be opened singly or altogether in a "gang" as this system is called. Some switches are mounted so that the blade opens downward and these may be provided with latches to keep the knife blade from jarring open.

(*ii*) **Oil Switches.** The oil switch has both the blade and the contact mounted in a oil-filled tank. The switch is usually operated from a handle on the outside of the case. As the switch opens, the arc formed between the blade and contact is quenched by the oil. Oil switches may be remote controlled as well as manually operated. These are used for capacitor switching, street lighting control and automatic disconnect in case of power failure.

(*iii*) **Isolators.** Since isolators (or isolating switches) are employed only for isolating circuit when the current has already been interrupted, they are simple pieces of equipment. They ensure that the current is not switched into the circuit until everything is in order.

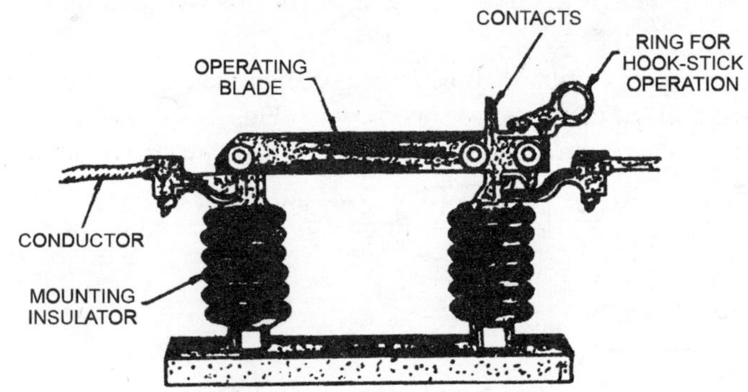

Fig. 20.14. ***Isolator***

Isolators or disconnect switches operate under no load condition. They are not equipped with arc-quenching devices. They do not have any specified current breaking capacity or current making capacity. The isolators in some cases are used for breaking charging current of transmission line.

Isolators are employed in addition to circuit breakers, and are provided on each side of every circuit breaker to provide isolation. While opening a circuit, the circuit breaker is opened first, then isolator. If an isolator is opened carelessly, when carrying a heavy current, the resulting arc could easily cause a flash-over to ground. This may shatter the supporting insulators and may even cause a fatal accident to the operator, particularly in *hv* circuits. While closing a circuit, the isolator is closed first, then circuit breaker. Isolators are necessary on the supply side of the circuit breakers in order to ensure isolation (disconnection) of the circuit breaker from the live parts for the purpose of maintenance. Automatic switching of isolators is preferred.

Isolators employed in power systems are usually 3-pole isolators, each having three identical poles. Each pole consists of two or three insulator posts mounted on a fabricated support. The fixed and moving conducting parts are of copper or aluminium rods. During the opening operation the conducting rods swing apart and isolation is obtained. The simultaneous operation of three poles is obtained by mechanical interlocking of the three poles.

To prevent the maloperation, the isolator is provided with the following interlockings :

(*a*) Interlocking between three poles for simultaneous operation.

(*b*) **Interlocking with circuit breakers.** *Isolator cannot be opened unless the circuit breaker is opened and circuit breaker cannot be closed unless the isolator is closed.*

(*iv*) **Load Interrupter Switches.** In addition to isolators and circuit breakers, there is one more device, called the **load-interrupting switch**, which combines the functions of the isolator and a switch. They are designed only for breaking and making the load currents.

These switches are designed and used to close and open high voltage circuits under normal working conditions (at normal load). The arc extinguishing device of the load interrupter is made in the form of a split, moulded plastic chute fitted with organic glass inserts. This chute surrounds the moving knife of the arc extinguishing system. The stationary arcing contact is located in the lower part of the chute.

Load interrupter switches are intended only for handling low-energy arcs resulting from the interruption of load current and altogether unsuitable for extinction of high-energy fault current arcs.

When the switch is opened, the working contacts between which the arc is drawn separate. Acted upon by the high temperature of arc, the walls of the organic material insert generate gases (mainly hydrogen), which create a longitudinal blast serving to extinguish the arc. Lever arm manually operating mechanisms are employed for closing and opening the load interrupter switches.

It is a wide practice to install load-interrupting switches in low-capacity installations like industrial shop, urban and rural, and like distribution substations where it is possible to provide short-circuit current protection with high-voltage fuses and where the only duty of load-interrupting switch is to make and break the load current.

The installation of a load-interrupting switch, including the high-voltage fuses, is not only cheaper, but also usually requires less space than the installation of a high-voltage circuit breaker.

(*v*) **Earthing Switch.** Earthing switch is connected between the line conductor and earth. Normally it is open and it is closed to discharge the voltage trapped on the isolated or disconnected line. When the line is disconnected from the supply end, there is some voltage on the line to which the capacitance between the line and earth is charged. This voltage is significant in *hv* systems. Before commencement of maintenance work it is necessary that these voltages are discharged to earth by closing the earthing switch. Normally, the earthing switches are mounted on the frame of the isolator.

Sequence of Operation During Opening/Closing of a Circuit.

While Opening : Open circuit breaker, open isolator and then close earthing switch, if any.

While Closing: Open earthing switch, close isolator and then close circuit breaker.

5. ***Power Transformers.*** Power transformers are used for stepping up the voltage for transmission at generating stations and for stepping down voltage for further distribution at main step-down transformer substations. Usually naturally cooled, oil immersed, known as ON type, two winding, three-phase transformers, are used up to the rating of 10 MVA. The transformers of rating higher than 10 MVA, are usually air blast cooled. For very high rating, the forced oil, water cooling and air blast cooling may be used. For regulating the voltage the transformers used are provided with on load tap changer.

They are put in operation during load hours and disconnected during light load hours *i.e.,* they are usually operated at approximately full load. This is possible because they are arranged in banks and can be thrown in parallel with other units or disconnected at will. So power transformers are designed to have maximum efficiency at or near full load (*i.e.,* with iron loss to full load copper loss ratio of 1 : 1). Power transformers are designed to have considerable leakage reactance than is permissible in distribution transformers because in power transformers inherent voltage regulation is not as much important as current limiting effect of the higher leakage reactance. Power transformers usually make use of flux density of 1.5 to 1.77, have percentage impedance ranging from 6-18% and regulation 6-10%.

Fig. 20.15. ***400/220 kV, 240 MVA Power Transformer***

The transformer specifications cover the following:

1. KVA rating; 2. Rated voltage; 3. Number of phases (single- or three-phase); 4. Rated frequency; 5. Connections (Δ or λ in case of 3-phase transformer); 6. Tappings if any; 7. Type of core (core or shell); 8. Type (power or distribution); 9. Ambient temperature (generally average 40ºC); 10. Type of cooling (*a*) cooling medium-air, oil or water (*b*) circulation type-natural or forced (*c*) simple or mixed cooling; 11. Temperature rise above ambient in ºC depending upon the class of winding insulation; 12. Voltage regulation [(*a*) Per cent or pu at full load at 75ºC unity pf or 0.8 pf lag (*b*) Impedance—per cent or pu (*c*) Reactance—per cent or pu]; 13. No-load current in amperes or per cent of rated current at rated voltage and rated

frequency; 14. Efficiency in per cent or pu at full load, 1/2 load, 3/4 load at unity power factor and 0.8 pf.

Power transformers are covered under IS 2026-1962.

The transformers are generally installed upon lengths of rails fixed on concrete slabs having foundation 1 to 1(1/2) metre deep.

6. Fuses. Fuses have already been described in detail in Chapter 6.

7. Instrument Transformers (CTs and PTs). Current transformers (CTs) and potential transformers (PTs) have already been described in Chapter 1.

8. Protective Relays. Protective relays have already been described in detail in Chapters 9 and 10.

9. Lightning Arresters. Lightning arresters have already been described in detail in Chapter 15.

10. *Indicating and Metering Instruments.* Ammeters, voltmeters, wattmeters, kWh meters, kVARh meters, power factor meters, reactive-volt-ampere meters are installed in substations to control and maintain a watch over the currents flowing through the circuits and over the power loads.

Detailed description of each and individual instrument is beyond the scope of the book and for it some book on instrumentation may be consulted.

11. *Batteries.* In electric power stations and large capacity substations, the operating and automatic control circuits, the protective relay systems, as well as emergency lighting circuits, are supplied by station batteries. The latter constitute independent sources of operative dc power and guarantee operation of the above mentioned circuits irrespective of any fault which has occurred in the station or substation, even in the event of complete disappearance of the ac supply in the installation.

Station batteries are assembled of a certain number of accumulator cells depending on the operating voltage of the respective dc circuits. Storage batteries are of two types viz lead acid and alkaline batteries. Lead-acid batteries are most commonly used in power stations and substations because of their higher cell voltage and low cost.

Brief description of lead acid batteries and its charger will be given later on (Art. 20.10).

12. *Carrier-Current Equipment.* Such equipment is installed in the substations for communication, relaying, telemetering or for supervisory control. This equipment is suitably mounted in a room known as *carrier room* and connected to the high voltage power circuit.

13. *Control Cables.* The control cables and conduit system are required for affecting automatic controls. The control system generally operates at 110 V or 220 V and the cables employed for this purpose are multi-core cables having 10 or 37 or 61 conductors according to requirement. For laying these cables generally ducts are run from control room basement to centrally located junction box from where the conduits are run to the required points.

14. *Switchyard.* The switchyard houses transformers, circuit breakers and switches for connecting and disconnecting the transformers and circuit breakers. It also has lightning arrestors for the protection of the power station against lightning strokes.

20.10. LEAD-ACID BATTERIES

20.10.1. Construction

Various parts of a lead-acid battery are shown in Fig. 20.16.

1. *Container.* The materials of which the battery containers are made should be resistant to sulphuric acid, should not deform or become porous, or contain impurities deterious to the electrolyte; of these iron and manganese are especially intolerable.

The containers of lead-acid cells are made of glass, lead lined wood, ebonite, hard rubber of bituminous compound, ceramic materials or moulded plastics and are sealed at the top (with three holes, one at each end for the posts and one in the middle for the vent plug and through which the electrolyte is poured and the gases escape out) to avoid spilling of electrolyte. At the bottom of the container there are four ribs (prisms), on two of them rest the positive plates, and on the other two the negative plates. These prisms serve as supports for the plates and at the same time protect them from short circuits that would otherwise occur as a result of fall of the active material from the plates on to the bottom of the container.

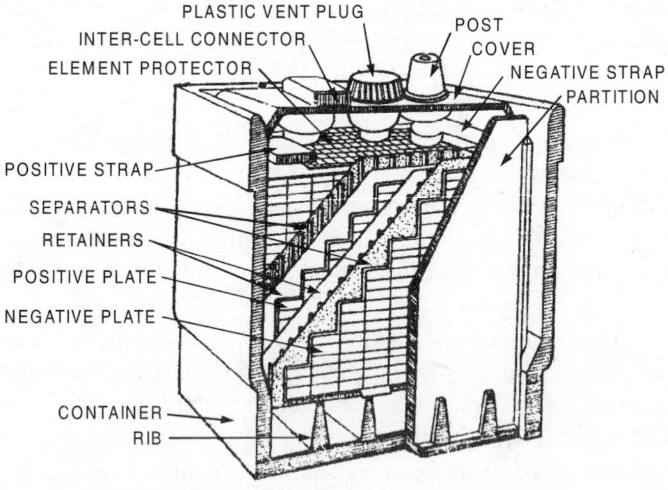

Fig. 20.16. *Lead Acid Battery*

2. *Plates.* The plates of lead-acid cells are of diverse designs, however, they all consist of some form of grid made of lead and the active material. The grid serves as a support for the active material and is also necessary for conducting the electric current, and as a means for distributing the current evenly over the active material. If the current is not distributed uniformly, an uneven change will take place in the volume of active material during charging and discharging processes, the result being that the active material will loosen and fall out and the plates warp. Lighter grids are employed for automotive (or SLI) batteries, designed mainly for short discharges. For motive power (traction type) and stationary batteries, which are designed for long term service, and in which the discharges are followed by idle intervals or are of long duration, heavier grids are employed.

The grids are cast of an alloy of lead and antimony. These are usually made with transverse ribs that cross the plate at right angle or diagonally. Grids for the positive and negative plates are often of the same design, but the grids serving the negative plates are made lighter, in as much they are not as essential to uniform conduction of the current and are less subject to corrosion than the grids of the positive plates.

Plates are of two types viz. Formed plates or Plante plates and pasted or Faure plates.

The positive plates are usually formed by the formation process. Plante plates are used largely for stationary batteries as these are heavier in weight and more costly than pasted plates but are more durable and less liable to lose active material by rapid charging and discharging. Plante plates have low capacity-weight ratio (12 to 21 Ah per kg of plate).

Faure process is much suitable for manufacturing of negative plates rather than positive plates. The negative active material (sponge lead) is quite tough (not hard and brittle as lead peroxide is) and moreover, it undergoes a comparatively negligible change in volume during charging and discharging processes of a cell. So it has no tendency to disintegrate or shed out of the grid but it tends to lose its porosity and become dense and, therefore, lose capacity. So in the manufacture of the negative plates by faure process, a small percentage of certain substances, such as powdered graphite, magnesium sulphate or barium sulphate is added in order to increase the material porosity.

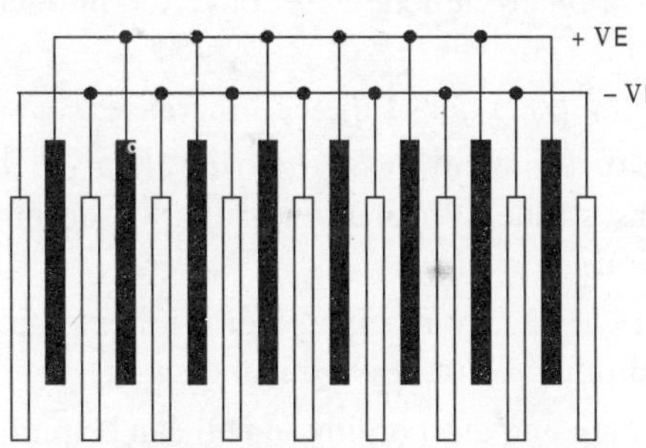

(*a*) *Arrangement of Plates in a Lead-acid Cell*

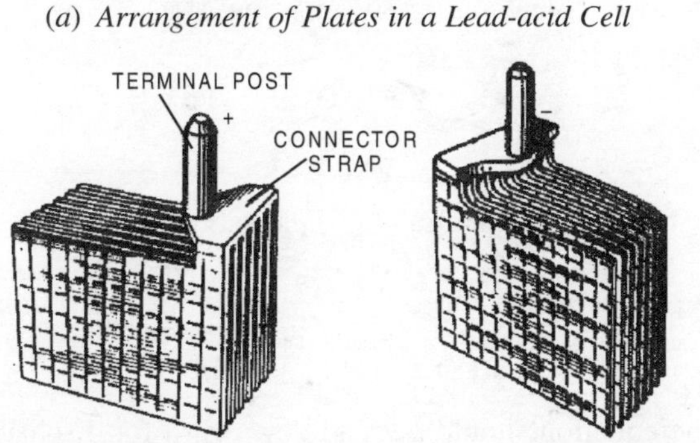

(*b*) *Positive Plates* (*c*) *Negative Plates*

Fig. 20.17. ***Grouping of Lead-acid Cell Plates***

In order to obtain large capacity from a battery, a large surface must be exposed to the electrolyte, and since the size of a single plate is limited so increased capacity is obtained by connecting a number of plates in parallel to form a group. By doing so internal resistance of the battery is also reduced. Two sets of plates are then sandwiched together, adjoining plates being separated from one another by perforated plastic, fiberglass, hard rubber or wooden separators. The – ve group of the plates contains one more plate than that of the + ve group so that when interleaved both the end plates become the – ve. This is done so that all the + ve plates can work equally on both sides. During charging lead sulphate ($PbSO_4$) on the + ve plate is changed to PbO_2 and the active material of the plate expands. If a positive plate were used at one end, one of its sides would remain inactive, when the other side expands and thus the plate is forced to buckle. The arrangement of plates in a cell is illustrated in Fig. 20.17 (*a*). Grouping of lead-acid cell plates is shown in Figs. 20.17 (*b*) and (*c*).

3. *Active Materials.* The materials, in a cell (or battery), taking active participation in chemical reaction (absorption or evolution of electrical energy) during charging or discharging are called the *active materials* of the cell.

The active materials of a lead-acid battery are :

1. **Lead peroxide** (PbO_2) dark chocolate brown in colour. It forms the positive active material.
2. **Sponge lead** (Pb) grey in colour. It forms the negative active material.
3. **Dilute sulphuric acid** (H_2SO_4) is used as electrolyte. The electrolyte of a fully charged battery contains about 31% sulphuric acid by weight and 21% by volume for specific gravity of 1.23 at 27ºC. This is the recommended value for the tropical climate. Slightly higher values can be used in winter, if desired. The acid should be made from sulphur and not from pyrites, as the latter is liable to contain injurious substances.

The mixture should be made by pouring the acid slowly into the water, *never the reverse*.

The lead peroxide and sponge lead, which form the positive and negative active materials respectively of a fully charged cell, as mentioned above, have little mechanical strength and, therefore, cannot be used alone. Supports of grids required for this purpose, in modern form, are made of either pure lead or an alloy of lead and antimony.

4. *Separators.* The separators, which are thin sheets of non-conducting porous material made of chemically treated wood, porous rubber or mats of glass fibres, are placed between the positive and negative plates to insulate them from each other. Separators are grooved vertically on one side and are smooth on the other. The grooved side is placed facing the positive plates to provide greater acid volume next to the surface for improved efficiency, to facilitate acid circulation within the cell, and to minimise the area of contact with the positive plate which has a highly oxidising effect on wood separators.

5. *Battery Terminals.* A battery has two terminals—the positive and the negative. The positive terminal with a diameter of 17.5 mm at the top is slightly larger than the negative terminal which is 16 mm in diameter.

20.10.2. Working Principle

When the sulphuric acid is dissolved, its molecules break up into hydrogen positive ions ($2H^+$) and sulphate negative ions (SO_4^{--}) and move freely. Now if two lead electrodes are immersed in this solution and connected to dc supply mains, the hydrogen ions being positively charged move towards the electrode connected to the negative terminal of the supply mains (*i.e.*, cathode) and SO_4^{--} ions being negatively charged move towards the electrode connected to the positive terminal of the supply main (*i.e.*, anode). Each positive hydrogen ion (H^+) reaching the cathode gets one electron from it, becomes atom of hydrogen and escapes out as hydrogen gas in the form of bubbles. Each of the negatively charged sulphate ion (SO_4^{--}) reaching the anode gives two electrons to it, reacts with water and forms sulphuric acid and oxygen according to chemical reaction $SO_4 + H_2O \rightarrow H_2SO_4 + O$. The oxygen produced attacks the lead anode and forms lead peroxide PbO_2. Thus during charging lead cathode remains as lead but lead anode gets converted into lead peroxide (PbO_2), chocolate in colour.

Now if the cell is disconnected from dc source of supply and voltmeter is connected between the electrodes, it will show potential difference between them and on connecting these electrodes, by a wire, current will flow from positive plate to the negative plate through external circuit, *i.e.,* the cell is now capable of supplying electrical energy.

Chemical Action During Discharging. When the cell is fully charged anode is of lead peroxide (PbO_2) and cathode is of sponge metallic lead (Pb). When the electrodes are connected through a resistance, the cell discharges and electrons flow in a direction opposite to that during charging. Now sulphuric acid molecules again break up into ($2H^+$) and (SO_4^{--}) ions. Each hydrogen ion now moves to the anode and reaching the anode receives one electron from the anode (coming from cathode through external circuit) and becomes hydrogen atom. Since it is directly in contact with anode of PbO_2, so it attacks and forms lead sulphate ($PbSO_4$), whitish in colour and water according to chemical equation.

$$PbO_2 + 2H = PbO + H_2O$$
$$PbO + H_2SO_4 = PbSO_4 + H_2O$$
$$\overline{PbO_2 + H_2SO_4 + 2H = PbSO_4 + 2H_2O}$$

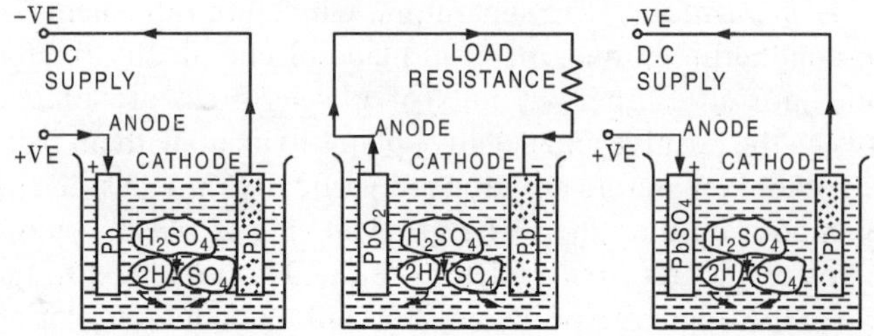

(a) Charging *(b) Discharging* *(c) Recharging*

Fig. 20.18. ***Charging and Discharging of Lead Acid Cells***

Each sulphate ion (SO_4^{--}) moves towards the cathode and reaching there gives up two electrons (moving to the anode through external circuit) becomes radical SO_4, attacks the metallic lead cathode and forms lead sulphate, whitish in colour according to chemical equation $Pb + SO_4 = PbSO_4$.

Thus during discharging (*i*) both of the electrodes are converted into lead sulphate, whitish in colour (*ii*) terminal potential of the cell falls (*iii*) density of the electrolyte decreases due to formation of water during chemical action and (*iv*) chemical energy stored is converted into electrical energy and is supplied to an external circuit.

Chemical Action During Recharging. When the cell gets discharged as explained above both of the electrodes are converted into lead sulphate ($PbSO_4$), whitish in colour. On recharging of the cell, the anode and cathode are connected to the positive and negative terminals, of the dc supply mains respectively. The molecules of sulphuric acid (H_2SO_4) in solution again breaks up into ($2H^+$) and (SO_4^{--}) ions. Hydrogen ions ($2H^+$) being positively charged move to the cathode, receives two electrons from there (coming from anode through external circuit), form hydrogen atom and react with lead sulphate cathode forming lead and sulphuric acid according to chemical equation,

$$PbSO_4 + 2H = H_2SO_4 + Pb$$

SO_4^{--} ion moves to the anode, gives up its two additional electrons, (moving to the cathode through external circuit) becomes radical SO_4, reacts with lead sulphate anode and forms lead peroxide (PbO_2) and sulphuric acid according to chemical equation

$$PbSO_4 + 2H_2O + SO_4 = PbO_2 + 2H_2SO_4$$

Hence during recharging

(*i*) Lead sulphate anode gets converted into lead peroxide (PbO_2) dark chocolate brown in colour.

(*ii*) Lead sulphate cathode gets converted into lead, grey in colour.

(*iii*) Terminal potential of the cell increases.

(*iv*) Density (concentration) of the electrolyte (H_2SO_4) increases (due to formation of H_2SO_4 during chemical reaction).

(*v*) Electrical energy received from the external source is stored in the form of chemical energy.

The charging and discharging of cell can be represented by a single reversible equation given below :

Positive plate Negative plate
$$PbO_2 + 2H_2SO_4 + Pb$$

Charge $\Uparrow\Downarrow$ Discharge

Positive plate Negative plate
$$PbSO_4 + 2H_2O + PbSO_4 + \text{Electrical energy}$$

The equation should be read downward for discharge and upward for recharge.

20.10.3. Charging Methods

Direct current is essential, and this may be obtained in some cases direct from the supply mains. In case the available source of supply is ac then it is converted into dc by some means such as motor-generator set, rotary convertor set or rectifier. The circuit shown in Fig. 20.19 may be used for charging the secondary batteries from dc supply main's.

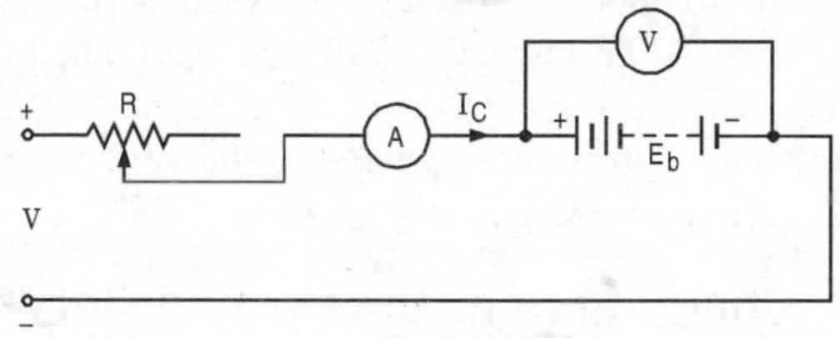

Fig. 20.19

The charging current is given by the expression

$$I_c = \frac{V - E_b}{R + r}$$

where V is supply voltage, E_b is battery voltage, R is current limiting resistance and *r* is the internal resistance of the battery.

If the source voltage is only slightly greater than the battery voltage, small source voltage or battery voltage variations will cause large variations in charging current. It is, therefore, better, whenever possible, to select a higher source voltage and a large resistance. This provides a more uniform charging current flow despite major changes in the battery voltage.

2. Charging of Batteries From AC Power Source. The basic requirements of common ac source chargers are like those of the dc power source, namely, the source voltage must be significantly greater than the battery voltage and the impedance must limit the current. The impedance of an ac power source can be reactive, resistive or combination of these. Limiting the current by reactance has the advantage that the reactive impedance component does not contribute to rectification. Rectification is used to convert ac into dc. Most chargers make use of a transformer which permits optimized matching of the source voltage to the battery charging voltage. Another very important feature of the transformer is that it isolates the charging output from the ac supply mains and thus avoids the hazard of electric shock. The rectifier used for this purpose may be mercury vapour rectifier (for medium and heavy charging loads only), copper oxide or selenium rectifiers or silicon or germanium rectifiers. The unit is usually provided with tapping points to give different low voltages *e.g.*, 6, 12, 18, 24 V. This is most economical and simple method of charging batteries. The circuit diagram is shown in Fig. 20.20.

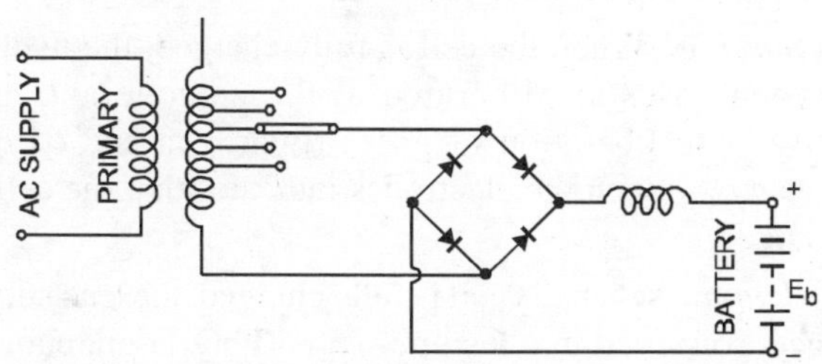

Fig. 20.20. ***Full-Wave Bridge Charging Circuit***

DC supply for battery charging from ac supply source can also be made available by employing a dc shunt wound generator coupled to an ac motor. In this method of charging batteries, the charging current can be controlled simply by regulating the generator excitation. In the event of failure of supply to the motor during charging, say because of blown fuses, the generator would continue to run as a shunt motor in the same direction of rotation, with no damage to battery.

Types of Charging. There are mainly two types of charging namely *constant voltage charging* and *constant current charging.*

(*a*) *Constant Voltage Charging.* In this method, the charging voltage is kept constant throughout the charging process. In this method the charging current is high in the beginning when a battery is in discharged condition, and it gradually drops off as the battery picks up charge resulting in increased back emf. Charging at constant voltage may be carried out only when the batteries have the same voltage, for example, 6 or 12 or 24 V. In this case source of current should have a voltage of 7.5, 15 or 30 V; these batteries are connected in parallel to the charging circuit.

The convenience of charging at constant voltage is that it allows cells with different capacities and at different degrees of discharge to be charged. The large charging current at the beginning of the charge is of relatively short duration and will not harm the cells. At the end of the charge the charging current drops to almost zero because the voltage of the battery becomes nearly equal to the voltage of the supply circuit. This method, is however, not very suitable for old, badly sulphated batteries which need prolonged charging at a slow rate.

This method is the most common method of charging lead-acid batteries and has been used successfully for over 50 years for different types of lead-acid batteries. With this method of charging, the charging time is almost reduced to half, capacity is increased by approximately 20% but efficiency is reduced by approximately 10%.

(*b*) *Constant Current Charging.* In this method of charging of batteries, the batteries are connected in series so as to form groups and each group is charged from the dc supply mains through loading rheostats. The number of batteries in each group depends on the charging circuit voltage which should not be less than 2.7 V per cell. The charging current is kept constant throughout the charging period by reducing the resistance in the circuit as the battery voltage goes up. This method is usually employed for initial charging of lead-acid batteries and for charging portable batteries in general.

In order to avoid excessive gassing or overheating, the charging may also be carried out in two steps, an initial charging of comparatively higher current and a finishing rate of low current. In this method the charge current is kept one-eighth of its ampere-hour rating.

The excess voltage of the supply circuit is absorbed in the series resistance. However, the groups of batteries to be charged should be so connected that the series resistance absorbs as little energy as possible. The series resistance used should be of current carrying capacity equal to or greater than the charging current required, otherwise the series resistance will overheat and burn out. The groups of batteries should be selected so that they all have the same capacity. If the batteries have the different capacities the charging current will have to be set according to the battery having the least capacity. This would both retard and complicate the charging.

Precautions During Charging. (*i*) During the charging period the temperature of the electrolyte should not exceed beyond 40 to 45 °C because of the danger of plate buckling. Therefore, the electrolyte temperature is carefully watched and if the temperature approaches the danger limit before the end of the charge, the charging current should be reduced or the charging discontinued until the batteries cool down.

(*ii*) While the charging is in process the fans in the battery room should be kept operating continuously. This is specially important during the last period of the charging in order to exhaust from the premises the dangerous oxygen and hydrogen gas evolved by the batteries during charging.

(*iii*) The gases, hydrogen and oxygen evolving from the batteries under charge can explode if spark or flame is brought too near. No metal tool should be used on batteries without switching off the battery charger.

(*iv*) Before the end of the charging it is necessary to examine all the batteries and make sure that gassing begins simultaneously in all the cells and with the same intensity. It is also necessary to measure the density of the electrolyte and the voltage of the battery cells.

Battery Charging Regulator. The basic components of the circuits are shown in Fig. 20.21. Diodes D_1 and D_2 are to establish a full-wave rectified signal across SCR_1 and the 12-V battery to be charged. When the battery is in discharged condition, SCR_2 is in the off state as will be clear after discussion.

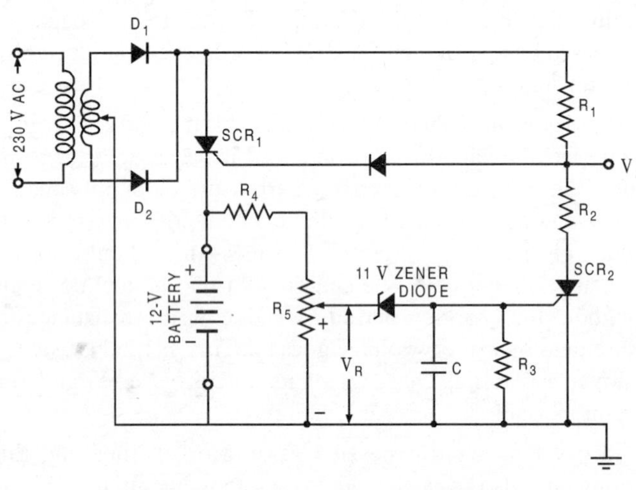

Fig. 20.21. ***Battery Charging Regulator***

When the full-wave rectified input is large enough to give the required turn-on gate current (controlled by resistor R_1), SCR_1 will *turn on* and the charging of the battery will commence. At the commencement of charging of battery, voltage V_R determined by the simple voltage-divider circuit R_4 and R_5 is too small to cause 11.0 V zener conduction. In the *off state*, zener diode is effectively an open circuit maintaining SCR_2 in the *off state* because of zero gate current. The capacitor C is included in the circuit to prevent any voltage transients in the circuit from accidentally turning *on* of the SCR_2. As charging continues, the battery voltage increases to a point where V_R is large enough to both turn *on* the 11.0 V zener diode and fire SCR_2. Once SCR_2 has fired, the short circuit representation for SCR_2 will result in a voltage-divider circuit determined by R_1 and R_2 that will maintain V_2 at a level too small to turn SCR_1 *on*. When this occurs, the battery is fully charged and the open-circuit state of SCR_1 will cut off the charging current. Thus the regulator recharges the battery whenever the voltage drops and prevents overcharging when fully charged.

20.10.4. Charging and Discharging Curves

Typical charge and discharge curves (variations in terminal voltage) of a lead-acid accumulator are shown in Fig. 20.22. When the cell is charged, the voltage of the cell increases from 1.8 V to 2.2 V during first two hours, then increases very slowly, rather remains almost constant for sufficient time and finally rises to 2.5 to 2.7 V.

When a charged storage cell has just been disconnected from a charging source , its terminal voltage falls rapidly to 2.2 V. On discharge, the voltage of the cell drops to 2.0 V in the beginning; remains constant for sufficient time and falls to 1.8 V finally, as illustrated in Fig. 20.22.

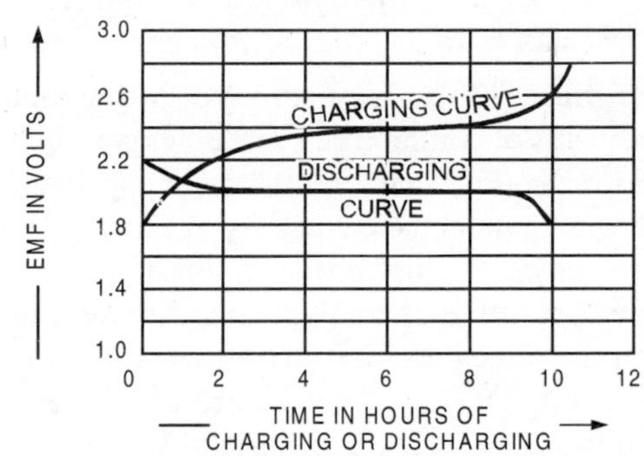

Fig. 20.22. ***Typical Charge and Discharge Curves For Lead-Acid Accumulator***

Caution. The cell should never be allowed to discharge beyond 1.75 V otherwise lead sulphate will be formed on the electrodes which is hard, insoluble and increases the internal resistance of the cell. The conversion of active material into lead sulphate is termed *sulphatization.*

20.10.5. Charging Indications

Full charging of lead-acid accumulator (or cells) can be judged from the following indications :

1. ***Gassing.*** When the cell is fully charged, the hydrogen and oxygen gases are liberated at the cathode and anode respectively, so liberation of gases (hydrogen and oxygen), known as *gassing*, on the electrodes indicates that the cells are fully charged.

2. ***Colour.*** When the cell is fully charged, the lead sulphate anode gets converted into lead peroxide (PbO_2) dark chocolate brown in colour and lead sulphate cathode gets converted into lead (Pb), grey in colour. It is considered one of the best tests for ascertaining the condition of a battery.

3. ***Voltage.*** When the cell is fully charged its terminal potential will be approximately 2.6 volts.

4. ***Specific Gravity of The Electrolyte.*** When the cell is fully charged, the specific gravity of the electrolyte will be approximately 1.21. When the cell is fully discharged its value falls to 1.17.

Cells are considered to be fully charged once three successive hourly readings of cell voltage and electrolyte gravity are found to be constant. However, the minimum total ampere-hours input, as mentioned by the manufacturers, must be provided to the cells even if the voltages and specific gravities are observed to be constant before that.

20.10.6. Electrical Characteristics

There are three important characteristics of an accumulator (or storage battery) namely (*i*) voltage (*ii*) capacity and (*iii*) efficiency.

(*i*) *Voltage.* Average emf of cell is approximately 2.0 volts. The value of emf of a cell does not remain constant but varies with the change in specific gravity of electrolyte, temperature and the length of time since it was last charged.

The emf of the cell increases with the increase in specific gravity of the electrolyte and vice versa but increase in specific gravity of the electrolyte also causes increase in internal resistance of the cell, therefore, its value should not go beyond 1.22. Best results are obtained with the electrolyte of specific gravity 1.21.

The emf of the cell, though not much, but slightly increases with the increase in temperature.

The terminal voltage of a battery is higher during charge than that during discharge due to the following reasons.

The internal voltage developed by chemical action, depends on the strength of electrolyte and increases slightly as the acid becomes stronger and the concentration of electrolyte increases due to formation of H_2SO_4 during charging and decreases due to formation of water during discharging. Moreover, since acid is formed in the pores of active material during charging and water is formed during discharging, and since it takes time for the acid or water to diffuse out, it follows that the strength of the electrolyte that is in actual contact with the active material is considerably greater during charging than the average strength of the electrolyte (acid) while during discharging it is considerably less than the average. Therefore, emf of the cell is greater during charging than that during discharging.

The terminal voltage of a battery is equal to $E + Ir$ while charging and to $E - Ir$ while discharging where I is the charging or discharging current and r is the internal resistance of the cell.

(*ii*) *Capacity.* The ability of an accumulator to last and give current is called *rated output* or the *capacity*. While the voltage of a cell is determined by its chemistry, the capacity of a cell is infinitely variable.

The capacity of a cell is essentially the number of electrons that can be obtained from it. Since the current is the number of electrons per unit time, cell capacity is the integration of current supplied by the cell over time. The capacity of the cell is, therefore, expressed in *ampere-hours* (A -h) and is equal to the product of the specified discharge current in amperes multiplied by the number of hours before the cell discharges to the specified extent. Thus, a rated output (or capacity) of 10 ampere-hours means that one ampere current can be drawn for 10 hours, or half an ampere current for 20 hours.

Thus the capacity of a battery may be defined as the useful quantity of electricity that can be drawn from a battery at the specified discharge rate before it falls to the specified value of voltage, which is equal to 1.75 V multiplied by the number of cells. The capacity of battery depends upon several factors, principal among which are area of plate surface; quantity, arrangement and porosity of the active material used in the manufacture of the plates; quantity and specific gravity of the electrolyte used; and the porosity of the separators. Rate of discharge and temperature also play important role.

The capacity of the cell increases with the increase in *plate surface area*. A rough rule for estimating the capacity of a battery is the surface area of positive plates in mm^2 multiplied by the number of such plates and divided by 1,000. For example the capacity of a battery having 5 positive plates each of dimensions of 100 mm and 50 mm, will be $\frac{100 \times 50 \times 5}{1,000}$ *i.e.*, 25 Ah.

Since electricity is produced by chemical action taking place within the cells, the capacity of the battery depends directly upon the *kind and amount of active material used.* Theoretically, roughly 4 grams of metallic lead on either element is required to be reduced to sponge lead or to lead peroxide to produce one ampere-hour of electricity. In practice, from four to six times this amount is required. The reason for this is that it is impossible to reduce all the active material, to bring every particle in contact with the electrolyte, or to cause every part to be penetrated by the current. Experiment shows that from 15 to 22 grams of sponge lead, and from 16 to 24 grams of metallic lead converted into PbO_2, are required on their respective elements to produce a discharge of one ampere-hour at ordinary commercial rates.

The capacity of the cell depends on the *concentration* or the *specific gravity* of the electrolyte as it affects the internal resistance and vigourisity of chemical reaction taking place in a cell. It increases with the increase in specific gravity of the electrolyte.

At a particular temperature, the capacity of the cell depends on its rate of discharging. For instance, a 100 ampere-hour battery capable of giving a continuous discharge of 10 A for 10 hours should theoretically give a discharge of 20 A continuously for 5 hours or 50 A for 2 hours or 100 A for one hour, but in reality, the *ampere-hour capacity decreases with the increase in discharging rate.* With the increase in the rate of discharge, voltage of the cell falls more rapidly due to internal resistance of the cell, the chemical reactions become brisk and so weaken the plates and reduce the capacity of the cell. If the cell is discharged too rapidly it may break the plates, and in case of pasted plates, a very sudden discharge will dislodge the paste. The capacities of chloride tubular stationary lead- acid cells at various rates of discharge, expressed as a percentage of the ampere-hours available at the 10 hour rate, are given in Fig. 20.23.

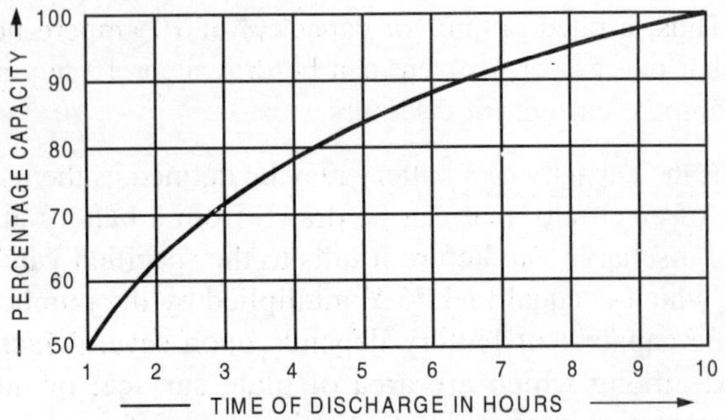

Fig. 20.23. ***Courtesy Chloride India Ltd.***

The capacity of a battery increases with the increase in temperature (Fig. 20.24) because at high temperature, the chemical reactions taking place within the cell become more vigorous, the acid resistance is reduced and diffusion of electrolyte is improved. However, at high temperature, the paste gets rapidly converted into lead sulphate which is always accompanied by expansion of paste particularly at positive plates causing in buckling and cracking of the grid. At high temperature the antimony-lead alloy grid, terminal posts and wooden separators are also attacked by the acid. So, it will not be advisable to operate lead-acid batteries beyond temperature of 40°C. With the fall in temperature, the chemical reactions become slow, cell internal resistance increases and diffusion of electrolyte becomes poor. Consequently, the capacity of the cell decreases with the fall in temperature till at freezing point (–35°C at specific gravity of 1.22 of electrolyte) the capacity is reduced to zero even though the battery otherwise be fully charged.

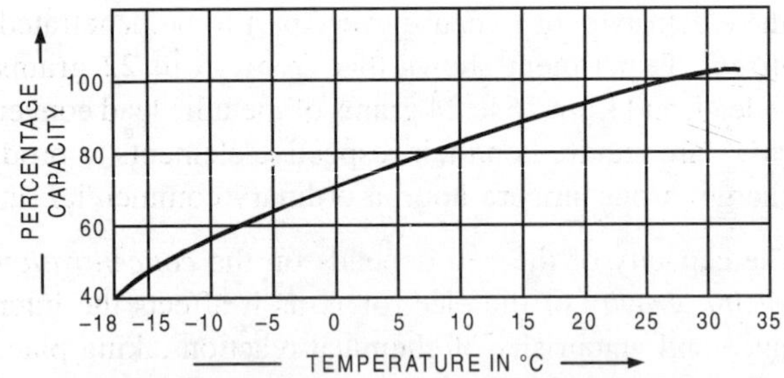

Fig. 20.24. ***Courtesy Chloride India Ltd.***

(*iii*) *Efficiency.* The efficiency of the cell can be given in two ways, enumerated and explained below.

1. *The Quantity or Ampere-Hour (A-H) Efficiency.* Since variations in terminal potential of the cell during charge and discharge are not taken into account while working out this efficiency and terminal potential of the cell during charge is higher than that during discharge, therefore, quantity efficiency is always higher than energy efficiency, in which variations of terminal potential of the cell are taken into account.

As generally efficiency is defined as the ratio of output to the input, similarly quantity efficiency or ampere-hour efficiency is defined as the ratio of ampere-hours of discharge and ampere-hours of charge.

i.e., Ampere-hour efficiency, η_{AH}

$$= \frac{\text{Ampere-hours of discharge}}{\text{Ampere-hours of charge}} \times 100 = \frac{I_d \times T_d}{I_c \times T_c} \times 100$$

The quantity efficiency of the lead acid cell varies from 90 to 95%. It would be 100 per cent if it were not for the gassing on charge, which represents a non-reversible chemical reaction. If the charging is discontinued each time as soon as the gassing becomes appreciable, the ampere-hour efficiency will be nearly 100 per cent but ampere-hour capacity will be reduced and it is advisable to give the battery a full charge from time to time in order to avoid deterioration of the otherwise unused lead sulphate. The quantity efficiency also decreases due to self discharge of the plates caused by local reactions and because of current leakage caused by faulty insulation between the cells and the battery.

2. *Energy or Watt-Hour Efficiency.* Energy efficiency is defined as the ratio of energy delivered in watt-hours by the cell during discharge and the energy drawn in watt-hours during charge.

i.e., Energy or watt-hour efficiency, η_{W-H}

$$= \frac{\text{Output in watt-hours}}{\text{Input in watt-hours}} \times 100$$

$$= \frac{\text{Current delivered} \times \text{time of discharge} \times \text{average pd during discharge}}{\text{Current drawn} \times \text{time of charge} \times \text{average pd during charge}} \times 100$$

$$= \frac{I_d \times T_d \times V_d}{I_c \times T_c \times V_c} \times 100 = \frac{I_d\, T_d}{I_c\, T_c} \times 100 \times \frac{V_d}{V_c} = \eta_{A-H} \times \frac{V_d}{V_c}$$

Operation at low rate of charge and discharge and at reduced ampere-hour capacity both tend to raise the watt-hour efficiency. Actual watt-hour or energy efficiency obtained in practice ranges from about 75 to 85 per cent.

20.10.7. Battery Ratings

The standards adopted, both by industry and government organisations, are given below :

1. *Ampere-hour Capacity Rating.* The ampere-hour rating of a battery is usually determined from its ability to deliver current continuously for 20 hours at 27°C. A battery that can deliver 5 amperes steadily for 20 hours then its rating will be 100 Ah.

2. *Reserve Capacity.* The reserve capacity of a battery is indicated in terms of minutes, a battery is capable of tolerating a drain of 25 A without dropping below 10.5 V (1.75 volts per cell). Higher this rating, better the battery is.

3. *Cold Rating.* The cold rating of a battery indicates the number of minutes a battery can deliver 300 A at – 18°C (0°F). This is applicable in relation to craft which ply in freezing weather.

4. *Cold Cranking Power Rating.* This rating is applicable to all 12 V batteries irrespective of their size. The battery is discharged at –18°C till its terminal voltage falls to 7.2 V. The output current is measured for 30 s. Higher the output, better the battery is.

20.10.8. Battery Testing

The condition of each cell should be tested at least once a week by a cell testing voltmeter. This test should be made at the end of discharge, when the voltage per cell should not fall below 1.85 V, the current being at the 10 hour discharge rate. The voltage of each cell should also be tested at the end of charge and the readings taken about 10 minutes before charging current is stopped.

The specific gravity of electrolyte in each cell should be tested once a week at the end of charge and, at the same time, the plates in every cell should be carefully examined. If any internal contacts are found between adjacent plates, they must be removed immediately.

20.10.9. Special Charges

The following types of special charges are usually required for proper maintenance of secondary batteries.

1. *Boosting Charge.* Though batteries are usually of sufficient capacity to work for the whole day without requiring any intermediate charge during the day. But in some cases such as in battery operated vehicles, however, the batteries may require an extra charge known as "*boosting charge*" in the middle of the day.

The current for this boosting charge is usually determined by an emperical rule according to which the *boosting charge current in amperes is numerically equal to battery discharge in ampere-hours up to that time when boosting charge is to be given.* For example for an 80 ampere-hour battery when discharged to the extent of 25 ampere-hours, the boosting charge current will be 25 amperes. Due to this charging current there is no gassing.

2. *Equalising Charge.* The equalising charge is a low rate charging and is given to the cells to ensure that every cell is brought into a fully charged condition at regular intervals. This charge is given until the voltage and specific gravity of the electrolyte remains constant for three successive hourly readings. This keeps the cell healthy for a longer period of time. The usual practice of giving equalising charge is as follows :

(*i*) For batteries subjected to "trickle charge"—not required.
(*ii*) For "floating" or idle batteries—once a month.
(*iii*) For batteries charged less frequently—once a fortnight.
(*iv*) For batteries charged every day—once a week.

3. *Trickle Charge.* The trickle charge is extremely a low rate charge and is applied to standby batteries for compensating the open-circuit losses. The charge current should be of such a magnitude that there is no gassing and when it is allowed to flow continuously through the fully charged batteries, specific gravity remains constant at the specific value.

The batteries subjected to 'trickle charge' can be maintained always in a fully charged condition for several years. The service life of the battery also increases.

20.11. SUBSTATION AUXILIARIES SUPPLY

In small unattended substations only a small amount of power for electric lighting during regular periods of inspection, maintenance and repair is required.

In regional substations the electric power is required for the auxiliaries—the lighting circuits, air blast fans of power transformers, battery charging sets, oil servicing facilities, compressor units in case of air blast circuit breakers, ventilating fans of the substation buildings, water supply and heating system equipment etc.

In substations incorporating synchronous condensers the supply is also required for the operation of auxiliary equipment of the synchronous condensers.

In large substations it is wide practice to connect two transformers to the 11 kV main bus-bars for supply of the auxiliaries at a voltage of 415 V/240 V.

An example of the auxiliary supply connections in a substation of medium capacity is shown in Fig. 20.25.

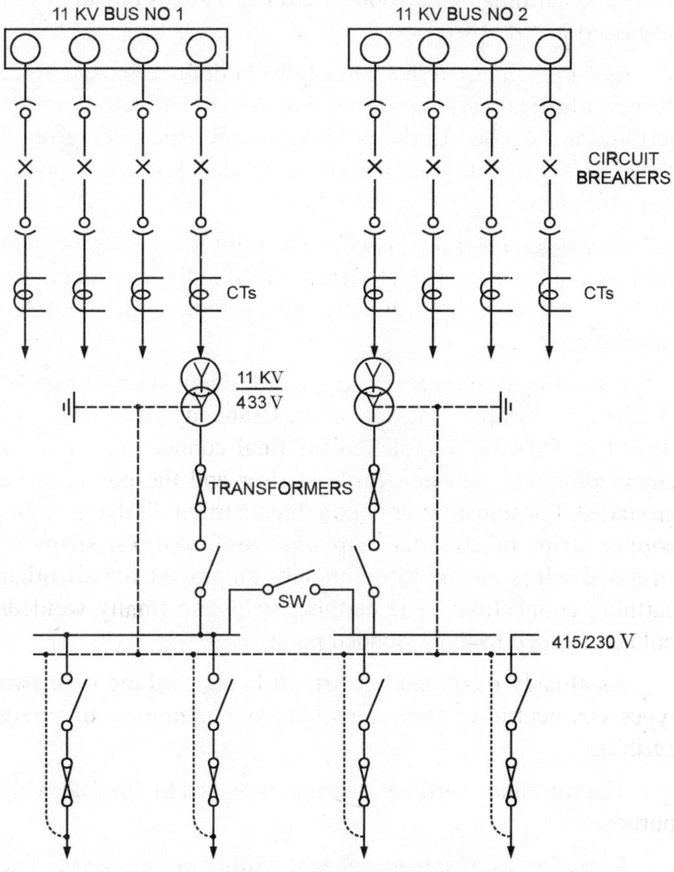

Fig. 20.25. ***Main Connection For Auxiliaries Supply***

20.12. SUBSTATION EARTHING

Modern substation earthing system has buried horizontal mesh of steel rods and vertical electrodes (spikes) welded to the mesh. Further, the vertical risers and the galvanised steel grounding strips or copper bars etc are connected between the grounding mesh and the points to be grounded.

Before 1960 s the design criterion of substation earthing system was 'low earth resistance' (below 0.5 Ω for hv installations). During 1960 s, the new criteria for the design and evaluation of substation earthing system were evolved particularly for EHV ac and *hv dc* substations. The new criteria is that the substation earthing system should have *low earth resistance, low touch potential* and *low step potential*.

The conventional criterion of "*low earth resistance*" and low current earth resistance measurement continues to be in practice for substations and power stations up to and including 220 kV.

The parts of the earthing system include the entire solid metallic conductor system between different earthed points and the underground earth mat. The earthed points are held near earth potential by low resistance conductor connections with earth mat, also called the *earthing grid* (see Art. 18.7).

The parts of the earthing system are:

An underground horizontal earth mesh, known as earth mat or earthing grid, earthing electrodes (or earthing spikes), earthing risers and earthing connections. Earthing mat has already been discussed in Art. 18.7.

Earthing electrode has already been defined in Art. 18.2. Several identical earth electrodes are driven vertically into the soil and are welded to the earthing rods of the underground mesh. Larger the number of earth electrodes, lower will be the earth resistance.

Earthing risers are usually the mild steel rods bent in vertical and horizontal shapes and welded to the earthing mesh at one end and brought directly upto equipment or structure foundation.

Earthing connections are galvanised steel strips or electrolytic copper flats or strips/stranded wires (cables)/ flexibles. These are employed for final connection (welded/ clamped/bolted) between earthing riser and the points to be grounded. For transformer neutral/high current discharge paths copper strips or stranded wires are preferred. GI strips or stranded wires are more commonly employed for all other earthing connections. The earthing strips are finally welded/ bolted/clamped to the grounded point.

As already mentioned in Art. 18.1, the earthing is of two types viz. neutral or system earthing and equipment or safety earthing.

The substation earthing system is provided for the following purposes.

1. *Safety of Operational and Maintenance Staff.* The earthing system is provided to ensure safety of the operational and maintenance staff working in a substation against electric shock. Deaths by shocks can be avoided completely by proper earthing of equipment. Before commissioning, check and certification of earthing system is essential. Before giving permit for carrying out maintenance work on the equipment it is essential that the said equipment is isolated and earthed from both ends.

2. *Discharge of Electrical Charges To Ground*. The ground system is provided to provide return path for discharging fault currents and discharge currents/voltages from the earthed points of lightning masts, lightning conductors, earthing switches, surge arresters etc, which are connected to the underground earthing system by solid or flexible earthing conductors of adequate short-time current carrying capability and low resistance.

3. *Grounding of Overhead Shielding Wires.* The overhead shielding wires and grounded-flanges of insulators and bushings are connected to the earthing system to obtain protection zone against lightning strokes for the outdoor and exposed conductors and equipment.

4. *Electromagnetic Interference.* The control rooms are provided with grounded screen in the walls and windows to ensure freedom from electromagnetic disturbances. Grounding of chassis of instruments, grounded screening of control room, computer room ensures freedom from electromagnetic disturbances on the operation of isolators, thyristors in main power circuits. This earthing system ensures freedom from electromagnetic interference in communication and data processing equipment in the substation.

Methods of Earthing. There are several methods of earthing viz wire or strip earthing, rod earthing, pipe earthing, plate earthing or earthing through a water mains. Most commonly used methods of earthing are pipe earthing and plate earthing.

1. *Pipe Earthing.* This is the most common and best system of earthing as compared to other systems suitable for the same earth and moisture conditions.

In this method of earthing, a galvanised steel and perforated pipe of approved length and diameter is placed up right in a permanently wet soil, as shown in Fig. 20.26.

The size of the pipe depends upon the current to be carried and type of soil. Usually the pipe used for this purpose is of diameter 40 mm and 2.5 metres in length for ordinary soil or of greater length in case of dry and rocky soil. The depth at which the pipe must be buried depends upon the moisture of the ground. The pipe is placed at a depth of 3.75 metres (minimum). The pipe is provided with a tapered casting at the lower end in order to facilitate the driving. The pipe at the bottom is surrounded by broken pieces of coke or charcoal for a distance of about 15 cm around the pipe. Generally alternate layers of coke and salt are used to increase the effective area of the earth and to decrease the earth resistance respectively. Another pipe of 19 mm diameter and minimum length 1.25 metres is connected at the top of GI pipe through reducing socket.

In our country in summer season the moisture in the soil decreases which causes increase in earth resistance. So a cement concrete work, as shown in Fig. 20.26, is done in order to keep the water arrangement accessible, and in summer to have an effective earth, 3 or 4 buckets of water are put through the funnel connected to 19 mm diameter pipe, which is further connected to GI pipe.

The earth wire (either GI wire or GI strip of sufficient cross section to carry faulty current safely) is carried in a GI pipe of diameter 12 mm at a depth of about 60 cm from the ground.

Care should be taken that earth wire is well protected from mechanical injury, when it is carried over from one machine to another.

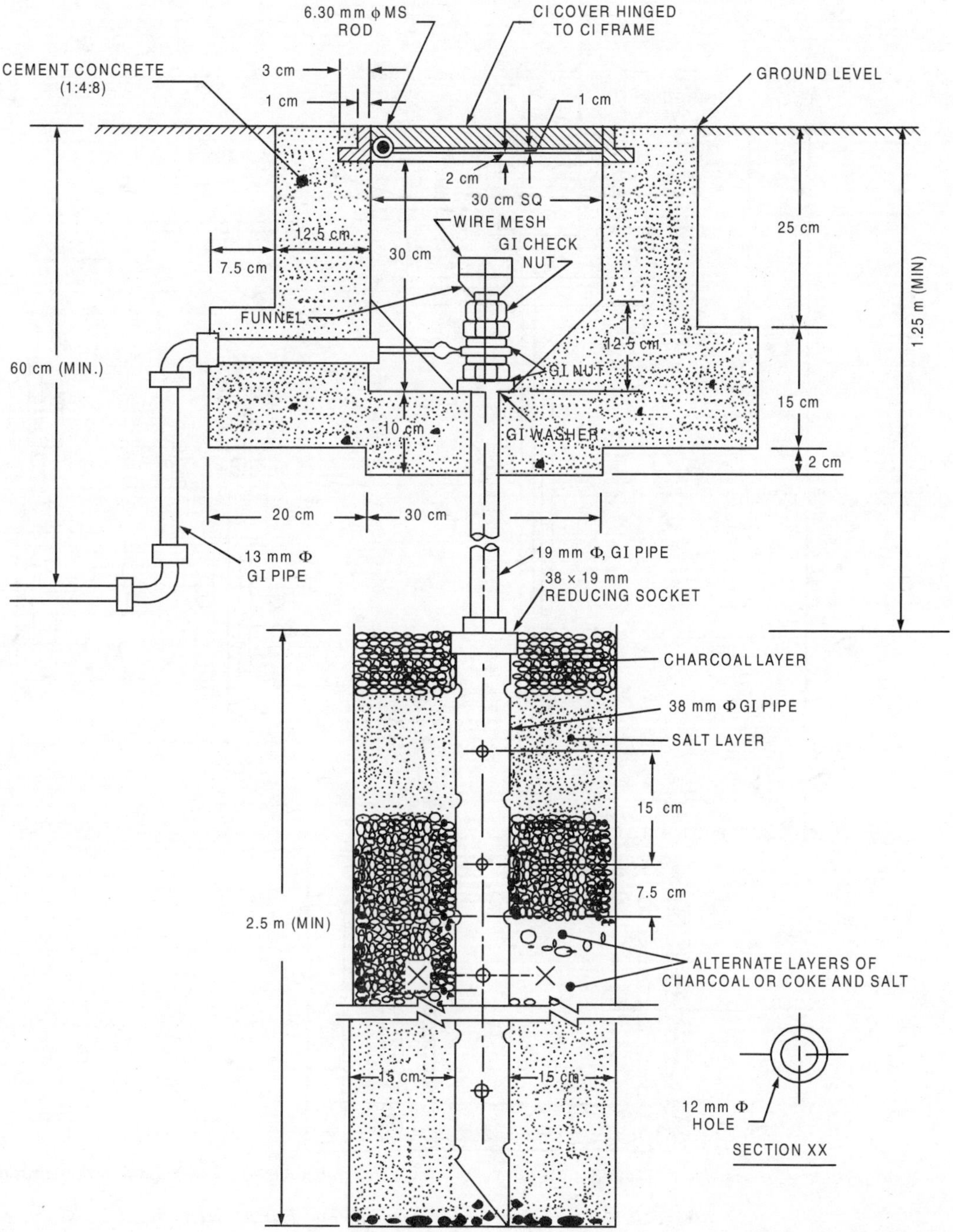

Note: Three or four buckets of water to be poured into sump every few days to keep the soil surrounding the earth pipe permanently moist.

Fig. 20.26. ***A Typical Illustration of Pipe Earth Electrode***

2. *Plate Earthing.* This is another common system of earthing. In plate earthing an earthing plate either of copper of dimensions 60 cm × 60 cm × 3 mm or of galvanised iron of dimensions 60 cm × 60 cm × 6 mm is buried into the ground with its face vertical at a depth of not less than 3 metres from ground level. The earth plate is embedded in alternate layers of coke and salt for a minimum thickness of 15 cm. The earth wire (GI wire for GI plate earthing and copper wire for copper plate earthing) is securely bolted to an earth plate with the help of a bolt, nut and washer made of material of that of earth plate (made of copper in case of copper plate earthing and of galvanised iron in case of GI plate earthing).

A small masonry brick wall enclosure with a cast iron cover on top or an RCC pipe round the earth plate is provided to facilitate its identification and for carrying out periodical inspection and tests.

The copper plate and copper wire are usually not employed for grounding purposes because of their higher cost.

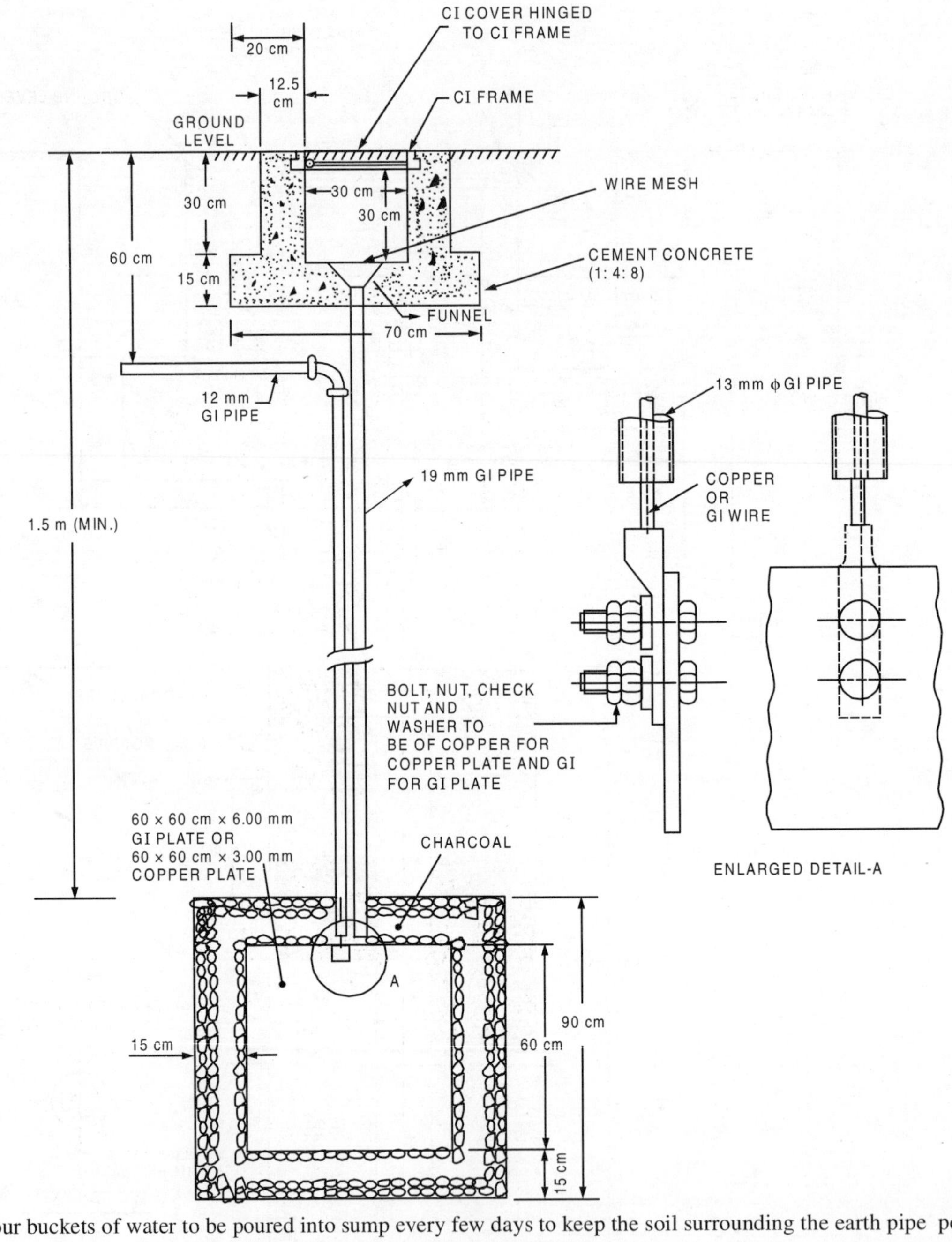

Note: Three or four buckets of water to be poured into sump every few days to keep the soil surrounding the earth pipe permanently moist.

Fig. 20.27. ***A Typical Illustration of Plate Earth Electrode***

Exercises

1. Explain the different types of the substations and draw layout for the same. Give the working of each component of a substation.
[U.P. Technical Univ. Power Station Practice 2006-07]
2. What are the functions of substation ? What are different types of substation ? Write short notes on each of them.
[M.D. Univ. Power System-I, May 2008]
3. Discuss different ways of classifying the substations. Give a comparison between indoor and outdoor substations. **[M.D. Univ. Power System-I, December 2008]**
4. Give layout of an indoor substation and describe briefly the function of each component used.
[M.D. Univ. Power System-I, December 2007]
5. What are the factors considered for the selection of site of an outdoor 11/0.4 kV substation. Describe advantages and disadvantages of outdoor substation over indoor type.
[M.D. Univ. Power System-I, May 2008]
6. Draw the key diagram of a typical 11 kV/400 V indoor substation. Explain functions of various equipments used.
[M.D. Univ. Power System-I, December 2008]
7. Draw and briefly explain the various parts of a typical 33 kV substation.
[U.P. Technical Univ. Power Station Practice 2007-08]
8. Draw and explain the layout of a substation. Also give different types of substations.
[G.B. Technical Univ. Power Station Practice 2011-12]
9. Make a list of the main equipment in a substation. Draw layout of a substation.
[Pb. Univ. Switchgear & Protection Sept. 1988]

10. What are the different types of substations ? Draw the single line diagram for any one type of substation and list down the various equipments used in that substation. **[A.M.I.E. Sec. B. Advanced Power Systems Summer 1996]**
11. Describe briefly the equipment that must be available in a substation. **[Pb. Univ. Electrical Power III (Switchgear and Protection) July 1992]**
12. Show the component diagram of 33 kV substation along with the various equipments used with rating. **[U.P. Technical Univ. Power Station Practice 2005-06]**
13. Draw the layout of typical 66/11 kV substation. **[M.D. Univ. Power System-I, December 2009]**
14. Explain the function of various components used in a substation. **[M.D. Univ. Power System-I, May 2009]**
15. Prepare a list of machineries and instruments necessary for 66 kV to 11 kV step-down outdoor type substation. Describe the functions of any four of them. **[M.D. Univ. Power System-I, December 2007]**
16. What is a substation ? Discuss the factors for designing a substation. **[M.D. Univ. Power System-I, December 2010]**
17. Differentiate between the functions of an isolator and a circuit breaker.
18. What is meant by switchgear ? Explain the difference between air-break switches, isolators and load interrupter switches.
19. Why is battery provided in a power stations ? Which type of battery is used in power substation ? Give construction, working principle, charging method and electrical characteristics of lead-acid batteries in brief.
20. Write short notes on the following :
   (*i*) Classification of substations. **[Agra Univ. 1982]**
   (*ii*) Indoor substations.
   (*iii*) Outdoor substations and their necessity.
   (*iv*) Pole-mounted substations.
   (*v*) Equipment for indoor substations.
   (*vi*) Substation layout.
   (*vii*) Power station batteries.
   (*viii*) Substation auxiliaries supply.
   (*ix*) Substation earthing—necessity and methods.
   (*x*) Grounding of a substation and its equipments.
   (*xi*) Earthing of substation.
   [W.B. Univ. of Technology 2009-10]

Short Answer Type Questions With Answers

Q. 1. State any four types of substations according to the service ?

Ans. The substations, according to the service rendered are: Transformer substations, Switching substations, Converting substations and Frequency changing substations.

Q. 2. State the relative merits of indoor and outdoor substations.

Ans. **Merits of Indoor Substations**. Less requirement of space, less maintenance and control cable length; protection from lightning; flexibility in installation, no dust and dirt.

Merits of Outdoor Substations. No building requirement, short erection time; Easy fault finding due to visibility of equipment; Easy repair work; Easier installation extension; Availability of sufficient space between equipments.

Q. 3. What are the factors governing the selection of site for substations ?

Ans. Factors governing the selection of site for a substation are : 1. Nearness to load centre 2. Availability of suitable and sufficient land 3. Availability of communication facilities 4. Drainage facilities 5. Availability of essential amenities to the staff. 6. Pollution free atmosphere 7. Facility for future extension 8. Distance from airport, military or police rifle training centre.

Q. 4. List out the equipment used in a substation.

Ans. Equipment used in a substation are Transformer(s); Bus-bars; Circuit breakers; CTs; PTs; Insulators; Isolators; Switchgear (fuses, air-break switches etc.,) Protective relays; Surge arresters; Cables, Firefighting equipment; Batteries; Earthing arrangement, Wave traps, Reactors etc.

Q. 5. What are the functions of panels and firefighting equipment in a substation ?

Ans. *Panels* or *cubicles* in a substation are provided to support the various equipment like meters, relays etc. It saves the horizontal space and facilitates the supervisory staff to record readings easily. All the panels/cubicles are properly grounded.

Special type of fire extinguishers are installed in a substation to quench electric fire in case of a fire accident because ordinary fire extinguishers or water cannot be used to quench electric fire as it is dangerous. They are provided in accessible position and every one in the substation must know its operation.

Q. 6. What is flexible bus ?

Ans. Flexible bus consists of flexible ACSR or all aluminium alloy stranded conductors supported by strain insulators from each end.

Q. 7. Name the interlockings provided with isolators.

Ans. The interlocking provided with isolators are :
1. Interlocking between three poles for simultaneous operation.
2. Interlocking with circuit breakers.

Q. 8. In what way is an isolator different from an air-break switch ?

Ans. Isolators are used to connect or disconnect the line only under no-load condition while air-break switches are used for making and breaking of the circuit under loaded condition. Air-break switches are provided with arc control devices.

Q. 9. Why are isolators necessarily provided on the supply side of the circuit breakers ?

Ans. Isolators are necessarily provided on the supply side of the circuit breaker in order to ensure isolation of the circuit breaker from the live parts for the purpose of maintenance.

Q. 10. Give the sequence of operation during opening and closing of a circuit.

Ans. Sequence of operation of circuit breaker, isolator and earthing switch during opening/closing of a circuit are given below.

While Opening: Open circuit breaker, open isolator and then close earthing switch if provided.

While Closing: Open earthing switch, close isolator and then close circuit breaker.

Q. 11. What do you understand by switchgear ?

Ans. The apparatus employed for switching, controlling and protecting the electrical circuits and equipment is known as switchgear.

Q. 12. What is a circuit breaker ?

Ans. Circuit breaker is a mechanical device designed to close or open contact members, thus closing or opening an electrical circuit under normal or abnormal conditions.

Q. 13. How does a circuit breaker differ from a switch ?

Ans. A circuit breaker is a device that interrupts the abnormal or fault currents and in addition performs the function of a switch.

Q. 14. What for isolating switches are provided along with circuit breakers ?

Ans. Isolating switches are provided for separating the disconnected element from the live portion of the system for the purposes of safety.

Q. 15. What is difference between primary and secondary batteries ?

Ans. Batteries are either primary or secondary. Primary batteries can be used only once because the chemical reactions that supply the current are irreversible. Secondary batteries, sometimes called *storage batteries* or *accumulators*, can be used, recharged and reused. In these batteries, the chemical reactions that provide current from the battery are readily reversed when current is supplied to the battery.

Q. 16. Lead acid battery can be charged where as primary cell cannot be charged. Why ?

Ans. Primary cells can be used only once because the chemical reactions that supply the electric current are irreversible but in lead acid battery, the chemical reactions that provide the current are readily reversed when current is supplied to the battery.

Q. 17. In what form is energy stored in a secondary battery ?

Ans. Energy is stored in the form of electrical energy in a secondary battery.

Q. 18. Why Faure process is much suitable for manufacturing of negative plates rather than positive plates?

Ans. Faure process is much suitable for manufacturing of negative plates because negative active material (sponge lead) is quite tough (not hard and brittle as lead peroxide is) and moreover, it undergoes a comparatively negligible change in volume during charging and discharging processes.

Q. 19. Why in grouping of lead acid cells the negative group is made to have one plate more than positive group ?

Ans. The negative group of the plates contains one more plate than that of the positive group so that when interleaved both the end plates become the negative. This is done so that all the positive plates can work equally well on both sides. If a positive plate were used at one end, one of its sides would remain inactive, when the other side expands, and thus the plate is likely to buckle.

Q. 20. Do we use ac or dc to charge a battery ?

Ans. For charging of a battery dc is essential.

Q. 21. Why should the lead acid cells not be allowed to stand completely discharged ?

Ans. Lead acid cells shall not be allowed to stand completely discharged because the lead sulphate formed on the plates crystallizes and the proper reforming of the plates becomes difficult.

Q. 22. What are the factors that affect the battery capacity.

Ans. The capacity of battery depends upon several factors, principal among which are area of plate surface, quantity, arrangement and porosity of the active material used in the manufacture of plates, quantity and specific gravity of the electrolyte used and the porosity of the separators, rate of charge and temperature.

Q. 23. Why the lead acid battery should not be allowed to discharge beyond 1.75 V ?

Ans. The lead acid battery should never be allowed to discharge beyond 1.75 V otherwise lead sulphate will be formed on the electrodes which is hard, insoluble and increases the internal resistance of the battery.

Q. 24. Why is the watt-hour efficiency of a cell lower than its ampere-hour efficiency ?

Ans. Watt-hour efficiency is given as

$$\eta_{AH} \times \frac{\text{average pd during discharge, } V_d}{\text{average pd during charge, } V_c}$$

and V_d is always lower than V_c, therefore η_{WH} is always lower than its η_{A-H}.

❑❑❑

CHAPTER 21

Power System Stability

INSIDE THIS CHAPTER

21.1. INTRODUCTION

In all power systems, the larger machines are of the synchronous type, these include substantially all the generators and condensers, and a considerable part of the motors. On such systems it is necessary to maintain synchronism, otherwise a standard of service to the consumers will not be achieved. The transient disturbances are caused by the changes in loads, switching operations, and, particularly, faults and loss of excitation. Thus, maintenance of synchronism during steady state conditions and regaining of synchronism or equilibrium after a disturbance are of prime importance to the electrical utilities. The term '*stability*' can be interpreted as '*maintenance of synchronism*'. These two terms are quite frequently used interchangeably.

The present trend is towards interconnection of the power systems; resulting into increased lengths and increased reactance of the system, this presents an acute problem of maintenance of stability of the system. The term '*power limit*' is also sometimes interpreted as '*stability*' because to have the maximum utility of the system it should be capable of supplying maximum power without causing instability. Power system stability, in general terms, may be defined as its ability to respond to a disturbance from its normal operation by returning to a condition where the operation is again normal.

The terms 'stability' and 'stability limit' are defined by A.I.E.E. as below :

Stability when used with reference to a power system, is that attribute of the system, or part of the system, which enables it to develop restoring forces between the elements thereof, equal to or greater than the disturbing forces so as to restore a state of equilibrium between the elements.

A *stability limit* is the maximum power flow possible through some particular point in the system or the part of the system to which the stability limit refers is operating with stability.

For the purpose of analysis there are three stability conditions that must be considered.

1. Steady state stability.
2. Transient stability.
3. Dynamic stability.

1. 'Steady state stability' may be defined as the capability of an electric power system to maintain synchronism between machines within the system and external tie lines following a small slow disturbance (normal load fluctuations, the action of automatic voltage regulators and turbine governors). In case the maximum power transfer exceeds under this condition, individual machines or groups of machines will cease to operate in synchronism, violent fluctuations of the voltage will occur and the *steady state limit* for the system as a whole would have been reached. The steady state stability limit refers to the maximum power which can be transferred through the system without loss of stability.

Loss of synchronism is, however, possible at loads (or disturbances) below the steady state limit, if they are suddenly applied or removed. When dealing with such large sudden disturbances a concept of *transient stability* is introduced. A sudden large disturbance includes application of faults, clearing of faults, sudden load changes, and inadvertent tripping of lines and generators. The maximum power which can be transferred through the system without the loss of stability under sudden disturbances is referred as *transient stability limit*.

Transient stability is the ability of the system to remain in synchronism during the period following a disturbance and prior to the time that the governors can act. Ordinarily the first swing of machine rotors will take place within about one second following the disturbance, but the exact time depends on the characteristics of the machines and the transmission system. Following this period, governors begin to take effect, and dynamic stability conditions are effective.

Dynamic stability is the ability of a power system to remain in synchronism after the 'initial swing' (transient stability period) until the system has settled down to the new steady state equilibrium condition. When sufficient time has elapsed after a disturbance, the governors of the prime movers will react to increase or reduce energy input, as may be required, to re-establish a balance between energy input and the existing electrical load. This usually occurs in about 1–1.5 seconds after the disturbance. The period between the time the governors begin to react and the time that steady state equilibrium is re-established is the period when *dynamic stability* characteristics of a system are effective. It is possible to have transient stable but dynamically unstable conditions. Immediately after a disturbance, the machine rotors will go through the first swing (before governor action) successfully, and then, after governor control is initiated, the oscillations will start increasing until the machine falls out of synchronism. Such action can occur if the time delays of the governor control are such that, following the sensing of necessity for increasing or reducing energy input, action is delayed sufficiently in time to augment rather than diminish the next swing. If such a condition exists, the oscillations of the machine rotor can continue to build up until the machine falls out of synchronism.

21.2. MAXIMUM STEADY STATE POWER

Line diagram of a 3-phase transmission line is shown in Fig. 21.1. The transmission line may supply power from a generating station, grid switching station or grid supply point. The ends of a transmission line are designated as buses. The concept of a bus in a one line diagram is essentially the same as that of a node in a circuit diagram.

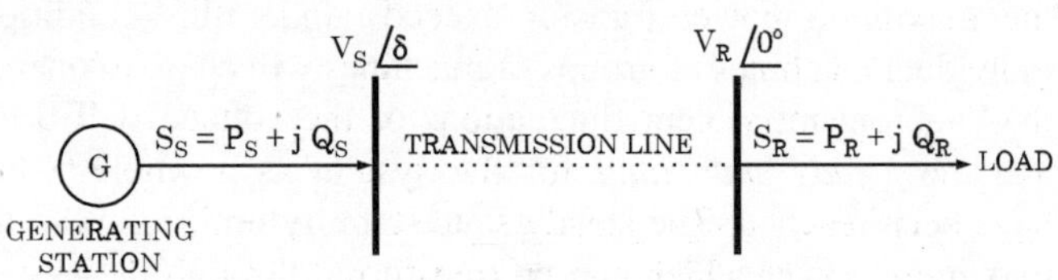

Fig. 21.1. *A Two-Bus Power System*

Let the receiving-end voltage be $V_R \angle 0°$ (taking V_R as reference phasor) and the sending-end voltage $V_S \angle \delta$ where δ is the phase angle between sending- and receiving-end voltages. Let the generalised line constants be expressed as $\mathbf{A} = A \angle \alpha$; $\mathbf{B} = B \angle \beta$; $\mathbf{D} = D \angle \Delta$.

The input voltage per phase (voltage at sending end) and the input current (current at sending end) of a transmission line can be expressed as

$$\mathbf{V_S} = \mathbf{AV_R} + \mathbf{BI_R} \quad \text{...(21.1)}$$

$$\mathbf{I_S} = \mathbf{CV_R} + \mathbf{DI_R} \quad \text{...(21.2)}$$

The complex power at the receiving end is given by the expression

$$\mathbf{S_R} = P_R + j\,Q_R = \mathbf{V_R I_{RC}} \quad \text{...(21.3)}$$

where $\mathbf{I_{RC}}$ is the conjugate of receiving-end current $\mathbf{I_R}$.

Receiving-end current can, however, be expressed in terms of receiving- and sending-end voltages [using Eq. (21.1)]

$$\mathbf{I_R} = \frac{\mathbf{V_S} - \mathbf{AV_R}}{\mathbf{B}}$$

$$= \frac{V_S \angle \delta - A \angle \alpha \times V_R \angle 0°}{B \angle \beta}$$

$$= \frac{V_S}{B} \angle \delta - \beta - \frac{A}{B} V_R \angle \alpha - \beta \quad \text{...(21.4)}$$

So conjugate of $\mathbf{I_R}$,

$$\mathbf{I_{RC}} = \frac{V_S}{B} \angle \beta - \delta - \frac{A}{B} V_R \angle \beta - \alpha$$

Substituting the value of $\mathbf{I_{RC}}$ from above equation in Eq. (21.3), we have

$$S_R = P_R + jQ_R = \frac{V_S V_R}{B} \angle \beta - \delta - \frac{A}{B} V_R^2 \angle \beta - \alpha \quad \text{...(21.5)}$$

Separating real and reactive components, we have

Receiving-end power,

$$P_R = \frac{V_S V_R}{B} \cos(\beta - \delta) - \frac{A}{B} V_R^2 \cos(\beta - \alpha) \quad \text{...(21.6)}$$

and receiving-end reactive power,

$$Q_R = \frac{V_S V_R}{B} \sin(\beta - \delta) - \frac{A}{B} V_R^2 \sin(\beta - \alpha) \quad \text{...(21.7)}$$

For fixed values of V_S and V_R the power received will be maximum when $\cos(\beta - \delta) = 1$ or when $\delta = \beta$

$$\therefore\ P_{R\,max} = \frac{V_S V_R}{B} - \frac{A}{B} V_R^2 \cos(\beta - \alpha) \quad \text{...(21.8)}$$

From Eq. (21.2),

$$\mathbf{I_R} = \frac{\mathbf{I_S} - \mathbf{CV_R}}{\mathbf{D}}$$

Substituting the value of $\mathbf{I_R}$ from above equation in Eq. (21.1), we have

$$\mathbf{V_S} = \mathbf{AV_R} + \mathbf{B}\,\frac{\mathbf{I_S} - \mathbf{CV_R}}{\mathbf{D}}$$

$$\text{or } \mathbf{DV_S} = \mathbf{ADV_R} + \mathbf{BI_S} - \mathbf{BCV_R}$$

$$\text{or } \mathbf{V_R}(\mathbf{AD} - \mathbf{BC}) = \mathbf{DV_S} - \mathbf{BI_S}$$

$$\text{or } \mathbf{V_R} = \mathbf{DV_S} - \mathbf{BI_S}$$

Since for a four-terminal network **AD** – **BC** = 1

and therefore,

$$\mathbf{I}_S = \frac{\mathbf{D}\mathbf{V}_S - \mathbf{V}_R}{\mathbf{B}}$$

$$= \frac{D\angle\Delta\ V_S\angle\delta - V_R\angle 0°}{B\angle\beta}$$

$$= \frac{D}{B}V_S\angle\Delta+\delta-\beta - \frac{V_R}{B}\angle-\beta \quad ...(21.9)$$

Conjugate of sending-end current,

$$\mathbf{I}_{SC} = \frac{D}{B}V_S\angle\beta-\Delta-\delta - \frac{V_R}{B}\angle\beta \quad ...(21.10)$$

Sending-end power,

$$\mathbf{S}_S = P_S + j\,Q_S$$

$$= \mathbf{V}_S\mathbf{I}_{SC} = \frac{D}{B}V_S^2\angle\beta-\Delta - \frac{V_R V_S}{B}\angle\beta+\delta \quad ...(21.11)$$

Separating real and reactive components, we have

Sending-end true power,

$$P_S = \frac{D}{B}V_S^2\cos(\beta-\Delta) - \frac{V_R V_S}{B}\cos(\beta+\delta) \quad ...(21.12)$$

and sending-end reactive power,

$$Q_S = \frac{D}{B}V_S^2\sin(\beta-\Delta) - \frac{V_R V_S}{B}\sin(\beta+\delta) \quad ...(21.13)$$

For fixed values of V_S and V_R, the power sent out will be maximum when $\cos(\beta+\delta) = -1$ or when $\beta+\delta = 180°$

$$\therefore P_{S\,max} = \frac{D}{B}V_S^2\cos(\beta-\Delta) + \frac{V_R V_S}{B} \quad ...(21.14)$$

The noteworthy points about Eqs. (21.5) and (21.11) are :

1. These equations give phase values of S_R and S_S in volt-amperes if V_R and V_S are phase values in volts (*i.e.,* line-to-neutral voltage).

2. The total 3-phase power is 3 times the power per phase. Since the line voltage is $\sqrt{3}$ times the phase voltage, these equations yield 3-phase power directly if V_R and V_S are expressed in line volts (line-to-line voltage).

3. If the voltages V_R and V_S are expressed in kV, the complex power is expressed in mega volt-amperes.

Now let us consider a loss-less transmission line having negligible shunt admittance.

For such a line

$$\mathbf{A} = \mathbf{D} = 1\angle 0°,\ \mathbf{B} = X\angle 90°;\ \text{and}\ \mathbf{C} = 0$$

and the expressions for receiving-end power, maximum receiving-end power, sending-end power and maximum sending-end power become

$$P_R = \frac{V_S V_R}{X}\sin\delta \quad ...(21.15)$$

$$P_{R\,max} = \frac{V_S V_R}{X} \quad ...(21.16)$$

$$P_S = \frac{V_S V_R}{X}\sin\delta \quad ...(21.17)$$

and $$P_{S\,max} = \frac{V_S V_R}{X} \quad ...(21.18)$$

From above expressions, we find that

(*i*) $P_R = P_S$; (*ii*) $P_{R\,max} = P_{S\,max}$; it is because of negligible line losses.

(*ii*) For fixed values of sending-end voltage V_S, receiving-end voltage V_R and line reactance X, the power sent out (or power received) varies directly as the sine of the load angle δ (phase shift between sending-end and receiving-end voltages).

Example 21.1. A 50 Hz transmission line 500 km long with constants given below ties up two large power areas.

R = 0.11 Ω/km, L = 1.45 mH/km, C = 0.009 μF/km, G = 0.

Find the steady state stability limit if | V_S | = | V_R | = 200 kV (constant). What will be the steady state stability limit if line capacitance is also neglected ? What will be the steady state limit if the line resistance is also neglected ? Comment on the results.

[*U.P. Technical Univ. Power System Analysis 2008-09*]

Solution : Total resistance of transmission line,

$$R = 0.11 \times 500 = 55\ \Omega$$

Total inductive reactance of transmission line,

$$X = 2\pi f L = 2\pi \times 50 \times 1.45 \times 10^{-3} \times 500 = 227.765\ \Omega$$

Total series impedance,

$$\mathbf{Z} = (R + jX) = (55 + j227.765)\ \Omega$$

$$= 234.31\angle 76.424°\ \Omega$$

Total shunt admittance,

$$\mathbf{Y} = (G + jB) = (0 + j2\pi f C)$$

$$= 0 + j2\times\pi\times 50\times 0.009\times 10^{-6}\times 500$$

$$= (j0.001414) = 0.001414\angle 90°\ S$$

$$\mathbf{YZ} = 0.001414\angle 90° \times 234.31\angle 76.424°$$

$$= 0.33125\angle 166.424° = (-0.322 + j0.0778)$$

$$\mathbf{Y^2Z^2} = \left(0.33125\angle 166.424°\right)^2 = 0.1097\angle 332.85°$$

$$= (0.0976 - j0.05)$$

$$\mathbf{Y^3Z^3} = 0.33125^3\angle 3\times 166.424$$

$$= 0.03635\angle 499.272° = 0.03635\angle 139.272°$$

$$= (-0.0275 + j0.0237)$$

Now auxiliary line constants

A = **D**

$$= 1 + \frac{\mathbf{YZ}}{2} + \frac{\mathbf{Y^2Z^2}}{24} + \frac{\mathbf{Y^3Z^3}}{720} + ...$$

$$= 1 + \frac{(-0.322 + j0.0778)}{2} + \frac{0.976 - j0.05}{24} + \frac{-0.0275 + j0.0237}{720}$$

$$= (1 - 0.161 + 0.0407 - 0.000038) + j(0.0389 - 0.00208 + 0.000033)$$

$$= (0.87966 + j0.036853) = 0.88 \angle 2.4°$$

$$\mathbf{B} = \mathbf{Z}\left\{1 + \frac{\mathbf{YZ}}{6} + \frac{\mathbf{Y}^2\mathbf{Z}^2}{120} + \frac{\mathbf{Y}^3\mathbf{Z}^3}{5,040}\right\} \Omega$$

$$= 234.31 \angle 76.424° \{1 + (-0.053666 + j0.01297)\}$$

higher terms being negligible

$$= 234.31 \angle 76.424° (0.946334 + j0.01297)$$

$$= 234.31 \angle 76.424° \times 0.946423 \angle 0.785°$$

$$= 221.756 \angle 77.21° \ \Omega$$

Steady-state stability limit

$$= P_{R\,max}$$

$$= \frac{V_S V_R}{B} - \frac{A}{B} V_R^2 \cos(\beta - \alpha)$$

$$= \frac{200 \times 200}{221.756} - \frac{0.88}{221.756} \times 200^2 \cos(77.21° - 2.4°)$$

$$= 180.4 - 41.6 = 138.8 \text{ MW} \quad \textbf{Ans.}$$

When shunt capacitance is neglected

$$\mathbf{A} = \mathbf{D} = 1 \angle 0° \text{ and } \mathbf{B} = \mathbf{Z} = 234.31 \angle 76.424° \ \Omega$$

Steady-state stability limit

$$= P_{R\,max} = \frac{V_S V_R}{B} - \frac{A}{B} V_R^2 \cos(\beta - \alpha)$$

$$= \frac{200 \times 200}{234.31} - \frac{1 \times 200^2}{234.31} \cos 76.424°$$

$$= 170.7 - 40 = 130.7 \text{ MW} \quad \textbf{Ans.}$$

When capacitance and resistance both are neglected.

Steady-state stability limit

$$= \frac{V_R V_S}{X} = \frac{200 \times 200}{227.765} = 175.62 \text{ MW} \quad \textbf{Ans.}$$

Comments on the Results. From above analysis it can be seen that when resistance is considered but shunt capacitance is neglected, the steady-state stability limit is reduced. If line resistance is also neglected steady-state stability limit is increased.

21.3. POWER ANGLE DIAGRAM

From Eqs. (21.15) and (21.17) it is obvious that for fixed values of sending-end voltage V_S, receiving-end voltage V_R and line reactance X the power received (or power sent out) varies as a sine function and, therefore, the curve drawn for power against load angle δ, called the *power angle diagram*, will obviously be a sine curve, as shown in Fig. 21.2 and the maximum power will occur for δ = 90°.

Now consider a line having resistance and reactance only (no shunt admittance)—a case of short transmission line.

For such a line, generalised line constants are:

$$\mathbf{A} = 1 \angle 0°; \ \mathbf{B} = Z \angle \beta \text{ where } \beta = \text{Tan}^{-1} \frac{X}{R}$$

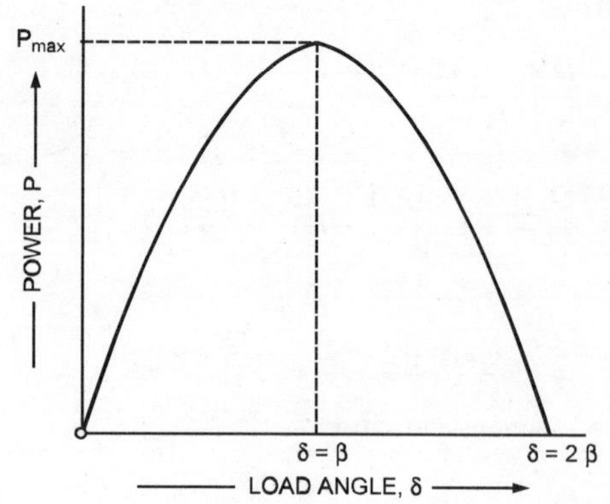

Fig. 21.2

and the power received will be

$$\mathbf{P}_R = \frac{V_S V_R}{Z} \cos(\beta - \delta) - \frac{V_R^2}{Z} \cos\beta$$

$$= \frac{V_R}{Z^2} [V_S Z \cos(\beta - \delta) - V_R Z \cos\beta]$$

$$= \frac{V_R}{Z^2} [V_S Z \cos(\beta - \delta) - V_R \cdot R] \quad ...(21.19)$$

Maximum power will occur when δ = β, since V_S, V_R and Z are of constant magnitude and

$$\therefore P_{R\,max} = \frac{V_R^2}{Z^2}\left(\frac{V_S}{V_R} Z - R\right)$$

$$\text{or } P_{R\,max} = \frac{V_R^2}{Z^2}\left[\frac{V_S}{V_R}\sqrt{R^2 + X^2} - R\right] \quad ...(21.20)$$

If the resistance of the line is kept constant and line reactance X is varied, the condition for P_{max} to be maximum is given as

$$\frac{d\,P_{Rmax}}{dX} = \frac{d}{dX}\left[\frac{V_R^2}{R^2 + X^2}\left(\frac{V_S}{V_R}\sqrt{R^2 + X^2} - R\right)\right] = 0$$

$$\text{or } -\frac{1}{2} \frac{2 V_S V_R X}{(R^2 + X^2)^{3/2}} + \frac{2 V_R^2 R X}{(R^2 + X^2)^2} = 0$$

$$\text{or } -XV_S V_R (R^2 + X^2)^{1/2} + 2V_R^2 RX = 0$$

$$\text{or } \frac{V_S}{V_R}\sqrt{R^2 + X^2} = 2R$$

$$\text{or } R^2 + X^2 = 4R^2\left(\frac{V_R}{V_S}\right)^2$$

$$\text{or } X = \sqrt{R^2\left[4\left(\frac{V_R}{V_S}\right)^2 - 1\right]} \quad ...(21.21)$$

For equal values of sending-end and receiving-end voltages (*i.e.*, $V_S = V_R$)

$$X = \sqrt{R^2 (4 - 1)} = \sqrt{3} R \quad ...(21.22)$$

We conclude that

(*i*) Maximum power will be transferred from the sending end to receiving end when the line reactance is $\sqrt{3}$ *times its resistance* (*i.e.,* $X = \sqrt{3}$ R). Normally the reactance is quite large as compared to the resistance.

(*ii*) It is not necessarily desirable to compensate by series capacitance for all the reactance.

(*iii*) Power can be transferred only if reactance is present. In case reactance X is zero transfer of power will not be possible.

(*iv*) Transfer of power will not be possible without phase shift if equal voltages are to be maintained at both ends.

21.4. STEADY STATE STABILITY

As stated earlier, *steady state stability* is the capability of the power system to restore to its initial condition after a small disturbance or to reach a condition very close to the initial one when the disturbance is still present and *steady state stability limit* refers to the maximum flow of power possible through a particular point without causing the loss of stability when the power is increased very gradually. The terms steady state stability and 'steady state stability limit' are frequently used interchangeably. Since the electrical system is always subject to small disturbances, the steady state stability requirement is essential for the system to operate properly.

The assessment of steady state stability is very important in planning and designing of electrical power systems, in developing special automatic control devices, putting into operation new elements of the system, or modifying its operating conditions (such as interconnection of the power systems and commissioning of new power plants, transformer substations, transmission lines etc.)

Furthermore, the assessment of the steady state stability may also be required in the power system analysis (including checking of stability of an electrical power system in a specified steady state, determination of its stability limit, qualitative estimation of transients) and synthesis (involving the choice of type of the excitation system and its control, the mode of control, the parameters of the excitation and automatic control systems). This choice is made on the basis of requirements to the stability limit or quality of electrical energy (*i.e.,* accuracy of maintaining the voltage) under steady-state conditions or during transients.

When all the machines in one part of a power system act together, they are treated as one large machine connected at that point, for the purpose of study of power system stability. Even if the machines are not connected on the same bus bars but are separated by large percentage reactances, they are considered as one machine. A large system in the power system stability problems is always supposed to have constant voltage and constant frequency and is treated like infinite bus. In this way, it is always possible to work out a complicated system on the basis of two machines or one machine connected to an infinite bus.

Let us consider a system consisting of a generator G, transmission line and a synchronous motor M in the form of load (Fig. 21.3).

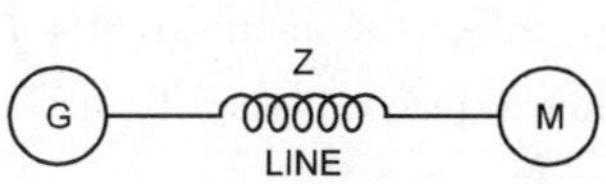

Fig. 21.3

Power developed by the generator G and synchronous motor M are given by the expressions

$$P_G = \frac{D}{B} V_G^2 \cos(\beta - \Delta) - \frac{V_G V_M}{B} \cos(\beta + \delta)$$

$$\text{and } P_M = \frac{V_G V_M}{B} \cos(\beta - \delta) - \frac{A}{B} V_M^2 \cos(\beta - \alpha)$$

Refer to Eqs. (21.12) and (21.6)

and the maximum power developed by the generator G and synchronous motor M are given by the expressions

$$P_{G\,max} = \frac{D}{B} V_G^2 \cos(\beta - \Delta) + \frac{V_G V_M}{B} \quad ...(21.23)$$

$$P_{M\,max} = \frac{V_G V_M}{B} - \frac{A}{B} V_M^2 \cos(\beta - \alpha) \quad ...(21.24)$$

[Refer to Eqs. (21.14) and (21.8)]

where A, B, and D are the generalised constants of a two terminal pair network. The above expressions will give power in watts per phase in case voltages are taken as phase voltages in volts. However, if the voltages are taken as line voltages in kV the above expressions will give total power (power for all 3 phases) in MW.

Now consider a synchronous motor connected to an infinite bus and running at a constant speed. The power input is equal to power output plus losses. If a small increment of the shaft load is added to the motor, the output power of the motor increases whereas the input to the motor remains unchanged. Hence, there is a net torque on the motor tending to retard it and its speed falls temporarily. As a result of momentarily reduction in motor speed, the phase angle between the internal voltage of the motor and the system voltage increases until the electrical power input again equals power output plus losses. During the transient interval electrical power input to the motor being less than mechanical load, the excess power required is supplied by the stored energy in the rotating system. The motor thus oscillates around the new point of equilibrium and may finally either come to rest or may lose synchronism. The latter will be the case when the load applied will be too large or will be applied too suddenly.

P_{Mmax} given by Eq. (21.24) is the maximum power that motor can develop. This value of power can be obtained only when load angle $\delta = \beta$. The load on the motor may, therefore, be increased till this condition is reached. After this condition if the load still keeps on increasing, load angle δ will increase further and the motor may lose synchronism and the excess of power required over power developed will come from the stored energy of the rotating system and the speed goes down. Larger and larger angle δ becomes, the smaller and smaller will be the power developed, till the motor comes to rest.

It is to be noted that $\delta = \beta$ is not the condition for maximum power developed by the generator but it needs not to be considered that condition since the motor will lose synchronism earlier (*i.e.,* at an angle $\delta = \beta$).

The difference between the generator and motor power developed for any value of δ is equal to the line losses. If the line resistance and shunt admittance are negligible, we get the following expression for the power transferred between the alternator and the motor

$$P = P_M = P_G = \frac{V_G V_M}{X} \sin \delta \qquad ...(21.25)$$

The power transferred will be maximum for $\delta = 90°$

$$\text{and } P_{max} = \frac{V_G V_M}{X} \qquad ...(21.26)$$

Omission of resistance gives a somewhat optimistic results (higher than actual value) for maximum steady state stability limit while in transient stability calculations the results obtained by neglecting resistance are pessimistic (lower than actual value) since resistance is an important damping element.

Methods of Improving Steady-State Stability Limit. From Eq. (21.26) it is obvious that the maximum power that can be transferred between an alternator and a motor is directly proportional to product of internal emfs, of the two machines and inversely proportional to the line reactance.

Hence the steady state stability limit can be increased by

1. increasing the excitation of a generator or motor or both so that the internal emfs are increased and consequently the maximum power transferred between the two machines increases. Further for a given power transfer and with the increased values of internal emfs, the load angle δ decreases.

2. reducing the transfer reactance—increasing the number of parallel lines between the transmission points is a common method of reducing the reactance. In general more power is transferred during a fault on one of the lines if there are two lines in parallel than would be transferred over a single faulted line. The use of bundle conductors is, of course, another method of reducing reactance. The bundle conductors are used mainly in EHV transmission lines to reduce corona loss and radio-interference. Another method of reducing the line reactance is by using capacitors in series with the lines. Series capacitors are only used in EHV lines to increase the power transfer and are most economical for transmission distances over 350 km.

Example 21.2. Find the steady-state stability limit of a system shown in Fig. 21.4, consisting of a generator of equivalent reactance 1.0 pu connected to an infinite bus through a series reactance of 0.6 pu. The terminal voltage of the generator (E_g) is held at 1.0 pu and the voltage of the infinite bus at 1.0 pu.

[*U.P.S.C., I.E.S. Electrical Engineering II 2000*]

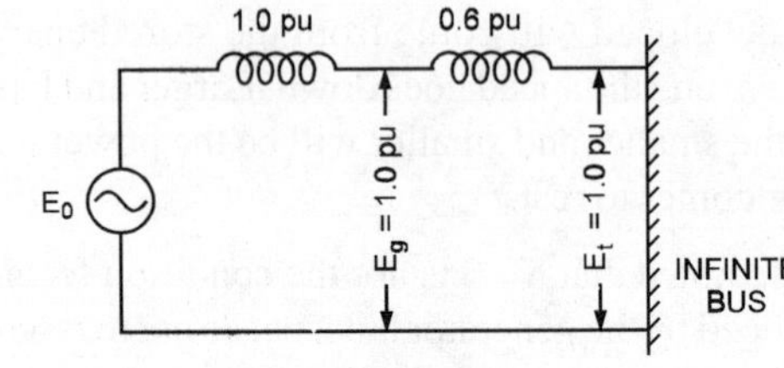

Fig. 21.4

Solution : The system is shown in Fig. 21.4.

Taking voltage of infinite bus as reference, we have

$$E_t = 1.0 \angle 0°$$

$$E_g = 1.0 \angle \theta$$

$$\text{Now, } I = \frac{E_g - E_t}{jX} = \frac{1.0\angle\theta - 1.0}{j0.6}$$

$$E_0 = E_g + jIX_d = 1.0 \angle\theta + j\left(\frac{1.0\angle\theta - 1.0}{j0.6}\right) \times 1.0$$

$$\text{or } E_0 = 2.6667 \angle\theta - 1.6667$$

$$= (2.6667 \cos\theta - 1.6667) + j\,2.6667 \sin\theta$$

Steady-state power limit is reached when E_0 has an angle of $\delta = 90°$, *i.e.*, its real part is zero

So $2.6667 \cos\theta - 1.6667 = 0$

$$\text{or } \theta = \text{Cos}^{-1}\frac{1.6667}{2.6667} = 51.32° \text{ (electrical)}$$

$$\text{Now } E_g = 1.0 \angle 51.32° = 0.625 + j\,0.7806$$

$$\text{Current } I = \frac{0.625 + j0.7806 - 1}{j0.6} = 1.3 + j\,0.625$$

$$E_0 = E_g + jIX_d = 0.625 + j\,0.7806 + j\,(1.3 + j\,0.625) \times 1.0$$

$$= j\,2.0806 = 2.0806 \angle 90°$$

Steady-state power limit is given by

$$P_{max} = \frac{E_0 \times E_t}{X_d + X} = \frac{2.0806 \times 1}{1.6} = 1.3 \text{ pu } \textbf{Ans.}$$

Example 21.3. Find the steady-state power limit of a system consisting of a generator equivalent reactance 0.5 pu connected to an infinite bus through a series reactance of 1.0 pu. The terminal voltage of the generator is held at 1.20 pu and the voltage of the infinite bus is 1.0 pu. [*Rajasthan Univ. Power System Engineering 2007*]

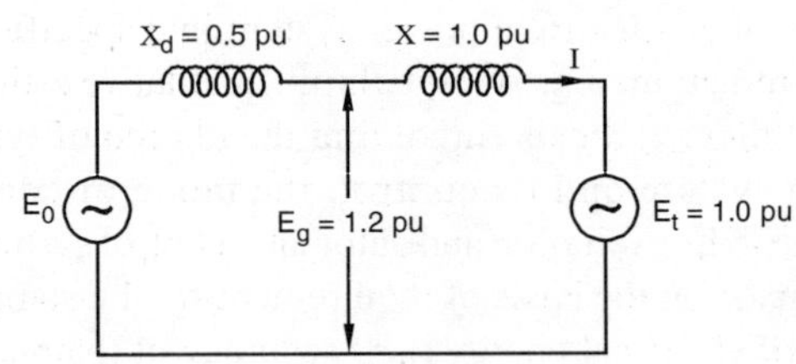

Fig. 21.5

Solution : The system is shown in Fig. 21.5.

Taking voltage at infinite bus as reference, we have

$$E_t = 1.0 \angle 0°$$

$$E_g = 1.2 \angle\theta$$

$$\text{Now, } I = \frac{E_g - E_t}{jX} = \frac{1.2\angle\theta - 1.0}{j1.0}$$

$$E_0 = E_g + jIX_d = 1.2 \angle\theta + j\,0.5 \times \frac{1.2\angle\theta - 1.0}{j1.0}$$

$$= 1.8 \angle\theta - 0.5$$

$$= (1.8 \cos\theta - 0.5) + j1.8 \sin\theta$$

Steady-state limit is reached when E_0 has an angle of $\delta = 90°$, *i.e.*, its real part is zero

so $1.8 \cos\theta - 0.5 = 0$

or $\theta = \cos^{-1}\frac{0.5}{1.8} = 73.87°$ (electrical)

Now $\mathbf{E}_g = 1.2\angle 73.87° = 0.3333 + j1.153$

Current $\mathbf{I} = \frac{0.3333 + j1.153 - 1}{j1} = 1.153 + j0.6667$

$$\mathbf{E}_0 = \mathbf{E}_g + j\mathbf{I}X_d = 0.3333 + j1.153 + j0.5(1.153 - j0.6667)$$

$$\simeq j1.73 = 1.73\angle 90°$$

Steady-state power limit is given by

$$P_{max} = \frac{E_0 \times E_t}{X_d + X} = \frac{1.73 \times 1}{0.5 + 1.0} = 1.153 \text{ pu } \textbf{Ans.}$$

If instead, generator emf is held fixed at a value of 1.2 pu, the steady state power limit would be

$$P_{max} = \frac{1.2 \times 1}{1.5} = 0.8 \text{ pu}$$

It is observed that regulating the generator emf to hold the terminal generator voltage at 1.2 pu raises the power limit from 0.8 pu to 1.153 pu, this is how voltage regulating loop helps in power system stability.

21.5. TRANSIENT STABILITY

A synchronous power system has transient stability if, after a large sudden disturbance, it can regain and maintain synchronism. A sudden large disturbance includes application of faults, clearing of faults, switching on and off the system elements (transmission lines, transformers, generators, loads etc.).

Usually, transient stability studies are carried out over a relatively short-time period that will be equal to the time of one swing. Normally, the time period will be one second or less. The analysis is carried out to determine whether the system loses stability during the first swing or not. In case the power system remains stable, it is assumed that subsequent swings will diminish and that power system will remain stable, as usually happens. However, there is a possibility of power system going unstable in some subsequent swing. For example negative damping is one cause. Control equipment improperly adjusted or applied can produce negative damping.

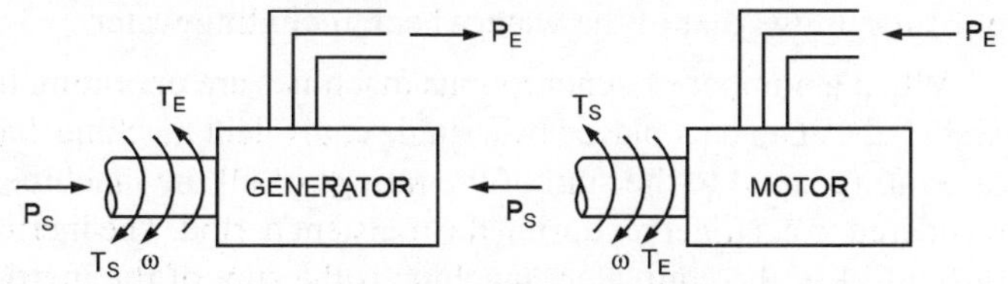

Fig. 21.6(*a*). *Flow of Mechanical and Electrical Powers in a Generator and a Motor*

In practice load is not applied gradually and the machine (motor or generator) is not in a position to meet the demand instantly. The limit of the system to meet such loads, called the *transient stability limit*, is obviously less than steady state stability.

As said earlier, the power systems have rotating synchronous machines, in order to know whether the system is stable or not, it is necessary to derive swing equation.

21.5.1. Swing Equation

Let angular displacement = θ radians

$$\text{Angular velocity, } \omega = \frac{d\theta}{dt} \text{ radians/second}$$

$$\text{Angular acceleration, } \alpha = \frac{d\omega}{dt} = \frac{d^2\theta}{dt^2} \text{ radians/second}^2$$

Power developed, P = Tω watts where T is torque in Nm

Angular momentum, M = Iω J-s/radian.

where I is moment of inertia in kg-m^2 or J-s^2/radian^2.

In case of a generator the input is a mechanical torque (shaft torque), T_S and the output is an electromagnetic torque, T_E and both of these torques are reckoned as positive. In case of a motor the input is electromagnetic torque, T_E and the output is shaft torque, T_S, both of these torques are reckoned as the negative. Torques due to rotational losses are assumed to be negligible in either case during the discussions.

Under normal working, the relative position of the rotor axis and the stator magnetic field axis is fixed. The angle between the two is called *load angle* (or torque angle) δ, which depends upon the loading of the machine. Larger the loading, larger is the load angle δ. In case load is added or removed from the shaft of the synchronous machine, the rotor will decelerate or accelerate accordingly with respect to the synchronously rotating stator field and a relative motion starts. Now rotor is said to be swinging with respect to the stator field. The equation giving the relative motion of the rotor (load angle δ) with respect to the stator field as a function of time is called the *swing equation*.

It is obvious that any difference between the input and output torques ($T_S \sim T_E$) will cause acceleration or retardation of the rotor depending whether the input torque is greater than output torque or otherwise. Accordingly for a generator

$$T_{AG} = T_S - T_E \quad \text{...(21.27)}$$

where T_{AG} is the net torque causing acceleration of the rotor and will be positive if $T_S > T_E$.

The above expression holds true for a motor also.

A similar relation holds good when expressed in terms of power, *i.e.*,

$$P_{AG} = P_S - P_E \quad \text{...(21.28)}$$

where P_{AG} is the accelerating power

$$\text{Further } P_{AG} = T_{AG}\omega = I\alpha\omega = M\alpha \quad \text{...(21.29)}$$

Since the angular position θ of the rotor is continually varying with time, it is more convenient to measure the angular position and velocity with respect to a synchronously rotating axis.

Angular displacement, θ with respect to time, *t* can be expressed as

$$\theta = \omega_S t + \delta \quad \text{...(21.30)}$$

where ω_S is the angular velocity of the reference axis rotating synchronously and δ is the angular displacement in electrical degrees from the synchronously rotating reference axis [Fig. 21.6(*b*)].

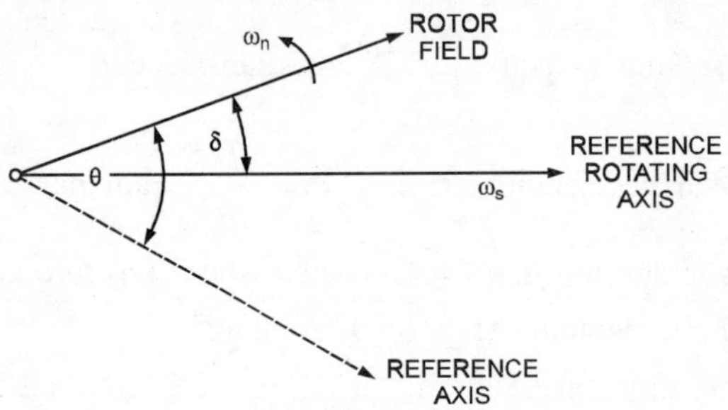

Fig. 21.6(*b*). *Angular Position of Rotor With Respect To Reference Axis*

Differentiating with respect to time *t*, we have

$$\frac{d\theta}{dt} = \omega_S + \frac{d\delta}{dt} \quad \text{...(21.31)}$$

and $$\frac{d^2\theta}{dt^2} = \frac{d^2\delta}{dt^2} \quad \text{...(21.32)}$$

From Eq. (21.29)

$$M\alpha = P_{AG}$$

Substituting $\alpha = \frac{d^2\theta}{dt^2} = \frac{d^2\delta}{dt^2}$ from Eq. (21.32) and $P_{AG} = P_S - P_E$ from Eq. (21.28) in above equation, we have

$$M\frac{d^2\delta}{dt^2} = P_S - P_E \quad \text{...(21.33)}$$

where $P_E = \frac{V_G \cdot V_M}{X} \sin\delta = P_{max}\sin\delta$

$$\therefore\ M\frac{d^2\delta}{dt^2} = P_S - P_{max}\sin\delta \quad \text{...(21.34)}$$

Equation (21.33) is called the *swing equation*. The angle δ is the difference between the internal angle of the machine and the angle of the synchronously rotating reference axis which in this case corresponds to the infinite bus. For a two-machine system two swing equations are necessary one for each machine but as illustrated in Art. 21.8, a two machine system can be reduced to a single machine connected to an infinite bus.

21.5.2. Inertia Constant

The stored kinetic energy of a rotating body is given as

$$KE = \frac{1}{2}I\omega^2 = \frac{1}{2}M\omega \quad \text{...(21.35)}$$

The above equation gives the unit of M in joule-seconds per radian (Js rad^{-1}) if the angular velocity ω is in radians per second. The energy is more conveniently expressed in MJ and angles are often measured in degrees. So M can be expressed in MJs per electrical degree.

From above equation M, being equal to $I\omega$, is not constant but varies somewhat during the swings because of variation in ω. In practice, the change in ω from the normal system angular velocity is not much during swing except of course when the machine falls out of step and, therefore, hardly any significant error is involved if M is assumed to remain constant and is equal to the value $I\omega_n$, where ω_n is the normal angular velocity of the machine. This value of M is known as the *inertia constant* of the machine, and is normally used in calculations for stability studies. The inertia constant is truely constant because it is the angular momentum at synchronous speed.

M, the inertia constant of the machine, depends on the type and the size of the machine. Another constant H having narrow range of values for each class of machine regardless of its capacity is defined as the ratio of the kinetic energy at rated speed to the rated apparent power of the machine, *i.e.*,

$$H = \frac{\text{Stored energy in MJ}}{\text{Machine rating in MVA, } i.e., \text{ G}} \quad \text{...(21.36)}$$

Thus stored energy in mega joules equals GH.

A relation between M and H is derived as follows:

From Eqs. (21.35) and (21.36)

$$GH = \frac{1}{2}M\omega = \frac{1}{2}M \times 360f \quad \text{...(21.37)}$$

where *f* is the frequency in Hz and ω is expressed as 360 *f* in terms of electrical degrees per second.

$$\text{or } M = \frac{GH}{180f} \text{ MJ-s/electrical degree} \quad \text{...(21.38)}$$

$$\text{and } M = \frac{GH}{\pi f} \text{ MJ-s/electrical radian} \quad \text{...(21.39)}$$

Equation (21.38) relates the two inertia constants of the machine. For stability studies it is necessary to determine M which depends upon the size and speed of the machine, but instead H has a characteristic value or a range of values for each class of machines. The average value of H in MJ per MVA for turbo-generators is from 3 to 9; for waterwheel generators between 2 to 4; for synchronous condensers between 1 and 1.25; while for synchronous motors about 2.

It is observed that the value of H is considerably higher for turbogenerators than for waterwheel generators. 30 to 60 per cent of the total inertia of a turbogenerator unit is that of the prime mover, while only 4 to 15% of the inertia of a waterwheel generator unit is that of the waterwheel, including water.

When a number of synchronous machines are operating in parallel, they can be replaced by a single equivalent machine that has a rating equal to the sum of the ratings of all the machines considered to act together during the transient period. The inertia constant M of the equivalent machine is the sum of the inertia constants of the individual machines, *i.e.*,

$$M_{eq} = M_1 + M_2 + M_3 + \ldots M_n$$

$$\text{or } H_{eq}G_{base} = H_1G_1 + H_2G_2 + H_3G_3 + \ldots H_nG_n$$

$$\text{or } H_{eq} = \frac{H_1G_1}{G_{base}} + \frac{H_2G_2}{G_{base}} + \frac{H_3G_3}{G_{base}} + \ldots \frac{H_nG_n}{G_{base}} \quad \text{...(21.40)}$$

It is evident from the above discussion that the value of H for a given rating varies inversely as the base.

For a multi-machine system solution of several swing equations is required. In fact for each machine a swing equation is required. The solution is tedious and normally the point by point or step by step method (discussed in Art. 21.9) is used for it. Even for a single machine system connected to infinite bus with resistance neglected, the formal solution of the swing equation is possible only when $P_S = 0$ and by use of elliptic integrals. The solution of Eq. (21.34) gives a graph of load angle δ versus time *t*, called the *swing curve*. This curve is used to determine the stability of the system. In case δ increases indefinitely it indicates instability whereas if it reaches a maximum and starts decreasing it shows that the system will not lose stability and the oscillations of the machine about the equilibrium point will become smaller and smaller and finally it reaches equilibrium. Many a time, in a multi-machine system one of the machines may stay in step on the first swing and yet go out of step in the subsequent swing. For a two-machine system under the assumption of constant input, no damping and constant voltage behind transient reactance, the machines either fall out of step in the first swing or never. Under this condition the two machines are said to be running at standstill with respect to each other.

For solution of swing equation certain assumptions and simplifications are usually made. These are listed below along with their justification and their effects on accuracy of results.

1. It is assumed that mechanical input for a generator and mechanical load on a motor remains constant during the entire period of analysis. This assumption is quite valid for a synchronous motor because the speed of the motor does not change significantly unless stability is lost. The validity of this assumption for a generator is based on the fact that the prime-mover governor, steam valve etc. are slow acting and act only after sensing a change in speed which does not change significantly during the transient period. This assumption in case of a generator makes the results somewhat pessimistic.
2. The resistances of the transmission lines and the synchronous machines are neglected. This assumption leads to a pessimistic result as resistance introduces damping term in swing equation and stability is improved.
3. The damping term contributed by synchronous machine damper windings is also ignored. This assumption leads to a pessimistic result for the transient stability limit.
4. All shunt capacitances are neglected. This assumption does not introduce any significant error.
5. Direct axis transient reactance of the synchronous machine is considered in the analysis and the saliency effects of the synchronous machine are neglected. This assumption is quite valid, because during a disturbance the period of mechanical oscillation may be of the order of one second. The sub-transient time constant will be approximately 0.3 second, so it is usually ignored in transient stabilities. A typical armature time constant would be about 0.2 s, so in many studies the direct component of armature current is ignored. The transient component of armature current may have a time constant of a few seconds, which is longer than the period of mechanical oscillation, so it is necessary to consider the transient reactance.
6. The voltage behind transient reactance is assumed to remain constant during the entire period of analysis. This means that the action of excitation control system is ignored and this leads to pessimistic stability results. This is a stronger assumption than assuming P_S constant as the excitation control system is faster than the governor system. However, with the advent of digital computer, the effects of voltage regulators and turbine governors can be incorporated in the transient stability analysis.

The above assumptions simplify the calculations to some extent. Even then the calculations are so lengthy that the use of digital computer is preferred. The solution for a multi-machine system is possible only through a computer simulation.

In case of only one machine connected to an infinite bus-bar, by application of equal area criterion (discussed in forthcoming Article), the transient stability of the system can be worked out without detailed solution of swing equation.

Modern power systems are so large that even after lumping of machines, the system remains a multi-machine one. Even then a simple two-machine system greatly aids the understanding of the transient stability problem. The two-machine system can be reduced to one machine connected to infinite bus [refer to Art. 21.8].

Example 21.4. A 2-pole, 50 Hz, 11 kV, turbogenerator has a rating of 60 MW, power factor 0.85 lagging. Its rotor has a moment of inertia of 8,800 kg-m². Calculate its inertia constant in MJ per MVA and its momentum in MJ-s/electrical degree.

[*Pb. Univ. Power System Analysis December 1995*]

Solution : Moment of inertia,

$$I = 8{,}800 \text{ kg-m}^2$$

Angular velocity, $\omega = 2\pi f = 2\pi \times 50 = 100\pi$ electrical radians/second

Rating of machine in MVA,

$$S = \frac{MW}{pf} = \frac{60}{0.85} = 70.588 \text{ MVA}$$

Kinetic energy stored in the rotor at synchronous speed,

$$KE = \frac{1}{2} I\omega^2 = \frac{1}{2} \times 8{,}800 \times (100\pi)^2$$

$$= 434.265 \times 10^6 \text{ J or } 434.26 \text{ MJ}$$

$$\text{Inertia constant, } H = \frac{\text{Kinetic energy stored}}{\text{S in MVA}} = \frac{434.26}{70.588}$$

$$= 6.152 \text{ MJ/MVA } \textbf{Ans.}$$

$$\text{Angular momentum, } M = \frac{\text{Kinetic energy stored in MJ}}{180 f}$$

$$= \frac{434.26}{180 \times 50} = 0.04825 \text{ MJ-s/electrical degree } \textbf{Ans.}$$

Example 21.5. A 100 MVA synchronous generator operates on full load at a frequency of 50 Hz. The load is suddenly reduced to 50 MW. Due to time lag in governor system, the steam valve begins to close after 0.4 second. Determine the change in frequency that occurs in this time. Given: H = 5 kW-s/kVA of generator capacity.

[*U.P.S.C. I.E.S. Electrical Engineering II 1995*]

Solution : The rating of the machine,

$$G = 100 \text{ MVA}$$

Inertia constant,

$$H = 5 \text{ kW–second/kVA} = 5 \text{ MJ/MVA}$$

Kinetic energy stored in the rotating parts of the generator and turbine at synchronous speed ($f = 50$ Hz)

$$= HG = 5 \times 100 = 500 \text{ MJ}$$

Excess power input to the generator shaft before the steam valve begins to close

= 100 – 50 = 50 MW

Excess energy transferred to rotating parts in 0.4 second

= 50 × 0.4 = 20 MJ

Since kinetic energy,

KE ∝ (speed)^2 ∝ (frequency)^2

So, frequency at the end of 0.4 second,

$$f_2 = f_1 \times \left(\frac{\text{Total energy stored in 0.4 second}}{\text{Energy stored at synchronous speed}}\right)^{\frac{1}{2}}$$

$$= 50 \times \sqrt{\frac{500+20}{500}} = 50 \times 1.02 = 51 \text{ Hz}$$

Change in frequency

$= f_2 - f_1 = 51 - 50 = 1$ Hz **Ans.**

Example 21.6. A 50 Hz, 4-pole turbogenerator rated 100 MVA, 11 kV has an inertia constant of 8.0 MJ/MVA. Determine (*i*) the stored energy in the rotor at synchronous speed, (*ii*) if the mechanical input is suddenly raised to 80 MW for an electrical load of 50 MW, find rotor acceleration neglecting mechanical and electrical losses.

Solution : The rating of the machine, G = 100 MVA

Inertia constant, H = 8 MJ/MVA

(1) Stored energy in the rotor at synchronous speed

= GH = 100 × 8 = 800 MVA **Ans.**

Accelerating power, $P_{AG} = P_S - P_E$ = 80 MW – 50 MW = 30 MW

Angular momentum,

$$M = \frac{GH}{180f} = \frac{100 \times 8}{180 \times 50} = \frac{4}{45} \text{ MJ-s/electrical degree}$$

$$\text{Acceleration } \alpha = \frac{P_{AG}}{M} = \frac{30}{4/45} = \frac{30 \times 45}{4}$$

= 337.5 electrical degrees/s^2 **Ans.**

Example 21.7. A 50 Hz 4-pole turbogenerator rated 20 MVA, 13.2 kV has H = 9 MW-s/MVA. Determine : (*i*) kinetic energy stored at N_S (*ii*) acceleration if the input less the rotational losses are 25,000 hp and the electric power developed is 15 MW, (*iii*) if the acceleration computed for the generator is constant for a period of 15 cycles, determine the change in torque angle (δ) in that period and speed at the end of 15 cycles. Assume that the generator is synchronised with large system and has no accelerating torque before the 15-cycle period begins.

[*Power System Engineering, Rajasthan Technical Univ. 2010; Rajasthan Univ. 2008*]

Solution : Rating of the machine, G = 20 MVA

Inertia constant, H = 9 MJ/MVA

Kinetic energy stored in the rotor at synchronous speed

= H × G = 9 × 20 = 180 MJ **Ans.**

Accelerating power, P_{AG}

= Mechanical power input – electrical power developed

$$= P_S - P_E = \frac{25{,}000 \times 0.7355 - 15{,}000}{1{,}000} = 3.3875 \text{ MW}$$

Angular momentum, M

$$= \frac{GH}{180f} = \frac{180}{180 \times 50} = 0.02 \text{ MJ-s/electrical degree}$$

Acceleration α

$$= \frac{P_{AG}}{M} = \frac{3.3875}{0.02} = 169.375 \text{ electrical degrees/s}^2 \textbf{ Ans.}$$

For a 50 Hz system, 15 cycles corresponds to $\frac{15}{50}$, *i.e.,* 0.3 second

Change in torque angle δ in the period of 0.3 second

$$= \frac{1}{2} \times \alpha \times t^2 = \frac{1}{2} \times 169.375 \times (0.3)^2$$

= 7.621875 electrical degrees **Ans.**

Synchronous speed,

$$N_S = \frac{120f}{P} = \frac{120 \times 50}{4} = 1{,}500 \text{ rpm}$$

Acceleration,

$$\alpha = \frac{169.375}{\text{Pair of poles}} \text{ mechanical degrees/s}^2$$

$$= \frac{169.375}{2} = 84.6875 \text{ mechanical degrees/s}^2$$

$$= \frac{84.6875 \times 60}{360} = 14.11458 \text{ rpm/second}$$

Speed at the end of 15 cycles

$= N_S + \alpha t$ = 1,500 + 14.11458 × 0.3 = 1,504.234375 rpm **Ans.**

Example 21.8. A 50 Hz four-pole turbogenerator rated 20 MVA, 13.2 kV has an inertia constant of H = 9.0 kW-sec/kVA. Determine the kinetic energy stored in the rotor at synchronous speed. Also calculate the acceleration if the input less the rotational losses is 25,000 HP and the electric power developed is 15,000 kW.

[*U.P. Technical Univ. Power System Analysis 2007-08*]

Solution : Kinetic energy stored in the rotor at synchronous speed

= H × G = 9 × 20 = 180 MJ **Ans.**

Acceleration, α = 169.375 electrical degrees/s^2 **Ans.**

As worked out in Example 21.7

Example 21.9. A 3-phase, 60 Hz, 500 MVA, 13.8 kV, 4-pole steam turbine-generating unit has an H constant of 5.0 pu-s. Determine : (*i*) ω_{syn} (*ii*) $\omega_{m\ syn}$, (*ii*) the kinetic energy in joules stored in the rotating masses at synchronous speed, (*iii*) the mechanical angular acceleration α_m and electrical angular acceleration α if the unit is operating at synchronous speed with an acceleration power of 500 MW. [*G.B. Technical Univ. Power System Analysis 2010-11*]

Solution : Rating of the machine, G = 500 MVA

Inertia constant, H = 5.0 pu-s

Number of poles, P = 4

Frequency, f = 60 Hz

Accelerating power, P_{AG} = 500 MW

(*i*) Synchronous angular velocity,

$\omega_{syn} = 2\pi f = 2\pi \times 60 = 120\pi$ electrical radians/second **Ans.**

Mechanical synchronous angular velocity

$$= \frac{\omega_{syn}}{P/2} = \frac{120\pi}{4/2} = 60\pi \text{ mechanical radians/second } \textbf{Ans.}$$

(*ii*) Kinetic energy stored in rotating masses at synchronous speed

= HG = 5 × 500 = 2,500 MJ **Ans.**

(*iii*) Angular momentum,

$$M = \frac{GH}{180f} = \frac{2{,}500}{180 \times 60} = 0.2315 \text{ MJ-s/electrical degree}$$

Electrical angular acceleration,

$$\alpha = \frac{P_{AG}}{M} = \frac{500}{0.2315} = 2{,}160 \text{ electrical degrees/second}^2 \textbf{ Ans.}$$

Mechanical angular acceleration,

$$\alpha_m = \frac{\alpha}{\text{Pair of poles}} \text{ mechanical degrees/second}^2$$

$$= \frac{2{,}160}{4/2} = 1{,}080 \text{ mechanical degrees/second}^2 \textbf{ Ans.}$$

Example 21.10. A 50 Hz synchronous generator is connected to an infinite bus through a line. The pu reactances of the generator and the line are *j* 0.3 pu and *j* 0.2 pu respectively. The generator no load voltage is 1.1 pu and that of the infinite bus is 1.0 pu. The inertia constant of the generator is 3 MW-sec/MVA.

(*a*) If the generator is loaded to 60% of its maximum power transfer capability and small perturbation is given, calculate resulting natural frequency of oscillations.

(*b*) If the generator is loaded to 75% of its maximum power transfer capability and a small perturbation is given, calculate the new natural frequency of oscillations.

[U.P.S.E. I.E.S. Electrical Engineering-II, 2007]

Solution : Inertia constant, H = 3 MW-s/MVA = 3 MJ/MVA

No-load voltage, V_1 = 1.1 pu

Infinite bus voltage, V_2 = 1.0 pu

Total reactance, $X = X_G + X_L = 0.3 + 0.2 = 0.5$ pu

Angular momentum, $M = \frac{GH}{\pi f} = \frac{1 \times 3}{50\pi} = 0.06/\pi$ MJ-s/radian

When the load is changed (say, increased by ΔP), the load angle changes (say by Δδ) and the swing equation can be linearized around the initial operating point (P_0, δ_0) of the power angle curve and is given as

$$M\frac{d^2}{dt^2}\Delta\delta = -\Delta P = -\left.\frac{\partial P}{\partial \delta}\right|_{\delta_0}.\Delta\delta \qquad ...(i)$$

Taking $\frac{d}{dt} = p$, above equation reduces to

$$Mp^2\Delta\delta + \left.\frac{\partial P}{\partial \delta}\right|_{\delta_0}.\Delta\delta = 0$$

or $$\left[Mp^2 + \left.\frac{\partial P}{\partial \delta}\right|_{\delta_0}\right]\Delta\delta = 0 \qquad ...(ii)$$

Above Eq. (*ii*) is the characteristic equation with the two roots

$$p = \pm\left[\frac{-\left.\frac{\partial P}{\partial \delta}\right|_{\delta_0}}{M}\right]^{1/2} \qquad ...(iii)$$

For positive value of $\frac{\partial P}{\partial \delta}$, the two roots are pure imaginary and conjugate, the rotor motion is oscillatory and undamped around δ_0 and for negative value of $\frac{\partial P}{\partial \delta}(\delta > 90°)$, both the roots are real (one positive and the other negative respectively) and so the system is unstable. For zero value of $\frac{\partial P}{\partial \delta}$, δ = 90° the system is critically stable. The frequency of oscillation of the system is given by the roots of the characteristic equation.

Since the system is operating initially under steady state condition, a small perturbation in power will make the rotor oscillate. The natural frequency of oscillation is given by

$$f_n^2 = \frac{\left.\frac{\partial P_e}{\partial \delta}\right|_{\delta_0}}{M}$$

(*i*) *For 60% loading*

$$\sin\delta_0 = \frac{60}{100} = 0.6$$

$$\cos\delta_0 = \cos\sin^{-1} 0.6 = 0.8$$

$$\frac{\partial P_e}{\partial \delta} = \frac{V_1V_2}{X}\cos\delta_0 = \frac{1.1\times 1}{0.5}\times 0.8 = 1.76$$

$$f_n = \sqrt{\frac{1.76}{0.06}\pi} = 9.6 \text{ radians/sec} = \frac{9.6}{2\pi}\text{ Hz} = 1.528 \text{ Hz} \textbf{ Ans.}$$

(*ii*) *For 75% loading*

$$\sin\delta_0 = \frac{75}{100} = 0.75$$

or $\delta_0 = \sin^{-1} 0.75 = 48.59°$

$\cos\delta_0 = \cos 48.59° = 0.6614$

$$\frac{\partial P_e}{\partial \delta} = \frac{V_1V_2}{X}\cos\delta_0 = \frac{1.1\times 1}{0.5}\times 0.6614 = 1.455$$

$$f_n = \sqrt{\frac{1.455}{0.06}\times\pi} = 8.728 \text{ radians/second}$$

$$= \frac{8.728}{2\pi}\text{ Hz} = 1.39 \text{ Hz} \textbf{ Ans.}$$

Example 21.11. A generator is delivering rated power of 1.0 per unit to an infinite bus through a lossless network. A 3-phase fault under this condition reduces P_{max} to 0 per unit. The value of P_{max} before fault is 2.0 per unit and 1.5 per unit after fault clearing. If the fault in cleared is 0.05 second, calculate rotor angles at intervals of 0.05 second from *t* = 0 second to 0.1 second. Assume H = 7.5 MJ/MVA and frequency to be 50 Hz. *[GATE 1996]*

Solution : The value of P_{max} before fault = 2.0 pu

Power delivered, P_E = 1.0 pu

Since $P_E = P_{max}\sin\delta$

Load angle $\delta_0 = \sin^{-1}\frac{P_E}{P_{max}} = \sin^{-1}\frac{1}{2} = 30$ electrical degrees

Angular momentum,

$$M = \frac{GH}{180 f} = \frac{7.5\times 1}{180\times 50} = \frac{1}{1{,}200}\text{ pu}$$

Average accelerating power during the first interval of 0.05 second

$$P_{AG} = \frac{1+0}{2} = 0.5 \text{ pu}$$

Average acceleration,

$$\alpha_1 = \frac{P_{AG}}{M} = \frac{0.5}{\frac{1}{1{,}200}} = 600 \text{ electrical degrees/s}^2$$

Change in angular velocity

$= \alpha_1\Delta t = 600 \times 0.05 = 30$ electrical degrees/second

So $\frac{d\delta_1}{dt} = 0 + \alpha_1\Delta t = 0 + 30 = 30$ electrical degrees/second

The corresponding change in rotor angle during first interval of 0.05 s,

$\Delta\delta_1 = 30 \times 0.05 = 1.5$ electrical degrees

So rotor angle at the end of period of 0.05 second,

$\delta_1 = \delta_0 + \Delta\delta_1 = 30 + 1.5 = 31.5$ electrical degrees **Ans.**

Since P_{max} after the fault is cleared = 1.5 pu

So $P_E = P_{max}\sin\delta_1 = 1.5\sin 31.5° = 0.78375$ pu

Since P_{max} during fault is zero,

Average accelerating power

$$= \frac{0+0.78375}{2} = 0.391875 \text{ pu}$$

Average acceleration,

$$\alpha_2 = \frac{0.391875}{\frac{1}{1{,}200}} = 470.25 \text{ electrical degrees/second}^2$$

Change in angular velocity during second interval of 0.05 second

$= \alpha_2 \Delta t = 470.25 \times 0.05 = 23.5125$ electrical degrees/second

$$\text{So } \frac{d\delta_2}{dt} = \frac{d\delta_1}{dt} + \alpha_2 \Delta t = 30° + 23.5125°$$

$$= 53.5125 \text{ electrical degrees/second}$$

So change in rotor angle during second interval of 0.05 second

$\Delta\delta_2 = 53.5125 \times 0.05 = 2.68$ electrical degrees

Rotor angle at the end of second interval of 0.05 second

$\delta_2 = \delta_1 + \Delta\,\delta_2 = 31.5 + 2.68 = 34.18$ electrical degrees **Ans.**

Example 21.12. A power system has a total load of 1,250 MW at 50 Hz. The load varies 1.5% for every 1% change in frequency. Find the steady-state frequency deviation when a 60 MW load is suddenly tripped. The speed regulation parameter R = 0.0025 Hz per MW. *[U.P.S.C. I.E.S. Electrical Engineering-II, 2010]*

Solution : Load, $P_D = 1{,}250$ MW

Change in load, $\Delta P_D = -60$ MW $\qquad \because$ load is reduced

Speed regulation parameter, R = 0.0025 Hz per MW

$$\text{Constant, } B = \frac{\partial P_D}{\partial f} = \frac{1.5\% \text{ of } P_D}{1\% \text{ of } f} = \frac{1.5 \times 1{,}250}{1 \times 50} = 37.5 \text{ MW/Hz}$$

Steady-state frequency deviation

$$\Delta f = -\left[\frac{1}{B + \frac{1}{R}}\right] \Delta P_D \text{ Hz} = -\left[\frac{1}{37.5 + \frac{1}{0.0025}}\right] \times (-60)$$

$$= 0.137 \text{ Hz} \quad \textbf{Ans.}$$

21.6. EQUAL AREA CRITERION

For one machine system and infinite bus-bars a method known as '*equal area criterion of stability*' is employed. The use of this method eliminates partially or wholly the calculations of swing curves and thus saves a considerable amount of work. The method is applicable to any two-machine system. This method is not applicable to multi-machine system directly.

The principle of this method consists of the basis that when δ oscillates around the equilibrium point with constant amplitude, transient stability will be maintained.

Starting with swing equation

$$\frac{M d^2\delta}{dt^2} = P_S - P_E = P_A$$

Multiplying both sides of the above equation by $\frac{d\delta}{dt}$, we have

$$M \frac{d^2\delta}{dt^2} \cdot \frac{d\delta}{dt} = P_S \frac{d\delta}{dt} - P_E \frac{d\delta}{dt} = \frac{d\delta}{dt} (P_S - P_E)$$

$$\text{or } \frac{1}{2} M \frac{d}{dt}\left(\frac{d\delta}{dt}\right)^2 = (P_S - P_E) \frac{d\delta}{dt}$$

Rearranging, multiplying by dt and integrating, we have

$$\left(\frac{d\delta}{dt}\right)^2 = \int_{\delta_0}^{\delta} \frac{2(P_S - P_E)}{M} d\delta$$

$$\text{or } \frac{d\delta}{dt} = \sqrt{\int_{\delta_0}^{\delta} \frac{2(P_S - P_E)}{M} d\delta} \qquad \text{...(21.41)}$$

where δ_0 is the torque angle at which the machine is operating while running at synchronous speed under normal conditions. Under the above conditions the torque angle was not changing, *i.e.*, before the disturbance $\frac{d\delta}{dt} = 0$. Also if the system has transient stability the machine will again operate at synchronous speed after the disturbance, *i.e.*, $\frac{d\delta}{dt} = 0$.

Hence the condition for the transient state stability is given by the equation

$$\sqrt{\int_{\delta_0}^{\delta} \frac{2(P_S - P_E)}{M} d\delta} = 0$$

$$\text{or } \int_{\delta_0}^{\delta} (P_S - P_E)\, d\delta = 0$$

$$\text{or } \int_{\delta_0}^{\delta} P_A \, d\delta = 0 \qquad \text{...(21.42)}$$

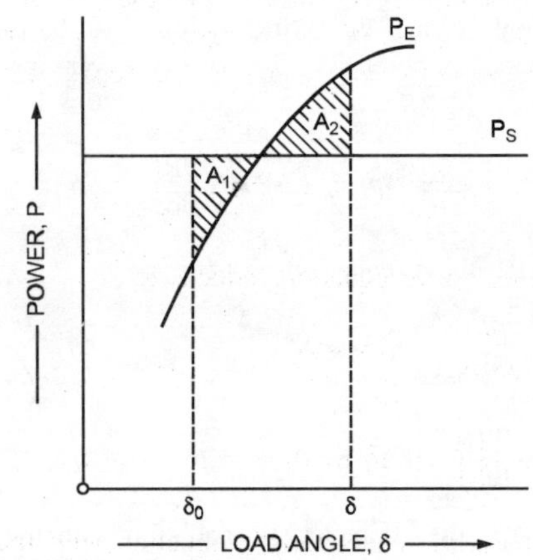

Fig. 21.7

This means that the area under the curve P_A should be zero which is possible only when P_A has both accelerating and decelerating powers, *i.e.*, for a part of the curve $P_S > P_E$ and for the other $P_E > P_S$ (Fig. 21.7). For a generator action $P_S > P_E$ for positive area A_1 and $P_E > P_S$ for negative area A_2 for stable operation. Hence the name *equal area criterion*. The area A_1 represents the kinetic energy stored by the rotor during acceleration and the area A_2 represents the kinetic energy given up by the rotor to the system and when it is all given up, the machine has returned to its original speed. It is to be noted that the kinetic energy involved in our explanation is fictitious as it is being calculated corresponding to relative speed rather than the actual speed.

The equal area criterion is also useful in determining the maximum limit on the load that the system can take without exceeding the stability limit. This can happen as long as area between P_S line and the P_E curve is equal to the area between the initial torque angle δ_0 and the P_S line. In case the area A_2 is less than the area A_1, the system will become unstable.

The main cases for the study of transient stability problems in a system are :

1. Sudden changes in load.

2. Switching of one of the lines which causes a change in the reactance of the system and hence a change in load conditions.
3. Sudden fault on the system which causes reduction in output, requiring arrangement for clearance of fault rapidly, and study of after-fault condition which may cause part of the system outage.

In each case, the procedure will be to determine the power angle curve for the initial condition of the system, for the condition under fault, and for the after-fault condition and plot the curves in per unit values. Then locate the points for the initial load conditions finding out δ_0. Then using equal-area criterion, determine the new angle of displacement δ. The maximum angle δ_{max} which may be allowed and the corresponding maximum permissible load can also be determined.

Now every case will be discussed individually.

21.6.1. Effect of Increase in Load

There are two possibilities of increasing load from the point of view of transient disturbance and system stability;

(*i*) Total load exceeds the steady-state limit of the system.

(*ii*) The rate of increase in load is so high as to set up oscillations which cross the critical point.

Consider a synchronous motor connected to an infinite bus-bar and operating at synchronous speed with mechanical output P_0. Let the load angle corresponding to output of P_0 be δ_0 determined from power angle diagram (Fig. 21.8). The power angle diagram is drawn using the relationship

$$P_E = \frac{V_G \times V_M}{X} \sin\delta$$

neglecting resistance of the system

where V_G is the voltage of the infinite bus-bars, V_M is the voltage behind the transient reactance of the synchronous motor, X is the transient reactance of the system (motor, transmission line, transformers etc) and δ is the load or torque angle.

Under these conditions the mechanical power output P_0 and electrical input P_E are equal

$$i.e.,\quad P_0 = P_E = \frac{V_G V_M}{X}\sin\delta$$

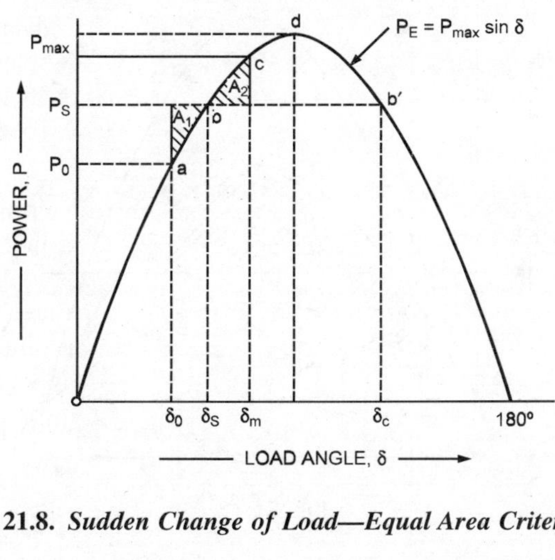

Fig. 21.8. ***Sudden Change of Load—Equal Area Criterion***

Now let the load on the motor be suddenly increased to P_S. Momentarily there is no change in load angle δ corresponding to electrical input to the motor $P_E < P_S$. Therefore, the motor decelerates supplying extra power required, as a result the load angle δ increases and the electrical input power P_E starts increasing. Thus electrical input keeps on increasing till it becomes equal to output power. This operating point is given by the point *b* on the diagram and the load angle corresponding to P_S is δ_S. At point *b* electrical input power P_E is equal to mechanical output power, P_S and therefore, decelerating force is zero but due to inertia of the rotor, the load angle goes on increasing. The speed of the motor beyond point *b* starts increasing. The speed is minimum at operating point *b*. When the speed of the motor becomes equal to synchronous speed beyond point *b*, say at point *c*, the load angle stops increasing. But between operating points *c* and *b*, electrical power input P_E > mechanical output power P_S; therefore the motor accelerates and the load angle δ starts decreasing till it reaches the point *b* where this time the speed is maximum and is more than the synchronous speed. At point *b* the acceleration ceases and the rotor moves towards point *a* with a finite deceleration. In this way the cycle keeps on repeating itself between points *a* and *c* till the oscillations are damped out. Applying 'equal area criterion of stability' and ignoring the losses, the area of the sectioned part above the curved portion *ab* must be equal to the sectioned part below the curved potion *bc*. The first area A_1 represents the rotational energy lost by the rotor which is due to the retarding torque given by

$$T_R = T_S - T_E$$

Also the energy gained by the rotor is represented by the sectioned area A_2, *i.e.*, area under the portion *bc*. This is due to accelerating torque given by

$$T_A = T_E - T_S$$

Since speeds at points *a* and *c* are synchronous, so the two areas are equal.

Let the load angle corresponding to the point *c*, known as the point of maximum swing of the rotor, be δ_m. This angle should be such that the stability of the system is maintained.

$$\text{Area over portion } ab,\ A_1 = \int_{\delta_0}^{\delta_s} (P_S - P_E)\, d\delta$$

$$\text{and area under portion } bc,\ A_2 = \int_{\delta_s}^{\delta_m} (P_E - P_S)\, d\delta$$

$$\text{Hence } A_1 - A_2 = \int_{\delta_0}^{\delta_m} (P_E - P_S)\, d\delta = 0$$

if the transient state stability condition given by the expression $\int_{\delta_0}^{\delta_m} (P_E - P_S)\, d\delta = 0$ to be satisfied.

$$\text{Also } \frac{d\delta}{dt} = 0 \text{ at the point } c$$

It follows from the above discussion that the point *c* should be so located that $A_1 = A_2$.

In fact if the point c can lie any where on the power angle diagram above the line $b\,b'$ the system will have transient stability. The point b' on the curve will give the load angle δ_c beyond which P_E will be less than P_S and an unstable operation will set in. At this instant the synchronism will be lost and the load angle δ will continue to increase till the motor stops.

The value of load angle beyond which the operation becomes unstable is called the *critical angle of load,* $\delta_{critical}$.

It is thus obvious that for maintaining transient stability it is necessary that rotor swing must not be beyond the critical load angle δ_c. This condition will reach either if the initial operating load angle δ_0 is too large or the sudden increase of power to be met was too high.

21.6.2. Effects of Switching Operation

Consider a generator supplying power to an infinite bus over two transmission lines operating in parallel, as shown in Fig. 21.9. The power angle diagram for the two lines operating in parallel is represented by curve A in Fig. 21.10. Curve B represents the

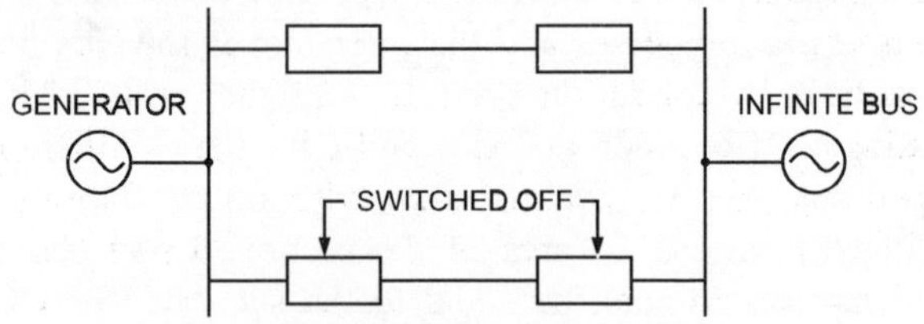

Fig. 21.9. ***Single Machine Tied To an Infinite Bus Through Two Parallel Lines***

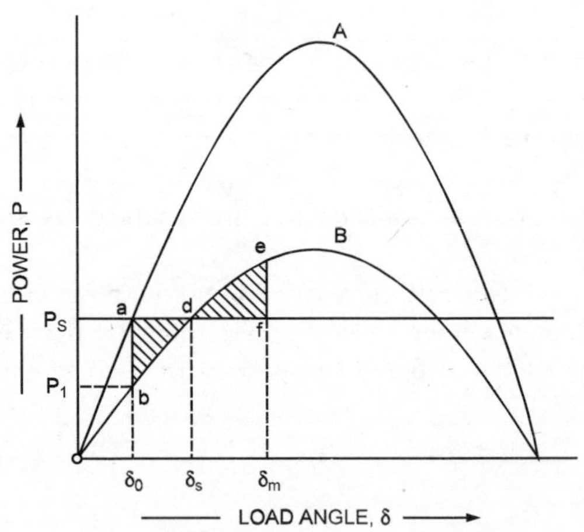

Fig. 21.10

power angle diagram after isolating one of the lines. Let the mechanical power input be P_S corresponding to point a on power angle curve A (power angle curve drawn for two lines), the load angle being δ_0. Now let one of the lines be opened instantaneously and simultaneously at both ends. As a result the equivalent impedance between the bus-bars is increased and hence curve B lies below curve A. As soon as one of the line is opened, the output of the generator is reduced, say to P_1 (say operating point b on curve B) and since input P_S remains constant which is higher than the output, the rotor accelerates and therefore, the torque angle δ increases and the operating point moves along curve B towards d from b. As it reaches d, the accelerating power becomes zero and the speed of the generator becomes more than that of infinite bus and the speed continues to increase. From d to e the rotor experiences deceleration but the speed is more than that of the infinite bus till it reaches point e where the relative speed is zero and the load angle ceases to increase. At point e, the output exceeds the input and hence the rotor decelerates and the speed goes down relative to the infinite bus till it reaches again point d where the speed is minimum. This load angle continues to decrease till it reaches point b where again the speed is equal to the speed of infinite bus. The cycle repeats itself in absence of damping. In practice due to damping present the rotor operates at point d on curve B and the torque angle is δ_S. In order to determine the transient stability limit for this case the input line P_S should be so raised that area below the line P_S and the curve B and above the line P_S and the curve B at the intersection of line P_S and curve B are equal. This is illustrated in Fig. 21.11. The value of P_S corresponding to this situation is known as the *transient stability limit.*

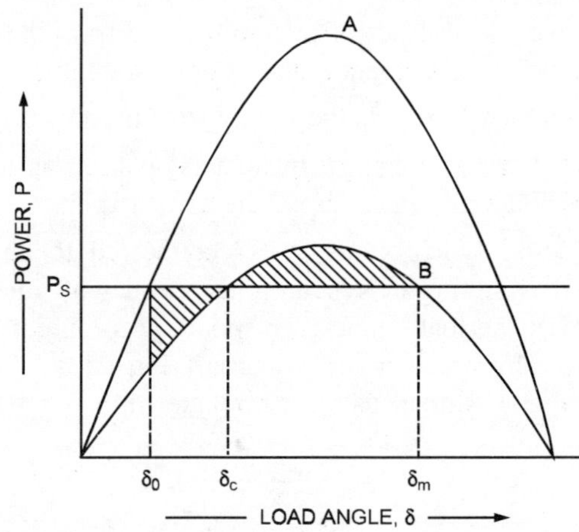

Fig. 21.11

The transient stability in this case depends upon (*i*) steady state limit of the remaining system and (*ii*) the difference in initial and final load angles.

21.6.3. Faults With Subsequent Circuit Isolation

Curve A in Fig. 21.12 represents the power angle diagram corresponding to healthy condition of the system shown in

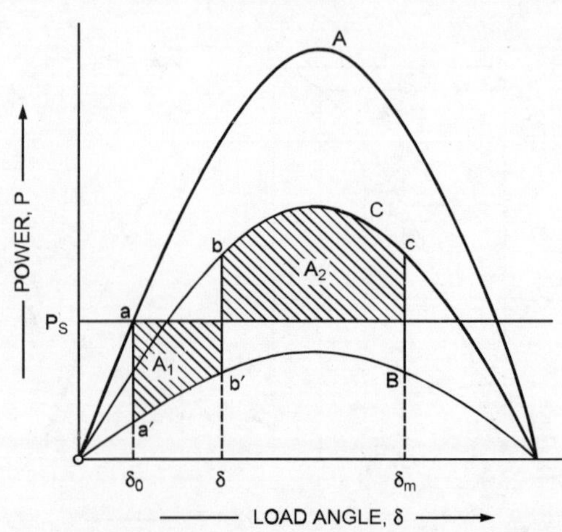

Fig. 21.12

Fig. 21.9. Curve B represents corresponding to fault on one of the two lines and fault allowed to exist for some time. Curve C corresponds to the situation when the faulted line is removed. Let P_S be the power supplied by the generator to the infinite bus in normal conditions and let δ_0 be the corresponding load angle. At the time of fault, the output of the generator reduces and becomes as at *a*'. Hence the rotor accelerates and rotor moves along the curve B to point *b*' when the faulty line is removed and the operating point becomes *b* on curve C where the output exceeds the input and the rotor decelerates till the speed becomes equal to the speed of the infinite bus and the load angle ceases to increase at point *c*.

It is thus clear that the transient stability limit not only depends upon the type of disturbance but it also depends upon the clearing time of the breaker. Faster the breaker operation, smaller will be the area A_1 and therefore, larger will be the transient stability limit.

21.6.4. Generator Connected To Infinite Bus Through a Line

Now let us consider another case of a generator connected to the infinite bus through a single line, as shown in Fig. 21.13. Let P_S be the power supplied by generator to the infinite bus in normal conditions and let δ_0 be the corresponding load angle. Now say there occurs a 3-phase fault on the line temporarily. The power angle curve (Fig. 21.14) will correspond to horizontal axis because output power becomes zero. If the breaker recloses after some time corresponding to clearing angle δ_c when the fault vanishes, the output will be more than the input and, therefore, the rotor decelerates. Finally, if the clearing angle δ_c is such that $A_1 = A_2$, the system becomes stable.

Fig. 21.13

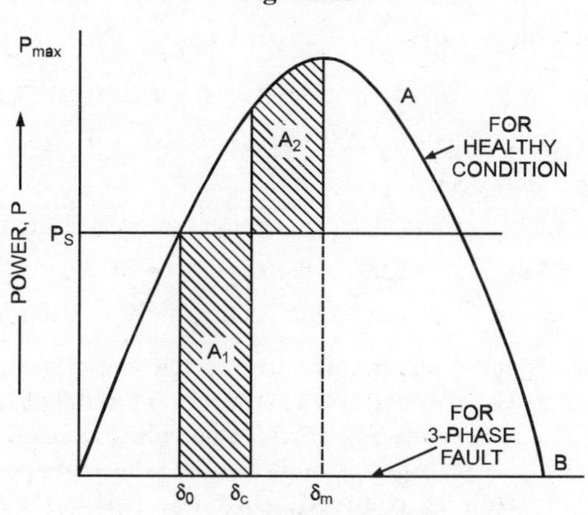

Fig. 21.14

Example 21.13. A 3-ϕ transmission line has a reactance of 12 Ω per phase. The voltage at each end is maintained at 132 kV (line to line). Determine the maximum steady state power that can be transmitted by the line. Also determine the limits of angular oscillations for transient stability, when the above line develops a sudden jerk at 2/5th of the steady state limit.

Solution : Line voltage at each end, $V_{SL} = V_{RL} = V_L = 132$ kV

Line reactance per phase, $X = 12\ \Omega$

Maximum steady state power that can be transmitted by the line (for all three phase),

$$P_{max} = \frac{V_{SL}\,V_{RL}}{X}\ \text{MW}$$

where V_{SL} and V_{RL} are line values in kV and X is in ohms

$$= \frac{132 \times 132}{12} = 1{,}452 \text{ MW} \quad \textbf{Ans.}$$

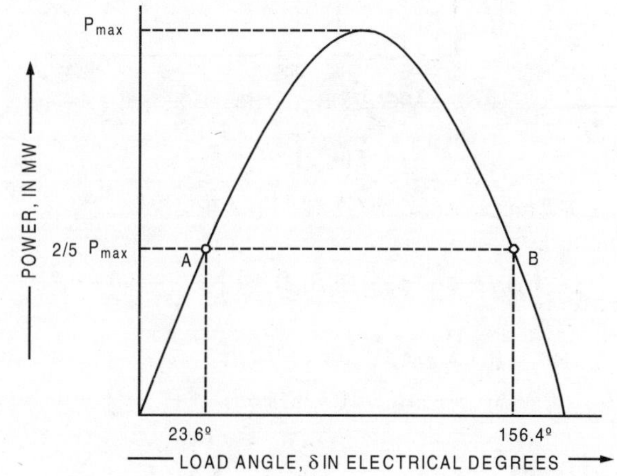

Fig. 21.15. ***Load Angle, δ in Electrical Degrees***

The power angle diagram is drawn, as illustrated in Fig. 21.15, and the operating point A, *i.e.*, $\frac{2}{5} P_{max}$ is marked on the graph and the corresponding load angle is read which is 23.6° (electrical). As soon as the jerk occurs on the system the rotor of the alternator at the sending end will tend to advance. If it advances up to point B transient stability will be maintained. B corresponds to 180° – 23.6° *i.e.*, 156.4°. This means that the limit of angular oscillations for transient stability is 23.6° – 156.4° (electrical) **Ans.**

Example 21.14. A synchronous motor is receiving 35% of the power that it is capable of receiving from an infinite bus. If the load on the motor is doubled, determine the maximum value of load angle δ during the swinging of the motor around its new equilibrium position.

Solution : Initial operating load,

$$P_0 = 0.35\,P_{max}$$

Operating load angle corresponding to this load,

$$\delta_0 = \text{Sin}^{-1}\frac{P_0}{P_{max}}$$

$$= \text{Sin}^{-1} 0.35 = 20.49° = 0.3576 \text{ radian (electrical)}$$

Increased load on the bus,

$$P_1 = 2 \times 0.35\,P_{max} = 0.7\,P_{max}$$

Operating load angle corresponding to this new load,

$$\delta_1 = \text{Sin}^{-1}\frac{P_1}{P_{max}}$$

$$= \text{Sin}^{-1} 0.7 = 44.3° \text{ or } 0.7754 \text{ radian (electrical)}$$

Let the maximum value of load angle during the swinging of the rotor be δ_m

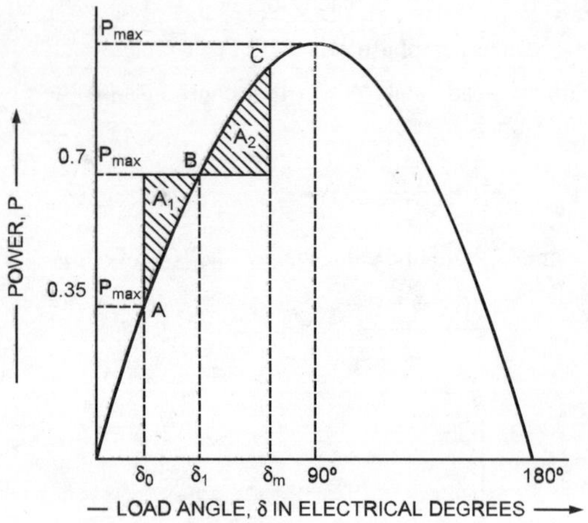

Fig. 21.16

For transient stability Area A_1 = Area A_2

$$\text{or} \int_{\delta_0}^{\delta_1} (0.7P_{max} - P_{max}\sin\delta)d\delta = \int_{\delta_1}^{\delta_m} (P_{max}\sin\delta - 0.7P_{max})d\delta$$

$$\text{or } 0.7(\delta_m - \delta_0) + \cos\delta_m - \cos\delta_0 = 0$$

$$\text{or } 0.7(\delta_m - 0.3576) + \cos\delta_m - 0.93675 = 0$$

$$\text{or } \cos\delta_m = 1.18707 - 0.7\,\delta_m$$

Solving above equation, we have δ_m = 1.2567 radians (electrical) or 72 electrical degrees **Ans.**

Example 21.15. A transmission line is acting as a connector between two constant voltage networks. Determine graphically or otherwise, the maximum additional load that can be suddenly applied to such an interconnector already carrying 50 MW if the power angle diagram is given by the equation

$$\mathbf{P = 100 \sin\alpha}$$

where P is the power transmitted in MW and α is the displacement between voltage phasors at the ends of the lines.

[Pb. Univ. Power System Analysis, Dec. 1993]

Solution : The maximum power that can be transferred,

$$P_{max} = 100 \text{ MW}$$

Initial operating load, P_0 = 50 MW

Operating load angle corresponding to this load angle,

$$\text{Then } \delta_0 = \text{Sin}^{-1}\frac{P_0}{P_{max}} = \text{Sin}^{-1}\frac{50}{100} = 30° \text{ or } 0.5236 \text{ radian (elec.)}$$

Let the maximum permissible suddenly increased load on the generator be P_1 MW and operating load angle corresponding this load be δ_1.

$$\text{Then } P_1 = P_{max}\sin\delta_1 = 100\sin\delta_1$$

As soon as the load is increased the rotor of the generator will advance beyond its new equilibrium point B and oscillate between original point A and the extreme point C and finally will reach new equilibrium point B due to damping present.

For transient stability, load angle δ_m corresponding to extreme point C must be less or equal to $\pi - \delta_1$.

For transient stability Area A_1 = Area A_2

$$\text{or} \int_{\delta_0}^{\delta_1} (P_{max}\sin\delta_1 - P_{max}\sin\delta)\,d\delta = \int_{\delta_1}^{\delta_m} (P_{max}\sin\delta - P_{max}\sin\delta_1)\,d\delta$$

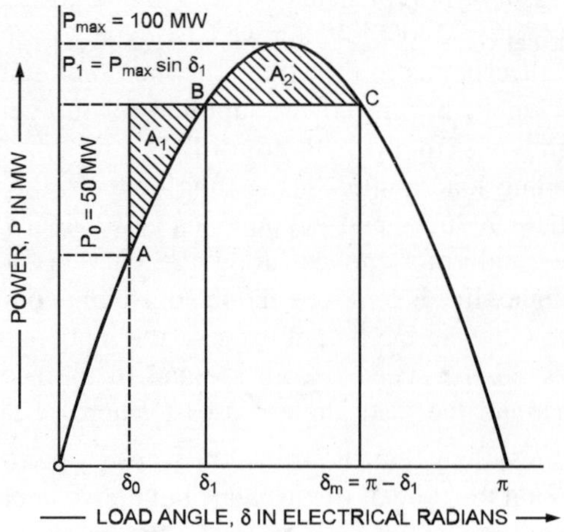

Fig. 21.17

$$\text{or} \int_{\pi/6}^{\delta_1} (100\sin\delta_1 - 100\sin\delta)\,d\delta = \int_{\delta_1}^{\pi-\delta_1} (100\sin\delta - 100\sin\delta_1)\,d\delta$$

$$\text{or} \int_{\pi/6}^{\delta_1} (\sin\delta_1 - \sin\delta)\,d\delta = -\int_{\delta_1}^{\pi-\delta_1} (\sin\delta_1 - \sin\delta)\,d\delta$$

$$\text{or} \int_{\pi/6}^{\delta_1} (\sin\delta_1 - \sin\delta)\,d\delta + \int_{\delta_1}^{\pi-\delta_1} (\sin\delta_1 - \sin\delta)\,d\delta = 0$$

$$\text{or} \int_{\pi/6}^{\pi-\delta_1} (\sin\delta_1 - \sin\delta)\,d\delta = 0$$

$$\text{or } \left[\delta\sin\delta_1 + \cos\delta\right]_{\pi/6}^{\pi-\delta_1} = 0$$

$$\text{or } \frac{5}{6}\pi\sin\delta_1 - \delta_1\sin\delta_1 - \cos\delta_1 - \cos\frac{\pi}{6} = 0$$

$$\text{or } 2.618\sin\delta_1 - \delta_1\sin\delta_1 - \cos\delta_1 - 0.866 = 0$$

$$\text{or } \sin\delta_1(2.618 - \delta_1) - \cos\delta_1 - 0.866 = 0$$

Solving the above equation, we have

$$\delta_1 = 1.0472 \text{ electrical radians or 60 electrical degrees}$$

$$\therefore P_1 = 100\sin 60° = 100 \times 0.866 = 86.66 \text{ MW}$$

∴ Sudden increase permissible = 86.66 – 50 = 36.66 MW **Ans.**

Example 21.16. An alternator is connected to an infinite bus, as shown in Fig. 21.18. It delivers 1.0 pu current at 0.8 pf lagging at V = 1.0 pu. The reactance is 1.2 pu. Determine the active power output and steady state power limit. Keeping the active power fixed if the excitation is reduced, find the critical excitation corresponding to operation at stability. *[GATE 1998]*

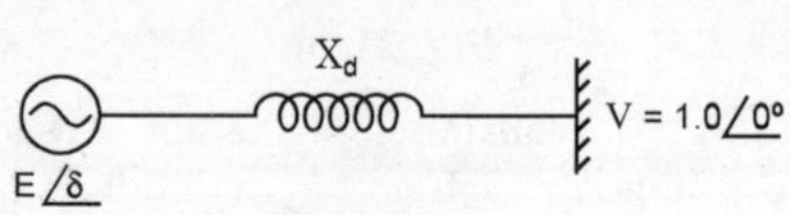

Fig. 21.18

Solution : I = 1.0 pu

Power factor, cos φ = 0.8 (lagging)

$\text{Sin } \phi = -0.6$

So current delivered, $\mathbf{I} = 1\,(0.8 - j\,0.6)$

$X_d = 1.2$ pu

Receiving-end voltage, $\mathbf{V} = 1.0\angle 0° = 1 + j\,0$

Sending-end voltage,

$$\mathbf{E} = \mathbf{V} + j\mathbf{I}X_d$$

$$= 1.0 + j\,(0.8 - j0.6) \times 1.2 = 1.72 + j\,0.96$$

$$= 1.97\,\angle 29.17\,° \text{ pu}$$

Power output $= VI \cos\phi = 1.0 \times 1 \times 0.8 = 0.8$ pu **Ans.**

Steady state power limit,

$$P_{max} = \frac{EV}{X_d} = \frac{1.97 \times 1.0}{1.2} = 1.64 \text{ pu } \textbf{Ans.}$$

For fixed active power output,

Generated emf,

$$E = \frac{\text{Power output} \times X_d}{V \times \sin 90°} = \frac{0.8 \times 1.2}{1.0 \times 1} = 0.96 \text{ pu } \textbf{Ans.}$$

$\because$ Power output $= \dfrac{EV}{X_d} \sin\delta$ and for stability limit, $\delta = 90°$

Example 21.17. A loss-free alternator supplies 50 MW to an infinite bus, the steady-state stability limit being 100 MW. Determine whether the alternator will remain stable if the input to the prime mover of the alternator is abruptly increased by 40 MW. *[GATE 1995]*

Solution : Since sudden increase permissible for stability = 36.66 MW (Refer to solution to Example 21.15), the alternator will not remain stable.

21.7. CRITICAL CLEARING ANGLE

We have seen that for any given initial load there is a critical clearing angle and if actual clearing angle exceeds critical clearing angle the system becomes unstable otherwise it will be stable.

Let the power angle curves drawn for the healthy condition, faulty condition and for the condition after isolation of fault be as follows and represented by curves A, B and C respectively in Fig. 21.19.

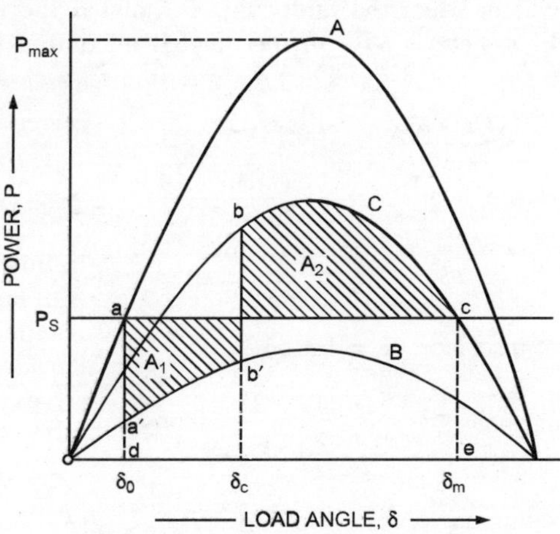

Fig. 21.19. *Critical Clearing Angle*

$A = P_{max} \sin\delta$ ...Before the fault

$B = \gamma_1 P_{max} \sin\delta$ ...During the fault

$C = \gamma_2 P_{max} \sin\delta$ ...After the fault

where γ_1 is the ratio of system reactance in healthy condition to that of during the fault and γ_2 is the ratio of steady state limit of the system after the isolation of fault and that of system under initial condition.

For transient stability limit, two areas $A_1 = A_2$

or equivalently the area under the curve *adec* (a rectangle)

= Area under the curve *da′b′bce*

$$\text{or } P_S(\delta_m - \delta_0) = \int_{\delta_0}^{\delta_c} \gamma_1 P_{max} \sin\delta\, d\delta + \int_{\delta_c}^{\delta_m} \gamma_2 P_{max} \sin\delta\, d\delta$$

$$= \gamma_1 P_{max}(\cos\delta_0 - \cos\delta_c) + \gamma_2 P_{max}(\cos\delta_c - \cos\delta_m)$$

Now substituting $P_S = P_{max} \sin\delta_0$, we have

$$(\delta_m - \delta_0) P_{max} \sin\delta_0 = \gamma_1 P_{max}(\cos\delta_0 - \cos\delta_c) + \gamma_2 P_{max}(\cos\delta_c - \cos\delta_m)$$

$$\text{or } (\delta_m - \delta_0)\sin\delta_0 = \text{Cos}\,\delta_c\,(\gamma_2 - \gamma_1) + \gamma_1\cos\delta_0 - \gamma_2\cos\delta_m$$

$$\text{or Cos}\,\delta_c = \frac{(\delta_m - \delta_0)\sin\delta_0 - \gamma_1\cos\delta_0 + \gamma_2\cos\delta_m}{\gamma_2 - \gamma_1} \quad ...(21.43)$$

Also from the curves

$$P_S = P_{max}\sin\delta_0 = \gamma_2 P_{max}\sin\delta_m = \gamma_2 P_{max}\sin(\pi - \delta_m)$$

$$\text{or Sin}\,\delta_0 = \gamma_2 \sin(\pi - \delta_m)$$

$$\text{or } \delta_m = \pi - \sin^{-1}\left(\frac{\sin\delta_0}{\gamma_2}\right) \quad ...(21.44)$$

Thus if γ_1, γ_2 and δ_0 are known, the critical clearing angle δ_c, can be determined .

Example 21.18. A 50 Hz, 3-phase alternator is supplying 60% of P_{max} to an infinite bus through a reactive network. A fault occurs which increases the reactance of the network between the generator internal voltage and the infinite bus by 400%. When the fault is cleared, the maximum power that can be delivered is 80% of the original maximum value. Determine the critical clearing angle for the condition described. *[U.P.S.C. I.E.S. Electrical Engineering-II, 2009]*

Solution : Let P_{max} be the maximum power that can be transferred.

$$\therefore P_S = P_{max} \sin\delta_0 = 0.6\,P_{max}$$

Initial loading angle,

$$\delta_0 = \text{Sin}^{-1}\frac{P_S}{P_{max}} = \text{Sin}^{-1}\,0.6 = 36.87° \text{ or } 0.6435 \text{ radian}$$

During fault the reactance is 400% of the value before the fault

$$\therefore \gamma_1 = \frac{\text{Reactance before the fault}}{\text{Reactance during the fault}} = \frac{100}{400} = 0.25$$

$$\text{and } \gamma_2 = \frac{\text{Maximum power that can be transferred after isolation of fault}}{\text{Maximum power that can be transferred initially}}$$

$$= 0.8$$

$$\delta_m = \pi - \sin^{-1}\frac{\sin\delta_0}{\gamma_2}$$

$$= \pi - \sin^{-1}\frac{0.6}{0.8} = \pi - \sin^{-1} 0.75$$

$$= \pi - 0.848 = 2.29353 \text{ radians or } 131.4°$$

Critical clearing angle,

$$\delta_c = \text{Cos}^{-1} \frac{\sin\delta_0 \times (\delta_m - \delta_0) - \gamma_1 \cos\delta_0 + \gamma_2 \cos\delta_m}{\gamma_2 - \gamma_1}$$

$$= \text{Cos}^{-1} \frac{0.6 \times (2.29353 - 0.6435) - 0.25 \times 0.8 + 0.8 \times (-0.6613)}{0.8 - 0.25}$$

$$= \text{Cos}^{-1}\, 0.4745 = 61.67° \textbf{ Ans.}$$

Example 21.19. A 50 Hz generator is delivering 50% of power that it is capable of delivering through a transmission line to an infinite bus. A fault occurs that increases the reactance between the generator and the infinite bus to 500% of the value before the fault. When fault is isolated, the maximum power that can be delivered is 75% of the original maximum value. Determine the critical clearing angle for the condition described.

[U.P. Technical Univ. Power System Analysis 2005-06]

Solution : Let P_{max} be the maximum power that can be transferred

$$\therefore \quad P_s = P_{max} \sin\delta_0 = 0.5\, P_{max}$$

Initial loading angle $\delta_0 = \sin^{-1} 0.5 = 30°$ or 0.5236 radian (elec)

During fault the reactance is 500% of the value before the fault

$$\therefore \gamma_1 = \frac{\text{Reactance before the fault}}{\text{Reactance during the fault}} = \frac{100}{500} = 0.2$$

$$\text{and } \gamma_2 = \frac{\text{Maximum power that can be transferred after isolation of fault}}{\text{Maximum power that can be transferred initially}}$$

$$= 0.75$$

$$\delta_m = \pi - \sin^{-1} \frac{\sin\delta_0}{\gamma_2}$$

$$= \pi - \sin^{-1} \frac{0.5}{0.75} = \pi - 0.7297 = 2.412 \text{ radians (electrical)}$$

and critical angle

$$\delta = \cos^{-1} \frac{\sin\delta_0 \times (\delta_m - \delta_0) - \gamma_1 \cos\delta_0 + \gamma_2 \cos\delta_m}{\gamma_2 - \gamma_1}$$

$$= \cos^{-1} \frac{0.5 \times (2.412 - 0.5236) - 0.2 \times 0.866 + 0.75 \times (-0.7454)}{0.75 - 0.2}$$

$$= \cos^{-1} \frac{0.9442 - 0.1732 - 0.559}{0.55}$$

$$= \cos^{-1} \frac{0.212}{0.55} = \cos^{-1} 0.3854 = 67.33° \text{ (elec) } \textbf{Ans.}$$

Example 21.20. A large generator is delivering 1.0 pu power to an infinite bus through a transmission network. The minimum power which can be transferred to pre-fault, fault and post-fault conditions are 1.8 pu, 0.4 pu and 1.3 pu respectively. Find critical clearing angle. *[U.P.S.C. I.E.S. Electrical Engineering-II, 2001]*

Solution : All of the three power angle curves are shown in Fig. 21.19.

$$P_{max_1} = 1.8 \text{ pu}$$

$$P_{max_2} = 0.4 \text{ pu}$$

$$P_{max_3} = 1.3 \text{ pu}$$

$$\text{Initial loading, } P_S = 1.0 \text{ pu}$$

Initial loading angle,

$$\delta_0 = \text{Sin}^{-1} \frac{P_S}{P_{max_1}} = \text{Sin}^{-1} \frac{1}{1.8}$$

$$= 33.75° \text{ (electrical) or } 0.589 \text{ radian (electrical)}$$

$$\gamma_1 = \frac{\text{Maximum power that can be transferred during fault}}{\text{Maximum power that can be transferred initially}}$$

$$= \frac{P_{max_2}}{P_{max_1}} = \frac{0.4}{1.8} = 0.2222$$

$$\gamma_2 = \frac{P_{max_3}}{P_{max_1}} = \frac{1.3}{1.8} = \frac{13}{18}$$

$$\delta_m = 180° - \sin^{-1} \frac{\sin\delta_0}{\gamma_2}$$

$$= 180° - \sin^{-1} \frac{\frac{1}{1.8}}{\frac{13}{18}}$$

$$= 180° - 50.285° = 129.715° \text{ or } 2.264 \text{ radians (electrical)}$$

Critical clearing angle,

$$\delta_c = \text{Cos}^{-1} \frac{\sin\delta_0 \times (\delta_m - \delta_0) - \gamma_1 \cos\delta_0 + \gamma_2 \cos\delta_m}{\gamma_2 - \gamma_1}$$

$$= \text{Cos}^{-1} \frac{\frac{1}{1.8} \times (2.264 - 0.589) - 0.2222 \times 0.8315 + \frac{13}{18} \times (-0.639)}{\frac{13}{18} - \frac{4}{18}}$$

$$= \text{Cos}^{-1}\, 0.5686 = 55.35° \text{ (electrical) } \textbf{Ans.}$$

Example 21.21. A two-machine power system delivers a load of 25 MW at 0.8 pf lag and has a double circuit. The system reactance is 150 per cent on 100 MVA base. A sudden symmetrical line-to-ground fault occurs in one of the circuits which reduced the power supplied to 40% which is subsequently cleared by simultaneous action of the circuit breakers on both sides of the faulted line. Calculate in the electrical degrees the "critical clearing angle". Assume that during fault conditions the system reactance attains such a value that the maximum power becomes 30% of the normal maximum value, when the faulty line is isolated the maximum power of the system is 60% of the normal maximum power.

[U.P.S.C. I.E.S. Electrical Engineering-II, 2006]

Solution : Base MVA = 100 MVA

$$\text{Power supplied, } P_s = \frac{\text{Load delivered in MVA}}{\text{Base MVA} \times \text{pf}}$$

$$= \frac{25}{100 \times 0.8} = 0.3125 \text{ pu}$$

System reactance, X = 1.5 pu

$$\text{Maximum power, } P_{max} = \frac{1}{X} = \frac{1}{1.5} = 0.6666 \text{ pu}$$

$$\text{Initial loading angle, } \delta_0 = \text{Sin}^{-1} \frac{P_s}{P_{max}}$$

$$= \text{Sin}^{-1} \frac{0.3125}{0.6666}$$

$$= 27.95° \text{ or } 0.488 \text{ radian (electrical)}$$

$$\gamma_1 = \frac{P_{max2}}{P_{max1}} = 0.3$$

$$\gamma_2 = \frac{P_{max3}}{P_{max1}} = 0.6$$

$$\delta_m = \pi - \sin^{-1}\frac{\sin\delta_0}{\gamma_2}$$

$$= \pi - \sin^{-1}\frac{0.4688}{0.6}$$

$$= \pi - 0.89666 = 2.245 \text{ radians (electrical) or } 128.63° \text{ (electrical)}$$

Critical clearing angle,

$$\delta_c = \text{Cos}^{-1}\frac{\sin\delta_0 \times (\delta_m - \delta_0) - \gamma_1\cos\delta_0 + \gamma_2\cos\delta_m}{\gamma_2 - \gamma_1}$$

$$= \text{Cos}^{-1}\frac{0.4688 \times (2.245 - 0.4688) - 0.3 \times \cos 27.95° + 0.6 \times \cos 128.63°}{0.6 - 0.3}$$

$= \text{Cos}^{-1}\ 0.6437 = 49.93°$ (electrical) **Ans.**

Example 21.22. For the system given below in Fig. 21.20, the numerical values for different components are E = 1.2 pu, V = 1.0 pu, X'_d = 0.2 pu, $X_1 = X_2$ = 0.4 pu. Initially the generator was delivering power of 1.5 pu. If one of the double circuit lines is tripped out, using equal area criterion, determine whether the system would be able to maintain its stability. If stability is maintained, determine the maximum swing (δ) attained by generator. *[U.P. Technical Univ. Power System Analysis 2006-07]*

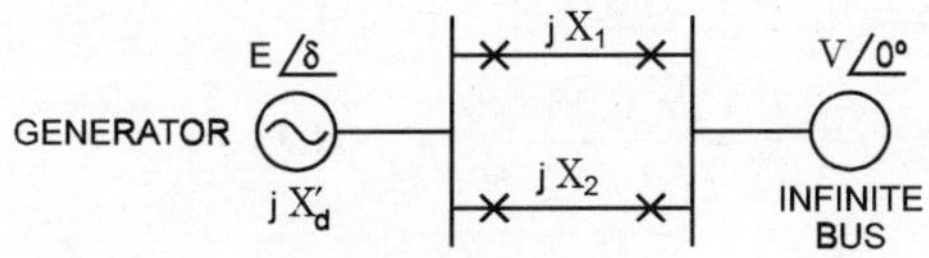

Fig. 21.20

Solution : The power angle curves are shown in Fig. 21.11.

For curve A, *i.e.,* when both the lines are in operation,

Transfer reactance,

$$X = X'_d + \frac{X_1}{2} = 0.2 + \frac{0.4}{2} = 0.4 \text{ pu}$$

Steady state power limit,

$$P_{max} = \frac{EV}{X} = \frac{1.2 \times 1}{0.4} = 3 \text{ pu}$$

For initial condition

$$P_0 = 1.5 \text{ pu}$$

Initial load angle,

$$\delta_0 = \text{Sin}^{-1}\frac{P_0}{P_{max}} = \text{Sin}^{-1}\frac{1.5}{3} = 30° \text{ or } 0.524 \text{ radian (electrical)}$$

For curve B (*i.e.,* when one line is in operation)

Transfer reactance,

$$X' = X'_d + X_1 \text{ (or } X_2) = 0.2 + 0.4 = 0.6 \text{ pu}$$

Steady state power limit,

$$P'_{max} = \frac{EV}{X'} = \frac{1.2 \times 1}{0.6} = 2 \text{ pu}$$

Electrical power developed,

$$P'_E = P'_{max}\sin\delta = 2\sin\delta$$

Load angle

$$\delta_c = \text{Sin}^{-1}\frac{P_S}{P'_{max}} = \text{Sin}^{-1}\frac{1.5}{2} = 48.6° \text{ or } 0.848 \text{ radian (electrical)}$$

and load angle

$$\delta_m = 180° - \delta_c = 180° - 48.6 = 131.4° \text{ or } 2.293 \text{ radian (electrical)}$$

The area A_1 is given as

$$A_1 = \int_{\delta_0}^{\delta_c}(P_S - P'_E)\,d\delta = \int_{0.524}^{0.848}(1.5 - 2\sin\delta)\,d\delta$$

$$= \left[1.5\delta + 2\cos\delta\right]_{0.524}^{0.848} = 0.0769$$

The system stability depends on whether there is enough negative area A_2 in the interval $\delta_c < \delta < \delta_m$ to match the area A_1

$$A_{2\,max} = \int_{\delta_c}^{\delta_m}(P_S - P'_E)\,d\delta = \int_{0.848}^{2.293}(1.5 - 2\sin\delta)\,d\delta$$

$$= \left[1.5\,\delta + 2\cos\delta\right]_{0.848}^{2.293} = -0.4777$$

Since $A_{2\,max}$ is greater than A_1 in magnitude, the system is stable.

The maximum value of rotor swing δ_m (Fig. 21.10) is given by condition that $A_2 = A_1$

$$\text{or} \quad \int_{\delta_s}^{\delta_m}(P_S - P'_E)\,d\delta = -0.0769$$

$$\text{or} \quad \int_{\delta_s}^{\delta_m}(1.5 - 2\sin\delta)\,d\delta = -0.0769$$

$$\text{or} \quad \left[1.5\,\delta + 2\cos\delta\right]_{0.848}^{\delta_m} = -0.0769$$

$$\text{or } 1.5\,\delta_m + 2\cos\delta_m = -0.0769 + 1.5 \times 0.848 + 2\cos 0.848$$

$$= 2.518.$$

The above equation is nonlinear and by solving this equation by iterative method, we have

$$\delta_m = 1.218 \text{ radian or } 69.8° \text{ (electrical)}$$ **Ans.**

Example 21.23. Explain the equal area criterion for stability. In the system given in following Fig. 21.21 where a three-phase fault is applied at the point P as shown. Find the critical clearing angle for clearing the fault with simultaneous opening of the breaker 1 and 2. The reactance values of various components are indicated on the diagram. The generator is delivering 1.0 pu power at the instant preceding the fault.

[U.P. Technical Univ. Power System Analysis 2008-09]

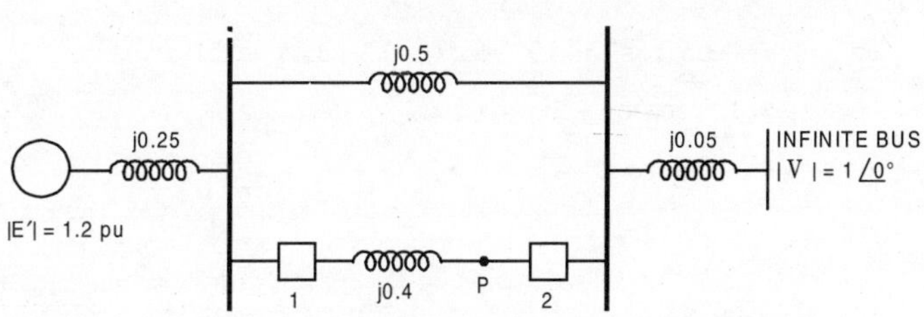

Fig. 21.21

Solution : The power angle curves are shown in Fig. 21.11.

For curve A *i.e.,* when both the lines are in operation

$$\text{Transfer reactance, } X = 0.25 + \frac{0.5 \times 0.4}{0.5 + 0.4} + 0.05 = 0.522 \text{ pu}$$

$$\text{Steady state power limit, } P_{max} = \frac{EV}{X} = \frac{1.2 \times 1}{0.522} = 2.298 \text{ pu}$$

For initial condition, P_0 = 1.0 pu

Initial load angle,

$$\delta_0 = \sin^{-1}\frac{P_0}{P_{max}} = \sin^{-1}\frac{1}{2.298} = 25.8° \text{ or } 0.45 \text{ radian (elec.)}$$

During fault, it is clear from Fig. 21.22 that no power is transferred during fault *i.e.*, $P_E = 0$

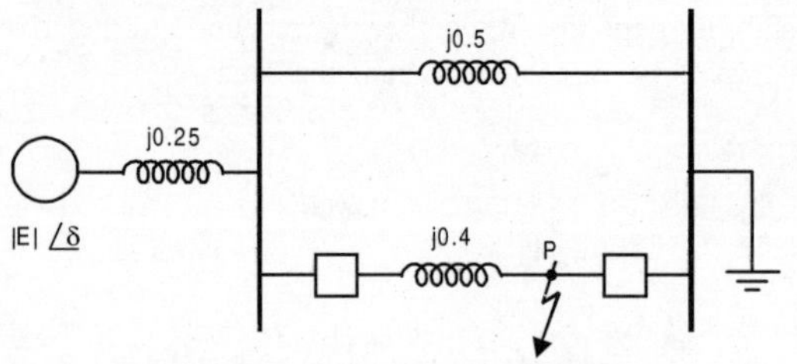

Fig. 21.22

For curve B *i.e.,* when one line is in operation

Transfer reactance,

$$X' = 0.25 + 0.5 + 0.05 = 0.8 \text{ pu}$$

Steady state power limit,

$$P'_{max} = \frac{EV}{X'} = \frac{1.2 \times 1.0}{0.8} = 1.5 \text{ pu}$$

Electrical power developed,

$$P'_E = P'_{max} \sin\delta = 1.5 \sin\delta$$

and load angle,

$$\delta_m = 180° - \sin^{-1}\frac{1}{1.5} = 180° - 41.81°$$

$$= 138.19° \text{ or } 2.41 \text{ radians (elec.)}$$

Applying equal area criterion for clearing angle δc

$$A_1 = P_0(\delta_c - \delta_0) = 1.0(\delta_c - 0.45) = \delta_c - 0.45$$

$$A_2 = \int_{\delta_c}^{\delta_m} (P'_E - P_0)\, d\delta$$

$$= \int_{\delta_c}^{138.19°} (1.5 \sin\delta - 1)\, d\delta$$

$$= \left[-1.5\cos\delta - \delta\right]_{\delta_c}^{138.19°}$$

$$= -1.5(\cos 138.19° - \cos\delta_c) - (2.41 - \delta_c)$$

$$= 1.5 \cos\delta_c + \delta_c - 1.292$$

Equating $A_1 = A_2$, we have

$$\delta_c - 0.45 = 1.5 \cos\delta_c + \delta_c - 1.292$$

$$\text{or} \quad \cos\delta_c = \frac{1.292 - 0.45}{1.5} = 0.561$$

$$\text{or} \quad \delta_c = \text{Cos}^{-1}\, 0.561 = 55.85° \text{ (electrical)}$$ **Ans.**

Example 21.24. The figure below shows a generator connected to a metropolitan system (infinite bus) through high-voltage lines. Per-unit reactances of generators, transformers and transmission lines are marked on the figure. Compute critical clearing angle for a three phase fault at point P when the generator is delivering 1.0 pu power. The voltage behind transient reactance is 1.2 pu for the generator and the voltage of infinite bus is 1.0 pu

[*Bharathiyar Univ. Power System Analysis April 1997*]

Solution : The equivalent reactance under healthy conditions is given as

$$X_1 = j\left[0.3 + \frac{0.15 + 0.25 + 0.15}{2} + j0.15\right]$$

$$= j\,(0.3 + 0.275 + 0.15) = j\,0.725 \text{ pu}$$

Under fault condition, the equivalent electrical circuit becomes, as illustrated in Fig. 21.24(*a*).

Simplified form of circuit shown in Fig. 21.24(*a*) is given in Fig. 21.24(*b*) or Fig. 21.24(*c*).

Delta ABP of Fig. 21.24(*c*) has been changed into an equivalent star, as shown in Fig. 21.24(*d*).

$$X_{OA} = j\frac{0.15 \times 0.55}{0.55 + 0.15 + 0.4} = j\,0.075 \text{ pu}$$

$$X_{OB} = j\frac{0.4 \times 0.55}{0.55 + 0.15 + 0.4} = j\,0.2 \text{ pu}$$

$$X_{OP} = j\frac{0.15 \times 0.4}{0.55 + 0.15 + 0.4} = j\,0.0545 \text{ pu}$$

Figure 21.24(*d*) is further simplified as shown in Fig. 21.24(*e*).

The star formed by reactances of $j\,0.375$, $j\,0.35$ and $j\,0.0545$ is again converted into equivalent delta. The circuit is shown in Fig. 21.24(*f*).

$$X'_{AB} = \frac{0.375 \times 0.35 + 0.35 \times 0.0545 + 0.0545 \times 0.375}{0.0545}$$

$$= j\,3.133 \text{ pu}$$

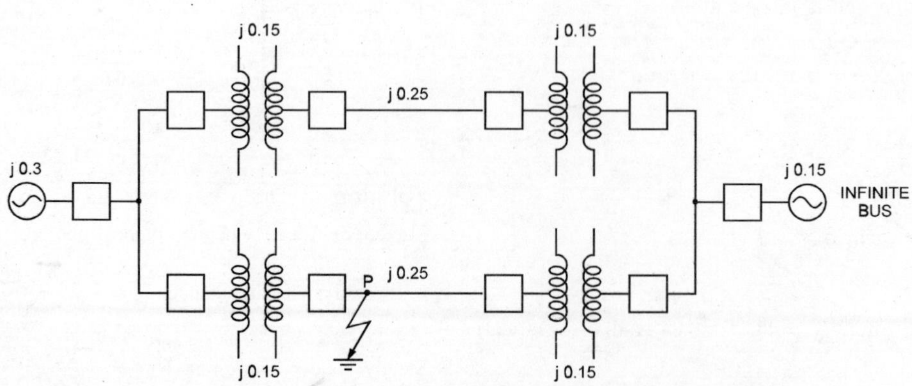

Fig. 21.23

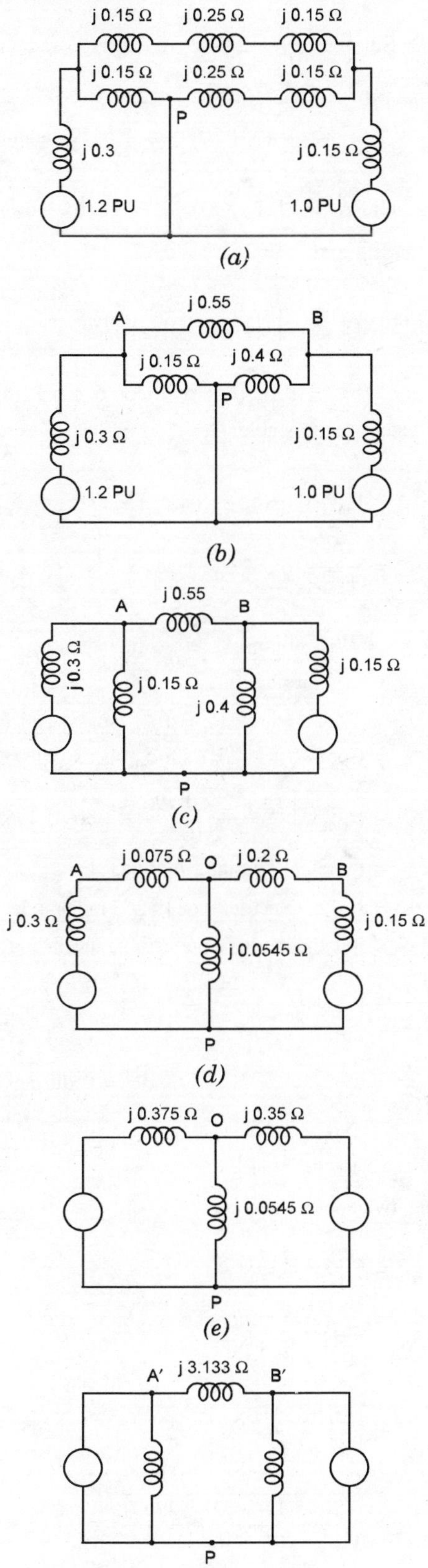

Fig. 21.24

Thus the reactance between generator and the infinite bus during fault, $X_2 = j\,3.133$ pu.

The reactance between generator and the infinite bus after the fault is cleared is given as

$$X_3 = j\,0.3 + j\,0.55 + j\,0.15 = j\,1.0 \text{ pu}$$

The maximum power output

$$\text{before the fault, } P_{max_1} = \frac{EV}{X_1} = \frac{1.2 \times 1}{0.725} = 1.655 \text{ pu}$$

$$\text{during the fault, } P_{max_2} = \frac{1.2 \times 1}{3.133} = 0.383 \text{ pu}$$

$$\text{after the fault, } P_{max_3} = \frac{1.2 \times 1}{1} = 1.2 \text{ pu}$$

$$\gamma_1 = \frac{P_{max_2}}{P_{max_1}} = \frac{0.383}{1.655} = 0.231 \text{ pu}$$

$$\gamma_2 = \frac{P_{max_3}}{P_{max_1}} = \frac{1.2}{1.655} = 0.725 \text{ pu}$$

Initial load angle,

$$\delta_0 = \text{Sin}^{-1} \frac{P_S}{P_{max_1}} = \text{Sin}^{-1} \frac{1}{1.655}$$

$$= 37.2^\circ \text{ or } 0.649 \text{ radian (electrical)}$$

Load angle,

$$\delta_m = 180^\circ - \sin^{-1} \frac{\sin \delta_0}{\gamma_2}$$

$$= 180^\circ - \sin^{-1} \frac{0.604}{0.725} = 123.58^\circ \text{ or } 2.157 \text{ radians (electrical)}$$

$$\text{Cos } \delta_C = \frac{(\delta_m - \delta_0) \sin \delta_0 - \gamma_1 \cos \delta_0 + \gamma_2 \cos \delta_m}{\gamma_2 - \gamma_1}$$

$$= \frac{(2.157 - 0.649) \times 0.604 - 0.231 \times 0.797 + 0.725(-0.553)}{0.725 - 0.231}$$

$$= -0.6888$$

or $\delta_c = \text{Cos}^{-1}(-0.6888) = 46.46$ electrical degrees **Ans.**

Example 21.25. The single line diagram of a three-phase power system is given below (Fig. 21.25). The generator is delivering 1.0 per unit power to the infinite bus. The pre-fault power angle equation is $P_e = 2.10 \sin \delta$. Calculate the critical clearing angle and critical clearing time when the system is subjected to a 3-phase fault at point P (*i.e.*, sending-end of the short line. The fault is cleared by opening the circuit breaker 5. The inertia constant H = 5 MJ/MVA.

[*U.P.S.C. I.E.S. Electrical Engineering-II, 2010*]

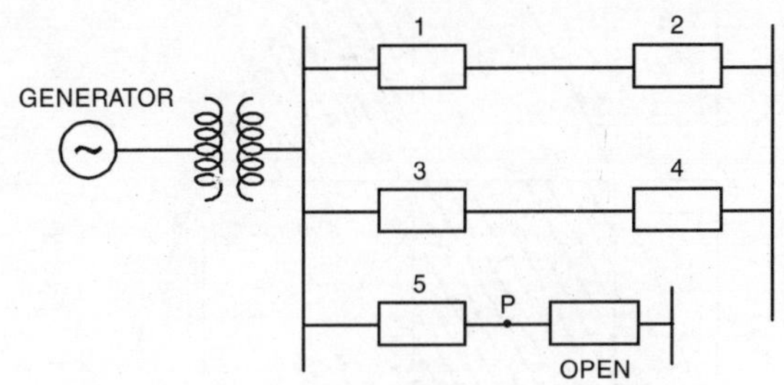

Fig. 21.25

Solution : Let P_S be the power supplied by generator to the infinite bus in normal condition and let δ_0 be the corresponding load angle, as shown by the point *a* on the power angle curve ($P_e - \delta$ diagram) of Fig. 21.26. When a three-phase fault occurs at point P of the outgoing radial line, the electrical output of the generator instantly reduces to zero *i.e.*, $P_E = 0$ and the operating point drops to *b*. The acceleration area A_1 begins to increase and so does the rotor angle while the operating point moves along *bc*. At time t_c corresponding to power angle δ_c the faulted line is cleared by the opening of circuit breaker 5. The values of δ_c and t_c are respectively known as the *clearing angle* and *clearing time.* The system once again becomes healthy and transmits power $P_e = P_{max} \sin \delta$, *i.e.*, the operating point shifts to *d* on the original power-angle curve. The rotor now decelerates and the decelerating area A_2 begins to increase while the operating point moves along *de*. If an angle δ_1 can be found such that $A_2 = A_1$, the system is found to be stable. The system finally settles down to the steady operating point *a* in an oscillatory manner because of inherent damping.

The value of clearing time corresponding to a clearing angle can be determined only by numerical integration except in this simple case. The equal area criterion therefore provides only qualitative answer to system stability as the time when the breaker should be opened is hard to establish.

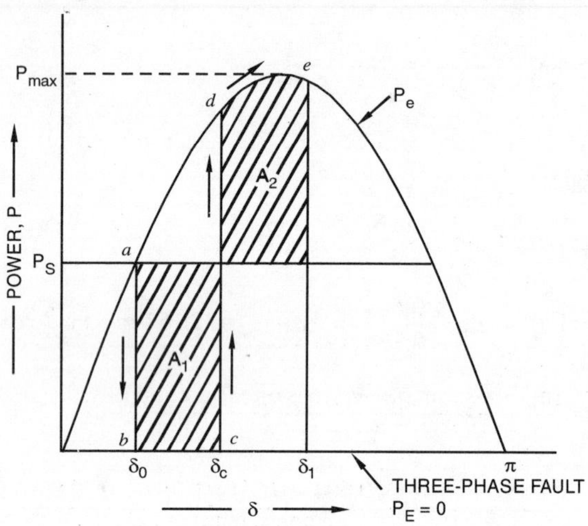

Fig. 21.26

With the delay in clearance of fault, area A_1 increases and so does power angled δ_1 to find $A_2 = A_1$, till $\delta = \delta_m$ as shown in Fig. 21.27. For a clearing time (or angle) larger than this value, the system would be unstable as $A_2 < A_1$. The maximum allowable value of the clearing time and angle for the system to remain stable are known respectively as critical clearing time t_c and angle δ_c.

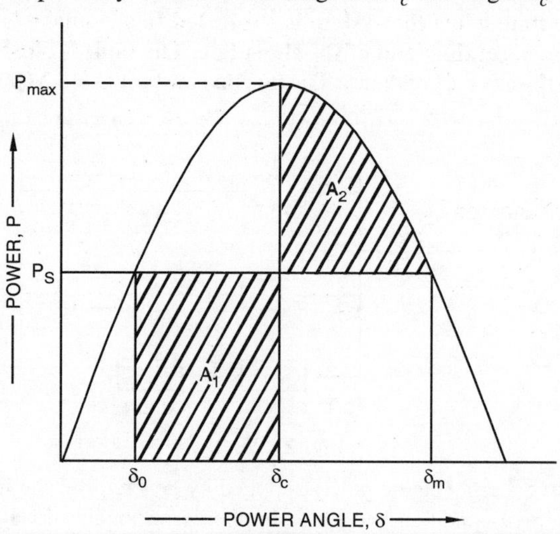

Fig. 21.27

$$P_e = P_{max} \sin \delta = 2.10 \sin \delta$$

$$P_S = 1.0 \text{ pu}$$

Before fault, $P_S = P_e = 2.10 \sin \delta_0$

or $\delta_0 = \sin^{-1} \dfrac{1}{2.10} = 28.44°$ or 0.4963 radian (electrical)

$$\delta_m = 180° - \delta_0 = 180° - 28.44° = 151.56° \text{ or } 2.6452 \text{ radians}$$

Accelerating area, $A_1 = \int_{\delta_0}^{\delta_c} (P_S - 0)\, d\delta = P_S(\delta_c - \delta_0)$

and area $A_2 = \int_{\delta_c}^{\delta_m} (P_{max} \sin \delta - P_S)\, d\delta$

$$= P_{max} (\cos \delta_c - \cos \delta_m) - P_S(\delta_m - \delta_c)$$

For the system to be stable

$$A_2 = A_1$$

i.e., $P_S(\delta_c - \delta_0) = P_{max} (\cos \delta_c - \cos \delta_m) - P_S(\delta_m - \delta_c)$

or $\cos \delta_c = \dfrac{P_S}{P_{max}} (\delta_m - \delta_0) + \cos \delta_m$

$$= \frac{P_{max} \sin \delta_0}{P_{max}} (\pi - \delta_0 - \delta_0) + \cos (\pi - \delta_0)$$

$\because P_S = P_{max} \sin \delta_0$ and $\delta_m = \pi - \delta_0$

or $\delta_c = \cos^{-1} [(\pi - 2\delta_0) \sin \delta_0 - \cos \delta_0]$

$$= \cos^{-1} \left[\frac{(\pi - 2 \times 0.4963)}{2.10} - \cos 0.4963 \right]$$

$$= \cos^{-1} (1.02333 - 0.87934)$$

$= 81.72°$ (electrical) or 1.4263 radian (electrical) **Ans.**

During the period the fault is persisting, the swing equation is

$$M \frac{d^2\delta}{dt^2} = P_S - P_{max} \sin \delta = P_S - P_e$$

or $\dfrac{d^2\delta}{dt^2} = \dfrac{P_S}{M}$ $\quad\because P_e = 0$ during fault period

$= \dfrac{P_S}{H/\pi f} = \dfrac{\pi f}{H} . P_S$ $\quad\because M = \dfrac{H}{\pi f}$ in pu system

Integrating twice, we have

$$\delta = \frac{\pi f}{2H} \times P_S t^2 + \delta_0$$

$$\delta_c = \frac{\pi f P_S}{2H} . t_c^2 + \delta_0$$

or $t_c = \left[\dfrac{2H(\delta_c - \delta_0)}{\pi f P_S} \right]^{1/2}$

$= \sqrt{\dfrac{2 \times 5(1.4263 - 0.4963)}{\pi \times 50 \times 1}} = 0.243$ second **Ans.**

Example 21.26. In the system shown in Fig. 21.28, the generator is delivering 1 pu power. The voltage behind transient reactance is 1.05 pu and infinite bus voltage is 1 pu. The number on the diagram are the pu reactance on a common system bus. Calculate δ_c and t_c when the system is subjected to a 3-φ fault on point P. Take H = 5 MJ/MVA and f = 50 Hz.

[Power System Engineering, Rajasthan Univ. 2002, 2003, 2005; Rajasthan Technical Univ. 2010]

Solution : System reactance

$$X = j\left(0.2 + 0.1 + \frac{0.4}{2}\right) = j\,0.5 \text{ pu}$$

$$P_S = \frac{V_S V_R}{X} \sin\delta = \frac{1.05 \times 1.0}{0.5} \sin\delta = 2.10 \sin\delta$$

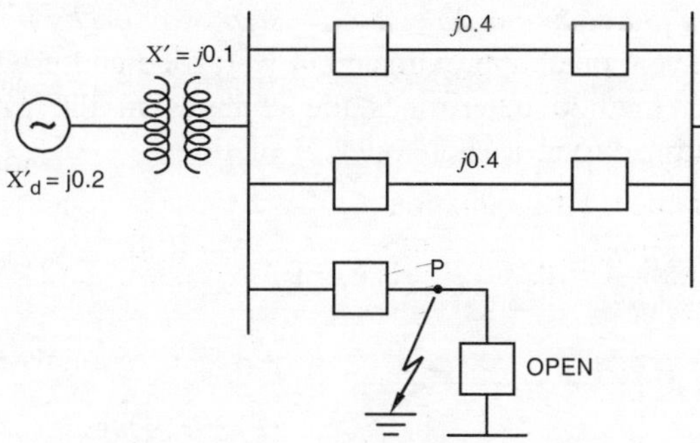

Fig. 21.28

δ = 28.44° or 0.4963 radian (electrical), as determined in Example 21.25

δ_m = 151.56° or 2.6453 radians (electrical)

Critical clearing angle,

$$\delta_c = \cos^{-1}[(\pi - 2\delta_0)\sin\delta_0 - \cos\delta_0]$$

= 81.72° (electrical) or 1.4263 radians (electrical) **Ans.** as determined in Example 21.25.

Critical clearing time,

$$t_c = \sqrt{\frac{2H(\delta_c - \delta_0)}{\pi f P_S}} = \sqrt{\frac{2 \times 5(1.4263 - 0.4963)}{\pi \times 50 \times 1}}$$

= 0.243 second **Ans.**

as determined in Example 21.25

21.8. TWO-MACHINE SYSTEM

A system having two machines can be replaced by an equivalent system having one machine connected to an infinite bus such that the swing equation and swing curves of angular displacement between the two machines are the same for both systems. This is shown below :

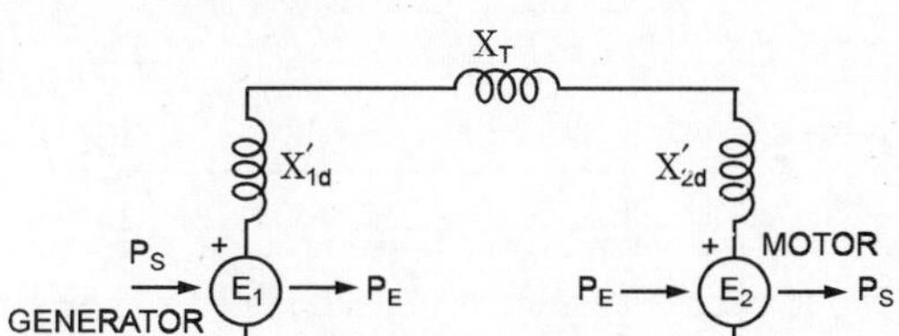

Fig. 21.29. *Two Machine System*

A simple two-machine system connected through a transmission line having reactance X_T is shown in Fig. 21.29. Each machine is represented by its transient reactance and voltage behind transient reactance. The mechanical power input to the generator is P_S and its electrical power output is P_E. Since both electrical and mechanical losses are neglected, the electrical power input to the motor is equal to the electrical output of generator, *i.e.*, P_E and its mechanical power output equals the power input to the generating machine, *i.e.*, P_S. Torque angles are measured positively in the direction of rotation. Let δ_1, M_1, P_{A_1} be the quantities referred to generator and δ_2, M_2 and P_{A_2} for the motor. The swing equation for the two machines can be written as

$$\frac{d^2\delta_1}{dt^2} = \frac{P_{A_1}}{M_1} = \frac{P_S - P_E}{M_1} \quad \text{...(21.45) For generating machine}$$

$$\frac{d^2\delta_2}{dt^2} = \frac{P_{A_2}}{M_2} = \frac{P_E - P_S}{M_2} \quad \text{...(21.46) For motoring machine}$$

The relative torque angle between the two rotor axes,

$$\delta = \delta_1 - \delta_2 \quad \text{...(21.47)}$$

Expressing the acceleration of generator relative to the motor as the algebraic difference of their accelerations which must be equal to the acceleration of the equivalent single generator relative to the infinite bus, the following equation can be obtained.

$$\frac{d^2\delta}{dt^2} = \frac{d^2\delta_1}{dt^2} - \frac{d^2\delta_2}{dt^2} \quad \text{...(21.48)}$$

Substituting the values of $\frac{d^2\delta_1}{dt^2}$ and $\frac{d^2\delta_2}{dt^2}$ from Eqs. (21.45) and (21.46) in above Eq. (21.48), we have

$$\frac{d^2\delta}{dt^2} = \frac{P_S - P_E}{M_1} - \frac{P_E - P_S}{M_2}$$

$$= (P_S - P_E)\left[\frac{1}{M_1} + \frac{1}{M_2}\right]$$

$$= (P_S - P_E)\frac{M_1 + M_2}{M_1 \times M_2} = \frac{P_S - P_E}{M} \quad \text{...(21.49)}$$

Equation (21.49) is the same as the swing equation for a single machine connected to an infinite bus with equivalent inertia constant.

$$M = \frac{M_1 M_2}{M_1 + M_2} \quad \text{...(21.50)}$$

Further, the electrical power transfer is given as

$$P_E = \frac{E_1 E_2}{X'_{1d} + X_T + X'_{2d}} \sin\delta = \frac{E_1 E_2}{X} \sin\delta \quad \text{...(21.51)}$$

Thus, the two-machine system shown in Fig. 21.29 is equivalent to a single generator of angular momentum M and voltage (behind transient reactance) E_1 connected to an infinite bus of voltage E_2 and the transfer reactance between the generator and infinite bus, *i.e.*, $X = X'_{1d} + X_T + X'_{2x}$.

Example 21.27. A 4-pole 50-Hz turboalternator is rated at 45 MW, 0.8 pf lag and has an inertia of 25,000 kg-m². It is connected via a transmission system to another set whose corresponding data is 2-pole, 50 Hz, 60 MW, 0.75 pf lag, 9,000 kg-m². Calculate the inertia constant of each set on its own rating and that of the single equivalent set connected to an infinite bus-bar and on a base rating of 100 MVA.

[Pb. Univ. Power System Analysis January 1993]

Solution : For Machine I

Moment of inertia, I_1 = 25,000 kg-m²

Rating of machine in MVA, $S_1 = \frac{MW_1}{pf_1} = \frac{45}{0.8} = 56.25$ MVA

Angular velocity, $\omega_1 = 2\pi f_1 = 2\pi \times 50 = 100\pi$ electrical radians/second

Kinetic energy stored in the rotor at synchronous speed,

$$KE_1 = \frac{1}{2} I\omega^2 = \frac{1}{2} \times 25{,}000 \times (100\,\pi)^2$$

$$= 1{,}233.7 \times 10^6 \text{ J or } 1{,}233.7 \text{ MJ}$$

Inertia constant,

$$H_1 = \frac{\text{Kinetic energy stored}}{S_1 \text{ in MVA}} = \frac{1{,}233.7}{56.25} = 21.932 \text{ MJ/.MVA } \textbf{Ans.}$$

Angular momentum,

$$M_1 = \frac{\text{Kinetic energy stored in MJ}}{180 f}$$

$$= \frac{1{,}233.7}{180 \times 50} = 0.1372 \text{ MJ-s per electrical degree } \textbf{Ans.}$$

For Machine II

Moment of inertia, $I_2 = 9{,}000$ kg-m^2

Rating of machine in MVA, $S_2 = \frac{MW_2}{pf_2} = \frac{60}{0.75} = 80$ MVA

Kinetic energy stored in the rotor at synchronous speed (*i.e.*, $2\pi f_2$ radians/s)

$$KE_2 = \frac{1}{2} I_2\omega^2 = \frac{1}{2} \times 9{,}000 \times (2\pi \times 50)^2 = 444.132 \times 10^6 \text{ J}$$

Inertia constant,

$$H_2 = \frac{KE_2}{S_2} = \frac{444.132}{80} = 5.55 \text{ MJ/MVA } \textbf{Ans.}$$

Angular momentum,

$$M_2 = \frac{KE_2}{180 f} = \frac{444.132}{180 \times 50} = 0.04935 \text{ MJ-s/electrical degree } \textbf{Ans.}$$

Angular momentum of single equivalent set,

$$M = \frac{M_1 M_2}{M_1 + M_2} = \frac{0.1372 \times 0.04935}{0.1372 + 0.04935}$$

$$= 0.0363 \text{ MJ-s/electrical degree}$$

Kinetic energy stored,

$$GH = 180 f M = 180 \times 50 \times 0.0363 = 326.65 \text{ MJ}$$

Inertia constant of equivalent machine on 100 MVA base,

$$H = \frac{GH}{G} = \frac{326.65}{100} = 3.2665 \text{ MJ/MVA } \textbf{Ans.}$$

21.9. NUMERICAL SOLUTION OF SWING EQUATION (POINT BY POINT SOLUTION)

As already mentioned the *equal area criterion* method can be employed in case of one-machine system and infinite bus-bar or two-machine power system but in most practical power systems after machine lumping has been done, there are still more than two machines to be considered from the point of view of system stability. Hence there is no choice but for solving the swing equation of each machine. Even in the case of a single machine tied to infinite bus-bar by equal area criterion method only the *critical clearing* angle can be determined where as we are interested in determining critical clearing time corresponding to the critical clearing angle, *i.e.*, the time when the rotor have moved to the critical clearing angle. The knowledge of critical clearing time helps in designing the operating times of the relay and circuit breaker so that these may operate within critical clearing time for stable operation of the system. The critical clearing time can be determined by solving swing equation by making certain approximations. There are several sophisticated methods for solving the swing equation. The *step by step* or *point by point method* is conventional, approximate but well tried and proven method. This method determines the changes in the rotor angular position during a short interval of time.

Consider the swing equation

$$M\frac{d^2\delta}{dt^2} = P_S - P_{max} \sin \delta = P_A \qquad \text{...(21.52)}$$

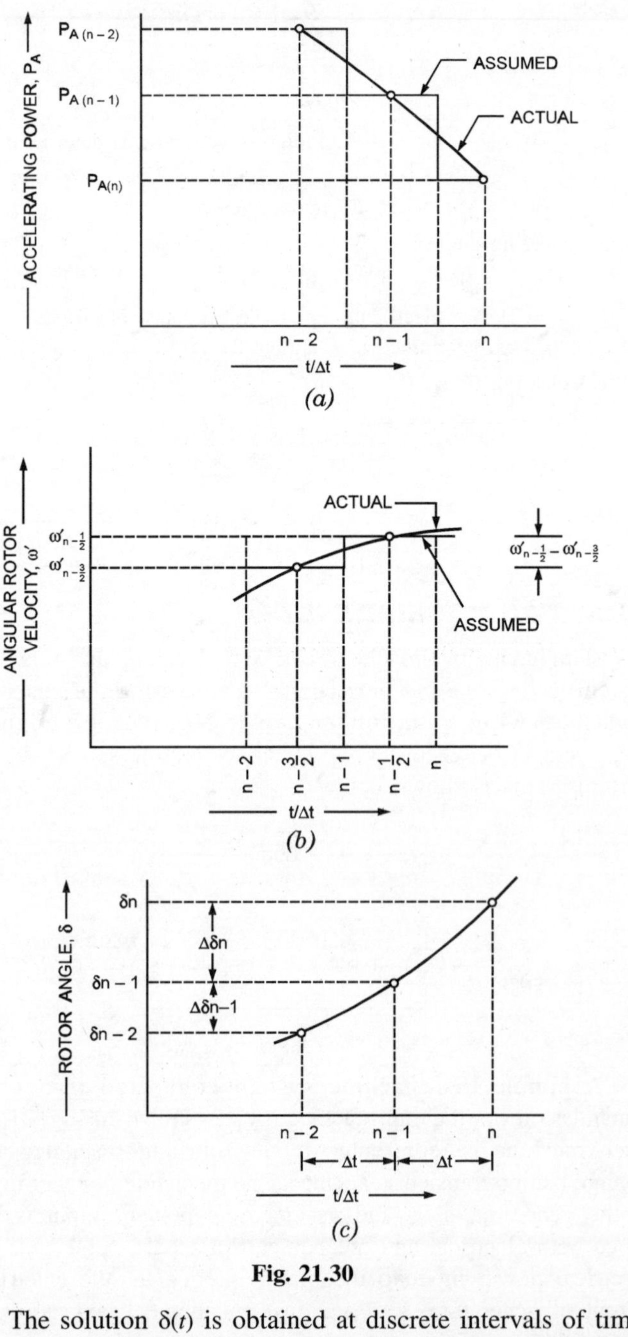

Fig. 21.30

The solution δ(*t*) is obtained at discrete intervals of time with interval spread of Δ*t* uniform throughout. Accelerating power, P_A and change in speed, which are continuous function of time and are described as below :

1. The accelerating power P_A computed at the beginning of an interval is assumed to remain constant from the middle of the preceding interval to the middle of the interval being considered, as illustrated in Fig. 21.30.
2. The angular rotor velocity ω', *i.e.*, $\frac{d\delta}{dt}$ (over and above synchronous velocity ω_0) is assumed to remain constant throughout any interval at the value computed for the middle of the interval, as illustrated in Fig. 21.30.

In Fig. 21.30 the numbering on $\frac{t}{\Delta t}$ axis pertains to the end of intervals.

The equation for accelerating power at the end of the (*n* – 1)th interval or for *n*th interval can be written as

$$P_{A\,(n-1)} = P_S - P_{max} \sin \delta_{n-1} \quad ...(21.53)$$

where δ_{n-1} has been earlier calculated.

The change in velocity caused due to $P_{A\,(n-1)}$ assumed to remain constant over Δ*t* from (*n* – 3/2) to (*n* – 1/2),

$$\Delta\omega'_{n-1/2} = \omega'_{n-1/2} - \omega'_{n-3/2}$$

$$= \frac{\Delta t}{M} P_{A(n-1)} \quad ...(21.54)$$

The change in rotor angle δ during (*n* – 1)th interval,

$$\Delta\delta_{n-1} = \delta_{n-1} - \delta_{n-2} = \Delta t\,\omega'_{n-\frac{3}{2}} \quad ...(21.55)$$

and during the *n*th interval,

$$\Delta\delta_n = \delta_n - \delta_{n-1} = \Delta t\,\omega'_{n-\frac{1}{2}} \quad ...(21.56)$$

Subtracting Eq. (21.55) from Eq. (21.56), we have

$$\Delta\delta_n - \Delta\delta_{n-1} = \Delta t\left(\omega'_{n-\frac{1}{2}} - \omega'_{n-\frac{3}{2}}\right) = \frac{(\Delta t)^2}{M} P_{A(n-1)}$$

$$\because \text{ from Eq. (21.54) } \omega'_{n-\frac{1}{2}} - \omega'_{n-\frac{3}{2}} = \frac{(\Delta t)}{M} P_{A(n-1)}$$

$$\text{or } \Delta\delta_n = \Delta\delta_{n-1} + \frac{P_{A(n-1)}}{M}(\Delta t)^2 \quad ...(21.57)$$

$$\therefore \delta_n = \delta_{n-1} + \Delta\delta_n \quad ...(21.58)$$

The above process of computation is repeated to obtain $P_A(n)$, $\Delta\delta_{n+1}$ and δ_{n+1}. The time solution in discrete form is thus carried out over the desired length of time, usually 0.05 second. Actual swing curve can be plotted by drawing a smooth curve through discrete values, as shown in Fig. 21.30. The accuracy of the solution depends upon the time duration of the intervals. As the time interval is reduced the computed swing curve approaches the true. Usually Δ*t* = 0.05 second provides good accuracy in results. The occurrence or removal of a fault or initiation of any switching action causes a discontinuity in accelerating power. There are three possibilities of occurrence of discontinuity.

(*i*) The discontinuity occurs at the beginning of the nth interval.

(*ii*) The discontinuity occurs at the middle of an interval.

(*iii*) The discontinuity occurs at some time other than the beginning or the middle of an interval.

In the first case, the average of the values of accelerating power P_A before and after discontinuity should be used. Thus in determining the increment of angle occurring during the first interval after the occurrence of fault at *t* = 0, Eq. (21.57) becomes

$$\Delta\delta_1 = \frac{P_{A\,o+}}{2M}(\Delta t)^2$$

whereas $P_{A\,o+}$ is the accelerating power immediately after the occurrence of the fault. Since immediately before the occurrence of the fault the system is in steady state, $P_{A\,o-} = 0$ and δ_0 is of known value.

If the fault is cleared at the beginning of the *n*th interval, in calculation for this interval the value of $P_{A\,(n-1)}$ should be taken as

$$\frac{1}{2}\left[P_{A(n-1)^-} + P_{A(n-1)^+}\right]$$

where $P_{A(n-1)^-}$ is the accelerating power immediately before clearing and $P_{A(n-1)^+}$ is that immediately after clearing the fault.

In second case, *i.e.*, when the discontinuity occurs at the middle of an interval, no special procedure is required. The increment of the angle during such an interval is computed, as usual, from the value of P_A at the beginning of the interval, *i.e.*,

$$P_A = P_S - \text{output during the fault}$$

where as at the beginning of the interval following clearing of the fault, P_A is given as

$$P_A = P_S - \text{output after clearance of fault.}$$

To compute accelerating power P_A in the third case, a weighted average value of P_A before and after the discontinuity may be used. It is found in practice that such a precise computation of accelerating power P_A is not required as the time interval used in computation is so short that it is sufficiently accurate to consider the discontinuity to occur at the beginning or at the middle of an interval and accelerating power P_A is computed as outlined above in the first two cases.

Example 21.28. A 50 MVA, 50 Hz generator is supplying 48 MW to the infinite bus over a double circuit line (Fig. 21.31). The generator has an inertia constant of 2.7 MJ/MVA at rated speed. The generator transient reactance is 0.2 pu. Each transmission line has a reactance of 0.4 pu on a 50 MVA base. Voltage at generating end is 1.04 pu and voltage at infinite bus-bar is 1.0 ∠0°. A 3-phase short circuit occurs at the mid point of one of the transmission lines. (*a*) Determine the critical clearing angle and the critical clearing time. (*b*) Plot the swing curve over the period of 0.5 second for a sustained fault. (*c*) Plot the swing curve if the fault is cleared in 0.1 second by simultaneous opening of the breakers at the both ends of the line.

Solution : Taking base MVA of 50

The rating of the generator, G = 1.0 pu

The inertia constant, H = 2.7 MJ/MVA

$$\text{Angular momentum, } M = \frac{GH}{180f} = \frac{1 \times 2.7}{180 \times 50} = 3 \times 10^{-4} \text{ pu}$$

During Normal Operation :

$$\text{Transient reactance of the system, } X = 0.2 + \frac{0.4}{2} = 0.4 = 0.4 \text{ pu}$$

$$\text{Maximum power transferred, } P_{max} = \frac{1.04 \times 1.0}{0.4} = 2.6 \text{ pu}$$

$$\text{Power transferred, } P_E = P_{max} \sin \delta$$

Initial power angle is given by the expression

$$2.6 \sin \delta_0 = \frac{48}{50} = 0.96$$

$$\text{or } \sin \delta_0 = \frac{0.96}{2.6} = 0.3692$$

$$\text{or } \delta_0 = \sin^{-1} 0.3692 = 21.7^\circ \text{ or } 0.3782 \text{ radian (electrical)}$$

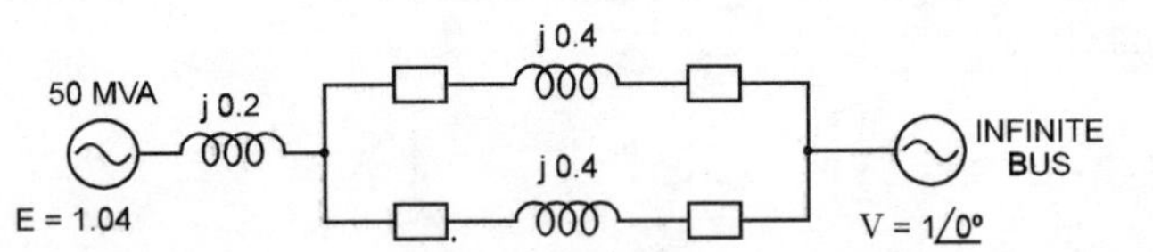

Fig. 21.31

(a) Under Fault Condition : A positive sequence reactance diagram is shown in Fig. 21.32 (*a*). Converting star to delta, we have the network, as shown in Fig. 21.32 (*b*).

Transient reactance of the system,

$$X' = \frac{0.2 \times 0.2 + 0.4 \times 0.2 + 0.2 \times 0.4}{0.2} = 1.0 \text{ pu}$$

$$\text{Power transferred, } P'_E = P'_{max} \sin \delta = \frac{1.04 \times 1.0}{1} \sin \delta = 1.04 \sin \delta$$

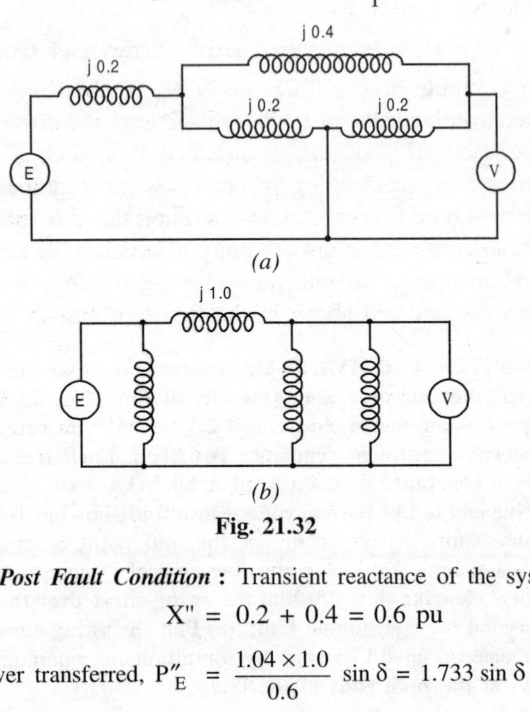

Fig. 21.32

Under Post Fault Condition : Transient reactance of the system,

$$X'' = 0.2 + 0.4 = 0.6 \text{ pu}$$

$$\text{Power transferred, } P''_E = \frac{1.04 \times 1.0}{0.6} \sin \delta = 1.733 \sin \delta$$

$$\gamma_1 = \frac{\text{Reactance before fault}}{\text{Reactance during fault}} = \frac{X}{X'} = \frac{0.4}{1.0} = 0.4$$

$$\gamma_2 = \frac{\text{Maximum power that can be transferred after isolation of fault}}{\text{Maximum power that can be transferred initially}}$$

$$= \frac{1.733}{2.6} = 0.6667$$

$$\delta_m = \pi - \sin^{-1} \frac{\sin \delta_0}{\gamma_2} = \pi - \sin^{-1} \frac{0.3692}{0.6667}$$

$$= 146.37^\circ \text{ or } 2.5546 \text{ radians (electrical)}$$

and $\cos \delta_C$

$$= \frac{\sin \delta_0 \times (\delta_m - \delta_0) - \gamma_1 \cos \delta_0 + \gamma_2 \cos \delta_m}{\gamma_2 - \gamma_1}$$

$$= \frac{0.3692 \times (2.5546 - 0.3782) - 0.4 \times 0.9291 + 0.6667 \times (-0.8326)}{0.6667 - 0.4}$$

$$= -0.4616$$

or Critical clearing angle,

$$\delta_C = \cos^{-1} (-0.4616) = 117.5 \text{ electrical degrees } \textbf{Ans.}$$

Taking $\Delta t = 0.05$ second, the recursive relationships for step by step swing curve calculation are given below:

$$P_{A(n-1)} = P_S - P_{max} \sin \delta_{n-1}$$

$$\Delta\delta_n = \Delta\delta_{n-1} + \frac{P_{A(n-1)}}{M} (\Delta t)^2 = \Delta\delta_{n-1} + \frac{P_{A(n-1)}}{3 \times 10^{-4}} \times (0.05)^2$$

$$= \Delta\delta_{n-1} + 8.33 P_{A(n-1)}$$

and $\delta_n = \delta_{n-1} + \Delta\delta_n$ electrical degrees.

(b) Sustained Fault : Since there is discontinuity at $t = 0$ and hence average value of P_A is used. At $t = 0^-$, $P_{max} = 2.6$ pu, $P_A = 0$

and At $t = 0^+$, $P_{max} = 1.04$ pu and

$$P_A = P_S - P_{max} \sin \delta_0 = \frac{48}{50} - 1.04 \sin 21.7^\circ$$

$$= 0.96 - 0.3692 = 0.5908 \text{ pu}$$

$$P_{A(0^+)} = \frac{0 + 0.5908}{2} = 0.2954 \text{ pu}$$

The details of the calculations are given in Table 21.1. The second column of the table shows that P_{max}, the maximum power that can be transferred at time t indicated in first column. P_{max} in case of a sustained fault undergoes a sudden change at $t = 0^+$ and remains constant thereafter. A representative calculation at $t = 0.20$ second is illustrated below.

$$t = 0.20 \text{ second}$$

$$\delta_{(0.15 \text{ second})} = 43.36^\circ$$

$$\sin \delta_{(0.15 \text{ second})} = 0.6866$$

$$P_{max} = 1.04$$

$$P_{E(0.15 \text{ second})} = P_{max} \sin \delta_{(0.15 \text{ second})}$$

$$= 1.04 \times 0.6866 = 0.714$$

$$P_{A(0.15 \text{ second})} = P_S - P_{E(0.15 \text{ second})} = 0.96 - 0.714 = 0.246$$

$$\frac{(\Delta t)^2}{M} P_{A(0.15 \text{ second})} = 8.33 P_{A(0.15 \text{ second})} = 8.33 \times 0.246 = 2.05$$

Table 21.1. Step By Step Solution of Swing Curve For Sustained Fault

t Seconds	P_{max} pu	Sin δ	$P_E = P_{max} \sin \delta$ pu	P_A $0.96 - P_E$ pu	$8.33\ P_A$ Electrical Degrees	Δ δ Electrical Degrees	δ Electrical Degrees
0^-	2.6	0.3692	0.96	0	—	—	21.7
0^+	1.04	0.3692	0.384	0.5908	—	—	21.7
$0_{average}$	—	—	—	0.2954	2.46	2.46	21.7
0.05	1.04	0.41	0.4264	0.5336	4.44	7.9	24.16
0.10	1.04	0.53	0.552	0.408	3.4	11.3	32.06
0.15	1.04	0.6866	0.714	0.246	2.05	13.35	43.36
0.20	1.04	0.8359	0.87	0.09	0.74	14.09	56.71
0.25	1.04	0.944	0.982	– 0.022	– 0.18	13.91	70.80
0.30	1.04	0.996	1.036	– 0.076	– 0.63	13.28	84.71
0.35	1.04	0.99	1.03	– 0.07	– 0.58	12.70	97.99
0.40	1.04	0.9355	0.973	– 0.013	– 0.10	11.70	110.69
0.45	1.04	0.8444	0.878	0.082	0.68	12.38	122.39
0.50	1.04	0.71	0.738	0.22	1.85	14.23	134.77

Table 21.2. Computation of Swing Curve For Fault Cleared in 0.10 second

t Seconds	P_{max} pu	Sin δ	$P_E = P_{max} \sin \delta$ pu	P_A $0.96 - P_E$ pu	$8.33\ P_A$ Electrical Degrees	Δδ Electrical Degrees	δ Electrical Degrees
0^-	2.6	0.3692	0.96	0	—	—	21.7
0^+	1.04	0.3692	0.384	0.5908	—	—	21.7
$0_{average}$	—	—	—	0.2954	2.46	2.46	21.7
0.05	1.04	0.41	0.4264	0.5336	4.44	7.9	24.16
0.10^-	1.04	0.53	0.552	0.408	—	—	32.06
0.10^+	1.733	0.53	0.92	0.04	—	—	32.06
$0.10_{average}$	—	—	—	0.224	1.866	9.766	32.06
0.15	1.733	0.667	1.156	– 0.196	– 1.63	8.136	41.826
0.20	1.733	0.766	1.327	– 0.367	– 3.0	5.136	49.96
0.25	1.733	0.822	1.425	– 0.465	– 3.88	1.26	55.32
0.30	1.733	0.8347	1.446	– 0.486	– 4.05	– 2.8	56.58
0.35	1.733	0.807	1.398	– 0.438	– 3.65	– 6.45	53.78
0.40	1.733	0.735	1.274	– 0.314	– 2.61	– 9.06	47.33
0.45	1.733	0.619	1.073	– 0.113	– 0.942	– 10.00	38.26
0.50	1.773	0.473	0.82	0.14	1.1662	– 8.83	28.26

$$\Delta\delta_{(0.20\text{ second})} = \Delta\delta_{(0.15\text{ second})} + \frac{(\Delta t)^2}{M} P_{A\,(0.15\text{ second})}$$

$$= 11.3 + 2.05 = 13.35$$

$$\delta_{(0.20\text{ second})} = \delta_{(0.15\text{ second})} + \Delta\,\delta_{(0.2\text{ second})}$$

$$= 43.36 + 13.35 = 56.71$$

The swing curve is plotted in Fig. 21.33 from which it is observed that the system is unstable.

(*c*) ***Fault Cleared in 0.10 Second*** **:** The swing curve remains the same as in case of sustained fault till 0.05 second. Owing to clearance of fault, P_{max} changes from 1.04 pu at t = 0.1 second to 1.733 pu at $t = 0.1^+$. Since the discontinuity occurs at the beginning of an interval we compute average value of P_A at t = 0.10 second and make use of this value in computing the change in δ at this time. The calculations are given in Table 21.2 and results are plotted in Fig. 21.33. From Fig. 21.33 it is clear that there is maximum swing of 56.58 electrical degrees and finally starts decreasing. So the system is stable.

From the swing curve plotted for sustained fault condition (Fig. 21.33) critical clearing time corresponding to critical clearing angle of 117.5 electrical degrees is 0.42 second. **Ans.**

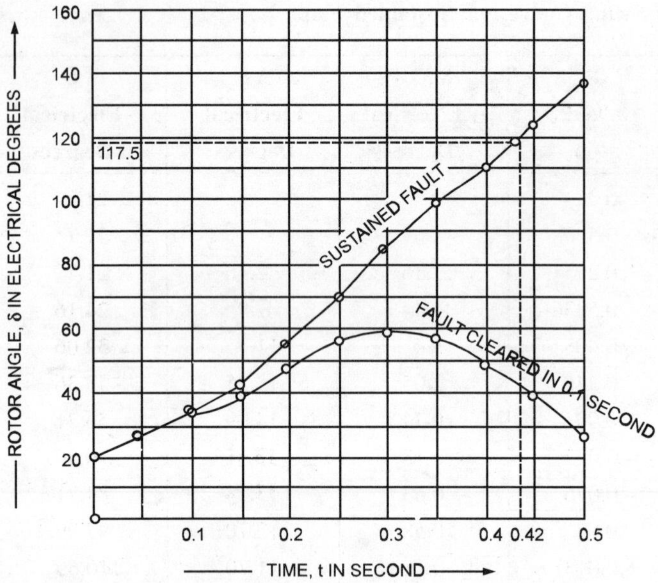

Fig. 21.33

Example 21.29. Consider the power system shown in Fig. 21.34. The values marked are per unit reactances and per unit voltages. The generator was delivering 1.0 pu power before a three phase fault occurs at P. The fault was cleared by opening the circuit breakers and isolating the faulty lines in 5 cycles. Generator has an inertia constant of 4.0 pu. Using point by point method, with time interval of 0.05 second, obtain the swing curve for a period of 0.2 second. Assume f = 50 Hz.

[U.P.S.C. I.E.S. Electrical Engineering II, 1999]

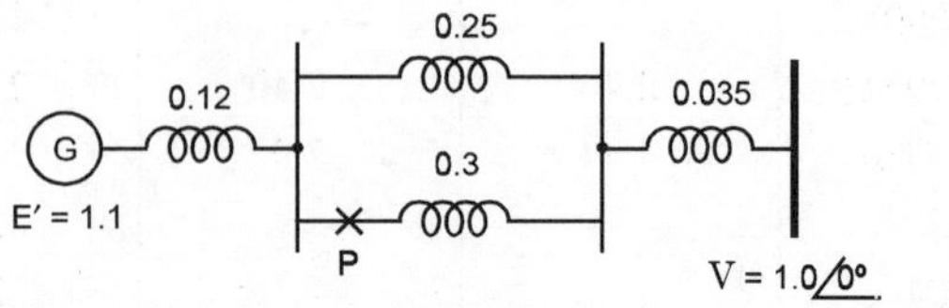

Fig. 21.34

Solution : ***During normal operation :***

Transient reactance of the system,

$$X_1 = j0.12 + j\frac{0.25 \times 0.3}{0.25 + 0.3} + j0.035 = j\,0.29136 \text{ pu}$$

Maximum power transferred,

$$P_{max_1} = \frac{E'V}{X_1} = \frac{1.1 \times 1.0}{0.29136} = 3.775 \text{ pu}$$

During fault,

$$P_{max_2} = 0$$

After the breakers operate, the reactance of the system

$$X_3 = j\,(0.12 + 0.25 + 0.035) = j\,0.405 \text{ pu}$$

$$P_{max_3} = \frac{1.1 \times 1.0}{0.405} = 2.716 \text{ pu}$$

Initial power angle

$$\delta_0 = \text{Sin}^{-1}\frac{P_E}{P_{max_1}}$$

$$= \text{Sin}^{-1}\frac{1.0}{3.775}$$

$$= \text{Sin}^{-1} 0.265 = 15.36^\circ \text{ or } 0.268 \text{ radian (electrical)}$$

For Swing Curve

Angular momentum,

$$M = \frac{GH}{180f} = \frac{1.0 \times 4}{180 \times 50} = 4.444 \times 10^{-4} \text{ pu}$$

Taking Δt = 0.05 second, the recursive relationships for step by step swing calculations are given below :

$$\frac{(\Delta t)^2}{M} = \frac{(0.05)^2}{4.444 \times 10^{-4}} = 5.625$$

Fault sustains for $\frac{5}{50}$ = 0.1 second

During sustained fault : Since there is discontinuity at t = 0 and hence average value of P_A is used.

At $t = 0^-$

P_{max} = 3.775 pu,

Sin δ = 0.265,

$P_E = 1.0,\ P_A = 0$

and δ = 15.36 electrical degrees

At $t = 0^+$, $P_{A(0^+)} = \frac{1.0 - 0.0}{2}$ = 0.5 pu because power transmitted is zero and shaft power is unity.

In view of Eq. (21.57)

$$\Delta\delta_n = \Delta\delta_{n-1} + \frac{P_{A(n-1)}}{M}(\Delta t)^2$$

$$= 0 + 0.5 \times 5.625 = 2.8125 \text{ electrical degrees}$$

So rotor angle, $\delta_1 = \delta_0 + \Delta\delta_1$ = 15.36 + 2.8125

= 18.1725 electrical degrees

Now consider the second interval *i.e.*, t = 0.05 s to t = 0.1 s

$P_{A_1} = 1 - 0 = 1.0$

For this interval $\Delta\delta_{n-1}$ = 2.8125 electrical degrees and so

$$\Delta\delta_2 = \Delta\delta_1 + \frac{P_{A_1}}{M}(\Delta t)^2$$

$$= 2.8125 + 1 \times 5.625 = 8.4375 \text{ electrical degrees}$$

So $\delta_2 = \delta_1 + \Delta\delta_2$ = 18.1725 + 8.4375

= 26.61 electrical degrees

Since at t = 0.1 second (or 5 cycles) the operation of the circuit breaker will isolate the faulty section of the system, therefore, the system operates on the power characteristic

$$P_E = 2.716 \sin\delta$$

To find out accelerating power after the breakers have operated, we take the average of the accelerating powers as there is a discontinuity. The accelerating power before the breakers operate is 1.0 pu.

Accelerating power

$$= P_S - P_E$$

$$= 1.0 - P_{max3} \sin\delta_2$$

$$= 1.0 - 2.716 \sin 26.61^\circ$$

$$= 1.0 - 1.2165 = -0.2165 \text{ pu}$$

So average accelerating power will be $\frac{1 - 0.2165}{2}$ = 0.39173 pu

$$\Delta\delta_3 = \Delta\delta_2 + \frac{P_A}{M}(\Delta t)^2$$

$$= 8.4375 + 0.39173 \times 5.625$$

$$= 10.64 \text{ electrical degrees}$$

and $\delta_3 = \delta_2 + \Delta\delta = 26.61 + 10.64 = 37.25$ electrical degrees

Now consider the next interval, *i.e.*, $t = 0.15$ s to $t = 0.20$ s

Accelerating power

$$= 1.0 - 2.716 \sin 37.25° = 1.0 - 1.644 = -0.644 \text{ pu}$$

$$\Delta\delta_4 = \Delta\delta_3 + \frac{P_A}{M}(\Delta t)^2 = 10.64 + (-0.644) \times 5.625 = 7.0175$$

$$\delta_4 = \delta_3 + \Delta\delta_4 = 37.25 + 7.0175 = 44.2675 \text{ electrical degrees}$$

For next interval $t = 0.20$ s to $t = 0.25$ s

Accelerating power $= 1 - 2.716 \sin 44.2675° = -0.8958$ pu

$$\Delta\delta_5 = \Delta\delta_4 + \frac{P_A}{M}(\Delta t)^2 = 7.0175 - 0.8958 \times 5.625$$

$$= 1.98 \text{ electrical degrees}$$

$$\delta_5 = \delta_4 + \Delta\delta_5 = 44.2675 + 1.98 = 46.2475 \text{ electrical degrees}$$

Swing curve is shown in Fig. 21.35.

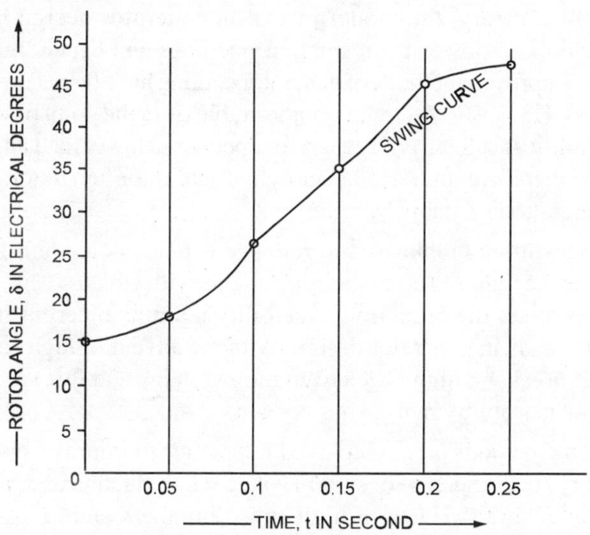

Fig. 21.35

21.10. DIGITAL SOLUTION OF SWING EQUATION

There are number of numerical techniques that can be used for the solution of swing equation. These techniques can be used for digital computations. In such techniques an *n*th order differential equation is written as *n* first order equations. These techniques include (*i*) Euler's method (*ii*) modified Euler's method and (*iii*) Runge-Kutta method. The last two methods are most popular methods used for solution of swing equation.

21.10.1. Euler's Method

This is a single step method. In case of a single step method for determining solution of x_{i+1}, we consider dependence at only one earlier point say x_i. In this case algorithm will be of the following form

$$y_{i+1} = y_1 \quad \underbrace{h\,\phi\,(x_i, y_i, h)}_{\text{Increment Function}} \qquad ...(21.59)$$

In the case of initial value problem, we know dependence at the earlier point from initial condition, and therefore, no difficulty arises in determination of solution at x_i for the first order differential equation. There is another advantage of single step method and it is that the step size can be changed at any stage.

$$\text{Let } \frac{dy}{dx} = y' = f(x, y)$$

with initial condition $y(x_0) = y_0$ ...(21.60)

Algorithm is $\quad y_{i+1} = y_i + h f(x_i, y_i)$ ...(21.61)

Geometrical integration of the algorithm (21.61) is clear from Fig. 21.36.

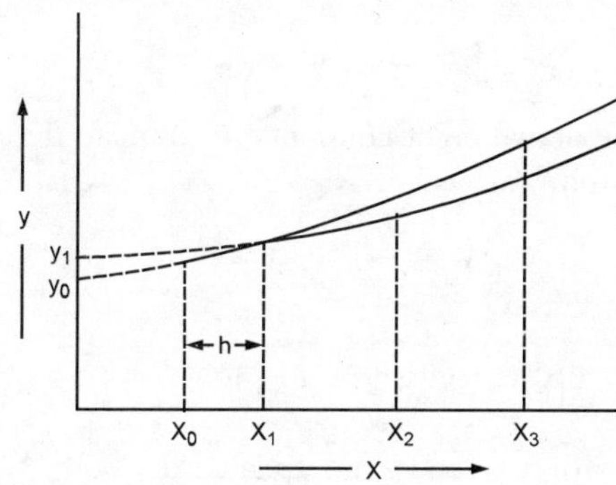

Fig. 21.36

Here the part of the curve in the interval *h* is being approximated by segment of the straight line whose slope is the same as that at the beginning of the subinterval. This method is only of theoretical importance.

21.10.2. Modified Euler's Method

This method is known as *predictor-corrector method*. Here the slope is approximated at the middle of the interval. The algorithm for predictor formula is

$$y_{i+1} = y_i + h f(x_i, y_i) \qquad ...(21.62)$$

and the algorithm for the corrector formula is

$$y_{i+1} = y_i + \frac{h}{2}[f(x_i, y_i) + f(x_{i+1}, y_{i+1})] \qquad ...(21.63)$$

From above it is obvious that, the predictor formula is the same as the Eulers' method where we obtain y_{i+1} at x_{i+1} with the initial value (or initial condition) of y_i at x_i. Then we correct the value of y_{i+1} thus obtained with the help of corrector formula.

21.10.3. Runge-Kutta Method

This is the most powerful method of solving swing equation on digital computer. Let us consider two first order differential equations in two variables *x* and *y* such that

$$\frac{dx}{dt} = f(x, y) \qquad ...(21.64)$$

$$\frac{dy}{dt} = g(x, y) \qquad ...(21.65)$$

Let us start with known initial conditions x^0 and y^0 and a time step Δt. We compute the following eight constants

$$h_1^0 = f(x^0, y^0)\,\Delta t \qquad ...(21.66\,a)$$

$$k_1^0 = g(x^0, y^0)\,\Delta t \qquad ...(21.66\,b)$$

$$h_2^0 = f\left(x^0 + 0.5h_1^0, y^0 + 0.5k_1^0\right)\Delta t \quad ...(21.66\ c)$$

$$k_2^0 = g\left(x^0 + 0.5h_1^0, y^0 + 0.5k_1^0\right)\Delta t \quad ...(21.66\ d)$$

$$h_3^0 = f\left(x^0 + 0.5h_2^0, y^0 + 0.5k_2^0\right)\Delta t \quad ...(21.66\ e)$$

$$k_3^0 = g\left(x^0 + 0.5h_2^0, y^0 + 0.5k_2^0\right)\Delta t \quad ...(21.66\ f)$$

$$h_4^0 = f\left(x^0 + h_3^0, y^0 + k_3^0\right)\Delta t \quad ...(21.66\ g)$$

$$k_4^0 = g\ (x^0 + h_3^0,\ y^0 + k_3^0)\ \Delta t \quad ...(21.66\ h)$$

We use the above eight constants to estimate the changes in x and y as follows:

$$\Delta x^0 = \frac{1}{6}\left(h_1^0 + 2h_2^0 + 2h_3^0 + h_4^0\right) \quad ...(21.67\ a)$$

$$\Delta y^0 = \frac{1}{6}\left(k_1^0 + 2k_2^0 + 2k_3^0 + k_4^0\right) \quad ...(21.67\ b)$$

The values of x, y and t are updated as

$$x^1 = x^0 + \Delta x^0 \quad ...(21.68\ a)$$

$$y^1 = y^0 + \Delta y^0 \quad ...(21.68\ b)$$

$$t^1 = 0 + \Delta t \quad ...(21.68\ c)$$

Then we replace x^0 and y^0 by x^1 and y^1 and recalculate h's, k's, Δx and Δy. In general we may write

$$x^{n+1} = x^n + \Delta x^n \quad ...(21.69\ a)$$

$$y^{n+1} = y^n + \Delta y^n \quad ...(21.69\ b)$$

$$t^{n+1} = (n + 1)\ \Delta t \quad ...(21.69\ c)$$

Application To Solution of Swing Equation. For using the Runge-Kutta method for solving the swing equation of one machine connected to infinite bus, let us substitute

$$x = \delta \quad ...(21.70\ a)$$

$$y = \alpha = \frac{d\delta}{dt} \quad ...(21.70\ b)$$

Then $$\frac{d\delta}{dt} = f(\delta, \alpha) = \alpha \quad ...(21.71\ a)$$

and $$\frac{d\alpha}{dt} = g\ (\delta, \alpha) = \frac{1}{M}\ (P_S - P_E) \quad ...(21.71\ b)$$

where $M = \frac{GH}{\pi f}$ MJ-s/electrical radian

The initial value of load angle δ_0 is determined using the prefault values.

$$P_S = P_E = P_{max} \sin \delta_0 \quad ...(21.72\ a)$$

$$\delta_0 = \text{Sin}^{-1} \frac{P_S}{P_{max}} \quad ...(21.72\ b)$$

The values of P_E used in Eq. (21.71b) are

For $0 < t < t_c$ $P_e = \gamma_1 P_{max} \sin \delta$ ...(21.73)

and for $t > t_c$ $P_e = \gamma_2 P_{max} \sin \delta$ ...(21.74)

where t_c is the fault clearing time, and γ_1 and γ_2 are as defined in Art. 21.7.

21.11. FACTORS AFFECTING TRANSIENT STABILITY AND METHODS OF ITS IMPROVEMENT

From the swing equation $\frac{d^2\delta}{dt^2} = \frac{P_A}{M}$ it is obvious that for given accelerating power the acceleration of the rotor is inversely proportional to the moment of inertia of the machine, *i.e.*, higher the moment of inertia the slower will be the change in the rotor angle of the machine and thus will allow a longer time for the operation of the breaker to isolate the fault before the machine passes through the critical clearing angle. Hence transient stability can be improved either by using machines of higher inertia or by connecting the synchronous motors to heavy flywheels. This method, however, cannot be used in practice because of economic reasons and reasons of excessive rotor weight.

On contrary, the modern trend in generator design is to achieve more power from smaller machines and hence lighter rotors. Improved methods of generator cooling have helped in this process. However, this trend is undesirable from the point of view of stability. A salient-pole generator operates at lower load angles and is, therefore, preferred over cylindrical rotor generators for consideration of stability.

Now more emphasis and reliance is placed on controls to provide the required compensating effects with which we may be able to offset the reduction in stability margins inherent from above trends in generator design. With the advent of high speed circuit breakers, high speed excitation systems and fast valving the loss in stability margin has been made up.

The methods often employed in practice to improve system stability are (*i*) increase of system voltage (*ii*) reduction of transfer reactance and (*iii*) use of high speed breakers and reclosing breakers.

1. Increasing System Voltage. Transient stability is improved by raising the system voltage profile, (*i.e.*, raising E and V). Increase in system voltage means the higher value of maximum power, P_{max} that can be transferred over the lines. Since shaft power, $P_S = P_{max} \sin \delta$, therefore, for a given shaft power initial load angle δ_0 reduces with the increase in P_{max} and thereby increasing difference between the critical clearing angle and initial load angle. Thus machine is allowed to rotate through large angle before it reaches the critical clearing angle which results in greater critical clearing time and the probability of maintaining stability.

2. Reduction in Transfer Reactance. Transient stability can also be improved by reducing the transfer reactance, as discussed in Art. 21.4. The effect of reducing the transfer reactance means increase of P_{max} resulting in increase in transient stability, as mentioned above.

The line reactance can be reduced by using more lines in parallel instead of a single line. In general, more power is transferred during a fault on one of the lines if there are two lines in parallel, than that would be transferred over a single faulted line. Increased power transfer means less available accelerating power, because the accelerating power is the difference between power input and power transfer. Lower accelerating power reduces the risk of instability. The use of bundled conductor lines also helps in reducing line reactance and improving stability.

The compensation of line reactance by series capacitors is another method of improving stability. For lines longer than 325 km this method has proved economical as a means of increasing stability.

The use of shunt reactors will reduce the danger of instability on lightly loaded lines. If saturable reactors are used, the regulation can be maintained steady over a wide range of loads.

Increasing the $\frac{X}{R}$ ratio increases the power limit of the line itself, which in turn aids stability.

3. Using High Speed Circuit Breaker. The best method of improving transient stability is the use of high speed circuit breakers. The quicker a breaker operates, the faster the fault is removed from the system and better is the tendency of the system to restore to normal operating conditions. The use of high speed breakers has materially improved the transient stability of the power systems and does not require any other method for the purpose.

Use of high speed breakers increases the decelerating area A_2 and decreases the accelerating area A_1 and so improves the stability.

4. Automatic Reclosing. As the majority of faults on the transmission lines are transient in nature and are self clearing, rapid switching and isolation of faulty lines followed by reclosing are quite helpful in maintaining stability. The modern circuit breaker technology has made it possible for line clearing to be done as fast as in 2 cycles. On occurrence of a fault on a transmission line, the faulted line is de-energized to suppress the arc in the fault and then the circuit breaker recloses, after a suitable time interval. Automatic reclosing increases the decelerating area A_2 and thus helps in improving stability.

5. Transient Stability can be improved by reducing the severity of faults which can be achieved by using lightning arresters for protection of the lines.

6. In systems where the stability is of prime importance, high neutral grounding impedance may be used.

The grounding is effective only for unbalanced faults. Zero-sequence impedance comes into picture to restrict the fault current only in case of faults like line-to-ground or line-to-line-to-ground. For an electrical system depicted in Fig. 21.37 (*a*), the equivalent impedance diagram is shown in Fig. 21.37 (*b*).

The power transfer is given by the expression

$$P = \left[\frac{E'V}{X_1 + X_2 + \frac{X_1 X_2}{X_f}}\right] \sin \delta \qquad \ldots(21.75)$$

where X_f is equivalent fault-shunt impedance.

Physically the resistance in the neutral of the transformer represents an absorption of electrical energy which in turn reduces the accelerating energy and thus improves the transient stability.

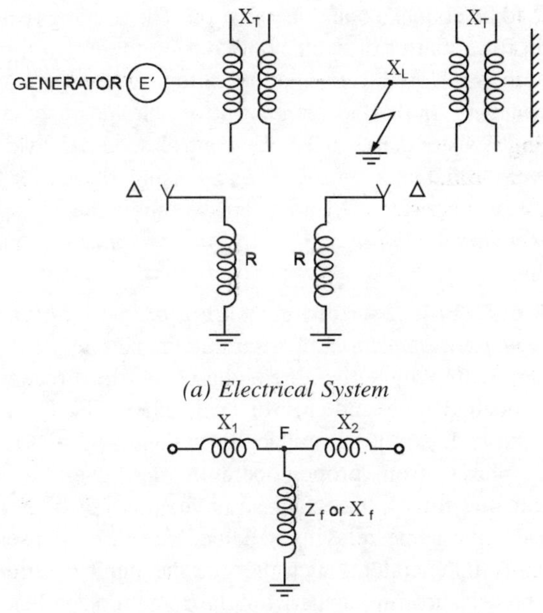

(b) Equivalent Circuit of System Shown in (a)

Fig. 21.37

The grounding resistor consumes power during a ground fault and thus exerts braking effect on the synchronous machine which is greater, the closer the fault is to the resistor and closer the machine is to the fault. A grounding resistor located near a generator is, therefore, beneficial. However, a grounding resistor should neither be employed near an actual or equivalent synchronous motor nor it should be employed near a synchronous condenser, as such machines already are retarded by faults. In a two-machine system, it is, therefore, advisable to have resistance grounding at the sending end and reactance grounding at the receiving end.

7. Turbine Fast Valving. One reason for power system instability is the excess energy supplied by the turbine during the disturbance period. Fast valving is a means of reducing turbine mechanical input power when a unit is under acceleration due to a transmission system fault. This can be initiated by load impedance relays, acceleration transducers or by relays that recognize only severe transmission system faults. For maximum stability gains with fast valving, the turbine input power should be reduced as fast as possible. During a fast valving operation, the interceptor valves are rapidly shut (in 0.1 to 0.2 second) and immediately reopened. This procedure increases the critical switching time long enough so that in most of the cases the unit will remain stable for faults with stuck-breaker clearing times. Presently some stations in USA have been put to use fast valving schemes.

8. Application of Braking Resistors. An alternative or supplement to fast turbine valving action is the application of braking resistors. Braking resistors, as employed in the context of electric power system stability, is the concept of applying an artificial electric load to a portion of the generator-transmission load complex to correct a temporary imbalance between power generated and power delivered. During a fault the resistors are connected to the terminals of the generator through circuit breakers by means of an elaborate control scheme. The control scheme determines the amount of resistance to be connected and its duration. The braking resistors remain on for a matter of cycles both during fault clearing and after system is restored to normal operation. A few cycles after the clearance of fault, the same control scheme disconnects the braking resistors. However, the control schemes available are not very reliable. Control schemes using thyristors have recently been suggested. The noteworthy point is that a *possible failure or maloperation of control scheme can make the matter still worse.*

9. Single-Pole Switching. *Single-pole switching* or *independent pole operation* of a circuit breaker refers to the mechanism with which the three phases of the breaker are closed or opened independently of each other. The failure of any one phase does not automatically prevent any of the two remaining phases from proper operation. However, for a 3-phase fault, the three phases are simultaneously activated for operation by the same relaying scheme. The three phases are mechanically independent, such that the mechanical failure of any one pole is not propagated for the remaining poles.

Single-pole switching is used at locations where the design criterion is to guard against a three-phase fault compounded with breaker failure. The successful independent pole operation of the failed breaker will reduce a three-phase fault to a single L-G fault (if one pole of the breaker is stuck), or to L-L-G fault (if two poles of the breaker are stuck). This criterion can be applied to the substation of a generating plant with multiple transmission outlets.

The advantages of single-pole switching under three-phase fault breaker failure contingency are twofold. First, they are among the cheapest stability aids. Single-pole switching operation is most efficient at high transmission voltages where equipments are costlier. Successful single-pole switching may allow the critical clearing time of a plant circuit breaker to increase by as much as 2 to 5 cycles. Second, it is relatively easy to install. Most EHV circuit breakers are equipped with separate pole mechanism due to the large component size and wide phase space requirements at high working voltages. The only additional complexity is to provide separate trip coils to activate each pole.

10. Use of Quick Acting Automatic Voltage Regulators. The satisfactory operation of synchronous generators of a complex power system at high power (or load) angles and during transient condition is very much dependent on the source of excitation for the generators and on the automatic voltage regulator. The power output of a generator is proportional to internal voltage E. Under fault conditions the terminal voltage V falls. A quick acting voltage regulator causes increase in E so that the terminal voltage V remains constant. A higher value of E means a higher generator output.

It has already been shown that the maximum value of a power angle curve is proportional to the per unit excitation. Field forcing can, therefore, cause the machine to operate on a higher power angle curve thereby allowing it to swing through a larger angle from its original position before it reaches the critical clearing angle.

As an example of the effectiveness of modern high speed automatic voltage regulators, consider Fig. 21.38. Two power-angle curves *a* and *b* have been drawn representing two different excitations. Curves of terminal voltage V against rotor angle for the same excitations have also been drawn. Consider working at point B on power angle curve *a*, where the voltage-rotor angle relation is normal. If we now wish to increase the power input and maintain the same rotor angle, we can do so by increasing excitation, *i.e.,* by working on power angle curve *b* at point C. Although C is in the static stability limit, the terminal voltage is above normal. However, the same power can be supplied by working at point D, in the inherently unstable region.

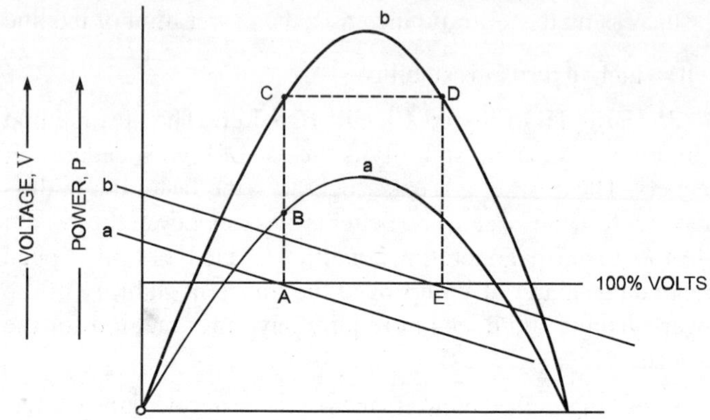

Fig. 21.38. ***Effectiveness of High Speed Automatic Voltage Regulators***

If the field current is constant, the slightest difference would cause instability. However, because of the high inertia, the swings are slow, and if an automatic voltage regulator is fitted, it can sense the drop in terminal voltage as the rotor angle is increased. Before the generator has swung an appreciable amount, sufficient boost is given to the field to make the power output greater than the input, thereby arresting the swing and maintaining stability.

Exercises

1. State the bad effects of instability. Distinguish between steady state and transient stabilities.
[Madras Univ. Power System Analysis April 1997]
2. Define the terms:
(*i*) Steady state stability (*ii*) Transient stability (*iii*) Steady state stability limit (*iv*) Transient stability limit.
[Pb. Univ. Power System Analysis December 1993]
3. Differentiate between steady state stability and transient state stability of a power system. Derive an expression for maximum power transfer between two nodes. Show that this power is maximum when $X = \sqrt{3}R$, where X is the reactance and R is the resistance of system.
[U.P. Technical Univ. Power System Analysis 2009-10]
4. Find out the conditions for maximum power to be transferred through a transmission line of impedance, $r + jx$.
[Rajasthan Univ. Power System Engineering 2005, 2006]
5. Distinguish between steady state, transient and dynamic stability. Derive the swing equation of a synchronous machine.
[Rajasthan Univ. Power System Engineering 2004, 2007]
6. Derive the relation for linearised swing equation for small incremental changes, and find out the steady state stability limits. What are the causes of instability ?
[Power System Engineering, Rajasthan Univ. 2005, 2008; Rajasthan Technical Univ. 2011]

7. Deduce the swing equation of a synchronous machine connected to an infinite bus.

[U.P. Technical Univ. Power System Analysis 2007-08]

8. Explain the steady state and transient stability of power systems. Derive the swing equation of a synchronous generator connected to an infinite bus-bar.

[Pb. Univ. Power System Analysis May 1995]

9. What is meant by the 'power angle' diagram ? Show how this may be used to determine the stability of a simple system under a transient fault condition.

[U.P.S.C. I.E.S. Electric Power System, 1974]

10. Derive the swing equation of a synchronous machine swinging against an infinite bus. Clearly state the assumptions made in deducing the swing equation.

[Madras Univ. Power System Analysis October 1996]

11. Define and derive swing equation for a finite machine connected to an infinite bus. Discuss the applications in the study of power system stability.

[U.P.S.C. I.E.S. Electrical Engineering-II, 2008]

12. Explain the effect of change of excitation on the steady state stability of a synchronous generator feeding an infinite bus-bar.

    Explain equal area criterion for stability by taking a suitable example of power system.

[Pb. Univ. Power System Analysis December 1995]

13. Explain the concept of equal area criterion. How it can be used to study transient stabilty ?

[Anna Univ. Chennai Power System Analysis May/June 2011]

14. Explain the equal area criterion for the stability of an alternator supplying infinite bus-bar via an inductive interconnector. Mention the limitations of the method.

[U.P.S.C. I.E.S. Electrical Engineering II, 1997]

15. How is transient stability affected by load increase and switching operations ?

[Rajasthan Technical Univ. Power System Engineering 2011]

16. Derive the equal area criterion for the stability of power systems using the above criterion show how the critical clearing angle could be found for the case of generator supplying power to an infinite bus through two parallel lines, one of which develops a dead three-phase short circuit.

[Rajasthan Technical Univ. Power System Engineering 2011]

17. Derive expression for critical clearing angle when sudden short circuit on one of two parallel lines, away from the alternator ends, occurs in the system.

[U.P. Technical Univ. Power System Analysis 2007-08]

18. Discuss the effect of clearing time on the stability of power system. Also derive the expression for critical clearing time and critical clearing angle.

[G.B. Technical Univ. Power System Analysis 2010-11]

19. An alternator with negligible damping is connected to an infinite bus bar. Write the swing equation in usual form and define the inertia constant (H) here. Deduce equal area criterion condition for stability analysis.

[U.P. Technical Univ. Power System Analysis 2006-07]

20. Using equal area criterion, discuss the concept of transient stability when sudden change in mechanical input of an alternator takes place.

[U.P. Technical Univ. Power System Analysis 2007-08]

21. Using equal area criterion, derive the expression for critical clearing angle for a system having a generator feeding a large system through a double circuit transmission line.

[U.P.S.C. I.E.S. Electrical Engineering, 2001]

22. Show that two synchronous generating sources of inertia constants M_1 and M_2 respectively and interconnected by means of a transmission line may be regarded for purpose of stability studies as a single generator of inertia constant $\dfrac{M_1M_2}{M_1+M_2}$ connected through the same transmission line to an infinite bus-bar.

[Pb. Univ. Power System Analysis December 1993]

23. Describe the Point by Point solution technique of swing equation for the transient stability of a power system.

[U.P. Technical Univ. Power System Analysis 2007-08]

24. Derive swing equation of a multi-machine power system. State assumptions and approximation if any. Is this a linear equation or nonlinear equation ? Justify your answer.

[Bharathiyar Univ. Power System Analysis November 1996]

25. Explain how transient stability analysis can be carried out using digital computer.

[U.P.S.C. I.E.S. Electrical Engineering II, 1992]

26. Discuss the procedure for solving swing equation using modified Euler method.

[Anna Univ. Chennai Power System Analysis May/June 2011]

27. Distinguish between the "transient" and "dynamic stability" of a power system. Discuss any three recent means of improving transient stability of a power system.

[U.P.S.C. I.E.S. Electrical Engineering II, 1995]

28. What are the methods of improving transient stability ?

[Anna Univ. Chennai Power System Analysis, May/June 2011]

29. Derive swing equation and discuss its application in the study of power stability. Discuss the methods for improving the transient state stability of a power system.

[U.P. Technical Univ. Power System Analysis 2009-10]

30. Discuss the methods of improving transient stability.

[Rajasthan Technical Univ. Power System Engineering 2011]

31. For improving the transient stability of power system, briefly discuss the following discrete supplementary control terms : (*i*) Dynamic braking, (*ii*) High speed circuit breaker reclosing, (*iii*) Independent control of excitation, (*iv*) Series capacitor insertion, (*v*) Generator tripping.

[G.B. Technical Univ. Power System Analysis 2010-11]

32. With the help of equal area criteria, explain how the 'ARCB' and turbine fast valving help to improve the stability.

[Power System Engineering, Rajasthan Univ. 2005, 2008; Rajasthan Technical Univ. 2010]

33. Derive equation of critical clearing angle and critical clearing time for the system given below, if sudden short circuit occurs on short transmission line near generator bus at point P.

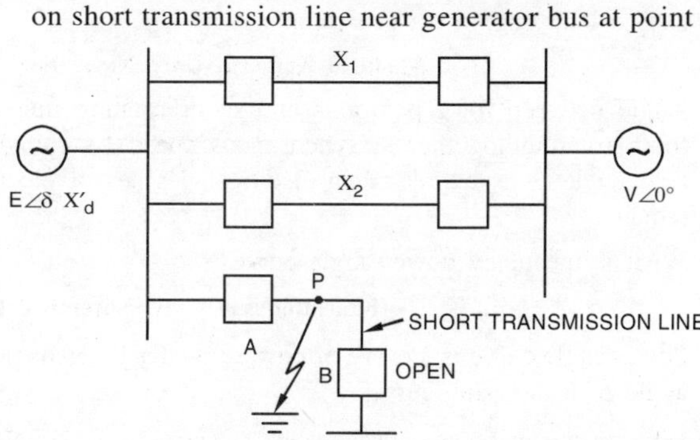

Fig. 1

[Rajasthan Univ. Power System Engineering 2008]

34. Write short notes on the following:

(*i*) Steady state and transient stability. **[Pb. Univ. December 1994]**

(*ii*) Power angle diagram and system stability.

(*iii*) Steady state stability and methods of its improvement.

(*iv*) Assumptions made in deducing equal area criterion and validity of the assumptions. **[Madras Univ. April 1997]**

(*v*) Equal area criterion applied to consider a step increase in load. **[Madras Univ. October 1996]**

(*vi*) Solution of swing equation by modified Euler method. **[Madras Univ. April 1997]**

(*vii*) Runge-Kutta method. **[Madras Univ. October 1996]**

(*viii*) Transient stability and methods of improvement.

Short Answer Type Questions With Answers

Q. 1. Define stability limit.

Ans. A stability limit is the maximum power flow possible through some particular point in the system or the part of the system to which the stability limit refers is operating with stability.

Q. 2. Define steady state stability. **[Madras Univ. April 1997]**

Ans. Steady state stability may be defined as the capability of an electric power system to maintain synchronism between machines within the system and external tie lines following a small slow disturbance (normal load fluctuations, the actions of automatic voltage regulators and turbine governors).

Q. 3. What is meant by steady state stability limit ?

Ans. The steady state stability limit refers to the maximum power which can be transferred through the system without loss of stability.

Q. 4. Distinguish between steady state stability and transient stability of a power system. **[Bharathiyar Univ. November 1996]**

Ans. Steady state stability refers to the conditions of stability in response to small and gradual changes of load while transient stability refers to the conditions of stability in response to large sudden disturbances. A sudden large disturbance includes application of faults, clearing of faults, sudden load changes, and inadvertent tripping of lines and generators.

Q. 5. Define dynamic stability of a power system. **[Bharathidasan Univ. November 1996]**

Ans. Dynamic stability is the ability of a power system to remain in synchronism after the initial swing (transient stability period) until the system has settled down to the new steady state equilibrium condition.

Q. 6. Define load angle of a generator. **[Madurai Kamaraj Univ. November 1996]**

Ans. Angle between rotor pole axis and stator rotating field axis (both rotating together at synchronous speed) is called the load angle. It is expressed in electrical degrees or electrical radians.

Q. 7. What is meant by power angle curve ? **[Bharathidasan Univ. November 1996]**

Ans. Power angle curve is a curve of power transfer P versus power angle δ. It is a sine curve.

Q. 8. What is meant by a swing curve ?

Ans. Swing curve is a graph of power angle δ versus time t.

Q. 9. On what basis do you conclude that a given synchronous machine has lost stability ? **[Bharathiyar Univ. April 1997]**

Ans. In case the power angle δ reduces after attaining maxima corresponding to $\frac{d\delta}{dt} = 0$ condition indicates stability. On contrary if the power angle exceeds critical angle of load, the system becomes unstable.

Q. 10. Define infinite bus in a power system. **[Bharathiyar Univ. April 1997]**

OR

What is an infinite bus ? **[Anna Univ. May/June 2011]**

Ans. Infinite bus has constant voltage, constant electrical angle of the voltage, constant frequency, infinite fault level and, therefore, the quantities of infinite bus do not get affected by connecting or disconnecting individual machines or transmission lines.

Q. 11. How can steady state limit be improved ?

Ans. Steady state limit can be improved either by increasing the excitation of the machine/machines or by reducing transfer reactance.

Q. 12. What do you know about swing equation ? **[M.D. Univ. Power Systems-I, 2010-11]**

OR

Write swing equation of a single machine. **[Bharathiyar Univ. April 1997]**

Ans. The equation giving the relative motion of rotor (load angle δ) with respect to the stator field as a function of time is called the swing equation.

The swing equation for a single-machine system can be written as $M\frac{d^2\delta}{dt} = P_S - P_E = P_S - P_{max}\sin\delta$

where M is angular momentum, P_S is shaft power (mechanical power), P_E is electromagnetic power, δ is load angle between E and V and P_{max} is the maximum steady state power.

Q. 13. Define inertia constants M and H. **[Bharathidasan Univ. Nov. 1996]**

Ans. M is the angular momentum of the machine and is expressed in joule-seconds per radian or in mega-joule-seconds per electrical degree. M, the inertia constant of the machine, depends on the type and size of the machine.

Another constant H having narrow range of values for each class of machine regardless of its capacity is defined as the ratio of the kinetic energy stored at rated speed to the rated power of the machine. H is expressed in MJ/MVA.

Q. 14. Give typical values of inertia constant H for turbogenerators and slow-speed waterwheel generators. **[Madras Univ. October 1996]**

Ans. The average value of H in MJ/MVA for turbogenerators is between 3 and 9 and for waterwheel generators it is between 2 and 4.

Q. 15. Write down the units of inertia constants M and H and their interrelationships. **[Madras Univ. April 1997]**

Ans. M is expressed in MJ-seconds/electrical degree or MJ-seconds/ electrical radian and H is expressed in MJ/MVA.

The interrelationship between M and H is given as

$$M = \frac{GH}{180f} = \frac{GH}{\pi f}$$

Q. 16. What is equal area criterion ? **[Bharathiyar Univ. April 1997]**

Ans. Equal area criterion is a simple graphical method of predicting the transient stability of two-machine systems or a single machine against infinite bus. This criterion does not require solution of swing equation for determination of stability conditions. The stability conditions are determined by equating the areas of segments on the power angle diagram between the P-curve and the new power transfer line for the given condition.

Q. 17. For the swing equation $M \frac{d^2\delta}{dt^2} = P_A$, what will be the value of P_A during steady state operation ?

[Madras Univ. October 1996]

Ans. During steady state operation the value of accelerating power P_A will be zero.

Q. 18. If the two machines with inertias M_1 and M_2 are swinging together, what will be the inertia of the equivalent machine ?

[Madras Univ. April 1997]

Ans. Equivalent machine inertia is given as $M_{eq} = \frac{M_1 M_2}{M_1 + M_2}$.

Q. 19. Sketch typical swing curves for a synchronous machine:

(*i*) showing that the machine is stable after a disturbance.

(*ii*) showing that the machine is unstable after a disturbance.

[U.P.S.C. I.E.S. Electrical Engineering-II, 2011]

Ans. If the system is unstable, $\delta(t)$ continues to increase indefinitely with time and the machine loses synchronism.

On the other hand, if the system is stable, $\delta(t)$ performs oscillations (nonsinusoidal) whose amplitude decreases in actual practice owing to damping terms.

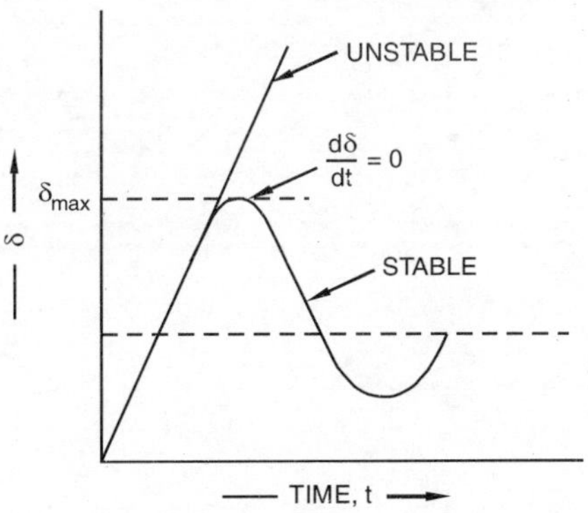

Fig. 2

Problems

1. A 4-pole, 50 Hz turbogenerator rated 30 MVA, 13.2 kV has an inertia constant H = 9.0 kW-s per kVA. Find out the kinetic energy stored in the rotor at synchronous speed. Also find the acceleration if the input less the rotational losses is 25,000 hp and the electric power developed is 15,000 kW.

[**Ans.** 270 MJ; 112.92 electrical degrees/s^2]

2. A 50 Hz, 4-pole turboalternator rotated 20 MVA has an inertia constant of H = 9 kW-s/kVA. Find the kinetic energy stored in the rotor at synchronous speed. Find the accelerating torque if the shaft input less the rotational losses is 26,000 hp (metric) and the electrical power developed is 16,000 kW.

[**Ans.** 180 MJ: 23,628 Nm]

3. In above Question 2 assuming the acceleration computed as constant for a period of 15 cycles, determine the change in torque angle in that period and the rpm at the end of 15 cycles. Assume that the generator is synchronised with a large system and has no accelerating torque before the beginning of 15 cycle period. [**Ans.** 8.36 electrical degrees; 1,504.6 rpm]

4. A 50-Hz, 4-pole turboalternator of rating 40 MVA, 11 kV has an inertia constant of 15 kW-s per kVA. Determine the kinetic energy stored in the rotor at synchronous speed. Determine the acceleration and accelerating torque if the shaft input less the rotational losses is 20 MW and the electrical power developed is 15 MW.

[**Ans.** 600 MJ, 37.5 degrees/second^2; 31,831 N-m]

5. A 50 Hz two-pole turboalternator rated 50 MVA, 13.2 kV has an intertia constant H = 5.0 MJ/MVA. Determine the kinetic energy stored in the rotor at synchronous speed. Determine the acceleration if the input less the rotational losses is 65,000 hp and the electrical power developed is 40 MW. If the acceleration computed for the generation is constant for a period of 10 cycles, determine the change in torque angle in that period and the rpm at the end of 10 cycles. Assume that the generator is synchronised with a large system and has no accelerating torque before the 10 cycles period begins.

[U.P.S.C. I.E.S. Electrical Engineering-II, 2005]

[**Ans.** 250 MJ, 280 electrical degrees per second^2; 0.9769 electrical radian per second, 3,009.328 rpm]

6. A synchronous generator feeds 1.0 pu power to an infinite bus via double circuit transmission line. A fault occurs on one line which reduces the maximum power transfer to 0.5 pu where as before the fault this power was 2.0 pu and after the clearance of the fault 1.5 pu. By the use of equal area criterion determine the critical clearing angle.

[Pb. Univ. Power System Analysis May 1994]

[**Ans.** 70.28 electrical degrees]

7. A 3-phase power system consists of a synchronous machine connected through a loss-less double-circuit transmission to an infinite bus-bar. A fault occurs on the transmission line. The maximum power transfer of this system when unfaulted is 5.0 pu and immediately prior to the instant of the fault the power transfer is 2.5 pu. The power angle curves during fault and post-fault conditions have peak values of 2.0 pu and 4.0 pu respectively. Determine the critical clearing angle using equal area criterion. **[Pb. Univ. Power System Analysis May 1995]**

[**Ans.** 90 electrical degrees]

8. A 50 Hz generator is supplying 40% of the power that it is capable of delivering through a transmission line to an infinite bus. A fault occurs that increases the reactance between the generator and the infinite bus to 600 per cent to the value before the fault. When the fault is isolated, the maximum power that can be delivered is 80% of the original maximum value. Determine the critical angle for the condition given.

[Rajasthan Technical Univ. Power System Engineering 2011]

[**Ans.** 86.56 electrical degrees]

9. A synchronous generator is operating at an infinite bus and supplying 0.45 pu of its maximum power capacity. A fault occurs and the reactance between the generator and the line becomes four time its value before fault. The maximum power that can be delivered after the fault is cleared, is 70% of the original maximum value. Determine the critical clearing angle.

[Rajasthan Univ. Power System Engineering, 2006]

[**Ans.** 72.2 electrical degrees]

10. The power system, shown below, is operating in steady-state with the generator delivering 1.5 pu power to the infinite bus. The circuit breaker A suddenly opens and remains open. Find (*i*) whether the system remains stable; (*ii*) the angle to which the rotor swings. [**Ans.** (*i*) remains stable (*ii*) 69.8° electrical]

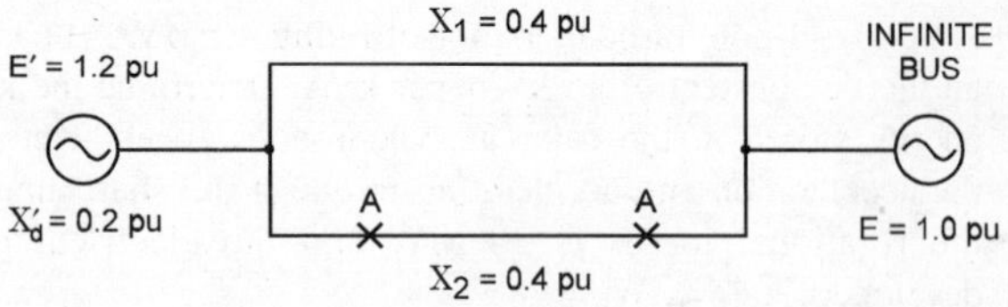

Fig. 1

11. A 3-ϕ transmission line has a reactance of 15 Ω per phase. The voltage at each end is maintained at 66 kV (line to line). Determine the maximum steady state power that can be transmitted by the line. Also determine the limits of angular oscillations for transient stability, when the above line develops a sudden jerk at 3/5th of the steady state limit.

[**Ans.** 290.4 MW; 36.9° – 143.1° (electrical)]

12. A motor is receiving 30% of the power that it is capable of receiving from an infinite bus. If the load on the motor is doubled, determine the maximum value of δ during the swinging of the motor around its new equilibrium portion.

[**Ans.** 58 electrical degrees]

13. A 50 Hz turbogenerator is delivering 50% of the power that is capable of delivering through a transmission line to an infinite bus. A fault occurs that increases the reactance between the generator and the infinite bus to 400% of the value before the fault. When the fault is isolated the maximum power that can be delivered is 75% of the original maximum value. Determine the critical clearing angle for the conditions given above.

[**Ans.** 70.3° (electrical)]

14. The steady state limit of a power system is 150 MW. A generator with constant excitation is supplying 60 MW to the system. Estimate the maximum permissible sudden increase in generator output without causing instability.

[**Ans.** 67.0 MW]

15. For the system shown below (Fig. 2), the per unit values of different quantities are :

$E_1 = 1.2$, $V = 1$, $X'_d = 0.2$, $X_1 = X_2 = 0.4$.

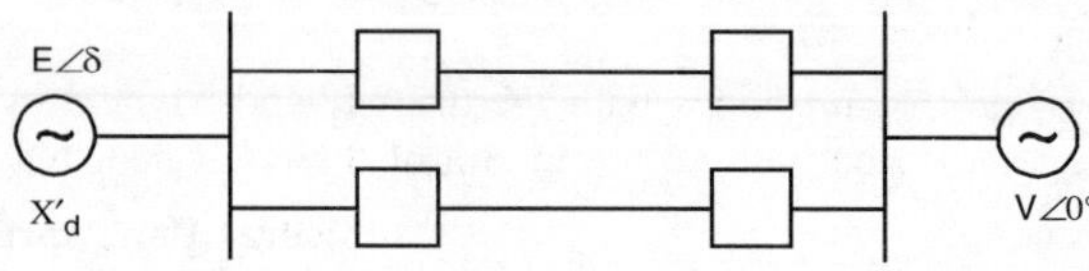

Fig. 2

The system is operating in equilibrium with $P_i = 1.5$ pu. When one of the line is suddenly switched out, predict whether the system will be stable or not. If the system is stable, find the maximum value which δ attains.

[Rajasthan Univ. Power System Engineering 2004]

[**Ans.** Stable; 69.8 electrical degrees]

❑❑❑

CHAPTER 22

Power Flow Studies

INSIDE THIS CHAPTER

22.1. INTRODUCTION

In a 3-phase ac power system, active and reactive power flow from the generating stations to the load through different network buses and branches (transmission lines). Active power P and reactive power Q is supplied by generators at generator buses. Active power is drawn by loads from load buses. Reactive power Q is supplied or drawn from the load buses by shunt compensation elements (shunt capacitors, reactor elements, static VAR system). The flow of active and reactive power is called the *power flow* or *load flow*. The voltages of buses and their phase angles are affected by the power flow and vice versa.

Power flow studies provide a systematic mathematical approach for determination of various bus voltages, their phase angles, active and reactive power flow through different branches, generators and loads under steady state conditions.

The power flow study in a power system constitutes a study of paramount importance.

Load flow studies are carried out to study short-circuit conditions for any interconnected power system. These are also required for planning the operation of power systems under existing conditions, its improvement and future expansion. Such studies facilitate us in determination of best size as well as the most favourable locations for the power capacitors both for power factor improvement and also for raising the network voltages. The load flow studies also help us in determination of the best location as well as optimal capacity of the proposed generating stations, substations and new lines. Thus load flow studies are very important for planning existing system as well as its future expansion.

For optimized operation of an interconnected system some informations, such as bus-bar voltage levels, machine excitation, tap-change and reactive compensation needs are required which are provided by load flow studies. The main information obtained from load flow studies comprises the magnitudes and phase angles of load bus voltages, reactive power at generator buses, active and reactive power flow in transmission lines, other variables being specified. Such information is essential for the continuous monitoring of the current state of the system and for analysis of the effectiveness of alternative plans for future system expansion to meet the increased load demand.

For many years, load flow studies were carried out by means of special purpose analog computer, called the *ac network analyzer*, but the advent of high speed digital computers has tended to replace their use for large system studies. This change from the ac network analyzer to the digital computer has resulted in greater flexibility, economy, accuracy and quicker operation. However, for system studies of a more local character, the network analyzer is still used, particularly in the initial planning stages.

Analysis of power system networks is possible by using either mesh current or nodal voltage techniques. The latter is more amenable to digital computer analysis, so that the nodal iterative approach to the solution of load or power flow problems has now become firmly established. The main advantages attributed to the use of nodal voltage methods are : (*i*) the number of equations is smaller, (*ii*) the system may be described in terms of its node numbers and the interconnecting impedances, (*iii*) parallel branches can be dealt with separately without adding to the number of equations and (*iv*) cross-over branches (if present) do not introduce any difficulties in respect of the formation of the admittance matrix.

The steps in the collection of system data are given below.

1. A single line diagram of the system is drawn.
2. Assuming a balanced 3-phase system, the transmission system is represented by its positive phase sequence network of linear lumped series and shunt branches. The line impedances and shunt admittances are then determined in per unit values, including transformer impedances, shunt capacitor ratings and transformer tappings.
3. Node self and mutual admittances are determined, using nodal analysis.
4. The shunt capacitance and per unit resistance and reactance between the terminating buses of the lines are found.
5. The operating conditions are selected. The static operating state of the system is then specified by the constraints on power and/or voltage at the network buses.

After having the above type of data, a suitable mathematical model of the system, adequately describing the relationships between voltages and powers in the interconnected system, is formed. Power and voltage constraints at various buses in the network are then specified and the load flow equations are solved numerically. On determination of different bus voltages, actual load flow in all transmission lines is computed.

The mathematical formulation of the load flow problem results in a system of nonlinear equations. These equations can be written in terms of either the bus admittance matrix or bus impedance matrix. The former is more amenable to digital computer analysis, because of the ease with which the bus admittance matrix could be formed and modified for network changes in subsequent cases. Further this approach (bus admittance matrix) is the most economical from the point of view of computer time and memory requirements.

22.2. BUS CLASSIFICATION

A *bus* is a node at which one or many lines, one or many loads and generators are connected. It is not necessary that all of these be connected at every bus. The bus is indicated by a vertical line at which several components are connected.

In a power system each node or bus is associated with four quantities, such as magnitude of voltage *i.e.*, $|V|$, phase angle of the voltage *i.e.*, δ, active or true power *i.e.*, P and reactive power, *i.e.*, Q. In a load flow problem two out of these four quantities are specified and the remaining two are required to be determined through the solution of equations. Depending upon which quantities have been specified, the buses are classified into three categories viz generation bus, load bus and slack bus.

1. Generation Bus or Voltage-Controlled Bus. This is also called the P-V bus, and on this bus the voltage magnitude corresponding to generation voltage and true or active power P corresponding to its ratings are specified. Voltage magnitude is maintained constant at a specified value by injection of reactive power. The reactive power generation Q and phase angle δ of the voltage are to be computed.

2. Load Bus. This is also called the P-Q bus and at this bus the total injected power is specified, *i.e.*, the active and reactive power injected into the network at this bus. Magnitude and phase angle of the voltage are to be computed. It is required to specify only P and Q at such a bus as at a load bus voltage can be permitted to vary within the permissible values, *i.e.*, 5%. Also phase angle of the voltage *i.e.*, δ is not very important for the load.

3. Slack, Swing or Reference Bus. One of the generation buses in a power system is chosen as *slack* or *swing bus*. At this bus, the magnitude and phase angle of the voltage are specified. The phase angle of the voltage is usually set equal to zero. The active and reactive powers at this bus are to be determined through the solution of equations. The swing bus is a fictitious concept in load flow studies and arises because the I^2R losses of the system are not known precisely in advance for the load flow calculations. Therefore, the total injected power cannot be specified at every single bus. At the swing bus, it is customary to regard the active power as unknown. The difference between the expected and solved output (MW) represents the error in the prior estimate of system I^2R losses. The generators at the swing bus supply the difference of the specified active power to be injected into the system at other buses and the total system output plus losses. If slack bus is not specified, the generation bus usually with maximum active power P is taken as reference bus. There can be more than one slack bus in a given network.

The above discussion can be summarised in the tabular form as below:

Table 22.1. Bus Classification

Type of Bus	Known or Specified Quantities	Unknown Quantities or Quantities to be Determined
Generation or P-V Bus	P, $|V|$	Q, δ
Load or P-Q Bus	P, Q	$|V|$, δ
Slack or Reference Bus	$|V|$, δ	P, Q

Phase angle of the voltage at the slack bus is usually taken as reference or zero.

Power system can be analysed using either mesh current method or nodal voltage method but latter is usually employed. This is already stated in Art 22.1. From the point of view of computer time and memory, the nodal admittance formulation, employing the nodal voltages as the independent variables is the most economical.

22.3. BUS ADMITTANCE MATRIX

A power system may comprise several buses interconnected through transmission lines. Power is injected into a bus from generators, while the loads are tapped from it. Of course, there may be buses with only generators, and there may be others with only loads. Some buses may have both generators and loads while some others may have static capacitors (or synchronous condensers) for reactive power compensation or voltage control. The surplus power at some of the buses is transported through transmission lines to the buses deficient in power.

Single line diagram of a simple 4-bus system with generators and loads at each bus is shown in Fig. 22.1. To arrive at the network model of a power system, a short line may be represented by a series impedance, long line by a nominal π model while very long lines by equivalent π. Line resistance is usually neglected with a small loss in accuracy but a great deal of saving in computation time. It is convenient to consider loads as negative generators and lump together the generator and load powers at the buses.

Let S_{Gi} denote the 3-phase complex generator power flowing into the ith bus and S_{Di} denote the 3-phase complex power demand at the ith bus. Let S_{Gi} and S_{Di} may be represented as

$$\mathbf{S}_{Gi} = P_{Gi} + j\,Q_{Gi}$$

$$\text{and} \quad \mathbf{S}_{Di} = P_{Di} + j\,Q_{Di}$$

Net complex power injected into the bus is given as

$$\mathbf{S}_i = P_i + j\,Q_i = (P_{Gi} - P_{Di}) + j\,(Q_{Gi} - Q_{Di})$$

The real and reactive powers injected into the ith bus are then

$$\left.\begin{aligned} P_i &= P_{Gi} - P_{Di} \\ Q_i &= Q_{Gi} - Q_{Di} \end{aligned}\right\} \; i = 1, 2, 3, \ldots, n \qquad \ldots(22.1)$$

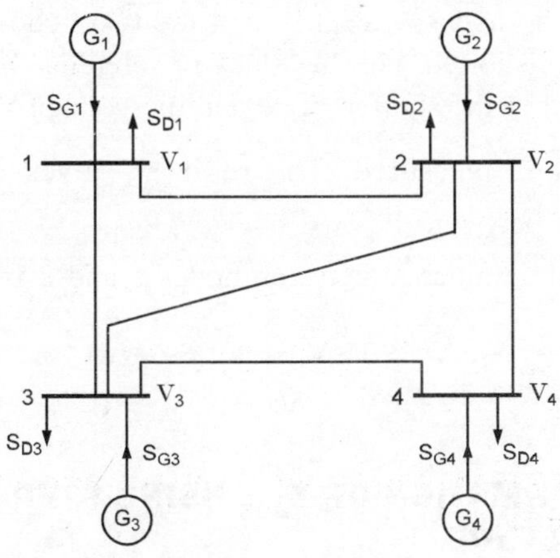

Fig. 22.1. ***One Line Diagram of a 4-Bus System***

Network model of the given power system, worked out on the above lines is shown in Fig. 22.2(a). S_1, S_2, S_3, S_4 denote the net 3-phase complex power flowing into the buses and I_1, I_2, I_3, I_4 denote the currents flowing into the buses. Each transmission line is represented by a π circuit.

The equivalent circuit of Fig. 22.2 (a) has been redrawn in Fig. 22.2 (b) where all the sources have been shown connected to a common reference at ground potential and the shunt admittances at the buses have been lumped. Besides the ground node, it has 4 other nodes (buses) at which the current from the sources is injected into the network. The line admittance between nodes i and k is depicted by $y_{ik} = y_{ki}$.

Further, the mutual admittance between lines is assumed to be zero.

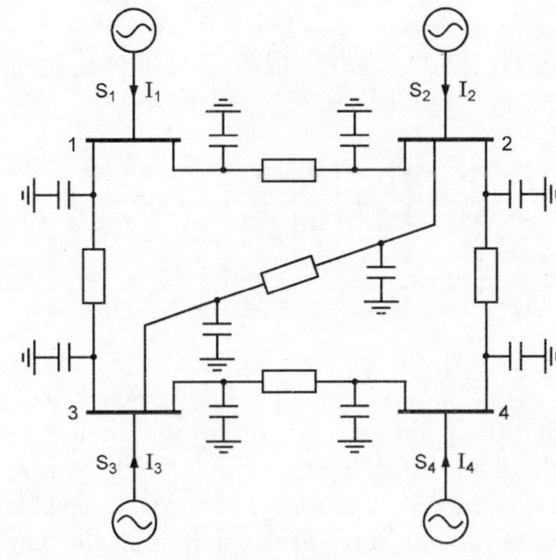

(a) Equivalent Circuit

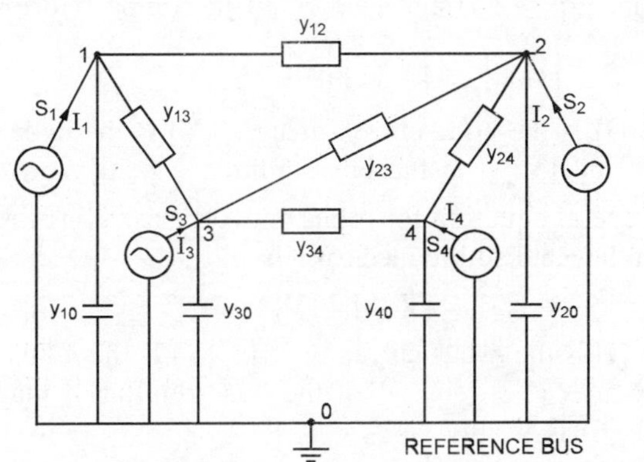

(b) Modified Equivalent Circuit (Admittance Network) of The Four Bus System Shown in Fig. 22.1

Fig. 22.2

Application of Kirchhoff's current law to the four nodes of Fig. 22.2 (b) gives the following equations

$$\begin{aligned} \mathbf{I}_1 &= \mathbf{V}_1 \mathbf{y}_{10} + (\mathbf{V}_1 - \mathbf{V}_2)\mathbf{y}_{12} + (\mathbf{V}_1 - \mathbf{V}_3)\mathbf{y}_{13} \\ \mathbf{I}_2 &= \mathbf{V}_2 \mathbf{y}_{20} + (\mathbf{V}_2 - \mathbf{V}_1)\mathbf{y}_{12} + (\mathbf{V}_2 - \mathbf{V}_3)\mathbf{y}_{23} + (\mathbf{V}_2 - \mathbf{V}_4)\mathbf{y}_{24} \\ \mathbf{I}_3 &= \mathbf{V}_3 \mathbf{y}_{30} + (\mathbf{V}_3 - \mathbf{V}_1)\mathbf{y}_{13} + (\mathbf{V}_3 - \mathbf{V}_2)\mathbf{y}_{23} + (\mathbf{V}_3 - \mathbf{V}_4)\mathbf{y}_{34} \\ \mathbf{I}_4 &= \mathbf{V}_4 \mathbf{y}_{40} + (\mathbf{V}_4 - \mathbf{V}_2)\mathbf{y}_{24} + (\mathbf{V}_4 - \mathbf{V}_3)\mathbf{y}_{34} \end{aligned} \qquad \ldots(22.2)$$

The above Eq. (22.2) can be rearranged and written in matrix form as below

$$\begin{bmatrix} I_1 \\ I_2 \\ I_3 \\ I_4 \end{bmatrix} = \begin{bmatrix} y_{10}+y_{12}+y_{13} & -y_{12} & -y_{13} & 0 \\ -y_{12} & y_{20}+y_{12}+y_{23}+y_{24} & -y_{23} & -y_{24} \\ -y_{13} & -y_{23} & y_{30}+y_{13}+y_{23}+y_{34} & -y_{34} \\ 0 & -y_{24} & -y_{34} & y_{40}+y_{24}+y_{34} \end{bmatrix} \begin{bmatrix} V_1 \\ V_2 \\ V_3 \\ V_4 \end{bmatrix} \qquad \ldots(22.3)$$

The admittance terms on RHS are redesignated as

Self Admittances	**Mutual Admittances**
$\mathbf{Y}_{11} = \mathbf{y}_{10} + \mathbf{y}_{12} + \mathbf{y}_{13}$	$\mathbf{Y}_{12} = \mathbf{Y}_{21} = -\,\mathbf{y}_{12}$
$\mathbf{Y}_{22} = \mathbf{y}_{20} + \mathbf{y}_{12} + \mathbf{y}_{23} + \mathbf{y}_{24}$	$\mathbf{Y}_{13} = \mathbf{Y}_{31} = -\,\mathbf{y}_{13}$
$\mathbf{Y}_{33} = \mathbf{y}_{30} + \mathbf{y}_{13} + \mathbf{y}_{23} + \mathbf{y}_{34}$	$\mathbf{Y}_{14} = \mathbf{Y}_{41} = -\,\mathbf{y}_{14} = 0$

$\mathbf{Y}_{44} = y_{40} + y_{24} + y_{34}$ $\quad \mathbf{Y}_{23} = \mathbf{Y}_{32} = -y_{23}$

$$\mathbf{Y}_{24} = \mathbf{Y}_{42} = -y_{24}$$

$$\mathbf{Y}_{34} = \mathbf{Y}_{43} = -y_{34}$$

Matrix Eq. (22.3) is written in terms of self bus admittance Y_i and mutual bus admittance Y_{ik} as follows

$$\begin{bmatrix} I_1 \\ I_2 \\ I_3 \\ I_4 \end{bmatrix} = \begin{bmatrix} Y_{11} & Y_{12} & Y_{13} & Y_{14} \\ Y_{21} & Y_{22} & Y_{23} & Y_{24} \\ Y_{31} & Y_{32} & Y_{33} & Y_{34} \\ Y_{41} & Y_{42} & Y_{43} & Y_{44} \end{bmatrix} \begin{bmatrix} V_1 \\ V_2 \\ V_3 \\ V_4 \end{bmatrix} \quad ...(22.4)$$

From network theory we know that each admittance $\mathbf{Y}_{ii}$ is known as *self-admittance* (or *driving point admittance*) of *i*th node and is equal to the sum of the admittances connected to *i*th node. Each off-diagonal term Y_{ik} is known as *mutual admittance* (or *transfer admittance*) between *i*th and kth nodes and is equal to the negative of the sum of all the admittances connected directly between *i*th and kth nodes. Further $\mathbf{Y}_{ik} = \mathbf{Y}_{ki}$.

The Eq. (22.4) can be written in compact form as

$$[\mathbf{I}_{bus}] = [\mathbf{Y}_{bus}][\mathbf{V}] \quad ...(22.5)$$

where **[I]** is the node current matrix, **[V]** is the node voltage matrix and $[\mathbf{Y}_{bus}]$ is the bus admittance matrix.

General equation for *n*-bus network based on Kirchhoffs' current law and admittance form is

$$[\mathbf{I}] = [\mathbf{Y}_{bus}] * [\mathbf{V}] \quad ...(22.6)$$

where **[I]** is the *n*-bus current matrix, **[V]** is the *n*-bus voltage matrix and, $[\mathbf{Y}_{bus}]$ is called the bus admittance matrix and Eq. (22.6) is written as

$$\mathbf{I} = \mathbf{Y}_{bus}\,\mathbf{V}$$

$$\text{where } \mathbf{Y}_{bus} = \begin{bmatrix} Y_{11} & Y_{12} & \cdots & Y_{1n} \\ Y_{21} & Y_{22} & \cdots & Y_{2n} \\ \cdots & \cdots & \cdots & \cdots \\ Y_{n1} & Y_{n2} & \cdots & Y_{nn} \end{bmatrix} \quad ...(22.7)$$

and is called the *bus admittance matrix*, and **V** and **I** are the *n*-element *node voltage matrix* and *node current matrix* respectively.

Y_{bus} matrix for *n*-bus network has *n* rows and *n* columns. Each of the Y terms in the rows and columns has two subscripts:

— The first subscript refers to the bus number on which the current is expressed.

— The second subscript refers to the bus number whose voltage has caused that current component.

— The terms on diagonal are self-admittances.

— All the non-diagonal terms are mutual admittances. It is seen that the current entering bus *i* is given as

$$\mathbf{I}_i = \sum_{k=1}^{n} \mathbf{Y}_{ik}\mathbf{V}_k \; ; \; i = 1, 2, ..., n \quad ...(22.8)$$

22.3.1. Self Admittance and Mutual Admittance Elements

1. Self Admittance of Node. The terms $\mathbf{Y}_{ii}$ (i = 1, 2, 3, 4) are *self admittances* of respective nodes and represent the *algebraic sum of all* the admittances connected to that node. Each diagonal term in the $\mathbf{Y}_{bus}$ matrix is a self admittance term.

If all the nodes, except node *i*, are shorted with the slack or reference bus and current I is injected in the particular node *i* bus and the voltage V across that node *i* and the reference bus is measured then the ratio $\frac{I}{V}$ gives the self admittance of that node.

Thus, self admittance of node 4 is (i = 4)

$$\mathbf{Y}_{44} = \frac{I_4}{V_4}\Bigg|_{\substack{V_4 \neq 0 \\ V_1 = V_2 = V_3 = V_5 = ...\, V_n = 0}} \quad ...(22.9)$$

2. Mutual Admittance Between Two Nodes. The mutual admittance terms (or transfer admittance terms) are the terms $\mathbf{Y}_{ik}$ in $\mathbf{Y}_{bus}$ matrix. All the non-diagonal terms in the $\mathbf{Y}_{bus}$ matrix are mutual admittance terms $\mathbf{Y}_{ik}$ (i; k = 1, 2, 3, 4, ... n)

Mutual admittance between two buses is the negative of the sum of all the admittances connected directly between those two buses.

Also $\mathbf{Y}_{ik} = \mathbf{Y}_{ki}$

For measuring the mutual admittance between the two nodes, all the nodes, except one of the two nodes (node *i* or node k), are shorted with the slack or reference bus. Current I is injected in the shorted node and the voltage V across the two nodes is measured. The ratio $\frac{I}{V}$ gives the mutual admittance between the two nodes. Thus,

Mutual admittance between nodes 1 and 2 is $\mathbf{Y}_{12} = \mathbf{Y}_{21}$

$$\mathbf{Y}_{21} = \mathbf{Y}_{12}\Bigg|_{\substack{V_1 \neq 0 \text{ (}V_1\text{ is not shorted)} \\ V_2 = V_3 = V_4 = ...\, V_n = 0}} \quad ...(22.10)$$

22.4. FORMATION OF Y_{BUS} USING STEP BY STEP METHOD

The admittance matrix can be formed from the parameters of the system components according to the discussion made above. A diagonal element $\mathbf{Y}_{ii}$ is the sum of all admittances connected to *i*th bus. An off-diagonal element $\mathbf{Y}_{ik}$ is the negative of the total admittance directly connected between *i*th and kth buses. The following step by step procedure for formation of $\mathbf{Y}_{bus}$ may be used. This procedure is very suitable.

Let us start with $[\mathbf{Y}_{bus}]$ array set to zero. The dimensions of the $\mathbf{Y}_{bus}$ matrix is ($n \times n$) where n is the number of buses. The total number of nodes are $m = n + 1$ including the ground or reference node. Consider an element having admittance $\mathbf{y}_{ik}$ connected between buses i and k. Four entries in $[\mathbf{Y}_{bus}]$ are affected: $\mathbf{Y}_{ii}$, $\mathbf{Y}_{ik}$, $\mathbf{Y}_{ki}$ and $\mathbf{Y}_{kk}$.

These entries can be modified as under

$$\begin{aligned} \mathbf{Y}_{ii,\,new} &= \mathbf{Y}_{ii,\,old} + \mathbf{y}_{ik} \\ \mathbf{Y}_{ik,\,new} &= \mathbf{Y}_{ik,\,old} - \mathbf{y}_{ik} \\ \mathbf{Y}_{ki,\,new} &= \mathbf{Y}_{ki,\,old} - \mathbf{y}_{ik} \\ \mathbf{Y}_{kk,\,new} &= \mathbf{Y}_{kk,\,old} + \mathbf{y}_{ik} \end{aligned} \quad ...(22.11)$$

We add the elements one by one and modify the entries of $[\mathbf{Y}_{bus}]$ according to Eq. (22.11). If an element is connected from ith bus to reference bus, only entry $\mathbf{Y}_{ii}$ is affected.

Example 22.1. Determine nodal admittance matrix for the power system represented by the single line diagram in Fig. 22.3.

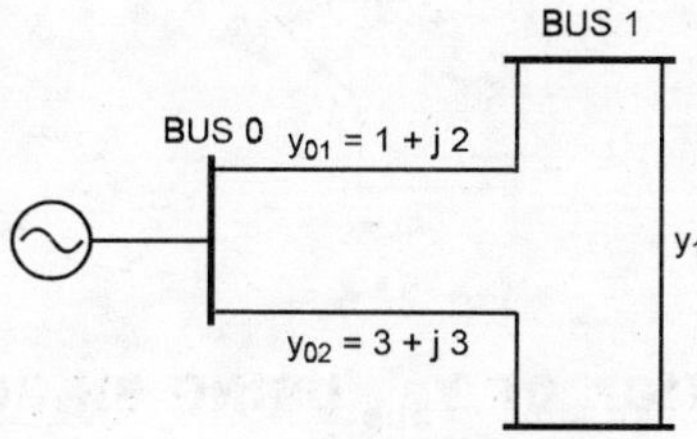

Fig. 22.3

Solution :

$\mathbf{Y}_{00}$ = Sum of all admittances terminating at bus 0

$= \mathbf{y}_{01} + \mathbf{y}_{02} = (1 + j\,2) + (3 + j\,3) = 4 + j\,5$

$\mathbf{Y}_{01} = -\mathbf{y}_{01} = -(1 + j\,2)$

$\mathbf{Y}_{02} = -\mathbf{y}_{02} = -(3 + j\,3)$

$\mathbf{Y}_{11}$ = Sum of admittances terminating at bus 1

$= \mathbf{y}_{01} + \mathbf{y}_{12} = (1 + j\,2) + (2 + j\,1) = 3 + j\,3$

$\mathbf{Y}_{12} = -\mathbf{y}_{12} = -(2 + j\,1)$

$\mathbf{Y}_{10} = -\mathbf{y}_{01} = -(1 + j\,2)$

$\mathbf{Y}_{22}$ = Sum of all admittances terminating at bus 2

$= \mathbf{y}_{12} + \mathbf{y}_{02} = (2 + j\,1) + (3 + j\,3) = 5 + j\,4$

$\mathbf{Y}_{21} = -\mathbf{y}_{12} = -(2 + j\,1)$

$\mathbf{Y}_{20} = -\mathbf{y}_{02} = -(3 + j\,3)$

Hence nodal admittance matrix is

$$\begin{bmatrix} \mathbf{Y}_{00} & \mathbf{Y}_{01} & \mathbf{Y}_{02} \\ \mathbf{Y}_{10} & \mathbf{Y}_{11} & \mathbf{Y}_{12} \\ \mathbf{Y}_{20} & \mathbf{Y}_{21} & \mathbf{Y}_{22} \end{bmatrix} = \begin{bmatrix} 4+j5 & -1-j2 & -3-j3 \\ -1-j2 & 3+j3 & -2-j1 \\ -3-j3 & -2-j1 & 5+j4 \end{bmatrix} \quad \textbf{Ans.}$$

Example 22.2. Formulate $\mathbf{Y}_{bus}$ matrix for the power network given in Fig. 22.4. The relevant pu line admittances are indicated on the diagram.

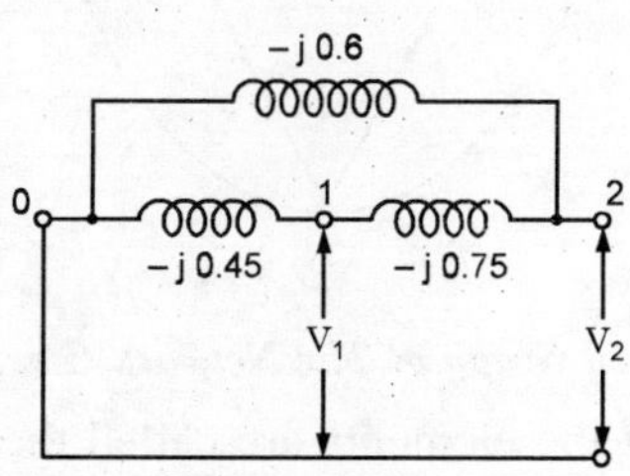

Fig. 22.4

Solution : The network has two buses,

Therefore $\mathbf{Y}_{bus}$ matrix is a 2 × 2 matrix

$\mathbf{Y}_{11}$ = Sum of all admittances terminating at bus 1

$= -j\,0.45 + (-j\,0.75) = -j\,1.2$

$\mathbf{Y}_{22}$ = Sum of all admittances terminating at bus 2

$= -j\,0.6 + (-j\,0.75) = -j\,1.35$

$\mathbf{Y}_{12} = \mathbf{Y}_{21} = -\mathbf{y}_{12} = -(-j\,0.75) = j\,0.75$

$$\text{So } [\mathbf{Y}_{bus}] = \begin{bmatrix} \mathbf{Y}_{11} & \mathbf{Y}_{12} \\ \mathbf{Y}_{21} & \mathbf{Y}_{22} \end{bmatrix} = \begin{bmatrix} -j1.2 & j0.75 \\ j0.75 & -j1.35 \end{bmatrix} \quad \textbf{Ans.}$$

By Using Step By Step Method. Let us start with $|\mathbf{Y}_{bus}| = \begin{bmatrix} 0 & 0 \\ 0 & 0 \end{bmatrix}$

According to Eq. (22.11), on adding $\mathbf{y}_{12} = -j\,0.75$, we have

$$[\mathbf{Y}_{bus}] = \begin{bmatrix} -j0.75 & -(-j0.75) \\ -(-j0.75) & -j0.75 \end{bmatrix} = \begin{bmatrix} -j0.75 & j0.75 \\ j0.75 & -j0.75 \end{bmatrix}$$

Now adding $-j\,0.45$, we have

$$[\mathbf{Y}_{bus}] = \begin{bmatrix} -j0.75 + (-j0.45) & j0.75 \\ j0.75 & -j0.75 \end{bmatrix}$$

$$= \begin{bmatrix} -j1.2 & j0.75 \\ j0.75 & -j0.75 \end{bmatrix} \quad \because \text{ Only } \mathbf{Y}_{11} \text{ is affected}$$

Now adding $-j\,0.6$, we have

$$[\mathbf{Y}_{bus}] = \begin{bmatrix} -j1.2 & j0.75 \\ j0.75 & -j0.75 + (-j0.6) \end{bmatrix}$$

$$= \begin{bmatrix} -j1.2 & j0.75 \\ j0.75 & -j1.35 \end{bmatrix} \quad \textbf{Ans. } \text{Only } \mathbf{Y}_{22} \text{ being affected.}$$

Example 22.3. Consider the network shown in figure (Fig. 22.5). The bus numbers and impedances are marked. Determine the bus impedance matrix of the network.

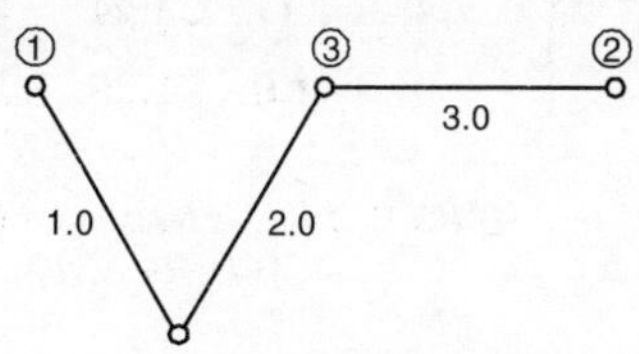

Fig. 22.5

[U.P.S.C. I.E.S. Electrical Engineering-II, 2008]

Solution : $\mathbf{Y}_{11}$ = Sum of all admittances terminating at bus 1

$= y_{10} + y_{12} + y_{13}$

$= \dfrac{1}{j1.0} + \dfrac{1}{\infty} + \dfrac{1}{\infty} = -j\,1$

$\mathbf{Y}_{22} = y_{20} + y_{12} + y_{23}$

$= j0 + j0 + \left(-\dfrac{1}{3}\right)j = -\dfrac{1}{3}j$

$\mathbf{Y}_{33} = y_{30} + y_{13} + y_{23}$

$= -\dfrac{1}{2}j + j0 - j\dfrac{1}{3} = \dfrac{-5}{6}j$

$\mathbf{Y}_{12} = \mathbf{Y}_{21} = -y_{12} = 0$

$\mathbf{Y}_{13} = \mathbf{Y}_{31} = -y_{13} = 0$

$$\mathbf{Y}_{23} = \mathbf{Y}_{32} = -y_{23} = \frac{1}{3}j$$

$$\mathbf{Y}_{bus} = \begin{bmatrix} \mathbf{Y}_{11} & \mathbf{Y}_{12} & \mathbf{Y}_{13} \\ \mathbf{Y}_{21} & \mathbf{Y}_{22} & \mathbf{Y}_{23} \\ \mathbf{Y}_{31} & \mathbf{Y}_{32} & \mathbf{Y}_{33} \end{bmatrix}$$

$$= \begin{bmatrix} -j1 & 0 & 0 \\ 0 & -\dfrac{j}{3} & \dfrac{j}{3} \\ 0 & \dfrac{j}{3} & \dfrac{-5}{6}j \end{bmatrix}$$

$$|\mathbf{Y}| = -j\left(\frac{5}{18}j^2 - \frac{j^2}{9}\right) = \frac{-j^3}{6} = \frac{j}{6}$$

$$\text{Adj } \mathbf{Y} = \begin{bmatrix} -\dfrac{1}{6} & 0 & 0 \\ 0 & -\dfrac{5}{6} & -\dfrac{1}{3} \\ 0 & -\dfrac{1}{3} & -\dfrac{1}{3} \end{bmatrix}^T = \begin{bmatrix} -\dfrac{1}{6} & 0 & 0 \\ 0 & -\dfrac{5}{6} & -\dfrac{1}{3} \\ 0 & -\dfrac{1}{3} & -\dfrac{1}{3} \end{bmatrix}$$

$$\mathbf{Z}_{bus} = \mathbf{Y}^{-1} = \frac{\text{Adj } \mathbf{Y}}{|\mathbf{Y}|}$$

$$= -j6 \begin{bmatrix} -\dfrac{1}{6} & 0 & 0 \\ 0 & -\dfrac{5}{6} & -\dfrac{1}{3} \\ 0 & -\dfrac{1}{3} & -\dfrac{1}{3} \end{bmatrix} = \begin{bmatrix} j1 & 0 & 0 \\ 0 & j5 & j2 \\ 0 & j2 & j2 \end{bmatrix} \quad \textbf{Ans.}$$

Example 22.4. Draw the reactance diagram of the system whose bus admittance matrix is given below. First, second, third and fourth rows refer to buses 1, 2, 3, 4 respectively.

$$\mathbf{Y}_{bus} = j\begin{bmatrix} \mathbf{-3.78} & \mathbf{1.25} & \mathbf{2.50} & \mathbf{0} \\ \mathbf{1.25} & \mathbf{-3.42} & \mathbf{1.11} & \mathbf{1.00} \\ \mathbf{2.5} & \mathbf{1.11} & \mathbf{-4.89} & \mathbf{1.25} \\ \mathbf{0} & \mathbf{1.0} & \mathbf{1.25} & \mathbf{-2.31} \end{bmatrix}$$

[*U.P.S.C. I.E.S. Electrical Engineering-II, 2001; M.D. Univ. Power Systems-I, 2001*]

Solution :

$$\mathbf{y}_{12} = -\mathbf{Y}_{12} = -\mathbf{Y}_{21} = -j1.25$$

$$\mathbf{y}_{13} = -\mathbf{Y}_{31} = -\mathbf{Y}_{13} = -j2.5$$

$$\mathbf{y}_{14} = -\mathbf{Y}_{41} = -\mathbf{Y}_{14} = 0$$

i.e., no line between bus 1 and bus 4

$$\mathbf{y}_{23} = -\mathbf{Y}_{23} = -\mathbf{Y}_{32} = -j1.11$$

$$\mathbf{y}_{24} = -\mathbf{Y}_{24} = -\mathbf{Y}_{42} = -j1.0$$

$$\mathbf{y}_{34} = -\mathbf{Y}_{34} = -\mathbf{Y}_{43} = -j1.25$$

$$\mathbf{Z}_{12} = \frac{1}{\mathbf{y}_{12}} = \frac{1}{-j1.25} = j0.8$$

$$\mathbf{Z}_{13} = \frac{1}{-j2.5} = j0.4$$

$$\mathbf{Z}_{14} = \infty$$

$$\mathbf{Z}_{23} = \frac{1}{-j1.11} = j0.9$$

$$\mathbf{Z}_{24} = \frac{1}{-j1.0} = j1.0$$

$$\mathbf{Z}_{34} = \frac{1}{-j1.25} = j0.8$$

The reactance diagram is shown in Fig. 22.6.

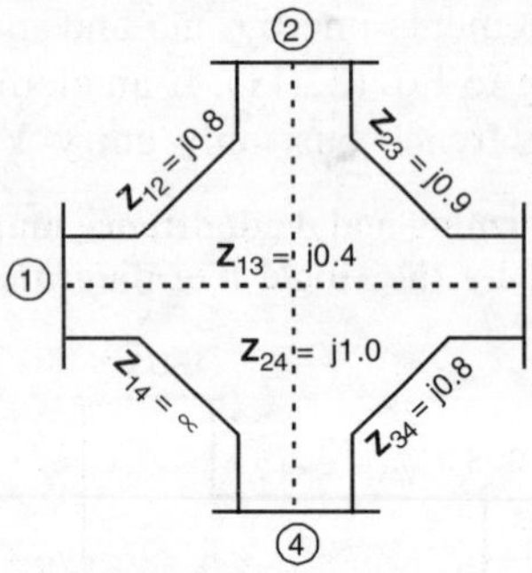

Fig. 22.6

22.5. FORMATION OF $\mathbf{Y}_{bus}$ USING SINGULAR TRANSFORMATION

The $\mathbf{Y}_{bus}$ can be alternatively formed by using singular transformation given by a graph theoretical approach. This alternative approach is of great theoretical and practical significance and is, therefore, discussed here.

To understand this alternative approach, the graph theory is briefly reviewed.

22.5.1. Graph Theory

In order to describe the geometrical structure of the network, it is sufficient to replace the different power system components (of the corresponding power system network) such as generators, transformers and transmission lines etc. by a single line element irrespective of the characteristics of the power system components. The geometrical interconnection of these line elements (of the corresponding power system network) is known as a *graph* (rather *linear graph* as the graph means always a linear graph). Each source and the shunt admittance across it are taken as a single element.

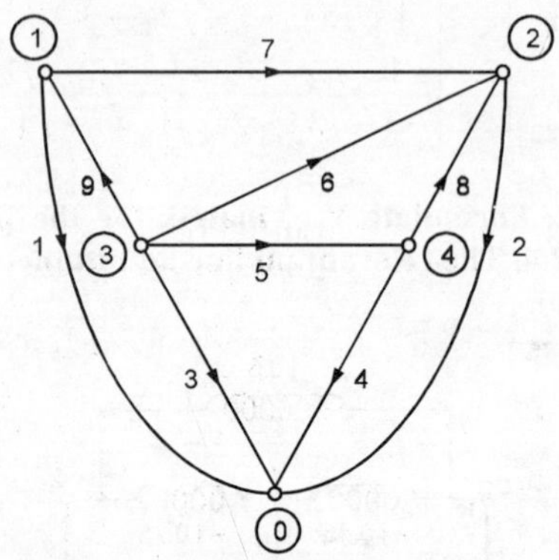

Fig. 22.7. ***Oriented Graph of The Network Given in Fig. 22.2 (b)***

The terminals of the elements are called the *nodes*. A *graph is connected* if, and only if, there exists a path between every pair of nodes. A single edge or a single node is a connected graph. If every edge of the graph is assigned a direction, the graph is termed as an *oriented graph*.

The direction is generally, so assigned as to coincide with the assumed positive direction of the current in the element.

Power networks are so structured that out of the m total nodes, one node (normally described by 0) is always at ground potential and the remaining $n = (m - 1)$ nodes are the buses at which the source power is injected. Figure 22.7 shows the oriented graph of the network given in Fig. 22.2 (*b*).

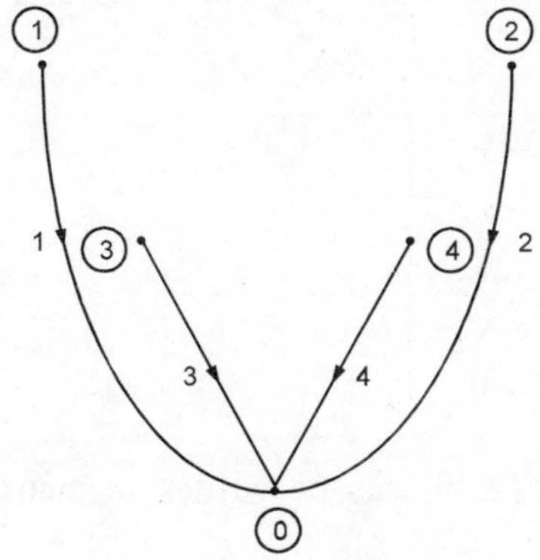

(a) Tree

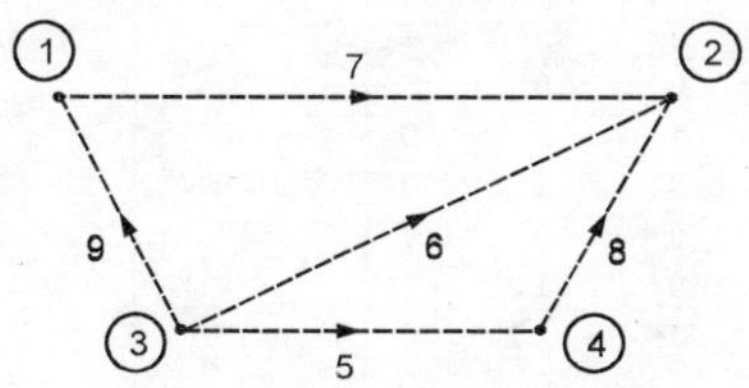

(b) Co-Tree

Fig. 22.8

A connected subgraph containing all the nodes of the original graph but no closed path is called a *tree*. The tree branches form a sub-set of the elements of the connected graph. The number of branches b required to form a tree is equal to the number of buses in the network (the total numoer of nodes, including the reference node, is one more than the number of buses), *i.e.*,

$$b = m - 1 = n \text{ (number of buses)} \qquad ...(22.12)$$

The elements of the original graph not included in the tree, form a subgraph which may not necessarily be connected, is known as *co-tree*. The co-tree is a complement of a tree. The elements of a co-tree are called the *links*. The number of links l of a connected graph with e elements is given as

$$l = e - b = e - m + 1 \qquad ...(22.13)$$

i.e., number of links equals number of elements less the number of tree branches.

A tree and the corresponding co-tree of graph shown in Fig. 22.7 are shown in Figs. 22.8 (*a*) and 22.8 (*b*) respectively.

If a link is added to the tree, the corresponding graph contains one closed path called a *loop*. Thus a graph has as many loops as the number of links.

The above system has 9 branches. So it has 18 variables (9 branch voltages and 9 branch currents). However, it can be easily seen that all these 18 variables are not independent. The number of independent variables is found from the concept of the tree. The number of tree branches gives the number of independent voltages. For any system the number of tree branches is equal to the number of buses. The number of links gives the number of independent current variables.

22.5.2. Primitive Network

A network is constituted by many branches and each branch consists of active and/or passive elements. Fig. 22.9(*a*) and 22.9(*b*) show a network branch, containing both active and passive elements in impedance and admittance representation. The impedance is a voltage source E_{rs} in series with and impedance, z_{rs}; while in admittance form there is a current source j_{rs} in parallel with an admittance y_{rs}. The element current is I_{rs} and element voltage, $V_{rs} = V_r - V_s$ where V_r and V_s are the voltages of the element nodes r and s, respectively.

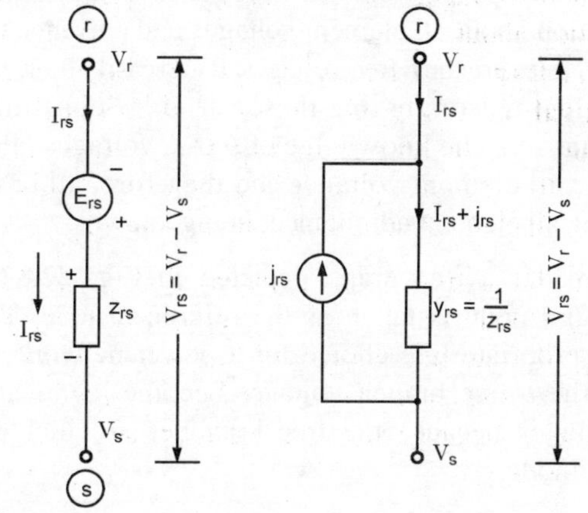

(a) Impedance Form *(b) Admittance Form*

Fig. 22.9. *Representation of a Network Element*

The noteworthy point is that for steady state ac performance, all element variables (V_{rs}, V_r, V_s, I_{rs}, J_{rs}) are phasors and element parameters z_{rs} and y_{rs} are complex numbers.

The performance equation for impedance representation, depicted in Fig. 22.9(*a*), can be written as

$$V_{rs} + E_{rs} = z_{rs} I_{rs} \qquad ...(22.14)$$

and for admittance representation depicted in Fig. 22.9(*b*)

$$I_{rs} + J_{rs} = y_{rs} V_{rs} \qquad ...(22.15)$$

The two representations shown in Figs. 22.9(*a*) and 22.9(*b*) are equivalent wherein the parallel source current in admittance form is related to the series voltage in impedance form by

$$\left.\begin{aligned} J_{rs} &= -y_{rs} E_{rs} \\ \text{and} \quad y_{rs} &= \frac{1}{z_{rs}} \end{aligned}\right\} \qquad ...(22.16)$$

A set of unconnected elements is known as *primitive network*. The performance equations in admittance (or impedance) form can be written for all branches. The set of these equations in impedance form is

$$\mathbf{V} + \mathbf{E} = \mathbf{ZI} \qquad ...(22.17)$$

and in admittance form

$$\mathbf{I}+\mathbf{J} = \mathbf{YV} \qquad ...(22.18)$$

where **V** and **E** are branch voltage and source voltage matrices, **I** and **J** are branch current and source current matrices, **Z** is primitive impedance matrix (*i.e.*, a matrix whose elements are branch self impedances) and **Y** is primitive admittance matrix (*i.e.*, matrix whose elements are branch self admittances). These are related as $\mathbf{Z} = \frac{1}{\mathbf{Y}}$. If there is no coupling between elements, **Z** and **Y** are diagonal matrices.

22.5.3. Network Variables in Bus Frame of Reference

Linear network graph is quite helpful in the systematic assembly of a network model. The main problem in the derivation of mathematical models for large and complex power networks is to select a minimum or zero redundancy (linearly independent) set of current or voltage variables which is sufficient to provide the information about all element voltages and currents. One set of such variables are the *b* tree voltages. It can easily be seen by using topological reasoning that these variables constitute a non-redundant set. The knowledge of *b* tree voltages allows us to compute all element voltages and therefore, all bus currents assuming all element admittances being known.

Consider a tree graph depicted in Fig. 22.8 (*a*) where the ground node is taken as the reference node. This is the most appropriate tree choice for a power network. With this choice the *b* tree branch voltages become identical with the bus voltages because the tree branches are incident to the ground mode.

22.5.4. Bus Incidence Matrix

If "G" is a graph with "*n*" nodes and "*e*" elements, then the matrix $\overline{A}$ whose *n* rows correspond to the "*n*" nodes (*i.e.*, vertices) and "*e*" columns correspond to the "*e*" elements, *i.e.*, edges, is known as an *incidence matrix*. The matrix elements are:

a_{ik} = 1 if *i*th element is incident to and directed away from the kth node (bus).

= – 1 if the *i*th element is incident to but directed towards the kth node

= 0 if the *i*th element is not incident to the kth node.

The dimension of this matrix is $n \times e$ and its rank is less than *n*.

Any node of the connected graph can be selected as the reference node and then the variables of the remaining $n - 1$ nodes which are termed as buses can be measured w.r.t. this assigned reference node. The matrix "A" obtained from the incidence matrix $\overline{A}$ by deleting the reference row (corresponding to the reference node) is termed as *reduced* or *bus incidence matrix* (the number of buses in the connected graph is equal to $n - 1$ where *n* is the number of nodes). The order of this matrix is $(n - 1) \times e$ and its rows are linearly independent with rank equal to $(n - 1)$.

For the specific system shown in Fig. 22.7, the 9-branch voltages ($V_{b1}, V_{b2}, V_{b3}, ... V_{b9}$) can be expressed in terms of 4-bus voltages (V_1 .... V_4) as below :

$$\left.\begin{aligned} V_{b1} &= V_1 \\ V_{b2} &= V_2 \\ V_{b3} &= V_3 \\ V_{b4} &= V_4 \\ V_{b5} &= V_3 - V_4 \\ V_{b6} &= V_3 - V_2 \\ V_{b7} &= V_1 - V_2 \\ V_{b8} &= V_4 - V_2 \\ V_{b9} &= V_3 - V_1 \end{aligned}\right\} \qquad ...(22.19)$$

Equation (22.19) can be written in matrix form as

$$\mathbf{V} = \mathbf{AV}_{\text{bus}} \qquad ...(22.20)$$

where the bus incidence matrix A is

A =

e \ Bus	1	2	3	4
1	1	0	0	0
2	0	1	0	0
3	0	0	1	0
4	0	0	0	1
5	0	0	1	– 1
6	0	–1	1	0
7	1	–1	0	0
8	0	–1	0	1
9	–1	0	1	0

$$= \begin{array}{c|c} e \backslash \text{Bus} & \text{Buses} \\ \hline \text{Tree Branches} & A_b \\ \hline \text{Link Branches} & A_l \end{array} \qquad ...(22.21)$$

This matrix is rectangular and, therefore, singular. Its elements a_{ik} are found as per rules given in the beginning of the sub-article 22.5.4.

22.5.5. Formulation of Y_{bus} and Z_{bus}

Substituting Eq. (22.20) into Eq. (22.18), we have

$$\mathbf{I} + \mathbf{J} = \mathbf{YAV}_{\text{bus}} \qquad ...(22.22)$$

Premultiplying Eq. (22.22) by $\mathbf{A}^T$ (*i.e.*, transpose of the bus incidence matrix), we have

$$\mathbf{A}^T\mathbf{I} + \mathbf{A}^T\mathbf{J} = \mathbf{A}^T\mathbf{YAV}_{\text{bus}} \qquad ...(22.23)$$

Each component of the *n*-dimensional vector $\mathbf{A}^T\mathbf{I}$ is the algebraic sum of the element currents leaving the nodes 1, 2, 3, ..., *n*.

Therefore, as per Kirchhoffs' current law

$$\mathbf{A}^T\mathbf{I} = 0 \qquad ...(22.24)$$

Similarly, each component of vector $\mathbf{A}^T\mathbf{J}$ can be recognized as the algebraic sum of all source currents injected into nodes 1, 2, ... *n*. These components are therefore the bus currents. Hence we can write

$$\mathbf{A}^T\mathbf{J} = \mathbf{J}_{\text{bus}} \qquad ...(22.25)$$

Equation (22.23) is then simplified to

$$\mathbf{J}_{\text{bus}} = \mathbf{A}^T\mathbf{YAV}_{\text{bus}} \qquad ...(22.26)$$

Comparing Eqs. (22.26) and (22.5), we have

$$\mathbf{Y}_{bus} = \mathbf{A}^T\mathbf{Y}\mathbf{A} \qquad ...(22.27)$$

Above Eq. (22.27) suggests the formulation of $\mathbf{Y}_{bus}$. Since matrix A is singular, $\mathbf{A}^T\mathbf{Y}\mathbf{A}$ is a singular transformation of **Y**. The bus incidence matrix can be obtained through a computer programme. Standard matrix multiplication and matrix transpose sub-routines can be employed to compute $\mathbf{Y}_{bus}$ using Eq. (22.27). As mentioned earlier $\mathbf{Z}_{bus}$ is the inverse of $\mathbf{Y}_{bus}$.

Example 22.5. Formulate Y_{bus} using singular transformation for the network shown in Fig. 22.2(*b*).

Solution :

$$\mathbf{Y} = \begin{bmatrix} y_{10} & & & & & & & & 0 \\ & y_{20} & & & & & & & \\ & & y_{30} & & & & & & \\ & & & y_{40} & & & & & \\ & & & & y_{34} & & & & \\ & & & & & y_{23} & & & \\ & & & & & & y_{12} & & \\ & & & & & & & y_{24} & \\ 0 & & & & & & & & y_{13} \end{bmatrix}$$

Using A from Eq. (22.21), we have

$$\mathbf{YA} = \begin{bmatrix} y_{10} & 0 & 0 & 0 \\ 0 & y_{20} & 0 & 0 \\ 0 & 0 & y_{30} & 0 \\ 0 & 0 & 0 & y_{40} \\ 0 & 0 & y_{34} & -y_{34} \\ 0 & -y_{23} & y_{23} & 0 \\ y_{12} & -y_{12} & 0 & 0 \\ 0 & -y_{24} & 0 & y_{24} \\ -y_{13} & 0 & y_{13} & 0 \end{bmatrix}$$

Finally, $\mathbf{Y}_{bus}$

$= \mathbf{A}^T\,\mathbf{Y}\,\mathbf{A}$

$$= \begin{bmatrix} (y_{10}+y_{12}+y_{13}) & -y_{12} & -y_{13} & 0 \\ -y_{12} & (y_{20}+y_{12}+y_{23}+y_{24}) & -y_{23} & -y_{24} \\ -y_{13} & -y_{23} & (y_{30}+y_{13}+y_{23}+y_{34}) & -y_{34} \\ 0 & -y_{24} & -y_{34} & (y_{40}+y_{24}+y_{34}) \end{bmatrix}$$

The elements of this matrix as obtained above are identical with elements of the matrix determined in Eq. (22.3).

Example 22.6. Compute the bus admittance matrix, primitive admittance matrix, and bus-incidence matrix in a 3-bus power system:

Bus Code	**Admittance**
1-Ground	y_1
2-Ground	y_2
3-Ground	y_3
1-2	$\mathbf{Y}_4$
2-3	$\mathbf{Y}_5$
3-1	$\mathbf{Y}_6$

Y_4, Y_5 and Y_6 are mutually coupled with each other.

[U.P.S.C. I.E.S. Electrical Engineering-II, 2000]

Solution : Assuming that the mutual admittance between elements is y_m

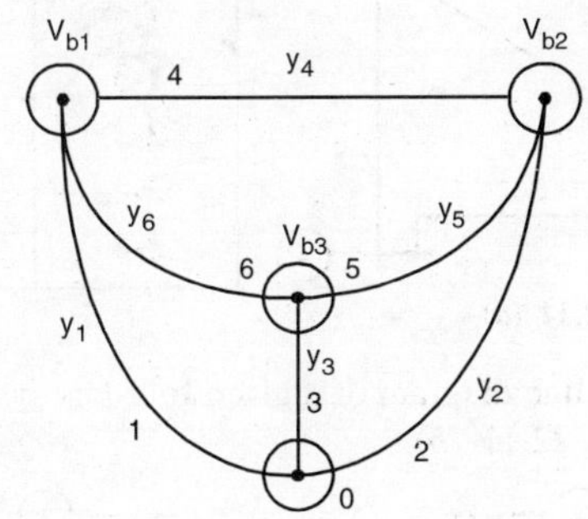

Fig. 22.10

$$\begin{aligned} V_{b1} &= V_1 \\ V_{b2} &= V_2 \\ V_{b3} &= V_3 \\ V_{l4} &= V_1 - V_2 \\ V_{l5} &= V_2 - V_3 \\ V_{l6} &= V_1 - V_3 \end{aligned}$$

$$\mathbf{Y} = \begin{array}{c} \\ 1 \\ 2 \\ 3 \\ 4 \\ 5 \\ 6 \end{array} \begin{array}{c} \begin{array}{cccccc} 1 & 2 & 3 & 4 & 5 & 6 \end{array} \\ \begin{bmatrix} y_1 & & & & & \\ & y_2 & & & & \\ & & y_3 & & & \\ & & & y_4 & y_m & y_m \\ & & & y_m & y_5 & y_m \\ & & & y_m & y_m & y_6 \end{bmatrix} \end{array}$$

$$A = \begin{bmatrix} 1 & 0 & 0 \\ 0 & 1 & 0 \\ 0 & 0 & 1 \\ 1 & -1 & 0 \\ 1 & 0 & -1 \\ 0 & 1 & -1 \end{bmatrix}$$

$$A^T = \begin{bmatrix} 1 & 0 & 0 & 1 & 1 & 1 \\ 0 & 1 & 0 & -1 & 0 & 1 \\ 0 & 0 & 1 & 0 & -1 & -1 \end{bmatrix}$$

where Y is primitive admittance matrix

A is bus incidence matrix

A^T is transpose of matrix

$$Y_{BUS} = A^T Y A$$

Example 22.7. Figure 22.11 (*a*) shows the one line diagram of a simple four-bus system. Table 22.2 gives the line impedance identified by the buses on which these terminate. The shunt admittance at all the buses is assumed negligible. (*i*) Find Y_{bus} assuming that line shown dotted is not connected; (*ii*) What modifications need to be carried out in Y_{bus} if the line shown is connected. [*Rajasthan Technical Univ. Power System Analysis 2011*]

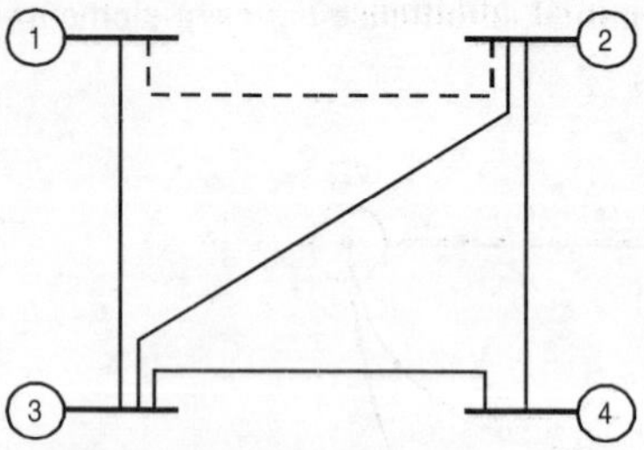

Fig. 22.11 (*a*)

Table 22.2

Line Bus To Bus	R in pu	X in pu
1-2	0.05	0.15
1-3	0.10	0.30
2-3	0.15	0.45
2-4	0.10	0.30
3-4	0.05	0.15

Solution : One line diagram of a given four-bus system is redrawn as shown in Fig. 22.11 (*b*).

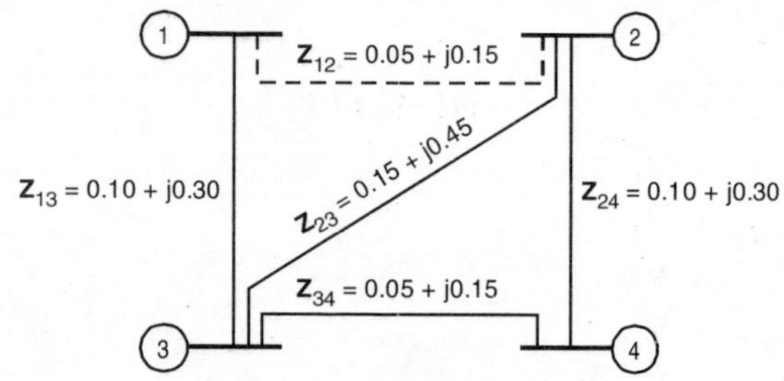

Fig. 22.11 (*b*)

The line admittances are worked out as below :

$$y_{12} = \frac{1}{0.05 + j0.15} = \frac{0.05 - j0.15}{0.025} = 2.0 - j6.0$$

$$y_{23} = \frac{1}{0.15 + j0.45} = \frac{0.15 - j0.45}{0.225} = 0.667 - j2.0$$

$$y_{13} = \frac{1}{0.1 + j0.30} = \frac{0.1 - j0.30}{0.1} = 1.0 - j3.0$$

$$y_{34} = \frac{1}{0.05 + j0.015} = 2.0 - j6.0$$

$$y_{24} = \frac{1}{0.10 + j0.30} = 1.0 - j3.0$$

$y_{10} = y_{20} = y_{30} = y_{40} = 0$ because shunt admittances at buses are negligible.

(*i*) When buses 1 and 2 are not connected, as shown by dotted line, *i.e.*, $y_{12} = 0$

Y_{bus}

$$= \begin{bmatrix} y_{10}+y_{12}+y_{13} & -y_{12} & -y_{13} & 0 \\ -y_{12} & y_{20}+y_{12}+y_{23}+y_{24} & -y_{23} & -y_{24} \\ -y_{13} & -y_{23} & y_{30}+y_{13}+y_{23}+y_{34} & -y_{34} \\ 0 & -y_{24} & -y_{34} & y_{40}+y_{24}+y_{34} \end{bmatrix}$$

$$= \begin{bmatrix} y_{13} & 0 & -y_{13} & 0 \\ 0 & y_{23}+y_{24} & -y_{23} & -y_{24} \\ -y_{13} & -y_{23} & y_{13}+y_{23}+y_{34} & -y_{34} \\ 0 & -y_{24} & -y_{34} & y_{24}+y_{34} \end{bmatrix}$$

$$= \begin{bmatrix} 1.0 - j3.0 & 0 & -1 + j3.0 & 0 \\ 0 & 1.667 - j5.0 & -0.667 + j2.0 & -1.0 + j3.0 \\ -1 + j3.0 & -0.667 + j2.0 & 3.667 - j11.0 & -2.0 + j6.0 \\ 0 & -1.0 + j3.0 & -2.0 + j6.0 & 3 - j9.0 \end{bmatrix} \quad \textbf{Ans.}$$

(*ii*) When buses 1 and 2 are connected, as shown by dotted line, *i.e.*, when $y_{12} = 2.0 - j\,6.0$

$$Y_{bus} = \begin{bmatrix} y_{12}+y_{13} & -y_{12} & -y_{13} & 0 \\ -y_{12} & y_{12}+y_{23}+y_{24} & -y_{23} & -y_{24} \\ -y_{13} & -y_{23} & y_{13}+y_{23}+y_{34} & -y_{34} \\ 0 & -y_{24} & -y_{34} & y_{24}+y_{34} \end{bmatrix}$$

$$= \begin{bmatrix} 3 - j9 & -20 + j6.0 & -1.0 + j3.0 & 0 \\ -2.0 + j6.0 & 3.667 - j11.0 & -0.667 + j2.0 & -1.0 + j3.0 \\ -1 + j3.0 & -0.667 + j2.0 & 3.667 - j11.0 & -2.0 + j6.0 \\ 0 & -1.0 + j3.0 & -2.0 + j6.0 & 3.0 - j9.0 \end{bmatrix}$$

Ans.

Example 22.8. The admittances between the various buses in a power system are given in the table below. Obtain the Y_{bus} matrix.

Buses	Admittances in pu
1-2	0.10
1-4	0.20
2-3	0.25
2-4	0.50
3-1	0.40
4-3	0.50

[*U.P.S.C. I.E.S. Electrical Engineering-II 2007*]

Solution :

$$Y_{bus} = \begin{bmatrix} Y_{11} & Y_{12} & Y_{13} & Y_{14} \\ Y_{21} & Y_{22} & Y_{23} & Y_{24} \\ Y_{31} & Y_{32} & Y_{33} & Y_{34} \\ Y_{41} & Y_{42} & Y_{43} & Y_{44} \end{bmatrix}$$

$$Y_{11} = y_{11} + y_{12} + y_{13} + y_{14} = j(-0.1 - 0.4 - 0.20) = -j\,0.7$$

$$Y_{12} = Y_{21} = -y_{12} = j0.1$$

$$Y_{31} = Y_{13} = -y_{13} = j0.4$$

$$Y_{41} = Y_{14} = -y_{14} = j0.2$$

$$Y_{22} = y_{21} + y_{22} + y_{23} + y_{24} = j\,(-0.1 - 0.25 - 0.5) = -j\,0.85$$

$$Y_{33} = y_{31} + y_{32} + y_{33} + y_{34} = j\,(-0.4 - 0.25 - 0.05) = -j\,1.15$$

$$Y_{44} = y_{41} + y_{42} + y_{43} + y_{44} = j\,(-0.2 - 0.5 - 0.5) = -j1.2$$

$$Y_{23} = Y_{32} = -y_{23} = j0.25$$

$$Y_{34} = Y_{43} = -y_{34} = j0.5$$

$$Y_{42} = Y_{24} = -y_{24} = j0.5$$

Thus, Y_{bus} matrix will be

$$Y_{bus} = \begin{bmatrix} -j0.7 & j0.1 & j0.4 & j0.2 \\ j0.1 & -j0.85 & j0.25 & j0.5 \\ j0.4 & j0.25 & -j1.15 & j0.5 \\ j0.2 & j0.5 & j0.5 & -j1.2 \end{bmatrix} \quad \textbf{Ans.}$$

Example 22.9. The impedance data for a sample power system is given below. Find the admittance matrix of the system in bus frame of reference.

Bus Code	Impedance	Line Charging Admittance
1-2	0.08 + *j*0.24	0.0
1-3	0.02 + *j*0.06	0.0
2-3	0.06 + *j*0.18	0.0

[U.P. Technical Univ. Power System Analysis 2005-06]

Solution : $y_{12} = \dfrac{1}{\mathbf{Z}_{12}} = \dfrac{1}{0.08 + j0.24} = \dfrac{1}{0.253 \angle 71.565}$

$$= 3.95285 \angle -71.565°$$

$$= 1.25 - j3.75$$

$$y_{13} = \frac{1}{\mathbf{Z}_{13}} = \frac{1}{0.02 + j0.06} = \frac{0.02 - j0.06}{0.004}$$

$$= 5 - j15$$

and $y_{23} = \dfrac{1}{\mathbf{Z}_{23}} = \dfrac{1}{0.06 + j0.18} = \dfrac{0.06 - j0.18}{0.036}$

$$= 1.666 - j5.0$$

Nodal admittance matrix is

$$\mathbf{Y}_{bus} = \begin{bmatrix} y_{12} + y_{13} & -y_{12} & -y_{13} \\ -y_{21} & y_{21} + y_{23} & -y_{23} \\ -y_{31} & -y_{32} & y_{32} + y_{31} \end{bmatrix}$$

$$= \begin{bmatrix} 6.25 - j18.75 & -1.25 + j3.75 & -5 + j15 \\ -1.25 + j3.75 & 2.916 - j8.75 & -1.666 + j5.0 \\ -5 + j15 & -1.666 + j5.0 & 6.666 - j20 \end{bmatrix} \quad \textbf{Ans.}$$

Example 22.10. From $\mathbf{Y}_{bus}$ for 4-bus system as shown in Fig. 22.12, if the line series impedances are as follows :

Line (bus to bus)	Impedance
1-2	0.15 + *j*0.6 pu
1-3	0.1 + *j*0.4 pu
1-4	0.15 + *j*0.6 pu
2-3	0.05 + *j*0.2 pu
3-4	0.05 + *j*0.2 pu

Neglect the shunt capacitance of line.

[U.P. Technical univ. Power System Analysis 2009-10]

Solution : $y_{12} = \dfrac{1}{0.15 + j0.6} = \dfrac{0.15 - j0.6}{0.15^2 + 0.6^2}$

$$= \frac{0.15 - j0.6}{0.3825} = 0.392 - j1.5686$$

$$y_{13} = \frac{1}{0.1 + j0.4} = \frac{0.1 - j0.4}{0.1^2 + 0.4^2} = \frac{0.1 - j0.4}{0.17}$$

$$= 0.588 - j2.353$$

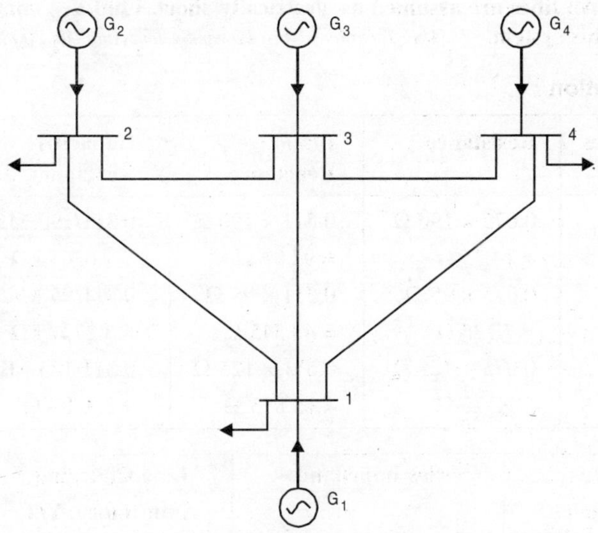

Fig. 22.12

$$y_{14} = \frac{1}{0.15 + j0.6} = \frac{0.15 - j0.6}{0.15^2 + 0.6^2} = \frac{0.15 - j0.6}{0.3825} = 0.392 - j1.5686$$

$$y_{23} = \frac{1}{0.05 + j0.2} = \frac{0.05 - j0.2}{0.05^2 + 0.2^2} = \frac{0.05 - j0.2}{0.0425} = 1.176 - j4.706$$

$$y_{34} = \frac{1}{0.05 + j0.2} = \frac{0.05 - j0.2}{0.0425} = 1.176 - j4.706$$

Nodal admittance matrix is

$$\mathbf{Y}_{bus} = \begin{bmatrix} y_{12} + y_{13} + y_{14} & -y_{12} & -y_{13} & -y_{14} \\ -y_{21} & y_{21} + y_{23} + y_{24} & -y_{23} & -y_{24} \\ -y_{31} & -y_{32} & y_{31} + y_{32} + y_{34} & -y_{34} \\ -y_{41} & -y_{42} & -y_{43} & y_{41} + y_{42} + y_{43} \end{bmatrix}$$

$$= \begin{bmatrix} 1.372 - j5.49 & -0.392 + j1.5686 & -0.588 + j2.353 & -0.392 + j1.5686 \\ -0.392 + j1.5686 & 1.568 - j6.2746 & -1.176 + j4.706 & 0 \\ -0.588 + j2.353 & -1.176 + j4.706 & 2.94 - j11.765 & -1.176 + j4.706 \\ -0.392 + j1.5686 & 0 & -1.176 + j4.706 & 1.568 - j6.2746 \end{bmatrix}$$

Ans.

Example 22.11. Consider the three bus system given in the following Fig. 22.13.

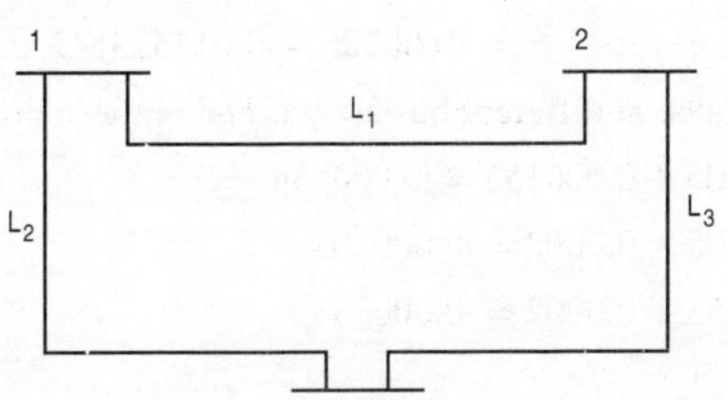

Fig. 22.13

All lines are assumed to have these parameter data.

R = 0.075 Ω/km and phase

ωL = 0.511 Ω/km and phase

1/ωC = 0.311 MΩ/km and phase

L_1 = 190 kms; L_2 = 95 km, L_3 = 125 kms.

All lines are assumed as electrically short. Find Y_{bus} matrix for this system. *[U.P. Technical Univ. Power System Analysis 2008-09]*

Solution :

Lines	Resistance	Inductive Reactance	Capacitive Reactance
1-2	0.075 × 190 Ω = 14.25 Ω	0.511 × 190 Ω = 97.09 Ω	0.311/190 MΩ = 1.6368 kΩ
1-3	0.075 × 95 Ω = 7.125 Ω	0.511 × 95 Ω = 48.545 Ω	0.311/95 MΩ = 3.2737 kΩ
2-3	0.075 × 125 Ω = 9.375 Ω	0.511 × 125 Ω = 63.875 Ω	0.311/125 MΩ = 2.488 kΩ

Bus code	Series impedance	Line Charging Admittance, Y/2
1-2	(14.25 + j97.09) Ω	j0.000305 ℧
1-3	(7.125 + j48.545) Ω	j0.000153 ℧
2-3	(9.375 + j63.875) Ω	j0.0002 ℧

Line series admittances are worked out as below :

$$y_{12} = \frac{1}{14.25 + j97.09} = \frac{14.25 - j97.09}{14.25^2 + 97.09^2}$$

$$= \frac{14.25 - j97.09}{9,629.53}$$

$$= (0.00148 - j0.01008)\ ℧$$

$$y_{13} = \frac{1}{7.125 + j48.545} = \frac{7.125 - j48.545}{7.125^2 + 48.545^2}$$

$$= \frac{7.125 - j48.545}{2,407.38}$$

$$= (0.00296 - j0.02016)\ ℧$$

$$y_{23} = \frac{1}{9.375 + j63.875} = \frac{9.375 - j63.875}{9.375^2 + 63.875^2}$$

$$= \frac{9.375 - j63.875}{4,167.9}$$

$$= (0.00225 - j0.015325)\ ℧$$

Shunt admittance at different bus are worked out as below :

$$y_{10} = 0.000305 + 0.000153 = j0.000458$$

$$y_{20} = 0.000305 + 0.0002 = j0.000505$$

$$y_{30} = 0.000153 + 0.0002 = j0.000353$$

$$Y_{bus} = \begin{bmatrix} y_{10} + y_{12} + y_{13} & -y_{12} & -y_{13} \\ -y_{12} & y_{20} + y_{12} + y_{23} & -y_{23} \\ -y_{13} & -y_{23} & y_{30} + y_{13} + y_{23} \end{bmatrix}$$

$$= \begin{bmatrix} 0.00444 - j0.02978 & -0.00148 + j0.01008 & -0.00296 + j0.02016 \\ -0.00148 + j0.01008 & 0.00373 + j0.0249 & -0.00225 + j0.015325 \\ -0.00296 + j0.02016 & -0.00225 + j0.015325 & 0.00521 - j0.03513 \end{bmatrix}$$

mho **Ans.**

Example 22.12. The parameters of a 4 bus system are as under :

Bus Code	Line Impedance	Charging Admittance (pu)
1-2	0.2 + j 0.8	j 0.02
2-3	0.3 + j 0.9	j 0.03
2-4	0.25 + j 1.0	j 0.04
3-4	0.2 + j 0.8	j 0.02
1-3	0.1 + j 0.4	j 0.01

Draw the network and find bus admittance matrix.

[Power System Analysis Rajasthan Univ. 2001, 2007; Rajasthan Technical Univ. 2011]

Solution : The network is shown in Fig. 22.14.

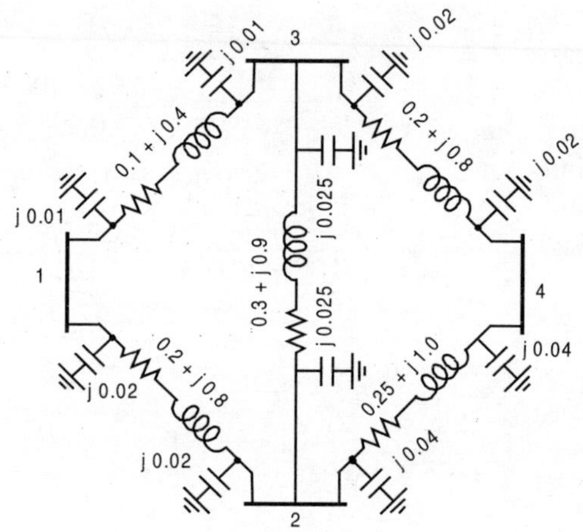

Fig. 22.14

The line series admittances are worked out as below :

$$\mathbf{y}_{12} = \frac{1}{0.2 + j0.8} = \frac{0.2 - j0.8}{0.68} = (0.294 - j1.176) \text{ pu}$$

$$\mathbf{y}_{13} = \frac{1}{0.1 + j0.4} = \frac{0.1 - j0.4}{0.17} = (0.588 - j2.352) \text{ pu}$$

$$\mathbf{y}_{23} = \frac{1}{0.3 + j0.9} = \frac{0.3 - j0.9}{0.9} = (0.333 - j1.0) \text{ pu}$$

$$\mathbf{y}_{24} = \frac{1}{0.25 + j1.0} = \frac{0.25 - j1.0}{1.0625} = (0.235 - j0.94) \text{ pu}$$

$$\mathbf{y}_{34} = \frac{1}{0.2 + j0.8} = \frac{0.2 - j0.8}{0.68} = (0.294 - j1.176) \text{ pu}$$

The shunt admittances at different buses are worked out as below:

$$\mathbf{y}_{10} = j\,0.01 + j\,0.02 = j\,0.03 \text{ pu}$$

$$\mathbf{y}_{20} = j\,0.02 + j\,0.04 + j\,0.03 = j\,0.09 \text{ pu}$$

$$\mathbf{y}_{30} = j\,0.01 + j\,0.02 + j\,0.03 = j\,0.06 \text{ pu}$$

$$\mathbf{y}_{40} = j\,0.02 + j\,0.04 = j\,0.06 \text{ pu}$$

$$\mathbf{Y}_{bus} = \begin{bmatrix} \mathbf{y}_{12}+\mathbf{y}_{13}+\mathbf{y}_{10} & -\mathbf{y}_{12} & -\mathbf{y}_{13} & 0 \\ -\mathbf{y}_{12} & \mathbf{y}_{12}+\mathbf{y}_{23}+\mathbf{y}_{24}+\mathbf{y}_{20} & -\mathbf{y}_{23} & -\mathbf{y}_{24} \\ -\mathbf{y}_{13} & -\mathbf{y}_{23} & \mathbf{y}_{13}+\mathbf{y}_{23}+\mathbf{y}_{34}+\mathbf{y}_{30} & -\mathbf{y}_{34} \\ 0 & -\mathbf{y}_{24} & -\mathbf{y}_{34} & \mathbf{y}_{24}+\mathbf{y}_{34}+\mathbf{y}_{40} \end{bmatrix}$$

$$= \begin{bmatrix} 0.882 - j3.498 & -0.294 + j1.176 & -0.588 + j2.352 & 0 \\ -0.294 + j1.176 & 0.862 - j3.026 & -0.333 + j1.0 & -0.235 + j0.94 \\ -0.588 + j2.352 & -0.333 + j1.0 & 1.215 - j4.468 & -0.294 + j1.176 \\ 0 & -0.235 + j0.94 & -0.294 + j1.176 & 0.529 - j2.056 \end{bmatrix}$$

Ans.

22.6. REPRESENTATION OF TRANSFORMERS

As we know that transformer is a static machine which is used to transform electric power from one circuit to another without change in frequency. It may raise or lower the voltage in a circuit but with a corresponding decrease or increase in current. Thus transformer plays a vital role in power systems. Practically every transformer is provided with taps for ratio control, and thus it is also used for controlling secondary voltage level. Most of the transformers are provided with *off-load tap-changers* while some of the transformers are provided with *on-load tap-changers*.

A power transformer with turn-ratio K can be represented by its impedance $\mathbf{Z}_L$ or admittance $\mathbf{Y}_L$ in series with an ideal transformer, as illustrated in Fig. 22.15. However, this model does not serve the purpose here. Though it may prove useful in other cases. We shall write the injected power equations in terms of the network bus admittance matrix, which is only possible if all branch elements (including transformer) are resolved into equivalent π-network; *i.e.*, transformer model depicted in Fig. 22.15 needs transformation into the model depicted in Fig. 22.16.

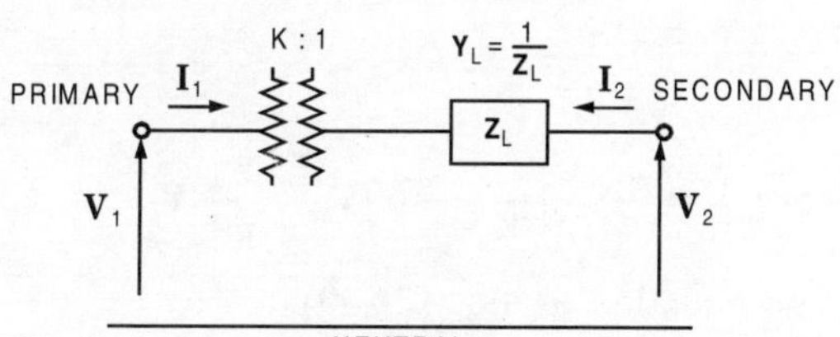

Fig. 22.15. ***Transformer Model***

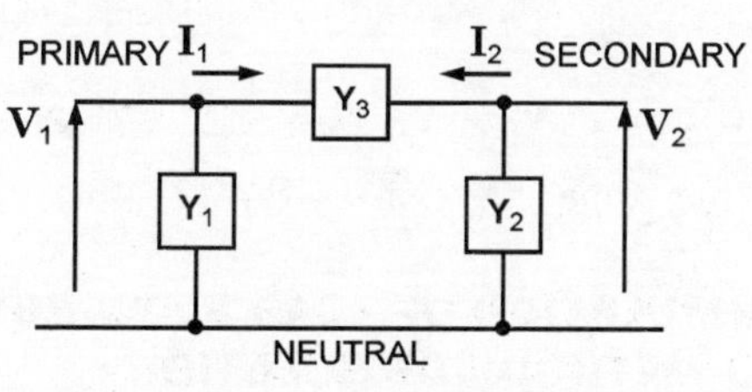

Fig. 22.16. ***π-Equivalent Model of Fig. 22.15***

The parameter of the π-equivalent model can be derived by equating the terminal currents of the transformer model shown in Fig. 22.15 with the corresponding currents of the π-equivalent model given in Fig. 22.16.

From the transformer model shown in Fig. 22.15, we have

$$\mathbf{I}_1 = -\frac{\mathbf{I}_2}{K} \qquad \text{...(22.28)}$$

and $$\mathbf{I}_2 = \left(\mathbf{V}_2 - \frac{\mathbf{V}_1}{K}\right)\mathbf{Y}_L = \mathbf{Y}_L\,\mathbf{V}_2 - \mathbf{Y}_L\frac{\mathbf{V}_1}{K} \qquad \text{...(22.29)}$$

From Eqs. (22.28) and (22.29), we have

$$\mathbf{I}_1 = -\frac{\mathbf{Y}_L}{K}\mathbf{V}_2 + \mathbf{Y}_L\frac{\mathbf{V}_1}{K^2} \qquad \text{...(22.30)}$$

where $\mathbf{Z}_L$ = Leakage impedance of transformer

$\mathbf{Y}_L = \frac{1}{\mathbf{Z}_L}$, Leakage admittance of transformer

K = Turn-ratio of transformer = $\frac{N_1}{N_2}$

From the π-equivalent model of transformer depicted in Fig. 22.16, we have

$$\mathbf{I}_1 = (\mathbf{Y}_1 + \mathbf{Y}_3)\mathbf{V}_1 - \mathbf{Y}_3\mathbf{V}_2 \qquad \text{...(22.31)}$$

$$\mathbf{I}_2 = -\mathbf{Y}_3\mathbf{V}_1 + (\mathbf{Y}_2 + \mathbf{Y}_3)\mathbf{V}_2 \qquad \text{...(22.32)}$$

Equating like coefficients, we have

$$\mathbf{Y}_1 + \mathbf{Y}_3 = \frac{\mathbf{Y}_L}{K^2} \qquad \text{...(22.33)}$$

$$\mathbf{Y}_3 = \frac{\mathbf{Y}_L}{K} \qquad \text{...(22.34)}$$

$$\mathbf{Y}_2 + \mathbf{Y}_3 = \mathbf{Y}_L \qquad \text{...(22.35)}$$

From the above equations, we have

$$\mathbf{Y}_1 = \frac{\mathbf{Y}_L}{K^2} - \frac{\mathbf{Y}_L}{K} = \frac{\mathbf{Y}_L}{K^2}(1 - K) \qquad \text{...(22.36)}$$

$$\mathbf{Y}_2 = \frac{\mathbf{Y}_L}{K}(K - 1) \qquad \text{...(22.37)}$$

and $$\mathbf{Y}_3 = \frac{\mathbf{Y}_L}{K} \qquad \text{...(22.38)}$$

So our mathematical model in matrix form comes out to be

$$\begin{bmatrix} \mathbf{I}_1 \\ \mathbf{I}_2 \end{bmatrix} = \begin{bmatrix} \frac{\mathbf{Y}_L}{K^2} & -\frac{\mathbf{Y}_L}{K} \\ \frac{-\mathbf{Y}_L}{K} & \mathbf{Y}_L \end{bmatrix} \begin{bmatrix} \mathbf{V}_1 \\ \mathbf{V}_2 \end{bmatrix} \qquad \text{...(22.39)}$$

With the above equation and the equivalent circuit depicted in Fig. 22.16, desired requirement in load flow equation is complete.

22.6.1. Auto-Transformer and Phase-Shifter in Tandem

Now let us make a model for a phase-shifting transformer connected in series with an auto-transformer, as shown in Fig. 22.17.

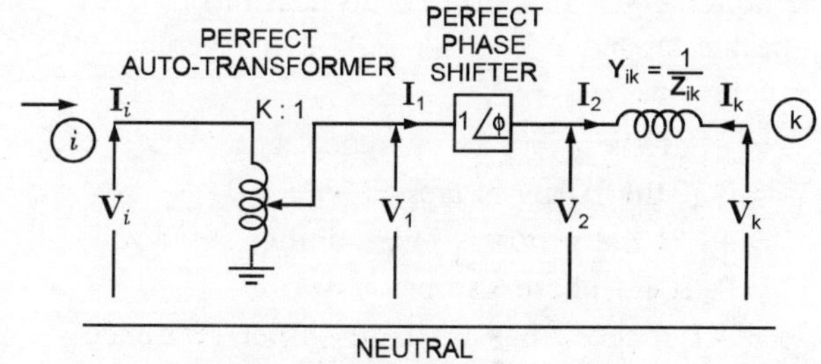

Fig. 22.17. ***Auto-Transformer and Phase-Shifter in Tandem***

Phase-shifter is used for regulating power flow through a line element or occasionally through a set of lines and also in the area interchange control.

For the circuit given in Fig. 22.17, we have

$$\mathbf{I}_k = \mathbf{Y}_{ik}(\mathbf{V}_k - \mathbf{V}_2) \quad \text{...(22.40)}$$

$$\mathbf{I}_2 = \mathbf{Y}_{ik}(\mathbf{V}_2 - \mathbf{V}_k) \quad \text{...(22.41)}$$

where $\mathbf{Y}_{ik} = \frac{1}{\mathbf{Z}_{ik}}$, equivalent line impedance (line impedance plus leakage impedances of auto-transformer and phase-shifting transformer) between nodes (or buses) *i* and k. Shunt admittance of the line is neglected.

At the phase-shifter

$$\mathbf{V}_2 = (1\angle\phi)\,\mathbf{V}_1 \quad \text{...(22.42)}$$

$$\mathbf{V}_1\mathbf{I}_1^* = \mathbf{V}_2\,\mathbf{I}_2^* \quad \text{...(22.43)}$$

At the auto-transformer

$$\mathbf{K}\mathbf{V}_1 = \mathbf{V}_i \quad \text{...(22.44)}$$

$$\mathbf{V}_1\mathbf{I}_1^* = \mathbf{V}_i\,\mathbf{I}_i^* \quad \text{...(22.45)}$$

From Eqs. (22.40), (22.42) and (22.44), we have

$$\mathbf{I}_k = \mathbf{Y}_{ik}\mathbf{V}_k - \mathbf{Y}_{ik}\,(1/\mathrm{K}\,\angle\phi)\,\mathbf{V}_i \quad \text{...(22.46)}$$

From Eqs. (22.41), (22.43), and (22.45), we have

$$\mathbf{I}_i^* = \frac{\mathbf{V}_1}{\mathbf{V}_i}\left[\frac{\mathbf{V}_2}{\mathbf{V}_1}(\mathbf{Y}_{ik}\mathbf{V}_2 - \mathbf{Y}_{ik}\mathbf{V}_k)\right] \quad \text{...(22.47)}$$

Using Eqs. (22.42) and (22.44), Eq. (22.47) may be reduced to

$$\mathbf{I}_i = (1/\mathrm{K}\,\angle{-\phi})\,(\mathbf{Y}_{ik}\mathbf{V}_2 - \mathbf{Y}_{ik}\mathbf{V}_k) \quad \text{...(22.48)}$$

Again using Eqs. (22.42) and (22.44), Eq. (22.48) is further reduced to

$$\mathbf{I}_i = -\,\mathbf{Y}_{ik}\,[1/\mathrm{K}\angle{-\phi}\,\mathbf{V}_k] + \mathbf{Y}_{ik}\mathbf{V}_i/\mathrm{K}^2 \quad \text{...(22.49)}$$

Now Eqs. (22.46) and (22.49) may be written in matrix form as below:

$$\begin{bmatrix}\mathbf{I}_i\\ \mathbf{I}_k\end{bmatrix} = \begin{bmatrix}\dfrac{\mathbf{Y}_{ik}}{\mathrm{K}^2} & -\mathbf{Y}_{ik}\left(\dfrac{1}{\mathrm{K}}\angle{-\phi}\right)\\ -\mathbf{Y}_{ik}\left(\dfrac{1}{\mathrm{K}}\angle\phi\right) & \mathbf{Y}_{ik}\end{bmatrix}\begin{bmatrix}\mathbf{V}_i\\ \mathbf{V}_k\end{bmatrix} \quad \text{...(22.50)}$$

where ϕ is phase shift angle between $\mathbf{V}_2$ and $\mathbf{V}_1$.

For the use of above Eq. (22.50) it is necessary to consider the reference bus for the phase shifter and the tap-side bus for the tap-changer. It is noteworthy that in Fig. 22.17 either could be the tapped side or phase-shifter reference. We may define here that

$\in = +1$ if the *p* bus is untapped side

$= 0$ if the *p* bus is tapped side

$\in' = +1$ if the *p* bus is phase-shifter reference

$= 0$, if no phase shifter

$= -1$ if the *q* bus is phase-shifter reference

Thus Fig. 22.18 represents complete model expressed in terms of equivalent admittances looking from the *p* bus or written in equation form as

$$\mathbf{I}_p = \mathbf{Y}_{pq}\left[1+\in\left(\frac{1}{\mathrm{K}^2}-1\right)\right]\mathbf{V}_p - \frac{1}{\mathrm{K}}\,\mathbf{Y}_{pq}\,(1\angle{-\in\phi})\,\mathbf{V}_q \quad \text{...(22.51)}$$

K = 1 in case there is no auto-transformer and only phase-shifter is present.

$\in' = 0$ or $\phi = 0$ if there is no phase-shifter and only auto-transformer is present.

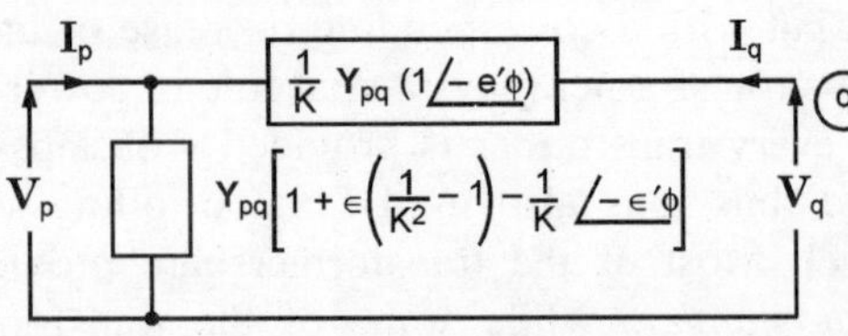

Fig. 22.18. *Model of Auto-Transformer and Phase-Shifter in Tandem*

It is to be noted that Eq. (22.50) in matrix form is not the π-equivalent model of our power system as is obvious from the admittance matrix elements. Mathematical model represented by Eq. (22.51) can be easily used in the construction of bus admittance matrix of the power system under power flow studies.

Let us derive Eqs. (22.46) and (22.49) from our general mathematical Eq. (22.51) as an illustration.

For determining $\mathbf{I}_i$, we substitute in Eq. (22.51)

$\in = 1 \qquad p \to i$ ...(22.52)

$\in' = 1 \qquad q \to \mathrm{k}$

and we have $\mathbf{I}_i = -\mathbf{Y}_{ik}\left(\dfrac{1}{\mathrm{K}}\angle{-\phi}\right)\mathbf{V}_k + \dfrac{\mathbf{Y}_{ik}}{\mathrm{K}^2}\mathbf{V}_i$ ...(22.53)

this is same equation as Eq. (22.49).

For determining $\mathbf{I}_k$, we substitute in Eq. (22.51)

$\in = 0 \qquad p \to \mathrm{k}$ ...(22.54)

$\in' = -1 \qquad q \to i$

and we have $\mathbf{I}_k = \mathbf{Y}_{ik}\mathbf{V}_k - \mathbf{Y}_{ik}\,(1/\mathrm{K}\,\angle\phi)\,\mathbf{V}_i$ ...(22.55)

This is the same equation as Eq. (22.46).

22.7. FORMULATION OF LOAD FLOW EQUATIONS AND METHODS OF SOLUTION

The complex power injected by the generating source into the *i*th bus of a power system is given as

$$\mathbf{S}_i = \mathrm{P}_i + j\,\mathrm{Q}_i = \mathbf{V}_i\,\mathbf{I}_i^* \quad i = 1, 2, \ldots n \quad \text{...(22.56)}$$

where V_i is the voltage at the *i*th bus with respect to ground and $\mathbf{I}_i^*$ is the complex conjugate of source current $\mathbf{I}_i$ injected into the bus.

It is convenient to handle load flow problem by using $\mathbf{I}_i$ rather than $\mathbf{I}_i^*$. So, taking the complex conjugate of Eq. (22.56), we have

$$\mathbf{S}_i^* = \mathrm{P}_i - j\mathrm{Q}_i = \mathbf{V}_i^*\,\mathbf{I}_i\,;\; n = 1, 2, 3, \ldots n \quad \text{...(22.57a)}$$

Substituting $\mathbf{I}_i = \sum_{\mathrm{k}=1}^{n}\mathbf{Y}_{ik}\mathbf{V}_k$ from Eq. (22.8) in above equation, we have

$$\mathbf{S}_i^* = \mathrm{P}_i - j\mathrm{Q}_i = \mathbf{V}_i^*\sum_{\mathrm{k}=1}^{n}\mathbf{Y}_{ik}\mathbf{V}_k\,;\; i = 1, 2, \ldots n \quad \text{...(22.57b)}$$

Equating real and imaginary parts, we have

$$\text{Real power} = P_i = R_e\left\{\mathbf{V}_i^* \sum_{k=1}^{n} \mathbf{Y}_{ik}\,\mathbf{V}_k\right\} \quad ...(22.58a)$$

$$\text{Reactive power} = Q_i = -I_m\left\{\mathbf{V}_i^* \sum_{k=1}^{n} \mathbf{Y}_{ik}\mathbf{V}_k\right\} \quad ...(22.58b)$$

In polar form $\mathbf{V}_i = V_i \angle\delta_i$, $\mathbf{V}_i^* = V_i \angle -\delta_i$

and $\mathbf{Y}_{ik} = Y_{ik} \angle\theta_{ik}$,

So real and reactive powers can now be expressed as

$$\text{Real power, } P_i = V_i \sum_{k=1}^{n} V_k\, Y_{ik} \cos(\theta_{ik} + \delta_k - \delta_i) \quad ...(22.59)$$

$$\text{Reactive power, } Q_i = -V_i \sum_{k=1}^{n} V_k Y_{ik} \sin(\theta_{ik} + \delta_k - \delta_i) \quad ...(22.60)$$

Above Eqs. (22.59) and (22.60) are known as *static load flow equations*. (SLFE). These equations are nonlinear equations and, therefore, only a numerical solution is possible. For each of the n system buses we have two such equations giving a total of $2n$ equations (n real flow power equations and n reactive power flow equations). Each bus is characterized by four variables P_i, Q_i, V_i and δ_i giving a total of $4n$ variables. To obtain a solution it is necessary to specify two variables at each bus so that the number of unknowns is reduced to $2n$. Evidently we should specify the variables over which we have physical control. The choice is influenced somewhat by the devices which are connected to a particular bus. Depending upon the quantities specified, the buses can be classified into three types, as already discussed in Art. 22.2.

The load buses are most common in power system. At these buses P_i and Q_i are known because P_{Di}, Q_{Di} are known from the load forecast data and P_{Gi} and Q_{Gi} are either zero (no generation at these buses) or specified. At the generator buses P_{Gi} and V_i are specified as these can be controlled by governor control and excitation control. P_i is known because P_{Di} is known from load forecast data. The need and importance of swing or slack bus is already explained in Art 22.2. Voltage at the slack bus is usually specified to be equal to 1 pu as the voltages throughout the system must be close to 1 pu. From elementary ac theory we know that any one phasor can be taken as reference and so the voltage of the slack or swing bus may be taken as reference phasor making its angle δ_i zero.

The solution to the power flow problem consists in assuming a certain initial bus load configuration, specifying the $2n$ known variables as discussed above and using some numerical method of determination of remaining $2n$ variables for the system with known Y_{bus} matrix. The final solution must satisfy some constraints such as (*i*) voltage magnitude at different buses must be within limits (*ii*) the real and reactive generator power at different buses must be within the minimum and maximum limits (*iii*) total generation must be equal to total load demand plus losses. In addition, from the point of view of system stability, δ_i cannot exceed a certain magnitude.

It has been demonstrated that load flow equations being essentially nonlinear algebraic equations, have to be solved through iterative numerical techniques. The most commonly used methods for power flow studies will be discussed in the forthcoming Arts.

At the cost of solution accuracy, it is possible to linearize power flow equations by making suitable assumptions and approximations so that fast and explicit solutions become possible.

22.7.1. Approximate Load Flow Study

Let us make the following assumptions and approximations in load flow Eqs. (22.59) and (22.60):

1. Line resistances being smaller are neglected. Shunt conductances of the overhead lines are always negligible. By making this assumption active power loss in the lines becomes zero and complexity of the equations is reduced because active power generation becomes equal to total active power demand. The effect of this assumption on Eqs. (22.59) and (22.60) is that $\theta_{ik} \simeq 90°$ and $\theta_{ii} = -90°$.

2. The angle δ_i is so small that $\sin\delta_i = \delta_i$. This approximation converts the nonlinear power flow equations into linear ones therefore, makes analytical solution possible.

3. All buses leaving the slack or swing bus (numbered as bus 1) are voltage controlled buses, *i.e.*, voltage magnitudes at all the buses including the slack bus are specified.

With the above assumptions and approximations, Eqs. (22.59) and (22.60) are modified to

$$P_i = V_i \sum_{k=1}^{n} V_k Y_{ik} (\delta_i - \delta_k) \quad ...(22.61)$$

$$Q_i = -V_i \sum_{\substack{k=1\\k\neq i}}^{n} V_k\, Y_{ik} \cos(\delta_i - \delta_k) + V_i^2 Y_{ii} \quad ...(22.62)$$

Since V_is are specified, Eq. (22.61) represents a set of linear algebraic equations. The only unknowns are angles δ. For slack or swing bus (bus no. 1) $\delta_i = 0$, therefore, we have $(n - 1)$ linear equations from which the values of δ at all buses can be determined. These values of δ when put in set of Eq. (22.62) provide the values of Q_i at all buses.

It is noteworthy that by making assumptions and approximations the Eqs. (22.61) and (22.62) have decoupled and there is no need of solving them simultaneously but can be solved sequentially [solution of Eq. (22.62) follows immediately upon simultaneous solution of Eq. (22.61)]. Since the solution is non-iterative and the dimension is reduced to $(n - 1)$ from $2n$, it is computationally highly economical.

Example 22.13. The one line diagram of a simple three-bus power system with generator at bus 1 is given in Fig. 22.19. The line impedances are marked in per unit on 100 MVA base. The voltage obtained by performing load flow are:

$$\mathbf{V_1 = 1.05 \angle 0° \text{ pu};}$$

$$\mathbf{V_2 = 0.98183 \angle -3.5035° \text{ pu};}$$

$$\mathbf{V_3 = 1.00125 \angle -2.8624° \text{ pu}}$$

Determine (*i*) slack bus power, (*ii*) complex power flow on line 1-2. [*U.P.S.C. I.E.S. Electrical Engineering-II, 2010*]

Solution : $V_1 = 1.05$ pu

$V_2 = 0.98183$ pu

$V_3 = 1.00125$ pu

$\delta_1 = 0°$

$\delta_2 = -3.5035°$

and $\delta_3 = -2.8624°$

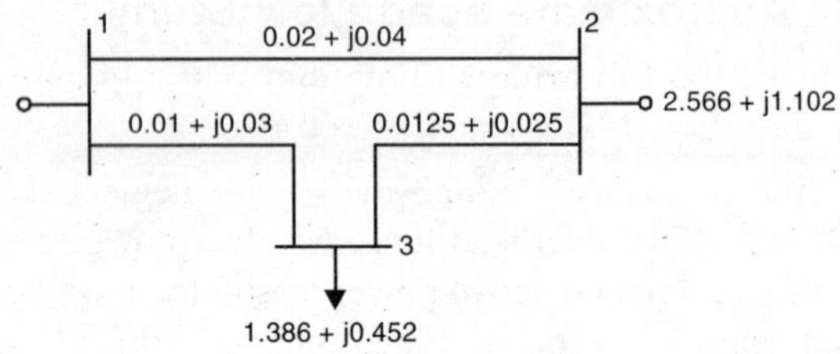

Fig. 22.19

Define impedances

$\mathbf{Z}_{12} = (0.02 + j0.04)$ pu

$\mathbf{Z}_{13} = (0.01 + j0.03)$ pu

$\mathbf{Z}_{23} = (0.0125 + j0.025)$ pu

Line admittances,

$$\mathbf{y}_{12} = \frac{1}{\mathbf{Z}_{12}} = \frac{1}{0.02 + j0.04} = 10 - j20 = 22.36 \angle -63.435° \text{ pu}$$

$$\mathbf{y}_{13} = \frac{1}{\mathbf{Z}_{13}} = \frac{1}{0.01 + j0.03} = (10 - j30) = 31.62 \angle -71.565° \text{ pu}$$

$$\mathbf{y}_{23} = \frac{1}{\mathbf{Z}_{23}} = \frac{1}{0.0125 + j0.025} = (16 - j32) = 35.78 \angle -63.433° \text{ pu}$$

$$Y_{bus} = \begin{bmatrix} \mathbf{Y}_{11} & \mathbf{Y}_{12} & \mathbf{Y}_{13} \\ \mathbf{Y}_{21} & \mathbf{Y}_{22} & \mathbf{Y}_{23} \\ \mathbf{Y}_{31} & \mathbf{Y}_{32} & \mathbf{Y}_{33} \end{bmatrix}$$

Elements of Y_{bus} are

$$\mathbf{Y}_{11} = \mathbf{y}_{12} + \mathbf{y}_{13} = (10 - j20) + (10 - j30)$$

$$= (20 - j50) = 53.85 \angle -68.2° \text{ pu}$$

$$\mathbf{Y}_{22} = \mathbf{y}_{12} + \mathbf{y}_{23} = (10 - j20) + (16 - j32)$$

$$= (26 - j52) = 58.13 \angle -63.435° \text{ pu}$$

$$\mathbf{Y}_{33} = \mathbf{y}_{13} + \mathbf{y}_{23} = (10 - j30) + (16 - j32)$$

$$= (26 - j62) = 67.23 \angle -67.25° \text{ pu}$$

$$\mathbf{Y}_{12} = \mathbf{Y}_{21} = -\mathbf{y}_{12} = 22.36 \angle 180° - 63.43° = 22.36 \angle 116.57° \text{ pu}$$

$$\mathbf{Y}_{13} = \mathbf{Y}_{31} = -\mathbf{y}_{13} = 31.62 \angle 180° - 71.565° = 31.62 \angle 108.435° \text{ pu}$$

$$\mathbf{Y}_{23} = \mathbf{Y}_{32} = -\mathbf{y}_{23} = 35.78 \angle 180° - 63.435° = 35.78 \angle 116.565° \text{ pu}$$

Now, power injected at ith bus

$$P_i = V_i \sum_{k=1}^{n} V_k Y_{ik} \cos(\theta_{ik} + \delta_k - \delta_i)$$

$$Q_i = -V_i \sum_{k=1}^{n} V_k Y_{ik} \sin(\theta_{ik} + \delta_k - \delta_i)$$

Slack bus power,

$$P_1 = V_1[V_1 Y_{11} \cos(\theta_{11} + \delta_1 - \delta_1) + V_2 Y_{12} \cos(\theta_{12} + \delta_2 - \delta_1) + V_3 Y_{13} \cos(\theta_{13} + \delta_3 - \delta_1)]$$

$$= 1.05[1.05 \times 53.85 \cos(-68.2°) + 0.98183 \times 22.36 \cos(116.57° + (-3.5035°) - 0 + 1.00125 \times 31.62 \cos(108.435° + (-2.8624°) - 0)]$$

$$= 1.05[20.998 - 8.597 - 8.499] = 4.097 \text{ pu}$$

$$Q_1 = -V_1[V_1 Y_{11} \sin(\theta_{11} + \delta_1 - \delta_1) + V_2 Y_{12} \sin(\theta_{12} + \delta_2 - \delta_1) + V_3 Y_{13} \sin(\theta_{13} + \delta_3 - \delta_1)]$$

$$= -1.05[1.05 \times 53.85 \sin(-68.2°) + 0.98183 \times 22.36 \sin(116.57° - 3.5035° - 0°) + 1.00125 \times 31.62 \sin(108.435° - 2.8624° - 0°)]$$

$$= -1.05[-52.4989 + 20.1985 + 30.4973] = 1.8932 \text{ pu}$$

Slack bus power $= \mathbf{S}_1 = P_1 + jQ_1 = (4.097 + j1.8932)$ pu

MVA base = 100 MVA

So, $\mathbf{S}_1 = (409.7 + j189.32)$ MVA **Ans.**

Complex power on line 1-2

$$\mathbf{S}_{12} = V_1 I_{12}^*$$

where
$$\mathbf{I}_{12} = \frac{\mathbf{V}_1 - \mathbf{V}_2}{\mathbf{Z}_{12}} = \frac{1.05 \angle 0° - 0.98183 \angle -3.5035}{0.02 + j0.04}$$

$$= 2.062 \angle -22.835° \text{ pu}$$

$$\mathbf{S}_{12} = \mathbf{V}_1 \mathbf{I}_{12}^* = 1.05 \angle 0° \times 2.062 \angle 22.835° \text{ pu}$$

$$= 2.165 \angle 22.835 \text{ pu} = (1.995 + j0.84) \text{ pu}$$

MVA base = 100

So, $\mathbf{S}_{12} = (199.5 + j84)$ MVA **Ans.**

Example 22.14. A 4-bus system is shown in Fig. 22.20. The magnitudes of the bus voltages at all the buses are 1 per unit each. The per unit values of line admittances, loads and active power generated at the different buses are indicated on the diagram. Determine (*i*) reactive power generation at buses (*ii*) active and reactive power losses and (*iii*) power flow in each of the transmission lines in the network.

Solution : As bus voltages are specified, it means that all the buses have controllable Q sources. It is also obvious that at buses 2 and 4 there is no active power generation but these buses have reactive power compensation equipment. Further since there is no active power losses (line resistances being negligible), the real power generation P_{G_1} at bus 1 can be determined straight way

$$P_{G_1} = P_{D_1} + P_{D_2} + P_{D_3} + P_{D_4} - P_{G_2} - P_{G_3} - P_{G_4}$$

$$= 1.5 + 2 + 0 + 2 - 0 - 3.5 - 0 = 2 \text{ pu}$$

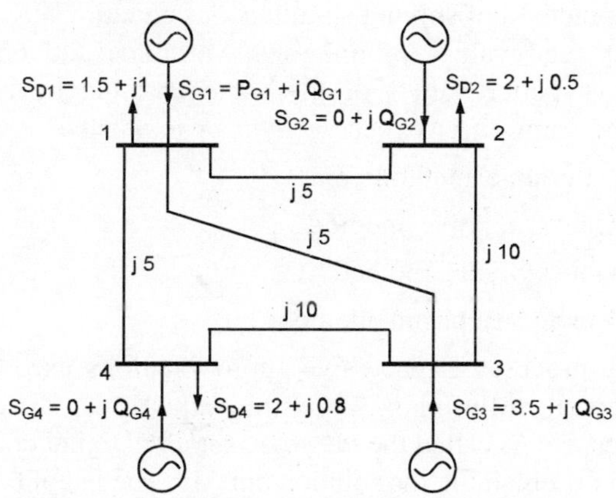

Fig. 22.20. ***4-Bus System For Example 22.14***

Assuming bus 1 as the reference bus, we have 7 unknowns. In the present problem the unknown state and controllable variables are δ_2, δ_3, δ_4, Q_{G_1}, Q_{G_2}, Q_{G_3} and Q_{G_4}.

Though the active power losses are zero, but there will be reactive losses. So total reactive generation will be more than the total reactive demand (*i.e.*, 1.0 + 0.5 + 0.8 = 2.3 pu)

The active power injected into the buses are

$$P_1 = P_{G_1} - P_{D_1} = 2 - 1.5 = 0.5 \text{ pu}$$

$$P_2 = P_{G_2} - P_{D_2} = 0 - 2 = -2 \text{ pu}$$

$$P_3 = P_{G_3} - P_{D_3} = 3.5 - 0 = 3.5 \text{ pu}$$

$$P_4 = P_{G_4} - P_{D_4} = 0 - 2 = -2 \text{ pu}$$

The admittances between different buses are

$$\mathbf{y}_{12} = \mathbf{y}_{13} = \mathbf{y}_{14} = -j\,5.0 \text{ pu}$$

$$\mathbf{y}_{23} = \mathbf{y}_{34} = -j\,10.0 \text{ pu}$$

Self admittances

$$\mathbf{Y}_{11} = \mathbf{y}_{12} + \mathbf{y}_{13} + \mathbf{y}_{14} = -j\,5 + (-j\,5) + (-j\,5) = -j\,15 \text{ pu}$$

$$\mathbf{Y}_{22} = \mathbf{y}_{12} + \mathbf{y}_{23} = -j\,5 - j\,10 = -j\,15 \text{ pu}$$

$$\mathbf{Y}_{33} = \mathbf{y}_{32} + \mathbf{y}_{31} + \mathbf{y}_{34} = -j\,10 - j\,5 - j\,10 = -j\,25 \text{ pu}$$

$$\mathbf{Y}_{44} = \mathbf{y}_{43} + \mathbf{y}_{41} = -j\,10 - j\,5 = -j\,15 \text{ pu}$$

$$\mathbf{Y}_{12} = \mathbf{Y}_{21} = \mathbf{Y}_{13} = \mathbf{Y}_{31} = \mathbf{Y}_{14} = \mathbf{Y}_{41} = j\,5 \text{ pu}$$

$$\mathbf{Y}_{23} = \mathbf{Y}_{32} = \mathbf{Y}_{34} = \mathbf{Y}_{43} = j\,10 \text{ pu}$$

Using Eq. (22.4), $\mathbf{Y}_{bus}$ can be written as

$$\mathbf{Y}_{bus} = \begin{vmatrix} -j15 & j5 & j5 & j5 \\ j5 & -j15 & j10 & 0 \\ j5 & j10 & -j25 & j10 \\ j5 & 0 & j10 & -j15 \end{vmatrix} \qquad \ldots(i)$$

Using the above $\mathbf{Y}_{bus}$ and bus powers as indicated in Fig. 22.20, approximately power flow Eq. (22.61) are expressed as (all voltage magnitudes are equal to 1.0 pu. Bus 1 being reference one, $\delta_1 = 0$)

$$P_2 = 5\delta_2 + 10(\delta_2 - \delta_3) = -2$$

$$\text{or } 15\delta_2 - 10\delta_3 = -2 \qquad \ldots(ii)$$

$$P_3 = 5\delta_3 + 10(\delta_3 - \delta_2) + 10(\delta_3 - \delta_4) = 3.5$$

$$\text{or } -10\delta_2 + 25\delta_3 - 10\delta_4 = 3.5 \qquad \ldots(iii)$$

$$\text{and } P_4 = 5\delta_4 + 10(\delta_4 - \delta_3) = -2$$

$$\text{or } 10\delta_3 - 15\delta_4 = 2 \qquad \ldots(iv)$$

From Eqs. (*ii*) and (*iv*), we have

$$\delta_2 = \delta_4$$

Substituting $\delta_2 = \delta_4$ in Eq. (*iii*), we have

$$-20\delta_2 + 25\delta_3 = 3.5 \qquad \ldots(v)$$

Solving Eqs. (*ii*) and (*v*), we have

$$\delta_4 = \delta_2 = -\frac{0.6}{7} = -0.08571 \text{ rad or } -4.91^\circ \text{ (electrical)}$$

and $\delta_3 = 0.07143$ rad or 4.0926° (electrical)

Substituting the values of δ_2, δ_3, δ_4 in Eq. (22.62), we have

$$Q_1 = -5\cos(4.91^\circ) - 5\cos(4.0926^\circ) - 5\cos(4.91^\circ) + 15$$

$$= 0.0495 \text{ pu} \qquad \ldots(vi)$$

$$Q_2 = -5\cos(4.91^\circ) - 10\cos[4.0926^\circ + 4.91^\circ] + 15 = 0.1415 \text{ pu}$$

$$Q_3 = -5\cos(4.0926^\circ) - 10\cos[4.0926^\circ + 4.91^\circ] - 10\cos[4.0926^\circ + 4.91^\circ] + 25$$

$$= 0.2591 \text{ pu}$$

and $Q_4 = -5\cos(4.91^\circ) - 10\cos(4.91^\circ + 4.0926^\circ) + 15 = 0.1415$ pu

(*i*) Reactive power generation at the 4 buses are

$$Q_{G_1} = Q_1 + 1 = 1.0495 \text{ pu}$$

$$Q_{G_2} = Q_2 + 0.5 = 0.6415 \text{ pu}$$

$$Q_{G_3} = Q_3 + 0 = 0.2591 \text{ pu}$$

$$Q_{G_4} = Q_4 + 0.8 = 0.1415 + 0.8 = 0.9415 \text{ pu } \textbf{Ans.} \qquad \ldots(vii)$$

(*ii*) Reactive line losses

$$= Q_1 + Q_2 + Q_3 + Q_4 = 0.0495 + 0.1415 + 0.2591 + 0.1415$$

$$= 0.5916 \text{ pu } \textbf{Ans.}$$

Since the line shunt capacitances have been neglected, the lines behave as short lines, the real power transferred over a line is given as

$$P_{ik} = \frac{V_i^2}{Z}\cos\theta - \frac{V_i V_k}{Z}\cos(\theta + \delta_i - \delta_k)$$

Substituting $Z = X_{ik}$ and $\theta = 90^\circ$, we have

$$P_{ik} = -P_{ki} = \frac{V_i V_k}{X_{ik}}\sin(\delta_i - \delta_k) \qquad \ldots(viii)$$

where P_{ik} is the real power flow from bus i to bus k.

(*iii*) Substituting the values of X and δ in above Eq. (*viii*), we have

$$\left.\begin{aligned} P_{12} &= -P_{21} = \frac{1}{0.2} \times 1 \times 1 \times \sin[0^\circ - (-4.91^\circ)] \\ &= 0.428 \text{ pu} \\ P_{13} &= -P_{31} = \frac{1}{0.2} \times 1 \times 1 \times \sin(0^\circ - 4.0926^\circ)] \\ &= -0.357 \text{ pu} \\ P_{14} &= -P_{41} = \frac{1}{0.2} \times 1 \times 1 \times \sin[0^\circ - (-4.91^\circ)] \\ &= 0.428 \text{ pu} \\ P_{23} &= -P_{32} = \frac{1}{0.1} \times 1 \times 1 \times \sin(-4.91^\circ - 4.0926^\circ) \\ &= -1.565 \text{ pu} \\ P_{34} &= -P_{43} = \frac{1}{0.1} \times 1 \times 1 \times \sin(4.91^\circ + 4.0926^\circ) \\ &= 1.565 \text{ pu } \textbf{Ans.} \end{aligned}\right\} \ldots(ix)$$

The reactive power transferred over the lines is given as

$$Q_{ik} = \frac{V_i^2}{Z}\sin\theta - \frac{V_i V_k}{Z}\sin(\theta + \delta_i - \delta_k)$$

Substituting $\theta = 90°$ and $Z = X_{ik}$, we have

$$Q_{ik} = \frac{V_i^2}{X_{ik}} - \frac{V_i V_k}{X_{ik}} \sin(\delta_i - \delta_k) \quad ...(x)$$

where Q_{ik} is the reactive power flow from bus *i* to bus k.

Substituting the values of X and δ in above Eq. (*x*), we have

$$Q_{12} = Q_{21} = \frac{1}{0.2} - \frac{1}{0.2}\cos(\delta_1 - \delta_2) = 5 - 5\cos(4.91°) = 0.01835 \text{ pu}$$

$$Q_{13} = Q_{31} = \frac{1}{0.2} - \frac{1}{0.2}\cos(\delta_1 - \delta_3) = 5 - 5\cos(-4.0926°)$$

$$= 0.01275 \text{ pu}$$

$$Q_{14} = Q_{41} = \frac{1}{0.2} - \frac{1}{0.2}\cos(\delta_1 - \delta_4) = 5 - 5\cos(4.91°)$$

$$= 0.01835 \text{ pu}$$

$$Q_{23} = Q_{32} = \frac{1}{0.1} - \frac{1}{0.1}\cos(\delta_2 - \delta_3) = 10 - 10\cos(-9.0026°)$$

$$= 0.1232 \text{ pu}$$

$$Q_{34} = Q_{43} = \frac{1}{0.1} - \frac{1}{0.1}\cos(\delta_3 - \delta_4) = 10 - 10\cos(9.0026°)$$

$$= 0.1232 \text{ pu } \textbf{Ans.}$$

Generation and load demands at all the buses and all the line flows are indicated in Fig. 22.21.

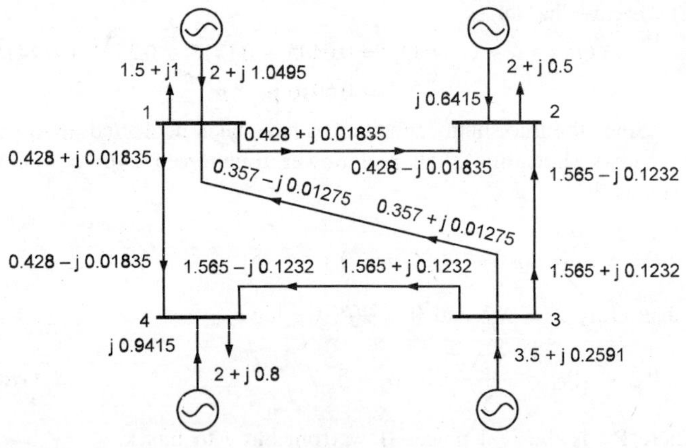

Fig. 22.21. ***Generation, Load and Line Flow for Example 22.14***

$P_{12} = -P_{21}$ means that bus 1 exports 0.428 pu active power to bus 2 or bus 2 imports 0.428 pu active power from bus 1.

$Q_{12} = Q_{21} = 0.01835$ pu.

It means both the buses export 0.01835 pu reactive power to one another. It means bus 1 is exporting 0.01835 pu reactive power (lagging) to bus 2 but bus 2 imports – 0.01835 pu reactive power (leading) from bus 1. Therefore, the complex power export from bus 1 to bus 2 has been indicated as (0.428 + *j* 0.01835) pu and complex power imported by bus 2 from bus 1 has been indicated as (0.428 – *j* 0.01835) pu. The same reasonings are followed for other lines.

22.8. ITERATIVE COMPUTATION OF NONLINEAR ALGEBRAIC EQUATIONS—GAUSS AND GAUSS-SEIDEL ITERATIVE TECHNIQUE

A set of simultaneous nonlinear algebraic equations cannot be solved by the conventional methods. For solving these equations, the method of numerical iteration is used.

The method of solving simultaneous equations by starting with assumed values of unknown variables and obtaining successive better values of the same variable by repeated cycles of solution is called *method of numerical iteration.*

The iterative methods are :

1. Gauss-iterative method.
2. Gauss-Seidel method.
3. Newton-Raphson method.

Any process of successive approximations used to have a numerical solution of algebraic equations, differential equations, etc. is called the *iterative method.* Digital computers are used for obtaining the solution quickly. The list of computer instructions stating the sequence of operations is an *algorithm.* The quality of an algorithm is decided by the speed of convergence.

Gauss-iterative method is essentially a trial and error method and provides a sufficiently accurate numerical solution. The steps in solving the problem by this method are :

1. Guess the initial solution approximately.
2. This guessed solution is used in conjunction with the original equation to compute a new and more accurate estimate for the solution.
3. The second estimate is used to find the third estimate.

The process is continued until the difference between successive estimates is extremely negligible.

The above method has two drawbacks:

1. It may require an extremely large number of iterations to reach sufficiently close to the solution, but with the use of computers it may not be thought of as a drawback.
2. The process may not converge, if the solution does not exist or if the iteration process is divergent.

Gauss-Seidel method is a modification to Gauss-iteration method. This modification reduces the number of iterations. In this method the values of unknowns immediately replace the previous values in the next step while in case of Gauss method as stated earlier, the calculated values replace the earlier values only at the end of the iteration. Because of it Gauss-Seidel method converges much faster than the Gauss method, *i.e.*, number of iterations required to obtain solution is much less in the Gauss-Seidel method compared to the Gauss method. This is why Gauss-Seidel method is very popular.

Gauss-Seidel method is of the simplest iterative methods and has been in use since early days of digital computer methods of analysis. It has got the following advantages :

(*i*) It is very simple method.

(*ii*) Computing costs are less.

Sometimes it is used to determine the initial solution for the other iterative processes.

The Gauss and Gauss-Seidel methods of iterative solution of nonlinear algebraic equations are illustrated in simple cases by the following examples.

Example 22.15. Find the value of x, accurate up to 4 decimal places in the equation $x^2 - 5x + 6 = 0$ using Gauss-iterative method.

Solution : Given equation is $x^2 - 5x + 6 = 0$

or $x = 0.2\,(x^2 + 6)$

Let the initial guess is such that

$x^{(0)} = 1.6$

Iteration 1: $x^{(1)} = 0.2\,[1.6^2 + 6] = 1.712$

Iteration 2: $x^{(2)} = 0.2\,[1.712^2 + 6] = 1.7861888$

Iteration 3: $x^{(3)} = 0.2\,[1.7861888^2 + 6] = 1.838094086$

Iteration 4: $x^{(4)} = 0.2\,[1.838094086^2 + 6] = 1.875717974$

Iteration 5: $x^{(5)} = 0.2\,[1.875717974^2 + 6] = 1.903663583$

Iteration 6: $x^{(6)} = 0.2\,[1.903663583^2 + 6] = 1.924787008$

Iteration 7: $x^{(7)} = 0.2\,[1.924787008^2 + 6] = 1.940961005$

Iteration 8: $x^{(8)} = 0.2\,[1.940961005^2 + 6] = 1.953465925$

Iteration 9: $x^{(9)} = 0.2\,[1.953465925^2 + 6] = 1.963205824$

Iteration 10: $x^{(10)} = 0.2\,[1.963205824^2 + 6] = 1.970835421$

Iteration 11: $x^{(11)} = 0.2\,[1.970835421^2 + 6] = 1.976838452$

Iteration 12: $x^{(12)} = 0.2\,[1.976838452^2 + 6] = 1.981578053$

Iteration 13: $x^{(13)} = 0.2\,[1.981578053^2 + 6] = 1.985330316$

Iteration 14: $x^{(14)} = 0.2\,[1.985330316^2 + 6] = 1.988307293$

Iteration 15: $x^{(15)} = 0.2\,[1.988307293^2 + 6] = 1.990673178$

Iteration 16: $x^{(16)} = 0.2\,[1.990673178^2 + 6] = 1.99255594$

Iteration 17: $x^{(17)} = 0.2\,[1.99255594^2 + 6] = 1.994055835$

Iteration 18: $x^{(18)} = 0.2\,[1.994055835^2 + 6] = 1.995251735$

Iteration 19: $x^{(19)} = 0.2\,[1.995251735^2 + 6] = 1.996205897$

Iteration 20: $x^{(20)} = 0.2\,[1.996205897^2 + 6] = 1.996967597$

Iteration 21: $x^{(21)} = 0.2\,[1.996967597^2 + 6] = 1.997575916$

Iteration 22: $x^{(22)} = 0.2\,[1.997575916^2 + 6] = 1.998061908$

Iteration 23: $x^{(23)} = 0.2\,[1.998061908^2 + 6] = 1.998450278$

Iteration 24: $x^{(24)} = 0.2\,[1.998450278^2 + 6] = 1.998760703$

Iteration 25: $x^{(25)} = 0.2\,[1.998760703^2 + 6] = 1.999008869$

Iteration 26: $x^{(26)} = 0.2\,[1.999008869^2 + 6] = 1.999207292$

Iteration 27: $x^{(27)} = 0.2\,[1.999207292^2 + 6] = 1.999365959$

Iteration 28: $x^{(28)} = 0.2\,[1.999365959^2 + 6] = 1.999492848$

Iteration 29: $x^{(29)} = 0.2\,[1.999492848^2 + 6] = 1.99959433$

Iteration 30: $x^{(30)} = 0.2\,[1.99959433^2 + 6] = 1.999675497$

Iteration 31: $x^{(31)} = 0.2\,[1.999675497^2 + 6] = 1.999740418$

Iteration 32: $x^{(32)} = 0.2\,[1.999740418^2 + 6] = 1.999792348$

Since the values of $x^{(31)}$ and $x^{(32)}$ are same up to 4 decimal places, the iteration need not be continued further and 1.999792348 can be taken as the value of x.

Example 22.16. Obtain the values of x and y by Gauss-iterative method for the following equations

$$3x + xy - 2 = 0$$

$$\text{and } 3y - xy + 2 = 0$$

Solution : The given equations can be written as

$$x = 0.66666 - 0.33333\,xy$$

$$\text{and } y = -0.66666 + 0.33333\,xy$$

Using the usual method, it is seen that the solution to the given equations is $x = 1$ and $y = -1$

Let us start iteration by assuming values of x and y such that

$x^{(0)} = 0.5$ and $y^{(0)} = -0.5$

Then

Iteration 1:

$x^{(1)} = 0.66666 - 0.33333 \times 0.5 \times (-0.5) = 0.75$

$y^{(1)} = -0.66666 + 0.33333 \times 0.5 \times (-0.5) = -0.75$

Iteration 2:

$x^{(2)} = 0.66666 - 0.33333 \times 0.75 \times (-0.75) = 0.854166667$

$y^{(2)} = -0.66666 + 0.33333 \times 0.75 \times (-0.75)$

$= -0.854166667$

Iteration 3:

$x^{(3)} = 0.66666 - 0.33333 \times 0.854166667 \times (-0.854166667)$

$= 0.909866898$

$y^{(3)} = -0.66666 + 0.33333 \times 0.854166667 \times (-0.854166667)$

$= -0.909866898$

Iteration 4:

$x^{(4)} = 0.66666 - 0.33333 \times 0.909866898 \times (-0.909866898)$

$= 0.942619256$

$y^{(4)} = -0.66666 + 0.33333 \times 0.909866898 \times (-0.909866898)$

$= -0.942619256$

Substitute the values obtained of x and y and continue steps $x^{(5)}$, $y^{(5)}$; $x^{(6)}$, $y^{(6)}$; $x^{(7)}$, $y^{(7)}$ etc. till final results obtained are quite steady.

Example 22.17. Obtain the values of x and y for the given equations in Example 22.16 by Gauss-Seidel (G-S) method.

Solution : As already stated, in the Gauss-Seidel method, the convergence is obtained faster by making use of upgraded iterates as soon as they are available, *i.e.*, use $x^{(k+1)}$ and not x^k, as soon as it is available.

Iteration 1: Same as in example 22.16.

Iteration 2:

$x^{(2)} = 0.66666 - 0.33333 \times 0.75 \times (-0.75) = 0.854166667$

$y^{(2)} = -0.66666 + 0.33333 \times 0.854166667 \times (-0.75)$

$= -0.880208332$

Iteration 3:

$x^{(3)} = 0.66666 - 0.33333 \times 0.854166667 \times (-0.880208332)$

$= 0.917281538$

$y^{(3)} = -0.66666 + 0.33333 \times 0.917281538 \times (-0.880208332)$

$= -0.935799616$

Iteration 4:

$x^{(4)} = 0.66666 - 0.33333 \times 0.917281538 \times (-0.935799616)$

$= 0.952797236$

$y^{(4)} = -0.66666 + 0.33333 \times 0.952797236 \times (-0.935799616)$

$= -0.963875762$

It will be observed that the speed of convergence in this method is increased or the number of iterations required are lesser.

22.8.1. Acceleration Factor

The convergence process can be significantly accelerated in the above methods by the application of acceleration factor. If $\Delta x^{(r)}$ and $\Delta y^{(r)}$ indicate the variable change in the $(r-1)$th and rth iteration, the difference variables can be written as:

$$\Delta x^{(r)} = x^{(r)} - x^{(r-1)} \quad ...(22.63)$$

$$\Delta y^{(r)} = y^{(r)} - y^{(r-1)} \quad ...(22.64)$$

If the difference is multiplied by the acceleration factor α which is larger than unity (usually between 1.5 and 1.7 for the solution of power flow equations), the accelerated iterates are obtained as follows:

$$x_{acc}^{(r)} = x_{acc}^{(r-1)} + \alpha \Delta x^{(r)} \quad ...(22.65)$$

$$y_{acc}^{(r)} = y_{acc}^{(r-1)} + \alpha \Delta y^{(r)} \quad ...(22.66)$$

The accelerated values of x and y are used in the iteration process instead of x, y.

22.9. GAUSS-SEIDEL METHOD FOR POWER FLOW STUDIES

For explaining the application of Gauss-Seidel method for power flow studies, let it be assumed that all buses other than the swing or slack bus are P-Q or load buses.* At slack bus both V and δ are specified and they remain fixed throughout. There are $(n-1)$ buses where P and Q are given. Initially we assume the magnitudes and angles at these $(n-1)$ buses and update these voltages at every step of iteration.

From Eq. (22.57*a*) $$\mathbf{I}_i = \frac{P_i - jQ_i}{\mathbf{V}_i^*} \quad ...(22.67)$$

From Eq. (22.8) $$\mathbf{V}_i = \frac{1}{\mathbf{Y}_{ii}}\left[\mathbf{I}_i - \sum_{\substack{k=1\\k\neq i}}^{n} \mathbf{Y}_{ik}\mathbf{V}_k\right] \quad ...(22.68)$$

Substituting $\mathbf{I}_i = \dfrac{P_i - jQ_i}{\mathbf{V}_i^*}$ from Eq. (22.67) in Eq. (22.68), we have

$$\mathbf{V}_i = \frac{1}{\mathbf{Y}_{ii}}\left[\frac{P_i - jQ_i}{\mathbf{V}_i^*} - \sum_{\substack{k=1\\k\neq i}}^{n} \mathbf{Y}_{ik}\mathbf{V}_k\right];\ i = 2, 3, ...\ n \quad ...(22.69)$$

Bus 1 is a slack bus with specified **V**, hence in above equation $\mathbf{V}_1$ is taken as specified.

Thus Eq. (22.69) represents a set of $(n-1)$ equations for $i = 2, 3, ..., n$ which are to be solved simultaneously for $\mathbf{V}_2, \mathbf{V}_3, \mathbf{V}_4 \ldots \mathbf{V}_n$.

In the **Gauss method**, we assume the voltage for all the buses except the slack bus where the voltage magnitude and phase angle are specified and remain fixed. Normally, we set *i.e.*, assume the voltage magnitude and phase angle of these buses equal to that of the slack bus and work in per unit system. The assumed bus voltage and the slack bus voltage along with P and Q are substituted in RHS of the Eq. (22.69) to obtain new set of bus voltages. After the entire iteration is complete, the new set of bus voltages is again substituted along with the specified slack bus voltage in the RHS of Eq. (22.69) to obtain a new set of bus voltages. The process is continued till

$$V_i^{r+1} - V_i^r \leq \varepsilon \text{ for } i = 2, 3 \ldots n$$

where r is an iteration count and ε is a very small number which depends upon the system accuracy and is normally equal to 0.0001 etc.

Thus the process is continued till the mod of the bus voltage obtained at the current iteration less the value of bus voltage at the previous iteration is smaller than a chosen very small number and in this way we obtain the solution, *i.e.*, magnitude and phase angle of voltage.

Gauss iterative method, explained above, is much slower to converge and may sometimes fail to do so.

In case of *Gauss-Seidel method*, the value of bus voltages calculated for any bus immediately replace the previous values in the next step while in case of Gauss method, as stated earlier, the calculated bus voltages replace the earlier value only at the end of the iteration. Due to this Gauss-Seidel method converges much faster than that of Gauss method compared to Gauss method.

Gauss-Seidel method was one of the most common methods employed for solving power flow equations. It has the following advantages and disadvantages.

Advantages. 1. Simplicity of technique.

2. Small computer memory requirement.

3. Less computational time per iteration.

Disadvantages. 1. Slow rate of convergence resulting in larger number of iterations.

2. Increase in number of iterations directly with the increase in the number of buses.

3. Effect on convergence due to choice of slack bus.

Because of the above drawbacks, use of Gauss-Seidel method is limited only to systems with smaller number of buses.

22.9.1. Gauss-Seidel Method of Solution of Power Flow Problems on Digital Computers

Let us continue to consider the case where all buses except slack bus are load or P-Q buses.

* We shall see later that the method can be easily applied to include generation or P-V buses as well.

Block 1. Reading of Primitive Y Matrix. With the load profile known at each bus (P_{Di} and Q_{Di} being known) P_{Gi} and Q_{Gi} are allocated to all generating stations.

While active and reactive generations are allocated to the slack bus, these are permitted to vary during iterative computation. This is necessary because the magnitude and angle of voltage are specified at this bus and only two variables can be specified at any bus.

With this step, bus injections ($P_i + jQ_i$) are known at all buses other than the slack bus.

Block 2. Formation of Bus Admittance Matrix $\mathbf{Y}_{bus}$. With the line and shunt admittance data stored in computer, bus admittance matrix $\mathbf{Y}_{bus}$ is formed by using the rules for self and mutual admittances explained in Art 22.3.1. $\mathbf{Y}_{bus}$ can also be formed by using Eq. (22.27) where input data is in the form of primitive matrix Y and singular connection matrix A.

Block 3. Iterative Computation of Bus Voltages. To start the iterations a set of initial voltage values is to be assumed. Since, in a power system the voltage spread is not too wide, it is normal practice to assume a flat voltage profile for all nodal voltages except the slack bus 1. Initially all voltages are set equal to $(1 + j0)$ except the voltage of the slack bus which is fixed. It is to be noted that voltages and admittances in Eq. (22.69) are complex quantities. In most of the computers the complex number operations are available. If complex number operations are not available in a computer, Eq. (22.69) can be converted into $2(n-1)$ equations in real unknowns. (e_i, f_i or V_i, δ_i) by writing

$$\mathbf{V}_i = e_i + jf_i = V_i \angle \delta_i \qquad ...(22.70)$$

It is possible to obtain significant reduction in computer time if all arithmetic operations are performed in advance when they do not change with iterations. Since P, Q and Y at a bus do not change with iterations, the term $\dfrac{P_i - jQ_i}{\mathbf{Y}_{ii}}$ can be evaluated before hand.

In order to save computer time it is more appropriate to write Eq. (22.69) in the following form:

$$\mathbf{V}_i = \frac{\mathbf{A}_i}{\mathbf{V}_i^*} - \sum_{\substack{k=1\\k\neq i}}^{n} B_{ik}\mathbf{V}_k \qquad ...(22.71)$$

where
$$\mathbf{A}_i = \frac{P_i - jQ_i}{\mathbf{Y}_{ii}} \quad \text{for } i_1 = 2, 3, ..., n \qquad ...(22.72)$$

and
$$\mathbf{B}_{ik} = \frac{\mathbf{Y}_{ik}}{\mathbf{Y}_{ii}} \quad \text{for } i = 2, 3, ..., n;$$

$$k = 1, 2, ..., n \text{ and } k \neq i \qquad ...(22.73)$$

The values of $\mathbf{A}_i$ and $\mathbf{B}_{ik}$ are computed once in the beginning and then used in every iteration.

Now for the $(r + 1)$th iteration, the voltage Eq. (22.71) becomes

$$\mathbf{V}_i^{(r+1)} = \frac{\mathbf{A}_i}{(\mathbf{V}_i^{(r)})^*} - \sum_{k=1}^{i-1} \mathbf{B}_{ik}\mathbf{V}_k^{(\ +1)} - \sum_{k=i+1}^{n} \mathbf{B}_{ik}\mathbf{V}_k^{(r)}$$

$$\text{for } i = 2, 3, 4, ..., n \qquad ...(22.74)$$

The second term on the RHS of the above Eq. (22.74) is obvious because the voltage prior to bus i should correspond to the value as computed during the current iteration.

The iterative process is continued till the change in magnitude of the bus voltages in successive iterations become less than a certain pre-specified tolerance level for all buses, *i.e.*,

$$\Delta V_i^{(r+1)} = V_i^{(r+1)} - V_i^{(r)} \leq \varepsilon \text{ for } i = 2, 3, 4 ..., n \qquad ...(22.75)$$

To reduce the number of iterations, an acceleration factor $\alpha = 1.6$ is considered to be a good value for power flow studies. A wrong choice of α may indeed slowdown convergence or even cause the method to diverge.

After computing bus voltages (magnitudes and phase angles) for all the buses, injected powers $\mathbf{S}_i$ and line flows are computed using the nodal voltages. Equation (22.57) is used to determine $\mathbf{S}_i$ and conjugate of $\mathbf{S}_i^*$ gives injected power $\mathbf{S}_i$.

The last step in power flow study is the calculation of *line flows* and line losses.

Computations of Line Flows and Line Losses. Figure 22.22 shows a line connecting ith and kth buses. Here we assume the normal π representation of transmission line.

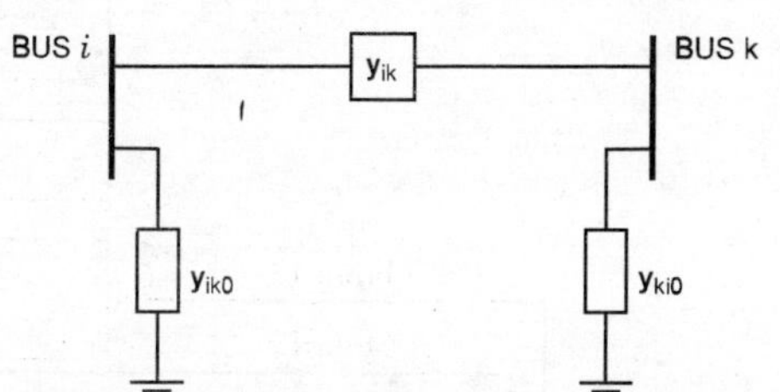

Fig. 22.22. ***π-Representation of a Line Between Two Buses***

Current flowing from bus i towards bus k,

$$\mathbf{I}_{ik} = [\mathbf{V}_i - \mathbf{V}_k]\mathbf{y}_{ik} + \mathbf{V}_i\mathbf{y}_{ik0} \qquad ...(22.76)$$

where $\mathbf{V}_i$ and $\mathbf{V}_k$ are the bus voltages at the buses i and k respectively which are already calculated from the power flow studies

The power flow in the line i-k at the bus i is given as

$$\mathbf{S}_{ik} = P_{ik} + jQ_{ik} = \mathbf{V}_i\mathbf{I}_{ik}^*$$

$$= \mathbf{V}_i\left(\mathbf{V}_i^* - \mathbf{V}_k^*\right)\mathbf{y}_{ik}^* + \mathbf{V}_i\mathbf{V}_i^*\mathbf{y}_{ik0}^* \qquad ...(22.77)$$

Similarly, the power flow in the line i-k at the bus k is given as

$$\mathbf{S}_{ki} = \mathbf{V}_k\left(\mathbf{V}_k^* - \mathbf{V}_i^*\right)\mathbf{y}_{ik}^* + \mathbf{V}_k\mathbf{V}_k^*\mathbf{y}_{kio}^* \qquad ...(22.78)$$

Thus power flows over all the lines can be computed.

The power losses in the (i-k)th line are given by sum of the power flows determined from above Eqs. (22.77) and (22.78) *i.e.*, power losses in the (i-k)th line = $S_{ik} + S_{ki}$. Total transmission losses can be computed by summing all the line flows (*i.e.*, $S_{ik} + S_{ki}$ for all *i*, k).

It may be noted that the slack bus power can also be determined by summing the power flows on the lines terminating at the slack bus. This concludes the power flow study for the case of P-Q buses only.

22.9.2. Treatment To Voltage Controlled Buses in Gauss-Seidel Method

In a power system, some of the buses are voltage controlled buses where P and V are specified while Q and δ are unknowns and are to be determined. However, usually limit of reactive power, *i.e.*, Q_{max} and Q_{min} to hold the generation voltage within limits are also given.

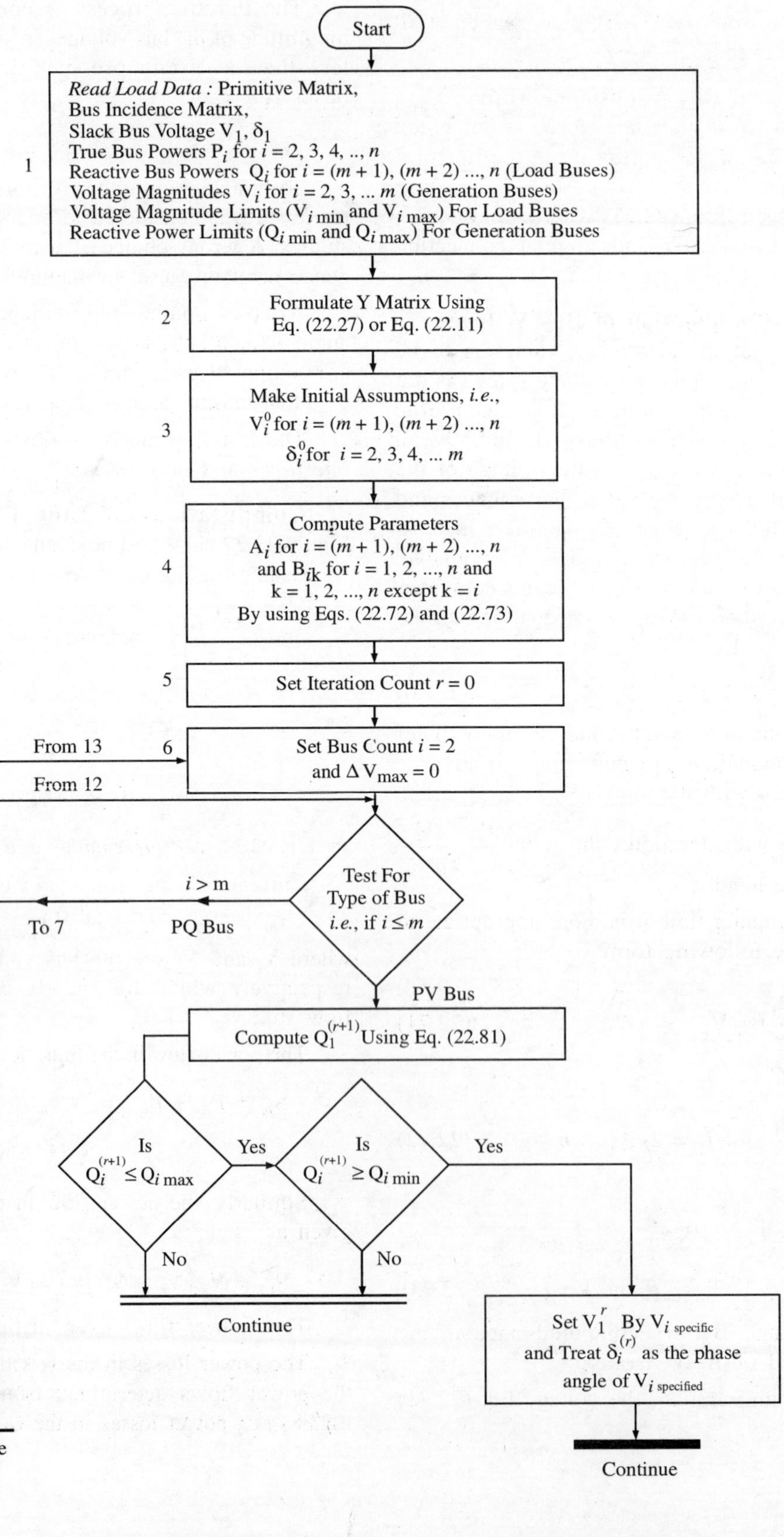

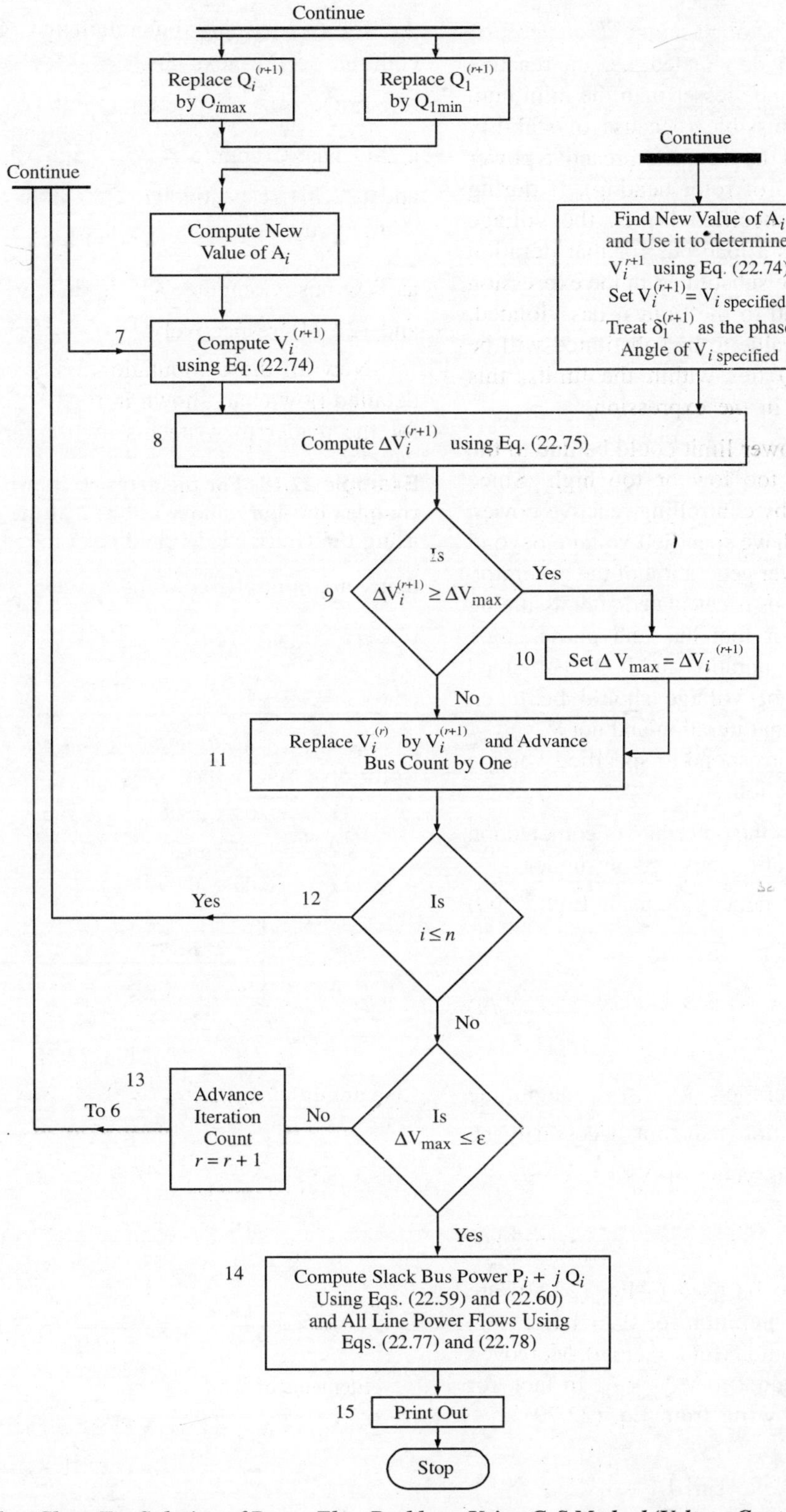

Fig. 22.23. *Flow Chart For Solution of Power Flow Problems Using G-S Method (Voltage Controlled Buses Included)*

So far power flow study in a system with generation or voltage controlled buses with G-S method the values of Q and δ are to be updated in every GS iteration through appropriate bus equations and therefore computational procedure needs some modifications.

Let the buses be numbered as

i = 1 slack or swing bus

i = 2, 3, ..., m voltage controlled or P-V buses

i = m + 1, m + 2, m + 3, ..., n load or P-Q buses

For the voltage controlled buses, the bus voltage V_i must be equal to the specified voltage $V_{i\ specified}$ in magnitude and the value of reactive power Q_i (for i = 2, 3, ..., m) must lie between the limits. Thus, the conditions to be satisfied are

1. $V_i = V_{i\ specified}$ for i = 2, 3, ..., m

2. $Q_{i\,min} < Q_i < Q_{i\,max}$ for $i = 2, 3, ..., m$

The second requirement may be violated, i.e., the reactive power generation at the bus may be lesser than the minimum reactive power generation permissible (because of stability problem) or it may be more than the maximum reactive power generation permissible (because of rotor heating). If during any iteration Q_i violates the specified limits, the voltage controlled bus is made to act as a load bus for that iteration only and the reactive power to be substituted in the expression for that iteration will correspond to the limit it has violated, *i.e.*, if it is less than Q_{min}, the value to be substituted will be Q_{min}. If calculated value of Q lies within the limits, this value of Q will be substituted in the expression.

The violation of reactive power limit could be due to the specified voltage either being too low or too high. Since $V_{i\,specified}$ can be obtained only by controlling reactive power, therefore, it is possible that we have specified voltage beyond the capability of the reactive power generation of the generator. If the value of Q for a generator bus is calculated after assuming magnitude equal to $V_{specified}$ for that bus and phase angle corresponding to the value in that iteration and if Q so obtained violates the limits, the value of voltage should be taken corresponding to the voltage in that iteration and not $V_{specified}$. Otherwise the voltage should correspond to specified voltage and phase angle as in that iteration.

In view of above observations the procedure of computation for voltage controlled or generation buses is an under.

(*i*) Calculate reactive power generation using Eq. (22.60) which is reproduced below:

$$Q_i = -V_i \sum_{k=1}^{n} V_k Y_{ik} \sin(\theta_{ik} + \delta_k - \delta_i) \qquad ...(22.79)$$

At the beginning of $(r+1)$th iteration, $V_i^{(r)}$, *i.e.*, magnitude of V_i obtained during rth iteration may not necessarily be equal to $V_{i\,specified}$. We have this value of $V_i^{(r)}$

$$V_i^{(r)} = V_i^{(r)} \angle \delta_i^{(r)} \qquad ...(22.80)$$

The values of V_k and δ_k to be used in Eq. (22.80) are those obtained during $(r+1)$th iteration for (k = 1 to $i-1$) and those obtained during rth iteration (for k = i to n). Moreover, for every iteration V_i must be equal to $V_{i\,specified}$. In fact, for the $(r+1)$th iteration one can write from Eq. (22.79)

$$Q_i^{r+1} = -V_{i\,specified} \sum_{k=1}^{n-1} Y_{ik} V_k^{(r+1)} \sin\left(\theta_{ik} + \delta_k^{(r+1)} - \delta_i^{(r)}\right)$$

$$= -V_{i\,specified} \sum_{k=1}^{n} Y_{ik} V_k^{r} \sin\left(\theta_{ik} + \delta_k^{r} - \delta_i^{r}\right) \qquad ...(22.81)$$

(*ii*) If $Q_i^{(r+1)}$ is found to lie within limits, *i.e.*, $Q_{i\,min} < Q_i^{(r+1)} < Q_{i\,max}$ calculate new value of A_i and then use Eq. (22.74) for computation of $V_i^{(r+1)}$ using $V_{i\,specified}$ and $\delta_i^{(r)}$ for the magnitude and phase angle of $V_i^{(r+1)}$. Reset $V_i^{(r+1)}$ to $V_{i\,specified}$ but retain the phase angle $\delta_i^{(r+1)}$ and continue to the next bus.

If $Q_i^{(r+1)} < Q_{i\,min}$, set $Q_i^{(r+1)} = Q_{i\,min}$ and treat bus i as a P-Q bus. Compute $A_i^{(r+1)}$ and $V_i^{(r+1)}$ from Eqs. (22.72) and (22.74) respectively.

If $Q_i^{(r+1)} > Q_{i\,max}$, set $Q_i^{(r+1)} = Q_{i\,max}$ and treat bus i as P-Q bus. Compute $A_i^{(r+1)}$ and $V_i^{(r+1)}$ from Eqs. (22.72) and (22.74) respectively.

Now all the computational steps are summarized in the detailed flow chart shown in Fig. 22.23 which serves as basis for the reader to write his own computer programme.

Example 22.18. For the network shown in Fig. 22.24, obtain the complex bus-bar voltage at bus 2 at the end of the first iteration using the Gauss-Siedel method. Line impedances shown in the figure are in pu. Given : Bus 1 is slack bus with $V_1 = 1.0 \angle 0°$, $P_2 + jQ_2 = -5.96 + j1.46$, $|V_3| = 1.02$ pu. Assume : $V_3^0 = 1.02 \angle 0°$ and $V_2^0 = 1.0 \angle 0°$.

[G.B. Technical Univ. Power System Analysis 2010-11]

Solution : Line impedances

$$Z_{12} = 0.04 + j0.06$$
$$Z_{23} = 0.02 + j0.03$$
$$Z_{13} = 0.06 + j0.09$$

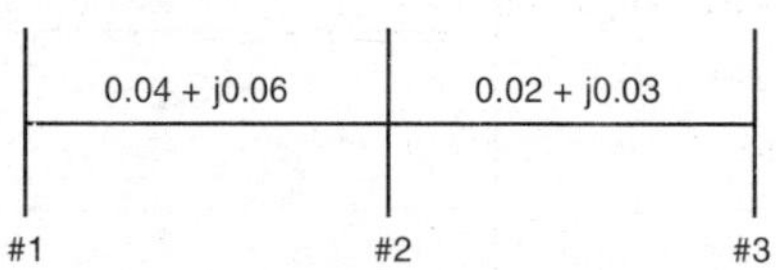

Fig. 22.24

Line admittances

$$y_{12} = \frac{1}{Z_{12}} = \frac{1}{0.04 + 0.06} = (7.69 - j11.54) \text{ pu}$$

$$y_{23} = \frac{1}{Z_{23}} = \frac{1}{0.02 + j0.03} = (15.38 - j23.077) \text{ pu}$$

$$y_{13} = \frac{1}{Z_{13}} = \frac{1}{0.06 + j0.09} = (5.128 - j7.692) \text{ pu}$$

Elements of Y_{BUS} are

$$Y_{11} = y_{12} + y_{13} = (7.69 - j11.54) + (5.128 - j7.692)$$
$$= (12.818 - j19.232) \text{ pu}$$
$$Y_{22} = y_{12} + y_{23} = (7.69 - j11.54) + (15.38 - j23.077)$$
$$= (23.07 - j34.617) \text{ pu}$$
$$Y_{33} = y_{13} + y_{23} = (5.128 - j7.692) + (15.38 - j23.077)$$
$$= (20.51 - j30.77)$$
$$Y_{12} = Y_{21} = -y_{12} = (-7.69 + j11.54) \text{ pu}$$
$$Y_{13} = Y_{31} = -y_{13} = (-5.128 + j7.692) \text{ pu}$$
$$Y_{23} = Y_{32} = -y_{23} = (-15.38 + j23.077) \text{ pu}$$
$$V_1^{(0)} = 1.0 \angle 0° \text{ pu}$$

$V_2^{(0)} = 1.0 \angle 0°$ pu

$V_3^{(0)} = 1.02 \angle 0°$ pu

$$V_2^{(1)} = \frac{1}{Y_{22}}\left[\frac{P_2 - jQ_2}{V_2^{(0)}} - Y_{21}V_1^{(0)} - y_{23}V_3^{(0)}\right]$$

$$= \frac{1}{(23.07 - j34.617)}\left[\frac{-5.96 - j1.46}{1.0 \angle 0°} - \right.$$

$$\left.(-7.69 + j11.54) \times 1.0 \angle 0° - (-15.38 + j23.077) \times 1.02 \angle 0°\right]$$

$$= \frac{-5.96 - j1.46 + 7.69 - j11.54 + 15.688 - j23.538}{23.07 - j34.617}$$

$$= \frac{17.418 - j36.538}{23.07 - j34.617} = \frac{40.4773 \angle -64.51°}{41.6 \angle -56.32°}$$

$$= 0.973 \angle -8.19° = (0.963 - j0.1386) \text{ pu}$$

$$V_{2\,acc}^{(1)} = V_2^{(0)} + \alpha[V_2^{(1)} - V_2^{(0)}]$$

$= (1 + j0) + 1.4[0.963 - j0.1386 - (1 - j0.0)]$

assuming acceleration factor to be 1.4

$= (1 + j0) + (-0.0518 - j0.19404)$

$= (0.9482 - j0.19404)$ pu volts **Ans.**

Example 22.19. Using Gauss-Seidel load flow method, find bus voltages at the end of one iteration for the following 2-bus system. Line reactances are shown in Fig. 22.25. Ignore resistance and line charging. Assume initial voltage at all buses to 1.0 $\angle 0°$. Use 1.0 as acceleration factor. The bus data is given in the table below :

Bus No.	Specified P (pu)	Injections Q (pu)	Specified Voltage (pu)
1	—	—	1.0
2	0.3	—	1.0
3	0.5	0.2	—

[GATE 1996]

Solution : Admittance of each line,

$$y_{12} = y_{23} = y_{13} = y_{21} = y_{32} = y_{31} = \frac{1}{j0.1} = -j\,10$$

The admittance matrix is given as

$$Y_{bus} = \begin{vmatrix} y_{12} + y_{13} & -y_{12} & -y_{13} \\ -y_{21} & y_{21} + y_{23} & -y_{23} \\ -y_{31} & -y_{32} & y_{32} + y_{31} \end{vmatrix}$$

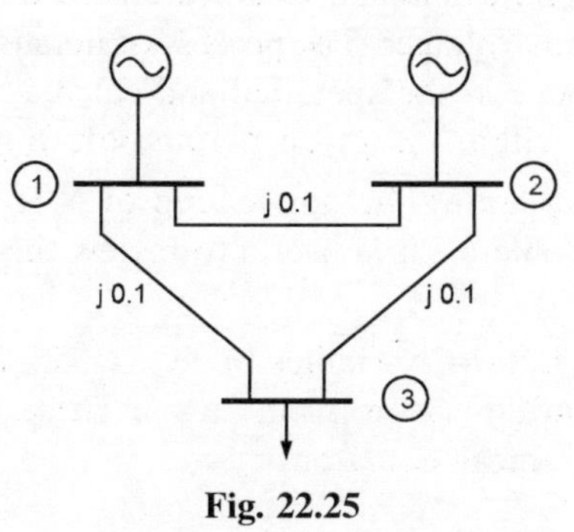

Fig. 22.25

$$= \begin{vmatrix} -j20 & j10 & j10 \\ j10 & -j20 & j10 \\ j10 & j10 & -j20 \end{vmatrix}$$

With the superfixes as iteration number, we have

$V_1^{(0)} = V_2^{(0)} = V_3^{(0)} = 1.0 \angle 0°$ assumed

$V_1^1 = 1 \angle 0°$ or $(1 + j0)$ **Ans.** ...Being specified

For the generator bus 2

$P_2 = 0.3$

$Q_2 = -g_m V_{20}(Y_{21}V_1 + Y_{22}V_2 + Y_{23}V_3)$

$= -g_m V_{20}(j\,10 \times 1 - j\,20 \times 1 + j\,10 \times 1) = 0$

$$V_2^{(1)} = \frac{1}{Y_{22}}\left[\frac{P_2 - jQ_2}{V_2^*} - Y_{21}V_1^{(0)} - Y_{23}V_3^{(0)}\right]$$

$$= \frac{1}{-j20}\left[\frac{0.3}{1 - j0} - j10 \times 1 - j10 \times 1\right]$$

$= j\,0.05\,(0.3 - j\,20) = (1 + j\,0.015)$

$$V_{2\,acc}^1 = V_2^{(0)} + \alpha\left[V_2^{(1)} - V_2^{(0)}\right]$$

$= (1 + j\,0) + 1\,(1 + j\,0.015 - 1)$

$= (1 + j\,0.015)$ pu **Ans.**

For the load bus 3

$$V_3^{(1)} = \frac{1}{Y_{33}}\left[\frac{P_3 - jQ_3}{V_3^*} - Y_{31}V_1^{(0)} - Y_{32}V_3^{(0)}\right]$$

$$= \frac{1}{-j20}\left[\frac{-0.5 - j0.2}{1 - j0} - j10 \times 1 - j10 \times 1\right]$$

$= j\,0.05\,[-0.5 - j\,20.2] = (1.01 - j\,0.025)$ pu

$$V_{3\,acc}^1 = V_3^{(0)} + \left[V_3^{(1)} - V_3^{(0)}\right]$$

$= (1 + j\,0) + 1\,[1.01 - j\,0.025 - 1]$

$= (1.01 - j\,0.025)$ pu **Ans.**

Example 22.20. The following is the system data for load flow solution.

The line admittances.

Bus Code	Admittance
1—2	2—j 8.0
1—3	1—j 4.0
2—3	0.666—j 2.664
2—4	1—j 4.0
3—4	2—j 8.0

The schedule of active and reactive powers :

Bus Code	P	Q	V	Remarks
1	—	—	1.06	Slack
2	0.5	0.2	1 + j 0.0	PQ
3	0.4	0.3	1 + j 0.0	PQ
4	0.3	0.1	1 + j 0.0	PQ

Determine the voltages at the end of first iteration using Gauss-Seidel method. Take α = 1.6.

[Pb Univ. Power System Analysis December 1993]

Solution : $Y_{11} = y_{12} + y_{13}$

$= (2 - j\,8.0) + (1 - j\,4.0) = (3 - j\,12.0)$

$$\mathbf{Y}_{21} = \mathbf{Y}_{12} = -\mathbf{y}_{12} = (-2 + j\,8.0)$$
$$\mathbf{Y}_{31} = \mathbf{Y}_{13} = -\mathbf{y}_{13} = (-1 + j\,4.0)$$
$$\mathbf{Y}_{41} = \mathbf{Y}_{14} = 0$$
$$\mathbf{Y}_{22} = \mathbf{Y}_{12} + \mathbf{y}_{23} + \mathbf{y}_{24} = (2 - j\,8.0) + (0.666 - j\,2.664) + (1 - j\,4.0) = (3.666 - j\,14.664)$$
$$\mathbf{Y}_{32} = \mathbf{Y}_{23} = -\mathbf{y}_{23} = -0.666 + j\,2.664)$$
$$\mathbf{Y}_{42} = \mathbf{Y}_{24} = -\mathbf{y}_{24} = (-1 + j\,4.0)$$
$$\mathbf{Y}_{33} = \mathbf{y}_{13} + \mathbf{y}_{34} + \mathbf{y}_{32} = (1 - j\,4.0) + (2 - j\,8.0) + (0.666 - j\,2.664) = (3.666 - j\,14.664)$$
$$\mathbf{Y}_{43} = \mathbf{Y}_{34} = -\mathbf{y}_{34} = -2 + j\,8.0$$
$$\mathbf{Y}_{44} = \mathbf{y}_{42} + \mathbf{y}_{43} = (1 - j\,4.0) + (2 - j\,8.0) = (3 - j\,12.0)$$

Thus the admittance matrix will be as given below:

$$\mathbf{Y}_{\text{bus}} = \begin{vmatrix} 3-j12.0 & -2+j8.0 & -1+j4.0 & 0 \\ -2+j8.0 & 3.666-j14.664 & -0.666+j2.664 & -1+j4.0 \\ -1+j4.0 & -0.666+j2.664 & 3.666-j14.664 & -2+j8.0 \\ 0 & -1+j4.0 & -2+j8.0 & 3-j12.0 \end{vmatrix}$$

The powers for generator bus are taken as positive while the powers for load buses are taken as negative.

For the given system

$$\mathbf{V}_2^{(1)} = \frac{1}{\mathbf{Y}_{22}}\left[\frac{\mathrm{P}_2 - j\mathrm{Q}_2}{\mathbf{V}_2^*} - \mathbf{Y}_{21}\mathbf{V}_1^{(0)} - \mathbf{Y}_{23}\mathbf{V}_3^{(0)} - \mathbf{Y}_{24}\mathbf{V}_4^{(0)}\right]$$
$$= \frac{1}{3.666 - j14.664}\left[\frac{-0.5 + j0.2}{1 - j0} - (-2 + j8)\times 1.06 - (-0.666 + j2.664)(1 + j0) - (-1 + j4.0)(1 + j0)\right]$$
$$= \frac{1}{3.666 - j14.664}[-0.5 + j\,0.2 + 2.12 - j\,8.48 + 0.666 - j\,2.664 + 1 - j\,4.0]$$
$$= \frac{1}{3.666 - j14.664}(3.286 - j\,14.944)$$
$$= (1.01187 - j\,0.02888)$$
$$\mathbf{V}_{2\,\text{acc}}^{(1)} = (1.0 + j\,0.0) + 1.6\,[(1.01187 - j\,0.02888) - (1 + j\,0)]$$
$$= (1.01899 - j\,0.04608)$$ **Ans.**

$$\mathbf{V}_3^{(1)} = \frac{1}{\mathbf{Y}_{33}}\left[\frac{\mathrm{P}_3 - j\mathrm{Q}_3}{\mathbf{V}_3^*} - \mathbf{Y}_{31}\mathbf{V}_1^{(0)} - \mathbf{Y}_{32}\mathbf{V}_2^{(1)} - \mathbf{Y}_{34}\mathbf{V}_4^{(0)}\right]$$
$$= \frac{1}{(3.666 - j14.664)}\left[\frac{-0.4 + j0.3}{1 - j0} - (-1 + j4.0)\times 1.06 - (-0.666 + j2.664)(1.01899 - j0.04608) - (-2 + j8)(1 + j0.0)\right]$$
$$= \frac{1}{3.666 - j14.664}[-0.4 + j\,0.3 + 1.06 - j\,4.24 + 0.5559 - j\,2.7453 + j\,2.742 + 2 - j\,8]$$
$$= \frac{3.2159 - j14.6853}{3.666 - j14.664} = [0.994145 - j\,0.02923]$$
$$\mathbf{V}_{3\,\text{acc}}^{(1)} = (1 + j\,0) + 1.6\,[(0.994145 - j\,0.02923) - (1 + j\,0)]$$
$$= [0.990632 - j\,0.046768]$$ **Ans.**

$$\mathbf{V}_4^{(1)} = \frac{1}{\mathbf{Y}_{44}}\left[\frac{\mathrm{P}_4 - j\mathrm{Q}_4}{\mathbf{V}_4^*} - \mathbf{Y}_{42}\mathbf{V}_2^{(1)} - \mathbf{Y}_{43}\mathbf{V}_3^{(1)}\right]$$
$$= \frac{1}{(3 - j12)}\left[\frac{-0.3 + j0.1}{1 - j0} - (-1 + j4.0)(-1.01899 + j0.04608) - (-2 + j8.0)(0.990632 - j0.046768)\right]$$
$$= \frac{1}{(3 - j12)}[-0.3 + j\,0.1 + 0.83465 - j\,4.122 + 1.6075 - j\,8.01852]$$
$$= \frac{2.14215 - j12.04052}{(3 - j12)} = (0.98636 - j\,0.06808)$$
$$\mathbf{V}_{4\,\text{acc}}^{(1)} = (1.0 + j\,0) + 1.6\,[(0.98636 - j\,0.6808) - (1 + j\,0)]$$
$$= (0.978176 - j\,10.893)$$ **Ans.**

22.10. NEWTON-RAPHSON METHOD

The power flow problem can also be solved by using Newton-Raphson method. In fact, among the numerous solution methods available for power flow analysis, the Newton-Raphson method is considered to be the most sophisticated and important. Many advantages are attributed to the Newton-Raphson (N-R) approach.

Gauss-Seidel (G-S) is a simple iterative method of solving n number load flow equations by iterative method. It does not require partial derivatives. Newton-Raphson method is based on *Taylor's* series and partial derivatives. The N-R method is recent, needs less number of iterations to reach convergence, takes less computer time hence computation cost is less and the convergence is certain. The N-R method is more accurate, and is insensitive to factors like slack bus selection, regulating transformers etc. and the number of iterations required in this method is almost independent of the system size.

The drawbacks of this method are difficult solution technique, more calculations involved in each iteration resulting in large computer time per iteration and the large requirement of computer memory but the last drawback has been overcome through a compact storage scheme.

Convergence can be considerably speeded up by performing the first iteration through the G-S method and using the values of voltages so obtained for starting the N-R iterations. These voltages are used to compute active power P at every bus except the swing bus and also reactive power Q wherever reactive power is specified. The difference between the specified and calculated values are used to determine the correction of bus voltages. The process of iteration is continued till the difference in the specified and calculated values of P, Q and V are within the given permissible limit.

Before explaining the application of N-R method to the power flow problems, it is useful to review this method in its general form.

Let the unknown variables be $(x_1, x_2, x_3, ..., x_n)$ and the specified quantities $(y_1, y_2, y_3, ..., y_n)$. These are related by the set of nonlinear equations :

$$\begin{aligned} y_1 &= f_1(x_1, x_2, x_3, ..., x_n) \\ y_2 &= f_2(x_1, x_2, x_3, ..., x_n) \\ y_3 &= f_3(x_1, x_2, x_3, ..., x_n) \\ &... \quad ... \quad ... \quad ... \\ y_n &= f_n(x_1, x_2, x_3, ..., x_n) \end{aligned} \qquad ...(22.82)$$

The solution of above nonlinear equations is started with an approximate solution $\left(x_1^0, x_2^0, x_3^0, ... x_n^0\right)$, the zero subscript defining the zeroth iteration in the process of solving the above nonlinear Eq. (22.82). The noteworthy point is that the initial solution for the nonlinear equations should not be very far from the actual solution, otherwise, there are chances of the solution diverging rather than converging and it may not be possible to have a solution whatever the computer time taken. At first glance it may appear to be a great drawback for the N-R method but the problem of initial guess for a power system is not at all difficult. A flat voltage profile, *i.e.*, $\mathbf{V}_i = (1.0 + j\,0)$ for $i = 1, 2, 3 ... n$ except the slack bus has been found to be satisfactory for almost all practical systems.

Let Δx_1^0, Δx_2^0, Δx_3^0, ... Δx_n^0 be the corrections, which on being added to the initial assumed values, give the actual solution. So

$$y_1 = f_1\left(x_1^0 + \Delta x_1^0, x_2^0 + \Delta x_2^0, x_3^0 + \Delta x_3^0, + ..., x_n^0 + \Delta x_n^0\right) \quad ...(22.83)$$

Expanding these equations in Taylor's series around the initial guess, we have

$$y_1 = f_1\left(x_1^0, x_2^0, x_3^0, ..., x_n^0\right) + \left[\Delta x_1^0\left(\frac{\partial f_1}{\partial x_1}\right)^0 + \Delta x_2^0\left(\frac{\partial f_1}{\partial x_2}\right)^0 + \Delta x_3^0\left(\frac{\partial f_1}{\partial x_3}\right)^0 + ... + \Delta x_n^0\left(\frac{\partial f_1}{\partial x_n}\right)^0\right] + \text{higher order terms} \quad ...(22.84)$$

where $\left(\frac{\partial f_1}{\partial x_1}\right)^0, \left(\frac{\partial f_1}{\partial x_2}\right)^0, \left(\frac{\partial f_1}{\partial x_3}\right)^0, ..., \left(\frac{\partial f_1}{\partial x_n}\right)^0$ are the derivatives of f_1 w.r.t. $x_1, x_2, x_3, ..., x_n$ evaluated at $\left(x_1^0, x_2^0, x_3^0, ..., x_n^0\right)$.

Partial derivatives of second and higher order are neglected according to N-R method. In fact it is this assumption that requires the initial solution to be close to the final solution. Linearizing all the equations and arranging them in matrix form, we have

$$\begin{bmatrix} y_1 - f_1\left(x_1^0, x_2^0, x_3^0, ..., x_n^0\right) \\ y_2 - f_2\left(x_1^0, x_2^0, x_3^0, ..., x_n^0\right) \\ y_3 - f_3\left(x_1^0, x_2^0, x_3^0, ..., x_n^0\right) \\ \cdots\cdots\cdots\cdots \\ y_n - f_n\left(x_1^0, x_2^0, x_3^0, ..., x_n^0\right) \end{bmatrix} = \begin{bmatrix} \frac{\partial f_1}{\partial x_1} & \frac{\partial f_1}{\partial x_2} & \frac{\partial f_1}{\partial x_3} & \cdots & \frac{\partial f_1}{\partial x_n} \\ \frac{\partial f_2}{\partial x_1} & \frac{\partial f_2}{\partial x_2} & \frac{\partial f_2}{\partial x_3} & \cdots & \frac{\partial f_2}{\partial x_n} \\ \frac{\partial f_3}{\partial x_1} & \frac{\partial f_3}{\partial x_2} & \frac{\partial f_3}{\partial x_3} & \cdots & \frac{\partial f_3}{\partial x_n} \\ \cdots & \cdots & \cdots & \cdots & \cdots \\ \frac{\partial f_n}{\partial x_1} & \frac{\partial f_n}{\partial x_2} & \frac{\partial f_n}{\partial x_3} & \cdots & \frac{\partial f_n}{\partial x_n} \end{bmatrix} \begin{bmatrix} \Delta x_1^0 \\ \Delta x_2^0 \\ \Delta x_3^0 \\ \cdots \\ \cdots \\ \cdots \\ \Delta x_n^0 \end{bmatrix} \quad ...(22.85)$$

In vector form above equation can be written as

$$\mathbf{B} = \mathbf{J}.\mathbf{C} \quad ...(22.86)$$

where **J** is the square matrix of the partial derivatives on the RHS and is known as **Jacobian matrix**. The solution of the equations needs calculation of left hand vector **B** which is the difference of the specified quantities and calculated quantities at $\left(x_1^0, x_2^0, x_3^0, ..., x_n^0\right)$. Similarly **J** is calculated at this guess. Solution of the matrix equation provides $\left(\Delta x_1^0, \Delta x_2^0, \Delta x_3^0, ..., \Delta x_n^0\right)$ and the better estimates of the solution are given by

$$\begin{aligned} x_1^1 &= x_1^0 + \Delta x_1^0 \\ x_2^1 &= x_2^0 + \Delta x_2^0 \\ x_3^1 &= x_3^0 + \Delta x_3^0 \\ x_n^1 &= x_n^0 + \Delta x_n^0 \end{aligned} \qquad ...(22.87)$$

Repeating the process of iterations, with these values, we get yet better estimated values. The (Δx_1, Δx_2, Δx_3, ..., Δx_n) becomes smaller and smaller with every iteration and finally the iteration process is stopped when (Δx_1, Δx_2, Δx_3, ..., Δx_n) are lesser than pre-specified values.

22.10.1. Newton-Raphson Method Applied to Power Flow Problem

N-R method can be applied to power flow problems in a number of ways, the most common being those using (*i*) rectangular coordinates and (*ii*) polar coordinates.

1. N-R Method Using Rectangular Coordinates. In this formulation the quantities are expressed in rectangular form. The general expression for power is given as

$$P_i - j\,Q_i = \mathbf{V}_i^* I_i = \mathbf{V}_i^* \sum_{k=1}^{n} \mathbf{Y}_{ik} \mathbf{V}_k \quad ...(22.88)$$

Let $\mathbf{V}_i = e_i + j f_i$

where e_i and f_i are the real and imaginary components of tne bus voltage V_i, and therefore

$$\mathbf{V}_i^* = e_i - j f_i \quad ...(22.89)$$

Similarly $\mathbf{V}_k = e_k + j f_k \quad ...(22.90)$

and $\mathbf{Y}_{ik} = G_{ik} - j\,B_{ik} \quad ...(22.91)$

where G_{ik} and B_{ik} are conductance and susceptance respectively. Substituting Eqs. (22.89), (22.90) and (22.91) in Eq. (22.88) for power, we have

$$P_i - j\,Q_i = (e_i - jf_i) \sum_{k=1}^{n} (G_{ik} - jB_{ik})(e_k + jf_k) \quad ...(22.92)$$

Separating the real and imaginary parts, we have

$$P_i = \sum_{k=1}^{n} e_i (e_k G_{ik} + f_k B_{ik}) + f_i (f_k G_{ik} - e_k B_{ik}) \quad ...(22.93)$$

and $$Q_i = \sum_{k=1}^{n} f_i (e_k G_{ik} + f_k B_{ik}) - e_i (f_k G_{ik} - e_k B_{ik}) \quad ...(22.94)$$

Also $V_i^2 = e_i^2 + f_i^2 \quad ...(22.95)$

Separating for ith bus, the power Eqs. (22.93) and (22.94) become

$$P_i = e_i(e_i G_{ii} + f_i B_{ii}) + f_i(f_i G_{ii} - e_i B_{ii}) + \sum_{\substack{k=1\\k\neq i}}^{n} e_i(e_k G_{ik} + f_k B_{ik}) + f_i(f_k G_{ik} - e_k B_{ik}) \quad ...(22.96)$$

and

$$Q_i = f_i(e_i G_{ii} + f_i B_{ii}) - e_i(f_i G_{ii} - e_i B_{ii}) + \sum_{\substack{k=1\\k\neq i}}^{n} f_i(e_k G_{ik} + f_k B_{ik}) - e_i(f_k G_{ik} - e_k B_{ik}) \quad ...(22.97)$$

Thus the above formulation results in a system of nonlinear algebraic equations, two equations (one for P_i and the other for Q_i) at each bus. So excluding the slack bus (bus 1) where V and δ are specified and remains fixed throughout, the total number of equations to be solved for n bus system will be $(2n - 1)$ equations.

With the help of the Newton-Raphson method, the above nonlinear algebraic equations of power is transferred into a set of linear algebraic equations inter-relating the changes in power (i.e., error in power) with the change in real and reactive components of bus voltages with the help of Jacobian matrix. Thus we have from Eqs. (22.93) and (22.94)

$$\begin{bmatrix} \Delta P_2 \\ \Delta P_3 \\ \vdots \\ \Delta P_n \\ \Delta Q_2 \\ \Delta Q_3 \\ \vdots \\ \Delta Q_n \end{bmatrix} = \left[\begin{array}{cccc|cccc} \frac{\partial P_2}{\partial e_2} & \frac{\partial P_2}{\partial e_3} & \cdots & \frac{\partial P_2}{\partial e_n} & \frac{\partial P_2}{\partial f_2} & \frac{\partial P_2}{\partial f_3} & \cdots & \frac{\partial P_2}{\partial f_n} \\ \frac{\partial P_3}{\partial e_2} & \frac{\partial P_3}{\partial e_3} & \cdots & \frac{\partial P_3}{\partial e_n} & \frac{\partial P_3}{\partial f_2} & \frac{\partial P_3}{\partial f_3} & \cdots & \frac{\partial P_3}{\partial f_n} \\ \vdots & \vdots & & \vdots & \vdots & \vdots & & \vdots \\ \frac{\partial P_n}{\partial e_2} & \frac{\partial P_n}{\partial e_3} & \cdots & \frac{\partial P_n}{\partial e_n} & \frac{\partial P_n}{\partial f_2} & \frac{\partial P_n}{\partial f_3} & \cdots & \frac{\partial P_n}{\partial f_n} \\ \hline \frac{\partial Q_2}{\partial e_2} & \frac{\partial Q_2}{\partial e_3} & \cdots & \frac{\partial Q_2}{\partial e_n} & \frac{\partial Q_2}{\partial f_2} & \frac{\partial Q_2}{\partial f_3} & \cdots & \frac{\partial Q_2}{\partial f_n} \\ \frac{\partial Q_3}{\partial e_2} & \frac{\partial Q_3}{\partial e_3} & \cdots & \frac{\partial Q_3}{\partial e_n} & \frac{\partial Q_3}{\partial f_2} & \frac{\partial Q_3}{\partial f_3} & \cdots & \frac{\partial Q_3}{\partial f_n} \\ \vdots & \vdots & & \vdots & \vdots & \vdots & & \vdots \\ \frac{\partial Q_n}{\partial e_2} & \frac{\partial Q_n}{\partial e_3} & \cdots & \frac{\partial Q_n}{\partial e_n} & \frac{\partial Q_n}{\partial f_2} & \frac{\partial Q_n}{\partial f_3} & \cdots & \frac{\partial Q_n}{\partial f_n} \end{array}\right] \begin{bmatrix} \Delta e_2 \\ \Delta e_3 \\ \vdots \\ \Delta e_n \\ \Delta f_2 \\ \Delta f_3 \\ \vdots \\ \Delta f_n \end{bmatrix} \quad ...(22.98)$$

In short form Eq. (22.98) can be written as

$$\left[\begin{array}{c} \Delta P \\ \hline \Delta Q \end{array}\right] = \left[\begin{array}{c|c} J_1 & J_2 \\ \hline J_3 & J_4 \end{array}\right] \left[\begin{array}{c} \Delta e \\ \hline \Delta f \end{array}\right] \quad ...(22.98\ a)$$

In case the system contains all types of buses, the set of equations can be written as

$$\left[\begin{array}{c} \Delta P \\ \hline \Delta Q \\ \hline \Delta V_i^2 \end{array}\right] = \left[\begin{array}{c|c} J_1 & J_2 \\ \hline J_3 & J_4 \\ \hline J_5 & J_6 \end{array}\right] \left[\begin{array}{c} \Delta e \\ \hline \Delta f \end{array}\right] \quad ...(22.98\ b)$$

The elements of the Jacobian matrix can be derived from the three power flow Eqs. (22.96), (22.97) and (22.95).

The off-diagonal elements of J_1 are

$$\frac{\partial P_i}{\partial e_k} = e_i G_{ik} - f_i B_{ik} \quad \text{for } k \neq i \quad ...(22.99)$$

and diagonal elements of J_1 are

$$\frac{\partial P_i}{\partial e_i} = 2e_i G_{ii} + f_i B_{ii} - f_i B_{ii} + \sum_{\substack{k=1\\k\neq i}}^{n} (e_k G_{ik} + f_k B_{ik})$$

$$= 2e_i G_{ii} + \sum_{\substack{k=1\\k\neq i}}^{n} (e_k G_{ik} + f_k B_{ik}) \quad ...(22.100)$$

The off-diagonal elements of J_2 are

$$\frac{\partial P_i}{\partial f_k} = e_i B_{ik} + f_i G_{ik} \quad \text{for } k \neq i \quad ...(22.101)$$

and the diagonal elements of J_2 are

$$\frac{\partial P_i}{\partial f_i} = e_i B_{ii} + 2f_i G_{ii} - e_i B_{ii} + \sum_{\substack{k=1\\k\neq i}}^{n} (f_k G_{ik} - e_k B_{ik})$$

$$= 2f_i G_{ii} + \sum_{\substack{k=1\\k\neq i}}^{n} (f_k G_{ik} - e_k B_{ik}) \quad ...(22.102)$$

The off-diagonal elements of J_3 are

$$\frac{\partial Q_i}{\partial e_k} = e_i B_{ik} + f_i G_{ik} \quad \text{for } k \neq i \quad ...(22.103)$$

and the diagonal elements of J_3 are

$$\frac{\partial Q_i}{\partial e_i} = f_i G_{ii} - f_i G_{ii} + 2e_i B_{ii} - \sum_{\substack{k=1\\k\neq i}}^{n} (f_k G_{ik} - e_k B_{ik})$$

$$= 2e_i B_{ii} - \sum_{\substack{k=1\\k\neq i}}^{n} (f_k G_{ik} - e_k B_{ik}) \quad ...(22.104)$$

The off-diagonal elements of J_4 are

$$\frac{\partial Q_i}{\partial f_k} = -e_i G_{ik} + f_i B_{ii} \quad \text{for } k \neq i \quad ...(22.105)$$

and the diagonal elements of J_4 are

$$\frac{\partial Q_i}{\partial f_i} = e_i G_{ii} + 2f_i B_{ii} - e_i G_{ii} + \sum_{\substack{k=1\\k\neq i}}^{n} (e_k G_{ik} + f_k B_{ik})$$

$$= 2f_i B_{ii} + \sum_{\substack{k=1\\k\neq i}}^{n} (e_k G_{ik} + f_k B_{ik}) \quad ...(22.106)$$

The off-diagonal and diagonal elements of J_5 are

$$\frac{\partial V_i^2}{\partial e_k} = 0 \quad \text{for } k \neq i \quad ...(22.107)$$

and

$$\frac{\partial V_i^2}{\partial e_i} = 2e_i \quad ...(22.108)$$

The off-diagonal and diagonal elements of J_6 are

$$\frac{\partial V_i^2}{\partial f_k} = 0 \quad \text{for } k \neq i \quad ...(22.109)$$

$$\frac{\partial V_i^2}{\partial f_i} = 2 f_i \quad ...(22.110)$$

Algorithm: The steps for solving power flow problem by the N-R method are given below :

1. For the load buses where P and Q are given, we assume the bus voltages magnitude and phase angle for all the buses except the slack bus where V and δ are specified. Normally we have the flat voltage start, *i.e.*, we set the assumed bus voltage magnitude and its phase angle (*i.e.*, the real and imaginary components *e* and *f* of the bus voltages) equal to the slack bus quantities.

2. Substituting this assumed bus voltages (*i.e.*, *e* and *f*) in Eqs. (22.96) and (22.97), we calculate the real and reactive components of power, *i.e.*, P_i and Q_i for all the buses $i = 2, 3, 4, ..., n$ except the slack bus (bus no. 1).

3. Since P_i and Q_i for any bus *i* is given, *i.e.*, specified, the error in power will be

$$\Delta P_i^r = P_{i\,\text{specified}} - P_i^r$$

$$\Delta Q_i^r = Q_{i\,\text{specified}} - Q_i^r \quad ...(22.111)$$

where *r* is an iteration count.

Here P_i^r and Q_i^r are the power calculated with the latest value of bus voltages at any iteration *r*.

4. Then the elements of Jacobian matrix (J_1, J_2, J_3 and J_4) are determined with the latest bus voltages and calculated power Eqs. (22.96) and (22.97).

5. After this the linear set of Eq. (22.98) is solved by iterative technique or by the method of elimination (normally by Gaussian elimination method) to determine the voltage correction, *i.e.*, Δe_i and Δf_i at any bus *i*.

6. This value of voltage correction is used to determine the new estimate of bus voltages as follows:

$$e_i^{r+1} = e_i^r + \Delta e_i^r \quad ...(22.112)$$

$$f_i^{r+1} = f_i^r + \Delta f_i^r$$

where *r* is an iteration count.

7. Now this new estimate of the bus voltage, *i.e.* e_i^{r+1} and f_i^{r+1} is used in Eqs. (22.96) and (22.97) for power to recompute the error in power and thus entire algorithm starting from step 3 as listed above is repeated.

Here in each iteration, the elements of Jacobian are computed as these depend upon the latest voltage estimate and calculated power. The process is continued till the error in power becomes very small.

$$\textit{i.e.}, \ \Delta P < \varepsilon \text{ and } \Delta Q < \varepsilon \quad ...(22.113)$$

where ε is very small number.

2. N-R Method Using Polar Coordinates. The Newton-Raphson method can also be applied to the solution of power flow problem when the bus voltages are expressed in polar form. In fact, only polar form is used in practice because the use of polar form results in a smaller number of equations than the total number of equations involved in rectangular form.

For any *i*th bus, we have

$$\mathbf{V}_i = V_i e^{j\delta i}, \text{ then } \mathbf{V}_i^* = V_i e^{-j\delta i},$$

$$\text{and } \mathbf{V}_k = V_k e^{j\delta k} \quad ...(22.114)$$

$$\text{and } \mathbf{Y}_{ik} = Y_{ik} e^{-j\theta ik}$$

when δ is the phase angle of the bus voltages and θ_{ik} is an admittance angle.

Then according to Eq. 22.57(*b*) for any *i*th bus

$$\mathbf{S}_i^* = P_i - jQ_i = \mathbf{V}_i^* \sum_{k=1}^{n} \mathbf{Y}_{ik}\mathbf{V}_k \,;\ i = 1, 2, ...\ n \quad ...(22.115)$$

Substituting the values of $\mathbf{V}_i^*$, $\mathbf{V}_k$ and $\mathbf{Y}_{ik}$ from Eq. (22.114) in Eq. (22.115), we have

$$P_i - jQ_i = \sum_{k=1}^{n} V_i V_k Y_{ik}\, e^{-j(\theta_{ik} + \delta_i - \delta_k)} \quad ...(22.116)$$

$$\text{Thus } P_i = \text{Real } \mathbf{V}_i^* \sum_{k=1}^{n} \mathbf{Y}_{ik}\mathbf{V}_k = \sum_{k=1}^{n} V_i V_k Y_{ik} \cos(\theta_{ik} + \delta_i - \delta_k)$$

$$= V_i V_i Y_{ii} \cos\theta_{ii} + \sum_{\substack{k=1\\k\neq i}}^{n} V_i V_k Y_{ik} \cos(\theta_{ik} + \delta_i - \delta_k) \quad ...(22.117)$$

$$\text{and } Q_i = \text{Imaginary } \mathbf{V}_i^* \sum_{k=1}^{n} \mathbf{Y}_{ik}\mathbf{Y}_k$$

$$= \sum_{k=1}^{n} V_i V_k Y_{ik} \sin(\theta_{ik} + \delta_i - \delta_k)$$

$$= V_i V_i Y_{ii} \sin\theta_{ii} + \sum_{\substack{k=1\\k\neq i}}^{n} V_i V_k Y_{ik} \sin(\theta_{ik} + \delta_i - \delta_k) \quad ...(22.118)$$

for $i = 2, 3, 4, ...\ n$ because bus 1 is slack bus

Now the linear equation in polar form becomes

$$\begin{bmatrix} \Delta P \\ \Delta Q \end{bmatrix} = \begin{bmatrix} J_1 & J_2 \\ J_3 & J_4 \end{bmatrix} \begin{bmatrix} \Delta\delta \\ \Delta V \end{bmatrix} \quad ...(22.119)$$

where J_1, J_2, J_3 and J_4 are the elements of Jacobian matrix and can be determined from power Eqs. (22.117) and (22.118) as follows:

The off-diagonal and diagonal elements of J_1 are

$$\frac{\partial P_i}{\partial \delta_k} = V_i V_k Y_{ik} \sin(\theta_{ik} + \delta_i - \delta_k) \text{ for } k \neq i \quad ...(22.120)$$

$$\text{and } \frac{\partial P_i}{\partial \delta_i} = -\sum_{\substack{k=1\\k\neq i}}^{n} V_i V_k Y_{ik} \sin(\theta_{ik} + \delta_i - \delta_k) \quad ...(22.121)$$

The off-diagonal and diagonal elements of J_2 are

$$\frac{\partial P_i}{\partial V_k} = V_i Y_{ik} \cos(\theta_{ik} + \delta_i - \delta_k) \text{ for } k \neq i \quad ...(22.122)$$

$$\text{and } \frac{\partial P_i}{\partial V_i} = 2V_i Y_{ii} \cos\theta_{ii} + \sum_{\substack{k=1\\k\neq i}}^{n} V_k Y_{ik} \cos(\theta_{ik} + \delta_i - \delta_k) \quad ...(22.123)$$

The off-diagonal and diagonal elements of J_3 are

$$\frac{\partial Q_i}{\partial \delta_k} = -V_i V_k Y_{ik} \cos(\theta_{ik} + \delta_i - \delta_k) \quad ...(22.124)$$

and $$\frac{\partial Q_i}{\partial \delta_i} = \sum_{\substack{k=1 \\ k \neq i}}^{n} V_i V_k Y_{ik} \cos(\theta_{ik} + \delta_i - \delta_k) \quad ...(22.125)$$

The off-diagonal and diagonal elements of J_4 are

$$\frac{\partial Q_i}{\partial V_k} = V_i Y_{ik} \sin(\theta_{ik} + \delta_i - \delta_k) \text{ for } k \neq i \quad ...(22.126)$$

and $$\frac{\partial Q_i}{\partial V_i} = 2V_i Y_{ii} \sin\theta_{ii} + \sum_{\substack{k=1 \\ k \neq i}}^{n} V_k Y_{ik} \sin(\theta_{ik} + \delta_i - \delta_k) \quad ...(22.127)$$

The elements of Jacobian matrix are computed with the latest voltage estimate and computed power. However, the procedure (*i.e.*, algorithm, here) is the same as that of the rectangular coordinates. The formulation in the polar coordinates take less computational efforts and also needs less memory space.

Approximation to N-R Method. It is well known that the flow of reactive power Q is not much affected by the change in phase angle δ and similarly the flow of active power remains almost inaffected by variation in magnitude of nodal voltage (ΔV). Keeping these facts in mind and using

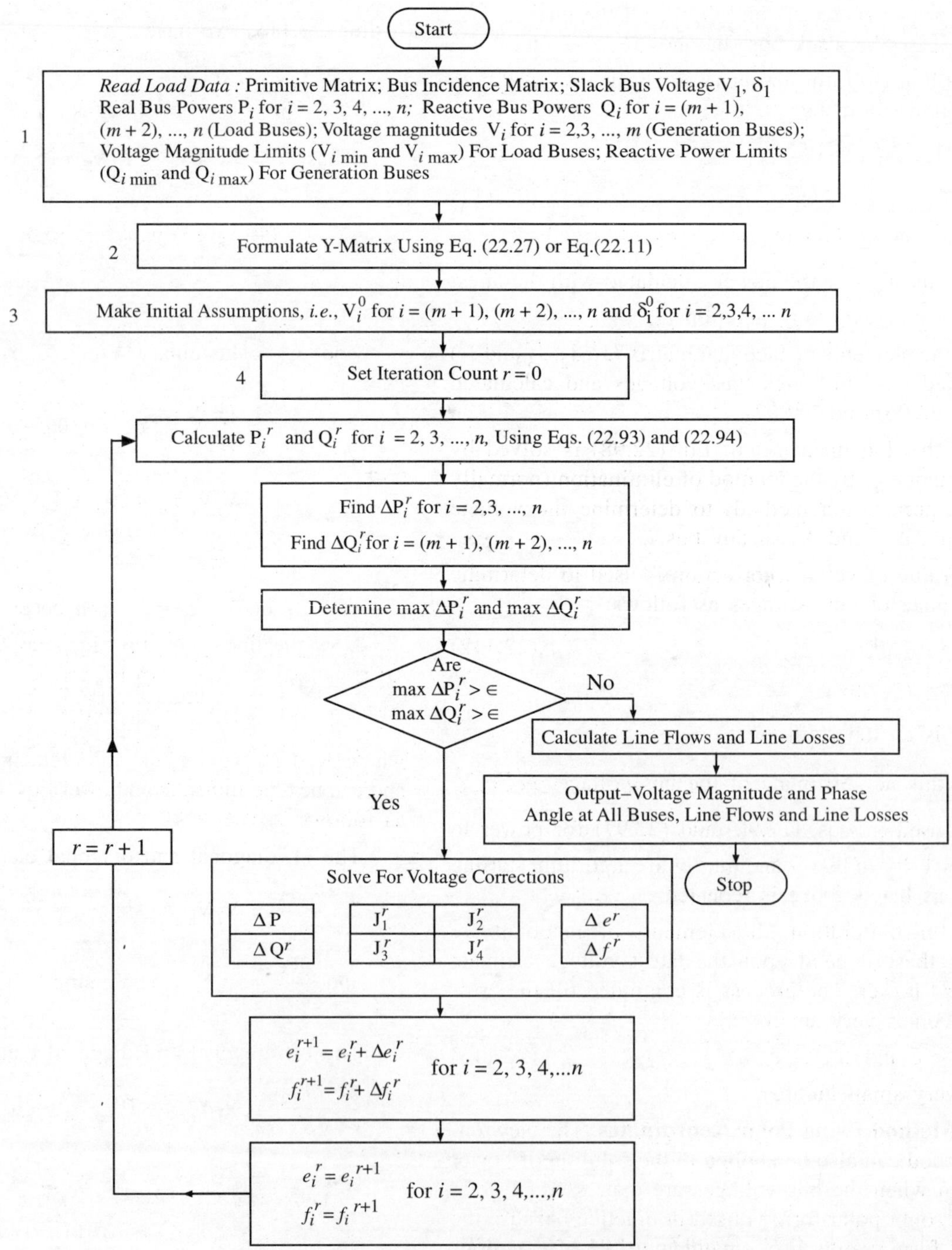

Fig. 22.26. ***Flow Chart For Power Flow Solution Using N–R Method***

the polar coordinates, the set of linear power flow equations can be written in matrix form as follows:

$$\begin{bmatrix} \Delta P \\ \Delta Q \end{bmatrix} = \left[\begin{array}{c|c} J_1 & 0 \\ \hline 0 & J_4 \end{array}\right] \begin{bmatrix} \Delta\delta \\ \Delta V \end{bmatrix} \quad \text{...(22.128)}$$

where J_1 corresponds to the elements $\frac{\partial P}{\partial \delta}$ which exist.

J_2 corresponds to the elements $\frac{\partial P}{\partial V}$ which do not exist and, therefore, are zero.

J_3 corresponds to the elements $\frac{\partial Q}{\partial \delta}$ which do not exist and, therefore, are zero.

J_4 corresponds to the elements $\frac{\partial Q}{\partial V}$ which exist.

This certainly simplifies the calculations and results in small computation time.

Voltage-Controlled or Generation Bus. Here P and the magnitude of voltage V are given. Now the real power P for any bus i is given as

$$P_i = \text{Real } V_i^* \sum_{k \neq 1}^{n} V_k Y_{ik} \quad \text{...(22.129)}$$

and also for ith bus, we have

$$V_i^2 = e_i^2 + f_i^2 \quad \text{...(22.130)}$$

where V_i is the voltage magnitude and e_i and f_i are its real and imaginary components.

The matrix equations interrelating the changes in bus powers and square of the bus voltage magnitude to the changes in the real and imaginary components of voltage are given by Eq. (22.98 b), where $\Delta (V_i^r)^2 = (V_{i\,specified})^2 - (V_i^r)^2$ and V_i^r is the calculated bus voltage after the rth iteration.

Elements of Jacobian matrix are calculated using Eqs. (22.107)...(22.110).

Here V_i^r is the bus voltage computed at the rth iteration and $V_{i\,specified}$ is the voltage specified at any bus i as it is the generation bus.

After obtaining bus voltages, power flow and lines losses are calculated using Eqs. (22.77) and (22.78).

The flow-chart for N-R method using polar coordinates is given in Fig. 22.26.

Example 22.21. A 3-bus system is shown in Fig. 22.27. The series impedance and shunt admittance of each line are (0.0197 + j 0.0788) pu and j 0.04 pu respectively. The bus specification and power input etc. at the bus is as under.

Bus	P_G	Q_G	P_L	Q_L	Bus Voltage
1	–	–	2.0	1.0	1.04 $\angle 0°$ pu
2	0.5	1.0	–	–	Unspecified
3	–	–	1.5	0.6	1.04 $\angle 0°$ pu

Form Y_{bus}, P_2^0, Q_2^0 and P_3^0 and voltage of bus 2 by the N-R method.

Solution : The series impedance of each line is (0.0197 + j 0.0788) pu

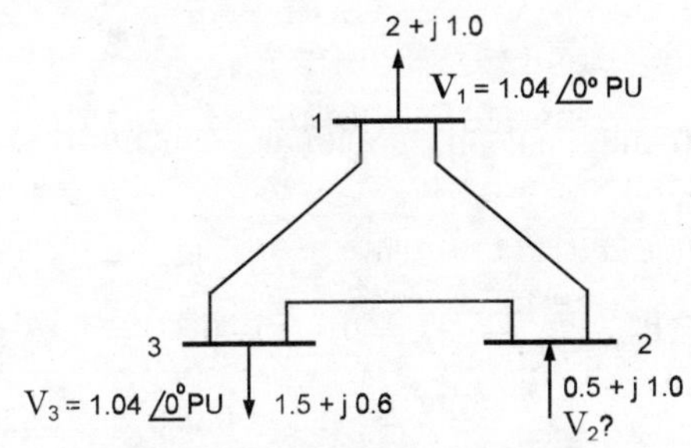

Fig. 22.27

The series admittance of each line

$$= \frac{1}{0.0197 + j0.0788}$$

$$= 12.31 \angle -75.96° = (0.298 - j\,11.94) \text{ pu}$$

So $Y_{12} = Y_{21} = Y_{13} = Y_{31} = Y_{23} = Y_{32}$

$$= -(0.298 - j\,11.94) \quad = -0.298 + j\,11.94$$

$$= 12.31 \angle 104.04° \text{ pu}$$

Each line can be represented by a nominal π circuit. The shunt admittance at each end of line is 0.5 × j 0.04 = j 0.02 pu. Since two lines are connected to each bus and shunt admittance between each bus and ground is j 0.04 pu. So

$Y_{11} = Y_{22} = Y_{33}$

$$= 2\,(0.298 - j\,11.94) + j\,0.04 \text{ pu}$$

$$= (0.596 - j\,23.84) = 24.23 \angle -75.95° \text{ pu}$$

Thus bus matrix

$$Y_{bus} = \begin{vmatrix} 24.23 \angle -75.95° & 12.31 \angle 104.04° & 12.31 \angle 104.04° \\ 12.31 \angle 104.04° & 24.23 \angle -75.95° & 12.31 \angle 104.04° \\ 12.31 \angle 104.04° & 12.31 \angle 104.04° & 24.23 \angle -75.95° \end{vmatrix}$$

Ans.

Taking Bus 1 as slack bus, $n = 3$ and $i = 2$ and using Eq. (22.117), we have

$$P_2^0 = V_2^0 V_1^0 Y_{21} \cos(\theta_{21} + \delta_1^0 - \delta_2^0) + V_2^0 V_3^0 Y_{23} \cos(\theta_{23} + \delta_3^0 - \delta_2^0) + (V_2^0)^2 Y_{22} \cos\theta_{22}$$

$$= 1 \times 1.04 \times 12.31 \cos(104.04°)$$

$$+ 1 \times 1.04 \times 12.31 \cos(104.4°) + 1^2 \times 24.23 \cos(-75.95°)$$

$= -0.33$ pu **Ans.**

$$P_3^0 = V_3^0 V_1^0 Y_{31} \cos(\theta_{31} + \delta_1^0 - \delta_3^0) + V_3^0 V_2^0 Y_{32} \cos(\theta_{32} + \delta_2 - \delta_3) + (V_3^0)^2 Y_{33} \cos\theta_{33}$$

$$= 1.04 \times 1.04 \times 12.31 \cos(104.04°)$$

$$+ 1.04 \times 1\times 12.31 \cos(104.04°) + (1.04)^2 \times 24.23 \cos(-75.95°)$$

$= 0.026$ pu **Ans.**

From Eq (22.118), we have

$$Q_2^0 = -V_2^0V_1^0Y_{21}\sin\left(\theta_{21}+\delta_1^0-\delta_2^0\right)-V_2^0V_3^0Y_{23}\sin\left(\theta_{23}+\delta_2^0-\delta_3^0\right)-\left(V_2^0\right)^2Y_{22}\sin\theta_{22}$$

$$= -1\times 1.04\times 12.31\sin(104.04°) - 1\times 1.04\times 12.31\sin(104.4°) - 1^2\times 24.23\cos(-75.95°)$$

$$= -1.33 \text{ pu}$$ **Ans.**

Using Eq. (22.111), we have

$$\Delta P_2^0 = P_{2\,\text{specified}} - P_2^0 = 0.5 - (-0.33) = 0.83 \text{ pu}$$

$$\Delta P_3^0 = P_{3\,\text{specified}} - P_3^0 = -1.5 - 0.026 = -1.526 \text{ pu}$$

$$\Delta Q_2^0 = 1.0 - (-1.33) = 2.33 \text{ pu}$$

For the given 3-bus system (with V_3 known) Eq. (22.119) becomes

$$\begin{bmatrix}\Delta P_2^0\\ \Delta P_3^0\\ \Delta Q_2^0\end{bmatrix} = \begin{bmatrix}\dfrac{\partial P_2}{\partial\delta_2} & \dfrac{\partial P_2}{\partial\delta_3} & \dfrac{\partial P_2}{\partial V_2}\\ \dfrac{\partial P_3}{\partial\delta_2} & \dfrac{\partial P_3}{\partial\delta_3} & \dfrac{\partial P_3}{\partial V_2}\\ \dfrac{\partial Q_2}{\partial\delta_2} & \dfrac{\partial Q_2}{\partial\delta_3} & \dfrac{\partial Q_2}{\partial V_2}\end{bmatrix}\begin{bmatrix}\Delta\delta_2^0\\ \Delta\delta_3^0\\ \Delta V_2^0\end{bmatrix} \quad \text{..(A)}$$

Differentiating P_i from Eqs. (22.117) and (22.118) and substituting numerical values in above Eq. (A), we have

$$\begin{bmatrix}0.83\\ -1.526\\ 2.33\end{bmatrix} = \begin{bmatrix}24.27 & -12.23 & 5.64\\ -12.23 & 24.95 & -3.05\\ -6.11 & 3.05 & 22.54\end{bmatrix}\begin{bmatrix}\Delta\delta_2^0\\ \Delta\delta_3^0\\ \Delta V_2^0\end{bmatrix} \quad \text{...(B)}$$

By matrix inversion principles to three matrices

$$A = BC \quad \text{then } B^{-1}A = C$$

i.e.,

$$\begin{bmatrix}\Delta\delta_2^0\\ \Delta\delta_3^0\\ \Delta V_2^0\end{bmatrix} = \begin{bmatrix}24.27 & -12.23 & 5.64\\ -12.23 & 24.95 & -3.05\\ -6.11 & 3.05 & 22.54\end{bmatrix}\begin{bmatrix}0.83\\ -1.526\\ 2.33\end{bmatrix} \quad \text{...(C)}$$

Thus, we have

$$\begin{bmatrix}\Delta\delta_2^0\\ \Delta\delta_3^0\\ \Delta V_2^0\end{bmatrix} = \begin{bmatrix}0.05179 & 0.02653 & -0.00937\\ 0.02666 & 0.05309 & 0.00051\\ 0.01043 & -0.00001 & 0.04176\end{bmatrix}\begin{bmatrix}0.83\\ -1.526\\ 2.33\end{bmatrix} \quad \text{...(D)}$$

Solving for ΔV_2^0, we have

$$\Delta V_2^0 = 0.01043\times 0.83 + 0.00001\times 1.526 + 0.04176\times 2.33 = 0.106 \text{ pu}$$

The $V_2^1 = V_2^0 + \Delta V_2^0 = 1 + 0.106 = 1.106$ pu

The process is repeated until convergence is had. Finally we have $V_2 = 1.081\angle{-1.36°}$ pu **Ans.**

22.11. COMPARISON BETWEEN G-S AND N-R METHODS OF POWER FLOW SOLUTION

Gauss-Seidel method is a well established and simple method while the Newton-Raphson method is a recent and most sophisticated method of power flow studies.

From the point of view of computer memory requirements, polar coordinates are preferred for solution based on N-R method and rectangular coordinates for the G-S method. N-R method uses polar coordinates preferably, because if rectangular coordinates are used, it needs more memory.

The time taken to perform one iteration of computation is relatively smaller in case of G-S method as compared to N-R method but the number of iterations required by G-S method for a particular system are more than that required by the N-R method and they increase with the increase in size of the system. In case of N-R method the number of iterations is more or less independent of the size of the system and vary between 3 and 5 iterations. The N-R method takes longer time to perform one iteration because elements of Jacobian are to be computed for each iteration while G-S method takes less time per iteration because it needs fewest number of arithmetic operations to complete one iteration and due to the sparsity of the network matrix and simplicity of solution technique.

In G-S method, the rate of convergence is slow and the convergence characteristic is linear while N-R method has quadrature convergence characteristics. Acceleration factors are used in G-S method for convergence. In N-R method convergence is certain while in G-S method it is not.

G-S method takes more computer time and costs more than N-R method.

The main advantage of G-S method as compared to N-R method is its ease in programming and most efficient use of core memory. However, for large systems the N-R method is faster, more accurate and more reliable than the G-S method.

G-S method is used to compute the solution of small system problems while N-R method is used with advantage for large power systems.

22.12. SPARSITY

In a realistic power system, there may be a large number of buses (say 100 or more) but each bus is connected to only a small number (usually two or three) of the remaining buses. It means Y_{bus} of a large power system is very sparse (*i.e.*, it has a large number of zero elements). In a large system of 100 nodes, the non-zero elements may be as small as 20% of the total elements. The sparsity feature of Y_{bus} minimizes the computer memory requirement (as only non-zero terms need to be stored) and results in faster computations.

22.13. FAST DECOUPLED POWER FLOW METHOD

The fast decoupled power flow method is a very fast and efficient method of obtaining power flow problem solution. In this method, both, the speed as well as the sparsity are exploited. This is actually an extension of Newton-Raphson method formulated in polar coordinates with certain approximations which result into a fast algorithm for power flow solution. This method exploits the property of the power system where in MW flow-voltage angle and MVAR flow-voltage magnitude are loosely coupled. In other words a small change in the magnitude of the bus voltage does not affect the real power flow at the bus and similarly a small change in phase angle of the bus voltage has hardly any effect on reactive power flow. Because of this loose physical interaction between MW and MVAR flows in a power system, the MW-δ and MVAR-V calculations can be decoupled. This decoupling results in a very simple, fast and reliable algorithm. As already said in the preceding article, the sparsity feature of admittance matrix minimizes the computer memory requirements and results in faster computations. The accuracy is comparable to that of the N-R method.

The earlier equation of power flow studies [*i.e.*, Eq. (22.119)] using N-R method can be written in polar coordinates as

$$\begin{bmatrix} \Delta P \\ \Delta Q \end{bmatrix} = \begin{bmatrix} H & N \\ M & L \end{bmatrix} \begin{bmatrix} \Delta \delta \\ \dfrac{\Delta V}{V} \end{bmatrix} \quad ...(22.131)$$

where H, N, M and L are the elements (viz J_1, J_2, J_3 and J_4) of the Jacobian matrix.

Since changes in real power (*i.e.*, Δ P) are less sensitive to the changes in voltage magnitude (*i.e.*, Δ V) and changes in reactive power (*i.e.*, Δ Q) are less sensitive to the changes in phase angle of voltage (*i.e.*, Δ δ), Eq. (22.131) can be reduced to

$$\begin{bmatrix} \Delta P \\ \Delta Q \end{bmatrix} = \begin{bmatrix} H & 0 \\ 0 & L \end{bmatrix} \begin{bmatrix} \Delta \delta \\ \dfrac{\Delta V}{V} \end{bmatrix} \quad ...(22.132)$$

The Eq. (22.132) is decoupled equation and can be expanded as

$$\Delta P = H\Delta\delta \quad ...(22.133)$$

$$\Delta Q = L\frac{\Delta V}{V} \quad ...(22.134)$$

The next step in deriving the algorithm is to make suitable assumptions in deriving the expressions for H and L.

Off-diagonal element of H is

$$H_{ik} = \frac{\partial P_i}{\partial Q_k} = V_i V_k Y_{ik} \sin(\theta_{ik} + \delta_i - \delta_k)$$

$$= V_i V_k Y_{ik} [\sin\theta_{ik} \cos(\delta_i - \delta_k) + \cos\theta_{ik} \sin(\delta_i - \delta_k)]$$

$$= V_i V_k [Y_{ik} \sin\theta_{ik} \cos(\delta_i - \delta_k) + Y_{ik} \cos\theta_{ik} \sin(\delta_i - \delta_k)]$$

$$= V_i V_k [-B_{ik} \cos(\delta_i - \delta_k) + G_{ik} \sin(\delta_i - \delta_k)] \quad ...(22.135)$$

Similarly off-diagonal element of L is

$$L_{ik} = \frac{\partial Q_i V_k}{\partial V_k} = V_i V_k Y_{ik} \sin(\theta_{ik} + \delta_i - \delta_k)$$

$$= V_i V_k [G_{ik} \sin(\delta_i - \delta_k) - B_{ik} \cos(\delta_i - \delta_k)] \quad ...(22.136)$$

From Eqs. (22.135) and (22.136), we have

$$H_{ik} = L_{ik} = V_i V_k [G_{ik} \sin(\delta_i - \delta_k) - B_{ik} \cos(\delta_i - \delta_k)]$$

The diagonal elements of H are given as

$$H_{ii} = \frac{\partial P_i}{\partial \delta_i} = -\sum_{\substack{k=1 \\ k \neq i}}^{n} V_i V_k Y_{ik} \sin(\theta_{ik} + \delta_i - \delta_k)$$

$$= -Q_i + V_i V_i Y_{ii} \sin\theta_{ii} = -Q_i - V_i^2 B_{ii} \quad ...(22.137)$$

Similarly diagonal elements for the matrix are given as

$$L_{ii} = \frac{\partial Q_i V_i}{\partial V_i}$$

$$= 2V_i^2 Y_{ii} \sin\theta_{ii} + \sum_{\substack{k=1 \\ k \neq i}}^{n} V_i V_k Y_{ik} \sin(\theta_{ik} + \delta_i - \delta_k)$$

$$= 2V_i^2 Y_{ii} \sin\theta_{ii} + Q_i - V_i^2 Y_{ii} \sin\theta_{ii}$$

$$= Q_i + V_i^2 Y_{ii} \sin\theta_{ii} = Q_i - V_i^2 B_{ii} \quad ...(22.138)$$

In the case of fast decoupled power flow method of power flow studies the following approximations are made for evaluating Jacobian elements.

$$\cos(\delta_i - \delta_k) \simeq 1$$

$$G_{ik} \sin(\delta_i - \delta_k) \leq B_{ik} \quad ...(22.139)$$

$$Q_i << B_{ii} V_i^2$$

With the above assumptions the Jacobian elements become

$$H_{ik} = L_{ik} = -V_i V_k B_{ik} \text{ for } k \neq i$$

$$\text{and } H_{ii} = L_{ii} = -V_i^2 B_{ii} \quad ...(22.140)$$

With these Jacobian elements Eqs. (22.133) and (22.134) become

$$\Delta P_i = H\Delta\delta = V_i V_k B'_{ik} \Delta\delta_k \quad ...(22.141)$$

$$\text{and } \Delta Q_i = L\frac{\Delta V}{V} = V_i V_k B''_{ik} \frac{\Delta V_k}{V_k} \quad ...(22.142)$$

where B'_{ik} and B''_{ik} are elements of $-B_{ik}$ matrix. Further decoupling and final algorithm for the fast-decoupled power-flow studies are obtained by

(*i*) Omitting from B′, the representation of those network elements that affect MVAR flows, *i.e.*, shunt reactances and off-nominal in phase transformer taps.

(*ii*) Omitting from B″, the angle shifting affects of phase shifters.

(*iii*) Dividing Eqs. (22.141) and (22.142) by V_i and assuming $V_k = 1.0$ pu and also neglecting series resistance in calculating the elements of B′.

With the above assumptions, Eqs. (22.141) and (22.142) for the power flow studies become

$$\frac{\Delta P_i}{V_i} = B'\Delta\delta \quad ...(22.143)$$

$$\text{and } \frac{\Delta Q_i}{V_i} = B''\Delta V \quad ...(22.144)$$

In the above equations B′ and B″ are real and sparse and have similar structures as those of H and L respectively. Since, they contain only network admittances, they are constant and do not change during successive iterations for solution of power flow problem, they need to be evaluated only once and inverted once during the first iteration and then used in all successive iterations. It is due to the nature of Jacobian matrices B′ and B″ and the sparsity of these matrices that the method is fast. In this method of power flow studies, each cycle of iteration consists of one solution for $\Delta\delta$ to update δ and one solution for ΔV to update V. The iterations are continued till ΔP and ΔQ at all load buses and ΔP at all generation buses are within prescribed (or assumed) tolerances.

22.14. IMPORTANCE AND OBJECTIVES OF POWER FLOW STUDIES

The power flow (or load flow) studies are the essential and vital part of power system studies. Data about active and reactive power flows through the branches and the bus voltage under steady state is required by power system engineers. Power flow study provides such data. Power flow studies are extremely important and essential for power system planning, designing, expansion design and for providing guidelines to control room operating engineers (power controllers) in the following activities.

1. Providing operating instructions to generating station and substation control rooms for loading, reactive power compensation, relay settings, tap-setting and switching sequence. Selecting the optimum settings of overcurrent relays.
2. Analysing the effect of rearranging the circuits on the power flows, bus voltages.
3. Preparing software for on-line operation, control and monitoring of the power system.
4. Analysing the effect of temporary loss of generating station or transmission path on the power flow.
5. Knowing the effect of reactive power compensation on bus voltages.
6. Calculation of line losses for different power flow conditions.
7. Evaluation of the operating performance of a power system under normal steady state.
8. Planning expansion of system. Introducing HVDC line, interconnection, EHV ac lines etc.
9. Obtaining initial input data for various other power system studies such as *economic load dispatch, reactive power and voltage, control, state estimation, fault calculations, generation planning,* transmission planning etc.

Exercises

1. Discuss the advantages of using Y_{bus} model of power system network for load flow analysis.
[U.P.S.C. I.E.S. Electrical Engineering II, 1997]
2. Discuss classification of buses in a power system.
[Rajasthan Univ. Power System Analysis 2001, 2005, 2008]
3. Explain the network formulation of bus admittance matrix of a power system and its uses.
[Bharathidasan Univ. Power System Analysis November 1996]
4. With the help of a suitable example explain process of obtaining nodal admittance matrix.
[Power System Analysis; Rajasthan Univ. 2005; Rajasthan Technical Univ. 2010]
5. Derive the equation
$Y_{bus} = A^t Y A$, where Y_{bus} – bus admittance matrix, A – bus incidence matrix, $A^t \rightarrow$ transpose of A and Y $\rightarrow$ primitive admittance matrix. **[U.P.S.C. I.E.S. Electrical Engineering II, 2000]**
6. Derive static load flow equations. Write it in the complex form and real form.
[Rajasthan Technical Univ. Power System Analysis 2011]
7. Explain the necessity of load flow studies in power system. Derive static load flow equations for 2 bus system. Develop the real power and reactive power balance equation. Hence comment on the real and reactive power losses in transmission line. **[W.B. Univ. of Technology Power System-II, 2010]**
8. Derive mathematical model of phase-shifting transformer to be used in power flow analysis.
[Madurai Kamaraj Univ. Power System Analysis Nov. 1996]
9. Discuss different types of tap changing transformers. Obtain partial Y_{Bus} of a tap changing transformer. Show that a tap changing transformer can be represented by a π model only if its turns ratio is real.
[U.P. Technical Univ. Power System Analysis 2007-08]
10. Discuss the classification of the buses for power flow analysis. What is Y_{bus} ? Explain the development of power flow equation.
[U.P. Technical Univ. Power System Analysis 2008-09]
11. Develop necessary equations and describe the load flow solution using Gauss-Seidel method.
[U.P.S.C. I.E.S. Electrical Engineering II, 1999]
12. Formulate the load flow equation and explain the method of solving equations using G-S method.
[Computer Applications To Power System Analysis May 2011]
13. Explain Gauss-Seidel iterative method for solving load flow equations. **[U.P.S.C. I.E.S. Electrical Engineering-II, 2009]**
14. Illustrate the Gauss Seidel iterative procedure for solving power flow equations.
[Rajasthan Univ. Power System Analysis 2005, 2008]
15. Explain how Q-limits at P-V buses and V-limits at P-Q buses are taken into account in Gauss-Seidel and Newton Raphson methods of load flow.
[U.P. Technical Univ. Power System Analysis 2004-05]
16. Classify various types of buses in power system for load flow studies. Discuss clearly with a flow chart the computational procedure for load flow solutions using Gauss-Seidel method when the system contains all types of buses.
[U.P. Technical Univ. Power System Analysis 2009-10]
17. Explain Algorithm for solution of load flow problem using Newton-Raphson method when the system consists of all types of buses (Flow chart not required)
[U.P.S.C. I.E.S. Electrical Engineering-II, 2008]
18. With neat flow chart, explain the load flow study using N-R power flow method. How does the method get modified when PV buses are also present.
[Anna Univ. Chennai Power System Analysis May/June 2011]
19. Develop load flow algorithm with flow chart by Newton - Raphson method using nodal admittance approach.
[U.P. Technical Univ. Power System Analysis 2009-10]
20. What is Newton-Raphson method ? How it is applied for the solution of power flow equations ? Explain with the help of an example.
[U.P. Technical Univ. Power System Analysis 2008-09]

21. Derive the load flow equations in polar form for Newton - Raphson method. Also write the algorithm for the solution of load flow problem using N-R method.

[U.P. Technical Univ. Power System Analysis 2008-09]

22. Formulate the mathematical model in rectangular form for the load flow analysis of a power system considering the voltage controlled buses into account using Newton - Raphson method. How the line flows and line losses are calculated ?

[U.P. Technical Univ. Power System Analysis 2007-08]

23. Draw flow chart of Newton-Raphson method for load flow studies including PV buses. Explain each block of the chart.

[U.P.S.C. I.E.S. Electrical Engineering II, 1997]

24. Draw the flow chart algorithm for Newton-Raphson load flow method. Also compare the Gauss-Siedel and Newton-Raphson methods of load flow.

[G.B. Technical Univ. Power System Analysis 2010-11]

25. Distinguish between voltage control bus and swing bus. What do you mean by Jacobian matrix ? Discuss the NR method of solving SLFE and develop the corresponding flow chart. **[W.B. Univ. of Technology Power System-II, 2010]**

26. Derive the equations of various elements of the Jacobian matrix in case of Newton-Raphson method.

[Rajasthan Technical Univ. Power System Analysis 2011]

27. Explain the comparison of various load flow methods.

[Pb. Technical Univ. Computer Applications To Power System Analysis, May 2007]

28. Write the algorithm and draw the flow chart for load flow studies using fast decoupled N-R method.

[Pb. Technical Univ. Computer Applications To Power System Analysis, December 2007]

29. What are the assumptions made in obtaining fast decoupled load flow equations ? Give steps to solve a load flow problem using Newton-Raphson load flow method.

[U.P. Technical Univ. Power System Analysis 2007-08]

30. Write short notes on the following :

(*i*) Importance of power flow studies.

(*ii*) Classification of bus-bars.

(*iii*) Gauss-Seidel method for solving the load flow equations.

(*iv*) Newton-Raphson method of load flow studies.

[Pb. Univ. December 1994].

(*v*) Fast decoupled method of power flow studies.

Short Answer Type Questions With Answers

Q. 1. What is the purpose served by power flow analysis ?

[Bharathiyar Univ. April 1997]

Ans. Power flow analysis provides a systematic mathematical approach for determination of various bus voltages, their phase angles, active and reactive power flows through different branches, generators and loads under steady state conditions.

Q. 2. Name the three categories of buses.

[Madras Univ. December 1996]

Ans. Depending upon which quantities have been specified, the buses are classified into following three categories;

1. Generation or voltage-controlled or P-V buses,
2. Load or P-Q buses and
3. Slack or swing or reference bus.

Q. 3. What are quantities specified at a generator bus ?

[Bharathidasan Univ. November 1996]

Ans. At a generator bus voltage magnitude and true or active power are specified.

Q. 4. Mention the items specified and not specified at a reference bus. **[Madras Univ. April 1997]**

Ans. At a reference bus, the magnitude and phase angle of the voltage are specified. The unknown quantities are active and reactive power flows.

Q. 5. Distinguish between voltage control bus and swing bus.

[W.B. Univ. of Technology 2010]

Ans. *Voltage-Controlled Bus.* This is also called the P-V bus, and on this bus the voltage magnitude corresponding to generation voltage and true or active power P corresponding to its ratings are specified. Voltage magnitude is maintained constant at a specified value by injection of reactive power. The reactive power generation Q and phase angle δ of the voltage are to be computed.

Slack, Swing or Reference Bus. One of the generation buses in a power system is chosen as *slack* or *swing bus*. At this bus, the magnitude and phase angle of the voltage are specified. The phase angle of the voltage is usually set equal to zero. The active and reactive powers at this bus are to be determined through the solution of equations. The swing bus is a fictitious concept in load flow studies and arises because the I^2R losses of the system are not known precisely in advance for the load flow calculations. Therefore, the total injected power cannot be specified at every single bus. At the swing bus, it is customary to regard the active power as unknown. The difference between the expected and solved output (MW) represents the error in the prior estimate of system I^2R losses. The generators at the swing bus supply the difference of the specified active power to be injected into the system at other buses and the total system output plus losses. If slack bus is not specified, the generation bus usually with maximum active power P is taken as reference bus. There can be more than one slack bus in a given network.

Q. 6. Why is one of the buses taken as slack bus in load flow studies ? **[U.P.S.C. I.E.S. E.E.-II, 2001; Rajasthan Univ. 2001, 2008; M.D. Univ. 2010-11]**

Ans. In a power system, there are mainly two types of buses viz. load and generator buses. For these buses, we have to specify real power P injections, which is positive for generator buses and negative for load buses. The losses remain unknown until the load flow solution is complete. Because of this, one of the generator buses is made to take the additional power to supply transmission losses. This bus is called the slack or swing bus. At this bus, the magnitude and phase angle of the voltage are specified and the active and reactive powers are to be determined through the solution of the equations.

Q. 7. What is necessity of designating one of the bus as a slack bus ?

Ans. The need to designate one of the buses as a slack bus is obviously due to the fact that the system power losses are not known initially and, therefore, the net power flow into the system cannot be fixed in advance.

Q. 8. How is the swing bus selected in a load flow study ? **[Bharathidasan Univ. November 1996]**

OR

What are the guidelines for choosing slack bus ? **[Bharthiyar Univ. Nov. 1996]**

Ans. One of the generation bus is selected as slack bus and the generators at this bus supply the difference between the specified real power injected into the system at other buses and the total output of the system plus system power losses.

Q. 9. Define the bus admittance matrix. **[Bharthiyar Univ. April 1997]**

Ans. General equation for n-bus network based on Kirchhoffs' current law and admittance form is $[\mathbf{I}] = [\mathbf{Y}_{bus}]^* [\mathbf{V}]$ where $[\mathbf{I}]$ is the n-bus current matrix, $[\mathbf{V}]$ is the n-bus voltage matrix and $[\mathbf{Y}_{bus}]$ is called the bus-admittance matrix and above equation is written as

$$\mathbf{I} = \mathbf{Y}_{bus}\,\mathbf{V}$$

where $\mathbf{Y}_{bus} = \begin{bmatrix} Y_{11} & Y_{12} & Y_{13} & \cdots & Y_{1n} \\ Y_{21} & Y_{22} & Y_{23} & \cdots & Y_{2n} \\ Y_{31} & Y_{32} & Y_{33} & \cdots & Y_{3n} \\ \cdots & \cdots & \cdots & \cdots & \cdots \\ Y_{n1} & Y_{n2} & Y_{n3} & \cdots & Y_{nn} \end{bmatrix}$ and is called the bus admittance matrix.

Q. 10. Define tree, co-tree and cut-set.

Ans. *Tree* is a subgraph containing all the nodes of the original graph but no closed path. The elements of the tree are known as branches.

Co-tree is a complement of a tree. The elements of the original graph not included in the tree, form a subgraph which may not necessarily be connected, is known as co-tree.

Cut-set is the minimal set of the elements in the connected graph which when removed, divide the graph in only two parts, *i.e.*, two connected subgraphs.

Q. 11. Give any two operating constraints imposed in load flow studies. **[Madras Univ. April 1997]**

Ans. Two operating constraints may be voltage and reactive power.

Q. 12. Write general expression for the complex power injected by the source into the ith bus of a power system. **[Madras Univ. April 1997]**

Ans. General expression for the complex power (conjugate) injected into ith bus is given as

$$\mathbf{S}_i^* = P_i - j\,Q_i = \mathbf{V}_i^* \sum_{k=1}^{n} \mathbf{Y}_{ik}\mathbf{V}_k \quad i = 1, 2, \ldots n$$

Q. 13. Name any two iterative methods used for solution of load flow problems.

Ans. The most commonly used iterative methods for solution of load flow problems are 1. Gauss-Seidel method and Newton-Raphson method.

Q. 14. Write Gauss-Siedel Iterative formula for solving flow equations. **[M.D. Univ. 2010-11]**

Ans. Gauss-Seidel load flow formula is given as

$$\mathbf{V}_i = \frac{1}{\mathbf{Y}_{ii}}\left[\frac{P_i - jQ_i}{\mathbf{V}_i} - \sum_{\substack{k=1 \\ k\neq i}}^{n} \mathbf{Y}_{ik}\mathbf{V}_k\right] \quad i = 2, 3, \ldots, n \text{ Bus 1 is}$$

a slack bus with specified V.

Q. 15. What is meant by acceleration factor in load flow solution and its best value ? **[Bharthidasan Univ. November 1996]**

Ans. Acceleration factor is used in Gauss-Seidel method of load flow solution to increase the rate of convergence. Acceleration factor = 1.6 is considered to be a good value for power flow studies.

Q. 16. Why an acceleration factor is commonly used in load flow studies using G-S method ?

Ans. In the Gauss-Seidel method, a large number of iterations are required to arrive at the specified (desired) convergence. The rate of convergence can be significantly increased by using accelerating factor to the solution obtained after each iteration. The acceleration factor is a multiplier that enhances correction between the values of voltage in two successive iterations.

If $\Delta x^{(r)}$ and $\Delta y^{(r)}$ indicate the variable change in the $(r-1)$th and rth iteration, the difference variables can be written as:

$$\Delta x^{(r)} = x^{(r)} - x^{(r-1)}$$
$$\Delta y^{(r)} = y^{(r)} - y^{(r-1)}$$

If the difference is multiplied by the acceleration factor α which is larger than unity (usually between 1.5 and 1.7 for the solution of power flow equations), the accelerated iterates are obtained as follows:

$$x_{acc}^{(r)} = x_{acc}^{(r-1)} + \alpha\,\Delta x^{(r)}$$
$$y_{acc}^{(r)} = y_{acc}^{(r-1)} + \alpha\,\Delta y^{(r)}$$

The accelerated values of x and y are used in the iteration process instead of x, y.

Q. 17. Why is Gauss-Seidel method considered superior to Gauss-iterative method ?

Ans. The Gauss-Seidel method uses upgraded iterates as soon as they are available and thus has an advantage over the Gauss-iterative method which takes longer time. Gauss-Seidel method converges much faster than the Gauss-iterative method, *i.e.*, the number of iterations required to obtain the solution is much less in Gauss-Seidel method compared to the Gauss-iterative method.

Q. 18. Which method out of Gauss-Seidel method and Newton-Raphson method is used for large power system ?

Ans. Newton-Raphson method as it needs less number of iterations to reach convergence, takes less computer time, convergence is certain, computer cost is less and it is more accurate and is insensitive to factors like slack bus selection, regulating transformer etc.

Problems

1. For the network shown in Fig. 1 give the total number of elements, nodes, buses and branches. Write the elements of Y_{bus} matrix directly by inspection. **[Madras Univ. Power System Analysis October 1996]**

[Ans. $e = 8$; $m = 4$, $n = 3$, branches = 6;

$$Y_{bus} = \begin{bmatrix} -j45 & j15 & j20 \\ j15 & -j40 & j15 \\ j20 & j15 & -j45 \end{bmatrix}$$ **]**

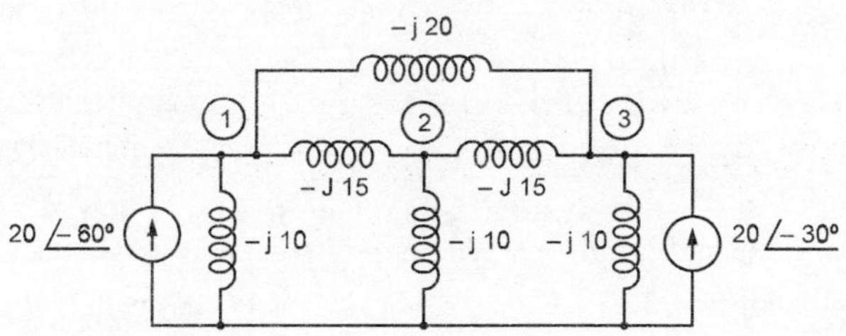

Fig. 1

2. For a system shown in Fig. 2 form the bus admittance matrix. [**Madras Univ. Power System Analysis October 1997**]

$$\text{[Ans. } Y_{bus} = \begin{bmatrix} -j7.5 & j2.5 & j5 \\ j2.5 & -j7.5 & j5 \\ j5 & j5 & -j10 \end{bmatrix}\text{]}$$

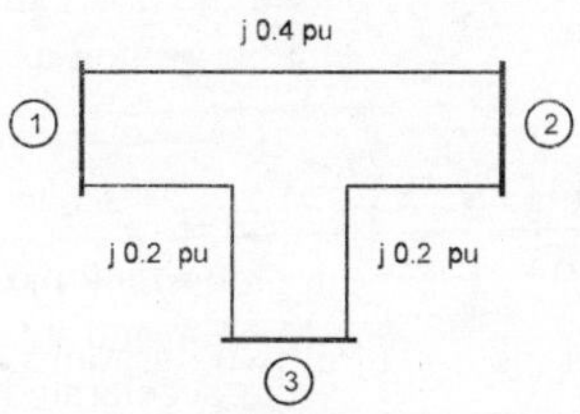

Fig. 2

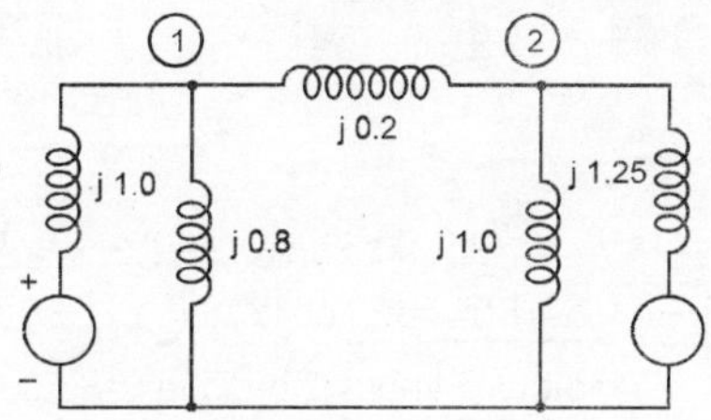

Fig. 3

3. Compute the bus admittance matrix for the system shown in Fig. 3. Per-unit reactances are marked on the figure.

[**Madurai Kamaraj Univ. Power System Analysis Nov. 1996**]

$$\text{[Ans. } Y_{bus} = \begin{bmatrix} -j7.25 & j5 \\ j5 & -j6.8 \end{bmatrix}\text{]}$$

4. One line diagram of a simple 4-bus system is given in Fig. 4. The relevant per unit line impedances are indicated on the diagram. The shunt admittances at the buses may be neglected.

(*a*) Determine Y_{bus} if buses 1 and 2 are not connected, as indicated by dotted line in Fig. 4

(*b*) What modifications will have to be carried out in Y_{bus} if the buses 1 and 2 are connected, as indicated by dotted line in Fig. 4.

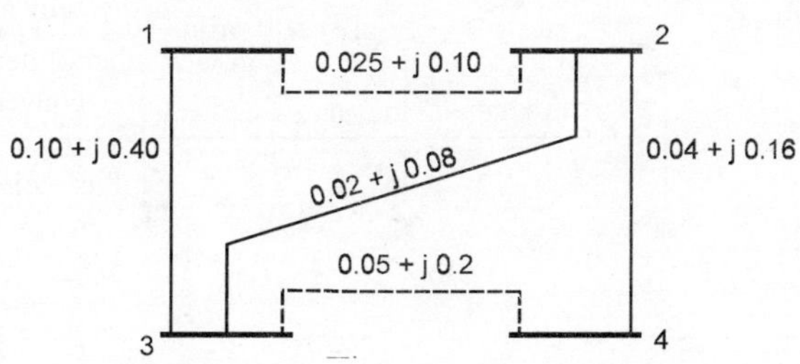

Fig. 4

Ans. (*i*) $Y_{bus} = \begin{bmatrix} 0.588 - j2.353 & 0 & -0.588 + j2.353 & 0 \\ 0 & 4.412 - j17.647 & -2.941 + j11.765 & -1.471 + j5.882 \\ -0.588 + j2.353 & -2.941 + j11.765 & 4.705 - j18.824 & -1.176 + j4.706 \\ 0 & -1.471 + j5.882 & -1.176 + j4.706 & 2.647 - j10.588 \end{bmatrix}$

(*ii*) $Y_{bus} = \begin{bmatrix} 2.941 - j11.765 & -2.353 + j9.412 & -0.588 + j2.353 & 0 \\ -2.353 + j9.412 & 6.765 + j27.059 & -2.941 + j11.765 & -1.471 + j5.882 \\ -0.588 + j2.353 & -2.941 + j11.765 & 4.705 - j18.824 & -1.176 + j4.706 \\ 0 & -1.471 + j5.882 & -1.176 + j4.706 & 2.647 - j10.588 \end{bmatrix}$

5. Determine Y_{bus} for the 4-bus system shown in Fig. 5. The line series impedances are as follows :

Line (Bus To Bus)	**Impedance (Per Unit)**
1-2	0.25 + *j*1.0
1-3	0.20 + *j*0.8
1-4	0.30 + *j*1.2
2-3	0.20 + *j*0.8
3-4	0.15 + *j*0.6

Neglect the shunt capacitances of the lines.

[**Rajasthan Univ. Power System Analysis 2002**]

$$\text{Ans. } Y_{bus} = \begin{bmatrix} 0.725 - j2.9 & -0.235 + j0.94 & -0.294 + j1.176 & -0.196 + j0.784 \\ -0.235 + j0.94 & 0.529 - j2.116 & -0.294 + j1.176 & 0 \\ -0.294 + j1.176 & -0.294 + j1.176 & 0.98 - j3.92 & -0.392 + j1.568 \\ -0.196 + j0.784 & 0 & -0.392 + j1.568 & 0.588 - j2.352 \end{bmatrix}$$

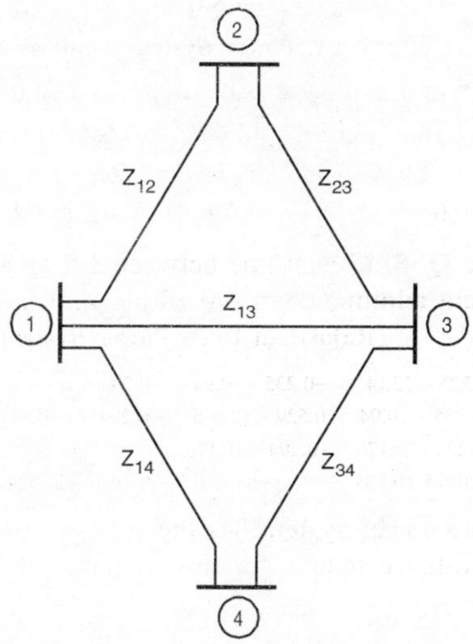

Fig. 5

6. Figure 6 shows a 4 bus system. The shunt admittances at the bus are negligible. The line impedances are as under :

Line (Bus To Bus)	1-2	2-3	3-4	1-4
R (pu)	0.025	0.02	0.05	0.04
X (pu)	0.10	0.08	0.20	0.16

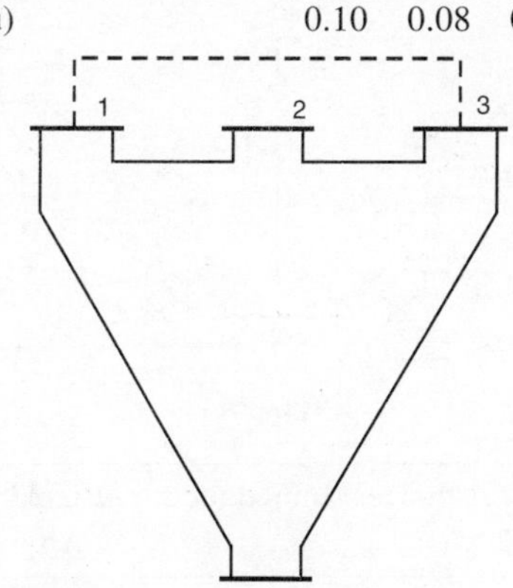

Fig. 6

(*a*) Assume that the line shown dotted (from bus 1 to bus 3) is not present. Formulate Y_{bus}.

(*b*) Which element of the Y_{bus} obtained above are affected when the line from bus 1 to bus 3 is added ? (The new line has no mutual coupling with the other lines).

If the pu impedance of this line is $0.1 + j0.4$, find the new Y_{bus}. **[Rajasthan Univ. Power System Analysis 2000]**

Ans. (*a*) $Y_{bus} = \begin{bmatrix} 3.82 - j15.29 & -2.35 + j9.41 & 0 & -1.47 + j5.88 \\ -2.35 + j9.41 & 5.29 - j21.17 & -2.94 + j11.76 & 0 \\ 0 & -2.94 + j11.76 & 4.116 - j16.466 & -1.176 + j4.706 \\ -1.47 + j5.88 & 0 & -1.176 + j4.706 & 2.646 - j10.586 \end{bmatrix}$

(*b*) $Y_{bus} = \begin{bmatrix} 4.408 - j17.642 & -2.35 + j9.41 & -0.588 + j2.352 & -1.47 + j5.88 \\ -2.35 + j9.41 & 5.29 - j21.17 & -2.94 + j11.76 & 0 \\ -0.588 + j2.352 & -2.94 + j11.76 & 4.704 - j18.818 & -1.176 + j4.706 \\ -1.47 + j5.88 & 0 & -1.176 + j4.706 & 2.646 - j10.586 \end{bmatrix}$

7. Write down the admittance matrix for the following 4-bus system data.

Bus Code	Admittance
1—2	2—*j* 8.0
1—3	1—*j* 4.0
2—3	0.666—*j* 2.664
2—4	1—*j* 4.0
3—4	2—*j* 8.0

[Pb. Univ. Power System Analysis December 1994]

$$\begin{bmatrix} 3 - j12.0 & -2 + j8.0 & -1 + j4.0 & 0.0 \\ -2 + j8.0 & 3.666 - j14.664 & -0.666 + j2.664 & -1 + j4.0 \\ -1 + j4.0 & -0.666 + j2.664 & 3.666 - j14.664 & -2 + j8.0 \\ 0.0 & -1 + j4.0 & -2 + j8.0 & 3 - j12.0 \end{bmatrix}$$

8. In above Q. 5, if each line between 1-2, 1-3 and 1-4 has a total shunt admittance of $-j0.16$ pu, find the modified Y_{bus} matrix. **[Rajasthan Univ. Power System Analysis 2002]**

Ans. $Y_{bus} = \begin{bmatrix} 0.725 - j3.14 & -0.235 + j0.94 & -0.294 + j1.176 & -0.196 + j0.784 \\ -0.235 + j0.94 & 0.529 - j2.196 & -0.294 + j1.176 & 0 \\ -0.294 + j1.176 & -0.294 + j0.176 & 0.98 - j4.0 & -0.392 + j1.568 \\ -0.196 + j0.784 & 0 & -0.392 + j1.568 & 0.588 - j2.432 \end{bmatrix}$

9. A sample power system has the following line data. Form bus admittance matrix for this system.

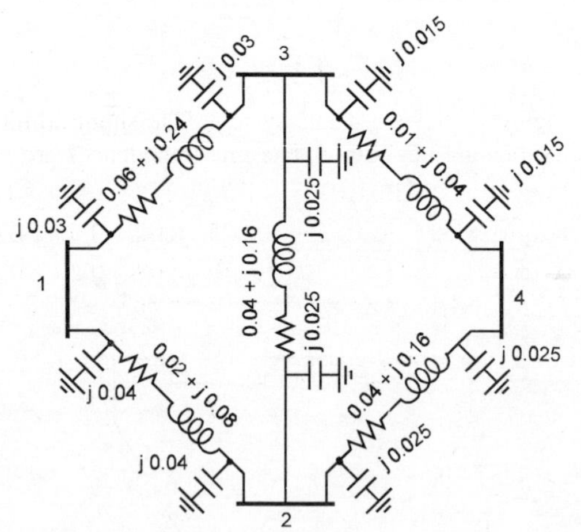

Fig. 7

Bus Code	Series Impedance	PU Line Charging Admittance Y/2
1—2	0.02 + *j* 0.08	0.0 + *j* 0.04
1—3	0.06 + *j* 0.24	0.0 + *j* 0.03
2—3	0.04 + *j* 0.16	0.0 + *j* 0.025
2—4	0.04 + *j* 0.16	0.0 + *j* 0.025
3—4	0.01 + *j* 0.04	0.0 + *j* 0.015

[Bharathiyar Univ. Power System Analysis November, 1996]

Ans. $Y_{bus} = \begin{bmatrix} 3.92 - j15.61 & -2.94 + j11.76 & -0.98 + j3.92 & 0 \\ -2.94 + j11.76 & 5.88 - j23.43 & -1.47 + j5.88 & -1.47 + j5.88 \\ -0.98 + j3.92 & -1.47 + j5.88 & 8.33 - j33.26 & -5.88 + j23.53 \\ 0 & -1.47 + j5.88 & -5.88 + j23.53 & 7.35 - j29.37 \end{bmatrix}$

10. A three-bus system is shown in Fig. 8 below. Line impedances are marked in pu. Treat bus 1 as slack bus with $E_1 = 1.05\angle 0°$ pu and bus # 2 as PQ bus with $P_2 + jQ_2 = -5.96 + j\,1.46$ pu and bus # 3 as PV bus with $E_3 = 1.02$ pu. Calculate the voltage of bus # 2 at the end of first iteration. Use Gauss-Seidel method with acceleration factor 1.4. Assume $E_2^0 = 1.0\angle 0°$ pu and $E_3^0 = 1.02\angle 0°$. Neglect line charging. **[U.P.S.C. I.E.S. Electrical Engineering II, 1998]**

[Ans. (0.9702 – j 0.1942) volt]

Fig. 8

11. A 2-bus system is shown in Fig. 9. $\mathbf{Y}_{11} = \mathbf{Y}_{22} = 1.6 = \angle -80°$ pu and $\mathbf{Y}_{12} = \mathbf{Y}_{21} = 1.9\angle 100°$ pu. Determine the voltages at bus 2 by Gauss-Seidel method after two iterations.

[Ans. $1.047\angle -8.6°$ pu]

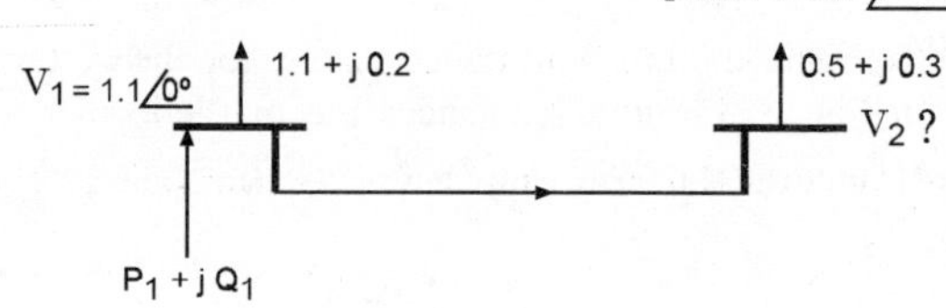

Fig. 9

12. A three-bus system is shown in Fig. 10 below, line impedances are marked in pu. Treat bus #1 at slack bus with $E_1 = 1.05\angle 0°$ pu, bus #2 as PQ bus with $P_2 + jQ_2 = -5.96 + j1.46$ pu and bus #3 as PV bus with $|F_3| = 1.02$ pu. Calculate the voltage of bus #2 at the end of first iteration. Use Gauss-Seidel method with acceleration factor 1.4. Assume $E_2(0) = 1.0\angle 0°$ pu and $E_3(0) = 1.02\angle 0°$ pu. Neglect line charging. **[U.P.S.C. I.E.S. Electrical Engineering-II, 1998]**

[Ans. $(0.9717 - j0.01932)$ pu volts]

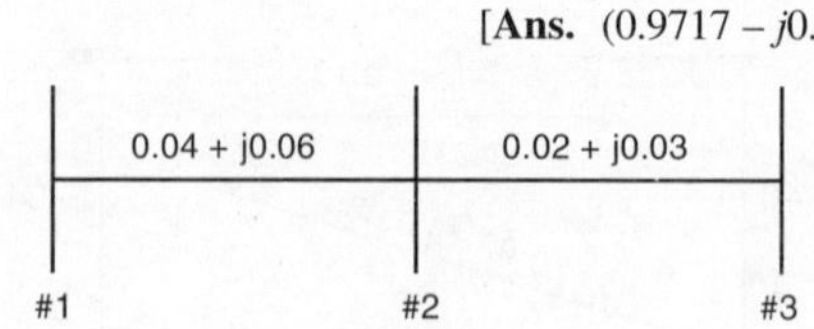

Fig. 10

❑❑❑

OBJECTIVE TYPE QUESTIONS

MULTIPLE CHOICE TYPE QUESTIONS

Note: Each of the following questions comprises multiple responses (two or more than two answers). Select the correct answers. In case you feel that there is more than one correct answer, select the answer which you consider the best. In any case choose ONLY ONE answer or response for each question.

1. Short-circuit currents are due to
(a) single-phase to earth fault.
(b) phase-to-phase fault.
(c) all the three phases-to earth fault.
(d) all the three phases short-circuited.
(e) any of the above.

2. The most serious consequence of a major uncleared short-circuit fault could be
(a) blowing of fuse. (b) fire.
(c) heavy voltage drop. (d) none of these.

3. Which of the following results in a symmetrical fault ?
(a) Single-phase-to earth. (b) Phase-to-phase.
(c) All the three phases-to earth. (d) Two phases-to earth.

4. Which portion of the transmission system is more prone to faults?
(a) Alternator. (b) Transformer.
(c) Overhead lines. (d) Underground cable.

5. Which portion of the power system is least prone to faults ?
(a) Alternators. (b) Switchgear.
(c) Transformers. (d) Overhead lines.

6. The magnitude of fault current depends upon
(a) total impedance up to fault. (b) voltage at the fault point.
(c) load current being supplied before occurrence of fault.
(d) both (a) and (b).

7. The most common type of fault is
(a) single-phase-to ground. (b) phase-to-phase.
(c) two-phase-to ground. (d) three-phase to ground.
[A.M.I.E. Sec B. Summer 1999]

8. The maximum short-circuit current occurs in the case of:
(a) three-phase bolted fault. (b) double-line-to-ground fault.
(c) line-to-line fault. (d) single-line-to-ground fault.
[M.D. Univ. Transmission and Distribution, May 2011]

9. For a fault at the terminals of synchronous generator, the fault current is maximum for a
(a) 3-phase fault. (b) 3-phase to ground fault.
(c) line-to-ground fault. (d) line-to-line fault.
[GATE E.E. 1997]

10. The load currents in short-circuit calculation are neglected because
1. short-circuit currents are much larger than load currents.
2. short-circuit currents are greatly out of phase with load currents.
Which of these statement(s) is/are correct ?
(a) neither 1 nor 2 (b) 2 alone
(c) 1 alone (d) 1 and 2
[U.P.S.C. I.E.S. E.E.-II, 2001]

11. For the power system shown in the figure below, the specifications of the components are the following
G1 : 25 kV, 100 MVA, X = 9%
G2 : 25 kV, 100 MVA, X = 9%
T1 : 25 kV/220 kV, 90 MVA, X = 12%
T2 : 25 kV/25 kV, 90 MVA, X = 12%
Line 1: 220 kV, X = 150 ohms

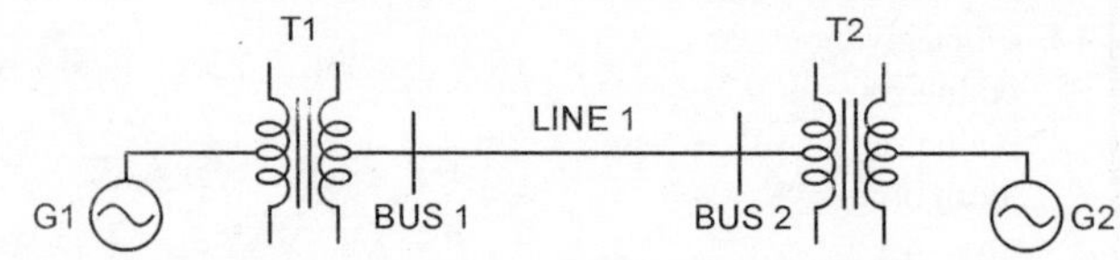

Choose 25 kV as the base voltage at the generator G1, and 200 MVA as the MVA base. The impedance diagram is

(a)
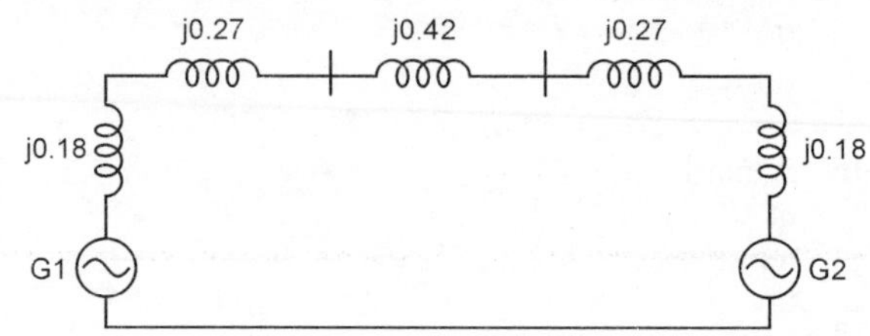

(b)
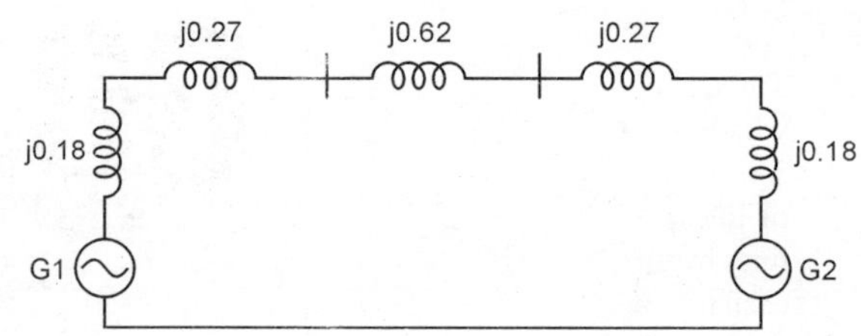

(c)
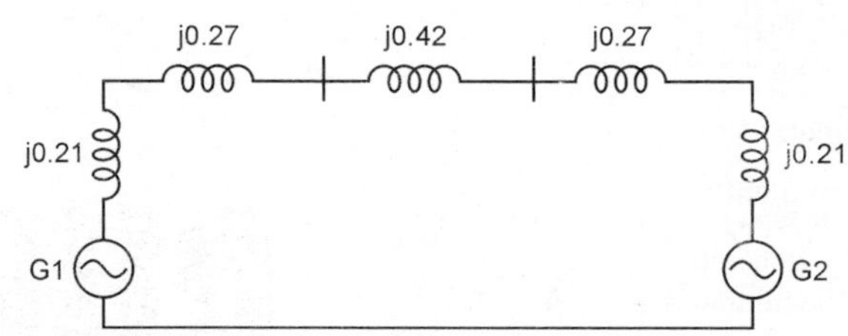

(d)
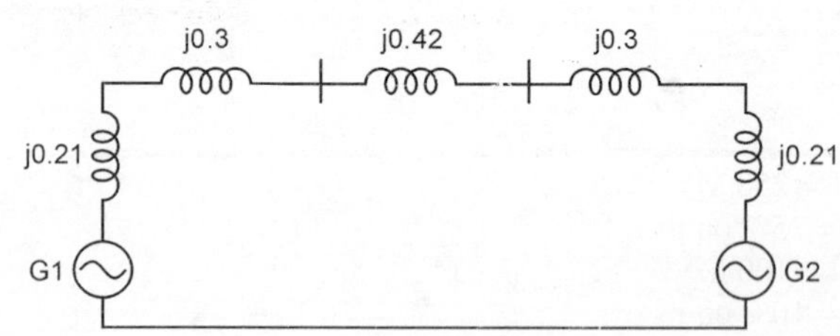

[GATE E.E. 2010]

12. The per unit impedance $Z_{(Pu)}$ in a 3-phase system is

(a) $\dfrac{Z_{(ohms)} \times (MVA)_B}{(KV)_B^2}$ (b) $\dfrac{Z_{(ohms)} \times (KV)_B^2}{(MVA)_B}$

(c) $\dfrac{1000 \times (KV)_B}{\sqrt{3} I_B}$ (d) $\dfrac{1000 \times (KV)_B}{I_B \times 10^6}$

[U.P.S.C. I.E.S. E.E.-II, 2004]

13. Z_{pu}^{old} is the per unit impedance on the power base S_B^{old} and voltage base V_B^{old}. What would be the per unit impedance on the new power base S_B^{new} and voltage base V_B^{new} ?

Answers: 1. (e), **2.** (b), **3.** (c), **4.** (c), **5.** (a), **6.** (d), **7.** (a), **8.** (a), **9.** (c), **10.** (c), **11.** (b), **12.** (a), **13.** (b),

(a) $Z_{pu}^{new} = Z_{pu}^{old} \dfrac{S_B^{old}}{S_B^{new}} \left(\dfrac{V_B^{new}}{V_B^{old}}\right)^2$. (b) $Z_{pu}^{new} = Z_{pu}^{old} \dfrac{S_B^{new}}{S_B^{old}} \left(\dfrac{V_B^{old}}{V_B^{new}}\right)^2$.

(c) $Z_{pu}^{new} = Z_{pu}^{old} \dfrac{S_B^{new}}{S_B^{old}} \left(\dfrac{V_B^{old}}{V_B^{new}}\right)$. (d) $Z_{pu}^{new} = Z_{pu}^{old} \dfrac{S_B^{old}}{S_B^{new}} \left(\dfrac{V_B^{new}}{V_B^{old}}\right)$.

[U.P.S.C. I.E.S. E.E.-II, 2004]

14. For a given base voltage and base volt-amp, the per-unit impedance value of an element is X. What will be the per-unit impedance value of this element when the voltage and a volt-amp bases are both doubled ?

(a) 4X (b) 2X (c) X (d) 0.5X

[U.P.S.C. I.E.S. E.E.-II, 2002]

15. A 10 kVA, 400 V/200 V single-phase transformer with 10% impedance, draws a steady short-circuit current of

(a) 50 A (b) 150 A
(c) 250 A (d) 350 A

[U.P.S.C. I.E.S. E.E.-II, 2010]

16. A transformer rated for 500 kVA, 11 kV/0.4 kV has an impedance of 10% and is connected to an infinite bus. The fault level of the transformer is

(a) 500 kVA (b) 5000 kVA
(c) 500 MVA (d) none of these.

[W.B. Univ. of Technology Power System-II, 2009-10]

17. Three individual 3-phase 400 kV power systems, each one of which consists of a generator transformer and a radial line having 3-ϕ symmetrical fault level of 20 GVA for fault at the remote end of lines, are now reinforced by interconnecting the remote end lines by three lines each of reactance $j12\ \Omega$ so that the fault level at each remote end of the radial lines on ignoring resistance becomes

(a) 10 GVA (b) (20/3) GVA
(c) 24 GVA (d) 20 GVA

[U.P.S.C. I.E.S. E.E.-II, 2010]

Linked Answer Questions

Statement for Linked Answer Questions 18-19:

Two generator units G1 and G2 are connected by 15 kV line with a bus at the midpoint as shown below.

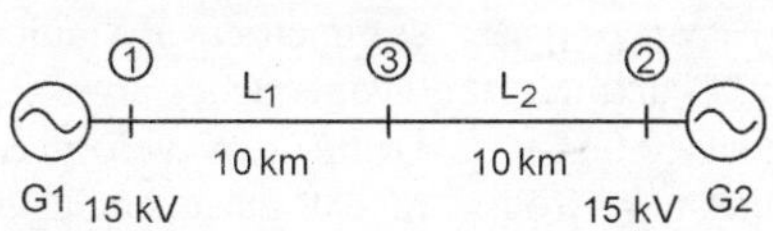

G1 = 250 MVA, 15 kV, positive sequence reactance X = 25% on its own base

G2 = 100 MVA, 15 kV, positive sequence reactance X = 10% on its own base

L_1 and L_2 = 10 km, positive sequence X = 0.225 Ω/km

18. For the above system, the positive sequence diagram with the pu values on the 100 MVA common base is

(a)
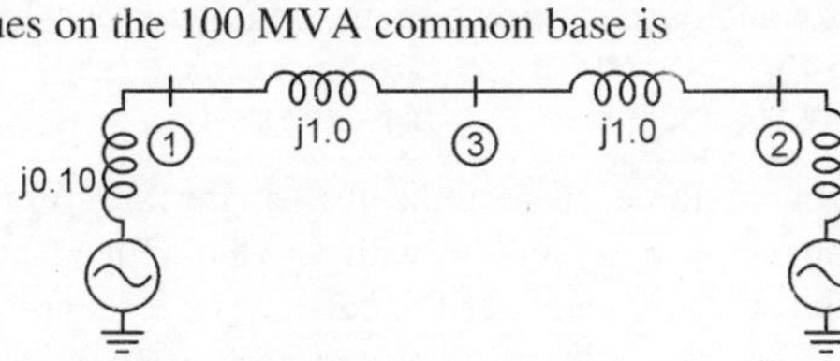

(b)
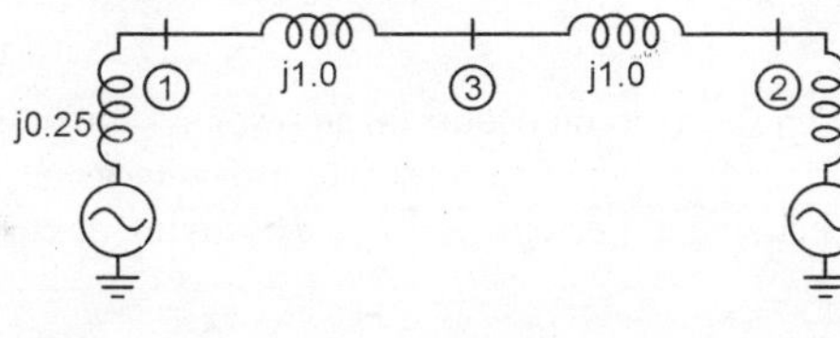

(c)
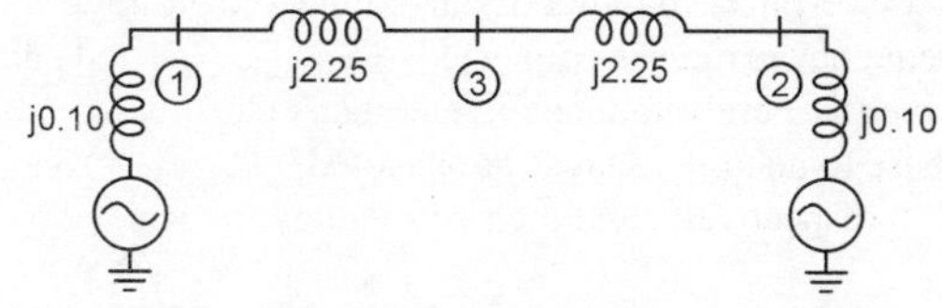

(d)
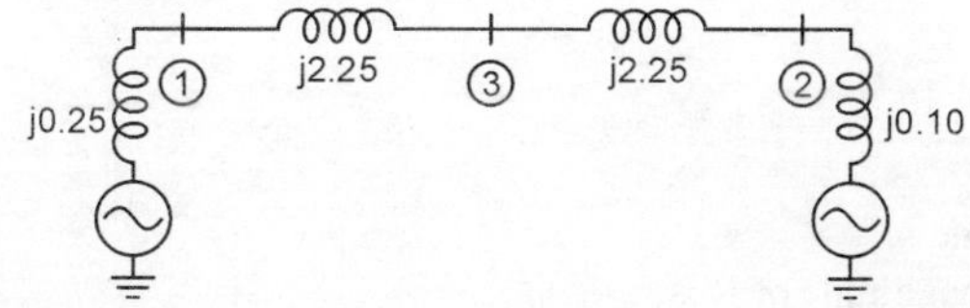

[GATE E.E., 2011]

19. In the above system, the three-phase fault MVA at the bus 3 is

(a) 82.55 MVA (b) 85.11 MVA
(c) 170.91 MVA (d) 181.82 MVA

[GATE E.E., 2011]

20. The per-unit impedance of an alternator corresponding to base values 13.2 kV and 30 MVA is 0.2 p.u. The p.u. value of the impedance for base values of 13.8 kV and 50 MVA in pu will be

(a) 0.131 (b) 0.226 (c) 0.305 (d) 0.364

[U.P.S.C. I.E.S. E.E.-II, 2002]

21. An isolated synchronous generator with transient reactance equal to 0.1 pu on a 100 MVA base is connected to the high voltage bus through a step-up transformer of reactance 0.1 pu on a 100 MVA base. The fault level at the bus is

(a) 1,000 MVA. (b) 500 MVA.
(c) 50 MVA. (d) 10 MVA.

[U.P.S.C. I.E.S. E.E.-II, 2000]

22. Four alternators, each rated at 5 MVA, 11 kV, with 20% reactance are working in parallel. The short-circuit level at bus-bars is

(a) 6.25 MVA. (b) 20 MVA.
(c) 25 MVA. (d) 100 MVA.

[U.P.S.C. I.E.S. E.E.-II, 1999]

23. A generator is connected through a 20 MVA, 13.8/138 kV step-down transformer, to transmission line. At the receiving end of the line a load is supplied through a step-down transformer of 10 MVA, 138/69 kV rating. A 0.72 pu load, evaluated on load side transformer ratings as base values, is supplied from the above system. For system base values of 10 MVA and 69 kV in load circuit, the value of the load (in per unit) in generator circuit will be

(a) 36 (b) 1.44 (c) 0.72 (d) 0.18

[GATE E.E. 2006]

24. A new generator having $E_g = 1.4\angle 30^\circ$ pu [equivalent to $(1.212 + j0.70)$ pu] and synchronous reactance 'X_s' of 1.0 pu on the system base, is to be connected to a bus having voltage V, in the existing power system. This existing power system can be represented by Thevenin's voltage $E_{th} = 0.9\angle 0^\circ$ pu in series with Thevenin's impedance $Z_{th} = 0.25\angle 90^\circ$ pu. The magnitude of the bus voltage V_t of the system in pu will be

(a) 0.990 (b) 0.973 (c) 0.963 (d) 0.900

[GATE E.E. 2004]

25. A 3-phase generator rated at 110 MVA, 11 kV is connected through circuit breakers to a transformer. The generator is having direct axis sub-transient reactance $X''_d = 19\%$, transient reactance $X'_d = 26\%$ and synchronous reactance = 130%. The generator is operating at no load and rated voltage when a three phase short circuit fault occurs between the breakers and the transformer. The magnitude of initial symmetrical rms current in the breakers will be

(a) 4.44 kA (b) 22.20 kA (c) 30.39 kA (d) 38.45 kA

[GATE E.E. 2004]

Answers: 14. (d), **15.** (c), **16.** (c), **17.** (c), **18.** (a), **19.** (d), **20.** (c), **21.** (b), **22.** (d), **23.** (a), **24.** (b), **25.** (c),

26. A three-phase balanced star-connected voltage source with frequency ω rad/s is connected to a star-connected balanced load which is purely inductive. The instantaneous line currents and phase to neutral voltages are denoted by (i_a, i_b, i_c) and (V_{an}, V_{bn}, V_{cn}) respectively, and their rms values are denoted by V and I.

If $R = [V_{an}, V_{bn}, V_{cn}]\begin{bmatrix} 0 & \frac{1}{\sqrt{3}} & -\frac{1}{\sqrt{3}} \\ -\frac{1}{\sqrt{3}} & 0 & \frac{1}{\sqrt{3}} \\ \frac{1}{\sqrt{3}} & -\frac{1}{\sqrt{3}} & 0 \end{bmatrix}\begin{bmatrix} i_a \\ i_b \\ i_c \end{bmatrix}$, then the magnitude of R is

(a) 3VI (b) VI (c) 0.7VI (d) 0

[GATE E.E. 2007]

27. The diagram shows the layout of a power station having two generators A and B, connected to the 11 kV buses which are also fed through two transformers C and D from a 132 kV grid. The 11 kV buses are interconnected through a reactor R.

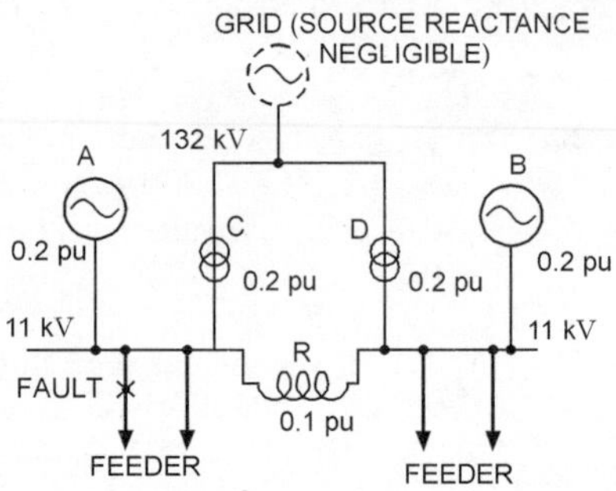

The reactances of A, B, C, D and R are in pu on a common MVA- and kV- base. All the generated voltages in A, B and grid are each 1.0 pu and assumed as in phase at the time of fault.

The steady state symmetrical fault-current for a 3-phase fault on the 11 kV feeders is

(a) 10 pu (b) 15 pu (c) 20 pu (d) 25 pu

[U.P.S.C. I.E.S. E.E.-II, 2001]

28. Four identical 100 MVA, 33 kV generators are operating in parallel, as shown below, in two bus-bar sections, interconnected through a current limiting reactor of x pu reactance on the generator-base. Each generator has a reactance of 0.2 pu. The value of the reactor x to limit a symmetrical short-circuit (a-b-c) current through the circuit breaker to 1,500 MVA is

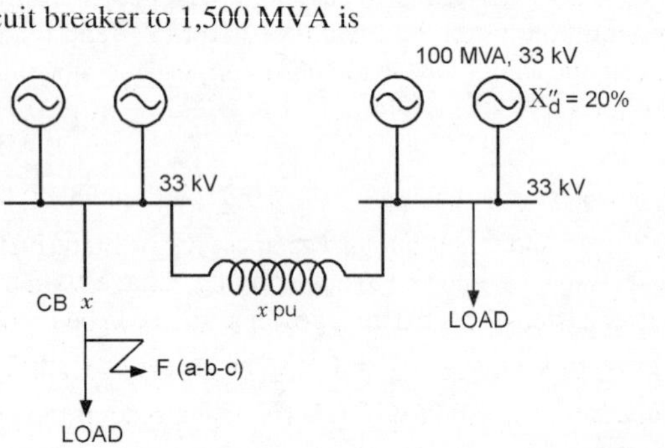

(a) 0.05 pu (b) 0.10 pu (c) 0.15 pu (d) 0.20 pu

[U.P.S.C. I.E.S. E.E.-II, 2003]

29. Two power stations S_1 and S_2 are interconnected through a transmission line of per unit (pu) reactance of 0.4. Station S_1 has one generator of pu reactance of 0.2, and station S_2 has two generators of pu reactance 0.4 each.

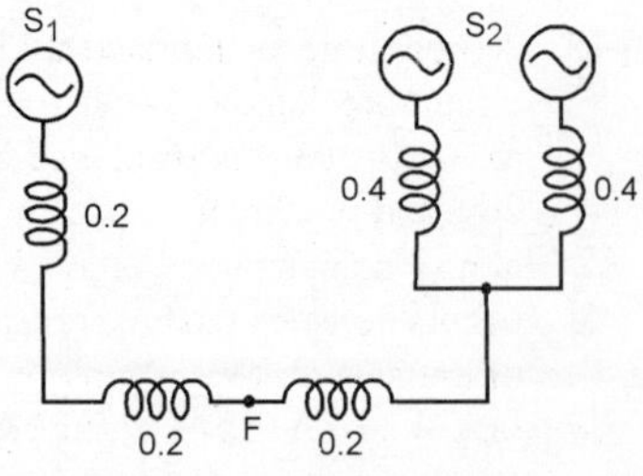

For a 3-phase symmetrical short circuit at the middle of the transmission line (point F), what is the change in the value of the equivalent fault reactance, with one of the generators of S_2 removed?

(a) 10% (b) 20% (c) 30% (d) 40%

[U.P.S.C. I.E.S. E.E.-II, 2007]

Data for Q. 30, and Q. 31 are given below. Solve the problems and choose the correct answers.

For a power system the admittance and impedance matrices for the fault studies are as follows.

$$Y_{bus} = \begin{bmatrix} -j8.75 & j1.25 & j2.50 \\ j1.25 & -j6.25 & j2.50 \\ j2.50 & -j2.50 & -j5.00 \end{bmatrix}$$

$$Z_{bus} = \begin{bmatrix} j0.16 & j0.08 & j0.12 \\ j0.08 & j0.24 & j0.16 \\ j0.12 & j0.16 & j0.34 \end{bmatrix}$$

The prefault voltages are 1.0 pu at all the buses. The system was unloaded prior to the fault. A solid 3 phase fault takes place at bus 2.

30. The post fault voltages at buses 1 and 3 in per unit respectively are

(a) 0.24, 0.63 (b) 0.31, 0.76
(c) 0.33, 0.67 (d) 0.67, 0.33 [GATE E.E. 2006]

31. The per unit fault fed from generators connected to buses 1 and 2 respectively are

(a) 1.20, 2.51 (b) 1.55, 2.61
(c) 1.66, 2.50 (d) 5.00, 2.50 [GATE E.E. 2006]

Data for Q. 32, Q. 33 and Q. 34 are given below. Solve the problems and choose the correct answers.

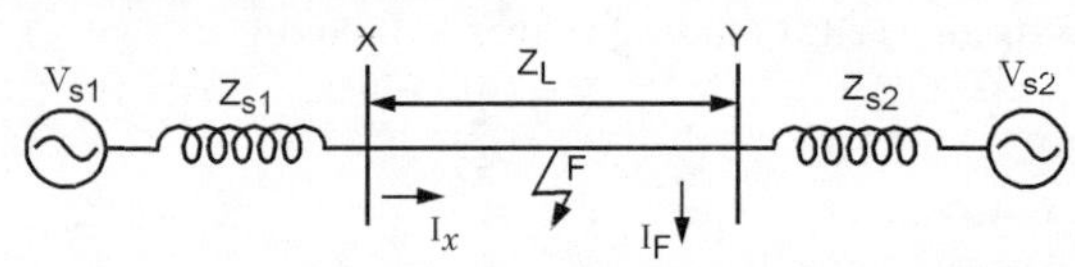

Given that: $V_{s1} = V_{s2} = 1 + j0$ pu, +ve sequence impedances are $Z_{s1} = Z_{s2} = 0.001 + j0.01$ pu and $Z_L = 0.006 + j0.06$ pu, 3ϕ. Base MVA = 100, voltage base = 400 kV(L – L).

Nominal system frequency = 50 Hz. The reference voltage for phase 'a' is defined as $V(t) = V_m \cos(\omega t)$. A symmetrical 3-ϕ fault occurs at centre of the line, i point 'F' at time t_0 the +ve sequence impedance from source S, to point 'F' equals (0.004 + j0.04) pu. The waveform corresponding to phase 'a' fault current from bus X reveals that decaying dc offset current is –ve and in magnitude at its maximum initial value. Assume that the negative sequence are equal to +ve sequence impedances and the zero sequence (Z) are 3 times +ve sequence (Z).

32. The instant (t_0) of the fault will be

(a) 4.682 ms (b) 9.667 ms
(c) 14.667 ms (d) 19.667 ms [GATE E.E. 2008]

33. The rms value of the ac component of fault current (I_x) will be

(a) 3.59 kA (b) 5.07 kA
(c) 7.18 kA (d) 10.15 kA [GATE E.E. 2008]

34. Instead of the three phase fault, if a single line-to ground fault occurs on phase 'a' at point 'F' with zero fault impedance, then the rms of the ac component of fault current (I_x) for phase 'a' will be

(a) 4.97 pu (b) 7.0 pu (c) 14.93 pu (d) 29.85 pu

[GATE E.E. 2008]

35. If a sudden short circuit occurs on a power system (considered as R-L series circuit), the current waveform consists of

1. a decaying ac current. 2. a decaying dc current.

Answers: **26.** (a), **27.** (b), **28.** (b), **29.** (b), **30.** (d), **31.** (d), **32.** (a), **33.** (c), **34.** (a), **35.** (a),

Let the alternator reactance be X and the power system resistance R. Which one of the following is correct ?
(a) The decay in (1) is caused by the increase in X but in (2) is caused by R.
(b) The decay in (1) is caused by R but in (2) is caused by increase in X.
(c) The decay in both (1) and (2) is caused by R.
(d) The decay in both (1) and (2) is caused by the increase in X.
[U.P.S.C. I.E.S. E.E.-II, 2004]

36. Series reactors are used to
(a) improve the transmission efficiency.
(b) improve the power factor of the power system.
(c) improve the voltage regulation.
(d) bring down the fault level within the capacity of the switchgear.

37. Current limiting reactors may be
(a) air-cored air-cooled.
(b) oil immersed magnetically shielded.
(c) oil immersed non-magnetically shielded.
(d) any of the above.

38. The symmetrical components are used in the fault analysis because
(a) the number of equations becomes smaller.
(b) the sequence networks do not have mutual couplings.
(c) the results are required in terms of symmetrical components.
[U.P.S.C. I.E.S. E.E.-II, 2000]

39. In a star-connected system without neutral grounding, zero-sequence currents are
(a) zero. (b) phasor sum of phase currents.
(c) same as rms value of phase currents.
(d) same as peak value of phase currents.

40. A balanced 3-phase system consists of
(a) zero-sequence currents only.
(b) positive-sequence currents only.
(c) negative- and zero-sequence currents.
(d) zero, negative and positive sequence currents.

41. The positive sequence current of a transmission line is:
(a) always zero.
(b) 1/3 of negative sequence current.
(c) equal to negative sequence current.
(d) 3 times negative sequence current.
[M.D. Univ. Transmission Distribution May 2011]

42. In case of an unbalanced star-connected load supplied from an unbalanced 3-phase, 3-wire system, load currents will consist of
(a) positive-sequence components.
(b) negative-sequence components.
(c) zero-sequence components. (d) only (a) and (b).

43. For an unbalanced fault, with paths for zero-sequence currents, at the point of fault
(a) the negative- and zero-sequence voltages are minimum.
(b) the negative- and zero-sequence voltages are maximum.
(c) the negative-sequence voltage is minimum and zero-sequence voltage is maximum.
(d) the negative-sequence voltage is maximum and zero-sequence voltage is minimum. **[GATE E.E. 1996]**

44. When a line-to-ground fault occurs, the current in the faulted phase is 100 A. The zero-sequence current in this case will be:
(a) zero (b) 33.3 A
(c) 66.6 A (d) 100 A
[M.D. Univ. Transmission and Distribution May 2011]

45. Zero-sequence fault current is absent when fault is
(a) single-line-to ground fault.
(b) line-to-line ground fault.
(c) double-line-to ground fault.
(d) line-to-line. **[W.B. Univ. of Technology Power System-II, 2010]**

46. The current of a single-phase load drawn from a 3-phase system has:
(a) Zero-sequence component of current as zero.
(b) Negative-sequence component of current more than positive sequence component.
(c) Positive-, negative- and zero-sequence components equal.
(d) Negative-sequence component of current less than positive-sequence component. **[U.P.S.C. I.E.S. E.E.-II, 2011]**

47. Zero-sequence currents can flow from a line to transformer bank if the windings are in
(a) grounded star/delta. (b) delta/star.
(c) star/grounded star. (d) delta/delta.
[U.P.S.C. I.E.S. E.E.-II, 1997]

48. If the positive-, negative- and zero-sequence reactances of an element of a power system are 0.3, 0.3 and 0.8 pu respectively, then the element would be a
(a) synchronous generator. (b) synchronous motor.
(c) static load. (d) transmission line.
[U.P.S.C. I.E.S. E.E.-II, 2000]

49. Negative-sequence reactance of a transformer is
(a) equal to the positive-sequence reactance.
(b) larger than the positive-sequence reactance.
(c) smaller than the positive-sequence reactance.
[A.M.I.E. Sec B. Summer 1998]

50. If all the sequence voltages at the fault point in a power system are equal, then the fault is a
(a) three-phase fault (b) line-to ground fault
(c) line-to-line fault (d) double-line-to ground fault
[U.P.S.C. I.E.S. E.E.-II, 2001]

51. The zero-sequence current of a generator for line-to ground fault is $j2.4$ pu. Then the current through the neutral during the fault is
(a) $j2.4$ pu (b) $j0.8$ pu (c) $j7.2$ pu (d) $j0.24$ pu
[U.P.S.C. I.E.S. E.E.-II, 20[illegible]]

52. The line currents of a 3-phase power supply are:

$$I_R = 3 + j5A$$
$$I_Y = 2 + j2A$$
$$I_B = -2 - j1A$$

The zero-sequence current will be:
(a) $1 + j2$ A (b) $5 + j7$ A
(c) $1 + j4$ A (d) $-2 - j1$ A
[U.P.S.C. I.E.S. E.E.-II, 2011]

53. The zero-sequence circuit of the three-phase transformer shown in the figure is

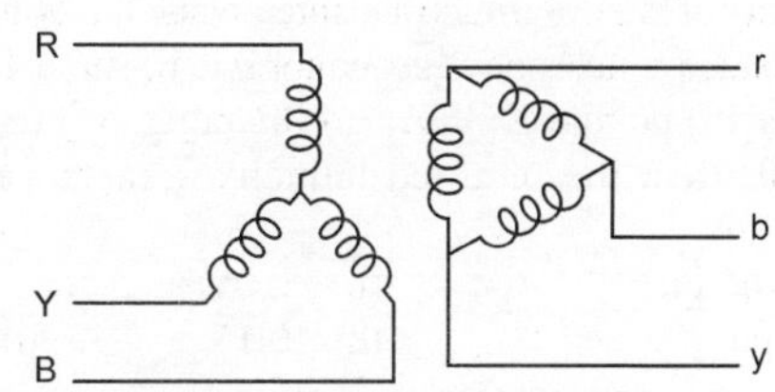

(a) 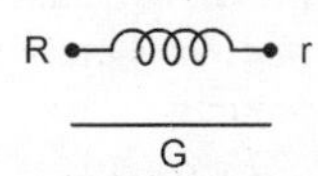

(b) 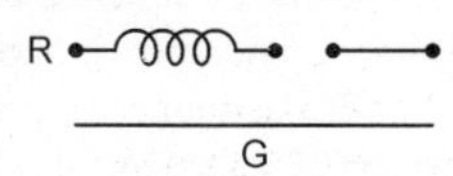

Answers: **36.** (d), **37.** (d), **38.** (b), **39.** (a), **40.** (b), **41.** (c), **42.** (d), **43.** (b), **44.** (b), **45.** (d), **46.** (c), **47.** (a), **48.** (d), **49.** (a), **50.** (d), **51.** (c), **52.** (a), **53.** (c),

(c) 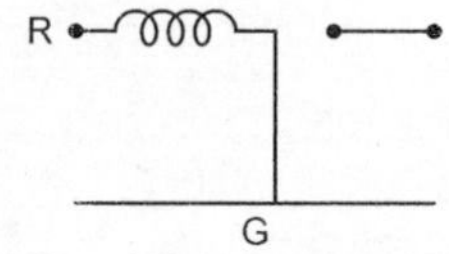

(d) 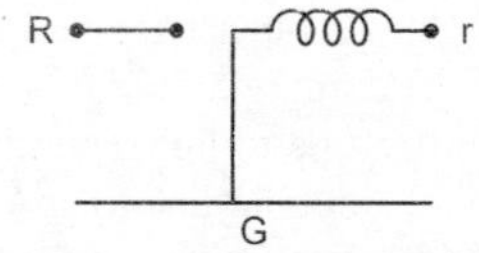

[GATE E.E. 2010]

54. One current transformer (CT) is mounted over a 3-phase 3-core cable with its sheath and armour removed from the portion covered by the CT. An ammeter placed in the CT secondary would measure
(a) the positive-sequence current.
(b) the negative-sequence current.
(c) the zero-sequence current.
(d) three times the zero-sequence current. [U.P.S.C. I.E.S. E.E.-II, 1998]

55. Consider the following statements regarding the fault analysis :
1. The neutral grounding impedance Z_n appears as $3Z_n$ in zero-sequence equivalent circuit.
2. For faults on transmission lines, 3-phase fault is the least severe amongst other faults.
3. The positive- and negative-sequence networks are not affected by method of neutral grounding.

Which of the statements given above are correct ?
(a) 2 and 3 (b) 1 and 2 (c) 1 and 3 (d) 1, 2 and 3
[U.P.S.C. I.E.S. E.E.-II, 2004]

56. A 3-phase transmission line supplies Δ-connected load Z. The conductor '*c*' of the line develops an open circuit fault as shown in figure. The currents in the lines are as shown on the diagram. The +ve sequence current component in line 'a' will be

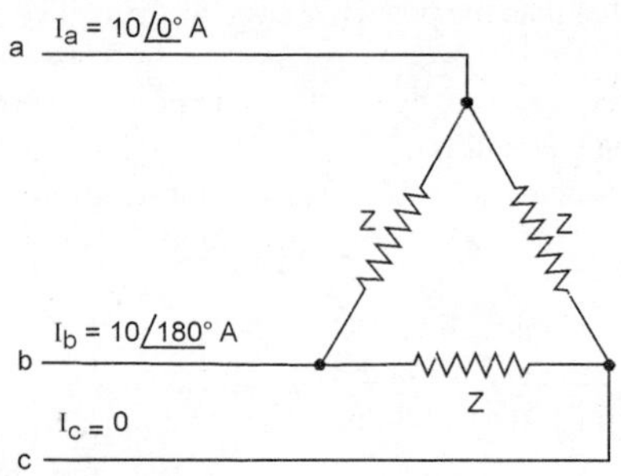

(a) $5.78\angle -30°$ (b) $5.78\angle 90°$
(c) $6.33\angle 90°$ (d) $10.00\angle -30°$
[GATE E.E. 2004]

57. A star-connected 3-phase 11 kV, 25 MVA alternator with its neutral grounded through a 0.033 pu reactance (based on the alternator rating) has positive-, negative- and zero-sequence reactances of 0.2 pu, 0.1 pu and 0.1 pu respectively. A single line to ground fault on one of its terminals would result in a fault MVA of
(a) 150 MVA (b) 125 MVA
(c) 100 MVA (d) 50 MVA
[U.P.S.C. I.E.S. E.E.-II, 1999]

58. The severity of line-to-ground and three-phase faults at the terminals of an unloaded synchronous generator is to be same. If the terminal voltage is 1.0 pu and $\mathbf{Z}_1 = \mathbf{Z}_2 = j\,0.1$ pu, $\mathbf{Z}_0 = j\,0.05$ pu for the alternator, then the required inductive reactance for neutral grounding is
(a) 0.0166 pu (b) 0.05 pu
(c) 0.1 pu (d) 0.15 pu [GATE E.E. 2000]

59. The positive-sequence current for a L-L fault of a 2 kV system is 1,400 A, and corresponding current for a L-L-G fault is 2,220 A. The zero-sequence impedance of the system is
(a) 62.75 Ω (b) 4.5275 Ω (c) 5.275 Ω (d) 0.5275 Ω
[U.P.S.C. I.E.S. E.E.-II, 1999]

60. A generator is connected to a transformer which feeds another transformer through a short feeder shown in the figure. The zero sequence impedance values are expressed in pu on a common base and are indicated in the figure. The Thevenin equivalent zero sequence impedance at point B is

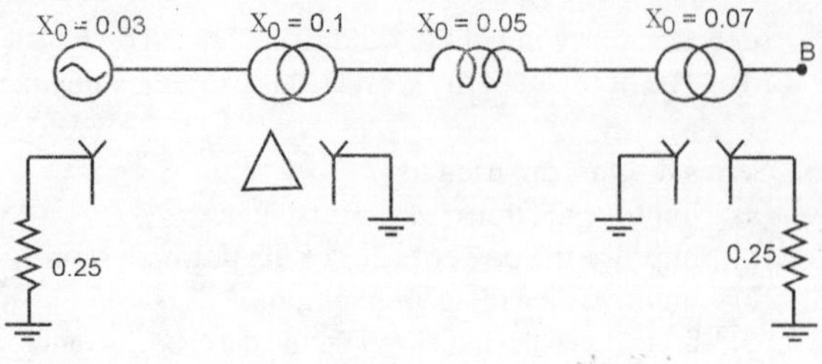

(a) $0.8 + j0.6$ (b) $0.75 + j0.22$
(c) $0.75 + j0.25$ (d) $1.5 + j0.25$ [GATE E.E. 2002]

61. The parameters of a transposed overhead transmission line are given as :
Self reactance $X_s = 0.4$ Ω/km and Mutual reactance $X_m = 0.1$ Ω/km. The positive sequence reactance X_1 and zero sequence reactance X_0, respectively, in Ω/km are
(a) 0.3, 0.2 (b) 0.5, 0.2
(c) 0.5, 0.6 (d) 0.3, 0.6 [GATE E.E. 2005]

62. An alternator (star connected) is connected to Delta-Star transformer with star neutral earthed. On the star side of the transformer a single line-to-ground fault occurs. This is equivalent to :
(a) A line-to-ground fault on the generator side of the transformer.
(b) A line-to-line fault on the generator side of the transformer.
(c) A 3-phase fault on the generator side of the transformer.
(d) A single-line-to-ground fault on the generator side of the transformer. [U.P.S.C. I.E.S. E.E.-II, 2003]

63. A 500 MVA, 50 Hz, 3-phase turbo-generator produces power at 22 kV. Generator is Y-connected and its neutral is solidly grounded. Its sequence reactances are $X_1 = X_2 = 0.15$ and $X_0 = 0.05$ pu. It is operating at rated voltage and disconnected from the rest of the system (no load). The magnitude of the sub-transient line current for single-line-to ground fault at the generator terminal in pu will be
(a) 2.851 (b) 3.333 (c) 6.667 (d) 8.553
[GATE E.E. 2004]

64. For the network shown in figure the zero-sequence reactances in pu are indicated. The zero-sequence driving point reactance of the node 3 is

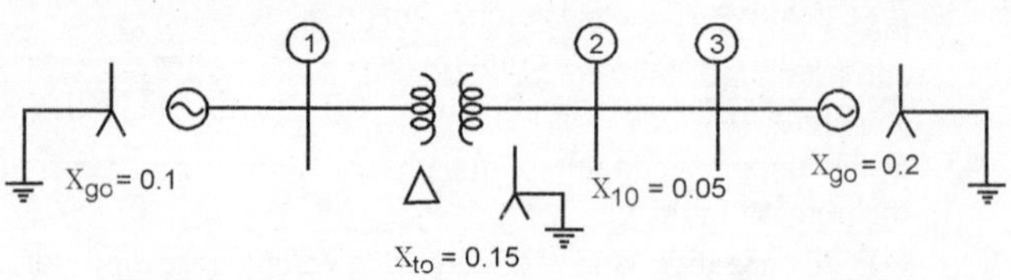

(a) 0.12 (b) 0.30 (c) 0.10 (d) 0.20
[GATE E.E. 1998]

65. A fault occurring at the terminals of an unloaded synchronous generator operating at its rate voltage has resulted in the following values of currents and voltages
$I_{a0} = j2.37$ pu, $I_{a1} = -j3.05$ pu, $I_{a2} = j0.68$ pu,
$V_{a0} = V_{a1} = V_{a2} = 0.237$ pu
Which one of the following faults has occurred ?
(a) L-L fault. (b) L-G fault.
(c) L-L-G fault. (d) L-L-L fault.
[U.P.S.C. I.E.S. E.E.-II, 2005]

Answers: 54. (d), **55.** (c), **56.** (a), **57.** (a), **58.** (a), **59.** (d), **60.** (b), **61.** (d), **62.** (b), **63.** (d), **64.** (c), **65.** (c),

66. For the system shown in the diagram given below, what is a line-to-ground fault on the line side of the transformer equivalent to ?

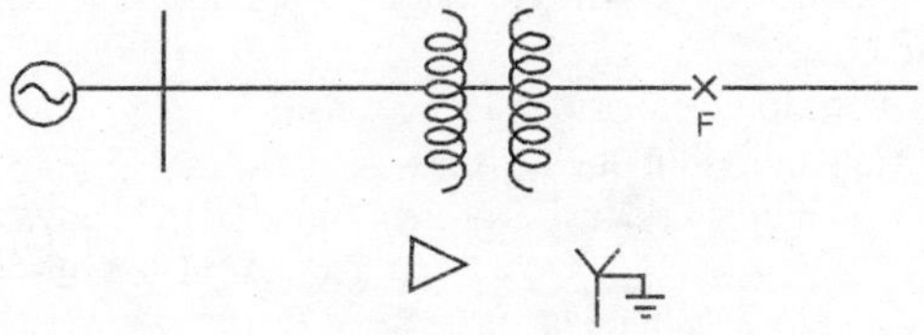

(a) A line-to-ground fault on the generator side of the transformer.
(b) A line-to-line fault on the generator side of the transformer.
(c) A double-line-to-ground fault on the generator side of the transformer.
(d) A 3-phase fault on the generator side of the transformer.
[U.P.S.C. I.E.S. E.E.-II, 2006]

Data for Q. 67 and Q. 68 are given below. Solve the problems and choose the correct answers.
At a 220 kV substation of a power system, it is given that the three-phase fault level is 4,000 MVA and single-line-to ground fault level is 5,000 MVA. Neglecting the resistance and the shunt susceptances of the system.

67. The positive sequence driving point reactance at the bus is
(a) 2.5 Ω (b) 4.033 Ω (c) 5.5 Ω (d) 12.1 Ω
[GATE E.E. 2005]

68. The zero sequence driving point reactance at the bus is
(a) 2.2 Ω (b) 4.84 Ω (c) 18.18 Ω (d) 22.72 Ω
[GATE E.E. 2005]

69. A 20 MVA, 6.6 kV, 3-phase alternator is connected to a 3-phase transmission line. The per unit positive-sequence, negative-sequence and zero-sequence impedances of the alternator are $j0.1$, $j0.1$ and $j0.04$ respectively. The neutral of the alternator is connected to ground through an inductive reactor of $j0.05$ pu. The per unit positive-, negative- and zero-sequence impedances of the transmission line are $j0.1$, j0.1 and j0.3, respectively. All per unit values are based on the machine ratings. A solid ground fault occurs at one phase of the far end of the transmission line. The voltage of the alternator neutral with respect to ground during the fault is
(a) 513.8 V (b) 889.9 V (c) 1,112.0 V (d) 642.2 V
[GATE E.E. 2003]

70. A two-machine power system is shown. Transmission line XY has positive sequence impedance of $Z_1\Omega$ and zero sequence impedance of $Z_0\Omega$.
An 'a' phase to ground fault with zero fault impedance occurs at the centre of the transmission line. Bus voltage at X and line current from X to F for the phase 'a', are given by V_a volts and I_a amperes, respectively. Then, the impedance measured by the ground distance relay located at the terminal X of line XY will be given by

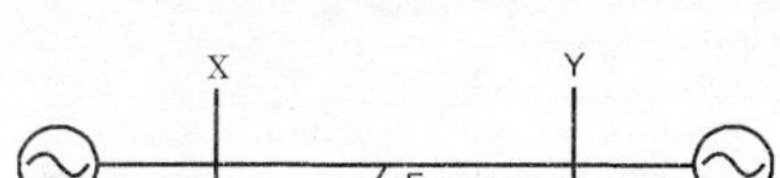

(a) $Z_1/2\ \Omega$ (b) $Z_0/2\ \Omega$
(c) $(Z_0 + Z_1)/2\ \Omega$ (d) $V_a/I_a\ \Omega$ [GATE E.E. 2008]

71. A three-phase alternator generating unbalanced voltages is connected to an unbalanced load through a 3-phase transmission line as shown in figure. The neutral of the alternator and the star point of the load are solidly grounded. The phase voltages of the alternator are $E_a = 10\angle 0°$ V, $E_b = 10\angle -90°$ V, $E_c = 10\angle 120°$ V. The positive sequence component of the load current is

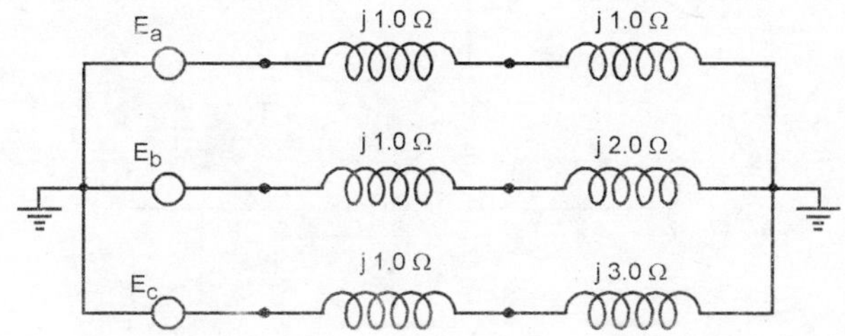

(a) $1.310\angle -107°$ A (b) $0.332\angle -120°$ A
(c) $0.996\angle -120°$ A (d) $3.510\angle -81°$ A [GATE E.E. 2003]

72. A 3-phase transmission line is shown in figure.

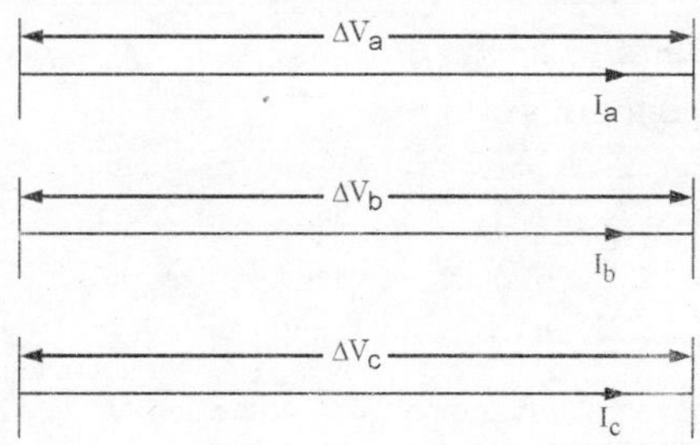

Voltage drop across the transmission line is given by the following equation :

$$\begin{bmatrix} \Delta V_a \\ \Delta V_b \\ \Delta V_c \end{bmatrix} = \begin{bmatrix} Z_s & Z_m & Z_m \\ Z_m & Z_s & Z_m \\ Z_m & Z_m & Z_s \end{bmatrix} \begin{bmatrix} I_a \\ I_b \\ I_c \end{bmatrix}$$

Shunt capacitance of the line can be neglected. If the line has positive sequence impedance of 15 Ω and zero sequence impedance of 48 Ω, then the values of Z_s and Z_m will be
(a) $Z_s = 31.5\ \Omega$; $Z_m = 16.5\ \Omega$
(b) $Z_s = 26\ \Omega$; $Z_m = 11\ \Omega$ (c) $Z_s = 16.5\ \Omega$; $Z_m = 31.5\ \Omega$
(d) $Z_s = 11\ \Omega$; $Z_m = 26\ \Omega$ [GATE E.E. 2008]

73. Suppose we define a sequence transformation between "a-b-c" and p-n-o" variables as follows :

$$\begin{bmatrix} f_a \\ f_b \\ f_c \end{bmatrix} = K \begin{bmatrix} 1 & 1 & 1 \\ \alpha^2 & \alpha & 1 \\ \alpha & \alpha^2 & 1 \end{bmatrix} \begin{bmatrix} f_p \\ f_n \\ f_0 \end{bmatrix} \quad \text{where } \alpha = e^{j\frac{2\pi}{3}} \text{ and } k \text{ is a constant}$$

Now, if it is given that : $\begin{bmatrix} V_p \\ V_n \\ V_0 \end{bmatrix} = \begin{bmatrix} 0.5 & 0 & 0 \\ 0 & 0.5 & 0 \\ 0 & 0 & 2.0 \end{bmatrix} \begin{bmatrix} i_p \\ i_n \\ i_0 \end{bmatrix}$

and $\begin{bmatrix} V_a \\ V_b \\ V_c \end{bmatrix} = Z \begin{bmatrix} i_a \\ i_b \\ i_c \end{bmatrix}$ then,

(a) $Z = \begin{bmatrix} 1.0 & 0.5 & 0.75 \\ 0.75 & 1.0 & 0.5 \\ 0.5 & 0.75 & 1.0 \end{bmatrix}$ (b) $Z = \begin{bmatrix} 1.0 & 0.5 & 0.5 \\ 0.5 & 1.0 & 0.5 \\ 0.5 & 0.5 & 1.0 \end{bmatrix}$

(c) $Z = 3k^2 \begin{bmatrix} 1.0 & 0.75 & 0.5 \\ 0.5 & 1.0 & 0.75 \\ 0.75 & 0.5 & 1.0 \end{bmatrix}$ (d) $Z = \dfrac{K^2}{3} \begin{bmatrix} 1.0 & -0.5 & -0.5 \\ -0.5 & 1.0 & -0.5 \\ -0.5 & -0.5 & 1.0 \end{bmatrix}$

[GATE E.E. 2007]

74. The given circuit shows a 3-phase system having a balanced self-inductive reactance of X_L per phase and a mutual reactance of X_m between pairs of phases. The line currents are I_a, I_b and I_c, and they may be unbalanced The series voltage-drop relations would be

Answers: 66. (b), **67.** (d), **68.** (d), **69.** (d), **70.** (d), **71.** (d), **72.** (b), **73.** (b), **74.** (a),

$$V_{aa'} = I_aX_L + I_bX_m + I_cX_m$$
$$V_{bb'} = I_aX_m + I_bX_L + I_cX_m$$
$$V_{cc'} = I_aX_m + I_bX_m + I_cX_L$$

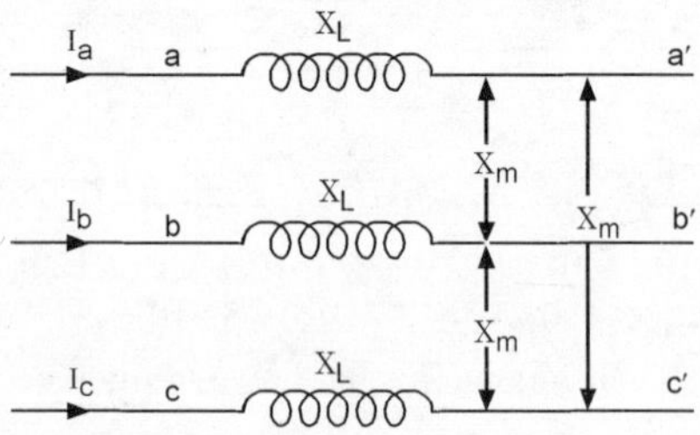

The positive, negative and zero-sequence reactances X_1, X_2, and X_0, respectively, would be
(a) $(X_L - X_m)$, $(X_L - X_m)$ and $(X_L - 2X_m)$.
(b) $(X_L + X_m)$, $(X_L - 2X_m)$ and $(X_L + 2X_m)$.
(c) $(X_L - 2X_m)$, $(X_L + X_m)$ and $(X_L - X_m)$.
(d) $(X_L + 2X_m)$, $(X_L + 2X_m)$ and $(X_L + X_m)$. **[U.P.S.C. I.E.S. E.E.-II, 2003]**

75. Fuses have got advantages of
(a) cheapest type of protection.
(b) inverse time-current characteristic.
(c) no maintenance.
(d) current limiting effect under short-circuit conditions.
(e) all of the above.

76. The primary function of a fuse is to
(a) open the circuit. (b) protect the appliance.
(c) protect the line.
(d) prevent excessive currents from flow through the circuit.

77. The material used in liquid fuses is
(a) SF_6 (b) distilled water.
(c) carbon tetra chloride. (d) transformer oil.

78. In comparison to rewirable fuses HRC fuses have the advantage(s) of
(a) high speed operation. (b) high rupturing capacity.
(c) no ageing effect. (d) all of the above.

79. In HRC fuse the time between cut-off and final current zero is called the
(a) pre-arcing time. (b) arcing time.
(c) total operating time. (d) none of these.

80. If a combination of HRC fuse and a circuit breaker is employed, the circuit breaker operates for
(a) short-circuit current. (b) low overload currents.
(c) under all abnormal currents.
(d) the combination is newer employed in practice.

81. The least expensive protection for overcurrent in low-voltage system is
(a) rewirable fuse. (b) isolator.
(c) circuit breaker. (d) air-break switch.

82. A circuit breaker normally operates
(a) when the power is to be supplied.
(b) when the line is to be tested.
(c) when the switch is to be put on.
(d) whenever fault occurs in the line.

83. When a fault occurs in a high voltage transmission line, first the
(a) circuit breaker operates then the relay.
(b) relay operates then the circuit breaker.
(c) relay operates, then successively the isolator and the circuit breaker.
(d) isolator operates, then successively the relay and the circuit breaker. **[A.M.I.E. Sec B. Winter 1999]**

84. The initiation of electric arc at the instant of contact separation is caused by
(a) thermionic emission of electrons.
(b) field emission of electrons.
(c) (a) and (b) both. (d) none of the above. **[A.M.I.E. Sec B. Summer 1998]**

85. Arcing voltage will be the least in case of
(a) carbon. (b) copper. (c) silver. (d) tungsten.

86. The heat produced at the contact point owing to flow of electric current is least affected by
(a) temperature of the surrounding medium.
(b) contact resistance.
(c) magnitude of electric current flowing.
(d) duration of flow of current.

87. Which of the following should have low value for the contacts and their material ?
(a) Thermal capacity. (b) Contact resistance.
(c) Thermal conductivity. (d) None of these.

88. In a CB contact wipe is necessary
(a) to decrease the speed of opening of the CB.
(b) to increase the speed of opening and reduce the arcing of contacts.
(c) to increase the dielectric strength across the CB contacts.
(d) none of the above. **[A.M.I.E. Sec B. Winter 1999]**

89. Interrupting medium in a contactor may be
(a) air at atmospheric pressure.
(b) SF_6 gas. (c) oil.
(d) compressed air. (e) any of these.

90. The fluids used in circuit breakers should be of
(a) high dielectric strength and thermal stability.
(b) noninflammability. (c) arc extinguishing ability.
(d) all of the above.

91. Sparking between contacts can be reduced by inserting
(a) a capacitor in parallel with the contacts.
(b) a capacitor in series with the contacts.
(c) a resistor in the line. (d) a reactor in the line.

92. For magnetic blow out of arc the magnetic field is produced
(a) in the load circuit.
(b) at right angles to the axis of the arc.
(c) in line with the axis of the arc.
(d) any of the above.

93. In circuit breakers the contact space is ionized by
(a) field emission from the contact surface.
(b) thermal emission from the contact surface.
(c) thermal ionization of gas.
(d) any of the above.

94. In a circuit breaker, ionization is facilitated by
(a) increase in field strength. (b) increase of mean free length.
(c) high temperature of surrounding medium.
(d) all of the above.

95. The stability of arc in vacuum depends on
(a) The contact material only.
(b) The contact material and its vapour pressure.
(c) The circuit parameters only.
(d) The combination of (b) and (c). **[U.P.S.C. I.E.S. E.E.-II, 2005]**

96. The interrupting time of a circuit breaker is the period between the instant of

Answers: **75.** (e), **76.** (d), **77.** (c), **78.** (d), **79.** (b), **80.** (b), **81.** (a), **82.** (d), **83.** (b), **84.** (c), **85.** (c), **86.** (a), **87.** (b), **88.** (c), **89.** (e), **90.** (d), **91.** (a), **92.** (b), **93.** (b), **94.** (d), **95.** (d), **96.** (b),

(a) initiation of short circuit and the arc extinction on an opening operation.
(b) energizing of the trip circuit and the arc extinction on an opening operation.
(c) initiation of short circuit and the parting of primary arc contacts.
(d) energizing of the trip circuit and the parting of primary arc contacts. **[GATE E.E. 2003]**

97. For a high speed circuit breaker the total clearing time is around
(a) few minutes. (b) few seconds.
(c) 1 to 2 cycles. (d) 5 to 20 cycles.

98. The resistance of an electric arc can be increased by
(a) increasing the concentration of ionized particles.
(b) reducing the arc length. (c) splitting the arc.
(d) increasing the arc x-section.

99. The arc voltage in a circuit breaker is
(a) in the phase with the arc current.
(b) lagging the arc current by 90°.
(c) leading the arc current by 90°.
(d) lagging the arc current by 180°. **[A.M.I.E. Sec B. Summer 1999]**

100. In a circuit breaker, the active recovery voltage depends upon
(a) circuit conditions. (b) power factor.
(c) armature reaction. (d) all of these.

101. In a circuit breaker the current that exists at the instant of contact separation is called the ______ current.
(a) restriking (b) breaking (c) arc (d) recovery

102. The rate of rise of restriking voltage (RRRV) depends upon
(a) system voltage. (b) circuit pf only.
(c) switching condition only. (d) both (b) and (c).

103. The RRRV depends upon the
(a) type of the circuit breaker.
(b) capacitance of the system only.
(c) inductance of the system only.
(d) inductance and capacitance of the system.

104. It is difficult to interrupt a capacitive circuit because
(a) the current has a leading power factor.
(b) the restriking voltage can be high.
(c) current magnitude is very small.
(d) stored energy in the capacitor is very high. **[A.M.I.E. Sec B. Summer 1993]**

105. A 50 Hz, 3-phase synchronous generator has inductance per phase of 15 mH. The capacitance of generator and circuit breaker is 0.002 μF. What is its natural frequency of oscillation ?
(a) 29 kHz (b) 2.9 kHz (c) 290 kHz (d) 29 MHz **[U.P.S.C. I.E.S. E.E.-II, 2008]**

106. In connection with the arc extinction in the circuit breaker, resistance switching is employed wherein a resistance is placed in parallel with the poles of the circuit breaker as shown in the figure. This process introduces damping in the L-C circuit. For critical damping, the value of '*r*' should be equal to

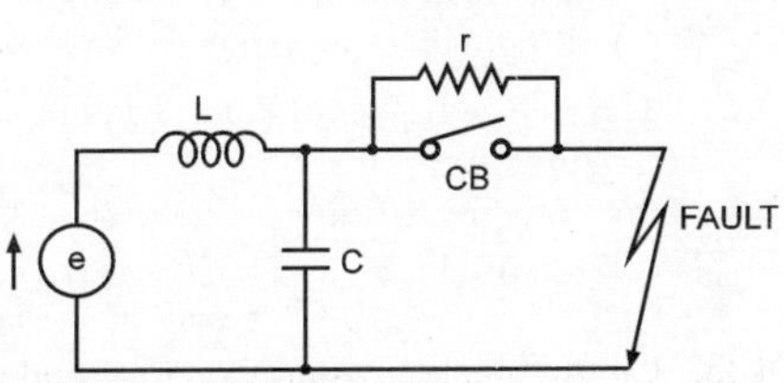

(a) $\sqrt{C/L}$ (b) $0.5\sqrt{C/L}$
(c) $0.5\sqrt{L/C}$ (d) $\frac{1}{2\pi}\sqrt{\frac{L}{C}}$ **[U.P.S.C. I.E.S. E.E.-II, 1998]**

107. If the inductance and capacitance of a power system are respectively 1 H and 0.01 μF and the instantaneous value of interrupted current is 10 A, then the voltage across the breaker contact will be
(a) 50 kV (b) 57 kV (c) 60 kV (d) 100 kV **[U.P.S.C. I.E.S. E.E.-II, 2000]**

108. A 50 Hz, 17.32 kV generator is connected to a power system. The system reactance and capacitance per phase are : 10 mH and 0.02 μF, respectively. What is the maximum voltage across the contacts of the circuit breaker at an instant when it passes through zero ?
(a) 28.28 kV (b) 29.28 kV
(c) 30.28 kV (d) 31.28 kV **[U.P.S.C. I.E.S. E.E.-II, 2005]**

109. In circuit breakers, oil, as quenching medium, has the advantage(s) of
(a) good cooling properties. (b) absorption of arc energy.
(c) high dielectric strength. (d) acting as an insulator.
(e) all of the above.

110. In circuit breakers, oil, as quenching medium, has the disadvantage(s) of
(a) need of periodical reconditioning or replacement.
(b) risk of formation of explosive mixture with air.
(c) possibility of causing fire hazards.
(d) all of the above.

111. Main purpose of oil in OCB is to
(a) provide insulation. (b) provide cooling of contacts.
(c) quenching arc. (d) none of above. **[A.M.I.E. Sec B. Summer 1994]**

112. Arc interruption in an oil circuit breaker takes place when
(a) contacts apart. (b) voltage goes through zero.
(c) current goes through zero. **[A.M.I.E. Sec B. Winter 1999]**

113. The chances of arc interruption in subsequent current zeros
(a) increase in case of ABCB but decrease in OCB.
(b) decrease in case of ABCB but increase in OCB.
(c) decrease in both cases.
(d) increase in both cases. **[U.P.S.C. I.E.S. E.E.-II, 2008]**

114. In a minimum oil circuit breaker the oil is used
(a) to act as circuit breaking medium only.
(b) for circuit breaking and providing insulation.
(c) for providing insulation only.
(d) for none of the above purposes. **[A.M.I.E. Sec B. Summer 1993]**

115. The air-blast circuit breakers for 400 kV system are designed to operate in
(a) 0.1 s (b) 0.5 s (c) 50 ms (d) 100 μs **[A.M.I.E. Sec B. Summer 1999]**

116. The minimum oil circuit breaker has less volume of oil because
(a) the oil between the breaker contacts has greater strength.
(b) there is insulation between the breaker contacts.
(c) solid insulation is provided for insulating the contacts from the earth.
(d) none of the above.

117. Air used in air blast circuit breaker must
(a) be ionized. (b) be free from moisture.
(c) have least CO_2. (d) have oil mist.

118. While using air-blast circuit breaker, current chopping is a phenomenon often observed when
(a) a long overhead line is switched off.
(b) a bank of capacitors is switched off.
(c) a transformer on no load is switched off.
(d) a heavy load is switched off. **[U.P.S.C. I.E.S. E.E.-II, 1994]**

Answers: 97. (c), **98.** (c), **99.** (a), **100.** (d), **101.** (b), **102.** (d), **103.** (d), **104.** (b), **105.** (a), **106.** (c), **107.** (d), **108.** (a), **109.** (e), **110.** (d), **111.** (c), **112.** (c), **113.** (b), **114.** (a), **115.** (c), **116.** (c), **117.** (b), **118.** (c),

119. In air-blast circuit breaker resistance switching is used to
(a) reduce the magnitude of fault current.
(b) control the CB operating time.
(c) damp out the fast transient.
(d) change the fault current power factor.
[A.M.I.E. Sec B. Summer 1993]

120. SF_6 gas has excellent heat transfer properties because of its
(a) low gaseous viscosity. (b) high dielectric strength.
(c) higher molecular weight. (d) both (a) and (b).
(e) both (a) and (c).

121. SF_6 circuit breakers have the advantages of
(a) very much reduced electrical clearance, reduced moisture problem and minimum current chopping problem.
(b) noiseless operation, less arcing time resulting in less contact erosion.
(c) no reduction in dielectric strength of SF_6.
(d) performance independent of ambient conditions.
(e) all of the above.

122. SF_6 circuit breakers have the drawback(s) of
(a) sealing problem of gas.
(b) ingress of moisture in the gas system–dangerous.
(c) deterioration of SF_6 with time.
(d) both (a) and (b).

123. Which of the following statements is not correct?
(a) SF_6 is non-toxic and noninflammable gas.
(b) SF_6 has very high dielectric strength roughly 24 times of that of air.
(c) SF_6 is yellow in colour.
(d) SF_6 is about 100 times more effective than air in arc extinction.
(e) SF_6 has density 5 times that of air at 20°C.

124. The single most important property that makes SF_6 a very efficient medium for circuit breaking is
(a) it is non-toxic and noninflammable.
(b) it has high dielectric constant.
(c) it has high breakdown strength.
(d) it is highly electronegative in character.
[A.M.I.E. Sec B. Summer 1993]

125. During arc extinction SF_6 gas gets
(a) decomposed into SF_4 and SF_2.
(b) decomposed into S and F ions.
(c) reduced to SF_2. (d) oxidised.

126. SF_6 gas is transported in
(a) air cylinders. (b) gas cylinders.
(c) liquid form in cylinders. (d) solid form.

127. The current chopping tendency is mimimised by using SF_6 gas at relatively
(a) low pressure and high viscosity.
(b) low pressure and low viscosity.
(c) high pressure and low viscosity.
(d) high pressure and high viscosity.

128. To limit current chopping in vacuum circuit breakers, the contact material employed should have the properties of
(a) low conductivity and high vapour pressure.
(b) low conductivity and low vapour pressure.
(c) high conductivity and high vapour pressure.
(d) high conductivity and low vapour pressure.

129. A thermal protection switch provides protection against
(a) overload. (b) temperature.
(c) short circuit. (d) overvoltage.

130. For remote operation, the circuit breaker must be equipped with
(a) inverse time trip. (b) shunt trip.
(c) time delay trip. (d) all of these.

131. Which of the following circuit breakers is generally used in railway electrification ?
(a) Air-blast circuit breaker. (b) Minimum oil circuit breaker.
(c) Bulk oil circuit breaker. (d) SF_6 circuit breaker.
[U.P.S.C. I.E.S. E.E.-II, 2006]

132. What is the preferred type of CB to be installed in extra high voltage ac system ?
(a) Bulk oil type CB (b) Air-blast CB
(c) Vacuum CB
(d) Sulphur hexafluoride (SF_6) CB **[U.P.S.C. I.E.S. E.E.-II, 2007]**

133. Keeping in view the cost and overall effectiveness, the following circuit breaker is best suited for capacitor bank switching
(a) vacuum. (b) air-blast.
(c) SF_6 (d) oil. **[GATE E.E. 2006]**

134. Which of the following circuit breakers has the lowest operating voltage ?
(a) SF_6 gas. (b) Air-break.
(c) Air-blast. (d) Minimum oil.

135. Which of the following circuit breakers produce the least arc energy?
(a) Plain oil. (b) Minimum oil.
(c) Air-blast. (d) Air break.

136. Which of the following circuit breakers has high reliability and negligible maintenance?
(a) Air-blast. (b) SF_6 (c) Oil. (d) Vacuum.

137. Which of the following circuit breakers take minimum time in installation?
(a) Air-blast. (b) Minimum oil.
(c) Bulk oil. (d) SF_6

138. Where voltages are high and current to be interrupted is low, the circuit breaker preferred is ... one.
(a) air-break (b) vacuum (c) oil (d) air-blast

139. For rural electrification in a country like India with complex network, the circuit breaker preferred is ... one.
(a) air-break (b) oil (c) vacuum (d) minimum oil

140. The most suitable circuit breaker for short line fault without switching resistor is ... one.
(a) minimum oil (b) air-blast
(c) SF_6 (d) air-break

141. The rating of a circuit breaker is usually determined on the basis of ... fault.
(a) symmetrical (b) line to line
(c) single line to ground (d) double line to ground

142. The transient phenomenon lasts in a power system for a period ranging from
(a) few ms to 1 s. (b) 1 s to 2 s.
(c) 2 s to 3 s. (d) greater than 3 s.
[W.B. Univ. of Technology Power System-III, 2009-10]

143. Circuit breakers usually operate under
(a) steady short-circuit current.
(b) sub-transient state of short-circuit current.
(c) transient state of short-circuit current.
(d) none of these. **[W.B. Univ. Technology Power Systems-II, 2010]**

144. The restriking voltage is measured in
(a) RMS value. (b) Peak value.
(c) Instantaneous value. (d) Average value.
[U.P.S.C. I.E.S. E.E.-II, 2010]

Answers: **119.** (c), **120.** (e), **121.** (e), **122.** (d), **123.** (c), **124.** (d), **125.** (a), **126.** (c), **127.** (b), **128.** (a), **129.** (a), **130.** (b), **131.** (a), **132.** (d), **133.** (a), **134.** (b), **135.** (c), **136.** (b), **137.** (d), **138.** (b), **139.** (c), **140.** (c), **141.** (a), **142.** (a), **143.** (b), **144.** (b),

145. The making and breaking currents of 3-phase ac circuit breakers in power system are respectively in what form?
(a) rms value, rms value.
(b) instantaneous value, rms value.
(c) rms value.
(d) instantaneous value, instantaneous value.
[U.P.S.C. I.E.S. E.E.-II, 2010]

146. The breaking capacity of a 3-phase circuit breaker is given as
(a) service line voltage in volts × rated symmetrical current in amperes $\times 10^{-6}$ MVA.
(b) $\sqrt{3}$ service line voltage in volts × rated symmetrical current in amperes $\times 10^{-6}$ MVA.
(c) 1.1 × service line voltage in volts × rated symmetrical current in amperes $\times 10^{-6}$ MVA.
(d) $\sqrt{2}$ service line voltage in volts x rated symmetrical current in amperes $\times 10^{-6}$ MVA.

147. Rated breaking capacity (MVA) of a circuit breaker is equal to
(a) the product of rated breaking current (kA) and rated voltage (kV).
(b) the product of rated symmetrical breaking current (kA) and rated voltage (kV).
(c) the product of breaking current (kA) and fault voltage (kV).
(d) twice the value of rated current (kA) and rated voltage (kV).

148. The making to breaking current ratio for an EHV circuit breaker is
(a) more than 1. (b) equal to 1.
(c) less than 1. (d) a negative number.
[A.M.I.E. Sec B. Winter 1995]

149. The making capacity of a circuit breaker is
(a) Less than the asymmetrical breaking capacity of the breaker.
(b) Greater than the asymmetrical breaking capacity of the breaker.
(c) Equal to the symmetrical breaking capacity of the breaker.
(d) Equal to the asymmetrical breaking capacity.
[U.P.S.C. I.E.S. E.E.-II, 2007]

150. A three-phase, 33 kV oil circuit breaker is rated 1200 A, 2,000 MVA, 3 s. The symmetrical breaking current is
(a) 1,200 A (b) 3,600 A (c) 35 kA (d) 104.8 kA
[GATE E.E., 2010]

151. A three-phase breaker is rated 2,000 MVA, 33 kV. Its making current will be
(a) 35 kA (b) 49 kA (c) 70 kA (d) 89 kA
[U.P.S.C. I.E.S. E.E.-II, 2000; W.B. Univ. of Technology Power System-II, 2009-10]

152. In a 220 kV system, the inductance and capacitance up to the circuit breaker location are 25 mH and 0.025 μF respectively. The value of resistor required to be connected across the breaker contacts which will give no transient oscillations, is
(a) 25 Ω. (b) 250 Ω. (c) 500 Ω. (d) 1000 Ω.
[U.P.S.C. I.E.S. E.E.-II, 2001]

153. Which of the following sequences of operations represents the rated operating duty cycle of a circuit breaker?
(O-open; C-close; t = 3s; T = 3 minutes)
(a) O-t-CO-T-CO. (b) O-CO-t-CO-T-C.
(c) O-C-t-OC-T. (d) O-CO-T-CO-T-C.
[U.P.S.C. I.E.S. E.E.-II, 1999]

154. Which of the following statements is not correct?
(a) Arc chutes are used in air break circuit breakers.
(b) Air-blast circuit breakers are employed for high voltage traction system.
(c) Resistance switching is employed for overcoming current chopping.
(d) Linear resistors are used in resistance switching.

155. Capacitor switching in 33 kV power system is better done with ... circuit breakers.
(a) air-blast (b) minimum oil
(c) vacuum
[A.M.I.E. Sec B Winter 1996]

156. The probable cause(s) for fall in insulation resistance between phase terminal and earthed frame could be
(a) dirty insulation surface.
(b) ingress of moisture.
(c) sticking of carbon or copper particles to the internal surface.
(d) all of the above.

157. The probable cause(s) for failure of a circuit breaker on electrical compound could be
(a) trip circuit open. (b) trip latch defective.
(c) spring defective. (d) any of the above.

158. Routine tests are conducted on
(a) bulk oil circuit breakers. (b) minimum oil circuit breakers.
(c) air-blast circuit breakers. (d) all of the above.

159. An isolator is installed
(a) to isolate one portion of the circuit from another.
(b) usually on both sides of a circuit breaker.
(c) as a substitute for a circuit breaker.
(d) both (a) and (b).

160. Current rating is not necessary in case of
(a) isolators. (b) circuit breakers.
(c) load break switches.
(d) circuit breakers and load break switches.
[A.M.I.E. Sec B. Power Systems Winter 1999]

161. An isolator is meant for
(a) breaking abnormal current.
(b) making under fault condition.
(c) breaking the circuit under no-load condition.
(d) none of the above.
[A.M.I.E. Sec B. Advanced Power Systems Summer 1998]

162. Isolators used in transmission lines are capable of breaking:
(a) fault current. (b) no current.
(c) charging current. (d) load current.
[M.D. Univ. Transmission and Distribution May, 2011]

163. The oil switches may be
(a) remote or manually controlled.
(b) used for capacitor switching, street lighting control and automatic disconnect on power failure.
(c) used for high voltage and large current circuits.
(d) all of the above.

164. Which of the following sequence of operation is correct for operation of circuit breaker, isolator and earthing switch while opening a circuit ?
(a) Open circuit breaker–open isolator–close earthing switch.
(b) Open isolator–open circuit breaker–close earthing switch.
(c) Open earthing switch–open isolator–then open circuit breaker.
(d) Open circuit breaker–close earthing switch–open isolator.

165. Which of the following sequence of operation is correct for operation of circuit breaker, isolator and earthing switch while closing a circuit?
(a) Ensure circuit breaker is open-close isolator–open earthing switch–close circuit breaker.
(b) Ensure circuit breaker is open–open earthing switch-close isolator–close circuit breaker.
(c) Close circuit breaker–open earthing switch–close isolator.
(d) Open earthing switch–close circuit breaker–close isolator.

Answers: **145.** (d), **146.** (b), **147.** (b), **148.** (a), **149.** (b), **150.** (c), **151.** (d), **152.** (c), **153.** (a), **154.** (d), **155.** (c), **156.** (d), **157.** (d), **158.** (d), **159.** (d), **160.** (a), **161.** (c), **162.** (c), **163.** (d), **164.** (a), **165.** (b),

166. Purpose of backup protection is
(a) to increase the speed. (b) to increase a reach.
(c) to leave no blind spot.
(d) to guard against failure of primary. [A.M.I.E. Sec B. Winter 1998]

167. Protective relays are the devices that detect abnormal conditions in electrical circuits by measuring
(a) voltage. (b) current.
(c) constantly the electrical quantities which differ during normal and abnormal conditions.
(d) none of the above.

168. The actuating quantity for the relays may be
(a) magnitude. (b) frequency.
(c) phase angle. (d) any of these.

169. Protective relays can be designed to respond to
(a) light intensity. (b) temperature.
(c) resistance, reactance or impedance.
(d) voltage or current. (e) all of the above.

170. The protective relay is provided to
(a) provide additional safety to the circuit breaker in its operation.
(b) close the contacts when the actuating quantity attains a certain predetermined value.
(c) limit the arcing current during the circuit breaker operation.
(d) earth any stray voltage.

171. The operating speed of a relay depends upon the
(a) rate of flux built up. (b) armature core air-gap.
(c) spring tension. (d) all of these.

172. Plug setting of an electromagnetic relay can be altered by varying
(a) number of ampere turns. (b) air gap of magnetic path.
(c) adjustable back stop. (d) none of these.
[W.B. Univ. of Technology Power System-II, 2009-10]

173. The torque produced in shaded pole structure induction type relay is
(a) proportional to the square of the current.
(b) proportional to the current.
(c) inversely proportional to the current.
(d) inversely proportional to the square of the current.
[A.M.I.E. Sec. B Summer 1999]

174. Given ϕ_{1m}, ϕ_{2m} = the fluxes produced by the two portions of the shaded pole,
θ = the angle between ϕ_{1m} and ϕ_{2m},
R = resistance of the disc, the torque developed in an induction relay would be proportional to which of the following
1. ϕ_{1m} and ϕ_{2m} 2. 1/R
3. R 4. sin θ
Select the correct answer using the codes given below
Codes:
(a) 1, 2 and 4 (b) 1, 3 and 4
(c) 1 and 2 (d) 2 and 4
[U.P.S.C. I.E.S. E.E.-II, 1993]

175. The most efficient torque-producing actuating structure for induction-type relays is
(a) shaded pole structure. (b) watt-hour meter structure.
(c) induction-cup structure.
(d) single induction loop structure. [U.P.S.C. I.E.S. E.E.-II, 1994]

176. A bimetal strip, in a thermal relay, consists of two metal strips of different
(a) specific heat. (b) thermal conductivity.
(c) coefficients of expansion. (d) all of the above.

177. Thermal relays are used for protection of motors against over-current owing to
(a) short circuits. (b) heavy loads.
(c) earth fault. (d) all of these.

178. Buchholz relay is
(a) located in the conservator tank.
(b) located in the transformer tank itself.
(c) connected in the pipe connecting main tank of transformer and conservator.
(d) installed in the circuit breaker.

179. Buchholz relays are used on
(a) air cooled transformers. (b) instrument transformers.
(c) distribution transformers.
(d) oil immersed power transformers of ratings above 500 kVA.

180. The Buchholz relay protects a transformer from
(a) types of internal faults. (b) a turn to turn fault.
(c) winding to winding fault. (d) none of these.
[W.B. Univ. of Technology Power System-II, 2009-10]

181. Which of the following relays is/are overload relays?
(a) Thermal. (b) Electromagnetic.
(c) Induction. (d) Solid state.
(e) All of the above.

182. In an overcurrent protection the setting of the earth fault relay is
(a) more than the phase fault relay.
(b) equal to the phase fault relay.
(c) less than the phase fault relay.
(d) the two settings are unrelated to each other.

183. Inverse time-current relays are used for the protection of
(a) feeders. (b) transformers.
(c) alternators. (d) both (a) and (b).

184. IDMT relays are used to protect the power transformers against
(a) external short circuits. (b) overloads.
(c) internal short circuits. (d) both (a) and (b).

185. Consider the following statements with reference to protective relays :
1. The minimum relay coil current at which the relay operates is called pick-up value.
2. The pick-up value of a relay is 7.5 A and fault current is 30 A. Therefore, its plug setting multiplier is 5.
3. An earth fault current is generally lesser than the short-circuit current.
4. Induction relays are used with both ac and dc quantities.
Which of these statements are correct ?
(a) 1 and 2. (b) 2 and 3.
(c) 1 and 3. (d) 1, 2 and 4.
[U.P.S.C. I.E.S. E.E.-II, 2003]

186. A fault current of 2,000 A is passing on the primary side of a 400/5 C.T. On the secondary side of the C.T., an inverse-time over current relay is connected whose plug setting is set at 50%. The plug setting multiplier will be :
(a) 25 (b) 50 (c) 10 (d) 30
[U.P.S.C. I.E.S. E.E.-II, 2003]

187. In an inverse definite minimum time, electromagnetic type over-current relay, the minimum time feature is achieved because of
(a) saturation of the magnetic circuit.
(b) proper mechanical design.
(c) appropriate time-delay element.
(d) electromagnetic damping. [GATE E.E. 2000]

Answers: **166.** (d), **167.** (c), **168.** (d), **169.** (e), **170.** (b), **171.** (d), **172.** (a), **173.** (a), **174.** (a), **175.** (c), **176.** (c), **177.** (b), **178.** (c), **179.** (d), **180.** (a), **181.** (e), **182.** (c), **183.** (d), **184.** (d), **185.** (c), **186.** (c), **187.** (a),

188. Earth fault relays are
(a) directional relays. (b) non-directional relays.
(c) short operate time relays. (d) none of these.

189. The domains of power system where directional overcurrent relay is indispensable are
(a) in case of parallel feeder protection.
(b) in case of ring main feeder protection.
(c) both (a) and (b). (d) none of the above.
[A.M.I.E. Sec. B Summer 1998]

190. Directional overcurrent relays have two exciting coils connected across
(a) CT secondaries of two different phases.
(b) VT secondaries of two different phases.
(c) CT and VT secondaries of the same phase.
(d) CT and VT secondaries of two different phases.
[A.M.I.E. Sec. B Winter 1995]

191. Directional element connections of directional overcurrent relays are popularly described in terms of the angle by which unity power factor balanced load current flowing in tripping direction leads the voltage applied to relay potential coil. A protection scheme having the following potential coil and current coil connection is termed.

Relay	*Potential Coil Voltage*	*Current Coil Current*
A	V_{ac}	I_a
B	V_{ba}	I_b
C	V_{cb}	I_c

(a) 0° connection. (b) 30° connection.
(c) 60° connection. (d) 90° connection.
[U.P.S.C. I.E.S. E.E.-II, 1993]

192. The distribution system shown in the figure is to be protected by overcurrent system of protection. For proper fault discrimination, directional overcurrent relays will be required at locations

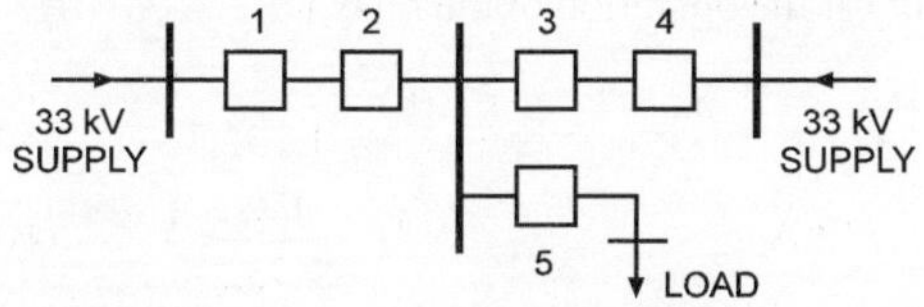

(a) 1 and 4. (b) 2 and 3.
(c) 1, 4 and 5. (d) 2, 3 and 5.
[Gate E.E. 1993; U.P.S.C. I.E.S. E.E.-II, 1999]

193. Consider the following statements. Distance relays used for protection of transmission lines might operate undesirably due to
1. magnetizing-current inrush.
2. power swings.
3. series compensating capacitors.
4. switching overvoltages.

Of these statements
(a) 1 and 2 are correct. (b) 2 and 3 are correct.
(c) 1, 2 and 4 are correct. (d) 1, 3 and 4 are correct.
[U.P.S.C. I.E.S. E.E.-II, 1993]

194. In distance protection, the relay measures
(a) negative sequence impedance of the line from relay up to the fault point.
(b) positive sequence impedance of the line from relay up to the fault point.
(c) self impedance of the line from relay up to the fault point.
(d) zero sequence impedance of the line from relay up to the fault point.
[A.M.I.E. Sec B. Winter 1996]

195. If the fault occurs near the relay, the V/I ratio will be
(a) lower than that of if the fault occurs away from the relay.
(b) constant for all distances.
(c) higher than that of it the fault occurs away from the relay.
(d) none of the above.
[A.M.I.E. Sec B. Summer 1998]

196. Impedance relay may use
(a) balance beam structure. (b) induction-cup structure.
(c) shaded pole structure. (d) either (a) or (b).

197. Impedance relay can be used for
(a) phase faults only. (b) earth faults only.
(c) both earth and phase faults.
(d) none of above.

198. An impedance relay is a
(a) voltage restrained overcurrent relay.
(b) voltage restrained directional relay.
(c) directional restrained overcurrent relay.
(d) directional restrained overvoltage relay.

199. Threshold characteristics of a plain impedance relay in a complex *Z* plane is a
(a) circle passing through origin.
(b) circle with the centre at the origin.
(c) straight line passing through origin.
(d) straight line offset from the origin. **[A.M.I.E. Sec B. Summer 1994]**

200. On transient fault OC and impedance relay.
(a) under reaches. (b) over reaches.
(c) reach unaffected. (d) none of these.
[A.M.I.E. Sec B. Winter 1998]

201. The structure used in reactance relays is
(a) induction cup. (b) double induction loop.
(c) single induction loop. (d) either (a) or (b).

202. A reactance relay is
(a) voltage restrained overcurrent relay.
(b) voltage restrained directional relay.
(c) directional restrained overcurrent relay.
(d) directional restrained overvoltage relay.

203. The operating characteristics of a reactance relay in the complex impedance plane is a
(a) circle with its origin at the centre of the R–X plane.
(b) circle passing through the origin.
(c) straight line passing through the origin.
(d) straight line parallel to the X-axis. **[A.M.I.E. Sec B. Summer 1993]**

204. Reactance relay is normally used for protection against
(a) earth fault only. (b) phase fault only.
(c) both earth and phase faults. (d) none of the above.

205. Admittance relay is
(a) non-directional relay. (b) directional relay.
(c) differential relay. (d) none of these.

206. A mho relay is a
(a) voltage restrained directional relay.
(b) voltage restrained overcurrent relay.
(c) directional restrained overcurrent relay.
(d) directional restrained overvoltage relay.

207. A Mho relay is a
(a) voltage restrained directional relay.
(b) voltage controlled overcurrent relay.
(c) directional restrained overcurrent relay.
(d) directional restrained overvoltage relay.
[W.B. Univ. of Technology Power System-II, 2009-10]

Answers: 188. (a), **189.** (c), **190.** (c), **191.** (b), **192.** (b), **193.** (b), **194.** (b), **195.** (a), **196.** (d), **197.** (a), **198.** (a), **199.** (b), **200.** (a), **201.** (d), **202.** (c), **203.** (d), **204.** (a), **205.** (b), **206.** (a), **207.** (a),

208. The operating characteristic of a distance relay in the R-X plane is shown in the figure. It represents operating characteristic of a

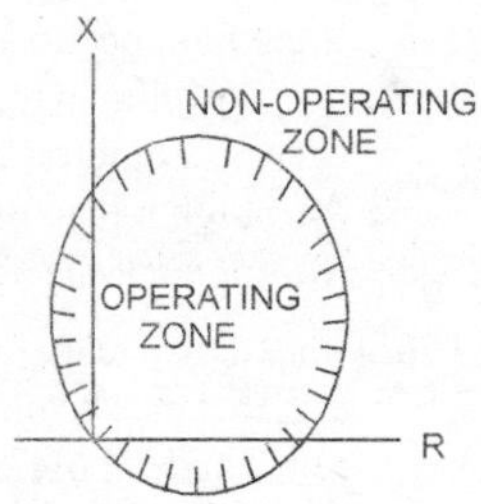

(a) reactance relay.
(b) directional impedance relay.
(c) impedance relay.
(d) mho relay.

[U.P.S.C. I.E.S. E.E.-II, 2001]

209. For ground fault use preffer:
(a) plain impedance relay. (b) directional relay.
(c) reactance relay. (d) over current relay.

[A.M.I.E. Sec B. Summer 1994; M.D. Univ. Transmission and Distribution May, 2011]

210. A distance relay is said to be inherently directional if its characteristics on R-X diagram
(a) is a straight line off-set from the origin.
(b) is a circle that passes through the origin.
(c) is a circle that encloses the origin.
(d) always a separate directional relay is required.

[W.B. Univ. of Technology Power System-II, 2009-10]

211. For the protection of lines against faults involving variable fault resistance, the preferred relaying scheme is a
(a) plain impedance relay. (b) directional overcurrent relay.
(c) mho relay. (d) reactance relay.

[A.M.I.E. Sec B. Summer 1995]

212. The relay used for phase fault protection of short transmission lines is
(a) reactance relay. (b) impedance relay.
(c) mho relay. (d) IDMT relay.

[A.M.I.E. Sec B. Summer 1999]

213. Mho relay is usually employed for the protection of
(a) short lines only. (b) medium lines only.
(c) long lines only. (d) any line.

[U.P.S.C. I.E.S. E.E.-II, 2008]

214. Where severe synchronising swing occur, the relay employed is
(a) impedance relay. (b) mho relay.
(c) reactance relay. (d) induction relay.

215. Which of the following relays has inherent directional characteristic?
(a) Mho. (b) Reactance.
(c) Impedance. (d) None of the above.

216. For the protection of a very long extra high voltage line, the protective relay used is
(a) overcurrent with extremely inverse characteristics.
(b) percentage differential relay.
(c) reactance type distance relay.
(d) mho type distance relay. [A.M.I.E. Sec B. Summer 1993]

217. A differential relay responds to
(a) algebraic difference of two or more similar electrical quantities.
(b) phasor difference of two or more similar electrical quantities.
(c) algebraic difference between two currents.
(d) algebraic difference between two voltages.

218. Differential relays are used for protection of equipment against
(a) internal faults. (b) overcurrent.
(c) reverse current. (d) reverse power.

219. Undervoltage relays are mainly used for
(a) motor protection. (b) transformer protection.
(c) transmission line protection.
(d) All of these. [U.P.S.C. I.E.S. E.E.-II, 2010]

220. In a biased differential relay the bias is defined as a ratio of
(a) number of turns of restraining and operating coil.
(b) operating coil current and restraining coil current.
(c) fault current and operating coil current.
(d) fault current and restraining coil current. [GATE E.E. 2006]

221. Differential relays are used for protection of
(a) feeders. (b) alternators.
(c) transformers. (d) all of these.

222. Both voltage and current signals are required for
(a) a plain overcurrent relay.
(b) a differential relay. (c) a directional relay.
(d) a biased differential relay. [A.M.I.E. Sec B. Summer 1994]

223. A 75 MVA, 10 kV synchronous generator has $X_d = 0.4$ pu. The X_d value (in pu) to a base of 100 MVA, 11 kV is
(a) 0.578. (b) 0.279.
(c) 0.412. (d) 0.44. [GATE E.E. 2001]

224. The relay used for feeder protection is
(a) undervoltage relay. (b) Translay relay.
(c) thermal relay. (d) Buchholz relay.

225. In the case of transmission line protection, overcurrent relay is used
(a) only up to 110 kV. (b) only up to 220 kV.
(c) only up to 50 kV. (d) only up to 1,100 kV.

[A.M.I.E. Sec B. Summer 1994]

226. Phase relays are used to provide protection against
(a) single line to ground fault. (b) three phase fault only.
(c) phase faults involving two or more phases.
(d) double phase to ground fault only. [A.M.I.E. Sec B. Summer 1994]

227. In a 3-phase circuit, residual current means the flow of ... current.
(a) negative-sequence (b) zero sequence
(c) dc transient [A.M.I.E. Sec B. Winter 1999]

228. The plug setting of a negative sequence relay is 0.2 A. The current transformer ratio is 5 : 1. The minimum value of line to line fault current for the operation of the relay is
(a) 1 A (b) $\frac{1}{1.732}$ A
(c) 1.732 A (d) $\frac{0.2}{1.732}$ A [GATE E.E. 2000]

229. Zero-sequence current is used for relaying purposes only in the case of
(a) phase overcurrent relay. (b) phase impedance relay.
(c) ground overcurrent relay. (d) ground impedance relay.

[A.M.I.E. Sec B. Summer 1994]

230. Over conventional electromechanical relays, the transistor relays have the advantages of
(a) low power consumption, less maintenance and greater sensitivity.
(b) quick operation and high reset to pick-up ratio.
(c) wide range of characteristics approaching more or less to ideal requirements.
(d) all of the above.

231. Over conventional electromechanical relays, the transistor relays have the limitations of
(a) characteristics varying with temperature and ageing.
(b) low short-time overload capacity.
(c) reliability dependent upon the large number of small components and their electrical connections.
(d) all of the above.

Answers: 208. (d), **209.** (c), **210.** (b), **211.** (d), **212.** (a), **213.** (c), **214.** (b), **215.** (a), **216.** (d), **217.** (b), **218.** (a), **219.** (d), **220.** (b), **221.** (d), **222.** (c), **223.** (d), **224.** (b), **225.** (c), **226.** (c), **227.** (b), **228.** (a), **229.** (c), **230.** (d), **231.** (d),

232. In a static overcurrent relay, inverse time characteristics are obtained by
(a) a transistor amplifier. (b) an integrating circuit.
(c) a transistor switch. (d) a differentiating circuit.
[A.M.I.E. Sec B. Summer 1993]

233. Two-input phase comparator in a static relay is made up of a
(a) transformer amplifier. (b) transistor logic circuit.
(c) rectifier bridge. (d) thyristor bridge.
[A.M.I.E. Sec B. Summer 1993]

234. The phase comparators in case of static relays and electro-mechanical relays normally are
(a) cosine and sine comparators respectively.
(b) sine and cosine comparators respectively.
(c) both are sine comparators. (d) both are cosine comparators.

235. Merz-Price protection is employed for protection of
(a) alternators. (b) transformers.
(c) transmission lines. (d) both (a) and (b).
[A.M.I.E. Sec B. Summer 1999]

236. Negative-sequence voltages in the stator induces.
(a) high frequency currents in the rotor.
(b) high frequency voltage in the rotor.
(c) low frequency currents in the rotor.
(d) high frequency currents in the stator.

237. The short-circuit current of an alternator, in case of line to line fault, depends on its
(a) short-circuit resistance. (b) transient reactance.
(c) synchronous reactance. (d) none of the above.

238. We do not require any protection against prime mover failure in case of
(a) turbo-generator sets. (b) hydrogenerator sets.
(c) diesel-engine driven alternators.
(d) back pressure turbo-generators.

239. A large ac generator supplying power to an infinite bus, has a sudden short circuit occurring at its terminals. Assuming the prime mover input and the voltage behind the transient reactance to remain constant immediately after the fault, the acceleration of the generator rotor is
(a) inversly proportional to the moment of inertia of the machine.
(b) inversly proportional to the square of voltage.
(c) directly proportional to the square of the short-circuit current.
(d) directly proportional to the short-circuit power.
[U.P.S.C. I.E.S. E.E.-I, 1994]

240. For the protection of stator winding of an alternator against internal fault involving ground, the relay used is a
(a) biased differential relay. (b) directional overcurrent relay.
(c) plain impedance relay. (d) Buchholz relay.
[A.M.I.E. Sec B. Summer 1993]

241. The bias factor S in unit protection synchronous generators
(a) lies between 0.05 and 0.1 pu.
(b) is less than 0.05 pu. (c) lies between 0.1 to 0.25 pu.
(d) is greater than 0.025 pu. **[A.M.I.E. Sec B. Summer 1994]**

242. Field failure in an alternator occurs due to
(a) failure of exciter. (b) faulty field circuit breaker.
(c) inrush of large currents. (d) both (a) and (b).

243. A soon as a fault develops in a generator stator, it is essential to suppress field excitation, otherwise
(a) terminal voltage will drop.
(b) it may lead to loss of synchronism.
(c) it will continue to supply power to the stator winding fault.
(d) all of the above.

244. If there is an open circuit in the field circuit of a generator
(a) the generator will continue to operate as an induction generator supplying load at a very low leading power factor.
(b) the supply voltage will drop, and the system stability will be affected.
(c) the circulating current at slip frequency may overheat the generator.
(d) all of the above.

245. The earth fault in stator causes
(a) arcing to core.
(b) severe heating in conductors and thereby damaging the insulation.
(c) open-circuit in the stator. (d) both (a) and (b).

246. Failure of insulation on stator winding of generator results in
(a) short circuit between turns.
(b) short circuit between one or more phases and earth.
(c) short circuit between phases.
(d) all of the above.

247. The magnitude of earth-fault current for a given fault position within a winding depends upon
(a) the winding connections.
(b) the method of neutral grounding.
(c) unmatched characteristics of CTs.
(d) both (a) and (b).

248. The neutral of 10 MVA, 11 kV alternator is earthed through a resistance of 5 ohms. The earth fault relay is set to operate at 0.75 A. The CT's have a ratio of 1,000/5. What percentage of the alternator winding is protected ?
(a) 85% (b) 88.2% (c) 15% (d) 11.8%
[GATE E.E. 1998]

249. Consider a stator winding of an alternator with an internal high-resistance ground fault. The currents under the fault condition are as shown in the figure. The winding is protected using a differential current scheme with current transformers of ratio $\frac{400}{5}$ A as shown. The current through the operating coil is

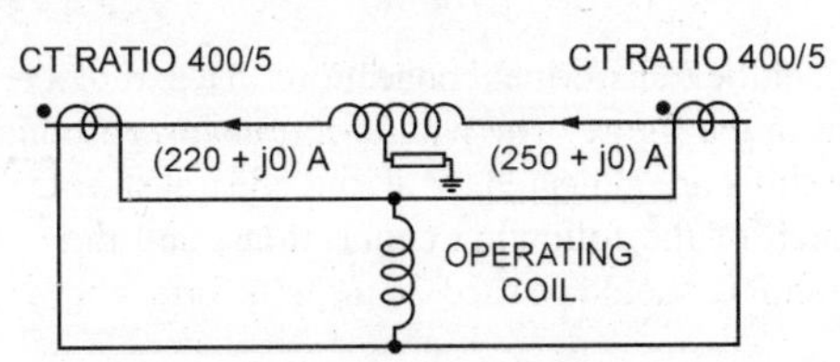

(a) 0.1875 A (b) 0.2 A
(c) 0.375 A (d) 60 kA **[GATE E.E., 2010]**

250. For the protection of a large squirrel-cage induction motor against single phase normally
(a) an overcurrent relay is used.
(b) a differential relay is used.
(c) a directional relay is used.
(d) negative sequence current sensitive relay is used.
[A.M.I.E. Sec B. Summer 1993]

251. The main function of undervoltage protective device generally employed with a motor starter is to
(a) open the supply circuit on failure of power supply.
(b) control the motor voltage.
(c) prevent the opening of supply circuit.
(d) none of the above.

Answers: **232.** (b), **233.** (c), **234.** (a), **235.** (d), **236.** (a), **237.** (c), **238.** (a), **239.** (a), **240.** (a), **241.** (a), **242.** (d), **243.** (c), **244.** (d), **245.** (d), **246.** (d), **247.** (d), **248.** (b), **249.** (c), **250.** (d), **251.** (a),

252. While adding extra stop buttons to an existing control system they must be in
(a) parallel to the load. (b) series with the contactor coil.
(c) series with the motor. (d) parallel to the start button.

253. The types of faults that occur in electric motors are
(a) stator. (b) rotor.
(c) overload. (d) undervoltage.
(e) unbalanced supply voltages and reverse or open phase starting.
(f) all of the above.

254. Negative-sequence currents flow in the stator windings of an ac motor due to
(a) unbalance in the supply voltage.
(b) unbalanced load. (c) phase reversal.
(d) either (a) or (b).

255. A negative sequence relay is commonly used to protect
(a) an alternator. (b) a transformer.
(c) a transmission line. (d) a bus bar. **[GATE E.E. 2011]**

256. Earth fault protection for an electric motor is provided by means of
(a) instantaneous overcurrent relay.
(b) instantaneous relay having a setting of aproximately 30% of motor rated current in the residual circuits of two CTs.
(c) ground wire. (d) both (b) and (c).

257. Any winding of a power transformer develops short-circuit owing to
(a) loose connection. (b) insulation failure.
(c) impulse voltage. (d) mechanical vibrations.

258. Which one of the following relays has the capability of anticipating the possible major fault in a transformer ?
(a) Overcurrent relay (b) Differential relay
(c) Buchholz relay (d) Overfluxing relay
[U.P.S.C. I.E.S. E.E.-II, 2001]

259. Which relay is used to detect and protect internal faults of a transformer ?
(a) Buchholz (b) Directional relay
(c) Thermal relay (d) Distance relay
[U.P.S.C. I.E.S. E.E.-II, 2007]

260. Consider the transformer connections in a part of a power system shown in the figure. The nature of transformer connections and phase shifts are indicated for all but one transformer

Which of the following connections, and the corresponding phase shift θ, should be used for the transformer between A and B?

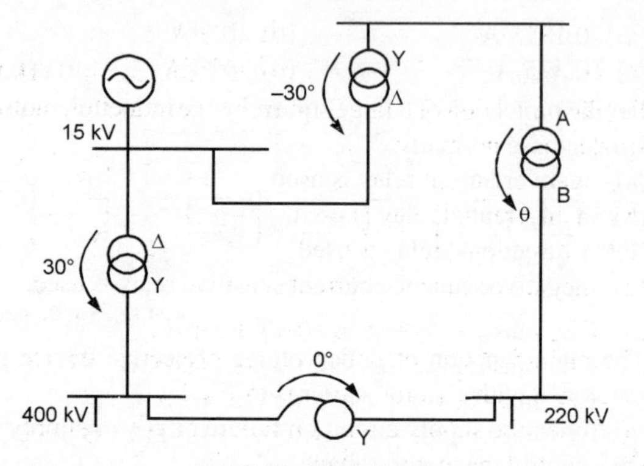

(a) Star-Star (θ = 0°) (b) Star-Delta (θ = –30°)
(c) Delta-Star (θ = 30°) (d) Star-Zigzag (θ = 30°)
[GATE E.E. 2007]

261. A three-phase transformer having zero-sequence impedance of Z_0 has the zero-sequence network as shown in the figure. The connections of its windings are

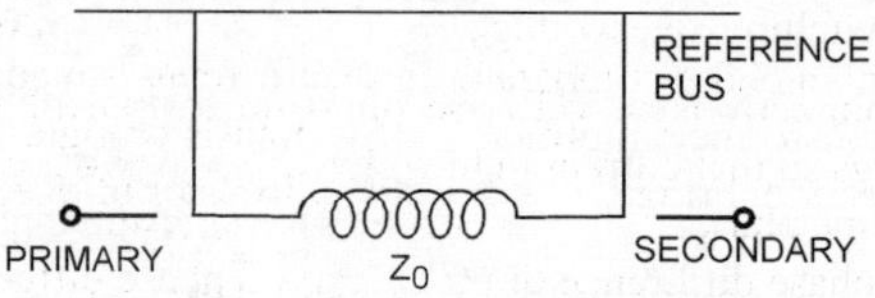

(a) star-star (b) delta-delta
(c) star-delta
(d) delta-star with neutral grounded **[U.P.S.C. I.E.S. E.E.-II, 2001]**

262. To protect the power transformer (Y-Y with neutral grounded) against fault, what type of connection do the CTs have ?
(a) Δ-Y connection (b) Δ-Δ connection
(c) Y-Y connection (d) Y-Δ connection
[U.P.S.C. I.E.S. E.E.-II, 2006]

263. In Merz-Price percentage differential protection of a Δ-Y transformer, the CT secondaries connection in the primary and secondary winding of the transformer would be in the form of
(a) Δ-Y. (b) Y-Δ. (c) Δ-Δ. (d) Y-Y.
[U.P.S.C. I.E.S. E.E.-II, 2000]

264. A 10 kVA, 400 V/200 V single phase transformer with 10% impedance draws a steady short-circuit line current of
(a) 50 A. (b) 150 A. (c) 250 A. (d) 350 A.
[GATE E.E. 1999]

265. A 3-phase 11/66 kV delta/star transformer, protected by Merz-Price scheme has CT ratio of 400/5 on *lt* side. Ratio of CT on *ht* side will be equal to
(a) 1 : 23 (b) 23 : 1 (c) $23 : \sqrt{3}$ (d) $\sqrt{3} : 23$
[U.P.S.C. I.E.S. E.E.-II, 1999]

266. To prevent mal operation of differentially connected relay while energising a transformer, the relay restraining coil is biased with
(a) second harmonic current. (b) third harmonic current.
(c) fifth harmonic current. (d) seventh harmonic current.
[U.P.S.C. I.E.S. E.E.-II, 1995]

267. The inrush current of a transformer at no load is maximum if the supply voltage is switched on
(a) at peak voltage value. (b) at zero voltage value.
(c) at half voltage value. (d) at 0.866 time voltage value.

268. Magnetising inrush phenomena occur in
(a) large transformer primary only, when the transformer is switched on.
(b) large transformer primary when a 3-phase fault occurs in the secondary.
(c) large transformer secondary only when the transformer is switched on in the primary.
(d) none of the above. **[A.M.I.E. Sec B. Winter 1996]**

269. In case of power transformer :
(a) positive-, negative- and zero-sequence impedances are all equal.
(b) positive- and zero-sequence impedances are equal but less than negative-sequence impedance.
(c) positive- sequence impedance is greater than both negative- and zero-sequence impedances.
(d) positive- sequence impedance is less than negative-sequence impedance but equal to zero-sequence impedance.
[A.M.I.E. Sec B. Advanced Power Systems Summer 1998]

Answers: 252. (b), **253.** (f), **254.** (d), **255.** (a), **256.** (d), **257.** (b), **258.** (c), **259.** (a), **260.** (a), **261.** (b), **262.** (b), **263.** (b), **264.** (c), **265.** (b), **266.** (a), **267.** (b), **268.** (a), **269.** (a),

270. For the protection of transformers, harmonic restraint is used to guard against
(a) magnetising inrush current.
(b) unbalanced operation. (c) lightning.
(d) switching overvoltage. [GATE E.E. 2001]

271. The connections of CTs are opposite to that of transformer windings so that current in the pilot wires of two are/have
(a) same phase. (b) opposite in phase.
(c) a phase difference of 60°. (d) a phase difference of 90°.

272. For preventing the maloperation of Merz-Price protection scheme on inrush of magnetizing current
(a) the relay restraining coil is biased with second harmonic current.
(b) time lag is provided in the relay.
(c) relay sensitivity is reduced by employing a shunt.
(d) all of the above.

273. Problems associated with differential protection is/are
(a) magnetizing inrush current.
(b) mismatching characteristics of CTs.
(c) charge of ratio as a result of tapping.
(d) all of the above.

274. Which of the following sections can be employed for bus-bars?
(a) Bars. (b) Rods.
(c) Tubes. (d) Any of the above.

275. The material used for bus-bars should have
(a) low resistivity. (b) low cost.
(c) high softening temperature.
(d) all of the above.

276. A bus-bar is rated by
(a) current only. (b) voltage only.
(c) current, voltage and frequency.
(d) current, voltage, frequency and short-circuit current.

277. Single bus-bar arrangement
(a) is cheapest in initial as well as maintenance cost.
(b) provides simple operation and relaying system.
(c) has the drawback that there will be complete shutdown when fault occurs on the bus itself.
(d) all of the above.

278. Double bus-bar arrangement makes use of two identical bus-bars and has the advantage(s) that
(a) it does not require any bus coupler and permits switchover from one bus to another whenever desired.
(b) it provides maximum flexibility and reliability.
(c) either bus-bar may be taken up for maintenance.
(d) the infeed and load circuits may be divided into two separate groups.
(e) all of the above.

279. A bus coupler circuit breaker is utilised in a substation for
(a) joining the transmission line with station bus-bar.
(b) joining main and transfer bus in a substation.
(c) joining the generator with transfer.
(d) joining the neutral of the generator with earth.
[A.M.I.E. Sec B. Summer 1993]

280. Zero-sequence impedance of the transmission line is
(a) equal to the positive-sequence impedance of the line.
(b) about 2.5-3 times the positive-sequence impedance of the line.
(c) less than positive-sequence impedance of the line.
(d) none of the above. [A.M.I.E. Sec B. Winter 1996]

281. Which is the pilotless protection method for feeder line?
(a) Differential protection. (b) Carrier current protection.
(c) Time graded protection. (d) None of these.

282. The most commonly used method for the protection of 3-phase feeder is
(a) time graded protection. (b) differential protection.
(c) reverse power protection.

283. A transmission line is protected by:
(a) distance protection.
(b) current graded overcurrent protection.
(c) time graded overcurrent protection.
(d) inrush protection.
[M.D. Univ. Transmission and Distribution May 2011]

284. In the case of transmission line protection, overcurrent used is:
(a) only up to 110 kV. (b) only up to 220 kV.
(c) only up to 50 kV. (d) only up to 1,100 kV.
[M.D. Univ. Transmission and Distribution May 2011]

285. Which is the main relay for protecting up to 90% of the transmission line-length in the forward direction ?
(a) Directional overcurrent relay. (b) Mho relay.
(c) Carrier-current protective relay. (d) Impedance relay.
[U.P.S.C. I.E.S. E.E.-II, 2006]

286. For complete protection of a 3-phase transmission line we require
(a) three-phase and three earth fault relays.
(b) three-phase and two earth fault relays.
(c) two-phase and two earth fault relays.
(d) two-phase and one earth fault relay.

287. For complete protection of a 3-phase line
(a) three-phase and three-earth fault relays are required.
(b) three-phase and two-earth fault relays are required.
(c) two-phase and two-earth fault relays are required.
(d) two-phase and one-earth fault relays are required.
[W.B. Univ. of Technology Power System-II, 2009-10]

288. How many relays are used to detect interphase fault of a three-line system ?
(a) One (b) Two (c) Three (d) Six
[U.P.S.C. I.E.S. E.E.-II, 2006]

289. The transmission line distance protection relay having the property of being inherently directional is
(a) impedance relay (b) MHO relay
(c) OHM relay (d) reactance relay [GATE E.E. 2004]

290. The sections of a feeder are provided with circuit breakers CB_1, CB_2, CB_3, CB_4, CB_5 and CB_6. For a fault at F as indicated in figure

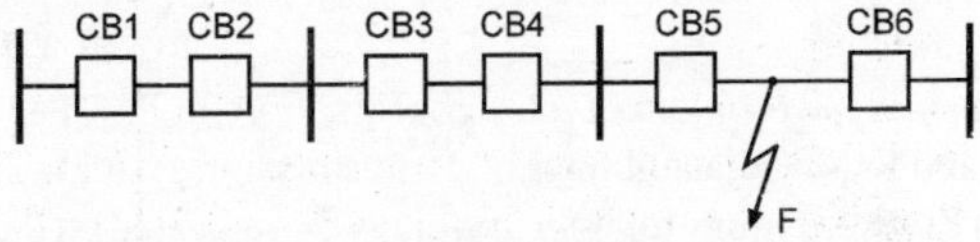

(a) CB_5 must be set to trip after CB_1 trips.
(b) CB_5 must be set to trip after CB_3 and CB_4 trip.
(c) CB_5 must be set to trip after CB_2 trips.
(d) CB_5 must be set to trip before CB_1, CB_2, CB_3 and CB_4 trip.
[GATE E.E. 1999]

291. Time graded protection of a radial feeder can be achieved by using
(a) definite time relays. (b) inverse time relays.

Answers: **270.** (a), **271.** (a), **272.** (a), **273.** (d), **274.** (d), **275.** (d), **276.** (d), **277.** (d), **278.** (e), **279.** (b), **280.** (b), **281.** (b), **282.** (b), **283.** (a), (b), (c), **284.** (c), **285.** (b), **286.** (d), **287.** (d), **288.** (c), **289.** (b), **290.** (d), **291.** (c),

(c) both definte and inverse time relays.
(d) none of the above. [A.M.I.E. Sec B. Summer 1999]

292. For protection of parallel feeders fed from one end, the relays required are
(a) Non-directional relays at the source end and directional relays at the load end.
(b) Non-directional relays at both ends.
(c) Directional relays at source end and non-directional relays at load end.
(d) Directional relays at both ends. [U.P.S.C. I.E.S. E.E.-II, 2005]

293. Pilot-wire protection scheme is most economical and provides high speed relaying for
(a) short length of lines up to 15 km.
(b) medium length of lines up to 60 km.
(c) long length of lines up to 200 km.
(d) all of the above.

294. Adjustable resistors are connected in the pilot wires in order to
(a) change the phase of the current flowing through the relay.
(b) get equipotential points on pilot wires.
(c) reduce current flowing through the relay.
(d) none of the above.

295. Signal mixing transformers are used for
(a) induction disc type overcurrent relay.
(b) direction sensitive distance relay.
(c) pilot-wire feeder protection.
(d) plain differential relay. [A.M.I.E. Sec B. Summer 1993]

296. In a balanced voltage pilot-wire protection scheme if the pilot-circuit gets opened, the relay will
(a) fail to trip on internal faults. (b) trip on full load.
(c) trip instantaneously on external faults.

297. In a balanced voltage pilot-wire protection scheme, if the pilot-circuit gets short-circuited, the relay will
(a) fail to trip on internal faults.
(b) fail to trip on external faults.
(c) trip on full-load. (d) trip instantaneously.

298. Voltage phasors at the two terminals of a transmission line of length 70 km have a magnitude of 1.0 per unit but are 180 degrees out of phase. Assuming that the maximum load current in the line is 1/5th of minimum 3-phase fault current. Which one of the following transmission line protection schemes will NOT pick up for this condition ?
(a) Distance protection using mho relays with zone-1 set to 80% of the line impedance.
(b) Directional overcurrent protection set to pick up at 1.25 times the maximum load current.
(c) Pilot relaying system with directional comparison scheme.
(d) Pilot relaying system with segregated phase comparison scheme. [GATE E.E. 2008]

299. Consider the problem of relay coordination for the distance relays R_1 and R_2 on adjacent lines of a transmission system. The Zone 1 and Zone 2 settings for both the relays are indicated on the diagram. Which of the following indicates the correct time setting for the Zone 2 of relays R_1 and R_2.

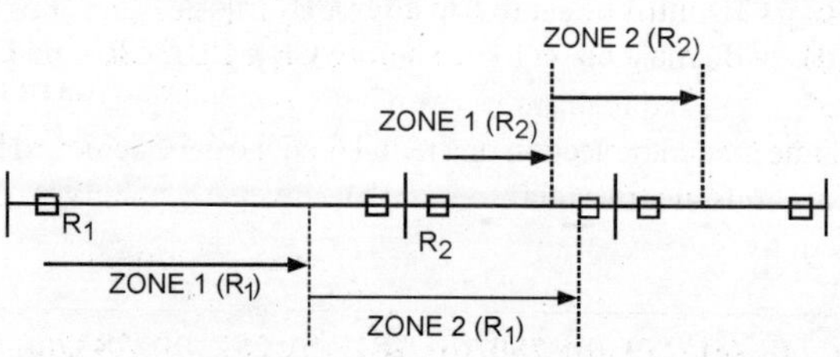

(a) $TZ2_{R1} = 0.6s$, $TZ2_{R2} = 0.3s$.
(b) $TZ2_{R1} = 0.3s$, $TZ2_{R2} = 0.6s$.
(c) $TZ2_{R1} = 0.3s$, $TZ2_{R2} = 0.3s$.
(d) $TZ2_{R1} = 0.1s$, $TZ2_{R2} = 0.3s$. [GATE E.E. 2002]

300. Consider the protection system shown in the figure. The circuit breakers numbered from 1 to 7 are of identical type. A single line to ground fault with zero fault impedance occurs at the midpoint of the line (at point F), but circuit breaker 4 fails to operate ("Stuck breaker"). If the relays are coordinated correctly, a valid sequence of circuit breaker operation is

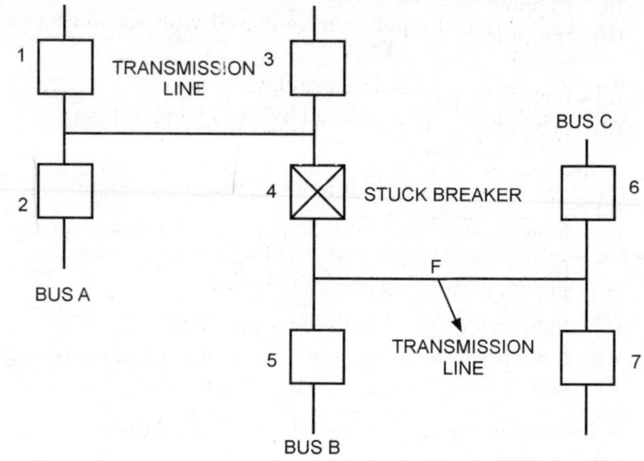

(a) 1, 2, 6, 7, 3, 5 (b) 1, 2, 5, 5, 7, 3
(c) 5, 6, 7, 3, 1, 2 (d) 5, 1, 2, 3, 6, 7 [GATE E.E. 2008]

301. The distribution system shown in figure is to be protected by overcurrent system of protection.
At which locations are directional overcurrent relays required for proper fault discrimination ?

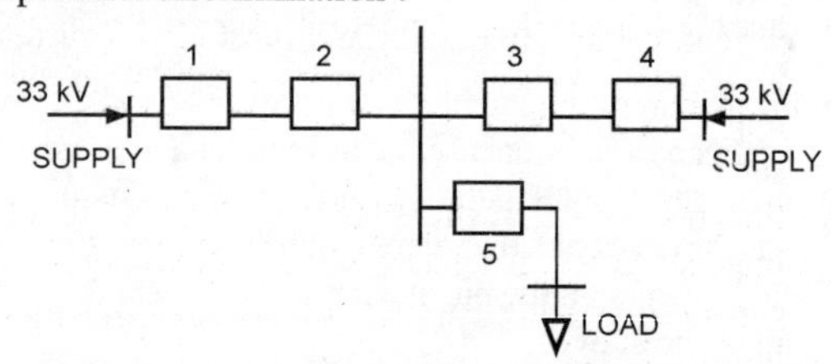

(a) 1 and 4 (b) 2 and 3
(c) 1, 4 and 5 (d) 2, 3 and 5
[U.P.S.C. I.E.S. E.E.-II, 2005]

302. The figure given below shows a schematic arrangement of a Distance Relay provided with a 'Replica Impedance' Z_r. The CT ratio = $\frac{I}{i}$ and VT ratio = $\frac{V}{v}$.
When a fault occurs on the line being protected, when would the relay operate ?

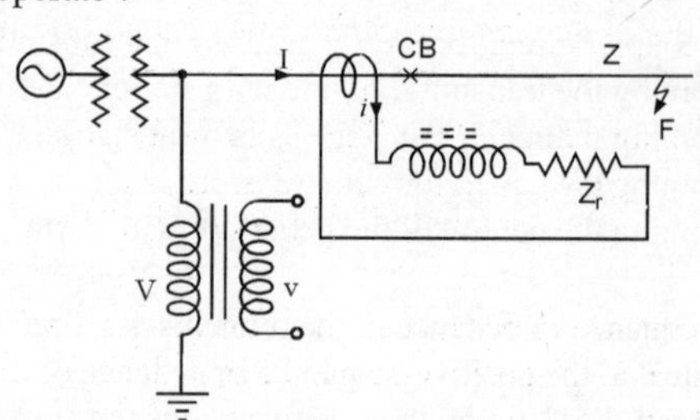

(a) $Z_r > Z$ (b) $Z_r < Z$
(c) $Z_r > Z \cdot \frac{I}{i}$ (d) $Z_r > Z \cdot \frac{V}{v}$
[U.P.S.C. I.E.S. E.E.-II, 2008]

Answers: **292.** (a), **293.** (a), **294.** (b), **295.** (c), **296.** (a), **297.** (c), **298.** (a), **299.** (a), **300.** (c), **301.** (b), **302.** (b),

303. The frequency of carrier in the carrier current pilot scheme is in the range of
(a) 1 kHz to 10 kHz. (b) 10 kHz to 25 kHz.
(c) 25 kHz to 50 kHz. (d) 50 kHz to 500 kHz.
[A.M.I.E. Sec B. Summer 1994]

304. The frequency of carrier transmitted by microwave pilot is in the range of
(a) 1,000 kHz to 1,500 kHz (b) 2,000 kHz to 5,000 kHz
(c) 900 MHz to 6,000 MHz (d) 10,000 MHz to 15,000 MHz
[A.M.I.E. Sec B. Summer 1994]

305. In carrier current protection the purpose of the wavetrap is for
(a) trapping power frequency waves.
(b) trapping high frequency waves entering into generators/ transformer unit.
(c) both (a) and (b).
(d) none of the above. **[A.M.I.E. Sec B. Summer 1998]**

306. A line trap in carrier current relaying tuned to carrier frequency presents
(a) high impedance to carrier frequency but low impedance to power frequency.
(b) low impedance to both carrier and power frequency.
(c) high impedance to both carrier and power frequency.
(d) low impedance to carrier frequency but high impedance to power frequency. **[U.P.S.C. I.E.S. E.E.-II, 1998]**

307. A line trap in a long transmission line is used to
(a) improve the power factor.
(b) dampen the overvoltage oscillations.
(c) confine the carrier signals in the line.
(d) protect the line against direct lightning stroke.
[A.M.I.E. Sec B. Advanced Power Systems Winter 1999]

308. Line trap and coupling capacitors are used for carrier current protection in which
(a) line trap has high impedance to 50 Hz signal but low impedance to carrier current signal whereas a coupling capacitor has low impedance to 50 Hz signal but high impedance to carrier signal.
(b) line trap has low impedance to 50 Hz signal but high impedance to carrier current signal, where as a coupling capacitor has high impedance to 50 Hz signal but low impedance to carrier signal.
(c) both the line trap and coupling capacitor have low impedance to 50 Hz signal but high impedance to carrier current signal.
(d) both the line trap and coupling capacitor have high impedance to 50 Hz signal but low impedance to carrier current signal.
[U.P.S.C. I.E.S. E.E.-II, 1999]

309. In a 3-step distance protection, the reach of the three zones of the relay at the beginning of the first line typically extends into
(a) 100% of the first line, 50% of the second line and 20% of the third line.
(b) 80% of the first line, 50% of the second line and 20% of the third line.
(c) 80% of the first line, 50% of the second line and 10% of the third line.
(d) 50% of the first line, 50% of second line and 20% of the third line. **[GATE E.E. 2000]**

310. Distance protection scheme is preferred over graded time-lag over-current protection in HV and EHV lines because
(a) it is faster in operation. (b) it is simple.
(c) it is cheaper in cost. (d) all of the above.

311. Three step time-distance characteristic of distance relay can be had by
(a) changing taps on voltage transformer.
(b) separate measuring elements for zones 2 and 3.
(c) switching resistance in relay restraint circuit at pre-set time intervals by means of timer element.
(d) any of the above.

312. Lightning is a huge spark caused by electrical discharge taking place between
(a) clouds. (b) within the same cloud.
(c) cloud and earth. (d) any of the above.

313. Switching overvoltages are more hazardous than lightning surges in case of
(a) low voltage system. (b) 33 kV system.
(c) EHV and UHV systems. (d) all of these.

314. The overvoltage surges in power systems may be caused by
(a) lightning. (b) resonance.
(c) switching. (d) all of these.

315. Requirements of protection power station buildings against direct strokes are
(a) interception. (b) conduction.
(c) dissipation. (d) reflection and convection.
(e) (a), (b) and (c).

316. The protection against direct lightning strokes and high voltage steep waves is provided by
(a) ground wires. (b) lightning arresters.
(c) lightning arresters and ground wires.
(d) earthing of neutral.

317. Which of the following factors should be considered in the design of a transmission line against lightning with ground wire?
(a) Mechanical strength of ground wire.
(b) Clearance between line conductor and ground wire.
(c) Clearance between line conductor and earth.
(d) All of the above.

318. For protection of rotating machines against lightning surges _______ is used
(a) lightning arrester. (b) capacitor.
(c) combination of lightning arrester and capacitor.
(d) lightning conductor and arrester.

319. Consider the following statements. To provide reliable protection for a distribution transformer against overvoltages using lightning arresters, it is essential that the
1. lead resistance is high.
2. distance between the transformer and the arrester is small.
3. transformer and the arrester have a common interconnecting ground.
4. sparkover voltage of the arrester is greater than the residual voltage.

Of these statements.
(a) 1, 3 and 4 are correct. (b) 2 and 3 are correct.
(c) 2, 3 and 4 are correct. (d) 1 and 4 are correct.
[U.P.S.C. I.E.S. E.E.-II, 1997]

320. Impulse ratios of insulators and lightning arresters should be
(a) both low. (b) high and low respectively.
(c) low and high respectively. (d) both high.
[U.P.S.C. I.E.S. E.E.-II, 1996]

321. Lighting arresters are used in power systems to protect electrical equipments against

Answers: 303. (d), **304.** (c), **305.** (b), **306.** (a), **307.** (c), **308.** (b), **309.** (b), **310.** (a), **311.** (d), **312.** (d), **313.** (c), **314.** (d), **315.** (e), **316.** (c), **317.** (d), **318.** (c), **319.** (b), **320.** (b), **321.** (c),

(a) direct strokes of lightning.
(b) overvoltages due to indirect lightning stroke.
(c) power frequency overvoltages.
(d) overcurrents due to lightning. [U.P.S.C. I.E.S. E.E.-II, 1995]

322. Which of the following is the protective device against lightning overvoltages?
(a) Rod gaps. (b) Surge absorbers.
(c) Horn gaps. (d) All of the above.
[A.M.I.E. Sec B. Power Systems Winter 1999]

323. Overhead ground wires are used to protect a transmission line against
(a) line-to-ground faults. (b) arcing earths.
(c) voltage surges due to direct lightning stroke.
(d) high voltage oscillations due to switching.
[A.M.I.E. Sec B. Advanced Power Systems Winter 1999]

324. An overhead transmission line is provided with earth wire for protection against
(a) switching surge. (b) lightning surge.
(c) power frequency overvoltage.
(d) none of these. [W.B. Univ. of Technology Power System-III, 2009-10]

325. In a thyrite lightning arrester the resistance
(a) varies linearly with the applied voltage.
(b) increases with the applied voltage.
(c) decreases linearly with the applied voltage.
(d) is high at low current and low at high current.
[A.M.I.E. Sec B. Advanced Power Systems Winter 1999]

326. An ideal surge diverter should have the characteristics
(a) its power frequency breakdown or sparkover must be above normal or abnormal fundamental frequency.
(b) when the value of voltage transient peak exceeds the spark-over value of the surge diverter, a conducting path to earth must be provided.
(c) after occurrence of breakdown, it must carry the resulting discharge current without any damage to it.
(d) after breakdown the power frequency current must be interrupted as soon as the transient voltage has dropped below the breakdown value.
(e) all of the above.

327. Consider the following statements:
1. Present-day surge diverters use nonlinear resistance elements.
2. A travelling wave is usually represented as a step wave in the analysis.
3. A travelling wave suffers reflection when it reaches a discontinuity.
4. The function ($f(vx \pm t)$) represents a travelling wave.

Which of the above statements is/are correct?
(a) 1 only. (b) 1 and 2 only.
(c) 1, 2 and 3. (d) 3 and 4 only.
[U.P.S.C. I.E.S. E.E.-II, 2009]

328. A lightning arrester provides
(a) low impedance path. (b) high impedance path.
(c) low resistance path.
(d) high resistance path between line and earth during operation.

329. Surge absorbers are used for protection against
(a) high voltage low frequency oscillations.
(b) high voltage high frequency oscillations.
(c) low voltage high frequency oscillations.
(d) low voltage low frequency oscillations.

330. Surge modifiers are employed for
(a) reducing the steepness of wavefront.
(b) reducing the current of wavefront.
(c) reducing the voltage of wavefront.
(d) modify the shape of the wavefront.

331. Which of the following are the important limitations of rod gap surge arresters ?
1. They are not capable of sealing off power frequency follow up current.
2. After a discharge, the rods are destroyed completely.
3. Performance is affected by climatic conditions.

Use the following codes for selecting the correct answer.
Codes :
(a) 1, 2 and 3 (b) 1 and 2 (c) 2 and 3 (d) 1 and 3
[U.P.S.C. I.E.S. E.E.-II, 1993]

332. A thyrite type lightning arrester
(a) blocks the surge voltage appearing in a line.
(b) absorbs the surge voltage appearing in a line.
(c) offers a low resistance path to the surge appearing in a line.
(d) returns the surge back to the source.
[U.P.S.C. I.E.S. E.E.-II, 1993]

333. A lightning arrester connected between the line and earth in a power system
(a) protects the terminal equipment against travelling surges.
(b) protects the terminal equipment against direct lightning stroke.
(c) suppresses high frequency oscillations in the line.
(d) reflects back the travelling waves approaching it.
[U.P.S.C. I.E.S. E.E.-II, 1994]

334. Which of the following is a non-linear diverter?
(a) Expulsion type arrester. (b) Valve type arrester.
(c) Electrolytic type arrester. (d) Rod gap arrester.

335. A valve type lightning arrestor in a substation should be placed
(a) close to the circuit breaker.
(b) close to the transformer.
(c) away from the transformer.
(d) none of these. [W.B. Univ. Technology Power System-III, 2009-10]

336. Surge absorber ______ the energy of travelling waves
(a) absorbs (b) reflects
(c) diverts (d) partly absorbs and partly diverts

337. Coupling factor of a ground wire can be increased by
(a) reducing the footing impedance.
(b) increasing the ground wire size.
(c) using cantilever rods on the crossing alongwith the areas of ground wire.
(d) all of the above.

338. In the presence of corona, electrostatic coupling ______ and electro-magnetic coupling ______
(a) decrease, increases. (b) increases, decreases.
(c) increases, remains the same.
(d) remains the same, decreases.

339. When a transmission line is energized, ______ propagate on it
(a) voltage wave only (b) current wave only
(c) both voltage and current waves
(d) none of the above.

340. When a wave propagate on a transmission line, it suffers reflection several times at
(a) load end. (b) sending end.
(c) sending end and other end. (d) tapping point.

Answers: 322. (d), **323.** (c), **324.** (b), **325.** (d), **326.** (e), **327.** (c), **328.** (a), **329.** (c), **330.** (a), **331.** (d), **332.** (c), **333.** (a), **334.** (b), **335.** (b), **336.** (d), **337.** (d), **338.** (c), **339.** (c), **340.** (c),

341. Travelling voltage wave and current wave have the same waveforms and travel together along the transmission line at a velocity
(a) of sound. (b) of light.
(c) slightly lesser than that of light.
(d) slightly lesser than that of sound.

342. The relation between travelling voltage wave and current wave is given as
(a) $ei = \sqrt{\frac{L}{C}}$ (b) $\frac{e}{i} = \sqrt{\frac{L}{C}}$
(c) $ei = \sqrt{LC}$ (d) $e/i = \sqrt{LC}$

343. For a transmission line the standing wave ratio is the ratio of
(a) peak voltage to rms voltage.
(b) maximum current to minimum current.
(c) maximum voltage to minimum voltage.
(d) maximum impedance to minimum impedance.

344. The steepness of the wavefront can be reduced by connecting
(a) an inductor in series with the line.
(b) a capacitor between line and earth.
(c) either (a) or (b).
(d) an inductor between line and earth or a capacitor in series with the line.

345. If a travelling wave travelling along a loss-free overhead line does not result in any reflection after it has reached the far end, then the far end of the line is
(a) open circuited. (b) short circuited.
(c) terminated into a resistance equal to surge impedance of the line.
(d) terminated into a capacitor. [U.P.S.C. I.E.S. E.E.-II, 2001]

346. A rectangular voltage wave is impressed on a loss-free overhead line, with the far end of the line being short-circuited. On reaching the end of this line
(a) The current wave is reflected back with positive sign, but the voltage wave with negative sign.
(b) The current wave is reflected back with negative sign, but the voltage wave with positive sign.
(c) Both the current and the voltage waves are reflected with positive sign.
(d) Both the current and the voltage waves are reflected with negative sign. [U.P.S.C. I.E.S. E.E.-II, 2003]

347. Consider the following statements for transmission lines :
1. When a transmission line is terminated by its characteristic impedance the line will not have any reflected wave.
2. For a finite line terminated by its characteristic impedance the velocity and current at all points on the line are exactly same.
3. For a lossless half wave transmission line the input impedance is not equal to load impedance.

Which of the statements given above are correct ?
(a) 1 and 2 only. (b) 2 and 3 only.
(c) 1 and 3 only. (d) 1, 2 and 3.
[U.P.S.C. I.E.S. E.E.-I, 2008]

348. For a loss-free long transmission line, the conventional line equations become,

$V_s = (\cos \beta l)V_R + (j \sin \beta l)I_R$ and $I_S = \left(\frac{j \sin \beta l}{Z_C}\right)V_R + (\cos \beta l)\, I_R$

Which one of the following statements is correct ?
If the line is terminated at the receiving end by its natural load impedance Z_c, then
(a) the voltage is constant in magnitude at all points along the line and, V_s and V_R always remain in phase.
(b) the voltage is constant in magnitude at all points along the line but V_s advances in phase relative to V_R by an angle β radians per km.
(c) the magnitude of the voltage along the line changes in proportion to the line-length and the V_s and V_R always remain in phase.
(d) the magnitude of the voltage along the line changes in proportion to the line-length, V_s advances in phase relative to V_R by β radians per km. [U.P.S.C. I.E.S. E.E.-I, 2006]

349. A cable has the following characteristics, L = 0.201 μH/m and C = 196.2 pF/m. The velocity of wave propagation through the cable is
(a) 32 m/s (b) 159.24 m/μs
(c) 0.0312 m/s (d) 159.24 m/s [GATE E.E. 1998]

350. An extra high voltage transmission line of length 300 km can be approximated by a lossless line having propagation constant β = 0.00127 radians per km. Then the percentage ratio of line length to wavelength will be given by
(a) 24.24% (b) 12.12% (c) 19.05% (d) 6.06%
[GATE E.E. 2008]

351. The reflection coefficient for the voltage wave in overhead lines is given as
(a) $\frac{R_0}{R_0 - R_L}$ (b) $\frac{R_L}{R_0 - R_L}$
(c) $\frac{R_L - R_0}{R_L + R_0}$ (d) $\frac{R_L + R_0}{R_0 - R_L}$

352. If a line of surge impedance Z_o is terminated in an impedance Z then the reflection for current and voltage surges at the termination are given respectively by
(a) $\frac{Z_o - Z}{Z_o + Z}, \frac{2Z}{Z_o + Z}$ (b) $\frac{Z_o - Z}{Z_o + Z}, \frac{Z - Z_o}{Z_o + Z}$
(c) $\frac{2Z_o}{Z_o + Z}, \frac{2Z}{Z_o + Z}$ (d) $\frac{2Z_o}{Z_o + Z}, \frac{Z - Z_o}{Z_o + Z}$
[U.P.S.C. I.E.S. E.E.-II, 2010]

353. If the load impedance is 100 ohm and input impedance is 25 ohm, then the characteristic impedance of the transmission line is
(a) 70 ohm. (b) 60 ohm.
(c) 50 ohm. (d) 40 ohm.
[U.P.S.C. I.E.S. E.E.-I, 2010]

354. A transmission line of characteristic impedance of 50 ohm is terminated by a load impedance of (15 – *j*20) ohm. What is the normalized load impedance?
(a) 0.6 – *j*0.8 (b) 0.3 – *j*0.6
(c) 0.3 – *j*0.4 (d) 0.3 + *j*0.4
[U.P.S.C. I.E.S. E.E.-I, 2009]

355. A loss-less transmission line with characteristic impedance of 600 ohms is terminated in a purely resistive load of 900 ohms. The reflection coefficient is
(a) 0.2 (b) 0.5 (c) 0.667 (d) 1.5
[U.P.S.C. I.E.S. E.E.-I, 2003]

356. For a line of characteristic impedance Z_0 terminated in a load of Z_R such that $Z_R = Z_0/3$, what is the reflection coefficient T_L ?
(a) 1/3 (b) 2/3 (c) –1/3 (d) –1/2
[U.P.S.C. I.E.S. E.E.-I, 2006]

Answers: 341. (c), **342.** (b), **343.** (c), **344.** (c), **345.** (c), **346.** (a), **347.** (a), **348.** (b), **349.** (b), **350.** (d), **351.** (c), **352.** (b), **353.** (c), **354.** (c), **355.** (a), **356.** (d),

357. The reflection coefficient for the transmission line shown in figure at point P is

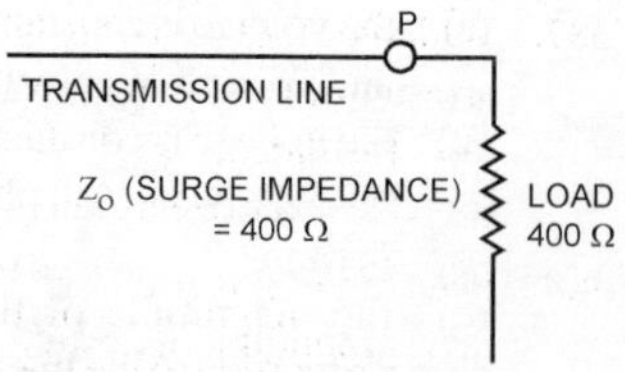

(a) +1 (b) 0.5
(c) 0 (d) −1

[U.P.S.C. I.E.S. E.E.-II, 2002]

358. Consider a step voltage wave of magnitude 1 pu travelling along a lossless transmission line that terminates in a reactor. The voltage magnitude across the reactor at the instant the travelling wave reaches the reactor is

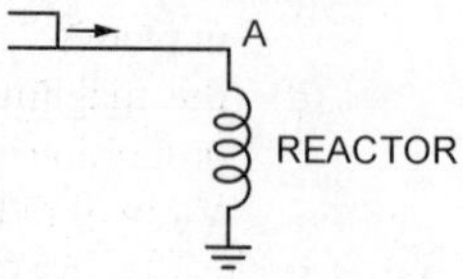

(a) −1 pu (b) 1 pu (c) 2 pu (d) 3 pu

[GATE E.E., 2010]

359. The reflection coefficient of a short-circuited line is
(a) −1 (b) +1 (c) 0.5 (d) zero

[U.P.S.C. I.E.S. E.E.-II, 1997]

360. The propagation constant of a transmission line is

$$0.15 \times 10^{-3}\ j\ 1.5 \times 10^{-3}$$

The wavelength of the travelling wave is

(a) $\dfrac{1.5\times10^{-3}}{2\pi}$ (b) $\dfrac{2\pi}{1.5\times10^{-3}}$

(c) $\dfrac{1.5\times10^{-3}}{5\pi}$ (d) $\dfrac{\pi}{1.5\times10^{-3}}$

[U.P.S.C. I.E.S. E.E.-II, 1997]

361. The reflection coefficient at the load end of a short-circuited line is
(a) zero. (b) $1\ \angle 0^\circ$ (c) $1\ \angle 90^\circ$ (d) $1\ \angle 180^\circ$

[U.P.S.C. I.E.S. E.E.-I, 1999]

362. A surge voltage of 1,000 kV is applied to an overhead line with its receiving end open. If the surge impedance of the line is 500 Ω, then the total surge power in the lıne is
(a) 2,000 MW (b) 500 MW
(c) 2 MW (d) 0.5 MW

[U.P.S.C. I.E.S. E.E.-II, 1996]

363. A surge of 260 kV travelling in a line of natural impedance of 500 Ω arrives at the junction with two lines of natural impedances of 250 Ω and 50 Ω respectively. The voltage transmitted in the branch lines is
(a) 400 kV (b) 260 kV (c) 80 kV (d) 40 kV

[U.P.S.C. I.E.S. E.E.-II, 1999]

364. An overhead line with surge impedance of 400 Ω is terminated through a resistance R. A surge travelling over the line will not suffer any reflection at the junction, if the value of R is
(a) 100 Ω (b) 200 Ω (c) 400 Ω (d) 800 Ω

[U.P.S.C. I.E.S. E.E.-II, 2000]

365. An overhead line with a surge impedance of 400 Ω is connected to a transformer by a short length of cable of surge impedance of 100 Ω. If a rectangular surge wave of 40 kV travels along the line towards the cable, then the voltage of the wave travelling from the junction of the overhead line through the cable towards the transformer would be
(a) 16 kV (b) 24 kV (c) 32 kV (d) 36 kV

[U.P.S.C. I.E.S. E.E.-II, 2000]

366. Two loss-less resistive transmission lines each of characteristic impedance Z are connected as shown in the circuits below. If the maximum voltage on the two lines is the same and the power transmitted by line A is W_1, then what is the power transmitted by the line B?

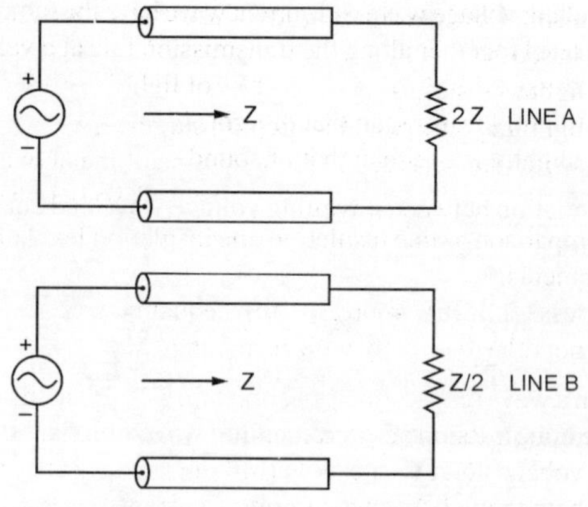

(a) $4\ W_1$ (b) $3\ W_1$ (c) $2\ W_1$ (d) $1\ W_1$

[U.P.S.C. I.E.S. E.E.-I, 2009]

367. A travelling wave due to lightning with an incident voltage V travels through the overhead line of surge impedance of 400 Ω enters a cable of surge impedance of 40 Ω. What is the voltage entering the cable at the junction ?
(a) V/11 (b) 4 V/11 (c) 2 V/11 (d) V

[U.P.S.C. I.E.S. E.E.-I, 2006]

368. A surge voltage rising at 100 kV/μs travels along a lossless open-circuited transmission line. It takes 10 μs to reach the open end. The reflected wave from the open end, will be rising at
(a) 1,000 kV/μs. (b) 200 kV/μs.
(c) 1,000 kV/μs. (d) 2,000 kV/μs.

[U.P.S.C. I.E.S. E.E.-I, 2002]

369. An overhead line having a surge impedance of 400 Ω is connected in series with an underground cable having a surge impedance of 100 Ω. If a surge of 50 kV travels from the line end towards the line-cable junctions, the value of the transmitted voltage wave at the junction is
(a) 30 kV (b) 20 kV (c) 80 kV (d) −30 kV

[GATE E.E. 1999]

370. A lightning stroke discharges impulse current of 10 kA (peak) on a 400 kV transmission line having surge impedance of 250 Ω. The magnitude of transient overvoltage travelling waves in either direction assuming equal distribution from the point of lightning strike will be
(a) 1,250 kV (b) 1,650 kV (c) 2,500 kV (d) 2,900 kV

[GATE E.E. 2004]

371. A surge of 100 kV travels along an overhead line towards its junction with a cabie. The surge impedances for the overhead line and cable are 400 ohms and 50 ohms respectively. The magnitude of the surge transmitted through the cable is
(a) 11.11 kV (b) 22.22 kV (c) 12.50 kV (d) 88.89 kV

[U.P.S.C. I.E.S. E.E.-II, 2002]

372. A short length of cable between the dead end tower and the power transformer
(a) reduces the steepness of the wave under certain conditions only.
(b) always reduces the steepness of the incident wave.
(c) steepness of the incident wave remains unaffected.

373. The insulation strength of an EHV transmission line is mainly governed by
(a) load power factor. (b) switching overvoltages.
(c) harmonics. (d) corona.

[GATE E.E. 2005]

Answers: **357.** (c), **358.** (a), **359.** (a), **360.** (b), **361.** (d), **362.** (a), **363.** (d), **364.** (c), **365.** (a), **366.** (a), **367.** (c), **368.** (c), **369.** (b), **370.** (a), **371.** (b), **372.** (b), **373.** (b),

374. The insulation coordination for UHV lines (above 500 kV) is done based on
(a) lightning surges.
(b) lightning surges and switching surges.
(c) switching surges. (d) none of the above.
[U.P.S.C. I.E.S. E.E.-II, 1999]

375. In comparison to line insulation, the insulation level of the station equipment is
(a) less. (b) more. (c) equal.
(d) not related directly with each other.

376. When a wave reaches an open circuit the ______ at the termination is/are double the incident value(s)
(a) voltage (b) current
(c) both the voltage and current

377. When a wave reaches a short-circuit the ______ at the termination is/are zero
(a) voltage (b) current
(c) both the voltage and current

378. A backward wave means a
(a) negative voltage wave. (b) negative current wave.
(c) a wave travelling in the negative direction.

379. The size of the earth wire is determined on the basis of
(a) voltage of the service line.
(b) current carrying capacity of the service line.
(c) atmospheric conditions. (d) none of the above.

380. Neutral earthing is provided for
(a) the safety of personnel from electric shock.
(b) the safety of equipment and personnel against lightning and voltage surges.
(c) reducing the voltage stress on lines and equipment with respect to earth under various operating and fault conditions.
(d) controlling the earth fault currents for protective relaying.
(e) both (c) and (d).

381. Isolated neutral system has the disadvantage(s) of
(a) voltage oscillations. (b) difficult earth fault relaying.
(c) persistent arcing ground. (d) all of the above.

382. Isolated neutral transmission system is not recommended as the
(a) system insulation is overstressed due to overvoltages.
(b) insulation overstress may lead to its failure resulting in phase-to-phase faults.
(c) system is not adequately protected against earth fault.
(d) all of the above.

383. Neutral earthing has the advantage(s) of
(a) elimination of arcing grounds.
(b) low maintenance and operating costs over isolated neutral systems.
(c) simplified design of earth fault protection.
(d) discharging of overvoltages due to lightning to earth.
(e) all of the above.

384. The advantage of neutral earthing is
(a) safety of personnel. (b) reduction of earth fault current.
(c) elimination of arcing ground.
(d) none of the above. [A.M.I.E. Sec B. Winter 1993]

385. Resistance earthing is employed for voltages between
(a) 3.3 and 11 kV. (b) 11 and 33 kV.
(c) 33 and 66 kV. (d) 66 kV and 132 kV.

386. Earthing of transformer neutral through reactance will improve its
(a) transient stability. (b) steady state stability.
(c) both of the above. (d) none of these.

387. Which of the following neutral systems will require the lightning arrester of least voltage rating ?
(a) Insulated. (b) Solidly earthed.
(c) Resistance earthed. (d) Reactance earthed.

388. Petersan coil is used for
(a) grounding of system neutral.
(b) to reduce fault current.
(c) connecting two interconnected systems.
(d) for shunt compensation of transmission lines.
[A.M.I.E. Sec B. Power Plant Engineering Summer 1999]

389. A 3-phase, 50 Hz transmission line has the following constants (line-to-neutral) Resistance = 10 Ω, Inductive reactance = 20 Ω; Capacitive susceptance = 4×10^{-4} ℧. The inductance of the arc suppressor coil to be used in the system is
(a) 0.06368 H (b) 0.127 H
(c) 1.1274 H (d) 2.654 H
[U.P.S.C. I.E.S. E.E.-II, 1994]

390. Which one of the following grounding methods is used to reduce the tower footing resistance where earth resistance is low?
(a) Single driven rod. (b) Multiple rods.
(c) Counter poises. (d) Plates.
[U.P.S.C. I.E.S. E.E.-II, 1993]

391. Tower footing resistance of a transmission tower should be
(a) as high as possible. (b) as low as possible.
(c) moderately high. (d) moderately low.
[W.B. Univ. of Technology Power System-I, 2009-10]

392. The method of neutral grounding affects the
(a) positive-sequence network. (b) negative-sequence network.
(c) zero-sequence network.
(d) both positive- and zero-sequence networks.
[A.M.I.E. Sec B. Advanced Power Systems Winter 1998]

393. The earth transformer is used to
(a) avoid the harmonics in the transformers.
(b) provide artificial neutral earthing where neutral point is not accessible.
(c) improve stability of the system.
(d) measure the voltage.
[A.M.I.E. Sec B. Advanced Power Systems Summer 1999]

394. The positive-, negative- and zero-sequence impedances of a solidly grounded system under steady-state condition always follow the relations
(a) $Z_0 < Z_1 < Z_2$ (b) $Z_1 > Z_2 > Z_0$
(c) $Z_1 < Z_2 < Z_3$ (d) $Z_0 > Z_1 > Z_2$

395. In an isolated neutral system, when a single line to ground fault occurs
(a) persistent arcing grounds will be developed.
(b) voltage in the healthy phases rise to full line value causing insulation breakdown.
(c) the capacitive current in the faulty phase rises to 3 times its normal value.
(d) all of the above.

396. The voltage of a transmission line can be controlled by
(a) excitation control. (b) using induction regulator.
(c) reactive VAR injection methods.
(d) any of the above.

397. The voltage of a particular bus is regulated by controlling the
(a) active power of the bus. (b) reactive power of the bus.
(c) phase angle.
(d) phase angle and reactive power. [U.P.S.C. I.E.S. E.E.-I, 2001]

Answers: **374.** (c), **375.** (d), **376.** (a), **377.** (a), **378.** (c), **379.** (b), **380.** (e), **381.** (d), **382.** (d), **383.** (e), **384.** (c), **385.** (a), **386.** (a), **387.** (b), **388.** (a), **389.** (d), **390.** (a), **391.** (b), **392.** (c), **393.** (b), **394.** (b), **395.** (d), **396.** (c), **397.** (d),

398. In ac distribution system the voltage can be controlled by using
(a) tap changing transformer. (b) booster transformer
(c) induction regulator. (d) any of the above.

399. Tap changing transformers are employed for
(a) stepping down the voltage. (b) stepping up the voltage.
(c) supplying low voltage current to instruments.
(d) both stepping up and stepping down the voltage.

400. The best location for use of a booster transformer in a transmission line is
(a) at the sending end. (b) at the receiving end.
(c) at the intermediate point. (d) anywhere in the line.

401. For voltage control in ac distribution system the induction regulators have the advantage(s) of
(a) reliability of operation.
(b) operation independent of load and power factor variations.
(c) stepless voltage variations without arcing or short-circuiting of turns as in the case of transformers.
(d) all of the above.

402. A line voltage regulator is to be used in a single phase 200 V, 5 kVA system to keep the voltage constant for voltage variations within ± 10%.
The rating (in kVA) of the voltage regulator is
(a) 0.05 (b) 0.5 (c) 5 (d) 50
[U.P.S.C. I.E.S. E.E.-II, 1993]

403. Series capacitor is used in a transmission line to
(a) compensate the voltage drop.
(b) reduce line losses. (c) limit short-circuit current.
(d) improve load power factor. **[A.M.I.E. Sec B. Winter 1993]**

404. Shunt compensation in an EHV line is used to improve
(a) stability and fault level. (b) fault level and voltage profile.
(c) voltage profile and stability.
(d) stability, fault level and voltage profile.
[U.P.S.C. I.E.S. E.E.-II, 2000]

405. ______ are used to provide compensation at the receiving end of a transmission line so as to improve its voltage profile.
(a) Condensers. (b) Resistors.
(c) Reactors.
(d) Condensers, resistors or reactors.

406. The combined effect of series and shunt compensation on transmission lines in terms of degree of series compensation (K_{se}), degree of shunt compensation (K_{sh}), and surge impedance of uncompensated line (Z_0) is given by which one of the following equations ?
(a) $Z'_0 = Z_0\sqrt{1-K_{se}}\cdot\sqrt{1-K_{sh}}$
(b) $Z'_0 = \left[\sqrt{1-K_{se}}\cdot\sqrt{1-K_{sh}}\,\right]/Z_0$
(c) $Z'_0 = Z_0\sqrt{(1-K_{se})/(1-K_{sh})}$
(d) $Z'_0 = Z_0\sqrt{(1-K_{sh})/(1-K_{se})}$ **[U.P.S.C. I.E.S. E.E.-II, 2004]**

407. Consider the following statements :
Addition of lumped capacitances in parallel to a loss-free transmission line increases
1. characteristic impedance. 2. propagation constant.
3. system stability. 4. charging current.
Which of these statements are correct ?
(a) 1 and 3 (b) 2 and 4 (c) 2, 3 and 4 (d) 1, 2 and 4
[U.P.S.C. I.E.S. E.E.-I, 2002]

408. What will happen if a short-circuit fault occurs in a switched capacitor controlled reactor?
(a) Oscillation. (b) Capacitor discharge.
(c) Over voltage. (d) Noise.
[U.P.S.C. I.E.S. E.E.-II, 2010]

409. An AC capacitor is to be switched in parallel with AC line using back to back connected thyristor. What is the firing angle of thyristor for first switching?
(a) 0° (b) 180° (c) 90° (d) 45°
[U.P.S.C. I.E.S. E.E.-II, 2009]

410. In between the generating station (power house) and consumers a number of transformation and switching stations exist, these are called the
(a) switchgears. (b) substations.
(c) intermediate substations. (d) transformation stations.

411. Which of the following is provided with arcing horns?
(a) Isolator. (b) Air-break switch.
(c) Oil switch. (d) None of the above.

412. Pole mounted substations are used in
(a) primary transmission system.
(b) secondary transmission system.
(c) primary distribution system.
(d) secondary distribution system.
[W.B. Univ. of Technology Power System-I, 2009-10]

413. With large reactance interconnector between two power stations
(a) the power stations may fall out of step due to large angular displacement between the power stations.
(b) power will be transferred with minimum power losses.
(c) power will be transferred with voltage fluctuations and noise.
(d) none of the above.

414. AC network analyser is employed for solving the problems of
(a) load flow. (b) load flow and stability.
(c) load flow and short circuit.
(d) load flow, stability and short circuit.

415. Consider the following statements :
Stability studies constitute
1. the major analytical approach to the study of power system electromechanical dynamic behaviour.
2. the involvement of one or just a few machines undergoing slow or gradual changes in operating conditions.
3. the determination of the locus of essentially steady-state operating points of the system.
4. the determination of whether or not the rotors of the machines being perturbed, return to the constant speed operation.
Which of the statements given above are correct ?
(a) 1, 2 and 3. (b) 2, 3 and 4.
(c) 1 and 4. (d) 1, 2, 3 and 4.
[U.P.S.C. I.E.S. E.E.-II, 2004]

416. Power system stability is defined as
(a) that attribute of the system or part of the system which enables it to develop restoring forces between the elements there of equal or greater than disturbing forces so as to restore a state of equilibrium between the elements.
(b) the maximum power flow possible through some particular point in the system when the entire or part of the system is disturbed.
(c) both (a) and (b). (d) neither (a) nor (b).

417. Stability limit of a power system is defined as
(a) that attribute of the system or part of the system which enables it to develop restoring forces between the elements there of equal or greater than disturbing forces so as to restore a state of equilibrium between the elements.

Answers: 398. (d), **399.** (d), **400.** (c), **401.** (d), **402.** (b), **403.** (a), **404.** (b), **405.** (a), **406.** (c), **407.** (b), **408.** (a), **409.** (c), **410.** (b), **411.** (b), **412.** (d), **413.** (a), **414.** (d), **415.** (c), **416.** (a), **417.** (b),

(b) the maximum power flow possible through some particular point in the system when the entire or part of the system is disturbed.
(c) both (a) and (b). (d) neither (a) nor (b).

418. Steady-state stability of a power system is the ability of the power system to
(a) maintain voltage at the rated voltage level.
(b) maintain frequency exactly at 50 Hz.
(c) maintain a spinning reserve margin at all times.
(d) maintain synchronism between machines and on external tie lines. **[GATE E.E. 1999]**

419. Transient disturbances are caused by
(a) sudden load changes. (b) switching operations.
(c) inadvertent tripping of lines and generators.
(d) faults in the power system.
(e) all of the above.

420. Transient stability of a 3-phase power systems having more than one synchronous generator is not affected by which one of the following specifications?
(a) Initial operating conditions of generators.
(b) Quantum of large power disturbance.
(c) Fast fault clearance and reclosure.
(d) Small changes in system frequency. **[U.P.S.C. I.E.S. E.E.-II, 2009]**

421. The power transmission capacity of a transmission line is
(a) proportional to transmission voltage.
(b) proportional to the square of transmission voltage.
(c) inversely proportional to transmission voltage.
(d) inversely proportional to the square of transmission voltage.

422. The load carrying capability of a long ac transmission line is
(a) always limited by the conductor size.
(b) limited by statbility considerations.
(c) reduced at low ambient temperatures.
(d) decreased by the use of bundled conductors of single conductors. **[GATE E.E. 1999]**

423. Maximum power will be transferred from the sending end to the receiving end by a transmission line when
(a) the line reactance is $\sqrt{3}$ times its resistance, *i.e.*, $X = \sqrt{3}$ R.
(b) the torque angle $\delta = 90^o$.
(c) both (a) and (b). (d) neither (a) nor (b).

424. A 100 km transmission line is designed for a nominal voltage of 132 kV and consists of one conductor per phase. The line reactance is 0.726 Ω/km. The static transmission capacity of the line, in Mega watts, would be
(a) 132 (b) 240 (c) 416 (d) 720 **[U.P.S.C. I.E.S. E.E.-II, 1993]**

425. An 800 kV transmission line is having per phase line inductance of 1.1 mH/km and per phase line capacitance of 11.68 nF/km. Ignoring the length of the line, its ideal power transfer capability in MW is
(a) 1,204 MW (b) 1,504 MW
(c) 2,085 MW (d) 2,606 MW **[GATE E.E. 2004]**

426. An 800 kV transmission line has a maximum power transfer capacity of P. If it is operated at 400 kV with the series reactance unchanged, then maximum power transfer capacity is approximately
(a) P (b) 2P (c) P/2 (d) P/4 **[GATE E.E. 2005]**

427. If a 500 MVA, 11 kV three-phase generator at 50 Hz feeds, through a transfer impedance of (0.0 + *j*0.605) Ω per phase, an infinite bus also at 11 kV, then the maximum steady-state power transfer on the base of 500 MVA and 11 kV is
(a) 1.0 pu (b) 0.8 pu (c) 0.5 pu (d) 0.4 pu **[U.P.S.C. I.E.S. E.E.-II, 2001]**

428. An alternator having an induced emf of 1.6 pu is connected to an infinite bus of 1.0 pu. If the bus bar has reactance of 0.6 pu and alternator has reactance of 0.2 pu, what is the maximum power that can be transferred ?
(a) 2 pu (b) 2.67 pu (c) 5 pu (d) 6 pu **[U.P.S.C. I.E.S. E.E.-II, 2007]**

429. A transmission line has a total series reactance of 0.2 pu. Reactive power compensation is applied at the midpoint of the line and it is controlled such that the midpoint voltage of the transmission line is always maintained at 0.98 pu. If voltage at both ends of the line are maintained at 1.0 pu, then the steady state power transfer limit of the transmission line is
(a) 9.8 pu (b) 4.9 pu (c) 19.6 pu (d) 5 pu **[GATE E.E. 2002]**

430. A generator with constant 1.0 pu terminal voltage supplies power through a step-up transformer of 0.12 pu reactance and a double-circuit line to an infinite bus bar as shown in the figure. The infinite bus voltage is maintained at 1.0 pu. Neglecting the resistances and susceptances of the system, the steady state stability power limit of the system is 6.25 pu. If one of the double-circuit is tripped, then resulting steady state stability power limit in pu will be

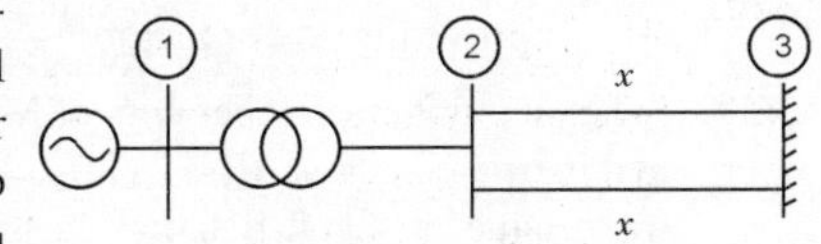

(a) 12.5 pu (b) 3.125 pu (c) 10.0 pu (d) 5.0 pu **[GATE E.E. 2005]**

431. The stability of the power system is not affected by
(a) generator reactance. (b) line reactance.
(c) line losses. (d) excitation of generators.
(e) output torque.

432. The power systems are operated with power angle around
(a) 10° (b) 30° (c) 70° (d) 80° **[U.P.S.C. I.E.S. E.E.-II, 2005]**

433. The steady state stability limits for round rotor and salient pole 3-phase synchronous generator are attained at the values of power angle δ
(a) $= \pi/2$, and $= \pi/2$, respectively.
(b) $< \pi/2$, and $< \pi/2$, respectively.
(c) $< \pi/2$, and $= \pi/2$, respectively.
(d) $= \pi/2$, and $< \pi/2$, respectively. **[U.P.S.C. I.E.S. E.E.-II, 2006]**

434. A round rotor generator with internal voltage $E_1 = 2.0$ pu and $X = 1.1$ pu is connected to a round rotor synchronous motor with internal voltage $E_2 = 1.3$ pu and $X = 1.2$ pu. The reactance of the line connecting the generator to the motor is 0.5 pu. When the generator supplies 0.5 pu power, the rotor angle difference between the machines will be
(a) 57.42° (b) 1° (c) 32.58° (d) 122.58° **[GATE E.E. 2003]**

435. Which one of the following statements is true?
(a) steady-state stability limit is greater than transient stability limit.
(b) steady-state stability limit is equal to transient stability limit.
(c) steady-state stability limit is less than the transient stability limit.

Answers: **418.** (d), **419.** (e), **420.** (d), **421.** (b), **422.** (b), **423.** (c), **424.** (b), **425.** (c), **426.** (d), **427.** (d), **428.** (a), **429.** (d), **430.** (d), **431.** (c), **432.** (b), **433.** (d), **434.** (c), **435.** (a),

(d) no generalisation can be made regarding the equality or otherwise of the steady, state stability limit and transient stability limit. **[U.P.S.C. I.E.S. E.E.-II, 1995]**

436. Steady-state stability of a power system is improved by
(a) reducing fault clearing time.
(b) using double circuit line instead of single circuit line.
(c) single pole switching.
(d) decreasing generator inertia. **[U.P.S.C. I.E.S. E.E.-II, 1998]**

437. The methods employed in improving the system stability are
(a) increasing the system voltage only.
(b) reducing the transfer reactance only.
(c) using high-speed auto-reclosing circuit breaker only.
(d) using all (a), (b), (c) together.
[A.M.I.E. Sec B. Advanced Power Systems Summer 1998]

438. Series capacitors are used to
(a) compensate for line inductive reactance and improve the stability of the power system.
(b) improve the voltage.
(c) reduce fault level.
(d) improve the power factor.

439. When a number of generators of MVA capacities $S_1, S_2, S_3 \ldots S_n$ and inertia constants $H_1, H_2, H_3 \ldots H_n$ respectively are connected to the same bus-bar in a station, the inertia constant of the equivalent machine on a base of S_b is given by

(a) $\sum_{i=1}^{n} H_i S_i$ (b) $\sum_{i=1}^{n} \frac{S_b}{S_i} H_i$

(c) $\sum_{i=1}^{n} \frac{S_i}{S_b} H_i$ (d) $\sum_{i=1}^{n} \frac{S_b}{S_i} \cdot \frac{1}{H_i}$

[U.P.S.C. I.E.S. E.E.-II, 1998]

440. If a generator of 250 MVA rating has an inertia constant of 6 MJ/MVA, its inertia constant on a 100 MVA base is
(a) 15 MJ/MVA (b) 10.5 MJ/MVA
(c) 6 MJ/MVA (d) 2.4 MJ/MVA
[U.P.S.C. I.E.S. E.E.-II, 2000]

441. If the inertia constant H = 8 MJ/MVA for a 50 MVA generator, the stored energy is:
(a) 50 MJ (b) 8 MJ (c) 400 MJ (d) 6.25 MJ
[U.P.S.C. I.E.S. E.E.-II, 2011]

442. For 800 MJ stored energy in the rotor at synchronous speed, what is the inertia constant H for a 50 Hz, four pole turbo-generator rated 100 MVA, 11 kV?
(a) 2.0 MJ/MVA (b) 4.0 MJ/MVA
(c) 6.0 MJ/MVA (d) 8.0 MJ/MVA
[U.P.S.C. I.E.S. E.E.-II, 2005]

443. The pu parameters for a 500 MVA machine on its own base are: inertia, M = 20 pu; reactance, X = 2 pu.
The pu values of inertia and reactance on 100 MVA common base, respectively, are
(a) 4, 0.4 (b) 100, 10 (c) 4, 10 (d) 100, 0.4
[GATE E.E. 2005]

444. A 500 MW, 21 kV, 50 Hz, 3-phase, 2-pole synchronous generator having a rated pf = 0.9, has a moment of inertia of 27.5×10^3 kg-m^2. The inertia constant (H) will be
(a) 2.44 MJ/MVA (b) 2.71 MJ/MVA
(c) 4.88 MJ/MVA (d) 5.42 MJ/MVA **[GATE E.E. 2009]**

445. Two 50 Hz generating units operate in parallel within the same power plant and have the following ratings ·
Unit 1 : 500 MVA, 0.85 power factor, 20 kV, 3000 rpm, H_1 = 5 MJ/MVA
Unit 2 : 200 MVA, 0.9 power factor, 20 kV, 1500 rpm, H_2 = 5 MJ/MVA
The equivalent inertia constant H in MJ/MVA on 100 MVA base is
(a) 2.5 (b) 5.0 (c) 10.0 (d) 35.0
[U.P.S.C. I.E.S. E.E.-II, 2001]

446. A power station consists of two synchronous generators *A* and *B* of ratings 250 MVA and 500 MVA with intertia 1.6 and 1.0 pu respectively on their own base MVA ratings. The equivalent pu inertia constant for the system on 100 MVA common base is
(a) 2.6 (b) 6.15 (c) 1.625 (d) 9
[GATE E.E. 1998]

447. The inertia constant of a 100 MVA 50 Hz, 4-pole generator is 10 MJ/MVA. If the mechanical input to the machine is suddenly raised from 50 MW to 75 MW, the rotor acceleration will be equal to
(a) 225 electrical degrees/s^2. (b) 22.5 electrical degrees/s^2.
(c) 125 electrical degrees/s^2. (d) 12.5 electrical degrees/s^2.
[U.P.S.C. I.E.S. E.E.-II, 1999]

448. In a multimachine interconnected system, subsequent to a 3-phase fault, the transient stability is examined by
(a) equal-area criterion.
(b) solution of swing equation.
(c) either by equal-area criterion or by solution of swing equation
(d) combination of equal-area criterion and solution of swing equation. **[U.P.S.C. I.E.S. E.E.-II, 2002]**

449. The equal area criterion of stability is used for
(a) no load on the bus-bar. (b) one machine and infinite bus-bar.
(c) more than one machine and infinite bus-bar.
(d) none of the above.

450. Principle of equal area criterion is to be applied to determine, for a given initial load P_1, the maximum amount of sudden increase in load ΔP, to maintain transient stability of a cylindrical rotor synchronous motor operating from an infinite bus. Applying this criterion (in each case the area A_1 = area A_2), which one of the following diagrams is correct ?

(a) 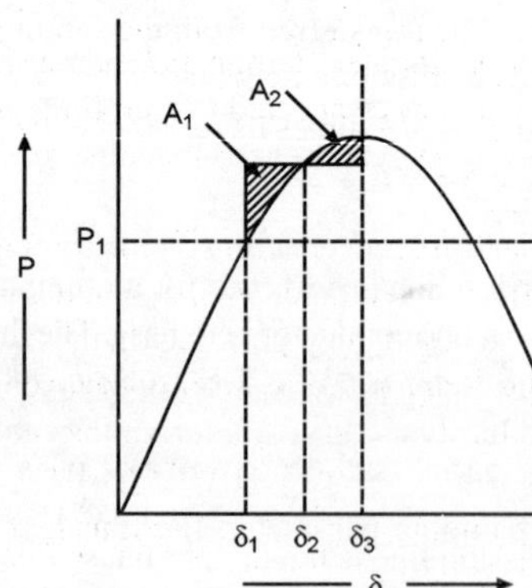

(b) 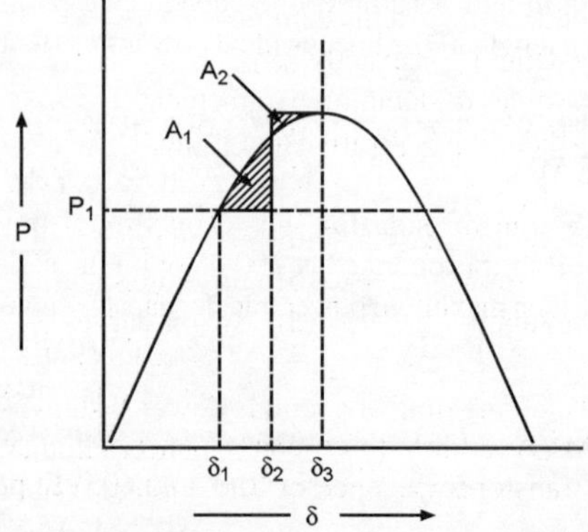

Answers: 436. (b), **437.** (d), **438.** (a), **439.** (c), **440.** (a), **441.** (c), **442.** (d), **443.** (d), **444.** (a), **445.** (d), **446.** (d), **447.** (a), **448.** (b), **449.** (b), **450.** (a),

(c) 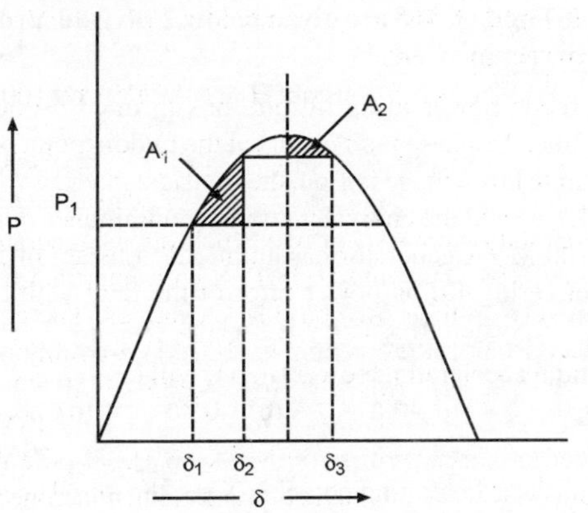

(d) 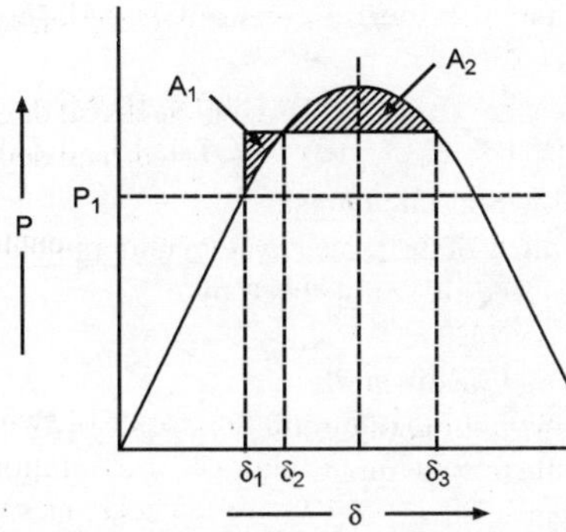

[U.P.S.C. I.E.S. E.E.-II, 1993]

451. The 'equal area criterion' for the determination of transient stability of a synchronous machine connected to an infinite bus
(a) ignores line as well as synchronous machine resistances and shunt capacitances.
(b) assumes accelerating power acting on the rotor as constant.
(c) ignores the effect of voltage regulator and governor but considers the inherent damping present in the machine.
(d) takes into consideration the possibility of machine losing synchronism after it has survived during the first swing.
[U.P.S.C. I.E.S. E.E.-II, 1996]

452. The 'Equal area criterion' for the determination of transient stability of the synchronous machine connected to an infinite bus:
(a) Ignores line as well as synchronous machine resistances and shunt capacitances.
(b) Assumes accelerating power acting on the rotor as constant.
(c) Ignores the effect of voltage regulator and governor but considers the inherent damping present in the machine.
(d) Takes into consideration the possibility of machine loosing synchronism after it has survived during the first swing.
[U.P.S.C. I.E.S. E.E.-II, 2011]

453. For what value of damping parameter, the transient stability is assured by equal area criterion?
(a) Independent of systems damping.
(b) If only damping is exactly zero.
(c) For all values of damping parameters.
(d) If only damping is positive and finite.
[U.P.S.C. I.E.S. E.E.-II, 2010]

454. Equal area criterion gives the information regarding
(a) stability region. (b) absolute stability.
(c) relative stability. (d) swing curves.
[U.P.S.C. I.E.S. E.E.-II, 1998]

455. The critical clearing time of a fault is power system is related to
(a) reactive power limit. (b) short-circuit limit.
(c) steady-state stability limit. (d) transient stability limit.
[U.P.S.C. I.E.S. E.E.-II, 1998]

456. With fault clearing time, the transient stability limit of a power system
(a) increases. (b) decreases.
(c) first increases and then decreases.
(d) first decreases and then increases.
[A.M.I.E. Sec B. Advanced Power Systems Winter 1999]

457. For which one of the following types of motors, is the equal-area criterion for stability applicable ?
(a) Three-phase synchronous motor.
(b) Three-phase induction motor.
(c) DC series motor. (d) DC compound motor.
[U.P.S.C. I.E.S. E.E.-II, 2006]

458. Figure 2 given below is the "equal area" criterion diagram for the determination of transient stability limit of the power system shown in Fig. 1 for a fault on the transmission line.
What is the type of fault, and the time of its clearing, from figure 2?

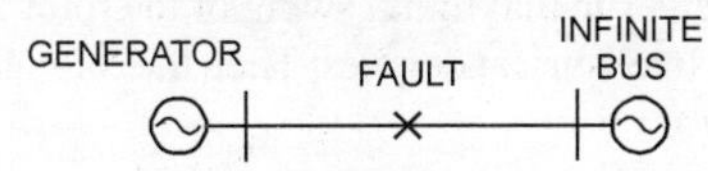

Fig. 1

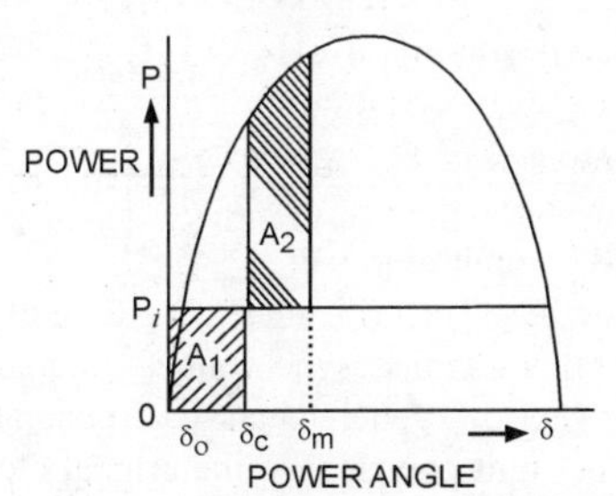

Fig. 2

(a) Three-phase fault with instantaneous clearing.
(b) Three-phase fault with subsequent clearing.
(c) Single-line-to ground fault with instantaneous clearing.
(d) Single-line-to ground fault with subsequent clearing.
[U.P.S.C. I.E.S. E.E.-II, 2007]

459. The inertia constants of two groups of machines which do not swing together are M_1 and M_2 such that $M_1 > M_2$. It is proposed to add some inertia to one of the two groups of machines for improving the transient stability of the system. It should be added to
(a) M_1 (b) M_2
(c) either to M_1 or to M_2. (d) neither of the above.

460. The transient stability limit of a power system can be appreciably increased by introducing
(a) series inductance. (b) shunt inductance.
(c) series capacitance. (d) shunt capacitance.
[U.P.S.C. I.E.S. E.E.-II, 1994]

461. A shunt reactor of 100 MVAr is operated at 98% of its rated voltage and at 96% of its rated frequency. The reactive power absorbed by the reactor is;
(a) 98 MVAr (b) 10.402 MVAr
(c) 96.04 MVAr (d) 100.04 MVAr **[GATE E.E. 1998]**

462. A 3-phase 11 kV generator feeds power to a constant power unity power factor load of 100 MW through a 3-phase transmission line. The line-to-line voltage at the terminals of the machine is maintained constant at 11 kV. The per unit positive sequence impedance of the line based on 100 MVA and 11 kV is $j0.2$. The line to line voltage at the load terminals is measured to be less than 11 kV. The total reactive power to be injected at the terminals of

Answers: **451.** (b), **452.** (a), **453.** (b), **454.** (b), **455.** (d), **456.** (b), **457.** (a), **458.** (b), **459.** (b), **460.** (c), **461.** (d), **462.** (d), the load to increase the line-to-line voltage at the load terminals to 11 kV is

(a) 100 MVAR (b) 10.1 MVAR
(c) –100 MVAR (d) –10.1 MVAR **[GATE E.E. 2003]**

463. A 50 Hz, 4-pole, 500 MVA, 22 kV turbo-generator is delivering rated magavolt-amreres at 0.8 power factor. Suddenly a fault occurs reducing its electric power output by 40%. Neglect losses and assume constant power input to the shaft. The accelerating torque in the generator in MNm at the time of the fault will be

(a) 1.528 (b) 1.018 (c) 0.848 (d) 0.509 **[GATE E.E. 2004]**

464. A generator delivers power of 1.0 pu to an infinite bus through a purely reactive network. The maximum power that could be delivered by the generator is 2.0 pu. A three-phase fault occurs at the terminals of the generator which reduces the generator output to zero. The fault is cleared after t_c second. The original network is then restored. The maximum swing of the rotor angle is found to be δ_{max} = 110 electrical degrees. Then the rotor angle in electrical degrees at $t = t_c$ is

(a) 55 (b) 70 (c) 69.14 (d) 72.4 **[GATE E.E. 2003]**

465. A lossless single machine infinite bus power system is shown below :

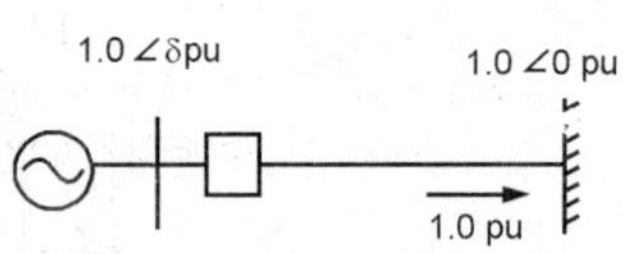

The synchronous generator transfers 1.0 per unit of power to the infinite bus. The critical clearing time of circuit breaker is 0.28 s. If another identical synchronous generator is connected in parallel to the existing generator and each generator is scheduled to supply 0.5 per unit power, then the critical clearing time of the circuit breaker will

(a) reduce to 0.14 s.
(b) reduce but will be more than 0.14 s.
(c) remain constant at 0.218 s.
(d) increase beyond 0.28 s. **[GATE E.E. 2008]**

466. Consider a synchronous generator connected to an infinite bus by two identical parallel transmission lines. The transient reactance X′ of the generator is 0.1 pu and the mechanical power input to it is constant at 1.0 pu. Due to some previous disturbance, the rotor angle (δ) is undergoing an undamped oscillation, with the maximum value of δ(*t*) equal to 130º. One of the parallel lines trip due to relay maloperation at an instant when δ(*t*) = 130º as shown in the figure. The maximum value of the per unit line reactance, X such that the system does not lose synchronism subsequent to this tripping is

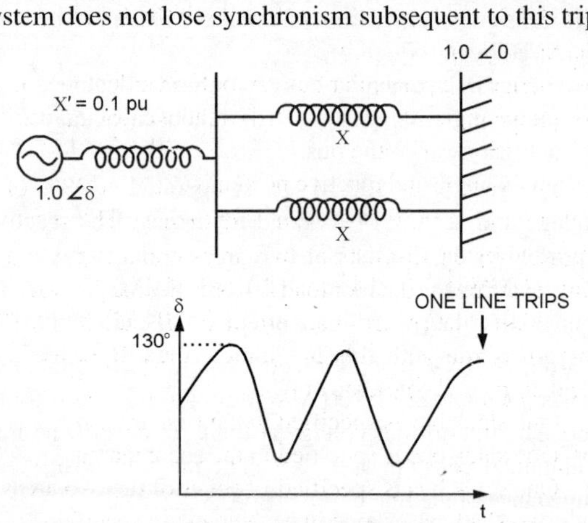

(a) 0.87 (b) 0.74 (c) 0.67 (d) 0.54 **[GATE E.E. 2007]**

Data for Q. 467 and Q. 468 are given below. Solve the problems and choose the correct answers.

A generator feeds power to an infinite bus through a double circuit transmission line. A3 phase fault occurs at the middle point of one of the lines. The infinite bus voltage is 1 pu, the transient internal voltage of the generator is 1.1 pu and the equivalent transfer admittance during fault is 0.8 pu. The 100 MVA generator has an inertia constant of 5 MJ/MVA and it was delivering 1.0 pu power prior of the fault with rotor power angle of 30º. The system frequency is 50 Hz.

467. The initial accelerating power (in pu) will be

(a) 1.0 (b) 0.6 (c) 0.56 (d) 0.4 **[GATE E.E. 2006]**

468. If the initial accelerating power is X pu, the initial acceleration in elect deg/sec, and the inertia constant in MJ-sec/elect deg respectively will be

(a) 31.4X, 18 (b) 1800X, 0.056
(c) X/1800, 0.056 (d) X/31.4, 18 **[GATE E.E. 2006]**

469. Consider the following statements:
The transient stability of the power system under unbalanced fault condition can be effectively improved by
1. excitation control.
2. phase shifting transformer.
3. single-pole switching of circuit breakers.
4. increasing the turbine input.

Of these statements
(a) 1 and 2 are correct. (b) 2 and 3 are correct.
(c) 3 and 4 are correct. (d) 1 and 3 are correct. **[U.P.S.C. I.E.S. E.E.-II, 1998]**

470. The use of high speed circuit breakers
(a) improves transient stability.
(b) decreases transient stability.
(c) has no effect on system stability. **[U.P.S.C. I.E.S. E.E.-II, 2000]**

471. The use of fast acting relays and circuit breakers for clearing a sudden short-circuit on a transmission line between a generator and the receiving end bus improves the transient stability of the machine because the
(a) short-circuit current becomes zero.
(b) post-fault transfer impedance attains a value higher than that during the fault.
(c) ordinate of the post-fault power-angle characteristic is higher than that of during fault characteristic.
(d) voltage behind the transient reactance increases to a high value. **[U.P.S.C. I.E.S. E.E.-II, 1998]**

472. A loss-less finite generator is connected directly to an infinite bus at rated voltage. Conside the following operating conditions.
A: Generator exciation is maintained constant at a value which gives rated voltage at its terminals at rated current.
B: Generator excitation is regulated to maintain rated voltage at its terminals at all loads.

The maximum steady-state power in a *A* and *B* will be

	A	*B*
(a)	finite	infinite.
(b)	infinte	finite.
(c)	finite	finite.
(d)	infinite	infinite. **[U.P.S.C. I.E.S. E.E.-II, 1993]**

473. A 50 Hz synchronous generator is initially connected to a long lossless transmission line which is open-circuited at the receiving end. With the field voltage held constant, the generator is disconnected from the transmission line. Which of the following may be said about the steady state terminal voltage and field current of the generator?

Answers: 463. (b), **464.** (c), **465.** (b), **466.** (c), **467.** (c), **468.** (b), **469.** (d), **470.** (a), **471.** (c), **472.** (c), **473.** (d),

(a) The magnitude of terminal voltage decreases, and the field current does not change.
(b) The magnitude of terminal voltage increases, and the field current does not change.
(c) The magnitude of terminal voltage increases, and the field current increases.
(d) The magnitude of terminal voltage does not change, and the field current decreases. [GATE E.E. 2010]

474. Which of the following statements is incorrect?
(a) The inertia constant of an alternator is the ratio of the stored energy in MJ to its machine rating in MVA.
(b) The angular momentum M of a machine is $\frac{GH}{180f}$ MJs per electrical degree.
(c) For moment of inertia J in kg.m^2, the swing equation is $J\frac{d^2\delta}{dt^2} = P_S - P_E$.
(d) In a two machine system, it is advisable to have resistance grounding at the sending end and the reactance grounding at the receiving end.
(e) For the transient stability of the power system the necessary condition is $\frac{d\delta}{dt} = 0$ and $\frac{dP}{dt} > 0$.

475. Load flow study is carried out for
(a) load-frequency control. (b) planning of power system.
(c) fault calculations.
(d) study of stability of the system.

476. Load flow studies must be made on a power system before
(a) making short-circuit studies but not for transient stability studies on the power system.
(b) making transient stability studies but not for short-circuit studies on the power system.
(c) making both short-circuit and transient stability studies on the power system.
(d) for neither making short-circuit studies nor transient stability studies on the power system. [U.P.S.C. I.E.S. E.E.-II, 2010]

477. Load flow studies involve solving simultaneous
(a) linear algebraic equations. (b) non-linear algebraic equations.
(c) linear differential equations.
(d) nonlinear differential equations. [U.P.S.C. I.E.S. E.E.-II, 2000]

478. In a power system, each bus or node is associated with four quantities, namely
1. real power. 2. reactive power.
3. bus-voltage magnitude. 4. phase angle of the bus voltage.
For load-flow solution, among these four, the number of quantities to be specified is
(a) any one. (b) any two.
(c) any three. (d) all the four.
[U.P.S.C. I.E.S. E.E.-II, 1998]

479. Consider the following quantities :
1. Real power 2. Reactive power
3. Power factor 4. Input current
5. Bus voltage magnitude 6. Bus voltage phase-angle
For the purpose of the load flow studies of a power system, each bus or node is associated with which one of the combinations of the above four quantities ?
(a) 1, 3, 4 and 5 (b) 1, 2, 3 and 4
(c) 2, 3, 5 and 6 (d) 1, 2, 5 and 6
[U.P.S.C. I.E.S. E.E.-II, 2004]

480. In load flow studies of a power system, a voltage control bus is specified by :
(a) Real power and reactive power.
(b) Reactive power and voltage magnitude.
(c) Voltage and voltage phase angle.
(d) Real power and voltage magnitude. [U.P.S.C. I.E.S. E.E.-II, 2006]

481. For a load-flow solution the quantities normally specified at a voltage controlled bus are
(a) P and Q (b) P and |V|
(c) Q and |V| (d) P and δ.
[W.B. Univ. of Technology Power System-II, 2009-10]

482. Load bus is specified by
(a) P and V. (b) P and δ. (c) P and |V|. (d) P and Q.

483. At slacks bus, which one of the following combinations of variables is specified ?
(a) | V |, δ (b) P, Q (c) P, | V | (d) Q, | V |
(The symbols have their usual meaning) [U.P.S.C. I.E.S. E.E.-II, 2004]

484. In a power system, the maximum number of buses are
(a) generator buses. (b) load buses.
(c) slack buses. (d) P-V buses.

485. Buses for load flow studies are classified as
1. Load bus 2. PV bus 3. Slacks bus
Which one of the following is the correct combination of the pair of quantities specified having their usual meaning for different buses ?

	Load bus	*PV bus*	*Slack bus*
(a)	P, | V |	P, Q	P, δ.
(b)	P, Q	P, | V |	| V |, δ.
(c)	| V |, Q	P, δ	P, Q.
(d)	P, δ	Q, | V |	Q, δ.

[U.P.S.C. I.E.S. E.E.-II, 2005]

486. If a voltage controlled bus is treated as a load bus, then which one of the following limits would be violated ?
(a) Voltage. (b) Active power.
(c) Reactive power. (d) Phase angle.
[U.P.S.C. I.E.S. E.E.-II, 1998]

487. In load-flow analysis, the load connected at a bus is represented as
(a) constant current drawn from the bus.
(b) constant impedance connected at the bus.
(c) voltage and frequency dependent source at the bus.
(d) constant real and reactive power drawn from the bus.
[GATE E.E. 1993]

488. The voltage of a particular bus can be controlled by controlling the
(a) active power of the bus. (b) reactive power of the bus.
(c) phase angle. (d) both (a) and (b).
(e) (b) and (c).

489. The voltage of a particular bus can be controlled by controlling
(a) phase angle. (b) reactive power of the bus.
(c) active power of the bus.
(d) phase angle and reactive power.
[W.B. Univ. of Technology Power System-II, 2009-10]

490. A power system consists of two areas connected via a tie line. While entering the data for load flow the tie line parameters and its connectivity data were inadvertently left out. If the load flow program is run with this incomplete data, then the load flow calculations will converge only if
(a) One slack bus is specified in the first area.
(b) One slack bus is specified in the second area.
(c) One slack bus is specified in either of the two areas.
(d) Two slack buses, one in each area, are specified.
[U.P.S.C. I.E.S. E.E.-II, 2003]

Answers: **474.** (c), **475.** (b), **476.** (c), **477.** (b), **478.** (b), **479.** (d), **480.** (d), **481.** (b), **482.** (d), **483.** (a), **484.** (b), **485.** (b), **486.** (a), **487.** (d), **488.** (b), **489.** (b), **490.** (d),

491. A power system consists of 2 areas (Area 1 and Area 2) connected by a single tie-line as shown in the figure. It is required to carry out a load flow study on this system. While entering the network data, the tie-line data (connectivity and parameters) is inadvertently left out. If the load flow program is run with this incomplete data

(a) The loadflow will converge only if the slack bus is specified in Area 1.
(b) The loadflow will converge only if the slack bus is specified in Area 2.
(c) The loadflow will converge if the slack bus is specified in either Area 1 or Area 2.
(d) The loadflow will not converge if only one slack bus is specified.

Tie Line

Area 1 Area 2

[GATE E.E. 2002]

492. The principal information obtained from load flow studies in a power system are

1. magnitude and phase angle of the voltage at each bus.
2. reactive and real power flows in each of the lines.
3. total power loss in the network.
4. transient stability limit of the system.

Select the correct answer from the codes given below :

(a) 1 and 2 (b) 3 and 4
(c) 1, 2 and 3 (d) 2 and 4

[U.P.S.C. I.E.S. E.E.-II, 2001]

493. Y_{BUS} as used in load flow study, and Z_{BUS} as used for short circuit study are :

(a) the same. (b) inverse of each other.
(c) are not related to each other. **[U.P.S.C. I.E.S. E.E.-II, 2003]**

494. Which one of the following statements is correct ?
The elements of each row of Y_{BUS} matrix for load flow studies in power system add up to zero,

(a) always.
(b) if the shunt admittances at the buses are ignored.
(c) if mutual couplings between transmission lines are absent.
(d) if both (b) and (c) are satisfied. **[U.P.S.C. I.E.S. E.E.-II, 2006]**

495. The Y_{BUS} matrix of a 100-bus interconnected system is 90% sparse. Hence the number of transmission lines in the system must be

(a) 450 (b) 500 (c) 900 (d) 1,000

[U.P.S.C. I.E.S. E.E.-II, 2002]

496. For a graph of power system network shown in figure, where bus numbers and impedances are marked, assuming equal R/X of impedances, find the bus impedance matrix element Z_{22}.

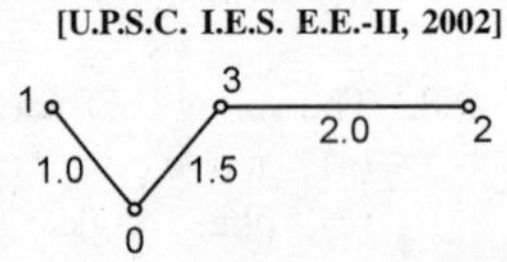

(a) 2.0 (b) 3.5 (c) 0.5 (d) 4.5

[U.P.S.C. I.E.S. E.E.-II, 2010]

497. In matrix form, the equation of a 4 terminal network representing a transmission line is

$$\begin{bmatrix} V_s \\ I_s \end{bmatrix} = \begin{bmatrix} A & B \\ C & D \end{bmatrix} \begin{bmatrix} V_g \\ I_R \end{bmatrix}$$

The two networks considered are

(A) 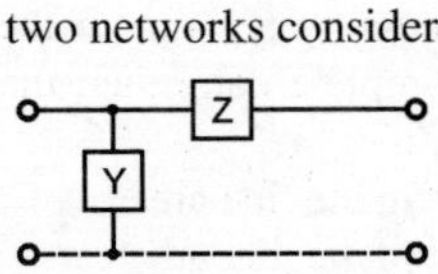

(B) 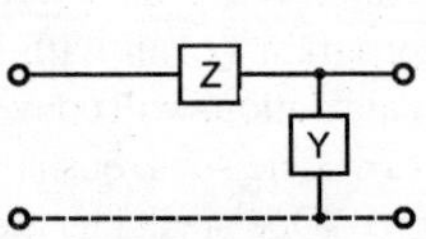

The possible transfer matrix for the networks (*A*) and (*B*) could be

(i) $\begin{bmatrix} 1 & YZ \\ 0 & Z \end{bmatrix}$ (ii) $\begin{bmatrix} Y & 0 \\ YZ & 1 \end{bmatrix}$

(iii) $\begin{vmatrix} 1 & Z \\ Y & (1+YZ) \end{vmatrix}$ (iv) $\begin{bmatrix} 1+YZ & Z \\ Y & 1 \end{bmatrix}$

The correct combinations for the two networks (*A*) and (*B*), could be

(a) (i) and (ii). (b) (i) and (iii).
(c) (ii) and (iv). (d) (iii) and (iv).

[U.P.S.C. I.E.S. E.E.-II, 1997]

498. The bus admittance matrix of a power systems is given as

$$\begin{array}{c} \\ 1 \\ 2 \\ 3 \end{array}\begin{array}{c} \begin{array}{ccc} 1 & 2 & 3 \end{array} \\ \begin{bmatrix} -j50 & +j10 & +j5 \\ +j10 & -j30 & +j10 \\ +j5 & +j10 & -j25 \end{bmatrix} \end{array}$$

The impedance of line between bus 2 and 3 will be equal to

(a) $+j0.1$ (b) $-j0.1$ (c) $+j0.2$ (d) $-j0.2$

[U.P.S.C. I.E.S. E.E.-II, 2003]

499. For the Y-bus matrix of a 4-bus system given in per, unit, buses having shunt elements are

$$Y_{BUS} = j\begin{bmatrix} -5 & 2 & 2.5 & 0 \\ 2 & -10 & 2.5 & 4 \\ 2.5 & 2.5 & -9 & 4 \\ 0 & 4 & 4 & -8 \end{bmatrix}$$

(a) 3 and 4 (b) 2 and 3 (c) 1 and 2 (d) 1, 2 and 4

[GATE E.E. 2009]

500. The network shown in the given figure has impedances in pu is indicated. The diagonal element Y_{22} of the bus admittance matrix Y_{BUS} of the network is

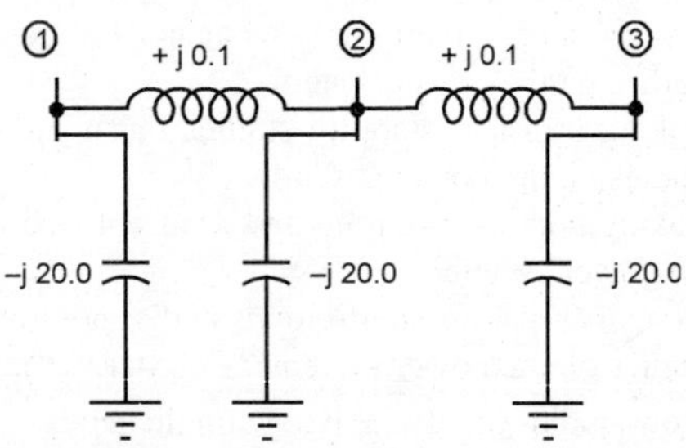

(a) $-j19.8$ (b) $+j20.0$ (c) $+j0.2$ (d) $-j19.95$

[GATE E.E. 2005]

501. A three-bus network is shown in the figure below indicating the pu impedance of each element.

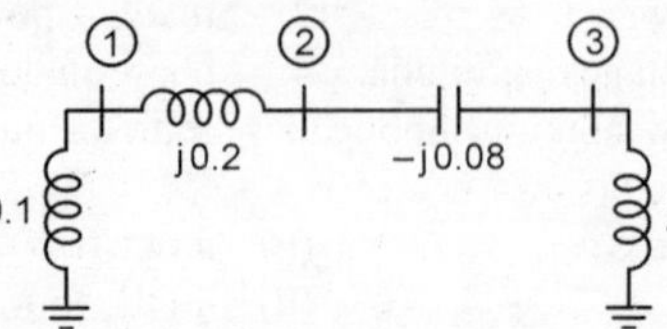

The Bus admittance matrix, Y-Bus, of the network is

(a) $j\begin{bmatrix} 0.3 & -0.2 & 0 \\ -0.2 & 0.12 & 0.08 \\ 0 & 0.08 & 0.02 \end{bmatrix}$ (b) $j\begin{bmatrix} -15 & 5 & 0 \\ 5 & 7.5 & -12.5 \\ 0 & -12.5 & 2.5 \end{bmatrix}$

(c) $j\begin{bmatrix} 0.1 & 0.2 & 0 \\ 0.2 & 0.12 & -0.08 \\ 0 & -0.08 & 0.10 \end{bmatrix}$ (d) $j\begin{bmatrix} -10 & 5 & 0 \\ 5 & 7.5 & 12.5 \\ 0 & 12.5 & -10 \end{bmatrix}$

[GATE E.E., 2011]

Answers: 491. (d), **492.** (a), **493.** (c), **494.** (d), **495.** (c), **496.** (a), **497.** (d), **498.** (a), **499.** (c), **500.** (d), **501.** (b),

502. The bus impedance matrix of a 4-bus power system is given by

$$Z_{bus} = \begin{bmatrix} j0.3435 & j0.2860 & j0.2723 & j0.2277 \\ j0.2860 & j0.3408 & j0.2586 & j0.2414 \\ j0.2723 & j0.2586 & j0.2791 & j0.2209 \\ j0.2277 & j0.2414 & j0.2209 & j0.2791 \end{bmatrix}$$

A branch having an impedance of $j0.2\ \Omega$ is connected between bus 2 and the reference. Then the values of $Z_{22,\ new}$ and $Z_{23,\ new}$ of the bus impedance matrix of the modified network are respectively

(a) $j0.5408\ \Omega$ and $j0.4586\ \Omega$ (b) $j0.1260\ \Omega$ and $j0.0956\ \Omega$
(c) $j0.5408\ \Omega$ and $j0.0956\ \Omega$ (d) $j0.1260\ \Omega$ and $j0.1630\ \Omega$

[GATE E.E. 2003]

503. In the network as shown in figure, the marked parameters are pu impedances. The bus-admittance matrix of the network is

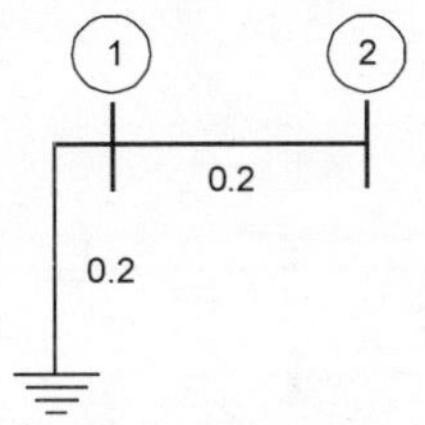

(a) $\begin{bmatrix} 10 & -5 \\ -5 & 5 \end{bmatrix}$ (b) $\begin{bmatrix} 5 & -5 \\ -5 & 10 \end{bmatrix}$

(c) $\begin{bmatrix} -10 & 5 \\ 5 & -5 \end{bmatrix}$ (d) $\begin{bmatrix} -5 & 5 \\ 5 & -10 \end{bmatrix}$

[U.P.S.C. I.E.S. E.E.-II, 2003]

504. A sample power system network is shown in the figure. The reactances marked are in pu. What is the pu value of Y_{22} of the Bus Admittance Matrix (Y_{Bus})?

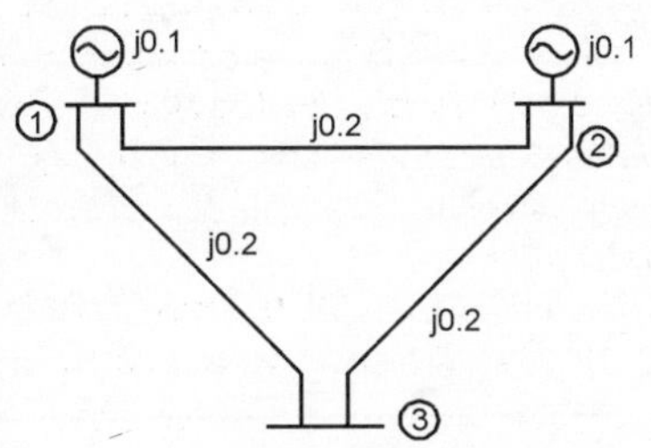

(a) $j10.0$ (b) $j0.5$
(c) $-j0.1$ (d) $-j20.0$

[U.P.S.C. I.E.S. E.E.-II, 2005]

505. Gauss-Seidel interative method can be used for solving a set of
(a) linear differential equations only.
(b) linear algebraic equations only.
(c) both linear and nonlinear algebraic equations.
(d) both linear and nonlinear differential equations.

[GATE E.E. 1997]

506. What is the form of the Y_{bus} matrix for carrying out load flow studies by Gauss-Seidal method of a power system having mesh connection of nodes?
(a) Symmetric but not diagonal matrix.
(b) Diagonal matrix.
(c) Antisymmetric matrix.
(d) Sparse asymmetric matrix. **[U.P.S.C. I.E.S. E.E.-II, 2009]**

507. The Gauss Siedel load flow method has following disadvantages. Tick the incorrect statement.
(a) Unreliable convergence (b) Slow convergence.
(c) Choice of slack bus affects convergence.
(d) A good initial guess for voltages is essential for convergence.

[GATE E.E. 2006]

508. Compared to Gauss-Seidel method, Newton-Raphson method takes
(a) less number of iterations and more time per iteration.
(b) less number of iterations and less time per iteration.
(c) more number of iterations and more time per iteration.
(d) more number of iterations and less time per iteration.

[U.P.S.C. I.E.S. E.E.-II, 1999]

509. Consider the following statements in respect of load flow studies in power systems
1. Bus admittance matrix is a sparse matrix
2. Gauss-Seidel method is preferred over Newton-Raphson method for load flow studies.
3. One of the buses is taken as slack bus in load flow studies.

Which of the statements given above are correct ?
(a) 1, 2 and 3. (b) 1 and 2.
(c) 1 and 3. (d) 2 and 3.

[U.P.S.C. I.E.S. E.E.-II, 2005]

510. In the solution of load-flow equation, Newton-Raphson (NR) method is superior to the Gauss-Seidel (GS) method, because
(a) the time taken to perform one iteration in the NR method is less when compared to the time taken in the GS method.
(b) number of iteration required in the NR method is more when compared to that in the GS method.
(c) number of iterations required is not independent of the size of the system in the NR method.
(d) convergence characteristics of the NR method are not affected by the selection of slack bus. **[U.P.S.C. I.E.S. E.E.-II, 1997]**

511. Let $x^2 - 117 = 0$. The iterative steps for the solution using Newton-Raphson's method is given by

(a) $x_{k+1} = \dfrac{1}{2}\left(x_k + \dfrac{117}{x_k}\right)$ (b) $x_{k+1} = x_k - \dfrac{117}{x_k}$

(c) $x_{k+1} = x_k - \dfrac{x_k}{117}$ (d) $x_{k+1} = x_k - \dfrac{1}{2}\left(x_k + \dfrac{117}{x_k}\right)$

[GATE E.E., 2009]

512. Regulating transformers are used in power systems for control of
(a) voltage. (b) power factor.
(c) power flow. (d) all of the above.

513. In a load-flow study a PV bus is treated as a PQ bus when
(a) voltage limit is violated. (b) active power limit is violated.
(c) phase angle is high.
(d) reactive power limit is violated. **[U.P.S.C. I.E.S. E.E.-II, 2000]**

514. For a 15-bus power system with 3 voltage controlled buses, the size of Jacobian matrix is
(a) 11 × 11 (b) 12 × 12 (c) 24 × 24 (d) 25 × 25

[U.P.S.C. I.E.S. E.E.-II, 1996]

515. A 12-bus power system has 3 voltage controlled buses. The dimensions of the Jacobian matrix will be
(a) 21 × 21 (b) 21 × 19 (c) 19 × 19 (d) 19 × 21

[U.P.S.C. I.E.S. E.E.-II, 2000]

516. A power system consists of 300 buses out of which 20 buses are generator bus, 25 buses are ones with reactive power support and 15 buses are the ones with fixed shunt capacitors. All the other buses are load buses. It is proposed to perform a load flow analysis in the system using Newton-Raphson method. The size of the Newton Raphson Jacobian matrix is
(a) 553 × 553 (b) 540 × 540
(c) 555 × 555 (d) 554 × 554 **[GATE E.E. 2003]**

517. If the reference bus is changed in two load flow buses with same system data and power obtained for reference bus taken as specified P and Q in the latter run
(a) the system losses will be unchanged but complex bus voltages will change.
(b) the system losses will change but complex bus voltages remain unchanged.
(c) the system losses as well as complex bus voltages will change.
(d) the system losses as well as complex bus voltages will be unchanged. **[GATE E.E. 1996]**

Answers: 502. (b), **503.** (a), **504.** (d), **505.** (b), **506.** (a), **507.** (a), **508.** (a), **509.** (c), **510.** (d), **511.** (a), **512.** (c), **513.** (d), **514.** (d), **515.** (b), **516.** (b), **517.** (a),

518. Consider the two power systems shown in figure A below, which are initially not interconnected, and are operating in steady state at the same frequency. Separate load flow solutions are computed individually for the two systems, corresponding to this scenario. The bus voltage phasors so obtained are indicated on figure A. These two isolated systems are now interconnected by a short transmission line as shown in figure B, and it is found that $P_1 = P_2 = Q_1 = Q_2 = 0$.

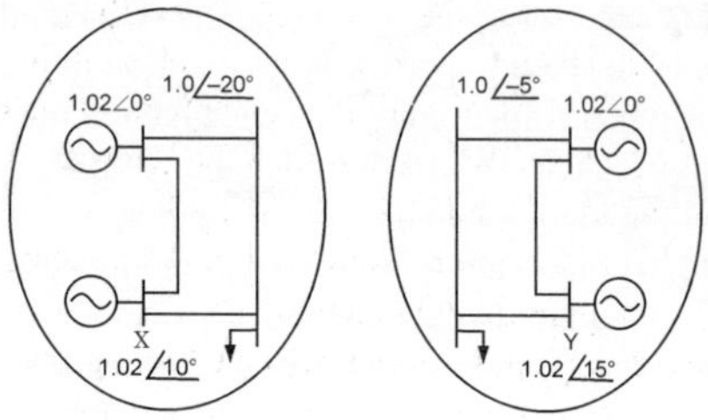

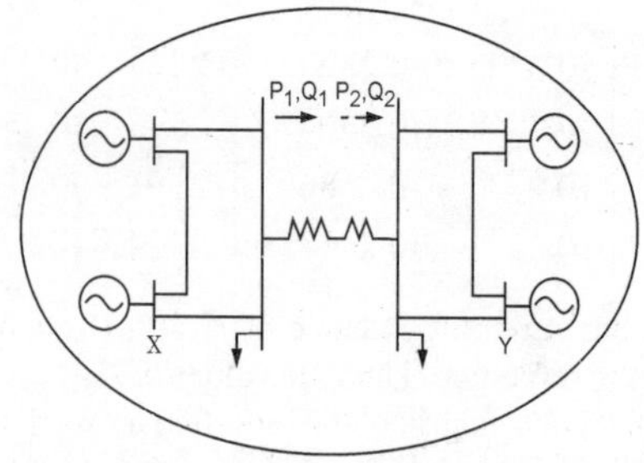

The bus voltage phase angular difference between generator bus X and generator bus Y after the interconnection is

(a) 10° (b) 25° (c) –30° (d) 30°

[GATE E.E. 2007]

Answers: 518. (a).

ASSERTION (A)/REASON (R) TYPE QUESTIONS

Note : Each of the following questions consists of two statements, one labelled the 'Assertion (A)' and the other labelled the 'Reason (R)'. Examine the two statements carefully and decide if the Assertion (A) and Reason (R) are individually true and if so, whether the reason (R) is correct explanation of the Assertion (A). Select your answers to these questions using the codes given below:

Codes: (a) Both A and R are true and R is a correct explanation of A.
(b) Both A and B are true and R is not a correct explanation of A.
(c) A is true but R is false. (d) A is false but R is true.

1. (A): In L.P.S. (Longitudinal Power Supply) system, the voltage regulation is generally poor.
(R): L.P.S. system has low level of short-circuit MVA. **[U.P.S.C. I.E.S. E.E.-II, 2000]**

2. (A): When a Line-to-Line (L—L) fault takes place at the terminals of an open-circuited generator, phase voltages are sometimes 'indeterminate', though line-to-line voltages are always determinable.
(R): During a Line-to-Line fault, zero-sequence voltage is always indeterminate. **[U.P.S.C. I.E.S. E.E.-II, 2002]**

3. (A): In comparison to making capacity of a circuit breaker, its breaking capacity is normally higher.
(R): The breaking capacity of a circuit breaker (3-phase) is expressed as $\sqrt{3}$ VI × 10^{-6} MVA where V is the rated service voltage in volts and I is rms value of symmetrical breaking current in amperes.

4. (A): For ground fault protection, the reactance relay is preferred.
(R): Ground fault involves variable fault resistance and the operation of reactance relay is not affected by fault resistance.

5. (A): Reactance relay is more suitable for the protection of short lines.
(R): Operation of reactance relay is independent of arc fault resistance. **[U.P.S.C. I.E.S. E.E.-II, 1994]**

6. (A): A capacitor is connected in the polarizing circuit of the mho relay to prevent its operation when three-phase faults occur very close to the relay.
(R): Capacitor provides memory action and thus maintains polarizing flux for several cycles following the fault. **[U.P.S.C. I.E.S. E.E.-II, 2009]**

7. (A): Residual earth fault relays cannot give satisfactory protection in th case of resonant earthed system.
(R): Earth fault currents are very large. **[U.P.S.C. I.E.S. E.E.-II, 1998]**

8. (A): Transistor static relays are the most widely used static relays.
(R): The transistor circuit not only perform the essential functions of a relay but also provide necessary flexibility to suit the various relay requirements.

9. (A): Testing of circuit breaker is more difficult than that of other electrical equipment like transformer or rotating machines.
(R): Short-circuit currents are very large and there is no satisfactory method for testing circuit breaker at reduced power.

10. (A): The core-balance leakage protection used for power transformers against earth fault on *hv* winding is inferior to combined leakage and overload protection.
(R): Relay does not operate in core-balance leakage protection when fault occurs between phases.

11. (A): Zero-sequence currents flow from a line into a transformer tank.
(R): Transformer windings are connected in grounded star/delta.

Answers: 1. (a), **2.** (c), **3.** (d), **4.** (a), **5.** (b), **6.** (a), **7.** (c), **8.** (a), **9.** (a), **10.** (a), **11.** (a),

12. (A): Merz-Price protection system is not suitable for protection of feeders beyond 33 kV.

(R): Difficulties are experienced in balancing the secondaries of the two CTs above 33 kV.

13. (A): Mho relay is normally used for protection of long transmission lines.

(R): Under power swing conditions, Mho relay is stable.

14. (A): Thyrite arrestor is most common and is mostly used for the protection against high dangerous voltages.

(R): The thyrite arrester discharges several thousand amperes without the slightest tendency to flash-over on the edges and there is absolutely no time lag in its performance.

15. (A): Present-day surge diverters use nonlinear resistance elements.

(R): A nonlinear resistance element has the property that its resistance increases sharply as the voltage across it increases. **[U.P.S.C. I.E.S. E.E.-II, 2006]**

16. (A): To provide effective protection to the terminal apparatus against lightning long lead length to the feeder line and to the ground are recommended for arrester applications.

(R): Long lead length offers high impedance path to lighting surge current. **[U.P.S.C. I.E.S. E.E.-II, 1995]**

17. (A): Shunt capacitors are used in conjunction with lighting arresters for the protection of station equipment against surges.

(R): Shunt capacitors reduce the crest of surges. **[U.P.S.C. I.E.S. E.E.-II, 1996]**

18. (A): A short length of cable between an overhead line and the substation is effective in reducing the incoming surge voltage.

(R): The surge impedance of the cable is quite low when compared with that of the incoming overhead line. **[U.P.S.C. I.E.S. E.E.-II, 1998]**

19. (A): Voltage-time insulation coordinations curves for the equipment to be protected and the lightning arrester to be employed are used for determining the required insulation levels.

(R): Voltage time insulation coordination curves indicate the time of arrival of the peak of the surge wave at the equipment location. **[U.P.S.C. I.E.S. E.E.-II, 2000]**

20. (A): The 'short-circuit capacity' of a bus in a large power grid is defined as the product of the prefault voltage and the 3-phase fault current at a point very close to the bus.

(R): The larger the short-circuit capacity, the larger would be the equivalent source impedance at the bus. **[U.P.S.C. I.E.S. E.E.-II, 2008]**

21. (A): The best method of improving transient stability is by use of high speed circuit breakers.

(R): The quicker the circuit breaker operates, the faster the fault is removed from the system and normal operating conditions are restored.

22. (A): Load flow studies are usually carried out using Y_{BUS}.

(R): Y_{BUS} is a sparse matrix and is generally matching with topology of the PS network. **[U.P.S.C. I.E.S. E.E.-II, 2010]**

23. (A): During the load-flow study, a voltage controlled bus is treated as a load bus when the magnitude of voltage is very much higher or very much lower than the scheduled voltage.

(R): A tap changing transformer is represented as a π-network in the load-flow model. **[U.P.S.C. I.E.S. E.E.-II, 1997]**

24. (A): Bus admittance matrix is a sparse matrix.

(R): In load flow studies, one of the buses is taken as a slack bus. **[U.P.S.C. I.E.S. E.E.-II, 2006]**

25. (A): The bus admittance matrix (Y_{Bus}) used in load-flow studies is a sparse matrix.

(R): The elements of Y_{Bus} exhibit the physical interconnection of nodes of the power system. **[U.P.S.C. I.E.S. E.E.-II, 1994]**

26. (A): For the load-flow solution of a large power system Newton-Raphson method is superior to the Gauss-Seidel method.

(R): In the Gauss-Seidel method, the number of iterations vary between 3 to 5, irrespective of the size of the system. **[U.P.S.C. I.E.S. E.E.-II, 1994]**

27. (A): The main advantage of decoupled load flow (DLF) method as compared to Newton Raphson method is its reduced memory requirements in storing the Jacobian.

(R): An important characteristic of any practical electric power transmission system operating in steady state is the strong interdependence between the real power and bus voltage angles, and between the reactive powers and voltage magnitudes. **[U.P.S.C. I.E.S. E.E.-II, 2004]**

Answers: 12. (a), **13.** (b), **14.** (a), **15.** (c), **16.** (c), **17.** (a), **18.** (a), **19.** (a), **20.** (c), **21.** (a), **22.** (a), **23.** (b), **24.** (b), **25.** (a), **26.** (c), **27.** (b).

MATCHING TYPE QUESTIONS

Note: Each of the following questions consists of two lists. Match the items given in list 1 and those in list 2. Select your answers to these questions using the codes given in each of the questions.

1.

List 1 (Symbol for 3-phase Transformer Connections)

A. 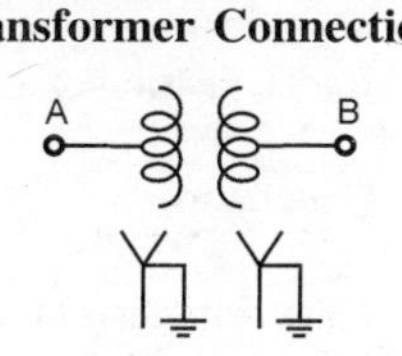

B. 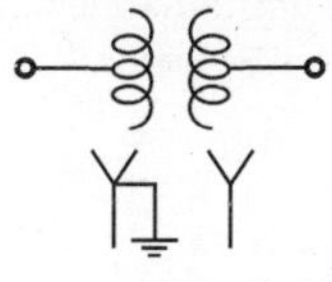

C. 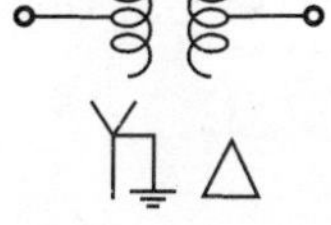

List 2 (Zero-Sequence Equivalent Circuits)

1. 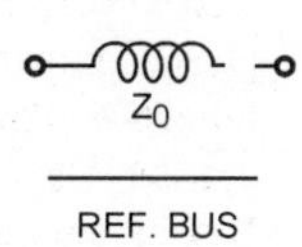

2. 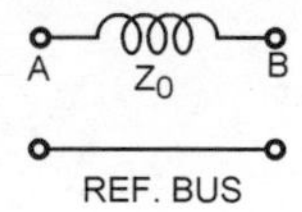

3. 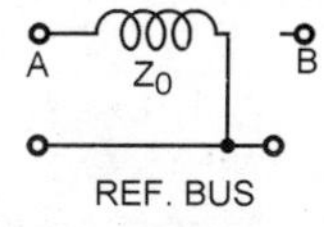

Answers: 1. (a),

Codes:	A	B	C
(a)	2	1	3
(b)	2	3	1
(c)	1	2	3
(d)	3	2	1

[U.P.S.C. I.E.S. E.E.-II, 2007]

2.

List 1	**List 2**
A. All types of faults.	1. Negative-sequence currents are present.
B. All unsymmetrical faults.	2. Zero-sequence currents are present.
C. Fault involving ground with neutral of the system grounded.	3. Zero-sequence currents are absent.
D. Fault involving ground with floating neutral.	4. Positive-sequence currents are present.

Codes:	A	B	C	D
(a)	3	1	2	4
(b)	4	1	2	3
(c)	3	2	1	4
(d)	4	2	1	3

[U.P.S.C. I.E.S. E.E.-II, 2010]

3. **List 1**

A. Arc voltage.
B. Recovery voltage.
C. Restriking voltage.
D. Short-circuit current.

List 2

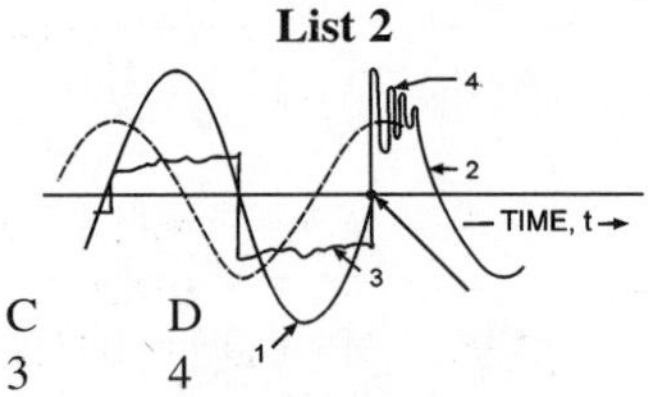

Codes :	A	B	C	D
(a)	1	2	3	4
(b)	2	3	4	1
(c)	3	2	4	1
(d)	3	4	2	1

4. The inverse characteristics of an induction-disc relay are shown below :

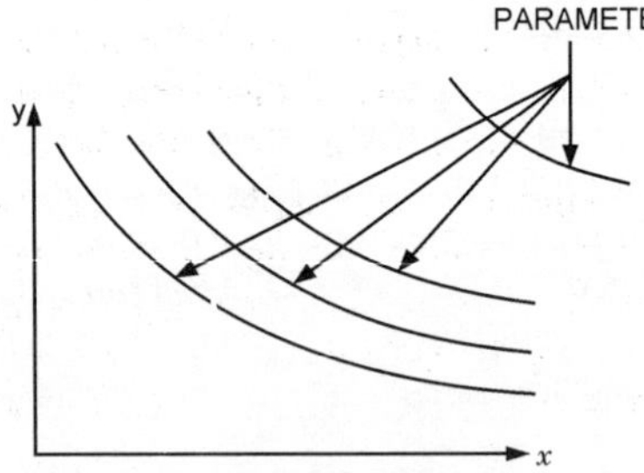

Match List 1 (*x*, *y* coordinates and parameter) with List 2 (Variables) and select the correct answer using the codes given below:

List 1 (*x*, *y* Coordinates and Parameter)	**List 2 (Variables)**
A. *x*-coordinate	1. Plug setting voltage.
B. *y*-coordiante	2. Current as multiplier of plug setting.
C. Parameter	3. Operating time.
	4. Time multiplier setting.
	5. Power factor.

Codes :	A	B	C
(a)	5	4	1
(b)	2	3	4
(c)	5	3	4
(d)	2	4	1

U.P.S.C. I.E.S. E.E.-II, 2002]

5. **List 1**

A. Mho relay.
B. Plain impedance relay.
C. Directional relay.
D. Angle impedance relay.

List 2

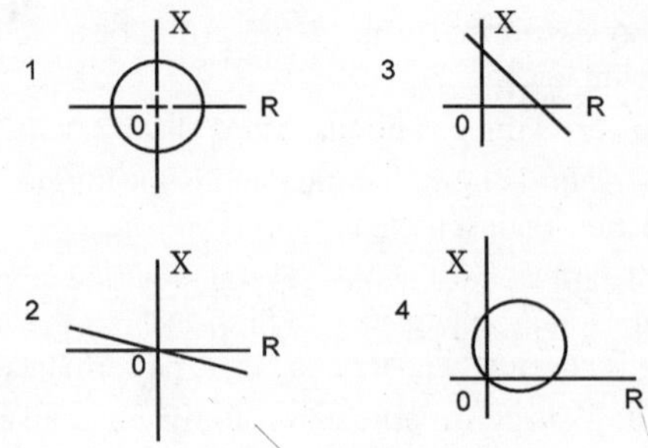

Fig. Q. 5.

Codes :	A	B	C	D
(a)	4	3	2	1
(b)	4	1	2	3
(c)	3	2	1	4
(d)	3	2	4	1

[U.P.S.C. I.E.S. E.E.-II, 1993; 2011]

6.

List 1 (Types of Relays)	**List 2 (Types of Protection)**
A. Directional relay.	1. Relay operates for fault within certain distacne of its location.
B. Impedance relay.	2. Relay will trip for fault in one location and block for all other locations.
C. Differential relay.	3. High speed protection for entire transmission line.
D. Pilot relay.	4. The principle of current continuity is used to devise a simple and effective relaying system over a small physical space.

[U.P.S.C. I.E.S. E.E. II, 2004]

Codes :	A	B	C	D
(a)	1	2	4	3
(b)	2	1	3	4
(c)	2	1	4	3
(d)	1	2	3	4

7.

List 1	**List 2**
A. Buchholz relay.	1. Feeder.
B. Translay relay.	2. Transformer.
C. Negative sequence relay.	3. Generator.
D. Directional over-current relay.	4. Long overhead line.

Codes:	A	B	C	D
(a)	1	2	3	4
(b)	2	1	3	4
(c)	2	1	4	3
(d)	1	2	4	3

[U.P.S.C. I.E.S. E.E.-II, 1998]

8.

List 1 (Type of Relay)	**List 2 (Most Suited For)**
A. Buchholz relay.	1. Feeder.
B. Translay relay.	2. Transformer.
C. Carrier current phase comparison relay.	3. Radial distributor.
D. Directional over-current relay.	4. Generator.
E. Negative sequence relay.	5. Ring main distributor.
	6. Long overhead transmission line.

Answers: **2.** (b), **3.** (c), **4.** (b), **5.** (b), **6.** (c), **7.** (b), **8.** (c),

Codes:	A	B	C	D	E
(a)	2	1	6	4	5
(b)	2	1	3	4	5
(c)	2	1	6	5	4
(d)	1	2	4	5	6

[GATE E.E. 1995]

9. **List 1 (Type of Realy)** — **List 2 (Components Protected)**

A. Distance realy
B. Under frequency relay
C. Differential realy
D. Buchholz relay

1. Transformers
2. Turbines
3. Bus-bars
4. Shunt capacitors
5. Alternators
6. Transmission lines

Codes :	A	B	C	D
(a)	6	5	3	1
(b)	4	3	2	1
(c)	5	2	1	6
(d)	6	4	5	3

[GATE E.E. 2003; U.P.S.C. E.E.-II, 2005]

10. **List 1 (Types of Relays)** — **List 2 (Protective Schemes)**

A. Negative-sequence relay.
B. Harmonic restraint differential relay.
C. Overcurrent relay with time-delay.
D. Mho relay.

1. Distribution feeder protection.
2. Long ht lines protection.
3. Transformer protection
4. Rotor protection of alternators.

Codes :	A	B	C	D
(a)	4	3	1	2
(b)	1	2	4	3
(c)	4	2	1	3
(d)	1	3	4	2

[U.P.S.C. I.E.S. E.E.-II, 2002]

11. **List 1 (Type of Transmission Line)** — **List 2 (Type of Distance Relay Preferred)**

A. Short Line
B. Medium Line
C. Long Line

1. Ohm Relay
2. Reactance Realy
3. Mho Relay

(a) $a \to 2, b \to 1, c \to 3$
(b) $a \to 3, b \to 2, c \to 1$
(c) $a \to 1, b \to 2, c \to 3$
(d) $a \to 1, b \to 3, c \to 2$

[GATE E.E. 2009]

12. **List 1 (Type of Fault)** — **List 2 (Type of Protection)**

A. External phase fault
B. Faults between phases
C. Phase to earth fault
D. Overheating

1. Differential protection
2. Earth fault relay
3. Overcurrent protection (inverse time delay o/c relay)
4. Thermal relay

Codes :	A	B	C	D
(a)	3	1	2	4
(b)	4	2	1	3
(c)	3	2	1	4
(d)	4	1	2	3

[U.P.S.C. I.E.S. E.E.-II, 2006]

13. **List 1 (Equipment)** — **List 2 (Function)**

A. Circuit breaker.
B. Lightning arrester.
C. Governor.
D. Exciter.

1. Voltage control.
2. Power control.
3. Overvoltage protection.
4. Overcurrent protection.

[GATE E.E. 1992]

Codes:	A	B	C	D
(a)	1	2	3	4
(b)	4	1	2	3
(c)	2	3	4	1
(d)	4	3	2	1

[U.P.S.C. I.E.S. E.E.-II, 1995]

14. **List 1 (Equipments)** — **List 2 (Applications)**

A. Metal oxide arrester.
B. Isolator.
C. Auto-reclosing CB.
D. Differential relay.

1. Protects generator against short circuit faults.
2. Improves transient stability.
3. Allows CB maintenance.
4. Provides protection against surges.

Codes :	A	B	C	D
(a)	4	3	2	1
(b)	3	4	1	2
(c)	4	3	1	2
(d)	3	4	2	1

[U.P.S.C. I.E.S. E.E.-II, 2001]

15. **List 1** — **List 2**

A. Thyrite arrester.
B. Sag template.
C. Cable sheaths.
D. Circuit breaker.

1. Tower location.
2. Cross bonding.
3. Restriking voltage.
4. Nonlinear resistor.

Codes:	A	B	C	D
(a)	3	1	2	4
(b)	4	1	2	3
(c)	3	2	1	4
(d)	4	2	1	3

[GATE E.E. 2011]

16. **List 1** — **List 2**

A. Transient stability. improvement
B. Economic dispatch.
C. Load frequency control.
D. Dynamic stability.

1. Incremental transmission loss.
2. Area control error.
3. Power system stabilizers.
4. Turbine fast valving.

Codes :	A	B	C	D
(a)	2	3	4	1
(b)	4	1	2	3
(c)	2	1	4	3
(d)	4	3	2	1

[U.P.S.C. I.E.S. E.E.-II, 2008]

17. **List 1 (Phenomenon)** — **List 2 (Dominant Features)**

A. Voltage stability.
B. Transient stability.
C. Oscillatory instability.
D. Steady-state Dynamics.

1. Power system stabilizer.
2. Damping power.
3. 'Angle' stability.
4. Reactive power.

Codes :	A	B	C	D
(a)	1	3	2	4
(b)	4	2	3	1
(c)	1	2	3	4
(d)	4	3	2	1

[U.P.S.C. I.E.S. E.E.-II, 2002]

18. **List 1** — **List 2**

A. Ryle's formula.
B. Gauss-Siedel method.
C. Stringing chart.
D. Line and machine charts.

1. Interactive solution of load flow.
2. Surge impedence.
3. Transmission lower.
4. Sag Vs temperature Vs tension.
5. Voltage drop and regulation calculation.

Answers: 9. (a), **10.** (a), **11.** (c), **12.** (a), **13.** (d), **14.** (a), **15.** (b), **16.** (b), **17.** (d), **18.** (b).

6. Line loss and machine loss calculation.
7. Determination of sending end voltages for a required receiving end voltage for various loads.

Codes:	A	B	C	D
(a)	1	7	4	6
(b)	7	1	4	6
(c)	7	1	5	6
(d)	5	7	6	4

[A.M.I.E. Sec B. Power Systems Winter 1997]

TRUE(T)/FALSE(F) TYPE QUESTIONS

Note : Identify whether the following statements are TRUE (T) or FALSE (F).

1. A short-circuit current is identified heavy current flow.

2. The overcurrent fault occurs mainly because of short-circuit or leakage owing to corona and sometimes due to overload on the supply system.

3. Reverse power fault occurs only in interconnected systems.

4. The first peak short-circuit current is $\sqrt{2}$ times the rms value of the periodic component of the short-circuit current 2-3 periods after the short-circuit is initiated.

5. The MVA at fault is equal to base MVA divided per unit equivalent fault impedance.

6. A sudden short-circuit in an ac power system causes a fall in current in the short-circuited phase.

7. The reactor limits the fault by limiting the current that can flow into it from other healthy parts of the system.

8. Oil immersed type reactors are used for voltages above 1.1 kV.

9. In a tie-bar system of bus-bar reactors the generators are connected to the common bus-bar through the reactors but the feeders are fed from the generator side of the reactors.

10. In delta-connected load, the line currents do not have zero sequence components.

11. In a power system, the 3-phase fault MVA is always higher than the single line-to-ground fault MVA at a bus. **[GATE E.E. 1994]**

12. For simulating a three-phase to ground fault positive and negative sequence networks are connected in series. **[A.M.I.E. Sec B Summer 1999]**

13. Zero-sequence currents flow in a transmission line when there is a double line to grounded fault.

14. Breaking capacity of a fuse is the rating corresponding to the rms value of the ac component of the maximum prospective current and the system voltage.

15. HRC fuses obey inverse time-current characteristic. **[A.M.I.E. Sec B. Winter 1994]**

16. HRC fuses are reusable after blowing off. **[A.M.I.E. Sec B. Winter 1995]**

17. A fuse has low braking capacity as compared to a circuit breaker. **[A.M.I.E. Sec B. Summer 1999]**

18. The relation between fusing current (I) and fuse wire diameter is $I \alpha d^{1.5}$.

19. Rewirable fuse has linear time-current characteristic.

20. For the same rupturing capacity, the actual current to be interrupted by an HRC fuse is much more than that of any circuit breaker.

21. A circuit breaker is a current interrupting device.

22. On occurrence of fault on the protected circuit, a circuit breaker operates manually through a control switch.

23. Current chopping phenomenon is absent in ABCB. **[A.M.I.E. Sec B. Winter 1998]**

24. RRRV of a system is expressed in kV/km. **[A.M.I.E. Sec B. Winter 1994]**

25. For preventing arc strike in a circuit breaker the rate of transient recovery voltage must be greater than the rate of build-up of dielectric strength across the contacts. **[A.M.I.E. Sec B. Summer 1999]**

26. Successful arc interruption is achieved when the system recovery is faster than dielectric recovery. **[A.M.I.E. Sec B. Winter 1998]**

27. Bulk oil circuit breakers utilise oil for arc quenching only. **[A.M.I.E. Sec B. Winter 1995]**

28. In small oil volume circuit breaker insulation are provided to provide sufficient insulation of the current carrying parts of the circuit breaker. **[A.M.I.E. Sec B. Winter 1996]**

29. Magnetic blow out coils in air circuit breaker helps arc interruption. **[A.M.I.E. Sec B. Winter 1998]**

30. Air-blast circuit breakers at 220 kV level make great noise while breaking a circuit. **[A.M.I.E. Sec B. Winter 1993]**

31. The chances of arc interruption in subsequent current zero increase in case of OCB but decrease in air blast circuit breaker. **[A.M.I.E. Sec B. Winter 1993]**

32. Chimney effect occurs in an air blast circuit breaker. **[A.M.I.E. Sec B. Winter 1996]**

33. Current chopping leads to dangerously high voltages. **[A.M.I.E. Sec B. Winter 1994]**

34. Air-blast circuit breakers for 400 kV power system are designed to operate in 50 ms. **[A.M.I.E. Sec B. Summer 1998]**

35. At normal temperature and pressure SF_6 is a colourless liquid. **[A.M.I.E. Sec B. Winter 1993]**

36. SF_6 gas is not an efficient quenching medium because of its electronegativity. **[A.M.I.E. Sec B. Summer 1998]**

37. SF_6 gas is an excellent arc quenching medium. **[A.M.I.E. Sec B. Winter 1998]**

38. SF_6 gas is used as an efficient arc quenching medium because of its electronegativity. **[A.M.I.E. Sec B. Winter 1996]**

39. The behaviour of arc in high vacuum is affected by contact material and contact surface conditions.

40. The insulation requirements of switchgear are determined on the basis of transient voltage.

41. Making capacity of a circuit breaker is greater than breaking capacity. **[A.M.I.E. Sec B. Winter 1998]**

Answers: **1.** (T), **2.** (T), **3.** (T), **4.** (F), **5.** (T), **6.** (F), **7.** (T), **8.** (F), **9.** (T), **10.** (T), **11.** (T), **12.** (F), **13.** (T), **14.** (T), **15.** (T), **16.** (F), **17.** (T), **18.** (T), **19.** (F), **20.** (F), **21.** (T), **22.** (F), **23.** (F), **24.** (F), **25.** (F), **26.** (F), **27.** (F), **28.** (T), **29.** (T), **30.** (T), **31.** (T), **32.** (T), **33.** (T), **34.** (T), **35.** (T), **36.** (F), **37.** (T), **38.** (T), **39.** (T), **40.** (T), **41.** (T),

42. The arcing contacts for a circuit breaker are made of stainless steel.

43. With the increase in mechanical force applied, the contact resistance decreases.

44. The fault clearing time of a circuit breaker is usually few cycles of supply voltage.

45. Bulk oil circuit breakers do use pneumatic operating mechanism.

46. Oil circuit reclosures are increasingly used in unattended substations and rural distribution schemes whereas the circuit breakers are installed in outlying areas.

47. Auto-recloser type breakers never close during a fault. **[A.M.I.E. Sec B. Winter 1994]**

48. A switch is used in an electric circuit as a device for making or breaking the electric circuit in a convenient way.

49. Isolators may be operated under short-circuit conditions.

50. While opening a transmission line under normal load condition the isolator should open first before the opening of the circuit breaker. **[A.M.I.E. Sec B. Winter 1996]**

51. Backup protection functions when primary protection fails to operate.

52. In the protection scheme relay functions as a switching device.

53. Relay contacts are usually made of silver.

54. By burden of the relay, we generally mean VA rating of the relay.

55. Holding ratio is the ratio of drop off value to pick up value.

56. The holding ratio of a relay is usually more than unity.

57. Bounce proof contacts are used in relays. **[A.M.I.E. Sec B. Summer 1998]**

58. Pressure at the relay contacts is kept low. **[A.M.I.E. Sec B. Summer 1998]**

59. The relay is designed for maximum torque/friction ratio. **[A.M.I.E. Sec B. Summer 1998]**

60. The cores of protection CT have high saturation level. **[A.M.I.E. Sec B. Winter 1995]**

61. Backup relays are more selective than primary relays. **[A.M.I.E. Sec B. Winter 1994]**

62. Interlocking prevents one set of contacts from closing while other set of contacts is closed.

63. In electromagnetic relays the restraining torque is produced by springs.

64. Induction disc relays are faster in operation than induction cup relays. **[A.M.I.E. Sec B. Winter 1994]**

65. Induction type overcurrent relays can be used for the protection of a dc distributor. **[A.M.I.E. Sec B. Winter 1995]**

66. Relays using induction disc principle operate both on ac and dc system. **[A.M.I.E. Sec B. Summer 1998]**

67. The rotation of disc of an induction disc relay under the poles is from unshaded pole to shaded pole.

68. Induction cup relays are basically phase comparators. **[A.M.I.E. Sec B. Winter 1993]**

69. Buchholz relay is a current actuated relay.

70. Buchholz relays are used for protection of power transformers against all types of internal and external faults.

71. Overcurrent relay plug settings are determined from maximum fault current rating. **[A.M.I.E. Sec B. Winter 1993]**

72. The plug setting multiplier for a relay with 15% current setting CT ratio of 500/5 and primary fault current of 600 A is 8. **[A.M.I.E. Sec B. Summer 1999]**

73. Induction type reverse power relays are very suitable for the protection of parallel feeders.

74. Operation of distance relay is based upon the measurement of impedance between the location of the relay and the fault point.

75. Plain impedance relays are direction sensitive **[A.M.I.E. Sec B. Winter 1994]**

76. Plain impedance relays are non-directional in nature. **[A.M.I.E. Sec B. Winter 1994]**

77. Reactance relay is the best for ground fault protection. **[A.M.I.E. Sec B. Winter 1998]**

78. Reactance relay is a voltage restrained overcurrent relay. **[A.M.I.E. Sec B. Winter 1994]**

79. Mho relay is a voltage restrained over-current relay. **[A.M.I.E. Sec B. Winter 1993]**

80. Mho relay is used for short transmission line protection. **[A.M.I.E. Sec B. Winter 1996]**

81. Protection against unbalanced supply voltage is provided by negative phase sequence relays.

82. Directional overcurrent relays are used for protection of ring main distribution lines.

83. Mho relay has inherent directional properties.

84. Mho relays have an R-X plane characteristic depicted by a circle passing through origin.

85. Static relays cause more VA burden on CTs than their electromagnetic counter part. **[A.M.I.E. Sec B. Winter 1993]**

86. Static relays pose lower burden on protective circuits. **[A.M.I.E. Sec B. Winter 1994]**

87. Static relays offer low burden than electromagnetic relays. **[A.M.I.E. Sec B. Winter 1995]**

88. Frequency relays are used in aternators for overspeed protection.

89. Loss of excitation protection of generator is rendered by offset mho relay. **[A.M.I.E. Sec B. Winter 1993]**

90. It is not necessary to switch off the field supply on occurrence of a fault in the generator stator winding.

91. Differential protection does not provide protection against/turn-to-turn fault on the same phase winding of the stator of an alternator as it does not cause the imbalance between the CT pairs.

92. The main drawback of Merz-Price differential protection relay is its response to severe external faults.

93. The function of the AVR of a generator connected to the grid is to control the power factor. **[A.M.I.E. Sec B. Winter 1998]**

94. The most commonly used system for the protection of generator is circulating current system. **[A.M.I.E. Sec B. Summer 1999]**

95. In the event of prime mover failure, the generator starts motoring, and this can be detected by a directional relay.

96. Whatever be the cause of unbalance in an alternator, obviously negative phase sequence current will result in heating of stator.

97. Inter-turn fault on the same phase of the stator winding can be detected by longitudinal differential protection.

98. In Merz-Price protection scheme, the relays are located adjacent to the CTs. An earth fault is the stator winding of an alternator causes an increase in current through one CT, as compared to current through the other CT, causing current imbalance and spilting of current through relay and therefore, tripping of circuit.

99. The boundary of a protective system is determined by the location of CTs.

Answers: 42. (F), **43.** (T), **44.** (T), **45.** (F), **46.** (T), **47.** (F), **48.** (T), **49.** (F), **50.** (F), **51.** (T), **52.** (F), **53.** (T), **54.** (T), **55.** (T), **56.** (F), **57.** (T), **58.** (F), **59.** (T), **60.** (T), **61.** (F), **62.** (T), **63.** (F), **64.** (F), **65.** (F), **66.** (F), **67.** (T), **68.** (T), **69.** (F), **70.** (F), **71.** (F), **72.** (T), **73.** (T), **74.** (T), **75.** (F), **76.** (T), **77.** (T), **78.** (F), **79.** (F), **80.** (F), **81.** (T), **82.** (T), **83.** (T), **84.** (T), **85.** (F), **86.** (T), **87.** (T), **88.** (T), **89.** (T), **90.** (F), **91.** (T), **92.** (T), **93.** (F), **94.** (T), **95.** (T), **96.** (T), **97.** (F), **98.** (T), **99.** (T),

100. Flow of unbalanced three-phase currents through the stator winding of an alternator will cause heating of the rotor.

101. A large-size generator is protected against overloads by providing overcurrent relays.

102. A single phasing relay is used with a large 3-phase squirrel cage induction motor.

103. For protection of motors against short circuits on stator winding thermal or dash pot type overcurrent tripping devices are used.

104. Heating effect of the positive phase sequence current is more than that of negative sequence current.

105. In a differential protection of star/delta transformer, the CTs on the star side are connected is delta and on delta side in star. **[A.M.I.E. Sec B. Summer 1999]**

106. Buchholz relay can detect faults above oil level in the transformer. **[A.M.I.E. Sec B. Summer 1999]**

107. Magnetising inrush current is transformers is rich in second harmonic contents. **[A.M.I.E. Sec B. Summer 1998]**

108. In a differential protection of a star/delta transformer the CTs on the star side are connected in star and on the delta side in delta. **[A.M.I.E. Sec B. Winter 1996]**

109. Magnetising inrush occurs in a motor under fault condition. **[A.M.I.E. Sec B. Winter 1996]**

110. Transformer, differential relays usually have high bias settings. **[A.M.I.E. Sec B. Winter 1995]**

111. Bus-bar term is used for a main bar or conductor carrying an electric current to which many connections may be made.

112. Bus-bar operates at variable voltage. **[A.M.I.E. Sec B. Summer 1999]**

113. Bus-coupler is very essential in main and transfer bus.

114. Overcurrent/distance relays of other zones cannot be employed for protection of bus-bars.

115. Single line opening in a transmission system is detected by ground fault relays. **[A.M.I.E. Sec B. Winter 1994]**

116. Differential protection scheme for longer lines is very costly. **[A.M.I.E. Sec B. Summer 1999]**

117. Pilot protection scheme is used for protection of long EHV transmission lines in interconnected power systems.

118. In a protective scheme of a feeder, a summation transformer is used. It has tappings on the primary and cannot operate with secondary open-circuited.

119. Unit type of protection responds to fault is protected zone alone.

120. Differential protection scheme compares the current entering and current leaving the protected circuit.

121. In a biased or percentage differential relay, the restraining coil provides a biasing feature and thus the flow of heavy out-of-balance current through the relay can be avoided.

122. Differential protection schemes employed for protection of feeders make use of pilot wires.

123. The relays used in balanced voltage schemes will behave in the same way as in circulating scheme on a pilot fault.

124. Power line carrier is a low frequency signal sent through the power line conductors.

125. Wavetraps are series resonating circuits. **[A.M.I.E. Sec B. Winter 1994]**

126. Wavetraps of carrier protection are series resonating circuits at carrier frequency. **[A.M.I.E. Sec B. Winter 1993]**

127. Line traps are used to restrict the flow of high frequency carrier signals between the local end transmitter and remote end receiver. **[A.M.I.E. Sec B. Winter 1996]**

128. Line traps are series resonant circuits. **[A.M.I.E. Sec B. Winter 1998]**

129. Microwave pilot scheme need both a line trap and coupling capacitor. **[A.M.I.E. Sec B. Summer 1998]**

130. Carrier blocking scheme uses healthy line for carrier transfer. **[A.M.I.E. Sec B. Winter 1998]**

131. First zone distance relays are never allowed to over reach. **[A.M.I.E. Sec B. Winter 1993]**

132. Distance protection schemes widely used in transmission lines provides discriminative protection without using pilot wires.

133. Micro-pilot scheme need both line trap and coupling capacitor.

134. The most severe cause of switching overvoltage is closing of charged line.

135. Switching overvoltages are proportional to the square of system voltages.

136. The electrostatically induced emf in a line due to discharge of one cloud over another one is called the direct lightning stroke.

137. A surge may be considered as high voltage of very high frequency.

138. The damage caused by surge depends on steepness of the wave front.

139. Nonlinear resistors are employed in lightning arresters.

140. Lightning arresters are used before the switchgear.

141. Magnetically blow out type lightning arresters are designed to discharge the lightning overvoltages as well as switching over-voltages.

142. In EHV line ground wire is used to provide a shield to phase conductors from direct lightning strokes.

143. A lightning arrester should operate only at abnormal frequencies.

144. A rod gap arrester is capable of stopping power frequency to earth.

145. Expulsion type arrester is also called the protector tube.

146. Protector tubes can be used for the protection of station insulation.

147. The angle of protection should not exceed 30° for good shielding of the line conductor.

148. Earthling lightning rods are provided at the substations.

149. Location of lightning arrester is near a transformer.

150. Travelling wave gets reflected from short-circuited line with reflected current wave of opposite polarity and voltage wave of the same polarity.

151. In long transmission lines as we move away from the receiving end the reflected component of voltage increases in magnitude and phase.

Answers: **100.** (T), **101.** (F), **102.** (T), **103.** (T), **104.** (F), **105.** (T), **106.** (F), **107.** (T), **108.** (F), **109.** (F), **110.** (T), **111.** (T), **112.** (F), **113.** (T), **114.** (F), **115.** (F), **116.** (T), **117.** (T), **118.** (T), **119.** (T), **120.** (T), **121.** (T), **122.** (T), **123.** (F), **124.** (F), **125.** (T), **126.** (T), **127.** (T), **128.** (F), **129.** (F), **130.** (T), **131.** (T), **132.** (T), **133.** (F), **134.** (T), **135.** (F), **136.** (F), **137.** (T), **138.** (T), **139.** (T), **140.** (F), **141.** (T), **142.** (T), **143.** (T), **144.** (F), **145.** (T), **146.** (F), **147.** (T), **148.** (F), **149.** (T), **150.** (F), **151.** (F),

152. In long transmission lines incident and reflected waves travel in opposite directions.

153. Voltage surge travelling on a transmission line enters the lightning arrester first.

154. Surge impedance depends on tower height and base.

155. Small protective angles are used for low towers and large angles for high towers.

156. A surge diverter should provide a large impedance path to the travelling wave reaching the diverter.

157. Proper coordination will insure that the volt-time curve of the equipment will lie below the volt-time curve of the protective device. **[A.M.I.E. Sec B. Advanced Power Systems Summer 1999]**

158. The crest voltage of the standard wave that will not cause flashover of the insulation is called the basic impulse insulation (BIL).

159. BIL are reference levels expressed in impulse crest voltage with a standard wave not longer than 1.2 × 50 wave.

160. Impulse insulation breakdown strength of the protected equipment must be greater than that of the protecting device.

161. Insulation coordination depends upon voltage-time characteristics.

162. Insulation strength of EHV line is governed by lightning over-voltage.

163. Tail time of a wave means time from beginning of wave to half the crest value on the tail.

164. In transmission lines, the ground wires run below the line conductors.

165. The insulation strength of EHV lines is mainly governed by power frequency overvoltages.

166. Capacitive voltage transformer (CVT) is used to reduce the incidence of overvoltage surges on transmission lines.

167. All ac power systems operate with solid earthing nowadays.

168. Arcing ground is also called the earth fault.

169. The arc-suppression coil method of neutral grounding is normally confined to medium voltage overhead transmission lines which are supplied through transformers.

170. Neutral grounding resistors are normally designed to carry their rated current for a short period (usually 30 s).

171. When there are more than two or three generators or power transformer banks at one station, it is usually desirable to use only one resistor for neutral earthing.

172. With solid-grounded system 3 overload and 1 earth-leakage scheme is preferred over 2 overload and 1 earth-leakage scheme for protection of a feeder.

173. For effective application of counter poise it can be buried into the ground to any depth but the depth should be enough to avoid theft.

174. The induction regulator is essentially a step-up transformer.

175. Quadrature booster injects the voltage lagging the system voltage by 90°.

176. The addition of synchronous compensator in a transmission system causes the decrease in fault current.

177. The addition of SVS in a power system has no effect on fault current.

178. Bulk supply and industrial substations are installed separately for individual consumer.

179. Higher excitation voltage reduces the stability of the system.

180. Shunt compensation of lines is resorted to improve stability of the system. **[A.M.I.E. Sec B. Winter 1993]**

181. From stability point of view it is desirable to ground the neutral of the generator through a reactor and that of the motor through a resistor. **[A.M.I.E. Sec B. Winter 1993]**

182. Series compensation of lines normally should not exceed 60%. **[A.M.I.E. Sec B. Winter 1994]**

183. For better transient stability it is desirable to have machines with larger SCR. **[A.M.I.E. Sec B. Winter 1994]**

184. In comparison to synchronous condensers shunt capacitors have more losses.

185. Initial cost of synchronous condensers in comparison to shunt capacitor is less.

186. Transient stability limit is larger than steady state stability limit. **[Bharathiyar Univ. April 1997]**

187. A power system remains stable as long as a $\frac{\partial P_e}{\partial \delta}$ is less than zero.

$$P_e = \frac{E_1 E_2}{X} \sin \delta$$

[A.M.I.E. Sec B. Advanced Power Systems Summer 1999]

188. The swing equation gives the variation of torque angle δ w.r.t. time t.

189. The higher the operating time of the circuit breaker, the smaller will be the critical clearing angle.

190. In a dynamically stable system, oscillations of synchrounous machines after load changes will increase with time.

191. The synchronising coefficient between two areas of a 2-area power system is $\frac{\partial P}{\partial \delta}$.

192. By decreasing excitation of an alternator, its steady-state limit can be improved.

193. The critical clearing time is related to its transient stability limit.

194. An alternator with induced emf E_g supplies a motor with back emf E_b.

195. The equivalent circuit series reactance is X. The maximum power transferred will be $P_{max} = \frac{E_g E_b}{X}$.

196. The higher the initial load, the lower the critical clearing angle.

197. Maximum power has optimistic value when resistance is neglected.

198. Slack control bus is specified by real power and voltage magnitude.

199. The bus admittance matrix (Y_{bus}) of a power system is a null matrix.

200. In load-flow analysis, the load connected to a bus is represented as constant real and reactive power drawn from the bus.

Answers: **152.** (T), **153.** (T), **154.** (T), **155.** (F), **156.** (F), **157.** (F), **158.** (T), **159.** (T), **160.** (T), **161.** (T), **162.** (F), **163.** (T), **164.** (F), **165.** (F), **166.** (T), **167.** (F), **168.** (T), **169.** (T), **170.** (T), **171.** (T), **172.** (F), **173.** (T), **174.** (F), **175.** (F), **176.** (F), **177.** (T), **178.** (T), **179.** (F), **180.** (F), **181.** (F), **182.** (F), **183.** (T), **184.** (F), **185.** (F), **186.** (F), **187.** (F), **188.** (T), **189.** (F), **190.** (F), **191.** (T), **192.** (F), **193.** (T), **194.** (T), **195.** (T), **196.** (F), **197.** (T), **198.** (F), **199.** (F), **200.** (T).

FILL IN THE BLANKS TYPE QUESTIONS

Note : Fill in the blanks with appropriate word or words.

1. The frequency of occurrence of three-phase fault is _______ than that of line-to-line fault. [A.M.I.E. Sec B. Summer 1994]
2. In power system maximum numbers of fault involve _______ . [A.M.I.E. Sec B. Winter 1995]
3. A 3-phase dead short circuit is _______ severe than a line-to-ground short circuit. [A.M.I.E. Sec B. Summer 1994]
4. The first peak short-circuit current is _______ times the rms value of the periodic component of short-circuit current 2-3 periods after the short-circuit is initiated. [A.M.I.E. Sec B. Winter 1993]
5. For a fault at generator, the fault current is maximum for _______ fault. [Bharathidasan Univ. Nov. 1996]
6. The synchronous reactance of a 200 MVA, 10 kV, 3-phase, 50 Hz generator is 1.0 pu at its own base. Its pu reactance at 100 MVA, 20 kV base will be _______ . [GATE E.E. 1997]
7. The percentage impedance of a 100 kVA, 11 kV/400 V, delta/wye, 50 Hz transformer is 4.5%. For the circulation of half the full-load current during short-circuit test, with low voltage terminals shorted, the applied voltage on the high voltage side will be _______. [GATE E.E. 1995]
8. For limiting the short circuits we use _______ . (resistors/reactors/capacitors)
9. Oil immersed type reactors are used for voltages _______ kV.
10. Zero-sequence component in 3-phase voltage of delta is _______.
11. In the case of balanced three-phase currents, the zero-sequence component of current is _______ . (minimum/maximum)
12. Zero-sequence fault currents are absent when the fault is _______.
13. The negative- and zero-sequence networks contain no _______ but include the impedance of generators to negative and zero-sequence currents respectively. [A.M.I.E. Sec B. Winter 1995]
14. The zero-sequence impedance of high voltage OH transmission lines is about _______ times as large as positive-sequence impedance. [A.M.I.E. Sec B. Winter 1995]
15. Zero-phase sequence component consists of three phasors equal in magnitude and with _______ phase displacement from each other. [A.M.I.E. Sec B. Winter 1995]
16. The positive-sequence component of the voltage at the point of fault in a power system is zero for a _______ fault. [GATE E.E. 1995]
17. Negative-sequence currents are present in all types _______ of faults. [A.M.I.E. Sec B. Winter 1994]
18. For simulation of L-G fault the three sequence networks are to be connected in _______ . [A.M.I.E. Sec B. Winter 1994]
19. For simulation of line-to-line fault, positive- and negative-sequence networks are to be connected in _______ .[A.M.I.E. Sec B. Winter 1993]
20. A fault occurring at the terminals of an unloaded synchronous generator operating at its rated voltage has resulted in the following values of current and voltage.

    $I_{a_0} = j\,2.37$ pu $\qquad I_{a_1} = -j\,3.05$ pu

    $I_{a_2} = j\,0.68$ pu $\qquad V_{a_0} = V_{a_1} = V_{a_2} = 0.237$ pu

    The fault that has ocurred is _______ . [GATE E.E. 1991]
21. When a line-to-ground fault occurs, the current in a faulted phase is 300 A. The zero-sequence current in this case will be _______ amperes.
22. _______ is essentially a small piece of metal inserted in a circuit which melts when a predetermined value of current flows through it. (circuit breaker, fuse) [A.M.I.E. Sec B. Summer 2000]
23. Current rating of a fuse element depends upon the permissible _______ of the contacts of the fuse holder, fuse and upon the deterioration of fuse caused by oxidation.
24. _______ is defined as the minimum value of current at which the fuse element or fuse wire melts.
25. The fusing current depends upon the type of material used, x-sectional area, _______ , size and location of the terminals, diameter of wire, type of enclosure used and type of surface.
26. If two stranded fuse wires are twisted together, then fusing current of this wire will be _______ the sum of fusing current of two wires. (equal to, less than) [A.M.I.E. Sec B. Summer 2000]
27. The ratio of minimum fusing current and the current rating of the fuse element is called the _______ .
28. The value of fusing factor is always _______ than 1. (more, less) [A.M.I.E. Sec B. Summer 2000]
29. Breaking capacity of a fuse is the rating corresponding to the _______ value of ac component of the maximum prospective current and the system voltage. (average/rms) [A.M.I.E. Sec B. Summer 2000]
30. The maximum value to which the fault current reaches before the fuse melts is called the _______ current.
31. HRC fuses are capable of clearing _______ currents. [A.M.I.E. Sec B. Summer 2000]
32. Operating time of in HRC fuse in the event of a fault is _______ than one cycle. (less, more) [A.M.I.E. Sec B. Winter 1999]
33. The operating time of HRC fuse is _______ cycle. [A.M.I.E. Sec B. Winter 1993]
34. HRC fuses provide best protection against _______ .
35. Heat generated in HRC fuses is dissipated mainly through _______. (radiation/conduction/convection)
36. The rating of fuse is expressed in terms of _______ . (amperes/volts/VA)
37. The delays fuses are used for the protection of _______ . (motors/transformers)
38. Arcing contacts of circuit breakers are normally made of _______.
39. Power factor of an arc is _______ .
40. The normal frequency _______ voltage that appears across the circuit breaker poles after final arc extinction has occurred is termed as the recovery voltage. (average, rms) [A.M.I.E. Sec B. Summer 2000]
41. The _______ voltage that appears across the contacts of CB at the instant of arc extinction is called the restriking voltage. (rms, transient) [A.M.I.E. Sec B. Summer 2000]
42. The rate of rise of restriking voltage of a circuit breaker defined by the maximum slope of the voltage curve is called _______ rate of rise of restriking voltage. (average, maximum)
43. If the inductance and capacitance of a power system network up to a circuit breaker location are 1 H and 0.01 μF respectively, the value of the shunt resistor across the circuit breaker, required for critical damping of the restriking voltage is _______ . [GATE 1995]

44. The use of oil in a circuit breaker offers _______ dielectric strength. **[A.M.I.E. Sec B. Summer 1993]**

45. MOCB stands for _______ circuit breaker. **[A.M.I.E. Sec B. Winter 1993]**

46. Minimum oil type circuit breakers use oil for _______ purpose. **[A.M.I.E. Sec B. Summer 1996]**

47. _______ air circuit breakers are designed for use on dc circuits and low voltage ac circuits.

48. The speed of operation of air circuit breakers is _______ than that of oil circuit breakers. **[A.M.I.E. Sec B. Winter 1999]**

49. Air blast circuit breakers and SF_6 circuit breakers are preferred for _______ voltage applications.

50. For long EHV/UHV transmission lines we use _______ circuit breaker. **[A.M.I.E. Sec B. Summer 1996]**

51. Resistance switching is normally restored in case of _______ circuit breakers. **[A.M.I.E. Sec B. Summer 1994]**

52. SF_6 provides better arc quenching because of its _______ property. **[A.M.I.E. Sec B. Winter 1993]**

53. The order of vacuum in vacuum circuit breakers is _______ . **[A.M.I.E. Sec B. Winter 1993]**

54. The rupturing capacity of a circuit breaker is expressed in _______. **[A.M.I.E. Sec B. Winter 1999]**

55. Short-circuit tests are conducted on circuit breakers to prove their _______ .

56. A 3-phase circuit breaker is rated at 200 MVA, 33 kV, its making current will be _______ kamp. **[A.M.I.E. Sec B. Summer 1994]**

57. An oil circuit _______ consists essentially of an oil switch or breaker actuated by relays which make it to open when predetermined current flows through it.

58. The contact resistance of a circuit breaker is about _______ Ω. (20, 2, 20 m, 20 μ)

59. Sparking occurs on switching off the load due to circuit high _______ . (resistance/inductance/capacitance)

60. For modern EHV circuit breakers the operating time between the instant of receiving trip signal and final contact separation is a about _______ second. (0.001/0.003/0.03/0.1)

61. The transient voltage that appears across the contacts at the instant of arc extinction is known as _______ voltage.

62. The normal frequency rms voltage that appears across the circuit breaker pole after final arc extinction has occurred is called the _______ voltage.

63. The recovery voltage will be maximum for power factor of _______ . (0, 0.5, 0.707, 1.0)

64. In a circuit breaker the current that exists at the instant of contact separation is called the _______ current.

65. A fault is more severe from the view point of RRRV if it is a _______ line fault. (short/medium/long)

66. The maximum value of restriking voltage in a circuit breaker in term of system voltage will always be _______ times the normal peak. (1.5, 2, 2.5, 3)

67. Magnetic circuit breaker has _______ trip action.

68. Thermal circuit breaker has _______ trip action.

69. Resistance switching is normally employed in _______ circuit breakers. (all, bulk oil, minimum oil, air-blast)

70. Plain break oil CB can be used for voltage up to _______ kV.

71. Bulk-oil CB can be used for voltage up to _______ kV. (0.4, 1.1, 11, 33, 66)

72. MOCBs use oil for _______ only.

73. LVCBs are used for voltage up to _______ volts. (200/400/600/1100)

74. The medium used for arc extinction in air circuit breaker is _______. (air/compressed air/SF_6)

75. An ABCB is usually used for _______ duty. (short/repeated/intermittent)

76. The pressure of air used in ABCB is around _______ kg/cm^2. (1, 10, 20–30, 100–200)

77. The pressure of SF_6 gas in a circuit breaker is around _______ kg/cm^2. (0.25, 1, 4, 20)

78. In a vacuum circuit breaker, the vacuum is of the order of _______ torr. ($1, 10^{-3}, 10^{-6}, 10^{-12}$)

79. Insulation resistance of hv CBs is more than _______ MΩ. (1, 10, 200/2000)

80. The making current of a CB is normally specified in terms of _______ value. (average/effective/peak)

81. Making capacity of a CB is equal to _______ times symmetrical breaking current. ($1.5, \sqrt{2}, 2.55, \sqrt{3}$)

82. Magnetic blow-out coils are generally used in _______ circuit breaker. (oil, vacuum, air-break)

83. A circuit breaker, under normal conditions should be inspected once in _______ months.

84. For testing hv ac circuit breakers the rated short-circuit current is passed for _______ seconds. (2, 3, 5, 15)

85. For testing a hv circuit breaker for endurance, it is tested for at least _______ opening-closing operations. (20/200/1,000/5,000)

86. Autorecloser improves _______ stability. **[A.M.I.E. Sec B. Summer 1994]**

87. Isolators are never used for _______ short circuit current. **[A.M.I.E. Sec B. Summer 1994]**

88. A _______ isolator has three identical poles, which operate simultaneously.

89. Isolators used in transmission lines are capable of breaking _______ current. (lead/fault/nil/charging)

90. The protective relay performs the function of _______ . **[A.M.I.E. Sec B. Summer 1996]**

91. Relay contacts are normally made up of _______ . **[A.M.I.E. Sec B. Summer 1996]**

92. The operating force is a force which tends to _______ the contacts of the relay. (open, close) **[A.M.I.E. Sec B. Summer 2000]**

93. A relay is said to over reach when it operates at a current which is _______ than its setting. (lower, higher) **[A.M.I.E. Sec B. Summer 2000]**

94. The pick-up current of a relay is usually _______ than the drop-out value. (larger, smaller) **[A.M.I.E. Sec B. Winter 1993]**

95. In definite-time relays, the operating time is _______ of the operating current magnitude. **[A.M.I.E. Sec B. Winter 1999]**

96. The unit of burden of a current transformer is _______ . (amperes, volt-amperes) **[A.M.I.E. Sec B. Winter 1999]**

97. 1 VA relay is _______ sensitive than 3 VA relay. **[A.M.I.E. Sec B. Summer 1996]**

98. _______ protection system can be employed for protection of any equipment or line against the current more than predetermined one.

99. The overcurrent relay operating coil is supplied through _______. **[A.M.I.E. Sec B. Summer 1996]**

100. In _______ relays the time of operation is approximately inversely proportional to the magnitude of the current or other quantity causing operation.

101. By placing a fuse in parallel with an instantaneous of definite time lag relay it can be made as _______ relay.

102. _______ resistor is placed across relay contacts in order to avoid arc jumping across them causing their pitting and burning.

103. The shape of the disc of an induction disc relay is _______ . (circular/spiral/elliptical)

104. The ratio of reset to pick-up current for an induction cup relay is approximately _______ . (0.7/0.99/1.02/1.25)

105. Thermal relays are commonly used in _______ .

106. Buchholz relay is connected between _______ and _______ . **[A.M.I.E. Sec B. Winter 1993]**

107. Buchholz relay is used for the protection of _______ . **[A.M.I.E. Sec B. Summer 1984]**

108. Buchholz relay is used is transformer for _______ fault. (external, internal) **[A.M.I.E. Sec B. Summer 2000]**

109. The ratio of pick-up current to set current of an IDMT induction disc relay is approximately _______ . **[A.M.I.E. Sec B. Winter 1996]**

110. Plug setting multiplier is defined as the ratio of _______ current referred to CT secondary to the relay setting current. (full load, fault) **[A.M.I.E. Sec B. Winter 1999]**

111. The setting of an earth fault relay will be usually _______ than that of phase fault relay. (lower, higher) **[A.M.I.E. Sec B. Winter 1999]**

112. The relay with inverse time characteristic will operate within _______ seconds. (5 – 10, 10 – 25, exceeding 25)

113. In case of electromagnetic relay, coil is normally designed for _______ and _______ amperes.

114. If the time of operation of a relay for unity TMS is 10 seconds, the time of operation for 0.4 TMS will be _______ seconds.

115. An over current relay, having a current setting of 125% is connected to a supply circuit through a CT of ratio 400/5. The pick-up value is _______ A. (5/6.25/10/12.5/15)

116. Undervoltage relays are mostly used for _______ protection.

117. Earth fault relays have _______ current setting than over-current relays. **[A.M.I.E. Sec B. Summer 1996]**

118. If the time of operation for a relay with TMS = 1 and PSM 0.5 is 10 seconds, the time of operation for the same current with TMS = 0.5 and PSM = 1 will be _______ seconds. **[A.M.I.E. Sec B. Winter 1996]**

119. A protective system responding to the ratio of voltage and current is called the _______ protection.

120. The performance of distance relays is governed by _______ . (current, ratio of voltage and current) **[A.M.I.E. Sec B. Summer 2000]**

121. Threshold characteristic of a plain impedance relay in a complex Z-plane is _______ . **[A.M.I.E. Sec B. Summer 1996]**

122. The tendency of an impedance relay to operate at impedances _______ than its setting value is known as over reach. (larger, smaller) **[A.M.I.E. Sec B. Winter 1999]**

123. Reactance relay has a _______ characteristic on R-X plane. **[A.M.I.E. Sec B. Summer 1996]**

124. The distance relay with inherent directional property is known as _______ relay. **[GATE 1995]**

125. Mho relay is normally used for the protection of _______ lines. **[A.M.I.E. Sec B. Winter 1994]**

126. The admittance (mho) relay is inherently _______ . **[A.M.I.E. Sec B. Summer 1996]**

127. The impedance relaying scheme is used for the protection of _______ .

128. In the R-X plane the characteristics of an impedance relay is represented by a _______ .

129. Reactance relays are employed for phase fault in _______ lines. (short/medium/long)

130. For phase fault _______ relay is used.

131. _______ relay is best suited for phase fault relaying for medium transmission lines.

132. _______ relay is used for the protection of EHV/UHV transmission lines.

133. A relay used for protection of motors against overload is _______ relay.

134. In a three-phase system, the minimum number of relays used to detect phase-to-phase fault is _______ . (1/2/3/6)

135. The protection from negative-sequence currents is provided for _______ . (motors/generators/transformers)

136. A relay which has no moving part and performs measurement with solid state circuit is called the _______ relay.

137. A static coincidence circuit is used for _______ comparison. (phase/amplitude)

138. A protective system responding to vector difference of two electrical quantities is called the _______ protection.

139. Relay used in differential protection should operate only for _______ faults. (external, internal) **[A.M.I.E. Sec B. Summer 2000]**

140. The differential protection trips the faulty feeder quickly _______ the necessity of time grading. (without, with) **[A.M.I.E. Sec B. Summer 2000]**

141. Translay scheme is used for the protection of _______ . **[A.M.I.E. Sec B. Summer 1994]**

142. _______ relays do not use moving parts or elements. (electromagnetic, static) **[A.M.I.E. Sec B. Summer 2000]**

143. An underexcited alternator operates at _______ power factor. **[A.M.I.E. Sec B. Winter 1994]**

144. Merz-Price differential relay is used for _______ protection. **[A.M.I.E. Sec B. Summer 1996]**

145. Differential protection scheme is used for alternators of capacity more than _______ MVA. (1/2/5/25/500)

146. The most common type of protection used for stator winding faults in case of alternators is _______ protection.

147. The percentage bias for a generator protection lies between _______ to _______ percent. **[A.M.I.E. Sec B. Winter 1996]**

148. Prime mover failure of an alternator connected to a grid can be detected using _______ relay. (differential/reverse power) **[A.M.I.E. Sec B. Winter 1999]**

149. Field failure of an alternator connected to grid will result is the machine running as _______ . (motor, induction generator) **[A.M.I.E. Sec B. Winter 1999]**

150. During loss of field excitation, the equivalent impedance of the generator drops, and this can be detected by a _______ relay.

151. In a motor starter, thermal relay provides _______ protection as well as single phasing protection.

152. Supply of an unbalanced voltage to a 3-phase, delta connected induction motor can be detected using _______ relay. (negative-sequence, zero-sequence) **[A.M.I.E. Sec B. Winter 1999]**

153. The overload protection is so designed that it closely matches the _______ curve of majority of motors.

154. Operation of an induction motor on undervoltage will cause _______ current.

155. Heating effect on the rotor windings due to negative sequence components is proportional to _______ times ac resistance value where s is the slip.

156. Heating effect on the rotor windings due to positive sequence component is proportional to _______ resistance value. (dc/ac)

157. The frequency of the negative sequence currents in the motor during unbalance is _______ where s is the slip and f is the supply frequency.

158. Phase fault protection is provided by _______ high set instantaneous relay elements.

159. Core-balance leakage protection is used to provide protection against _______ faults.

160. A Buchholz relay has _______ float switches.

161. The way in which transformers are connected in 3-phase circuits does not affect the _______ impedance in the equivalent circuit. **[A.M.I.E. Sec B. Winter 1995]**

162. Magnetising inrush current is a transformer is rich in _______ harmonic component. **[A.M.I.E. Sec B. Winter 1999]**

163. Inter-turn fault in a power transformer can be detected using _______ relay. [differential, Buchholz] **[A.M.I.E. Sec B. Winter 1999]**

164. Magnetising current of transformer contains even harmonics due to _______ . (saruation, hysteresis) **[A.M.I.E. Sec B. Winter 1999]**

165. The differential protection scheme of an alternator will have a _______ bias setting than that of transformer. (larger, smaller) **[A.M.I.E. Sec B. Winter 1999]**

166. To protect a Δ-Y connected power transformer against internal fault, CTs should be connected in _______ . (star-delta delta-star) **[A.M.I.E. Sec B. Summer 2000]**

167. For differential protection of power transformers (delta-delta), the CTs will have _______ connections.

168. The percentage differential protection of a transformer protects the transformer against _______ fault.

169. The connections of CTs are opposite to that of transformer windings so that current in the plot wires of the two are/have _______ phase. (same/opposite)

170. Differential protection scheme is employed for power transformers of rating exceeding _______ . **[A.M.I.E. Sec B. Winter 1993]**

171. Bus-bars are usually made of _______ .

172. The diameter of aluminium pipe for 400 kV bus-bar is _______ mm.

173. All bus-bar protection schemes are designed to respond to _______ only. (overcurrents, ground faults)

174. In a 4 bus system the circuit breakers and the various zones of protection are shown in figure. If the circuit breakers, *E*, *F* and *G* trip, the location of the fault is on _______ .

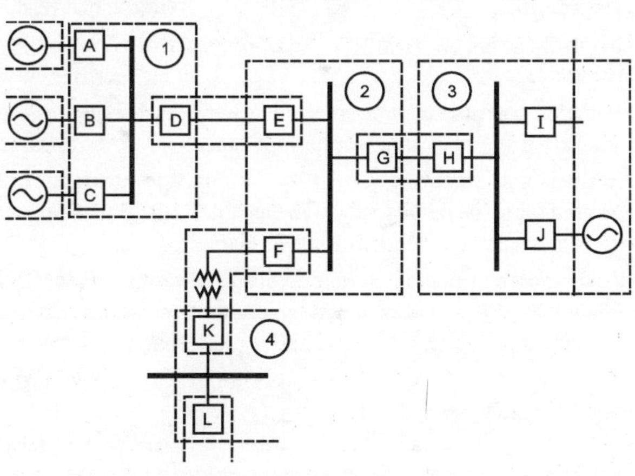

175. Tuned power lines have the property _______ and are used _______ . Tuning is done by _______ . **[A.M.I.E. Sec B. Power Systems Winter 1997]**

176. The protection system for transmission lines having no pilot wires are known as _______ type.

177. When an interconnector between substations is used in a ring main system the relays used on the interconnector must of the _______ type. (directional, non-directional)

178. Translay protection scheme used for protection of feeders is an example of _______ balance differential protection. (voltage, current)

179. Distance relays are adjusted on the basis of the _______ phase sequence impedance between the relay location and the fault location. (negative, positive)

180. _______ protection used for protection of transmission lines is a unit type protection.

181. For 3-phase feeder protection in a distribution network the number of earth fault relays required is _______ . (1/2/3/more)

182. The number of pilot wires required for protecting 3-phase transmission lines using Translay system of protection is _______.

183. Summation transformers are used in _______ relays.

184. Carrier current protection scheme is normally used for protection of _______ transmission lines only.

185. The frequency of carrier in the carrier-current pilot scheme is in the range _______ Hz to _______ Hz. **[A.M.I.E. Sec B. Summer 1996]**

186. The line trap in the case of carrier current pilot scheme consists of _______ . **[A.M.I.E. Sec B. Summer 1996]**

187. The frequency of carrier transmitted by microwave pilot is in the range of _______ Hz to _______ Hz. **[A.M.I.E. Sec B. Summer 1996]**

188. Line traps used in carrier protection schemes are parallel resonant circuits tuned to _______ frequency. (carrier, power) **[A.M.I.E. Sec B. Winter 1999]**

189. Coupling capacitors used in carrier current protection give very _______ impedance to 50 Hz power currents. (low, high) **[A.M.I.E. Sec B. Summer 2000]**

190. Line traps in carrier current protection are _______ resonant circuit. (series, parallel) **[A.M.I.E. Sec B. Summer 2000]**

191. In an OH transmission line, wavetraps are used for the detection of _______ signals. (carrier/fault power)

192. Coupling capacitors used in carrier protection have values usually in the range of _______ to _______ pF.

193. On _______ transmission lines carrier pilot relaying is cheaper and more reliable than pilot-wire relaying. (short, long) **[A.M.I.E. Sec B. Summer 2000]**

194. Time graded overcurrent protection is used in _______ feeders. (radial/ring-main) **[A.M.I.E. Sec B. Winter 1999]**

195. In a _______ pilot-wire scheme, if the pilot circuit gets open-circuited, the relay will trip on full load. (circulating current/balanced voltage)

196. In a circulating current pilot-wire scheme, if pilot circuit gets short-circuited the relay will _______ for internal faults. (trip, fail to trip)

197. First zone distance relay protects a transmission line section up to _______ of its full length. (25/50/800/100%)

198. In distance protection of transmission line, the voltage at the relay point _______ with the increase of distance of fault point from the relay. (increases/decreases)

199. Switching overvoltage arising from energising of uncharged line can be reduced by using high voltage shunt _______ .

200. In horn gaps the arc is produced at the bottom during _______ voltage surge.

201. _______ type lightning arresters are employed for the protection of distribution transformers.

202. The _______ resistor in lightning arrester offers low resistance to the flow of discharge currents and high resistance to power frequency voltage.

203. The impulse ratio of rod gap is around _______ . **[A.M.I.E. Sec B. Winter 1996]**

204. Lightning conductor is used for the protection of _______ .
[A.M.I.E. Sec B. Winter 1993]

205. The insulation requirement of lightning arrester is _______ than that of a power transformer. [A.M.I.E. Sec B. Winter 1993]

206. The lightning discharge current is of the order of _______ . (10 kA, 10 mA)

207. The gap length of rod type surge diverter used for 132 kV line should be _______ . (65 cm, 125 cm)

208. Most of the damaged caused due to lightning discharge depends on _______ of discharge. (magnitude, duration)

209. A lightning stroke is actually _______ flash(es).

210. A lighting stroke rises to its maximum _______ but decays _______.

211. Flash-over voltage of the gap should be as _______ as possible. (low, high)

212. Transmission lines can be protected against direct lightning strokes by providing _______ . (auxiliary discharge path, drainage facility)

213. The coupling factors of earth wire should be _______ . (low, high)

214. The main purpose of providing a lightning arrester in a substation is to protect _______ . (transformers, circuit breakers)

215. Lighting stroke attains its maximum in _______ .

216. The order of energy in a lightning stroke is of _______ units.
[A.M.I.E. Sec B. Winter 1993]

217. In a lightning arrestor, flash-over occurs _______ times the arrestor rating.

218. The velocity of propagation of electromagnetic waves is _______ on overhead lines and on cables it depends on _______ .

219. Time taken for surge to travel a 1,200 km long overhead transmission line is _______ .

220. Ground wire causes _______ in the capacitance between conductor and earth. (decrease, increase)

221. Capacitance of line and equipment _______ the wavefront of the surge. (decreases, increases)

222. BIL is expressed as the impulse crest voltage of a standard wave not larger than _______ μs wave.
[A.M.I.E. Sec B. Advanced Power Systems Winter 1999]

223. For a 220 kV line, the impulse withstand kV crest voltage and one minute power frequency test voltage kV_{rms} are respectively _______. (900, 395/1550, 660)

224. For a 400 kV transmission line the switching overvoltage crest is _______ . (825, 1155)

225. A suitable value of shielding angle for EHV transmission lines is _______ . (20°, 45°)

226. A suitable value of shielding angle for high voltage lines is _______. (30°, 20°)

227. If a wave is represented by a step wave the analytical results are _______ . (optimistic, pessimistic)

228. In a cable with the insulation permittivity 4, the velocity of propagation of electromagnetic waves is _______ $\times 10^{-8}$ m/p. (1.5, 6)

229. A voltage wave + 200/1/50 means a positive wave having a crest of _______ front of _______ ms and tail of _______ ms.

230. For an open-circuited line, the resulting current will be _______ .

231. For a short-circuited line, the resulting voltage will be _______ .

232. Steepness of the travelling waves is attenuated by line _______ . (resistance/inductance/capacitance)

233. For an open-circuited transmission line, the reflection coefficient is _______ .

234. The real part of propagation constant of a transmission line is _______ constant. (attenuation/phase)

235. Two transmission lines each having an impedance of 200 Ω is separated by a cable like, for zero reflection, the impedance of the the cable should be _______ Ω. (100/200/400/600)

236. The earth wire should not be of size smaller than _______ SWG copper. (4/6/8/10)

237. A system is said to be effectively grounded if $\frac{X_0}{X_1}$ is _______ and $\frac{R_0}{X_1}$ is _______ .

238. The arcing ground phenomenon is commonly experienced with _______ neutral systems.

239. Peterson coil has _______ reactance.

240. Grounding by Peterson coil _______ arcing ground.
[A.M.I.E. Sec B. Summer 1994]

241. Peterson coil is used for _______ grounding.
[A.M.I.E. Sec B. Summer 1994]

242. Tower footing resistance should be _______ (low, high)

243. The footing resistance of a tower depends on _______ . (soil conditions, material of tower)

244. Top most wire of a transmission line is _______ wire. (ground, neutral)

245. Grounding is done generally at _______ end. (supply/load)

246. Solid earthing is done for voltage below _______ kV. (11/33/66/0.4)

247. Grounding transformer in a system is usually connected in _______. (delta-delta/star-delta/delta-star/star-star)

248. Arcing on transmission lines is prevented by connecting a/an _______ in the neutral. (resistor/inductor/capacitor)

249. During arcing ground conditions, the phase voltage of the system rises to _______ times its normal value. ($\sqrt{3}, 5-6, 15, 20$)

250. Electromechanical voltage regulators are used in _______ .

251. The voltage regulator having dead zero effect is _______ .

252. The electromechanical type of voltage regulators are of _______ type.

253. The magnetic amplifier type voltage regulators affect the VR by varying the _______ . (resistance/reactance)

254. In tap changing transformers, the tappings are provide on _______ winding. (hv//lv/tertiary)

255. For effective use of a counterpoise wire, its leakage resistance should be _______ than surge impedance. (less, more)

256. In an isolated neutral system, on occurrence of single line-to-ground fault, the persistency of the arc due to the flow of capacitive current give rise to a condition called the _______ .

257. To restrict the value of circulating third-harmonic current flow in a generator _______ are used in the neutral grounding connections of generators, from which the harmonics emanate.

258. The long and high voltage lines arching earths are caused if the power system neutral is kept _______ .
[A.M.I.E. Sec B. Advanced Power Systems Winter 1999]

259. Booster transformers are required for _______ .
[A.M.I.E. Sec B. Power Systems Winter 1997]

260. Induction regulators are used for voltage control in _______ system. (transmission/distribution)

261. The isolators used on both sides of a breaker are for _______ (isolating the breaker, backup protection of the breaker)
[A.M.I.E. Sec B. Winter 1999]

262. Shunt compensation in an EHV line is used to _______ .

263. Step-up substations are associated with _______ substations.

264. For cost and safety the outdoor substations are employed for voltages ______ .

265. Shunt capacitors in a substations ______ lagging VARs. (delivers/consumes)

266. A device used for connecting a conductor to the earth is known as ______ switch. (earthing/limiting/isolator)

267. Earthting switch is usually installed on ______ .

268. The angular displacement between two interconnected power stations is mainly due to ______ of the interconnector. (resistance/reactance)

269. The transfer of power between two stations is maximum when the phase displacement between the voltages of the two stations is ______ . (0°/90°/120°/180°)

270. The steady-state stability of the power system can be increased by connecting lines in ______ . (series/parallel)

271. The inertia of two groups of machines, which swing together are M_1 and M_2. The inertia constant of the system is ______ .

272. The inertia constant H of a machine of 200 MVA is 2 pu, its value corresponding to 400 MVA is ______ .

[A.M.I.E. Sec B. Summer 1994]

273. The transient stability of a transmission systems may be better improved by using fast acting ______ . (circuit breaker, voltage regulator) **[A.M.I.E. Sec B. Winter 1996]**

274. Automatic fast acting voltage regulators for synchronous generators ______ the dynamic stability of such machines.

[A.M.I.E. Sec B. Advanced Power Systems Winter 1995]

275. The relative velocity between the rotor axis and the stator magnetic axis for transient stability is ______ .

[A.M.I.E. Sec B. Advanced Power Systems Winter 1994]

276. As compared to the stability of a cylindrical rotor alternator, the stability of a salient pole alternator is ______ .

277. For stable operation of an interconnected system, the passive element used as an interconnecting element is ______ . (reactor, capacitor)

278. Maximum power condition of $X = \sqrt{3}R$ for an under ground system is obtained by inserting ______ . (series reactors, shunt capacitors)

279. The addition of a synchronous compensator in the power system network ______ the stability of the system. (improves, reduces)

280. Series compensation ______ stability limit. (increases, decreases)

281. A series compensated transmission line has better ______ stability. (transient, steady-state)

282. Slack bus is a ______ bus. (load/feeder/generator)

283. In load flow studies of a power system, the quantities specified at a voltage controlled bus are ______ . **[GATE E.E. 1994]**

284. In load-flow analysis, a voltage controlled bus is treated as a load bus in subsequent iteration when its ______ limit is violated.

[GATE E.E. 1995]

285. The diagonal elements of a nodal admittance matrix are strengthened by adding ______ . (shunt inductances, shunt capacitor)

286. For accurate load-flow calculations on large power systems ______ method is the best one. (GS/NR/decoupled)

287. For fast load-flow calculations, the best method is ______ method. (GS/NR/decoupled)

288. An acceleration factor is used in ______ method of load-flow studies. (GS/NR/decoupled)

ANSWERS

1. lower	**2.** earth	**3.** more	**4.** $1.8\sqrt{2}$
5. L-G	**6.** 0.125	**7.** 247.5 V	**8.** reactors
9. above 33	**10.** zero	**11.** minimum	**12.** line-to-line
13. internal voltage	**14.** 2-4	**15.** 0°	**16.** three-phase
17. unsymmetrical	**18.** series	**19.** opposition	**20.** L-L-G fault
21. 100	**22.** Fuse	**23.** temperature rise	**24.** Fusing current
25. length	**26.** less than	**27.** fusing factor	**28.** more
29. rms	**30.** cut-off	**31.** high as well as low	**32.** less
33. one-fourth of a	**34.** short circuit	**35.** radiation	**36.** amperes
37. motors	**38.** copper tungsten alloy	**39.** unity	**40.** rms
41. transient	**42.** average	**43.** 5 kΩ	**44.** high
45. minimum-oil	**46.** arc quenching	**47.** low-voltage	**48.** less
49. extra high	**50.** SF_6	**51.** air-blast	**52.** electronegativity
53. 10^{-5} to 10^{-6} torr	**54.** kilo amperes	**55.** ratings	**56.** 89
57. reclosure	**58.** 20 μ	**59.** inductance	**60.** 0.03
61. restriking	**62.** recovery	**63.** zero	**64.** breaking
65. short	**66.** 2	**67.** instantaneous	**68.** delayed
69. air-blast	**70.** 11	**71.** 66	**72.** arc quenching
73. 400	**74.** air	**75.** repeated	**76.** 20-30
77. 4	**78.** 10^{-6}	**79.** 200	**80.** peak
81. 2.55	**82.** air-break	**83.** 3 or 6	**84.** 3
85. 1,000	**86.** transient	**87.** making or breaking	**88.** triple-pole

89. nil
90. watch and ward
91. silver
92. close
93. lower
94. larger
95. independent
96. volt-amperes
97. more
98. overload
99. current transformer
100. Inverse-time lag
101. inverse time lag
102. discharge
103. circular
104. 0.99
105. motor starters
106. main oil tank, conservator
107. power transformers
108. internal
109. unity
110. fault
111. lower
112. 5-10
113. 1, 5
114. 4
115. 6.25
116. motor
117. lower
118. 2.5
119. distance
120. ratio of voltage and current
121. circle with centre at origin
122. larger
123. straight line parallel to R-axis
124. admittance
125. long
126. directional relay
127. transmission line
128. circle
129. short
130. distance
131. Impedance
132. Mho
133. thermal
134. 2
135. generators
136. static
137. phase
138. differential
139. internal
140. without
141. feeders
142. static
143. leading
144. generator
145. 1
146. Merz-Price
147. 5; 10
148. reverse-power
149. induction generator
150. mho
151. overload
152. negative sequence
153. heating
154. over-
155. (2-s)
156. dc
157. (2-s) *f*
158. two
159. ground
160. two
161. per unit
162. second
163. differential
164. saturation
165. smaller
166. star-delta
167. star-star
168. internal
169. same
170. 2 MVA
171. aluminium
172. 100
173. ground faults
174. bus 2
175. of adjustable inductance; in power carrier system; connecting variable inductance
176. non-unit type
177. non-directional
178. voltage
179. positive
180. Carrier current
181. 1
182. 2
183. pilot-wire
184. HV
185. 50 k; 500 k
186. a parallel tuned L-C circuit
187. 450×10^6; $10{,}000 \times 10^6$
188. carrier
189. high
190. parallel
191. carrier
192. 1,000, 5,000
193. long
194. radial
195. circulating current
196. fail to trip
197. 80%
198. increases
199. reactors
200. high
201. secondary
202. nonlinear
203. 1.3
204. tall buildings
205. lesser
206. 10 kA
207. 65 cm
208. duration
209. number of
210. quickly; slowly
211. low
212. auxiliary discharge path
213. high
214. transformers
215. 1-10 ms
216. 250
217. 3 to 4.5
218. 3×10^8 m/s; relative permittivity of dielectric
219. 4 ms
220. increase
221. decreases
222. 1.2×50
223. 900, 395
224. 1155
225. 20°
226. 30°
227. pessimistic
228. 1.5
229. 200 kV; 1, 50
230. zero
231. zero
232. resistance
233. 1
234. attenuation
235. 200
236. 10
237. not greater than 3; not greater than 1
238. insulated
239. inductive
240. eliminates
241. resonant
242. low
243. soil conditions
244. ground
245. supply
246. 33
247. star/delta
248. inductor
249. 5-6
250. alternators
251. electromagnetic
252. rheostatic or vibrating
253. reactance
254. hv
255. less
256. arcing ground
257. harmonic suppressor
258. isolated
259. injecting the voltage in the line
260. distribution
261. isolating the breaker
262. improve the voltage profile
263. generating
264. 33 kV and above
265. delivers
266. earthing
267. isolator frame
268. reactance
269. 90°
270. parallel
271. $\dfrac{M_1M_2}{M_1 + M_2}$
272. one pu
273. circuit breaker
274. improves
275. zero
276. better
277. reactor
278. series reactors
279. improves
280. increases
281. transient
282. generator
283. *P* and |*V*|
284. reactive power
285. shunt inductances
286. NR
287. decoupled
288. GS

VOLUME-II

UTILIZATION OF ELECTRIC POWER AND ELECTRIC TRACTION

[ELECTRIC DRIVES AND INDUSTRIAL APPLICATIONS AND ELECTRIC POWER UTILIZATION]

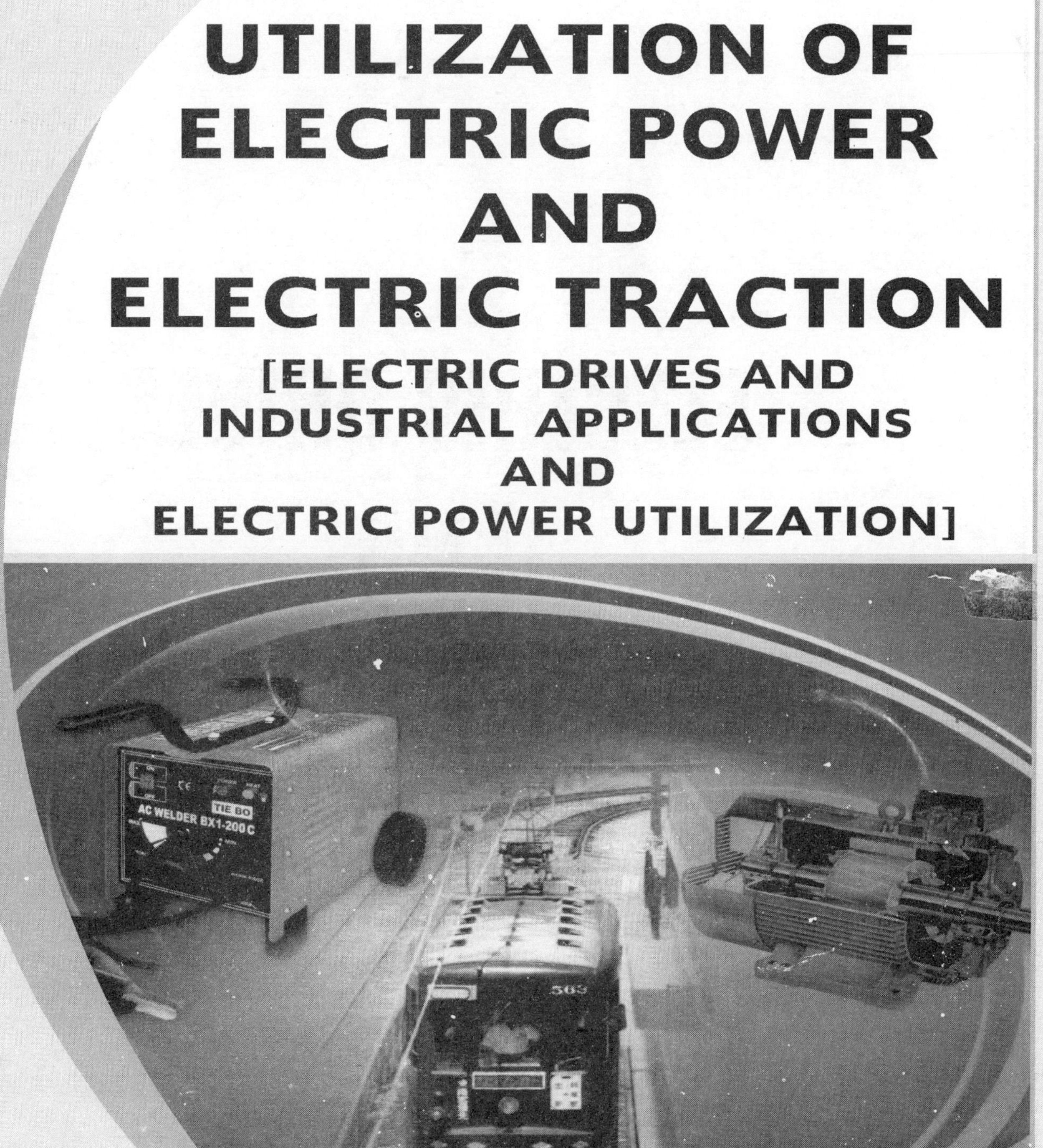

SECTION - A

ELECTRIC DRIVES AND INDUSTRIAL APPLICATIONS

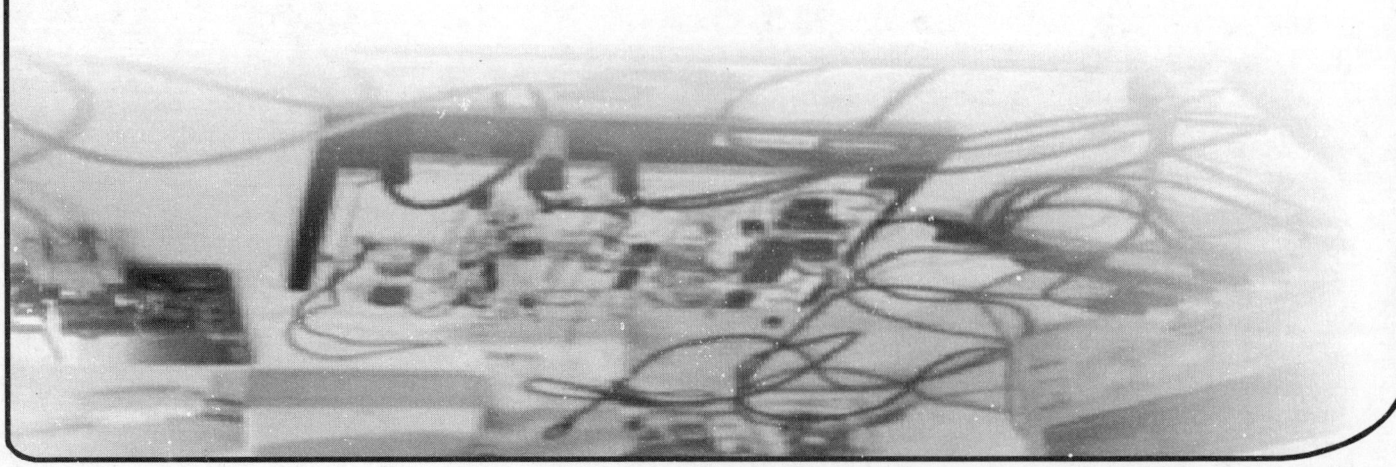

CHAPTER 1

Electric Drives

INSIDE THIS CHAPTER

1.1. INTRODUCTION

Every well developed machine may be considered to consist of three important different parts : the prime mover (engine or motor), the transmission system and the working machine proper. The function of the first two parts, the prime mover with its system of control, and the transmission system, which may comprise a set of shafts, pulleys, belts, gears etc., is to impart motion to and operate the working machine. Because of their function of maintaining the working machine in motion the first two parts (prime mover and the transmission system) are embraced by the common term *drive*. The simplest example of a drive is a hand-powered device. Devices of this class are still used in the form of household appliances and some types of agricultural implements, though industrially they are of no importance today. Horse-powered drive which at one time replaced hand-powered drive by using animals for providing motive power also has no industrial importance today. These two means of motive power were eventually replaced by mechanical drives powered by windmills, waterwheels and turbines, steam engines and internal combustion engines.

Nowadays modern water turbines and steam turbines are extensively used in electric power stations generating large amounts of electrical energy for centralised supply of industry, electrified traction facilities, agriculture and domestic needs. In utilising this source of energy, the principal means of driving all kinds of working machines is the electric motor. This makes it the predominate type of drive in modern conditions, the term *electric drive* being applied to it.

An electric drive is defined as a form of machine equipment designed to convert electrical energy into mechanical energy and provide electrical control of this process.

The basic elements of an electric drive are the electric motor, the transmission, and the electrical control system. Electric drives often include conversion equipment (frequency changers, rectifier units, motor-generator sets etc.) to attain smoother and more versatile control and make the motor operate on specific speed-torque characteristic.

Electric drive is becoming more and more popular for its simplicity, cleanliness, easy and smooth control, flexibility in layout, compactness, reliability, long life, easy starting, low cost (initial as well as maintenance cost), facility for remote control etc. Modern electric drive with a system of automatic control fulfills several functions other than of imparting motion to the working machine so as to carry out the production processes involved in the most efficient manner. Using automatically controlled electric drives, machinery and production units may be operated with high productivity, ensuring at the same time high quality. Sometimes electric drive is employed although the prime mover is already existing there, such as in diesel electric traction system and for ship propulsion. In diesel electric traction system the electricity is used for driving the locomotive. If the diesel engine had to drive the locomotive directly, the operation would have required a complex gear system, since the diesel engine provides best performance only at one speed. But in diesel electric traction the desired output can be had by controlling the dc motors and running the diesel engine-dc generator set at a constant speed, all the time. In electric ship propulsion the prime mover (steam turbine or diesel engine) drives an ac or dc generator, which supplies ac or dc motors mounted on the propeller shafts. Thus the power which used to be transmitted by the shafts previously is now carried through a cable coming from the engine room and use of long shafts going from engine room to the propellers is avoided.

Its main disadvantage is non-availability of drive on failure of electrical power supply. It also cannot be employed in distant places where electric power supply is not available.

1.2. FACTORS GOVERNING SELECTION OF ELECTRIC MOTORS

Parallel to the growth of machine tools, an equally rapid progress has been made in the field of electric motors. The electric motors of various designs are available to meet the needs of various classes of industry. So the selection of electric motor itself has become an important and tedious process. The conditions under which an electric motor has to operate and the type of load it has to handle, determine its selection. The various factors that are to be considered in the selection of an electric motor for a particular service are :

1. Nature of Electric Supply.
2. Types of Drives.
3. Nature of Load.
4. Electrical Characteristics.
   (*a*) Operating or Running Characteristics.
   (*b*) Starting Characteristics.
   (*c*) Speed Control.
   (*d*) Braking Characteristics.
5. Mechanical Considerations
   (*a*) Types of Enclosures.
   (*b*) Types of Bearings.
   (*c*) Types of Transmission For Drive.
   (*d*) Noise Level.
   (*e*) Heating and Cooling Time Constants of the Motor.
6. Service Capacity and Rating.
   (*a*) Requirements For Continuous, Intermittent or Variable Load Cycle.
   (*b*) Pull-Out Torque and Overload Capacity.
7. Appearance.
8. Cost Considerations.
   (*a*) Capital or Initial Cost.
   (*b*) Running or Operating Cost–power factor, losses, maintenance and depreciation etc.

From the above it is seen that a large number of factors are to be considered in making the choice of an electric motor for a given drive. Although in the list of various considerations the cost has been placed at the end, yet the final choice has to be governed by it. The motor selected must fulfil all the necessary load requirements and at the same time it should not be very costly if it has to be a commercial success. In any particular case some of the factors may be definitely fixed by existing circumstances while others will probably be conflicting and, therefore, for making most satisfactory choice understanding of various possibilities is necessary. The basic problem is of matching the mechanical output of the electric motor with the load requirements, *i.e.,* to provide a motor with the correct speed-torque characteristics as demanded by the load.

The factors mentioned above and their effects on the various types of electric motors available will be individually discussed in this chapter, so that the reader may understand the importance of each.

1.3. NATURE OF ELECTRIC SUPPLY

The electric supply available may be 3-phase ac, single-phase ac or dc.

In case three-phase ac supply is available, polyphase induction motor, squirrel cage type for small ratings and slip-ring type for higher ratings may be used, provided this suits the requirements of the load. In the cases where speed variation is required these cannot be conveniently used, so pole changing motors or motors with stepped pulleys may be used. Where accurate control of speed is required, Schrage motors may be used. Use of single-phase motors is limited to small loads only owing to their limited outputs.

DC motors are not used so widely as ac ones. There are several reasons for this, the most important of which are given below :

1. At present a large percentage (in fact, the whole) of the electrical energy used for domestic and commercial purposes is generated in ac form because of economic and technical reasons, *i.e.,* supply available is usually ac. Additional equipment is, therefore, required for converting existing ac supply into dc supply.
2. DC motors have commutators that are subject to trouble resulting from sparking, brush wear, arc-over and the presence of moisture and destructive fumes in the surrounding air.
3. DC motors are generally more expensive than ac motors for similar working conditions.

On contrary, ac motors have the advantages of possessing few working parts, requiring less maintenance and replacement of spares and providing uninterrupted long service life. Three-phase induction motors have high efficiency and provide high starting and uniform operating torques and balanced loading of supply.

DC motors are nevertheless used in a large number of applications, specially when their excellent torque and speed operating characteristics cannot be duplicated by ac motors. Moreover, the speed control that can be provided for dc motors is far more flexible and satisfactory than that which can be provided for ac motors. These requirements are particularly significant in connection with services such as operation of traction equipment, hoists and elevators. In some cases, such as in electric excavators, steel mills and cranes, the speed control is so important that existing ac supply is converted into dc supply in order to employ dc motors.

With electrical utilities transmitting energy on ac, the usual arrangement used to be a motor-generator set (a dc generator driven by an ac motor) supplying power to a dc motor. However, the advent of solid state diodes and silicon controlled rectifiers (SCRs) eliminated the need for the motor-generator set and made dc motor drives more versatile and more popular than even before.

Nature of dc supply available, to some extent affects the performance, and therefore, size of the motor required. DC supply available from a dc generator will be more or less smooth but the output voltage from thyristor converter consists of a dc component and ac harmonic components. Torque or mechanical power is developed by the dc component of the current where as armature heating is developed by the effective or rms value of current. The form factor for half-wave three-phase thyristors may be taken as 1.2 while for full-wave three-phase thyristors it is 1.1. This increases electrical losses, and, therefore, heating is 5 to 7% more for three-phase full-bridge converters while for three-phase half-bridge convertors it is from 15 to 20%. Thus when a dc motor is driven from thyristor convertor, slightly large size motor will be required for a given kW power requirement.

The commercial ac power is generated at a frequency of 50 Hz but this frequency is not capable of meeting the requirement of certain industries, defence services and railway signalling. As such, there has been recent trend towards development of high frequency power generation. Relative merits and demerits of high-frequency and power-frequency equipment are enumerated below.

Merits of High Frequency. 1. The output of the equipment increases with the increase in frequency or speed, as obvious from the following equations

The output of an alternator per phase,

$$P = 4K_f K_p K_d \phi f T \times I \cos \phi \qquad ...(1.1)$$

where K_f, K_p and K_d are the form factor, pitch factor and distribution factor, ϕ is the useful flux per pole in webers, f is the frequency in hertz, T is the number of turns in series per phase, I is the output current per phase and cos ϕ is the power factor.

The output of a motor is given as

$$P = KD^2LN \qquad ...(1.2)$$

where D is diameter, L is the length and N is the speed of motor in rpm.

For a given power output, high frequency machine may be of one-eighth of the weight of the 50 Hz machine. Because of high power-weight ratio, relative cost of the equipment is reduced with the increase in frequency. Reduction in weight of equipment makes also transportation and handling of equipment easy and convenient at high altitude for defence services.

2. For the generation of high frequency power, along with the small sized generator, the prime mover will also be of smaller size. This will result in reduction of high noise problem.

3. High frequency equipment requires less maintenance and has low operating cost, higher efficiency and long life span requiring minimum repairing cost.

4. More reliable service is provided, in arduous conditions of high altitude, low temperature and high humidity, by high frequency equipment.

The *drawback of high frequency equipment* is that it has a problem of increased power losses due to hysteresis and eddy currents. However, this problem of large power losses can be overcome by using low loss materials, proper flux density and adequate cooling arrangements.

1.4. TYPES OF DRIVES

The various types of electric drives used in industry may be divided into three types : *the group, individual* and *multi-motor* drives.

Group Drive. By group drive is meant a drive in which a single electric motor drives a line shaft by means of which an entire group of working machines may be operated. It is also sometimes called the *line shaft drive.* The line shaft is fitted with multi-stepped pulleys and belts that connect these pulleys and the shafts of the driven machines and serve to vary their speed. This drive is economical in consideration of the first cost of the motors and control gear. A single motor of large capacity costs less than the total cost of a number of small motors of the same total capacity viz a single 100 kW motor costs much less than that of ten motors of 10 kW each. Since all the machines may not be used on full load at the same time, therefore, the kW rating of the motor for group drive can often be less than the aggregate of kW output ratings of all the individual motors, and further causes reduction in cost. The efficiency and power factor of a large group drive motor will be higher, provided it is operated fairly near its rated load. If the machines are liable to short but sharp overloads, group drive is again advantageous, because 100% overload on an individual machine will cause hardly 10% overload when being driven by group drive.

Owing to its following drawbacks and objectionable features, this form of drive has become obsolete nowadays and the modern trend is to employ individual and multimotor drives:

1. In group drive, speed control of individual machine is very cumbersome using stepped pulleys, belts etc.
2. Owing to use of line shafting, pulleys and belts group drive does not provide good appearance and is also less safe to operate.
3. In group drive since machines have to be installed to suit the layout of the line shafting, as such flexibility of layout of the various machines is lost. Also, it is not possible to instal any machine at a distant place.
4. The possibility of installation of additional machines in an existing industry is limited.
5. If at a certain instance all the machines are not in operation, the motor will operate at low capacity and, therefore, operation efficiency will be low.
6. The breakdown of large single motor causes all the operations to be stopped. In some processes, where stoppage of one operation causes stoppage of whole sequence such as in textile industry or flour mills, this drawback is not effective and group drive, therefore, can be used.
7. Considerable power loss takes place in the energy transmitting mechanism.
8. The level of the noise produced at the work site is quite large.

Group drive is adopted, when existing factories are changed over from engine drive to electric motor drive simply by replacing the oil or steam engine by an electric motor of corresponding output retaining all the old shafts and belts.

Individual Drive. In an individual drive, a single electric motor is used to drive one individual machine. Example of such a drive are single-spindle drilling machines, various types of electrical hand tools and simple types of metal working machine tools and mechanisms. Though it costs more than group drive, but each operator has complete control of his machine, which enables him to vary its speed if necessary and stop while not is use, thus eliminating no-load losses. The machines can be placed in any desired position and can be moved very easily. The motor and its control unit can be built as an integral part of the machine which results in good appearance, cleanliness and safety. For driving heavy machines such as for lifts, cranes, shapers, lathes etc. and for the purposes where constancy of speed and flexibility of control is required, such as in paper mills and textile industry, individual drive is essential.

For new factories individual drive is preferred, as it causes some saving in the cost of superstructure because of being much lighter and less expensive.

In case of individual drive too, the energy is transmitted to the different parts of the same mechanism by means of mechanical parts (such as gears, pulleys etc.). Thus some power loss occurs in the energy transmission mechanism. This drawback is overcome in the case of *multimotor drives.*

Multimotor Drive consists of several individual drives each of which serves to operate one of many working members or mechanisms in some production unit. For example, in travelling cranes, there are three motors : one for hoisting, another for long travel motion and the third for cross travel motors. Such drive is essential in complicated metal-cutting machine tools, paper making machines, rolling mills, rotary printing machines and similar types of other machinery. The use of multimotor drive is continuously expanding in modern industry as their advantages outweigh the increase in capital cost as compared to the group drives.

The use of individual and multimotor drives has enabled introduction of automation in production processes, which in turn has considerably increased the productivity of various industrial organisations. Complete or partial automation helps to operate various mechanisms at optimum conditions and to increase reliability and safety of operations.

Example 1.1. Compare the total annual cost of a group drive with a motor costing Rs. 60,000 with that of ten individual motors, each costing Rs. 18,750. With group drive annual consumption is 75,000 kWh. With separate drives annual consumption is 60,000 kWh. Electrical energy costs Rs. 4.00 per kWh. Depreciation, maintenance and other fixed charges amount to 12% in the case of group drive and 15% in the other case.

Discuss possible reasons for which you may decide in favour of the drive with a higher capital cost.

Solution : Capital cost on group drive = Rs. 60,000

Annual depreciation, maintenance and fixed charges

$$= \text{Rs. } 60{,}000 \times \frac{12}{100} = \text{Rs. } 7{,}200$$

Energy consumption per annum = 75,000 kWh

Annual energy charges = Rs. 4.00 × 75,000 = Rs. 300,000

Total annual cost in case of group drive

= Rs. 7,200 + Rs. 300,000 = Rs. 307,200

Capital cost on individual drive

= Rs. 18,750 × 10 = Rs. 187,500

Annual depreciation, maintenance and fixed charges

$$= \text{Rs. } 187{,}500 \times \frac{15}{100} = \text{Rs. } 28{,}125.00$$

Annual energy consumption = 60,000 kWh

Annual energy charges = Rs. 4.00 × 60,000 = Rs. 240,000

Total annual cost in case of individual drive

= Rs. 28,125 + Rs. 240,000 = Rs. 268,125

Though the capital cost is higher in case of individual drive, but because of lower annual cost and inherent advantages of individual drive over group drive such as flexibility of control and convenience it is preferred.

1.5. TYPES OF LOADS

There are two types of loads—those which provide active torques and those which provide passive torques. Active torques are due to either gravitational force (as in case of hoists, lifts or elevators and railway locomotive operating on gradients) or deformation in elastic bodies. Such torques are also developed during compression or release of springs. Since functioning of hoisting mechanism, operation of locomotives on gradients and compression/release of springs are all associated with a change in potential energy of the drive, active torques are also closely connected to the potential energy. In upward movement of load/compression of a spring, the stored potential energy increases and the active torque developed opposes the action that takes place *i.e.,* the torque is acted against the upward movement/compression. On the other hand in the downward movement of load/release of a spring the stored potential energy decreases and torque associated with it helps the action. Thus, it can be seen that the active torque continues to act in the same direction even after the direction of the drive has been reversed.

Passive torques are those due to friction or due to shear and deformation in inelastic bodies (lathes, pumps, fans etc.). They always oppose the motion, retarding the rotation of the driven machine. Moreover, with change in direction of motion, the sense of torque also changes. For example, when a weight is raised upward, the friction torque adds to the useful torque but when lowered down it subtracts from the latter.

1.6. SINGLE QUADRANT OPERATION OF ELECTRIC MOTORS

Single quadrant operation means that load and the motor shall be required to run only in one specified direction under the action of developed torque.

When an electric motor is switched to electric supply mains and drives mechanical devices such as line shaft and machine tools, the machine is said to be operating in forward motoring mode or simply motoring mode. In this mode of operation armature current flows in opposition to the emf induced in the armature. The direction of emf induced in the armature is in direct opposition to the applied voltage. Electromagnetic torque developed is in the direction of armature rotation.

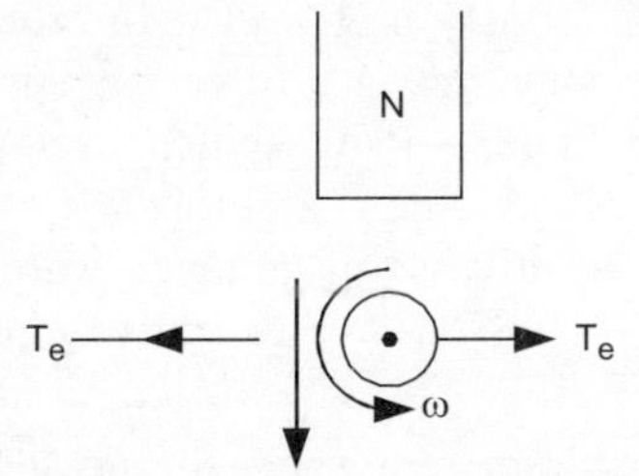

Fig. 1.1. *Single Quadrant Operation of Motor*

1.7. FOUR QUADRANT OPERATION OF ELECTRIC MOTORS

In view of the fact that both active and passive load torques can be present in general, in a drive system, the motor driving the load may operate in different regimes—not only as a motor, but for specific periods, also as a generator and as a brake. Further, in many applications, the motor may be required to run in both directions. Therefore, in sketching the speed-torque characteristics of either the load or the motor, it is preferable to make use of all four quadrants of the speed-torque plane for plotting, rather than to confine, the characteristics to the first quadrant alone.

The conventions used for positive and negative values of speed, motor torque and load torque in a diagram of this type are as follows :

The speed is assumed to have a positive sign, if the direction of rotation is counter-clockwise or is in such a way to cause an 'upward' or 'forward' motion of the drive. In case of reversible drives the positive sign for speed may have to be assigned arbitrarily either to counter-clockwise or clockwise direction of rotation. The motor torque is taken to be positive when it causes an increase in speed in the positive sense. The load torque is assigned a positive sign when it acts against the motor torque.

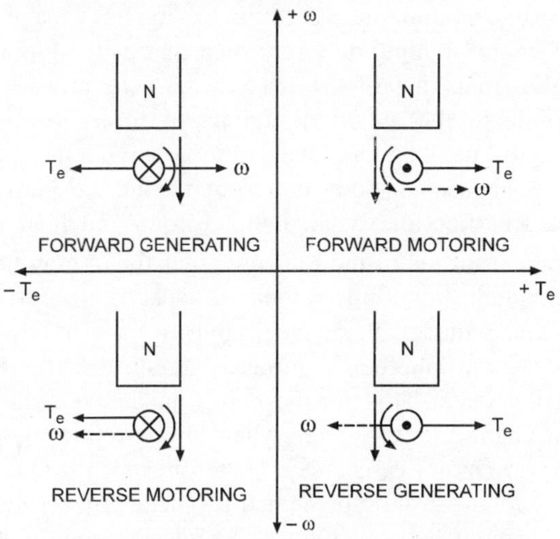

Fig. 1.2. *Four Quadrants of Operation of Motor*

Figure 1.2 illustrates the four different modes of operation of an electric machine. Although the diagram indicates that the field polarity is maintained and the armature current is reversed to obtain negative torque, the same effect is obtained by reversing the field polarity and maintaining the armature

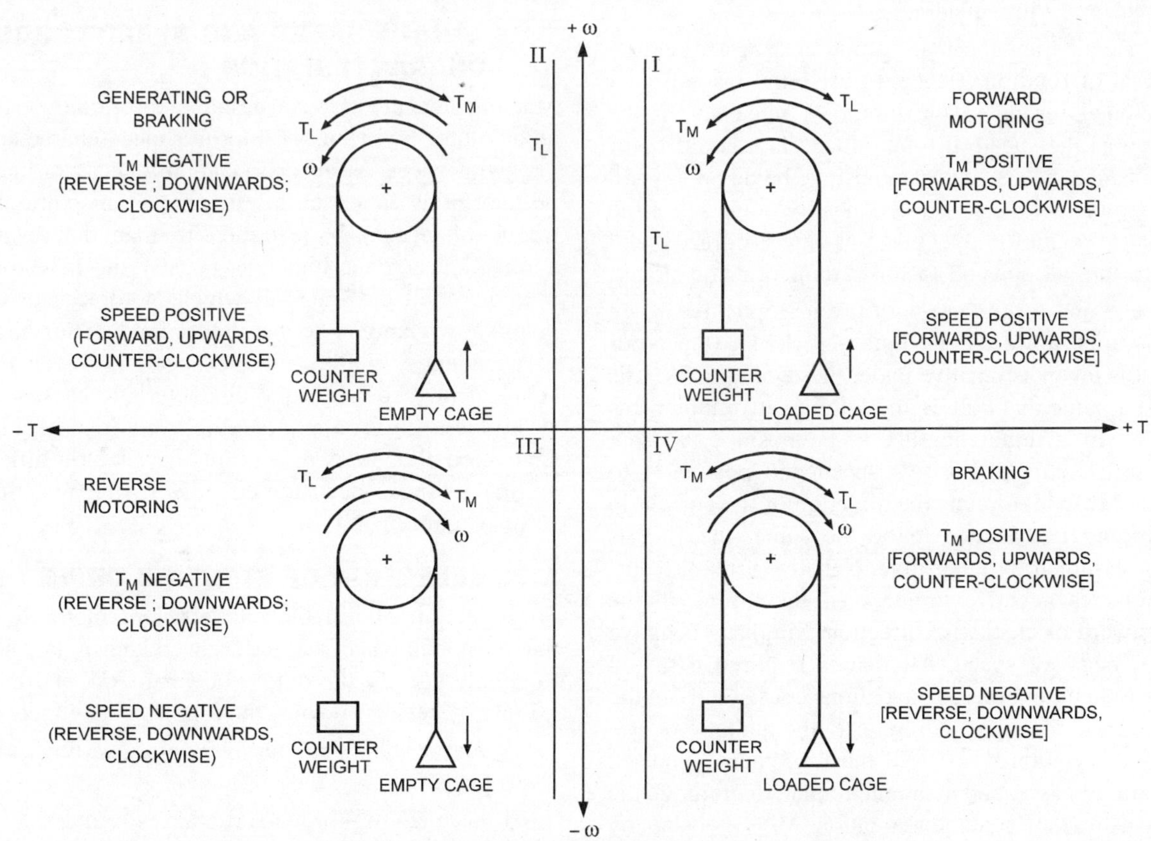

Fig. 1.3. *Four Quadrant Operation of a Motor Driving a Hoist Load*

current direction. Field reversal is necessary with some forms of rectifier control. These four quadrants of operation are feasible for any dc or ac rotating machine but the dc machine is much freer to transfer its operation between quadrants and operates satisfactorily at any point within the envelope.

Figure 1.3 illustrates the four quadrant operation of a motor driving a hoist consisting of a cage with or without load, a rope wound onto a drum to hoist the cage and a balance weight magnitude larger than that of the empty cage but smaller than that of the loaded one. The arrows in this figure indicate the actual directions of motor (or electromagnetic) torque, load torque and motion in the four quadrants. It can be easily seen that they correspond to the sign conventions mentioned earlier for speed, motor torque and load torque.

The load torque of the hoisting mechanism may be taken to be constant (*i.e.,* independent of speed); forces due to friction and windage being negligible in case of low speed hoists and the torque being primarily due to the gravitational pull on the cage. This torque being an active load torque does not change its sign even when the direction of rotation of the driving motor is reversed. Therefore speed-torque curves of a hoist load can be represented by means of vertical lines passing through two quadrants. The speed-torque characteristic of a loaded hoist is shown in Fig. 1.3 by means of the vertical line passing the first and fourth quadrants. Since the counterweight is assumed to be heavier than the empty cage, the inherent tendency of the load viz the empty cage is to move in an opposite direction to that of load presented by the loaded cage and hence the speed-torque curve of the unloaded hoist is represented by the vertical line passing through second and third quadrants.

In the first quadrant the load torque acts in a direction opposite to that of rotation. Hence, to drive the loaded hoist up, the developed torque in the motor M must act in the same direction as the speed of rotation *i.e.,* T_M should be of positive sign. Since the speed is also positive being an upward motion, the power will also have a positive sign, *i.e.,* the drive is said to be motoring. Quadrant I is arbitrarily and conventionally, thus, designated as 'forward motoring quadrant'.

Since speed and torque are both taken as positive when motoring forwards, upwards or counter-clockwise, the product of E and I is taken as positive under these conditions. It is convenient to assume that both E and I are positive under this condition, *i.e.,* to assume the flux to be positive, so that positive E corresponds to positive speed and positive I to positive torque. If in all circuits the direction of the current in the field winding is assumed to be the same, the flux is automatically assumed to be positive. Negative armature current then indicates negative torque, *i.e.,* the torque in the reverse, downward or clockwise direction. Similarly negative emf indicates negative speed. All that is required to avoid difficulties is to maintain these conventions consistently in the circuit analysis and the signs will indicate the mode of operation. The motor equation $V = E + IR$ must, of course, be used consistently for all cases, and no attempt made to prejudge the behaviour by using the "generator equation" $V = E - IR$ at any time.

The hoisting up of the unloaded cage is represented in the second quadrant. Since the counterweight is heavier than the empty cage, the speed at which the hoist is moved upwards may attain a dangerously high value. In order to avoid this, the motor torque must act in a direction opposite to that of rotation, *i.e.,* the motor should switch over to a braking or generator regime. Note that T_M will have a negative sign and speed still has a positive sign, being forwards, upwards and counter-clockwise, giving power a negative sign, corresponding to the generating or braking operation.

The third quadrant represents the downward motion of the empty cage. The downward journey of the cage opposed by the torque due to the counterweight and friction at the transmitting parts. So for moving the cage downwards, the motor torque must act in the same direction as the motion of the cage. The electrical machine acts as a motor as in the first quadrant, but in the reverse direction. Thus quadrant-III becomes 'reverse motoring'. The motor torque has a negative sign as it causes an increase in speed in the negative sense and the speed also has a negative sign being a downward motion. Power, thus, has a positive sign.

The downward motion of the loaded cage is shown in the IVth quadrant. The motion can take place under the action of load itself, without the use of any motor. But in order to limit the speed of the downward motion of the hoist, the electrical machine must act as a brake. The motor torque has a positive sign as it causes a decrease in speed in the downward motion. The speed, of course, has a negative sign, being a downward journey. The power, thus, acquires a negative sign, corresponding to the braking operation of the motor.

1.8. TORQUE-SPEED AND SPEED-TORQUE CHARACTERISTICS

Machine specialists or designers express the performance of an ac motor by means of a torque-speed characteristic curve, *i.e.,* with speed as abscissa and torque as ordinate. This is because it is much easier to obtain an expression for the torque in terms of speed or slip than the reverse. But in application engineering, it is usually the load torque which is known and the speed at which motor can drive this load is unknown and is to be determined. Thus the torque is independent variable and the speed is dependent variable consequently speed-torque characteristic curves with torque as abscissa are commonly employed in utilization studies. Characteristic curves drawn thus may look a little unfamiliar at first to those accustomed only to the machine designer form of curve.

1.9. SELECTION OF ELECTRIC DRIVE

For selection of reliable and efficient motor it is essential that the conditions of service are well known. It is not sufficient to simply specify the output power in kW and the speed but it is also necessary to know the following additional particulars :

(*i*) Torque at the shaft during running, starting and at different loads.

(*ii*) Accelerating torque and braking torque.

(*iii*) Switching frequency.

(*iv*) Efficiency of motor at different loads.

(*v*) Other working requirements.

In studying the behaviour of a motor selected for a particular driven unit, one of the first problems involved is to determine whether the speed-torque characteristic of the motor suits the requirements imposed by the speed-torque characteristic of the driven unit. Drive behaviour during the transient period of a start-up, braking, or speed change-over also depends upon how the speed-torque characteristics of the motor and the driven unit vary with speed. It is, therefore, imperative to study these characteristics in order to be able to select correctly the motor and obtain an economical drive.

1.9.1. Speed-Torque Characteristics of Machines or Mechanisms

The speed-torque characteristic of a machine or mechanism given by the relation $\omega = f(T_L)$ is defined as the relationship between the speed at which it is operated and the resisting or load torque it develops.

Different kinds of mechanisms and machines exhibit different speed-torque characteristics. However, several general conclusions may be drawn if we use the following empirical equation for the speed-torque characteristic of some driven unit of industrial equipment

$$T_L = T_0 + (T_{rn} - T_0)\left(\frac{\omega}{\omega_n}\right)^x \qquad ...(1.3)$$

where T_L is gross load (or resisting) torque developed by the unit at speed ω, T_0 is the resisting torque developed by the unit due to friction in its moving parts, T_{rn} is the resisting torque developed by the unit when it is driven at its nominal rated speed ω_n and x is the exponential coefficient characterising the change in resisting torque with the change in speed.

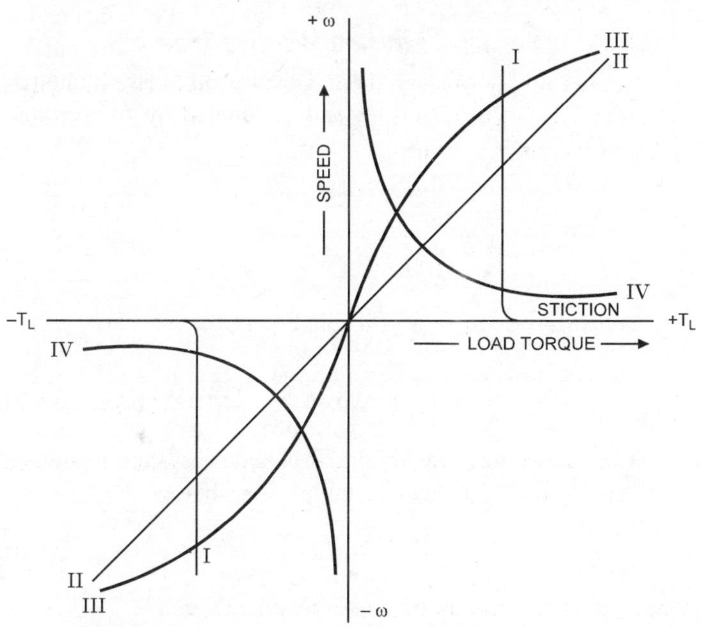

Fig. 1.4. ***Speed-Torque Characteristics of Typical Different Types of Machines and Mechanisms***

The above Eq. (1.3) permits the speed-torque characteristics of different kinds of machines and mechanisms to be roughly divided into the following categories :

1. *Loads Requiring Constant Torque at All Speeds.* Such a load presents to the motor a passive torque which is essentially independent of speed. It is characterized also by the requirement of an extra torque at very near zero speed. For this characteristic $x = 0$ and the load torque T_L does not depend upon speed. The speed-torque characteristic for such loads is illustrated by vertical line in Fig. 1.4. Such loads are dry friction, cranes during hoisting, hoist winches, machine tool feed mechanism, piston pumps operating against a constant pressure head, and conveyors handling a constant weight of material per unit time. In power applications it is usually called the breakaway torque and in control systems, it is referred to as *stiction* (derived from sticking friction).

Since it changes sign with reversal of rotation, the dry friction torque characteristic is discontinuous, as illustrated in Fig. 1.4.

2. *Loads With Linear-Rising Characteristic.* Such speed-torque characteristics, illustrated by straight line II in Fig. 1.4, are exhibited by calendering machines, eddy current brakes, separately excited dc generators supplying fixed ohmic resistance loads and fluid or viscous friction. In this case $x = 1$ and the load torque T_L rises in direct proportion to the speed.

3. *Loads With Non-Linear-Rising (Parabolic) Characteristic.* For such a characteristic $x = 2$ and the load torque T_L is proportional to the square of the speed. Such a characteristic is illustrated by curve III in Fig. 1.4. A load with considerable windage, of which a fan is the extreme example, has a torque which varies almost as the square of the speed. Blowers, centrifugal pumps, propellers in ships or aeroplanes, water wheels, pipe friction, velocity head of pumps etc. also have the same type of speed-torque characteristics.

4. *Loads With Non-Linear Falling (Hyperbolic) Characteristic (or Constant Power Load).* For such a characteristic $x = -1$ and the load torque T_L is inversely proportional to the speed, while the power required to drive the given unit remains unchanged. Such a characteristic is illustrated by curve IV in Fig. 1.4. Certain types of lathes, boring machines, milling machines and other kinds of metal-cutting machine tools, steel-mill coilers fall under this category of loads.

The categories of loads listed above do not cover all the cases that may be met with in practice but give a good idea of the characteristics typical of a great many kinds of industrial equipment. In actual practice we may come across loads which are combination of these basic types of loads.

1.9.2. Load Torque-Time Characteristics

Perhaps, the variation of load torque with time is of equal or greater importance in selection of motor. This variation in certain applications, can be periodic and repetitive, one cycle of variation being called a duty cycle. Different types of loads from the point of view of load torque characteristics, may be classified as follows.

(*i*) **Continuous, constant loads** such as paper making machines, centrifugal pumps or fans operating for a long time under the same conditions.

(*ii*) **Continuous, variable loads** such as hoisting winches, metal-cutting lathes, conveyors etc.

(*iii*) **Pulsating loads** such as reciprocating pumps and textile looms and, in general, all machines having crank shaft.

(*iv*) **Impact load** such as rolling mills, shearing machines, presses, forging hammers etc. Apparent, regular and repetitive load peaks or pulses occur in such loads.

(*v*) **Short time loads** such as motor-generator sets for charging of batteries; servo-motors used in remote control of drilling machine clamping rods.

(*vi*) **Short time intermittent loads** such as cranes and hoisting mechanisms, excavators, roll trains etc.

Certain machines (such as ball mills) do not strictly belong to any category mentioned above. If such loads (ball mills, stone crushers etc.) were characterized by frequent impacts of comparatively small peaks, it would be more appropriate to place them in the category of continuous variable loads rather than impact loads. Sometimes, it is quite cumbersome to make distinction between pulsating loads and impact loads, because both of them are periodic in nature.

One and same drive can be represented by a load torque which varies either with speed or with time. The most appropriate example is a fan load whose load torque T_L is proportional to the square of speed, is also a continuous constant load.

1.9.3. Load Torques Varying With Shaft Displacement Angle

In all machines having crankshafts, such as reciprocating pumps and compressors, frame-saws etc. load torque varies with the angular displacement of the shaft or rotor of the motor. For all such machines, the load torque T_L can be resolved into two components—one of constant magnitude T_{av} and the other a variable T_L' which changes periodically in magnitude depending on the angular position of the shaft. Such load torque characteristics, can, for simplicity, be represented by a Fourier series as a sum of oscillations of fundamental and harmonic frequencies, *i.e.*,

$$T_L' = \sum_{r=0}^{m} T_{Lr}' \sin (r\theta + \phi_r) \qquad \text{...(1.4)}$$

where $\theta = \omega t$, ω being the angular speed of the motor shaft driving the compressor.

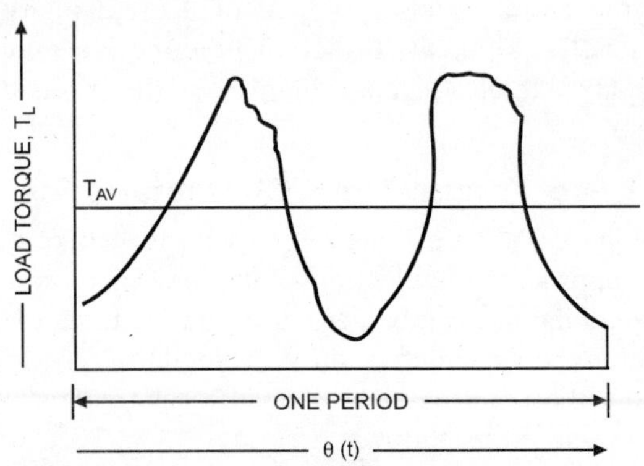

Fig. 1.5

During variations in speed, only small deviations from a fixed value of speed ω_a occur, therefore, displacement can be represented by $\theta = (\omega_a + \Delta\omega)t$. Thus the variable portion of the load torque may be given as

$$T_L' = \sum_{r=0}^{m} T_{Lr}' [\sin (r\,\omega_a t + \phi_r) + r\Delta\omega t] \qquad \text{...(1.5)}$$

The term $r\Delta\omega t$ being very small in magnitude may be neglected. Thus, restricting to small deviations in angle from the equilibrium position, a load torque varying with angular displacement of shaft can be transformed to one which varies periodically w.r.t. time.

1.9.4. Load Torques Depending on Path or Position of Load During Motion

In Art 1.9.1. load torques varying with speed were considered. However, load torques, that depend not only speed but also on the nature of the path traced out by the load during its motion, do exist both in hoisting mechanisms and transport systems. For example, the resistance to motion of a train moving upgradient or taking a turn depends on the magnitude of the gradient or the radius of curvature of the track respectively.

Force due to the gradient is given as

$$F_g = 1{,}000 \text{ W} \sin \theta \text{ kg.} \qquad \text{...(1.6)}$$

where W is the weight of train in tonne.

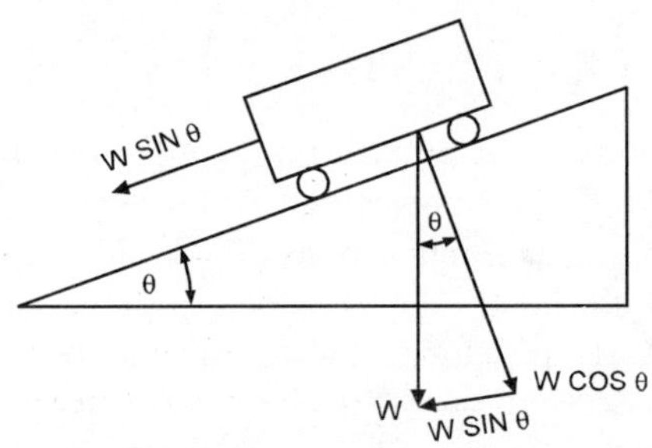

Fig. 1.6. ***Upgradient Motion of Train***

But in railway work gradient is expressed as rise in metres in a track distance of 100 m, and is denoted by *'percentage gradient'* (G%)

$$i.e., \quad G = \text{Sin } \theta \times 100$$

$$\text{or } \sin \theta = \frac{G}{100}$$

Substituting $\sin \theta = \dfrac{G}{100}$ in Eq. (1.6), we have

$$F_g = 1{,}000 \text{ W} \times \frac{G}{100} = 10 \text{ WG kg.} \qquad \text{...(1.7)}$$

The tractive force required to overcome curvature resistance is given by the empirical formula given below

$$F_c = \frac{700{,}000 \text{ W}}{\text{R}} \text{ kg} \qquad \text{...(1.8)}$$

where R is the radius of curvature in metres.

In hoisting mechanisms in which tail ropes or balancing ropes are not employed (Fig. 1.7), the load torque is not only due to the weight of the unloaded or the loaded cage but also due to that of the lifting ropes or cables, which depends on the position of two cages. When cage 1 is at the bottom most position and is to be lifted upwards, the entire weight of the rope is also to be moved up. When both the cages are at the same height, the weight of the rope to be lifted becomes zero, since the weight of the ropes on both sides balance each other, being equal in length. When cage 1 is at a higher position than cage 2, a portion of the weight of the rope acts in such a way as to assist the upward motion of the cage 1. Ultimately when cage 1 reaches the top most position, the whole weight of the rope assists the upward movement.

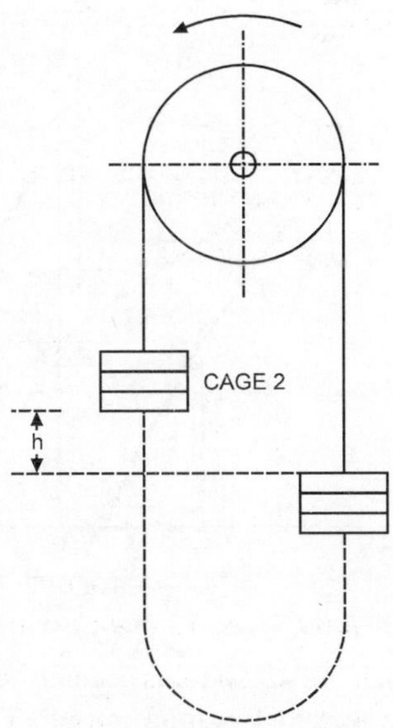

Fig. 1.7. *Hoisting Mechanism*

The force resisting the upward movement of the load, F_r due to varying weight of the rope depending on the position of the load is given as

$$F_r = W_r\left(1 - \frac{2x}{h}\right) \text{ kg} \qquad ...(1.9)$$

where W_r is the total weight of the rope in kg, h is the desired maximum height to which the cage is to be moved upwards, in metres and x is the height of the cage at any arbitrary position from the bottom most position in metres.

For large values of h, the force F_r affects to a large extent, the performance of the drive used in hoisting mechanisms because in such a situation the weight of the rope may be considerably larger than that of the load to be lifted upwards. If we use tail ropes, as illustrated by means of dotted lines in Fig. 1.7, the weight of the connecting rope can be balanced and almost smooth movement of the cages can be had.

1.10. MOTOR SPEED-TORQUE CHARACTERISTICS

A motor speed-torque characteristic is defined as the relationship between the speed at which it operates and the torque it developes, *i.e.,* $\omega = f(T)$.

Practically all electric motors—shunt wound, series wound, compound dc motors, squirrel-cage and slip-ring induction motors, and ac commutator motors, have drooping speed-torque characteristics, *i.e.,* their speed falls off as the load torque increases. However, the degree by which the speed changes with the change in torque differs for various types of motors, it being characterized by the so-called hardness of their speed-torque characteristics.

Electric motor speed-torque characteristics may be classified into three main groups.

1. ***Absolutely hard (flat) speed-torque characteristic,*** a characteristic exhibiting no change in speed with change in load torque. Synchronous motors operate with such a characteristic (horizontal straight line I in Fig. 1.8).

2. ***Hard speed-torque characteristic,*** a characteristic showing a speed which drops only slightly with increase in torque. A hard characteristic is exhibited by a shunt wound dc motor, this also being true of induction motors over the operating region of the speed-torque characteristic (curve II in Fig. 1.8).

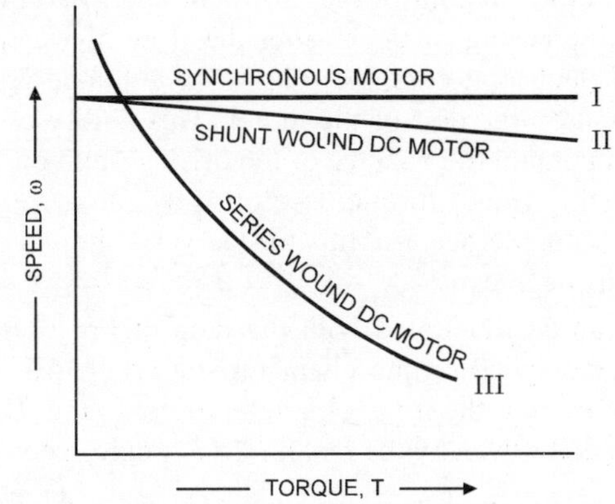

Fig. 1.8. *Speed-Torque Characteristics of Different Types of Motors*

The speed-torque characteristic of an induction motor exhibits a "hardness" that differs according to what point along the characteristic is taken into consideration (Fig. 1.9). Between the points of the maximum torque in motor operation $T_{max\ M}$ and the maximum torque in generator operation $T_{max\ G}$ an induction machine will exhibit a rather hard characteristic.

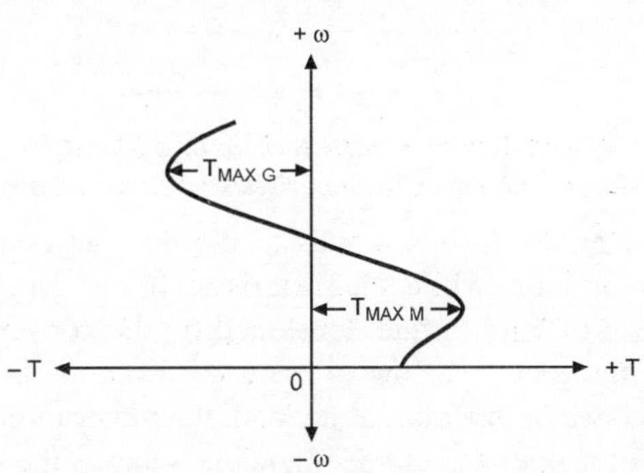

Fig. 1.9. *Speed-Torque Characteristic of an Induction Motor*

3. ***Soft speed characteristic,*** a characteristic showing considerable drop in speed with rise in torque. Series wound dc motor possesses such characteristic, especially along the low-torque portion of the characteristic (curve III in Fig. 1.8). For such motors, the degree of hardness of the characteristic varies all along the curve.

Compound wound dc motors, depending on the degree of hardness their speed-torque characteristics display, may be considered as hard- or soft-characteristic motors.

1.11. JOINT SPEED-TORQUE CHARACTERISTIC OF AN ELECTRIC MOTOR AND A DRIVEN UNIT

The joint operation of an electric motor and the unit it drives, when the speed has a steady value, corresponds to a condition of balance between the driving torque of the motor and the resisting torque developed by the driven unit at a given speed. When the resisting or load torque developed at motor shaft by the driven unit undergoes some change, the speed and torque developed by the motor wiil change automatically so as to restore stable operation at a new value of speed and load torque.

In case of non-electric prime movers (water turbine, steam turbine or diesel/petrol engine) the balance between the resisting torque and the driving torque is achieved by using respective type of governor to control the inflow of energy to the prime mover by increasing or decreasing the flow of water, steam or fuel. In electric motors, the role of the automatic governor is performed by the emf of the motor. This ability of electric motors to maintain the balance of the drive system on changes in the resisting (load) torque developed by the driven unit is extremely valuable because this torque very often is unstable to a certain degree.

This can be illustrated with the help of Fig. 1.10 which illustrates the speed-torque characteristic (curve III) of a dc shunt motor and the two characteristics I and II of the production unit driven by the motor (for example, a conveyor).*

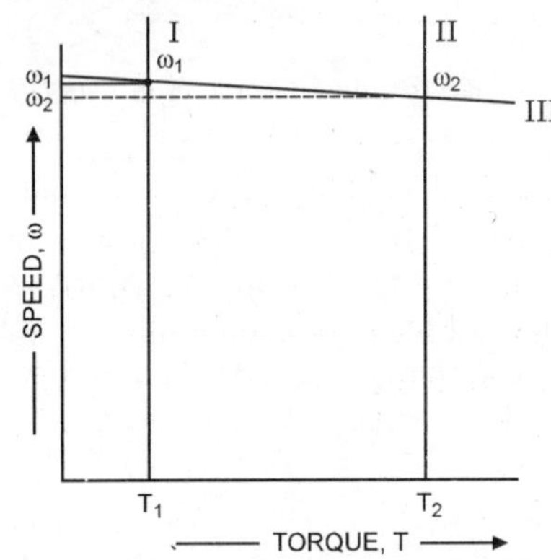

Fig. 1.10. ***Speed-Torque Characteristic of a Shunt Wound DC Motor and Two Characteristics of Driven Unit***

Characteristic I corresponds to the no-load condition of the conveyor unit, while characteristic II corresponds to a greater level of load torque developed by the conveyor when it handles the required flow of material. Initially, at the time when the conveyor operates at no load, the motor torque $T = T_1$ and the motor operates at speed ω_1. As soon as the conveyor starts to maintain a flow of material, the increase in load on the motor acts to brake the motor and reduce its speed. This makes the motor to develop a smaller emf. Consequently the armature current increases and the motor begins to develop a larger driving torque. The motor torque grows until a point of balance is reached at which the torque developed by the motor = resisting torque of the driven unit, *i.e.*, $T = T_2$ (where the speed is ω_2). This new point also constitutes the one that is common both to the speed-torque characteristic II of the conveyor and the speed-torque characteristic III of the motor.

In studying the operation of the motor and the unit it drives, it is sometimes convenient to make use of the so-called joint speed-torque characteristic of the electric drive, a curve representing the algebraic sum of the speed-torque characteristic of the driven unit and that of the drive motor.

Speed-torque characteristics of a fan and the drive motor and the joint speed-torque characteristic of a motor fan unit are represented by curves I, II and III respectively in Fig. 1.11.

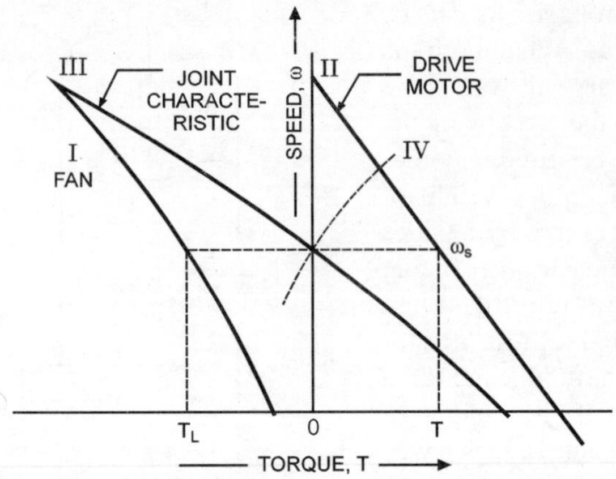

Fig. 1.11. ***Joint Speed-Torque Characteristic of a Motor-Fan Drive***

When the unit attains a steady speed ω_s, the motor operates with the torque T = load torque, T_L. The torque indicated on the joint characteristic will be zero under such a condition. Operation of the unit at the steady speed ω_s will be seen to be stable in this case because any increase in speed leads to a negative change (drop) in torque, while any drop in speed results in a positive change (increase) in torque. Curve III is, therefore, an example of a joint speed-torque characteristic of a drive that will be able to operate with stability. If the joint characteristic had the form of curve IV, operation would not be stable because slight increase in speed gives rise to acceleration as the motor torque exceeds the load torque. On the other hand slight decrease in speed gives rise to retardation because the motor torque becomes less than the load torque.

The conditions of drive operation in the steady state discussed above constitute those required for static stability of the drive, and are only applicable when speed and torque vary slowly. During periods of transient (rapid) change involving the dynamic stability, the conditions of drive stability will be different.

Usually, when an electric drive is designed for a specific drive, its speed-torque characteristic is known before hand. The problem of attaining stable operation in steady-state condition at known speeds and load torques of the driven unit therefore consists in selecting a motor whose speed-torque characteristic will be compatible with that of the driven unit. This may be achieved by first selecting a suitable type of motor and then changing accordingly the electrical parameters of its circuits. Sometimes, in order to provide the required speed-torque characteristics it becomes imperative to set up special power and control circuits for involved switching of the drive motor and the control apparatus.

1.12. DYNAMICS OF MOTOR-LOAD COMBINATION

In translational motion, an active or driving force F_d is counterbalanced by a resisting force F_r developed by the driven

* For convenience, the two speed-torque characteristics of the production unit are drawn in the first quadrant although the resisting torque is actually negative in the sense with respect to the driving torque.

machine and by an inertia force $m\frac{dv}{dt}$ arising from the change in speed. When the body involved has a mass *m* expressed in kg and a velocity *v* expressed in m/s, the inertia force, like other forces, will be expressed in newtons (kg-m/s^2).

The equation for force equilibrium in translation of a body may accordingly, be written in the following form.

$$F_d - F_r = m\frac{dv}{dt} \text{ newtons} \quad ...(1.10)$$

The equation for torque equilibrium in translation of a body may, accordingly, be written in the following form.

$$T_M - T_L = J\frac{d\omega}{dt} \text{ Nm} \quad ...(1.11)$$

The above Eq. (1.11) shows that the torque T_M developed by the motor is counterbalanced by a resisting or load torque T_L exerted at its shaft and by an inertia or dynamic torque $J\frac{d\omega}{dt}$. In above Eqs. (1.10) and (1.11) it is assumed that the mass *m* of the bodies involved and the polar moment of inertia J of the drive remain constant, an assumption which holds good for a large number of industrial machines and mechanisms. In certain drives, it becomes necessary to deal with a variable polar moment of inertia as in the case of crank drives.

From analysis of Eq. (1.11), it is possible to determine the different states at which an electric drive causing rotational motor can remain.

1. When $T_M > T_L$, $\frac{d\omega}{dt} > 0$, *i.e.,* the drive will undergo acceleration, in particular, picking up speed to attain rated speed.

2. When $T_M < T_L$, $\frac{d\omega}{dt} < 0$, *i.e.,* the drive will undergo deceleration and, particularly, coming to rest.

   Deceleration will evidently also occur at negative* values of motor torque. A motor develops a negative torque when it passes over to braking operation.

3. When $T_M = T_L$, $\frac{d\omega}{dt} = 0$, *i.e.,* the drive will run at a steady-state speed.

The above statements, namely, that when $T_M > T_L$ the drive accelerates and that when $T_M < T_L$ the drive decelerates, are valid only when load or restraining torque T_L happens to be a passive torque. The reverse may occur with active torque loads. For example, if a motor is switched on for hoisting up a winch, while it is coming down on its own weight, until the direction of rotation changes, deceleration of the drive takes place when $T_M > T_L$. In case $T_M < T_L$ in the above situation when the motor has been switched on for moving the winch up, the load will continue to come down and the motor will accelerate instead of decelerating.

The inertia or dynamic torque $J\frac{d\omega}{dt}$ appears only during transient conditions, *i.e.,* when the speed of the drive varies. During acceleration of the drive, the inertia torque opposes the drive motion, but during braking it maintains the motion of the drive. The inertia torque, both in magnitude and in sign, is determined as the algebraic sum of the motor torque and the resisting or load torque.

In view of the above, the signs for T_M and T_L in Eq. (1.11) corresponding to motoring operation of the driving machine and to passive load torque (or an active braking torque) respectively. In the general form, the torque equation can be written as

$$\pm T_M \mp T_L = J\frac{d\omega}{dt} \quad ...(1.12)$$

Selection of sign to be placed before each of the torques in above Eq. (1.12) depends upon the operating condition and on the nature of the resisting or load torques. The equation of motion for a drive, makes it possible to determine the dependence of torque, current, speed and path upon the time of operation under transient conditions. All the torques in the equation for motion have to be referred to some given element in the system. Most frequently, both the load torque and the dynamic torque are referred to the motor shaft.

Example 1.2. A motor is coupled to a load having the following characteristics :

Motor : $T_m = 15 - 0.5\omega_m$

Load : $T_l = 0.5\omega_m^2$

Find out the stable operating point for this combination.

[*U.P. Technical Univ. Electric Drives 2004-05*]

Solution : The stable operation will be obtained when

$$T_m = T_l$$
$$\text{or } 15 - 0.5\omega_m = 0.5\omega_m^2$$
$$\text{or } \omega_m^2 + \omega_m - 30 = 0$$
$$\text{or } \omega_m = 5 \text{ or } -6$$

Discarding minus figure, we have

$$\omega_m = 5 \text{ and } T = 12.5$$

So stable operating point is (12.5, 5) **Ans.**

1.13. REFERRED LOAD TORQUES AND MOMENTS OF INERTIA

A motor generally drives an industrial machine through some transmission system whose individual parts operate at different speeds. In carrying out practical calculations, a necessity arises to refer the torques and masses of individual parts to some convenient element, say, a certain shaft.

Load torques may be referred from one shaft to another on the basis of the power balance of the system. In this case account is taken of the power losses in the intermediate links of the transmission by introducing the respective efficiency values.

Let the speed of the motor shaft be ω_M and the speed of shaft of the given industrial machine ω_L.

On the basis of equality of power flow, we have

$$T_L\omega_L\frac{1}{\eta_T} = T'_L\omega_M$$

or load torque referred to the motor shaft,

* A motor torque is considered to be positive in sense when it acts in the direction necessary for normal operation of the drive. When the motor develops torque opposing this direction of motion, the torque is taken to be negative in sense.

$$T_L' = T_L \frac{\omega_L}{\omega_M} \times \frac{1}{\eta_T} = \frac{T_L}{i\eta_T} \qquad \text{...(1.13)}$$

where T_L is the load torque, η_T is the efficiency of transmission and i is speed transmission ratio and is equal to ω_M/ω_L.

When there are several stages in transmission between the drive motor and the driven machine, as illustrated schematically in Fig. 1.12, with the transmission ratios i_1, i_2, ..., i_n and the respective transmission efficiencies η_{T_1}, η_{T_2}, ..., η_{T_n}, the load torque referred to the motor shaft is given as

$$T_L' = T_L \times \frac{1}{i_1 i_2 \ldots i_n} \times \frac{1}{\eta_{T_1}\eta_{T_2}\ldots\eta_{T_n}} \qquad \text{...(1.14)}$$

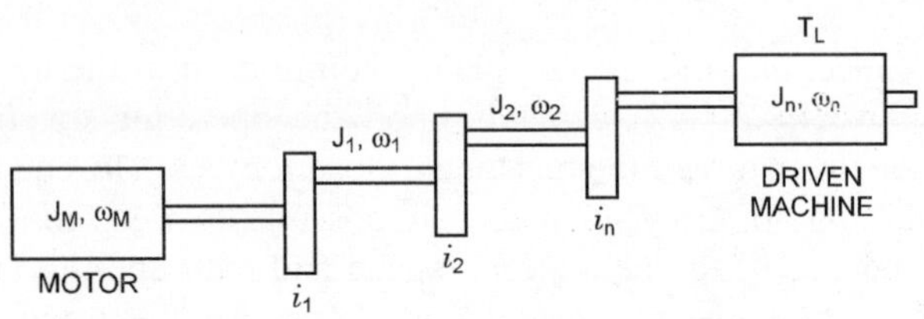

Fig. 1.12. *Motor and Driven Machine With Gears*

Moments of inertia are referred to a given shaft on the basis that the total amount of kinetic energy stored in the moving parts and referred to the given shaft remains unchanged. With the rotating parts having polar moments of inertia of J_M, J_1, J_2, ..., J_n and angular velocities of ω_M, ω_1, ω_2, ... ω_n (Fig. 1.12), their dynamic action may be replaced by that of a single polar moment of inertia referred, say, to the motor shaft, and we may write the following equations.

$$\frac{J\omega_M^2}{2} = J_M\frac{\omega_M^2}{2} + J_1\frac{\omega_1^2}{2} + J_2\frac{\omega_2^2}{2} \ldots + J_n\frac{\omega_n^2}{2}$$

or the polar moment of inertia referred to the motor shaft

$$J = J_M + J_1\left(\frac{\omega_1}{\omega_M}\right)^2 + J_2\left(\frac{\omega_2}{\omega_M}\right)^2 + \ldots J_n\left(\frac{\omega_n}{\omega_M}\right)^2$$

$$= J_M + \frac{J_1}{i^2} + \frac{J_2}{(i_1 i_2)^2} + \ldots \frac{J_n}{(i_1 . i_2 \ldots i_n)^2} \qquad \text{...(1.15)}$$

Example 1.3. A motor drives a rotational load through a reduction gear of teeth ratio a = 0.1 and efficiency as 90%. The load has a moment of inertia of 10 kg-m^2 and a torque of 50 N-m. The motor has inertia of 0.4 kg-m^2 and runs at a constant speed of 1,400 rpm. Determine the equivalent inertia of the motor and load combination referred to the motor side and the power developed by the motor. [*U.P. Technical Univ. Electric Drives 2004-05*]

Solution : Motor inertia, J_M = 0.4 kg-m^2

Load inertia, J_1 = 10 kg-m^2

Teeth ratio of gear, $a = \frac{1}{i} = 0.1$

Total moment of inertia referred to the motor shaft (or side)

$$J = J_M + \frac{J_1}{i^2} = 0.4 + \frac{10}{\left(\frac{1}{0.1}\right)^2}$$

$$= 0.4 + 10 \times (0.1)^2 = 0.5 \text{ kg-m}^2 \quad \textbf{Ans.}$$

Total equivalent torque referred to motor shaft,

$$T_L' = \frac{a_1 T_L}{\eta_T} = \frac{T_L}{i\eta_T} = \frac{50}{\frac{1}{0.1}\times 0.9} = 5.55 \text{ N-m}$$

Power developed by motor

$$= T_L' \times \frac{2\pi N}{60} = 5.55 \times \frac{2\pi \times 1{,}400}{60} = 815 \text{ W} \quad \textbf{Ans.}$$

1.14. REFERRING FORCES AND MASSES HAVING TRANSLATIONAL MOTION TO A ROTATING SHAFT

In many machines (such as in hoists, cranes, shaping machines etc.) some of the moving parts rotate while others go through a translational motion. A motion of translation will have to be referred to a rotating shaft drive involving such machines.

The resisting force developed by a translating body is referred to a rotating shaft from the condition of unchanged kinetic energy balance in much the same way as torques are referred.

Figure 1.13 illustrates the forces and torques acting on a hoist drive. If a translating body (or moving mass) has a velocity of v m/s and the motor shaft has the angular velocity ω_M rad/second, then

$$F_r\, v \times \frac{1}{\eta_T} = T_L'\,\omega_M \qquad \text{...(1.16)}$$

where F_r is the resisting force developed by the given machine due to the translating motion of a load or body of weight W.

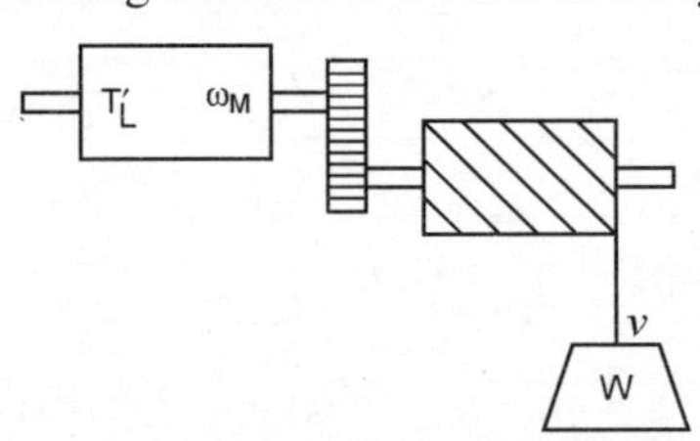

Fig. 1.13. *Motor-Hoist Drive*

The referred load torque at the motor shaft will be, therefore, given as

$$T_L' = \frac{F_r\, v}{\omega_M \eta_T} \qquad \text{...(1.17)}$$

A **translating mass** is referred to a rotating shaft on the basis of equality of kinetic energy

$$\frac{mv^2}{2} = J\frac{\omega_M^2}{2}$$

The polar moment of inertia, referred to the motor shaft, is therefore, given as

$$J = m\left(\frac{v}{\omega_M}\right)^2 = \frac{W}{g}\left(\frac{v}{\omega_M}\right)^2 \qquad \text{...(1.18)}$$

Example 1.4. A motor drives two loads one has a rotational motion. It is coupled to the motor through a reduction gear of teeth ratio a = 0.1 and efficiency as 90%. The load has a moment of inertia 10 kg-m^2 and a torque of 50 N-m. Other load has a translational motion and consists of 600 kg weight to be lifted at a uniform speed of 2.0 m/s. Coupling between this load and the motor has an efficiency of 88%. Motor has inertia of

0.4 kg-m^2 and runs at a constant speed of 1,500 rpm. Determine the equivalent inertia referred to the motor shaft and the power developed by the motor.

Solution : Motor inertia, J_M = 0.4 kg-m^2

Load inertia, J_l = 10 kg-m^2

Teeth ratio of gear, a = 0.1

$$\text{Motor speed, } \omega_M = \frac{2\pi N}{60} = \frac{2\pi \times 1{,}500}{60}$$

$$= 157 \text{ radians/speed}$$

Uniform speed of weight lifting, v = 2.0 m/s

Total moment of inertia referred to the motor shaft,

$$J = J_M + a_1^2 J_1 + M_1 \times \left(\frac{v}{v_m}\right)^2$$

$$= 0.4 + (0.1)^2 \times 10 + 600 \times \left(\frac{2}{157}\right)^2 = 0.6 \text{ kg-m}^2 \quad \textbf{Ans.}$$

Total torque referred to motor shaft

$$T'_L = \frac{a_1 T_L}{\eta_T} + \frac{F_r v}{\eta'_T \omega_M} = \frac{0.1 \times 50}{0.9} + \frac{600 \times 9.81 \times 2}{0.88 \times 157} = 90.75 \text{ N-m}$$

Power developed by motor

$= T'_L \times \omega_m = 90.75 \times 157 = 14{,}250$ W or 1.425 kW **Ans.**

1.15. REFERRING TORQUES AND TRANSLATING MASSES WHEN SPEED IS VARIABLE

In certain types of machines which convert the motion of rotation into a translational one with the aid of a crank gear, the speed and acceleration of the moving masses vary both in magnitude and in sign throughout each revolution of the crankshaft.

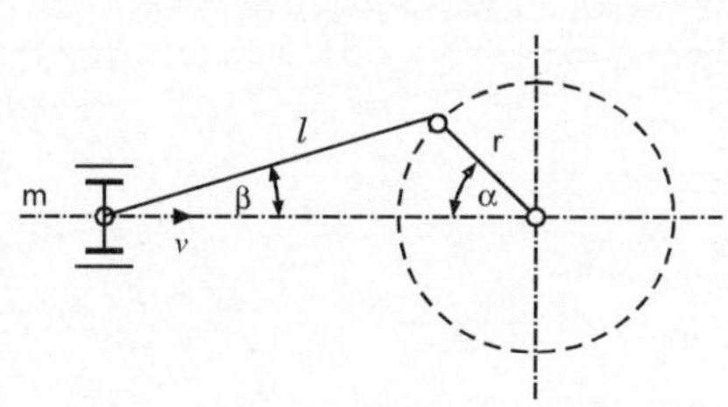

Fig. 1.14. ***Crankshaft Mechanism***

The kinetic energy stored in these masses will, accordingly, change from zero to maximum. The MOI of these masses referred to the crankshaft is given as

$$J = \frac{mv^2}{\omega^2} \quad ...(1.19)$$

where m is the mass of given translating part moving with velocity v and ω is the angular velocity of the crankshaft.

Using the geometrical relationships of the crank gear shown in Fig. 1.14, after certain transformations, we have the following expression for the velocity of the translating part

$$v = \frac{r\omega \sin(\alpha + \beta)}{\cos \beta} \quad ...(1.20)$$

where $\beta = \sin^{-1}\left(\frac{r}{l}\sin\alpha\right)$; the value of α depends on the position of the crank pin.

Substituting the value of v from Eq. (1.20) in Eq. (1.19) for moment of inertia, we have

$$J = \frac{mr^2 \sin^2(\alpha + \beta)}{\cos^2 \beta} \quad ...(1.21)$$

The total moment of inertia will be the sum of the moment of inertia J determined from above Eq. (1.21) and that of all other moving parts referred to the crankshaft. If there is some intermediate stage of transmission between the crankshaft and the driving motor, the moments of inertia referred to the crankshaft should then be referred to the motor shaft by making use of Eq. (1.17).

Thus static load torque referred to the motor shaft,

$$T'_L = \frac{F_r\, r \sin(\alpha + \beta)}{i\, \eta_T \cos \beta} \quad ...(1.22)$$

where F_r is the resisting force developed by the translating part involved, i is the motor shaft-to-crankshaft transmission ratio, $\left(i.e., \frac{\omega_M}{\omega}\right)$ and η_T is the efficiency of transmission.

With variable moment of inertia as in the case of crank gears, the equation of motion for the electric drive becomes more complex since $J = f(\alpha)$. The kinetic energy stored in the system, at the crankshaft, amounts to

$$(KE)_c = \frac{1}{2} J \omega^2 \quad ...(1.23)$$

and the dynamic power corresponding to this energy,

$$P_{dynamic} = \frac{d(KE)_c}{dt}$$

$$= J\omega \frac{d\omega}{dt} + \frac{1}{2}\omega^2 \frac{dJ}{d\alpha}\cdot\frac{d\alpha}{dt}$$

$$= J\omega \frac{d\omega}{dt} + \frac{1}{2}\omega^3 \frac{dJ}{d\alpha} \quad \therefore \omega = \frac{d\alpha}{dt} \quad ...(1.24)$$

Taking into account that the dynamic torque $T_{dynamic} = \frac{P_{dynamic}}{\omega}$, the equation of motion becomes

$$T_M - T_L = J\frac{d\omega}{dt} + \frac{1}{2}\omega^2 \frac{dJ}{d\alpha} \quad ...(1.25)$$

The additional term, that has appeared is obviously owing to the variable moment of inertia.

1.16. OPERATING OR RUNNING CHARACTERISTICS OF ELECTRIC MOTORS

While studying electrical behaviour of a machine under normal operating conditions, the speed-torque characteristic, speed-current characteristic, losses, efficiency, magnetising current and power factor at various loads are kept in view. The last two factors, *i.e.,* magnetising current and power factor are to be considered in case of ac motors only.

Other features, such as temperature rise and insulation strength, are taken care of by Indian Standard Specifications, therefore if a motor is purchased built by a reputed manufacturer in accordance with the appropriate specifications, it will prove satisfactory in these respects.

1.16.1. DC Motors

In all dc machines, the speed and torque are given by expressions

$$N = \frac{V - I_a R_a}{\phi Z} \times \frac{60\,A}{P} = K_1 \frac{V - I_a R_a}{\phi} = K_1 \frac{E_b}{\phi} \quad ...(1.26)$$

∵ Z, A and P are constant for a particular machine

$$T = \frac{\phi}{2\pi} \times Z \times \frac{P}{A} I_a = K_2 \phi I_a \quad ...(1.27)$$

Torque and Speed of a DC Motor. Equations (1.26) and (1.27) reveal that

$$N \propto \frac{E_b}{\Phi} \text{ and } T \propto \Phi\, I_a$$

The above relations reveal that increase in flux would cause increase in torque and decrease in speed. It cannot be so because torque is the cause of rotation and increase in torque should cause increase in speed rather than decrease. The apparent inconsistency between the above two relations may be reconciled as below.

Let the motor be running at a steady speed and the field strength be weakened, which is possible by inserting an additional resistance in the field circuit in case of a dc shunt motor connected across constant voltage supply. The immediate effect is that, owing to the momentary reduction in back emf, the current flowing through armature circuit increases out of all proportion to the reduction in flux. As a result, inspite of the weakened field, the torque increases momentarily tremendously and will exceed considerably the value corresponding to load. The surplus torque, thus available causes the motor to accelerate, and the back emf to increase. The motor will attain steady speed again when the back emf has attained such a value that the current flowing in the armature develops with the slight weakened flux, only just sufficient torque to drive the load.

For avoiding undue rush of current the field strength should be reduced slowly.

The manner in which the motor slows down on strengthening of the field is also of equally importance. When the field is strengthened, the immediate effect is increase in back emf. Quite a small increase in the flux will normally be sufficient to increase the back emf beyond supply voltage, and in consequence, will not merely reduce the value of armature current, but will cause it to reverse its direction of flow. The machine will, therefore, act temporarily as a generator, and feed energy to the supply system, this energy being available at the expense of the kinetic energy associated with the machine and its load, with the result that a rapid reduction in speed takes place. The decrease in speed will cease when the back emf has fallen to a value such that the motor takes just sufficient current to develop the torque the load requires. By thus strengthening the field the motor speed is reduced by a form of "regenerative braking".

From the above discussion it is obvious that torque developed in a motor is certainly a function of flux and armature current and is independent of speed and the speed depends upon the torque.

(a) DC Shunt Motors

1. ***Speed-Armature Current Characteristic.*** If applied voltage V is kept constant, the field current will remain constant, hence flux will have maximum value on no load but will slightly decrease due to armature reaction as the load increases but for most purposes the flux is considered to be constant, neglecting armature reaction effect.

From speed equation, speed N is directly proportional to back emf E_b or $(V - I_a R_a)$ and inversely proportional to the flux Φ. Since flux is considered to be constant as mentioned above, so with the increase in armature current the speed slightly falls due to increase in voltage drop in armature and the speed-armature current curve coincides with the back emf-armature current curve. Since voltage drop in armature at full load is very small as compared to applied voltage so drop in speed from no load to full load is very small. If demagnetising effect of armature reaction is considered, the drop in speed due to voltage drop in armature is compensated for up to some extent due to decrease in flux with the increase in armature current and speed-armature current characteristic is less drooping, as shown dotted in Fig. 1.15 or even be rising if demagnetisation is high. Thus the drop in speed from no load to full load is very small and for all practical purposes the dc shunt motor is taken as a *constant speed motor*.

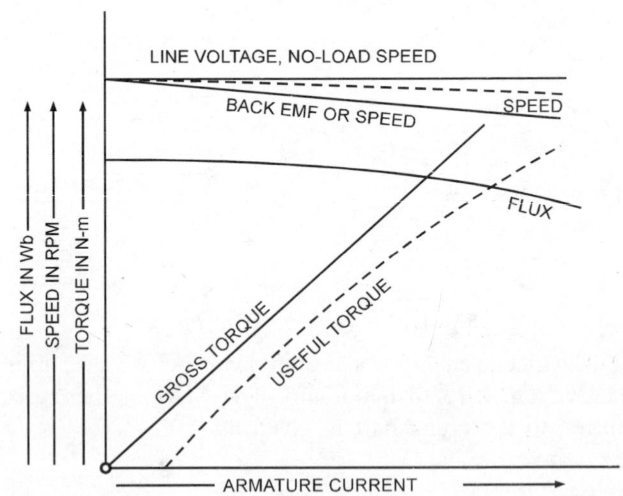

Fig. 1.15

Since there is a slight variation in speed of the shunt motor from no load to full load and this slight variation in speed can be made up by inserting resistance in the shunt field and so reducing the flux. Therefore, shunt motors can be used for the loads which are totally and suddenly thrown off without resulting in excessive speed. Shunt motors being constant speed motors are best suited for driving of line shafts, machine lathes, milling machines, conveyors, fans and for all purposes where constant speed is required. It is not suitable for use with flywheel or with fluctuating loads or for parallel operation due to its constant speed characteristic.

It is also useful where a moderate degree of speed control is required.

2. ***Torque-Armature Current Characteristic.*** From the expression for the torque of a dc motor, torque is directly proportional to the product of flux per pole ϕ and armature current, I_a. Since in case of a dc shunt motor the flux per pole ϕ is considered to be constant, therefore, torque increases with the increase in load current following linear law, *i.e.*, torque-armature current characteristic is a straight line passing through origin O (Fig. 1.15). But the weakening of field due to armature reaction causes the torque line to droop slightly, and the iron and friction losses cause it to be slightly lower than the line representing the electromagnetic developed torque corresponding to $T = 0.159\phi\, ZP \frac{I_a}{A}$ Nm.

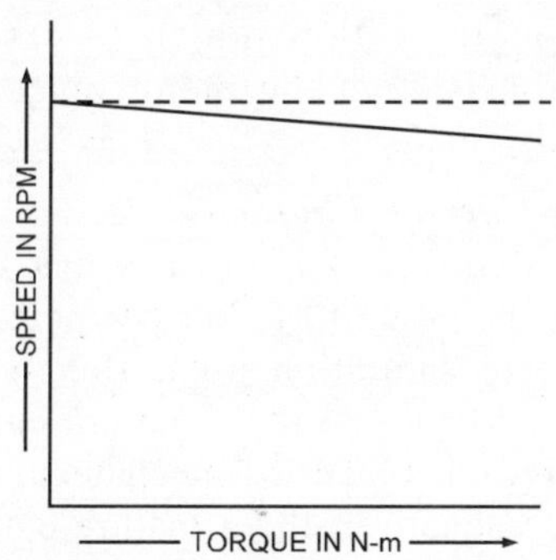

Fig. 1.16. *Speed-Torque Characteristic of DC Shunt Motor*

3. ***Speed-Torque Characteristic.*** Substituting $I_a = \frac{T}{K_2 \phi}$ from Eq. (1.27) in Eq. (1.26), we have

$$N = K_1 \frac{V}{\phi} - K_1 \frac{\frac{T}{K_2 \phi} R_a}{\phi} = K_1 \frac{V}{\phi} - \frac{K_1}{K_2} \frac{T R_a}{\phi^2} \quad ...(1.28)$$

For a dc shunt motor, ϕ remains more or less constant and, therefore speed of the motor is given as

$$N = C - DT \quad ...(1.29)$$

where C and D are constants for the machine, *i.e.*, the characteristic will be linear in nature, as illustrated in Fig. 1.16.

This type of motor is used in applications requiring medium starting torque such as centrifugal pumps, blowers, fans, conveyors, boring mills, shapers, woodworking machines, spinning and weaving machines, printing presses, machine tools etc.

DC shunt motors should never be started on heavy loads because such loads need heavy starting current.

4. ***Break in Field Circuit of a DC Motor.*** It is imperative that the field circuit of a dc motor never be opened while it is running. If the field circuit is opened, the flux will drop to practically zero and, therefore, the speed would increase to an extremely high value. The centrifugal force developed at the high speed may cause the windings to be pulled out of the slots or may cause the commutator segments to buckle, resulting in damage to the motor.

Proper use of overload devices would protect the motor. Since back emf would be reduced due to reduction in flux and cause increase in armature current or current drawn from the supply mains by the motor. This excessive current would operate the overload device, disconnect the motor from the supply, and thus protect the motor.

5. ***Performance Curves of a DC Shunt Motor.*** In Fig. 1.17, four important characteristics of a dc shunt motor, namely, torque, speed, armature current, and efficiency, each plotted against the useful output power, are shown. These curves are also known as *performance curves* of a motor.

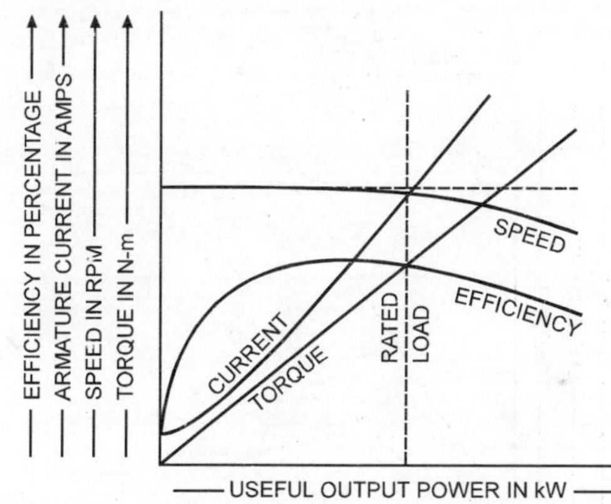

Fig. 1.17. *Performance Curves of a DC Shunt Motor*

From the speed–useful output power curve it is seen that the shunt motor has a definite speed at no load. Hence it does not 'run away' when the load is suddenly thrown off provided that the field circuit remains closed. The fall in speed from no load to full load is small, hence, these motors are used where a substantially constant speed is required as in machine shop drives.

From the efficiency–output power curve it is noted that the efficiency of a shunt motor increases rapidly in the beginning, reaches its maximum value and then falls. In shunt wound motors the field, friction and iron losses are approximately constant but at light loads their percentage is high as compared with the load. The armature copper loss varies as the square of the current so this loss, termed as variable loss, increases rapidly with the increase in load. The efficiency of the shunt machine becomes maximum when variable loss is equal to the constant loss.

From armature current–output power curve it is noted that a certain value of current is required even when the output is zero. The motor input power under no-load conditions is consumed in meeting the various losses occurring within the machine.

A shunt wound motor is said to have a low starting torque as compared to other motors but it does not mean that a shunt motor is incapable of starting a heavy load. A shunt motor is capable of starting heavy load also but it would require more excess of current input over normal values than the series and compound motors. For instance, if twice full-load torque is required at start the shunt motor will draw twice full-load current (since $T \propto I_a$) but a series motor would draw only approximately 1.414 times the full-load current (Since $T \propto I_a^2$). Due to this reason, shunt motors are always started at no load or light load and then the load is applied. When a motor is required to start heavy loads series or compound motor is used.

(b) DC Series Motors

1. *Speed-Current Characteristic.* In case of a dc series motor, the mmf due to the exciting coils increases in direct proportion to the line or armature current, so (neglecting armature reaction effects) the value of flux varies with the load current according to the ordinary magnetization curve. Owing to armature reaction, the actual curve representing the useful flux falls below the open-circuit magnetization curve, as illustrated in Fig. 1.18. With larger currents the magnetic circuit gets saturated and flux Φ tends to approach a constant value.

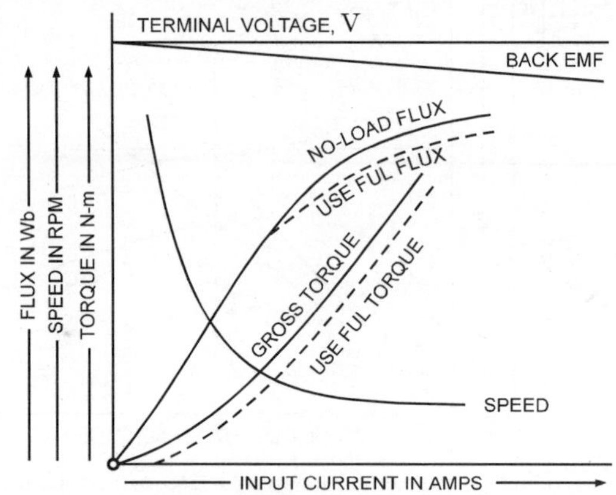

Fig. 1.18

From the speed equation, it is obvious that speed is proportional to back emf E_b and inversely proportional to flux per pole Φ. With the increase in armature current voltage drop in armature circuit and series field $[I(R_a + R_{se})]$ increases and, therefore, back emf E_b decreases, as shown in Fig. 1.18. However, under normal conditions $I(R_a + R_{se})$ drop is quite small and may be neglected. Thus if the applied voltage remains constant, speed N is inversely proportional to flux Φ. If a curve is drawn between speed and input (or line) current I, it will be a rectangular hyperbola before magnetic saturation as up to saturation point the magnetisation curve is a straight line. In this region, the speed decreases abruptly with the increase in input current. After magnetic saturation, the flux Φ tends to become constant and speed-current characteristic becomes a straight line and speed decreases slightly due to voltage drop in armature and series field, as shown in Fig. 1.18. The speed becomes zero when the input current is the normal short-circuit current of the motor *i.e.,* equal to applied voltage divided by the motor resistance $(R_a + R_{se})$. Normally this current is many many times full-load value.

From speed-current characteristic curve it is obvious that the series motor is a *variable speed motor i.e.,* speed varies with the variation in line current. With the decrease in load on a dc series motor, the speed increases and may become dangerously high at very light loads.

Since on no load the speed is dangerously high, the machine may get damaged due to heavy centrifugal forces set up in the rotating parts. *This is the reason that series motors are never started on no load.*

To start a dc series motor, mechanical load is first put and then the motor is started.

Since on no load the series motor attains dangerously high speed, which causes heavy centrifugal force resulting in the damage of machine, series motors are not suitable for the services:

(*i*) where the load may be entirely removed and

(*ii*) for driving by means of belts because mishap to the belt would cause the motor run on no load.

The motors are suitable for gear drive, because gears provide some load on account of the frictional resistance of the gear teeth in case of sudden release of load.

However, very small series motors may be used with belts, since in case of mishap to belt, the comparatively large frictional resistance would represent an appreciable load on it.

The minimum load on a dc series motor should be great enough (not below 15% of full load) to keep the speed of the motor within limits. In case the speed becomes dangerously high the motor must be disconnected from the supply main.

2. *Torque-Current Characteristic.* From the expression of mechanical torque T it is obvious that torque is directly proportional to the product of flux per pole Φ and armature current I_a. Up to saturation point flux is proportional to field current and hence to the armature current, because $I_a = I_f$. Therefore, on light load mechanical torque T is proportional to the square of the armature current *i.e.,* $T \propto I_a^2$ and hence curve drawn between torque and armature current up to saturation point is a parabola, as shown in Fig. 1.18.

After saturation point, flux Φ is almost independent of excitation current and so the torque is proportional to the armature current *i.e.,* $T \propto I_a$. Hence the characteristic becomes a straight line. The useful (or shaft) torque is, of course, less than the total (or gross) torque developed. This is due to torque lost in iron and friction and windage losses.

From the torque-armature current curve it is evident that so long as the field of the motor is not saturated, the series motor exerts a torque proportional to the square of current *i.e.,* starting torque is very high. Hence series motors are used where large starting torque is required for accelerating heavy masses quickly such as in hoists, electric railways, trolleys and electric vehicles.

From torque-current characteristic it is evident that series motor develops large starting torque to accelerate the heavy masses and from speed-current characteristic it is also evident that speed falls as the load increases, so series motor is automatically relieved from heavy excessive load, therefore, series motor is best suited for electric traction work.

3. *Speed-Torque Characteristic.* For low values of load current,

$$I \propto \phi \text{ and } T = K_3 \phi^2$$

$$\phi = \sqrt{\frac{T}{K_3}} \qquad \text{...(1.30)}$$

Substituting $\phi = \sqrt{\dfrac{T}{K_3}}$ in Eq. (1.28) we have

$$N = K_1 \frac{V}{\sqrt{T/K_3}} - \frac{K_1 . K_3}{K_2} \frac{\phi^2 R_a}{\phi^2} = \frac{C_1}{\sqrt{T}} - D_1 \quad ...(1.31)$$

where C_1 and D_1 are two constants, *i.e.,* the shape of the speed-torque characteristic will be hyperbolic.

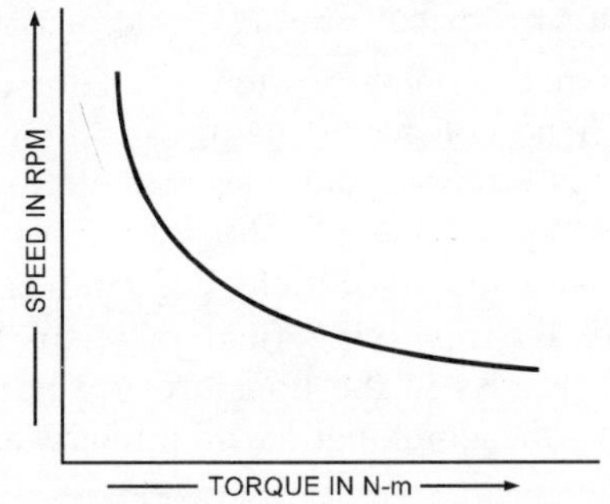

Fig. 1.19. ***Speed-Torque Characteristics of DC Series Motor***

For large values of load current, ϕ remains almost constant and therefore, speed will be given as

$$N = C_2 - D_2 T \quad ...(1.32)$$

where C_2 and D_2 are constants for the machine, *i.e.,* speed-torque characteristic will be linear in nature.

Speed-torque characteristic, is also known as *mechanical characteristic*. Speed sharply falls with the increase in torque for smaller values of load. But at higher loads, the speed drops linearly but slowly with increasing torque.

Hence series motors are best suited for services where the motor is directly coupled to the load such as fans whose speed falls with the increase in load torque.

4. ***Performance Curves of a DC Series Motor.*** In Fig. 1.20, four important characteristics of a dc series motor, namely torque, speed, current and efficiency each plotted against useful output power are shown.

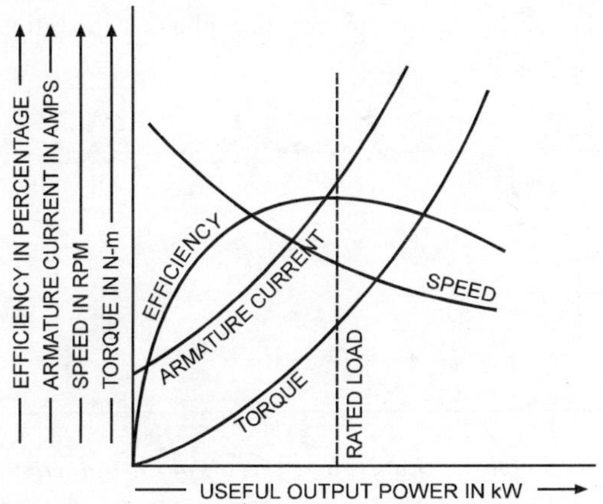

Fig. 1.20. ***Performance Curves of a DC Series Motor***

From the performance curves for a dc series wound motor it is noted that;

(*i*) The speed of series motor falls rapidly with the increase in load, so a series motor is not suitable for services requiring a substantially constant speed.

(*ii*) The efficiency increases rapidly in the beginning, reaches its maximum value and then decreases. This is due to the fact that at light loads the friction and iron losses are large compared with the load and the effect of these losses becomes less with the increase in load. The armature and field copper loss varies as the square of the current, so these losses increase rapidly with the increase in load. The efficiency becomes maximum when friction and iron losses are practically equal to the copper loss.

(*iii*) Series motor develops a starting torque comparatively greater than that developed by a shunt motor for a given current. Hence series motors are best suited where huge starting torque is required *i.e.,* for streetcars, cranes, hoists and locomotives.

(*iv*) In addition to the large starting torque, there is another unique characteristic of series motors which makes them best suited for traction services. For instance let the streetcar ascend a gradient. In case the shunt wound motor is used for driving the above streetcar, it will maintain the speed of the car at approximately the same value that it had on the level ground and, therefore, will tend to draw an excessive current. A series motor, on the other hand, automatically slows down on such a gradient because of increased current demand and, therefore, it develops more torque at reduced speed. This drop in speed allows the motor to develop a large starting torque with but a moderate increase of power. Hence under the same load conditions, the rating of a series motor would be less than for a shunt motor.

(c) Compound Wound DC Motors

Operating Characteristics of Cumulative Compound Wound DC Motors. The characteristics of cumulative compound wound motor are combination of shunt and series characteristics and lie between those of shunt and series motors.

As the load is increased, the flux due to series field winding increases and causes the torque greater than it would have been with shunt field winding alone for a given machine and for a given current. The increase in flux due to series field winding on account of increase in load cause the speed to fall more rapidly than it would have done in shunt motor. The characteristics are shown in Fig. 1.21.

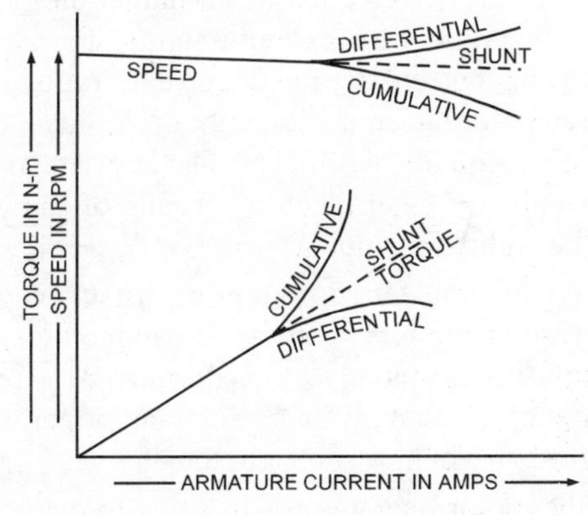

Fig. 1.21. ***Characteristics of Compound Wound DC Motors***

Like the series motor, the cumulative compound wound motor has excellent characteristics for starting heavy loads and yet there is no danger of "running away" at light loads when used without a speed controller.

By varying the proportions of shunt field mmf and series field mmf, compound wound motors may be designed to give speed and torque characteristics anywhere intermediate between those for shunt and series motors. In many cases only enough shunt field is provided to guarantee a safe no-load speed, in which case the motor be called a *stabilised series motor*. Compound wound motors are usually designed so that there will be a drop in speed of about 20% between no load to full load.

Cumulative compound wound motors are used in driving machines which are subject to sudden applications of heavy loads, such as occur in rolling mills, punching and shearing machines, lifts, mine-hoists etc. This type of motor is used also where a large starting torque is required but series motor cannot be employed conveniently such as in cranes and elevators.

Another advantage of the cumulative compound wound motor for suddenly applied loads is that the motor automatically undergoes a substantial drop in speed when the load is applied. Accordingly much of its stored kinetic energy becomes available for supplying a part of the increased load thus reducing the electrical load on the motor as well as the peaks on the power system. The available kinetic energy is frequently increased by the use of a flywheel, particularly with rolling mill motors.

Cumulative compound wound motors are not practical for applications requiring adjustable speed for field control. With a weakened shunt field, the series field becomes dominating hence the changes in load may produce unstable speeds.

1.16.2. Three-Phase AC Motors

(a) Three-Phase Induction Motors. 3-phase induction motors have gained extremely wide application in industry by virtue of the advantages they have over other types of motors. Induction motors are simple in design, rugged in construction and reliable in service, as they have no commutator. Besides this, they have low initial cost, easy operation and simple maintenance, high efficiency and simple control gear for starting and speed control.

The speed-torque characteristics of induction motors are quite important in the selection of an induction motor drive. In addition, the ratio of maximum torque to rated torque, ratio of starting current to rated current, ratio of starting torque to rated torque and the ratio of no-load current to rated current are equally significant. The above characteristics can be conveniently determined by means of the equivalent circuit of the induction motor.

The energy is transferred from primary (stator) winding to secondary (rotor) winding entirely by induction, therefore, induction motor is essentially a transformer. At standstill, the induction motor is actually a static transformer having its secondary (rotor) winding short circuited.

When the motor operates at slip s, the frequency of rotor currents is s times the frequency of the stator currents, therefore, the revolving field produced by the rotor currents revolves with respect to rotor itself at speed

$$\frac{120sf}{P} = \frac{120f}{P} \times s = N_s \times s,$$

Mechanical speed of the motor, $N = N_s(1 - s)$. The speed of the revolving field of the rotor with respect to stator or space is obtained by combining the rotational speed of rotor field with respect to rotor with mechanical speed of the rotor. Hence speed of rotor revolving field with respect to stationary stator or space $= sN_s + N_s(1 - s) = N_s$.

Hence from the point of view of stator, the induction motor still can be considered as a static transformer, even when its rotor is rotating, and it is possible to represent the performance of an induction motor by a transformer phasor diagram. Actually the rotor field does not exist alone but combines with the revolving field of stator to produce a resultant field, just as in the transformer, the resultant field is produced by the combination of primary and secondary ampere-turns.

In the transformer the load on secondary is electrical and in an induction motor the load is mechanical which can be replaced by an equivalent electrical load of resistance R'_L given by

$$R'_L = \frac{R_2}{K^2}\left(\frac{1}{s}-1\right)$$

where R_2 is the resistance per phase of rotor and K is the turn-ratio of the secondary (rotor) to primary (stator).

If R'_2, the effective rotor resistance referred to stator, is combined with R'_L, the fictitious resistance representing electrically the mechanical power developed in rotor, we have

$$R'_2 + R'_L = R'_2 + R'_2\left(\frac{1}{s}-1\right) = \frac{R'_2}{s}, \quad \text{...(1.33)}$$

The simplified equivalent circuit for an induction motor is shown in Fig. 1.22 where V is applied voltage per phase, R_1 and X_1 are stator resistance and leakage reactance per phase, R'_2 and X'_2 are rotor resistance and standstill leakage reactance per phase referred to stator, R_0 and X_0 are the resistance and inductive reactance per phase of the magnetizing branch, I'_2 is the rotor current per phase referred to stator and s is the slip.

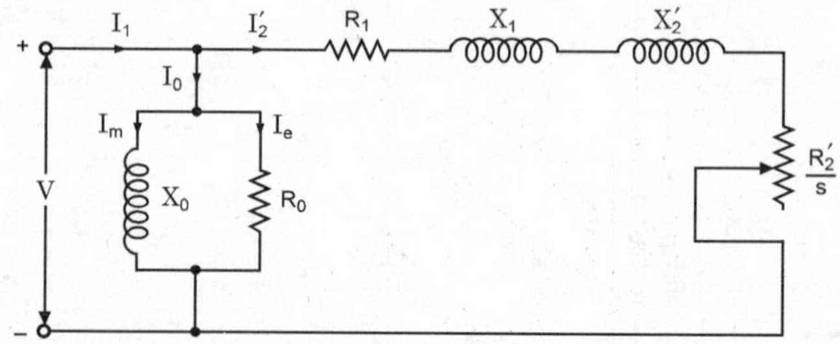

Fig. 1.22. ***Simplified Equivalent Circuit of an Induction Motor***

In accordance with the equivalent circuit, rotor current referred to stator comes out to be

$$I'_2 = \frac{V}{\sqrt{\left(R_1 + \frac{R'_2}{s}\right)^2 + (X_1 + X'_2)^2}} \quad \text{...(1.34)}$$

Using the expression for losses, $\frac{2\pi N_s}{60} Ts = 3(I'_2)^2R'_2$, we have the expression for the torque of the induction motor. The expression for the torque is, therefore, given as

$$T = \frac{3(I'_2)^2 R'_2}{s\frac{2\pi N_s}{60}} \quad ...(1.35)$$

Substituting the value of I'_2, from Eq. (1.34) in Eq. (1.35), we have

$$T = \frac{3V^2R'_2/s}{\frac{2\pi N_s}{60}\left[\left(R_1+\frac{R'_2}{s}\right)^2 + (X_1+X'_2)^2\right]}$$

Newton-metres ...(1.36)

Typical speed-torque curves are illustrated in Fig. 1.23. The two curves are identical; one being drawn for the forward direction of rotation and the other for the reverse. Extension of motoring curve from the first quadrant into the fourth quadrant indicates that the positive torque is developed even though the motor is rotating in the reverse direction. The extension of the same curve into the second quadrant indicates that a negative torque is developed when the machine operates at speeds above synchronous one. Thus regeneration is possible if the system to which the motor is connected is capable of supplying required reactive power for excitation.

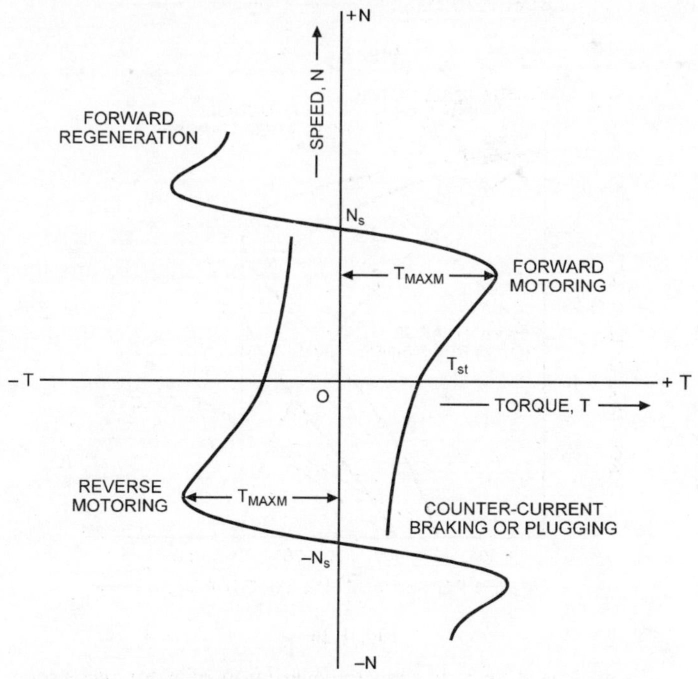

Fig. 1.23. ***Speed-Torque Characteristics of Induction Machine***

By differentiating the torque Eq. (1.36) w.r.t. slip s and equating to zero we get slip corresponding to maximum torque, as

$$s_{max\,T} = \pm\frac{R'_2}{\sqrt{R_1^2 + (X_1+X'_2)^2}} \quad ...(1.37)$$

Substituting the value of slip corresponding to maximum torque from Eq. (1.37) in Eq. (1.36), the maximum torque is obtained as,

$$T_{max} = \frac{3V^2}{\frac{4\pi N_s}{60}\left[R_1 \pm \sqrt{R_1^2 + (X_1+X'_2)^2}\right]}$$

Newton-metres ...(1.38)

Plus and minus sign in Eqs. (1.37) and (1.38) corresponds to motoring (or braking) and generating operation respectively.

Dividing Eq. (1.36) by Eq. (1.38), we have

$$\frac{T}{T_{max}} = \frac{2\left[R_1 + \sqrt{R_1^2 + (X_1+X'_1)^2}\right] \times \frac{R'_2}{s}}{\left(R_1+\frac{R'_2}{s}\right)^2 + (X_1+X_2)^2} \quad ...(1.39)$$

For motor operation, R'_2 in the above Eq. (1.39) can be replaced by its value in terms of $s_{max\,T}$ from Eq. (1.37). After simplification, we have

$$\frac{T}{T_{max}} = \frac{1+\sqrt{\left(\frac{X_1+X_2}{R_1}\right)^2+1}}{1+\frac{1}{2}\sqrt{\left(\frac{X_1+X_2}{R_1}\right)^2+1}\left(\frac{s}{s_{max\,T}}+\frac{s_{max\,T}}{s}\right)} \quad ...(1.40)$$

Neglecting stator resistance R_1, the above expression becomes more simple and approximate one, which is as

$$\frac{T}{T_{max}} = \frac{2}{\frac{s}{s_{max\,T}}+\frac{s_{max\,T}}{s}} \quad ...(1.41)$$

From above Eq. (1.41) it is obvious that if the maximum torque T_{max} and the slip corresponding to maximum torque, $s_{max\,T}$ are specified, the speed-torque characteristic is almost fixed throughout the entire speed range provided that the motor parameters are constant. The above statement is not true for motors with variable rotor resistance.

Effect of Variation in Supply Voltage on Speed-Torque Characteristics. From the viewpoint of practice, it is of prime importance to note here the influence that variations in supply voltage will have on the speed-torque characteristics of induction motor. As is obvious from Eq. (1.36), for any given value of slip the torque developed by the motor will be proportional to the square of applied voltage. This means that the motor will be sensitive to fluctuations in supply voltage.

Since at full load, slip s is very small, therefore, the terms R_1, X_1 and X'_2 can be neglected in comparison with term $\frac{R'_2}{s}$ and the equation for torque may be written as

$$T = \frac{3V^2 s}{\frac{2\pi N_s}{60}R'_2} \quad ...(1.42)$$

or $T \propto sV^2$

If the supply voltage is changed, it changes the torque under running conditions also. With the decrease in supply voltage, torque under running condition decreases. As the motor now develops a reduced torque, the motor slows down (*i.e.,* slip increases) so as to cause increase in torque to meet the load torque and the motor may come to a standstill (if the load torque exceeds the pull-out torque of the motor corresponding to reduced supply voltage) unless the load is removed. Figure 1.24 shows the variation of torque with slip for three different values of line voltage.

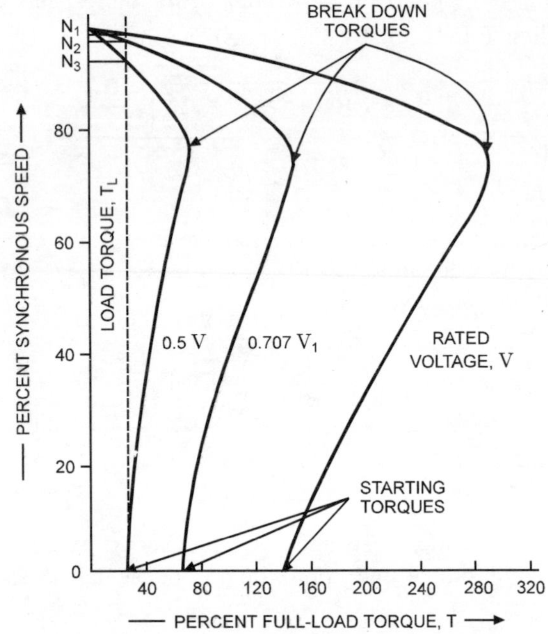

Fig. 1.24

Effects of Varying Supply Voltage and Supply Frequency Together on Speed-Torque Characteristics. This kind of variation can occur for example, if the generator supplying the power to the induction motor (or motors) is subject to speed variations, either due to momentary overloads or because of a noticeable speed regulation. Both output voltage and frequency will vary as the speed if no automatic correction is provided for it. On some large marine drives, the propellor motors are of induction type and are speed-controlled from such a local supply which is provided by synchronous generators coupled to variable speed turbines.

Since emf induced in stator, E_1 is proportional to the product of stator flux ϕ and supply frequency f (*i.e.,* $E_1 \alpha \phi f$), a change of frequency will give rise to a change of flux level unless the induced emf is changed in the same ratio. An imbalance in this ratio will cause either an excessive flux and saturation, or diminished flux and reduced torque per ampere of current. Constant flux per pole is approximated if the supply volts per Hz (V/f) is maintained constant because E_1 is approximately equal to V, the supply voltage. However, an idealized variable-frequency control scheme might aim to ensure that any particular torque should be obtained at the same flux.

With this criterion of controlling $\frac{E_1}{N_s}$ to a constant value for a particular torque, the speed-difference (N_s – N) is also fixed for the same torque, since (N_s – N) in turn determines the slip frequency f', the rotor emf E_2, impedance Z_2 and current I_2, which remain unchanged. From this it follows that as the frequency is reduced, the shape of the torque-slip (or torque-speed) curves do not change, but displaced horizontally, as illustrated in Fig. 1.25.

Effect of Rotor Resistance on Speed-Torque Characteristics. From torque Eq. (1.36) it is obvious that for a constant supply voltage when the rotor resistance R_2 is very small, the term $\frac{R'_2}{s}$ is very small as compared to X'_2 and the torque for a given slip s is directly proportional to R'_2 (neglecting stator impedance). But when the rotor resistance R_2 is very large, the terms R_1, X_1 and X'_2 can be neglected and the torque for a given slip s is inversely proportional to R_2.

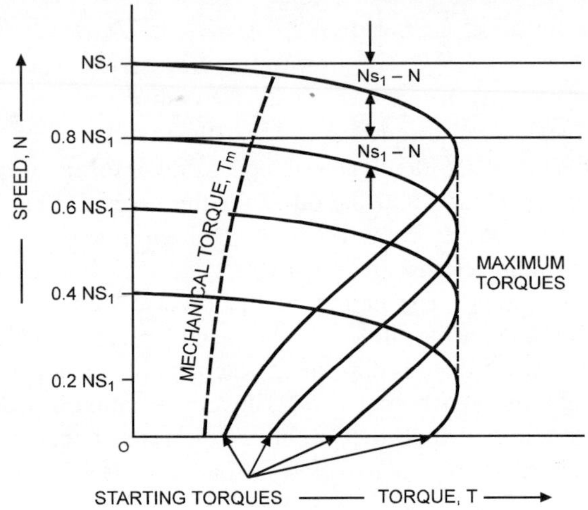

Fig. 1.25

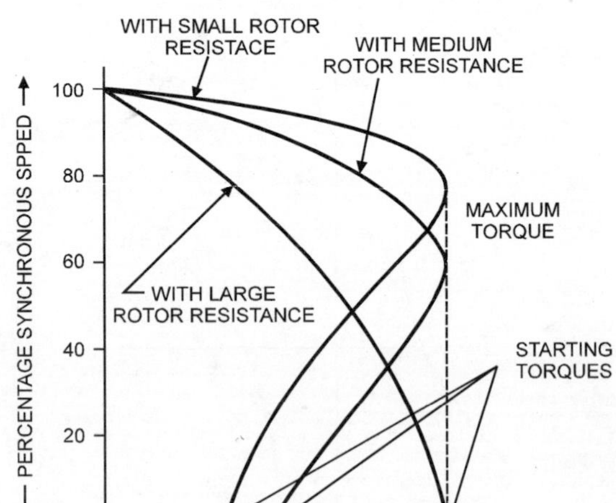

Fig. 1.26

From Fig. 1.26 it is seen that variation of rotor resistance does not change the magnitude of maximum torque T_{max}; merely changes the value of slip at which it occurs. This fact is also obvious from maximum torque Eq. (1.38) and Eq. (1.37) which gives condition for maximum torque. Larger the rotor resistance, the greater is the slip (or smaller is speed) at which the maximum torque occurs.

For a given value of torque T, slip s is proportional to rotor resistance R_2, so addition of external resistance in the rotor circuit does not make the curve narrower but stretches it so that the same torque values occur at lower speeds, as illustrated in Fig. 1.26.

Speed-Torque Curve and Operating Region. The speed-torque curve for an induction motor is given in Fig. 1.27. From Fig. 1.27 it is evident that for any load torque there are two operating points (say B and D). But at position B, the operation is unstable because if there is tendency of speed rise the developed torque also increases than the load torque causing further rise in speed. Thus at position B the operation is unstable. At position D, operation is stable because a tendency to rise in speed will be opposed by the decrease in developed torque. Similarly when there is a tendency to fall in speed, there will be increase in developed torque to bring the motor to operating position D. Thus the region AC of the torque-speed characteristic is *unstable region* of operation while the region CF is stable region of operation.

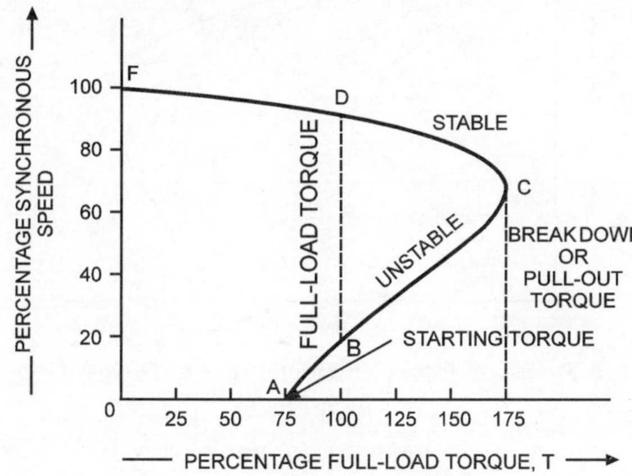

Fig. 1.27

At full load, the motor runs at a speed of N rpm. If the mechanical load applied to the motor is increased, motor speed drops till the torque developed by the motor equals the load torque. As long as the two torques are in balance, the motor will run at a constant speed. However, if the load torque exceeds the breakdown torque of the motor, the motor will immediately stop.

The value of breakdown torque varies with the design of the motor but ranges from 200 to 300 per cent of full-load torque in standard squirrel cage motors.

The speed-torque characteristics of an induction motor in the operating region are quite similar to those of a dc shunt motor. Even at full load the drop in speed from no load does not exceed 5 per cent.

Performance Curves of Induction Motors. Steady-state operating characteristics of an induction motor provides graphically the variation of speed (or slip), power factor, stator current and efficiency with the variation in power output from no load to full load. The performance curves of a 75 kW, 2200 V, 3-phase, 50 Hz squirrel cage induction motor are shown in Fig. 1.28.

1. ***Speed- Output Characteristic.*** The three-phase, squirrel cage induction motor like a dc shunt motor, operates at substantially constant speed from no load to full load. As the rotor cannot attain the speed of the rotating magnetic field (*i.e.,* synchronous speed), it must at all times operate with a certain amount of slip. Because of the extremely low impedance of the rotor, only a slight decrease in speed is enough to cause a large increase in rotor current to develop the torque necessary to drive the load. The per cent slip at no load is less than 1 per cent while at full load it is usually 2 to 5 per cent. This small percentage change in slip (or speed) from no load to full load likewise indicates why a squirrel cage induction motor is considered a constant speed motor.

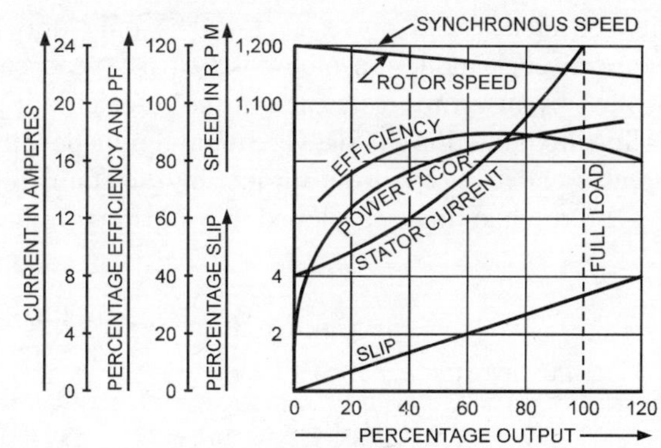

Fig. 1.28. ***Performance Curves of a 75 kW, 2,200 V, 3-phase, 50 Hz Squirrel Cage Motor***

2. ***Power Factor Characteristic.*** Because of the air gap between the stator and rotor of an induction motor, the reluctance of the magnetic circuit is high. As a result an induction motor draws a large magnetising current (I_m) to produce the required flux in the air gap.

The current drawn by an induction motor operating at no load is largely a magnetising current, so no-load current lags behind the applied voltage by a large angle and thus the power factor of an induction motor at no load is quite low (about 0.1 lagging). As the load is increased, the active or power component of the current increases, but the magnetising component remains practically the same. Consequently, the power factor improves. However, because of the large value of magnetising current, which is present regardless of load, the power factor of an induction motor, even at full load, seldom exceeds 90 per cent.

3. ***Efficiency Characteristic.*** At all loads, there are certain fixed losses (such as core loss, friction and windage loss) in addition to variable losses that include stator and rotor copper losses varying nearly as the square of the load. At light loads, the efficiency is quite low because the fixed losses are a relatively large part of the input. With the increase in load, the efficiency, at first, increases rapidly and attains a maximum value, the fixed and variable losses being equal at this point. Beyond this point, the copper losses (I^2R losses) become relatively large, causing the efficiency to decrease.

The maximum efficiency occurs at about 80 to 95% of rated output, the higher values being applicable to larger motors.

4. ***Stator Current Characteristic.*** The no-load current is about 30 to 40 per cent of the rated current, the higher values being applicable to smaller motors. With increase in load, slip increases in a straight line characteristic, the rotor current will likewise increase in practically a direct proportion and cause the stator current to increase accordingly.

Double Cage Rotor Motor. There are three types of 3-phase induction motors namely plain or squirrel cage induction motor, slip-ring or wound rotor induction motor and double cage rotor induction motor.

The simple squirrel cage induction motor is superior to slip-ring induction motor on account of motor output, frame size, initial, maintenance and repair costs, weight, life, reliability, efficiency and power factor. The only drawback of the squirrel cage induction motor is that its speed cannot be controlled by inserting resistance in the rotor circuit as in case of slip-ring induction motor. Maximum torque developed, which is also called the pull-out torque, by the squirrel cage motor, is more than that developed by slip-ring induction motor.

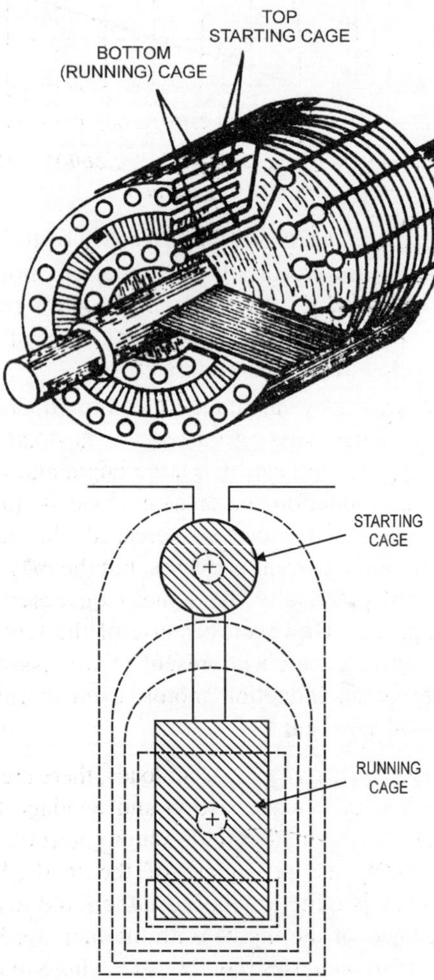

Fig. 1.29

The *double cage rotor motor* is designed to provide a high starting torque with a low starting current. The rotor is so designed that the motor operates with the advantages of high-resistance rotor circuit during starting and a low-resistance rotor circuit under running conditions. The starting torque of a double cage rotor motor ranges from 200 to 250 per cent of full-load torque with a starting current of 400 to 600 per cent of full-load value. The rotor of a double squirrel cage motor carries two squirrel cage windings embedded in two rows of slots (Fig. 1.29). The outer slots contain a high resistance and low leakage reactance winding and the inner slots contain a low resistance and high leakage reactance winding. At starting, the rotor current has the same frequency as the supply current, and the high reactance winding carries very little current, so that the performance approximates to that of high-resistance low-reactance cage alone. At normal speed, however, the rotor current frequency is quite low so that the reactances of the two rotor cages are insignificant compared with the resistances, and the performance approximates to that of the low-resistance cage. The cost of double squirrel motor is 20 to 30% higher than that of an ordinary squirrel cage motor. The speed-torque characteristics of individual cages as well as resultant are shown in Fig. 1.30.

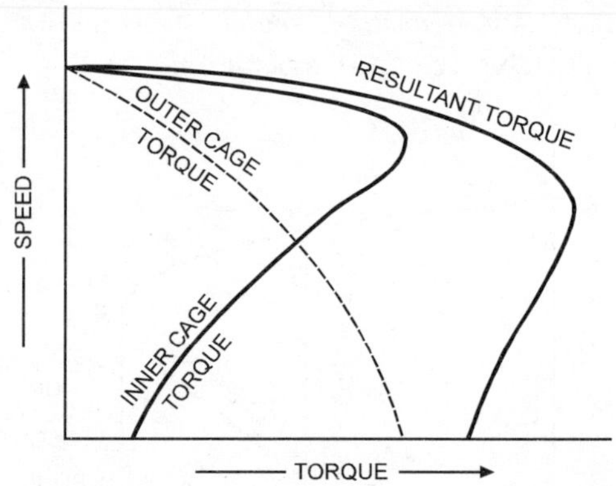

Fig. 1.30. ***Speed-Torque Characteristics of Double Cage Induction Motor***

The choice of an induction motor of any type depends on the requirements of starting torque. *Squirrel cage* induction motor is suitable for constant speed industrial drives of small power where speed control is not required and where starting torque requirements are of medium or low value, such as for printing machinery, flour mills and other shaft drives of small power. *Wound rotor* or *slip-ring* induction motors are used for loads requiring severe starting conditions or for load requiring speed control such as for driving line shafts, lifts, pumps, generators, winding machines, cranes, hoists, elevators, compressors, small electric excavators, printing presses, turntables, strokers, large ventilating fans, crushers etc. *Double squirrel cage motor* is particularly suited where both high starting torque and a small slip on full load are required.

(b) Three-Phase Synchronous Motors. The synchronous motor is a constant speed motor whose speed is fixed by the supply frequency and number of poles and, therefore, is independent of load. Since this motor is not self starting, therefore, some special arrangement is required to be made for making it self starting. The special arrangement may be such as embedding of squirrel cage winding in the pole faces (as used in case of plain synchronous motor) or by arranging the field winding in the form of an ordinary induction motor rotor winding (as used in case of synchronous-induction motor). In both cases the motor starts as a plain induction motor and when it attains speed near the synchronous speed the dc excitation is switched on and the motor pulls into synchronism.

On no load synchronous motor draws very little current from the supply to meet the internal losses. With fixed excitation the input current increases with the increase in load. After the input current reaches maximum no further increase in load is possible. If the motor is further loaded, the motor will stop.

It can operate under a wide range of power factor both lagging and leading. This enables the motor to perform phase compensation action in addition to driving the load.

Torque developed by the synchronous motor varies directly as the voltage where as in an induction motor it varies as square of the applied voltage, therefore, synchronous motors are best suited to withstand large voltage variations.

These motors can be constructed with wider air gaps than induction motors, which make them better mechanically. These motors usually operate at higher efficiencies (92 to 96% as against 87 to 90% for induction motors). Such motors can be made to take large overloads.

Synchronous motors are rarely used below 40 kW output in the medium speed range because of their much higher initial cost in comparison to that of induction motors. In addition they need a dc excitation source, and the starting and control devices which are usually more expensive. Where low speeds and high kW outputs are involved the induction motors are no long cheaper. The various classes of services for which synchronous motors are employed may be classified as (*i*) power factor correction (*ii*) voltage regulation and (*iii*) constant-speed, constant-load drives.

Because of the higher efficiency possible with synchronous motors, they can be employed advantageously for the loads where constant speed is required. Typical applications of high-speed synchronous motors (above 500 rpm) are drives such as fans, blowers, dc generators, line shafts, centrifugal pumps and compressors, reciprocating pumps and compressors, constant speed frequency changers, rubber and paper mills etc.

The fields of applications of low-speed synchronous motors (below 500 rpm) are drives such as reciprocating compressors when started unloaded, dc generators, centrifugal and screw type pumps, vacuum pumps, electroplating generators, line shafts, rubber and band mills, ball and tube mills, chippers, metal rolling mills etc. Flywheel is used for pulsating loads.

High power electronic convertors generating very low frequencies made the operation of synchronous motors at ultra low speeds. Synchronous motors in very large sizes (as high as 10 MW size) operating at ultra low speeds are employed to drive crushers, rotary kilns and variable speed ball mills.

(c) Synchronous Induction Motors. The synchronous induction motor, as its name implies, is a machine which is capable of running both as an induction motor and as a synchronous motor, the former being its mode of operation during the starting period, and the latter its mode of operation during normal running.

The secondary winding of a synchronous induction motor consists of a polyphase winding, almost invariably of the three phase type. Since the rotor is uniformly slotted around the whole periphery, there is no saliency, and thus it is a true round rotor machine.

For the synchronous mode of operation it is necessary to supply the rotor winding with direct current. When the rotor windings of a wound rotor induction motor are excited by a direct current, the mmf distribution develops alternate N and S poles in the same way as does a 3-phase current, but the essential difference is that the dc excitation being fixed, the pole axes due to dc excitation are fixed in space and do not shift as in case when the rotor winding carries ac. These fixed rotor poles get magnetically locked with the rotating magnetic field developed by the 3-phase stator windings carrying ac and the motor runs at a constant speed equal to the synchronous speed.

The synchronous induction motor is provided with a large air gap as provided in a plain synchronous motor. A long air gap gives a stiffer machine with a larger overload capacity. The rotor slots are made fewer in number and larger in size. These machines are provided with a heavy rotor winding in order to have a low slip, which facilitates in pulling it into synchronism. Also in order that the induced emf in the field at the starting may not be too high, the field turns provided are few in number and the excitation voltage is kept low. As the exciter winding serves the purpose of damper winding, so there is no need of providing separate damper winding with synchronous induction motors.

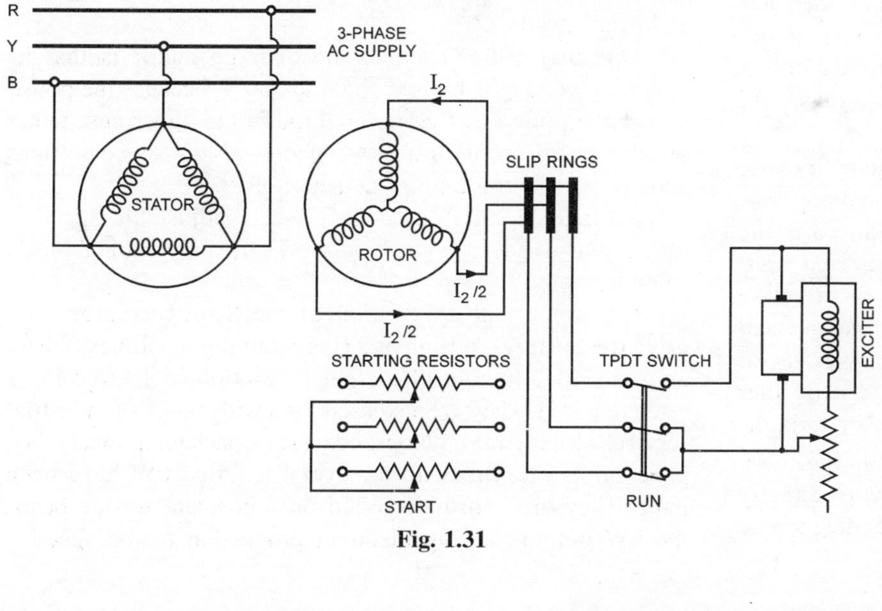

Fig. 1.31

The schematic arrangement of starting a synchronous induction motor is shown in Fig. 1.31. The synchronous induction motor is started as a slip-ring induction motor by inserting resistances in the rotor circuit. When the additional resistances are completely cut out and the motor attains the normal induction motor speed, the rotor is disconnected from the starting resistances and connected to the exciter which is usually mounted on the same shaft. Motor will now be operating as a synchronous motor.

With secondary rheostats for starting, such a motor gives low starting current and high starting torque of the wound rotor induction motor and an improved power factor under load. Hence the synchronous induction motor is essentially a motor, having the induction motor features like high starting torque with low starting current combined with the synchronous motor features like constant speed and power factor control.

The interesting point about this motor is that its peak torque as an induction motor exceeds that as a synchronous motor. If it is momentarily overloaded as a synchronous motor, it may continue to operate as an induction motor-albeit with considerable pulsations of current and torque due to dc until the load falls and it is able to re-synchronise.

However, the extra cost, low efficiency as compared to that of standard types, the small copper space available with a distributed rotor winding, and the necessary compromise between excessive ring voltage at start, or excessive rotor current in operation are sever handicaps.

The synchronous induction motor is rarely built for ratings below 25 kW because of the relatively higher cost of the exciter. Synchronous induction motors are used where a high starting torque and constant speed operation are required. Typical applications of synchronous induction motors include fans, pumps, blowers, generators, air-compressors, ammonia compressors, machinery and line shafting in industrial works such as cement mills, rolling mills, flour mills, paper mills, rubber works and textile mills.

Synchronous induction motors are very often installed alongwith other induction motors, so that they may improve the overall power factor of the system. Their leading kVAR capacity is designed to offset the lagging kVAR demand from the induction motors. They have been made for ratings as high as 30 MW.

(d) Linear Induction Motor. It will be discussed in chapter 12 (Art. 12.7 page 343).

(e) Variable Speed Commutator Type 3-Phase Induction Motor (Schrage Motor). This is also an improvement of the plain induction motor designed to provide a variable speed. It also consists of a commutator winding placed on the rotor in addition to primary winding, whose emf is collected and injected into the secondary winding, placed on the stator. It is provided with two sets of brushes, as illustrated in Figs. 1.32 and 1.33.

The speed can be controlled over 3 to 1 range by moving the brush sets relatively to one another in opposite directions. The speed which depends upon the total emf induced in the secondary winding, will be affected by the injected emf. The speed will decrease if the injected emf is in phase opposition to the induced emf and will increase if the injected emf is in phase with the induced emf. Some power factor control can be had by moving both the brush sets together in the same direction. The operating power factor is high.

Speed-torque characteristic curves are shown in Fig. 1.34. The output power available from the machine is directly proportional to the operating speed. Maximum operating torque ranges from 140 to 250% of full-load speed.

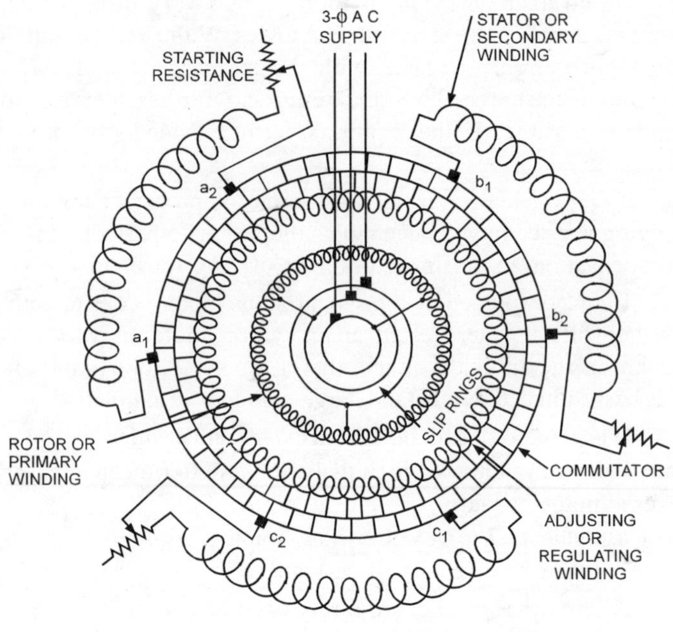

Fig. 1.32. ***Connection Diagram of The Schrage Motor***

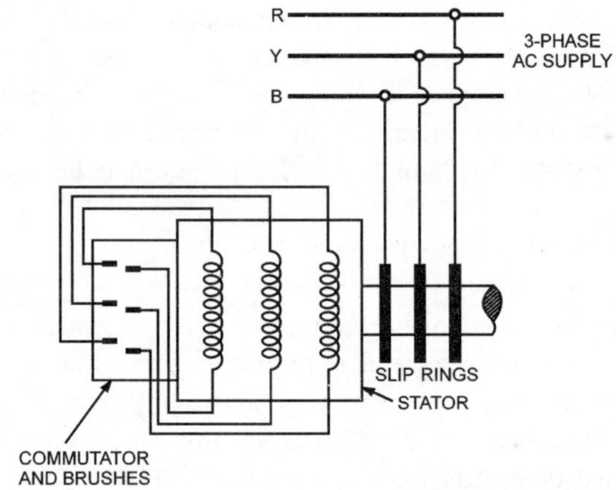

Fig. 1.33. ***General Arrangement of The Schrage Motor***

The main advantages of the Schrage motor are (*i*) continuous speed regulation within the required range (*ii*) high power factor for high speed settings (*iii*) high efficiency at all speeds except synchronous speed and very much high at lower speeds.

One major drawback of the Schrage motor is that its operating voltage is limited about to 700 V, because the power is to be supplied to the motor through the slip rings. Other disadvantages are its low power factor at low speed settings and its poor commutation and high cost.

Schrage motor is used only where adjustable speed is required as in bakery, machinery, stokers, printing machines, calenders etc.

One important application of the Schrage motor is to drive the hosiery-knitting and ring-spinning machines, where an automatically controlled speed variation of 10 to 30% is often required. They are also used for a wide range of industrial services, driving fans, pumps, conveyors, packaging machinery, paper mills etc. Sizes up to several hundred kW have been built. They are normally rated on a constant torque basis, the kW output varying in direct proportion to the speed.

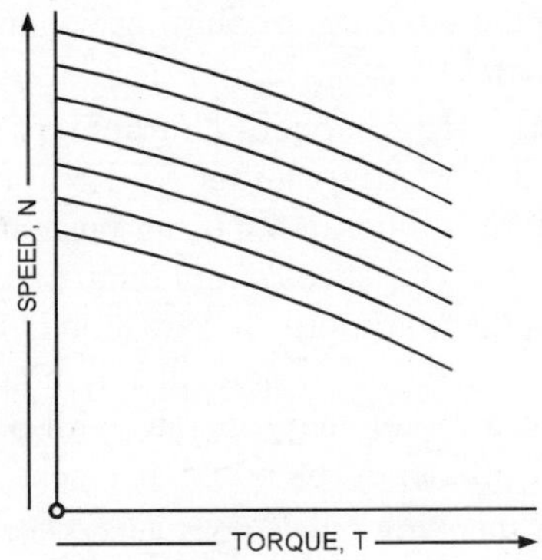

Fig. 1.34

(f) Compensated Induction Motor. This is an improvement of the plain induction motor designed to operate at approximately unity or slightly leading power factor over the whole range of loads. The usual type of motor is *no-lag type* motor, which consists of a primary winding placed on the rotor and the secondary winding on the stator, as shown in Fig. 1.35. The rotor consists of an additional winding, known as *commutator winding,* whose emfs are collected by the brushes from the commutator and injected into the secondary winding in such a way that power factor improvement is accomplished.

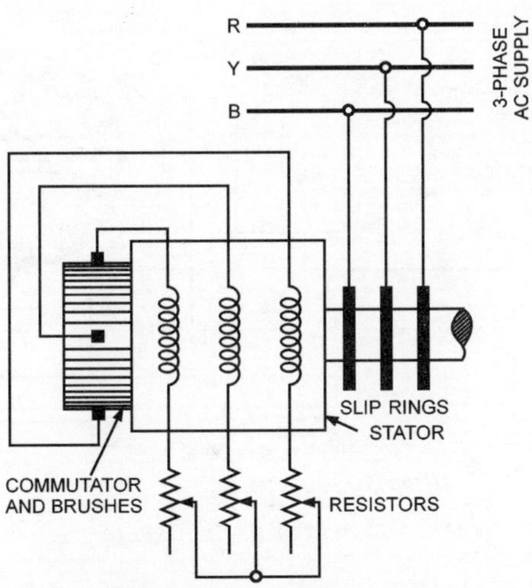

Fig. 1.35. ***General Arrangement of The No-Lag Motor***

No-lag motor was developed earlier and Schrage motor is actually modification form of the no-lag motor. In this motor phase of the emf can be varied but not the magnitude.

(g) Three-Phase AC Series Commutator Motor. This machine has a 3-phase stator winding similar to that of an ordinary induction motor and may be wound for high voltage if required. The rotor carries a winding similar to that of a dc machine and its commutator is provided with three brush sets per pair of poles spaced 120 electrical degrees apart. Connections are shown in Fig. 1.36. In effect, the stator and rotor windings are in series, so that the working flux is dependent on the current providing series speed characteristic. The speed can be controlled by moving the brush gear—the backward movement (*i.e.,* in a direction opposite to that of rotation of the rotor) of brushgear increases the speed and vice versa. For any given position of brushes, the speed falls with the increase in load, as in case of a dc series motor. The power factor is high approaching unity for speeds near and above synchronous speeds. The speed range is limited between 40 and 150 per cent of synchronous speed owing to commutation difficulties. Such a motor is useful when large starting torque is required such as in haulage and hoisting work if only ac supply is available.

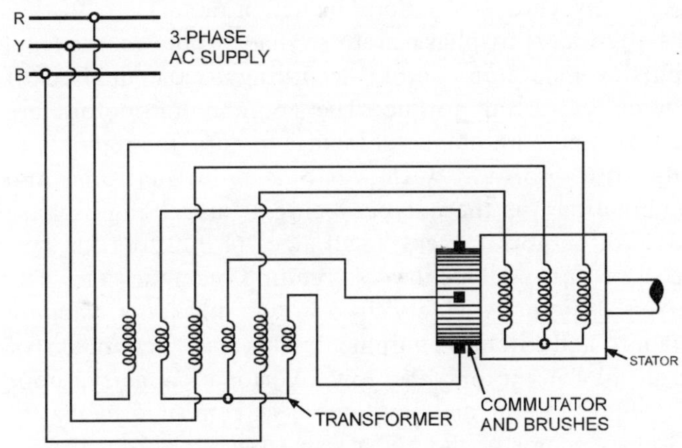

Fig. 1.36. ***Connection Diagram For a 3-Phase AC Series Commutator Motor***

(h) Three-Phase AC Shunt Commutator Motor. The connection diagram of a 3-phase ac shunt commutator motor is shown in Fig. 1.37. In this motor, the stator is connected directly to the supply while the rotor brushes are connected to the supply through a transformer. Speed variation is obtained by changing the tapping points on the transformer. It is to be noted here that brushes are fixed in position in the motor of this type. Such a motor provides more or less constant speed at all loads, for a given tapping on the transformer.

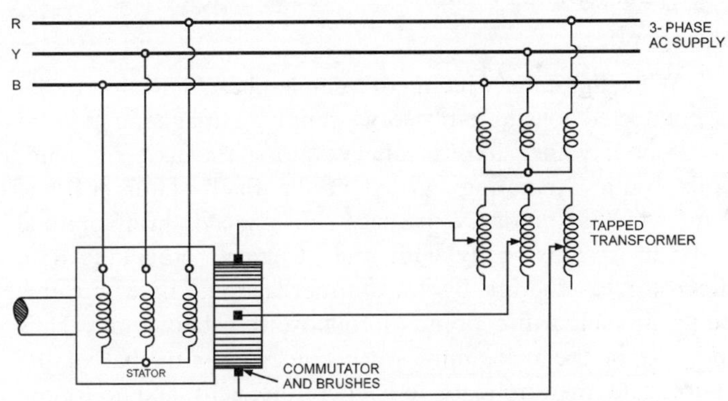

Fig. 1.37. ***Connection Diagram For a 3-Phase AC Shunt Commutator Motor***

1.16.3. Single-Phase AC Motors

Single-phase motors are more widely used than 3-phase motors because of two following basic reasons.

First for reasons of economy, most houses, offices and also rural areas are supplied with single-phase ac and the second factor is the economics of the motor and its branch circuit. Fixed loads requiring not more than 0.5 kW can generally be served most economically with single phase power and a single phase motor. Single-phase motors are simple in construction, reliable, easy to repair and comparatively cheaper in cost and, therefore, find wide use in fans, refrigerators, vacuum cleaners, washing machines, other kitchen equipment, tools, blowers, centrifugal pumps, small farming appliances etc.

Single-phase ac motors may be divided in three general classes namely (*i*) induction motors (*ii*) commutator motors and (*iii*) synchronous motors. Induction motors are further classified as split-phase motors, shaded-pole motors and repulsion-induction motors according to the method of producing starting torque. The commutator motors are the series motors, universal (ac/dc) motors, the repulsion–induction motors, with various modifications and combinations of these types being in use. Single-phase induction motors in very small sizes (1/400 to 1/25 kW) are used in toys, hair dryers, vending machines etc. The universal motor is widely used in portable tools, vacuum cleaners and kitchen equipment. The main drawbacks of single-phase motors are low overload capacity, poor efficiency, low *pf* and low output as compared to that of a 3-phase motor of the given frame size.

(a) Single-Phase Induction Motors. A single-phase induction motor is similar to a 3-phase squirrel cage induction motor in physical appearance. The rotor of a single phase squirrel cage induction motor is essentially the same as that employed in 3-phase induction motors. There is uniform air gap between stator and rotor but no electrical connection between them (stator and rotor). Except for shaded-pole types, the stator core is also very similar. A single-phase motor can be wound for any even number of poles, two, four, and six being most common. Like three-phase machines, adjacent poles have opposite magnetic polarity and synchronous speed equation ($N_s = \dfrac{120f}{P}$) also applies.

When the stator winding of a single-phase induction motor is connected to single-phase ac supply, a magnetic field is developed, whose axis is always along the axis of stator coils. With alternating current in the fixed stator coil the mmf wave is stationary in space but pulsates in magnitude and varies sinusoidally with time. Currents are induced in the rotor conductors by transformer action, these currents being in such a direction as to oppose the stator mmf. Thus the axis of the rotor mmf wave coincides with that of the stator field, the torque angle is, therefore, zero, and no torque is developed at starting. However, if the rotor of such a motor is given a push by hand or by another means in either direction, it will pick up the speed and continue to rotate in the same direction developing operating torque. Thus a single-phase induction motor is not self starting and needs special starting means.

Commercial single-phase induction motors employ the principle of 'phase split' and are, therefore, known as *split–phase motors.* Split-phase motors are of the following types.

1. *Resistance–Start Single-Phase Induction Motor.* It is a form of split-phase motor having a high resistance connected in series with the auxiliary or starting winding, as illustrated in Fig. 1.38 (*a*).

A typical torque-speed characteristic is shown is Fig. 1.38 (*b*). The starting torque is 1.5 times to twice the full-load starting torque and starting current is 6 to 8 times full-load current. The speed regulation of a resistance-start single phase induction motor is very good. The per cent slip in most fractional kilowatt split-phase motors is about 4-6 per cent. Such a motor may operate with a power factor of 0.55-0.65 and efficiency of 60-65 per cent.

Owing to their low cost, resistance start split-phase motors are most popular single phase motors. These motors are made in fractional kilowatt $\left(\dfrac{1}{20} \text{ to } \dfrac{1}{4} \text{ kW}\right)$ ratings with speed ranging from 2,875 to 700 rpm. Such motors find wide applications for low inertia loads, continuous operating loads and applications requiring moderate starting torque such as for driving washing machines, fans, blowers, centrifugal pumps, domestic refrigerators, duplicating machines, office appliances, food processing machines, woodworking tools, grinders, oil burners etc. Because of low starting torque these motors are seldom employed in sizes larger than $\dfrac{1}{4}$ kW.

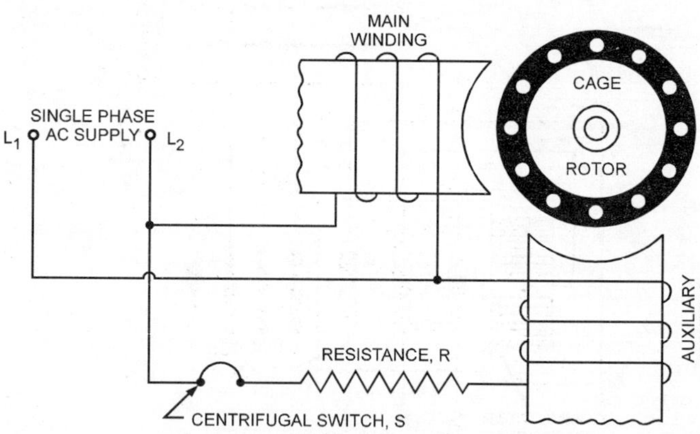

(a) Schematic Diagram

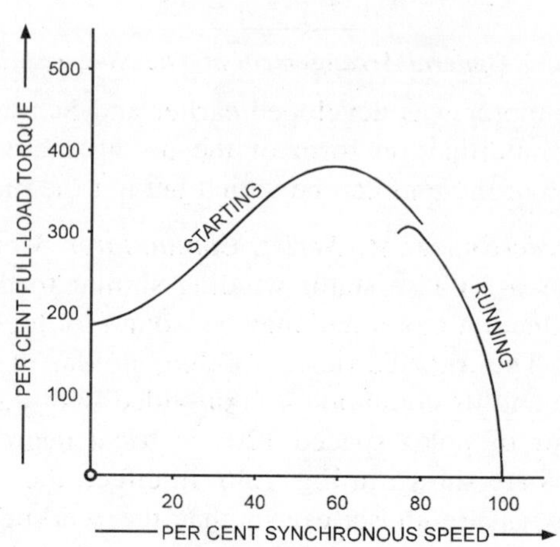

(b) Torque-Speed Characteristic

Fig. 1.38

The direction of rotation of a capacitor-start induction motor may be reversed by inter-changing the connections to the supply of either the main or starting winding. The noteworthy point is that reversal of a split-phase induction motor takes place only when it is at rest and the starting switch is in its normally closed de-energized position.

2. ***Capacitor-Start Single-Phase Induction Motor.*** Fundamentally this motor is very similar to the resistance-start split-phase motor except that the starting winding has a few more turns and consists of a heavier wire than the starting winding of a resistance-start motor; there is also a large electrolytic capacitor in place of resistor. The capacitor-start motor develops a greater starting torque than does an equally rated resistance-start motor. The line current of a resistance-start motor was found to be two-third higher than line current of a corresponding capacitor-start motor, while the starting torque of the capacitor-start motor was twice of that of resistance-start motor.

It is to be noteworthy that the capacitor-start motor differs from the resistance-start motor only during the starting period, for it develops more starting torque (about 3.5-4.5 times full-load torque) with a lower inrush current. After the machine attains normal speed and the starting winding is cut out, its operating characteristics are identical to those of a corresponding resistance-start motor.

Capacitor-start motors are manufactured in rating ranging from $\frac{1}{10}$ kW to $\frac{3}{4}$ kW, but larger sizes are also available. They are widely used for heavy duty general purpose applications like refrigerator units, air-conditioners, compressors, jet pumps, sump pumps, centrifugal pumps, compressors, conveyors, fans, blowers, farm and home workshop tools, oil burners etc.

3. ***Capacitor-Start Capacitor-Run Single-Phase Induction Motor.*** The capacitor start capacitor–run motor is similar to capacitor-start, induction-run motor, except that the auxiliary winding and a capacitor remain connected in the circuit at all times. Of course, such as motor consists of two different capacitors–one small value oil impregnated paper continuous rating capacitor C_R and another much larger value electrolytic short-duty capacitor C_S, as illustrated in schematic diagram given in Fig. 1.39 (*a*).

Such a motor operates as a two-phase motor from single phase supply, thereby, producing a constant torque and not a pulsating torque, as in other single-phase motors. Besides their ability to start heavy loads they are extremely quiet in operation, have better efficiency (55-65%) and power factor (0.8-0.95) when loaded and develop up to 25% greater overload capacities. They are indeed splendid machines where the load requirements are severe as in case of compressors, refrigerators and fire stokers, pumps, conveyors etc. The disadvantage of such machines is only high cost. These motors are often employed requiring a quiet operating motor such as in hospitals, studios etc.

In recent years, home air-conditioners have become extremely popular. In many cases, the electrical system in the home may not have the capacity to handle the electrical load of the air-conditioner(s). For this reason, capacitor-start capacitor-run motor have found more wide spread use in the compressors. Both the starting and running currents are low—full-load current may be $\frac{3}{4}$th of that for an equivalent capacitor-start motor.

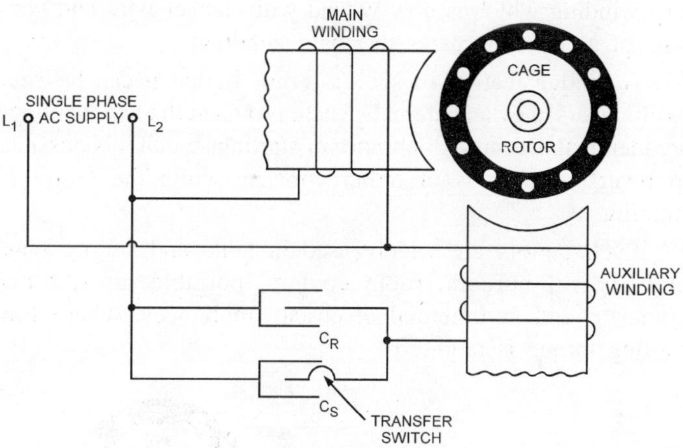

(*a*) *Schematic Diagram*

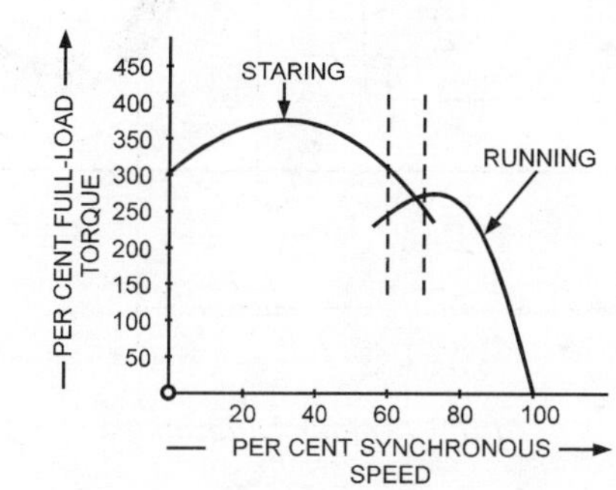

(*b*) *Torque-Speed Characteristic*

Fig. 1.39. ***Capacitor-Start, Capacitor-Run Single-Phase Induction Motor***

Such motors have become increasingly popular and are now manufactured in the larger sized single-phase motors where previously only repulsion-start motors were available. A typical torque-speed characteristic is shown in Fig. 1.39 (*b*). The direction of rotation of motor may be reversed by interchanging the connections to the supply of either the main or auxiliary winding.

4. ***Permanent Capacitor Single Phase Induction Motor.*** Permanent capacitor single-phase induction motor, also sometimes called the *single-value capacitor–run motor*, has two stator windings placed mutually 90 electrical degrees apart. The main or running winding is connected directly across the single phase ac supply. A capacitor in series with the auxiliary winding is also connected across the supply lines.

The auxiliary winding is always in the motor circuit, and therefore, the operation of the motor when loaded resembles more closely to that of a 2-phase motor. One advantage of such operation is that the rotating field developed by the two stator windings is a more nearly uniform, and therefore, the motor is quieter in operation. The other advantages of such a motor are (*i*) increased pull-out torque (*ii*) a higher power factor (0.8-0.95), (*iii*) higher operation efficiency (55-65 per cent) and (*iv*) smaller full-load line current.

The permanent capacitor induction motor is more expensive than the equivalent resistance-start or capacitor—start induction motor mainly owing to the increased cost of auxiliary winding which is now wound with thicker wire and consists of as many turns as the main winding.

A unique feature of such a motor is that it can be very easily reversed by an external switch, provided the two windings are identical. Because of absence of starting switch it is possible to reverse the leads of either winding while the motor is running.

Such motors are widely used in table and ceiling fans, blowers, oil burners, room coolers, portable tools, other domestic and commercial electrical appliances, where low starting torque is required.

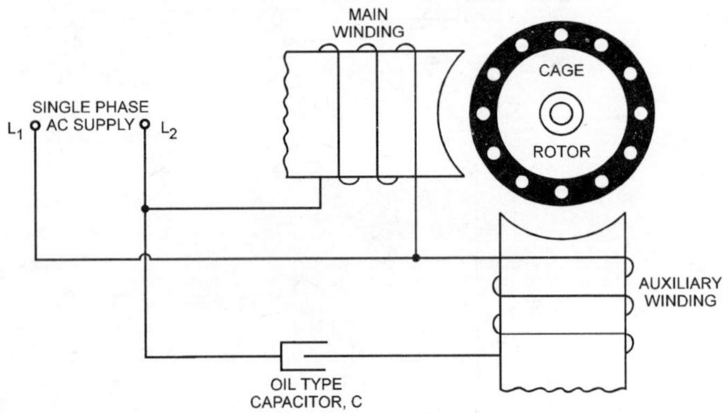

(a) Schematic Diagram

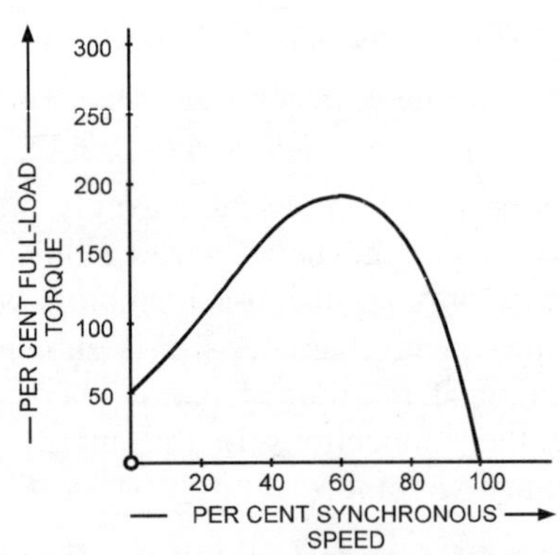

(b) Torque-Speed Characteristic

Fig. 1.40. ***Permanent Capacitor Single-Phase Induction Motor***

Because of the simplicity of reversal of motor, such motors are often used to operate devices that must be moved back and forth frequently such as induction regulators, rheostats, furnace controls, valves and arc-welding controls.

Because of the absence of starting switch, it is possible to run a permanent (or single value) capacitor-run motor over a wide range of speeds, and such a motor has many applications where adjustable speed is required. Such a motor can be arranged for an adjustable speed by the use of a tapped winding or an auto-transformer regulator. It is known that the torque developed in an induction motor is proportional to the square of the applied voltage. For a given load, therefore, if the applied voltage is reduced, the motor will operate with increased slip to develop the required torque. Thus the speed can be controlled by varying the voltage applied to the motor.

5. ***Shaded-Pole Motor.*** As illustrated schematically in Fig. 1.41 (*a*) the shaded-pole motor usually has salient poles with one portion of each pole surrounded by a short-circuited turn of copper called a *shading coil.* Induced currents in the shading coil cause the flux in the shaded portion of the pole to lag the flux in the other portion. The result is like a rotating field moving in the direction from the unshaded portion to the shaded portion of the pole, and a low starting torque is developed. A typical torque-speed characteristic is shown in Fig. 1.41 (*b*). Starting torque is very small (about 40-80% of full-load torque). The efficiency is quite low (20 to 50%). Power factor is also very poor ranging from 0.5 to 0.6. It has little overload capacity. Speed falls with the decrease in applied voltage. Speed can be varied by varying the voltage applied across its stator winding (by using an auto-transformer, tapped reactance coil or by using tapped exciting winding). Shaded-pole motors are least expensive type of small power motors.

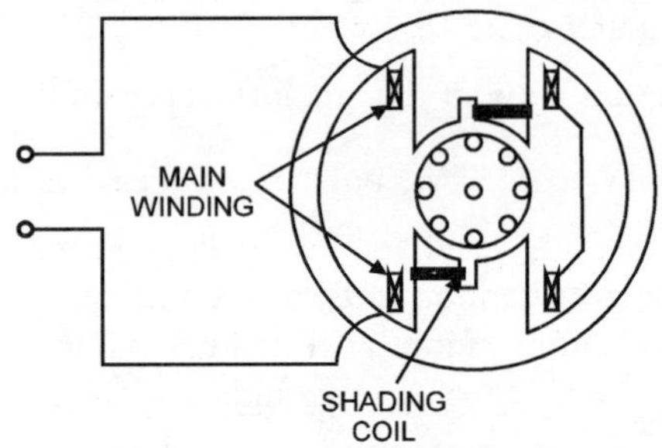

(a) Schematic Diagram

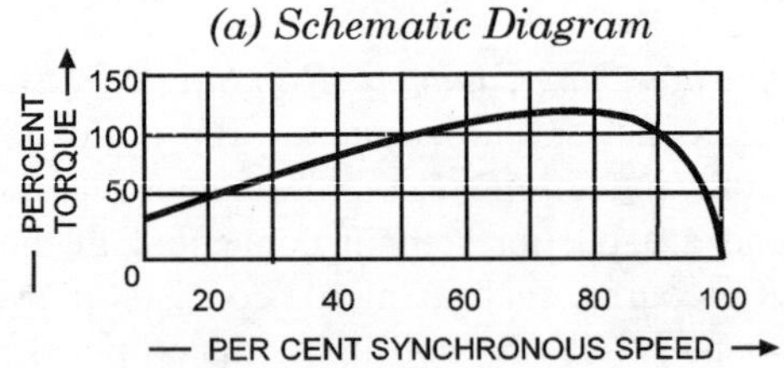

(b) Torque-Speed Characteristic

Fig. 1.41. ***Shaded Pole Motor***

Because of their poor starting torque, low power factor and poor efficiency, such motors are only suitable for low power applications such as for toys, small fans, electric clocks, hair dryers, time phonographs, slide projectors, humidifiers, small business machines (photocopying machines, vending machines, advertising displays etc.) and other similar applications. They are manufactured in small sizes ranging from below 5 W to about 100–150 W. They are available with built-in gear reducers to give almost any speed (as low as below 1 rpm.)

(*b*) ***AC Series Motor.*** DC shunt or series motor rotates in the same direction regardless polarity of supply, *i.e.,* if the line terminals be reversed, the motor continues to rotate in a direction it was rotating before the line terminals were reversed. From this it seems that any dc motor would operate satisfactorily when connected to single phase ac supply.

However, it is not true. Some modifications are necessary in a dc series motor that is to operate satisfactorily on single phase ac supply. So the construction of an ac series motor is very similar to a dc series motor except that some modifications (such as whole magnetic circuit laminated, series field with as few turns as possible, large number of armature conductors, use of high-resistance carbon brushes, numerous poles with lesser flux per pole, very short air gap etc.) are incorporated. The machine is provided with compensating winding and interpoles to improve commutation. A schematic diagram of a single-phase series motor with interpole and compensating windings is given in Fig. 1.42.

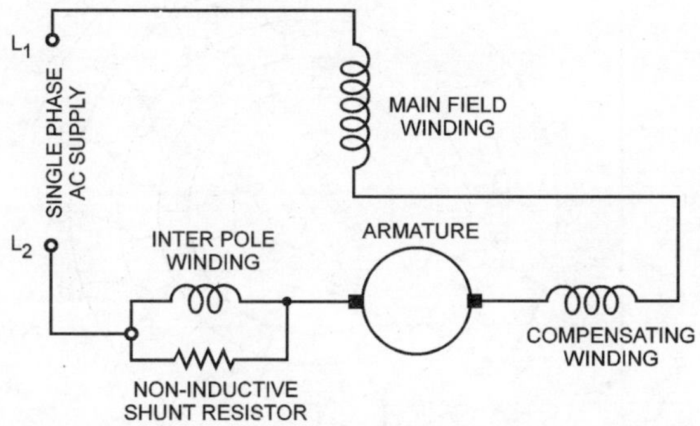

Fig. 1.42. ***Series Motor With Interpole and Compensating Winding***

The average value of the torque on the motor shaft is given as

$$T_{av} = \frac{1}{2\pi\sqrt{2}} Z \frac{P}{A} \phi_{max} \text{ I} \cos \theta \qquad ...(1.43)$$

where I is the effective value of current, ϕ_{max} is the peak value of the flux per pole and θ is the phase angle between phasors ϕ and I.

For a given value of torque T and applied voltage, the armature current is same but voltage drop in case of an ac series motor is much more than that in case of a dc series motor and so speed of an ac series motor for a given developed torque is less than that of a dc series motor, as illustrated in Fig. 1.43.

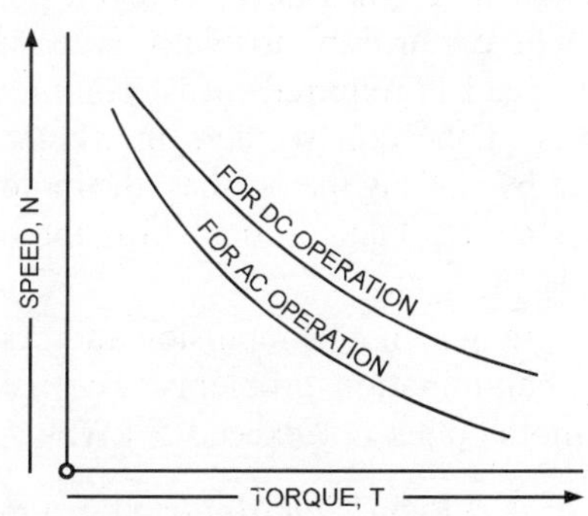

Fig. 1.43. ***Speed-Torque Characteristics of Series Motors***

The single-phase ac series motor has practically the same operating characteristics as the dc series motor. The torque, or tractive effort, varies nearly as the square of the current and the speed varies inversely as the current. This is illustrated in Fig. 1.44.

However, in case of an ac series motor (*i*) power factor is very low at starting and on overloads on account of high inductive nature of the series field and armature circuits (*ii*) efficiency is not as good as in a corresponding dc machine due to eddy current losses and effects of power factor and (*iii*) starting torque is low due to poor power factor at starting.

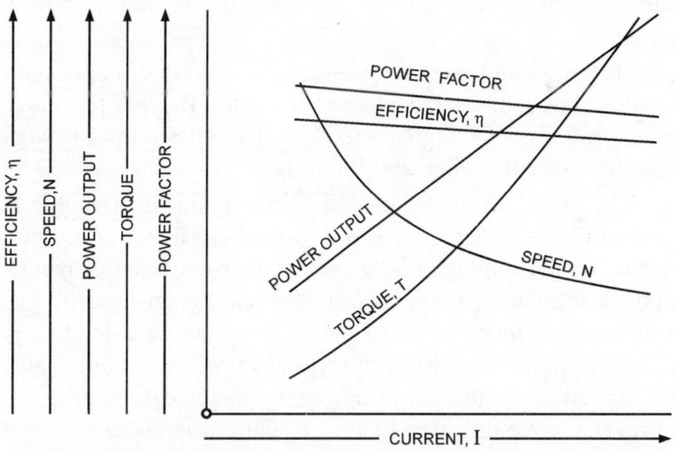

Fig. 1.44. ***Performance Characteristics of AC Series Motors***

For a given kilowatt rating ac series motor is 1.5-2 times in size and weight of the corresponding dc series motor. The construction cost of an ac series motor is much more than that of a dc series motor.

The speed of an ac series motor may be controlled efficiently by taps on a transformer, which is not possible in case of a dc series motor.

The torque-speed characteristic of the single phase series motor is similar to that of the dc series motor *i.e.*, high starting torque and decrease in speed with increase in load making it to have self-relieving property from heavy excessive load, so such a machine is particularly useful for traction services.

(c) Universal Motor. A universal motor is a specifically designed series wound motor, that operates at approximately

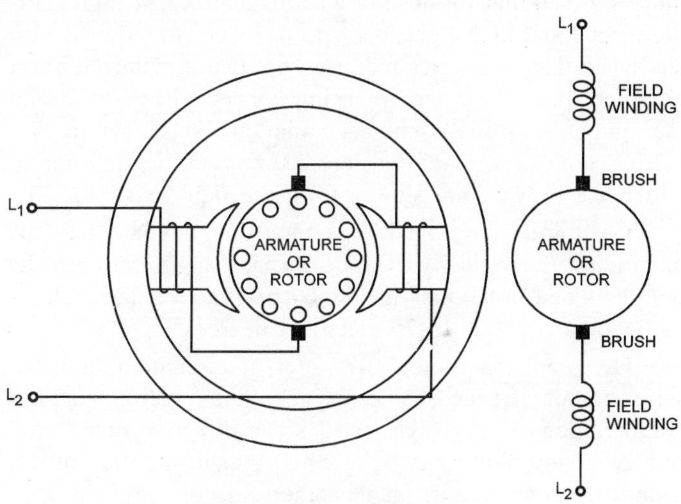

Fig. 1.45. ***Universal Motor***

the same speed and output on either dc or ac of approximately same voltage. Because of the difficulty in obtaining like performance on dc and ac at low speeds, most universal motors are designed to operate at speeds exceeding 3,500 rpm. Motors operating at speeds 8,000 to 10,000 rpm are common. Universal motor is constructed with few series field turns, laminated armature and field circuits, low-reluctance magnetic path, increased armature conductors and commutator segments and using low flux densities so as to minimize the adverse effects caused by high field reactances, eddy current and hysteresis losses.

Universal motors may be either compensated (distributed field) or uncompensated (concentrated field) type, the latter type being used for higher speeds and smaller output ratings (usually not exceeding 200 W) only.

The speed-torque characteristic of a universal motor is quite similar to that of a series wound dc motor, *i.e.,* high starting torque and high no-load speed. Universal motors beings high speed motors, are smaller in size and lighter in weight as compared to other motors of the same output. Full-load pf is high (approximately 0.9) but it is poor at start and on overloads. The direction of rotation of any series motor can be reversed by reversing the direction of flow of current in either the field or armature circuit (but not through both). The speed of universal motor, for any given load, can be changed by changing either flux or applied voltage or both.

Very small power output rating universal motors, which usually do not exceed 5 or 10 watts are employed in such equipment as sewing machines, fans, portable hand tools, hair dryers, motion-picture projectors, and electric shavers. The higher rating (5-500 W) universal motors are used in vacuum cleaners, electric typewriters, food mixers and blenders, motion-picture projectors, cameras, adding machines, and calculating machines.

The small series motors are frequently furnished as motor parts, *i.e.,* consisting of bare stators and rotors (with shaft) but without bearings or feet. They can then be compactly "built in" to the power using devices.

(d) Repulsion Motors. The characteristics of repulsion motors are similar to those of a series motor, *i.e.,* high starting torque and high light-load speed. Its construction is also similar to that of a series motor except that armature is short circuited on itself instead of being connected in series with the stator. Simplified schematic diagram is shown in Fig. 1.46. A repulsion motor develops a torque in the direction in which the brushes are shifted from the field axis.

The torque developed by a repulsion motor should be maximum, theoretically, when the space angle between the pole axis and brush axis is 45°, but in practice the angle of inclination is about 15-25 electrical degrees.

The repulsion motor has better commutation than the series motor at speeds below synchronous speed and poor commutation at very high speeds. The direction of rotation of a repulsion motor can be reversed by shifting the brushes round the commutator, on the other side of the field axis. Speed control can be affected by varying the applied voltage to motor or by mounting the brushes on a rocker which can be rotated by a lever handle mounted on the motor end-shield.

The repulsion motor has high starting torque (about 3-5 times full-load torque) and moderate starting current (about 3-4 times full-load current) but poor speed regulation. Shifting the brushes during operation gives a wide range of speed control, as high as 6:1 ratio, and yet provides a continuous variation. The upper speed is not limited by frequency. The motor is a reversing type, and the direction may be changed during rotation.

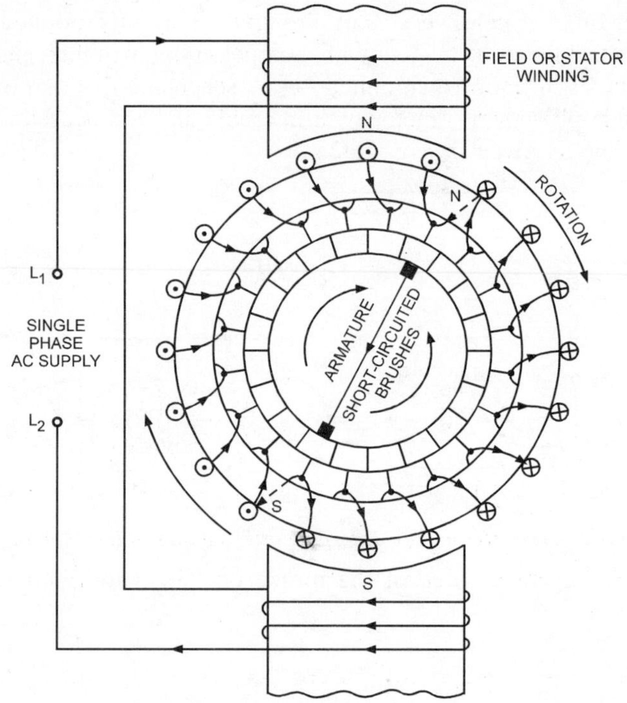

Fig. 1.46. ***Simplified Schematic Diagram of a Repulsion Motor***

The drawbacks of repulsion motors are (*i*) speed variations with the variations in loads-dangerously high at no load (*ii*) low power factor, except at high speeds, (*iii*) tendency to spark at the brushes-sparking at the brushes is negligible at the rated speed which usually occurs near synchronous speed (*iv*) higher cost and (*v*) more attention and maintenance is required.

The repulsion motor has never enjoyed popularity. The motor is used where sturdy motor with large starting torque and adjustable speed is required. Most common use of this type of motor is in the coil winders in which the operator adjusts the speed by shifting the brushes; the motor is equipped with a special lever mechanism, that shifts the brushes when a foot treadle is pressed.

The ratings in which the repulsion motors are built is limited due to commutation problems. The usual rating of the repulsion motor does not exceed 5 kW.

1. ***Compensated Repulsion Motor.*** It is a modified form of the repulsion motor discussed above. It is provided with an additional winding, called the *compensating winding*, to improve power factor and provide better speed regulation. This winding is much smaller than the stator winding and is usually wound in the inner slots of each main pole and is connected in series with rotor through an additional set of brushes placed midway between the usual short-circuited brushes, as illustrated in Fig. 1.47.

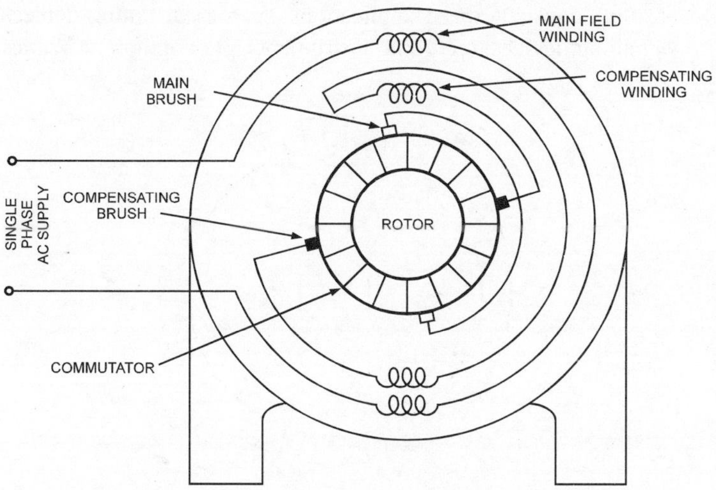

Fig. 1.47. ***Compensated Repulsion Motor***

2. *Repulsion-Start Induction Motor.* As the name implies, the repulsion-start induction motor starts as a repulsion motor and runs as an induction motor.

The general construction of a repulsion-start induction motor is quite similar to that of a repulsion motor (*i.e.,* a laminated stator core complete with one winding similar to the main winding of a split-phase motor and a rotor, also called the *armature,* consisting of slots in which a winding is placed and connected to a commutator). The only difference is that in addition to the basic repulsion-motor construction it is equipped with a centrifugal device which operates at about 75-80 per cent of synchronous speed and short circuits all of the commutator segments.

The repulsion-start induction motor combines the desirable starting characteristics of the repulsion motor (*i.e.,* high starting torque with moderate starting current) with operating characteristic (constant speed running) of the induction motor. A typical performance characteristic of a repulsion-start induction motor is given in Fig. 1.48. The starting torque is 3 to 6 times of full-load torque, and the starting current is approximately 3-4 times of full-load current.

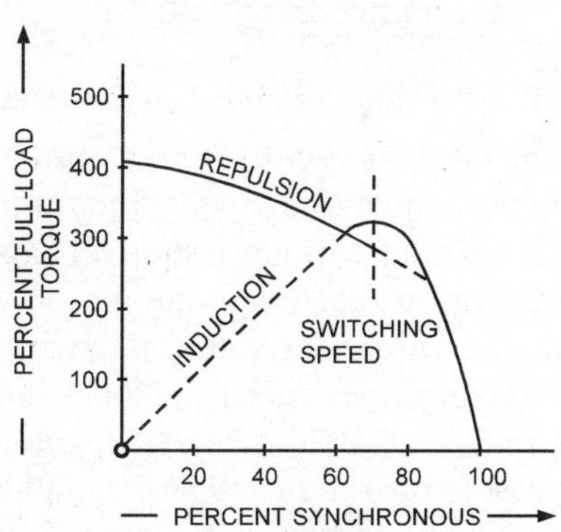

Fig. 1.48. ***Torque-Speed Characteristic For Repulsion-Start Induction Motor***

The direction of rotation of such a motor may be reversed by shifting the brushes, but the brush rigging is not readily accessible, and the method is poorer than with any of the capacitor type motors.

The efficiency and the maximum operating torque of the repulsion-start induction motor are usually less than those of a squirrel cage induction motor of comparative size *i.e.,* a repulsion-start induction motor must be larger than a squirrel cage motor of the same rating to give the same performance.

As compared to capacitor motors, these motors (*i*) need more maintenance because of commutator, brushes and centrifugal switch, (*ii*) are relatively more expensive, (*iii*) cause radio interference during starting, and (*iv*) are noiser, especially when the brushes continue to ride on commutator.

However, despite all the drawbacks, these motors are still used in integral kilowatt ratings because they (*i*) develop higher starting torque with low starting current and (*ii*) can withstand longer starting periods than capacitor type motors.

Such motors are suitable for commercial refrigerators, compressors, pumps, hoists and other constant speed drives, particularly those which have a high inertia and prolonged starting period. The usual rating is from $\frac{1}{3}$ kW to 12 kW but for special applications, ratings as high as 30 kW are available.

3. *Repulsion-Induction Motor.* The repulsion-induction motor has a single phase stator winding, as a respulsion-start

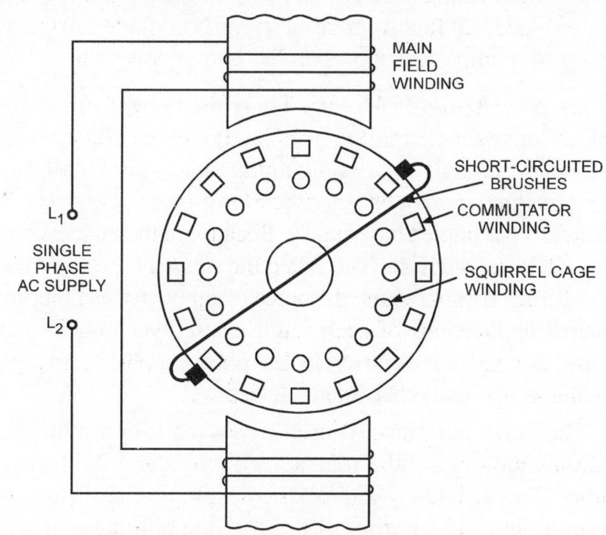

(*a*) *Schematic Diagram*

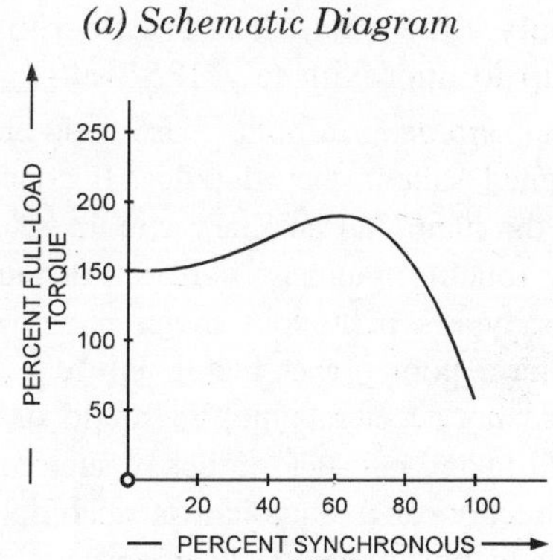

(*b*)

Fig. 1.49. ***Torque-Speed Characteristic of a Repulsion-Induction Motor***

induction motor has, but it has two separate and independent windings on rotor in common slots. The inner winding is a squirrel cage winding with rotor bars permanently short-circuited while outer winding is a repulsion or commutator winding similar to a dc armature winding. When the motor starts, the squirrel cage winding, due to its high reactance, has no effect and the motor starts as a repulsion motor. As the motor speeds up, the current shifts from the outer winding to the inner winding owing to decreasing reactance of the inner winding with the increasing speed and the squirrel cage winding comes into action. At normal speed both of the windings develop torque, and the output of the motor is the combined output of both of the rotor windings.

The starting torque is about 2.5-3 times the full-load torque and its starting current is about 3-4 times the full-load current. The commutation of the motor is good at all speeds. No-load speed is above synchronous speed but normal speed is somewhat below synchronous one. No-load current is rather high, sometimes greater than the full-load current. The motor besides being costlier, has brushes always on the commutator involving sparking and maintenance problems. This motor is particularly suitable where the load can be removed entirely by the declutching or by loose pulley.

Such a motor is used for applications requiring a high starting torque with essentially a constant running speed. The common ratings are 1/6 to 4 kW although in large ratings are also made. Common applications are household refrigerators, garage air pumps, gasoline pumps, compressors etc.

*(e) **Synchronous Motors.*** There are many timing-device applications where small motors having exact constant speed characteristics will be very advantageous. Very small motors with constant speed characteristics have been developed. They operate from a single phase supply. Because of their exact constant speed characteristics, they are called the *single phase synchronous motors.* They do not require dc source of supply for excitation. The principal applications of such single phase synchronous motors are for driving of electric clocks, phonographs, record-player turntables, tape and other timing devices.

The most commonly used types of single-phase synchronous motors are the reluctance motors and the hysteresis motors. The efficiency and torque developing ability of these motors is low. The output of most of the commercial motors available is only a few watts. It is practical to design hysteresis motors up to approximately 125 watts.

Reluctance motor is a split-phase induction motor with properly designed salient shaped poles. It consists of a stator carrying both the main and auxiliary windings for developing a synchronously rotating magnetic field. Rotor punching for a 4-pole reluctance type synchronous motor is shown in Fig. 1.50. Such motors have poor power factor, low efficiency and poor torque. They cannot accelerate high inertia loads to synchronous speed. The pull-in and pull-out torques of such motors are small. Reversal of direction of rotation can be accomplished as in any single phase induction motor. Such motors are widely used for absolute constant speed applications, such as in timing devices, signalling devices, recording instruments, phonograph turntables, control apparatus etc.

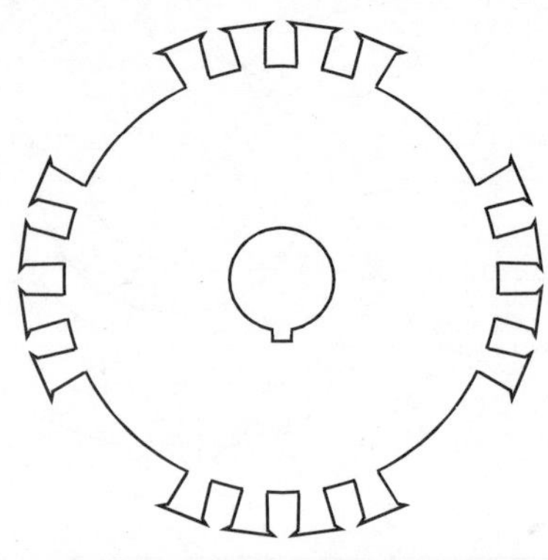

Fig. 1.50. ***Reluctance Motor Rotor***

Hysteresis motor is a synchronous motor with uniform air gap but with no dc excitation. Its operation depends upon the effect of hyteresis. A 2-pole shaded-pole type hysteresis motor employed for driving ordinary clocks is shown in Fig. 1.51. Because of noiseless operation and ability to drive high-inertia loads, hysteresis motors are particularly well suited to drive timing devices, electric clocks, tape decks, turntables and other precision audio-equipment. Commercial motors, being 2-pole motors, run at 3,000 rpm and so for driving an electric clock and other indicating devices, gear train is connected to the motor shaft for reducing the speed. By changing number of stator poles through pole-changing connections, a set of synchronous speeds can be obtained for the motor.

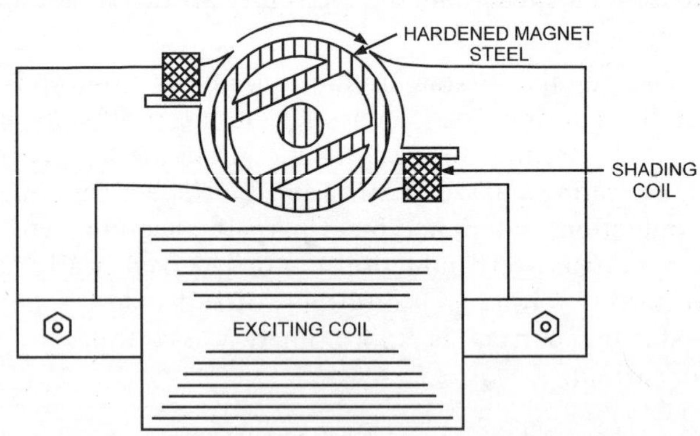

Fig. 1.51. ***2-Pole Shaded Pole Type Hysteresis Motor***

Permanent magnet synchronous motor is constructed by embedding permanent magnets in the rotor, as illustrated in Fig. 1.52. The rotor itself has a squirrel cage construction to provide starting torque. When the motor is connected to a single phase ac supply, it starts as an induction motor, attains nearly synchronous speed and locks into synchronism with the rotating stator field near synchronous speed. Such a motor is quieter in operation and has high pf and efficiency–approaching that of a polyphase motor. It is thus finding more use, even in the low – integral kW ratings (0.5–1.5 kW).

*(f) **Stepper Motor.*** The stepper motor* is a form of synchronous motor which is designed to rotate through a

* A stepper motor is an incremental motion machine *i.e.,* the motor which turns in discrete movement (called the steps) is known as the stepper motor. It does not rotate continuously, as a conventional motor does.

specific number of degrees for each electrical pulse received by its control unit. Typical steps are 2, 2.5, 5, 7.5, and 15° per pulse. The stepper motor is used in digital control systems, where the motor receives open-loop commands as a train of pulses to turn a shaft or move a plate by a specific distance. A typical application for the motor is positioning a worktable in two dimensions for automatic drilling in accordance with hole–location instructions on tape. With a stepper motor a position sensor and feedback system is normally not required to make the output member follow input instructions. Stepper motors are built to follow signals as rapid as 1,200 pulses per second and with equivalent power ratings up to several kilowatts.

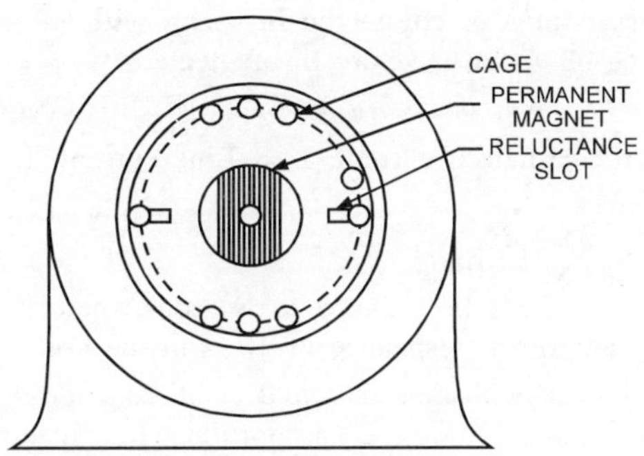

Fig. 1.52. ***Permanent Magnet Synchronous Motor***

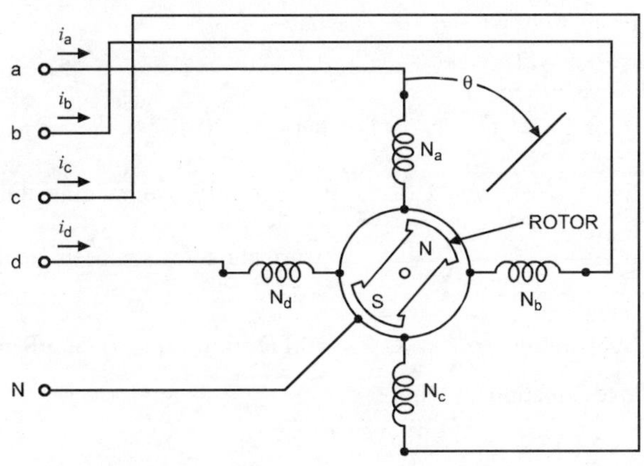

Fig. 1.53. ***Elementary Diagram of a 4-Phase Stepper Motor***

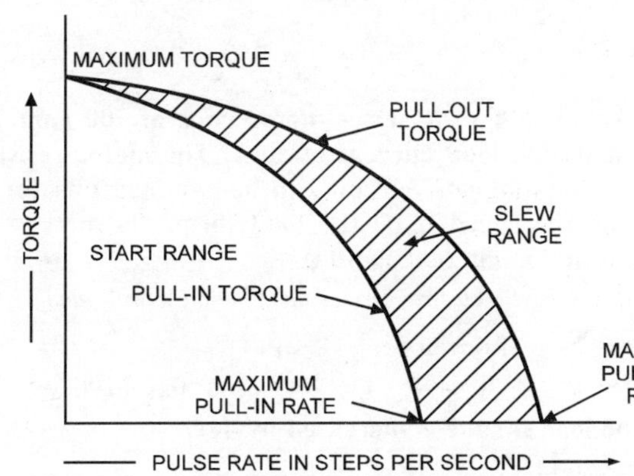

Fig. 1.54. ***Torque Versus Pulse-Rate Characteristics of a Stepper Motor***

Stepper motors are usually designed with a multi-pole, multiphase stator winding that is not unlike the windings of conventional machines. They typically use 3- and 4-phase windings, the number of poles being determined by the desired angular change per input pulse. The rotors are either of the variable reluctance or permanent magnet type. Stepper motors operate with an external drive logic circuit; as a train of pulses is applied to the input of the drive circuit, the circuit delivers appropriate currents to the stator windings of the motor to make the axis of the air-gap field step around in coincidence with the input pulses. Depending upon the pulse rate and the load torque, including inertia effects, the rotor follows the axis of the air-gap magnetic field by virtue of the reluctance torque and/or the permanent-magnet torque.

The elementary operation of a 4-pole stepper motor with a 2-pole, rotor is illustrated in the sequence of Fig. 1.53. The rotor assumes the angles $\theta = 0, 45°, 90°$ ... as the windings are excited in the sequence N_a, $N_a + N_b$, N_b, ... The stepper motor shown in Fig. 1.53 may also be employed for 90° steps by exciting the coils singly. In the latter case, only the permanent-magnet rotor can be employed.

The characteristics of a stepper motor are frequently presented as the torque versus stepping rate of the pulses applied to the drive unit, as illustrated in Fig. 1.54. As the stepping rate is increased, the motor can provide less torque because the rotor has less time to drive the load from one position to the next as the stator-winding-current pattern is shifted.

The stepper motor is essentially a position control device and has the following advantages over a conventional machine :

1. The angular displacement can be precisely controlled without any feedback arrangement.
2. It can be readily interfaced with microprocessor/computer based controller.

Stepper motors have a wide range of applications—paper feed motors in typewriters and teleprinters, positioning of print heads, pens in X-Y graphical plotters, recording heads in computer disk drives and in positioning of worktables and tools in numerically controlled machining equipment.

The stepper motor is also employed to perform many other functions such as metering, mixing, cutting, blending, stirring etc. in many commercial, military and medical applications, usually along, with microprocessor and controlled switches.

Example 1.5. A 220 V, 1,000 rpm, 100 A separately excited dc motor has an armature resistance of 0.1 Ω. The motor is driving a constant torque load equal to the rated torque. Calculate the motor speed if the voltage drops to 200 V.

[*U.P. Technical Univ. Electric Drives 2004-05*]

Solution : At rated load and rated supply voltage

Supply voltage, $V = 220$ V

Current, $I = 100$ A

Back emf, $E_b = V - IR_a = 220 - 100 \times 0.1 = 210$ V

At reduced voltage, $V' = I = 100$ A $\because$ Load torque $T \propto \phi I_a$ is constant and also ϕ is constant

Back emf, $E'_b = V' - I'R_a = 200 - 100 \times 0.1 = 190$ V

$$\text{Motor speed, } N' = N \times \frac{E'_b}{E_b} = 1{,}000 \times \frac{190}{210} = 905 \text{ rpm } \textbf{Ans.}$$

Example 1.6. A 5 kW, 230 V, shunt motor has an armature resistance of 0.5 Ω and a field resistance of 230 Ω. At no load

the motor runs at a speed of 1,000 rpm and draws a current of 3 A. At full load and rated voltage, the current drawn is 23 A and the armature reaction causes a drop of 2% in flux. Determine (*i*) full-load speed (*ii*) full-load torque.

[*U.P.S.C. I.E.S. Electrical Engineering-II, 1998*]

Solution : Shunt field current, $I_{sh} = \frac{V}{R_{sh}} = \frac{230}{230} = 1$ A

At No Load Line current, $I_{L0} = 3$ A

Armature current, $I_{a0} = I_{L0} - I_{sh} = 3 - 1 = 2$ A

No-load back emf, $E_{b0} = V - I_{a0} R_a$
$= 230 - 2 \times 0.5 = 229$ V

No-load speed, $N_0 = 1{,}000$ rpm

At Full Load

Armature current, $I_{af} = I_{Lf} - I_{sh} = 23 - 1 = 22$ A

Back emf, $E_{bf} = V - I_{af} R_a$
$= 230 - 22 \times 0.5 = 219$ V

Flux $\Phi_f = 0.98\ \Phi_o$

$\because$ Reduction in flux at full load is 2%

$$(i)\ \text{Speed, } N_f = N_0 \times \frac{E_{bf}}{E_{b0}} \times \frac{\Phi_0}{\Phi_f}$$

$$= 1{,}000 \times \frac{219}{229} \times \frac{\Phi_0}{0.98\ \Phi_0}$$

$= 976$ rpm **Ans.**

$$(ii)\ \text{Full-load torque, } T_f = \frac{9.55 \times E_{bf} \times I_{af}}{N_f}$$

$$= \frac{9.55 \times 219 \times 22}{976}$$

$= 47.15$ Nm **Ans.**

Example 1.7. A 440 V dc shunt motor has a no-load ($I_0 = 0$) speed of 2,000 rpm. It is running at 1,000 rpm at full-load torque, reduced armature voltage and full field. If load torque is reduced to 50% of rated value with armature voltage and field voltage held constant at previous values, the speed increases to 1,050 rpm. Find the armature voltage drop at full load. Neglect effect of armature reaction. [*GATE 1994*]

Solution : Supply voltage, V = 440 V

No-load speed, $N_0 = 2{,}000$ rpm

No-load back emf, $E_{b0} = V = 440$ V

$\because$ armature current on no load = 0

Full-load speed, $N_f = 1{,}000$ rpm

Half-full load speed, N' = 1,050 rpm

$$\text{Full-load back emf, } E_{bf} = E_{b0} \times \frac{N_f}{N_0} = 440 \times \frac{1{,}000}{2{,}000} = 220 \text{ V}$$

$\because$ Flux is unchanged

$$\text{Back emf at half full load, } E_b = E_{b0} \times \frac{N'}{N_0} = 440 \times \frac{1{,}050}{2.000} = 231 \text{ V}$$

Let the reduced armature voltage be V_1 volts and armature voltage drop at full load be v volts.

Since flux remains unchanged, when load torque is reduced to 50%, armature current would reduce to 50%, torque T being proportional to the product of flux per pole and armature current. Since voltage drop in armature is proportional to armature current, armature voltage drop will also reduce to 50% *i.e.* become $\frac{v}{2}$ when load torque is reduced to 50%.

So $E_{bf} = V - v$

and $E'_b = V - \frac{1}{2} v$

or $220 = V - v$ ...(*i*)

and $231 = V - \frac{1}{2} v$ ...(*ii*)

Solving Eqs. (*i*) and (*ii*), we have

Armature voltage drop at full load, $v = 22$ V **Ans.**

Example 1.8. A 230 V dc shunt motor with constant field drives a load whose torque is proportional to its speed. When running at 750 rpm it takes 30 A. Find the speed at which it will run if a 10 Ω resistance is connected in series with its armature. The resistance of armature may be neglected.

[*A.M.I.E. Sec. B. Electrical Drives and Their Control Winter 1993*]

Solution : Armature current, I_a = Line current, $I_L = 30$ A

neglecting shunt field current

Speed, $N_1 = 750$ rpm

Let the speed be reduced to N_2 rpm where $N_2 = KN_1$ after inserting additional resistance of 10 Ω in the armature circuit.

Since with constant field excitation, torque developed is proportional to armature current *i.e.,*

$T \propto I_a$ ...(*i*)

Load torque, $T \propto N$ ...(*ii*) given

From Eqs. (*i*) and (*ii*), we have

$I_a \propto T \propto N$

$$\text{or } I_{a2} = I_{a1} \times \left(\frac{N_2}{N_1}\right) = K I_{a_1} = 30\ K \qquad ...(iii)$$

Since $N \propto [V - I_a(R_a + R)]$ flux ϕ remaining the same

$$\text{or } \frac{N_2}{N_1} = \frac{V - I_{a2} R}{V} \qquad \because \text{ armature resistance is negligible}$$

Substituting $\frac{N_2}{N_1} = K$, $I_{a2} = 30$ K from Eq. (*iii*), and R = 10 Ω in above equation, we have

$$K = \frac{230 - 30 K \times 10}{230}$$

$$\text{or } K = \frac{230}{230 + 300} = 0.434$$

New speed, $N_2 = KN_1 = 0.434 \times 750 = 325$ rpm. **Ans.**

Example 1.9. A 200 V dc series motor runs at 700 rpm when operating at its full-load current of 20 A. The motor resistance is 0.5 Ω and the magnetic circuit can be assumed unsaturated. What will be the speed if (*i*) the load torque is increased by 44% (*ii*) the motor current is 10 A ?

[*A.M.I.E. Sec. B. Electric Drives and Their Control Summer 1994*]

Solution : Back emf developed,

$E_{b1} = V - I_1 (R_a + R_{se}) = 200 - 20 \times 0.5 = 190$ V

(*i*) When the load torque is increased by 44%

i.e., $T_2 = 1.44\ T_1$

or $\Phi_2 I_2 = 1.44\ \Phi_1 I_1$ $\because T \propto \Phi I_a$

or $I_2^2 = 1.44\ I_1^2$ $\because \Phi \propto I$

or $I_2 = \sqrt{1.44} \cdot I_1 = 1.2 \times 20 = 24$ A

Back emf developed,

$E_{b2} = V - I_2 (R_a + R_{se}) = 200 - 24 \times 0.5 = 188$ V

Since $E_b \propto \Phi N$

$$\text{Speed } N_2 = \frac{E_{b2}}{E_{b1}} \times \frac{\Phi_1}{\Phi_2} \times N_1 = \frac{188}{190} \times \frac{20}{24} \times 700$$

$$= 577.2 \text{ rpm } \textbf{Ans.}$$

(*ii*) **When motor current is 10 A**

Back emf developed,

$$E_{b3} = V - I_3(R_a + R_{se}) = 200 - 10 \times 0.5 = 195 \text{ V}$$

$$\text{Speed, } N_3 = \frac{E_{b3}}{E_{b1}} \times \frac{\Phi_1}{\Phi_3} \times N_1 = \frac{195}{190} \times \frac{20}{10} \times 700$$

$$= 1{,}437 \text{ rpm } \textbf{Ans.}$$

Example 1.10. A 230-V dc series motor has an armature resistance of 0.08 Ω and field resistance of 0.05 Ω. The magnetisation characteristic of the machine at 700 rpm is as follows:

Field current	30	60	90	120	150 A
EMF	70	125	180	210	230 V

The motor is run on a 230 V supply. An additional resistance of 1.5 Ω is connected in series with the armature. Determine the torque and speed when the motor draws a current of 90 A. [*A.M.I.E. Sec. B. Electric Drives and Their Control Winter 1994*]

Solution : Total resistance between motor terminals,

$$R_m = R_a + R_{se} = 0.08 + 0.05 = 0.13\ \Omega$$

Field current, I_f = Line current $I_1 = 90$ A

Additional resistance inserted in armature circuit, R = 1.5 Ω

Back emf developed,

$$E_{b_1} = V - I_1(R_m + R) = 230 - 90\,(0.13 + 1.5) = 83.3 \text{ V}$$

Let speed be N_1

From given datas for magnetisation curve when field current is 90 A

EMF developed, E_2 = 180 V

Speed, N_2 = 700 rpm

Since $N \alpha \dfrac{E_b}{\Phi}$

$$\therefore \frac{N_1}{N_2} = \frac{E_{b1}}{E_2} \times \frac{\Phi_2}{\Phi_1} = \frac{E_{b1}}{E_2}$$

$\because \Phi_2 = \Phi_1$, field current being the same

$$\text{or } N_1 = N_2 \times \frac{E_{b1}}{E_2} = 700 \times \frac{83.3}{180} = 324 \text{ rpm } \textbf{Ans.}$$

$$\text{Torque developed, } T = \frac{9.55 \times E_b I_a}{N} = \frac{9.55 \times 83.3 \times 90}{324}$$

$$= 221 \text{ Nm } \textbf{Ans.}$$

Example 1.11. A dc series motor has been connected as shown in the figure below, for obtaining different speed-torque characteristics.

(*i*) **Prove that the motor torque is maximum when the speed of the motor is given by**

$$\omega = \frac{RR_1 - RR_2 - R_1R_2}{K_f R} \quad \text{where } K_f = \frac{E_b}{I_f\omega} \text{ is the back}$$

emf constant.

(*ii*) **Derive the expression for the maximum torque in terms of voltage, resistances and the torque constant.**

[*A.M.I.E. Sec. B. Electric Drives and Their Control Summer 1996*]

Solution : From circuit diagram shown in Fig. 1.55, we have

$$V = E_b + RI + R_2I_a = K_f\omega I_f + RI + R_2I_a \quad ...(i)$$

where K_f is the back emf constant.

$$\text{Also } V = R_1I_f + RI \quad ...(ii)$$

Comparing Eqs. (*i*) and (*ii*), we have

$$R_1I_f = K_f\omega I_f + R_2I_a$$

$$\text{or } I_a = \frac{I_f}{R_2}(R_1 - K_f\omega) \quad ...(iii)$$

From Eq. (*ii*), we have

$$V = R_1I_f + R(I_a + I_f)$$

$$= I_f(R + R_1) + \frac{R}{R_2}I_f(R_1 - K_f\omega)$$

$$= I_f\left[R + R_1 + \frac{RR_1}{R_2} - \frac{R}{R_2}K_f\omega\right]$$

$$= \frac{I_f}{R_2}\left[RR_2 + R_1R_2 + RR_1 - RK_f\omega\right]$$

$$= \frac{I_f}{R_2}\left[A - RK_f\omega\right] \quad ...(iv)$$

$$\text{or } I_f = \frac{VR_2}{A - RK_f\omega} \quad ...(v) \quad \text{where } A = RR_1 + R_1R_2 + R_2R$$

Also we know that $T = K\phi I_a = K_tI_aI_f$ where K_t is a torque constant.

Substituting $I_a = \dfrac{I_f}{R_2}(R_1 - K_f\omega)$ from Eq. (*iii*), we have

$$T = K_t\frac{I_f^2}{R_2}(R_1 - K_f\omega)$$

Substituting $I_f = \dfrac{VR_2}{A - RK_f\omega}$ from Eq. (*v*) in above equation we have

$$T = \frac{K_tV^2R_2(R_1 - K_f\omega)}{(A - RK_f\omega)^2} \quad ...(vi)$$

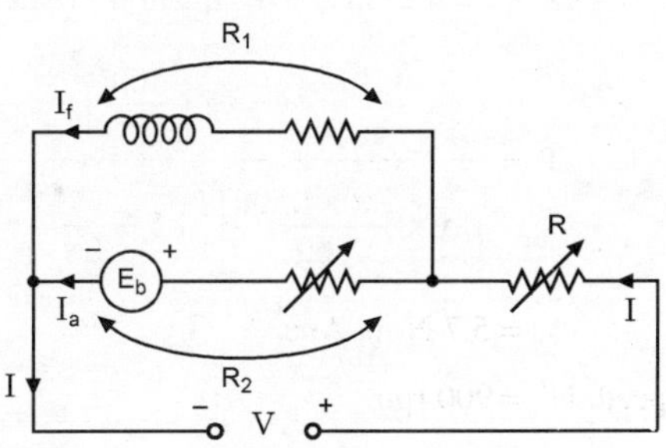

Fig. 1.55

For the torque to be maximum, the differential of T w.r.t. ω is set to zero.

$$\frac{dT}{d\omega} = R_2K_tV^2[(A - RK_f\omega)^2 \times (-K_f) - (R_1 - K_f\omega) \times 2(A - RK_f\omega)(-RK_f)] = 0$$

$$\text{or } (A - RK_f\omega)^2 \times K_f = 2RK_f(A - RK_f\omega)(R_1 - K_f\omega)$$

$$\text{or } RK_f\omega = 2RR_1 - A = 2RR_1 - [RR_1 + R_1R_2 + R_2R]$$

$$= RR_1 - R_1R_2 - R_2R$$

$$\text{or } \omega = \frac{RR_1 - RR_2 - R_1R_2}{K_fR} \quad \textbf{Proved.}$$

Substituting the value of ω in the Eq. (*vi*), we have

$$T = \frac{K_t V^2 R_2 \left[R_1 - \frac{RR_1 - RR_2 - R_1R_2}{R} \right]}{[RR_1 + RR_2 + R_1R_2 - (RR_1 - RR_2 - R_1R_2)]^2}$$

$$= \frac{K_t V^2 R_2 (RR_2 + R_1R_2)}{R[2RR_2 + 2R_1R_2]^2} = \frac{K_t V^2}{4R(R + R_1)} \quad \textbf{Ans.}$$

Example 1.12. A 400 V, 50 Hz, 4 pole, 1,350 rpm star-connected induction motor has the following parameters referred to stator. R_s = 1.5 Ω, R'_r = 4 Ω, X_s = X'_r = 4 Ω, X_m can be neglected for calculation purpose. The motor is controlled using stator voltage controller. When driving a fan load, it runs at a speed of 1,350 rpm at rated voltage. Assuming the load torque to be proportional to speed square for a fan load, calculate the motor torque and terminal voltage at 900 rpm.

[*U.P. Technical Univ. Electric Drives 2004-05*]

Solution : Synchronous speed,

$$N_s = \frac{120f}{P} = \frac{120 \times 50}{4} = 1{,}500 \text{ rpm}$$

Actual speed of rotation,

$$N = 1{,}350 \text{ rpm}$$

$$\text{Slip, } s = \frac{N_s - N}{N_s} = \frac{1{,}500 - 1{,}350}{1{,}500} = 0.1$$

Torque developed by a 3-phase induction motor is given by an expression

$$T = \frac{3V^2 R'_2 / s}{\frac{2\pi N_s}{60}\left[\left(R_1 + \frac{R'_2}{s}\right) + (X_1 + X'_2)^2\right]} \text{ N-m}$$

Refer to Eq. (1.36)

Substituting $V = \frac{400}{\sqrt{3}}$ V, R'_2 = 4 Ω, s = 0.1, N_s = 1,500 rpm, R_1 = 1.5 Ω, X_1 = 4 Ω, X'_2 = 4 Ω in above equation, we have

$$T = \frac{3 \times \left(\frac{400}{\sqrt{3}}\right)^2}{0.1 \times \frac{2\pi \times 1{,}500}{60}\left[\left(1.5 + \frac{4}{0.1}\right)^2 + (4+4)^2\right]}$$

$$= 5.7 \text{ N-m} \quad \textbf{Ans.}$$

When speed, $N' = 900$ rpm

$$\text{Slip, } s' = \frac{1{,}500 - 900}{1{,}500} = 0.4$$

$$\text{Torque developed, } T' = T \times \left(\frac{N'}{N}\right)^2$$

$$= 5.7 \times \left(\frac{900}{1{,}350}\right)^2 = 2.22 \text{ N-m} \quad \textbf{Ans.} \because T \propto N^2$$

$$\text{Torque develope} \propto \frac{V^2 R'_2}{s\left(R_1 + \frac{R'_2}{s}\right)^2 + (X_1 + X'_2)^2}$$

$$\text{So } \frac{T'}{T} = \frac{(V')^2 \times s\left(R_1 + \frac{R'_2}{s}\right)^2 + (X_1 + X'_2)^2}{V^2 \times s'\left(R_1 + \frac{R'_2}{s'}\right) + (X_1 + X'_2)^2}$$

$$\text{or } V' = V \times \sqrt{\frac{T'}{T} \times \frac{s'\left(R_1 + \frac{R'_2}{s'}\right)^2 + (X_1 + X'_2)^2}{s\left(R_1 + \frac{R'_2}{s}\right)^2 + (X_1 + X'_2)^2}}$$

$$= 400 \times \sqrt{\frac{4}{9} \times \frac{0.4}{0.1} \times \frac{\left(1.5 + \frac{4}{0.4}\right)^2 + (4+4)^2}{\left(1.5 + \frac{4}{0.1}\right)^2 + (4+4)^2}}$$

$$= 400 \times \frac{2}{3} \times 2 \times \sqrt{\frac{(11.5)^2 + (8)^2}{(41.5)^2 + 8^2}} = 176.8 \text{ V} \quad \textbf{Ans.}$$

Example 1.13. A 75 kW, 415 V, 4-pole, 3-φ, 50 Hz induction motor is delivering rated torque at 5% slip driving a pump load. The motor delivers maximum torque at 10% slip. A 3-phase ac regulator controls the rms supply voltage to effect limited speed control. Assume the torque-speed characteristics of the motor in the stable region to be linear. Calculate the rms voltage at which the motor runs at 1,350 rpm.

[*A.M.I.E. Sec. B. Electric Drives and Their Control Summer 1996*]

Solution : Full-load slip, s_f = 5% or 0.05

Full-load speed, $N = N_s(1 - s)$

$$= \frac{120f}{P}(1 - s) = \frac{120 \times 50}{4}(1 - 0.05)$$

$$= 1{,}425 \text{ rpm}$$

Slip corresponding to maximum torque, s_{maxT} = 10% or 0.1

Ratio of rotor resistance to standstill reactance,

$$\frac{R_2}{X_2} = s_{maxT} = 0.1$$

or $X_2 = 10 R_2$

When speed, N' = 1,350 rpm

$$\text{Full-load slip, } s'_f = \frac{\frac{120f}{P} - N'}{\frac{120f}{P}} = \frac{\frac{120 \times 50}{4} - 1{,}350}{\frac{120 \times 50}{4}} = 0.1$$

For pump load assuming that the torque varies as the speed so torque corresponding to 1,350 rpm.

$$T' = T \times \frac{1{,}350}{1{,}425}$$

$$\text{Torque developed} = \frac{KR_2 sE_2^2}{R_2^2 + s^2X_2^2}$$

$$\text{So } \frac{T'}{T} = \frac{s'_f (E'_2)^2 / \left[R_2^2 + (s'_f X_2)^2\right]}{s_f E_2^2 / R_2^2 + s_f^2 X_2^2}$$

$$= \frac{s'_f (V')^2 \left[R_2^2 + s_f^2 X_2^2\right]}{s_f V^2 \left[R_2^2 + (s'_f X_2)^2\right]} \quad \because E_2 \alpha V, \text{ the supply}$$

$$\text{So } V' = V \times \sqrt{\frac{T'}{T} \times \frac{s_f}{s'_f} \times \frac{R_2^2 + (s'_f X_2)^2}{R^2 + (s_f X_2)^2}}$$

$$= 415 \times \sqrt{\frac{1,350}{1,425} \times \frac{0.05}{0.1} \times \frac{R_2^2 + (0.1 \times 10 R_2)^2}{R_2^2 + (0.05 \times 10 \times R_2)^2}}$$

$$= 415 \times \sqrt{\frac{1,350}{1,425} \times 0.5 \times \frac{8}{5}} = 361.3 \text{ V } \textbf{Ans.}$$

Example 1.14. An induction motor runs at a slip frequency of 2 Hz when supplied from a 3-phase, 400 V, 50 Hz supply. For the same developed torque, find the slip frequency at which it will run when supplied from a 3-phase, 340 V, 40 Hz system. Slip at which the machine develops maximum torque using 50 Hz supply is 0.1. Neglect stator impedance and assume linear torque-slip characteristic between zero torque and maximum torque in the working region. *[GATE 1992]*

Solution : At 400 V, 50 Hz supply

$$\text{Slip, } s_1 = \frac{f'}{f} = \frac{2}{50} = 0.04$$

Ratio of rotor resistance to standstill reactance,

$$\frac{R_2}{X_2} = s_{\max T} = 0.1$$

$$\text{or } R_2 = 0.1\, X_2$$

Torque developed at slip s_1 of 0.04,

$$T_1 = \frac{K s_1 E_2^2}{R_2^2 + s_1^2 X_1^2} = \frac{K_1 s_1 V_1^2}{R_2^2 + s_1^2 X_1^2} \quad \because E_2 \alpha V$$

$$= \frac{K_1 \times 0.04 \times (400)^2}{(0.1 X_2)^2 + (0.04 X_2)^2} = \frac{6,400\text{ K}}{0.0116\, X_2^2}$$

At 340 V, 40 Hz supply; Let the machine develop torque equal to T_1 at slip s_2

$$\text{So } \frac{K s_2 V_2^2}{R_2^2 + s_2^2 X_2^2} = T_1$$

$$\text{or } \frac{K s_2 \times 340^2}{(0.1 X_2)^2 + s_2^2 X_2^2} = \frac{6,400\text{K}}{0.0116 X_2^2}$$

$$\text{or } \frac{1,15,600 s_2}{0.01 + s_2^2} = \frac{6,400}{0.0116}$$

$$\text{or } 6,400 s_2^2 - 1,340.96 s_2 + 64 = 0$$

$$\text{or } s_2^2 - 0.2095 s_2 + 0.01 = 0$$

$$\text{or } s_2 = \frac{0.2095 \pm \sqrt{(0.2095)^2 - 0.04}}{2}$$

$$= 0.07356 \text{ or } 0.1359$$

Rejecting higher value, slip $s_2 = 0.07356$

Slip frequency $f' = s_2 f = 0.07356 \times 40 = 2.94$ Hz **Ans.**

Example 1.15. A 3-phase, 50 Hz induction motor has $R_s = R'_r = 0.02\ \Omega$. $X_{ls} = X'_{lr} = 0.1\ \Omega$ as the equivalent circuit parameters. The motor is operated at one-half of its rated-voltage and 25 Hz frequency. Calculate the maximum torque at this operating condition in terms of its rated voltage, rated-frequency maximum torque value.

[A.M.I.E. Sec. B. Electric Drives and Their Control Winter 1998]

Solution : Maximum torque at rated voltage and frequency,

$$T_{\max} = \frac{3 V_1^2}{\frac{4\pi N_s}{60}\left[R_1 + \sqrt{R_1^2 + (X_1 + X'_2)^2}\,\right]} \quad \ldots(i)$$

Refer to Eq. (1.38)

Maximum torque at one-half voltage and 25 Hz frequency (*i.e.,* half the rated frequency)

$$T'_{\max} = \frac{3 V_1^2 / 4}{\frac{4\pi}{60} \times \frac{N_s}{2}\left[R_1 + \sqrt{R_1^2 + \frac{(X_1 + X'_2)^2}{4}}\right]} \quad \ldots(ii)$$

From Eqs. (*i*) and (*ii*) we have

$$T'_{\max} = T_{\max} \frac{R_1 + \sqrt{R_1^2 + (X_1 + X'_2)^2}}{2\left[R_1 + \sqrt{R_1^2 + \frac{(X_1 + X'_2)^2}{4}}\right]}$$

$$= \frac{0.02 + \sqrt{0.02^2 + (0.1 + 0.1)^2}}{2\left[0.02 + \sqrt{0.02^2 + \frac{(0.1+0.1)^2}{4}}\right]} T_{\max}$$

$$= 0.9058\ T_{\max} \textbf{ Ans.}$$

1.17. STARTING OF ELECTRIC MOTORS

The most important processes associated with controlled electric drives are (*i*) starting, (*ii*) speed control (*iii*) braking, (*iv*) reversal of direction of rotation. All these are transient* processes that occur in a drive system. Starting of an electric drive involves a change in its state from rest to a steady-state speed of rotation. The process of starting is the most important phenomenon in the entire operation of the drive.

An electric motor is started by switching on the supply. There are, however, three possibilities which have to be considered :

1. Interference with the supply in the form of an excessive voltage drop which is more than that can be tolerated by other equipment or other consumers connected to the same power supply circuit.
2. The starting currents will add to the motor heating by an amount that depends upon their rms values and the frequency of starting. The excessive currents may cause damage to the motor itself.

   In a dc motor the limitation may be good commutation rather than heating, as dc machine have a certain maximum limit for the current dictated by the commutation process.

* The operation of an equipment between two steady-state conditions is known as transient operation and during this period, for an electric drive, the speed, torque and current undergo change.

3. Damage to the connected load through too rapid acceleration. This may also be taken to include unacceptable discomfort to passengers in lifts and trains.

In some cases none of the above is applicable and full-voltage or direct-on-line (DOL) starting is permissible, but in many cases one or more have to be guarded against, usually at the expense of additional equipment.

Necessity of Starting Equipment. If a motor is switched directly on to the supply it will, except in the small sizes, draw a current much above the permissible limit. The main purpose of the starting equipment is, therefore, to limit the starting current to a safe value without, at the same time, reducing the starting torque to a value less than that required. This limitation of current may be accomplished by reducing the voltage applied to the motor by means of a resistance in series with the motor or by some other suitable means as briefed below.

1. *Direct-on-Line Starting or Full-Voltage Starting.* This involves the application of full line voltage to the motor terminals. Whether or not this method of starting is used, depend, upon the following factors :

(*i*) Size and design of the motor (*ii*) kind of application (*iii*) location of the motor in the distribution system and (*iv*) capacity of the power supply circuit and the rules governing such installations as established by power supply companies.

As a general rule, dc motors up to 2 kW and squirrel-cage induction motors as well as certain small synchronous motors up to 4 or 5 kW are usually started by this method.

2. *Reduced-Voltage Starting.* The starting of a dc motor is usually accomplished by inserting a suitable external resistance in the armature circuit and as the motor speeds up the starting resistance is cut out in steps. For details refer to Art. 1.17.4.

Reduced voltage for starting of 3-phase induction motors is achieved by (*i*) stator resistance starting (*ii*) stator reactor starting (*iii*) star-delta starting and (*iv*) auto-transformer starting. For details refer to Art. 1.17.5. The above methods are applicable equally to the synchronous motors. With reduced voltage starting the transition to full voltage may be done either before or after synchronisation although the former method is usually preferred.

Slip-ring induction motors, though can be started by using the starting methods that are used in case of squirrel-cage induction motors, are usually started with full-line voltage across the stator terminals and by introducing variable resistance in each phase of the rotor circuit. The external resistance introduced in each phase of the rotor circuit not only reduces the current at the starting instant but increases the starting torque also. As the motor accelerates, the external resistance is cut out in steps so that the available electromagnetic torque remains maximum during the accelerating period. Ultimately when the machine attains the normal speed, the rotor winding is short circuited automatically.

3. *Starting By Means of Smooth Variation of Voltage or Frequency.* With ac motor-dc generator sets, dc motors can be started by smooth variation of applied voltage and with variable frequency sources both induction and synchronous motors can be started by smooth variation of supply frequency, simultaneously varying proportionally the voltage applied to the motors.

4. *Starting of AC Commutator Motors.* AC commutator motors may be started by applying a reduced voltage or by shifting the brushes. Both of these methods are also employed for speed control, and in most cases starting can be satisfactorily affected by adjusting the control equipment to the low-speed position and then switching on full-line voltage. In some cases it may be necessary to provide in addition special starting tappings on an auto-transformer or a series resistance.

Special Features of Starting Equipment (Starters). There are a number of features common to starters of motor of different types. Firstly, it is desirable to provide some protection against the flow of excessive currents for a prolonged time duration, even though currents of the same magnitude are permissible for a short period during the initial acceleration. Such 'overcurrent' protection can be had by providing means to return the starter to the 'off ' position or by disconnecting the supply by other means. The operation of the device may be achieved electromagnetically by energization of solenoid connected in series with the motor, or thermally by a bimetallic strip heated by the motor current. In the former case a "dashpot" is required to prevent operation during acceleration period; in the latter case, the inherent delay caused by the thermal capacity of the device may be sufficient.

Secondly, it would be pointless to provide starting equipment which remained in the "full-on" position in the event of failure of power supply, since the restoration of supply would cause the very result for the prevention of which the starting equipment was provided. Furthermore, unexpected restarting of the motor and connected load could endanger equipment and personnel. A "no-volt release" is, therefore, provided in all starters, to return them to the "off" position in the event of failure of supply. This applies even to D-O-L starters for the second reason mentioned above.

A third common feature is a matter of definition. Most starters operate in a number of discrete steps and not by continuous smooth variation. It is necessary to define the number of steps, n, as the number of accelerating positions, including "full-on". Since the starter has an "off" position, there are $(n + 1)$ positions in all.

1.17.1. Starting Characteristics of Electric Motors

The study of starting characteristics of a motor is essential to know whether the starting torque that the motor is capable of developing is sufficient to start and accelerate the motor and its load to the rated speed in a reasonable time or not. This feature is very important particularly when the motor is to be selected for the services in which either it is to start against full-load torque (as for driving grinding mills or expellers) or it is to be stopped and started very frequently (as in case of lifts and hoists), *i.e.*, where high starting torque is essential.

The torque developed by the motor at the starting is required. (*i*) to overcome the initial static friction and (*ii*) accelerate the motor and load to the full speed. The torque required to overcome the initial static friction cannot be easily determined and in certain cases as for starting a machine, which has been out of use for some time, it may be much more than the full-load torque.

The torque for accelerating depends upon the load torque. The loads which are usually met with, may be divided according to accelerating torque requirements into the following categories.

(*i*) Load requiring very small accelerating torque in comparison with full-load torque such as when the motor is to be run light.

(*ii*) Load requiring the torque which may increase with speed and it may be proportional to $(\text{speed})^2$ as in case of centrifugal pumps or fans.

(*iii*) Load having constant load torque at all speeds as in case of a hoists, lifts etc.

(*iv*) When the motors have to start and accelerate against full-load torque and in addition to accelerate some heavy moving parts as in case of rolling mills or with motors equipped with a flywheel.

It is obvious that starting of a motor in category (*i*) is quite a different proposition from starting in category (*iv*), even though the full-load torque and speed may be the same in each case; a study of the load characteristic is thus necessary in order to be able to choose the most suitable type of motor and starter.

Another factor of importance is the time taken to run the motor up to speed, since this governs the heating of the motor and control equipment because of the heavy starting currents.

1.17.2. Acceleration Time

The study of behaviour of an electric drive during transient period (*i.e.*, the duration of starting, braking, and transition from one period to another) is of great practical importance, as it helps in determining the requisite control equipment for starting, braking etc. so as to have minimum energy loss during such operation and also to ensure that the machine operates stably after the transients die out.

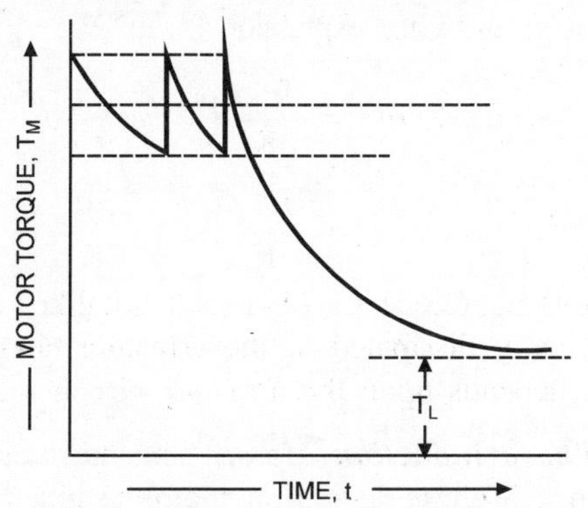

Fig. 1.56. *Motor Torque During Starting*

Determination of the time during which a transient will continue to exist is based on intergration of the equation of motion for the drive, *i.e.*, Eq. (1.11). Solving the equation for time, we have

$$dt = \frac{J\,d\omega}{T_M - T_L} \quad ...(1.44)$$

The time required to change the speed of drive from ω_1 to ω_2 will amount to

$$t = J\int_{\omega_1}^{\omega_2} \frac{d\omega}{T_M - T_L} \quad ...(1.45)$$

In the above Eq. (1.45) the inertia J is assumed to be constant. In case J varies with time or speed, the equation may be rewritten suitably.

For the case, where ω_1 is zero at t equal to zero and ω_2 is equal to a steady-state speed ω_n attained by the drive, accelerating time is given as

$$t = J\int_{0}^{\omega_n} \frac{d\omega}{T_M - T_L} \quad ...(1.46)$$

The above Eq. (1.46) shows that the acceleration time duration depends upon the area under the curve relating speed ω and $\frac{1}{T_M - T_L}$. Also, as the steady-state speed of the motor is reached, the term $(T_M - T_L)$ tends to be zero and hence $\frac{1}{T_M - T_L}$ tends to be infinite and this would lead to an infinite accelerating time. This difficulty is overcome, in practice, by computing the time required to reach say 95 to 98 per cent of the final steady-state speed. The desired value of accelerating time for any motor-load combination can be achieved by suitably modifying the speed-torque characteristic of the driving motor.

Acceleration Time For Specific Nature of Motor and Load Torques. For determination of accelerating time it is necessary to know the motor torque T_M, load torque T_L, and inertia constant J. We now will derive expressions for the acceleration time for different motor and load torque conditions.

(*i*) Constant Motor and Load Torques. Such a situation may arise, for example, during starting of a hoisting mechanism with a constant load torque. If the motor also develops a constant torque during starting, which is possible with shunted armature connections of a dc series motor or rotor resistance starting of a wound rotor induction motor or a double cage induction motor, the net torque available for acceleration $(T_M - T_L)$ will also be constant, and acceleration time, will therefore, be given as

$$t_1 = \frac{J\omega_n}{T_M - T_L} \text{ seconds} \quad ...(1.47)$$

Also, if the motor is started on no load, *i.e.*, $T_L = 0$, the time required to bring the motor from rest to no-load speed is given as

$$t_1 = \frac{J\omega_n}{T_M} \quad ...(1.48)$$

The time t_1 given by the above Eq. (1.48) is called the mechanical time constant of the motor and denoted by t_m. For dc motors and induction motors rated at 1,000 rpm, t_m varies from 0.4 to 0.6 second.

(*ii*) Linearly Varying Motor Torque and Constant Load Torque. For a particular motor (such as for a slip-ring induction motor with high resistance rotor circuit) let the motor starting torque be T_{st} (*i.e.*, when $\omega = 0$, $T_M = T_{st}$) and be zero when speed is synchronous, ω_s (*i.e.*, when $\omega = \omega_s$; $T_M = 0$). The relationship between motor torque and slip s is expressed as

$$\frac{T_{st}}{1} = \frac{T_M}{s}$$

$$\text{or } s = \frac{T_M}{T_{st}} \quad \text{...(1.49)} \quad \text{where } s \text{ is slip.}$$

$$\text{Also } s = \frac{\omega_s - \omega}{\omega_s}$$

$$\text{or } \omega_s ds = -d\omega \quad \text{...(1.50)}$$

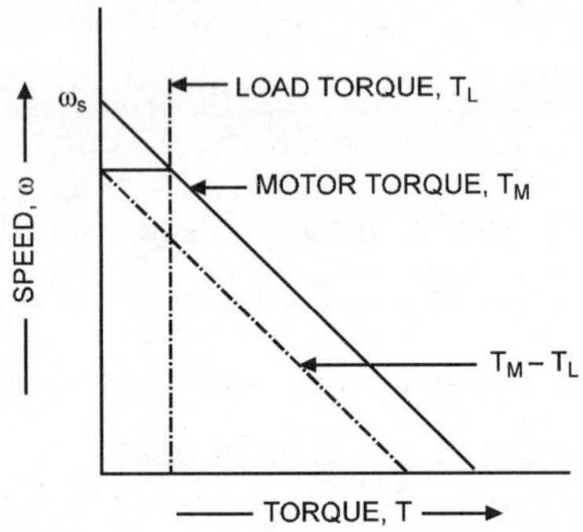

Fig. 1.57. ***Linear Motor Speed-Torque and Constant Load Curves***

Substituting $d\omega = -\omega_s ds$ and $T_M = sT_{st}$ in Eq. (1.46), we have

$$t = J\int \frac{-\omega_s ds}{sT_{st} - T_L} = -J\omega_s \int \frac{ds}{sT_{st} - T_L} \quad \text{...(1.51)}$$

Similarly the acceleration time duration for different motor torque and load torque conditions can be determined using Eq. (1.46).

1.17.3. Energy Consumption During Starting

An expression for the energy lost in the armature circuit of a *dc shunt motor* can be derived as follows.

The voltage equations of a dc shunt motor is

$$V = E_b + I_a R_a = K\omega + I_a R_a$$

$$\text{or } I_a R_a = V - K\omega \quad \text{...(1.52)}$$

The above equation is also valid for a separately excited dc motor.

The equation of motion on no load will be

$$T_M = K I_a = J\frac{d\omega}{dt} \quad \text{...(1.53)}$$

Multiplying Eqs. (1.52) and (1.53), we have

$$I_a^2 R_a = \frac{JV}{K}\frac{d\omega}{dt} - J\omega\frac{d\omega}{dt} \quad \text{...(1.54)}$$

On no load, $I_a^2 R_a$ will be negligibly small and hence $V = K\omega_n$ where ω_n is the no-load speed of the shunt motor.

Substituting $\frac{V}{K} = \omega_n$ in Eqs. (1.52) and (1.54), we have

$$I_a R_a = K(\omega_n - \omega) \quad \text{...(1.55)}$$

$$I_a^2 R_a\, dt = J\omega_n d\omega - J\omega\, d\omega = J(\omega_n - \omega)d\omega \quad \text{...(1.56)}$$

Now energy consumption during starting is given by the expression

$$W_{st} = \int_0^t I_a^2 R_a\, dt = J\int_0^{\omega_n} (\omega_n - \omega)d\omega$$

$$= J\omega_n^2 - \frac{J\omega_n^2}{2} = \frac{J\omega_n^2}{2} \quad \text{...(1.57)}$$

Thus, energy lost in the armature circuit of a dc shunt motor or a separately excited motor during starting on no load will be equal to the kinetic energy stored by the armature and is thus independent of the armature circuit resistance.

If the motor were started with a constant load torque T_L, the energy lost in the armature circuit during starting could be determined as follows.

The equation of motion will, now, be of the form

$$T_M = KI_a = J\frac{d\omega}{dt} + T_L \quad \text{...(1.58)}$$

From Eqs. (1.55) and (1.58), we have

$$I_a^2 R_a = K(\omega_n - \omega)I_a = (\omega_n - \omega)\left[J\frac{d\omega}{dt} + T_L\right] \quad \text{...(1.59)}$$

Now energy consumption during starting is given by the expression

$$W_{st} = \int_0^t I_a^2 R_a\, dt = \int_0^t (\omega_n - \omega)\left[J\frac{d\omega}{dt} + T_L\right]dt$$

$$= J\int_0^{\omega_n} (\omega_n - \omega)d\omega + T_L\int_0^t (\omega_n - \omega)dt \quad \text{...(1.60)}$$

DC Series Motor. Since a dc series motor cannot be started on no load, let us assume that it is started with a constant load torque T_L. The equation of motion will be

$$T_M = KI_a^2 = J\frac{d\omega}{dt} + T_L \quad \text{...(1.61)}$$

Therefore, energy consumed in the armature circuit of a dc series motor is given by the expression

$$W_{st} = \int_0^t I_a^2 R_a\, dt = \frac{JR_a}{K}\int_0^{\omega_n} d\omega + \frac{T_L R_a}{K} t_{st}$$

$$= \frac{JR_a}{K}\omega_n + \frac{T_L R_a}{K} t_{st} \quad \text{...(1.62)}$$

From above Eq. (1.62) it is obvious that unlike in a dc shunt motor, the energy dissipated in the armature circuit of a dc series motor depends upon the armature circuit resistance.

Three-Phase Induction Motor. The mechanical torque developed in a 3-phase induction motor is given as

$$T_M = \frac{3I_2^2 R_2}{\omega_s s} \quad \text{...(1.63)}$$

and the equation of motor under no-load condition, neglecting friction torque becomes

$$T_M = J\frac{d\omega}{dt} = -J\omega_s \frac{ds}{dt} \quad \because \omega = \omega_s(1 - s) \quad \text{...(1.64)}$$

The total energy loss in the rotor circuit of an induction motor during starting period is given as

$$W_{st} = \int_0^{t_s} 3 I_2^2 R_2 \, dt$$

$$= \int_0^{t_s} s\omega_s T_M \, dt = -J\omega_s^2 \int_1^0 s \, ds$$

$$= \frac{J\omega_s^2}{2} \text{ joules} \quad ...(1.65)$$

From above Eq. (1.65) it is revealed that the energy lost in the rotor circuit during starting is equal to the kinetic energy stored by the rotor and is independent of rotor circuit resistance. The total energy loss including the loss in the stator circuit, however, depends upon the rotor and stator resistances. An expression for the total energy lost in the motor during a change in speed can be given as

$$W_M = 3\int_{t_1}^{t_2} I_2^2 (R_1 + R_2) dt \quad \because I_2 \simeq I_1 \quad ...(1.66)$$

During starting, the energy lost in the motor will, therefore, be given as

$$(W_M)_{st} = \frac{J\omega_s^2}{2}\left(1 + \frac{R_1}{R_2}\right) \quad ...(1.67)$$

where J is the moment of inertia of rotor.

The energy loss during starting of a motor can be reduced by adopting the following methods :

(*i*) reduction in moment of inertia of rotor.

(*ii*) smooth variation of applied voltage in case of dc shunt motors.

(*iii*) smooth variation of supply frequency (V/*f* control) in case of induction motors.

1.17.4. Starting of DC Motors

As already stated [refer to Art 1.16.1 Eq. (1.27)], the torque of a dc motor is directly proportional to the product of flux and armature current and is quite independent of speed, *i.e.*, $T \propto \phi I_a$. Hence in order to have high starting torque for a given armature current, the flux must be increased to the maximum value possible.

In case of a *dc shunt motor,* the flux ϕ remains constant as field is connected directly across the constant voltage supply mains and the armature current is controlled by connecting a starting resistance in series with the armature. The torque, which is directly proportional to armature current (as shown in Fig. 1.58) is limited by the maximum starting current–full-load current will develop full-load torque, twice full-load current will provide twice full-load torque, and so on. As the motor speeds up the starting resistance is cut out.

In case of a dc shunt motor reduced-voltage starting method cannot be used because a reduction in supply voltage would reduce the flux proportionately and hence the starting torque.

In the case of a dc series motor as the field winding is connected in series with the armature therefore, the current in the series field winding and the armature is the same. Since up to saturation point the flux is directly proportional to the current flowing through the field and after saturation point the flux is independent of current and remains almost constant, the torque, therefore, varies as the square of the armature current up to saturation point and varies directly as the armature current beyond saturation point. The torque-armature current characteristic is shown in Fig. 1.58.

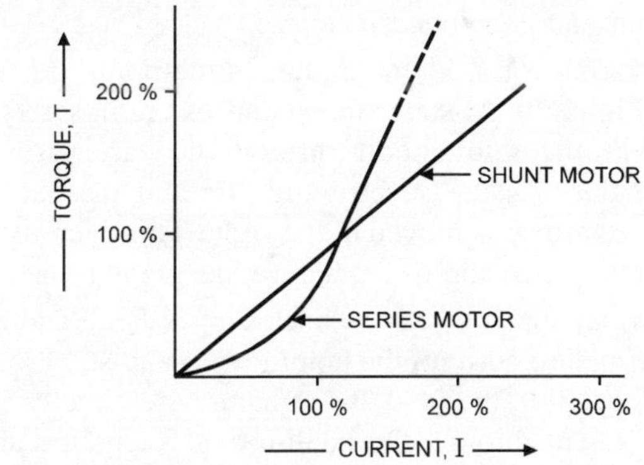

Fig. 1.58. ***Torque-Current Curves For DC Motors***

In case of series wound motor since torque is independent of the applied voltage and the speed, hence reduction of voltage applied across the series motor is a very suitable method for its starting, which is achieved by connecting a resistance in series with the motor.

When the series motors operate in pairs, series-parallel method of voltage control (as described in chapter 13) can be employed.

Starters For Shunt and Compound Wound DC Motors. As mentioned above, starters are used to limit the starting current to a safer value and obtain the required starting torque.

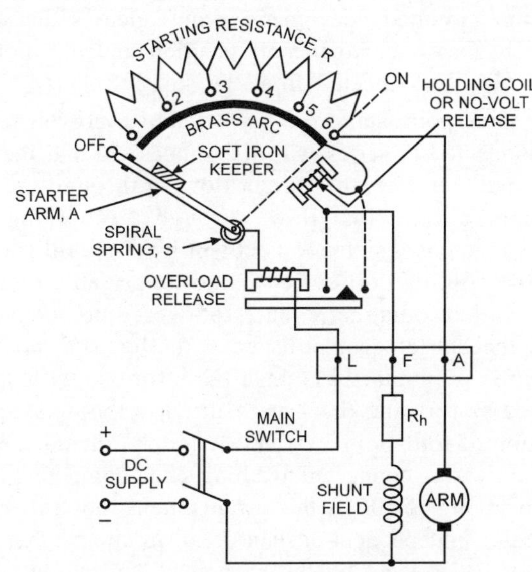

Fig. 1.59. ***Three-Point Starter***

In this context it is pertinent to note that while starting dc shunt and compound motors, it is advantageous to keep the field excitation at its maximum value. A large field current, therefore, a higher value of flux will result in a low operating speed and in higher motor torque for a particular value of starting current because motor torque is proportional to the product of flux per pole and armature current. Thus for a given load torque, the motor will accelerate quickly and take less time to reach the lower operating speed from the starting instant. This will result in less heating of the armature during starting. Thus the rheostat, in series with the shunt field winding, should be at zero resistance position at the time of starting of the dc shunt and compound motors.

The starter for a shunt motor (three-point starter) is shown in Fig. 1.59. At start all resistances are in series with the armature, the holding coil (through the starter arm A) is across the line. As the speed builds up and the back emf increases the arm A is moved to the right. This decreases the armature resistance and the armature has greater emf.

At normal speed when the back emf is large enough to limit the armature current, the holding coil and starter arm A make contact. Both the field and armature are directly across the line. Current through the holding coil keeps the arm in this position. If for any reason the line voltage should fail, the magnetic field of the holding coil ceases and under the action of the spring the arm goes back to its off position. This will keep the line voltage open. Because of this action the starter protects the motor. Restarting after a line voltage failure can only be done by using the protective series resistance in the armature circuit. The starter is also provided with a overload release coil to protect the motor against the flow of excessive current due to overload. This coil is connected in series with motor and so carries full-load current. When the motor is overloaded it draws heavy current, which also flows through the coil and magnetises it to such an extent, that it pulls its armature upwards and so short circuit the no-voltage release coil, as shown in Fig. 1.59. The no-volt release coil being short circuited demagnetises and releases the starter arm which goes back to the off position and the motor is automatically disconnected from the supply mains.

In a three-point starter, discussed above, no-volt release coil is connected in series with the shunt field and the field rheostat, and therefore, the current flowing through the field is the same current that flows through the holding coil. Thus if sufficient resistance is cut in by the field rheostat so that the holding coil current is no longer able to create sufficient electromagnetic pull to overcome the spring tension, the starter arm pulls back to the 'off' position. Hence three-point starter is unsuitable for use with speed-controlled motors and that has resulted in the wide spread application of four-point starters. A four-point starter with its internal wiring connected to a long shunt compound motor is shown in Fig. 1.60. In this arrangement 'no-volt release coil' circuit is independent of shunt field circuit and, therefore, it will not be affected by the change of the current in the shunt field circuit.

DC Series Motor Starter. The series motor starter serves the same purpose as the three- and four-point starters employed with shunt and compound wound motors. However, series motor starter has different internal and external connections.

A series motor starter with holding coil and overload release is illustrated in Fig. 1.61. Holding coil protects the motor against "no volt" and "no load" whereas overload release protects the motor against the flow of excessive current.

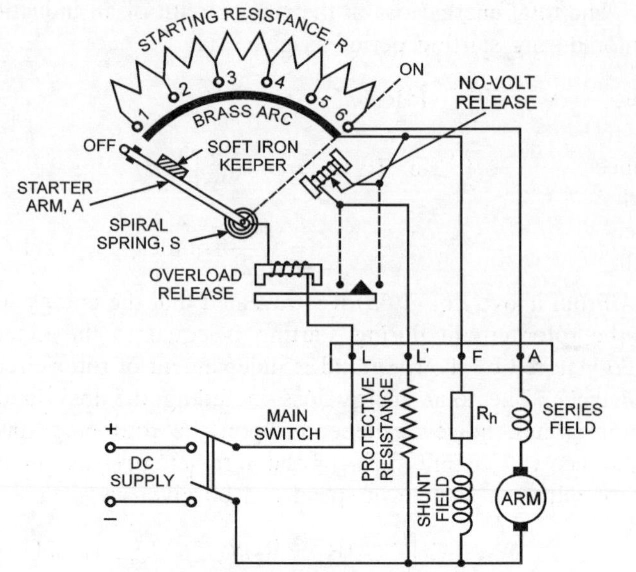

Fig. 1.60. ***Four-Point Starter***

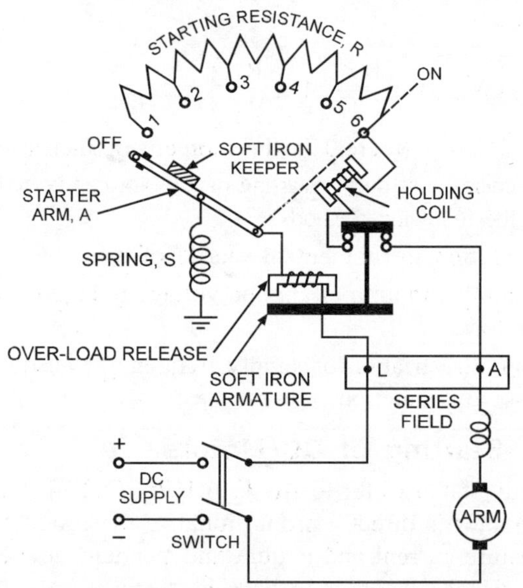

Fig. 1.61. ***Face-Plate Starter For DC Series Motor***

For switching off the dc series motor, the line switch should always be opened rather than throwing back the starting arm, as in case of dc shunt and compound motors. If it is done, then heavy sparking is caused at the stud 1, because here the field circuit is broken and the entire energy stored in the magnetic field is dissipated in the form of heavy spark.

However, there is one significant difference. In case of a dc series motor, the flux does not remain constant but varies with the current because the line current, field current and armature current are the same. Thus the back emf at any given speed varies as the current varies between the upper and lower limits. The practical result of this is that a series motor starter has a smaller number of steps than that required for a starter of a dc shunt motor of the same rating with the same current limits. This is because an increase in current causes increase in back emf and thus the current rise tends to be self-limiting. As a result of the fewer steps, the resistance of each section of the starter resistor is greater than for the shunt motor starter.

Since flux varies during the starting operation and its relation to the field current is non-linear, the determination of the number of steps is rather complicated.

Drum Type Series Motor Speed Controller. DC series motors are often employed on cranes, elevators, tramcars and other applications, where the motor is under the direct control of an operator. In these applications, frequent starting, variations of speed, stopping and reversing may be necessary. A manually operated controller, more rugged than a starting rheostat, called a *drum controller,* is employed.

The controller is in the form of rotating drum having segments which makes contact with the fixed points. Speed controller of this type is shown in Fig. 1.62 in which K is an arc braking coil: EM is a braking electromagnet; FLS is a switch for rotation in one direction and RLS is a switch for rotation in reverse direction. The controller has six positions for forward and six positions for reverse rotation of the motor. The working positions of the controller are shown by vertical dotted lines. The electromagnet EM, is connected in parallel with the motor and releases the motor at starting. When the motor is disconnected it is braked mechanically.

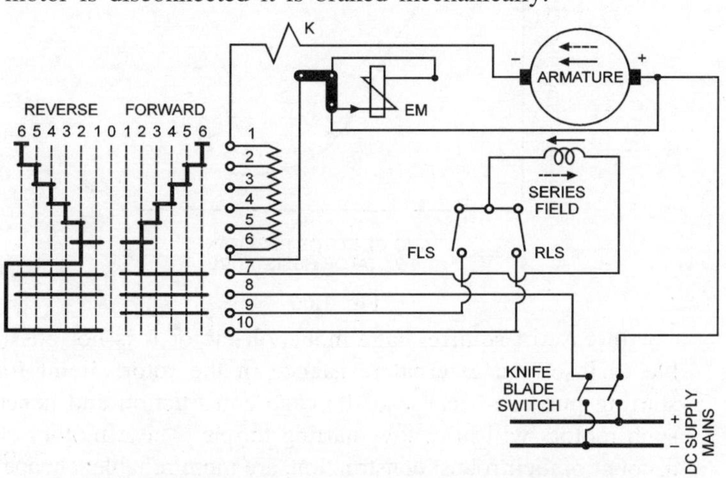

Fig. 1.62. *Drum Type Series Motor Speed Controller*

Now when the knife blade switch is closed and the controller is set in forward position 1, the connections are along vertical line one. In this position segments make contacts with fixed points 6 and 7, 8 and 9. In this position current flows from +ve bus-bar through armature winding of the motor, arc braking coil K, all the starting resistances 1–6, fixed points 6 and 7 through controller segments, series field winding, the forward limit switch FLS, fixed points 9 and 8 through controller segments and returns back to the negative bus-bar.

In the second forward position, the segments of the controller come in contact with fixed points 5 and 6 thereby bringing some of the starting resistance out of circuit, so speed of the series motor increases. In the subsequent forwarding positions 3rd, 4th and 5th, additional steps of the starting resistances are brought out of the circuit and finally in 6th position all the starting resistances are short circuited and motor attains maximum speed.

In all the six forward positions the direction of current in armature as well as in series field winding is same (from right to left), as shown in Fig. 1.62.

But in first reverse position the current flows from + ve bus-bar through armature winding, arc braking coil K, all the starting resistances 1–6, fixed points 6 and 10 through controller segments, the reverse limit switch RLS, the series field winding, fixed points 7 and 8 through controller segments and returns back to the – ve bus-bar. Hence it is obvious that in reversing positions direction of flow of current in armature winding remains unchanged, while reverses through the series field winding, thereby reversing the direction of rotation of motor.

Automatic Starters. Push-button type of automatic starters are quite common in use in industry. Even an inexperienced operator, with the help of such starters, called the *auto-starters*, can start and stop the motor without any difficulty.

The operation of such starters depends upon, either the time delay or the counter emf developed across the armature terminals. Here only the basic principle of the counter emf starter will be discussed.

Counter-EMF Starter. This type of starter depends upon the build up of the back or counter emf to operate the contactors (a heavy duty relay designed to open or close an electrical power circuit), which in turn causes the starting resistors to be shorted. The connections of a counter-emf starter are shown in Fig. 1.63.

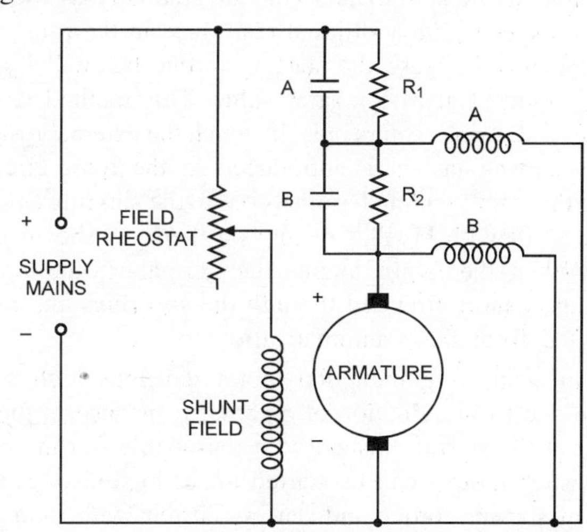

Fig. 1.63

When the motor is switched on, the counter emf developed by the motor is zero and the voltage across contactor coil A or B is insufficient to energize the relays. Consequently the contactors A and B are normally open and, therefore, starting resistances R_1 and R_2 are in series with the armature. As the motor picks up speed and builds up a counter emf the voltage across coils A and B increases until the operating voltage of coil A is reached. Contactor A is then energized, closing the normally open contact A, shorting out the starting resistor R_1. The motor continues to pick up speed building up the counter emf until the operating voltage of coil B is reached. Contactor B is energized, closing normally open contactor B, which in turn shorts out the starting resistor R_2 and places the armature directly across the supply mains.

The drawback of counter-emf automatic starter is that if the motor fails to start, the counter emf remains zero, the voltage sensitive relays cannot operate and so the starting resistances may burn out. Such occurrences can, however, be avoided by using definite time-limit starters.

1.17.5. Starting of 3-Phase Induction Motors

At start, speed of motor is zero and, therefore, slip $s = 1.0$. Substituting $s = 1.0$ in Eqs. (1.34) and (1.36), we have

Starting current per phase,

I_{st} = Rotor current referred to stator,

$$I'_2 = \frac{V}{\sqrt{(R_1+R'_2)^2+(X_1+X'_2)^2}} \text{ amperes} \qquad ...(1.68)$$

(neglecting magnetizing current)

and starting torque,

$$T_{st} = \frac{3V^2R'_2}{\frac{2\pi N_s}{60}\left[(R_1+R'_2)^2+(X_1+X'_2)^2\right]} \text{ Nm} \qquad ...(1.69)$$

The starting torque becomes maximum when the rotor resistance is made equal to its leakage reactance. Since the rotor resistance is not more than 1 or 2 per cent of its leakage reactance otherwise efficiency will be low therefore, in order to obtain high starting torque resistance must be inserted in the rotor circuit at start and cut out gradually as the motor picks up speed. The additional resistance in the rotor circuit will not only increase the starting torque but will limit the starting current also to a safer value. This method is useful in case of slip-ring motors only, in which the external resistance at the starting instant is introduced in the rotor circuit by taking the rotor winding terminals out to the slip rings mounted on the shaft with brushes resting on them, as shown in Fig. 1.64. When the motor attains the normal speed, the rotor winding is short circuited through the slip rings and brushes are lifted from them automatically.

Since slip-ring induction motor develops high starting torque due to introduction of external resistance in the rotor circuit at the starting instant with reasonable starting current hence such motors can be started under load also. The variation of starting torque and starting current with rotor resistance are depicted in Figs. 1.65 and 1.66 respectively.

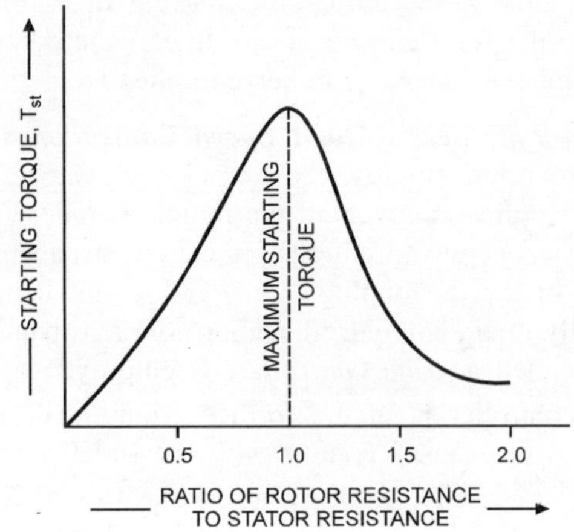

Fig. 1.65

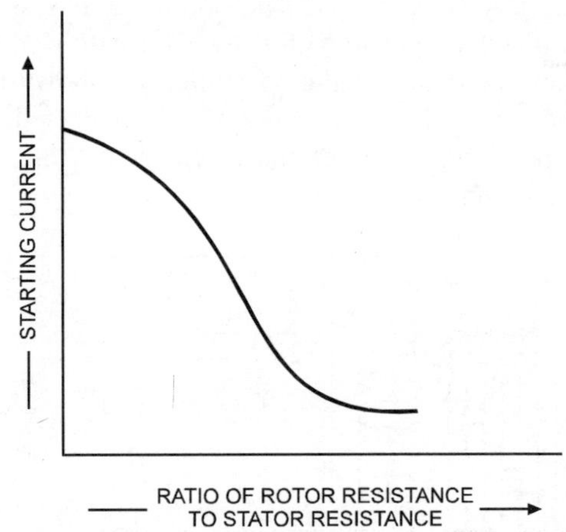

Fig. 1.66

In case of squirrel cage induction motor it is not possible to insert an external resistance in the rotor circuit for starting purpose because of its cage construction and hence such motors will have low starting torque. These motors on account of their robust construction, are more reliable; cheaper and easier to maintain as compared to slip-ring induction motors. These motors are employed where constant torque is required, *e.g.*, in paper mills, textile mills, compressors and conveyers etc.

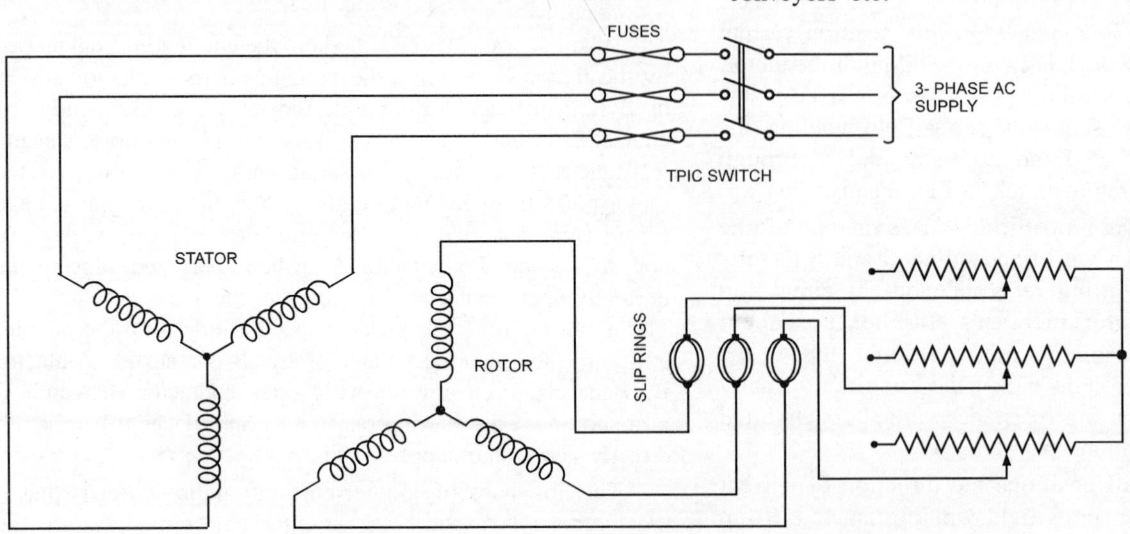

Fig. 1.64. ***Slip-Ring Induction Motor With Starting Rheostat***

Starting of Squirrel-Cage Induction Motors

1. ***Direct-on-Line Starting or Full-Voltage Starting.*** As already mentioned, squirrel-cage induction motors, as a general rule, up to 4 or 5 kW are started by this method. As the name suggests, this method involves direct switching of polyphase squirrel cage induction motor to supply mains, as illustrated in Fig. 1.67(*a*).

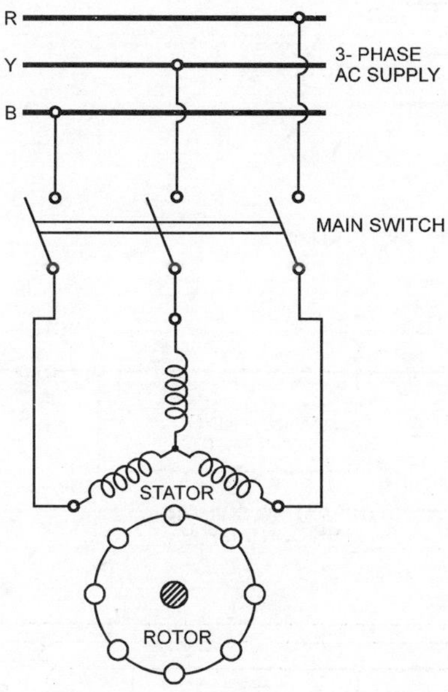

(a)

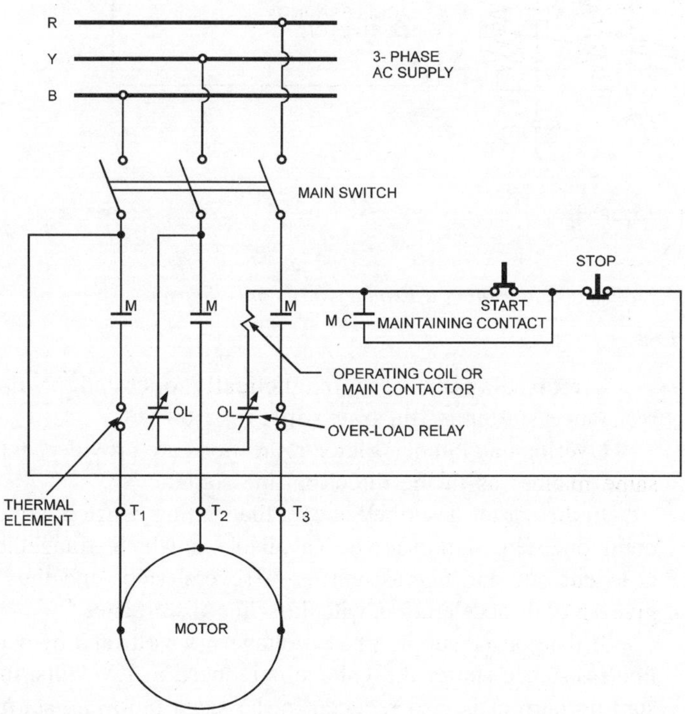

(b) Wiring Diagram of a Direct-on-Line Starter With Protective Devices

Fig. 1.67

The push-button type direct-on-line starter, which is very common in use, is shown in Fig. 1.67(*b*). It is simple, inexpensive and easy to instal and maintain. It consists of a set of 'start' and 'stop' push buttons, a contactor (an electromagnet) with its associated contacts and usually overload and under-voltage protection devices. The start button is momentary contact that is held normally open by a spring. The stop button is held normally closed by a spring. Thermal overload relays are commonly used for motor overload protection.

Under-voltage protection is provided by the contactor (that is controlled by a three-wire control circuit) maintains the interruption of circuit even after the supply is restored.

Torque Developed on Starting of Induction Motor By Direct Switching.

$$\text{Input power to rotor} = T\omega$$

$$\text{Rotor copper loss} = s \times \text{input power to rotor}$$

$$\text{or } 3\, I^2 R_2 = sT\omega$$

where I_2 is the rotor current per phase, R_2 is rotor resistance per phase, T is the torque developed, s is the slip and ω is the supply angular velocity.

$$\text{or Torque developed, } T = \frac{3I^2 R_2}{s\omega}$$

$$\textit{i.e.}\ T \propto \frac{I_2^2}{s} \text{ if rotor resistance is constant}$$

Since rotor current is proportional to stator current I_1, so

$$T \propto \frac{I_1^2}{s}$$

$$\text{or } T = K\frac{I_1^2}{s} \text{ where K is any constant}$$

At starting slip, $s = 1$

So starting torque,

$$T_{st} = K(I_{st})^2 \text{ where } I_{st} \text{ is the starting current}$$

$$\text{Full-load torque, } T_f = \frac{KI_f^2}{s_f}$$

where I_f is the full-load motor current and s_f is the full-load slip

$$\text{So } \frac{T_{st}}{T_f} = \frac{KI_{st}^{\ 2}}{KI_f^2/s_f} = \left(\frac{I_{st}}{I_f}\right)^2 s_f$$

$$\text{or } T_{st} = T_f \times \left(\frac{I_{st}}{I_f}\right)^2 s_f \qquad ...(1.70)$$

When the motor is connected directly across the supply mains, the starting current is equal to the short-circuit current, I_{sc}

$$\therefore \text{ Starting torque, } T_{st} = T_f \left(\frac{I_{sc}}{I_f}\right)^2 s_f \qquad ...(1.71)$$

2. *Primary Resistor (or Reactor) Starting*. This method of starting of a polyphase cage motor is very simple and provides smooth acceleration of the motor. In this method of starting of a 3-phase squirrel cage induction motor reduced voltage is obtained by means of resistors (or reactors) that are connected in series with each stator lead, as shown in Fig. 1.68, during the starting period. The voltage drop in resistors (or reactors) causes a reduced voltage across the motor terminals. As the motor picks up the speed, the resistors (or reactors) are cut out in steps and finally short circuited when the motor attains the operating speed.

The primary resistor method is quite similar to the dc shunt motor starter with its series armature resistor, because both provide a voltage drop between the voltage supply and the motor terminals. There are, however, some very important differences. In a dc shunt motor the starting torque is directly proportional to the starting armature current whereas in case of

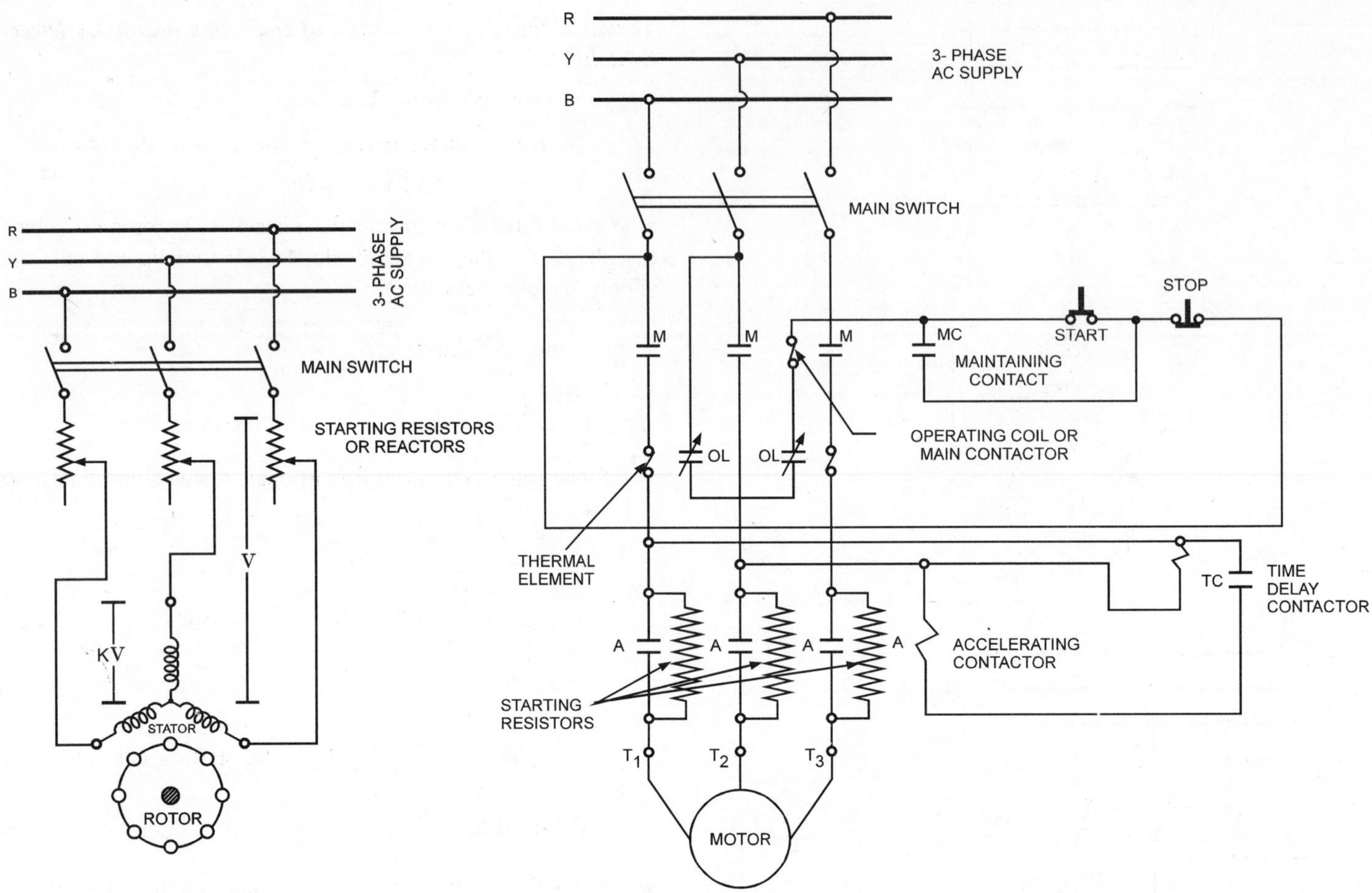

(*a*) (*b*) *Wiring Diagram of a Primary Resistance Starter*

Fig. 1.68

a polyphase induction motor the torque varies as the square of the motor terminal voltage *i.e.*, there is a far greater reduction of starting torque for an equivalent reduction of starting current.

Although the initial cost of reactors is high in comparison to that of resistors, reactor starting is preferred because this method incurs small power losses and is more effective* in reducing the voltage applied to the stator at starting.

The advantages and disadvantages of this method of starting of cage motors are given below:

Advantages. 1. Smooth acceleration.
2. High power factor during start.
3. Less expensive than auto-transformer starter in lower output ratings.
4. Closed transition starting.
5. Available with as many as 7 accelerating points.

Disadvantages 1. Resistors give off heat.
2. Low torque efficiency.
3. Starting duration usually exceeds 5 seconds, so needs expensive resistors.
4. Starting voltage is difficult to adjust to meet varying conditions.

Circuit diagram of a magnetically operated primary resistance starter is illustrated in Fig. 1.68 (*b*).

Overload and undervoltage protections are provided in the same manner as in the direct-on-line starter.

In the starter described above, the starting resistors are cut out in one step. Starters are also available, in which arrangement is to cut out starting resistances in several steps and thus to give smooth acceleration with less line disturbance.

If the normal supply phase voltage = V volts and by using line resistance starter the voltage is reduced to KV volts, then starting current is also reduced in the same ratio *i.e.*, starting current $I_{st} = K\, I_{sc}$ where I_{sc} is the short-circuit current.

Starting torque,

$$T_{st} = T_f \times \left(\frac{I_{st}}{I_f}\right)^2 s_f$$

$$= T_f \times K^2 \left(\frac{I_{sc}}{I_f}\right)^2 s_f \qquad ...(1.72)$$

$= K^2 \times$ Torque obtained by switching the motor directly

This method finds its application in some cases where only very low starting torques are required. A curve plotted between

* The power factor of an induction motor is quite low and if reactors are used for reducing voltage across the motor terminals, the voltage drop across the reactors will be nearly in phase with the supply voltage and so the net reduction in the applied voltage to the stator will be more. On the other hand in case of resistance starting, the voltage drop across the resistors will be nearly in quadrature with the applied voltage and so the net reduction in the applied voltage to the stator terminals will be less.

the starting-current/full-load current and the starting-torque/full-load torque in case of stator resistance method of starting of induction motor is depicted in Fig. 1.69.

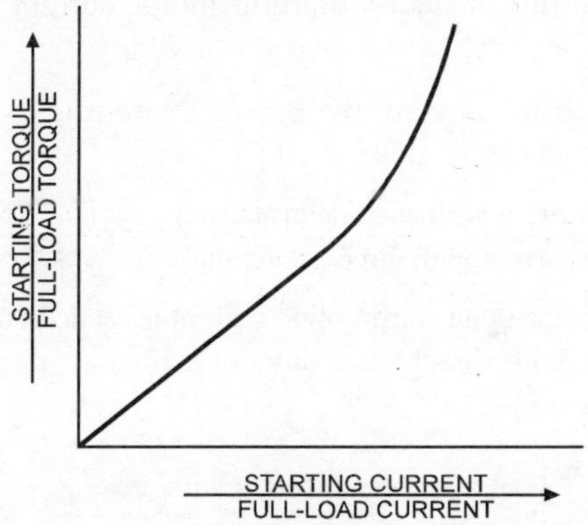

Fig. 1.69

3. *Auto-Transformer Starting*. In auto-transformer starting method the reduced voltage is obtained by taking tappings at suitable points from a three phase auto-transformer, as shown in Fig. 1.70(*a*). The auto-transformers are generally tapped at the 50, 60 and 80 per cent tapping points, so that adjustment at these voltages may be made for proper starting torque requirements. Since the contacts frequently break large values of current, arcing is sometimes quenched effectively by having them assembled to operate in an oil bath.

Auto-transformer starters may be either manually or magnetically operated. The wiring diagram of an auto-transformer starter is shown in Fig. 1.70(*b*).

Let the motor be started by an auto-transformer having transformation ratio, K. If I_{sc} is the starting current when normal voltage is applied, and applied voltage to stator winding at starting = KV

then motor input current,

$$I_{st} = KI_{sc} \quad \text{....(1.73)}$$

Supply current

$$= \text{Primary current of auto-transformer}$$

$$= K \times \text{secondary current of auto-transformer} = K^2 I_{sc}$$

Starting torque,

$$T_{st} = T_f \left(\frac{I_{st}}{I_f}\right)^2 \times s_f$$

$$= T_f K^2 \left(\frac{I_{sc}}{I_f}\right)^2 \times s_f$$

$$= K^2 \times \text{Torque obtained by direct switching.} \quad \text{...(1.74)}$$

Hence the line current and the starting torque are reduced in the square ratio.

The advantages of this method of starting of cage induction motors lies in the fact that the voltage is reduced by transformation and not by dropping the voltage in the resistors (or reactors) and, therefore, the current and power drawn from the supply mains are also reduced in comparison to primary resistor (or reactor) starting. The internal losses of the starter itself are small during long starting periods. Other advantages of this method are (*i*) availability of highest torque per ampere of supply current (*ii*) adjustment of starting voltage by selection of proper tap on the auto-transformer (*iii*) suitability for long starting periods (*iv*) closed transition starting and (*v*) motor current larger than supply current.

The drawbacks of such starters are low power factor and higher cost in case of lower output rating motors.

This method can be employed for starting of star-connected as well as delta-connected motors.

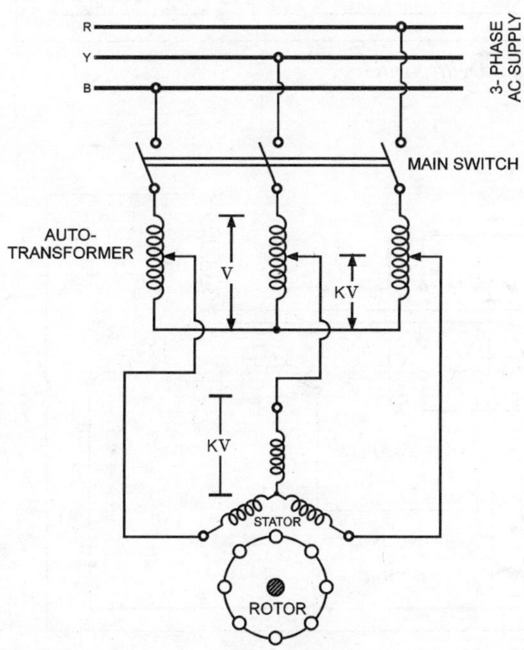

Fig. 1.70(*a*)

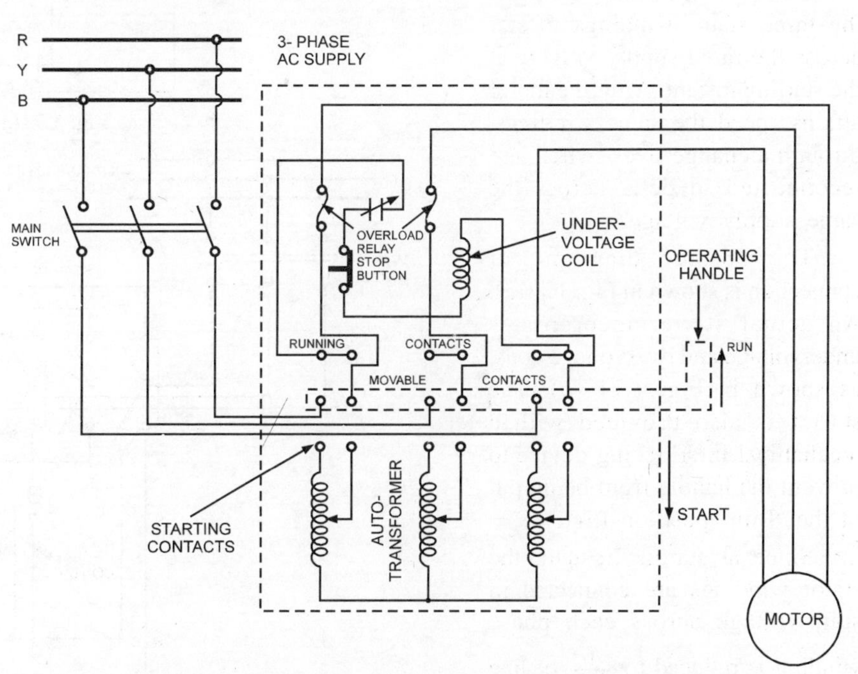

Fig. 1.70(*b*). *Circuit Diagram of an Auto-Transformer Starter*

For starting of large cage motors (of output rating exceeding 20 kW) this method of starting is often used.

The values of starting torque available, together with the corresponding values of the voltage applied to the stator and the motor and line currents expressed as percentages of the normal full-load values, for different tap positions are shown in Fig. 1.70 (*c*).

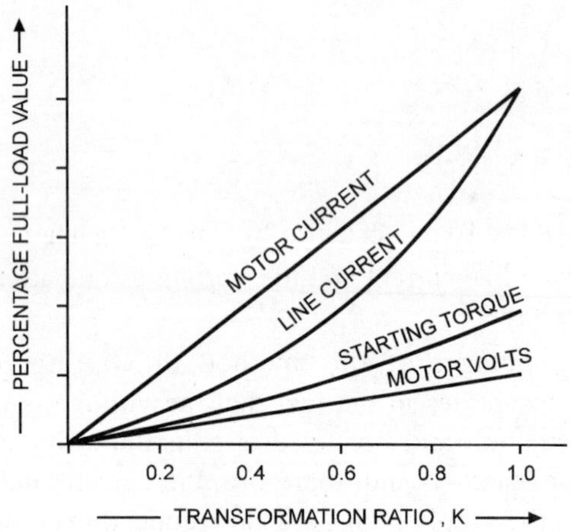

Fig. 1.70(*c*)

4. *Star-Delta Starting.* This method of starting of cage induction motors is based upon the principle that with 3 windings connected in star, the voltage across each winding is $1/\sqrt{3}$ *i.e.* 57.7 % of the line-to-line voltage whereas the same winding connected in delta will have full line-to-line voltage across each.

The star-delta starter connects the three stator windings in star across the rated supply voltage at the starting instant. After the motor attains speed the same windings, through a change-over switch, are reconnected in delta across the same supply voltage.

The basic diagram of connection is shown in Fig. 1.71(*a*). An actual starter incorporates undervoltage and overvoltage coils, as shown in Fig. 1.71 (*b*). The starter is also provided with a mechanical interlocking device to prevent the handle from being put in the 'Run' position first.

Since at starting instant, the stator windings are connected in star, voltage across each phase winding is reduced to $\frac{1}{\sqrt{3}}$ of line voltage and, therefore, starting current per phase becomes equal to $I_{sc}/\sqrt{3}$.

Starting line current by connecting the stator windings in star at the starting instant = Starting motor current per phase $= I_{sc}/\sqrt{3}$.

Starting line current by direct switching with stator windings connected in delta $= \sqrt{3}\, I_{sc}$.

$$\therefore \quad \frac{\text{Line current with star - delta starting}}{\text{Line current with direct switching}} = \frac{I_{sc}/\sqrt{3}}{\sqrt{3}\, I_{sc}} = \frac{1}{3}$$

Hence by star-delta starting line current is reduced to one-third of line current with direct switching.

Starting torque,

$$T_{st} = T_f \left(\frac{I_{st}}{I_f}\right)^2 s_f = T_f \left[\frac{I_{sc}/\sqrt{3}}{I_f}\right]^2 s_f$$

$$= \frac{1}{3} T_f \left(\frac{I_{sc}}{I_f}\right)^2 s_f \qquad \text{...(1.75)}$$

Hence with star-delta switching, the starting torque is also reduced to one-third of starting torque obtained with direct switching.

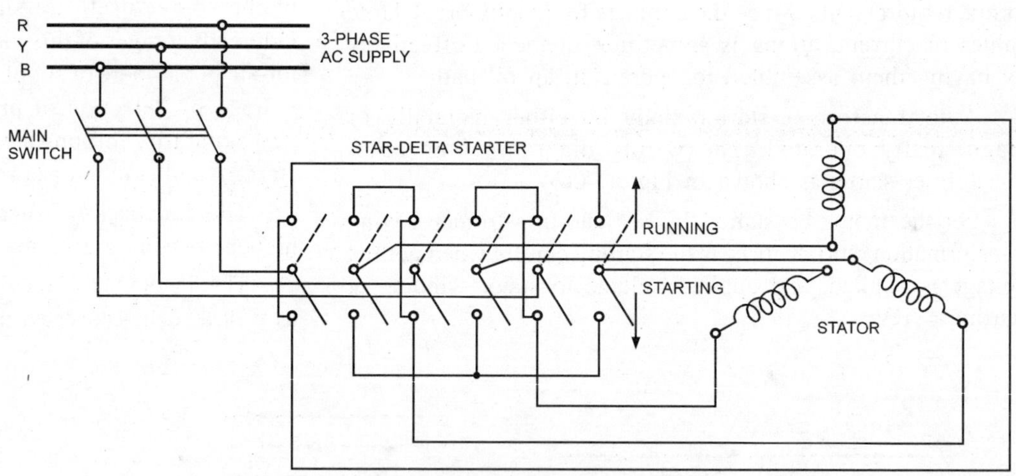

Fig. 1.71(*a*). *Star-Delta Switch*

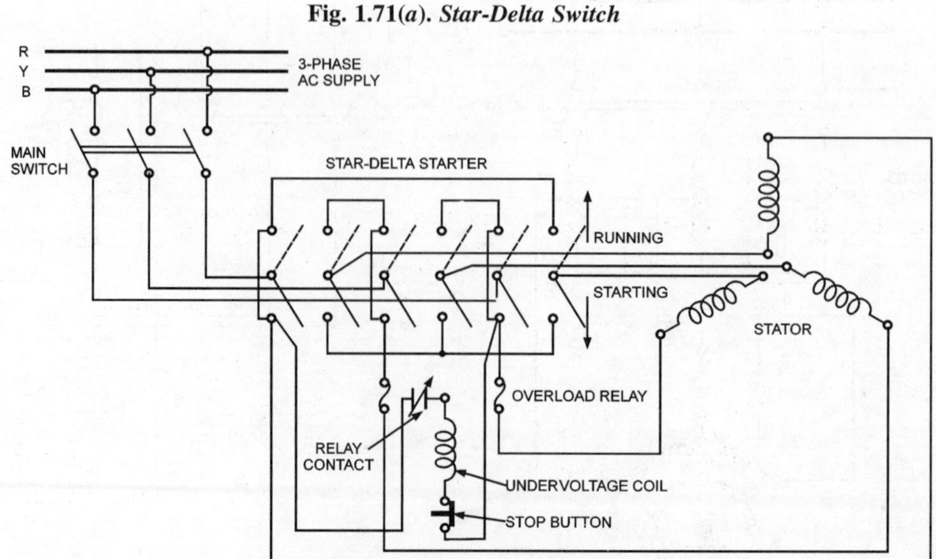

Fig. 1.71(*b*). *Star-Delta Starter*

This method of starting of cage motors is simple, cheap, effective and efficient since no power is lost in auxiliary components. This method is also suitable for high inertia and long acceleration loads.

This method needs a motor to be delta-connected for normal operation and all the six terminals of the 3-phase stator windings are to be brought out. The reduction in voltage is fixed and starting torque is also low. So this method is limited to application where high starting torque is not the essential requirement e.g. machine tools, pumps, motor-generator sets etc. This method is unsuitable for line voltage exceeding 3,000 V, because of excess number of stator turns required for delta connection.

Such starters are employed for starting 3-phase squirrel cage induction motors of rating between 4 and 20 kW.

Precaution With Star-Delta Starting. The initial current flowing when the motor is started in star is 57.7% of the short-circuit current in delta together with a transient in each phase. The transient currents decay rapidly but the steady state is not reached until the motor has attained 70 per cent of its synchronous speed. The change-over from star to delta connection should not be made until the motor attains about 90 per cent of synchronous speed, otherwise there will be a current surge considerably greater than full-load current which may even be greater than the standstill current with star-connection.

1.17.6. Starting Methods of Synchronous Motors

As already stated the synchronous motor has got no self starting torque and the motor must first be brought up to synchronous speed by some external means before it is capable of operating with load angle δ. Various methods of starting of synchronous motors are described as follows:

1. *From DC Source.* If the dc supply and dc compound motor is available, the synchronous motor is coupled and started by means of a dc compound motor. The speed of dc motor is adjusted by the speed regulator. The synchronous motor is then excited and synchronised with ac supply mains. At the moment of synchronising the synchronous motor is switched on with the ac mains, and either the dc motor is disconnected from the dc supply mains or the field of the dc machine is strengthened until it begins to function as a generator. Now the synchronous machine is operating, as a motor, from ac supply mains and dc machine acts as load on it.

The synchronous motor can also be started by the exciter mounted on an overhung synchronous motor bracket and shaft extension. Here again, an available dc source operates the exciter as a motor during the starting period; then, after the synchronous machine is brought up to speed and synchronised, the exciter assumes its normal function.

2. *By Means of AC Motor.* A small direct-coupled induction motor, called the *pony motor*, may be used for starting the synchronous motor unless the motor is required to start against full-load torque. The induction motor frequently has two poles less than the synchronous motor and so is capable of raising the speed of the latter to synchronous speed. Before switching the ac supply to the synchronous motor, it must be synchronised with the bus-bars. After normal operation is established, the pony motor is sometimes uncoupled from the synchronous motor. This method is not very satisfactory and not suited to industrial needs. Modern machines are usually of the self starting type and are arranged to start as induction motors.

3. *By Means of Damper Grids in The Pole Faces.* The synchronous motor is made self starting by providing a special winding on the rotor poles, known as *damper winding* or *squirrel cage winding*. The damper winding consists of short-circuited copper bars embedded in the face of the field poles. AC supply given to the stator produces a rotating magnetic field which causes the rotor to rotate, therefore, in the beginning synchronous motor provided with damper winding starts as a *squirrel cage induction motor*. The exciter moves along the rotor. When the motor attains about 95% of synchronous speed, the rotor winding is connected to exciter terminals and the rotor is magnetically locked by the rotating field of the stator and the motor runs as a synchronous motor.

Advantages. When the motor is overloaded it does not stop, continues running as a squirrel cage induction motor. When the speed falls slightly below the synchronous speed due to overloading an emf is induced in the rotor short-circuited winding, called the *damper winding*. Due to this emf a torque is produced which keeps the motor running even on no load.

Disadvantages. Since damper winding resistance is low so at start it takes large current from supply mains.

To avoid large starting current drawn from the supply mains the synchronous motor is started usually as slip-ring induction motor. The rotor winding is connected in series with triplex rheostat at the time of starting by throw-over switch. Rotor circuit is completed and an emf is induced in the rotor circuit as in the rotor of a slip-ring induction motor.

As the motor attains synchronous speed the throw-over switch is put on the side of the exciter terminals and the dc supply passes through the rotor winding. While running, the rotor acts like rotor of a synchronous motor. Hence the motor runs as a synchronous motor. If the starting and excitation is done automatically the motor is known as *auto-synchronous motor*.

Precautions. *In starting a synchronous motor the field winding must be shorted through a suitable resistance so that the induced voltage is distributed throughout the whole winding and no part can be subjected to the high voltage induced in the entire winding.*

Another method of eliminating the hazard of breakdown of insulation of the field winding due to induced high voltage in it is by means of dividing the field winding into several sections during the starting period. While emf is still induced in the field coils, the number of coils in series is reduced, so that the magnitude of emf induced in any section is kept within safe limits. Sectionalizing of field winding may be accomplished by a centrifugally operated switch that keeps the field circuit sectionalized until the rotor revolves at near synchronous speed.

Example 1.16. A 3-phase squirrel-cage induction motor has a starting current of 6 times its full-load current. The motor has a full-load slip of 4 per cent. Calculate (*i*) the starting torque (*ii*) the slip at which maximum torque occurs and (*iii*) the ratio of maximum torque to full-load torque.

[*A.M.I.E. Sec. B. Electric Drives and Their Controls Summer 2000*]

Solution : Starting current, $I_{st} = 6I_f$

Full-load slip, $s_f = 4\%$, *i.e.*, 0.04

(*i*) Starting torque, $T_{st} = T_f \times \left(\frac{I_{st}}{I_f}\right)^2 \times s_f$

$$= T_f \times (6)^2 \times 0.04 = 1.44\ T_f \quad \textbf{Ans.}$$

From Eq. (1.34), neglecting stator resistance and magnetising current, we have

$$\text{Full-load current, } I_f = \frac{V}{\sqrt{\left(\frac{R_2'}{s_f}\right)^2 + (X_1 + X_2')^2}} \quad ...(i)$$

$$\text{and starting current, } I_{st} = \frac{V}{\sqrt{(R_2')^2 + (X_1 + X_2')^2}} \quad ...(ii)$$

Dividing Eq. (*ii*) by Eq. (*i*), we have

$$\frac{I_{st}}{I_f} = \sqrt{\frac{\left(\frac{R_2'}{s_f}\right)^2 + (X_1 + X_2')^2}{(R_2')^2 + (X_1 + X_2')^2}}$$

or $$\frac{\left(\frac{R_2'}{0.04}\right)^2 + (X_1 + X_2')^2}{(R_2')^2 + (X_1 + X_2')^2} = \left(\frac{I_{st}}{I_f}\right)^2 = 6^2 = 36$$

or $625(R'_2)^2 + (X_1 + X'_2)^2 = 36(R'_2)^2 + 36(X_1 + X'_2)^2$

or $$\frac{R_2'}{(X_1 + X_2')} = \sqrt{\frac{36 - 1}{625 - 36}} = \sqrt{\frac{35}{589}} = 0.244$$

(*ii*) From Eq. (1.37), slip corresponding to maximum torque

$$s_{max\,T} = \frac{R_2'}{X_1 + X_2'} = 0.244 \text{ or } 24.4\% \textbf{ Ans.}$$

Neglecting stator resistance R_1

(*iii*) From Eq. (1.41), we have

Ratio of maximum torque to full-load torque

$$= \frac{1}{2}\left(\frac{s_f}{s_{max\,T}} + \frac{s_{max\,T}}{s_f}\right)$$

$$= \frac{1}{2}\left[\frac{0.04}{0.244} + \frac{0.244}{0.04}\right] = 3.132. \textbf{ Ans.}$$

Example 1.17. A 3-phase squirrel cage induction motor has a starting current 8 times the full-load value. The full-load slip is 4%. Compute the starting torque as a percentage of full-load torque if the motor is started (*i*) direct-on-line (*ii*) by a star-delta starter, (*iii*) using an auto-transformer to limit the starting current to three times the full-load value.

[*A.M.I.E. Sec. B. Electric Drives and Their Control Winter 1999*]

Solution : Short-circuit current, $I_{sc} = 8\ I_f$

Full-load slip, $s_f = 4\% = 0.04$

(*i*) Starting torque, when started by means of direct switching,

$$T_{st} = \left(\frac{I_{sc}}{I_f}\right)^2 s_f\, T_f = 8^2 \times 0.04\ T_f$$

$$= 2.56 \text{ times full-load torque. } \textbf{Ans.}$$

(*ii*) Starting torque, when started by means of a star-delta starter,

$$T_{st} = \frac{1}{3}\left(\frac{I_{sc}}{I_f}\right)^2 s_f\, T_f = \frac{1}{3} \times 8^2 \times 0.04\ T_f$$

$$= 0.853 \text{ times full-load torque. } \textbf{Ans.}$$

(*iii*) If K is the transformation ratio of auto-transformer, then

Starting current, $I_{st} = K^2 I_{sc}$

$$\text{or Transformation ratio, } K = \sqrt{\left(\frac{I_{st}}{I_{sc}}\right)} = \sqrt{\frac{3I_f}{8I_f}} = \sqrt{\frac{3}{8}}$$

$$\text{Starting torque, } T_{st} = K^2\left(\frac{I_{sc}}{I_f}\right)^2 s_f\, T_f$$

$$= \frac{3}{8} \times (8)^2 \times 0.04\, T_f$$

$$= 0.96 \text{ times full-load torque. } \textbf{Ans.}$$

Example 1.18. Determine the ratio of starting to full-load current of a 10 hp, 400 V, 3-phase induction motor with star-delta starter. The motor has a full-load power factor of 0.8 lagging and an efficiency of 0.9. The short-circuit current of motor is 7.2 A at 160 V. Neglect the effect of magnetising current.

[*Pb. Univ. Electric Drives and Utilization of Electrical Energy August 1984*]

Solution : Output = 10 hp = 10 × 0.7355 = 7.355 kW

$$\text{Full-load line current, } I_f = \frac{\text{Output in kW} \times 1{,}000}{\sqrt{3} \times V_L \times \cos\phi \times \eta}$$

$$= \frac{7.355 \times 1{,}000}{\sqrt{3} \times 400 \times 0.8 \times 0.9} = 14.75 \text{ A}$$

Short-circuit current at 160 V

= 7.2 A

Short-circuit current at normal voltage of 400 V,

$$I_{sc} = 7.2 \times \frac{400}{160} = 18.0 \text{ A}$$

Starting current with star-delta starter,

$$I_{st} = \frac{I_{sc}}{3} = \frac{18.0}{3} = 6.0 \text{ A}$$

Ratio of starting current to full-load current

$$= \frac{I_{st}}{I_f} = \frac{6.0}{14.75} = 0.407. \textbf{ Ans.}$$

Example 1.19. A 50 kVA, 400 V, 3-phase, 50 Hz squirrel cage induction motor has full-load slip of 5%. Its standstill impedance is 0.866 Ω/phase. It is started using a tapped auto-transformer. If the maximum allowable supply current at the time of starting is 100 A, calculate the tap position and the ratio of starting torque to full-load. [*GATE 1997*]

Solution : Full-load current drawn by a 3-phase induction motor,

$$I_f = \frac{\text{Output in kVA} \times 1{,}000}{\sqrt{3} \times \text{line voltage}} = \frac{50 \times 1{,}000}{\sqrt{3} \times 400} = 72.2 \text{ A}$$

Short-circuit current,

$$I_{sc} = \frac{\text{Supply voltage (line-value)}/\sqrt{3}}{\text{Standstill impedance per phase}} = \frac{400/\sqrt{3}}{0.866} = 266.7 \text{ A}$$

Supply current at start, I_{st} = 100 A

Tap position of auto-transformer,

$$K = \sqrt{\frac{I_{st}}{I_{sc}}} = \sqrt{\frac{100}{266.7}} = 0.6124 \text{ or } 61.24\ \% \textbf{ Ans.}$$

Full-load slip, s = 5 % or 0.05

Ratio of starting torque to full-load torque,

$$\frac{T_{st}}{T_f} = K^2\left[\frac{I_{sc}}{I_f}\right]^2 s_f = 0.6124^2\left[\frac{266.7}{72.2}\right]^2 \times 0.05 = 0.256 \textbf{ Ans.}$$

Note : 3-phase induction motor has been assumed to be star-connected and losses have been neglected.

Example 1.20. A star-connected cage induction motor has the following data : 440 V, 50 Hz, 4-pole, 1460 rpm, $R_s = 2\ \Omega$, $R'_r = 2\ \Omega$, $X'_s = X_r = 3.0\ \Omega$. Calculate starting torque and starting current of this motor at 50 Hz and 10 Hz for V/*f* control.

[A.M.I.E. Sec. B. Electric Drives and Their Control Winter 1996]

Solution : At 50 Hz supply

Supply voltage per phase, $V = \frac{440}{\sqrt{3}} = 254$ V

Equivalent resistance of motor as referred to stator,

$$R_{01} = R_s + R'_r = 2 + 2 = 4\ \Omega,$$

Equivalent reactance of motor as referred to stator,

$$X_{01} = X_s + X'_s = 3 + 3 = 6\ \Omega$$

Starting current,

$$I_{st} = \frac{V}{\sqrt{(R_{01})^2 + (X_{01})^2}} = \frac{254}{\sqrt{4^2 + 6^2}} = 35.23 \text{ A } \textbf{Ans.}$$

Total copper losses in rotor

$$= 3 I_{st}^2 \times R'_r = 3 \times (35.23)^2 \times 2 = 7{,}446 \text{ watts}$$

Rotor power input at standstill,

$$P_2 = \text{Total copper losses} = 7{,}446 \text{ watts} \qquad \because s = 1$$

Synchronous speed, $N_s = \frac{120f}{P} = \frac{120 \times 50}{4} = 1{,}500$ rpm

Starting torque, $T_{st} = \frac{9.55 P_2}{N_s} = \frac{9.55 \times 7{,}446}{1{,}500} = 47.41$ N-m **Ans.**

At 10 Hz supply

Supply voltage per phase, $V' = V \times \frac{f'}{f} = 254 \times \frac{10}{50} = 50.8$ V

Equivalent resistance of motor referred to stator,

$R'_{01} = 4\ \Omega$, the same as before

Equivalent reactance of motor as referred to stator,

$$X'_{01} = X_{01} \times \frac{f'}{f} = 6 \times \frac{10}{50} = 1.2\ \Omega$$

Starting current,

$$I'_{st} = \frac{V'}{\sqrt{(R'_{01})^2 + (X_{01})^2}} = \frac{50.8}{\sqrt{4^2 + 1.2^2}} = 12.166 \text{ A } \textbf{Ans.}$$

Power input to rotor at standstill,

$$P'_2 = \text{Total copper losses}$$
$$= 3(I'_{st})^2 R'_r = 3 \times (12.166)^2 \times 2 = 888.1 \text{ watts}$$

Synchronous speed, $N'_s = \frac{120 f'}{P} = \frac{120 \times 10}{4} = 300$ rpm

Starting torque, $T'_{st} = \frac{P'_2 \times 9.55}{N'_s} = \frac{888.1}{300} \times 9.55 = 28.27$ N-m **Ans.**

Example 1.21. At particular instant during starting up a drive, following data are recorded.

Motor torque = 100 Nm; Load torque = 30 Nm; Acceleration = 10 revolutions/second^2. Find the moment of inertia of the drive and specify its units.

[A.M.I.E. Sec B. Electric Drives and Their Control Winter 1996]

Solution : Motor torque, T_M = 100 Nm

Load torque, T_L = 30 Nm

Angular acceleration, $\frac{d\omega}{dt}$ = 10 revolutions per second^2

$= 2\pi \times 10 = 20\pi$ rad/s^2

From Eq. (1.44).

Moment of inertia of the drive,

$$J = \frac{T_M - T_L}{d\omega/dt} = \frac{100 - 30}{20\pi} = 1.114 \text{ kg-m}^2 \textbf{ Ans.}$$

Example 1.22. A dc shunt motor delivers rated output of 37.5 kW at 500 rpm. Its moment of inertia is 20 kg-m^2. Find the time to attain full-load speed when the motor is started from rest against full-load torque. The starting current fluctuates between 1.1 and 1.4 times the full-load current.

[A.M.I.E. Sec. B. Electric Drives and Their Control Summer 1993]

Solution : Full-load torque,

$$T_L = \frac{\text{Output in kW} \times 1{,}000}{2\pi \times \frac{N}{60}} = \frac{37.5 \times 1{,}000 \times 60}{2\pi \times 500} = 716.2 \text{ Nm}$$

The motor is required to be accelerated against full-load. The starting current varies from 1.1 to 1.4 times full-load current, therefore, starting torque fluctuates between 1.1 to 1.4 times full-load torque (in a dc shunt motor torque is proportional to current).

Average starting torque, $T_{st} = \frac{1.1 + 1.4}{2} \times 716.2 = 895.25$ Nm

Torque available for acceleration, $T_a = T_{st} - T_L$

$= 895.25 - 716.2$

$= 179.05$ N-m

Time in attaining full-load speed,

$$t_1 = \frac{J\omega_n}{T_M - T_L} \text{ seconds} \qquad \text{Refer to Eq. (1.47)}$$

$$= \frac{J \times 2\pi N/60}{T_a} = \frac{20 \times 2\pi \times 500}{60 \times 179.05} = 5.85 \text{ seconds. } \textbf{Ans.}$$

Example 1.23. A 37.5 kW, 500 rpm shunt motor is used for shop. The stored energy of the machinery averages 660 kg-m per kW. Assuming the load torque during starting to be equal to the full-load torque and the starting current to be twice the full-load current, estimate the time taken to start the motor.

Solution : Load torque during starting,

$$T_L = \text{Full-load torque, } T_f$$

$$= \frac{\text{Output in kW} \times 1{,}000}{2\pi N/60} = \frac{37.5 \times 1{,}000 \times 60}{2\pi \times 500} = 716.2 \text{ N-m}$$

Torque developed by motor during starting,

$$T_M = \frac{\text{Starting current}}{\text{Full-load current}} \times \text{Full-load torque}$$

$= 2 \times 716.2 = 1{,}432.4$ N-m $\qquad \because$ shunt motor $T \propto I_a$

Torque available for acceleration,

$$T_a = T_M - T_L = 1,432.4 - 716.2 = 716.2 \text{ N-m}$$

Stored energy of machinery

$$= 37.5 \times 660 \times 9.81 = 242,800 \text{ N-m}$$

Since kinetic energy stored $= \frac{1}{2} J\omega^2$

∴ Moment of inertia of the rotating system,

$$J = \frac{\text{KE stored}}{\frac{1}{2}\omega^2} = \frac{242,800 \times 2}{\left(\frac{2\pi \times 500}{60}\right)^2} = 177 \text{ kg-m}^2$$

$$\text{Angular acceleration, } \alpha = \frac{d\omega}{dt} = \frac{T_a}{J} = \frac{716.2}{177} = 4.04 \text{ rad/s}^2$$

$$\text{Starting period, } t = \frac{\omega}{\alpha} = \frac{2\pi N/60}{\alpha} = \frac{2\pi \times 500}{60 \times 4.04}$$

$= 12.95$ seconds **Ans.**

Example 1.24. A 220 V dc series motor driving a constant load torque runs at 200 radians/second and draws 20 A from the supply. The total resistance of armature and field is 1.0 Ω. The MI of the motor together with load is 5 kg-m². Calculate total energy dissipated in the armature circuit if the motor starts from rest and attains the steady speed of 200 radians/second is 2.5 seconds. *[U.P. Technical Univ. Electric Drives 2008-09]*

Solution : Motor output

$= V \times I -$ Power losses in armature and field

$= 220 \times 20 - 20^2 \times 1.0 = 4,000$ W

Speed, $\omega_n = 200$ rad/s

$$\text{Load torque, } T_L = \frac{\text{Motor output in watts}}{\omega} = \frac{4,000}{200} = 20 \text{ N-m}$$

Starting time, $t_{st} = 2.5$ seconds

$$\text{So} \quad \frac{d\omega}{dt} = \frac{200}{2.5} = 80 \text{ rad/second}^2$$

From Eq. (1.61), we have

$$K = \frac{J\frac{d\omega}{dt} + T_L}{I_a^2} = \frac{5 \times 80 + 20}{20^2} = 1.05$$

From Eq. (1.62), total energy dissipated in the motor circuit

$$W_{st} = \frac{JR_a}{K}\omega_n + \frac{T_L R_a}{K} t_{st}$$

$$= \frac{5 \times 1}{1.05} \times 200 + \frac{20 \times 1}{1.05} \times 2.5 = 1,000 \text{ W} \quad \textbf{Ans.}$$

Also, current drawn from supply is constant during starting period load torque being constant, and is equal to 20 A.

So, total energy dissipated,

$$W_{st} = I_{st}^2 R_a \times t_{st} = 20^2 \times 1 \times 2.5 = 1,000 \text{ W, same as above.}$$

1.18. SPEED CONTROL OF ELECTRIC MOTORS

Speed control means intentional change of the drive speed to a value required for performing the specific work process. This concept of speed control or adjustment should not be taken to include the natural change in speed which occurs due to change in the load on the drive shaft. The desired change in speed is accomplished by acting accordingly on the drive motor or on the transmission connecting it to the unit it serves to drive. This may be done manually by the operator or by means of some automatic control device.

Any given piece of industrial equipment may have its speed changed or adjusted mechanically by means of stepped pulleys, sets of change gears, variable speed friction clutch mechanism, and other mechanical devices. Historically, this proved to be the first step in transition from non-adjustable speed to adjustable speed drive. The electrical speed control has many economical as well as engineering advantages over mechanical speed control.

The nature of the speed control requirement for an industrial drive depends upon its type. Some drives may require continuous variation of speed for the whole of the range from zero to full speed, or over a portion of this range; while the others may require two or more fixed speeds. Some machines may require creeping speed for adjusting or setting up the work. Such a speed is of the order of few rpm. For most of the drives, however, a control of speed within the range of ± 20% may be suitable.

Some of the important terms related to speed are defined below.

Constant Torque Drive. It is a drive in which the motor shaft torque remains constant over a given speed range. In such a drive, shaft power varies as the speed.

Constant Power Drive. It is a drive in which the motor shaft power, given by the product of shaft torque and speed, remains constant over a given speed range. In a constant power drive, higher torques are available at lower speeds and vice versa. The motor size is always determined by the highest torque requirement at the lowest speed.

$$\text{Percentage speed regulation} = \frac{N_0 - N_f}{N_f} \times 100 \quad ...(1.76)$$

where N_o and N_f are no-load and full-load speeds respectively.

1.18.1. Speed Control of DC Motors

One of the attractive features the dc motor offers over all other types is the relative ease with which speed control can be achieved and, therefore, dc motors are indispensable for many adjustable speed drives. The various schemes available for speed control can be deduced from the expression of speed for a dc motor which is repeated here with one modification

$$N = K\frac{V - I_a(R + R_a)}{\Phi}$$

The modification involves the inclusion of an external resistance in the armature circuit. The above expression reveals that the speed can be controlled by adjusting any one of the three factors appearing on the right hand side of the expression: (*i*) applied voltage to the armature terminals, V (*ii*) external resistance in the armature circuit, R and (*iii*) flux per pole, Φ. The first two possibilities involve adjustment affecting the armature circuit, whereas the third involves change in the magnetic field. Therefore, speed control methods are broadly classified as *armature control methods* and *field control methods.* Sometimes a combination of the two methods is employed. With armature control the speed decreases as the voltage applied to the armature terminals is reduced, whereas with field control the speed increases as the flux is reduced.

1. *Field Control Methods.* In case of *dc series motors,* the flux can be varied by an one of the following methods.

(i) Field Divertor Method. The field flux can be reduced by shunting a portion of motor current around the series field, thus reducing the excitation mmf and weakening of field. This method is illustrated in Fig. 1.72 (*a*). This method provides speeds above normal because flux is reduced by this method. Lesser the divertor resistance, less the field current, less flux and, therefore, more the speed. This method is convenient as well as economical and provides the speed control range usually not exceeding 2 : 1. This method of speed control is used in electric drives in which the speed should rise sharply as soon as the load falls.

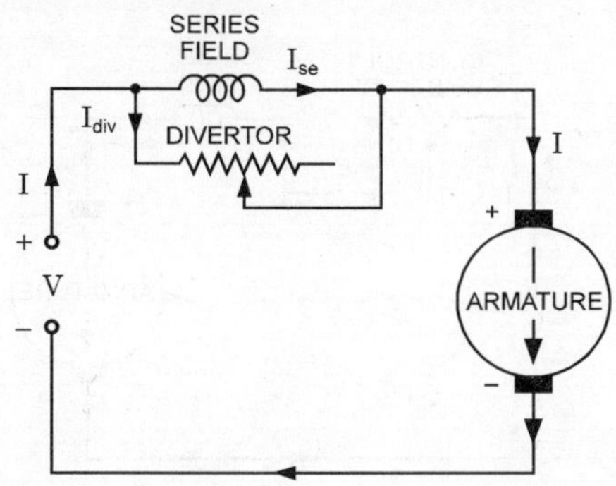

Fig. 1.72 *(a) Field Divertor Method of Speed Control For DC Series Motors*

(ii) Tapped Field Control. This is another method of increasing the speed by reducing the flux and it is accomplised by reducing the number of turns of the field winding through which the current flows. In this method of speed control of dc series motors a number of tappings from the field winding are brought outside, as illustrated in Fig. 1.72(*b*). A number of series field turns can be short circuited according to the requirement. When all field turns are in circuit, the motor runs at the lowest speed and speed increases with cutting out some of the series field turns. This method is usually employed in electric traction.

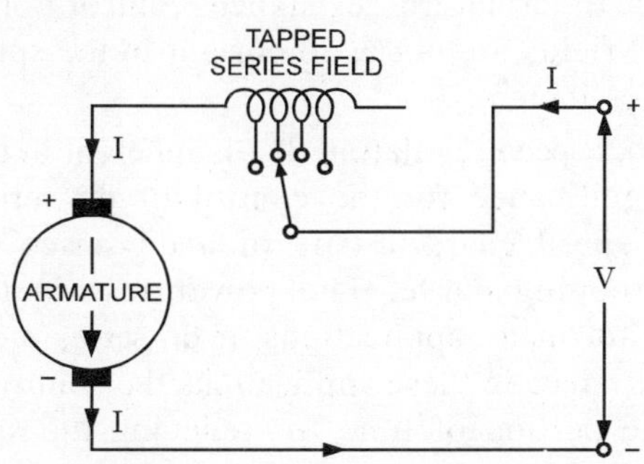

Fig. 1.72 *(b) Tapped Field Control For DC Series Motors*

(iii) Paralleling Field Coils Method. In this method, used in fan motors, several speeds can be obtained by regrouping of field coils, as illustrated in Fig. 1.72 (*c*).

In case of dc **shunt motors** the flux can be varied by inserting a variable resistance in series with the field winding (Refer to Fig. 1.73). Since this resistance has to carry only a small current, so it is made up of slide-wire type of resistor to have continuously variable speed over the range.

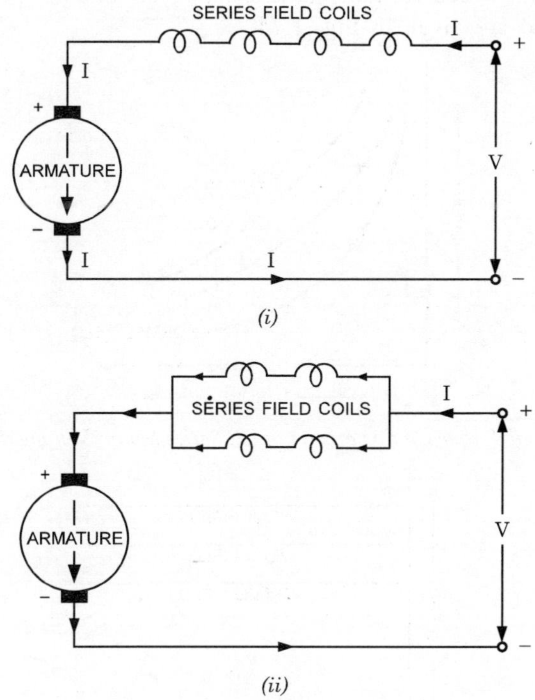

Fig. 1.72(*c*)

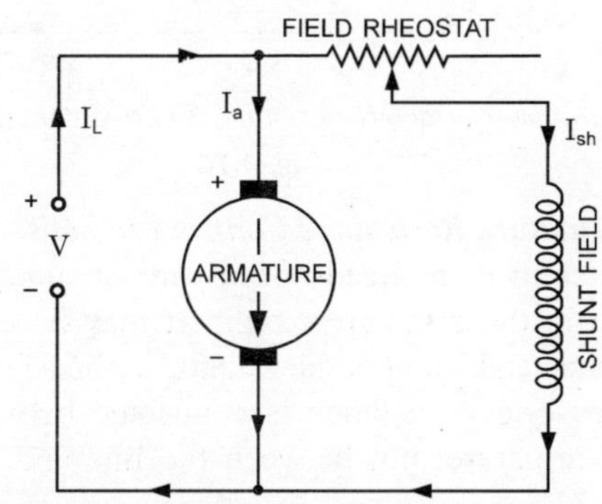

Fig. 1.73. ***Field Rheostat Control For a DC Shunt Motor***

Since the flux can be only decreased (not increased) so the speeds only above normal one can be obtained by this method. The speed is minimum at the maximum value of flux and depends upon the design of the field and its saturation point. The speed is maximum at the minimum value of flux, which is governed by the demagnetizing effect of armature reaction on the field as at higher speeds the motor tends to be unstable and difficulties in commutation arise. The high speed is also restricted due to mechanical considerations as the centrifugal forces are set up at high speeds. Speed variation by this method is limited to ratio of 4 or 5 to 1. Creeping speeds cannot be obtained by this method. The power output being proportional to TN or VI_a remains constant. This method is, therefore, suitable only where power of the load remains constant. This method of speed control is very simple, convenient and most economical and is, therefore, extensively used in electric drives. The power wasted in the controlling resistance is very little, field current being very small. This method of speed control is independent of load on the motor and permits the remote control of speed.

The speed-torque characteristics obtainable with this method are given in Fig. 1.74.

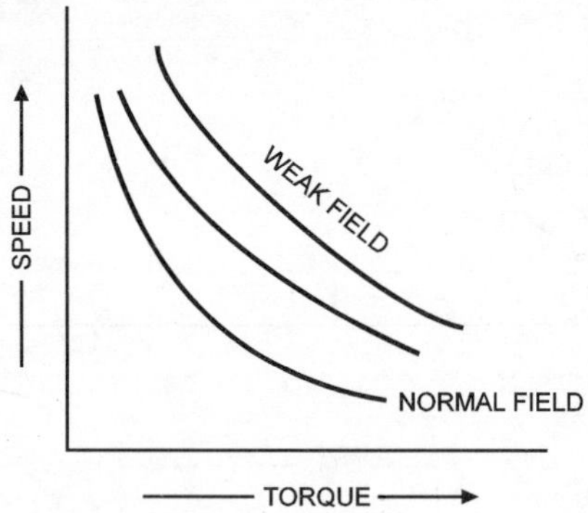

(a) Speed-Torque Characteristics of DC Series Motor With Field Control

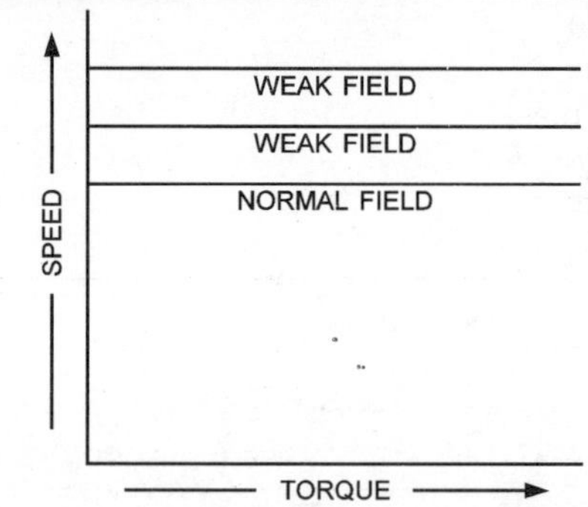

(b) Speed-Torque Characteristics of DC Shunt Motor With Field Control

Fig. 1.74

2. ***Armature Resistance Control Methods.*** In this method of speed control, reduced speeds are obtained by inserting resistance in the armature circuit. It may be employed with series, shunt and compound wound motors. For the last two types, the series resistance is connected between the shunt field and armature, not between the line and motor.

In case of a dc *shunt motor* an increase in armature circuit resistance will cause more voltage drop in the armature circuit, so the speed will be reduced. Field current will remain unaffected, as the shunt field is directly connected across the supply voltage. For a constant torque load, the armature current remains the same so input to the motor remains the same but the output decreases in proportion to the speeds. Operating costs are, therefore, comparatively high for long time running at reduced speed. In case of fans and centrifugal pumps where the load torque decreases with the decrease in speed, the losses are considerably low and because of its low initial cost and simplicity this method may be quite convenient and economical for short-time or intermittent slowdowns. Wide range of speed (below normal one) can be obtained by this method and at the same time motor will develop any desired torque over its operating range. The main advantage of this method is that speeds below base speed down to creeping speeds of only a few rpm are easily available. This method of speed control, therefore, is employed where speeds lower than rated one are required for a short period only and also occasionally as in printing machines, cranes and hoists where the motor is frequently started and stopped. This method of speed control is also employed where the load drops off rapidly with the decrease in speed, as in fans and blowers.

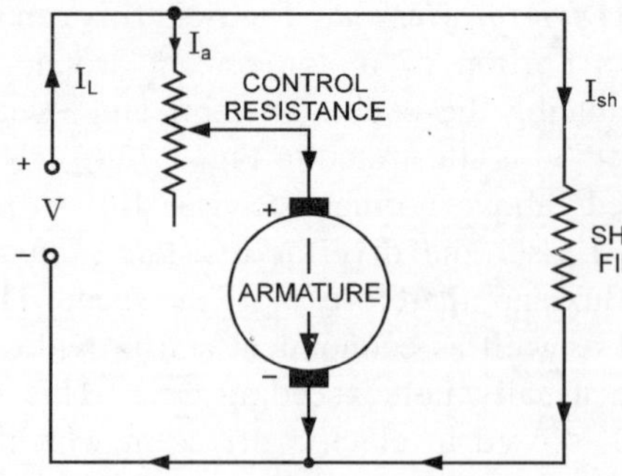

(a) For DC Shunt Motor

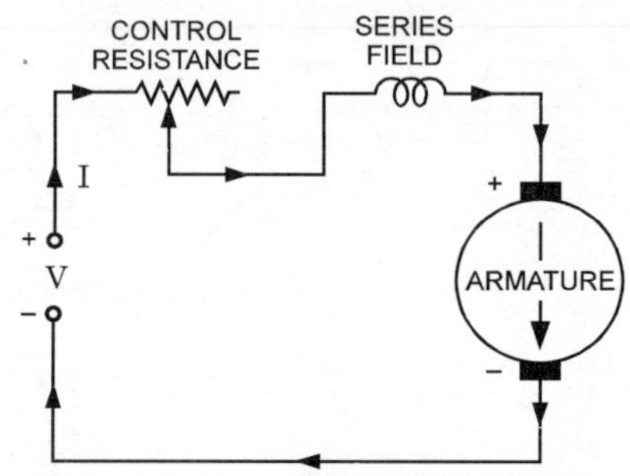

(b) For DC Series Motor

Fig. 1.75

In case of a dc *series motor* an increase in armature circuit resistance will cause more voltage drop in the armature circuit, so speed will be reduced. The torque which is proportional to the product of the flux and armature current will also be reduced with the increase in armature circuit resistance as increase in armature circuit resistance will reduce the armature current and the flux which also depends upon the armature current. In order to obtain different speeds for constant load torque, the armature current is kept constant so the flux. The resistance in the armature circuit is increased to reduce the speed. The drawbacks of armature resistance control for machines with shunt fields are not as important in the speed control of dc series motors.

The poor speed regulation that is inherent in this method has no significance for the control of dc series motors, since the speed characteristic of a dc series motor is a rapidly drooping curve. The power loss in the control resistance for many applications of dc series motors is not too serious, since in these applications the control is utilized for a large portion of time for reducing the speed under light-load conditions and is only employed intermittently when the motor is carrying full load. The speed-torque characteristics of dc series motors with resistance control are shown in Fig. 1.76 (*b*). The maximum range of speed control of about 3 : 1 will be available depending on the load. This method of speed control is most economical for constant torque drives. This method of speed control is employed chiefly for dc series motors driving cranes, hoists, trains etc. because such drives operate on intermittent duty.

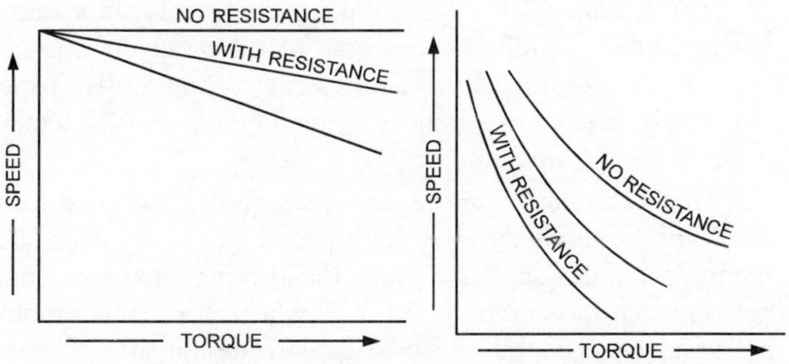

(a) For DC Shunt Motor *(b) For DC Series Motor*

Fig. 1.76. ***Speed-Torque Characteristics With Armature Resistance Control***

The speed-torque characteristics of armature resistance control as applied to dc shunt and series motors are shown in Figs. 1.76 (*a*) and 1.76 (*b*) respectively.

3. *Shunted Armature Control.* The series resistance arrangement described above suffers from a drawback that the speed varies with the load, *i.e.,* the speed increases with the fall in load. This drawback is overcome by shunting the armature with a variable resistance, as shown in Fig. 1.77.

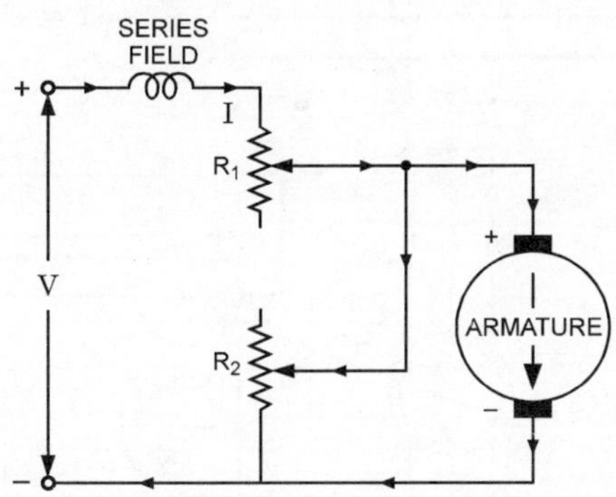

(a) For Series Motor

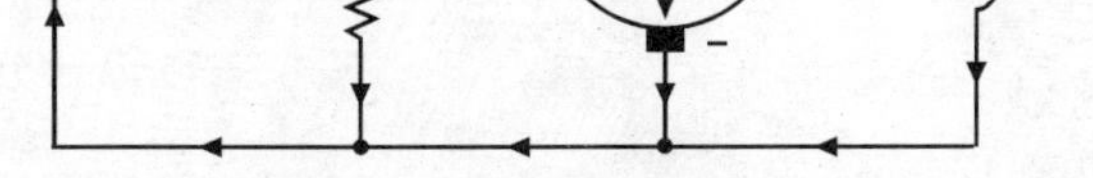

(b) For DC Shunt Motor

Fig. 1.77. ***Shunted Armature Control***

In effect resistors, R_1 and R_2 act as a voltage divider applying a reduced voltage to the armature. Greater flexibility is possible because two resistors may now be adjusted to provide the desired performance. For series motors, no-load speed may be adjusted to a finite, reasonable value, and the scheme is, therefore, applicable to the production of low speeds at light loads. For shunt motors the speed regulation in the low speed range is appreciably improved because the no-load speed is definitely lower than the value with no controlling resistor. The main disadvantages of this method are its low efficiency and the fact that a heavy current is drawn from the supply under certain load conditions.

In Fig. 1.78, natural speed-torque characteristic, speed-torque characteristic with conventional series resistance and shunted armature control are represented by curves I, II and III respectively.

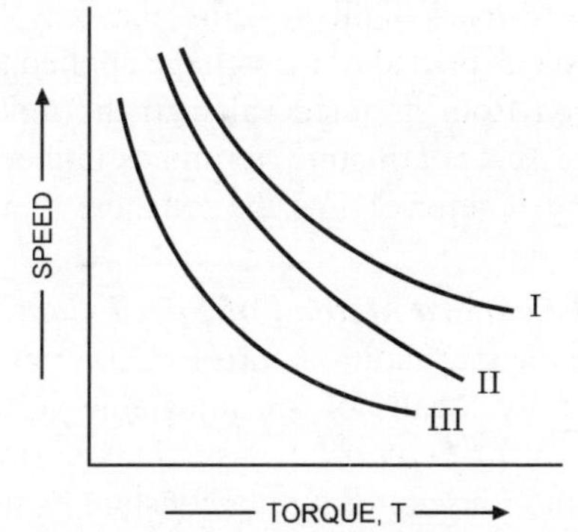

(a) For Series Motor

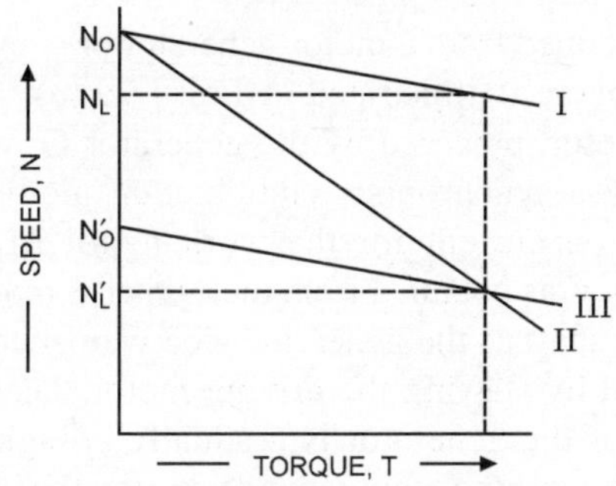

(b) For Shunt Motor

Fig. 1.78. ***Speed-Torque Characteristics***

The shunted armature control is generally employed in drives of comparatively small power rating. This method is also not economical for continuous operation as there is a relatively large power loss in the speed controlling resistances. It is usually employed as a means for sharply reducing the speed so as to achieve accurate stopping.

4. *Armature Voltage Control.* This method of speed control requires a variable source of voltage separate from the source supplying the field current. This method avoids the disadvantages of poor speed regulation and low efficiency which are characteristics of the armature–resistance control method but it is more expensive in initial cost. The adjustable voltage for the armature is obtained from an adjustable voltage generator or from an adjustable electronic rectifier*. This method gives a large speed range with any desired number of speed points. It is essentially a constant–torque system, because the output delivered by the motor decreases with a decrease in applied voltage and a corresponding decrease in speed.

* For most applications dc motors are fed by rectification of the available ac supply. It is usually necessary to insert a transformer immediately ahead of the rectifier.

This particular system has a further advantage that can be employed to provide excellent starting characteristics by bringing the generator voltage gradually up from zero, starting and bringing the motor up to speed with a comparatively slowly increasing voltage. Because of the excellent starting characteristics, this system is used largely for modern high-speed elevators, and on account of the combination of excellent starting characteristics and the wide speed range available, it is used to some extent for reversing planer installations. This method is not applied to any great extent, generally on account of higher initial cost of the generating equipment.

Another drawback of this system is that only the speed below the base speed can be had as the voltage applied to the armature can be reduced from its rated value. If the motor is operated with a voltage across armature terminals higher than the rated value for a long duration of time, the armature winding insulation may get damaged.

5. *Ward-Leonard Method of Speed Control.* The basic adjustable-voltage armature control method of speed control accomplished by means of an adjustable-voltage generator is called the *Ward-Leonard system.* This system consists of simply working the motor with a constant excitation and applying a variable voltage to its armature to provide the required speed. The variable voltage supply is obtained from a motor-generator or convertor set. The scheme is illustrated in Fig. 1.79 (*a*). M_1 is the work motor, powered by the generator G, which is driven by a synchronous or induction motor M_2. The excitation current for the work motor M_1 and the generator G is obtained from the exciter E mounted on the same shaft as the generator. The Ward-Leonard set is started by starting the driving motor. The field rheostat R of the generator is gradually brought out of the circuit as the generator picks up the speed and the work motor begins to rotate. The variable voltage across the terminals of the generator or across the motor is obtained by varying the exciting current of the generator G by means of shunt regulator R. The direction of rotation of motor armature can be reversed by reversing the direction of exciting current of the generator G with the help of reversing switch RS. The converter set runs always in the same direction.

Braking of motor M_1 may be carried out by reducing the generator excitation so that its emf is less than the counter emf of motor M_1. Under these conditions, motor M_1 begins to operate momentarily as a generator, generator G as a motor and ac driving machine M_2 as a generator. As a result kinetic energy of motor M_1 and its load is returned to the supply mains and braking action on the motor M_1 takes place.

Advantages. 1. Very fine speed control over the whole range from zero to normal speed in both directions can be obtained.

2. Uniform acceleration can be obtained.

3. Speed regulation is good.

Disadvantages. 1. Two extra machines are required, so arrangement is costly.

2. Low overall efficiency of the system, especially at light loads.

Applications. This system of speed control is best suited where almost unlimited speed control in either direction of rotation is required such as in steel rolling mills, paper machines, elevators, cranes, mine hoists, diesel-electric propulsion of ships etc.

Ward Leonard Ilgner System, which is a modified form of Ward Leonard system, incorporates a heavy flywheel mounted on the shaft coupling the driving motor M_2 and generator G. In operation with a flywheel, the driving motor has to have a drooping speed-load characteristic *i.e.,* its speed must drop with the increase in load on the shaft. The function of the flywheel is to reduce the fluctuations in the power demand from the supply mains as explained below:

An increase in the load on the shaft causes the work motor to draw more current from the generator so more power is required to drive the later and if there were no flywheel, the driving motor would take all the additional power from the supply line, thus causing sharp fluctuations in it. The heavy flywheel, however, stores a large amount of kinetic energy. When an increase in the generator load causes the driving motor to slow down, some of the kinetic energy of

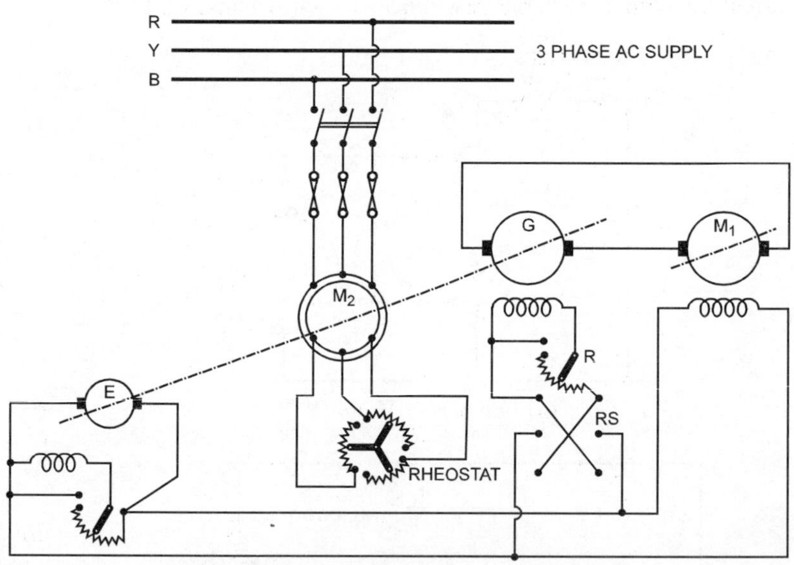

Fig. 1.79 (*a*) *Ward-Leonard Control System*

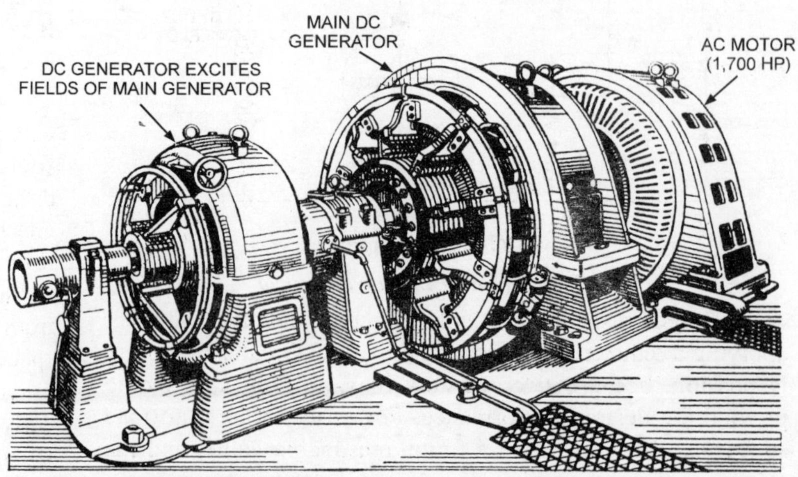

Fig. 1.79 (b). GEC Ward-Leonard Set Used in Mines 1,200 kW Output Generator Controlling a Large Hoist Motor

the flywheel goes to sustain the peak load on the shaft of the work motor. When the load on the work motor decreases, the driving motor picks up speed and the flywheel stores up kinetic energy.

This system of speed control is employed where the load on the motor shaft sharply varies such as in mine hoists, rolling mills etc.

6. *Booster Speed Control.* A booster in series with the armature of the main motor may be employed for varying the voltage across the driving motor. The booster is a separately excited dc generator, which adds or subtracts the voltage in the motor feeder. The adding (boosting) or subtracting (bucking) depends upon the field current and its direction, which is achieved by parallel resistance arrangement. By this method the speed can be varied over a wide range depending upon the size of the booster. Since booster generator is to provide only the variation in the voltage for boosting and bucking its output, therefore, need not to be as much as that of the main motor inspite of the fact that it has to carry the main motor current. The exact size of the booster however, depends upon the range of speed control required. The chief disadvantages of this method is that it requires two extra machines which is expensive, Generally this method is employed where only a small variation in speed is required.

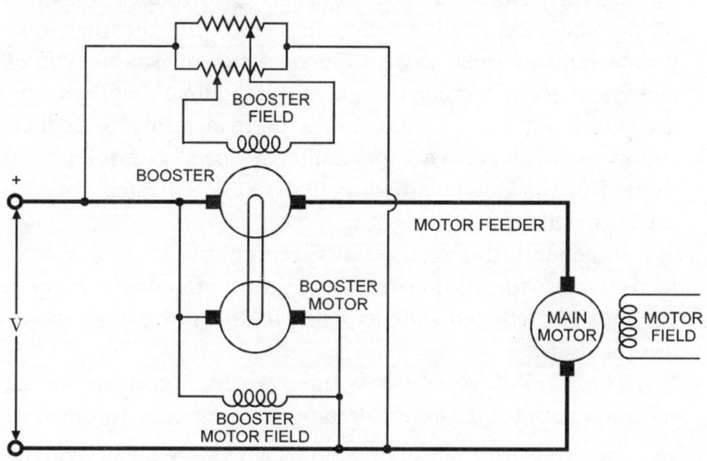

Fig. 1.80. ***Booster Control***

7. *Metadyne Control.* This method will be discussed in details in Chapter 13 (Art. 13.9).

8. *Series-Parallel Control.* This method will be also discussed in details in Chapter 13 [Arts. 13.2 (2) and 13.4]

9. *Grid-Controlled Mercury-Arc Rectifier Method.* In this method, grid-controlled mercury-arc rectifier consisting of a multi-anode steel tank or glass-tube unit or a group of single-anode (thyratron or ignitron) tubes is employed to supply variable voltage to the driving motor.

The output voltage can be varied from zero to maximum value by means of grid control and direction of rotation can only be changed by reversing the connection at the motor armature. The advantage of this method is that the control equipment is static, has a higher operating efficiency and occupies less floor area. The disadvantages of this method are that the power factor on the input side of the rectifier decreases approximately in proportion to speed of the main motor, the control and auxiliary equipment are complicated and it has got no storage capacity so it cannot be employed where large fluctuations of load are involved.

10. *Constant Current System.* For certain services, where there is a possibility of the motors being stalled such as with excavators, ship's windlasses etc. the constant current system is advantageous and is used. In this system a dc generator delivering a constant current by means of a special exciter under all load conditions is employed. The current remains the same through out and the voltage varies with the variation in load. The connections are shown in Fig. 1.81.

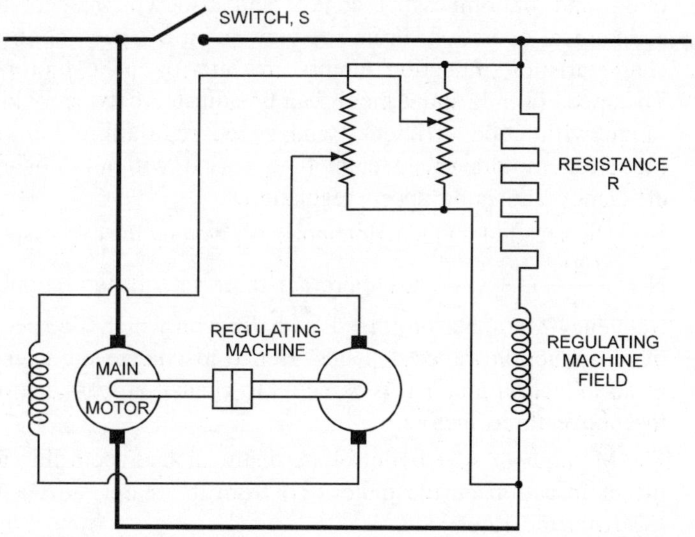

Fig. 1.81. ***Connections For Constant Current Motor***

The field of the main motor is excited from tappings across the resistance R and field winding of the regulating machine so that variable voltage in either direction can be applied to the field system. The regulating machine is mounted on the motor shaft and is so connected in series with the motor field that the emf of the regulating machine opposes the main field current. Any decrease in speed of the motor which may occur due to increase in load, would reduce opposite emf, (emf induced in the regulating machine) and thus field current and torque would increase. The torque would reach maximum when the motor stalls. Speed-torque characteristics for the various positions of the potential divider are shown in Fig. 1.82.

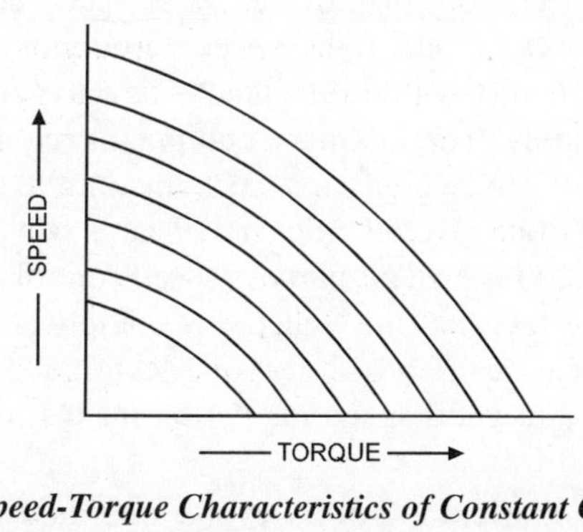

Fig. 1.82. ***Speed-Torque Characteristics of Constant Current Motor***

A full-range of speed control can be had with the maximum torque when the motor stalls without any excessive current.

Any number of motors can be connected in series with each other as the current is independent of load. Thus cable cost and copper losses are reduced. The motor can be stopped simply by closing switch S.

1.18.2. Speed Control of 3-Phase Induction Motors

The problem of speed control of electrical motors in general and of induction motors in particular is of great practical importance.

In a number of industries motors must satisfy very strict speed characteristic requirements, both with respect to the range and smoothness of control and also with respect to economical operation. From the view point of speed control characteristics, induction motors are inferior to dc motors. The speed of a dc shunt motor can be adjusted between wide range with good efficiency and speed regulation, but in induction motors speed cannot be varied without loosing efficiency and good speed regulation.

The speed of an induction motor is given by the expression $N = \frac{120f}{P}(1 - s)$. Thus there are three factors viz, supply frequency f, number of poles P and slip s on which the speed of an induction motor depends. Hence to change the speed of an induction motor it is essential to change at least one of the above three factors.

Methods of speed control are distinguished according to the main action on the motor : (*i*) from the stator side, and (*ii*) from the rotor side.

Various methods of speed control from stator side are (*a*) variation of supply frequency (*b*) variation of applied voltage and (*c*) by changing the number of poles. From the rotor side the speed may be controlled (*a*) by changing the resistance in the rotor circuit and (*b*) by introducing into the rotor circuit an additional emf of the same frequency as the fundamental emf of the rotor.

For the latter method of speed control of induction motors, an additional electrical machine or several such machines are required. A set consisting of a regulated induction motor and one or more additional electrical machines connected to it electrically or mechanically is called *cascade*. Commutator machines are commonly used as the additional machines.

1. *Speed Control By Variation of Supply Frequency*. This method of speed control provides wide speed-control range with gradual variation of the speed throughout this range. The major difficulty with this method is how to get the variable frequency supply. The auxiliary equipment required for this purpose results in a high first cost, increased maintenance and lowering of the overall efficiency. That is why, this method is not employed for general-purpose speed control applications. Inspite of the fact that this scheme is complicated, there are certain applications in which its wide, continuously variable, speed range and good speed regulation makes its use highly desirable.

If an induction motor is to be operated at different frequencies with practically constant values of efficiency, power factor, overload capacity and a constant absolute slip, then, with the iron unsaturated, it is essential that the supply voltage be varied in proportion to supply frequency.

This kind of variation can occur for example, if the supply generator is subjected to the speed variations, either due to momentary overloads or because of a noticeable speed regulation. Both output voltage and frequency vary as the speed if no automatic correction is provided. On some large marine drives, the propeller motors are induction type and are speed controlled from such a local supply which is provided by synchronous generators coupled to variable speed turbines. Even in such cases the range of speed variation is limited, because the efficiency of the prime movers falls rapidly with the change in speed from that for what they are designed.

Typical speed-torque curves for four different frequencies are shown in Fig. 1.25.

2. *Speed Control by Variation of Supply Voltage*. This is a slip-control method with constant frequency variable supply voltage. In this method of speed control of induction motors, the voltage applied to the stator is varied for varying the speed.

This method of speed control is simple, low in first cost and has low maintenance cost but it has limited use because (*i*) the operation at voltages exceeding rated voltage is restricted by magnetic saturation, (*ii*) a large change in voltage is required for a relatively small change in speed (*iii*) the developed torque reduces greatly with the reduction in supply voltage and the motor will come to standstill if the load torque exceeds the pull-out torque of the motor corresponding to reduced supply voltage and (*iv*) the range of speed control is very limited in the downward direction *i.e.,* from rated speed to lower speeds.

In general, this method of speed control is used only on loads where the torque required drops off considerably as the speed is reduced such as with small squirrel cage motors driving fans.

The variable voltage may be obtained by means of either saturable reactors, variac, or tap-changing transformers.

The speed-torque characteristics with reduced voltages are illustrated in Fig. 1.24.

3. *Speed Control By Changing The Number of Poles*. This method is easily applicable to squirrel cage motors because a cage winding automatically reacts to create the same number of poles as the stator. This method of speed control is generally not practicable with wound rotor motors as in such machines this method would involve considerable complications of design and switching, since the interconnections of both primary and secondary would have to be changed simultaneously in a manner to produce the same number of poles in both windings. Otherwise, negative torque will be developed by certain of the rotor conductor belts.

The number of pole pairs in the stator can be changed as follows :

***(a) By Using Multiple Stator Windings*.** In this method of speed control two or more completely independent windings, each wound for different number of poles, are placed in the same stator slots. The number of stator winding poles are in this case in no way interrelated and can be arbitrarily chosen depending on the operating conditions of this motor. For example, a two speed motor may have two stator windings, one wound for 4 poles and another for 6 poles which will give synchronous speeds of 1,500 rpm and 1,000 rpm with a supply frequency of 50 Hz. Motors with four independent stator windings are also used sometimes and they give four different synchronous (and hence running) speeds. Of course, one winding is used at a time, the others being entirely inoperative. Change-over from one speed to the another may be done by a mechanical switch or by contactors. With such an arrangement, the winding or windings which is/are not being used must be kept open circuited by the switch, or at least left in star. Otherwise, because of transformer action, the winding which is connected to the supply would induce voltages in the idle winding(*s*) and cause overheating due to the subsequent circulating currents.

The regulation itself amounts to changing the motor speed in steps as one or another stator winding is connected to the supply mains. At each change, the motor finds itself under conditions essentially analogous to the starting conditions, *i.e.,* a large current and reduced torque. Because of inactive conductors, this method of pole changing requires a large stator than a single-speed motor of the same rating. This method has been used for elevator motors, traction motors and also for small motors driving machine tools.

(b) Consequent Pole Method. The method of pole changing for speed control of 3-phase induction motor described above in (*a*) *i.e.,* multiple stator winding method, is employed only in small motors. In large motors, it is more expedient to employ *consequent pole technique* which eliminates the necessity of having multiple stator windings for changing the number of stator poles. This method of speed control makes use of the entire winding for each speed. Normally, for a single speed motor, successive coil groups are wound so as to create poles of opposite polarity in succession. This is illustrated in Fig. 1.83, where one phase of a four-pole winding is developed.

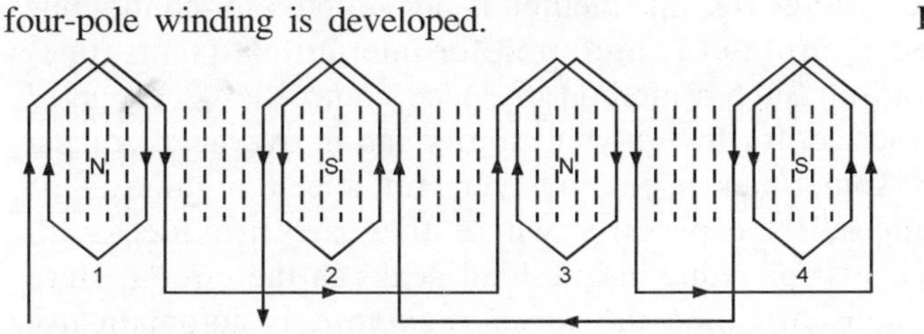

Fig. 1.83. ***Four-Pole Conventional Winding, Series Arrangement***

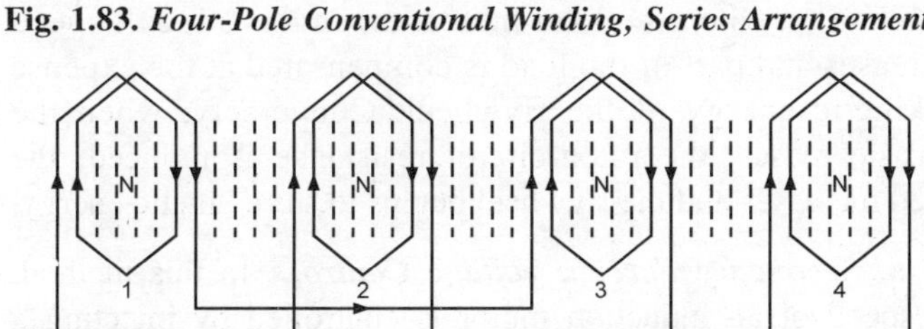

Fig. 1.84. ***Consequent-Pole Method For 8-pole Winding, Series-Parallel Arrangement***

It should be noted that all the coils are in series but that the end connections are made to alternate groups, *i.e.,* group 1 to group 3, and group 4 to group 2. Now the end connections of the groups, with the same coils, can be changed so as to have successive coils of the same polarity, as illustrated in Fig. 1.84. By paralleling the series connections of groups 2-4 with that groups 1-3, the polarity of group 2 and 4 has changed, so that there are now four successive north poles. As a consequence of this south poles are created between each of the north poles, and the motor now has an 8-pole winding. Thus for a 50 Hz supply, the series arrangement gives a synchronous speed of 1,500 rpm while the series-parallel arrangement gives a synchronous speed of 750 rpm. It is also possible to use a series parallel arrangement for the 4-pole winding and then switch to the series arrangement for 8-pole winding. A circuit for switching a three phase winding from 2 to 4 pole winding is shown in Fig. 1.85.

By reconnecting all the coil groups for the same polarity, the stator acts as though it has twice as many poles as pole groups. The additional poles so created are called the *consequent poles,* and the winding is known as a *consequent pole winding.* By an unique switching arrangement, it is possible to have the standard (or conventional) connection on one side of a double-throw switch and the consequent pole connection on the other side of the switch. Thus two speeds

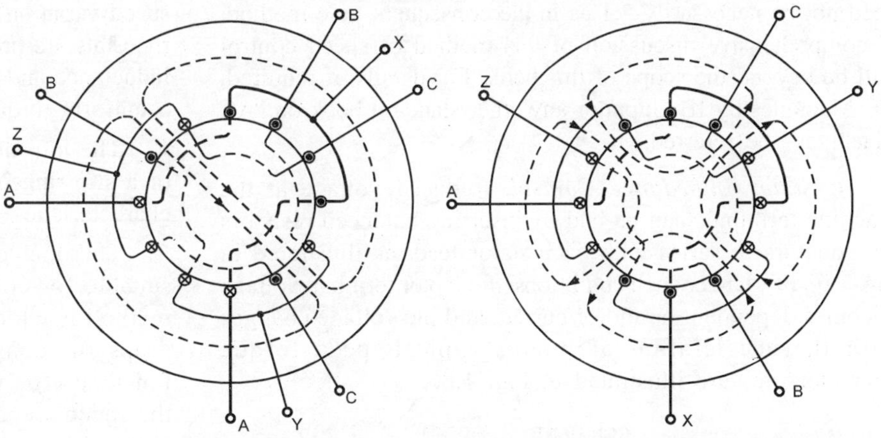

Fig. 1.85. ***Switching Stator Winding From Two to Four Poles***

can be obtained, the higher speed with the conventional connection and half speed with consequent-pole connection.

In the motors using consequent-pole technique for speed control, the best possible design is not usually obtainable at both speeds. That is, desirable characteristics, such as high power factor, are sacrificed at one speed in order that a reasonably good power factor be obtained at the other speed. Sometimes the stator connections are changed from delta to star at the same time that the pole connections are changed. This changes the voltage and makes possible a better motor at each speed.

Two-speed motors are usually made with one winding on the stator, the number of poles being changed in the ratio 1:2. Three- and four-speed motors are provided with two windings on the stators, one or both of which are made with changing the number of poles. For example if it is desired to obtain a motor for four synchronous speeds; 1500, 1,000, 750 and 500 rpm, two windings should be placed on the stator, one of which giving 4 and 8 poles and the other giving 6 and 12 poles.

Employing a series-parallel connection of phase groups of individual phases, the phases can themselves be connected in star/delta resulting in two-speed operation with the three types of torque-speed characteristics viz *constant torque, constant power output* and *variable torque.*

Owing to complications in the design and switching of interconnection of the stator windings, it is not feasible to obtain more than four speeds for any motor through this method of speed control. This method also does not provide gradual speed control. That is why, this method is not employed for general-purpose speed control applications. However, this method is very satisfactory for applications, such as ventilating fans, conveyors, machine tools, or other applications which require operation at only two or four approximately constant speeds which it will provide. This method has the advantage of simplicity, good speed regulation for each setting, high efficiency and moderate initial cost and maintenance.

The number of pole pairs on the stator can also be changed by using *pole amplitude modulation technique*–a suppressed-carrier modulation technique. This method of pole changing has the advantage that the ratio of the two speeds obtainable need not be necessarily 2:1 as in the consequent pole method. A comprehensive discussion of this method of speed control will be beyond the scope of this book. For details, if required, AC Machines by JB Gupta or any other standard book on "AC Machines" be referred.

4. *Stator Impedance Control.* Reduced voltage at the machine terminals can be had by inserting balanced resistors or inductors in series with each stator lead, as illustrated in Fig. 1.68 (*a*). Under such conditions, the motor terminal voltage becomes dependent on motor current and the voltage changes with the acceleration of motor. Typical-speed torque characteristics are illustrated in Fig. 1.86.

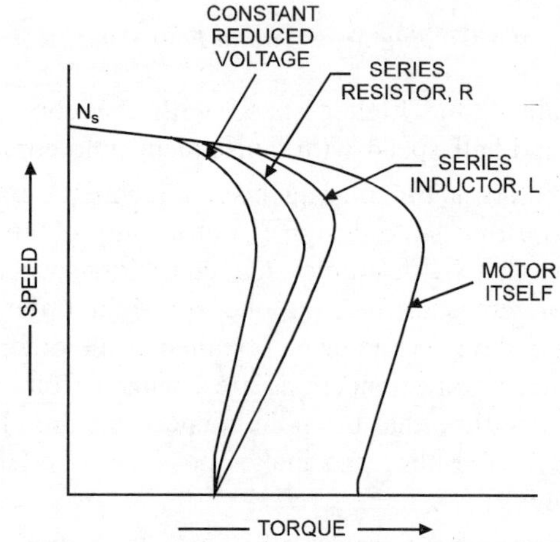

Fig. 1.86. ***Speed-Torque Characteristics with Stator Impedance Control***

If impedance (R or X) is adjusted so as to provide the same starting torque, the speed-torque characteristic in case of addition of inductor would have larger torque than with additional resistance. Besides, both of these characteristics enable us to have torques larger than that obtainable with the characteristic with reduced applied voltage, for same starting torque. Insertion of resistor in the stator circuit improves the power factor, but at the cost of greater energy losses.

5. *Speed Control by Variation of Rotor Resistance (or Rotor Resistance Control).* Wound rotor motors are usually started by connecting starting resistances in the secondary circuit, which are shorted out in steps as the motor speeds up. If the ohmic values of these resistances are properly chosen and if these resistances are designed for continuous operation, they can serve dual purpose, starting and speed control. As obvious this method can be applied to wound rotor motors alone. Smoothness of speed control depends upon the number of steps of resistances that are available. Control is provided in the direction downward from the base speed. The range of speed control is not constant and depends on the value of the load. This method of speed-control has characteristics similar to those of dc shunt motor speed control by means of resistance in series with the armature. Its main disadvantages are lower efficiency and poor speed regulation due to increase in rotor resistance at low-speed settings. For instance, at 50 per cent of synchronous speed the efficiency of rotor is only 50 per cent and the efficiency of the motor would be somewhat less. The speed variation is limited to a range of 50-100% only. Liquid rheostats are preferred over metallic rheostats as they have no inductance and the resistances can be cut out smoothly giving a uniform torque.

The introduction of external resistors in the rotor circuit of a slip-ring induction motor will modify the speed-torque characteristics as illustrated in Fig. 1.26.

This method of speed control as such is, therefore, not suitable for controlling speed at constant torque. But this method is widely used for loads where the torque required drops off considerably as the speed is reduced such as fan loads for which the power input drops noticeably as the speed is reduced which in turn reduces the rotor copper losses. Moreover, this method is not adopted for continuous speed control but is preferred for intermittent (short time) operation. Such a method of speed control is widely used in practice, for low-power motors and in overhead cranes. It is sometimes used however, for speed regulation of rolling mills especially where they are provided with flywheels for reducing the load peaks in the circuit. Here the rheostat, called the *speed regulator,* is automatically switched on when the load increases, and as a result speed decreases and part of the load is compensated at the expense of kinetic energy of the flywheel. Conversely, when the load decreases, the rotor circuit resistance is reduced; the speed increases and the flywheel begins to store kinetic energy.

6. *Secondary Foreign Voltage Control.* In this method, the speed of an induction motor is controlled by injecting a voltage of slip frequency in the secondary circuit. If injected voltage is in phase with the induced emf in the secondary circuit, it is equivalent to decreasing of resistance in the secondary circuit, therefore, slip decreases or speed increases. It is possible to achieve super-synchronous speeds by injecting a voltage of slip frequency in phase with the induced emf in the rotor circuit. If injected voltage is in phase opposition to the induced emf in the secondary circuit, it is equivalent to increasing of resistance in the secondary circuit, hence resulting in increase in slip or reduction in speed. Thus a large range of speed control is possible by this method by including frequency-converting equipment in the rotor circuit.

In this method the disadvantages of lower efficiency and poor speed regulation are overcome, but, since it requires one and usually more auxiliary machines to inject emf of slip frequency in the secondary circuit, therefore, it is costlier and used with motors of very large rating, such as for motors in steel mills.

Several methods have been developed to control speed of induction motors by injecting a voltage of slip frequency in the secondary circuit. The different methods vary in their requirements for auxiliary equipment; from the incorporation of the auxiliary equipment in the design of induction motor itself to a fairly complicated system of auxiliary rotating machines and adjustable ratio transformers. The basic creteria that such a system must fulfill are : (*i*) that the variable foreign voltage have a slip frequency, *i.e.,* frequency equal to s_f and (*ii*) that the energy associated with the foreign voltage be recoverable. A comprehensive discussion of this method of speed control is beyond the scope of this book. The most common systems based on this principle are Kramer system, Leblanc system and Scherbius system of speed control of polyphase induction motors. For details of these speed control systems, AC Machines by J.B. Gupta or any other standard book on AC Machines be referred.

7. *Speed Control by Concatenation*. In this method, two motors are required, at least one of which must have a wound rotor. The two motors may be mechanically coupled together to drive a common load. If the stator winding of one of them, a wound rotor motor, is connected to 3-phase ac supply, and its rotor winding is connected to the stator winding of the second motor, the speed of the combination will be determinated by the sum or difference of the number of poles in the two machines. In practice, it is customary, to connect the rotor output of the first machine to the stator of the second machine in such a way that the revolving fields of both are in the same direction; under this condition, the resulting speed will be determined by the following expression.

$$N = \frac{120f}{P_1 + P_2} \quad ...(1.77)$$

where f is supply frequency, P_1 and P_2 are the number of poles on machines I and II respectively.

For obtaining speed, above normal the torque of the second motor is reversed by simply changing two of the leads of the second motor. This is called the *differential cascading* and under this condition speed will be given by the expression.

$$N = \frac{120f}{P_1 - P_2} \quad ...(1.78)$$

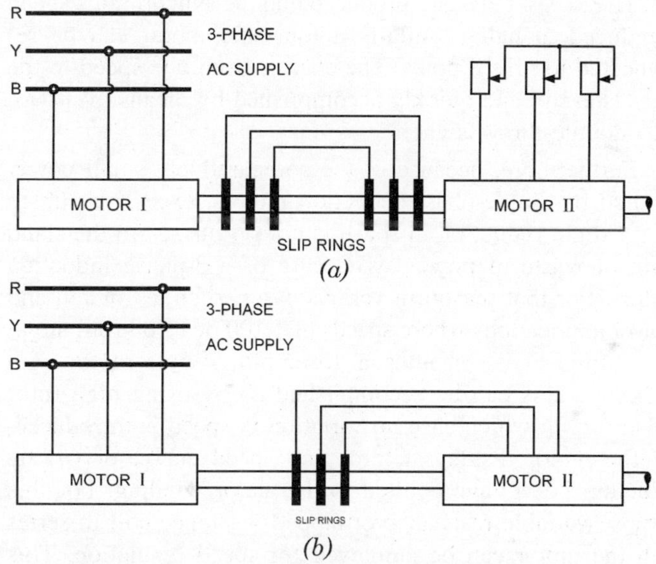

Fig. 1.87. *Cascade Connections of Induction Motors*

Further speed control, if desired, can be obtained by having second machine also of wound rotor type and inserting control resistance in the rotor circuit of the second machine.

If the two motors have the same number of poles and are connected in cumulative cascade [Eq. (1.77)], they will operate at practically half speed. Since the shaft torque is largely reduced in differential cascade, it is not employed in practice.

The mechanical output of the two machines are approximately in the ratio of $(1 - s_1): s_1$ or $\left(1 - \frac{P_2}{P_1 + P_2}\right) : \frac{P_2}{P_1 + P_2}$ or $P_1 : P_2$, *i.e.,* in the ratio of number of poles on the machines.

1.18.3. Speed Control of Single-Phase Induction Motors

The stator winding of a single-phase induction motor can easily be arranged to give two synchronous speeds, one double the other. In Fig. 1.88 (*a*) the connections for two pole motors

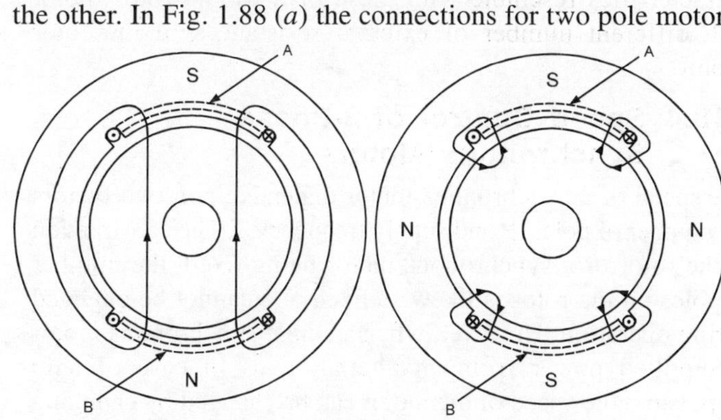

(*a*) *2-Pole Connection* (*b*) *4-Pole Connection*

Fig. 1.88

are shown in which two coils A and B have been connected in series with their mmfs helping each other. If the connections to coil B are reversed so that both the coils are again in series but their mmfs opposing each other, as shown in

Fig. 1.88 (*b*), 4 poles are produced and the synchronous speed is reduced to half. Similarly a four-pole motor can be re-connected to give 8 poles. The change from one speed to the other is easily and quickly accomplished by means of a two-pole double-throw switch.

Furthermore, because of the comparatively small power handled by single-phase induction motors, excessive slip to give a slight reduction in speed is not serious, from the stand point of waste of power, as in case of a 3-phase induction motors. For that reason in various cases, such as for fan and blower applications where speeds of 1,100 or 1,300 rpm might be required, it is not unusual to employ 4-pole motor with high slip. This can be accomplished (*i*) by using high rotor resistance, in which case no variation in speed is introduced, but the motor displays a drooping speed curve; or (*ii*) by reducing the voltage applied to the stator winding. For this purpose variable resistance or tapped reactance coil in series with the motor can be employed for speed regulation. This method is inefficient, gives large variations in speed with the variations in load, but it is much used inspite of these draw-backs. Alternatively, the voltage applied to the motor stator winding can be varied either by means of a transformer with taps on its secondary or by means of a variac, which provides the maximum possible number of taps and thus gives finest speed adjustment. This is one of the simplest methods of speed control of fractional kilowatt motors. The speed of the single phase-induction motor can also be controlled by employing tapped exciting winding, so that the constant voltage source may be impressed across the entire winding or some part of it. By this means, a normal speed and a single reduced speed are provided.

Speed control because of its more general use, has brought about the standardization of fan speeds, allowing the use of high slip motors. A squirrel cage motor with 8-10 per cent slip and with low pull-out torque operates satisfactorily not only for constant speed drives, but also for adjustable speed drives through the use of voltage control on the stator winding. With this method of speed control, the torque-speed characteristics resemble with those of a wound rotor motor with different number of external resistances in the rotor circuit.

1.18.4. Speed Control of 3-Phase Synchronous Motors

The speed of a synchronous motor depends upon two factors viz number of poles, P and supply frequency, *f*. The construction of the rotor of a synchronous motor being fixed, the number of poles on the rotor is likewise fixed and cannot be changed during operation. However, when a single synchronous motor is supplied power from an alternator, as in case of ship propulsion, the speed of the motor can be changed by changing the speed of the alternator–the speed of the motor changes exactly in the same proportion as that of the alternator supplying power to it. It is to be noted here that the voltage and frequency are directly proportional to the speed at which the alternator is driven. So no special control is required for maintaining V/*f* ratio constant. Field excitation control, however, is required as voltage applied to the terminals of stator is changed. If not, power factor is greatly affected. Assuming the motor is operating at near unity power factor and normal voltage, increase in speed causes increase in applied voltage across the motor stator terminals and the motor is badly under-excited (*i.e.*, low lagging power factor) for fixed dc excitation. On the other hand, lowering of speed causes overexcitation (*i.e.*, low leading power factor). Thus field excitation control is essential.

1.18.5. Speed Control of AC Commutator Motors

Speed control of ac commutator motors can be had either by varying the applied voltage or by movement of the brushes round the commutator. The former method makes use of tappings on a transformer or, occasionally, a series resistor or inductor so as to provide variation of voltage in steps, while brush shifting provides a gradual control throughout the range.

In case of a **single phase ac series motor,** the speed control can be accomplished, when required, only by voltage variation because it operates with fixed brush position. Transformer tappings provide variation between zero and maximum speed, as depicted in Fig. 1.89, without any appreciable additional loss such as would occur with series resistance. The latter method is employed, only when speed control is required very occasionally, and where capital cost is the prime importance.

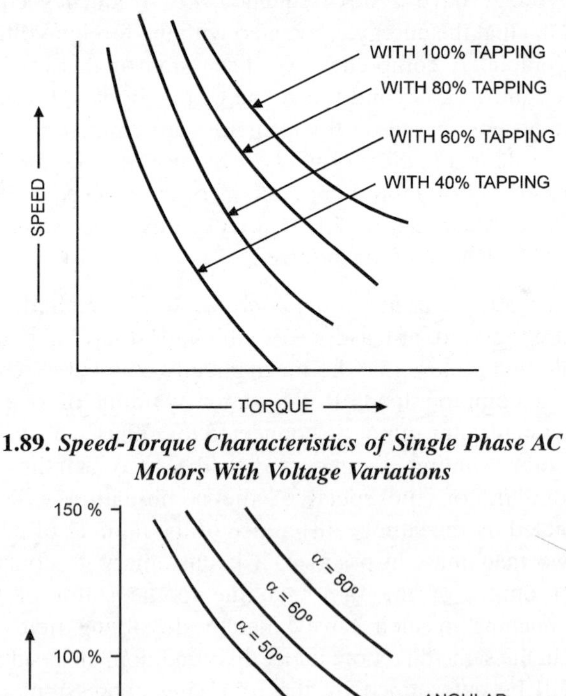

Fig. 1.89. ***Speed-Torque Characteristics of Single Phase AC Series Motors With Voltage Variations***

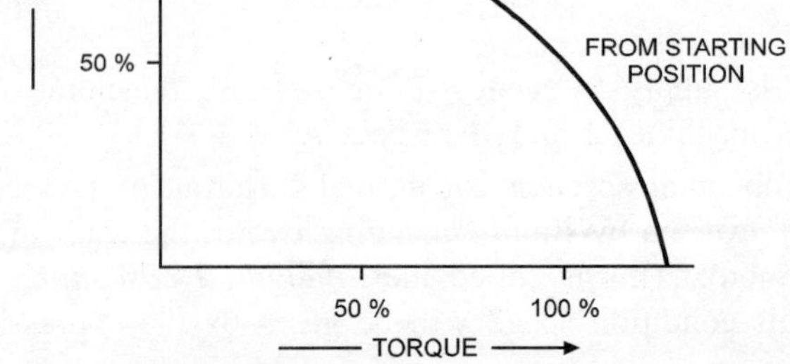

Fig. 1.90. ***Speed-Torque Characteristics of Repulsion Motor With Brush Shifting Speed Control***

In case of *repulsion motors,* brush shifting is the simplest method of speed control for a speed range between about 0.5 and 1.1 times synchronous speed, but beyond this range voltage variation control is preferred due to commutation difficulties. Speed-torque characteristics with different brush positions are shown in Fig. 1.90.

In case of 3-phase series motors, brush shifting is the most economical method of speed control and provides a range of 3 or 4 to 1 with characteristics similar to those of repulsion motors.

Example 1.25. A 220 V dc shunt motor takes 22 A at rated voltage and runs at 1,000 rpm. Its field resistance is 100 Ω and armature circuit resistance is 0.1 Ω. Compute the value of additional resistance required in the armature circuit to reduce the speed to 800 rpm when (*i*) the load torque is proportional to speed (*ii*) the load torque varies as the square of the speed.

[U.P.S.C. I.E.S. Elec. Engineering II, 1993]

Solution : In normal conditions

Line current, $I_{L_1} = 22$ A

Shunt field current, $I_{sh} = \frac{V}{R_{sh}} = \frac{220}{100} = 2.2$ A

Armature current, $I_{a_1} = I_{L_1} - I_{sh} = 22 - 2.2 = 19.8$ A

Back emf, $E_{b_1} = V - I_{a_1} R_a = 220 - 19.8 \times 0.1 = 218.02$ V

Speed, $N_1 = 1{,}000$ rpm

(*i*) Let the additional resistance required in the armature circuit be of R ohms to reduce the speed to 800 rpm when the load torque is proportional to speed.

$$T_2 = T_1 \times \frac{N_2}{N_1} = T_1 \times \frac{800}{1{,}000} = 0.8\ T_1$$

$$\text{or } I_{a_2}\Phi_2 = 0.8\ I_{a_1}\Phi_1 \quad \because T\ \alpha\ \Phi I_a$$

$$\text{or } I_{a_2} = 0.8 \times 19.8 = 15.84 \text{ A} \quad \because \Phi_2 = \Phi_1$$

$$E_{b_2} = V - I_{a_2}(R + R_a)$$

$$= 220 - 15.84(R + 0.1) = 218.416 - 15.84R$$

$$\text{Since } \frac{E_{b_2}}{E_{b_1}} = \frac{N_2}{N_1} \text{ flux being constant}$$

$$\therefore \frac{218.416 - 15.84R}{218.02} = \frac{800}{1{,}000}$$

or R = 2.778 Ω **Ans.**

(*ii*) Let the additional resistance required in the armature circuit be of R' ohms to reduce the speed to 800 rpm when no load torque varies as the square of the speed

$$T_3 = T_1 \times \left(\frac{N_2}{N_1}\right)^2 = T_1 \times \left(\frac{800}{1{,}000}\right)^2 = 0.64T_1$$

$$\text{or } I_{a_3}\Phi_3 = 0.64\ I_{a_1}\Phi_1 \quad \because T\ \alpha\ \Phi I_a$$

$$\text{or } I_{a_3} = 0.64 \times 19.8 = 12.672 \text{ A} \quad \because \Phi_3 = \Phi_1$$

$$E_{b_3} = V - I_{a_3}(R' + R_a)$$

$$= 220 - 12.672\ (R' + 0.1)$$

$$= 218.7328 - 12.672R'$$

$$\text{Since } \frac{E_{b_3}}{E_{b_1}} = \frac{N_3}{N_1} \text{ flux being constant}$$

$$\therefore \frac{218.7328 - 12.672R'}{218.02} = \frac{800}{1{,}000}$$

or R' = 3.497 Ω **Ans.**

Example 1.26. A dc series motor, with unsaturated magnetic circuit and negligible resistance, when running at a certain speed on a given load, takes 50 A at 500 V. If the load torque varies as the cube of the speed, find the resistance to be inserted to reduce the speed by 50%.

[A.M.I.E. Sec. B. Elec. Drives and Their Control Summer 1993]

Solution : Motor input current, $I_1 = 50$ A

Speed, N_1 = N (Say)

Speed, N_2 = 0.5 N

Back emf, $E_{b_1} = V = 500$ V

Since motor resistance is negligible

Since $T \propto N^3$

$$\therefore \frac{T_2}{T_1} = (0.5)^3 = 0.125$$

and since $T \propto \Phi I$

$$\text{or } \frac{\Phi_2 I_2}{\Phi_1 I_1} = 0.125$$

$$\text{or } \left(\frac{I_2}{I_1}\right)^2 = 0.125 \quad \because \Phi \propto I, \text{ magnetic circuit being unsaturated}$$

$$\text{or } I_2 = I_1\sqrt{0.125} = 50\sqrt{0.125} = 17.678 \text{ A}$$

Let the resistance of R ohms be connected in series with the motor circuit

$$\text{Back emf, } E_{b_2} = V - I_2R = 500 - 17.678R$$

$$\text{Since } E_b \propto \Phi N$$

$$\therefore \frac{E_{b_2}}{E_{b_1}} = \frac{\Phi_2}{\Phi_1} \times \frac{N_2}{N_1}$$

$$\text{or } \frac{500 - 17.678R}{500} = \frac{17.678}{50} \times 0.5$$

or R = 23.283 Ω **Ans.**

Example 1.27. A series motor is required to produce constant full-load torque whilst its speed is raised by 20% above its full-rated value by means of a divertor. Determine the divertor resistance as a percentage of field resistance. Neglect motor losses.

The magnetisation curve of the motor is as follows :

Current	40	60	80	100%	of full-load value
Flux per pole	62	80	92	100%	of full-load value

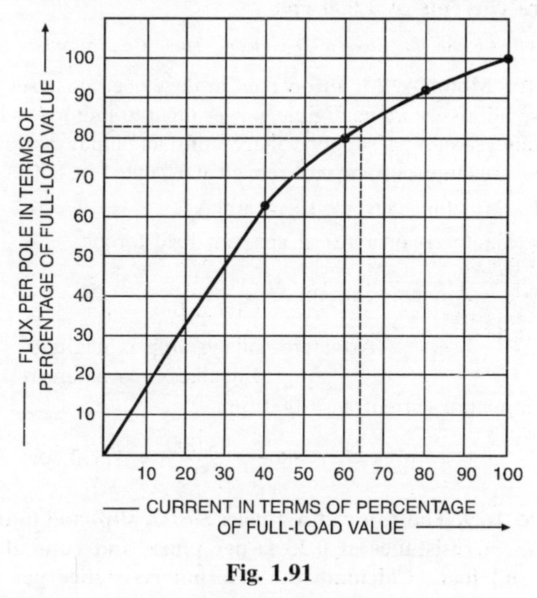

Fig. 1.91

Solution : Let the motor be connected across a supply of V volts and draw current of I amperes while running at normal speed of N rpm and developing full load torque.

$$\text{Since speed, } N \propto \frac{E_b}{\Phi} \propto \frac{V - IR}{\Phi} \propto \frac{V}{\Phi} \quad \because \text{ losses are negligible}$$

After connecting a divertor, the speed,

$$N_2 = 1.2\ N$$

$$\text{and since } \frac{N_2}{N_1} = \frac{V/\Phi_2}{V/\Phi_1} = \frac{\Phi_1}{\Phi_2}$$

$$\text{or } \frac{\Phi_1}{\Phi_2} = \frac{1.2\ N}{N} = 1.2$$

$$\text{or } \Phi_2 = \frac{\Phi_1}{1.2} = 83.3\ \%\ \text{of full load value of flux}$$

From saturation curve the field current corresponding to 83.3% of full-load value of flux is 65% of full-load value of field current.

$$i.e.,\ I_{se_2} = 0.65\ I_{se_1} = 0.65\ I_{a_1}$$

Since load torque is constant

$$\therefore I_{a_2}\Phi_2 = I_{a_1}\Phi_1$$

$$\text{or } I_{a_2} = I_{a_1}\frac{\Phi_1}{\Phi_2} = 1.2 I_{a_1}$$

Since series field current,

$$I_{se_2} = \frac{R_{div}}{R_{se} + R_{div}} \times I_{a_2}$$

$$\text{or } 0.65\ I_{a_1} = 1.2 I_{a_1}\frac{R_{div}}{R_{se} + R_{div}}$$

$$\text{or } R_{div} = \frac{0.65}{0.55} R_{se} = 1.182\ R_{se}$$

i.e., Resistance of divertor is 118.2% of series field resistance. **Ans.**

Example 1.28. Two adjustable speed dc separately excited motors have maximum speeds of 1,000 and 500 rpm. Speed is controlled by variation of field flux. Motor A drives a load requiring constant developed power over the speed range and motor B drives a load requiring constant torque. All losses, including armature resistance and armature reaction may be neglected and magnetic circuit is assumed linear. If the developed power is equal at 500 rpm and armature currents are each 800 A, what will be the armature currents at 1,000 rpm ?

[*A.M.I.E. Sec B. Electric Drives and Their Control Summer 1999*]

Solution : Motor A. Motor operates at developed constant power and, since all losses are negligible, power input to motor A will also be constant. Assuming supply voltage constant, input current and, therefore, armature current will remain unchanged at 800 A. **Ans.**

Motor B. Armature current, I_{ab} = 800 A

Since motor B operates at constant load torque,

$$\text{Armature current} \propto \frac{1}{\text{Field flux}} \propto N$$

$\because$ Armature voltage drop is negligible owing to negligible armature resistance

So armature current at, 1,000 rpm,

$$I'_{ab} = I_{aa} \times \frac{N'}{N} = 800 \times \frac{1,000}{500} = 1,600\ \text{A} \quad \textbf{Ans.}$$

Example 1.29. The rotor of a 4-pole, 50 Hz, slip-ring induction motor has a resistance of 0.25 Ω per phase and runs at 1,440 rpm at full load. Calculate the external resistance per phase which must be added to lower the speed to 1,200 rpm, the torque being the same as before.

[*A.M.I.E. Sec B. Advanced Elec. Machines Winter 1993; Elec. Machines II Bharathiyar Univ. Nov. 1996; Madurai Kamarajar Univ. Nov. 1996*]

Solution : Synchronous speed of motor,

$$N_s = \frac{120 f}{P} = \frac{120 \times 50}{4} = 1,500\ \text{rpm}$$

Full-load speed, N = 1,440 rpm

$$\text{Full-load slip, } s_1 = \frac{N_s - N}{N_s} = \frac{1,500 - 1,440}{1,500} = 0.04$$

After inserting resistance (say R ohms per phase) in the rotor circuit.

Motor speed, N_2 = 1,200 rpm

$$\text{Slip, } s_2 = \frac{1,500 - 1,200}{1,500} = 0.2$$

$$\text{Since slip, } s = \frac{\text{Rotor copper loss}}{\text{Input power to rotor}} = \frac{3I_2^2 R_2}{\text{Input power to rotor}}$$

For constant power input to rotor and rotor current, $s \propto R_2$

$$\frac{s_2}{s_1} = \frac{R_2 + R}{R_2}$$

$\because$ for constant load torque power input to rotor and rotor current remains the same

$$\text{or } \frac{0.2}{0.04} = \frac{0.25 + R}{0.25}$$

or R = 1 Ω **Ans.**

Example 1.30. Two 50 Hz, 3-phase induction motors having six and four poles respectively are cumulatively cascaded, the 6-pole motor being connected to the main supply. Determine the frequency of the rotor currents and slips referred to each stator field if the set has a slip of 2%. [*Madras Univ. Elec. Machinery II, 1987*]

Solution : Synchronous speed of the set,

$$N_{sc} = \frac{120 f}{P_1 + P_2} = \frac{120 \times 50}{6 + 4} = 600\ \text{rpm}$$

Actual speed of the set, $N = N_{sc}(1 - s') = 600\ (1 - 0.02) = 588$ rpm

Synchronous speed of 6-pole motor stator field,

$$N_s = \frac{120 \times f}{P_1} = \frac{120 \times 50}{6} = 1,000\ \text{rpm}$$

Slip referred to the stator field,

$$s = \frac{N_s - N}{N_s} = \frac{1,000 - 588}{1,000} = 0.412 \text{ or } 41.2\%\ \textbf{Ans.}$$

The 6-pole motor rotor current frequency or the impressed frequency on 4-pole motor stator field,

$$f' = sf = 0.412 \times 50 = 20.6\ \text{Hz}\ \textbf{Ans.}$$

The synchronous speed of 4-pole induction motor with frequency f',

$$N'_s = \frac{120 f'}{P_2} = \frac{120 \times 20.6}{4} = 618\ \text{rpm}$$

The slip as referred to the 4-pole induction motor,

$$s'' = \frac{N'_s - N}{N'_s} = \frac{618 - 588}{618} = 0.0485 \text{ or } 4.85\%\ \textbf{Ans.}$$

Frequency of rotor current of 4-pole induction motor,

$$f'' = s'' f' = 0.0485 \times 20.6 = 1.0\ \text{Hz}\ \textbf{Ans.}$$

Example 1.31. A 10 hp, 4-pole, 50 Hz, 220 V induction motor is connected in cascade with another 15 hp, 6-pole, 50 Hz, 220 V motor. What are the possible speeds obtainable with this combination and what is the maximum load in hp which may be delivered without overloading either machine. ? Also determine the ratio of mechanical power outputs of the two machines at this load.

Solution : Synchronous speed of machine I (6-pole machine),

$$N_{s_1} = \frac{120f}{P_1} = \frac{120 \times 50}{6} = 1{,}000 \text{ rpm}$$

Synchronous speed of machine II (4-pole machine),

$$N_{s_2} = \frac{120f}{P_2} = \frac{120 \times 50}{4} = 1{,}500 \text{ rpm}$$

Concantenated speed of the set when cumulatively connected,

$$N_{sc_1} = \frac{120f}{P_1 + P_2} = \frac{120 \times 50}{6+4} = 600 \text{ rpm}$$

Concantenated speed of the set when differentially connected,

$$N_{sc_2} = \frac{120f}{P_1 - P_2} = \frac{120 \times 50}{6-4} = 3{,}000 \text{ rpm}$$

So speeds available are 1,000, 1,500, 600 and 3,000 rpm **Ans.**

In cascade connection the rating of the main motor should be such that it covers the power output of the auxiliary motor also. So the maximum load which can be delivered is 15 hp **Ans.**

The ratio of the mechanical power outputs of the main and auxiliary induction motors are $P_1 : P_2 :: 6 : 4$ or $1.5 : 1$ **Ans.**

Example 1.32. A 3-phase, 440 V, 50 Hz, 110 kW, 24-pole, 245 rpm slip-ring induction motor has both its stator and rotor windings connected in star. The ratio of stator turns to rotor turns is 1.25. The resistance measured between each pair of slip-rings is 0.04 Ω. This motor drives a fan which requires 110 kW at the full speed of the motor. The torque required to drive the fan varies as the square of speed. What resistance should be connected in series with each slip-ring so that the fan will run at 175 rpm. Neglect stator resistance and leakage reactance and rotational losses. [*A.M.I.E. Sec. B. Electrical Drives and Their Control Winter 1993*]

Solution : Synchronous speed of motor,

$$N_s = \frac{120 \times 6}{P} = \frac{120 \times 50}{24} = 250 \text{ rpm}$$

$$\text{Full-load slip, } s_f = \frac{N_s - N}{N_s} = \frac{250 - 245}{250} = 0.02$$

$$\text{Torque at full-load, } T_f = \frac{\text{Output is kW} \times 1{,}000}{2\pi N/60}$$

$$= \frac{110 \times 1{,}000 \times 60}{2\pi \times 245} = 4{,}287.44 \text{ N-m}$$

Rotor resistance per phase,

$$R_2 = \frac{\text{Resistance between a pair of slip-rings}}{2}$$

$$= \frac{0.04}{2} = 0.02\ \Omega$$

Rotor resistance referred to stator,

$$R'_2 = R_2 \times K^2 = 0.02 \times (1.25)^2 = 0.03125\ \Omega$$

Torque developed by rotor is given as

$$T_f = \frac{3}{2\pi N_s/60} \times \frac{V_1^2 R_2' s_f}{(R_2')^2 + s_f^2 (X_2')^2} \qquad \text{Refer to Eq. (1.36)}$$

Substituting $V_1 = \frac{440}{\sqrt{3}}$ V (stator being connected in star), $s_f = 0.02$, $R_2' = 0.03125\ \Omega$, $T_f = 4{,}287.44$ N-m, and $N_s = 250$ rpm in above equation, we have

$$4{,}287.44 = \frac{3 \times (440)^2 \times 0.03125 \times 0.02 \times 60}{3 \times 2\pi \times 250\left[(0.03125)^2 + (0.02 X_2')^2\right]}$$

$$\text{or} \quad (X'_2)^2 = \left[\frac{440^2 \times 0.03125 \times 1.2}{4{,}287.44 \times 500\,\pi} - (0.03125)^2\right] \times \left(\frac{1}{0.02}\right)^2$$

$$= 0.2536$$

$$\text{or} \quad X'_2 = 0.5036\ \Omega$$

$$\text{Slip at 175 rpm, } s = \frac{250 - 175}{250} = 0.3$$

Let the rotor resistance referred to stator per phase corresponding to speed of 175 rpm be R''_2 and torque developed be T

$$\text{So} \quad \frac{T}{T_f} = \left[\frac{N}{N_f}\right]^2 = \left(\frac{175}{245}\right)^2$$

or Torque at 175 rpm,

$$T = T_f \times \left(\frac{175}{245}\right)^2 = 4{,}287.44 \times \left(\frac{175}{245}\right)^2$$

$$= 2{,}187.47 \text{ Nm.}$$

Substituting T = 2,187.47 Nm, $V_1 = \frac{440}{\sqrt{3}}$ V, $s = 0.3$ and $X_2 = 0.5036\ \Omega$ in torque equation, we have

$$2{,}187.47 = \frac{(440)^2 \times R_2'' \times 0.3 \times 60}{2\pi \times 250[(R_2'')^2 + (0.3 \times 0.5036)^2]}$$

$$\text{or} \quad (R''_2)^2 + 0.022825 = 1.0142 R''_2$$

$$\text{or} \quad R''_2 = 0.023\ \Omega \text{ or } 0.991\ \Omega$$

The first value, being lesser than original value, is rejected. So resistance to be added to each slip-ring $= \frac{0.991 - 0.03125}{(1.25)^2}$ $= 0.6142\ \Omega$ **Ans.**

1.19. ELECTRIC BRAKING

If the load is removed from an electric motor and supply to it be disconnected, it will continue to run for some time due to inertia. The time elapsing before it stops will be especially long if the motor is massive and has run at high speed. It is essential, however, in many cases that the motor and its driven machine be stopped quickly (in machine tools, cranes, hoists etc). In fact, quick stopping of a motor is more essential than quick starting. Delay in starting up a motor only causes the machinery to stand idle; a delay in stopping a motor may result in heavy damage to equipment or to the manufactured products and even the loss of human life.

Based on the purpose for which braking is employed, it is of the two types, viz, braking while bringing the drive to rest and braking while lowering loads. In the first type, the device employed for braking absorbs the kinetic energy of the moving parts while in the second one, it absorbs, in addition to the kinetic energy, potential energy, usually gravitational which can drive the system at an excessively high speed.

Braking, while stopping, may be used for any one of the following objectives:

(*i*) Reducing the time taken to stop.

(*ii*) Stopping exactly at specified points, for example in lifts; sometimes such precise stops are necessary for reasons of safety.

(*iii*) Feeding back, atleast a portion of the power, to the supply system.

Braking, while lowering loads, may be employed for any of the following purposes:

(*i*) Controlling the speed at which the load comes down and limiting it to a safe value.

(*ii*) Feeding power back to the supply system.

The main features of a satisfactory brake are that

(*i*) the braking should be quick and reliable in action,

(*ii*) the braking torque must be controllable,

(*iii*) some suitable means must be provided for dissipation of kinetic energy of the moving parts of the motor and its driven machines and

(*iv*) failure of any part of the braking system must result in application of brakes.

Braking torque can be applied either by mechanical (or friction) brakes or electrodynamically. In mechanical braking, the stored energy of rotating parts is dissipated in the form of heat by a brake shoe or band rubbing on a wheel or a brake drum. In electrical (or electrodynamic) braking the stored energy of rotating parts is converted into electrical energy and dissipated in the resistance in the form of heat or returned to the supply. Here we will deal with electric (or electrodynamic) braking only.

Advantages and Disadvantages of Electric Braking Over Mechanical Braking. Before we deal with electric braking it will be desirable to have the knowledge of advantages and limitations of electric braking over mechanical braking.

Advantages. (*i*) Since mechanical braking, due to excessive wear on the brake blocks or brake lining, needs frequent and costly replacement but no such replacement is required in electric braking, hence there is saving on this account.

(*ii*) Frequent adjustments are required to compensate for the wear in mechanical braking thereby making maintenance costly. In electric braking there is no cost by way of maintenance of the brake shoes or lining.

(*iii*) By employing electric braking the capacity of the system can be increased by way of higher speeds and haulage of heavy loads.

(*iv*) In some cases of electric braking, a part of energy is returned to the supply thereby affecting a considerable saving in the operating cost. This is not possible with mechanical braking.

(*v*) In electric braking, heat producd, in no way, is harmful to the braking system where as in mechanical braking heat is produced at brake blocks or brake lining which may result in failure of brake.

(*vi*) Mechanical braking produces metal dust which, if finds ingress to bearings, will cause over-wearing of the bearings. No such dust is produced in electric braking.

(*vii*) Incorrect adjustment of mechanical brakes may result in very harsh and sudden braking being applied which may produce shock loading of ropes and mechanical parts or in case of passenger lifts and trains it may result in the discomfort to the occupants. Electric braking on the other hand is smooth and without shock.

Disadvantages and Limitations. (*i*) In electric braking, the driving motor operates as a generator during the period of braking, and motor ceases to operate as a generator at standstill so that although an electric brake can almost stop a machine or load, but it cannot hold it stationary, therefore, a friction brake is required in addition.

(*ii*) Since the motor has to function as a generator during the braking period, therefore, it must have suitable braking characteristics *i.e.,* choice of motor is limited.

(*iii*) Additional complications, high initial cost, special motors capable of generating electrical energy make electric braking costly.

Economic considerations have a considerable bearing on the use of electric braking.

Types of Electric Braking. There are three types of electric braking, all of which are applicable to the usual types of electric motors, viz plugging (or counter-current braking), dynamic (or rheostatic) braking and regenerative braking.

1. *Plugging or Counter-Current Braking.* This is the simplest type of braking. Plugging or counter-current braking occurs when the motor windings are connected for reverse direction of rotation at a time when the armature is still rotating in the forward direction either under the action of an external torque or due to inertia. Plugging involves reconnection of power supply to the motor so that it tends to drive in the opposite direction. It is obvious that, left to itself, the system will come to rest and then accelerate in the reverse direction. In case, it is required to bring the drive system to rest, it is necessary to include a special device to cut off the supply exactly at the instant when the motor stops.

During the braking period, the energy is drawn from the supply. The energy drawn from the supply and stored or kinetic energy of the rotating parts of motor and its driven machines are dissipated in the series current limiting resistor. Thus this method is wasteful of energy. Other drawbacks of this method of electric braking are the shock caused to the motor and equipment and heavy inrush of current at the time of braking.

This method provides greater braking torque than that provided by rheostatic (or dynamic) braking. This method is used to get either a quick reversal or to get a rapid stop. This method is commonly used in controlling rolling mills, elevators, printing presses, machine tools etc.

This method can be applied to direct current, alternating current induction and synchronous motors.

2. *Dynamic (or Rheostatic) Braking*. In this method of braking the motor is disconnected from the supply and operated as a generator driven by the kinetic energy of the rotating parts of the motor and its driven machines. Thus the kinetic energy of rotation is converted into electrical energy, which is dissipated in the external resistance connected across the motor at the braking instant. By this method, energy required from the supply to brake the motor, has been eliminated as compared to the previous method (plugging) and this is marked advantage. This method of braking can be applied to brake the direct current motors, synchronous motors, and induction motors (with a separate source of dc excitation during braking).

3. *Regenerative Braking*. In the previous two methods of electric braking namely plugging and rheostatic braking stored energy of the rotating parts of the motor and its driven machine is wasted whilst in plugging extra energy is drawn during the braking period and is wasted. In regenerative braking, mechanical energy is converted into electrical energy, part of which is returned to the supply and the rest of the energy is lost as heat in the windings and the bearings of the electrical machines. Regeneration does not, in general, involve any switching operation, unless it is required to change the speed at which it becomes effective. Most of the electrical machines pass smoothly from motoring to generating operation, when overdriven by load.

For regenerative braking it is necessary for (*i*) supply voltage to drop, (*ii*) the motor to be overexcited or (*iii*) the motor to be running at a speed higher than no-load speed. In all of these cases the armature current is reversed and a retarding torque is produced, slowing down the motor until the back emf of the armature becomes equal to the supply voltage. Regenerative braking will not stop the motor. It is effective only for braking hauling loads. They may occur in a crane motor due to action of the lowering load or in the motor of an electric locomotive moving down grade.

1.20. ELECTRIC BRAKING WITH DC MOTORS

1.20.1. Plugging or Counter-Current Braking

In a dc motor a reversed torque is obtained by reversing the current either in the armature or in the field (not both), Polarity reversal of field winding is rarely used because it results in longer braking time due to relatively large inductance of the field winding in comparison to that can be obtained by polarity reversal of armature winding.

The connections for dc series and shunt motors during normal running and braking conditions are shown in Figs. 1.92 and 1.93 respectively. In case of a dc series motor it should be ensured that the direction of flow of current in the field winding remains unchanged when the current flow in the armature winding is reversed.

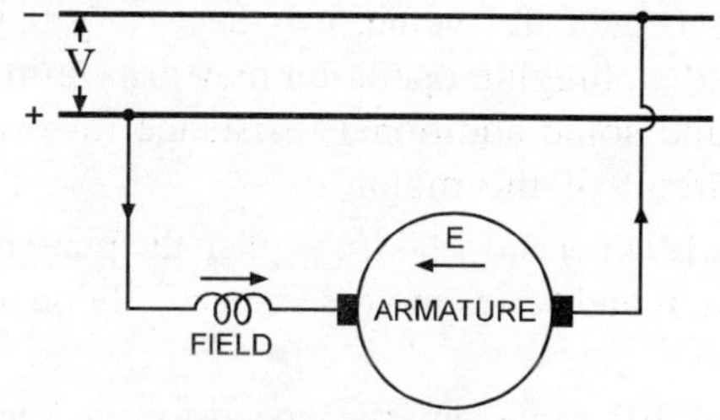

(*a*) *Normal Running*

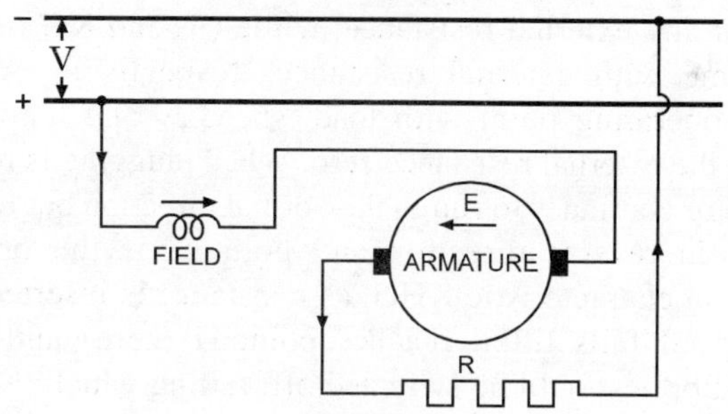

(*b*) *Plugging*

Fig. 1.92

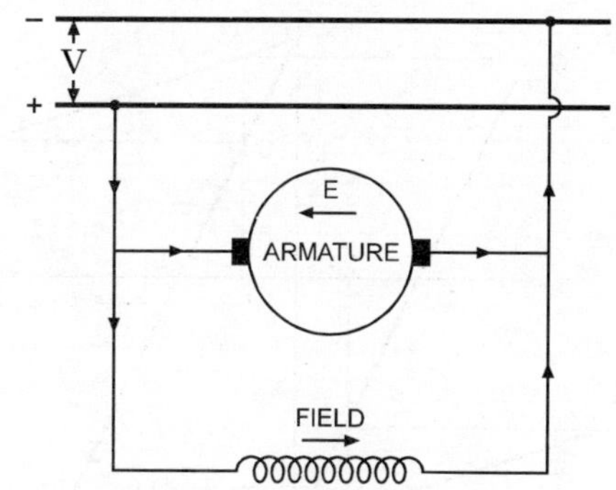

(*a*) *Normal Running*

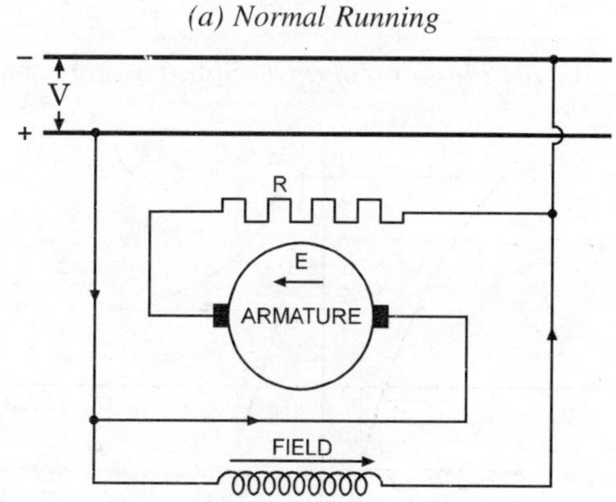

(*b*) *Plugging*

Fig. 1.93

In the normal running position the back emf is nearly equal to applied voltage and opposite in direction, so that a small voltage acts across the armature circuit to drive the normal current through a small resistance of the motor. In the plugging position, the induced emf acts in the direction of applied voltage, therefore, at the instant of switching the motor to the plugging position, twice of supply voltage acts across the machine circuit and heavy current would flow (about twice the current drawn by the stationary motor on normal rated voltage). Hence to avoid flow of heavy current and limit it to the safer value, it is necessary that switching performing the plugging operation may also re-insert starting resistance and some additional resistance in series with the armature circuit of the motor.

Figures 1.94 (*a*) and 1.94 (*b*) explain the plugging operation of the dc shunt and series motors respectively on a *quadrantal diagram.*

AB and EF represent the speed-torque characteristics of the motor in normal and reverse direction of rotation without any external resistance, while CL and KG represent the same with external resistances respectively. A is the initial operating point with load, speed N and load torque T_L and the external resistance zero. When plugging is resorted, the motor continues to run in the normal direction but develops torque in reverse direction and point G is the operating point (on characteristic KHG, as resistance is inserted here). The speed falls till it reaches point H (zero) and at this stage supply should be switched off, failing which the motor attains speed in reverse direction under motoring action and operates at point E on switching out external resistance.

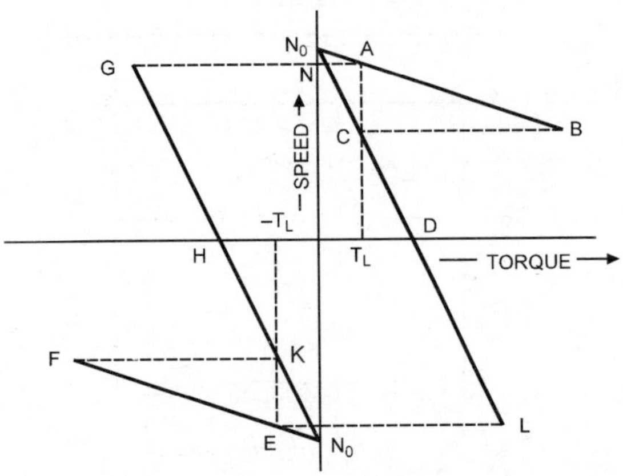

(*a*) *Plugging of a DC Shunt Motor*

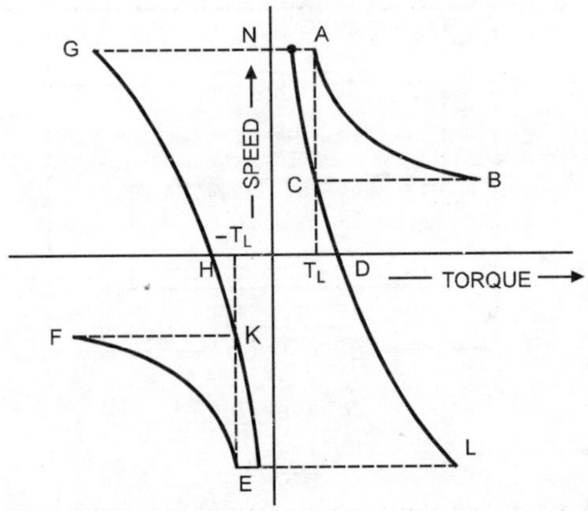

(*b*) *Plugging of a DC Series Motor*

Fig. 1.94. ***Quadrantal Diagrams***

1.20.2. Dynamic (or Rheostatic) Braking

1. ***Rheostatic Braking With DC Shunt Motors.*** The armature of a dc shunt motor is disconnected from the supply and connected across a braking resistance, keeping the field connected to the supply, as shown in Fig. 1.95 (*b*). The machine then acts as a generator loaded with a resistance. Since the field current flows as before but the armature current in the reversed direction, therefore, a retarding torque is produced. As the motor speed falls during rheostatic braking, the retarding effect decreases. In order to maintain the retarding torque, some sections of the braking resistance may be then cut out of circuit. If, during rheostatic braking, supply fails, the magnetic flux of the motor will collapse, and there will be no braking effect. This drawback is overcome by providing a series winding in the armature circuit which is connected during the braking period only. The braking effect can be controlled by varying the braking resistance R.

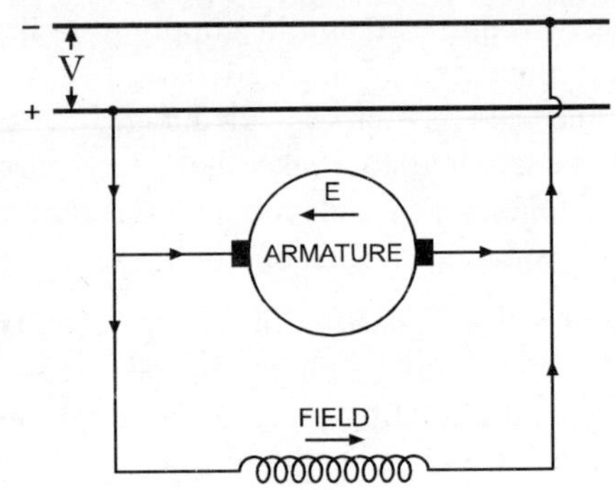

(*a*) *Normal Running*

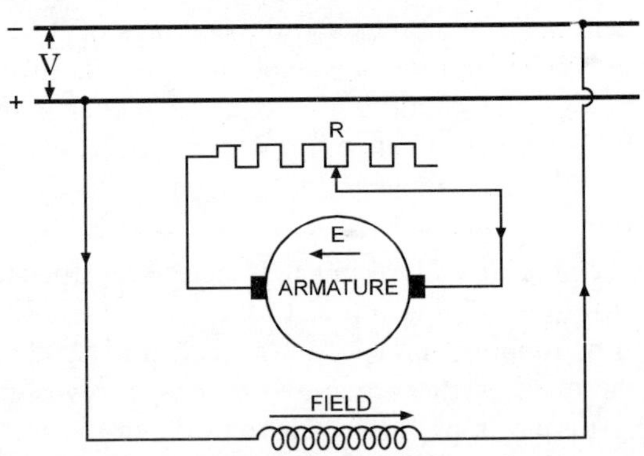

(*b*) *Rheostatic Braking*

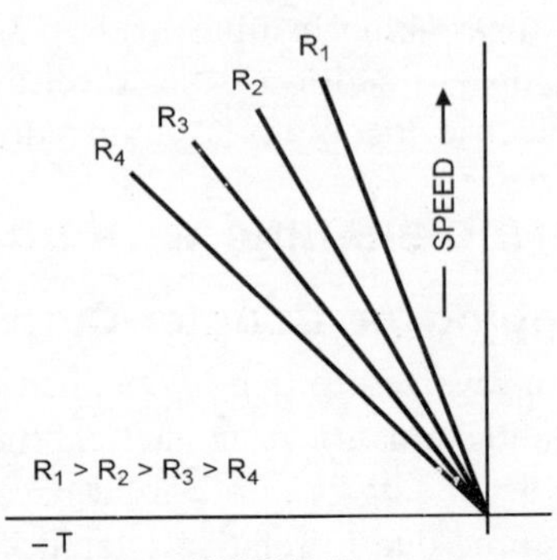

(*c*) *Speed-Torque Characteristics*

Fig. 1.95. ***Rheostatic Braking With DC Shunt Motors***

2. ***Rheostatic Braking With DC Series Motors.*** In the case of a series wound direct current motor, when it is disconnected from the supply and is connected across a braking resistance, it is essential that the connections of the field winding are changed over. If this is not done the current through the field will also be reversed, and no braking will occur. To avoid this, the field connections are so changed that the current through the field winding will flow in the same direction as before. The resistance inserted in the circuit must be less than the critical resistance, otherwise the generator will not be self-exciting.

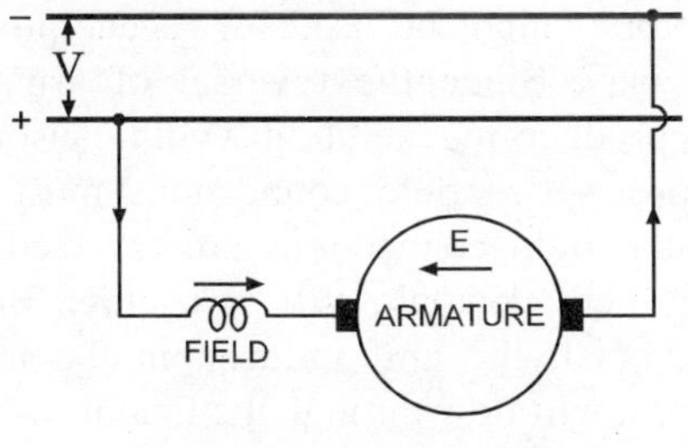

(a) Normal Running

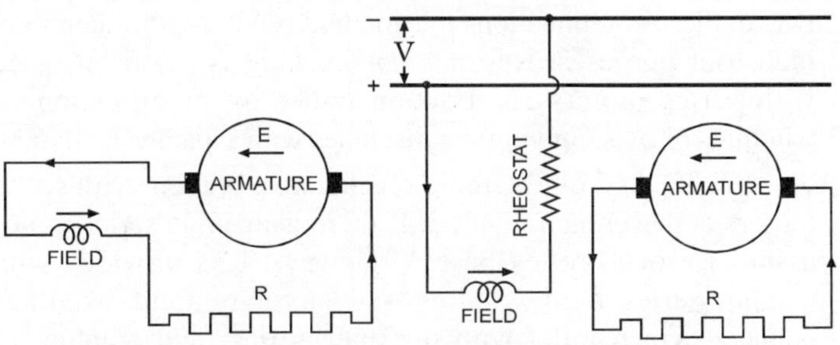

(b) Rheostatic Braking With Self Excitation *(c) Rheostatic Braking With Separate Excitation*

Fig. 1.96

Rheostatic braking of a dc series motor can also be affected by feeding its field winding from an independent source and connecting a resistance across the armature terminals *i.e.*, by reconnecting it as a separately excited machine, as shown in Fig. 1.96 (*c*). An additional resistance is also connected in series with the field winding to limit the field current to a safe value. Since the machine operates as a separately excited generator, the braking characteristics will be same as those for dc shunt motor during dynamic braking.

Dynamic braking with self excitation is normally not used as the field of the machine may build up suddenly and cause the braking torque to jump to high value and the driven unit may experience sudden jerks.

3. *Rheostatic Braking With DC Compound Motors*. For direct current compound wound motors rheostatic braking is performed in the same way as for dc shunt wound motors.

The relation between the braking torque and speed for direct current motors can be determined as below.

$$\text{Electric braking torque} = K_1\Phi I = K_1\Phi.\frac{E}{R}$$

since braking current, $I = \frac{E}{R}$, where E is the induced emf and R is the total resistance in the motor circuit.

$$\text{Again } E \propto \Phi N$$

$$\text{or } E = K_2\Phi N$$

$$\therefore \text{Electric braking torque} = \frac{K_1\Phi}{R} \times K_2\Phi N = K_3\Phi^2 N \quad ...(1.79)$$

In case of series wound motors since flux Φ varies with the variation in current the value of torque therefore, can only be determined graphically from the magnetisation curve.

In case of shunt wound motors since flux remains constant, the electric braking torque, therefore, for any given value of R is directly proportional to speed.

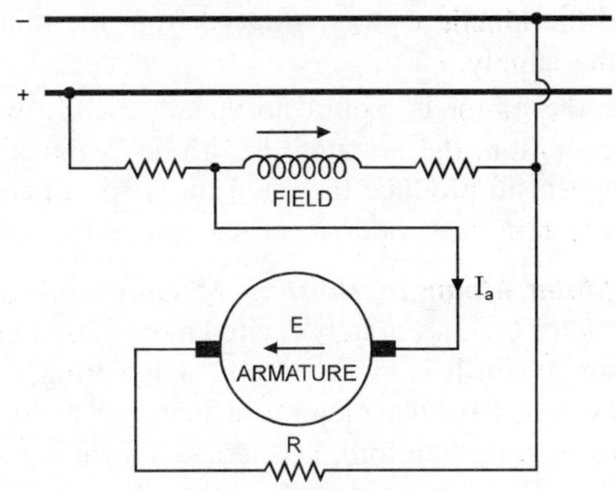

Fig. 1.97

In actual practice, in order to have maximum braking torque resistance in the armature circuit is so reduced with speed that armature current and, therefore, braking torque remains constant till external resistance becomes zero. The braking torque falls linearly with the speed and becomes zero at zero speed. In case of dc series motors with self-excitation speed-torque curve is not linear because magnetic flux changes with the change in armature current which changes with the variations in speed.

In practice rheostatic braking is preferred for non-reversing drives as this simplifies the drive circuit. For reversing drives, plugging is preferred for braking and starting of the motor in the opposite direction in a short time.

Potentiometric Lowering Circuit. Such a circuit is shown in Fig. 1.97 which is just the same as shown in Fig. 1.93 except that the armature carrying reverse current is supplied from a potential divider of which field winding forms a part (not across full supply voltage).

1.20.3. Regenerative Braking

1. *Regenerative Braking With DC Shunt Motors*. Regenerative braking is an inherent characteristic of shunt wound motors and does not require any changing of connections.

Let us consider a shunt motor running in normal condition. Now if due to, say, hauling load, speed increases above normal, due to increase in speed the induced emf will increase and if it exceeds the line voltage, the machine will start supplying current to the line, thus have tendency to prevent any further increase in speed.

Similarly if the field current is increased so that motor emf exceeds the line voltage and it may start supplying current to the line, the motor will slow down to the speed corresponding to this new value of field current.

Regenerative braking can be easily applied to dc shunt motors, particularly in cases where it is required to hold a load at a certain speed for instance lowering a hoist, it is not, however, possible to obtain regenerative braking down to very low speeds, since a sufficient increase in field current cannot be obtained.

Before bringing the motor to rest, the field excitation is increased to the permissible maximum value, as a result of which the speed of the motor is reduced to the minimum value and the kinetic energy released from the motor is fed back to the supply.

Since the motor is requird to operate with a weak field at rated condition, the armature has to be designed to carry a large current to produce the rated torque and hence motor will be larger in size, poor in efficiency and costlier.

DC Shunt Motor in Hoisting Mechanism. Consider a dc shunt motor (or a separately excited motor) used in hoisting mechanism. When it is switched on for lowering a load, the torque developed by the motor and that due to load torque act in unison and, therefore, accelerates the motor. With the increase in speed, induced emf in the armature (*i.e.,* back emf, E) increases till E = applied voltage V. At this moment the speed becomes the ideal no-load speed, the armature current and, therefore, the developed electromagnetic torque is zero, so that the downward motion of the hoist is sustained only by the downward moving load. When the speed exceeds the ideal no-load speed, induced emf E exceeds the applied voltage V and, therefore, armature current becomes negative. The drive, then acts as a generator and provides the braking torque. Under such condition

$$\text{Current } I = \frac{E - V}{R}$$

and electric braking torque,

$$T = K_1 \phi I = K_1 \phi \frac{E - V}{R}$$

$$= K_1 \phi \frac{K_2 N \phi - V}{R}$$

where K_1 and K_2 are constants.

$$\text{or braking torque, } T = \frac{K_1 K_2 \phi^2 N}{R} - \frac{K_1 \phi V}{R}$$

$$= K_3 \phi^2 N - K_4 \phi \qquad ...(1.80)$$

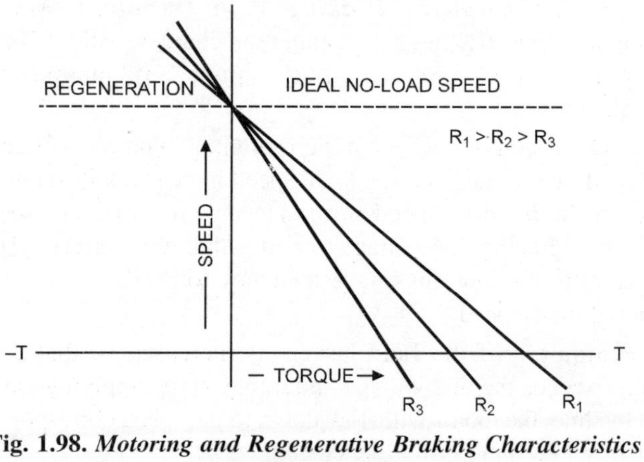

Fig. 1.98. ***Motoring and Regenerative Braking Characteristics of a DC Shunt Motor***

The speed-torque characteristics of motor during motoring and braking operation are shown in Fig. 1.98. From Fig. 1.98 it is obvious that braking or regeneration can occur only at speeds greater than the ideal no-load speed of the motor. From Fig. 1.98 it can also be noted that higher the resistance of the armature circuit, to develop a particular braking torque, higher is the speed required.

2. *Regenerative Braking With DC Series Motors.* The dc series motors cannot be used for regenerative braking in an ordinary way. Since the reversal of armature current necessary to produce regeneration would cause a reversal of field, therefore, series field connections must be reversed. But even if the field connections are reversed at the exact moment, this method would still be useless. Because at the instant of reversal, the emf induced in the motor will be small, so current will flow through the field in wrong direction, which will reverse the field and cause the motor emf to help the supply voltage. This will result in short circuit of supply. Due to these complications this method is not used for common industrial purposes. Regenerative braking is, however, used with series motors for traction either by modification of windings or by supplying the machines with separate excitation.

One method of obtaining regenerative braking with series motors is the French method. If there is a single series motor as in case of a trolley buses, tramways, it is provided with a main series field winding and auxiliary field windings connected in parallel with the main series field winding as shown in Fig. 1.99 (*a*). During regeneration (braking period) the auxiliary field windings are put in series with each other and are switched across the supply, as shown in Fig. 1.99 (*b*). The machine acts as a compound generator slightly differentially compounded. Such an arrangement is quite stable. Any change in line voltage causes a change in excitation which produces a corresponding change in the induced emf of the machine so that inherent compensation is provided. For example, if the line voltage increases beyond the emf of the generator the increased voltage across the generator's field will send a large exciting current through it causing the emf of the generator to rise. The reversal of this will happen when the line voltage decreases.

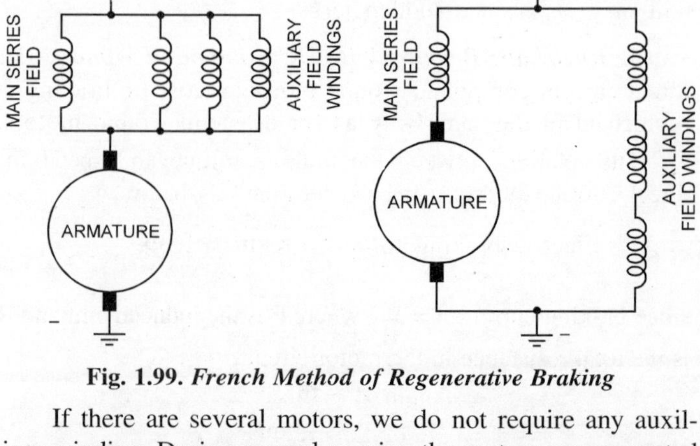

Fig. 1.99. ***French Method of Regenerative Braking***

If there are several motors, we do not require any auxiliary winding. During normal running the motors are connected in parallel with the field winding connected in series with their respective armatures, as shown in Fig. 1.100 (*a*).

But during regeneration the motors are connected, as shown in Fig. 1.100 (*b*), *i.e.,* all armatures are connected in parallel and series field windings of all motors but one are connected in series and placed across the supply. Suitable resistance is also connected in series with the series field windings, as illustrated in the Fig. 1.100 (*b*).

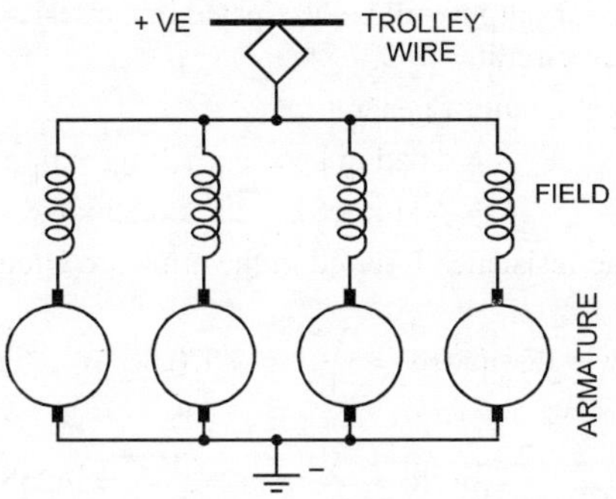

(*a*)

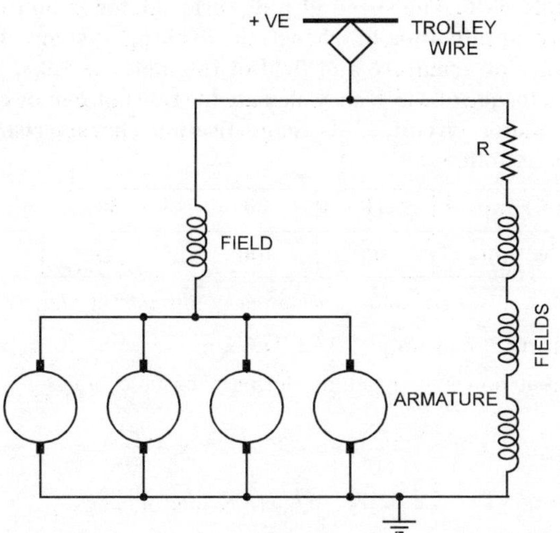

(*b*)

Fig. 1.100

Another alternative method of regenerative braking, which obviates the losss of power in the resistances in the field circuit, is by using a separate exciter for controlling the excitation of the field windings during regeneration. The exciter may either be axle driven or driven by a motor operated from the auxiliary supply. Many devices have been used to secure compensation against variations in line voltage, most of which make use of some from of differential winding on the exciter. One such method is illustrated in Fig. 1.101 (*a*) in which the exciter is differentially compounded. The shunt field winding of the exciter is separately excited from an auxiliary supply, the series field winding is connected in the main-motor circuit in such a way that it opposes the separately excited winding during regeneration. It can be seen that a rise in line voltage during regeneration, which will tend to reduce the regenerated current, will strengthen the exciter field and counteract the change. The stabilising resistance R also assists in this action.

Another method used for regenerative braking is depicted in Fig. 1.101 (*b*). Here the exciter armature along with the field windings of the series motors are connected across the stabilizing resistance in the main circuit. The current through the stabilizing resistance is the sum of the exciter current and the regenerated current. Any increase in regenerated current, due to fall in line voltage, causes larger voltage drop in the stabilizing resistance and, therefore, less current through the exciter circuit, again counter-acting the effect of line voltage variations.

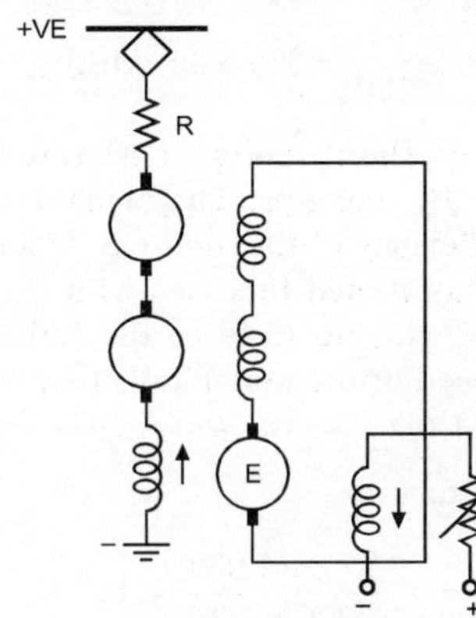

(*a*) *Differentially Compounded Exciter*

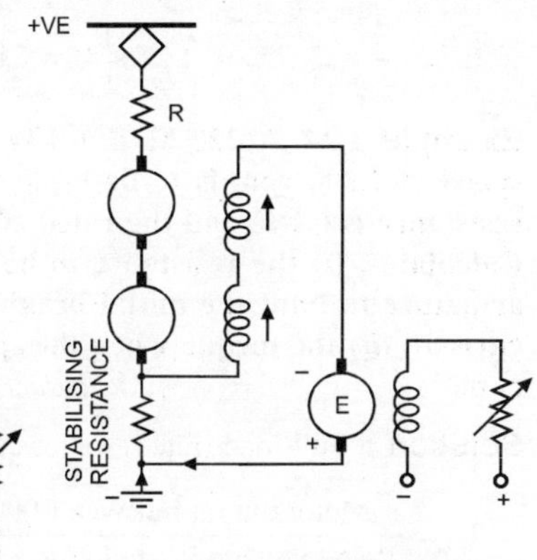

(*b*) *Separately Excited Exciter*

Fig. 1.101. *Regenerative Braking With Separate Exciter*

3. *Regenerative Braking With DC Compound Motors.* Compound wound motors can be regeneratively retarded by an hauling load only if their shunt field is stronger than their series field.

Example 1.33. A 220 V separately excited motor running at 1,000 rpm draws 100 A from the source. The motor is braked by plugging. Calculate (*i*) resistance to be inserted in the armature circuit to limit the braking current to twice the full-load current. (*ii*) Initial braking torque. (*iii*) The braking torque when speed has reduced to 500 rpm.

[*U.P. Technical Univ. Electric Drives 2005-06*]

Solution : Full-load motor current,

$$I_f = 100 \text{ A}$$

Supply voltage, $V = 220$ V

$$\text{Full-load torque, } T_f = \frac{V \times I_f}{\frac{2\pi N}{60}} = \frac{220 \times 100 \times 60}{2\pi \times 1000} = 210 \text{ N-m}$$

Neglecting motor losses

Back emf, $E_{b_f} = V = 220$ V Neglecting armature resistance

Voltage across the armature at the instant of braking

$$= V + E_{b_f} = 220 + 220 = 440 \text{ V}$$

(*i*) Resistance to be connected in series with the armature to limit the initial braking current to twice the full-load current (*i.e.* $I_b = 2I_f = 2 \times 100 = 200$ A),

$$R = \frac{V + E_{b_f}}{I_b} - R_a = \frac{440}{200} - 0 = 2.2\ \Omega \quad \textbf{Ans.}$$

(*ii*) Initial braking torque,

$$T_b = T_f \times \frac{I_b}{I_f}$$ $\because T \propto \phi I_a$ and flux in a separately excited motor is assumed to remain constant

$$= 210 \times 2 = 420 \text{ N-m} \quad \textbf{Ans.}$$

(*iii*) When the speed has reduced to 500 rpm, the induced emf falls in the ratio of speed because the flux is assumed to remain constant.

So induced emf, $E'_b = E_{b_f} \times \frac{N'}{N_f} = 220 \times \frac{500}{1,000} = 110$ V

Armature braking current,

$$I'_b = \frac{V + E'_b}{R + R_a} = \frac{220 + 110}{2.2} = 150 \text{ A}$$

Electric braking torque,

$$T'_b = T_f \times \frac{I'_b}{I_f} = 210 \times \frac{150}{100} = 315 \text{ N-m } \textbf{Ans.}$$

Example 1.34. A 220 V, 17.6 kW dc shunt motor at its rated speed of 1,200 rpm is to be braked by plugging. The armature resistance is 0.1 Ω and the rated efficiency of the motor is 80%. Calculate : (*i*) the resistance to be connected in series with the armature to limit the initial braking current to twice the rated current, (*ii*) the torque when the speed of the motor falls to 400 rpm. [*U.P. Technical Univ. Electric Drives 2008-09*]

Solution : Full-load (rated) motor current,

$$I_f = \frac{\text{Motor output in kW} \times 1,000}{\text{Rated voltage} \times \text{rated efficiency}} = \frac{17.6 \times 1,000}{220 \times 0.8} = 100 \text{ A}$$

Full-load torque,

$$T_f = \frac{\text{kW output} \times 1,000}{2\pi N / 60} = \frac{17.6 \times 1,000}{2\pi \times 1,200 / 60} = 140 \text{ N-m}$$

Full-load armature current,

$I_{af} = I_f = 100$ A, neglecting shunt field current

Induced emf in motor,

$E_{b_f} = V - I_{af}R_a = 220 - 100 \times 0.1 = 210$ V

Voltage across the armature at the instant of braking

$= V + E_{b_f} = 220 + 210 = 430$ V

(*i*) Resistance required to be connected in series with the armature to limit the initial current to twice the rated current (*i.e.* $I_b = 2I_f = 2 \times 100 = 200$ A),

$$R = \frac{V + E_b}{I_b} - R_a = \frac{430}{200} - 0.1 = 2.05 \ \Omega \ \textbf{Ans.}$$

(*ii*) When the speed has fallen to 400 rpm, the induced emf falls in the ratio of speed because in case of a dc shunt motor flux remains constant,

$$\text{So, induced emf, } E'_b = E_{b_f} \times \frac{N'}{N} = 210 \times \frac{400}{1,200} = 70 \text{ V}$$

$$\text{Armature braking current, } I'_b = \frac{V + E'_b}{R + R_a} = \frac{220 + 70}{2.05 + 0.1} = 134.9 \text{ A}$$

$$\text{Electric braking torque, } T'_b = T_f \times \frac{I'_b}{I_f} = 140 \times \frac{134.9}{100}$$

$$= 188.84 \text{ N-m } \textbf{Ans.}$$

Example 1.35. A 220 V, dc shunt motor having an efficiency of 85% drives a hoist having an efficiency of 80%. Calculate the current drawn from the supply to raise a load of 400 kg at 2.5 m/s. What resistance must be added in the armature circuit in order to lower the load at 2.5 m/s using rheostatic braking, assuming that the efficiency of the motor and load will remain the same. [*U.P. Technical Univ. Electric Drives 2007-08*]

Solution : Work done per second

$= \text{Load in kg} \times g \times v = 400 \times 9.81 \times 2.5 = 9,810$ N-m/s or watts

$$\text{Power input} = \frac{\text{Output}}{\eta_m \times \eta_h} = \frac{9,810}{0.85 \times 0.8} = 14,426.5 \text{ W}$$

Current drawn from supply mains,

$$I = \frac{\text{Power input}}{\text{Rated voltage}} = \frac{14,426.5}{220} = 65.6 \text{ A } \textbf{Ans.}$$

When the load is lowered, the machine operates as a generator and the generator output will be dissipated in the resistance connected in the armature circuit.

So, power output of generator,

$P_{out} = \text{Load in kg} \times g \times v \times \eta_m \times \eta_h$

$= 400 \times 9.81 \times 2.5 \times 0.85 \times 0.8 = 6,670.8$ watts

If R is the resistance inserted in the armature circuit,

$$\text{Then power dissipated} = \frac{V^2}{R} = 6,670.8$$

$$\text{or } R = \frac{V^2}{P_{out}} = \frac{220^2}{6,670.8} = 7.255 \ \Omega \ \textbf{Ans.}$$

Example 1.36. The speed of a dc series motor is limited to 250 rpm by application of rheostatic braking system. The total resistance of armature and field of the motor is 1.0 Ω and load exerts a torque of 245 N-m. Calculate the resistance to be connected in the motor circuit, if its magnetisation characteristic at 250 rpm is as follows:

I in Amps	10	20	30	40
E in Volts	90	190	248	295

[*Agra Univ. Traction and Utilization of Elec. Power 1982*]

Solution : Load torque, T = 245 N-m

Input power to machine during rheostatic braking

$$= T \times \frac{2\pi N}{60} = 245 \times 2\pi \times \frac{250}{60}$$

$= 6,414$ N-m/s or watts

During rheostatic braking the machine operates as a dc series generator and the electromagnetic power developed by the generator (neglecting rotational losses), *i.e.*, E I should be 6,414 watts.

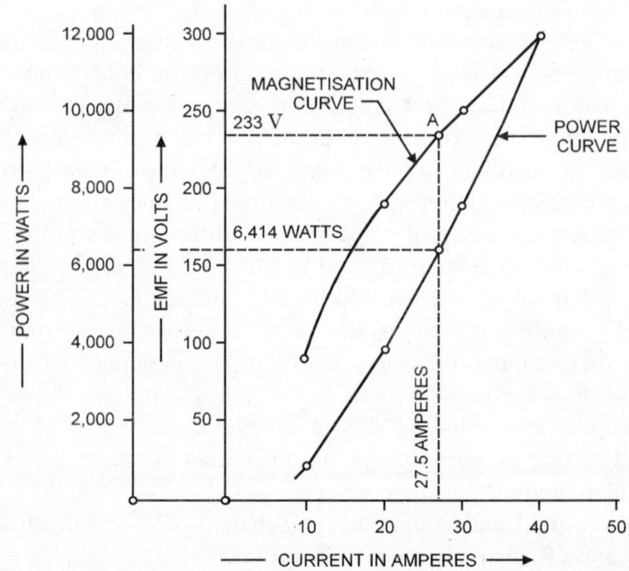

Fig. 1.102

In order to determine E and I separately, plot curves between E and I and power P and I, as shown in Fig. 1.102 from the datas given below :

I in Amps	10	20	30	40
E in Volts	90	190	248	295
Power, P (EI) in watts	900	3,800	7,440	11,800

Current corresponding to power of 6,414 watts is found to be 27.5 A and corresponding induced emf E is 233 V.

Hence total resistance of the circuit = $\frac{E}{I} = \frac{233}{27.5} = 8.47\ \Omega$.

External resistance to be inserted in the motor circuit

$= 8.47 - 1.0 = 7.47\ \Omega$ **Ans.**

Example 1.37. A 500 V, 45 kW, 600 rpm dc shunt motor has a full-load efficiency of 90%. The field resistance is 200 Ω and the armature resistance is 0.2 Ω. Find the speed under each of the following conditions at which the machine will develop an electromagnetic torque equal to rated value—(*i*) regenerative braking: no limiting resistance, (*ii*) plugging: external limiting resistance of 5.5 Ω inserted, (*iii*) dynamic braking: external limiting resistance of 2.6 Ω inserted. The field current is maintained constant and armature reaction and the brush drop may be neglected. [*A.M.I.E. Sec B. Elec. Drives and Their Control Summer 1994*]

Solution : Rated line current,

$$I_{Lf} = \frac{\text{Output in kW} \times 1{,}000}{\text{Supply voltage} \times \text{full-load efficiency}} = \frac{45 \times 1{,}000}{500 \times 0.9} = 100 \text{ A}$$

Shunt field current,

$$I_{sh} = \frac{V}{R_{sh}} = \frac{500}{200} = 2.5 \text{ A}$$

Armature current on full load, $I_{af} = I_{Lf} - I_{sh} = 100 - 2.5 = 97.5$ A

Induced emf on full load, $E_f = V - I_{af} R_a = 500 - 97.5 \times 0.2 = 480.5$ V

Rated speed, $N_f = 600$ rpm.

Since field current remains constant and the armature reaction and brush drop are negligible so electromagnetic torque developed is directly proportional to armature current and speed is directly proportional to induced emf.

As torque to be developed in each case is equal to rated torque so armature current in each case is equal to armature current at full load *i.e.*, 97.5 A.

(*i*) Induced emf to give armature current of 97.5 A for regenerative braking

$$E_1 = V + I_{af} R_a = 500 + 97.5 \times 0.2 = 519.5 \text{ V}$$

and speed, $N_1 = \frac{E_1}{E_f} \times N_f = \frac{519.5}{480.5} \times 600 = 648.7$ rpm **Ans.**

(*ii*) Induced emf to give armature current of 97.5 A for plugging with external limiting resistance of 5.5 Ω

$$E_2 = I_{af} \times (R + R_a) - V = 97.5 \times (5.5 + 0.2) - 500 = 55.75 \text{ V}$$

and speed, $N_2 = \frac{E_2}{E_f} \times N_f = \frac{55.75}{480.5} \times 600 = 69.6$ rpm **Ans.**

(*iii*) Induced emf to give armature current of 97.5 A for dynamic braking with external limiting resistance of 2.6 Ω

$$E_3 = I_{af} \times (R + R_a) = 97.5 \times (2.6 + 0.2) = 273 \text{ V}$$

and speed, $N_3 = \frac{E_3}{E_f} \times N_f = \frac{273}{480.5} \times 600 = 340.9$ rpm **Ans.**

Example 1.38. A 230 V, 870 rpm, 100 A separately excited dc motor has an armature resistance of 0.05 Ω. It is coupled to an overhauling load with a torque of 400 N-m. Determine the speed at which motor can hold the load by regenerative braking.

[*U.P. Technical Univ. Electric Drives 2009-10*]

Solution : Supply voltage, V = 230 V

Armature current, $I_a = 100$ A

Back emf, $E_b = V - I_a R_a = 230 - 100 \times 0.05 = 225$ V

Torque developed by motor, $T = \frac{E_b I_a}{2\pi N / 60} = \frac{225 \times 100 \times 60}{2\pi \times 870} = 247$ N-m

Overhauling load torque, $T_L = 400$ N-m

Assuming field current constant and armature reaction and brush drop negligible electromagnetic torque developed will be directly proportional to armature current and speed will be directly proportional to induced emf

So armature current, $I'_a = I_a \times \frac{T_L}{T} = 100 \times \frac{400}{247} = 162$ A

Induced emf to provide armature current of 162 A for regenerative braking

$$E' = V + I'_a R_a = 230 + 162 \times 0.05 = 238.1 \text{ V}$$

and speed, $N' = \frac{E'}{E_b} \times N = \frac{238.1}{230} \times 870 = 900$ rpm **Ans.**

Example 1.39. A 230 V, 500 rpm, 100 A separately excited dc motor has an armature resistance of 0.1 Ω. The motor is driving, under rated conditions, a load whose torque is constant and independent of speed.

Find the amount by which the flux should be reduced to get the motor speed of 800 rpm.

If the above motor is now coupled with a load torque of 800 N-m, find the speed at which the motor can hold the load by regenerative braking. Neglect motor rotational losses.

[*G.B. Technical Univ. Electric Drives 2010-11*]

Solution : Armature current, $I_{a_1} = 100$ A

Back emf, $E_{b_1} = V - I_{a_1} R_a = 230 - 100 \times 0.1 = 220$ V

Speed, $N_1 = 500$ rpm

Now let the flux and armature current be ϕ_2 and I_2 respectively when the speed is raised from $N_1 = 500$ rpm to $N_2 = 800$ rpm

Back emf, $E_{b_2} = V - I_{a_2} R_a = 230 - 0.1 I_{a_2}$ ...(*i*)

Since load torque is constant

$I_{a_2} \phi_2 = I_{a_1} \phi_1$ $\because T \propto \phi I_a$

or $I_{a_2} = I_{a_1} \times \frac{\phi_1}{\phi_2} = 100 \frac{\phi_1}{\phi_2}$ ...(*ii*)

and $\frac{N_2}{N_1} = \frac{E_{b_2}}{E_{b_1}} \times \frac{\phi_1}{\phi_2}$

or $\frac{800}{500} = \frac{230 - 0.1 I_{a_2}}{220} \times \frac{\phi_1}{\phi_2}$ $\because N \propto \frac{E_b}{\phi}$

or $352 = \left(230 - 0.1 \times 100 \frac{\phi_1}{\phi_2}\right) \frac{\phi_1}{\phi_2}$

$\because$ From Eq. (*ii*), $I_{a_2} = 100 \times \frac{\phi_1}{\phi_2}$

or $10\left(\frac{\phi_1}{\phi_2}\right)^2 - 230 \frac{\phi_1}{\phi_2} + 352 = 0$

$$\text{or} \quad \frac{\phi_1}{\phi_2} = \frac{230 \pm \sqrt{(230)^2 - 4 \times 10 \times 352}}{2 \times 10}$$

$$= 11.5 \pm 9.85 = 21.35 \text{ or } 1.65$$

$$\text{or } \phi_2 = \frac{\phi_1}{1.65} = 0.6066 \textbf{ Ans.}$$

$$\text{Armature current, } I_{a_2} = 100 \times \frac{\phi_1}{\phi_2} = 100 \times 1.65 = 165 \text{ A}$$

$$\text{Back emf, } E_{b_2} = V - I_{a_2} R_a = 230 - 165 \times 0.1 = 213.5 \text{ V}$$

$$\text{Speed, } N_2 = 800 \text{ rpm}$$

$$\text{Torque developed, } T_2 = \frac{E_{b_2} \times I_{a_2}}{2\pi N_2 / 60} = \frac{213.5 \times 165}{2\pi \times 800} \times 60 = 420 \text{ N-m}$$

When motor is coupled to a load torque T_3 of 800 N-m

$$\text{Armature current, } I_{a_3} = I_{a_2} \times \frac{T_3}{T_2} \qquad \text{assuming constant flux}$$

$$= 165 \times \frac{800}{420} = 314.3 \text{ A}$$

Induced emf to provide armature current of 314.3 A for regenerative braking,

$$E_{b_3} = V + I_{a_3} R_a = 230 + 314.3 \times 0.1 = 261.43 \text{ V}$$

$$\text{Speed, } N_3 = \frac{E_{b_3}}{E_{b_2}} \times N_2 = \frac{261.43}{213.5} \times 800 = 980 \text{ rpm } \textbf{Ans.}$$

1.21. ELECTRICAL BRAKING OF POLYPHASE INDUCTION MOTORS

1.21.1. Plugging (or Counter-Current Braking)

Plugging can be achieved in an induction motor merely by reversing two of the three phases which causes a reversal of the direction of the rotating magnetic field. At the instant of switching the motor to the plugging position the motor runs in the opposite direction to that of the field and the relative speed is approximately twice [(2 – *s*) times] of synchronous speed *i.e.*, the slip is very nearly equal to two, being equal to (2 – *s*). So voltage induced in the rotor will be twice of normally induced voltage at standstill and the winding must be provided with the additional insulation to withstand this much voltage.

During plugging period, the motor acts as a brake and it absorbs kinetic energy from the still revolving load causing its speed to fall. The associated power P_{mech} is dissipated as heat in the motor. At the same time, the rotor also continues to draw power from the stator which is also dissipated as heat. Due to increased rotor frequency there are additional iron losses in the rotor. Thus the heat developed in the rotor during braking period are about three times the heat developed during starting period (or under blocked condition of rotor).

The size of the induction motor where plugging is to be applied, is therefore, decided not by loading conditions alone but by braking condition also.

In case of a squirrel cage motor, energy is dissipated wholly within the machine; whereas in case of wound rotor motor this energy is dissipated also in the external resistance added in the rotor circuit for this purpose.

The conditions for braking an induction motor can be studied by considering the torque-slip curves of the motor when extended beyond the point of 100 % slip, as shown in Fig. 1.103(*a*). The ordinate at the point B represents the torque at the instant of plugging and we can see that the torque increases gradually as the motor approaches the standstill, after which, of course, if the supply line is not disconnected it will start up again in the reverse direction in accordance with the ordinary torque-speed curves from A to O. Thus we see that for bringing the motor to rest, it is essential to disconnect the supply to the motor at zero speed.

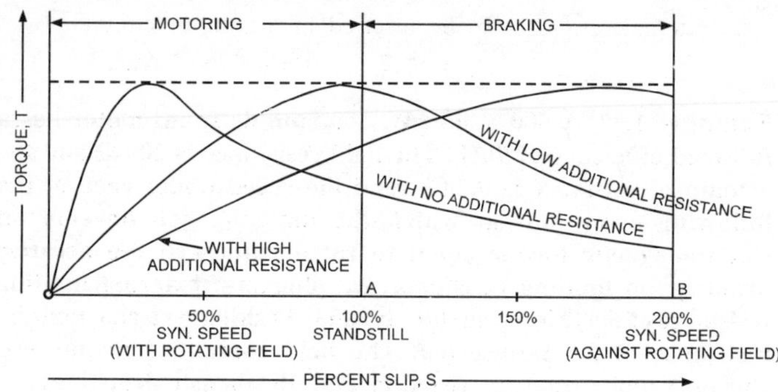

Fig. 1.103(*a*). ***Torque-Slip Curves of Polyphase Induction Motor***

From the torque-slip curves shown in Fig. 1.103 (*a*), it is observed that magnitude of the braking torque is very much low as compared to the maximum torque exerted by the motor. This is specially so for low rotor resistances. As more and more resistance is inserted in the rotor circuit, maximum torque occurs at increasing values of the slip. This is the reason that squirrel cage induction motors, which are primarily designed for maximum efficiency and, therefore, for low resistance are not much suitable for such type of braking. In case of wound rotor motors, addition of resistance in the rotor circuit gives rise to more and more braking torque and, therefore, wound rotor motors are more suitable for plugging. If the change in rotor resistance is so adjusted that maximum braking torque is exerted throughout, optimum braking will be achieved in minimum time.

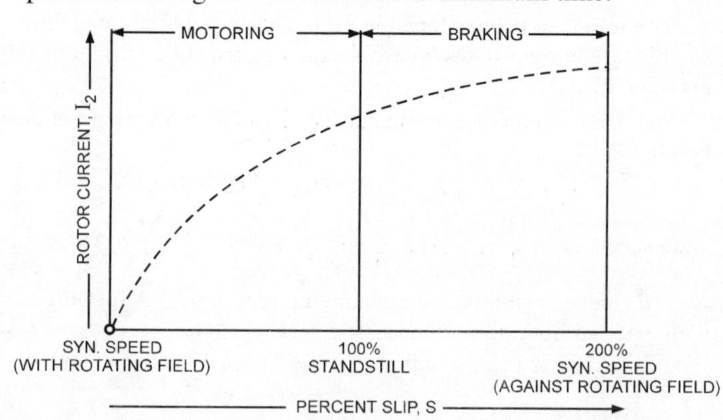

Fig. 1.103 (*b*). ***Rotor Current-Slip Curve of Polyphase Induction Motor***

The expression for the braking torque neglecting stator impedance and magnetizing reactance, can be deduced, and is given as

$$T_b = \frac{3 \times 60}{2\pi N_s} \cdot \frac{R_2 s E_2^2}{R_2^2 + s^2 X_2^2} \qquad \text{...(1.81)}$$

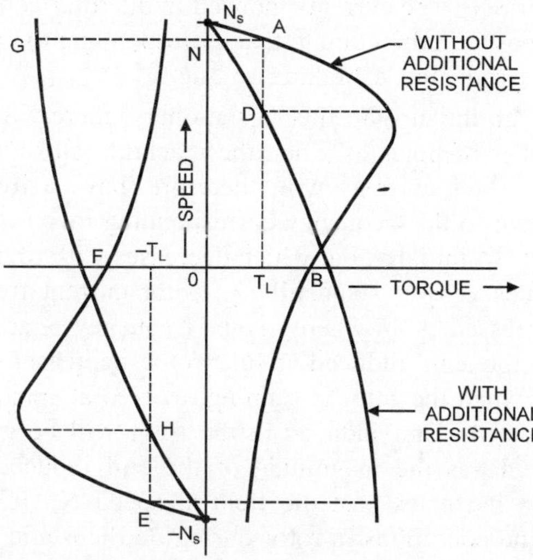

Fig. 1.103(*c*). ***Quadrantal Diagram of Induction Motor During Plugging***

The rotor current can be determined during the braking period from the following relation and is plotted, as shown in Fig. 1.103 (*b*).

$$I_2 = \frac{sE_2}{\sqrt{R_2^2 + s^2X_2^2}} \qquad ...(1.82)$$

where E_2 is the emf induced in each phase of the rotor at standstill, R_2 is the rotor resistance per phase and X_2 is the rotor standstill reactance per phase and s is the slip.

During plugging period, since the induced emf in the rotor is very high, as already stated above, so the rotor current and thereby stator current are very high. However, braking current can be reduced by inserting external resistance in the rotor circuit. From this point also, the wound rotor motors are beneficial as compared to squirrel cage rotor motors.

Cage motors of about 20 kW output are plugged directly, using the star connection if a star-delta switch is provided. Larger motors need stator resistors. Wound rotor motors employ rotor resistors for current limitation as well as developing higher braking torques. Problems of thermal rating may arise for frequent braking duty, as already explained above, and contactors may need frequent servicing.

Simultaneous reversal of stator connections and inclusion of rotor resistance in the wound rotor motors is quite practicable, but the same effect can be achieved by including saturistors in the rotor windings. Alternatively a deep bar rotor cage may give rise in effective rotor resistance following the plugging operation as the slip frequency is suddenly increased from sf to $(2 - s)f$ where f is the supply frequency.

In practice, for reversing drives where braking and starting up of induction motor in reverse direction comprises stages of the same continuous process, plugging is advantageously employed.

The plugging operation can also be explained with the help of quadrantal diagram shown in Fig. 1.103 (*c*). Speed-torque characteristics shown in Fig. 1.103 (*c*) are actually extension of the motor characteristics in the second and fourth quadrant. The explanation of Fig. 1.103 (*c*) is exactly identical to one for Fig. 1.94 (*a*) or 1.94 (*b*).

1.21.2. Dynamic (or Rheostatic) Braking

The rheostatic braking with a polyphase induction motor can be obtained by disconnecting the stator winding from the ac supply and exciting it from a dc source to produce a stationary dc field. In rheostatic braking, the stator winding is employed

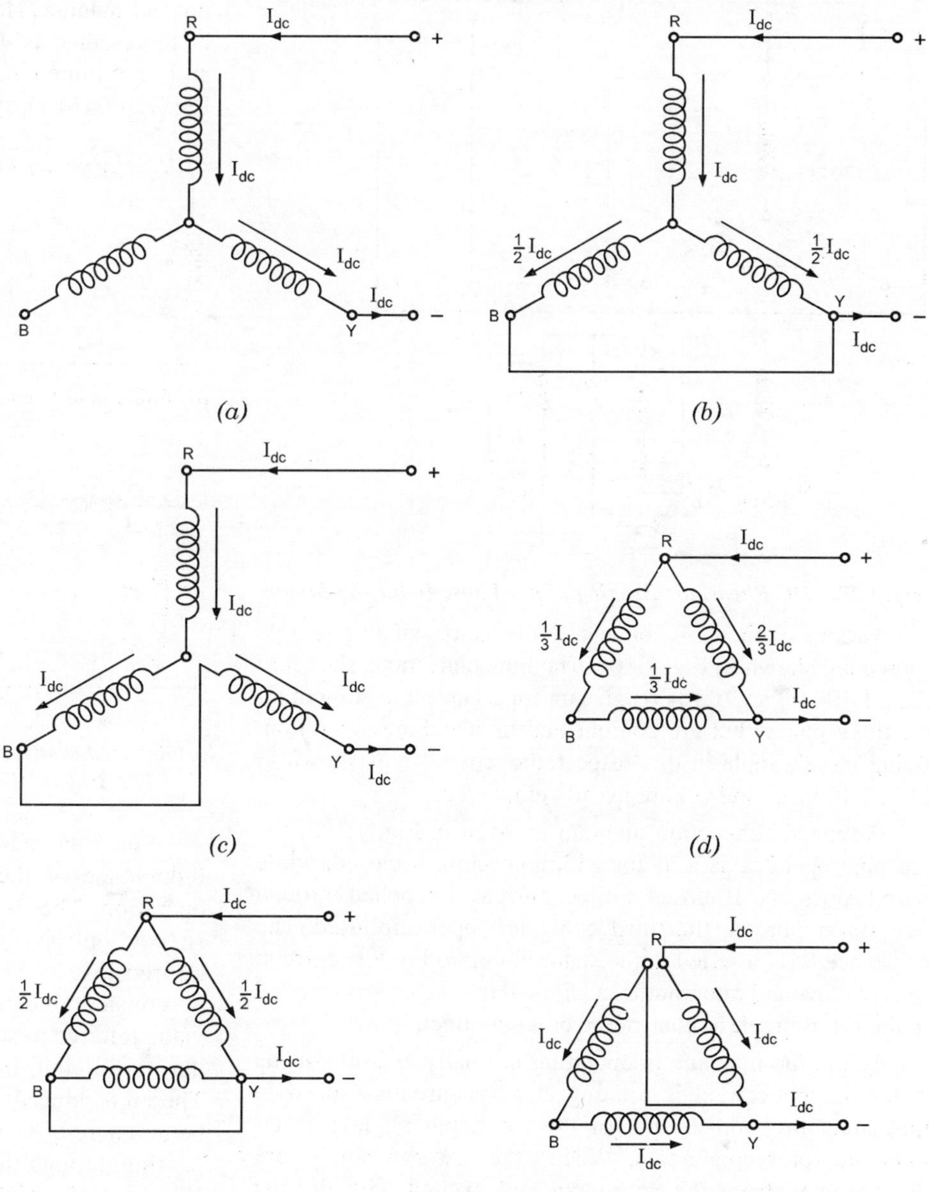

Fig. 1.104. ***Connection of Stator Windings of 3-Phase Induction Motors For Dynamic Braking***

as a dc field winding and the rotor winding as an armature winding. With a wound rotor machine, external resistors can be inserted into the rotor circuit to provide a load. With squirrel cage machines, however, the rotor winding itself has to form the load.

The source of excitation may be provided either by an independent dc source or from the ac mains through a transformer rectifier set, as shown in Fig. 1.105.

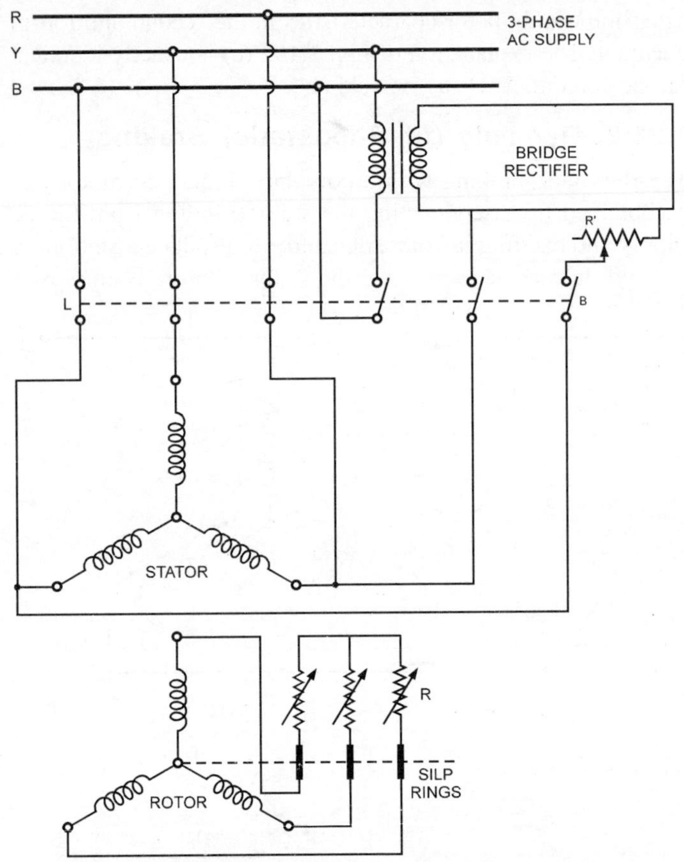

Fig. 1.105. ***DC Rheostatic Braking of a 3-Phase Induction Motor***

Various methods of connecting the stator winding to a dc source are shown in Fig. 1.104. Though connections shown in Figs. 1.104 (*c*) and 1.104 (*f*) give uniform current loading of all the three phases but are complicated in switching operation. Being more simple in this respect, the connections shown in Figs. 1.104. (*a*) and (*b*) are usually employed.

A typical connection diagram is given in Fig. 1.105; the machine operates as a motor with contactors L closed, while with L open and B closed a direct current is supplied through two stator phases, the third being left open circuited. The resistance R′ is inserted in the stator circuit to limit its current. In case of wound rotor motors additional resistance R is inserted in the rotor circuit to control the braking effect.

While the machine is operating normally as a motor, its stator magnetic field is rotating at a synchronous speed in the same direction as that of the rotor, but slightly faster than the rotor conductors. When the stator windings are disconnected from the ac supply and excited with dc, the magnetic field produced will be stationary in space, thus making the rotor conductors move past the field with a speed $(1 - s)N_s$ or SN_s. The currents induced in the rotor conductors will be opposite in direction to that corresponding to motoring operation, producing a braking torque.

Although the air-gap flux is stationary, there will be the same number of poles as when the machine is excited with ac and the rotor currents will, therefore, have a frequency which is near to the frequency corresponding to synchronous speed (*i.e.*, N_s) initially, but which decreases with rotor speed and becomes zero at standstill *i.e.*, rotor current frequency can be expressed as Sf, where f is the frequency of ac supply. Likewise, the emf induced in the rotor decreases from a maximum when the rotor is running, to zero at standstill. In other words, the emf induced in the rotor will be given by SE_2 where E_2 is the magnitude of the emf induced in the rotor when it rotates past the field at speed N_s. It is, thus seen that the conditions in rotor during dc rheostatic braking with the speed falling from synchronous to standstill are very much the same as when the motor accelerates in the normal manner. Hence the equivalent circuit of the rotor can be represented as shown in Fig. 1.106 (*a*). Dividing the voltage and the impedance by S, we have the circuit shown in Fig. 1.106 (*b*) carrying the same current I_2.

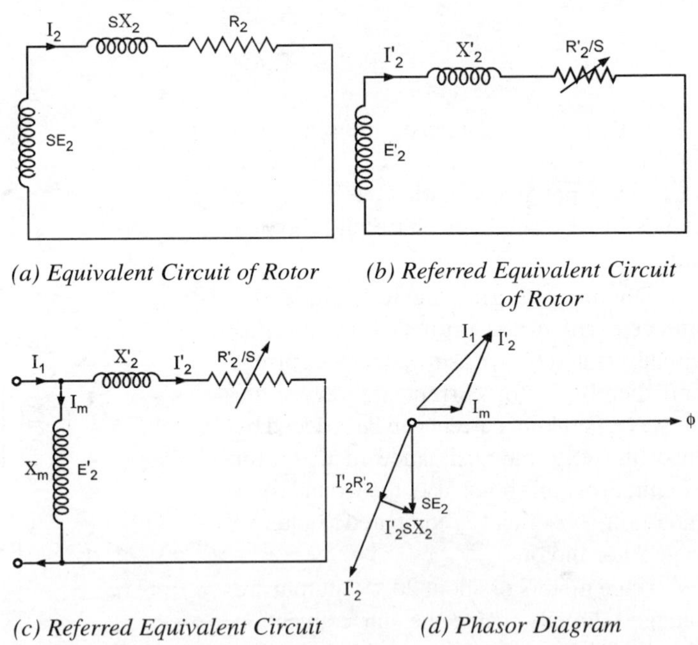

(*a*) *Equivalent Circuit of Rotor* (*b*) *Referred Equivalent Circuit of Rotor*

(*c*) *Referred Equivalent Circuit For Braking Operation* (*d*) *Phasor Diagram*

Fig. 1.106

Since the stator winding carries only direct current, the inductance of the stator has no effect under steady-state operation. The dc voltage applied across the stator winding is fixed only by the stator winding resistance. There will be no iron loss in the stator core. But the rotor core loss will be of considerable magnitude and hence the rotor resistance value referred to stator, which determines the torque developed during braking, must be corrected to take this into account. The emf induced in the rotor, at synchronous speed, E'_2 will be given by $I_m X_m$ where I_m represents the magnetizing current passing through the magnetizing reactance. The circuit shown in Fig. 1.106 (*b*) can be modified into a circuit shown in Fig. 1.106 (*c*). The phasor diagram corresponding to this modified circuit is shown in Fig. 1.106 (*d*).

It may be noted that the rotor current is alternating in nature although it is produced by a constant flux, caused by a magnetizing mmf $I_m N_1$, which remains stationary in space. However, when viewed from the rotor, both the magnetizing mmf and the resultant flux in the stator will appear as alternating. Also, since the rotor ampere-turns $I_2 N_2$ $(= I'_2 N_1)$ must be balanced by the stator ampere-turns so as to maintain the magnetizing ampere-turns $I_m N_1$, the stator ampere-turns $I_1 N_1$ as seen from the rotor, represents the phasor sum of $I_m N_1$ and $I'_2 N_1$. The effective alternating current I_1 depends on the magnitude of direct current carried by the stator and the nature of the stator winding connections, as explained above.

From above discussion, it can be concluded that the operation of the motor during dc rheostatic braking can be analysed by using the equivalent circuit (the type of which we are already familiar) shown in Fig. 1.106 (*c*).

The braking torque can be determined from the relation

$$T_b = \frac{60}{2\pi N_s} \cdot 3(I'_2)^2 \frac{R'_2}{S} \text{ N-m}$$

$$= \frac{60}{2\pi N_s} \cdot 3 I_1^2 \frac{X_m^2 \left(\frac{R'_2}{S}\right)}{\left(\frac{R'_2}{S}\right)^2 + (X_m + X'_2)^2} \text{ N-m} \quad ...(1.83)$$

The above equation enables us to determine the braking torque at any speed of a given induction motor, when excited by direct current, I_1 corresponding to an equivalent ac, which would have produced the same mmf as that by the actual direct current carried by the stator windings.

Differentiating Eq. (1.83) w.r.t. S and equating it to zero for a maximum we have slip corresponding to maximum braking torque

$$S_{\max T} = \frac{R'_2}{X_m + X'_2} \quad ...(1.84)$$

and maximum torque,

$$T_{\max} = \frac{60}{2\pi N_s} \cdot 3 I_1^2 \frac{X_m^2}{2(X_m + X'_2)} \quad ...(1.85)$$

Typical performance characteristics of an induction motor are illustrated in Fig. 1.107. The magnitude of the braking torque developed by the motor depends upon the excitation (strength of the field developed by the stator winding), the rotor circuit resistance, and the speed of the motor. Effect of variation of excitation I_1 and rotor resistance R_2 are also illustrated in Fig. 1.107 (*a*). For design purposes, the torque-rotor resistance curves illustrated in Fig. 1.107 (*b*) are sometimes used. With the increase in rotor resistance R_2, the speed at which maximum braking torque occurs also increases. But, the maximum torque would not increase in proportion to the square of current (I_1^2), as pointed out by Eq. (1.85), because magnetising reactance X_m decreases due to saturation caused by increase in current I_1.

Stable braking operation is on the steeper portion of torque-speed curve [*i.e.,* on the left hand side of the maximum torque position in Fig. 1.107 (*a*)]. If inadvertently the speed rises above that at which the torque is a maximum, further rise of speed is caused by a decrease in braking torque, which may lead to a further rise in speed in the case of an overhauling load. At the start of applying braking, therefore, we require high resistance in the rotor circuit, which has to be gradually reduced with decrease in speed. Braking torque can be controlled by any or both of the following methods: (*i*) By controlling the dc excitation (*ii*) By varying the rotor resistance.

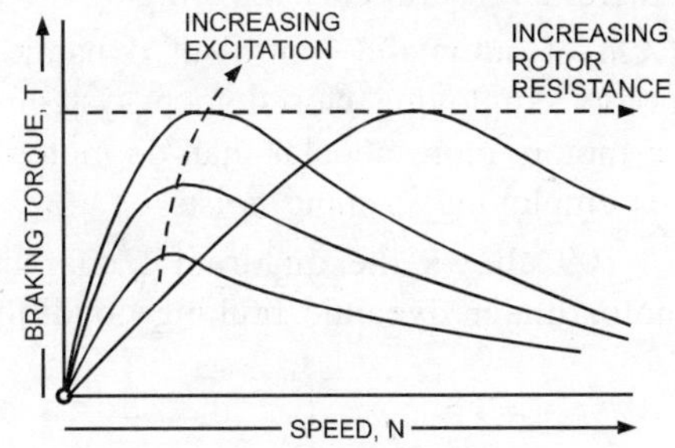

(*a*)

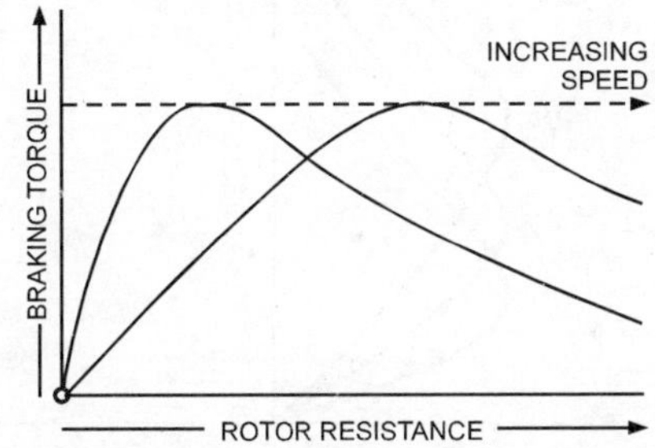

(*b*)

Fig. 1.107. ***DC Rheostatic Braking Characteristics***

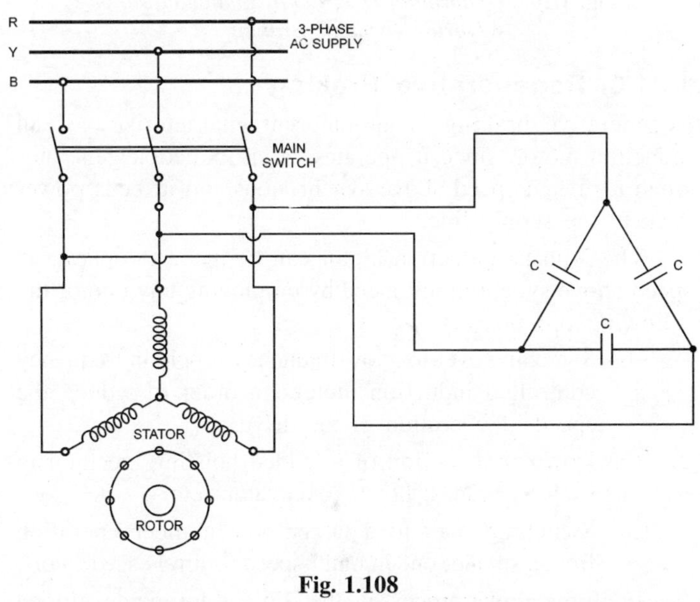

Fig. 1.108

AC dynamic braking may be obtained with self excitation by connecting the stator windings to a bank of static capacitors as shown in Fig. 1.108. In this case the induction machine operates as an induction generator receiving its ac excitation from the static capacitors. The generated electrical energy is dissipated as heat in the rotor circuit. AC dynamic braking is not popular owing to relatively high cost of capacitor banks.

DC dynamic braking takes little power from the supply and provides smooth braking torque, useful for mine winders and high inertia loads. Its advantages over plugging are the absence of the reverse rotation air-gap field [and, therefore, no tendency for the machine to run backwards] and a lower rotor I^2R loss. By increasing rotor resistance, the braking torque can be made more effective at higher speeds. This method is, therefore, very much suitable for retarding over-hauling loads. In case of a squirrel cage motor, since rotor resistance is small, it will apply braking torque at very low speeds and, therefore, useful only for bringing the load to rest.

The advent of automatic control of dynamic braking of induction motors employing closed loop systems, has made the induction motors more popular than dc motors, especially for the drives employing in mine hoists.

Figure 1.109 shows the quadrantal characteristics of induction motor under dynamic braking condition.

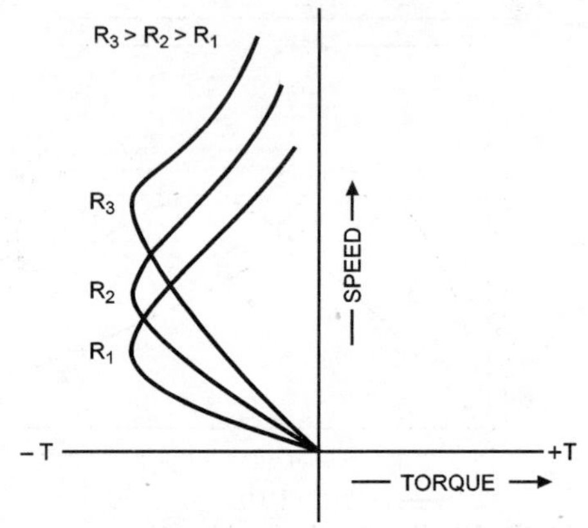

Fig. 1.109. ***Quadrantal Diagram of Induction Motor During Dynamic Braking***

1.21.3. Regenerative Braking

Regenerative braking is an inherent characteristic of an induction motor, since it operates as an induction generator when it runs at speed above synchronous and it feeds power back to the supply line.

The 3-phase induction motor can be made to operate at speed above synchronous speed by employing any one of the following processes.

(*i*) Switching over to a low frequency supply in frequency controlled induction motors in order to reduce the speed of operation of the drive.

(*ii*) Downward motion of a loaded hoisting mechanism such as cranes, hoists, excavators etc.

(*iii*) Switching over to a larger pole number operation from a smaller one in multi-speed squirrel cage motors.

In all the above processes, the slip and torque developed become negative, as shown in Fig. 1.110, and thus the machine acts as a generator, receiving mechanical energy and giving it back to the supply system in the form of electrical energy.

If the load drives the motor above synchronous speed, no switching operation is required. Once the machine is driven above synchronous speed, the braking operation automatically starts. The operating point will depend upon the magnitude of load torque and the nature of torque-speed characteristic of the machine during generating operation. By varying the resistance in the rotor circuit, it is possible to operate at any speed above synchronous speed during braking. In case the driving torque of the load exceeds the maximum braking torque, of which the machine is capable, the system will become unstable and the speed will rise further, probably to a disastrous value, since, the faster the machine runs, the lesser will be the braking torque developed.

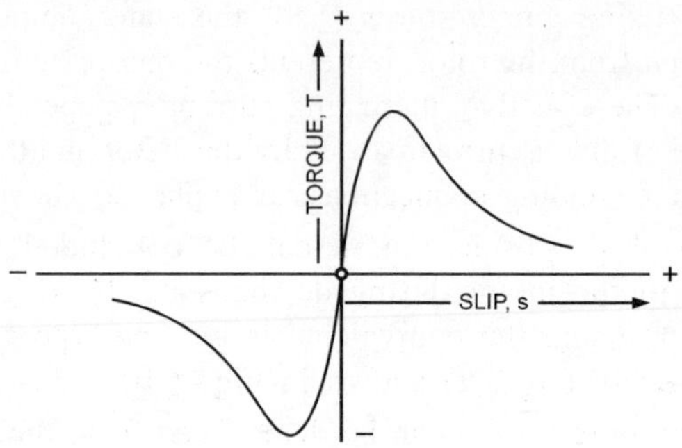

Fig. 1.110

In the case of a squirrel cage induction motor, stable speed is obtained at a speed considerably in excess of the synchronous speed and the regenerative braking cannot be applied unless the motor is specially designed to withstand the excessive speed.

Regenerative braking has the disadvantage of the possibility of braking only at super-synchronous speeds and, therefore, is seldome used for braking. This method can be used only in hoisting type of mechanism or with a multi-speed squirrel cage motor. It is advantageous on mountain railways too. It returns about 20% of the total energy on certain railway runs and saves a great deal of brake shoe wear.

Example 1.40. A 30 kW, 400 V, 3-phase, 4-pole, 50 Hz induction motor has full-load slip of 5%. If the ratio of standstill reactance to resistance per rotor phase is 4, estimate the plugging torque at full speed.

[*Pb. Technical Univ. Electric Drives and Utilization December 2008*]

Solution : Synchronous speed, $N_s = \frac{120f}{P} = \frac{120 \times 50}{4} = 1{,}500$ rpm

Full-load slip, $s_f = 5\%$ or 0.05

Full-load speed, $N_f = N_s(1 - s_f) = 1{,}500\,(1 - 0.05) = 1{,}425$ rpm

$$\text{Full-load torque, } T_f = \frac{\text{Power developed in kW} \times 1{,}000}{2\pi N/60} = \frac{30 \times 1{,}000}{2\pi \times 1{,}425/60} = 201.04 \text{ N-m}$$

Since torque developed,

$$T \propto \frac{sR_2E_2^2}{R_2^2 + (sX_2)^2}$$

$$\text{So} \quad \frac{T_2}{T_1} = \frac{s_2R_2E_2^2/(R_2^2 + s_2^2X_2^2)}{s_1R_2E_2^2/(R_2^2 + s_1^2X_2^2)}$$

$$= \frac{s_2(R_2^2 + s_1^2X_2^2)}{s_1(R_2^2 + s_2^2X_2^2)}$$

$$= \frac{s_2}{s_1}\,\frac{1 + s_1^2\left(\frac{X_2}{R_2}\right)^2}{1 + s_2^2\left(\frac{X_2}{R_2}\right)^2} = \frac{s_2}{s_1}\,\frac{1 + 16s_1^2}{1 + 16s_2^2} \quad \because \frac{X_2}{R_2} = 4$$

Slip, $s_1 = s_f = 0.05$

Slip, $s_2 =$ Slip at plugging $= 2 - s_f = 2 - 0.05 = 1.95$

So plugging torque, $T_P = \dfrac{1.95}{0.05} \times \dfrac{1+16(0.05)^2}{1+16(1.95)^2} \times T_f$

$$= 39 \times \frac{1.04}{61.84} \times 201.04 = 131.86 \text{ N-m } \textbf{Ans.}$$

Example 1.41. A 3-phase, 400 V, 50 Hz, 8-pole, star-connected induction motor has the following parameters: $R_1 = 0.1\ \Omega$, $R_2 = 0.5\ \Omega$, $X_1 = 0.6\ \Omega$, $X_2 = 2.4\ \Omega$, $s_f = 0.05$ and the ratio of effective stator to rotor turns $\frac{1}{2}$. The motor is to be braked at rated speed and an external resistance of 2 Ω per phase (referred to stator) has been inserted into the rotor circuit. Determine the initial braking torque in (*i*) plugging and (*ii*) dc dynamic braking.

Solution : Rotor resistance referred to stator,

$$R_2' = 0.5 \times \left(\frac{1}{2}\right)^2 = 0.125\ \Omega$$

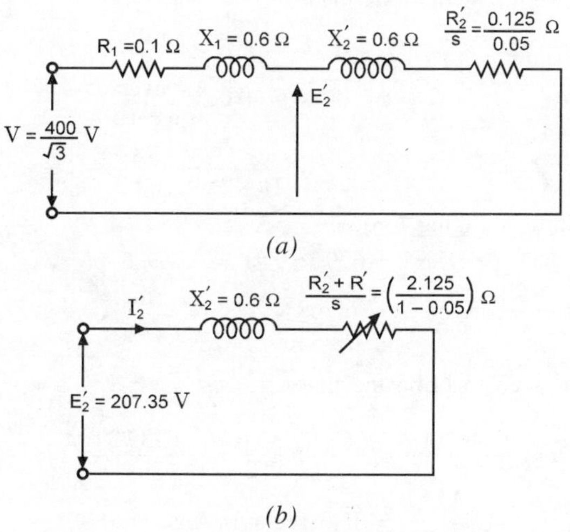

Fig. 1.111

Rotor leakage reactance referred to stator,

$$X_2' = 2.4 \times \left(\frac{1}{2}\right)^2 = 0.6\ \Omega$$

(*i*) **During plugging**

Slip, $s = 2 - sf = 2 - 0.05 = 1.95$

Synchronous speed, $N_s = \dfrac{120 f}{P} = \dfrac{120 \times 50}{8} = 750$ rpm

Initial rotor current during plugging,

$$I'_2 = \frac{V}{\sqrt{\left[R_1 + \dfrac{(R_2' + R_L')}{s}\right]^2 + (X_1 + X_2')^2}}$$

$$= \frac{400/\sqrt{3}}{\sqrt{\left(0.1 + \dfrac{0.125+2}{1.95}\right)^2 + (0.6+0.6)^2}}$$

$= 136.7$ A

Initial braking torque, $T_b = \dfrac{3 \times 60}{2\pi N_s} \times (I_2')^2 \times \dfrac{R_2' + R'}{s}$

$$= \frac{3 \times 60}{2\pi \times 750} \times (136.7)^2 \times \frac{2.125}{1.95}$$

$= 777$ Nm **Ans.**

(*ii*) ***During dc dynamic braking.*** The equivalent circuit of the motor corresponding to slip of 0.05 and the equivalent-circuit of the machine during dynamic braking at rated speed are given in Figs. 1.111 (*a*) and 1.111 (*b*) respectively,

$$E'_2 = \frac{\sqrt{\left(\dfrac{0.125}{0.05}\right)^2 + 0.6^2}}{\sqrt{\left(\dfrac{0.125}{0.05} + 0.1\right)^2 + (1.2)^2}} \times \frac{400}{\sqrt{3}} \text{ V} = 207.35 \text{ V}$$

During dc dynamic braking, $S = 1 - 0.05 = 0.95$

$$I'_2 = \frac{207.35}{\sqrt{0.6^2 + \left(\dfrac{2.125}{0.95}\right)^2}} = 89.53 \text{ A}$$

Initial braking torque,

$$T_b = \frac{3 \times 60}{2\pi N_s} \times (I_2')^2 \times \frac{R_2' + R'}{S} = \frac{3 \times 60}{2\pi \times 750} \times (89.53)^2 \times \frac{2.125}{0.95}$$

$= 684.9$ Nm **Ans.**

1.22. ELECTRIC BRAKING OF SYNCHRONOUS MOTORS

In principle, a synchronous motor may be stopped by either plugging or rheostatic braking. But the former method has certain limitations, as explained below.

If the direction of dc excitation is reversed, dc and ac fields will rotate in opposite directions and no braking torque will be developed in case of plain synchronous motors (*i.e.*, synchronous motors provided with no damper windings). The braking effect will be produced, however, due to eddy currents induced in damper windings, if provided for starting purposes. Plugging as applied to synchronous motors provided with damper windings, does not prove satisfactory. Because of usually large size of such motors it results in heavy surges of stator current causing line disturbances beyond permissible limits and also it calls for additional control equipment to switch off the motor when the speed falls to zero. Also, since the plugging torque arises from the damper windings and unless it is specially designed for the purpose, the braking torque developed during plugging are of smaller magnitudes. Hence, *rheostatic or dynamic braking* is almost exclusively used for braking synchronous motors.

As already mentioned above, synchronous motors are usually braked with rheostatic braking. In rheostatic braking, the stator of synchronous motor is disconnected from the ac supply and is short circuited through a set of balanced 3-phase resistors, keeping the field excitation unchanged. In case field excitation is increased, greater braking torques will be developed. During dynamic braking, the machine acts as an alternator and dissipates the stored energy in the resistors. The terminal voltage and frequency drop with the fall in speed and both become zero at standstill. The maximum torque occurs at approximately that speed for which the machine reactance becomes equal to the braking resistance. The appearance of high torque immediately prior to stopping has a very positive effect on the braking process.

1.23. DYNAMICS OF BRAKING

Let electrical and mechanical torque in N-m be T_E and T_F respectively.

Electrical torque,

$T_E \propto$ speed of rotation if excitation is constant.

Braking torque,

$$T_B = T_E + T_F = K\omega + T_F$$

Angular retardation,

$$\beta = \frac{T_B}{J} = \frac{K\omega + T_F}{J} \text{ rad/s}^2;\ \beta \text{ is negative being retardation.}$$

Time required to bring the motor from an initial speed ω_1 to final speed ω in the same direction is given as

$$t = \int_{\omega_1}^{\omega} -\frac{d\omega}{\beta} = -J\int_{\omega_1}^{\omega} \frac{d\omega}{K\omega + T_F} = \frac{-J}{K}\left[\log_e (K\omega + T_F)\right]_{\omega_1}^{\omega}$$

$$\text{or } t = \frac{-J}{K}\left[\log_e (K\omega + T_F) - \log_e (K\omega_1 + T_F)\right]$$

$$= \frac{J}{K}\log_e \frac{K\omega_1 + T_F}{K\omega + T_F} \quad ...(1.86)$$

Time required to bring the motor to standstill is obtained by substituting $\omega = 0$ in above equation,

$$i.e.,\ t = \frac{J}{K}\log_e \frac{K\omega_1 + T_F}{T_F} \quad ...(1.87)$$

From Eq. (1.86)

$$K\omega + T_F = (K\omega_1 + T_F)\, e^{\frac{-Kt}{J}}$$

$$\text{or } \omega = \frac{1}{K}\left[(K\omega_1 + T_F)e^{\frac{-Kt}{J}} - T_F\right] \quad ...(1.88)$$

Number of revolutions made,

$$n = \frac{1}{2\pi}\int \omega\, dt = \frac{1}{2\pi}\int_0^t \left[(K\omega_1 + T_F)e^{\frac{-Kt}{J}} - T_F\right] dt$$

$$= \frac{1}{2\pi K}\left[\frac{(K\omega_1 + T_F)e^{-Kt/J}}{-\frac{K}{J}} - T_F\, t\right]_0^t$$

$$\text{or } n = \frac{1}{2\pi K}\left\{\left[\frac{K\omega_1 + T_F}{-\frac{K}{J}}e^{-Kt/J} - T_F\, t\right] - \left[\frac{K\omega_1 + T_F}{-\frac{K}{J}} - 0\right]\right\}$$

$$= \frac{1}{2\pi K}\left[\frac{J}{K}(K\omega_1 + T_F)(1 - e^{-Kt/J}) - T_F\, t\right] \quad ...(1.89)$$

Example 1.42. A 400 V, 8-pole, 3-phase synchronous motor and its load have a total moment of inertia of 630 kg-m^2. Determine time taken and the number of revolutions made for it to come to standstill if (*i*) it is braked by plugging, which produces a constant braking torque of 165 kg-m. (*ii*) rheostatic braking is employed which gives an initial electric braking torque of 690 kg-m. Assume a frictional torque 1.4 kg-m and supply frequency 50 Hz.

Solution : Total moment of inertia of rotating parts,

$$J = 630 \text{ kg-m}^2$$

Frictional torque, $T_F = 1.4 \times 9.81 = 13.734$ N-m

(*i*) Electrical torque produced by plugging,

$$T_E = 165 \times 9.81 = 1{,}618.65 \text{ N-m}$$

Total braking torque, $T_B = T_E + T_F = 1{,}618.65 + 13.734 = 1{,}632.4$

$$\text{Angular retardation, } \beta = \frac{T_B}{J} = \frac{1{,}632.4}{630} = 2.591 \text{ rad/s}^2$$

Synchronous speed of motor,

$$N_s = \frac{120 f}{P} = \frac{120 \times 50}{8} = 750 \text{ rpm}$$

$$\omega_1 = \frac{2\pi \times 750}{60} = 78.54 \text{ radians/second}$$

Time taken to stop the motor,

$$t = \frac{\omega_1}{\beta} = \frac{78.54}{2.591} = 30.3 \text{ seconds. } \textbf{Ans.}$$

$$\text{Now } t = -\int \frac{d\omega}{\beta}$$

$$\text{or } dt = -\frac{d\omega}{\beta}$$

and number of revolutions made,

$$n = \frac{1}{2\pi}\int \omega\, dt = \frac{-1}{2\pi}\int_{\omega_1}^{0} \frac{d\omega}{\beta} = \frac{-1}{2\pi\beta}\left[\frac{\omega^2}{2}\right]_{\omega_1}^{0}$$

$$= \frac{1}{2\pi \times 2.591}\left[\frac{78.54^2}{2}\right] = 189.5 \textbf{ Ans.}$$

(*ii*) Initial braking torque,

$$K\omega_1 = 690 \times 9.81 = 6{,}769 \text{ N-m}$$

$$\text{and } K = \frac{6{,}769}{78.54} = 86.2$$

Time taken to bring the motor to rest,

$$t = \frac{J}{K}\log_e \frac{K\omega_1 + T_F}{T_F} = \frac{630}{86.2}\log_e \frac{6{,}769 + 13.734}{13.734}$$

$$= 45.3 \text{ seconds } \textbf{Ans.}$$

Number of revolutions made,

$$n = \frac{1}{2\pi K}\left[\frac{J}{K}(K\omega_1 + T_F)(1 - e^{-Kt/J}) - T_F\, t\right]$$

$$= \frac{1}{2\pi \times 86.2}\left[\frac{630}{86.2}(6{,}769 + 13.734)(1 - e^{-86.2 \times 45.3/630}) - 13.734 \times 45.3\right]$$

$$= \frac{1}{541.8}[49{,}555\ (1 - 0.00203) - 622] = 90 \text{ revolutions } \textbf{Ans.}$$

Example 1.43. A 37.5 kW, 400 V, 750 rpm synchronous motor having a total moment of inertia of 20 kg-m^2 employs rheostatic braking for obtaining rapid stopping in cases of emergency. When the motor is operating on full load, estimate the time taken and the number of revolutions made before the motor is stopped after switching over to a star-connected braking resistor of 2.5 Ω per phase. Assume full-load efficiency of 90% and full-load power factor to be 0.85.

Solution : Full-load torque,

$$T_f = \frac{\text{Output in kW} \times 1{,}000}{2\pi N/60} = \frac{37.5 \times 1{,}000 \times 60}{2\pi \times 750} = 477.46 \text{ N m}$$

Full-load current,

$$I_L = \frac{\text{Output in kW} \times 1{,}000}{\sqrt{3}\, V_L \times \cos\phi \times \eta} = \frac{37.5 \times 1{,}000}{\sqrt{3} \times 400 \times 0.85 \times 0.9} = 70.75 \text{ A}$$

Approximate line current at the start of braking period (neglecting impedance drop)

$$I_b = \frac{400}{\sqrt{3} \times 2.5} = 92.376 \text{ A}$$

Assuming the excitation to remain constant

Initial electric braking torque

$$= 477.46 \times \frac{92.376}{70.75} = 623.4 \text{ Nm}$$

Total initial braking torque = 477.46 + 623.4 = 1,100.86 Nm

If T_F is the mechanical braking torque due to load (assumed to be constant) and T_E is the electric braking torque, we have

Total braking torque, $T_B = T_F + T_E = T_F + K\omega$

$$\text{Now } \omega = \frac{2\pi N}{60} = \frac{2\pi \times 750}{60} = 78.54 \text{ rad/s}$$

$$T_E = 623.4 \text{ N/m}$$

$$\text{and, therefore, } K = \frac{623.4}{78.54} = 7.94$$

$$\text{Thus, retardation, } \beta = \frac{T_B}{J} = \frac{T_F + K\omega}{J}$$

Time taken to bring the motor to standstill,

$$t = \int_0^{\omega_1} \frac{d\omega}{\beta} = J\int_0^{\omega_1} \frac{d\omega}{K\omega + T_F} = \left[\frac{J}{K}\log_e (T_F + K\omega)\right]_0^{\omega_1}$$

$$\text{or } t = \frac{J}{K}\log_e \frac{T_F + K\omega_1}{T_F} = \frac{20}{7.94}\log_e \frac{1{,}100.86}{477.46} = 2.1 \text{ seconds } \textbf{Ans.}$$

Number of revolutions made,

$$n = \frac{1}{2\pi K}\left[\frac{J}{K}(K\omega_1 + T_F)(1 - e^{-Kt/J}) - T_F t\right]$$

$$= \frac{1}{2\pi \times 7.94}\left[\frac{20}{7.94} \times 1{,}100.86\left(1 - e^{\frac{-7.94 \times 2.1}{20}}\right) - 477.46 \times 2.1\right]$$

$$= \frac{1}{49.8885} [2{,}772.95 (1 - 0.43444) - 1{,}002.666]$$

$= 11.34$ revolutions.

Example 1.44. Determine the time taken by a dc shunt motor to fall in speed from 1,500 rpm to 600 rpm with the excitation maintained constant. The armature is connected across a resistance of 15 Ω. The emf induced in the armature at 1,500 rpm is 240 V. The moment of inertia of the armature is 20 kg-m^2. Neglect the effect of the friction in motor. Also determine the time taken for the same fall in speed if there existed a frictional torque of 1.5 kg-m.

Solution : The emf induced in the armature at 1,500 rpm,

$E = 240$ V

Power supplied to the resistance connected across the armature.

$$P = \frac{E^2}{R} = \frac{240^2}{15} = 3{,}840 \text{ W}$$

Electric braking torque,

$$T_B = \frac{P}{2\pi N/60} = \frac{3{,}840 \times 60}{2\pi \times 1{,}500} = 24.45 \text{ N-m}$$

Since rheostatic braking is being used keeping field excitation constant and connecting the armature to a resistance of 15 Ω. At any state the braking effect will be due to the current supplied which in turn depends upon the emf induced in the armature at that speed. The electric braking torque is thus proportional to rotational speed or angular velocity.

$T_E = K\omega = 24.45$ N-m

$$\omega_1 = \frac{2\pi N}{60} = \frac{2\pi \times 1{,}500}{60} = 157 \text{ radians/second}$$

$$K = \frac{T_E}{\omega_1} = \frac{24.45}{157} = 0.1557$$

Time taken to bring the motor speed from 1,500 rpm to 600 rpm (*i.e.*, 157 rad/s to 62.832 rad/s),

$$t = J\int_{\omega_2}^{\omega_1} \frac{d\omega}{K\omega + T_F} = \left[\frac{J}{K}\log_e (T_F + K\omega)\right]_{\omega_2}^{\omega_1}$$

$$= \frac{J}{K}\log_e \frac{T_F + K\omega_1}{T_F + K\omega_2}$$

When frictional torque $T_F = 0$

$$t = \frac{J}{K}\log_e \frac{K\omega_1}{K\omega_2} = \frac{20}{0.1557}\log_e \frac{157}{62.832}$$

$= 111.7$ seconds **Ans.**

When frictional torque, $T_F = 1.5 \text{ kg-m} = 1.5 \times 9.81 = 14.715$ Nm

$$\text{So } t = \frac{20}{0.1557}\log_e \frac{14.715 + 24.45}{14.715 + 24.45 \times \frac{600}{1{,}500}}$$

$$= \frac{20}{0.1557}\log_e \frac{39.165}{24.495} = 60.3 \text{ seconds } \textbf{Ans.}$$

1.24. TYPES OF ENCLOSURES

In order to protect machines against the ingress of dirt and dust or larger foreign bodies like spanners, vermin etc. into it, it is desirable that some suitable enclosure be provided. But the enclosure provided for this protection will interfere with the free entry of cool air into the machine with the result that an enclosed machine does not give the same output as does an open type machine of the same dimensions for a given temperature rise.

The purpose of providing enclosure is two fold—firstly providing protection to persons against contact with live or moving parts inside the enclosure and protection to machines against the ingress of solid foreign bodies, and secondly providing protection to machines against harmful ingress of water. Various types of enclosures giving various degrees of protection and cooling facilities for use in different working conditions can be built. ISI 4691 provides an ingress protection code which consists of the letter IP followed by two numbers—the first characteristic numeral designates the extent of protection of persons against contact and of the machine against ingress of solid foreign bodies, while the second characteristic numeral indicates the extent of protection to the machine against harmful ingress of water. Further addition of letter S or M indicates that the degree of protection against ingress of water is applicable only while the machine is stationary or rotating respectively. Absence of both the letters means that protection is provided under both the conditions. For example enclosure marked IP 22 will provide protection against ingress bodies of diameter more than 12 mm and against ingress of water falling up to 15° from vertical under both the conditions. Similarly enclosures marked IP 44 and IP 56 will provide protection against ingress of bodies of diameter more than 1 mm and against water splashed under both the conditions and protection against ingress of bodies in powder form and water projected by a nozzle under both the conditions respectively. For further details refer to ISI code 4691.

The various types of enclosures, as specified by relevant ISI are as follows:

1. *Open Type.* The bearings are supported on simple pedestals or end brackets. This type of enclosure does not protect the machines against dust, dirt and larger foreign bodies but facilitates in cooling. The machines of this type are usually large dc generators and dc motors, which are installed in a neat and clean separate room such as in case of power stations, substations, collieries, rolling mills. Because all the moving parts are exposed, hand rail and other suitable rail guards are provided to prevent unauthorized persons coming into contact specially with moving parts.

2. *Protected Type.* Protected type machines are such in which access to live and rotating parts is so limited as to prevent accidental contact with such parts or ingress of foreign bodies while ventilation is not materially obstructed. Protected type motors are used mostly for industrial purposes.

3. *Screen Protected Type.* In this type of machines openings provided for ventilation are covered with wire screen, expanded metal or other suitable perforated covers having apertures not exceeding 3.25 cm^2 in area but not less than 0.6 cm^2 in area. Though such an enclosure does not protect the motor against dirt and dust but larger bodies and big insects cannot enter into the machine . Such motors are very widely employed for most of industrial applications where dry clean conditions prevail and gas or chemical fumes are not present. Fig. 1.112 (*a*) illustrates a screen-protected motor in which no attempt has been made to exclude dust or fluff from the windings. It is suitable for most forms of electric drive and permits ample ventilation.

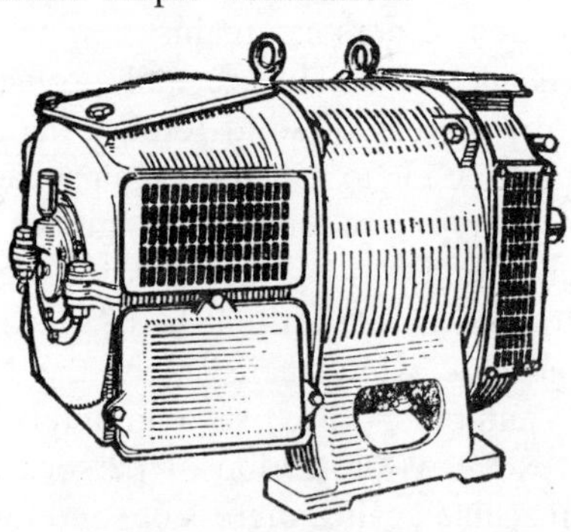

(a) Screen-Protected Motor

(b) Drip-Proof Motor

Fig. 1.112

4. *Drip Proof Type.* In such machines openings provided for ventilation are so protected by overhanging cowls that liquid drops or dust particles falling on the machine vertically cannot enter the machine. Such motors are used in damp situations such as for pumping sets. Fig. 1.112 (*b*) illustrates a drip-proof motor, similar to the screen-protected type, but having protection against dripping moisture.

5. *Splash Proof Type.* In such machines the ventilating openings are so constructed that liquid drops or dust particles falling on the machine or coming towards it in a straight line at any angle not greater than 100° from the vertical cannot enter the machine. Such motor can be used in rains.

6. *Totally Enclosed Type.* Such machines have solid frames and end shields but no opening for ventilation. They get cooled by surface radiation only. In such machines no dirt or foreign matter can enter the motor and clog the air passage. These machines are used in situations where dust, water, oil, chemicals, gases or other injurious substances are likely to enter the machine but natural cooling is sufficient such as for driving coal handling plant, saw mills etc. In these machines due to cooling difficulties the output obtainable is only about 50 to 70% of that obtainable from a protected type of same dimensions and for same temperature rise. Totally enclosed motors are seldom furnished in sizes above 1.5 or 2 kW because higher rating motors require frames of much larger sizes than fan-cooled motors of corresponding rating.

7. *Totally Enclosed, Fan Cooled Type.* In order to assist in heat dissipation, most of the manufacturers have developed totally enclosed fan cooled motors in which an outer casing is built around the motor and air is blown by means of a fan mounted on the shaft between the casing and external surface of the motor. To increase cooling rate, motor is provided with external fans. Such motors are used in saw mills, flour mills, cement works etc. These motors are not water proof and should not be used in places likely to be submerged. Fig. 1.112 (*c*) illustrates a totally enclosed motor in which the main means of cooling is by radiation and internal air circulation.

(c) Totally-Enclosed Motor

Fig. 1.112

8. *Pipe Ventilated Type.* These are totally enclosed type machines in which to overcome the heat dissipation difficulties, a duct or pipe is provided through which clean air is led to the motor from outside the dust laden building or area. The cost of the piping may be overweighed by the saving in cost of motor, as motor of smaller size can be used with same output.

Pipe ventilated motor is shown in Fig. 1.112 (*d*). Fresh clean air is brought to the motor by piping and drawn in through a big inlet, as illustrated in (*i*), and air is then expelled through louvres or taken away by pipe, as illustrated in (*ii*).

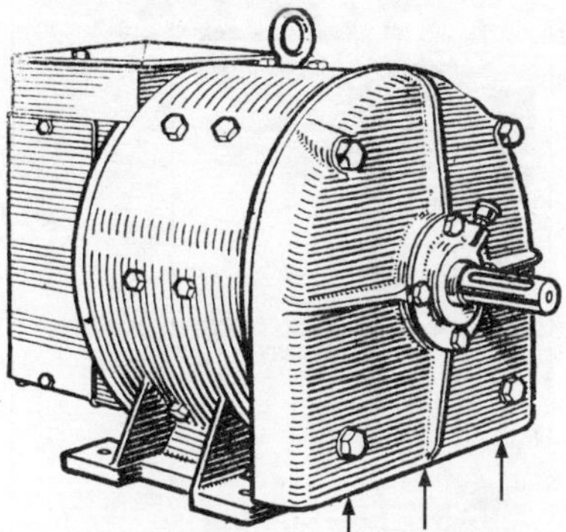

(*i*)

(*ii*)

(*d*) *Pipe-Ventilated Motor*

Fig. 1.112

9. *Flame Proof Type.* The flame proof type machine is provided enclosure designed and constructed to withstand an explosion of the gas or vapour which may occur within it and to prevent the ignition of the gas or vapour surrounding the machine. The ventilating openings in the machine are made long enough for the flame to be contained within the frame. Flame proof type motors are used in collieries, petroleum stations or chemical plants.

1.25. BEARINGS

Bearings are those parts of a machine which house supports and restrain the rotating parts. The moving part of the bearing is known as *journal* and the stationary part that supports the moving part is known as *bearing*.

The main functions of a bearing, besides supporting the rotating parts, are (*i*) to permit free rotation of the moving components with minimum friction and (*ii*) to maintain the rotating member of the machine in a fixed physical location relative to the stationary member.

The bearings are of two types, namely (*i*) ball or roller bearings and (*ii*) sleeve or bush bearings.

1. *Ball or Roller Bearings.* Ball or roller bearings are those bearings in which shafts are supported by balls or rollers which themselves are movable. The action in ball or roller bearings is not that of rubbing but of rolling between the balls and or rollers and their races, the inner surface being fixed to the shaft while the outer surface is fixed to the housing supporting the bearing. Such bearings are supplied and filled with grease. The period between the renewal of grease depends upon the speed of rotation of the motor. These are also known as *antifriction bearings.*

Up to about 75 kW use of ball or roller bearings is made. Very small motors are usually provided with ball bearings. In vertical shaft drives the ball bearings are essential as they are capable of taking an axial thrust.

Though the initial cost of bearings of this type is high but their life is longer, occupy less space, friction loss is smaller and their maintenance cost is low. These have made use of induction motors with small air gap possible as the roller bearing does not wear appreciably. However, these cannot be recommended where noise is to be avoided.

2. *Sleeve or Bush Bearings.* Sleeve or bush bearings are those bearings in which the rotating shaft is supported by a bearing component rigidly fixed to the frame of the machine. They may be (*i*) journal bearing offering support to the shafts at right angles to the shaft axes (*ii*) foot-step bearings supporting shafts parallel to the shaft axes and (*iii*) thrust or collar bearings offering support to shafts subjected to end or axial thrust.

Sleeve bearings are normally of bronze but for universal and fan-cooled motors, sintered powder or porous metal bearings may be used because of their self lubricating properties due to capillary action.

Sleeve bearings on very small motors are lubricated with an oil wick pressed lightly against the shaft by a spring. The wick must be kept saturated with a fairly light oil and will remain moist through several weeks of heavy use. Large sleeve bearings are ring lubricated which consists of a ring freely rotating on the shaft, carrying oil to the bearings. Lubricating oil, which is free from acid and resin, is renewed every six months. Temperature of bearing (either sleeve or roller) is the indication of the bearing condition. ISS 325 provides for maximum temperature rise at sleeve or ball bearings to be 50°C.

Bearings of this type are necessary to reduce the noise level for certain types of couplings and for the positions subject to external vibrations. They are also used where there is a danger of individual balls or rollers indenting the race during long rest periods.

Motors with sleeve bearings should always be used with shaft horizontal and fixed coupling between motor and the driven machine. Because of larger wear of the bearings, relatively large air gap in the induction motor has to be introduced.

1.26. TRANSMISSION OF DRIVE

Mechanical power available at the motor shaft has to be transmitted to the driven machine or machines. There are various methods of doing so. Among these may be mentioned the following :

1. *Direct Drive*. In this type of drive the driving member is connected direct to the driven member, without any interposed gearing, by means of solid or flexible coupling. It is the simplest method, space required is less than with belt drive and efficiency is 100 per cent. Solid coupling requires very accurate aligning otherwise there is possibility of damage to shafts. Another shortcoming of solid coupling is that sudden jerks of load are transmitted to the motor. By employing flexible coupling both of the above shortcomings are overcome to some extent. Flexible coupling can accommodate some angular, lateral or vertical misalignment. The other disadvantages of direct drive are that (*i*) end thrust is exerted on the motor by the load if it happens to be a centrifugal pump and (*ii*) the driving and driven members must run at the same speed, which is usually not desirable. This type of drive is employed where there is a possibility of arranging the driving member in line wiιh the driven member and where speed of driven member is same as that of driving member.

2. *Belt Drives*. These are the least expensive drives. These are employed where a speed change is desired in the transmission of power and where it is not absolutely essential to maintain a fixed speed ratio between the driving and driven shaft. The belt drives are of two types namely (*i*) flat belt drive and (*ii*) V-belt drive.

(i) Flat Belt Drive. It is a long distance drive with a limited degree of slip. The belts used for drive are made of hair, woven cotton, rubberized cotton ply, nylon, terylene and leather. The leather belt drive is the most economical type of drive. Vertical belt drive should be avoided to the possible extent. In horizontal belt drive the slack side of the belt should be kept preferably on the top, so that the natural sag increases the arc of contact on two pulleys.

Two motors of 15 kW and 10 kW arranged for flat belt drive are illustrated in Fig. 1.113 (*a*).

The advantages, disadvantages and limitations of flat belt drive are :

Advantages. (*i*) Greater flexibility in the original design of a plant is possible and new arrangements of old apparatus can be made at any time.

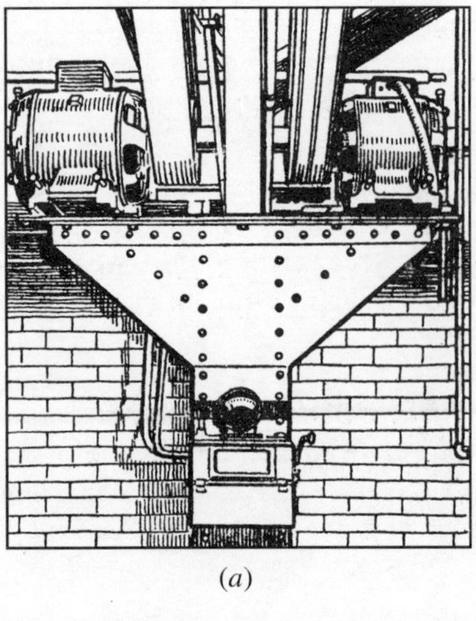

(*a*)

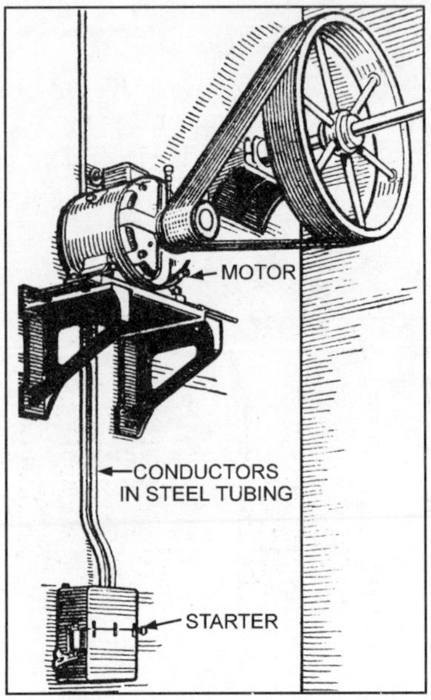

(*b*)

(*c*)

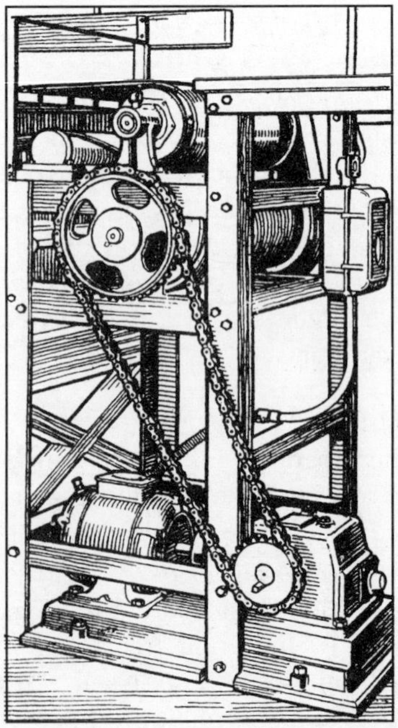

(*d*)

Fig. 1.113. ***Transmission of Drive***

(*ii*) It provides conveniently any desired speed ratio and permits the use of high-speed motors, which are cheaper than low-speed motors.

(*iii*) The tendency of slipping, especially under heavy overloads, is often beneficial in that it will absorb a portion of the shock of suddenly applied loads, and thus protect, to some extent, both the driving and driven machines.

Disadvantages and Limitations. (*i*) Considerable space is required and action is not positive. Distance between drive and driven pulley should not be less than four times the diameter of larger pulley. Typical figures for the minimum distance between pulley centres for motors of 22 to 30 kW are about 6 m for a 6 : 1 ratio and about 1.5 m for a $\frac{3}{2}$: 1 ratio.

(*ii*) Belts exert a side pull on the bearings which results in wear, also loss of power by friction.

(*iii*) The maximum power which can be conveniently transmitted by means of belts is about 150 to 220 kW.

(*iv*) The most economical speed has been found to be from 1,000 to 1,500 metres per minute, though practical limitations such as speed of shafting and pulley diameters, often limit belt speeds to much lower values.

Makers of belts and some motor makers publish tables from which the recommended pulley sizes, driving centres and section of belts or ropes can be conveniently calculated. In arriving at the final speed of the driven machine it must not be forgotten that there is always some slip (3 or 4 per cent) in a belt drive.

(*ii*) *V-Belt Drive.* It is essentially a short centre, silent drive with a negligible slip (say about 1 per cent). Such type of drive is suitable for motors of ratings up to 450 kW. The power transmitted depends upon the size of the belt, the speed of the belt (limited between 2.5 to 25 m/s), the size of the sheaves, (the larger the sheave, the greater the kW power transmitted) and the arc of contact on the smaller sheave.

The drives should be so designed that the arc of contact on the smaller V-pulley is not less than 90° and more than 130°, as far as possible 12 kW motor with multiple V-belt drive is shown in Fig. 1.113 (*b*).

3. *Rope Drive.* Rope drive, the drive by means of ropes running over pulleys having a number of grooves, is used for power ranges beyond the limit of V-belt drive. It is a long centre drive. Main advantages of rope drive are negligible slip and ability of taking sudden loads. Where the speed ratio is very high rope drive is indispensable. This type of drive is coming into use on small and medium sized machines, nowadays.

Coupled motors driving through multiple ropes working in grooved pulleys is shown in Fig. 1.113 (*c*).

4. *Chain Drive.* Though chain drive is more costlier than belt and rope drives but it is more efficient, can be used for high speed ratio (the limit being 7:1 with a distance between pulley centres of about 1.5 to 2 times the diameter of the larger pulley) and has no slip. It can be used conveniently for drives involving up to 12 or more parallel shafts; the driving centres can be shorter than with any drive except the direct or gear drive. It is particularly suitable for damp and dirty conditions because the chain case protects the drive from outside influences. With chain drive, it must be ensured that shafts are parallel and chains run centrally on their sprockets at right angle to the shafts to give long life and avoid end thrust on the motor. This method is suitable for speeds up to 900 m/minute needing positive drives without any slipping. Chain drive, through reduction gearing, from a 2.5 kW motor is shown in Fig. 1.113 (*d*).

5. *Gear Drive.* It is a short centre positive drive. It is very important to have proper alignment in the absence of which, otherwise, we may have to pay in the form of bent motor shafts. It was the first form of power transmission in the early steam driven mills. Modern straight cut gears, and the particular variation found in the worm reduction drive, are common features in many heavy machines such as cloth calenders. A typical 35 kW worm gear with a speed reduction of 15 : 1 will have a mechanical efficiency of about 98 per cent.

6. *Vertical Drive.* In this type of drive motor is arranged with its axis vertical. This arrangement frequently proves to be convenient.

1.27. CHOICE OF DRIVE

Choice of drive is governed by the speed of driving and driven machines, convenience, space available, clutching arrangement required and cost. Choice of speed of motor is most important factor as it not only affects the performance of the motor but also overall cost. The dimensions and, therefore, the first cost of a motor for a given output are approximately inversely proportional to the speed, so for the same output kW the cost of high-speed motor is less than that of a low-speed motor. The efficiency and, in case of induction motors, power factor also decreases with the decrease in speed. Thus for a low-speed drive high-speed motor using a reduction gear is usually found cheaper than a low-speed direct coupled motor. These days there is a trend to make motors for low-speed drives by incorporating gear in the motor frame, the motor being designed for a high speed. The overall cost of such a motor including the gear is less than that of the corresponding low-speed motor; gear efficiency ranging from about 96 per cent; for a 5 : 1 ratio to about 80 per cent for a 50 : 1 ratio can be obtained

1.28. NOISE

A feature of importance in choosing a motor, as well as any other apparatus, is that of noise; extreme silence is, of course, essential in motors for domestic purposes, use in hospitals, theatres, another similar institutions. Even for ordinary industrial purposes it is essential to keep motor noise to a minimum because it causes fatigue to the workers and, therefore, affects the overall efficiency and output of the factory adversely. The increasing use of steel-framed buildings, which transmit noise readily, tends to make the problem of greater importance than hitherto.

The level of sound is measured in bels (B), which is defined as

Difference in loudness

$$= \log_{10} \frac{W_2}{W_1} = \log_{10} \frac{P_2^2}{P_1^2} \text{ bels} \qquad \because W \propto P^2$$

$$= 2 \log_{10} \frac{P_2}{P_1} \qquad ...(1.90)$$

where W_1 and W_2 are the powers of sound waves under consideration and P_1 and P_2 are the pressures corresponding to the sound waves concerned.

In practice the decibel (one-tenth of a bel) is more convenient unit and represents approximately the smallest difference in sound which the ear can detect.

Thus the difference in loudness between two sounds

$$= 10 \log_{10} \frac{W_2}{W_1} \text{ dB} = 20 \log_{10} \frac{P_2}{P_1} \text{ dB} \qquad ...(1.91)$$

Noise may be due to purely mechanical causes, such as resonance or vibration of laminations, or it may be due to magnetic pulsations caused by an incorrect choice of the number of slots. The elimination of such features is a matter of the manufacturer, but their effects can be minimized by preventing transmission of vibrations to other parts of the building. This can be accomplished by resilient or spring mounting the motor. Resilient mounting makes use of cushioning material such as felt mat, cork and rubber slabs which is effective in the audible frequency range. Spring mounting on the other hand can absorb all vibration frequencies. To avoid transmission of vibrations electrical connections to the motor are made by means of flexible pipe rather than by conduit pipe. It is, however, impossible to exterminate the noise.

1.29. INSULATING MATERIALS

Insulating materials are employed for providing electrical insulation between parts at different potentials. These materials begin to deteriorate at relatively small temperatures. For reliable operation, it is essential that the temperature rise in electrical machines and equipments do not exceed the permissible maximum temperature of the insulating materials used therein.

The three fundamental electrical properties of insulating materials of utmost importance in the operation of electrical machines or equipment are resistivity (insulation resistance), dielectric strength and dielectric loss angle. In addition to the above, other important properties of insulating materials are mechanical strength, heat resistance, hygroscopicity.

An insulating material should have high resistivity, high dielectric strength, low dielectric loss, good heat conductivity, sufficient mechanical strength to withstand vibrations etc., capability of withstanding a repeated heat cycle without deterioration and should be non-hygroscopic.

The materials used for insulation in electrical machines and apparatus are numerous (such as insulating varnishes, mineral waxes, synthetic waxes, varnish, cloth, silk, cotton, rubber, vulcanized rubber, hard rubber, bakelite, paper, wood, mica, asbestos, glass, porcelain, marble, slate, insulating oils etc.). Application of some of the most important insulating materials are given below:

Cotton (cotton tape, empire cloth etc.) is mainly employed for covering bare wire used in winding of small magnet coils, armature windings of small machines, chokes and coils of small transformers. **Silk** is coated with a waxy form of lubricant having good insulating properties and is mainly used for winding of small machines. In electrical machines, **glass** is mainly used as a glass tape and varnished glass cloth. **Glass tapes** are used for insulating the windings and winding ends of electrical machines operating at high temperatures. **Varnished glass cloth** is used mainly for insulating the stator windings in hv electrical machines. **Impregnating varnishes** serve to treat porous fibrous or solid insulators used in electrical machines and apparatus. **Bakelite** is used as an insulating cylinder in transformers. **Mica** is largely used in commutators for separating its segments.

1.29.1. Classification of Insulating Materials

The temperature rise which electric motor can safely withstand is determined by the thermal stability (limiting temperature) of the insulating materials used in them. By using materials of higher thermal stability, motors of the same physical size may be rated for larger power output.

Different insulating materials have different limiting temperatures. According to their limiting temperatures all insulating materials employed in electrical machines are classified as per IS 1271 as below.

1. **Class Y-** Cotton yarn and fabrics, paper and fibrous materials of cellulose or silk neither impregnated with nor immersed in liquid dielectric. The limiting temperature is 90° C.
2. **Class A-** The same materials as mentioned above, but impregnated with or immersed in liquid dielectric. Limiting temperature is 105° C.
3. **Class E-** Certain synthetic organic films and other materials having the same thermal stability. Limiting temperature is 120° C.
4. **Class B-** Mica, asbestos or glass-fibre base materials combined with an organic binding agent, The limiting temperature is 130° C.
5. **Class F-** Mica, asbestos or glass-fibre base materials combined with suitable synthetic binding and impregnating agents. The limiting temperature is 155° C.
6. **Class H-** Mica, asbestos or glass fibre combined with silicon binding and impregnating agents. The limiting temperature is 180° C.
7. **Class C-** Mica, ceramic materials, glass, or quartz used without or with inorganic binding agents. For this class of insulating materials, the limiting temperature is above 180° C and is restricted only by the specific physical, chemical or electrical properties of the material concerned.

The limiting temperatures are never exceeded in practice, otherwise it will result in deterioration and breakdown of the insulation and will shorten the life of the motors.

1.30. TEMPERATURE RISE IN ELECTRICAL MACHINES

Heat is developed in all electrical machines due to the losses in the various parts, principally copper loss (I^2R loss) in conductors, eddy current and hysteresis loss in iron and mechanical losses (in rotating machines only) due to friction of the bearing, air friction or windage causing the temperature of that part to rise. This temperature rise continues until all the heat generated is dissipated to the surroundings by one or more of the natural modes of heat transfer, viz. conduction, convection and radiation.

Ultimately then, under steady load, each part achieves a final temperature, the magnitude of which depends on the balance between the rate at which heat is developed in that part (or received by conduction from a hotter part), and the rate at which the heat can be dissipated which is determined by effectiveness of the cooling method.

The temperature rise depends upon (*i*) the amount of heat produced and (*ii*) the amount of heat dissipated per 1°C rise of the surface of a machine.

According to Newton's law of cooling the rate of loss of energy of a hot body is proportional to the difference in temperature between that body and its surroundings. This law is approximately true for moderate temperature difference (up to 100° C) and for bodies dissipating heat by radiation and natural convection. It means that the amount of heat dissipated per 1° C rise of the surface of a machine depends on the surface area of cooling.

The size of motor for any service is governed by the maximum temperature rise when operating under the given load conditions and the maximum torque required. The former is more important because if the motor operates satisfactorily for maximum temperature rise, it will usually provide the required maximum torque, except in special cases where the load consists of heavy peaks followed by relatively long intervals of no load. Electrical machines are, therefore, designed for a limited temperature rise. In fact, the continuous rating of a machine is that rating for which the final temperature rise is equal to or just below the permissible value of temperature rise for the insulating material used in protection of motor windings. When the machine is overloaded for such a long time that its final temperature rise exceeds the permissible limit, it is likely to be damaged. In worst cases, it will result in an immediate thermal breakdown of the insulating material which will cause a short circuit in the motor, thus putting a stop to its functioning. The short circuit may also lead to a fire. In less severe cases immediate thermal breakdown of the insulating material may not occur, but the quality of the insulation will deteriorate such that thermal breakdown with future overloads or even normal loads might soon occur, thus shortening the useful life of the machine.

The temperature rise to which a motor be allowed to rise is limited by the insulation. The maximum temperature rises which should not be exceeded by different types of motor are fully set out in the relevant ISS. Since temperature rise is one of the chief features in fixing the size of a motor, its calculations become a matter of importance and may be carried out as detailed below.

Heating-Time Curve. For determination of an expression for the temperature rise of an electrical machine after time *t* seconds from the instant of switching it on, let

Power converted into heat = P joules/s or watts

Mass of active parts of machine = m kg

Specific heat of material= C_p joules/kg/°C

Surface area of cooling = S metres^2

Coefficient of cooling = α in watts per metre^2 of surface per °C of difference between surface and ambient cooling temperatures.

Assumptions Made. (*i*) Losses or heat produced remains constant during the temperature rise.

(*ii*) Temperature of the cooling medium remains unchanged.

and (*iii*) Heat dissipated is directly proportional to the difference in temperature of the motor and the cooling medium.

Suppose a machine attains a temperature rise of θ°C above ambient temperature after t seconds of switching on the machine and further rise of temperature by $d\theta$ in very small time dt.

$$\text{Energy converted into heat} = Pdt \text{ joules}$$

$$\text{Heat absorbed} = mC_p d\theta \text{ joules}$$

$$\text{Heat dissipated} = S\theta\alpha dt \text{ joules}$$

Since energy converted into heat

$$= \text{Heat absorbed} + \text{heat dissipated}$$

$$\therefore\ Pdt = mC_p d\theta + S\theta\alpha dt \qquad \text{...(1.92)}$$

$$\text{or } (P - S\alpha\theta)dt = mC_p\, d\theta$$

$$\text{or } \frac{dt}{mC_p/S\alpha} = \frac{d\theta}{\frac{P}{S\alpha} - \theta} \qquad \text{...(1.93)}$$

When final temperature is reached, there is no absorption of heat. Whatever heat is generated, has to be dissipated.

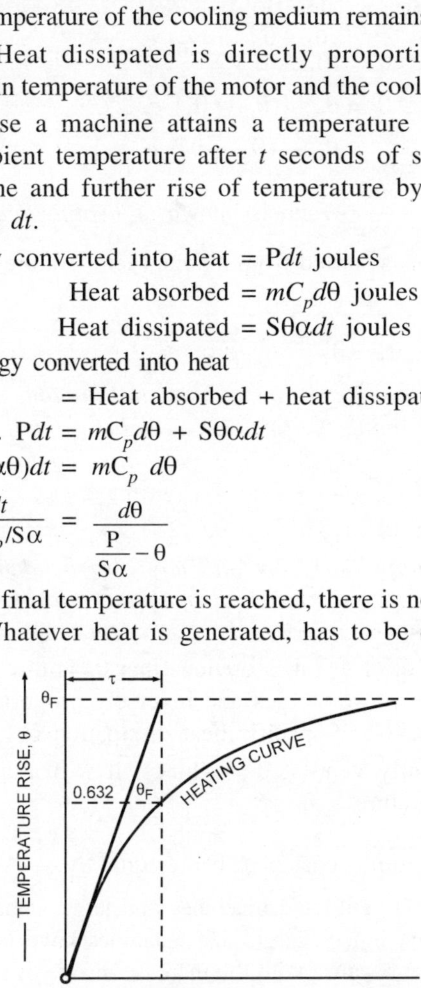

Fig. 1.114(*a*). *Temperature Rise-Time Curve*

$$\therefore\ Pdt = S\theta_F\alpha\, dt$$

or Final temperature rise,

$$\theta_F = \frac{P}{S\alpha} \qquad \text{...(1.94)}$$

Substituting $\frac{P}{S\alpha} = \theta_F$ in Eq. (1.93), we get

$$\frac{S\alpha\, dt}{mC_p} = \frac{d\theta}{\theta_F - \theta}$$

Integrating both sides, we get

$$\frac{S\alpha}{mC_p}t = -\log_e(\theta_F - \theta) + K_1 \qquad ...(1.95)$$

where K_1 is a constant of integration.

Substituting $t = 0$, $\theta = \theta_1$, initial temperature rise from initial conditions, we have

$$0 = -\log_e(\theta_F - \theta_1) + K_1$$

or $K_1 = \log_e(\theta_F - \theta_1)$

Substituting $K_1 = \log_e(\theta_F - \theta_1)$ in Eq. (1.95), we get

$$\frac{S\alpha}{mC_p}t = -\log_e(\theta_F - \theta) + \log_e(\theta_F - \theta_1)$$

$$= \log_e\left(\frac{\theta_F - \theta_1}{\theta_F - \theta}\right)$$

or $e^{\frac{S\alpha t}{mC_p}} = \frac{\theta_F - \theta_1}{\theta_F - \theta}$

or $\theta_F - \theta = (\theta_F - \theta_1)\, e^{-S\alpha t/m C_p}$

or $\theta = \theta_F - (\theta_F - \theta_1)\, e^{-S\alpha t/m C_p}$

$$= \theta_F - (\theta_F - \theta_1)\, e^{-t/\tau} \qquad ...(1.96)$$

where $\tau = \frac{mC_p}{S\alpha}$ and is known as *heating time constant.*

If motor is started from ambient temperature, $\theta_1 = 0$, we have

$$\theta = \theta_F - \theta_F^{-t/\tau} = \theta_F(1 - e^{-t/\tau}) \qquad ...(1.97)$$

Heating time constant is defined as the time taken by the motor in attaining the final steady value if the initial rate of rise of temperature were maintained throughout the operation. Substituting $t = \tau$ in Eq. (1.97), we have

$$\theta = \theta_F(1 - e^{-1}) = 0.632\,\theta_F \qquad ...(1.98)$$

So *heating time constant may also be defined as the time duration during which the machine will attain 63.2% of its final temperature rise above ambient temperature.*

From Eq. (1.94) it is obvious that θ_F is directly proportional to the power losses and inversely proportional to the surface area S and specific heat dissipation α.

For poorly ventilated machines, it will attain a higher final temperature rise.

Heating time constant τ, being equal to $\frac{mC_p}{S\alpha}$, has small value for well ventilated machines and large value for poorly ventilated machines. Large size machines have large heating time constant because with the increase in size of machine, the volume and hence the mass increases in proportion to the third power of linear dimensions and surface area S in proportion to second power. Typical values of heating time constant lie between about $1\frac{1}{2}$ hours for small motors (7.5 kW to 15 kW) up to about 5 hours for motors of several hundred kW.

Cooling-Time Curve. Let the machine be switched off after reaching final steady-temperature rise of θ_F. When the machine is switched off, no heat is produced, therefore,

0 = Heat absorbed + heat dissipated

or $0 = mC_p d\theta + S\theta\alpha' dt$ where α' is the rate of heat dissipation during cooling

or $dt = \frac{-mC_p\, d\theta}{S\alpha'\theta}$

Integrating both sides, we have

$$t = \frac{-mC_p}{S\alpha'}\log_e\theta + K_2 \qquad ...(i)$$

where K_2 is a constant of integration

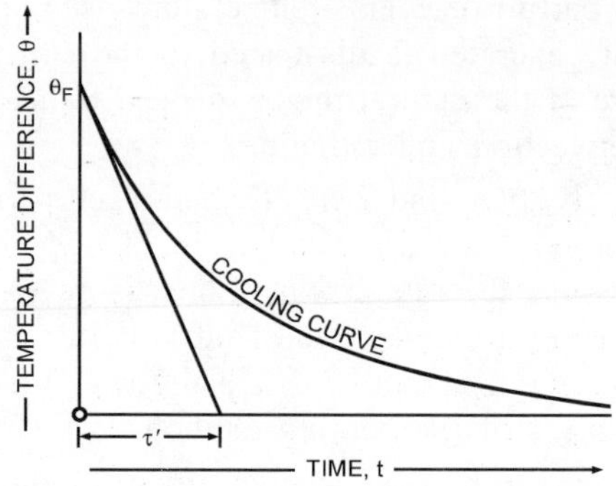

Fig. 1.114(*b*)

Substituting $t = 0$, $\theta = \theta_F$ from initial conditions in above equation, we have

$$K_2 = \frac{mC_p}{S\alpha'}\log_e\theta_F$$

Substituting $K_2 = \frac{mC_p}{S\alpha'}\log_e\theta_F$ in Eq. (*i*), we have

$$t = \frac{-mC_p}{S\alpha'}\log_e\theta + \frac{mC_p}{S\alpha'}\log_e\theta_F = \frac{-mC_p}{S\alpha'}\log_e\frac{\theta}{\theta_F}$$

or $\theta = \theta_F\, e^{-S\alpha' t/mC_p} = \theta_F\, e^{-t/\tau'} \qquad ...(1.99)$

where $\tau' = \frac{mC_p}{S\alpha'}$, called the cooling time constant.

Substituting $t = \tau'$ in Eq. (1.99), we have

$$\theta = \theta_F e^{-1} = 0.368\,\theta_F \qquad ...(1.100)$$

i.e., cooling time constant may also be defined as the time required to cool the machine down to 0.368 times the initial temperature rise above ambient temperature.

The heating and cooling curves follow an exponential law. Heating time constant and cooling time constant may be different for the same machine, as the ventilation conditions in the two cases may not be the same. The cooling time constant of a rotating machine is usually larger than its heating time constant owing to poor ventilation conditions when the machine cools. In self-cooled rotating machines the cooling time constant is about 2-3 times greater than its heating time constant because cooling conditions are worse at standstill.

1.31. DUTY CYCLES

The nominal duty of a drive motor is the duty corresponding to the service conditions and performance marked on its name plate.

There are three types of duties viz continuous duty, short-time duty and intermittent duty.

The heating and cooling curves for *continuous-duty motor* are shown in Fig. 1.115 (*a*). Continuous duty is that duty when the on-period is so long that the motor attains a steady-state temperature rise.

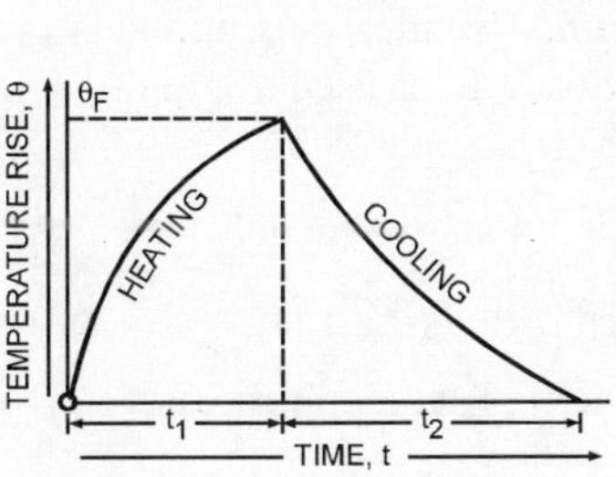

(a) For Continuous Duty Motor

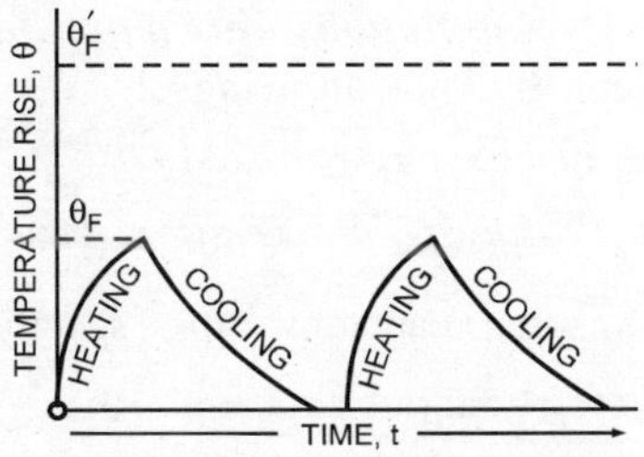

(b) For Short-time Duty Motor

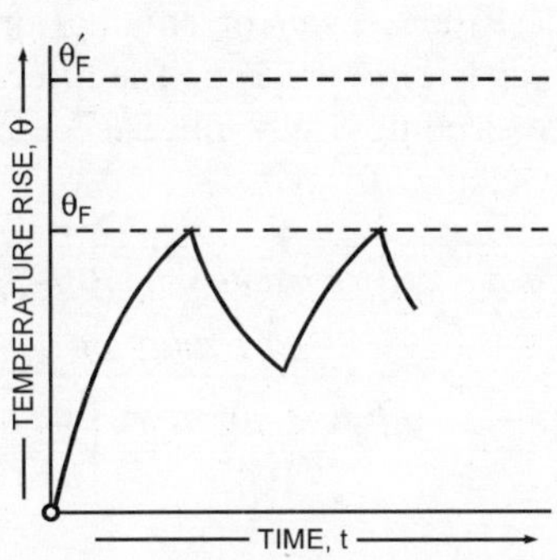

(c) For Intermittent Periodic Duty Motor

Fig. 1.115. ***Heating and Cooling Curves***

The heating and cooling curves for *short-time duty motor* are given in Fig. 1.115 (*b*). The short-time duty motor operates at a constant load for some specified periods which is then followed by a period of rest. The period of run (or load) is so short that machine cannot attain its steady temperature rise while the period of rest is too long that the motor temperature drops to the ambient temperature.

The heating and cooling curves for *intermittent periodic duty motors* are illustrated in Fig. 1.115 (*c*). On intermittent duty the periods of constant load and rest with machine de-energized alternate. The loading periods are too short to allow the motor to attain its final steady-state value while periods of rest are too small to allow the motor to cool down to the ambient temperature. *Intermittent rating of a machine is defined as the load which is applied during a certain fraction of time of a load cycle and the temperature rise limit is not exceeded.*

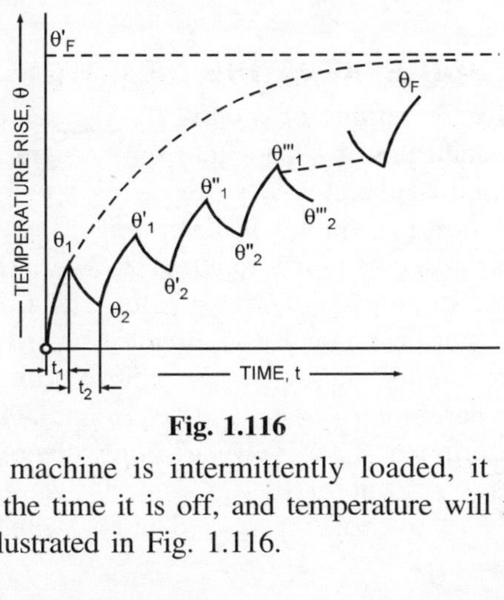

Fig. 1.116

When a machine is intermittently loaded, it will cool down during the time it is off, and temperature will rise when it is on, as illustrated in Fig. 1.116.

Continuous duty motors are employed to drive fans, compressors, generators etc. They may be in operation for many hours and even days in succession. *Short-time duty motors* are used in navigation-lock gates, railway turntables, bascule bridges, and the like. *Intermittent-duty motors* are employed in cranes, hoists, lifts, rolling mills, some metal-working machines, etc.

The duty cycles for various motor duties are shown in Fig. 1.117.

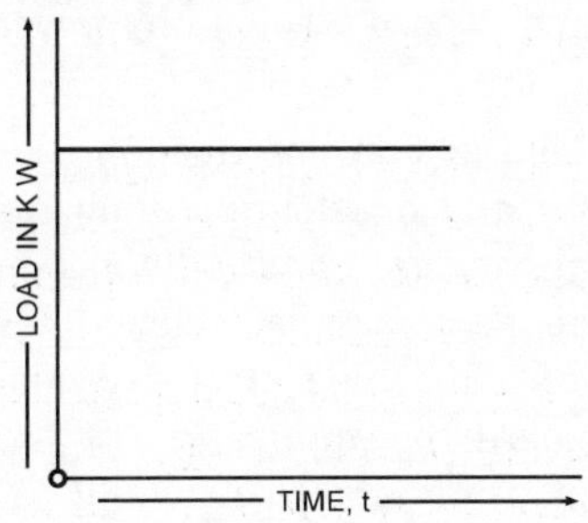

(a) Continuous Duty Cycle

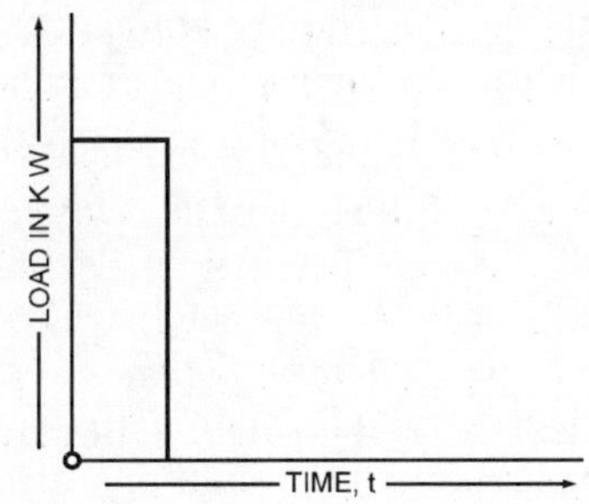

(b) Short-Time Duty Cycle

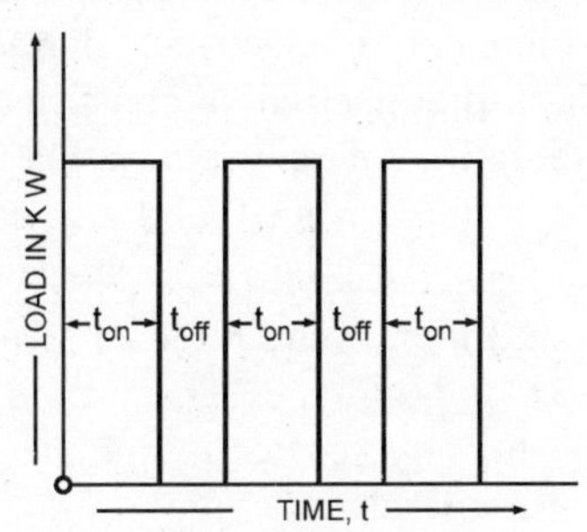

(c) Intermittent Duty Cycle

Fig. 1.117

1.31.1. Maximum Temperature Attained With Intermittent Loading

When a machine is intermittently loaded it will cool down during the time duration it is off and its temperature will rise when it is on, as illustrated in Fig. 1.116. Let θ_1, θ'_1, θ''_1, be θ'''_1 be the temperature rise after heating and θ_2, θ'_2, θ''_2, θ'''_3 be the temperature rise after cooling, as illustrated in Fig. 1.116.

$$\text{Now } \theta_1 = \theta'_F(1 - e^{-t_1/\tau}) = \theta_F'(1 - e^x) \text{ where } x = \frac{-t_1}{\tau}$$

$$\text{and } \theta_2 = \theta_1 e^{-t_2/\tau'} = \theta_1 e^y = \theta'_F(1 - e^x)e^y \text{ where } y = \frac{-t_2}{\tau'}$$

$$\begin{aligned}
\theta'_1 &= \theta'_F - (\theta'_F - \theta_2)e^x \\
&= \theta'_F(1 - e^x) + \theta_2 e^x \\
&= \theta'_F(1 - e^x) + \theta'_F(1 - e^x)\, e^y e^x \\
&= \theta'_F(1 - e^x)\,(1 + e^x e^y) \\
\theta'_2 &= \theta'_1 e^y = \theta'_F(1 - e^x)\,(1 + e^{x+y}).e^y \\
\theta''_1 &= \theta'_F(1 - e^x) + \theta'_2 e^x \\
&= \theta'_F(1 - e^x) + \theta'_F e^x e^y (1 - e^x)\,(1 + e^x e^y) \\
&= \theta'_F(1 - e^x)\,[1 + e^x e^y + e^{2x}.e^{2y}]
\end{aligned}$$

For n times intermittency, we have

$$\theta_1^{1-n} = \theta'_F(1 - e^x)[1 + e^x e^y + e^{2x} e^{2y} + \dots e^{(n-1)x} \times e^{(n-1)y}]$$

$$= \theta'_F(1 - e^x)\left[\frac{1 - e^{nx} e^{ny}}{1 - e^x e^y}\right]$$

When $n = \infty$ both e^{nx} and e^{ny} will be zero as x and y are negative. If θ_F is the maximum temperature rise with intermittent load then

$$\theta_F = \theta_F' \left(\frac{1-e^x}{1-e^x e^y} \right) = \theta_F' \frac{1-e^x}{1-e^{x+y}} = \theta_F' \frac{1-e^{-t_1/\tau}}{1-e^{-(t_1/\tau + t_2/\tau')}} \quad ...(1.101)$$

1.32. RATING OF MACHINES

A name plate fixed to the outside frame of an electrical machine records the data pertaining to its rating. A machine rating specifies the voltage, current, speed, excitation, power factor, efficiency, power output etc.

The rating of a machine should give all information necessary so that, if the machine is operated within the limits of all factors specified in its rating, the machine will operate satisfactorily and safely and will give reasonable length of service. Therefore, the rating of a machine must give the necessary information to safeguard the application of the machine from conditions of operation which (*i*) would result in unsafe mechanical or electrical strains upon any part of its structure or (*ii*) would result in excessive deterioration of the mechanical or electrical characteristics of the materials of which the machine is constructed, To give this information, the rating of an electric machine should include the output, voltage, speed, and any other information that may necessary for the proper operation of the machine.

Generators and motors are rated in terms of kW output at a given speed and voltage. The size and rating of an electric machine for any service is mainly governed by the factor 'temperature rise'. The maximum temperature, to which an electric machine is allowed to reach, is limited by the type of insulation used.

The maximum temperature rise permissible with insulation A and insulation B are 40°C and 50°C respectively. Overloads are generally permissible for short period of time but when machines are required to carry greater loads than those specified, they must be kept under inspection to see that the temperature does not rise too much and that severe sparking at the commutator does not occur.

The type of service to which a machine is subjected is of great importance. The machines operating continuously at rated (or near rated) load are physically larger than those working at intermittent loads. Also electric machines that are not enclosed and are, in addition, well cooled by fans are likely to have higher ratings than covered up machines or those machines located where air does not circulate freely through and over them. High-speed machines and machines employing mica, glass tapes, and the new silicons as insulation can generally be physically smaller, in given ratings, than the low-speed machines employing standard insulations. The output of dc machines is also limited by the factor 'commutation sparking'. This factor often limits the output, even though heating may not have proceeded to permissible values.

The machines according to ISI specifications are classified as follows.

Continuous Rating. This is an output, which a machine delivers continuously without exceeding the permissible temperature rise. It can deliver 25 per cent overload for two hours.

Continuous Maximum Rating. Similar to continuous rating but not allowing any overload. It is used for motors of capacity larger than 1.84 kW (2.5 hp) per rpm.

Short-Time Rating. This is an output which an electric machine can deliver for a specified period (say 1 hr., $\frac{1}{2}$ hr., $\frac{1}{4}$ hr. etc.) without exceeding the specified temperature rise.

1.32.1. Effect of Altitude on Rated Output

At high altitudes, the density of cooling air decreases. As a result, cooling capacity is reduced. The effect is negligible for elevations not exceeding 1,000 m. The output is reduced as stated below in tabular form.

Altitude above sea level in metres	*Permissible output in percentage of the rated value for motor speeds*	
	up to 1,000 rpm	*above 1,000 rpm*
0–1,000	100	100
2,000	95	92
3,000	90	85
4,000	87	80
5,000	85	77

1.32.2. Overload Capacity of Induction Motors

Motors are designed to carry overloads, without any adverse effect, as mentioned below.

Sustained overloads	*Time duration*	*Motor kW output per 1,000 rpm*
25% overload in torque	2 hours	7.5 and above
	30 minutes	3 to 6.8
	15 minutes	0.75 to 3
Momentary overloads		
50% overload in torque	1 minute for all sizes	
100% overload in torque	15 seconds for all sizes	

Single phase induction motors are usually not designed to carry overload. However, 20-25% overload in torque intermittently may be permitted in split-phase ac motors and 40-45% overload in torque intermittently for capacitor-start motors.

1.33. CHOICE OF RATING OF MOTORS

The choice of ratings of motors for the services requiring fairly constant power such a fans, air-compressors, blowers, pumps, motor-generator sets etc. is every simple. In such cases the power required is determined and motor having continuous rating of power required is chosen. It is, however, worthwhile to take considerable trouble to find out exactly how much power is actually required by the driven machinery, since if the motor chosen be too small it will be overloaded and will, therefore, overheat and deteriorate, while if it is too large the efficiency and power factor will probably be less than those of a motor of the correct size. If there is no possibility of overloading then a motor with a continuous maximum rating may be selected, which will be comparatively smaller in size and cheaper in cost. Such a motor may operate at a higher temperature (55°C in place of 40°C).

For the constant torque loads, the size of the motor can be determined as follows :

In case of rotary motion, rating of motor required is given as

$$P = T \times \frac{2\pi N}{60} \times \frac{1}{\eta} \times \frac{1}{1{,}000} \text{ kW} = \frac{TN}{9{,}550\eta} \text{ kW} \quad ...(1.102)$$

where T is the load torque in N-m, N is the speed in rpm and η is the product of the efficiency of the driven load and that of the transmitting device. Efficiency varies considerably with the type of drive, bearings etc. and its value can best be dictated by experience in applications.

In case of linear motion, rating of the motor required is given as

$$P = \frac{F \times 9.81 \times v}{\eta} \times \frac{1}{1{,}000} \text{ kW} = \frac{F \times v}{102\,\eta} \text{ kW} \quad ...(1.103)$$

where F is the force caused by the load in kg and v is the velocity of motion of load in m/s.

The above expression [Eq. (1.103)] can be applied in case of hoisting mechanism, lifts or elevators with slight modification and that is due to counterweight (balancing the weight of the cage or car as well as one-half of the useful load). The modified expression is

$$P = \frac{Fv}{2 \times 102\,\eta} \text{ kW} \quad ...(1.104)$$

v in case of normal passenger lift cabins lies in the range of 0.5–1.5 m/s.

In case of pumps, the rating of the motor required,

$$P = \frac{Q \times \rho \times h}{102\,\eta} \text{ kW} \quad ...(1.105)$$

where Q is the delivery of pumps in m^3/s, ρ is the density of liquid pumped in kg/m^3 and h is the gross head (static head + friction head) in m. Efficiency η lies in the range of 80 to 90% in case of reciprocating pumps and 40 to 80% in case of centrifugal pumps.

Similarly, the rating of a fan motor is given as

$$P = \frac{Qh}{102\eta} \text{ kW} \quad ...(1.106)$$

where Q is the volume of air or any other gas in m^3/s and h is the pressure in mm of water or in kg/m^2. Efficiency may be taken to be 0.8 for larger power fans and 0.6 in case of smaller power fans.

The selection of size of motor is also simple for the services where the motor is run for a short time so that the temperature rise is less than the maximum permissible value and then it is allowed to cool down to the ambient temperature. For such services a motor rated at the required output power for the given period is chosen.

In cases where the load fluctuates over a given cycle, as in rolling mills or colliery winders, the size of the motor is determined accurately by finding the heating and cooling curves of motors under consideration and plotting heating curve for each motor, when working on the given cycle. The smallest motor which delivers the load according to given cycle without exceeding the specified temperature rise may be choosen.

Other methods for determination of rating of a motor for continuous duty and variable load are given below.

1. *Method of Average Losses.* In this method it is assumed that the temperature rise attained by the motor with fluctuating loading conditions over a certain period of time will be the same as that attained by the motor with a certain load of constant magnitude. This will be true provided that the average losses in the motor for the same time are the same for both the conditions of operation.

This method consists of determining average losses in the motor when it operates according to the given load diagram and then comparing with the losses corresponding to the continuous duty of the machine when operated at its nominal rating.

2. *Equivalent Current Method.* This method is based upon the assumption that the actual variable current may be replaced by an equivalent current I_{eq} which produces the same losses in the motor as the actual current.

$$\therefore \quad I_{eq} = \sqrt{\frac{I_1^2 t_1 + I_2^2 t_2 + I_3^2 t_3 + \ldots + I_n^2 t_n}{t_1 + t_2 + t_3 + \ldots + t_n}} \quad ...(1.107)$$

The heating and cooling conditions in self ventilated machines depend upon its speed. At low speeds the cooling conditions are poorer than at normal speeds. Therefore, if the work cycle involves slow-speed operation, it must be taken into consideration when using Eq. (1.107).

The equivalent current is compared with the rated current of the motor selected. The equivalent current may be less than or equal to the rated current of the machine.

3. *Equivalent Torque Method.* This method is based upon the assumption that the motor current is proportional to the torque and heating is proportional to the square of current (*i.e.,* heating is proportional to the square of torque).

The equivalent torque is determined in the same manner as the equivalent current *i.e.,*

$$T_{eq} = \sqrt{\frac{T_1^2 t_1 + T_2^2 t_2 + T_3^2 t_3 + \ldots + T_n^2 t_n}{t_1 + t_2 + t_3 + \ldots + t_n}} \quad ...(1.108)$$

A motor having torque nearly equal to T_{eq} is selected.

4. *Equivalent Power Method.* At constant speed or where the variations in speed are small, the equivalent power is given by the relationship,

$$P_{eq} = \sqrt{\frac{P_1^2 t_1 + P_2^2 t_2 + P_3^2 t_3 + \ldots + P_n^2 t_n}{t_1 + t_2 + t_3 + \ldots + t_n}} \quad ...(1.109)$$

A motor having power rating P_{eq} is selected.

The equivalent current method is the most accurate out of the four methods discussed above. This method may be employed for determining the capacity for all applications except where it is necessary to take into account the variations in so called 'constant losses'.

The equivalent torque method cannot be employed for applications where equivalent current method cannot be applied or in cases in which flux does not remain constant like dc series motors.

The equivalent power method cannot be used for motors whose speed varies considerably under load, especially when dealing with starting and braking conditions.

Example 1.45. The outside surface of a 12 hp (metric) motor is equivalent to a cylinder of 65 cm and 1 metre length. The motor weighs 400 kg and has a specific heat of 700 J/kg/°C. The outer surface is capable of heat dissipation of 12 W per metre square per degree centigrade. Find the final temperature rise and thermal constant of motor when operating at full load with an efficiency of 90 %. *[J.N. University Hyderabad April 2003]*

Solution : Power output of machine = 12 × 735.5 = 8,826 watts

$$\text{Power input} = \frac{\text{Power output}}{\eta} = \frac{8,826}{0.9} = 9,807 \text{ watts}$$

Power losses, P = Power input – power output

= 9,807 – 8,826 = 981 watts

Mass of active parts of machine, m = 400 kg

Specific heat of material, C_p = 700 J/kg/°C

Coefficient of cooling, α = 12 watts/m²/°C

Surface area of cooling $= \pi dl = \pi \times 0.65 \times 1 = 2.042 \text{ m}^2$

$$\text{Final temperature rise, } \theta_m = \frac{P}{S\alpha} = \frac{981}{2.042 \times 12} = 40° \text{ C } \textbf{Ans.}$$

Heating time constant,

$$\tau = \frac{mC_p}{S\alpha} = \frac{400 \times 700}{2.042 \times 12}$$

= 11,427 seconds or 3 hours 11 minutes **Ans.**

Example 1.46. The mean temperature rise of a machine to run on full load is :

Time in hours	0.2	0.4	0.6	1.0	1.5	2.0	3.0
Temperature rise in °C	7.0	13.5	19.3	29.1	39.1	46.7	57.3

Estimate the final steady-state temperature and heating time constant. *[Agra Univ. Traction & Utilization of Elec. Power 1979]*

Solution : For determination of steady-state temperature plot temperature-rise against time from the given datas, as shown in Fig. 1.118. Now plot increments of temperature for equal intervals of time against temperature rise to the left of Y-axis, as shown in Fig. 1.118.

Straight line drawn through the plotted points cuts the Y-axis, where the rate of change of temperature rise is zero. Therefore the point of intersection of the line drawn, as mentioned above, with the Y-axis gives steady-state temperature. From Fig. 1.118, θ_F = 74°C. **Ans.**

From definition, heating time constant is the time at which temperature rise is 0.632 times of final steady state temperature, *i.e.*, 0.632 × 74 = 46.7°C.

From Fig. 1.118, time corresponding to temperature rise of 46.7°C is 2 hours. Hence heating time constant is 2 hours. **Ans.**

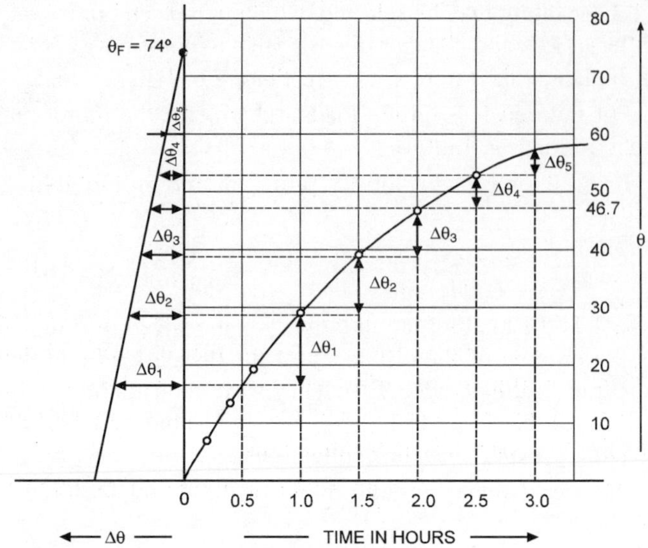

Fig. 1.118

Example 1.47. The full-load temperature rise of totally enclosed motor is 20°C after one hour and 34°C after two hours. Find (*i*) final steady-temperature rise on full load, (*ii*) the heating time constant, (*iii*) one hour rating of the motor for the same final temperature rise. Take no-load losses = copper losses at rated continuous load.

Solution : From Eq. (1.97)

$$\theta = \theta_F(1 - e^{-t/\tau}) \quad ...(i)$$

Now from given datas when t = 1 hour, θ = 20° C and when t = 2 hours, θ = 34° C. Substituting the above values of t and θ in Eq. (*i*) we get

$$20 = \theta_F(1 - e^{-1/\tau}) \quad ...(ii)$$

$$\text{and } 34 = \theta_F(1 - e^{-2/\tau}) \quad ...(iii)$$

Dividing Eq. (*iii*) by Eq. (*ii*) we get

$$\frac{34}{20} = \frac{1 - e^{-2/\tau}}{1 - e^{-1/\tau}}$$

$$\text{or } 1 + e^{-1/\tau} = 1.7$$

$$\text{or } e^{-1/\tau} = 0.7$$

$$\text{or Heating time constant, } \tau = \frac{-1}{\log_e 0.7} = 2.8 \text{ hours } \textbf{Ans.}$$

Substituting τ = 2.8 hours in Eq. (*ii*) we get

$$\text{Final steady temperature rise, } \theta_F = \frac{20}{1 - e^{-1/2.8}} = 66.67° \text{ C } \textbf{Ans.}$$

If temperature rise after one hour is 66.67°, final temperature rise θ_F will be given as,

$$\theta'_F = \frac{\theta_F}{1 - e^{-t/\tau}} = \frac{66.67}{1 - e^{-1/2.8}} = 222° \text{ C}$$

Total losses at full load

= Iron loss + copper loss = 2 × copper loss

∵ Iron loss = no-load loss = full-load copper loss

= 2P_c where P_c is full-load copper loss

Let one hour rating be x times full load then

Total losses = Iron losses + $x^2 P_c$

$= P_c + x^2 P_c = (1 + x^2) P_c$

$$\therefore \quad \frac{\text{Total losses at one-hour rating}}{\text{Total loss at full-load}} = \frac{\theta_F'}{\theta_F}$$

$$\text{or} \quad \frac{(1+x^2)P_c}{2P_c} = \frac{222}{66.67}$$

$$\text{or} \quad x = 2.38$$

So one-hour rating of the motor is 2.38 times full-load rating. **Ans.**

Example 1.48. Determine the half-hour rating of a 25 kW motor having a time constant of 1.5 hours. Assume that the motor cools down completely between each load period.

[*U.P. Technical Univ. Electric Drives 2005-06*]

Solution : Heating time constant, $\tau = 1.5$ hours

Let P kW be the half-hour rating, θ_F be the final temperature rise in case of continuous rating and W be full-load copper loss.

Final temperature rise in case of half-hour loading

$$\theta_F' = \frac{\theta_F}{1-e^{-t/\tau}} = \frac{\theta_F}{1-e^{-0.5/1.5}} = \frac{\theta_F}{1-e^{-1/3}} = 3.528\,\theta_F$$

Losses at 25 kW

$= 0.9W + W = 1.9$ W assuming iron loss to be 90 per cent of copper loss at full load

$$\text{Losses at P kW} = 0.9\text{ W} + \left(\frac{P}{25}\right)^2 W = W\left(0.9 + \frac{P^2}{625}\right)$$

Since copper loss varies as the square of the load

$$\frac{\text{Total losses at P kW}}{\text{Total losses at 25 kW}} = \frac{\theta_F'}{\theta_F}$$

$$\text{or} \quad \frac{W\left(0.9+\frac{P^2}{625}\right)}{1.9\text{ W}} = 3.528$$

or $P = \sqrt{(3.528 \times 1.9 - 0.9) \times 625} = 60.2$ kW say, 60 kW **Ans.**

Example 1.49. A motor running continuously on full load has a temperature rise of 20 °C. The heating time constant is 60 minutes. How long the motor should be run at twice the continuously rated output without overheating ? The maximum efficiency of motor at full load is 80%.

[*Pb. Technical Univ. Electric Drives and Utilization May 2008*]

Solution : Heating time constant, $\tau = 60$ minutes

Final temperature rise, $\theta_F = 20$°C in case of continuous rating

Let P kW be the continuous rating, and t be the time duration in minutes for which motor can be run at twice the continuously rated output without overheating.

Since maximum efficiency occurs 80% of full-load *i.e.*, 0.8 P, therefore at 0.8 P load copper loss is equal to constant loss, let each be equal to W watts.

$$\text{Total losses at P kW} = W + \left(\frac{P}{0.8P}\right)^2 W = 2.5625\text{ W}$$

$$\text{Total losses at 2P kW} = W + \left(\frac{2P}{0.8P}\right)^2 W = 7.25\text{ W}$$

Since copper loss varies as the square of load

$$\frac{\text{Total losses at 2P kW}}{\text{Total losses at P kW}} = \frac{\theta_F'}{\theta_F}$$

$$\text{or} \quad \theta_F' = \theta_F \times \frac{\text{Total losses at 2P kW}}{\text{Total losses at P kW}}$$

$$= 20 \times \frac{7.25}{2.5625} = 56.585\text{ °C}$$

$$\text{Also} \quad \theta_F' = \frac{\theta_F}{1-e^{-t/\tau}}$$

$$\text{or } t = \tau \log_e \frac{1}{1-\frac{\theta_F}{\theta_F'}} = 60 \log_e \frac{1}{1-\frac{20}{56.585}}$$

= 26.17 minutes say, 26 minutes **Ans.**

Example 1.50. If one hour rating of a machine is $\sqrt{3}$ times the continuous rating and if the steady temperature rise for one hour rating is twice for normal load, find the ratio of iron losses to I^2R losses at full load.

Solution : Losses on continuous rating $= P_c + P_i$

Losses on one-hour rating $= x^2 P_c + P_i = 3P_c + P_i$

$\because$ one hour rating is $\sqrt{3}$ times the continuous rating *i.e.*, $x = \sqrt{3}$

Ratio of final steady temperature rises with 1 hour and continuous rating is given as

$$\frac{\theta_F'}{\theta_F} = \frac{P_i + x^2 P_c}{P_i + P_c}$$

$$\text{or } 2 = \frac{3P_c + P_i}{P_c + P_i}$$

$$\text{or } P_i = P_c$$

Thus iron losses are equal to full-load copper losses at full load. **Ans.**

Example 1.51. The 15 minute rating of a motor is 400 watts. The heating time constant is 60 minutes. Determine the continuous rating of the motor if the maximum efficiency occurs at 80 per cent of full load.

[*A.M.I.E. Sec. B. Electric Drives and Their Control Winter 1996*]

Solution : Since maximum efficiency occurs at 80% of full load so at 80% of full-load, iron loss is equal to copper loss, let each equal to P_c watts

$$\text{Losses on full load} = P_c + \left(\frac{1}{0.8}\right)^2 P_c = 2.5625\,P_c$$

$$\text{Losses at load of 400 watts} = P_c + \left(\frac{400}{0.8 \times \text{full load}}\right)^2 P_c$$

Ratio of final temperature rises with 15 minute rating and continuous rating is given as,

$$\frac{\theta_F'}{\theta_F} = \frac{1}{1-e^{-t/\tau}} = \frac{\text{Total loss on 15 - minute rating}}{\text{Total loss on full load}}$$

$$\text{or} \quad \frac{1}{1-e^{-15/60}} = \frac{P_c + \left(\frac{400}{0.8 \times \text{full load}}\right)^2 P_c}{2.5625\,P_c}$$

or Continuous rating = 153.7 watts **Ans.**

Example 1.52. A motor has a thermal time constant of 45 minutes. When the motor runs continuously on full load, its final temperature rise is 80° C. (*i*) What is the temperature rise after 1 hour if the motor runs continuously on full load? (*ii*) If the temperature on one-hour rating is 80° C, find the maximum steady-state temperature at this rating. (*iii*) How much time does the motor take for its temperature to rise from 50° C to 80° C if it is working at its 1 hour rating ?

[*A.M.I.E. Sec. B. Electric Drives and Their Control Winter 1993*]

Solution : (*i*) From given datas :

Final temperature rise, $\theta_F = 80$° C

Thermal time constant, τ = 45 minutes = 0.75 hour

Temperature rise after 1 hour (*i.e* when t = 1) can be had by substituting $\theta_F = 80°$ C; τ = 0.75 hour and t = 1 hour in Eq. (1.97)

$$i.e.,\ \theta = \theta_F\,(1 - e^{-t/\tau}) = 80\,(1 - e^{-1/0.75}) = 58.9°\text{ C } \textbf{Ans.}$$

(*ii*) Let the maximum steady-temperature rise at one hour rating be θ'_F

Now from given datas τ = 0.75 hour, θ = 80° C when t = 1 hour

$$\therefore\ \theta'_F = \frac{\theta_F}{1-e^{-t/\tau}} = \frac{80}{1-e^{-1/0.75}} = 108.64°\text{ C } \textbf{Ans.}$$

(*iii*) Time taken to attain temperature of 80° C (assuming initial temperature as 0° C) in one hour or 60 minutes,

Final temperature rise, θ'_F = 108.64° C

Now let the time taken to attain temperature of 50° C from 0° C be t hours then

$$50 = 108.64\,(1 - e^{-t/0.75})$$

or t = 0.4625 hour or 27.75 minutes

Hence time taken to increase temperature from 50° C to 80° C = 60 – 27.75 = 32.25 minutes **Ans.**

Example 1.53. A motor has a continuous rating of 100 kW. The heating and cooling time constants are 50 and 70 minutes respectively. The motor has a maximum efficiency at 80% of full-load and is employed in an intermittent periodic load cycle consisting of a load period of 10 minutes followed by no-load period of 10 minutes. Calculate the value of load in kW during the load period. [*U.P. Technical Univ. Electric Drives 2009-10*]

Solution : Heating time constant, τ = 50 minutes

Cooling time constant, τ' = 70 minutes

On-load period, t_1 = 10 minutes

Off-load period, t_2 = 10 minutes

From Eq. (1.101)

$$\frac{\theta'_F}{\theta_F} = \frac{1-e^{-\left(\frac{t_1}{\tau}+\frac{t_2}{\tau'}\right)}}{1-e^{-t_1/\tau}} = \frac{1-e^{-\left(\frac{10}{50}+\frac{10}{70}\right)}}{1-e^{-10/50}} = \frac{1-0.70974}{1-0.81873} = 1.6$$

Let P be rating of motor when subjected to a duty cycle of 10 minutes on certain load followed by 10 minutes on no load.

Since maximum efficiency occurs at 80 per cent of full load *i.e.* at a load of 80 kW, therefore, at 80 kW load copper loss is equal to constant loss, let each be equal to W watts.

$$\text{Losses at 100 kW load} = W + \left(\frac{100}{80}\right)^2 W = 2.5625\ W$$

$$\text{Losses at P kW load} = W + \left(\frac{P}{80}\right)^2 W = W\left(1+\frac{P^2}{6{,}400}\right)$$

$$\text{Thus}\quad \frac{\text{Total losses at P kW output}}{\text{Total losses at 100 kW output}} = \frac{\theta'_F}{\theta_F}$$

$$\text{So}\quad \frac{W\left(1+\frac{P^2}{6{,}400}\right)}{2.5625\ W} = 1.6$$

$$\text{or } P = \sqrt{(2.5625\times1.6-1)\times6{,}400} = 140.85\text{ kW } \textbf{Ans.}$$

Example 1.54. The temperature rise of a motor after operating for 30 minutes on full load is 20°C. After another 30 minutes on the same load the temperature rise becomes 30°C. Assuming that the temperature increases according to an exponential law, determine the final temperature rise and the heating time constant. [*U.P. Technical Univ. Electric Drives 2007-08*]

Solution : From Eq. (1.97)

$$\theta = \theta_F(1-e^{-t/\tau})$$

Now from given datas when t_1 = 30 minutes, θ_1 = 20°C and when t_2 = 30 + 30 = 60 minutes, θ_2 = 30°C. Substituting the values for t and θ in above equation, we have

$$20 = \theta_F(1-e^{-30/\tau}) \quad ...(i)$$

$$30 = \theta_F(1-e^{-60/\tau}) \quad ...(ii)$$

Dividing Eq. (*ii*) by Eq. (*i*), we have

$$\frac{1-e^{-60/\tau}}{1-e^{-30/\tau}} = \frac{30}{20} = 1.5$$

$$\text{or } 1+e^{-30/\tau} = 1.5$$

$$\text{or Heating time constant, } \tau = \frac{-30}{\log_e 0.5} = 43.28\text{ minutes}$$

Substituting τ = 43.28 minutes in Eq. (*i*), we have

$$\theta_F = \frac{20}{1-e^{-30/43.28}} = 40\ °\text{C } \textbf{Ans.}$$

Example 1.55. Calculate the maximum overload that can be carried by a 25 kW motor if the temperature rise is not to exceed 50°C afte: one hour on overload. The temperature rise on full load after one hour is 30°C and after 2 hours is 40°C. The losses vary as square of the load.

Solution : The expression for temperature rise is given as

$$\theta = \theta_F(1 - e^{-t/\tau}) \quad ...(i)$$

From given datas when t = 1 hour, θ = 30°C and when t = 2 hours, θ = 40°C.

Substituting these values of t and θ above expression (*i*), we have

$$30 = \theta_F(1 - e^{-1/\tau}) \quad ...(ii)$$

$$\text{and } 40 = \theta_F(1 - e^{-2/\tau}) \quad ...(iii)$$

Solving above Eqs. (*ii*) and (*iii*), we have

$$e^{-1/\tau} = \frac{40}{30} - 1 = \frac{1}{3}$$

Substituting $e^{-1/\tau} = \frac{1}{3}$ in Eq. (*ii*), we have

$$\theta_F = \frac{30}{1-\frac{1}{3}} = 45°\text{C}$$

If temperature rise after 1 hour is not to exceed 50°, final temperature rise, θ'_F will be given as

$$\theta'_F = \frac{50}{1-e^{-1/\tau}} = \frac{50}{1-\frac{1}{3}} = 75°\text{C} \quad \because t = 1 \text{ and } e^{-1/\tau} = \frac{1}{3}$$

Let the maximum overload that can be carried by a 25 kW motor be P kW.

Since losses vary as the square of load

$$\frac{\theta'_F}{\theta_F} = \frac{\text{Losses at P kW}}{\text{Original losses}}$$

$$\frac{75}{45} = \left(\frac{P}{25}\right)^2$$

$$\text{or } P = 25\sqrt{75/45} = 32.3\text{ kW } \textbf{Ans.}$$

Example 1.56. A totally enclosed motor has a temperature rise of 20°C after half an hour and 35°C after 1 hour on full load. Determine the temperature rise after 2 hours on load. If a fan is fitted driving cool air over the external surface of the machine which reduces the final steady-temperature rise of the machine to 80 per cent of the original value, determine the temperature rise from cold after 1 hour on full load.

Solution : From Eq. (1.97)

$$\theta = \theta_F (1 - e^{-t/\tau}) \quad ...(i)$$

Now from given datas when $t = \frac{1}{2}$ hour, $\theta = 20°$ C and when $t = 1$ hour, $\theta = 35°$ C

Substituting the above values for t and θ in Eq. (i) we get

$$20 = \theta_F (1 - e^{-0.5/\tau}) \quad ...(ii)$$

$$35 = \theta_F (1 - e^{-1/\tau}) \quad ...(iii)$$

Dividing Eq. (iii) by Eq. (ii) we get

$$\frac{1 - e^{-1/\tau}}{1 - e^{-0.5/\tau}} = \frac{35}{20} = 1.75$$

or $1 + e^{-0.5/\tau} = 1.75$

$$\text{or } \tau = \frac{-0.5}{\log_e 0.75} = 1.738 \text{ hours}$$

Substituting $e^{-0.5/\tau} = 1.75 - 1 = 0.75$ in Eq. (ii) we get

$$\theta_F = \frac{20}{1 - 0.75} = 80°\text{C}$$

Temperature rise after 2 hours,

$$\theta' = \theta_F (1 - e^{-t/\tau}) = 80(1 - e^{-2/1.738}) = 54.7° \text{ C } \textbf{Ans.}$$

When a fan is fitted driving cooling air over the external surface of the machine, coefficient of cooling is increased resulting in decrease in steady final temperature rise θ_F and heating time constant τ. Let the new coefficient of cooling, steady final temperature rise and heating time constant be α_1, θ_{F_1} and τ_1 respectively.

$$\theta_{F_1} = 0.8\theta_F$$

$$\text{Since } \theta_{F_1} = \frac{P}{S\alpha}$$

$$\text{So } \frac{\alpha_1}{\alpha} = \frac{\theta_F}{\theta_{F_1}}$$

$$\text{or } \alpha_1 = \frac{\theta_F}{\theta_{F_1}} \alpha = \frac{\alpha}{0.8} = 1.25\,\alpha$$

Heating time constant is given as

$$\tau = \frac{mC_p}{S\alpha}$$

$$\text{So } \frac{\tau_1}{\tau} = \frac{\alpha}{\alpha_1}$$

$$\text{or } \tau_1 = \frac{\alpha}{1.25\alpha} \tau = 0.8\,\tau = 0.8 \times 1.738 = 1.39 \text{ hours}$$

$$\theta_{F_1} = 0.8\,\theta_F = 0.8 \times 80 = 64° \text{ C}$$

Temperature rise from cold after 1 hour at full load

$$\theta_1 = \theta_{F_1} (1 - e^{-t/\tau_1}) = 64(1 - e^{-1/1.39}) = 32.8° \text{ C } \textbf{Ans.}$$

Example 1.57. A certain motor has to perform the following duty cycle: 100 kW for 10 minutes, no load for 5 minutes, 50 kW for 8 minutes and no load for 4 minutes. The duty cycle is repeated indefinitely. Draw the curve for the load cycle. Assuming that the heating is proportional to the square of the load, determine the suitable size of a continuously rated motor.

Solution : The load cycle is illustrated in Fig. 1.119.

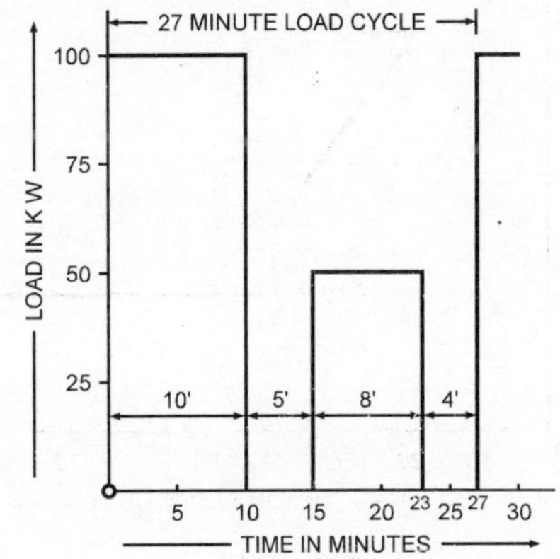

Fig. 1.119

Since heating is proportional to the square of the load,

Rating of continuously rated motor = RMS kW

$$= \sqrt{\frac{(100)^2 \times 10 + (50)^2 \times 8}{27}}$$

$$= 66.67 \text{ kW}$$

i.e. A motor of 70 kW will be adequate.

Example 1.58. An electric motor has load variation as given below: Torque 240 N-m for 20 minutes, 140 N-m for 10 minutes, 300 N-m for 10 minutes, 200 N-m for 20 minutes. If the speed of the motor is 720 rpm, find the power rating of motor.

[A.M.I.E. Sec. B. Electric Drives and Their Control Winter 1993]

Solution : Equivalent load torque, from Eq. (1.108),

$$T_{eq} = \sqrt{\frac{T_1^2 t_1 + T_2^2 t_2 + T_3^2 t_3 + T_n^2 t_n}{t_1 + t_2 + t_3 + t_n}}$$

$$= \sqrt{\frac{(240)^2 \times 20 + (140)^2 \times 10 + (300)^2 \times 10 + (200)^2 \times 20}{20 + 10 + 10 + 20}}$$

$$= 225.39 \text{ Nm}$$

Speed of motor N = 720 rpm

Power rating of motor,

$$P = T_{eq} \times \frac{2\pi N}{60} = 225.39 \times 2\pi \times \frac{720}{60}$$

$$= 16{,}994 \text{ Watts or } 17 \text{ kW } \textbf{Ans.}$$

Example 1.59. Based on the rms torque, estimate the kW rating of a 750 rpm motor used for driving an equipment having the following load torque curve :

(i) For first 10 seconds, the torque is constant at 40 kg-m;

(ii) For next 30 seconds the torque varies linearly with time from 35 kg-m to 15 kg-m;

(iii) For the last 50 seconds, the torque is constant and equal to 10 kg-m.

[A.M.I.E. Sec B. Electric Drives and Their Control Summer 2000]

Solution : The load cycle is shown in Fig. 1.120.

$$(T_{eq})^2 \times \text{Time of cycle} = \int T^2\, dt \text{ over the cycle}$$

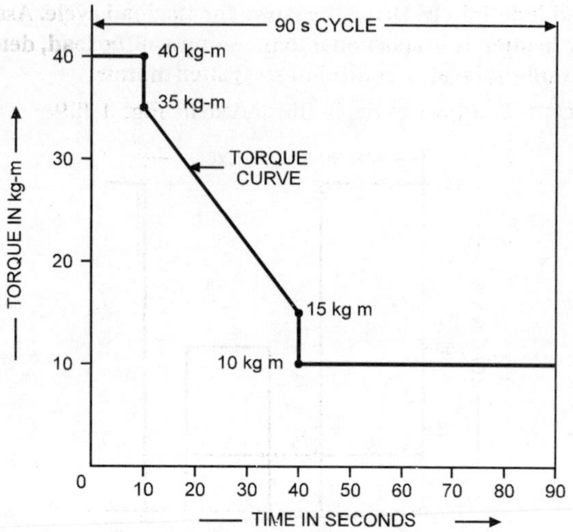

Fig. 1.120

$$= \int_0^{10} 40^2\,dt + \int_{10}^{40}\left[35 - \frac{(35-15)}{30}(t-10)\right]^2 dt + \int_{40}^{90} 10^2\,dt$$

$$= \int_0^{10} 1{,}600\,dt + \int_{10}^{40}\left[1{,}225 - \frac{140}{3}(t-10) + \frac{4}{9}(t-10)^2\right]dt + \int_{40}^{90} 100\,dt$$

$$= \left[1{,}600t\right]_0^{10} + \left[1{,}225t - \frac{140}{6}(t-10)^2 + \frac{4}{27}(t-10)^3\right]_{10}^{40} + \left[100t\right]_{40}^{90}$$

$$= 16{,}000 + 1{,}225 \times 30 - \frac{70}{3}\times 900 + \frac{4}{27} \times 27{,}000 + 5{,}000$$

$$= 16{,}000 + 36{,}750 - 21{,}000 + 4{,}000 + 5{,}000 = 40{,}750$$

or $T_{eq} = \sqrt{\dfrac{40{,}750}{90}} = 21.278$ kg-m

Power rating of motor,

$$P = T_{eq} \text{ in kg-m} \times 9.81 \times \frac{2\pi N}{60} = 21.278 \times 9.81 \times \frac{2\pi \times 750}{60}$$

$$= 16{,}400 \text{ W or } 16.4 \text{ kW } \textbf{Ans.}$$

Example 1.60. A motor driving a colliery winder has the following duty cycle :

(*i*) Acceleration period 0–15 s – Load rising uniformly from zero to 1,050 hp

(*ii*) Full-speed period 15–85 s – Load constant at 600 hp

(*iii*) Deceleration period 85–95 s – HP returned 200 to zero uniformly

(*iv*) Idle period 95–120 s – Motor stationary.

Estimate continuous rating of the motor for this load cycle.

[U.P. Technical Univ. Electric Drives 2008-09]

Solution : The load cycle is shown in Fig. 1.121.

Assuming losses proportional to $(hp)^2$, we have

(HP rating of motor)^2 × cycle duration

$= \int (\text{HP})^2\,dt$ over the complete cycle

$$= \int_0^{15}\left(\frac{1{,}050}{15}t\right)^2 dt + \int_{15}^{85} 600^2\,dt + \int_{85}^{95}\left[\frac{200}{95-85}(t-85)\right]^2 dt$$

$$= \left[\frac{4{,}900t^3}{3}\right]_0^{15} + \left[360{,}000t\right]_{15}^{85} + \left[\frac{400(t-85)^3}{3}\right]_{85}^{95}$$

$$= 5{,}512{,}500 + 25{,}200{,}000 + 133{,}333 = 30{,}845{,}833$$

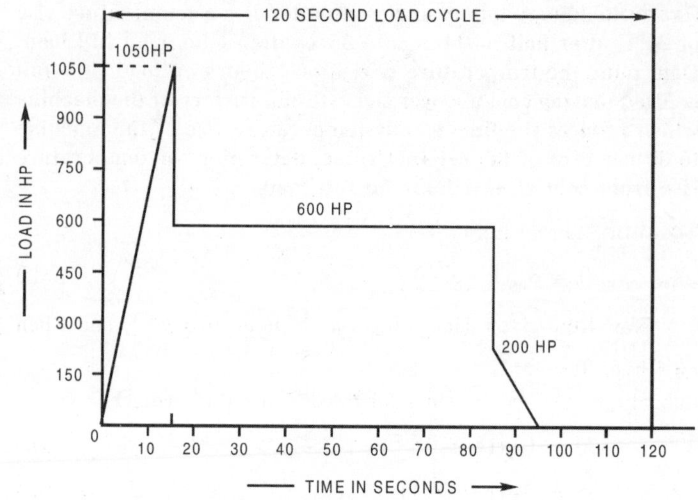

Fig. 1.121

or HP rating of motor $= \sqrt{\dfrac{30{,}845{,}833}{120}} = 507$ hp **Ans.**

Hence the rating of continuously rated motor on the basis of rms value should be 507 hp in order to supply the above load without exceeding the specified temperature.

Note : During regenerative braking, even though the power is returned to the line, the machine will be carrying current. So far as heating is concerned, it is immaterial whether machine is drawing current or returning it to the line.

1.34. METHODS OF VENTILATION AND COOLING OF MACHINES

It is necessary to provide suitable ventilation and cooling for the machines so that the temperature rise at any part of the machine does not exceed the permissible limit governed by the type of insulation employed in the construction. The cooling of electrical machine by means of an air stream is called *ventilation of the machine.*

The circulation of air in the machine can be arranged by the use of a built-in fan or fan with a separate drive. Either a centrifugal type fan or an axial type fan is employed for rotating electrical machines. A centrifugal fan forces air from the centre to flow outwards and operates the same way irrespective of the direction of rotation of the machine. An axial or propeller type fan moves the air in the opposite direction when the direction of rotation of the machine reverses.

According to IS : 4722-1968, the cooling system for rotating electrical machines are classified into three types depending up the origin of cooling.

1. *Natural Cooling.* The machine is cooled by air movements in the machine due to its rotation or due to temperature difference between inside parts and the outside temperature of the air. The machine thus is cooled without the use of a fan, by the movement of air and radiation. This method of cooling is employed only in small fractional kW machines with rating within several hundred watts, since the condition of cooling are comparatively favourable.

2. *Self Cooling.* The machine is cooled by air, blown by fan integrally built with the motor or mounted on the shaft.

3. ***Separate Cooling.*** The machine is cooled either by a fan not driven by its shaft, or it is cooled by a cooling medium other than air put into motion by means not belonging to machine.

According to the manner of cooling the ventilation systems are classified as

1. ***Open-Circuit Ventilation.*** The heat is given up directly to the cooling air flowing through the machine which is being continuously replaced. The open-circuit ventilation can be further divided into two types in accordance with how the air is brought into the machine.

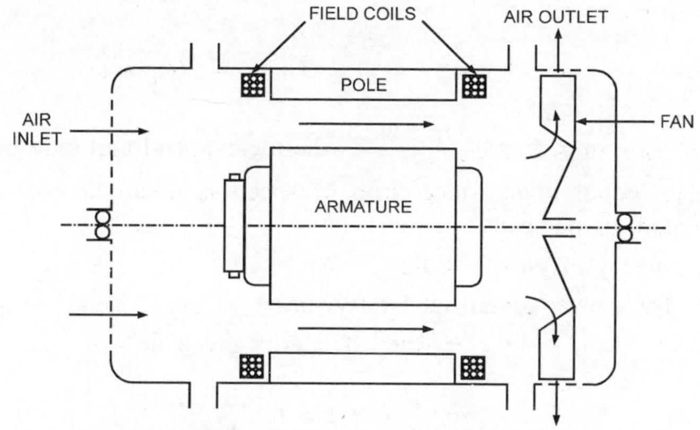

Fig. 1.122. ***Induced Ventilation in Electrical Machine***

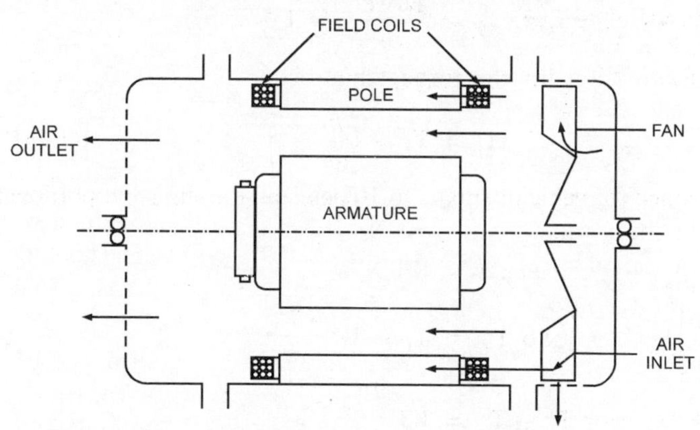

Fig. 1.123. ***Forced Ventilation in Electrical Machine***

(a) Induced Ventilation. Induced ventilation method used in an electrical machine is illustrated in Fig. 1.122. In this arrangement, a fan produces a reduced pressure of air inside the machine and air from outside is sucked into the machine owing to the atmospheric pressure. The air is circulated through the machine and then pushed out by the fan into the atmosphere. The fan is normally arranged inside the machine, as illustrated in the figure. The fan can be provided externally also.

(b) Forced Ventilation. In this method of ventilation, illustrated in Fig. 1.123, the fan sucks the air from outside and forces it into the machine and finally pushed out into the atmosphere. For forced ventilation, the pressure of air inside the machine is greater than the atmospheric pressure outside. The fan may be arranged internally, as illustrated in the figure or externally,

2. ***Closed Circuit Ventilation.*** In this system of cooling, heat is transferred to the cooling medium such as air or hydrogen through an intermediate cooling medium circulating in a closed circuit between the interior of the machine and the cooling device, *e.g.*, a water cooler can be used to cool the cooling medium such as air.

For increasing the surface of contact for cooler, stator and rotor cores of electrical machines are provided with ducts. These ducts may be radial or axial type depending on the direction of air flow through them.

3. Surface Ventilation. The heat is transferred from inside of the machine, by cooling medium, to the external surface of a totally enclosed machine. The external surface is being cooled by natural means or mainly by air blown by fan.

1.35. FRAME SIZE

A frame is the mechanical structure required to house a stator along with its bearings, end covers and terminal box. The standardization of frame size leads to economy. All modern motors of small size are built with standard frame sizes as specified in IEC 72 which lists a coherent range of main structural dimensions with centre heights between 56 and 1,000 mm.

A frame size is designated by a number which is its centre height (H) expressed in mm. Thus frame designated 132 has a centre height-(H) of 132 mm. The frame sizes with heights between 56 and 1,000 mm as per recommendations of IEC are 56, 63, 71, 80, 90, 100, 112, 132, 160, 180, 200, 225, 250, 280, 315, 355, 400, 450, 500, 630, 710, 800, 900, 1,000 mm.

The outputs from the standard frames are periodically assessed to take into account the latest technological advancements.

The frames may be diecast or fabricated. Machines of ratings up to 50 kW have their frames diecast in a strong silicon aluminium alloy and in some cases with the stator core cast in. The frames of larger machines are fabricated by welding steel plates. The advantage of fabrication is its adaptability to new designs and modifications.

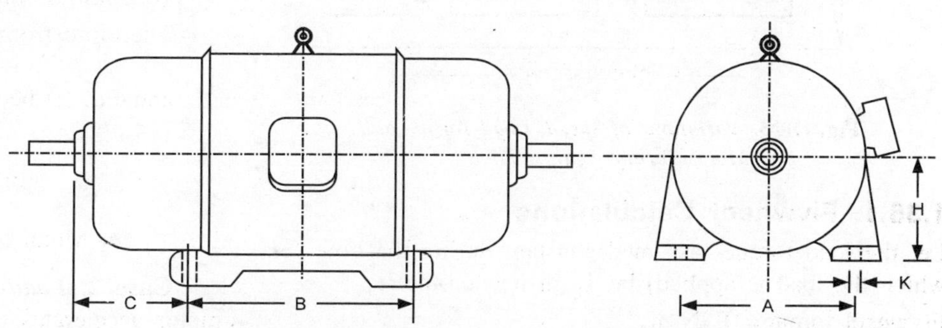

Fig. 1.124. ***Structural Dimensions of Standard Frames***

1.36. LOAD EQUALIZATION

In many industrial drives, such as in rolling mills, planning machines, electric hammers, presses, reciprocating pumps, the load fluctuates over a wide range in the space of a few seconds. It is desirable to smoothen out the fluctuations in load, otherwise during interval of peak load, it will draw a heavy current from the supply either producing large voltage drop in the distribution system or requiring cables and wires of heavy section. The process of smoothing out of fluctuations in the load is known as *load equalization.* In this process, energy is stored during the interval of light load and given out during the interval of peak load. Thus the power drawn from the supply mains remains almost constant.

The most common method of load equalization is by use of flywheel. During the light load period the flywheel accelerates and stores the excessive energy drawn from the supply and during peak load period the flywheel decelerates and supplies some of its stored energy to the load in addition to the energy supplied from the supply. Thus the load demand is reduced.

The motors used for such load should have drooping speed-torque characteristics, so that the speed may fall with the increase in load and enables the flywheel to give up its stored energy. For the load in which the motor have to run in the same direction and is not to be stopped and started frequently, flywheel may be mounted on the motor shaft. For a reversing drive, such as for colliery winder, the Ward Leonard control system is generally used for reversing and speed control, so flywheel can be mounted on the shaft of the motor-generator set. The load torque required and motor torque developed as well as speed variations with time are shown in Fig. 1.125.

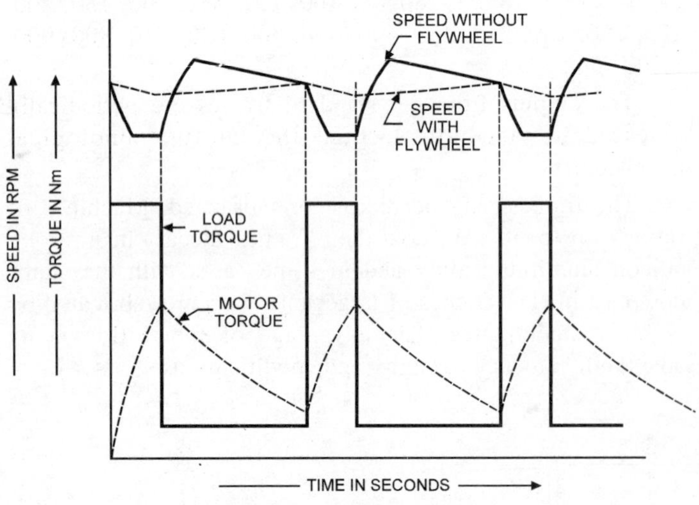

Fig. 1.125. ***Variations of Speed, Load Torque and Motor Torque Against Time***

1.36.1. Flywheel Calculations

Let the load torque (assumed constant during the time for which the load is applied) be T_L in newton-metres.

Flywheel torque $= T_F$ N-m

No-load torque $= T_0$ N-m (say, due to friction and windage etc.)

Motor torque at any instant $= T_m$ N-m

Motor speed on no load $= \omega_0$ radians/s

Motor speed at any instant $= \omega$ radians/s

Motor slip, $s = \omega_0 - \omega$

Moment of inertia of flywheel $=$ J kg-m^2

Time $= t$ seconds.

Case 1. ***Load Increasing.*** During this period the flywheel decelerates and gives up a part of stored energy in it. The torque required to be supplied by the motor, $T_m = T_L - T_F$.

Energy given out by flywheel when its speed is reduced from ω_0 to ω

$$= \frac{1}{2}J(\omega_0^2 - \omega^2) = J\frac{\omega_0 + \omega}{2}(\omega_0 - \omega)$$

but $\omega_0 - \omega = s$ and $\frac{\omega_0 + \omega}{2}$ is the mean speed and may be taken equal to ω, since drop of speed is assumed not to exceed 10 per cent.

$\therefore$ Energy given out $= J\omega s$

The power given out by flywheel

$= $ Rate of energy given up

$$= \frac{d}{dt}(J\omega s) = J\omega\frac{ds}{dt}$$

$$\text{Flywheel torque, } T_F = \frac{\text{Power}}{\omega} = J\frac{ds}{dt}$$

Expression for the torque thus becomes

$$T_m = T_L - J\frac{ds}{dt} \quad \text{...(1.110)}$$

Since for value of slip up to 10 per cent, the slip is proportional to the torque,

i.e., $s = KT_m$

$$\text{So } T_m = T_L - JK\frac{dT_m}{dt}$$

$$\text{or } T_L - T_m = KJ\frac{dT_m}{dt}$$

$$\text{or } \frac{dT_m}{T_L - T_m} = \frac{1}{KJ}dt$$

$$\text{or } -\log_e(T_L - T_m) = \frac{1}{KJ}t + C_1 \quad \text{...}(i)$$

From initial conditions when $t = 0$, $T_m = T_0$

Therefore from Eq. (*i*) $C_1 = -\log_e(T_L - T_0)$

$$\text{and Eq. } (i) \text{ becomes } \frac{-1}{KJ}t = \log_e\frac{T_L - T_m}{T_L - T_0}$$

$$\text{or } \frac{T_L - T_m}{T_L - T_0} = e^{-t/KJ}$$

$$\text{or Motor torque, } T_m = T_L - (T_L - T_0)e^{-t/KJ} \quad \text{...(1.111)}$$

Case 2. ***Load Decreasing.*** While load decreases, the motor accelerates the flywheel to its normal speed, slip is therefore decreased and $\frac{ds}{dt}$ is –ve.

Motor torque, $T_m = T_0 + T_F$

$$\text{or } T_m = T_0 - J\frac{ds}{dt} \text{ since } \frac{ds}{dt} \text{ is negative}$$

By solving the above expression, we get

$$T_m = T_0 + (T'_m - T_0)e^{-t/KJ} \quad ...(1.112)$$

where T'_m is the motor torque at the instant when the load is removed.

Example 1.61. A 4-pole. 50 Hz induction motor has a flywheel on its shaft. Total inertia at the motor shaft is 1,000 kg-m². Load torque is 100 kg-m for 10 seconds followed by a no-load period long enough for the flywheel to regain its full speed. Motor has a slip of 6% at a torque of 50 kg-m. Calculate the speed at the end of deceleration period. Assume motor speed-torque characteristic to be a straight line in the region of interest and neglect friction and windage.

Solution : Load torque, $T_L = 100 \text{ kg-m} = 100 \times 9.81 = 981$ N-m

Load period, $t = 10$ seconds

Moment of inertia, $J = 1{,}000$ kg-m^2

No-load torque, $T_0 = 0$

$$\text{Synchronous speed, } N_s = \frac{120f}{P} = \frac{120 \times 50}{4} = 1{,}500 \text{ rpm}$$

$$\text{Slip speed} = 0.06 \times 1{,}500 = 90 \text{ rpm or } 2\pi \times \frac{90}{60} = 9.42 \text{ rad/s}$$

$$K = \frac{\text{Slip speed}}{\text{Torque}} = \frac{9.42}{50 \times 9.81} = 0.0192$$

$$\therefore T_m = T_L - (T_L - T_0)\, e^{-t/KJ}$$

$$= 981 - (981 - 0)\, e^{\frac{-10}{0.0192 \times 1{,}000}} \qquad \because T_0 = 0$$

$$\text{or } T_m = 981(1 - e^{-0.52}) = 981\,(1 - 0.594) = 398 \text{ N-m}$$

$$\text{Slip speed} = K \times T_m = 0.0192 \times 398 \text{ rad/s}$$

$$= \frac{0.0192 \times 398}{2\pi} \times 60 = 73 \text{ rpm}$$

$\therefore$ Speed at the end of deceleration period

$= 1{,}500 - 73 = 1{,}427$ rpm **Ans.**

Example 1.62. The following data refers to a 500 hp rolling mill induction motor equipped with a flywheel.

No-load speed – 40 rpm; Slip at full load – 12%; load torque during actual rolling – 41,500 kg-m. Duration of each rolling period – 10 seconds.

Determine the inertia of the flywheel required in the above case to limit the motor torque to twice its full-load value. Neglect the no-load losses and assume that the rolling mill torque falls to zero between each rolling period. Assume that the motor slip is directly proportional to the torque.

[*Pb. Univ. Elec. Drives and Utilization of Elec. Energy Nov. 1980*]

Solution : Output of motor = 500 hp = 500 × 735.5 W

No-load speed, $N_0 = 40$ rpm

Full-load slip, $s_f = 12\%$ or 0.12

Full-load speed, $N_f = 40(1 - 0.12) = 35.2$ rpm

$$\text{Full-load torque, } T_f = \frac{500 \times 735.5}{2\pi \times \frac{35.2}{60}} = 99{,}766 \text{ N-m}$$

Motor torque, $T_m = 2 \times T_f$

$= 2 \times 99{,}766 = 199{,}532$ N-m

Load torque, $T_L = 41{,}500 \times 9.81 = 407{,}115$ N-m

No-load torque, $T_0 = 0$

Load period, $t = 10$ seconds

Slip speed = Full-load slip × no-load speed

$= 0.12 \times 40 = 4.8$ rpm

$$\text{or } \frac{2\pi \times 4.8}{60} = 0.5026 \text{ rad/s}$$

$$K = \frac{\text{Slip speed}}{\text{Torque}} = \frac{0.5026}{99{,}776} = 5.0383 \times 10^{-6}$$

$\because s \;\alpha$ torque

$$\because T_m = T_L - (T_L - T_0)\, e^{-t/KJ} = T_L(1 - e^{-t/KJ}) \qquad \because T_0 = 0$$

$$\therefore J = \frac{-t}{K \log_e\left(1 - \frac{T_m}{T_L}\right)}$$

$$= \frac{-10}{5.0383 \times 10^{-6} \log_e\left(1 - \frac{199{,}532}{407{,}115}\right)}$$

$= 2.947 \times 10^6$ kg-m^2 **Ans.**

Example 1.63. A 50 kW, 3-phase, 6-pole, 50 Hz induction motor fitted with a flywheel supplies a constant load torque of 300 N-m and at wide intervals an additional load torque of 1,500 N-m for 10 seconds. The full-load slip of the motor is 4%. Assuming that the stator losses are negligible and the motor operates on the linear portion of the torque-slip curve, determine :

(*i*) Weight of the flywheel if the radius of gyration is 0.9 m and the motor torque is not to exceed twice rated torque.

(*ii*) Time taken after removal of extra load before motor torque is 700 N-m.

Solution : Output of motor = 50 kW

$$\text{Synchronous speed, } N_s = \frac{120f}{P} = \frac{120 \times 50}{6} = 1{,}000 \text{ rpm}$$

Full-load speed of motor,

$$N_f = N_s(1 - s_f) = 1{,}000(1 - 0.04) = 960 \text{ rpm}$$

$$\text{Full-load torque, } T_f = \frac{50 \times 1{,}000}{2\pi \times \frac{960}{60}} = 497.36 \text{ N-m}$$

Motor torque, $T_m = 2 \times T_f = 2 \times 497.36 = 994.72$ N-m

Load torque, $T_L = 300 + 1{,}500 = 1{,}800$ N-m

No-load torque $T_0 = 300$ N-m

Load period, $t = 10$ seconds

$$\text{Slip speed} = \frac{0.04 \times 1{,}000 \times 2\pi}{60} = 4.1888 \text{ rad/s}$$

$$K = \frac{\text{Slip speed}}{T_f} = \frac{4.1888}{497.36} = 8.422 \times 10^{-3}$$

$$\because T_m = T_L - (T_L - T_0)\, e^{-t/KJ}$$

$$\therefore e^{-t/KJ} = \frac{T_L - T_m}{T_L - T_0} = \frac{1{,}800 - 994.72}{1{,}800 - 300} = 0.5368$$

$$\text{or } J = \frac{-t}{K \log_e 0.5368}$$

$$= \frac{-10}{8.422 \times 10^{-3} \log_e 0.5368} = 1{,}909 \text{ kg-m}^2$$

$$\text{Weight of flywheel} = \frac{J}{r^2} = \frac{1{,}909}{(0.9)^2} = 2{,}357 \text{ kg } \textbf{Ans.}$$

(*ii*) Let the time taken be t_1 seconds.

The expression for the motor torque is given as

$$T_m = T_0 + (T'_m - T_0)e^{-t_1/KJ}$$

$$\text{or } t_1 = KJ \log_e \frac{T'_m - T_0}{T_m - T_0}$$

$$= 8.422 \times 10^{-3} \times 1{,}909 \log_e \left(\frac{994.72 - 300}{700 - 300}\right)$$

$$= 8.876 \text{ seconds. } \textbf{Ans.}$$

∵ Motor torque at peak speed,

T'_m = 994.72 Nm and T_m = 700 N-m given

Example 1.64. A motor equipped with a flywheel has to supply a load torque of 600 N-m for 10 seconds followed by a no-load period long enough for the flywheel to regain its full speed. It is desired to limit the motor torque to 450 N-m. What should be the moment of inertia of the flywheel ? The no-load speed of the motor is 600 rpm and it has a slip of 8% at torque 400 N-m. Assume the motor speed-torque characteristic to be a straight line in the range of operation. Motor has an inertia of 10 kg-m².

[U.P. Technical Univ. Electric Drives 2009-10]

Solution : Full-load torque, T_L = 600 N-m

No-load torque, T_0 = 0

Motor torque, T_m = 450 N-m

$$\text{No-load speed, } \omega_0 = \frac{2\pi N}{60} = \frac{2\pi \times 600}{60} = 20\pi$$

$$\text{Slip speed} = 0.08 \times 20\pi = 1.6\pi$$

$$K = \frac{\text{Slip speed}}{\text{Torque}} = \frac{1.6\pi}{450} = 0.01117$$

From Eq. (1.111)

$$\text{Motor torque, } T_m = T_L - (T_L - T_0)\, e^{-t/KJ}$$

Substituting the values, we have

$$450 = 600 - (600 - 0)e^{\frac{-10}{0.01117J}}$$

$$\text{or } e^{\frac{-895}{J}} = 0.25$$

$$J = \frac{-895}{\log_e 0.25} = \frac{-895}{-1.3863} = 645.6 \text{ kg-m}^2 \textbf{ Ans.}$$

Moment of inertia of flywheel,

$$J_F = J - J_M = 645.6 - 10 = 635.6 \text{ kg-m}^2 \textbf{ Ans.}$$

Example 1.65. A 3-phase 100 kW, 6-pole, 960 rpm induction motor has a constant load of 600 N-m at wide intervals and is subjected to additional torque of 2,400 N-m for 10 seconds. Calculate the moment of inertia of the flywheel used for load equalization, if motor torque is not to exceed twice the rated torque. The moment of inertia of the motor is 200 kg-m².

[U.P. Technical Univ. Electric Drives 2005-06]

Solution : Output of motor, P = 100 kW

$$\text{Synchronous speed, } N_s = \frac{120f}{P} = \frac{120 \times 50}{6} = 1{,}000 \text{ rpm}$$

Full-load speed, N = 960 rpm

$$\text{Slip speed} = N_s - N = 1{,}000 - 960$$

$$= 40 \text{ rpm} = 40 \times \frac{2\pi}{60} = 4.1888 \text{ rad/s}$$

$$\text{Full-load torque, } T_F = \frac{100 \times 1{,}000}{2\pi \times \frac{960}{60}} = 994.72 \text{ N-m}$$

Motor torque, T_m = 2 × 994.72 = 1,989.44 N-m

Load torque, T_L = 600 + 2,400 = 3,000 N-m

No-load torque, T_0 = 600 N-m

$$K = \frac{\text{Slip speed}}{T_f} = \frac{4.1888}{994.72} = 0.004211$$

From Eq. (1.111), $T_m = T_L - (T_L - T_0)e^{-t/KJ}$

$$\text{or } J = \frac{-t}{K \log_e \frac{T_L - T_M}{T_L - T_0}}$$

$$= \frac{-10}{0.04211 \log_e \frac{3{,}000 - 1{,}989.44}{3{,}000 - 600}} = 2{,}745.5 \text{ kg-m}^2$$

Moment of inertia of flywheel,

$$J_F = J - J_M = 2{,}745.5 - 200 = 2{,}545.5 \text{ kg-m}^2 \textbf{ Ans.}$$

1.37. APPEARANCE

Pleasing appearance is a good aspect of design, which can be obtained by harmonious blending of the motor finish, mounting and dimensions in line with those of the driven machine.

1.38. COST

Although the cost has been placed as last factor in the above list, but it is the major factor for the choice of motor. Sometimes we have to choose a motor which may not be an ideal choice from engineering consideration but cheap in cost—because special features are not worth the additional investment. While considering the cost, the capital as well as running cost (cost of losses and cost due to maintenance, interest and depreciation) of a motor should be considered. In most of the cases, we find that equipment very cheap in first cost is costlier in running cost due to poor efficiency, low power factor and short life. In a nutshell the motor choosen should give the desired service at minimum overall cost.

As compared to plain squirrel cage 3-phase induction motors, which are the cheapest, the cost of dc motors range from 1.5 to 3 times its cost. The cost of synchronous motors, 3-phase slip-ring induction motors and commutator motors range from 1.5 to 4 times the cost of a plain squirrel cage induction motor of the same output. For single phase motors it is 1.5 to 2.5 times the cost of plain squirrel cage induction motor of the same size.

Exercises

1. What is an electric drive ? What are its advantages ?
[G.B. Technical Univ. Electric Drives 2010-11]
2. Discuss criteria for selection of electric motor for a particular duty. **[U.P. Technical Univ. Electric Drives 2005-06]**
3. Discuss various factors which are required to be considered for selecting a motor.
[Pb. Technical Univ. Electric Drives & Utilization Dec. 2006]
4. What are the disadvantages of dc drives due to which the three-phase induction motor drive is fastly replacing it ?
[G.B. Technical Univ. Electric Drives 2010-11]
5. Enlist the classification of electric drives and explain.
[G.B. Technical Univ. Electric Drives 2010-11]

6. Differentiate bewteen group drive and individual drive and give their merits.
[U.P. Technical Univ. Electric Drives 2005-06]

7. What is meant by "group drive" and "individual drive" ? Discuss their merits and demerits.
[U.P. Technical Univ. Electric Drives 2008-09]

8. Write advantages and disadvantages of group drives.
[Pb. Technical Univ. Electric Drives & Utilization Dec. 2006]

9. Explain the various types of loads. What are the components of the load torque ?
[G.B. Technical Univ. Electric Drives 2010-11]

10. What is meant by nature of load torque ? Explain difference between "passive load" and "active load" and give two examples of each category.
[U.P. Technical Univ. Electric Drives 2008-09]

11. Explain active load and passive load giving two examples for each. **[U.P. Technical Univ. Electric Drives 2005-06]**

12. How do you define passive load torques and active load torques ? What are the differences between the two ?
[U.P. Technical Univ. Electric Drives 2009-10]

13. What are the components of load torque in a drive ? How each component is dependent on speed ?
[U.P. Technical Univ. Electric Drives 2007-08]

14. Discuss the motoring and braking operations of an electric drive in both directions, forward and reverse taking the hoist as an example. **[A.M.I.E. Sec. B. Electric Drives & Their Control Summer 1999]**

15. Explain four quadrant operation of an electric drive.
[G.B. Technical Univ. Electric Drives 2009-10, 2010-11]

16. Describe four quadrant operation of a motor with illustrative figures for each. Show clearly the direction of motor torque, load torque and speed for each quadrant.
[U.P. Technical Univ. Electric Drives 2007-08]

17. A motor is driving a hoist load. Discuss four quadrant operation of the motor and show that in the speed-torque plane. Indicate the polarity of torque, speed and power in each quadrant. **[U.P. Technical Univ. Electric Drives 2004-05]**

18. Discuss factors which decide selection of electric drive for a particular industrial application.
[U.P. Technical Univ. Electric Drives 2008-09]

19. Classify various mechanical loads on the basis of their offered torque-speed characteristics.
[U.P. Technical Univ. Electric Drives 2005-06]

20. Give typical example for following types of loads encountered in a drive.
(*i*) Continuous constant load.
(*ii*) Continuous variable load.
(*iii*) Pulsating load.
(*iv*) Impact load.
(*v*) Short time intermittent loads.
[U.P. Technical Univ. Electric Drives 2007-08]

21. What are constant torque and constant power operations of an electric drive ?
[G.B. Technical Univ. Electric Drives 2010-11]

22. What is meant by "constant torque drive" and "constant power drive" ? Explain them and mention one application of each such drive. **[U.P. Technical Univ. Electric Drives 2005-06]**

23. Derive equation of motion and explain under what conditions motor will be accelerating, decelerating or at steady state. **[U.P. Technical Univ. Electric Drives 2008-09]**

24. Draw the torque-speed characteristics, giving examples with justification, of the following types of loads:
(*i*) Constant torque (*ii*) Constant power (*iii*) Torque directly proportional to speed.
[A.M.I.E. Sec. B. Electric Drives & Their Control Winter 1996]

25. A motor running at a speed of N rpm is driving a rotational load L_1 directly coupled to its shaft and another rotational load L_2 through a gear to reduce the speed by a factor K the inertia of motor, L_1 and L_2 are respectively J_m, J_1 and J_2. The load torque of L_1 and L_2 are τ_1 and τ_2 respectively. Derive an expression for effective inertia and load torque on the same motor in terms of the above parameters.
[U.P. Technical Univ. Electric Drives 2007-08]

26. Derive the necessary relationship between the variation of electromagnetic torque and load torque with respect to speed for a stable operation of an electric drive. Comment on the stability of the equilibrium points A, B, C and D in Fig. 1.
[G.B. Technical Univ. Electric Drives 2010-11]

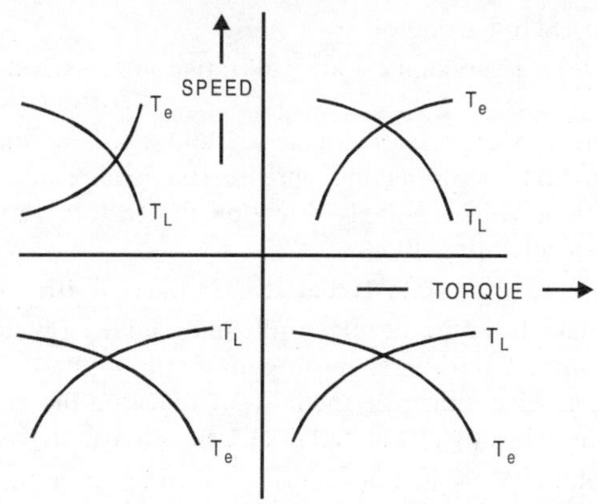

Fig. 1

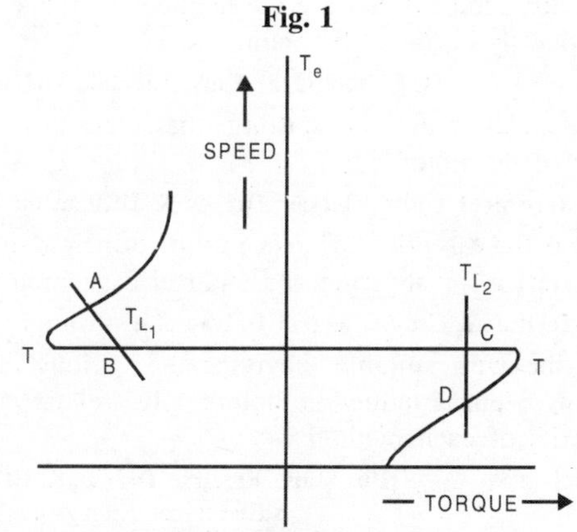

Fig. 2

27. Figure 2 shows plots of speed vs motor and load torques. Comment on the stability of operating points A, B, C and D (Fig. 2). **[U.P. Technical Univ. Electric Drives 2009-10]**

28. Give the characteristics of dc shunt motors. Why are such motors not suitable for traction purposes ?

29. Draw speed-torque characteristics of a dc series motor. What modifications are expected in above characteristics when armature reaction is considered ?
[A.M.I.E. Sec. B. Electric Drives & Their Control Summer 1998]

30. Explain that the characteristic of a dc series motor is suitable for traction application.
[A.M.I.E. Sec. B. Electric Drives & Their Control Winter 1993]

31. Define steady state stability of electric drive. Derive a condition for steady state stability. Using this condition determine steady state operation point on torque-slip characteristics of three phase induction motor supplying constant load torque. **[U.P. Technical Electric Drives 2005-06]**

32. Discuss, with the help of curves, the characteristics and applications of the following electric motors:

(*a*) 3-phase, cage type induction motor (*b*) 3-phase wound rotor induction motor (*c*) single phase induction motor (*d*) synchronous motor (*e*) dc shunt motor.

[A.M.I.E. Sec. B. Electric Drives & Their Control Summer 1996]

33. What do you understand by "series" and "shunt" speed-torque characteristics of electric motors used in industrial drives ? Name (at least one of each type) ac and dc motor with such characteristics.

Name the characteristics and applications of the following type of motors :

(*a*) DC compound wound motor.

(*b*) Synchronous-induction motor.

(*c*) Double-squirrel cage induction motor

(*d*) Schrage motor.

[Gorakhpur Univ. Utilization of Electrical Power & Traction Sept. 1983]

34. Derive an expression for accelerating time of three-phase induction motor having pure inertia load, started from a constant voltage supply. Mention the factors which affect the accelerating time.

[U.P. Technical Univ. Electric Drives 2005-06]

35. Explain how the accelerating time and energy losses are calculated during a starting of a three-phase induction motor. Also enlist the methods of reducing the energy loss during starting. **[G.B. Technical Univ. Electric Drives 2010-11]**

36. Explain the dynamics during starting of a three-phase induction motor. What are the methods of reducing energy loss during starting ? Explain.

[U.P. Technical Univ. Electric Drives 2009-10]

37. Draw and explain the starting characteristics of various types of dc motors.

[Pb. Technical Univ. Electric Drives & Utilization Dec. 2009]

38. What is the essential difference in the principle of operation of a starter for slip-ring and squirrel cage motor ?

[Pb. Technical Univ. Electric Drives & Utilization May 2007]

39. Describe with suitable diagrams the various methods of starting 3-phase induction motors. Give relative merits and demerits of each method.

[Pb. Univ. Electric Drives & Utilization of Electrical Energy August, 1984]

40. Discuss, with circuit diagram, the star-delta starter for a squirrel cage induction motor. Explain factors which decide the starter-switch change-over from star-to-delta position.

[A.M.I.E. Sec. B. Electric Drives & Their Control Winter 1998]

41. What is the drawback of a star/delta starter for an induction motor ? How is it overcome ?

For a three-phase induction motor with negligible stator resistance and no-load current, show that the ratio of starting current to its stator current at any slip can be expressed as

$$\frac{I_{st}}{I} = \sqrt{\frac{s^2 + s_{max}^2}{s^2(1 + s_{max}^2)}}$$

where I_{st} = starting current, I = stator input current at any slip s and s_{max} = slip at which maximum torque occurs.

[A.M.I.E. Sec. B. Electric Drives & Their Control Summer 1999]

42. How do you start a synchronous motor ?

[A.M.I.E. Sec. B. Electric Drives & Their Control Winter 1999]

43. Explain why field (or flux) control is employed for obtaining speeds higher than rated value and armature voltage control for speeds lower than rated in the case of dc separately excited motor.

[A.M.I.E. Sec. B. Electric Drives & Their Control Summer 1999]

44. Explain which properties make series motors more suitable for heavy torque applications. Also discuss various methods for controlling speed of dc series motors.

[Pb. Technical Univ. Electric Drives & Utilization May 2007]

45. Draw the speed-torque characteristics for dc series motor. Show the effect of increase in series resistance on above characteristics. What is base speed in dc motors? How these motors can be run above the base speed ?

[A.M.I.E. Sec. B. Electric Drives & Their Control Summer 2000]

46. Describe with the help of connection diagram the Ward-Leonard method of speed control and explain its working. What are the merits and demerits of this method ? State its field of application.

47. Name the different methods of speed control of a 3-phase induction motor. Describe in brief speed control of induction motor by pole changing. **[Rajasthan Univ. 2008]**

48. Describe the variable voltage, variable-frequency method of speed control of induction motors. Explain, with reasons, the voltage-frequency relationship for full-range speed control.

[A.M.I.E. Sec. B. Electric Drives & Their Control Winter 1993]

49. Derive the induction motor torque-speed characteristics under V/*f* control. Explain why E/*f* control is superior to V/*f* control. **[A.M.I.E. Sec. B. Electric Drives and Their Control Winter 1993]**

50. What is electric braking ? What are the advantages and disadvantages of electric braking.

[Rajasthan Univ. Utilization of Electric Power 2003, 2004, 2005, 2007, 2008]

51. What do you understand by electric braking? Why is it necessary to have mechanical brakes in electric drives? What are the advantages of electric brakes over mechanical brakes? Describe in brief various methods of electric braking and mention their relative merits and demerits.

52. Explain various methods of electric braking. State the conditions to be fulfilled for each method of braking.

[M.D. Univ. Utilization of Elec. Power & Traction Dec. 2008]

53. What are the advantages and disadvantages of electric braking ?

[Pb. Technical Univ. Electric Drives & Utilization May 2007]

54. Explain the types of braking of dc motors. Which one is usually employed and why ?

[G.B. Technical Univ. Electric Drives 2010-11]

55. Explain the various types of braking as used in a dc motor. What necessary modifications are required while braking a dc series motor ?

[U.P. Technical Univ. Electric Drives 2009-10]

56. How can plugging be applied to a separately excited dc motor ? **[U.P. Technical Univ. Electric Drives 2008-09]**

57. Describe different methods of braking used for dc shunt and 3-phase squirrel cage induction motor.

[U.P. Technical Univ. Electric Drives 2007-08]

58. Explain dc dynamic braking of 3-phase slip-ring induction motor and derive an expression for braking torque. How can the magnitude and the point of occurrence of the maximum braking torque be varied ?

[U.P. Technical Univ. Electric Drives, 2008-09]

59. Explain dc dynamic braking of 3-phase induction motor. Discuss effects of excitation and rotor resistance on the torque-slip characteristics of the motor.

[U.P. Technical Univ. Electric Drives 2005-06]

60. Explain in 4-quadrant torque-speed plane the reversal of direction of a loaded dc shunt motor by plugging using a single resistor in the motor armature circuit.

[A.M.I.E. Sec. B. Electric Drives & Their Control Summer 1993]

61. Explain, with diagram, the Ward-Leonard dc drive using rotating machines. How is the regenerative braking achieved in this drive ? **[A.M.I.E. Sec. B. Electric Drives and Their Control Summer 1993]**

62. What is regenerative braking? Why it cannot be accomplished with a series wound dc motor without modifications in the circuit diagram. Give its merits and demerits.

[Gorakhpur Univ. Utilization of Electric Power and Traction Sept. 1984]

63. Specify the different types of enclosures of motors, highlighting the special feature of each type. Mention the area of application of each type.

[A.M.I.E. Sec. B. Utilization of Elec. Power Summer 1985]

64. Discuss the different types of drives required for transmission of power from the driving machines to the loads.

[A.M.I.E. Sec. B. Utilization of Elec. Power Winter 1983]

65. Discuss the following features pertaining to the selection of a motor:

Types of enclosures, bearings, transmission of drive and noise.

[A.M.I.E. Sec. B. Utilization of Elec. Power Summer 1982]

66. What are the mechanical features to be considered in the selection of a motor for a particular service ?

What do you understand by the type of enclosure indicated as IP 44 ?

List the advantages of using roller or ball bearings for an induction motor.

State the factors that contribute to mechanical vibrations and noise.

Describe (*i*) flat belt drive (*ii*) V-belt drive (*iii*) rope drive (*iv*) chain drive (*v*) gear drive and (*vi*) direct drive.

[A.M.I.E. Sec B. Utilization of Elec. Power Summer 1980]

67. What are the limiting temperatures of motors when the class of insulation used is :

(*i*) Y, (*ii*) A, (*iii*) E, (*iv*) B, (*v*) F, (*vi*) H and (*vii*) C ?

Give typical examples of loads demanding continuous duty, intermittent duty and short-time duty.

68. From basic principles of heat transfer, develop an expression for temperature rise of an electric motor delivering fixed rated power at a fixed (rated) speed.

[U.P. Technical Univ. Electric Drives 2007-08]

69. What is heating time constant ? Explain how the rating of a motor is affected by the temperature rise.

[Pb. Technical Univ. Electric Drives & Utilization Dec. 2008]

70. Explain what do you understand by 'continuous', 'short time', intermittent rating of an electric machine.

[Pb. Technical Univ. Electric Drives & Utilization May 2009]

71. Define continuous duty, short time duty and periodic intermittent duty. Draw typical variations of output power losses, temperature rise with time in each case.

[U.P. Technical Univ. Electric Drives 2007-08]

72. Derive an expression for the temperature rise of an electric motor. State the assumptions made.

[A.M.I.E. Sec. B. Electric Drives & Their Control Winter 1994]

73. A motor of smaller rating can be selected for a short time duty. Why ? **[U.P. Technical Univ. Electric Drives 2009-10]**

74. Discuss different classes of motor duty cycles. Give industrial application for each class of duty cycle.

Discuss the equivalent power method for determining the rating of an electric motor which is subjected to intermittent periodic duty.

[A.M.I.E. Sec. B. Electric Drives & Their Control Summer 1993]

75. Draw the pattern of temperature rise characteristic under steady-state for (*i*) short-time duty (*ii*) intermittent duty.

What is the equivalent current method of estimating motor rating ? Explain. Can this method be used if motor speed varies during a load cycle ?

[A.M.I.E. Sec. B. Electric Drives & Their Control Winter 1996]

76. Explain the load diagram. Why is it important to know for selection of drive rating ?

[A.M.I.E. Sec. B. Electric Drives & Their Control Summer 1998]

77. Discuss 'rms current', 'rms torque' and 'rms power' methods for evaluation and rating of motor for continuous duty, variable load conditions. Give their merits and limitations.

[U.P. Technical Univ. Electric Drives 2005-06]

78. What are the problems faced in case a motor of wrong rating is chosen ?

[A.M.I.E. Sec. B. Electric Drives & Their Control Winter 1999]

79. What is load equalization ? How it is achieved ?

[A.M.I.E. Sec. B. Electric Drives & Their Control Summer 2000]

80. What are the reasons for load equalization in electric drive ?

[U.P. Technical Univ. Electric Drives 2007-08]

81. Derive expression of torque equation in case of load increasing or flywheel releasing energy.

[U.P. Technical Univ. Electric Drives 2009-10]

82. Derive an expression for the motor torque driving a rolling mill when equipped with flywheel.

[Pb. Technical Univ. Electric Drives & Utilization May 2009]

83. What do you understand by load equalization ? Where do you find its applications ? Derive the expression for instantaneous motor torque, MOI of the flywheel, and the motor slip. State any assumptions made.

[Pb. Univ. Elec. Drives and Utilization of Electric Energy, August 1984]

84. Explain the following statements.

(*a*) Pole changing method of speed control is used only in squirrel cage induction motors.

(*b*) Ward-Leonard method of speed control facilitates speed variation below the base speed, and is suited for constant torque drives.

(*c*) Motors used in domestic fans should not have low rotor resistance.

(*d*) By feeding energy to the rotor of an induction motor the speed variation can be achieved above the synchronous speed.

[A.M.I.E. Sec. B. Utilization of Elec. Power Winter 1984]

85. Justify the following statements.

(*i*) DC series motors are best suited where high starting torque is required.

(*ii*) Starting torque of a dc series motor is proportional to the square of its armature current.

(*iii*) DC series motors should never be started on no load.

(*iv*) DC series motor is able to produce a far greater torque compared to a dc shunt motor. **[Gorakhpur Univ. Sept. 1984]**

(*v*) A differential compound motor tends to run in reverse direction when started and is found to have a tendency to race on overload.

(*vi*) 3-phase induction motors are most commonly used.

(*vii*) The speed of the rotor cannot be equal to that of the revolving field.

(*viii*) Schrage motor can run above and below synchronous speed.

(*ix*) Synchronous motors are not self-starting.

(*x*) Synchronous motors will operate at constant speed irrespective of load.

(*xi*) In case of fans and centrifugal pumps where the load torque decreases with the fall in speed armature control is quite convenient and economical. **[Gorakhpur Univ. Sept. 1984]**

(*xii*) Ward-Leonard system of speed control is used where fine speed adjustment and smooth acceleration is required. **[Gorakhpur Univ. 1982]**

(*xiii*) Quick stopping of motor is more essential than quick starting. **[Gorakhpur Univ. 1982]**

(*xiv*) Friction brake is required in addition to electric braking system. **[Gorakhpur Univ. Sept. 1984]**

86. Write short notes on the following :

(*i*) Group and individual drives. **[Pb. Univ. April 1979]**

(*ii*) Ward-Leonard method for speed control. **[Rajasthan Univ. 2005]**

(*iii*) Cascade connections of induction motors. **[Pb. Univ. April 1974]**

(*iv*) Different methods of electric braking. **[M.D. Univ. May 2009]**

(*v*) Regenerative braking as applied to traction system.

(*vi*) Types of enclosures for electrical machines. **[Pb. Univ. April 1981]**

(*vii*) Rating of electric motors for different use with examples.

(*viii*) Load equalization. **[Pb. Univ. 1983; Pb. Univ. Nov. 1983; April 1981]**

Short Answer Type Questions With Answers

Q. 1. What is an electric drive ?

Ans. An electric drive is defined as a form of machine equipment designed to convert electrical energy into mechanical energy and provide electrical control of this process.

The basic elements of an electric drive are the electric motor, the transmission, and the electrical control system. Electric drives often include conversion equipment (frequency changers, rectifier units, motor-generator sets etc.) to attain smoother and more versatile control and make the motor operate on specific speed-torque characteristic.

Q. 2. Why electric drive is preferred over mechanical drive ? **[Gorakhpur Univ. 1982]**

Ans. Electric drive is preferred over mechanical drive because of its simplicity, cleanliness, easy and smooth control, flexibility in layout, compactness, reliability, long life, easy starting, low cost (initial as well as maintenance cost), facility for remote control etc. Modern electric drive with a system of automatic control fulfills several functions other than imparting motion to the working machine so as to carry out the production process involved in the most efficient manner. Using automatically controlled electric drives, machinery and production units may be operated with high productivity, ensuring at the same time high quality.

Q. 3. Enumerate the factors governing selection of electric motors.

Ans. The various factors that govern the selection of electric motor for a particular service are :

1. Nature of electric supply available—dc or ac.
2. Types of drives—individual, group or multimotor.
3. Nature of load—active or passive.
4. Electrical characteristics (operating, starting, speed control and braking).
5. Mechanical considerations (Types of enclosures, types of bearings, types of transmission for drive, noise level etc.)
6. Service capacity rating
7. Appearance and
8. Cost

Q. 4. Write classification of electric drives. **[P.T.U. December 2006]**

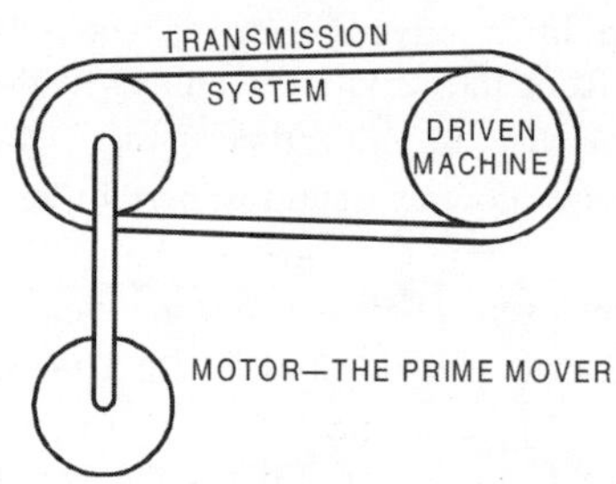

Fig. 1

Ans. The various types of electric drives used in industry may be divided into three types : The group drive, the individual drive and multimotor drive.

Q. 5. What is group drive ?

Ans. Group drive, also sometimes called the line shaft drive, means a drive in which a single electric motor drives a line shaft by means of which an entire group of working machines may be operated.

Q. 6. What is individual drive ?

Ans. In an individual drive, a single motor is used to drive one individual machine. Example of such drives are single-spindle drilling machines, various types of electrical hand tools and simple types of metal working machine tools and mechanisms.

Q. 7. What is multimotor drive ?

Ans. Multimotor drive consists of several individual drives each of which serves to operate one of many working members or mechanisms in some production unit.

Q. 8. Why individual drive is preferred over group drive nowadays ? **[Gorakhpur Univ. 1981]**

Ans. Though the individual drive costs more than group drive, but each operator has complete control of his machine, which enables him to vary its speed if necessary and stop while not is use, thus eliminating no-load losses. The machines can be placed in any desired position and can be moved very easily. The motor and its control unit can be built as an integral part of the machine which results in good appearance, cleanliness and safety. For driving heavy machines such as for lifts, cranes, shapers, lathes etc. and for the purposes where constancy of speed and flexibility of control is required, such as in paper mills and textile industry, individual drive is essential.

For new factories individual drive is preferred, as it causes some saving in the cost of superstructure because of being much lighter and less expensive.

Q. 9. How does active torque differ from passive torque ?

Ans. Active torques are due to either gravitational force or deformation in elastic bodies whereas passive torques are due to friction or due to shear and deformation in inelastic bodies. Active torque continues to act in the same direction even after the direction of drive has been reversed while passive torque always opposes the motion, retarding the rotation of the driven machine.

Q. 10. What is meant by two-quadrant operation of a dc motor ?

Ans. The operation of an electric motor in forward motoring mode and in regenerating (or braking mode) is called the two-quadrant operation of the motor.

Q. 11. What factors govern selection of motor for a drive ? **[P.T.U. December 2007]**

Ans. For selection of reliable and efficient motor it is essential that the conditions of service are well known. It is not sufficient to simply specify the output power in kW and the speed but it is also necessary to know the following additional particulars :

(*i*) Torque at the shaft during running, starting and at different loads.

(*ii*) Accelerating torque and braking torque.

(*iii*) Switching frequency.

(*iv*) Efficiency of motor at different loads.

(*v*) Other working requirements.

In studying the behaviour of a motor selected for a particular driven unit, one of the first problems involved is to determine whether the speed-torque characteristic of the motor suits the requirements imposed by the speed-torque characteristic of the driven unit. Drive behaviour during the transient period of a start-up, braking, or speed change-over also depends upon how the speed-torque characteristics of the motor and the driven unit vary with speed.

Q. 12. Why speed-torque (not torque-speed) characteristic curves are commonly used in utilization studies ?

Ans. In application engineering, it is usually the load torque which is known and the speed at which the motor can drive this load is unknown and is to be determined. Thus the torque is independent variable and speed is dependent variable. Consequently speed-torque characteristic curves with torque as abscissa are commonly used in utilization studies.

Q. 13. What will be the effect on direction of rotation of a dc motor if the supply terminals are reversed ?

Ans. No effect. Motor will rotate in the same direction, as before the reversal of the line terminals.

Q. 14. Why should a dc series motor not be run without load ?

Ans. On no load dc series motor would attain tremendously high speed and the motor may get damaged due to heavy centrifugal forces set up in rotating parts.

Q. 15. Why is series motor preferred for traction purposes ?

Ans. Because dc series motor develops large starting torque and slows down with the increase in load and so gets automatically relieved from heavy excessive load.

Q. 16. Why the dc motors are mostly gear motors ?

Ans. Because the low-speed motors are much more expensive than high-speed motors of the same kW output rating.

Q. 17. Why are induction motors called asynchronous motors ?

Ans. Because the rotor does not turn in synchronous with the rotating field developed by the stator currents.

Q. 18. Is the maximum torque of a 3-phase induction motor dependent on the rotor resistance ?

Ans. No, it is independent of rotor resistance.

Q. 19. Why the power factor of an induction motor is low at starting ?

Ans. The rotor frequency and rotor reactance are high under starting conditions and, therefore, rotor currents lag the rotor emf by a large angle. This results in low power factor at starting.

Q. 20. The starting torque of a squirrel cage induction motor cannot be altered, when the applied voltage is constant, why ? **[M.K. Univ. Nov. 1996]**

Ans. The starting torque of a squirrel cage induction motor cannot be increased as there is no provision for inserting resistance in the rotor circuit.

Q. 21. How will you improve the starting torque of a 3-phase induction motor ? **[Madras Univ. Elec. Machines II October 1995]**

Ans. The starting torque of a 3-phase induction motor can be improved by inserting external resistance in each phase of the rotor circuit, which is possible only in case of wound rotor motors.

Q. 22. What are the main advantages of a cage motor ? **[M.S. Univ. April 1996]**

Ans. The main advantages of a cage motor are simple and rugged construction, low initial as well as maintenance cost, better efficiency and power factor, better cooling, more pull-out torque, greater maximum power output, nearly constant speed and simple starting arrangement.

Q. 23. What are the main disadvantages of a cage motor ? **[M.S. Univ. April 1996]**

Ans. The main drawbacks of a cage motor are low starting torque with large starting current and no possibility of a speed control.

Q. 24. What are the advantages and disadvantages of wound rotor over squirrel cage rotor in case of induction motor ? **[P.T.U. December 2002]**

Ans. In wound rotor external resistance can be inserted and, therefore,

(*i*) high starting torque can be obtained by inserting an external resistance of proper value in each phase of the rotor circuit and

(*ii*) adjustable speed can be obtained simply by adding external resistors in the rotor circuit.

The drawback of wound rotor over squirrel cage rotor are initial and maintenance costs are more because wound rotor needs slip rings, brushes, short-circuiting devices etc.

Q. 25. Why in some induction motors double cage rotors are provided ? **[A.M.I.E. Sec. B. Summer 1994]**

Ans. In some induction motors double cage rotors are provided so as to get high starting torque with a low starting current as well as high operating efficiency. Such motors operate with the advantages of high-resistance rotor circuit during starting and low-resistance rotor circuit under running conditions.

Q. 26. State the advantages of squirrel cage induction motors over dc motors. **[Rajasthan Univ. 2008]**

Ans. The important advantages of squirrel cage induction motors over dc motors are: (*i*) Ruggedness (*ii*) Lower maintenance (*iii*) Better reliability, (*iv*) Lower cost, weight, volume and interia, (*v*) Higher efficiency, (*vi*) Ability to operate satisfactorily with sharp voltage fluctuations and in dirty environment.

Q. 27. Why the V/f is kept constant while controlling the speed of a 3-phase induction motor ?

Ans. In order to maintain flux level constant.

Q. 28. How the power factor of a synchronous motor is changed keeping the shaft load undisturbed ?

Ans. By varying the field excitation.

Q. 29. For which type of loads, induction motor with voltage controller is suitable and why ? **[A.M.I.E. Sec. B. Summer 1996]**

Ans. Induction motor with voltage controller is suitable only for loads where the torque required drops off considerably with the drop in speed such as for driving fans because a large change in voltage is required for a relatively small change in speed and developed torque reduces greatly with the reduction in supply voltage.

Q. 30. How can the speed of a linear induction motor be controlled ? **[Bharathiyar Univ. Nov. 1995]**

Ans. By varying both frequency and voltage simultaneously.

Q. 31. Is the torque developed by a single phase induction motor at synchronous speed zero ?

Ans. No, it is negative.

Q. 32. Why single phase induction motors are usually set on rubber spring mounts ?

Ans. In a single phase induction motor, pulsating torque is developed as a direct consequence of the pulsating power in a single phase circuit.

Q. 33. Does a shaded-pole motor have large starting torque ?

Ans. No, it has very little starting torque.

Q. 34. What is a shaded-pole motor ?

Ans. Shaded-pole motor is a salient-pole, split-phase type, squirrel cage rotor single-phase induction motor provided with one or more auxiliary short-circuited stator windings displaced in magnetic position from the main winding.

Q. 35. State the applications of shaded-pole single phase induction motors. **[Madras Univ. April 1996]**

Ans. Shaded-pole motor develops very little starting torque and, therefore, it is used only for small fans, electric clocks, hair dryers, time phonographs, advertising displays, toys, instruments and such other applications.

Q. 36. Why is the normal full-load slip of a single phase induction motor higher than that of a three-phase motor ? **[A.M.I.E. Sec. B. Summer 1993]**

Ans. The normal full-load slip of a single phase induction motor is higher than that of a 3-ϕ motor, owing to development of backward rotating field. The power is to be delivered to the backward field from the power converted into mechanical power by the forward field.

Q. 37. What are the applications of electric drives ? **[P.T.U. December 2009]**

Ans. Squirrel cage induction motor is suitable for constant speed industrial drives of small power where speed control is not required and where starting torque requirements are of medium or low value, such as for printing machinery, flour mills and other shaft drives of small power.

Wound rotor (or slip-ring) induction motors are used for loads requiring severe starting conditions or for loads requiring speed control such as for driving line shafts, lifts, pumps, generators, winding machines, cranes, hoists, elevators, compressors, small electric excavators, printing presses, turn tables, strokers, large ventilating fans, crushers etc.

Single phase induction motors find wide use in fans, refrigerators, air conditioners, vacuum cleaners, washing machines, other kitchen equipment, record players, small farming appliances etc.

Q. 38. State whether repulsion motor has series speed-torque characteristics, or shunt speed-torque characteristics.

Ans. Series speed-torque characteristics.

Q. 39. What is a stepper motor ?

Ans. A motor which turns in discrete movement is called a stepper motor.

Q. 40. For stopping the motor, the line switch should always be opened rather than throwing back the starting arm. Why ?

Ans. If the starter arm is thrown back the field circuit is broken at the last contact button and owing to inductive nature of field, this will cause a hot arc and burn the contact.

Q. 41. Why do you require starters for starting a 3-phase induction motor ?

Ans. Starter is required to limit the initial rush of current to a predetermined value at the starting instant.

Q. 42. Which is the cheapest method of starting a 3-phase induction motor ?

Ans. Direct-on-line (D-O-L) starting.

Q. 43. What type of protection is provided in the starters meant for 3-phase induction motors ?

Ans. Overload and undervoltage protection.

Q. 44. Which method we must adopt to control the speed of a dc shunt motor above the base speed ?

Ans. Field control method.

Q. 45. While controlling the speed of a dc shunt motor what should be done to achieve a constant torque drive ? **[A.M.I.E. Sec. B. Summer 1995]**

Ans. Applied voltage should be maintained constant so as to maintain field strength constant.

Q. 46. If the number of poles on a motor is increased to lower the speed, how will the power factor be affected ?

Ans. Power factor will be reduced.

Q. 47. How is speed control affected in a 3-phase ac series commutator motor ?

Ans. In a 3-phase ac series commutator motor, the speed control is affected by moving all brushes simultaneously round the commutator by means of a hand-wheel.

Q. 48. How is speed control affected in a 3-phase ac shunt commutator motor ?

Ans. In a 3-phase ac shunt commutator motor, the speed is adjusted by changing the tapping points on a transformer inserted between the supply mains and the brushes, which are in fixed position.

Q. 49. How speed variation is affected in a repulsion motor ?

Ans. Speed variation in a repulsion motor is affected either by varying the impressed voltage or by changing the position of brushes.

Q. 50. What is meant by plugging ?

Ans. Plugging means stopping of an electric motor by instantaneously reversing its torque till it stops.

Q. 51. How plugging of a dc motor is done ?

Ans. Plugging is obtained in a dc motor by reversing the supply terminals to the armature of the motor.

Q. 52. What is the most economical method of electric braking ?

Ans. Regenerative braking.

Q. 53. Electric braking is accomplished in an induction motor by interchanging two stator terminals. Name the method.

[A.M.I.E. Sec. B. Winter 1995]

Ans. Plugging or counter-current braking.

Q. 54. When it is required to control the induction motor speed during deceleration what kind of braking is preferable to employ ? **[A.M.I.E. Sec. B. Winter 1996]**

Ans. DC dynamic braking is highly suitable for controlling the speed of an induction motor during deceleration by controlling the dc excitation.

Q. 55. How is choice of drive governed ?

Ans. Choice of drive is governed by the speed of driving and driven machines, convenience, space available, including clutching arrangement required and cost.

Q. 56. What is meant by "rating of motor" ? **[P.T.U. December 2009]**

Ans. A name plate fixed to the outside frame of an electrical machine records the data pertaining to its rating. A machine rating specifies the voltage, current, speed, excitation, power factor, efficiency, power output etc.

The rating of a machine should give all information necessary so that, if the machine is operated within the limits of all factors specified in its rating, the machine will operate satisfactorily and safely and will give reasonable length of service. Therefore, the rating of a machine must give the necessary information to safeguard the application of the machine from conditions of operation which (*i*) would result in unsafe mechanical or electrical strains upon any part of its structure or (*ii*) would result in excessive deterioration of the mechanical or electrical characteristics of the materials of which the machine is constructed, To give this information, the rating of an electric machine should include the output, voltage, speed, and any other information that may necessary for the proper operation of the machine.

Q. 57. On what factors does the rate of rise of temperature depend ?

Ans. The rate of temperature rise depends upon (*i*) amount of heat produced and (*ii*) the amount of heat dissipated per °C rise of the surface of the machine.

Q. 58. Why the cooling time constant of a rotating machine is usually larger than its heating time constant ?

[A.M.I.E. Sec. B. Winter 1996]

Ans. The cooling time constant of a rotating machine is usually larger than its heating time constant owing to poor ventilation conditions when the machine cools.

Q. 59. What is meant by load equalization ?

Ans. The process of smoothing out of fluctuations in load is known as load equalization.

Q. 60. How is load equalization achieved ?

Ans. The most common method of load equalization is by use of flywheel. During the light load period the flywheel accelerates and stores the excessive energy drawn from the supply and during peak load period the flywheel decelerates and supplies some of its stored energy to the load in addition to the energy supplied from the supply. Thus the load demand is reduced.

Problems

1. Compare the total annual cost of a group drive with a motor costing Rs. 65,000 with that of 8 individual motors, each costing Rs. 25,000. With group drive annual consumption is 60,000 kWh. With individual drive the consumption is 45,000 kWh. Electrical energy costs Rs. 1.20 per kWh. Annual interest, depreciation and maintenance charges may be taken as 18% in either case.

[**Ans.** Rs. 83,700 per annum in case of group drive and Rs. 90,000 per annum in case of individual drive]

2. A drive has the following equations for motor and load torques

$T = 1 + 2\omega_m$ and $T_L = 3\sqrt{\omega_m}$.

Obtain equilibrium points and determine their steady state stability. **[U.P. Technical Univ. Electric Drives 2008-09]**

3. A 220 V dc series motor is running at a speed of 800 rpm and draws 100 A. Calculate at what speed the motor will run when developing half the torque. Total resistance of the armature and field is 0.1 Ω. Assume that the magnetic circuit is unsaturated. [**Ans.** 1,147 rpm]

4. A series motor takes 20 A at 400 V and runs at 250 rpm. The armature resistance is 0.6 Ω and the field resistance 0.4 Ω. The torque required by the device varies as the square of the speed. Find the applied voltage and current to run the device at 350 rpm. [**Ans.** 772.8 V; 28 A]

5. A 3-phase induction motor at rated voltage and frequency exerts a maximum torque of 275 per cent and a starting torque of 125% of full-load torque. Determine the slip (*i*) at maximum torque and (*ii*) at full-load. Assume constant rotor resistance and neglect stator resistance and rotational losses.

[**Ans.** (*i*) 24% (*ii*) 4.52%]

6. A 3-phase, 50 Hz, 8-pole induction motor has a full-load slip 4%. The rotor resistance is 0.001 Ω per phase and standstill reactance 0.005 Ω per phase. Find the ratio of maximum to full-load torque and the speed at which the maximum torque occurs.

[Madras Univ. Elec. Machines II, April 1997]

[**Ans.** 2.6 ; 600 rpm]

7. A 3-phase star-connected 6.5 kV, 20-pole, 50 Hz induction motor has rotor resistance of 0.12 Ω and standstill reactance of 1.12 Ω. The motor has speed of 292.5 rpm at full load. Calculate slip at maximum torque and ratio of maximum torque to full load torque. **[Rajasthan Univ. Utilization of Elec. Power 2008]**

[**Ans.** 0.107; 9.68]

8. The star-connected stator winding of a 3-phase, 6-pole induction motor is supplied from a 240 V, 50 Hz source. The rotor resistance and standstill reactance are 0.12 Ω and 0.85 Ω per phase respectively. The stator to rotor turn-ratio is 1.8. Find the pull-out torque and the corresponding speed.

[**Ans.** 99.86 Nm; 859 rpm]

9. Determine approximately the starting torque of an induction motor in terms of its full-load torque when started by means of (*i*) star-delta starter and (*ii*) by an auto-transformer starter with 50 % taps. The motor draws 5 times full-load current when switched directly on line and has a full-load slip of 4 %.

[**Ans.** One-third and one-fourth of full-load torque]

10. A squirrel cage induction motor takes 5 times full-load current and develops 2.5 times full-load torque when switched directly on the line. Determine the torque developed and the initial value of the current drawn from the supply mains when started through an auto-transformer with 60 % tapping.

[**Ans.** 1.8 times of full-load current; 81% of full load torque]

11. A 250 V dc shunt motor has an armature resistance of 0.5 Ω and a field resistance of 250 Ω. When driving a load of constant torque at 600 rpm, the armature current is 20 A. If it is desired to raise the speed from 600 to 800 rpm, what resistance should be inserted in the shunt field circuit? Assume that the magnetic circuit is unsaturated. [**Ans.** 88 Ω]

12. A 220 V shunt motor has an armature resistance of 0.5 Ω and takes an armature current of 40 A on a certain load. By how much the main flux be reduced to raise the speed by 50% if the developed torque is constant ? Neglect saturation and armature reaction. [**Ans.** 37.3%]

13. A 200 V dc shunt motor running at 1,000 rpm takes an armature current of 17.5 A. It is required to reduce the speed to 600 rpm. What must be the value of resistance to be inserted in the armature circuit if the original armature resistance is 0.4 Ω ? Take armature current to be constant during this process.

[**Ans.** 4.41 Ω]

14. A 250 V shunt motor with armature resistance of 0.5 Ω runs at 600 rpm on full load and takes an armature current of 20 A. If resistance of 1.0 Ω is placed in the armature circuit, find the speed at (*i*) full load torque (*ii*) half-full load torque.

[**Ans.** (*i*) 550 rpm (*ii*) 587.5 rpm]

15. A 230 V dc shunt motor takes an armature current of 20 A on a certain load. Resistance of armature is 0.5 Ω. Find the resistance required in series with the armature to halve the speed if (*i*) the load torque is constant (*ii*) the load torque is proportional to the square of the speed. [**Ans.** (*i*) 5.5 Ω (*ii*) 23.5 Ω]

16. A 250 V dc series motor takes 40 A when developing full-load torque at 1,500 rpm. Its total resistance is 0.5 ohm. If the load torque varies as the square of the speed, determine the resistance to be connected in series with the armature to reduce the speed to 1,200 rpm. Assume a straight line magnetisation curve.

[**Ans.** 2.7125 ohms]

17. A series motor working on a 500 V dc supply mains at a speed of 800 rpm when the load is such that the current is 120 A. The motor armature resistance is 0.11 ohm and field resistance is 0.04 ohm. Assuming an unsaturated field calculate the speed of the motor when the torque is one half of the full-load value and the field winding is connected in parallel with divertor resistance of 0.08 ohm. [**Ans.** 1,395 rpm]

18. A 4-pole, 250 V dc series motor takes 20 A and runs at 900 rpm. Each field coil has resistance of 0.025 Ω and the resistance of the armature is 0.1 Ω. At what speed will the motor run developing the same torque if (*i*) a divertor of 0.2 Ω is connected in parallel with the series field (*ii*) rearrranging the field coils in two series and parallel groups. Assume unsaturated magnetic operation.

[Nagpur Univ. Electric Drives & Their Control 1993]

[**Ans.** (*i*) 1,101.6 rpm (*ii*) 1,275.4 rpm]

19. The rotor of a 6 pole, 50 Hz slip-ring induction motor has a resistance of 0.2 Ω per phase and runs at 960 rpm on full load. Calculate the approximate resistance per phase of a rotor rheostat such that speed is reduced to 800 rpm for full-load torque.

[**Ans.** 0.8 Ω/phase]

20. The voltage across two slip rings of a stationary star-connected 100 hp, 415 V induction motor is 273 V with full voltage across stator. If full-load slip is 2 per cent, find the resistance to be added in series with each slip ring to reduce the speed by 20 %. Assume load torque to be constant and rotor reactance negligible.

[Pb. Univ. Elec Machines II Dec. 1993]

[**Ans.** 0.196 Ω]

21. What synchronous speeds can be obtained from the combination of 50 Hz induction motors having 8 poles and 4 poles.

[**Ans.** 1,500 rpm; 750 rpm; 500 rpm]

22. A cascade set consists of two motors A and B with 4 poles and 6 poles respectively. The motor A is connected to 50 Hz supply. Determine (*i*) the speed of the set (*ii*) the electric power transferred to motor B when the input to motor A is 25 kW. Losses may be neglected. [**Ans.** (*i*) 600 rpm (*ii*) 10 kW]

23. A 50 hp, 440 V dc shunt motor is braked by plugging. Calculate the value of the resistance to be placed in series with the armature circuit to limit the initial braking current to a double of the full-load current. Armature resistance is 0.1 Ω. Full-load armature current is 100 A. [**Ans.** 4.25 Ω]

24. A 25 hp, 220 V dc shunt motor with a full-load speed of 600 rpm is to be braked by plugging. Estimate the value of the resistance which should be placed in series with it to limit the current to 130 A. What should be the initial value of the electric braking torque and value when the speed has fallen to half its full load value ? Armature resistance of motor is 0.1 Ω. Full-load armature current is 95 A.

[**Ans.** 3.211 Ω ; 400.5 Nm; 302.57 Nm]

25. A 400-V,25 hp, 450 rpm dc shunt motor is braked by plugging when running on full-load. Determine the braking resistance necessary if the maximum braking current is not to exceed twice the full-load current. Determine also the maximum braking torque and the braking torque when the motor is just reaching zero speed. The efficiency of the motor is 74.6% and the armature resistance is 0.2 Ω. [**Ans.** 6.1 Ω; 1,028 Nm; 522 Nm]

[**Hint :** Maximum braking torque will be developed initially when motor speed is maximum *i.e.* at a speed of 450 rpm]

26. A 250 V dc shunt motor has an armature resistance of 0.1 Ω and with full field has an emf of 240 V at a speed of 1,200 rpm. The motor is driving an overhauling load with a torque of 180 Nm.

Calculate the minimum speed at which the motor can hold the load by means of regenerative braking. [**Ans.** 1,297 rpm.]

27. Rheostatic braking is used to limit the speed of a series type dc crane motor to 500 rpm when the supply is cut off. The motor has a resistance of 1.35 Ω and the descending load exerts a torque of 442 N-m. Determine the resistance to be connected across the motor terminals.

The magnetisation characteristic of the machine at 500 rpm is as follows:

Current (A)	20	30	40	50
EMF (V)	360	440	490	520

State assumptions made if any.

[Gorakhpur Univ. Utilization of Elec. Power and Traction Sept. 1984] [**Ans.** 10.005 Ω]

28. A 400 V, 50 Hz, 4-pole cage type induction motor is braked by plugging, while working at full load delivering 50 hp; the full-load slip being 4%. The ratio of rotor resistance to standstill reactance = 1:5.

Calculate the initial braking torque.

[Gorakhpur Univ. Utilization of Electric Power and Traction 1983] [**Ans.** 128 N-m]

29. A 40 hp, 400 V, 3-ϕ, 4-pole, 50 Hz induction motor has full-load slip of 5%. If the ratio of standstill reactance to resistance per motor phase is 4, estimate the plugging torque at full speed.

[Pb. Univ. Elec. Drives and Utilization of Elec. Energy April 1981] [**Ans.** 129.3 N-m]

30. A 50 kW, 415 V, 3-phase, 4-pole, 50 Hz induction motor has a full-load slip of 5%. If the ratio of standstill reactance to resistance per phase of rotor is 5, calculate the torque developed at the start of reverse current braking at rated speed. Neglect stator impedance and magnetizing current.

[U.P. Technical Univ. Electric Drives 2007-08]

[**Ans.** 144.5 N-m]

31. The outside of a 12.5 kW dc motor is equivalent to a cylinder of a 100 cm diameter and 140 cm in length. The machine may be considered to be a homogenous material weighing 650 kg and of specific heat 700 J/kg/°C. Assuming that the outer cylindrical surface dissipates heat at 0.0012 W/cm^2/°C, estimate the thermal time constant of the machine and its final temperature rise when operating on full load with an efficiency of 85%.

[**Ans.** 2.394 Hours; 41.8° C]

32. In a temperature rise test on a dc machine at full load, the following temperature rises were observed; 16° C after one hour and 28° C after 2 hours. Find (*i*) final steady temperature rise and the time constant (*ii*) the steady temperature rise after one hour at 50% overload from cold. Assume that the final temperature rise on 50% overload is 92° C.

[**Hint :** Heating time constant of a motor is independent of rate of heat production so the heating time constant at 50% overload is the same as on full load.]

[**Ans.** (*i*) 64° C; 3.481 hours (*ii*) 123° C]

33. The heating time constant of a 22 kW motor is 90 minutes. If the motor is to be run continuously on full load, it attains a maximum steady temperature rise of 40° C. Calculate the half-hour rating of the motor for this temperature rise. Assume that losses are proportional to the square of the load and the motor cools down completely between each load cycle.

[BHU Electric Drives & Power Electronics 1978]

[**Ans.** 41.32 kW]

34. A motor driving some haulage equipment for a mine has to deliver a load which follows the following duty cycle: 100 kW for 10 minutes, no load for 4 minutes, 50 kW for 10 minutes, no load for 6 minutes. The cycle being repeated indefinitely. Estimate the suitable size of continuously rated motor for the purpose.

[**Ans.** 65 kW]

35. A machine on a short test run gave the following mean temperature rise :

Time hours	0	0.25	0.5	0.75	1.0	1.25	1.5
Temp. rise °C	42.5	45.0	47.4	49.5	51.4	53.0	54.4

Estimate graphically the final temperature rise and the time constant. Justify the procedure adopted by you.

[**Ans.** 62°C, 1.6 hours]

36. The temperature rise of a motor when operating for 25 minutes on full load is 25°C and becomes 40°C when operates for another 25 minutes on the same load. Determine the heating time constant and steady-state temperature rise.

[A.M.I.E. Sec. B. Electric Drives & Their Control Winter 1998]

[**Ans.** 48.94 minutes; 62.5°C]

37. A motor driving some haulage equipment for a mine has to deliver a load which follows the following duty cycle :

100 hp for 10 minutes, no load for 4 minutes, 50 hp for 10 minutes, no load for 6 minutes.

The cycle being repeated indefinitely. Estimate the suitable size of continuously rated motor for the purpose.

[Agra Univ. Traction and Utilization of Elec. Power 1982]

[**Ans.** 65 hp]

38. A constant speed drive has the following duty cycle :

Load rising from 0 to 400 kW : 5 min

Uniform load of 500 kW : 5 min

Regenerative power of 400 kW returned to supply : 4 min

Remains idle for : 2 min

Estimate the power rating of the motor. Assume losses to be proportional to (power)^2.

[A.M.I.E. Sec. B. Electric Drives & Their Control Summer 1998]

[**Ans.** 330 kW]

39. A 500 rpm synchronous motor and its load have a total moment of inertia of 420 kg-m^2 and a frictional torque of 140 N-m. The motor is braked by rheostatic method, field current and the braking resistance being kept constant at value which gives an initial electric braking torque of 5,400 N-m. Find the time taken to bring the motor to test. [**Ans.** 15 seconds]

40. A motor fitted with a flywheel supplies a load torque of 100 N-m for 20 seconds. During the no-load period, the flywheel regains its original speed. The motor torque is required to be limited to 500 N-m. Determine the moment of inertia of the flywheel. The no-load speed of the motor is 500 rpm and it has a slip of 10% on full load.

[Pb. Univ. Elec. Drives & Utilization of Elec. Energy April 1980]

[**Ans.** 2,755 kg-m^2]

41. A dc shunt motor 30 hp, 400 V, 500 rpm draws a current of 40 A when running at full load. The moment of inertia of the rotating system is 12.4 kg-m^2. If the current during the starting period varies from 1.05 to 1.25 times of full-load current, calculate (*i*) the full-load torque, and (*ii*) the time required for the motor to attain the rated speed against full load.

[**Ans.** (*i*) 421.4 N-m (*ii*) 10.27 seconds]

42. A 6-pole, 50 Hz induction motor has a flywheel of 1,200 kg-m^2 moment of inertia. Load torque is 100 kg-m for 10 seconds. No-load period is long enough for the flywheel to regain its full-speed. Motor has a slip of 6% at a torque of 50 kg-m. Calculate (*i*) maximum torque exerted by motor (*ii*) speed at the end of deceleration period.

[Pb. Univ. Utilization of Elec. Energy 1973 Supp.]

[**Ans.** (*i*) 469 N-m (*ii*) 925 rpm]

❑❑❑

CHAPTER 2

Control Devices For Industrial Motors

INSIDE THIS CHAPTER

2.1. INTRODUCTION

The apparatus such as suitable switches required for opening and closing the necessary circuits and for operating rheostats required for starting, reversing, speed control and braking of electric motors together with the protective equipment for guarding against faults or careless usage is known as the *control gear.*

The simplest method of motor control is manual control, which is accomplished by plain knife switches, rotary switches, starting and speed control rheostats, push buttons and controllers. The use of manual control devices involves a certain additional expenditure of time for accomplishing control and hence lowers the productivity of the equipment concerned, especially when its operation calls for frequent starting or speed control. The use of manual control devices also precludes the possibility of remote control. This makes it altogether unsuitable for many kinds of modern automatic equipment. In large capacity electric drives, manual control is difficult to provide, and in some cases is altogether impracticable because the operators would be unable to exert the effort required for switching the apparatus.

The need to eliminate the shortcoming of manual control has led to the development of semi-automatic and fully-automatic control equipment. Semi-automatic control makes use of control apparatus which is energized by pressure on a push button. In the case of automatic control, the driven machinery itself may start up the drive motor, or this can be done by suitable controllers.

Automatic electric drive control is of great importance to the national economy because it permits the productivity of labour to be raised, makes for better working conditions, improves the quality of the products being produced, reduces the expenditure of electrical energy, and raises the reliability of operation of the equipment.

Automation of the drive simplifies the operation of the equipment and allows it to be remotely controlled. This is particularly valuable where local control of drive motors is inconvenient by reason of position or impossible because of the nature of the production process.

The above advantages are tending to increase the applications of fully-automatic control equipment, inspite of the fact that it costs about three times as much as the corresponding manual control equipment.

2.2. AUTOMATIC REGULATING EQUIPMENT

In industrial processes a certain quantity, such as the speed of a motor, the temperature of an oven, the tension in a wire, output voltage of the generator or output current is required to be held at some specified value, usually constant, although in certain cases a predetermined cycle of variation may be required.

Electrical devices used for providing such regulation operate on the closed cycle or feedback principle. Such an arrangement requires essentially (*i*) an actuating device for providing a control signal dependent on the deviation of the controlled quantity from its specified value (*ii*) an operating device for controlling the power supply feeding the apparatus to be regulated in such a way as to counteract any deviation and (*iii*) a connecting link for transferring the actuating signal in an appropriate manner to the operating device.

The connecting link used for this purpose should be of high power gain and of a very rapid response so that the smallest departure from the specified value will be sufficient to initiate the correcting effect of the operating device and there is no time delay between the occurrence of the actuating signal and the correcting effect.

The thermionic valves, usually of gas filled type, rotating amplifiers (amplidynes) and magnetic amplifiers are extensively used as connecting link.

2.2.1. Thyratron

Thyratron is a hot cathode gas filled triode. Normally it is filled with a small amount of inert gas such as argon, hydrogen, neon or mercury vapour. The fast acting devices do not use mercury vapour as the mercury vapour has the disadvantage of characteristics varying with temperature and it takes time in deionization. The characteristics of hydrogen thyratron are stable and the devices are fast. Plate is a disc of graphite. The special construction oxide coated cathode is used so that its surface is not damaged due to bombardment of positively charged particles. The control grid has a metal cylinder surrounding the cathode with one or more perforated discs known as *grid baffles* near the centre. Thus the control grid acts as an electrostatic shield between cathode and anode except for the holes in the grid baffles.

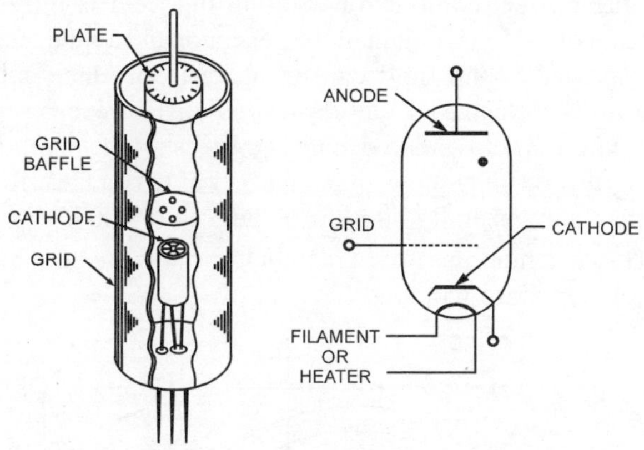

(a) Construction *(b) Symbolic Representation*

Fig. 2.1. ***Thyratron***

Before the tube fires, its characteristics are similar to those of a vacuum triode *i.e.,* the plate current (small in magnitude) can be varied by variation of potential applied to the grid. After firing of the tube the grid loses control over plate current completely and plate current is limited by the external resistance in the plate circuit *i.e.,* the thyratron behaves like a diode. The introduction of grid in between the plate and cathode makes it possible to control the conditions under which the discharge should start.

Since the only function of the grid is to delay or prevent the starting of the current, the grid is naturally designed to achieve its purpose with as little grid voltage as possible. The grid, therefore, is quite an extensive structure. The critical grid voltage varies with the plate voltage to some extent, and also with the design of the grid.

Plate Voltage-Plate Current (or Plate) Characteristics. These characteristics are drawn for various grid voltages.

For a given grid voltage when the plate voltage is increased from zero, initially there is no plate current but when the plate voltage is further increased, a voltage is reached, called the *breakdown voltage*, the current at once shoots up and is limited only by the external resistance in the plate circuit. If the grid is made more negative, the tube will break down at higher plate voltage, as illustrated in Fig. 2.2.

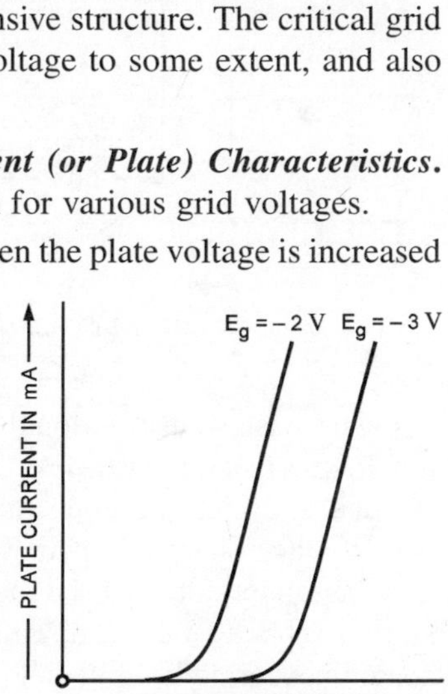

Fig. 2.2

Grid Control Characteristics. The curve drawn between plate voltage and critical grid voltage is called *the grid control characteristic or starting characteristic or breakdown characteristic*. This characteristic can be determined by determining the value of critical grid voltage for different values of plate voltage and plotting them, as shown in Fig. 2.3. The slope of this characteristic is known as *control ratio.*

Applications. Thyratron has two main fields of applications :

(*i*) as an electronic switch to handle heavy currents and (*ii*) as a grid controlled rectifier when voltage control is required. The thyratron, when used as electronic switch, keeps the circuit open from the plate to cathode so long as the grid is kept at high negative potential. As soon as the grid potential is brought to such a value that the tube fires, the circuit is closed.

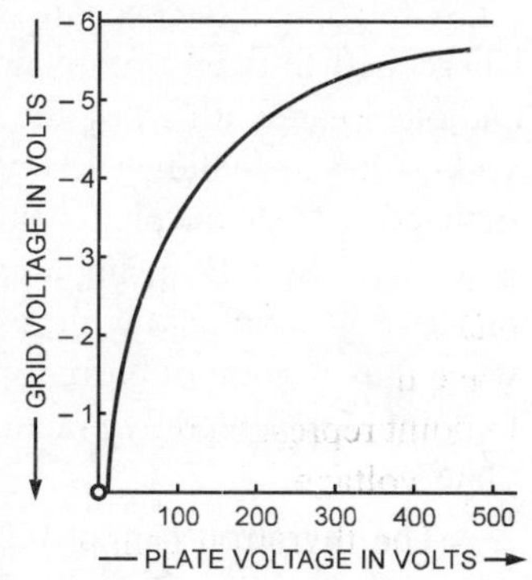

Fig. 2.3

The thyratron switch has two great advantages over mechanical switches : (*i*) it is electrically operated and requires only a minute amount of energy to close it, so that it leads itself readily to automatic operation; (*ii*) operation is very fast. However, it has got some disadvantages also such as (*i*) voltage drop of 8 to 18 volts in the circuit; (*ii*) it cannot open dc circuit; (*iii*) a single thyratron will pass only the positive half waves. A second thyratron connected in parallel with the first one but with opposite polarity may be used to pass the negative half waves. A thyratron can open ac circuit, with a delay that never exceeds one half cycle, so long as frequency is not so high that the time of one half cycle is less than the deionizing time.

The thyratron finds wide use in timing and other control applications.

The fact that its time of firing may be controlled makes the thyratron adaptable to heat control in resistance welding. By delaying or advancing the time during the cycle at which the tube fires, the average amount of welding current allowed to flow per cycle may be increased or decreased. In addition to this, the number of cycles that the welding current is allowed to flow may be controlled very accurately.

Thyratron rectifiers are also used in electronic methods of electric motor control for both small and large powers. The main advantages of electronic motor control are that the control can be made to depend on very minute signals such as those obtained from a photocell or even a thermocouple and that where necessary an extremely precise control can be obtained..

Control may be affected in several ways such as (*i*) with direct voltage applied to the plate as well as to the grid (*ii*) with alternating voltage to the plate circuit and direct voltage to the grid and (*iii*) alternating voltage applied on both the grid and plate. In the first case the tube will fire and conduct whenever the grid voltage is made less negative than critical grid voltage and will again stop to conduct when the plate voltage is reduced below striking voltage or the plate circuit is opened. In the second case the grid will regain control when the plate voltage becomes negative, hence conduction will be stopped every half cycle and the tube breakdown will occur each half cycle as the plate voltage becomes positive, the point at which the tube fires depends upon the grid voltage. The average current flow can be controlled, therefore, by controlling the length of time during the half cycle that the tube is conducting *i.e.,* by controlling the breakdown point. In the last case breakdown point is determined by the phase relation of the two voltages as well as by their magnitudes. There will thus be a point in each positive half cycle of the cathode–plate path when the tube will break down, if the grid voltage has a small negative value than the critical grid voltage with reference to the plate voltage, which will result in ionization and conduction. By using a phase shifting device to control the phase of grid voltage with respect to the cathode-plate voltage wave the duration of current flow may be controlled from zero to point represented by practically all the positive half cycles of plate voltage.

The thyratron cannot be employed as an amplifier like a vacuum triode because the grid voltage has no control over the magnitude of plate current after firing of the tube.

2.2.2. Ignitron

Ignitron is a gaseous tube and consists of carbon anode, boron carbide ignitor tube and mercury pool cathode. It is a cold cathode tube and does not have any heating element or grid control as in thyratron. Therefore, electrons are to be made available before the tube starts conducting. This is achieved by the ignitor which remains dipped in the mercury pool but being made of boron carbide does not get wet and, therefore, gives high resistance, say 10 to 500 Ω, between the mercury pool and ignitor. On closing of switch S full voltage is impressed between ignitor and mercury pool. This causes intense electric field which pulls out electrons from mercury pool to start with. The electrons of the arc ionize the mercury vapours if the ignitor current is of the magnitude between 20 and 40 amperes. Positive ions thus released are attracted towards cathode mercury pool and momentum of these mercury ions releases large number of electrons from the cathode pool by secondary emission. These electrons are attracted towards the carbon anode. Now resistance of main path of the current is very low as compared to that of ignitor path and, therefore, ignitor current stops. This process is to be repeated every time anode becomes positive in ac voltage cycle.

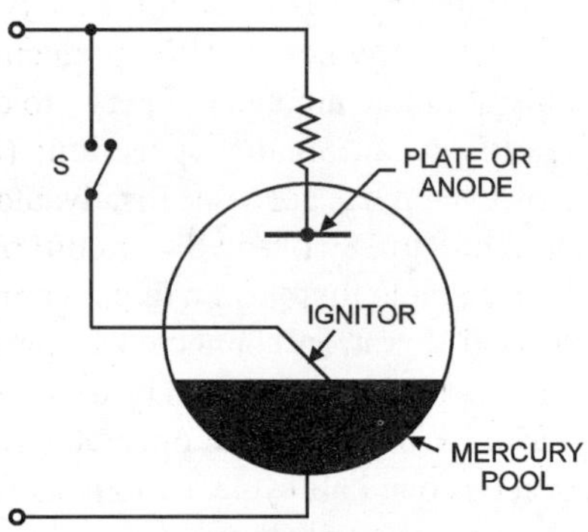

Fig. 2.4

Ignitrons can handle current from 40 to 10,000 A. It may also be noted that electronic current flow from ignitor to cathode, spoils the ignitor and, therefore, should not be permitted.

The power output can be controlled by timing the firing by means of the ignitor, usually by shifting the phase of the pulse to the ignitron.

The single units may be connected in 3-phase, 6-phase, or 12-phase. In smaller ratings, sealed-in glass tubes are used; in the intermediate ratings, sealed-in water-cooled metal tanks are used. In the larger power units, pumped all steel water-cooled tanks are used.

2.2.3. Amplidyne (Rotary Amplifier)

The amplidyne is a rotary amplifier in which a small amount of power supplied to excite the field is amplified in the generator output. Though the ordinary separately excited generator operated at low flux densities can be used as a power amplifier since the power required for exciting the field is only 1 to 3 per cent of the rated output of the generator and a small change in the power to the field can be made to produce a large proportionate change in the power output. However, a much larger power amplification can be had by using the amplidyne, since in this machine power is amplified in two stages. Rotary amplifiers are generally built for outputs up to 100 kW.

This machine uses the cross-field principle and a single two-pole armature winding.

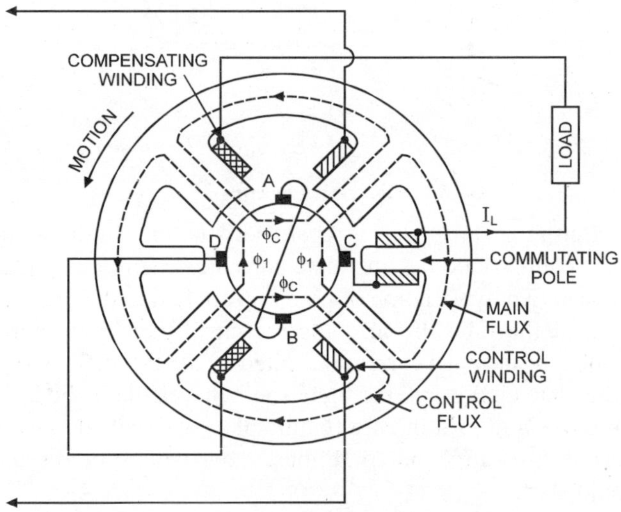

(a) Constructional Features

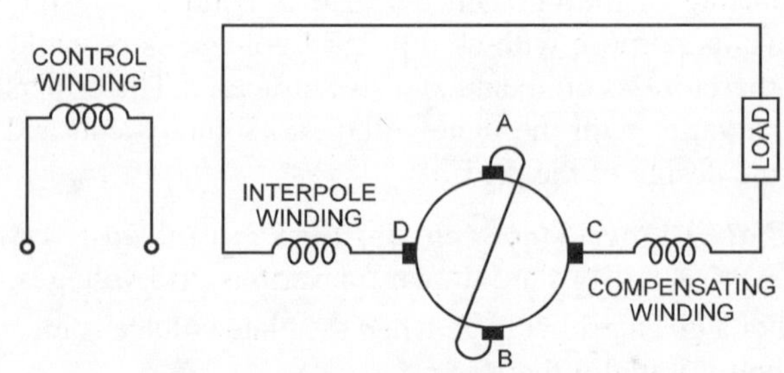

(b) Connection Diagram

Fig. 2.5. ***Amplidyne***

The basic constructional features for a 2-pole armature amplidyne are shown in Fig. 2.5 (*a*). The poles, however, are divided and give the appearance of a four-pole field though actually the poles are in pairs with the same polarity. The poles may be salient ones as used in ordinary dc machines or may be slotted so as to take a distributed winding. The machine is sometimes termed a split-field machine because it has divided poles. Commutating poles may be arranged in all the interpolar gaps. Control windings and compensating windings are arranged, as illustrated in Fig. 2.5 (*a*). Compensating winding in the form of a concentrated coil wound around the field cores, as shown in Fig. 2.5 (*a*) is used in many cases but such a winding does not completely neutralize the armature load current mmf and some distortion of the flux, therefore, exist. For better compensation a distributed type of compensating winding, properly located in slots in the pole faces must be used. The magnetic circuit may then consist of slotted punchings. This form does not provide as much space for a variety of field windings as the form with polar projections, and the various control inputs are more usually combined in an electronic or magnetic amplifier providing the control excitations instead of in the machine itself. Magnetic circuit should be of laminated construction throughout.

The action of the machine is now explained. The control windings produce a small control flux Φ_c in the horizontal axis. When the machine is driven, a voltage is developed between the short-circuited brushes A and B and, therefore, current flows between them (the control winding is provided with few ampere-turns in order to limit the current through the short-circuited armature winding). Now the current I_1 through the short-circuited armature winding produces armature reaction flux at right angles to the control flux Φ_c which in turn develops emf between the output brushes C and D. In the figure separate flux paths have been shown. Although the fluxes are superimposed in the poles and parts of the armature and yoke, in the absence of saturation they are mutually independent.

When the amplidyne is loaded, load current I_L flows which in turn sets up the flux opposing the control flux Φ_c. It is essential, therefore, for the armature reaction mmf set up by the output current I_L to be neutralised if the machine is to operate as a high-gain amplifier. A compensating winding is, therefore, connected in series with the power output brush terminals. The machine operated in this manner is called the amplidyne. This device is virtually two machines in one. Current through the short-circuited brushes which form the load for the first stage, acts also as the field excitation for the second stage. Mechanical input at the shaft is converted into an electrical output, modulated by the control flux. With full compensation, gains of the order of 20,000 : 1 are possible and the power supplied to the control winding may be fraction of a watt even in a 10-20 kW rated machine. Hence the amplidyne can be fed directly from an electronic amplifier in an automatic control scheme.

The basic machine, described above, had only one control winding, which was separately excited. Actual amplidynes are provided with additional field windings separately excited as well as self excited to provide desirable characteristics. Separately excited field windings are called the control windings since the energy or signals, as they are generally called, which are supplied to the control windings control the machine output. Since these windings are of very small power capacity, a small change of input power to a control field winding will develop a large change in the output voltage across brushes C and D and, therefore, a large change in power output. It is feasible to construct amplidynes with four separate control field windings so that the output of the machine may be controlled from several independent signals. Commutating poles are fitted at the brush axis to assist commutation, but it is not usual to fit them at the cross axis, except in large machines owing to the small value of I_1, the exciting current.

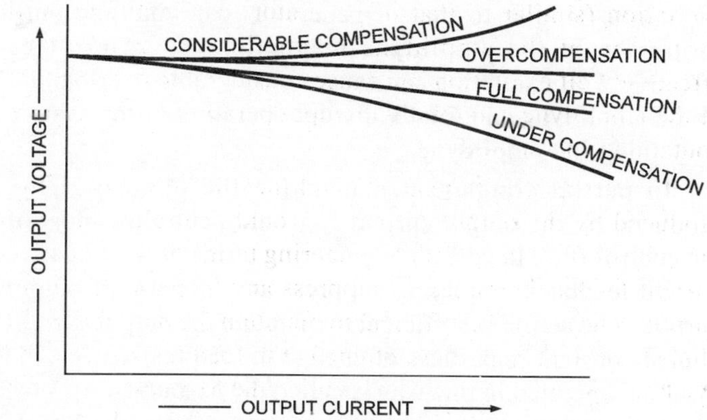

(a) For Various Degrees of Compensation Amplidyne External Characteristics

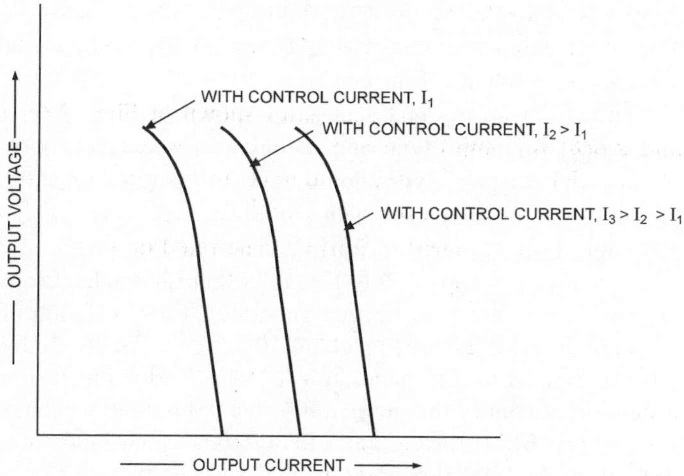

(b) For Various Control Currents Metadyne External Characteristics

Fig. 2.6

The external characteristic of an amplidyne is similar to that of an ordinary separately excited generator, but its slope can be regulated by acting on the compensating winding of the amplidyne. This winding is usually designed to provide magnetizing force somewhat greater than the demagnetising force of armature reaction of the amplidyne *i.e.,* to over-compensate for the armature reaction, which is necessary under certain amplidyne operating conditions. But usually the compensating winding is shunted by a rheostat to reduce the magnetising force. This allows the slope of the external characteristic to be regulated. In the case of under compensation, the voltage drop, with increasing load, will be due not only to the voltage drop in the armature, but also to armature reaction. In the case of full compensation of armature reaction a fall in the amplidyne voltage with increasing load will be due only to voltage drop in the armature. With slight overcompensation, an increase in flux with the load can be made to compensate exactly for the increase in voltage drop in the amplidyne armature, in which case the voltage of the amplidyne will not vary with the load. However, in the case of extreme overcompensation, undesirable self excitation of the amplidyne *i.e.*, spontaneous rise in voltage and load current at constant control current (or with the control circuit open) can occur. Such self-excitation (similar to that of generator) can continue until saturation of the amplidyne magnetic system becomes effective. Self excitation can cause considerable overloading of the amplidyne and totally disrupt operation of the system containing the amplidyne.

In partial compensated machine the opposing flux produced by the output current I_L would neutralize most of the control flux. In control engineering terms, it is a negative current feedback tending to suppress any increase of output current. The action is sufficient to maintain the output current almost constant regardless of change in load resistance. The machine operated in this way is called the Metadyne. In such cases electrical input and output powers are the same, neglecting copper losses and the only mechanical power that is required to overcome the mechanical losses. The output current is controlled by the current through control windings. The output characteristics of Metadyne with different control currents are shown in Fig. 2.6(*b*).

Note : From the characteristics shown in Figs. 2.6(*a*) and 2.6(*b*) for amplidyne and metadyne respectively it is obvious that an amplidyne should never be operated on short circuit and a metadyne on open circuit.

Metadyne Generator Rating. The rated output power of a metadyne is appreciably less than that obtainable from the same armature in a normal generator. First; the output current has to be reduced by about 10 per cent, owing to the heating caused by the quadrature current I_1 flowing in the armature. Secondly the provision of the commutating zones for two sets of brushes reduces the effective pole-face area, and hence the total flux and terminal voltage by about one-third. The rated output power of the armature is thus about $0.9 \times \frac{2}{3} = 0.6$ of that of the same armature in an ordinary dc generator. For this reason it is not economical to use metadynes for large output power, and, above 15 to 20 kW separately excited dc generators are employed with metadynes as exciters to provide similar results as from a metadyne generator.

Applications of Amplidynes. An amplidyne is widely used for various purposes, particularly in feedback control systems where the regulated power requirement is in the range of 1 to 50 kW. It may sometimes be found economical to employ it as a voltage regulator in the excitation system of a large alternator to buck or boost voltage in series with the field winding of the main exciter. It can also be employed as a generator in a Ward-Leonard system of speed or position control. In modern control systems these functions of an amplidyne have been taken over by SCR circuitry.

Amplidynes are employed in installations for controlling reversing rolling mills, cold rolling mills, mine hoists, metal cutting and paper machines. However, amplidynes posses a number of shortcomings: poor commutation, which sometimes disrupts the operating conditions of the installation, and a low efficiency of operation. There is a trend in industry to replace amplidynes by static devices, such as magnetic amplifiers operating in conjunction with transistor amplifiers.

2.2.4. Magnetic Amplifiers

In a magnetic amplifier, the basic component is a steel-cored coil with an additional winding energized by direct current. It is intended for controlling alternating current of relatively large power by means of a low power dc.

The simplest practical magnetic amplifier has two identical steel cores with identical ac windings, called the *load winding,* on each. In addition they have a common dc winding called the *control winding.* The ac windings are connected together in series or in parallel, and in series with them there is a load. The series connection is used when a short-time response and a high voltage are required, usually in controlling loads requiring low power. The parallel connection, on the other hand, has a slow response of approximate 1 to 3 seconds, but permits control of large current. The control winding is energized by dc.

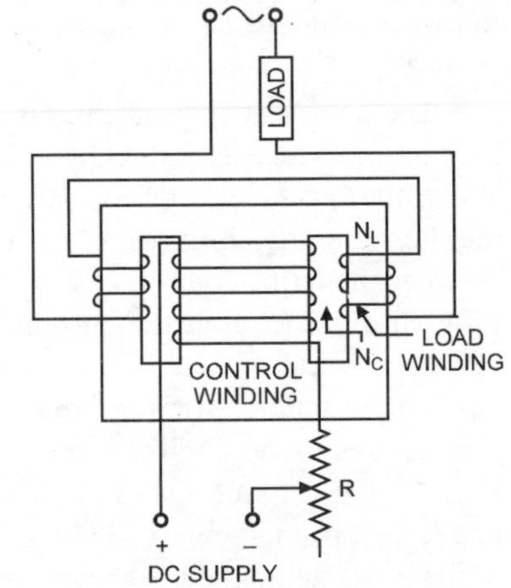

Fig. 2.7. ***Simplest Magnetic Amplifier Circuit***

With no current in control winding, the ac winding has a large inductance, and a high reactance is offered to an ac potential. As a result alternating current to the load is limited to a low value by the inductive reactance, and the load voltage is small. When a dc voltage is now applied to control winding the dc magnetic flux passes through the core and the core approaches magnetic saturation, even when a small value of dc voltage is applied. This causes the effective inductance of the ac winding to decrease, so that the ac winding impedance drops. Consequently the alternating current flowing through the ac winding (and the load voltage) increases as the flow of direct current through the control winding increases. Thus, varying the magnitude of control current over a small range causes the load voltage to vary over a wide range.

If a magnetic amplifier consisted only of one core with two windings (ac and control), the load current would be noticeably non-sinusoidal. This is one of the reasons why the simplest practical magnetic amplifier is made double. A second reason is that for one core, the control winding would be linked with an alternating magnetic flux, and an alternating emf would be induced in it. This emf could become very significant for a large value of the ratio $\frac{N_C}{N_L}$. With two cores, the total flux linking the control winding does not have an alternating component and, hence, no alternating emf. is induced in it. At present there are a large number of different magnetic amplifier circuits and types of construction.

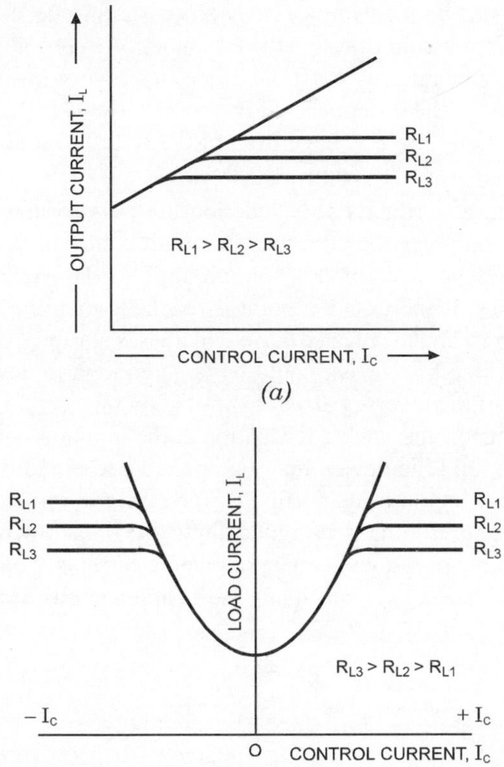

(a)

(b)

Fig. 2.8. ***Dynamic Characteristic of Non-Polarised Magnetic Amplifier***

The output or transfer characteristic curves of a magnetic amplifier are shown in Fig. 2.8 for various load resistances. From the curves it is obvious that with the exception of a small region near the origin the load current is directly proportional to the control current up to a value determined by the load resistance, and secondly that the characteristic curve does not begin from the origin but intersects the ordinate axis above the origin. This is due to the current flowing in ac winding which is determined by the applied voltage and the inductive reactance of the winding. Thus for zero control current, load current, I_L will be equal to $\frac{E}{X_L}$, known as *quiescent current.*

Magnetic amplifiers have got following advantages over electronic ones :

1. Mechanically sturdy and inherently shockproof.
2. Relatively inexpensive amplifiers can be designed to have very high gain, with low input power requirements, and to be compact and light in weight.
3. It does not have any filament to be heated as is in the case of vacuum tube amplifiers.
4. It can be designed for a wide range of input and output impedances and since the control and load circuits are not conductively coupled, they provide somewhat more flexibility than vacuum tube circuits.

On the other hand magnetic amplifiers have limited frequency response and produce an output that is a modulated ac carrier. They have a much lower input impedance than vacuum tubes and provide only a finite power gain where as vacuum tube may theoretically at least, give an infinite power gain. Other disadvantage is that magnetic amplifier, in general, is required to be designed specifically for each application, since both the supply voltage and the load impedance critically affect the operation.

Magnetic amplifiers are used in electrical systems of control, measurement and automatic regulation as power amplifiers (for example for automatic control of ac and dc motors or in smooth regulation of lighting).

The schematic diagram of a magnetic amplifier control circuit for squirrel cage motor starting as a function of voltage is shown in Fig. 2.9.

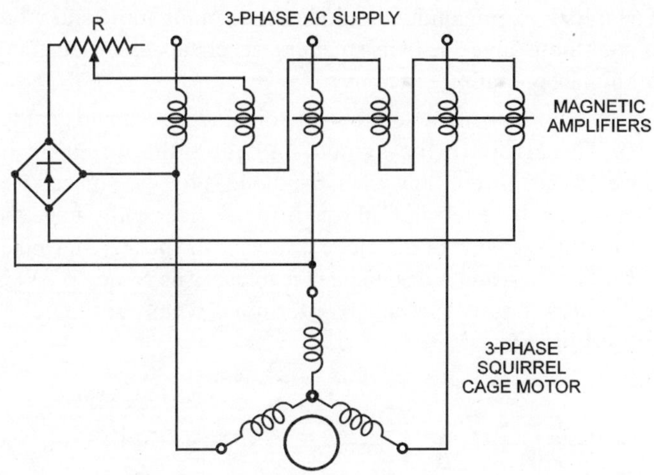

Fig. 2.9. ***Magnetic Amplifier Control Circuit For Squirrel Cage Motor Starting***

When the ac supply is applied to the motor for starting, the load windings of the magnetic amplifier have a high inductive reactance and only a reduced voltage acts across the stator terminals of the motor. As the motor begins to gain speed, the starting current drops and allows the voltage at the stator terminals to rise. Two of the stator winding terminals are connected to a semiconductor bridge rectifier which serves to excite the control windings of all the three magnetic amplifiers. As the voltage across the stator terminals rises during starting, the direct current through the control windings also rises. As a result, the inductive reactance of the ac windings is reduced and the voltage at the stator terminals rises. The control winding current, due to rise in stator terminal voltage, increases until it finally saturates the amplifiers. By this time the inductive reactance of the ac winding drops practically to zero and the motor operates at full voltage.

The character of motor current and torque change with time during starting may be adjusted by setting up a suitable ratio of the electromagnetic time constant of the magnetic amplifier control winding circuit to the electromechanical time constant of the drive.

2.3. SOLID STATE DEVICES*

Solid state devices find more and more applications in industry and, therefore, it becomes imperative to have preliminary knowledge of such devices. The main devices usually used for motor control are diode, thyristor or silicon controlled rectifier, triac, diac, power transistors etc. Every device has certain maximum voltage, current and time response (speed) capabilities. These can very rarely be achieved simultaneously in a single device. Only one of these is realized in one device.

2.3.1. Diode

'The term *diode* denotes a two-electrode device. A *semiconductor diode* is simply a P-N junction with connecting leads or terminals on the two sides of the P-N junction. A diode is a unidirectional device permitting the easy flow of current in one direction but restraining the flow in opposite direction. A major application of diodes is in *rectification i.e.,* in conversion of ac into dc. Semiconductor diode is becoming more and more popular these days due to its smaller size, cheapness, robustness and higher operating efficiency.

The symbol for a semiconductor diode is shown in Fig. 2.10. The arrow in the symbol indicates the direction of conventional current flow when the diode is forward biased *i.e.,* from the positive terminal through the device to the negative terminal. The P-side of the diode is always the positive terminal for forward bias and is designated the *anode*. The N-side is called the *cathode* and is the negative terminal when the device is forward biased.

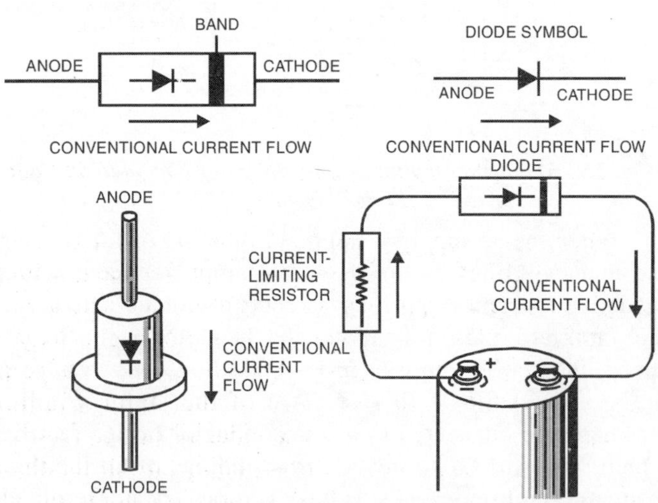

Fig. 2.10. *Semiconductor Diode*

The semiconductor diode may be either silicon one or germanium one. The silicon junction diode is similar in appearance to the germanium diode but differs in properties. Silicon diodes have, in general, higher peak inverse voltage (PIV) and current ratings and wider temperature ranges than germanium diode.

The volt-ampere characteristic of a diode are illustrated in Fig. 2.11. When the junction is in forward biased very little current, called the *forward current,* flows until the forward voltage exceeds the junction barrier potential (0.3 V for Ge and 0.7 V for Si). The characteristic follows an exponential law. With the increase in forward voltage forward current increases almost linearly and the P-N junction starts behaving as a resistor and when the applied voltage exceeds a certain value, extremely large current would flow and the P-N junction may get damaged due to overheating. A practical diode has a very low forward resistance (of the order of 75 Ω in case of Ge and 150 Ω in case of silicon) and has a voltage drop of 0.3 to 0.7 V at all current levels in forward direction.

When an ordinary P-N junction diode is reverse biased, normally only very small reverse saturation current flows. It is almost independent of the voltage applied. However, if the reverse bias is increased, a point is reached when the junction breaks down and the reverse current increases abruptly, as shown in Fig. 2.11. This current could be large enough to destroy the junction. If the reverse current is limited by means of a suitable series resistor, the power dissipation at the junction will not be excessive, and the device may be operated continuously in its breakdown region to its normal (reverse saturation) level. It is found that for a suitably designed diode, the breakdown voltage is very stable over a wide range of reverse currents. This quality gives the *breakdown diode* many useful applications as a *voltage reference source.*

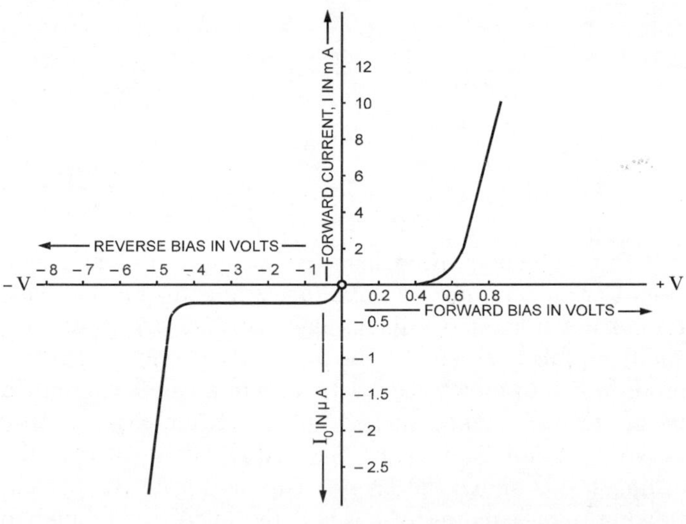

Fig. 2.11. *Volt-Ampere Characteristics of Semiconductor Diode*

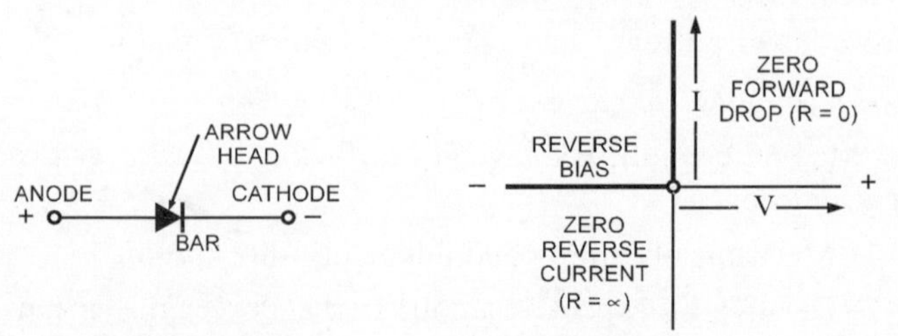

(a) Ideal Diode *(b) Ideal Diode Characteristics*

Fig. 2.12

An *ideal diode* presents zero impedance to current flow in one direction and infinite impedance in opposite direction. The symbol and V-I characteristic of an ideal diode are shown in Fig. 2.12. An ideal diode acts like an *automatic switch.*

* For details refer Electronic Devices and Circuits by J.B. Gupta or any other standard book on the subject.

The switch is closed when the diode is forward biased and is opened when reverse biased.

The main parameters of a diode are blocking or peak inverse voltage or PIV (about 1,000 V for Si and 400 V for Ge), average forward current and maximum operating junction temperature (nearly 125°C for Si and 60° for Ge). Protection is necessarily provided against voltage surges, overcurrents and excessive temperature rise.

A "freewheeling" diode is commonly employed in motor control systems to provide an alternate path for continuity of current in the inductive circuit following the switching off of some power device between the motor and the supply source.

Zener Diode. Zener diode, also sometimes called the *breakdown diode,* is a P-N junction diode specially designed for operation in the breakdown region in reverse bias condition. The breakdown diode may be silicon or germanium one but silicon is preferred over germanium because of higher operating temperature and current capability. The knee point is also more sharp in case of a silicon diode. These diodes are analogous to gas diodes in which large current appears on reaching the breakdown potential.

The symbol of a zener diode is given in Fig. 2.13. This is similar to that of a normal diode except that the line representing the cathode is bent at both ends *i.e.,* the bar is turned into Z-shape; stands for zener.

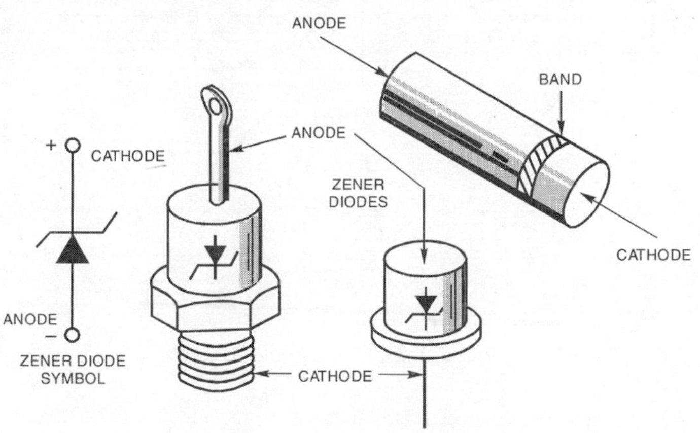

Fig. 2.13. ***Zener Diode***

The zener diode, though not a true power controlling device, is quite often used as a voltage control and sensing device in many motor controllers.

Power Diode. Diodes find many applications in electronics and electrical engineering circuits. Power diodes play a significant role in power electronics for conversion of electric power. A diode acts as a switch to perform various functions, such as switching in rectifiers, freewheeling in switching regulators, charge reversal of capacitor and energy transfer between components, voltage isolation, energy feedback from load to the power source, and trapped energy recovery.

Power diodes can be assumed as ideal switches for most applications but practical diodes differ from the ideal characteristics and have certain limitations. The power diodes are similar to P-N junction signal or low-power diodes. However, the power diodes have large power–, voltage–, and current-handling capabilities than those of ordinary diodes. The frequency response (or switching speed) is low in comparison to that of ordinary diodes.

Power diodes are of three types: General purpose, high speed (or fast-recovery), and Schottky. General-purpose diodes are available up to 6000 V, 4500 A, and the rating of fast-recovery diodes can go up to 6000 V, 1100 A. The reverse recovery time varies between 0.1 and 5 μs. The fast-recovery diodes are essentially required for high-frequency switching of power converters. Schottky diodes have low on-state voltage and very small recovery time, typically nanoseconds. The leakage current increases with the voltage rating and their ratings are limited to 100 V, 300 A. A diode conducts when its anode voltage is higher than that of the cathode; and the forward voltage drop of a power diode is very small, typically 0.5 and 1.2 V.

2.3.2. Thyristor

Thyristor is the general name given to a family of semiconductor devices having four layers with a control mechanism, although this term is most commonly applied to the SCR (silicon controlled rectifier). This term is derived from *thyratron* and *transistor* because the device combines the rectification action of thyratron and control action of transistor.

Since its inception the thyristor has come to stay as a basic building block in many industrial and power system applications. Its ability to be controlled, compactness, fast response, high reliability, better efficiency, large power handling capacity, high voltage and current ratings, good trigger sensitivity, static operation, large power gain, sturdy construction, long life, very little maintenance and low cost of fabrication—due to advancement in the field of fabrication—have given the thyristor a colourful reception in every field. Today thyristors are finding applications in the control of dc/ac motors; for the improvement of power factor; and as switching devices. Thyristors have helped in further cost reduction and in the development of drive system by changing the emphasis from dc motors to ac motors. With cycle converters and inverters, the speed of an ac motor can also be controlled with ease and reliability. Apart from these main applications it finds use as a switching device, particularly in the improvement of power factor of transmission lines and mains. Thyristor can be used as a power switching device with a power handling capacity ranging from a few watts to as high as 4 MW (2,500 A at 1,600 V). Some thyristors have a rating as high as 35,000 A, 10,000 V for use in hv dc transmission lines.

Thyristors, with its large number of advantages and tremendous control capabilities, have numerous applications and have completely replaced the electromagnetic control systems. Thyristor basically serves two functions viz electronic switching and electronic control. Some of the applications of thyristors are listed below:

1. Speed control of dc and ac motors.
2. As rectifier for conversion of ac into dc.
3. As inverter for conversion of dc into ac.
4. As dc chopper or dc to dc converter for converting dc at one level to dc at another level.
5. As cycloconverter for converting ac of one frequency into ac of another frequency.
6. Control of temperature, level, position and illumination.
7. Power switches (dc and ac circuit breakers).

8. As static switches.
9. Control of induction heating.
10. Relay control.
11. Phase control.

SCRs and triacs having high voltage and current ratings are widely employed for power control applications whereas other members of thyristor family are employed for small power applications and for switching in control and digital circuits.

2.3.3. Silicon Controlled Rectifier (SCR)

The SCR is a four-layer, three-junction and a three-terminal device and is shown in Fig. 2.14 (*a*). The end P-region is the anode, the end N-region is the cathode and the inner P-region is the gate. The anode to cathode is connected in series with the load circuit. Essentially the device is a switch. Ideally it remains off (voltage blocking state), or appears to have an infinite impedance until both the anode and gate terminals have suitable positive voltages with respect to the cathode terminal. The SCR then *switches on* and current flows and continues to conduct without further gate signals. Ideally the SCR has zero impedance in conduction state. For switching off or reverting to the blocking state, there must be no gate signal and the anode current must be reduced to zero. Current can flow only in one direction. Now SCRs of voltage rating 10 kV and an rms current rating of 35,000 A with speed of 1 μs corresponding to a power-handling capacity of 30 MW are available. It can be switched on by a low voltage supply of about 1 A and 10 W. Obviously it has tremendous power amplification (of the order of 3 × 10^6). Because of its compactness, high reliability and low loss, the SCR has almost replaced the earlier power switching devices—thyratron and a magnetic amplifier.

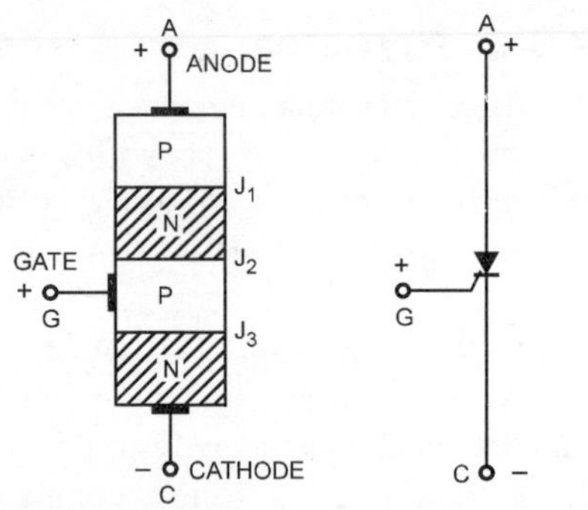

(a) Schematic Diagram *(b) Symbolic Representation*

Fig. 2.14. ***Silicon Controlled Rectifier***

Volt-Ampere Characteristics. As mentioned above, the SCR is a four layer device with three terminals, namely, the anode, the cathode and the gate. When the anode is made positive with respect to the cathode, junctions J_1 and J_3 are forward biased and junction J_2 is reverse biased and only the leakage current will flow through the device. The SCR is then said to be in the *forward blocking state* or in the *forward mode* or *off state*. But when the cathode is made positive with respect to the anode, junctions J_1 and J_3 are reverse biased, a small reverse leakage current will flow through the SCR and the SCR is said to be in the *reverse blocking state* or *in reverse mode*.

When the anode is positive with respect to cathode *i.e.,* when the SCR is in forward mode, the SCR does not conduct unless the forward voltage exceeds certain value, called the *forward breakover voltage*, V_{FBO}. In non-conducting state, the current through the SCR is the leakage current which is very small and is negligible. If a positive gate current is supplied, the SCR can become conducting at a voltage much lesser than forward breakover voltage. The larger the gate current, lower the breakover voltage. [Fig. 2.15(*c*)] With sufficiently large gate current, the SCR behaves identical to PN rectifier. Once the SCR is switched on, the forward voltage drop across it is suddenly reduced to very small value, say about 1 volt. In the *conducting* or *on-state*, the current through the SCR is limited by the external impedance.

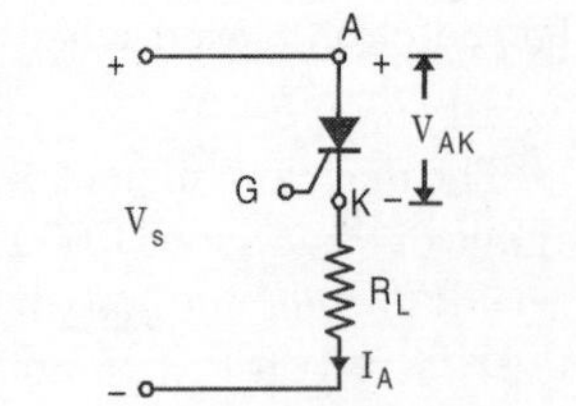

(a) Circuit Diagram

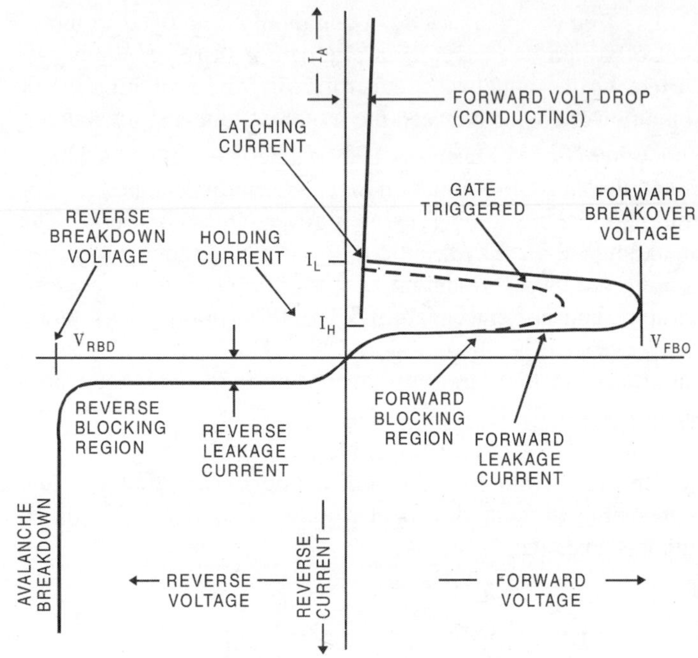

(b) V-I Characteristics

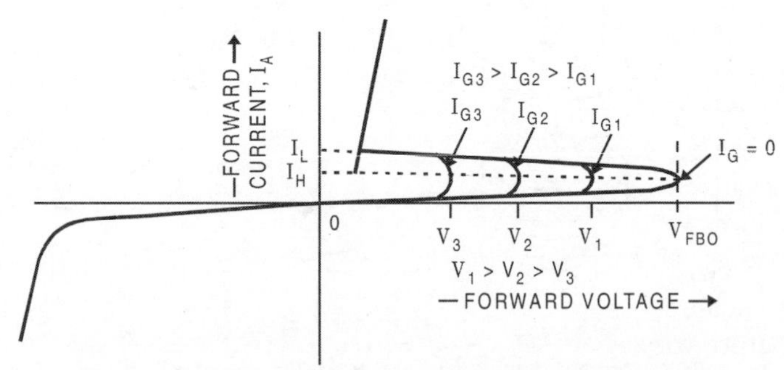

(c) Effects of Gate Current On Forward Blocking Voltage

Fig. 2.15

When the anode is negative with respect to cathode *i.e.,* when the SCR is in reverse mode or in blocking state no current flows through the SCR except very small leakage current of the order of few microamperes. But if the reverse voltage is increased beyond a certain value, called the *reverse breakover voltage*, V_{RBO} *avalanche breakdown* takes place. Forward breakover voltage V_{FBO} is usually higher than reverse breakover voltage, V_{RBO}.

From the foregoing discussion, it can be seen that the SCR has two stable and reversible operating states. The change over from *off-state* to *on-state*, called *turn-on*, can be achieved by increasing the forward voltage beyond V_{FBO}. A more convenient and useful method of turning on the device employs the gate drive. If the forward voltage is less than the forward breakover voltage, V_{FBO}, it can be turned on by applying a positive voltage between the gate and the cathode. This method is called the *gate control*. Another very important feature of the gate is that once the SCR is triggered to on-state the gate loses its control.

The switching action of gate takes place only when

(*i*) SCR is forward biased *i.e.,* anode is positive with respect to cathode, and

(*ii*) suitable positive voltage is applied between the gate and the cathode.

Once the SCR has been switched on, it has no control on the amount of current flowing through it. The current through the SCR is entirely controlled by the external impedance connected in the circuit and the applied voltage. There is, however, a very small, about 1 V, potential drop across the SCR. The forward current through the SCR can be reduced by reducing the applied voltage or by increasing the circuit impedance. There is, however, a minimum forward current that must be maintained to keep the SCR in conducting state. This is called the *holding current* rating of SCR. If the current through the SCR is reduced below the level of holding current, the device returns to off-state or blocking state.

The SCR can be switched off by reducing the forward current below the level of holding current which may be done either by reducing the applied voltage or by increasing the circuit impedance.

Note : *The gate can only trigger or switch on the SCR, it cannot switch off.*

Alternatively the SCR can be *switched off* by applying negative voltage to the anode (reverse mode), the SCR naturally will be *switched off.*

Here one point is *noteworthy*, the SCR takes certain time to switch off. The time, called the *turn-off time,* must be allowed before forward voltage may be applied again otherwise the device will switch on with forward voltage without any gate pulse. The turn-off time is about 15 microseconds, which is immaterial when dealing with power frequency, but this becomes important in the inverter circuits, which are to operate at high frequency.

There are mainly two types of SCRs for use in motor controllers—*inverter grade* and *converter grade.* Former ones are used in inverters, cycloconverters, and brushless dc motor systems while the latters are used in choppers, phase-controlled rectifiers, ac regulators etc. These two grades vary in their time responses. Inverter grade SCRs are costlier than the converter grade.

Typical SCR Parameters. Commercially available SCRs have a wide range of voltage and current ratings, turn-on and turn-off times and dynamic resistances in cut-off and in saturation regions. Typical values of these parameters for SCRs are given below in tabular form.

Table 2.1. Typical SCR Parameters

S. No.	Parameters	Typical values
1	Forward Breakover voltage, V_{FBO}	50 to 500 volts
2	Maximum on-state voltage	About 1.5 V
3	Peak reverse voltage, PRV	Up to 2.5 kV
4	Holding voltage, V_H	0.5 to 20 volts
5	Forward breakover current	Less than a few hundred μA
6	Peak forward current	30 A to over 100 A
7	Holding current	A few mA to few hundred mA
8	Turn-on and turn-off times	A few tenths of μs for fast acting SCRs
		A few μs for slow acting SCRs
9	Dynamic resistance in cut-off region	A few MΩ to a few hundred MΩ
10	Dynamic resistance in saturation region	Lesser than 1 Ω for currents of several amperes Lesser than 10 Ω for large currents

2.3.4. Triac

The triac is another three-terminal ac switch that is triggered into conduction when a low-energy signal is applied to its gate terminal. Unlike the SCR, the triac conducts in either direction when turned on. The triac also differs from the SCR in that either a positive or negative gate signal triggers it into conduction. Thus the triac is a three terminal, four layer bidirectional semiconductor device that controls ac power whereas an SCR controls dc power or forward biased half cycles of ac in a load. Because of its bidirectional conduction property, the triac is widely used in the field of power electronics for control purposes. Triacs of 16 kW rating are readily available in the market.

"Triac" is an abbreviation for three terminal ac switch. 'Tri'-indicates that the device has three terminals and 'ac' indicates that the device controls alternating current or can conduct in either direction.

Though the triac can be turned *on* without any gate current provided the supply voltage becomes equal to the breakover voltage of the triac but the normal way to turn on the triac is by applying a proper gate current. As in case of SCR, here too, the larger the gate current, the smaller the supply voltage at which the triac is turned *on*.

Triac can conduct current irrespective of the voltage polarities of terminals MT_1 and MT_2 with respect to each other and that of gate and terminal MT_2.

Typical V-I characteristic of a triac is shown in Fig. 2.16 (*b*). The triac has on- and off-state characteristics similar to SCR but now the characteristic is applicable to both positive and negative voltages. This is expected because triac consists of two SCRs connected in parallel but opposite in directions.

Next to SCR, the triac is the most widely used member of the thyristor family. In fact, in many of control applications, it has replaced SCR by virtue of its bidirectional conductivity. Motor speed regulation, temperature control, illumination control, liquid level control, phase control circuits, power switches etc. are some of its main applications.

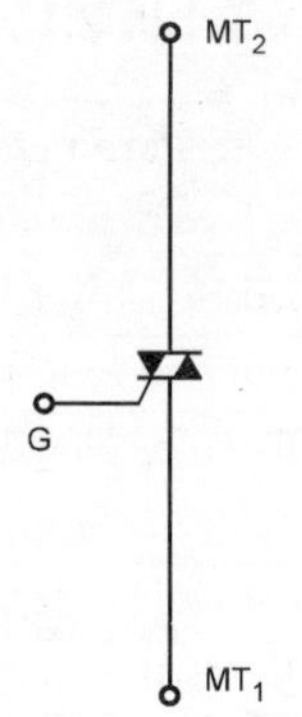

(a) Schematic Symbol

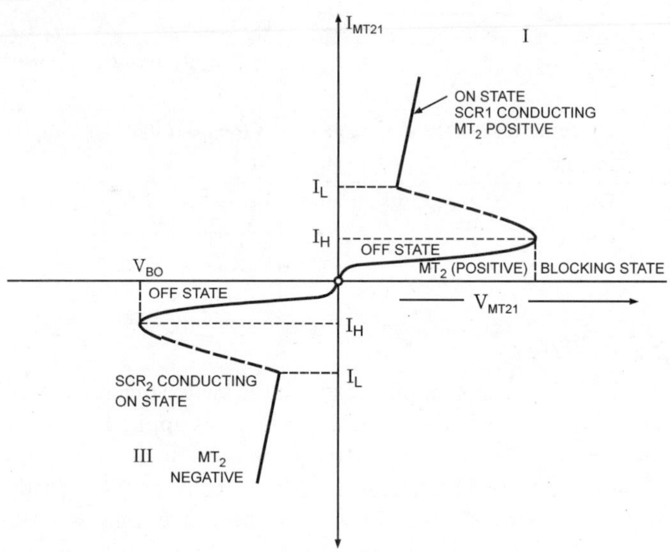

Fig. 2.16. ***Triac***

2.3.5. Diac

A diac is an important member of the thyristor family and is usually employed for triggering triacs. A diac is a two-electrode bidirectional avalanche diode which can be switched from off-state to the on-state for either polarity of the applied voltage. This is just like a triac without gate terminal. Its equivalent circuit is a pair of inverted four layer diodes. Two schematic symbols are shown in Fig. 2.17 (*a*). Again the terminal designations are arbitrary since the diac, like triac, is also a bilateral device.

The switching from off-state to on-state is achieved by simply exceeding the avalanche breakdown voltage in either direction. Volt-ampere characteristic is shown in Fig. 2.17 (*b*).

The diacs, because of their symmetrical bidirectional switching characteristics, are widely used as triggering devices in triac phase control circuits used in heat control, universal motor speed control etc.

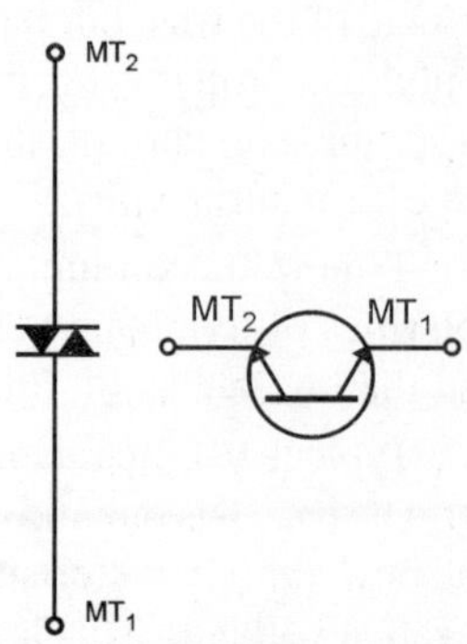

(a) Schematic Symbols

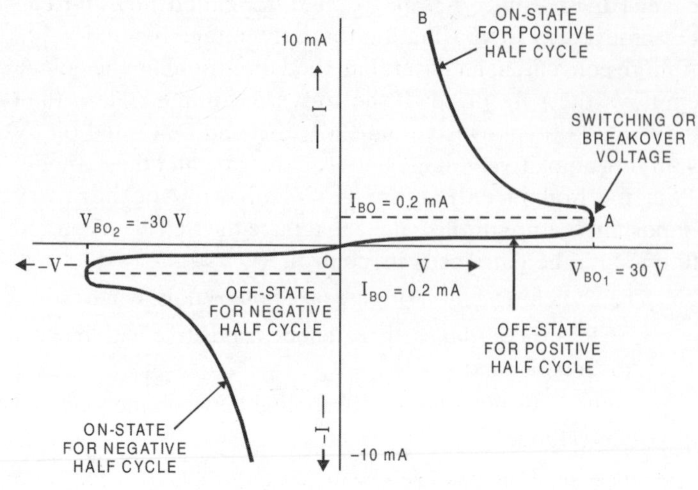

(b) V-I Characteristics

Fig. 2.17. ***Diac***

2.3.6. Gate Turn-Off (GTO) Thyristors

Conventional thyristors are used as ideal switches in power electronic applications. They have the capability of blocking several thousand volts in the *off*-state and conducting several thousand amperes in the *on*-state with only small *on*-state voltage drop. The only drawback of conventional thyristor is that when the gate turns into *on*-state, the gate loses the control over the device and, therefore, it cannot be turned *off* without the technique called commutation. The circuitry required for commutation makes the device costlier and bulky. This can be avoided by using a device of thyristor family, known as the gate turn-off thyristor (GTO).

A GTO like an SCR can be *turned on* by applying a positive gate signal. However, a GTO can be *turned off* by a negative gate signal. A GTO is a non-latching device and can be built with current and voltage ratings similar to those of an SCR. The device incorporates the character of both of a thyristor and a transistor.

The GTOs have the advantages over SCRs : (*i*) elimination of commutating components in forced commutation, resulting in reduction in cost, weight and volume (*ii*) faster turn-off permitting high switching frequencies (*iii*) improved efficiency of converters (*iv*) reduction in accoustic and electromagnetic noise due to elimination of commutation chokes (*v*) high blocking voltage and large current capability.

In low-power applications, GTOs have the following advantages over BJTs :

1. Higher blocking voltage capability.
2. High ratio of peak controllable current to average current.
3. High ratio of peak surge current to average current, typically 10 : 1.
4. High on-state gain (anode current and gate current) typically 600.
5. Pulsed gate signal of short duration.

Under surge conditions, a GTO goes into deeper saturation due to regenerative action. On the other hand, a BJT tends to come out of saturation.

However, the GTO has the following disadvantages :

1. High latching and holding currents.
2. More on-state voltage drop and associated loss.

3. Higher triggering gate current due to multi-cathode structure as compared to conventional thyristor.

The symbols of GTO are shown in Fig. 2.18.

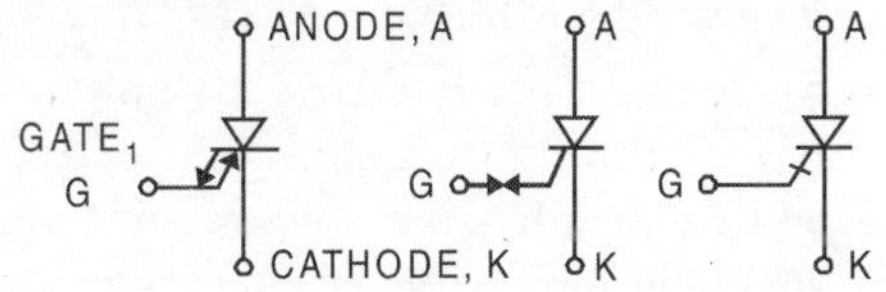

(*i*) (*ii*) (*iii*)

Fig. 2.18. ***Circuit Symbols For GTO***

Applications : GTOs are mostly used in voltage-source converters in which a fast recovery antiparallel diode is required across each GTO. Thus, GTOs normally do not require reverse voltage capability. Such GTOs are known as *asymmetric* GTOs. This is achieved by a so-called buffer layer, a heavily doped N^+ layer at the end of the N layer. Asymmetric GTOs have lower voltage drop and higher voltage and current ratings.

GTO devices are now used in (*i*) high performance drive systems, such as the field-oriented control scheme used in rolling mills, robotics and machine tools, (*ii*) traction purposes because of their lighter weight and (*iii*) adjustable-frequency inverter drives. At present, GTOs with ratings up to 5,000 V and 3,000 A are available.

2.3.7. Power Transistors

The transistor is a two junction, three terminal, current controlled solid state device. This device is a 3-layer P-N-P (or N-P-N) structure as illustrated in Fig. 2.19. The transistor is capable of amplification and in most respect it is analogous to a vacuum triode. The transistor is only six decade old, yet it has replaced vacuum tubes in almost all applications. The reasons are obviously its advantages over vacuum tubes such as compact size, light weight, rugged construction more resistive to shocks and vibrations, instantaneous operation (no heating required), low operating voltage, high operating efficiency and long life with essentially no ageing effect if operated within permissible limits of temperature and frequency.

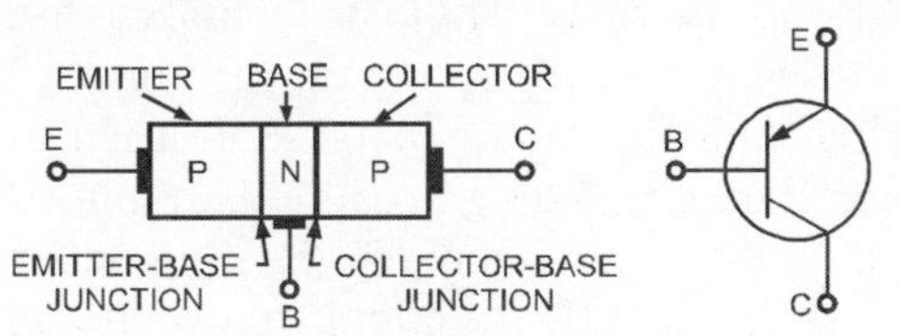

(a) PNP Transistor

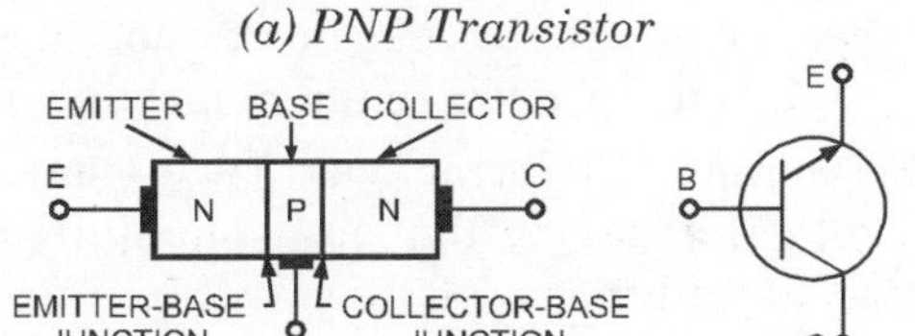

(b) NPN Transistor

Fig. 2.19

In normal operation the emitter-base junction is forward biased and collector-base junction is reverse biased. The collector current is almost equal but slightly less than the emitter current. The collector current can be controlled by base current. Except for small motors, transistors are used in the switching (on/off) mode. The common-emitter configuration is normally employed as it has high power gain.

Power transistors are now available in ratings suitable for motor control. The common-emitter saturation voltage $V_{CE\,(Sat)}$ for typical power transistors varies from 0.3 to 0.7 V. This is sufficiently less than the on-state anode-cathode voltage drop of an SCR, and therefore, the average power loss in a transistor is less than that of an SCR of an equivalent power rating.

The switching of a power transistor is controlled through the base current. A continuous base current is required to keep the transistor in the saturated state, *i.e.,* the on-state. In a high-current device, a base current of several amperes is required to keep the device in the on-state owing to its low current gain (about 10) and, therefore, the power loss in the base circuit may be appreciable.

Power transistors have no surge current capacity and can withstand only a low rate of change of current. The switching operations of power transistors are generally faster than those of SCRs, and the commutation problems of SCR are not present in transistors. Power transistors are becoming more and more popular in low-to-medium-power applications, where they compete successfully with thyristors and GTOs. Power transistor applications are in the range of a few kW to several hundred kWs size in voltage-fed choppers and inverters with switching frequency up to 10-15 kHz.

2.4. PILOT DEVICES AND ACCESSORIES

Motor control devices may be classified as primary control devices and pilot control devices. Pilot devices control the primary control devices which in turn controls (connects or disconnects) the motor or load. For example in D-O-L (direct-on-line) starter main contactors and tripple pole switch, which connect the motor to the line are classified as *primary control devices* where as start or stop push buttons which energizes or de-energizes the operating coil are called the *pilot devices.* Pilot devices usually have sensing element to sense temperature, liquid level, fluid pressure or mechanical pressure applied etc. Pilot devices convert what they sense into the control of primary control devices. The most important of pilot devices are push buttons, master switches, float switches, thermostats, pressure regulators, plugging switches, limit switches.

1. *Push Button.* This is the most simple pilot device and provides control by pressing a button. The push-button switches are used in remotely operated electromagnetic apparatus and signalling circuits. Push buttons are usually of two types namely *maintained contact type* in which on pushing of the button the contacts close and remain closed until stop button is pushed and the *momentary contact type* in which contacts remain closed till push button is held pushed. The contacts of push buttons are usually double break, actuated by plastic buttons. Two sets of contacts are usually provided namely normally open contacts 'NOC' and normally closed contacts 'NCC.' so that when the button is

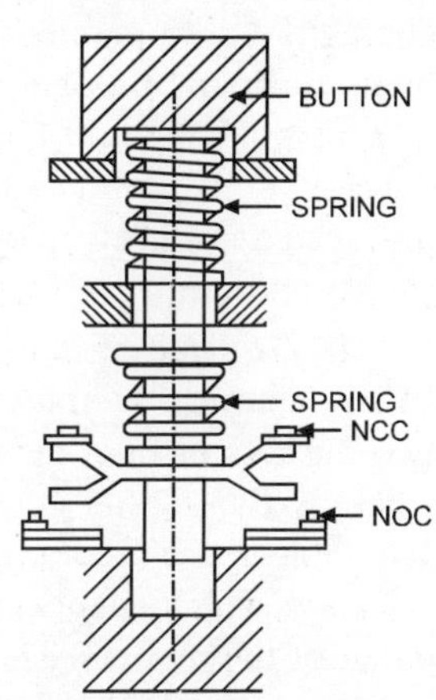

Fig. 2.20. ***Push Button***

pressed one set is opened and the other set is closed. The push button station usually consists of a green button to start and red button to stop. Start push button is provided with normally open contact (NOC) and stop push button is provided with normally closed contact (NCC). The contacts are made silver tipped to withstand numerous of switching operations. Since push buttons are used in control circuits, not used in power circuits, current rating of push button station is, therefore, usually very small (2 A, 500 V for ac working). On pressing start push-button contactor coil is energized, which closes the main contactor in the circuit.

2. *Limit Switches.* Limit switches are ordinarily used as pilot devices in control circuits of magnetic starters to govern starting, stopping, or reversing of electric motors. These may be used either as control devices for regular operation or as emergency switches to prevent improper functioning of machinery. These are actuated by the travel of the driven machinery and automatically disconnect the machine after it has travelled through its entire path. These switches can be actuated by cams, levers, push rods and the like, and also through gears. Lever type actuators may be either of normal roller type or angular type. The switch is either stationary and actuated by some moving member or it moves with the machine and is actuated by a stationary member. These are usually provided with one NOC (normally open contact) and one NCC (normally closed contact). These switches are usually air break type. These are also used for opening and closing of doors, cinema curtains, lifts and elevators etc.

3. *Master Controllers.* A device or a group of devices that governs in a predetermined manner the supply of electric power to an electric apparatus with a view to accelerate, decelerate, start, stop or reverse it according to requirement is called the *controller.* The device performing the above functions in respect of many units at a time is called the *master controller.* Master controllers can employ different methods for closing or opening the contactors namely all-electric method, electropneumatic method and the camshaft method. In the all-electric method, the contactors are operated by solenoids and the drawback of this method is that supply variations may cause uncertainty of action of the contactors. In the electropneumatic method solenoids operate the valves that admit air into cylinders in which movable piston close or open the contactors and in the cam shaft method, the contactors are operated by cams mounted on a shaft driven by an electric or a pneumatic motor. In electropneumatic or in cam shaft method large mechanical forces can be put on the contactors, so that vibration can cause no harm : moreover the systems are electrically simple and mechanically very robust and are, therefore, favoured.

4. *Transfer Switches.* Sometimes operation and control of an electrical equipment becomes necessary from either of two masters of pilot devices, such as control of multiple unit stock by two master controllers at the ends of a train. This is accomplished by making use of switches, called the *transfer switches,* which enable electrical apparatus to be connected to either of the masters. These are merely double throw switches having required number of poles.

5. *Speed Governors.* These are employed for protection of rotating machinery against overspeed, under-speed or to perform some functions at a given speed of the machine. These are actuated by centrifugal force.

6. *Float Switches.* This is a type of limit switch and is actuated by a float through a chain, one end of which is connected to the float and the other to the counterweight. Stops connected to the chain move a lever which turns and tilts the mercury tube forming a switch contact. These are designed and used for maintaining a certain level of liquid. The raising or lowering of a float mechanically attached to electrical contacts may start motor driven pump to empty or to fill tanks, whichever is desired. They are also used for opening or closing piping valves for the control of fluids.

7. *Pressure Switches and Regulators.* The pressure switches and regulators are widely used in industry. These are designed to cover the wide range of requirements encountered in the control of pneumatic or hydraulic machines. These controls are most commonly used in the control circuits of welding equipment, high pressure lubricating system, machine tools and motor driven pumps and compressors. These are available in various pressure ranges. A diaphragm actuated device is designed for applications requiring sensitive response to small pressure changes at low pressure ranges. A seamless metal-bellows actuated control is used for pressures as high as 1.5 kg/mm^2 and the piston operated hydraulic switches are used for pressures up to 10 kg/mm^2.

Gauge type pressures regulators provide accurate control of pressure or vacuum conditions on the systems on which they are used. When used as pilot control devices with magnetic starters, they govern operation of pump or compressor motor similar to pressure switches. Reverse action regulators may also be employed on pressure system interlocks to prevent an operation until the pressure in that system has increased to the point desired.

Pressure regulators consist of a Bourdon type pressure gauge and control relay. Delicate contacts on the gauge energise the relay, causing it to be closed or opened, as the case may be. Thus the relay contacts are used to control a large motor starter, avoiding possible burning of the gauge contacts. Regulators for standard operation open the circuit at high pressure and close the circuit at low pressure. Special regulators operate on reverse operation to close the circuit at high pressure and to open the circuit at low pressure.

8. *Thermostats.* The thermostat is a two or three (or more) wire control pilot device that is temperature actuated. Thermostats indirectly control large motors in air-conditioning systems and in many industrial applications.

General industrial temperature controllers are recommended for heating applications where the temperature to be controlled is higher than the normal or ambient temperature. In general, the application would be in connection with the controlled liquids rather than gases because of the relatively greater conductivity between bulb and liquid when compared to conductivity between bulb and gas. Generally, where air or gas temperature is to be controlled, the sensitivity will decrease and the difference between "on and off" will widen.

Gas vapour or liquid pressure expands a metal bellows against the force of a spring to give a motion to an operating pin, which snaps a precision switch to its operating position when the setting is reached. This is very similar to industrial pressure switches. The operating pressure is generated in proportion to the temperature of liquid in a closed bulb. The precision switch snap back to its normal position when the pressure in the bellows drops enough to allow the main spring to compress the bellows. Thermostats are of many different designs and are used in control circuits to operate heaters, blowers, fans, pumps, solenoid, valves and other devices.

9. *Plugging Switches*. Plugging switches, also called *zero-speed switches* or speed relays, were primarily designed for incorporation into control circuits as pilot devices to provide quick automatic stopping of machines driven by squirrel cage motors. Properly adjusted plugging switches are capable of preventing reversal of rotation of the controlled drive after it has been brought to standstill by the reversal of motor connections. These are often used on machine tools that must come to sudden stop at some point in their cycle of operation to prevent inaccuracies in the work or damage to the machine. Another application is found in processes where the machine must come to a complete stop before the next step of work is started, here the reduced stopping time saves time for a greater total output.

The shaft of the plugging switch, is mechanically coupled to the motor shaft or to a suitable shaft on the driven machine. Motion of the motor or driven machine is transmitted to the switch contacts by a centifugal mechanism or a magnetic induction (eddy current disk) arrangement within the switch. The contacts are wired to the reversing starters which control the motor. Acting as a link between the motor and reversing starter, the switch allows the starter to apply just enough 'reverse' power to bring the motor to a quick stop.

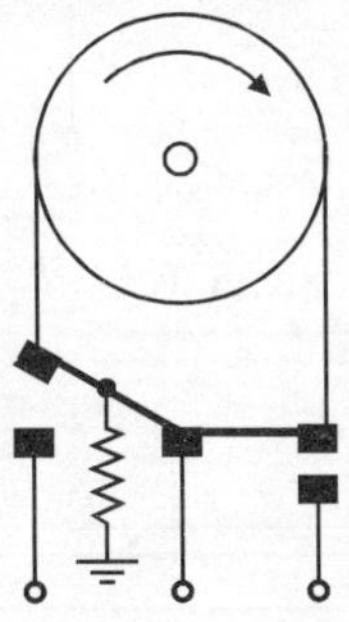

Fig. 2.21. ***Plugging Switch***

2.5. CONTACTORS

Magnetic contactors are electromagnetically operated switches that provide a safe and convenient means for repeatedly making and breaking branch circuits. The principal difference between a contactor and a motor starter is that contactor does not contain overload relays. Contactors are used to switch, with pilot control device, such loads as lighting, heating and controlling electric motors where overload protection is provided separately. The larger sizes are used extensively for remote control of relatively high current circuits where expense prohibits running the power leads to the remote controlling location. This is one of the main advantages of electromagnetic control over manual control. Pilot devices such as push buttons, pressure switches, float switches, limit switches or thermostats are used to provide the necessary control for operating contactors.

The contactor consists of an electromagnet (having an armature, a core and a pick-up coil connected in its control circuit), main contacts (one stationary and the other movable) that make and break the power circuit, electrical interlocks that are used for blocking and control of automation and signal circuits, a blowout structure (a chute with magnetic blow-out or with special grids) that provides quick arc suppression when breaking a circuit.

Contactors are manufactured for ac and dc circuits and may have one or several poles. Besides they may have normally open and (or) normally closed contacts that are switched instantaneously or with some delay. DC contactors have long pick-up coils with a small diameter and a large outside surface, which enhances cooling and enables larger currents to be carried. AC contactors are manufactured for rated voltages of 220, 380 and 500 volts and for rated currents from 20 A to 600 A. They can have from one to five main contacts and from two to four interlocks. AC contactors consist of the same elements as dc contactors. Their magnetic circuit, however, is built up from varnished electrical steel laminations to reduce eddy current losses. Three-pole ac contactors with normally open contacts are the most widespread.

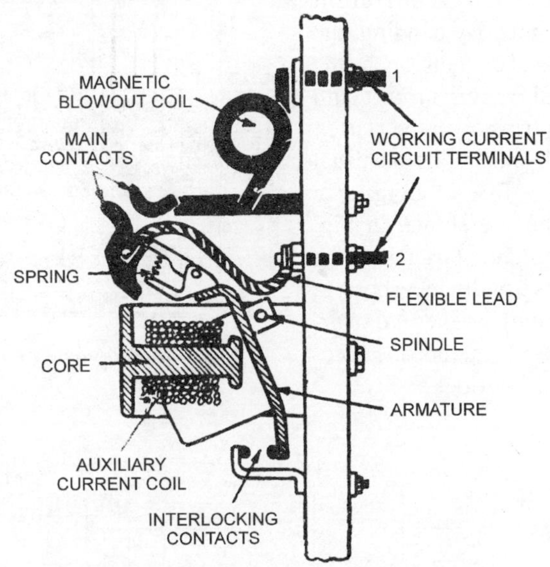

Fig. 2.22 (*a*) *DC Electromagnetic Contactor*

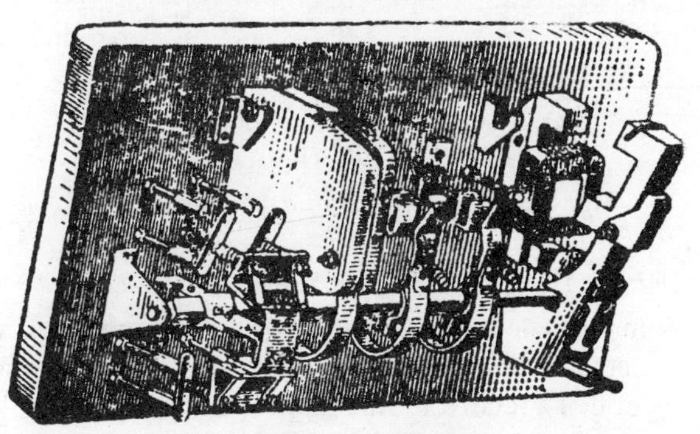
Fig. 2.22 (*b*) *AC Contactor*

Figure 2.22 (*a*) shows the construction of a dc electromagnetic contactor. The auxiliary current passes through the contactor coil and attracts a steel armature to its core; the armature rotates about an axis and closes the main contacts in the working-current circuit. The spring ensures proper pressure of the movable contact. The main circuit is connected to the terminals 1 and 2; its current passes through the magnetic blowout coil, the main contacts and the flexible connecting wire. This contactor closes the contacts in the auxiliary current circuit so-called interlocking contacts which serve to perform the auxiliary control operations in addition to the main circuit. Figure 2.22(*b*) shows a diagram of an ac contactor.

2.5.1. Magnetic Starters

Magnetic starters are special forms of magnetic contactors designed to include thermal overload relays and electrical interlocks. Like contactors, magnetic starters are also used for remote control of electric motors. However, unlike contactors they will disconnect the motor from supply when overload arise. A magnetic starter, shown in Fig. 2.23, essentially comprises a contactor and a set of thermal relays. Each relay consists of two closely bonded strips of different alloys with different temperature coefficients of expansion. When the current flowing through such a relay exceeds its setting, it heats the bimetal element and makes it bend because the two strips increase in length by different amounts. By bending, the bimetal element opens the holding coil circuit and thus trips open the starter. The connections of a usual three-pole magnetic starter are shown in Fig. 2.24. The starter in such cases usually incorporates two built-in thermal overload relays and one electrical interlock.

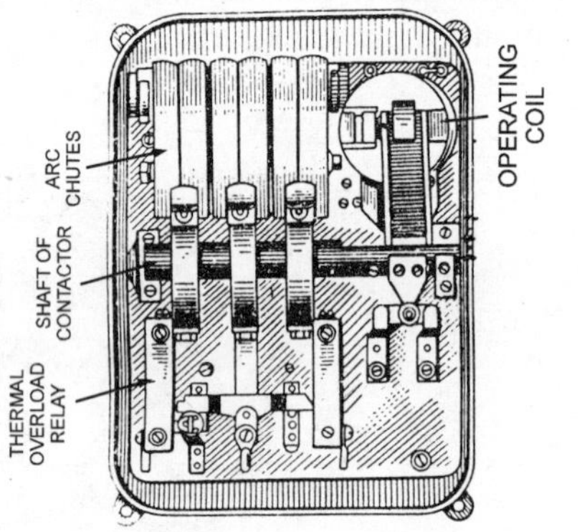

Fig. 2.23. ***Magnetic Starter***

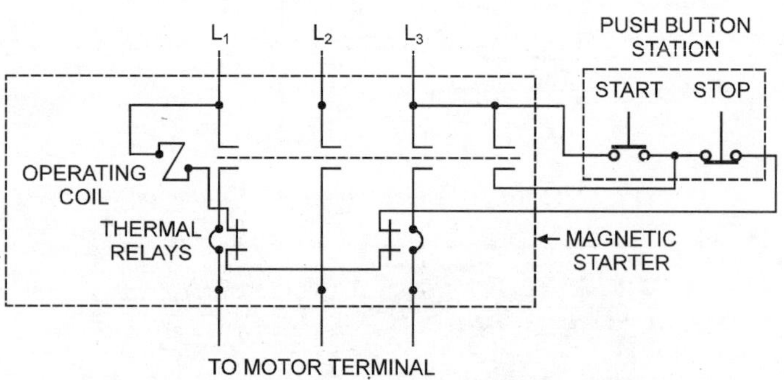

Fig. 2.24. ***Connections of a Magnetic Starter***

2.6. ELECTROMAGNETIC RELAYS

A relay, in the general sense of the word, is any apparatus which serves to actuate or control the regime of a high power system by the action of a relatively small power on the relay. An electric relay is a mechanism which actuates auxiliary electric circuits or mechanical devices, when electrical factors act on it. The relay mechanism consists of a sensing member and an operating member; the former "perceives" the action of the electrical factors and actuates the operating member, which performs some operation.

Relays are widely used for protection against circuit overload, keying remote switching, reversing, control etc. The uses of relays in industrial control circuits are numerous and it will not be possible to describe all of them here. It can generally be said, however, that the relay is used in any industrial circuit in which a large amount of load power is to be controlled by a small amount of control power.

Basic Construction. Figure 2.25 (*a*) shows the basic construction of a typical electromagnetic relay. The electromagnetic relay essentially consists of a coil to which a voltage is applied, a core upon which the coil is wound, relay contacts, and a movable steel armature which is held against one of the contacts by means of the spring. Before the voltage is applied to the coil the relay is said to be in its *normal* or *de-energized* position. Its action depends on the interaction of the magnetic field set up by the coil carrying the current and the movable steel armature.

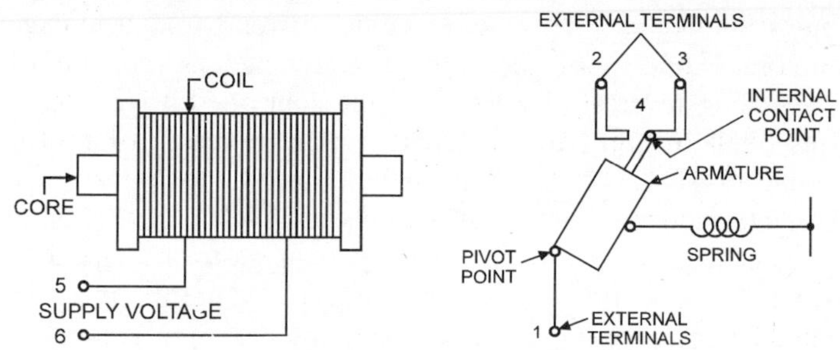

(a) Basic Construction

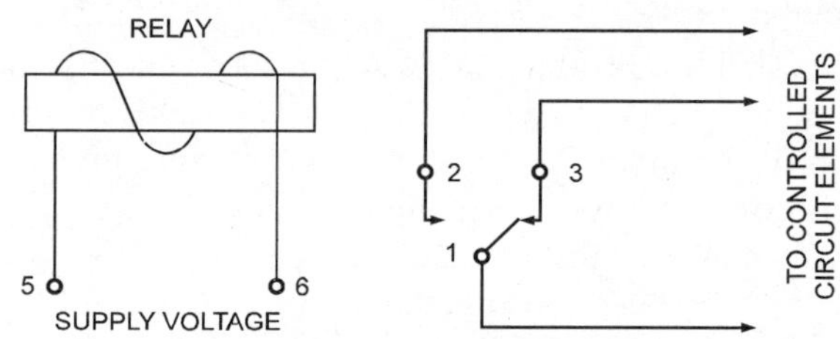

(b) Schematic Interpretation

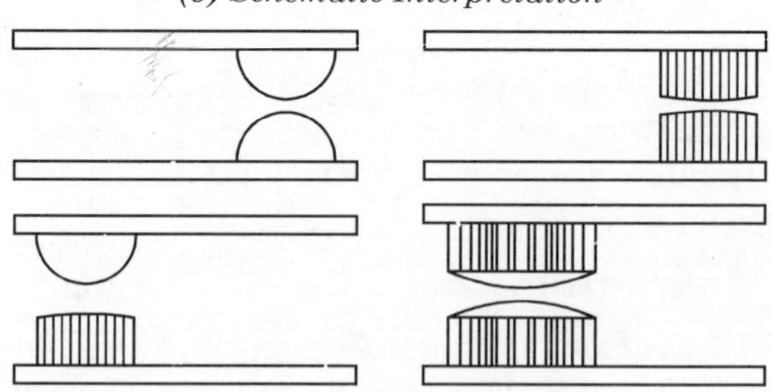

(c) Relay Contacts

Fig. 2.25. ***Electromagnetic Relay***

Operation. In the relay shown in Fig. 2.25(*a*) any external circuit connected between terminals 1 and 3 is in a closed position at this instant and the external circuit connected between terminals 1 and 2 is in a open position. When the voltage is applied to the coil, the relay is energized and the magnetic field set up around the coil makes the armature to be attracted toward it. Contact 4 breaks with terminal 3 but makes with terminal 2, and any external (controlled) circuit connected between terminals 1 and 2 is now completed. The relay remains in the position till the coil remains energized.

The schematic interpretation of this basic relay is illustrated in Fig. 2.25 (*b*). Terminals 2 and 3 are the fixed contacts, and terminal 1 is connected to the movable armature (in some relays the armature remains isolated from all contacts). The contacts of the basic relay acts as single pole double throw (SPDT) switch. Although switch contacts are manually controlled, but relay contacts are remotely controlled.

Relay Contacts. Relay contacts are made of metals and alloys of low resistance in order to prevent overheating. In addition, the metal used in the manufacture of the contacts must be able to resist both the welding action that occurs when the contacts close and the arcing that forms when the contacts open. The shapes of the contacts should be such as to resist the pitty action. Some common shapes are shown in Fig. 2.25 (*c*). Metal used in the manufacture of relay contacts include silver, tungsten, palladium and certain alloys. The current at which any given pair of contacts will interrupt successfully (without arcing) is generally smaller for dc than ac voltages, and is smaller at higher voltages.

AC Relay. Such relays have a *shading coil* placed over a portion of the pole face at the air gap to reduce vibration and chattering of the contacts. The shading coil (a single turn of wire fastened over approximately half of the pole face area) shifts the phase of the flux in the shaded portion of the pole face. This prevents the total force of the flux (taken over the entire pole) from decreasing to zero at any point in the cycle. As a result, chattering is minimized. In certain ac relays, a capacitor is connected in series with the relay coil, producing a frequency-sensitive device. The relay is then energized only when the applied ac signal is at or near the resonant frequency of the coil-capacitor combination.

2.7. PROTECTIVE DEVICES AND INTERLOCKS

Protective devices must be incorporated in the control circuit (or circuits) to interrupt the flow of current to the motor to guard against overloads or other conditions dangerous to the motor or the driven machine, and also to guard against faulty or careless operation of the equipment itself.

The most frequent causes of abnormal operation of electric motors are overloads, open-field, open-phase, reversed phase, overspeed, overheating of bearings or other mechanical parts, low voltage, or voltage failure.

Overload Protection. An overload is an increase in motor current above its safe limit and may be caused by an overload on driven machinery, by a low line voltage, or by an open line polyphase system that results in single phase operation. Overload protection is an essential feature of any controller that is designed to properly protect a motor and still obtain its maximum available power under a range of overload and temperature conditions.

An excess current lasting for only a second or two will not normally cause any serious damage hence protective device should not operate on overloads of short duration and limited magnitude. On the other hand, the protective device must respond to overloads of long duration and of high magnitude which are dangerous to the motor in that they may burn the windings and insulation. So whatever overload protective device is employed, it should possess a time lag to ensure that it operates only for a sustained overload.

Apparatus having an inverse time lag will trip the motor out after a very short interval on a heavy overload but after a relatively long interval on a small overload. Such protection is obviously more satisfactory for most purposes, because the motor temperature will rise more quickly on a heavy overload. This inverse-time characteristic is possessed by certain special types of relays, by a fuse, and also by an ordinary solenoid relay provided with an oil dashpot.

Overload or overcurrent protection may be affected by fuses, electromagnetic and thermal overload relays. Overload relays or fuses need only be placed in two of the three phases of a 3-phase machine if the neutral of the system is not earthed, but should be placed in all three phases if it is earthed. Although an overload relay is more expensive than a fuse but it has the advantage that it automatically resets when the switch is reclosed. The selection of a particular protective device depends on the kW rating, type and purpose of the motor, starting duty and nature of overload.

Open-Field Protection. DC motors of large size may attain dangerously high speed with the loss of field excitation while those of small size may not race due to friction etc. Field loss relays are available to protect dc shunt or compound wound motors against loss of field excitation. Different arrangements exist with starting equipment for dc and ac synchronous motors.

Open-Phase Protection. This condition may be caused by a blown fuse, an open connection, or a broken line. If the motor is at standstill at the time of occurring phase failure, stator current will rise to and remain at a very high value, but the motor will remain stationary and thus the windings may get damaged due to poor ventilation. Dangerous conditions may exist also while the motor is running.

Reversed Phase Protection. In case two phases of the supply of a 3-phase induction motor are interchanged, the direction of rotation of the motor will reverse. This is known as *phase reversal.* In elevator operation and industrial applications, this might result in serious damage. Phase failure and phase reversal relays are available. These protect the motor, machines, and personnel from the hazards of open-phase or reversed phase sequence conditions.

Overspeed Protection. Excessive speeds are possible with certain motors that might damage a driven machine, materials in the industrial process, or the motor. Overspeed protection may involve a fine selection and use of control equipment for such applications as paper and printing plants, steel mills, processing plants and the textile industry.

A centrifugal device operates at a certain predetermined speed, usually about 20 per cent above normal, and causes the motor to be shut down.

Protection Against Overheating of Bearings. Protection against overheating of bearings or other mechanical parts may be provided by placing thermocouples in the desired situations and arranging so that when the temperature reaches a certain value, relays are operated and the motor is caused to shut down or alarm bells are caused to be rung.

Low-Voltage Protection. This protection is necessary, partly to avoid damage to the motor and its driven machine due to heavy currents on switching on full rated voltage after a temporary shutdown, and partly to protect the operator of the equipment. The low-voltage release consists usually of a simple solenoid held up by the voltage of the supply and which causes the starting equipment to return to the "off" position on the failure of the supply.

In addition to the features mentioned above to guard against faults on the motor or driven machine, protection must also be provided against careless manipulation of hand-operated gear or faulty operation of automatic equipment owing to sticking of relays or contactors or to dirty contacts.

The main danger with hand-operated gear is the too fast movement of the starting arm and it can be avoided by providing a mechanical ratchet to prevent the arm being moved more than one step at a time. The other possibility is starting of dc machines with a weak field resulting in an unnecessarily heavy current to provide a given torque. For protection against such possibility mechanical or electrical interlocks may be provided. The mechanical interlock may consist of a metal plunger, operated by the field rheostat arm, that prevents the movement of the starting arm except when the rheostat is in the full-field position. In an electrical interlock, the plunger is controlled by a solenoid, which is energized only when the field is fully on. Similar interlocks are to be provided in the control equipment for ac commutator motors used in speed control by brush shifting; the main switch should be prevented from closing unless the brushes are in the low-speed position.

With fully automatic equipment, there is always possibility that one or more of the accelerating contactors remain closed after being de-energised, due to dirt having clogged their movement. However, this can be avoided by placing auxiliary contacts on the moving arm which are closed when the contactor is open. These contacts are connected in series with the start button, so that the latter is inoperative if any of the contactors have remain closed.

Exercises

1. What is meant by control gear ? Give the advantages of automatic control over manual control.
2. What is thyratron? How does it differ in construction and properties from triode ? Give it uses too.
3. What is an ignitron ? What is its usual application ?
4. How does an amplidyne differ from an ordinary dc generator ? Give its construction, principle of operation and uses.
5. Describe drawing neat sketches the principle of working and construction of magnetic amplifier. Why ac windings are provided in pair ? Mention a few applications of magnetic amplifiers.
[A.M.I.E. Sec B. Utilization of Elec. Power Summer 1983]
6. Explain with diagram the principle of operation of magnetic amplifiers, rotating amplifiers and thyristors.
[Pb. Univ. Elec. Drives and Utilization of Elec. Energy April, 1983]
7. Why are solid state devices becoming more and more popular in industry nowadays ? What is diode? Give its V-I characteristic and uses.
8. What is Zener diode ? Give its schematic symbol and uses.
9. What do you understand by the word "Thyristor" ? Give its applications.
10. What is an SCR ? Give its volt-ampere characteristic and typical parameters.
11. With a neat diagram explain the constructional features of a triac. Sketch its V-I characteristics. How does it differ from an SCR ? Mention its applications.
12. What is diac ? Give its schematic symbol, V-I characteristic and uses.
13. What is GTO ? Give its schematic symbol and its advantages over SCR.
14. Describe the construction of a dc electromagnetic contactor.
15. What is a relay ? What are their applications ?
16. What are the differences between an amplidyne and an ordinary dc generator ?
How does a saturable reactor differ from an ordinary reactor ? Explain.
Distinguish clearly the primary control devices and pilot control devices of a motor.
Where do (*i*) limit switch (*ii*) pressure switch (*iii*) plugging switch and (*iv*) transfer switch find applications ? Explain.
[A.M.I.E. Sec B. Utilization of Elec. Power Winter 1981]
17. Explain the action of (*i*) push buttons (*ii*) limit switches (*iii*) float switches (*iv*) pressure switches (*v*) thermostats (*vi*) speed governors (*vii*) plugging switches and (*viii*) master controllers.
[A.M.I.E. Sec B. Utilization of Elec. Power Summer 1980]
18. Write short notes on the following :
(*i*) Amplidyne
(*ii*) Magnetic amplifier
(*iii*) Thyratron and its applications
(*iv*) SCR
(*v*) Triac, diac and GTO switch.

Short Answer Type Questions With Answers

Q. 1. What are the two main applications of a thyratron ?
Ans. Two main applications of thyratron are (*i*) switching and (*ii*) rectification.

Q. 2. Why is series compensating winding employed in amplidyne ?
Ans. Compensating winding is connected in series with the power output brush terminals of the amplidyne to neutralize the armature reaction set up by the load current and increase amplification factor.

Q. 3. Why are amplidynes provided with split-poles ?
Ans. Amplidynes are provided with split-poles to provide space for interpoles.

Q. 4. What is the effect of temperature on the reverse current of a P-N junction ?
Ans. Reverse current of a P-N junction increases with the increase in junction temperature. It has been found that reverse current of a P-N junction approximately doubles for each 10°C rise in temperature in case of germanium and for each 8°C in temperature in case of silicon.

Q. 5. What is Zener diode ?
Ans. Zener diode is a P-N junction diode specially designed for operation in its breakdown region in reverse bias condition.

Q. 6. How is current limited in conducting state of an SCR ?
Ans. The current in conducting state of an SCR is controlled by an external impedance.

Q. 7. What is false triggering ?
Ans. False triggering is unintended turn-on of an SCR either through gate due to noise pick-up or excessive anode voltage.

Q. 8. Why is it necessary to keep supply voltage much less than breakover voltage of an SCR ?
Ans. The supply voltage is kept much less than breakover voltage of an SCR otherwise the device will get damaged.

Q. 9. Where do you recommend the use of freewheeling diode ?
Ans. Use of freewheeling diode is recommended in motor control circuits to provide an alternate path for continuity of current in the inductive circuit following the switching-off of some power device between the motor and the supply source.

Q. 10. Where are float switches used ?
Ans. Float switches are used for opening or closing pipe valves for the control of fluids.

❑❑❑

CHAPTER 3

Thyristor Control of Electric Motors

INSIDE THIS CHAPTER

3.1. INTRODUCTION

Electric motors are employed in a very large power range, from a few watts to thousands of kilowatts. Many applications require very precise position adjustments (as in robotics). In many applications optimum performance and efficiency are the main concern. Variable speed drive (VSD) systems help in optimisation of process so as to reduce investment costs, operational costs and maintenance costs. Energy saving is another big advantage of variable speed drives.

The advent of thyristors capable of handling large currents has revolutionized the field of electric power control. Thyratrons, ignitrons, mercury-arc rectifiers, magnetic amplifiers motor-generator sets have all been replaced by solid state circuits employing semiconductor diodes and thyristors. Thyristor controlled drives employing both dc and ac motors find wide application in industry as variable speed drives. In the 1960's, ac power used to be converted to dc power for direct control of drive motors with solid state devices (high power silicon diodes and silicon controlled rectifiers). Initially saturable reactors were employed in conjunction with high power silicon rectifiers for dc drives. Nowadays, thyristors are extensively used for ac-to-dc conversion.

Thyristorized control has the following advantages and disadvantages:

Advantages. 1. The response of the control device is faster as it eliminates the time lag introduced by the inductances of the generator field and the armature.

2. Due to low voltage drop across the thyristor, the efficiency of the control system is high.
3. The control device is smaller in size, lighter in weight, cheaper in cost, requiring less space and minimal maintenance.
4. Simple and reliable operation.

Disadvantages. 1. Because of higher ripple content in the converter output, motor heating and commutation problems are serious.

2. Due to switching action of thyristors and non-sinusoidal nature of current, there is more possibility of interference with the communication networks.

In all variable speed drive systems power electronic converter acts as an interface which accepts electric power from the existing source and converts it in a controlled manner into a suitable form compatible with the particular load or the process for which it is employed. The main sources for electric power are: Single or three-phase 50 Hz ac from utility systems and dc from storage batteries or solar cells. The four basic forms of power conversion required are: ac to dc, dc to dc, dc to ac and ac to ac. The modern converters are compact, cheap, reliable, durable, flexible, completely controllable. They need reduced maintenance too. They are suitable for all the four basic forms of power conversion mentioned above through rectifiers (ac/dc), choppers (dc/dc), inverters (dc/ac) and cycloconverters or ac regulators (ac/ac).

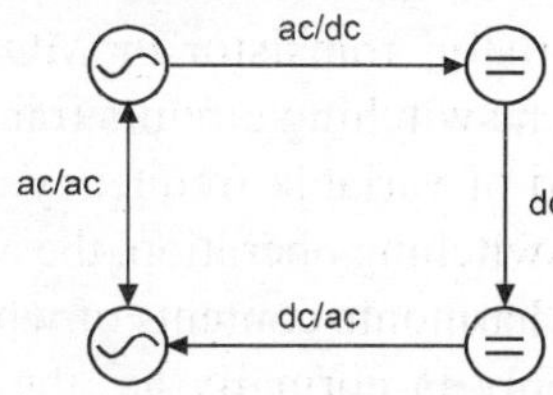

Fig. 3.1. ***Four Basic Forms of Power Conversion***

For dc motor control, controlled dc power from a constant voltage ac supply is obtained by means of *controlled rectifiers* (usually termed as *converters*) using thyristors and diodes. The control of dc voltage is achieved by varying the phase angle at which the thyristors are fired relative to the applied alternating voltage waveform. This scheme of control is known as the *phase control*. In another control system, known as the *integral cycle control*, the current is gated to flow from the ac supply for a number of complete cycles and is then quenched for a further few cycles, the process being repeated continuously. Control is applied by adjusting the ratio of *on* and *off* durations. This method is suitable for the control of fractional kW output dc motors.

Phase controlled converters are simple to operate and are less expensive as they do not require additional circuitry for commutation process. In such converters natural commutation is achieved, *i.e.,* when an incoming thyristor is turned on, it immediately reverse biases the outgoing thyristor and turns it off.

Phase control and integral cycle control methods are also applicable for ac motors wherein the converter circuitry is not required.

Control of dc motors supplied from dc supply is achieved by means of a thyristor switching circuit called the *chopper*. In chopper circuits, the control of the average voltage is achieved by changing the on-to-off time duration ratio for which the dc supply voltage is applied to the motor. This provides an efficient and stepless control of dc motors. The motor can also be made to operate in the regenerative braking mode. Instead of converter circuits, it is possible to use an uncontrolled rectifier, which provides a constant direct voltage, followed by a chopper to provide a variable mean direct output voltage. The chopper controller needs forced commutation of the thyristor.

For the control of ac motors supplied from dc supply, thyristor, transistor or MOSFET based inverters are used. Such switching circuits transfer energy from dc supply to ac load of variable frequency and/or variable voltage. Because of switching operation, the ac voltage waveforms are stepped, the harmonic contents of which are filtered out. As the power supply is normally ac, the complete scheme for obtaining variable-voltage and frequency power involves the use of both an inverter and a converter.

A *cycloconverter* is a control unit for providing variable-voltage and frequency power directly from a fixed frequency supply without the necessity of an intermediate dc stage. The mechanism of voltage and frequency control is a combination of those employed in the phase-controlled converter and the pulse-modulated inverters. Cycloconverters inspite of their attraction of direct ac to ac conversion, suffer from certain drawbacks because of which they did not become popular. Some of these drawbacks are : cycloconverters can generate only a sub-frequency output; they produce output with large harmonic content and have a low input power factor. Cycloconverters are employed for low-speed drives and for controlling linear motors in high-speed transportation systems.

3.2. GENERAL CONFIGURATION OF A MOTOR DRIVE

Figure 3.2 shows a block diagram illustrating the control of a motor drive. The main constituents are power electronic converter, motor, processes, process control computer and controller.

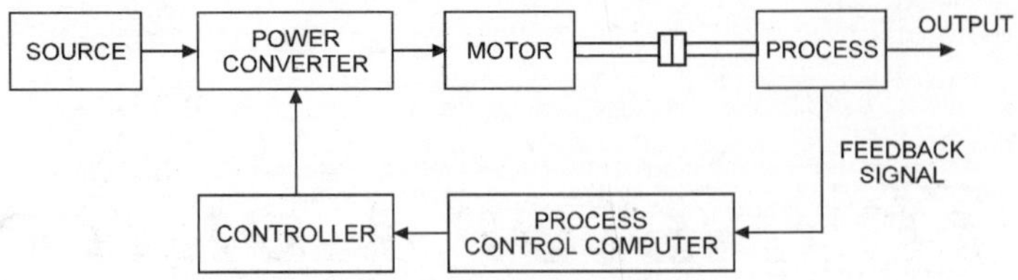

Fig. 3.2. ***Block Diagram of a Motor Drive***

The required characteristics of the motor drive are determined by the process. The process may need a variable-speed drive or a servo drive etc. The motor is selected on the basis of these requirements. The power electronic converter converts ac single phase or 3-phase input into a supply suitable to provide desired characteristics of the motor. The process control computer receives a feedback from the process as regards the extent to which the requirements are met. This feedback may be of speed, position etc. The computer feeds an error signal to the controller which, in turn, does the corrective action. The corrective action may be, say, the change in the firing angle of thyristors of the power electronic converter. In some situations the accuracy and response time of the motor to adjust the new speed is of utmost importance. In some other situations the requirements may not be so critical.

3.3. CLASSIFICATION OF POWER CONVERTERS

Solid state power converters are employed for obtaining the appropriate form of electrical energy such as direct current or adjustable-frequency alternating current (required to operate most electronic circuits and motor drives) from fixed-frequency alternating current.

According to functions, the static power converters may be classified in the four basic categories namely ac/dc converters, ac/ac converters, dc/dc converters and dc/ac converters, already discussed in Art. 3.1 [refer to Fig. 3.1].

The inverters (dc/ac converters) may be further categorised as per their source nature as voltage source and current source inverters. A voltage source inverter (VSI) is fed by a constant voltage source system while a current source inverter (CSI) is supplied from a constant current source. A voltage source inverter is one in which dc source has small or negligible impedance. Because of a low internal impedance, the terminal voltage of a VSI remains almost constant with variations in load. It is, therefore, equally suitable to single motor and multimotor drives. On the other hand, CSI is supplied with a controlled current from a dc source of high impedance. Because of a large impedance, the terminal voltage of a CSI varies substantially with a change in load. CSIs are not suitable for multimotor drives because a change in load on any motor affects other motors.

The static power converters may also be categorised into two categories viz. transistor converters and thyristor converters. The transistor converters, which do not require additional commutation circuitry as needed in thyristor converters, have almost completely replaced thyristor converters below 75 kW. The availability of low-cost high-power BJTs has also made possible the use of pulse width modulation (PWM) for improving the motor current waveform.

According to the method of commutation the thyristor converters can mainly be categorised in two types viz. naturally commutated and force commutated converters. In case of ac circuits, ac line voltage is available across the device. When the current in the SCR passes through a natural zero, the device is turned off. This process is known as natural commutation process and the converters based on this principle are known as *line commutated* or *naturally commutated converters*. AC/DC converters, cycloconverters, and ac/ac voltage regulators using SCRs are of the naturally commutated type.

In case of dc circuits, since the supply voltage does not pass through the zero point, some external source is required to commutate the device. This process is known as forced commutation process and the inverters based on this principle are known as force commutated inverters. As the device is to be commutated forcefully, such inverters need complicated commutation circuitries.

Converters may also be grouped into one-quadrant, two-quadrant and four-quadrant converters. A semiconverter is a one-quadrant converter and it has one polarity of output voltage and current. A full converter is a two-quadrant converter and polarity of its output voltage can be either positive or negative. However, its output current has one polarity only. A dual converter is a four-quadrant converter and both its output voltage and current can be either positive or negative.

3.3.1. AC/DC Converters

Rectification is a process of converting alternating current or voltage into a direct current or voltage. This conversion can be achieved by a variety of circuits based on and using switching devices. The widely used switching devices are diodes, thyristors, power transistors, power MOS etc. The rectifier circuits using only diodes as the rectifying elements are known as *uncontrolled rectifiers* because for a fixed value of ac input voltage, the output dc voltage is fixed and cannot be controlled. For some low-power devices where control of output voltage is not required, rectifier circuits using only diodes may be employed. Such circuits have the advantages of low cost and simple circuitry.

Rectifier circuits using thyristors are known as *controlled rectifiers* or *ac-dc converters*. The basic principle of a controlled rectifier is to control the point in time at which the thyristors are allowed to conduct during each cycle. Thus, it is possible to select the time segments of the ac voltage waves which appear at the dc terminals, and the mean output voltage is controlled continuously. This process of control is called the phase control and the controlled rectifiers are, therefore, also called the phase-controlled rectifiers. Since thyristors are available in high voltage and large current ratings, controlled rectifiers using thyristors are employed in many high-power devices. A controlled rectifier may be a single-phase or 3-phase depending on whether the input is single-phase or 3-phase ac.

Controlled rectifiers are of two types namely fully-controlled and half-controlled. The fully-controlled rectifier uses thyristors as the rectifying elements and the dc output voltage is a function of the amplitude of the ac supply voltage and the point-on-wave at which the thyristors are triggered (called firing angle α). The half-controlled rectifiers consist of a mixture of diodes and thyristors, allowing a more limited control over the dc output voltage level than the fully-controlled rectifier. The half-controlled rectifier is cheaper than a fully-controlled rectifier of the same ratings but has operational limitations.

Uncontrolled and half-controlled rectifiers permit power to flow only from the ac system to the dc load and are, therefore, referred to as unidirectional, one-quadrant or semiconverters. However, with a fully-controlled rectifier it is possible, by control of the point-on-wave at which switching occurs, to allow power to be transferred from the dc side of the converter back into the ac system. When this occurs, operation is said to be in the inverting mode. The fully-controlled converters may, therefore, be referred to as bidirectional or two-quadrant converters. Thus it is possible for the phase-controlled converters to provide either a one-quadrant, two-quadrant or four-quadrant operation at its dc terminals. In a four-quadrant operation, two fully-controlled converters, connected back-to-back with one another, are used. Four-quadrant converters or dual converters provide the facility for bidirectional current flow through the load.

The converter type depends on the power to be handled and how much voltage ripple is tolerable. For low powers (below 20 kW), single-phase circuits are adequate but they themselves can take different forms. For high powers (above 20 kW), three-phase circuits are invariably employed.

3.3.2. AC/AC Converters

Thyristor-based ac voltage controllers using the phase control principle convert fixed alternating voltage directly to variable alternating voltage without a change in frequency. Applications of ac voltage controllers or regulators include domestic and industrial heating, transformer tap changing, lighting control, speed control of single-phase and three-phase induction motors etc. Earlier, the devices used for such applications were auto-transformers, tap-changing transformers, magnetic amplifiers, saturable reactors etc. But these devices are now being replaced by ac regulators using thyristors and triacs because of their high efficiency, flexibility in control, compact size and less maintenance. However, ac regulators using thyristors and triacs introduce objectionable harmonics in the circuit. AC regulators are classified as single phase or three phase. Each of these may be half-wave (unidirectional) or full-wave (bidirectional). Because of ac input, ac regulators are always line or naturally commutated and as such no complex commutation circuitry is required in these controllers.

Cycloconverter is defined as a power controller which converts input power at one frequency to output power at a different frequency with one-state conversion. Thus it is an ac/ac converter without any intermediate dc link. Cycloconverters can be classified as single phase-to-single phase, three phase-to-single phase, 3-phase-to-three phase devices. They may also be classified as step-down and step-up. In step-down cycloconverters, the output frequency is lower than the supply or input frequency. In step-up cycloconverters, output frequency is higher than supply frequency. Step-down cycloconverters use line or natural commutation while step-up cycloconverters need forced commutation. Generally step-down cycloconverters are in much use. The desired amplitude of output voltage can be had by changing firing angle of thyristors used in the circuit. With the advent of high-power thyristors cycloconverters are becoming more and more popular. At present, the applications of cycloconverters include speed control of high-power ac drives, induction heating, static VAR compensation, conversion of variable-speed alternator voltage to constant frequency output voltage for use as power supply in aircraft or shipboards.

3.3.3. DC/DC Converters

A dc chopper, also known as dc-dc converter, is a static device (switch) used to obtain variable dc voltage from a source of constant dc voltage. Thus, chopper may be thought of as dc equivalent of an ac transformer as it behaves in an identical manner. Besides the saving in power, the dc chopper offers greater efficiency, fast response, lower maintenance, small size, smooth control, and, for many applications, lower cost than motor-generator sets or gas tubes approaches.

Solid state choppers due to their inherent advantages are widely employed in trolley cars, battery-operated vehicles, traction motor control, control of a large number of dc motors from a common dc bus, marine hoists, forklift trucks, mine haulers etc. They can also be used for regenerative braking in electric traction.

3.3.4. DC/AC Converters

The dc to ac power converters are known as inverters. In other words, an inverter is a circuit which converts a dc power into an ac power at desired output voltage and frequency. The ac output voltage could be fixed at a fixed or variable frequency. The dc input supply may come from a battery, solar cells, fuel cells etc.

The conversion can be achieved either by transistors or thyristors. For low and medium power outputs transistorised inverters are used but for high power outputs, thyristors should be used. For low power outputs, self-oscillating transistorised inverters are suitable but for high power outputs, driven inverters are more common than self-oscillating ones. Moreover for multiphase ac output, there is no alternative other than driven inverters. The driven inverters have better frequency stability because a separate master oscillator is used for this purpose. For the applications in inverters, transistors have some edge over thyristors regarding the switching speed, simplicity in control circuits, higher efficiency and greater reliability. This is mainly due to the fact that thyristor inverters need extra circuits to turn thyristors off, moreover, additional complex logic circuits may be required to avoid false triggering and provide proper commutation timings. Thyristors can handle much higher load currents than transistors and consequently, for high power outputs, thyristors become more desirable than transistors.

The output voltage waveforms of an ideal inverter should be sinusoidal, but the output voltage waveforms of practical inverters are non-sinusoidal and contain certain harmonics. Square wave or quasi-square wave voltages may be acceptable for low and medium power applications, but for high power applications low-distorted sinusoidal waveforms are required. The output frequency of an inverter is determined by the rate at which the semiconductor devices are switched on and off by the inverter control circuitry and consequently, an adjustable frequency ac output is readily provided. The harmonic contents of output voltage can be reduced significantly by switching techniques of available high speed power semiconductor devices. The filtering of harmonics is not feasible when the output frequency varies over a wide range, and the generation of ac waveform with low harmonic content is important. When the ac output voltage of an inverter is supplied to a transformer or ac motor, this output voltage must be varied in conjunction with frequency to maintain the proper magnetic conditions. Some of the important applications of inverters are: variable speed ac motor drives, induction heating, aircraft power supplies uninterruptable power supplies (UPS), high-voltage dc transmission lines etc.

3.4. MATCHING POWER ELECTRONIC CONVERTER AND MOTOR

The motor has to be selected to match the load requirements. The power electronic converter has to match the motor. Some important considerations in matching power electronic converter and motor are as under.

1. ***Current Rating.*** The torque developed by an electric motor depends on the current drawn by it. During most of the operating duration a motor is required to deliver its rated torque, and therefore, the current drawn is also equal to its rated current. But a motor is capable of developing much higher torque (as compared to its rated torque) provided the duration of the torque is small as compared to thermal constant of the motor. Since torque depends on the current, a motor for developing a large torque, requires a large current which has to be supplied by the converter. However, the overcurrent capability of electronic devices is very small. A large current leads to high junction temperature which may destroy the electronic converter. The thermal time constants of the power electronic devices are very small in comparison to those of motors. The current of the converter must be selected properly on the basis of average, rms and peak currents of the motor.

2. ***Voltage Rating.*** Every motor develops a back emf e_b that opposes the applied voltage v. The rate of change of torque depends on the rate of change of current which, in turn, can be expressed as

$$\frac{di}{dt} = \frac{v - e_b}{L} \qquad ...(3.1)$$

where $\frac{di}{dt}$ is the rate of change of current and L is the inductance of the motor. For a reasonably large value of $\frac{di}{dt}$, the output voltage v of converter should be much larger than e_b. Since e_b depends on the motor speed, it is necessary that the voltage rating of the converter is selected on the basis of maximum speed of the motor.

3. ***Switching Frequency and Motor Inductance.*** The steady-state ripple content in the motor current depends on motor inductance L. Large ripple content leads to large motor losses. So the steady-state ripple content in the motor current should be small. This is possible if inductance L is large. But large inductance means that the motor cannot respond quickly to the change in load demand. To meet these opposing requirements, the switching frequency of converter should be high (so that the steady-state ripple content is small). However, a high switching frequency leads to large power losses in the converter. Therefore, a proper selection of switching frequency and inductance is essential.

3.5. DC MOTOR CONTROL THROUGH CONVERTERS

DC motors are widely used in adjustable-speed drives and position control applications. Their speeds below base speed can be controlled by armature voltage control while speeds above base speed are obtained by field flux control. DC motors are preferred over ac motors because these are inherently most suitable for smooth, efficient, wide-range speed control and quick reversals.

Phase-controlled converters provide an adjustable dc output voltage from a fixed ac input voltage. DC choppers also provide adjustable dc output voltage from a fixed dc input voltage. When available supply is ac, an ac-dc converter (phase-controlled converter) is ideally suited to convert the constant ac voltage into a variable dc voltage. Phase-controlled converters are simple to operate and are less expensive as they do not require additional circuitry for commutation process. In such converters natural commutation is achieved, *i.e.,* when an incoming thyristor is turned on, it immediately reverse biases the outgoing thyristor and turns it off. The power conversion efficiency of these converters is more than 95% because of low losses in thyristors. Thus, dc motors can be controlled conveniently and efficiently through phase-controlled converters.

The dc motors used in conjunction with power converters are dc separately excited motors and dc series motors. The power converters are used in speed control of fractional kW dc motors as well as in large sized motors used in variable-speed reversing drives for rolling mills with motor ratings as large as several MWs.

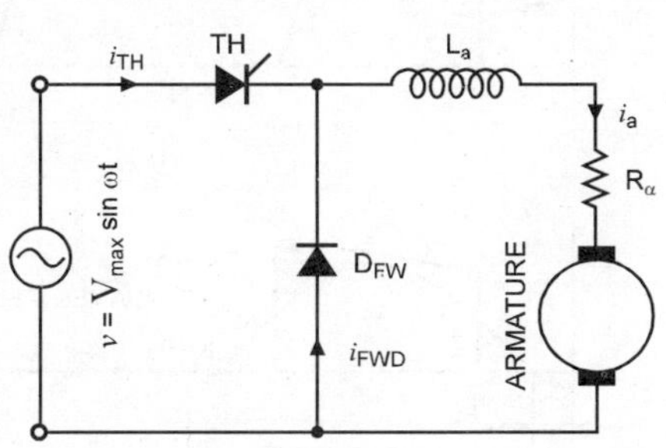

(a) Circuit Diagram

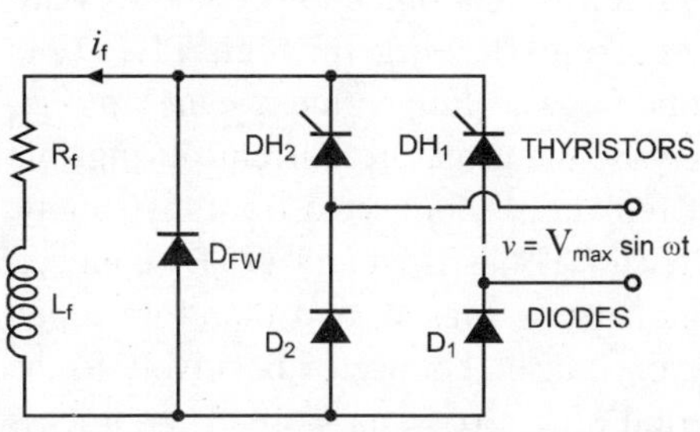

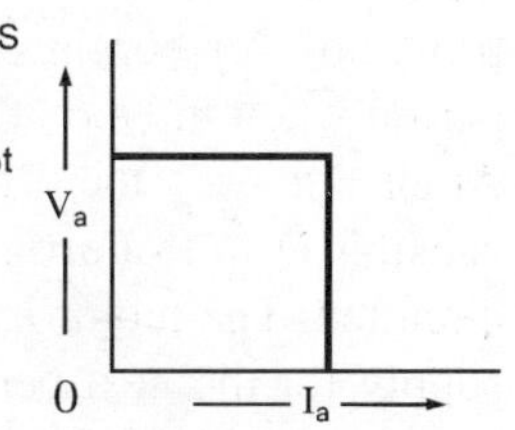

(b) Quadrant of Operation

Fig. 3.3. ***Separately Excited DC Motor Fed From Single Phase AC Supply Through Half-Wave Converter***

3.6. THYRISTOR CONTROL OF SEPARATELY EXCITED DC MOTOR DRIVE

3.6.1. Single Phase Drives

The single phase ac is converted into dc by a controlled rectifier or converter and the dc is supplied to the dc motor. By varying the firing angle the voltage applied to the motor can be varied and thus the speed of the motor can be controlled. Single phase drives can be further classified as

1. Single phase half-wave converter drives.
2. Single phase semiconverter drives.
3. Single phase full converter drives.
4. Single phase dual converter drives.

1. *Single Phase Half-Wave Converter Drives.* Figure 3.3 (*a*) shows a single phase half-wave converter for controlling a separately excited dc motor. It needs a single thyristor and a freewheeling diode (D_{FW}). Freewheeling diode, also sometimes known as *bypass diode* or *commutating diode*, is used to improve the waveshape of load current and power factor. Freewheeling diode is connected across the motor terminals to allow for dissipation of energy stored in the motor inductance and to provide for continuity of motor current when the thyristors are blocked. It also provides protection against transient overvoltages. Separate converters are employed for the armature and field circuits. It is desirable that the supply to the field winding is provided through a semiconverter or full converter. If half-wave converter is used for supply to field circuit, the high ripple content in the field circuit would cause increase in the iron losses of the machine. It is one quadrant drive, as illustrated in Fig. 3.3 (*b*).

Because of inductance of the field and armature, the thyristor would not turn off at $\omega t = \pi$. Therefore it is desirable to have freewheeling diode (D_{FW}), as shown.

Average value of voltage applied across armature is given by

$$V_a = \frac{1}{2\pi}\left[\int_{\alpha_a}^{\pi} V_{max} \sin \omega t \, d(\omega t) + 0\right]$$

$$= \frac{V_{max}}{2\pi}(1 + \cos \alpha_a) \quad \text{for } 0 \le \alpha_a \le \pi \qquad ...(3.2)$$

where α_a is the firing angle of converter in the armature circuit.

In this circuit the motor current is always discontinuous, resulting in poor performance. This type of drive is used only for small dc motors of rating up to 500 W or so.

2. *Single Phase Semiconverter Drives.* Figure 3.4 (*a*) shows the circuit of a semiconverter feeding a separately excited dc motor. Both armature and field circuits are supplied from single phase ac supply through semiconverters. Because of inductance in the armature and field circuits, freewheeling diodes

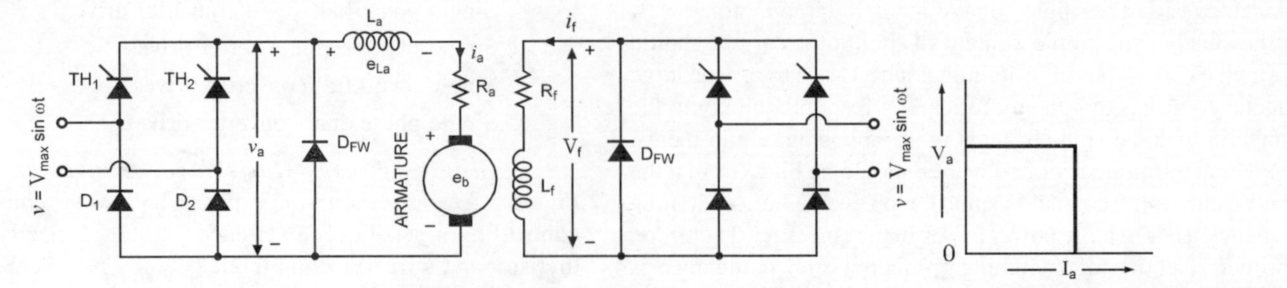

(a) Circuit Diagram *(b) Quadrant of Operation*

Fig. 3.4. ***Separately Excited DC Motor Fed From Single Phase AC Supply Through Semiconverters***

are required in both the circuits. The voltages applied to the armature and field are

$$V_a = \frac{V_{max}}{\pi}(1 + \cos \alpha_a) \text{ for } 0 \le \alpha_a \le \pi \qquad ...(3.3)$$

$$V_f = \frac{V_{max}}{\pi}(1 + \cos \alpha_f) \text{ for } 0 \le \alpha_f \le \pi \qquad ...(3.4)$$

where α_a and α_f are the firing angles of converters in the armature and field circuits respectively.

This is also a one quadrant drive [Fig. 3.4 (*b*)] which provides voltage and current of one polarity at dc terminals. It, therefore, does not provide for regenerative braking, *i.e.,* power flow from dc motor to the ac supply. Where regeneration is not required, this converter is used for reasons of economy. This drive is used for motors up to 15 kW rating.

Typical steady-state voltage and current waveforms are illustrated in Fig. 3.5. The thyristor TH_1 is fired at angle α_a while TH_2 at angle $\pi + \alpha_a$ with respect to supply voltage v and process is repeated continuously.

Under steady conditions, as thyristor TH_1 is fired ($\omega t = \alpha_a$), TH_1 and diode D_2 conduct and the motor gets supply, *i.e.,* v_a. At $\omega t = \pi$, v_a tends to become negative as the input voltage reverses in polarity. This makes freewheeling diode to become forward biased and armature current flowing through TH_1 D_2 is transferred to D_{FW} for the freewheeling period $\pi < \omega t < \pi + \alpha_a$ providing for continuity of the armature current during this period when the motor remains disconnected from the supply. At $\omega t = \pi + \alpha_a$, the thyristor TH_2 is fired and TH_2D_1 conduct, causing D_{FW} to become reverse biased, and therefore, open circuited. The motor is once again connected positively to the supply for the next period of $\pi + \alpha_a < \omega t < 2\pi$. This process repeats continuously.

Different voltage and current waveforms of a separately excited dc motor supplied through a semiconverter are illustrated in Fig. 3.5. Though the voltage across motor terminals [Fig. 3.5 (*c*)] contains harmonics over and above a steady dc value, it is rightly assumed here that the motor does not respond to these harmonics and thus runs at constant speed of *n rps* and has constant induced back emf e_b. As TH_1 fires at $\omega t = \alpha_a$, the motor current is given as

$$\frac{1}{L_a}\int_{\alpha_a}^{\omega t}(v - e_b)\, d(\omega t);$$

assuming R_a negligibly small

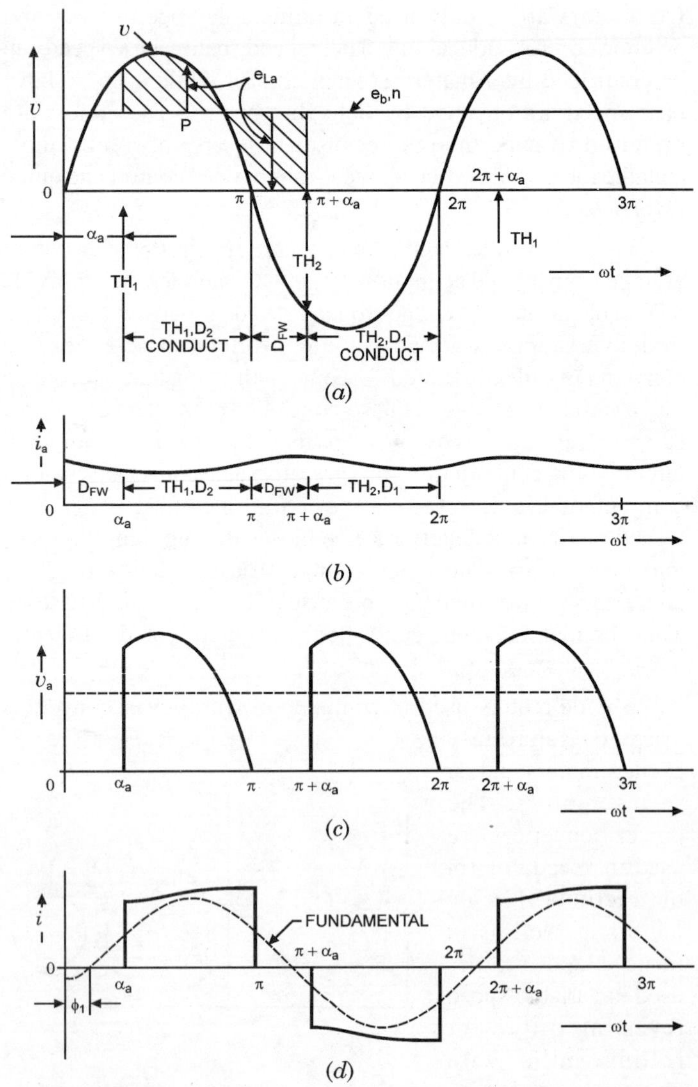

Fig. 3.5. ***Voltage and Current Waveforms of Semiconverter Feeding Separately Excited DC Motor; Continuous Armature Current***

up to the point P illustrated in Fig. 3.5 (*a*) ; $v > e_b$ so that the motor current increases. So does the motor back emf e_b. During this period, apart from energy being delivered to the load, energy is also being stored in the motor armature inductance L_a. Beyond the point P, voltage v becomes lesser than induced back emf e_b and the motor current starts to drop. This also implies the reversal of voltage across motor armature inductance which now feeds energy into the system. During the freewheeling period ($\pi < \omega t < \pi + \alpha_a$), the diode continues to be forward biased by the reversal of the inductive voltage. During this period a part of the energy stored in armature inductance is consumed in supplying the mechanical load. The motor current, speed, and emf, therefore, all decrease. This process then repeats over the next period ($\pi + \alpha_a < \omega t < 2\pi + \alpha_a$) through $TH_2 D_1$, and later through D_{FW}. The current drawn from the supply shown in Fig. 3.5 (*d*) is that part of the armature current which flows over the periods (α_a, π), ($\pi + \alpha_a$, 2π), when the motor is connected to the supply. It is not necessary to employ the freewheeling diode. In its absence at $\omega t = \pi$, diode D_1 becomes forward biased so that freewheeling occurs through $TH_1 D_1$ till TH_2 is fired. At $\omega t = 2\pi$ freewheeling occurs through $TH_2 D_2$ and so on.

Discontinuous Armature Current. The armature current becomes discontinuous for large values of the firing angle, high speed and low values of torque. Discontinuous armature current results in deterioration of motor performance. The ratio of peak to average and rms to average value of armature current increases. Thus it is desirable to operate the motor in the continuous current mode. This can be achieved by using an external armature circuit choke, which reduces the rate of decay of current during the freewheeling operation.

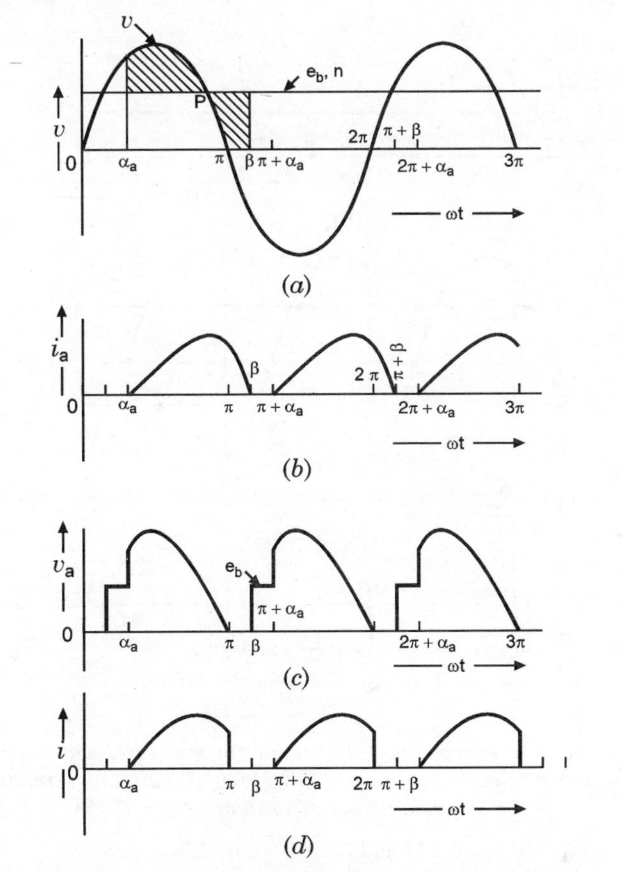

Fig. 3.6. ***Voltage and Current Waveforms of Semiconverter Feeding Separately Excited DC Motor: Discontinuous Armature Current***

The voltage and current waveforms for semiconverter with discontinuous current are illustrated in Fig. 3.6. The motor is connected to the supply through TH_1 and D_2 for the period $\alpha_a < \omega t < \pi$. Beyond π, the motor is shorted through the freewheeling diode D_{FW}. The armature current falls to zero at angle β (extinction angle) $< \pi + \alpha_a$ *i.e.*, before the thyristor TH_2 is fired, thereby making the armature current discontinuous, During α_a to π, the conduction period through TH_1 and D_2, the motor terminal voltage is the same as the input voltage. During π to β, the motor terminal voltage is zero, motor terminals being shorted by the freewheeling diode D_{FW}. From β to $\pi + \alpha_a$, the motor coasts and, therefore, its terminal voltage is the same as its induced emf. It should be noted here that the fundamental of the current drawn from the mains lags behind the voltage by an angle ϕ, ($< \alpha_a$).

Torque-Speed Characteristics. The emf induced in armature conductors of a motor, known as back emf, is given by

$$E_b = k_e \phi N \qquad ...(3.5)$$

where K_e is emf constant and is equal to $\frac{PZ}{60A}$.

In a separately excited dc motor, the field winding is excited from a separate source. So voltage applied to the armature is given by

$$V_a = E_b + I_a R_a \qquad ...(3.6)$$

where I_a is armature current in amperes and R_a is the armature resistance in ohms.

Thus, from Eqs. (3.5) and (3.6), we have

$$\text{Back emf, } E_b = V - I_a R_a = k_e N \phi$$

$$\text{or speed of motor, } N = \frac{V - I_a R_a}{k_e \phi} \qquad ...(3.7)$$

Since field is excited separately, field current I_f and flux ϕ are constant. Moreover voltage drop in armature, $I_a R_a$ is negligible. So motor speed N is constant if voltage applied to the armature is constant, and speed varies directly in proportion to the applied voltage *i.e.*, $N \propto V_a$.

Electromagnetic torque developed in a dc motor is given by

$$T_e = k_T \phi I_a \qquad ...(3.8)$$

where k_T is torque constant and is equal to $\frac{1}{2\pi}\frac{PZ}{A}$

$$\text{and torque constant, } k_T = \frac{1}{2\pi}\frac{PZ}{A} = \frac{60}{2\pi} \times \frac{PZ}{60\,A} = \frac{60}{2\pi} k_e \qquad ...(3.9)$$

From Eq. (3.8), we have

$$\text{Armature current, } I_a = \frac{T_e}{k_T \phi} \qquad ...(3.10)$$

From Eqs. (3.7) and (3.10), we have

$$N = \frac{V_a}{k_e \phi} - \frac{T_e R_a}{k_T k_e \phi^2} \qquad ...(3.11)$$

With a semiconverter,

$$V_a = \frac{V_{max}}{\pi}(1 + \cos \alpha_a) \qquad \text{... Refer to Eq. (3.3)}$$

$$\text{So } N = \frac{V_{max}(1 + \cos \alpha_a)}{\pi k_e \phi} - \frac{T_e R_a}{k_T k_e \phi^2} \qquad ...(3.12)$$

The first term of RHS of Eq. (3.12) represents the theoretical speed and the second term represents the speed drop due to voltage drop in armature circuit.

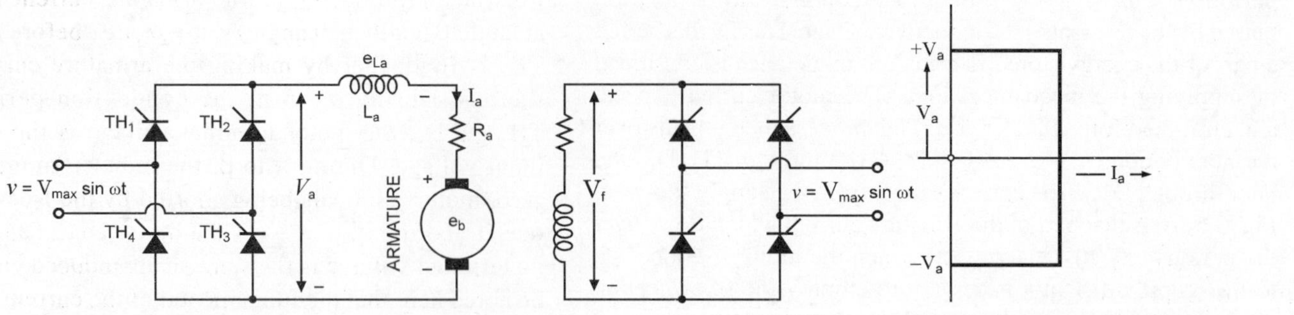

(a) Circuit Diagram (b) Quadrant of Operation

Fig. 3.7. ***Separately Excited DC Motor Supplied From Single Phase AC Supply Through Full Converters***

The theoretical no-load speed can be controlled by varying the firing angle α_a.

3. ***Single Phase Full-Converter Drives.*** The circuit diagram is illustrated in Fig. 3.7 *(a)*. A fullconverter needs four thyristors but no freewheeling diode. Full converters are employed for both the armature and field supply. As illustrated in Fig. 3.7 *(b)*, two quadrant operation is possible. The converter in the armature circuit provides $+V_a$ or $-V_a$ thus allowing operation in first and fourth quadrants. Current remains unidirectional because of the unidirectional thyristors. When operating in the fourth quadrant regenerative braking is possible and motor feeds back energy to the source. If α_a and α_f are the firing angles for the converters used in the armature and field circuits, the armature voltage V_a and field voltage V_f are given as

$$V_a = \frac{2V_{max}}{\pi} \cos \alpha_a \text{ for } 0 \le \alpha_a \le \pi \qquad ...(3.13)$$

$$V_f = \frac{2V_{max}}{\pi} \cos \alpha_f \text{ for } 0 \le \alpha_f \le \pi \qquad ...(3.14)$$

This drive is also employed for motors of rating up to 15 kW.

The voltage and current waveforms are illustrated in Fig. 3.8; armature current i_a being assumed to be almost constant. During the time interval $\alpha_a < \omega t < \pi + \alpha_a$, thyristors TH_1 and TH_3 conduct and connect the motor to supply. At $\pi + \alpha_a$, thyristors TH_2 and TH_4 are triggered. Immediately the supply voltage appears in reverse bias across thyristors TH_1 and TH_3 and turns them off. This is *natural* or *line commutation*. The motor current is transferred from thyristors TH_1 and TH_3 to thyristors TH_2 and TH_4. During the time interval α_a to π energy is supplied to the motor (both v and i are positive and, therefore, are v_a and i_a). But during time interval π to $\pi + \alpha_a$, some of the motor energy is fed back to the supply (v and i and so the v_a and i_a being of opposite polarity). The noteworthy point is that the fundamental of the current drawn from the supply mains lags behind the applied voltage by angle $\phi_1 = \alpha_a$.

The voltage and current waveforms for $\alpha_a > 90°$ are illustrated in Fig. 3.9. The average motor terminal voltage is now negative. On reversing of the motor terminals, it will operate as a generator supplying power back to the ac supply. This is the *inversion operation* of the converter and is employed in regenerative braking of the motor. The noteworthy point here is that during the conduction period of either TH_1 TH_3 or TH_2 TH_4 as the supply voltage becomes negative, the armature current begins to drop, causing the inductance polarity to reverse and thus conducting thyristors continue to be forward biased.

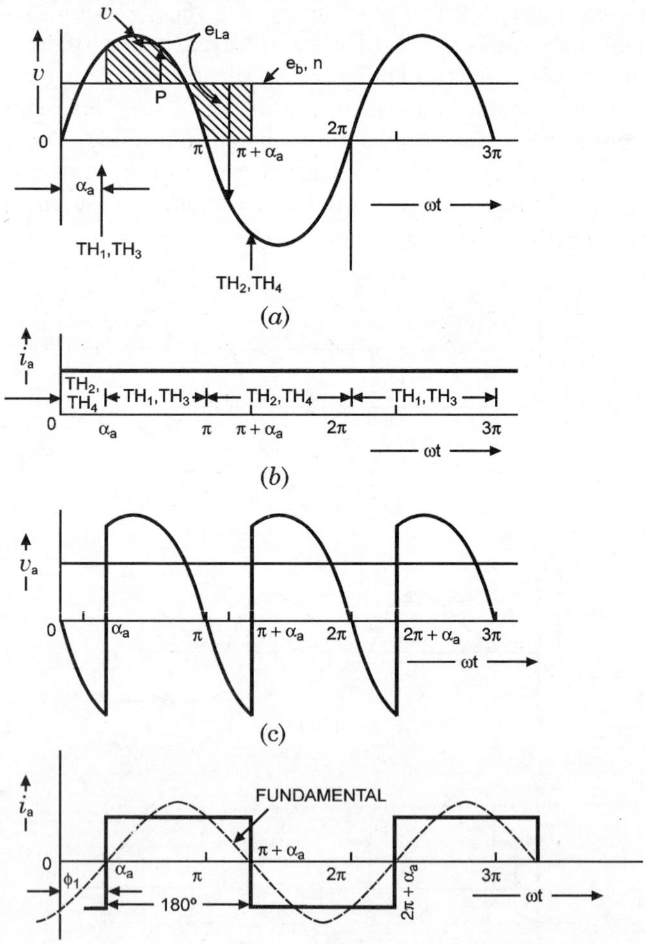

(d)

Fig. 3.8. ***Voltage and Current Waveforms of Full converter Supplying Separately Excited DC Motor; Continuous Armature Current (Almost Constant): α < 90°***

From Eqs. (3.11) and (3.13) we have

$$\text{Speed N} = \frac{2V_{max} \cos \alpha_a}{\pi k_e \phi} - \frac{T_e R_a}{k_T k_e \phi^2} \qquad ...(3.15)$$

In the case of discontinuous current, the average voltage at motor terminals depends upon the extinction angle β which itself depends on the average motor speed N, average armature current I_a and firing angle α_a. We need not to go in its analytical treatment.

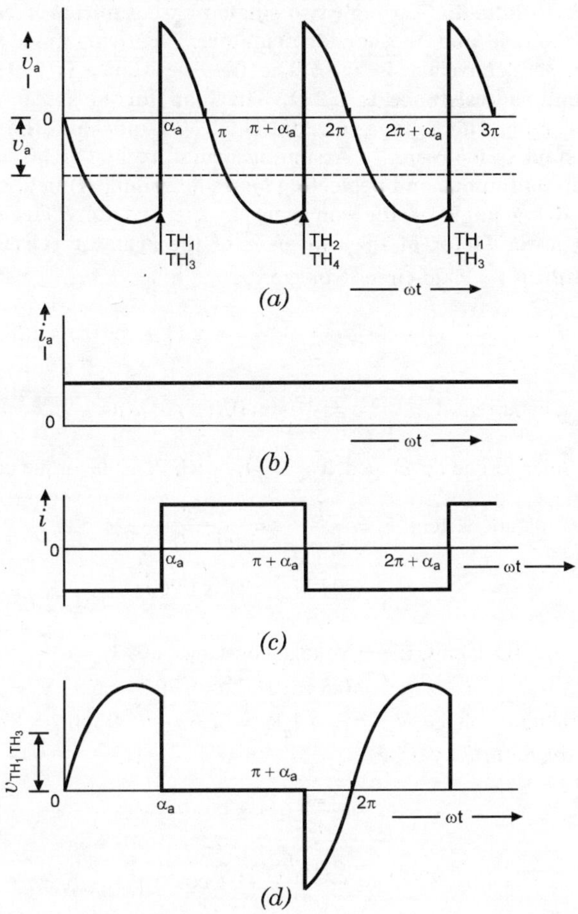

Fig. 3.9. *Voltage and Current Waveforms For Full Converter:* $\alpha_a > 90°$

4. ***Single Phase Dual Converter Drives.*** The circuit diagram is illustrated in Fig. 3.10 (*a*). In this drive, the armature is supplied from two full converters while the field is supplied from another full converter. Converter 1 provides operation in the first and fourth quadrants while converter 2 provides operation in second and third quadrants. Thus it is a *four quadrant drive* and allows all four modes of operation, *i.e.*, forward motoring, forward regenerative braking, reverse motoring and reverse regenerative braking. When converter 1 is operating, a positive voltage V_a is applied to the armature. When converter 2 is operating a negative voltage $-V_a$ is applied to the armature. It is employed for motors of rating up to 15 kW. If α_{a_1} is the firing angle of converter 1, the armature voltage V_a is given as

$$V_a = \frac{2V_{max}}{\pi}\cos\alpha_{a_1},\ \text{for } 0 \le \alpha_{a_1} \le \pi \qquad ...(3.16)$$

The firing angle α_{a_2} of converter 2 will be equal to $(\pi - \alpha_{a_1})$ and when converter 2 is operating, armature voltage is given as

$$V_a = \frac{2V_{max}}{\pi}\cos\alpha_{a_2}\ \ \text{for } 0 \le \alpha_{a_2} \le \pi \qquad ...(3.17)$$

The voltage applied to the field is given as

$$V_f = \frac{2V_{max}}{\pi}\cos\alpha_f\ \ \text{for } 0 \le \alpha_f \le \pi \qquad ...(3.18)$$

where α_f is the firing angle of converter supplying the field. Figure 3.10 (*b*) illustrates the four quadrants of operation.

Example 3.1. A single-phase half-wave converter is operated from a 120 V, 60 Hz supply. If the load is resistive, R = 10 Ω and the delay angle $\alpha = \frac{\pi}{3}$, determine:

***(i)* the efficiency *(ii)* the form factor *(iii)* the ripple factor *(iv)* the transformer utilisation factor ; and *(v)* peak inverse voltage of the thyristor.**

[A.M.I.E. Sec. B. Electric Drives and Their Control Summer 2000]

Solution : In a half-wave converter, negative half waves are eliminated. The average voltage across load is

$$V_{dc} = \frac{1}{2\pi}\int_{\alpha}^{\pi} V_{max}\sin\omega t\, d(\omega t) = \frac{V_{max}}{2\pi}(1+\cos\alpha)$$

$$= \frac{120\sqrt{2}}{2\pi}\left(1+\cos\frac{\pi}{3}\right)$$

$$= \frac{120\sqrt{2}}{2\pi}\times 1.5 = 40.5 \text{ V}$$

$$I_{dc} = \frac{V_{dc}}{R} = \frac{40.5}{10} = 4.05 \text{ A}$$

RMS voltage across the load,

$$V_{rms} = \left[\frac{1}{2\pi}\int_{\alpha}^{\pi}(V_{max}\sin\omega t)^2\, d(\omega t)\right]^{1/2}$$

$$= V_{max}\left[\frac{1}{4\pi}\int_{\alpha}^{\pi}(1-\cos 2\omega t)\, d(\omega t)\right]^{1/2}$$

$$= V_{max}\left[\frac{1}{4\pi}\left|\left(\omega t - \frac{\sin 2\omega t}{2}\right)\right|_{\alpha}^{\pi}\right]^{1/2}$$

$$= V_{max}\left[\frac{\pi-\alpha}{4\pi}+\frac{\sin 2\alpha}{8\pi}\right]^{1/2}$$

$$= 120\sqrt{2}\left[\frac{\pi-\pi/3}{4\pi}+\frac{\sin 2\pi/3}{8\pi}\right]^{1/2} = 76.1 \text{ V}$$

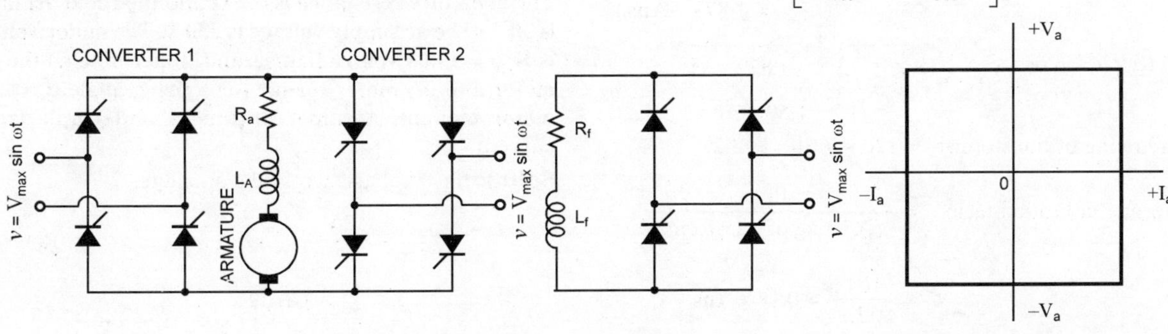

(a) Circuit Diagram *(b) Quadrant of Operation*

Fig. 3.10. *Separately Excited DC Motor Supplied From Single Phase AC Supply Through Dual Converters*

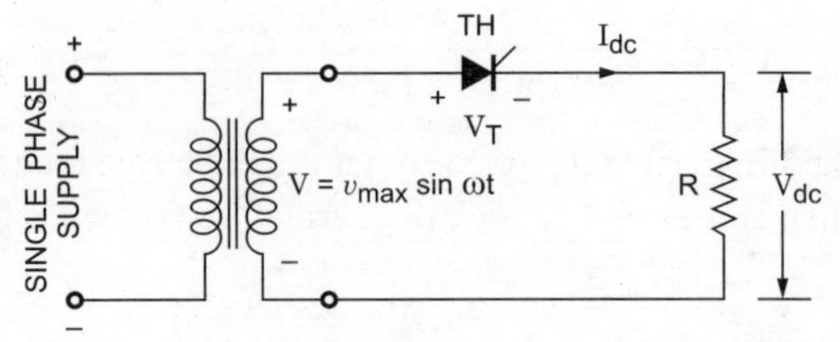

(*a*) *Circuit Diagram*

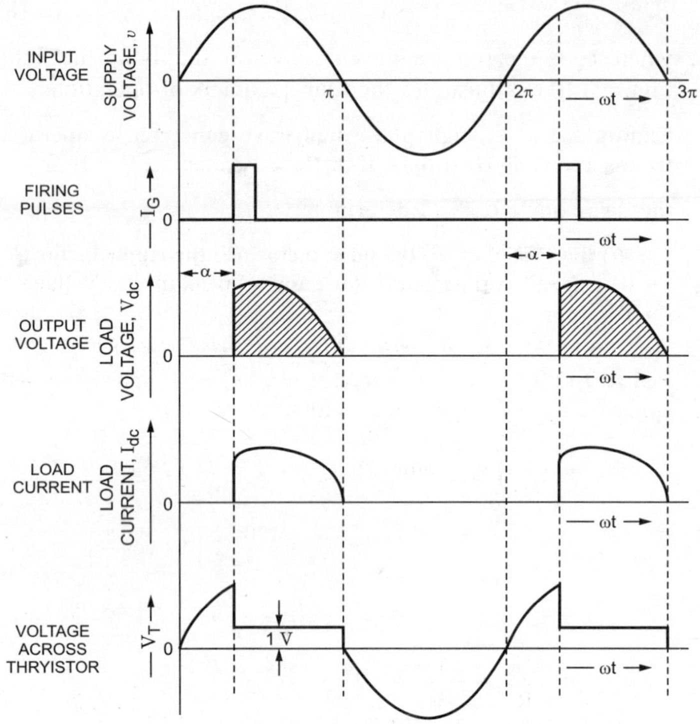

(*b*) *Waveforms of Input Voltage, Firing Pulses, Output Voltage, Current and Voltage Drop Across Thyristor*

Fig. 3.11. ***Single Phase Half-Wave Converter***

$$I_{rms} = \frac{V_{rms}}{R} = \frac{76.1}{10} = 7.61 \text{ A}$$

DC output, $P_{dc} = V_{dc} I_{dc} = 40.5 \times 4.05 = 164$ W

AC input, $P_{ac} = V_{rms} I_{rms} = 76.1 \times 7.61 = 580$ W

(*i*) Rectification efficiency, $\eta = \frac{P_{dc}}{P_{ac}} = \frac{164}{580} = 0.283$ **Ans.**

(*ii*) Form factor, $K_f = \frac{\text{RMS voltage}}{\text{Average voltage}} = \frac{76.1}{40.5}$

$= 1.879$ **Ans.**

(*iii*) Ripple factor, $\gamma = \sqrt{(K_f^2 - 1)} = \sqrt{1.879^2 - 1}$

$= 1.59$ **Ans.**

Volt-ampere rating of transformer $= 120 \times 7.61 = 913.2$

(*iv*) Transformer utilisation factor $= \frac{P_{dc}}{\text{VA rating of transformer}}$

$= \frac{164}{913.2} = 0.18$ **Ans.**

(*v*) Peak inverse voltage, PIV $= V_{max} = 120\sqrt{2} = 169.7$ V **Ans.**

Example 3.2. A single phase, 220 V, 50 Hz supply feeds a separately excited dc motor through two single phase semiconverters, one for the field and another for armature. The firing angle for the field semiconverter is zero. The field resistance is 250 Ω and armature resistance is 0.2 Ω. The load torque is 50 N-m at 1,000 rpm. The voltage constant is 0.8 V/A-rad/s and the torque constant is 0.8 N-m/A². Assuming armature and field currents to be continuous and neglecting losses determine (*i*) field current (*ii*) firing angle of the converter in the armature circuit and (*iii*) power factor of the converter of the armature circuit.

Solution : Field circuit voltage,

$$V_f = \frac{V_{max}}{\pi}(1 + \cos\alpha_f) = \frac{220\sqrt{2}}{\pi} \times (1 + \cos 0°) = 198 \text{ V}$$

(*i*) Field current, $I_f = \frac{V_f}{R_f} = \frac{198}{250} = 0.792$ A **Ans.**

(*ii*) Since torque developed, $T_d = K_f I_f I_a$ where K_f is torque constant

So armature current, $I_a = \frac{T_d}{K_f I_f} = \frac{50}{0.8 \times 0.792} = 78.9$ A

$$\omega = \frac{2\pi N}{60} = \frac{2\pi \times 1,000}{60} = 104.72 \text{ rad/s}$$

Back emf, E_b = Voltage constant × ω × I_f

$= 0.8 \times 104.72 \times 0.792 = 66.35$ V

Armature voltage, $V_a = E_b + I_a R_a = 66.35 + 78.9 \times 0.2 = 82.13$ V

Since from Eq. (3.3),

$$V_a = \frac{V_{max}}{\pi}(1 + \cos\alpha_a)$$

So $82.13 = \frac{220\sqrt{2}}{\pi}(1 + \cos\alpha_a)$

or $\alpha_a = 99.83°$ **Ans.**

(*iii*) Power output of converter in the armature circuit,

$= V_a I_a = 82.13 \times 78.9 = 6,480$ W

$$\text{Input current} = \left[\frac{2}{2\pi}\int_{\alpha_a}^{\pi} I_a^2 \, d(\omega t)\right]^{1/2}$$

$$= I_a\left[\frac{\pi - \alpha_a}{\pi}\right]^{1/2} = 78.9\left[\frac{180 - 99.83}{180}\right]^{1/2}$$

$= 52.66$ A

Input volt-amperes $= 220 \times 52.66 = 11,585.2$

Input power factor $= \frac{6,480}{11,585.2} = 0.56$ lagging **Ans.**

Example 3.3. A single phase full converter is used to control the speed of 5 kW, 220 V, 1,000 rpm separately excited dc motor. The armature resistance is 0.2 Ω and the rated armature current is 30 A. The ac supply voltage is 230 V. The motor voltage constant is $K_a\phi$ = 1.668 volt/radian/second. Determine (*i*) the speed of the motor and (*ii*) motor torque for a firing angle α = 45° and rated motor current. Assume continuous and ripple free armature current.

Solution : Voltage across the armature,

$$V_a = \frac{2V_{max}}{\pi}\cos\alpha_a \qquad \text{Refer to Eq. (3.13)}$$

$$= \frac{2 \times 230\sqrt{2}}{\pi}\cos 45° = 146.42 \text{ V}$$

Back emf, $E_b = V_a - I_a R_a = 146.42 - 30 \times 0.2 = 140.42$ V (*i*) So speed, $\omega = \frac{E_b}{K_a\phi} = \frac{140.42}{1.668} = 84.186$ radians/second

or $N = \frac{\omega}{2\pi} \times 60 = \frac{84.186}{2\pi} \times 60 = 804$ rpm **Ans.**

(*ii*) Motor torque, $T = K_a\phi I_a = 1.668 \times 30 = 50.04$ Nm **Ans.**

Example 3.4. A 200 V, 875 rpm, 150 A separately excited dc motor has an armature resistance of 0.06 Ω. It is fed from a single phase fully controlled rectifier with an ac source voltage of 220 V, 50 Hz. Assuming continuous conduction calculate the firing angle for rated motor torque at 750 rpm.

[*A.M.I.E. Sec. B. Electric Drives and Their Control Winter 1998*]

Solution : Back emf of motor when operated from 200 V dc supply and runs at 875 rpm,

$$E_{b_1} = V_{dc} - I_aR_a = 200 - 150 \times 0.06 = 191 \text{ V}$$

Back emf of motor when fed from 220 V 50 Hz single phase supply through fully controlled rectifier and runs at 750 rpm (assuming flux ϕ to remain constant),

$$E_{b_2} = E_{b_1} \times \frac{N_2}{N_1} = 191 \times \frac{750}{875} = 163.71 \text{ V}$$

Armature current $I'_a = I_a = 150$ A

∵ torque remains the same and flux is assumed to remain constant

Armature voltage, $V_a = E_b + I_aR_a = 163.71 + 150 \times 0.06 = 172.71$ V

Since for a fully controlled single phase rectifier voltage across armature is given as

$$V_a = \frac{2V_{max}}{\pi} \cos\alpha_a$$

So $172.71 = \frac{2 \times 220\sqrt{2}}{\pi} \cos\alpha_a$

or $\alpha_a = 29.31°$ **Ans.**

Example 3.5. A single phase fully controlled thyristor converter is connected to a load comprising 2 Ω resistance and 0.3 H inductance. The supply voltage is 230 V at 50 Hz. Estimate the average load voltage, average load current and input power factor for triggering angle of 30º. Assume continuous and ripple free load current. [*U.P. Technical Univ. Energy Conversion 2005-06*]

Solution : Supply voltage $V_{(\text{rms value})} = 230$ V

$$V_{max} = 230\sqrt{2} \text{ V}$$

Average load voltage, $V_{L\,av} = \frac{2V_{max}}{\pi}\cos\alpha = \frac{2 \times \sqrt{2} \times 230 \cos 30°}{\pi} = 179.33$ V **Ans.**

Average load current, $I_{L\,av} = \frac{V_{L\,av}}{R} = \frac{179.33}{2} = 89.665$ A **Ans.**

As load current is ripple free,

RMS value of load current, $I_{rms} = I_{L\,av} = 89.665$ A

Active power of load, $P = V_{L\,av} \times I_{L\,av} = 179.33 \times 89.665 = 16{,}080$ W

Reactive power of load, $Q = \frac{2V_{max}}{\pi} I_{L\,av} \sin\alpha$

$$= \frac{2 \times \sqrt{2} \times 230}{\pi} \times 89.665 \sin 30° = 9{,}283.6 \text{ VAR}$$

Input power factor

$$= \cos\tan^{-1}\frac{Q}{P} = \cos\tan^{-1}\frac{9283.6}{16{,}080} = \cos 30° = 0.866 \text{ (lagging)}$$ **Ans.**

Example 3.6. A separately excited dc motor is controlled by varying its armature voltage using a single phase full-converter bridge as shown in Fig. 3.12. The field current is kept constant at the rated value. The motor has an armature resistance of 0.2 Ω, and the motor voltage constant is 2.5 V/rad/sec. The motor is driving a mechanical load having a constant torque of 140 Nm. The triggering angle of the converter is 60º. The armature current can be assumed to be continuous and ripple free.

(*a*) Calculate the motor armature current.

(*b*) Evaluate the motor speed in rad/sec. [*GATE 2001*]

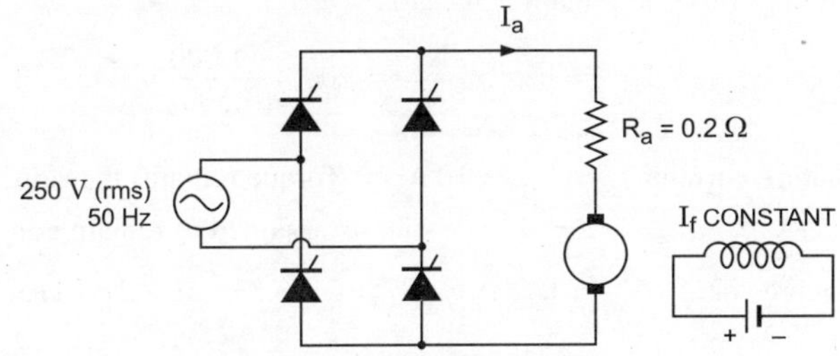

Fig. 3.12

Solution : (*a*) Voltage across the motor armature

$$V_a = \frac{2V_{max}}{\pi}\cos\alpha = \frac{2 \times 250\sqrt{2}}{\pi}\cos 60° = 112.5 \text{ V}$$

∵ $V_{max} = 250\sqrt{2}$ V and $\alpha = 60°$

Now $T\omega = E_bI_a = K_a\omega I_a$ ∵ $E_b = K_a\omega$ ...(*i*)

or Armature current, $I_a = \frac{T}{K_a} = \frac{140}{2.5} = 56$ A **Ans.**

(*b*) Back emf, $E_b = V - I_aR_a = 112.5 - 56 \times 0.2 = 101.34$ V

From Eq. (*i*) Speed, $\omega = \frac{E_bI_a}{T} = \frac{101.34 \times 56}{140} = 40.536$ radians/second **Ans.**

Example 3.7. A 220 V, 1500 rpm, 10 A separately excited dc motor is fed from a single phase fully controlled rectifier with an ac source voltage of 230 V, 50 Hz, R_a = 2 Ω. Assuming continuous conduction, find the firing angles for :

(*i*) Half the rated torque 500 rpm

(*ii*) Rated motor torque and (–1000) rpm.

[*U.P. Technical Univ. Electric Drives 2009-10*]

Solution : Back emf of motor when operated from 220 V dc supply and runs at 1,500 rpm

$$E_{b_1} = V_{dc} - I_{a_1}R_a = 220 - 10 \times 2 = 200 \text{ V}$$

(*i*) Back emf of motor when fed from 230 V, 50 Hz single phase supply through fully controlled rectifier and runs at 500 rpm (assuming flux ϕ to remain constant)

$$E_{b_2} = E_{b_1} \times \frac{N_2}{N_1} = 200 \times \frac{500}{1{,}500} = 66.67 \text{ V}$$

Armature current, $I_{a_2} = \frac{1}{2} I_{a_1} = \frac{1}{2} \times 10 = 5$ A

∵ Torque is half and flux is assumed to remain constant

Armature voltage, $V_{a_1} = E_{b_2} + I_{a_2} R_a = 66.67 + 5 \times 2 = 76.67$ V

Since for a fully controlled single phase rectifier voltage across armature is given as

$$V_a = \frac{2V_{max}}{\pi} \cos \alpha_a$$

$$\text{or } 76.67 = \frac{2 \times \sqrt{2} \times 230}{\pi} \cos \alpha_{a_1}$$

$$\text{or } \cos \alpha_{a_1} = \frac{76.67 \times \pi}{2\sqrt{2} \times 230} = 0.37$$

$$\text{or } \alpha_{a_1} = \cos^{-1} 0.37 = 68.27° \text{ **Ans.**}$$

(*ii*) Back emf of motor when fed from 230 V, 50 Hz single phase supply through fully controlled rectifier and runs at –1,000 rpm (assuming flux *f* to remain constant)

$$E_{b_3} = E_{b_1} \times \frac{N_3}{N_1} = 200 \times \frac{-1{,}000}{1{,}500} = -133.33 \text{ V}$$

Armature current, $I_{a_3} = I_{a_1} = 10$ A $\because$ Torque remains the same and flux is assumed to remain constant

Armature voltage, $V_{a_2} = E_{b_3} + I_{a_3} R_a = -133.33 + 10 \times 2 = -113.33$ V

$$\text{and } V_{a_2} = \frac{2V_{max}}{\pi} \cos \alpha_{a_2}$$

$$\text{So } -113.33 = \frac{2 \times \sqrt{2} \times 230}{\pi} \cos \alpha_{a_2}$$

$$\text{or } \cos \alpha_{a_2} = \cos^{-1}(-0.5473) = 123.18° \text{ **Ans.**}$$

Example 3.8. A 220 V single 50 Hz supply feeds the armature and field circuits of a separately excited dc motor through two full converters. The firing angle of field circuit converter is zero. The armature and field resistances are 0.2 Ω and 200 Ω respectively. The torque and voltage constants are 1.12. The firing angle of the armature circuit converter is 30° and armature current is 40 A. Find torque developed and motor speed. Assume that field and armature currents are continuous and constant and neglect brush contact drop.

Solution : Field circuit voltage,

$$V_f = \frac{2V_{max}}{\pi} \cos \alpha_f = \frac{2 \times 220\sqrt{2}}{\pi} \cos 0° = 198 \text{ V}$$

Field circuit current,

$$I_f = \frac{V_f}{R_f} = \frac{198}{200} = 0.99 \text{ A}$$

Armature circuit voltage,

$$V_a = \frac{2V_{max} \cos \alpha_a}{\pi} = \frac{2 \times 220\sqrt{2}}{\pi} \cos 30° = 171.53 \text{ V}$$

Back emf, $E_b = V_a - I_a R_a = 171.53 - 40 \times 0.2 = 163.53$ V

Torque developed,

$$T_d = K_t I_a I_f = 1.12 \times 40 \times 0.99 = 44.352 \text{ N-m} \quad \textbf{Ans.}$$

$$\text{Speed, } N = \frac{60\omega}{2\pi} = \frac{60}{2\pi} \times \frac{\text{back emf}}{K_t I_f} = \frac{60}{2\pi} \times \frac{163.53}{1.12 \times 0.99}$$

$$= 1{,}408 \text{ rpm} \quad \textbf{Ans.}$$

3.6.2. Three Phase Drives

For large sized (rating exceeding 15 kW) motors 3-phase drive is employed. The output voltage of a 3-phase converter has less ripple contents than that of single phase converter. So the filtering requirements for smoothing out the armature current are less and the current is mostly continuous. Three-phase converters could be half-wave, semiconverter, full-converter and dual converter. The 3-phase half-wave converter drives are only of theoretical importance and are generally not employed in industrial applications. This is because of presence of dc component in supply currents. Semi- and full-converters are widely used. A dual converter is employed in case of reversible drives with a power rating up to several MW.

1. *Three-Phase Semiconverter Drives.* It is a one quadrant drive and is employed for motors of rating up to about 100 kW. The field converter may be 3-phase or single-phase. The armature voltage V_a is given as

$$V_a = \frac{3\sqrt{3}V_{max}}{2\pi} (1 + \cos \alpha_a) \text{ for } 0 \le \alpha_a \le \pi \qquad ...(3.19)$$

where V_{max} is the peak value of line-to-neutral voltage and α_a is the firing angle of the converter in the armature circuit.

The circuit diagram for a separately excited dc motor supplied from 3-phase ac supply through a three-phase semiconverter is shown in Fig. 3.13.

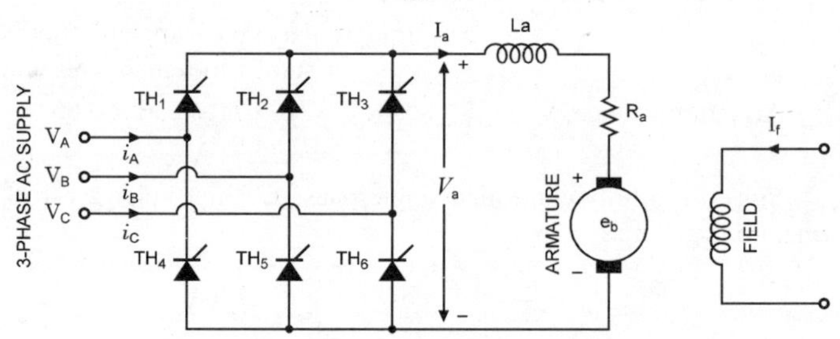

Fig. 3.13. ***Three-Phase Semiconverter Supplying Separately Excited DC Motor***

2. *Three-Phase Full Converter Drives.* Three-phase full converter drive is a two quadrant drive and is employed up to 1,500 kW motors. The converter in the field circuit may be single-phase or 3-phase. When 3-phase full converters are used for both armature and field, the armature and field voltages are given as

$$V_a = \frac{3\sqrt{3}V_{max}}{\pi} \cos \alpha_a \text{ for } 0 \le \alpha_a \le \pi \qquad ...(3.20)$$

$$V_f = \frac{3\sqrt{3}V_{max}}{\pi} \cos \alpha_f \text{ for } 0 \le \alpha_f \le \pi \qquad ...(3.21)$$

3. *Three-Phase Dual Converter Drives.* 3-phase dual converter consists of two converters and can be operated in all the four quadrants. Either converter 1 is operated to supply armature voltage V_a or converter 2 is operated to supply armature voltage $-V_a$. It is used for motors of rating up to 1,500 kW. The armature voltage supplied by converter 1 is given as

$$V_a = \frac{3\sqrt{3}\,V_{max}}{\pi} \cos \alpha_{a_1}, \text{ for } 0 \le \alpha_{a_1} \le \pi \qquad ...(3.22)$$

where α_{a_1} is the firing angle for converter 1.

For converter 2, the voltage V_a is given as

$$V_a = \frac{3\sqrt{3}\,V_{max}}{\pi} \cos \alpha_{a_2} \text{ for } 0 \le \alpha_{a_2} \le \pi \qquad ...(3.23)$$

where $\alpha_{a_2} = \pi - \alpha_{a_1}$.

If the field is also supplied by a 3-phase full converter,

$$V_f = \frac{3\sqrt{3}\ V_{max}}{\pi} \cos \alpha_f \text{ for } 0 \le \alpha_f < \pi \qquad ...(3.24)$$

Example 3.9. A 400 V, 3-phase ac supply feeds a separately excited dc motor through two 3-phase full converters, one for armature and the other for the field. The firing angle of the field circuit converter is zero. The armature and field resistances are 0.2 Ω and 250 Ω respectively. The armature current is 60 A. The motor voltage constant is 1.3 V/A-rad/second. The speed is 1,200 rpm. Determine the firing angle of the armature circuit converter.

Solution : Field circuit voltage,

$$V_f = \frac{3\sqrt{3}}{\pi} V_{max} \cos \alpha_f = \frac{3\sqrt{3}}{\pi} \times \frac{400}{\sqrt{3}} \times \sqrt{2} \cos 0° = 540.2 \text{ V}$$

...Refer to Eq. (3.21)

Field circuit current,

$$I_f = \frac{V_f}{R_f} = \frac{540.2}{250} = 2.16 \text{ A}$$

Back emf developed,

$$E_b = KI_f\omega = 1.3 \times 2.16 \times \frac{2\pi \times 1{,}200}{60} = 352.86 \text{ V}$$

Armature circuit voltage,

$$V_a = E_b + I_aR_a = 352.86 + 60 \times 0.2 = 364.86 \text{ V}$$

Also armature circuit voltage

$$= \frac{3\sqrt{3}}{\pi} V_{max} \cos \alpha_a \qquad \text{...Refer to Eq. (3.20)}$$

$$\text{So } 364.86 = \frac{3\sqrt{3}}{\pi} \times \frac{400}{\sqrt{3}} \times \sqrt{2} \cos \alpha_a$$

or $\alpha_a = 47.52°$ **Ans.**

Example 3.10. A 3-phase half-controlled thyristor bridge with 400 V, 3-phase, 50 Hz supply is feeding a separately excited dc motor. Armature resistance is 0.2 Ω, armature rated current is 100 A and back emf constant is 0.25 V/rpm. Determine no-load speed if no-load armature current is 5 A and firing angle is 45°. Also determine firing angle to obtain a speed of 1,500 rpm at rated current.

Solution : Armature circuit voltage at no load,

$$V_{a_0} = \frac{3\sqrt{3}}{2\pi} V_{max} (1 + \cos \alpha_a) \qquad \text{...Refer to Eq. (3.19)}$$

$$= \frac{3\sqrt{3}}{2\pi} \times \frac{400\sqrt{2}}{\sqrt{3}} (1 + \cos 45°) = 461 \text{ V}$$

Back emf at no load, $E_{b_0} = V_{a_0} - I_{a_0} R_a = 461 - 5 \times 0.2 = 460$ V

Speed at no load, $N_0 = \frac{460}{0.25} = 1{,}840$ rpm **Ans.**

Back emf developed at a speed of 1,500 rpm,

$$E_b = 1{,}500 \times 0.25 = 375 \text{ V}$$

Input voltage to motor, $V_a = E_b + I_aR_a = 375 + 100 \times 0.2 = 395$ V

$$\text{Also } V_a = \frac{3\sqrt{3}}{2\pi} V_{max} (1 + \cos \alpha_a)$$

$$\therefore\ 395 = \frac{3\sqrt{3}}{2\pi} \times \frac{400\sqrt{2}}{\sqrt{3}} (1 + \cos \alpha_a)$$

or $\alpha_a = 62.45°$ **Ans.**

Example 3.11. A separately excited dc motor is supplied from 3-phase, 400 V, 50 Hz supply through half controlled 3-phase bridge rectifier. The motor parameters are inductance 0.012 H, resistance 0.8 Ω, armature constant 2 V/rad/s (N-m/A). Neglecting rectifier losses and supply impedance, determine speed-torque characteristics up to 80 N-m at fixed firing angle of 90°.

Solution : Because of armature inductance, the armature current can be assumed to remain constant.

Armature circuit voltage,

$$V_a = \frac{3\sqrt{3}}{2\pi} V_{max} (1 + \cos \alpha_a)$$

$$= \frac{3\sqrt{3}}{2\pi} \times \frac{400}{\sqrt{3}} \times \sqrt{2}\ (1 + \cos 90°) = 270 \text{ V}$$

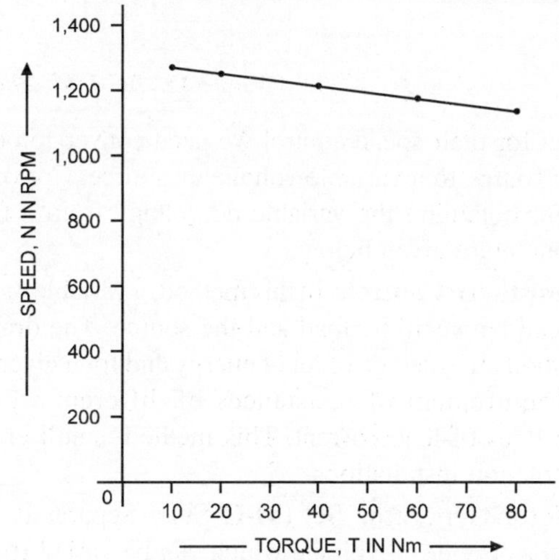

Fig. 3.14

Let us determine torque and speed for armature current of 5, 10, 20, 30, 40 A, as tabulated below :

Armature Current in Amperes, I_a	*Torque in N-m $= K_tI_a$ $= 2 \times I_a$*	*Back emf, $E_b = V_a - I_aR_a$ $= 270 - I_a \times 0.8$ in volts*	*Speed, N in RPM $= \frac{E_b}{K_t} \times \frac{60}{2\pi}$ $= \frac{E_b}{2} \times \frac{30}{\pi}$*
5	2 × 5 = 10	270 – 5 × 0.8 = 266	$\frac{266}{2} \times \frac{30}{\pi} = 1{,}270$
10	2 × 10 = 20	270 – 10 × 0.8 = 262	$\frac{262}{2} \times \frac{30}{\pi} = 1{,}250$
20	2 × 20 = 40	270 – 20 × 0.8 = 254	$\frac{254}{2} \times \frac{30}{\pi} = 1{,}213$
30	2 × 30 = 60	270 – 30 × 0.8 = 246	$\frac{246}{2} \times \frac{30}{\pi} = 1{,}175$
40	2 × 40 = 80	270 – 40 × 0.8 = 238	$\frac{238}{2} \times \frac{30}{\pi} = 1{,}136$

Speed-torque characteristic is shown in Fig. 3.14.

3.7. CHOPPER CONTROL OF DC MOTORS

In many cases, conversion of dc supply source voltage to different levels is required. For example, subway cars, trolley buses, or battery operated vehicles are fed from a fixed voltage dc supply

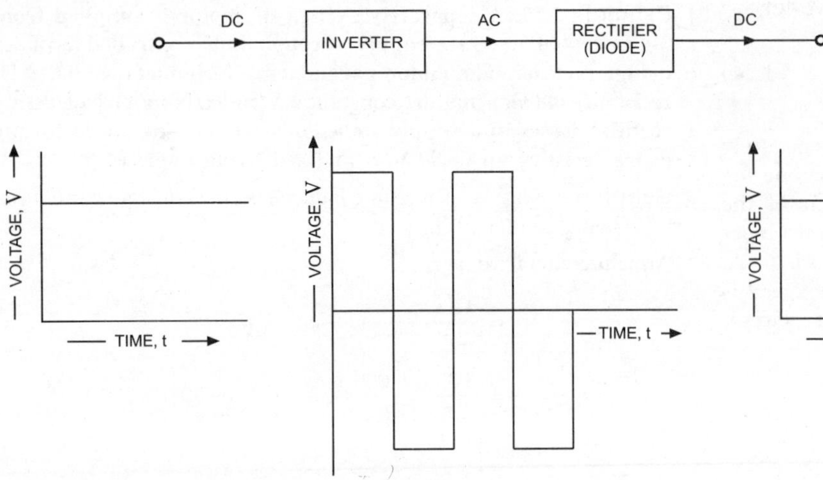

Fig. 3.15. *AC-Link Chopper*

source. But for their speed control we need conversion of fixed voltage dc source to a variable voltage dc source. Conventional methods for obtaining the variable dc voltage from a fixed dc voltage source are given below:

1. Resistance Control. In this method, a variable resistance is introduced between the load and the source. The drawbacks of this method are wastage of lot of energy and for a given output voltage, requirement of resistances of different values for different values of load current. This method is still employed for older traction installations.

2. Motor-Generator Set (M-G Set). Separately excited dc generator provides a voltage which can be varied from zero to rated value with either polarity. It has disadvantages of higher cost, large bulk, low efficiency and sluggish response because of the generator field time constant.

3. AC Link Chopper (Inverter-Rectifier). In this method, the dc supply is first converted to ac supply by an inverter (dc to ac converter). The obtained ac supply is then stepped up or down by a transformer and then rectified back to dc by a rectifier. Because of two stages of conversion, the set-up is costly, bulky and less efficient.

4. DC Chopper (DC to DC Power Converter). As already mentioned in Art. 3.3.3, a dc chopper is a static switch to provide variable dc voltage from a source of constant dc voltage. Choppers are now being used all over the world for rapid transit systems. These are also employed in trolley cars, marine hoists, forklift trucks and mine haulers. The future electric automobiles are likely to use choppers for their speed control and braking. Chopper systems offer smooth control, high efficiency, fast response and regeneration.

3.7.1. Principle of Operation of Chopper

A *dc chopper* is a high speed on/off semiconductor switch that connects the load to and disconnects it from the supply and produces a chopped load voltage from a constant input supply voltage. As it is an on/off switch between source and load, it can be symbolically represented by a switch S inside a dotted rectangle, as shown in Fig. 3.16. Actually, a chopper consists of main power semiconductor device together with their turn-on and turn-off mechanisms. In low-power chopper circuits, power transistors, GTOs etc. are used widely. In high-power chopper circuits, however, thyristors are in common use.

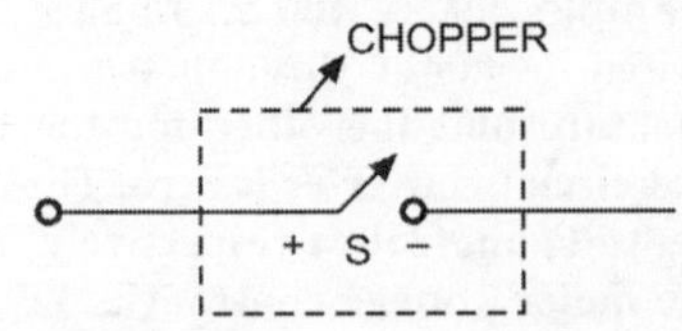

Fig. 3.16. *Symbol of Chopper*

Figure 3.17 illustrates the principle of operation of a chopper. The switch may be closed or opened by supplying on/off pulses to it. Here the mechanism of sending pulses to

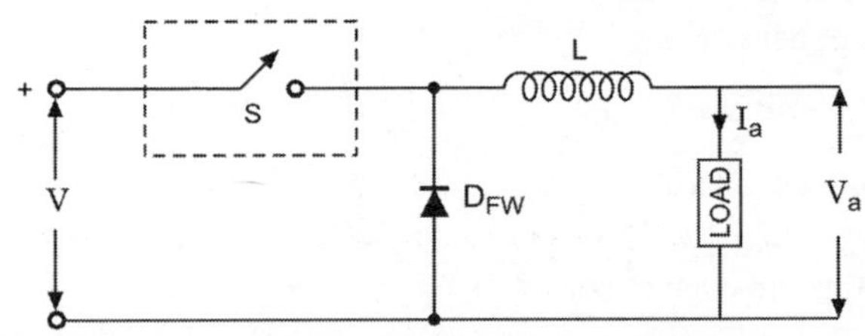

Fig. 3.17. *Elementary Chopper Circuit*

a switch has not been shown. A shunting diode D_{FW} is provided across the load for freewheeling the load current when the switch S is off. When the switch S is closed, the source voltage V, assuming that voltage drop across switch S in on-state is negligible, appears across the load for the duration the switch is closed. This duration is designated T_{on}. After the pulse to switch is removed the load is disconnected from the source. The energy stored in the inductor circulates a current through the freewheeling diode D_{FW}. As a result the load is short circuited and voltage across it falls to zero but the current through the load falls exponentially, as illustrated in Fig. 3.18. The duration for which switch is off is called off time T_{off}. The total time of a pulse is, therefore, given by $T = T_{on} + T_{off}$.

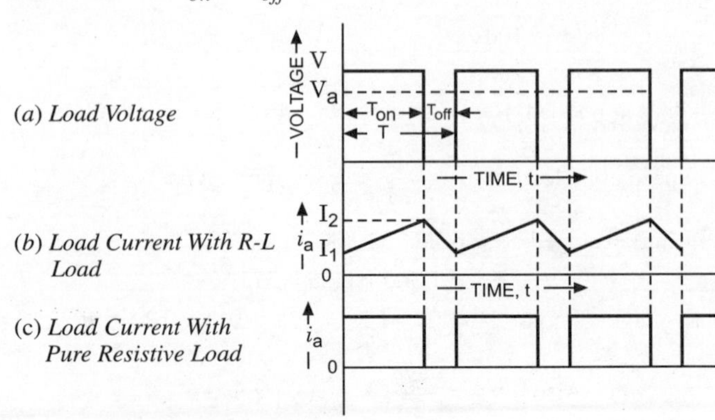

Fig. 3.18. *Voltage and Current Waveforms*

Thus, a dc chopped voltage is produced at the load terminals [Fig. 3.18 (*a*)] and the load current is continuous as shown in Figs. 3.18 (*b*) and (*c*).

From Fig. 3.18 (*a*), the average load voltage V_a is given by

$$V_a = V\frac{T_{on}}{T_{on}+T_{off}} = V\frac{T_{on}}{T} = \alpha V \qquad ...(3.25)$$

where $\alpha = \frac{T_{on}}{T}$ is called *duty cycle* of the circuit. It is analogous to transformation ratio of ac transformer. The value of α ranges from 0 to 1. Hence the output voltage varies between zero and supply voltage V. In the present case the output voltage is less than the supply voltage. It is, therefore, called **step-down** chopper. In choppers, called step-up choppers, the output voltage is more than supply voltage.

The load current and voltage waveforms, as shown in Figs. 3.18 (*b*) and (*a*) are for R-L load. However, in case, resistive load is used in the output, the current and voltage waveforms have similar variations as illustrated in Fig. 3.18 (*c*) and (*a*).

In case of R-L load, the inductor absorbs energy from the source during on-period while same is released by it during off-period. We can write, by using conservation of energy principle;

$$\text{Energy absorbed} = \text{Energy released}$$

$$\text{or} \quad VI = V_a I_a$$

$$\text{or} \quad I = \frac{V_a}{V} I_a = \alpha I_a$$

$$\text{or} \quad \frac{I}{I_a} = \alpha = \frac{V_a}{V} \qquad ...(3.26)$$

Above relation is similar to that for an ac transformer. Here α, the duty cycle, corresponds to transformation ratio of transformer. The output of a chopper, as can be observed from Eqs. (3.25) and (3.26), can be controlled by a variation in α, the duty cycle.

Control Strategies. It is seen from Eq. (3.25), that, average value of output voltage, V_a can be controlled by periodic opening and closing of the semiconductor switch. There are two types of control strategies for operating the switches that can be used in dc choppers. These control strategies are time-ratio control (TRC) and current limit control.

1. Time-Ratio Control. In this control strategy, the value of $\frac{T_{on}}{T}$ is varied. This is affected in two ways namely variable frequency control and constant frequency control.

(*a*) **Variable Frequency Control (FM).** Under such scheme the chopping frequency *f* is kept variable and for this purpose either (*a*) *on* time, T_{on} or (*b*) *off* time, T_{off} is kept constant. This scheme is, therefore, called frequency modulation scheme and is illustrated in Fig. 3.19.

In Fig. 3.19 (*a*) and (*b*) *on* time T_{on} is kept constant while *off* time T_{off} of the switch is variable. The duty cycles in two figures are $\alpha_1 = \frac{T_{on}}{T_{on}+T_{off}} = \frac{3}{4}$ and $\alpha_2 = \frac{1}{2}$ respectively.

Figure 3.19 (*c*) and (*d*) are for variable on time T_{on} and constant off time T_{off}. The duty cycle in the two cases are respectively $\frac{1}{4}$ and $\frac{2}{5}$. This technique is suitable for switches which require forced commutation to turn off.

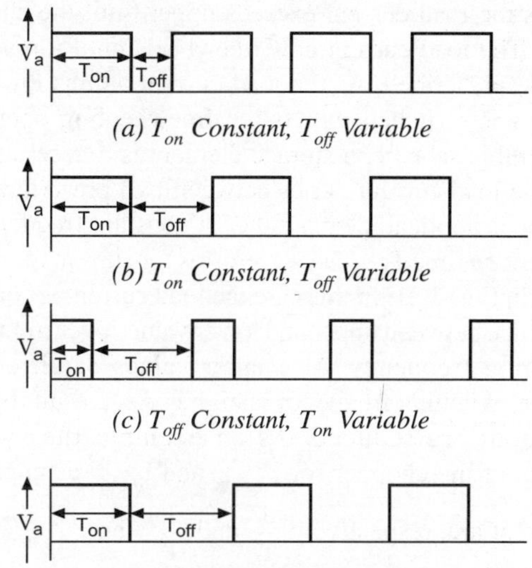

(*a*) T_{on} *Constant*, T_{off} *Variable*

(*b*) T_{on} *Constant*, T_{off} *Variable*

(*c*) T_{off} *Constant*, T_{on} *Variable*

(*d*) T_{off} *Constant*, T_{on} *Variable*

Fig. 3.19. ***Output Voltage Waveforms For Variable Frequency Control***

(*b*) **Constant Frequency Control.** Under this control scheme the chopping period T remains constant but the pulse width T_{on} is varied to control the output. This is, therefore, called pulse width modulation (PWM) control. The scheme is illustrated in Fig. 3.20. In the figure, time $T = T_{on} + T_{off}$ is constant but T_{on} and T_{off} both are variable. The duty cycles of α = 0.75 and 0.50 are shown, respectively, in Figs. 3.20 (*a*) and (*b*).

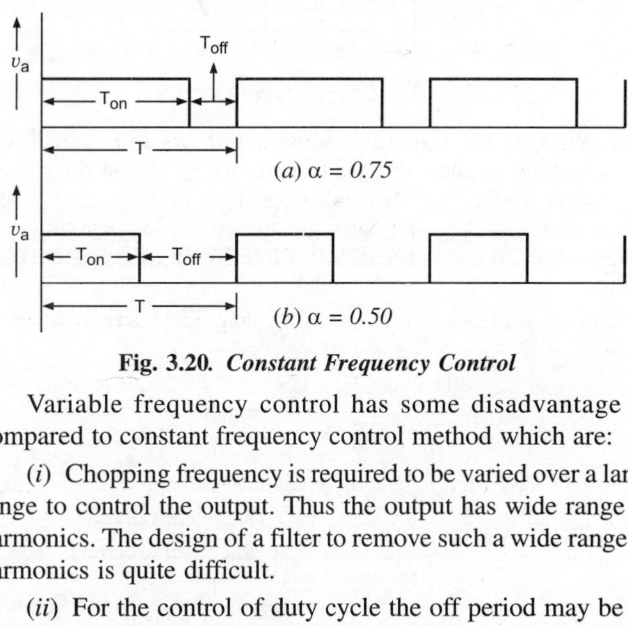

(*a*) α = 0.75

(*b*) α = 0.50

Fig. 3.20. ***Constant Frequency Control***

Variable frequency control has some disadvantage as compared to constant frequency control method which are:

(*i*) Chopping frequency is required to be varied over a large range to control the output. Thus the output has wide range of harmonics. The design of a filter to remove such a wide range of harmonics is quite difficult.

(*ii*) For the control of duty cycle the off period may be so large that load current falls to zero between two consecutive pulses. This leads to discontinuity in the load current which is not desirable.

(*iii*) For the control of duty cycle α, frequency variations would be wide. As such, there is a possibility of interference with signalling and telephone lines in frequency modulation control.

For above reasons PWM control is preferred over FM control in chopper circuits.

2. Current Limit Control. Under this scheme the chopper is switched *on* and *off* between upper and lower limits of load current. As the load current exceeds upper limit the chopper is turned *off*. The load current now freewheels until it reaches the lower value prescribed. As the load current attains lower limit, the chopper is switched *on*. As the freewheeling operation is involved in this scheme, a storage element is a necessity in the load. As the load current varies between two prescribed limits, this scheme is applicable only either at constant frequency or at constant turn *on* time T_{on}. The waveforms for current limit control are shown in Fig. 3.21. In this case the load current is continuous and difference between upper and lower values of current decides the switching frequency. At smaller values of difference in current, the switching frequency is higher. As a result the ripples in the output are reduced. As an example the switching frequencies of the chopper for $I'_{a\,max}$ and $I_{a\,max}$ as upper limits of the current are, respectively $\frac{1}{T'}$ and $\frac{1}{T}$ where $I'_{a\,max} < I_{a\,max}$ and $T' < T$. By further reducing the difference between upper and lower limits of the current the switching frequency can be increased correspondingly which leads to reduction in the ripple content in the output.

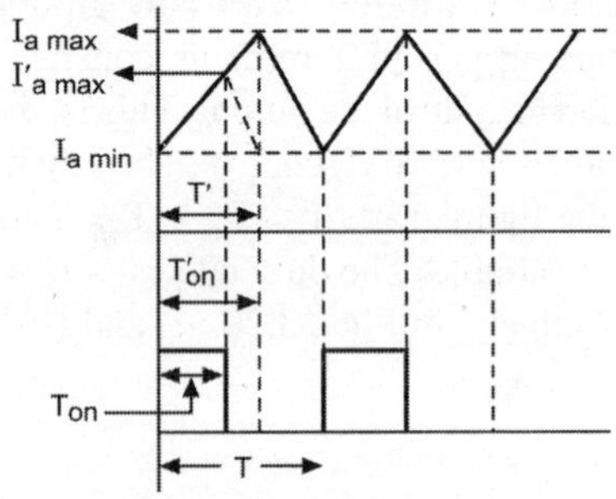

Fig. 3.21. *Current Limit Control*

Example 3.12. For a dc to dc converter shown in Fig. 3.17, the source voltage is 200 volts, load resistance 10 Ω and duty cycle α = 0.4. If load is resistive, calculate the following (*a*) Average load voltage (*b*) Thyristor average current (*c*) Average and rms freewheeling diode currents (*d*) Effective input resistance of chopper.

Solution : For resistive load, the load voltage and load current are in phase:

(*a*) Average load voltage, $V_a = \alpha V_{dc} = 0.4 \times 200 = 80$ volts **Ans.**

(*b*) Average thyristor current $= V_a/R = \frac{80}{10} = 8$ Amps

(*c*) Diode currents are zero in case of resistive load. **Ans.**

(*d*) Effective input resistance $= \frac{\text{DC source voltage}}{\text{Average source current}}$

$$= \frac{V}{\alpha \frac{V}{R}} = \frac{R}{\alpha} = \frac{10}{0.4} = 25\ \Omega \text{ Ans.}$$

3.7.2. Step-up Choppers

In a step-down chopper the output voltage lies between zero and supply voltage. However, there are techniques by which the output of a chopper obtained is higher than the supply voltage. Such types of choppers are called step-up choppers. The circuit diagram of a step-up chopper is shown in Fig. 3.22 (*a*). In figure, the chopper is represented by a switch S.

In this chopper, a large inductor in series with the supply voltage source is essential, as shown in Fig. 3.22 (*a*). When the chopper is on, the current flows through inductance L, chopper S and power supply source, as shown in Fig. 3.22 (*b*). During this on-period T_{on}, the inductor L stores energy.

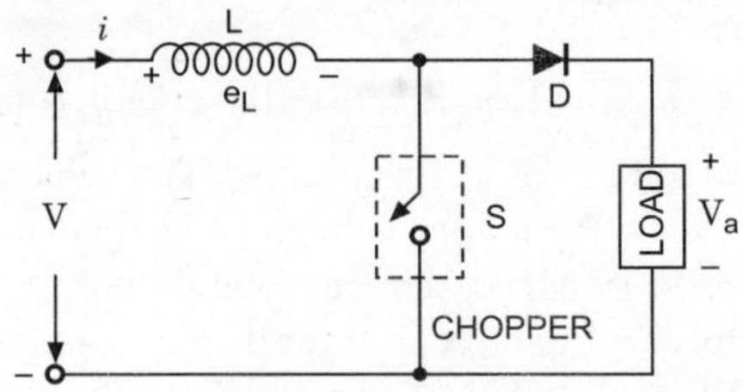

(a) Step-up Chopper

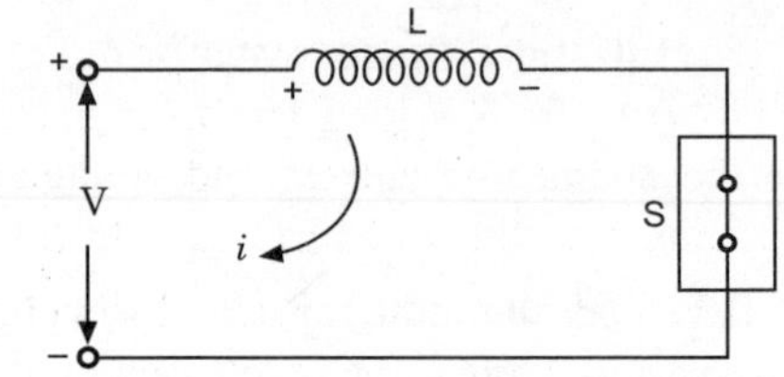

(b) Equivalent Circuit During On-Period

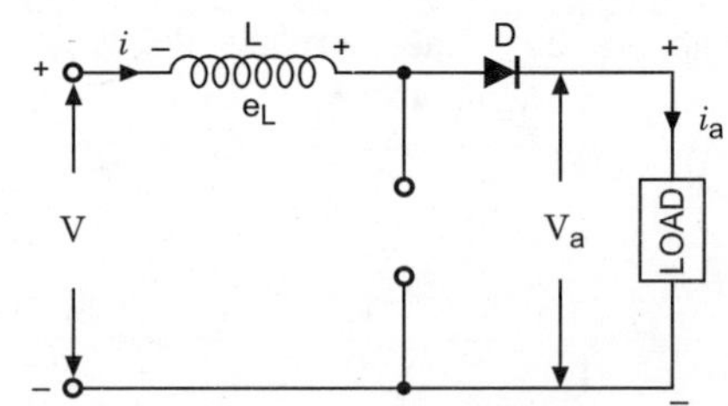

(c) Equivalent Circuit During off-Period

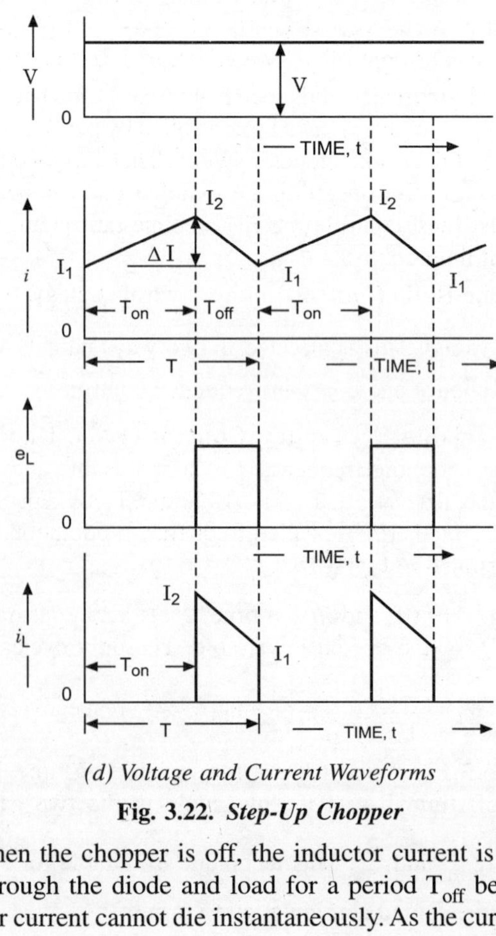

(d) Voltage and Current Waveforms

Fig. 3.22. *Step-Up Chopper*

When the chopper is off, the inductor current is forced to flow through the diode and load for a period T_{off} because the inductor current cannot die instantaneously. As the current tends to decrease, polarity of the emf induced in the inductor is reversed, as illustrated in Fig. 3.22 (*c*).

Assuming negligible voltage drop across diode D, the voltage equation for outer loop may be written as

$$-V_a + e_L + V = 0$$

$$\text{or } V_a = V + e_L \quad ...(3.27)$$

If the rate of fall of current through inductor is $\frac{di}{dt}$, Eq. (3.27) may be written as

$$V_a = V + L\frac{di}{dt} \quad ...(3.28)$$

It means that the inductor voltage adds to the source voltage to force the inductor current into the load. In this manner, the energy stored in inductor is released to load. Here, higher value of inductance L is preferred for getting lesser ripple in the output.

During the time T_{on}, when the chopper is on, the energy stored in the inductor is given by

$$W_1 = \text{Voltage} \times \text{average current through L} \times T_{on}$$

$$= V \times I_a \times T_{on} \quad ...(3.29)$$

Above Eq. (3.29) is based on the assumption that the variation of output current is linear.

During the period T_{on}, the current through the inductor rises to peak value I_2 while during off period T_{off} it falls to value I_1 and rises again to I_2 in the next on state of the chopper. It is shown in Fig. 3.22 (*d*). Thus, in a cycle the average value of current is given by $I_a = \frac{I_1 + I_2}{2}$. Hence the energy stored in the inductor L is given by

$$W_1 = V \times \frac{I_1 + I_2}{2} \times T_{on} \quad ...(3.30)$$

and this energy is released by the inductor to the load during the off-period, T_{off}. Thus energy released by inductor L during off period is given by

$$W_0 = \text{Voltage across inductor} \times \text{average current through inductor} \times T_{off}$$

$$= (V_a - V)\left(\frac{I_1 + I_2}{2}\right) T_{off} \quad ...(3.31)$$

Assuming the system to be lossless, in the steady-state the two energies will be equal *i.e.*,

$$W_1 = W_0$$

$$\text{or } V \times \frac{I_1 + I_2}{2} \times T_{on} = (V_a - V)\left(\frac{I_1 + I_2}{2}\right) T_{off}$$

$$\text{or } V(T_{on} + T_{off}) = V_a T_{off}$$

$$\text{or } V_a = \frac{V(T_{on} + T_{off})}{T_{off}}$$

$$= \frac{V \times T}{T - T_{on}} = \frac{V}{1 - \frac{T_{on}}{T}} = \frac{V}{1 - \alpha} \quad ...(3.32)$$

The value of α, the duty cycle lies between zero and unity. Corresponding average values of load voltage V_a lies between V and infinity. Thus the output voltage of step-up chopper is never less than the supply voltage V *i.e.*, for $0 \le \alpha \le 1$ the output voltage of step-up chopper lies in the range $V \le V_a \le \alpha$.

The step-up chopper can be employed for regenerative braking of dc motors even at low operating speeds. Let V represent the motor armature voltage and V_a the dc source voltage in Fig. 3.22 (*a*). Regenerative braking takes place when $\left(V + L\frac{di}{dt}\right)$ exceeds V_a. Even at decreasing motor speeds, duty cycle α can be so adjusted that $\left(V + L\frac{di}{dt}\right)$ exceeds fixed supply voltage V_a.

Example 3.13. A step-up chopper shown in Fig. 3.22 (*a*) is fed by a dc source of 220 volts. The inductance has value of 30 mH. If load voltage is 250 volts across a resistance of 10 ohms, calculate average current and duty cycle.

Solution : Average value of load voltage is

$$V_a = \frac{V}{1-\alpha} = 250 \text{ volts}$$

$$\text{or } 250 = \frac{220}{1-\alpha}$$

$$\text{or } \alpha = 1 - \frac{220}{250} = 0.12 \textbf{ Ans.}$$

$$\text{Average load current} = \frac{V_a}{R} = \frac{250}{10} = 25 \text{ A } \textbf{Ans.}$$

3.7.3. Step-up/down Chopper

A chopper can also be used for stepping up or down the supply voltage by varying its duty cycle. The principle of operation is illustrated in Fig. 3.23. As shown, the output voltage polarity is opposite to that of input voltage V.

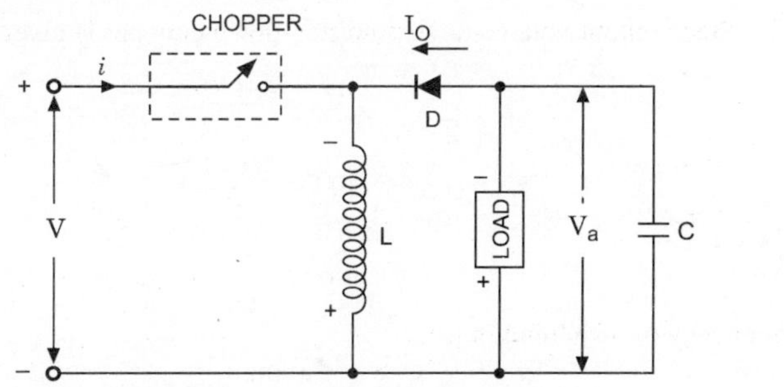

Fig. 3.23. ***Step-Up/Down Chopper***

When the chopper is on, the supply current flows through the path V_+ – chopper – inductor L – V_-. As a result energy is stored in the inductor. When the chopper is made off, the current through the inductor L decreases. As a result an emf is induced in the inductor with lower end positive and the energy stored in the inductor is released through the load and diode and the voltage across the load is negative as illustrated. It is thus used to reverse the polarity of load voltage w.r.t. source voltage.

The energy stored in inductor L during on-period T_{on} is given by

$$W_1 = V \times \text{average current} \times T_{on}$$

But the average current through the inductor, as in case of step-up chopper discussed in Art. 3.7.2, is

$$I_a = \frac{I_1 + I_2}{2} \text{ giving}$$

$$W_1 = V \times \frac{I_1 + I_2}{2} T_{on} \qquad ...(3.33)$$

During *off* period T_{off}, the energy released by inductor to the load is

$$W_0 = V_a \times \frac{I_1 + I_2}{2} T_{off} \qquad ...(3.34)$$

Under steady-state condition, assuming no loss of power.

Energy stored in inductor = Energy released to load

$$\text{or } V \times \frac{I_1 + I_2}{2} T_{on} = V_a \times \frac{I_1 + I_2}{2} T_{off}$$

$$\text{or } V_a = V \times \frac{T_{on}}{T_{off}} = \frac{V \times T_{on}}{T - T_{on}} = \frac{V}{\frac{T}{T_{on}} - \frac{T_{on}}{T_{on}}}$$

Substituting $\frac{T_{on}}{T} = \alpha$ in above equation, we have

$$V_a = \frac{V}{\frac{1}{\alpha} - 1} = \frac{\alpha}{1-\alpha} V \qquad ...(3.35)$$

Equation (3.35) can be used for step-up and step-down operation of chopper. In the range $0 < \alpha < 0.5$ the circuit operates in step-down mode while for duty cycle in the range $0.5 < \alpha < 1$ it acts as a step-up chopper.

Example 3.14. A step-up/step-down chopper has an input voltage of 125 V and output voltage of 200 V. If the period of conduction is 1 ms, determine the frequency of switching pulse.

Solution : Average output voltage, $V_a = 200$ V

Supply voltage, $V = 125$ V

Period of conduction, $T_{on} = 1 \text{ ms} = 0.001$ s

Since output voltage of step-up/step-down chopper is given by

$$V_a = \frac{\alpha V}{1-\alpha},$$

$$200 = \frac{\alpha \times 125}{1-\alpha}$$

$$\text{or } \alpha = \frac{8}{13}$$

Frequency of switching pulse,

$$f = \frac{1}{T} = \frac{1}{T_{on}/\alpha} = \frac{\alpha}{T_{on}} = \frac{8/13}{1 \times 10^{-3}} = \frac{8,000}{13} \text{ Hz} = 615.4 \text{ Hz} \qquad \textbf{Ans.}$$

3.7.4. Chopper Control of Separately Excited DC Motor Drive

Chopper drives can be operated in the following modes :

1. Motoring mode.
2. Regenerative braking mode.
3. Rheostatic braking mode.

1. *Motoring Mode.* Figure 3.24 (*a*) illustrates this mode of operation with a one quadrant chopper feeding the motor from dc supply of voltage V, the field being excited from a separate source. Figure 3.24 (*b*) shows the quadrant of operation. When the chopper conducts, voltage across the motor armature is supply voltage V, and when it blocks, voltage across the motor armature is zero. Therefore, both average voltage and current are positive and power can flow from supply source to the load, and thus, represents *motoring operation*. The voltage across the motor armature is given as

$$V_a = \alpha V \qquad ...(3.36)$$

where α is the duty cycle of the chopper.

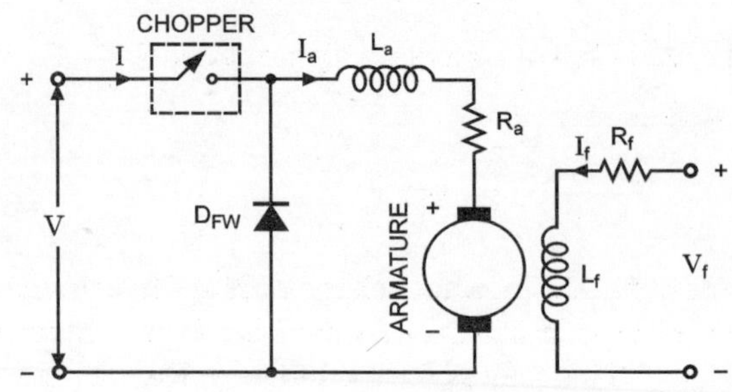

(*a*) *Circuit Diagram*

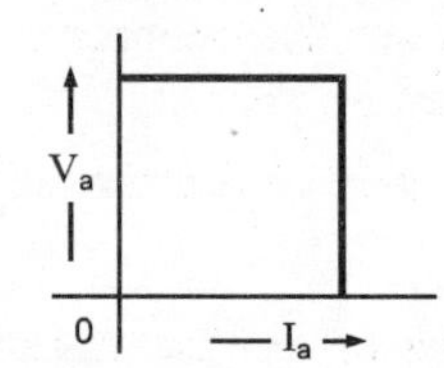

(*b*) *Quadrant of Operation*

Fig. 3.24. ***Separately Excited DC Motor Supplied by a Chopper***

Freewheeling diode D_{FW} is necessary so that the armature current may flow when the chopper is off. The power supplied to the motor is given as

$$P = V_a I_a = \alpha V I_a \qquad ...(3.37)$$

The current drawn from the supply is

$$I = \alpha I_a \qquad ...(3.38)$$

By varying the duty cycle α, the speed of the motor can be controlled.

2. *Regenerative Braking Mode.* When the motor is to be stopped, brakes are to be applied. In regenerative braking the motor is made to operate as a generator and the kinetic energy of the motor is converted into electrical energy and returned to the supply. Fig. 3.25 (*a*) shows the circuit diagram for this mode of operation. During regenerative braking mode, the chopper is gated so that the motor armature is short circuited through the chopper. Kinetic energy of the motor is partly dissipated in armature resistance R_a and partly stored in inductance L_a. When chopper is turned off, freewheeling diode D_{FW} is turned on and the energy stored in inductance L_a is returned to the supply source (provided the source is receptive).

The operation is in second quadrant, as illustrated in Fig. 3.25 (*b*).

The average voltage V_{ch} is given as

$$V_{ch} = (1 - \alpha) V \qquad ...(3.39)$$

Regenerated power P_g is given as

$$P_g = V_{ch} \times I_a = (1 - \alpha) V I_a \qquad ...(3.40)$$

where I_a is the armature current

Generated voltage E_g is given as

$$E_g = V_{ch} + I_a R_a = (1 - \alpha) V + I_a R_a \qquad ...(3.41)$$

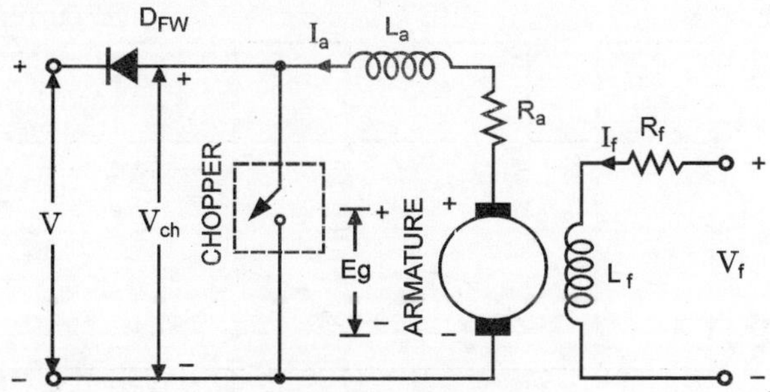

(a) Circuit Diagram

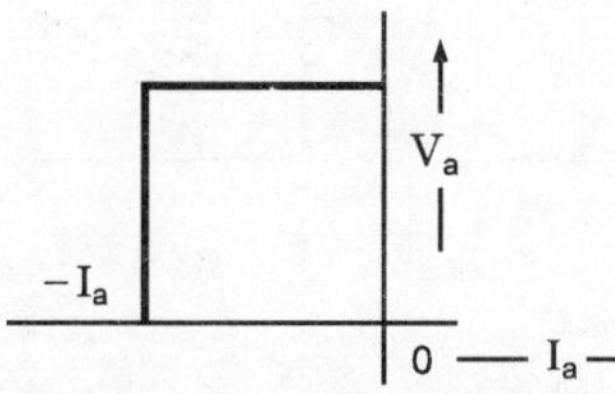

(b) Quadrant of Operation

Fig. 3.25. ***Regenerative Braking of Separately Excited DC Motor***

3. ***Rheostatic Braking Mode.*** Figure 3.26 (*a*) illustrates the circuit for this mode of operation. In this mode also the motor is made to operate as a generator and the kinetic energy of the motor is converted into electrical energy. This electrical energy is dissipated in braking resistance R_b connected to the terminals, as illustrated in the figure. When this mode is to be used, chopper is turned on by a gate pulse.

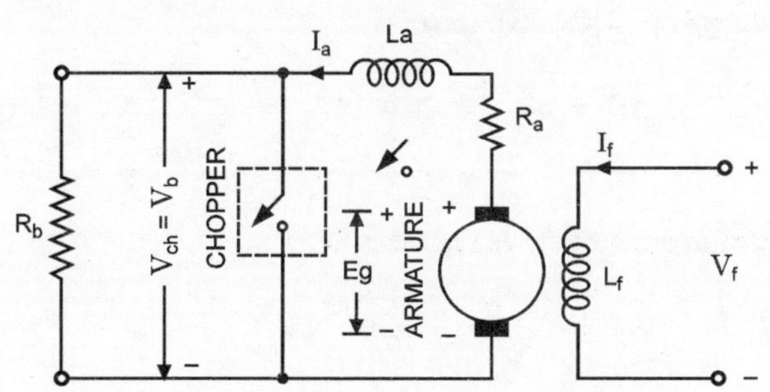

(a) Circuit Diagram

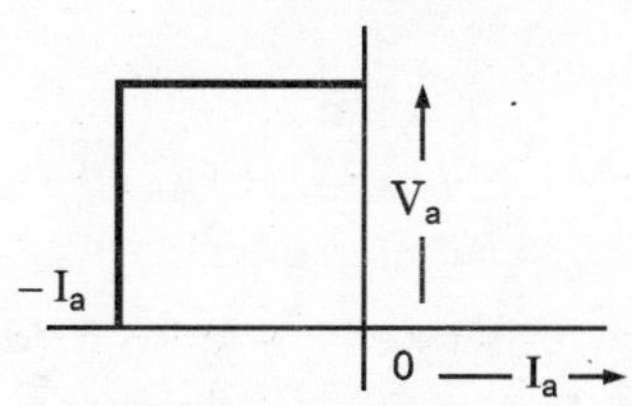

(b) Quadrant of Operation

Fig. 3.26. ***Rheostatic Braking of Separately Excited DC Motor***

The kinetic energy is partly dissipated in armature resistance R_a and partly stored in armature inductance L_a. When chopper is turned off, the energy stored in inductance L_a is transferred to braking resistance R_b and dissipated as heat.

The average current in braking resistance R_b is given as

$I_b = (1 - \alpha)\, I_a$ ...(3.42) where I_a is the armature current.

Voltage across braking resistance R_b,

$$V_b = I_b R_b = R_b I_a (1 - \alpha) \quad ...(3.43)$$

Power dissipated in braking resistance R_b,

$$P_b = I_a^2 R_b (1 - \alpha)^2 \quad ...(3.44)$$

In this mode also the operation is in second quadrant, as illustrated in Fig. 3.26 (*b*).

It is also possible to combine the regenerative and rheostatic braking modes. Regenerative braking is used when the supply system is receptive. During remaining period the rheostatic braking is used.

3.7.5. Step-Down Chopper With R-L-E_b Load

The choppers are widely employed for speed control of dc motors in industrial and traction drives. Figure 3.27 (*a*) shows the basic

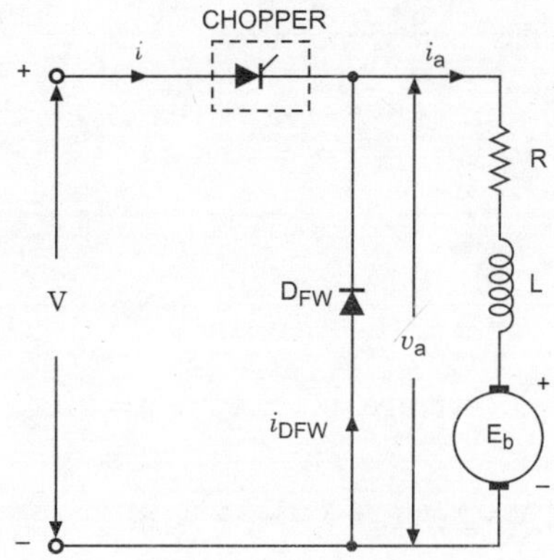

(a) Circuit Diagram

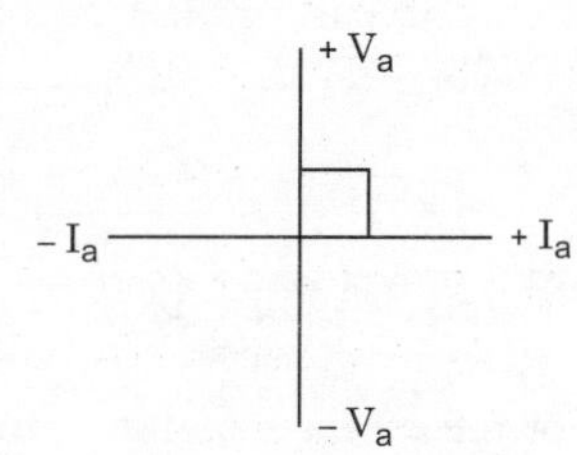

(b) Quadrant Diagram

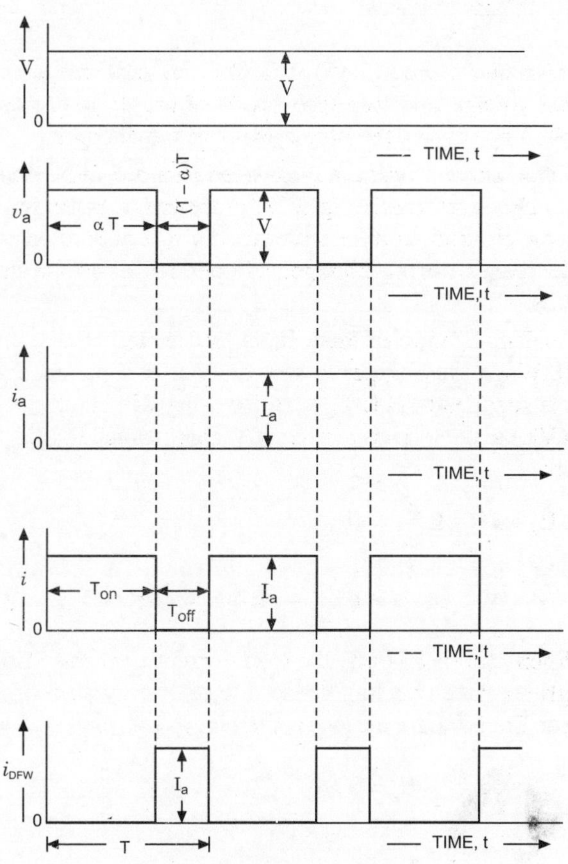

(c) Waveforms

Fig. 3.27. ***Step-Down Chopper With R-L-E_b Load***

chopper circuit for the control of a dc series motor. The chopper is shown to consist of a force-commutated thyristor, it could well be a transistor switch. It offers one-quadrant drive [Fig. 3.27 (*b*)]. Armature current is assumed continuous and ripple free. The waveforms for the source voltage V, armature terminal voltage v_a, armature current i_a, supply current i and freewheeling-diode current i_{DFW} are shown in Fig. 3.27 (*c*). From these waveforms we have

Average value of voltage across motor,

$$V_a = \frac{T_{on}}{T} V = \alpha V = f T_{on} V \quad \text{...Refer to Eq. (3.25)}$$

where α = Duty cycle = $\frac{T_{on}}{T}$ and f = Chopping frequency = $\frac{1}{T}$

Power delivered to load (motor)

= Average value of voltage across motor × average motor current

$= V_a I_a = \alpha V I_a$

Average supply current

$= \frac{T_{on}}{T} I_a = \alpha I_a$

Input power to chopper

= Average supply voltage × average source current = $V\alpha I_a$

For the motor armature circuit,

$$V_a = \alpha V = E_b + I_a (R_a + R_{se}) = k_e \omega_m + I_a (R_a + R_{se})$$

or Motor speed,

$$\omega_m = \frac{\alpha V - I_a (R_a + R_{se})}{k_e} \quad \text{...(3.45)}$$

From above Eq. (3.45) it is obvious that the armature terminal voltage and therefore, speed of the dc motor can be controlled by varying the duty cycle α of the chopper.

So far, armature current i_a has been assumed ripple free and accordingly waveforms in Fig. 3.27 (*c*) are drawn. But in practice, the motor armature current increases during chopper on period and decreases during chopper off period, as illustrated in Fig. 3.28.

For R-L-E_b type of load, E_b is load voltage (back emf of motor). When the chopper is switched on for a period T_{on} and power is supplied to R-L-E_b load. By using Kirchhoff's voltage law (KVL) in the loop consisting of V, chopper and R-L-E_b load, we have

$$V - E_b - iR - L\frac{di}{dt} = 0$$

$$\text{or } V = L\frac{di}{dt} + iR + E_b \text{ for } 0 \le t \le T_{on} \quad \text{...(3.46)}$$

When chopper is off, the load current continues flowing through the freewheeling diode due to energy stored in the inductor. The KVL for the loop containing R-L-E_b and D_{FW} yields

$$-iR - L\frac{di}{dt} - E_b = 0$$

$$\text{or } L\frac{di}{dt} + iR + E_b = 0 \quad \text{for } T_{on} \le t \le T \quad \text{...(3.47)}$$

The above Eqs. (3.46) and (3.47) can be solved by using Laplace transform. It is seen from Fig. 3.28 that initial value of load current is I_{min} for Eq. (3.46) and I_{max} for Eq. (3.47).

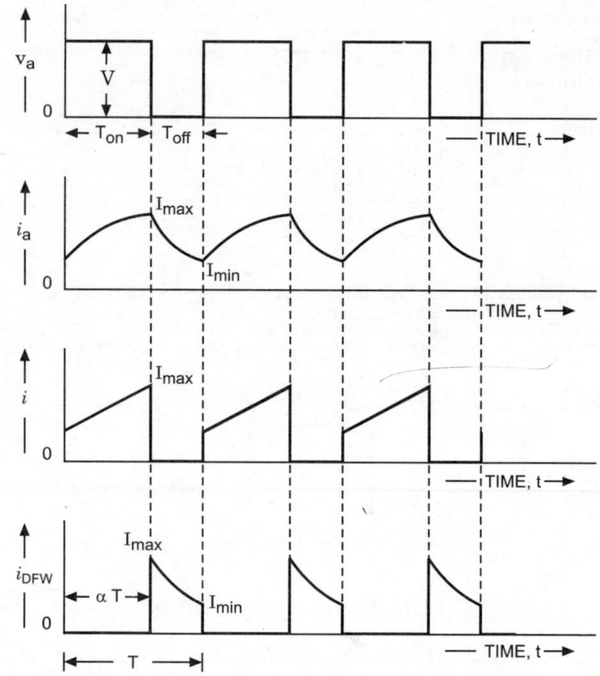

Fig. 3.28. ***Waveforms For DC Chopper Drive of Fig. 3.27 (a)***

So Laplace transform of Eqs. (3.46) and (3.47) we have

$$RI(s) + L[s\,I(s) - I_{min}] = \frac{V - E_b}{s} \quad \text{...(3.48)}$$

$$\text{and } RI(s) + L[s\,I(s) - I_{max}] = \frac{-E_b}{s} \quad \text{...(3.49)}$$

From Eq. (3.48), we have

$$I(s) = \frac{V - E_b}{s(R + Ls)} + \frac{L I_{min}}{R + sL} = \frac{V - E_b}{sL\left(s + \frac{R}{L}\right)} + \frac{I_{min}}{s + \frac{R}{L}}$$

Laplace inverse of above equation is

$$i(t) = \frac{V - E_b}{R}\left(1 - e^{\frac{-R}{L}t}\right) + I_{min}\, e^{\frac{-R}{L}t} \quad \text{...(3.50)}$$

Similarly the time-domain expression for current from Eq. (3.49) is

$$i(t') = -\frac{E_b}{R}\left(1 - e^{\frac{-R}{L}t'}\right) + I_{max}\, e^{\frac{-R}{L}t'} \quad \text{...(3.51)}$$

Here R = Armature resistance R_a + series field resistance R_{se}

and L = Armature inductance + series field inductance L_{se}.

The values of maximum current I_{max} and minimum current I_{min} can be determined by using the fact that $i(t) = I_{min}$ when $t = 0$ and $i(t') = I_{max}$ when $t' = 0$. The values are found to be

$$I_{max} = \frac{V}{R}\left[\frac{1 - e^{\frac{-R\alpha T}{L}}}{1 - e^{\frac{-RT}{L}}}\right] - \frac{E_b}{R} \quad \text{...(3.52)}$$

$$\text{and } I_{min} = \frac{V}{R}\left[\frac{e^{\alpha TR/L} - 1}{e^{RT/L} - 1}\right] - \frac{E_b}{R} \quad \text{...(3.53)}$$

Example 3.15. An R-L-E_b load is supplied from 400 V through a chopper circuit. If R = 0, L = 0.5 H and α = 0.25, calculate the chopping frequency to limit the ripple current to 10 A.

Solution : Average load voltage, $V_a = \alpha V = 0.25 \times 400 = 100$ V

In R-L-E_b load, the difference between input voltage and average voltage is responsible for creating magnetic field in the coil. So

$$L\frac{di}{dt} = V - V_a = 400 - 100 = 300 \text{ V}$$

$$L = 0.5 \text{ H}$$

$$\text{and } \frac{di}{dt} = \frac{10\text{ A}}{T_{on}}$$

$$\text{So } \frac{0.5 \times 10}{T_{on}} = 300$$

$$\text{or } T_{on} = \frac{0.5 \times 10}{300} = \frac{1}{60} \text{ second}$$

$$\text{and } T = \frac{T_{on}}{\alpha} = \frac{1/60}{0.25} = \frac{1}{15} \text{ second}$$

$$\text{Chopping frequency } f = \frac{1}{T} = 15 \text{ Hz.} \quad \textbf{Ans.}$$

Example 3.16. The speed of a separately excited dc motor is controlled by a chopper. The supply voltage is 120 V, armature circuit resistance = 0.5 Ω, armature circuit inductance = 20 mH and motor constant 0.05 V/rpm. The motor drives a constant load torque requiring an average current of 20 A. Assume motor current is continuous. Calculate (*a*) the range of speed control (*b*) the range of duty cycle. [*Pb. Univ., Nov. 1990*]

Solution : The motor will operate at maximum speed (or base speed) when the ratio of time-on (T_{on}) to cycle (T), α = 1, In this case

Voltage across the motor armature, $V_a = V = 120$ V

$$\text{Back emf, } E_b = V_a - I_a R_a = 120 - 20 \times 0.5 = 110 \text{ V}$$

$$\text{Motor speed, } N = \frac{E_b}{\text{Motor constant}} = \frac{110}{0.05} = 2{,}200 \text{ rpm}$$

The minimum speed is zero and at zero speed

$$\text{Back emf, } E_{b_0} = 0$$

Voltage across the motor armature,

$$V_a = E_b + I_a R_a = 0 + 20 \times 0.5 = 10 \text{ V}$$

$$\text{Since, } V_a = V \times \frac{T_{on}}{T}$$

$$\therefore \text{ Duty cycle, } \alpha = \frac{T_{on}}{T} = \frac{V_a}{V} = \frac{10}{120} = \frac{1}{12}$$

Hence (*a*) Range of speed control is from 0 to 2,200 rpm **Ans.**

(*b*) Range of duty cycle is from $\frac{1}{12}$ to 1 **Ans.**

Example 3.17. A 200 V, 940 rpm, 100 A separately excited dc motor has an armature resistance of 0.02 Ω. The motor is fed from a single quadrant step-down chopper for motoring operation. The input to the chopper is 200 V dc. Assuming continuous load current, calculate the duty cycle of the chopper for motoring at rated torque and 500 rpm.

[*U.P. Technical Univ. Electric Drives 2004-05*]

Solution : Back emf induced in the motor armature when it is drawing a current of 100 A from 200 V supply and running at 940 rpm

$$E_{b_1} = V - I_a R_a = 200 - 100 \times 0.02 = 198 \text{ V}$$

Back emf induced at a speed of 500 rpm

$$E_{b_2} = E_{b_1} \times \frac{N_2}{N_1} = 198 \times \frac{500}{940} = 105.32 \text{ V}$$

Voltage across armature when running at 500 rpm

$$V_a = E_{b_2} + I_a R_a = 105.32 + 100 \times 0.02$$
$$= 107.32 \text{ V} \qquad \because \text{Torque is rated one}$$

$$\text{Duty ratio, } \alpha = \frac{T_{on}}{T} = \frac{V_a}{V} = \frac{107.32}{200} = 0.5366 \quad \textbf{Ans.}$$

Example 3.18. A dc chopper fed from 400 V supply runs a separately excited motor. The armature resistance R_a = 0.1 Ω. The motor voltage constant is 4 Vs/radian, the average armature current I_a = 150 A. The armature current is continuous and has negligible ripple. Determine (*i*) the input power, (*ii*) the motor speed and (*iii*) the developed torque for a duty cycle of 60%. Neglect the losses in chopper. If the duty cycle of the chopper varies between 10% and 90%, find the minimum and maximum speeds. [*A.M.I.E. Sec. B. Electric Drives and Their Control Summer 1999*]

Solution : (*i*) Power input $= \frac{\alpha \times V \times I_a}{1{,}000} = \frac{0.6 \times 400 \times 150}{1{,}000}$

$= 36$ kW **Ans.**

Voltage across armature circuit, $V_a = \alpha V = 0.6 \times 400 = 240$ V

$$\text{Back emf developed, } E_b = V_a - I_a R_a$$
$$= 240 - 150 \times 0.1 = 225 \text{ V}$$

$$(ii) \text{ Speed, } N = \frac{60}{2\pi}\omega = \frac{60}{2\pi} \times \frac{E_b}{\text{voltage constant}} = \frac{60}{2\pi} \times \frac{225}{4}$$
$$= 537 \text{ rpm} \quad \textbf{Ans.}$$

$$(iii) \text{ Torque, } T = \frac{E_b I_a}{2\pi N/60} = \frac{225 \times 150 \times 60}{2\pi \times 537} = 600 \text{ Nm} \quad \textbf{Ans.}$$

For duty cycle of 10%,

$$E_b = \alpha V - I_a R_a = 0.1 \times 400 - 150 \times 0.1 = 25 \text{ V}$$

$$\text{Speed, } N = \frac{60}{2\pi} \times \frac{25}{4} = 60 \text{ rpm}$$

For duty cycle of 90%,

$$E_b = \alpha V - I_a R_a = 0.9 \times 400 - 150 \times 0.1 = 345 \text{ V}$$

$$\text{Speed, } N = \frac{60}{2\pi} \times \frac{345}{4} = 824 \text{ rpm}$$

Minimum speed will be at 10% duty and will be equal to 60 rpm while the maximum speed will be at 90% duty and will be equal to 824 rpm. **Ans.**

Example 3.19. A one quadrant chopper is used for rheostatic braking of a separately excited dc motor. Armature resistance is 0.1 Ω, braking resistance is 7.5 Ω, voltage constant 1.6 V/A–rad/s, armature current 150 A and field current is 1.5 A. The duty cycle of the chopper is 0.4. Determine (*i*) voltage across the chopper (*ii*) power dissipated in the braking resistance and (*iii*) motor speed.

Solution : (*i*) Average voltage across chopper,

$$V_{ch} = V_b = (1 - \alpha) R_b I_a \qquad \text{... Refer to Eq. (3.43)}$$
$$= (1 - 0.4) \times 7.5 \times 150 = 675 \text{ V} \quad \textbf{Ans.}$$

(*ii*) Power dissipated in braking resistance,

$$P_b = I_a^2 R_b (1 - \alpha)^2 \qquad \text{... Refer to Eq. (3.44)}$$
$$= 150^2 \times 7.5 (1 - 0.4)^2 = 60{,}750 \text{ W or } 60.75 \text{ kW} \quad \textbf{Ans.}$$

Generated emf, $E_g = V_a + I_a R_a = 675 + 150 \times 0.1 = 690$ V

(*iii*) Speed, $N = \frac{60}{2\pi} \times \omega$

$= \frac{60}{2\pi} \times \frac{E_g}{\text{voltage constant} \times \text{field current}}$

$= \frac{60}{2\pi} \times \frac{690}{1.6 \times 1.5} = 2{,}745$ rpm **Ans.**

Example 3.20. The motor given in Example 3.8 is to be employed for regenerative braking by reversing the direction of field current. Find the firing angle of the armature circuit converter for the given value of armature current. Also determine the power supplied back to the supply source.

Solution : When the field current is reversed, the direction of back emf is reversed. The motor is now acting as a generator.

Generated emf, $E_g = -163.53$ V

Voltage across armature circuit,

$V_a = E_g + I_a R_a = -163.53 + 40 \times 0.2 = -155.53$ V

Since $V_a = \frac{2V_{max}}{\pi} \cos \alpha_a$

$\therefore \quad -155.53 = \frac{2 \times 220 \times \sqrt{2}}{\pi} \cos \alpha_a$

or $\alpha_a = 141.74°$ **Ans.**

Power supplied back to the source

$= V_a I_a = 155.53 \times 40$

$= 6{,}221.2$ W or 6.2212 kW **Ans.**

Example 3.21. A constant frequency system is used to control the speed of a dc traction motor from 220 V dc supply. The combined armature and field resistance is 0.2 Ω. The average current in the circuit is 25 A and chopper frequency is 200 Hz. Determine the pulse width, if the average value of back emf is 100 V. [*A.M.I.E. Sec B. Electric Drives and Their Control Winter 1995*]

Solution : Average voltage across chopper,

$V_{ch} = V_a = E_b + I_a (R_a + R_{se}) = 100 + 25 \times 0.2 = 105$ V

Also $V_a = V \times \frac{T_{on}}{T} = V \times T_{on} \times f$

So $105 = 220 \times T_{on} \times 200$

or $T_{on} = \frac{105}{220 \times 200} = 0.00239$ second or 2.39 ms **Ans.**

3.8. COMPARISON OF CONVERTER FED AND CHOPPER FED DC DRIVES

For speed control of a dc motor it is necessary that it is supplied with an adjustable voltage which can be obtained either from a controlled converter or from a chopper. The relative advantages and disadvantages of these two are given below :

1. When available supply is ac, an ac-dc converter is ideally suited to convert the constant ac voltage into a variable dc voltage. If we use chopper, which converts a constant dc voltage into a variable dc voltage, diode rectifier will be required to convert constant ac voltage into a constant dc voltage for feeding it to chopper.
2. The ripple in the output of a chopper is much less than that in the output of a converter. Less ripple means lesser losses.
3. The chances of discontinuous operation in a chopper drive are lesser than that in a converter drive. So speed regulation and transient response of chopper drive are better than that in converter drive.
4. For reducing ripple in the output a converter with higher pulse number is required and a converter with higher pulse number has lower utilisation factor and more cost.

3.9. SPECIAL FEATURES OF THYRISTOR DRIVE MOTORS

The most commonly used dc motors for thyristor drive are separately excited and series excited dc motors. The thyristor drive motors usually differ from conventional dc motors in construction. Before discussing the special features incorporated in thyristor drive motors, discussion on the effects of thyristor power supply on performance of a dc motor is necessary.

3.9.1. Effects of Thyristor Power Supply on The Performance of a DC Motor

1. The output voltage from thyristor converter consists of a dc component and ac harmonic components.
2. The output voltage can change very rapidly in comparison to that of a motor-generator set owing to the absence of field time constants associated with the generator.
3. An abnormally high value of armature current may rise in the event of thyristor fault when operating in the inverter mode.
4. The magnitude of harmonics decreases with the increase in frequency of harmonic for all converters. The magnitude of the harmonics increase as the firing angle is increased and dc voltage is reduced for all bidirectional converters. The higher the armature inductance, the lesser will be the harmonic currents.
5. Torque is developed by the dc component of the current whereas heating is developed by the effective (or rms) value of current. The form factor (ratio of effective value to average value) for half-wave three-phase thyristors may be taken as 1.2 while for full-wave three-phase thyristors it is 1.1. This increases electrical losses and, therefore, heating is 5 to 7 per cent more for three-phase full-bridge converters while for three-phase half-bridge converters it is from 15 to 20 per cent.
6. The commutating ability is seriously affected by the presence of harmonic currents. The peak value of the current is increased, interpole flux will be reduced in magnitude and a time lag will be introduced between the interpole current and the flux due to eddy currents generated in the iron path of the inter polar flux.

The other effects of thyristor power supply on motor performance are heating of interpole winding, saturation of interpole magnetic circuit, transformer voltage at the brushes and increase in voltage commutator segment.

3.9.2. Special Features of Thyristor Drive Motors

The thyristor drive dc motors are designed with the following special features in order to improve their performance.

1. The thyristor drive dc motors are made with larger diameter armature and larger size poles of reduced height.
2. The commutators are made larger in order to provide extra insulation to withstand larger and rapid voltage fluctuations.
3. The yoke as well as the main and commutating poles are laminated to reduce the eddy current effects.
4. Low inertia armatures are employed for improving the response.
5. Compensating windings are used in large motors to reduce armature reaction effect.
6. Split brushes of good commutating quality are used for reducing the effect of transformer voltage in the coils undergoing commutation.
7. The use of a laminated yoke instead of a solid yoke improves the commutation to a greater extent.
8. The use of a large number of commutator bars reduces the voltage between commutator segments and improves the commutation.
9. The use of an octagonal, rather than circular, shape for the frame accommodates more material and gives a larger rating for the same frame size.
10. The current densities used for the armature and interpole winding are reduced compared to conventional dc motors of the same frame size and rating in order to reduce the effect of heating armature and interpoles.
11. A better class of insulation (class F materials as standard insulation) is used to allow higher temperature rise and dissipation of more losses from a given frame.
12. The ratio pole arc/pole pitch is reduced in order to reduce the ratio of commutating zone to neutral zone.
13. Armature inductance is increased to reduce the ripple current. It may also increase reactance voltage. The number of turns per armature coil is kept at the minimum since the reactance voltage is proportional to the square of the turns per coil. Use of dummy coils is avoided.
14. Forced cooling by an auxiliary ac motor is widely used for improving cooling of the motor at reduced speeds.
15. Great care is required to be taken for accurate construction : spacing of brushes, poles and manufacture of commutators.

3.10. SPEED CONTROL OF DC MOTOR USING FEEDBACK LOOPS

The systems of speed control discussed in Art. 3.6 are open-loop systems. An open-loop system can be converted into a closed-loop system by employing a feedback loop, as illustrated in Fig. 3.29. The tacho-generator is a permanent magnet dc generator whose output emf is directly proportional to the speed. The difference of input speed setting and speed feedback provides the error signal. The error signal is amplified by the amplifier and supplied to the firing circuit of the converter. Thus the firing of converter thyristors is advanced or delayed depending on whether the actual speed is less than or more than the desired speed.

Let the desired speed of motor be 1,200 rpm, error signal be 0.1 V, converter output be 240 V and output of tacho-generator be 10 V per 1,000 rpm. If the input setting is 12.1 V, the converter output will be 240 V and motor speed will be 1,200 rpm. If the load torque causes the decrease in speed and let the speed become 1,190 rpm, the output of tacho-generator will be 11.9 V increasing the error signal to 0.2 V. The converter output voltage will increase causing the speed to increase to 1,200 rpm. If the speed tends to increase beyond 1,200 rpm, the error signal will become less than 0.1 V and the output voltage of the converter will decrease, thus reducing the motor speed. Thus negative feedback is employed for control of motor speed.

A change in motor speed causes a temporary increase in motor and thyristor currents. Excessive current, however, can be avoided by an automatic current limiting circuit using another feedback loop, as shown in Fig. 3.30. The dc current transformer (dc CT), measures the converter output current and develops a voltage proportional to current. This signal is fed back, as illustrated at a position between the two amplifiers. If the converter current tends to increase, the output of dc CT increases thus reducing the current error. Thus the output current of the converter is reduced to the safe value.

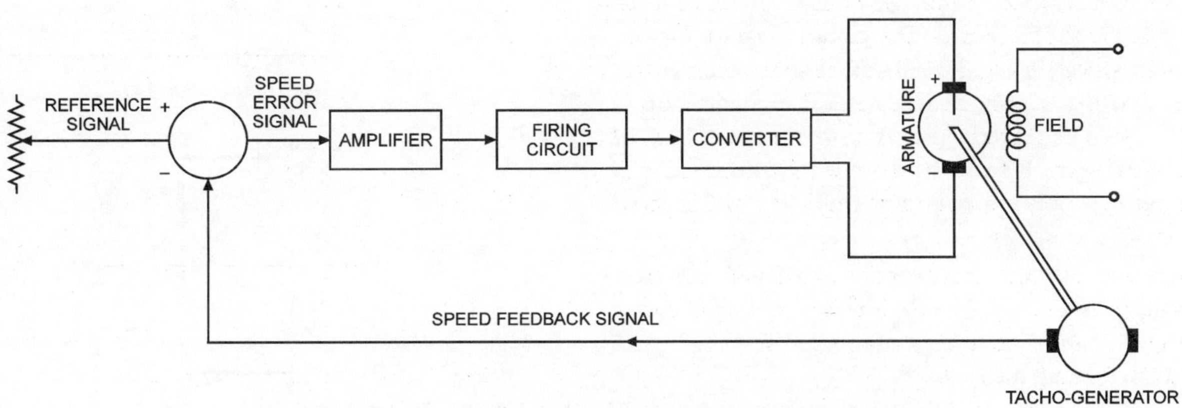

Fig. 3.29. *Speed Control of DC Motor With Negative Feedback*

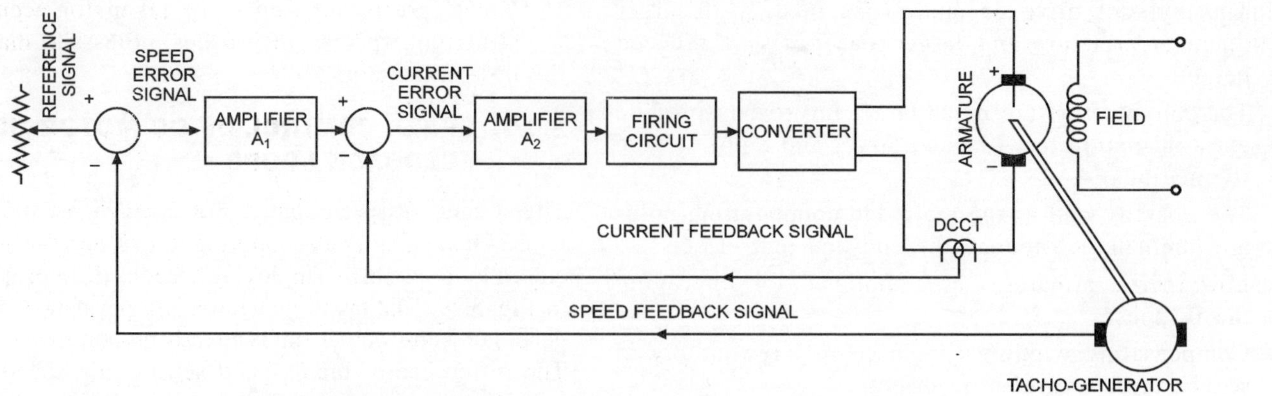

Fig. 3.30. ***Speed Control of DC Motor With Current Limiting***

3.11. ELECTRIC BRAKING OF RECTIFIER CONTROLLED SEPARATELY EXCITED DC MOTOR

Figure 3.31 (*a*) illustrates a simplified representation of a single phase fully controlled rectifier supplying a separately excited dc motor. The polarities of voltages and directions of currents are also marked. The firing angle of the rectifier is in the range of $0 \le \alpha \le 90°$.

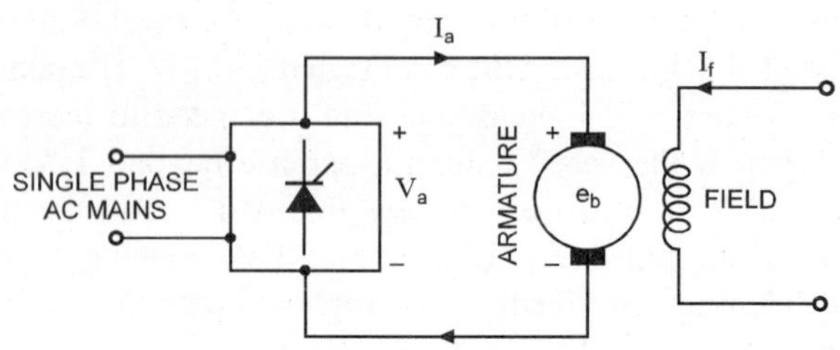

(a) Motor Action $0 \le \alpha \le 90°$, $V_a > E_b$

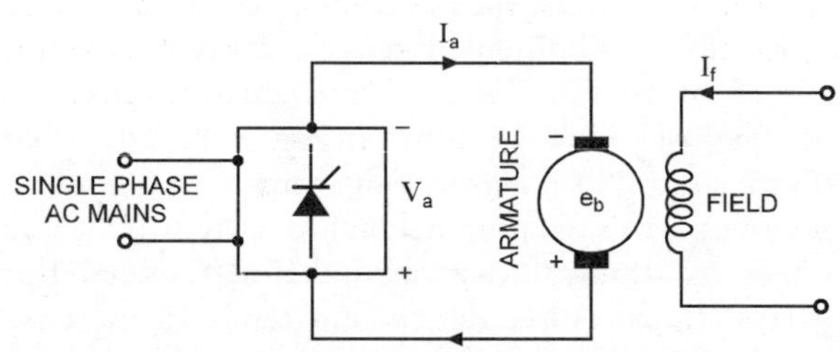

(b) Generator Action $90° \le \alpha \le 180°$, $E_b > V_a$

Fig. 3.31. ***Electric Braking of Separately Excited DC Motor***

For regenerative braking, the direction of power flow is to be reversed. The motor has to operate as a generator and supply power to converter which now operates as an inverter and feeds back power to the supply mains. The circuit diagram for this operation is illustrated in Fig. 3.31 (*b*). Since the inverter output is connected to the supply mains, natural commutation is possible. It is seen that direction of flow of current in both of the circuits is the same. It has to be so since the current through a thyristor can flow only from anode to cathode. For this mode of operation $90° \le \alpha \le 180°$.

It is seen from Fig. 3.31 that direction of induced emf (back emf) in circuit shown in Fig. 3.31 (*b*) is reverse of that in circuit shown in Fig. 3.31 (*a*). The reversal of induced emf can be had by any of the following methods :

1. By reversing the direction of flow of field current. This provides forward regeneration.
2. By reversing the armature terminals (keeping direction of rotation unchanged). This also provides forward regeneration.
3. By driving the motor by an active load coupled to it in reverse direction. No change in connection is required. This provides reverse regeneration.

If the motor is supplied by a semiconverter regenerative braking is not possible because a semiconverter operates only in a first quadrant (*i.e.,* cannot operate as an inverter). However, plugging is possible with both fully controlled and semiconverter. In this case it is necessary to reverse direction of induced emf by any of the above three methods. However, plugging needs an external resistor to be connected to the machine (to dissipate energy) and results in wastage of energy. Therefore, plugging is generally not employed in converter fed dc motors.

3.12. THYRISTOR CONTROL OF DC SERIES MOTORS

3.12.1. Fully Controlled Rectifier

Figure 3.32 (*a*) illustrates the circuit diagram for a dc series motor supplied from single-phase ac mains through a fully

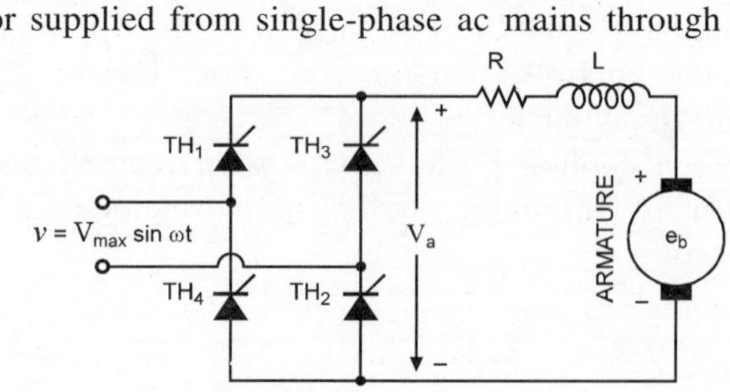

(a) Circuit Diagram

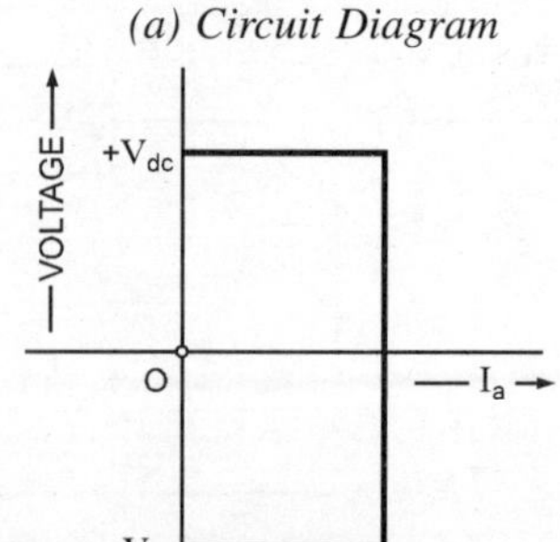

(b) Quadrants of Operation

Fig. 3.32

controlled rectifier. The motor armature and field windings have resistance as well as inductance. Since the output current of a rectifier is not a perfect dc, the role of inductance also comes in. R is resistance of armature including that of field and L is inductance of armature including that of field. During positive half cycle, thyristors TH_1 and TH_2 are forward biased and start conducting at $\omega t = \alpha$. The load current flows through TH_1, motor and TH_2. At $\omega t = \pi$, the supply voltages reverses. Because of inductance L thyristors TH_1 and TH_2 continue to conduct beyond $\omega t = \pi$. From $\omega t = \pi + \alpha$ to $\omega t = 2\pi$, thyristors TH_3 and TH_4 are forward biased. When TH_3 and TH_4 are triggered at $\omega t = \pi + \alpha$, thyristors TH_1 and TH_2 are subjected to reverse bias and are turned off by natural commutation. Load current is transferred from TH_1 and TH_2 to TH_3 and TH_4. This mode of operation continues till TH_1 and TH_2 are triggered in the next positive half cycle. The quadrants of operation and waveshapes are shown in Figs. 3.32 (*b*) and 3.32 (*c*). If inductance L is sufficiently large, the motor current is more or less constant. The value of α should be such that when thyristors are triggered the instantaneous value of ac input voltage $V_{max} \sin \omega t$ is more than back emf E_b. This sets a lower limit on firing angle α.

Average output voltage of converter,

$$V_a = \frac{2}{2\pi} \int_{\alpha}^{\pi+\alpha} V_{max} \sin \omega t \, d(\omega t)$$

$$= \frac{2\,V_{max}}{2\pi} \left| -\cos \omega t \right|_{\alpha}^{\pi+\alpha} = \frac{2\,V_{max}}{\pi} \cos \alpha$$

For continuous motor current I_a, we can write

$$V_a = \frac{2 V_{max}}{\pi} \cos \alpha = RI_a + E_b$$

Since the field current is also equal to I_a, the back emf E_b can be written as $K_1 I_a N$ where K_1 is a constant. In writing E_b as $K_1 I_a N$ we have neglected the residual magnetism in the motor.

$$\text{So} \quad \frac{2 V_{max}}{\pi} \cos \alpha = RI_a + K_1 I_a N$$

$$\text{or } N = \frac{\dfrac{2 V_{max}}{\pi} \cos \alpha - RI_a}{K_1 I_a} \quad ...(3.54)$$

$$\text{and torque, } T \propto \phi I_a = K_2 I_a^2 \quad ...(3.55)$$

Substituting value of I_a from Eq. (3.54) in Eq. (3.55), we have

$$T = K_2 \left[\frac{\dfrac{2\,V_{max} \cos\alpha}{\pi}}{R + K_1\,N} \right]^2$$

$$\text{or } T = \frac{4 K_2 V_{max}^2 \cos^2 \alpha}{\pi^2 (R + K_1 N)^2} \quad ...(3.56)$$

Using above Eq. (3.56) speed-torque characteristics for different values of α can be plotted, as illustrated in Fig. 3.32 (*d*).

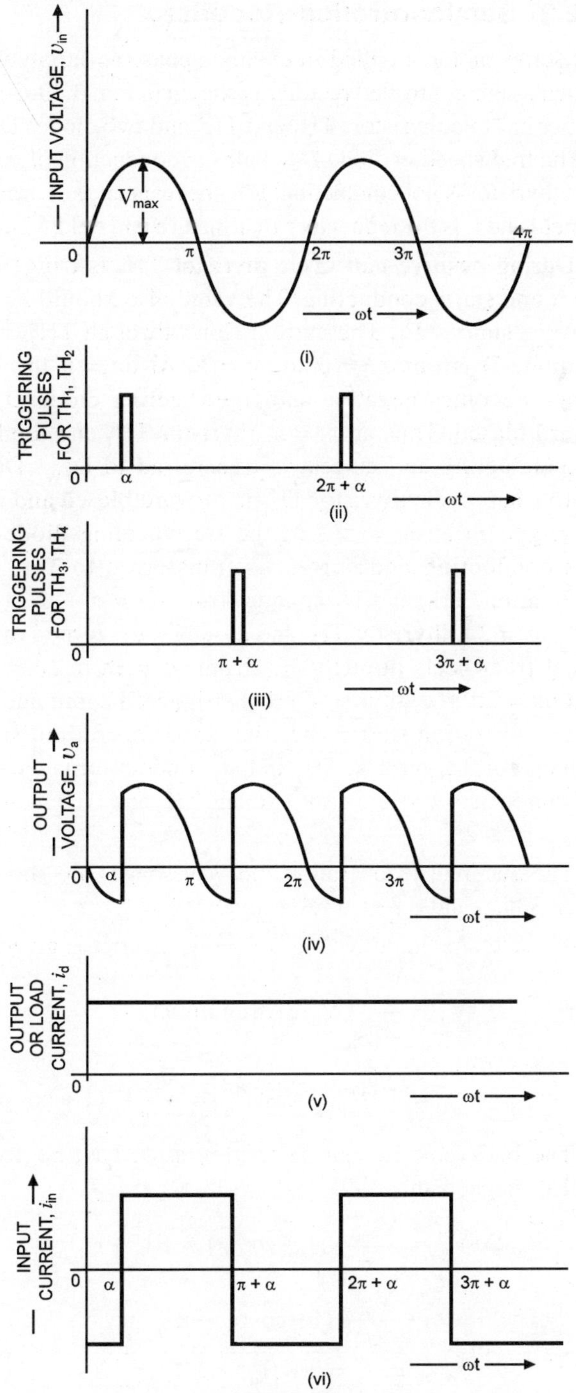

(*c*) *Waveshape*

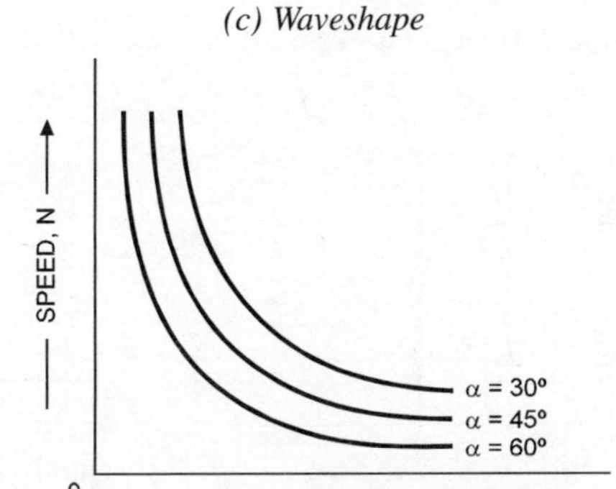

(*d*) *Speed-Torque Characteristics*

Fig. 3.32. ***DC Series Motor Supplied From Single Phase AC Supply Through Fully Controlled Rectifier***

3.12.2. Semicontrolled Rectifier

A dc series motor supplied from single phase ac supply mains through a semicontrolled rectifier is shown in Fig. 3.33 (*a*). The rectifier has two thyristors TH_1 and TH_2 and two diodes D_1 and D_2. The freewheeling diode D_{FW} helps in conduction of current when thyristor is not conducting. R is the resistance of armature and field and L is the inductance of armature and field windings.

During positive half cycle thyristor TH_1 is triggered at $\omega t = \alpha$ and starts conducting. The value of α should be such that $V_{max} \sin \alpha > E_b$. The current flows through TH_1, motor and diode D_1 from $\omega t = \alpha$ to $\omega t = \pi$. At $\omega t = \pi$ the input voltage becomes negative and freewheeling diode D_{FW} is forward biased. Thus at $\omega t = \pi$, thyristor TH_1 and diode D_1 stop conducting and current is transferred to D_{FW}. During negative half cycle thyristor TH_2 is forward biased and when it is triggered at $\omega t = \pi + \alpha$, the freewheeling diode D_{FW} stops conducting and current is transferred to $TH_2 - D_2$ combination. TH_2 and D_2 conduct from $\omega t = \pi + \alpha$ to $\omega t = 2\pi$. At $\omega t = 2\pi$, thyristor TH_2 and diode D_2 get turned off and current freewheels through D_{FW} from $\omega t = 2\pi$ to $\omega t = 2\pi + \alpha$. At $\omega t = 2\pi + \alpha$, thyristor TH_1 is triggered again and next cycle of operation starts. Thus the circuit operation is: D_{FW} conducts for $0 \le \omega t \le \alpha$; TH_1 and D_1 conduct for $\alpha \le \omega t \le \pi$; D_{FW} conducts for $\pi \le \omega t \le \pi + \alpha$; and TH_2 and D_2 conduct for $\pi + \alpha \le \omega t \le 2\pi$.

The quadrant of operation and waveshapes are shown in Figs. 3.33 (*b*) and 3.33 (*c*) respectively.

The average output voltage of semiconverter is given as

$$V_a = \frac{2}{2\pi} \int_{\alpha}^{\pi} V_{max} \sin \omega t \, d(\omega t)$$

$$= \frac{2 V_{max}}{2\pi} \left| -\cos \omega \right|_{\alpha}^{\pi} = \frac{V_{max}}{\pi} (1 + \cos \alpha)$$

The back emf E_b can be written as $K_1 I_a N$, neglecting residual magnetism

$$\text{So } V_a = \frac{V_{max}}{\pi} (1 + \cos \alpha) = RI_a + K_1 I_a N$$

$$\text{or } N = \frac{\frac{V_{max}}{\pi}(1 + \cos\alpha) - RI_a}{K_1 I_a} \quad ...(3.57)$$

and torque $T \propto \phi I_a = K_2 I_a^2$

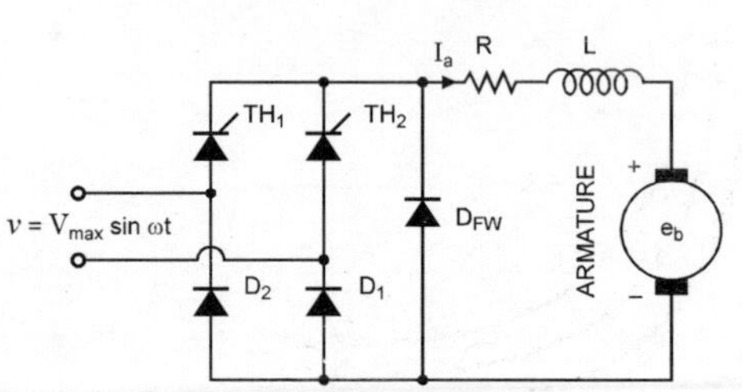

(*a*) Circuit Diagram

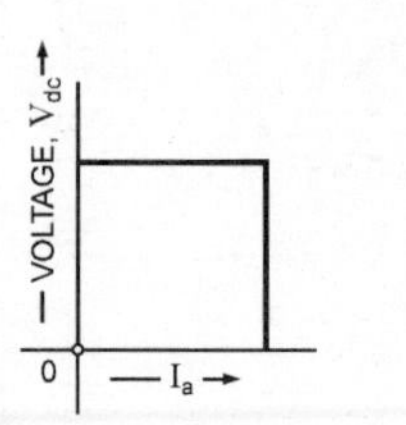

(*b*) Quadrant of Operation

Fig. 3.33

Substituting the value of I_a from Eq. (3.57) in above equation, we have

$$T = \frac{K_2 V_{max}^2 (1 + \cos\alpha)^2}{\pi^2 (R + K_1 N)^2} \quad ...(3.58)$$

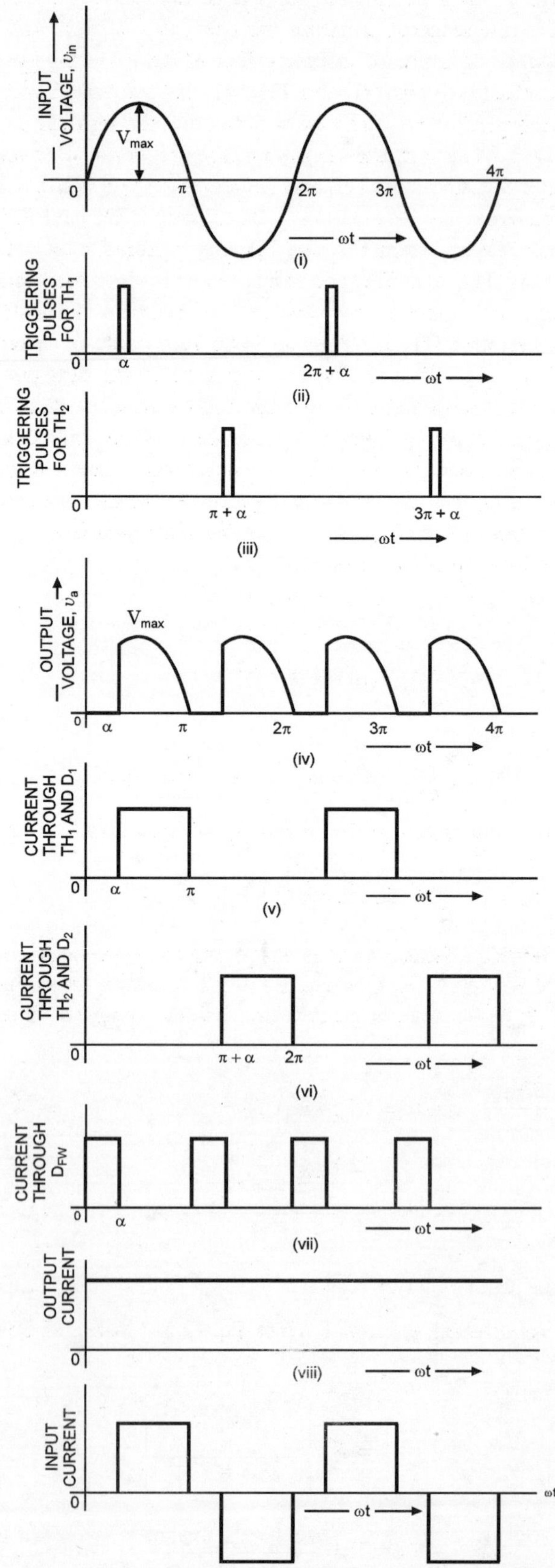

(*c*) Waveshapes

Fig. 3.33

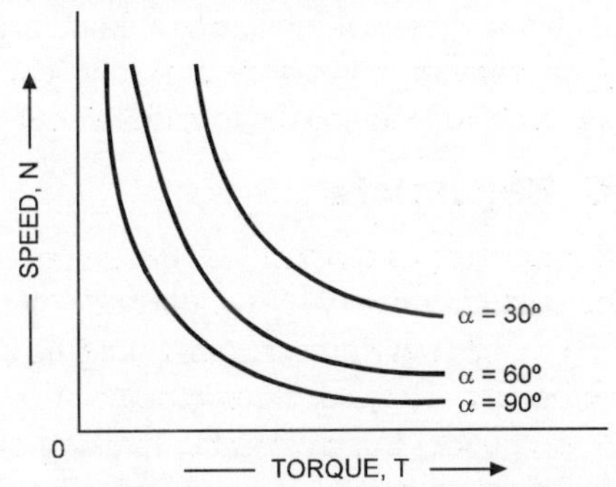

(d) Speed-Torque Characteristics

Fig. 3.33. ***DC Series Motor Supplied From Single Phase AC Supply Through Single Phase Semiconverter***

Using above Eq. (3.58) speed-torque characteristics for different values of α are plotted, as illustrated in Fig. 3.33 (*d*).

During operation of dc separately excited and dc series motors, it is observed that the region of discontinuous current is smaller in case of dc series motors. The use of semiconverter provides continuous current over almost the entire operating range except when the load torque is very small. If inductance is employed in the armature circuit, continuous armature current can be obtained over the entire range of operation.

DC series motors are normally employed for constant output power applications. However, speed-torque characteristics do not correspond to the constant output power for a given firing angle and, therefore, in order to obtain constant output power over the entire range of speeds, the firing angle α has to be adjusted. The semiconverter system, because of its freewheeling action, helps in maintaining continuous current and thus provides better performance to motors in comparison to full-converter system. It has also been observed that dc series motor along with semiconverter provides better performance.

Phase controlled converters have poor power factor especially when the output voltage is less than the maximum, *i.e.*, when firing angle α is large. Semiconverters provide better power factor in comparison to full converters even though the improvement is marginal.

Example 3.22. The speed of a 10 kW, 220 V, 1000 rpm dc series motor is controlled using a single phase (*i*) semiconverter (*ii*) full converter. The combined armature and field circuit resistance is 0.25 Ω. Determine motor current and motor torque for a firing angle of α = 30°. Assume that motor current is continuous and ripple free, motor speed, N = 1,000 rpm, K = 0.025 N-m/A², and ac supply voltage = 240 V.

Solution : Speed, $\omega = \frac{2\pi N}{60} = \frac{2\pi \times 1{,}000}{60} = 104.72$ radians/s

(*i*) When controlled through semiconverter

$$\text{Voltage across armature, } V_a = \frac{V_{max}}{\pi}(1 + \cos\alpha)$$

$$= \frac{\sqrt{2} \times 240}{\pi}(1 + \cos 30°) = 201.6 \text{ V}$$

$$\text{Armature current, } I_a = \frac{V_a}{R + K\omega} = \frac{201.6}{0.25 + 0.025 \times 104.72}$$

$$= 70.3 \text{ A} \quad \textbf{Ans.}$$

$$\text{Motor torque, } T = KI_a^2 = 0.025 \times 70.3^2 = 123.53 \text{ N-m} \quad \textbf{Ans.}$$

(*ii*) When controlled through full converter

$$\text{Voltage across armature, } V_a = \frac{2V_{max}}{\pi}\cos\alpha = \frac{2 \times \sqrt{2} \times 240}{\pi}\cos 30°$$

$$= 187.13 \text{ V}$$

$$\text{Armature current, } I_a = \frac{V_a}{R + K\omega} = \frac{187.13}{0.25 + 0.025 \times 104.72}$$

$$= 65.25 \text{ A} \quad \textbf{Ans.}$$

$$\text{Motor torque, } T = KI_a^2 = 0.025 \times 65.25^2 = 106.43 \text{ N-m} \quad \textbf{Ans.}$$

Example 3.23. A dc series motor is supplied from a single phase, 230 V, 50 Hz supply through a full-wave diode rectifier. The armature and field resistance is 1.0 Ω and torque constant K_t is 0.3 Nm/A². If load torque is 30 N-m, find average motor current and speed. Assume that motor current is constant and emf constant is 0.3 V/A-rad/second.

Solution : DC output voltage,

$$V_{dc} = \frac{1}{\pi}\int_0^{\pi} V_{max} \sin\omega t \, d(\omega t) = \frac{\sqrt{2} \times 230}{\pi}\int_0^{\pi} \sin\omega t \, d(\omega t)$$

$$= \frac{\sqrt{2} \times 230}{\pi} \times 2 = 207 \text{ V}$$

Applying Kirchhoff's second or voltage law to the motor circuit, we have

$$RI_a + (L_a + L_f)\frac{dI_a}{dt} + K_e\omega I_a = V_{max} \sin\omega t$$

where I_a is the armature current, L_a is the armature winding inductance, L_f is the field winding inductance, R is the combined resistance of field and armature windings, ω is the motor angular speed and K_e is the emf constant.

Since armature current I_a is constant, $\frac{dI_a}{dt} = 0$

$$\text{So } 1.0 \times I_a + 0.3\,\omega I_a = 230\sqrt{2}\sin\omega t$$

Integrating above equation over one half cycle, we have

$$I_a + 0.3\omega I_a = 207 \quad ...(i)$$

$$\text{Moreover, load torque, } T_L = K_t I_a^2$$

$$\text{So } 30 = 0.3\, I_a^2$$

$$\text{or } I_a = \sqrt{\frac{30}{0.3}} = 10 \text{ A} \quad \textbf{Ans.}$$

Substituting I_a = 10 A in Eq. (*i*), we have

$$\omega = \frac{207 - I_a}{0.3\, I_a} = \frac{207 - 10}{0.3 \times 10} = 65.667 \text{ rad/second}$$

$$\text{and speed, } N = \frac{\omega \times 60}{2\pi} = \frac{65.667 \times 60}{2\pi} = 627 \text{ rpm} \quad \textbf{Ans.}$$

Example 3.24. A dc series motor is coupled to a load whose torque varies as square of the speed and is supplied from a transformer through a semiconverter. The secondary voltage of the transformer is 675 V. When the firing angle is 100°, the motor speed and armature current are 400 rpm and 36 A. It is desired to raise the speed to 600 rpm. Determine the armature current and firing angle of semiconverter. Armature and field resistance is 0.4 Ω. Assume that magnetic circuit is unsaturated.

Solution : Torque developed by a dc motor is given as

$$T = K_t\,\phi\, I_a$$

or $T \;\alpha\; I_a^2$ $\because$ magnetic circuit is unsaturated and, therefore, $\phi \;\alpha\; I_a$

Also load torque, $T \propto N^2$

Under steady-state conditions

$I_a^2 \propto N^2$

or $I_a \propto N$

$$\text{So } I_{a_2} = I_{a_1} \times \frac{N_2}{N_1} = 36 \times \frac{600}{400} = 54 \text{ A} \quad \textbf{Ans.}$$

Voltage across armature when firing angle $\alpha_1 = 100°$

$$V_{a_1} = \frac{V_{max}}{\pi}(1 + \cos\alpha_1) = \frac{675\sqrt{2}}{\pi}(1 + \cos 100°) = 251.09 \text{ V}$$

Back emf, $E_{b_1} = V_{a_1} - I_{a_1}(R_a + R_f) = 251.09 - 36 \times 0.4 = 236.69$ V

Let V_{a_2} be the output voltage of semi-converter when speed is 600 rpm

Back emf, $E_{b_2} = V_{a_2} - I_{a_2}(R_a + R_f) = V_{a_2} - 54 \times 0.4 = (V_{a_2} - 21.6)$ volts

Since $E_b \propto \phi N \propto I_a N$

$$\text{So } E_{b_2} = E_{b_1} \times \frac{I_{a_2} N_2}{I_{a_1} N_1} = \frac{236.69 \times 54 \times 600}{36 \times 400} = 532.56 \text{ V}$$

Also $E_{b_2} = V_{a_2} - 21.6$

or $V_{a_2} = E_{b_2} + 21.6 = 532.56 + 21.6 = 554.16$ V

$$\text{Also } V_{a_2} = \frac{V_{max}}{\pi}(1 + \cos\alpha)$$

$$\text{or } 554.16 = \frac{675\sqrt{2}}{\pi}(1 + \cos\alpha)$$

or $\alpha = 34.54°$ **Ans.**

3.13. THYRISTOR CONTROL OF THREE PHASE INDUCTION MOTORS

As already stated in Art. 1.18.2, the speed of an induction motor is given as

$$N = \frac{120 f}{P}(1 - s)$$

Thus the speed of an induction motor with fixed number of poles depends upon the supply frequency f and the slip s which in turn depends upon the voltage or current supplied to the motor. Following methods are available for speed control of 3-phase induction motors using thyristors.

1. Stator voltage control or variable voltage constant frequency control.
2. Variable voltage and variable frequency control.
3. Rotor resistance control.
4. Secondary foreign voltage control.

Stator voltage variation is accomplished by means of *ac regulators*, which control the rms value of the ac voltage applied to the motor by introducing thyristors connected back-to-back in each supply line. Variable frequency power is obtained by means of a *cycloconverter*, which directly converts fixed frequency ac to variable frequency ac or by an inverter which converts dc to ac. The effective value of an external resistance introduced into the rotor circuit can be controlled by connecting a high frequency chopper across the resistance and varying the time for which the chopper is *on* during the cycle. Static frequency converters are being employed for replacing the auxiliary machines in the Scherbius system. The Kramer scheme has also been changed by making use of a diode bridge rectifier in place of rotor converter, but a dc motor is still required for converting the rectified slip power to mechanical power.

3.13.1. AC Regulators

An ac regulator converts a constant ac voltage into a variable ac voltage of the same frequency. No doubt, ac voltage level can be changed by auto-transformer, tap-changing transformer, saturable reactors etc. These devices have been in use for long time and still are in use. But ac regulators using thyristors and triacs are becoming more and more popular because of their high efficiency, fast control and compact size. However, ac regulators using thyristors and triacs introduce objectionable harmonics in the circuits. AC regulators are classified as single-phase or three-phase. Each of these may be half-wave (*i.e.,* unidirectional) or full-wave (*i.e.,* bidirectional).

Since the input to an ac regulator is ac, it is always line commutated. Therefore, forced commutation is not required. As such the circuits of ac regulators are quite simple. Two types of controls are used in ac regulators. These are known as *integral cycle control* and *phase control.*

In an *integral cycle control*, also known as *on-off control*, the thyristors are used as switches to connect the motor to the supply source for a certain number of cycles of the source voltage and then to disconnect it for another certain number of cycles. Each of the on- and off-times consists of an integral number of cycles. Thyristors are switched on by gate pulses at zero voltage crossing of input voltage.

In *phase control*, the thyristors are employed as switches for connecting the motor to the supply for a certain portion of each cycle of the supply voltage. Most ac regulators use phase control. The power circuit configurations for integral cycle control and phase control do not differ in any manner.

Smooth variation of three-phase ac voltage can be realized by means of different configurations of the power circuit.

The three-phase regulators, as already said, may be half-wave or full-wave. Circuit of *three-phase half-wave regulator* for delta-connected motors or star-connected motors in which neutral point is not accessible is shown in Fig. 3.34. This circuit uses three thyristors and three diodes. Though half-wave ac regulator illustrated in Fig. 3.34 affects a saving in the cost of semiconductor devices and does not give rise to dc components in any part of the system but it introduces more harmonics into the line current than does the full-wave regulator. Half-wave circuit is not used in actual practice.

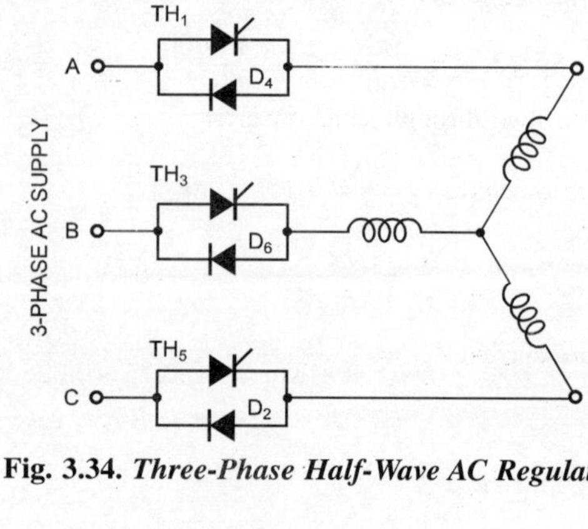

Fig. 3.34. ***Three-Phase Half-Wave AC Regulator***

Figure 3.35 shows a 3-phase full-wave regulator. It uses 6 thyristors, 2 for each phase. The input transformer may or may not be employed. As regards the heating of motor windings are concerned, star-connected motor supplied through a full-wave ac regulator is preferred over a delta-connected motor supplied through a full-wave ac regulator. This is so, because any third harmonic voltages generated by the motor back emf can cause circulating currents in case of delta-connected motor.

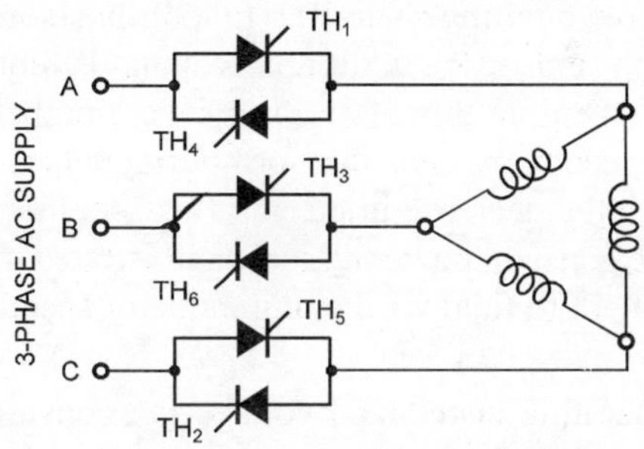

Fig. 3.35. *Three-Phase Full-Wave AC Regulator*

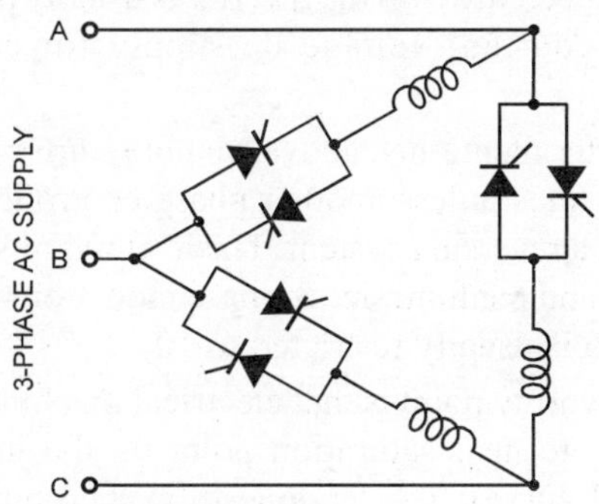

Fig. 3.36. *Delta-Connected Speed Controller*

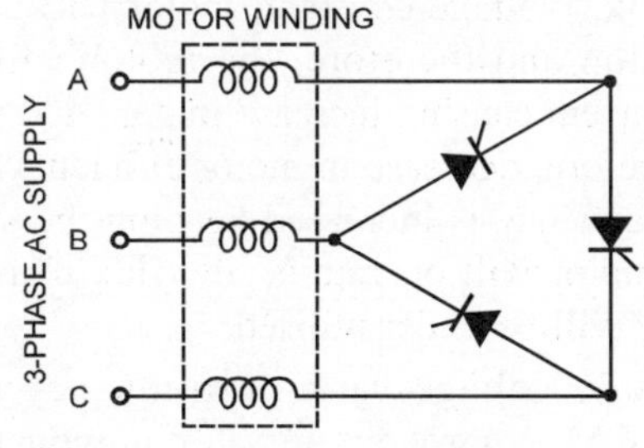

Fig. 3.37. *Neutral Connected Regulator*

For delta-connected load circuits in which each end of each phase is accessible, the arrangement illustrated in Fig. 3.36 is employed. This arrangement has the advantage of reducing the current of the device as it has to carry now $1/\sqrt{3}$ of the current if they were connected in the line of the delta winding. Once the phase current wave is known, the line current wave can be constructed by superposition.

For star-connected load circuits in which neutral point is accessible and can be opened, the arrangement illustrated in Fig. 3.37 may be used. In this arrangement, the number of thyristors required reduces to three and control circuitry becomes considerably simplified. The power consumption of the motor may be as high 100% greater than that with sine wave voltage control especially at reduced speeds.

3.13.2. Variable Voltage and Variable Frequency Control

If only frequency is changed and stator voltage is kept constant, the stator flux will not be at its rated value. The operation with flux below or above the rated value is not desirable. For constant flux operation, it is necessary that the induced emf increases or decreases linearly with applied frequency. At higher voltages and at high frequency operation stator drops are very small and thus constant flux operation is obtained by keeping V/f ratio constant. Speed-torque characteristics of a 3-phase induction motor with variable voltage variable frequency (constant V/f control) are illustrated in Fig. 1.25. It may be seen from the Fig. 1.25 that the maximum torque does not alter because the flux per pole remains constant. This mode of operation ensures the ability of developing the highest possible torque per unit stator current and, therefore, results in the best possible utilisation of the available current capability of the drive.

The variable stator voltage and frequency can be obtained from the systems illustrated in Fig. 3.38 or in Fig. 3.41, known as square-wave inverter and pulse width modulated (PWM) inverter respectively.

A square-wave inverter power circuit is illustrated in Fig. 3.38. The three-phase ac supply is converted into dc by a controlled rectifier. The output of the rectifier is supplied to the filter circuit to remove the harmonics. The dc output from filter is fed to a controlled inverter which provides variable voltage variable frequency output. This supply is fed to the stator of the 3-phase induction motor whose speed is to be controlled.

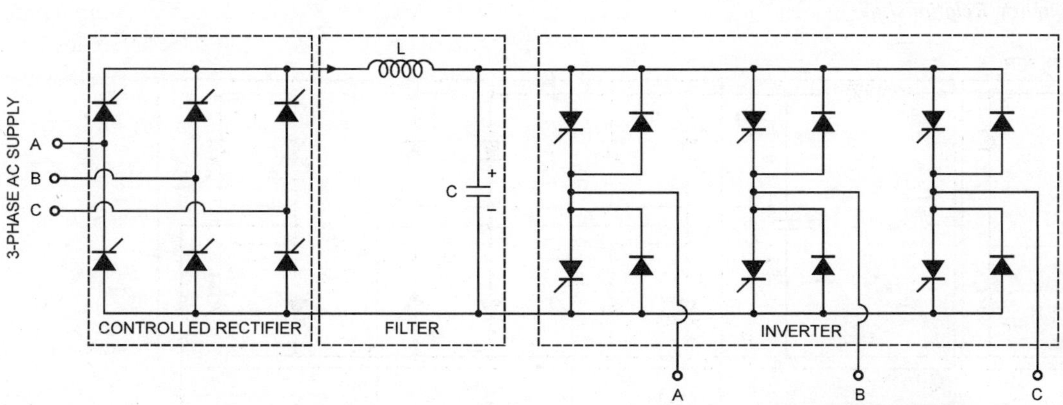

Fig. 3.38. *Variable Voltage Variable Frequency Control of 3-Phase Induction Motor (Square-Wave Inverter Power Circuit)*

Figure 3.39 shows the phase voltage waveforms V_{AN}, V_{BN}, V_{CN} and line- to-line voltage waveforms V_{AB}, V_{BC} and V_{CA}. Each line voltage waveform is displaced in time phase by 120 electrical degrees from one another and is of quasi-square wave with 120° width. The noteworthy point is that the inverter thyristors are force commutated because

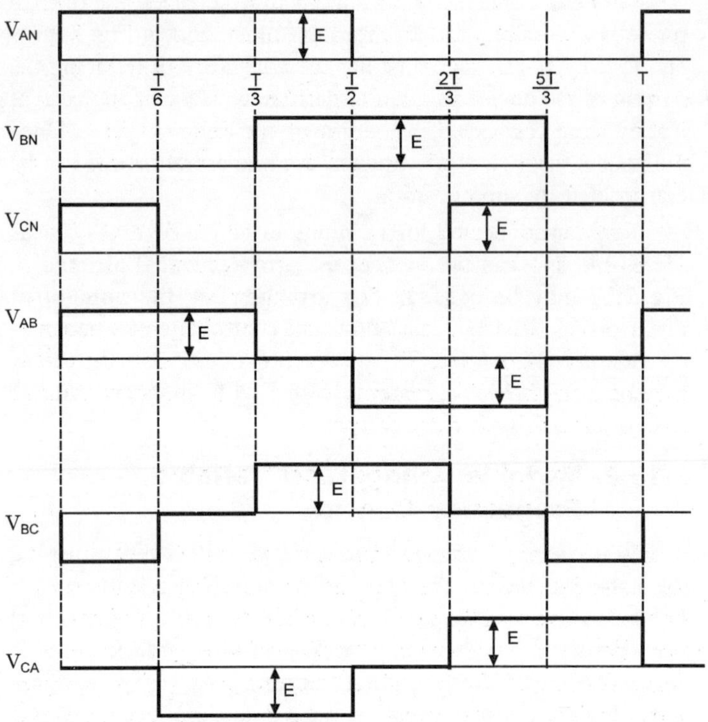

Fig. 3.39. ***Phase and Line Voltage Waveforms at the Motor Terminals***

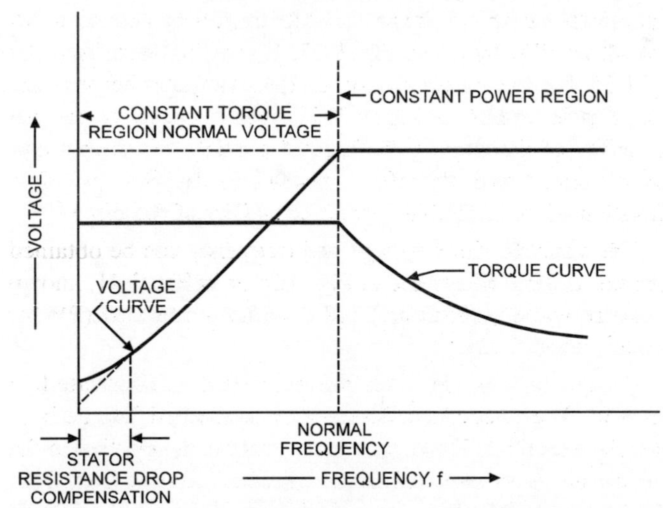

Fig. 3.40. ***Voltage-Frequency Relationship***

induction motor is a lagging power factor load. The feedback diodes help in circulation of load reactive power with filter capacitor and maintain the output voltage waves fixed to the level of dc link voltage.

Desired voltage-frequency relationship of induction motor is illustrated in Fig. 3.40. When the frequency is less than the normal frequency, the voltage is reduced by the same proportion so as to keep V/*f* constant. At very low frequencies, as the reactance drop becomes smaller in comparison to stator resistance drop ($\omega L < R$), additional voltage is required to be applied to compensate this effect. It means a higher V/*f* ratio. When the frequency exceeds the normal frequency, the torque decreases with the decrease in air-gap flux and the motor now operates in the constant power region, as illustrated in Fig. 3.40. This is equivalent to field weakening mode of speed control of a dc motor.

When operating at reduced voltage the converter voltage decreases and, therefore, the commutation capability of the capacitor also decreases. So the inverter is usually provided with an auxiliary constant voltage dc supply for commutation purposes.

The inverter mentioned above cannot return power back to the ac supply lines unless another phase controlled rectifier is added to form a reversing system. This method of speed control is used in low and medium size 3-phase induction motors where the speed ratio is usually restricted to 10 : 1.

The noteworthy point is that electrical machine is designed corresponding to near saturation point on the magnetization curve (or B-H curve). This is done from the point of view of full utilisation of the core. If the stator frequency is reduced, keeping the stator voltage constant, motor will operate in the saturation region and therefore, the motor will draw large magnetising current causing increase in core and stator copper losses and therefore, decrease in motor efficiency. However, if only supply frequency is increased keeping the stator voltage constant, the motor will operate at low flux density and thus motor capacity will be under utilised.

The variable voltage variable frequency pulse width modulation (PWM) control circuitry for an induction motor is illustrated in Fig. 3.41. This is recent method and supersedes the square-wave inverter circuit discussed above.

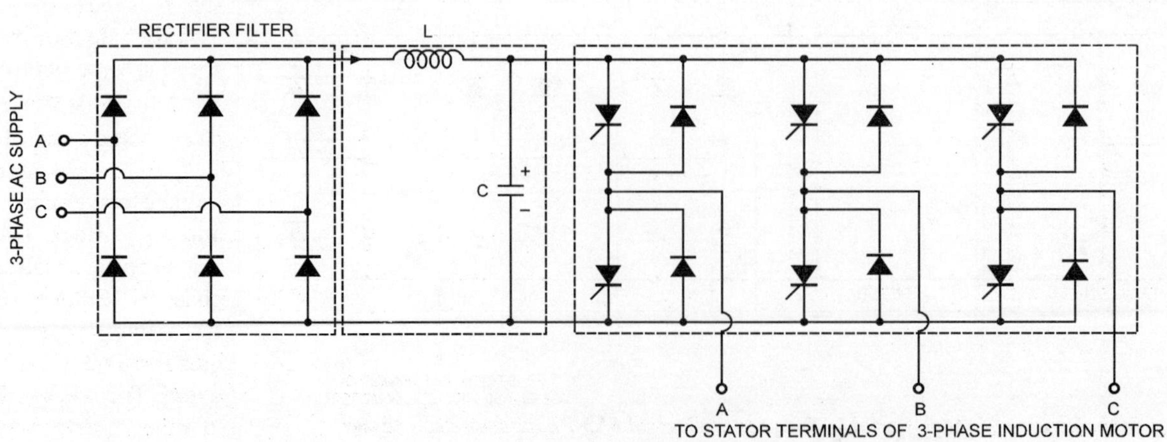

Fig. 3.41. ***Variable Voltage Variable Frequency Control of 3-Phase Induction Motor (PWM Control)***

Pulse width modulated (PWM) inverters use the chopping or pulsing technique for control of alternating voltage output of a static inverter. The dc link voltage is kept uncontrolled by a diode rectifier. The square or step-wave output voltage is rapidly switched on and off several times during each half cycle so that a number of pulses of equal amplitude are formed. Each pulse has the amplitude of the inverter input voltage V_{dc}. The magnitude of the fundamental output voltage is controlled by variation of the total on-time during a half cycle. By commutating one side of the bridge several times during a half cycle, the output voltage of the waveform shown in Fig. 3.42 (*a*) can be obtained. Simple PWM inverters can be easily made to produce a waveform with only two pulses per half cycle within a six step envelope, as illustrated in Fig. 3.42 (*b*). Significant fifth and seventh harmonics are present in such waveform and they cause appreciable deterioration in the low-speed performance of the ac motor. For elimination of low order harmonics, more refined techniques of PWM are employed, in which high frequency pulsing occurs throughout the half cycle.

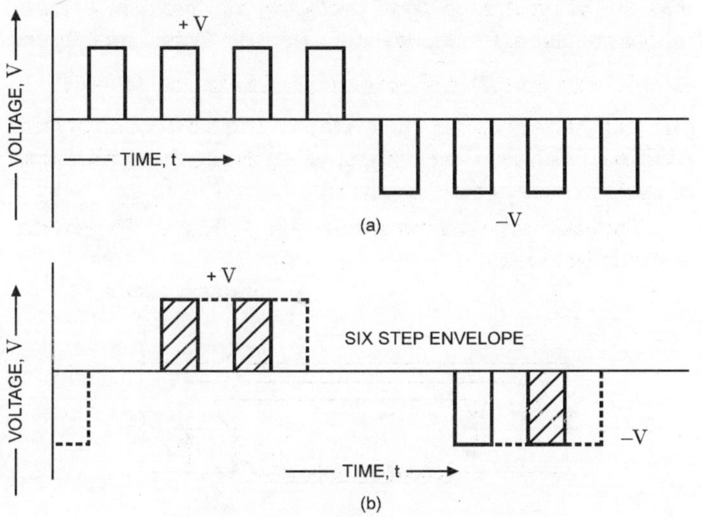

Fig. 3.42. ***Basic Waveform of PWM***

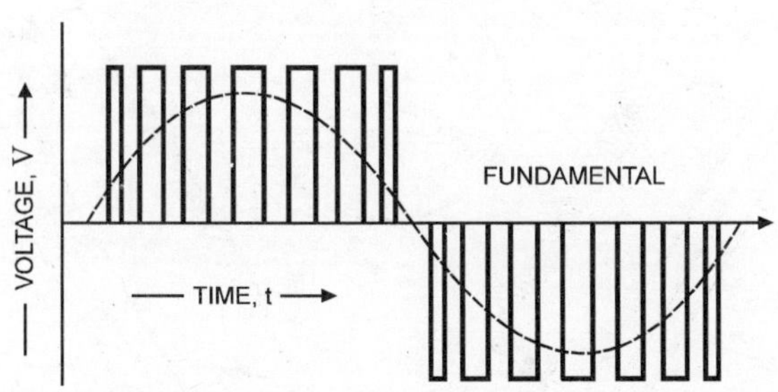

Fig. 3.43. ***Output Voltage Waveform of a PWM Inverter With Sinusoidal Modulation of The Pulse Width***

In sophisticated PWM systems, the pulse width is varied throughout the half cycle in a sinusoidal manner, as illustrated in Fig. 3.43. Actually, the pulses should be regularly spaced and the pulse width at a particular position should be proportional to the area under the sine wave at that position. In the PWM waveform, the lowest harmonic frequency is at the pulse repetition frequency and, if this is much higher than the fundamental frequency, adequate filtering is provided by the inductance of the machine. Such waveforms are usually produced by means of a control circuit in which a high frequency triangular waveform is mixed with a sinusoidal waveform of the desired frequency. Voltage control is obtained by varying the widths of all pulses without affecting the sinusoidal relationship.

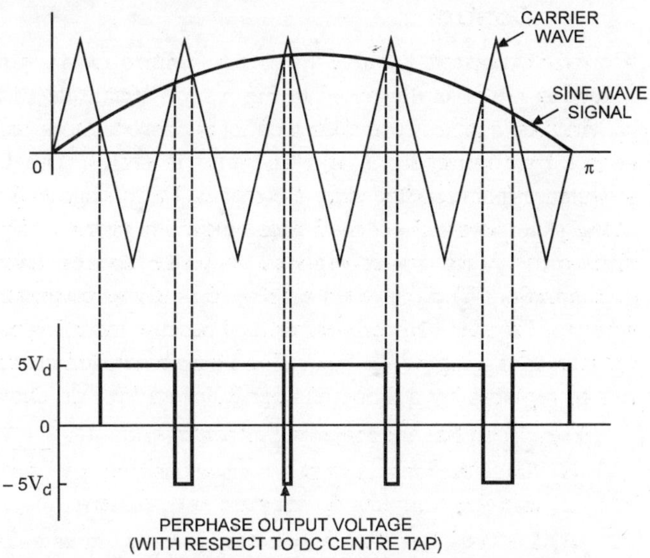

Fig. 3.44. ***Principle of Pulse Width Modulation (PWM)***

Figure 3.44 illustrates sinusoidal PWM technique where an isosceles triangular wave is compared with the sinusoidal wave signal and the points of commutation are determined by the crossover points. If modulation index is found to be less than unity, only carrier frequency harmonics with fundamental frequency related side bands appears at the output. Such waveform generates less harmonic heating and torque pulsation in comparison to that of the square wave. When the modulation index exceeds unity, maximum voltage is

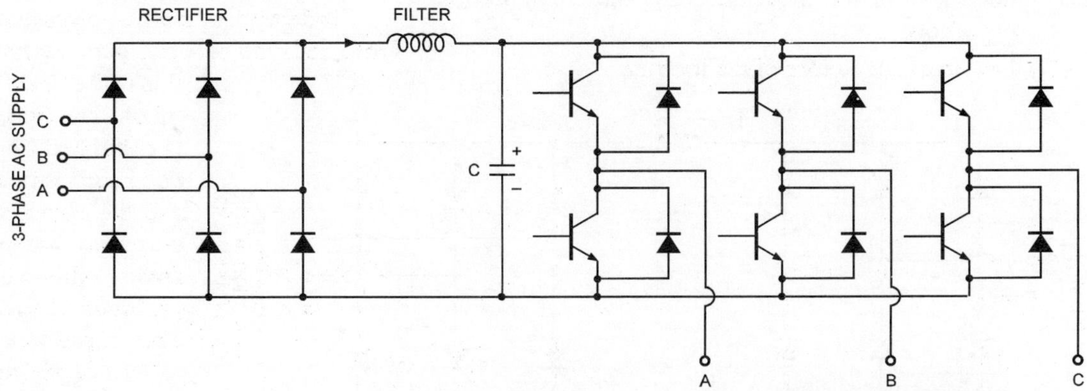

Fig. 3.45. ***Transistorized PWM Control***

obtained in the square-wave mode. Thus, PWM mode is applicable in the constant torque region while in the constant power region the operation is similar to that of the square-wave mode.

The transistorized PWM control, shown in Fig. 3.45, is used for control of low to medium size motors. No doubt power transistors cost much more than that of the thyristors of the same capability but savings due to elimination of commutation circuit and corresponding commutation losses, the circuit proves to be more economical and efficient. Also, the transistors operate faster, PWM is possible at higher frequency. This cuts down further the machine losses.

3.13.3. Variable Current Variable Frequency Control

A variable current variable frequency control circuit for an induction motor is illustrated in Fig. 3.46. Variable dc voltage provided by a phase controlled rectifier is converted to a current source by connecting a large inductor in series. The large inductance maintains the current constant. The voltage available at the stator terminals of the 3-phase induction motor is almost sinusoidal with superimposed voltage spikes due to commutation. The converter employed is a line commutated while the inverter is forced commutated because induction motor operates at lagging power factor. The phase controlled converter can be replaced by a diode rectifier followed by a dc chopper.

The circuit has the following ***advantages*** :

(*i*) Since the input current is constant, misfiring of devices and short circuits do not pose any problem.

(*ii*) Less number of components in inverter circuit and less commutation loss.

(*iii*) Rugged and reliable power circuit.

(*iv*) Simpler and more reliable control circuit. This is because only 6 thyristors are to be controlled.

(*v*) Peak current of devices is limited.

(*vi*) It can handle reactive or regenerative loads without freewheeling diodes.

Disadvantages : (*i*) Somewhat sluggish response of the drive.

(*ii*) Somewhat bulky and expensive inverter. This is due to large size of the inductance and the commutation capacitors.

(*iii*) Low frequency range of the inverter.

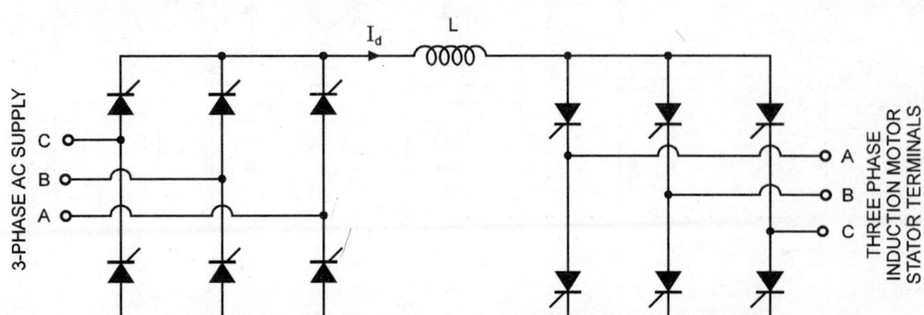

Fig. 3.46. ***Variable Current Variable Frequency Control***

(*iv*) It cannot operate under no-load condition. This is because some minimum load current is necessary for the satisfactory commutation of the inverter.

3.13.4. Cycloconverters

A cycloconverter converts ac at one frequency to an ac of another frequency. Cycloconverters can be classified as single-phase to single-phase, three-phase to single-phase and three-phase to 3-phase devices. They can also be classed as step-up and step-down cycloconverters. A step-up cycloconverter provides an output whose frequency is higher than that of input while a step-down cycloconverter provides an output of frequency lower than that of input. Step-down cycloconverter uses line or natural commutation. Cycloconverters were developed initially for electric traction systems operating at 25 Hz and $16\frac{2}{3}$ Hz. In their early stages of use, the cycloconverters used mercury-arc rectifiers. With the development of thyristors the applications of cycloconverters have increased.

The basic power circuit scheme of a 3-phase cycloconverter is shown in Fig. 3.47.

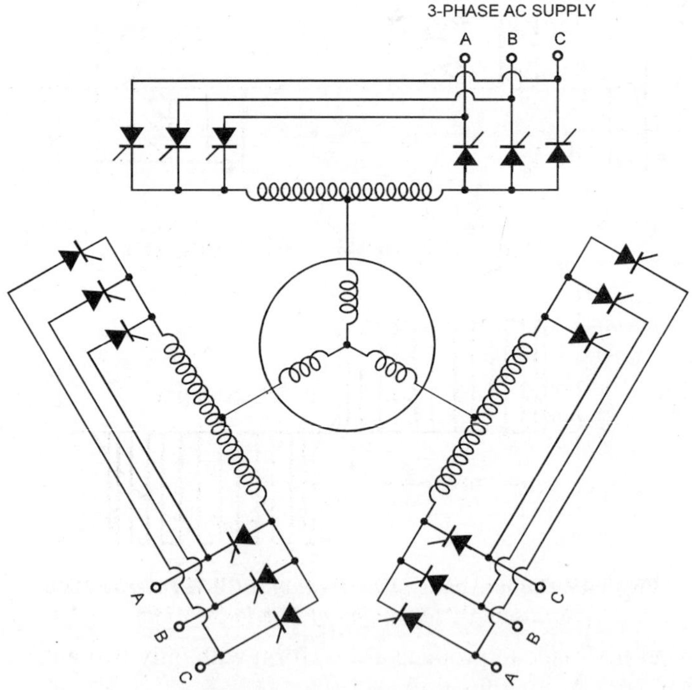

Fig. 3.47. ***3-Phase Cycloconverter***

Independent control of output frequency and voltage is obtained with only one parameter variation, *i.e.,* viz., by variation of the firing points of the controlled rectifiers. The frequency of the output voltage is controlled by the rate at which the firing points are varied about the quiescent point and the output voltage is controlled by the maximum excursion of the firing points from the quiescent point. The cycloconverter, with its associated firing circuit, produces an output voltage that is replica of the reference voltage.

The operation of cycloconverter is characterized by several features. They are, generally, employed as step-down frequency converters. There is no fixed minimum ratio of input to output frequency; however, the output frequency is restricted typically to one-third or one-half of the input or line frequency. Below these ratios, the efficiencies of both the cycloconverters and motors supplied by them start falling significantly.

Reversibility is another feature of cycloconverter drive systems. A cycloconverter fed ac motor drive will respond to a change in polarity of the input signals by changing the direction of rotation of the motor without the use of contactors to reverse phase sequence.

The ability of cycloconverter to handle power flow in either direction is another important feature. This, together with the above mentioned reversibility feature, provides an induction motor drive capable of operating in any of the four quadrants of the motor's speed-torque curve.

While the cycloconverter has many attractive features from a theoretical point of view, there are several limitations because of which they have not gained popularity. It needs more power semiconductors than an inverter. For example, the three-phase cycloconverter needs 18 thyristors, whereas the rectifier-inverter combination (Fig. 3.38) needs only 12 thyristors. Cycloconverters can produce only a sub-frequency output. Line pollution with harmonics and low power factor can also be problems with cycloconverters of high power rating. However, recent advancements in fast switching devices have resulted into devices known as *forced commutated direct frequency changers* (FCDFC) which operate at high efficiency and have low harmonic content.

Cycloconverters drives are normally used for large size motors as the cost and complexity of the power and control circuits prohibit their use for general applications. Cycloconverters have been employed in diesel electric locomotives where a high frequency alternator coupled to the engine shaft provides power at the input. These have also been employed in gearless cement mill or ball mill drives.

3.13.5. Reduced Voltage Starting (Soft Start)

The starting line current at full voltage of an induction motor may be about 6 times the rated full-load current. Such a high current may cause severe voltage dip in the network supplying the induction motor.

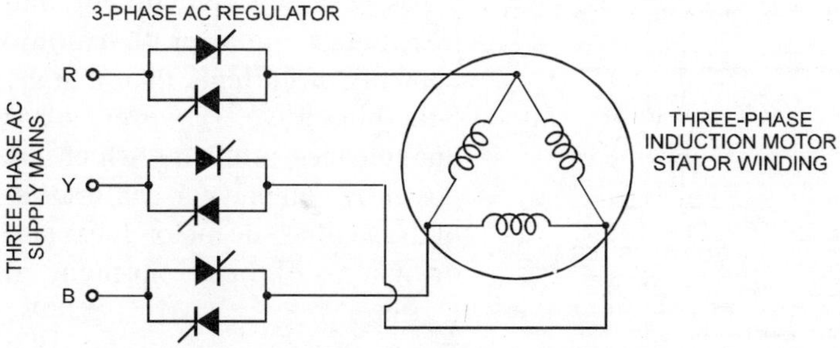

Fig. 3.48. ***Reduced Voltage Starting of a 3-Phase Induction Motor***

The circuit illustrated in Fig. 3.48 can be used for feeding a reduced voltage at the start. As seen it is a 3-phase ac regulator. By proper control of firing angle the regulator provides a low output voltage which is supplied to the induction motor. When the motor attains the full or rated speed, the regulator can be short circuited by mechanical contactor so that the motor operates normally at rated voltage. Moreover, if the motor is employed for a constant speed drive it is possible to operate the motor at reduced voltage when the mechanical load is small. Operation at reduced voltage causes decrease in power losses in the motor and thus results in energy saving.

3.13.6. Rotor Resistance Control

The conventional method of connecting resistances across the slip rings of a wound rotor induction motor is a form of rotor voltage control [Art. 1.18.2 (5)]. The main drawback of this method of speed control is its poor efficiency because of power wasted in the external resistors.

Figure 3.49 (*a*) shows a 3-phase diode rectifier and gate-turn-off thyristor (GTO)* connected in the rotor circuit of a wound rotor induction motor. The GTO operating as a chopper varies the resistance R as per the duty cycle α. The effective resistance R_e is given by

$$R_e = R(1 - \alpha) \qquad ...(3.59)$$

Thus the speed of a wound rotor induction motor is controlled by varying the ratio of on-time to off-time.

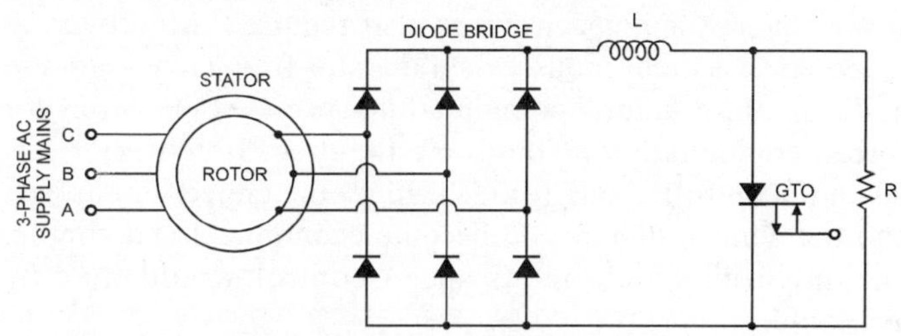

Fig. 3.49 (*a*). ***Rotor Voltage Control of a 3-Phase Wound Rotor Induction Motor (Chopper Control)***

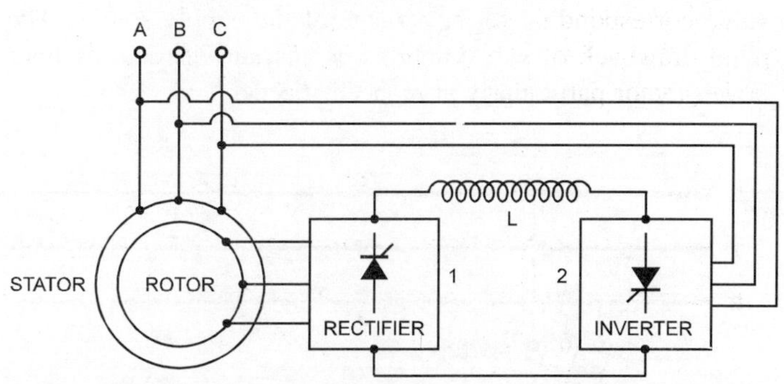

Fig. 3.49 (*b*). ***Rotor Voltage Control of a 3-Phase Wound Rotor Induction Motor (Slip-Power Recovery Scheme)***

3.13.7. Slip-Power Recovery Scheme

Figure 3.49 (*b*) shows a scheme known as *slip-power* recovery scheme. The rotor terminals are connected to the 3-phase ac

* A gate turn-off thyristor (GTO) is a special thyristor which can be turned on by a positive gate signal and can be turned off by a negative gate signal. Evidently the use of a GTO in a power electronics circuit eliminates the need of forced commutation circuit because turn-off is achieved by applying a negative gate signal.

supply mains through two fully controlled thyristor bridges. Bridge 1 acts as a rectifier (or converter) and bridge 2 acts as an inverter. Power output from rotor can be supplied back to the supply source. Since the frequency of rotor currents is slip frequency, this method is known as *slip-power recovery scheme.* By controlling the firing angles of the two bridges, the rotor power output can be varied. Thus the motor slip and speed (for the same torque) will also change. However, the drawback of this scheme is that both the bridges draw reactive power from the supply mains. Therefore, the overall power factor of the motor is poor. If speeds only below synchronous are desired, bridge 1 may be uncontrolled and, thus, may consists of diodes. If both bridges are controlled, the operation of the two bridges can be reversed also to get speeds above synchronous one.

Infact, the slip-power is either returned to the supply network as in Scherbius scheme or used to drive an auxiliary motor which is mechanically coupled to the induction motor shaft, as in Kramer scheme.

1. *Static Scherbius Drive.* The static Scherbius drive also uses the principle of slip-power recovery. The schematic diagram is shown in Fig. 3.50. For achieving both sub-sychronous and supersynchronous speed control, converters 1 and 2 must be fully-controlled thyristor bridges, one functioning at slip frequency as a rectifier or inverter while the other operating at supply system frequency as an inverter or rectifier. The cost of converters is quite appreciable and a slip-frequency gating circuit is also required. Moreover, at speeds near synchronous, when the slip-frequency emfs are quite small for natural commutation, special connections for forced commutation methods are required. If converter 1 is taken uncontrolled one (diode bridge) the converter cascade and the control unit would become economical and simple, but only sub-synchronous speed control would then be available.

The 3-phase transformer between the supply source and inverter 2 is meant to bring the rotor circuit voltage up to the value corresponding to the voltage of the supply source. The main drawback of sub-synchronous cascade drive is its poor power factor particularly at reduced speeds.

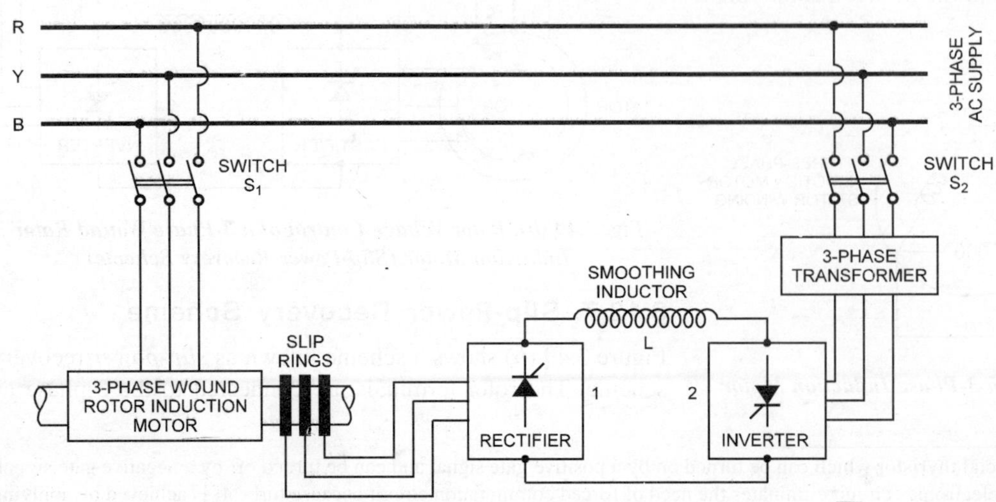

Fig. 3.50. ***Static Scherbius Drive***

This drive has applications in large power fan and pump drives which need speed control in a narrow range only. Power ratings of diode bridge, inverter and transformer is just maximum slip times the power rating of the motor resulting in a low cost drive. This drive provides a constant torque control. Constant power control can be obtained from Kramer drive discussed below.

2. *Static Kramer Drive.* Figure 3.51 shows the schematic diagram of the Kramer cascade with static converter. The rotor circuit of a slip-ring induction motor feeds the slip power, rectified by a diode bridge, to the armature of a separately excited dc motor mechanically coupled to the induction motor. Speed control is achieved by varying the field current of the dc motor. An emf proportional to the back emf of the dc motor may be considered to be injected into the rotor circuit of the induction motor to cause variation in the speed of the system.

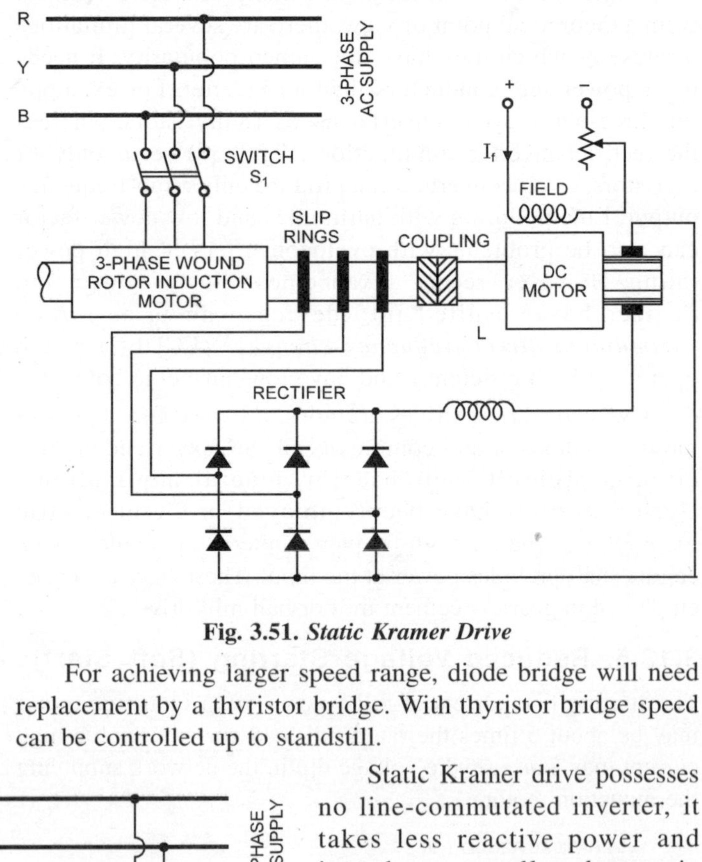

Fig. 3.51. ***Static Kramer Drive***

For achieving larger speed range, diode bridge will need replacement by a thyristor bridge. With thyristor bridge speed can be controlled up to standstill.

Static Kramer drive possesses no line-commutated inverter, it takes less reactive power and introduces smaller harmonic contents of currents than a static Scherbius drive. However, it has maintenance problems which are posed by commutator and brushes of the auxiliary dc motor. It has also drawback of large moment of inertia.

Static Kramer drive systems are used in large power pumps and compressor type loads where speed control is within narrow range and below synchronous speed.

3.14. THYRISTOR CONTROL OF A 3-PHASE SYNCHRONOUS MOTOR

The synchronous motor is a constant speed motor and it develops torque only at synchronous speed which is directly proportional to supply frequency. Therefore, variation of frequency of ac supply is a convenient method to control the speed of a synchronous motor. Before the advent of power electronic devices change of frequency was a very difficult task. However, the power electronic circuits provide a simple method of changing the frequency. The 50 Hz ac is converted into dc by a controlled rectifier. A variable frequency inverter converts dc into variable frequency ac. To keep the flux density in the motor constant voltage and frequency must be varied in the same ratio so that voltage/frequency ratio is constant.

Figure 3.52 shows a system of speed control of a synchronous motor. The 3-phase 50 Hz ac is rectified by a controlled 3-phase rectifier. The filter removes harmonics from the output of the rectifier. The variable frequency inverter feeds a variable frequency ac to the synchronous motor. Thus the speed of a synchronous motor can be controlled. The dc output from filter is also used to excite the rotor winding.

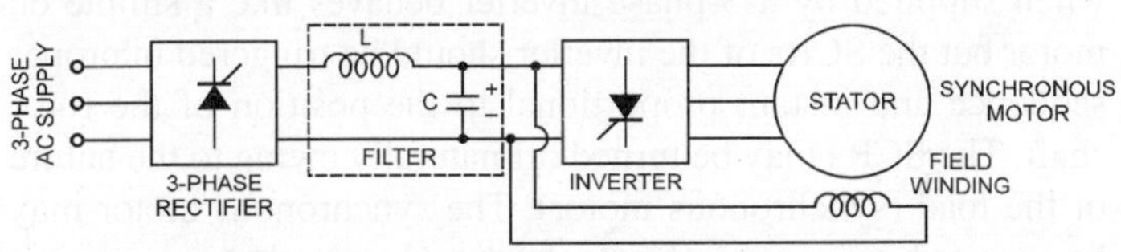

Fig. 3.52. *Speed Control of a Synchronous Motor*

In low-speed high power applications it is also possible to use a cycloconverter to convert 50 Hz ac into variable frequency ac. The cycloconverter does the conversion without an intermediate dc link. The maximum output frequency is limited to about one-third of the line frequency so that the output waveform has low harmonic content.

3.14.1. Load Commutated Inverter Drive For Synchronous Motor

For large sized motors (motors having output rating of more than 750 kW), load commutated synchronous motor drive becomes competitive with induction motor drive in adjustable speed requirements. Figure 3.53 shows the circuit for a load

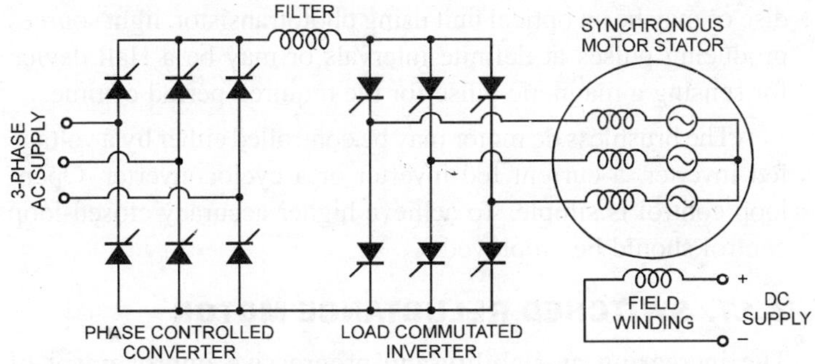

Fig. 3.53. *Load Commutated Inverter Drive*

commutated inverter drive. The three-phase ac supply is rectified by phase controlled 3-phase bridge converter. The dc supply is passed through a filter to improve waveform and then fed to the load commutated inverter. Each phase of the synchronous motor is represented by an internal induced emf in series with an inductance of stator winding. The frequency and phase of stator currents are synchronised to the rotor position. The current commutation in the load commutated inverter to supply currents to the stator phases in an appropriate sequence is provided by the stator induced emfs. The presence of 3-phase induced emf in the motor stator winding causes the commutation of the thyristor. A control of firing angle of converter controls its dc output voltage and, therefore, the current.

A complete circuit of load commutated inverter drive controller is shown in Fig. 3.54. The input voltage to stator of a synchronous motor is measured to calculate the rotor field position as function of time. The measured voltage is rectified to provide a dc signal proportional to instantaneous speed of synchronous motor. The turn-off time T_{off} available to thyristor in the inverter is kept constant. Keeping field current and T_{off} constant, the actual speed is compared with the reference speed. The error signal is amplified to provide reference I_d. If actual current I_d is less than reference, the rectifier increases the voltage supplied to inverter, thus increasing I_d and the motor torque. Based on current I_d and measured voltage, the fire pulses to the gates of thyristors of inverter are provided to keep T_{off} constant.

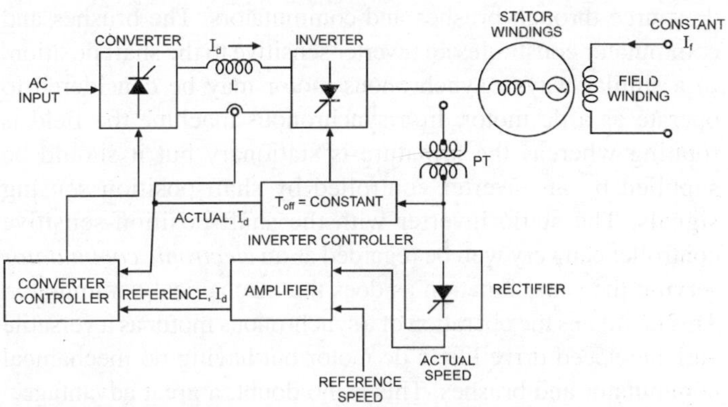

Fig. 3.54. *Load Commutated Inverter Drive For a Synchronous Motor*

3.15. SPEED CONTROL OF SINGLE PHASE INDUCTION MOTOR USING TRIAC (FAN REGULATOR)

The conventional speed regulator for a fan uses a resistance regulator. The regulator resistance is in series with the fan motor. The speed of the fan is reduced, whenever desired, by increasing the regulator resistance. Thus the voltage drop across the regulator resistance increases and, therefore, voltage applied to the motor is reduced and so the speed. The resistance regulator causes loss of energy. This loss of energy becomes significant at low speeds.

Figure 3.55 shows a speed control circuit using a diac-triac pair. The diac is meant to trigger the triac into conduction. It is preferable to use a matched diac-triac pair. The R-C circuit forms the triggering network. By adjusting the resistor R, the voltage across capacitor C can be adjusted. When voltage across capacitor C exceeds breakdown voltage of diac, it is triggered into conduction and sends a triggering signal to the triac gate. When triac is turned on the motor starts. By varying resistance R, The firing angle of triac can be altered and thus the voltage applied to the motor is changed and consequently the motor speed is changed. Since resistance R, carries very small current, the loss of energy is very small. Sometimes an R-C snubber circuit is added in parallel with the triac to protect it from high voltage transients and high $\frac{dv}{dt}$.

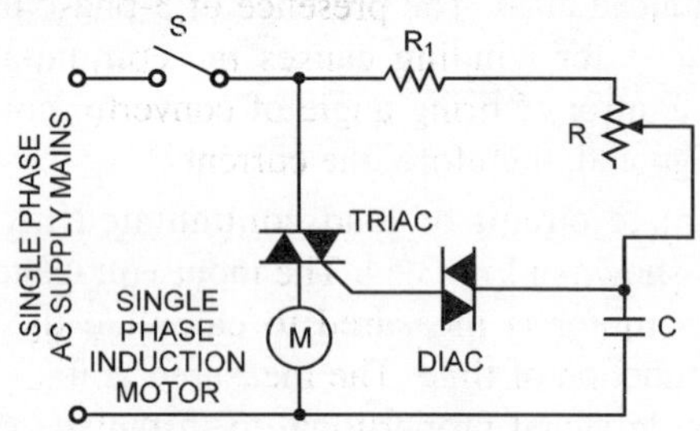

Fig. 3.55. ***Speed Control of Single Phase Induction Motor Using Triac***

3.16. BRUSHLESS (OR COMMUTATORLESS) DC MOTOR

A dc motor can be thought of as an ac synchronous machine in which the field is stationary and the armature with its multiphase ac winding is rotating. The armature receives ac supply from a dc source through brushes and commutators. The brushes and commutator constitutes an inverter sensitive to the shaft position. In a similar way, a synchronous motor may be considered to operate as a dc motor. In a synchronous machine the field is rotating whereas the armature is stationary but it should be supplied by an inverter controlled by shaft-position sensing signals. The static inverter with the shaft-position-sensitive controller can very well be regarded as an *electronic commutator* serving the same function as does the mechanical commutator. This facilitates the operation of a synchronous motor as a versatile variable-speed drive like a dc motor but having no mechanical commutator and brushes. This is, no doubt, a great advantage.

No doubt, the brushless dc motors are generally more expensive for the same kW rating that conventional (commutator and brush) motors but possess certain advantages too such as little or no maintenance, much longer operating life, no risk of explosion or possibility of RF radiation due to arcing, higher efficiency, rapid response etc. These motors produce no brush or commutator particles or gases as by- products of operation. These motors are capable of operation submerged in fluids, combustible gases and may even be hermetically sealed. They provide a fairly linear output torque vs input current characteristic, which lends itself to servo applications.

The stator winding of a brushless dc motor may be conventional three-phase winding of a synchronous motor or the conventional armature winding of a dc motor. However, in both cases, the stator winding has to be supplied from a static inverter triggered by shaft-position sensitive signals so that the supply frequency is proportional to the shaft speed. The dc field winding is placed on the rotor and supplied from a static dc source through slip rings mounted on the motor shaft. The equivalent block diagram of a brushless dc motor is shown in Fig. 3.56.

It has already been explained that static inverter together with the shaft-position sensitive trigger circuit is equivalent to the mechanical commutator of a dc machine. A 3-phase inverter is equivalent to a three-segment commutator. If the stator winding is similar to the armature winding of an ordinary dc machine, six symmetrical tappings from the winding can be taken out and the stator winding may be supplied from a six-phase SCR bridge inverter. This is equivalent to six segment commutator. There may be many other variations of the electronic commutation arrangement. For instance, as individual stator coils may be controlled by series SCR switches. However, three segment commutation with a three-phase static inverter is widely used for drives.

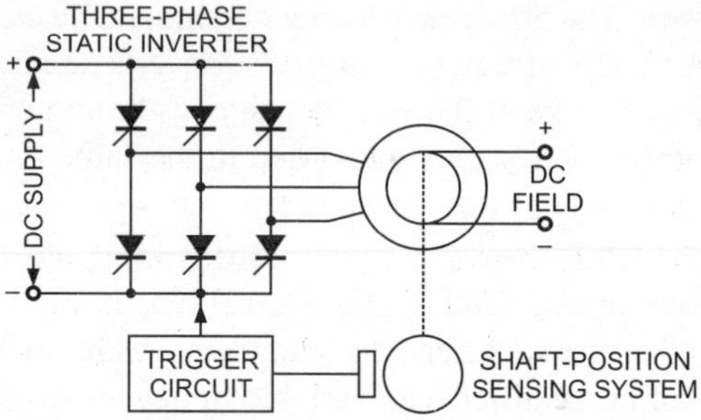

Fig. 3.56. ***Equivalent Block Diagram of Brushless DC Motor***

It is, therefore, obvious that a 3-phase synchronous motor when supplied by a 3-phase inverter behaves like a simple dc motor but the SCRs of the inverter should be triggered in proper sequence and instant proportional to the position of the rotor shaft. The SCRs may be turned off naturally owing to the nature of the load (synchronous motor). The synchronous motor may be operated at any power factor (leading, lagging behind or unity) by variation of the dc field excitation. When the motor is operated at leading power factor, the inverter can be operated with load commutation *i.e.,* the SCRs are turned off naturally. Otherwise, a force commutated inverter should be employed.

Since the system behaves like a conventional separately excited dc motor, the speed can be controlled by the variation of the dc supply to the inverter or to the field. The speed is inversely proportional to the field strength. The torque-speed characteristics are similar to those of a separately excited dc motor but are slightly more drooping in this case.

The shaft-position sensing unit normally consists of a toothed disc on a free motor shaft end and two sets of three position detecting sensors. The detecting sensor may be an oscillator triggered into action for a definite period by the toothed disc or may be an optical unit using phototransistor, light sources producing pulses at definite intervals or may be a Hall device for sensing a magnetic pulse for the required period of time.

The brushless dc motor may be controlled either by a voltage fed inverter, a current fed inverter or a cycloconverter. Open-loop control is simple. To achieve higher accuracy closed-loop control should be employed.

3.17. SWITCHED RELUCTANCE MOTOR

The increasing availability and progressive improvement of power semiconductor devices as well as intensive computer aided study of the geometry of conventional reluctance motors for optimum torque production have led to the introduction of yet another type of drive called the *switched reluctance motor drive.* It is becoming a serious competitor to converter supplied dc and ac variable speed drive systems. The switched reluctance motor is being considered, nowadays, for applications ranging from low power servomotors to high power traction drives. Motors of power ratings varying from 4 to 22 kW are commercially available at present for many applications.

Construction. The switched reluctance motor (SRM) has both salient pole stator and rotor like a variable reluctance stepper motor. Figure 3.57 shows a one phase winding of a 4-phase switched reluctance motor having 8 poles on stator and 6 poles on rotor. While the rotor has no windings, each stator pole has a concentrated winding around it and each pair of diametrically opposite coils comprise one phase of the motor. These motors are designed for applications different from those for which stepper motors are designed. A stepper motor is designed suitable for open-loop position and speed control in small power applications, where efficiency is not significant. On the other hand a switched reluctance motor (SRM) is employed in variable speed drives and designed to operate efficiently for wide range of speed and torque and needs rotor position sensing. It is, also, quite different from a standard synchronous reluctance motor, described in Art. 1.16.3 (*e*), in two aspects. A synchronous reluctance motor has the same number of poles on stator and rotor but the rotor of a SRM has lesser poles than the stator, which is an essential feature to provide self-starting capability and bidirectional control. They are also different in stator construction—the synchronous reluctance motor has a cylindrical stator with distributed winding, while the SRM has a salient pole stator with concentrated coils like a dc motor.

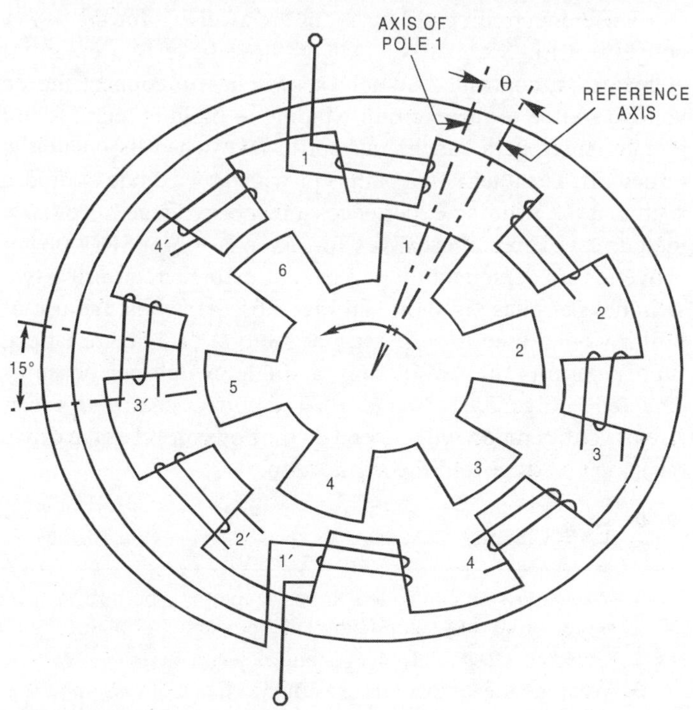

Fig. 3.57. ***One Phase Winding of a 4-Phase 8/6 Pole Switched Reluctance Motor***

Though various combinations of stator and rotor pole numbes are possible in SRMs, the commonly used are 8/6 and 6/4. As already mentioned, the stator has concentrated coils and diametrically opposite coils are connected in series or parallel to provide one phase. Thus, motors with pole numbers 6/4 and 8/6 will have three and four phases respectively.

Operation. Reluctance torque is produced when a stator phase is excited by means of unidirectinal currents. This results in the magnetic attraction of an adjacent rotor pole as it tends to align into a position of minimum reluctance. When the number of stator and rotor poles differ, the sequential switching of the excitation from one set of stator poles to the next, in synchronism with the rotor position, produces an almost constant torque resulting in an uniform rotation of rotor. The synchronisation of the switching on the excitation with rotor position can be accomplished with simple rotor position feedback.

Neglecting nonlinearity of the magnetic circuit, the instantaneous torque produced in such machines may be given as

$$T = \frac{1}{2} i^2 \frac{dL}{d\theta} \qquad ...(3.60)$$

where *i* is the instantaneous current in the exciting winding and L is the self inductance of that winding varying as a function of the angular position of the rotor. It may be noted that the torque developed is independent of the direction of current flow in the windings, so that unidirectional currents can be used to control the motor. The direction of rotation can be reversed by exciting the phases in the reverse sequence. The motor can also provide regenerative braking. If a phase is excited after the rotor has crossed the position of minimum reluctance, the rotor will experience a torque in opposition to its motion, it will decelerate, mechanical energy drawn from it will be converted into electrical energy and supplied to the source. In fact, the possibility of operating in all of the four quadrants of the speed-torque plane and obtaining flexible speed-torque characteristics simply by appropriate switching of current pulses makes the motor very versatile.

Figure 3.58 (*a*) and (*b*) illustrates the ideal variation of inductance of the exciting winding with respect to the angular position of the rotor over a periphery of one rotor pole pitch and the corresponding torque developed, for an assumed value of constant current [using Eq. (3.60)].

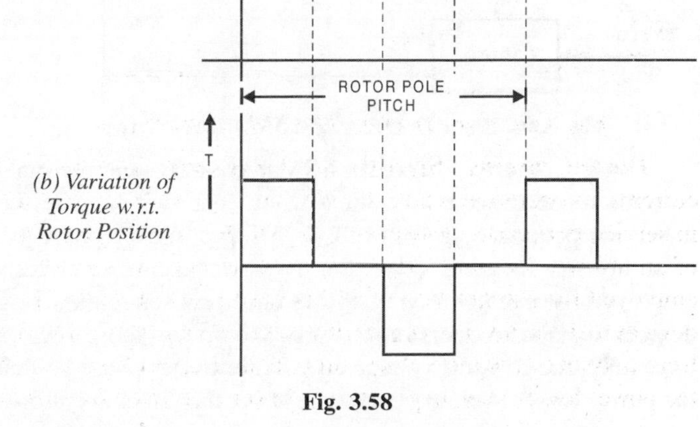

Fig. 3.58

Modes of Operation of Motor. There are two distinct modes of operation corresponding to low or high speed. Monitoring of exciting current during low speed operation is essential because of long duration of each phase period and needs chopping of energization to restrict each phase current within the semiconductor ratings. Moreover, the developed torque is controlled by varying the average phase current. Hence, accurate monitoring of the exciting current is required for obtaining high degree of controllability possible. During high speed operation, current control is not essential because the inductance of the winding and the motional back emf induced restrict the excitation to single pulses of current. Torque is controlled by optimal positioning of these pulses rather than the current level. Current monitoring, however, is retained for the sake of protection.

Typical phase current waveforms of the two modes (low speed and high speed) of operation are shown in Figs. 3.59 (*a*) and (*b*).

The complete drive system, comprising a SRM coupled to a load, a power converter and a control system involving rotor position transducer and current sensor is shown in Fig. 3.60.

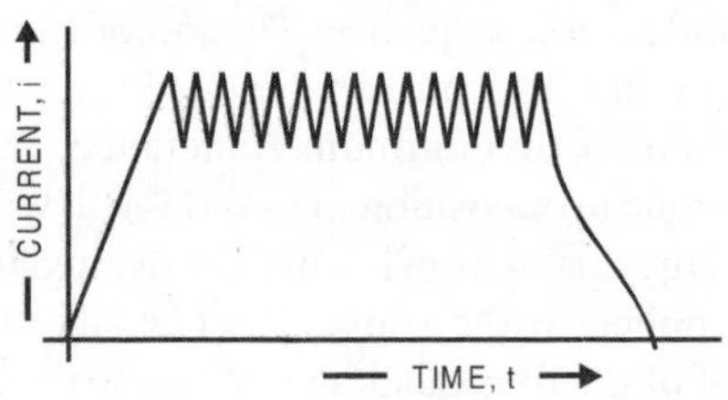

(*a*) *Low Speed Operation*

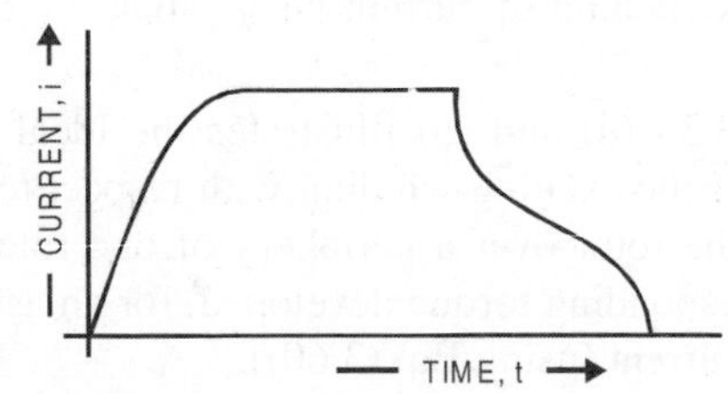

(*b*) *High Speed Operation*

Fig. 3.59. ***Current Waveforms***

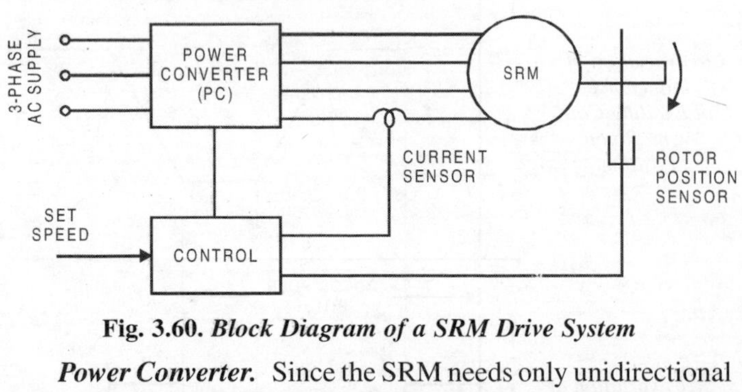

Fig. 3.60. ***Block Diagram of a SRM Drive System***

Power Converter. Since the SRM needs only unidirectional currents, its operation is possible with only one switching device in service per phase, instead of two in series in each phase leg of an inverter for an ac drive. So, the power converter circuits employed for energization of SRMs have few semiconducting devices than the inverters supplying ac motors and those devices have only one forward voltage drop in series per phase so that the power losses may, in general, be lower than in conventional inverters. Because of these facts, other factors being the common, results in reduction in the physical size of converter and increase in its reliability.

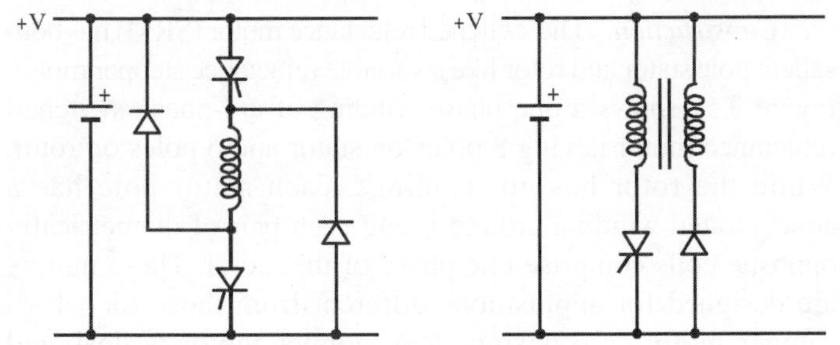

(*a*) (*b*)

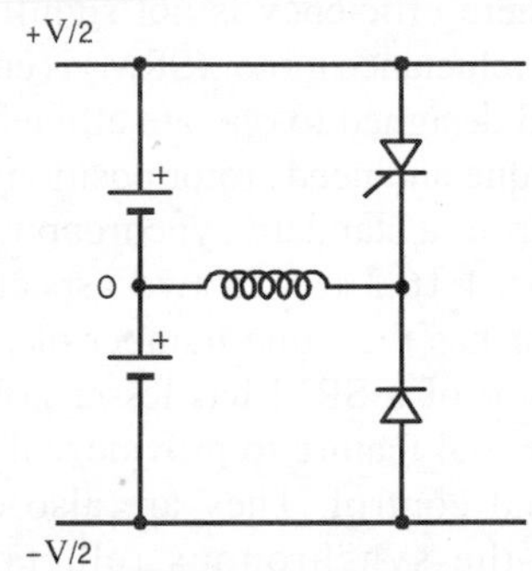

(*c*)

Fig. 3.61. ***Different Power Circuit Configurations***

Different circuit configurations are available for the power converters for SRM drive. All of these have two essential elements—a controlled switch or switches to connect the dc voltage source to the exciting winding to build up current and an alternative path for the current to take when the switch is turned off. The alternative path is provided by a diode or diodes such that the winding experiences a reverse voltage to collapse the current. Three alternatives for one phase of a SRM power converter are depicted in Fig. 3.61. A flexible circuit using two switches per phase is shown in Fig. 3.61 (*a*), whilst the use of bifilar wound motor (as in stepper motors) or a centre tapped supply permits the use of only a single switch per phase as depicted in Figs. 3.61 (*b*) and (*c*). The correct choice of power circuit configuration will depend on the power level of the drive, the supply voltage and the application.

Exercises

1. Why thyristor controlled drives employing both dc and ac motors find wide application in industry ? Explain.
2. Give the classification of static power converters.
3. What are the factors that govern the choice of current rating, voltage rating and switching frequency of the power electronic converter for electric motor drive ?
4. Explain working principle of single phase half controlled bridge converter fed separately excited dc motor and draw waveforms of motor terminal voltage, motor current and supply current.
   Compare its merits and demerits with respect to when fed by fully controlled bridge converter.

[U.P. Technical Univ. Electric Drives 2005-06]

5. Explain how the speed control of a separately excited dc motor can be achieved by a fully controlled single phase thyristor bridge converter.
[Madurai Kamaraj Univ. November 1996]

6. Draw the complete drive circuit of a single phase fully controlled rectifier controlling a separately excited dc motor. The circuit should include the single phase supply, controlled rectifier (including the thyristors) and the motor. Draw typical waveforms of supply voltage, armature voltage and current at a fixed speed under continuous conduction. Derive an expression for motor speed in terms of supply voltage, firing angle and motor constants.
[U.P. Technical Univ. Electric Drives 2007-08]

7. Compare and contrast a half-controlled converter with a fully controlled converter for speed control of a separately excited dc motor. In which case supply power factor is better and why ? **[A.M.I.E. Sec. B. Electric Drives & Their Control Summer 1996]**

8. Compare the performance of a separately excited dc motor when fed from a 1-phase fully controlled bridge converter with that of the same motor fed from a 1-phase half controlled bridge. Draw the relevant waveforms in each case and sketch the variation of average output voltage *vs* firing angle. **[A.M.I.E. Sec. B. Electric Drives & Their Control Summer 1994]**

9. Describe the operation of a single phase half controlled bridge converter fed separately excited dc motor and draw waveforms of motor terminal voltage, armature and supply currents assuming continuous current conduction. List its merits and demerits in comparison to that of with fully controlled bridge rectifier.
[U.P. Technical Univ. Electric Drives 2008-09]

10. How do you change from motoring to braking operation in forward direction of a dc separately excited motor when two fully controlled thyristor converters are used in back-to-back connection? Explain in detail.
[A.M.I.E. Sec. B. Electric Drives & Their Control Summer 1999]

11. Draw the circuit arrangements for a dual converter and explain its operation. **[Madras Univ. April 1997]**

12. Describe, with power circuit diagram, the dual converter. What is the relationship between firing angles of the two full converters of the dual converter under continuous current conduction mode ? Explain Why ?
[A.M.I.E. Sec. B. Electric Drives & Their Control Summer 1993]

13. Explain two modes in which a dual converter is used to control the dc motor speed. Which of the two methods is better and why ? Explain.
[U.P. Technical Univ. Electric Drives 2009-10]

14. A three phase fully controlled bridge converter is feeding a separately excited dc motor. Assuming continuous load current, derive the equation relating speed and torque of a motor as a function of triggering delay 'α'.
[U.P. Technical Unv. Electric Drives 2004-05]

15. Enumerate the different methods of obtaining variable dc voltage from a fixed dc voltage source. Give their advantages and disadvantages.

16. What is a dc chopper ? Explain the principle of operation of a chopper giving its circuit diagram and output voltage waveforms.

17. Explain the time ratio control (TRC) and current limit control used for choppers.

18. What is the analogy between a chopper and transformer ? Describe the working and obtain an expression for average load voltage of step-up chopper.

19. Discuss in full, the chopper control of a dc separately excited motor.
[Bharathiyar Univ. Electrical Drives and Control Oct. 1996]

20. Draw the block diagram for chopper fed dc motor drive. Explain the working. **[A.M.I.E. Sec. B. Electrical Drives & Their Control Summer 2000]**

21. Draw the power circuit of a single quadrant dc chopper and explain its working.
[A.M.I.E. Sec. B. Electric Drives & Their Control Winter 1998]

22. Draw the circuit for the chopper control of a separately excited dc motor and derive an expression for speed in terms of motor constants, supply voltage, duty ratio and load torque. **[U.P. Technical Univ. Electric Drives 2007-08]**

23. Regenerative braking is energy efficient braking, explain in view of chopper drives.
[A.M.I.E. Sec. B. Electric Drives & Their Control Summer 1998]

24. Discuss a chopper circuit for regenerative braking of a separately excited dc motor. Explain the operating principle.
[U.P. Technical Univ. Electric Drives 2004-05]

25. State the advantages of chopper controlled dc motors.
[A.M.I.E. Sec. B. Electric Drives & Their Control Winter 1994]

26. How is the performance of a dc motor affected by thyristor power supply ? Explain.

27. What are the special problems associated with the operation of dc drives where speed is controlled by phase controlled converters and choppers ?
[A.M.I.E. Sec. B. Electric Drives & Their Control Summer 1993]

28. Explain the speed control of a dc series motor using thyristors.

29. Explain in brief, on the importance of supply harmonics, power factor and ripples in the motor current, in the performance of a converter fed dc motor drive.
[G.B. Technical Univ. Electric Drives 2010-11]

30. Explain the chopper control techniques for each of a separately excited dc motor and dc series motor.
[G.B. Technical Univ. Electric Drives 2010-11]

31. How ac motor speed control is achieved by using voltage source inverter ?
[A.M.I.E. Sec. B. Electric Drives & Their Control Summer 2000]

32. Explain, giving block diagram, the operation of a dc link voltage source inverter for speed control of 3-phase induction motors.
[A.M.I.E. Sec. B. Electric Drives & Their Control Winter 1998]

33. What is 'pulse width modulation (PWM)' ? Explain the purpose. **[A.M.I.E. Sec. B. Electric Drives & Their Control Summer 2000]**

34. State the advantages of ac voltage control by PWM technique. Explain the sinusoidal PWM technique together with relevant waveforms.
[A.M.I.E. Sec. B. Electric Drives & Their Control Winter 1998]

35. What is meant by VVVF control of induction motors ? Explain with a block diagram and with reference to both constant torque and constant power modes of operation.
[A.M.I.E. Sec. B. Electric Drives & Their Control Winter 1999]

36. Draw the power circuit of a 3-phase current source inverter and explain its working. State also its inherent advantages over voltage source inverter for speed control of ac motor.
[A.M.I.E. Sec. B. Electric Drives & Their Control Winter 1998]

37. Distinguish between voltage source and current source inverters. Compare the two inverters for speed control of 3-phase induction motor.
[A.M.I.E. Sec. B. Electric Drives & Their Control Summer 1996]

38. Discuss the advantages and disadvantages of cycloconverter for speed control of 3-phase induction motors.
[A.M.I.E. Sec. B. Electric Drives & Their Control Summer 1996]

39. What is a cycloconverter ? Enumerate some of its industrial applications.
[A.M.I.E. Sec. B. Electric Drives & Their Control Summer 1995]

40. What is meant by "soft start" ? State and explain the soft start schemes (including electronic means, if any) employed for induction motors. **[A.M.I.E. Sec. B. Electric Drives and Their Control Winter 1999]**

41. Draw neat diagram for static rotor resistance control scheme using diode bridge and switch controlled resistor.
[A.M.I.E. Sec. B. Electric Drives & Their Control Summer 1998]

42. With neat circuit diagram, explain the speed control of a 3-phase slip-ring induction motor by static control of rotor resistance through chopper.
[A.M.I.E. Sec. B. Electric Drives & Their Control Summer 1996]

43. Discuss a static slip-power recovery scheme for speed control of slip-ring induction motor.
[A.M.I.E. Sec. B. Electric Drives & Their Control Winter 1998]

44. Explain the use of chopper for rotor resistance control in induction motors. Compare this scheme with slip-power recovery scheme, specially at lower speeds, stating the merits and demerits. Include circuits as needed.
[A.M.I.E. Sec. B. Electric Drives & Their Control Summer 1999]

45. Explain thyristor equivalent Kramer system of controlling speed of a slip-ring induction motor and regenerative operation. **[A.M.I.E. Sec. B. Electric Drives & Their Control Winter 1993]**

46. Explain any one slip power recovery scheme for the control of three-phase induction motor drive. Can supersynchronous speeds be achieved through it ? Explain.
[G.B. Technical Univ. Electric Drives 2010-11]

47. Explain the slip power recovery scheme of a three-phase induction motor control. What are its advantages and disadvantages ? **[U.P. Technical Univ. Electric Drives 2009-10]**

48. Explain working of static slip power recovery control scheme returning power to supply mains for speed control of three phase slip-ring induction motor. Why is supply power factor poor and how can it be improved ?
[U.P. Technical Univ. Electric Drives 2008-09]

49. Discuss static Scherbius scheme for speed control of a slip-ring induction motor. Draw a neat circuit diagram of the complete scheme. Mention one main advantage of this scheme compared to rotor resistance control.
[U.P. Technical Univ. Electric Drives 2004-05]

50. Discuss with the help of a suitable schematic diagram, the operation of a load commutated-fed synchronous motor drive. Why is this drive suitable for high speed and high power applications?
[A.M.I.E. Sec. B. Electric Drives & Their Control Winter 1998]

51. Explain the self controlled scheme for the synchronous motor control. **[G.B. Technical Univ. Electric Drives 2010-11]**

52. Describe self-controlled and load commutated inverter controlled synchronous motor drives in detail and compare them. **[U.P. Technical Univ. Electric Drives 2009-10]**

53. Explain static speed control of a single phase induction motor. **[U.P. Technical Univ. Electric Drives 2005-06]**

54. What are the similarities between a brushless dc motor and a conventional dc motor? Give the advantages of brushless dc motor over conventional dc motor.

55. Briefly explain the construction and principle of operation of a switched reluctance motor.
[G.B. Technical Univ. Electric Drives 2010-11]

56. Describe the construction and principle of operation of a switched reluctance motor.
[U.P. Technical Univ. Electric Drives 2009-10]

57. Describe working principle of switched reluctance motor. Discuss any one converter circuit used. What are its advantages over other ac motor drives.
[U.P. Technical Univ. Electric Drives 2005-06

58. Write short notes on the following :
(*i*) Half controlled converters.
[Madurai Kamaraj Univ. Nov. 1996]
(*ii*) Full controlled converters.
(*iii*) Dual converter drives.
(*iv*) Chopper drives.
(*v*) Comparison of converter fed and chopper fed dc drives.
(*vi*) Thyristor control of dc series motors.
(*vii*) Control circuits for ac drives.
[Bharathidasan Univ. Nov. 1996]
(*viii*) Cycloconverters.
(*ix*) Slip-power recovery scheme for speed control of slip-ring induction motor.
(*x*) Load commutated inverter drive for synchronous motor.
(*xi*) Brushless dc motor.
(*xii*) Switched reluctance motor.

Short Answer Type Questions With Answers

Q. 1. Why the thyristor control is preferred over Ward Leonard system of speed control ?

Ans. The thyristor control has some distinct advantages over Ward Leonard system of speed control such as low installation cost, ease of maintenance, higher accuracy, greater reliability, quick response, higher efficiency of operation, lesser floor space requirements, operation at a wide range of temperature etc.

Q. 2. What are freewheeling diodes ?
[Bharathiyar Univ. April 1997]

Ans. Freewheeling diode is a bypass or commutating diode. It is connected across motor terminals to allow for dissipation of energy stored in motor inductance and to provide for continuity of motor current when the thyristors are blocked. It also provides protection against transient overvoltages.

Q. 3. Why half-wave converter is not used for supply to the field circuit of a dc motor ?

Ans. If half-wave converter is used for supply to the field circuit of a dc motor, the high ripple content in the field circuit would cause increase in iron losses of the machine.

Q. 4. What is the relationship of V_{dc} and V_{ac} in terms of firing angle α for half-controlled and full-controlled converters?
[Madurai Kamaraj Univ. Nov. 1996]

Ans. For half-controlled converter,

$$V_{dc} = \frac{\sqrt{2}\,V_{ac}}{\pi}(1 + \cos\alpha) \text{ for } 0 \le \alpha \le \pi.$$

For full-controlled converter,

$$V_{dc} = \frac{2\sqrt{2}\,V_{ac}}{\pi}\cos\alpha \text{ for } 0 \le \alpha \le \pi.$$

Q. 5. What are the drawbacks of half-wave rectifier circuits ?

[Bharathiyar Univ. April 1997]

Ans. The output voltage wave of a half-wave rectifier is not a perfect dc. It is a discontinuous current and contains various harmonics. So half-wave converter drives have poor performance and their use is limited to small drives (rating up to 500 W).

Q. 6. What is the necessity of dc choke coil and freewheeling diode in a converter circuit ?

[Madurai Kamaraj Univ. Nov. 1996]

Ans. Freewheeling diode allows for dissipation of energy stored in the motor inductance and dc choke coil which reduces the rate of decay of current during the freewheeling operation when thyristors are blocked and thus motor current is made continuous. Discontinuous armature current results in deterioration of motor performance.

Q. 7. Why three-phase half-wave drives are generally not employed in industrial applications ?

Ans. Because of presence of dc component in supply currents.

Q. 8. What is a chopper ?

Ans. A chopper is a static switch to provide variable dc voltage from a source of constant dc voltage source.

Q. 9. Give the applications of dc choppers.

Ans. DC choppers are employed in trolley cars, marine hoists, forklift trucks, mine haulers and electric traction.

Q. 10. What is meant by duty ratio of a dc chopper ?

[Madras Univ. April 1997]

Ans. The ratio of on-period to the time period is known as duty ratio, *i.e.*,

$$\text{Duty ratio, } \alpha = \frac{T_{on}}{T} \text{ or } T_{on} \times \text{frequency or } \frac{T_{on}}{T_{on} + T_{off}}.$$

Q. 11. What are the advantages of chopper fed drive over converter fed drive ?

Ans. The advantages of chopper fed drive over converter fed drive are:

1. The ripple in the output of chopper is much less than that in the output of a converter. Lesser ripple means lesser losses.
2. The chances of discontinuous operation in chopper drive are lesser than that in converter drive. So speed regulation and transient response of chopper drive are better than that in converter drive.

Q. 12. Enumerate the control strategies used in dc choppers.

Ans. There are two types of control strategies for operating the switches that can be used in dc choppers. These control strategies are time ratio control (TRC) and current limit control.

Q. 13. What is TRC ? In how many ways can it be affected ?

Ans. TRC (time-ratio control) is a control strategy in which the value of $\frac{T_{on}}{T}$ is varied. This is affected in two ways namely variable frequency control (FM control) and constant frequency control (PWM).

Q. 14. What quantity is varied in PWM and frequency modulation ?

Ans. In pulse width modulation (PWM), the chopping period T is kept constant but the pulse width T_{on} is varied to control the output.

In frequency modulation the chopping frequency is varied and for this purpose either T_{on} or T_{off} is varied.

Q. 15. What is a step-up chopper ?

Ans. A step-up chopper, like a step-up transformer, raises the level of input dc voltage.

Q. 16. On what principle does a step-up chopper operate ?

Ans. A step-up chopper operates on the principle of storing energy in an inductance to provide output voltage higher than source voltage.

Q. 17. What is steady-state ripple ?

Ans. When chopper feeds R-L or R-L-E_b load, the current varies between I_{max} and I_{min}. The difference of I_{max} and I_{min} is known as steady-state ripple.

Q. 18. When is steady-state ripple maximum.

Ans. When $\alpha = 0.5$, steady-state ripple is maximum.

Q. 19. What factors can make the current discontinuous in a chopper feeding an R-L-E_b load ?

Ans. Higher L/R ratio, low duty cycle and higher back emf can make the current discontinuous.

Q. 20. The commutators of the thyristor controlled dc motors are made larger than those for ordinary dc motors. Why ?

Ans. The commutators of the thyristor controlled dc motors are made larger in order to provide extra insulation to withstand larger and rapid voltage fluctuations.

Q. 21. Semiconverter system is preferred for dc series motor drive. Why ?

Ans. The semiconverter system, because of its freewheeling action helps in maintaining continuous current and thus provides better performance to dc series motors in comparison to full converter system.

Q. 22. Why are ac regulators preferred over conventional methods of providing ac variable voltage of constant frequency for ac motor drives ?

Ans. AC regulators using thyristors are becoming popular because of high efficiency, fast control and compact size.

Q. 23. What is difference between on-off control and phase control in the case of ac voltage controller ?

[Madras Univ. October 1996]

Ans. In on-off control, also known as integral cycle control, the thyristors are used to connect the motor to the supply source for a certain number of cycles and then to disconnect it for another certain number of cycles; each of the on and off times consist of an integral number of cycles. On the other hand in phase control, the thyristors are employed as switches for connecting the motor to the supply for a certain portion of each cycle of the supply voltage.

Q. 24. What is the significance of maintaining the ratio of supply voltage to frequency constant in variable frequency drives of induction motors. **[U.P.T.U. Electric Drives 2007-08]**

Ans. If only frequency is changed and stator voltage is kept constant, the stator flux will not be at its rated value. The operation with flux below or above the rated value is not desirable.

The noteworthy point is that electrical machine is designed corresponding to near saturation point on the magnetization curve (or B-H curve). This is done from the point of view of full utilisation of the core. If the stator frequency is reduced, keeping the stator voltage constant, motor will operate in the saturation region and therefore, the motor will draw large magnetising current causing increase in stator core and copper losses and therefore, decrease in motor efficiency. However, if only supply frequency is increased keeping the stator voltage constant, the motor will operate at low flux density and thus motor capacity will be under utilised.

For constant flux operation, it is necessary that the induced emf varies linearly with applied frequency. At higher voltages and at high frequency operation stator voltage drops are very small and constant flux operation may be obtained by keeping V/*f* ratio constant.

Q. 25. State two advantages of current source inverter.

[Madurai Kamaraj Univ. Nov. 1996]

Ans. Current source inverter has the following two main advantages:

1. It can handle reactive or regenerative loads without freewheeling diodes.
2. Misfiring of devices and short circuits, because of constant input current, do not pose any problem.

Q. 26. Mention any two drawbacks of cycloconverters.

[Bharathiyar Univ. April 1997]

Ans. The two drawbacks of cycloconverters are :

1. It needs more power semiconductors than an inverter.
2. They can produce only a sub-frequency output.

Q. 27. Give some applications of cycloconverter.

[Madurai Kamaraj Univ. Nov. 1996]

Ans. Cycloconverter drives are normally used for large sized motors. They have also been employed in gearless cement mill or ball mill drives.

Q. 28. What is meant by GTO ? **[Madras Univ. April 1997]**

Ans. Gate-turn-off switch is a special thyristor which can be turned on by a positive gate signal and can be turned-off by a negative gate signal.

Problems

1. A single phase 230 V, 50 Hz supply feeds a separately excited dc motor through single-phase semiconverters. The firing angle of the field semiconverter is zero. The armature and field resistances are 0.1 Ω and 150 Ω respectively. The load torque is 150 N-m at 750 rpm. The voltage constant is 0.9 V/A radian/s and torque constant is 0.9 N-m/A^2. Determine (*i*) field current (*ii*) firing angle of converter in armature circuit and (*iii*) power factor of armature circuit converter. Assume continuous current operation and neglect losses. [**Ans.** (*i*) 1.38 A (*ii*) 96.6° (*iii*) 0.585 lagging]
2. A 230 V single phase ac supply is used to supply a separately excited dc motor through two full converters. The firing angle of converter in the field circuit is kept at zero. The field and armature resistances are 150 Ω and 0.15 Ω respectively. The firing angle of converter in the armature circuit is 30° and armature current is 80 A. The torque and voltage constants are 1.2 N-m/A^2 and 1.2 V/A-radian/second respectively. Determine torque developed and motor speed. [**Ans.** 132.48 N-m; 964.7 rpm]
3. A 220 V, 1500 rpm, 10 A, separately excited dc motor with an armature resistance of 3 Ω is fed from a single phase fully controlled rectifier connected to an ac source voltage of 230 V, 50 Hz. Assuming continuous conduction, calculate the firing angle for half the motor torque at 600 rpm. **[U.P. Technical Univ. Electric Drives 2007-08]** [**Ans.** 63.93°]
4. A 400 V 3-phase supply feeds a separately excited dc motor through two three-phase full converters, one for armature and the other for field. The firing angle of the field converter is zero. The armature and field resistances are 0.2 Ω and 200 Ω respectively. The armature current is equal to rated value of 60 A. The motor voltage constant is 1.05 V/A-rad/second. The speed is 1,000 rpm. Find firing angle of armature circuit converter. [**Ans.** 56.65°]
5. A 3-phase fully controlled bridge converter is used to control the speed of a 40 kW, 400 V, 1,000 rpm separately excited dc motor. The converter is fed from a 3-phase, 400 V, 50 Hz ac power supply. The induced emf constant is 0.3 V/rpm and the armature resistance is 0.5 Ω. Determine (*i*) rated torque (*ii*) firing angle when the motor operates at full load and rated speed. (*iii*) firing angle when the motor should operate at half load and at 500 rpm. **[U.P. Technical Univ. Electric Drives 2008-09]** [**Ans.** (*i*) 382 N-m (*ii*) 42.23° (*iii*) 68.27°]
6. A 230 V ac supply feeds the armature and field circuits of a separately excited dc motor through two full converters. The firing angle of the field converter is zero. The armature and field resistances are 0.5 Ω and 300 Ω respectively. The torque and voltage constants are 1.35 each. Determine motor speed and torque for armature current of 60 A and firing angle of armature circuit converter of 45°. [**Ans.** 1,173 rpm; 55.89 N-m]
7. A dc chopper is supplying a resistive load. If the input and output voltages are 200 V and 128 V and chopping frequency is 800 Hz, find the periods of conduction and blocking in each cycle. [**Ans.** 0.8 × 10^{-3} s and 0.45 × 10^{-3} s]
8. A chopper circuit is operating on TRC principle at a frequency of 4 kHz on 200 V dc supply. If the load voltage is 160 V, compute the conducting and blocking period of thyristor in each cycle. [**Ans.** 0.2 ms and 0.05 ms]
9. A step-up chopper has an ac input voltage of 100 V and an output voltage of 250 V. The period of blocking in each cycle is 0.6 ms. Find period of conduction and chopping frequency. [**Ans.** 0.36 ms; 1.6667 kHz]
10. A step-up chopper has input voltage of 200 V and output voltage of 600 V. If the conducting period of chopper is 50 ms, compute the pulse width of output voltage. In case output voltage pulse width is halved for constant frequency operation, determine the average value of new output voltage. [**Ans.** 25 ms; 1,200 V]
11. A separately excited dc motor is supplied from a 500 V source through a chopper. Armature current is 200 A and armature resistance is 0.1 Ω. The voltage and torque constants are 1.4 V/A rad/s and 1.4 Nm/A^2. The field current is 2.5 A. The duty cycle of the chopper is 0.5. Determine (*i*) input power (*ii*) speed and (*iii*) torque. [**Ans.** (*i*) 50 kW (*ii*) 627.5 rpm (*iii*) 700 N-m]
12. The chopper used for control of a dc separately excited motor has supply voltage of 220 V, on-period is 10 ms and off-period is 15 ms. Determine the average load current when the speed of the motor is 1,200 rpm and has a voltage constant of 0.5 V/rad per second. The armature resistance is 2.5 Ω. [**Ans.** 10.07 A]
13. A one quadrant chopper is used for rheostatic braking of a separately excited dc motor. Armature and braking resistances are 0.2 Ω and 6 Ω respectively. Armature and field currents are 160 A and 1.9 A respectively. Duty cycle of chopper is 0.4 and voltage constant is 1.1 V/A-rad/second. Determine (*i*) average voltage across chopper (*ii*) speed of motor and (*iii*) power dissipated in braking resistance. [**Ans.** (*i*) 576 V (*ii*) 2,778 rpm (*iii*) 92.16 kW]
14. A dc series motor is coupled to a load whose torque varies as the square of speed. The motor is supplied through a semiconverter with input voltage of 700 V. For firing angle of 90° the armature current is 25 A and motor speed is 400 rpm. Determine the firing angle and armature current if the motor speed is to be raised to 500 rpm. Assume unsaturated magnetic circuit and the combined resistance of armature and field to be 0.4 Ω. [**Ans.** 56.46°; 31.25 A]

❑❑❑

CHAPTER 4

Motors For Particular Services

INSIDE THIS CHAPTER

4.1. INTRODUCTION

We have so far discussed dynamics of electrical drives, electrical characteristics (operating or running characteristics, starting characteristics, speed control and braking characteristics), mechanical features, service capacity and ratings of different types of electrical motors available in the market. In this chapter we will discuss the requirement of individual application and suggest possible choice of motor for particular application, keeping in mind its requirement. Space does not permit of a discussion of more than a few of applications, so only the important applications will be included here for discussion.

4.2. TYPES OF MOTORS USED IN INDUSTRIAL DRIVES

While selecting a motor for a particular service the following points must be known about it.

(*i*) Speed-torque characteristics—whether they are series, shunt or constant.

(*ii*) The power factor at which the motor normally operates.

(*iii*) Speed control range.

(*iv*) Starting torque and starting current.

(*v*) Maintenance cost—whether low, medium or high.

(*vi*) Initial cost.

The characteristics including all the above six points and applications of various types of motors in common use are given in Table 4.1.

4.3. APPLICATIONS OF ELECTRIC DRIVE

1. ***Domestic Uses.*** Requirements are low torque, nearly constant speed, frequent start and comparatively small load duration. Small universal motors, usually series type, are used for domestic uses such as for driving sewing machines, vacuum cleaners, mixers, electric shavers, hair dryers, small fans, cloth-washing machines etc. Single phase capacitor-start induction motor with direct-on-line start is used for domestic refrigerators, window air-conditioners etc.

2. ***Machine Tool Drives.*** Many machine tools require a simple more or less constant speed drive. For this squirrel-cage induction motors together with simple manual controls are adequate. For transmission of power to cutting tools or work pieces at different finite number of speeds, we use gear boxes and stepped pulleys. The use of gear mechanism gives rise to vibrations and noise, and thus affects the machining accuracy. Stepless speed control, if used, will provide better machine timings and surface finishes. Electro-hydraulic and electromagnetic controls are used for providing desired smooth speed control. Though thyristor controlled drives have replaced these methods up to some extent. At present, in our country, the use of thyristorized drives is limited to special machine tools only, as the cost of such drives is sufficiently high.

In some cases pole changing is preferred over gear changing. If more than two speeds are required, two pole changing windings may be used in the same stator slots, *e.g.*, 4/8 and 6/12 poles providing speeds in the ratio of 1 : 1.5 : 2 : 3. Such motor is convenient for small vertical drilling machines.

In some cases, speeds in excess of 3,000 rpm are required as for example in woodworking machinery. A direct drive is preferred over a geared drive, but with a 50 Hz supply this can be achieved only by commutator machines. If there are several such drives, it is worthwhile to install an induction type frequency-changer to provide high frequencies (100 Hz and/or 150 Hz). This permits the relatively cheap and robust squirrel cage motors to be employed for speeds approaching 6,000 and 9,000 rpm respectively.

Other drives need a variable speed. For a wider speed range the Ward-Leonard system is supreme but now it has been replaced by a cheaper alternative such as a rectifier-dc motor combination. For a smaller speed range, field-controlled

Table 4.1

Type of Motor	*Characteristics and Costs*	*Applications*
1. DC Motors		
(*a*) Series wound	(*i*) Very high starting torque (3 times full-load torque with twice full-load current at start). (*ii*) Torque varies nearly as the square of the current. (*iii*) Varying speed; dangerously high at no load. (*iv*) Self adjustable speed–decrease in speed with increase in load. (*v*) Speed control possible, maximum range of 3 : 1. (*vi*) Maintenance cost — medium. (*vii*) Relative cost — 1.5.	For heavy duty applications such as electric railways, rolling mills; metallurgical works, mine hoists, continuous conveyors, cranes, valve operation etc.
(*b*) Shunt Wound	(*i*) Nearly constant speed. (*ii*) Torque varies nearly as the current. (*iii*) Medium starting torque (twice full-load torque with twice full-load current at start). (*iv*) (*a*) Continuous speed range of 4 : 1 maximum. (*b*) Continuous speed range of 100 : 1 and more by means of Ward Leonard control. (*v*) Maintenance cost — medium. (*vi*) Relative cost — 1.5 to 3.0.	For driving constant speed line shafts, lathes, vacuum cleaners, pressure blowers, constant head centrifugal pumps, compressors, reciprocating pumps, fans, wood-working machines such as circular saws, wood planners, laundry washing machines, milling machines, passenger elevators, continuous conveyors, grinders, polishers, small printing presses, paper making machines, metal cutting machines etc.
(*c*) Cumulatively Compound Wound	(*i*) High starting torque (2 to 3 times full-load torque with twice full-load current at start). (*ii*) Varying speed but within limits. (*iii*) Self adjustable speed with changing load. (*iv*) Speed control possible 2 to 1 range. (*v*) Maintenance cost — medium. (*vi*) Relative cost — 1.5 to 3.0.	For driving compressors, stamping machines, pressure blowers, variable head centrifugal pumps, door lifts, rotary presses, flat bed presses, circular saws, metal planner, shearing machines, punch presses, rolling mills, freight elevators, passenger elevators, continuous conveyors, hydro-extractors etc.
(*d*) Differentially Compound Wound	(*i*) Low starting torque. (*ii*) Constant speed but dangerously high at no load. (*iii*) Maintenance cost — medium. (*iv*) Relative cost — 1.5 to 3.0	Seldom used.
2. AC Motors (*a*) *Three-Phase Motors* (*i*) *Synchronous Motors*		
Plain Synchronous Motor	(*i*) Constant speed at varying load. (*ii*) No self-starting torque. (*iii*) Speed control—not possible. (*iv*) Capacity of working over wide range of power factor both leading and lagging. (*v*) Maintenance cost — medium. (*vi*) Relative cost — 2.	(*i*) For driving continuously operating equipment at constant speed such as ammonia and air compressors, motor-generator sets etc. (*ii*) For main drives of continuous rolling mills, medium and large drives in the paper and cement industries. (*iii*) For voltage control at the end of transmission line. (*iv*) For improving power factor like capacitors in large power systems in which considerable lagging power factor equipment is installed.
Synchronous Induction Motor.	(*i*) Constant speed. (*ii*) Twice full-load torque with twice full-load current at start. (*iii*) Capability of working over wide range of power factor both leading and lagging. (*iv*) Speed control — not possible. (*v*) Maintenance cost — medium. (*vi*) Relative cost — 2.	It is used when some starting torque is to be developed but load torque remains fairly steady, such as for large fans, compressors, line shafts, pumps etc.

Type of Motor	*Characteristics and Costs*	*Applications*
(*ii*) *Induction Motors*		
Squirrel Cage Type	(*i*) Nearly constant speed. (*ii*) Poor starting torque (full-load torque with five times full-load current at start). (*iii*) High overload capacity. (*iv*) Simple starting arrangement. (*v*) Speed control — not possible. (*vi*) Power factor — 0.75 to 0.9 (lagging). (*vii*) Maintenance cost — very low. (*viii*) Relative cost —unity.	For low and medium power drives where speed control is not required such as for water pumps, tube wells, lathes, drills, line shafts, circular saws, grinders, polishers, wood planers, compressors, laundry washing machines, fans, blowers etc.
Double Squirrel Cage Type	(*i*) Nearly constant speed. (*ii*) High starting torque with low starting current (2 to 2.5 times of full-load torque with a starting current of 4 to 6 times of full-load current). (*iii*) High overload capacity. (*iv*) No need of starting equipment. (*v*) Speed control — not possible. (*vi*) Power factor — 0.75 to 0.9 (lagging). (*vii*) Maintenance cost — very low. (*viii*) Relative cost — 1.5.	For driving loads requiring high starting torque such as for compressors, pumps, large refrigerators, crushers, conveyor equipment, boring mills, woodworking equipment, textile machinery, cranes, punches, lathes etc.
Multiple-Speed Motor	(*i*) Nearly constant speed for a given setting. (*ii*) Poor starting torque (full-load torque with a starting current of 5 times of full-load current). (*iii*) Power factor — 0.75 to 0.9 (lagging). (*iv*) Speed control by changing number of poles of stator winding; speed range 6 : 1; no control between steps; maximum 1/6th of top speed. (*v*) Maintenance cost — very low. (*vi*) Relative cost — 1.5 to 3.0.	For applications requiring several definite speeds but not necessarily a continuously adjustable speed such as for ventilating fans, conveyors, machine tools, elevators, traction etc.
Slip-Ring Type or Wound Rotor Type	(*i*) Nearly constant speed. (*ii*) High starting torque (twice full-load torque with a starting current of twice full-load current). (*iii*) High overload capacity. (*iv*) Speed control can be affected as to 50-75%. (*v*) Lower efficiency. (*vi*) Power factor — 0.75 to 0.9 (lagging). (*vii*) Maintenance cost — low. (*viii*) Relative cost — 1.5.	For industrial drives of high power where high starting torque and speed control are required such as for driving line shafts, lifts, pumps, generators, winding machines, auxiliary fans, smoke exhausters, printing presses, elevators, compressors etc.
(*iii*) *Commutator Motors*		
3-Phase Series Motor	(*i*) Speed torque characteristic—series. (*ii*) High starting torque (2.5 times full-load torque with a starting current of twice full-load current). (*iii*) Power factor — 0.9 to 1.0 (lagging). (*iv*) Speed control over the full range. (*v*) Maintenance cost — high. (*vi*) Relative cost — 2 to 3.	For drives where load increases with the speed such as fans, centrifugal pumps and compressors or where large starting torque is required such as for haulage and hoist work.
Schrage Motor	(*i*) Nearly constant speed for a setting. (*ii*) High starting torque (2.5 times full-load torque with a starting current of twice full-load current). (*iii*) Power factor — 0.7 to 1.0. (*iv*) Speed control range 3:1. (*v*) Maintenance cost — high. (*vi*) Relative cost — 3 to 4.	For drives where adjustable speed is required such as for driving paper making machines, printing presses, calenders and textile works. It is also employed for lifts, pumps, machine tools, belt conveyors etc. because of its high starting torque and convenient speed control.

Type of Motor	Characteristics and Costs	Applications
(b) Single Phase Motors		
(i) Synchronous Motor	*(i)* Constant speed. *(ii)* Power factor — 0.9 to 1.0. *(iii)* Speed control — not possible. *(iv)* Maintenance cost — medium. *(v)* Relative cost — 2.5.	For applications requiring constant speed but very low torque such as teleprinters, clocks, control apparatus, all kinds of timing devices, recording instruments, sound recording and reproducing apparatus.
(ii) Commutator Motors		
Single Phase Series Motor	*(i)* High starting torque (3 times full-load torque with a starting current of twice full-load current). *(ii)* Torque varies nearly as the square of the current. *(iii)* The speed varies inversely as the current nearly. *(iv)* Speed control — full range with transformer. *(v)* Power factor — 0.8 to 0.9 (lagging). *(vi)* Cost of maintenance — high. *(vii)* Relative cost — 2.5.	In small size it is used in driving small domestic appliances like vacuum cleaners, refrigerators etc. In large size it is employed for traction work.
Repulsion Motor	*(i)* High starting torque (3 times full-load torque with a starting current of twice full-load current). *(ii)* High speed at light load. *(iii)* Speed control possible by shifting of brushes–full range. *(iv)* Power factor–0.8 to 0.9 (lagging). *(v)* Maintenance cost–high. *(vi)* Relative cost–2.5.	For services requiring large starting torque and adjustable but constant speed as in coil winder.
Repulsion Induction Motor	*(i)* Constant speed. *(ii)* High starting torque (2.5 to 3 times full-load torque with a starting current of 3 to 4 times full-load current). *(iii)* Power factor — 0.85 to 0.9 (lagging). *(iv)* Speed control — not possible. *(v)* Maintenance cost — medium. *(vi)* Relative cost — 2.	For applications requiring high starting torque with an essential constant running speed such as for driving compressors, lifts etc.
(iii) Induction Motors		
Capacitor-Start Type	*(i)* High starting torque (1.5 times full-load torque with a starting current of twice full-load current). *(ii)* Constant running speed. *(iii)* Power factor — 0.7 to 0.9 (lagging). *(iv)* Speed control — not possible. *(v)* Maintenance cost — low. *(vi)* Relative cost — 1.5.	For compressors, refrigerators, and other applications involving high starting torque such as in small portable hoists.
Capacitor-Start, Capacitor-Run Type	*(i)* High starting torque (1.5 times full-load torque with a starting current of twice full-load current). *(ii)* Constant speed. *(iii)* Quiet in operation. *(iv)* Speed control — not possible. *(v)* High efficiency. *(vi)* Power factor — 0.9 to 1.0 (lagging). *(vii)* Maintenance cost — low. *(viii)* Relative cost — 2.0.	For applications requiring quiet operating motor and replacing repulsion start motor.
Repulsion-Start Type	*(i)* High starting torque (2.5 to 4.0 times of full-load torque with a starting current of 3.5 times full-load current). *(ii)* Constant running speed. *(iii)* Speed control — not possible. *(iv)* Power factor — 0.8 to 0.9 (lagging). *(v)* Maintenance cost — medium. *(vi)* Relative cost — 2.	For refrigerators, compressors, and other drives particularly those which have a high inertia and prolonged starting period.

dc shunt motor is adequate, and if dc supply is not available a free-firing rectifier may be employed. AC commutator motors with induction regulators or brush-shifting control of speed may also be employed for variable-speed drives.

For use such as internal grinding spindles for horological applications, drilling of printed circuit boards etc., high frequency, three-phase induction motors operating up to speeds of 180,000 rpm and providing fairly high output powers in small sizes are used. Typical ratings are 1.2 kW at 120,000 rpm for high-speed grinder drives and 0.5 kW at 150,000 rpm for high precision horological applications.

Some applications need braking torques. Almost all the methods described in chapter 1 are applicable. Many motors associated with machine-tool drives are of sufficiently small rating to allow capacitors to be employed for braking purposes without the size and cost being prohibitive. Where considerable inertia is involved, as in planing-machine drives, the regenerative braking of the Ward-Leonard set is a further advantage in addition to its others.

3. *Cranes and Hoist Work.* Essential requirement of a crane or hoist motor is that it should develop high starting torque and should withstand large number of switching operations. DC series or compound wound motors are preferred for cranes on account of their high starting torque and smooth speed control. The motors used for lifting, travelling and reversing, and for conveying and hoisting are dc compound wound, dc series wound and ac slip-ring induction motors (with rotor resistance control) respectively.

Hoist motors are provided with special electromechanical brakes, which by means of springs, are made to hold the load as soon as supply fails. On restoration of supply, solenoid connected across the terminals of motor is energized and brakes are released. Such motors are half-, or one-hour rated, developing starting torque twice of full-load torque, and robust to withstand severe strains to which these are exposed.

Drum-type controllers are used, operated by contactors, which are controlled by press buttons.

4. *Lifts.* High smooth accelerating torque (twice full-load torque at start), high overload capacity and pull-out torque, high degree of silence and moderate speed are essential requirements. Motors are normally one-hour rated for duty cycle of 150 to 200 starts per hour. DC compound wound and ac slip-ring induction motors are mostly used. Induction repulsion and variable speed ac commutator motors are also used. In single phase installation a shunt type commutator motor is also used. In case of dc motors speed control is affected by shunt field variation or by variable voltage control (Ward Leonard). In case of polyphase slip-ring induction motor the speed is controlled by rotor resistance and involves copper loss (I^2R loss).

5. *Lathes, Milling and Grinding Machines.* Lathes are usually driven by constant speed squirrel cage induction motors. Multi-speed ac motors or adjustable speed dc motors are sometimes used.

Milling machines are usually driven by constant speed squirrel cage induction motors. Larger machines such as planer type milling machines have separate motors for each milling head and separate motors for each of feed motions. The general practice is to use adjustable speed motors for headstock drive and either constant or adjustable-speed motor for the wheel drive. Separate motors are often used to feed the wheel with respect to the work.

Grinding machinery varies considerably in the design of individual machines. The general practice is to employ adjustable speed dc motors for the headstock drive, constant speed squirrel cage or adjustable speed dc motors for wheel drive and constant speed squirrel cage motors for traverse drive.

Constant speed squirrel cage induction motors with the grinding wheels mounted directly on the motor shaft extension are used for bench, pedestal and centreless grinders.

6. *Planers.* Planer consists of a bed having attached to it a platen which moves forwards and backwards on the bed. The job is clamped to the platen and is planed by a stationary clamped tool; at the end of the stroke the tool moves slightly in order to make a new cut. Cutting stroke is slow and return stroke is quick, therefore adjustable speed, reversing motors are required. In case of dc supply shunt or compound wound motor and in case of ac supply slip-ring induction motor is used.

7. *Punches, Presses and Shears.* Special features are high peak loads and heavy starting torque. These are usually equipped with flywheels to supply momentarily demand of power. Motors used for these machines should have drooping speed-torque characteristics to allow the motor to slow down when heavy loads are encountered. High-slip squirrel cage induction motors, slip-ring induction motors or dc cumulative compound motors are used for such applications.

8. *Shapers and Slotters.* Constant speed squirrel cage induction motors are used for shapers and slotters.

9. *Blowers and Fans.* Squirrel cage induction motors and synchronous motors are used for driving blowers and fans.

10. *Frequency Converters.* Squirrel cage induction motors and synchronous motors are used for driving converters. Nowadays owing to the recent advances in the field of silicon controlled rectifiers, static frequency changers are being increasingly used.

11. *Drilling Machines.* Constant speed squirrel cage or dc shunt or multi-speed ac motors are used for this purpose.

12. *Air Compressors.* Slip-ring induction motors and synchronous motors are usually employed in large size units. The latter type can be used for power factor correction in addition to driving air compressors if so designed. Squirrel cage induction motors are used for small compressors only.

13. *Electric Traction.* DC series motors, which are simple and robust in construction, provide high starting torque and smooth speed control, are suitable for all types of services but more particularly for suburban services where high rate of acceleration is essential.

Single phase ac compensated series motors, which have speed-current and speed-torque characteristics similar to those of dc series motors, are extensively used for the main line work. Due to poor starting torque these motors are not suitable for suburban services where stops are frequent.

14. *Pumps.* Drip proof or totally enclosed surface cooled polyphase induction motors are usually employed, often mounted on a common bed plate and direct coupled to the pump. Sometimes motor is flange mounted. In case the speed does not come within the range of the fixed motor speed, V-belt drive is employed. Squirrel cage motors with reduced voltage starters are used for centrifugal pumps, where starting torque required is about 40 to 50 per cent of full-load torque. Slip-ring induction motors are used for reciprocating pumps in which starting torque required may be 100 to 200% of full-load torque.

15. *Refrigeration and Air Conditioning.* In vapour compression system of refrigeration, motor employed to drive the compressor is thermostat controlled. On re-starting the motor has to drive compressor against high head pressure. Motor used, therefore, should be capable of developing starting torque 2 to 2.5 times full-load torque.

Single phase, 230 V, capacitor-type induction motors with D-O-L starters are usually employed for small units. For large installations high torque squirrel cage induction motors or slip-ring induction motors are employed. For very large plants, synchronous motor, driving turbo-compressor, may suit specially if power factor correction is also required.

16. *Breweries.* Totally enclosed surface cooled type motors are used because working atmosphere is damped one.

17. *Excavators.* Totally enclosed motors are used because of arduous and rugged work.

18. *Belt Conveyors.* Belt conveyors are used for moving sand and gravel. Heavy loads are required to be accelerated. Normal starting current and high starting torque double cage induction motor with direct-on-line starter or wound rotor induction motors are used. These motors must be totally enclosed, surface cooled type as grit and dust remain present in the atmosphere.

19. *Flour Mills.* Pipe-ventilated or totally enclosed type motors are used. Speed of the motor varies from 600-1,000 rpm.

20. *Quarrying.* Totally enclosed surface cooled type motors are used. Such motors must be of robust construction.

21. *Woodworking Machinery.* Screen-protected motors, except where there are chances of motor being burried in saw dust are usually used. In places where there are chances of motor being buried in sawdust totally enclosed surface cooled motor is used. Requirement of speed above 3,000 rpm, if required, can be met with induction motors in conjunction with induction type frequency converter or by using commutator motors.

22. *Printing Machinery.* In printing machinery for constant speed work such as guillotines, driving platens or other small machines squirrel cage induction motors are used. Where line capacity and moderate starting torque permits single phase capacitor motors may be used. In case of rolling presses where a slow forward stroke and a quick return stroke is required, two squirrel cage motors–one for each stroke, are used. For other purposes; where variable speed motors are required such as for rotary presses, dc compound or three-phase slip-ring induction motors with rotor resistance control or ac commutator motors may be used. For obtaining the steady crawling speed for inching the press, a pony or barring motor is used.

23. *Petrochemical Industries.* In petroleum and chemical industries, fluid handling equipment is extensively used. Pumps have been driven by induction motors with flow control, when desired, accomplished by throttling valves. As per estimates 30% or more of electrical energy used by a typical refinery is wasted due to throttling effect of control valves. However, adjustable-speed pumping utilizing multi-speed induction motors or inverter-fed induction motors are quite economical because of reduced energy consumption and maintenance costs.

Other important tasks within the production processes of the petrochemical industry are gas liquefaction, compression, refrigeration and heat recovery. Compressor drive systems constitute important components in plants for such processes. The majority of the drives operate at a constant speed, employing a 4- or 6-pole motor with step-up gear. For large ratings, synchronous motors with cylindrical rotors having both excitation and starting windings (damper bars) are preferred. Such motors are specially designed considering the oscillating torques developed during asynchronous acceleration owing to the magnetic and electrical anisotropy in the rotor and special cooling conditions.

24. *Sugar Mills.* In sugar mills, a centrifuge is employed for separating out crystalized sugar from the syrup obtained from steam evaporator and for drying it out by the action of centrifugal force. The duty cycle of a centrifuge motor is illustrated in Fig. 4.1. The functions involved are charging, intermediate spinning, spinning, regenerative and reverse current

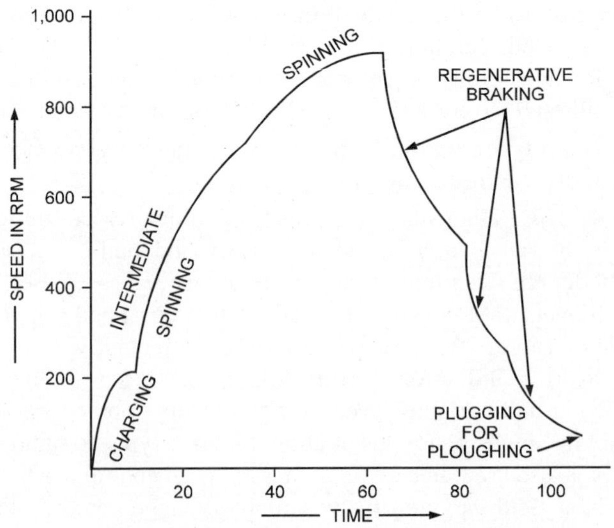

Fig. 4.1. ***Duty Cycle of a Sugar Centrifuge***

braking and plugging for ploughing. All these functions are to be performed at different speeds, variation of which may be as much as 1 : 30. The motors used for this purpose are usually four-speed, pole-changing motors having two sets of stator windings that enable us to obtain synchronous speeds 1,500/750/214/107 rpm or 1,000/500/214/107 rpm. They not only are capable of providing the desired fixed speeds of operation, but also of returning a portion of energy back to the supply mains during regenerative braking accomplished by switching over to a higher pole operation from a lower pole one.

In order to satisfy the duly cycle illustrated in Fig. 4.1, first 28 pole winding is switched on to provide a speed of approximately 200 rpm. During charging supply is cut off and then intermediate spin speed of about 450 rpm is obtained by energizing the 12 pole winding. The final spin speed of about 950 rpm is achieved by switching over to the 6 pole winding. After centrifuging operation is complete, supersynchronous regenerative braking is applied by connecting the motor for 12, 28 and 56 poles successively which brings the speed down to 500, 214 and 107 rpm respectively. Ploughing speed of about 50 rpm is obtained by the application of reverse current braking. Automatic control of the entire duty cycle is achieved by feeding the control equipment from the output of a tachogenerator directly coupled to centrifuge motor.

Motors employed for driving centrifuges have their own special features in construction. They are vertically mounted so as to be coupled with the centrifuge shaft. The motors may be made with larger air gap so as to take care of the possible rotor oscillations about the vertical axis. Insulation used in motors must be humid proof so that it can operate successfully in a humid environment. For protection of motor against overheating, thermal elements are embedded in the windings. These elements, which are also called sensotherms by trade name, operate few degrees below the maximum permissible temperature of the winding. Operation of thermal elements will either directly trip the motor or give visible and audible warning signals so that the particular duty cycle may be completed. Following this, next cycle cannot be commenced until temperature of the motor has come down to normal.

25. *Cement Mills*. The driving motors used in cement works can be broadly classified as (*i*) raw mill and cement mill drives (*ii*) kiln drives (*iii*) crusher drives (*iv*) waste gas fan drives and (*v*) compressor drives etc.

The starting torque for the mill motors for large cement plants is limited to 125% of the full-load torque and the pull-out torque is restricted to nearly 240% of the full-load torque. Slip-ring induction motors of 6.6 kV with liquid resistance starters are usually employed. Gearboxes are employed in order to get the desired mill speed of about 15 rpm and high voltage capacitors of adequate reliability and automatic capacitor control switchgear and circuit breakers are used for power factor correction. Owing to large ratings (above 3,000 kW) required for the raw and cement mill drives and due to the limitations in the availability of large size gearboxes and motors, twin drives are used–the two motors employed for twin drives have to be more or less identical to each other and so also their liquid resistance starters.

The rotary kiln is an indispensable part of the cement plant. The rating of the motors employed for driving the kilns vary from 100–1,000 kW. The maximum speed of the kiln is about 1 rpm and speed range required is of the order of 1 : 10. The starting torque required may be between 200% and 250% of rated torque. In early years, variable speed ac commutator motors to be employed for kiln drives, which has been superseded by Ward-Leonard drives as ac commutator motors proved expensive and had maintenance problems. Nowadays use of dc motors with static supply is made. To cope up with increasing kiln capacities, the modern trend is to employ twin motor dc drives for kiln application.

Usually, the starting torque for crusher drives is limited to 160% of the full-load torque and pull-out torque is limited to 200–250% of the full-load torque. The slip-ring induction motors are usually used in crushers. Normally the motors are designed to withstand locked rotor current during running, without any external resistance inserted in the rotor circuit, for one minute–very important feature owing to frequent chances of getting crushers jammed.

The starting torque required for fan drives is about 120% of the full-load torque and the pull-out torque is from 200% to 250% of the full-load torque. The slip-ring induction motors with a speed variation are used for such drives. For starting and controlling the speed, the cast iron grid resistance controllers are normally used. The slip-rings and brush gears are totally enclosed and kept external to the motor enclosures for convenience in maintenance.

The air compressor motors are usually of smaller ratings, say 300-450 kW. Squirrel cage or wound rotor induction motors can be used as per power needs. The enclosures are usually of TEFC type and the speeds vary between 1,000 and 750 rpm.

26. *Mining Work*. The electric motors employed in coal mines can be classified into two categories namely the auxiliary motors, used for driving auxiliaries such as compressors, fans, conveyors, pumps, hoists etc., and the mine face motors, used on continuous miners, drills, shuttle cars, cutting machines, loaders etc. The motors of the two categories differ considerably from each other. The auxiliary motors are normally modifications of general purpose industrial motors, while the face motors are specially designed motors as per specific needs. The duty of the auxiliary motors is, often, well defined and steady, while that of face motors consists of random loading and has a number of high-shock loads.

The motors employed for coal drills are usually short-time rated induction motors of rating 1 kW, 125 V, 2860 rpm, 50 Hz with a flame proof enclosure. Sometimes, high frequency motors (150 or 200 Hz) with the help of a frequency changer housed in the drill panel are also used.

Slip-ring induction motors with water rheostats are, usually employed for driving the colliery winder used for raising the coal, raising or lowering the persons and other loads. DC rheostatic braking is used initially and mechanical braking is employed only towards the fag end of the winding period. For higher capacities dc motors supplied from a Ward-Leonard generator are used for driving the mine winders as precise control of drum speed and position is necessary for accurate decking. The recent trends are to use thyristor convertors for supplying variable dc voltage supply.

The motors employed for haulage (mine-winches) work are subjected to frequent starting and stopping and also to braking and reversing. For this work the loads, often, require different speeds at different states of haulage. Hence slip-ring induction motors with drum resistance controllers are used. The drum controller and the rotor resistances are separately mounted in flame proof enclosures.

There are two types of ventilation fans used in mines namely axial flow type and radial flow type. Although squirrel cage induction motors having large number of poles can be used directly but high speed motors having 4 to 6 poles with a reduction gear are usually employed.

There are two types of colliery pumps–triple ram type driven by a high speed motor with a gear and the centrifugal type directly coupled to the driving motor. The ram type pump requires a starting torque of about 200% of rated torque, and, therefore wound rotor motors are used. For driving centrifugal pump, which requires only about 40% of rated torque squirrel cage induction motors are used.

Working surroundings is the chief factor which distinguishes electrical motors used in mining from those used in general industry. The motors employed in mining work must be flame proof, and all commutators and slip-rings must be in flame proof chambers with an inch metal to metal flange.

27. *Textile Mills.* In early years the textile mills were usually equipped with diesel generators, steam turbines. The aspect of boiler, engine etc. made it essential to consider the electric drive. Then the ac power supply was preferred. This was on account of power cost to drive the mill. Thereafter, the group drive of machines were designed, but with the recent developments, the need for individual drive was recognised.

The standard motors in open type or in totally enclosed type cannot be used in a textile mills but specially designed motors are required on account of environmental (presence of fluff and dust fly, high humidity, restricted ventilation etc.), operating (wide voltage fluctuations, frequent starting and stopping, intermittent operation resulting in rapid change in torque and power requirement) and drive (restricted space, special mountings and special shaft extension requirement) conditions.

The special features of textile motors are :

Electrical : Extra high or moderate torque with smooth acceleration characteristics.

Thermal : Normally high thermal withstand capacity for frequent starting, prolonged starting time or restricted ventilation or inching.

Environmental : Ability to withstand high humidity, temperature, dust and fluff.

Mounting : Foot, Cradle and Flat Base.

Construction : Totally enclosed with axial or circumferential ribs with steam-lined shapes for smooth floor for fluff.

Cooling : Surface cooled or fan cooled.

Miscellaneous : Terminal box or loose leads and special shapes for restricted overall dimensions, special shaft extensions.

Loom Motors. In order to achieve a short pick-up time, the starting torque of the loom motor should be high (2–2.5 times of the rated torque). The loom machine operates on a reciprocating mechanism, which causes both torque and current pulsations. Also the loom motor is subjected to frequent starts and stops. This results in a high temperature rise and is taken care by having a liberal thermal dissipation capacity of the motor. These motors are totally enclosed surface cooled. A high degree of fluff in the atmosphere calls for a smooth surface finish of the housing and shields so that the fluff do not get accumulated on the motor surface. The insulation of the motor must be able to withstand high moisture content. These motors have normally a higher slip or lower speed than that of a standard motor in order to give flywheel effect to smoothen the current pulsations. The ratings of the motors employed for driving looms for light fabrics such as silk, rayon, cotton, nylon etc. are 0.37, 0.55, 0.75, 1.1 and 1.5 kW; while those of the motors used for making heavy fabrics (wool and canvas) are 2.2 and 3.7 kW. The motors are usually of 6 or 8 poles.

Carding Motors. The inertia of the carding drum is high, as such the starting of such motors have to be designed to withstand prolonged starting time. Once the drum is started, the operation is continuous and uninterrupting. The commonly used motors are totally enclosed and totally enclosed fan cooled, 3-ϕ, high torque squirrel cage induction motors of rating 1.1 and 1.5 kW for light fabrics and 2.2, 3, 3.7 and 5.5 kW for heavy fabrics. Here again, the motors are usually of 6 or 8 poles.

Spinning Motors. In spinning, the moderate starting torque and the smooth acceleration is essential. This calls for a low dip in the speed-torque curve of the motor. Totally enclosed fan cooled motors of ratings 5–30 kW are used for spinning frame operation. In general, three types of drives are employed for spinning frame operation namely single-speed motor, two-speed motor and two-motor drives. Usually a 4 pole or 6 pole squirrel cage induction motor is employed for single-speed drive. Two-speed motors (4/6 or 6/8 poles) are used in order to give maximum production with minimum breakage. The increased production may compensate for the additional outlay owing to larger size and higher cost. In case of two-motor drive, two separate motors are employed to drive the common pulley of the ring frame. Although this drive is costlier and requires more space, it has several advantages such as (*i*) availability of any desired speed by adjustment of speed ratios, (*iii*) facility of adjustment of yarn tension independently, (*ii*) continued production even in case of failure of one of the motors.

The most common electrical method of controlled-torque starting involve the use of standard squirrel cage induction motors and different methods of applying reduced voltage to the motors during a selected starting period. One of the most effective methods of controlling the stator voltage of motors employed in textile mills is that of using series reactors. This method provides stepless, closed transition increase to approximately full speed. Variable iron-core inductors offer almost infinite choices of starting torques. Reactors with fixed tappings may be used to provide variety of starting torques. With the ever increasing use of solid state devices, nowadays ac regulators are being used to provide control of starting torque.

28. *Woollen Mills*. Loom motors are totally enclosed type, capable of developing 3 times full-load torque at start if direct driven or alternatively started light and controlled through clutch. Motors including switchgear in dyeing section should be of totally enclosed type, and should be treated with acid proof paint as atmosphere is laden with fumes.

29. *Paper Mills*. There are two main processes, that take place in a paper mill namely pulp making and paper making. The drives used are quite different for each one of these processes.

There are two methods of pulp making–purely by mechanical means and by both mechanical and chemical processing. The former method involves grinding logs of wood of about a metre length on large grind-stones. Grinders operate at almost constant speeds of 200–300 rpm and can be started under light load conditions. Hence constant speed synchronous motors and geared drives are most suitable. Usually, pulp making by purely mechanical means consumes more than half of the total power requirement of a paper mill, therefore, large size grinders driven by 3,000–4,000 kW motors are, usually, considered as economical. Pulp can also be made by cutting wood logs into chips of several cms length and treating them with alkalies alongwith other raw materials such as grass, rags etc. During the chemical treatment the material is continuously beaten. Wood choppers have random load characteristics and their inertia is large, depending upon the size of the disc, on which the knives of the chopper are mounted. Beaters, usually are required to start with large load. Depending upon the size of the mill, the ratings of motor used for chipping, beating, refining and storing range from several hundreds to thousand killowatts. Except for beaters, synchronous motors are employed in these drives. Since beaters, very often, require speeds less than 200 rpm and large starting torque, wound rotor induction motors are, therefore, most suitable for such drives.

The paper making machine has to perform the job of forming sheets, removing water from sheets, drying of sheets, pressing of sheets and reeling up of sheets.

The basic requirements of paper making machine drive are :

(*i*) For the actual formation and production of sheets and from the point of view of economy, constancy of the speed of the drive is necessary.

(*ii*) For the paper machine to be multipurpose, its speed should be adjustable over a range as large as 10 : 1.

(*iii*) In the calendar section and reel section, variations in tension in sheet can occur even if the correct relative speeds are maintained due to uneven drying and other factors. It is, therefore, imperative to augment the speed control circuit by an overriding tension control.

(*iv*) While cleaning the wire, it has to be moved forward a few cms at a time for its proper cleaning and inspection. Hence the motor employed should be capable of running at inching speeds of 10-25 m/minute as long as a particular button in pressed.

(*v*) Each section should be able to run at the crawling speed of 10–25 m/minute for running in felts, wire and heating up of dryer cylinders.

(*vi*) Smooth and quick starting of the sections, without causing excessive starting current, are required.

(*vii*) Control system employed should be flexible in nature.

There are two types of drives used for making paper from pulp, namely *line shaft drive* and *sectional drive.* In line shaft drive, the various sections of the paper machine are driven from a line shaft running the full length of paper machine. Cone pulleys and belt-combinations drive the various sections of the machine from line shaft through right angled gear reductions. Electric motors are usually employed for driving the transmission shaft. Both ac and dc drives can be employed for practically loss-less speed control. In ac drive, only the ac commutator motor with shunt characteristic is suitable to provide an economic speed control system. However, the speed of an ac commutator motor depends on load and, therefore, its use as a paper making machine drive with stringent requirements of constant speed is not advisable.

Also, its speed range (usually of the order of 1 : 3) as well as the power required greatly affects the size of the motor. The open-loop speed control of the ac commutator motor is sluggish in comparison with a dc drive, as speed is varied by adjusting an induction regulator and shifting the brush rocker. In dc drives, the speed of the paper making machine is controlled by varying the armature voltage of a separately excited dc motor. The variable dc voltage is obtained from ac by means of either rotary converters or static converters.

In *sectional drive* each section of the paper making machine has its own electrical motor. The speed of the paper making machine can be controlled by varying the supply voltage. By adjusting the field excitation of any motor, it is possible to vary the speed of that particular motor with respect to other motors.

The requirements of a line shaft drive are not as stringent as those of the sectional drive. It has a number of major limitations as compared to the sectional drive system. The lower cost of the electrical equipment of a line shaft drive system is offset significantly by the additional cost of the mechanical equipment required and its maintenance.

30. *Ship-Propulsion*. Except for small vessels, electric drive for ship-propulsion is universal. In electric ship-propulsion the prime mover (steam turbine or diesel engine) drives an ac or dc generator, which feeds power to ac or dc motor mounted on the propeller shafts. For large ships requiring power above 2,500 kW steam turbines with ac equipment and for smaller vessels diesel engines with dc equipment are generally employed.

In case of ac system 3-phase induction or synchronous motors are used. Speed is controlled partly by variation of applied voltage and partly by frequency variation. Speed of induction motors can also be changed by pole-changing method which gives two economical speeds. Reversal is achieved by changing the phase sequence of the supply. Power on the ship is generated by turbo-alternators at 2,200 or 6,600 volts.

In case of dc system, the speed control and reversing are carried out by varying field excitation of generator field in Ward-Leonard system. Power on the ship is generated by diesel-electric sets comparatively at low voltage of 650 volts.

Instead of constant voltage system, constant current system can be used to feed propeller motor armatures while their fields are supplied from constant voltage supply. Such a supply system is inherently proof against overloads, provides maximum safe torque without any risk of interruption of current. Motor is controlled through field control. This is quite suitable for small vessels requiring maximum manoeuvrability in restricted waters, such as tugs and ferries.

In case of very small vessels, propeller shafts are directly coupled to diesel engines through slip couplings. Speed of propeller shaft can be controlled by varying dc excitation, and hence the slip, of the coupling.

31. *Rolling Mills*. Rolling of steel, the process during which the cross section of the metal gets reduced, while increasing its length proportionately, is the major function of steel mills. Rolling mills are usually classified according to the material produced by them. The principal types of mills, classified according to their products rolled are :

Blooming mills (the mills used for producing blooms–any length of rolled metal having almost a square cross section), *slabbing mills* (the mills used to produce slab—any length of metal having a rectangular cross section), *billet mills* (the mills used for producing only a limited range of sizes which are further reduced in finishing mills), *plate mills* (the mills used to produce plates from slabs previously rolled by a blooming or slabbing mill), *structural* mills (the mills used for rolling beams, heavy angles, channels etc. from blooms or billets), *rail mills* (the mills used for rolling rails only from blooms or billets), *merchant mills* (the mills used for rolling small angles, channels, rounds, squares, etc. from billets), *rod mills* (the mills used for rolling small rounds, which are later drawn into wire), *hot-strip mills* (the mills used for rolling strips from slabs), *cold-strip mills* (the mills used for cold-rolling strips previously rolled on hot-strip mills to obtain sheets of high quality or to obtain gauges thinner than those produced on a hot mill), *sheet mills* (the mills used for rolling sheet bars or packs prepared from strips rolled on hot strip mills), *skinpass or temper mills* (the mills used to produce strips, previously rolled on a cold mill, of the desired temper, flatness, surface, and luster by using rolling pressure or a combination of pressure and tension and *tube mills* (the mills used for producing tubes, *e.g.*, pipe and conduit, by either the seamless or the butt or lap-welded processes).

Direct drive is preferred whenever the speed of the mill permits. In general, it is not desirable to use direct drive (direct-connected motor) for speed less than 150 rpm and outputs less than 750 kW. In cases, the speed of the mills is not suitable for direct drive, gear drive is employed. It is common practice to use machine-cut, double helical gears for such work, although sometimes single helical gears are used, and provision is made to carry the thrust that is produced by such gears. Spur gearing is used only for the lower speed drives where the pitch line speeds are 450 metres per minute or less. The use of gears introduces losses and offsets the increased efficiency of high-speed motors, but in many cases the total cost of installation is less, and there are other advantages, *e.g.*, higher power factor, with induction motors and sometimes a reduction of space required.

The flywheel where necessary, should be mounted between the gear and the mill, if the speed of the mill permits, so that gear does not have to transmit the peak loads which are carried by the flywheel. Where the mill shaft runs at a speed too low to permit economic flywheel design, the flywheel should be mounted on the pinion shaft of the gear unit in which case the peripheral speed of the wheel can be relatively high, and the weight correspondingly less, but the gears must be large enough to transmit the peak loads to the mill.

The dc motors, because of their inherent characteristics, are best suited for the rolling mills. Motors for reversing mills must have high starting torque, wide speed range, precise speed control, be capable to withstand overload and pull-out torque which may be as three times the rated value and have good commutation throughout. Acceleration from zero to base speed and then to top speed and subsequent reversal from top speed forward to top speed backward must be achieved in a couple of seconds. The moment of inertia of the armature must be as small as possible. Requirement of higher constant torque at low speed needed for affecting heavy reduction in first few passes can be achieved by varying the applied voltage to the armature, keeping field constant. On the other hand low torque and higher speed required for rolling operation in final passes can be achieved by weakening the field. The motors employed for steel mill duty are, therefore, of totally enclosed, forced ventilated, higher class of insulation, lesser diameter and larger length. In general, the use of dc motors increases the initial cost and introduces losses owing to the necessity of providing converting equipment, but the advantages obtainable by use of such motors greatly outweigh these additional losses and costs. The supply to dc motors for rolling mill drive can be made by means of grid controlled mercury-arc rectifier, silicon controlled rectifier or Ward-Leonard system. The main advantages of grid-controlled mercury-arc rectifier are higher efficiency, higher overload capacity, stepless voltage control from 0–100% and quick response to very rapid changes in operating variables but suffers from the drawback of causing higher current peaks in the power system accompanied by wide voltage fluctuations. The silicon controlled rectifier has the advantage of small size but suffers from the drawback that it does not have overload capacity. Hence grid controlled mercury-arc rectifiers or silicon controlled rectifiers are used for supplying the rolling mills where ac supply network is sufficiently strong. The main advantages of employing Ward-Leonard set are possibility of having speed variation within ± 100% of rated speed, possibility of electric braking by working the mill motor as a generator and of reducing the fluctuations in the power demand from the supply system. Ward-Leonard system, however, suffers from having low efficiency, longer response time and requires more maintenance.

Synchronous motors are used for driving constant speed mills and are advantageous where power factor correction is desirable or where slow-speed motors are necessary. These are particularly adaptable to continuous mills where the load is maintained for relatively long periods or for drives where the length of the material is such that the load exists for such long periods that a flywheel cannot be economically used and consequently the motor must carry the entire load. For slow speeds, synchronous motors can be built more economically than induction motors.

Slip-ring type induction motors are suitable for roughing and re-rolling mills where very precise speed control is not required. Their efficiency is low because of wastage of energy taking place in the rotor resistance. There is also abrupt rise in the motor speed when the material leaves the rolling stands. The disadvantage of the slip-ring induction motor with variable rotor resistance can be overcome by employing cascaded induction motor. The low frequency rotor output is rectified by silicon rectifier and the rectified power is made to operate against an adjustable back emf provided by a separately excited dc motor mounted on the same shaft, which converts the rectified power into mechanical power. The slip-ring induction motor now has a shunt speed-torque characteristic and speed control is affected without power loss in the secondary resistances. The starting torque is about 4 times of the rated torque when the motor is started with full dc excitation on. The cascading arrangement becomes cheaper as compared to dc motors where speed variation desired is not more than 25%.

Exercises

1. Give comparative costs and applications of various types of electric motors.
2. State with reasons, the type of ac motor suitable for following loads:
   (*i*) Table fan (*ii*) Domestic mixie.
   [A.M.I.E. Sec. B. Electric Drives & Their Control Summer 1993]
3. Explain the selection of electric motors for machine tool application regarding (*i*) electrical features (*ii*) mechanical features. (*iii*) How are frequent starting and reversing requirements are handled ? (*iv*) Methods of braking used for machine tool motors.
   [A.M.I.E. Sec B. Electric Drives and Their Control Winter 1993]
4. What are the desirable characteristics of the drives for use in cranes and hoists ? Make a comparative study of dc and ac motors for crane applications in the light of the above points.
   [A.M.I.E. Sec B. Electrical Drives and Their Control Winter 1999]
5. Explain the factors that must be taken into account while selecting a motor for a particular work. On the basis of these factors suggest and justify motors for the following :
   (*a*) Paper mill drive (*b*) Cranes
   (*c*) Marine drive (*d*) Textile mill drive.
   [Pb. Univ. Elec. Drives & Utilisation of Elec. Energy Nov. 1983]
6. State, giving reasons, what type of motor you would suggest for the following drives :
   (*i*) Rolling mills (*ii*) Centrifugal pumps
   (*iii*) Textile machinery (*iv*) Passenger lifts.
   [Gorakhpur Univ. Utilisation of Elec. Power & Traction Sept. 1984]
7. State, giving reasons, what type of motor you would consider suitable for the following drives :
   (*a*) Air compressor (*b*) Reciprocating pump
   (*c*) Paper mill drive and (*d*) Elec. traction.
   [Gorakhpur Univ. Utilisation of Elec. Power & Traction Sept. 1983]
8. Give an account of the special conditions imposed upon the electrical equipment of rolling mills in a steel works. Describe fully how these conditions are fulfilled in the case selected. Assume an ac supply, the driving motors being either direct or alternating.
   [Pb. Univ. Elec. Drives & Utilisation of Elec. Energy August 1984; April 1983]
9. What type of drives do you suggest for paper mill and steel rolling mill ? Explain in brief.
   [A.M.I.E. Sec. B. Electric Drives & Their Control Summer 2000]
10. Suggest suitable type of motors with reasons for spinning mills and cement mills.
    [Pb. Technical Univ. Electric Drives & Utilization Dec. 2008]
11. Suggest suitable type of motors with reasons for textile mill and paper mill drives.
    [Pb. Technical Univ. Electric Drives & Utilization May 2009]
12. Write typical motors and controllers for
    (*i*) Pump drives, (*ii*) Celing fan, (*iii*) Kiln drive of cement mills, (*iv*) Rolling mill drives and (*v*) Electric traction.

❑❑❑

SECTION - B

ELECTRIC POWER UTILIZATION

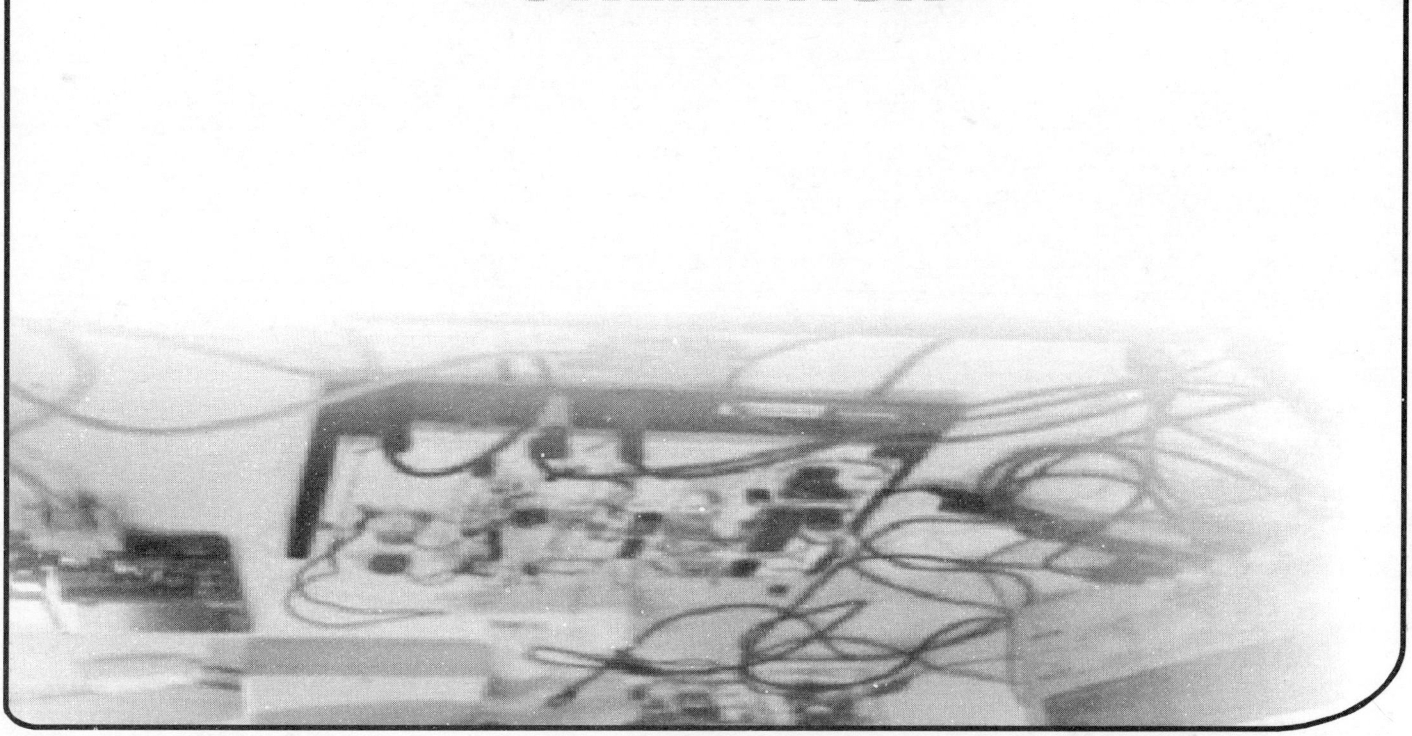

CHAPTER 5

Electric Heating

INSIDE THIS CHAPTER

5.1. INTRODUCTION

Heating is required for domestic purposes such as cooking and heating of buildings, as well as for industrial purposes such as melting of metals, hardening and tempering, case-hardening, drying and welding. Practically all the heating requirements can be met by some form of electric heating equipment. The main advantages of electric heating over other systems of heating (coal, oil or gas heating) are given below :

1. ***Economical.*** Electric heating is economical as electric furnaces are cheaper in initial cost as well as maintenance cost. It does not require any attention so their is a considerable saving in labour cost over other systems of heating. Electrical energy is also very cheap as it is being produced on large scale.

2. ***Cleanliness.*** Since dust and ash are completely eliminated in electric heating system, it is a clean system and cleaning costs are rendered to a minimum.

3. ***Absence of Flue Gases.*** Since no flue gas is produced in this system, there is no risk of atmosphere or objects being heated and operation is, therefore, hygienic.

4. ***Ease of Control.*** Simple, accurate and reliable temperature control can be had either by hand operated or by fully automatic switches. Desired temperature or temperature cycle can be had accurately in electric heating system, which is not convenient in other heating systems.

5. ***Automatic Protection.*** Automatic protection against overcurrents or overheating can be provided through suitable switchgears in the electric heating system.

6. ***Upper Limit of Temperature.*** There is no upper limit to the temperature obtainable except the ability of the material to withstand heat.

7. ***Special Heating Requirements.*** Certain requirements of heating such as uniform heating of material or heating of one particular portion of the job without effecting others, heating of non-conducting materials, heating with no oxidation, can be met only in electric heating system.

8. ***High Efficiency of Utilization.*** The overall efficiency of electric heating is comparatively higher, since in this system of heating the source can be brought directly to the point where heat is required, thereby reducing the losses. Further there is no product of combustion in which heat losses are involved.

It has been practically ascertained that 75 to 100% of heat produced by electric heating can be successfully utilised whereas in cases of gas, solid fuel and oil heating the efficiencies are 60%, 30% and 60% respectively.

9. ***Better Working Conditions.*** Electric heating system produces no irritating noise and also the radiating losses are low. Thus working with electric furnaces is convenient and cool.

10. ***Safety.*** Electric heating is quite safe and responds quickly.

5.2. MODES OF TRANSFER OF HEAT

A heated substance will give off heat to another substance at a lower temperature. The different modes by which heat is transferred are conduction, convection and radiation.

1. ***Conduction.*** In this mode of transfer of heat, one molecule of the substance gets heated and transfer the heat to the adjacent one and so on. Thus heat is transferred through a substance from one part to another, or between two substances in contact.

The rate of conduction of heat along a substance depends upon the temperature gradient. It may be expressed in MJ per hour per square metre per metre or in watts per square centimetre per centimetre while dealing electric heating.

In a plate of thickness t metres, having x-sectional area of its two parallel faces A square metres and temperature of its two faces T_1 and T_2 °C absolute the quantity of heat passed through it during T hours is given by

$$Q = \frac{kA}{t}(T_1 - T_2)\,T \qquad ...(5.1)$$

where k is the coefficient of thermal conductivity for the material in MJ/m^2/m/°C/hour.

2. *Convection.* Heat is transferred by convection in case of immersion type water heater or in case of low temperature heating equipment for buildings. The air in contact with a heated radiator element in a room receives heat from contact with the element. The heated air expands and rises, cold air flowing into take its place. Thus there is a constant flow of air upwards across the heating element. This process is called *convection.* These convection currents give up some of their heat to the colder parts of the room. The room and its contents are thus gradually heated by this means. A similar action takes place in an electric water heater, a continuous flow of water passing upwards across the immersed heater element, with the result that the whole of the water in the tank becomes hot.

The quantity of heat absorbed from the heater by convection depends mainly upon the temperature of the heating element above the surroundings and upon the size of the surface of the heater. It also depends partly on the position of the heater. Heat dissipation is given by the following expression.

$$\text{Heat dissipation, } H = a(T_1 - T_2)^b \text{ W/m}^2 \quad ...(5.2)$$

where a and b are constants, whose value depend upon the heating surface facilities for heating etc. T_1 and T_2 are the temperatures of the heating surface and the fluid in °C absolute respectively.

For vertical surfaces in air, the above expression becomes

$$\text{Heat dissipation, } H = 3.875\,(T_1 - T_2)^{1.25} \text{ watts/m}^2 \quad ...(5.3)$$

In furnaces heat transfer by convection is not to any great extent.

3. *Radiation.* In this mode of heat transfer the heat reaches the substance to be heated from the source of heat without heating the medium in between. Rate of heat radiation is given by Stefan's law, according to which

Heat dissipation,

$$H = 5.72 \times 10^4\, k\, e \left[\left(\frac{T_1}{1{,}000}\right)^4 - \left(\frac{T_2}{1{,}000}\right)^4\right] \text{W/m}^2 \quad ...(5.4)$$

where T_1 is temperature of source of heat in °C absolute, T_2 is temperature of substance to be heated in °C absolute, k is a constant known as *radiant efficiency,* k is unity for a single element and between 0.5 and 0.8 for several elements placed side by side and e is emissivity, which is unity for black body and is equal to 0.9 for resistance heating elements.

Since radiation is proportional to the difference of fourth powers of temperatures, we can have very efficient heating at high temperatures.

5.3. CLASSIFICATION OF ELECTRIC HEATING METHODS

Electric heating can be broadly classified as (*i*) power frequency heating and (*ii*) high frequency heating.

Power frequency heating can be further classified as (*i*) resistance heating and (*ii*) arc heating.

Resistance heating can be further classified as (*i*) direct resistance heating, (*ii*) indirect resistance heating and (*iii*) infrared or radiant heating.

Similarly arc heating can be further classified as (*i*) direct arc heating and (*ii*) indirect arc heating.

High frequency heating can be classified into (*i*) induction heating and (*ii*) dielectric heating. Induction heating can further be classified as (*i*) direct induction heating and (*ii*) indirect induction heating.

1. *Direct Resistance Heating.* Electric current is made to pass through the body to be heated. This principle of heating is employed in resistance welding and electrode boiler for heating water.

2. *Indirect Resistance Heating.* Electric current is made to pass through a wire or other high resistance material forming a *heating element;* heat so developed is transferred from the heating element to the body by the agency of radiation or convection. Normally this method is used in immersion heaters, resistance ovens, domestic and commercial cooking and heat treatment of metals.

3. *Infrared or Radiant Heating.* Heat energy from an incandescent lamp is focused upon the body to be heated up in the form of electromagnetic radiations. This is employed to dry the wet paints on an object.

4. *Arc Heating.* The arc drawn between two electrodes develops high temperature (about 3,000–3,500°C) depending upon the electrode material. The electric arc may be used in the following different ways:

(*i*) By striking the arc between the charge and the electrode or electrodes. In this method the heat is directly conducted and taken by the charge. The furnaces operating on this principle are known as *direct arc furnaces.*

(*ii*) By striking the arc between the two electrodes. In this method the heat is transferred to the charge by radiation. The furnaces operating on this principle are known as *indirect arc furnaces.*

(*iii*) By striking an arc between an electrode and the two metallic pieces to be joined, as in arc welding.

5. *Direct Induction Heating.* In this method of heating the currents are induced by electromagnetic action in the body to be heated. The induced currents when flowing through the resistance of the body to be heated develop the heat and thus raise the temperature. In induction furnace heat is used to melt the charge and eddy current heaters used for heat treatment of metals are other forms of direct induction heating.

6. *Indirect Induction Heating.* In this method of electric heating the eddy currents are induced in the heating element by electromagnetic action. Eddy currents set up in the heating element produce the heat which is transferred to the body to be heated up, by radiation and convection. Certain types of induction ovens used for heat treatment of metals operate on this principle.

7. *Dielectric Heating.* In this method of electric heating use of dielectric losses is made to heat the non-metallic materials. Non-metallic material to be heated is placed between two metal electrodes across which a high voltage having high frequency is applied, the heat is developed owing to the dielectric losses taking place.

5.4. RESISTANCE HEATING

Resistance heating is based upon the I^2R effect. This method of heating has wide applications such as heat treatment of metals (annealing, hardening etc.), drying and baking of potteries, stoving of enamelled ware and commercial and domestic cooking. Temperature up to about 1,000°C can be obtained in ovens employing wire resistances for heating elements. There are two methods of resistance heating.

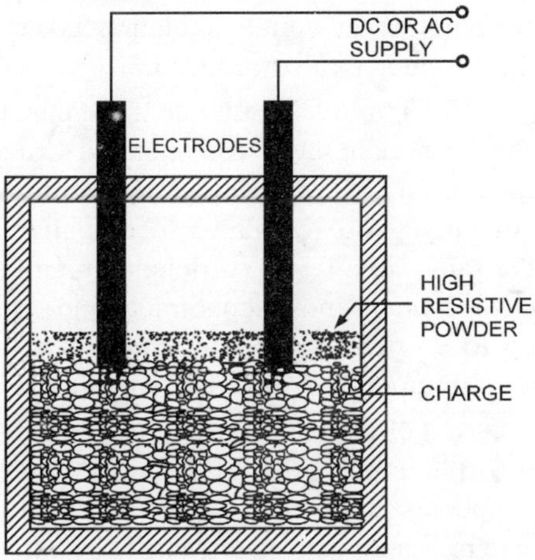

Fig. 5.1. ***Direct Resistance Heating***

1. *Direct Resistance Heating.* In this method of heating, the material or charge to be heated is taken as resistance and current is passed through it. The charge may be in the form of powder, pieces or a liquid. Two electrodes are immersed in the charge and connected to the supply in case of availability of direct current or single phase ac supply and three electrodes are immersed in the charge and connected to supply in case of availability of 3-ϕ ac supply. When some pieces of metals are to be heated some highly resistive powder is sprinkled over the surface of pieces to avoid direct short circuit. The current flows through the charge and heat is produced. This method has high efficiency since heat is produced in the charge itself. As the current in this case is not easily variable, therefore, automatic temperature control is not possible. However, uniform and high temperature can be obtained. This method of heating is used in salt bath furnaces and in the electrode boiler for heating water.

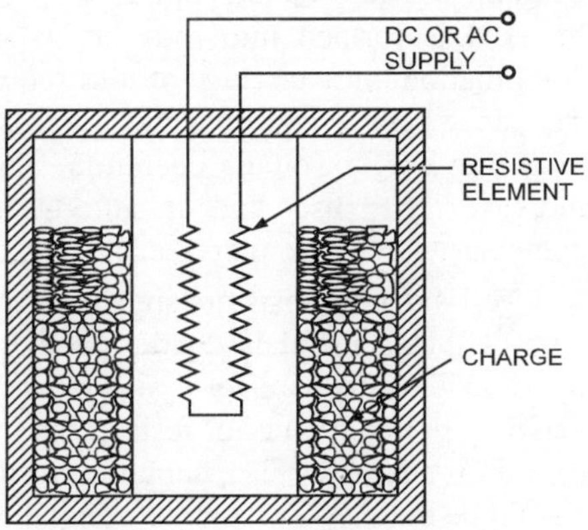

Fig. 5.2. ***Indirect Resistance Heating***

2. *Indirect Resistance Heating.* In this method of heating, the current is passed through a wire or other high resistance material forming a heating element. The heat proportional to I^2R loss produced in the heating element is delivered to the charge by one or more of the modes of transfer of heat viz. conduction, convection and radiation. If the heat transfer is by conduction the resistor must be in contact with the charge. An enclosure, known as *heating chamber,* is required for heat transfer by radiation and convection for the charge. For industrial purposes, where a large amount of charge is to be heated, the heating element is kept in a cylinder surrounded by jacket containing the charge.

This arrangement provides a uniform temperature. Automatic temperature control can be provided in this case. This method of heating is used in room heaters, immersion water heaters and in various types of resistance ovens employed in domestic and commercial cooking, and salt bath furnaces.

Resistance Ovens and Furnaces. Resistance furnaces may be classified according to their operating temperature. Low temperature heating chamber with provision for ventilation is termed as *oven.* Resistance ovens are used for drying and baking potteries, drying varnish coatings, vulcanising and hardening of synthetic materials and commercial and domestic heating. They are also employed for tempering hardened steel pieces. *Medium temperature furnaces* having operating temperatures between 300°C and 1,050°C are employed for annealing, normalising of steel and non-ferrous metals, melting of non-ferrous metals and stove enamelling. *High temperature furnaces having* operating temperatures between 1,050°C and 1,350°C are employed for hardening applications. Typical furnace using heating elements and having provision of carrying out heating in particular atmosphere is depicted in Fig. 5.3.

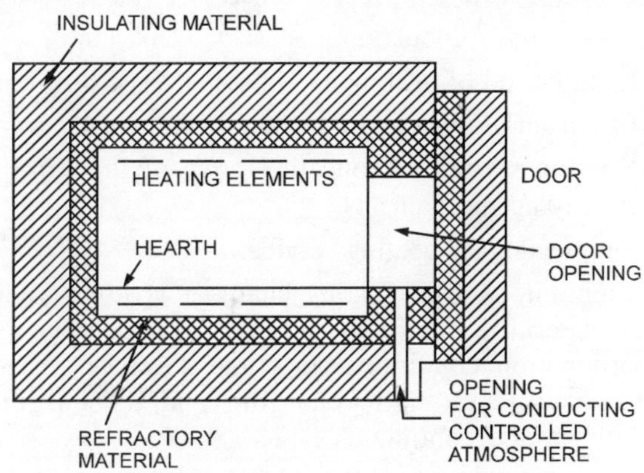

Fig. 5.3. ***Resistance Furnace***

The resistance oven is constructed of firebricks or other heat insulating material supported on a metal framework. The heating elements are mounted on the top, sides or bottom of the oven, according to circumstances.

The resistance furnace is an enclosure with a refractory lining : a surrounding layer of heat insulation and outer casing of steel plate, bricks or tiles. The inside proportions of a heating chamber are made to suit the character of the charge and type of furnace or oven. The nature of material required for the insulation is determined by the maximum temperature of the inner face of the layer of insulation of a heating chamber. The heating elements are mounted on top, sides or bottom of the oven as the circumstances permit.

Heating Chambers. An enclosure of a charge to be heated by radiation or convection or by the joint effect of these modes of heat transfer is known as *heating chamber.* Apart from mechanical considerations the design of the chamber is related primarily to temperature and the major mode of heat transfer to be used. The essential features are : the enclosing chambers, the resistors, and the atmosphere within the chamber. The functions of the heating chambers are :

(*a*) to control the distribution of heat within the chamber;

(*b*) to control the cooling rate of charge, if required;

(*c*) to confine the atmosphere around the charge; and

(*d*) to store as much of the heat supplied as may be practicable and economical.

A cylinder with an inside diameter equal to its height represents an ideal shape and proportions of a heating chamber, keeping only the utilisation of material and rate of heat loss from the exterior surfaces in view. This ideal is seldom attained because of many practical considerations in heating service. Electric heating permits wide latitude in the proportions and shape of a heating chamber, and factory planning takes this into account in developing the sequence of movements of material through manufacturing operations.

Heating chambers may be of 'batch' or 'continuous' type. In the former type, the charge remains stationary during the heat application. The cycle may include cooling the charge in the chamber. In case of continuous type chambers, the charge is heated as it moves through the chamber. In some cases the chamber is extended for more or less cooling of the charge before it leaves the chamber.

The application of the batch type heating chamber is defined by the following conditions.

(*i*) Intermittent production.

(*ii*) Long period of heating and in some cases slow cooling.

(*iii*) For material that cannot be handled on conveyors.

(*iv*) Heating service beyond the range of the capacity of conveying mechanisms.

(*v*) Supplementary heating service.

A continuous type heating chamber is used where the flow of material is reasonably uniform and continuous, *i.e.*, mass production conditions. In some cases batch heating chambers with automatic charging and discharging are essentially of the continuous type.

5.4.1. Heating Element Materials

A material which is used as a heating element must possess the following properties :

(*i*) ***High resistivity.*** The material to be used for heating element should be of high specific resistance so that a small length of wire may be sufficient to produce the required amount of heat.

(*ii*) ***High melting point.*** Melting point of material to be used for heating element should be high so that charge can be heated to a high temperature.

(*iii*) ***Low temperature coefficient.*** The material for heating element should have low temperature coefficient so that resistance may not vary with the change in temperature otherwise starting current would be high.

(*iv*) ***Free from oxidation.*** The material for heating element should be such that it may withstand the required temperature without getting oxidised, otherwise it would have to be replaced frequently.

The materials commonly used for heating elements for low and medium temperature services are either alloy of nickel and chromium (Nickel 80% and chromium 20%) or alloy of nickel, chromium and iron (Ni 65%, Cr 15% and Fe 20%). The addition of iron of the alloy reduces the temperature at which oxidation takes place but the cost of the product is also reduced.

Ni-Cr alloy is suitable for temperatures up to 1,150°C and for work in severe conditions Ni-Cr-Fe alloy is recommended for use up to 950°C. Nickel-chromium (Nichrome) alloy has not only good resistance to oxidation up to 1,150°C but at the same time it has sufficient strength. Ni-Cr-Fe alloy is the cheapest, and most economical and stronger for temperatures up to 950°C. Alloys containing iron, chromium, cobalt and aluminium can withstand temperatures as high as 1,350°C.

Nickel-copper alloy (Eureka or constantan by trade name) is often employed for heating elements operating at low temperatures. Its most important property is that it has virtually zero resistance temperature coefficient. Iron-chromium-aluminium alloy (Kanthal by trade name) has very good resistance to oxidation at high temperatures but are weaker in mechanical strength and, therefore, such elements need more support.

Resistors for operating temperatures above 1,150°C are made of silicon carbide, molybdenum, tungsten and graphite.

Silicon carbide is the basis of a resistor material for operating in air for temperatures up to about 1,500°C. The material is formed into rods of diameters and lengths for combination into circuits of the required electrical rating. The range between the cold and hot resistances is small enough to permit single voltage operation. The resistance gradually increases with use, and an auto-transformer with taps to compensate for this increase of resistance is desirable.

Molybdenum resistors are suitable for temperatures up to 1,650°C. This metal is ductile enough at room temperature for drawing into wire for resistor windings. The wide range between the cold and hot resistances makes multiple voltage operation necessary. The supports (insulators) of the winding should be magnesia or zirconia. Molybdenum resistors cannot be operated in air and also must be protected against reactions with silicon and carbon. The metal is immune from reactions with sulphur, nitrogen, hydrogen, and water vapour. A hydrogen atmosphere is usually used for the protection of these resistors. Molybdenum is not suitable for resistors of vacuum furnaces because of its high vapour pressure.

Tungsten resistors can be used for temperatures up to 2,000°C. The maximum temperature is limited by the refractory supports of the resistor. The metal must be heated for drawing into shapes. The low vapour pressure of tungsten makes it useful for resistors of vacuum furnaces.

Graphite resistors are suitable for any temperature that can be used. The resistors require protection against oxidation above about 600°C. Also because of the chemical activity of carbon, special consideration must be given to the surrounding atmosphere. The material is available in a wide range of shapes and dimensions.

5.4.2. Maximum Operating Voltage

Maximum operating voltage is limited, by electrical insulation at high temperatures and from safety consideration, to 600 volts. This value may exceed in special cases. Resistor winding may be in one, two or more circuits and may be connected either to single or to polyphase supply system.

5.4.3. Design of Heating Element

Knowing the electrical input and its voltage the size and length of the wire required as the heating element to produce the given temperature can be calculated. The wire employed may be circular or rectangular like a ribbon. The latter permit the use of higher wattage per unit area and are used in ovens, vulcanizers, toasters etc. The heating element on reaching a steady temperature will dissipate the heat from its surface equivalent to electrical input. Since generally the heat will be dissipated from the heating elements at high temperatures, it is reasonable to assume that the whole of the heat energy is dissipated solely by radiation.

Heat dissipated according to Stefan's law,

$$H = 5.72ke\left[\left(\frac{T_1}{100}\right)^4 - \left(\frac{T_2}{100}\right)^4\right] \text{ watts/m}^2 \qquad ...(5.5)$$

$$\text{Electrical input, } P = \frac{V^2}{R}$$

where V is the supply voltage and R is resistance of heating element and is given by the expression,

$$R = \rho\frac{l}{a} = \frac{\rho l}{\frac{\pi d^2}{4}} = \frac{4\rho l}{\pi d^2}$$

for a circular wire of diameter d, length l and resistivity ρ.

Electrical power input,

$$P = \frac{V^2}{R} = \frac{V^2}{4\rho l/\pi d^2} = \frac{\pi d^2 V^2}{4\rho l}$$

$$\text{or} \quad \frac{l}{d^2} = \frac{\pi V^2}{4\rho P} \qquad ...(5.6)$$

Surface area, $S = \pi dl$

Heat dissipated $= \pi dl\, H$

Since at steady temperature

Power input = Heat dissipated

$\therefore \quad P = \pi dl\, H$

$$\text{or} \quad \frac{\pi d^2 V^2}{4\rho l} = \pi dl\, H$$

$$\text{or} \quad \frac{d}{l^2} = \frac{4\rho H}{V^2} \qquad ...(5.7)$$

Solving Eqs. (5.6) and (5.7) length and diameter of wire can be determined.

For ribbon type of conductor let w be the width and t be the thickness.

Electrical power input,

$$P = \frac{V^2}{R} = \frac{V^2}{\rho l/a} = \frac{V^2}{\rho l/wt} = \frac{V^2 wt}{\rho l}$$

$$\text{or} \quad \frac{l}{wt} = \frac{V^2}{\rho P} \qquad ...(5.8)$$

Since at steady temperature

Power input = Heat dissipated

$\therefore \quad P = 2wlH$ neglecting the side area $2tl$ as the thickness is negligible

$$\text{or} \quad \frac{V^2 wt}{\rho l} = 2wlH$$

$$\text{or} \quad \frac{t}{l^2} = \frac{2H\rho}{V^2} \qquad ...(5.9)$$

So by solving the two Eqs. (5.8) and (5.9), length l and width w for a ribbon of thickness t will be evaluated.

5.4.4. Efficiency and Losses

The heat produced in the heating elements is also to overcome the losses occurring due to

(*i*) heat used in raising the temperature of oven or furnace;

(*ii*) heat used in raising the temperature of the containers or carriers;

(*iii*) heat conducted through the walls;

(*iv*) escapement of heat due to opening of door; in addition to heat required to raise the temperature of the charge to the required value.

The efficiency of the oven

$$= \frac{\text{Heat required to raise the temperature of the charge to the required value}}{\text{Heat required to raise the temperature of the charge to the required value + losses}}$$

The heat required to raise the temperature of the charge to the required value, W = Mass of charge in kg × specific heat of charge in J/kg/ °C × temperature rise = mst Joules.

The heat used for raising the temperature of the oven or furnace can be calculated in a similar way by knowing the weight of the refractory material and its specific heat. In case the oven is used continuously this loss becomes negligible.

Heat used for raising the temperature of container is calculated in exactly the same way as for oven or furnace. The container usually has to be heated up afresh for each charge.

Since the heat is continuously conducted through the walls, this source of heat loss is most important. It can be determined knowing the mean value of the inside and outside surface areas A in square metres; the thickness of walls t in metre; inside and outside temperatures T_1 and T_2 in °C and thermal conductivity of walls k in MJ per hour per square metre, per metre, from the following expression

Heat loss by conduction through walls

$$= \frac{kA}{t}(T_1 - T_2) \text{ MJ/hour} \qquad ...(5.10)$$

Though there is no specific formulae for determination of loss occurring due to opening of door for inspection of the charge, however, this loss may be taken as 0.575 to 1.15 MJ/ m^2 of the door area if the door is opened for a period of 20 to 30 seconds.

The efficiency defined above lies between 60 and 80 per cent. A more convenient figure for expressing the behaviour of an oven, however, is the energy required per tonne of charge treated. Typical values for different processes are given in Table 5.1.

Table 5.1.

Process	*Energy Consumption in kWh per Tonne of Charge*
Baking of bread	50-100
Annealing copper	100-200
Annealing steel	200-250
Carburising	250-500
Vitreous enamelling of sheet steel	400-700

5.4.5. Temperature Control of Resistance Ovens/Furnaces

Temperature control is necessary in resistance ovens/furnaces–temperature may have to be kept constant or varied according to requirements. Control may be manual or automatic.

In resistance ovens/furnaces heat developed depends upon $I^2 R t$ or $\frac{V^2}{R} t$. So there are three ways in which the temperature can be controlled. Firstly, by varying the applied voltage to the elements or current flowing through the element; secondly, by varying the resistance of elements and thirdly, by varying the ratio of on and off times of supply.

Voltage across the oven can be controlled by changing the transformer tappings. This is economical and most suitable if the transformer is to be used for stepping down the voltage for supply to ovens or furnaces, but such conditions do not arise usually. Auto-transformer or induction regulator can also be used for variable voltage supply. In case of large furnace, there may be an independent generating set for the supply to the furnace and in that case variable voltage supply can be provided.

Alternatively voltage across the oven or furnace can be controlled by varying the impedance connected in series with the circuit. But this method is not economical as power is continuously wasted in the controlling resistance. Its use is, therefore, limited to small furnaces.

Temperature can also be controlled by switching the various combinations of groups of resistances used in the ovens or furnaces in the following ways.

1. ***Use of Variable Number of Elements.*** In this method, the number of heating elements in working are changed; so total power input or heat developed is changed. This method does not provide uniform heating unless the number of heating elements in the circuit at any particular instant are distributed over the surface area, which requires complicated wiring.

2. ***Change of Connections.*** In this method the elements are arranged to be connected either all in series or all in parallel or combination of both or in star or in delta by means of switches at different instants according to the requirements. This is the simplest and most commonly used method of control.

An *on-off switch* can also be employed for temperature control but its use is restricted to small ovens. The time duration for which the oven is connected to the supply and the time duration for which it remains cut off from supply will determine the temperature. Here an oven is supplied through a thermostat switch which makes and breaks the supply connections at particular temperature. The ratio of $\frac{\text{Time duration during which supply remains on}}{\text{Total time duration of an on-off cycle}}$ is an indication of temperature. The higher the ratio, the larger will be the temperature of the oven. Advantage of this method is that it is more efficient than series impedance method.

Normally the temperature is required to be maintained between limits. This necessitates some type of automatic control of the thermostat type. When the temperature of the oven exceeds or falls below certain predetermined values, the thermostat connected in the circuit operates a relay which varies the current in the elements by any one of the methods given above. The relay may even switch the oven out of service if so required. The actual switching is carried out by contactors.

For large capacity ovens, some protection is to be provided against overloading, excessive temperatures, and in some cases against the possibility of workmen loading or unloading the furnace with the power on.

The only possibility of overload is due to a part of a metallic charge falling against and short circuiting the elements. An instantaneous overload relay is provided and set to trip the circuit at 10 or 15 per cent, above normal current. This adequately guards against damage due to such a contingency. In addition to above, fuses are also provided either in the main oven circuit or in the hold-on coil of the energizing contactors to provide protection in case of failure of automatic control system. For the protection of nickel-chrome wire in this way a gold fuse is employed, because it has a sharp melting point of 1,060°C; and does not oxidise. Fuses melting at a lower temperature can be had, but although of some service are not completely reliable.

5.4.6. Causes of Failure of Heating Elements

The chief causes of failure of heating elements are given below.

1. ***Formation of Hot Spots.*** Hot spots are the points in the heating element which are at a higher temperature than the main body of the element. One of the reasons of formation of hot spot in the heating element is high rate of local oxidation that may reduce the element wire x-section, thereby increasing the resistance at that spot and producing more heat locally and causing breakdown of the element. Another reason may be due to shielding of element by supports etc. that will reduce the local heat loss by radiation and cause an increase of temperature of shielded portion of the element. However, it can be minimized by using minimum number of supports without producing distortion of the element. Other causes are due to too high element temperature, insufficient support for the element or selection of wrong fuse material, may result in sagging and warping of the material. The sagging and warping may cause uneven spacing of sections, thereby developing hot spots where sections are closest together or even there may be actual shorting of adjacent sections of an element.

2. ***Oxidation and Intermittency of Operation.*** Continuous and tenacious oxide scale is formed on the surface of the heating element when it attains high temperature. This oxide layer is so strong that it prevents further oxidation of the inner metal of the heating element. But in case of frequent use of the element, the scale is subjected to thermal stresses developed owing to frequent heating and cooling. The oxide scale, therefore, cracks and flakes off, exposing further fresh metal to oxidation and thus resulting in increased local oxidation and producing hot spots.

3. ***Embrittlement Due to Grain Growth.*** All heating alloys containing iron tend to form large brittle grains at high temperatures. When cold, the heating elements are very brittle and liable to rupture easily on slighest handling and jerks.

4. ***Contamination and Corrosion.*** Gases of the controlled atmosphere prevailing in annealing furnaces or fumes from flux used in brazing furnaces, or oil fumes caused by heat treatment of components contaminated with lubricant contaminate the elements and produce dry corrosion.

Of all the conditions stated above, which determine the element life, most critical are the temperature of the hottest point and ratio of intermittent to continuous working.

5.4.7. Special Types of Resistance Furnaces

The most important applications of resistance ovens are air-circulation oven and bright annealing furnace.

1. ***Air-Circulation or Convection Type Furnaces.*** The charge may not get uniformly heated in an ordinary furnace in which heat is transferred to the charge almost entirely by radiation. If uniform distribution of heat is necessary and the temperature is not to exceed about 600°C, such furnaces are used. The common applications are drawing and hardening of steel, heat treatment of aluminium and light metals. In such furnaces, air is passed over the heating elements and this heated air is circulated over or through the charge imparting its heat uniformly to all parts of the charge by conduction. The direction of flow in air-circulation furnaces is sometimes reversed periodically in order to make the distribution of heat more uniform. If this were not done, the temperature of that part of the charge which comes across the hot air first will be higher than that which comes across it afterwards. However, this difference can be minimised by circulating large volume of air. Blowers and fans are used for circulating the gas or air. The advantage of the circulating air or gas is to impart all the parts of the charge a uniform application of the atmosphere. Generally the charge is to be cooled in the furnace itself and to achieve this gases are circulated after being cooled. Gases are let through cooling pipes during the cooling process. Nitriding is an important application. In this process the steel after being heated by hot air, is exposed to ammonia. The steel absorbs nitrogen and gets hardened. An important item in the design of such an oven is the characteristic of the fan. A centrifugal fan is used, but it must be of such a design that a change in the resistance of the air path due to charges of different size and shape does not appreciably affect the volume of air delivered or kW required by the motor.

2. ***Resistance Furnace For Bright Annealing.*** The process of cooling a heated job slowly for removing its brittleness is known as *annealing*. In the usual process of annealing the job is left covered with a scale of oxide which must be removed in order to have bright finish. The formation of oxide scale can be prevented if annealing process is carried out in an atmosphere free from oxygen, water vapour and carbon dioxide. Such a condition can be achieved by heating the job in an air tight furnace to which is connected a non-return air valve. During the period of heating air and any oil absorbed by the job are expelled through the non-return valve due to their expansion. During cooling the non-return valve closes, thereby preventing access of oxygen or other gases to the job. Thus the surface of the job under annealing process is maintained bright.

5.4.8. Salt-bath Heating

Salt-bath furnaces are used for the purposes of tempering, quenching and hardening of steel tools. The characteristics of salt-bath heating are rapidity and uniformity and selective localised heating combined with protection from oxidation.

This type of furnace consists of a bath of some salt such as molten sodium chloride and two electrodes immersed in it. When the current between electrodes immersed in the salt is passed, heat is developed and the temperature of the salt bath may be in the range between 1,000°C and 1,500°C depending upon the type of salt used. In this bath the material to be heated is dipped and the necessary heat treatment, as required for tool steel, is given to it. The heat transfer in this case is by conduction and by better contact of the body with the heated salt, which is in molten state.

As dc would cause electrolysis of the salt, therefore, alternating current is used. The voltages up to 20 volts and currents up to 3,000 amperes depending on the size of the furnace are used. Voltage on the secondary side is changed by tap changing gear included in the primary of the transformer.

The resistance of the salt decreases with the increase of temperature of the salt therefore, in order to maintain constant power input it is necessary that the salt furnaces are started with highest tap and as the temperature rises, this is gradually brought to lower taps. Control of power input is also affected by varying the depth of immersion and distance between electrodes. It must be ensured that the current flows through the salt and not through the job being heated.

5.4.9. Infrared or Radiant Heating

In the ordinary resistance furnaces the heat is transferred from heating elements to the charge partly by radiation and partly by convection, the latter predominating at low and medium temperatures but in this form of heating suitable for low and medium temperature heating purely radiant or infra-red heating is used. In this method of heating, heating elements consist of tungsten filament lamps together with reflectors to direct the whole of the heat emitted on to the charge. The lamps are operated at 2,300°C instead of 3,000°C giving greater proportion of infrared radiation and a longer life. The reflectors are plated with rhodium which, being harder, has a longer life and is easier to maintain. The lamps employed are usually of rating between 250 and 1,000 watts operating at 115 volts. The operation at low voltage results in a robust filament.

Plant sizes range from a single lamp to chambers containing several hundred kW of lamps. With this arrangement charge temperature between 200°C and 300°C can be obtained. Heat emission intensities up to 7,000 watts/square metre of the chamber surface can be obtained. These are much higher than those obtained with ordinary resistance ovens (1,500 watts/m^2). In radiant heating heat absorption remains practically constant whatever the charge temperature is where as it falls off rapidly as the temperature of charge rises in the ordinary resistance furnace.

In this process the lamps with their reflectors are grouped close together to form the walls and possibly the top of a heating chamber; the lamps being mounted on a light angle-iron framework, no heat insulation is required as all heat is directed inwards. Individual reflectors for each lamp or trough reflectors may be used, the latter enabling a greater number of lamps to be installed in the given area. The precise shape of the chamber depends on the size and shape of the charge to be heated and the framework can easily the made adjustable. Continuous processes can be had, by carrying the articles to be heated through the chamber on a conveyor belt.

Infrared heating possesses the advantages of (*i*) rapid heating, (*ii*) compactness of heating units, (*iii*) flexibility and (*iv*) safety and finds application in (*i*) paint stoving (*ii*) drying of wood furniture (*iii*) preheating of plastics prior to moulding (*iv*) softening of thermoplastic sheets and (*v*) drying of pottery, paper, textiles etc. where moisture content is not large.

For obtaining best results, the infrared lamps should be located at a distance of 25-30 cm from the object to be heated.

Example 5.1. Six resistances each of 80 Ω are used in resistance heating furnace. How much power is drawn with following connections and supply voltages ?

(*a*) 400 V single phase
- **(*i*) Six groups in parallel, each of one resistance.**
- **(*ii*) Three groups in parallel, each of two resistances in series.**
- **(*iii*) Two groups in parallel, each of three resistances in series.**

(*b*) 400 V three-phase star-connections
- **(*i*) Two elements in parallel in each phase.**
- **(*ii*) Two elements in series in each phase.**

(*c*) 400 V three-phase delta-connections
- **(*i*) Two elements in parallel in each branch.**
- **(*ii*) Two elements in series in each branch.**

(*d*) Same as in case of (*b*) and (*c*) but with 50% tapping on auto-transformer.

[Rajasthan Univ. Utilization of Elec. Power 2000]

Solution : (*a*) (*i*) *Six groups in parallel, each of one resistance*

Equivalent resistance,

$$R_1 = \frac{1}{\frac{1}{80}+\frac{1}{80}+\frac{1}{80}+\frac{1}{80}+\frac{1}{80}+\frac{1}{80}} = \frac{80}{6} \text{ or } \frac{40}{3}\ \Omega$$

$$\text{Power drawn, } P_1 = \frac{V^2}{R_1} = \frac{400^2}{\frac{40}{3}} = 12{,}000 \text{ W or } 12 \text{ kW} \quad \textbf{Ans.}$$

(*ii*) *Three groups in parallel, each of two resistances in series*

Equivalent resistance,

$$R_2 = \frac{1}{80+80} + \frac{1}{80+80} + \frac{1}{80+80} = \frac{160}{3}\ \Omega$$

$$\text{Power drawn, } P_2 = \frac{V^2}{R_2} = \frac{400^2}{\frac{160}{3}} = 3{,}000 \text{ W or } 3 \text{ kW} \quad \textbf{Ans.}$$

(*iii*) *Two groups in parallel, each of three resistances in series*

Equivalent resistance,

$$R_3 = \frac{1}{\frac{1}{80+80+80}+\frac{1}{80+80+80}} = 120\ \Omega$$

$$\text{Power drawn, } P_3 = \frac{V^2}{R_3} = \frac{400^2}{120} = 1{,}333 \text{ W or } 1.333 \text{ kW} \quad \textbf{Ans.}$$

(*b*) (*i*) *Two elements in parallel in each phase*

Equivalent resistance,

$$R_4 = \frac{1}{\frac{1}{80}+\frac{1}{80}} = 40\ \Omega$$

$$\text{Phase voltage, } V_P = \frac{400}{\sqrt{3}} \text{ V}$$

Power drawn per phase,

$$P_4 = \frac{V_P^2}{R_4} = \frac{\left(\frac{400}{\sqrt{3}}\right)^2}{40} = \frac{4{,}000}{3} \text{ W or } \frac{4}{3} \text{ kW}$$

$$\text{Total power drawn} = 3P_4 = 3 \times \frac{4}{3} = 4 \text{ kW} \quad \textbf{Ans.}$$

(*ii*) Two elements in series in each phase

Equivalent resistance, R_5 = 80 + 80 = 160 Ω

$$\text{Total power drawn, } P_5 = 3 \times \frac{\left(\frac{400}{\sqrt{3}}\right)^2}{160} = 1{,}000 \text{ W or } 1 \text{ kW} \quad \textbf{Ans.}$$

(*c*) (*i*) This is the same case as (*a*) (*i*),

Power drawn = 12 kW **Ans.**

(*ii*) This is the same case as (*a*) (*ii*), Power drawn = 3 kW **Ans.**

or Power drawn will be three times that have been drawn under (*b*) (*i*) and (*b*) (*ii*).

(*d*) Power drawn will be of one-fourth of the values obtained under (*b*) (*i*), (*b*) (*ii*), (*c*) (*i*) and (*c*)(*ii*) when applied voltage is 50 per cent.

Example 5.2. A 2.5 kW, 240 V, single-phase resistance oven is to have nichrome wire heating elements. If the wire temperature is to be 1,500°C and that of the charge 450°C, estimate the diameter and length of wire. The resistivity of nichrome alloy is 42.5 μΩ-cm. Assume the radiating efficiency and the emissivity of the element as 1.0 and 0.9 respectively.

[G.B. Technical Univ. Utilization of Electrical Energy and Electric Traction 2010-11]

Solution : Power supplied, P = 2.5 kW = 2,500 W

Supply voltage, V = 240 V

Radiating efficiency, K = 1.0

Emissivity, *e* = 0.9

Resistivity, $\rho = 42.5 \times 10^{-6}\ \Omega\text{-cm} = 0.425 \times 10^{-6}\ \Omega\text{-m}$

From Eq. (5.6), $\frac{l}{d^2} = \frac{\pi V^2}{4\rho P} = \frac{\pi \times 240^2}{4 \times 0.425 \times 10^{-6} \times 2{,}500}$

$= 42.5778 \times 10^6$ ...(*i*)

From Eq. (5.5),

$$H = 5.72\, ke\left[\left(\frac{T_1}{100}\right)^4 - \left(\frac{T_2}{100}\right)^4\right] \text{ W/m}^2$$

$$= 5.72 \times 1 \times 0.9\left[\left(\frac{1{,}500 + 273}{100}\right)^4 - \left(\frac{450 + 273}{100}\right)^4\right] \text{ W/m}^2$$

$$= 5.72 \times 1 \times 0.9\ [(17.73)^4 - (7.23)^4] \text{ W/m}^2 = 494{,}647$$

Since heat dissipated = Electrical power input

$\therefore$ $494{,}647 \times \pi \times dl = 2{,}500$

or $d^2 l^2 = \left(\frac{2{,}500}{\pi \times 494{,}647}\right)^2 = 0.000002588$ ...(*ii*)

Multiplying Eqs. (*i*) and (*ii*), we have

$l^3 = 110.2 \text{ m}^3$

or $l = 4.79$ m **Ans.**

and diameter of wire, $d = \frac{\sqrt{0.000002588}}{4.79}$

$= 0.000336$ m or 0.336 mm **Ans.**

Example 5.3. In a three-phase 440 V, 50 Hz, star-connected 20 kW oven, the temperature of wire is 1,200°C and that of charge is 700°C. If radiating efficiency is 0.6 and emissivity is 0.9, design the heating element. A strip of thickness 0.025 mm having resistivity 1.05×10^{-6} Ω-m is used.

[*U.P. Technical Univ. Utilization of Electrical Energy and Electric Traction 2006-07*]

Solution : Power per phase, $P = \frac{20 \times 1{,}000}{3} = 6{,}667$ W

Phase voltage, $V_P = \frac{440}{\sqrt{3}} \text{ V} = 254.034 \text{ V}$

Resistance of strip, $R = \frac{V_P^2}{P} = \frac{254.034^2}{6{,}667} = 9.68\ \Omega$

If l is the length of the strip and w is the width of strip, then

Resistance of the strip, $R = \frac{\rho \times l}{w \times t} = \frac{1.05 \times 10^{-6} \times l}{w \times 0.025 \times 10^{-3}}$

or $\frac{l}{w} = \frac{9.68 \times 0.025 \times 10^{-3}}{1.05 \times 10^{-6}} = 230.476$ ...(*i*)

Heat dissipated from the surface,

$$H = 5.72\, ke\left[\left(\frac{T_1}{100}\right)^4 - \left(\frac{T_2}{100}\right)^4\right] \text{ W/m}^2$$

$$= 5.72 \times 0.6 \times 0.9\left[\left(\frac{1{,}200 + 273}{100}\right)^4 - \left(\frac{700 + 273}{100}\right)^4\right]$$

$$= 117{,}727.4 \text{ W/m}^2$$

Surface area available for heat radiation = $2wl$ m^2

Since heat dissipated = Power input

$117{,}727.4 \times 2wl = 6{,}667$

or $wl = \frac{6{,}667}{2 \times 117{,}727.4} = 0.0283 \text{ m}^2$ ...(*ii*)

Dividing Eq. (*i*) by Eq. (*ii*), we have

$w = \sqrt{\frac{0.0283}{230.476}} = 0.011084 \text{ m} = 11.084 \text{ mm}$ **Ans.**

Length of strip, $l = 230.476 \times 11.084 \times 10^{-3} = 2.555$ m **Ans.**

Example 5.4. A cubic water tank has surface area of 6.0 m^2 and is filled to 90% capacity six times daily. The water is heated from 20 °C to 65 °C. The losses per square metre of tank surface per 1 °C temperature difference is 6.3 watts. Find the loading in kW and the efficiency of the tank. Assume specific heat of water = 4,200 J/kg/°C and one kWh = 3.6 MJ.

Solution : Let the one side of the tank be l metres

Surface area of the tank, $6l^2 = 6.0 \text{ m}^2$

or $l = 1.0$ m

Volume of the tank $= 1.0 \times 1.0 \times 1.0 = 1.0 \text{ m}^3$

Water to be heated daily $= 6 \times 0.9 \text{ m}^3 = 5{,}400$ kg.

$\because$ 1 cubic metre water weights 1,000 kg

Heat required to raise the temperature of water

$= ms(t_2 - t_1) = 5{,}400 \times 4{,}200 \times (65 - 20)$

$= 1{,}020 \text{ MJ} = \frac{1{,}020}{3.6} = 283.5$ kWh

Losses from the surface of the tank

$= \frac{6.3 \times 6.0 \times (65 - 20) \times 24}{1{,}000} \text{ kWh} = 40.8$ kWh

Energy supplied $= 283.5 + 40.8 = 324.3$ kWh

Loading in kW $= \frac{324.3}{24} = 13.5$ kW **Ans.**

Efficiency of the tank $= \frac{\text{Output}}{\text{Input}} \times 100 = \frac{283.5}{324.3} \times 100$

$= 87.4\%$ **Ans.**

5.5. ARC FURNACES

When a high voltage is applied across an air gap, the air in the gap gets ionised under the influence of electrostatic forces and becomes conducting medium. Current flows in the form of a continuous spark, called the *arc*. It is to be noted that a very high voltage is required to establish an arc across an air gap but to maintain an arc small voltage may be sufficient. Alternatively an arc can also be produced by short circuiting the two electrodes momentarily and then withdrawing them back. In this method of striking an arc, high voltage is not required. Arc drawn between two electrodes produces heat and has a temperature between 1,000 °C and 3,500 °C depending on the material of the electrodes used. *The use of this principle is made in electric arc furnaces.*

Usually arc furnaces are of cylindrical shape but recently conical shaped shells have been used. Even in conical shapes the horizontal cross section is cylindrical. The conical shape has the advantage of a large surface area per unit bath volume. It consumes less power and the radiation losses are reduced. Melting time is also reduced. Bottom, side walls and roof of the furnace are lined with firebricks, magnesite bricks and silica bricks. The bottom is lined with magnesite mix or ground ganister mix depending upon whether the basic or acidic lining is required. The electrodes project through the top or sides of the chamber and are arranged for easy replacement and adjustment. Each furnace is provided with charging door and tap hole for introducing the charge and taking out the molten metal. These doors and tap holes are also to be suitably lined. To minimize heat losses the size of the doors/tap holes and the opening time has to be reduced. The furnace rests on platform which can be titled for pouring the molten metal through the door opening in the side of the shell. Tilting mechanism consists of a motor which drives a pinion which meshes with a semicircular toothed rack at the bottom of the platform.

The furnaces may be *door-charge type* or the *top-charge type.* Large furnace installations are often equipped with side door charging. Top charging is growing in favour for medium size furnaces. In this method for charging the roof of the furnace is removed, and a complete charge is placed in the chamber by a drop bucket handled by an overhead crane. This saves labour and time.

The electrodes used in arc furnaces are of three types namely carbon electrodes, graphite electrodes and self-baking electrodes. Material of the electrodes namely carbon and graphite has been selected on account of their electrical conductivity, insolubility, infusibility, chemical inertness, mechanical strength and resistance to thermal shock. The size of these electrodes may be 18 cm to 27 cm in diameter. With graphite or carbon electrode, the temperature obtainable from the arc is between 3,000 °C and 3,500 °C. The trend is toward the general use of graphite electrodes. Carbon electrodes are used with small furnaces for manufacture of ferro-alloys, aluminium, calcium carbide, phosphorous etc. Self-baking electrodes are employed in ferro-alloys and electrochemical furnaces and in electrolytic production of aluminium. Owing to lower resistivity of graphite (one fourth that of carbon), graphite is required half in size for the same current resulting in easy replacement. Graphite begins to oxidise at about 600 °C where as carbon at about 400 °C. Under average conditions the consumption of graphite electrodes is about one half that of carbon electrodes. Electrode consumption for steel-melting furnaces varies between about 4.5 and 9 kg of electrodes per tonne of steel with carbon electrodes, and between about 2.27 and 6.8 kg for graphite electrodes. Carbon electrodes are very cheap and cost less than one half as much for same weight as graphite electrodes. The large area of carbon electrodes allow for more uniform heating. However, the arc has to be brought nearer the surface of the refractory lining and, therefore, the life of the refractory is shortened. Carbon electrodes are made of anthracite coal and coke where as graphite electrodes are obtained by heating the carbon electrodes to a very high temperature. Self-baking electrodes are made of special paste, whose composition depends upon the type of process for which it is used, contained in thin steel cylinder. The flow of current produces heat and the paste is baked and formed into an electrode.

5.5.1. Types of Arc Furnaces

Arc furnaces are of two types namely (*i*) direct arc furnaces and (*ii*) indirect arc furnaces. In a direct arc furnace, shown in Fig. 5.4, the arc is formed between the electrodes and the charge, while in the indirect arc furnace the arc is formed between the electrodes, as shown in Fig. 5.5, and heat is transmitted to the charge solely by radiation.

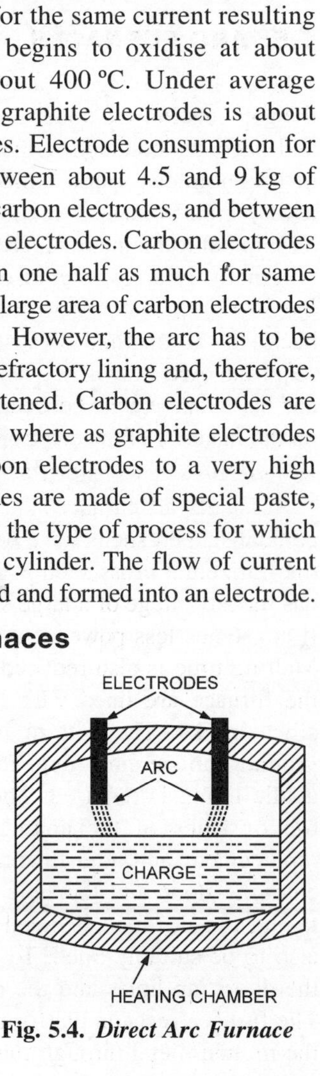

Fig. 5.4. ***Direct Arc Furnace***

1. ***Direct Arc Furnace.*** In a direct arc furnace charge acts as one of the electrodes and the charge is heated by producing arc between the electrodes and the charge. Since in a direct arc furnace, the arc is in direct contact with the charge and heat is also produced by flow of current through the charge itself, the charge can be, therefore, heated to highest temperature. In case of a single phase arc furnace two electrodes are taken vertically downward through the roof of the furnace to the surface of the charge and in a 3-phase furnace three electrodes put at the corners of an equilateral triangle, project on the charge through the roof and three arcs are formed. The current passing through the charge develops electromagnetic field and necessary stirring action is automatically obtained by it. Thus uniform heating is obtained.

It is commonly used for production of steel. The usually size of such a furnace is between 5 and 10 tonnes, though 50 and 100 tonne arc furnaces have also been developed. The main advantage of direct arc furnace over cupola method for production of steel is that purer production is obtained and the composition can be exactly controlled during refining process. Another advantage is that arc furnace can operate on 100% steel scrap which is cheaper than pig iron where as the cupola requires a proportion of pig iron in cupola charge. This is the reason, that direct arc furnaces even being costlier in initial as well as operating costs are preferred. Though such furnace is employed both for melting and refining, but due to its higher cost it is used for refining than melting. The power factor is about 0.8 lagging. For 1 tonne furnace power required is about 200 kW and energy consumed is 1,000 kWh/tonne.

2. ***Submerged Arc Furnace.*** These are cylindrical furnaces in which arc is formed between the carbon electrodes (from the top) and hearth electrodes. The hearth lining is of magnesite which becomes comparatively good electrical conductor when hot. It is also mixed with coke or graphite. Sometimes a conducting hearth is used as electrode. The number of electrodes, taken from the roof depends on the type of supply. One for single phase, two or four for two phase and three for three phase supply, bottom conductor being connected to the neutral. Usually 3-phase supply is used for dealing with large power, usually required. The current from the top electrode passes through the arc to the charge and returns through the electrode at the bottom of the charge. Better distributed resistance heating is obtained since charge behaves as the resistance. Similarly better mixing of charge takes place. The current under short circuit is limited due to charge, which otherwise in indirect furnace is very high. Power is controlled by varying distance between the electrodes or by varying the voltage applied to the electrodes. In a direct arc furnace the length of arc is varied for this purpose. These furnaces are used for the manufacture of ferro-alloys like ferro-chrome and ferro-manganese. The power factor is about 0.8 lagging. The stirring action is inherent due to electromagnetic forces set up by the current resulting in uniform heating of charge.

3. ***Indirect Arc Furnace.*** In this case arc is formed between two electrodes above the charge, as shown in Fig. 5.5

and heat is transmitted to the charge solely by radiation. In this case the temperature of the charge is, therefore, lower than that in case of direct arc furnace. Since in this furnace current does not flow through the charge, there is no stirring action and the furnace is required to be rocked mechanically. That is why the furnace is made of cylindrical shape, with the electrodes projecting through the chamber from each end and along the horizontal axis. By rocking action there is thorough mixing of the charge. The life of the refractory lining also increases since the molten metal comes in contact with the lining and takes away some of its heat thus preventing it from attaining excessive temperature. The efficiency is increased because the charge is heated not only by radiation from the arc but by conduction from the heated refractory during rocking action. Its construction limits the number of electrodes to two, so single-phase supply is required. The size of the furnace is thus limited by the amount of single-phase load, which can be taken from one point. The arc is produced by bringing the electrodes into solid contact and then withdrawing them. Power input is regulated by adjusting the arc length by moving the electrodes.

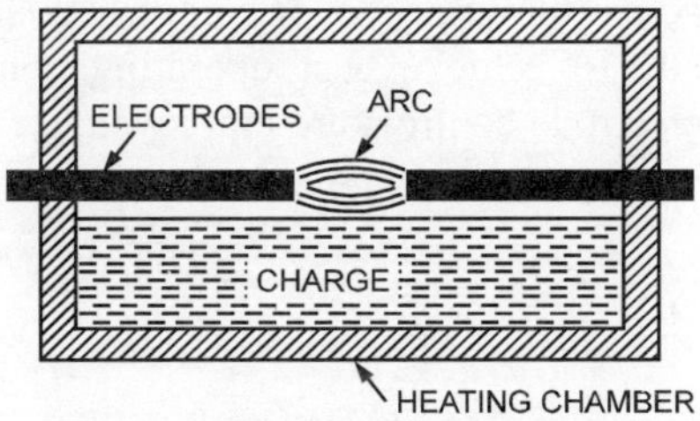

Fig. 5.5. ***Indirect Arc Furnace***

An electric motor is employed for operating suitable grinders and rollers to provide rocking action to the furnace. At start, the rocking action is carried through an angle of 15 to 20º and as the melt proceeds the angle is increased to about 200º at a frequency of about 2 cycles per minute.

Such furnaces have got following advantages :

(i) ***Low Metal Losses.*** Since furnace chamber is closed and a reducing atmosphere above the metal is produced due to carbon arc, therefore, metal losses due to oxidation and volatilisation are quite low.

(ii) ***Economical.*** Melting is rapid and takes place in a completely closed chamber resulting in small heat losses and low power consumption. Higher production rate gives low labour costs resulting in lower overall production cost per tonne of molten metal.

(iii) ***Sound Castings.*** The agitation produced by the rocking action of the furnace and the absence of combustion gases ensure products free from blow holes, inclusion and segregations. Higher pouring temperature may readily be attained. Thus sound castings in thin and intricate designs can be produced.

(iv) ***Flexible.*** Single furnace is capable of handling small or large heat of widely differing analysis. Interchangeable furnace shells can be used for different alloys and can be substituted in a few minutes.

Due to indirect heating the furnace is suitable for comparatively low melting point. So these furnaces are used for melting non-ferrous metals. These are also used to a considerable extent in iron foundries where relatively small quantity of metal is required intermittently. The capacity of furnace varies from 0.25 tonne to 3 tonnes. The power factor is about 0.85. The energy consumption is about 275 kWh/tonne for brass; 325 kWh/tonne for bronze and 400 kWh per tonne for copper.

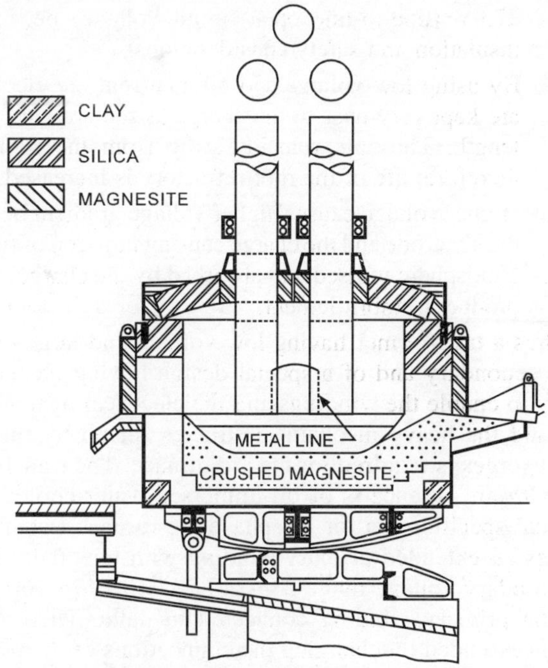

Fig. 5.6. ***Three-Phase Arc Furnace***

Even though Ajax Wyatt induction furnace is its chief rival in the field of the non-ferrous metals, owing to its higher efficiency, it is more suitable in situations where the charge has to be varied frequently or where heating is intermittent.

The three-phase arc furnace having a steel bowl chamber with a refractory lining is shown in Fig. 5.6. Single-phase rocking arc furnace is shown in Fig. 5.7.

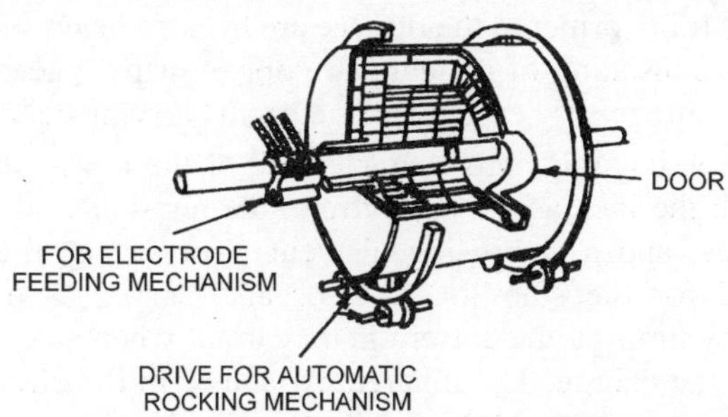

Fig. 5.7. ***Single-Phase Rocking Arc Furnace***

5.5.2. Power Supply and Control

An arc furnace used for melting and refining of steel requires power of about 500 kW per tonne for small furnaces (say for 1/2 tonne capacity) and of about 200 kW per tonne for very large furnaces (say for 50–100 tonne capacity). The energy required is usually between 600 and 800 kWh per tonne. For the arc furnaces used for refining only electrical energy consumption is about 100–120 kWh/tonne. Thus it is seen that the power consumption of the arc furnace is very high. The arc voltage lies between 50 and 150 volts, the current required to give the above mentioned power is, therefore, of the order of several hundred or thousand amperes. The reasons for employing low voltage high current power supply for arc furnaces are as follows :

(*i*) Heating effect is proportional to the square of the current, therefore, to achieve higher temperatures heavy currents are essential.

(*ii*) The maximum secondary voltage is also limited to 275 V (line-to-line open-circuit voltage) because of insulation and safety considerations.

(*iii*) By using low voltage and high current the electrodes are kept very near to the charge as the arc is of small length. Thus arc remains away from the roof and, therefore, life of the roof refractory is increased.

(*iv*) Higher voltage causes higher voltage gradient between the electrode and the charge causing nitrogen of furnace atmosphere ionised and absorbed by the charge, which produces embrittlement.

Thus a transformer having low voltage and large current on the secondary and of a special design having mechanical rigidity to enable the windings and the electrical insulation to withstand the heavy mechanical stresses set up by the high current surges is required for the arc furnace. The transformer used with arc furnace is of oil-immersed water-cooled type. A typical specification for a 3-phase arc furnace transformer includes an extended primary winding with taps there in for the secondary voltage range 235-220-205-190-175-160 volts, with the primary winding connected in delta. This voltage range is extended by changing the connections of the primary windings from delta to star giving 58% voltage from each tap. Both core type and shell type transformers, can be used, but the latter type is preferred because it facilitates the bringing out and bracing of the heavy current leads.

It is desirable to arrange the furnace and transformer in such a way that the current leads are of shorter length. Also in order to keep inductance of these leads low they should be placed as close together as possible and they must be of such a shape as to reduce skin effect to a minimum. Since the leads from the transformer to the furnace are to carry heavy currents, so leads consisting of rectangular copper strips spaced a few mm apart are mostly employed as current carrying conductors.

The voltage-ampere characteristic of the arc is negative, *i.e.*, with the increase in arc current, the resistance of the arc decreases, and a stabilising element (*reactor*) is, therefore, required for the stability. Reactor also serves as a safety device by limiting the current in the circuit when an electrode touches the charge. The inherent reactance in the circuit of a large furnace 5,000 kVA and larger may be and usually is sufficient for the purpose but the smaller installations require more or less supplemental reactance.

The equivalent circuit of an electric arc furnace is shown in Fig. 5.8 where R_T and X_T are the equivalent resistance and reactance of the furnace transformer referred to the secondary side. R_L and X_L are the resistance and reactance of leads. R_E is the resistance of electrodes and $E_A = IR_A$, the voltage drop across the arc. The arc current depends upon the resistance and reactance of transformer (R_T and X_T), resistance and reactance of the leads (R_L and X_L), resistance of the electrodes (R_E) and the arc resistance R_A.

The power input can be controlled by raising or lowering the electrodes which results in variation of the arc resistances R_A. The power input can be also controlled by changing the tapping of transformer which results in variation of voltage across the furnace. For complete control of furnace temperature and to achieve best operating conditions both voltage and electrode controls are employed.

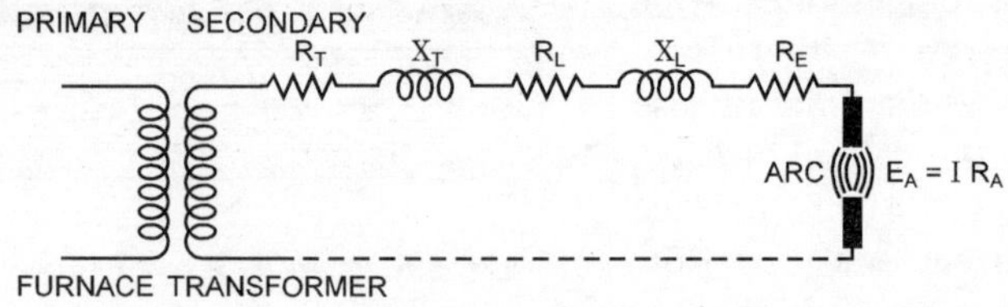

Fig. 5.8. ***Equivalent Circuit of An Arc Furnace***

Figure 5.9 illustrates the performance characteristics of a typical arc furnace. Initially when the electrodes are touching the charge, *i.e.*, arc resistance is zero, the current is limited by the impedance of the furnace circuit and the power input to the furnace is almost zero (copper losses in the transformer windings). Again when the electrodes are raised so that the arc is extinguished, the input to the furnace is zero. In between two these two limits there is a particular loading when the power input to the furnace is maximum.

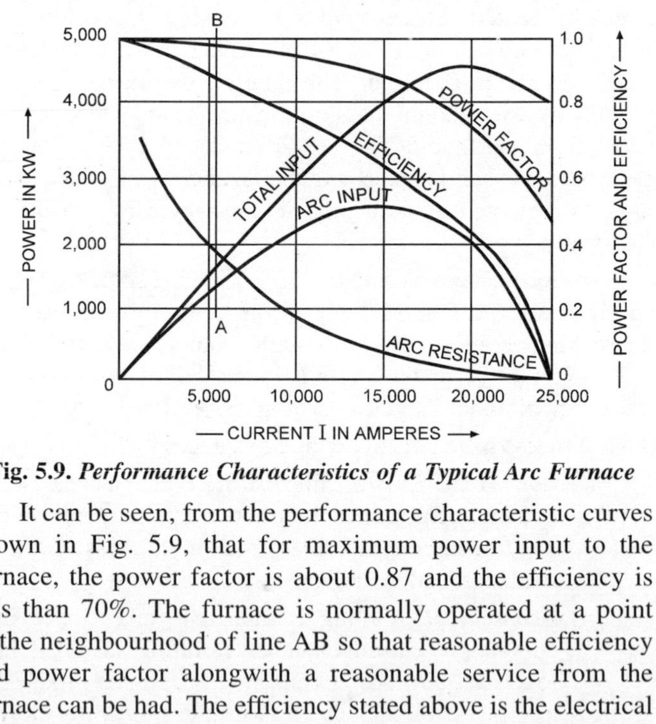

Fig. 5.9. ***Performance Characteristics of a Typical Arc Furnace***

It can be seen, from the performance characteristic curves shown in Fig. 5.9, that for maximum power input to the furnace, the power factor is about 0.87 and the efficiency is less than 70%. The furnace is normally operated at a point in the neighbourhood of line AB so that reasonable efficiency and power factor alongwith a reasonable service from the furnace can be had. The efficiency stated above is the electrical efficiency (taking into account only the ohmic losses occurring in the furnace circuit), overall efficiency is about 50-60 per cent taking into account all other losses such as used in heating up the refractory material and furnace chamber, losses through walls, losses through escaping gases, losses through water cooling. Water cooling is applied to the openings in the roof through which electrodes are placed, the roof ring and the door frame.

Condition For Maximum Output

$$\text{Arc current, } I = \frac{V}{\sqrt{(R_T + R_L + R_A)^2 + (X_T + X_L)^2}}$$

$$\text{Power loss in the arc} = I^2 R_A$$

$$= \frac{V^2}{(R_T + R_L + R_A)^2 + (X_T + X_L)^2} \times R_A$$

$$= \frac{V^2 R_A}{R_A^2 + 2R_A(R_T+R_L) + (R_T+R_L)^2 + (X_T+X_L)^2}$$

$$= \frac{V^2}{R_A + 2(R_T+R_L) + \frac{(R_T+R_L)^2}{R_A} + \frac{(X_T+X_L)^2}{R_A}}$$

Power loss will be maximum when denominator will be minimum

$$\therefore \quad \frac{d}{dR_A}\left[R_A + 2(R_T+R_L) + \frac{(R_T+R_L)^2 + (X_T+X_L)^2}{R_A}\right] = 0$$

$$\text{or } 1 + 0 - \frac{(R_T+R_L)^2 + (X_T+X_L)^2}{R_A^2} = 0$$

$$\text{or } R_A = \sqrt{(R_T+R_L)^2 + (X_T+X_L)^2}$$

i.e., Power loss in the furnace will be maximum when the arc resistance R_A will be numerically equal to the impedance of the whole electrical circuit referred to the secondary excluding arc resistance R_A.

Power factor at maximum power loss,

$$\text{Cos } \phi = \frac{R_A + R_T + R_L}{\sqrt{(R_A+R_T+R_L)^2 + (X_T+X_L)^2}}$$

$$= \frac{R_A + R_T + R_L}{\sqrt{(R_A^2 + 2R_A(R_T+R_L)) + (R_T+R_L)^2 + (X_T+X_L)^2}}$$

Substituting $(R_T + R_L)^2 + (X_T + X_L)^2 = R_A^2$, we get

$$\text{Cos } \phi = \frac{R_A + R_T + R_L}{\sqrt{2R_A^2 + 2R_A(R_T+R_L)}}$$

$$= \sqrt{\frac{R_A + R_T + R_L}{2R_A}}$$

$$= \frac{1}{\sqrt{2}}\sqrt{1 + \frac{R_T+R_L}{R_A}} = \frac{1}{\sqrt{2}} = 0.707$$

if we neglect $R_T + R_L$ in comparison with R_A. However, it is not economical to operate an arc furnace with primary side power factor under 0.8.

Example 5.5. A 3-phase arc furnace is used to melt 4.3 tonnes of steel in one hour. Determine the average kW and kVA input to the furnace, arc voltage, arc resistance and pf of the current drawn from the supply. Given:

Specific heat of steel = 444 J/kg/°C

Latent heat of steel = 37.25 kJ/kg

Melting point of steel = 1,370°C

Initial temperature of steel = 19.1°C

Overall efficiency of furnace = 50% = 0.5

Input current = 5,700 A

Resistance of transformer referred to secondary and reactance of the transformer referred to secondary are 0.008 and 0.014 ohm respectively. [*Pb. Technical Univ. Electric Drives and Utilization May 2009*]

Solution : Energy required to melt 4.3 tonnes of steel

$= ms(t_2 - t_1) + mL$

$= m[s(t_2 - t_1) + L]$

$= 4.3 \times 1,000[444(1,370 - 19.1) + 37.25 \times 1000]$

$= 2,739,313,280$ J

$$= \frac{2,739,313,280}{3.6 \times 10^6} = 760.92 \text{ kWh}$$

Time taken in melting of steel = 1 hour

$$\text{Average output} = \frac{\text{Total energy required in kWh}}{\text{Time of melting in hours}} = \frac{760.92}{1} = 760.92 \text{ kW}$$

$$\text{Average input} = \frac{\text{Average output}}{\text{Overall efficiency}} = \frac{760.92}{0.5} = 1,521.84 \text{ kW} \textbf{ Ans.}$$

Current input, I = 5,700 A

Voltage drop due to resistance of furnace leads (including transformer)

$= 5,700 \times 0.008 = 45.6$ V

Voltage drop due to reactance of furnace leads (including transformer)

$= 5,700 \times 0.014 = 79.8$ V

Let the arc drop be V_A volts, resistive in nature

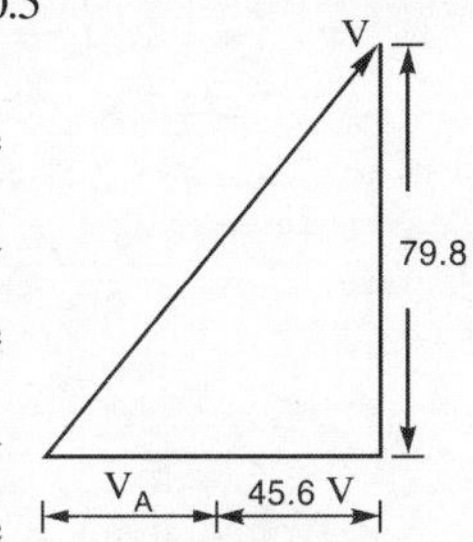

Fig. 5.10

Open-circuit phase voltage of trans-former secondary

$$= \sqrt{(V_A + 45.6)^2 + (79.8)^2}$$

$$\text{Power factor, cos } \phi = \frac{V_A + 45.6}{\sqrt{(V_A + 45.6)^2 + 79.8^2}}$$

Total power input = 3 × power drawn per phase

= 3 × current drawn per phase × secondary phase voltage × pf

$$\text{or } 1,521.84 \times 1,000 = 3 \times 5,700 \times \sqrt{(V_A + 45.6)^2 + (79.8)^2} \times \frac{V_A + 45.6}{\sqrt{(V_A + 45.6)^2 + (79.8)^2}}$$

$$\text{or } V_A = \frac{1,521.84 \times 1,000}{3 \times 5,700} - 45.6 = 43.4 \text{ V} \textbf{ Ans.}$$

$$\text{Arc resistance, } R_A = \frac{\text{Arc voltage}}{\text{Arc current}} = \frac{43.4}{5,700} = 0.0076\ \Omega \textbf{ Ans.}$$

Open-circuit phase voltage of transformer (secondary)

$$= \sqrt{(V_A + 45.6)^2 + (79.8)^2}$$

$$= \sqrt{(43.4 + 45.6)^2 + (79.8)^2} = 119.54 \text{ V}$$

Total kVA drawn from the supply

= 3 × current drawn per phase × secondary phase voltage × 10^{-3}

$= 3 \times 5,700 \times 119.54 \times 10^{-3}$

= 2,044 kVA **Ans.**

Power factor of current drawn from supply

$$= \frac{V_A + 45.6}{\sqrt{(V_A + 45.6)^2 + (79.8)^2}}$$

$$= \frac{43.4 + 45.6}{\sqrt{(43.4 + 45.6)^2 + (79.8)^2}} = 0.7445 \text{ (lagging)} \textbf{ Ans.}$$

Example 5.6. A 3-phase arc furnace is used to melt 4.3 tonnes of steel in one hour. Determine the average kW and kVA input to the furnace, arc voltage, arc resistance and pf of the current drawn from the supply. Given:

Specific heat of steel = 0.12 kcal/kg/°C

Latent heat of steel = 8.89 kcal/kg/°C

Melting point of steel = 1,370°C

Initial temperature of steel = 19.1°C

Overall efficiency of furnace = 50%

Input current = 5,700 A

Resistance of transformer referred to secondary and reactance of the transformer referred to secondary are 0.008 and 0.014 ohm respectively. [*Pb. Technical Univ. Electric Drives and Utilization Dec. 2008*]

Solution : Energy required to melt 4.3 tonnes of steel

$$= m[s(t_2 - t_1) + L]$$

$$= 4.3 \times 1000[0.12(1,370 - 19.1) + 8.89]$$

$$= 735,291.4 \text{ kcals}$$

$$= \frac{735,291.4}{860} = 855 \text{ kWh} \qquad \because 1 \text{ kWh} = 860 \text{ kcals}$$

$$\text{Average output} = \frac{855}{1} = 855 \text{ kW}$$

$$\text{Average input} = \frac{855}{0.5} = 1710 \text{ kW} \quad \textbf{Ans.}$$

Current input, I = 5,700 A

Voltage drop due to resistance of furnace leads (including transformer)

$$= 5,700 \times 0.008 = 45.6 \text{ V}$$

Voltatge drop due to reactance of furnace leads (including transformer)

$$= 5,700 \times 0.014 = 79.8 \text{ V}$$

Let the arc drop be V_A volts, resistive in nature

Open-circuit phase voltage of transformer secondary

$$= \sqrt{(V_A + 45.6)^2 + (79.8)^2}$$

$$\text{Power factor, } \cos\phi = \frac{V_A + 45.6}{\sqrt{(V_A + 45.6) + (79.8)^2}}$$

Total power input = 3 × I × secondary phase voltage × pf

$$\text{or } 1,710 \times 1,000 = 3 \times 5,700 \times \sqrt{(V_A + 45.6)^2 + (79.8)^2} \times \frac{V_A + 45.6}{\sqrt{(V_A + 45.6)^2 + (79.8)^2}}$$

$$\text{or Arc voltage, } V_A = \frac{1,710 \times 1,000}{3 \times 5,700} - 45.6 = 54.4 \text{ V} \quad \textbf{Ans.}$$

$$\text{Arc resistance, } R_A = \frac{VA}{I} = \frac{54.4}{5,700} = 0.00954\ \Omega$$

Open-circuit voltage of transformer secondary

$$= \sqrt{(V_A + 45.6)^2 + (79.8)^2}$$

$$= \sqrt{(54.4 + 45.6)^2 + (79.8)^2} = 127.94 \text{ V}$$

Total kVA drawn = 3 × 5,700 × 127.94 × 10^{-3} = 2,187.7 kVA **Ans.**

$$\text{Power factor, } \cos\phi = \frac{\text{kW}}{\text{kVA}} = \frac{1,710}{2,187.7} = 0.7816 \text{ (lagging)} \quad \textbf{Ans.}$$

5.6. INDUCTION HEATING

In induction heating effect of currents induced by electromagnetic action in the charge is employed. The heat developed depends on the power drawn by the charge. The power consequently depends upon the voltage and resistance of the charge, because power drawn is equal to $\frac{V^2}{R}$. So to develop heat sufficient to melt the charge, the resistance of the charge must be low, which is possible only with metals, and voltage must be higher, which is obtained by employing higher flux and higher frequency. Magnetic materials, therefore, can be easily treated than non-magnetic materials because of their higher permeability.

The various types of induction furnaces are described as follows.

5.6.1. Core Type Furnaces

The core type furnace is just like a transformer having primary connected to the supply and the charge to be heated as secondary, as shown in Fig. 5.11.

1. *Direct Core Type Induction Furnace.* Direct core type induction furnace is shown in Fig. 5.11 which consists of an iron core, crucible of some insulating material and primary winding connected to an ac supply. The charge is kept in the crucible, which forms a single turn short-circuited secondary circuit. The current in the charge is very high, of the order of several thousand amperes.

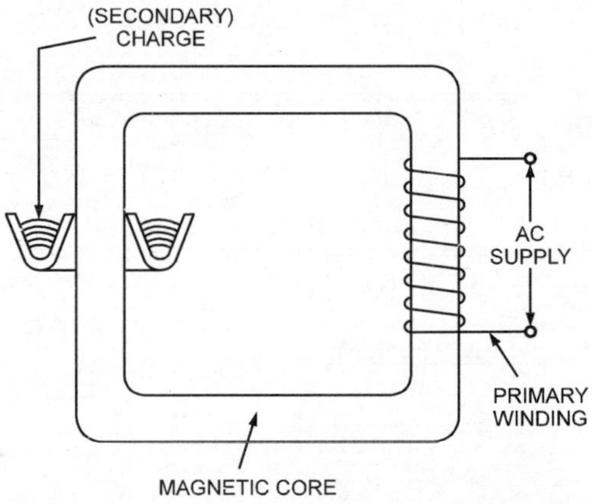

Fig. 5.11. ***Direct Core Type Induction Furnace***

This type of furnace has following drawbacks :

1. As magnetic coupling between the primary and secondary circuit is poor, therefore, leakage reactance is high and power factor is low. This difficulty, however, is overcome by employing supply of frequencies as low as 10 Hz for operation of such furnaces. For obtaining low frequency supply motor-generator set or frequency changer is required, which involves extra cost.
2. If normal frequency supply is employed for operation of such furnaces, the electromagnetic forces cause severe stirring action in the molten metal. Low frequency supply, therefore, is also necessary from this point of view.
3. If the current density exceeds about 5 amps per mm^2 the *pinch effect* (formation of bubbles and voids etc.) due to electromagnetic forces may cause complete interruption of the secondary circuit and so of supply.
4. The crucible for the charge is of odd shape and inconvenient from the metallurgical point of view.
5. For functioning of the furnace the closing of the secondary circuit is essential which necessitates the formation of

complete ring of the charge around the core. For starting the furnace, either molten metal is poured into the crucible or sufficient molten metal is allowed to remain in the crucible from the previous operation. This is required otherwise, the secondary will remain open circuited and no current will circulate and no heating will take place. However, once the melting starts additional metal can be added and the furnace tapped periodically. Also in order to close the secondary circuit, an iron ring may be placed in the crucible or the lining may be of graphite.

Such furnaces are not suitable for intermittent services or where different types of charges are to be melted.

On account of the drawbacks mentioned above such furnaces have become obsolete nowadays.

2. *Vertical Core Type Induction Furnace.* This is an improved form of furnace described above. The furnace of this type, known as Ajax Wyatt vertical core type furnace (shown in Fig. 5.12) employs a vertical channel instead of horizontal one for the charge. The convection currents keep the circulation of molten metal round the V portion. As V-channel is narrow, so even a small quantity of charge is sufficient to keep the secondary circuit closed. Hence the chances of discontinuity of the circuit are less. Due to pinch effect the adjoining molecules carrying current in same direction will try to repel each other, but because of the weight of the charge they will remain in contact and chances of interruption will be reduced.

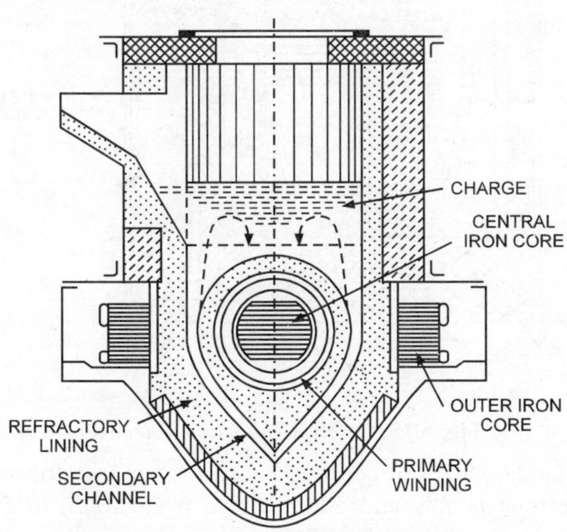

Fig. 5.12. *Ajax Wyatt Vertical Core Furnace*

The output of the furnace depends upon the type and dimensions of the channels used. In certain furnaces instead V-shaped channels U-shaped channels or rectangular channels are employed.

The inside of the furnace is lined depending upon the charge. Clay lining is used for yellow brass. For red brass and bronze an alloy of magnesia and alumina or corrundum having high contents of alumina is employed.

The shell of the furnace is of heavy steel. The top of the furnace is covered with an insulated cover which can be removed for charging. Necessary hydraulic arrangements are usually made for tilting the furnace to take out the molten metal.

Advantages. 1. Highly efficient heat, low operating costs and improved production.

2. Accurate temperature control, uniform castings, reduced metal losses and reduction of rejects.
3. Absence of crucibles.
4. Consistent performance and simple control.
5. Ideal working conditions in a cool atmosphere with no dirt, noise or fuel.
6. Absence of combustion gases resulting in elimination of the most common source of metal contamination.
7. Comparatively high power factor (the average power factor between 0.8 and 0.85) with normal supply frequency since primary and secondary are both on the same central core.

However, it is to be noted that the Vee must be kept full of charge in order to maintain continuity of the secondary circuit. For this reason, this type of furnace is suitable only for continuous operation. These furnaces are widely used for melting and refining of brass and other heavy non-ferrous metals. Its efficiency is about 75 per cent. Standard sizes of these furnaces range 60 to 300 kW, all single phase, 50 Hz in this country for standard voltage up to 600 V.

3. *Tama Furnace.* The main drawback of Ajax Wyatt furnace is that it cannot be used for melting aluminium and its alloys. This is mainly because of tendency of ahderence, of any aluminium oxide formed, to the walls of V channel. This makes the operation of furnace impossible any longer. This difficulty, however, can be overcome by using a ring type furnace, known as the *Tama furnace,* depicted in Fig. 5.13.

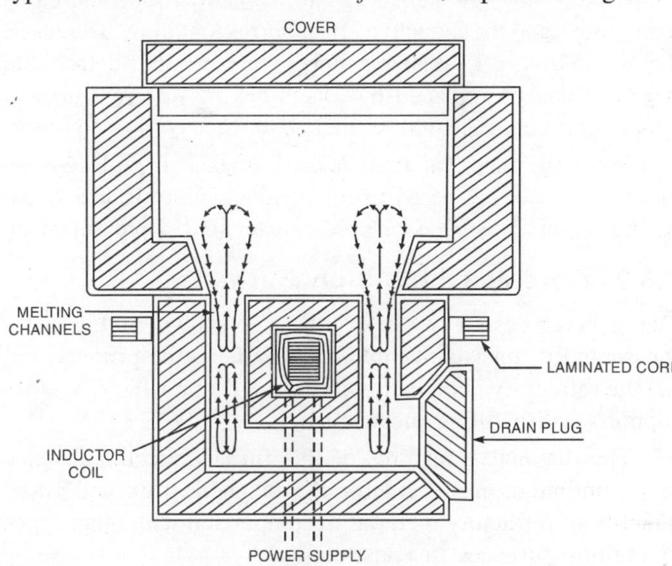

Fig. 5.13. *Tama Furnace*

In this furnace inductor channels are built vertical which facilitates removal of the deposits with a simple and cheap type cleaning tool. Vertical channels can be cleaned without emptying the furnace and without taking long time. The horizontal channel through the base of furnace does not require frequent cleaning. No doubt, it needs emptying of furnace before removal of drain plug but the operation is speedy. The pf of the Tama furnace is reduced because of its straight sided channels and therefore installation of additional capacitors is required which makes the Tama furnace costlier than the Ajax Wyatt furnace. Stirring action is also set up due to electromagnetic forces in this furnace like Ajax Wyatt furnace.

4. *Indirect Core Type Induction Furnace.* In such a furnace an inductively heated element is made to transfer its heat to the charge by radiation. In this type of furnace the principle of induction has been utilised for providing heat treatment of metallic and other charges in addition to its use for melting metals. Such a furnace is shown in Fig. 5.14.

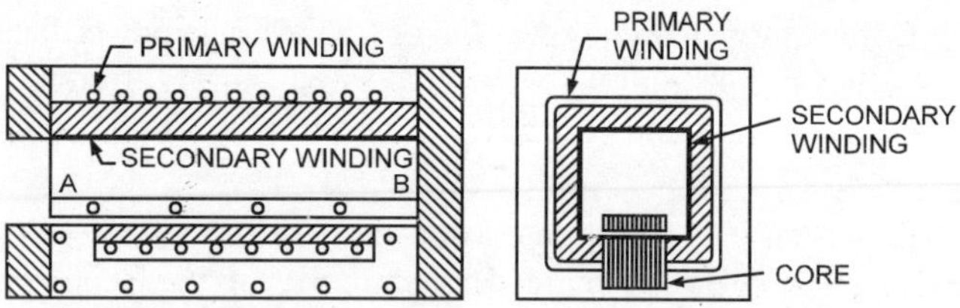

Fig. 5.14. ***Indirect Core Type Induction Furnace***

It consists of an iron core linking with the primary winding and secondary also. In this case secondary consists of a metal container forming the walls of the oven proper. Primary winding is connected to the ac supply, inducing currents and heating the metal container. Heat is transmitted to the charge by radiation. It is advantageous in respect of temperature control without use of external control equipment. It consists of part AB of the magnetic circuit situated in the oven chamber and consisting of a special alloy which looses its magnetic properties at a particular temperature and regains them when cooled to the same temperature. As soon as the oven attains the critical temperature, the reluctance of the magnetic circuit increases many times and the inductive effect correspondingly decreases, thereby cutting off the heat supply. The bar AB is detachable type and can be replaced by others having different critical temperatures between 400 °C and 1,000 °C, according to needs.

From the mode of transmission of heat it will be seen that this furnace is directly in competition with resistance oven; but has comparatively poor power factor (0.8 approximately).

5.6.2. Coreless Induction Furnace

The general design of this furnace is shown in Fig. 5.15. It essentially consists of three main parts (*i*) the primary coil (*ii*) the refractory container and (*iii*) the frame which includes supports and a tilting mechanism.

The distinctive features of this furnace are the absence of a continuous iron path for the magnetic flux and small quantity of refractory material in comparison with other types of melting furnaces in construction.

Standard preformed crucibles are employed for small furnaces (up to about 200 kg holding capacity). The base and the wall around the crucible are made by ramming granular refractory material. The top of the wall is sealed with refractory cement. The containers of the large furnaces are made in place, the procedure being same as for the smaller furnaces except that a hollow collapsible form is substituted for the crucible to form the receptacle. Acid or basic materials are used as per requirements.

For floor level mounting, the electrical connections are made by knife contacts in order to make the handling of the furnace as a ladle for pouring. For platform mounting the electrical connections are made with flexible cables. This is done as with this arrangement, power can be left on the furnace while pouring—a feature often desirable when a number of castings are poured from one heat. A variation is the lift-coil furnace made in the smaller sizes. The primary coil is lowered and raised over the load crucible.

The charge is put into the crucible and primary winding coil is connected to high frequency ac supply. The flux created by primary winding sets up eddy currents in the charge which tend to flow concentrically with those in the inductor. These eddy currents heat up the charge to its melting point and also set up electromagnetic forces producing stirring action which is essential for obtaining uniform quality of metal. Because of high frequency employed, which is necessary to induce the required voltage in the secondary, the skin effect produces heat in the primary winding coils. The primary winding coils are, therefore, made from hollow tube and are cooled by circulation of water through it. Insulated supporting structure is employed for such furnaces, otherwise stray magnetic field outside the primary will set up emf in it, which will result in circulation of eddy currents in it and so reduction of efficiency.

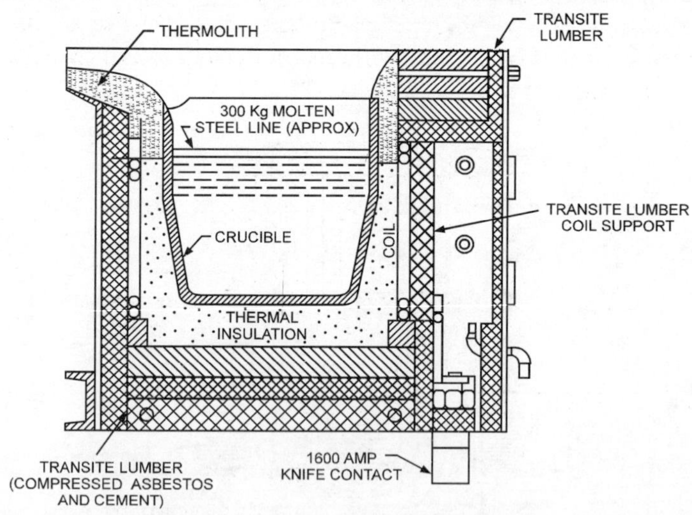

Fig. 5.15. ***Coreless Induction Furnace***

Standard sizes of coreless induction furnaces for melting non-ferrous metals and alloys range from 50 kg to 500 kg holding capacity.

The choice of frequency of operation plays vital role. It is governed by the factors, material to be heated and thickness of cylinder layer at the outside edge of the crucible. The frequency of the primary current can be ascertained by using penetration formula. Accordingly,

$$t = \frac{1}{2\pi}\sqrt{\frac{\rho \times 10^7}{\mu f}} \text{ m} \qquad ...(5.11)$$

where ρ is the resistivity of molten metal in Ω-m, f is supply frequency in Hz and μ is relative permeability (for molten steel $\mu = 1$).

The exact theory also shows that for efficient operation the ratio of radius of piece of material in the charge to the thickness should be greater than 3. If it is taken as 4, then the expression for the frequency, for efficient operation, becomes

$$f = \frac{1}{4\pi^2} \times \frac{\rho \times 10^7 \times 16}{\mu \times r^2} \quad ...(5.12) \quad \because \frac{r}{t} = 4$$

From above Eq. (5.12) it can be seen that by increasing r efficient operation can be obtained with lower frequencies. In most of the modern coreless induction furnaces the frequencies in the range of 500 to 1,000 Hz are used. However, for smaller units for melting small quantities of finely divided metal frequencies up to 100 kHz or even 1,000 kHz are used.

The refractory container makes necessary a large air gap (loose coupling) with consequent low power factor (between 0.1 and 0.3). Static capacitors are, therefore, invariably employed in parallel with such a furnace in order to improve the pf. Since in case of coreless induction furnace operation pf does not remain constant, capacitance in the circuit during heat cycle is varied to maintain power factor approximately unity.

The choice of frequency is influenced also by the cost of the capacitors, used for power factor improvement, which decreases with increase of frequency, and the cost of the converting apparatus, which increases with frequency.

Advantages of Coreless Induction Furnaces. The advantages of a coreless induction furnace over other types are given below :

(*i*) Low operating cost.
(*ii*) Low erection cost.
(*iii*) Automatic stirring action produced by eddy currents.
(*iv*) Absence of dirt, smoke, noise etc.
(*v*) Simple charging and pouring.
(*vi*) Possibility of operating the furnace intermittently, as no time is lost in warming up.
(*vii*) Precise control of power.
(*viii*) Less melting time.
(*ix*) Possibility of employing vacuum heating necessary for precious metal melting.
(*x*) No contamination of charge and very accurate control of composition so most suitable for production of high grade alloy steels.

The coreless induction furnace is mainly employed as a metal melting unit. An important application of this furnace is the production of carbon free ferrous alloys. Various special uses are : vacuum melting, duplexing steel, heating of charges of non-conducting materials (with or without melting) by the use of conducting crucibles etc. The energy consumption is 600 to 1,000 kWh per tonne of steel.

Example 5.7. Estimate the rating of an induction furnace to melt two tonnes of zinc in one hour if it operates at an efficiency of 80%. Specific heat of zinc = 0.1, latent heat of fusion of zinc = 26.67 kcals per kg. Melting point is 555°C. Assume initial temperature to be 35°C.

[G.B. Technical Univ. Utilization of Electrical Energy and Electrical Traction 2010-11]

Solution : Quantity of zinc to be melt, m = 2 tonnes = 2,000 kg

Initial temperature, t_1 = 35°C

Melting temperature, t_2 = 555°C

Specific heat of zinc = 0.1 kcal/kg/°C

Latent heat of fusion, L = 26.67 kcals/kg

Heat energy required to melt 2 tonnes of zinc

$$= m[s(t_2 - t_1) + L]$$
$$= 2{,}000[0.1(555 - 35) + 26.67]$$
$$= 157{,}340 \text{ kcals}$$
$$= \frac{157{,}340}{860} \text{ kWh} = 182.95 \text{ kWh}$$

$$\text{Energy input} = \frac{\text{Energy required}}{\eta} = \frac{182.95}{0.8} = 228.7 \text{ kWh}$$

$$\text{Rating of furnace} = \frac{\text{Energy input}}{\text{Time of operation in hours}}$$
$$= \frac{228.7}{1} = 228.7 \text{ kW} \quad \textbf{Ans.}$$

Example 5.8. Estimate the efficiency of a high frequency induction furnace which takes 10 minutes to melt 1.8 kg of aluminium. The input to the furnace being 5 kW and initial temperature 15ºC.

Given : Specific heat of aluminium = 880 J/kg/ºC

Melting point of aluminium = 660ºC

Latent heat of fusion of aluminium = 32 kJ/kg;

1 J = 2.78 × 10⁻⁷ kWh

[Pb. Univ. Electric Drives and Utilization of Electric Energy August 1984]

Solution : Quantity of aluminium to be melt, m = 1.8 kg

Initial temperature, t_1 = 15ºC

Melting temperature, t_2 = 660ºC

Specific heat of aluminium, s = 880 J/kg/ºC

Latent heat of aluminium, L = 32,000 J/kg

Heat required to melt 1.8 kg of aluminium

$$= m\,[s\,(t_1 - t_1) + L]$$
$$= 1.8\,[880\,(660 - 15) + 32{,}000 = 1{,}079{,}280 \text{ J}$$
$$= \frac{1{,}079{,}280}{3.6 \times 10^6} = 0.3 \text{ kWh} \qquad \because 1 \text{ kWh} = 3.6 \text{ MJ}$$

$$\text{Energy input} = 5 \times \frac{10}{60} = 0.8333 \text{ kWh}$$

$$\text{Efficiency of furnace, } \eta = \frac{\text{Output}}{\text{Input}} \times 100 = \frac{0.3}{0.8333} \times 100 = 36\% \textbf{ Ans.}$$

Example 5.9. A low frequency induction furnace operating at 10 volts in secondary circuit takes 500 kW at 0.5 pf when the hearth is full. If the secondary voltage be maintained at 10 volts, estimate the power absorbed and the power factor when the hearth is half-full. Assume the resistance of the secondary circuit to be thereby doubled and the reactance to remain the same.

[U.P. Technical Univ. Utilization of Electrical Energy and Traction, 2009-10]

Solution : Secondary voltage, V = 10 volts

Power drawn, P = 500 kW

$$\text{Secondary current, I} = \frac{P}{V\cos\phi} = \frac{500 \times 1{,}000}{10 \times 0.5} = 100{,}000 \text{ A}$$

Impedance of the secondary circuit when hearth is full

$$Z = \frac{V}{I} = \frac{10}{100{,}000} = 1 \times 10^{-4}\ \Omega$$

Resistance of secondary circuit when hearth is full

$R = Z\cos\phi = 1 \times 10^{-4} \times 0.5 = 0.5 \times 10^{-4}\ \Omega$

Reactance of secondary circuit when hearth is full

$X = Z\sin\phi = 1 \times 10^{-4} \times 0.866 = 0.866 \times 10^{-4}\ \Omega$

When the hearth is half-full

Resistance of secondary circuit, $R' = 2R = 2 \times 0.5 \times 10^{-4}\ \Omega = 1 \times 10^{-4}\ \Omega$

Reactance of secondary circuit, $X' = X = 0.866 \times 10^{-4}\ \Omega$

Impedance of secondary circuit,

$$Z' = \sqrt{(R')^2 + (X')^2} = \sqrt{(1\times10^{-4})^2 + (0.866\times10^{-4})^2} = 1.323 \times 10^{-4}\ \Omega$$

$$\text{Power factor} = \frac{R'}{Z'} = \frac{1\times10^{-4}}{1.323\times10^{-4}} = 0.756 \textbf{ Ans.}$$

$$\text{Secondary current, } I' = \frac{V}{Z'} = \frac{10}{1.323\times10^{-4}} = 75{,}593\text{ A}$$

$$\text{Power drawn, } P = (I')^2R = (75{,}593)^2 \times 1.0 \times 10^{-4} = 571{,}428\text{ W or } 571.428\text{ kW} \textbf{ Ans.}$$

Example 5.10. A low frequency induction furnace, whose secondary voltage is maintained constant at 10 volts, takes 400 kW at 0.6 power factor when the hearth is full. Assuming the resistance of the secondary circuit to vary inversely as the height of the charge and reactance to remain constant, find the height up to which hearth should be filled to obtain maximum heat.

[*Gorakhpur Univ. Utilization of Electrical Power & Traction Sept., 1983*]

Solution : Secondary voltage, V = 10 volts

Power drawn, P = 400 kW = 400,000 watts

$$\text{Secondary current, } I = \frac{P}{V\cos\phi} = \frac{400{,}000}{10\times0.6} = 6.667 \times 10^4\text{ A}$$

Impedance of secondary circuit when hearth is full

$$Z = \frac{V}{I} = \frac{10}{6.667\times10^4} = 1.5 \times 10^{-4}\ \Omega$$

Resistance of secondary circuit when hearth is full,

$R = Z\cos\phi = 1.5 \times 10^{-4} \times 0.6 = 0.9 \times 10^{-4}\ \Omega$

Reactance of secondary circuit, X

$= Z\sin\phi = 1.5 \times 10^{-4} \times 0.8 = 1.2 \times 10^{-4}\ \Omega$

Let the height of charge be x times of full hearth, *i.e.*, $h = x$H

Since resistance varies inversely as the height of the charge

∴ Resistance of the charge with height of charge as

$$x\text{H} = \frac{R}{x} = \frac{0.9\times10^{-4}}{x}\ \Omega$$

Power drawn or heat produced will be maximum when resistance of secondary circuit will be equal to reactance of secondary circuit

$$\textit{i.e.,}\quad \frac{0.9\times10^{-4}}{x} = 1.2 \times 10^{-4}$$

$$\text{or } x = \frac{0.9}{1.2} = \frac{3}{4}$$

i.e., Maximum heat will be obtained with the height of charge as three-fourth of height of hearth. **Ans.**

5.7. HIGH FREQUENCY EDDY CURRENT HEATING

The articles to be heated are placed within a high frequency current carrying coil, alternating magnetic field is set up, eddy currents are induced in the article and heating is, therefore, affected (Fig. 5.16). The high frequency current carrying coil is known as *heater coil* or *work coil,* the material which is to be heated is known as the *charge* or *load* and the process employed is referred to as *high frequency eddy current heating.*

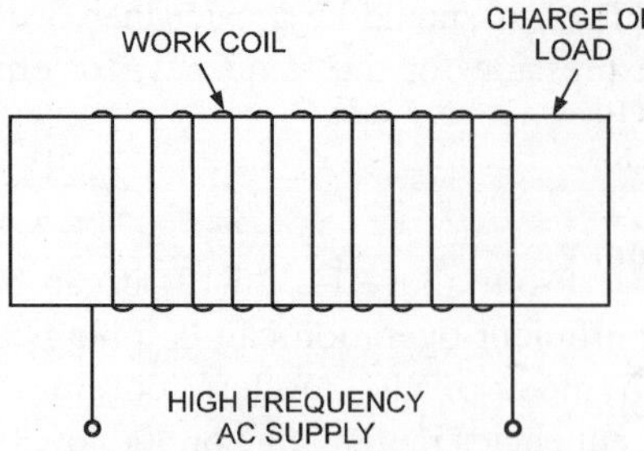

Fig. 5.16. ***High Frequency Eddy Current Heating***

The eddy current loss is primarily responsible for the production of heat, but hysteresis loss also contribute to it, though to a little extent, in the case of magnetic materials.

Since the eddy current loss is proportional to the square of the product of supply frequency and flux density, therefore, by controlling the frequency and the flux density the amount of heat can be controlled.

However, it is observed that higher the frequency employed in induction heating, the greater will be the tendency of the induced heating currents to remain at the surface of the material being heated. This property is called *skin effect.* This has a strong effect upon the uses and limitations of high frequency eddy current heating.

Induced eddy current is of greatest magnitude at the surface of the material to be heated and its value decreases with the distance normal to the surface as we go inside the material.

Since the depth of penetration of the eddy currents into the charge is inversely proportional to the under root of supply frequency $\left(\delta = \frac{1}{2\pi}\sqrt{\frac{\rho\times10^7}{\mu f}}\text{ m}\right)$, therefore eddy current heating can be restricted to any desired depth of the material to be heated by judicious selection of frequency of the heating current. The supply frequency is usually employed between 10 to 400 kHz.

To be effective the coil must be so shaped to provide as tight a magnetic coupling as possible between itself and the work (or load), the turns of the coil being parallel to the paths in which it is desired that the eddy currents shall flow. With the higher frequencies it is possible to achieve power inputs to the charge of 1.6 kW or more per square centimetre of surface.

Advantages of Eddy Current Heating. (*i*) It is quick, clean and convenient method.

(*ii*) There is little wastage of heat, as heat is produced in the body to be heated up directly.

(*iii*) It can easily take place in vacuum or other special atmosphere, where as other conventional types of heating are not possible in such places.

(*iv*) The control of temperature is very easy.

(*v*) The heat can be made to penetrate into the metal surface to any desired depth.

(*vi*) Unskilled labour can also operate the equipment.

(*vii*) The area of surface over which heat is produced can be accurately controlled.

(*viii*) The amount of heat produced can be accurately controlled by suitable timing devices.

(*ix*) The work coils are not required to fit closely around the object being treated. It enables the same coil to heat many different objects of different shapes and sizes and in fact often obviates the necessity of requiring odd shaped coils to heat irregular shaped objects.

From the economic point of view, the generation of heat is costly, efficiency of equipment is quite low (less than 50%) and initial cost of equipment is also high.

Applications of High Frequency Eddy Current Heating. The important applications of high frequency eddy current heating are :

(*a*) *Surface Hardening.* It is very important application of high frequency eddy current heating. The bar whose surface is to be hardened, by heat treatment, is surrounded by the work coil. An alternating current of high frequency is made to pass through the work coil. The desired depth of penetration is obtained by judicious selection of the frequency of the heating current in the work coil. After about a few seconds, when the surface has reached the proper temperature, the ac supply is cut off and the bar is at once dipped in water. In the case of hardening done by other methods, the heat treatment of the inner metal is disturbed which produces scales on the surface and also warps a long piece. This is avoided in this method of surface hardening. Such treatment due to induction heating reduces the cost, labour and time considerably.

(*b*) *Annealing.* In conventional method of annealing the process takes long time, which results in scaling of metal—a great shortcoming. But in eddy current heating method, the time taken in the process is less, thus, there is no scale formation. By this method a temperature of the order 750°C can be attained in one minute (approximately) up to depth of 25 mm.

(*c*) *Soldering.* For soldering it is necessary that required amount of heat is produced at the soldering point and the remaining portion of the solder may remain cold. Eddy current heating can be economically employed for soldering precisely for high temperature soldering, Silver, copper and their alloys are used as solders. By induction heating heat develops very rapidly melting the solder which runs into the joint sealing it properly.

Other applications are welding, drying of paints, melting of precious metals, sterilization of surgical instruments and forging of bolt-heads and rivet heads.

Example 5.11. The depth of penetration, in the case hardening of a steel pulley, required is 1.5 mm. The relative permeability is unity and the specific resistance for steel is 5 × 10^{-7} Ω-m. Determine the frequency required.

Solution : Specific resistance, $\rho = 5 \times 10^{-7}$ Ω-m

Relative permeability, $\mu = 1$

Depth of penetration, $\delta = 1.5$ mm $= 0.0015$ m

$$\text{Since } \delta = \frac{1}{2\pi}\sqrt{\frac{\rho \times 10^7}{\mu f}}$$

Frequency required,

$$f = \frac{\rho \times 10^7}{\mu \delta^2 \times 4\pi^2} = \frac{5 \times 10^{-7} \times 10^7}{1 \times (0.0015)^2 \times 4\pi^2}$$

$= 56{,}290$ Hz or 56.3 kHz **Ans.**

5.8. HIGH FREQUENCY POWER SUPPLY SOURCES

The power supply for coreless induction furnace is usually obtained from ordinary supply system and its frequency is converted to higher value either by means of motor-generator set with salient pole alternator (suitable for frequencies up to about 1,000 Hz and for any output required) or by means of motor-generator set with inductor alternator (suitable for frequencies up to about 10,000 Hz and output of 1 tonne) or by means of valve oscillators (suitable for very small furnaces which necessitates frequencies up to 1 million Hz). The oscillator which is employed in induction heating differs from the oscillator used in radio transmitters. The induction heaters are rugged, compact, portable and are almost automatic in operation. They are to be handled by unskilled operators and in comparatively dirty atmosphere. They are also subjected to overload and vibration. Hence while designing an oscillator for high frequency current supply for induction heating, the above factors should be given a proper consideration.

The round about value of frequency used for induction heating is 400 kHz, the circuit being used is shown in Fig. 5.17.

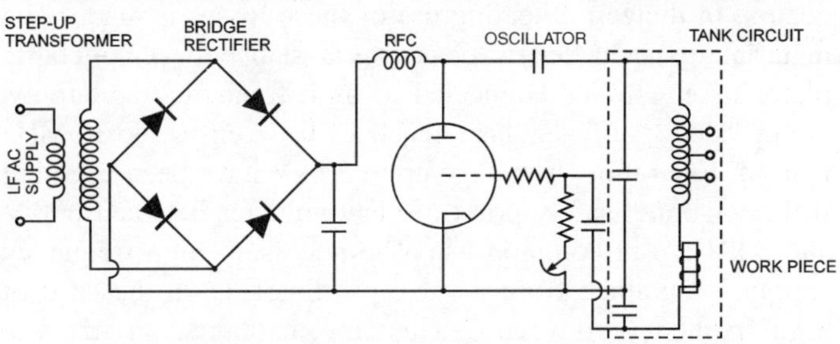

Fig. 5.17. ***Simplified Circuit of a Vacuum Tube Oscillator***

The ac supply is stepped up by the transformer initially, the stepped up voltage is rectified by using a bridge rectifier circuit, the rectified voltage is applied to the oscillator to obtain high frequency currents. The capacitances and inductances (including that of work piece) decide the frequency of supply to work piece. The vacuum tube oscillators are less efficient as compared to SCR, the voltage drop across the SCR being very small (of the order of 1 V). The operating efficiency of SCR is about 90%. The SCRs are fired by pulses of gate current produced by a UJT. The circuit is shown in Fig. 5.18. The frequency of supply to work piece depends upon the value of R and C. The smaller the product of R and C, higher is the frequency across the work piece.

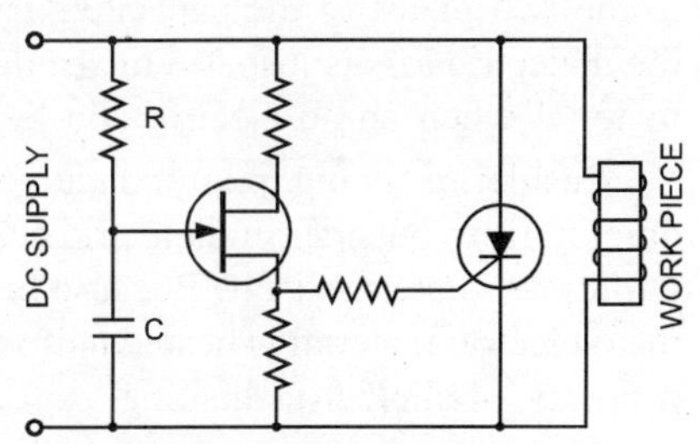

Fig. 5.18. ***SCR Inverter Circuit***

Sometimes spark gap oscillators are also employed to provide high frequency supply. The basic principle of operation of a spark gap converter is the alternate charging and discharging of a capacitor.

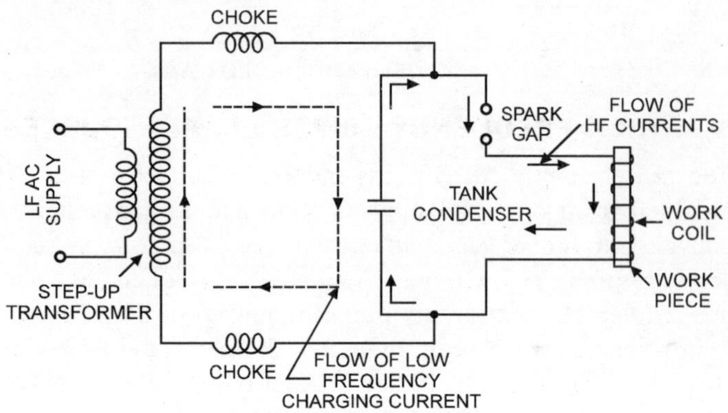

Fig. 5.19. ***Basic Spark-Gap Converter Circuit***

The transformer step ups the voltage, say to about 500 V, the tank condenser gets a voltage build-up till the spark gap breaks down and flashover takes place. The value of inductance and capacitance in the discharge circuit decides the frequency of current through the gap. Frequencies up to 1 MHz can be obtained.

5.9. DIELECTRIC HEATING

When non-metallic parts such as wood, plastics, bones are subjected to an alternating electrostatic field dielectric loss occurs. In dielectric heating use of these losses is made. The material to be heated is placed as a slab between metallic plates or electrodes connected to high frequency ac supply. For producing sufficient heat frequency between 10 and 30 MHz is used. Even though voltage up to 20 kV have been used but from personnel safety point of view voltages between 600 V and 3 kV are in common use. The necessary high-frequency supply is obtained from a valve oscillator, as in the case of high frequency eddy current heating, explained in Art. 5.8. An overall efficiency is about 50%.

The current drawn by the capacitor, when an ac supply voltage is applied across its two plates, does not lead the supply voltage by exactly 90º and there is always an in-phase component of the current. Due to this in-phase (or active) component of current, heat is always produced in the dielectric material placed in between the two plates of the capacitor. The electric energy dissipated in the form of heat energy in the dielectric material is known as *dielectric loss.* The dielectric loss is directly proportional to the frequency of ac supply given to the two plates of the capacitor. The physical conception of the dielectric loss is just as a molecular friction in the dielectric material when an ac electrostatic field is applied to it.

Insulators being poor conductors cannot be heated up quickly from outside. In dielectric heating the heat is produced within the material itself. Because heat generation is uniform, the dielectric material is heated uniformly. This is the important property of dielectric heating.

In insulators or non-conducting materials, the amount of heat produced by dielectric heating can be calculated as follows :

The material to be heated may be considered as the imperfect dielectric of a condenser and may be, therefore, represented as a capacitance placed in parallel with a resistance, as shown in Fig. 5.20 (*b*). The phasor diagram of the circuit is shown in Fig. 5.20 (*c*). If V is supply voltage in volts, f is supply frequency in Hz, C is the capacitance of the condenser in farads and cos ϕ is power factor of the load or charge,

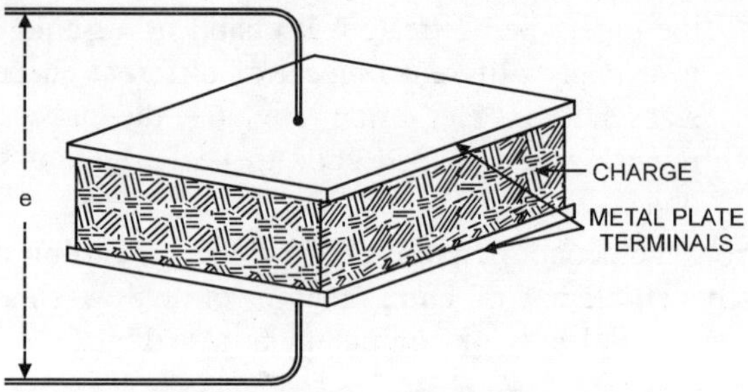

(a) Charge Between Parallel Metal Plates

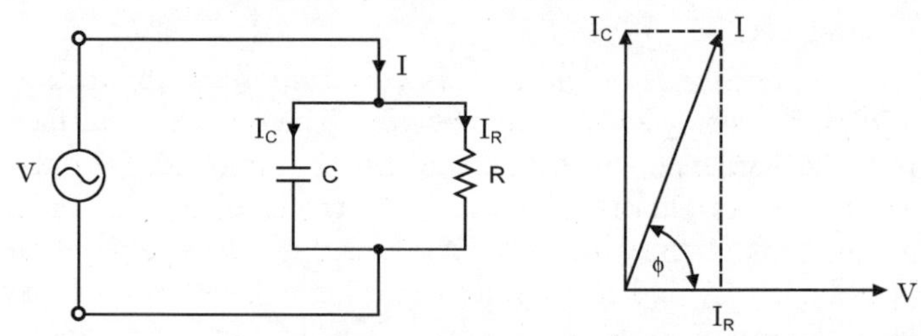

(b) Equivalent Circuit *(c) Phasor Diagram*

Fig. 5.20. ***Dielectric Heating***

Current through the capacitor,

$$I_C = \frac{V}{X_C} = \frac{V}{1/2\pi fC} = 2\pi fCV \text{ amperes}$$

where C is in farads and V is in volts.

The current drawn from supply,

$$I \simeq I_C = 2\pi fCV \text{ amperes}$$

$$\text{Power produced, } P = VI\cos\phi = V \times 2\pi fCV \times \cos\phi \text{ watts}$$
$$= 2\pi fCV^2 \cos\phi \text{ watts} \quad ...(5.13)$$

The value of power factor for a particular non-conducting material is constant, and the capacity is determined from the dimensions of plates, the charge which serves as the dielectric medium and the dielectric constant. Therefore, the dielectric heating depends upon the values of the frequency and the voltage. By varying one of these two quantities, the rate of dielectric heating can be varied. The insulation problem limits the voltage to be used; hence to achieve more heat, high frequency is used. However, there is a limit also, on the high frequency, imposed by cost involved in getting a circuit for obtaining very high frequencies and other difficulties.

The capacity of the condenser can be calculated from the following relation

$$C = \frac{\epsilon_r \epsilon_0 A}{t} \text{ farads} \quad ...(5.14)$$

where ϵ_r is relative permittivity of dielectric, ϵ_0 is absolute permittivity of vacuum and equals 8.854×10^{-12} F/m, t is the thickness of dielectric in metres and A is the surface area of plates in m^2.

Advantages : (*i*) If the material to be heated is homogeneous, and the alternating (or varying) electric field is uniform, heat is developed uniformly and simultaneously throughout the entire mass of the charge.

(*ii*) As materials heated by this process are non-conducting, so by other methods heat cannot be conducted to inside so easily.

Applications. The cost of the equipment required for dielectric heating is so high that it is employed only where other methods are impracticable or too slow. Some of the applications of dielectric heating are given below.

1. *Preheating of Plastic Preforms.* The raw material in the form of tablets or biscuits, commonly called plastic preforms, is required to be heated uniformly before putting them into the hot moulds so that whole mass becomes fluid at a time, otherwise if the raw material is put directly into the moulds, usually heated by steam, the outer skin of the preforms will become hot and start curing while the core of the material has not reached fluid temperature resulting in unequal hardening of the plastic and improper filling of moulds corners. Difficulty arises due to the fact that plastic raw material once cured cannot be softened again satisfactorily. Any method of heating depending upon conduction of heat from surface to the core would miserably fail because plastic is bad conductor of heat. Dielectric heating is the only method which can be used for preheating of plastic preforms to proper temperature uniformly.

2. *Gluing of Wood.* Dielectric heating is most commonly used for gluing of wooden sheets or boards as in this method of gluing the moisture contents of the wooden sheets remain unaltered. It is due to the fact that heat can be applied to the desired surface. Main difficulty in using animal glues is of long curing time and that parts to be joined are to be kept under mechanical pressure after application of glue for a period of about 24 hours. Mechanical pressure may be applied in gluing of wood by dielectric heating in order to secure better adhesion. Because of higher loss factor of glue as compared to that of wood most of the heat developed goes into the glue and very little heat is wasted.

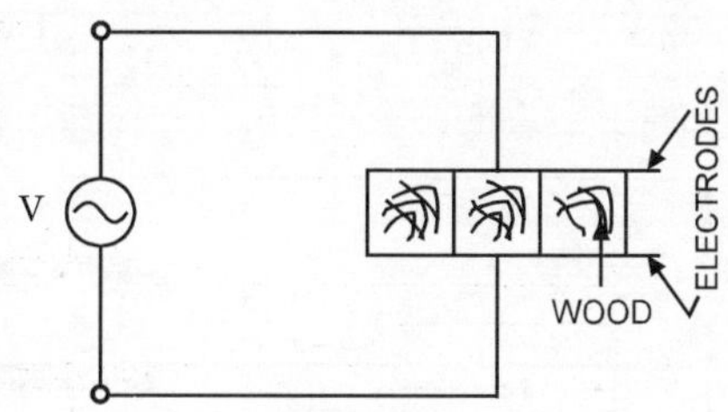

Fig. 5.21

High frequency dielectric heating is very economical for obtaining curved wood sections such as furniture. The curves obtained by this method are stable.

3. *Baking of Foundary Cores.* In foundaries resin type thermosetting binders are employed as they set almost instantaneously when brought to polymerizing temperature. The dielectric heating evaporates water rapidly from the core mix and at the same time raises the temperature of the core material to polymerization point. Hence dielectric heating is most suitable for baking foundary cores mixed with thermosetting resin type core binders.

4. *Diathermy.* Dielectric heating is also employed for heating tissues and bones of the body required for the treatment of certain types of pains and diseases.

5. *Sterlization.* The dielectric heating is quite suitable for sterilization of bandages, absorbent cotton, sterile gauge, instruments etc.

6. *Textile Industry.* In textile industry the dielectric heating is employed for drying purposes.

7. *Electronic Sewing.* Nowadays rain coats, umbrellas, food containers, medicine containers etc. are made from plastic film materials. In case of ordinary stitching by thread they will be no longer water tight and also become weak at seam. With adhesives curing times will be longer. In case of electronic sewing the films to be stitched are rolled in between cold rollers to which radio-frequency voltage is applied. The heat produced in the material seals it all along the line where mechanical pressure is applied. The cold rollers prevent the outer surface of the films from being softened.

8. *Food Processing.* The use of dielectric heating for food processing is one of the most modern method. It has brought many advantages for the food processing industry and has set forth such processes which are outside the realm of cooking. The dielectric heating can be appreciably employed for the following purposes :

(*a*) Pasteurising of milk and beer inside bottles.

(*b*) Dehydrating of fruits, milk, cream, vegetables and eggs etc.

(*c*) Cooking of foods without removing the outer shells.

(*d*) Defrosting of frozen foods such as meat and vegetables.

(*e*) Germicidal heating–In dielectric heating process the products do not loose flavour.

Disinfestation of cereals and grains–Dielectric heating may be used to destroy the eggs of insects and pests saving grains and cereals from damage and also to prevent these eggs, larva or moth etc. to pass on to cereal processing equipment and develop into dangerous worms later on under favourable conditions of temperature and humidity.

Example 5.12. A laminated wooden board 30 cm × 16 cm × 25 cm thick is to be heated from 20°C to 170°C in 10 minutes by dielectric heating using 30 MCPS (30 MHz) supply source. Specific heat of wood is 0.35 and density 0.55 gm/cc, relative permittivity = 5, pf = 0.05. Determine the voltage across the workpiece and current during heating. Assume loss of energy by conduction, connection and radiation as 15%.

[Rajasthan Univ. Utilization of Elec. Power 2006]

Solution : Volume of plywood to be heated

$$= 30 \times 10^{-2} \times 16 \times 10^{-2} \times 25 \times 10^{-2} = 0.012 \text{ m}^3$$

Density of plywood, $d = 0.55 \text{ gm/cc} = \dfrac{0.55}{1{,}000} \times 10^6 = 550 \text{ kg/m}^3$

Weight of plywood $= 550 \times 0.012 = 6.6$ kg

Specific heat of wood $= 0.35$ calorie per gm per °C

$= 0.35 \times 4.18 \times 1{,}000$ joule per kg per °C

$= 1{,}463$ J/kg/°C

Heat required to raise the temperature of plywood from 20°C to 170°C

$$= 1{,}463 \times 6.6 \times (170 - 20) = 1{,}448{,}370 \text{ joules}$$

$$= \frac{1{,}448{,}370}{3{,}600} \text{ kWh} = 402{,}325 \text{ Wh}$$

Since it is to be done in 10 minutes or 1/6 hour

$$\text{Power utilised} = \frac{402.325}{1/6} = 402.325 \times 6 = 2{,}413.95 \text{ watts}$$

$$\text{Power input required, P} = \frac{2{,}413.95}{(100-15)} \times 100 = 2{,}840 \text{ W}$$

Power factor, $\cos\phi = 0.05$

Capacitance of capacitor formed,

$$C = \frac{\varepsilon_0 \varepsilon_r A}{t} = \frac{8.854 \times 10^{-12} \times 5 \times 30 \times 10^{-2} \times 16 \times 10^{-2}}{25 \times 10^{-2}}$$

$$= 8.5 \times 10^{-12} \text{ F}$$

Voltage required,

$$V = \sqrt{\frac{P}{2\pi f C \cos\phi}} = \sqrt{\frac{2{,}840}{2\pi \times 30 \times 10^6 \times 8.5 \times 10^{-12} \times 0.05}}$$

$$= 5{,}954 \text{ V} \quad \textbf{Ans.}$$

Current flowing through the material,

$$I = \frac{P}{V\cos\phi} = \frac{2{,}840}{5{,}954 \times 0.05} = 9.54 \text{ A} \quad \textbf{Ans.}$$

Example 5.13. A slab of insulating material 150 cm² in area and 1 cm thick is to be heated by dielectric heating. The power required is 400 W at 30 MHz. Material has relative permittivity of 5 and pf 0.05. Absolute permittivity = 8.854 × 10⁻¹² F/m. Determine the necessary voltage.

[*M.D. Univ. Utilization of Elec. Power and Traction December 2009*]

Solution : Power absorbed, P = 400 W

Power factor, $\cos\phi = 0.05$

Supply frequency, f = 30 MHz = 30×10^6 Hz

Capacitance of capacitor formed,

$$C = \frac{\varepsilon_0 \varepsilon_r A}{t} = \frac{8.854 \times 10^{-12} \times 5 \times 150 \times 10^{-4}}{1 \times 10^{-2}}$$

$$= 66.405 \times 10^{-2} \text{ F}$$

Voltage required,

$$V = \sqrt{\frac{P}{2\pi f C \cos\phi}} \qquad \because P = 2\pi f C V^2 \cos\phi$$

$$= \sqrt{\frac{400}{2\pi \times 30 \times 10^6 \times 66.405 \times 10^{-12} \times 0.05}} = 800 \text{ W } \textbf{Ans.}$$

Example 5.14. The power required for dielectric heating of a slab of resin 150 cm² in area and 2 cm thick is 200 watts, frequency of 30 MHz. The material has a relative permittivity of 5 and pf of 0.05. Determine the voltage necessary and current flowing through the material. If the voltage is limited to 600 V, what will be the value of the frequency to obtain the same heating ?

[*Gorakhpur Univ. Utilization of Electrical Power With Traction 1980*]

Solution : Power absorbed, P = 200 watts

Power factor, $\cos\phi = 0.05$

Supply frequency, $f = 30 \times 10^6$ Hz

Capacitance of the capacitor formed,

$$C = \frac{\epsilon_0 \epsilon_r A}{t} = \frac{8.854 \times 10^{-12} \times 5 \times 150 \times 10^{-4}}{0.02} = 33.2 \times 10^{-12} \text{ F}$$

∴ Voltage required,

$$V = \sqrt{\frac{P}{2\pi f C \cos\phi}} = \sqrt{\frac{200}{2\pi \times 30 \times 10^6 \times 33.2 \times 10^{-12} \times 0.05}}$$

$$= 800 \text{ V } \textbf{Ans.}$$

Current flowing through the material,

$$I = \frac{P}{V\cos\phi} = \frac{200}{800 \times 0.05} = 5.0 \text{ A } \textbf{Ans.}$$

Since heat produced $\alpha\ f V^2$

$$\therefore f_2 V_2^2 = f_1 V_1^2$$

or Frequency required with 600 V supply to obtain the same heat, *i.e.*, of 200 watts,

$$f_2 = f_1 \left(\frac{V_1}{V_2}\right)^2 = 30 \times 10^6 \times \left(\frac{800}{600}\right)^2 = 53.3 \text{ MHz } \textbf{Ans.}$$

Example 5.15. A small piece of plywood having dimensions : length 5 cm, width 2 cm and thickness 1 cm is placed in between two electrodes having dimensions, length 25 cm, width 2 cm and with 2 cm distance between them, Electric current at frequency of 10 MHz is passed through electrodes for heating of wood by the process of dielectric heating. If the power consumption for heating the piece of wood is 1 kW, determine (*i*) voltage across the electrodes and (*ii*) current through it during the process of heating. Assume that the plywood is an imperfect dielectric and has relative permittivity of 5 and power factor 0.04.

[*A.M.I.E. Sec. B. Utilization of Electric Power Summer 1983*]

Solution : The arrangement of the heating material is shown in Fig. 5.22. With this configuration the capacitance of the parallel plate condenser formed with two dielectrics, plywood and air is given by the expression

$$C = \epsilon_0 \left[\frac{A_1 \epsilon_{r2}}{t} + \frac{A_2}{\frac{t_1}{\epsilon_{r1}} + \frac{t_2}{\epsilon_{r2}}}\right] \text{ Farads}$$

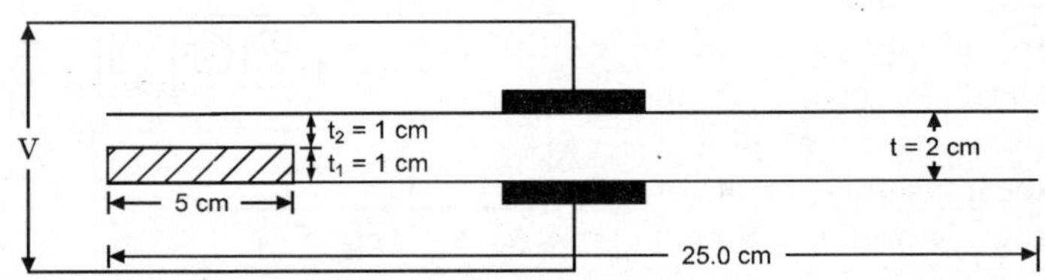

Fig. 5.22

where $A_1 = (25 - 5) \times 2 = 40 \text{ cm}^2 = 0.004 \text{ m}^2$

$A_2 = 5 \times 2 = 10 \text{ cm}^2 = 0.001 \text{ m}^2$

$t = 2 \text{ cm} = 0.02 \text{ m}$

$t_1 = 1 \text{ cm} = 0.01 \text{ m}$

$t_2 = 2 - 1 = 1 \text{ cm} = 0.01 \text{ m}$

$\epsilon_0 = 8.854 \times 10^{-12}$ F/m

ϵ_{r1} = Relative permittivity of plywood = 5

ϵ_{r2} = Relative permittivity of air = 1

∴ Capacitance of the capacitor,

$$C = 8.854 \times 10^{-12} \left[\frac{0.004 \times 1}{0.02} + \frac{0.001}{\frac{0.01}{5} + \frac{0.01}{1}}\right]$$

$$= 2.509 \times 10^{-12} \text{ Farads}$$

(*i*) Voltage across the electrodes,

$$V = \sqrt{\frac{P}{2\pi f C \cos\phi}} = \sqrt{\frac{1{,}000}{2\pi \times 10 \times 10^6 \times 2.509 \times 10^{-12} \times 0.04}}$$

$$= 12{,}594 \text{ V } \textbf{Ans.}$$

(*ii*) Current through plywood,

$$I = \frac{P}{V\cos\phi} = \frac{1{,}000}{12{,}594 \times 0.04} = 1.985 \text{ A } \textbf{Ans.}$$

5.10. CHOICE OF FREQUENCY

The selection of frequency for heating purposes is an important factor. Although, the selection of frequency has a great bearing on the work to be heated and the method of heating to be employed (induction heating or dielectric heating). Power frequency (50 Hz) furnaces can be of 1 MW capacity where as medium frequency (500 Hz to 1,000 Hz) furnaces have capacity of 500 kW and high frequency (100 kHz to 2 MHz) furnaces have capacity from 200 kW to 500 kW.

Induction Heating. For making a choice of frequency for induction heating the following factors are to be considered.

1. The thickness of surfaces to be heated. Higher the frequency, thinner the surface will get heated.
2. The time of continuous heating. Longer the duration, deeper the penetration of heat owing to conduction.
3. The temperature to be obtained, and duration of heating. If higher temperatures are to be obtained in shorter time, higher capacities of generator should be selected.

Dielectric Heating. The rate of heat produced in dielectric heating is given by the expression $P = 2\pi f CV^2 \cos\phi$ watts where V is the supply voltage, f is the supply frequency and C is the capacitance of the condenser formed, which depends upon the relative permittivity ϵ_r of the material.

Thus the rate of heat production $\alpha\ V^2 \times f \times \cos\phi \times \epsilon_r$

The voltage across any specimen is limited by its thickness, *i.e.*, potential gradient, breakdown voltage, insulation and safety consideration. The voltages used for dielectric heating are usually from 600 V to 3,000 V, however voltage up to 20,000 volts is also sometimes used. Alternatively higher rate of production of heat can be obtained by employing higher frequency but this is also limited due to the following considerations.

(*i*) Necessity of incorporating a special matching circuit at higher frequencies due to the fact that maximum power output is obtained from the supply oscillator only when the impedance of the oscillator match with that of load.

(*ii*) Possibility of existing of a standing wave across the surface of electrodes having wavelength approximately equal to or more than one quarter of the frequency wavelength at any time.
It causes a variation in the voltage across the electrodes which results into uneven heating. This is avoided by imposing limits on the length of the electrode at a particular frequency.

(*iii*) At very high frequency it is difficult for tuning inductance to resonate with the charge capacitance.

(*iv*) At higher frequencies uniform voltage distribution cannot be obtained.

(*v*) The higher frequencies may disturb nearby radio station services due to radiation. So special care should be taken that there is no radiation etc.

Exercises

1. What are the advantages of electric heating ? Give classification of various electric heating methods alongwith brief account of their working principles.
[Gorakhpur Univ. Utilization of Elec. Power & Traction Sept., 1984]
2. Write the Stefan's law of heat radiation and label all constants.
[A.M.I.E. Sec B. Utilization of Elec. Power Winter 1983]
3. Discuss various methods of electric heating and explain any one of them.
[Pb. Technical Univ. Electric Drives and Utilization December 2009]
4. Explain the different methods of electric heating and their field of applications.
[Rajasthan Univ. Utilization of Electrical Power 2007]
5. What are the desirable properties the materials for heating element should have ? Name a few materials.
Discuss the methods of temperature control of resistance ovens.
Explain the design procedure of the heating elements when the power and voltage of the oven are known.
[Gorakhpur Univ. Utilization of Elec. Power & Traction 1983]
6. What are the essential properties of resistance heating elements ? How will you design such elements in case of resistance heating.
[Pb. Technical Univ. Electric Drives and Utilization May 2008]
7. What are the requirements of a good heating material. Describe different type of heating materials. Also describe causes of failure of heating elements.
[Rajasthan Technical Univ. Utilization of Elec. Power 2010]
8. Describe with the help of neat sketches salient features of (*i*) air circulation ovens (*ii*) bright annealing (*iii*) radiant heating and (*iv*) salt bath heating and give application of each.
[A.M.I.E. Sec B. Utilization of Elec. Power Summer 1983]
9. What are the different methods of heating ? Describe briefly the methods of direct and indirect resistance heating. List the applications of these two methods.
[U.P. Technical Univ. Utilization of Electrical Energy and Electric Traction, 2005-06]
10. Describe various methods of controlling the temperature in resistance ovens.
[West Bengal Univ. of Technology Utilization of Electric Power 2009]
11. Describe the construction and operation of direct type arc furnace.
[U.P. Technical Univ. Utilization of Electrical Energy and Electric Traction 2006-07]
12. Describe advantages of electrical heating and functioning of indirect arc furnace.
[U.P. Technical Univ. Utilization of Elect. Energy and Traction 2009-10]
13. Describe the construction and working of a 3-phase arc furnace.
[A.M.I.E. Sec B. Utilization of Electric Power Winter 1982]
14. With neat sketches, describe the construction, principle of operation, applications and control methods of direct and indirect arc furnaces.
[M.D. Univ. Utilization of Elec. Power and Traction May 2009]
15. Explain construction, operation, performance and application of arc furnace.
[Rajasthan Univ. Utilization of Electrical Power 2007]
16. Explain Principle of induction heating.
[U.P. Technical Univ. Utilization of Elect. Energy and Traction 2009-10]
17. Explain different methods of induction heating. Give some applications of induction heating.
[Rajasthan Univ. Utilization of Elec. Power 2008]

18. With neat sketches, describe the construction, principle of operation, applications and control methods of Ajax-Wyatt vertical core furnace.
[G.B. Technical Univ. Utilization of Elect. Energy and Traction 2010-11]

19. Explain with neat sketches and describe the construction, principle of operation and application of Ajax Wyatt furnace. **[Rajasthan Univ. Utilization of Elec. Power Dec. 2008]**

20. Explain the working of Tama furnace with the help of neat diagram. **[U.P. Technical Univ. Utilization of Elect. Energy and Traction 2009-10]**

21. With the help of diagram describe the working of a coreless type induction furnace. Mention advantages of coreless type furnace.
[Pb. Technical Univ. Electric Drives & Utilization Dec. 2008]

22. Describe a coreless induction furnace.
[West Bengal Univ. of Technology Utilization of Elec. Power 2009]

23. Explain the method of induction heating and describe coreless type of induction furnace.
[M.D. Univ. Utilization of Electric Power and Traction Dec. 2009]

24. Describe the construction and working of an induction furnace suitable for melting non-ferrous metals. What are the various sources of generating high frequency energy for induction heating ? Describe in brief.
[U.P. Technical Univ. Utilization of Elect. Energy and Traction 2005-06]

25. Explain the principle of high frequency induction heating. Discuss the factors affecting the depth of penetration of heat.
[M.D. Univ. Utilization of Electric Power and Traction Dec. 2008; May 2009]

26. Describe with the help of a neat sketch, the working of a vertical core type induction furnace. Give the applications of high frequency eddy current heating.
[Rajasthan Technical Univ. Utilization of Electrical Power 2011]

27. Describe the principle of induction heating at high frequency and highlight applications of the eddy current heating.
[M.D. Univ. Utilization of Electric Power & Traction May 2010]

28. What is dielectric heating ? Explain the factors on which the dielectric loss in a dielectric material depend.
[D.E.I. Agra Utilization of Electric Power 1989]

29. Explain dielectric heating. What is the rate of rise of temperature in dielectric heating ? Discuss the advantages of dielectric heating.
[Pb. Technical Univ. Electric Drives and Utilization May 2007]

30. State the advantages of electric heating. Explain the principle of induction heating. What are the applications of induction heating ?
What is dielectric heating and dielectric loss ?
[West Bengal Univ. of Technology Utilization of Electric Power 2009]

31. Explain the principle of 'dielectric heating'. Derive the mathematical expression of power consumed in such process. State important applications of dielectric heating.
[R.T.U. Utilization of Electrical Power 2010]

32. What is dielectric heating ? Explain the factors on which the dielectric loss in a dielectric material depends.
[M.D. Univ. Utilization of Electric Power and Traction Dec. 2008; Dec. 2009]

33. What is dielectric heating ? Explain why very high frequency should not be used for dielectric heating.
[M.D. Univ. Utilization of Electric Power & Traction May 2009]

34. Compare induction heating and dielectric heating. Discuss different applications where dielectric heating and induction heating are used. **[U.P. Technical Univ. Utilization of Elect. Energy and Traction 2006-07]**

35. Discuss various applications of dielectric heating.
[U.P. Technical Univ. Utilization of Elect. Energy and Traction 2009-10]

36. Write short notes on the following :
(*i*) Electric heating. **[M.D. Univ. May 2010]**
(*ii*) Resistance heating. **[G.B.T.U. 2010-11]**
(*iii*) Arc furnace.
(*iv*) Induction heating. **[G.B.T.U. 2010-11]**
(*v*) Coreless induction furnace.
(*vi*) Dielectric heating. **[W.B. Univ. of Technology 2009]**

Short Answer Type Questions With Answers

Q. 1. Why electric heating is preferred over other forms of heating ?

Ans. Electric heating is preferred over other forms of heating because of its numerous inherent advantages like economy, cleanliness, absence of flux gases, easy of control, automatic protection, upper limit of temperature, special heating needs.

Q. 2. What is an oven ?

Ans. Low temperature heating chamber with provision for ventilation is termed as oven.

Q. 3. Write and explain Joules law of electric heating.
[P.T.U. December 2006]

Ans. According to Joule's law of electric heating, the energy expended or heat generated in joules is given as

Work done or heat generated

$$= I^2Rt = \frac{V^2}{R}t = VIt = Wt \text{ joules.}$$

i.e., According to Joules law of heating, heat produced in an electric circuit of resistance R and carrying current I for t seconds is

(*i*) proportional to the square of the current *i.e.*, $H \propto I^2$

(*ii*) proportional to the resistance of the circuit *i.e.*, $H \propto R$, and

(*iii*) proportional to the time duration for which the current flows through the circuit *i.e.*, $H \propto t$

Heat produced in kcals,

$$H = \frac{I^2 Rt}{4,180} \simeq \frac{I^2 Rt}{4,200} \qquad [\because 1 \text{ kcal} = 4,180 \text{ J} \simeq 4,200 \text{ J}]$$

Q. 4. What properties are considered for selecting material for heating element ?

Ans. The properties to be considered in selection of material for heating elements are resistivity, melting point, temperature coefficient of resistance, and freedom from oxidation.

Q. 5. What are the essential properties of resistance heating elements ? **[P.T.U. Dec. 2008]**

Ans. The essential properties of resistance heating elements are : High resistivity, high melting point, low temperature coefficient of resistance and freedom from oxidation.

Q. 6. Name few materials, that are widely used as heating element materials.

Ans. The materials widely used as materials for heating elements are nickel-chromium (Nichrome) alloy, Ni-Cr-Fe alloy, nickel-copper alloy (Eureka or constantan), iron-chromium-aluminium alloy (Kanthal) and alloys containing iron, chromium, cobalt and aluminium.

Q. 7. What are the causes of failure of heating elements ?

Ans. The chief causes of failure of heating elements are formation of hot spots, oxidation and intermittency of operation, embrittlement due to grain growth, contamination and corosion.

Q. 8. How can you control the temperature inside the resistance furnace ? **[P.T.U. Dec. 2006]**

Ans. Refer to Art. 5.4.5.

Q. 9. Name few applications of infrared heating.

Ans. Infrared heating finds applications in (*i*) paint stoving (*ii*) drying of radio-cabinets and wood furniture (*iii*) preheating of plastics prior to moulding (*iv*) softening of thermo-plastic sheets, (*v*) drying of pottery, paper, textiles etc. where moisture content is not large.

Q. 10. Name the types of electrodes used in arc furnaces.

Ans. The electrodes used in arc furnaces are of three types namely carbon, graphite and self-baking electrodes.

Q. 11. Why are indirect arc furnaces not built in large sizes ? **[P.T.U. May 2009]**

Ans. The arc is produced between two electrodes and heat is transferred to the charge by radiation only. Such construction limits the use of single phase supply. The size of the furnace is thus limited by the amount of single-phase load, which can be taken from one point.

Q. 12. How is power input to submerged arc furnace controlled ?

Ans. The power input to submerged arc furnace is controlled by varying distance between electrodes or by varying the voltage applied to the electrodes.

Q. 13. How is control of power affected in electric arc furnaces ?

Ans. The power input to arc furnaces can be controlled by raising or lowering the electrodes or by varying the voltage applied to the arc furnace (by changing of tapping on supply transformer); preferably both voltage and electrode controls.

Q. 14. At what power factor would you like to operate the electric arc furnace ?

Ans. At power factor of about 0.9, but not below 0.8 in any case.

Q. 15. What type of supply is used for direct core type induction furnace in our country ?

Ans. Single-phase, 50 Hz, 600 V supply is used for direct core type induction furnaces.

Q. 16. How amount of heating is controlled in high frequency eddy current heating ?

Ans. The amount of heating is controlled by controlling the supply frequency and the flux density in high frequency eddy current heating.

Q. 17. What are the applications of high frequency eddy current heating ?

Ans. The applications of high frequency eddy current heating are surface hardening, annealing, soldering, welding, drying of paints, melting of precious metals, sterilization of surgical instruments and forging of bolt-heads and rivet heads.

Q. 18. How can the rate of dielectric heating be varied ?

Ans. Rate of dielectric heating can be varied by varying either supply voltage or supply frequency.

Q. 19. Why dielectric heating is employed only where other methods are impracticable or too slow ?

Ans. Because of very high cost of equipment required for dielectric heating.

Problems

1. A 4.5 kW, 200 V, 1-phase resistance oven is to have nichrome wire heating elements. If the wire temperature is to be 1,000°C and that of charge 500°C, estimate the diameter and length of wire. Resistivity of nichrome alloy = 42.5 μΩ-cm, radiant efficiency = 1.0 and emissivity of the element = 0.9.

[M.D. Univ. Utilization of Electric Power & Traction May 2009]

[**Ans.** 0.9072 mm; 13.5176 m]

2. Determine the length and diameter of the nichrome wire in the resistance element of a single phase electric furnace rated at 20 kW and 220 V. The wire temperature is not to exceed 1,170°C and the temperature of the charge is to be 500 °C. Take k = 0.57, e = 0.95 and ρ = 10 μΩ-cm.

[Pb. Univ. Drives and Utilization of Elec. Energy April 1981]

[**Ans.** 16.70 m; 3.1 mm]

3. A 40 kW, 3-phase, 400 V resistance oven is to employ Ni-Cr strip of 0.3 mm thickness. The heating elements are star-connected. If the wire temperature is to be 1,127°C and that of the charge is to be 727°C, estimate the suitable width and length of wire required. Take radiation efficiency = 60%; specific resistance of Ni-Cr = 1.03×10^{-6} Ω-m; emissivity = 0.9.

[Gorakhpur Univ. Utilization of Elec. Power & Traction 1982]

[**Ans.** 8,075 mm, 9.4 m]

4. A 30 kW, 3-phase, 400 V resistance oven is to employ nickel-chrome strip 0.025 cm thick for the 3-phase star-connected heating elements. If the wire temperature is to be 1,100 °C and that of charge is to be 700°C, estimate a suitable width for the strip. Assume radiating efficiency as 0.6 and emissivity as 0.9. The specific resistance of the nichrome-alloy is 1.03×10^{-6} Ω-m. State any assumption made.

[A.M.I.E. Sec B. Utilisation of Elec. Power Summer 1982]

[**Ans.** 6.86 mm]

5. In a 3-phase arc furnace to melt 10 tonne steel in 2 hours, estimate the average input to the furnace, if overall efficiency is 50%. If the current input is 9,000 A with the above kW input and the resistance and reactance of furnace leads (including transformer) are 0.003 Ω and 0.005 Ω respectively, estimate the arc voltage and total kVA taken from the supply.

Specific heat of steel = 444 J kg^{-1} °C^{-1}

Latent heat of fusion of steel = 37.25 kJ/kg

Melting point of steel = 1,370 °C

[Pb. Univ. Elec. Drives & Utilization of Electric Energy Nov., 1983]

[**Ans.** 38.5 V; 2,146 kVA]

6. Estimate the energy required to melt 0.5 tonne of brass in a single-phase induction furnace. If the melt is to be carried out in 0.5 hour, what must be the average power input in the furnace. **[West Bengal Univ. of Technology Utilization of Electric Power 2009]**

[**Hint:** Assume room temperature to be 20°C, specific heat of brass 393.6 J/kg/°C, latent heat of fusion 163,000 J/kg, melting point of brass 920° and furnace efficiency 70%]

[**Ans.** 102.625 kWh; 205.25 kW]

7. Estimate the efficiency of high frequency induction furnace which takes 12 minutes to melt 2.5 kg of aluminium, the input to the furnace being 3.75 kW. The initial temperature is 20°C, specific heat of aluminium 880 J kg^{-1} C^{-1}, melting point 650°C, latent heat of fusion 32 kJ/kg.

[B.H.U. Traction & Utilization 1978] [**Ans.** 54.3%]

8. A low frequency induction furnace operating at 12 V in the secondary circuit takes 48 kW at 0.5 pf when the hearth is full. If the secondary voltage be maintained at 12 V estimate the power factor and power absorbed when hearth is half-full. Assume the resistance of the secondary circuit to be thereby doubled and the reactance to remain the same.

[M.D. Univ. Utilization of Elec. Power and Traction May 2010]

[**Ans.** 0.756 (lagging); 548.584 kW]

9. A low frequency induction furnace, whose secondary voltage is maintained constant at 10 volts, takes 400 kW at 0.6 power factor when the hearth is full. Assuming the resistance of the secondary circuit to vary inversely as the height of the charge and the reactance to remain constant, determine the height up to which the hearth should be filled to obtain maximum heat.

[Gorakhpur Univ. Utilization of Electrical Power & Traction Sept., 1981]

[**Ans.** Three-fourth of height of hearth]

10. An insulating slab 2 cm thick and 150 cm^2 in area is to be heated by dielectric heating. The power required is 300 watts at a frequency of 30 MHz. The material has a relative permittivity of 5 and pf of 0.05. Determine the voltage necessary and current that will flow through the material. If the voltage is limited to 600 V, calculate the frequency that will produce the same loss.

[Pb. Univ. Electric Drives & Utilization of Electric Energy Nov., 1980]

[**Ans.** 980 V; 6.12 A; 80 MHz]

11. A piece of plastic material of size 4 × 2 × 1 cm is heated by being placed between two electrodes, each having an area of 20 × 2 cms and the distance of separation between two being 1.6 cm. The frequency of voltage impressed across the electrodes is 20 MHz. If the power consumed is 80 watts, find the voltage applied across the electrodes and current through the material. Assume relative permittivity as 5 and power factor 0.05.

[Gorakhpur Univ. Utilization of Electrical Power & Traction 1981]

[**Ans.** 2,189 V; 0.73 A]

12. A 3-phase electric arc furnace has the following data :

Current drawn = 5,000 A

Arc voltage = 50 V

Resistance of transformer referred to secondary = 0.002 Ω

Reactance of transformer referred to secondary = 0.004 Ω

(*i*) Calculate the pf and the kW drawn from the supply.

(*ii*) If the overall efficiency of the furnace is 65% find the time required to melt 2 tonnes of steel when latent heat of steel = 8.99 kcal/kg; specific heat of steel = 0.12, melting point of steel = 1,370°C and initial temperature of steel = 20°C.

[A.M.I.E. Sec. B. Utilization of Elec. Power Winter 1976]

[**Ans.** 0.9487; 900 kW; 40 minutes 46 seconds]

13. A 3-phase arc furnace takes 7,500 A when the arc voltage is 75 V. The circuit comprising the transformer and the electrodes excluding the arc has a reactance of 0.003 Ω and a resistance of 0.002 Ω. The steel has a specific heat of 278 J/kg/°C; a latent heat of fusion 37,000 J/kg and melting point 1,370°C.

Assuming furnace overall efficiency as 50% and initial temperature of steel as 15°C, determine

(*i*) the time taken to melt 5 tonnes of steel, (*ii*) the kVA taken from the supply and (*iii*) the electrical efficiency.

[**Ans.** (*i*) 2,040 seconds (*ii*) 2,088 kVA (*iii*) 83.3%]

14. Determine the efficiency of a high frequency induction furnace, which takes 10 minutes to melt 1.815 kg of aluminium, the input to the furnace being 5 kW and the initial temperature is 15°C.

Specific heat of Al = 0.212

Melting point of Al = 660°C

Latent heat of Al = 76.8 kcal/kg. [**Ans.** 54%]

15. A slab of insulating material 130 cm^2 in area and 1 cm thick is to be heated by dielectric heating. The power required is 380 W at 30 MHz. Material has a relative permittivity of 5 and pf of 0.05. Absolute permittivity = 8.854×10^{-12} F/m. Determine the necessary voltage.

[Rajasthan Univ. Utilization of Electrical Power 2008]

[**Ans.** 837 V]

❑❑❑

CHAPTER 6

Electric Welding

INSIDE THIS CHAPTER

6.1. INTRODUCTION

Welding is a materials joining process that produces coalescence* of materials by heating them to the welding temperature with or without the application of pressure or by the application of pressure alone, and with or without the use of filler metal. It is used to make welds. A weld is "a localized coalescence of metals or non-metals produced either by heating materials to the welding temperature, with or without the application of pressure, or by the application of pressure alone and with or without the use of filler metal."

All metals can be joined by one welding process or another. There is a saying, "If it's metal, weld it," and it is certainly true. This should be qualified by stating that all metals commercially employed for structural or strength parts are weldable. Some metals are easy to weld, and others are difficult to weld. The metals that are easily weldable can be welded in thickness from the very thinnest, about the thickness of this paper page, to the thickest or heaviest produced. The difficult-to-weld metals require special procedures and techniques that must be developed for specific applications. Some metals may never be welded or joined. For example, mercury is a liquid at room temperature and cannot be welded, while sodium and potassium melt just below the temperature of boiling water and are of no use as a strength member and cannot be welded. In general, metals that have a low melting point or low strength would not be welded. Some metals are so scarce or expensive that they would not be used where welding would be required. The physical and mechanical properties (such as melting temperature, density, thermal conductivity, tensile strength, ductility etc.), availability, and price all help in determining if a metal will be used in applications where welding is required.

6.1.1. Importance of Welding

Welding is the most efficient way to join metals. It is the only way to join two or more pieces of metal to make them act as one piece. Welding is widely employed to manufacture or repair all products made of metal. Look around, almost every thing made of metal is welded; the world's tallest building, moon rocket engines, nuclear reactors, home appliances, and automobiles barely start the list.

The use of welding is still increasing. If a joint is welded, it is permanent joint. Obviously, if the joint must be disassembled occasionally, it should not be welded. Thus we should change our statement to "welding is the most economical method to permanently join metal parts." To join two members by bolting or riveting requires holes in the parts to accommodate the bolts or rivets. These holes reduce the x-sectional area of the members to be joined by up to 10 per cent. The joint may also require the use of one or two gusset plates, thus increasing the weight of material required and the cost. This expense can be eliminated by the use of a weld. The greatest economy of a welded design will be obtained if the x-sectional area of the entire structural member is reduced by the amount of the bolt holes. This can be done since the entire cross section of a member of a welded design is utilized to carry the load. The amount of material required is reduced as well as its cost. This same design concept applies to joining plates used to build a ship or a container. In view of this material savings, ships and storage tanks are no longer riveted.

Pipes joined by welding offer similar economies. The wall thickness of a pipe should be heavy enough to carry the required load. However, if the pipe is joined by screw threads a heavier wall thickness is used to allow for cutting away a portion of the thickness for the threads. A thinner pipe wall thickness is used for the entire welded pipe system. This reduces the amount of metal required and the cost. The inside surface of the welded joint is smoother. Larger diameter pipes are no longer connected together with screw threads and pipe fittings.

Converting castings to weldments allows the designer to reduce weight by reducing metal thickness. Welding is a design concept which allows freedom and flexibility not possible with

* Coalescence means the growing together or growth into one body of the materials being welded.

cast construction. Heavy plates can be used where strength is required and thin ones can be used where possible. The uniform thickness rule and minimum thickness required for foundry practice are not necessary for weldments. Additionally, high-strength materials can be employed in specific areas, while normal-strength materials are used where required.

Welding is the best way to protect and conserve materials by protecting their surface with special metal overlays. Corrosion and wear of metals account for losses running into billions of dollars annually. Together they are responsible for an untold loss of lives. Waste from both of these destructive forces can be largely reduced by welding. Special alloys are weld deposited on base metals to provide corrosion resistance surfaces. Hard surfacing overlays can be made by welding to provide special alloys with wear resistant surfaces. A typical application is the resurfacing of a cinder crusher roll with hard weld metal. Weld surfacing is used to reduce the costly abrasive and corrosive wear of machinery.

Thus in short we can say, "today, the science and art of welding encompasses a wide range of processes and procedures applicable to materials of any thickness and shape—from tiny electronic components to mammoth machines and structures. Welding has opened up prospects for radical improvements in the manufacture of a large variety of machines, devices and structures. It has promoted mechanization and automation in their manufacture. The use of automatic, semi-automatic and mechanized welding equipment has relieved a large number of workers for employment elsewhere in the national economy. Welding is nowadays extensively used in automobile industry, aircraft machine frames, structural work, tanks, machine repair work, shipbuilding, pipeline fabrication in thermal power plants and refineries, fabrication of metal structures."

6.1.2. Advantages and Disadvantages of Welding

Some of the advantages of welding are given below :

1. Welding is the lowest-cost joining method.
2. It affords lighter weight through better utilization of materials.
3. It joins all commercial metals.
4. It can be used anywhere.
5. It provides design flexibility.

The limitations of welding are :

1. Some welding depends on the human factor.
2. It often needs internal inspection.

Most of these limitations can be overcome by means of good controls and supervision.

Welding is also an economical manufacturing method wherever there is a need to join parts permanently. In the high-volume production industries it is common to see welding operations intermixed with bending, machining, forming, assembly and so on. Welding is an important manufacturing process taking its place with other metal working operations to help bring us good-quality products at economical prices.

6.1.3. Personnel Protection and Safety Rules

There are a number of safety and health problems associated with welding. When correct precautionary measures are followed, welding is a safe occupation. The hazards that are more or less peculiar to welding are electrical shock, arc radiation, air contamination, fire and explosion, compressed gases, welding cleaning and other hazards related to specific processes or occupations.

Combustible materials must not be allowed to collect in or near the welding workplace. Good housekeeping practices should always be employed in the welding shop. Adequate safety devices should be provided, such as fire extinguishers, life-saving and support equipment, first-aid kits, and so on, plus the training of personnel to utilize this equipment properly. Only approved equipment be used, and it must be properly installed and maintained in good working conditions.

6.1.4. Classification of Welding Processes

There are many ways to make a weld and there are many different kinds of welds. The welder behind the hood making sparks is using one of the more popular welding processes, known as *arc welding*. Some welding processes do not cause sparks; in some cases electricity is not used, and in some cases there is not even extra heat. Welding has become complex and technical. It needs considerable knowledge to select the proper welding process for critical work.

Modern methods of welding may be classified under two broad headings—plastic welding and fusion welding. They are also called pressure welding, and non-pressure welding respectively. In the *plastic welding* or *pressure welding*, the pieces of metal to be joined are heated to a plastic state and then forced together by external pressure. This includes forge welding, resistance welding, thermit welding and gas welding, in which pressure is required. In the *fusion welding* or *non-pressure* welding, the material at the joint is heated to a molten state and allowed to solidify. This includes gas welding, arc welding and thermit welding without pressure.

The various welding processes used in general engineering are given below :

1. Gas Welding — (*i*) Oxyacetylene (*ii*) Air-acetylene and (*iii*) Oxy-hydrogen.
2. Resistance Welding — (*i*) Butt (*ii*) Spot (*iii*) Projection (*iv*) Seam and (*v*) Percussion.
3. Arc Welding — (*i*) Carbon arc (*ii*) Metal arc (*iii*) Gas metal arc (*iv*) Gas tungsten arc (*v*) Atomic hydrogen arc (*vi*) Plasma arc (*vii*) Submerged arc (*viii*) Flux-cored arc and (*ix*) Electro-slag.
4. Thermit Welding.
5. Solid State Welding — (*i*) Friction (*ii*) Ultrasonic (*iii*) Diffusion and (*iv*) Explosive.
6. Newer Welding — (*i*) Electron beam and (*ii*) Laser.

Here we will discuss only about electric welding (Resistance welding and arc welding) and some introduction to other modern welding techniques.

6.2. RESISTANCE WELDING

By definition, *resistance welding is that process in which a sufficiently strong electric current is sent through the two metal pieces in contact to be welded which melts the metals by the resistance they offer to the flow of the electric current.*

Resistance welding includes butt welding, spot welding, projection welding, seam welding and percussion welding. All are alike in the principle of resistance heating but differ in the details of application.

In resistance welding a heavy current (above 100 A) at a low voltage is passed directly through the workpiece and heat developed by the resistance to the flow of current, given by the expression I^2Rt (where I is the current in amperes, R is the resistance in ohms and t is the time or duration of flow of current in seconds) is utilised. The heat developed at the contact area between the pieces to be welded reduces the metal to a plastic state; the pieces are then pressed together to complete the weld. In this process, preferably two copper electrodes are incorporated in a circuit of low resistance and the metals to be welded are pressed between the electrodes. The electrical voltage required ranges from 4 to 12 volts depending on the composition, area, thickness etc. of the metal pieces to bc welded. The amount of power supplied to the weld usually ranges from about 60 watts to 180 watts for each sq. mm of area. *Alternating current is found to be most suitable for resistance welding as it can provide any desired combination of current and voltage by means of a suitable transformer.*

In order to avoid the surface distortion, the portion of the metal adjacent to the weld or joint should not be allowed to be overheated.

Resistance to the flow of current is made of (*i*) resistance of the current path in the work (*ii*) resistance between the contact surfaces of the parts being welded and (*iii*) resistance between the electrodes and the surface of the parts being welded. In order to develop higher temperature between the interfaces of the work to be welded rather than at the surface of the work in contact with the electrodes it is necessary to keep the resistance between the electrodes and the surface of the body being welded to minimum.

In order to obtain a good weld it is necessary to maintain the contact resistance uniform which depends upon the surface condition.

For welding thin materials the resistance of the current path in the work is kept minimum. For welding thick materials of low conductivity the resistance of the current path have a comparatively greater value and the control of contact resistance is not necessary. For welding thick materials of high conductivity either reduced pressure or high resistance electrodes having melting point higher than that of metal to be welded, can be used. For welding two dissimilar metals having different conductivity, low conductivity electrodes on high conductivity metal side and vice versa are used in order to prevent overheating on the low conductivity metal and to develop sufficient heat to melt high conductivity metal side.

The pressure which is to be applied on the weld is also an important factor. At high pressure, low temperature plastic welds can be obtained and where as if the pressure is lowered the resistance to the welding current is to be increased. There is a limit up to which the resistance can be increased and after that there will be surface burning. The pressure necessary to effect the weld varies from 2.5–5.5 kgf/mm^2.

The magnitude of current is controlled by varying either the primary voltage of the welding transformer (by using auto–transformer between supply and the welding transformer) or changing the primary turns of the welding transformer. Alternative method of controlling the current to weld is to vary the magnitude and wave of the primary as well as secondary current by using Thyratron or Ignitron tubes in the primary circuit.

In resistance welding, the time for which current flows is very important. Usually automatic arrangements are devised which switch off the supply after a predetermined time from applying of pressure (starting of weld). The pressure may be applied manually, by air pressure, by springs or by hydraulic means. After switching off the supply, the pressure is maintained on the electrodes until the weld cools. In machines which are operated continuously, the electrodes are cooled by water circulating through hollow electrodes.

Electrical circuit diagram for resistance welding is shown in Fig. 6.1. The machine employed for resistance welding contains a transformer provided with necessary taps, a clamping device for holding the metal pieces, and a mechanical means for forcing the pieces, to be welded, together to complete the weld.

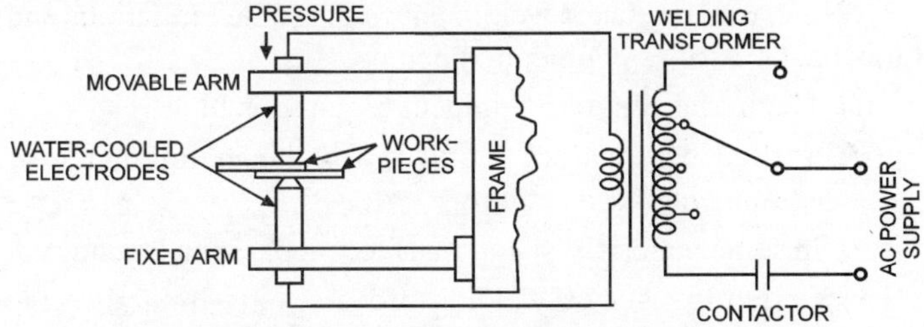

Fig. 6.1. ***Electrical Circuit For Resistance Welding***

Resistance welding has the advantage of producing a large volume of work at high speeds that are reproducible with high quality. Resistance welds are made very quickly; however, each process has its own time cycle. Resistance welding operations are automatic. Good-quality welds do not depend on welding operator skill but more on proper set-up and adjustment of the equipment and adherence to weld schedules.

Resistance welding is employed mainly for mass production. It is easily adapted to those components which can be moved to the machine and are light. The operation is extremely rapid and simple. This is the only process where heat can be controlled and which permits a pressure action at the weld. Metals of medium and high resistance, such as steel, stainless steel, monel metal and silicon bronze are easy to weld. Special control gear is required, however, in the case of high-carbon steels and special equipment providing very high current impulses (stored energy welding) is used in case of materials of low electrical resistance.

The automotive industry is the major user followed by the appliance industry. It is used by many industries manufacturing a variety of products made of thinner-gauge metals and for manufacturing pipes, tubing, and smaller structural sections.

When specifying the material intended to be resistance welded, consideration must be given to the state in which this to be supplied to the welding shop. Whilst slight rust, mill scale etc., on the material may not affect the efficiency of arc welds to a considerable degree, lack of cleanliness will be fatal to resistance welded joints. Pickling or shot-blasting immediately prior to the resistance welding operation is essential for making this latter method a success.

Material up to 5 mm thickness which is to be used on resistance-welded jobs is usually purchased in a pickled and slightly oiled condition, and should be carefully stored in order to keep it clean. It can then be used without removing the oil film provided that the oil is clean. Material above 5 mm thickness should be shot-blasted before being taken to the resistance welding machines. No long delay should occur between the shot blasting and the welding, in order to avoid new corrosion which might eliminate the advantage gained by the former. Sand-blasting is not recommended, as particles of the siliceous material may be embedded in the steel surface and influence its electrical resistance.

High frequency resistance welding is done with 400 to 450 kHz current usually supplied by an oscillator. The high frequency current readily breaks through oxide film barriers and produces a thin heat-affected zone because it travels on the surface of the material.

However, resistance welding have got some limitations and drawbacks also, as enumerated below :

(*i*) The initial cost of equipment required is high.

(*ii*) Skilled persons are required for the maintenance of equipment and its controls.

(*iii*) In some materials, special surface preparation is required.

(*iv*) Certain resistance welding processes are limited to lap joints. A lap joint has an inherent device between the two metal pieces, which cause stress concentration in applications where fatigue is present. The device may also cause trouble when corrosion is present.

6.2.1. Butt Welding

Welding of two pieces of metal together with a butt weld can be carried out by different resistance welding processes. The butt weld consists of joining of two pieces of metal together either on face or on edge. It is of two types namely (*a*) upset butt welding and flash butt welding.

(*a*) *Upset Butt Welding.* It is a resistance welding process that produces coalescence over the entire area of faying surfaces or progressively along a butt joint by the heat obtained from the resistance to the flow of welding current through the area where those surfaces are in contact. It can be used only if the parts to be welded are equal in cross-sectional area.

End to end welds, lap welds, and butt welds are included. The faces of the metal pieces to be joined are prepared for even contact. The metal pieces to be joined are fixed in clamps and butted squarely against each other, as shown in Fig. 6.2 and heavy current is passed through them. In this case heating is obtained by the contact resistance of the metals to be welded. When sufficient heat is developed, the pieces are rammed by force to complete the weld. Pressure is applied either manually or with toggle mechanism. The pressure applied upsets or forges the parts together. The joint is cooled under pressure. After cooling, the force is released and the weld is completed. Upset butt welding, which has the advantage of creating no flash spatter and a smooth and symmetrical upset, is used principally on non-ferrous materials for welding bars, rods, wire, tubing, formed parts etc. The voltage required for welding is 2–8 volts and current varies from 50 A to several hundred amperes, depending upon the material and the area to be welded at a time. The current, pressure and time may be manual or automatic controlled. The latter method is preferred in production machines.

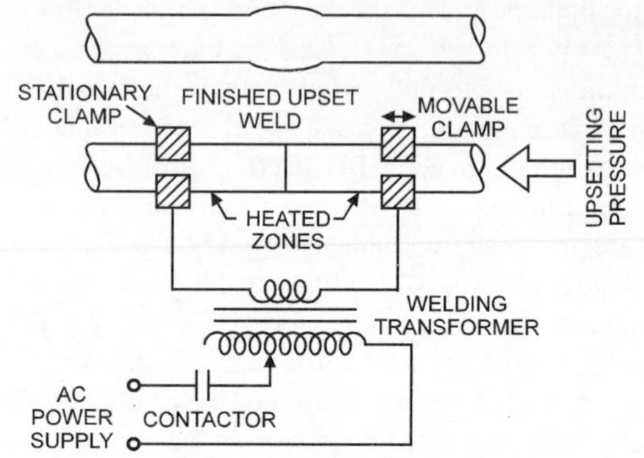

Fig. 6.2. ***Upset Welding Process***

If intimate contact is not obtained because of improper joint preparation, the weld will be defective.

(*b*) *Flash Butt Welding.* Even though it is not a resistance welding process, it is desirable to consider it here, as in this method the two metal pieces are butted together and a current is passed through the joint which heats the ends to red hot. Flash welding may be defined as a resistance welding process that produces coalescence at the faying surfaces of a butt joint by flashing action and by application of pressure after heating is substantially completed (Fig. 6.3).

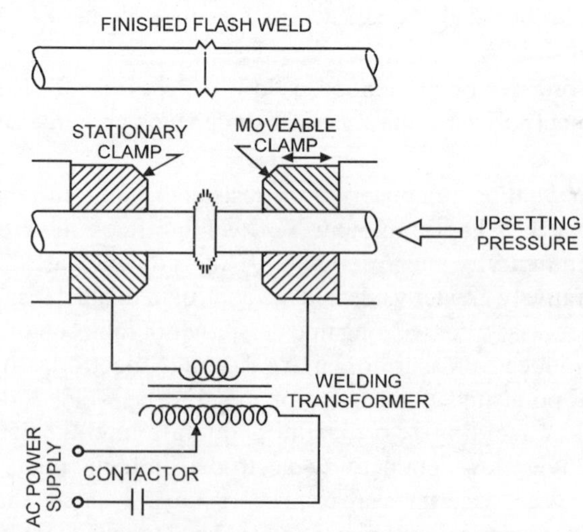

Fig. 6.3. ***Flash Butt Welding Process***

In this process of welding no special preparation of the faces to be welded is necessary. In flash butt welding the voltage to the metal parts to be joined together is applied before putting them together. Protruding small parts of the faces to be welded are brought into contact; the heat developed owing to high resistance of the contact areas raises the temperature of the material at these points and causes them to melt. The molten metal is blown out, and a small arc is formed which raises the temperature of adjoining parts of the abutting surfaces. The ejection of molten metal from high spots continues until the whole of the surfaces are heated to the welding temperature. Some additional heat may be created by the combustion of molten particles, the forceable action of which is probably due to the pressure of metallic vapour. The flashing action is aided by an oscillating movement of one of the components. As soon as the whole of the surfaces to be welded reach forging temperature, the pressure is applied, and the power supply is cut off.

Flash butt welding creates a joint which has practically the full strength of the parent metal under static loading conditions and a strength slightly less than that of the parent metal under dynamic loads. The shape and area of the two components must be identical at the point to be joined, and a suitable upset allowance must be made.

Flash butt welding can be used on most metals (low carbon steels, tool steels, stainless steels, aluminium alloys with thickness exceeding 1.25 mm, copper alloys with high zinc content, magnesium alloys, molybdenum alloys, nickel alloys, titanium alloys, medium strength and high strength low alloy steels) leaving cast iron, lead, zinc, antimony alloys and bismuth. Flash butt welding is considered superior to upset butt welding on the following accounts.

(*i*) Many dissimilar metals with different melting temperatures can be flash welded.

(*ii*) Flash weld provides a smaller upset.

(*iii*) Flash welding is faster than other methods.

(*iv*) All the foreign metals appearing on the butting surfaces is burnt due to first few flashes preceding application of welding pressure. Thus clean and pure weld is obtained.

(*v*) Power requirement is less.

(*vi*) Flash welding can be used for the welding of those highly alloyed steels which cannot be welded satisfactorily by any other process. This is because, in flash welding, under correctly controlled conditions, the heating is not only even but extremely local; so that the cooling stresses are maintained at a minimum; this avoids hardening and cracking in highly alloyed steels.

Flash welding have some limitations and drawbacks also, as enumerated below :

(*i*) The most undesirable feature of flash welding is the flashing operation during which it is impossible to protect the welding machine and the surrounding area from these particles, which can burn into sideway bearings, insulation etc. This necessitates more frequent maintenance.

(*ii*) The process presents a considerable fire hazards. Operators need to be protected from flying particles.

(*iii*) Metal is lost during flashing and upsetting.

(*iv*) Concentricity and straightness of the workpieces during welding process is often difficult to maintain.

Applications of Flash Welding. (*i*) Flash welding is applied primarily in the butt welding of metal sheets, tubing, bars, rods, forgings, fittings etc.

(*ii*) Flash welding finds applications in automotive and aircraft products, household appliances, refrigerators and farm implements.

(*iii*) The process is also used for welding the band-saw blades into continuous loops, and joining of tool steel drill, tap and reamer bodies to low carbon steel and alloy steel shanks.

(*iv*) Flash welding is used to produce assemblies that otherwise would require more costly forgings or castings.

Difference Between Flash Welding and Upset Butt Welding. (*i*) In upset butt welding no arcing (hence flashing) occurs between the surfaces being joined. Heat is produced solely by the electrical resistance at the abutting surfaces to the passage of an electric current.

(*ii*) In flash welding, unlike upset butt welding, the movable platen keeps on moving constantly toward the stationary platen.

(*iii*) Flash welding consumes much less welding current than consumed by upset butt welding process; the time allowed for weld to be completed is, however, more.

(*iv*) In flash welding heat application precedes the pressure where as in upset butt welding constant pressure is applied during the heating process which eliminates flashing.

6.2.2. Spot Welding

By definition, *spot welding is a form of resistance welding in which the parts or pieces are joined in spots, accompanied by heated relatively small sections of the parts or pieces between suitable electrodes under pressure.*

Spot welding is perhaps the most widely used form of resistance welding. It is characterized by low cost, speed, and dependability making it a common electrical resistance welding process.

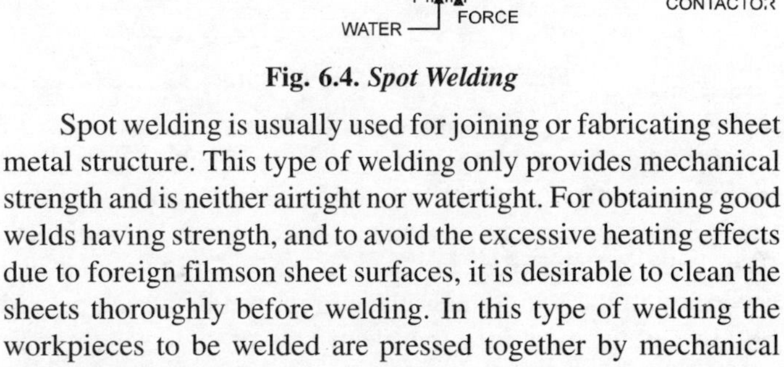

Fig. 6.4. ***Spot Welding***

Spot welding is usually used for joining or fabricating sheet metal structure. This type of welding only provides mechanical strength and is neither airtight nor watertight. For obtaining good welds having strength, and to avoid the excessive heating effects due to foreign filmson sheet surfaces, it is desirable to clean the sheets thoroughly before welding. In this type of welding the workpieces to be welded are pressed together by mechanical pressure exerted through electrodes (made of pure copper or alloy of copper) as shown in Fig. 6.4 and then supply is switched on for a particular period. The mechanical pressure may be developed by a foot lever or by air pressure or by hydraulic cylinders. The metal in the zone of pressure gets up heated to fusion and the joint thus made gets cooled under pressure. The pressure remains applied till the weld cools and regains sufficient strength. Current required is above 5,000 A and the voltage between the electrodes is usually less than 2 volts. The open-circuit voltage is less than 12 volts. The time period of flow of current varies widely depending upon the thickness of sheets, kinds of metal etc. For thin sheets this period is about 1/50 second for each 0.25 mm of total thickness of the two sheets to be joined.

In order to avoid electrode sticking it is necessary to avoid heating of junction faces between electrodes and work. This is achieved (*i*) by using water-cooled electrodes and (*ii*) by employing electrodes made of material having high electrical as well as thermal conductivity. In order to minimise the wear and tear at the tip of the electrodes it is necessary to employ electrodes made of material having high mechanical strength.

Efficient spot welds are possible only by a correct combination of current, pressure, and welding time. The heat developed being proportional to the product of the welding time and the square of the current, it is possible to obtain good welds by employing either low currents for a longer duration or higher currents for a shorter duration. Longer welding time usually produces stronger welds but this involves lot of distortion of workpieces, electrode maintenance and high energy expenditure. The welding pressure required is determined on the basis of a specific electrode pressure of 7 kg per sq. mm over the electrode tip area. In view of the fact, however, that plates will not always be perfectly flat, a certain additional pressure is usually recommended, which will overcome slight distortion of the plates without the resultant necessary welding pressure.

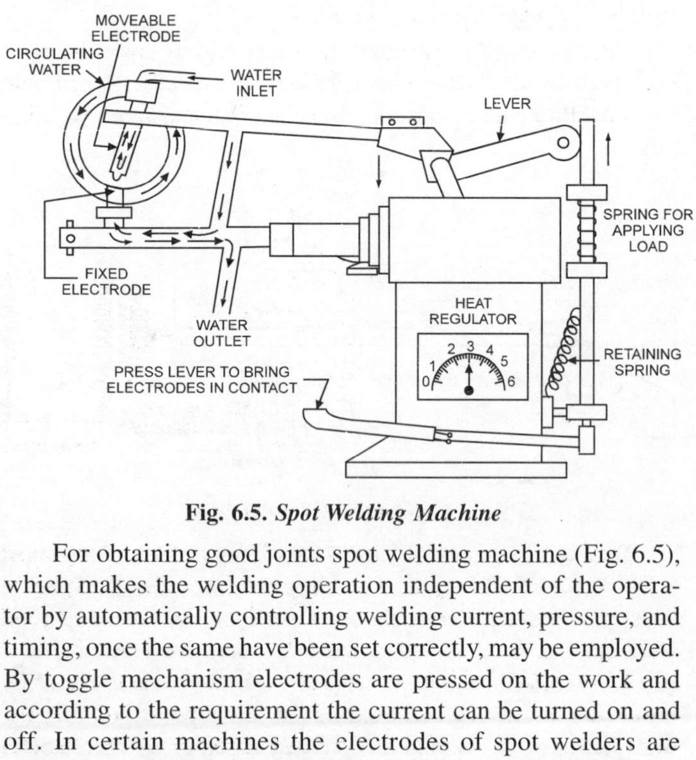

Fig. 6.5. ***Spot Welding Machine***

For obtaining good joints spot welding machine (Fig. 6.5), which makes the welding operation independent of the operator by automatically controlling welding current, pressure, and timing, once the same have been set correctly, may be employed. By toggle mechanism electrodes are pressed on the work and according to the requirement the current can be turned on and off. In certain machines the electrodes of spot welders are equipped with water cooling arrangement. In such cases the electrodes are made hollow and the water is circulated through them. Spot welding machines are of capacities ranging from 10 to 150 kVA.

Spot welding is applicable for joining components made from plate material, the plate thickness being limited (usually to 10 or 12 mm) by the pressure and current capacity of the available spot welding machine. It has an advantage over projection welding in cases where the number of welds required is relatively small, where the size of the job is too large for the capacity of projection welders, and where the design allows the necessary spacing and arrangement of the welds to be obtained. The lower cost of the equipment makes spot welding economical for small quantities.

6.2.3. Projection Welding

Projection welding is a resistance welding process that produces coalescence by the heat obtained from the resistance to the flow of the welding current. The resulting welds are localized at predetermined points by projections, embossments or intersections.

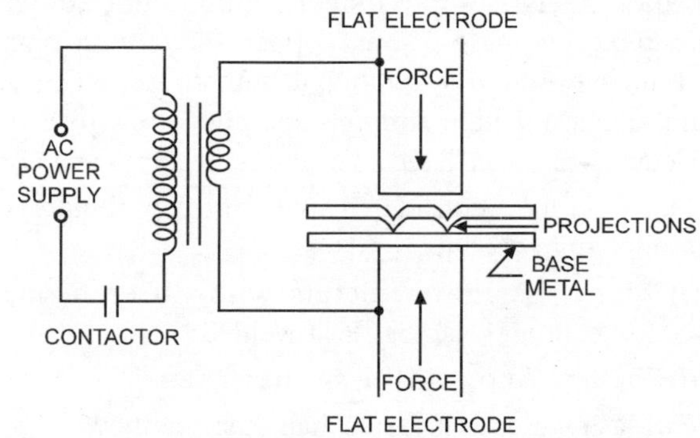

Fig. 6.6. ***Projection Welding***

Projection welding uses the same equipment as spot welding. The only difference is that the electrodes used are flat on the ends and larger in diameter than the flat electrodes used in spot welding. Successful projection welding depends greatly on the surface preparation of the pieces to be welded. Projections, small deformations that will touch the surface of the material to be welded, are made on the weld areas. One of the main advantages of these projected points is that welding area can be located easily, which makes projection welding a high-production welding technique. As the current flows through the two parts to be welded, the projected points are the main contact area. Whenever a point or projection touches the metal, a weld nugget will start. These weld points soon reach the plastic state, and the force applied by the electrodes finishes the weld nugget. The cycle tine is the same as the spot-weld time. As the points reach their plastic state, the metal is compressed so that the finished weld is similar to the spot weld with the exception of the small indentations created by projections. Projection welding, as spot welding, needs no protective atmosphere in order to produce successful results. It is also useful in securing small circular component, to larger pieces of metal. For projection welding, a large flat-surfaced electrode is used, which covers the areas of the projections. The areas of projection, one or more, and the spots where they touch on the other piece of metal will be the only places where fusion takes place, even with a large electrode. Projection welding has been used by auto manufacturers for many years. Auto bodies have many areas that are projection welded. Another area that uses projection welding is stud welding, which has also been utilized for car bodies.

Projection welding reduces the amount of current and pressure required in order to form a good bond between two surfaces. With this reduction there is less chance of shrinkage and distortion in the areas surrounding the weld zone, one reason why it has been incorporated into many manufacturing processes.

Projection welding has the following advantages over spot welding :

(*i*) Projection makes the welding process simple.

(*ii*) It is possible to join several welding points in projection welding.

(*iii*) More output is obtained since more than one weld are done at a time.

(*iv*) Electrode life is increased due to use of low current density and low pressure.

(*v*) Good finished appearance is obtained as surface remains unidented by the electrodes.

(*vi*) Welds are automatically located by the position of projections.

(*vii*) As the electrodes are flat, therefore the contact area over the projection is sufficient. It facilitates the easier maintenance of electrodes, alignments of parts and construction of various jigs and fixtures.

(*viii*) With the projection welding, it is easy to weld certain parts which cannot be welded by spot welding.

When the two plates to be welded are of different cross-sectional area then in order to obtain desirable strength it becomes necessary to have the projections on the thicker plate.

When the two plates to be welded are of different material then the projections should be made on the high conductivity materials.

The thickness of plates to be jointed is limited to 2 × 6 mm in case of steel, 2 × 4 mm in case of aluminium, and 2 × 2 mm in case of copper. This reduction in maximum thickness of different metals is due to increase in conductivity.

The initial cost of the equipment required for projection welding is high as press type machines are required for this type of welding.

This type of welding is usually employed on punched, formed or stamped parts, where the projections automatically exist.

Small fasteners, nuts etc. can be welded to larger components.

Projection welding is particularly applicable to mass production work, *e.g.*, the welding of refrigerator condensers, crossed-wire welding, refrigerator racks, gratings, grills etc.

6.2.4. Seam Welding

Seam welding can be defined as series of continuous spot welds. This process is employed for making a continuous joint between two overlapping pieces of sheet metal. In this process the work to be welded is placed between the two wheels which apply sufficient mechanical pressure and also carry sufficient current for producing continuous welds. The mechanical pressure applied is kept constant and the current is regulated by a timer (the current is kept on for a definite time and then off for another fixed interval). As the pressure is applied, the drive is started, the welding current switched on and simultaneously the overlapping surfaces of the metal are forced together as fast as they are heated. The heat is generated due to flow of current through the resistance in the welding circuit. The heat generated can be controlled by either varying the current or pressure between the sheets which varies the contact resistance. If the rate of heat generation is high then the speed of the rollers is increased thereby reducing the welding time and vice versa. The electrodes are made of copper alloys and are water circulated in order to dissipate heat from them. In high speed seam welding using continuous current the frequency of the current acts as an interrupter. In order to localise the current and pressure to the welding point, the contact area of the electrodes is kept small. Usually number of spots obtained per metre of the joint are between 200 and 400 depending upon the nature of the joint required. Welding speed ranges from 0.25 metre to 10 metres per minute. This type of welding is very important as it provides pressure tight or leak proof joint.

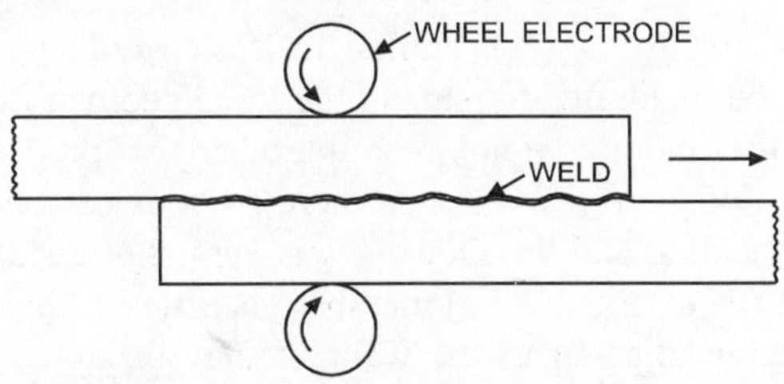

Fig. 6.7. ***Seam Welding***

It is very essential in case of seam welding that the surfaces to be welded are cleaned and made dust and dirt free. This is not necessary in other forms of resistance welding, *e.g.*, in spot welding the detrimental effect of scale and surface dirt can be compensated for by adjustment of time, pressure, current or by other means but this is not true of seam welding. Materials which can be seam welded after they are cleaned include high-carbon, stainless and coated steels and alloys of aluminium, nickel and magnesium. Seam welding of copper and high copper alloys is not recommended. Seam welding is employed for welding pipes, conduits, tanks, transformers, refrigerators, gasoline tanks, aircrafts and various types of containers. This process is best adopted for metal thickness ranging from 0.25 mm to 3 mm.

6.2.5. Percussion Welding

It is a recent development in welding. It is a resistance welding process where in coalescence is produced simultaneously over the entire area of abutting surfaces by heat obtained from an arc produced by a rapid discharge of stored electrical energy. This is a self-timing spot welding method.

In this process, the pieces to be welded are held apart, one in a stationary holder and the other in a clamp mounted in a slide and backed up against heavy spring pressure. When the movable clamp is released, it moves rapidly carrying with it the piece to be welded. When the two pieces are about 1.5 mm apart, there is a sudden discharge of electrical energy, causing intense arcing between the two surfaces and heating them to high temperature. As pieces come in contact with each other under heavy pressure the arc is extinguished due to the percussion blow of the two parts and the force between them effects the weld.

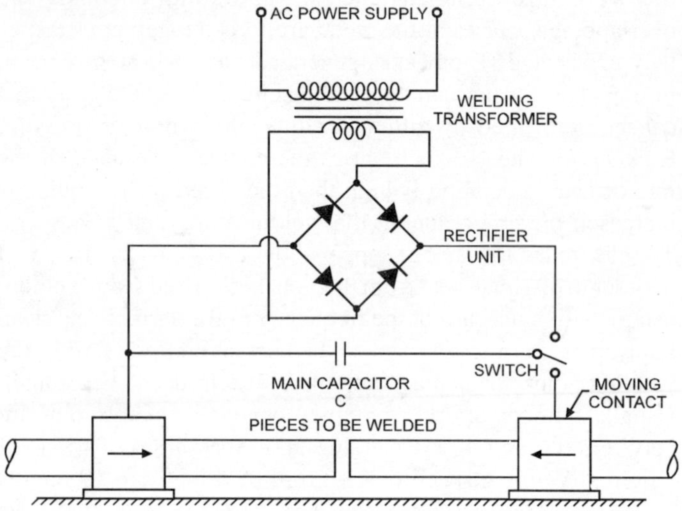

Fig. 6.8. ***Percussion Welding***

This process is quite similar to flash welding and upset welding, but is limited to parts of the same geometry and cross section. It is more complex than the other two processes. The advantage of the process is that there is an extremely shallow depth of heating and the time cycle is very short. This process is so fast (takes about 0.1 *s*) that there is little heating effect in the material adjacent to the weld. It may be noted that this process is limited to only small areas (up to 3.2 cm^2) of nearly regular sections. Thin sheets of equivalent area cannot be joined by this process. It can be used for welding a large number of dissimilar metals. It is used for very specialized applications and the process is entirely automatic. Metals which can be percussion welded include, copper alloys, aluminium alloys, nickel alloys, low-carbon steel, medium-carbon steel and stainless steels. This process is used for welding satellite tips to tools, copper to aluminium or stainless steel, silver contact tips to copper, cast iron to steel, lead-in-wires on electric lamps and zinc to steel. Gold, silver, copper-tungsten, silver-tungsten and silver-cadmium oxide percussion welded to copper alloys for commonly used assemblies for electric contacts.

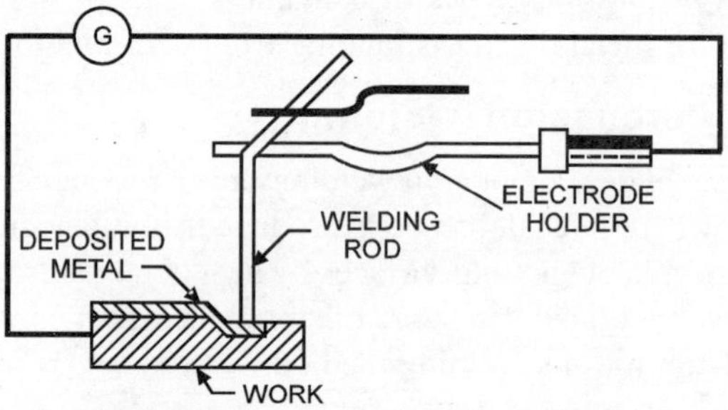

Fig. 6.9. ***Electric Arc Welding Circuit***

6.3. ELECTRIC ARC WELDING

By definition, *arc welding is that process in which the pieces of metal to be welded are brought to the proper welding temperature at point of contact by the heat liberated at the arc terminals and in the arc stream so that the metals are completely fused into each other, forming a single solid homogeneous mass, after it solidifies.*

In this process an electric arc is produced by bringing two conductors (electrode and metal piece) connected to a suitable source of electric current, momentarily in contact and then separating by a small distance. The current continues to flow across the small gap and gives intense heat. The heat developed is utilised to melt the part of the workpiece, and the filler metal and thus forms the joint. So arc welded joint is a union of metal parts made by localised heating without any pressure. That is why sometimes this type of welding is known as the *non-pressure welding*. The heat developed by arc is also used for cutting of metal.

The voltage necessary to strike the arc is higher than that required to maintain it. The temperature is of the order of 3,600°C, at which mechanical pressure is not required for jointing. According to BSS 638 maximum voltage specified for welding is about 100 volts for the safety of the operator. The arc voltage varies from 20 to 40 volts and current from 50 A in sheet metalwork to 1,000 A in heavy automatic welding.

The welding circuit for an electric arc welding process is shown in Fig. 6.9. The welding circuit consists of a welding machine, two leads, an electrode holder, an electrode and the work itself.

The electric arc welding is widely used for joining of metal parts, the repair of fractured castings and the fillings by the deposition of new metal on worn-out parts.

Arc welding is again subdivided into five groups namely carbon arc welding, metal arc welding, atomic hydrogen arc welding, inert gas metal arc welding and submerged arc welding.

6.3.1. Carbon Arc Welding

Carbon arc welding differs from metal arc welding in that it is a puddling process and is somewhat similar to the gas welding process. The carbon arc is very stable and easy to maintain. The length of the arc can be varied over wide limits without causing the arc to go out. There is no tendency for the electrode to freeze or stick, as in the case of the metallic electrode. Accordingly the arc can be struck without difficulty at any point, and rapidly moved over the surface of the work to the point where the weld is to be made. High melting speed and efficient welds are the main advantages claimed.

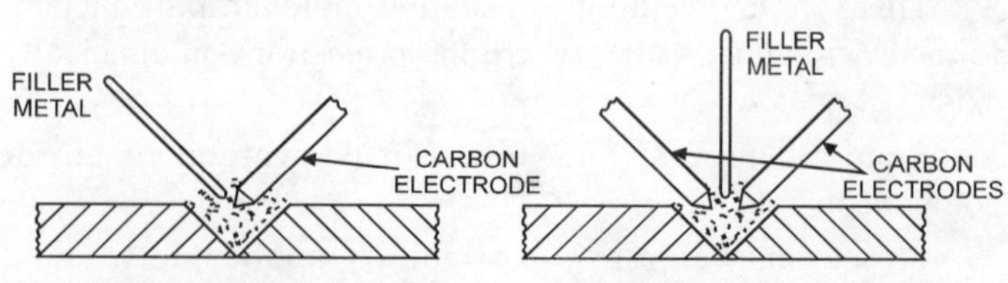

(a) Welding With Single Carbon Electrode *(b) Welding With Double Carbon Electrodes*

Fig. 6.10. ***Carbon Arc Welding***

In this process, a carbon or graphite rod is used as a negative electrode and the work being welded as positive. Mostly graphite electrodes are used as they yield longer life and have low resistance and thus capable of conducting more current. The arc produced between the two electrodes heats the metal to the melting temperature. This is about 3,200°C on the negative electrode, and 3,900°C on the positive electrode. The reason to use carbon rod as negative electrode is that less heat will be generated at the electrode tip than that at the workpiece, and carbon from the electrode will not fuse and mix up with the job. If this so happens the resultant weld will be rich in carbon, and consequently very much brittle and unsound. For this type of welding only dc can be used. The use of ac is not recommended because no fixed polarity can be maintained. This process is normally adopted where addition of filler material is not required, *e.g.*, flange or edge joints, but if filler metal is required, it is provided by the welding rod made of the similar metal to that of the metal to be welded. Some protection for the molten weld metal may be provided by using a long arc which produces a carbon monoxide gas envelope. There are two methods of carbon arc welding. In one method no flux is used and in the other method flux either in the form of powder or paste is used to prevent the weld from oxidation. Former method is confined to non-ferrous metals and the later method is usually employed for ferrous metals.

Electrodes up to 25 mm in diameter with currents up to 600 or 800 A are used, so that the carbon arc process is well suited to rapid work, or cases such as arise in the repair of castings, where large quantities of filler metals have to be deposited.

Carbon arc welding is employed for welding sheet steel, copper alloys, brass, bronze, and aluminium. It is not suitable for vertical and overhead welding.

In another process of carbon arc welding, known as *twin carbon arc welding,* two carbon electrodes are used between which the arc is formed, as illustrated in Fig. 6.10 (*b*). In this way the arc generates heat which is applied close to the parts to be welded, causing the edges to fuse. In heavy welds, a filler rod is melted by the arc and deposited in the weld. An ac supply is recommended for twin carbon arc welding. In case dc supply is used, the positive electrode will disintegrate and consume at a much faster rate as compared to the –ve electrode because two-thirds of the total heat is generated at the positive pole. This will produce an unstable arc and need more frequent adjustment of the electrodes. The electrodes used for twin carbon arc welding are approximately of the same diameter as the thickness of workpiece. The magnitude of arc current required depends upon both, electrode diameter and plate thickness. Twin carbon arc welding, though, more complex than single carbon arc welding, possesses the advantage that arc is independent of the job and can be moved anywhere without getting extinguished. Moreover, the workpiece is not a part of the electrical circuit.

6.3.2. Metal Arc Welding

In this type of welding, a metal rod of the same metal as being welded forms one of the electrode and also serves as a filler and no filler rod is used separately. The arc struck between the work being welded and the electrode causes the melted rod to flow across the arc into the metal pool of the parent metal. This deposition of metal is accomplished by contact made between the molten metal and the globules formed on the end of the electrode filler rod. The temperature produced is about 2,400°C and 2,600°C on the negative electrode and positive electrode respectively. The concentration of heat energy at the terminals of the electrode causes a small part of the work being welded to melt almost instantaneously and an intermittent flow of metal across the arc stream. The metal in the arc stream is in both the liquid and gaseous forms, the liquid metal being transferred across the arc by molecular attraction, adhesion, cohesion, surface tension or a combination of these. For this type of welding both ac and dc can be used. For dc supply 50-60 volts and for ac supply 70-100 volts are used for welding. For current above 750 A ac equipment is preferred as it has high efficiency, negligible loss at no load and minimum maintenance and initial cost. Welding with dc also gives rise to "arc blow", owing to the magnetic forces created by the current, a trouble which does not exist with ac welding as each magnetic force is immediately followed by one in opposite direction and the arc-deflecting forces are automatically cancelled. Metal in the molten state has an affinity for foreign elements such as oxygen, nitrogen etc.; and in order to prevent impurities from entering the weld modern arc welding electrodes are covered with a coating which, under the welding heat, generates a gas shield round the arc, and also forms a slag which is deposited on and around the molten metal, protecting it against foreign matter during the cooling stages. A metal arc welding circuit is shown in Fig. 6.11.

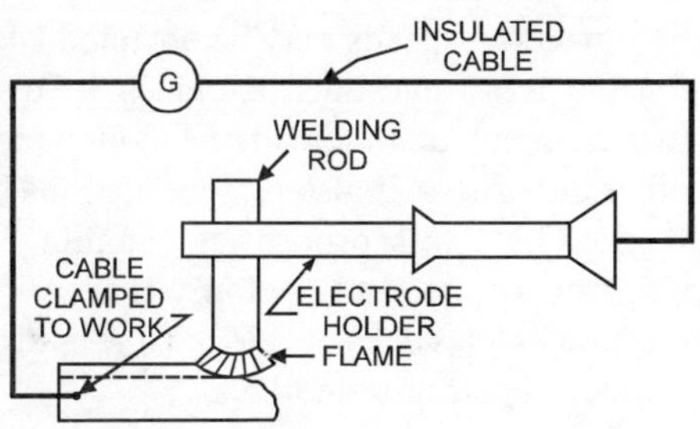

Fig. 6.11. ***Metal Arc Welding Circuit***

The correct welding current, voltage and welding speed are very important. Various welding currents can have a deciding effect on the forming of proper beads. A welding shows bead characteristics under different conditions : (*i*) welding current too low excessive piling of the metal (*ii*) welding current too high causing excessive spatter, (*iii*) voltage too high bead too small (*iv*) welding speed too low excessive piling up of weld metal. Proper current and timing create a smooth, regular, well-formed bead. Undercutting is also a result of too much current. No enough current results in overlapping and a lack of fusion with the metal.

6.3.3. Atomic Hydrogen Arc Welding

The principle of atomic hydrogen welding is based on (*i*) the possibility of obtaining atomic hydrogen by means of an electric arc between the two tungsten electrodes, in an atmosphere of hydrogen at atmospheric pressure, (*ii*) the very high temperature produced by the recombination of the atoms, which occurs in the cooler regions immediately outside the arc, and (*iii*) the very great heat conductivity of hydrogen at high temperatures, owing to the dissociation of the hydrogen molecules into the atomic state, and resulting in an extremely high rate of delivery of heat to the surfaces to be welded–roughly twice that for the oxyacetylene welding flame.

In this method of arc welding an ac arc is maintained between the two non-consumable tungsten electrodes while a stream of hydrogen gas, under a pressure of about 0.5 kg/cm^2, is passed through the arc and around the electrodes. AC supply is used in order to obtain equal consumption of the electrodes. The hydrogen is usually supplied from steel cylinders and serves two-fold purpose—firstly as protective screen for the arc, secondly as cooling agent for the glowing tungsten electrode points. As the molecules of hydrogen pass through the electric arc, they are changed into the atomic state and thus absorb a considerable amount of energy—resulting in cooling of glowing tungsten electrodes. But when the atoms of hydrogen recombine into molecules just outside the arc, a large amount of heat is liberated. This extra heat, added to the intense heat of the arc itself, produces a temperature of about 4,000ºC, as compared to 2,000ºC produced by the combination of normal hydrogen and oxygen. This heat is used in making fusion welds. When additional metal is required, filler rods are melted into the joint. Due to the decarbonizing character of the welding flame (with the formation of hydrocarbons) the filler rod material should have a carbon content higher than that of the material to be welded, if the finished weld metal is to have the same carbon content as the parent material. Owing to the concentrated heat, high welding speeds may be obtained keeping distortion small. Further more hydrogen excludes all oxygen and other gases which might combine with the molten metal to form oxides and other impurities. It also removes oxides from the surface of the work. Thus, this process is capable of producing smooth, uniform, strong, and ductile welds. Automatic equipment for this process has been developed, the hydrogen being fed through a water-cooled nozzle block.

Arc currents up to 150 A can be used and the power supply arrangements are similar to those for ordinary ac arc welding except that the transformer voltages are higher, being about 300 V on open circuit for striking the arc and between 80 and 100 V while operating.

This method is very flexible—practically any metal or alloy, ferrous and non-ferrous, can be welded. Its greatest usefulness is for the fusion welding of certain special ferrous alloys, such as chrome nickel steels, aluminium, and duralumin, but it must use flux to weld some of these materials. Welding of thin sheets, production of tubing, and the repairing of expensive tools and dies are some of the common uses of this method.

6.3.4. Inert Gas Metal Arc Welding

It is a gas shielded, metal arc welding process which uses the intense heat of an electric arc between a continuously fed, consumable electrode wire and the material to be welded. A consumable bare electrode wire producing the filler metal is automatically fed from a reel on to the job through a welding gun which also carries a nozzle. Through this nozzle helium or argon is blown around the arc and on to the weld. Because of low arc voltage for a given welding current argon is suitable for thin materials as tendency to burn through is reduced. Because of low arc voltage and low power for given welding current in case of argon gas welding tendency for metal fusion and metal sag or run is reduced and, therefore, it is preferred for inposition welding. On the other hand, because of high arc voltage and high power in case of helium gas welding, it is used for welding of thick materials and metals with high heat conductivity. Mixture of the two gases is useful to get intermediate characteristics. AC and dc both can be used. This type of welding has the following advantages :

(*a*) In this method concentration of heat is possible and thus distortion is minimised.

(*b*) In this method no flux is required since atmosphere is inert and air does not come in contact with the molten metal.

This process is particularly employed for welding light alloys, stainless steel and non-ferrous metals such as copper, aluminium and their alloys. For welding aluminium and aluminium alloys ac supply and argon gas is used where as for welding magnesium and magnesium alloys either ac supply and argon gas or dc supply, reversed polarity and either argon or helium gas is used. For the welding of stainless steel, mild steel, copper and copper alloys, dc and either argon or helium or ac high frequency stabilized with argon or helium used.

6.3.5. Submerged Arc Welding

It is an arc welding process that uses an arc between a bare metal electrode and the weld pool. The arc and molten metal are shielded by a blanket of granular flux on the workpieces. The process is used without pressure and with the filler metal from an electrode, and sometimes from a supplemental source such as a welding rod, flux, or flux with metal granules.

It is an automatic process developed primarily for the production of high quality butt welds in thicker steel plate than is normally suited to other manual arc welding processes.

As in open arc welding, the source of heat in the submerged arc welding process is an electric arc or arcs between a metal electrode, or electrodes and the work. The welding zone is shielded by a blanket of flux, so that the arc is not visible. Hence the name "*submerged arc welding*". The arc melts the parent metal, the electrode, and the flux. The fused flux produces a blanket of liquid slag which forms a protective envelope both around the arc and the welding zone. Right beneath the electrode tip a weld pool is formed, holding an amount of molten metal. On melting, the electrode metal forms globules which go over into the puddle and mix with the molten parent metal. The pressure of the arc forces the mix out of the puddle so that it forms the weld (Fig. 6.12). This action continues as long as the arc advances. The shape and size of the puddle have a strong effect on those of the weld.

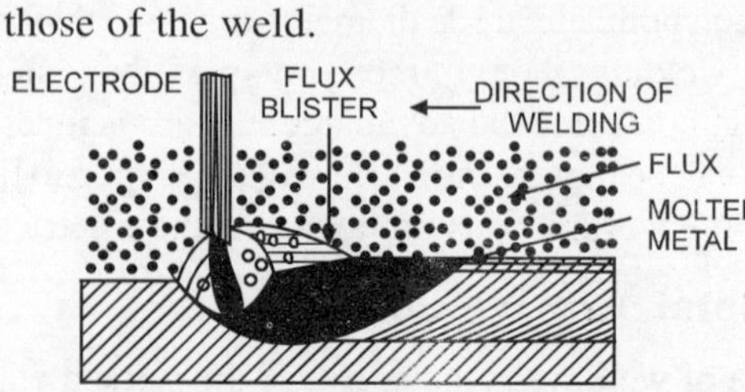

Fig. 6.12. ***Cross Section Through The Welding Zone in Submerged Arc Welding***

The method of making a weld by submerged arc welding is illustrated diagrammatically in Fig. 6.13. Flux is fed continuously onto the work just ahead of the advancing arc. Since the arc is not visible, the operator can wear goggles with ordinary glasses, using neither a hand screen nor a helmet. Since molten flux is lighter than molten metal, it floats to the surface and solidifies as a crust of brittle slag which is readily removable from the weld surface. Much of the flux spread on the work remains unfused. It is recovered by suction into the hopper for re-use.

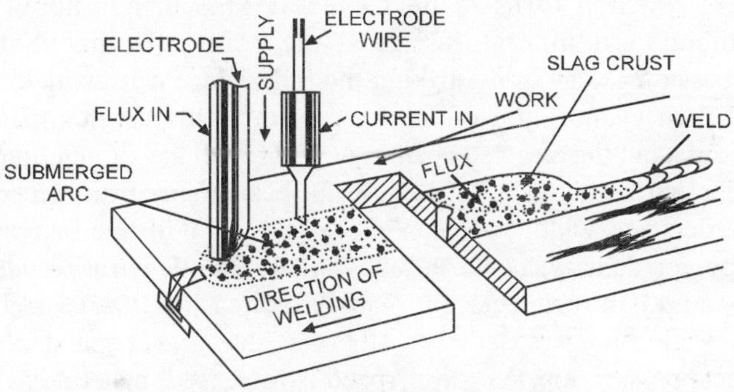

Fig. 6.13. ***Submerged Arc Welding***

The flux may be made of silica, metal oxides, and other compounds fused together and then crushed to proper size. Another group of fluxes is made of similar material "bonded" and formed into *granules.*

The bare electrode is fed from a reel down through the gun or nozzle. The operator can move it slowly, a little at a time for the start of the arc and then set the proper feed rate on the control box. Wire of several alloys for steel, stainless steel, copper etc. is available in diameters from 3 to 6 mm. Voltage employed is from 25 to 40 volts. Current employed depends considerably on workpiece thickness. Normally dc is used employing 600 to 1,000 A current for welding alloy and stainless steel, although ac is preferable particularly for low carbon steel. The current for ac is usually 2,000 A.

The advantages of submerged arc welding are :

(*i*) Partly because it is often automated, it is much faster than regular arc welding. Speeds up to 3,800 mm/minute are possible on 3 mm thick steel at 100% efficiency.

(*ii*) Deep penetration with high quality weld is possible.

(*iii*) Less distortion occurs from high speed and uniform heat input, specially when automated.

(*iv*) No edge preparation is required.

(*v*) Operator is not exposed to usual spatter and can work more easily without helmet and other safety equipment.

(*vi*) The welds obtained have good ductility, impact strength, uniformity, low nitrogen content and high corrosion resistance.

(*vii*) Manipulative skills normally not involved.

(*viii*) High utilisation of electrode wire.

(*ix*) Easily automated for high operator factor.

(*x*) Little or no smoke.

The submerged arc welding is widely used in heavy steel plate fabrication work. This includes the welding of structural shapes, the longitudinal seam of larger diameter pipe, the manufacture of machine components for all types of heavy industry, the manufacture of vessels and tanks for pressure and storage use. It is widely used in the ship building industry for splicing and fabricating subassemblies, and by many other industries where steels are used in medium to heavy thickness. It is also used for surfacing and build-up work, maintenance, and repair.

6.4. ULTRASONIC WELDING

Ultrasonic welding is a solid state welding process that produces coalescence of materials by the local application of high-frequency (20-60 kHz) vibratory energy as the workpieces are held together under pressure. No flux or filler metals are used, no electrical current flows through the weld metal, and usually no heat is applied. Since vibratory action in ultrasonic welding breaks up and disperses moisture, oxide and other (*e.g.*, insulation) coatings, normally, degreasing may be the only cleaning required before welding (especially with aluminium).

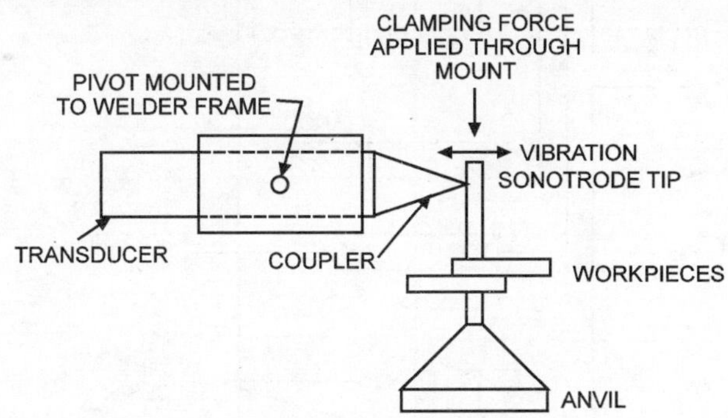

Fig. 6.14. ***Ultrasonic Welding***

Welding occurs when the ultrasonic tip or electrode, the energy coupling device, is clamped against the workpieces and is made to oscillate in a plane parallel to the weld interface. The combined clamping pressure and oscillating forces introduce dynamic stresses in the base metal. This produces minute deformations which create a moderate temperature rise in the base metal at the weld zone. This coupled with the clamping pressure provides for coalescence across the interface to produce the weld.* Ultrasonic energy will aid in cleaning the weld area by breaking up oxide films and causing them to be carried away. The vibratory energy that produces the minute deformation comes from a transducer which converts high-frequency alternating electrical energy into mechanical energy. The transducer is coupled to the work by various types of tooling which can range from tips similar to resistance welding tips to resistance roll welding electrode wheels.

The temperature at the weld is not raised to the melting point and, therefore, there is no nugget similar to resistance welding. Weld strength is equal to the strength of the base metal. Most ductile metals can be welded together and there are many combinations of dissimilar metals that can be welded. The process is restricted to relatively thin materials normally in the foil or extremely thin gauge thicknesses. The maximum thickness by these processes ultrasonically may vary from 0.38 to 2.5 mm depending upon the metal.

This process is used extensively in the electronics, aerospace, and instrument industries. It is also used for producing packages and containers and for sealing them. The process can also be employed for joining plastics and is finding wider use in this field than in joining metals.

* The coupling system and welding tip form a unit called a sonotrode.

6.5. ELECTRON BEAM WELDING

Basically, electron beam welding in vacuum utilises the kinetic energy of electrons travelling with high velocity in a high vacuum (10^{-3} to 10^{-5} mm Hg). When the electrons strike the surface of the metal, they give up the bulk of their energy as heat, and this goes to melt the metal. If the work is dore in a high vacuum, no electrodes, gases, or filler metals need contaminate it and pure welds can be made. Moreover, high vacuum is necessary around the filament so that it will not burn up anJ will also produce and focus a stable beam (Fig. 6.15).

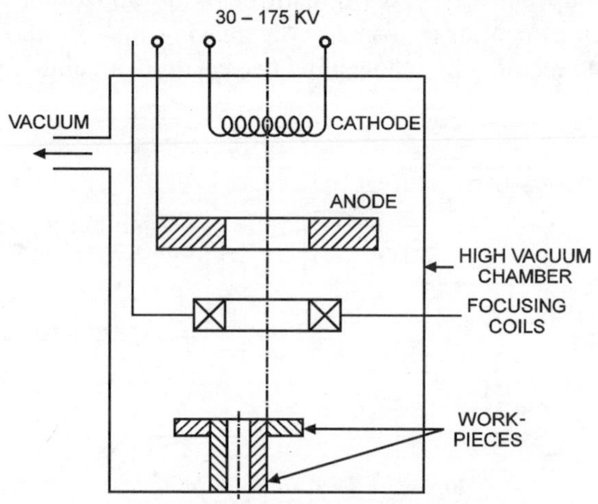

Fig. 6.15. *Electron Beam Welding*

In all types of electron beam machines, a tungsten filament which serves as a cathode emits a mass of electrons that are accelerated and focused to a 0.25–1 mm diameter beam of high energy density up to 0.5–10 kW/mm^2. The temperature produced is about 2,500°C. This is sufficient to melt and vaporize the workpiece material and thus fills a narrow weld gap even without a filler rod (although filler rods may be employed).

The speed of the beam is stepped to one-half to two-third of the speed of light by passing it through a high-voltage electrostatic field. An electromagnetic lens is used to obtain correct focusing of the beam.

A similar process, known as *non-vacuum electron beam welding* is rapidly coming into use today, *i.e.*, many welds are now made without elaborate arrangement required for vacuum electron beam welding. In this case, the vacuum chamber in which the beam is created is evacuated to a lower pressure. In situation where contamination of the workpiece must be held to a minimum, the beam should be passed through argon or helium. To make the chamber high vacuum, it takes about 5 to 30 minutes to evacuate the air, depending on the size of the chamber. A medium size electron beam welder operates below 60 kV. The welding head or the work is moved by numerical control or by hand.

The advantages of electron beam welding are that the welds are clean, with no porosity since there is no air ; no shielding gas is required ; and as the energy input is in a narrow, concentrated beam, distortion is almost eliminated. The speed may be as fast as 2,500 mm per minute, and it will weld or cut any metal or ceramic, diamond, sometimes as thick as 150 mm.

The major advantage of electron beam welding is its tremendous penetration, which occurs when the highly accelerated electron hits the base metal. It will penetrate slightly below the surface and at that point release the bulk of its kinetic energy, which turns to heat energy. This brings about a tremendous temperature rise at the point of impact. The succession of electrons striking the same place causes melting and then evaporation of the base metal. This creates metal vapours, but the electron beam travels through the vapour much easier than solid metal. This causes the beam to penetrate deeper. The depth-to-width ratio can exceed 20 : 1. With the increase in power density, penetration is increased. Penetration also depends on the beam current. With the increase in beam current, penetration is increased. The other variable, travel speed, also affects penetration. As travel speed is increased, penetration is reduced.

Since the electron beam has tremendous penetrating characteristics, with the lower heat input, the heat affected zone is much smaller than of any arc welding process. In addition, because of almost parallel sides of the weld nugget, distortion is greatly reduced. The cooling rate is much higher, and for many metals this is advantageous; however, for high carbon steel this is a drawback and cracking may occur.

Almost all metals can be welded with the electron beam welding process. The metals that are most often welded are super-alloys, the refractory metals, the reactive metals, and the stainless steels. Many combinations of dissimilar metals can also be welded.

One of the disadvantages of the electron beam process is its high capital cost. The price of equipment is very high, and it is expensive to operate, due to the requirement for vacuum pumps. In addition, fitup must be precise and locating the parts with respect to the beam must be perfect.

Today automobile, airplane, aeroscape, farm and other types of equipments including ball-bearing over 100 mm are being welded by the electron beam process.

6.6. LASER BEAM WELDING

Laser beam welding is a welding process that uses the heat from a laser beam impinging on the joint. The process is used without a shielding gas and without the application of pressure. The laser is a device that produces a concentrated coherent light beam by stimulating electronic or molecular transitions to lower energy levels. The word *laser* is an acronym for "*light amplification by stimulated emission of radiation*". The laser beam is a highly concentrated source of energy which has many applications. It can be used for welding, for cutting metals and non-metals, for surface heat treating of metals, and for cladding by fusing powders to base materials. It can also be used for brazing and soldering, and for drilling, machining, and marking. It is also used in other fields : medicine, communications, marking and surveying.

There are two basic types of lasers employed in metal working. The original types are the solid state lasers, which use a solid sodium. The second type are the gas lasers, which normally use a mixture of helium, nitrogen, and CO_2 gas in a tube. In either case, when the medium is excited, it emits photons, which become the laser beam.

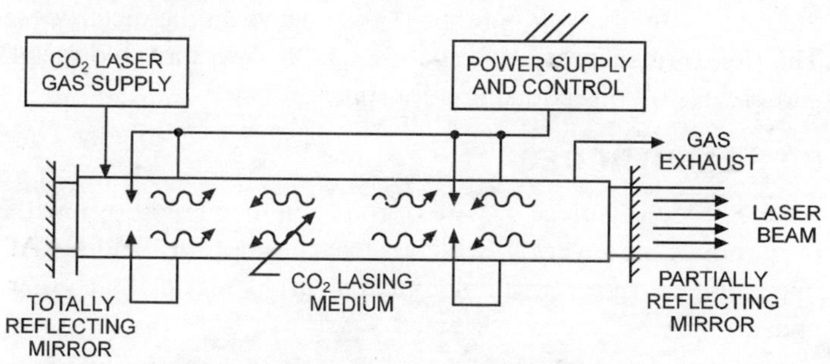

Fig. 6.16. *Basic CO_2 Laser System*

A gas laser beam is shown in Fig. 6.16. Excitation of gas laser is by means of high-voltage low-current electric power. The electrical discharge excites the CO_2 molecules, which on returning to their original state, emit photons. Mirrors are placed on both ends of the tube, one entirely reflective and the other with small partially transmissive area to permit the beam to exit. This forms a cavity in which photons builds up. The freed photons travel between the mirrors and excite the CO_2 molecules, starting a chain reaction of photon emissions. A stream of photons, the laser beam, exits through the unsilvered section of the one mirror. The wavelength of the laser beam is 10.6 μm. This wavelength is longer and does not have the safety problems of the short wavelength beam. The CO_2 gas lasers can be operated in the continuous wave mode or in the pulsed mode.

There are several types of CO_2 lasers. The early or low powered type used a "sealed tube" which has a power output of from 3 to 100 W. The axial flow type, which is more complex method of producing laser beams, will produce from 50 to 2,000 W average output power. Another type, known as the transverse flow type, has an average output power of from 2 to 15 kW. The laser power supply system is called an oscillator. These different types of CO_2 lasers involve different designs of the beam-producing apparatus. The size of the apparatus and the efficiency of the system are different.

The laser beam is very intense and unidirectional but can be focused and reflected in the same way as an ordinary light beam. The focus size is controlled by the choice of lenses and mirrors and the distance to the workpiece. The spot size can be made as small as 1 mm to up to 12 mm. The smaller focus spot size is used for cutting and welding, and the larger spot size is used for heat treating. The laser beam can be used in open air and can be transmitted long distances with only minimal loss of power.

A block diagram of a laser beam welding system is shown in Fig. 6.17. This shows the major components.

When using a laser beam for welding, the beam impinges on the surface of the base metal with such a concentration of energy that the surface is melted and volatilized. When the metal is raised to its melting temperature, the surface conditions have a minor effect on reflecting the beam. The distance from the optical cavity to the workpiece has little effect on the laser, because it can be focused to the proper spot size at the work with the same amount of energy available whether it is close or far away.

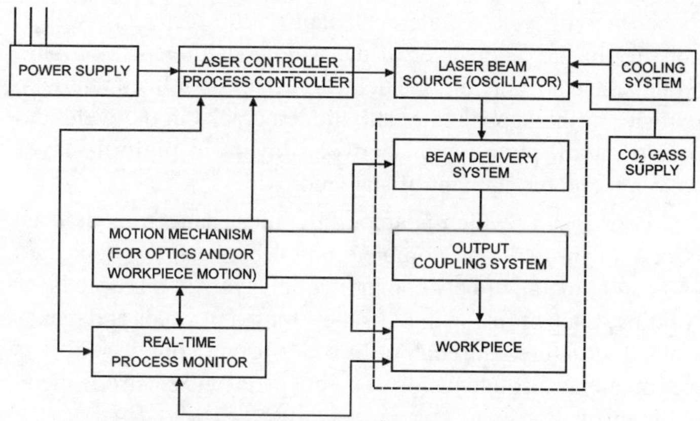

Fig. 6.17. *Block Diagram of CO_2 Laser System*

With laser beam welding the molten metal takes on a radial configuration similar to plasma arc welding*, known as "melt-in" welding. When the power density rises above a certain threshold level, keyholing occurs, the same as with plasma arc or electron beam welding. Keyholing provides for extremely deep penetration, which provides the weld a high depth-to-width ratio. For some applications, inert gas is used to shield the molten metal from the atmosphere.

The welding characteristics of the laser are similar to those of the electron beam. The laser can weld the same joint types. The concentration of energy by both beams is similar, with the laser having a power density of the order of 10^6 W/cm^2. The power density of electron beam is slightly greater.

The location of the focal point of the beam with respect to the surface of the workpiece is important. Maximum penetration occurs when the beam is focused slightly below the surface. Penetration is less when the beam is focused on the surface or deep within the surface. As the power is increased, the depth of penetration increases.

Heating and cooling rates are much higher in laser beam welding than in arc welding, and the heat-affected zones are much smaller. Rapid cooling rates can create problems, such a cracking in high-carbon steels.

The laser beam has been used to weld carbon steels, high-strength low-alloy steels, aluminium, stainless steel, and titanium. Laser welds are similar in quality to welds made by electron beam process. Filler metal is used to weld metals that tend to show porosity when welded with either electron beam or laser beam welding.

Materials 12 mm thick are being welded at a speed of 250 mm per minute. The efficiency of laser beam welding equipment is 5 to 10%, but as new equipment becomes available this will increase. The cost of the equipment will decrease in the future. More welding applications for laser welding are being found daily, and its use will increase.

6.7. REQUIREMENTS OF GOOD WELD

The weld is said to be good weld if it is free from external defects such as irregular width and heights of beads, deviation of the weld from the prescribed dimensions, unfilled craters on

* Plasma arc welding is an arc welding process that uses a constricted arc between a non-consumable electrode and the weld pool (transferred arc) or between the electrode and the constricting nozzle (non- transferred arc).

the surface of welds, undercuts, slag on the surface of welds, porosity in the outer layer of the welds, visible cracks in and around the weld and internal defects such as poor fusion, hidden porosity in the deposited metal, hidden cracks in the weld and parent metal, poor fusion between layers in multiple layer welding, and entrapment of slag and oxides.

Poor fusion, which is among the most dangerous defects, is lack of through and complete union between the deposited and parent metal, or between individual layers in multiple layer welding. Lack of fusion impairs the strength of welds and makes welded structures unreliable. In order to get sound weld it is essential that absolute fusion of parent metal and electrode metal is obtained.

Porosity is entirely ruled out in weldments where tight seams are critical (pressure vessels, storage tanks, pipelines etc.). Porosity may be brought about by insufficient puddling, as it does not give time for the trapped gas to escape from the molten metal before it solidifies. The gas remaining in the molten metal forms gas pockets. They may be scattered about or link up into strings of pores, giving the welded metal a spongy appearance. Porosity makes welds leaky, pervious to liquids and gases and weak.

Cracks in welds may arise from locked-up stresses up by non-uniform heating and cooling, excess sulphur or phosphorus in the weld metal, and some other causes. Cracks most often occur in butt welds and may be either longitudinal or transverse. Apart from visible cracks, there may be hair cracks, usually appearing at the line of fusion between the parent and deposited metal. They may be brought about by lack of fusion, large gas pockets, or slag inclusions where thermal stresses concentrate, giving rise to hair cracks which may develop into large cracks. That is why hair cracks are intolerable.

6.8. PREPARATION OF WORK

There are several factors which must be considered while preparing work for welding in order to obtain good results.

For obtaining good results it is desirable that provision may be made for expansion and contraction wherever possible. The strength of the weld will depend on the correct bevelling and spacing of parts to be welded. Uniform fusion is directly dependent on the proper bevelling and spacing.

The cleaning of the surfaces is another important factor. The cleaning can be done by wire brushing, machining or sand blasting. Impurities if present tend to make the joint weaker as the welded portion is filled with gas and slag inclusions and metal becomes brittle and the cohesion between the metals is, therefore, poor.

Another fundamental of good preparation or joint design is to keep the cross section of the added metal as small as practicable, so as to localise the effective heat in adjacent sections.

Another important point in welding is to overcome the oxidation at high temperatures. The oxides of metals formed at high temperatures encountered in welding have serious weakening effects. This influence is counteracted by using a flux which removes the oxides and permits perfect cohesion of the metals. In arc welding the flux is coated on the electrodes. The flux forms a protective coating of slag over the weld metal and creates a non-oxidising atmosphere.

6.9. ELECTRODES

An electrode is a piece of wire or rod (of a metal or alloy), with or without flux covering, which carries current for welding. At one end it is gripped in a holder and an arc is set up at the other end.

Electrodes are made of materials having high electrical conductivity as well as thermal conductivity so that heat developed at junction faces between electrodes and work is minimum in the first instance and secondly, whatever heat is produced is conducted away efficiently. In addition to high electrical and thermal conductivity, electrode material should be of very high mechanical strength so that the wear and tear at the electrode tip is minimum.

The materials used for electrodes are : Hard drawn copper (softening temperature 150°C and for welding of aluminium alloys); Cadmium copper (softening temperature 250°C and for joining MS sheets); Chromium copper (softening temperature 500°C and for joining of steel) and Tungsten copper (softening temperature 1,000°C and for welding of steel) and copper alloys.

The diameter of the electrode is approximately given by the expression $d = 5\sqrt{t}$

Where d is the diameter of tip in mm and t the thickness of sheet in mm.

Either non-consumable or consumable electrodes may be employed in arc welding.

Non-consumable electrodes may be of carbon, graphite or tungsten which do not consume during the welding operation. Consumable electrodes may be made of various metals depending upon their purpose and the chemical composition of the metals to be welded.

In general, the electrodes can be divided into three categories depending upon outer aspects as well as their technological properties.

Bare electrodes are most commonly used in automatic and semi-automatic welding. In using the plain or bare electrodes, as the globules of the metal pass from the electrode to the work, they (globules) are exposed to the oxygen and nitrogen in the surrounding air. This causes the formation of some non-metallic constituents which are trapped in the rapidly solidifying weld metal and thereby reduces the strength and ductility of the metal.

Lightly coated (or washed) electrodes have a coating layer several tenths of a millimetre thick. This coating usually consists of lime mixed with soluble glass which serves as a binder. The primary purpose of the light coating is to increase the arc stability, so they are also called the stabilising coatings. No attempt is made to prevent oxidation and no slag is formed on the weld, nor are the mechanical properties of the weld metal improved. For this reason lightly coated electrodes may only be used for welding non-essential jobs.

Heavy coated electrodes, sometimes referred to as *shielded arc electrodes*, are used to obtain a weld metal of high quality, comparable with, and even superior to, the parent metal in terms of mechanical properties. Heavy coatings employed are composed of ionising (chalk), deoxidising (aluminium, ferro-manganese etc.), gas generating (starch), slag forming (kaolin), alloying and binding materials. These electrodes have the following distinct advantages :

(*i*) The physical and metallurgical properties of the weld can be influenced by adding alloying components to the covering.

(*ii*) Basic salts of silicon, magnesium and calcium in the covering form slag which floats on the surface of metal and prevents rapid cooling of the weld. Thus weld does not become brittle.

(*iii*) The weld metal is protected from oxidising action of atmospheric oxygen and nitrifying action of nitrogen of air due to gases formed from the covering material.

(*iv*) In case of ac supply arc cools at zero current and there is a tendency of deionizing the arc path. Covering gases keep the arc space ionized.

(*v*) During welding the covering extends beyond the core wire. This directs the arc and concentrates the arc stream, reduces thermal losses and causes increase in temperature of electrode tip.

There may also be *powder-cored electrodes* which have a good portion of the source of metal for depositing in the joint located as powdered iron mixed in with the flux coating. This makes the electrodes much larger on the outside for the same diameter core wire as standard flux coated electrodes. These electrodes are well suited to down-hand welding and are said to deposite more metal in a given period time than standard flux covered electrodes.

Both bare and coated electrodes, for manual arc welding, are made in the shape of rods up to 12 mm in diameter and 450 mm long. Semi–automatic and automatic welding use electrode wire in coils.

6.10. POWER SUPPLY FOR RESISTANCE WELDING

AC supply is used for resistance welding because of the ease and covenience with which the required high current at a low voltage can be obtained by means of a transformer. The kVA required for resistance welding, when actually making a weld, ranges from a few kVA to as much as 1 MVA. The power factor will be about 0.25 or 0.3 lagging. The power factor is low mainly due to the high ratio of reactance to resistance of the loop formed by the jaws of the welding machine. Such heavy intermittent single-phase loads may cause serious voltage drop difficulties in the supply network. Such problems can be overcome to some extent by connecting capacitors of suitable capacity in parallel with the welding transformer so as to improve the power factor. But with this arrangement the power factor will become leading when welding current is not being drawn. This problem can be avoided by connecting the capacitors in series with the welding transformer to neutralize the reactance drop in the supply circuit.

6.11. MACHINES FOR RESISTANCE WELDING

The machine for resistance welding incorporates a transformer, suitable electrodes for supplying current to the weld and arrangement for controlling the mechanical pressure, and finally, means for controlling the duration of weld current flow. The mechanical pressure may be exerted through levers and clutch by an electric motor or by compressed air. The magnitude of pressure required depends upon the type of work and may vary from a few kg for thin sheets or wires up to a tonne or more for heavy work.

In the older type of welding machines, the electrodes were brought on to the work and the electrical circuit closed by the operation of a pedal. Thus application of pressure and the time duration of current flow used to be controlled by the operator and for this operator needs to be experienced and skilled. The modern practice is to pass heavy currents for shorter time durations (ranging from 10 ms to 100 ms). The equipment used for this purpose may be *constant time, current-actuated,* or *energy-actuated* types.

Constant time equipment is employed in high speed production where the work has a consistently clean surface. Constant time equipment may be provided with mechanical control or electrical control. In mechanical control providing up to 300 welds per minute, the device employed is a cam-operated switch, connected in the primary circuit of a welding transformer, driven from the welding machine. For a large number of welds per minute the mechanical arrangement becomes unsuitable because it is not capable of providing consistently accurate timing, due to wear of the cam and operating mechanism, arcing and burning of the contacts and irregularities caused by closing of switch at different instants in the cycle.

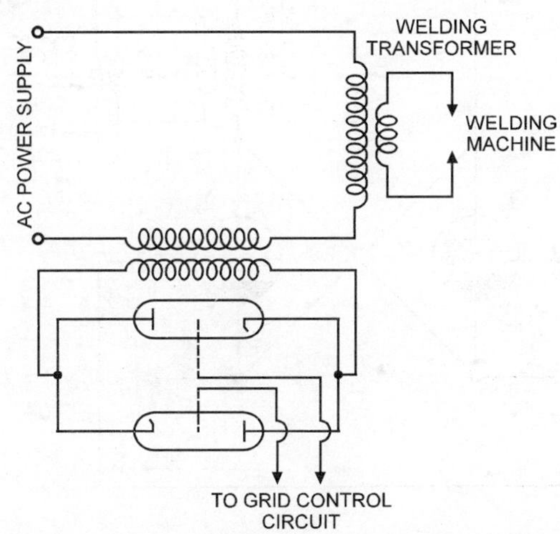

Fig. 6.18. ***Valve Controlled Resistance Welding Circuit***

An alternative arrangement is to control the timing through grid controlled ignitrons or thyratrons. It is easier to build tubes for a high voltage and small current than with a low voltage and high current. An arrangement using valves in the secondary circuit of a series transformer is shown in Fig. 6.18. When the tubes conduct, the series transformer secondary is almost short-circuited, and whole of the supply voltage is available across the welding transformer primary. But when the tubes are not conducting, the series transformer primary winding offers a high impedance in the circuit of the welding transformer and the current is reduced to a negligible value. Auxiliary valves are employed for controlling the timing of the negative potential applied to the grids of the main tubes.

Constant-time method of control does not yield consistently good results when there may be variations in the conditions under which successive welds are made, due to variations in supply voltage or mechanical pressure, wearing of electrodes, surface irregularities etc. The energy-actuated control, wherein definite amount of energy is supplied to the weld, is used. Constant-time method of control did not prove to be successful, particularly with modern high-speed welding. The energy-actuated control, which permits the current to flow until a predetermined amount of energy has been supplied to the weld, is theoretically an ideal method. However, the control equipment is quite complicated.

6.12. ELECTRONIC CONTROL IN RESISTANCE WELDING

As there are several factors, *i.e.*, current, pressure, heat, time, to be considered, manual control does not yield good results in case of resistance welding. For the precise control of these factors electronic control welding circuits are used. Some of the electronic control circuits are given here.

6.12.1. Ignitron Contactor

Theory of ignitron has already been dealt in Art. 2.2.2. We will now discuss its use as a contactor for controlling heavy currents.

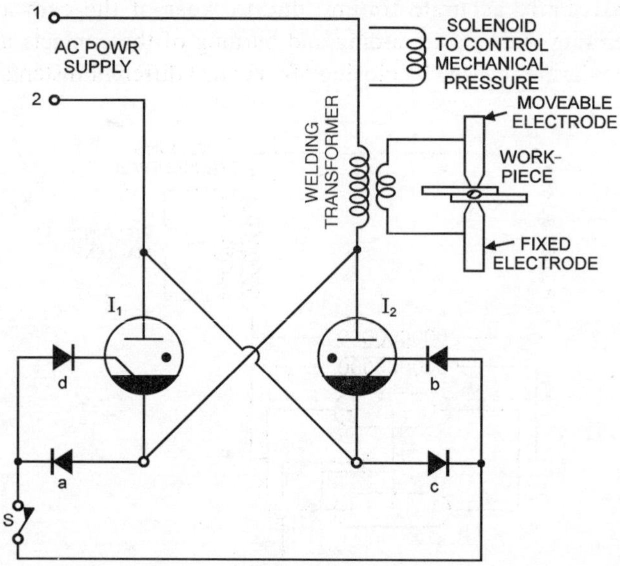

Fig. 6.19

A simple line contactor using two ignitrons is shown in Fig. 6.19. If the switch S is closed at the instant the line 1 is positive, current will rush through the primary of welding transformer, rectifier *a*, switch S, rectifier *b*, the ignitron I_2 and back to the line 2. The current will strike an arc in ignitron I_2 and the tube starts conducting. Now voltage drops across I_2 to a low value, causing a voltage drop in the ignitron circuit. Therefore, the ignitron will conduct just long enough to strike an arc. Similarly, during next half cycle, the line 2 will be positive and the current will flow from line 2 through rectifier *c*, switch S, rectifier *d* to ignitron I_1. During this half cycle, as the anode of ignitron I_2 becomes – ve, it stops conducting. Metallic rectifiers are used in this circuit. They conduct the current in proper direction, thus, preventing the application of negative voltages to the electrodes and saving the ignitrons from inverse current damages. A solenoid is used to apply suitable pressure through the upper movable electrode. Manual control of the contactor switch is possible only if the welds are of long duration. But for the accurate time control of the welds of short duration, thyratrons are used to ignite the ignitrons, as shown in Fig. 6.20. The grid circuits of the thyratrons are controlled by a suitable timing control circuit.

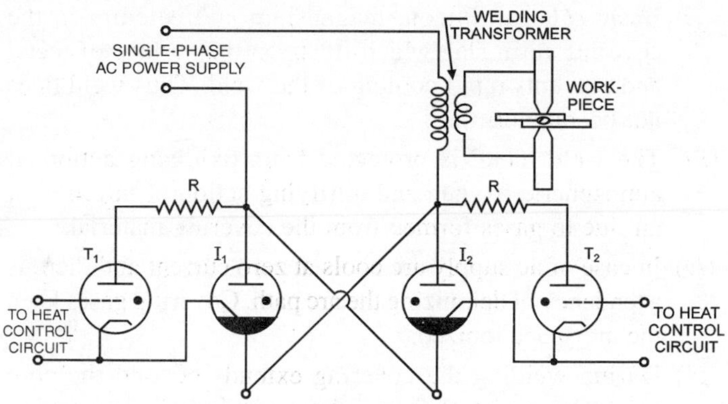

Fig. 6.20

Since very heavy current flows through the ignitrons, say 1,000 amperes, and arc drop be taken constant at 10 volts, therefore, losses to the tune of 10 kW will take place. So the ignitrons are always water cooled. In case of very heavy load, temperature of water becomes too high, normally closed thermostat contacts open and the ignitrons stop conducting.

6.12.2. Heat Control Unit

It is an electronic circuit which helps the delay in the firing of the ignitrons by a definite, predetermined angle in each cycle and operates in conjunction with the line contactor. A typical circuit employed for the heat control is shown in Fig. 6.21. This is essentially a phase shift control circuit and delays the firing of the ignitrons, thus reducing the magnitude of welding current as per requirements.

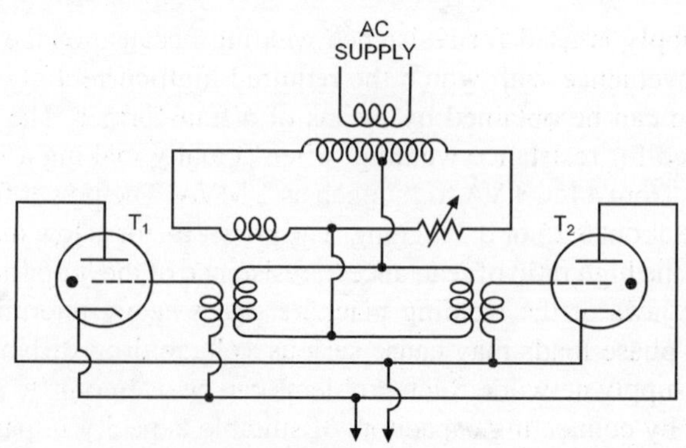

Fig. 6.21. *Heat Control Unit*

6.12.3. AC Timer Circuit

When the condenser C is discharged through a resistor R the voltage across the condenser falls exponentially as given by the expression

$$v = Ve^{-t/CR}$$

$$\text{or} \quad t = RC \log_e \frac{V}{v}.$$

From the above expression it is clear that greater the capacity of condenser and value of resistor, greater will be the time required for voltage to fall by given amount. In all timer circuits provision is, therefore, made for charging a condenser to a particular value of voltage and then discharge by short-circuiting switch till condenser is discharged to a particular value when relay will operate and particular contact will open or close.

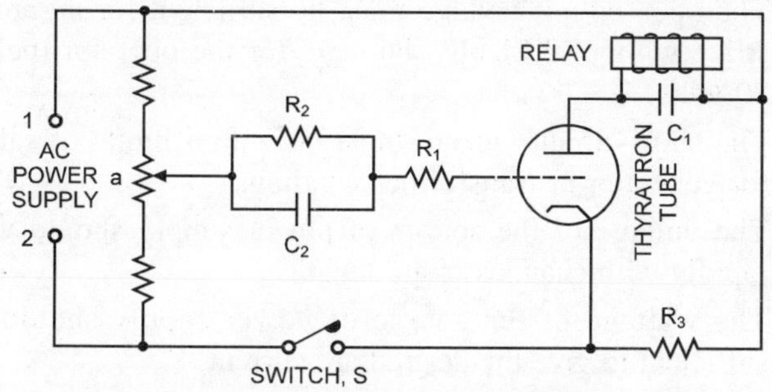

Fig. 6.22. *AC Timer Circuit*

Typical ac timer circuit is shown in Fig. 6.22. Such a timer circuit is used to control the number of cycles for which power may be supplied into the weld. The action of such a timer circuit is explained below.

When switch S is open and supply terminal 1 is positive w.r.t. terminal 2, the cathode and anode of thyratron are at the same potential, and grid is – ve w.r.t. cathode, and, therefore no current flow between cathode and grid. When terminal 1 is negative w.r.t. terminal 2, the potential at *a* is positive, grid becomes positive w.r.t. cathode and anode and electronic current flows through R_2, R_1 from grid to cathode, through R_3 and to terminal 1. In a few cycles the capacitor C_2 will get charged to the maximum voltage between *a* and 1. This is due to the large value of time constant. The capacitor does not discharge much during the negative half cycle of the grid voltage. Resistance R_1 limits the grid-cathode circuit current to a safe value and also determines the number of cycles in which C_2 will be fully charged. So long switch S remains open, capacitor C_2 remains charged by the grid rectification action. As soon as switch S is closed, the grid becomes very much negative w.r.t. cathode and there is no capacitor charging current through grid rectification. Capacitor will, therefore, start discharging through R_2 and –ve bias* of the grid will gradually decrease depending upon the time constant R_2C_2 of the discharge circuit. Conduction in thyratron tube will start during positive half cycle of anode voltage when the grid voltage instantaneously rises to critical grid voltage. Current through relay coil is rectified half waves. As such to avoid relay terminals chatter, a capacitor C_1 is connected across relay coil.

6.12.4. Energy Storage Welding Processes

To meet the demand of heavy current of very high conductivity metals such as aluminium and magnesium *energy storage welding circuits* are used. There are basically two such circuits namely electrostatically stored energy circuits and electromagnetically stored energy circuits.

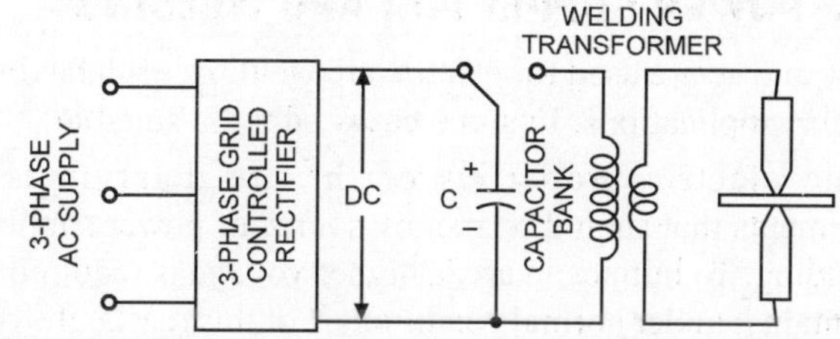

Fig. 6.23. *Capacitor Discharge Welding Circuit*

1. *Capacitor Discharge Welding Circuit.* As shown in Fig. 6.23, condenser C (capacitor bank of capacity of 2,000 to 3,000 μF) is charged to about 3,000 volts from grid controlled rectifier. When the condenser is connected to the primary of welding transformer by ignitron contactor, it will discharge and thus high transient current will be produced in the secondary to weld the material. The noteworthy points in connection with this circuit are :

(*i*) As the voltage of condenser approaches the voltage of the source of supply, charging rate becomes lower, therefore to charge condenser to about 3,000 V at high charging rate voltage of about 5,000 V to 6,000 V will be required. A voltage regulating circuit cuts off the rectifier from the bank when the voltage of the bank becomes 3,000 V.

(*ii*) If there is residual magnetism near saturation, it will result in low rate of change of flux linkages in the secondary and, therefore, in production of low heat. Hence in the welding transformer core flux should not be present.

2. *Magnetic Energy Storage Welding Circuit.* In this type of welding, energy stored in magnetic circuit is used in the welding operation. The dc voltage of the rectifier is suitably controlled so that the current in the primary of the transformer rises gradually without inducing large current in the secondary. This is necessary to avoid preheating of metals at the weld joint. Preheating in aluminium, magnesium etc. is undesirable as it causes deformation.

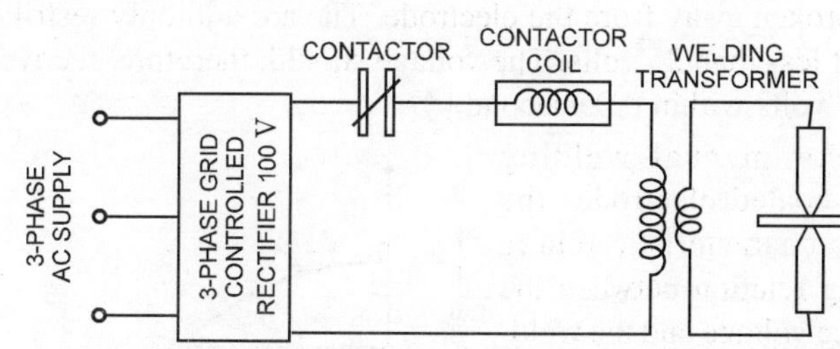

Fig. 6.24. *Magnetic Energy Storage Welding Circuit*

When sufficient energy has been stored up in the transformer core, the contactor opens, dc flow ceases and there is a rapid collapse of magnetic field. The decay of flux induces heavy currents in the secondary of the transformer for welding.

The kVA demand on the line in magnetic energy storage welding is higher as compared to that in capacitor discharge welding but a high voltage rectifier and costly capacitor bank are not required.

* Because of the combination of R_2 and C_2, the grid has two voltages impressed on it : one is the ac component (ac voltage between *a* and 2) and the other is the dc component due to the voltage to which the capacitor C_2 is charged. The dc component dies out at exponential rate.

6.13. POWER SUPPLY FOR ARC WELDING

Both dc and ac are used for electric arc welding, each having its particular applications; in some cases either is suitable.

The electric properties of the arc determines the requirements that should be met by source of power supply for arc welding. To initiate an arc, a higher voltage is required than to maintain it under normal conditions. For this reason the open-circuit voltage of the power supply source (when there is no arc drawn between the electrode and the work) must be higher than the arc, or closed-circuit voltage (when there is an arc drawn between the electrode and the work).

With dc the open-circuit voltage must be at least 30 or 35 volts, while with ac it should be not lower than 50 or 55 V. An open arc will be sustained at 18 to 25 volts. Open-circuit voltage usually ranges from 50 to 90 volts.

To initiate an arc, the electrode is touched to the work and is then withdrawn a short distance. On contact, the electrode shorts the welding circuit and an inadvertent short circuit may give rise to an excessive current in the circuit, which may dangerously heat and even burn the insulation. A power supply source for welding should, therefore, be able to limit the surge of current, when the arc is initiated, to a small percentage above the desired value.

In metal arc welding the globules of molten electrode metal are carried across the arc, thus changing the resistance of the arc and consequently the arc voltage and current. In fact the arc resistance is continually varying within broad limits in very short periods of time. The point is that the molten globules of metal pass from the electrode to the plate at the rate of 30 or more per second, continually causing approximate short circuits from the electrode to the work. If the arc is not to go out, the power supply source must be able to quickly vary its voltage with the resistance of the arc.

Of special importance is the ability of a power supply source to restore the voltage quickly after a globule of molten metal has broken away from the electrode. The arc will only restrike at not less than 25 volts. The voltage should, therefore, recover to 25 volts within 0.05 second.

For manual welding with coated electrodes the static characteristic (curve giving relation between the source voltage and the welding current) is represented by curve I in Fig. 6.25. This is a drooping or negative characteristic. OA represents the open-circuit voltage, OB represents the arc or closed circuit voltage, OD represents the arc current and OE represents the short-circuit current. The point C on the curve I shows the instant of striking the arc. With this type of volt-ampere characteristic the short-circuit current in the welding circuit is reduced to a limit safe enough for the windings of the welding generator or transformer.

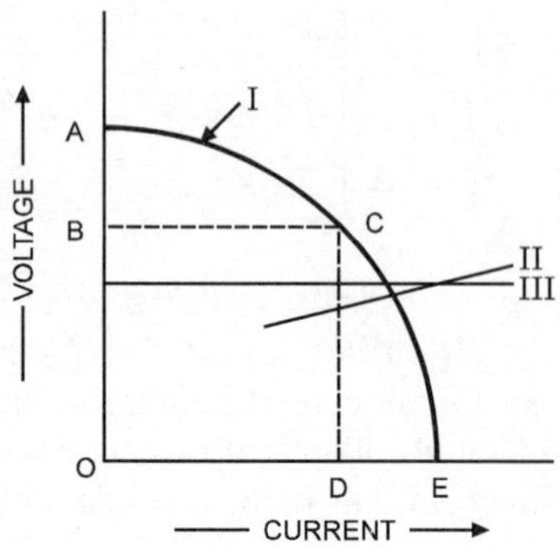

Fig. 6.25

If the arc has a flat (curve III) or a rising volt-ampere characteristic (curve II) such as in shielded arc welding, automatic welding with large currents the source of power supply should also have a flat or a rising characteristic respectively.

To sum up, the power supply for manual arc welding should meet the following requirements :

1. The open-circuit voltage must be sufficient for an arc to strike without difficulty, but safe for the operator (below 80 volts).
2. The short-circuit current must be within limits of safety for generator or transformer windings.
3. The voltage of the source of power supply should vary rapidly with changes in arc length.
4. The wattage of the source of power supply should be sufficient to give the desired arc current.

6.14. ELECTRIC WELDING EQUIPMENT

The electric welding sets may be either dc or ac type.

1. *DC Welding Sets.* Such sets are of two types namely (*i*) generator type and (*ii*) rectifier type.

Generator type welding set consists of a differential compound wound dc generator, giving drooping volt-ampere characteristic, driven by any type of prime mover (a squirrel cage induction motor or a petrol or diesel engine). In a differential compound wound dc generator the terminal voltage falls automatically with the increase in load current. The control may be obtained by tapping the series field or by providing a suitable shunt across the series field winding. The open-circuit voltage is adjusted from shunt field.

If supply from existing dc distribution system is to be used for welding then a ballast (resistance) is put in series with the equipment and control is obtained by varying this external resistance. This method is also suitable when a number of operators are working on the same supply system. In such cases each operator is provided with a separate ballast. The special field of this method is where the service of each arc welding circuit is infrequently in use (say 25% of the time or less). In such cases the loss in the series resistor is less than the no-load loss of a generator type unit.

Another type of dc welding set is a *dry type rectifier* used in conjunction with a multi-phase, high leakage reactance transformer. Many of these rectifier type welders use selenium rectifiers which are forced air cooled. Rectifier type welders are said to combine some of the desirable arcing characteristics of dc welding, such as easy arc starting, with those of welding transformers, such as reduced no-load losses. DC voltage is controlled by regulating the transformer output in this case.

2. *AC Welding Sets.* Single phase or 3-phase step-down transformers which provide low voltage (80-100 voltage on open-circuit) power for welding with some means of output control. One or two taps are provided on the primary side to take care of the voltage variations. The secondary winding of an air-cooled set is generally made of bare wound-on-edge copper strip. The welding transformers may be air or oil-cooled type. The air-cooled sets are lighter in weight and permit the use of insulating materials of the highest thermal classification. They are also cheaper, less hazardous and easier to maintain. On the other hand, the oil-cooled sets are more compact. Being heavy, they are more suitable for stationary applications in workshops. Synthetic liquid filled sets may be employed in hazardous locations. Some machines are provided with an *arc booster* that provides a momentary surge of current to give an arc a good start when it is struck.

In transformer type welding machines the current control is achieved by using (*a*) magnetic shunt or (*b*) a choke coil or reactor placed in series with primary or secondary winding or (*c*) tap changing switch in the primary winding.

In the magnetic shunt type, an adjustable gate of iron laminations is placed between the primary and the secondary windings. Part of the flux is diverted through this gate without linking with the secondary winding, thus providing a simple and reliable means of stepless voltage control.

The choke coil type regulator is a variable reactance with a movable iron plunger. The reactor is such that it operates well below the saturation point. Though it improves the arc stability but reduces the power factor.

The tap-changer type regulator is prone to give trouble on account of arcing on the contacts and the moving finger.

The use of series resistance can also be made for current control but efficiency is reduced.

6.15. ARC WELDING WITH DC AND AC

AC and dc both can be used in arc welding, each having its distinctive merits and applications.

The advantages of arc welding with dc lie in the higher arc stability and the degree to which the work is heated. It is best suited for thinner sheet metal (below 6 mm) and also for welding of non-ferrous metals.

AC welding, due to the absence of 'arc blow', is considered superior for production welding involving large size electrodes. AC welding has a number of economical advantages such as given below :

1. As power supply in factories is usually ac so a rotating dc generator or a rectifier is required for dc welding whilst for ac welding only transformer is required which is relatively cheaper in initial cost.
2. Maintenance of a transformer is less difficult and expensive than that of a dc generator.
3. Operating costs of ac equipment are also lower than those of dc equipment.
4. The electric energy consumption per kg of deposited metal in ac welding ranges from 3 to 4 kWh while for dc welding it is as high as 6 to 10 kWh.

One disadvantage of ac welding is the comparatively low power factor (about 0.4 lagging) which can be corrected by the use of capacitors.

Comparison between dc welding and ac welding is given in tabular form :

S.No.	Particulars	DC Welding	AC Welding
1.	Equipment	Motor–generator set or rectifier is required in case of availability of ac supply, otherwise oil engine-generator set is required.	In case of availability of ac supply only a transformer is required Transformer costs less and its maintenance cost is also low.
2.	Prime cost	Two to three times as compared to that of a transformer.	Comparatively low.
3.	Operating efficiency	Low; consequently high cost of electrical energy.	High (about 85%).
4.	No-load voltage	Low (safer).	Frequently too high (dangerous).
5.	No-load requirements	Very high.	Low (advantageous).
6.	Power factor	Comparatively higher.	Low, requiring capacitors for correction.
7.	Arc blow	Pronounced.	Not so pronounced with ac (advantageous).
8.	Electrodes	Both bare (non-coated) and thus cheap electrodes can be used.	Only coated electrodes—expensive ones.
9.	Heating	Uniform	Not so uniform as in case of dc.
10.	Connected load (cross-sectional area of conductors and fuses)	Normal.	Considerably higher because of low power factor.
11.	Welding of non-ferrous metals	Suitable.	—
12.	Arc stability	Higher.	—

6.16. COMPARISON BETWEEN RESISTANCE AND ARC WELDING

S.No.	Particulars	Resistance Welding	Arc Welding
1.	Supply.	Usually ac only.	AC or dc.
2.	Voltage.	Very low.	The striking voltage is high so requires voltage control.
3.	Power factor.	Very low.	Poor.
4.	Additional material requirement.	No material is added in any form to get the two pieces joined.	Suitable filler metal electrodes are necessary to get proper strength.
5.	External pressure.	External pressure is required.	No external pressure is required hence the equipment is more simple and easy to control.

S.No.	Particulars	Resistance Welding	Arc Welding
6.	Development of heat.	Heat is developed due to flow of current through the contact resistance mainly.	Heat is developed due to arc between electrode and the work piece.
7.	Temperature.	The temperature attained is not very high as in case of arc welding.	The temperature of the arc is very high and so likely to damage the work if not properly handled.
8.	Applications.	It cannot be used for repair work. It is most suitable for mass production.	It is not suitable for mass production. It is most suitable for repair work and where more metal is to be deposited.

Exercises

1. Describe with neat sketches the various methods of electric resistance welding. Give its merits or demerits with respect to arc welding.
[Pb. Univ. Electric Drives and Utilization of Electric Energy August 1984]
2. Name and describe various types of resistance welding processes.
[M.D. Univ. Utilization of Electric Power & Traction May 2009]
3. Why projection welding is considered superior to spot welding ?
[Pb. Technical Univ. Electric Drives and Utilization May 2009]
4. Explain with neat sketches, the construction, principle of operation and applications of three different types of resistance welding. Give its demerits with respect to arc welding.
[U.P. Technical Univ. Utilization of Electrical Energy and Electric Traction 2005-06]
5. Explain the principle of electric spot welding and seam welding.
Discuss the merits and demerits of electric resistance and arc welding. Indicate when each find its use.
[Gorakhpur Univ. Utilization of Elect. Power & Traction 1982]
6. Explain various types of electric welding. Explain seam welding in detail.
[Rajasthan Technical Univ. Utilization of Electrical Power 2011]
7. Explain the role of ignition contactor in welding process. What are its advantages over mechanical switches ?
[U.P. Technical Univ. Utilization of Electrical Energy & Traction 2009-10]
8. What is the fundamental difference between electric arc welding and resistance welding ? Explain with neat sketch how the spot welding is carried out by spot welding machine.
[Gorakhpur Univ. Utilization of Electrical Power & Traction Sept. 1983]
9. Explain the various types of resistance and arc welding. How the mechanical pressure, amount of current and its duration is controlled in case of resistance welding ? What is atomic hydrogen welding ? Explain. Give its relative merits. **[Pb. Univ. Electric Drives & Utilization of Electric Energy Nov. 1983]**
10. Describe briefly various types of arc welding processes used in industry.
[Pb. Technical Univ. Electric Drives & Utilization 2003]
11. Why a drooping characteristics of supply voltage is essential for maintaining steady arc in electric welding and how this characteristic is obtained in case of a dc source.
[Pb. Technical Univ. Electric Drives & Utilization May 2009]
12. Differentiate between carbon arc and metal arc welding.
[Rajasthan Technical Univ. Utilization of Electrical Power 2011]
13. What are the advantages of using coated electrodes ? Mention types of coating used.
[U.P. Technical Univ. Utilization of Electrical Energy and Electric Traction 2005-06]
14. Compare the performance of various electrodes used in electric arc furnaces.
[M.D. Univ. Utilization of Electric Power & Traction Dec. 2009]
15. Describe various welding schemes specifying the equipment for controlling temperature, pressure etc.
Draw a neat sketch of welding transformer showing primary and secondary circuits.
[Pb. Univ. Electric Drives & Utilization of Electric Energy April 1982]
16. Discuss laser welding and electron beam welding.
[Pb. Technical Univ. Electric Drives & Utilization Dec. 2007]
17. Explain electron beam welding and give its disadvantages.
[U.P. Technical Univ. Utilization of Electrical Energy & Traction 2009-10]
18. Explain different types of electric welding. What are qualities of good weld.
[M.D. Univ. Utilization of Electric Power & Traction May 2010]
19. With suitable example explain the requirements of good welding. **[U.P. Technical Univ. Utilization of Electrical Energy & Traction 2009-10]**
20. Compare dc and ac welding methods.
[Pb. Technical Univ. Electric Drives & Utilization May 2008]
21. What are the basic components of dc and ac arc welding sets and explain their working.
[Pb. Technical Univ. Electric Drives & Utilization May 2008]
22. Discuss the need of dc supply for electric welding. Why dc supply is preferred over ac supply for welding ?
[U.P. Technical Univ. Utilization of Electrical Energy 2009-10]
23. What is welding ? What are the different types of electric welding ? What type of electric supply is suitable for electric welding ? Explain.
[Rajasthan Univ. Utilization of Electrical Power 2003, 2005, 2006]
24. What is the fundamental difference between electric arc welding and resistance welding ? Give their merits and demerits.
[M.D. Univ. Utilization of Electric Power & Traction Dec. 2009]
25. Compare in details, the electric arc welding with resistance welding. Explain with the help of a neat sketch the process of spot welding.
[Rajasthan Technical Univ. Utilization of Electrical Power 2011]
26. Explain the basic difference between electric arc welding and resistance welding.
[Pb. Technical Univ. Electric Drives and Utilization Dec. 2009]
27. Write short notes on the following :
   (*i*) Resistance welding. **[Rajasthan Univ. 2007]**
   (*ii*) Arc welding.
   (*iii*) Resistance and arc welding. **[Agra Univ. 1983]**
   (*iv*) Atomic hydrogen welding.
   (*v*) Ultrasonic welding. **[G.B.T.U. 2010-11]**
   (*vi*) Laser welding. **[W.B. Univ. of Technology 2009]**

(*vii*) Advantages of coated electrodes. **[Pb. Univ. Nov. 1980]**

(*viii*) Power control in arc welding. **[G.B.T.U. 2010-11]**

(*ix*) Electric supply for arc welding.

(*x*) Welding transformer. **[Rajasthan Univ. 2004, 2007]**

(*xi*) Welding generator and welding transformer. **[M.D. Univ. Dec. 2008]**

Short Answer Type Questions With Answers

Q. 1. What is welding ?

Ans. Welding is a materials-joining process that produces coalescence of materials by heating them to the welding temperature, with or without application of pressure or by the application of pressure alone, and with or without the use of filler metal.

Q. 2. "All metals can be joined by one welding process or another". Comment on this statement.

Ans. The above statement needs slight modification and may be stated as follows.

All metals commercially employed for structural or strength parts are weldable. In general, metal that have a low melting point or low strength would not be welded.

Q. 3. What are the different types of electric welding ? **[P.T.U. May 2007]**

Ans. Electric welding may be resistance welding or arc welding. Resistance welding includes butt welding, spot welding, projection welding seam welding and percussion welding. Arc welding is subdivided into five groups namely carbon arc welding, metal arc welding, atomic hydrogen arc welding, inert gas metal arc welding and submerged arc welding.

Q. 4. What is the difference between plastic welding and fusion welding ? **[Gorakhpur Univ. 1983]**

Ans. In the plastic welding, also called the pressure welding, the pieces of metal to be joined are heated to a plastic state and then forced together by external pressure whereas in fusion welding, also called the non-pressure welding, the material at the joint is heated to a molten state and allowed to solidify. Former one includes forge welding, resistance welding, thermit welding and gas welding while the latter includes gas welding, arc welding and thermit welding without pressure.

Q. 5. What is resistance welding ?

Ans. Resistance welding is a group of welding processes that produce coalescence of the faying surfaces with the heat obtained from resistance of the work to the flow of welding current in a circuit of which the work is a part and by the application of pressure.

Q. 6. What is the fundamental difference between electric arc welding and resistance welding ? **[Gorakhpur Univ. Sept., 1983]**

Ans. The resistance welding processes differ from arc welding in that pressure is used but filler metal or fluxes are not.

Q. 7. Why alternating current is found most suitable for resistance welding ? **[Gorakhpur Univ. 1980]**

Ans. Alternating current is found most suitable for resistance welding as it can provide any desired combination of current and voltage by means of a transformer.

Q. 8. Explain the theory of arc production. **[P.T.U. Dec. 2006]**

Ans. An electric arc is produced when two conductors (electrode and metal piece) are connected to a suitable source of electric current momentarily in contact and then separated by a small distance. Both dc and ac can be used for producing an electric arc, however, dc provides more steady arc.

Basically an arc is nothing but an electric discharge through a path of ionised particles of air. The electrical energy is converted into heat at a temperature of about 5,000°C.

Q. 9. What is meant by negative resistance characteristic of an electric arc and what precautions are taken for ill effects due to this ?

Ans. Negative resistance characteristic of an arc means decrease in arc resistance with the increase in temperature or current. The reduced resistance causes further increase in current and, therefore, steady arc cannot be obtained. For obtaining steady arc, supply source should have a drooping characteristic (*i.e.,* voltage decreasing with the increase in current).

Steady arc for dc welding can be obtained by using differential compound wound generator. The control may be obtained by tapping the series field or by providing a suitable shunt across the series field winding. The open-circuit voltage is adjusted from shunt field.

In an ac welding sets, welding transformer is used and drooping characteristic is obtained by using external reactor.

Q. 10. What are the fluxes used in arc welding ? **[P.T.U. Dec. 2009]**

Ans. In arc welding, flux coated electrodes are used. The fluxes used are silicon oxiders, graphite, sodium silicate, calcium oxide and aluminium in right proportions. Sometimes, the electrodes are washed (lightly coated) of several tenths of millimetres thickness, lime mixed with soluble glass.

The heavy coated electrodes have coating composed of ionising (chalk) deoxidising (aluminium, ferro-manganese etc.), gas generating (starch), sag forming (kaoline), alloying and binding materials.

Q. 11. What are the qualities of good weld ? **[P.T.U. Dec. 2007; Dec. 2008]**

Ans. The weld is said to be good weld if it is free from external defects such as irregular width and heights of beads, deviation of the weld from the prescribed dimensions, unfilled craters on the surface of welds, undercuts, slag on the surface of welds, porosity in the outer layer of the welds, visible cracks in and around the weld and internal defects such as poor fusion, hidden porosity in the deposited metal, hidden cracks in the weld and parent metal, poor fusion between layers in multiple layer welding, and entrapment of slag and oxides.

Q. 12. What is main drawback of electron beam process ?

Ans. The main drawback of electron beam process is its high capital cost. The price of the equipment required is very high, and it is expensive to operate, due to the requirement of vacuum pumps. In addition, fitup must be precise and locating the parts w.r.t. the beam must be perfect.

Q. 13. The location of the focal point of the beam w.r.t. the surface of the work piece is very important in laser beam welding. Why ?

Ans. Maximum penetration occurs when the beam is focused slightly below the surface. Penetration is less when the beam is focused on the surface or deep within the surface.

Q. 14. What is meant by welding electrode ?

Ans. An electrode is a piece of wire or rod (of a metal or alloy), with or without flux covering, which carries current for welding.

Q. 15. What type of dc generator is used in electric arc welding ?

Ans. Differential compound wound dc generator.

❑❑❑

CHAPTER 7

Illumination Engineering

INSIDE THIS CHAPTER

7.1. INTRODUCTION

Light is the prime factor in the human life as all activities of human beings ultimately depend upon the light. Where there is no natural light, use of artificial light is made. Artificial lighting produced electrically, on account of its cleanliness, ease of control, reliability, steady output, as well as its low cost, is playing an increasingly important part in modern everyday life. Apart from its aesthetic and decorative aspects, good lighting has a strictly utilitarian value in increasing production, reducing workers fatigue, protecting their health, eyes and nervous system, and reducing accidents. The science of illumination engineering is, therefore, becoming of major importance.

Nature of Light. Light is a form of radiant energy. Various forms of incandescent bodies are the sources of light and the light emitted by such bodies depend upon the temperature of bodies. Heat energy is radiated into the medium by a body which is hotter than the medium surrounding it. The heat of the body, as seen, can be classified as red-hot or white-hot. While the body is red-hot the wavelength of radiated energy will be sufficiently large and the energy available is in the form of heat. When the temperature increases the body changes from red-hot to white-hot state, the wavelength of the energy radiated becomes smaller and enters into the range of the wavelength of light. Actually the hot bodies emit heat as well as light energy, the velocity of these being equal to 3×10^8 m/s. The light waves have wavelengths varying from 0.00075 mm to 0.0004 mm. The wavelength of light is usually expressed in Angstrom ; 1 Angstrom being equal to 10^{-10} metre. The ratio of the energy emitted by the body in the form of light to the total energy emitted by the body is known as the '*radiant efficiency*' of the body, which depends upon the temperature. Higher the temperature of the body ; lower the wavelength of the radiant energy and higher the radiant efficiency. The radiant efficiency will be maximum when the temperature of the body will be such that the wavelength of the shortest wave radiated by the body is 0.0004 mm (4,000 Å). It is found that maximum radiant efficiency would occur at about 6,250°C and even then it would be 20%. Since this temperature is far above the highest that has yet been obtained in practice, it is obvious that the actual efficiency of all artificial light sources, *i.e.*, those depending upon the temperature of incandescence is quite low.

Colour. The sensation of colour is due to the difference in the wavelengths of the light radiations. Visible light can have a wavelength between 4,000 Å and 7,500 Å and the colour varies in the way, as shown in Figs. 7.1 and 7.2.

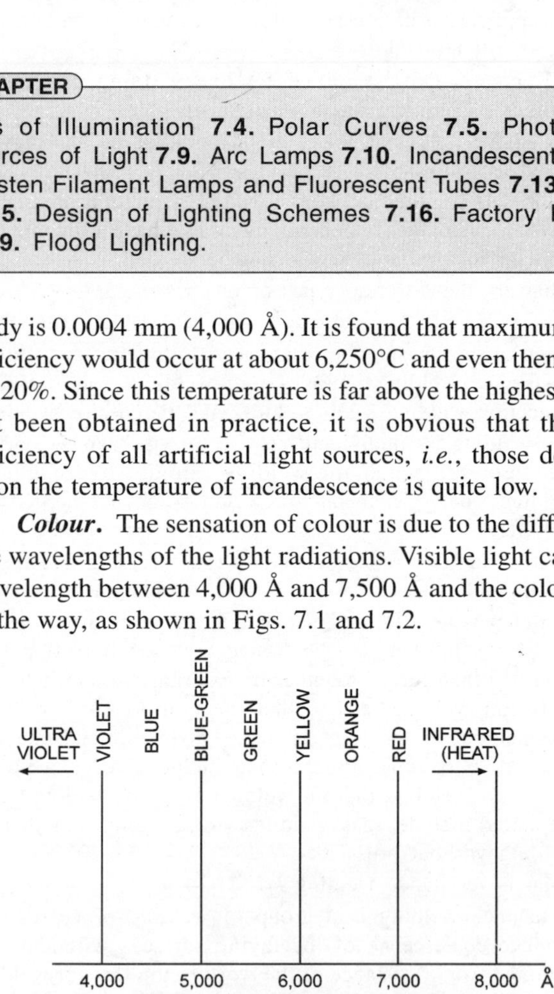

Fig. 7.1. ***Wavelength and Colour***

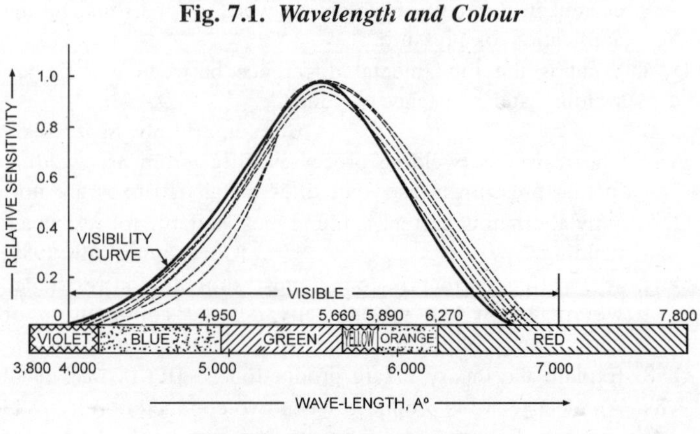

Fig. 7.2. ***Eye Sensitivity Curve***

Relative Sensitivity. The sensitivity of the eye to the lights of different wavelengths varies from person to person and according to the age. The average relative sensitivity is shown in Fig. 7.2. The eye has greatest sensitivity for wavelengths of about 5,500 Å ; that is, yellow-green can be seen under poor conditions of illumination when blue or red cannot be seen under dim illumination, the sensitive curve tends to shift, as shown by the shaded region in Fig. 7.2. Therefore, violet disappears first and red remains visible. Yellow disappears last as the illumination becomes very dim. As each colour disappears, it becomes a grey shade, and finally black. The sensitivity of the eye to yellow-green radiation is taken as unity or 100% and the sensitivity to other wavelengths is expressed as a fraction or percentage of it. The relative sensitivity at any particular wavelength λ is written $K\lambda$ and is known as *relative luminosity factor.*

Illumination differs from light very much, though generally these terms are used more or less synonymously. Strictly speaking light is the cause and illumination is the result of that light on surfaces on which it falls. Thus the illumination makes the surface look more or less bright with a certain colour and it is this brightness and colour which the eye sees and interprets as something useful, or pleasant or otherwise.

Light may be produced by passing electric current through filaments as in the incandescent lamps, through arcs between carbon or metal rods, or through suitable gases as in neon and other gas tubes. In some forms of lamps the light is due to fluorescence excited by radiation arising from the passage of electric current through mercury vapour.

Some bodies reflect light in some measure, and when illuminated from an original source they become secondary sources of light. A good example is the Moon, which illuminates the earth by means of the reflected light originating in the Sun.

7.2. TERMS USED IN ILLUMINATION

The devising of modern lighting schemes and the selection of fittings and types of lamps requires a knowledge of the terms and quantities in general use for such purposes. Therefore, the following definitions are given in simple form to facilitate easy identification and reference.

(*i*) Light. It is defined as the radiant energy from a hot body which produces the visual sensation upon the human eye. It is usually denoted by Q, expressed in lumen-hours and is analogous to watt-hours.

(*ii*) Luminous Flux. It is defined as the total quantity of light energy emitted per second from a luminous body. It is represented by symbol F and is measured in lumens (or *cd-sr*). The conception of luminous flux helps us to specify the output and efficiency of a given light source.

(*iii*) Luminous Intensity. Luminous intensity in any given direction is the luminous flux emitted by the source per unit solid angle, measured in the direction in which the intensity is required. It is denoted by symbol I and is measured in candela (*cd*) or lumens per steradian.

If F is the luminous flux radiated out by source within a solid angle of ω steradians in any particular direction then $I = F/\omega$ lumens/steradian or candela (*cd*).

(*iv*) Lumen. The lumen is the unit of luminous flux and is defined as the amount of luminous flux given out in a space represented by one unit of solid angle by a source having an intensity of one candle power in all directions.

$$i.e., \quad \text{Lumens} = \text{Candle power} \times \text{solid angle} = CP \times \omega$$

or total lumens given out by source of one candela is 4π lumens.

(*v*) Candle Power. Candle power is the light radiating capacity of a source in a given direction and is defined as the number of lumens given out by the source in a unit solid angle in a given direction. It is denoted by symbol CP.

$$i.e., \quad CP = \frac{\text{Lumens}}{\omega}$$

(*vi*) Illumination. When the light falls upon any surface, the phenomenon is called the *illumination.* It is defined as the number of lumens, falling on the surface, per unit area. It is denoted by symbol E and is measured in lumens per square metre or lux or metre-candle.

If a flux of F lumens falls on a surface of area A, then the illumination of that surface is

$$E = \frac{F}{A} \text{ lumens/m}^2 \text{ or lux } (lx)$$

(*vii*) Lux or Metre-Candle. It is the unit of illumination and is defined as the luminous flux falling per square metre on the surface which is everywhere perpendicular to the rays of light from a source of one candle power and one metre away from it.

(*viii*) Foot-Candle. It is also the unit of illumination and is defined as the luminous flux falling per square foot on the surface which is everywhere perpendicular to the rays of light from a source of one candle power and one foot away from it.

$$i.e., \quad 1 \text{ Foot-candle} = 1 \text{ lumen/ft}^2$$

$$= \frac{1 \text{ lumen}}{\left(\frac{1\text{ m}}{3.28}\right)^2} = 10.76 \text{ metre-candle or lux.}$$

(*ix*) Candela. It is the unit of luminous intensity. It is defined as 1/60th of the luminous intensity per cm^2 of a black body radiator at the temperature of solidification of platinum (2,043 K).

(*x*) Mean Horizontal Candle Power (MHCP). It is defined as the mean of candle powers in all directions in the horizontal plane containing the source of light.

(*xi*) Mean Spherical Candle Power (MSCP). It is defined as the mean of candle powers in all directions and in all planes from the source of light.

(*xii*) Mean Hemispherical Candle Power (MHSCP). It is defined as the mean of candle powers in all directions above or below the horizontal plane passing through the source of light.

(*xiii*) Reduction Factor. Reduction factor of a source of light is the ratio of its mean spherical candle power to its mean horizontal candle power.

$$i.e., \quad \text{Reduction factor} = \frac{MSCP}{MHCP}$$

(*xiv*) Lamp Efficiency. It is defined as the ratio of the luminous flux to the power input. It is expressed in lumens per watt.

(*xv*) Specific Consumption. It is defined as the ratio of the power input to the average candle power. It is expressed in watts per candela.

(*xvi*) Brightness or Luminance. When the eye receives a great deal of light from an object we say it is bright, and 'brightness' is an important quantity in illumination. It is all the same whether the light is produced by the object or merely reflected from it. The object sends out light as though each small piece of its surface were of a certain luminous intensity. Generally, the brightness of an object is not the same from all points of view. When brightness is considered in terms of measurable quantity, the term *luminance* is employed.

Brightness or *luminance* is defined as the luminous intensity per unit projected area of either a surface source of light or a reflecting surface and is denoted by L.

If a surface of area A has an effective luminous intensity of I candelas in a direction θ to the normal, then the luminance (brightness) of that surface is

$$L = \frac{I}{A\cos\theta} \text{ candela/m}^2 \text{ or nits}$$

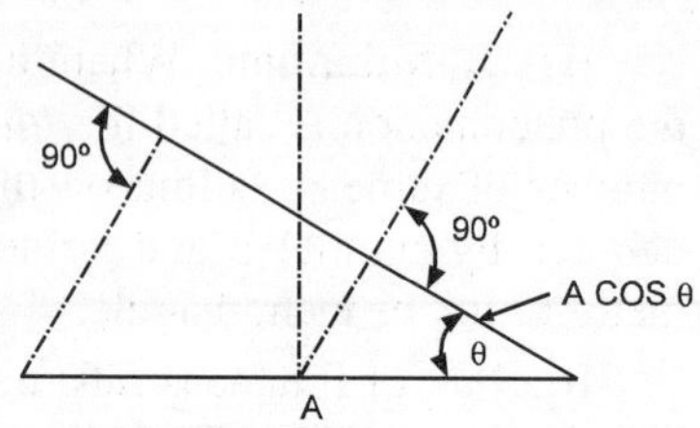

Fig. 7.3

i.e., nit is defined as candela per square metre. Bigger unit of luminance is *stilb* which is defined as candelas per square cm. Lambert is also the unit of brightness which is lumens/cm^2. Foot-lambert is lumens/ft^2.

***(xvii)* Glare.** The size of the opening of the pupil in the human eye is controlled by its iris. If the eye is exposed to a very bright source of light, the iris automatically contracts in order to reduce to amount of light admitted and prevent damage to retina ; this reduces the sensitivity, so that other objects within the field of vision can be only imperfectly seen. This effect is referred to 'glare' and is familiar in connection with motor-car headlights. *In other words glare may be defined as the brightness within the field of vision of such a character as to cause annoyance, discomfort, interference with vision or eye fatigue.*

***(xviii)* Space-height Ratio.** It is defined as the ratio of horizontal distance between adjacent lamps and height of their mountings.

$$i.e., \text{ Space-height ratio} = \frac{\text{Horizontal distance between two adjacent lamps}}{\text{Mounting height of lamps above working plane}}$$

***(xix)* Utilization Factor or Coefficient of Utilization.** It is defined as the ratio of total lumens reaching the working plane to total lumens given out by the lamp.

$$i.e., \text{ Utilization factor} = \frac{\text{Total lumens reaching the working plane}}{\text{Total lumens given out by the lamp}}.$$

***(xx)* Maintenance Factor.** Due to accumulation of dust, dirt and smoke on the lamps, they emit less light than that they emit when they are new ones and similarly the walls and ceilings, etc., after being covered with dust, dirt and smoke do not reflect the same output of light, which is reflected when they are new.

The ratio of illumination under normal working conditions to the illumination when the things are perfectly clean is known as maintenance factor.

i.e., Maintenance factor

$$= \frac{\text{Illumination under normal working conditions}}{\text{Illumination when everything is perfectly clean}}.$$

***(xxi)* Depreciation Factor.** This is merely the inverse of the maintenance factor and is defined as the ratio of initial metre-candles to the ultimate maintained metre-candles on the working plane. Its value is more than unity.

***(xxii)* Waste Light Factor.** Whenever a surface is illuminated by a number of sources of light, there is always a certain amount of waste of light on account of overlapping and falling of light outside the edges of the surface. The effect is taken into account by multiplying the theoretical value of lumens required by 1.2 for rectangular areas and 1.5 for irregular areas and objects such as statues, monuments etc.

***(xxiii)* Absorption Factor.** In the places where atmosphere is full of smoke fumes, such as in foundries, there is a possibility of absorption of light. The ratio of total lumens available after absorption to the total lumens emitted by the source of light is called the absorption factor. Its values varies from unity for clean atmosphere to 0.5 for foundries.

***(xxiv)* Beam Factor.** The ratio of lumens in the beam of a projector to the lumens given out by lamps is called the *beam factor.* This factor takes into account the absorption of light by reflector and front glass of the projector lamp. Its value varies from 0.3 to 0.6.

***(xxv)* Reflection Factor.** When a ray of light impinges on a surface it is reflected from the surface at an angle of incidence, as shown in Fig. 7.4. A certain portion of incident light is absorbed by the surface. The ratio of reflected light to the incident light is called the '*reflection factor*'. It is always less than unity.

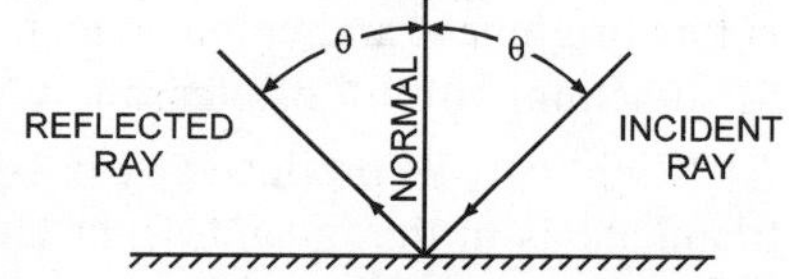

Fig. 7.4

***(xxvi)* Solid Angle.** Plane angle is subtended at a point in a plane by two converging straight lines and its magnitude is given by

$$\theta = \frac{\text{Arc}}{\text{Radius}} \text{ radians}$$

The largest angle subtended at a point is 2π radians.

Solid angle is the angle generated by the surface passing through the point in space and the periphery of the area. Solid angle is denoted by ω, expressed in steradians and is given by the ratio of the area of the surface to the square of the distance between the area and the point.

$$i.e., \ \omega = \frac{\text{Area}}{(\text{Radius})^2} = \frac{A}{r^2}$$

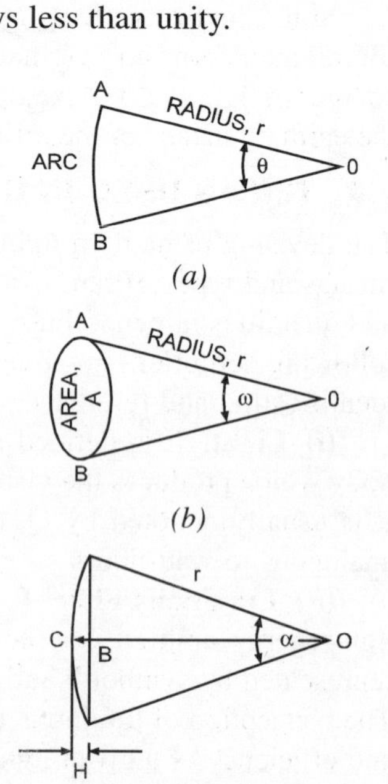

(a)

(b)

(c) Sectional View for Solid Angle

Fig. 7.5

The largest solid angle subtended at a point is that due to a sphere at its centre. If r is the radius of any sphere, its surface area is $4\pi r^2$ and the distance of its surface area from the centre is r, therefore, solid angle subtended at its centre by its surface,

$$\omega = \frac{4\pi r^2}{r^2} = 4\pi \text{ steradians.}$$

Relationship between plane and solid angle can be obtained as follows.

Consider a curved surface of a spherical segment of height H and radius r [Fig. 7.5. (c)].

Surface area of such a segment = $2\pi r\text{H}$

From Fig. 7.5 (c)

$$\text{H} = \text{OC} - \text{OB} = r - r\cos\frac{\alpha}{2} = r\left(1 - \cos\frac{\alpha}{2}\right)$$

Surface area of the surface under consideration

$$= 2\pi r\text{H} = 2\pi r \times r\left(1 - \cos\frac{\alpha}{2}\right) = 2\pi r^2\left(1 - \cos\frac{\alpha}{2}\right)$$

$$\text{Solid angle, } \omega = \frac{\text{Surface area}}{(\text{Radius})^2} = \frac{2\pi r^2\left(1 - \cos\frac{\alpha}{2}\right)}{r^2} = 2\pi\left(1 - \cos\frac{\alpha}{2}\right)$$

Thus solid angle subtended at a point can be determined directly from the above relation knowing the plane angle subtended by its section.

(*xxvii*) Steradian. It is the unit of solid angle and is defined as the solid angle that subtends a surface on the sphere equivalent to the square of the radius.

7.3. LAWS OF ILLUMINATION

There are two laws of illumination (i) Law of inverse squares and (ii) Lambert's cosine law.

1. *Law of Inverse Squares*. If a source of light which emits light equally in all directions be placed at the centre of a hollow sphere, the light will fall uniformly on the inner surface of the sphere, that is to say, each square mm of the surface will receive the same amount of light. If the sphere be replaced by one of the larger radius, the same total amount of light is spread over a larger area proportional to the square of the radius. The amount which falls upon any square mm of such a surface will, therefore, diminish as the radius increases, and will be inversely proportional to the square of the distance.

A similar relation holds if we have to deal with a beam of light in the form of a cone or pyramid, as shown in Fig. 7.6 (a). If we consider parallel surfaces which cut the pyramid at different distances from the source, the areas of these surfaces are proportional to the square of these distances, and, therefore, the amount of light which falls on one unit of the area of these surfaces, is inversely proportional to the square of the distance from the source. This relationship is referred to as the *law of inverse squares.*

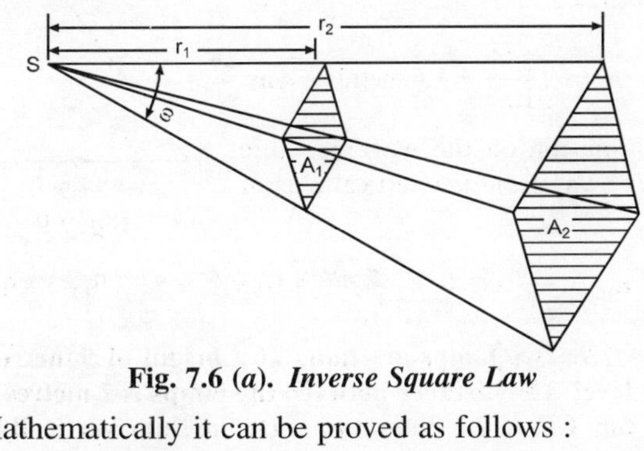

Fig. 7.6 (*a*). *Inverse Square Law*

Mathematically it can be proved as follows :

Let us consider surface area A_1 and surface area A_2 at distances r_1 and r_2 respectively from the point source S of luminous intensity I and normal to the rays, as shown in Fig. 7.6 (a).

Let the solid angle subtended be ω steradians

Luminous flux radiated per steradians = I

Total luminous flux radiated = Iω lumens

Illumination on the surface of area A_1

$$= \frac{\text{I}\omega}{\text{A}_1} \text{ lumens/unit area}$$

$$\text{and area } \text{A}_1 = \omega r_1^2$$

∴ Illumination on the surface of area A_1,

$$\text{E}_1 = \frac{\text{I}\omega}{\omega r_1^2} = \frac{\text{I}}{r_1^2} \text{ lumens per unit area}$$

Similarly illumination on the surface of area A_2,

$$\text{E}_2 = \frac{\text{I}\omega}{\text{A}_2} = \frac{\text{I}\omega}{\omega r_2^2} = \frac{\text{I}}{r_2^2} \text{ lumens per unit area.}$$

Hence the illumination of a surface is inversely proportional to the square of the distance between the surface and the light source provided that the distance between the surface and the source is sufficiently large so that the source can be regarded as a point source.

2. Lambert's Cosine Law. Very often the illuminated surface is not normal to the direction of light as AC in Fig. 7.6 (b) but is inclined as AB. The area over which the light is spread is then increased in the ratio

$$\frac{\text{AB}}{\text{AC}} = \frac{1}{\cos\theta}$$

and the illumination decreases in the ratio $\dfrac{\cos\theta}{1}$.

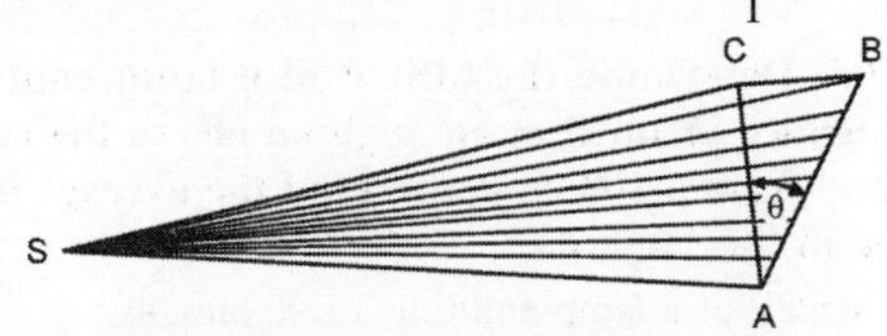

Fig. 7.6 (*b*). *Lambert's Cosine Law*

The expression for the illumination then becomes

$$\text{E} = \frac{\text{I}\cos\theta}{r^2}$$

According to this law the illumination at any point on a surface is proportional to the cosine of the angle between the normal at that point and the direction of luminous flux.

Example 7.1. The flux emitted by a lamp in all directions is 1,000 lumens. Calculate its MSCP.

[*A.M.I.E. Sec. B. Utilization of Elec. Power Winter 1984*]

Solution : Total luminous flux emitted, F = 1,000 lumens

$$\text{MSCP of the lamp} = \frac{\text{F}}{4\pi} = \frac{1{,}000}{4\pi} = 80 \textbf{ Ans.}$$

Example 7.2. State laws of illumination. A 250 V lamp has a total flux of 1,500 lumens and takes a current of 0.4 A. Calculate (*i*) Lumen per watt (*ii*) MSCP per watt.

[*M.D. Univ. Utilization of Electric Power and Traction Dec. 2009*]

Solution : Wattage of lamp = V × I = 250 × 0.4 = 100 watts

$$\text{Mean spherical candle power of lamp, MSCP} = \frac{\text{F}}{4\pi} = \frac{1{,}500}{4\pi} = 120$$

$$(i) \text{ Lumens per watt} = \frac{\text{Output of lamp in lumens}}{\text{Wattage of lamp in watts}} = \frac{1{,}500}{100} = 15 \textbf{ Ans.}$$

$$(ii) \text{ MSCP per watt} = \frac{\text{MSCP of lamp}}{\text{Wattage of lamp}} = \frac{120}{100} = 1.2 \textbf{ Ans.}$$

Example 7.3. A 0.4 metre diameter diffusing sphere of opal glass (20 per cent absorption) encloses an incandescent lamp with a luminous flux of 4,850 lumens. Calculate the average luminance of the sphere.

Solution : Flux emitted by the globe, F = 0.8 × 4,850 = 3,880 lumens

Surface area of the globe $= 4\pi r^2 = 4\pi\,(0.2)^2 = 0.16\,\pi$

$$\text{Average luminance of sphere} = \frac{\text{Flux emitted}}{\text{Surface area}}$$

$$= \frac{3{,}880}{0.16\pi} = 7{,}720 \text{ lumens/m}^2 \quad \textbf{Ans.}$$

Example 7.4. Filament lamp of 200 W is suspended at a height of 5 metres above working plane and gives uniform illumination over an area of 4 metre diameter. Assume efficiency of reflector as 50%. Determine the illumination on the working plane. Efficiency of lamp is 0.89 watt/CP.

[*West Bengal Univ. of Technology Utilization of Electric Power 2009*]

Solution : Candle power of lamp,

$$CP = \frac{\text{Wattage of lamp}}{\text{Efficiency of lamp in watt / CP}} = \frac{200}{0.89} = 224.72$$

Lumens output of lamp $= 4\pi CP = 4\pi \times 224.72 = 2{,}823.9$ lumens

Flux emitted by reflector

= Reflector efficiency × total output lumens

= 0.5 × 2,823.9 = 1,411.95 lumens

$$\text{Area of working plane, A} = \frac{\pi}{4}d^2 = \frac{\pi}{4} \times 4^2 = 4\pi \text{ m}^2$$

Illumination on the working plane

$$= \frac{\text{Lumens output}}{\text{Working plane area, A}}$$

$$= \frac{1{,}411.95}{4\pi^2} = 112.36 \text{ lumens/m}^2 \quad \textbf{Ans.}$$

Example 7.5. Determine the MSCP of a lamp emitting 1,000 lumens. A surface inclined at an angle of 60° to the rays is kept 5 metres away from a 100 cp lamp. Find the average intensity of illumination on the surface.

Solution : MSCP of a lamp emitting 1,000 lumens

$$= \frac{1{,}000}{4\pi} = 80 \quad \textbf{Ans.}$$

Average intensity of illumination on the surface inclined at an angle 60° to the rays and kept 5 metres away from a 100 CP lamp.

$$= \frac{CP}{d^2}\cos\theta = \frac{CP}{d^2}\cos(90° - 60°) = \frac{100}{(5)^2} \times 0.866$$

$$= 3.464 \text{ lux} \quad \textbf{Ans.}$$

Example 7.6. Deduce the relation to find illumination at any point on the surface due to light source suspended at a height *h* from the plane surface.

Solution : Let us consider a point P on the plane surface AB where illumination due to light source S of candle power CP at a height *h* from the plane AB is to be determined.

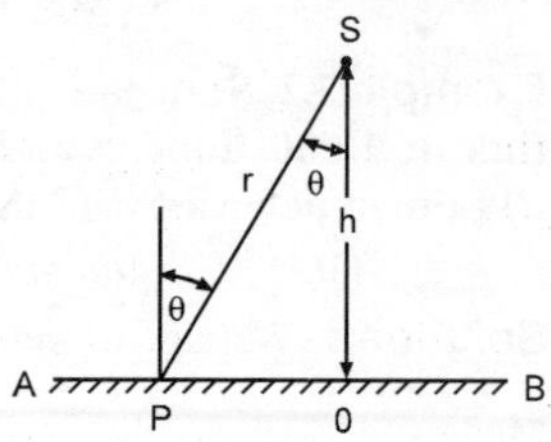

Fig. 7.7

Let the distance between point P and light source S be *r*.

By laws of illumination,

Illumination at point P

$$= \frac{CP}{r^2}\cos\theta \quad \text{where } \cos\theta = \frac{h}{r}$$

$$= \frac{CP}{(h/\cos\theta)^2}\cos\theta = \frac{CP}{h^2}\cos^3\theta$$

and illumination at any point O, vertically below the source of light $= \frac{CP}{h^2}$.

Hence illumination at any point on a plane is $\cos^3\theta$ *times of illumination at a point just vertically below the light source, where* θ *is the angle between the normal to the surface at the point and rays of light.*

Example 7.7. Define and explain 'illumination' and 'luminous flux intensity'.

A lamp of 500 W having a mscp of 1,250 is suspended 2.7 metres above the working plane. Calculate :

(*i*) Illumintion directly below the lamp at the working plane

(*ii*) Lamp efficiency

(*iii*) Illumination at a point 3 m away on the horizontal plane from vertically below the lamp.

[*U.P. Technical Univ. Utilization of Electrical Energy and Electric Traction 2005-06*]

Solution : Luminous flux intensity of lamp, I = 1,250

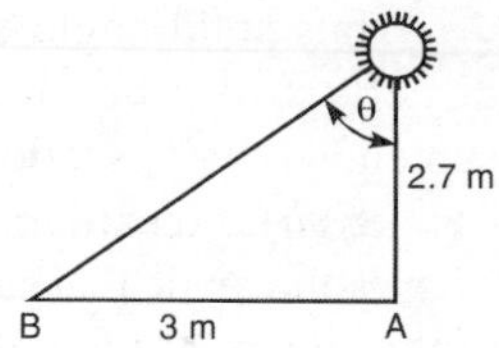

Fig. 7.8

Height of lamp, *h* = 2.7 m

(*i*) Illumination directly below the lamp at the working plane,

$$E_A = \frac{1{,}250}{(2.7)^2} = 171.47 \text{ lux} \quad \textbf{Ans.}$$

$$(ii) \text{ Lamp efficiency} = \frac{\text{Luminous flux}}{\text{Wattage of lamp}}$$

$$= \frac{4\pi \times \text{mscp}}{W} = \frac{4\pi \times 1{,}250}{500} = 31.42 \text{ lumens/watt} \quad \textbf{Ans.}$$

(*iii*) Illumination at a point 3 metres away on the horizontal plane from vertically below the lamp,

$$E_B = \frac{I}{h^2}\cos^3\theta = \frac{1{,}250}{(2.7)^2} \times \frac{2.7^3}{[3^2 + 2.7^2]^{3/2}}$$

$$= 51.33 \text{ lux} \quad \textbf{Ans.}$$

Example 7.8. The illumination at a point on a working plane directly below the lamp is to be 100 lumens/m². The lamp gives 256 CP, uniformly below the horizontal plane. Determine the height at which the lamp is suspended.

Also find illumination at a point on the working table 1.2 metres away from the vertical axis of the lamp.

Solution : Luminous intensity of the lamp, I = 256 cd

Illumination at a point on a working plane directly below the lamp,

$E = 100$ lumens/m²

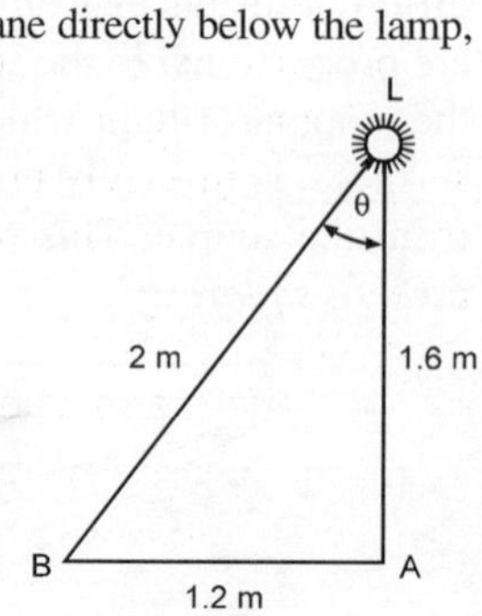

Fig. 7.9

Let the height at which the lamp is suspended be *h* metres

$\because\ E = I/h^2$

$\therefore\ 100 = 256/h^2$

$$\text{or } h = \sqrt{\frac{256}{100}} = 1.6 \text{ metres} \quad \textbf{Ans.}$$

Illumination on the working table 1.2 metres away from the vertical axis of the lamp,

$$E_B = \frac{I}{h^2}\cos^3\theta = \frac{256}{(1.6)^2} \times (0.8)^3 = 51.2 \text{ lux} \quad \textbf{Ans.}$$

Example 7.9. Two lamps are hung at a height of 9 metres from the floor level. The distance between the lamps is 2 metres. Lamp one is of 500 CP. If the illumination on the floor vertically below this lamp is 20 lux, find the candle power of the lamp two.

Solution : Candle power of lamp L_1 = 500 CP

Let the candle power of lamp L_2 be CP_x

Illumination at the point A on the floor level vertically below lamp L_1

= Illumination due to lamp L_1

+ Illumination due to lamp L_2

$$20 = \frac{500}{(9)^2} + \frac{CP_x}{(9)^2}\cos^3\theta$$

$$\text{or } 20 = \frac{500}{81} + \frac{CP_x}{81}\left(\frac{9}{\sqrt{85}}\right)^3$$

$$\text{or } CP_x = \frac{1{,}620 - 500}{\left(\frac{9}{\sqrt{85}}\right)^3}$$

$$= \frac{1{,}120}{0.93} = 1{,}204 \text{ CP} \quad \textbf{Ans.}$$

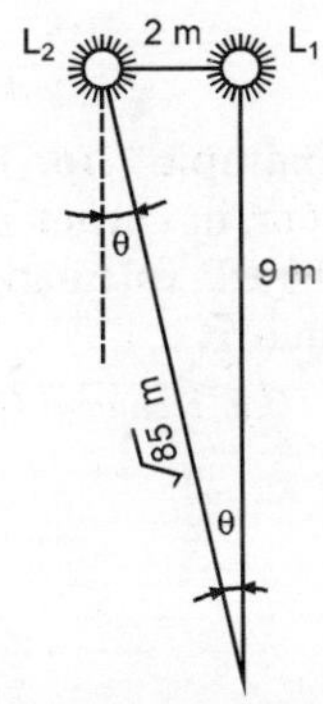

Fig. 7.10

Example 7.10. A small light source with intensity uniform in all directions is mounted at a height of 10 metres above a horizontal surface. Two points A and B both lie on the surface with point A directly beneath the source. How far is point B from point A if the illumination at point B is only 1/10 as great as at point A ?

Solution : Let the intensity of the lamp L be I candela and the distance between points A and B be x metres.

Illumination at point A, $$E_a = \frac{I}{(10)^2} = \frac{I}{100}$$

Illumination at point B, $$E_B = \frac{I}{(10)^2} \times \left(\frac{10}{\sqrt{(10)^2 + x^2}}\right)^3$$

$$= \frac{10I}{(100 + x^2)^{3/2}}$$

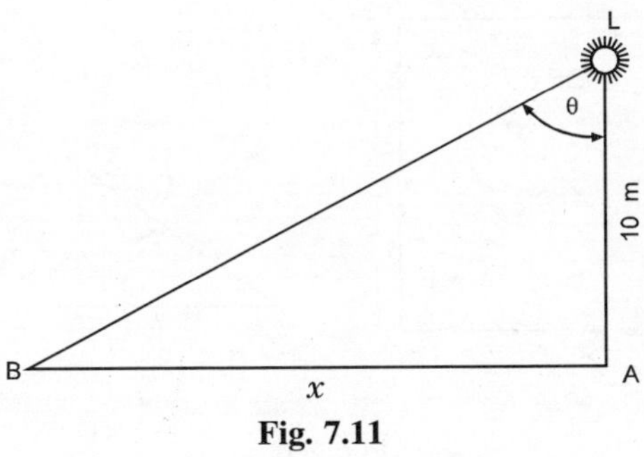

Fig. 7.11

$$\because \quad E_B = \frac{1}{10} E_A$$

$$\therefore \quad \frac{10I}{(100 + x^2)^{3/2}} = \frac{1}{10} \times \frac{I}{100}$$

$$\text{or } (100 + x^2)^{3/2} = 10{,}000$$

$$\text{or } (100 + x^2) = (10{,}000)^{2/3} = 464$$

$$\text{or } x = \sqrt{464 - 100}$$

$$= 19.1 \text{ metres } \textbf{Ans.}$$

Example 7.11. A light source having an intensity of 400 CP in all directions is fitted with a reflector so that it directs 80% of its light along a beam having a divergence of 15°. Determine the total light flux emitted along the beam. Also determine the average illumination produced on a surface normal to the beam direction at a distance of 8 metres.

[Gorakhpur Univ. Utilization of Electrical Power and Traction Sept. 1984]

Solution : Intensity of light source, I = 400 cd

Total light flux emitted along the beam $= 0.8 \times 4\pi \times 400$

$= 4{,}020$ lumens

Beam angle, $\theta = 15°$

Distance of sphere from light source, $l = 8$ m

Radius of circle of the surface to be illuminated,

$$r = l \tan\frac{\theta}{2} = 8 \tan\frac{15°}{2} = 1.05322 \text{ m}$$

Area of surface to be illuminated,

$$A = \pi r^2 = \pi \times (1.05322)^2 = 3.485 \text{ m}^2$$

$$\text{Average illumination} = \frac{4{,}020}{3.485} = 1{,}153 \text{ lux} \quad \textbf{Ans.}$$

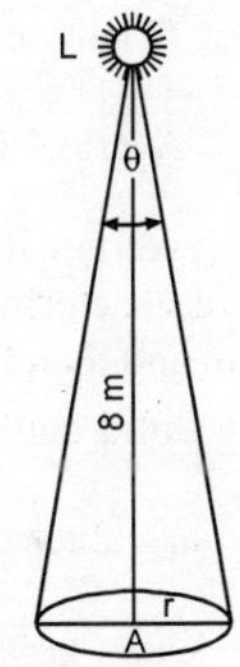

Fig. 7.12

Example 7.12. A lamp with a reflector is mounted 12 m above the centre of a circular area of 24 metres diameter. If the combination of the lamp and reflector gives a uniform CP of 1,000 over the circular area, determine the maximum and minimum illumination produced on the area.

[Pb. Univ. Elec. Drives and Utilization of Elec. Energy Nov. 1983]

Solution : Candle power of the lamp, CP = 1,000

The maximum illumination will occur directly below the lamp, *i.e.*, at point A and

$$= \frac{CP}{h^2} = \frac{1{,}000}{(12)^2} = 6.94 \text{ lux} \quad \textbf{Ans.}$$

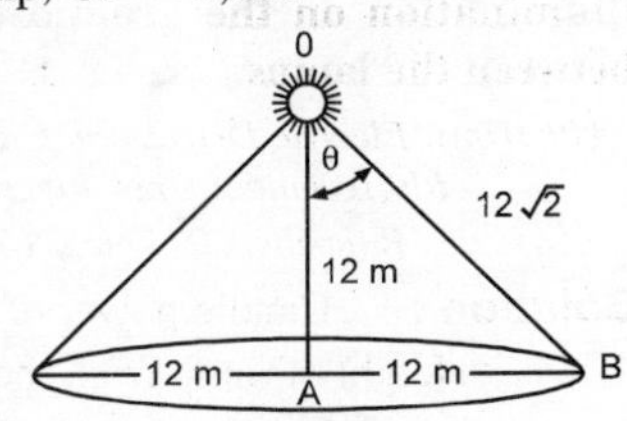

Fig. 7.13

Minimum illumination will occur at the periphery of the circular area, *i.e.*, at point B and

$$= \frac{CP}{h^2}\cos^3\theta = \frac{1{,}000}{(12)^2} \times \left(\frac{12}{\sqrt{12^2 + 12^2}}\right)^3$$

$$= \frac{1{,}000}{(12)^2} \times \frac{1}{(\sqrt{2})^3} = 2.46 \text{ lux} \quad \textbf{Ans.}$$

Example 7.13. A lamp having a uniform CP of 200 in all directions is provided with a reflector which directs 60 per cent of the total light uniformly on to a circular area of 10 m diameter. The lamp is hung 6 m above the area. Calculate the illumination

(*i*) at the centre and (*ii*) at the edge of the surface with and without the reflector.

Determine also the average illumination over the area without the reflector.

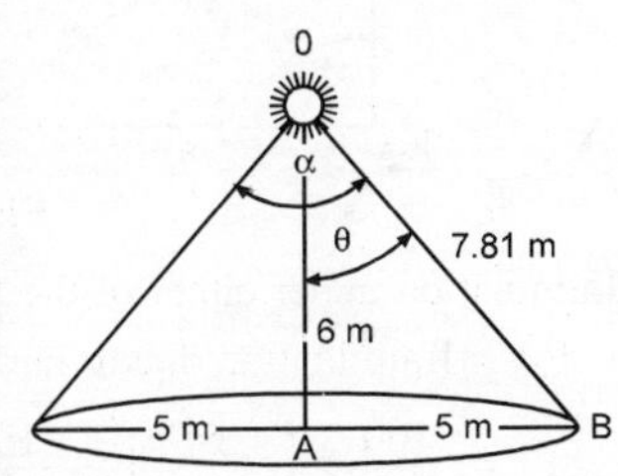

Fig. 7.14

Solution : Candle power of the lamp, CP = 200

(*i*) Illumination at the centre of the area without reflector

$$= \frac{CP}{h^2} = \frac{200}{(6)^2} = 5.56 \text{ lux} \quad \textbf{Ans.}$$

(*ii*) Illumination at the edge of the surface without reflector

$$= \frac{CP}{h^2}\cos^3\theta \text{ where } \cos\theta = \frac{6}{\sqrt{5^2 + 6^2}} = 0.768$$

$$= \frac{200}{(6)^2} \times (0.768)^3 = 2.52 \text{ lux} \quad \textbf{Ans.}$$

With the reflector the illumination at the edge and at the centre will be the same since the reflector directs the light uniformly on the surface.

Total lumens given out, $\phi = 4\pi\, CP = 4\pi \times 200 = 800\,\pi$

Total lumens reaching the surface $= 0.6 \times \phi = 0.6 \times 800\,\pi$

Total surface area, A = $\pi r^2 = \pi \times (5)^2 = 25\,\pi\ m^2$

Average illumination $= \dfrac{0.6 \times 800\pi}{25\pi} = 19.2$ lux **Ans.**

Average illumination of the surface without reflector will be found by first determining the solid angle subtended by the surface at the lamp and then finding out the luminous flux emitted in that solid angle.

Solid angle subtended by the area at the lamp,

$$\omega = 2\pi\left(1 - \cos\frac{\alpha}{2}\right) = 2\pi\,(1 - \cos\theta) = 2\pi\,(1 - 0.768)$$

$$= 0.464\,\pi \text{ steradians}$$

Total flux reaching the surface = $I\omega = 200 \times 0.464\,\pi$

$= 92.8\,\pi$ lumens

Average illumination $= \dfrac{92.8\pi}{\pi(5)^2} = 3.7$ lux **Ans.**

Example 7.14. To lamp posts are 16 m apart and are fitted with a 100 CP lamp each at a height of 6 m above ground. Calculate the illumination on the ground (*a*) under each lamp (*b*) midway between the lamps.

[*Pb. Univ. Electric Drives and Utilization of Elec. Energy December 1982; Pb. Technical Univ. Electric Drives & Utilization December 2008; Rajasthan Technical Univ. Utilization of Electrical Power 2011*]

Solution : Candle power of each lamp = 100 CP

Height of lamp from the ground, h = 6 metres

Distance between the two lamp posts = 16 metres

Illumination midway between the lamps,

E_C = Illumination due to lamp L_1 + Illumination due to lamp L_2

$$= \frac{100}{(6)^2} \times (0.6)^3 + \frac{100}{(6)^2} \times (0.6)^3 = 1.2 \text{ lux} \quad \textbf{Ans.}$$

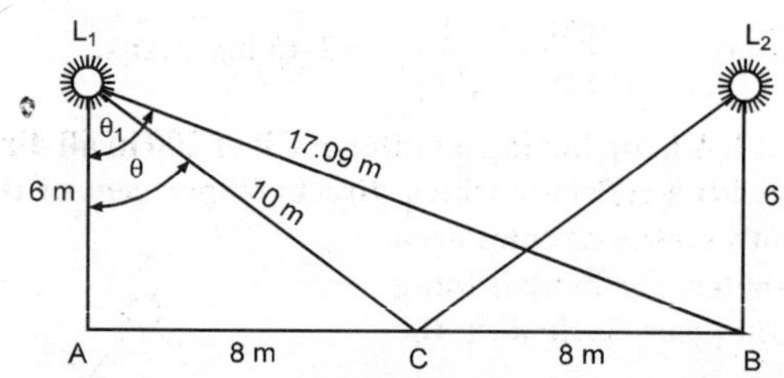

Fig. 7.15

Illumination under either of the lamps, say under lamp L_2,

E_B = Illumination due to lamp L_1 + Illumination due to lamp L_2

$$= \frac{100}{(6)^2} \times \left(\frac{6}{17.09}\right)^3 + \frac{100}{(6)^2} = 2.9 \text{ lux} \quad \textbf{Ans.}$$

Example 7.15. Explain laws of illumination:

In a street lighting scheme, lamps having uniform CP of 800 are hung at a height of 10 m. Distance between consecutive lamp posts is 12 metres. Find illumination under lamp and at the centre in between the lamp posts.

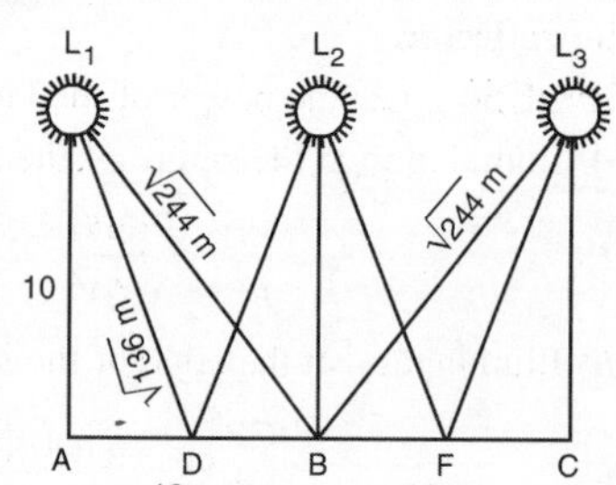

Fig. 7.16

[*U.P. Technical Univ. Utilization of Electrical Energy & Traction 2009-10*]

Solution : Illumination under lamp L_2

E_B = Illumination due to lamp L_2 + Illumination due to lamp L_1 + Illumination due to lamp L_3

$$= \frac{800}{10^2} + \frac{800}{10^2} \times \frac{10^3}{(244)^{3/2}} + \frac{800}{10^2} \times \frac{10^3}{(244)^{3/2}}$$

$= 8 + 2.1 + 2.1 = 12.2$ lux **Ans.**

Illumination at the centre in between the lamp posts

$$E_D \text{ or } E_F = \frac{800}{10^2} \times \frac{10^3}{(136)^{3/2}} + \frac{800}{10^2} \times \frac{10^3}{(136)^{3/2}}$$

$= 5.044 + 5.044 = 10.088$ lux **Ans.**

Example 7.16. Four lamps are hung at a height of 10 m from the floor in corner of a square of 20 m × 20 m. If each lamp is of 400 CP, calculate the illumination on the floor at the centre of the square.

[*U.P. Technical Univ. Utilization of Electrical Energy & Traction 2006-07*]

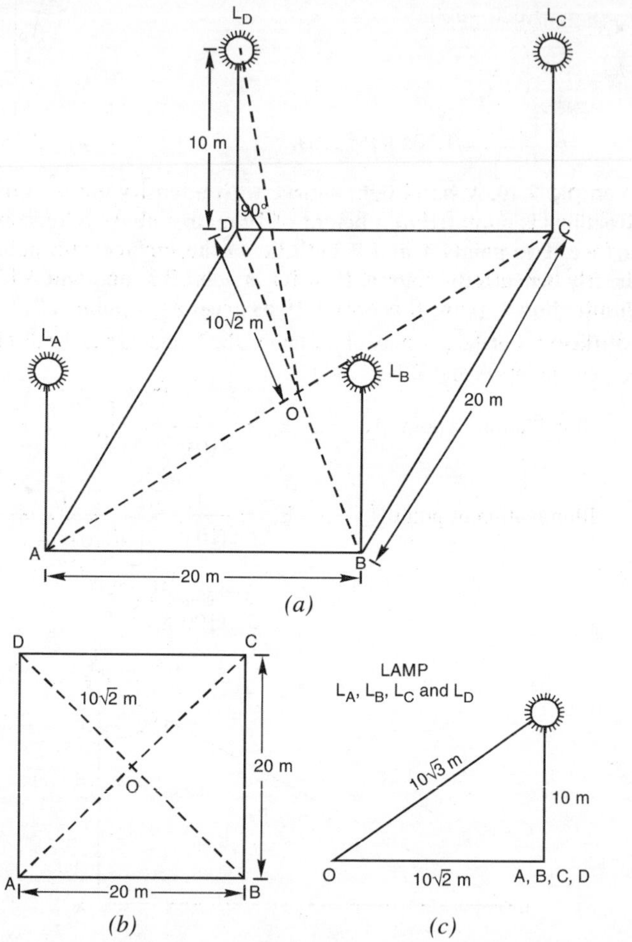

Fig. 7.17

Solution : Candle power of lamp = 400 CP

Height of lamp from ground, h = 10 m

Distance between lamp posts = 20 m

Distance of the centre of square from either corner or lamp post

$$= \frac{1}{2}\sqrt{20^2 + 20^2} = 14.142 \text{ m}$$

Illumination on the floor at the centre of the square,

E_D = Illumination due to lamp L_A + illumination due to lamp L_B + illumination due to lamp L_C + illumination due to lamp L_D

$$= \frac{400}{10^2} \times \left(\frac{10}{\sqrt{300}}\right)^3 + \frac{400}{10^2} \times \left(\frac{10}{\sqrt{300}}\right)^3 + \frac{400}{10^2} \times \left(\frac{10}{\sqrt{300}}\right)^3 + \frac{400}{10^2} \times \left(\frac{10}{\sqrt{300}}\right)^3$$

$\because$ Distance of each lamp from the centre of the square is same

$= 0.7698 + 0.7698 + 0.7698 + 0.7698 = 3.08$ lux **Ans.**

Example 7.17. Two similar lamps having uniform intensity of 500 candle power in all directions below the horizontal are mounted at a height of 4 metres. What must be the maximum spacing between the lamps so that the illumination on the ground midway between the lamps shall be at least one half the illumination directly under the lamps.

[*Gorakhpur Univ. Utilization of Elec. Power and Traction 1982*]

Solution : Candle power of lamp = 500 CP

Height of lamps from the ground, $h = 4$ metres

Let the maximum spacing between the lamps be of d metres

Illumination midway between the lamps,

$E_C = 2 \times$ Illumination due to either lamp

$$= 2 \times \frac{500}{(4)^2} \times \frac{(4)^3}{\left(4^2 + \frac{d^2}{4}\right)^{3/2}} = \frac{4{,}000}{\left(4^2 + \frac{d^2}{4}\right)^{3/2}}$$

Illumination under either of the lamps, say under lamp L_2,

E_B = Illumination due to lamp L_1 + Illumination due to lamp L_2

$$= \frac{500}{(4)^2} \times \frac{(4)^3}{(4^2 + d^2)^{3/2}} + \frac{500}{(4)^2} = \frac{2{,}000}{(4^2 + d^2)^{3/2}} + 31.25$$

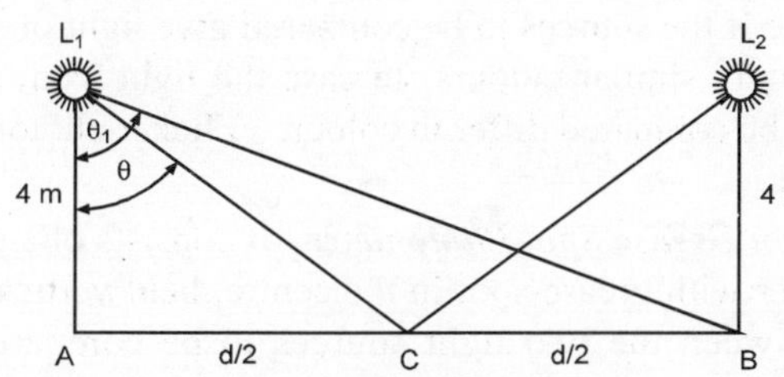

Fig. 7.18

Since $E_C = \frac{1}{2} E_B$

$$\therefore \quad \frac{4{,}000}{\left(4^2 + \frac{d^2}{4}\right)^{3/2}} = \frac{1}{2}\left[31.25 + \frac{2{,}000}{(4^2 + d^2)^{3/2}}\right]$$

or $$\frac{4{,}000}{\left(4^2 + \frac{d^2}{4}\right)^{3/2}} = 15.625 + \frac{1{,}000}{(4^2 + d^2)^{3/2}}$$

or $d = 9.56$ m **Ans.**

Example 7.18. A small area of 5 metres in diameter is to be illuminated by a lamp at a height of 4.5 m above the centre of the area. A lamp having an efficiency of 24 lumens/watt is fitted with a reflector which directs the light output only over the surface to be illuminated. If the utilization coefficient is 0.65 and illumination 800 lux, determine the wattage of the lamp.

[*M.D. Univ. Utilization of Electric Power and Traction May 2009*]

Solution : Area of working plane,

$$A = \frac{\pi}{4} d^2 = \frac{\pi}{4} \times (5)^2 = 6.25\pi \text{ m}^2$$

Illumination on working plane, E = 800 lux

Luminous flux reaching the working plane,

$$F = EA = 800 \times 6.25\pi \text{ lumens}$$

Total luminous flux emitted by lamp

$$= \frac{F}{\text{Utilization coefficient}} = \frac{800 \times 6.25\pi}{0.65}$$

$$= 24{,}166 \text{ lumens}$$

Wattage of the lamp $= \frac{24{,}166}{24} = 1{,}006.92 \text{ W} \simeq 1 \text{ kW}$ **Ans.**

7.4. POLAR CURVES

All discussions made so far were based on the assumption that the luminous intensity or the candle power from a source is uniformly distributed over the surrounding surface. But none of the practical type of lamp provides light uniformly distributed in all directions because of its un-symmetrical shape. It is often necessary to know the distribution of light in various directions to ascertain how the candle power of a light source varies in different directions. The luminous intensity in all the directions can be represented by *polar curves*. If the luminous intensity in a horizontal plane passing through the lamp is plotted against angular position, a curve known as *horizontal polar curve is obtained.* If the luminous intensity in a vertical plane is plotted against the angular position, a curve known as *vertical polar curve* is obtained. The typical polar curves for an ordinary filament lamp are shown in Fig. 7.19.

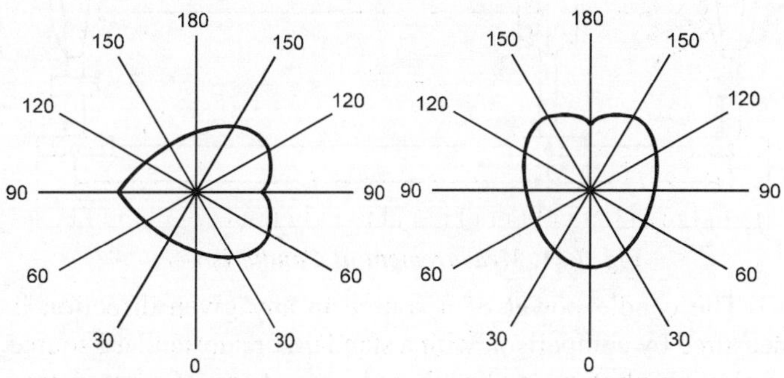

(a) Polar Curve For Horizontal Plane *(b) Polar Curve For Vertical Plane*

Fig. 7.19

The polar curves are used to determine the mean horizontal candle power (mhcp) and mean spherical candle power (mscp). These are also used to determine the actual illumination of a surface by employing the candle power in that particular direction as read from the vertical polar curve in illumination calculations.

The mean horizontal candle power of a lamp can be determined from the horizontal polar curve by taking the mean value of the candle power in a horizontal direction.

Mean spherical candle power can be determined from the vertical polar curve by Rousseau's construction.

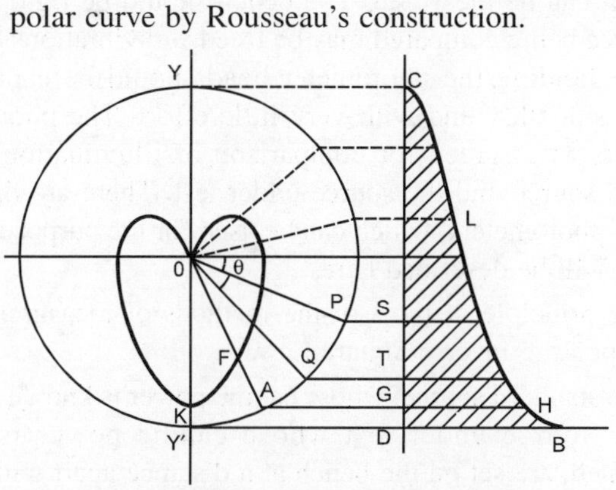

Fig. 7.20. ***Rousseau's Construction***

Rousseau's Construction. The construction is illustrated in Fig. 7.20. A semicircle of any convenient radius is drawn with the pole of the polar diagram as centre. The line CD is drawn equal and parallel to the vertical diameter YY′. Now from this line CD ordinate equal to corresponding radius on the polar curve are set up such as DB = OK, GH = OF and so on. The curve obtained by joining the ends of these ordinates is known as *Rousseau's curve*. The mean ordinate of this curve gives the mscp of the lamp having polar curve given in the Fig. 7.20.

$$\text{The mean ordinate of the curve} = \frac{\text{Area CSTGDBHLC}}{\text{Length of CD}}$$

The area under the curve can either be determined on a graph paper or found by Simpson's rule.

7.5. PHOTOMETRY

We shall now discuss the comparison and measurement of candle powers.

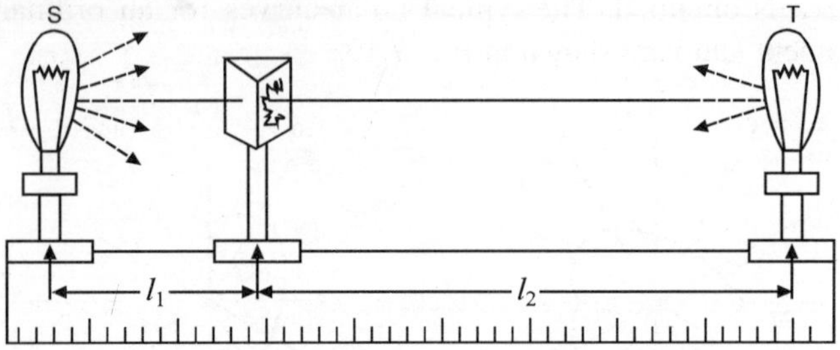

Fig. 7.21. ***Measurement of Candle Power***

The candle power of a source in any given direction is measured by comparison with a standard or substandard source employing photometer bench and some form of photometer. The experiment is performed in a dark room with dead black walls and ceiling in order to eliminate errors due to reflected light.

The photometer bench consists essentially of two steel rods which carry the stands or saddles for holding the two sources, the carriage for the photometer head and for any other apparatus employed in making measurements. One of the bar carries a bar strip, graduated in centimetres and millimetres. The carriages which slides upon the bench have except that carrying the photometer head, a circular table which can be rotated in a horizontal plane and clamped in any position. The circular table is provided with a scale graduated in degrees round its edge so that the angle of rotation of lamp from the direction of the axis of bench can be measured. The bench should be rigid so that the source being compared may be free from vibrations and the carriage holding the photometer head should be capable of moving smoothly and with very little effort. The photometer head acts as a screen for comparison of illumination of the standard source and the source under test. There are different types of photometers, which can be used for the purpose. Some of them will be described here.

The principle of most of the methods of measurement is based upon the Inverse Square Law.

The standard source, whose candle power is known (say S) and the source under test whose candle power is to be determined, are set on the bench at a distance apart with some type of screen in line with, and between, them, as shown in Fig. 7.21. The photometer head or screen is moved in between the two fixed sources until the illumination on both sides of the screen is same. If the distances of the standard source S and source under test T from photometer head are l_1 and l_2 respectively then according to inverse square law

$$\frac{\text{Candle power of source under test}}{\text{Candle power of standard source}} = \frac{l_2^2}{l_1^2}$$

$$\text{or} \quad \text{Candle power of source under test} = S \times \frac{l_2^2}{l_1^2}$$

Since the square of distances are involved in measurement, distances should be measured accurately. In order to obtain distance exactly two points are determined at which there is a perceptible difference in illumination from the two sides and the point half way between them is then taken as the position of equal illumination.

Most of the photometer heads consist of some device by means of which the illumination of two surfaces, side by side– one illuminated by standard source and other by the source whose candle power is to be determined may be compared under exactly similar conditions without movement of the eye. The photometers which are most common in use are the Bunsen and the Lummer Brodhun photometers. These photometers are best for use if the sources to be compared give light of same or approximately similar colours. In case the light from the two sources to be compared differ in colour, a Flicker photometer is best suited.

Bunsen Grease Spot Photometer. It consists of a piece of tissue paper with grease spot in the centre, held vertically in a carrier between the two light sources to be compared. The position of the carrier is then adjusted until the semi-transparent spot and the opaque parts of the paper are equally bright, *i.e.*, the grease spot is invisible. The distances of the tissue paper from both light sources are measured. The candle power of the source under test is then determined from the relation.

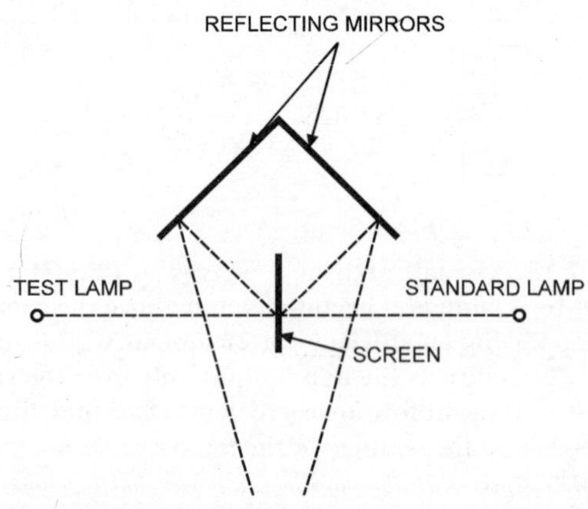

Fig. 7.22. ***Bunsen Grease Spot Photometer Head***

$$\frac{\text{Candle power of source under test}}{\text{Candle power of standard source}}$$

$$= \left(\frac{\text{distance of source under test from photometer}}{\text{distance of standard source from photometer}}\right)^2$$

The Bunsen Grease spot photometer can be used in other way also. In this method of use, the photometer is fixed at one position and one side of the paper is illuminated first by test lamp and then by a standard lamp and distance of the lamp from the photometer head is measured in each case on disappearing of grease spot.

If l_1 and l_2 are the distances then the candle power of source under test

$$= \text{Candle power of standard lamp} \times \frac{l_1^2}{l_2^2}$$

The simplest and best method of use of Bunsen Grease spot photometer for comparison of illumination on either side of the screen is to use two mirrors placed behind the paper and view the illumination on both the sides through the mirrors simultaneously by standing in a plane passing through the photometer head and perpendicular to the lines joining the lamps. The position of the photometer head for equal contrast in illumination between the opaque and transparent portions of the paper on the two sides is determined. Now the candle power of the lamp under test can be calculated as before.

There are two types of Lummer Brodhun heads—one is the equality of brightness type and second is the contrast type. Contrast type is more accurate and, therefore, more used in photometric measurements.

Equality of Brightness Type Photometer Head. The photometer head consists of a plaster of paris screen, two mirrors M_1 and M_2, a compound prism P and a telescope. The compound prism is made up of two right angled glass prisms, one of which has principal surface as spherical one with a small flat portion at the centre. This small flat portion at the centre makes optical contact with the flat surface of the other prism, as shown in Fig. 7.23.

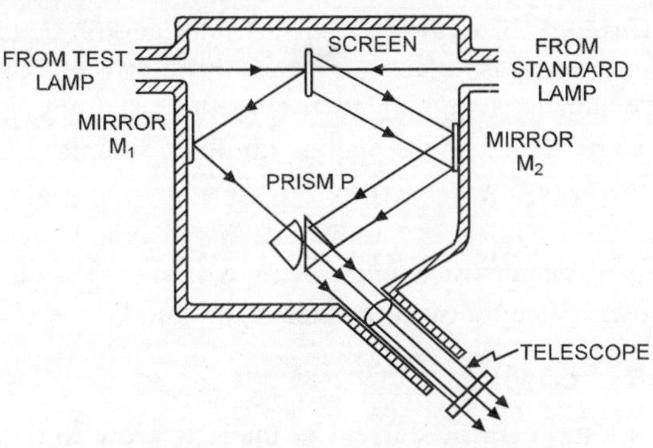

Fig. 7.23. ***Lummer Broadhun Photometer (Quality of Brightness Type)***

Each side of the screen S is illuminated by light entering from one of the sources to be compared. The light rays coming from the test lamp and standard lamp are reflected by the screen on to the mirrors M_1 and M_2 respectively which in turn reflects the same to the compound prism P. Only that portion of light reflected by mirror M_1 which falls on the flat surface of the spherical prism, passes straight into the telescope while the rest of it is reflected back. The portion of light reflected by mirror M_2, which falls on the surface of contact between the two prism passes straight through the compound prism while the rest of it is reflected through the telescope. Thus the observer sees the centre portion of the circular area illuminated by the test lamp and outer ring illuminated by the standard lamp.

While carrying out measurements, the photometer head is moved until the dividing line between the centre portion and surrounding disappears. The disappearance of this dividing line indicates the equality of brightness. Now the distance of the photometer head from the two lamps is noted and the candle power of lamp under test can be calculated by using *inverse square law principle*.

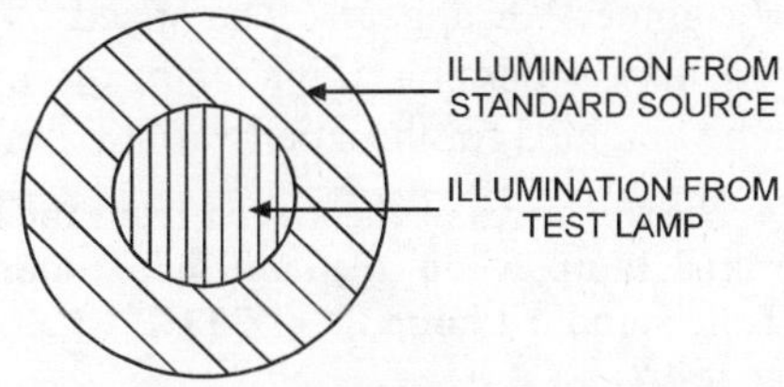

Fig. 7.24

Contrast Type of Lummer Brodhun Photometer Head. The compound prism, in the contrast type photometer head, is made up of right angled prisms with their principal surfaces face to face. One of the prism [the left hand one in Fig. 7.25 (*a*)] has its hypotenuses surface etched away at X, Y and Z so that it forms a pattern similar to that shown in the figure.

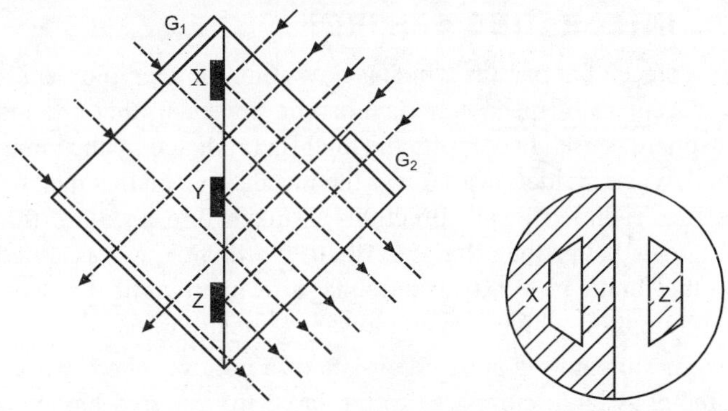

Fig. 7.25. ***Lummer Brodhun Photometer Head (Contrast Type)***

The light falling on the compound prism from both sides of the screen will pass through the un-etched portions of the prism at the junction but will be reflected by the etched surfaces. The etched portion will have difference in illumination as compared with the un-etched portion in case the illumination of the surfaces of reflections are different, as shown in Fig. 7.25 (*b*). G_1 and G_2 are two glass plates which give a little reflected light in order to maintain some difference between illumination of the etched and un-etched areas for all the positions of the photometer head. The difference is about 8% in balance position of photometer head. The photometer head will be in balance position when the difference in illumination of etched and the un-etched portions will be same on each half of the circular area. In case the balance position is disturbed, the difference in illumination of area Z and its surrounding area will decrease while the difference in illumination of area XY and the inner trapezium will increase. The photometer head is moved until the equal contrast is obtained. This type of head gives accuracy within 1% when lights of similar colours are compared.

Flicker Photometers. These photometers are employed when the two sources giving light of different colours are to be compared. The working of these photometers depend upon the fact that if two illuminated surfaces are presented to the eye with rapid alternations, the flicker disappears when the surfaces are of equal brightness. These photometers are not affected by the difference in colour of two lights to be compared as much as photometers of the steady comparison type are affected since the colour difference between two alternating fields of light disappears at a lower speed of alternation than the speed at which difference in brightness disappears. The speed of alternation should be kept as low as possible at which the disappearance of the flicker can be obtained for the small variation in brightness.

Example 7.19. The distances of the screen from the lamp under test and standard lamp when adjusted are 1.6 m and 0.9 m respectively. If the standard lamp be of 80 CP what is the CP of the lamp under test ?

Solution : Candle power of standards lamp, $CP_s = 80$

Distance of test lamp from screen, $l_1 = 1.6$ m

Distance of standard lamp from screen, $l_2 = 0.9$ m

Candle power of lamp under test, CP_t

$$= \frac{CP_s}{l_2^2} \times l_1^2 = \frac{80}{(0.9)^2} \times (1.6)^2 = 252.84 \text{ CP} \quad \textbf{Ans.}$$

7.6. INTEGRATING SPHERE

It is a piece of apparatus which is now commonly employed for measurement of mean spherical candle power. In this method the sphere is used to measure the total flux radiated by the lamp, which when divided by 4π gives the mscp. Since in this method the flux radiated in all directions is taken into account, this method is better than that described in which it was assumed that the candle power distribution is same in all vertical planes, an assumption which may not always be justifiable.

The integrating sphere consists of a hollow sphere whose diameter is large compared to the lamp to be tested having a smooth inner surface with a uniform coating of white paint. If the lamp is hung inside the sphere, the light is so diffused that an uniform illumination is produced over the whole surface. A small window of translucent glass provided at one side of the sphere is illuminated by reflected light from the inner surface of the sphere. The small screen is inserted in between the lamp and window in order to prevent the light from the lamp reaching the window directly.

The measurement of mscp is done as follows :

The lamp, whose mscp is to be determined, is placed at the centre of sphere and brightness of window is measured with the help of some form of illuminometer. The test lamp is replaced by a standard lamp whose mscp is known and brightness of the window is again measured. Since the mscp of the two lamps are proportional to the respective brightness of the window, therefore, the mscp of the lamp under test can be determined by knowing the mscp of standard lamp and brightness of window in the two cases.

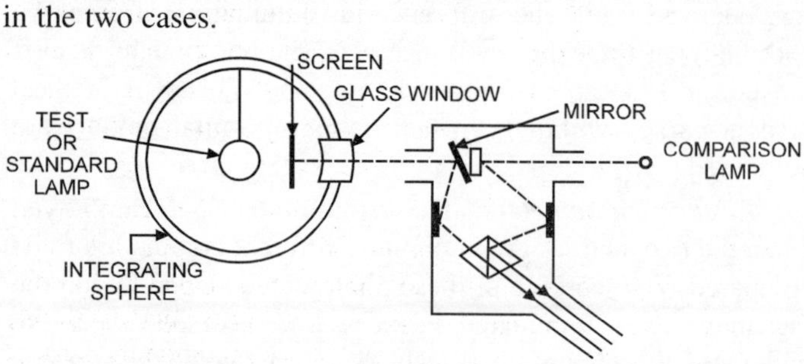

Fig. 7.26

A Lummer Brodhun photometer head can also be employed for the measurement of brightness of window with some modifications. A small mirror is placed along with the screen, as shown in Fig. 7.26 in order to reflect the light from the window of the sphere into compound prism of the photometer and balance is obtained using a lamp on the other side of the photometer. Balance is obtained by moving the comparison lamp first with the standard or substandard lamp in the sphere, and then with the lamp test inside the sphere. The squares of the corresponding distance of the comparison lamp from the photometer head is inversely proportional to the mscp of lamps.

7.7. MEASUREMENT OF ILLUMINATION

The best and simplest method of measuring illumination is by a portable type illumination photometer calibrated to read directly in lux. No shadow should be allowed to fall on the meter during measurements, and the meter should be held perfectly horizontal and stationary for accurate results. The meter should be colour-corrected type so that it can be used for all kinds of light sources.

7.8. SOURCES OF LIGHT

According to principle of operation the light sources may be grouped as follows :

1. ***Arc Lamps.*** Electric discharge through air provides intense light. This principle is utilised in arc lamps.

2. ***High Temperature Lamps.*** Oil and gas lamps and incandescent filament type lamps, which emit light when heated to high temperature.

3. ***Gaseous Discharge Lamps.*** Under certain conditions, it is possible to pass electric current through a gas or metal vapour, which is accompanied by visible radiations. Sodium and mercury vapour lamps operate on this principle.

4. ***Fluorescent Type Lamps.*** Certain materials, when exposed to ultraviolet rays, transform the absorbed energy into radiations of longer wave length lying within the visible range. This principle is employed in fluorescent lamps.

7.9. ARC LAMPS

In an arc lamp electric current is made to flow through two electrodes in contact with each other which are drawn apart. The result is an arc being struck. The arc maintains the current, and is very efficient source of light. There are various forms of arc lamps such as carbon-arc, flame-arc or magnetic-arc lamps.

7.9.1. Carbon-Arc Lamp

This is earliest type and is still used in cinema projectors and search lights. When two hard carbon rods are placed end to end and connected to the terminals of dc supply mains of not less than 45 volts, the current flows through them and the ends of carbon rods soon become incandescence due to high resistance. If they are slightly pulled (say about 2 or 3 mm) an arc will be formed between two carbon rods and white light will be produced. The arc is maintained by transfer of carbon particles from one rod to another one. It is found that these particles travel from the positive carbon rod to negative one.

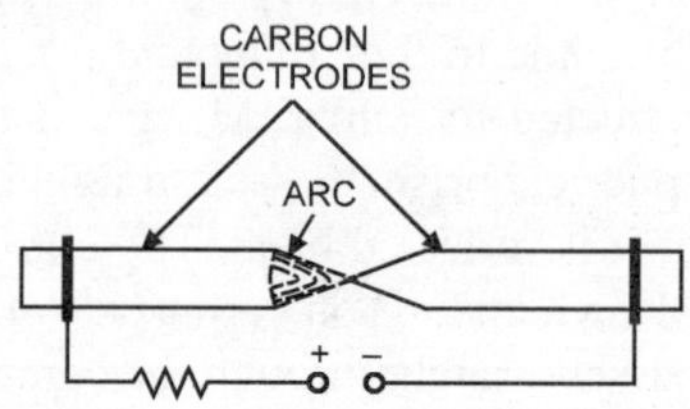

Fig. 7.27. ***Carbon-Arc Lamp***

That is why, the positive rod after sometime of use becomes hollow and negative becomes a pointed pencil.

Due to transfer of particles and oxidation both rods burn away by and by. The rate of burning of positive rod is twice of that of negative rod, that is why, the x-section of the positive rod is made twice of that of negative one. In case of ac supply the rate of burning of both rods is same and, therefore, they are made of equal x-section. Since the carbon is consumed during operation, therefore, requires adjustment of distance between two rods now and then. Except in small lamps, an automatic device is required for feeding the carbon into the arc as they burn away. The voltage required for maintaining the arc is given approximately by $V = (39 + 2.8\, l)$ where l is the length of the arc in mm.

The carbon arc is unstable, since as the current increases the amount of vaporised carbon increases, and the resistance decreases so much that the product of current and resistance, *i.e.*, the voltage drop across the arc falls. A series resistance is, therefore, used for stabilising the arc, and this leads to a low overall luminous efficiency. The voltage drop across the arc is about 45 to 60 volts and the supply voltage is about 70 to 100 volts. In case of ac supply an inductor is used in place of series resistance for stabilising the arc.

In a carbon-arc lamp about 85% of light is given out by the positive carbon, 10% by the negative carbon and 5% by the air. The temperature of positive carbon is between 3,500°C and 4,000°C and that of negative one is about 2,500°C. The luminous efficiency of such a lamp is 12 lumens per watt.

7.9.2. Flame-Arc Lamp

The principle of operation is similar to that of carbon-arc lamp. The electrodes of such a lamp has 5 to 15% fluoride (called the flame material) and 85 to 95% carbon. Generally core type carbon electrodes are used and the cavities are filled with fluoride. The fluoride has a characteristic which radiates light energy efficiently from a very high heated arc stream. Fluoride turns into vapour along with the carbon and these fluoride

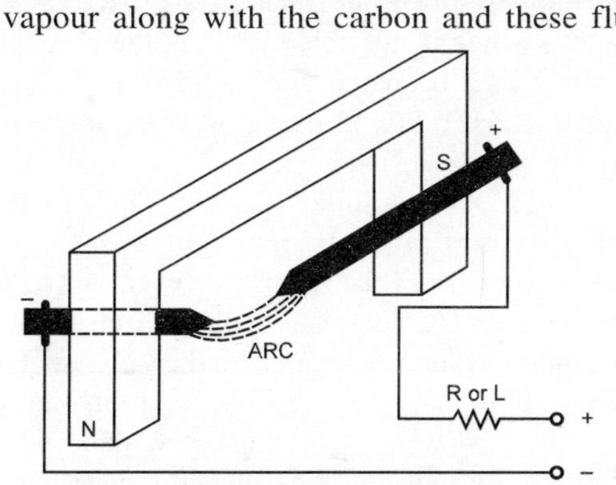

Fig. 7.28. *Flame-Arc Lamp*

vapours cause a very high luminous intensities. In addition to fluoride, there are also other some flame materials. Different flame materials will produce different colours. Most of these colours do not appeal to eyes and at the same time they produce strain on them. The arc can be drawn out to one side with the help of a magnet. Though the arc is very efficient but owing to objection to its colours it has now been superseded by the electric discharge lamps. The luminous efficiency of such a lamp is 8 lumens per watt.

7.9.3. Magnetic-Arc Lamp

In such a lamp positive electrode made of copper and negative electrode made of magnetic oxide of iron are used. The arc is struck in the similar way as in case of carbon-arc lamp. Such lamps are rarely used.

7.10. INCANDESCENT LAMPS

For sometimes the arc lamps were used for general lighting purpose but as the carbon-arc lamps were complicated so these have been superseded by filament lamps for general lighting.

When an electric current is passed through a fine metallic wire, heat is produced and the temperature of the wire increases. At low temperatures the wire radiates heat energy, as the temperature of the wire increases due to heating it radiates heat as well as light energy. The higher the temperature of the wire, higher is the amount of light energy radiated. A black body when heated to 6,250°C emits the maximum energy in the visible spectrum range.

The incandescent or filament type lamp consists of a glass globe completely evacuated and a fine wire, known as *filament,* within it. The glass globe is evacuated to prevent the oxidization and convection currents of the filament and also to prevent the temperature being lowered by radiation.

The material, which can be used for the filaments of incandescent lamps, must possess the properties of high melting point, low vapour pressure, high resistivity, low temperature coefficient, ductility and sufficient mechanical strength to withstand vibrations during use.

The materials which can be used for the filaments in incandescent lamps are carbon, osmium, tantalum and tungsten. Though the carbon has a melting point of 3,500°C but its main drawback is that it starts vaporizing at very fast rate if it is operated beyond 1,800°C, moreover its temperature coefficient of resistance is negative, *i.e.*, its resistance decreases with the rise in temperature and, therefore, it takes more current from mains. Because of its low operating temperature its efficiency is quite low and is of the order of 3.5 lumens per watt. Osmium is very rare and expensive metal. The melting point of osmium is 2,600°C. Average efficiency of osmium lamp is of the order of 5 lumens per watt. The melting point of tantalum is only 2,800°C and, therefore, it cannot be operated at higher temperature to give more efficiency. The efficiency of tantalum filament lamps is about 5 lumens per watt. Nowadays tungsten is the most commonly used metal for filament due to its high melting point (3,400°C), high resistivity, low temperature coefficient (0.0051), low vapour pressure, being ductile and mechanically strong to withstand vibration during use. The hot resistance of tungsten filament is about 15 times the cold resistance and, therefore, there is a inrush of current at the switching instant. However, the inrush current is not beyond 15 times the normal current. Since the filament heats up immediately and resistance increases almost instantaneously, so the inrush current attains a maximum value in 0.003 second and falls to normal operating value in 0.2 second. The average efficiency of tungsten filament lamp is about 10 lumens per watt. The light spectrum of an incandescent lamp is continuous, *i.e.*, it contains all the colours, but contains relatively excess of red and yellow radiations and less of blue and violet radiations.

Aging Effects. The light output of an incandescent lamp decreases gradually. As a tungsten lamp ages, its light output decreases for two reasons. Evaporation of the filament tends to cause the bulb to blacken. Also, evaporation makes the filament slowly decrease in diameter, which means that the resistance of the filament increases. Therefore, an old filament draws less current and operates at a lower temperature, which reduces its light output. In turn the efficiency of the lamp (lumens output/watt input) also decreases with the passage of time. The current drawn and the power consumed by the filament decrease at the same rate as the lamp ages. However, the efficiency decreases about four times as fast, and the light output decreases approximately five times as fast. The total depreciation of light output is roughly 15% over the useful life range.

Effects of Voltage Variations. The operating characteristics of an incandescent lamp are materially affected by departure from its normal operating voltage. An increase of 5% in operating voltage increases the lumens output by 20% but shortens the life of the lamp by 50%. On the other hand, reduction of 5% in operating voltage causes a reduction of approximately 20% in lumens output but doubles the life of the lamp. The efficiency of a lamp (lumens/watt) increases with the increase in voltage owing to increase in temperature and is proportional to the square of the voltage. 1% change in applied voltage modifies lamp wattage by 1.5%, efficiency by 2%, lumens output by 3.5%. The normal life of a filament lamp is about 1,000 working hours. The variations in power consumption, lumens output, efficiency and life of incandescent lamps with the variation in voltage are shown in Fig. 7.29.

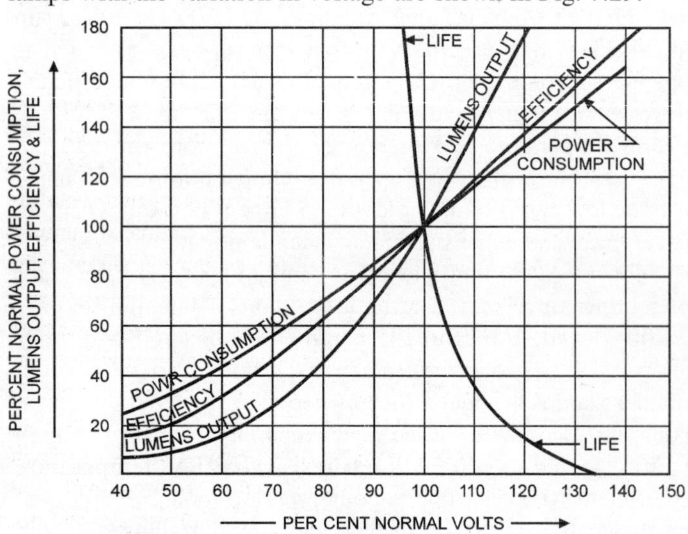

Fig. 7.29

Analytically these relationships are given as below :

Lumens output $\propto (V)^{3.55}$

Power consumption $\propto (V)^{1.55}$

Luminous efficiency $\propto (V)^2$

Life $\propto (V)^{-13}$ for vacuum lamps;

$\propto (V)^{-14}$ for gas-filled lamps

Advantages. 1. Operating power factor unity.

2. Direct operation on standard distribution voltage.
3. Availability in various shapes and shades.
4. Good radiation characteristic in the luminous range.
5. No effect of surrounding air temperature.

Filament Dimensions. Depending upon the voltage and wattage the diameter of a tungsten lamp filament may be as small as 10 microns (about one-sixth of diameter of human hair).

The influence of the diameter of the filament on the heating is found easily if it is assumed that the heat lost by convection is negligible compared with that lost by radiation.

The power absorbed by the filament

$$= I^2R = I^2\rho\frac{l}{a} = I^2\rho\frac{l}{\frac{\pi}{4}d^2} = \frac{4\rho l}{\pi d^2}I^2$$

where I is filament current in amperes, ρ is the resistivity of filament material at the operating temperature, l is length of the filament and d is diameter of the filament.

The power radiated by the filament

$$= eK \times \text{surface area} \times (T_1^4 - T_0^4) = eK \times \pi dl \times (T_1^4 - T_0^4)$$

where e is the emissivity of the surface, K is a constant, T_1 is the temperature of filament and T_0 is the temperature of the surrounding medium.

Since the power absorbed = Power radiated

$$\frac{4\rho l}{\pi d^2}I^2 = eK \times \pi\, dl \times (T_1^4 - T_0^4)$$

$$\text{or} \quad I = d^{3/2}\sqrt{\frac{eK\pi^2}{4\rho}(T_1^4 - T_0^4)}$$

For two filaments of same material operating at the same temperature the diameter d is proportional to $I^{2/3}$.

Gas Filled Lamps. A metal filament can work in an evacuated bulb up to 2,000°C without oxidation and if it is worked beyond this temperature it vaporizes quickly and blackens the lamp. For higher efficiency it is necessary to use working temperature more than 2,000°C keeping down the evaporation, which is possible by filling the bulb with an inert gas–argon with a small percentage of nitrogen. Nitrogen is added to reduce the possibility of arcing. Krypton is the best gas for this purpose but it is so expensive that it is used only in special purpose lamps, such as miners lamps. The tungsten filament can safely be burn at temperature of 2,400°C to 2,750°C according to the size of the lamp. However, due to presence of gas there is heat loss due to convection currents. This loss depends upon the surface area of filament. As such coiled coil filaments, which take much less space compared with coiled filaments are used with such lamps. A coiled coil filament is made by winding tungsten wire on a fine iron wire to form a spiral which is again wound on to a thick iron wire to form a coiled coil. (iron is later on dissolved out by acid). Helical filaments also have a slower rate of tungsten evaporation. This evaporation eventually causes bulb blackening because the tungsten vapour condenses as a black film on the inner surface of the bulb. In a gas filled lamp, the hot gas carries the tungsten vapour upward. Therefore, a black spot forms at the top of the bulb instead of spreading over the entire inner surface, as in a high-vacuum bulb. Chemicals called '*getters*' are often placed inside the bulb to capture tungsten vapour and thereby reduce the rate of blackening. A piece of wire mesh called a *collector grid* may also be attached to each lead-in-wire to attract the particles of tungsten vapour. Efficiency of coiled-coil lamp is about 12 lumens/watt.

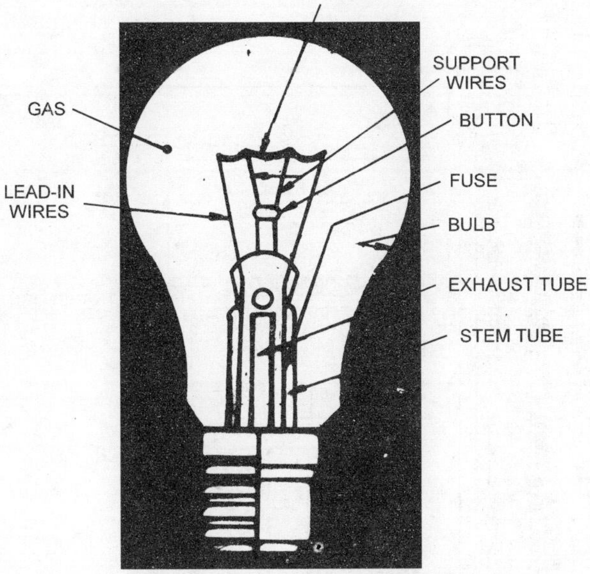

Fig. 7.30. *Gas Filled Lamp*

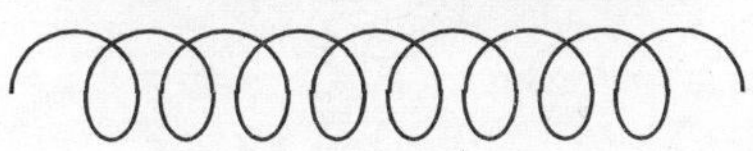

(a) Coiled Filament

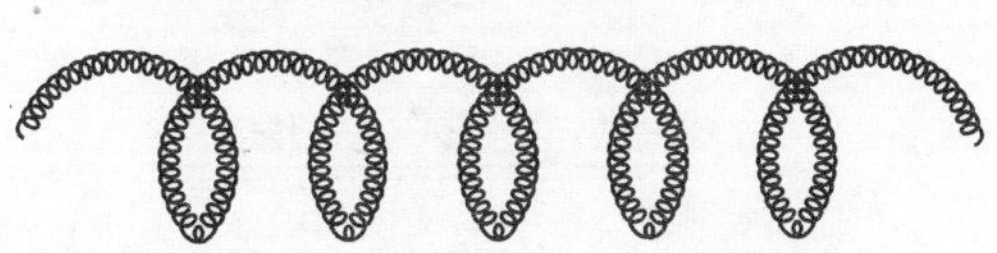

(b) Coiled Coil Filament

Fig. 7.31

For low wattage lamps, however, the heat loss due to introduction of gas is more than in medium wattage lamps, so for low wattage (up to 40 watts) vacuum type lamps are used.

Halogen Lamp. The halogen lamp is the latest member in the family of incandescent lamps. It possesses numerous advantages over the ordinary incandescent lamp. As already stated, the life and efficiency of an incandescent lamp fall off with use—partly due to slow evaporation of the filament and partly due to black deposit formed on the inside of the bulb. The addition of a small amount of halogen vapour to the filling gas restores part of the evaporated tungsten vapour back to the filament by means of a chemical reaction, *i.e.*, there is a sort of '*regenerative cycle*'.

Halogen lamps possess the following advantages :

(*i*) No blackening of lamp, hence no depreciation of lumens output.
(*ii*) High operating temperature with increased luminous efficiency varying from 22 lumens per watt to 33 lumens per watt.
(*iii*) Reduced dimensions of lamps–miniature size.
(*iv*) Long life—2,000 hours.
(*v*) Better colour rendition.

Halogen lamps, which are being manufactured in sizes up to 5 kW, are suitable for outdoor illumination of buildings, playing fields, large gardens, fountains, car parks, airport runways etc. and for lighting of public halls, factories, sport-halls, photo film and TV studios etc.

7.11. GASEOUS DISCHARGE LAMPS

Incandescent lamp suffers from two disadvantages—low efficiency and coloured light. The gaseous discharge lamps have been developed to overcome these drawbacks.

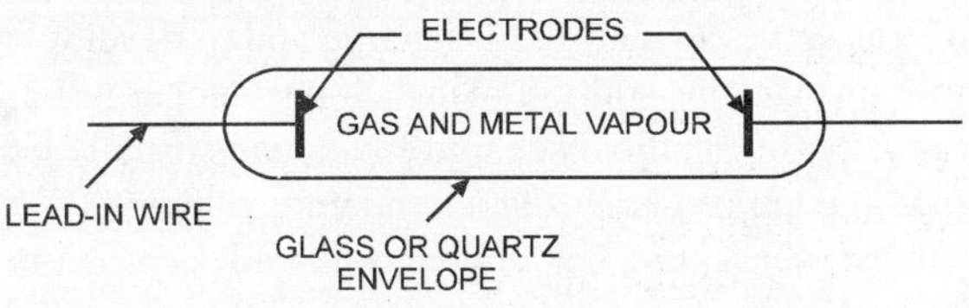

Fig. 7.32. *Gaseous Discharge Lamp*

The basic principle of a gaseous discharge lamp is illustrated in Fig. 7.32. Gases are normally poor conductors, specially at atmospheric and higher pressures, but application of suitable voltage, *called the ignition voltage*, across the two electrodes can result in a discharge through the gas, which is accompanied by electromagnetic radiation. The wavelength of this radiation depends upon the gas, its pressure, and the metal vapour used in lamp. Argon gas and sodium and mercury vapours are commonly employed in the manufacture of gaseous discharge lamps.

Once the ionisation has commenced in the gas, it has a tendency to increase continuously accompanied by a fall in the circuit resistance, *i.e.*, gaseous discharge lamp possesses a *negative resistance* characteristic. In order to limit the current to a safe value use of a choke or a ballast is made. The choke performs the dual functions of providing the ignition voltage initially, and limiting the current subsequently. Since due to use of choke the power factor becomes poor (0.3—0.4), therefore, in order to improve the power factor of the gaseous discharge lamp use of a condenser is made. The light spectrum obtained is, however, discontinuous (*i.e.*, it consists of one or more coloured lines). The colour of the light obtained depends upon the nature of the gas or vapour used. Discharge lamps are of two types:

(*i*) Those which give the light of the same colour as produced by the discharge through the gas or vapour such as sodium vapour, mercury vapour and neon gas lamps.

(*ii*) Those which use the phenomenon of fluorescence and are known as *fluorescent lamps*. In these lamps, the discharge through the vapour produces ultraviolet waves which cause fluorescence in certain materials called as *phosphors*. The inside of a fluorescent lamp is coated with a phosphor which absorbs invisible ultraviolet rays and radiates visible rays. Example is fluorescent mercury vapour tube.

The gaseous discharge lamps are, in general, considered superior to metal filament lamps. However, they suffer from the following drawbacks:

(*i*) High initial cost and poor power factor.

(*ii*) Starting is somewhat complicated requiring starters in some cases and transformer in others.

(*iii*) These take time to attain full brilliancy.

(*iv*) Ballasts are necessary for stabilising the current since such lamps have negative resistance characteristic.

(*v*) Light output fluctuates at twice the supply frequency. The flicker causes *stroboscopic* effect.

(*vi*) These lamps can be used only in particular position.

7.11.1. Sodium Vapour Discharge Lamp

Principally the sodium vapour discharge lamp consists of a bulb containing a small amount of metallic sodium, neon gas, and two sets of electrodes connected to a pin type base. The presence of neon gas serves to start the discharge and to develop enough heat to vaporize the sodium. Since long discharge paths are necessary, therefore, the discharge envelope is usually bent into U shape. The lamp operates at a temperature-like 300°C and in order to conserve the heat generated and assure the lamp operating at normal air temperatures the discharge envelope is enclosed in a special vacuum envelope designed for this purpose. The lamp must be operated horizontally, or nearly so, to keep the sodium well spread out along the tube, although some small lamps may be operated vertically, lamp cap up. Care should be taken in handling these lamps, particularly when replacing inner U-tube, for if it is broken and sodium comes in contact with moisture fire will result.

The sodium vapour lamp is only suitable for alternating current, and therefore, requires choke control. This requirement is met by operating the lamp from a stray field, step-up, tapped auto-transformer with an open-circuit secondary voltage of 470 to 480 volts. The uncorrected power factor is very low, about 0.3, and a capacitor must be used to improve the power factor to about 0.8.

When the lamp is not in operation, the sodium is usually in the form of solid deposited on the side walls of the tube, therefore, at first when it is connected across the supply mains the discharge takes place in the neon gas and gives red-orange glow. The metallic sodium gradually vaporizes and then ionizes, thereby producing the characteristic monochromatic yellow light, which makes objects appear as grey. The lamp will come up to its rated light output in approximately 15 minutes. It will restart immediately should the power supply be momentarily interrupted since the presence of vapour is quite low and the voltage is sufficient to restrike the arc.

The wiring diagram of sodium vapour lamp(s) is given in Fig. 7.33 (*c*).

The efficiency of a sodium vapour lamp under practical conditions is about 40—50 lumens/watt. The major application of this type of lamp is for highway and general outdoor lighting where colour discrimination is not required, such as street lighting, parks, rail yards, storage yards etc. Such lamps are manufactured in 45, 60, 85 and 140 watt ratings. The average life is about 3,000 hours and is not affected by voltage variations. At the end of this period the light output will be reduced by 15% due to aging.

The lamp fails to operate when

(*i*) the filament breaks or burns out,

(*ii*) the cathode stops to emit electrons,

(*iii*) the sodium particles may concentrate on one side of the tube,

(*iv*) the lamp tube is blackened owing to sodium vapour action on the glass, in which case the output will be reduced.

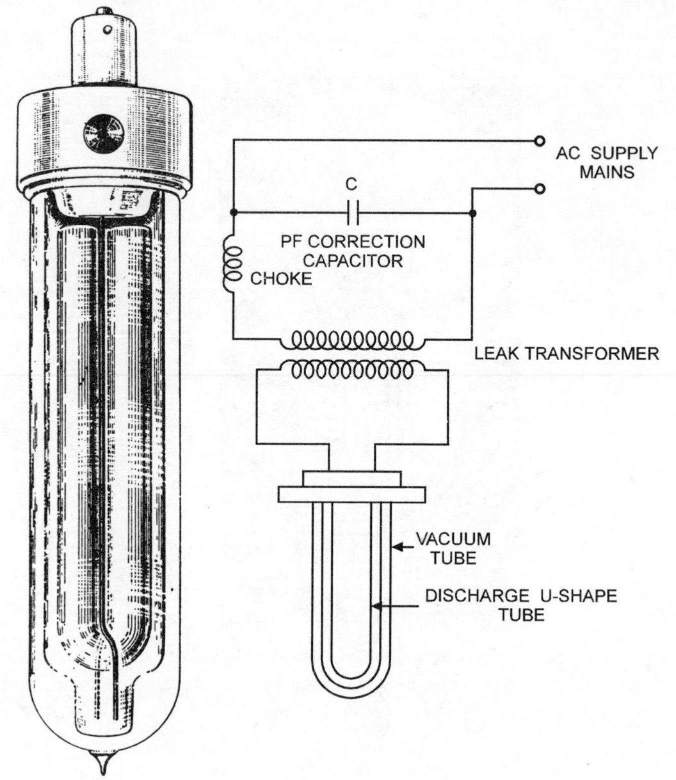

(*a*) (*b*)

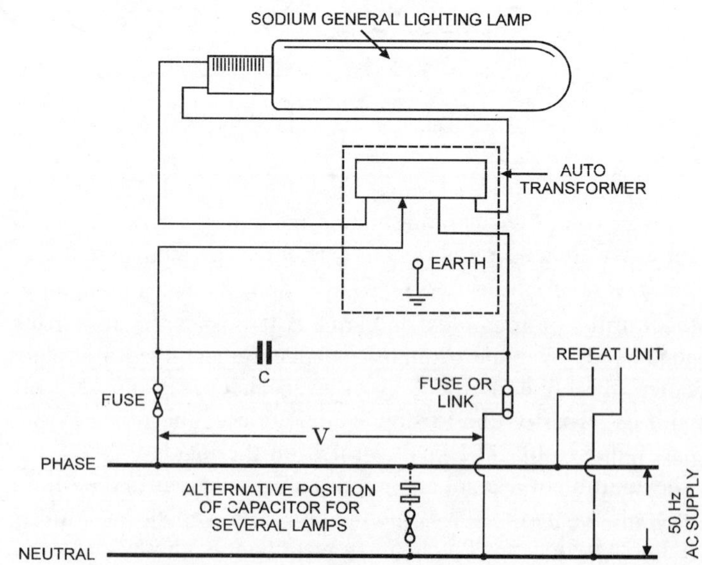

(*c*) *Wiring Diagram*

Fig. 7.33. ***Sodium Vapour Discharge Lamp***

7.11.2. High Pressure Mercury Vapour Discharge Lamp

The extensive use of the mercury vapour discharge lamp depends entirely upon the versatility of the mercury vapour as regards pressure, temperature, voltage and other characteristics, each change resulting in a lamp of different spectral quality and efficiency.

The mercury vapour discharge lamp is similar in construction to the sodium vapour discharge lamp. It consists of a discharge envelope enclosed in an outer bulb of ordinary glass. The discharge envelope may be of hard glass or quartz. The space between the bulb is partially or completely evacuated to prevent heat loss by convection from the inner bulb. The outer bulb absorbs harmful ultraviolet rays. The inner bulb contains argon and a certain quantity of mercury. In addition to two main electrodes a starting (auxiliary) electrode is connected through a high resistance (as shown in Fig. 7.34) is also provided. The main electrodes are made of tungsten wire in the shape of helix. In this case no separate heater is required for the cathode which is heated by the constant bombardment of the heavy mercury ions.

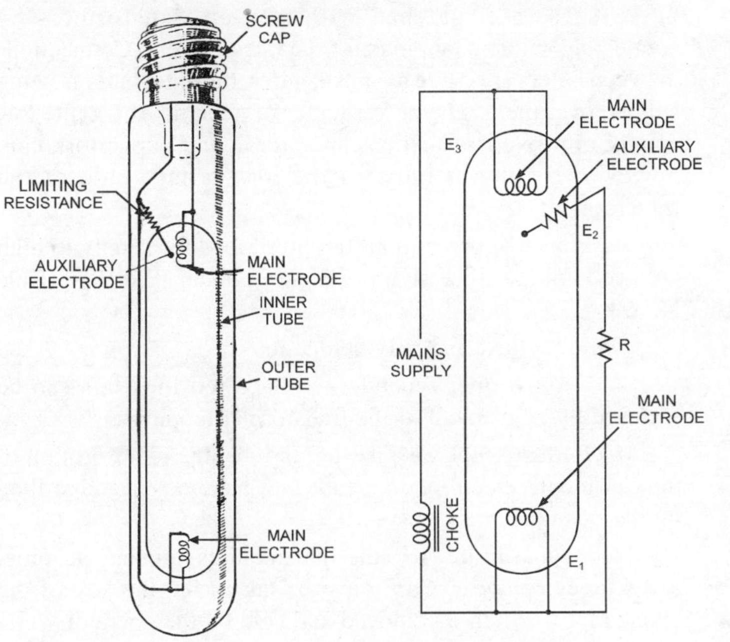

Fig. 7.34. ***High Pressure Mercury Vapour Discharge Lamp***

The lamp has to have auxiliary equipment for use with standard mains voltage, and the necessary connections are shown in Fig. 7.35. The choke is provided to limit the current to a safe value. This choke lowers the power factor, so a capacitor is connected across the circuit to improve the power factor. These lamps must be operated vertically, since if they are used horizontally convection will cause the discharge to touch the glass bulb, which will fail. Lamps which are intended to operate horizontally are fitted with a magnetic device which will hold the luminous column central.

Complete wiring diagram for high pressure mercury vapour discharge lamp is shown in Fig. 7.35.

When the supply is switched on, full mains voltage is applied between the auxiliary electrode and neighbouring main electrode; this breaks down the gap and a discharge through the argon gas takes place. This enables the main discharge to commence. As the lamp warms up, mercury is vaporised, increasing the vapour pressure and the luminous column becomes brighter and narrower. The lamp requires 4 or 5 minutes to attain full brilliancy. If the supply is interrupted, the lamp must cool down and the vapour pressure be reduced before it will start. This takes 3 or 4 minutes. The temperature of operation inside the inner bulb is about 600°C. It gives greenish blue colour light, which causes colour distortion. The efficiency is about 30—40 lumens/watt. These lamps are manufactured in 250 and 400 watt ratings for use on 200—250 volts ac supply mains. The pressure of vapour in these lamps is 2-3 atmosphere. Lamps of this type are used for general industrial lighting, railway yards, ports, work area, shopping centres etc., where greenish blue colour light is not objectionable.

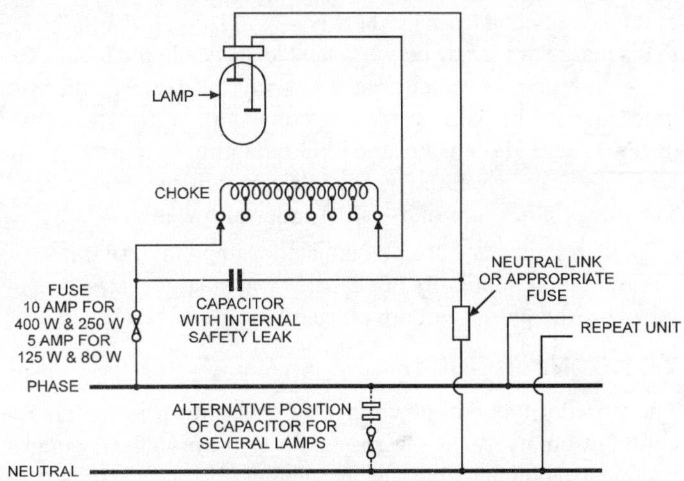

Fig. 7.35. ***Complete Wiring Diagram of HPMV Lamp***

The lamp described above is MA type. Another type, which is manufactured in 300 and 500 watt rating for use on ac as well as dc supply mains, is MAT type. This is similar to MA type except that choke is not used as a ballast. Space between two tubes instead of being evacuated consists of a tungsten filament in series with a discharge tube which acts as a ballast. When the supply is switched on, it operates as a filament lamp, its full output being given by the outer tube. At the same time the discharge or inner tube begins warming up and at a particular temperature a thermal switch operates cutting a part of the filament and thereby increasing the voltage across the discharge tube. The filament contributes a considerable portion of red rays. The combination of the rays from the filament and the blue radiations from the discharge tube produce a useful colour. As the filament acts as a resistance, the overall power factor of the lamp is about 0.95 and therefore, capacitor is not required.

Lower wattage lamps, such as 80 and 125 watts, are manufactured in a different design and using high vapour pressure of about 5-10 atmosphere. These are known as MB types. These operate in a manner similar to MA type except that resistance in series with starting electrode is large and outer bulb is of quartz, (not of ordinary glass) in order to withstand high temperature so that these lamps can be used in any position.

7.11.3. Mercury Iodide Lamps

These lamps are similar in construction to high pressure mercury vapour lamps but in addition to mercury, a number of iodides are added which fill the gaps in the light spectrum, and thus, improve the colour characteristic of the light. Their efficiency is also higher (75—90 lumens/watt). A separate ignition device, in addition to the choke, is required for the mercury iodide lamp. Such lamps are suitable for application in the fields of floodlighting, industrial lighting and public lighting.

7.11.4. Neon Lamp

It is a cold cathode lamp and consists of a glass bulb filled with neon gas with a small percentage of helium. These lamps give orange pink coloured light. Electrodes are of pure iron and are spaced only few mm apart so that lamps can be made for voltages as low as 110 volts ac or 150 dc. For use on ac the electrodes are of equal size. On dc the gas glows near the negative electrode, therefore, negative electrode is made larger in size. The efficiency of neon lamp lies between 15-40 lumens/watt. Owing to discharge of the gas between the electrodes in the form of an arc, it may cause the current drawn by the lamp to increase indefinitely. This is prevented by connecting a high resistance of few thousand ohms in series and mounting it in the cap. The lamp of this type is of the size of an ordinary incandescent lamp. The power consumption is of the order of 5 watts.

Neon lamps are used as indicator lamps, night lamps, for determination of polarity of dc mains and in larger sizes as neon tubes for the purpose of advertising.

7.11.5. Neon Tubes

The popularity of high voltage neon lighting arose almost entirely from its use in advertising, for signs, or in the decorative treatment of buildings, but later the lighting field became important. The neon tube, which is used in varying lengths up to about 9 metres, may be bent into almost any desired shape during manufacture. It consists of a length of glass tubing containing two electrodes, normally cylindrical in shape, of iron, steel, or copper.

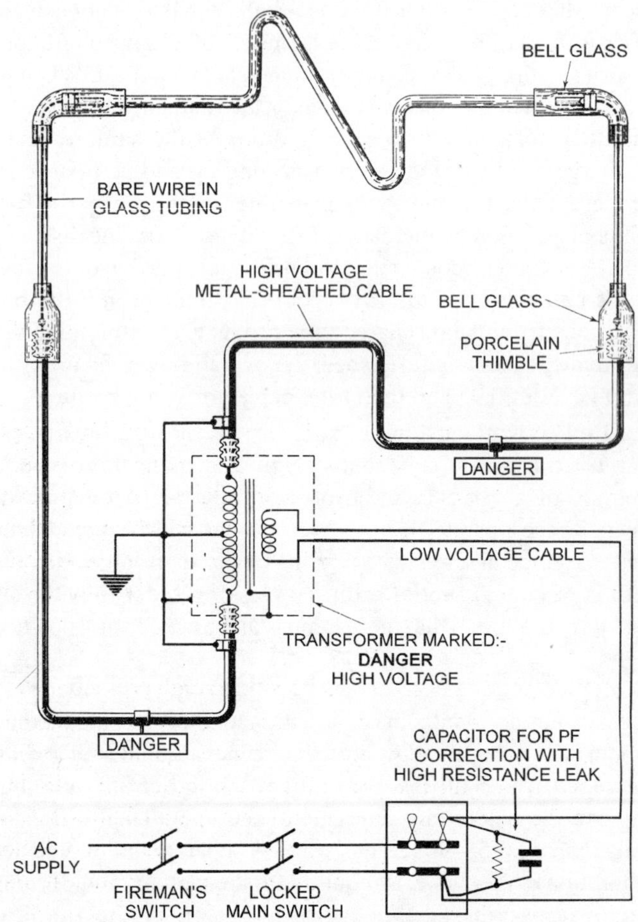

Fig. 7.36. ***Neon Tube Installation***

The true neon tube contains neon, but the term is now used also for tubes with fillings of other rare gases. By varying the composition of glass and adding different substances to neon gas different colours such as orange, red, yellow, green etc. are obtained. The diameters of the tube vary, and common sizes of 10, 15, 20 and 30 mm carry currents of 25, 35, 60 and 150 mA respectively. Voltage required may vary from 300 V to 1,000 V per metre of tube length and for starting the discharge a striking voltage, about 1½ times this value is required. Such voltage is obtained by making use of a step-up transformer having a high leakage reactance so that it gives a drooping characteristic. The usual operating voltage is 6,000 volts.

The tubes are mounted either on a wooden frame or a metal base. These are matched with step-up transformers by connecting suitable tappings for the rated current. Connections between letters are made by nickel wires, the glass tubings being slipped over them. The power factor of neon tubes is quite low and is improved by using capacitors. The capacitors can, however, be placed only on the low voltage side of the transformer.

A number of precautions are prescribed for supply to high voltage luminous tube sign installations under IER 1956 (Rule No. 71).

Faults. Flickering may be due to;

(*i*) Transformer secondary voltage too low; this can be rectified by adjustment of the transformer tappings.

(*ii*) Reduction of gas pressure in tube due to absorption of the gas into the electrodes, the tube may be removed and refilled by the manufacturers.

Neon tubes installed in the open requires frequent cleaning, say 4 times per year. Care must be taken that the key of the opened lock switch is removed and held by the operator while work is being done.

7.11.6. Fluorescent Tubes

Fluorescent lighting has a great advantage over other light sources in many applications. The tubes can be obtained in a variety of lengths, with illumination in a variety of colours. It is possible to achieve quite high lighting intensities without excessive temperature rise and, owing to the nature of light sources, the danger of glare is minimised. It must not, however, be thought that fluorescent lamps can be used indiscriminately without giving careful consideration to the type of reflector to be employed or to the correct positioning of the light source. An exposed tube in the line of vision is usually as bad as a tungsten lamp would be. The efficiency of fluorescent lamp is about 40 lumens per watt, about three times the efficiency of an equivalent tungsten filament lamp. The fluorescent tube consists of a glass tube 25 mm in diameter and 0.38 m–1.52 m in length. The inside surface of the tube is coated with a thin layer of fluorescent material in the form of a powder.

The coating materials used depend upon the colour effect desired and may consist of zinc silicate, cadmium, silicate or calcium tungstate. These organic chemicals are known as *phosphors* which transforms short-wave invisible radiations into visible light. By mixing the various powders light of any desired colour including day light can be obtained. The tube contains small quantity of argon gas at a pressure of 2.5 mm of mercury and one or two drops of mercury. It is provided with two electrodes coated with electron emissive material. A starting switch is provided in the circuit, which puts the electrodes directly across the supply mains at the time of starting, so that electrodes may get heated and emit sufficient electrons. A stabilising choke is connected in series with it, which acts as a ballast in running condition and provides a voltage impulse for starting. A capacitor is connected across the circuit to improve the power factor.

Fluorescent tubes are available in the following sizes :

Length	Wattage
38 cm	14 W
46 cm	15 W
61 cm	20 W
100 cm	25 W
122 cm	40 W
152 cm	65 W
152 cm	80 W

The fluorescent lamp in common with most electrical devices may cause a certain amount of radio interference. This interference may be caused due to one of the following factors :

1. Direct radiation from the bulb to the antenna. This effect diminishes rapidly as the radio is separated from the lamp. Thus, for example, at a radius of 2.8 metres interference from this cause is negligible.

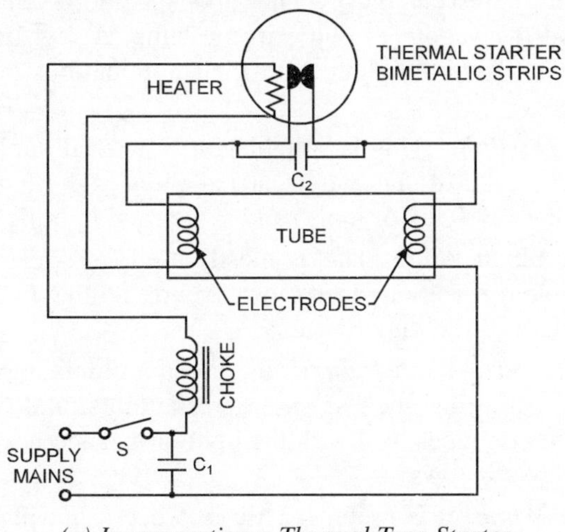

(a) Incorporating a Thermal Type Starter

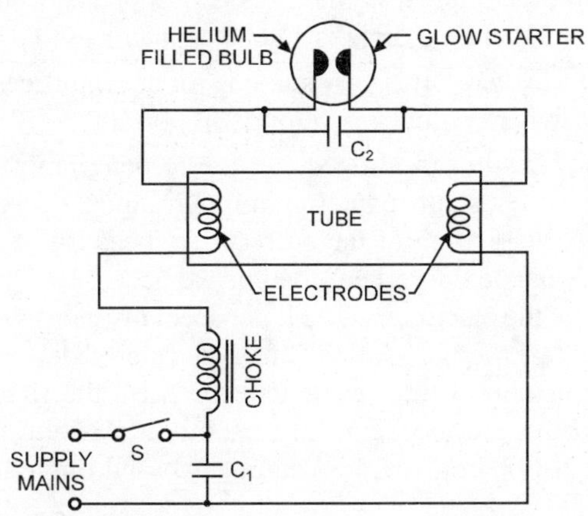

(b) Incorporating a Glow Type Starter

Fig. 7.37. ***Connection Diagram For a Fluorescent Tube***

2. Line radiation from the electric supply line to the antenna.
3. Line feedback from the lamp through the line to the radio.

The radio interference effect may be reduced to a minimum by connecting a small capacitor (0.05 μF) across starter terminals, as shown in Figs. 7.37 (*a*) and 7.37 (*b*).

The starting switches are of two types, namely the thermal type and the glow type.

The connections of a fluorescent tube incorporating a thermal type starter are shown in Fig. 7.37 (*a*). The thermal starter is a current operated device and consists of two metallic strips and a heater coil. The bimetallic strips are in contact with each other when the lamp is not in operation. When the supply is switched on, the two electrodes get connected in series through the thermal switch, the relatively large current rising them to incandescence. The current also flows through the heater element as a result of which bimetallic strips break contact. This causes interruption in the current flowing through the circuit, which further results in a high voltage surge across the electrodes of the tube, which is enough to strike the arc between the electrodes. This arc is then maintained by the normal lamp voltage. Bombardment of the electrode's surface by the positive mercury ions maintains their temperature so that they continue to emit electrons. Thus the tube is put in operation. The thermal switch is now generally obsolete because of its more complicated construction, greater cost and greater power loss.

The connections of a fluorescent tube incorporating a glow type starter are shown in Fig. 7.37 (*b*). The glow type starter is a voltage operated device and consists of two bimetallic electrodes enclosed in a glass bulb filled with a mixture of helium and hydrogen. Normally the contacts are open. When the supply is switched on, the potential across bimetallic electrodes causes a small glow discharge at a small current not enough to heat up the tube electrodes (filament). This discharge is enough, however, to heat the bimetallic strips of the switch causing them to bend and make contact. The result is a large current through the electrodes, their temperature being raised to incandescence and the gas in the immediate neighbourhood is ionised. After one or two seconds the bimetallic strips cool down and the contacts open. This opening of contacts in series with the choke causes a momentary high voltage, which is sufficient to start the discharge in the main tube. The starter ceases to glow as the voltage is now too low. A small capacitor is placed inside the starter to suppress arcing and radio interference.

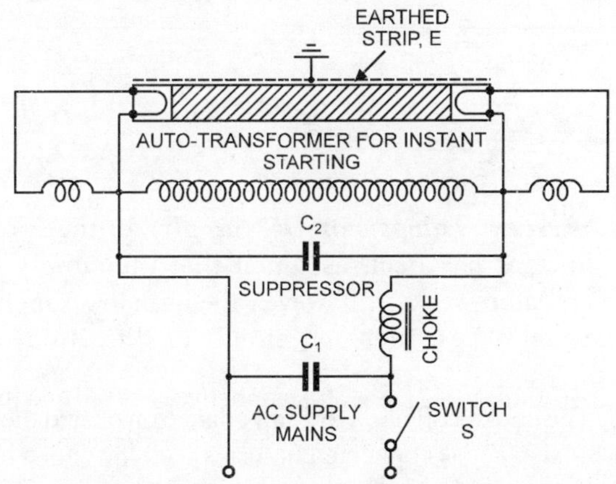

Fig. 7.38. ***Circuit For Instant Starting of Fluorescent Tube***

Nowadays starterless circuits, as instant start, rapid start etc. for such tubes are also employed (Fig. 7.38). The primary winding of an auto-transformer is connected in parallel with the tube and receives practically full mains voltage. The filaments are provided with current from secondary tappings. When the lamp has started the transformer receives the normal tube voltage and the filament currents are correspondingly reduced.

Stroboscopic Effect. At the usual alternating current supply frequency of 50 Hz, a discharge lamp will be extinguished 100 times per second. Although this effect is seldom noticeable in normal conditions. It is possible that this may, in some instances, give rise to a stroboscopic effect, that is moving objects such as rotating parts of machinery, illuminated by this light, may appear to be moving in disunity or jerks, or rotating more slowly than their actual speed.

In actual applications, however, where this effect might cause annoyance, it can be practically eliminated in a three lamp unit by connecting each lamp on a separate phase of a 3-phase system and it can be greatly reduced in a two lamp unit by the use of a two lamp control unit, which employs a condenser in the ballast of one of the lamps. (Fig. 7.39). The current through the lamps is almost 90° out of phase and under these conditions light output of one of the lamps is at a maximum. This method has an additional advantage of giving an overall power factor of nearly unity for the unity of two lamps. In this arrangement one circuit remains at a low power factor at about 0.5 lagging, while the other circuit incorporating a series capacitor C remains at a power factor of about 0.5 leading.

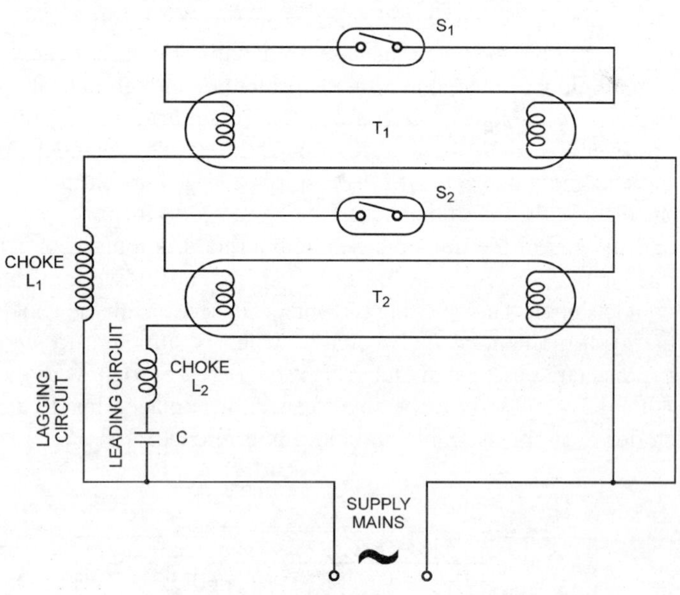

Fig. 7.39

Fluorescent Lamps For DC Supply. In the foregoing discussions, it has been assumed that the supply to the fluorescent lamps is ac. If however, the available supply is dc some special accessories and circuit modification will be required.

(*i*) The choke coil has a low impedance on dc and therefore, a ballast resistance is connected in series with the choke in order to limit the current.

(*ii*) On systems below 220 V, starting becomes less certain on dc. Only thermal type starters should be used.

(*iii*) The positive end becomes relatively dark on account of the tendency of the mercury vapour to migrate towards the negative end of the tube. In order to overcome this defect a reversing switch is included in the circuit between the supply and the fitting, as shown in Fig. 7.40.

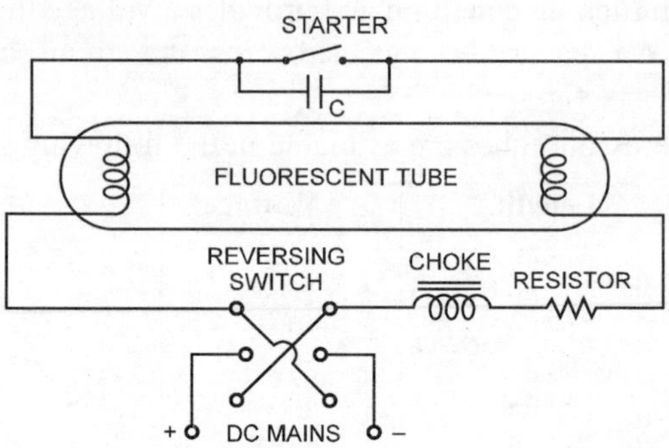

Fig. 7.40. ***Operation of Fluorescent Tube on DC***

In dc operation of fluorescent tube there is no problem of power factor correction and no stroboscopic effect. Its disadvantages are—low efficiency due to power loss in ballast series resistance, increased cost of the ballast resistance and reversing switch and less life of the tube (about 80% of that with ac operation).

Colour of Fluorescent Light. As already mentioned the colour of fluorescent light depends upon the fluorescent powder used and the vapour pressure in the lamp. A few important variations of white colour with their applications are given below.

(i) Daylight. This is suitable for industrial applications where checking of material is carried out at high illumination levels (say 1,000 lux or more) and for normal lighting service.

(ii) Warm-White. This is mostly used for street lighting where colour rendition, due to low level of lighting (say 10 lux or less), is of little importance.

(iii) Warm-White Deluxe. It is used for illuminating rooms of large gathering such as shops, restaurants, and dwelling houses. It combines well with the light of incandescent lamps, as in show-windows etc.

(iv) White. It is useful for use in offices, drawing offices, schools and factories. It harmonizes exceptionally well with the daylight, and therefore, light is a good supplement to daylight.

White-Deluxe. It is very suitable for use in offices, schools and shops where colour rendition is important.

Useful Lamp Life. In general, fluorescent lamps lose their usefulness because of reduction in light output before they fail to operate. Darkening of the bulb occurs because of the effect of mercury on the fluorescent coating and because of the material given off by the electrodes. The latter specially causes darkening at the ends of the bulb late in life. The rate of depreciation in light output diminishes throughout the life ; the first hundred hours produce approximately as much darkening as the following 1,000 hours. Rated output is based on conditions at 100 hours.

Frequent starting of lamps may take more life out of the electrodes than long hours of burning because momentarily there is a higher than normal voltage drop at the electrodes which causes the active material to sputter or evaporate off. If a lamp be started once a minute, for example, the hours of burning will be shorter than normal, but if it be switched on and burned continuously, its life will be longer than normal. When the active material on the electrodes is nearly exhausted, the voltage required for starting will rise and may equal or exceed the available supply. This may occur after the lamp has been started thousands of times or burned beyond its rated life. Sometimes the end of life is indicated by the lamp flashing momentarily and then going out.

The normal life of a fluorescent lamp is 7,500 hours. The average life is for three burning hours per switching operation. The actual life may vary from 5,000 to 10,000 hours depending upon the operating conditions. Light output is reduced by 15-20% after 4,000 hours of operation and it is, therefore, a good practice to replace the fluorescent lamp after 4,000-5,000 hours burning on economical grounds.

Performance Curves. Performance curves are shown in Fig. 7.41.

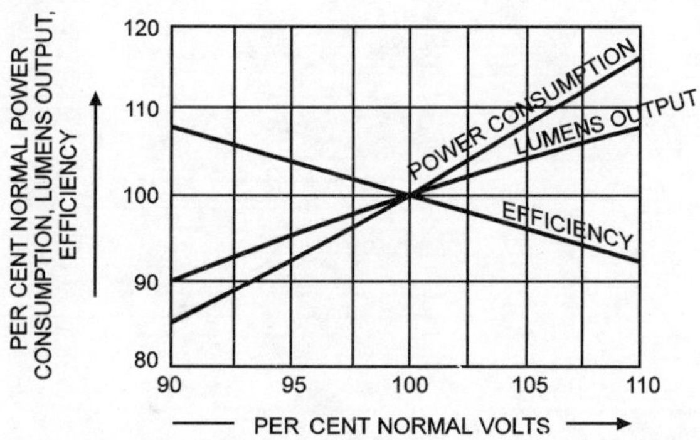

Fig. 7.41. ***Performance Curves of Fluorescent Lamps***

The effect of voltage variation in the case of fluorescent lamps is less marked as compared to the incandescent lamps. However, their life and performance are adversely affected both by low and high voltage. With increased voltage there is a greater heating of electrodes and they lose emissive material by evaporation. With reduced voltage, the current reduces causing sputtering at the electrodes shortening their life.

The best performance of fluorescent lamps is obtained at 20-25°C operating temperature. It decreases rapidly when a lamp is operated at a lower temperature or is exposed to cold wind drafts. The fluorescent lamps may be enclosed in outer envelops for their protection from cold weather in exposed locations. For operation at high temperature, the fittings with provision for air circulation should be employed.

Operating Instructions. (*i*) A fluorescent lamp should start up with little blinking. Blinking indicates a defective starter, low voltage or a defective tube. Both the choke and the starter will get damaged due to overheating, if this condition is allowed to persist. The tube or the starter must be replaced immediately.

(*ii*) Correct voltage must be ensured.

(*iii*) The starter should be replaced every time a tube is changed. A bad starter shortens the life of the tube.

(*iv*) Frequent switching operations should be avoided.

Merits and Demerits. As already stated the fluorescent lamps have got merits of high luminous efficiency, long life, low running cost, low glare level and less heat output. The demerits are stroboscopic effect, small wattage requiring large number of fittings and magnetic hum associated with the choke causing disturbance. The problem of noise can be solved to some extent by mounting chokes on resilient pads.

Trouble-Shooting. Various faults occurring in fluorescent tubes, their possible causes and remedies are given below :

Symptoms	Possible Causes	Remedies
1. Blinking on and off.	1. Low circuit voltage. 2. Loose contact. 3. Defective starter. 4. Defective tube. 5. Wrong connections.	Change ballast tapping if required. Check all connections. Replace starter. Replace tube. Check all connections.
2. End of a tube remain lighted.	1. Short-circuited starter.	Replace starter.
3. Failure of lamp to light up.	1. Defective tube. 2. Defective starter. 3. Defective choke. 4. Defective and loose holders.	Replace tube. Replace starter. Replace choke. Turn tube through 90°. Try another starter and check connections.
4. Blackening of the end (early life) (normal life)	1. Too low or too high voltage. 2. Mercury deposits at the ends.	Adjust ballast tapping. It will evaporate as the lamp operates.
5. Burn out electrodes	1. Control unit in circuit or choke short circuited.	Test with a new choke.
6. Dark streaks along lamp	1. Mercury globules.	Rotate tube through 180°.
7. Snaking spiralling (with glow type starter)	1. Insufficient heating due to either starter opening quickly or wrong tappings made on the ballast.	Replace starter or ballast. In case of new lamp, it cures after few days.

7.11.7. Compact Fluorescent Lamps (CFLs)

A compact fluorescent lamp (CFL), also known as a *compact fluorescent light* or energy saving light, is a type of fluorescent lamp which combines the energy efficiency of fluorescent lighting with the convenience and popularity of incandescent fixtures. CFLs can replace incandescents that are roughly 3-4 times their wattage, saving up to 75% of the initial light energy. Although CFLs costs 3-10 times more than comparable incandescent lamps, they last about 10 times as long (10,000 hours).

CFLs are most cost effective in areas where lights are ON *for long periods of time.*

Because of the potential of reducing electric consumption and pollution, some organizations have encouraged the adoption of CFLs. Some electric utilities and local governments have subsidized CFLs as a means of reducing electric demand. In India also almost all the states are encouraging the adoption of CFLs in place of incandescent lamps and are providing incentives or even distributing CFLs to people free of cost to save electricity.

CFLs radiate a light spectrum different from that of incandescent lamps. Improved phosphor formulations have improved the perceived colour of the light emitted by CFLs such that some sources rate the best "soft white" CFLs as subjectively similar in colour to standard incandescent lamps.

Like all fluorescent lamps, CFLs contain mercury, which makes their disposal complicated.

Operation. CFLs work much like standard fluorescent lamps. They consist of two components : a gas-filled tube, and a magnetic or electronic ballast. The gas in the tube glows with ultraviolet light when switched ON and electric current from the ballast flows through it. This in turn excites a white phosphor coating on the inside of the tube, which emits visible light throughout the tube surface.

CFLs with magnetic ballasts flicker slightly at start. They are also heavier than those with electronic ballasts. This may make them too heavy for some light fixtures. Electronic ballasts are more expensive, but light immediately (especially at low temperature). They are also more efficient than magnetic ballasts. The tubes will last about 10,000 hours and the ballast about 50,000 hours. Most currently available CFLs have electronic ballasts.

Electronic ballasts contain a small circuit board with rectifiers, a filter capacitor and usually two switching transistors connected as a high-frequency resonant series dc to ac inverter. The resulting high frequency (about 40 kHz or higher) is applied to the lamp tube. Since the resonant converter tends to stabilize lamp current and so the light produced over a range of input voltages, standard CFLs do not respond well in dimming applications and special lamps are required for dimming services.

CFLs are designed to operate within a specified temperature range. Temepratures below the range cause reduced output. Most of the CFLs are for indoor use, but there are models available for outdoor use. Outdoor CFLs need installation in enclosed fixtures so as to minimize the adverse effects of low temperatures.

Types of CFLs. CFLs are available in a variety of styles or shapes. Some have two, four, or six tubes while others have circular or spiral-shaped tubes. The size or total surface area of the tube(s) determines how much light it produces.

Some CFLs have the tubes and ballast permanently connected. Other CFLs have separate tubes and ballasts which facilitates in replacement of tubes without changing the ballast. They are also of types enclosed in a glass tube. These look somewhat similar to conventional incandescent lamps, except they are larger.

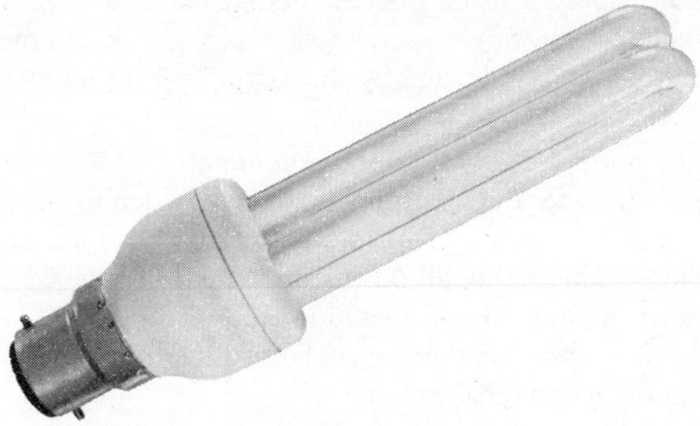

(a) The Tubular Type Compact Fluorescent Lamp

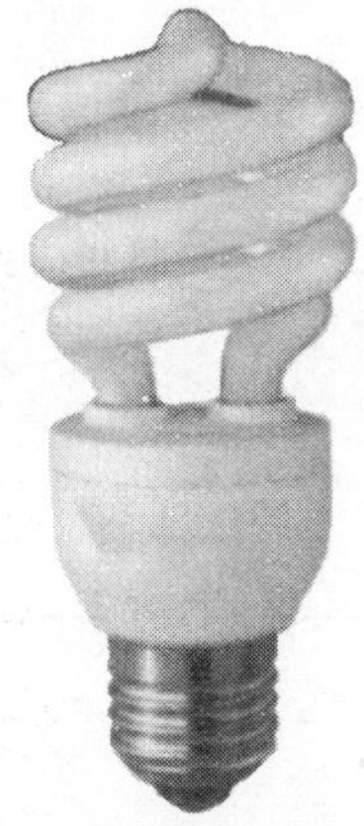

(b) Helical Integrated CFL

Fig. 7.42

CFLs are of two types : integrated and non-integrated lamps.

Integrated lamps combine a tube, an electronic ballast and either an Edison screw or a bayonet in a single unit. The incandescent lamps can be easily replaced with CFLs. Integrated CFLs work well in many standard incandescent light fixtures, reducing the cost of conversion.

Non-integrated CFLs have the ballast permanently installed in the luminaire, and only lamp is usually replaced at the end of life. Since the ballasts are placed in the light fixture, they are larger and last longer in comparison to the integrated ones. Non-integrated CFL housings can be both more expensive and sophisticated. They have two types of tubes : bi-pin tubes and quad-pin tubes. Bi-pin tubes are designed for conventional ballasts, and quad-pin tubes are designed for electronic ballasts or conventional ballasts with external starters. A bi-pin tube contains an integrated starter which obviates the need for external heating pins but causes incompatibility with electronic ballasts.

CFLs are produced for operation on both dc and ac supply. DC CFLs are popular for use in recreational vehicles and off-the-grid housing. CFLs can also be operated with solar powered street lights, using solar panels located on the top or sides of a pole and light fixtures that are specially wired to use the lamps.

Advantages of CFLs Over Incandescent Lamps

1. *Life Span.* As already mentioned, the average rated life of a CFL is about 10 times that of incandescent lamps.

The lifetime of any lamp depends on various factors, such as operating voltage, exposure to voltage spikes, mechanical shock, frequency of cycling on and off, lamp orientation, ambient temperature, manufacturing quality etc. The life of a CFL is significantly reduced if it is turned on and off frequently. In the case of a 5 minute on/off cycle the life span of a CFL can be reduced to "close to that of incandescent lamp".

CFLs provide less light in their lives than when they are new. The light output decay is exponential, with the fastest losses being soon after the lamp is first used. By the end of their lives, they can be expected to produce 70-80% of their original light output.

2. *Energy Efficiency.* For a given light output, CFLs consume 20 to 33 per cent of the power of equivalent incandescent lamps. Electrical power equivalents for different lamps are indicated below.

Electrical Power Consumption in Watts		Minimum Light Output in Lumens
CFLs	Incandescent Lamps	
8-9	40	450
9-15	60	800
15-20	75	1,100
20-25	100	1,600
25-45	150	2,600

3. *Heating and Cooling.* If in a building, indoor incandescent lamps are replaced by CFLs, the heat produced due to lighting will be significantly reduced. In warm climates office/ industrial buildings where air-conditioning is usually requried, CFLs would reduce the load on the cooling system resulting in saving in electrical consumption, in addition to the energy efficiency savings of using CFLs instead of incandescent lamps. However, in cooler climates in which buildings need heating, the heating system would need more electrical energy. While the CFLs are still saving electricity, total greenhouse gas emissions may increase in certain scenarios, such as the operation of natural gas furnace to replace unintended heating from CFLs running on low-GHG electricity. Thus overall energy cost saving depends on the climate, increased heating demand offsets some of the lighting energy saved.

4. *Efficiency.* The typical luminous efficiency of CFLs is 60 to 72 lumens/watt, and that of normal domestic incandescent lamps is 13 to 18 lumens/watt. Compared to a theoretical 100% efficient lamp, these figures are equivalent to lighting efficiency ranges of 9 to 11% for CFLs (60/680 to 72/680), and 1.9 to 2.6% for incandescent (13/680 to 18/680).

5. *Embodied Energy.* No doubt, CFLs need more energy in manufacturing than the incandescent lamp but this embodied energy is offset by their longer life and small energy consumption than equivalent incandescent lamps.

6. *Cost.* While the purchase price of an integrated CFL is typically 3 to 10 times higher than that of an equivalent incandescent lamp, the extended lifetime and lower energy consumption will more than compensate for the higher initial cost. CFLs are extremely cost-effective in commercial buildings when used to replace incandescent lamps.

7. *Starting Time.* Incandescent lamps attain full brilliancy a fraction of second after being switched on, CFLs turn on within a second, but many still take time to warm up to full brightness. The light colour may be slightly different immediately after being switched on. Some CFLs are marketed as "instant on" and have no noticeable warm-up period, but others can take up to a minute to attain full brightness, or longer in very cold temperatures. Some that make use of a mercury amalgam can take up to three minutes to attain full brightness. This and shorter life of CFLs when turned on and off for short periods may make CFLs less suitable for applications like motion-activated lighting.

Disadvantages of CFLs

1. *Health Issues.* If individuals are exposed to the light produced by some single-envelope CFLs for long periods of time at distances of less than 20 cm, it could lead to ultraviolet exposures approaching the current workplace limit set to protect workers from skin and retinal damage. The ultraviolet radiation received from CFLs is too small to contribute to skin cancer and the use of double-envelope CFL lamps "largely or entirely" mitigates any other risks.

2. *Environmental Issues.* CFLs, like all fluorescent lamps, contain small amounts of mercury as vapour inside the glass tubing. Because mercury is poisonous, even these small amounts are a concern for landfills and waste incinerators where the mercury from lamps may be released and contribute to air and water pollution. Health and environmental concerns about mercury have promoted many jursdictions to require used lamps to be properly disposed or recycled rather than being included in the general waste stream sent to land fills. So disposal of used CFLs required proper care.

7.12. COMPARISON BETWEEN TUNGSTEN FILAMENT LAMPS AND FLUORESCENT TUBES

	Tungsten Filament Lamps	Fluorescent Tubes
1.	Voltage fluctuation has comparatively more effect on the light output.	Voltage fluctuation has comparatively low effect on light output as the variations in voltage are absorbed in the choke.
2.	Luminous efficiency increases with the increase in voltage of the lamp.	Luminous efficiency increases with the increase in wattage and increase in length of tube.
3.	It gives light close to natural light, therefore, objects are properly seen.	It does not give light close to natural light, therefore, colour rendering is defective.
4.	Luminous efficiency of coloured filament lamps is poor because coloured glasses are used for this purpose.	Different colour lights can be obtained by using different composition of fluorescent powder. Hence efficiency is high and better colours are obtained.

	Tungsten Filament Lamps	Fluorescent Tubes
5.	Due to comparatively high working temperature heat radiations are also present.	Due to low working temperature heat radiation is low.
6.	Its brightness is more.	Its brightness is less.
7.	With the time light output is reduced.	With the time light output is gradually reduced.
8.	No stroboscopic effect.	It has objectionable stroboscopic effect.
9.	Though the life of filament lamps varies with the working voltage, however, its normal life is 1,000 working hours.	Life of fluorescent tubes is not affected so much by variations in voltage but it depends on the frequency of starting. The life of the tube is about 7,500 working hours.
10	The initial cost per lamp is quite low.	The initial cost per tube is more.
11	For same lumens output more lamps are required and wiring cost is more. Life of lamp is also low. Hence overall cost of maintenance is more.	For same lumens output, lesser number of tubes is required, the wiring cost is low. Since its life is comparatively more, replacement cost is low. Hence overall maintenance cost is low.

Example 7.20. Two 220 V lamps, one of 60 W and the other of 75 W are connected in series across a 440 V supply. Calculate the pd across each lamp, neglecting any variation in resistance. Assuming the candle power to be proportional to the fourth power of the voltage, calculate the candle power of each lamp under this condition as a percentage of its value under normal operation at 220 V.

Solution : Resistance of 60 watt, 220 V lamp,

$$R_1 = \frac{V^2}{W_1} = \frac{(220)^2}{60} = 806.67\ \Omega$$

Resistance of 75 watt, 220 V lamp,

$$R_2 = \frac{V^2}{W_2} = \frac{(220)^2}{75} = 645.33\ \Omega$$

Current flowing through the lamps when connected in series across 440 V,

$$I = \frac{440}{806.67 + 645.33} = 0.303\ \text{A}$$

PD across 60 watt lamp, $V_1 = IR_1 = 0.303 \times 806.67 = 245$ V **Ans.**

PD across 75 watt lamp, $V_2 = IR_2 = 0.303 \times 645.33 = 195$ V **Ans.**

Normal operating voltage, V = 220 V

Candle power of 60 watt lamp under this condition

$$= \left(\frac{V_1}{V}\right)^4 \times \text{normal candle power}$$

$$= \left(\frac{245}{220}\right)^4 \times 100 \text{ per cent of candle power under normal operation at 220 V}$$

= 154% of candle power under normal operation at 220 V **Ans.**

Similarly candle power of 75 watt lamp under this condition

$$= \left(\frac{195}{220}\right)^4 \times 100\% \text{ of CP under normal operation at 220 V}$$

= 62% of CP under normal operation at 220 V **Ans.**

Example 7.21. A lamp of 84 watts operates at 240 volts and power factor 0.7. Its power factor is to be corrected to unity. Determine the capacitance required for the condenser.

Solution : Wattage of lamp = 84 watts

Uncorrected pf, $\cos\phi = 0.7$

$$\text{Current draw by lamp, } I = \frac{\text{Wattage of lamp}}{V \times \cos\phi} = \frac{84}{0.7 \times 240} = 0.5\ \text{A}$$

Relative volt-amperes drawn by the lamp

$$= VI \sin\phi = 240 \times 0.5 \times 0.714 = 85.7$$

Corrected power factor, $\cos\phi' = 1$

Reactive volt-amperes drawn by the lamp after pf correction

$$= VI \sin\phi' = 0$$

Reactive volt-amperes to be compensated by the capacitor,

$2\pi f CV^2 = 85.7$

$$\text{or}\quad C = \frac{85.7}{2\pi f V^2} = \frac{85.7}{2\pi \times 50 \times (240)^2} = 4.74\ \mu\text{F} \quad \textbf{Ans.}$$

Note : Supply frequency be assumed as 50 Hz.

Example 7.22. A 110-V lamp develops 16 CP and lamp of the same material and working at the same efficiency develops 25 CP on 220 V. Compare the diameter and length of the filaments.

[Pb. Technical Univ. Electric Drives and Utilization May 2009]

Solution : As the efficiency are the same, their operating temperatures must be the same.

Let d_1 and d_2 be the diameters of the filaments, l_1 and l_2 their respective lengths and I_1 and I_2 their respective currents.

Assuming that the power intakes of the two lamps are proportional to their output, we have

$$16 \ \alpha\ 110\, I_1 \quad \text{and} \quad 25\ \alpha\ 220\, I_2$$

$$\text{or}\quad \frac{I_1}{I_2} = \frac{16}{25} \times \frac{220}{110} = \frac{32}{25}$$

Since the material and operating temperatures are same

$$\therefore\quad \frac{d_1}{d_2} = \left(\frac{I_1}{I_2}\right)^{2/3} = \left(\frac{32}{25}\right)^{2/3} = 1.18 \quad \textbf{Ans.}$$

Since output $\alpha\ l_1 d_1$

$$\therefore\quad \frac{16}{25} = \frac{l_1 d_1}{l_2 d_2}$$

$$\text{or}\quad \frac{l_1}{l_2} = \frac{16}{25} \times \frac{d_2}{d_1} = \frac{16}{25} \times \frac{1}{1.18} = 0.543 \quad \textbf{Ans.}$$

Example 7.23. A 60 CP, 250 volt metal filament lamp has a measured candle power of 71.5 CP at 260 volts and 50 CP at 240 volts.

(*a*) Find the constant for the lamp in the expression $C = aV^b$ where C = candle power and V = voltage.

(*b*) Calculate the change of candle power per volt at 250 V. Determine the percentage variation of candle power due to a voltage variation of ± 4% from the normal value.

Solution : (*a*) Substituting C = 71.5 CP when V = 260 volts and C = 50 CP when V = 240 volts in expression $C = aV^b$ we get

$$71.5 = a\,(260)^b \quad ...(i)$$

$$\text{and}\quad 50 = a\,(240)^b \quad ...(ii)$$

Dividing Eq. (*i*) by Eq. (*ii*), we get

$$\left(\frac{260}{240}\right)^b = 1.43$$

or $b = 4.5$ **Ans.**

Substituting $b = 4.5$ in Eq. (*ii*), we get

$$a = \frac{50}{(240)^{4.5}} = 0.98 \times 10^{-9} \quad \textbf{Ans.}$$

Thus expression becomes $C = 0.98 \times 10^{-9}\, V^{4.5}$ ...(*iv*)

(*b*) Differentiating Eq. (*iii*) w.r.t. V for V = 250 volts, we get

$$\frac{dC}{dV} = 0.98 \times 10^{-9} \times 4.5\, V^{3.5} = 0.98 \times 10^{-9} \times 4.5 \times (250)^{3.5}$$

$$= 4.4\ V \quad \textbf{Ans.}$$

When voltage increases by 4%, $\frac{C_2}{C_1} = (1.04)^{4.5}$

and percentage change in candle power

$$= \frac{C_2 - C_1}{C_1} \times 100 = [(1.04)^{4.5} - 1] \times 100$$

$$= (1.193 - 1) \times 100 = 19.3\% \quad \textbf{Ans.}$$

When voltage falls by 4 percentage, $\frac{C_2}{C_1} = (0.96)^{4.5}$

and percentage change in candle power

$$= \frac{C_2 - C_1}{C_1} \times 100 = [(0.96)^{4.5} - 1] \times 100$$

$$= (0.832 - 1) \times 100 = -16.8\% \quad \textbf{Ans.}$$

7.13. BASIC PRINCIPLES OF LIGHT CONTROL

When light falls on a surface, depending upon the nature of the surface, some portion of light energy is reflected, some portion is transmitted through the medium of the surface and the rest is absorbed. *The ratio of reflected light energy to the incident light energy is known as reflection factor.*

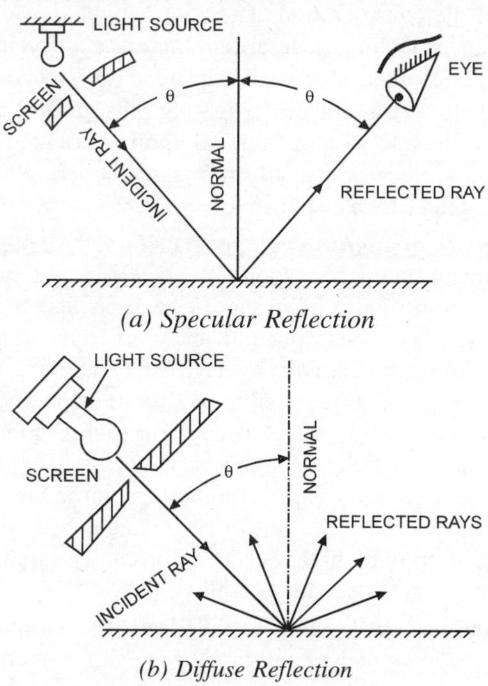

Fig. 7.43. ***Basic Types of Reflections***

There are two basic types of reflection (*i*) mirror or specular reflection and (*ii*) diffuse reflection. In case of specular reflection a beam of light is reflected but not scattered, and unless the eye is placed in the path of the reflected beam, the viewer is unaware of the existence of light. Moreover, if his eye is placed in the path of the reflected beam, he sees, not the illuminated surface, but the light source. Surfaces causing specular reflection are silvered mirrors, highly polished metals etc. With diffuse reflection the reflected light is scattered in all directions, and the viewer sees the illuminated surface, not the light source. Surfaces causing diffuse reflection are paper, frosted glass, chalk, dry earth, plaster etc. If a surface that is uniformly illuminated by a beam of light appears to be equally bright when viewed from all possible angles the reflection is said to be perfectly diffused. Perfect mirror surfaces and perfect diffusing surfaces are ideals that do not exist in nature. The reflection from any actual surface is partly specular and partly diffuse, the proportional varying widely. A surface that is almost free from mirror reflection is called a *mat* surface.

7.14. TYPES OF LIGHTING SCHEMES

The distribution of the light emitted by lamps is usually controlled to some extent by means of reflectors and translucent diffusing screens, or even lenses.

The interior lighting schemes may be classified as (*i*) direct lighting (*ii*) semi-direct lighting (*iii*) semi-indirect lighting (*iv*) indirect lighting and (*v*) general lighting.

1. Direct Lighting. It is the most commonly used type of lighting scheme. In this lighting scheme more than 90 per cent of total light flux is made to fall directly on the working plane with the help of deep reflectors. Though it is most efficient but causes hard shadows and glare. It is mainly used for industrial and general outdoor lighting.

2. Semi-Direct Lighting. In this lighting scheme 60 to 90 per cent of the total light flux is made to fall downwards directly with the help of semi-direct reflectors, remaining light is used to illuminate the ceiling and walls. Such a lighting scheme is best suited to rooms with high ceilings where a high level of uniformly distributed illumination is desirable. Glare in such units is avoided by employing diffusing globes which not only improve the brightness towards the eye level but also improve the efficiency of the system with reference to the working plane.

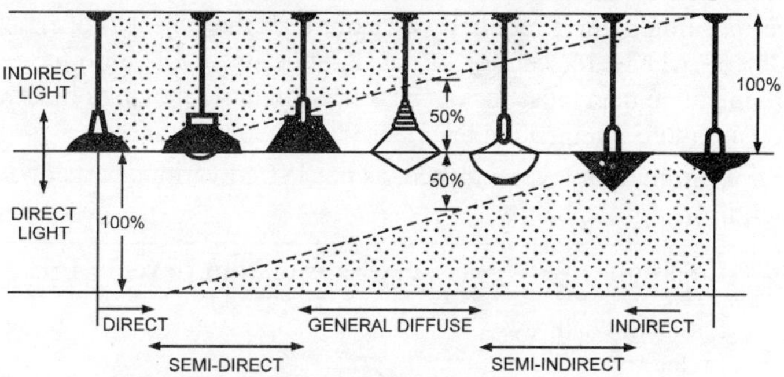

Fig. 7.44. ***General Classification of Luminaires***

3. Semi-Indirect Lighting. In this lighting scheme 60 to 90 per cent of total light flux is thrown upwards to the ceiling for diffuse reflection and the rest reaches the working plane directly except for some absorption by the bowl. This lighting scheme is with soft shadows and glare free. It is mainly used for indoor light decoration purposes.

4. Indirect Lighting. In this lighting scheme more than 90 per cent of total light flux is thrown upwards to the ceiling for diffuse reflection by using inverted or bowl reflectors. In such a system the ceiling acts as the light source, and the glare is reduced to minimum. The resulting illumination is softer and more diffused, the shadows are less prominent and the appearance of the room is much improved over that which results from direct lighting. It is used for decoration purposes in cinemas, theatres and hotels etc. and in workshops where large machines and other obstructions would cause troublesome shadows if direct lighting is employed.

5. General Lighting. In this scheme lamps made of diffusing glass are used which give nearly equal illumination in all directions.

7.15. DESIGN OF LIGHTING SCHEMES

The lighting scheme should be such that it may,

(*i*) provide adequate illumination,

(*ii*) provide light distribution all over the working plane as uniform as possible,

(*iii*) provide light of suitable colour, and

(*iv*) avoid glare and hard shadows as far as possible.

The following factors are required to be considered while designing the lighting scheme.

1. *Illumination Level.* This is the most vital factor because a sufficient illumination is the basic means where by we are able to see our surroundings, unless they are themselves light sources, since only when illuminated do the objects take on the necessary brightness. It is the task of illumination to give objects a distributed brightness. Body colours have property of reflecting light in different degrees. It is this differential brightness which gives essential perception of details. For each type of work there is a range of brightness most favourable to output, *i.e.*, which causes minimum fatigue and gives maximum output in terms of quality and quantity. Degree of illumination, to give necessary brightness to the objects depends upon (*i*) the size of the object to be seen and its distance from the observer–greater the distance of the object from observer and smaller the size of the object, greater will be the illumination required for its proper perception and (*ii*) contrast between the object and back ground—greater the contrast between the colour of the object and its back ground, greater will be the illumination required to distinguish the object properly. Objects which are seen for longer duration of time require more illumination than those for casual work. Similarly moving objects require more illumination than those for stationary object.

Illumination level required, as per ISI, in various parts of a building is given below:

Location	Illumination Level in Lux
Entrances, hallways	100
Living room	300
Dining room	150
Bed room General	300
Dressing tables, bed heads	200
Tables games	300
Games or recreation room	100
Kitchen	200
Kitchen sink	300
Laundry	200
Bathroom	100
Bathroom mirror	300
Sewing	700
Workshop	200
Stairs	100
Garage	70
Study	300

Illumination level required, as per ISI, for various types of traffic routes is given below :

Classification of Lighting Installation	Type of Road	Average Level of Illumination on Road Surface
Group A_1	Important traffic routes carrying fast traffic.	30
Group A_2	Other main roads carrying mixed traffic like main city streets, arterial roads, through ways etc.	15
Group B_1	Secondary roads with considerable traffic like principal local traffic routes, shopping streets etc.	8
Group B_2	Secondary roads with light traffic.	4

2. *Uniformity of Illumination.* The human eye adjusts itself automatically to the brightness within the field of vision. If there is a lack of uniformity, pupil or iris of the eye has to adjust more frequently and thus fatigue is caused to the eye and productivity is reduced. It has been found that visual performance is best if the range of brightness within the field of vision is not greater than 3 : 1, which can be achieved by employing general lighting in addition to localised lighting. Apart from the consideration of causing fatigue, local lighting without using matching general lighting creates psychological feeling of loneliness, gloom and unfriendliness. The modern trend is thus towards 'localised lighting plus general lighting' and towards the adoption of "general lighting oriented towards the working surface" especially in mass production factories, offices, drawing offices, shops etc.

3. *Colour of Light.* The appearance of the body colour entirely depends upon the colour of the incident light. In general the composition of the light should be such that the colour appears natural, *i.e.*, its appearance by artificial light is not appreciably different from that by daylight. Daylight fluorescent tubes nowadays make it possible to illuminate economically even large spaces with artificial daylight giving good colour rendering and at sufficiently high level. For certain applications such as street lighting, colour of light does not matter much if different components have not to be distinguished from each other by their colours, highly efficient discharge lamps, which cause colour distortion, can be used.

4. *Shadows.* In lighting installations, formation of long and hard shadows causes fatigue to eyes and, therefore, is considered to be a shortcoming. Complete absence of shadows altogether again does not necessarily mean an ideal condition of lighting installations. Contrary, perhaps to popular opinion, a certain amount of shadow is desirable in artificial lighting as it helps to give shape to the solid objects and makes them easily recognised. Objects illuminated by shadowless light appear flat and uninteresting, contours are lost, and it is difficult for the eye to form a correct judgement of the shape of an object. However, there is one exception to this, *i.e.*, in drawing offices, where we are to see flat surfaces, shadowless light is essential otherwise shadows will hinder the work. Hard and long shadows can be avoided by (*i*) using large number of small luminaries mounted at a height not less than 2.5 metres and (*ii*) by using wide surface sources of light using globes over filament lamps or by using indirect lighting system.

5. *Glare.* It may be direct or reflected, *i.e.*, it may come directly from the light source or it may be reflected brightness such as from a desk top, nickelled machine parts, or calendered paper.

Direct glare from a source of light is the more common, and is more often a hindrance to vision. A glance at the Sun proves that an extremely bright light source causes acute eye discomfort. Light sources of far less brilliancy than the Sun, such as the filament of an incandescent lamp, or the incandescent mantle of a gas lamp, also cause discomfort by a direct glare. Reflected glare is a glare which comes to the eyes as glint or reflection of the light source in some polished surface.

Toleration of bright light sources in the intermediate vicinity is made possible by locating them at such a height as to place them above the ordinary range of vision.

Metal reflectors for industrial lighting are ordinarily provided with a skirt around the rim of the reflector.

6. *Mounting Height.* The mounting height will largely be governed by the type of the building and type of lighting scheme employed. In the case of direct lighting, in rooms of large floor area, the luminaries should be mounted as close to the ceiling as possible. Lowering them not only will make the illumination less uniform, but will also bring them more into the field of vision, thus increasing the glare, without causing an appreciable increase in the coefficient of utilization. In the unusual case of small rooms with high ceilings, there is something to be gained by lowering the luminaries, but even here a better solution might be to use filament lamps with focusing reflectors and to mount them high.

In the case of indirect and semi-indirect lighting, it would of course be desirable to suspend the luminaries far enough down from the ceiling in order to give reasonably uniform illumination on the ceiling. In practice this is usually taken to mean that the length of the suspension tubes should be one-quarter to one-third the horizontal spacing between the rows of luminaries. Unfortunately it is often impossible to suspend them as far down as this, because it is quite generally recognized that there should be a minimum clearance of 2.5 metres between the luminaries and the floor. It follows that in rooms having low ceilings, the illumination on the ceiling is unavoidably non-uniform. This is not objectionable, because the ceiling is not a working plane.

7. *Spacing of Luminaries.* Correct spacing is of great importance to provide uniform illumination over the whole area and thus do away with comparatively dark areas which are so often found when the fittings are badly spaced.

In the case of direct and semi-direct luminaries the ratio of the horizontal spacing between rows to the height of the luminaries above the working plane depends to quite an extent on the candle power-distribution curve of the luminaire. With fluorescent luminaries it is good practice to aim at a value of unity for this ratio, and to set an upper limit of 3/4. In the case of tungsten lamps combined with focussing reflectors, the ratio of spacing to height should be about 0.6.

In the case of indirect and semi-indirect luminaries, it is good practice to aim at a horizontal spacing between rows approximately equal to a height of the ceiling above the working plane, and in no case should the horizontal spacing exceed 1½ times this height.

In the case of fluorescent luminaries, it is a common practice to join two or more luminaries end to end so that they can share a common outlet. In fact it often works out well to use continuous rows of luminaries, especially when the specified illumination is fairly high.

8. *Colour of Surrounding Walls.* The illumination in any room depends upon the light reflected from the walls and ceilings. White walls and ceilings reflect more light as compared to coloured ones.

7.16. FACTORY LIGHTING

Adequate lighting of factories is of vital importance, as it provides improved amenities for the employees, increased production and has a definite economic value in reducing accidents with their consequent loss of time and compensation payments.

General Requirements and Types of Installations. A factory lighting installation, in common with other indoor equipments should provide an adequate illumination on the working plane and give a good distribution of light, employ simple and easily cleaned fittings and avoid glare. It is essential not only to avoid glare from the lamp itself but also reflected glare from any polished surface, which may be within the line of vision.

General Lighting. The usual scheme in factories and workshops is to mount a number of lamps at a sufficient height so that uniform distribution of light over the working plane is obtained. In large machine shops the height is governed by the necessity of keeping the lamps above the travelling crane. In such cases it is often desirable to supplement the main lighting by side lighting in order to give additional illumination on a vertical plane. Since light coloured walls and ceiling add to the effectiveness of an installation, therefore; it is necessary to get whitewashing or painting done.

Local Lighting. On some points fairly intense illumination is required. For this purpose local lighting can be provided by means of adjustable fittings attached to the machine or bench in question or mounted on portable floor standards. Such lamps should be mounted in deep reflectors so that glare is avoided.

Portable hand lamps attached to wall plugs by means of trailing leads are used for maintenance work and emergency lighting. Low voltage lamps of not more than 50 volts are recommended for use as portable hand lamps because such lamps have thicker filament and are, therefore, more robust than those for normal voltage and danger of shock is also avoided in these few volt lamps. The supply for such lamps can be obtained from a special low voltage distribution system running through out the factory or by means of a small transformer for each individual lamp.

Local lighting should never be employed alone, good general lighting is essential so that the dark places between the local lighting units are avoided. Dark places between the local lighting units cause fatigue to the eyes on account of its continually having to adjust itself to new conditions.

Emergency Lighting. Some lights, such as for (*i*) internal pilot lighting required for safe and speedy evacuation of personnel after main lighting circuit is off (*ii*) external pilot lighting, provided with careful shades leading to shelters required for evacuation of personnel (*iii*) for control posts, first aid centres etc. (*iv*) dials and gauges in important plants required to be watched regularly are required during an air raid when all the factory lights are off as a matter of air-raid precaution. The circuit supplying the above emergency lights should be independently controlled. It is very desirable to provide auxiliary lighting from the source other than the main electric supply preferably from batteries or from small petrol driven generator set. If, however, emergency lighting circuits are operated from main electric supply, these should be completely separated from main lighting circuit.

7.16.1. Industrial Lighting Fittings

Reflectors for industrial purposes must be simple in design and easily cleaned. The requirements of most of the installations can be met by one of the following types of fittings.

Standard Reflectors. These reflectors are made to accommodate lamps of ratings from 40 to 1,500 watts and designed so that they give adequate and uniform illumination when they are mounted at a spacing equal to about 1.5 times their mounting height above the working plane.

Diffusing Fittings. When more diffused light is required than that given by the standard reflector a diffusing glass screen may be fixed across a standard type of reflector. Such fittings are used where highly polished articles are dealt with.

Concentrating Reflectors. A reflector with a concentrated beam is employed in large machine shops and foundaries, where the fittings are to be mounted on a considerable height above the working plane. In such places an ordinary reflector would have too wide angle of divergence and would waste a great deal of light on the walls.

Enclosed Diffusing Fittings. An opal globe completely enclosing the lamp giving a very even and well diffused light is used when light coloured walls and ceiling are there.

Angle Reflectors. Angle reflectors are used to provide illumination in a vertical plane where concentrating type reflectors are used. These can be mounted on suitable stanchions or the walls.

7.16.2. Maintenance

In order to maintain the fittings in a condition of reasonable efficiency it is necessary to clean the light fittings periodically. The frequency of cleaning depends on the conditions in the particular factory under consideration and varies from once or twice a week for very dirty surroundings to every four or six weeks under the best conditions.

7.16.3. Types of Lamps

The discharge lamps have been used in where colour rendering is not important. The fluorescent lamps are widely employed on account of its natural day—light colour, its even illumination and absence of glare and in some cases, the fact that it gives rise to considerably less heat than filament lamps of the same light output.

7.17. METHODS OF LIGHTING CALCULATIONS

A number of methods have been employed for lighting calculations, among which may be mentioned :

1. Watts per square metre method.
2. Lumen or light flux method.
3. Point to point or inverse-square law method.

1. *Watts Per Square Metre Method.* This is principally a "rule of thumb" method, very handy for rough calculation or checking. It consists in making an allowance of watts per square metre of area to be illuminated according to the illumination desired on the assumption of an average figure of overall efficiency of the system.

2. *Lumen or Light Flux Method.* This method is applicable to those cases where the sources of light are such as to produce an approximate uniform illumination over the working plane or where an average value is required.

From the size of lamp or lamps employed and from their efficiency total lumens output are determined. Multiplying the total lumens output from the source by coefficient of utilization, the lumens received on the working plane are determined. If the lamps and surroundings are not perfectly clean, then in determination of lumens received on working plane, the depreciation factor or maintenance factor should be included, *i.e.*,

Lumens received on working plane

= Number of lamps × wattage of each lamp × efficiency of each lamp in terms of lumens/watt × $\dfrac{\text{coefficient of utilization}}{\text{depreciation factor}}$

Also Lumens received on working plane

= Number of lamps × wattage of each lamp × efficiency of each lamp in terms of lumens per watt × coefficient of utilization × maintenance factor.

Coefficient of Utilization or Utilization Factor. The whole light radiated by the lamps does not reach the working plane. The ratio of lumens reaching the working plane to the total lumens given out by the lamp or lamps, is known as *utilization factor* or *coefficient of utilization*. Higher the value of utilization factor, more lumens will reach the working plane for the given lumens output of the lamps. The value of utilization factor depends upon (*i*) the mounting height of lamps—utilization factor decreases with the increase in mounting height of lamps (*ii*) area to be illuminated—for given height, proportion of direct light becomes more and more if floor area increases, *i.e.*, utilization factor increases with the increase in area to be illuminated (*iii*) type of lighting—more for direct lighting and low for indirect lighting and (*iv*) colours of surroundings etc.—more for light colours and less for dark colours. Its value varies from 0.25 to 0.5 and from 0.1 to 0.25 for direct and indirect lighting schemes respectively.

Maintenance Factor. The illumination produced by a lighting installation is considerably less after a year or two of use than it was initially. The loss is due partly to the aging of the lamps and partly to the accumulation of dust on the lamps, on the reflecting and transmitting surfaces of the fixtures and on the ceiling and walls. This fact is taken into account by including the maintenance factor, which is defined as the ratio of the ultimate maintained metre-candles on the working plane to the initial metre-candles. Its value is more if the lamp fittings are cleaned regularly, say 0.8, and less if there is much dust etc. say 0.6.

Depreciation Factor. This is merely the inverse of the maintenance factor and is defined as the ratio of the initial metre-candles to the ultimate maintained metre-candles on the working plane. Its value is more than unity.

3. *Point to Point or Inverse-Square Law Method.* This method is applicable where the illumination at a point due to one or more sources of light is required, the candle powers of the sources in the particular direction under consideration being known.

If a polar curve of lamp and its reflector giving candle powers of the lamp in different directions is known, the illumination at any point within the range of the lamp can be calculated from the inverse square law. If two and more than two lamps are illuminating the same working plane, the illumination due to each can be calculated and added. This method is not much used because of its complicated and cumbersome applications. It is employed only in some special problems, such as flood lighting, yard lighting etc.

Example 7.24. A small assembly shop 16 m long, 10 m wide and 3 m up to trusses is to be illuminated to a level of 200 lux. The utilization and maintenance factors are 0.74 and 0.8 respectively. Calculate the number of lamps required to illuminate the whole area if the lumen output of the lamp selected is 3,000 lumens.

[M.D. Univ. Utilization of Electric Power & Traction Dec. 2009]

Solution : Working area, $A = 16 \times 10 = 160 \text{ m}^2$

Required illumination level, $E = 200$ lux

Total lumens required $= E \times A = 200 \times 160 = 32{,}000$ lumens

Total lumens to be given out by the lamp

$$= \frac{\text{Total lumens required on working area}}{\text{Utilization factor} \times \text{Maintenance factor}} = \frac{32{,}000}{0.74 \times 0.8} = 54{,}000 \text{ lumens}$$

$$\text{Number of lamps required} = \frac{\text{Total lumens given out by lamps}}{\text{Lumen output of each lamp}} = \frac{54{,}000}{3{,}000} = 18 \textbf{ Ans.}$$

Example 7.25. A minimum illumination of 80 lumens/m^2 is required in a factory shed of 100 m × 10 m. Calculate the number, location and the wattage of the units to be used. Assume that the depreciation factor is 0.8, coefficient of utilization is 0.4 and efficiency of the lamp is 40 lumens/watt.

[Pb. Technical Univ. Electric Drives & Utilization May 2008]

Solution : Area to be illuminated, $A = 100 \times 10 = 1{,}000 \text{ m}^2$

Illumination required, $E = 80$ lumens/m^2

Total lumens required $= A \times E = 1{,}000 \times 80 = 80{,}000$

Utilization factor = 0.4

Depreciation factor = 0.8

$$\text{Gross lumens required, } \phi = \frac{A \times E}{UF \times DF} = \frac{80{,}000}{0.4 \times 0.8} = 250{,}000$$

$$\text{Total wattage required} = \frac{\phi}{\text{Lamp efficiency}} = \frac{250{,}000}{40} = 6{,}250$$

42 lamps of 150 W rating in 2 rows, each row having 21 lamps, can be used giving spacing of 4.76 m in length and 5 metres in width.

Disposition of lamps is shown in Fig. 7.45.

Example 7.26. An illumination on the working plane of 75 lux is required in a room 72 m × 15 m in size. The lamps are required to be hung 4 m above the work bench. Assuming a suitable space-height ratio, a utilization factor of 0.5, a lamp efficiency of 14 lumens per watt and a candle power depreciation of 20%, estimate the number, rating and disposition of lamps.

[Pb. Univ. Elec. Drives and Utilization of Elec. Energy August 1984; Pb. Technical Univ. Electric Drives & Utilization May 2009]

Solution : Area to be illuminated, $A = 72 \times 15 = 1{,}080 \text{ m}^2$

Illumination required, $E = 75$ lux

Total lumens required $= A \times E = 1{,}080 \times 75 = 81{,}000$

Utilization factor = 0.5

Maintenance factor = 1 – candle power depreciation = 1 – 0.2 = 0.8

$$\text{Gross lumens required, } \phi = \frac{A \times E}{UF \times MF} = \frac{81{,}000}{0.5 \times 0.8} = 202{,}500$$

$$\text{Total wattage required} = \frac{\phi}{\text{Lamp efficiency}} = \frac{202{,}500}{14} = 14{,}464$$

80 lamps in 4 rows, each row having 20 lamps, can be used giving spacings of 3.6 metres in length and 3.75 metres in width and space-height ratios of 0.9 and 0.9375 respectively.

$$\text{Wattage of each lamp} = \frac{14{,}464}{80} = 180.8 \simeq 200 \text{ W (say)} \quad \textbf{Ans.}$$

Disposition of lamps is shown in Fig. 7.46.

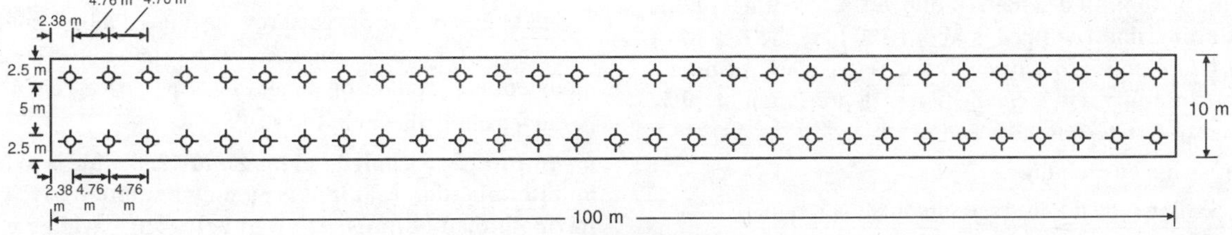

Fig. 7.45

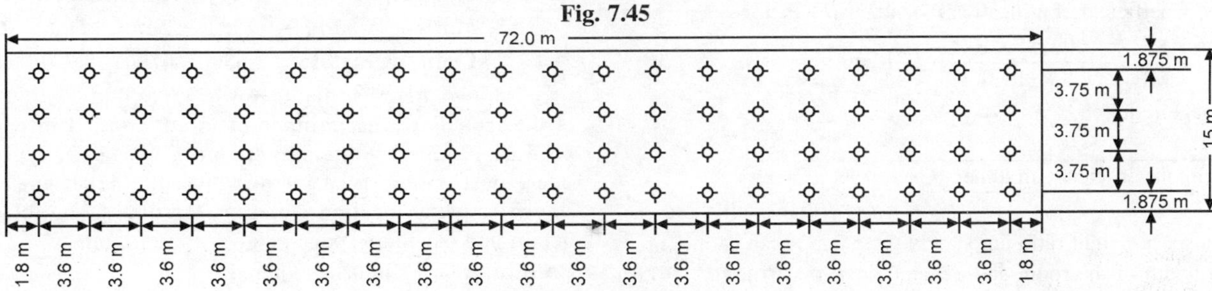

Fig. 7.46

Example 7.27. A class room of size 30 m × 30 m is to be illuminated with 75 lux. The lamps are required to be hung 5 m above the workbench. Assume a space-height ratio of 0.9-1.0, a utilization factor of 0.5, lamp efficiency of 15 lumens/watt, candle power depreciation of 20%. Estimate number, rating and disposition of lamps.

[*U.P. Technical Utilization of Electrical Energy and Electric Traction 2006-07*]

Solution : Gross lumens required,

$$\phi = \frac{A \times E}{UF \times DF} = \frac{30 \times 30 \times 75}{0.5 \times (1-0.2)} = 168{,}750$$

Total wattage required,

$$W = \frac{\phi}{\text{Lamp efficiency}} = \frac{168{,}750}{15} = 11{,}250$$

Total number of lamps required, assuming wattage rating of each lamp to be 300 W

$$N = \frac{11{,}250}{300} = 37.5 \text{ (say 42)}$$

6 lamps can be mounted along one side of class room (giving space-height ratio unity) and 7 lamps can be mounted along the other side of the room (giving space-height ratio of $\frac{30/7}{5}$ i.e. 0.857).

Disposition of lamps is shown in Fig. 7.47.

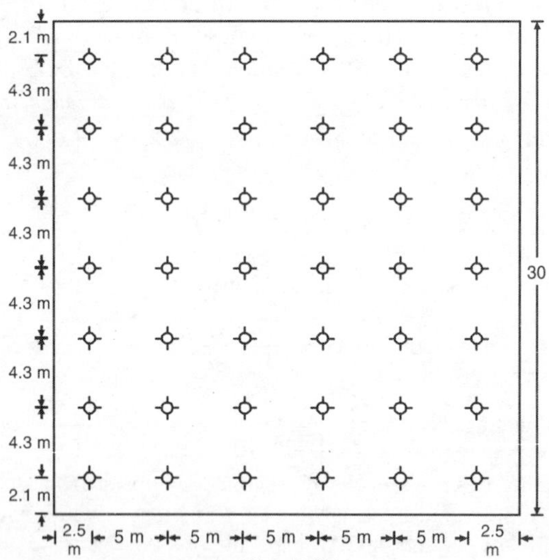

Fig. 7.47

Example 7.28. Estimate the number and wattage of lamps which would be required to illuminate a workshop space 60 × 15 metres by means of lamps mounted 5 metres above the working plane. The average illumination required is about 100 lux, coefficient of utilization = 0.4, luminous efficiency 16 lumens per watt. Assume a space-height ratio of unity and a candle power depreciation of 20%.

[*Gorakhpur Univ. Utilization of Elec. Power and Traction 1983*]

Solution : Gross lumens required,

$$\phi = \frac{\text{Area to be illuminated} \times \text{average illumination required}}{\text{Coefficient of utilization} \times \text{maintenance factor}}$$

$$= \frac{A \times E}{UF \times MF} = \frac{60 \times 15 \times 100}{0.4 \times (1-0.2)} = 281{,}250$$

Total wattage required,

$$W = \frac{\phi}{\text{Luminous efficiency in lumens / watt}} = \frac{281{,}250}{16}$$

$$= 17{,}578 \text{ watts}$$

For a space-height ratio unity, only three lamps can be mounted along the width of the room and 12 lamps can be arranged along the length of the room. Thus total number of lamps required = 12 × 3 = 36 **Ans.**

$$\text{Wattage of each lamp} = \frac{\text{Total wattage required}}{\text{Number of lamps}} = \frac{17{,}578}{36} = 488.3$$

We will take the nearest standard lamp of 500 watts. Location of the lamps is shown in Fig. 7.48.

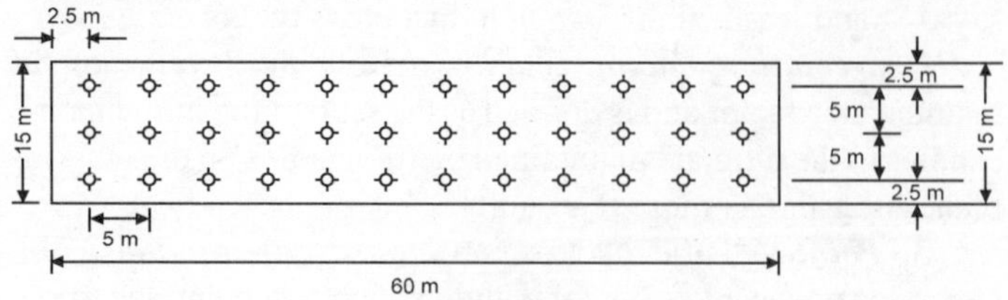

Fig. 7.48

Example 7.29. A hall 30 metres by 15 metres with a ceiling height of 5 metres is to be provided with a general illumination of 120 lumens/m². Taking a coefficient of utilization of 0.5 and depreciation factor of 1.4, determine the number of fluorescent tubes required, their spacing, mounting height and total wattage. Take luminous efficiency of fluorescent tube as 40 lumens per watt for 80 watt tubes.

Solution : Gross lumens required, $\phi = \frac{A \times E \times DF}{UF}$

$$= \frac{30 \times 15 \times 120 \times 1.4}{0.5}$$

$$= 151{,}200$$

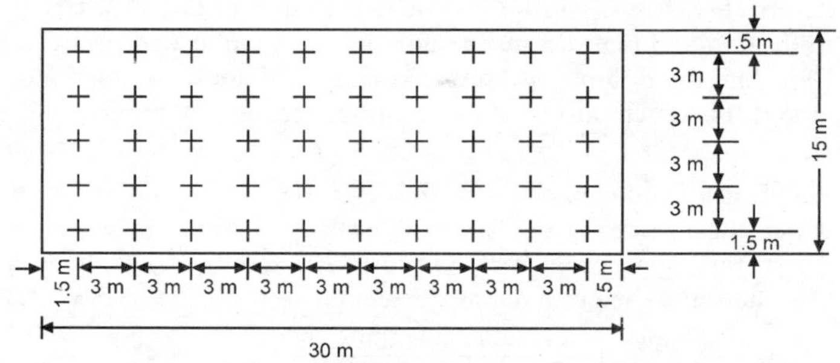

Fig. 7.49

$$\text{Total wattages required} = \frac{151{,}200}{40} = 3{,}780 \text{ watts} \quad \textbf{Ans.}$$

$$\text{Number of tubes required} = \frac{\text{Total wattage required}}{\text{Wattage of each tube}}$$

$$= \frac{3{,}780}{80} = 47.25 \cong 48 \text{ (say)}$$

50 tubes in 5 rows, each row having 10 tubes, can be used giving spacings of 3 metres in length as well as in width. These tubes may be mounted at an height of 3 metres giving space-height ratio of unity. Arrangement is shown in Fig. 7.49. **Ans.**

Example 7.30. A hall 30 m long and 12 m wide is to be illuminated and illumination required is 50 metre candles. Five types of lamps having lumen outputs, as given below, are available :

Watts	**100**	**200**	**300**	**500**	**1,000**
Lumens	**1,615**	**3,650**	**4,700**	**9,950**	**21,500**

Taking a depreciation factor of 1.3 and utilization coefficient of 0.5, calculate the number of lamps needed in each case to produce required illumination. Out of above five types of lamps select most suitable type and design a suitable scheme and make a sketch showing location of lamps. Assume a suitable mounting height and calculate space-height ratio of lamps.

Solution : Gross lumens required,

$$\phi = \frac{A \times E \times DF}{UF} = \frac{30 \times 12 \times 50 \times 1.3}{0.5} = 46{,}800$$

If 100 W lamps are used, number of lamps required

$$= \frac{46,800}{1,615} \cong 29 \quad \textbf{Ans.}$$

If 200 W lamps are used, number of lamps required

$$= \frac{46,800}{3,650} \cong 13 \quad \textbf{Ans.}$$

If 300 W lamps are used, number of lamps required

$$= \frac{46,800}{4,700} \cong 10 \quad \textbf{Ans.}$$

If 500 W lamps are used, number of lamps required

$$= \frac{46,800}{9,950} \cong 5 \quad \textbf{Ans.}$$

If 1,000 W lamps are used, number of lamps required

$$= \frac{46,800}{21,500} \cong 2 \quad \textbf{Ans.}$$

Let the mounting height be of 5 metres.

Most suitable type of lamps will be 300 W lamps requiring 10 lamps in two rows, each row having 5 lamps giving spacing of 6 metres in length as well as in width and space-height ratio of 6/5, *i.e.*, 1.2.

Lamps of lower wattage will require more number of fittings and, therefore, will increase the cost where as lamps of higher wattage will be few in requirement giving space-height ratio much more than desirable. Arrangement is shown in Fig. 7.50.

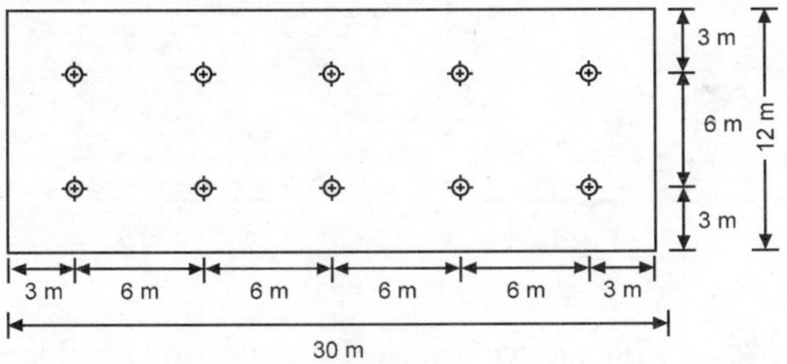

Fig. 7.50

7.18. STREET LIGHTING

The main objects of street lighting are :

(*i*) To make the traffic and obstructions on the road clearly visible in order to promote safety and convenience.

(*ii*) To make the street more attractive.

(*iii*) To increase the community value of the street.

The principle employed for street lighting is different from that of interior lighting. There are no walls and ceiling which reflect or diffuse light, hence only direct lighting scheme can be employed and hard shadows and high contrast cannot be avoided. Because of areas to be illuminated being large, the value of illumination for economic reasons, is very low compared to that for indoor lighting and the question of colour rendering is also of minor importance. High level of illumination is not necessary, because firstly one has seldom to look continuously at nearby objects, secondly it is the impression of whole scene that is required. In fact in case of interior lighting the objects are seen by the light reflected by them but in case of street lighting the objects are seen in relatively bright back ground *i.e.*, silhouetted against a relatively bright back ground. Further more, owing to the low illumination, the eye is in its most sensitive state and, therefore, glare must be avoided.

Two general principles are usually employed in the design of street lighting installations, namely (*i*) the diffusion principle and (*ii*) the specular reflection principle.

1. *Diffusion Principle*. In this case the lamps fitted with suitable reflectors are used. The reflectors are so designed that they may direct the light downwards and spread as uniformly as possible over the road surface. In order to avoid glare the reflectors are made to have a cut-off of between 30° to 45° so that the filament is not visible except from underneath it. The diffusing nature of the road surface causes the reflection of a certain proportion of the incident light in the direction of the observer, and therefore, the road surface appears bright to the observer. The illumination at any point on the road surface is calculated by applying point to point or inverse-square law method. Over certain proportions of the road the surface is illuminated from two lamps and the resultant illumination is the sum of the illuminations due to each lamp.

2. *Specular Reflection Principle*. In this case the reflectors are curved upwards so that the light is thrown on the road at a very large angle of incidence. It is observed that a motorist requires to see objects about 30 metres away. Thus in Fig. 7.51 the observer is shown about 30 metres from the object. Much of the light from the lamp L_3 is not reflected towards the observer, where as most of the light from lamps L_1 and L_2 is reflected towards him. Thus the object will appear silhouetted against the bright road surface due to lamps at a longer distance. The requirement of a pedestrian, who requires to see objects in his immediate neighbourhood, is also fulfilled in this method as some light from the lamps falls directly downwards. This method of street lighting is only suitable for straight sections of the road. This method is more economical also as compared to the diffusion method of lighting but it suffers from the disadvantage that it produces glare for the motorists.

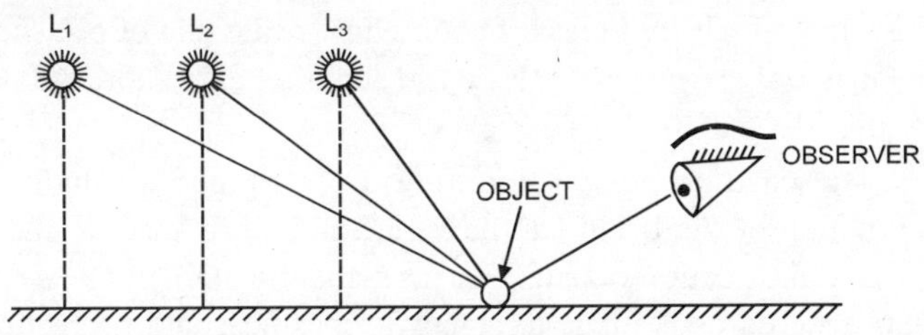

Fig. 7.51

***Illumination Level For Street Lighting and Mounting Height of Lamps*.** The illumination required depends upon the class of street lighting installations. In class A installations, *i.e.*, in important shopping centres and road junctions, illumination level of 30 lumens/m^2 is required where as a in poorly lighted suburban streets, illumination level of 4 lumen/m^2 is sufficient. An average well-lighted street, will require illumination level between 8 to 15 lumens per square metre. Excellent illumination is considered when the distance apart is not more than 8 times the height of the luminaires ; that is, with the luminaires 8 metres above street level, they should be spaced not more than 64 metres (normal spacing for standard lamps is 50 metres with a mounting height of 8 metres). Lamp posts should always be fixed at the junction of roads and as far as possible lamps near large trees should be avoided.

Types of Lamps For Street Lighting. Mercury vapour and sodium discharge lamps have been found to have certain particular advantages for street lighting purposes ; the most important of these is the lower power consumption for a given amount of light, which, inspite of the higher cost of the lamps, makes the overall cost of an installation with discharge lamps less than that employing filament lamps. The colour and monochromatic nature of the light produced by discharge lamps do not matter much in street lighting installations.

7.19. FLOODLIGHTING

Floodlighting means flooding of large surfaces with light from powerful projectors. It is employed to serve one or more of the following purposes.

1. ***Aesthetic Floodlighting.*** For enhancing beauty of building at night such as public places, ancient buildings and monuments, religious buildings on important festive occasions etc.

2. ***Industrial and Commercial Floodlighting.*** For illuminating railway yards, sports stadiums, car parks, construction sites, quarries etc.

3. ***Advertising.*** For illuminating advertisement boards and showcases.

For floodlighting it is necessary to concentrate the light from the light source into a relatively narrow beam. The particular type of reflector and its housing used for concentrating the light into narrow beam is known as *floodlight projector.* Since it is usually installed in remote positions, therefore, it must be robust and weatherproof in construction. The reflecting surface is the most important part in a projector. This may be made of silvered glass or chromium plate or stainless steel ; the efficiency of silvered glass is about 90% while that of polished metal is only about 70%. Metal reflectors being more robust are usually preferred.

The casing and its mounting are arranged so that the inclination of the beam can be varied in both a vertical and a horizontal direction on site. For permanent installations use of cast-metal cases is made to achieve robustness and protection against weather and for temporary installations or those in sheltered situations, use of sheet metal casing is made. When 500 or 1,000 watt lamps are used in projectors, ventilation may have to be provided for adequate cooling. This may be achieved by providing sufficient radiating surface. The front of the projector is usually of clear glass, often bowed outwards to protect it from the heat of the lamp ; use of diffusing glass is made when a diffuse beam is required. As far as possible the projectors should not be visible to the passers-by. In some cases the projectors may be housed in ornamental stand standards.

Projectors are classified according to the beam spread.

(*i*) *Narrow beam projectors* with beam spread between 12-25°. These are used for distance beyond 70 metres.

(*ii*) *Medium angle projectors*—Projectors with beam spread between 25-40°. These are used for distance between 30 to 70 metres.

(*iii*) *Wide angle projectors*—projectors with beam spread between 40-90°. These are used for distance below 30 metres.

For economic reasons use of wide angle projector with high wattage lamp is encouraged over narrow beam projector with low wattage lamp. High wattage lamp is more efficient than low wattge lamp used in narrow beam projectors. Standard gas filled tungsten filament lamps of 250, 500 or 1,000 watts are used in medium and wide angle projectors, although for narrow angle beams which require accurate location of the light source at the focal point of the reflector, special lamps having bunched filaments and known as *projector lamps* are generally employed.

Location and Mounting of Projectors. One of the most important factor which affects the choice of projector is the location of the projector. There are two possible locations of projectors in practice. Fig. 7.52 (*a*) shows symmetric projector kept 20 to 35 metres away from the surface to be floodlighted and providing approximately parallel beam having beam spread of 25° to 30°. Fig. 7.52 (*b*) shows the case when the projector cannot be located away from the building. In such a case, an unsymmetric reflector mounted in a basement area or on a bracket attached to the building is used which directs more intense light towards the top of the building.

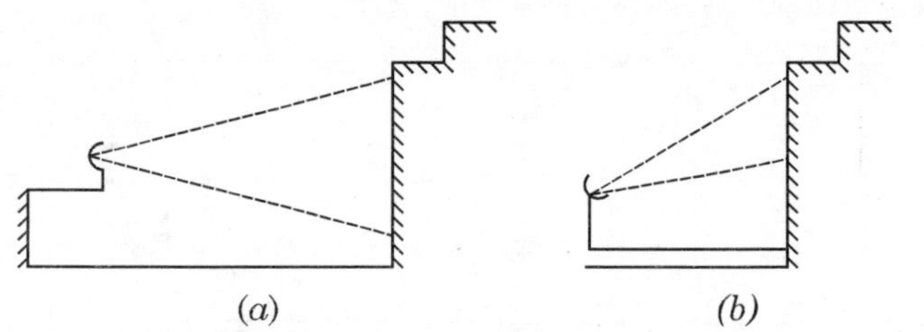

Fig. 7.52

Floodlighting Calculations. The problem of flood-lighting calculations may be roughly separated into three steps.

First Step : Illumination Level Required. The illumination level (lumens/m^2) required depends upon the type of building, the purpose of the floodlighting, the amount of conflicting light in the vicinity etc. Typical figures are given below :

Building Surfaces	Reflection Factor	Lux Recommended For Brightness of Locality		
		Low	Medium	High
White terra cotta	60—80	50	100	150
Cream terra cotta				
Light marble				
Light gray lime stone	40—60	70	120	200
Bed Ford lime stone				
Buff lime stone				
Smooth buff face brick				
Briar hill sand stone	20—40	100	150	250
Smooth gray brick				
Medium gray lime stone				
Common tan brick				
Dark field gray brick	10—20	120	180	300
Common red brick				
Brown stone				

Second Step : Type of Projector. Two considerations enter into the choice of a projector, viz., beam size and light output. The former determines the area covered by the beam and the latter the illumination provided. Beam angle of the projector is decided keeping in view the distance of projector from the surface.

Third Step : Number of Projectors. For any desired intensity over a definite surface the number of projectors required is obtained from the following relation

$$N = \frac{A \times E \times \text{depreciation factor} \times \text{waste light factor}}{\text{Utilization factor} \times \text{wattage of lamp} \times \text{lumious efficiency of lamp}}$$

where N = Number of projectors
A = Area of surface to be illuminated in square metres
E = Illumination level required in lumens/m^2

The terms used in flood lighting calculations are described below :

1. *Waste Light Factor.* Whenever a surface is illuminated by a number of light sources, there is always a certain amount of waste light on account of overlapping and falling of light beyond the edges of the area to be illuminated. This effect is taken into account by multiplying the theoretical value of lumens required by 1.2 for rectangular areas and 1.5 for irregular areas and objects such as statues and monuments etc.

2. *Depreciation Factor.* Dirt and dust depositing on the reflectors surface and on the source of light reduces the effectiveness of the projector and 50 to 100 per cent more light must be provided so that the illumination shall be adequate at the end of the interval between the cleaning periods. Depreciation factor is defined as the ratio of illumination under ideal conditions to the illumination under normal conditions.

3. *Coefficient of Utilization.* It is also known as beam factor and is defined as the ratio of beam lumens to the lamp lumens. Its value lies between 0.3 and 0.5. This factor is due to the fact that all the light emitted by the projector is not along the direction of the beam but some of it is absorbed by the reflector and by the front glass. When the above losses are considerable the value of utilization is low. The coefficient of utilization factor goes up with deep reflectors having a good reflecting surface, such as silvered glass.

Example 7.31. The front of a building 50 m × 16 m is illuminated by sixteen 1000-W lamps arranged so that uniform illumination on the surface is obtained. Assuming a luminous efficiency of 17.4 lumens/watt and a coefficient of utilization of 0.4, determine the illumination on the surface. [*Madras Univ., Utilization of Elec. Energy 1970*]

Solution : Area to be floodlighted A = 50 × 16 = 800 m^2

Total lumens given out by 16 lamps

ϕ = Number of lamps × wattage of each lamp × luminous efficiency of each lamp

= 16 × 1,000 × 17.4 = 278,400 lumens

Total lumens reaching the surface to be floodlighted

$$= \frac{\phi \times \text{utilization factor}}{\text{Depreciation factor} \times \text{waste light factor}}$$

$$= \frac{278{,}400 \times 0.4}{1.3 \times 1.2} = 71{,}400 \text{ lumens}$$

Assuming depreciation factor = 1.3 and waste light factor = 1.2

Illumination on the surface, $E = \dfrac{71{,}400}{800} = 89.25$ lumens/m^2 **Ans.**

Example 7.32. The following table gives the watts and lumens

Watts	300	500	1,000	1,500
Lumens	5,000	9,000	18,000	27,000

The front of the building measuring 60 m and 15 m is to be floodlighted by means of projectors placed at a distance of 8 m from the wall. The average illumination required is 50 lux. Determine the number and size of projectors required. Assume waste light factor of 1.2, depreciation factor 0.8 and coefficient of utilization 0.5.

[*G.B. Technical Univ. Utilization of Electrical Energy and Traction 2010-11*]

Solution : Area to be illuminated, A = 60 × 15 = 900 m^2

Total lumens required on the surface = A × E = 900 × 50

= 45,000 lumens

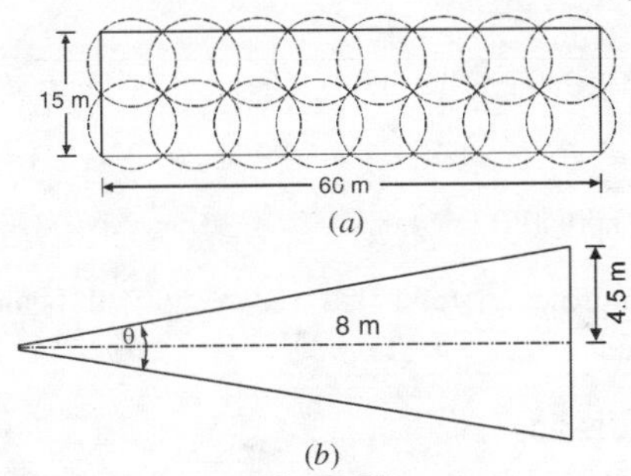

Fig. 7.53

Total lumens required to be given out by lamps

$$\phi = \frac{\text{Total lumens required on the surface} \times \text{waste light factor}}{\text{Utilization factor} \times \text{depreciation factor}}$$

$$= \frac{45{,}000 \times 1.2}{0.5 \times 0.8} = 135{,}000 \text{ lumens}$$

Assuming that 500 watt lamps having a lumens output of 9,000 lumens are used.

Number of projectors required

$$= \frac{\phi}{\text{Lumens output of each 500 W lamp}} = \frac{135{,}000}{9{,}000} = 15; \text{ say (16)}$$

$$\because \text{Tan}\ \frac{\theta}{2} = \frac{r}{l}$$

where r is the radius of illuminated circle at distance l from light source whose beam angle is θ

So, angle of spread, $\theta = 2\tan^{-1}\dfrac{4.5}{8} = 60°$ **Ans.** [Fig. 7.53 (*b*)]

Hence 16 projectors of 500 watts each with beam angle of 60° will be required.

Example 7.33. A 50 m × 15 m of a concrete building in an industrial town is to be illuminated by floodlighting projectors 25 m away. If the required illumination is 100 lux, coefficient of utilization 0.5, depreciation factor 1.5, waste light factor 1.2, estimate the number and size of projectors.

[*Rajasthan Technical Univ. Utilization of Electric Power 2011*]

Solution : Area to be illuminated, A = 50 × 15 = 750 m^2

Total lumens required on the surface = A × E = 750 × 100

= 75,000 lumens

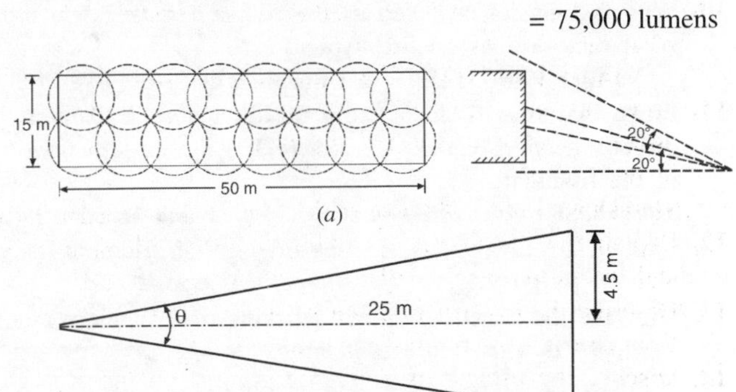

Fig. 7.54

Total lumens required to be given out by lamps

$$\phi = \frac{\text{Total lumens required on the surface} \times \text{depreciation factor} \times \text{waste light factor}}{\text{Utilization factor}}$$

$$= \frac{75{,}000 \times 1.5 \times 1.2}{0.5} = 270{,}000$$

Assuming that 1,000 watt lamps having a luminous efficiency of 17 lumens/watt are used.

Number of projectors required

$$= \frac{\phi}{\text{Wattage of lamp} \times \text{luminous efficiency of lamp}} = \frac{270{,}000}{1{,}000 \times 17}$$

$= 16$ **Ans.**

Angle of spread $\theta = 2 \tan^{-1} \frac{4.5}{25} = 20°$ **Ans.** [Fig. 7.54 (*b*)]

Hence 16 projectors of 1,000 watts each with beam angle of 20° will be required.

Exercises

1. Explain the basic nature of light.
2. Define and explain the following terms :
   (*i*) Mean horizontal candle power (*ii*) Mean spherical candle power (*iii*) Mean hemi-spherical candle power (*iv*) Flux (*v*) Solid angle (*vi*) Specific consumption (*vii*) Glare (*viii*) Absorption factor (*ix*) Beam factor and (*x*) Waste light factor.
3. Explain laws of illumination.
   [U.P. Technical Univ. Utilization of Electrical Energy & Traction 2009-10; M.D. Univ. Utilization of Electric Power and Traction May 2010]
4. Derive relation between E, I and L for a uniform diffuse source.
   [M.D. Univ. Utilization of Electric Power & Traction Dec. 2008]
5. Derive expressions for the illumination on a surface (*i*) when it is normal and (*ii*) when it is inclined to the axis of a beam of incident light.
   [Gorakhpur Univ. Utilization of Elec. Power & Traction 1981]
6. Define : (*i*) plane angle and solid angle, (*ii*) Depreciation factor, (*iii*) Cosine inverse law, (*iv*) Polar curve.
   [Rajasthan Technical Univ. Utilization of Electrical Power 2011]
7. What are polar curves ? Mention the uses of polar curves.
   [W.B. Univ. of Technology Utilization of Electric Power 2009]
8. What are polar curves ? Explain Rousseau diagram and its importance in illumination engineering ?
   [Rajasthan Technical Univ. Utilization of Electrical Power 2009]
9. Explain Rousseau diagram and its importance in illumination engineering.
   [M.D. Univ. Utilization of Electric Power & Traction Dec. 2008]
10. Why is tungsten selected as the filament material and on what factors does its life depend ?
    [Agra Univ. Traction & Utilization of Elec. Power 1983]
11. Prove that in a filament lamp the diameter of filament is directly proportional to $I^{2/3}$, where I is the current flowing in the filament.
    [Gorakhpur Univ. Utilization of Elec. Power and Traction 1983]
12. Explain the advantages of using inert gas in filament lamps and the purpose to get the filament as coiled coil.
13. Describe the construction and working of a filament lamp. Compare it with fluorescent lamp.
14. Discuss the effects of voltage variation on the life and illumination as regards tungsten filament lamp and fluorescent lamp.
15. Explain the working of halogen lamp.
    [U.P. Technical Univ. Utilization of Electrical Energy & Electric Traction 2006-07]
16. Describe with a neat sketch the principle of electric discharge lamp. State the advantages and disadvantages of discharge lamps over the filament lamp and give their applications.
    [B.H.U. Elec. Traction & Utilization 1978]
17. Explain with a neat diagram the principle of operation of a sodium vapour lamp. Mention its use.
    [M.D. Univ. Utilization of Electric Power & Traction Dec. 2009]
18. Describe the construction and principle of operation and application of mercury vapour lamp.
    [M.D. Univ. Utilization of Electric Power & Traction Dec. 2008]
19. Explain the operation of the following with circuit diagrams: (*a*) LP sodium vapour lamp, (*b*) HP mercury vapour lamp.
    [W.B. Univ. of Technology Utilization of Electric Power 2009]
20. Write construction and working of Neon lamp.
    [Pb. Technical Univ. Electric Drives & Utilization Dec. 2006]
21. Explain the working of a fluoescent tube with the help of the circuit giving the function of various parts. How stroboscopic effect is eliminated in fluorescent tube lighting.
    [Pb. Technical Univ. Electric Drives and Utilization May 2009]
22. Describe the working of fluorescent tube. What are the functions of starter and choke.
    Explain street lighting. Mention its principles and explain in detail.
    [Pb. Technical Univ. Electric Drives and Utilization May 2008]
23. With the help of circuit diagrams explain the working of the following light sources :
    (*a*) High pressure mercury vapour lamp (*b*) Fluorescent tube (*c*) Carbon-arc lamp.
    What are the usual values of power factors for the above lamps ? Indicate the systems of illumination where above lamps are used. **[Bombay Univ. 1970]**
24. Describe the construction and principle of operation and application of fluorescent lamp.
    [M.D. Univ. Utilization of Electric Power May 2009]
25. Discuss the construction and principle of operation and application of (*a*) filament lamp (*b*) sodium vapour lamp, (*c*) mercury vapour lamp (*d*) fluorescent lamp.
    [Rajas. Technical Univ. Utilization of Electric Power 2010, 2011]
26. Compare metal filament lamp with discharge lamp.
    [M.D. Univ. Utilization of Electric Power & Traction May 2010]
27. Compare the merits of alternative lighting schemes using incandescent lamps, sodium vapour lamps or fluorescent lamps for industrial lighting.
    [Gorakhpur Univ. Utilization of Elec. Power & Traction 1982]
28. Why CFL lamps are becoming more popular, nowadays ? Discuss technically.
    [U.P. Technical Univ. Utilization of Electrical Energy and Electric Traction 2006-07]
29. What do you understand by direct, indirect, and semi-indirect lighting ?
    [Agra Univ. Traction and Utilization of Elec. Power, 1982]
30. State and describe various types of lighting schemes.
    [Pb. Univ. Elec. Drives and Utilization of Elec. Energy April, 1981]
31. Describe in brief the requirements of good lighting. Enumerate the factors to be considered while designing an indoor lighting scheme.
    [Gorakhpur Univ. Utilization of Elec. Power & Traction Sept., 1984; Pb. Technical Univ. Electric Drives & Utilization Dec. 2008]
32. How does designing of indoor and outdoor system differ ? Explain method for designing an indoor system.
    [U.P. Technical Univ. Utilization of Electrical Energy and Electric Traction 2005-06]

33. What are the requirements of good street lighting ?

[Pb. Univ. Elec. Drives and Utilization of Elec. Power Nov. 1980; Pb. Technical Univ. Electric Drives & Utilization May 2009]

34. What are the general principles that are usually employed in the design of street lighting explain.

[Pb. Univ. Elec. Drives and Utilization of Elec. Energy April, 1983; Pb. Technical Univ. Electric Drives & Utilization Dec. 2008]

35. Explain the specular reflection principle of street lighting.

[Allahabad Univ. Utilization of Elec. Energy and Traction, 1975]

36. Enumerate the various factors to be considered while designing street lighting.

[Gorakhpur Univ. Utilization of Elec. Power and Traction 1983; Pb. technical Univ. Electric Drives & Utilization May 2009]

37. What are the aims of street lighting and how are they achieved ? List types of lamps used for street lighting.

[U.P. Technical Univ. Utilisation of Electrical Energy and Traction 2006-07]

38. Explain the various factors to be taken into account for designing schemes for (*i*) street lighting (*ii*) floodlighting.

[M.D. Univ. Utilization of Electric Power & Traction May 2009]

39. Explain the factors to be taken into account for designing schemes for (*i*) street lights, (*ii*) floodlighting and (*iii*) highway lighting.

[W.B. Univ. of Technology Utilization of Electric Power 2009]

40. Write short notes on the following:

(*i*) Laws of illumination. **[P.T.U. December 2009; W.B. Univ. of Technology 2009]**

(*ii*) Polar curves. **[Pb. Univ. Nov. 1980]**

(*iii*) Rousseau's diagram. **[Agra Univ. 1976]**

(*iv*) Sodium vapour lamp. **[M.D. Univ. Dec. 2008; May 2009; G.B.T.U. 2010-11]**

(*v*) Mercury vapour lamp. **[M.D. Univ. Dec. 2009]**

(*vi*) Halogen lamp.

(*vii*) Fluorescent lamps. **[G.B.T.U. 2010-11]**

(*viii*) CFL

(*ix*) Street lighting. **[Pb. Univ. April, 1978]**

(*x*) Essential requirement of good lighting.

(*xi*) Floodlighting. **[P.T.U. Dec. 2009]**

Short Answer Type Questions With Answers

Q. 1. What are the advantages of artificial lighting produced electrically ?

Ans. Artificial lighting produced electrically have the advantages of cleanliness, ease of control, reliability, steady output as well as low cost.

Q. 2. What is Angstrom unit (AU) ?

Ans. Angstrom unit (AU) is unit of wavelength of light and equals to 10^{-10} m.

Q. 3. What is difference between illumination and light ?

Ans. Light is the cause and illumination is the result of that light on surface on which it falls.

Q. 4. Explain candle power and mean horizontal candle power. **[U.P.T.U. 2005-06]**

Ans. Candle power. Candle power is the light radiating capacity of a source in a given direction and is defined as the number of lumens given out by the source in a unit solid angle in a given direction. It is denoted by symbol CP.

$$i.e., \quad CP = \frac{\text{Lumens}}{\omega}$$

Mean horizontal candle power. It is defined as the mean of candle powers in all directions in the horizontal plane containing the source of light.

Q. 5. Define and explain "illumination" and "luminous intensity". **[U.P.T.U. 2005-06]**

Ans. Illumination. When the light falls upon any surface, the phenomenon is called the *illumination.* It is defined as the number of lumens, falling on the surface, per unit area. It is denoted by symbol E and is measured in lumens per square metre or lux or metre-candle.

If a flux of F lumens falls on a surface of area A, then the illumination of that surface is

$$E = F/A \text{ lumens/m}^2 \text{ or lux } (lx)$$

Luminous Intensity. Luminous intensity in any given direction is the luminous flux emitted by the source per unit solid angle, measured in the direction in which the intensity is required. It is denoted by symbol I and is measured in candela (*cd*) or lumens per steradian.

If F is the luminous flux radiated out by source within a solid angle of ω steradians in any particular direction then $I = F/\omega$ lumens/steradian or candela (*cd*).

Q. 6. Differentiate between candela and lux. **[U.P.T.U. 2006-07]**

Ans. Candela or lumens per steradian is the unit of luminous intensity while lux is the unit of illumination.

Q. 7. Define luminous intensity. **[P.T.U. Dec. 2006]**

Ans. Please refer to Answer of Q. 5.

Q. 8. Define illumination. Mention properties of good illumination. **[P.T.U. May 2008]**

Ans. When the light falls upon any surface, the phenomenon is called the *illumination.* It is defined as the number of lumens, falling on the surface, per unit area. It is denoted by symbol E and is measured in lumens per square metre or lux or metre-candle.

If a flux of F lumens falls on a surface of area A, then the illumination of that surface is

$$E = F/A \text{ lumens/m}^2 \text{ or lux } (lx)$$

Good illumination means adequate illumination level, uniform distribution of light all over the working plane, suitable light colour and absence of glare and hard shadows.

Q. 9. What do you understand by foot candle and luminance ? **[P.T.U. Dec. 2008]**

Ans. Foot-Candle. It is also the unit of illumination and is defined as the luminous flux falling per square foot on the surface which is every where perpendicular to the rays of light from a source of one candle power and one foot away from it.

$$i.e., \quad 1 \text{ Foot-candle} = 1 \text{ lumen/ft}^2 = \frac{1 \text{ lumen}}{\left(\frac{1 \text{ m}}{3.28}\right)^2}$$

$$= 10.76 \text{ metre candle or lux.}$$

Luminance. When the eye receives a great deal of light from an object we say it is bright, and 'brightness' is an important quantity in illumination. It is all the same whether the light is produced by the object or merely reflected from it. The object sends out light as though each small piece of its surface were of a certain luminous intensity. Generally, the brightness of an object is not the same from all points of view. When brightness is considered in terms of measurable quantity, the term *luminance* is employed.

Brightness or *luminance* is defined as the luminous intensity per unit projected area of either a surface source of light or a reflecting surface and is denoted by L.

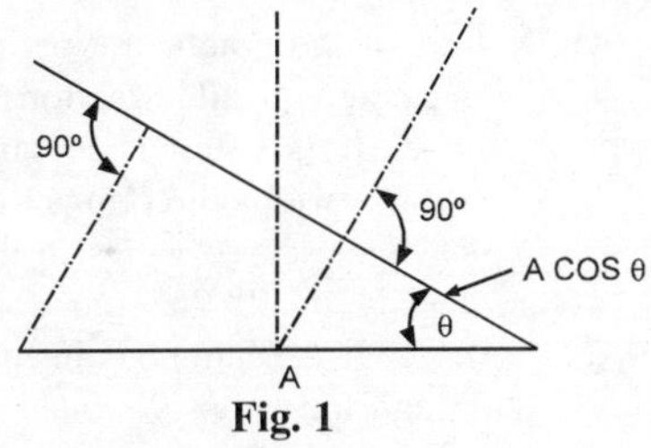

Fig. 1

If a surface of area A has an effective luminous intensity of I candelas in a direction θ to the normal, then the luminance (brightness) of that surface is

$$L = \frac{I}{A \cos \theta} \text{ candela/m}^2 \text{ or nits}$$

i.e., nit is defined as candela per square metre. Bigger unit of luminance is *stilb* which is defined as candelas per square cm. Lambert is also the unit of brightness which is lumens/cm^2. Foot-lambert is lumens/ft^2.

Q. 10. What is luminance ? **[P.T.U. Dec. 2009]**

Ans. Please refer to Answer of Q. 9.

Q. 11. Differentiate between illumination and luminous intensity. **[P.T.U. December 2009]**

Ans. Please refer to Answer of Q. 5.

Q. 12. What is relation between candela (cd) and lumens per steradian ?

Ans. Candela (cd) and lumens per steradian, both, are units of luminous intensity. 1 Candela = 1 lumen per steradian.

Q. 13. What is relation between lux and lumens/m^2 ?

Ans. Lux and lumens/m^2, both, are units of illumination.
1 lux = 1 lumen/m^2.

Q. 14. Differentiate between lamp efficiency and specific consumption.

Ans. Lamp efficiency is the ratio of the luminous flux in lumens to the power input in watts while the specific consumption is the ratio of the power input in watts to the average candle power in candela.

Q. 15. What is difference between maintenance factor and depreciation factor ?

Ans. Maintenance factor is defined as the ratio of illumination under normal working conditions to the illumination when every thing is perfectly clean. Depreciation factor is merely inverse of the maintenance factor.

Q. 16. What are the two laws of illumination ?

Ans. (*i*) Law of inverse squares (*ii*) Lambert's cosine law.

Q. 17. What are laws of illumination ? **[P.T.U. May 2007]**

Ans. There are two laws of illumination (*i*) Law of inverse squares and (*ii*) Lambert's cosine law.

According to law of inverse square the illumination of a surface is inversely proportional to the square of the distance between the surface and the light source provided that the distance between the surface and the source is sufficiently large so that the source can be regarded as a point source.

According to Lambert's cosine law, the illumination at any point on a surface is proportional to the cosine of the angle between the normal at that point and the direction of luminous flux.

Q. 18. Why none of the practical type of lamp provides uniformly distributed light in all directions ?

Ans. Because of unsymmetrical shape of the practical types of lamps.

Q. 19. What do you understand by polar curves ? **[P.T.U. May 2009; Pb. Univ. April 1980]**

Ans. None of the practical type of lamp provides light uniformly distributed in all directions because of its un-symmetrical shape. It is often necessary to know the distribution of light in various directions to ascertain how the candle power of a light source varies in different directions. The luminous intensity in all the directions can be represented by *polar curves*. If the luminous intensity in a horizontal plane passing through the lamp is plotted against angular position, a curve known as *horizontal polar curve is obtained.* If the luminous intensity in a vertical plane is plotted against the angular position, a curve known as *vertical polar curve* is obtained.

Q. 20. How is polar curve useful to an illumination engineer ?

Ans. The polar curve are used to determine the mhcp and mscp of a lamp. They are also used to determine the actual illumination of a surface by using the cp in that particular direction as read from the vertical polar curve in illumination calculations.

Q. 21. What for an integrating sphere is used ?

Ans. Integrating sphere is a piece of apparatus which is commonly used for measurement of mscp.

Q. 22. What are various methods of producing light ? **[P.T.U. Dec. 2007]**

Ans. Artificial light can be produced by any one of the following methods :

1. By passing electric current through a wire raising its temperature to incandescence.
2. By establishing an arc between two electrodes.
3. By maintaining an electric discharge through a vapour or gas.

Q. 23. List various light sources ? **[P.T.U. May 2008]**

Ans. Refer to Art. 7.8.

Q. 24. Why the glass globe of a filament lamp is completely evacuated ?

Ans. The glass globe of a filament of incandescent lamp is completely evacuated to prevent the oxidization and convection currents of the filament and also to prevent the temperature being lowered by radiation.

Q. 25. Why is tungsten selected as the filament material ? **[Agra Univ. 1983]**

Ans. Nowadays tungsten is the most commonly used metal for filament due to its high melting point (3,400°C), high resistivity, low temperature coefficient (0.0051), low vapour pressure, being ductile and mechanical strong to withstand vibration during use.

Q. 26. Why does the efficiency of a filament lamp increase with the increase in operating voltage ?

Ans. The efficiency (lumens/watt) of a lamp increases with the increase in operation voltage owing to increase in temperature and is proportional to the square of the operating voltage.

Q. 27. Why sodium vapour discharge lamps are not used for general lighting ? **[Gorakhpur Univ. 1981]**

Ans. Sodium vapour discharge lamps are not used for general lighting because they have got drawback of colour discrimination.

Q. 28. Why thermal type starter for use in fluorescent tubes have become obsolete nowadays ?

Ans. Because of more complicated construction, greater cost and greater power loss.

Q. 29. What is the role of condenser and choke used in fluorescent lamp ? **[P.T.U. May 2009]**

Ans. When the supply is switched on to a fluorescent lamp, the starter provides a path through the electrodes and choke for providing preheating. The starter switch then opens automatically thereby interrupting the current. Since this heating current reduces to zero instantaneously in the choke circuit also, the choke field collapses releasing stored energy and so developing a high voltage between the electrodes and thus causes the preionized tube to strike.

Thus choke performs two functions : (*i*) it provides inductive voltage surge to start the discharge of electrons and (*ii*) it limits the current in the lamp circuit during normal working. Power factor of fluorescent lamp is about 0.5 (lagging) due to presence of choke, so condenser is connected across the supply to improve the power factor. Overall pf of the fluorescent tube circuit is about 0.95 (lagging).

Q. 30. What is meant by stroboscopic effect ? How this effect is eliminated in fluorescent tube lightings ?
[Gorakhpur Univ. Sept. 1983]

Ans. At the usual alternating current supply frequency of 50 Hz, a discharge lamp will be extinguished 100 times per second. Although this effect is seldom noticeable in normal conditions. It is possible that this may, in some instances, give rise to a stroboscopic effect, that is moving objects such as rotating parts of machinery, illuminated by this light, may appear to be moving in disunity or jerks, or rotating more slowly than their actual speed.

In actual applications, however, where this effect might cause annoyance, it can be practically eliminated in a three lamp unit by connecting each lamp on a separate phase of a 3-phase system and it can be greatly reduced in a two lamp unit by the use of a two lamp control unit, which employs a condenser in the ballast of one of the lamps. The current through the lamps is almost 90° out of phase and under these conditions light output of one of the lamps is at a maximum. This method has an additional advantage of giving an overall power factor of nearly unity for the unity of two lamps. In this arrangement one circuit remains at a low power factor at about 0.5 lagging, while the other circuit incorporating a series capacitor C remains at a power factor of about 0.5 leading.

Q. 31. What are the drawbacks of direct lighting system ?
[P.T.U. May 2007]

Ans. The drawbacks of direct lighting scheme are that hard shadows and glare are caused.

Q. 32. What system of lighting should be recommended for drawing office ? **[P.T.U. Dec. 2007]**

Ans. Direct lighting scheme may be adopted on the working table in drawing office. Light should be concentrated by reflector and shadow and glare should be avoided.

Q. 33. What is floodlighting ?

Ans. Floodlighting means flooding of large surfaces with light from powerful projector(s).

Q. 34. How projectors are classified according to the beam spread ?

Ans. Projectors, according to beam spread, are classified as narrow beam angle projectors (beam spread 12-25°); medium beam angle projectors (beam spread 25-40°) and wide beam angle projectors (beam spread 40-90°)

Problems

1. The efficiency of a certain filament lamp is stated to be 15 lumens per watt. What is the mean spherical power of the lamp if its intake is 100 watts? **[Ans.** 119.4]

2. Find the candle power of a lamp to produce an illumination of 8 lux on a surface 5 m from the lamp when the light rays are (*i*) normal to the surface; (*ii*) inclined to normal at 60° (*iii*) parallel to the surface.
[Pb. Univ. Elec. Drives & Utilization of Elec. Energy Nov. 1977]
[Ans. (*i*) 200 (*ii*) 400 (*iii*) ∞]

3. A filament lamp of 20 W is suspended at a height of 5 metres above working plane and gives uniform illumination over an area of 4 m diameter. Assume efficiency of reflector as 50%. Determine the illumination on the working plane. Efficiency of lamp is 0.89 W per candle power.
[Pb. Technical Univ. Electric Drives & Utilization May 2007; Dec. 2007; Dec. 2008] **[Ans.** 11.236 lumens/m^2]

4. A corridor is lighted by four lamps spaced 10 metres apart and suspended at a height of 5 metres above the centre line of the floor. If each lamp gives 200 cp in all directions below the horizontal, find the illumination at the point on the floor midway between the second and third lamp.
[Bombay Univ. October 1971] **[Ans.** 6.16 lumens/m^2]

5. Two lamps A and B of 240 candela and 400 candela respectively are situated 100 metres apart. The height of A above the ground level is 10 metres and that of B is 20 metres. If a photometer is placed at the centre of the line joining the two lamp posts, calculate its reading.
[Gujarat Univ. 1970] **[Ans.** 0.066 lumens/m^2]

6. It is desired to illuminate a corridor. 100 CP lamps, three number, 8 metres apart are suspended 5 metres above floor level. Calculate the intensity of illumination exactly vertically below the lamp position on the surface of the floor. **[Ans.** 4.7 lux; 5.19 lux; 4.7 lux]

7. Two lamps, each emitting a luminous flux of 3,000 lumens below the horizontal plane, are suspended 6 metres above the centre line of a horizontal roadway and are 30 metres apart. If the road is 12 metres wide between kerbs, calculate the illumination at the middle of the kerb between the lamps. Assume that the luminous intensity of each lamp is uniform in all directions below the horizontal plane. **[Ans.** 7 lux]

8. A 100 cp lamp is hung 2 metres above the centre of a circular area of 3 metre diameter. Determine the illumination at (*i*) the centre of the area, (*ii*) circumference of the area, and (*iii*) average illumination. Also calculate the average illumination, if a reflector of 50% efficiency is used.
[Ans. (*i*) 25 lux (*ii*) 12.8 lux (*iii*) 17.8 lux (*iv*) 88.0 lux]

9. A lamp having a uniform candle power of 200 in all directions is provided with a reflector which directs 60% of the total light uniformly on to a circular area of 10 metre diameter. The lamp is hung 7 metres above the area. Calculate the illumination at the edge of the surface with and without the reflector. Determine also the average illumination over the area without the reflector.
[Gorakhpur Univ. Utilization of Elec. Power With Traction 1980]
[Ans. 19.2 lux; 2.2 lux; 2.98 lux]

10. A light source having an intensity of 700 CP in all directions is fitted with a reflector so that it directs 70% of its light along a beam having a divergence of 20°. What would be the average illumination produced on a surface normal to the beam direction at a distance of 25 metres ? Calculate also the CP of the lamp and the reflector as viewed from the illuminated surface.
[U.P.S.C. Engineering Service Examination 1974]
[Ans. 100.8 lux; 64,480 CP]

11. In a street lighting scheme lamps with candle power of 500 are hung at a height of 5 m. The distance between the posts is 10 m. Determine the illumination under lamps and at the mid point between the posts.
[M.D. Univ. Utilization of Electric Power & Traction Dec. 2008]
[Ans. 23.58 lux; 14.14 lux]

12. Two lamp posts are 14 m apart and are fitted with 200 CP lamp each at a height of 5 m above the ground. Calculate :
    (*i*) illumination mid way between them.
    (*ii*) illumination under each lamp.
[M.D. Univ. Utilization of Electric Power & Traction May 2010]
[Ans. 3.142 lux; 8.304 lux]

13. In a street lighting scheme, lamps having uniform candle power of 520 candela are hung at a height of 6 m. The distance between consecutive lamp posts is 8 m. Find the illumination under lamp and a distance midway between the lamp posts.

[Rajasthan Technical Univ. Utilization of Electrical Power 2010]

[**Ans.** 20.684 lux; 16.64 lux]

14. The relation between the voltage (V) and candle power (C) of a metal filament lamp is given by $C = aV^b$. Calculate the percentage change in candle power for a change of voltage of ± 5% from the rated voltage if $b = 4.5$.

[**Ans.** 24.6%; –20.6%]

15. Find the illumination due to 20 bulbs of 200 W in a room of 15 m × 8 m assuming following data.

Coefficient of utilization 0.6, Depreciation factor 1.2, MSCP of 200 W lamp is 250. [**Ans.** 262 lux]

16. It is required to provide an illumination of 100 lux in a factory hall 30 m × 12 m. Assume that the depreciation factor is 0.8, the coefficient of utilization 0.4, and the efficiency of proposed lamps 11 lumens per watt. Calculate the number of lamps and their disposition.

[Pb. Univ. Elec. Drives & Utilization of Elec. Energy April 1982]

[**Ans.** 40 lamps of 200 W each in 4 rows, each row having 10 lamps]

17. A minimum illumination of 80 lumens/m^2 is required in the factory shed of 50 m × 12 m. Calculate the number, the location and wattage of the units to be used. Assume that the depreciation factor is 0.8, coefficient of utilization is 0.4 and the efficiency of lamp units is 14 lumens/watt.

[M.S. Univ. of Baroda, Power System I, 1971]

[**Ans.** 36 lamps of 300 watts in 3 rows of 12 lamps in each row]

18. A uniform illumination of 50 lux is to be obtained on the floor of a room measuring 10 m × 10 m by arranging electric lights suitably. The system to be employed should be shadowless and glare free. Calculate the number of lamps and watt rating of each lamp if the efficiency of lamp is 15 lumens/watt. Assume suitable values of coefficient of utilization and depreciation. Indicate the layout.

[Bombay Univ. Elec. Power System I October 1972]

[**Ans.** 4 lamps of 250 W rating]

19. A building measuring 30 m × 20 m is to be floodlit on the front side with brightness of 25 lumens/m^2. Coefficient of reflection of building surface is 0.25. Lamps of 500 W having lumens output of 8,000 each are used. Assume beam factor as 0.6, waste light factor 1.2 and maintenance factor 0.75, determine the number of lamps required.

[Rajasthan Univ. Utilization of Electrical Power, 2006]

[**Ans.** 20 lamps]

20. Estimate the number of 1,000 W floodlight projectors required to illuminate the upper 75 metres of one face of a 96 metre tower of width 13 metres and approximate initial average luminance is to be 6.85 cd/m^2. The projectors are mounted at ground level 51 metres from base of tower. Utilization factor is 0.2, reflection factor of wall is 25% and efficiency of lamp is 18 lumens per watt. [**Ans.** 24]

21. A building frontage 50 m × 15 m is to be illuminated by floodlighting projectors situated 25 metres away. If the illumination is 100 lux, coefficient of utilization 0.5, depreciation factor 1.5, waste light factor 1.2, estimate the number and size of projectors. Sketch the projectors recommended indicating the usual adjustments provided.

[A.M.I.E. Sec B. Utilisation of Elec. Power. May, 1968]

[**Ans.** 16 projectors of 1,000 W each with beam angle of 20°]

22. A factory requires 3,20,000 lumens for proper illumination and the total lighting period in a year is 3,000 hours. Compare the annual cost of energy and the cost of lamp renewals for the following schemes.

(*a*) 100 watt filament lamps having a life of 1,000 hours, a luminous efficiency of 13.4 lumens/watt and renewal cost of Rs. 20 per lamp.

(*b*) Twin 40 watt fluorescent lamps having a life of 3,000 hours, a luminous efficiency of 30 lumens/watt and renewal cost of Rs. 200 per lamp.

The factory gets its power supply at the rate of Rs. 2,500 per kW per year plus Rs 1.20 per unit.

[**Ans.** Scheme A : Annual cost of energy Rs. 1,46,400
and cost of lamp renewals Rs. 14,400.00
Scheme B : Annual cost of energy Rs. 65,400
and cost of lamp renewals Rs. 53,600.00]

❑❑❑

CHAPTER 8

Electrolytic Processes

INSIDE THIS CHAPTER

8.1. INTRODUCTION

The fact that electrical energy can produce chemical changes and the processes based on it, called the '*electrolytic processes*', are widely used for the extraction of pure metals from their ores (such as aluminium, zinc, copper, magnesium, sodium etc.), manufacturing of various chemicals (such as caustic soda, potassium permanganate, hydrogen, oxygen, chlorine etc.), electrodeposition of metals including electroplating, electrotyping, electroforming, building up of worn-out parts in metallurgical, chemical and other industries. Though the various processes mentioned are different in apparent detail but fundamentally they are alike, being based on the principle of electrolysis.

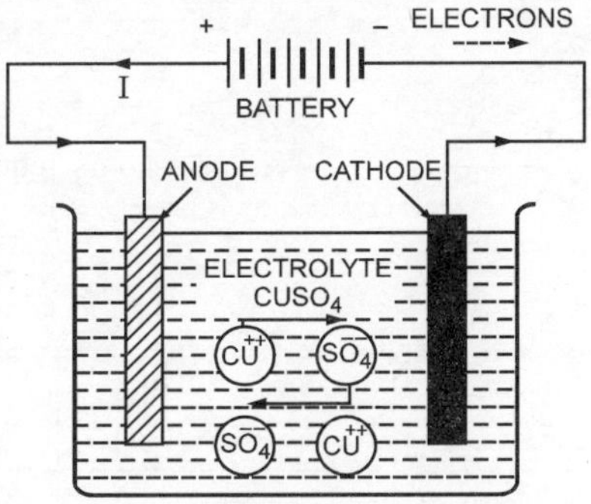

Fig. 8.1

8.2. ELECTROLYSIS—BASIC PRINCIPLE

In nature atoms of electrolyte (chemical compound) are closely bound together but bond becomes weaker, when dissolved and the molecules of the electrolyte split up into two types of ions carrying electric charges, called the *cations* and *anions*, and moving freely in the solution. Now if two electrodes are dipped into the electrolyte and connected to the dc supply, ions associated with positive charge (cations) and moving freely in the solution are attracted by the cathode (electrode connected to the negative terminal of the supply) and the ions associated with negative charge (anions) and moving freely in the solution are attracted by the *anode* (electrode connected to the positive terminal of the supply).

For example, when copper sulphate ($CuSO_4$) is dissolved in water, immediately it gets dissociated into +vely charged copper ions (Cu^{++}) and negatively charged sulphions (SO_4^{--}) moving freely in the solution and if a potential difference is applied between the two electrodes immersed in the solution the +vely charged copper ions will move towards the cathode and the –vely charged sulphions will move towards the anode. Each of the positively charged copper ions reaching the cathode will take two electrons from it and become a metallic atom of copper, and similarly each of the negatively charged sulphions reaching the anode will give up two electrons to it and cease to be anion. Thus the electrons will move from anode to cathode in the external circuit and constitute flow of current from anode to cathode in the electrolyte. Thus the function of source of supply seems only to serve as an electron pump pumping electrons from +ve side of the supply and supply to the –ve side of the supply.

As mentioned above each +vely charged copper ion on reaching the cathode takes two electrons from it, becomes atom of metallic copper and deposite there. Similarly each –vely charged sulphion on reaching the anode gives up two electrons to it and becomes SO_4 radical but since SO_4 radical cannot exist in the electrical neutral state so it will attack the anode and will form the corresponding sulphate of the material of the anode; for example if the anode is of copper then copper sulphate ($CuSO_4$) will be formed, but if the anode is of such a material which cannot be attacked by SO_4, for example if the anode is of carbon the SO_4 will react with water and form sulphuric acid and liberate oxygen according to chemical reaction

$$2SO_4 + 2H_2O \rightarrow 2H_2SO_4 + O_2$$

The whole process described above is called the *electrolysis* and the effect is that the copper gets dissolved from the anode and deposited on the cathode. During the process there is no accumulation of charge at any point in the circuit and the mass of copper deposited at the cathode is exactly equal to that removed from the anode.

8.3. FARADAY'S LAWS OF ELECTRODEPOSITION

The laws governing the electrolytic processes were formulated by Michael Faraday, an English Scientist, and are known after his name. These may be stated as below :

Faraday's First Law. According to this law the chemical deposition due to flow of current through an electrolyte is directly proportional to the quantity of electricity (coulombs) passed through it.

i.e., mass of chemical deposition,

$$m \propto \text{Quantity of electricity, Q}$$

$$\text{or} \quad m \propto It \qquad \because \quad Q = It$$

$$\text{or} \quad m = ZIt \qquad \text{...(8.1)}$$

where I is the steady current in amperes flowing through the electrolyte for t seconds and Z is a constant of proportionality and is known as the *electrochemical equivalent* of the substance.

If $I = 1$ A and $t = 1$ second, then $Z = m$

Thus electrochemical equivalent, Z, of a substance is defined as the amount of the substance deposited on passing a steady electric current of 1A for one second through its solution. It is usually expressed in terms of milligrams per coulomb. The SI unit of electrochemical equivalent Z is the kilogram per coulomb (kg C^{-1}).

Faraday's Second Law. This law states that when the same quantity of electricity is passed through several electrolytes, the mass of the substances deposited are proportional to their respective chemical equivalents or equivalent weights.*

From this law it follows that the constant of proportionality Z in Eq, (8.1) is proportional to the chemical equivalent.

Table 8.1 gives the values of atomic weight, valency, chemical equivalent and electrochemical equivalent of various elements.

8.4. CALCULATION OF CURRENT REQUIRED FOR DEPOSITING GIVEN AMOUNT OF METAL

The theoretical value of current required for depositing a given quantity of metal and the time for which this current should be passed through the electrolyte, can be calculated from the Faraday's laws, if the electrochemical equivalent of the metal is known. The following example will make the method of calculation clear.

Example 8.1. Calculate the ampere-hours required to deposit a coating of silver 0.05 mm thick on a sphere of 5 cm radius. Assume electrochemical equivalent of silver = 0.001118 and density of silver to be 10.5.

Solution : Surface area of the sphere, $S = 4\pi r^2 = 4\pi \times (5)^2 = 314 \text{ cm}^2$

Thickness of coating, $t = 0.05 \text{ mm} = 0.005 \text{ cm}$

Mass of silver to be deposited, $m = S \times t \times \text{density of metal}$

$= 314 \times 0.005 \times 10.5$

$= 16.493$ gm

$= 0.016493$ kg

ECE of silver, $Z = 0.001118$ gm/coulomb

$$= \frac{0.001118 \times 3{,}600}{1{,}000}$$

$= 0.0040248$ kg/A-h

$$\text{Ampere-hours required} = \frac{m}{Z} = \frac{0.016493}{0.0040248}$$

$= 4.12$ **Ans.**

Table 8.1

Elements	Symbol	Atomic Weight	Valency	Chemical Equivalent	Electrochemical-Equivalent Kg/Coulomb
Aluminium	Al	27.00	3	9.00	9.4×10^{-8}
Cadmium	Cd	112.4	2	56.2	58.2×10^{-8}
Calcium	Ca	40.08	2	20.04	20.75×10^{-8}
Chlorine	Cl	35.45	1	35.37	36.74×10^{-8}
Chromium	Cr	52.01	3,6	17.3, 8.6	17.9×10^{-8}, 9.1×10^{-8}
Copper (Cupric)	Cu	63.54	2	31.8	32.9×10^{-8}
Gold	Au	197	3	65.4	67.68×10^{-8}
Hydrogen	H	1.008	1	1.0	1.045×10^{-8}
Ferrous	Fe	55.9	2	27.95	29.02×10^{-8}
Ferric	Fe	55.5	3	18.63	19.35×10^{-8}
Lead	Pb	207.21	2	103.2	107.16×10^{-8}
Magnesium	Mg	24.32	2	11.97	12.43×10^{-8}
Nickel	Ni	58.7	2	29.3	30.43×10^{-8}
Nitrogen	N	14	3	4.67	4.85×10^{-8}
Oxygen	O	16	2	7.98	8.296×10^{-8}
Potassium	K	39.1	1	39.04	40.54×10^{-8}
Silver	Ag	107.88	1	107.67	111.81×10^{-8}
Sodium	Na	22.99	1	22.99	23.87×10^{-8}
Tin	Sn	118.7	2	59.3	61.4×10^{-8}
Zinc	Zn	65.38	2	32.45	33.88×10^{-8}

* The chemical equivalent or equivalent weight of a substance is defined as the weight of that substance which will combine with or displace unit weight of hydrogen. The chemical equivalent of hydrogen is, thus, unity. Since valency of a substance is equal to the number of hydrogen atoms, which it can replace or with which it can combine, the chemical equivalent of a substance, therefore, may be defined as the ratio of its atomic weight to its valency.

$$\text{Thus chemical equivalent} = \frac{\text{Atomic weight}}{\text{Valency}}.$$

Example 8.2. If 18.258 gm of nickel is deposited by 100 A flowing for 10 minutes, how much copper would be deposited by 50 A for 6 minutes? Atomic weight of nickel = 58.6 and that of copper 63.18. Valency of both is 2.

[Rajasthan Technical Univ. Utilization of Electrical Power 2010]

Solution : ECE of nickel, $Z = \frac{m}{I \times t} = \frac{18.258 \times 10^{-3}}{100 \times 10 \times 60}$

$= 3.043 \times 10^{-7}$ kg / C

ECE of copper, $Z' = Z \times \frac{\text{Chemical equivalent of copper}}{\text{Chemical equivalent of nickel}}$

$= 3.043 \times 10^{-7} \times \frac{63.18/2}{58.6/2}$

$= 3.2808 \times 10^{-7}$ kg/C

Mass of copper deposited, $m' = Z' \times I' \times t'$

$= 3.2808 \times 10^{-7} \times 50 \times 6 \times 60$

$= 0.0059$ kg or 5.9 gm **Ans.**

Example 8.3. A steady current of 40 A is passed through a solution of copper sulphate for 1 hour. Find the weight of copper deposited, if ECE of copper is 0.01044 mgm/C.

[Pb. Technical Univ. Electric Drives and Utilization Dec. 2006]

Solution : ECE of copper, $Z = 0.01044$ mg/C $= 1.044 \times 10^{-8}$ kg/C

Current strength, I = 40 A

Time for which current is passed,

$t = 1 \times 60 \times 60 = 3{,}600$ seconds

Mass of copper deposited, $m = Z \times I \times t$

$= 1.044 \times 10^{-8} \times 40 \times 3{,}600$

$= 0.0015$ kg or 1.5 gm **Ans.**

Example 8.4. Find the thickness of copper deposited on a plate area of 0.00025 m^2 during electrolysis if a current of one ampere is passed for 100 minutes. Density of copper is 8,900 kg/m^3 and ECE of copper = 32.95 × 10^{-8} kg/coulomb.

[Gorakhpur Univ. Utilization of Elec. Power and Traction, Sept., 1983; Pb. Technical Univ. Electric Drives & Utilization May 2009]

Solution : ECE of copper, $Z = 32.95 \times 10^{-8}$ kg C^{-1}

Current strength, I = 1 A

Copper density, D = 8,900 kg/m^3

Time for which current is passed, $t = 100 \times 60 = 6{,}000$ seconds

Mass of copper deposited, $m = Z I t$

$= 32.95 \times 10^{-8} \times 1 \times 6{,}000$

$= 0.001977$ kg

Volume of copper deposited, $v = \frac{m}{D} = \frac{0.001977}{8{,}900}$

$= 0.222 \times 10^{-6}$ m^3

Thickness of copper deposited, $t = \frac{v}{A} = \frac{0.222 \times 10^{-6}}{0.00025}$

$= 0.888 \times 10^{-3}$ m

$= 0.888$ mm **Ans.**

Example 8.5. Find the thickness of copper deposited on a plate area of 0.00025 m^2 during electrolysis, if a current of 1.5 A is passed for one hour. Density of copper is 8,900 kg/m^3 and ECE of copper = 32.95 × 10^{-8} kg/C.

[U.P. Technical Univ. Utilization of Electrical Energy and Traction 2006-07]

Solution : Mass of copper deposited,

$m = Z \times I \times t = 32.95 \times 10^{-8} \times 1.5 \times 60 \times 60 = 0.0017793$ kg

Volume of copper depsotied,

$v = \frac{m}{D} = \frac{0.0017793}{8{,}900} = 0.199 \times 10^{-6}$ m^3

Thickness of copper deposited,

$t = \frac{v}{A} = \frac{0.199 \times 10^{-6}}{0.00025} = 0.0008$ m or 0.8 mm **Ans.**

Example 8.6. In a copper-sulphate voltameter, the copper cathode is increased in weight by 50 gm in two hours, when the current maintained was constant. Calculate the value of this current. Given that

Atomic weight of copper = 63.5

Atomic weight of hydrogen = 1

Atomic weight of silver = 108

Electrochemical equivalent of silver = 0.001118 gm

Solution : Mass of copper deposited, $m = 50$ gm

Time of flow of current, $t = 2$ hours

$= 2 \times 60 \times 60$

$= 7{,}200$ seconds

ECE of silver = 0.001118 gm

$= 111.8 \times 10^{-8}$ kg C^{-1}

Chemical equivalent of silver $= \frac{\text{Atomic weight}}{\text{Valency}}$

$= \frac{108}{1} = 108$

Chemical equivalent of copper $= \frac{63.5}{2} = 31.75$

ECE of copper, Z = ECE of silver × $\frac{\text{Chemical equivalent of copper}}{\text{Chemical equivalent of silver}}$

$= 111.8 \times 10^{-8} \times \frac{31.75}{108} = 32.867 \times 10^{-8}$ kg C^{-1}

Strength of current, $I = \frac{m}{Z \times t} = \frac{50 \times 10^{-3}}{32.867 \times 10^{-8} \times 7{,}200}$

$= 21.13$ A **Ans.**

Example 8.7. A copper refining plant, using 500 electrolytic cells, carries a current of 6,000 A, voltage per cell being 0.25 volt. If the plant were to work 40 hours/week, calculate the energy consumption per tonne, assuming ECE of copper as 0.3281 mg/ coulomb of electricity.

Solution : ECE of copper, Z = 0.3281 mg/C

$= 32.81 \times 10^{-8}$ kg C^{-1}

$= 32.81 \times 10^{-8} \times 3{,}600$ kg/A-h

$= 0.00118116$ kg/A-h

Total number of ampere-hours per annum

$= 500 \times 6{,}000 \times 40 \times 52 = 624 \times 10^{7}$ Ah

Annual output of plant $= 0.00118116 \times 624 \times 10^{7}$ kg

$= 7{,}370{,}438$ kg $= 7{,}370.438$ tonnes

Energy consumed per annum = A-h per annum × voltage per cell

$= 624 \times 10^{7} \times 0.25$

$= 156 \times 10^{7}$ watt-hours

$= 156 \times 10^{4}$ kWh

Energy consumption per tonne $= \frac{\text{Annual consumption in kWh}}{\text{Annual output in tonnes}}$

$= \frac{156 \times 10^{4}}{7{,}370.438}$

$= 211.66$ kWh/tonne **Ans.**

Example 8.8. Determine the maximum voltage required for electrolysis of water if one kg of hydrogen on oxidation to water liberates 14.212 × 10^7 joules and electrochemical equivalent of hydrogen is 1.0384 × 10^{-8} kg C^{-1}.

[*G.B. Technical Univ. Utilization of Electrical Energy & Traction 2010-11*]

Solution : During electrolysis of water the energy is required to decompose water into hydrogen and oxygen and this is equal to the energy expanded in the circuit in forcing the quantity of electricity through the electrolyte.

$$\text{Energy expanded during electrolysis} = \frac{1}{Z} \times V \text{ watt-second per kg}$$

$$= \frac{V}{Z} \text{ joules per kg}$$

Energy liberated by 1 kg of hydrogen when it combines with oxygen

$$= 14.212 \times 10^7 \text{ joules}$$

$$\therefore \frac{V}{Z} = 14.212 \times 10^7$$

$$\text{or } V = 14.212 \times 10^7 \times Z = 14.212 \times 10^7 \times 1.0384 \times 10^{-8}$$

$$= 1.476 \text{ volts } \textbf{Ans.}$$

8.5. CURRENT EFFICIENCY

Owing to impurities, which causes secondary reactions, the quantity of substance or substances liberated is slightly less than that calculated from Faraday's laws. This is taken into account by employing a factor, called the *current efficiency*.

The current efficiency is defined as the ratio of the actual quantity of substance liberated or deposited to the theoretical quantity, as calculated from Faraday's laws.

i.e., Current efficiency

$$= \frac{\text{Actual quantity of substance liberated or deposited}}{\text{Theoretical quantity of substance liberated or deposited}}$$

Its value usually lies between 90 and 98%.

In certain cases this efficiency is very low. For example in chromium plating it is roughly 12 to 15 per cent. It is because only 15 per cent of the total current passed through the electrolyte consisting of some chromium acid solution, is used in depositing chromium and the rest is wasted in producing oxygen and hydrogen gases, which for the purpose in hand, are useless.

Example 8.9. How much aluminium will be produced from aluminium oxide in 24 hours if the average current is 3,000 A and current efficiency is 92%. Aluminium is trivalent and atomic weight is 27. The chemical equivalent weight of silver is 107.98 and 0.00111 gm of silver is deposited by one coulomb.

[*Gorakhpur Univ. Utilization of Elec. Power and Traction 1981; M.D. Univ. Utilization of Electric Power and Traction May 2010*]

Solution : ECE of silver = 0.00111 gm/C or 111 × 10^{-8} kg C^{-1}

Chemical equivalent weight of silver = 107.98

$$\text{Chemical equivalent weight of aluminium} = \frac{27}{3} = 9$$

∴ ECE of aluminium,

$$Z = \frac{\text{ECE of silver} \times \text{chemical equivalent weight of aluminium}}{\text{Chemical equivalent weight of silver}}$$

$$= 111 \times 10^{-8} \times \frac{9}{107.98} = 9.252 \times 10^{-8} \text{ kg } C^{-1}$$

Current efficiency = 92% = 0.92

Average current, I = 3,000 A

Duration of flow of current, t = 24 × 60 × 60 = 86,400 seconds

Mass of aluminium produced, m = Z I t × current efficiency

$$= 9.252 \times 10^{-8} \times 3{,}000 \times 86{,}400 \times 0.92$$

$$= 22.062 \text{ kg } \textbf{Ans.}$$

Example 8.10. A worn-out shaft of 10 cm diameter and 25 cm long is to be coated with 2 mm thick layer of nickel. Determine the quantity of electricity required and time taken if current density of 160 A/sq metre is assumed. Assume current efficiency of 90%. Density of nickel may be taken as 8.9 gm/CC.

[*U.P. Technical Univ. Utilization of Electrical Energy & Traction 2009-10*]

Solution : Surface area of circular shaft portion to be coated,

$$S = \pi \times d \times l = \pi \times 0.1 \times 0.25 = 0.07854 \text{ m}^2$$

Mass of nickel to be deposited,

$$m = S \times \text{thickness of coating} \times \text{density of metal}$$

$$= 0.07854 \times 2 \times 10^{-3} \times 8.9 \times 10^3 = 1.398 \text{ kg}$$

Electrochemical equivalent of nickel,

$Z = 30.43 \times 10^{-8}$ kg/C ...Refer to Table 8.1

Theoretical value of quantity of electricity required,

$$Q = \frac{m}{Z} = \frac{1.398}{30.43 \times 10^{-8}} = 0.045942 \times 10^8 \text{ C or A-s}$$

$$= \frac{0.045942 \times 10^8}{3{,}600}$$

$$= 1{,}276 \text{ A-h}$$

Actual value of quantity of electricity required

$$Q' = \frac{Q}{\text{Current efficiency}} = \frac{1{,}276}{0.9} = 1{,}418 \text{ A-h } \textbf{Ans.}$$

Since the current density is 160 A/sq meter and surface area of sheet is 0.07854 m^2, therefore current used

$$I = 160 \times 0.07854 = 12.5664 \text{ A}$$

Time required,

$$t = \frac{1{,}418}{12.5664} = 112.84 \text{ hours } \textbf{Ans.}$$

8.6. VOLTAGE

The voltage necessitated to pass the current through any electrolytic cell is equal to the sum of voltage drop in the resistance of the electrolyte and the voltage drops at electrodes (anode and cathode). From the resistivity of the electrolyte and the x-sectional area and the length of the electric current path, the resistance of electrolyte can be determined. For economy the electrolyte resistance should be reduced to minimum and to achieve it in many cases special conducting agents are added to the electrolyte. The addition of sulphuric acid to copper sulphate solution in copper plating is an example of this. There is some potential difference between the cathode and the electrolyte and between the electrolyte and anode. This potential difference is a measure of the tendency of the metal to go into the solution and is known as *electrode potential.* The electrode potential depends upon the exact conditions (*i.e.,* temperature and concentration), and also upon the nature of the metal and the electrolyte. Under ideal conditions the value of electrode potential for most of the substances lie between 0.5 to 1.0 volt. The total potential difference required to pass the necessary current through electrolytic cell is about 1 or 2 volts.

8.7. ENERGY EFFICIENCY

On account of various secondary effects and reactions the substance deposited by a given quantity of electricity is less than that determined theoretically from Faraday's laws. Voltage required is also higher than that determined theoretically. Hence actual energy consumption will be higher than that determined theoretically for depositing a given quantity of the substance.

The ratio of theoretical energy required to the actual energy required for depositing a given quantity of metal is known as *energy efficiency*.

$$i.e., \quad \text{Energy efficiency} = \frac{\text{Theoretical energy required}}{\text{Actual energy required}}$$

8.8. EXTRACTION OF METALS

Extraction of metal is an electrochemical process used for the production of metal with commercially acceptable purity. There are two methods of extraction of metals depending upon the physical state of the ore. In one of the processes the ore is treated with a strong acid to obtain a salt and the solution of such a salt is electrolysed to liberate the metal. The second process is employed when the ore is available in molten state or can be fused and in this process the ore, which is in a molten/fused state, is electrolysed in a furnace.

The details of the processes adopted for some of the metals are given in Table 8.2.

Table 8.2. Extraction of Metals

Metal to be Extracted	Treatment of Ore	Solution	Energy Consumption in kWh/Tonne
Aluminium	—	Fused cryolite	20,000—25,000
Copper	Roasted and treated with sulphuric acid	Copper sulphate	2,000—2,500
Magnesium	—	Fused magnesium chloride or carnalite	17,000—20,000
Sodium	—	Fused sodium hydrate or sodium chloride and sodium nitrate	10,000—20,000
Zinc	Treated with concentrated sulphuric acid	Zinc chloride and zinc sulphate	3,000—5,000

The methods adopted for extracting zinc and aluminium are explained below :

1. *Extraction of Zinc.* This is an example in which an aqueous solution of the salt is used. The ore, consisting largely of zinc oxide, is treated with concentrated sulphuric acid, roasted, and passed through various chemical processes in order to remove impurities (such as cadmium, copper etc.) by precipitation. The zinc sulphate solution so obtained is then electrolysed. The electrolysis of zinc sulphate is accomplished in large lead-lined wooden boxes having a number of aluminium cathodes and lead anodes. Zinc gets deposited on the cathodes and is removed periodically (once or twice a day). The current density on the cathodes is about 1,000 A/m^2. The voltage per cell is about 3.5 V and usually 100 to 150 cells are used in series requiring a pressure of roughly 500 V. Energy consumption is 3,000 to 5,000 kWh/tonne.

2. *Extraction of Aluminium.* This is an example of fused electrolyte process. Aluminium is produced from bauxite containing aluminium oxide or alumina (70% in case of high grade bauxite), silica (silicon oxide) and iron oxide. The bauxite ore is first reduced to aluminium oxide by chemical treatment and then it is dissolved in fused cryolite. Cryolite is a solution of aluminium fluoride and fluoride of either of sodium, potassium or calcium. The mixture thus obtained is electrolysed. The fusion and electrolysis are accomplished in a large shallow rectangular steel bath lined with carbon; carbon anodes projecting downwards into the bath and the bottom of the bath forms the cathode. The charge is melted by the arc struck between the carbon anodes and cathode and is then maintained in a molten state by the heating action of the electric current flowing through the charge. The liquid metal deposits at the cathode and settles at the bath bottom and is periodically siphoned out into large ladles from which it is poured into pig or ingot moulds. Fresh alumina is fed into the bath at short intervals to replace that which has been decomposed by the current and the process is, therefore, a continuous one. The aluminium obtained by this process is 99.5% pure. A furnace having an area of about 15 square metres will need a voltage of about 6 volts and a current of about 40,000 amperes. Energy consumption is 20,000 to 25,000 units/tonne. Almost the whole of the aluminium required in the present day's industry is produced in this way. As the electrolytic process requires large amount of electric power and process is continuous, so such plants are installed near hydroelectric power stations. The high temperature (1,000°C) necessary to keep the ores in a fused state is maintained by the ohmic losses due to the current flowing through the electrodes and electrolyte.

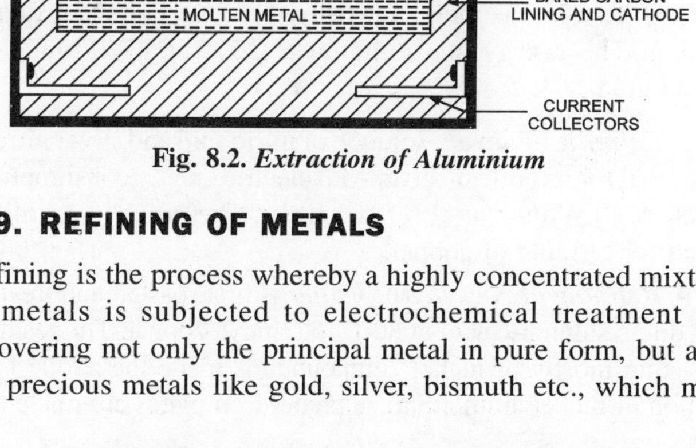

Fig. 8.2. ***Extraction of Aluminium***

8.9. REFINING OF METALS

Refining is the process whereby a highly concentrated mixture of metals is subjected to electrochemical treatment for recovering not only the principal metal in pure form, but also the precious metals like gold, silver, bismuth etc., which may be present in the form of minute traces. Some of the processes are explained below.

1. ***Refining of Copper.*** The impure copper is cast into plates which are made the anodes in a bath of copper sulphate acidified with sulphuric acid. The cathodes consist of sheets of pure copper which are coated with graphite so that the electrolyte deposit of copper may be easily stripped off. When the current is passed, copper ions are discharged at the cathode and are deposited as pure copper while hydroxyl ions discharged at the anode bring an equivalent amount of cupric ions into the solution. Thus there is a transfer of pure copper from the anode to the cathode while the impurities like nickel, zinc and iron, pass into the solution as sulphates and impurities like silver and gold, which are not affected by sulphuric acid—copper sulphate solution settles down as the *anode sludge or mud*. At regular intervals the deposit is stripped from the cathode and the anode is replaced. Copper extracted from its ore, known as blister copper, is 98 to 99 per cent pure. Copper of purity as high as 99.95% is obtained from blister copper by electrolytic process. Energy consumption in copper refining by electrolytic process is 150—300 kWh/tonne of refined copper.

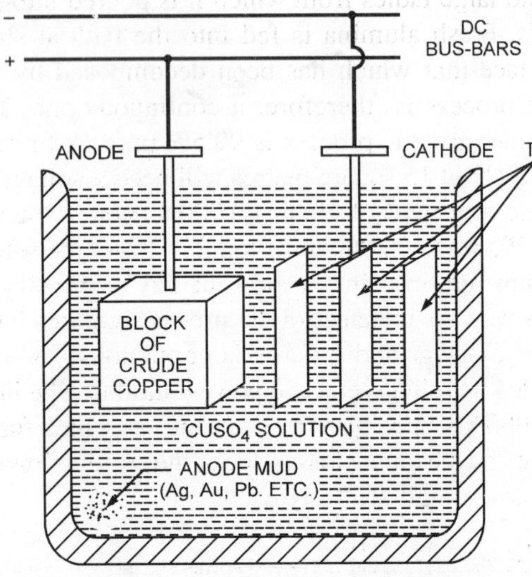

Fig. 8.3. ***Electrolytic Refining of Copper***

2. ***Refining of Gold.*** Blocks of impure gold are made the anodes in an electrolytic cell containing a solution of gold chloride ($AuCl_3$) in hydrochloric acid. Thin plates of pure gold constitute the cathodes. On electrolysis pure gold is deposited on the cathode. Gold thus obtained is 99.9% pure. For a higher purity, if desired, the metal may be put to a second electrolytic refining. The energy consumption is 300-350 kWh/tonne of refined gold.

3. ***Refining of Silver.*** Solution of nitric acid and silver nitrate is required for refining of silver with electric energy consumption of 400-425 kWh/tonne of refined silver. The process is similar to that for refining of copper.

4. ***Refining of Nickel.*** The *matter* is first roasted and treated with dilute sulphuric acid at 80°C to remove copper. The residue consisting mostly of nickel compounds is made the anode in a solution of nickel ammonium sulphate. Iron plates are made the cathode. On electrolysis nickel deposits on these plates. The energy consumption is 2,500-4,000 kWh/tonne of refined nickel.

5. ***Refining of Lead.*** The desilverised lead can be purified further by electrolysing a solution of lead fluorosilicate ($PbSiF_6$) and hydrofluorosilicic acid (H_2SiF_6) containing a little gelatin. The desilverised lead is cast into anodes while the cathodes are sheets of pure lead. Gelatin facilitates the formation of a coherent deposit of lead at the cathode. The metallic impurities which are more electropositive than lead, such as iron and tin, go into solution while the rest of the impurities are thrown down as anode mud. The energy consumption is 100-120 kWh per tonne of refined lead.

6. ***Refining of Zinc.*** The crude ore is roasted and leached with dilute sulphuric acid. The resulting solution is purified from elements present such as copper, cadmium, silver, cobalt, antimony, arsenic by treatment with the powdered zinc and with ferric hydroxide. The solution of purified zinc sulphate is then electrolysed in large concrete tank lined with lead or with rubber, using chemically pure lead as anode and aluminium cathode. The electrolysis is carried at 35-40°C using a current density of 300-400 A/m^2 and a cell voltage of 3.25-3.5 volts. The zinc is stripped from the aluminium cathodes and then melted down to give a material of high purity (99.99 per cent zinc).

7. ***Refining of Iron.*** In case of purification of iron, the iron-ammonium sulphate is used as electrolyte and energy consumption 1,000—1,600 kWh per tonne of refined iron.

The summary of the some of the important refining processes is given in Table 8.3.

Table 8.3. Refining of Metals

Metal to be Refined	Solution	Energy Consumption in kWh/Tonne
Copper	Copper sulphate	150—300
Gold	Gold chloride	300—350
Iron	Iron-ammonium sulphate	1,000—1,600
Lead	Lead fluosilicate	100—120
Nickel	Nickel-ammonium sulphate	2,500—4,000
Silver	Nitric acid and silver nitrate	400—425

8.10. PRODUCTION OF CHEMICALS

There are various industrial applications of electrolysis in the production of chemicals such as production of caustic soda, chlorine, potassium permanganate, ammonium per sulphate, hydrogen and oxygen by electrolysis.

The most important processes of production caustic soda by electrolysis of brine and production of hydrogen and oxygen gases by electrolysis of water are explained below.

1. ***Production of Caustic Soda.*** The oldest process is *Diaphragm process.* There are a number of variations, but all essentially consist of an anode compartment separated from a cathode compartment by a porous diaphragm which prevents the mechanical mixing of the two solutions. Chlorine is formed at the anode, and most of it is evolved as a gas, a small part going into solution. Sodium is discharged at the cathode and reacts with the hydroxyl ions to form sodium hydroxide and

hydrogen gas is liberated at the cathode. Usually, the brine is fed into the anode compartment to resist the flow of hydroxyl ions towards the anode.

Another method of production of caustic soda by electrolytic process is *mercury-cathode* process. Mercury cells may be built in small (1,000 A) to very large sizes (3,000 A) or even as high as 50,000 A or more per unit. Purified saturated brine is fed to the cells and commonly circulated through them with a reduction in salt concentration. Brine is resaturated for another cycle. Each unit consists of two sections ; one which is closed and has a chlorine outlet, graphite anodes, and a mercury cathode; and a second, the denuder section, containing a mercury amalgam anode and iron cathodes. Circulation of mercury between the two sections constitutes the electrical connection. The current efficiencies depend upon the concentration of the amalgam formed in the first section and the effectiveness of the circulation of the amalgam. Current densities in mercury cells are considerably higher than in diaphragm units, being of the order of 1.5 mA or more per square mm of anode surface.

Energy efficiencies are of the order of 50-60%. Anode consumption, when graphite anodes are employed, is of the order of 3-3.5 kg per tone of chlorine. The voltage requirement is about 4 volts per cell.

Other productions, that can be obtained from electrolysis of brine by interaction of the cathodic and the anodic products are sodium hydrochloride, bleaching powder and hydrogen when the solution is cold and sodium chlorate ($NaClO_3$) and hydrogen when the solution is hot.

2. *Production of Hydrogen and Oxygen Gas By Electrolysis of Water.* Gas obtained by this process are of high purity and at a cheap cost because of the low energy consumption. The electrolyte consists of 15-20% solution of caustic soda or its equivalent caustic potash, and electrodes are of iron. Sulphuric acid is no longer used. With caustic soda as electrolyte the chemical reactions are :

Cathode reaction $2Na + 2H_2O = 2NaOH + H_2$

Anode reaction $2OH = H_2O + \frac{1}{2} O_2$

Table 8.4. Production of Chemicals

Solution	Product	Energy Consumption in kWh/Tonne
Brine	Caustic soda	3,000—4,000
Brine	Chlorine, compounds of sodium (chlorates, per-chlo-rates, hypochlorates)	3,000—7,000
Potassium-manganate	Potassium-permanganate	70—80
Water	Hydrogen and oxygen	About 6 kWh per cubic metre of hydrogen and $\frac{1}{2}$ cubic metre of oxygen
Ammonium-sulphate	Ammonium persulphate	2,000—2,500

Thus hydrogen and oxygen gases are liberated at cathode and anode respectively and water disappears while the quantity of caustic soda remains constant. It is, therefore, necessary to add water to the solution periodically.

The voltage requirement is 2-2.2 V per cell during operation and 2.3-2.5 V per cell during starting period. Energy consumption is about 6 kWh per cubic metre of hydrogen and 1/2 cubic metre of oxygen.

Various industrial applications of electrolysis in the production of chemicals are summarized is Table 8.4.

8.11. ELECTRODEPOSITION

The process of depositing a coating of one metal over an other metal or non-metal electrically is called the *electrodeposition.* It is used for protective, decorative and functional purposes and includes such processes as electroplating, electroforming, electrotyping, electrofacing, electrometallisation, electrodeposition of rubber and building up of worn-out parts for repairs.

The factors, on which the quality of electrodeposition depends, are given below.

1. *Nature of Electrolyte.* The formation of smooth deposit largely depends upon the nature of electrolyte employed. The electrolyte from which complex ions can be obtained, such as cyanides, provides a smooth deposit.

2. *Current Density.* Electrodeposition depends upon the rate at which crystals grow and the rate at which fresh nuclei are formed, therefore, at low current densities the deposits are coarse and crystalline in nature. The deposit of metal will be uniform and fine-grained if the current density is used at rate higher than that at which the nuclei are formed. In case the rate of formation of nuclei is very high due to very high current density, there is a chance that the limiting value of the electrolyte is exceeded. At such instances, the deposit will be spongy and porous.

3. *Temperature.* A low temperature of the solution favours formation of small crystals of metal; and a high temperature, large crystals. In some cases this is very marked, a difference of only 15°C resulting in a 50% decrease in strength of the metal deposited. On the other hand, high temperature may give beneficial results due to (*a*) increased solubility of the salts, permitting greater metal concentration and higher current densities ; (*b*) increased conductivity, which also permits higher current densities and reduces the tendency to form trees ; (*c*) decreased occlusion of hydrogen in the deposited metal, which in many cases is the case of bad deposits. Since both (*a*) and (*b*) tend to decrease crystal size, they may in some cases counteract the tendency of temperature alone to increase the crystal size.

4. *Conductivity.* The use of a solution of good conductivity is important from the standpoint of view of economy in power consumption and also because it reduces the tendency to form trees and rough deposits.

5. *Electrolytic Concentration.* Higher current density, which is necessary to obtain uniform and fine-grain deposit, can be achieved by increasing the concentration of the electrolyte.

6. *Additional Agents.* The addition of acids or other substances to the electrolyte reduces its resistance, as already mentioned. There is another class of additional agents which takes little or no direct part in the chemical reactions but influences the nature of deposit, sometimes even making an otherwise unworkable process into one of practical importance. Such additional agents are glue, gums, dextrose, dextrin, gelatine, agar, alkaloids, albumen, phenol, glycrin, sugar, glucose, rubber etc. The crystal nuclei absorbs the additional agent added in the electrolyte. This prevents it to have large growth and thus deposition will be fine-grained. For obtaining satisfactory deposit of zinc from zinc sulphate solution addition of glucose or certain types of sugar is necessary.

7. *Throwing Power.* This is the ability of electrolyte to produce uniform deposit on an article of irregular shape and is one of the most important characteristics of plating or deposition bath. The distance between the various portions of cathode and anode will be different due to irregular shape of the cathode. Due to unequal distance, the resistance of the current path through the electrolyte for various portions of the cathode will be different but the potential difference between the anode and any point on the article to be plated (cathode) will, of course be the same and the result will be that the current density will be more on the portion nearer to anode and it will cause uneven deposit of the metal. Throwing power can be improved in two ways—firstly by increasing the distance between the anode and cathode and secondly by reducing the voltage drop at the cathode surface. In some cases decrease of current density causes a decrease in voltage drop at cathode, leaving more voltage available for overcoming the resistance of the electrolyte, thus tending to counteract any change in current concentration. This is the reason that solutions of the cyanides of metals usually have a better throwing power than solutions of the sulphates.

8. *Polarization.* The rate of deposition of metal increases with the increase in electroplating current density up to a certain limit after which electrolyte surrounding the base metal becomes so much deplated of metal ions that the increase in current density does not cause increase in rate of deposition. Use of current density beyond this limit causes electrolysis of water and hydrogen liberation on the cathode. This hydrogen evolved on the cathode blankets the base metal which reduces the rate of metal deposition. This phenomenon is called the *polarization.* Blanketing effect can be reduced by agitating the electrolyte. With reverse current electroplating, in which at regular intervals plating current is reversed for a second or so, sufficient electron concentration is established around the base metal and the polarization effect becomes negligible even with very high overall speed of plating.

The other advantages of *reverse current plating* are (*i*) unsound and inferior metal is deplated during reverse current period and flat level surfaces are produced. (*ii*) Metal surface is brightened causing elimination of buffing or polishing operation.

8.12. ELECTROPLATING

Electroplating is an art of depositing a superior or a more noble metal on an inferior or a base metal by means of electrolysis of an aqueous solution of a suitable electrolyte. For example, metals like iron which are easily corroded by atmospheric air, moisture and carbon dioxide, are coated electrolytically with deposits of nickel or chromium which are more resistant to chemical attack. Picture frames and machinery parts are often chromium-plated to protect them from corrosion and at the same time to give them a good polish.

Sometimes, electroplating is done, with a view to repair worn-out parts of a machinery. In such cases the suitable material is deposited electrolytically on the effected parts of the machinery.

Electroplating is also done occasionally for ornamentation and decoration purposes. For example, several articles made of copper or its alloys, such as tablewares, decoration pieces, are coated with silver or gold.

The electrolytic deposits are crystalline in nature. The crystals must be very fine in order to get firm, coherent and uniform deposits. For this purpose, suitable electrolytes should be used in the electrolytic bath and current density used should have an appropriate value. The temperature should also be maintained at a proper level. By experiments, a certain optimum value of current density and temperature have been worked out for each electrolyte. For example, for Na_3 $[Cu(CN)_4]$ bath, the optimum current density lies between 32-65 A/m^2 and optimum temperature is 45°C. The optimum conditions for $[Ag(CN)_2]$ bath, are 42 A/m^2 of current density and 20°C as the temperature. The articles to be coated with the noble metals should be in as high a state of purity as possible. These conditions are briefly discussed below.

Preparation For Plating. The preparation of an object for plating may involve any or all of the following operations :

1. Removal of oil, grease, or other organic material.
2. Removal of rust, scale, oxides, or other inorganic coatings adhering to the metal.
3. Mechanical preparation of the surface of the metal to receive the deposited metal, by polishing, buffing etc.

For the first, soaps, hot alkali solutions, or organic solvents such as gasoline or carbon tetrachloride are used, for the second, various acids, alkali and salt solutions, mechanical abrasion, and electrolytic cleaning, and for the third mechanical abrasion and polishing are used.

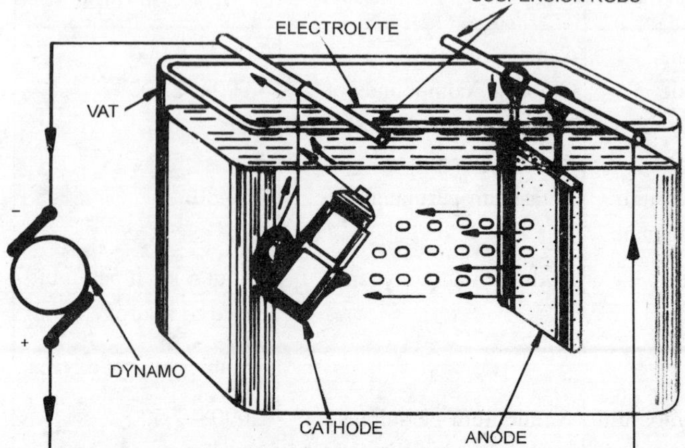

Fig. 8.4

Cleaning Methods. In case the object to be electroplated is not cleaned, polished and degreased, the deposit formed may not be well adherent to the base metal and is likely to peel off. For smooth, bright and strong deposit, the surface upon which a layer of a noble metal is required, should be thoroughly cleaned first mechanically by grinding or scratching (against a rough surface) or sandblasting and then chemically by treatment with hot alkalies (if the surface is greasy) or with dilute acids (if the surface has oxide films) or with organic solvent.

Electrolytic Bath. The electrolyte used in the electrolytic bath depends upon the nature of the metal to be deposited.

For *copper plating* two types of electrolytic baths are used. In acid type bath, solution is made of 150-200 gm of copper sulphate and 25-35 gm of sulphuric acid per 1,000 cc of solution. Current density used is 150-400 A/m^2 and temperature of 25 to 50°C. Deposit obtained is thick and rough requiring polishing. In cyanide bath solution is made of 25 gm of copper cyanide, 28 gm of sodium cyanide, 6 gm of sodium carbonate and 6 gm of sodium bisulphate per 1,000 cc of solution. Current density used is 50-150 A/m^2 and the temperature required is 25-40°C. It provides thin and smooth deposits. Copper anodes are used in both of the baths.

For *silver plating* solution consisting of 24 gm of silver cyanide, 24 gm of potassium carbonate and 36 gm of potassium cyanide per 1,000 cc is used. The required current density and temperature are 50-150 A/m^2 and temperature of 20-35°C respectively.

For *gold plating* solution used consists of 18 gm of potassium gold cyanide, 12 gm of potassium cyanide, 6 gm of potassium sulphate and 12 gm of caustic potash per 1,000 cc Anode employed is of stainless steel. Current density of 50—150 A/m^2 and temperature of 50–70°C are used.

For *chromium plating* solution most commonly used consists of 180-300 gm of chromic acid and 2-3 gm of sulphuric acid per 1,000 cc. Current density employed is 1,500-2,500 A/m^2 and the working temperature is 35-50°C. Current density used is higher for hard chromium plating than for decorative plating. Anodes are of antimonial lead. Vats used for chromium plating are of steel with lead lining. Chromic acid is added in the solution when required. Arrangements for removal of the fumes are also to be provided.

For *nickel plating* the solution consists of 180-240 gm of nickel sulphate, 36 gm of nickel chloride and 24 gm of boric acid per 1,000 cc. The current density used is 100-200 A/m^2. The working temperature is 25-40°C. Anode is of pure nickel.

Summary of the various processes is given in Table 8.5.

Table 8.5. Summary of Electroplating Practice

Metal Deposited	Principal Uses	Solution	Current Density in A/m^2	Voltage	Temperature of Solution	Throwing Power
Cadmium	Protection	Cadmium oxide and sodium cyanide	100—400	1—4	Room temperature	Good
Chromium	Decorative Engineering (hard) cylinder liners (porous)	Chromic acid	1,500—2,500	6—8	35—50°C	Poor
Copper	Electroforming under coat for other metals	Copper sulphate	150—400	1—2	25—50°C	Fair
		Cuprous cyanide and sodium cyanide	50—150	1.5—3	25—40°C	Good
Gold	Decorative	Double cyanide of potassium and gold	50—150	5—15	50—70°C	Good
Iron	Electroforming	Iron chloride	600	—	80–90°C	—
	Repair	Iron sulphate	200	—	Room temperature	—
Lead	Protection	Fluoborate	100—800	0.5	Room temperature	Good
	Bearing surfaces	Sulphate	50—400	3—8	25—50°C	Poor
Nickel	Protection decorative Electroforming under coat for Cr etc.	Nickel chloride and Nickel sulphate	100—200	0.5—3	25—40°C	Fair
Silver	Decorative Protective Bearing surfaces	Double cyanide of silver and sodium	50—150	1	20—35°C	Good
Tin	Protection Food and dairy Bearings Electrical industry	Sodium stannate	400	4—8	65—90°C	Excellent
		or Stannous Stannate	400	1—3	Room	Fair
Zinc	Protection	Zinc sulphate	150—4,000	—	25—40°C	Fair
		Zinc cyanide	100—500	—	40°C	Good
Brass	Rubber bonding decorative	Double cyanide of zinc and potassium	30—100	2—3	25—40°C	Good
Bronze	Decorative under coat for chromium	Cyanide-stannate	200—1,000	3—6	70°C	Excellent

8.13. DEPOSITION OF ALLOYS

It is possible to deposite alloys also provided the electrode potentials of the constituent metals are not much different. Electroplating of brass and bronze are example of it. For deposition of alloys the anode is made of an alloy to be deposited and the electrolyte consists of a mixture of electrolytes which would have been employed for separate deposition of the metals consumed. For brass plating solution of double cyanides of zinc and potassium and of copper and potassium is used as electrolyte and current density of 25-40 amperes/m^2 is used.

8.14. ELECTROFORMING

Reproduction of objects by electrodeposition on some sort of a mould or form is known as *electroforming*. This is another application of electrodeposition.

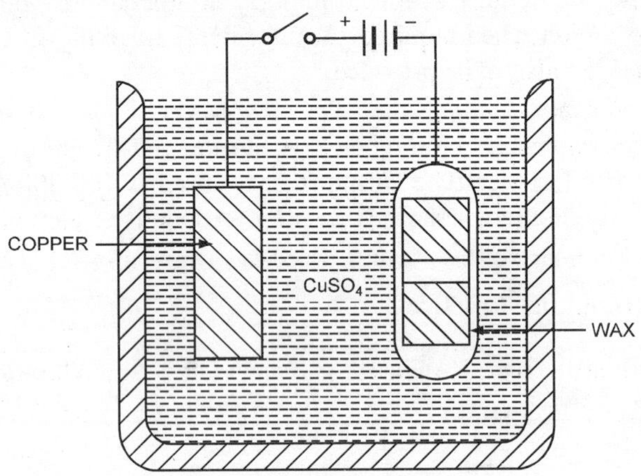

Fig. 8.5. *Electroforming*

In the reproduction of coins, medals, engravings and the like, a mould is first made by impressing the object, say, in wax. The surface of the wax, which bears exact impressions of the object, is coated by powdered graphite in order to make it conducting. The mould is then dipped in an electroforming cell as a cathode. After obtaining coating of desired thickness, the article is removed and the wax core is melted out of the metal shell.

This process is employed in the manufacture of gramophone records, in which the original recording is done on a record made of wax composition. This mould is then coated with powdered gold to make it conducting and dipped into a blue vitriol electrolyte as the cathode. The solution is kept saturated by using a copper anode. The plating of copper on to the wax record produces a negative master plate which is used to stamp a large number of shellac discs.

Production of seamless tubes is another application of electroforming. Ferric chloride solution is used as an electrolyte, cast iron as anodes and revolving mandrel as cathode for producing iron tubes. For producing copper tubes copper sulphate acid solution is used as an electrolyte and cast copper is made as anode. A thin coating of material, which renders easy withdrawl of tube of material deposited, is given to the mandrel.

8.15. ELECTROTYPING

This is a special application of electroforming and it is used to reproduce printers type, engraving and medals etc. The process is same as for electroforming.

8.16. ELECTROFACING

It is a process of coating of metallic surface with a harder metal by electrodeposition in order to increase its durability.

8.17. ELECTROMETALLISATION

It is a process of depositing metal on conducting base for protection and decoration. Non-conductive base is made conductive by a coating of graphite employed as cathode.

8.18. ELECTRODEPOSITION OF RUBBER

Rubber latex, as obtained from the tree, consists of very fine colloidal particles of rubber suspended in water. Like other colloidal solutions, particles of rubber are negatively charged. On electrolysis of the solution the rubber particles migrate towards the anode and deposit on it. Current density of roughly 100 A/m^2 is used.

8.19. ANODIZING

Anodizing is a process of anodic oxidation in which a thin uniform passive (tough thin, uniform and impermeable) film is produced artificially by the passage of electric current. Passive film is formed to protect the base metal from further corrosion. In the process of anodizing electric current of 10 to 30 amperes per square metre is passed through an electrolyte of 3% solution of chromic acid for a period of half an hour to one hour, the article to be anodized is thoroughly cleaned and made anode and carbon rod is made cathode. Aluminium and magnesium have capacity of producing such passive films and, therefore, anodizing is applicable to them. In case of aluminium great feature of these coatings is their submicroscopic porosity. This makes the passive films to take dyes which are absorbed and affect the entire firm length.

8.20. ELECTROPOLISHING

Closely allied to bright dipping is electropolishing, which utilises anodic treatment in specially formulated electrolytes to bring up a polished surface on such metals as stainless steel. Electropolishing is also useful as a tool in preparing metallic surfaces for microscopic examination. Both bright dipping and electropolishing depend on the more rapid eating away by the solution of micro-projections on the metal, so that a smoother surface results.

Electropolishing has one advantage over buffing. Unlike buffing electropolished surfaces are free from cold worked surface cemented layer embedding abrasive particles of mop in the metal.

The process of polishing, in principle, consists of making the work as anode in a suitable electrolytic solution. This develops insoluble compounds that are broken down by more anodic action on the hills than on valleys of surface. Thus smooth surface is produced.

8.21. POWER SUPPLY FOR ELECTROLYTIC PROCESSES

Power supply required for electrolytic processes is direct current and at very low voltage. The power required for electrodeposition is usually very small (between 100 and 200 amperes at 10 or

12 volts) and can be obtained either by employing a motor-generator set consisting of a standard induction motor driving a heavy-current low-voltage dc generator (preferably separately excited) or by employing the copper oxide rectifier. The latter is preferred because of low maintenance cost, occupying less space and higher operating efficiency. Mercury-arc rectifier cannot be used because it has low efficiency at low output dc voltage on account of constant voltage drops at cathode and anodes. The plate rectifier unit is usually placed along with its transformer in the oil so that it may be protected from the corrosive fumes of the electrolyte. Recently the solid state rectifying devices employing germanium and silicon diodes have been developed for use. These solid state devices occupy very small space even as compared to metal rectifiers. Output dc voltage can be controlled by controlling the excitation of the generator in case of motor-generator set and by means of continuously variable auto-transformer in case of rectifier supply. This method of control is suitable where only one bath is being supplied. In case, more than one bath are supplied, a variable resistance is connected in series with each bath so that the supply to each bath can be controlled independently. With the development of SCRs or thyristors, which can control output voltages of power supplies, their use in power supplies for electrochemical processes has increased. They are also compact and light in weight, even cooling attachments are included. Voltage control is by using output transformers.

Power supply required for extraction and refining of metals and large scale manufacture of chemicals is in very large amounts. Since most of the processes are continuous, therefore, have a load factor of 100 per cent. Because of power requirements in huge amount and at 100 per cent load factor, such plants are located near the hydroelectric power stations or atomic power stations even if extra transportation of raw material is necessitated. The advantage of a high load factor is greater with such stations than with steam stations and also transmission costs are eliminated. Nangal fertilizer factory producing calcium ammonium nitrate and heavy water and utilizing power of 180 MW from left bank Bhakra power house and Sri Ram Fertilizer factory located at Kota (Rajasthan) are instances in the support of the above statement.

The voltage of each cell is about 10 volts, but if many cells are connected in series, current of the order of several thousand amperes will be required at the voltage of the order of 500 to 800 volts. Thus by employing heavy current motor-generators, rotary convertors or even mercury-arc rectifiers, the required supply may be obtained from the modern grid network.

Exercises

1. State Faraday's laws of electrolysis and explain them clearly.
[U.P. Technical Univ. Utilization of Electrical Energy & Electric Traction 2005-06; M.D. Univ. Utilization of Electric Power & Electric Traction Dec. 2009; Pb. Technical Univ. Electric Drives & Utilization Dec. 2009]
2. Discuss various laws of electrolysis. What are different applications of electrolysis.
[Pb. Technical Univ. Electric Drives & Utilization May 2007]
3. What is electrolysis? State Faraday's laws of electrolysis and explain them clearly.
[M.D. Univ. Utilization of Electric Power & Traction Dec. 2008]
4. What are the various applications of electrolysis ? Explain extraction of metals. (Take example of any metal).
[Gorakhpur Univ. Utilization of Electrical Power & Traction September 1984]
5. Briefly describe the process of extraction of aluminium from the ore.
[U.P. Technical Univ. Utilization of Electrical Energy and Electric Traction 2006-07]
6. Explain with the help of a neat sketch the extraction process of aluminium.
[Rajasthan Technical Univ. Utilization of Electrical Power 2011]
7. Explain the terms 'polarisation'; 'throwing power' and 'electrodeposition'. How are zinc and copper refined from their base metals electrically ?
[Gorakhpur Univ. Utilization of Electrical Power and Traction September, 1983]
8. What is electrolysis? Explain the following terms used in electrolytic process :
(*i*) Current efficiency (*ii*) Throwing power (*iii*) Electro-chemical equivalent ?
[M.D. Univ. Utilization of Electric Power & Traction May 2009]
9. Define and explain 'current efficiency', 'voltage required' and 'energy efficiency' when referred to electrolytic processes. Also state and explain Faraday's laws of electrolysis.
[G.B. Technical Univ. Utilization of Electrical Energy & Traction 2010-11]
10. Describe principle of electrode position in electrolytic process.
[U.P. Technical Univ. Utilization of Electrical Energy & Traction 2009-10]
11. What is electrodeposition ? Explain the various factors on which the quality of deposition depends.
[Pb. Technical Univ. Electric Drives & Utilization Dec. 2008]
12. Explain the basic principle of electrodeposition. Give its applications.
[U.P. Technical Univ. Utilization of Electrical Energy & Electric Traction 2005-06]
13. State and explain Faraday's laws of electrolysis. What factors govern the rate of electrodeposition process?
[Rajasthan Technical Univ. Utilization of Electrical Power 2010]
14. What is electroplating and what for is it done? Describe various operations involved in electroplating.
[M.D. Univ. Utilization of Electric Power & Traction Dec. 2008]
15. Why is it necessary to clean a job before electroplating and how it is performed?
[Rajasthan Technical Univ. Utilization of Electrical Power 2010]
16. Describe in detail the process of nickel electroplating in industry.
[M.D. Univ. Utilization of Electric Power & Traction May 2009]
17. What do you understand by 'reverse current process' in electroplating? Discuss its advantages.
[M.D. Univ. Utilization of Electric Power and Traction Dec. 2009, May 2010]
18. Give the detailed description of equipment and process used for chromium plating. What is the composition of electrolyte used ?
[U.P. Technical Univ. Utilization of Electrical Energy & Electric Traction 2005-06]
19. List the major applications of electrolysis. What do you understand by anodising ?
[U.P. Technical Univ. Utilization of Electrical Energy and Electric Traction 2005-06]

20. Explain briefly : (*i*) Electrodeposition of rubber (*ii*) Anodic oxidation of aluminium.
[G.B. Technical Univ. Utilization of Electrial Energy & Traction 2010-11]
21. What is meant by anodising ? Explain the process of anodising and describe the equipment used for it.
22. List the major applications of electrolysis. Explain the basic principle of electrodeposition. Discuss in detail the power supply requirements for different electrolytic processes.
23. State various types of power supplies used for electrolytic process. Discuss any one in detail.
[Rajasthan Technical Univ. Utilization of Electrical Power 2010]
24. Why it is preferred to supply large chemical works from hydroelectric or atomic power stations?
[Rajasthan Technical Univ. Utilization of Electrical Power 2011]
25. Write short notes on the following :
    (*i*) Electrolysis and its laws.
    (*ii*) Applications of electrolysis. **[UPTU 2006-07]**
    (*iii*) Nickel plating. **[UPTU 2006-07]**
    (*iv*) Extraction of metals using electrolytic processes.
    (*v*) Anodizing and its applications.
    [W.B. Univ. of Technology 2009]
    (*vi*) Reverse current process of electroplating.
    [UPTU 2006-07]
    (*vii*) Electroforming. **[UPTU 2006-07]**
    (*viii*) Manufacture of chemicals by electrolytic processes.
    (*ix*) Power supply requirements for electrolytic processes.
    (*x*) Extraction and refining of metals by electrolysis.

Short Answer Type Questions With Answers

Q. 1. What is electrolysis ? **[P.T.U. December 2009]**

Ans. The breaking of a substance (known as electrolyte) into ions by flow of electric current through it is known as electrolysis.

Q. 2. What happens when an electric current is passed through solution of copper sulphate ? **[P.T.U. December 2009]**

Ans. When an electric current is passed through solution of copper sulphate, the electrolyte ($CuSO_4$) is decomposed into cations (Cu^{++}) and anions (SO_4^{--}) according to chemical reaction

$$CuSO_4 \longrightarrow Cu^{++} + SO_4^{--}$$

The cations move, toward the cathode (or negative electrode) and each of the positively charged copper ions reaching the cathode takes two electrons from it and becomes a metallic atom of copper. The negatively charged sulphions (SO_4^{--}) move towards the anode (or positive electrode) and each of the negatively charged sulphions reaching the anode gives up two electrons to it and ceases to be anion. Thus, metallic copper is deposited on the cathode, while the sulphions are collected at the anode and combine with water forming H_2SO_4 and O_2 according to chemical reaction

$$2SO_4 + 2H_2O \longrightarrow 2H_2SO_4 + O_2$$

The oxygen gas so formed is liberated at the anode.

Q. 3. Write the applications of electrolysis. **[P.T.U. Dec. 2006]**

Ans. Electrolytic processes are widely used for the extraction of pure metals from their ores (such as aluminium, zinc, copper, magnesium, sodium etc.), manufacturing of various chemicals (such as caustic soda, potassium permanganate, hydrogen, oxygen, chlorine etc.), electrodeposition of metals including electroplating, electrotyping, electroforming, building up of worn-out parts in metallurgical, chemical and other industries.

Q. 4. What is meant by electrochemical equivalent of a substance ?

Ans. Electrochemical equivalent (ECE) of a substance is the amount of substance deposited on passing a steady electric current of 1 A for one second through its solution.

Q. 5. What is relation between chemical equivalent, atomic weight and valency of any substance ?

Ans. Chemical equivalent of any substance may be defined as the ratio of its atomic weight to its valency

i.e., $$\text{Chemical equivalent} = \frac{\text{Atomic weight}}{\text{Valency}}.$$

Q. 6. Define current efficiency in electrolytic process.
[P.T.U. Dec. 2006]

Ans. The current efficiency is defined as the ratio of the actual quantity of substance liberated or deposited to the theoretical quantity, as calculated from Faraday's laws.

i.e., Current efficiency

$$= \frac{\text{Actual quantity of substance liberated or deposited}}{\text{Theoretical quantity of substance liberated or deposited}}$$

Its value usually lies between 90 and 98%.

Q. 7. What is electrodeposition ?

Ans. The process of depositing a coating of one metal over another metal or non-metal electrically is called the electrodeposition.

Q. 8. What for electrodeposition process is used ?

Ans. The electrodeposition process is used for protective, decorative and functional purposes and includes electroplating, electroforming, electrotyping, electrofacing, electrometallisation, electrodeposition of rubber, building up of worn-out parts for repairs etc.

Q. 9. Enumerate the factors on which the quality of electrodeposition depends.

Ans. The factors on which the quality of electrodeposition depends are (*i*) nature of electrolyte (*ii*) current density (*iii*) temperature (*iv*) conductivity and concentration of electrolyte (*v*) additional agents (*vi*) throwing power and (*vii*) polarisation.

Q. 10. What is meant by 'throwing power' and how it can be improved ? **[Gorakhpur Univ. Sept. 1981]**

Ans. Throwing power is the ability of electrolyte to produce uniform deposit on an article of irregular shape and is one of the important characteristics of plating or depositing bath. It can be improved in two ways—firstly, by increasing the distance between the anode and cathode, and secondly, by reducing the voltage drop at the cathode surface.

Q. 11. What is polarization and how can it be made negligible ?

Ans. Use of current density, in electrodeposition process, beyond a certain limit causes electrolysis of water and hydrogen liberation on the cathode. This hydrogen evolved on the cathode blankets the base metal which reduces the rate of metal deposition. This phenomenon is called polarization.

Polarization can be reduced by agitating the electrolyte and can be made negligible by reverse current process.

Q. 12. What do you understand by "reverse current process" in electroplating ?

Ans. In reverse current process in electroplating, at regular intervals plating current is reversed for a second or so.

Q. 13. What is electroplating ?

Ans. Electroplating is an art of depositing a superior or more noble metal on an inferior or a base metal by means of electrolysis of an aqueous solution of a suitable electrolyte.

Q. 14. What is use of electroplating ? **[P.T.U. May 2007; Dec. 2007]**

Ans. Electroplating is done occasionally for ornamentation and decoration purposes. Sometimes, it is done, with a view to repair worn-out parts of a machinery.

Q. 15. Why the article to be electroplated is thoroughly cleaned, polished and degreased before putting it for electroplating ? **[Gorakhpur Univ. 1980]**

Ans. In case the object to be electroplated is not cleaned, polished, and degreased, the deposit formed may not be well adherent to the base metal and is likely to peel off.

Q. 16. What is electroforming ? **[P.T.U. Dec. 2008]**

Ans. Reproduction of objects by electrodeposition on some sort of a mould or form is known as *electroforming.* This is another application of electrodeposition.

Q. 17. What is meant by anodising ?

Ans. Anodizing is a process of anodic oxidation in which a thin uniform passive (tough, thin, uniform and impermeable) film is produced artificially by the passage of electric current.

Q. 18. What is anodizing and Faraday constant ? **[P.T.U. May 2009]**

Ans. For anodizing refer to Answer of Q. 17.

According to Faraday's second law of electrolysis; the chemical equivalent weight is proportional to electrochemical equivalent

i.e. $e \propto Z$ or $e = FZ$

where F is called the Faraday constant. The quantity of electricity which liberates one *gm* equivalent of an element is known as Faraday constant *i.e.* 1 F = 96,540 coulombs.

Q. 19. What type of power supply is required for electrolytic processes ?

Ans. Power supply required for electrolytic processes is dc and at very low voltage ; usually very small in magnitude (between 100 and 200 A at 10 or 12 V)

Q. 20. Why it is preferred to instal large chemical plants near hydroelectric or atomic power plants ?

Ans. Because of huge power requirement and 100 per cent load factor, large chemical plants are located near the hydroelectric or atomic power plants. The advantage of a high load factor is greater with such plants than with steam power plants.

Problems

1. A coating of silver 0.005 mm thick is to be deposited on a vessel whose surface area is 0.8 m^2. The current is maintained at 5 A. The weight of the silver per cc is 10.5 gm. The ECE of silver is 0.001118 gm per coulomb. Calculate the time taken for finishing the job.
[Ans. 2 hours 5 minutes and 15 seconds]

2. State Faradays laws of electrolysis and explain them clearly. If 20 gm of nickel is deposited by 120 A current flowing for 10 minutes, how much copper would be deposited by 100 A current in 5 minutes ? Atomic weight of nickel and copper are 58.6 and 63.18 respectively and valency of both is two. **[U.P. Technical Univ. Utilization of Electrical Energy and Electric Traction 2005-06]**
[Ans. 9 gm]

3. Find the thickness of copper deposited on a plate of 2.25 cm^2 during electrolysis if a current of one ampere is passed for 150 minutes. Density of copper is 8.9 gm per cc and electrochemical equivalent of copper is 0.0003295.
[Pb. Technical Univ. Electric Drives & Utilization May 2008]
[Ans. 1.48 mm]

4. Find the thickness of copper deposited on a plate of 2.25 cm^2 during electrolysis if a current of one ampere is passed for 100 minutes. Density of copper is 8.9 gm per cc and electrochemical equivalent of copper is 0.0003295.
[Ans. 0.9872 mm]

5. A group of 400 cells in a copper refining plant carries a current of 5,000 A. The voltage per cell is 0.25 V. If the plant working time per week is 100 hours, determine the annual output of refined copper from the cells and the energy consumption per tonne. ECE of copper = 0.327 mg per coulomb. **[Ans.** 12,243 tonnes; 212.4 kWh per tonne]

6. A copper refining plant employs 5,000 cells carrying a current of 5,000 A. The voltage per cell is 0.28 volt. If the plant is worked for 48 hours per week, determine the annual output of refined copper and energy consumption per tonne. ECE of copper is 32.95 × 10^{-8} kg C^{-1}.
[Gorakhpur Univ. Utilization of Elec. Power & Traction 1982]
[Ans. 74,019 tonnes; 236 kWh]

7. A copper refining plant, employing 500 cells carries a current of 4,000 A. The voltage per cell is 0.25 V. Determine the annual output of the plant and energy consumption per tonne, if the plant works for 80 hours per week and ECE of copper is 32.81 × 10^{-8} kg C^{-1}.
[Gorakhpur Univ. Utilization of Elec. Power & Traction 1982]
[Ans. 9,827 tonnes; 212 kWh/tonne]

8. A 20 cm long portion of a circular shaft having 10 cm diameter is to be coated with a layer of 1.5 mm nickel. Determine the quantity of electricity in Ah and the time taken for the process. Assume a current density of 195 A/m^2 and a current efficiency of 92%. Specific gravity of nickel is 8.9 and its ECE is 1.0954 kg per 1,000 Ah.
[Ans. 832.33 Ah; 68 hours]

❑❑❑

CHAPTER 9

Refrigeration and Air Conditioning

INSIDE THIS CHAPTER

9.1. INTRODUCTION

Refrigeration and air conditioning has become a necessity in every sphere of human utility and industrial set-ups whether it is in houses, schools, colleges, departmental stores, cinema houses, dancing halls, night clubs, restaurants, cars, buses, railways, manufacturing and measurement or inspection departments/shops of the industry, power houses. In the early years of the twentieth century it used to be considered a luxury but with the change in the atmospheric set-ups, changing environments and standards of living it has become necessity. It is employed for the comforts of human beings both as domestic appliances and industrial equipment for producing low air temperature and for cooling of storage chambers in which perishable foods, fruits, medicines, chemicals etc. are preserved. In industry it is employed to maintain the temperature of the factories, tool rooms, and working place below that of surroundings so that the workers, various types of materials and equipments can perform their duties efficiently. It is also used for production of ice, for liquifying of gases and vapours in chemical and pharmaceutical industries and in air conditioning of buildings, coaches, milk vans etc.

9.2. TERMINOLOGY

The important terms used in refrigeration and air conditioning are defined below:

1. ***Refrigeration.*** The term '*refrigeration*' may be defined as the process of removing heat from a substance under controlled conditions. It also includes the process of reducing and maintaining the temperature of a body below the general temperature of its surroundings. In other words, the refrigeration means a continued extraction of heat from a body whose temperature is already below the temperature of its surroundings.

2. ***Refrigeration System.*** The mechanism used for lowering or producing low temperature in a body or a space, whose temperature is already below the temperature of its surroundings, is called the refrigeration system. Here the heat is being generally pumped from the low level to the higher one and is rejected at high temperature.

3. ***Refrigerator.*** Theoretically, a refrigerator is a reversed heat engine or a heat pump which takes out heat from a cold body and delivers it to a hot body.

4. ***Refrigerant.*** The refrigerant is a heat carrying medium which during their cycle (*i.e.*, compression, condensation, expansion, and evaporation) in the refrigeration system absorbs heat from a low temperature system (cold body) and delivers it to a higher temperature system (hot body).

5. ***Air Conditioning.*** 'Air conditioning' involves the control of the temperature, humidity (moisture content), and motion of air in an enclosure. Air conditioners usually incorporate filters for removing lint and dust and may include equipment for services such as removing odours or killing bacteria.

6. ***Tonne of Refrigeration (TR).*** It is a practical unit of refrigeration and is defined as the amount of refrigeration effect produced by the uniform melting of one tonne (1,000 kg) of ice from and at 0ºC in 24 hours. It is equivalent to heat extraction rate of 50 kcal/minute. While large commercial plants are designated in tonnes, domestic refrigerators are designated by the capacity (volume) of the cabinet. One tonne capacity of a refrigeration plant needs approximately 3/4 kW output motor.

9.3. REFRIGERATION CYCLE

For lowering the temperature of a body below that of its surroundings and then maintaining its temperature at this lower level, a continuous extraction of the heat from the body is required. The lowering or producing of low temperature in a body or a space is accomplished by a mechanism called the *refrigeration system.* Here the heat is being generally pumped from the low level to the higher one and is rejected at that high temperature. This rejection of heat from the low level to the high level of temperature can only be performed with the help of external work according to second law of thermodynamics. The total amount of heat being rejected to the outside body consists of two parts—heat extracted from the body to be cooled and the heat equivalent to the mechanical work required for extracting it.

A refrigerator in the theoretical sense is nothing but a reverse heat engine run in the reverse direction by means of external aid. Every type of refrigeration system used for producing cold must have the following four basic units.

(*i*) Low temperature thermal sink to which the heat is rejected for cooling the space.

(*ii*) Means of extracting the heat energy from the sink, raising its level of temperature before delivering it to heat receiver.

(*iii*) A receiver is a storage to which the heat is transferred from the high temperature, high pressure refrigerant.

(*iv*) Means of reducing the pressure and temperature of the refrigerant before it returns to the sink.

The refrigeration cycle (reversed heat engine cycle) is shown in Fig. 9.1 in which the four basic units or processes of the cycle are opposite to those of the heat engine cycle's evaporation, expansion, condensation and compression *i.e.*, evaporation, compression, condensation and expansion. Thus we see that by reversing a heat engine cycle completely and by changing the working agent also to the refrigerant, a refrigeration cycle is obtained.

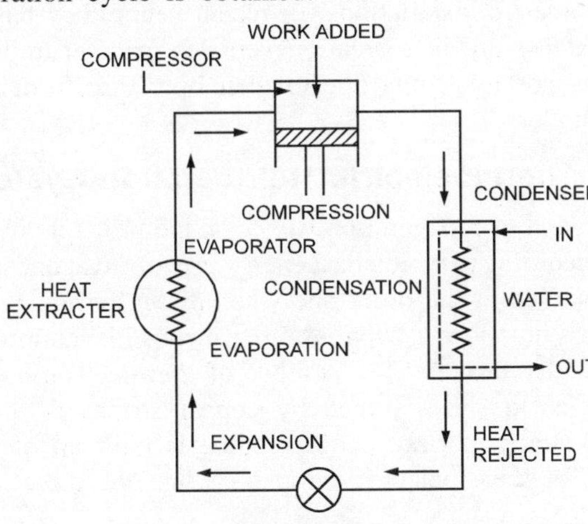

Fig. 9.1. ***Referigeration Cycle***

9.4. REFRIGERATION SYSTEMS

There are three refrigeration systems :

(*a*) Vapour compression refrigeration system.

(*b*) Vapour absorption refrigeration system.

(*c*) Thermoelectric refrigeration system.

9.4.1. Vapour Compression Refrigeration System

This is the most important system from the standpoint of commercial and domestic utility and most practical form of refrigeration. The working fluid, known as a *refrigerant*, used in this refrigeration system readily evaporates and condenses or changes alternately between the vapour and liquid phases without leaving the refrigerating plant. During evaporation it absorbs heat from the cold body and in condensing or cooling or liquifying it rejects heat to the external hot body. The heat absorbed from cold body during evaporation is used as its latent heat for converting it from liquid to vapour. Thus a cooling effect is created in the working fluid. This system of refrigeration, thus act as a *latent heat pump* since it pumps its latent heat from the cold body or brine and rejects it or deliver it to the external hot body or the cooling medium. According to the laws of thermodynamics, this can be done only on the expenditure of energy which is supplied to the system in the form of electrical energy driving the compressor.

The main advantages of this system of refrigeration over the other systems are the smaller size for a given refrigerating capacity (except for some recent development in air refrigeration system in the aircraft cooling), higher coefficient of performance or lower power requirements for a given capacity and less complexity in both design and operation. It can be used over a large range of temperatures.

The main disadvantages like toxity, inflammability and leakage of vapours have largely been eliminated by improvements in design resulting in greater safety and prevention of leakages and by development of non-toxic, nonflammable vapours to be used as refrigerants.

The vapour compression cycle is used in most of the modern refrigeration systems in large industrial plants requiring the rapid removal of large quantities of heat and in most of the domestic refrigerators. The vapour in this cycle is circulated through the various components of the system, where it undergoes a number of changes in its state or condition. Each cycle of operation consists of the four fundamental changes of state or processes (*i*) expansion (*ii*) vaporisation (*iii*) compression and (*iv*) condensation.

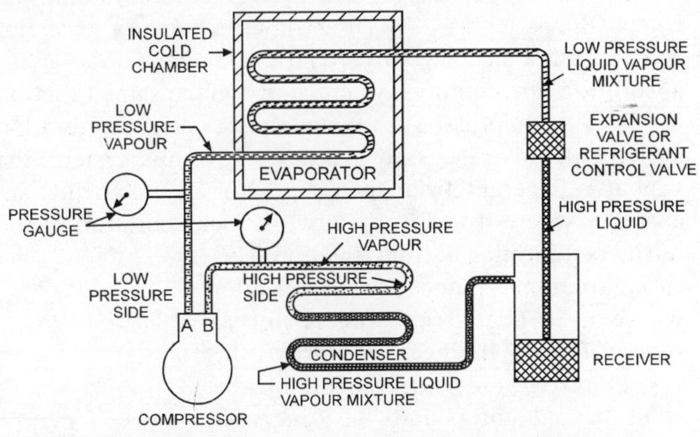

Fig. 9.2. ***Simple Vapour Compression Refrigeration System***

The schematic diagram of a vapour compression refrigeration system is shown in Fig. 9.2. It consists of the following five essential parts.

1. ***Compressor.*** The low pressure and temperature vapour refrigerant from evaporator is drawn into the compressor through the inlet or suction valve A, where it is compressed to a high pressure and temperature. The high pressure and temperature vapour refrigerant is discharged into the condenser through the delivery or discharge valve B.

2. *Condenser.* The condenser or cooler consists of coils of pipe in which the high pressure and temperature vapour refrigerant is cooled and condensed. The refrigerant, while passing through the condenser, rejects its latent heat to the surrounding condensing medium which is normally air or water. Thus hot refrigerant vapour received from compressor is converted into liquid form in condenser.

3. *Receiver.* The condensed liquid refrigerant from the condenser is stored in a vessel, known as *receiver,* from where it is supplied to the expansion valve or refrigerant control valve.

4. *Expansion Valve or Throttle Valve.* The function of this valve is to allow the liquid refrigerant under high pressure and temperature to pass at a controlled rate after reducing its pressure and temperature. Some of the liquid refrigerant evaporates as it passes through the expansion valve, but the greater portion is vaporised in the evaporator at the low pressure and temperature.

5. *Evaporator.* An evaporator consists of coils of pipe in which the liquid vapour refrigerant at low pressure and temperature is evaporated and changed into vapour refrigerant at low pressure and temperature.

During evaporation process, the liquid vapour refrigerant absorbs its latent heat of vaporization from the medium (air, water or brine) which is to be cooled.

9.4.2. Vapour Absorption Refrigeration System

The vapour absorption refrigeration system is one of the oldest method of producing refrigerating effect. The refrigerant, commonly used in this system, is ammonia. This system uses heat energy, instead of mechanical energy as in vapour compression systems, in order to change the conditions of the refrigerant required for the operation of the refrigeration cycle.

In the vapour absorption refrigeration system, the compressor is replaced by an absorber, a pump, a generator and a pressure reducing valve. These components in vapour absorption refrigeration system, perform the same function as that of a compressor in vapour compression refrigeration system, already discussed in Art 9.4.1. In this system, the vapour refrigerant from the evaporator is drawn into an absorber where it is absorbed by the weak solution of the refrigerant forming a strong solution. This strong solution is pumped to the generator where it is heated by some external source. During the heating process, the vapour refrigerant is driven off by the solution and enters into the condenser where it is liquified. The liquified refrigerant then flows into the evaporator and thus the cycle is completed.

The simple vapour absorption refrigeration system is shown in Fig. 9.3. It essentially consists of an absorber, a pump, a generator and a pressure reducing valve, in addition to condenser, receiver, expansion valve and evaporator, as in the vapour compression refrigeration system.

In this system, the low pressure ammonia vapour leaving the evaporator, enters the absorber where it is absorbed by the cold water. The water has the ability to absorb very large quantity of ammonia vapour and the solution so formed, is known as *aqua-ammonia.* The absorption of ammonia vapour in water lowers the pressure in the absorber which in turn draws more ammonia vapour from the evaporator and thus the temperature of the solution is raised. Some form of cooling arrangement (usually water cooling) is used in the absorber for removal of heat from the solution evolved there. This is necessary so that the absorption capacity of the water is increased. The strong solution thus formed in the absorber is pumped to the generator by the liquid pump. The pump increases the pressure of the solution up to 10 kg/cm^2.

The strong solution of ammonia in the generator is heated by some external source such as gas or steam. During the heating process, the ammonia vapour is driven off the solution at high pressure leaving behind the hot weak ammonia solution in the generator. This weak ammonia solution flows back to the absorber at low pressure after passing through the pressure reducing valve. The high pressure ammonia vapour from the generator is condensed in the condenser to a high pressure liquid ammonia. The liquid ammonia is passed to the expansion valve through the receiver and then to the evaporator. Thus the simple absorption cycle is completed.

This system has a least number of moving parts. They are quite in operation and are used both in commercial as well as in domestic installations. In recent years, they have been used extensively in recreation vehicles, camper-trailers and boats as portable units due to their operation from a small gas cylinder.

9.4.3. Thermoelectric Refrigeration System

This system of refrigeration works on the principle of *Peltier effect* according to which heat energy is evolved at one junction and absorbed at the other one when direct current is passed through a junction of two dissimilar metals like antimony and bismuth. It consists of a number of thermoelectric module assemblies in series joined by copper strips, as shown in Fig. 9.4. Each thermoelectric module is built up of a large number of thermoelectric couples. When the direct current

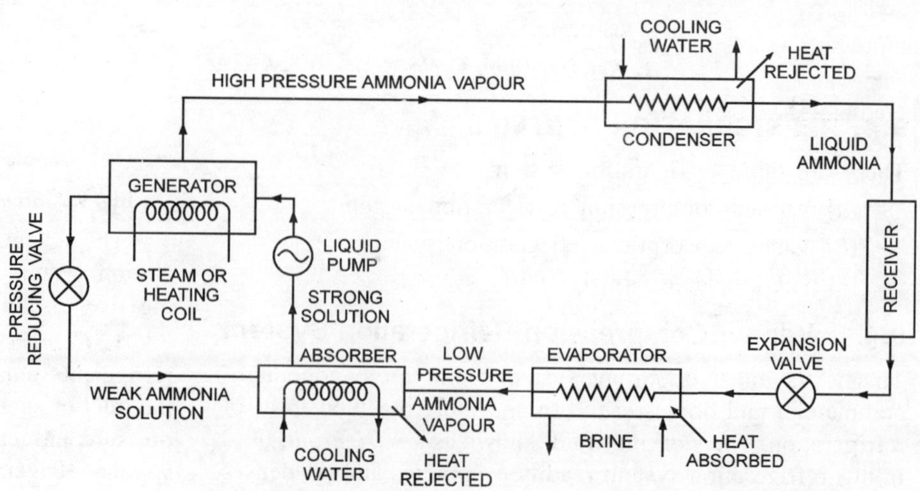

Fig. 9.3. ***Simple Vapour Absorption Referigeration System***

is passed through the thermoelectric module assemblies in the direction shown in Fig. 9.4, junctions at the top of the assembly are cooled and those at the bottom are heated up. Thus top junctions abstract heat from the surroundings and produce refrigerating effect and bottom junctions require cooling by water. Module assembly, therefore, abstracts heat from the medium at top and rejects the same to the medium at bottom. If the direction of flow of current is reversed, heat production at the junctions is reversed, *i.e.*, now bottom of the modules will be cooled and top ones get heated up. If it is an air conditioner, air will get heated up instead of getting cooled. Thus summer and winter operating of air conditioner can easily be obtained just by reversing the direction of flow of current. The main advantages of this system of refrigeration are absence of moving parts and ease of automatic control by controlling the magnitude of current. The only drawback of this system of refrigeration is its very high initial cost.

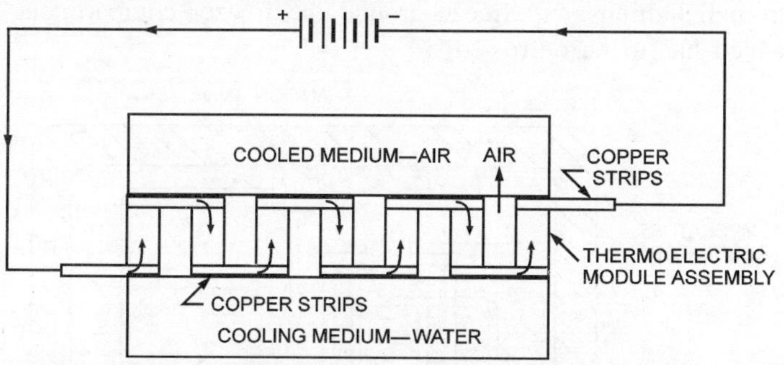

Fig. 9.4. *Thermoelectric Refrigeration System*

9.5. COMPOUND OR MULTI-STAGE VAPOUR COMPRESSION REFRIGERATION SYSTEM

In Art. 9.4.1, we have discussed the simple vapour compression refrigeration system in which the low pressure vapour refrigerant from the evaporator is compressed in single stage (or in a single compressor) and then delivered to a condenser at a high pressure. But sometimes, the vapour refrigerant is required to be delivered at a very high pressure as in the case of low temperature refrigerating systems. In such cases either the vapour refrigerant should be compressed by employing a single stage compressor with a very high pressure ratio between the condenser and evaporator or it should be compressed in two or more compressors placed in series. The compression carried out in two or more compressors is called *compound* or *multi-stage compression.*

In vapour compression refrigeration systems, the major operating cost is the energy supplied to the system in the form of mechanical work. Thus any method of increasing coefficient of performance is advantageous so long as it does not involve too heavy increase in other operating costs, as well as initial plant cost and consequent maintenance, Since the coefficient of performance of a refrigeration system is the ratio of refrigerating effect to the compression work, the coefficient of performance can be increased either by increasing the refrigerating effect or by decreasing the compression work. A little consideration will show that in a vapour compression system, the compression work is largely reduced if the refrigerant is compressed very close to the saturated vapour line. This can be achieved by compressing the refrigerant in more stages with intermediate intercooling. But it is economical only where the pressure ratio is considerable as would be the case when very low evaporator temperatures are desired or when high condenser temperature may be required. The compound compression is generally economical in large plants.

The refrigerating effect can be increased by maintaining the condition of the refrigerant in more liquid state at the entrance to the evaporator. This can be achieved by expanding the refrigerant very close to the saturated liquid line. It may be noted that by subcooling the refrigerant and by removing the flashed vapour, as they are during multi-stage expansion, the expansion can be brought close to the liquid line.

9.6. REFRIGERANTS

The suitability of a refrigerant for a certain application is determined by its physical, thermodynamic, chemical properties and by various practical factors. There is no one refrigerant which can be used for all types of applications, *i.e.,* there is no ideal refrigerant. If one refrigerant has certain advantages, it will have some drawbacks also. Hence, a refrigerant is chosen which has more advantages and less disadvantages.

For safe, economical and thermodynamically efficient production of cold, refrigerant should possess as may as practicable of the following desirable properties.

1. Low boiling point.
2. High critical temperature.
3. High latent heat of vaporisation.
4. Low specific volume of vapour.
5. Low specific heat of liquid.
6. Low cost.
7. Easy to liquify at moderate pressure and temperature.
8. Easy of locating leaks by odour or suitable indicator.
9. It should be noncorrosive.
10. It should be noninflammable.
11. It should be non-toxic.
12. It should be free from objectionable odour.

Various types of refrigerants used for domestic purposes and various industrial utilities in modern refrigerating equipment and plants are classified as primary refrigerants and secondary refrigerants. Primary refrigerants are those working mediums or heat carriers which actually traverse the whole cycle of evaporation, compression, condensation and liquidification *e.g.*, ammonia, carbon dioxide, sulphur dioxide, methyl chloride, methylene chloride, ethyl chloride and freon (mixture of chlorine, fluorine and carbon). Secondary refrigerants are those circulating cold substances which only transfer heat from a remote point to the evaporator of the refrigeration system, *e.g.*, ice, solid carbon dioxide etc. Ammonia is used for commercial and industrial refrigeration for ice manufacture, food processing, cold storage etc., as it is cheap and readily available and has the highest efficiency.

9.7. DOMESTIC REFRIGERATOR

The applications of refrigeration for domestic purposes are mainly in the form of domestic refrigerators and home freezers.

The main purpose of this type of refrigeration is to provide low temperature for storage and distribution of foods and drinks. It represents a significant portion of the refrigeration industry due to the use of these units in large number.

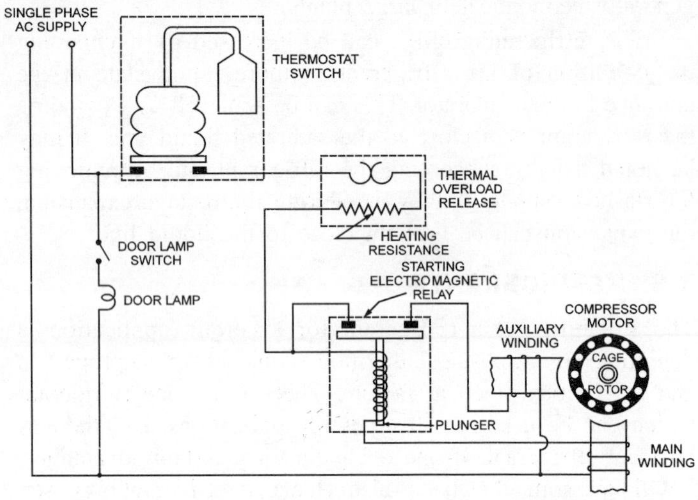

Fig. 9.5. *Electrical Circuit of a Refrigerator*

The growth of food destroying micro-organisms is quite fast (about 100 times) at 10ºC than at a temperature lower than 4ºC. In order to preserve the fruits, vegetables, fish, meat, milk, medicines etc. against spoilage at higher temperature (generally during summer season), the use of refrigerators is very much emphasized. The storage of perishable products at low temperature greatly reduces the activity of both the enzymes and micro-organisms.

For domestic preservation, the storage is generally short-term or temporary storage. The domestic refrigerators and freezers used for the purpose are usually small in sizes with ratings ranging from 1/20 to 1/2 tonne. The units are usually self-contained and hermetically sealed. Due to short-term storage the domestic refrigerator load is intermittent. The requirements of domestic refrigerator is that it should be simple in construction, automatic in action, nominal in initial cost, dependable, and without any necessity of expert inspection and repair. Non-irritant and non-toxic refrigerant should be used. Generally methylene chloride, Freon-12 and Freon-11 are used as the refrigerants.

The common type of domestic refrigerator have a cabinet shaped with compressor motor-fan assembly, the condenser and receiver fitted in their basement. The expansion valve and evaporator coils are exposed in the storage cabinet with the pipings, carrying liquid refrigerant passing through the body. The heat of the bodies to be cooled is carried to the evaporator coils by means of air trapped in the cabinet. The refrigerator used may be the open compressor type; hermetically sealed compressor type or absorption type. Usually hermetically sealed type refrigerators are used. The main advantages are that the problem of gas leakage is minimised as no moving part extends through the sealed housing, it has relatively silent operation and it does not require any maintenance. In open type compressor, the motor drives the compressor using pulley and belt arrangement. This is employed in large commercial refrigerators.

Refrigeration is not only provided with double walled cabinet packed with materials having high thermal insulation such as fibreglass, cork or expanded rubber but also all round the inside of door flap soft rubber seal is used which makes the cabinet airtight. Also the door is provided with automatic closing mechanism–door hinges are provided in such a way that door flap when left in open position automatically comes to closing position due to gravity and as it approaches closing position it is attracted by the magnetic strip fitted behind the sealing rubber ring and thus the door is closed with snap action. All this is done to prevent leakage of atmospheric heat inside the refrigerator.

Main precaution to be taken is that very hot things should not be put in refrigerator, if it is done it will quickly evaporate the refrigerant in evaporator coils producing a large vapour pressure resulting in increase in the duty of the compressor and as such motor has to work for longer duration in order to bring down the temperature in the freezer compartment. It may damage the motor, which is short-time rated. Also moisture released from hot things will freeze around the freezer compartment which has to be defrosted.

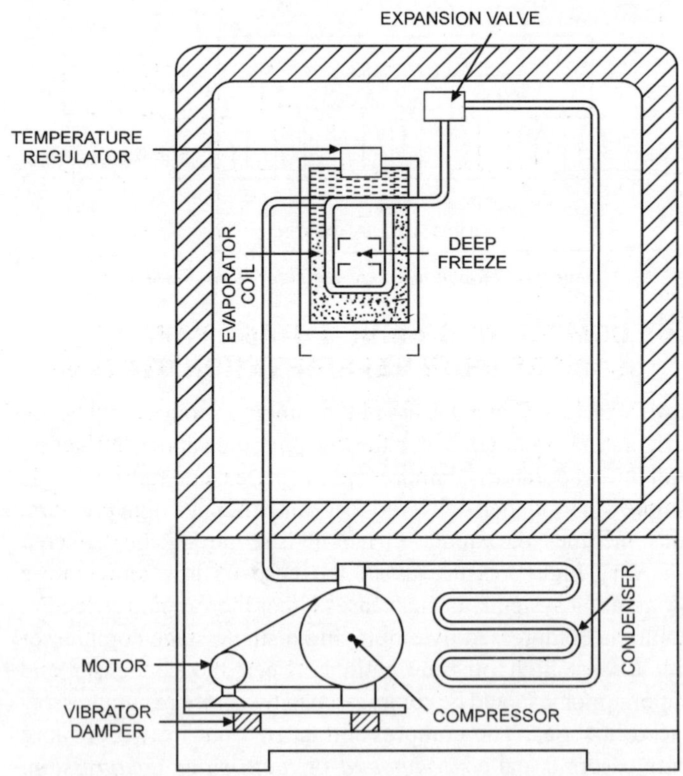

Fig. 9.6. *Domestic Refrigerator*

Electrical circuit of a refrigerator is shown in Fig. 9.5. Refrigerator is provided with a door push switch, which closes on opening of refrigerator and puts the lamp on. Capacitor-start single phase induction motor is used in open type refrigerators and split-phase single phase induction motor is used in sealed unit refrigerators. Electromagnetic relay is provided to connect auxiliary winding on the start and disconnect it when the motor picks up the speed. Thermal overload release is provided to protect the motor from damage against flow of overcurrent. Thermostat switch is provided to control the temperature inside the refrigerator. Temperature inside the refrigerator can be adjusted by means of temperature control screw. Greater the distance between contacts, greater will be the temperature inside the refrigerator and vice versa. To protect the motor against undervoltage use of automatic voltage regulator is essential since in case of fall in applied voltage, motor will draw heavy current to develop the required torque and will become hot, thermal overload relay will, therefore, repeatedly disconnect and connect the motor to supply, eventually burning it out.

Domestic refrigerator employing vapour compression refrigeration system is shown in Fig. 9.6.

9.7.1. Trouble Shooting of Refrigerator

Various troubles often found in a refrigerator are listed below.

Symptoms	Causes
1. Motor fails to start on giving supply.	(*i*) Blown fuses in the main switch. (*ii*) Thermostat contacts open. (*iii*) Overload release open. (*iv*) Open circuit in the main winding. (*v*) Open circuit in the auxiliary winding. (*vi*) Grounded winding. (*vii*) Burnt or shorted winding (*viii*) Excessive overload (*ix*) Worn-out or tight bearings. (*x*) Bent rotor shaft.
2. Motor runs slower than normal speed.	(*i*) Low voltage. (*ii*) Overload. (*iii*) Shorted main winding. (*iv*) Defective electromagnetic relay. (*v*) Worn-out bearings.
3. Motor runs hot.	(*i*) Low voltage. (*ii*) Overload. (*iii*) Shorted or grounding winding. (*iv*) Worn-out bearings.
4. Motors does not start and gives humming noise.	(*i*) Low voltage. (*ii*) Overload. (*iii*) Open circuit in the auxiliary winding. (*iv*) Defective electromagnetic (starting) relay-contacts being not closed.
5. Motor runs with noise	(*i*) Shorted winding. (*ii*) Improperly connected poles. (*iii*) Loose rotor bar. (*iv*) Worn-out bearings. (*v*) Too much end play. (*vi*) Foreign material in the rotor.
6. Motor keeps on running but (*a*) cooling is nil	(*a*) (*i*) No refrigerant gas in the system (indicated by the unit, tubing and condenser being at ambient temperature). (*ii*) Complete choking of capillary.
(*b*) cooling is insufficient	(*b*) (*i*) Less refrigerant gas in the system. (*ii*) Partial choking of capillary.
7. Motor keeps on running even though it is very cold inside the refrigerator.	(*i*) Wrong setting of thermostat. (*ii*) Defective thermostat—shorted wiring or sticking contacts.
8. Motor starts very frequently.	(*i*) Bad door seal. (*ii*) Wrong setting of thermostat.
9. Motor operates normal with normal cooling in freezer but cooling in the rest portion is unsatisfactory.	(*i*) Bad door seal. (*ii*) Opening of door too frequent.
10. Motor operates normal with good cooling but defrosting starts all of a sudden. Cooling again starts after some time	(*i*) Presence of moisture in the refrigerant cycle.
11. Too much frosting around the freezer.	(*i*) High atmospheric humidity or steaming hot liquids stored.

9.8. WATER COOLER

Water is one of the most needed thing for a person. In summer season cold water gives life to a thirsty person. At 10°C water is most refreshing. Thus cooling of water in summer season becomes necessary. Water coolers are used to produce cold water at about 7 to 13°C. The temperature of water is controlled with the help of a thermostatic switch.

Water coolers may be classified as follows:

(*i*) Bottle type water cooler,

(*ii*) Pressure type water cooler,

(*iii*) Self-contained or remote type water cooler.

9.8.1. Instantaneous Type Water Coolers

In this type of coolers the cooling coil is wrapped round the pipeline such that by the time water reaches the tank it is cooled to desired temperature. Various types of instantaneous type water coolers are described below.

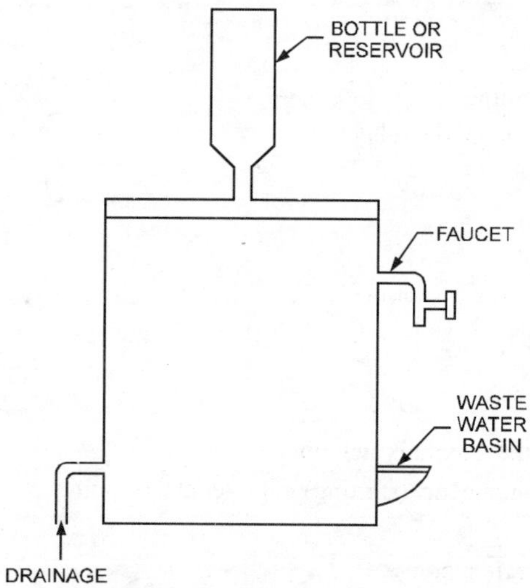

Fig. 9.7 (*a*). *Bottle Type Water Cooler*

1. *Bottle Type Water Cooler*. In this type of water cooler, water to be cooled is stored in a bottle or reservoir. For filling glass tumblers or containers faucet or similar means are provided. The dripping water from the faucet is collected in the waste water basin or water drip, as illustrated in Fig. 9.7 (*a*). The usual size is 25 litres and is suitable for places where plumbing installation is expensive and drains are available.

2. *Pressure Type Water Cooler*. In this type of cooler shown in Fig. 9.7 (*b*), water is supplied under pressure. For filling glass tumblers or containers faucets or similar means are provided. A valve is employed to control an appropriate flow of water or projected stream of water from a bubbler. An arrangement should be made to collect water and allow complete collection of water spreading from the bubbler. The temperature of waste water is low, it is used for cooling the supply water by passing through a pipe coil wrapped round the drainage line. By doing so, the cooling load for cooler is reduced. Since the water is supplied under pressure the cold water can be obtained from the top mounted at any height of the cooler. In case of bottle type, faucet has to be at a height up to which syphoned water can be obtained from the tank of the cooler. The refrigeration system is usually mounted at the bottom structure of the cooler body and a cooling coil is wrapped round the water tank. To ensure good surface contact between the evaporative tube and tank, either the tank surface is corrugated to accommodate pipe or pipes are secured using soft solder to provide metal contact. Sometimes a helical or U-type coil is immersed in the water tank. Although such as arrangement provides high heat transfer from water to the coil but formation of undesirable salt due to chemical reaction between water contaminant and the copper surface proves to be a great drawback in this system.

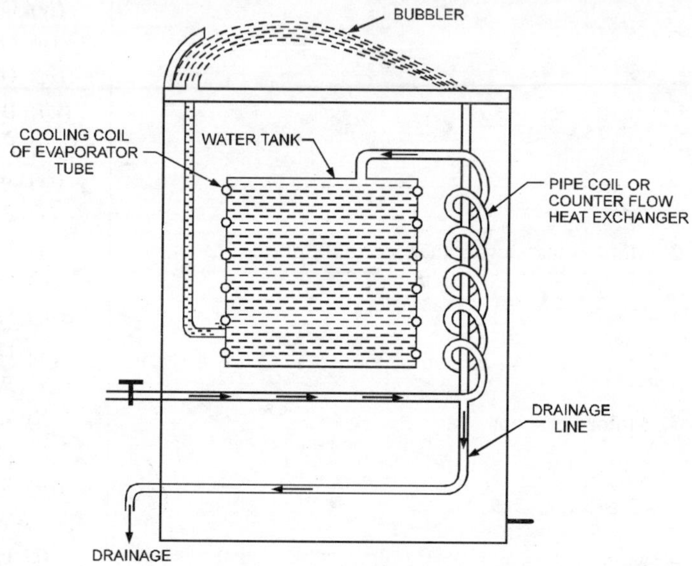

Fig. 9.7 (*b*). *Pressure Type Water Cooler*

3. *Self-Contained or Remote Type Water Cooler.* This type of cooler [Fig. 9.7 (*c*)] employs mechanical refrigeration system and is a factory assembled unit. A remote type water cooler cools the water which is supplied to the desired drinking place (away from the system). It is quite a useful unit as it does not need extra space near the place of work.

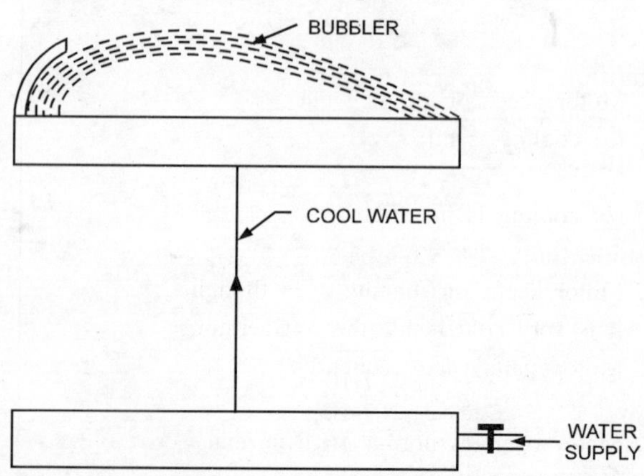

Fig. 9.7(*c*). *Self-Contained or Remote Type Water Cooler*

9.8.2. Storage Type Water Cooler

Such type of water coolers are used where continuous supply of water is not available. Fig. 9.8 shows a schematic storage type water cooler which is a self-explanatory. Here water is filled in the storage tank and level of water is kept the same by the use of a float valve. The storage tank is surrounded by an evaporator coil through which flows a low pressure liquid refrigerant which takes away the heat of water and thus makes it cold. When the water attains the desired temperature the thermostat operates and disconnects the power supply to the motor. The motor employed is capacitor-start capacitor-run single phase induction motor.

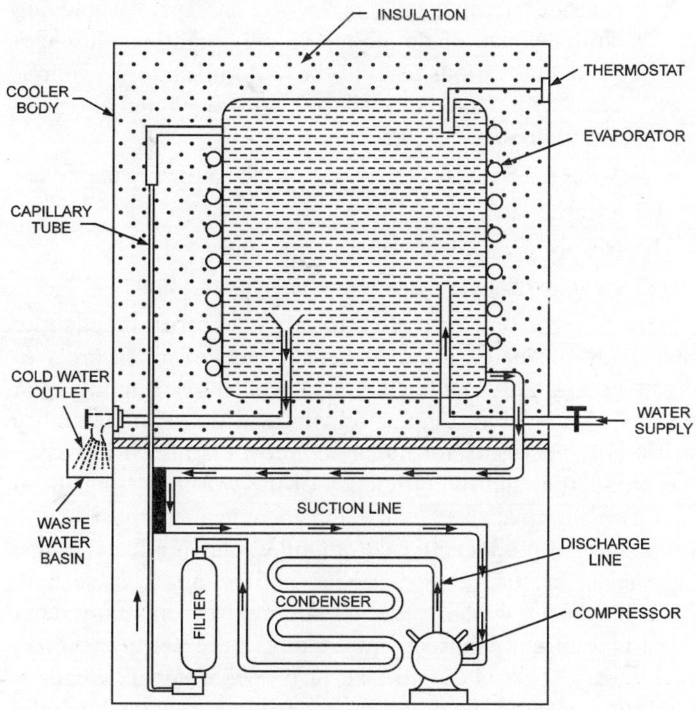

Fig. 9.8. *Storage Type Water Cooler*

9.8.3. Capacity of Water Coolers

For determination of cooling load for the water cooler (Q) the following relation may be used

$$Q = m_w \times C_p (T_i - T_0)$$

where m_w is rate of water consumption in kg, C_p is the specific heat of water in kJ/kg/K, T_i is the temperature of inlet water and T_0 is the temperature of outlet water.

The amount of cold water requirements under different conditions is given in Table 9.1 (figures based on extensive statistical survey).

Table 9.1. Amount of Drinking Water Requirements Under Different Conditions

Applications	Temperature in °C	Amount of Water Used including Drainage in Litres Per Hour
Restaurants	7-10	0.5 per person
Schools	7-10	0.56 per student
Offices	7-10	0.56 per person
Stores	7-10	4.5 per 100 customers
Hotels	7-10	0.4 per room
Hospitals	7-10	0.4 per bed and per attendent
Theatres	7-10	4.5 per 100 seats
Light Duty Manufacturing Unit	10-13	0.9 per person
Heavy Duty Manufacturing Unit	10-14	1.1 per person

* R-12 is the most common refrigerant up to 1 tonne refrigeration capacity and R-22 for 2 tonnes refrigeration capacity, and appropriate combinations for units of larger size.

** While estimating the amount of water consumption, the amount of wastage of cold water should be taken into consideration.

*** For ensuring significant heat transfer through insulation, usually heavy insulation around 4 cm to 6 cm thick glass wool or thermocole is provided.

9.9. DESERT COOLER

As the name implies, it is suitable for places where the humidity is quite low and temperature quite high. Such conditions are in conformity with desert areas. Hence the coolers are called the desert coolers. The principle on which a desert cooler operates is "*evaporative cooling*". Evaporative cooling is a process is which sensible heat is removed and moisture added to the air. When the air passes through a spray of water it gives up heat to water, some of water evaporates and picks up heat from the air equivalent to its latent heat. The vapours thus formed are carried along in air stream. In this way air is cooled and humidified.

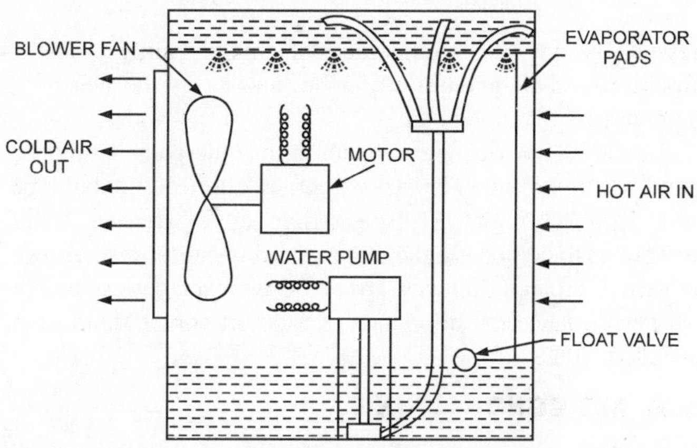

(a)

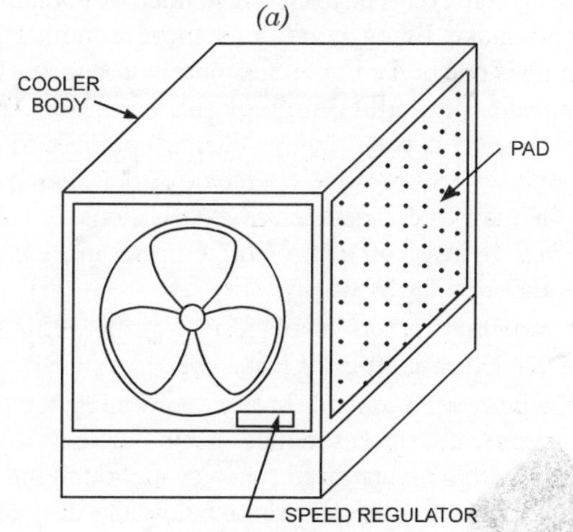

(b)

Fig. 9.9. *Desert Cooler*

Construction and Working. A desert cooler consists of the following main parts:

1. Blower/Fan 2. Water circulating pump 3. Water wetted pads. 4. Water tank 5. Float valve 6. Water traps.

The water is filled in the sump of the cooler from water supply mains, the level of which is controlled by a float valve. A water pump lifts the water and supplies it at the top of the cooler to the water distribution system which consists of small branches of copper pipe or so equipped with orifices which deliver equal amount of water to the troughs which in turn supply water to the wetted pads. The water which drops back from the pads is recirculated. The pump may be made of brass, stainless steel or even plastic.

The blower pulls the air through the wetted pads and delivers it to the space to be cooled through and opening in the fourth side of the desert cooler cabinet. The air sucked through the pads is cooled by the *principle of evaporative* cooling. The blower provides adequate velocity to the air before it is delivered to the space to be cooled.

To have long life of the desert cooler and better performance, pads should be changed every year and holes for water distribution system should be cleaned. The tank should be cleaned just after the season and coated with the corrosion resisting paint like red oxide and some other paints.

The efficiency of desert cooler is given as

$$\eta_{desert} = \frac{t_{db} - t_{dbg}}{t_{db} - t_{wb}}$$

where t_{db} is dry bulb temperature of ambient air, t_{dbg} is the temperature of air leaving the grille, and t_{wb} is the wet bulb temperature of ambient air.

According to ISS, the velocity across the pads should be around 40 m/minute. The desert cooler efficiency should be about 75%. Though the high efficiency of desert cooler corresponds to cooled air, the large humidity causes discomfort. The desert cooler efficiency above 80% is not desirable. For good performance the efficiency of a desert cooler should not fall below 70%.

9.10. AIR CONDITIONING

In this modern age everybody is trying to improve his surroudings in order to make living conditions more comfortable. Air conditioning is one of the important contribution in this respect. The technique of maintaining of air condition inside a building in such a way that produces comfort to human beings, ensures trouble free operation of delicate equipment or helps in industrial manufacturing operations is called *air conditioning*. It involves control of temperature, cleanliness, humidity and circulation of air.

The important factors involved in a proper, efficient and complete air conditioning systems are :

1. *Temperature Control.* In air conditioning, temperature control means the maintenance of a particular desired temperature within an enclosed space even though the outside temperature may be either above or below the desired inside temperature. This requires either the addition or removal of heat from the enclosed space as and when necessary. For this purpose the outside air to be supplied in the room is either cooled or heated.

In summer, air is cooled by one of the following methods :

(*i*) Circulation of cold water.

(*ii*) Evaporative system using water.

(*iii*) Ice activated system.

(*iv*) System using absorption principle.

(*v*) Vapour compression system employing a compressor, evaporator and a suitable refrigerant. This is commonly used method. Three arrangements are possible under it namely (*a*) individual window type or floor mounted package type air conditioners (*b*) a central plant employing chilled water in combination with unit conditioners located at various parts of the building and (*c*) a central plant with a ducting system to convey cooled air to the different parts of the building.

In winter, air is warmed by one of the following methods :

(*i*) Electric heaters

(*ii*) Electromechanical heat pumps and

(*iii*) Heat exchangers using waste system.

2. *Humidity Control.* Humidity control means the increasing or decreasing of water contents of air in order to create comfortable and healthy environments depending upon a particular season–summer or winter. Control of humidity is not only necessary for human comfort but also it increases the production and the efficiency of the workers. It has been found to improve the quality and to reduce the cost of the production of products like automobiles, clay products, drugs, chemicals, electric goods, foods, textiles etc. It is seen in factories during winter season or summer season, that though there is a normal environmental temperature yet the workers feel uneasy. That is due to lack of proper water contents or humidity of air. The humidity control is achieved by the process of humidification (addition of humidity or moisture) or dehumidification (reduction of humidity or water vapour). The humidification is accomplished either by supplying or spraying steam or hot water or cold water into the air. Dehumidification may be accomplished by using an air washer or by using absorbents. In the air washer system, the outside air is cooled below dew point temperature so that it loses moisture by condensation. The moisture removal is also accomplished when the spray water is chilled water and its temperature is lower than the dew point of the outside air. The absorbent materials are activated alumina, silicagel and calcium chloride. These are generally used for window type small room coolers. In general, for summer air conditioning, the relative humidity should not be less than 60% while for winter air conditioning it should not exceed 40%.

3. *Air-movement and Circulation.* In order to keep constant temperature throughout the conditioned rooms, it is necessary that there should be equi-distribution of air throughout the space to be air conditioned. For this purpose the air movement and its circulation should be controlled.

4. *Air-filtration, Cleaning and Purification.* Proper filtration, cleaning and purification of air is necessary to keep it free from dust, dirt and other impurities which are very injurious to health. The dust particles are liable to clog the nose, throat and lungs of the worker seriously affecting their health. The dirt present in air acts as a carrier of germs and is responsible for disease spreading. In case of certain industries like chemicals the dirt produced as a byproduct is very harmful for health if inhaled. The presence of dirt in the air used for air cooling of electrical machines and diesel sets in power plants is liable to clog their air passage resulting in reduced efficiency and life. Various types of air impurities other than gas or vapours called the Aerosols are also present. These are dust (earthly wet soil dust, dust of coal, cement, wheat, lint—a string like dust from cotton or wool); fumes (solid particles due to condensation and solidification of gaseous substances, zinc oxide fumes from zinc vapours); smoke (due to combustion of organic materials like tobacco, wood, oil, coal etc.); pollen (from weed, tree, flower and other vegetation—may be harmless or harmful); bacterial (living micro-organisms) etc.

There are various methods of cleaning, filtering and purifying the air from the above mentioned impurities. Mists or fogs in the air are usually removed by raising the temperature of air. This results in the formation of small suspended water particles into vapours making them invisible. The impurities like gases and vapours are removed by the process of condensation. Pollen grains responsible for hay fever, rose fever and other respiratory difficulties are removed by filters or by using a wet space. The particle of pollen grains will adhere to these wetted surfaces.

Bacteria present in the air is removed either by filter chemicals or by using a sterilising light of ultraviolet rays. For proper air sterilisation special ultraviolet ray lamps of radiation 2,600 Angstrom units are used. These are generally placed near the outlet grille on its inside end. They are shielded properly to avoid exposure of skin or eyes. Ultraviolet lamps are largely used in meat packing houses and storage rooms to avoid the growth of bacteria. Fine powder of propylene or triethylene glycol is also sprayed in the air to kill bacteria.

To remove dust particles from the air four methods using equipments like centrifugal devices, water filters or air washers, adhesive impregnated filters, electrostatic precipitators are used.

9.11. COMFORT AIR CONDITIONING

This is the type of air conditioning whose objective is to produce comfortable and healthy conditions for occupants. In other words, it is the process of treating air so as to control simultaneously the prevailing conditions of temperature, humidity, cleanliness and distribution in the conditioned space to meet the comfort and health requirements of the occupants. It includes the air conditioning of spaces, places and surroundings which are occupied by human beings.

Some of the important places intended for human occupancy and requiring comfort air conditioning are the residences, departmental stores, public buildings including theatres, cinema halls, dancing halls, skating rings, restaurants and night clubs, hospitals etc. In order to understand the comfort air conditioning the human requirements should be known.

Human Requirements. Our body is thermostat with inner temperature of 98.4°F (37°C) and outer temperature of about 26.5°C. Under normal conditions our body emits 2,500 k cals per 24 hours; 45% of which is dissipated by radiation, 30% by convection and remainder 25% by evaporation. Heat dissipated by radiation, convection and evaporation is governed by surrounding temperature, air temperature and humidity of air. It is necessary for our body to dissipate the above proportion of heat in order that we may feel convenient and comfortable. For producing comfort it is necessary to maintain surrounding atmosphere at a temperature between 22°C to 25°C during summer and 17°C to 20°C during winter and humidity between 40 to 60 per cent. It is not desirable however, to maintain more than a 8 to 11°C difference between indoor and outdoor temperatures since a higher temperature difference will cause discomfort when a person goes out of a room and comes back.

In *Indian Railway air-conditioned coaches the temperature is maintained at 24°C during summer and 20°C during winter, as these are most accepted temperature by passengers.*

9.12. INDUSTRIAL AIR CONDITIONING

A large number of manufacturing, inspection, measuring processes require a close control of temperature and relative humidity. Industrial air conditioning is generally defined as the process of simultaneously controlling the temperature, relative humidity, ventilation, cleaning and movement of air within an enclosure intended for the manufacture, inspection and measurement of a product which is sensitive to atmospheric conditions.

It covers a large number of industries requiring controlled air conditions for their efficient and accurate manufacture as well as healthy and comfortable environments for the workers. Air conditioning is very essential for the manufacture of rayon and various plastics, paper, colour printing, in tobacco industry, in the precision manufacturing and measurement works for year-round production, textiles, felt goods, leather, electrical equipment, photographic material manufacturing, candy etc.

9.13. EFFECTIVE TEMPERATURE

Comfort requirements for human beings excluding radiation, air odours and cleanliness generally depend upon the dry- and wet-bulb temperatures of the air and its movement rate. An arbitrary index which combines in a single value the degree of sensation warmth or coolness felt by the human body due to the effect of air, temperature, humidity and air movement is called the *effective temperature.* Its numerical value cannot be measured directly but is taken as that temperature of the saturated still air (at a velocity of 4.5 to 7.5 metres per minute) which induces the same sensation of warmth or coolness as those created by the air surrounding a person. An atmospheric condition will have an effective temperature of 20°C if it induces the same sensation of warmth or coolness as created by saturated still air at 20°C.

From the investigations and results of various studies it has now been proved that the effect of humidity shown by effective temperature scale are not great if the air temperature and the human activities are such that very little heat is lost by evaporation of the perspiration. For an individual doing an activity in sitting position with air temperature between 23°C and 25°C a relative humidity in the range of 30 to 65% will have no effect on his comfort. At the same temperature, a higher rate of activity may result in a sensation of warmth but a relative humidity of 40% would be more comfortable than 60% relative humidity under similar conditions. In case of a large number of persons doing different activities like sitting, playing, dancing etc. in the same space, a 23.5°C temperature with 40% or less relative humidity generally provides most comfortable condition for all.

9.14. SUMMER AIR CONDITIONING SYSTEM

It is the most important type of air conditioning, in which air is cooled and generally dehumidified. Figure 9.10 shows a schematic arrangement of a typical summer air conditioning system.

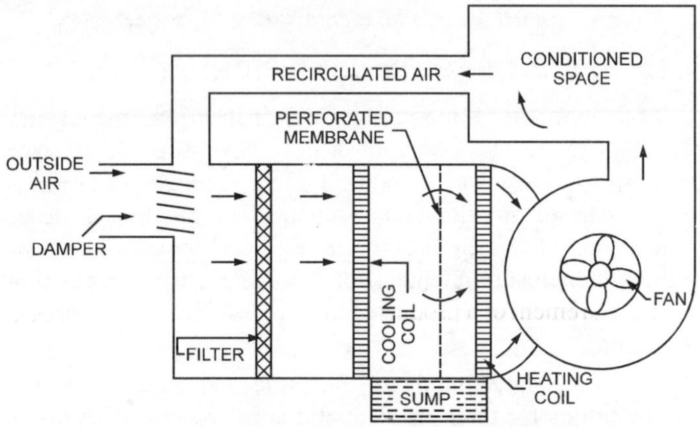

Fig. 9.10. *Summer Air Conditioning System*

The outside air flows through the damper, and mixes up with recirculated air (which is had from the conditioned space). The mixed air passes through a filter for removing dirt, dust and other impurities. The air now passes through a cooling coil, which has a temperature much below the required dry bulb temperature of air in the conditioned space. The cooled air passes through a perforated membrane and loses its moisture in the condensed form which is collected in a sump. After that, air is made to pass through a heating coil to heat up the air slightly. This is accomplished for bringing the air to the designed dry bulb temperature and relative humidity.

Now the conditioned air is supplied to the conditioned space by a fan. From the conditioned space, a part of the used air is exhaused to atmosphere by the exhaust fans or ventilators. The remaining part of the used air, known as *recirculated* air, is again conditioned as illustrated in Fig. 9.10. The outside air is sucked and made to mix with the recirculated air in order to make up for the loss of conditioned (or used) air through exhaust fans or ventilators from the conditioned space.

9.15. WINTER AIR CONDITIONING SYSTEM

In winter air conditioning, the air is heated, which is usually accompanied by humidification. The schematic arrangement of winter air conditioning system is shown in Fig. 9.11.

In this system, the outside air flows through a damper and mixes up with the recirculated air coming from the conditioned space. The mixed air is passed through a filter for removing dirt, dust and other impurities. The air is now passed through a preheater so as to avoid the possible freezing of water and to control the evaporation of water in the humidifier. After that, the air is made to pass through a reheat coil to bring the air to the designed dry bulb temperature. Now, the conditioned air is supplied to the conditioned space by fan. From the conditioned space, a part of the used air is exhausted to atmosphere by the exhaust fans or ventilators. The remaining used air is again conditioned as shown in Fig. 9.11. The outside air is sucked and made to mix with recirculated air, in order to make up for the loss of conditioned (or used) air through exhaust fans or ventilators from the conditioned space.

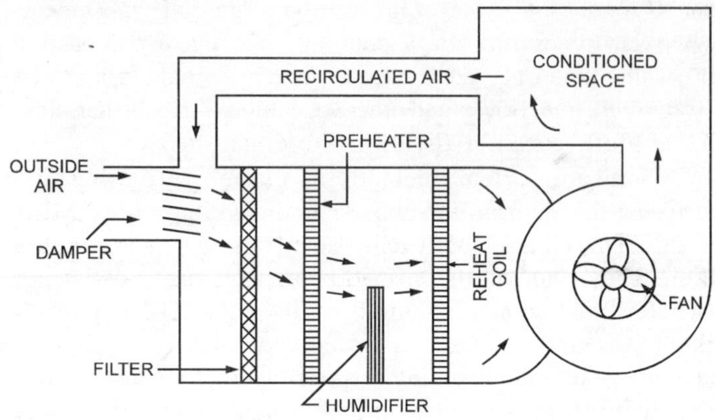

Fig. 9.11. *Winter Air Conditioning System*

9.16. YEAR-ROUND AIR CONDITIONING SYSTEM

The year-round air conditioning system has equipment for both the summer and winter air conditioning. The schematic arrangement of a modern year-round air conditioning system is shown in Fig. 9.12.

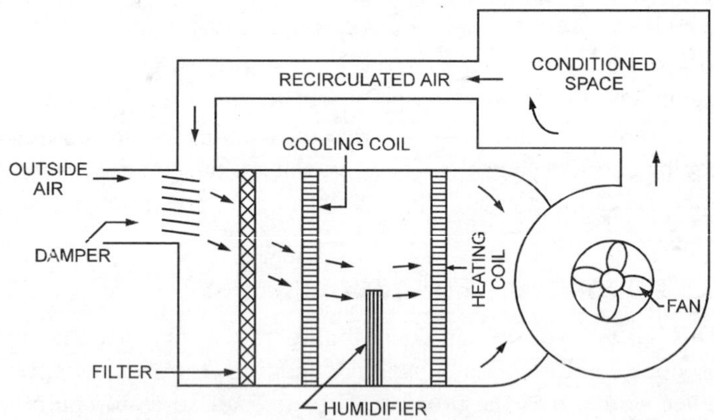

Fig. 9.12. *Year-around Air Conditioning System*

The outside air flows through the damper and mixes up with the recirculated air coming from the conditioned space. The mixed air is passed through a filter so as to remove dirt, dust and other impurities. In summer, the cooling coil is operated to cool the air to the desired value. The dehumidification is obtained by operating the cooling coil at a temperature below than the dew point temperature (apparatus dew point). In winter, the cooling coil is made inoperative and the heating coil is operated to heat the air. The spray type humidifier is also used in the dry season to humidify the air.

9.17. TYPES OF AIR CONDITIONING SYSTEMS

There are three types of air conditioning systems.

1. ***Unitary Air Conditioning System.*** In this system, factory assembled air conditioners are installed in or adjacent to the space to be conditioned. The unitary air conditioning systems are of the following two types.

(a) Window Units. These are self-contained units of small capacity of 1 tonne to 3 tonnes, and are mounted in a window or through the wall. They are employed to condition the air of one room only. If the room is of bigger size, two or more units are installed. Now smaller units of 1 tonne, even 3/4 tonne are also available.

(b) Vertical Packed Units. These are also self-contained units of larger capacity of 5 to 20 tonnes and are installed adjacent to the space to be conditioned. This is very useful for conditioning the air of a restaurant, bank or small office.

The unitary air conditioning system may be adopted for winter, summer or year-round air conditioning.

2. *Central Air Conditioning System.* This is most important type of air conditioning system, which is adopted, when the cooling capacity required is 25 tonnes or more. This system is also adopted when the air flow exceeds 300 m^3 per minute or different zones in a building are to be air conditioned.

In this type of air conditioning various essential components, or units of air conditioning are situated at some central place usually in basement from where conditioned air is led through sheet metal ducts to the various rooms or space of building. There are return ducts for carrying air from these rooms back to the central place where it is dehumidified, cooled and recharged with fresh air. The circulation of air is achieved either naturally or by a fan. Advantages of this type of air conditioning system are high efficiency of operation and robustness of the equipment. The main disadvantages are absence of any adjustment of individual room temperature, necessity of costly ducting system and mixing of odours, smoke and bacteria present in the return air from infected rooms and redistributing them to healthy rooms. This type of air conditioning is best suited for big installations, factories and industrial concerns.

3. *Unitary Central Air Conditioning System.* The main disadvantage of the central system is that it needs large sized ducts which are costly and occupy large space. The ducts have to carry full volume of air (outdoor plus return) to be supplied to all the rooms. The size of ducts is, therefore, made large enough to carry the whole supply of air, Similarly for return ducts, the same remarks apply. To overcome this problem, *unitary central system* is used.

In this type of air conditioning the features of both the central type and unitary type air conditioning systems have been combined in order to use it for all purposes and to make greater life expectancy. In this type of air conditioning heating and cooling equipments are installed in the central room. The various unit air conditioners in different rooms are supplied with heat energy (in the form of steam or hot water) or chilled water through the pipes. The air conditioning unit in this case comprises of heating and cooling coils, filters, blowers, dampers, valves, control and other associated equipment. All the disadvantages of central type are done away with. There will be no return air duct, no mixing of odours and bacterial of one room and redistribution of the same among other rooms. Temperature of each space unit can be controlled by room unit. Fresh air duct required to supply only 15% fresh air, rest of 85% air being recirculated, will be of very much small size. Fresh air introduced will displace corresponding quantity of air which will escape through openings etc.

9.18. ROOM AIR CONDITIONERS

It is a common type of air conditioner used to condition the air of a particular space occupied by human beings *i.e.*, office room, residential room etc. It has automatic operation to cool and humidify the air.

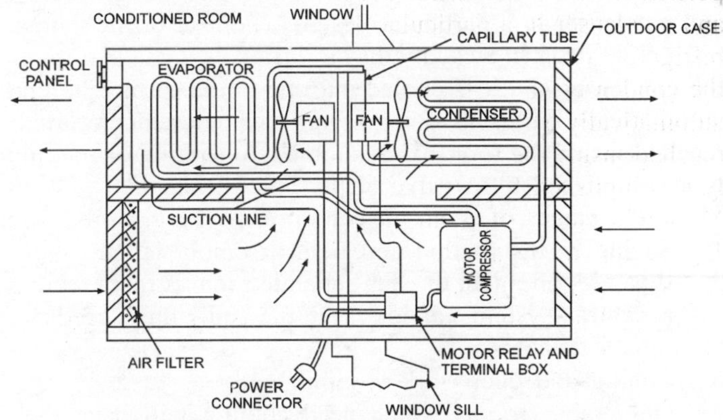

Fig. 9.13. ***Room Air Conditioner (Window Type)***

A window type air conditioner is shown in Fig. 9.13. It consists of a case divided into two parts by a partition with a small opening at the top as the outdoor part and indoor part. The outdoor portion consists of a hermetically sealed motor compressor unit, condenser, motor driven fan and a tray. This portion partitioned by a portion L into two parts is provided with a left hand side opening.

The indoor portion consists of evaporator, motor driven fan, remote bulb, refrigerant, control, a control panel, an air filter, power connector and a tray. This portion is further subdivided into two parts by opening on the right hand side. A pipeline connects the two trays in the inner and outer parts. A capillary line control through a refrigerant filter connects the condenser to the evaporator. Evaporator is connected to the compressor by a suction pipeline. These units are usually fitted flush with window still on inside having attractive frontage. On outside it is supported on brackets and projects beyond the wall. The front and back of the inner and outer portions of the cooler is fitted with shutters. These shutters are adjustable at different inclinations according to the requirements.

The working substance is a vapour which readily evaporates and condenses. The most popular substances are ammonia, freon, carbon dioxide etc.

As the unit is put into operation, the low pressure vapour through the suction pipeline is drawn from the evaporator and passed to the compressor. The compressor delivers it at high pressure to the condenser. In the condenser, the vapour gets condensed and the heat is removed from the refrigerant vapour. The liquid refrigerant collected at its lower coils is passed through the filter into the capillary tube control before flowing back at low pressure into the evaporator coils.

The low pressure liquid refrigerant inside the evaporator coils quickly picks up from its surface. The air is drawn from inside the room through a filter from the lower portion of the unit by a motor driven fan. This air is passed over the evaporator coils. The cooled air from the evaporator coils surface is supplied back into the room. Due to dehumidification of moisture from the circulating air, the moisture after flowing downward from the surface of the evaporator coils is collected at the bottom of the evaporator into a pan. This moisture in the pan flows due to gravity into another pan placed in the outdoor portion at the bottom of the condenser. The moisture helps to cool the compressor and condenser due to evaporation.

In the compressor and condenser compartment, the fan draws in the outside air from the lower portion of the unit, circulates it over the condenser and discharges from the upper portion of unit to the outside due to mounting of the compressor and condenser in a particular design. The heat of the vapour refrigerant is taken away by the air during its passage through the condenser and it is cooled into the liquid form. The unit automatically stops as soon as the desired temperature is reached inside the room. The automatic operation of the unit is accomplished by a valve at the control panel.

Main advantages of room air conditioners are given below :

1. Saving in installation and field assembly labour.
2. Exact requirement of each separate room is met whereas a central system cannot meet the individual needs of separate rooms.
3. Zoning and duct work is eliminated.
4. Only those rooms which need cooling will have their units running, whereas the central plant will have to operate all the time for the sake of only a few rooms.
5. Failure of unit affects a single room where as all the rooms are affected when failure occur in the central system.
6. Low initial cost.
7. Flexibility of operation.
8. Starting with only a few rooms, the air conditioning may be progressively extended to other rooms without spending a large sum from the very beginning.

9.18.1. Comparison Between Desert Cooler and Window Type Air Conditioner

Particulars	Desert Cooler	Window Type Air Conditioner
Initial Cost	Low	Quite high
Main Components	Motor and fan	Complete refrigeration unit: compressor, condenser, evaporator, and controls etc.
Humidity Control	Not possible	Possible
Power Required Per Unit Cooling	Less	More
Location	Inside or in the window of the room.	Fitted either in the window or wall.
Utility	Can be used to cool the air and that too in the dry season.	Can be used for both cooling and heating purposes.

9.19. CENTRAL AIR CONDITIONING SYSTEMS

The central air conditioning systems serving the needs of large building or spaces are generally called the year-round air conditioning systems. They have a cooling capacity of 25 tonnes and circulate about 300 m^3/minute of conditioned air. Depending upon its type such a system may contain some of the following equipments :

(*a*) Heating coils supplied with steam or hot water.

(*b*) Cooling and dehumidifying coils—both for chilled water and for direct expansion of the refrigerants.

(*c*) Blower and driving motor.

(*d*) Sprays for cooling and dehumidifying or for washing the air.

(*e*) Air-cleaning equipment containing filters, electrostatic precipitators, odour removing equipment and germicidal lamps.

(*f*) Various types of control devices.

The devices assembled as an air-conditioning unit are housed and enclosed in sheet metal housings or plenums. The plenum is usually covered with insulating board over its entire outer surface and is stoutly braced. The refrigeration condensing units and steam or hot water boilers for air conditioning in this system are either located in the same room or in the basement or in an adjoining room.

The various factors affecting the choice of the equipment for design and installation of the central air-conditioning systems are capacity of the plant, filters and mixing plenums, refrigeration and heating equipment, insulation, noise and vibration.

1. *Capacity of the plant.* For each 30 m^3/minute of air handled, heating and cooling coils handling air must provide 0.185 m^2 of face area (four row coil) at face velocities of about 150 m/minute. One of the limiting factors influencing the design of this system is the physical size of the coil and filter banks.

The horizontal space between flows of a building is usually kept between 3.65 and 4.87 metres because the operating expenses of the system depends upon the floor area utilised. Thus for a single air conditioning unit a practical limit of 1,200 m^3/minute or 100 tonnes of cooling capacity is set. For the spaces or buildings requiring more than this capacity two or more zones may be planned with an air conditioning system installed to handle each unit.

The extent of the zoning to be specified is generally determined by the nature of the building. A theatre or departmental store installation requiring spaces of large cubical structure are most economically conditioned by only a few (preferably one) central station air conditioning units. Office buildings, schools, colleges and similar buildings requiring many rooms necessitate greater zoning. Air conditioning units of 300—900 m^3/minute are usually employed. Considering all the factors discussed above, it seems more economical to supply chilled water, hot water, steam or refrigerant through pipes to various smaller air conditioning units than erect massive supply and return ducts for serving an extremely large unit.

2. *Filters and mixing plenums.* Before filtering, the return air and fresh outside air should be brought into a plenum for proper mixing. The amount of outside air to be introduced is usually controlled by motorised dampers. Depending upon the nature of occupancy, the dampers are controlled by a thermostat to admit less outside air during extreme outdoor temperature or by smoke and odour sensing devices to increase the amount of ventilation air according to requirement in the conditioned space.

Vertically assembled and cleanable types of filter banks are generally used. For central systems of less than 300 m^3/minutes capacity disposable filters are used. The various types of filters and air cleaning devices commercially available accomplish all the following cleaning operations.

(*a*) Removal of solid particles of visible size (dust, dirt, lint, pollen, cinders etc.)

(*b*) Removal of particles of microscopic size (smoke, haze, etc). Electrostatic precipitators are used for this purpose.

(*c*) Proper control of odour— for this purpose activated charcoal is usually used.

(*d*) Killing of bacteria—for this purpose germicidal lamps and certain chemicals are used.

3. *Refrigeration and Heating Equipment.* The design of central station system is greatly affected by these equipments. The direct-expansion (DX) refrigerant coils and hot water coils are unusually employed in systems requiring only one or two zones. The air conditioning units in these systems are generally located close to the cooling and heating equipments. The chilled water is used and circulated as a cooling medium in the air conditioning coils for many zones at remote locations which are served by a single cooling and heating plant. The heating medium is generally the hot water or steam. For serving the various zones water is generally pumped to the mechanical rooms.

4. *Insulation.* If the supply ducts are not running through the conditioned spaces, they should be properly insulated. Generally 25 mm thick insulating board is used to insulate the outer surface of the entire housing for the central station air conditioning unit. In order to prevent the heat losses and gains, all the steam, hot water, chilled water and suction refrigerant piping should be properly insulated.

5. *Noise and Vibration.* The noise and vibrations produced in the air conditioning units are very important factors for considerations while designing and selecting equipments. The undue noise and vibrations are considered undesirable.

9.20. HEATING OF BUILDINGS

The purpose of most forms of space-heating installation is to provide comfortable conditions in which people can work or relax. Space heating is far from being an exact science because opinions vary considerably as to what constitutes a comfortable environment. It is, however, reasonable to say that in order to be comfortable the feet and head should be at about the same temperature. The head should not be subjected to intense radiation and there should be as little draught as possible. Under these circumstances an ambient temperature between 20°C and 22°C will be acceptable to most people.

9.20.1. Factors Affecting The Choice of Heating Systems

The type of heating equipment used is influenced by several considerations. Probably the most significant of these is whether heat is to be required for long or short periods. *Short-term heating* generally relies on radiation which can be very effective in no time. Convected heat is best suited to *long-term* heating in which the time taken to 'warm up' is not so important.

The construction of building and the conditions existing in it are important. Where ceilings are low and changes of air are limited it may be practical to heat all the air in the room. If on the other hand the air is changing continually or there is a very high ceiling some sort of directional heating may be required.

Some forms of heating can only be installed effectively and economically during the construction of a new building whilst other may be used to improve an existing building.

Complete or Partial Heating Systems. In the past when coal or wood fires were the normal heat source it was customary to treat each room as a separate room. This persisted for many years after electricity was first used for space heating and in many cases a portable electric heater simply took the place of an open fire. This approach to heating produced result that tended to be both uncomfortable and uneconomical. Heat being continually lost from warm parts of the building to the cold and the wide differences of temperature experienced preventing all parts being used to equal advantage. Better results are obtained by designing a heating installation to suit an entire building. With this is done, a comfortable temperature can be achieved throughout the building and the points at which heat losses occur can be readily identified and the necessary steps taken. These steps would normally include double glazing of windows, draught-sealing doors and insulating the roof space with glass wool or some other thermal insulant.

One- and Two-Level Heating. Having decided to design a comprehensive heating system for a building another choice must be made. This is whether to settle upon a single value of temperature and heat the entire building to it or not. In office blocks and factories where every effort is made to get the maximum use out of the available floor space there is little waste involved in heating to the same temperature throughout.

In domestic installations the amount of use given to different parts varies largely and also there are considerable differences in levels of activity from one time of the day to another. To allow for such variations a two-level heating system is often employed. In such a case general heating is provided to a fairly low comfort level in the region of 15°C. This is called the *background heating* and keeps all the house or flat comfortable for housework or similar activities. Additional appliances are available, controlled by the consumer, to raise the temperature of specific areas a further 5° or 6° when he or she is sitting still for any length of time. This is called *topping up*. Background heating is usually provided by convection heaters and radiant heaters are used for topping up.

9.20.2. Types of Heating Equipments Used For Space Heating

The different types of equipments used for heating of buildings are given below :

1. *High Temperature Radiators.* The high temperature radiator essentially consists of a high resistance element wound on porcelain or refractory former and mounted into decorative frames. Resistance elements get heated to high temperature (1,400°C to 1,600°C) due to flow of electric current through them and 50% to 70% of heat so produced is dissipated by radiation and remaining by convection. Advantages of this type of heater are that it is portable, cheap in initial cost, simple in construction and attains its highest temperature quickly. The main drawback of this type of heater is that heat dissipated is localised. These are used for domestic purposes or where intermittent heating is required. The usual sizes are from 0.5 kW to 3 kW.

2. *Tubular Heaters or Low Temperature Convectors.* The most common type of low temperature heater is 5 cm diameter steel tubes of length ranging from 0.6 to 5 metres and containing resistance elements arranged on mica, fire clay, or porcelain former. Such tubes are mounted around the skirting board of room to be heated. The normal loading is 200 watts per metre length of the convector. The temperature of the external surface is maintained at about 93° C. About 90 per cent of the heat is transferred by convection. These are spaced to produce convection currents in the room so as to heat whole of it. These are specially suitable for thermostatic control, by which means they are switched on and off automatically as the room temperature falls below or rises above the desired value. When fixed at the base of a cold wall or underneath a window they prevent cold down draughts, which cause cold feet to the occupants of the room.

3. *Convector Heaters*. In these heaters, heating elements are mounted within sheet-metal cases which admits cool air from bottom. Air after getting heated up, comes out from grills at the top. The front and top may be of moulded plastic material for the sake of pleasing appearance. These heaters may be either portable or fixed, and may be handled without fear of burns. In these type of heaters, heating element attains low temperature of 66° to 93°C and proportion of heat transmitted by convection is large as compared to heat transmitted by radiation.

A very satisfactory method of heating a room is to use convector heaters to warm the room to about 13°C, together with a small radiant heater for local heating.

4. *Panel Heaters*. In this system of heating large panels of fire clay or some other suitable heat resisting materials are used along the walls or ceiling of a room. The resistance elements are embedded in these panels. The heat produced is utilised in the form of convection or radiations. This system of heating of buildings is advantageous from the decoration point of view and gives a uniform distribution of heat. Loadings are of the order of 400 watts/m^2 for the low temperature panels and 6,000 watts/m^2 for the high temperature panels. Low temperature panels operating at temperature from 27°C to 65°C may be fixed in walls or ceiling of a room and plastered over. High temperature panels operating at temperature up to 288°C are used on walls and ceiling but are not fixed in the wall surface. They are sometimes suspended from the ceiling.

5. *Underfloor Heating*. This is one of the systems that may be installed in a building whilst it is being built.

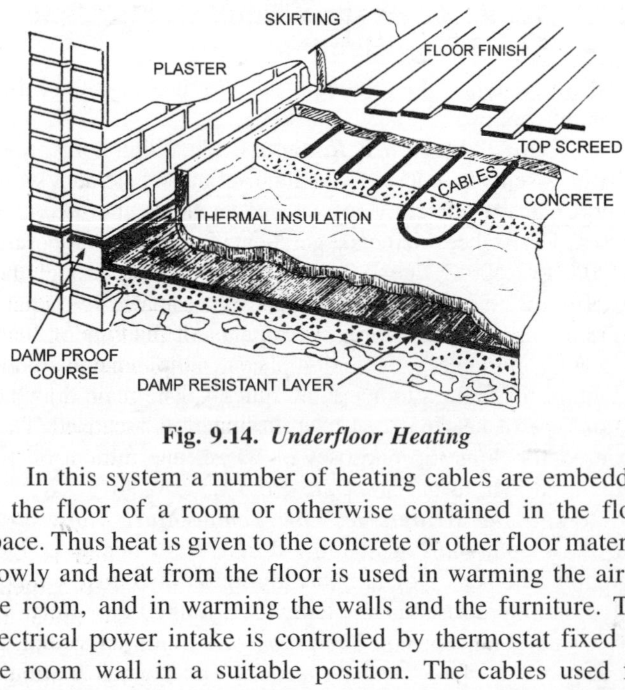

Fig. 9.14. *Underfloor Heating*

In this system a number of heating cables are embedded in the floor of a room or otherwise contained in the floor space. Thus heat is given to the concrete or other floor material slowly and heat from the floor is used in warming the air in the room, and in warming the walls and the furniture. The electrical power intake is controlled by thermostat fixed on the room wall in a suitable position. The cables used for heating are made from alloys of nickel, chromium, iron or other resistance materials and are insulated with asbestos mineral insulation, butyl rubber, or other heat-resisting insulating materials. For mechanical protection lead, copper, aluminium, or PVC sheathing can be used. Loading of from 100 to 150 watts per metre^2 of floor area are generally satisfactory.

In this system also the advantage of cheaper power available during night hours can be taken but in such a case it will be desirable that one or more convectors or radiant heaters for use towards the end of the floor-warming output cycle are installed.

6. *Thermal Storage System*. Roughly this system is similar to well-known central-heating system employing a coal- or coke-fired boiler except that the heat is supplied to the storage tank electrically. Hot water is circulated through the heating pipes and radiators situated throughout the building. It must be noted that in electrical heating system the tank and piping must be lagged to prevent loss of heat, a feature which is usually omitted with fuel-fired systems, since with such systems the efficiency is so low that the additional loss due to lack of lagging is relatively insignificant. There are two types of heaters used in this system : the immersion heater for a load up to 100 to 200 kW and the electrode boiler type for loads exceeding 200 kW.

The *immersion heater* used for heating the water in a storage tank consists of a spiral of resistance wire embedded in refractory insulating medium like magnesium oxide inside a copper or steel tube, which is bent to the desired shape. The tube is provided with a flange at one end, so that it can be bolted to the vertical side of the tank with the heating element projecting horizontally into the water. The loading of such a tube is usually between 16 and 24 kW/m^2 of face, so that a tube of length 305 mm and diameter 51.0 mm will be rated at about 1 kW. Single tubes in ratings of 4-5 kW are available. For higher ratings group of tubes are used.

The *electrode boiler* is only suitable for use on ac, usually three phase, since dc would electrolyse the water. In electrode boilers water is heated by placing cast iron electrodes directly in the water. For 3-ϕ supply three electrodes are placed in the water and the tank is earthed so as to form a star-connected system. The resistance of average cold tap water varies between about 20 and 50 Ω-m. Sodium carbonate (soda) is, therefore, added for increasing the conductivity of water. Another noteworthy point is that resistance of water decreases with the increase of its temperature, the value at 100°C being one-third of that at 20°C. Automatic control of the power supply is, therefore, required if a constant input is required. Control of power input may be carried out by varying the height of the electrodes in the water, by means of a movable shield surrounding the electrodes, by moving one electrode relative to the other, or in large units by varying the water level by means of a pump. A low-voltage boiler operating at 400 V is usually employed for a requirement up to 500-700 kW while for larger requirement a high-voltage boiler operating at voltages up to 11,000 V or even higher is used.

The advantage of this system is that the water can be heated during night (off-peak periods) and is supplied throughout the day, thus use of cheap power available during off-peak period can be made. This system of heating also provides hot water service for the building.

9.21. CALCULATION OF RATING OF ELECTRICAL EQUIPMENT

The amount of heat required for personal comfort in an enclosed space such as a room or office depends upon the following factors :

1. Number of changes of air per hour.
2. Area of windows.
3. Situation of walls, external or internal (most rooms have at least one external wall).
4. The exposure of the ceiling (open to the roof or with heated bedroom over).

5. Material of which walls, floor and ceilings are composed.
6. The temperature of outside air and
7. Whether the building is to be heated continuously or intermittently.

The method of calculations will be clear from the following examples.

Example 9.1. It is proposed to condition 3,000 m^3 of air per hour from a temperature of 5° C to 20° C. It is further necessary to evaporate 5 kg of moisture per 1,000 m^3 of air per hour to control humidity. Estimate the power required if heat required to raise the temperature of 1 m^3 of air through 1° C is 1,220 J and latent heat of evaporation is 2,450 × 10^3 J/kg.

Solution : Heat required to raise the temperature of air,

H_1 = Volume of air to be conditioned × heat required to raise the temperature of 1 m^3 of air through 1°C × difference of temperatures

$= 3,000 \times 1,220 \times (20 - 5)$ Joules $= 54.9 \times 10^6$ J

The moisture present in the air $= \frac{5 \times 3,000}{1,000} = 15$ kg

Latent heat required to evaporate the moisture

$H_2 = 15 \times 2,450 \times 10^3 = 36.75 \times 10^6$ J

Total heat required, $H = H_1 + H_2$

$= (54.9 + 36.75) \times 10^6 = 91.65 \times 10^6$ J/hour

Power required $= \frac{91.65 \times 10^6}{3,600 \times 1,000} \simeq 25$ kW **Ans.**

Example 9.2. A drawing office 6 m × 5 m × 4 m size is to be heated. The temperature of the room is to be kept at 25°C when the outside temperature is 5°C. The air is to be renewed every 15 minutes. Find the necessary rating of the electric heater. The heat loss from the walls is 836 × 10^3 J/°C/hour. Assume heat required to raise temperature of 1 m^3 air is 1,220 J.

[*Pb. Technical Univ. Electric Drives & Utilization Dec. 2008*]

Solution : Volume of air to be conditioned

$= 6 \times 5 \times 4 \times \frac{60}{15} = 480$ m^3/hour

Heat required to raise the temperature of 1 m^3 of air through 1°C = 1,220 J

Heat required to raise the temperature of 480 m^3 air through (25 – 5), *i.e.*, 20°C

$H_1 = 480 \times 1,220 \times 20 = 11.712 \times 10^6$ J/hour

Heat losses from the walls $= 836 \times 10^3 \times 20 \times 1$

$= 16.72 \times 10^6$ J/hour

Total heat required, $H = H_1 + H_2$

$= 11.712 \times 10^6 + 16.72 \times 10^6$

$= 28.432 \times 10^6$ J/hour.

Rating of heater $= \frac{28.432 \times 10^6}{3,600 \times 1,000} \simeq 8$ kW **Ans.**

Exercises

1. Some people have an idea that refrigeration is luxury but others feel it is a necessity. What is your opinion in respect of this particular controversial point. Give some concrete examples in support of your opinion.
2. Give a diagram describing a basic refrigeration system. **[Pb. Technical Univ. Electric Drives & Utilization May 2008]**
3. Write about vapour compression cycle in refrigerator. **[Pb. Technical Univ. Electric Drives & Utilization Dec. 2006]**
4. Enumerate the successive operations to be performed on a refrigerant in (*i*) a compression system and (*ii*) an absorption system. Why is ammonia the only commonly refrigerant used in the absorption system ?
5. Discuss in details
   (*a*) Vapour compression refrigeration system.
   (*b*) Vapour absorption refrigeration system.
   (*c*) Thermoelectric refrigeration system.
   [Pb. Technical Univ. Electric Drives & Utilization Dec. 2009]
6. Give the merits and demerits of vapour absorption system over vapour compression system.
7. Enlist the main requirements of a good refrigerant. What are the primary and secondary refrigerants ? Name the refrigerants generally used.
8. Explain the working principle of thermoelectric refrigeration system. List out the merits and demerits of the thermoelectric refrigeration system over other refrigeration systems. What are the field of its application ? **[Pb. Univ. Electric Drives and Utilization of Electric Energy August 1984]**
9. Draw the electric circuit of a refrigerator and explain its working. How can temperature inside the refrigerator be adjusted ? **[Pb. Technical Univ. Electric Drives & Utilization May 2009]**
10. Explain vapour compression refrigeration cycle and show how cooling effect is produced ? Describe electric circuit of a refrigerator. **[Pb. Technical Univ. Electric Drives & Utilization Dec. 2007]**
11. With the help of a circuit diagram, explain the working of a water cooler.
12. What is air conditioning ? On what factors it depends ? **[Pb. Technical Univ. Electric Drives & Utilization Dec. 2008]**
13. Define air conditioning. On what factors does the air conditioning depend ? Explain in detail. **[Pb. Univ. Elec. Drives & Utilization of Elec. Energy April 1981]**
14. Discuss the role of air conditioning in our day-to-day life.
15. What are the different conditions which constitute human comfort ? **[Pb. Univ. Elec. Drives & Utilization of Elec. Energy April 1981]**
16. Differentiate between comfort air conditioning and industrial air conditioning. **[Pb. Univ. Elec. Drives & Utilization of Elec. Energy April 1982]**
17. What is air conditioning ? How is air from micro-organism, gaseous contaminants and odours purified ? **[Pb. Univ. Elec. Drives & Utilization of Elec. Energy Nov. 1990]**
18. Describe in detail functioning of window air conditioner. **[U.P. Technical Univ. Utilization of Electrical Energy & Traction 2009-10]**
19. Explain with the help of diagram the working of window type room air conditioner. **[Pb. Technical Univ. Electric Drives & Utilization Dec. 2009]**
20. Explain the working of central air conditioning system. **[Pb. Technical Univ. Electric Drives & Utilization May 2009]**
21. Describe complete arrangement of centrally air conditioning plant. **[Pb. Univ. Elec. Drives & Utilization of Elec. Energy April 1982]**
22. Discuss the effects of temperature and humidity in the atmosphere on the human body from the view point of comfort. Define the effective temperature in air conditioning. Specify suitable values of effective temperature and relative humidity for summer as well as winter air conditioning of temperature and tropical climate assuming approximately still air.
23. Discuss different methods of heating of buildings. **[Agra Univ. Traction & Utilization Elec. Power 1982]**
24. What are different space heating systems ? Discuss them briefly. **[Pb. Univ. Elec. Drives & Utilization of Elec. Energy Nov. 1981; April 1981]**
25. What are different space heating systems ? Discuss them. **[Pb. Technical Univ. Electric Drives & Utilization May 2008]**
26. Write short notes on the following :
   (*i*) Thermoelectric refrigeration. **[Pb. Univ. Nov. 1983, April 1983]**
   (*ii*) Refrigerator **[G.B.T.U. 2010-11]**
   (*iii*) Water cooler **[G.B.T.U. 2010-11]**
   (*iv*) Air conditioning

(v) Effective temperature [G.B.T.U. 2010-11]
(vi) Window air conditioner [G.B.T.U. 2010-11]
(vii) Room air conditioner [Pb. Univ. Nov. 1983, April 1983]
(viii) Heating of buildings. [Pb. Univ. Nov. 1983, April 1983]
(ix) Heating of rooms and buildings [Agra Univ. 1983]
(x) Space heating systems. [Pb. Univ. April 1980]

Short Answer Type Question With Answers

Q. 1. What is practical unit of refrigeration ? Define it.
Ans. Tonne of refrigeration (TR) is the practical unit of refrigeration and is defined as the amount of refrigeration effect produced by the uniform melting of one tonne (1,000 kg) of ice from and at 0°C in 24 hours.

Q. 2. Draw the neat sketch of refrigeration cycle. **[P.T.U. Dec. 2009]**
Ans. Refer to Fig. 9.1.

Q. 3. What are the applications of vapour compression refrigeration cycle ?
Ans. The vapour compression refrigeration cycle is used in most of the modern refrigeration systems in large industrial plants requiring rapid removal of large amount of heat and in most of the domestic refrigerators.

Q. 4. On what principle does thermoelectric refrigeration system operate ?
Ans. Thermoelectric refrigeration system operates on the principle of 'Peltier effect'.

Q. 5. What are the desirable properties of refrigerants ? **[P.T.U. Dec. 2009]**
Ans. For safe, economical and thermodynamically efficient production of cold, refrigerant should possess as many as practicable of the following desirable properties.

1. Low boiling point.
2. High critical temperature.
3. High latent heat of vaporisation.
4. Low specific volume of vapour.
5. Low specific heat of liquid.
6. Low cost.
7. Easy to liquify at moderate pressure and temperature.
8. Easy of locating leaks by odour or suitable indicator.
9. It should be noncorrosive.
10. It should be noninflammable.
11. It should be non-toxic.
12. It should be free from objectionable odour.

Q. 6. Why is ammonia the only common refrigerant used in the absorption system ?
Ans. Because ammonia is cheap, readily available and has the highest efficiency.

Q. 7. Why thermostat is used in a domestic refrigerator ?
Ans. Thermostat switch is provided in a domestic refrigerator to control the temperature inside a refrigerator.

Q. 8. With the aid of a sketch of a thermostat explain how the temperature inside a refrigerator is adjusted. **[P.T.U. Dec. 2006]**
Ans. Thermostat is a thermal switch that is provided to control the temperature inside the refrigerator. Temperature inside the refrigerator can be adjusted by means of temperature control screw. Greater the distance between contacts, greater will be the temperature inside the refrigerator and vice versa.

Q. 9. Why it is not advisable to put very hot things into the refrigerator ?
Ans. Main precaution to be taken is that very hot things should not be put in refrigerator, if it is done it will quickly evaporate the refrigerant in evaporator coils producing a large vapour pressure resulting in increase in the duty of the compressor and as such motor has to work for longer duration in order to bring down the temperature in the freezer compartment. It may damage the motor, which is short-time rated. Also moisture released from hot things will freeze around the freezer compartment which has to be defrosted.

Q. 10. Why is an automatic door closing mechanism provided in a modern refrigerator ? **[P.T.U. Dec. 2006]**
Ans. The door is provided with automatic closing mechanism–door hinges are provided in such a way that door flap when left in open position automatically comes to closing position due to gravity and as it approaches closing position it is attracted by the magnetic strip fitted behind the sealing rubber ring and thus the door is closed with snap action. All this is done to prevent leakage of atmospheric heat inside the refrigerator.

Q. 11. What will happen if automatic starter relay stops to operate in refrigerator ? **[P.T.U. Dec. 2007]**
Ans. Capacitor-start single phase induction motor is used in open type refrigerators and split-phase single phase induction motor is used in sealed unit refrigerator. Electromagnetic relay is provided to connect auxiliary winding on the start and disconnect it when the motor picks up the speed.
If the automatic starter relay keeps the auxiliary winding open after the motor has been operating, little harm should occur since the motor is provided with overcurrent protection. The motor simply won't start on the next attempt. On the other hand, should the switch remain closed, the motor will start and run. The auxiliary winding will get quickly overheated, and if not detected, or cut out by a protective relay, this winding may cause damage to the main winding due to their proximity.

Q. 12. Mention the requirements of air conditioning. **[P.T.U. May 2008]**
Ans. In this modern age every body is trying to improve his surroudings in order to make living conditions more comfortable. Air conditioning is one of the important contribution in this respect. The technique of maintaining of air condition inside a building in such a way that produces comfort to human beings, ensures trouble free operation of delicate equipment or helps in industrial manufacturing operations is called *air conditioning.* It involves control of temperature, cleanliness, humidity and circulation of air.

Q. 13. What is comfort air conditioning ? **[P.T.U. May 2009]**
Ans. This is the type of air conditioning whose objective is to produce comfortable and healthy conditions for occupants. In other words, it is the process of treating air so as to control simultaneously the prevailing conditions of temperature, humidity, cleanliness and distribution in the conditioned space to meet the comfort and health requirements of the occupants. It includes the air conditioning of spaces, places and surroundings which are occupied by human beings.

Q. 14. Define the effective temperature in air conditioning ?
Ans. An arbitrary index which combines into a single value the degree of sensation warmth or coolness felt by a human body due to the effect of air, temperature, humidity and air movement is called the effective temperature.

Problems

1. In an air-conditioning plant it is required to raise the temperature of 2,000 m^3 of air per hour from 0°C. It is further necessary to evaporate 0.5 kg of moisture per 100 m^3 of air per hour to control the humidity. The density of air may be taken as 1.3 kg/m^3, its specific heat 0.25 and latent heat of evaporation as 500 calories per gram. Estimate the power required.
[Pb. Univ. Elec. Drives & Utilization of Elec. Energy Nov. 1977]
[Ans. 13.4 kW**]**
2. A room measures 4 m × 4 m × 9 m. The air in it is to be renewed every 30 minutes and maintained at a temperature of 10°C above that of incoming air. Find the necessary rating of an electric heater ignoring losses. Find the rating if the heat loss from the walls etc. is 3,00,000 calories per °C per hour. Take for air : density 1.28 g per litre, specific heat 0.24.
[Pb. Univ. Elec. Drives & Utilization of Elec. Energy Nov. 1979]
[Ans. 4.5 kW**]**

❑❑❑

SECTION - C

ELECTRIC TRACTION

CHAPTER 10

Traction Systems

10.1. INTRODUCTION

Act of drawing or state of being drawn; propulsion of vehicle is called the *traction* and the system of traction involving the use of electricity is called the *electric traction system.* There are various systems of traction prevailing in our country such as steam engine drive, internal combustion engine drive, diesel electric drive, battery electric drive, straight electric drive. These systems of traction may be classified broadly into two groups namely (*i*) the traction systems which do not involve the use of electricity at any stage and called the non-electric traction systems such as steam engine drive, internal combustion engine drive etc. (*ii*) the traction systems which involve the use of electricity at some stage or the other, and called the electric traction systems such as diesel electric drive, straight electric drive etc. Systems of electric traction may further be subdivided into two main groups: (*i*) the group consisting of vehicles which receive electric power from a distributing network fed at suitable points from either a central power station or substations suitably spaced such as tramways, trolley buses, electric railways etc. (*ii*) the group consisting of self-contained locomotives such as diesel electric trains and ships, petrol electric trucks and lorries, battery-driven road vehicles etc.

Each system of traction has got its own merits, demerits and fields of applications and no single system fulfills the requirements of an ideal traction system which are given below:

1. High starting tractive effort in order to have rapid acceleration.
2. Self-contained and compact locomotive or train unit so that it may be able to run on any route.
3. Equipment capable of withstanding large temporary overloads.
4. Minimum wear on the track.
5. Braking should be such that minimum wear is caused on the brake shoes, and if possible the energy should be regenerated and returned to the supply during braking period.
6. Equipment required should be minimum, of high efficiency, and low initial and maintenance cost.
7. No interference to the communication lines (telephone and telegraph lines) running along the track.
8. Easy speed control.
9. It should be pollution free.

10.2. DIFFERENT SYSTEMS OF TRACTION

The various traction systems commonly used are described below.

1. *Direct Steam Engine Drive.* The steam engine drive used to be widely employed for railway work. In this type of drive, the reciprocating steam engine is invariably used for getting the necessary motive power because of its inherent simplicity, operational dependability, simplified maintenance, the simplicity of connections between the cylinders and the driving wheels and easy speed control. The locomotive and train unit is self-contained, therefore, it is not tied to a route. It causes no interference to the communication lines running along the track. The capacity is very high in comparison to direct internal combustion engine drive, and battery electric drive etc. Initial investment required is low in comparison to that of electric drive as no overhead structure, distribution system and power generation is required. It is cheap for low density traffic areas and in initial stages of communication by rail.

Steam engine drive is losing ground because of the following drawbacks:

1. Because of difficulty of installing a condenser on a locomotive, the steam engines employed are of non-condensing type, therefore, efficiency is very poor (6 to 8 per cent). Further fires have to be banked even when locomotive is idle.
2. Adequate supplies of feed water at regular intervals is required.

3. The overload capacity of steam locomotive is limited.
4. Steam locomotive has to carry sufficient quantity of coal and water which correspondingly reduces the payload that can be hauled.
5. Limited availability and cruising range which are influenced not only due to quantity of water and coal to be carried in the locomotive but also due to periodic stops necessary to clean the ash pan. For standard passenger locomotives this cruising range is about 160 km.
6. Performance of steam locomotive is governed by the rate of firing of the coal. Single fireman can maintain a firing rate of about 2 tonne/hour for short periods and with two firemen it may be about 2.7 tonne/hour.
7. More number of crew (one driver and two firemen) are required, therefore, wage bill is increased.
8. Extensive and costly ancilliary equipment, such as coaling cranes, ash and fire cleaning pits, turntable, water supply plants, is required. The maintenance of these facilities also costs money.
9. Steam locomotive requires more repair and maintenance. The repair costs of steam locomotive are heavy specially for the boiler. Steam locomotive requires large number of unskilled and semi-skilled staff for maintenance.
10. Larger size of running sheds and workshops are required.
11. Owing to coal dust, cinder and ash the steam engine drive is not clean drive.
12. Steam locomotive cannot be put into service at any moment as time is required for steam raising.
13. Since driving wheels are very close, more concentrated adhesive weight is required.
14. Steam locomotive causes considerable wear on the track owing to unbalanced reciprocating parts and corrosion of steel structures by the smoke emitted.
15. Because of unbalanced reciprocating masses of steam engine riding qualities are not good.
16. Because of unbalanced forces produced by reciprocating masses, coefficient of adhesion of steam locomotive is low (0.25 as compared with that of electric ranging from 0.3 to 0.45).
17. Due to low coefficient of adhesion power-weight ratio of steam locomotive is low.
18. Owing to high centre of gravity of steam locomotive, speed is limited.
19. Owing to smoke, steam engine drive cannot be employed for underground railways.
20. Steam locomotive is only available for hauling for about 60 per cent of its working day, the remainder of the time is spent in coaling, fire cleaning, watering and servicing.
21. Because of low coefficient of adhesion it is not possible to have acceleration and retardation with steam engine drive as high as with electric drive. Steam engine drive is, therefore, not suitable for urban or suburban services where distance between stations is small. It is not also suitable for hauling heavy loads on steep gradients.

A few steam turbine locomotives have been built having a condenser and reduction gearing between the turbine and driving wheels for experimental purposes. Though the efficiency of such locomotives is between 10 and 15 per cent but due to their high initial and maintenance costs they could not be put into general use.

Gas-turbine locomotives have also been built on an experimental basis.

2. *Direct Internal Combustion Engine Drive.* Direct internal combustion engine drive is widely employed for road transport (buses, trucks, cars etc.). The efficiency of internal combustion engine at its normal speed is about 25 per cent. It is self-contained unit and therefore, it is not tied to any route. Initial investment required, being only the cost of vehicle and garage, is very low. Speed control and braking system employed is very simple. It is cheap drive for the outer suburbs and country districts.

The disadvantages of this type of drive are:

1. Its overload capacity is limited on account of the speed-torque characteristics of the engine and even 10% of overloading may result in stalling of the engine.
2. Its operation at any but the normal speed is very uneconomical.
3. As torque exerted by an internal combustion engine is approximately constant, hence an engine capable of giving high starting torque required is not used to its full capacity when running at full speed unless some form of gear is employed.
4. Speed control is possible only by employing a gearbox.
5. It is not suitable for heavy railway work as the arrangement possible for speed control and utilizing the torque of engine to full capacity when running at full speed is not altogether satisfactory for it.
6. The life of propulsive equipment is much shorter than that of electrical equipment of a tramcar or a trolley bus.
7. The maintenance and running costs are higher.
8. Since the fuel oil is to be imported in our country, therefore, there is drain on foreign exchange which is vital for expanding economy of the country. In wartime this system may get completely paralysed.
9. The oil engined bus accommodate slightly less passengers than the trolley bus.
10. It has no starting torque and must be started by some auxiliary means such as compressed air or electrically.

3. *Steam Electric Drive.* A few locomotives employing steam turbine for driving a generator used for supplying current to electric motors have been built for experimental purposes. Such locomotives have not been put into general use because of some mechanical difficulties and complications.

4. *Internal Combustion Engine Electric Drive.* In this drive the reduction gear and gearbox are eliminated as the diesel engine is to drive the dc generator coupled to it at a constant speed. This type of drive has found considerable favour for railway work and locomotives of this type are becoming widely used. The advantages and disadvantages of this drive are given below.

Advantages. 1. No modification of existing track is required for conversion from steam traction to diesel engine electric traction.

2. Initial investment required is low as no overhead structure distribution system and equipment is required.
3. The diesel engine electric vehicle can accommodate more passengers than steam engine vehicle and because of its higher acceleration and braking retardation the schedule speed over a given route will be higher than that of the latter vehicle, so that fewer diesel-electric locos will be required for operating a given service.
4. Power loss in speed control is very low because it can be carried out by the field control of generator.
5. The locomotive and train is a self-contained unit and, therefore, it is not tied to any route.
6. Though the maintenance cost is fairly heavy but the time spent in maintenance and overhaul is very less (only 5 to 10 per cent of its working day). Hence such a locomotive is available for hauling for almost $6\frac{1}{2}$ days in a week for 24 hours duty.
7. It can be put into service at any moment.
8. Overall efficiency is greater than that of steam locomotives (about 25 per cent in comparison of 8 per cent or so in case of steam locomotives).

Disadvantages. 1. Its overload capacity is limited as diesel engine cannot be overloaded.

2. Life of the diesel engine is comparatively shorter.
3. Special cooling system is required for cooling the diesel engine in addition to motor generator set.
4. Running and maintenance costs are high.
5. The dead weight of such locomotives is more. It requires more axles (six for diesel electric locomotive in comparison with four for electric locomotive).
6. Since the fuel oil is to be imported in our country, therefore, there is drain on foreign exchange which is vital for expanding economy of the country. In wartime this system may be completely paralysed.
7. For the same power output diesel electric locomotive is costlier than steam or electric locomotive because it carries diesel engine, main generator and traction motors, all capable of handling full locomotive power. This also makes diesel electric locomotive considerably heavier than plain electric locomotive.
8. Regenerative braking cannot be used with such drives but there is no bar in making use of rheostatic braking.

5. *Petrol Electric Traction.* This system has been used in heavy lorries and buses. Due to electric conversion it provides a very fine and continuous control which makes the vehicle capable of moving slowly at an imperceptible speed and creeping up the steepest slope without throttling the engine.

6. *Battery Electric Drive.* In this drive the locomotive carries the secondary batteries which supply power to dc motors employed for driving the vehicle. Such a drive is well suited for frequently operated service such as for local delivery of goods in large towns with maximum daily run of 50 to 60 km, shunting and traction in industrial works and mines. Battery driven vehicle is easy to control and very convenient to use. Other advantages are low maintenance cost and absence of fumes. The major limitation of this type of drive is the small capacity of the batteries and the necessity for frequent charging, speed range is also limited. Battery vehicles are started by series-parallel grouping of batteries—in parallel for starting and running at the speed up to half maximum speed and in series for running at full maximum speed.

7. *Electric Drive.* The drive of this type is most widely used. In this system of traction the vehicle draws electrical energy from the distribution system (from a contact wire suspended above the track or from an additional rail laid along side it) fed at suitable points from either a central power station or substations. The advantages and disadvantages of electric drive are as follows:

Advantages. 1. It is the cleanest of all other types of systems of traction. Due to this only it is ideally suitable for the underground and tube railways.

2. No water and coaling depots are required in electric drive.
3. Ancilliary equipment, such as coaling cranes, ash and fire cleaning pits, water supply plants, is not required. Size of running sheds and workshops is comparatively smaller.
4. The electrical energy required for lights and fans of the train can be drawn from the lines directly and, therefore, there is no need of providing Rosenberg generators.
5. Owing to better speed control in electric drive locomotion on steep gradients at slow speed is possible.
6. It has got another advantage of rapid acceleration and braking retardation. An electric locomotive has an acceleration of 1.5 to 3.5 kmphps for urban service. Trolley buses have acceleration of 4 to 6 kmphps. The steam locomotive has an acceleration of only 0.8 kmphps. Hence the schedule speed may be from 50 to 100 per cent higher than that corresponding to steam operation over short runs and with frequent stops.
7. Because of *larger* passenger carrying capacity and higher schedule speed the electric railway can handle the traffic of any amount—up to double the amount possible with steam railway.
8. The electric train can be divided and run in sections during the periods of light traffic, thereby enabling a frequent service to be maintained.
9. Less terminal space is required on account of speedy movement and lesser number of units required for operating a given service.
10. An electric locomotive requires much less time for maintenance and repairs than a steam locomotive and such vehicles can, if desired, be kept in service for 95 per cent or more of the working day.
11. Maintenance and repair cost is about 50 per cent of that of steam locomotive.
12. The electric locomotive can be put into service immediately whereas steam locomotive takes about two hours to get up steam.

13. There is no damage to the plant, equipment and building due to corrosive smoke fumes because of absence of unwanted gases and fire hazards.
14. This system is healthier from the hygienic point of view owing to complete absence of smoke and fumes.
15. Electric traction is most economical in areas of high traffic density particularly if electrical energy is cheap.
16. Centre of gravity of electric locomotive is lower than that of steam locomotive and, therefore, electric locomotive is able to negotiate curves at comparatively higher speeds.
17. The vibrations in electrically operated vehicles are less as the torque exerted by the electric motors is continuous.
18. In electric traction it is possible to instal power units in two or more power cars in the train and to control them from one point whereas in steam traction each locomotive is to be manned by its own crew.
19. The coefficient of adhesion is better in electric traction than that in steam traction. This is due to continuous torque of electric motors and distribution of driving wheels along the length of the train. Higher coefficients of adhesion not only reduces the ratio of weight to output kW of the locomotive but also results in improvement in riding qualities and reduction in wear and tear of the track.
20. Electric equipment can withstand large temporary overloads and can draw relatively large power from the distributing system.
21. Electric traction helps in saving of high grade coal which is limited in quantity in our country.
22. Power requirement for railway electrification has been of the order of 50 kW/track km. This is a sizeable load which tends to increase power development schemes. Traction load has high load factor of the order of 60 to 70%. Electric traction, therefore, provides a most important base load. This, therefore, enables the use of large generator units having high thermal efficiency possible in thermal stations. High base load also affords economic development of hydroelectric potential.
23. Railway electrification encourages rural electrification as no special transmission lines have to be run for this purpose.
24. Electric braking is superior to mechanical braking because of less wear on brake shoes, wheels and track. In case of regenerative braking a part of energy is returned to the supply instead of being wasted as heat in the brake shoes, thus saving in overal consumption is affected. Goods traffic on gradients is more safe and speedy.
25. In case of steam traction brake blocks and wheels are likely to overheat on continuous grades and sometimes trains have to be stopped for allowing these brake blocks and wheels to cool down but in case of electric traction it is not so. In electric traction dynamic braking is applied to hold the train under control and brake blocks are applied for stopping the train and thus wheels and brake blocks are always cool whenever required for stopping the train.

Disadvantages. 1. The greatest disadvantage of electric traction is that it involves a heavy initial expenditure for the power supply system and, in the case of tramways, for the track rails and electrical return system, in addition to the cost of vehicles, depots etc.

2. Number of somewhat severe statutory regulations, concerning leakage currents from overhead conductors and the voltage drop in the track rails, have to be complied with.
3. Failure of power supply for a few minutes may paralyse the whole system. Such an occurrence is, however, rare nowadays due to increased reliability of power supply.
4. Steam locomotives can use their steam for heating the compartments in cold weather, very cheaply whereas the electric locomotive have to do it at extra cost.
5. Electric traction system is tied up to only electrified routes.
6. In case of ac traction the communication lines running along the track experience considerable interference from the power lines. The communication lines therefore must, either be removed away from the track, or replaced by special expensive cables. The cost incurred for doing this may be as high as 15% of the capital outlay.
7. Additional equipment is required for regeneration. In case of dc series motors regeneration is not simple process.
8. In cold countries a service locomotive is required to run up and down the line in order to prevent the formation of a layer of ice on the conductor rails.

The main consideration in the electrification of railways or other traction services is one of the economics. If operation by electric traction proves to be cheaper than any other method (*i.e.,* the electrical energy is available at cheaper rates and the density of the traffic is high) and necessary capital can be raised then no other factor or disadvantage would prevent the conversion of steam railway to electric railway. There is no technical difficulty in adoption of electric traction.

So far as Indian Railways are concerned the route kilometrage on which electrification is justified is quite considerable. The reasons are:

1. About 72% of route kilometrage comprises of single line section and average traffic density even at sixties level of traffic is very much higher than that necessary for justification of electrification of single line sections.
2. The traffic densities for justification of electrification of railways in case of graded sections are much less due to relative increase in fuel energy charges and savings in the energy affected by employing regenerative braking and in our country nearly 30% of railway track is having gradients steeper than 0.33%.
3. In our country electric energy is available at cheaper rates comparatively because of availability of substantial sources of hydropower generation and nuclear power generation. By electrification of railways high grade coal, which is limited in quantity in our country, will be saved which can be used for some other useful purposes.

Efficiences of different types of locomotives are given below in tabular form.

Table 10.1

S.No.	Traction System	Efficiency
1.	Steam locomotive	6–8%
2.	Gas turbine electric locomoive	10%
3.	Diesel electric locomotive	25–30%
4.	Electric locomotive with thermal power plant	About 35%
5.	Electric locomotive with hydroelectric power plant	About 40%

10.3. SYSTEMS OF ELECTRIC TRACTION

As already mentioned, systems of electric traction may be divided into two main groups.

(1) The group consisting of vehicles which receive electric power from a distributing network fed at suitable points from either a central power station or substations suitably spaced. This group is further subdivided into:

(*a*) Systems operating with dc such as trolley buses, tramways and railways.

(*b*) Systems operating with ac, such as railways.

(2) The group consisting of self-contained locomotives. This group is further subdivided into:

(*a*) Diesel electric trains and ships.

(*b*) Petrol electric trucks and lorries.

(*c*) Battery driven road vehicles.

Out of these self-contained electric vehicles, only the diesel electric traction system is widely employed and hence keeping in view the limitations discussion will be limited to it.

10.3.1. Tramways

The power is supplied to the tramcar, usually at 600 dc volts, from a single overhead conductor (of positive polarity) fed at suitable points from a central power station or substations, the track rail forms the return conductor. The tramcar is provided with at least two driving axles so that necessary adhesion is secured and control is possible from either end. The use of two or more driving axles lends itself to the use of two motors with series-parallel control. Field weakening method may be employed to obtain higher running speeds for suburban service. Two drum type controllers, one at each end, usually employed for control of vehicle, are connected in parallel but are provided with some suitable interlocking arrangements so that operation is possible from one end at a time. With the increasing outputs required by present-day tramcars the controller is tending to become so large that its operation causes fatigue to the driver, and master controller equipment similar to that used on railways is employed in recent designs. The great disadvantage of the tramway is that it needs laying of overhead supply system and track for its use which is costly to maintain and constitutes a source of danger to other road users. For very dense traffic in large cities the tramway is most economical means of transportation. Nowadays the trolley bus or internal combustion engine omnibus is preferred on account of its greater manoeuvrability. The equipment required for tramcars is similar to that used in railways but of considerably smaller output (not exceeding 45 or 55 kW). Life of tramcar equipment is much more than that of internal combustion engine omnibus. The conditions for regenerative braking are not very favourable but it may be used in hilly areas and on level track up to a speed of 20 kmph. Rheostatic and mechanical brakings are employed for normal service, Mechanical braking is employed when the speed becomes too low that electrical braking becomes ineffective. For mechanical braking, electromechanical drum brakes are employed. Magnetic brakes are also used for providing better retardation. The magnetic brakes are in fact electromagnets suspended on springs. These are attracted to the rail track, where they exert sufficient braking force. Thus dual braking effect is obtained. Tramcars, if provided with magnetic brakes, can be excited by the rheostatic braking current.

10.3.2. Trolley Buses

Serious drawback of tramway is lack of manoeuvrability in congested areas and noise. These are overcome by trolley bus drive. The trolley bus is an electrically operated pneumatic tyred vehicle and is fed usually at 600 dc volts from two overhead contact wires by means of two collectors. Although the overhead equipment required by the trolley buses is more expensive and more unsightly than that for the tram, this drawback is usually more than offset by the elimination of the necessity of the track in the roadway, which is costly to maintain and also constitutes a source of danger to the other road users.

Another advantage of trolley bus is that it is able to manoeuvre in a traffic over a width of several metres on each side of the centre line of the trolley wires. As the adhesion between a rubber-tyred wheel and the roadways is sufficiently high, therefore, only a single driving axle and so a single motor is normally used. A dc compound motor of output of 50 to 100 kW is usually employed.

Speed control is obtained by field weakening (by providing a resistance in the shunt field or the tappings on series field or divertors with the series field) method. Foot operated master controllers are used so that driver may have his hands free to steer the vehicle and apply handbrake. One pedal controls the starting, speed control and regenerative braking if any, and second pedal controls rheostatic and compressed air brakes. Regenerative braking is usually not employed in trolley bus drive because of difficulty of ensuring that supply system is always in a position to absorb the energy regenerated. In order to avoid damage to the power transmission gear, which may occur, in case plain rheostatic braking is used, due to high adhesion between road and rubber tyres, stabilised rheostatic braking is used. The lighting system in the car is low-voltage dc supplied from a motor-generator set connected in parallel with a battery. The vehicles are usually provided with secondary batteries so that the vehicles can be manoeuvered in case of emergency. As the body of the vehicle is insulated from earth because of tyred wheels, care must be taken to ensure adequate insulation resistance. The insulation resistance is checked at the end of the day when it is hot and damp and, therefore, in its worst condition. The trolley bus can accommodate slightly more passengers than an oil engined bus, and because of its high acceleration and braking retardation the schedule speed over a given route will be higher than that of the latter vehicle, so that fewer trolley buses will be required for operating a given service. Trolley buses are, therefore, used for the inner suburbs with medium traffic density.

Although oil-engined buses are more expensive to maintain and operate than a trolley bus, urban and rural services operating with a service interval of more than 15 or 20 minutes are most economically dealt with by these vehicles, as the cost of overhead equipment for a trolley bus service would be too large relative to the number of passengers carried and the revenue earned.

10.3.3. Diesel Electric Traction

The high initial cost of any system of track electrification renders it uneconomical except where the traffic density is high or where heavy grades and loads preclude any other type of traction. Under other conditions self-contained motive power units are required, and these may employ either steam or internal combustion engines as the prime movers. No doubt for road transport the IC engine is supreme but for railway the choice lies between steam and heavy-oil (diesel) engines. The factors governing the choice of traction system are, therefore, largely economic and depend on the natural resources of the country/region concerned. For instance countries (like Denmark, South America and Siam) where fuel has to be imported and water is sometimes scarce, will like to adopt diesel electric traction. The diesel electric traction system is quite economical where the cost of diesel oil is low as in USA. On the other hand countries rich in coal, like Great Britain, may like to have electric traction for urban areas and steam traction for suburban areas. This system is also adopted in the areas where coal for steam engines is not available in plenty and water is also scarce as in Rajasthan. This avoids transportation of large quantities of coal over long distances. Diesel locomotive consumes very small amount of water as compared to steam locomotives.

Diesel locomotives were introduced for the first time in our country in 1945 for shunting purposes on BG sections and in 1956 for main line services on MG sections. Indian Railway started using diesel locomotives extensively for main line services since 1958.

The diesel-electric locomotives employed in practice are of following four types:

(*i*) The main-line diesel electric locomotive having engines of output not exceeding, 1,500 kW and speeds 160 kmph.

(*ii*) The shunting diesel electric locomotive having an engine of 225-375 kW output and speed between 25 to 50 kmph.

(*iii*) The diesel-electric multiple unit stock of which each motor has an engine of 135-150 kW output and the train is capable of having speeds between 80 to 110 kmph.

(*iv*) The diesel-electric rail car having an engine of 75-450 kW output which may operate as a single car or with one or more trailer coaches.

In diesel electric traction system the electric motors employed for driving the locomotive are fed from a dc generator driven by the diesel engine put on the same locomotive. Thus diesel electric locomotive is self-contained motive power unit and, therefore, heavy expenditure on track electrification and substations etc. is eliminated.

The diesel engine employed for diesel electric locomotives is usually high speed type with three or four different running speeds with different output power. The generator may be self-excited type or separately excited. In case of a separately excited generator excitation is obtained from an auxiliary generator connected in parallel with the battery, so that standby supply is available. Similarly diesel engine is made self starting employing the supply from the battery. The voltage control of main generator becomes easier.

The characteristic of diesel engine is that the torque obtainable is approximately constant and independent of speed and the output power increases fairly uniformly with the speed, as illustrated in Fig. 10.1. It has no starting torque and must be started by some auxiliary means such as compressed air or electrically, as in the ordinary motor car. The overload capacity of the diesel engine is also very small (not exceeding 10%).

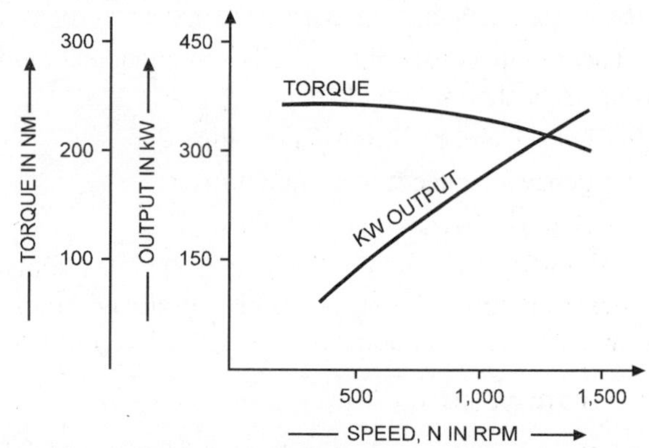

Fig. 10.1. ***Diesel Engine Characteristics***

The torque required in traction service is roughly inversely proportional to the speed being a maximum at start and at low speeds so that the constant torque characteristic of a diesel engine is not entirely suitable; if such an engine is directly coupled to the driving wheels and made of such a size that it provides the required torque at very low speeds, then at full load it would be loaded up to 20 to 25 per cent of its full load.

Some form of torque converter must, therefore, be included in the transmission between the engine shaft and the driving axles so as to convert the constant torque at full engine speed into a torque which decreases with the increase in speed, maintaining the kilowatts output (proportional to the product of torque and speed) approximately constant and, therefore, enabling the available output of the diesel engine to be used to the full at all speeds.

There are two types of transmission systems which are usually employed.

1. ***Mechanical transmission system*** employs a variable gear between the engine shaft and driving wheels as in the ordinary motor car. This method is almost universally employed for road and light railway vehicles but it is not suitable for railway work where output required, exceeds 300 kW due to the mechanical difficulties with the gears and gear changing equipment. This system of transmission of power is light and more efficient than electrical transmission system but it is not so flexible as the latter one.

2. *Electrical transmission system* employs a diesel engine coupled to a dc generator which supplies current to ordinary dc series motors on the driving axles. This system makes use of the simple speed control properties and desirable speed-torque characteristic of the dc series motors.

Assuming that the diesel engine is running at a constant speed and the generator is giving a constant voltage, any increase in the tractive effort required (say, due to its meeting an up gradient) will reduce the speed of the traction motors owing to their characteristics. The reduction in speed will not, however, be sufficient to compensate for the increased tractive effort, and, therefore, output power required will increase and the engine will be overloaded.

In order that the power delivered by the diesel remains the same, the product of tractive effort and speed *i.e.*, $T_E \times N$ must remain constant.

$$\text{Hence speed} \quad N \propto \frac{1}{T_E}$$

$$\text{But also} \quad N \propto \frac{V}{I} \qquad \text{If } \phi \propto I$$

$$\text{and} \quad T_E \propto I^2$$

$\therefore$ For constant output

$$\frac{V}{I} \propto N \propto \frac{1}{T_E} \propto \frac{1}{I^2}$$

$$\text{or} \quad V \propto \frac{1}{I}$$

Thus the generator voltage must be varied inversely as the current in order to keep the output power required constant.

This is done by employing a series winding on the generator or its exciter, which is so connected in the circuit of traction motor that the increase in traction motor current demagnetises the field. This method does not, however fulfil the above condition very closely and a large number of modifications and additions to it have been developed and employed by different manufacturers to obtain the constant power output curve.

In diesel electric traction system the series motors are supplied power directly from the generator instead of through dropping resistance in series as employed in straight electric traction system. The starting and speed control is obtained by varying the excitation of the generator.

Diesel engine driven Ward Leonard drive is depicted in Fig. 10.2. Such drive employs a non-electrical prime mover (diesel engine or a gas turbine) for driving a dc generator that feeds a dc motor. The generator-motor combination operates as a torque converter, like a stepless gear, to provide to the motor speed-torque characteristics required by the load. The motor runs at variable speed, the prime mover and the generator coupled to it runs at a fixed higher speed resulting in reduced their size and cost and optimized efficiency. Regenerative breaking is not possible bcause the flow of energy in reverse direction is not possible but the rheostatic braking can be employed if required.

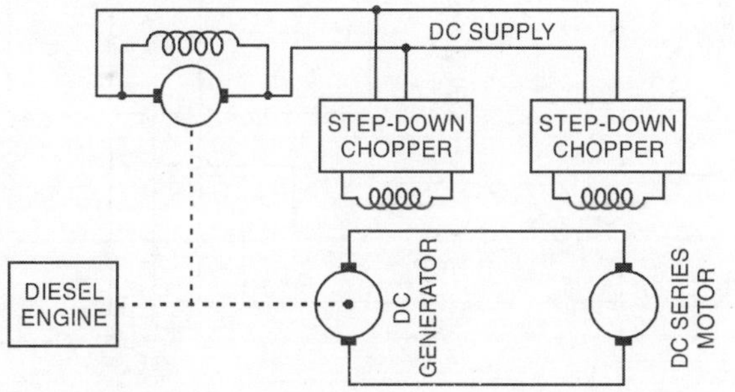

Fig. 10.2. *Diesel Engine Driven Ward Leonard Drive*

The dc generator speed is restricted to a value because of commutation difficulties. This may not permit the prime mover to be driven at an optimum speed. Further, commutator also imposes restrictions on the maximum power rating of a dc generator. In some large power applications, the generator is to feed a number of motors. The generator, therefore, should be of a size larger than what a dc generator can be. Furthermore, a dc generator needs frequent maintenance because of commutator. All these limitations and problems can be avoided if a synchronous generator and an uncontrolled rectifier bridge are used instead of a dc generator. The voltage across the dc series motor is controlled by varying the excitation of the synchronous generator.

Several electric transmission have been employed, the most common one are :

(*a*) Diesel engine driven dc generator feeding traction motors (dc series motors).

(*b*) Diesel engine driven three phase synchronous generator feeding traction motors through semiconductor diode rectifier.

(*c*) Diesel engine driven three phase synchronous generator feeding three phase squirrel-cage induction motors through semiconductor diode rectifier followed by a 3-phase VSI (voltage source inverter).

(*a*) Diesel Engine Driven DC Generator Feeding DC Series Motors. Block diagram of a diesel electric traction using a dc generator and dc series motors is shown in Fig. 10.3. Diesel engine does not have starting torque. It is made self starting employing the supply from the battery to feed the generator.

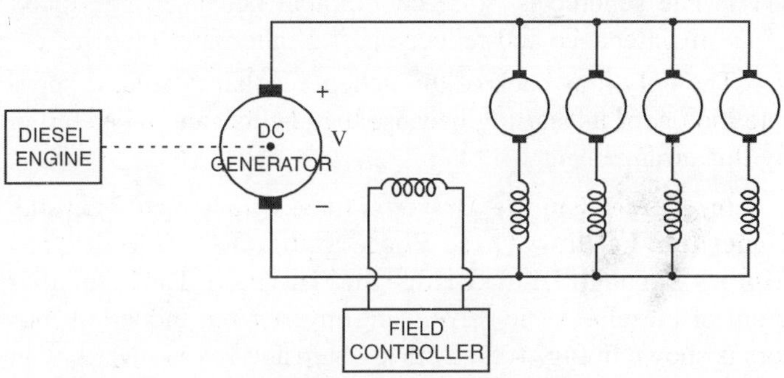

Fig. 10.3. *Diesel Engine Driven DC Generator Feeding Traction Motors (DC Series Motors)*

(*b*) Diesel Engine Driven Three Phase Synchronous Generator Feeding DC Series Motors Through Semiconductor Diode Rectifier. Block diagram of diesel electric traction

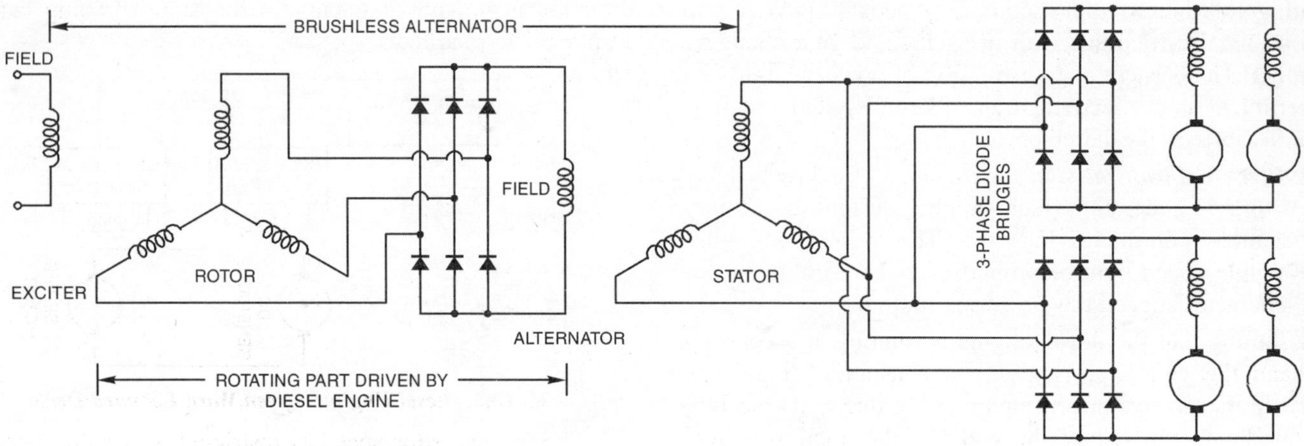

Fig. 10.4. *Diesel Engine Driven Brushless Alternator Feeding DC Series Traction Motors*

using brushless alternator and dc series traction motors is shown in Fig. 10.4.

3-Phase synchronous generator is excited by supplying dc into the rotor by means of slip-rings and brushes which need frequent maintenance. The dc series traction motors are fed through two diode bridges. In the scheme shown in Fig. 10.4, slip rings and brushes are eliminated by making use of brushless excitation system. It employs an exciter which is an alternator with inverted rotor and stator functions *i.e.,* field on the stator and three phase winding on the rotor. Exciter is mounted on the alternator shaft. Three phase voltages induced in the exciter rotor are rectified by a diode bridge which is mounted on the rotor shaft, and therefore, revolves with it. The rectified voltage is supplied to the field of synchronous generator. The excitation to alternator is adjusted by controlling exciter field current. The advantages of this scheme over dc generator scheme shown in Fig. 10.3 are :

(*i*) Because of restricted maximum speed, the maximum power rating of a dc generator is low. But with this scheme there is no restriction.

(*ii*) Because of brushless alternator, maintenance is reduced.

(*iii*) Transmission efficiency is improved by 2-4%.

(*iv*) Advantage of good adhesion is retained due to parallel connection of all motors.

(*v*) Reduced weight of generating unit (around 50%) brings down the volume and cost of generating unit.

(*vi*) The scheme is more economical because of reduced maintenance and reduced cost of alternator rectifier.

The only drawback of this scheme is that it needs a separate motor for its starting because the engine cannot be started by this arrangement.

(*c*) Diesel Engine Driven Three Phase Synchronous Generator Feeding Three Phase Squirrel-Cage Induction Motors Through Diode Bridge and Inverter. The block diagram of diesel electric drive using inverter fed induction motors is shown in Fig. 10.5. Both six-step and PWM inverter can be employed.

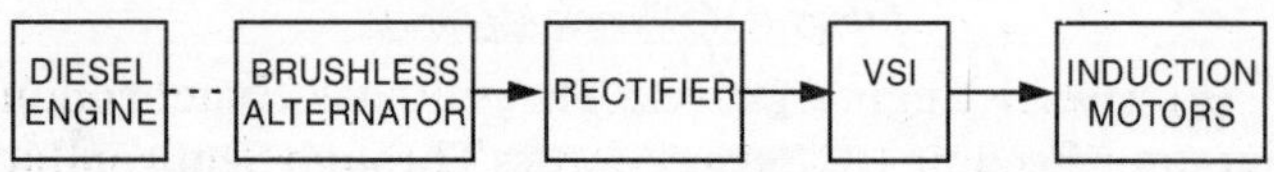

Fig. 10.5. *Diesel Electric Drive Using Inverter Fed Induction Motors*

With the use of 6-step inverter, voltage is controlled by adjusting field current of alternator and with the use of PWM inverter, voltage is controlled by the inverter itself and the rectifier output voltage can be maintained constant at the rated value. This allows reduction in the current rating of alternator and diode rectifier; further resulting in reduction in the cost, weight and size of the generator unit. A cycloconverter can also be employed instead of diode rectifier and inverter. But the alternator must be driven at a higher speed. This scheme has the following advantages :

(*i*) High power rating, high efficiency due to small power losses and reduced maintenance.

(*ii*) High ratio of power output to locomotive weight. So heavier trains on faster schedules can be pulled by the locomotive.

(*iii*) All motors can be connected in parallel providing best condition from the point of view of adhesion.

10.4. SYSTEMS OF TRACK ELECTRIFICATION

These comprise (*i*) the dc system (*ii*) the single phase ac system (*ii*) the three phase ac system and (*iv*) the composite system (*a*) single phase-three phase system and (*b*) single phase-direct current system.

1. *DC System.* In this system the electric motors employed for getting the necessary motive power are usually dc series motors, although the compound motors are coming into favour for tramways and trolley buses where regenerative braking is desired. The operating voltage is from 600 volts to 750 volts for tramways and for many suburban railways and from 1,500 volts to 3,000 volts for main line railways. The driving motors receive power from distribution system, (in dc as well as single phase ac system). The distribution system consists of one contact wire in case of tramways and two contact wires in case of trolley buses (the running rail acts as current returning conductor in case of tramways) which is fed from substations (except in case of very short railways where a direct supply from a dc generating station may be possible) suitably spaced. The spacing of the substations depends upon the operating voltage and the traffic density of the route. For suburban railways where the operating voltage is 600 volts the substations are spaced 3 to 5 km apart and for main line railways having operating voltage of 1,500 to 3,000 volts they may be spaced 15 to 40 km. These substations receive ac power from a 3-phase high voltage (33 to 132 kV) transmission lines and convert it into direct current either by rotary convertors or by mercury-arc rectifiers or thyristor converters with suitable transformation ratio. These substations are usually unattended type using supervisory control system. This arrangement reduces the cost of transmission lines and copper losses in transmission lines. As regards the merits and demerits of dc system the dc motors have better characteristics, low maintenance cost and a smaller weight per kW output, better speed control, and efficient regenerative braking but because of low operating voltage conductor of large x-section for the distribution system and more number of substations spaced at shorter distances for feeding the distribution system are required for a given service. The cost of substations is also more due to heavy cost of converting machinery required for converting ac into dc and of additional equipment such as negative boosters etc. required to keep the voltage of return rail within limits.

The dc system is preferred for suburban services and road transport where stops are frequent and distances of run are small.

2. *Single Phase AC System.* In single phase ac system ac series motors are used for getting the necessary motive power. The voltage employed for distribution network is 15,000 to 25,000 volts at $16\frac{2}{3}$ or 25 Hz, which is stepped down on the locomotive to a low voltage (300 to 400 V) suitable for supplying to single phase ac series motors by means of a step-down transformer carried on the locomotive. The distribution network may be fed directly, at high voltage, from a generating station when the extent of electrification is within a radius of about 30 km from the generating station. For longer distances the economic voltage for the power transmission system is higher than that desirable for traction system, and therefore, transformer substations are employed for feeding the distribution network. The spacing of substations is 50 to 80 km because of low current requirement at high voltage. The substations receive power at high voltage either from the existing extra high voltage transmission lines or from a separate system employing single phase generating stations and steps down the voltage to the required value and/or change the supply frequency through frequency convertors. The change of supply frequency becomes necessary because of better performance of the ac series motors at low frequency. Low frequency operation of ac series motor improves its commutation properties, power factor and efficiency. Low frequency operation of overhead transmission line reduces the line reactance and hence the voltage drop directly.

In case the supply at low frequency is obtained directly from a generating station exclusively meant for traction purposes, there is no problem as power corresponding to this frequency can always be generated. If, however, electric supply is taken from industrial frequency network, which is usually the case, substations in addition to transformers have necessarily the frequency converters to convert 3-phase, 50 Hz into single phase $16\frac{2}{3}$ or 25 Hz as required.

AC single phase system is invariably adopted for main line service where the cost of overhead structure is of more importance and rapid acceleration and retardation is not so important as for suburban railways.

3. *Three Phase AC System.* In this system 3-phase induction motors operating at 3,000 to 3,600 volts and normal frequency of $16\frac{2}{3}$ Hz are employed for getting the required motive power. The three phase induction motors are simple and robust in construction and have a high operating efficiency and property of automatic regenerative braking without requiring any additional equipment. However, the induction motors have few drawbacks such as low starting torque, high starting current and absence of speed control. The distribution system consists of two overhead wires and track rail for the third phase and receives power either directly from the generating station or through the transformer substations. The substations receive power from high voltage transmission lines at power frequency and step down the voltage and change the frequency. Since two trolley wires per track are necessary, the line voltage of the distribution systems has to be limited to values below the highest voltage used for single phase railways, and as the 3-phase induction motors can be built economically for operation at moderately high voltage, the line voltage is usually chosen so that the motors may be supplied directly from the trolley wires. This system is employed in some hilly areas, where output power required is high and regeneration on large scale is possible. On account of disadvantages of two overhead contact wires and constant speed characteristic of induction motors this system is no longer likely to be adopted.

4. *Composite Systems.* Efforts have been made to combine the advantages of the dc/ac and 3-phase/single-phase systems and composite systems have been evolved. The single phase ac sysem is preferable from the viewpoint of the distribution and contact wire system. It can be converted into either three-phase ac or dc at low voltage in the locomotive. At present there are two composite systems in vogue.

(*i*) Kando system or single phase-to-three phase system.

(*ii*) Single phase ac-to-dc system.

(a) Single Phase-To-Three Phase System or Kando System. In this system single phase high voltage ac system is employed for distribution purposes and three phase induction motors for getting the necessary motive power in order to have the advantage of low cost of single phase overhead distribution system together with desirable characteristics of three phase induction motors. The voltagc used for distribution network is 15,000 volts at 50 Hz. The locomotive carries a phase convertor for converting single phase supply into 3-phase supply at relatively low frequency. The development of silicon controlled rectifier (SCR) used as invertor has made possible to get variable low frequency supply ($\frac{1}{2}$ to 10 Hz) at which induction motor develops high starting torque without drawing excessive current from the system and speed control is also conveniently obtained by varying the supply frequency. There exists good possibility of this traction system to develop for further adoption.

(b) Single Phase-To-Direct Current System. In this system the advantages of the two systems are combined together by employing high voltage ac single phase system for overhead distribution and dc series motors for the necessary drive so that high efficiency of the ac distribution system is obtained together with the desirable characteristics of dc series motors. The locomotive carries transformer and converting machinery to step down the voltage and convert into dc. The voltage employed for overhead distribution system is 25 kV at normal supply frequency of 50 Hz. This system is adopted in all the areas of major traffic density on main lines. It has been decided to adopt this system of traction employing 25 kV, industrial frequency (50 Hz) ac supply for track electrification in our country and this is being installed on sections which are being electrified.

This system has got numerous advantages over dc system as enumerated below:

1. The line current for a given demand of power is reduced on account of high system voltage. It causes reduction in the cross section of overhead conductor which leads to a saving in the cost of supporting structures and their foundations being lighter in weight and simpler. This fact will be obvious from the following datas:

   The weight of copper required per track km for the 1,500 V dc system is 3,700 kg and for the 3,000 V dc system is 3,300 kg whereas for ac system it is 1,600 kg. The weight of steel required per track km in ac system is only about half of that required for dc system.
2. On account of high voltage the substations can be spaced at longer distances (50 to 80 km) whereas the substations are spaced at 12 to 30 km in case of 3,000 V dc system and at 5 to 12 km in case of 1,500 V dc system. It affects substantial saving in the capital expenditure on traction substations and transmission lines on account of reduced number of substations to be erected and flexibility offered in the location of site for substations. AC substations are simpler and cheaper than dc substations as they only have step-down transformers and associated switchgear.
3. A very special feature of the ac locomotive in comparison with the dc locomotive is the greater amount of adhesion obtained with ac locomotive. In the case of ac rectifier locomotive employing dc traction motors, coefficient of adhesion is obtained as much as 45% against the value of 27% in case of dc locomotives. The greater adhesion is obtained because of possibility of connecting of all dc series motors in parallel permanently and controlling their speeds by varying the voltage across each motor by tap changing auto-transformer on the ac side. In this system slipping of a particular motor does not affect the voltage across the remaining motors and so adhesion is maintained but in case of dc traction employing resistance and series—parallel control slipping of a particular motor throws its load on the other motor which may reduce the voltage across the other motor and may, therefore, cause slipping of this motor also. The increase in coefficient of adhesion increases the hauling capacity of the locomotives (75 tonne ac locomotive can haul the same load as the 110 tonne dc locomotive).
4. The starting efficiency of an ac locomotive is higher than that of a dc locomotive as in case of ac locomotive supply voltage for starting is reduced by means of 'on-load tap changer' installed on either the primary or the secondary side of the transformer where as in case of dc locomotive it is obtained by using starting resistors. Elimination of starting bulky resistors not only increases the starting efficiency but also makes the torque-speed curve more steep and thus improves adhesion. This in turn increases the power-weight ratio in the case of ac locomotives.
5. Since the dc series motors having ideal traction characteristics are employed in this system for getting the required propelling power, therefore, this system have got the advantages of the dc system.
6. AC locomotive has less kW demand at starting than that of a dc locomotive owing to elimination of starting resistances in ac locomotives. This is of particular importance at peak traffic hours specially when a number of locomotives are to be started simultaneously after an interruption of supply.

There are, however, some drawbacks with the single phase ac system such as unbalance imposed on the power supply system and interference with neighbouring communication lines. Fortunately both of the shortcomings can be minimised by taking suitable precautions such as

(*i*) Taking of supply for the traction substations at very high voltage (say 132 or 220 kV) and from a supply system having high capacity.

(*ii*) Balancing of traction loads equally on all the three phases which is possible by connecting different traction substations across different phases in rotation.

(*iii*) Employing the Scott or T-connected three phase/two phase transformers in the traction substations.

(*iv*) Replacing of open aerial communication lines by lead or aluminium sheathed underground cables and earthing the sheathing of the cables at regular intervals.

(*v*) Providing of isolating transformers which reduce the induced voltages and keep them below 60 V under operating conditions.

(*vi*) Making use of booster transformer and return conductor for suppressing interference at the source in case of thickly populated areas.

10.5. COMPARISON BETWEEN DC AND AC SYSTEMS OF RAILWAY ELECTRIFICATION FROM THE POINT OF VIEW OF MAIN LINE AND SUBURBAN LINE RAILWAY SERVICES

1. *Main Line Railway Service.* The most important requirements of main line railway service are:

1. Higher maximum speed.
2. Minimum cost of overhead structure.
3. Acceleration and retardation are not so important for main line service as in suburban service.

The cost of overhead structure can be reduced by employing single phase ac system.

Single phase ac system is preferred for main line railway service on account of special advantages enumerated below:

(*a*) Because of possibility of employing high voltage the cost of distribution system employing conductor of smaller section to carry the small current required is reduced.

(*b*) Also because of possibility of employing high voltage for distribution purposes, the spacing of the substations is higher and, therefore, number of substations required is less.

(*c*) The initial, maintenance and operating costs of ac substations are less as the substation equipment required in single phase ac system is less, cheap and efficient.

Single phase-to-three phase composite system can also be employed for approximately same results.

2. ***Suburban Line Railway Service.*** The most important requirements of suburban line railway service are:

1. Rapid acceleration and retardation as frequent stopping and starting is required.
2. The working of motors should not be affected by voltage fluctuations.
3. There should be no interference to the communication lines (telephone and telegraph lines) running along the track on account of power system adopted.

The dc system fulfills all of the requirements mentioned above for suburban service, and therefore, invariably adopted. The advantages and disadvantages of dc system are as follows.

Advantages. 1. In similar conditions the energy consumption in dc system is less as compared with that in ac system.

2. For exerting the same torque, current required in dc system is less than that in ac system.
3. The dc locomotive and motor coach equipment is lighter in weight, cheaper in initial as well as in maintenance cost and more efficient.

Disadvantages. 1. Due to low voltage employed for distribution purpose, the current required is high, therefore, the section of conductor required is large, the spacing of substations required is less and the number of substations required is more.

2. Substations are more costly as converting machinery is required.
3. Efficiency of substations is less.
4. Additional equipment such as negative booster etc. is required for keeping the voltage of return rail within limits.

The above disadvantages are partly eliminated by adopting composite system single phase ac-to-dc system.

Exercises

1. Explain the requirements for ideal traction and show which drive satisfy almost all the requirements.
[M.D. Univ. Utilization of Electric Power & Traction May 2009]
2. State the main requirements of an ideal traction system. Explain different traction systems.
[M.D. Univ. Utilization of Electric Power & Traction May 2010]
3. Discuss various types of traction systems.
[Pb. Technical Univ. Elecric Drives & Utilization Dec. 2009]
4. State merits and demerits of electric traction with reference to electric power crisis in India.
[Pb. Technical Univ. Electric Drives and Utilization Dec. 2006]
5. State the main requirements for an ideal traction system. What factors govern the operation of dieselisation or electrification of a steam operated railway section ? Which system of electric traction is now being preferred for electrification of main line sections and why ?
[Gorakhpur Univ. Utilization of Electrical Power & Traction Sept. 1981]
6. What are the advantages and disadvantages of electric traction over other types of traction system.
[Rajasthan Univ. Utilization of Electrical Power 2005, 2006, 2007, 2008]
7. What are the different traction systems ? Discuss them in brief.
Indicate the general trend in the adoption of traction systems for city road and rail transport, and discuss their merits and demerits. **[Gorakhpur Univ. Utilization of Electrical Power & Traction 1983]**
8. Describe briefiy a battery powered vehicle.
[A.M.I.E. Sec B. Electrical Drives and Their Control Winter 1998]
9. Briefly discuss the general considerations for a tramway.
[Pb. Univ. Electric Traction December 1992]
10. What are the disadvantages of a tramcar (tramway) system of electric traction ? How are these overcome in the trolley bus system ? **[Pb. Univ. Electric Traction December 1991]**
11. Explain the trolley bus system of electric traction.
[Pb. Technical Univ. Electric Drives & Utilization Dec. 2008]
12. The diesel engine has an approximately constant torque at all speeds. Explain how it is electrically transmitted to motors so that their tractive effort increases as the speed decreases.
13. What are the main characteristics of diesel engine with special reference to its application for traction purposes ?
[Pb. Univ. Electric Traction December 1991]
14. Discuss the problems associated with diesel electric traction. Suggest ways to overcome these in actual system.
[Pb. Univ. Electric Drives and Utilization May 1989]
15. Discuss diesel electric traction system. Write the situation, when diesel electric traction is preferred over electric traction.
[U.P. Technical Univ. Utilization of Electrical Energy & Electric Traction 2005-06]
16. Explain working of a diesel engine driven dc generating feeding dc series motors systems.
[U.P. Technical Univ. Utilization of Electrical Energy & Traction 2009-10]
17. What are the different systems of railway electrification? Which one is being commonly adopted in our country and why ?
Under what conditions is diesel electric traction preferred to track electrification ?
[Gorakhpur Univ. Utilization of Electrical Power & Traction Sept. 1983]
18. Compare the following systems of track electrification as regards operating voltages, economy in operation and suitability to the topography of regions, where they are best suited.
(*i*) DC system (*ii*) AC system (single phase) and (*iii*) Composite system. **[Pb. Univ. Electric Traction Nov. 1973]**
19. Discuss the merits and demerits of the dc and single phase ac systems for the main and suburban line electrification of the railways.

Which system you consider to be the best for the suburban railways in the vicinity of large cities ? Give reasons for your answer. **[Bombay Univ. Elec. Power 1971]**

20. Compare dc and ac systems of railway electrification from the point of main line and suburban line railway service. **[Rajasthan Univ. Utilization of Electrical Power 2008]**

21. Explain the factors that determine the choice of supply system for main line traction and give voltage and frequency of the supply as adopted by Indian Railways. **[A.M.I.E. Sec. B. Electric Drives & Their Control Winter 1996]**

22. What are the systems of electric traction ? Explain why single phase ac electrification is superseding the other types of system. **[Pb. Technical Univ. Electric Drives and Utilization 2003]**

23. Give the nature and specifications of supply sysetms for electric traction used in India. **[U.P. Technical Univ. Utilization of Electrical Energy & Electric Traction 2005-06]**

24. Discuss the advantages and disadvantages of the system of traction using 25 kV, 50 Hz supply and dc series motor. **[Pb. Technical Univ. Electric Drives & Utilization May 2008]**

25. State and explain the track electrification system generally employed for the following services.
(*i*) Tramway service in a busy town area.
(*ii*) Electric suburban railway service.
(*iii*) Main line services.
[Rajasthan Technical Univ. Utilization of Electrical Power 2011]

26. Write advantages of 25 kV ac system over dc system. **[Pb. Technical Univ. Electric Drives & Utilization Dec. 2006]**

27. Write disadvantages of 25 kV ac traction systems. **[Pb. Technical Univ. Electric Drives & Utilization Dec. 2007]**

28. Compare ac and dc systems of traction. **[Pb. Technical Univ. Electric Drives & Utilization May 2007; Dec. 2007]**

29. Compare dc and ac systems of railway electrification from the view of main line and suburban line railway services. **[Pb. Technical Univ. Electric Drives & Utilization May 2009]**

30. Describe the following systems of railway electrification and write down the relative merits of each:
(*i*) DC system (*ii*) Single phase ac system, (*iii*) 3-phase ac system, (*iv*) Composite system.
[Rajasthan Technical Univ. Utilization of Electrical Power 2010]

31. Write short notes on the following:
(*i*) Traction systems in India.
(*ii*) Battery driven vehicles and their applications.
(*iii*) Trolley bus system of traction. **[Pb. Univ. July 1992]**
(*iv*) Diesel electric traction system. **[Pb. Univ. Nov. 1988]**
(*v*) Comparison of various systems of track electrification.
(*vi*) Relative merits of ac single phase and dc systems of supply in electric traction.

Short Answer Type Questions With Answers

Q. 1. Why steam engine drive is not suitable for urban and suburban services where distance between stations is small ? **[Gorakhpur Univ. 1980]**

Ans. Because of low coefficient of adhesion it is not possible to have acceleration and retardation with steam engine drive as high as with electric drive. Steam engine drive is, therefore, not suitable for urban or suburban services where distance between stations is small.

Q. 2. Why steam engine drive is not suitable for underground railway?

Ans. Owing to smoke, steam engine drive cannot be employed for underground railways.

Q. 3. How is speed controlled in direct IC engine drive ?

Ans. By using a gearbox.

Q. 4. What is the operational efficiency of IC engine electric drive ?

Ans. About 25 per cent.

Q. 5. What are the limitations of battery electric drive ?

Ans. The major limitations of the battery electric drive are the small capacity of the batteries and the necessity for frequent charging and limited speed range.

Q. 6. What is the major drawback of electric traction ?

Ans. The greatest drawback of electric traction is that it involves a heavy initial expenditure for the power supply system and, in the case of tramways, for the track rails and electrical return system, in addition to the cost of vehicles, depots etc.

Q. 7. What are the major disadvantages of the tramway ?

Ans. The great disadvantage of tramway is that it needs laying of overhead supply system and track for its use which is costly to maintain and constitutes a source of danger to other road users. The other main drawback of tramway is lack of manoeuvrability in congested areas and noise.

Q. 8. Why electric traction system is considered superior over other systems of traction in India, even when there is a power crisis in India.

Ans. In our country electric energy is available at cheaper rates comparatively because of availability of substantial sources of hydropower generation and nuclear power generation. By electrification of railways high grade coal, which is limited in quantity in our country, will be saved which can be used for some other useful purposes.

Q. 9. Which of the system (dc or ac) is preferred for suburban services and road transport where stops are frequent and distances of run are small ?

Ans. DC system.

Q. 10. What is the value of voltage used in overhead supply for trains in India ? **[A.M.I.E. Sec. B. Winter 1999]**

Ans. 25 kV.

Q. 11. Why single phase system is preferred for main line railway service ? **[P.T.U. Dec. 2008; May 2009]**

Ans. AC single phase system is invariably adopted for main line service where the cost of overhead structure is of more importance and rapid acceleration and retardation is not so important as for suburban railways.

Q. 12. What are the most important requirements of the main line railway service ?

Ans. High maximum (crest) speed and minimum cost of overhead structure.

Q. 13. Which of the system is preferred for main line railway service ?

Ans. Single phase ac system.

Q. 14. What are the most important requirements of suburban line railway service ?

Ans. Rapid acceleration and retardation, no interference to nearby communication lines and operation of motors free from adverse effects of voltage fluctuations.

❑❑❑

CHAPTER 11

Train Movement and Energy Consumption

INSIDE THIS CHAPTER

11.1. INTRODUCTION

The movement of trains and their energy consumption can be most conveniently studied by means of speed-time and speed-distance curves, which show respectively the speed at different time instants after the start of run and the speed at different distances from the starting point. Of the two, the speed-time curve is generally the more useful.

The curve drawn between speed and time, taking speed (in km/hour) on the Y-axis and time (in seconds or minutes) on X-axis, is known as *speed-time curve*. The speed-time curve provides complete information of the motion of the train. This curve gives the speed at various time instants after the start of run directly. Slope of the curve at any point gives the acceleration at the corresponding instant or speed. The area covered by the curve, the time axis and the ordinates through the instants between which the time is taken, represents the distance covered in the corresponding time.

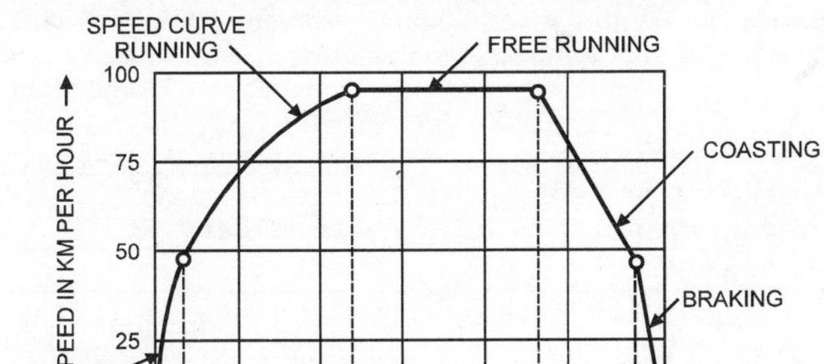

Fig. 11.1. *Typical Speed-Time Curve For Main Line Service*

Speed-time curve (Fig. 11.1) mainly consists of (*i*) initial acceleration (*ii*) constant speed run or free run (*iii*) coasting and (*iv*) retardation. Typical speed-time curve for main line service is shown in Fig. 11.1.

(*i*) *Acceleration.* It consists of two parts known as (*a*) constant acceleration or acceleration while notching up and (*b*) speed curve running or acceleration on the speed curve.

(a) Constant Acceleration or Acceleration During Notching Up. During notching up period (0 to t_1) the current is maintained approximately constant and the voltage across the motor is gradually increased by cutting out the starting resistance. Thus tractive effort is constant and, therefore, acceleration remains constant during this period.

(b) Speed Curve Running or Acceleration on Speed Curve. During speed curve running (t_1 to t_2) the voltage acting across the motor remains constant and current starts decreasing with the increase in speed according to the characteristics of the motor and finally the current taken by the motor becomes constant. During this period, though the train accelerates but acceleration decreases with the increase in speed and finally becomes zero at the speed at which the tractive effort developed by the motor becomes exactly equal to the resistance to motion of the train.

(ii) Free Run or Constant Speed Run. At the end of speed curve running *i.e.*, at t_2 the train attains the maximum speed. During this period the train runs with constant speed attained at t_2 and constant power is drawn.

(iii) Coasting. At the end of free running period (*i.e.*, at t_3) power supply is cut off and the train is allowed to run under its own momentum. The speed of train starts decreasing on account of resistance to the motion of train. The rate of decrease of speed during coasting period is known as *coasting retardation*.

(iv) Retardation or Braking Period. At the end of coasting period (*i.e.*, at t_4) the brakes are applied to bring the train to rest. During this period speed decreases rapidly and finally reduces to zero.

11.2. TYPICAL SPEED-TIME CURVES

1. *Urban or City Service.* In urban or city service the distance between the two stops is comparatively very short (say 1 km or so). The time required for this run is very small (few minutes). The acceleration as well as retardation is required to be high so that high average speed and short time of run is obtained. The acceleration and retardation for urban service is between 1.5 and 4 kmphps and between 3 and 4 kmphps respectively. Free run is not present in this run. The coasting retardation is about 0.15 kmphps. The coasting period is also small. Typical speed time curve for urban service is shown in Fig. 11.2.

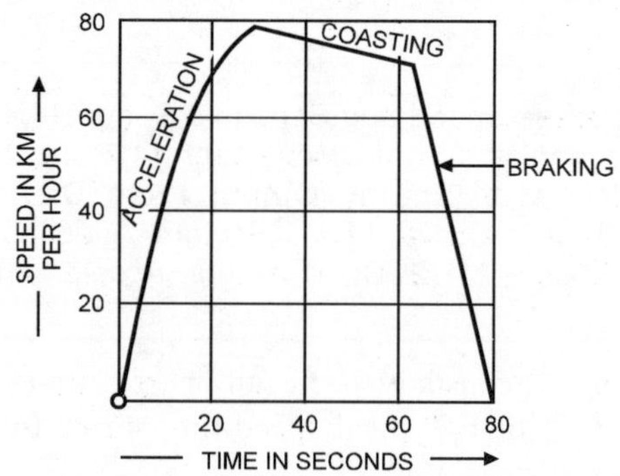

Fig. 11.2. *Typical Speed-Time Curve For Urban Service*

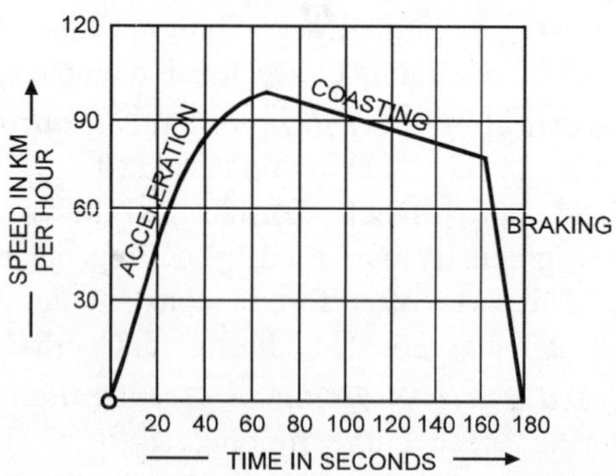

Fig. 11.3. *Typical Speed-Time Curve For Suburban Service*

2. *Suburban Service.* In this service the distance between the stops is little longer than urban service but smaller than main line service (say between 2 and 5 km). Free run is still not possible. Coasting is for a comparatively longer period. Acceleration and retardation required are as high as for urban service.

3. *Main Line Service.* The distance between two stops in main line service is considerably more (usually more than 10 km). In this service free run is of longer duration. The duration of acceleration and retardation is a small fraction of total running time. Since rate of acceleration and retardation does not affect the average and schedule speeds, therefore, these are of little importance in main line service.

Characteristics of various types of services are given in Table 11.1.

The acceleration and free-running periods are governed by the speed-torque characteristics of the electric traction motors, and these must be available before this part of the curve can be determined. Coasting and braking are governed by the train resistance and the allowable braking retardation respectively. The actual shape of any particular speed-time curve may be somewhat different from the above due to the existence of gradients and with different lengths of run.

11.3. CREST SPEED, AVERAGE SPEED AND SCHEDULE SPEED

Crest Speed. The maximum speed attained by the vehicle during the run is known as *crest speed*.

Average Speed. The mean of the speeds from start to stop *i.e.*, the distance covered between two stops divided by the actual time of run is known as *average speed*.

$$\text{Mathematically average speed} = \frac{\text{Distance between stops}}{\text{Actual time of run, T}}$$

Schedule Speed. The ratio of distance covered between two stops and total time of run including time of stop is known as *schedule speed*.

Mathematically schedule speed

$$= \frac{\text{Distance between stops}}{\text{Actual time of run + stop time}}$$

From above it is obvious that the schedule speed is always smaller than the average speed. The difference is large in case of urban and suburban services and is negligibly small in case of main line service.

11.4. FACTORS AFFECTING SCHEDULE SPEED

The schedule speed of a given train when running on a given service (*i.e.*, with a given distance between stations) is affected by the following factors:

Table 11.1

S.No.	Type of Service	Acceleration in kmphps	Retardation in kmphps	Maximum Speed in kmph	Distance between Stations in km	Special Remarks
1.	Urban	1.5 to 4.00	3 to 4	120	1	No free running period, coasting period—small
2.	Suburban	1.5 to 4.00	3 to 4	120	2 to 5 km	No free running period, coasting period—long
3.	Main Line	0.6 to 0.8	1.5	160	More than 10 km	Free running and coasting periods—long. Acceleration and braking periods—comparatively small.

1. Acceleration and braking retardation.
2. Maximum or crest speed.
3. Duration of stop.

1. *Effect of Acceleration and Braking Retardation.* For a given run and with fixed crest speed the increase in acceleration will result in decrease in actual time of run and, therefore, increase in schedule speed. Similarly increase in braking retardation will affect the schedule speed. Variation in acceleration and retardation will have more effect on schedule speed in case of shorter distance run in comparison to longer distance run.

2. *Effect of Maximum Speed.* For a constant distance run and with fixed acceleration and retardation the actual time of run will decrease, and therefore, schedule speed will increase with the increase in crest speed. The effect of variation in crest speed on schedule speed is considerable in case of long distance run.

3. *Effect of Duration of Stop.* For a given average speed the schedule speed will increase by reducing the duration of stop. The variation in duration of stop will affect the schedule speed more in case of shorter distance run as compared to longer distance run. It is because stops in case of urban and suburban services are kept very small (say 15 to 20 seconds). When the distance between stations is much larger and duration of actual run is long as in case of main line service, station stop has less conspicuous effect on the schedule speed.

11.5. SIMPLIFIED SPEED-TIME CURVES

In order to study the performance of a service at different schedule speeds the speed-time curves are replaced by simple geometric shaped curves. From these simplified curves the relationships between acceleration, retardation, average speed and distance can be easily worked out. These can have either quadrilateral or trapezoidal shape.

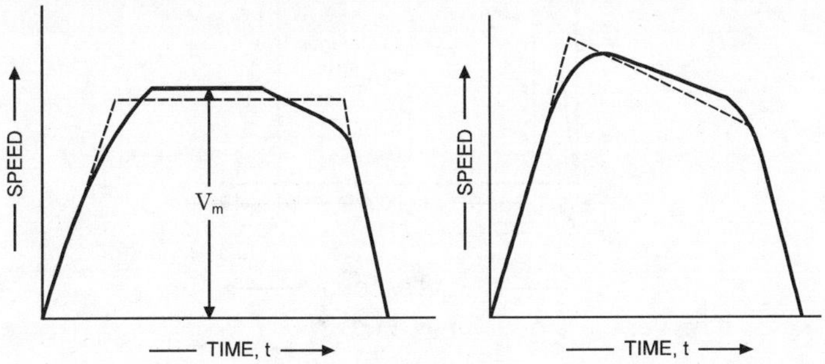

(a) *(b)*

Fig. 11.4. *Approximate Speed-Time Curves*

The speed-time curve of an urban service can be replaced by an equivalent speed-time curve of simple quadrilateral shape. The speed-time curve of a main line service is best and most easily replaced by a trapezoid. Since the area of the speed-time curve represents the total distance travelled hence the areas of the two curves should be same. The values of acceleration and retardation are also kept the same as those of the original speed-time curve. In case of simplified trapezoidal speed-time curve, speed curve running and coasting periods are replaced by constant speed period, as illustrated in Fig. 11.4 (*a*). While in case of quadrilateral speed-time curve, initial acceleration and coasting periods are extended, as illustrated in Fig. 11.4 (*b*).

The following examples illustrate the methods of calculations.

1. *Calculations By Trapezoidal Speed-Time Curve*

Let α = Acceleration in kmphps

β = Retardation in kmphps

V_m = Crest speed in kmph

T = Total time of run in seconds.

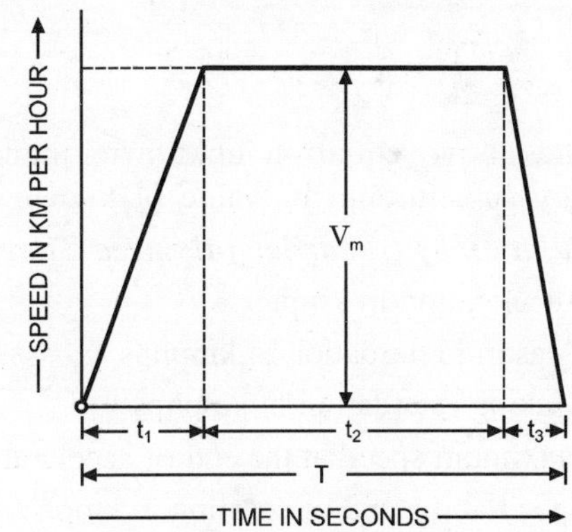

Fig. 11.5. *Trapezoidal Speed-Time Curve*

Time for acceleration in seconds, $t_1 = \dfrac{V_m}{\alpha}$

Time for retardation in seconds, $t_3 = \dfrac{V_m}{\beta}$

Time for free running in seconds, $t_2 = T - (t_1 + t_3)$

$$= T - \left(\frac{V_m}{\alpha} + \frac{V_m}{\beta}\right)$$

Total distance of run in km,

S = Distance travelled during acceleration + distance travelled during free run + distance travelled during braking

$$= \frac{1}{2} V_m \frac{t_1}{3{,}600} + V_m \frac{t_2}{3{,}600} + \frac{1}{2} V_m \frac{t_3}{3{,}600}$$

Substituting $t_1 = \dfrac{V_m}{\alpha}$, $t_3 = \dfrac{V_m}{\beta}$ and $t_2 = T - \left(\dfrac{V_m}{\alpha} + \dfrac{V_m}{\beta}\right)$ we have

$$S = \frac{V_m^2}{7{,}200\alpha} + \frac{V_m}{3{,}600}\left[T - \left(\frac{V_m}{\alpha} + \frac{V_m}{\beta}\right)\right] + \frac{V_m^2}{7{,}200\,\beta}$$

or

$$S = \frac{V_m^2}{7{,}200\alpha} + \frac{V_m}{3{,}600}T - \frac{V_m^2}{3{,}600\alpha} - \frac{V_m^2}{3{,}600\beta} + \frac{V_m^2}{7{,}200\,\beta}$$

$$= \frac{V_m T}{3{,}600} - \frac{V_m^2}{7{,}200\alpha} - \frac{V_m^2}{7{,}200\beta} \quad ...(11.1)$$

or

$$\frac{V_m^2}{3{,}600}\left(\frac{1}{2\alpha} + \frac{1}{2\beta}\right) - \frac{V_m T}{3{,}600} + S = 0$$

or

$$V_m^2\left(\frac{1}{2\alpha} + \frac{1}{2\beta}\right) - V_m T + 3{,}600\,S = 0$$

This is a quadratic equation for V_m. Substituting $\dfrac{1}{2\alpha} + \dfrac{1}{2\beta} = K$, we get

$$KV_m^2 - V_m T + 3{,}600\,S = 0$$

$$\text{or} \quad V_m = \frac{T \pm \sqrt{T^2 - 4K \times 3{,}600 S}}{2K}$$

$$= \frac{T}{2K} \pm \sqrt{\frac{T^2}{4K^2} - \frac{3{,}600\,S}{K}}$$

The +ve sign cannot be adopted, as value of V_m obtained by using +ve sign will be much higher than that is possible in practice. Hence –ve sign will be used and, therefore, we have

$$V_m = \frac{T}{2K} - \sqrt{\frac{T^2}{4K^2} - \frac{3{,}600\,S}{K}} \quad ...(11.2)$$

From the above equation unknown quantity can be determined by substituting the value of known quantities.

2. *Calculation by Quadrilateral Speed-Time Curve.* Let

α = Acceleration in kmphps

β_c = Coasting retardation in kmphps

β = Braking retardation in kmphps

V_1 = Maximum speed at the end of acceleration in kmph

V_2 = Speed at the end of coasting in kmph

T = Total time of run in seconds

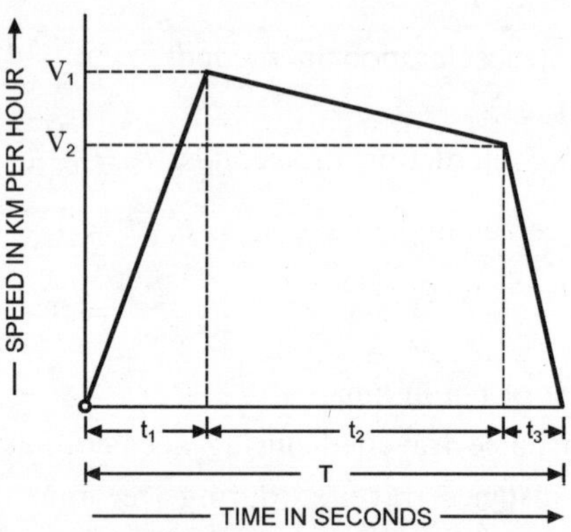

Fig. 11.6. ***Quadrilateral Speed-Time Curve***

Time of acceleration in seconds, $t_1 = \dfrac{V_1}{\alpha}$

Time of coasting in seconds, $t_2 = \dfrac{V_1 - V_2}{\beta_c}$

Time of braking in seconds $t_3 = \dfrac{V_2}{\beta}$

Total distance travelled in km,

S = Distance travelled during acceleration + distance travelled during coasting + distance travelled during retardation

$$= \frac{1}{2}V_1 \times \frac{t_1}{3{,}600} + \frac{V_1 + V_2}{2} \times \frac{t_2}{3{,}600} + \frac{1}{2}V_2 \times \frac{t_3}{3{,}600}$$

$$= \frac{V_1 t_1}{7{,}200} + \frac{V_1 t_2}{7{,}200} + \frac{V_2 t_2}{7{,}200} + \frac{V_2 t_3}{7{,}200}$$

$$= \frac{V_1}{7{,}200}(t_1 + t_2) + \frac{V_2}{7{,}200}(t_2 + t_3)$$

$$\text{or } S = \frac{V_1}{7{,}200}(T - t_3) + \frac{V_2}{7{,}200}(T - t_1)$$

Since $t_1 + t_2 + t_3 = T$

$$\text{or } S = \frac{T}{7{,}200}(V_1 + V_2) - \frac{V_1 t_3}{7{,}200} - \frac{V_2 t_1}{7{,}200}$$

$$\text{or } S = \frac{T}{7{,}200}(V_1 + V_2) - \frac{V_1}{7{,}200} \times \frac{V_2}{\beta} - \frac{V_2}{7{,}200} \times \frac{V_1}{\alpha}$$

$$= \frac{T}{7{,}200}(V_1 + V_2) - \frac{V_1 V_2}{7{,}200\,\beta} - \frac{V_1 V_2}{7{,}200\,\alpha}$$

$$\text{or } 7{,}200S = T(V_1 + V_2) - V_1 V_2\left(\frac{1}{\alpha} + \frac{1}{\beta}\right) \quad ...(11.3)$$

We have $V_2 = V_1 - \beta_c t_2 = V_1 - \beta_c(T - t_1 - t_3)$

$$= V_1 - \beta_c\left(T - \frac{V_1}{\alpha} - \frac{V_2}{\beta}\right)$$

$$\text{or} \left(V_2 - \frac{\beta_c}{\beta}V_2\right) = V_1 - \beta_c\left(T - \frac{V_1}{\alpha}\right)$$

$$\text{or} \quad V_2 = \frac{V_1 - \beta_c T + \dfrac{\beta_c}{\alpha}V_1}{1 - \dfrac{\beta_c}{\beta}} \quad ...(11.4)$$

Solving Eqs. (11.3) and (11.4) values of S, V_1, V_2 etc. can be obtained.

Example 11.1. A train service consists of following :

Uniform acceleration of 5 km/hr/sec for 30 seconds followed by free running for 10 minutes, then uniform braking at 5 km/hr/sec to stop followed by a stop of 5 minutes.

Draw the speed Vs time curve and calculate

(*i*) Distance between the stations

(*ii*) Average speed and (*iii*) Scheduled speed.

[U.P. Technical Univ. Utilization of Electrical Energy & Electric Traction 2005-06]

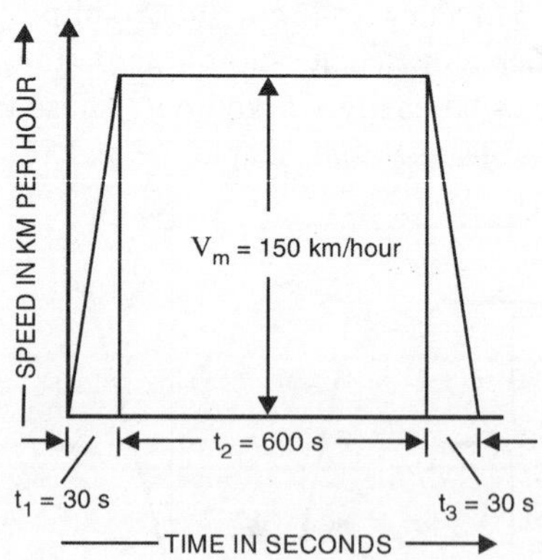

Fig. 11.7

Solution : Acceleration, α = 5 kmphps

Accelerating period, t_1 = 30 seconds

Maximum speed, $V_m = \alpha t_1 = 5 \times 30 = 150$ kmph

Time for free running, $t_2 = 10 \times 60 = 600$ seconds

Retardation, β = 5 kmphps

Time for retardation, $t_s = \dfrac{V_m}{\beta} = \dfrac{150}{5} = 30$ seconds

Distance travelled during acceleration period,

$$S_1 = \frac{1}{2}\frac{V_m t_1}{3{,}600}$$

$$= \frac{1}{2} \times \frac{150 \times 30}{3{,}600} = 0.625 \text{ km}$$

Distance travelled during free running,

$$S_2 = \frac{V_m \times t_2}{3,600} = \frac{150 \times 600}{3,600} = 25 \text{ km}$$

Distance travelled during braking period,

$$S_3 = \frac{1}{2} \times \frac{V_m \times t_3}{3,600} = \frac{1}{2} \times \frac{150 \times 30}{3,600} = 0.625 \text{ km}$$

(*i*) Total distance between stations, S

$$= S_1 + S_2 + S_3 = 0.625 + 25 + 0.625$$
$$= 26.25 \text{ km} \quad \textbf{Ans.}$$

$$\text{Average speed, } V_a = \frac{S \times 3,600}{T} = \frac{26.25 \times 3,600}{30 + 600 + 30}$$
$$= 143.18 \text{ kmph} \quad \textbf{Ans.}$$

$$\text{Scheduled speed, } V_s = \frac{S \times 3,600}{T + \text{stop time}} = \frac{26.25 \times 3,600}{660 + 5 \times 60}$$
$$= 98.44 \text{ kmph} \quad \textbf{Ans.}$$

Example 11.2. An electric train has an average speed of 42 kmph on a level track between stops 1,400 m apart. It is accelerated at 1.7 kmphps and is braked at 3.3 kmphps. Draw the speed-time curve for the run.

[*Pb. Technical Univ. Electric Drives & Utilization Dec. 2008*]

Solution : Acceleration, α = 1.7 kmphps

Retardation, β = 3.3 kmphps

Distance of run, S = 1,400 m = 1.4 km

Average speed, V_a = 42 kmph

$$\text{Time of run, } T = \frac{S}{V_a} \times 3,600 = \frac{1.4}{42} \times 3600$$
$$= 120 \text{ seconds}$$

$$\text{Using equation } V_m = \frac{T}{2K} - \sqrt{\frac{T^2}{4K^2} - \frac{3,600\,S}{K}}$$

$$\text{where } K = \frac{1}{2\alpha} + \frac{1}{2\beta} = \frac{1}{3.4} + \frac{1}{6.6} = 0.4456$$

Maximum speed,

$$V_m = \frac{120}{2 \times 0.4456} - \sqrt{\frac{(120)^2}{(2 \times 0.4456)^2} - \frac{3600 \times 1.4}{0.4456}}$$
$$= 52 \text{ kmph}$$

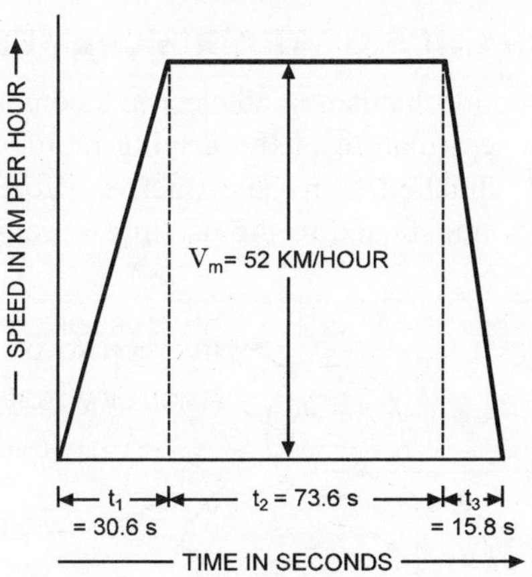

Fig. 11.8

$$\text{Acceleration period, } t_1 = \frac{V_m}{\alpha} = \frac{52}{1.7} = 30.6 \text{ seconds}$$

$$\text{Braking period, } t_3 = \frac{V_m}{\beta} = \frac{52}{3.3} = 15.8 \text{ seconds}$$

$$\text{Free running period, } t_2 = T - (t_1 + t_2) = 120 - (30.6 + 15.8)$$
$$= 73.6 \text{ seconds.}$$

The desired speed-time curve is shown in Fig. 11.8.

Example 11.3. A schedule speed of 45 km per hour is required between two stops 1.5 km apart. Find the maximum speed over the run if the stop is of 20 second duration. The values of acceleration and retardation are 2.4 kmphps and 3.2 kmphps respectively. Assume a simplified trapezoidal speed-time curve.

[*Pb. Technical Univ. Electric Drives and Traction 2003*]

Solution : Acceleration, α = 2.4 kmphps

Retardation, β = 3.2 kmphps

Distance of run, S = 1.5 km

Schedule speed, V_s = 45 kmph

$$\text{Schedule time, } T_s = \frac{S \times 3,600}{V_s} = \frac{1.5 \times 3,600}{45}$$
$$= 120 \text{ seconds}$$

$$\text{Actual time for run, } T = T_s - \text{stop duration}$$
$$= 120 - 20 = 100 \text{ seconds}$$

Using Eq. (11.2) we have

$$\text{Maximum speed, } V_m = \frac{T}{2K} - \sqrt{\frac{T^2}{4K^2} - \frac{3,600\,S}{K}}$$

$$\text{where } K = \frac{1}{2\alpha} + \frac{1}{2\beta} = \frac{1}{2 \times 2.4} + \frac{1}{2 \times 3.2} = 0.3646$$

$$\text{So, } V_m = \frac{100}{2 \times 0.3646} - \sqrt{\frac{100^2}{4 \times (0.3646)^2} - \frac{3,600 \times 1.5}{0.3646}}$$
$$= 74 \text{ kmph} \quad \textbf{Ans.}$$

Example 11.4. An electric train is to have acceleration and braking retardation of 0.8 km/h/s and 3.2 km/h/s respectively. If the ratio of maximum to average speed is 1.3 and time for stops 26 seconds, find schedule speed for a run of 1.5 km. Assume simplified trapezoidal speed-time curve.

[*Gorakhpur Univ. Utilization of Electrical Power & Traction Sept. 1984; M.D. Univ. Utilization of Electric Power & Traction May 2010; Rajasthan Technical Univ. Utilization of Electrical Power 2008, 2010*]

Solution : Let the actual time of run be T seconds

$$\text{Average speed, } V_a = \frac{3,600S}{T} = \frac{3,600 \times 1.5}{T} = \frac{5,400}{T} \text{ km per hour}$$

$$\text{Maximum speed, } V_m = 1.3V_a = 1.3 \times \frac{5,400}{T} = \frac{7,020}{T} \text{ km per hour}$$

$$\text{Since } V_m^2\left(\frac{1}{2\alpha} + \frac{1}{2\beta}\right) - V_m T + 3,600\,S = 0$$

$$\therefore V_m^2 = \frac{V_m T - 3,600\,S}{\frac{1}{2\alpha} + \frac{1}{2\beta}}$$

$$= \frac{\frac{7,020}{T} \times T - 3,600 \times 1.5}{\frac{1}{2 \times 0.8} + \frac{1}{2 \times 3.2}} = \frac{7,020 - 5,400}{5/6.4}$$

$$\text{or } V_m = \sqrt{1,620 \times \frac{6.4}{5}} = 45.53 \text{ kmph}$$

$$\text{and } V_a = \frac{V_m}{1.3} = \frac{45.53}{1.3} = 35.028 \text{ kmph}$$

$$\text{Actual time of run } T = \frac{3,600\,S}{V_a} = \frac{3,600 \times 1.5}{35.028} = 154 \text{ seconds}$$

Schedule time, T_s = Actual time of run + time of stop
= 154 + 26 = 180 seconds

Schedule speed, $V_s = \frac{S \times 3{,}600}{T_s} = \frac{1.5 \times 3{,}600}{180} = 30$ kmph **Ans.**

Example 11.5. A train has schedule speed of 30 kmph over a level track, distance between stations being 1 km. Station stopping time is 20 seconds. Assuming braking retardation of 3 kmphps and maximum speed 25 per cent greater than average speed, calculate acceleration required to run the service if the speed-time curve is approximated by a trapezoidal curve.

[*Rajasthan Univ. Utilization of Electrical Power 2005*]

Solution : Schedule time of run, $T_s = \frac{S \times 3{,}600}{V_s} = \frac{1 \times 3{,}600}{30}$
= 120 seconds

Actual time of run, T = T_s – duration of stop
= 120 – 20 = 100 seconds

Average speed, $V_a = \frac{S \times 3{,}600}{T} = \frac{1 \times 3{,}600}{100}$
= 36 kmph

Maximum speed, $V_m = 1.25\, V_a = 1.25 \times 36$
= 45 kmph

Using Eq. (11.1) we have

$$V_m^2\left(\frac{1}{2\alpha}+\frac{1}{2\beta}\right) - V_m T + 3{,}600\, S = 0$$

$$\text{or} \quad \frac{1}{2\alpha}+\frac{1}{2\beta} = \frac{V_m T - 3{,}600\, S}{V_m^2}$$

$$= \frac{45 \times 100 - 3{,}600 \times 1}{(45)^2} = \frac{4}{9}$$

$$\text{or} \quad \frac{1}{2\alpha} = \frac{4}{9} - \frac{1}{2\beta} = \frac{4}{9} - \frac{1}{2 \times 3} = 0.2777$$

$$\text{or} \quad \alpha = \frac{1}{2 \times 0.2777} = 1.8 \text{ kmphps} \quad \textbf{Ans.}$$

Example 11.6. A suburban electric train has a maximum speed of 70 kmph. The schedule speed including a station stop of 30 seconds is 45 kmph. If the acceleration is 1.5 kmphps, find the value of retardation when the average distance between stops is 4 km. [*A.M.I.E. Sec B. Utilization of Elec. Power Winter 1984; Pb. Technical Univ. Electric Drives & Utilization May 2009*]

Solution : Schedule time of run, $T_s = \frac{3{,}600 \times S}{V_s} = \frac{3{,}600 \times 4}{45}$
= 320 seconds

Actual time of run, T = T_s – duration of stop
= 320 – 30 = 290 seconds

$$\text{Since} \quad V_m^2\left(\frac{1}{2\alpha}+\frac{1}{2\beta}\right) - V_m T + 3{,}600\, S = 0$$

$$\text{or} \quad \frac{1}{2\alpha}+\frac{1}{2\beta} = \frac{V_m T - 3{,}600\, S}{V_m^2}$$

$$= \frac{70 \times 290 - 3{,}600 \times 4}{(70)^2} = 1.204$$

$$\text{or} \quad \frac{1}{2\beta} = 1.204 - \frac{1}{2\alpha} = 1.204 - \frac{1}{2 \times 1.5} = 0.87$$

$$\text{or} \quad \beta = \frac{1}{2 \times 0.87} = 0.575 \text{ kmphps} \quad \textbf{Ans.}$$

Example 11.7. A train is required to run between two stations 1.6 km apart at an average speed of 40 kmph. The run is to be made to a simplified quadrilateral speed-time curve. If the maximum speed is to be limited to 64 kmph, acceleration to 2.0 kmphps and coasting and braking retardation to 0.16 kmphps and 3.2 kmphps respectively, determine the duration of acceleration, coasting and braking periods.

[*Gorakhpur Univ. Utilization of Electrical Power with Traction 1980; Pb. Univ. Electric and Utilization January 1991; Rajasthan Technical Univ. Utilization of Electrical Power 2011*]

Solution : Distance of run, S = 1.6 km

Average speed, V_a = 40 km per hour

Maximum speed, V_1 = 64 km per hour

Acceleration, α = 2.0 kmphps

Coasting retardation, β_c = 0.16 kmphps

Braking retardation, β = 3.2 kmphps

$$\text{Duration of acceleration, } t_1 = \frac{V_1}{\alpha} = \frac{64}{2.0} = 32 \text{ seconds} \quad \textbf{Ans.}$$

$$\text{Actual time of run, } T = \frac{3{,}600\, S}{V_a} = \frac{3{,}600 \times 1.6}{40} = 144 \text{ seconds}$$

Let the speed before applying brakes be V_2

$$\text{then duration of coasting, } t_2 = \frac{V_1 - V_2}{\beta_c} = \frac{64 - V_2}{0.16} \text{ seconds}$$

$$\text{Duration of braking, } t_3 = \frac{V_2}{\beta} = \frac{V_2}{3.2} \text{ seconds}$$

Since actual time of run, T = $t_1 + t_2 + t_3$

$$\therefore \quad 144 = 32 + \frac{64 - V_2}{0.16} + \frac{V_2}{3.2}$$

$$\text{or} \quad V_2\left(\frac{1}{0.16} - \frac{1}{3.2}\right) = 32 + 400 - 144$$

$$\text{or} \quad V_2 = \frac{288}{6.25 - 0.3125} = 48.5 \text{ km per hour}$$

$$\text{Duration of coasting, } t_2 = \frac{V_1 - V_2}{\beta_c} = \frac{64 - 48.5}{0.16} = 96.85 \text{ seconds} \quad \textbf{Ans.}$$

$$\text{Duration of braking, } t_3 = \frac{V_2}{\beta} = \frac{48.5}{3.2} = 15.15 \text{ seconds} \quad \textbf{Ans.}$$

11.6. MECHANICS OF TRAIN MOVEMENT

Essential driving mechanism of an electric locomotive is shown in Fig. 11.9. The armature of the driving motor has a pinion of diameter d' attached to it. The tractive effort at the edge of the pinion is transferred to the driving wheel by means of a gearwheel.

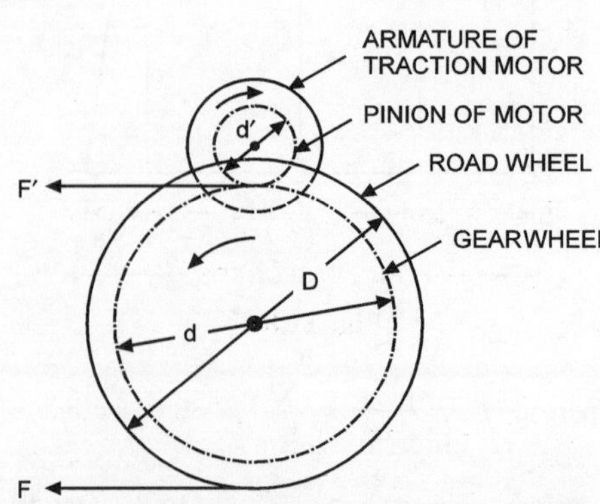

Fig. 11.9. ***Transmission of Tractive Effort***

Let the driving motor exert a torque T in Nm.

Tractive effort at the edge of pinion is given by the equation

$$T = F'\frac{d'}{2} \quad \text{or} \quad F' = \frac{2T}{d'}$$

Tractive effort transferred to the driving wheel

$$F = \eta F'\left(\frac{d}{D}\right) = \eta\frac{2T}{d'}\left(\frac{d}{D}\right) = \eta T\left(\frac{2}{D}\right)\left(\frac{d}{d'}\right)$$

$$= \eta T\frac{2\gamma}{D} \qquad ...(11.5)$$

where d is diameter of gearwheel in metres, D is diameter of driving wheel in metres and η is the efficiency of transmission, γ is the gear ratio and is equal to $\frac{d}{d'}$.

The maximum frictional force between the driving wheel and the track = μW where μ is the *coefficient of adhesion* between the driving wheel and the track and W is the weight of the train on the driving axles (called *adhesive weight*). Slipping will not take place unless tractive effort $F > \mu W$. For motion of trains without slipping tractive effort F should be less than or at the most equal to μW but in no case greater than μW.

The magnitude of the tractive effort that can be employed for propulsion, therefore, depends upon the weight coming over the driving wheels and the coefficient of adhesion between the driving wheel and the track. The coefficient of adhesion is defined as

i.e., Coefficient of adhesion,

$$\mu = \frac{\text{Maximum tractive effort that can be applied without slipping of wheels}}{\text{Adhesive weight (Weight on the driving wheels)}} \qquad ...(11.6)$$

The coefficient of adhesion reduces with the increase in speed as shown below in Table 11.2.

Table 11.2. Variation of coefficient of adhesion with speed on dry rails.

Speed in kmph	0	15	30	45	60	75
Coefficient of Adhesion	0.25	0.18	0.14	0.12	0.10	0.09

The normal value of coefficient of adhesion with clean dry rails is 0.25 and with wet or greasy rails the value may be as low as 0.08.

Electric traction has a very important advantage over steam traction and that is on account of greater adhesive weight - in a motor coach 100 per cent of the weight is on the driving wheels, in an electric locomotive 70 per cent whereas in a steam locomotive less than 50 per cent. The coefficient of adhesion in electric traction is also greater than that in steam traction due to the following two reasons.

(*i*) In electric traction the torque exerted is continuous whereas in steam traction the torque is pulsating which causes jolting and skidding.

(*ii*) In electric traction the driving wheels are distributed over a much greater length whereas in steam traction they are close to each other.

Since higher value of tractive effort can be used in electric traction, therefore, an electric train can be made to accelerate at a faster rate. This results in saving of time, especially when the distance between stops is small *e.g.* in urban and suburban areas.

11.7. TRACTIVE EFFORT FOR PROPULSION OF TRAIN

The effective force, necessary to propel the train, at the wheels of locomotive is called the *tractive effort*. It is tangential to the driving wheels and measured in newtons.

Total tractive effort required to run a train on track

= Tractive effort required for linear and angular acceleration + tractive effort to overcome the effect of gravity + tractive effort to overcome the train resistance.

$$\text{or } F_t = F_a \pm F_g + F_r \qquad ...(11.7)$$

1. *Tractive Effort For Acceleration.* According to laws of dynamics force is required to accelerate the motion of the body and is given by the expression

Force = Mass × acceleration

Consider a train of weight W tonnes being accelerated at α kmphps

The weight of train = 1,000 W kgf

Mass of train, m = 1,000 W kg

Acceleration = α kmphps

$$= \alpha \times \frac{1{,}000}{3{,}600} \text{ m/s}^2$$

$$= 0.2778\ \alpha \text{ m/s}^2$$

Tractive effort required for linear acceleration,

$$F_a = m\alpha$$

$$= 1{,}000\ W \times 0.2778\ \alpha$$

$$= 277.8\ W\ \alpha \text{ newtons} \qquad ...(11.8)$$

With the linear acceleration of the train, the rotating parts of the train such as wheels and motors also accelerate in an angular direction, and therefore, the tractive effort required is equal to the arithmetic sum of tractive effort required to have the angular acceleration of rotating parts and tractive effort required to have the linear acceleration. The tractive effort required to have the angular acceleration depends upon the individual weight, radius of gyration etc. of the rotating parts requiring angular acceleration. Hence the equivalent or accelerating weight of the train is taken as W_e which is higher than the dead weight W requiring linear acceleration to consider the tractive effort for the angular acceleration. In practice W_e is higher than W by 8 to 15%. The normal value lies between 10 and 12 per cent.

Hence tractive effort required for acceleration

$$F_a = 277.8\ W_e\ \alpha \text{ newtons} \qquad ...(11.9)$$

2. *Tractive Effort For Overcoming The Effect of Gravity.* When a train is on a slope, a force of gravity equal to the component of the dead weight along the slope acts on the train and tends to cause its motion down the gradient or slope. Hence force due to gradient, $F_g = 1{,}000\ W \sin\theta$ kg ...(11.10)

But in railway work gradient is expressed as rise in metres in a track distance of 100 metres and is denoted as '*percentage gradient*' (G%).*

* Some authors define percentage gradient as rise in metres in a track distance of 1,000 metres.

i.e., $G = \sin\theta \times 100$

or $\sin\theta = \frac{G}{100}$

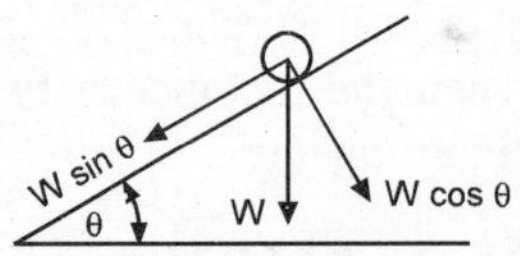

Fig. 11.10

Substituting $\sin\theta = \frac{G}{100}$ in Eq. (11.10) we have,

$$F_g = 1{,}000\ W \times \frac{G}{100}$$
$$= 10\ WG\ kg = 10\ WG \times 9.81$$
$$= 98.1\ WG\ newtons \quad ...(11.11)$$

When the train is going up a gradient, the tractive effort will be required to balance this force due to gradient but while going down the gradient, the force will add to the tractive effort.

3. *Tractive Effort For Overcoming Train Resistance.* Train resistance consists of all the forces resisting the motion of a train when it is running at uniform speed on a straight and level track. Under these circumstances the whole of the energy output from the driving axles is expended against train resistance. Train resistance is due to (*i*) the friction at the various parts of the rolling stock (*ii*) friction at the track and (*iii*) air resistance. The first two components constitute the *mechanical resistance* component of train resistance. The train resistance depends upon various factors, such as shape, size and condition of track etc., and is expressed in newtons per tonne of the dead weight. For a normal train the value of specific resistance has been 40 to 70 newtons/tonne.

The general equation for train resistance is given as

$$R = k_1 + k_2V + k_3V^2 \quad ...(11.12)$$

where k_1, k_2 and k_3 are constants depending upon the train and the track, R is the resistance in newtons and V is the speed in kmph. The first two terms represent the mechanical resistance and the last term represents air resistance.

Tractive effort required to overcome the train resistance,

$$F_r = W \times r \text{ newtons} \quad ...(11.13)$$

where r is the specific resistance in newtons per tonne of the dead weight.

Total tractive effort required,

$$F_t = F_a \pm F_g + F_r = 277.8\ W_e\alpha \pm 98.1\ WG + Wr \quad ...(11.14)$$

+ ve sign for the motion up the gradient and

– ve sign for the motion down the gradient.

11.8. POWER OUTPUT FROM THE DRIVING AXLES

Power, P = Rate of doing work

$$= \text{Tractive effort} \times \frac{\text{distance}}{\text{time}}$$
$$= \text{Tractive effort} \times \text{speed}$$
$$= F_t \times v \quad \text{where } F_t \text{ is in newtons and } v \text{ is in m/s}$$

$$\text{or } P = F_t \times V \times \frac{1{,}000}{3{,}600} \text{ watts} = \frac{F_t \times V}{3{,}600} \text{ kW} \quad ...(11.15)$$

where V is in kmph.

Example 11.8. An electric train weighing 200 tonne has eight motors geared to driving wheels, each wheel is 90 cm diameter. Determine the torque developed by each motor to accelerate the train to a speed of 48 kmph in 30 seconds up a gradient of 1 in 200. The tractive resistance is of 50 newtons per tonne, the effect of rotational inertia is 10% of the train weight, the gear ratio is 4 to 1 and gearing efficiency is 80 per cent.

[*Agra Univ. Traction & Utilization of Elec. Power 1983*]

Solution : Weight of train, W = 200 tonnes

Diameter of driving wheels, D = 90 cm = 0.9 m

Percentage gradient, $G = \frac{1}{200} \times 100 = 0.5\%$

Tractive resistance, r = 50 newtons/tonne

Gear ratio, $\gamma = 4$

Gearing efficiency, $\eta = 0.80$

Equivalent accelerating weight of the train

$$W_e = 1.10\ W = 1.1 \times 200 = 220 \text{ tonnes}$$

Maximum speed, V_m = 48 kmph

Time taken in accelerating the train to a speed of 48 kmph,

$$t_1 = 30 \text{ seconds}$$

$$\text{Acceleration, } \alpha = \frac{V_m}{t_1} = \frac{48}{30} = 1.6 \text{ kmphps}$$

Tractive effort required,

$$F_t = 277.8\ W_e\alpha + 98.1\ WG + Wr$$
$$= 277.8 \times 220 \times 1.6 + 98.1 \times 200 \times 0.5 + 200 \times 50$$
$$= 117{,}596 \text{ N}$$

Total torque developed,

$$T = \frac{F_t \times D}{\eta \times 2\gamma} = \frac{117{,}596 \times 0.9}{0.8 \times 2 \times 4} = 16{,}537 \text{ N-m}$$

Torque developed by each motor $= \frac{16{,}537}{8} = 2{,}067$ N-m **Ans.**

Example 11.9. A 200 tonne motor coach having 4 motors each developing 6,000 N-m torque during acceleration, starts from rest. If the gradient is 30 in 1,000; gear ratio 4, gear transmission efficiency 90%, wheel radius 45 cm, train resistance 50 N/tonne; addition of rotational inertia 10%, calculate the time taken to attain a speed of 50 kmph. If the line voltage is 3,000 V dc and efficiency of motors 85%, find the current during notching period.

[*Pb. Univ. Electric Drives and Utilization Dec. 1989; Rajasthan Univ. Utilization of Electrical Power 2006*]

Solution : Weight of train, W = 200 tonnes

Diameter of driving wheels, D = 2 × 45 = 90 cm = 0.9 m

Percentage gradient, $G = \frac{30}{1{,}000} \times 100 = 3\%$

Tractive resistance, r = 50 N/tonne

Gear ratio, $\gamma = 4$

Gear transmission efficiency, η = 90% or 0.9

Equivalent accelerating weight of the train,

$$W_e = 1.10 \times 200 = 220 \text{ tonnes}$$

Total torque developed, T = 4 × 6,000 = 24,000 N-m

Tractive effort,

$$F_t = \frac{\eta T 2\gamma}{D} = \frac{0.9 \times 24{,}000 \times 2 \times 4}{0.9} = 192{,}000 \text{ N}$$

Let the acceleration of the train be α kmphps

Then $F_t = 277.8\ W_e\alpha + 98.1\ WG + Wr$

or $192{,}000 = 277.8 \times 220\alpha + 98.1 \times 200 \times 3 + 200 \times 50$

$$\alpha = \frac{123{,}140}{277.8 \times 220} = 2.015 \text{ kmphps}$$

Time taken for the train to attain a speed of 50 kmph

$$t = \frac{V_m}{\alpha} = \frac{50}{2.015} = 24.82 \text{ seconds}$$ **Ans.**

Power output from the driving axles

$$= \frac{F_t \times V}{3,600} = \frac{192,000 \times 50}{3,600} = 2,666.7 \text{ kW}$$

$$\text{Power input} = \frac{\text{Power output}}{\eta_m} = \frac{2,666.7}{0.85} = 3,137 \text{ kW}$$

$$\text{Total current drawn} = \frac{\text{Power input in kW} \times 1,000}{V}$$

$$= \frac{3,137 \times 1,000}{3,000} = 1,046 \text{ A}$$

Current drawn per motor $= \frac{1,046}{4} = 261.5$ A **Ans.**

Example 11.10. A train weighing 120 tonnes is to be driven up an incline of 2 per cent at a speed of 36 kmph. If the train resistance at this speed is 2 kg per tonne, find the current required at 1,500 V dc if the efficiency of the motors and gearing is 88 per cent. If the current were cut off, how long would the train take to come to rest

Solution : Tractive effort,

F_t = 98.1 WG + W*r*

= 98.1 × 120 × 2 + 120 × 2 × 9.81 = 25,900 N

Speed, V = 36 kmph

$$\text{Power output of driving axles} = \frac{F_t \times V}{3,600} = \frac{25,900 \times 36}{3,600} = 259 \text{ kW}$$

$$\text{Power input} = \frac{\text{Power output}}{\text{Efficiency of motors and gears}} = \frac{259}{0.88} = 294.3 \text{ kW}$$

$$\text{Current required} = \frac{\text{Power input in kW} \times 1,000}{\text{Voltage of the line}}$$

$$= \frac{294.3 \times 1,000}{1,500} = 196.2 \text{ amperes}$$ **Ans.**

Let the coasting retardation be β_c

then $F_t = 277.8\, W_e(-\beta_c) + 98.1\, WG + Wr$

Assuming W_e = 110 per cent of dead weight, we have

$0 = -277.8 \times 1.1\, W \beta_c + 98.1 \times 2W + 2 \times 9.81\, W$

$$\text{or } \beta_c = \frac{98.1 \times 2.2}{277.8 \times 1.1} = 0.7 \text{ kmphps}$$

Time taken to come to rest $= \frac{V}{\beta_c} = \frac{36}{0.7} = 51$ seconds **Ans.**

Example 11.11. An electric train accelerates uniformly from rest to a speed of 48 km/hour in 24 seconds. It then coasts for 69 seconds against a constant resistance of 58 N/tonne and is braked to rest at 3.3 km/hour/second in 11 seconds. Calculate (*i*) the acceleration (*ii*) coasting retardation and (*iii*) the schedule speed, if the station stops are of 20 second duration. What would be the effect on schedule speed of reducing the station stops to 15 second duration, other conditions remaining same? Allow 10% for rotational inertia.

Solution : Time of acceleration, t_1 = 24 seconds

Time of coasting, t_2 = 69 seconds

Time of braking, t_3 = 11 seconds

Maximum speed, V_1 = 48 kmph

(*i*) Acceleration, $\alpha = \frac{V_1}{t_1} = \frac{48}{24} = 2$ kmphps **Ans.**

Let the coasting retardation be β_c

$F_t = 277.8\, W_e(-\beta_c) + 98.1\, WG + Wr$

$= -277.8 \times 1.1\, W \beta_c + 98.1\, W \times 0 + 58\, W$

(*ii*) or $\beta_c = \frac{58}{277.8 \times 1.1} = 0.19$ kmphps **Ans.**

Speed at the end of coasting period V_2

$= V_1 - \beta_c \times t_2 = 48 - 0.19 \times 69 = 34.9$ kmph

Distance travelled,

$$S = \frac{V_1 t_1}{7,200} + \frac{(V_1 + V_2) t_2}{7,200} + \frac{V_2 t_3}{7,200} \text{ km}$$

$$= \frac{48 \times 24}{7,200} + \frac{(48 + 34.9) \times 69}{7,200} + \frac{34.9 \times 11}{7,200} = 1.0 \text{ km}$$

Schedule time, $T_s = t_1 + t_2 + t_3$ + duration of stop = 24 + 69 + 11 + 20 = 124 seconds

(*iii*) Schedule speed, $V_s = \frac{S \times 3,600}{T_s} = \frac{1.0 \times 3,600}{124}$

= 29 kmph **Ans.**

When the duration of stop is 15 seconds

Schedule time, T_s' = 24 + 69 + 11 + 15 = 119 seconds

and schedule speed, $V_s' = \frac{1.0 \times 3,600}{119} = 30.25$ kmph **Ans.**

Example 11.12. An electric train has quadrilateral speed-time curve as follows:

(*i*) Uniform acceleration from rest at 2 kmphps for 30 seconds;

(*ii*) Coasting for 50 seconds; (*iii*) Uniform braking to rest for 20 seconds.

If the train is moving a uniform upgradient of 10/1,000, train resistance is 40 N/tonne, rotational inertia effect 10% of dead weight and duration of stop 30 seconds, find the schedule speed. [*Pb. Technical Univ. Electric Drives & Utilization May 2009*]

Solution : Time of acceleration, t_1 = 30 seconds

Time of coasting, t_2 = 50 seconds

Time of braking, t_3 = 20 seconds

Acceleration, α = 2 kmphps

Maximum speed, $V_1 = \alpha t_1 = 2 \times 30 = 60$ km/hour

$$\text{Gradient, } G = \frac{10}{1,000} \times 100 = 1\%$$

Let the coasting retardation be β_c

$F_t = 277.8\, W_e(-\beta_c) + 98.1\, WG + Wr$

or $0 = -277.8 \times 1.1\, W \beta_c + 98.1 \times 1 \times W + 40\, W$

$$\text{or } \beta_c = \frac{98.1 + 40}{277.8 \times 1.1} = \frac{138.1}{277.8 \times 1.1} = 0.452 \text{ kmphps}$$

$V_2 = V_1 - \beta_c t_2 = 60 - 0.452 \times 50 = 37.4$ km per hour

$$\text{Braking retardation, } \beta = \frac{V_2}{t_3} = \frac{37.4}{20} = 1.87 \text{ kmphps}$$

$$\text{Distance travelled, } S = \frac{V_1 t_1}{7,200} + \frac{(V_1 + V_2) t_2}{7,200} + \frac{V_2 t_3}{7,200} \text{ km}$$

$$= \frac{60 \times 30}{7,200} + \frac{60 + 37.4}{7,200} \times 50 + \frac{37.4 \times 20}{7,200}$$

= 1.03 km

Schedule time, $T_s = t_1 + t_2 + t_3$ + duration of stop

= 30 + 50 + 20 + 30 = 130 seconds

$$\text{Schedule speed, } V_s = \frac{3,600 \times S}{T_s} = \frac{3,600 \times 1.03}{130}$$

= 28.52 km per hour **Ans.**

Example 11.13. The speed-time curve of an electric train on a uniform rising gradient of 1 in 100 comprises

(*i*) Uniform acceleration from rest at 2 kmphps for 30 seconds

(*ii*) Coasting with power off for 70 seconds

(*iii*) Braking at 3 kmphps to standstill.

The weight of the train is 250 tonnes, the train resistance on level track being 5 kg/tonne, and allowance for rotary inertia 10%.

Calculate the maximum power developed by traction motors and total distance travelled by the train. Assume transmission efficiency as 97%.

[*Agra Univ. Traction & Utilization of Elec. Power 1979*]

Solution : Weight of train, W = 250 tonnes

Effective weight of train, $W_e = \left(1+\frac{10}{100}\right)W = 1.1\ W$

Train resistance, r = 5 kg/tonne = 5 × 9.81 = 49.05 N/tonne

Gradient, G = 1 per cent

Maximum velocity, $V_m = \alpha t_1 = 2 \times 30 = 60$ km/hour

Tractive effort required,

$$F_t = 277.8\ W_e\ \alpha + 98.1\ WG + Wr$$
$$= 277.8 \times 1.1 \times 250 \times 2 + 98.1 \times 250 \times 1 + 250 \times 49.05$$
$$= 189{,}577 \text{ newtons}$$

Maximum power output from the driving axles

$$= \frac{F_t V_m}{3{,}600} = \frac{189{,}577}{3{,}600} \times 60 = 3{,}160 \text{ kW}$$

Maximum power developed by traction motors

$$= \frac{3{,}160}{0.97} = 3{,}258 \text{ kW} \quad \textbf{Ans.}$$

Let the coasting retardation be β_c then

$$F_t = 277.8\ W_e(-\beta_c) + 98.1\ WG + Wr$$

or $0 = -277.8 \times 1.1\ W \times \beta_c + 98.1\ W \times 1 + W \times 49.05$

or $\beta_c = \frac{98.1 + 49.05}{277.8 \times 1.1} = 0.4815$ kmphps

$$V_2 = V_1 - \beta_c t_2 = 60 - 70 \times 0.4815 = 26.3 \text{ km/hour}$$

Braking period, $t_3 = \frac{V_2}{\beta} = \frac{26.3}{3} = 8.77$ seconds

Total distance travelled by the train,

$$S = \frac{V_1 t_1}{7{,}200} + \frac{(V_1 + V_2)t_2}{7{,}200} + \frac{V_2 t_3}{7{,}200}$$

$$= \frac{60 \times 30}{7{,}200} + \frac{60 + 26.3}{7{,}200} \times 70 + \frac{26.3 \times 8.77}{7{,}200} = 1.12 \text{ km} \quad \textbf{Ans.}$$

11.9. ENERGY OUTPUT FROM DRIVING AXLES

Assuming the run according to trapezoidal speed-time curve Total energy required for the run

= Energy required during acceleration + energy required during free run

= Average power during acceleration × acceleration period + average power during free run × duration of free run

$$= \frac{1}{2}\frac{F_t V_m}{3{,}600} \times \frac{t_1}{3{,}600} + \frac{F_t' V_m}{3{,}600} \times \frac{t_2}{3{,}600} \text{ kWh} \quad ...(11.16)$$

where V_m is the maximum speed in kmph, t_1 is the time of acceleration in seconds, t_2 is the time of free run in seconds, F_t is the tractive effort required during acceleration in newtons and F_t' is the tractive effort required during free run in newtons. Instead of expressing the energy in kWh, it is more convenient for the purpose of comparison to introduce the weight of the train and the distance of run and to express the energy in watt-hours per tonne-km.

i.e.,
$$\frac{\text{Energy output in watt-hours}}{\text{Weight of the train in tonnes} \times \text{distance of run in km}} \quad ...(11.17)$$

This quantity is called *specific energy output*, and is used for comparing the dynamical performances of trains operating to different schedules.

The energy input to the motors is called the *energy consumption* of the train, since it is the energy consumed for propelling the train. The total energy drawn from the distribution system will be greater than this by the quantity required for lighting, heating, control and braking.

The energy consumption can be expressed in watt-hours per tonne-km.

i.e.,
$$\frac{\text{Energy consumption of train in watt-hours}}{\text{Weight of train in tonnes} \times \text{distance of run in km}} \quad ...(11.18)$$

The above quantity is called the *specific energy consumption*.

11.10. DETERMINATION OF SPECIFIC ENERGY OUTPUT USING SIMPLIFIED SPEED-TIME CURVE

Let the track have a gradient of G per cent through out its run.

Energy output to accelerate the train from rest to a speed V_m

$$= \frac{1}{2}\frac{F_t V_m}{3{,}600} \times \frac{t_1}{3{,}600} \text{ kWh} = \frac{1}{2}\frac{F_t V_m}{3{,}600} \times \frac{V_m}{3{,}600\alpha} \text{ kWh}$$

since $t_1 = \frac{V_m}{\alpha}$

$$= \frac{1}{2}\frac{V_m^2}{(3{,}600)^2 \alpha}[277.8\ W_e\alpha + 98.1\ WG + Wr] \text{ kWh} \quad ...(11.19)$$

since $F_t = 277.8 W_e \alpha + 98.1 WG + Wr$

Energy output to run the train at the speed V_m against the gradient and resistance to motion

$$= \frac{F_t' V_m}{3{,}600} \times \frac{t_2}{3{,}600} \text{ kWh} = \frac{F_t' \times S'}{3{,}600} \text{ kWh}$$

$$= [Wr + 98.1WG] \times \frac{S'}{3{,}600} \text{ kWh} \quad ...(11.20)$$

since $\frac{V_m t_2}{3{,}600} = S'$, the distance travelled during free run

Total energy output for the run

$$= \frac{V_m^2}{2\alpha(3{,}600)^2}[277.8W_e\alpha + 98.1WG + Wr] + [Wr + 98.1WG] \times \frac{S'}{3{,}600} \text{ kWh}$$

$$= \frac{V_m^2 \times 1{,}000}{2\alpha \times (3{,}600)^2}[277.8W_e\alpha + 98.1WG + Wr] + [Wr + 98.1WG] \times \frac{S' \times 1{,}000}{3{,}600} \text{ Wh}$$

$$= \frac{V_m^2}{2\alpha} \times \frac{1{,}000}{(3{,}600)^2} \times 277.8W_e\alpha + \frac{V_m^2}{2\alpha} \times \frac{1{,}000}{(3{,}600)^2}[98.1WG + Wr] + \frac{1{,}000S'}{3{,}600}[Wr + 98.1WG]$$

$$= 0.01072 V_m^2 W_e + [98.1WG + Wr]\left[\frac{V_m^2}{2\alpha \times 3{,}600} + S'\right] \times \frac{1{,}000}{3{,}600}$$

$$= 0.01072 V_m^2 W_e + \frac{1{,}000}{3{,}600}[98.1WG + Wr][S'' + S']$$

where $\frac{V_m^2}{2\alpha \times 3{,}600} = S''$, distance travelled during accelerating period

$$= 0.01072 V_m^2 W_e + 0.2778\,(98.1WG + Wr)S_1 \quad ...(11.21)$$

where S_1 is the distance travelled during acceleration and free run in km

Specific energy output

$$= \frac{\text{Energy output for the run in watt-hours}}{\text{Weight of train in tonnes} \times \text{distance of run in km}}$$

$$= \frac{0.01072 V_m^2 W_e + 0.2778\,(98.1WG + Wr)S_1}{W \times S}$$

$$= \frac{0.01072\, V_m^2}{S} \times \frac{W_e}{W} + 0.2778\,\frac{S_1}{S}(98.1\,G + r) \text{ watt-hours per tonne-km} \quad ...(11.22)$$

If the track is level one, G = 0 then specific energy output

$$= \frac{0.01072\, V_m^2}{S} \times \frac{W_e}{W} + 0.2778\,\frac{S_1}{S}\, r \text{ watt-hours/tonne-km} \quad ...(11.23)$$

Specific energy consumption

$$= \frac{\text{Specific energy output at driving wheels}}{\text{Overall efficiency of the motors and gearing}}$$

$$= \frac{0.01072 V_m^2}{S\eta} \times \frac{W_e}{W} + 0.2778\,\frac{S_1}{S} \times \frac{r}{\eta} \text{ watt-hours/tonne-km} \quad ...(11.24)$$

11.10.1. Factors Affecting Energy Consumption

Energy consumption in propelling the train is required for

1. having the linear and angular acceleration;
2. working against gravity while moving up the gradient;
3. working against the resistance to motion;
4. overcoming the losses due to gearing system and
5. supplying losses in motors and other electrical system.

Energy required being the product of power and time depends upon the power required and the duration for which the power is taken.

1. The power (in watts) required in accelerating the train on level track depends upon the rate of acceleration α, the effective weight of the train W_e and the speed at the end of acceleration V_m, being equal to 77.17 $V_m W_e \alpha$ whereas the energy (in watts-hours) depends upon the rate of acceleration, effective weight of train, maximum speed and the time for which the train accelerates, being equal to 0.01072 $V_m W_e \alpha\, t$.
2. The power (in watts) required in overcoming the force of gravity while going up the gradient being equal to 27.334 WG V_m depends upon the gradient. While going down the gradient the energy is returned back. The energy required or returned (in watt-seconds) is given by the expression 27.334 WG $V_m t$ and, therefore, depends upon the gradient, dead weight of train, maximum speed and duration of run.
3. The power (in watts) required in overcoming the resistance to motion, is equal to 0.2778 Wr V_m and energy required = 0.2778 Wr $V_m t$ watt-seconds. The expression for energy required indicates that it depends upon the resistance to motion, maximum speed and time of run.

The energy output of train for a given speed-time curve is independent of the type of drive employed, but energy consumption depends upon the type of drive employed.

11.11. FACTORS AFFECTING SPECIFIC ENERGY CONSUMPTION OF AN ELECTRIC TRAIN OPERATING ON A GIVEN SCHEDULE SPEED

The specific energy consumption of a train operating at a given schedule speed depends upon the following factors

(*a*) distance between stops (*b*) acceleration (*c*) retardation (*d*) maximum speed (*e*) nature of route and (*f*) the type of train equipment.

Specific energy output is given by the expression
Specific energy output

$$= \frac{0.01072\, V_m^2 W_e}{SW} + \frac{0.2778\,(98.1\,G + r)S_1}{S} \text{ watt-hours/tonne-km} \quad ...(11.25)$$

Specific energy output is independent of locomotive overall efficiency but the specific energy consumption being equal to specific energy output divided by locomotive overall efficiency depends upon the overall efficiency of the locomotive. Greater the overall efficiency lesser will be the specific energy consumption for a given specific energy output at axles.

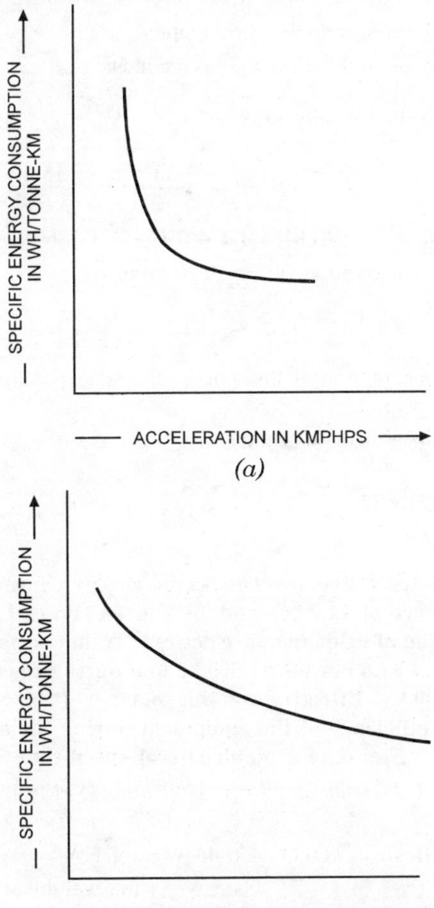

(*a*)

(*b*)

Fig. 11.11

The Eq. (11.25) clearly shows that specific energy consumption depends upon the maximum speed V_m, the distance travelled by the train while power is on, S_1, the specific resistance r, gradient G and distance between stops. Greater the distance between stops lesser will be the specific energy consumption. For a given run at a given schedule speed, greater the value of acceleration and retardation, more will be the period of coasting and, therefore, lesser the period during which power is on *i.e.*, S_1 will be small and, therefore, specific energy consumption will accordingly be less. Steep gradient will involve more energy consumption even if regenerative braking is used. Similarly more the train resistance, greater will be the specific energy consumption.

How the specific energy consumption falls with the increase in value of acceleration (retardation) and distance of run is illustrated in Figs. 11.11 (*a*) and 11.11 (*b*) respectively.

Typical values of specific energy consumption are (*i*) 50 – 75 watt-hours per tonne-km for suburban services and (*ii*) 20 – 30 watt-hours per tonne-km for main line service.

Example 11.14. An electric train has an average speed of 42 kmph on a level track between stops 1,400 m apart. It is to be accelerated at 1.7 kmphps and is braked at 3.3 kmphps. Draw the speed-time curve of the run. Estimate the energy consumption at the axles of the train per tonne-km. Take tractive resistance constant at 50 N per tonne and allow 10 per cent for rotational inertia. [*W.B. Univ. of Technology Utilization of Electric Power 2009*]

Solution : From solution of Example 11.2 we have

Maximum speed, V_m = 52 kmph

Duration of braking, t_3 = 15.8 seconds

$$\text{Distance travelled during braking} = \frac{1}{2} \times V_m \times \frac{t_3}{3{,}600}$$

$$= \frac{1}{2} \times 52 \times \frac{15.8}{3{,}600} = 0.114 \text{ km}$$

S_1 = S – 0.114 = 1.4 – 0.114 = 1.286 km

Resistance to motion, r = 50 N/tonne

$$\frac{W_e}{W} = 1.1$$

So energy consumption at the axle of the train per tonne-km

$$= \frac{0.01072 V_m^2}{S} \times \frac{W_e}{W} + 0.2778 r \frac{S_1}{S}$$

$$= \frac{0.01072 \times 52^2}{1.4} \times 1.1 + 0.2778 \times 50 \times \frac{1.286}{1.4} = 35.534 \text{ Wh} \quad \textbf{Ans.}$$

Example 11.15. Calculate the specific energy consumption if a maximum speed of 12.2 m/s and for a given run of 1,525 metres an acceleration of 0.366 m/s² are desired. Train resistance during acceleration is 52.6 newtons/1,000 kg and during coasting is 6.12 newtons/1,000 kg. Effective rotating mass is 10% of the dead weight. The efficiency of the equipment during the acceleration period is 65%. Assume a quadrilateral speed-time curve.

[*U.P. Technical Univ. Utilization of Electrical Energy and Electric Traction 2005-06*]

Solution : Effective weight of train W_e = 1.1 W

where W is the weight of train in kg

Acceleration α = 0.366 m/s²

Maximum speed, V_m = 12.2 m/s

Distance of run, S = 1,525 metres

$$\text{Acceleration period, } t_1 = \frac{V_m}{\alpha} = \frac{12.2}{0.366} = 33.3 \text{ seconds}$$

Train resistance during acceleration,

r = 52.6 N/1,000 kg = 0.0526 N/kg

Tractive effort required during acceleration,

$F_t = W_e \alpha + Wr$ = 1.1 W × 0.366 + 0.0526 W = 0.4552 W newtons

Total energy required for the run

= Energy required during acceleration as there is no free run

= Average power during acceleration × acceleration period

$$= \frac{1}{2} F_t V_m \times t_1 = \frac{1}{2} \times 0.4552 \text{ W} \times 12.2 \times 33.3 \text{ joules}$$

= 92.56 W joules or watt-seconds

$$= \frac{92.56 \text{ W}}{3{,}600} = 0.0257 \text{ W watt-hours}$$

$$\text{Specific energy output} = \frac{\text{Energy required for the run}}{W \times S}$$

$$= \frac{0.0257 \text{ W}}{W \times 1{,}525}$$

= 1.685 × 10^{-5} Wh/kg-m

$$\text{Specific energy consumption} = \frac{\text{Specific energy output}}{\eta}$$

$$= \frac{1.685 \times 10^{-5}}{0.65}$$

= 2.59 × 10^{-5} Wh/kg-m **Ans.**

Example 11.16. An electric train has quadrilateral speed-time curve as follows:

(*i*) Uniform acceleration from rest at 2 kmphps for 30 seconds

(*ii*) Coasting for 50 seconds

(*iii*) Braking period of 20 seconds.

The train is moving a uniform upgradient of 1%, tractive resistance is 40 newtons per tonne, rotational inertia effect 10% of dead weight, duration of station stop 15 seconds and overall efficiency of transmission gear and motor as 75%. Calculate the value of its schedule speed and specific energy consumption of run. [*Pb. Univ. Electric Drives and Utilization Nov. 1988; Rajasthan Univ. Utilization of Electric Power 2003*]

Solution : Maximum speed, $V_1 = \alpha t_1$ = 2 × 30 = 60 kmph

Refer to Fig. 11.7 and also solution to Example 11.12.

Coasting retardation, β_c = 0.452 kmphps

V_2 = 37.4 kmph

Braking retardation, β = 1.87 kmphps

Total distance travelled, S = 1.03 km

Schedule time, $T_s = t_1 + t_2 + t_3$ + stop duration

= 30 + 50 + 20 + 15 = 115 seconds

$$\text{Schedule speed, } V_s = \frac{3{,}600 \times S}{T_s} = \frac{3{,}600 \times 1.03}{115}$$

= 32.24 kmph **Ans.**

Distance travelled when power is on,

S_1 = Distance travelled during acceleration period

$$= \frac{V_1 t_1}{7{,}200} = \frac{60 \times 30}{7{,}200} = 0.25 \text{ km}$$

Specific energy output

$$= \frac{0.01072 V_m^2}{S} \cdot \frac{W_e}{W} + 0.2778\,(98.1\,G + r)\frac{S_1}{S}$$

$$= \frac{0.01072 \times 60^2}{1.03} \times 1.1 + 0.2778\,(98.1 \times 1 + 40) \times \frac{0.25}{1.03}$$

= 50.5 Wh/tonne-km

$$\text{Specific energy consumption} = \frac{50.5}{0.75} = 67.34 \text{ Wh/tonne-km} \quad \textbf{Ans.}$$

Example 11.17. An electric train has quadrilateral speed-time curve as follows :

(*i*) Uniform acceleration from rest at 2 kmphps for 30 seconds.

(*ii*) Coasting for 50 seconds.

(*iii*) Braking period of 20 seconds.

The train is moving a uniform down gradient of 1%, tractive resistance 40 newtons per tonne, rotational inertia effect 10% of dead weight, duration of stop 15 seconds and overall efficiency of transmission gear and motor as 75%. Calculate its schedule speed and specific energy consumption of run.

[*Pb. Univ. Electric Drives and Utilization May 1989*]

Solution : Maximum speed, $V_1 = \alpha t_1 = 2 \times 30 = 60$ kmph

Let the coasting retardation be β_c

Tractive effort, $F_t = 277.8\ W_e\ (-\beta_c) - 98.1\ WG + Wr$

or $0 = -277.8 \times 1.1\ W\ \beta_c - 98.1 \times W \times 1 + 40\ W$

or $\beta_c = \dfrac{-58.1}{277.8 \times 1.1} = -0.19$ kmphps

$V_2 = V_1 - \beta_c t_2 = 60 - (-0.19) \times 50 = 69.5$ kmph

Distance travelled, $S = \dfrac{V_1 t_1}{7{,}200} + \dfrac{V_1 + V_2}{7{,}200} \times t_2 + \dfrac{V_2\, t_3}{7{,}200}$ km

$= \dfrac{60 \times 30}{7{,}200} + \dfrac{60 + 69.5}{7{,}200} \times 50 + \dfrac{69.5 \times 20}{7{,}200}$

$= 0.25 + 0.899 + 0.193 = 1.342$ km

Schedule time, $T_s = t_1 + t_2 + t_3 +$ stop duration

$= 30 + 50 + 20 + 15 = 115$ seconds

Schedule speed, $V_s = \dfrac{3{,}600 \times S}{T_s} = \dfrac{3{,}600 \times 1.342}{115}$

$= 42$ kmph **Ans.**

Specific energy output

$$= \frac{0.01072 V_1^2}{S} \times \frac{W_e}{W} + 0.2778\,(98.1G + r)\frac{S_1}{S}$$

$$= \frac{0.01072 \times 60^2}{1.342} \times 1.1 + 0.2778\,[98.1(-1) + 40] \times \frac{0.25}{1.342}$$

$= 28.6$ Wh/tonne-km

Specific energy consumption $= \dfrac{28.6}{0.75} = 38.14$ Wh/tonne-km **Ans.**

Example 11.18. An electric train weighing 100 tonnes has a rotational inertia of 10%. This train while running between two stations which are 2.5 km apart has an average speed of 50 km/hour. The acceleration and retardation during braking are respectively 1 km/hour/second and 2 km per hour per second. The percentage gradient between these two stations is 1% and the train is to move up the incline. The track resistance is 40 N/tonne. If the combined efficiency of the electric train is 60%, determine (*i*) maximum power at the driving axle (*ii*) total energy consumption and (*iii*) specific energy consumption. Assume that journey estimation is being made on simplified trapezoidal speed-time curve.

Solution : Effective weight of train,

$$W_e = \left(1 + \frac{10}{100}\right) W = 1.1\ W = 1.1 \times 100 = 110 \text{ tonnes}$$

Distance of run, $S = 2.5$ km

Average speed, $V_a = 50$ kmph

Duration of run $= \dfrac{3{,}600 \times S}{V_a} = \dfrac{3{,}600 \times 2.5}{50} = 180$ seconds

$$K = \frac{1}{2\alpha} + \frac{1}{2\beta} = \frac{1}{2 \times 1} + \frac{1}{2 \times 2} = 0.75$$

Maximum speed, $V_m = \dfrac{T}{2K} - \sqrt{\left(\dfrac{T}{2K}\right)^2 - \dfrac{3{,}600\ S}{K}}$

$$= \frac{180}{2 \times 0.75} - \sqrt{\left(\frac{180}{2 \times 0.75}\right)^2 - \frac{3{,}600 \times 2.5}{0.75}}$$

$= 71$ kmph

Acceleration period, $t_1 = \dfrac{V_m}{\alpha} = \dfrac{71}{1} = 71$ seconds

Braking period, $t_3 = \dfrac{V_m}{\beta} = \dfrac{71}{2} = 35.5$ seconds

Time of free run, $t_2 = 180 - (71 + 35.5) = 73.5$ seconds

Tractive effort required,

$F_t = 277.8\ W_e\alpha + 98.1\ WG + Wr$

$= 277.8 \times 110 \times 1 + 98.1 \times 100 \times 1 + 100 \times 40 = 44{,}368$ N

(*i*) Maximum power at the driving axle

$= \dfrac{F_t \times V_m}{3{,}600} = \dfrac{44{,}368 \times 71}{3{,}600} = 875$ kW **Ans.**

Total energy output

$$= \frac{1}{2} F_t \times \frac{V_m}{3{,}600} \times \frac{t_1}{3{,}600} + \frac{F'_t V_m}{3{,}600} \times \frac{t_2}{3{,}600}$$

where F'_t is effort required during free run and is equal to $(98.1WG + Wr)$

i.e., $98.1 \times 100 \times 1 + 100 \times 40 = 13{,}810$ N

∴ Total energy output

$$= \frac{1}{2} \times 44{,}368 \times \frac{71}{3{,}600} \times \frac{71}{3{,}600} + 13{,}810 \times \frac{71}{3{,}600} \times \frac{73.5}{3{,}600}$$

$= 14.19$ kWh

(*ii*) Total energy consumption $= \dfrac{14.19}{0.6} = 23.65$ kWh **Ans.**

(*iii*) Specific energy consumption

$$= \frac{\text{Total energy consumption in watt-hours}}{\text{Weight of train in tonnes} \times \text{distance of run in km}}$$

$= \dfrac{23.65 \times 1{,}000}{100 \times 2.5} = 94.6$ Wh/tonne-km **Ans.**

Example 11.19. A train weighing 203 tonnes accelerates uniformly from rest to a speed of 45 kmph up a gradient of 1 in 500, the time taken being 30 seconds. The power is then cut off and train coasts down a uniform gradient of 1 in 1,000 for a period of 40 seconds when brakes are applied for a period of 15 seconds so as to bring the train uniformly to rest on this gradient. Calculate (*i*) the maximum power output from the driving axles (*ii*) the energy taken from the conductor rails in kWh, assuming an efficiency of 60%.

Assume train resistance to be 44 newtons per tonne at all speeds, and allow 10% for rotational inertia.

[*Agra Univ. Traction & Utilization of Elec. Power 1978*]

Solution : Weight of train, $W = 203$ tonnes

Effective weight of train, $W_e = 1.1\ W = 1.1 \times 203$

$= 223.3$ tonnes

The train resistance, $r = 44$ newtons/tonne

Gradient, $G = \dfrac{1}{500} \times 100 = 0.2\%$

Maximum speed, $V_m = 45$ kmph

Acceleration period, $t_1 = 30$ seconds

Acceleration, $\alpha = \dfrac{V_m}{t_1} = \dfrac{45}{30} = 1.5$ kmphps

Tractive effort required,

$F_t = 277.8\ W_e\alpha + 98.1\ WG + Wr$

$= 277.8 \times 223.3 \times 1.5 + 98.1 \times 203 \times 0.2 + 203 \times 44$

$= 105{,}964$ newtons

(*i*) The maximum power output from driving axles

$$= \frac{F_t \times V_m}{3,600} = \frac{105,964}{3,600} \times 45 = 1,324.55 \text{ kW} \quad \textbf{Ans.}$$

Total energy required for the run

= Energy required during acceleration as there is no free run

$$= \frac{1}{2}\frac{F_t V_m}{3,600} \times \frac{t_1}{3,600} \text{ kWh} = \frac{1}{2} \times \frac{105,964 \times 45}{3,600} \times \frac{30}{3,600}$$

$$= 5.52 \text{ kWh}$$

(*ii*) The energy taken from conductor rails $= \frac{5.52}{0.6} = 9.2$ kWh **Ans.**

Example 11.20. It is proposed to put an electric trolley service in a city. The schedule speed is to be 45 kmph. The distance between stops is 2.8 km. The track is assumed to be level. Each stop is of 30 seconds duration. Using simplified speed-time curve, calculate the maximum speed assuming the acceleration to be 2 kmphps and retardation to be 3.2 kmphps. The dead weight of the car is 16 tonnes, rotational inertia is 10 per cent of the dead weight and track resistance is 40 newtons per tonne. If the overall efficiency is 80%, calculate (*a*) maximum power output from driving axles (*b*) the specific energy consumption in watt-hours per tonne-km.

Solution : Schedule speed, V_s = 45 kmph

$$\text{Schedule time of run, } T_s = \frac{S \times 3,600}{V_s} = \frac{2.8 \times 3,600}{45}$$

$$= 224 \text{ seconds}$$

Actual time of run, T = 224 – 30 = 194 seconds

$$K = \frac{1}{2\alpha} + \frac{1}{2\beta} = \frac{1}{4} + \frac{1}{6.4} = \frac{13}{32} = 0.40625$$

Maximum speed,

$$V_m = \frac{T}{2K} - \sqrt{\left(\frac{T}{2K}\right)^2 - \frac{3,600\,S}{K}}$$

$$= \frac{194}{2 \times 0.40625} - \sqrt{\left(\frac{194}{2 \times 0.40625}\right)^2 - \frac{3,600 \times 2.8}{0.40625}}$$

= 58 kmph **Ans.**

$$\text{Acceleration time, } t_1 = \frac{V_m}{\alpha} = \frac{58}{2} = 29 \text{ seconds}$$

$$\text{Duration of braking, } t_3 = \frac{V_m}{\beta} = \frac{58}{3.2} = 18.125 \text{ seconds}$$

Time of free run, $t_2 = T - (t_1 + t_3)$

$= 194 - (29 + 18.125) = 146.875$ seconds

Tractive effort during acceleration,

$F_t = 277.8\, W_e \alpha + Wr$

$= 277.8 \times 16 \times 1.1 \times 2 + 16 \times 40 = 10{,}418.56$ Newtons

(*a*) Maximum power output

$$= \frac{F_t V_m}{3,600} \text{ kW} = \frac{10,418.56 \times 58}{3,600} = 167.855 \text{ kW} \quad \textbf{Ans.}$$

(*b*) Distance travelled during braking period

$$= \frac{1}{2}\frac{V_m \times t_3}{3,600} = \frac{1}{2} \times \frac{58 \times 18.125}{3,600} = 0.146 \text{ km}$$

Distance travelled with power on S_1 = 2.8 – 0.146 = 2.654 km

Specific energy output

$$= \frac{0.01072\, V_m^2}{S} \times \frac{W_e}{W} + 0.2778 r \frac{S_1}{S} \text{ Wh/tonne-km}$$

$$= \frac{0.01072 \times 58^2}{2.8} \times 1.1 + 0.2778 \times 40 \times \frac{2.654}{2.8}$$

= 24.7 Wh/tonne-km

$$\text{Specific energy consumption} = \frac{24.7}{0.8} = 30.9 \text{ Wh/tonne-km} \quad \textbf{Ans.}$$

Example 11.21. An electric train weighing 300 tonnes runs 1% upgradient with following speed-time curve:

(*i*) Uniform acceleration of 1.5 kmphps for 30 seconds

(*ii*) Constant speed for 40 seconds

(*iii*) Coasting for 30 seconds

(*iv*) Braking at 2.5 kmphps

Calculate the specific energy consumption if the tractive effort is 45 N/tonne, rotational inertia effect 10% and overall efficiency of transmission and motor 75%.

[*Utilization of Electrical Power Rajasthan Univ. 2009; Rajasthan Technical Univ. 2010*]

Solution : Speed at the end of accelerating period,

$V_1 = \alpha t_1 = 1.5 \times 30 = 45$ kmph

Let the coasting retardation be β_c, then

$F_t = 277.8\, W_e(-\beta_c) + 98.1\, WG + Wr$

or $0 = -277.8 \times 1.1\, W \times \beta_c + 98.1 \times 1\, W + 45\, W$

$$\text{or } \beta_c = \frac{143.1}{277.8 \times 1.1} = 0.4683 \text{ kmphps}$$

Speed at the end of coasting period,

$V_2 = V_1 - \beta_c t_3 = 45 - 0.4683 \times 30 = 30.95$ kmph

$$\text{Braking duration, } t_4 = \frac{V_2}{\beta} = \frac{30.95}{2.5} = 12.38 \text{ seconds}$$

Total distance travelled,

$$S = \frac{V_1 t_1}{7,200} + \frac{V_1 t_2}{3,600} + \frac{V_1 + V_2}{7,200} \times t_3 + \frac{V_2 t_4}{7,200}$$

$$= \frac{45 \times 30}{7,200} + \frac{45 \times 40}{3,600} + \frac{45 + 30.95}{7,200} \times 30 + \frac{30.95 \times 12.38}{7,200}$$

$= 0.1875 + 0.5 + 0.3165 + 0.0532 = 1.0572$ km

Distance travelled when power is on,

$$S_1 = \frac{45 \times 30}{7,200} + \frac{45 \times 40}{3,600} = 0.1875 + 0.5 = 0.6875 \text{ km}$$

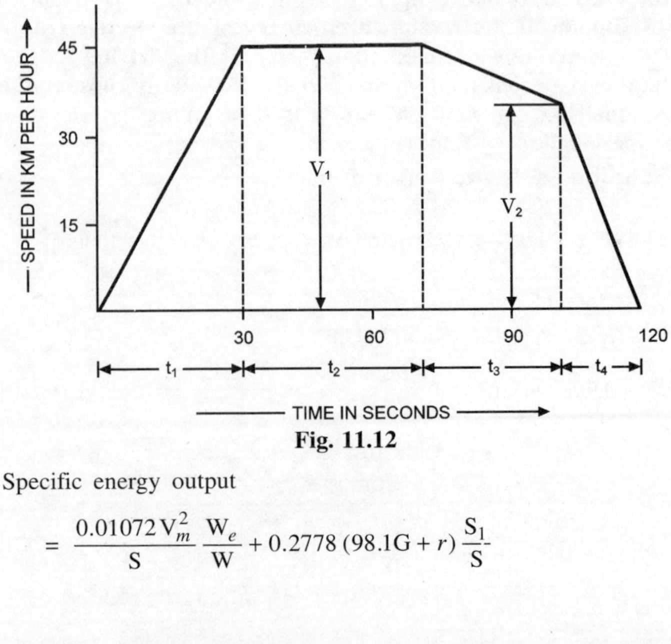

Fig. 11.12

Specific energy output

$$= \frac{0.01072\, V_m^2}{S}\frac{W_e}{W} + 0.2778\,(98.1G + r)\frac{S_1}{S}$$

$$= \frac{0.01072 \times 45^2 \times 1.1}{1.0572} + 0.2778\,(98.1 \times 1 + 45) \times \frac{0.6875}{1.0572}$$

$= 22.587 + 25.852 = 48.44$ Wh/tonne-km

Specific energy consumption $= \frac{48.44}{0.75} = 64.587$ Wh/tonne-km **Ans.**

Example 11.22. An electric train weighing 500 tonnes climbs up gradient with G = 8 and following speed-time curve.

(*i*) uniform acceleration of 2.5 km/hr/sec for 60 seconds.

(*ii*) Constant speed for 5 minutes.

(*iii*) Coasting speed for 3 minutes.

(*iv*) Dynamic braking at 3 kmphps to rest.

The train resistance is 25 N/tonne, rotational inertia effect 10% and combined efficiency of transmission and motor is 80%. Calculate the specific energy consumption.

[*U.P. Technical Univ. Utilization of Electrical Energy & Traction 2009-10*]

Solution : Speed at the end of accelerating period,

$V_1 = \alpha t_1 = 2.5 \times 60 = 150$ kmph

Let the coasting retardation be β_c, then

$F_t = 277.8\, W_e\, (-\beta_c) + 98.1 WG + Wr$

$$\text{or } 0 = -277.8 \times 1.1\, W \times \beta_c + 98.1\, W \times \frac{8}{1{,}000} \times 100 + 25\, W$$

$$\text{or } \beta_c = \frac{103.48}{277.8 \times 1.1} = 0.3386 \text{ kmphps}$$

Speed at the end of coasting period,

$V_2 = V_1 - \beta_c \times t_3 = 150 - 0.3386 \times 3 \times 60 = 89.052$ kmph

Braking retardation,

$$t_4 = \frac{V_2}{\beta} = \frac{89.052}{3} = 29.684 \text{ seconds}$$

Total distance travelled,

$$S = \frac{V_1 t_1}{7{,}200} + \frac{V_1 t_2}{3{,}600} + \frac{V_1 + V_2}{7{,}200} \times t_3 + \frac{V_2 t_4}{7{,}200}$$

$$= \frac{150 \times 60}{7{,}200} + \frac{150 \times 5 \times 60}{3{,}600} + \frac{150 + 89.052}{7{,}200} \times 3 \times 60 + \frac{89.052 \times 29.684}{7{,}200}$$

$= 1.25 + 12.5 + 5.9763 + 0.3671 = 20.0934$ km

Distance travelled when power is on,

$$S_1 = \frac{150 \times 60}{7{,}200} + \frac{150 \times 5 \times 60}{3{,}600} = 1.25 + 12.5 = 13.75 \text{ km}$$

Specific energy output

$$= \frac{0.01072\, V_m^2}{S} \times \frac{W_e}{W} + 0.2778(98.1G + r)\frac{S_1}{S}$$

$$= \frac{0.01072 \times 150^2}{20.0934} \times 1.1 + 0.2778\left(98.1 \times \frac{8}{1{,}000} \times 100 + 25\right) \times \frac{13.75}{20.0934}$$

$= 13.204 + 19.67 = 32.874$ Wh per tonne-km

Specific energy consumption

$$= \frac{32.874}{0.8} = 41.09 \text{ Wh/tonne-km} \quad \textbf{Ans.}$$

11.12. DEAD WEIGHT, ACCELERATING WEIGHT AND ADHESIVE WEIGHT

Dead Weight. The total weight of locomotive and train to be pulled by the locomotive is known as *dead weight*.

Accelerating Weight. The dead weight of the train *i.e.*, the weight of locomotive and train can be considered to be divided into two parts.

1. The weight, which requires angular acceleration such as weight of wheels, axles, gears etc. and
2. The weight, which requires linear acceleration.

Hence the effective weight, which is greater than dead weight is called the *accelerating weight*. Accelerating weight is taken 5 to 10 per cent more than dead weight.

Adhesive Weight. The total weight to be carried on the driving wheels is known as the *adhesive weight*.

Example 11.23. 400 tonne goods train is to be hauled by a locomotive up a gradient of 2% with acceleration of 1 kmphps. Coefficient of adhesion is 20%, track resistance 40 N/tonne and effective rotating masses 10% of the dead weight. Find the weight of the locomotive and number of axles if the axle load is not to increase beyond 22 tonnes.

[*Agra Univ. Traction & Utilization of Elec. Power 1979*]

Solution : Let the weight of the locomotive be W_L tonnes

Dead weight of train and locomotive, $W = (400 + W_L)$ tonnes

Gradient, $G = 2$ per cent

Coefficient of adhesion, $\mu = 0.2$

Effective weight of train and locomotive, $W_e = 1.1\,(400 + W_L)$

Tractive effort,

$F_t = 277.8\, W_e\, \alpha + 98.1\, WG + Wr$

$$= \left[277.8 \frac{W_e}{W} \alpha + 98.1\, G + r\right] W$$

$= [277.8 \times 1.1 \times 1 + 98.1 \times 2 + 40]$ W

$= 541.78$ W newtons

$$= \frac{541.78\, W}{9.81 \times 1{,}000} = 0.05523 \text{ W tonnes}$$

$= 0.05523\,(400 + W_L)$ tonnes ...(*i*)

Maximum value of tractive effort that can be possible with W_L adhesive weight of locomotive

$F_t = \mu\, W_L = 0.2\, W_L$ ...(*ii*)

Comparing Eqs. (*i*) and (*ii*) we get

$0.2\, W_L = 0.05523\,(400 + W_L)$

$$\text{or } W_L = \frac{0.05523 \times 400}{0.2 - 0.05523} = \frac{22.092}{0.14477} = 152.6 \text{ tonnes} \quad \textbf{Ans.}$$

$$\text{Number of axles required} = \frac{W_L}{\text{Allowable load}} = \frac{152.6}{22} = 7 \quad \textbf{Ans.}$$

Example 11.24. An electric locomotive is required to haul a train of 12 coaches each weighing 30 tonnes on the main line service requiring an initial acceleration of 0.8 kmphps up a gradient of 1 in 100. Estimate the adhesive weight and hence the number of driving axles the locomotive must have if the permissible axle loading is 20 tonnes per axle assuming rotational inertia to be 4% for the coaches and 15% for the locomotive. Maximum coefficient of adhesion is 0.2 and the tractive resistance 5 kg/tonne. [*B.H.U. Elec. Traction & Utilization 1978*]

Solution : Dead weight of train,

$W = 12 \times 30 = 360$ tonnes

Accelerating weight of train,

$W_e = 1.04 \times 360 = 374.4$ tonnes

Let adhesive weight of locomotive $= W_L$

Accelerating weight of locomotive $= 1.15\, W_L$

Tractive resistance, $r = 5$ kg per tonne $= 5 \times 9.81 = 49.05$ N/tonne

Coefficient of adhesion, $\mu = 0.2$

Tractive effort,

$F_t = (277.8\ \alpha)(W_e + 1.15\ W_L) + (98.1\ G + r)(W + W_L)$

$= (277.8 \times 0.8)(374.4 + 1.15\ W_L) + (98.1 + 49.05)(360 + W_L)$ newtons

$= \dfrac{222.24\,(374.4 + 1.15\,W_L) + 147.15\,(360 + W_L)}{1{,}000 \times 9.81}$ tonnes

$= (13.882 + 0.041\ W_L)$ tonnes ...(*i*)

Maximum value of tractive effort that can be possible with W_L adhesive weight of locomotive

$= \mu W_L = 0.2\ W_L$ ...(*ii*)

Equating Eqs. (*i*) and (*ii*), we get

$0.2\ W_L = 13.882 + 0.041\ W_L$

or $W_L = \dfrac{13.882}{0.159} = 87.3$ tonnes

Number of driving axles$= \dfrac{87.3}{20} = 4.37 \simeq 5$ **Ans.**

Example 11.25. A locomotive of 100 tonnes can just accelerate train of 500 tonnes with an acceleration of 1 kmphps up gradient of 1 %. Adhesive weight of locomotive is 70% of total dead weight, tractive resistance 45 newtons/tonne and inertia 10%. If this locomotive is helped by another locomotive of 130 tonnes with 100% adhesive weight, find out :

(*i*) Trailing weight that can be hauled up the same gradient under the same conditions.

(*ii*) The maximum gradient, trailing hauled load remaining unchanged.

[*U.P. Technical Univ. Utilization of Electrical Energy & Traction 2009-10*]

Solution : Dead weight of train and locomotive,

$W = 100 + 500 = 600$ tonnes

Effective weight,

$W_e = 1.1\ W = 1.1 \times 600 = 660$ tonnes

Tractive effort required,

$F_t = 277.8 W_e \alpha + 98.1 WG + Wr$

$= 277.8 \times 660 \times 1 + 98.1 \times 600 \times 1 + 600 \times 45$

$= 183{,}348 + 58{,}860 + 27{,}000 = 269{,}208$ newtons ...(*i*)

Maximum tractive effort for the first locomotive

$= \mu W_L \times \dfrac{\text{Adhesive weight}}{\text{Dead weight}}$

$= \mu \times 100 \times 1{,}000 \times 9.81 \times 0.7 = 686{,}700\ \mu$ newtons ...(*ii*)

Equating Eqs. (*i*) and (*ii*), we have

$686{,}700\ \mu = 269{,}208$

or $\mu = \dfrac{269{,}208}{686{,}700} = 0.392$

Adhesive weight of second locomotive

$= 130 \times 1 = 130$ tonnes

Total adhesive weight with both locomotives

$= 100 \times 0.7 + 130 = 200$ tonnes

Total tractive effort developed

$= \mu \times$ adhesive weight of both locomotive

$= 0.392 \times 200$

$= 78.4$ tonnes or $78.4 \times 9.81 \times 1{,}000$ *i.e.*, 769,104 newtons ...(*iii*)

Let the trailing weight that can be hauled, up the same gradient and under the same condition be W_1 tonnes

Now the total dead weight,

$W' = (100 + 130 + W_1) = (230 + W_1)$ tonnes

Tractive effort required,

$F'_t = (277.8 \times 1.1 \times 1 + 98.1 \times 1 + 45)W'$

$= 448.68\ W'$ ...(*iv*)

Equating Eqs. (*iii*) and (*iv*), we have

$448.68W' = 769{,}104$

or $W' = \dfrac{769{,}104}{448.68} = 1{,}714$ tonnes

or Trailing weight,

$W_1 = W' - 230 = 1{,}714 - 230 = 1{,}484$ tonnes **Ans.**

(*ii*) Total hauled load, $W'' = 100 + 500 + 130 = 730$ tonnes

Tractive effort required,

$F''_t = \left[277.8 \times \dfrac{W_e}{W''} \times \alpha + 98.1G + g\right] W''$

$= (277.8 \times 1.1 \times 1 + 98.1G + 45) \times 730$

$= (350.8 + 98.1G) \times 730$ newtons ...(*v*)

Equating Eqs. (*iii*) and (*v*), we have

$(350.8 + 98.1G) \times 730 = 769{,}104$

or $G = \left(\dfrac{769{,}104}{730} - 350.8\right) \times \dfrac{1}{98.1} = 7.164\%$ **Ans.**

Example 11.26. Determine acceleration in Example 11.25 if gradient and hauled load remains the same.

Solution : Tractive effort required,

$F''_t = \left[277.8 \times \dfrac{W_e}{W''} \times \alpha' + 98.1G + g\right] W''$

$= [277.8 \times 1.1 \times \alpha' + 98.1 \times 1 + 45] \times 730$

$= [305.58\alpha' + 143.1] \times 730$ newtons ...(*vi*)

Equating Eqs. (*iii*) and (*vi*), we have

$(305.58\alpha' + 143.1) \times 730 = 769{,}104$

or $\alpha' = \left(\dfrac{769{,}104}{730} - 143.1\right) \times \dfrac{1}{305.58} = 2.98$ kmphps **Ans.**

Example 11.27. A train with a locomotive having 4 motors has a total mass of 250 tonnes. Starting from rest the train attains a speed of 40 kmph in 20 seconds on a 1% up gradient. The gear ratio is 3, the gear efficiency 95%, the wheel diameter 95 cm, train resistance (average) is 40 N per tonne and rotational inertia is 10%. Find the torque developed by each of the motors and the minimum weight of the locomotive, given the adhesive coefficient is 0.25. [*Pb. Univ. Electric Drives and Utilization June 1991*]

Solution : Total mass of train and locomotive, W = 250 tonnes

Diameter of driving wheels, D = 95 cm = 0.95 m

Percentage gradient, G = 1%

Train resistance, r = 40 N/tonne

Gear ratio, $\gamma = 3$

Gear efficiency, η = 95% or 0.95

Equivalent accelerating weight of train,

$W_e = \dfrac{100 + 10}{100} \times 250 = 275$ tonnes

Time taken in accelerating the train to a speed of 40 kmph,

$t_1 = 20$ seconds

Acceleration, $\alpha = \dfrac{V_m}{t_1} = \dfrac{40}{20} = 2$ kmphps

Tractive effort required,

$F_t = 277.8\ W_e\ \alpha + (98.1\ G + r)W$

$= 277.8 \times 275 \times 2 + (98.1 \times 1 + 40) \times 250$

$= 1{,}87{,}315\ \text{N} = \dfrac{1{,}87{,}315}{9.81 \times 1{,}000} = 19.0943$ tonnes

Total torque developed,

$$T = \frac{F_t \times D}{\eta \times 2\gamma} = \frac{1{,}87{,}315 \times 0.95}{0.95 \times 2 \times 3} = 31{,}220 \text{ Nm}$$

Torque developed by each motor

$$= \frac{31{,}220}{4} = 7{,}805 \text{ Nm} \quad \textbf{Ans.}$$

Coefficient of adhesion, $\mu = 0.25$

$$\text{Adhesive weight, } W_L = \frac{F_t}{\mu} = \frac{19.0943}{0.25} = 76.4 \text{ tonnes}$$

Assuming an adhesive weight of locomotive to be 75% of its dead weight, we have

$$\text{Dead weight of locomotive} = \frac{W_L}{0.75} = \frac{76.4}{0.75} = 102 \text{ tonnes} \quad \textbf{Ans.}$$

Exercises

1. What do you understand by speech-time curves ? What is its use in practice ? **[Gorakhpur Univ. Utilization of Electrical Power & Traction 1983]**
2. Draw the speed-time curves for each of the following traction systems:
   (*i*) Tramcar (*ii*) Diesel electric (*iii*) Trolley buses (*iv*) Suburban electric train and (*v*) Urban services.
3. What is a speed-time curve ? Discuss briefly the distinct periods in a typical speed-time curve of a train running on main line. **[Pb. Univ. Electric Traction Dec. 1991]**
4. Draw a typical speed-time curve for train movement, and explain (*a*) acceleration (*b*) free running (*c*) coasting (*d*) retardation.
   Draw the speed-time curve for urban service and suburban service train movement and simplify these curves by approximation.
5. Draw and explain general speed-time curve of a train running between two stations. How can this curve be approximated for (*a*) main line service (*b*) suburban service ? **[U.P. Technical Univ. Utilization of Electrical Energy & Traction 2009-10]**
6. Sketch the speed-time curves for urban, suburban and main line services. Give the approximate values of acceleration, braking and coasting periods in each case. **[Rajasthan Univ. Utilization of Electrical Power 2004, 2005]**
7. Discuss the main features of the various train services. What type of train services corresponds to trapezoidal and quadrilateral speed-time curves ? **[M.D. Univ. Utilization of Electric Power & Traction Dec. 2008; May 2009; G.B. Technical Univ. Utilization of Electrical Energy & Traction 2010-11]**
8. Draw and explain a typical speed-time curve for an electric train and explain what do you understand by crest speed, average speed and schedule speed. **[Agra Univ. Traction & Utilization of Electric Power 1982; M.D. Univ. Utilization of Electric Power & Traction Dec. 2009]**
9. Define (*i*) crest speed, (*ii*) average speed, (*iii*) schedule speed. **[Rajasthan Technical Univ. Utilization of Electrical Power 2010]**
10. Draw a main line speed-time curve and label its various parts. Derive an expression for maximum speed of a suburban train in terms of total distance travelled, acceleration, retardation and average speed. Assume the speed-time curve to be trapezoidal. **[Utilization of Electrical Power Rajasthan Univ. 2009; Rajasthan Technical Univ. 2010]**
11. Define 'crest speed' and 'schedule speed' and discuss the factors which affect the schedule speed of a train. **[B.H.U. Elec. Traction & Utilization 1978]**
12. The speed-time curve of a locomotive going down gradient is represented by a quadrilateral speed-time curve. The locomotive is accelerated to V_1 speed in time t_1 seconds coasted for V_2 speed in t_2 seconds and braked to standstill in t_3 seconds. Obtain expressions for total distance covered and speed V_2. Choose acceleration during starting and coasting as α and β_c and retardation during braking as β. **[U.P. Technical Univ. Utilization of Electrical Energy and Electric Traction 2006-07]**
13. Define the terms crest speed, average speed and schedule speed. What are different methods of approximation of speed-time curves ? Derive expressions for distance travelled using quadrilateral approximation method of $V(t)$ curve. **[Gorakhpur Univ. Utilization of Electrical Power & Traction 1984]**
14. Explain speed-time curve for a main line service and derive the expression
    $\frac{1}{2}\left[\frac{1}{\alpha}+\frac{1}{\beta}\right] = \frac{3{,}600\text{ D}}{V_m^2}\left[\frac{V_m}{V_a}-1\right]$ where the symbols have their usual meanings. **[Pb. Univ. Electric Traction December 1992]**
15. Explain mechanics of train movement. **[Rajasthan Univ. Utilization of Electric Power 2003]**
16. Derive an expression for tractive effort required for propulsion of a train during (*a*) acceleration (*b*) free running (*c*) coasting (*d*) moving down a gradient. **[W.B. Univ. of Technology Utilization of Electric Power 2009]**
17. Derive an expression for :
    (*i*) The tractive effort for propulsion of a train on level track.
    (*ii*) The tractive effort for propulsion of a train up and down a gradient.
    [G.B. Technical Univ. Utilization of Electrical Energy & Traction 2010-11]
18. Explain various functions that the tractive effort exerted by traction unit is supposed to perform and derive an expression for total tractive effort. **[U.P. Technical Univ. Utilization of Electrical Energy & Traction 2009-10]**
19. Deduce the expressions for:
    (*i*) The tractive effort exerted by road wheel in terms of wheel diameter, motor torque, gear ratio and efficiency of transmission of power through gears.
    (*ii*) The tractive effort for the propulsion of train up and down the gradient.
    [Gorakhpur Univ. Utilization of Electrical Power & Traction 1982]
20. What is coefficient of adhesion ? How does it affect slipping of the driving wheels of the traction unit ? **[M.D. Univ. Utilization of Electric Power & Traction May 2010]**
21. Comment on the following statements:
    (*i*) The adhesion coefficient of a locomotive employing dc shunt motor will be higher than the using dc series motor.
    (*ii*) A higher ratio of adhesive weight to total weight is desirable for suburban trains, where the distance between the consecutive stops is very small.
22. Explain the terms (*i*) coefficient of adhesion (*ii*) tractive effort and (*iii*) gradient. **[Pb. Univ. Electric Traction December 1992]**

23. What do you mean by specific energy consumption ? What are the factors that affect specific energy consumption of a train ?
[W.B. Univ. of Technology Utilization of Electric Power 2009]

24. What are the factors which affects the specific energy consumption in electric trains ?
[Utilization of Electric Power Rajasthan Univ. 2000, 2004, 2006; Rajasthan Technical Univ. 2010]

25. Discuss various factors affecting specific energy consumption in the traction.
[G.B. Technical Univ. Utilization of Electrical Energy & Traction 2010-11]

26. Derive equation of motion of a vehicle. How does the effective mass of a vehicle differ with its physical (actual) mass ? Explain significance of equation of motion in energy consumption calculations.
[Gorakhpur Univ. Utilization of Electrical Power & Traction September 1984]

27. Derive a simple expression for the maximum power output of traction motors in terms of tractive effort, maximum speed and efficiency of transmission gear.

28. Derive an expression for specific energy consumption of a suburban train for a typical inter-station run.
[B.H.U. Elec. Traction & Utilization 1978]

29. Differentiate between adhesive weight and total weight of an electric locomotive.
[Gorakhpur Univ. Utilization of Electrical Power & Traction 1983]

30. Explain the terms 'dead weight', 'effective weight' and 'adhesive weight' in a locomotive.
[B.H.U. Elec. Traction & Utilization 1978]

31. Define (*i*) adhesive weight (*ii*) coefficient of adhesion (*iii*) Tractive effort. **[Rajasthan Technical Univ. Utilization of Electrical Power 2008, 2010]**

32. Write short notes on the following
(*i*) Speed-time curve.
(*ii*) Mechanics of train movement.
[Rajasthan Univ. 2003; M.D. Univ. May 2009; Dec. 2009]
(*iii*) Tractive effort for propulsion of train.
(*iv*) Factors affecting energy consumption in propelling a train.
(*v*) Specific energy consumption. **[M.D. Univ. May 2010]**
(*vi*) Dead weight, accelerating weight and adhesive weight of a train.
[Agra Univ. Traction & Utilization of Elec. Power 1982]

Short Answer Type Questions With Answers

Q. 1. What is meant by speed-time curve ?
Ans. The curve drawn between speed and time taking speed on the Y-axis and time on the X-axis is known as speed-time curve.

Q. 2. Define terms 'coasting' and 'braking'. **[U.P.T.U. 2006-07]**
Ans. ***Coasting.*** At the end of free running period (*i.e.,* at t_3) power supply is cut off and the train is allowed to run under its own momentum. The speed of train starts decreasing on account of resistance to the motion of train. The rate of decrease of speed during coasting period is known as *coasting retardation.*

Braking. At the end of coasting period (*i.e.,* at t_4) the brakes are applied to bring the train to rest. During this period speed decreases rapidly and finally reduces to zero.

Q. 3. What is meant by crest speed ?
Ans. The maximum speed attained by the vehicle during the run is known as crest speed.

Q. 4. What is meant by schedule speed of a train ?
[Gorakhpur Univ. 1975]
Ans. The ratio of distance covered between two stops and total time of run including time of stop is known as schedule speed.

Q. 5. What are those factors which affect the average speed of a train ?
Ans. The average speed of a given train when running on a given service (*i.e.,* distance between two stops) is affected by the (*i*) acceleration and retardation and (*ii*) crest speed.

Q. 6. What are the factors which affect the schedule speed of a train ?
Ans. The schedule speed of a given train when running on a given service is affected by (*i*) acceleration and retardation (*ii*) maximum or crest speed and (*iii*) time duration of stop.

Q. 7. How does increase in acceleration and braking retardation affect the schedule speed of a train for a given run with fixed crest speed ?
Ans. For a given run and with fixed crest speed the increase in acceleration will result in decrease in actual time of run, and therefore, increase in schedule speed. Similarly increase in braking retardation will affect the schedule speed.

Q. 8. How does increase in crest speed affect the schedule speed for a given run with fixed acceleration and retardation ?
Ans. For a given run with fixed acceleration and retardation the actual time of run will decrease, and therefore, schedule speed will increase with the increase in crest speed.

Q. 9. How does increase in duration of stop affect the schedule speed ?
Ans. For a given average speed, the schedule speed will increase with the decrease in duration of stop.

Q. 10. What is meant by coefficient of adhesion ?
Ans. The coefficient of adhesion is defined as the ratio of the tractive effort just necessary to slip the wheels on the track and adhesive weight.

Q. 11. How does coefficient of adhesion vary with the increase in speed ?
Ans. The coefficient of adhesion decreases with the increase in speed.

Q. 12. What is coefficient of adhesion ? Why a train driven by separately excited motors has better adhesion than a train driven by series dc motors ? **[U.P.T.U. 2005-06]**
Ans. Coefficient of adhesion (μ) provides a quantitative measure of the tendency of wheels to slip and is defined as the ratio of the tractive effort just necessary to slip the wheels on the track to the adhesive weight (weight on the driving axles).

The coefficient of adhesion depends on the nature of speed-torque characteristics of driving motor and has a higher value for motor with low regulation of speed *i.e.,* when for a given increase of speed drop in torque is large. When a wheel slips, the speed of the driving motor increases. The torque drops by a large amount in motors with low regulation of speed and the wheels regain their grip on rails immediately avoiding slip.

Since separately excited dc motors have low speed regulation than dc series motors, a train driven by separately excited dc motors has better adhesion than a train driven by dc series motors.

Q. 13. What do you understand by train resistance and on what factors does this depend ? **[Rajasthan Univ. 2006]**

Ans. ***Tractive Effort For Overcoming Train Resistance.*** Train resistance consists of all the forces resisting the motion of a train when it is running at uniform speed on a straight and level track. Under these circumstances the whole of the energy output from the driving axles is expended against train resistance. Train resistance is due to (*i*) the friction at the various parts of the rolling stock (*ii*) friction at the track and (*iii*) air resistance. The first two components constitute the *mechanical resistance* component of train resistance. The train resistance depends upon various factors, such as shape, size and condition of track etc., and is expressed in newtons per tonne of the dead weight. For a normal train the value of specific resistance has been 40 to 70 newtons/tonne. The general equation for train resistance is given as

$$R = k_1 + k_2V + k_3V^2$$

where k_1, k_2 and k_3 are constants depending upon the train and the track, R is the resistance in newtons and V is the speed in kmph. The first two terms represent the mechanical resistance and the last term represents air resistance.

Q. 14. What is meant by dead weight ?

Ans. The total weight of locomotive and train to be pulled by the locomotive is known as dead weight.

Q. 15. Why the accelerating weight of an electric train is greater than its dead weight ? **[Gorakhpur Univ. 1981]**

Ans. With the linear acceleration of the train, the rotating parts of the train such as wheels and motors also accelerate in an angular direction, and therefore, the tractive effort required is equal to the arithmetic sum of tractive effort required to have the angular acceleration of rotating parts and tractive effort required to have the linear acceleration. Hence the equivalent or accelerating weight of the train is taken higher than the dead weight.

Q. 16. What is meant by the term 'specific energy consumption' ?

Ans. Specific energy consumption is the total energy consumption of the train divided by the product of the weight of the train and distance of run.

Q. 17. What are those factors which affect the specific energy consumption of a train ?

Ans. The specific energy consumption of a train operating at a given schedule speed depends upon (*i*) distance between stops (*ii*) acceleration (*iii*) retardation (*iv*) crest speed (*v*) nature of route and (*vi*) the type of train equipment.

Q. 18. How is specific energy consumption affected by variation in distance between stops ?

Ans. Greater the distance between stops lesser will be the specific energy consumption.

Q. 19. How does the value of acceleration and retardation affect the specific energy consumption for a given run at a given schedule speed ?

Ans. For a given run at a given schedule speed, greater the value of acceleration and retardation more will be the period of coasting and, therefore, lesser the period during which power is on and, therefore specific energy consumption will accordingly be less.

Q. 20. What is meant by the term 'adhesive weight' ?

Ans. The total weight to be carried on the driving wheels is known as adhesive weight.

Problems

1. A train runs at an average speed of 50 kmph between stations situated 2.5 km apart. It accelerates at 2 kmphps and retards at 3 kmphps. Draw the speed-time curve for the run and find its maximum speed assuming a trapezoidal speed-time curve. **[Ans.** 57.75 kmph]

2. A train runs at an average speed of 45 kmph between stations 2.5 km apart. The train accelerates at 2 kmphps and retards at 3 kmphps. Find its maximum speed assuming a trapezoidal speed-time curve. Calculate also the distance travelled by it before the brakes are applied. Derive the formula used for calculation of maximum speed.
[B.H.U. Elec. Traction & Utilization 1978]
[Ans. 50.263 kmph; 2.383 km]

3. The speed-time curve of a train consists of :
(*i*) Uniform acceleration of 6 km/h/s for 25 s;
(*ii*) Free running for 10 minutes;
(*iii*) Uniform deceleration of 6 km/h/s to stop the train;
(*iv*) A stop of 5 minutes.
Find the distance between the stations, the average and schedule speeds. **[A.M.I.E. Sec. B. Electric Drives and Their Control Winter 1996]**
[Ans. 26.042 km; 144.23 kmph and 98.68 kmph]

4. A train runs with an average speed of 72 kmph. Distance between stops is 6 km. The train accelerates at 2 kmphps and retards at 3 kmphps. Determine the crest speed and schedule speed of the train. Duration of stops is 60 seconds. Assume trapezoidal speed-time curve. **[Ans.** 81.2 kmph; 60 kmph]

5. An electric train is to have a braking retardation of 3.2 kmphps. If the ratio of maximum speed to average speed is 1.3, the time for stops is 26 seconds, and acceleration is 0.8 kmphps, find its schedule speed for a run of 1.5 km. Assume trapezoidal speed-time curve.
[Agra Univ. Traction & Utilization of Elec. Power 1982]
[Ans. 30 kmph]

6. The schedule speed with a 200 tonne train on an electric railway with stations 777 metres apart is 27.3 kmph and the maximum speed is 20 per cent higher than the average running speed. The braking rate is 3.22 kmphps and the duration of stop is 20 seconds. Find the acceleration required. Assume a simplified speed-time curve with free running at the maximum speed.
[Agra Univ. Traction & Utilization of Elec. Power 1976 Supp.]
[Ans. 2.73 kmphps]

7. An electric train operating on a suburban service has a maximum speed of 60 kmph. The average distance between stops is 4 km and the schedule speed including a station stop of 30 seconds is 40 kmph. Assuming a simplified speed-time curve, calculate the necesary acceleration if the retardation is 4 kmphps. **[Ans.** 0.364 kmphps]

8. A train is required to run between two stations 2 km apart at an average speed of 40 km per hour. The run is to be made according to a simplified quadrilateral speed-time curve. If the maximum speed is to be limited to 60 km per hour, acceleration to 2 kmphps, coasting retardation to 0.15 kmphps and braking retardations 3 kmphps, determine the duration of accelerating, coasting and braking periods.
[Rajasthan Univ. Utilization of Electrical Power 2004]
[Ans. 30 seconds; 137 seconds, 13 seconds]

9. A 200 tonne motor coach with train has four motors each delivering a shaft torque of 300 N-m during acceleration from rest. Find out the time taken for the train to attain a speed of 40 kmph starting from rest on 1/200 gradient with a wheel diameter of 1,140 mm and gear ratio 4.5. Assume train resistance 0.5 kg/tonne, rotary inertia as 10% and gear efficiency as 95%.
[Gorakhpur Univ. Utilization of Elec. Power & Traction 1981]
[Ans. 23.7 seconds]

10. An electric train is accelerated uniformly from rest to a speed of 40 kmph, the period of acceleration being 20 seconds. If it coasts for 60 seconds against a constant resistance

of 50 N/tonne and is brought to rest in a further period of a 10 seconds by braking, determine (*i*) the acceleration (*ii*) the coasting retardation (*iii*) the braking retardation (*iv*) distance travelled and (*v*) the schedule speed with station stops of 10 second duration. Allow 10% for rotational inertia.

[Pb. Univ. Elec. Traction 1973]

[**Ans.** (*i*) 2 kmphps (*ii*) 0.1636 kmphps (*iii*) 3 kmphps (*iv*) 0.736 km and (*v*) 27.5 kmph]

11. An electric train weighing 300 tonnes is accelerated from rest up a gradient 1 in 200 at a uniform rate of 0.8 kmphps until a speed of 64 kmph is attained. Power is then cut off and the train coasts for 1 minute along level track. The brakes are now applied and brought to rest with retardation of 3.2 kmphps. Find specific energy consumption. Allow 10% rotational inertia and train resistance 5 kg/tonne.

[Gorakhpur Univ. Utilization of Elec. Power & Traction 1981]

[**Ans.** 46.5 Wh per tonne-km]

12. A 250 tonne train with 10% rotational inertia effect is started with uniform acceleration and reaches a speed of 50 kmph in 25 seconds on level road. Find the specific energy consumption if the journey is to be made according to a simplified trapezoidal speed-time curve, the acceleration is 2 kmphps, braking retardation 3 kmphps and the distance between the two stations is 2.4 km, efficiency of motors is 0.9, track resistance 5 kg/tonne.

[Pb. Univ. Elec. Traction April 1975]

[**Ans.** 28.3 Wh/tonne-km]

13. An electric train weighing 200 tonnes runs a uniform upgradient of 1% with the following speed-time curve:

(*i*) Uniform acceleration 2 kmphps for 30 seconds
(*ii*) Constant speed for 40 seconds
(*iii*) Coasting for 30 seconds
(*iv*) Braking at 2.5 kmphps to rest
(*v*) Stop at station 15 seconds.

If the tractive resistance is 40 N/tonne, rotational inertia effect 10% of dead weight and overall efficiency of transmission and motor is 75% determine:
(*i*) schedule speed (*ii*) specific energy consumption (*iii*) total energy consumption (*iv*) distance between two stations.

[Pb. Univ. Electric Traction June 1993]

[**Ans.** (*i*) 39.88 kmph (*ii*) 69.93 Wh/tonne-km (*iii*) 20.7 kWh (*iv*) 1.48 km]

14. An electric train weighing 200 tonnes accelerates uniformly on the level track to a speed of 43 kmph in 20 seconds. Power is then cut off and the train coasts until the speed is 34 kmph when the brakes are applied and the train is brought to rest in 10 seconds. Allowing a train resistance of 5 kg per tonne during acceleration and 6.5 kg per tonne during coasting, and 12% for rotational inertia, calculate (*i*) the maximum horse power developed at the driving axles, and (*ii*) the specific energy output.

[**Ans.** (*i*) 2,340 hp (metric); 37.4 Wh/tonne-km]

15. A 40 tonne train starts a gradient of 1 in 75 at the rate of 1.6 kmphps. The tractive resistance is 66.75 N per tonne and allowance for rotational inertia is 10%. Calculate:

(*a*) the energy in kWh usually employed in attaining a speed of 48 kmph from rest,

(*b*) the specific energy consumption in Wh per tonne-km when running at a steady speed of 56 kmph. up this gradient if the overall efficiency of equipment is 70%.

[Gorakhpur Univ. Utilization of Elec. Power & Traction Sept. 1982]

[**Ans.** (*a*) 2.18 kWh; 78.4 Wh/tonne-km]

16. An electric train weighing 400 tonnes has to maintain an average speed of 40 km per hour between two stations 3 km apart on an incline of 1 in 200. The train accelerates at 2 kmphps and retards at 4 kmphps. The tractive resistance is 5 kg per tonne, and allowance for rotational inertia 10%. Assuming a trapezoidal speed-time curve, find the energy consumption for the run while going up the incline. The overall efficiency may be taken as 65%. [**Ans.** 64.58 kWh]

17. What is meant by 'adhesive weight' of a locomotive? Calculate the weight of a locomotive to haul a train weighing 500 tonnes up an incline of 1 in 75 with an acceleration of 1 kmphps. The tractive resistance is 5 kg per tonne and allowance for rotational inertia 10%. Coefficient of adhesion 0.25. [**Ans.** 120 tonnes]

18. A 1,500 V dc locomotive is required to start up a train weighing 1,000 tonnes (including loco) on an up gradient of 1/200 and accelerates it at the rate of 0.5 kmphps, to a speed of 60 kmph. The train resistance is 7 kg/tonne and rotational inertia 8%. Making your own assumptions for any missing data, determine the minimum weight of the locomotive and its horse power.

[**Ans.** 213 tonnes; 47,333 hp (metric)]

19. Differentiate between adhesive and total weight of an electric locomotive. Determine the minimum adhesive weight of a loco required to start a 2,000 tonnes (inclusive of loco) on 1 : 150 gradient and accelerate it at 0.1 kmphps. Assume coefficient of adhesion as 0.25, train resistance 4 kg/tonne and rotary inertia as 8%.

[Gorakhpur Univ. Utilization of Elec. Power & Traction 1983]

[**Ans.** 110 tonnes]

20. A 500 tonne goods train is to be hauled by a locomotive up a gradient of 2% with an acceleration of 1.2 kmphps. Coefficient of adhesion is 25%, track resistance 40 N/tonne and effecting rotating masses 10% of dead weight. Find the weight of the locomotive and number of axles if the axle load is not to exceed 20 tonnes. [**Ans.** 160 tonnes; 8]

21. A locomotive of 100 tonnes can just accelerate a train of 400 tonnes with an acceleration of 1.5 kmphps up a gradient of 1 in 100. Assume an adhesive weight of locomotive 70% of total dead weight. The tractive resistance is 40 N/tonne and rotational inertia increases the dead weight by 10%. If the above locomotive is aided by another locomotive of 150 tonnes with 80% adhesive weight, determine:

(*i*) The additional trailing weight that can be hauled up the same gradient under the same conditions;
(*ii*) The maximum gradient, trailing hauled load remaining unchanged;
(*iii*) The acceleration, if both gradient and hauled load remain unchanged.

[**Ans.** (*i*) 1,106 tonnes, (*ii*) 7.6%, (*iii*) 3.62 kmphps]

22. An electric locomotive is required to haul a train of 12 coaches, each weighing 30 tonnes on the main line service requiring an initial acceleration of 0.8 kmphps up a gradient of 1 in 100. Estimate the adhesive weight needed and hence the number of driving axles the locomotive must have if the permissible axles load is 20 tonnes. Allowance for rotational inertia to be 4% for the coaches and 15% for the locomotive. Maximum coefficient of adhesion is 0.2 and the tractive resistance is 3.5 kg per tonne.

[Gorakhpur Univ. Utilization of Elec. Power & Traction 1980, 1982]

[**Ans.** 80 tonnes; 4]

❑❑❑

CHAPTER 12

Electric Traction Motors

INSIDE THIS CHAPTER

12.1. GENERAL FEATURES OF TRACTION MOTORS

The chief requirements of electric motors used for traction work are:

1. *Mechanical Features.* (*i*) A traction motor must be robust and capable to withstand continuous vibrations since service conditions are extremely severe.

(*ii*) The weight of the traction motor should be minimum in order to increase the payload capacity of the vehicle. This is achieved by using high speed motors, upper limit being fixed by excessive centrifugal stresses.

(*iii*) The traction motor is located underneath a motor coach (except in few electric locomotives). The space underneath a motor coach is limited by the size of driving wheels and the track gauge. The traction motor, therefore, must be small in overall dimensions specially in its overall diameter.

(*iv*) The traction motor must be totally enclosed type, particularly when mounted beneath the locomotive or motor coach, to provide protection against ingress of dirt, dust, water, mud etc.

Thus, for magnetic circuit of a traction motor cast iron, which cannot withstand continuous vibrations, is not suitable. Use of cast steel or fabricated steel, which gives more mechanical strength, is made in place of cast iron. Those parts of the motor, which are not highly stressed; must be made of pressed or fabricated steel plates and light alloys.

2. *Electrical Characteristics.* (*i*) *High starting torque*. A traction motor must be capable of developing high starting torque, specially when the train is to be accelerated at a reasonably high rate such as in case of urban or suburban services.

(*ii*) *Simple speed control.* The traction motor should be amenable to simple speed control methods as an electric train has to be started and stopped very often.

(*iii*) *Self-relieving property.* The speed-torque characteristics of the motor should be such that the speed may fall with the increase in load. The motors having such speed-torque characteristics are self protective against excessive overloading as power output of a motor is proportional to the product of torque and speed.

(*iv*) *Possibility of dynamic or regenerative braking.* The traction motor should be amenable to easy and simple methods of rheostatic and/or regenerative braking.

(*v*) *Capability of withstanding voltage fluctuations.* Traction motor should be capable of withstanding rapid fluctuations in supply voltage without undue effect on its performance since in traction work rapid voltage fluctuation owing to heavy current inrush at start, is common feature.

(*vi*) *Capability of withstanding temporary interruption of supply.* Traction motor should be capable of withstanding temporary interruptions of supply without undue rush of current, since it occurs at the instant of crossing over the cross-overs and section insulators.

(*vii*) *Overload capacity.* Traction motor should be capable of taking excessive load as it is subjected to very arduous duty.

(*viii*) *Parallel running.* In traction work, usually more than one motor (two or four motors per motor car) are required. Traction motors, therefore, should be of such speed-torque and current-torque characteristics that when they are operated in parallel and mechanically coupled, they share the load almost equally.

No motor meets all the requirements mentioned above. Most suitable motors for dc systems are the series and compound motors whereas for ac systems the single phase series motors and 3-phase induction motors are employed.

12.2. OPERATING CHARACTERISTICS OF DC MOTORS

The performance and, therefore, suitability of a dc motor is determined from its characteristics, known as *performance characteristics*. The important characteristics of dc motors are:

1. ***Torque-Armature Current Characteristic.*** This characteristic curve gives relation between torque developed in the armature, T and armature current I_a. This is also known as *electrical characteristic*.

2. ***Speed-Armature Current Characteristic.*** This characteristic curve gives relation between speed N and armature current I_a.

3. ***Speed-Torque Characteristic.*** This characteristic gives relation between speed N and torque developed in armature, T. This is also known as *mechanical characteristic*. This curve may be derived from the two characteristic curves mentioned above.

The important relations to be kept in mind while discussing motor characteristics are:

$$(i)\ I_a = \frac{V - E_b}{R_a} \quad (ii)\ N \alpha \frac{E_b}{\phi} \quad \text{and} \quad (iii)\ T \alpha \Phi I_a.$$

1. Torque developed by a dc motor is found as follows:

Let T_e be the electromagnetic torque developed in newton-metres by the motor running at n rps.

Power developed = Work done per second

$$= T_e\omega = T_e \times 2\pi n \text{ watts} \quad ...(12.1)$$

Electrical equivalent of mechanical power developed by the armature, also

$$= E_b I_a \text{ watts} \quad ...(12.2)$$

Comparing Eqs. (12.1) and (12.2), we have

$$T_e \times 2\pi n = E_b I_a$$

$$\text{or } T_e = \frac{E_b I_a}{2\pi n} = 0.159\frac{E_b I_a}{n} \text{ N-m} \quad ...(12.3)$$

$$\text{Also } T_e = \frac{E_b I_a}{2\pi \frac{N}{60}} = 9.55\frac{E_b I_a}{N} \text{ N-m} \quad ...(12.4)$$

where N is speed in rpm

Substituting $E_b = \Phi Z \frac{N}{60} \times \frac{P}{A}$ in Eq. (12.4), we have

$$T_e = 9.55 \times \Phi \times Z \times \frac{N}{60} \times \frac{P}{A} \times \frac{I_a}{N}$$

$$= 0.159\Phi ZP\frac{I_a}{A} \text{ newton-metres} \quad ...(12.5)$$

2. The relation between speed and armature current is found as follows :

As the back emf developed in a dc motor is nothing but a generated emf, therefore, it is given by the expression

$$E_b = \frac{\phi ZN}{60} \times \frac{P}{A} \text{ volts} \quad ...(12.6)$$

where ϕ is the flux per pole in webers, Z is the number of armature conductors, N is the speed of armature in rpm, P is the number of poles and A is the number of parallel paths.

$$\text{Back emf also} = V - I_a R_a \quad ...(12.7)$$

Comparing Eqs. (12.6) and (12.7) we get

$$\frac{\phi ZN}{60} \times \frac{P}{A} = V - I_a R_a$$

$$\text{or } N = \frac{V - I_a R_a}{\phi Z} \times \frac{60 A}{P} \quad ...(12.8)$$

or $N \propto \dfrac{V - I_a R_a}{\phi}$ since Z, P and A, for a particular machine, are constant.

or $N \propto \dfrac{V}{\phi}$ since $I_a R_a$ is very small as compared to applied voltage.

or $N \propto \dfrac{1}{\phi}$ if applied voltage V is constant.

3. The relation between speed and torque is determined as follows.

Torque developed, $T = 0.159\phi Z I_a \dfrac{P}{A}$ [Refer to Eq. (12.5)]

$$\text{or } \frac{A}{\phi ZP} = \frac{0.159 I_a}{T}$$

Substituting $\dfrac{A}{\phi ZP} = \dfrac{0.159 I_a}{T}$ in Eq. (12.8), we get

$$N = \frac{(V - I_a R_a) \times 60 \times 0.159 I_a}{T}$$

$$= \frac{9.55(V - I_a R_a) I_a}{T} \quad ...(12.9)$$

The operating characteristics of dc motors have already been dealt in chapter 1 (Art. 1.16.1)

Example 12.1. The torque-armature current characteristics of a series traction motor are given as:

Armature Current (Amp)	5	10	15	20	25	30	35	40
Torque (N-m)	20	50	100	155	215	290	360	430

The motor resistance is 0.3 Ω. If this motor is connected across 230-V dc supply deduce the speed-armature current characteristic.

Solution : Supply voltage, V = 230 V

Resistance of motor, $R_m = R_a + R_{se} = 0.3\ \Omega$

Armature current, I_a in amperes	5	10	15	20	25	30	35	40
Torque, T in N-m	20	50	100	155	215	290	360	430
Back emf, E_b = $(V - I_a R_m)$ in volts	228.5	227.0	225.5	224.0	222.5	221.0	219.5	218.0
Speed, N = $\frac{9.55(V - I_a R_m)}{T} \cdot I_a$ in RPM	545	434	323	276	247	218	204	194

The deduced speed-armature current characteristic is shown in Fig. 12.1.

Example 12.2. The following figures give the magnetisation curve of a dc series motor when working as a separately excited generator at 600 rpm:

Field Current (amperes)	**20**	**40**	**60**	**80**
EMF (volts)	**215**	**381**	**485**	**550**

The total resistance of the motor is 0.8 ohm. Deduce the speed-torque curve for this motor when operating at a constant voltage of 600 volts.

Solution : Voltage applied across the motor, V = 600 volts

Resistance of the motor, $R_m = (R_a + R_{se}) = 0.8\ \Omega$

Speed, $N_1 = 600$ rpm

Field current (amperes)	20	40	60	80
Back emf, E_1 (volts) at speed N_1 (600 rpm) = EMF generated by the armature (given)	215	381	485	550
Back emf, E (volts) at speed N (to be determined) = $V - I_a R_m$	584	568	552	536
Speed, N (to be determined in RPM) = $\frac{N_1}{E_1} \times (V - I_a R_m)$	1,630	895	683	585
Torque, T = $\frac{9.55(V - I_a R_m)I_a}{N}$ in N-m	68.4	240	462	700

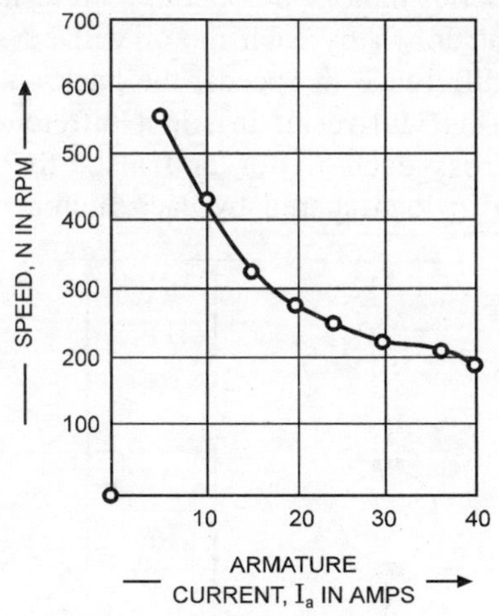

Fig. 12.1

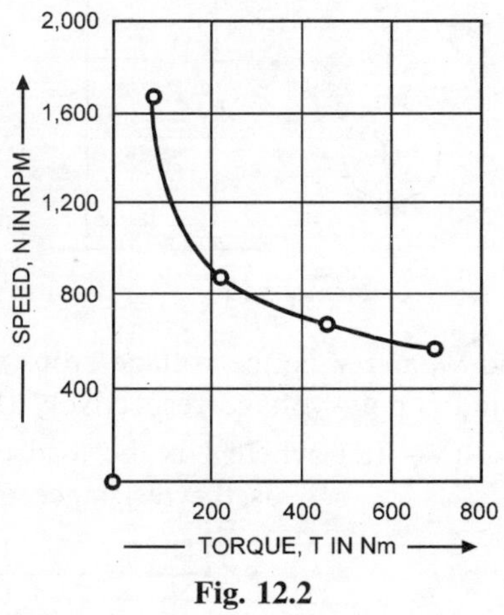

Fig. 12.2

The deduced speed-torque curve for the motor is shown in Fig. 12.2.

Example 12.3. A dc series motor drives a load. The torque of the motor varies as the square of the speed. The motor takes a current of 15 A when the speed is 640 rpm. Calculate the motor speed and current drawn when the field winding is shunted by a diverter of the same resistance as that of the field winding. Neglect all motor losses and assume that the magnetic circuit is unsaturated.

Solution : Speed, $N_1 = 640$ rpm

Series field current, $I_{se_1} = I_{ai} = I_1 = 15$ A

Let the speed be N_2 rpm, line current I_2 when the field winding is shunted by a diverter of the same resistance as that of field winding.

Series field current, $I_{se_2} = \frac{1}{2} I_2$

∵ because line current I_2 divides equally between series field and diverter, being of equal values.

Since torque developed, $T \propto \phi I_a \propto \phi I$ $\quad$ ∵ $I_a = I$

$$\therefore \quad \frac{T_2}{T_1} = \frac{\phi_2 I_2}{\phi_1 I_1} = \frac{\frac{1}{2} I_2^2}{I_1^2}$$

∵ $\phi \propto I_{se}$; magnetic circuit is unsaturated

$$\text{or} \quad \frac{T_2}{T_1} = \frac{I_2^2}{2 I_1^2} \quad \text{...}(i)$$

$$\text{Also} \quad \frac{T_2}{T_1} = \frac{N_2^2}{N_1^2} \quad \text{...}(ii) \quad ∵ \text{ Given that } T \propto N^2$$

Comparing Eqs. (*i*) and (*ii*) we have

$$\frac{N_2^2}{N_1^2} = \frac{I_2^2}{2 I_1^2}$$

$$\text{or} \quad \frac{N_2}{N_1} = \frac{1}{\sqrt{2}} \frac{I_2}{I_1} \quad \text{...}(iii)$$

Neglecting voltage drop in armature and series field and assuming applied voltage constant

$$N \propto \frac{1}{\phi}$$

$$\frac{N_2}{N_1} = \frac{\phi_1}{\phi_2} = \frac{I_1}{\frac{1}{2} I_2} = \frac{2 I_1}{I_2} \quad \text{...}(iv)$$

Comparing Eqs. (*iii*) and (*iv*) we have

$$\frac{1}{\sqrt{2}} \frac{I_2}{I_1} = \frac{2 I_1}{I_2}$$

Current drawn, $I_2 = \sqrt{2\sqrt{2} I_1^2} = \sqrt{2\sqrt{2} \times 15^2} = 25.227$ A **Ans.**

Speed, $N_2 = \frac{2 I_1}{I_2} N_1 = \frac{2 \times 15}{25.227} \times 640 = 761$ rpm **Ans.**

12.3. DC SERIES MOTOR

The characteristics of dc series motors have already been discussed in chapter 1 [Art. 1.16.1 (*b*)]. In this chapter dc series motor will be analysed from the point of view of requirements of traction service.

1. The dc series motor develops high torque at start which is an essential requirement of traction service.

2. The series motor is amenable to simple speed control methods. Various methods of speed control of dc series motors will be discussed in forthcoming chapter *i.e.*, chapter 13.
3. In case of dc series motors

$$\text{Speed } N \propto \frac{V}{\phi} \propto \frac{1}{I} \qquad ...(i)$$

assuming flux proportional to exciting current and applied voltage constant.

$$\text{Also } T \propto \phi I \propto I^2 \qquad ...(ii)$$

Comparing Eqs. (*i*) and (*ii*) we have

$$N \propto \frac{1}{I} \propto \frac{1}{\sqrt{T}}$$

But since power output of a motor is proportional to the product of torque and speed

$$\text{Power output} \propto TN \propto T \times \frac{1}{\sqrt{T}} \propto \sqrt{T}$$

Thus power drawn from supply mains varies as the square root of the load torque in case of dc series motors *i.e.*, the series motors to some extent are self-relieving ones.

4. Series motors, as discussed in chapter 1 (Art. 1.20.3) are not suitable for regenerative braking as these are not electrically stable unless certain measures are taken.
5. In case of dc series motors commutation is excellent up to twice full load, therefore, replacement of brushes etc. is not required frequently.
6. In case of dc series motors the flux varies as, the armature current, torque corresponding to a given armature current, therefore, is independent of line voltage and thus unaffected by variations in the supply voltage.

   With the sudden increase in supply voltage, the armature current tends to increase, but with the increase in armature current the flux also increases which causes increase in back emf and thus armature current restores to its initial value. Thus series motors are less susceptible to sudden change in supply voltage.
7. In case of dc series motors, up to the point of magnetic saturation, torque developed is proportional to the square of the armature current. Therefore, dc series motor requires comparatively less increased power input with the increase in load torque. Thus the series motors are capable of withstanding excessive loads.
8. Because of low time constant of field in case of series motor, the field flux dies away in very short time with the result that back emf of the motor ceases with temporary interruption of supply. In case of dc series motors, therefore, there is undue rush of initial current on temporary interruption of supply.
9. The dc series motors, when operated in parallel to drive a vehicle by means of different driving axles, share load almost equally even there is unequal wear of different driving wheels. This is explained as below.

   Consider two identical motors of series type (say of speed-current characteristic shown in Fig. 12.3) connected to different driving axles of the vehicle. For a given speed of vehicle, the peripheral speed of all the driving wheels will be same, but should there be a slight difference in the diameters of the driving wheels (which is possible on account of difference in wear) there will be a corresponding difference in the speeds of the driving axles to which they are connected and also, therefore, a difference in speed of the motors. The speed of the motors will be given by the relation $\frac{N_2}{N_1} = \frac{D_1}{D_2}$.

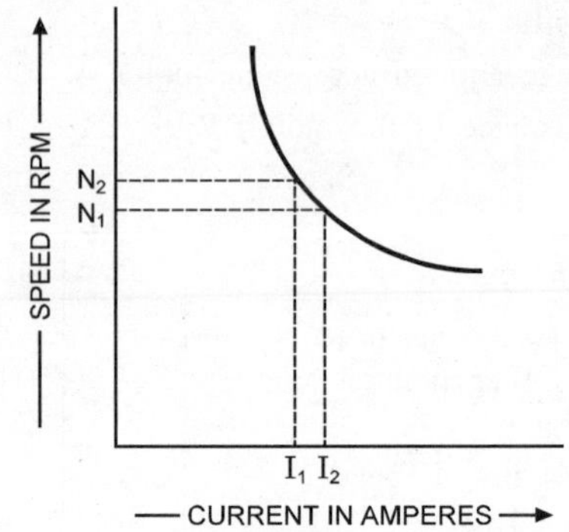

Fig. 12.3

When the motors are working in parallel the current drawn by each motor (determined from speed-current characteristic) due to slight difference in the speed does not differ much. As voltage across each motor is same, therefore motors share the load almost equally.

10. When the series motors are operating in series (Fig. 12.4), the current drawn by each motor will be same but due to slight difference in speeds, the back emfs will differ slightly. This will result in slight difference in voltage acting across each motor and also, therefore, slight difference in load shared by each motor.

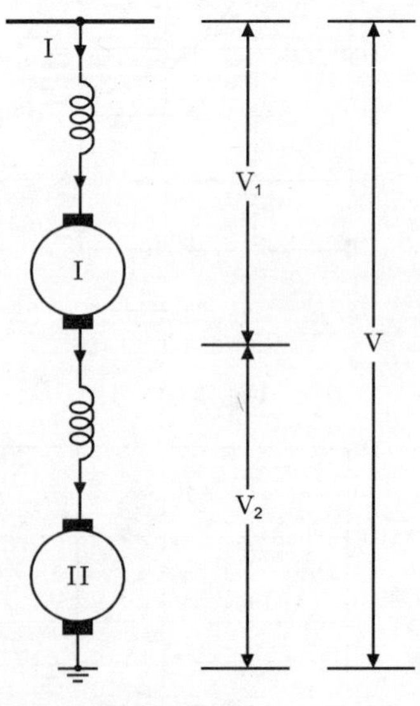

Fig. 12.4.

Let V_1 and V_2 and V be the voltage drops across motors I and II and supply voltage respectively (Fig. 12.4).

Since $N \propto (V - IR)$ where I is the load current and R is the resistance of each motor

$$\therefore \frac{N_2}{N_1} = \frac{V_2 - IR}{V_1 - IR} \qquad ...(12.10)$$

Also $V_1 + V_2 = V$ ...(12.11)

Solving above Eqs. (12.10) and (12.11) we have

$$V_1 = \frac{\frac{N_1}{N_2}(V - IR) + IR}{1 + \frac{N_1}{N_2}} \quad ...(12.12)$$

$$\text{and } V_2 = \frac{\frac{N_2}{N_1}(V - IR) + IR}{1 + N_2 / N_1} \quad ...(12.13)$$

11. The dc series motor is simple and robust in construction.

The dc series motor, owing to its characteristics, is most suitable for all types of traction services but more particularly for suburban and urban services where high rate of acceleration is essential. Motors of 45 kW and 75 kW, mass 5.4 to 7.2 kg per kW are available for trams and trolley buses although larger motors required for railway work are usually of masses of 12 to 18 kg per kW.

In 1,500 V dc system the dc series motors insulated for 1,500 V may be operated either at 1,500 V or at 750 V by connecting two of them permanently in series. In 3,000 V dc system the dc motors are insulated for 3,000 V but operated at 1,500 V by connecting two in series, as shown in Fig. 12.4.

The main drawbacks of dc series motors are owing to commutator such as restricted speed, voltage and current. There is a risk of flash-over and the brush gear requires considerable maintenance. The speed is also limited due to centrifugal forces acting on the rotor winding.

12.4. DC SHUNT MOTOR

The characteristics of dc shunt motors have already been discussed in chapter 1 [Art 1.16.1 (*a*)]. DC shunt motors are not suitable for traction purpose, the reasons are given below:

1. The dc shunt motor is a constant speed motor and its speed-torque characteristic does not meet with the requirements of traction.
2. In case of dc shunt motors, where the speed is approximately constant and independent of load torque, the power output $\propto TN \propto T$

   i.e., the power drawn from supply mains varies directly as the load torque. Hence a given overload will require a greater amount of power when shunt motors are employed than when dc series motors are employed.
3. With the increase in load, the dc shunt motor besides getting overloaded, develops commutation troubles also.
4. In case of dc shunt motors the flux is not independent of supply voltage, torque developed corresponding to a given armature current, therefore, not independent of supply voltage. Thus the torque developed is considerably affected by variation in supply voltage.

   Due to large inductance of field winding, the rate of change of current in this winding is low. Thus the current increases with the increase in supply voltage. With the build up of current in the field winding due to increase in supply voltage, the back emf increases and hence the value of increased current starts decreasing. The current inrush in case of shunt motors will, therefore, be of long duration and high initial value in comparison to that in series motors.
5. In case of dc shunt motors, the field flux remains constant and torque developed is directly proportional to the armature current. Hence for a given increase in torque developed a greater amount of current will be drawn from supply mains when dc shunt motor is used than when dc series motor is used.
6. Because of the large time constant of field winding of dc shunt motor and because it is connected across the armature, the back emf during interruption of supply is maintained even though it is decreasing. Now when the supply voltage is restored, the inrush current in the shunt motor will be determined by the resultant voltage in the armature circuit and impedance of armature circuit. Thus, inrush current in the shunt motor due to interruption of supply will be less in comparison to dc series motor.
7. In case of two identical shunt motors connected in parallel, the difference in currents drawn from supply mains due to small difference in speeds will be considerable, as obvious from Fig. 12.5, and therefore, motors will be loaded very unequally.

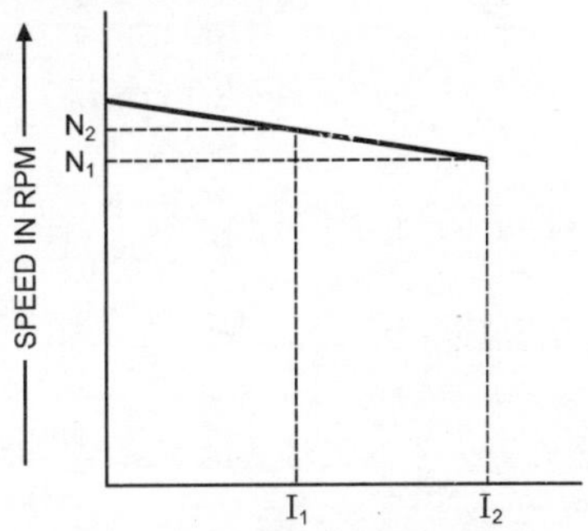

Fig. 12.5

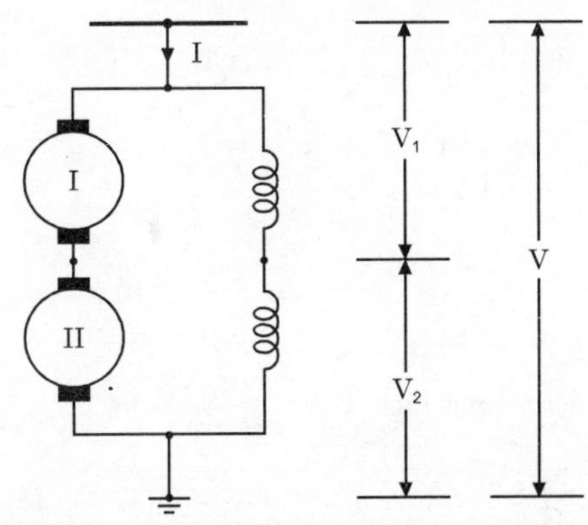

Fig. 12.6. ***Series Operation of DC Shunt Motors***

In case of two identical shunt motors connected in series (Fig. 12.6) there will be a slight difference in the voltage acting across each motor due to slight difference in the speeds and thus the motors will share the load almost equally.

From the above discussions it is obvious that shunt motor is not suitable for traction purposes.

Example 12.4. Two dc traction motors run at speed 700 rpm and 750 rpm respectively when each takes a current of 50 A from 500 V mains. Each motor has an effective resistance of

0.3 Ω. Calculate the speed and voltage across each machine when mechanically coupled and electrically connected in series and taking a current of 50 A from 500 V mains, the resistance of each motor being unchanged.

Solution : Let the two motors be 1 and 2 of speed N_1 = 700 rpm and N_2 = 750 rpm, respectively.

Resistance of each motor R_m = 0.3 Ω

Applied voltage across each motor, V = 500 V

Back emf of motor A when taking a current of 50 A

$$E_{b_1} = V - IR_m = 500 - 50 \times 0.3 = 485 \text{ V}$$

Back emf of motor B when taking a current of 50 A

$$E_{b_2} = V - IR_m = 500 - 50 \times 0.3 = 485 \text{ V}$$

When the machines are mechanically coupled and connected in series, the speed of each motor will be same, say N, current will be same and equal to 50 A (given) and the sum of voltage across the two motors will be equal to 500 V.

Let the voltage across motors 1 and 2 be V_1 and V_2 respectively

$$\text{Now } V_1 + V_2 = 500 \qquad ...(i)$$

$$\text{Back emf of motor 1, } E'_{b_1} = E_{b_1} \times \frac{N}{N_1} = \frac{485}{700} \times N$$

$$\text{Voltage across motor 1, } V_1 = E'_{b_1} + IR_m$$

$$= \frac{485}{700} N + 50 \times 0.3$$

$$= \frac{485}{700} N + 15$$

$$\text{Back emf of motor 2, } E'_{b_2} = E_{b_2} \times \frac{N}{N_2} = \frac{485}{750} N$$

$$\text{Voltage across motor 2, } V_2 = E'_{b_2} + IR_m = \frac{485}{750} N + 15$$

Substituting $V_1 = \frac{485}{700} N + 15$ and $V_2 = \frac{485}{750} N + 15$ in Eq. (*i*) we get

$$\frac{485}{700} N + 15 + \frac{485}{750} N + 15 = 500$$

$$\text{or} \quad \left(\frac{485}{700} + \frac{485}{750}\right) N = 470$$

$$\text{or} \quad N\left(\frac{1}{700} + \frac{1}{750}\right) = \frac{470}{485}$$

$$\text{or } N = \frac{470}{485} \times \frac{10{,}500}{15 + 14}$$

$$= 351 \text{ rpm} \quad \textbf{Ans.}$$

$$\text{PD across machine 1, } V_1 = \frac{485 \text{ N}}{700} + 15$$

$$= \frac{485 \times 351}{700} + 15$$

$$= 258.0 \text{ V} \quad \textbf{Ans.}$$

$$\text{PD across machine 2, } V_2 = \frac{485 \text{ N}}{750} + 15$$

$$= \frac{485 \times 351}{750} + 15$$

$$= 242.0 \text{ V} \quad \textbf{Ans.}$$

Example 12.5. A tramcar is equipped with two motors which are operating in parallel. Calculate the current drawn from the supply mains at 500 volts when the car is running at a steady speed of 40 kmph and each motor is developing a tractive effort of 1,800 newtons. The resistance of each motor is 0.4 ohm. The friction, windage and other losses may be assumed, as 3,200 watts per motor.

Solution : Voltage across each motor, V = 500 volts

Maximum speed, V_m = 40 kmph

Tractive effort, F_t = 1,800 newtons

Motor resistance, R_m = 0.4 Ω

Losses per motor = 3,200 watts

Power output of each motor

$$= \frac{F_t \times V_m \times 1{,}000}{3{,}600} \text{ watts}$$

$$= \frac{1{,}800 \times 40 \times 1{,}000}{3{,}600} = 20{,}000 \text{ watts}$$

Constant losses = 3,200 watts

Copper losses = $I^2 R_m = 0.4 I^2$

where I is the current drawn from supply mains.

Input to motor = Motor output + constant losses + copper losses

$$VI = 20{,}000 + 3{,}200 + 0.4 I^2$$

$$\text{or } 0.4 I^2 - 500 I + 23{,}200 = 0 \qquad \because V = 500 \text{ volts}$$

$$\text{or} \quad I = \frac{500 \pm \sqrt{(500)^2 - 4 \times 0.4 \times 23{,}200}}{2 \times 0.4}$$

or I = 48.25 A or 1,201.75 A

or Current drawn by each motor = 48.25 A

∵ 1,201,75 A , being unreasonably high, cannot be accepted

Total current drawn from supply mains

= 48.25 × 2 = 96.5 A **Ans.**

Example 12.6. A locomotive exerts a tractive effort of 35,300 N in hauling a train at 48 kmph on the level track. If the motor is to haul the same train on a gradient of 1 in 50 and the tractive effort required is 55,180 N, determine the power delivered by the locomotive if it is driven by (*i*) dc series motors (*ii*) induction motors. [*Pb. Technical Univ. Electric Drives & Utilization, Dec. 2008*]

Solution : Tractive effort, F_t = 35,300 N

Maximum speed, V = 48 kmph

Power output required on the level track

$$= \frac{F_t \times V \times 1{,}000}{3{,}600}$$

$$= \frac{35{,}300 \times 48 \times 1{,}000}{3{,}600} = 470{,}667 \text{ W or } 470.667 \text{ kW}$$

Power delivered by locomotive on upgradient track with dc series motors

$$= 470.667 \sqrt{\frac{55{,}180}{35{,}300}} = 588.5 \text{ kW} \quad \textbf{Ans.}$$

$$\because \text{Power output} \propto \sqrt{T} \propto \sqrt{F_t}$$

Power delivered by locomotive on upgradient track with induction motors

$$= 470.667 \times \frac{55{,}180}{35{,}300} = 735.7 \text{ kW} \quad \textbf{Ans.}$$

$$\because \text{Power output} \propto T \propto F_t$$

Example 12.7. A dc traction motor has speed-tractive effort and armature current characteristics as given below with driving wheels having 85 cm wheel diameter and gear ratio as 71 : 21.

Armature current (amperes)	60	120	180	240	300	360
Speed (kmph)	80	50	45	42	38	35
Tractive Effort in thousand newtons	1.7	5	10	14	16	20

Determine new characteristics if driving wheels are replaced by 90 cm diameter wheels and gear ratio changed to 74 : 19.

[Pb. Univ. Electric Traction July 1992]

Solution : Diameter of driving wheel, $D_1 = 85$ cm = 0.85 m

Diameter of new driving wheel, $D_2 = 90$ cm = 0.9 m

Gear ratio, $\gamma_1 = 71 : 21$

Gear ratio, $\gamma_2 = 74 : 19$

$$\text{New speed, } V_2 = \frac{D_2}{D_1} \times \frac{\gamma_1}{\gamma_2} \times V_1 = \frac{0.9}{0.85} \times \frac{71/21}{74/19} = 0.919\, V_1$$

$$\text{New tractive effort, } F_{t_2} = \frac{D_1}{D_2} \times \frac{\gamma_2}{\gamma_1} F_{t_1} = \frac{0.85}{9} \times \frac{74/19}{71/21} = 1.088\, F_{t_1}$$

New characteristics are:

Armature current (amperes)	60	120	180	240	300	360
Speed (kmph), $V_2 = 0.919\, V_1$	73.52	45.95	41.36	38.6	34.92	32.17
Tractive effort in thousand Newtons $F_{t_2} = 1.088\, F_{t_1}$	1.85	5.44	10.88	15.23	17.41	21.76

Example 12.8. A motor coach is being driven by two identical dc series motors. First motor is geared to driving wheel having diameter of 90 cm and other motor to driving wheel having diameter of 86 cm. The speed of the first motor is 500 rpm when connected in parallel with the other across 600 V supply. Find the motor speeds when connected in series across the same supply. Assume armature current to remain same and armature voltage drop of 10% at this current.

Solution : Speed of first motor, $N_1 = 500$ rpm

$$\text{Back emf, } E_{b_1} = 600 - \frac{10}{100} \times 600 = 540 \text{ volts}$$

When the motors are connected in series across 600 V supply, as shown in Fig. 12.7.

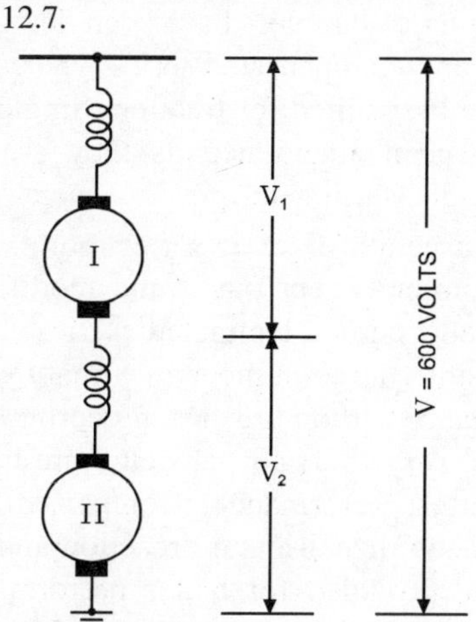

Fig. 12.7

Let the supply voltage across motors I and II be V_1 and V_2 volts and speed N'_1 and N'_2 respectively.

$$\text{Since speed, } N \propto \frac{V - IR}{\phi}$$

Current through the motors remains the same, therefore, flux produced by it also remains the same and $N \propto (V - IR)$

$$\therefore \quad \frac{N'_1}{N'_2} = \frac{V_1 - IR}{V_2 - IR} = \frac{V_1 - \frac{10}{100} \times 600}{V_2 - \frac{10}{100} \times 600} = \frac{V_1 - 60}{V_2 - 60} \quad ...(i)$$

And also $N'_1 D_1 = N'_2 D_2$

$$\text{or } \frac{N'_1}{N'_2} = \frac{D_2}{D_1} = \frac{86}{90} \quad ...(ii) \text{ Since peripheral speed is equal}$$

$$\therefore \quad \frac{V_1 - 60}{V_2 - 60} = \frac{86}{90}$$

or $90V_1 - 5{,}400 = 86V_2 - 5{,}160$

or $90V_1 - 86V_2 = 5{,}400 - 5{,}160 = 240$ ...(*iii*)

and also $V_1 + V_2 = V = 600$ V ...(*iv*)

Solving Eqs. (*iii*) and (*iv*) we get

$V_1 = 294.55$ V and $V_2 = 305.45$ V

Now the speeds of the motors can be calculated as follows:

$$\frac{N'_1}{N_1} = \frac{E'_{b_1}}{E_{b_1}}$$

$$\text{or } N'_1 = N_1 \times \frac{E'_{b_1}}{E_{b_1}} = 500 \times \frac{294.55 - 60}{600 - 60} = 217 \text{ rpm} \quad \textbf{Ans.}$$

$$\text{and } N'_2 = N'_1 \times \frac{D_1}{D_2} = 217 \times \frac{90}{86} = 277 \text{ rpm} \quad \textbf{Ans.}$$

Example 12.9. The characteristics of a dc series motor when operating at 600 volts and driving 107 cm wheels are as follows :

Current (amperes)	500	400	300	200
Train speed (km/hour)	35.6	38.5	42.5	50.5
Tractive effort (kg)	2,580	1,960	1,380	725

Two of such motors are mounted on a motor coach. The wheels driven by one motor A are 107 cm in dia and other motor B are 102 cm in dia.

Calculate the power input and tractive efforts of each motor when

(*i*) motors are connected in parallel and train speed is 40 kmph.

(*ii*) motors are in series and current is 400 A.

Assume resistance of each motor as 0.08 Ω.

[Pb. Univ. Elec. Traction July, 1974]

Solution : The characteristics of a dc series motor when operated at 600 V and driving 107 cm diameter wheels are drawn from the given datas (refer to Fig. 12.8(*a*)).

The characteristic of a dc series motor when operated at 600 V and driving 102 cm diameter wheel are derived as follows and drawn (Refer to Fig. 12.8 (*b*)).

Amperes	500	400	300	200
Train speed km/hr, $V_2 = \frac{V_1 \times 1.02}{1.07}$	33.9	36.7	40.5	48.2
Tractive effort kg $F_{t_2} = F_{t_1} \times \frac{1.07}{1.02}$	2,710	2,060	1,450	760

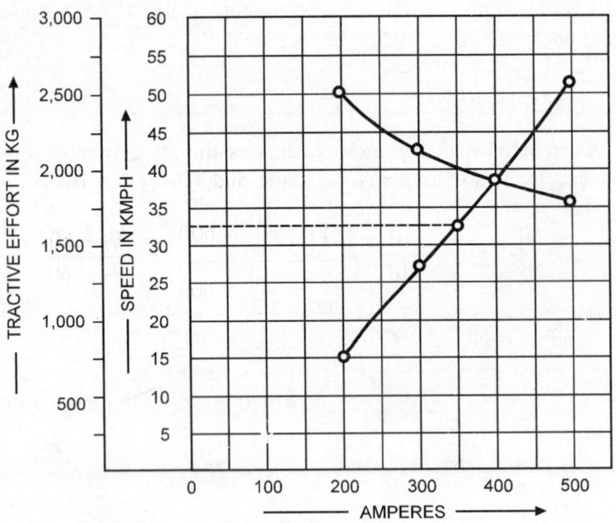

(a)

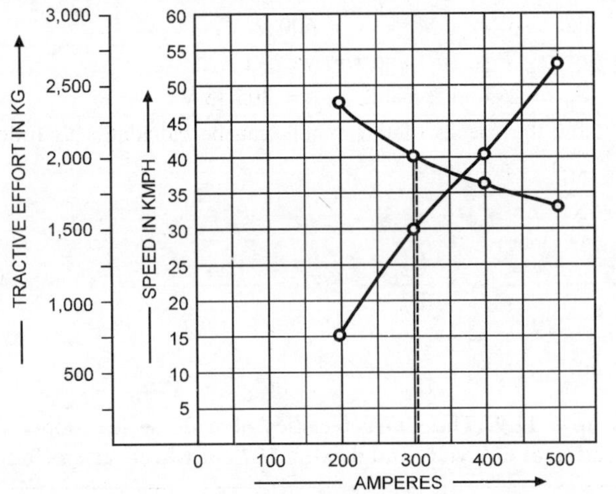

(b)

Fig. 12.8

Let the two motors be A and B of wheel diameters 107 and 102 cm respectively.

(i) When motors are connected in parallel and train speed is 40 kmph

(*a*) From speed-current characteristic corresponding to speed of 40 kmph

Current drawn by motor A, $I_A = 350$ A

Current drawn by motor B, $I_B = 305$ A

$$\text{Power input to motor A, } P_A = \frac{VI_A}{1{,}000} = \frac{600 \times 350}{1{,}000} = 210 \text{ kW}$$

$$\text{Power input to motor B, } P_B = \frac{600 \times 305}{1{,}000} = 183 \text{ kW}$$

(*b*) From tractive effort-current characteristics

Tractive effort of motor A, $F_{t_A} = 1{,}625$ kg **Ans.**

Tractive effort of motor B, $F_{t_B} = 1{,}480$ kg **Ans.**

(ii) When the motors are connected in series and current is 400 A

Resistance of each motor, $R_m = 0.08\ \Omega$

Applied voltage across each motor, V = 600 volts

Back emf of motor A when taking current of 400 A and at a speed of 38.5 kmph

$$E_{b_A} = 600 - 400 \times 0.08 = 568 \text{ V}$$

Back emf of motor B when taking current of 400 A and at a speed of 36.7 kmph

$$E_{b_B} = 600 - 400 \times 0.08 = 568 \text{ V}$$

After motors being connected in series, let the peripheral speed be *v* kmph, which will be same for both motors.

$$\text{Now back emf of motor A, } E'_{b_A} = E_{b_A} \times \frac{v}{v_A} = \frac{568}{38.5} \times v$$

$$\text{Voltage across motor A, } V_A = E'_{b_A} + IR_m = \frac{568}{38.5} \times v + 400 \times 0.08 = \frac{568v}{38.5} + 32$$

$$\text{Back emf of motor B, } E'_{b_B} = \frac{568}{36.7} \times v$$

$$\text{Voltage across motor B, } V_B = E'_{b_B} + IR_m = \frac{568}{36.7}v + 32$$

$$\text{Since } V_A + V_B = 600$$

$$\therefore \quad \frac{568v}{38.5} + 32 + \frac{568v}{36.7} + 32 = 600$$

$$\text{or} \quad 14.8v + 15.5v = 536$$

$$\text{or } v = \frac{536}{30.3} = 17.7 \text{ kmph}$$

$$\text{PD across motor A, } V_A = \frac{568}{38.5} \times 17.7 + 32 = 294 \text{ volts}$$

$$\text{PD across motor B, } V_B = \frac{568}{36.7} \times 17.7 + 32 = 306 \text{ volts}$$

$$(a) \text{ Power input to motor A, } P_A = \frac{V_A \times I}{1{,}000} = \frac{294 \times 400}{1{,}000} = 117.6 \text{ kW}$$

$$\text{Power input to motor B, } P_B = \frac{V_B \times I}{1{,}000} = \frac{306 \times 400}{1{,}000} = 122.4 \text{ kW}$$

(*b*) From tractive effort-current characteristics corresponding to 400 A of current,

Tractive effort of motor, A, $F_{t_A} = 1{,}960$ kg **Ans.**

Tractive effort of motor B, $F_{t_B} = 2{,}060$ kg **Ans.**

12.5. AC SERIES MOTOR

Many single phase ac motors have been developed for traction purposes but only compensated series type commutator motor is found to be best suited for traction. Single phase induction motors have been abandoned as they are not capable of developing high starting torque.

The construction of an ac series motor is very similar to a dc series motor except that some modifications (such as whole magnetic circuit laminated, series field with as few turns as possible, large number of armature conductors, use of high resistance carbon brushes, numerous poles with lesser flux per pole, very short air gaps etc.) are incorporated so as to obtain better performance. Compensating windings are provided to neutralize armature reaction and commutating or interpoles are provided for better performance in terms of higher efficiency and a greater output from a given size of armature core.

A schematic diagram of a single phase series motor with interpole and compensating windings is illustrated in Fig. 12.9.

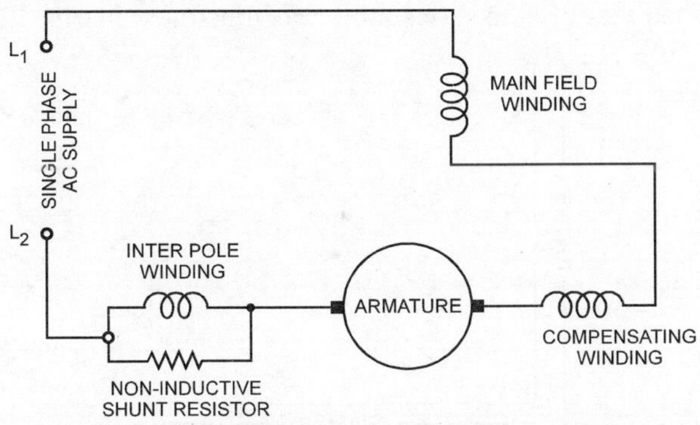

Fig. 12.9. ***AC Series Motor With Interpoles and Compensating Windings***

The average value of the torque on the motor shaft is given as

$$T = \frac{1}{2\sqrt{2}\pi} Z \frac{P}{A} \phi_{max} I \cos\theta \qquad ...(12.14)$$

where I is the effective value of current, ϕ_{max} is the peak value of the flux per pole and θ is the phase angle between phasors ϕ and I.

The phasor diagram of a single phase ac series motor is given in Fig. 12.10. Flux ϕ is produced by current I. Flux ϕ lags behind the current I by an amount too small to be important, so it has been shown in phase with current I. The resistive drops IR_{se}, IR_p, IR_c and IR_a due to resistances of series field, interpole circuit, compensating winding and of armature respectively are in phase with current I. The reactive drops IX_{se}, IX_p, IX_c and IX_a due to reactances of series field, interpole circuit, compensating winding and of armature respectively are in quadrature with current I and leading. Here it is noteworthy that reactive drop due to series field is much greater than that of either armature or the compensating field even all the measures to reduce series field inductance, are taken. Unlike the armature cross-flux ϕ′, the flux that causes this reactive drop cannot be neutralized because it is the necessity for the development of torque. The interpole circuit consists of the interpole winding in parallel with the non-inductive shunt, as illustrated in Fig. 12.9. R_p and X_p are the respective equivalent resistance and reactance of the interpole parallel circuit.

The phase angle of E, in the phasor diagram is most easily determined by considering instantaneous values. When the alternating field flux ϕ is at its peak value, the armature conductors cut the maximum flux, and the rotational or speed emf is, therefore, a maximum. When the field flux is zero, the rotational or speed emf, E is zero. Since the graphs of flux ϕ and generated emf E pass through zero at the same instant, and also through maximum at the same instant, the phasor E must be either exactly in phase with the phasor of flux ϕ or exactly in phase opposition. The first possibility is readily eliminated by noting that in case of a motor the generated voltage is in opposition to the direction of flow of current, and, therefore, E is in phase opposition with current and is called the counter emf.

The terminal voltage V is the phasor sum of the counter-emf reversed and all the resistive and reactive volt drops.

Series Motor Characteristics

1. *Power-Output Characteristic.* Mechanical power developed is given by the product of counter emf E and the armature current I . Though counter emf E decreases slightly with the increase in current, but if it is neglected in comparison with the magnitude of counter emf E, mechanical power developed increases almost in proportion with current I.

Power available at shaft or power output is equal to the mechanical power developed less rotational losses. Power output characteristic is shown in Fig. 12.14.

2. *Power Factor Characteristic.* The cosine of the phase angle ϕ is the power factor of the motor.

From phasor diagram shown in Fig. 12.10

$$\sin\phi = \frac{I(X_{se} + X_p + X_c + X_a)}{V} = \frac{IX}{V}$$

where $X = X_{se} + X_p + X_c + X_a$ and = effective reactance of the motor

$$\text{or Power factor, } \cos\phi = \sqrt{1 - \left(\frac{IX}{V}\right)^2} \qquad ...(12.15)$$

In order to have high power factor the reactance drops must be low and counter emf high. The reactance drops are lowest and the counter emf is highest at light loads, and therefore, the power factor of the single phase series motor is highest (nearly unity) at light loads and falls with the increase in current. This is the reason that power factor of a series motor is low at overloads and at starting. This is the reverse of the power factor relation that exists in case of the induction motor and transformer.

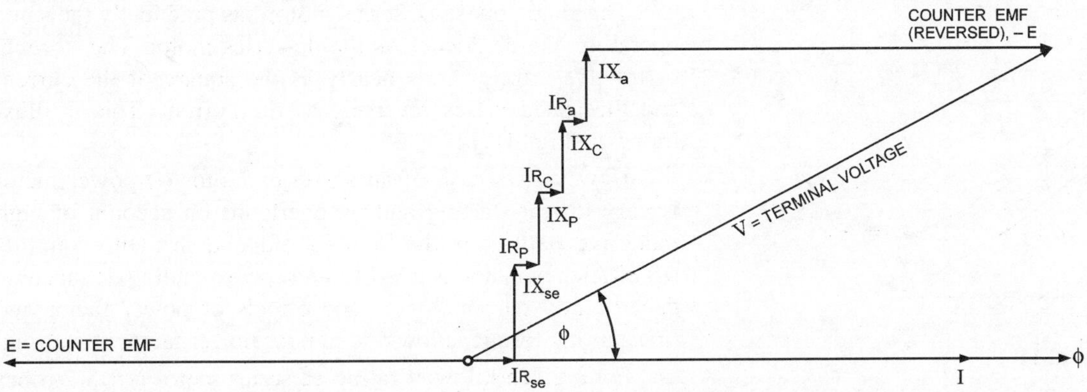

Fig. 12.10. ***Phasor Diagram For Single Phase Series Motor***

3. *Speed-Current Characteristic.* Speed of a commutator motor is proportional to counter emf E (E_r or E_b) or proportional to supply voltage less voltage drops because

$$E \; \alpha \; \phi N \text{ and } E = V - \text{voltage drops}$$

The current causes resistive drops in series field and armature in case of dc operation but in case of ac operation it also causes large reactance drops in series field, armature, compensating winding and interpole circuit in addition to resistive drops in these windings. Hence with ac operation, counter emf E developed is much less than that with dc operation. It means speed-current characteristic for ac series motors are more drooping as compared to those for dc series motors, as illustrated in Fig. 12.11.

Let the same series commutator motor be operated on dc as well as on ac and with same current I.

The back emf developed in case of dc operation is given

as $E_b = \dfrac{\phi Z N_{dc}}{60} \times \dfrac{P}{A}$ volts

and $E_r = \dfrac{\phi_{max}}{\sqrt{2}} \dfrac{Z N_{ac}}{60} \times \dfrac{P}{A}$ volts

Neglecting saturation, $\phi_{max} = \sqrt{2}\,\phi$ as the peak value of ac flux ϕ_{max} is set up by the peak value of current $\sqrt{2}$ while dc flux ϕ is set up by current I. So from above two equations

$$\frac{N_{ac}}{N_{dc}} = \frac{E_r}{E_b} = \frac{V\cos\phi - IR}{V - IR} = \text{Cos}\,\phi \qquad ...(12.16)$$

if the supply voltage is same in both of the cases and resistive drops IR are neglected in comparison with supply voltage V

where ϕ is the phase angle between supply voltage and current.

4. ***Torque-Current Characteristic.*** From Eq. (12.14) $T \,\alpha\, \phi I$ if phase angle between flux ϕ and current I is neglected or $T \alpha\, I^2$ neglecting saturation. Under saturated condition, flux ϕ becomes constant and, therefore, torque becomes directly proportional to current I. Torque-current characteristic for an ac series motor is shown in Fig. 12.12.

5. ***Torque-Speed Characteristic.*** The torque-speed characteristic for an ac series motor can be derived from speed-current and torque-current characteristics given in Figs. 12.11 and 12.12 respectively.

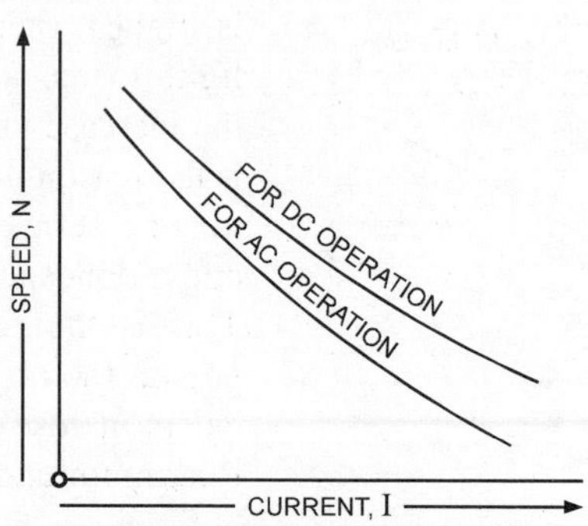

Fig. 12.11

For given value of torque T and applied voltage V, the armature current is same but voltage drop in case of ac series motor is much more than that in case of a dc series motor so speed of an ac series motor for a given developed torque is less than that of a dc series motor, as illustrated in Fig. 12.13.

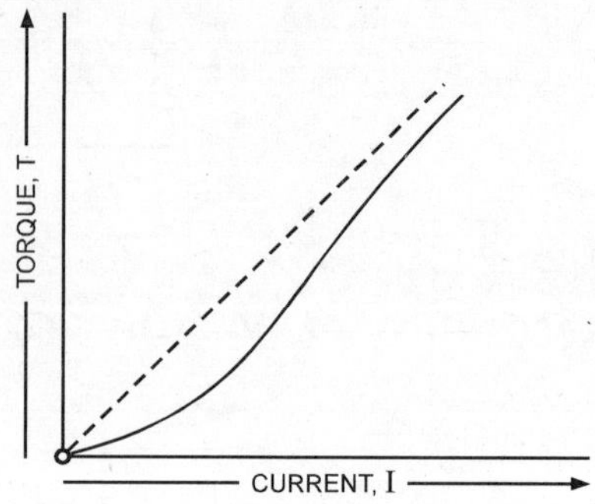

Fig. 12.12

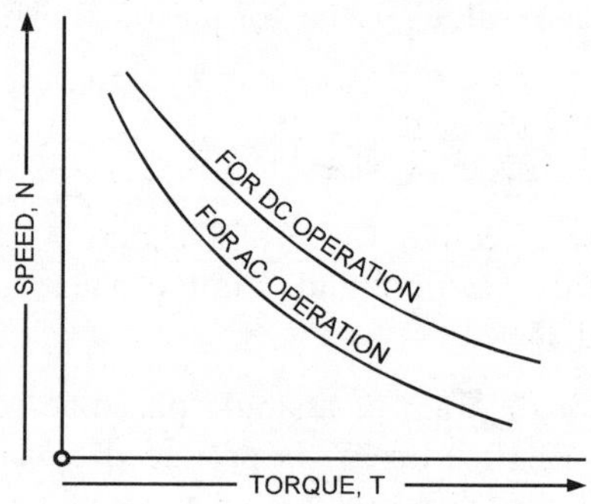

Fig. 12.13

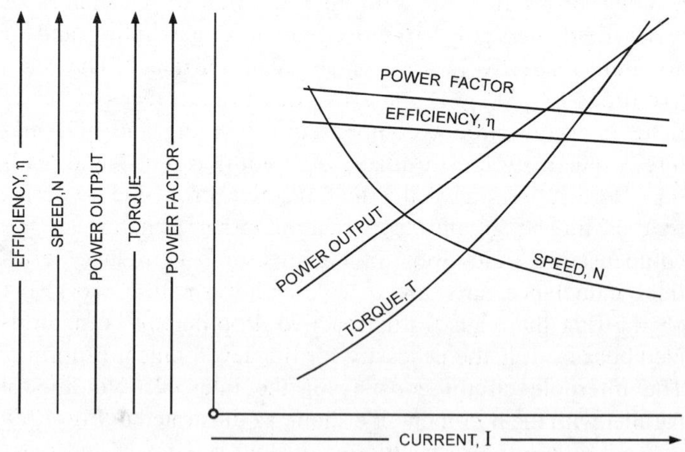

Fig. 12.14

The single phase ac series motor has practically the same operating characteristics as the dc series motor. The torque, or tractive effort, varies nearly as the square of the current and the speed varies inversely as the current. This is illustrated in Fig. 12.14.

However, in case of an ac series motor (*i*) power factor is very low at starting and on overloads on account of high inductive nature of the series field and armature circuits (*ii*) efficiency is not as good as in a corresponding dc machine due to eddy current losses and effects of power factor and (*iii*) starting torque is low due to poor power factor at starting.

For a given kilowatt rating ac series motor is 1.5-2 times in size and weight of the corresponding dc series motor. The construction cost of an ac series motor is much more than that of a dc series motor.

The speed of an ac series motor may be controlled efficiently by taps on a transformer, which is not possible in case of a dc series motor.

The torque-speed characteristic of the single phase series motor is similar to that of the dc series motor *i.e.,* high starting torque and decrease in speed with increase in load making it to have self-relieving property from heavy excessive load, so such a machine is particularly useful for traction services.

These motors are used for main line services. These are not suitable for urban and suburban services because of low starting torque and poor power factor at start.

However, single phase series motors have better performance (improved power factor, higher efficiency, improved commutation, better starting and fewer poles for a given output) on reduced supply frequency (say $16\frac{2}{3}$ or 25 Hz). The higher frequency results in higher leakage reactances and hence a relatively poor power factor. The weight per kW output is also greater for the higher frequency because of larger dimensions of the motor.

The 50 Hz motors have steeper speed-torque characteristics as they operate at lower flux densities than a $16\frac{2}{3}$ or 25 Hz motors. This results in fewer voltage steps at starting and the division of load between parallel connected motors will be less affected by wear of the driving wheels.

The operating voltage is kept low (300 to 400 V) in order to reduce the inductance. Because high voltage motor with proportionately low current would require a large number of turns to produce the given flux.

12.6. THREE PHASE INDUCTION MOTOR

The three phase induction motors have the advantages of simple and robust construction, trouble free operation, less maintenance, high voltage operation consequently requiring reduced amount of current and automatic regeneration. But due to their flat speed-torque characteristics, constant speed operation, developing low starting torque, drawing high starting current, complicated speed control system and complicated overhead feeding systems they are not suitable for electric traction work. It is also impossible to employ 3-phase induction motors for a multiple unit system in which two or more locomotives are used for propelling a heavy train. In traction system employing 3-ϕ induction motors it becomes necessary to couple all the driving axles through a connecting rod so that there is no possibility of a difference in speeds.

With the development of thyristorised inverter circuits, it has now been possible to invert the supply and obtain a variable frequency supply which could be used for the 3-phase induction motor and a very smooth speed control can be obtained.

Induction motor gives good power factor and good efficiency at the speed near to the synchronous speed of the motor. Now in conventional methods of speed control, change of slip method is normally used. But in this method at low speeds of motor or at high value of slip, rotor losses will be more so good efficiency cannot be obtained. To get good efficiency and good power factor, synchronous speed of the motor itself can be brought to lower speed near to the desired actual speed of the motor. This can be done by supplying the motor with the help of variable frequency supply.

Starting current of motor also decreases when motor is started at low frequency.

When 3-phase induction motor is used, distribution system consists of two overhead wires and track rail for the third phase to feed power to locomotive. This gives a complicated overhead structure and also a person who comes in contact of third live rail might be in danger. This drawback is removed by employing Kando system. In this system single phase HV supply is given to locomotive with the help of single overhead wire. In locomotive, this single phase supply is converted into three phase supply with the help of phase converters and fed to the motor. Kando system is in use in Hungary and in some sections of Italian State Railway.

12.7. LINEAR INDUCTION MOTOR

It is a special type of induction motor which gives linear motion instead of rotational motion as in the case of a conventional induction motor. It operates on the same principle on which a conventional induction motor operates *i.e.,* "whenever there occurs a relative motion between the field and the short-circuited conductors, currents are induced in them which results in electromagnetic forces and under the influence of these forces, according to Lenz's law, the conductors try to move in such a way as to eliminate the induced currents". In case of a conventional induction motor, movement of field is rotary about an axis so the movement of the conductors is also rotary. But is case of a linear induction motor, the movement of the field is rectilinear and so the movement of conductors.

In its simplest form, a linear induction motor consists of field system having a 3-phase distributed winding placed in slots as shown in Fig. 12.15. The field system may be a single primary system or double primary system (Fig. 12.15). The secondary of the linear induction motor is normally a conducting plate made of either copper or aluminium in which interaction currents are induced. Either member can be the *stator*, the other being the *runner* in accordance with the particular requirements imposed by the duty for which motor is intended.

In a single primary system a ferromagnetic plate is usually placed on the other side of the conducting plate to provide a path of low reluctance to the main flux. However, the ferromagnetic plate gets attracted towards the primary on energisation of the field and this causes unequal gap length on two sides of the conducting plate. This problem can be overcome by employing double primary system. [Fig. 12.15 (*b*)].

Which of the two primary and secondary will be shorter in length compared to the other depends upon the use of the motor. When the operating distance is large, the primary is made shorter than the secondary because it is uneconomical to wind a very long 3-phase primary. The short secondary form [Fig. 12.15 (*c*)] is useful with limited operating distance.

When the three phase primary winding of a linear induction motor is energized from a balanced three phase source, a magnetic field moving in a straight line from one end to the other at a linear synchronous speed v_s is developed. The linear synchronous speed is given as

$$v_s = 2\tau f \text{ metres/second} \quad \text{...(12.17)}$$

where τ is the pole pitch in metres and f is the supply frequency in Hz. *It is to be noted here that the synchronous speed does not depend on the number of poles, but only on the pole pitch and the stator supply frequency.*

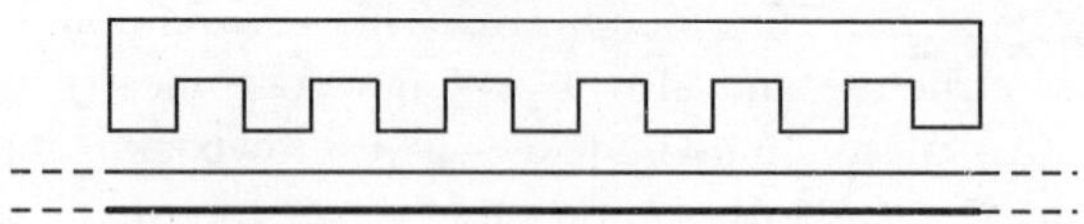

(i)

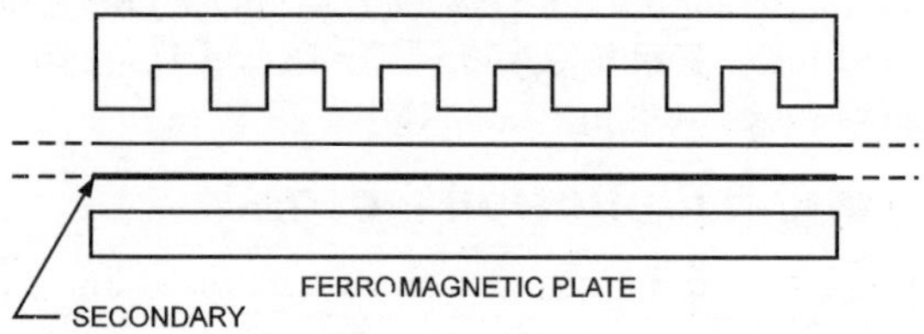

(ii)

(a) Short Single Primary

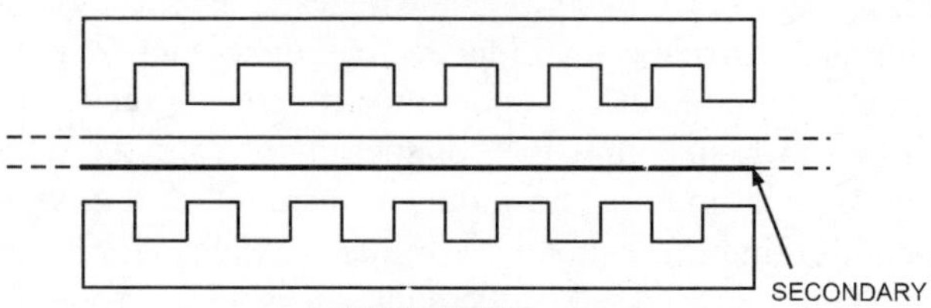

(b) Short Double Primary

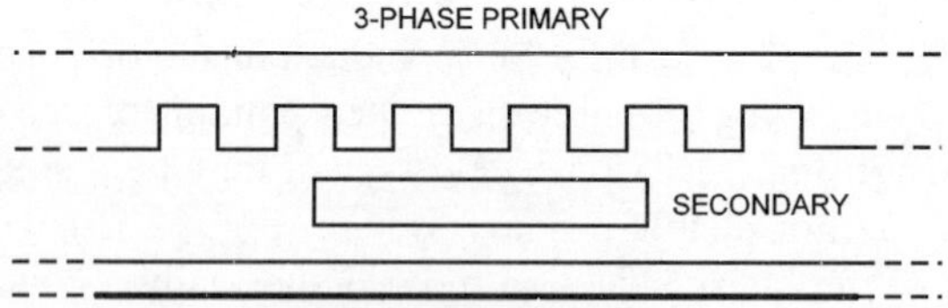

(c) Short Secondary

Fig. 12.15

As the flux moves linearly, it drags rotor plate alongwith it in the same direction. This reduces the relative speed of travel of the flux with respect to rotor plate. If the speed of the rotor plate is equal to that of the magnetic field, latter would be stationary when viewed from rotor plate. This is corresponding to the synchronous speed of induction motor. If the rotor plate is moved faster than this speed, the direction of the force would be reversed and a form of regenerative braking based on the principle of induction generator will come into being.

Slip of a linear induction motor is given as

$$s = \frac{v_s - v}{v_s} \quad \text{...(12.18)}$$

where v is the actual speed of rotor plate.

Thrust or force or tractive effort is given as

$$F = \frac{P_2}{v_s} \quad \text{... (12.19)}$$

where P_2 is the actual power supplied to the rotor

Active power flow is similar to that in a conventional rotary induction motor *i.e.,*

$$\text{Copper losses in rotor} = sP_2 \quad \text{...(12.20)}$$

and mechanical power developed, $P_{mech} = (1 - s)\, P_2$ ...(12.21)

Tractive effort will be a function of slip s *i.e.,* $(v_s - v)$. The thrust (or tractive effort)-speed characteristics of a linear induction motor, as shown in Fig. 12.16, are similar to the torque-speed characteristics of a conventional rotary induction motor.

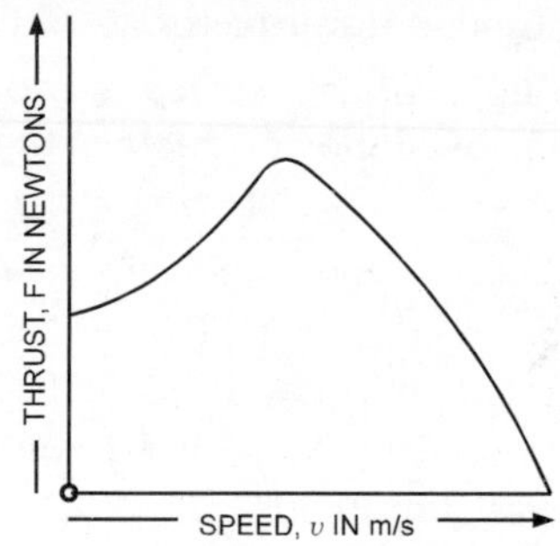

Fig. 12.16

Tractive effort, F can be controlled by varying both frequency and voltage simultaneously so that induction density remains constant.

There are two peculiar effects, which are encountered in a linear induction motor but not in a conventional rotary induction motor. These effects are *transverse edge effect* and *end effect.*

The paths of the induced currents in the secondary are not well defined because the secondary of a linear induction motor is a solid conducting plate. The portion of the current paths parallel to the direction of motion of secondary does not make any contribution towards the production of useful thrust but only contributes towards losses. This effect reduces the effective thrust and increases the losses and is known as the *transverse edge effect* because the current paths parallel to the direction of motion are more towards the edges of the conducting plate.

In the case of linear induction motors with short primary, the current paths towards the end of field structure on the conducting plate go beyond the field structure and such portions of current paths do not contribute to useful thrust but only towards motor losses. This is called the *end effect*. The end effect can be effectively reduced by increasing the number of poles on the motor.

The **advantages** of linear induction motors are (*i*) low initial cost (*ii*) low maintenance cost because of absence of rotating parts (*iii*) simplicity (*iv*) no limitation of tractive effort due to adhesion between the wheel and the rail (*v*) no limitation of maximum speed due to centrifugal forces (*vi*) no overheating of rotor because the motor moves continuously over cool rotor plate leaving behind heated rotor portion and (*vii*) better power to weight ratio.

The **disadvantages** of linear induction motors are (*i*) poor utilisation of motor due to transverse edge effect and end effect (*ii*) larger air gap and non-magnetic reaction rail (rotor plate) need more magnetizing current resulting in poor efficiency and low pf (*iii*) very high capital cost of reaction rail fixed along the centre line of the track (*iv*) complications and high cost involved in providing three phase collector system along the track (*v*) difficulties encountered in maintaining adequate clearances at points and crossings.

Applications. Linear induction motors can be used in conveyors, travelling cranes, haulers, electromagnetic pumps, high-speed rail traction etc. Such motors may be employed in applications in which the field system moves and the conducting plate remains stationary such as travelling crane motors. These motors can also be used in applications where the field system remains stationary and the conducting plate moves such as in automatic sliding doors in electric trains, metallic belt conveyors etc. It can be used on trolley cars for internal transport in workshop, as booster accelerator for moving heavy trains from rest or up the inclines or on curves or as a propulsion unit in marshalling yards in place of shunting locomotives. Linear induction motor has superiority over conventional rotary motor for speeds over 200 kmph. Linear induction motor provides excellent source of motive power for magnetically suspended trains where conventional rotary motor fails because the latter depends for torque conversion to linear tractive force upon the adhesive weight on driving wheels. The use of linear induction motor is limited to only a few applications till now owing to design difficulties and economic considerations.

Example 12.10. A single-sided linear induction motor has 98 poles, and its pole pitch is 50 cm. The motor is used for propelling an electric vehicle. Determine the linear synchronous velocity and the vehicle speed in km per hour if the frequency is 50 Hz and the slip is 0.25.

Solution : Supply frequency, $f = 50$ Hz

Pole pitch, $\tau = 50 \text{ cm} = 0.5 \text{ m}$

Slip, $s = 0.25$

Linear synchronous velocity $v_s = 2\tau f = 2 \times 0.5 \times 50$

$$= 50 \text{ m/s} = \frac{50 \times 60 \times 60}{1{,}000}$$

$= 180$ kmph **Ans.**

Vehicle speed, $v = v_s(1 - s) = 180(1 - 0.25)$

$= 135$ kmph **Ans.**

12.8. RATING AND VENTILATION

The motors are designed for intermittent loading as well as continuous loading. The motors employed in traction work are subject to intermittent loading particularly in case of tramway and suburban services, therefore, traction motors are given one-hour rating as well as continuous rating. The ratings have been fully defined in Indian Standard Specifications (ISS). The continuous rating of an ordinary self ventilated motor is approximately one-third to one-fourth of its one hour rating while the continuous rating of a forced ventilated motor is about four-fifth of its one hour rating.

The motors employed in traction work are usually totally enclosed so that protection against the ingress of dust, dirt, mud and water may be provided. Since the weight and overall dimensions of the motor are to be kept to the minimum, efficient cooling is, therefore, essential. For cooling purpose either self ventilation (by using a fan mounted on the motor shaft which blows air through suitable ducts in the motor) or forced ventilation (by employing an external fan) is used. In either case care must be taken in designing the ducts to prevent their becoming choked with dirt and axial rather than radial ducts are always used.

Forced ventilation is invariably employed in case of locomotives, as it is operative even when the vehicle is stationary but for motor coaches and road vehicles, self ventilation is usually employed as it does not cause noise due to fan employed for this purpose.

Exercises

1. Briefly discuss the special design features of traction motors. **[Pb. Univ. Electric Traction Dec. 1992]**
2. Discuss briefly the desirable properties of traction motors. **[M.D. Univ. Utilization of Electric Power & Traction Dec. 2009]**
3. Explain various characteristics that an ideal traction motor should possess. Discuss suitability of electric motors for main line, suburban and urban traction.
4. Discuss the suitability of dc series motor for its application in electric locomotive for traction duty. **[Rajasthan Technical Univ. Utilization of Electrical Power 2010]**
5. Discuss the advantages of series motors for electric traction on suburban railways. **[Rajasthan Univ. Utilization of Electrical Power 2006]**
6. Explain the merits of dc series motor that make it a more suitable drive for electric traction than other drives. **[Pb. Univ. Electric Traction Dec. 1992]**
7. Explain load sharing with respect to traction motors (*i*) separately excited dc motors (*ii*) dc series motors. **[A.M.I.E. Sec. B. Electric Drives and Their Control Winter 1999]**
8. What is the effect of changing wheel diameter and gear ratio on the characteristics of a motor ? **[Pb. Univ. Electric Traction April 1975]**
9. Explain how the difference in driving wheel diameters due to unequal wear affects the sharing of load by two similar series motors, working in parallel, driving an electric train. **[Agra Univ. Traction & Utilisation of Elec. Power 1976]**
10. What type of ac motor is usually employed for single phase electric traction ? Discuss briefly the principal features in the construction of the motor and analytically how good commutation and high power factor are obtained. For what frequency and voltage are such motors usually built and why ? **[Pb. Univ. Electric Traction Nov. 1973]**
11. Explain briefly the construction and characteristics of ac series motor, pointing out how they differ from the dc type. In what way is the good commutation and high power factor assured ? **[Agra Univ. Traction & Utilisation of Elec. Power 1976 Supp.; 1978]**
12. How a good power factor and commutation can be obtained in ac series motors ? **[Agra Univ. Traction & Utilisation of Elec. Power 1982]**

13. Compare the use of dc series motor and ac series motor in electric traction. **[Pb. Univ. Electric Traction June 1993]**

14. Compare the suitability of dc series motor for traction duty versus dc shunt motor and single phase ac series motor.

15. Describe with neat sketches the construction and working principle of high acceleration linear induction motor. Discuss its advantages and disadvantages.

[Pb. Univ. Electric Traction April 1975]

16. Explain the theory, working and characteristics of linear induction motor for traction purposes.

[Pb. Univ. Electric Traction June 1993]

17. Write short notes on the following:
    (*i*) Design features of traction motors. **[Pb. Univ. Dec. 1991]**
    (*ii*) Traction motors.
    [M.D. Univ. May 2010; Rajasthan Univ. 2003, 2008]
    (*iii*) Series motor for electric traction. **[Pb. Univ. May 1989]**
    (*iv*) Suitability of ac and dc series motors for traction service.
    (*v*) Suitability of squirrel-cage induction motor for traction drives. **[U.P.T.U. 2005-06]**
    (*vi*) Linear induction motor. **[Pb. Univ. Dec. 1991; Dec. 1992]**
    (*vii*) Selection of traction motors.

Short Answer Type Questions With Answers

Q. 1. Why the magnetic circuit of a traction motor is not made of cast iron ?

Ans. Because cast iron cannot withstand continuous vibrations but traction motors operate in extremely severe service conditions.

Q. 2. Why the traction motor must be small in overall dimensions specially in its overall diameter ?

Ans. The traction motor is located underneath a motor coach (except in few electric locomotives) and the space underneath a motor coach is limited by the size of the driving wheels.

Q. 3. Why the traction motor must be of totally enclosed construction ?

Ans. The traction motor must be totally enclosed type, particularly when mounted beneath the locomotive or motor coach, to provide protection against ingress of dirt, dust, water, mud etc.

Q. 4. Why the traction motor must be light in weight, robust and capable to withstand continuous operation ?

Ans. The traction motor must be light in weight in order to increase the payload capacity of the vehicle.

A traction motor must be robust and capable to withstand continuous vibrations since service conditions are extremely severe.

Q. 5. Why a dc series motor is ideally suited for traction purposes ? **[A.M.I.E. Sec. B Winter 1995]**

Or

Why series motors are preferred for electric traction ?

[Pb. Univ. November 1988]

Ans. The series motors are ideally suited for traction purposes because of the following reasons :

(*i*) The dc series motor develops high torque at start which is an essential requirement of traction service.

(*ii*) Simple speed control.

(*iii*) Power drawn from supply mains varies as the square of the load torque and so series motors are self relieving ones.

(*iv*) Commutation is excellent up to twice full load.

(*v*) The series motors are less susceptible to sudden change in supply voltage.

(*vi*) The series motors are capable of withstanding excessive overloads.

(*vii*) The dc series motors, when operated in parallel to drive a vehicle by means of different driving axles, share load almost equally even there is unequal wear of different driving wheels. This is explained as below.

Consider two identical motors of series type (say of speed-current characteristic shown in Fig. 12.3) connected to different driving axles of the vehicle. For a given speed of vehicle, the peripheral speed of all the driving wheels will be same, but should there be a slight difference in the diameters of the driving wheels (which is possible on account of difference in wear) there will be a corresponding difference in the speeds of the driving axles to which they are connected and also, therefore, a difference in speed of the motors. The speed of the motors will be given by the relation $\frac{N_2}{N_1} = \frac{D_1}{D_2}$.

When the motors are working in parallel the current drawn by each motor (determined from speed-current characteristic) due to slight difference in the speed does not differ much. As voltage across each motor is same, therefore motors share the load almost equally.

(*viii*) The dc series motor is simple and robust in construction.

Q. 6. Why are the ac single phase series motors not suitable for urban and suburban services ?

Ans. Single phase ac series motors develop low starting torque and poor power factor at start and for urban and suburban services large starting torque is an essential feature to provide high acceleration and frequent starting is involved.

Q. 7. Why ac series motors used for traction services are designed to operate at low frequency *i.e.*, $16\frac{2}{3}$ or 25 Hz ?

Ans. Single phase ac series motors have better performance (improved pf, higher efficiency, improved commutation, better starting etc.) if operated on low frequency ($16\frac{2}{3}$ or 25 Hz).

Q. 8. Why ac series motors used in traction services are designed to operate at low voltage (300 to 400 V) ?

Ans. The operating voltage is kept low (300 to 400 V) in order to reduce inductance. Because high voltage motor with proportionately low current would require a large number of turns to produce the required flux.

Q. 9. Why 3-phase induction motors are not used in electric traction ?

Ans. Because of flat speed-torque characteristics, constant speed operation, developing low starting torque, drawing higher current at start, complicated speed control system and complicated overhead feeding system, 3-phase induction motors are not suitable for electric traction.

Q. 10. Traction motors are given one-hour rating as well as continuous rating ?

Ans. The motors are designed for intermittent loading as well as continuous loading. The motors employed in traction work are subject to intermittent loading particularly in case of tramway and suburban services, therefore, traction motors are given one-hour rating as well as continuous rating.

Problems

1. The relation between the current and torque of a series motor determined by a static test is as follows:

Current (amps.)	20	40	60	80	100	120	140	160
Torque (N-m)	67.5	195	346	525	705	918	1,152	1,366

Deduce the speed-current curve of the motor when supplied at a constant voltage of 600 volts, having given the resistance of the main circuit of motor 0.3 ohm.

2. The following figures give the magnetisation curve of a dc series motor when working as a separately excited generator at 500 rpm.

Field current (amps)	10	20	30	40
EMF (volts)	144	254	324	366

The total resistance of the motor is 1.2 Ω. Deduce the speed-torque curve for this motor when operating at a constant voltage of 400 volts.

3. The following figures refer to the speed-current and tractive effort-current characteristic of a dc series traction motor:

Current (amps)	40	80	120	160	200
Speed (kmph)	51.5	33.6	28.8	26.0	24.6
Tractive effort (N)	1,200	4,200	7,400	10,700	14,000

The diameter of a car wheel is 96 cm, the gear ratio is 4 : 1. The car wheel is changed to 92 cm, the gear ratio to 3.5 : 1. Give the new characteristics.

4. Two dc series wound traction motors run at 800 and 850 rpm when each is connected to 700 volt dc mains and takes 100 A. The effective resistance of each motor is 0.5 Ω. If the two motors are connected in series to a 1,500 volt supply and draw 100 A and are also mechanically coupled, at what speed will they run and what will be the voltage across each motor ? **[Ans.** 887 rpm; 771 V; 729 V]

5. A locomotive exerts a tractive effort of 28,200 N in hauling a train at 48 kmph on the level track. If the motor is to haul the same train on a gradient and the tractive effort required is 44,000 N. Determine the power delivered by the locomotive in case it is driven by (*i*) dc series motors (*ii*) induction motors. **[Ans.** 375 kW; 470 kW]

6. A motor coach is equipped with two dc series motors, the characteristics of each at 675 V with 107 cm wheels are as follows:

Speed (kmph)	55.5	46.3	41	37.4
Tractive effort (kg)	610	1,120	1,680	2,240

If the wheels driven by one motor (A) are 107 cm in diameter and those driven by the other motor (B) are 101.5 cm in diameter, determine the total tractive effort at a train speed of 43 kmph when the motors are operating in parallel at normal voltage.

[Agra Univ. Traction & Utilisation of Elec. Power 1976]

[Ans. 2,625 kg]

7. A coach is driven by two similar series motors A and B. A drives 105 cm wheels and B drives 100 cm wheels. Characteristics of motor A are as follows:

Current (amperes)	200	300	400	500
Speed (kmph)	50.5	42.5	38.3	35.3
Tractive effort (kg)	730	1,340	1,970	2,580

Calculate the input to and tractive effort of each motor when (*a*) motors are in parallel and speed is 40 km per hour (*b*) motors are in series taking 400 A. For each motor, total resistance in the armature circuit is 0.8 W. Assume the motors are connected to 600 V dc supply system.

[Ans. (*a*) 213 kW; 183 kW; 1,675 kg, 1450 kg.
(*b*) 117.4 kW, 122.6 kW; 1,970 kg; 2,068 kg]

CHAPTER 13

Control of Traction Motors

INSIDE THIS CHAPTER

13.1. INTRODUCTION

If the traction motors are connected across the supply when the vehicle is at standstill, an enormous current would flow through them and they would develop an immense tractive effort. The result would be disastrous both electrically and mechanically; in fact, satisfactory operation would be impossible. Some means, therefore, is necessary to limit the starting current. It is also desirable to provide more than one economical running speed and electrical braking—rheostatic or regnerative. Thus control equipment is required for (*i*) connecting the motor or motors to supply without taking excessive current at start and provide smooth acceleration without causing sudden shock so as to avoid damage to the couplings and inconvenience to the passengers, (*ii*) adjusting the speed as per requirement, and (*iii*) providing rheostatic or regenerative braking.

13.2. STARTING AND SPEED CONTROL OF DC TRACTION MOTORS

As already discussed, only series and compound dc motors are suitable for traction work. With a dc series motor, the current and torque produced at standstill may be reduced by strengthening the field or lowering the terminal voltage or both. Motors may be placed in series, reducing the terminal voltage of each without loss in external resistance. External resistance may be placed in series with the motors to limit the starting current to any desired value, and by varying the resistance the current may be kept constant during notching up period as desired, as the back emf is being built up. Since maximum torque while starting demands full field strength, any shunt or reduced field connections are usually thrown out of action in starting.

With the dc compound motor the start may be made with full armature current in the series field coils and maximum current in the shunt field coils. A starting resistance inserted in the armature circuit is reduced in steps until the armature and series field are connected across the line. Further speed increase is affected by reducing the shunt field current in steps to the point where the shunt field winding is disconnected, and the action is then identical with that of a plain series motor.

Various methods used for starting and controlling of speed of dc series motors are discussed below.

1. *Rheostatic Control.* A series motor can be started by connecting an external resistance (starter) in series with the main circuit of the motor. At the starting instant, since the back emf developed by the motor is zero, therefore, the resistance connected in series with the motor is maximum and is of such a value that the voltage drop across it with full load rated current is equal to line voltage. As the motor speeds up, the back emf developed by the motor increases, therefore, the external resistance is gradually reduced in order to maintain the current constant throughout the starting or accelerating period. Basic traction motor circuit with rheostatic starting is shown in Fig. 13.1. In this method there is a considerable loss of energy in the external circuit.

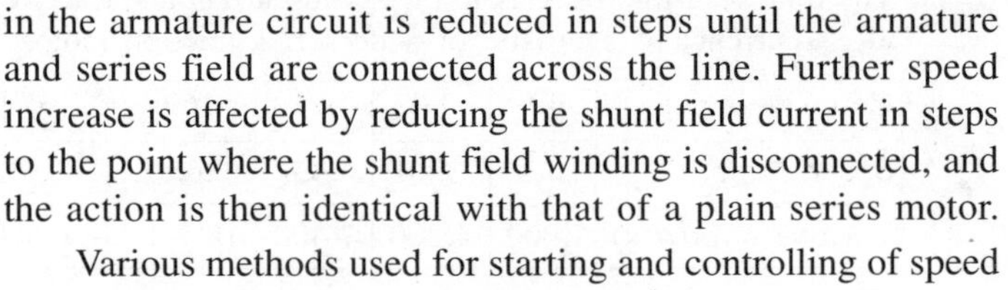

Fig. 13.1

The resistors employed are designed for short-time rating and not for continuous rating as they are required to carry current only during starting of motors. The motor can, therefore, have only one speed characteristic.

2. *Series-Parallel Control.* The main disadvantage of wastage of electrical energy in rheostatic control is partly overcome in this method when there are two or more motors.

In case of two motors, the motors are first connected in series with each other and a starting or control resistance as illustrated in Fig. 13.2 (*a*). The additional resistance is gradually cut out by the controller as the motors attain the speed and finally the control resistance is totally removed, then each motor has one half of the line voltage across it, as shown in Fig. 13.2 (*b*). This is the first running position. In this position for any given value of armature current, each motor will run at half of its normal speed.

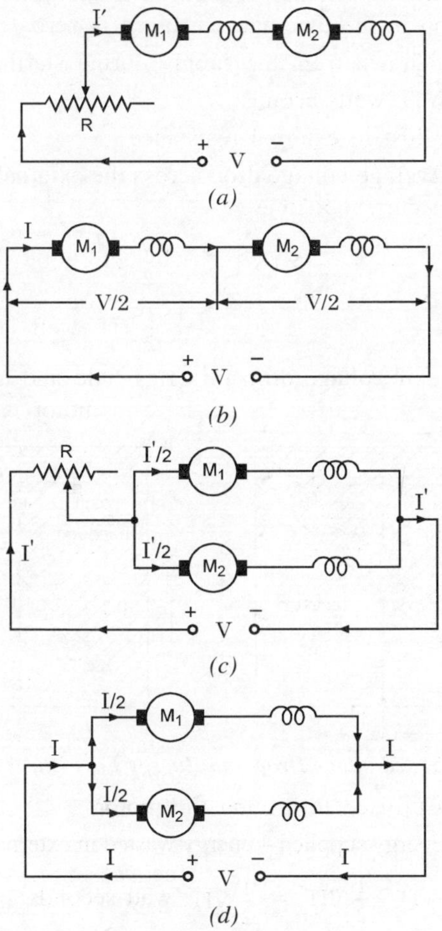

Fig. 13.2

Since there is no external resistance in the circuit, there is no wastage of energy and so motors operate at an efficiency nearly equal to that obtainable with full line voltage across the terminal of each motor.

In the next step the two motors are connected in parallel and in series with a variable resistance R, as shown in Fig. 13.2 (*c*). This resistance is gradually cut out as the motors attain the speed and finally when this resistance is totally removed from the circuit, as illustrated in Fig. 13.2 (*d*), the second running position is obtained. In this position each motor is connected across the full line voltage. A diagram of connections illustrating the switching sequence is given in Fig. 13.3.

3. *Field Control.* As the speed of the motor is inversely proportional to the flux (assuming line voltage constant), therefore, the speed can be varied by varying the flux. In case of series motors the flux can be varied either (*i*) by connecting a variable resistance known as *diverter* in parallel with the series field winding (refer to Fig. 13.4) or (*ii*) by cutting out some of the series field turns (refer to Fig. 13.5). Since in both the cases the flux can be only reduced, this method is known as *field weakening method* and speeds above normal can be obtained. By this method speed can be raised to the extent of 15 to 30 per cent of normal speed owing to design difficulties arising with traction motors.

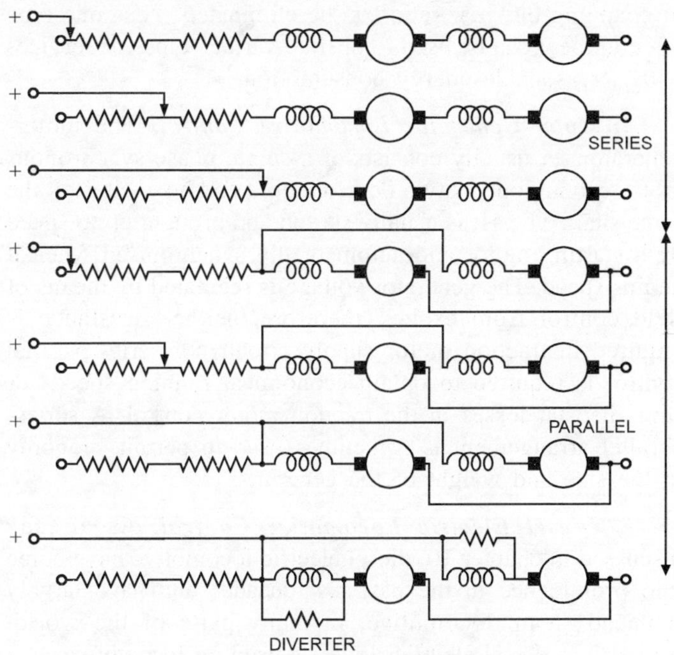

Fig. 13.3. ***Combination of Series-Parallel and Resistance Control of DC Series Motors***

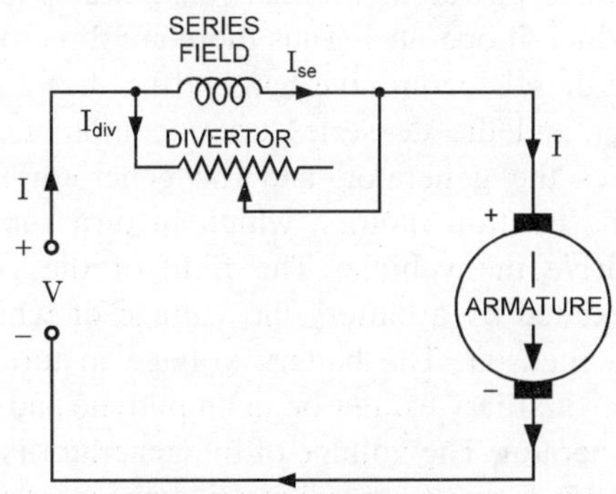

Fig. 13.4. ***Field Divertor Method of Speed Control For DC Series Motors***

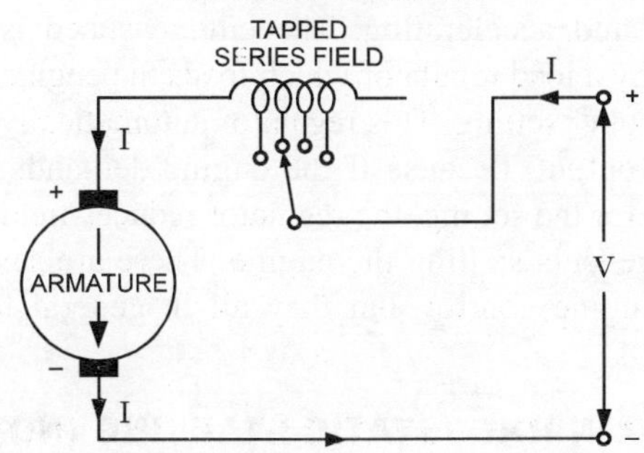

Fig. 13.5. ***Tapped Field Control For DC Series Motors***

The field weakening method is of no use for starting purpose. This method is used for increasing the speed of traction motors up to the extent of 10 or 15 per cent when they have attained maximum possible speed by series-parallel control system. The advantage of this system is that it increases the flexibility of the train utility.

For instance for city service, the speed required is low and frequent starting and stopping are required, the equipment can operate with the full field. When the same vehicle has to be employed for suburban and inter-urban lines where higher speeds are required, it would need a change in the gear ratio. By using field weakening method of speed control, the necessity of changing the gear ratio can be eliminated. Thus one type of equipment can be used to operate various types of services with a reasonable energy consumption.

4. ***Motor-Generator Locomotive Control.*** The motor-generator set usually consists of a single phase synchronous motor driving one or two dc generators and an exciter on the same shaft. The set is usually started and brought up to speed by a starting motor and automatically synchronised when it attains speed. The generator voltage is regulated by means of field control from exciter, therefore, neither resistance is required in traction motor supply circuit nor series-parallel control is required to obtain economical running speeds or save rheostat losses in the traction motor control. A series-parallel arrangement is sometimes used to permit economy in the size and weight of the generator.

5. ***Diesel Electric Locomotive Control.*** As already discussed in chapter 10, diesel electric locomotive have come into prominence in the past few decades and have largely replaced steam locomotive, in many parts of the world. Primarily a diesel-electric drive, as used on locomotive, is a self-contained system of power conversion in which a diesel engine supplies power to the traction motors through a dc generator which it operates. Thus there are three fundamental parts in the diesel locomotive namely the diesel engine, the dc generator and the dc series traction motors. The diesel engine drives the generator, and the generator supplies dc power to the traction motors, which in turn supply motive power to drive the vehicle. The field of the generator is indirectly excited by a battery the voltage of which may be varied by a rheostat. The battery voltage in turn varies the voltage of an auxiliary exciter or an amplidyne and this excites the main generator. The voltage of the generator is thus under control and this form of control makes unnecessary resistance grids in the main motor circuits to protect the traction motors in starting and accelerating. The engine speed is normally controlled by a load regulator which loads the engine according to the controller setting. This regulator automatically maintains a constant output, because if the engine demands more fuel than called for the setting, the regulator reduces field excitation and thus prevents stalling the engine. There are several types of diesel engine control, but they all in general follow this principle.

13.3. PLAIN RHEOSTATIC STARTING (NOTCHING)

As already mentioned in this method the voltage across the motor is gradually increased from zero to full (normal) voltage and the current is kept constant at normal rated value by reducing the external resistance gradually. Let the starting period be T seconds, constant current I amperes and line voltage V volts.

At the instant of switching on the supply, the back emf is zero and it starts building up immediately the motor starts rotating. At any instant the supply voltage is equal to emf developed by motor plus voltage drop in the series field and armature of motor plus voltage drop in starting rheostat.

$$i.e., \quad V = E_b + IR_m + IR \qquad ...(13.1)$$

At the end of starting period, the back emf becomes approximately equal to line voltage and is given as.

$$E_b = V - IR_m \qquad ...(13.2)$$

$\because$ external resistance is reduced to zero, and therefore, voltage drop in the external resistance is zero.

Energy drawn from the supply during starting

$= VIT$ watt-seconds

Energy lost in external resistance

= Average voltage drop across the external resistance × current × time

$$= \frac{V - IR_m + 0}{2} \times I \times T \quad = \frac{V}{2} \times I \times T$$

$$= \frac{1}{2} VIT \text{ watt-seconds}$$

if voltage drop in series field and armature of motor is neglected.

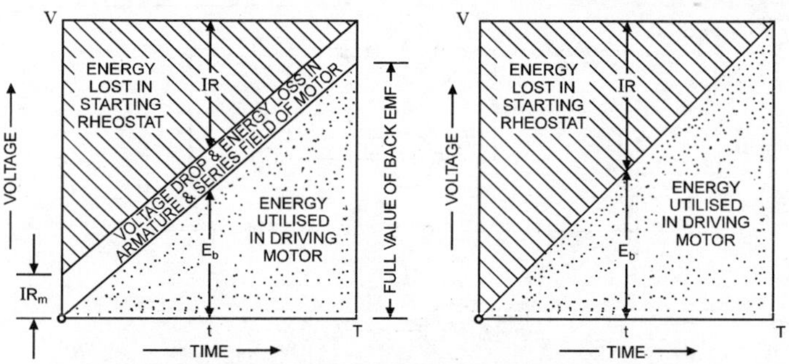

Fig. 13.6. ***Voltage Drop and Energy Loss Illustration***

Energy utilised in driving the motor.

= Energy supplied – energy wasted in external resistance

$$= VIT - \frac{1}{2} VIT = \frac{1}{2} VIT \text{ watt-seconds} \qquad ...(13.3)$$

Starting effeciency,

$$\eta = \frac{\text{Energy utilised}}{\text{Energy drawn}} \times 100 = \frac{VIT/2}{VIT} \times 100$$

$$= 50\% \textbf{ Ans.} \qquad ...(13.4)$$

The energy lost and the voltage drop can be illustrated with the help of illustrations shown in Fig. 13.6.

13.4. SERIES-PARALLEL STARTING

This method is usually employed in electric traction as motors employed for traction work are usually two, four, six or more than this even.

First consider two motors. Let the current be constant and equal to I amperes throughout, the line voltage be V volts and accelerating period be T seconds.

At the starting instant, both the motors are connected in series along with the starting resistance, the voltage across the motors (neglecting voltage drop in armature and series field resistance) is approximately equal to zero and voltage drop across external resistance is maximum and equal to V volts. As the motors speed up, the external resistance is gradually reduced to zero. During this period the voltage acting across each motor gradually increases from 0 to $\frac{V}{2}$ volts. Let the time required be half of total accelerating period *i.e.*, $\frac{T}{2}$ seconds.

Energy utilised by each motor

= Average voltage acting across each motor × current × time.

$$= \frac{0+\frac{V}{2}}{2} \times I \times \frac{T}{2} = \frac{VIT}{8} \text{ watt-seconds}$$

Energy utilised by two motors

$$= 2 \times \frac{VIT}{8} = \frac{VIT}{4} \text{ watt-seconds} \quad ...(13.5)$$

Energy drawn during this period

$$= VI \times \frac{T}{2} = \frac{VIT}{2} \text{ watt-seconds} \quad ...(13.6)$$

When the motors are changed from series grouping to parallel grouping external resistance is again inserted in the motors circuit. The voltage across each motor is equal to $\frac{V}{2}$ volts, the current per motor is I amperes and current drawn from the line is 2I amperes. As the motors speed up, the external resistance is gradually reduced to zero and voltage across each motor increases gradually from $\frac{V}{2}$ to V volts.

Energy utilised in driving the motors

= Number of motors × average voltage across each motor × current × time

$$= 2 \times \frac{\frac{V}{2}+V}{2} \times I \times \frac{T}{2} = \frac{3VIT}{4} \text{ watt-seconds} \quad ...(13.7)$$

Energy drawn during this period

$$= V \times 2I \times \frac{T}{2} = VIT \text{ watt-seconds} \quad ...(13.8)$$

Total energy utilised in driving the motors

$$= \frac{VIT}{4} + \frac{3VIT}{4} = VIT \text{ watt-seconds} \quad ...(13.9)$$

Total energy drawn from the line

$$= \frac{VIT}{2} + VIT = \frac{3}{2} VIT \text{ watt-seconds} \quad ...(13.10)$$

Overall starting efficiency,

$$\eta = \frac{\text{Energy utilised}}{\text{Energy input}} \times 100 = \frac{VIT}{\frac{3}{2}VIT} \times 100 = 66.67\% \quad ...(13.11)$$

In the case of 4 motors all the four motors are connected in series with starting resistance first. The voltage across each motor is gradually increased from 0 to $\frac{V}{4}$ volts by reducing the external resistance gradually to zero. The current per motor is equal to I amperes. The current drawn from the line is also equal to I amperes. The time required for increasing the voltage across each motor from 0 to $\frac{V}{4}$ volts is one-fourth of the total accelerating period *i.e.*, $\frac{T}{4}$ seconds.

In second step two groups, each consisting of two motors connected in series are put in parallel. The voltage across each motor increases from $\frac{V}{4}$ to $\frac{V}{2}$ volts. The current per motor is equal to I amperes and current drawn from the line is 2I amperes. The period is $\frac{T}{4}$ seconds.

In third step all the four motors are placed in parallel. The voltage across each motor gradually increases from $\frac{V}{2}$ to V volts. The current per motor is I amperes. Line current is 4I amperes. The period is $\frac{T}{2}$ seconds.

Energy utilised in driving the motors

$$= 4 \times \frac{\left(0+\frac{V}{4}\right)}{2} \times I \times \frac{T}{4} + 4\frac{\left(\frac{V}{4}+\frac{V}{2}\right)}{2} \times I \times \frac{T}{4} + 4\frac{\left(\frac{V}{2}+V\right)}{2} \times I \times \frac{T}{2}$$

$$= \frac{VIT}{8} + \frac{3VIT}{8} + \frac{3VIT}{2} = 2VIT \text{ watt-seconds} \quad ...(13.12)$$

$$\text{Total energy input} = V \times I \times \frac{T}{4} + V \times 2I \times \frac{T}{4} + V \times 4I \times \frac{T}{2}$$

$$= 2.75 \text{ VIT watt-seconds} \quad ...(13.13)$$

Overall starting efficiency

$$= \frac{\text{Energy utilised}}{\text{Energy input}} \times 100 = \frac{2VIT}{2.75VIT} \times 100 = 72.72\% \quad ...(13.14)$$

In case of 6 motors all the six motors are connected in series with starting resistance. The voltage across each motor is gradually increased from 0 to $\frac{V}{6}$ volts by reducing the external resistance gradually to zero. The current drawn per motor is equal to I amperes. The time required for increasing the voltage across each motor from 0 to $\frac{V}{6}$ volts is one-third of the total accelerating period *i.e.*, $\frac{T}{3}$ seconds.

In second step two groups, each consisting of three motors connected in series are put in parallel. The voltage across each motor increases from $\frac{V}{6}$ to $\frac{V}{3}$ volts. The current per motor is equal to I amperes and current drawn from the line is 2I amperes. The time required will be $\frac{T}{3}$ seconds.

In third step three groups each consisting of two motors connected in series are put in parallel. The voltage across each motor increases from $\frac{V}{3}$ to $\frac{V}{2}$ volts. The current per motor is equal to I amperes and current drawn from the line is 3I amperes. The time required will be again $\frac{T}{3}$ seconds.

Energy utilised in driving the motors

$$= 6 \times \frac{\left(0+\frac{V}{6}\right)}{2} \times I \times \frac{T}{3} + 6\frac{\left(\frac{V}{6}+\frac{V}{3}\right)}{2} \times I \times \frac{T}{3} + 6\left(\frac{\frac{V}{3}+\frac{V}{2}}{2}\right) \times I \times \frac{T}{3}$$

$$= \frac{VIT}{6} + \frac{VIT}{2} + \frac{5VIT}{6} = 1.5VIT \text{ watt-seconds} \quad ...(13.15)$$

Total energy input

$$= \frac{VIT}{3} + V \times 2I \times \frac{T}{3} + V \times 3I \times \frac{T}{3} = 2VIT \text{ watt-seconds} \quad ...(13.16)$$

∴ Starting efficiency

$$= \frac{\text{Energy utilised}}{\text{Energy input}} \times 100 = \frac{1.5VIT}{2VIT} \times 100 = 75\% \quad ...(13.17)$$

The system of double series-parallel control is used on heavy locomotives requiring a large number of rheostatic steps so as to provide smooth starting. With suburban trains, however, the slight saving in energy of double series-parallel control compared with series-parallel control is entirely offset by the additional cost, weight and maintenance of the double series parallel control equipment.

Advantages of Sereis-Parallel Starting. ***(i) Higher starting efficiency.*** 66.67% with two motors, 72.72% with four motors and 75% with six motors whereas only 50% efficiency is obtained with plain rheostat starting.

(ii) Different economical speeds. Two with two motors in the ratio of 1 : 2 : three with four motors in the ratio of 1 : 2 : 4 and three with six motors in the ratio of 1 : 2 : 3.

(iii) For the same power input the torque of different magnitude is obtained. Torque in parallel combination is 1/4th of that in series combination but speed is four times of that with series combination. For the same motor current, the torque remains the same but speed in parallel combination becomes double the speed of that of series combination.

(iv) Owing to low energy loss in the starting rheostats, they are not of cumbersome size.

The main difficulty associated with series-parallel control is in obtaining a satisfactory method of transition from series to parallel without interrupting the torque or permitting any heavy rush of current. Leaving smaller units it is quite inadmissible to open the circuit to affect the change-over.

13.4.1. Practical Requirements

As already stated above, starting resistance is inserted in series with the motors connected in series or series-parallel control at the start and then gradually reduced to zero with the increase in speed of motors. In practice it is not possible to cut off the resistance from the circuit gradually but it is affected with a limited number of steps in the starting rheostats so as to avoid excessive cost and complication of the controller. Hence the motor current during starting cannot be maintained absolutely constant as assumed but it varies between two definite upper and lower limits as the sections of the rheostat are cut out (Fig. 13.7). The number of steps to be provided

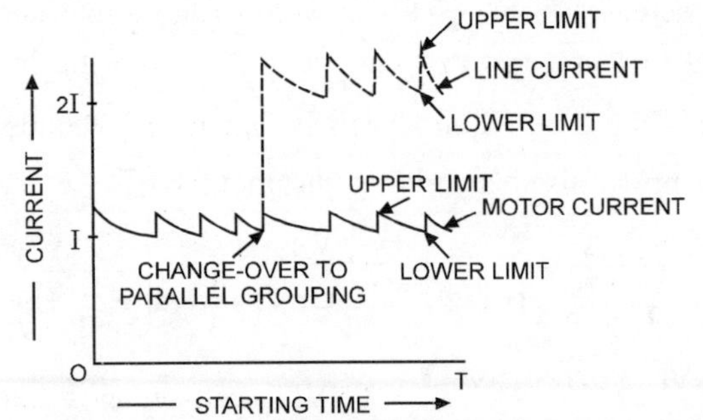

Fig. 13.7. ***Representation of Current Variation During Starting Period***

and the value of resistance cut out depends upon the ratio of maximum to minimum current values which also decides the proportional increase in the tractive effort taking place when a section of resistance is cut out. Because of heavy loads involved, shock tolerance limit is lower in the cases of locomotive trains than in the case of multiple unit trains. This, therefore, requires more rheostat steps in the case of controllers for locomotive trains than for multiple unit trains.

In order to maintain the average value of starting current constant it becomes necessary to change-over from series grouping to parallel grouping when all the starting rheostat has been cut out and the starting current is at its minimum value or at the lower limit. It is also essential to reinsert the starting resistance in the circuit while changing-over from series grouping to parallel grouping so that the motor current is limited to the prescribed value. Transition can be affected by any of the following methods.

13.5. TRANSITION METHODS

The methods of changing-over the connections from one grouping to another are known as *transition methods.* The change-over from *full series* (when the two motors are connected in series and voltage across each motor is equal to half of the supply voltage) to first parallel (when the two motors are connected in parallel and part of the starting resistance is reinserted in the circuit to cause the voltage drop equal to half of the supply voltage) is possible by three methods.

1. Open-circuit transition 2. Shunt transition and 3. Bridge transition.

1. *Open-Circuit Transition.* The various steps involved in open-circuit transition are shown in Fig. 13.8. In open-circuit transition first the series connection between the motors is disconnected and power is switched off with full load current. Then some of the starting resistance is reinserted in the circuit, the two motors are connected in parallel and supply is restored. It causes interruption of torque, and therefore, its use is limited to very small units.

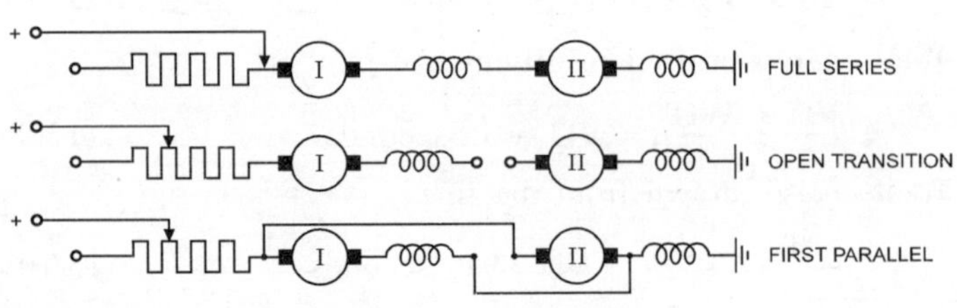

Fig. 13.8. ***Open-Circuit Transition***

2. *Shunt Transition.* In this method first the motors are run up and brought to full series position by cutting out the external resistance gradually. Then some resistance is re-inserted, one of the motor is short circuited, one end of the short circuited motor is opened and finally connected across in such a way that motors are placed in first parallel. The series resistance is now gradually reduced to zero and the motors are placed in full parallel. There is a jerk in this system as one motor is shorted and ceases to act and then other jerk when it is reinserted. This method is employed in tramways, industrial locomotives and main line locomotives. It is preferred for voltages above 600 volts.

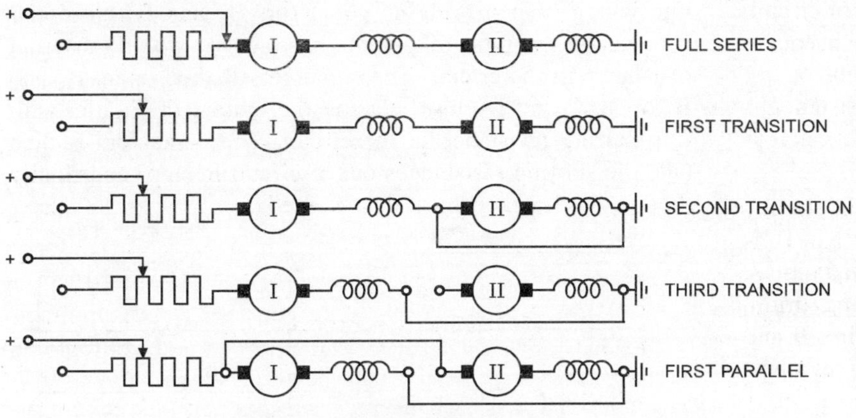

Fig. 13.9. *Shunt Transition*

3. ***Bridge Transition.*** In this method the starting resistance is split up into two equal parts. The motors and starting resistances are connected in series, as shown in Fig. 13.10 (first series). The starting resistances are gradually cut out and the motors come in full series, as shown in the figure. In the transition each motor have a section of the starting resistance in parallel with it and two such combinations are put in series with each other, then the bridge link is removed and the motors are placed in first parallel. Now the external resistance is gradually reduced to zero and the motors come in full parallel.

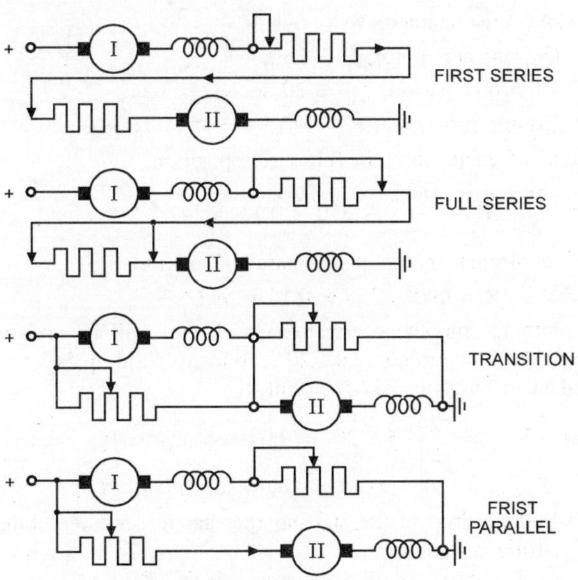

Fig. 13.10. *Bridge Transition*

The advantage of the bridge transition method is that during transition, the motors always remain connected to the supply and as the resistances are so adjusted that the value of current remains the same, the torque remains constant, and therefore, uniform acceleration is obtained without causing inconvenience to the passengers. This method is used for railway traction. In case this method is employed for voltage above 600 V, double contacts at series links are required in order to avoid heavy arcing due to heavy current.

13.6. DRUM CONTROLLER

A series-parallel controller (with shunt transition) for two series motors is shown in Fig. 13.11.

In addition to its primary function of providing series-parallel operation of dc series traction motors it also controls (*i*) speed and direction of motion of the vehicle (*ii*) magnitude of retardation during braking period and (*iii*) provides means to remove faulty motor if any fault occurs during the operation and operating the vehicle with the remaining motor and (*iv*) if possible it must also provide means to prevent its moving backward when stopped on a steep gradient.

The controller is in the form of a rotating drum having insulated and interconnected segments in the form of strips, which makes contact with the fixed points (*known as fingers*). The controller has 8 positions, 4 positions for series and four for parallel running. The working positions of the controller (*known as notches*) are shown by vertical dotted lines. The segments on the rotating drum are shown by black rectangles. Fixed contacts, *known as fingers*, are shown by vertical row of large circular dots on the left. Across these fingers, starting resistances and reversing contacts are connected. The reversing arrangement is provided by providing an additional rotating drum known as *reversing drum.*

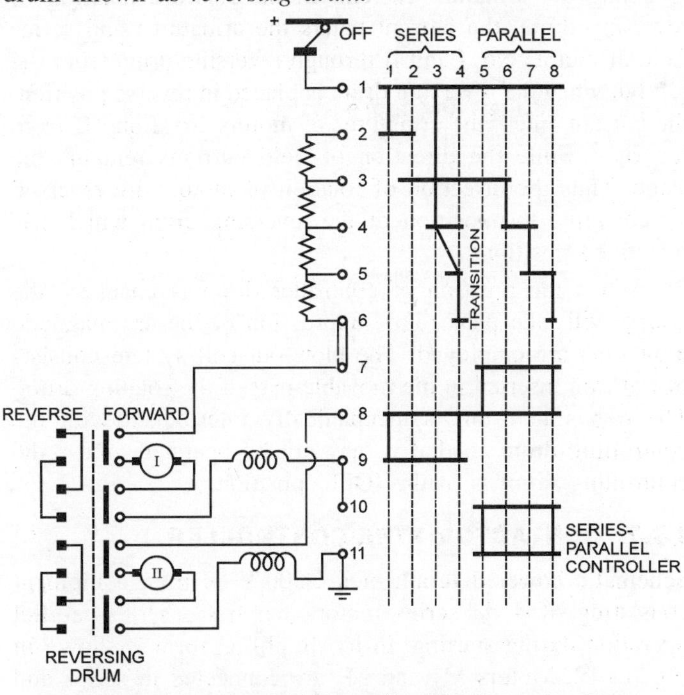

Fig. 13.11. *Series-Parallel Controller With Reversing Drum*

Now when the reversing drum is put in forward position and controlling drum is set in position 1, the connections with fixed points are along vertical line 1. In this position segments make contact with fixed points 1 and 2, 8 and 9. In this position current flows from + ve fixed point 1 through the drum segment, fixed point 2, all the starting resistances 2 – 6, fixed point 7, reversing drum to armature of motor No. I, series field of motor No. I, fixed points 9 and 8, armature and series field of motor No. II and returns back to the –ve terminal.

In the second position the segments of the controller drum come in contact with fixed points 1 and 3, 8 and 9 and thereby bringing some of the starting resistance out of circuit, therefore, speed of the motors increases. In the subsequent 3rd position additional step of resistance is brought out of the circuit and finally in fourth position starting resistance is reduced to zero and half of the supply voltage acts across each motor.

Further movement of the drum reinsert some of the starting resistance in the circuit and short circuit the motor No. II as the current from +ve fixed point 1 passes to fixed point 3, starting resistance, fixed points 6 and 7, reversing drum, armature and series field of motor No. 1, fixed points 9 and 10 and returns to the –ve terminal through lower two segments on the controller drum (on lines 5 to 8).

As the controller drum reaches vertical line No. 5 the motor No. II is again inserted in the circuit and operates in parallel with motor No. I. As the controller drum moves further the starting resistance is reduced and finally in vertical position 8 of controlling drum, both the motors are directly connected across the supply mains and are in full parallel.

For reversing the direction of rotation of motor, it is essential that the direction of current of either of the two (armature or field) is changed. Generally it is achieved by reversing the armature current. In the forward position of reversing drum, the current enters the armatures and series field of motors No. I and II through reversing drum from the left but when the reversing drum is placed in reverse position, the current enters the armatures of motors No. I and II from the right while the direction of field currents remains the same. Thus the direction of rotation of motors are reversed by changing the position of the reversing drum which has two fixed positions.

When the position of controller drum is changed, the arcing will take place. For suppression of the arc magnetic blow-outs are employed. The blow-out coil system consists of flat iron inserted in the movable part of the rotating drum. The reversing drum is mechanically interlocked with the controlling drum so that it may not be operated unless the controlling drum is in the 'OFF' position.

13.7. CONTACTOR TYPE CONTROLLER

Schematic power diagram of a 1,500 V dc train equipment consisting of 4 dc series motors wired for series-parallel operation during starting, in its simplified form is shown in Fig. 13.12. Motors M_1 and M_2 are connected in series and have three breaks L_1 and L_2 and D. Motors M_3 and M_4 are also connected in series and have three breaks L_3, L_4 and C. Bridging contactors B_1 and B_2 have full line voltage of 1,500 V across them when the motors are in parallel and for this reason there are two breaks. Resistance X is inserted in the circuit to protect the system at switch on in case of fault in any motor.

In the starting position line switches L_3, L_4 and bridging contractor B close and then line switches L_1 and L_2. Now motors M_1, M_2, M_3 and M_4 are in series with full starting resistance in the circuit. Then contactors R operate on the following notches and takes the starting resistances out of circuit in steps, the motors are running in series on the full line voltage. Then bridging contactors B_1 and B_2 close and B opens, the motors still operate in series but with no external resistance in the circuit. The contactors R then open, B_1 and B_2 open, C and D close, placing the motors in parallel with full starting resistance in the circuit. The contactors R then take the starting resistances out of circuit in steps and finally the required arrangement is obtained.

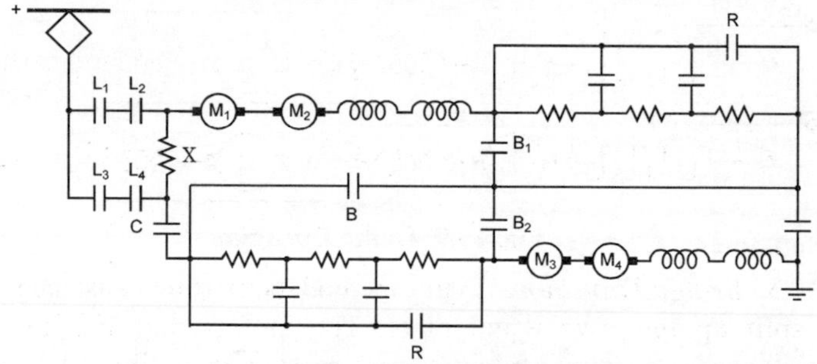

Fig. 13.12. ***Power Diagram of 1,500 V DC Train Equipment (Bridge Transition)***

Example 13.1. Two 600 V dc series motors of coach are started by series-parallel method with the current per motor being 350 A during starting time of 20 seconds and each motor has 0.15 Ω resistance. Calculate (*i*) energy lost in starting rheostat (*ii*) energy lost in motor (*iii*) total energy input from line.

[*U.P. Technical Univ. Utilization of Electrical Energy and Electric Traction 2006-07*]

Solution : Line voltage, V = 600 V

Current per motor, I = 350 A

Starting period, T_s = 20 second

Motor resistance, R = 0.15 Ω

Back emf of each motor in full series position,

$$E_{bse} = \frac{V}{2} - IR = \frac{600}{2} - 350 \times 0.15 = 247.5 \text{ V}$$

Back emf of each motor in full parallel position

$$E_{bp} = V - IR = 600 - 350 \times 0.15 = 547.5 \text{ V}$$

Assuming smooth acceleration, back emf will be built up at constant rate. Since motors take 20 seconds to build up 547.5 V, the time taken to build up 247.5 V emf,

$$i.e., \; T_{series} = \frac{247.5}{547.5} \times 20 = 9.04 \text{ seconds}$$

$$T_{parallel} = 20 - 9.041 = 10.96 \text{ seconds}$$

(*i*) Voltage drop in the starting rheostat in series combination at the starting instant

$$= V - 2IR = 600 - 2 \times 350 \times 0.15 = 495 \text{ V}$$

which reduces to zero in full series position

Energy dissipated in starting rheostat during series combination

$$= \frac{(V - 2IR) + 0}{2} \times I \times \frac{T_{series}}{3,600} = \frac{495 + 0}{2} \times 350 \times \frac{9.04}{3,600}$$

$$= 217.525 \text{ Wh}$$

Voltage drop across starting resistance in first parallel position is equal to $\frac{600}{2}$ *i.e.*, 300 V which is gradually reduced to zero.

Energy dissipated in starting resistance during parallel combination

$$= \frac{\frac{V}{2} + 0}{2} \times 2I \times \frac{T_{parallel}}{3600} = \frac{\frac{600}{2} + 0}{2} \times 2 \times 350 \times \frac{10.96}{3,600}$$

$$= 319.667 \text{ Wh}$$

Energy lost in starting rheostat

$= 217.525 + 319.667 = 537.2$ Wh or 0.5372 kWh **Ans.**

(*ii*) Energy lost in motors

$$= \frac{2I^2RT_s}{3,600 \times 1,000} = \frac{2 \times 350^2 \times 0.15 \times 20}{3,600 \times 1,000} = 0.102 \text{ kWh } \textbf{Ans.}$$

(*iii*) Total energy input from line

$$= \frac{VIT_{se} + 2\,VIT_P}{3,600 \times 1,000} = \frac{600 \times 350 \times 9.04 + 2 \times 600 \times 350 \times 10.96}{3,600 \times 1,000}$$

$$= 1.806 \text{ kWh } \textbf{Ans.}$$

Example 13.2. Two 600 V motors each having a resistance of 0.1 Ω are started on the series-parallel system, the mean current per motor throughout the starting period being 300 A. The starting period is 15 seconds and the train speed at the end of this period is 29 km per hour. Calculate (*i*) the rheostatic losses (in kWh) during (*a*) the series and (*b*) the parallel combinations of motors (*ii*) the train speed at which transition from series to parallel must be made.

[*Agra Univ. Traction & Utilization of Elec. Power 1976 Supp.*]

Solution : Line voltage, V = 600 volts

Current per motor, I = 300 amperes

Starting period, T_s = 15 seconds

Motor resistance, R = 0.1 Ω

Maximum speed, V_m = 29 kmph

Back emf of each motor in full series position,

$$E_{bse} = \frac{V}{2} - IR = \frac{600}{2} - 300 \times 0.1 = 270 \text{ volts}$$

Back emf of each motor in full parallel position,

$$E_{bp} = V - IR = 600 - 300 \times 0.1 = 570 \text{ volts}$$

Assuming acceleration smooth, back emf will be built up at constant rate.

Since motors take 15 seconds to build up 570 volts, therefore, time taken to build up 270 volts emf

i.e., $T_{series} = 15 \times \frac{270}{570} = 7.1$ seconds

$T_{parallel} = 15 - 7.1 = 7.9$ seconds

(*i*) (*a*) Voltage drop in the starting rheostat in series combination at the starting instant

$= V - 2IR = 600 - 2 \times 300 \times 0.1 = 540$ volts,

which reduces to zero in full series position.

Energy dissipated in starting resistance during series combination

$$= \frac{(V - 2IR) + 0}{2} \times I \times \frac{T_{series}}{3,600}$$

$$= \frac{540 + 0}{2} \times 300 \times \frac{7.1}{3,600} = 160 \text{ watt-hours}$$

$$= 0.16 \text{ kWh } \textbf{Ans.}$$

(*b*) Voltage drop across the starting resistance in first parallel position is equal to $\frac{V}{2}$ *i.e.*, 300 volts which gradually reduces to zero.

Energy dissipated in starting resistance during parallel combination

$$= \frac{\frac{V}{2} + 0}{2} \times 2I \times \frac{T_{parallel}}{3,600} = \frac{\frac{600}{2} + 0}{2} \times 2 \times 300 \times \frac{7.9}{3,600}$$

$$= 198 \text{ watt-hours}$$

$$= 0.198 \text{ kWh } \textbf{Ans.}$$

(*ii*) Acceleration, $\alpha = \frac{\text{Maximum speed}}{\text{Starting period}} = \frac{V_m}{T_s} = \frac{29}{15}$ kmphps

Speed at the end of series period

$$= \alpha T_{se} = \frac{29}{15} \times 7.1 = 13.7 \text{ km/hour } \textbf{Ans.}$$

Example 13.3. Of the four dc series motors of an electric train each has the resistance of 0.1 Ω and takes 200 A from 600 V supply during the starting period of 20 seconds. Calculate (*a*) the starting efficiency and (*b*) the speed at which the series connections are to be changed. Assume that the acceleration during the starting period is uniform and the speed just after the starting period is 80 kmph.

[*Allahabad Univ. Utilization of Elec. Energy & Traction 1975*]

Solution : Line voltage, V = 600 V

Current per motor, I = 200 A

Motor resistance, R = 0.1 Ω

Starting period, T_s = 20 seconds

Back emf of each motor during series combination, (assuming two motors in series and two such motor circuits in parallel)

$$E_{bse} = \frac{V}{2} - IR = \frac{600}{2} - 200 \times 0.1 = 280 \text{ volts}$$

Back emf of each motor during parallel combination when all the four motors are in parallel giving 4 circuits

$$E_{bp} = V - IR = 600 - 200 \times 0.1 = 580 \text{ volts}$$

Time of operation in series grouping,

$$T_{series} = T_s \times \frac{\frac{V}{2} - IR}{V - IR} = 20 \times \frac{280}{580} = 9.66 \text{ seconds}$$

Time of operation in parallel grouping,

$$T_{parallel} = T_s - T_{series} = 20 - 9.66 = 10.34 \text{ seconds.}$$

Motor input per motor

$$= \frac{VIT_{se}}{2 \times 3,600} + \frac{VIT_p}{3,600} = \frac{600 \times 200 \times 9.66}{2 \times 3,600} + \frac{600 \times 200 \times 10.34}{3,600}$$

$$= 505.7 \text{ watt-hours}$$

Energy lost in rheostat per motor

$$= \frac{1}{2} \times \frac{(V - 2IR) + 0}{2} \times I \times \frac{T_{series}}{3,600} + \frac{\frac{V}{2} + 0}{2} \times I \times \frac{T_{parallel}}{3,600}$$

$$= \frac{1}{2} \times \frac{600 - 2 \times 200 \times 0.1}{2} \times 200 \times \frac{9.66}{3,600} + \frac{\frac{600}{2} + 0}{2} \times 200 \times \frac{10.34}{3,600}$$

$$= 75.1 + 86.2 = 161.3 \text{ watt-hours}$$

Energy lost per motor

$$= \frac{I^2RT_s}{3,600} = \frac{(200)^2 \times 0.1 \times 20}{3,600} = 22.2 \text{ watt-hours}$$

Output per motor

= Input to each motor – energy lost in rheostat per motor – energy lost per motor

= 505.7 – 161.3 – 22.2 = 322.2 watt-hours

Starting efficiency,

$$\eta = \frac{\text{Output}}{\text{Input}} \times 100 = \frac{322.2}{505.7} \times 100 = 63.7\% \quad \textbf{Ans.}$$

(*b*) Acceleration, $\alpha = \frac{V_m}{T_s} = \frac{80}{20} = 4$ kmphps

Speed at which series connections are to be changed

$$= \alpha T_{series} = 4 \times 9.66 = 38.64 \text{ km per hour } \textbf{Ans.}$$

Example 13.4. A motor coach weighing 150 tonnes is equipped with four 600-V motors for series-parallel control. If during series-parallel starting the current per motor is maintained at 300 A.

Calculate:

(*a*) The durations of the starting period.

(*b*) The speed of the train at transition.

(*c*) The rheostatic losses during (*i*) series (*ii*) parallel steps and start.

At 300 A 600 V, the tractive effort is 15,000 N per motor and the train speed is 30 kmph. Assume the train is started up a gradient of 1 per cent and train resistance is 10 N per tonne. Allow 10 per cent for the effect of rotational inertia. Each motor has a resistance of 0.1 Ω.

[*Gorakhpur Univ. Utilization of Elec. Power and Traction 1975*]

Solution : Dead weight of train, W = 150 tonnes

Accelerating weight of train, $W_e = 1.1\ W = 1.1 \times 150 = 165$ tonnes

Average current per motor, I = 300 A

Maximum speed, V_m = 30 kmph

Supply voltage, V = 600 volts

Train resistance, r = 10 N per tonne

Motor resistance, R = 0.1 Ω

Tractive effort, $F_t = 4 \times 15{,}000 = 60{,}000$ N

$\because\ F_t = 277.8 \times W_e \times \alpha + Wr + 98.1\ WG$

$\therefore\ 60{,}000 = 277.8 \times 165 \times \alpha + 150 \times 10 + 98.1 \times 150 \times 1$

$$\text{or } \alpha = \frac{43{,}785}{277.8 \times 165} = 0.955 \text{ kmphps}$$

(*a*) Duration of starting period,

$$T_s = \frac{V_m}{\alpha} = \frac{30}{0.955} = 31.4 \text{ seconds. } \textbf{Ans.}$$

Back emf of each motor in series (2 circuits, each consisting of two motors in series)

$$E_{bse} = \frac{V}{2} - IR = \frac{600}{2} - 300 \times 0.1 = 270 \text{ V}$$

Back emf of each motor in full parallel combination,

$$E_{bp} = V - IR = 600 - 300 \times 0.1 = 570 \text{ V}$$

Duration for series running,

$$T_{\text{series}} = T_s \times \frac{E_{bse}}{E_{bp}} = 31.4 \times \frac{270}{570} = 14.9 \text{ seconds} \quad \textbf{Ans.}$$

Duration for parallel running,

$$T_{\text{parallel}} = T_s - T_{\text{series}} = 31.4 - 14.9 = 16.5 \text{ seconds}$$

(*b*) Speed of the train at transition

$$= \alpha \times T_{\text{series}} = 0.955 \times 14.9 = 14.23 \text{ kmph} \quad \textbf{Ans.}$$

(*c*) (*i*) Rheostat losses during series operation

$$= \frac{(V - 2IR) + 0}{2} \times 2I \times \frac{T_{\text{series}}}{3{,}600}$$

$$= \frac{(600 - 2 \times 300 \times 0.1) + 0}{2} \times 2 \times 300 \times \frac{14.9}{3{,}600}$$

$$= 270 \times 600 \times \frac{14.9}{3{,}600} = 670.5 \text{ watt-hours} \quad \textbf{Ans.}$$

(*ii*) Rheostat losses during parallel operation

$$= \frac{\frac{V}{2} + 0}{2} \times 4I \times \frac{T_{\text{parallel}}}{3{,}600} = \frac{300 + 0}{2} \times 4 \times 300 \times \frac{16.5}{3{,}600}$$

$$= 825 \text{ watt-hours} \quad \textbf{Ans.}$$

Rheostat losses during start = 670.5 + 825 = 1,495.5 watt-hours **Ans.**

Example 13.5. A series motor is required to produce constant full-load torque whilst its speed is raised 25% above the full-load value by means of a diverter. Calculate the diverter resistance required as a percentage of the field resistance. Neglect losses in the motor. The relation between field current and flux is as follows:

Current	**50**	**75**	**100%**	**of full-load value**
Flux	**60**	**86**	**100%**	**of full-load value**

Solution : Let the full-load speed be N_f; flux ϕ_f and armature current I_{af}

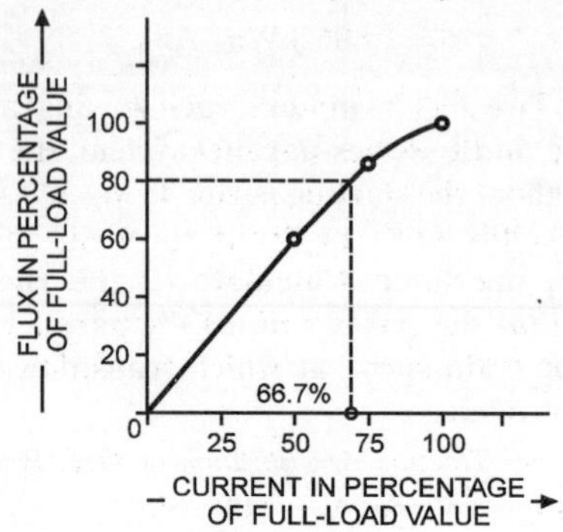

Fig. 13.13

After connecting diverter speed, $N_2 = 1.25\ N_f$

Since speed is inversely proportional to flux, therefore flux required,

$$\phi_2 = \frac{\phi_f N_f}{N_2} = \frac{\phi_f \times N_f}{1.25\ N_f} = 0.8\ \phi_f$$

From magnetic characteristics when $\phi_2 = 0.8\ \phi_f$

$$I_{se_2} = 0.667\ I_{sef}$$

Since torque is constant,

$$\therefore\ I_{a_2} I_{se_2} = I_{af} I_{sef}$$

$$\text{or } I_{a_2} = I_{af} \frac{I_{sef}}{I_{se_2}} = \frac{I_{af}}{0.667} = 1.5\ I_{af}$$

or $I_{a_2} = 1.5\ I_{sef}$ Since without diverter $I_a = I_{se}$

and $I_{se_2} = 0.667\ I_{sef}$

While using diverter, the armature current is equal to the sum of field current plus diverter current.

$$\therefore\ I_{a_2} = I_{se_2} + I_{div}$$

$$\text{or } 1.5\ I_{sef} = 0.667\ I_{sef} + I_{div}$$

$$\text{or } I_{div} = 0.883\ I_{sef}$$

Since current is inversely proportional to the resistance

$$\therefore\ I_{div} R_{div} = I_{se_2} R_{se}$$

$$0.883\ I_{sef} \times R_{div} = 0.667\ I_{sef} \times R_{se}$$

$$\text{or } R_{div} = \frac{0.667}{0.883} R_{se} = 75\% \text{ of series field resistance.} \quad \textbf{Ans.}$$

Example 13.6. A motor has characteristics as follows:

Current in amperes	**40**	**60**	**80**	**100**	**120**	**160**	**200**
Speed in kmph	**58.1**	**47.4**	**40.3**	**35.8**	**33.9**	**29.8**	**28.5**
Drawbar pull in kg	**185**	**440**	**700**	**970**	**1,245**	**1,800**	**2,360**

Deduce the corresponding approximate characteristics for the motor working with two-thirds of the normal field turns in the circuit.

Solution : Speed-armature current and draw bar pull-current curves are shown in Fig. 13.14.

Now when the field turns in the circuit are made two-third, the flux produced by armature current I_a will be equal to that, created by armature current $\frac{2}{3} I_a$ with all turns in series. Thus when the armature current is 120 amperes, the flux is that produced by a normal field coil with $\frac{2}{2} \times 120$ = 80 amperes.

Assuming supply voltage as constant and ignoring voltage drop in armature, speed is inversely proportional to the flux. The speed with the reduced field and armature current of 120 amperes is that which occured previously at 80 amperes. Thus point P_1 on the original speed-current curve becomes the point P_2. Similarly other points on the new speed-current curve can be located.

Tractive effort is proportional to ϕI_a *i.e.*, proportional to $\frac{I_a}{N}$ at constant voltage.

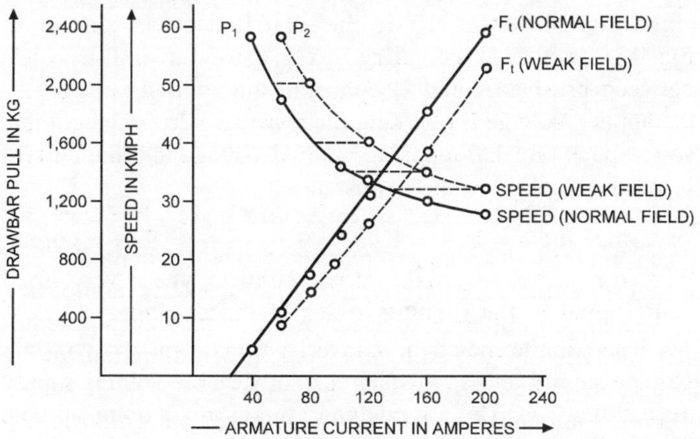

Fig. 13.14

The tractive effort with weak field

$$= \frac{\text{Tractive effort with normal field} \times \text{speed with normal field}}{\text{Speed with weak field}}$$

$$= F_{t_1} \times \frac{N_1}{N_2} \text{ for any armature current}$$

When armature current I_a = 200 A

F_{t_1} = 2,360 kg; N_1 = 28.5 kmph; N_2 = 32 kmph

$$\therefore F_{t_1} = F_{t_2} \times \frac{N_1}{N_2} = 2{,}360 \times \frac{28.5}{32} = 2{,}100 \text{ kg}$$

Similarly,

$$F_{t_2} \text{ for armature current of 160 A} = \frac{1{,}800 \times 29.8}{35} = 1{,}530 \text{ kg}$$

$$F_{t_2} \text{ for armature current of 120 A} = \frac{1{,}245 \times 33.9}{40.3} = 1{,}050 \text{ kg}$$

$$F_{t_2} \text{ for armature current of 100 A} = \frac{970 \times 35.8}{45} = 770 \text{ kg}$$

$$F_{t_2} \text{ for armature current of 80 A} = \frac{700 \times 40.3}{50} = 564 \text{ kg}$$

$$F_{t_2} \text{ for armature current of 60 A} = \frac{440 \times 47.4}{58.1} = 359 \text{ kg}$$

The corresponding approximate characteristics for the motor working with two-thirds of the normal field turns in the circuit are as follows:

Current in amperes	60	80	100	120	160	200
Speed in kmph	58.1	50	45	40.3	35	32
Drawbar pull in kg	359	564	770	1,050	1,530	2,100

13.8. BUCK AND BOOST METHOD OF SPEED CONTROL

The circuit diagram is shown in Fig. 13.15. The armatures of both the traction motors I and II and M-G set are connected in series. The whole series combination is connected across the supply. When the generator terminal voltage is equal to supply voltage in magnitude but of opposite polarity and the main contactor MC is closed, the voltage across the traction motors is zero and thus their speed is also zero. On reducing the generator voltage, voltage across the traction motors will start increasing and so their speed too. When the generator voltage becomes equal to zero, full line voltage will appear across both the motors *i.e.*, one-half of the supply voltage across each motor. In case polarities of both supply source and generator are same and generator voltage is equal to supply voltage, voltage across each traction motor will be equal to supply voltage. Thus the voltage across traction motors can be adjusted (bucked or boosted up) by adjusting the excitation of the generator.

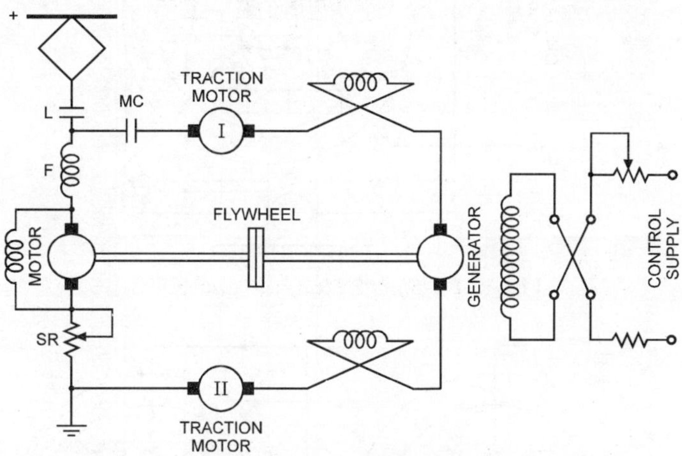

Fig. 13.15. ***Buck and Boost Method of Speed Control***

The above method of speed control have the following advantages:

(*i*) Traction motors can be operated on any speed, while in case of resistance control only a few speeds are possible.

(*ii*) There is no energy loss in the starting resistance of the traction motors. However, there is a loss of energy in the starting resistance of the M-G set.

(*iii*) In case of temporary interruption in the supply, the kinetic energy of the flywheel can be utilized in generating energy from the M-G set and supplied to the traction motors.

The main drawback of this method of speed control is requirement of M-G set. There is a loss of energy in M-G set.

13.9. METADYNE CONTROL

The metadyne control system is based on constant current system of speed control. In resistance control or series- parallel control a great deal of energy is dissipated in the starting resistance and jerks are experienced when the controller of the starter moves from one position to another position. In metadyne control, current throughout the accelerating period remains constant, therefore, uniform tractive effort is developed and very smooth control, without causing any wastage of energy in the starting resistance, is achieved.

The essential part of the metadyne control is metadyne converter. The metadyne converter is a cross-field machine which behaves like a transformer on direct current. The transformation ratio of a metadyne can be varied continuously. It takes power at constant voltage and variable current and delivers the same at constant current and varying voltage.

The metadyne converter essentially consists, in its simplest form, of a 2 pole dc armature with two pairs of brushes and a four pole field magnet, as shown in Fig. 13.16. One pair of the brushes (say A and C) are connected across constant voltage dc supply while the other pair (B and D) are connected to the load (normally a dc series motor).

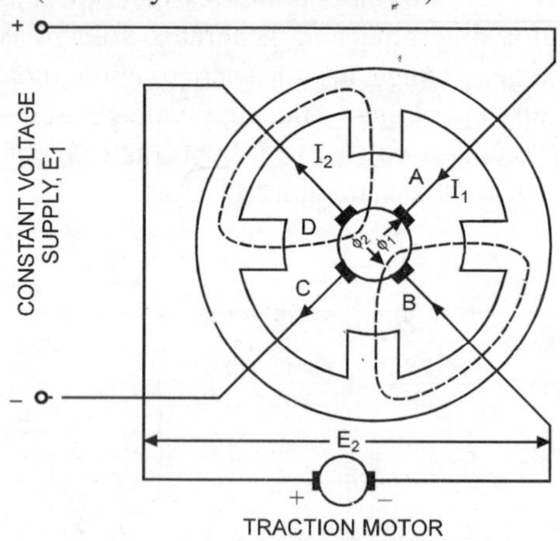

Fig. 13.16. *Simple Metadyne Converter*

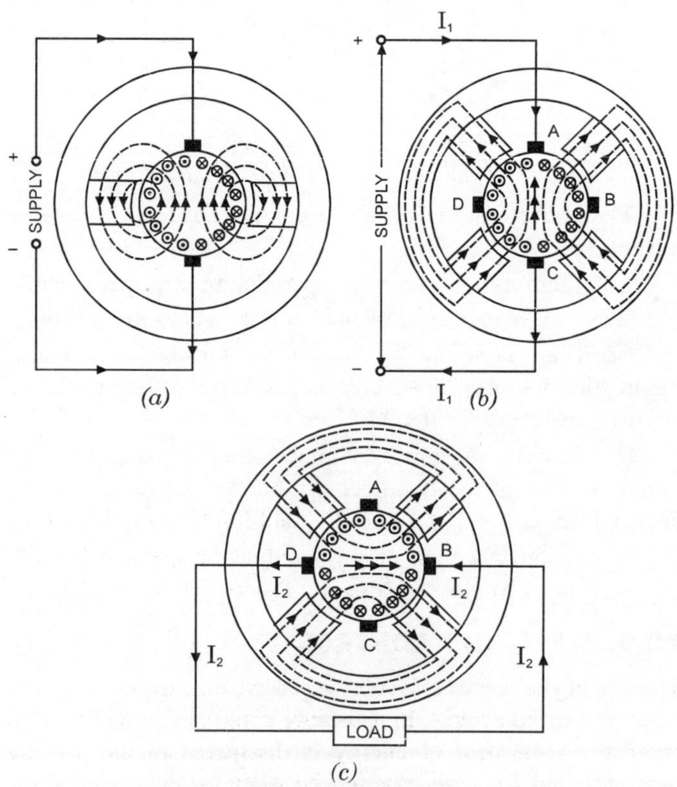

Fig. 13.17. *Illustration of Metadyne Principle*

For understanding the working of a metadyne converter consider first an ordinary dc machine with two poles and two brushes supplied with a current flowing in the direction shown in Fig. 13.17 (*a*). It will cause armature current distribution, as illustrated in the figure with corresponding cross flux, mainly confined to the poles.

Now consider that metadyne converter (a dc machine with two pairs of brushes and two pairs of poles) is running at constant speed and drawing a current I_1 from the dc supply main, which flows through the armature conductors via the brushes A and C, as shown in Fig. 13.17 (*b*). An armature reaction flux ϕ_1 set up in usual way is provided with a fairly low reluctance path through the yoke by the four poles, as shown in the figure. Due to rotation of armature conductors in this primary flux, and emf $E_2 = KI_1$ is set up between the brushes B and D. When these brushes B and D are connected to a load, a current I_2 flows through the load. The load current I_2 sets up another flux ϕ_2 known as *secondary flux*, at the right angles to the first, the distribution is shown in Fig. 13.17 (*c*). This secondary flux ϕ_2 causes an emf, $E_1 = KI_2$ between brushes A and C opposing the applied voltage. As the applied voltage is constant, the resistance drop is negligible so the back emf E_1 opposing applied voltage and the current I_2 producing E_1 are also constant.

Since input $= E_1I_1 = KI_2I_1 = KI_2 \times \frac{E_2}{K} = E_2I_2 =$ output, therefore, power required to drive the metadyne is very small being equal to the running losses of the machine.

This simple metadyne converter transforms the constant voltage supply into a constant current variable voltage supply to feed the load. The arrangement, therefore, is quite suitable for starting dc traction motors. With this arrangement the load current I_2 and supply voltage V remain constant and as the load increases on account of building of back emf in dc traction motor E_2 and I_1 increases to meet with the increased load.

The metadyne described above has no winding on the poles and is capable of delivering only a single value of constant current but for supplying dc traction motors, after the motor has gained speed, the load current I_2 has to be reduced to the running value. For this purpose the field magnet poles are provided with variator and regulator windings, as shown in Fig. 13.18.

The variator winding sets up a flux in the same direction as that set up by the load current I_2. Total flux ϕ_2 required in this axis being constant in order to produce a back emf E_1 equal to the constant supply voltage, therefore if some of this flux is set up by a separate winding, known as *variator winding*, the load current I_2 will decrease and can be of a smaller constant value. Similarly the load current can be increased by causing the current to flow in the variator winding in the opposite direction.

If the output current I_2 is, say, reduced in this way, the voltage remaining the constant, the total output will be reduced but input will remain the same and therefore, the set will speed up. In order to keep the speed of the metadyne converter constant, an additional winding known as *regulator winding* is provided. By adjusting the current in the regulator winding, the input current can be varied and therefore, input power can be adjusted equal to output power, the speed of the converter remaining the same.

The regulator winding is supplied from a small dc shunt generator mounted on the shaft of the metadyne. Any tendency towards a change in speed of the metadyne will cause corresponding change in the emf set up by the shunt machine and as it acts in the opposite direction of the supply voltage, so corresponding change in the regulator winding current will result in.

The variator winding is supplied excitation from an exciter mounted on the same shaft, as shown in Fig. 13.18.

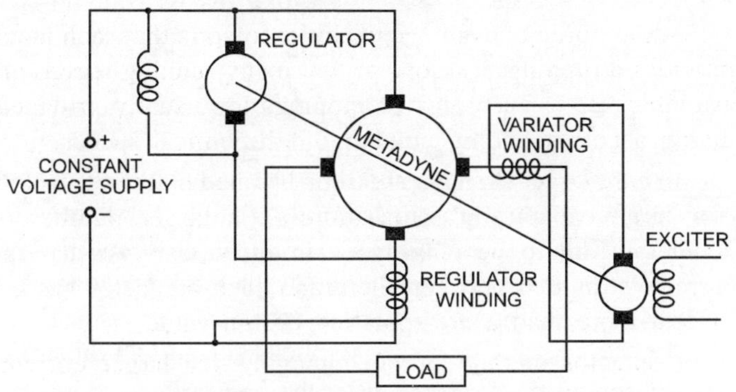

Fig. 13.18. ***Metadyne Control System***

With metadyne converter, regenerative braking can be accomplished very easily by reversing the field of the traction motor. This causes the reversal of direction of induced emf E_2 which in turn will change the direction of current I_1. Thus current I_1 can be supplied back to the supply source. By controlling the magnitude of reversed excitation of traction motors supplied by metadyne, the magnitude of regenerative braking can be regulated.

The metadyne is employed whenever control of dc motors is required. The control provided by the metadyne is smooth and does not require any switching. Thus switchgear and arcing are avoided. In some cases it is cheaper than the Ward Leonard system in initial cost. In traction it provides smooth acceleration without skill on the part of driver and regenerative braking down to very slow speeds. The savings due to these items may easily counterbalance the additional cost of the more complicated equipment required and its additional maintenance cost. It is already being employed in the underground railway.

13.10. THYRISTOR CONTROL OF TRACTION MOTORS

The modern trend is towards the use of dc motors (both separately excited and dc series motors) equipped with thyristor chopper control. The operating voltages are 600 V or 1,000 V. Brakings employed are mechanical, rheostatic and regenerative. Thyristorised converters provide accurate control and fast response. Main advantages of thyristor control are the absence of bulky on-load tap changer and electromagnetic devices, saving of energy, notchless control, incease in pulling ability of the motive power, and minimum wear and tear because of absence of conventional moving parts in the motor control circuits.

Various methods of speed control and electric braking employing thyristors have already been discussed in chapter 3.

In addition to ordinary phase control methods, *cycle selection methods* of control of SCR for varying the voltage applied to the traction motors are also employed. In this method the required average voltage is obtained by accepting or rejecting a certain number of complete half cycles. In practice, at the start only one half cycle out of eight is accepted and as the speed builds up, it is gradually raised to $\frac{2}{8}, \frac{3}{8}$ and finally $\frac{8}{8}$ for full-power operation. This method is advantageous due to low frequency harmonics, low rate of rise of current, better power factor etc.

In *chopper control* of traction motors, at start, the '*on*' period of pulse is kept very short which lengthens during the period of controlled acceleration. Thus the average voltage applied across the traction motors is gradually increased keeping the mean value of the input current close to the desired value.

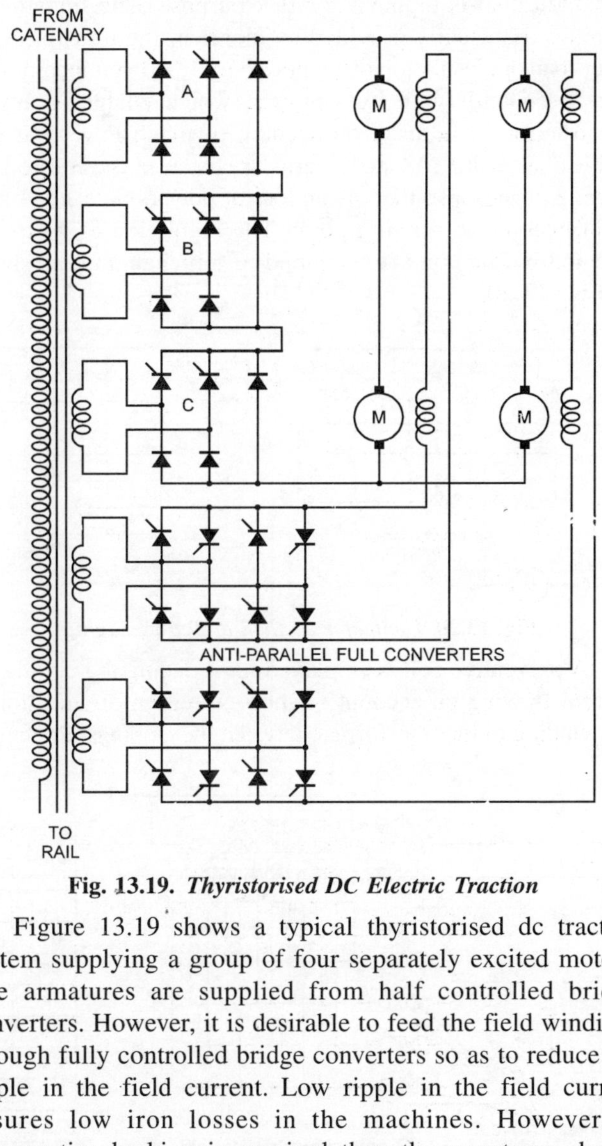

Fig. 13.19. ***Thyristorised DC Electric Traction***

Figure 13.19 shows a typical thyristorised dc traction system supplying a group of four separately excited motors. The armatures are supplied from half controlled bridge converters. However, it is desirable to feed the field windings through fully controlled bridge converters so as to reduce the ripple in the field current. Low ripple in the field current ensures low iron losses in the machines. However, if regenerative braking is required then the armatures should be supplied from fully controlled bridges. Freewheeling diodes are connected as illustrated to ensure good waveform of armature current. The armatures are connected in series-parallel arrangement to ensure good starting and running characteristics. It is seen that armatures are supplied by three bridges connected in series. For starting first only bridge A is triggered and firing angle is advanced as speed builds up. When bridge A is fully conducting (*i.e.,* when $\alpha = 0$), bridge B is triggered and then bridge C is triggered. During starting field currents are set to maximum to provide high starting torque. The use of three bridges ensures better power factor than would be possible with a single bridge.

13.11. SPEED CONTROL AND STARTING OF SINGLE PHASE SERIES MOTORS

The control of speed and torque in case of single phase ac series motor is carried out by varying the applied voltage. As the operating voltage for a single phase series motor is very low (about 300 volts), thus, transformer forms an essential part of the locomotive equipment and, therefore, variable voltage for both starting and speed control can be obtained by means of tappings on the transformer without requiring any additional equipment for this purpose. The transformer tappings are usually provided on *lt* side of the transformer in order to avoid insulation difficulties but 11,000 volt locomotives have been built with hv tappings, which enables a lighter construction to be used on account of much lower currents to be dealt with. As no external resistance is required for starting purpose, so there is no loss of power and each control point becomes a running point. The regulation of voltage is affected by the contactor method of tap changing, as shown in Fig. 13.20.

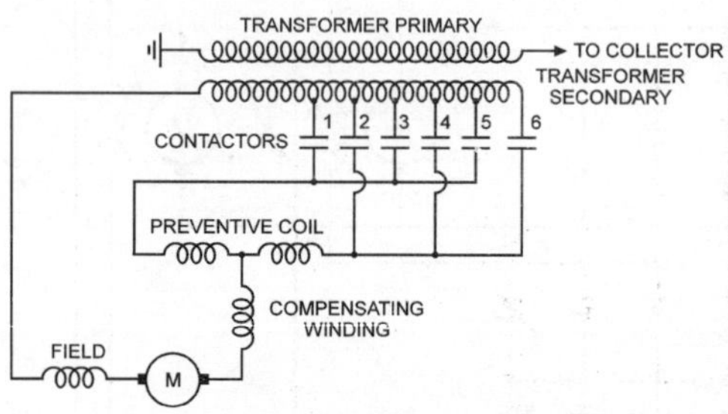

SEQUENCE OF CONTACTORS

NOTCH	CONTACTORS 1	2	3	4	5	6
1	•	•				
2		•	•			
3			•	•		
4				•	•	
5					•	•

Fig. 13.20. ***Contactor Method of Tap Changing***

A preventive coil is employed for reducing the circulating current flowing on account of short circuiting of a section of the winding of the transformer between the tappings. Preventive coil (choking coil having high resistance) is inserted between the terminals of two groups consisting of tappings 1, 3, 5 and 2, 4, 6 as shown in the Fig. 13.20. The motor is connected to the centre point of the preventive coil. Two contactors, connected to the adjacent tappings (1 and 2 or 2 and 3 or 3 and 4 or 4 and 5 or 5 and 6) are closed on each notch of the controller. Each contactor, therefore, carries approximately one half of the total current required. For changing-over voltage from one value to another one contactor is opened and another contactor belonging to the same group is closed.

An important advantage of this method is that each notch provides a running position, so that many running speeds are available. Between 6 and 12 tappings are usually provided, giving a corresponding number of economical speeds.

In case larger currents are to be handled by the contactors (for large capacity traction motors), a larger number of contactors are to be employed simultaneously so that the current is divided between them and each contactor has not to deal more than a pre-specified design value.

Contactor method of tap changing for larger currents (over 1,000 A) is shown in Fig. 13.21. In this method four adjacent contactors are connected to the same notch of the controller. There are three preventive coils, also called '*bridging coils*' arranged in such a fashion as to facilitate the correct division of current among the contactors.

The ac contactors are quite different from dc contactors. In general, dc contactor is much lighter than the ac contactor and needs less power for its operation. Other special features of ac contactors are (*i*) laminated magnetic circuit (*ii*) provision of shading coil in order to avoid chattering of contactors.

13.12. STARTING AND SPEED CONTROL OF 3-PHASE INDUCTION MOTORS

The method employed for speed control of 3-phase induction motors are (*i*) rheostatic control (*ii*) pole changing control (*iii*) cascade control and (*iv*) combination of cascading and pole changing control.

Different methods of speed control of 3-phase induction motors have already been discussed in chapter 1 (Art. 1.18.2). Thyristor control of 3-phase induction motors have been discussed in chapter 3 (Art. 3.13).

Rheostatic control is the simplest but the least efficient method of speed control of 3-phase induction motors. This method of speed control is employed in light locomotives and motor coaches where a single economical speed is sufficient and energy consumption is of no importance.

Pole-changing control is the simplest of the multi-speed control methods. This method has the advantage of simplicity, good speed regulation for each setting, high operation efficiency, and

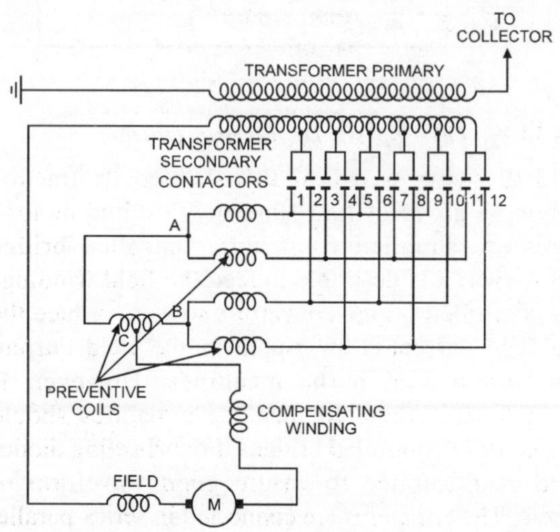

SEQUENCE OF CONTACTORS

STEP	CONTACTORS 1	2	3	4	5	6	7	8	9	10	11	12
1	•	•	•	•								
2		•	•	•	•							
3			•	•	•	•						
4				•	•	•	•					
5					•	•	•	•				
6						•	•	•	•			
7							•	•	•	•		
8								•	•	•	•	
9									•	•	•	•

Fig. 13.21. ***Contactor Method of Tap Changing For Larger Currents***

moderate first cost and maintenance. The choice of the number of poles on a pole-changing winding is in the ratio of 2 : 1, 3 : 2, 4 : 3.

Cascade control is employed for the two-speed goods locomotives (24 kmph and 50 kmph).

In *combination of cascading and pole-changing control method*, both of the motors employed are wound for the same number of poles and windings so arranged that the number of poles can be changed. With equal number of poles on both the machines one speed is obtained and then more speeds are made available by changing the number of poles. This method has been employed for the four-speed passenger locomotives (37, 51, 74 and 102 kmph) in some types of locomotives.

Exercises

1. Describe with the aid of diagrams following methods of speed control for dc traction motors, pointing out the merits of each:
(*i*) series-parallel control. (*ii*) control by weakening field.
Explain also the conditions under which the speed control by weakening of field is advantageous.

2. Explain with suitable connection diagrams, the resistance control method employed in dc traction drives.
[A.M.I.E. Sec. B. Electric Drives & Their Control Winter 1996]

3. Under what circumstances is the field weakening adopted for controlling speed of traction motors and what are the merits and demerits of this method of control ? Illustrate your answer with the help of tractive effort-speed characteristics of the locomotive.

4. Explain the method of series-parallel starting of traction motors. Mention its advantages.
[Pb. Univ. Electric Traction Dec. 1992]

5. Discuss the advantages of series-parallel control of starting as compared to the rheostatic starting for a pair of dc traction motors.
[W.B. Univ. of Technology Utilization of Electric Power 2009]

6. Explain the starting of dc series motor type traction motor and prove that with the series-parallel method of starting the efficiency of starter is increased from 50% in case of normal rheostatic starting to 66.67%.

7. Distinguish between shunt and bridge transition methods of starting a pair of dc traction motors.
[Rajasthan Univ. Utilization of Electrical Power 2005]

8. Draw and explain the complete view of drum controller giving shunt transition between two series traction motors.
[Pb. Univ. Electric Traction 1975]

9. Explain the short-circuit transition method of control of a pair of tram motors and give the developed view of the drum of a controller arranged for this method of control, indicating also the connections of the reversing barrel.

10. Give a schematic power diagram of 1,500 V dc train equipment of 2 dc series motors wired for series-parallel operation during starting and explain it.
[B.H.U. Electric Traction & Utilization 1978]

11. Explain with relevant diagrams the method of metadyne control of electric traction motors.
[B.H.U. Elec. Traction & Utilization 1978, Agra Univ. Traction & Utilization of Elec. Power 1983]

12. Describe with a neat connection diagram the principle of operation of a metadyne set to be used for starting the series traction motor. What are the advantages of variator and regulator windings ?
[Agra Univ. Traction & Utilization of Elec. Power 1979]

13. Describe with a neat diagram the principle of operation of a metadyne set to be used for starting of the series traction motor. **[Rajasthan Univ. Utilization of Electrical Power 2004, 2005, 2006, 2007]**

14. Discuss the advantages and disadvantages of thyristor control of electric traction.
[M.D. Univ. Utilization Electric Power and Traction Dec. 2008]

15. Discuss the advantages and disadvantages of thyristor control of traction motors.
Explain the chopper method of control of dc motors. What are the merits and demerits of the above method of control ?
[Pb. Univ. Electric Traction June 1993]

16. Describe the contactor method of tap changing for the control of speed and torque of a single phase series motor.

17. Compare the use of dc series motor and ac series motor in electric traction. Discuss the controllers used for such motors.
[Pb. Univ. Electric Traction June 1993]

18. Explain any one method of speed control of induction motor. What speed range is obtainable by this method ?
[Pb. Univ. Electric Traction Dec. 1991]

19. What are the requirements of electric drives for traction ? Explain a static control scheme used in ac traction employing ac motors. **[A.M.I.E. Sec. B. Electric Drives & Their Control Winter 1998]**

20. Write short notes on the following:
(*i*) Speed control of dc series motors.
(*ii*) Series-parallel control of starting for pair of dc traction motors. **[G.B.T.U. 2010-11]**
(*iii*) Series-parallel control of dc series motors.
[U.P.T.U. 2009-10]
(*iv*) Energy saving with series-parallel starting.
[Rajasthan Univ. 2000, 2004, 2006; Rajasthan Technical Univ. 2010]
(*v*) Tramway controller.
(*vi*) Metadyne control. **[D.E.I. Agra 1989]**
(*vii*) Thyristor control of electric traction.
[M.D. Univ. December 2009]
(*viii*) Speed control and starting of single phase ac series motors.
(*ix*) Drum controller. **[Pb. Univ. July 1974]**

Short Answer Type Questions With Answers

Q. 1. The resistors employed for control of traction motors are designed for short-time rating. Why ?

Ans. Because control resistors are required only during starting of traction motors.

Q. 2. Under what conditions is field weakening adopted for controlling the speed of a dc series traction motor ?

Ans. The field weakening method of speed control is used for increasing the speed of traction motors when they have attained maximum possible speed by series-parallel control system.

Q. 3. Under what circumstances series-parallel method of speed control of dc series traction motors adopted ?

Ans. The sereis-parallel method of speed control of dc series motors is adopted when motors employed are two, four, six or more than this even.

Q. 4. What is the main difficulty associated with series-parallel control ?

Ans. The main difficulty associated with series-parallel control is in obtaining a satisfactory method of transition from series to parallel without interrupting the torque or permitting any heavy rush of current.

Q. 5. What are the advantages of thyristor control of traction motors ?

Ans. The advantages of thyristor control of traction motors are accurate control and fast response, elimination of bulky on-load tap changer and electromagnetic devices, saving of energy, notchless control, increase in pulling ability of the motive power, and minimum wear and tear.

Q. 6. Enumerate the methods employed for speed control of 3-phase induction motors used in electric traction.

Ans. The methods employed for speed control of 3-phase induction motors used in electric traction are (*i*) rheostatic control (*ii*) pole-changing control, (*iii*) cascade control (*iv*) combination of cascading and pole-changing control method and (*v*) thyristorised control.

Q. 7. In ac locomotive tap-changer is provided on high voltage side of transformer. Give suitable reason. **[U.P.T.U. 2006-07]**

Ans. The transformer tappings are usually provided on low voltage side of the transformer to avoid insulation difficulties but 11,000 V ac locomotives have been built with high voltage tappings, which enables a lighter construction to be used on acount of much lower currents to be dealt with.

Problems

1. Two motors of a motor coach are started in series-parallel, the current per motor being 300 A (considered as being maintained constant during the starting period of 24 seconds). If the acceleration is uniform, the line voltage is 750 volts and the resistance of each motor is 0.1 Ω, determine
   (*a*) the time during which the motors are operating in series.
   (*b*) the time during which the motors are operating in parallel.
   (*c*) the energy drawn from the supply mains.
   (*d*) the energy dissipated in starting resistances.
   (*e*) the starting efficiency and
   (*f*) the speed at which series connections should be changed to parallel if the speed at the end is to be 48 kmph.
   [**Ans.** (*a*) 11.5 seconds (*b*) 12.5 seconds (*c*) 2.28125 kWh (*d*) 0.72125 kWh (*e*) 68.4% (*f*) 23 kmph]

2. Two motors of a motor coach are started on series-parallel system, the current per motor being 350 A (considered as being maintained constant during the starting period which is 18 seconds). If the acceleration is uniform, the line voltage is 600 volts and the resistance of each motor is 0.1 Ω, calculate
   (*a*) the time during which motors are operating in series
   (*b*) loss in the rheostats during starting period.
   [DEI Agra Utilization of Electric Power 1989; Rajasthan Univ. 2007]
   [**Ans.** (*a*) 8.44 seconds (*b*) 496 watt-hours]

3. A train equipped with two 600 V dc series motors is started on series-parallel system, the main accelerating current being 300 amperes. The train speed at the end of starting period of 15 seconds is 30 kmph. Determine the train speed at the time of transition and rheostatic losses in kWh during (*a*) series period (*b*) parallel period. Take resistance of each motor as 0.1 Ω. **[P.U. B.Sc. (Engg.) Nov. 1974]**
   [**Ans.** (*i*) 14.2 kmph (*ii*) (*a*) 0.16 kWh (*b*) 0.18 kWh]

4. Two 600 V dc series motors are started by series-parallel control. Each motor takes 500 A during starting time for 20 seconds and has 0.1 Ω resistance. Calculate (*i*) energy lost in starting rheostat (*ii*) energy lost in motor (*iii*) motor output and (*iv*) efficiency of starting.
   [**Ans.** (*i*) 0.77 kWh (*ii*) 0.139 kWh (*iii*) 0.7635 kWh (*iv*) 59.3%]

5. Two 600 volt dc traction motors each having a resistance of 0.15 Ω, are started with series-parallel control. The starting time is 20 seconds and the current in each motor during starting is maintained constant at 300 A. Calculate the energy in kWh dissipated in the starting rheostats.
   [**Ans.** 0.466 kWh]

6. A car driven by two dc traction motors takes current of 200 A from a 600 V line with motors in full series. Each motor has a resistance of 0.2 Ω. The motors are to be switched into parallel through a resistance. If the transition is to be affected without shock, calculate the value of this resistance.
   [**Ans.** 1.5 Ω]

7. A series traction motor has following characteristics:

Current	40	80	120	160	200 amperes
Speed	51.5	33.6	28.8	26	24.6 kmph
Tractive effort	120	420	740	1,070	1,400 kg

   Draw the new characteristics when 25% of the field turns are cut off. **[Pb. Univ. Electric Traction April, 1975]**

8. The tractive effort and speed characteristics of a motor are given below. Find the new characteristics if one third of the field turns are cut off.

Current (amperes)	50	75	100	150	200	250
Speed (kmph)	73.4	59.6	50.7	42.7	38.7	36
Tractive effort (kg)	207	496	794	1,400	2,030	2,660

   [Pb. Univ. Electric Traction July 1974]

❑❑❑

CHAPTER 14

Braking; Mechanical Considerations and Control Equipment

INSIDE THIS CHAPTER

14.1. BRAKING

Electrical and mechanical, both types of braking are used in electric traction. In electric braking the braking energy is converted into electrical energy instead of converting it into heat energy at the break shoes and either dissipated in the resistances mounted on the vehicle or returned to the supply system. Electric braking reduces the wear of the brake shoes and wheel tyres considerably and gives higher rate of braking retardation, thus brings the vehicle quickly to rest and shortens the running time to a considerable extent. In case regenerative braking is employed, the braking energy is returned to the supply system thereby a considerable saving is affected in the running cost, higher speeds are possible while going down the gradients as more braking power is available and heavier trains can be propelled down the gradients with safety and speed without dividing it into sections. Electric braking cannot replace the ordinary mechanical brakes, as the vehicle cannot be held stationary by it.

The desirable requirements of a braking system are given below:

1. The braking system should be simple, robust, quick, and reliable in action.
2. Maintenance needs should be minimum and braking system must be easy for driver to control and operate.
3. The system should apply brakes simultaneously over all the vehicles.
4. Normal service application of brakes should be very gradual and smooth so as to avoid damage to the goods and discomfort to the passengers.
5. The braking force applied to each axle should be proportional to the axle load so as to obtain uniform deceleration.
6. In case of emergency braking, safety consideration is the main consideration. As such retardation rate would be maximum consistent with the safety, so as to make unfailing halt in the minimum possible distance.
7. Kinetic energy of the train should as far as possible be storable during braking which could subsequently be used during acceleration of the train.
8. The braking system should be inexhaustible.
9. There should be automatic slack adjustment for constant piston stroke as a result of wear on the rim and the brake blocks in the case of mechanical braking.

The requirements of a braking system on a main line locomotive differ from those of motor coaches—the former usually needs braking to hold the train at the steady speed on a long down grade, whereas with the latter braking is primarily required for stopping the train. Rheostatic braking is employed in both the cases but regenerative braking is mainly confined to main line locomotives due to the complications involved in providing regenerative braking with dc series motors.

As already discussed in Art. 1.19, there are three methods of applying electric braking namely (*i*) plugging (*ii*) rheostatic braking and (*iii*) regenerative braking.

14.1.1. Plugging

This is the simplest type of braking. In this method of braking the torque of the motor is reversed, which brings the motor and its driven machine to standstill. This method of applying electric braking with dc motors, 3-phase induction motors and synchronous motors has already been discussed in chapter 1. [Arts. 1.20.1, 1.21.1 and 1.22].

14.1.2. Rheostatic or Dynamic Braking

In this method of braking, the motor is disconnected from the supply and operated as a generator driven by the kinetic energy of the rotating parts of the motor and its driven machines. Thus the kinetic energy of rotation is converted into electrical energy, which is dissipated in the external resistance connected across the motor at the braking instant. This method of applying electric braking with dc motors, 3-phase synchronous motors and induction motors (with a separate source of dc excitation during braking) has already been discussed in chapter 1. [Arts. 1.20.2, 1.21.2 and 1.22].

In traction work where two or more motors are employed, they are put in parallel across a resistance for braking, as shown in Fig. 14.1, as the series connection would produce too high voltage. During the braking period the motors are driven as generators owing to the kinetic energy of the train and the electrical energy so generated is dissipated in the form of heat in the resistances connected across them. An equaliser connection, as shown in Fig. 14.1 (*a*), is used in order to ensure that the two machines share the load equally. If the equaliser connection were not done, the machine that would build up first would send a large current through the other in the opposite direction, causing it to excite with reversed voltage and so the two machines would be short circuited on themselves. A large braking torque would be developed in this way, but it would be quite uncontrollable and the currents would be dangerously large. An alternative method of avoiding this situation is cross-connection of the two machines, as illustrated in Fig. 14.1 (*b*).

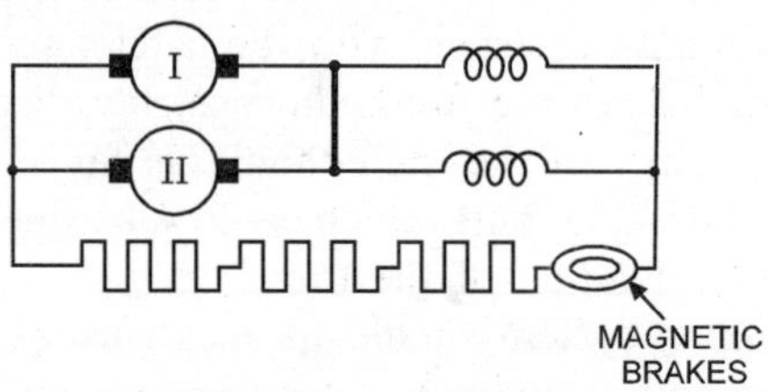

(a) Equaliser Connection

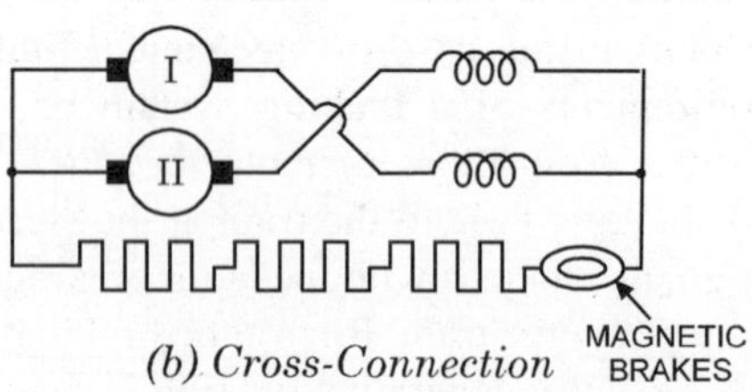

(b) Cross-Connection

Fig. 14.1. ***Rheostatic Braking***

In case the voltage of one of the machines becomes larger than that of the other, the first machine will cause flow of a higher current through the field winding of the second machine. As a result the second machine will have higher voltage while the field current in the first machine due to second one is smaller and hence it will cause lower voltage across first machine. Thus, an automatic compensation of the unbalancing is achieved.

The second method is advantageous to the first one as if the direction of rotation of the machine armatures reverses (may be on account of run back due to upgradient), the machines will fail to excite with equaliser connection and, therefore, no braking effect will be produced which may prove fatal for the passengers whereas with cross-connection the machines will build up in series and being short circuited on themselves will provide an emergency braking and the train will not be allowed to run back on the gradient.

Rheostatic braking cannot be employed with 3-phase induction motors.

In case of ac series motors the rheostatic braking is obtained by operating the machines as generators excited from the supply or as self-excited dc generators supplying power to resistance load. In the former case the fields are energised at low voltage from a suitable tapping on the main transformer while in the latter case the fields of the motors are excited from one of the motors acting as a series generator and in this case dc will be generated in rotors of the motors and the kinetic energy of the rotors will be dissipated as dc power in the loading resistors.

14.1.3. Regenerative Braking

In the previous two methods of electric braking namely plugging and rheostatic braking, stored energy of the rotating parts of the motor and its driven machines is wasted whilst in plugging extra energy is drawn during the braking period and wasted. In the regenerative braking the motors remain connected to the supply and return the braking energy to the supply.

For regenerative braking it is necessary that traction motors must generate power at a voltage higher than the supply voltage and at a reasonable constant voltage.

Regenerative braking is an inherent characteristic of *dc shunt wound motors* and does not require any change of connections. DC shunt motors are electrically stable also *i.e.,* braking torque is independent of line voltage fluctuations. Regenerative braking with dc shunt motors has already been discussed in Art. 1.20.3(1).

The dc series motors cannot be used readily for regenerative braking, as already discussed in Art. 1.20.3(2), and some modification of shunt winding or separate excitation of the series field at low voltage is necessary to be employed to enable the motor to act as a generator. One method of obtaining regenerative braking with series motors is *French method*, which has already been discussed in Art. 1.20.3(2).

Inherently a dc series generator has no stability of voltage, and therefore, an external means is required. In Fig. 14.2 (*a*) a separate motor driven exciter has a separately excited field winding F_1 and also another field winding F_2 connected in the main motor circuit in such a fashion that the field created by it opposes the field created by separately excited field winding F_1 during regeneration. The stabilising resistance R is employed to prevent current surges when the tram crosses from one section of the supply to another, and to compensate for variable line voltage. In case the line voltage falls, regenerated current will tend to increase resulting in strengthening of field F_2 which will weaken the field F_1 and, therefore, reduction in emf generated by the exciter. Thus the field of the traction motors will be weakened resulting in reduction of emfs generated by the traction motors operating as dc series generators. Thus compensation for a decrease in the line voltage is automatically provided.

The arrangement shown in Fig. 14.2 (*b*) has the exciter armature connected in the circuit of the field windings of

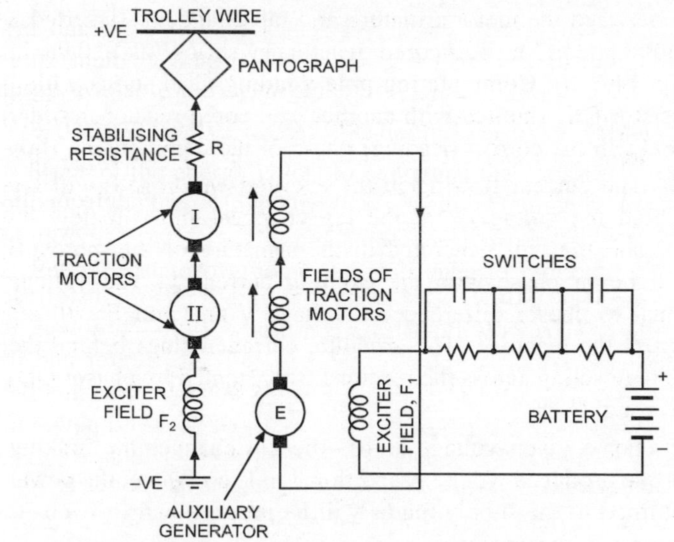

(a) *Differentially Compounded Exciter*

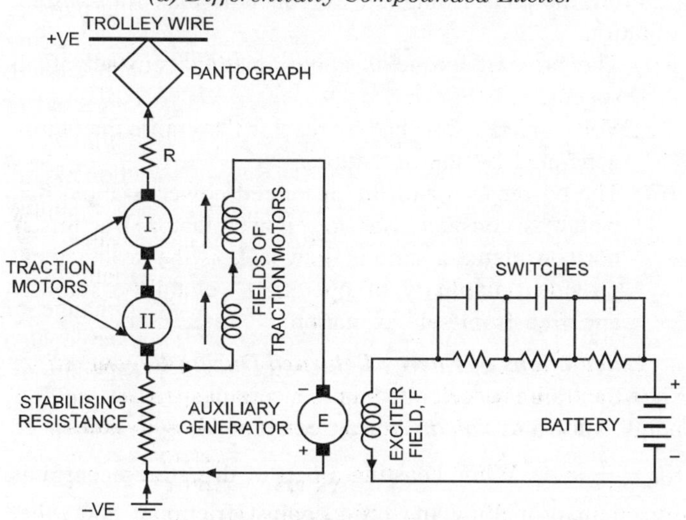

(b) *Separately Excited Exciter*

Fig. 14.2. *Regenerative Braking With Separate Exciter*

traction motors and the stabilising resistance. The current through the stabilising resistance is the sum of the exciter current and the regenerated current. The voltage of the exciter circuit can be regulated by either varying its field strength or manipulating resistances in series with the armature. In case the line voltage falls the regenerated current will tend to increase resulting in increase in voltage drop across the stabilising resistance and, therefore, reduction in the voltage available in the exciter armature circuit causing a reduction in the excitation current of the traction motors operating as dc series generators. This reduces the emf's generated and thus compensation is provided automatically.

Regenerative braking is an inherent characteristic of an *induction motor* since an induction motor operates as an induction (non-synchronous) generator when run at speeds above synchronous speed and it feeds power back to the supply line. The machine, however, is not self-exciting as a generator and is required to be connected to a system supplied from synchronous generators. This system supplies the excitation and determines the frequency at which the induction generator operates. Torque-speed characteristics of an induction machine are shown in Fig. 14.3. From the torque-speed characteristics of an induction machine it is obvious that without any extra resistance in the rotor circuit, there is only a slight variation of speed with the torque whereas with the external resistance inserted in the rotor circuit speed increases for a particular braking torque. Thus we see that with no extra resistance in the rotor circuit the speed during braking remains almost constant and independent of the gradient and the weight of the train. This is a great advantage with the induction motor when used for traction. But if increased speeds are necessary with light loads, these can be obtained by inserting external resistance in the rotor circuit. It is advantageous on mountain railways. It returns about 20% of the total energy on certain railway run and saves a great deal of brake shoe wear.

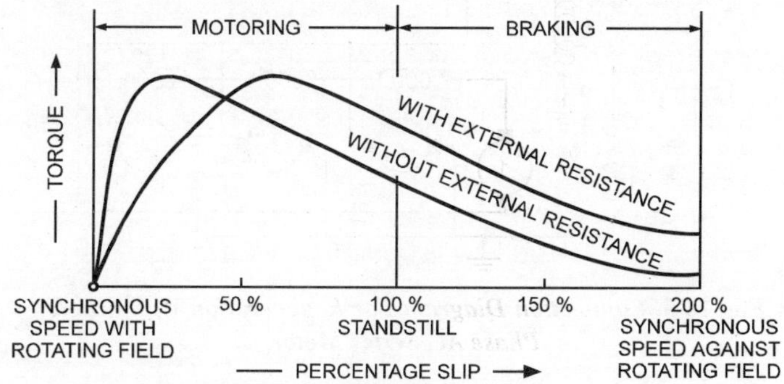

Fig. 14.3

Regenerative braking with ac series motors is more difficult than that with the dc series motor. The difficulties encountered are as follows.

During the regeneration period, the machine should not operate as a self-excited generator. Generally the high power factor is the main consideration in order to achieve a reasonable braking torque. Circuit for regeneration at high power factor usually suffers from self excitation, and those which are inherently stable and free from self excitation usually operate at a poor power factor.

For regenerative braking the regenerated power should have the same frequency as that of the main supply. This necessitates the energizing of the motor field winding from the ac mains. The regenerated current must be in phase opposition to the applied voltage and also the flux ϕ, so that power may be supplied back into the supply system.

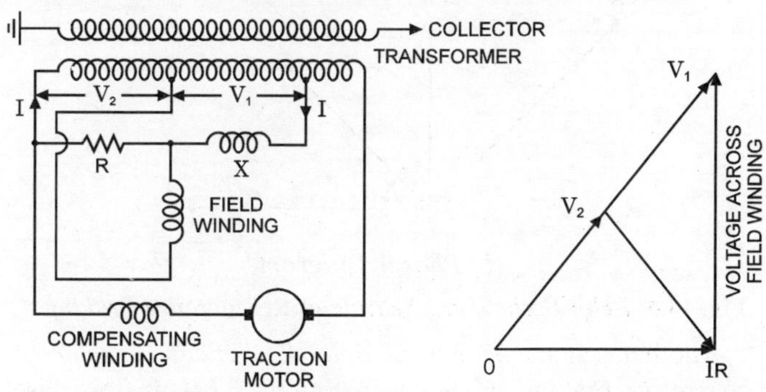

(a) *Connection Diagram* (b) *Phasor Diagram*

Fig. 14.4. *Regenerative Braking With Single Phase AC Series Motor*

The arrangement to provide the above conditions is illustrated in Fig. 14.4.

In another satisfactory arrangement one of the traction motors is used as a generator to supply current to excite the fields of the remaining motors. As the speed of the locomotive falls, the voltage of the motors can be controlled by increasing the excitation current. This is accomplished by increasing the voltage on the exciting motor by employing another transformer tap. Further flexibility can be obtained by changing the tap from the generating motor (Fig. 14.5).

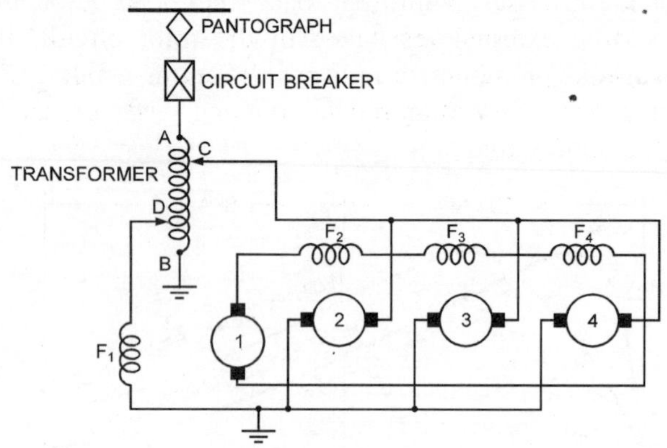

Fig. 14.5. ***Connection Diagram For Regeneration With Single Phase AC Series Motor***

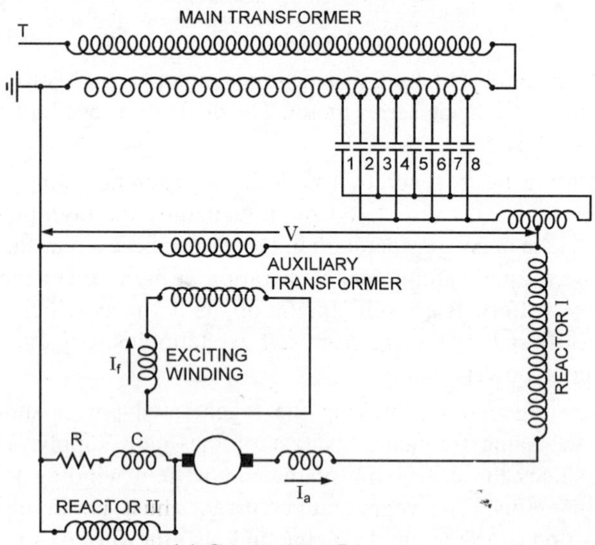

(a) Connection Diagram

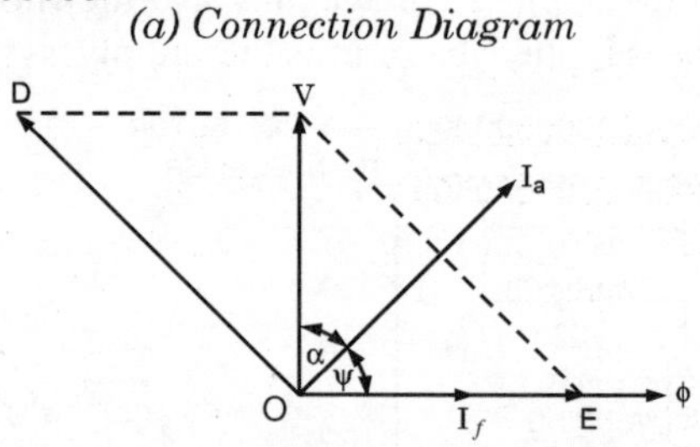

(b) Phasor Diagram

Fig. 14.6. ***Behn Eschenburg Scheme of Regenerative Braking***

Another scheme known as *Behn Eschenburg scheme* of regenerative braking is shown in Fig. 14.6 (*a*). In this case an auxiliary transformer is used to excite the exciting winding of the traction motor. The armature of the traction motor is connected to the main transformer through tap changer. In between the motor armature and tap changer is inserted a choking coil or iron-cored reactor in series, as shown in Fig. 14.6 (*a*). Commutating pole winding C in series with a resistor R is shunted with another iron-cored reactor in order to obtain the correct (leading) phase of the commutating flux.

The current flowing in the exciting winding, I_f will lag behind the voltage V at the tap changer approximately by 90° and the emf generated in the armature of the motor, E will be in phase with the exciting current I_f. The voltage equal to phasor difference of voltage V and emf E will act across the reactor I. The armature current I_a lags behind the voltage acting across the reactor I (represented by phasor OD) by roughly 90°.

For a given voltage across the tap changer the braking torque produced will be proportional to $I_a \cos \psi$ and the power returned to the supply mains will be proportional to $I_a \cos \alpha$ if losses are neglected.

From the phasor diagram we conclude that, for constant excitation,

(*i*) The braking torque is approximately constant at all speeds.

(*ii*) Volt-amperes absorbed by reactor I are more than those generated by the armature.

(*iii*) The power factor of the generated power is about 0.7, which is considered as low power factor. But this is not considered a serious drawback as the arrangement provides simplicity of operation, reliability, stability and free from self excitation.

Calculations of Energy Returned During Regeneration. When the train is accelerated up to a certain speed, it acquires energy, known as *kinetic energy,* corresponding to that speed $(KE = \frac{1}{2} mv^2)$. While coasting a part of this stored energy is utilised in propelling the train against frictional and other resistances to motion and, therefore, the speed falls. Under ideal conditions (no resistance to motion of the train) the speed of the train would have not decreased.

Similarly while the train going down the gradient or moving on level track, the speed remaining the same or reduced, this stored energy can be converted into electrical energy and returned back to the lines.

The amount of energy returned to the line depends upon (*i*) the initial and final speeds during regenerative braking (*ii*) the train resistance and gradient of the track also in case the train is moving down the gradient and (*iii*) efficiency of the system.

Let the initial and final speeds of the train be V_1 and V_2 kmph respectively. The kinetic energy stored in the train at a speed of V_1 kmph

$$= \frac{1}{2} \times \frac{1{,}000\, W_e}{9.81} \left(\frac{1{,}000\ V_1}{3{,}600} \right)^2 \text{ kg-m}$$

$$= \frac{1}{2} \times 1{,}000\, W_e \left(\frac{1{,}000\, V_1}{3{,}600} \right)^2 \text{ N-m or watt-seconds}$$

$$= \frac{1}{2} \times 1{,}000\, W_e \left(\frac{1{,}000\, V_1}{3{,}600} \right)^2 \times \frac{1}{3{,}600} \text{ watt-hours}$$

$$= 0.01072\ W_e\ V_1^2 \text{ watt-hours}$$

Similarly kinetic energy at speed

$V_2 = 0.01072\ W_e V_2^2$ watt-hours

Energy available during regeneration

$= 0.01072\ W_e (V_1^2 - V_2^2)$ watt-hours ...(14.1)

Some of the energy is lost to overcome the resistance to motion and the losses in the traction system including traction motors.

Energy lost to overcome the resistance to motion

$= \frac{W \times r \times S \times 1{,}000}{3{,}600}$ watt-hours $= 0.2778\ WrS$ Wh ...(14.2)

where r is the specific resistance in newton/tonne.

While going down the gradient in the hilly track service, energy is supplied as tractive effort due to the gradient and energy is added up to the energy available during regeneration. Energy available due to motion down the gradient.

$= \frac{98.1 GW \times S \times 1{,}000}{3{,}600} = 27.25$ GSW ...(14.3)

Hence total energy available during regeneration

$= [0.01072 W_e (V_1^2 - V_2^2) + 27.25\ GSW - 0.2778\ WrS]$ watt-hours ...(14.4)

Taking η as the efficiency of the system, Energy returned to the line

$= [0.01072 W_e (V_1^2 - V_2^2) + 27.25\ GSW - 0.2778 WrS] \times \eta$ watt-hours ...(14.5)

Advantages and Disadvantages of Regenerative Braking

Advantages. 1. A part of energy is returned to the supply system, so energy consumption for the run is considerably (about 20 to 30 per cent) reduced thereby affecting a considerable saving in the operating cost.

2. The wear of the brake shoes and wheel tyres is reduced to considerable extent, therefore, their life is increased and replacement cost is reduced.
3. Higher value of braking retardation is obtained so that the vehicle can be brought to rest quickly and running time is considerably reduced.
4. Small amount of brake dust is produced when the mechanical brakes are applied.
5. Higher speeds are possible while going down the gradients because the high braking retardation can be obtained with regenerative braking.
6. Propulsion of heavier trains on gradients is possible without dividing them into sections with speed and safety.

Disadvantages. 1. Additional equipment is required for control of regeneration and for protection of equipment and machines, hence initial as well as maintenance cost is increased.

2. The dc machines required in case of regenerative braking are of large size and cost more than those ordinarily employed, therefore, the weight of the locomotive and thus the required mechanical strength and cost increase.
3. Owing to recuperated energy the operation of the substations becomes complicated and difficult.
4. In case of substations employing mercury-arc rectifiers for conversion purpose, additional equipment is required either to deal with regenerated energy separately or to change one or more of the ordinary rectifiers over to inverted operation. No such difficulty is experienced in case of substations employing rotary convertors or motor-generator sets for converting purpose.
5. Regenerative braking is employed down to a speed of 16 kmph, then rheostatic braking to about 6.5 kmph and then mechanical braking is required to bring the locomotive to rest.

In most of the cases, however, and especially with motor-coach trains the increased cost of the train equipment and the additional features required in order to obtain regenerative braking, combined with the increase in the maintenance cost of the electrical equipment, may entirely offset the economics in the energy consumption and the other items.

For tramways, trolley buses, the regenerative braking is not recommended as it will unnecessarily increase the initial cost as well as increase the operating problems. Generally regenerative braking is desirable and necessary for the service lines having long gradients exceeding 0.6%.

Example 14.1. The following figures refer to the speed-current and torque-current characteristics of a 600 V dc series traction motor.

Current, amperes	**50**	**100**	**150**	**200**	**250**
Speed, kmph	**73.6**	**48**	**41.1**	**37.3**	**35.2**
Torque, N-m	**150**	**525**	**930**	**1,335**	**1,750**

Determine the braking torque at a speed of 48 kmph when operating as self-excited dc generator. Assume resistance of motor and braking rheostat to be 0.6 Ω and 3.0 Ω respectively.

Solution : **As Motor.** Terminal voltage, V = 600 volts

The motor current at a speed of 48 kmph (from speed-current characteristic curve), I = 100 A

Back emf developed by the motor,

$$E_b = V - I R_m = 600 - 100 \times 0.6 = 540 \text{ V}$$

***As Generator*.** At the instant of applying rheostatic braking at a speed of 48 kmph the terminal voltage of the machine will be equal to the emf developed by the machine *i.e.,* 540 V.

Total resistance in the circuit $= R_m + R_{rheostat} = 0.6 + 3 = 3.6\ \Omega$

Current delivered by the machine, $I = \frac{540}{3.6} = 150$ A

The braking torque (the torque corresponding to a current of 150 A from torque-current characteristic) is 930 N-m **Ans.**

Example 14.2. The following figures give the magnetisation curve of a dc series motor when working as a separately excited generator at 640 rpm :

Field current, amperes	**20**	**40**	**60**	**80**
EMF volts	**215**	**381**	**485**	**550**

The total resistance of the motor is 0.8 Ω. When connected for rheostatic braking against a load giving a torque of 40 kg-m it is desired to limit the speed to 600 rpm. What resistance must be connected across the motor during braking ?

Solution : Magnetisation characteristic corresponding to 600 rpm will be as follows :

Field current in amperes	20	40	60	80
EMF in volts	202	357	455	516

Load torque, T = 40 kg-m = 40 × 9.81 = 392.4 N-m

$$\text{Input to load} = T \times \frac{2\pi N}{60} = 392.4 \times \frac{2\pi \times 600}{60} = 24{,}650 \text{ W}$$

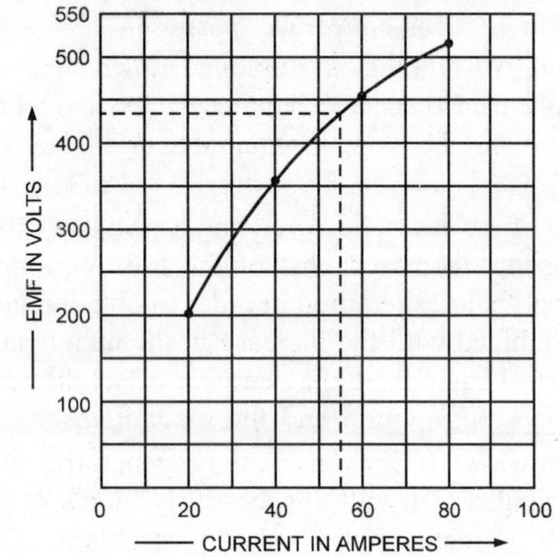

Fig. 14.7

During rheostatic braking the machine operates as a series generator. Assuming no mechanical losses, the generated power will be equal to the input to the load *i.e.,* 24,650 watts.

From curve shown in Fig. 14.7 we find that for the excitation current of 56 A, the induced voltage is about 440 V, which provides generated power of 24,650 watts.

$$\text{Total resistance of the circuit} = \frac{440}{56} = 7.857\ \Omega$$

External resistance to be connected across the motor during braking = 7.857 – 0.8 = 7.057 Ω **Ans.**

Example 14.3. A train weighing 400 tonnes has speed reduced by regenerative braking from 40 to 20 kmph over a distance of 2 km on a down gradient of 2 %. Calculate the electrical energy and average power returned to the line. Tractive resistance is 40 newtons/tonne and allow rotational inertia of 10% and efficiency of conversion 75%.

[*Agra Univ. Traction and Utilization of Elec. Power 1976*]

Solution : Accelerating weight of train,

W_e = 1.1 W = 1.1 × 400 = 440 tonnes

Distance travelled during period, S = 2 km

Gradient, G = 2%

Efficiency of conversion, η = 75% = 0.75

Energy available due to reduction in speed

$= 0.01072\, W_e (V_1^2 - V_2^2)$

$= 0.01072 \times 440 \times (40^2 - 20^2)$

= 5,660 watt-hours = 5.66 kWh

Tractive effort required while going down the gradient

$F_t = Wr - 98.1\, WG$

= 400 × 40 – 98.1 × 400 × 2 = – 62,480 newtons

i.e., tractive effort of 62,480 newtons is available.

Hence energy available on account of moving down the gradient a distance of 2 km

$$= \frac{F_t \times S \times 1{,}000}{1{,}000 \times 3{,}600} \text{ kWh} = \frac{62{,}480 \times 2 \times 1{,}000}{1{,}000 \times 3{,}600} = 34.7 \text{ kWh}$$

Total energy available = 5.66 + 34.7 = 40.37 kWh

Energy returned to the line = 0.75 × 40.37 = 30.28 kWh **Ans.**

$$\text{Average speed} = \frac{40 + 20}{2} = 30 \text{ kmph}$$

$$\text{Average time taken to cover 2 km} = \frac{2}{30} = \frac{1}{15} \text{ hours}$$

$$\text{Average power, } P = \frac{\text{Energy returned to the line}}{\text{Time}} = \frac{30.28}{1/15}$$

= 454 kW **Ans.**

Example 14.4. A 2,340 tonne train (including loco) proceeds down a gradient of 1 in 80 for 5 minutes, during which period, its speed gets reduced form 60 kmph to 36 kmph by application of regenerative braking. Find the energy returned to the lines, if the tractive resistance is 5 kg/tonne, rotational inertia 10% and the overall efficiency of the motors during regeneration as 70%.

Solution : Accelerating weight of train,

W_e = 1.1 × W = 1.1 × 2,340 = 2,574 tonnes

Track resistance, *r* = 5 × 9.81 N per tonne

$$\text{Gradient, } G = \frac{100}{80} = 1.25\%$$

Regenerative period, *t* = 5 minutes = 5 × 60 = 300 seconds

Overall efficiency, η = 0.7

Energy available due to reduction in speed

$= 0.01072\, W_e (V_1^2 - V_2^2)$

$= 0.01072 \times 2{,}574 \times (60^2 - 36^2)$ = 63,575 Wh or 63.575 kWh

Tractive effort required while going down the gradient

= W*r* – 98.1 WG

= 2,340 × 5 × 9.81 – 98.1 × 2,340 × 1.25 = – 172,165.5

i.e., tractive effort of 172,165.5 newtons is available.

Distance moved during regeneration

$$= \frac{V_1 + V_2}{2} \times \frac{1{,}000}{3{,}600} \times t = \frac{60 + 36}{2} \times \frac{1{,}000}{3{,}600} \times 300$$

= 4,000 metres

Energy available on account of moving down the gradient over distance of 4,000 metres.

$$= \frac{172{,}165.5 \times 4{,}000}{1{,}000 \times 3{,}600} = 191.3 \text{ kWh}$$

Total energy available = 63.575 + 191.3 = 254.875 kWh

Energy returned to the line = 0.7 × 254.875 = 178.4 kWh **Ans.**

Example 14.5. A train weighing 500 tonnes is going down a gradient of 20 in 1,000. It is desired to maintain train speed at 40 kmph by regenerative braking. Calculate the power fed into the line. Tractive resistance is 40 N/tonne and allow rotational inertia of 10% and efficiency of conversion of 75%.

Solution : Dead weight of train, W = 500 tonnes

Train speed, V = 40 kmph

Tractive resistance, *r* = 40 N/tonne

$$\text{Gradient, } G = \frac{20 \times 100}{1{,}000} = 2 \text{ per cent}$$

Efficiency of conversion, η = 75% = 0.75

Tractive effort required while going down the gradient

= W*r* – 98.1 WG = 500 × 40 – 98.1 × 500 × 2

= –78,100 newtons

i.e., tractive effort of 78,100 newtons is available

$$\text{Power available, } P = \frac{F_t \times V}{3{,}600} \text{ kW} = \frac{78{,}100 \times 40}{3{,}600} = 867.8 \text{ kW}$$

Power fed into the line = Power available × efficiency of conversion

= 867.8 × 0.75 = 651 kW **Ans.**

Example 14.6. The emf generated between brushes of a single phase series traction motor is 500 V and brushes are connected through a 0.5 Ω reactor to a 400 V tapping on the supply transformer. Neglecting losses, calculate the power generated.

[Pb. Univ. Electric Traction June 1993]

Solution : Voltage across the reactor

= Voltage difference of V and E

= Voltage represented by phasor OD = 640 V

Armature current, $I_a = \dfrac{\text{Voltage across reactor}}{\text{Ohmic value of reactor}} = \dfrac{640}{0.5} = 1{,}280$ A

Phase angle between V and I_a,

$\alpha = 38.66°$

Power generated, $P = VI_a \cos\alpha$

$= 400 \times 1{,}280 \cos 38.66°$

$= 399{,}804$ watts $= 399.804$ kW **Ans.**

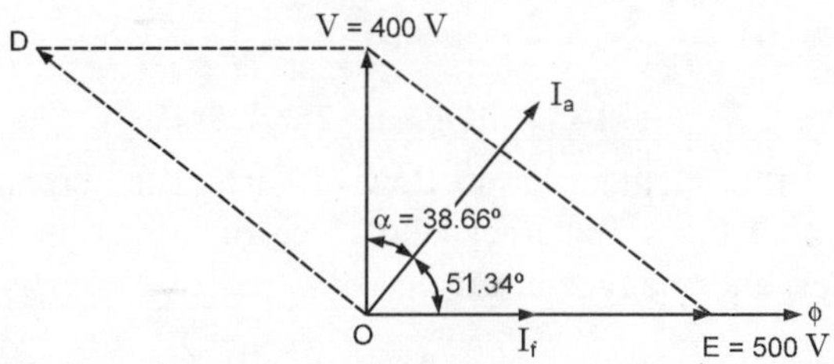

Fig. 14.8

14.2. MECHANICAL REGENERATIVE BRAKING

When the train is accelerated up to a certain speed it acquires energy, known as *kinetic energy,* corresponding to that speed (KE = $\frac{1}{2}mv^2$). While coasting a part of this stored energy is utilised in propelling the train against frictional and other resistances to motion and, therefore, the speed falls. Thus coasting may be considered as a form of *mechanical regenerative braking*. The saving in energy consumption can be considerably affected by increasing the coasting but it will reduce the schedule speed. If, however, the original schedule speed is to be maintained, an increase in the acceleration is required which is possible with large motors. Thus the saving in energy consumption may be neutralised by the increased train weight and the additional cost of the equipment.

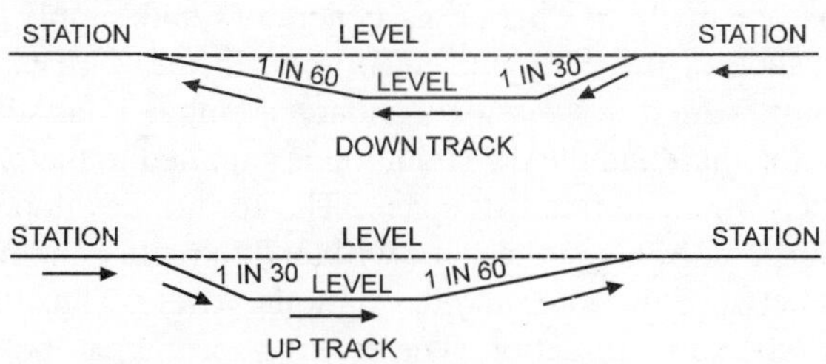

Fig. 14.9. ***Graded Track Construction***

With modern city and suburban traffic conditions, the coasting period is about 20 to 50 per cent of the total running period and consequently a large percentage of kinetic energy has to be wasted in brakes. One of the way of tackling this problem based on the principle of mechanical regenerative braking is adoption of graded track systems, as shown in Fig. 14.9. It facilitates the saving of energy consumption to about 25%.

In the graded track system the stations are situated at a certain height above level track, slope of track while leaving station is 1 in 30 and while approaching station is 1 in 60. In this system the kinetic energy of the train instead of being wasted in brakes is utilised and stored as potential energy as the train has to climb up a slope of 1 in 60 while approaching the station. This potential energy is utilised in moving the train down gradient and imparting acceleration to it. By using this system small motors can be used for driving the train and specific energy consumption can be reduced to 75 per cent of that when the level track was employed. This system can be employed only in underground railways because on surface railway difficulties will be experienced in the construction of the required track.

14.3. MECHANICAL BRAKING

For stopping the train and holding it stationary, in addition to electrical brakes, mechanical brakes are also required. The modern tendency is to employ regenerative braking down to a speed of 16 km per hour, then rheostatic braking down to about 6.5 km per hour, and finally mechanical braking to a standstill. Mechanical brakes are usually applied by brake block shoes pressed with force against the tread of the wheels. Braking force is caused by the movement of the piston which is transmitted to the brake blocks through a system of levers.

Mechanical brakes are of two types namely the *compressed air brakes* and *vacuum brakes*, the former is extensively used in electric railways whereas the latter is used in steam railways. The compressed air brakes are little advantageous to the vacuum brakes as compressed air can conveniently be stored up and released for quick action whereas in the vacuum brakes the necessary vacuum is to be created by the pump. However, this drawback is overcome by use of vacuum reservoir.

1. *Compressed Air Brakes.* The compressed air brake system consists of a reservoir of compressed air, a brake cylinder, compressed air pipe, a valve, springs, piston and a piston rod, as shown in Fig. 14.10. The piston is connected

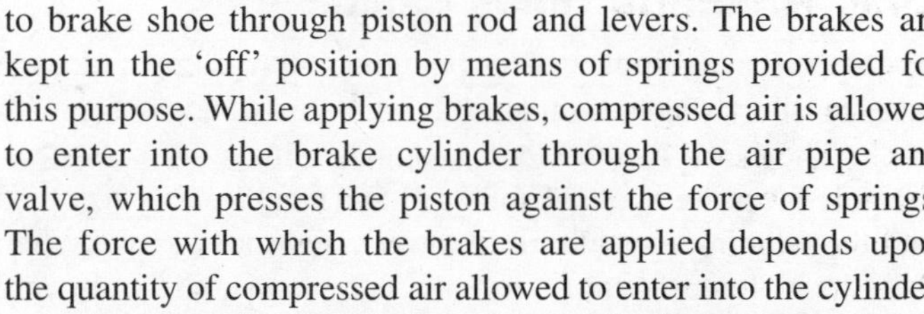

Fig. 14.10. ***Action of Compressed Air Brakes***

to brake shoe through piston rod and levers. The brakes are kept in the 'off' position by means of springs provided for this purpose. While applying brakes, compressed air is allowed to enter into the brake cylinder through the air pipe and valve, which presses the piston against the force of springs. The force with which the brakes are applied depends upon the quantity of compressed air allowed to enter into the cylinder. A valve placed at the inlet of brake cylinder controls the flow of air into it. The brakes are released by exhausting the air. The piston alongwith the brakes return to its original position under the influence of spring force on exhausting the air. The compressed air is used at about 5.5 kg/cm^2 pressure. The compressed air brakes are extensively employed on electric railways.

2. *Vacuum Brakes.* The vacuum brake system consists of a vertical cyclinder with a piston inside, the piston is an easy fit in the cylinder and is provided with rolling rubber ring. The piston is connected to the piston rod which operates the braking arrangement through a system of levers. Vacuum is always maintained on the top and the under side of the piston so that in normal conditions the piston rests at the bottom of the cylinder.

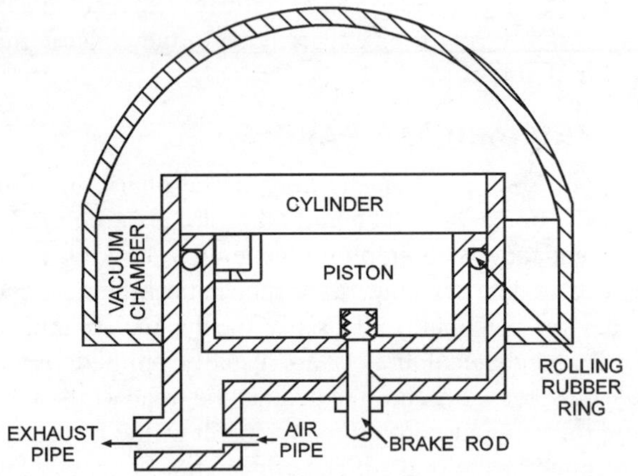

Fig. 14.11. ***Vacuum Brakes***

While applying brakes air at atmospheric pressure is gradually admitted beneath the piston through air pipe and valve the piston moves up and applies the brakes. The force with which the brakes are applied depends upon the rapidity with which the vacuum is destroyed.

The brakes are released by recreating the vacuum. The vacuum is recreated by drawing air from lower side of the piston through the pipe. The piston falls to bottom of the cylinder on account of its own weight.

Vacuum brakes are applied with steam locomotives.

14.4. HYDRAULIC BRAKING

This system consists of an assembly similar to a hydraulic coupling, the rotor being keyed to the axle and the stator being keyed to the bogie frame. The circulation of fluid (water or oil) through the assembly by the rotor vanes produces a braking torque. The brake is applied by filling the coupling by means of a pump.

14.5. MAGNETIC TRACK BRAKES

It comprises a bipolar electromagnet with elongated pole faces a short distance apart (about 6 mm) and along with track. The body of the electromagnet is made of cast steel, the pole faces are made of soft steel and can be easily replaced and the winding is enclosed in a water tight metal case. When the line voltage is impressed on these coils, the poles are strongly attracted to the rails and produce a retarding effect due partly to the magnetic pull (which increases the effective weight of the coils) and partly to mechanical drag.

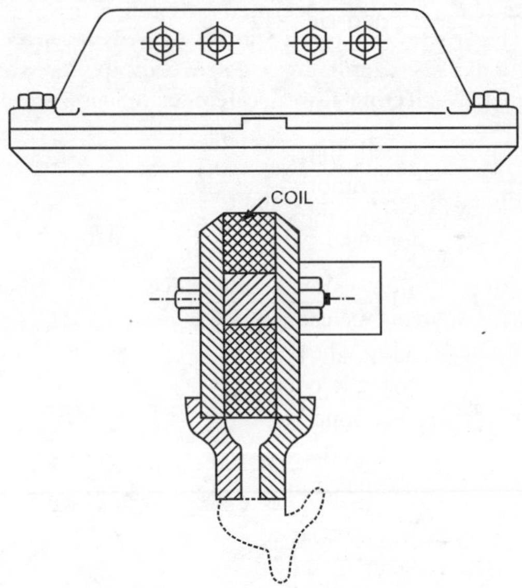

Fig. 14.12. ***Magnetic Track Brake***

The flux created due to flow of current is perpendicular to the pole faces and track and force of attraction between the track and magnet is given as

$$F = \frac{B^2 A}{2 \times 4\pi \times 10^{-7}} \text{ newtons} \qquad ...(14.6)$$

where B is flux density in Wb/m^2 and A is the area of the pole face in square metres. The braking force or the drag = μF where μ is the coefficient of friction. Magnetic brakes are usually employed on trolley buses and cars.

14.6. ELECTROMECHANICAL DRUM BRAKES

In this type the brake drum is mounted on the motor shaft. The brake shoe is pressed against it by springs. The brakes are released by means of a solenoid energised from a battery. This type of brake is usually employed on trolley buses and cars.

14.7. EDDY CURRENT BRAKES

Eddy current brakes are of two types viz linear type and rotary type. In the *linear type brakes*, shoe carrying the excitation current acts as a primary and the rail itself forms the secondary element of the system. During braking, traction motors are made to operate as generators and supply power to the shoe of the eddy current brakes. In the *rotary system*, the rotor, which acts as a secondary member is attached to the motor shaft and the dc excitation is supplied to the primary member which is fixed in space. The torque developed due to induced currents in the rotor will tend to reduce the relative speed between the rotor and the stationary field. Thus braking effect will come in action. The braking torque can be varied by the varying the dc excitation of the primary member (stator).

The main advantage of eddy current braking is the same as that of the disc brake and is used satisfactorily in conjunction with the friction brakes for bringing the vehicle to rest.

14.8. MECHANICAL CONSIDERATIONS

The main considerations in connection with the locomotive are the wheel arrangement and general disposition of equipment in order to have good riding properties and the transmission of power from motors to the driving wheels.

14.8.1. Wheel Arrangement

As already described, the maximum tractive effort which can be exerted by a locomotive without slipping of the driving wheels depends upon the coefficient of adhesion (which is about 0.15 to 0.25) and the weight on the driving wheels. Hence for propulsion of train, certain maximum weight on the driving wheels is essential. The maximum permissible weight on the driving wheels depends on the number of driving axles and the strength of the track, bridges etc. (which is between 15 and 30 tonnes). The minimum number of driving axles required, for a certain tractive effort F, given coefficient of adhesion μ and maximum permissible weight per driving axle.

$$= \frac{F}{\mu \times \text{maximum permissible weight per driving axle}} \quad ...(14.7)$$

For example, if the maximum tractive effort required is 30,000 kg, coefficient of adhesion is 0.2 and maximum allowable weight per driving axle is 19 tonnes, then number of driving axles required $= \frac{30,000}{0.2 \times 1,000 \times 19} = 8.$

Such a train will be hauld by two locomotives each having 4 driving axles making 8 in all.

In case of dc locomotives since the electrical equipment is lighter in weight, so whole of the locomotive weight is placed on the driving axles so that required tractive effort may be obtained.

In case of ac locomotives, since the electrical equipment is comparatively heavy, so the required tractive effort can be obtained by placing only a portion of the weight on the driving wheels and supporting the rest on trailing axles. The riding qualities of high speed locomotives are improved by the trailing axles, which are often used on dc locomotives for this purpose.

14.8.2. Drive

There are two methods of driving the electric train. These are individual drive and collective drive. In individual drive separate motor of smaller size is employed for each driving axle whereas in collective drive a single large sized motor is employed for driving all the axles through connecting rods as in steam locomotives. The collective drive has the following advantages and disadvantages over individual drive.

Advantages. 1. The wheels when coupled have got less tendency to slip.

2. The motor can be mounted well up in main body of the locomotive, therefore, the centre of gravity is raised thereby reducing the wear on track.
3. The motor for collective drive need not be paid so much attention in designing with regard to dimensions as much that for individual drive.

Disadvantages. Collective drive is not suitable for high speed locomotives because of stresses and vibrations in the connecting rods at high speeds.

From the above it can be concluded that individual drive is advantageous only for high-speed locomotives whereas for low-speed locomotives, which are usually employed for passenger and goods trains, collective drive is preferred.

14.8.3. Transmission of Drive

The electric locomotives are specifically designed to have springs between the driving axles and the main body and it is desirable to have as much of locomotive weight spring-borne as possible. This reduces not only vibrations in the locomotive but also reduces substantially the damage to the track owing to hammer blows. In case the motors are fixed on the locomotive or bogie frame in order to have maximum locomotive weight spring-borne, there will be a relative motion between the motor armature and the driving axle owing to compression and extension of the springs. Hence flexible drive between the two is required, which is one of the most difficult problems of the electric locomotive*. There are several methods of transmitting power from the armatures of the electric motors to the driving wheels and only a few will be described here.

1. *Gearless Drive.* (*a*) *Direct Drive.* The simplest way is to employ bipolar motors, whose armatures are mounted directly on the driving axles with the field attached to the frame of the locomotive, as shown in Fig. 14.13. The pole pieces of the motors are approximately flat so that the armature can move relatively to them without affecting the operation. The size of motor armature is limited by the diameter of driving wheels. In case driving wheels of large diameters are employed, the speed of the motor will have to be limited to lower one. Because of these facts and relatively large unsprung weight, poor utilization of material and the low centre of gravity of the locomotives the applications of this method are limited.

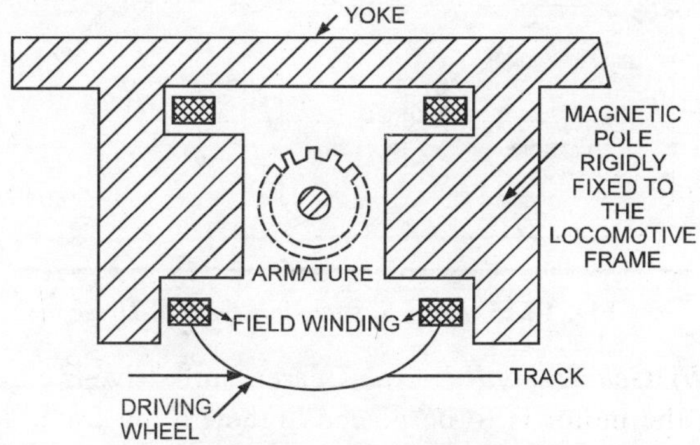

Fig. 14.13. ***Direct Drive***

(***b***) ***Direct Quill Drive.*** The quill is a hollow shaft which surrounds the driving axle and is connected to it by springs or some other flexible device, as shown in Fig. 14.14. The armature of the motor is the directly mounted on the quill and, therefore, it is entirely spring-borne. Thus the unsprung weight is reduced to a minimum and motor of normal design can be employed. The speed and size of armature are still limited by the diameter of the driving wheels.

* The problem does not arise in the steam locomotive, as the piston (attached to the driving axle) is free to move in the cylinder (attached to the frame).

One class of devices for obtaining a flexible connection between the quill flange and the driving wheel uses springs; other manufacturers avoid the use of springs and possible troubles owing to breakages by employing a link motion.

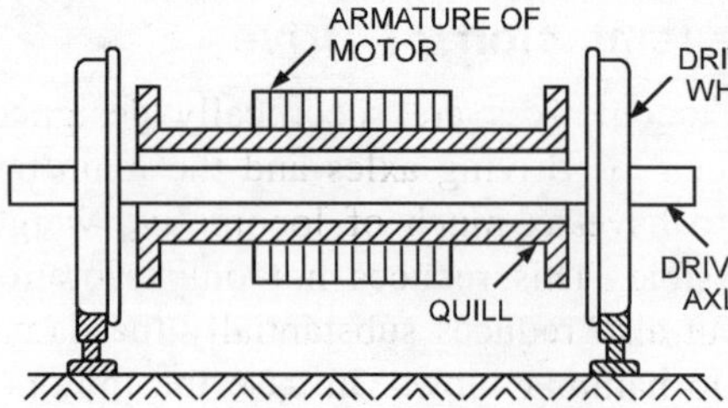

Fig. 14.14. ***Direct Quill Drive***

2. *Geared Drive.* The size of the motor is reduced with the increase in speed for a given output, so the gear drive is necessary in order to reduce the size of the motor for given output by running it at a higher speed. The gearing ratio usually employed is 3 to 5 : 1.

(*a*) *Nose-Suspension Geared Drive*. The most common type of geared drive is nose suspended motor shown in Fig. 14.15. In nose suspension the motor is partly supported from the driving axles and partly between springs which are put on the frame of the vehicle. More than 50 per cent of the total weight of the motor and gear is thus spring-borne. The motor is coupled to the driving axle by means of gear, one being on the same axle and the other on the shaft of the motor armature. Because of axle support the space available is short and limits the size of the motor to about 337.5 kW for normal gauge vehicles. The method is usually employed for motor coaches and tramways. Because of the relatively large unsprung weight this system is not used on large locomotives. This limitation is also imposed by the available space which will not accommodate motor of larger size.

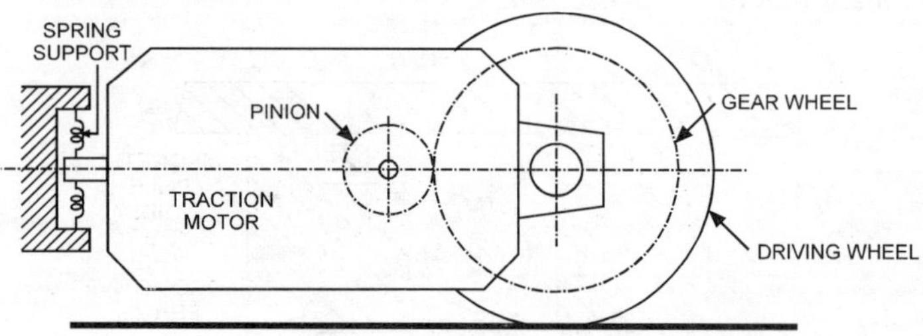

Fig. 14.15. ***Nose-Suspension Geared Drive***

(*b*) *Geared Quill Drive.* Geared quill drive is employed when the motor is to be placed higher in the frame than is possible with the nose suspension so that a motor of larger size (output) can be employed, centre of gravity can be raised and when unsprung weight is to be reduced. In this case quill surrounds the driving axle, as in direct quill drive and gear is mounted on the quill instead of motor armature.

One of the methods already discussed can be used for providing a flexible connection between the quill and the axle.

3. *Brown Boveri Individual Drive.* In this drive, flexibility is achieved by providing a special link arrangement between the gearwheel and the driving wheel, as illustrated in Fig. 14.16. This arrangement can be employed with considerable success on both high- and low-speed locomotives.

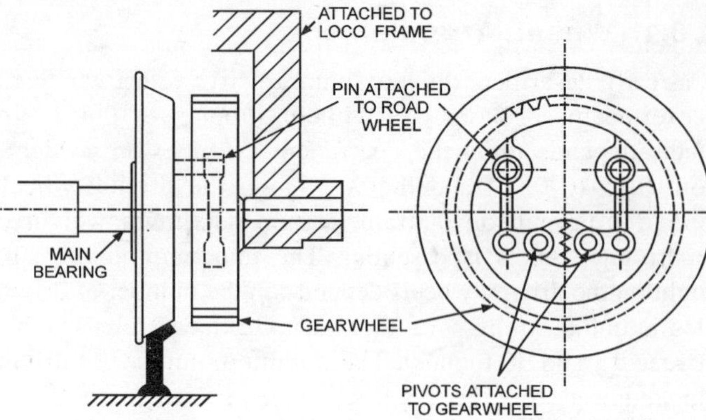

Fig. 14.16. ***Brown Boveri Individual Drive***

4. *Transmission Arrangement in Collective Drive.*

(*a*) *Collective Drive With Connecting rod.* Common method of collective drive is to drive with coupled wheels by means of a connecting rod attached to a crank mounted on the motor shaft or on a "jack" shaft geared to the motor shaft. The connecting rod is fixed in such a way as to remain nearly horizontal as in the steam locomotive. Such an arrangement eliminates the chances of any relative motion between the motor shaft and the driving wheels. Such an arrangement is illustrated in Fig. 14.17 (*a*).

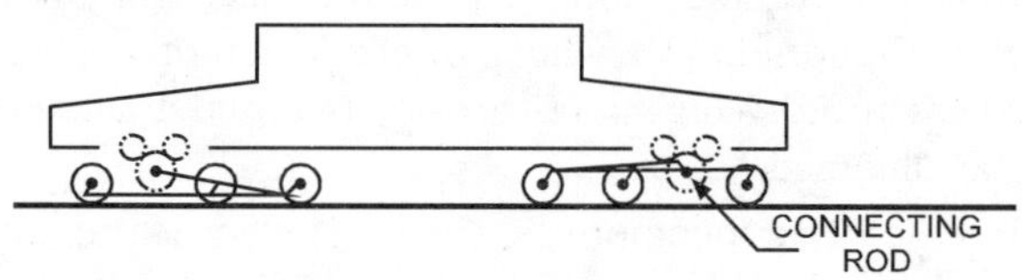

(*a*) *Collective Drive With Connecting Rod*

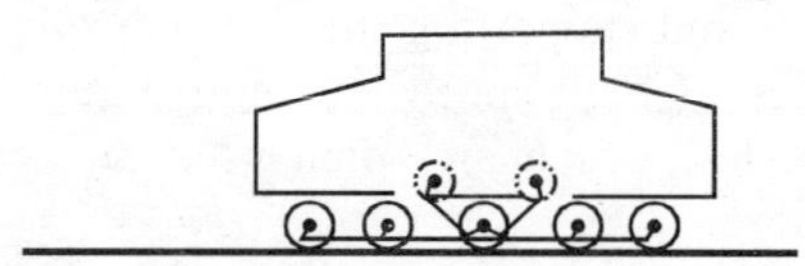

(*b*) *Collective Drive With Scotch Yoke*

Fig. 14.17. ***Transmission Arrangement in Collective Drive***

(*b*) *Collective Drive With Scotch Yoke.* In this drive also, known as Scotch Yoke, or Kando type of drive, a triangular frame work having two sides projecting from the apex is used. The apex of the triangular framework is attached to the coupling rod and the other two corners are attached to the cranks mounted on two motor shafts. Vertical play is allowed in the bearings to secure the required flexibility between motor shaft and driving wheels. Such an arrangement is shown in Fig. 14.17 (*b*).

14.9. CONTROL EQUIPMENT

As already mentioned in Art. 13.1, the conrol equipment is required for (*i*) connecting the motor or motors to the supply mains without taking excessive current at start and provide smooth acceleration without causing sudden shock so as to avoid damage to the coupling (*ii*) adjusting the speed as per requirement, and (*iii*) providing rheostatic or regenerative braking.

The operation may be *by hand* in which the closing of the accelerating contactors is governed by the position of the controller handle and so the rate of acceleration remains completely under the control of driver or *semi-automatic* in which the position of the controller handle governs the number of contactors which will close, but the sequence of operation and the time interval between their individual operations are governed by time-lag devices, as in fully automatic equipment. The hand-operation is preferred in locomotives, where there is a probability of divergences in the hauling loads and possibility of slipping of wheels. Semi-automatic operation is sometimes employed for suburban services in order to ensure uniform acceleration with a minimum energy consumption.

Plain drum controllers of the type shown in Fig. 13.11 are employed by the trams and small industrial locomotives but in railway work, where the currents to be dealt are too high for this type of equipment, *master controller system* is employed. In master controller system the contact instead of being made or broken by hand are closed or opened by the switches known as *contactors* operated by the solenoids.

The master conroller is usually equipped with '*dead man's handle*' the function of which is to stop the train in case the driver faints or becomes incapacitated. This consists of a contact attached to the knob of the controller handle. The driver is to hold it always while the train is running. As soon as the dead man's handle is released, circuit of the operating solenoid will get opened. This will lead to the opening of the motor circuit and bringing the train to a stop by automatic application of brakes.

There are various methods for closing and opening of the contactors by the master controller and the methods, usually employed are:

(*i*) All electric method (*ii*) Electropneumatic method and (*iii*) Cam shaft method.

1. ***All Electric Method.*** In this method the contactors are operated (closed or opened) by the solenoids energised from the master controller. The main disadvantages of this method is that the wide fluctuations in the supply voltage, which are liable to occur, may cause uncertainty of action of the contactors. The difficulty however, can be overcome by employing a special low-voltage generator to provide a supply at 100 V or lower for the operating circuits.

2. ***Electropneumatic Method.*** In this method, contactors are operated by the compressed air. The valves, which admit the air into the operating cylinders, are operated electrically by the solenoids energised from the master controller. This system is preferred to all electric operation system because

1. The contactor groups are more compact and robust.
2. Each valve magnet requires only a few watts for operation whereas the solenoids in all electric operation system may require from 100 to 200 watts.
3. A low voltage control supply can be employed, which simplifies the interlocking and auxiliary contacts on the contactors and their circuits, reduces the size of the master controller and gives reliability to the whole of the control apparatus and circuits.

As electric trains are almost universally equipped with compressed air brakes, the supply of air for electropneumatic operation is, therefore, always available.

3. ***Cam Shaft Drive.*** In this method the contactors are closed or opened directly by cams, which are mounted on a cam shaft driven by an electric or pneumatic motor. The electrical connections are very simple, therefore, this method is widely used. The position of the master controller handle governs the angle through which the cam shaft is turned and therefore, fixes the number of contactors, which are closed.

This method has also the advantages of (*i*) less wear and tear (*ii*) light in weight (*iii*) easy maintenance and (*iv*) occupying less space.

14.10. MULTIPLE UNIT CONTROL

For city and suburban services motor coach units are usually employed. Each motor coach unit consists of one motor coach and two or three trailing coaches for passengers. Motor coach carries driving electric motors and other controlling equipment. The number of motor coaches employed at different hours of the day are different, which depends upon the traffic to be handled at the various hours. In peak traffic hours number of motor coach units are combined to clear of the traffic. The units are so combined that they can easily be split

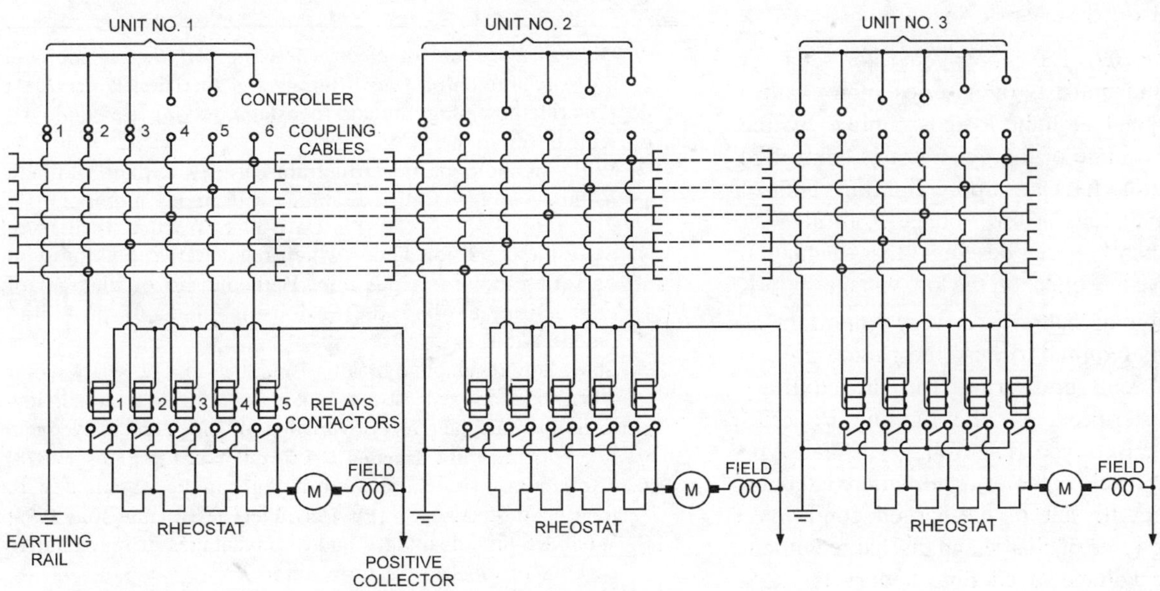

Fig. 14.18. *Multiple Unit Control*

up and run individually during the light traffic hours. The major coach unit (having number of motor coach units combined) thus formed must be capable of control from a single point. This is achieved in the *multiple unit control*.

In multiple unit control method each motor coach unit consists of two or more traction motors provided with a master controller, series-parallel controller (with rheostat and reverser), control-circuit multi-core cables (control bus line) with coupler sockets, acclerating relay; also a motor-generator set, battery or tapped line connected resistor (potentiometer) if the control circuit is to be supplied at low voltage. The trailer coach comprises master controller (when required), control-circuit multi-core cables with coupler sockets. Each motor coach is also equipped with overcurrent relay with driver's control switch, isolating switch for master controller, cut-out switch and fuses for isolating the control circuit of the motor controller from the control bus line.

All these individual coach units are put in parallel and are controlled from a single point by means of a master controller. Control circuits of various coach units are put in parallel by means of multi-core cables known as coupling cables. There being more than one master controller, it becomes necessary to incorporate interlocks so that only one master controller can be operated at a time.

Multiple unit control for three motor coach units is shown in Fig. 14.18. The controller of any one unit can be employed for master control over the whole train. If control of unit no. 1 is operated to start the motor, it is seen from Fig. 14.18 that relay 1 of all units are energised, whole of the starting resistance is placed in series with the motors and all the motors start simultaneously. In next step some of the starting resistance will be taken out of the circuits of all the motors by controller of unit no. 1. In the same way other operations will be carried out by controller of unit no. 1 which will be same and simultaneous for each unit.

14.11. AUXILIARY EQUIPMENT

Auxiliary equipment, in addition to main traction motors, required to be installed on the electric locomotives are:

1. *Motor-Generator Set.* The motor-generator set is of the two bearing type and consists of a series motor and a shunt generator. The speed at light load is limited by the ventilating fan and the voltage of the generator is controlled by an automatic regulator. It is to supply lighting, control system and other power circuits at low voltage from 30 V to 100 volts. The set must be of special design so that fluctuation of line voltage do not have any effect on the low voltage supply.

2. *Battery.* Provision of batteries is very important as some source of energy is required to raise the pantograph, to operate the air blast circuit breaker, to run the auxiliary compressor. Arno-converter provides cabin lighting. Batteries may be lead type or alkaline type. The voltage of lead type battery is 2 volts/cell. Lead acid batteries require more attention and maintenance, cannot be left in discharged conditions, and cannot withstand high rate of charge and discharge without reducing their life. The voltage of alkaline bettery is 1.2 V per cell. Alkaline cells can be left in discharged conditions and can withstand the short-circuit conditions even without any effect on their life. The battery is either charged by a separate rectifier or from a dc generator. The capacity of battery depends on the vehicle and may be between 175 and 375 AH at the 5 hours. The batteries are usually connected in parallel with the motor-generator set.

3. *Air Compressor.* The compressor is usually of two cylinder type and is directly connected to the motor. With compressors for motor coaches, a moderate speed motor is employed so that the weight is reduced and the compressor is driven through double-helical spur gearing.

Automatic starting and stopping of the motor to maintain the air pressure within desired limits is achieved by employing a pressure operated switch, the moving contact of which is actuated, through a system of levers and a toggle mechanism by the air pressure on a spring loaded diaphram. It is for supplying compressed air for operating brakes, pantograph operating gear and electropneumatic control gear. At 1,500 V or below the compressor motor can be supplied directly from the line, otherwise, it is usually fed from the auxiliary generator.

4. *Heating Equipment.* It is required for heating the locomotive and train if desired. This is supplied directly from the line up to 3,000 V. Where an electric locomotive has to haul steam-heated rolling stock an electrically heated boiler is often provided.

5. *Pantograph Operating Equipment.* Raising and lowering of pantograph is usually carried out by compressed air and suitable switches for controlling the solenoid-operated valves of the air supply are required to be provided.

6. *Traction Motor Blowers.* These are required on locomotives when forced ventilated motors are used. In some cases a single blower is used which delivers air at a low pressure into a central duct built into the under frame of the locomotive body from which it is distributed to the motors. In other cases with large frame mounted motors, each motor is provided with a separate blower. The blowers are then mounted on the motor frames and two or more blowers are coupled together and are driven by a single motor.

Exercises

1. Discuss a method of electric braking for traction motors. **[Rajasthan Technical Univ. Utilization of Electrical Power 2011]**
2. Describe how plugging and rheostatic braking are employed with dc motors.
3. With the help of neat diagram, briefly explain 'French method of regenerative braking' with series motors. **[Pb. Univ. Elec. Traction April 1975]**
4. State the condition under which regenerative braking with dc series motors is possible. With the aid of diagram of connection, explain the various methods of providing regeneration.
5. Show how regenrative braking is applied to a dc series motor when there are at least two motors working in parallel. How to compensate the effect of variation of voltage in the overhead line ? **[Agra Univ. Traction & Utilization of Elec. Power 1978]**
6. Explain how regenerative braking can be obtained with traction motors. **[Pb. Univ. Electric Traction June 1993]**
7. What are the advantages and disadvantages of regenerative braking of electric traction motor ? **[M.D. Univ. Utilization of Electric Power & Traction May 2010]**

8. Explain briefly how regenerative braking differs from other methods of braking traction motors and also write the usefulness of this method of braking with (*a*) dc series motors and (*b*) ac series motors in traction.
[M.D. Univ. Utilization of Electric Power & Traction Dec. 2009]
9. What are the conditions necessary to achieve electric regenerative braking ?
Explain the suitability of (*a*) dc series motor and (*b*) three-phase induction motor for the purpose of electric regenerative braking. **[Pb. Univ. Electric Traction July 1992]**
10. What are the advantages of regenerative braking ? Explain how regenerative braking can be obtained in dc locomotives.
[B.H.U. Elec. Traction and Utilization 1978]
11. Explain how hydraulic and eddy current brake works. What are their advantages ?
What is mechanical regenerative braking and what are its advantages ?
12. Explain with proper diagram the following braking systems used in traction:
(*a*) Vacuum brake (*b*) Compressed air brake (*c*) Magnetic track brake (*d*) Electromechanical drum brake (*e*) Regenerative braking with 3-ϕ induction motors.
13. Describe the various methods of drives which have been applied for transmitting motive power from motor shaft to driving wheels. **[Pb. Univ. Elec. Traction 1973]**
14. Explain the method of Quill drive of transmitting power from the motor to the driving wheels.
[B.H.U. Electric Traction and Utilization 1978]
15. State with sketches, the three important methods of transmission of mechanical power from traction motor to the driving wheels. **[Agra Univ. Traction & Utilization 1976 Supp.]**
16. Explain multiple unit control of dc motors.
[M.D. Univ. Utilization of Electric Power & Traction May 2009]
17. Describe the auxiliary electrical equipment for an electric locomotive. **[Pb. Univ. Elec. Traction Dec. 1992]**
18. Write short notes on the following:
(*i*) Electric braking in traction motors. **[U.P.T.U. 2005-06]**
(*ii*) Plugging.
(*iii*) Reverse current braking. **[Pb. Univ. July 1992]**
(*iv*) Regenerative braking.
(*v*) Vacuum brakes.
(*vi*) Quill drive.
(*vii*) Nose suspended motor.
(*viii*) Methods of transmission of power from motors to driving wheels.
(*ix*) Multiple unit control. **[M.D. Univ. Dec. 2008; Dec. 2009]**
(*x*) Auxiliary equipment for electric locomotives.

Short Answer Type Questions With Answers

Q. 1. Why ordinary mechanical braking cannot be totally replaced by electrical braking ?
Ans. Because electric braking cannot hold the vehicle stationary.
Q. 2. On what factors does the amount of energy returned to the line depend ?
Ans. The amount of energy returned to the line depends upon (*i*) the initial and final speeds during regenerative braking (*ii*) the train resistance and gradient of the track also in case the train is moving down the gradient and (*iii*) efficiency of the system.
Q. 3. Why is regenerative braking not recommended for tramways and trolley buses ?
Ans. Because it will unnecessarily increase the capital cost as well as enhance the operating problems.
Q. 4. What is meant by mechanical regenerative braking ?
Ans. During coasting a part of the stored energy (kinetic energy) is utilized in propelling the train against frictional and other resistances to motion and, therefore, speed falls. Thus coasting may be considered as a form of mechanical regenerative braking.
Q. 5. Why gear drive is necessary ?
Ans. The size of the motor is reduced with the increase in speed for a given output, so gear drive is necessary to reduce the size of the motor for given output by operating it at higher speed.
Q. 6. Why is master controller usually equipped with 'dead man's handle' ?
Ans. So as to stop the train in case the driver faints or becomes incapacitated.

Problems

1. A train weighing 300 tonnes has its speed reduced from 80 kmph to 30 kmph while going down an incline of 1 in 100 through a distance of 3 km by employing regenerative braking. Taking tractive resistance of 4 kg per tonne and allowance for rotational inertia as 8%, calculate the kWh returned to the line assuming an overall efficiency of 75%.
[Ans. 25.34 kWh]
2. A 300 tonne train on a down grade of 1 in 75 has its speed reduced from 60 kmph to its steady value 40 kmph. Calculate the energy returned to the line, in a distance of 1,000 metres by adopting regenerative braking. Assume overall efficiency of 85 per cent, track resistance of 5 kg per tonne and allowance for rotary inertia as 10 per cent.
[Ans. 13.28 kWh]
3. A 400 tonne train has its speed reduced by regenerative braking from 90 to 50 kmph in 2,000 metres on a down gradient of 1 in 100. Assume overall efficiency 80 per cent and allow 10 per cent for rotary inertia and 5 kg per tonne for tractive resistance. Find the energy in kWh returned to the line. **[Pb. Univ. Elec. Traction 1973] [Ans.** 29.81 kWh]
4. Rheostatic braking is used to limit the speed of a dc crane motor (series motor) to 500 rpm when the supply is cut off. The motor has a resistance of 1.2 Ω and the descending load exerts a torque of 45 kg-m. Calculate the value of the resistance to be connected across the motor terminals. State assumption made.
The magnetisation characteristics of the motor at 500 rpm is given by

Current amperes	20	30	40	50
EMF volts	360	440	490	520

[Ans. 10.155 Ω]
5. A dc series motor is to be used for rheostatic braking on a tram car. With the field respectively excited on test, the open-circuit voltage across armature at a speed corresponding to 24 kmph was:

Field amperes	10	20	40	60	80
Armature volts	145	280	500	680	700

When the machine is being used for braking a tramcar, what resistance would have to be put in the motor circuit to obtain a current of 70 A at a speed of 16 kmph ? The resistance of the motor armature and field windings may be neglected. **[Ans.** 6.4 Ω]
6. A 525 V series traction motor has the following data:

Current (amperes)	50	70	80	100
Speed (kmph)	33.8	26.9	25.1	23.8
Torque (N-m)	216	344	422	500

What will be the braking torque at a speed of 26 kmph when operating as a self-excited series generator, the resistance of the braking rheostat being 7.5 ohms, and that of motor being 0.5 ohm ? **[Ans.** 300 N-m]

CHAPTER 15

Power Supply For Electric Traction

INSIDE THIS CHAPTER

15.1. INTRODUCTION

As already discussed in chapter 10, the various systems of railway electrification are (*i*) dc system (*ii*) single phase ac system operating at low frequency (*iii*) three phase ac system operating at low frequency and (*iv*) composite system (*a*) Kando system and (*b*) single phase—dc system.

Electric traction was first introduced in Britian in the year 1890 at 600 V dc using third rail for urban and suburban services. For its application to long distance main-line service the developments were towards increasing the operating voltage and as a result the line rail system utilising 1,200 V dc was tried in Manchester-Bury line. This system of supply was not satisfactory due to inherent drawbacks, such as insufficient protection provided by the wooden planks and choking up of grooves by snow during winter season, and therefore, overhead contact system was adopted employing 1,500 V dc later on. This system is still used for suburban and main lines in many countries including our country.

In India electric traction was first introduced for suburban service in Bombay in 1925 and in Madras in 1931. It was later extended to Poona and Igatpuri. The system adopted was 1,500 V dc. The power used to be supplied to traction motors at 1,500 V dc from rotary converter substations. For 1,500 V supply two 750 V rotary converters, each of 2,500 kW capacity, were employed in series. Mercury-arc rectifiers were introduced for the first time for supplying dc power when Madras–Tambaram route was electrified. The rectifiers employed were multianode continuously evacuated steel tank type. Later on grid controlled mercury-arc rectifiers were used. Multianode air-cooled sealed type steel tank rectifiers having on-load tap changing arrangement for variation of voltage have also been employed. Nowadays solid state devices are finding increasing use in these dc substations.

After about 20 years, electric traction was introduced on the Howrah—Burdwan and Sheoraphuli-Tarakeshwar branch lines on 3,000 V dc, as at that time there was a general trend towards adopting of 3,000 V dc for electric traction.

The early applications of the three phase system were to Swiss mountain railways, and the first application to main-line railways (operating at 3,000 V) was in 1902. Subsequent development of the 3-ϕ system have been entirely for main-line electrification, and the principal applications are in Northern Italy.

The single phase ac system operating at low frequency ($16\frac{2}{3}$ Hz) found extensive applications in Switzerland, Germany, Austria and Scandinavia for suburban as well as main-line services. In America about 800 km (on the Pennsylvania and New Haven lines) are electrified on the single phase system, but at 25 Hz frequency and 1,100 V voltage. The single phase 50 Hz system has also been adopted for the electrification of the Katanga lines in the Belgian Congo.

Kando system (single phase-three phase system) was developed in Hungary in 1932.

From the reports obtained from French National Railways on relative merits and demerits of ac and dc systems it was found that ac system employing commercial frequency of 50 Hz was advantageous both financially and operationally. In 1957 Railway Board decided to adopt 25 kV, 50 Hz single phase system of electrification for all future schemes. This system gives considerable saving in the cost of overhead equipment, substations and locomotives. The size of the overhead system is reduced because of adoption of high voltage leading to saving in the cost of supporting structures and their foundations. On account of the high voltage, the traction substations can be spaced at longer distances. AC substations are simpler than dc substations because they have only step-down transformers and necessary switchgear. However, ac traction systems have some drawbacks also like induced voltage in aerial signalling equipment and interference with neighbouring communication lines, and therefore, it will need some remedial devices to overcome such drawbacks.

In this system 25 kV, 50 Hz single phase system is employed for supplying power to the locomotives. The locomotive draws power from the overhead equipment (OHE) through sliding contact by means of pantograph. The locomotive essentially consists of (*i*) air circuit breaker, provided on the roof of the locomotive, mainly to disconnect the locomotive from hv supply in the event of fault in the equipment and to isolate the locomotive equipment at a voltage change point or phase change point in the overhead equipment such as before entry into the neutral section (*ii*) step-down transformer to step down the supply voltage to utilization level (*iii*) on-load tap changer for providing variable voltage (*iv*) converting machinery for converting ac into dc, (*v*) smoothing chokes for reducing the magnitude of ac component of rectified undulating dc and (*vi*) dc traction motors.

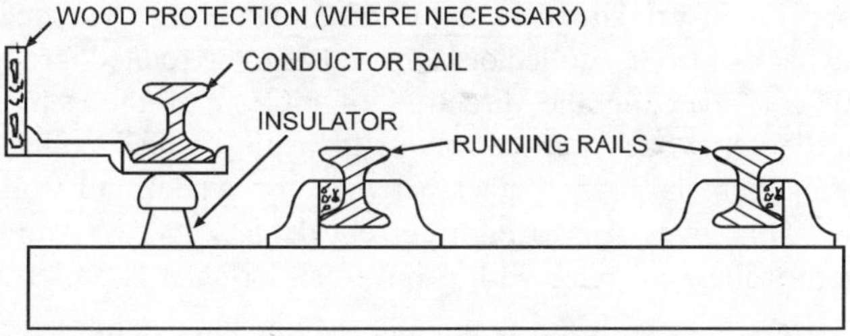

Fig. 15.1. *Conductor Rail System*

15.2. CURRENT COLLECTION SYSTEMS

There are mainly two systems of current collection for locomotives, tramways or trolley buses viz (*i*) conductor rail system and (*ii*) overhead system. Current collection from overhead system is far superior to that from the conductor rail system. This is because both theoretically and experimentally current collection is more difficult from a rigid body than from an elastic one. Further the insulation of the third rail at high voltages used on single phase ac traction would also be impracticable and endanger the safety of the personnel.

15.2.1. Conductor Rail System

The conductor rail system is employed at 600 V for suburban services on account of its relative cheapness and easier inspection and maintenance. The supporting structure does not interfere in the visibility of the signals. Also, it is not necessary to protect the rail from accidental contact by plate players etc., except at stations and in sidings. In this system, current is supplied to the electrical operated vehicle either through one rail conductor or through two rail conductors. In case of one rail conductor the track rail is employed as the return conductor. The advantage of using insulated return rail is elimination of electrolytic action due to return currents on other public services buried in the vicinity of the railway tunnels. The rails are mounted on insulators parallel with the track rails at a distance of 0.3 to 0.4 m from the running rail with their upper most surface acting as contact surface and are fed at suitable points from substations. Conductor rail with a side running contact, as shown in Fig. 15.2 (*a*), is also employed. Some railways employ an under-running contact. Latter systems are supposed to be more protective against accidental contact. The wearing of the rail conductor is due only to the friction of the collector shoes and pitting which usually occurs at starting when the current drawn is very large. The rails are designed from electrical considerations rather than mechanical. The main considerations are (*i*) electrical conductivity (*ii*) cost, (*iii*) wearing qualities (*iv*) contact surface available for the collector shoes and (*v*) shape and size of the conductor rail, keeping in view the type of insulators used. A special steel alloy is used for the rails for economy reasons. The composition is such that it provides high conductivity and at the same time satisfies all other requirements. A typical composition is iron 99.63%, carbon 0.05%, manganese 0.2%, phosphorous 0.05%, silicon 0.02% and sulphur 0.05%. It has a resistance of about 12.0×10^{-8} Ω-m at 18°C (about seven times of that of copper and about one quarter or one-fifth of that of the running rails). To reduce the voltage drop at joints, conductor rails are bonded together by short length flexible copper conductors riveted or welded to the rails. The conductor rail is not fixed rigidly to the insulators in order to take care of the contraction and expansion of rails. To prevent creepage due to friction of the collector shoes, however, it is anchored at intervals of 100 to150 metres.

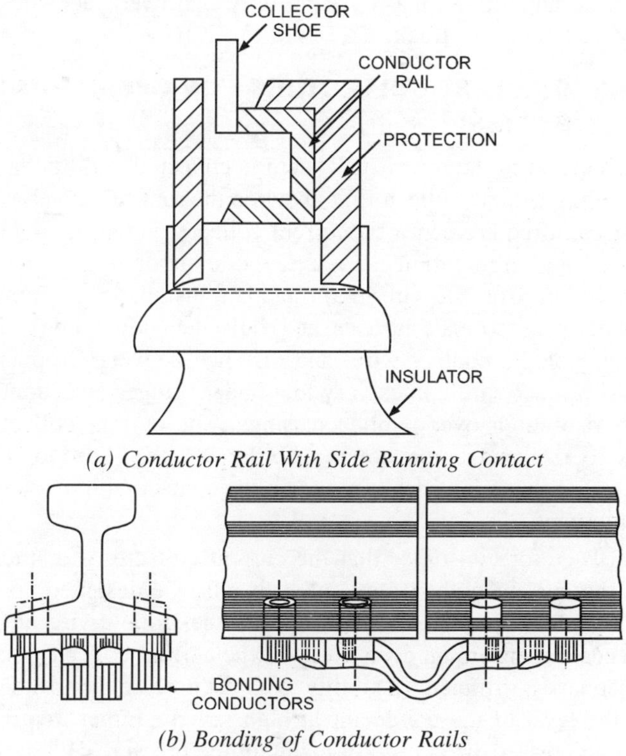

(*a*) *Conductor Rail With Side Running Contact*

(*b*) *Bonding of Conductor Rails*

Fig. 15.2

The current is conveyed from the conductor rail to the train equipment by means of collector shoe (flat steel shoe about 20 cm in length and 7.6 cm in width) which presses on to the rail with a force of about 15 kg. In case of top contact the necessary contact force is obtained by gravity and in case of side and under-running contacts springs are employed to obtain necessary contact force. The current which can be collected is about 300 to 500 A. Sometimes it may not be convenient to have the conductor rail on the same side of the track, therefore, it is desirable to provide shoes on both sides of the locomotive or train. Also, there are going to be gaps in the rails at points and crossings, at least two shoes must be provided on each side so as to avoid discontinuity in the current flow. This system is suitable for heavy current collection, top contact system for voltage up to 750 V and side contact system up to 1,200 V.

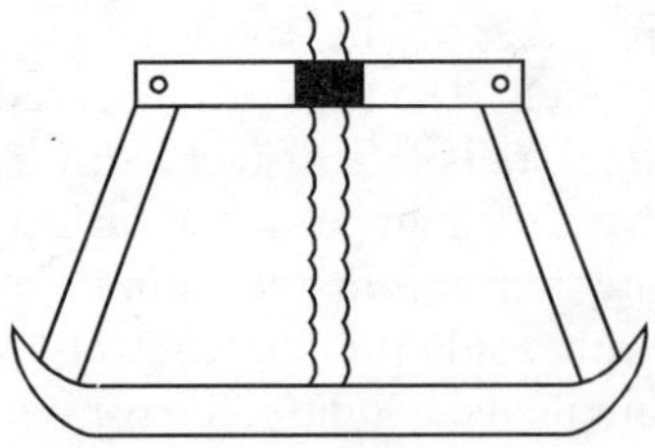

Fig. 15.3. ***Current Collecting Shoe***

15.2.2. Overhead System

This system is adopted when the trains are to be supplied at high voltage (1,500 volts or above). In this system with high operating voltage the trains requiring high power may be supplied through conductors of relatively small cross section and the collection of the current required by a train can be done by a collector with a sliding contact. Overhead construction is universal for all ac railways and is also used with dc tramways, trolley buses and locomotives operating at voltages of 1,500 volts and above. In all these cases the running rails are utilized as the return conductor, therefore, with dc and single phase system only one overhead wire is required for each track.

15.3. CURRENT COLLECTORS FOR OVERHEAD SYSTEM

The current in the overhead system is collected with the help of sliding contact collector mounted on the roof of the vehicle. The main requirement of a current collector is that it should, under no circumstances, leave the contact of overhead equipment. Contact wire in all practical installations is never perfectly horizontal, it rises and falls depending upon the weight of the contact wire, and distance between droppers. Also contact wire comes very low under bridges and tunnels and rises high over public crossings. The current collector has to rise and come down according to the speed of the electric vehicle in order to maintain the contact with overhead equipment.

Therefore, in order that the current collector gear picks up the current without any spark, it must be designed so as to maintain an even pressure against the trolley wire at all speeds. To ensure an even pressure the current collector gear must have a sufficient flexibility to adjust itself to the variation of the level of the conductor at high speeds. Either a spring or compressed air is used to maintain the pressure. When the contact wire rises the current collector has to move upwards so as to maintain contact, and the pressure exerted on the contact wire is the pressure of the spring or compressed-air piston less the force required to accelerate the current collector to its new position and any force necessary to overcome the frictional resistance. To design a satisfactory current collector therefore, particularly for high speed work, the frictional resistance must be reduced to a minimum and the moment of inertia must be kept low. Furthermore, means must be provided for lowering when idle and reasonable precautions taken to prevent the collector leaving the wire at points and crossings.

As the tramcars and trolley buses draw their power from overhead conductors suspended about 6.7 m (from 4.7 m to 7.5 m) above the road surface, the current collector must, therefore, be mounted on the vehicle. Three types of current collectors are in common use, viz, the trolley collector, bow collector and the pantograph collector.

1. ***Trolley Collector.*** The trolley collector is universally employed with tramways and trolley buses. This consists of a grooved gun metal wheel or grooved slider shoe with carbon insert carried at the end of a long pole. The other end of this pole is hinged to a swivelling base fixed to the roof of the vehicle. Necessary upward pressure for the pole and current collector is achieved by means of springs. As two trolley wires are required for a trolley bus a separate trolley collector is provided for each wire, the bases being mounted side by side. The pressure for wheel is approximately 10 kg and for a carbon insert slider is approximately 17 kg. The main drawback of trolley collector is that it has to be rotated through 180° for reversing the direction of motion of the vehicle. Another drawback particularly with trolley wheel collector is that there is poor contact between the wheel and trolley wire. This gives rise to high current density in gassy mines where trolley collector with grooved slider shoe is preferred.

The trolley collector is suitable for comparative low speeds (say 24 to 32 kmph) beyond this speed there is every possibility of its jumping off the trolley wire.

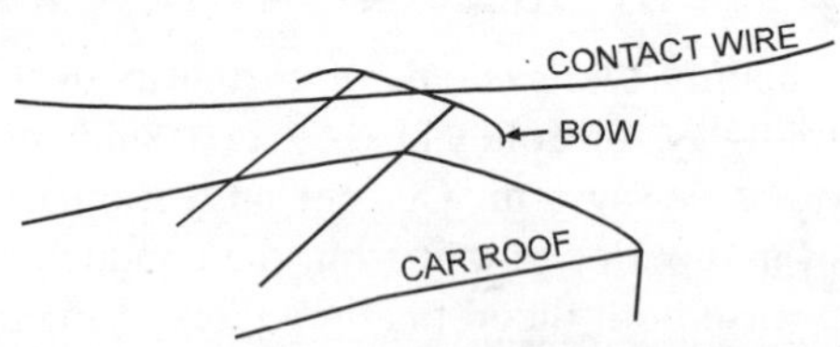

Fig. 15.4. ***Bow Collector***

2. ***Bow Collector.*** The bow collector is also employed for collecting the current with tramways. The bow collector consists of light metal strip or bow 0.6 or 0.9 metre wide pressing against the trolley wire and attached to a framework mounted on the roof of vehicle. Although provision of metal strip enables vehicle to run at higher speeds without running the risk of leaving the contact of trolley wire, its manoeuvrability is lost. It is as such not suitable for trolley buses. In order to avoid jumping off of the trolley wire at high speeds, it is desired that the wire be accurately located above the track and staggered about 15 cm to each side of centre line to avoid the wearing of the groove in the contact strip. Collection strip is made of soft material such as copper, aluminium or carbon so that it should wear instead of trolley wire as it is easy to replace worn out collection strip than trolley wire. Necessary upward pressure is obtained by springs. The length of the poles is kept sufficient to have lateral movement along the mean path of trolley bus. This arrangement enables vehicle to run at higher speed without danger of the jumping off wire. Bow collector has always to run trailing just like trolley collector. It, therefore, requires either provision of duplicate bows or an arrangement of reversing the bows for motion in the reverse direction. The bow collector has smaller inertia but is not so readily adoptable for collection of large currents as is a pantograph collector.

3. *Pantograph Collector*. The pantograph is employed in railways for collection of current where the operating speed is as high as 100 or 130 kmph and the currents to be collected are as large as 2,000 or 3,000 amperes. Pantographs are mounted on the roof of the vehicles and usually carry a sliding shoe for contact with the overhead trolley wire. These consist of a jointed frame usually of steel tubing. The contact shoes are usually about 1.2 metres long. There may be a single shoe or two shoes on each pantograph. The shoes may be straight throughout their lengths or cambered slightly or may be in the form of an auxiliary bow. Material used for pantograph is often steel with sometimes, wearing plates of copper or bronze inserted. The pressure varies from 5 to 15 kg.

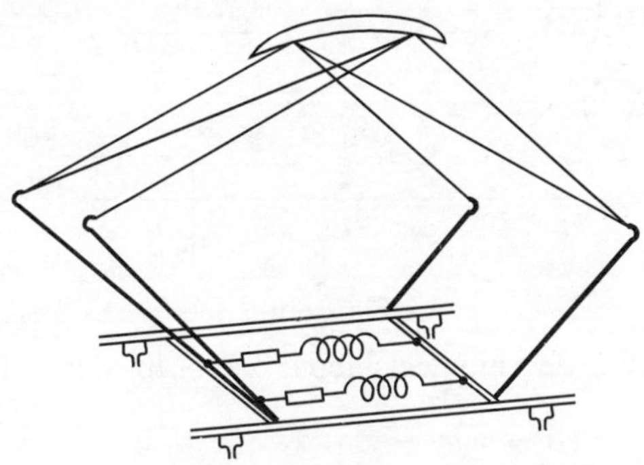

Fig. 15.5. ***Pantograph Collector***

The pantograph is raised or lowered from the driver cab by one of the following methods or with some modification of it.

(*i*) air raised, gravity lowered. (*ii*) air raised, spring lowered. (*iii*) spring raised, air lowered.

The pantographs should be of such construction that they maintain continuous contact with the overhead wire at reasonably constant pressure at any height. They must be light enough in order to follow the overhead wire as it changes height above the rails to pass under overhead crossing etc., with minimum inertia effects even at high speeds they must be sufficiently strong and rigid to resist air pressures both head on, due to speed of train, and transverse, due to wind; also they must resist stresses due to the sway of the locomotive at high speeds and to some extent, stresses due to blows from the overhead system at deflectors or turnouts etc. and from striking birds. If, however, there is a serious trouble on the overhead system, the pantograph must be sufficiently pliable so that it will be damaged rather than pulling down the overhead structure.

The above arrangement has the following advantages over other types of collectors.

(*i*) It can operate in either direction of motion.

(*ii*) There is no risk of leaving wire junction etc.

(*iii*) The erection of the overhead network is very simple due to absence of points and grooved crossings required for bows.

(*iv*) Its height can be varied from the drivers' cabin by carrying out simple operations.

15.4. OVERHEAD CONSTRUCTION FOR TRAMWAYS AND TROLLEY BUSES AND RAILWAYS

As already mentioned in Art. 15.2, for trams, trolley buses and for railways operating at 1,500 V and above overhead contact wire is used. The design of overhead system is more complex in comparison to conductor rail system because of greater difficulty of ensuring that the collector and contact wire remain in contact at a fairly even pressure.

15.4.1. Tramways and Trolley Buses

The overhead construction for tramways has to conform the following regulations:

1. The trolley wire is to be erected at a minimum height of 5.185 metres above the street surface (except under bridges).
2. The trolley wire is to be supported at intervals not exceeding 36.6 metres.
3. Each trolley wire must be divided into sections not exceeding 0.8 km in length with an emergency switch between every two sections.
4. The trolley wire should be capable of withstanding stresses due to variation in temperature.

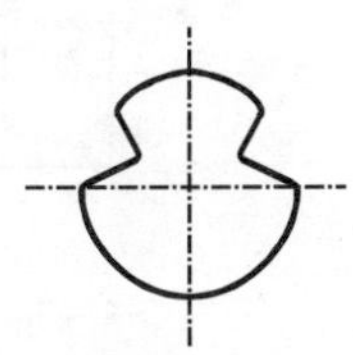

Fig. 15.6

The materials employed for trolley wire are hard drawn copper or alloys of copper (cadmium copper or silicon bronze). The alloy is preferred because it has uniform hardness throughout the x-section whereas in case of hard drawn copper the hardness is confined to the outer surface *i.e.,* the skin of the trolley wire only. The uniform hardness gives uniform wearing of conductor, which is highly desirable in electric traction. The trolley wire is usually of circular grooved cross section, as illustrated in Fig. 15.6. The cross-sectional area of a standard trolley wire is 80 mm^2.

The trolley wire is supported and insulated from a transverse span wire by means of a steel bolt screwed into a gunmetal ear. The gunmetal ear is attached to the trolley wire, and the bolt is insulated from and held in a hanger attached to the span wire. The span wire is supported from poles erected facing each other on both sides of the road and insulated there from.

On straight runs the span wire is above the trolley wire but at curves the span wire is in level with trolley wire. Hangers employed for curved paths are known as '*pull-offs*'. The span wire is of stranded galvanised steel. The span wires may be extended over the whole width of the road (in which case these are attached to poles on each side of the road) or may be of a shorter length (in which case these are attached to brackets carried from horizontal arms attached to the poles at one side of the road). The former arrangement is known as *span wire construction* and the latter *bracket arm construction.*

15.4.2. Railways

Since in railways a sliding current collector (bow or pantograph) is employed, therefore, the contact wire is required to be suspended with very small sag so that the contact between the wire and collector may be maintained at high speeds. This is achieved without excessive tension in the wire by employing spans of comparatively shorter length (3 to 4.5 metres). For such short spans an indirect method of suspension is adopted, in which the trolley wire is supported by another wire known as the "*Catenary*" or "*Messenger*" and is fixed with considerable sag (2 to 3 per cent of the span) between supports at suitable distance. The wire is usually made of more than seven strands of steel. To avoid the use of insulated droppers the whole of the messenger wire is usually insulated, and it may therefore be used if required for carrying current. The position of contact wire remains practically constant at all temperatures which may cause large amount of sag in the catenary wire. Nowadays the catenary construction employed is of either the single catenary or compound catenary.

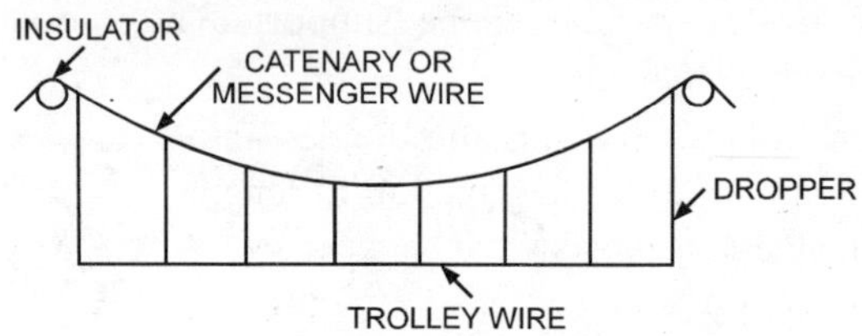

Fig. 15.7. ***Single Catenary Construction***

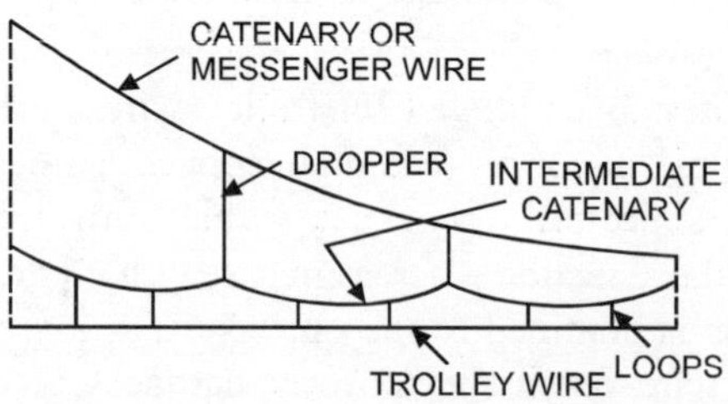

Fig. 15.8. ***Compound Catenary Construction***

In *single catenary construction,* a steel wire with 7 or more strands supports the trolley wire by means of droppers clipped to catenary and trolley wire at equidistant horizontal intervals. The span of catenary wire may be from 45 to 90 metres with a sag of from 1 to 2 metres respectively for straight track. The distance of droppers is between 3 and 5 metres. On curved track the spans of shorter lengths are employed and the trolley wire is kept in position by pull-off wires. Such construction provides sufficient flexibility for speeds up to 120 kmph without any problem of current collection. Besides this it is relatively cheaper, easy to erect and does not require heavy maintenance. Single catenary construction is considered suitable for most light weight ac systems and for branch and loop lines and secondary routes where traffic density is low and operating speeds are low.

The compound catenary construction is employed for heavy current *i.e.,* where traffic density is high. It consists of three wires, all in the same vertical plane. The upper wire is catenary wire, which is insulated from the supporting structures. From this wire another wire, known as *intermediate catenary,* is supported by droppers clipped to both wires. The third wire *i.e.,* trolley wire is maintained in definite and constant tension by means of automatic tightening gear. Both contact and catenary wires are divided in sections of 150 metres and each section is anchored at the centre. The purpose of intermediate catenary besides increasing the current carrying capacity is to provide more uniformity in elasticity.

15.5. SAG AND TENSION CALCULATION FOR A TROLLEY WIRE

A flexible wire when suspended horizontally between two supports it would assume the form of catenary. But in case of tramways the sag in the trolley wire rarely exceeds 1 or $1\frac{1}{2}$ per cent of the span length and, therefore, the wire may be considered to hang in the form of a parabola.

If $2l$ is the span length in metres, w is the weight in kg per metre length of wire and T is the tension of wire in kg.

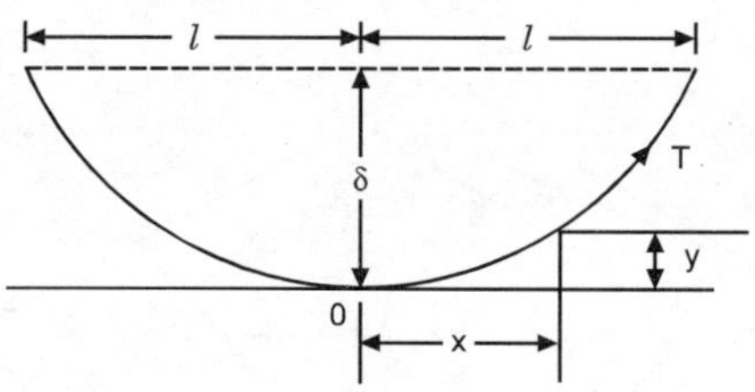

Fig. 15.9

Then by taking moments about one of the supports, we get

$$T\delta = wl \times \frac{l}{2}$$

$$\text{or} \quad \delta = \frac{wl^2}{2T} \qquad \text{...(15.1)}$$

The equation of parabola is given by the expression

$$y = \frac{\delta x^2}{l^2} = \frac{w}{2T}x^2 \qquad \text{...(15.2)}$$

We know that

$$\frac{ds}{dx} = \sqrt{1+\left(\frac{dy}{dx}\right)^2}$$

$$\text{or} \quad \frac{ds}{dx} = \sqrt{1+\left(\frac{wx}{T}\right)^2} \qquad \text{Since } y = \frac{wx^2}{2T}$$

$$\text{or} \quad \int ds = \int\left(1+\frac{w^2x^2}{T^2}\right)^{1/2} dx \qquad \therefore \frac{dy}{dx} = \frac{wx}{T}$$

$$\text{or } s = \int\left(1+\frac{w^2x^2}{T^2}\right)^{1/2} dx$$

$$= \int\left(1+\frac{w^2x^2}{2T^2}+\frac{3w^4x^4}{8T^4}+\ldots\right) dx$$

$$= x+\frac{w^2x^3}{6T^2}+\ldots$$

When $x = l$ *i.e.*, half of the span, the length of conductor in half of the span,

$$s = l + \frac{w^2 l^3}{6T^2} = l + \left(\frac{2T\delta}{l^2}\right)^2 \times \frac{l^3}{6T^2}$$

$$= l + \frac{2}{3} \cdot \frac{\delta^2}{l} \quad ...(15.3)$$

$$\text{Since } \delta = \frac{wl^2}{2T}$$

The change in temperature causes a change in length which will cause the change in sag and tension in a span. It should be noted that decrease in temperature reduces the sag and increases the tension. The reduction in the sag causes stretching which on the contrary tends to decrease the sag. In view of this combined effect the elasticity of the conductor material play an important role in determining the exact values of the sag at different temperatures.

Example 15.1. A trolley wire of a tramway is suspended from two poles 40 metres apart. If the diameter of the wire is 1 cm and weight per metre is 0.5 kg, find its sag if the tension applied is 500 kg. Find also the total length of wire required. Derive the necessary formula used.

[*Pb. Univ. Elec. Traction Nov. 1973; Pb. Technical Univ. Electric Drives & Utilization, May 2009*]

Solution : Span length, $2l = 40$ metres

or $l = 20$ metres

Weight per metre, $w = 0.5$ kg

Tension applied, $T = 500$ kg

$$\text{Sag, } \delta = \frac{wl^2}{2T} = \frac{0.5 \times 20^2}{2 \times 500} = 0.2 \text{ metre} \quad \textbf{Ans.}$$

Total length of wire required

$$2s = 2\left[l + \frac{2}{3}\frac{\delta^2}{l}\right] = 2\left[20 + \frac{2}{3} \times \frac{(0.2)^2}{20}\right]$$

$$= 40.00267 \text{ m} \quad \textbf{Ans.}$$

Example 15.2. Calculate the sag to be given to a trolley wire 1 cm in diameter and weighing 0.72 kg per metre when erected with a span of 60 metres and if the stress in the wire erected is 640 kg/cm^2.

Solution : Span length, $2l = 60$ metres

or $l = 30$ metres

Weight per metre, $w = 0.72$ kg

$$\text{Tension of wire, } T = Ea = 640 \times \frac{\pi}{4} \times (1)^2 = 503 \text{ kg}$$

$$\text{Sag, } \delta = \frac{wl^2}{2T} = \frac{0.72 \times (30)^2}{2 \times 503}$$

$$= 0.644 \text{ m} = 64.4 \text{ cm} \quad \textbf{Ans.}$$

Example 15.3. In a catenary system the catenary wire is of steel with a permissible tension of 1,000 kg. If the span is 60 metres, what shall be the minimum sag of the catenary ? The weight of the catenary wire is 0.9 kg per metre run, the weight of the trolley wire is 1.2 kg per metre run, and the allowance for droppers and fittings may be taken as 20% of the trolley wire weight.

Solution : Span length, $2l = 60$ metres

or $l = 30$ metres

Average weight of catenary wire, $w_1 = 0.9$ kg/metre run

Average weight of trolley wire, $w_2 = 1.2$ kg/metre run

Average weight of droppers and fittings,

$$w_3 = \frac{20}{100} \times 1.2 = 0.24 \text{ kg/metre run,}$$

Average weight of catenary wire, trolley wire and dropper per metre run,

$$w = w_1 + w_2 + w_3 = 0.9 + 1.2 + 0.24 = 2.34 \text{ kg}$$

Sag of the catenary,

$$\delta = \frac{wl^2}{2T} = \frac{2.34 \times (30)^2}{2 \times 1{,}000} = 1.053 \text{ metres} \quad \textbf{Ans.}$$

15.6. DC TRACK ELECTRIFICATION

The substations are necessary for control of electric supply to a considerably long network for electric traction.

15.6.1. DC Substations

These substations are required for tramways, trolley buses and railways. Their functions are to (*i*) step down the transmission voltage to the utilization level (*ii*) convert it into dc and (*iii*) sectionalizing the overhead equipment. Such substations, therefore, consist of step-down transformers, converting machinery, ac circuit breaker, dc circuit breaker and smoothing and metering equipment.

Mercury-arc rectifiers used to be employed as converting machinery because of their much higher all-day efficiency, 100% overload capacity for few minutes, low maintenance cost, noiseless and reliable operation at all voltages and adaptability to remote control with simplicity of operations. Nowadays solid state devices are finding increasing use in these substations.

Since the rectifier used to be non-reversible, hence suitable arrangements were required to be made to receive the energy returned during regenerative braking. In cases where regenerated energy was considerable, inverters (grid - controlled rectifiers arranged to operate with reversed voltage) were employed. In other cases the energy used to be dissipated in a load resistor connected to the dc bus bars.

15.6.2. Transmission System For Traction Substations

The substations lying along the railway track are fed by means of double transmission line running along the track. Double transmission line is employed to ensure continuity of supply. Normally, the current carried by each line is half of the total current but in case of fault on any line, the total current is carried by the healthy line.

Line diagram of duplicate transmission system is shown in Fig. 15.10. In case of fault on any section, the faulty section is opened out by the line circuit breaker and the voltage drop in the other line of this section is doubled because it has to carry double the current. For each single unit substation there are four line circuit breakers and one transformer circuit breaker and since the cost of high voltage equipment increases very rapidly with higher voltages, it is clear that transmission lines operating at 220 or 132 kV must be tapped as seldom as possible.

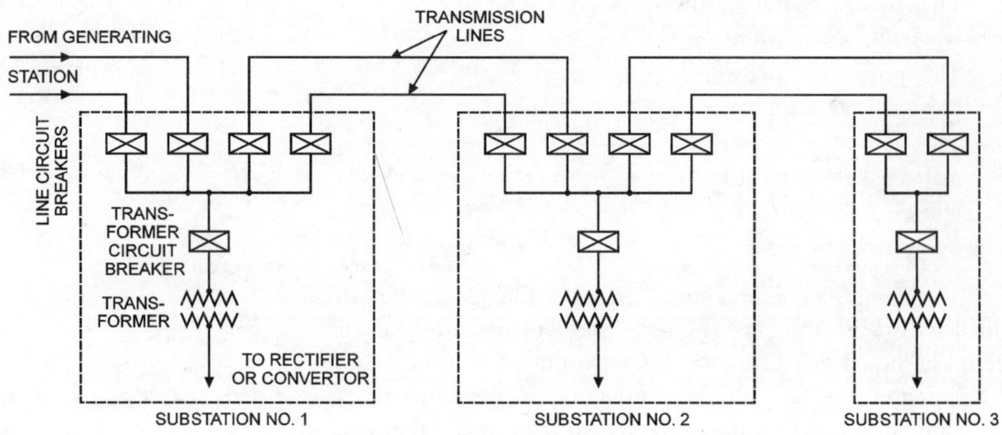

Fig. 15.10. *Duplicate Transmission System*

In case the ht supply is available from the grid, the substations are fed from the grid supply through step-down transformers. An arrangement for feeding the substations from the grid is shown in Fig. 15.11. The EHT supply is tapped at suitable points A, B and C, its voltage is stepped down by the transformers at substations at A, B and C. Now either a single ht transmission line may be taken to the substations from the transformer substations, as shown in Fig. 15.11(*a*) or duplicate transmission lines are provided between substations 1-2-3, 4-5-6, 7-8-9, as shown in Fig. 15.11(*b*) in order to avoid the risk of discontinuity of supply due to one of the faulty line.

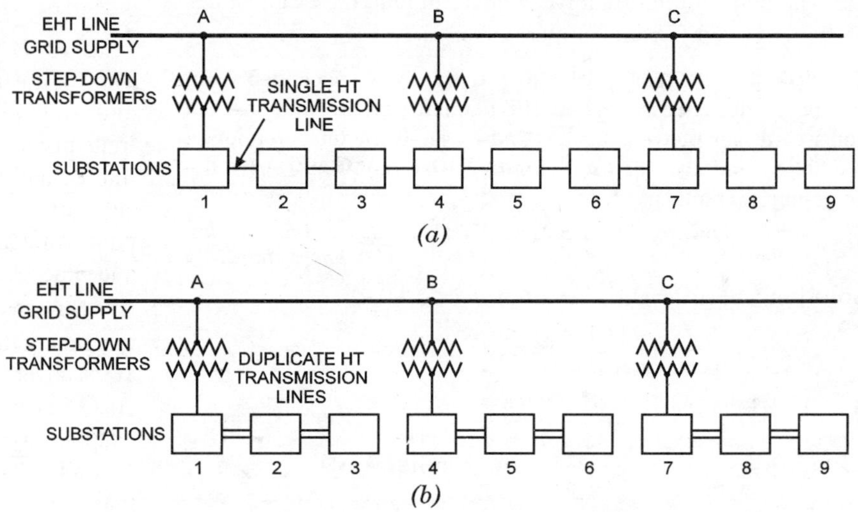

15.6.3. Location of DC Substations

Since in case of trolley bus both the conductors are trolley wires, therefore, more voltage drop takes place in it and the spacing between substations is reduced in comparison to tramway substations for similar capacity. Hence the substations required are more in number and of lesser capacity. Spacing of substations is kept 4.8 km for single track and 3.2 km for double track (up and down). Each section of 0.8 km length is individually fed, as shown in Fig. 15.12.

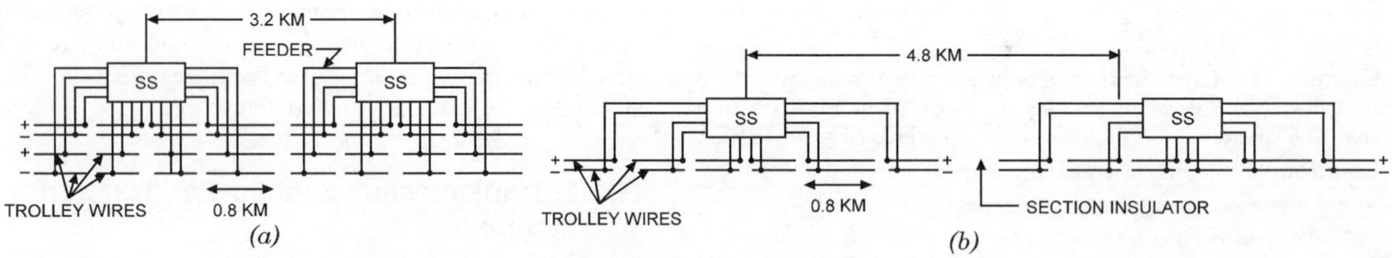

Fig. 15.12

For suburban railways the spacing of substations is usually determined from consideration of permissible variation of voltage at the trains and economy. The length of section depends upon the traffic density and substations spacing. The length of section may be from 3 to 5 km in general and distance between substations may be from 13 to 16 km. The sections can be isolated by sectional switches.

15.6.4. Feeding and Distributing Systems

The electrical energy is supplied from a power station (or a substation) to a number of circuits at a constant voltage through distributor cables (known as *distributors*), which are fed at suitable points (known as feeding points) by other cables known as *feeders*. The feeders connect the feeding points to the station bus-bars and maintain these points at a definite voltage. The distributors supply the circuits at practically constant voltage. Thus the choice of cross section of distributor depends on permissible variation of voltage along the distributor while that for the feeder on the minimum annual operating expenses. In electric traction systems the trolley wires and track rails (or two sets of conductor rails) form the distributors.

The tramways and trolley bus systems must conform to certain statutory regulations, chief among which are:

1. The voltage at the trolley wire is not to exceed 550 volts and that at generating stations (or substations) 650 volts.

2. The trolley wire shall be divided into sections not exceeding 1.6 km in length.
3. The potential difference between any two points of the track rail return systems shall not exceed 7 volts.
4. The potential difference between any point on the rail and earth should not exceed 4 volts.

Hence separate feeding systems are required for the trolley wire and the track rails in case of tramways in order to comply with these regulations.

In dc railways employing track rails as a return path, the voltage drop in the track rails is limited to such values that the working of telegraphic apparatus is not affected by earth currents. The positive portion of the distributing system need not to conform to statutory regulations with respect to voltage drop and sectionalising at every 0.8 km.

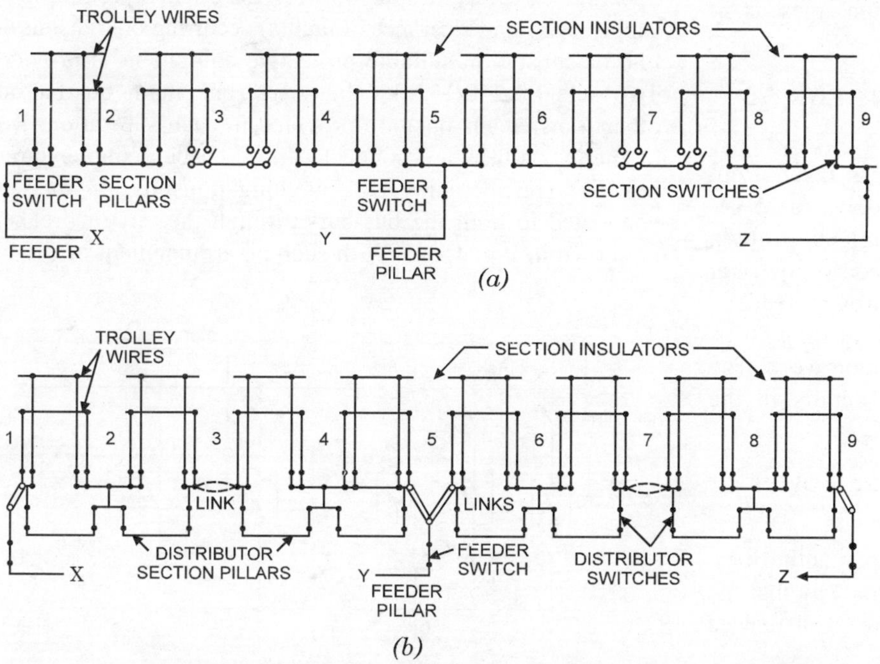

Fig. 15.13. ***Methods of Feeding and Sectionalizing Trolley Wires For Tramways***

The variation of voltage at the tramcars depends mainly upon the voltage drop in the trolley wire. The permissible variation in voltage depends upon service conditions.

The choice of length of the trolley wire between two feeding points depends upon the permissible voltage drop, current required in that section and particular of trolley wire.

The method of feeding the trolley wire in case of tramways is largely influenced by traffic density and economical considerations. For heavy traffic a separate feeder supplied from substation is required for each section of 0.8 km length while with light traffic, several 0.8 km sections of the trolley wire are fed from a single feeder because of economical considerations. The two methods are shown diagrammatically in Fig. 15.13.

In one case [Fig. 15.13 (*a*)] the trolley wire is used as a distributor and each distribution section comprises two or more sections of trolley wires. In other case [Fig. 15.13 (*b*)] feeders of large x-section are run to a few points in the system from which current is distributed to the various sections of trolley wire by means of graded distributing cables. The second method is more economical and the variation of voltage in the adjacent sections is less.

The voltage along the track is kept within the limits of 7 volts above earth by employing *negative booster* (Refer to Art. 15.17). Sometimes more number of negative feeders of comparatively low resistance are used.

15.7. LOW FREQUENCY SINGLE PHASE SYSTEM

If the supply available is from a National grid the transformers and frequency changers are required to convert the supply available (high voltage, commercial frequency supply) into low voltage (15,000 volts) and low frequency $\left(16\frac{2}{3}\text{ Hz}\right)$ supply in order to suit the traction requirements in addition to switchgear and protective and metering equipment.

If the primary supply system is designed solely for the purpose of feeding a traction system on single phase as is with most of the European railways, the generators employed at the generating stations are single phase, low frequency $\left(16\frac{2}{3}\text{ Hz}\right)$ and 15,000 volt machines. The substations are fed from generating stations through high voltage (220 kV/132 kV) transmission lines and step down the voltage to 15,000 volts by means of step-down transformers. Modern substations are of outdoor type because of low cost.

15.8. AC TRACK ELECTRIFICATION

As even in single phase electrification, the substations are necessarily required to keep the voltage drop within permissible limits. The type of substations required depends upon the nature of primary supply and system of electrification.

15.8.1. High Voltage AC Supply

Arrangements of power supply to the various electric traction substations in India are as follows:

1. Power is purchased from supply authorities who are responsible for the operation and maintenance of 220/132 kV transmission lines and grid substations up to 25 kV outgoing terminals of substations. However, 25 kV feeder circuit breaker is controlled from remote control centre by traction power controller (TPC).
2. All the 220/132 kV and 25 kV equipment of substation is owned, installed and maintained by the supply authorities whereas 25 kV circuit breakers are owned, installed, operated and maintained by the railway authorities. This

arrangement ends the dual control of 25 kV circuit breakers (operation by railways and installation and maintenance by supply authorities.)

3. For better control and expeditious execution of electrification programme Railways considered necessary to own, install, operate and maintain 220 kV/132 kV equipment at the substation themselves. The Railway authorities take high voltage supply of 220 kV/132 kV at the substations from the supply authorities. This arrangement exists on Madras-Villupuram, Rourkela-Durg and Virar-Sabarmati sections.
4. In another arrangement 220/132 kV 3-ϕ supply is taken at one point from the supply authorities and then it is distributed by the railway authorities to different traction substations through their own transmission lines as in case of Kanpur-Tundla section.

15.9. POWER SUPPLY ARRANGEMENT FOR AC TRACK ELECTRIFICATION

As mentioned in Art. 15.8.1, electric power is taken from grid supply at 220/132 kV and then stepped down to 25 kV single phase at traction substations through step-down transformers and suitable switchgear. The electric power at 25 kV ac is transmitted to feeding posts through feeders. Feeding posts are located opposite the track. At feeding posts, this power is fed to OHE above track for traction work. The power supply arrangement is described in details in the forthcoming sub-articles.

15.9.1. Sectioning Arrangement For Power Feeding

Power is generated and transmitted, by supply authorities, by 3-phase system and it is distributed in such a way that the loading of all the three phases remain almost same. Unbalanced

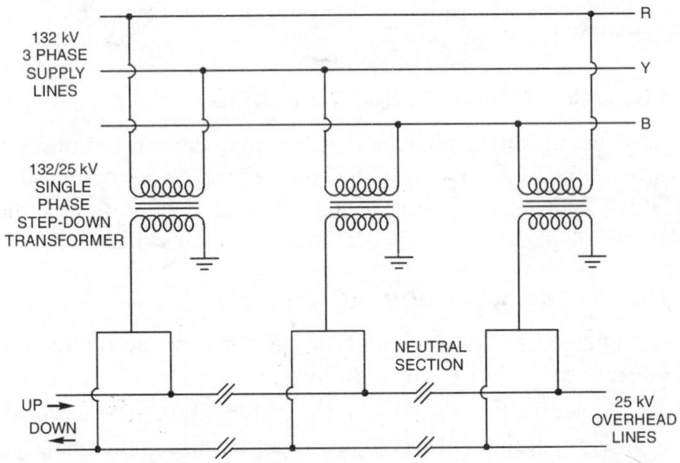

Fig. 15.14. *Sectioning Arrangement For Power Feeding*

loading of the three phase system results in phase shift and also causes rotor heating of alternators. Electric traction is a single phase load so in order to avoid unbalanced loading consecutive substations are supplied by different phases in rotation as depicted in Fig. 15.14. In dc track electrification, all the substations feed OHE in parallel but in ac track electrification, section supplied by one substation works as an independent unit and never work in parallel. As different sections fed by different substations work on different phases, a neutral section is provided approximately in the middle of two feeding posts. The purpose of providing neutral section is to isolate the two sections supplied by different phases. Thus, the pantograph of the locomotive does not bridge two different live phases of 25 kV supply while passing from one section to another one. Neutral section will be discussed in detail later on (refer to Art. 15.14.4)

15.9.2. Substations

Figure 15.15 depicts the layout diagram of a typical traction substation. Traction substation is supplied by 220/132 kV ac supply either from two sources of supply or by double circuit 220/132 kV transmission line for the sake of reliability of supply. For rapid clearance of fault(s) occurring on transmission line or substation suitable protective equipment is provided. Two sets of 220/132 kV bus-bars (one main bus-bar and other transfer bus-bar) are provided in each substation. Both of these bus-bars are connected through bus coupler circuit breaker. The incoming and outgoing transmission lines are connected to both the bus-bars through the circuit breakers, as shown in Fig. 15.15. With such an arrangement of the two

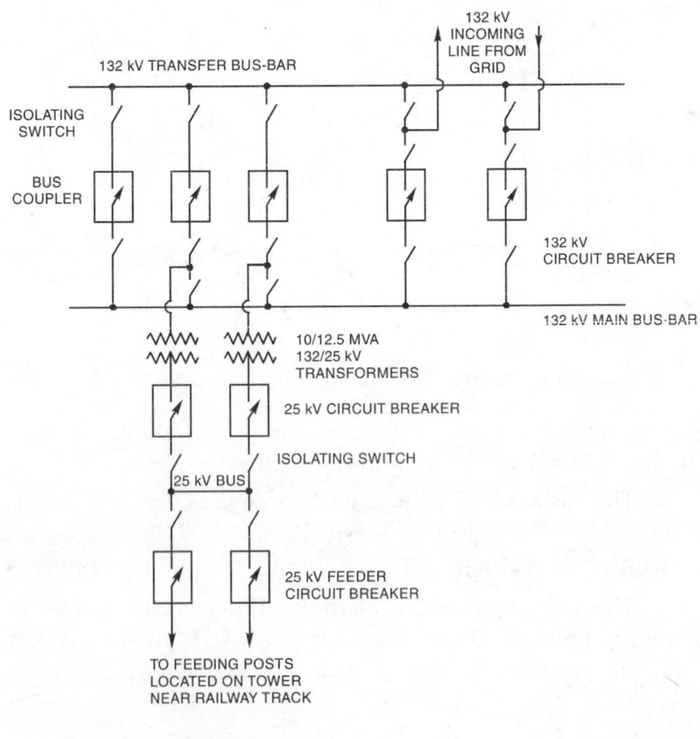

Fig. 15.15. *Layout Diagram of a Typical Traction Substation*

bus-bars and isolators on both sides of circuit breakers continuity of supply can be maintained while doing maintenance work on any bus-bar or circuit breaker on primary side of transformer. Normally two single phase transformers of rating 10 or 12.5 MVA, 132/25 kV are installed in a traction substation. Of the two transformers only one will be in operation normally, the other one acting as standby. Output voltage at 25 kV is maintained within –5% to +10% with the help of tappings provided on the transformer. One 25 kV transformer circuit breaker and one 25 kV feeder circuit breaker equipped with an overload relay and an impedance relay are provided on the output side of transformer for clearance of all faults on OHE. One terminal of 25 kV winding of transformer is connected to earth solidly at substation and return feeders from all the rails of electrified track opposite to feeding post are also connected to this earthed end solidly. Provision of 25 kV bus-bar on the output side facilitates the maintenance of any transformer or its associated circuit breaker without any interruption of power supply. These substations are unattended type and are controlled by remote control operation. Spacing of these substations are 50 to 90 km and they are located closed to feeding posts, maximum distance being 2 km.

15.9.3. Feeding Post

Figure 15.16 depicts the layout diagram of feeding post. It is located opposite the electrified track on towers. Two 25 kV incoming feeders from traction substation are connected to two sets of bus-bars in feeding post through circuit breakers. The circuit breakers are controlled from a remote control centre of the railway. Each feeder consists of two conductors, one insulated for 25 kV connected to the bus-bar and the other insulated for 3 kV connected to the track for return currents. Two sets of feeders and bus-bars along with one bus coupler make it possible to maintain continuity of supply in case of fault or carrying out maintenance work on any feeder or circuit breaker. In the case of two track lines there are four interruptors, two for each feeder and feeding two tracks on one side of feeding post only. Interruptors are single pole, low oil content outdoor type oil switches and are meant for connecting different sections of OHE to feeding post. Interruptors are normally arranged for remote control operation but can also be operated manually locally in case of emergency. Interruptors are not supposed to clear faults automatically like circuit breakers as these are not provided with any protective relays. These are supposed to operate (open or close) in normal load conditions only.

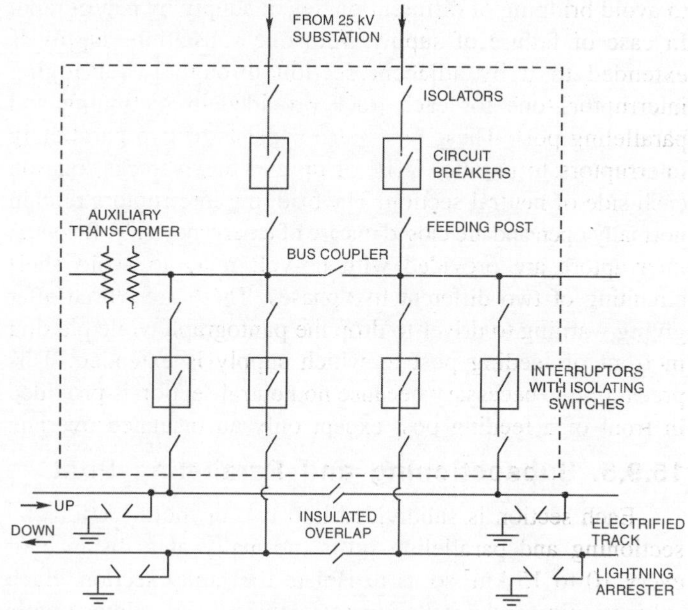

Fig. 15.16. *Feeding Post*

One auxiliary transformer of rating 25 kV/230 V, 10 kVA is also installed for 230 V supply at feeding post for charging batteries. Batteries are installed for providing uninterrupted power supply for remote control equipment, signalling and lighting circuits.

15.9.4. Sectioning and Paralleling Post

Zone covering up and down tracks between feeding post and neutral section is called one section and so each feeding post supplies two sections. The supply to each section is controlled by a circuit breaker. Figure 15.17 illustrates the power supply

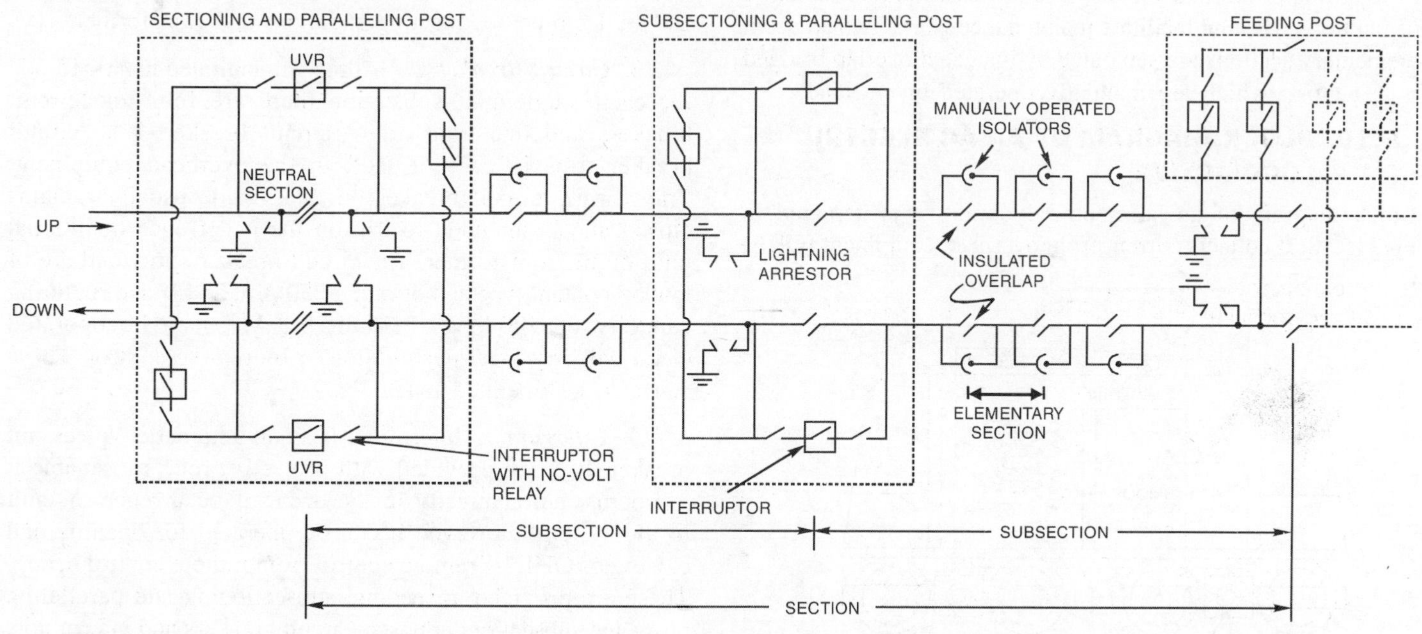

Fig. 15.17. *Power Supply Arrangement For a Two Track Line in Between a Feeding Post and a Neutral Section*

arrangement for a double track line in between a feeding post and a neutral section. Two adjacent sections supplied by different substations are connected through neutral section to avoid bridging of different phases of supply by pantograph. In case of failure of supply from one substation, supply is extended to it by adjacent section through two bridging interruptors–one for each track provided in sectioning and paralleling post. These posts also consists of two paralleling interruptors to parallel OHE of up and down tracks, one on each side of neutral section. The bridging interruptors remain normally open and are closed in case of emergency only. Bridging interruptors are provided with no-volt relay to avoid short circuiting of two different live phases. These are closed after giving warning to driver to drop the pantograph while passing in front of feeding post to which supply is extended. This precaution is necessary because no neutral section is provided in front of a feeding post except only an insulated overlap.

15.9.5. Subsectioning and Paralleling Post

Each section is subdivided into two or more sections by sectioning and paralleling posts, normally at a distance of every 10 to 15 km, so as to isolate the faulty section. Each subsectioning and paralleling post consists of two interruptors for bridging adjacent subsections and one interruptor for paralleling up and down tracks.

Sometimes only subsectioning post is provided and so there is no interruptor for paralleling up and down tracks. This is provided very rarely.

At some stations with large yards, change-over switches are sometimes provided so that the supply may be derived either from OHE of one section or the other. Normally, it is locked in one position but in an emergency it can be changed over after taking necessary precautions.

15.9.6. Elementary Section

Each subsection is further divided into elementary sections of length approximately 1 km so as to ensure rapid isolation of fault on OHE and facilitate maintenance work. Two adjacent elementary sections are separated by insulated overlap bridged by isolators, which are manually operated on no load.

15.10. BLOCK DIAGRAM OF AN AC ELECTRIC LOCOMOTIVE

Block diagram of an ac electric locomotive is depicted in Fig. 15.18. It collects current from overhead equipment (OHE)

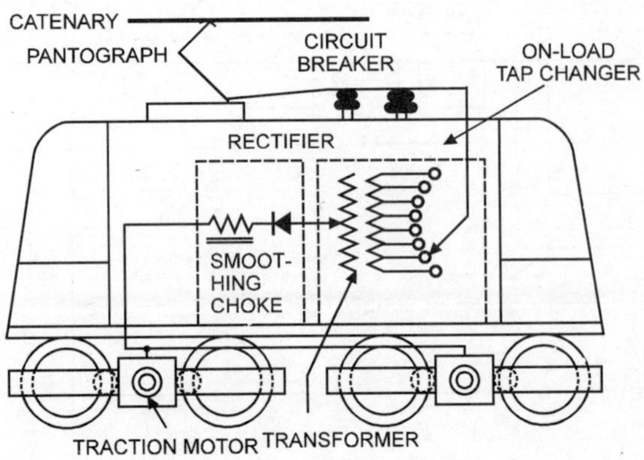

Fig. 15.18. ***Block Diagram of AC Electric Locomotive***

by means of a pantograph collector which usually carries a sliding shoe for contact with the overhead trolley wire. The locomotive essentially consists of (*i*) air circuit breaker (*ii*) on-load tap changer (*iii*) step-down transformer (*iv*) converting machinery (*v*) smoothing chokes and (*vi*) dc traction motors, as already mentioned in Art. 15.1.

15.11. MAJOR SUBSTATION EQUIPMENT

The major equipment provided at traction substations is (*i*) traction transformer (*ii*) circuit breakers and (*iii*) interruptor.

1. ***Traction Transformer.*** Transformers are required to step down the grid supply voltage (220 or 132 kV) to 25 kV. These are specially designed so as to withstand severe operating conditions and have large short-time peak capacity. Its windings are designed with class A insulation and are specially reinforced to withstand the stresses of frequent short circuits on 25 kV overhead equipment. Transformers are designed to have large per cent impedance (not less than 8.5% in case of 10 MVA and 10% for 12.5 MVA transformers) so as to limit the effect of short circuit. Transformers need not to be provided with on-load tap changer as voltage at substations is maintained within permissible limits by the supply authorities. However, off-load tappings are provided on the hv side at – 10%, – 5%, 0% and +10% of rated voltage so as to provide voltage between 27,500 V and 23,750 V. This has the advantage of less copper in the taps and smaller switches, which also may be all immersed in the oil used for cooling the transformers. These transformers are provided with standard accessories such as conservator tank, Buchholz relay, breather, oil level indicator, drain valve, dial gauge thermometer, relief pipe and two thermostats to indicate the oil temperature rise. One is meant to give an alarm indication while the other is for tripping of circuit breaker in case of excessive temperature rise. The cooling arrangement is obviously oil cooling, which is provided by radiators. The radiators can be easily removed for maintenance and repair simply by closing valves in the inlet and outlet pipes connecting radiators and transformer.

2. ***Circuit Breakers.*** As already mentioned in Art. 15.6.2, for each single unit substation there are four line circuit breakers and one transformer circuit breaker. Line circuit breaker trips in case of fault in the overhead equipment. Transformer circuit breaker on 25 kV side usually remains closed and is intended to trip on the occurrence of internal fault on the transformer. The circuit breakers provided are of low oil content type and of rating 750 A at 25 kV and rupturing capacity of 500 MVA. Circuit breakers can be operated electrically by local push button or by remote control. These can also be operated manually.

3. ***Interruptor.*** Interruptor is a non-automatic type circuit breaker. It is not provided with protective relays to enable it to operate automatically in the event of occurrence of fault in the system. However, it can be operated for opening and closing of OHE by remote control from remote control centre. These are provided at feeding, subsectioning and paralleling posts and subsectioning posts. Circuit breakers and interruptors are provided with the following interlocks.

1. Double pole switches cannot be operated unless the interruptor or circuit breaker is open so that sparking at the switches can be avoided.
2. The door, if any, into the bay of circuit breaker or interruptor, cannot be opened unless the double pole switch is locked in open position.
3. The circuit breaker/interruptor cannot be operated from remote control when its key is placed for operation locally.

15.12. REMOTE CONTROL CENTRE

It is provided to have a comprehensive picture of supply conditions and arrangement to operate switchgear with minimum delay so that continuity of electric supply to the track, which is of vital importance, can be maintained. Incharge of remote control centre, known as *traction power controller* (TPC), looks after it all the 24 hours and is directly responsible for carrying out all the switching operations on the electric supply system. Each remote control centre is connected to grid substations, feeding posts, sectioning and subsectioning posts, station master offices, important signalling cabins, divisional office and traffic controller office. It controls all the interruptors provided at different control posts and circuit breakers provided at different substations by remote control system that operates on transmission of voice frequency signals. This signalling system is accomplished by pilot wires with one pair for taking the operating signal and another for incoming indicating signalling. This is to avoid magnetically induced disturbance due to transient currents of switching operations. A mimic diagram depicting diagrammatically the entire OHE system as well as different substations and control posts is provided at each remote control centre to provide actual visual position of supply system to traction power controller. The mimic diagram is generally not lit up but whenever there is any change in the interruptor position at the switching post, corresponding point on the diagram gets illuminated and there is an alarm also to draw the TPC attention. The availability of the electrical supply to each and every section of OHE is monitored by the PTs at switching stations. Thus entire electrical supply system remains under the control of TPC who is fully equipped with facilities for instantaneous control of any switching operation for quick isolation of the faulty sections and restoration of power supply to the adjoining healthy sections and also for the isolation of sections for maintenance work anywhere in the system that may extend up to 250 km in every direction of the centre.

15.13. SIGNALLING

The purpose of railroad signals is to keep trains apart but permit spacing as close as safety permits. For optimum utilization of track capacity with utmost safety, automatic power-operated signalling systems are employed. Power operation also facilitates effective interlocking which reduces reliance upon human factors to the minimum. The signalling may be with semaphores arms, coloured lights or a combination of both. With semaphores or movable arms a lamp is usually so placed that, as the arm moves to one or two more positions, coloured lenses are shifted in front of the lamp to provide colour indications to the passing engine crew to correspond to the indication of the arms. Colour light signals use lamps that are brilliant enough to be seen in the brightest daylight.

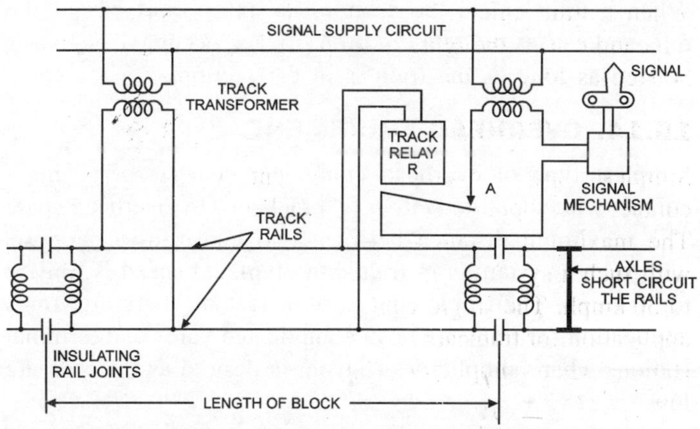

Fig. 15.19. *Simplified Diagram of Automatic Signal Track Circuits*

The rail track is divided in blocks. A block is a length of track which is controlled or protected by a signal at the entering end. A block system is a series of consecutive blocks and each block is insulated from the adjacent block.

Automatic signals are operated by the presence of a train in the block. Automatic signals may employ either battery current or ac supply. The track circuits are energized by power between 2 and 8 volts either alternating or direct current. The rails at each end of the block are cut and connected by insulated joints. There is sufficient insulation in the wood ties so that the rails constitute conductors in a closed circuit with signal power supply at one end and a relay at the other end to actuate the signal (Fig. 15.19). When a train passes the insulating joints, the axles short circuit the rails and causes the relay to drop. An advantage of this arrangement is that a broken rail or a failure of power will also cause the relay to drop and provide a stop-and-proceed indication to the signal. Track relay R holds contact A closed when the relay is energised and allows it to open by gravity if for any reason the relay is de-energized. If ac power is used, the relay is a frequency type, usually 50 or 100 Hz, so if it should be energized at a lower frequency than that for which it is designed to operate, it will open its contacts, causing the signal to go to stop position. The source of energy is of as low voltage as practicable, to prevent operation in the event of short circuit by the train is not complete— and if there should be stray currents present.

DC is used where tracks are free from stray currents from adjacent traction systems. AC operated system is employed when there is a possibility of stray currents or where the track is employed for return current. In the latter case, the blocks are coupled across the insulated joints by impedance bonds. Their function is to permit the flow of traction current around the insulated joints by offering low resistance to it and offer high impedance to the signalling current. An *impedance bond* consists of a massive laminated iron core carrying a few turns of heavy copper. Two end terminals are connected to the two rails and the mid-point or neutral is brought out which is connected to a similar bond beyond the insulated joint.

When the track is unoccupied, the relay circuit is completed, via the two rails and a clear signal is hoisted. When a train enters the section, its axles short circuit the rails and causes the relay to drop. "Track occupied" signal is hoisted as long as the train is in the section.

15.14. OVERHEAD EQUIPMENT

Simplest type of overhead equipment consists of a single contact wire supported either by bracket or by overhead span. The maximum distance between two consecutive supports with such a system is restricted to 30 m and speed is limited to 30 kmph. The single contact wire system, therefore, finds application for tramcars or in complicated yards and terminal stations where simplicity of layout is desired and speeds are low.

Catenary construction (either single catenary or compound catenary), as described in Art. 15.4.2, is used for railways.

15.14.1. Automatic Weight Tension and Temperature Compensation

For sparkless current collection by pantograph collector under high speeds it is necessary that the contact wire should not only remain horizontal at the time of erection but should remain so under all conditions of temperatures and wind pressures likely to be encountered in the service. In an unregulated OHE, in which there is no provision for temperature compensation, sag and tension in catenary and contact wire are so adjusted that contact wire remains approximately horizontal at the time of erection. This type of unregulated OHE is not suitable for current collection at high speeds as encountered on main lines as tension varies inversely as temperature. It affects the stiffness of the line and its dynamic behaviour.

In order that the contact wire remains at the same level under all weather conditions, it is necessary to keep the tension in the catenary and contact wire constant under all temperatures likely to be encountered in service and allow movements of wires along the track which result due to changes in temperature. This is achieved by using regulated OHE at either end of the tension length (Fig. 15.20). The tensioning device consists of a pulley block or a winch with suitable reduction ratio (about 5). Increase in tension reduces the static elasticity of the OHE and makes it more uniform and thus improves the dynamic behaviour and current collection. Contact and catenary wires are given tension of about 1,000 kg each so as to limit the lateral stresses exerted on the masts and foundations due to increase in tension. Cast iron weights of 400 kg are required by the automatic tensioning device (Fig. 15.20) in order to develop tension of 1,000 kg in each of contact and catenary wire. In order to prevent longitudinal to and fro oscillations of OHE in the running condition of train, it is fixed in the centre, which is known as *anticreep* (Fig. 15.21). For the temperature range experienced in electrified sections in India, maximum distance between the anticreep and the end of tension length *i.e.*, half tension length is restricted to 750 metres.

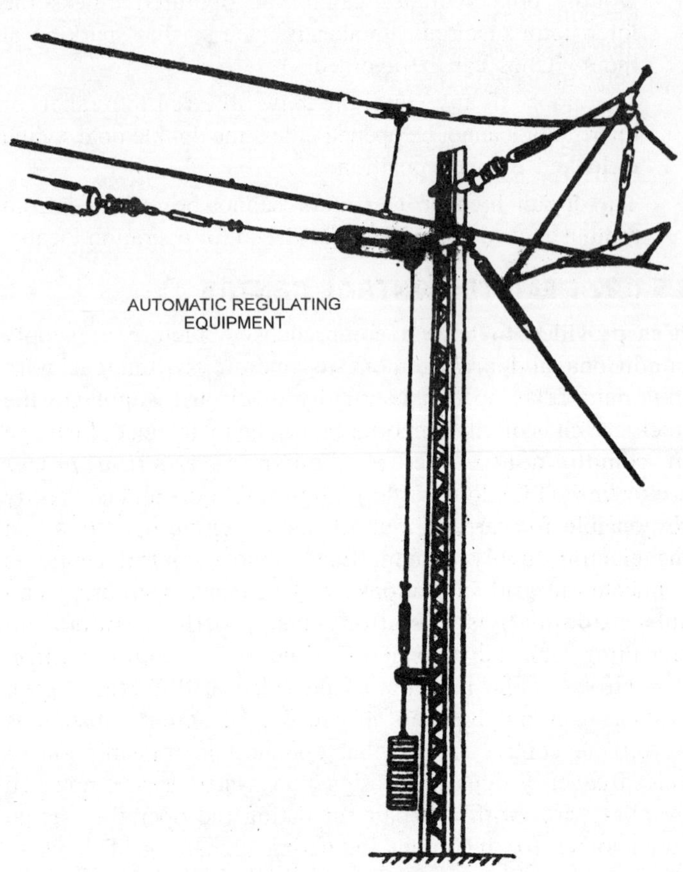

Fig. 15.20. ***Automatic Tensioning Device***

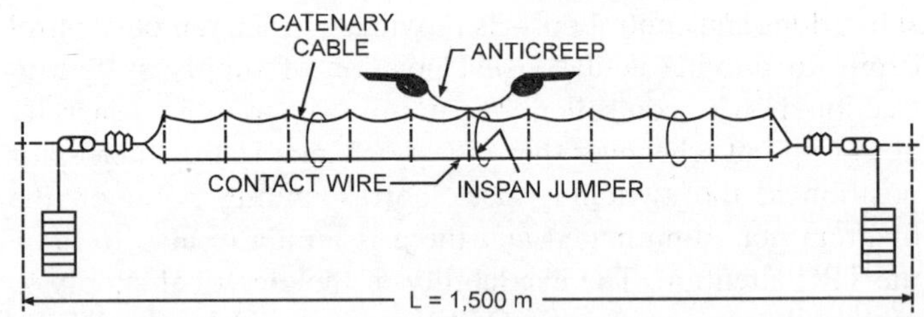

Fig. 15.21

Weight tensioning has the following advantages:

1. Sparkless current collection under all atmospheric conditions with consequent reduction of wear and tear of both contact wire and collecting strips of pantograph.
2. Automatically taking up of creep of the conductors by the tensioning device with the passage of time.
3. Some economy in the design of the supports and foundations because below 40°C temperature, the highest tensions attained are considerably less than with unregulated OHE.
4. Alleviation of problem of providing adequate clearances under overline structures and in tunnels as with fixed tension in the OHE, fluctuations in its height are eliminated.
5. Owing to contact tension in the conductors, there is only one critical velocity of propagation of waves in

the contact wire and this velocity is usually greater than that of unregulated OHE. In high speed traction, it is very important that there should be only one critical velocity of propagation and this should be higher than the speed of the train. From this point of view, the regulated OHE is advantageous because it has only one critical velocity much higher than the maximum speed of the train.

15.14.2. Uninsulated Overlap

In case OHE is made continuous, which is possible by splicing of conductors, long lengths of it would be affected whenever there is a fault or damage. It will also need long time for rectification of damage. It is, therefore, essential to divide the length of elementary section into a number of sub-elementary sections each of which is separated from its adjacent one by means of what is known as an *uninsulated overlap.* In an uninsulated overlap, two contact wires belonging to two adjacent sub-elementary sections are run parallel to each other for at least one span and electrically connected to each other by means of jumpers. The distance between the overlaps is restricted owing to the following reasons:

1. Limited size of the conductor and the weight of the drum on which conductor is rolled. This is to avoid difficulties encountered in handling in transportation.
2. Necessity of isolation of portions of OHE affected by the fault for effective operation and easy maintenance. It makes necessary that overlaps are provided at lesser intervals.
3. The friction of components used on swivelling type of bracket assemblies. This makes difficult to maintain uniform tension in the conductors in case the length of the OHE between anticreep and regulating equipment is long.

Length of regulated OHE is, therefore, never exceeded 1.5 km. Three span overlap is normally provided on regulated OHE with swivelling type bracket assemblies as standard overlap for 25 kV ac traction. The gap provided between two contact wires, in parallel running portion is about 20 cm which is maintained by mechanical linkage of the conductors.

15.14.3. Insulated Overlap

Insulated overlaps are installed chiefly for providing isolating facilities required for the operation and maintenance of OHE. In case two lengths of OHE belong to two adjacent elementary sections, there will be insulated overlap span. Insulated overlaps are also provided at feeding posts, subsectioning posts and booster stations. Insulated overlaps are similar to uninsulated overlaps except for the following:

1. In order to avoid accidental contact between the contact wires, gap between them is kept more in case of insulated overlap.
2. Electrical continuity between the two sections is maintained by means of jumper connections in series with isolator or interruptor. With such an arrangement, a section can be made dead and work can be carried out on the other equipment away from the overlap.
3. The anchoring portion of the catenary and contact wire is separated from the rest of the portion by means of insulators provided at a distance of 2.5 m from the supports and within overlap span. The tail ends of the anchoring OHEs are connected to the other nearby equipment so as to prevent it from attaining any other potential.

In dc traction, single span insulated overlap is employed. Lateral separation between the two equipments is kept as 0.3 m for 1,500 V and 0.4 m for 3,000 V. This separation is maintained by means of insulated rods clamped on to the wires. In 25 kV ac traction system, this lateral separation is increased to 0.5 m. Three span overlap is the normal type of insulated overlap provided on 25 kV ac OHE with swivelling type of bracket assemblies.

15.14.4. Neutral Section

In dc traction, supply to OHE can be provided from both ends so that voltage drop can be reduced and in such a case substations operate in parallel. In case it is felt necessary, say for sake of convenience in maintenance, to have some boundary of supply between two substations, it can be had simply by means of insulated overlap. There will be no damage in this case even if pantograph bridges the OHE. But in single phase ac traction above arrangement is not permissible because adjacent substations tap different phases of the three-phase system so as to equalise load on the three phases. So momentary passing of pantograph under insulated overlap will result in short circuit between two phases of the supply system and damage the pantograph and OHE. This situation is avoided by interposing a small length of OHE, called the *neutral section.* It is insulated from both sides and is not connected to any source of power supply. It's main function is to allow physically smooth and electrically sparkless passage of pantograph from one section to the other. As per Indian Railway practice, neutral section is located in the middle* of two feeding posts and in front of sectioning and paralleling post, as illustrated in Fig. 15.17.

Important precautions in connection with neutral section are given below:

1. Neutral section should be located far away from stopping signals so that the train can pick up enough momentum to enable it to coast through the neutral section without difficulty because there is no power supply in the neutral section.
2. Neutral section should neither be located on upgradient nor on curves because neutral section has to be negotiated with power off.

   The best position for its location would be at mid sections between two feeding posts, away from stopping signals and on down gradient in case of single track or straight level section in case of multi-track.

* In some countries neutral section is provided in front of feeding post also so as to avoid short circuit between the phases in the case of emergency when there is shutdown on the feeding post and power supply is extended from adjacent feeding posts. But Indian Railways did not consider it necessary as it is required only during emergency and additional expenditure will have to be incurred.

3. It is essential that the length of the neutral section should be as small as possible but not of length less than the spacing between extreme pantographs. This is because the train is to negotiate the neutral section with power off and if it gets stalled in this section, there will be dislocation in the traffic and arrangements will have to be made to feed the neutral section. In case its length is shorter than the spacing between extreme pantographs, they will short circuit the OHEs supplied from different phases.

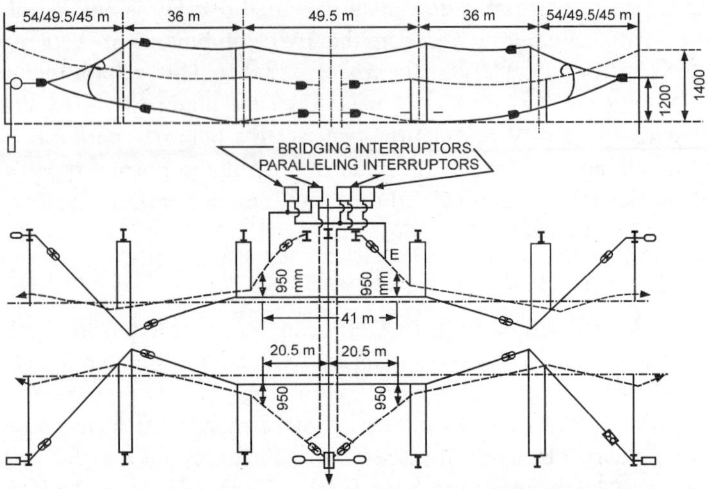

Fig. 15.22. ***Standard Neutral Section Adopted By Indian Railways***

When pantographs at the extreme ends of multiple head trains are raised they are about 40 m apart. Since both the air gaps at the extreme ends of the neutral section have not to be bridged simultaneously, the minimum length of the neutral section adopted by Indian Railways is 41 metres. Standard neutral section as adopted by Indian Railways is illustrated in Fig. 15.22.

Long neutral sections are usually provided in suburban sections and steep graded sections. These neutral sections make use of section insulators in place of air gaps at the extreme ends of standard neutral section. These section insulators are fixed within the span immediately after supports in the same direction in case of double tracks and in opposite direction to each other in case of single tracks, as illustrated in Fig. 15.23.

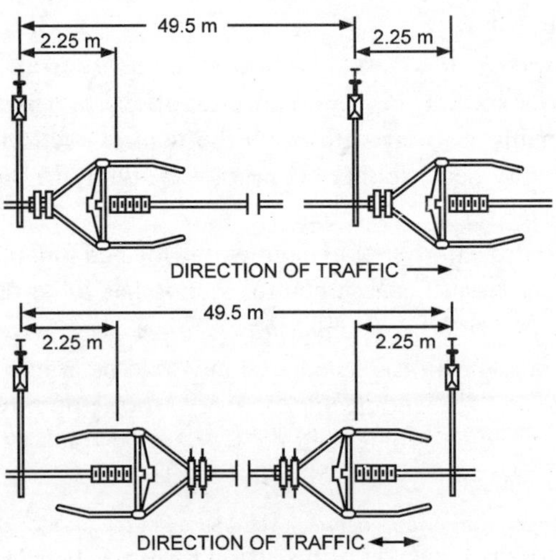

Fig. 15.23

15.14.5. Section Insulators

Section insulators are provided for insulating OHE of one elementary section from the OHE of another adjacent elementary section such that at cross-overs from one track to the other, from main line to siding etc. as shown in Fig. 15.17. Section insulator, as illustrated in Fig. 15.24, consists of straining insulator with two runners connected to one of the contact wires. Runners are at the same height as the contact wire on the other side and so shaped as to permit smoother passage of pantograph as the train passes underneath. Runners extend and overlap the contact wire on the other side so that the locomotive draws interruption free current.

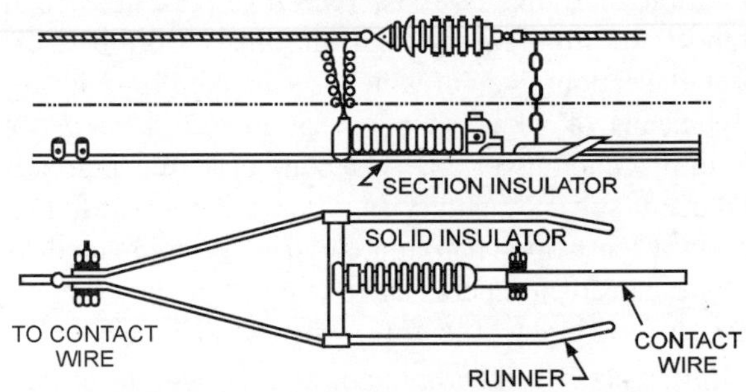

Fig. 15.24. ***Section Insulator***

15.15. INTERFERENCE IN TELECOMMUNICATION CIRCUITS

In all traction systems track is used as a return path for the current. Some part of the current flows through the track and the remaining portion of the current flows through earth, as shown in Fig. 15.25. Current flowing through earth causes interference with communication lines. Now to constrain the return current to flow through restricted path *i.e.,* along the track some steps are required to be taken. In ac traction system use of booster transformer, described in Art. 15.16, is made while in dc traction system negative booster, described in Art. 15.17, is used to reduce the potential drop across the track.

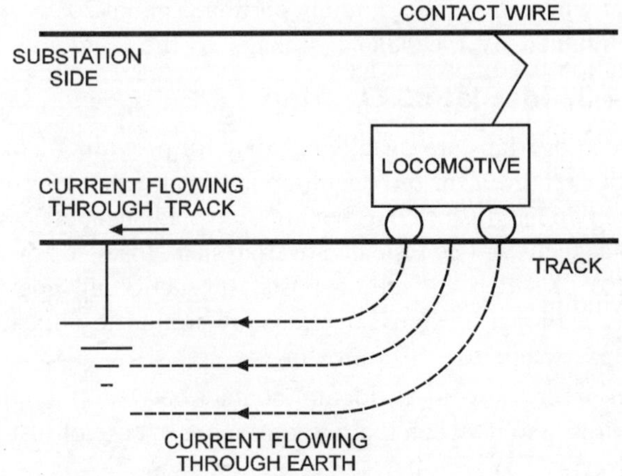

Fig. 15.25

Path of return current is also constrained by reducing the track resistance. The resistance per km length of continuous track rail is of order of 0.05 Ω but the resistance of fish-plate joints employed in jointing the commercial length rail mechanically is more than that of a rail. Hence in order to reduce the resistance of the track as a whole the joints are supplemented with a good conductor or "bond". Modern bonds are of the flexible type and are designed to be as short as the conditions of locations permit. The bonds are usually fixed under the fish plates as they are protected from damage while carrying out the repairs to the paving.

In single phase ac traction system presence of harmonics in supply system also causes interference in telecommunication systems.

Interference in telecommunication and signalling lines is also caused due to electromagnetic induction. Current flowing through contact wire sets up a magnetic field around itself. This magnetic field links with nearby circuits and induces emf in circuits.

Electrostatic induction in signalling and telecommunication circuits also causes interference in these circuits. There occurs a voltage gradient between hv contact wire and earth. Now statically induced emf in communication and singnalling lines is produced because of its positioning voltage gradient.

The interference effects on telecommunication circuits owing to reasons mentioned above (last three paras) can be reduced by taking following steps.

(*i*) All telecommunication circuits are installed underground in cables instead of in overhead lines.

(*ii*) Telecommunication or signalling lines are provided with aluminium sheath instead of lead.

(*iii*) Use of PVC or rubber covering is made to prevent sheath corrosion.

(*iv*) Telecommunication circuits may be removed to the distant place, to the extent practically possible, from the railway track.

(*v*) Long telecommunication circuits are broken by providing isolating transformers.

(*vi*) Booster transformer employing return feeder is also used to avoid magnetic linking of telecommunication lines with power lines.

15.16. BOOSTER TRANSFORMER

In ac traction system return current mainly flows to the substation through track leaving a small portion, which flows through ground because of leakage of current from track to the ground. It has already been described that this leakage current causes a heavy interference with communication lines which are running near the track. So it is very necessary to avoid flow of leakage current through ground. For this purpose booster transformer is used. Booster transformer consists of the windings of unity turn-ratio. Primary winding of booster transformer is inserted in series with the contact wire and the secondary is inserted in series with the return current circuit. Now current flowing through contact wire also passes through primary winding which induces current in the secondary winding in equal amount. This induced current in secondary winding forces the return current to flow through only return circuit, and so avoids leakage current.

There are two methods of connecting booster transformers.

1. ***Rail-Connected Booster Transformer.*** Circuit diagram for such connections is shown in Fig. 15.26. Primary winding of the booster transformer is connected in series with the contact wire and the secondary is connected in series with the rails. Now induced voltage in the secondary winding due to contact wire current flowing through primary winding causes flow of return current only through the rails. The main drawback of this system is that rail joints are insulated at the location of booster transformer and there may occur an insulation puncture of insulated rail joint which makes the booster transformer ineffective because of short circuiting the secondary winding. There may be also a voltage rise between rails and between rail and earth which may be dangerous for human life.

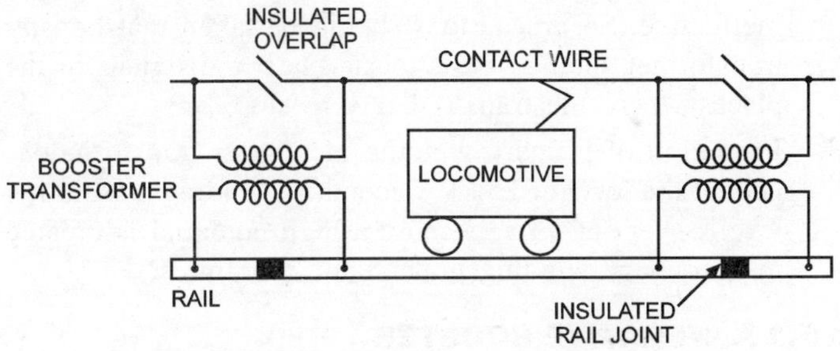

Fig. 15.26. ***Rail-Connected Booster Transformer***

2. ***Booster Transformer Using Return Feeder.*** Connection diagram for booster transformer using return feeder is shown in Fig. 15.27. In this method track is not used for return current flow to substations, but the return current flows through a return feeder which is provided in addition to contact wire. This system is more effective than rail-connected booster transformer system and, therefore, used more extensively. Turn ratio of booster transformer employed in this system also is unity. Primary winding of booster transformer is connected in series of contact wire and the secondary winding is connected in series of return feeder.

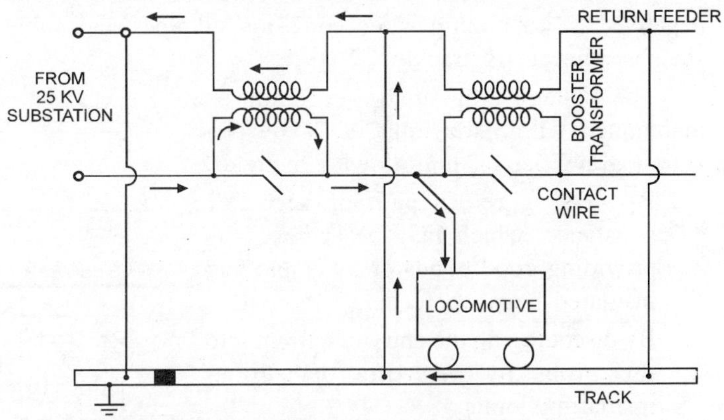

Fig. 15.27. ***Booster Transformer Using Return Feeder***

Return feeder is connected to rail at mid points between two booster transformers. Now current flowing through primary winding induces a voltage in the secondary winding which forces the current to flow only through return feeder.

Drawback of booster transformer is that it is connected in series of traction circuit causing increase in impedance of the circuit. Thus regulation of voltage in locomotive is increased. To improve the voltage regulation it is necessary to locate feeding posts closer, otherwise a capacitor of proper value may also be introduced in series of traction circuit. Former is costlier method and, therefore, the later is mostly used.

Precautions in Connecting Booster Transformers.

1. Booster transformers should be able to withstand severe mechanical stresses which will develop under short-circuit conditions when current of thousands of amperes flows.
2. Primary of booster transformer is connected at the two sections of insulated overlap. When the train passes under the insulated overlap, pantograph short circuits the primary winding for a moment and makes booster transformer ineffective. So precautions should be taken that booster transformer should not be located near the signals or the places where the train is likely to stop.
3. Insulation of primary winding of booster transformer is kept at the level of 25 kV. Secondary windings insulation level is kept at 3 kV because a high potential is created in secondary winding under fault conditions.

15.17. NEGATIVE BOOSTER

In case of dc traction system, return current flows through the rails, to the substations. So due to this current flow through rails some voltage drop exists along the track. Due to potential difference along the track some of the return currents leak out to the earth and follow low resistance paths provided by water pipes, gas mains, cable sheaths etc. as shown in Fig. 15.28. These leakage currents cause electrolytic corrosion to occur at places where the current leaves the pipes as they carry the metal of the pipe (acting as anode) to the electrolytic solution (moisture and salts in the surrounding soil).

The statutory regulations, according to which the voltage drop in the rails is limited to 7 volts and the voltage between any pipe and the rails is limited to 1 volt positive and 3 volts negative, are to minimize the corrosion of pipes and cable sheaths adjacent to track.

The voltage drop in the track rails is maintained within the limit of 7 volts by adopting any one of the following methods.

1. By making return path of very low resistance, which may be achieved by providing good bonds or by employing insulated negative feeders.
2. By discouraging the entry of current into the pipes by inserting occasional insulating joints.
3. By using negative booster.

As shown in Fig. 15.29, the negative bus-bar is connected to the track rail through negative feeder and the positive bus-bar is connected to the trolley wire near the generating stations. There are two boosters, one is positive booster which adds voltage to the line and the other is the negative booster which lowers the potential of the point to which it is connected. Both the boosters are mechanically coupled together and driven by a dc motor. Negative booster is separately excited dc generator whose armature is connected with negative busbar at the far end of the track through negative feeder. Field of negative booster is connected in series with positive feeder supplying the corresponding sections of the trolley wire. The boost on the feeder is adjusted by means of a diverter rheostat connected in parallel with the field winding.

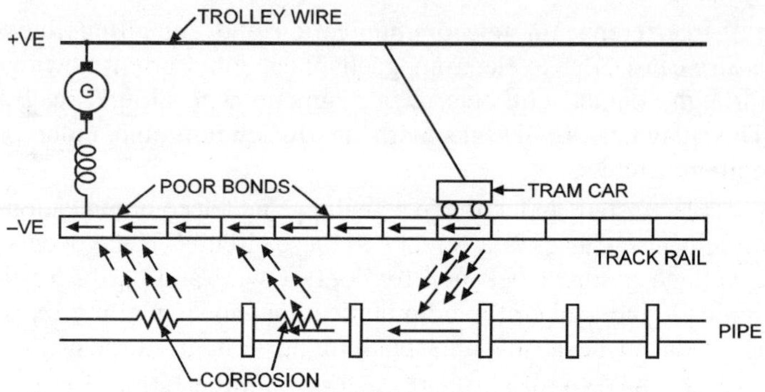

Fig. 15.28. ***Electrolysis by Earth Currents***

Sometimes a positive booster is also connected to the + ve side in addition to – ve booster on – ve side in order to compensate the voltage drop in trolley wire. Positive booster is a self excited series generator which is connected across the + ve busbar and at the far end of trolley wire in series with the field of the negative booster.

Example 15.4. In a section of tramway system the track resistance is 0.08 Ω and the total current which may be considered to be taken uniformly along the length of the section is 300 amperes. Calculate the current to be carried by a negative feeder connected to the distant end of the section to reduce the voltage drop between the ends of the section to 6 volts.

Solution : Let the current to be carried by a negative feeder connected to the distant end of the section to reduce the voltage drop between the ends of the section to 6 volts be of I′ amperes.

Total current, I = 300 amperes

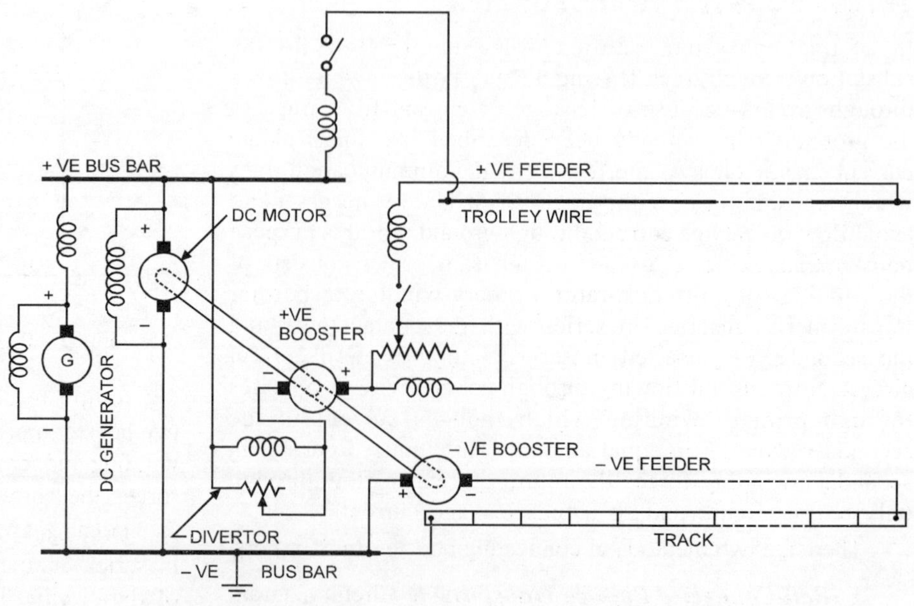

Fig. 15.29. ***Negative Booster Connection Diagram***

Total track resistance, R = 0.08 Ω

Since voltage drop between the ends of the section $= \frac{IR}{2} - I'R$

$$\therefore R\left(\frac{I}{2} - I'\right) = 6$$

$$\text{or} \quad 0.08\left(\frac{300}{2} - I'\right) = 6$$

$$\text{or} \quad I' = 150 - \frac{6}{0.08} = 75 \text{ A} \quad \textbf{Ans.}$$

Example 15.5. A section of a tramway 3 km long has a track resistance of 0.02 Ω per km and a uniformly distributed load of 125 A per km. The earth potential of the track at the far end of the section is zero. Find the potential of the track at the far end of the section.

If a negative feeder and booster are employed to reduce the far end voltage to 7 volts, find the current carried by the negative feeder.

Solution :

Current per km length of the track, i = 125 A
Resistance per km length of the track, r = 0.02 Ω
Length of track, l = 3 km

Potential of the track at the far end
= Potential drop along the track

$$= \frac{irl^2}{2} = \frac{125 \times 0.02 \times (3)^2}{2} = 11.25 \text{ volts} \quad \textbf{Ans.}$$

Let the current carried by the negative feeder be of I amperes

Since the potential drop along the track $= \frac{irl^2}{2} - Irl$

$$\therefore \frac{125 \times 0.02 \times (3)^2}{2} - I \times 0.02 \times 3 = 7$$

$$\text{or } I = 70.833 \text{ amperes} \quad \textbf{Ans.}$$

Example 15.6. A section MN of a tramway 5 km long has a resistance of 0.02 Ω/km and a loading of 200 A/km which may be assumed to be uniformly distributed. A negative feeder is connected to the track at a point X, 3 km from M. If the end M is earthed and the potential of X is 1 volt above M, calculate the current carried by the negative feeder.

Solution : Let the point P at a distance of x km from point M, be the point of highest potential.

Since the point P is the point of highest potential on the section MX, therefore, current at this point is zero and increases towards M and X. Direction of currents in different sections of track are shown in Fig. 15.30.

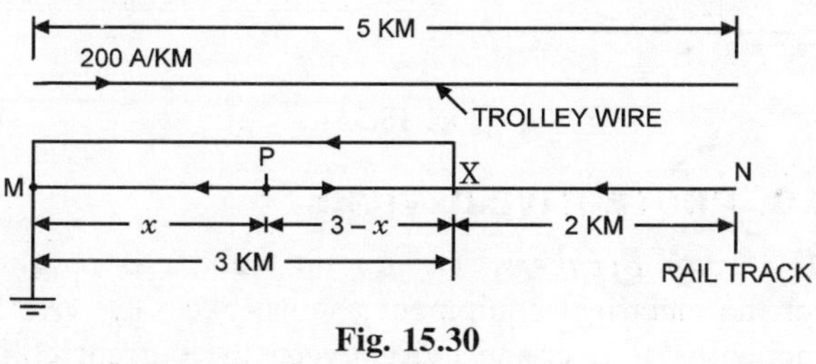

Fig. 15.30

The voltage drop along section PM,

$$V_{PM} = \frac{ix \times rx}{2} = \frac{irx^2}{2} = \frac{200 \times 0.02 x^2}{2} = 2x^2$$

Similarly the voltage drop along section PX,

$$V_{PX} = \frac{ir(3-x)^2}{2} = \frac{200 \times 0.02\,(3-x)^2}{2} = 2(3-x)^2$$

The voltage drop along section XM,

$$V_{XM} = V_{PM} - V_{PX}$$

$$\text{or } 1 = 2x^2 - 2(3-x)^2$$

$$\text{or } x = \frac{19}{12} \text{ km}$$

$$\therefore I_{PX} = 200 \times \left(3 - \frac{19}{12}\right) = 283.33 \text{ A}$$

and $I_{NX} = 200 \times 2 = 400$ A

∴ Current through negative booster

$$= I_{PX} + I_{NX} = 283.43 + 400 = 683.33 \text{ A} \quad \textbf{Ans.}$$

Example 15.7. A section ABC of a tramway rail system is 4.8 km long. A is earthed and B is 3.2 km from A. The resistance of the rails is 0.02 Ω/km. The loading is 250 A per km and may be assumed uniformly distributed.

A negative feeder (connected to a booster) is tapped to the rails at B and maintains B at 2 volts above earth potential. Calculate the maximum pd across any two points on the rails. Also determine the voltage and kW capacity of the booster used. Resistance of negative feeder is 0.02 Ω.

Solution : The rail track is earthed at point A *i.e.,* the potential of point A, $V_A = 0$

Potential of point B, V_B = 2 volts
Current per km length of track, i = 250 A
Resistance per km length of track, r = 0.02 Ω

Let the point D at a distance of x km from point A be point of highest potential on the section AB.

Since point D is the point of highest potential on the section AB, current at this point is zero and increases towards A and B. Directions of currents in different sections are shown in Fig. 15.31.

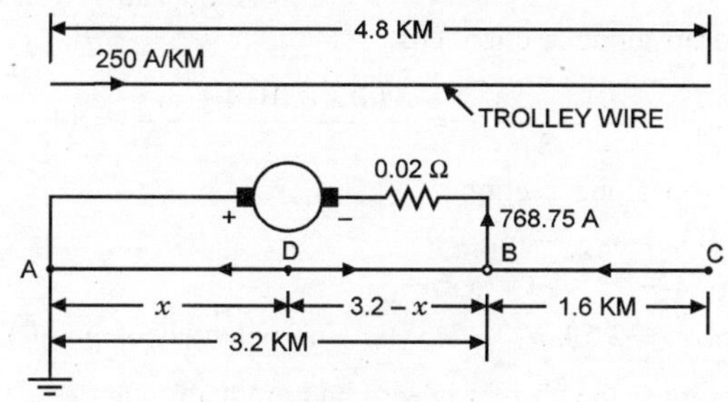

Fig. 15.31

The voltage drop along section DA;

$$V_{DA} = \frac{irx^2}{2} = \frac{250 \times 0.02 x^2}{2} = 2.5x^2$$

The voltage drop along section DB,

$$V_{DB} = \frac{ir(3.2-x)^2}{2} = \frac{250 \times 0.02(3.2-x)^2}{2}$$

$$= 2.5\,(3.2 - x)^2$$

The voltage drop along section BA = $V_{DA} - V_{DB}$

$$\therefore 2 = 2.5x^2 - 2.5\,(3.2 - x)^2$$

$$\therefore x = 1.725 \text{ km}$$

The current distribution will be as follows :

At point C the current is zero and gradually increases till point B is reached, where it is 1.6 × 250 *i.e.,* 400 amperes. At point D current is zero and gradually increases towards points A and B. At point A the current is 250 × 1.725 = 431.25 amperes and at point B the current is 250 (3.2 – 1.725) *i.e.,* 368.75 amperes.

The points D and C, where the current is zero, are at maximum potential.

Potential of point C
= Potential of point B + voltage drop is section BC

$$= 2 + \frac{250 \times 0.02 \times (1.6)^2}{2} = 8.4 \text{ volts}$$

Potential of point D

$$= \frac{irx^2}{2} = \frac{250 \times 0.02 \times (1.725)^2}{2} = 7.439 \text{ volts}$$

Thus the maximum potential difference on any two points on the rails = 8.4 volts **Ans.**

The current carried by the negative booster

= 400 + 368.75 = 768.75 A

Voltage drop in negative feeder = 768.75 × 0.02 = 15.375 V

Boost to be given by negative booster = 15.375 – 2 = 13.375 V

$$\text{Capacity of the booster} = \frac{13.375 \times 768.75}{1{,}000} = 10.28 \text{ kW} \quad \textbf{Ans.}$$

Example 15.8. A section of tramway ABC is 6 km long and earthed at A. Its resistance is 0.01 Ω per km and loading of 200 A/km uniformly distributed. Negative feeder with booster is taken off from point B, 4 km from A. If potential of point B is 4 V above earth, calculate the rating of the booster, negative feeder resistance being 0.02 Ω. [*Electric Traction Pb. Univ. December 1991; Rajasthan Univ. Utilization of Electrical Power 2001*]

Solution : Let the point D, at a distance of x km from point A, be the point of highest potential on section AB.

Since point D is the point of highest potential on the section AB, current at this point is zero and increases towards A and B. Directions of currents in different sections are shown in Fig. 15.32.

Voltage drop along section DA,

$$V_{DA} = \frac{irx^2}{2} = \frac{200 \times 0.01 \times x^2}{2} = x^2$$

$\because i = 200$ A/km and $r = 0.01$ Ω/km

Voltage drop along section DB,

$$V_{DB} = \frac{ir(4-x)^2}{2} = \frac{200 \times 0.01(4-x)^2}{2} = (4-x)^2$$

Voltage drop along section

BA = $V_{DA} - V_{DB}$

or $4 = x^2 - (4 - x)^2$

or x = 2.5 km $\because$ potential of point B is 4 V

At point C the current is zero and gradually increases till point B is reached where it is 2 × 200 = 400 A. At point D current is zero and gradually increases towards points A and B. At point A current is 2.5 × 200 = 500 A and at point B current is (4 – 2.5) × 200 = 300 A.

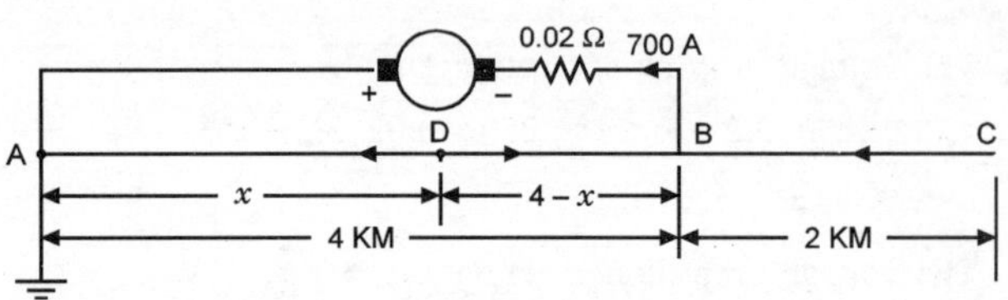

Fig. 15.32

Current carried by negative booster = 400 + 300 = 700 A

Voltage drop in negative feeder = 700 × 0.02 = 14 V

Boost to be given by negative booster = 14 – 4 = 10 V

$$\text{So rating of the booster} = \frac{700 \times 10}{1{,}000} = 7 \text{ kW} \quad \textbf{Ans.}$$

Example 15.9. A section of tramway by using positive and negative booster arrangements is supplied by the dc generator of voltage rating 600 V output. When trolley is at the far end of the section it should get voltage difference between trolley wire and track of at least 578 V. Voltage drop across the trolley wire is 60 V and that across the track is 12 V when the trolley is at far end.

(*a*) Determine the voltage available to the trolley when it is at the far end, and boosters are not used.

(*b*) How much would positive booster boost to keep the voltage at the far end within limit if negative booster provides boost of 10 volts.

Solution : Voltage drop along the trolley wire up to the far end, V_W = 60 V

Voltage drop along the track up to the far end, V_T = 12 V

Total voltage drop up to the far end = 60 + 12 = 72 V

(*a*) So voltage available to trolley when it is at the far end, without using boosters.

$V_A = V_S - (V_W + V_T) = 600 - (60 + 12) = 528$ V **Ans.**

(*b*) Boost given by negative booster, V_N = 10 V

Let the boost given by the positive booster be V_P volts

Recommended voltage to the trolley when it is at the far end, V_R = 578 V

Supply voltage at the generator side, V_S = 600 V

Voltage drop along the track with – ve booster in use

$V_{TN} = V_T - V_N = 12 - 10 = 2$ V

Voltage drop along the trolley with + ve booster in use

$V_{WP} = V_W - V_P = (60 - V_P)$ volts

Total voltage drop at the far end with boosters in use

$= V_{TN} + V_{WP} = 2 + (60 - V_P) = (62 - V_P)$ volts

Admissible maximum voltage drop = 600 – 578 = 22 V

$\therefore 62 - V_P = 22$ V

or $V_P = 62 - 22 = 40$ V

So positive booster should provide a boost of 40 V. **Ans.**

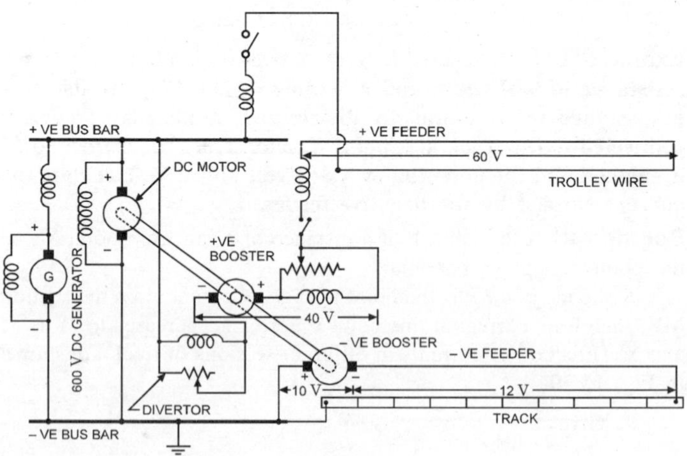

Fig. 15.33

15.18. PROTECTIVE DEVICES

1. *Circuit Breakers.* The circuit breaker is required to protect the electrical equipment against excessive overloads automatically. It is connected between the current collector and the main wiring. One circuit breaker is required in case of tramcar and two circuit breakers are required in case of trolley bus, one in each trolley circuit. The main contacts of the circuit breaker are located in an arc chute and are provided with a magnetic blow-out. The blow-out coil actuates also the tripping mechanism. It is also provided with a handle to trip the circuit breaker when it is moved to the "off" position by hand.

2. *Radio Interference Suppressor.* The emfs of radio frequency may be induced in the wiring due to arcing of sparking at the collector, controller, motors etc. and radiated out from the trolley wires. The most serious interference occurs at the contacts of the master controller in systems with electromagnetic contactors operated at line voltage. Radio-interference suppressor consists of two radio-frequency chokes, one connected in each supply lead to the master controller, a capacitor filter comprising three star-connected capacitors connected between the trolley wires and chassis. The capacity of each capacitor is about 0.1 μF.

3. *Lightning Arrestor.* Only the vehicles operating in the area liable to lightning storms are provided with lightning arrester or surge diverter. One type consists of a small cylindrical block of carborundum held between two electrodes. One electrode with a small air gap in series is connected to the positive trolley wire and the other one to the – ve trolley wire or earth.

Exercises

1. What are the requirements that a current collector gear of electric vehicle should fulfil ?
   Describe with a neat diagram the pantograph collector.
2. What are different types of current collectors? Give their merits and demerits. **[Rajasthan Univ. Utilization of Electrical Power 2005, Rajasthan Technical Univ. 2011]**
3. Discuss and compare various arrangements of current collection used in electric traction.
   [M.D. Univ. Utilization of Electric Power & Traction Dec. 2008]
4. Discuss different types of catenary construction for traction system.
   [Pb. Technical Univ. Electric Drives and Utilization May 2007]
5. Discuss the limitations of single catenary construction and explain some alternative construction to remove these limitations. **[Pb. Univ. Electric Traction June 1993]**
6. Describe with suitable diagram the compound catenary in overhead lines.
7. Describe a typical traction substation with the help of a sketch of substation equipment layout.
   [Rajasthan Univ. Utilization of Electrical Power 2002, 2006, 2007, 2008]
8. Discuss the location of substations with regard to ac and dc systems of electric traction.
   [Pb. Univ. Electric Traction Dec. 1991]
9. Give layout of a traction substation and describe it.
   [Pb. Technical Univ. Electric Drives & Utilization May 2008]
10. What are the functions of dc traction substation ? What is the major requirement for dc substation ?
    [Pb. Technical Univ. Electric Drives & Utilization Dec. 2007]
11. Discuss the various feeding and distributing systems used in electric traction.
    [Gorakhpur Univ. Utilization of Electric Power & Traction 1983]
12. State the various 'statutory regulations' that should be observed in cases of feeding and distributing systems for tramways. Why are such regulations imposed ?
13. Explain the working of a remote control centre. What are its advantages ? What is mimic diagram ?
14. Briefly discuss the salient design features of OHE (overhead equipment) giving suitable diagrams.
    Differentiate between insulated and uninsulated overlaps. Also explain the importance of neutral section in OHE.
    [Pb. Univ. Electric Traction Dec. 1992]
15. Explain the importance of neutral section in traction system.
    [Rajasthan Technical Univ. Utilization of Electrical Power 2008, 2010]
16. Explain why neutral section is provided in the OHE of ac traction system but not in dc traction system. What factors determine the span length of the OHE ?
17. What is critical velocity, discuss its significance.
    [Pb. Univ. Electric Traction Dec. 1991]
18. With the help of neat diagrams, briefly explain the following :
    (*a*) Pantograph (*b*) Negative booster
    (*c*) Interference effect of railway electrification on communication circuits. **[Pb. Univ. Electric Traction April 1975]**
19. What steps can be taken to reduce the interference in the telecommunication circuits ? Explain the use of booster transformer. **[Pb. Univ. Electric Traction July 1992]**
20. In traction system employing the running rails as return conductors, describe measures and precautions that may be necessary to keep the voltage within reasonable limits.
21. Describe the negative booster for use on a traction system.
    [D.E.I. Agra 1989]
22. Explain with relevant diagram the necessity of using positive and negative boosters in dc electric traction.
    [B.H.U. Elec. Traction & Utilization 1978]
23. Explain the functions of (*i*) feeder booster (*ii*) negative booster in an electric tramway system.
    [Agra Univ. Traction & Utilization of Elec. Power 1982]
24. Draw and describe the complete connection diagram of a negative booster along with the motor starter set in a 650 V dc tramway line. If the potential at the point of connection is zero, show the distribution of current and voltage in the rail at least in two successive sections.
    [Agra Univ. Traction & Utilization of Elec. Power 1978]
25. Discuss the protection of underground installation against corrosion due to railway electrification. Discuss the application of negative booster. **[Pb. Univ. Electric Traction June 1993]**
26. Write short notes on the following:
    (*i*) Comparison of various arrangements of current collection used in electric traction. **[U.P.T.U. 2005-06]**
    (*ii*) Conductor rail system. **[W.B. Univ. of Technology 2009]**
    (*iii*) Collector gear for overhead equipment. **[Agra Univ. 1976]**
    (*iv*) Overhead collector gear employed on trams and electric trains. **[Agra Univ. 1978]**
    (*v*) Pantograph. **[Allahabad Univ. 1975]**
    (*vi*) Methods of supporting overhead control contact wire equipment for electric tramways and railways. **[Agra Univ. 1978]**
    (*vii*) Compound catenary. **[Agra Univ. 1976; Pb. Univ. Dec. 1991]**
    (*viii*) Effect of temperature on level of contact wires. **[Pb. Univ. Nov. 1974]**
    (*ix*) Traction substations.
    (*x*) System of remote control.
    (*xi*) Booster transformer.
    (*xii*) Protection of underground installations against corrosion. **[Pb. Univ. July 1992]**
    (*xiii*) Phenomenon of electrolysis by earth currents. **[G.B.T.U. 2010-11]**
    (*xiv*) Negative booster.
    [Pb. Univ. December 1992; Rajasthan Univ. 2003, 2005]

Short Answer Type Questions With Answers

Q. 1. What is the arrangement of power supply on Kanpur-Tundla section ?

Ans. 132 kV supply is taken at one point from the supply authorities and then it is distributed by the railway authorities.

Q. 2. Name the current collection systems employed for electric locomotives, tramways or trolleys.

Ans. There are mainly two systems of current collection for locomotives, tramways or trolleys viz. (*i*) conductor rail system and (*ii*) overhead system.

Q. 3. The design of overhead system is more complex as compared to conductor rail system. Why ?

Ans. Because great difficulty is experienced in ensuring that the current collector and contact wire remain in contact at a fairly even pressure.

Q. 4. Why current collection from overhead system is considered superior over conductor rail system ?

Ans. Current collection from overhead system is considered superior over conductor rail system because both theoretically and experimentally current collection is more difficult from a rigid body than from an elastic one. Further the insulation of the third rail at high voltages used on single phase ac traction would also be impracticable and endanger the safety of the personnel.

Q. 5. What is intermediate category ? **[P.T.U. December 2009]**

Ans. Compound catenary construction consists of three wires, all in the same vertical plane. The upper wire is catenary wire, which is insulated from the supporting structures. From this wire another wire, known as *intermediate catenary,* is supported by droppers clipped to both wires. The third wire *i.e.,* trolley wire is maintained in definite and constant tension by means of automatic tightening gear.

Q. 6. What are the functions of dc traction substations ? **[P.T.U. December 2007]**

Ans. The function of dc traction substations are to (*i*) step down the transmission voltage to the utilization level (*ii*) convert it into dc and (*iii*) sectionalizing the overhead equipment. Such substations, therefore, consist of step-down transformers, converting machinery, ac circuit breaker, dc circuit breaker and smoothing and metering equipment.

Q. 7. Why separate feeding systems are required for the trolley wire and the track rails ?

Ans. Separate feeding systems are required for the trolley wire and track rails in case of tramways in order to comply with the following regulations.

1. The voltage at the trolley wire is not to exceed 550 volts and that at generating stations (or sub-stations) 650 volts.
2. The trolley wire shall be divided into sections not exceeding 1.6 km in length.
3. The potential difference between any two points of the track rail return systems shall not exceed 7 volts.
4. The potential difference between any point on the rail and earth should not exceed 4 volts.

Q. 8. What for TPC stands ?

Ans. TPC stands for traction power controller.

Q. 9. How mimic diagram enables the TPC to visualize the whole of the power supply system ?

Ans. A mimic diagram depicting diagrammatically the entire OHE system as well as different substations and control posts is provided at each remote control centre to provide actual visual position of supply system to traction power controller. The mimic diagram is generally not lit up but whenever there is any change in the interruptor position at the switching post, corresponding point on the diagram gets illuminated and there is an alarm also to draw the TPC attention. The availability of the electrical supply to each and every section of OHE is monitored by the PTs at switching stations. Thus entire electrical supply system remains under the control of TPC who is fully equipped with facilities for instantaneous control of any switching operation for quick isolation of the faulty sections and restoration of power supply to the adjoining healthy sections and also for the isolation of sections for maintenance work anywhere in the system that may extend upto 250 km in every direction of the centre.

Q. 10. Why are insulated overlaps installed ?

Ans. Insulated overlaps are installed mainly to provide isolating facilities required for the operation and maintenance of OHE.

Q. 11. How insulated overlap is different from uninsulated overlap ?

Ans. **Uninsulated overlap.** In case OHE is made continuous, which is possible by splicing of conductors, long lengths of it would be affected whenever there is a fault or damage. It will also need long time for rectification of damage. It is, therefore, essential to divide the length of elementary section into a number of sub-elementary sections each of which is separated from its adjacent one by means of what is known as an *uninsulated overlap.* In an uninsulated overlap, two contact wires belonging to two adjacent sub-elementary sections are run parallel to each other for at least one span and electrically connected to each other by means of jumpers.

Insulated Overlap. Insulated overlaps are installed chiefly for providing isolating facilities required for the operation and maintenance of OHE. In case two lengths of OHE belong to two adjacent elementary sections, there will be insulated overlap span. Insulated overlaps are also provided at feeding posts, sub-sectioning posts and booster stations. Insulated overlaps are similar to uninsulated overlaps except for the following:

1. In order to avoid accidental contact between the contact wires, gap between them is kept more in case of insulated overlap.
2. Electrical continuity between the two sections is maintained by means of jumper connections in series with isolator or interruptor. With such an arrangement, a section can be made dead and work can be carried out on the other equipment away from the overlap.
3. The anchoring portion of the catenary and contact wire is separated from the rest of the portion by means of insulators provided at a distance of 2.5 m from the supports and within overlap span. The tail ends of the anchoring OHEs are connected to the other nearby equipment so as to prevent it from attaining any other potential.

Q. 12. Why neutral section is provided in the OHE of ac traction system but not in dc traction system ?

Ans. In dc traction, supply to OHE can be provided from both ends so that voltage drop can be reduced and in such a case substations operate in parallel. In case it is felt necessary, say for sake of convenience in maintenance, to have some boundary of supply between two substations, it can be had simply by means of insulated overlap. There will be no damage in this case even if pantograph bridges the OHE. But in single phase ac traction above arrangement is not permissible because adjacent substations tap different phases of the three-phase system so as to equalise load on the three phases. So momentary passing of pantograph under insulated overlap will result in short circuit between two phases of the supply system and damage the pantograph and OHE. This situation is avoided by interposing a small length of OHE, called the *neutral section.* It is insulated from both sides and is not connected to any source of power supply. It's main function is to allow physically smooth and electrically sparkless passage of pantograph from one section to the other. As per Indian Railway practice, neutral section is located in the middle of two feeding posts and in front of sectioning and paralleling post, as illustrated in Fig. 15.17.

❑❑❑

OBJECTIVE TYPE QUESTIONS

MULTIPLE CHOICE TYPE QUESTIONS

Note: Each of the following questions comprises multiple responses (two or more than two answers). Select the correct answers. In case you feel that there is more than one correct answer, select the answer which you consider the best. In any case choose ONLY ONE answer or response for each question.

1. The basic elements of an electric drive are :
(a) electric motor and the transmission system.
(b) electric motor, the transmission and control system.
(c) the transmission and control system.
(d) electric motor and conversion equipment.

2. Electric drive is becoming more and more popular because
(a) it is simple, clean, compact and reliable.
(b) it provides easy and smooth control, flexibility in layout, easy starting and facility for remote control.
(c) it is cheaper in initial as well as in maintenance cost.
(d) all of the above.

3. The main drawback of electric drive is that
(a) it is cumbersome drive.
(b) it is costlier in initial as well as in maintenance cost.
(c) electrical power supply failure makes the drive standstill.
(d) all of the above.

4. An existing workshop is to be changed over from an engine drive to an electric drive. The type of drive likely to be adopted is
(a) individual drive. (b) group drive.
(c) multimotor drive. (d) any of the above.

5. The type of drive used for a paper mill requiring constant speed operation and flexibility of control is
(a) group drive. (b) multimotor drive.
(c) individual or multimotor drive.
(d) individual drive.

6. Introduction of automation in production processes has become possible only because of use of
(a) an electric drive. (b) group drives.
(c) individual and multimotor drives.
(d) individual drive.

7. The selection of an electric motor is governed by
(a) nature of load to be handled.
(b) environmental conditions.
(c) nature of electric supply available. (d) all of the above.

8. DC motors are still preferred for use in
(a) electric excavators, steel mills and cranes.
(b) lathes and machine tools.
(c) flour mills and jaw crushers.
(d) paper industry.

9. The least significant electrical characteristic in selection of electric motor for a flour mill is
(a) starting characteristics. (b) braking.
(c) running characteristics. (d) efficiency.

10. The least significant feature while selecting a motor for centrifugal pump is
(a) speed control. (b) power rating of motor.
(c) operating speed. (d) starting characteristics.

11. A typical active load is
(a) hoist. (b) blower. (c) pump. (d) lathe.
[A.M.I.E. Sec B. Elec. Drives & Their Control Summer 1998]

12. An elevator drive is required to operate in
(a) one quadrant only. (b) two quadrants.
(c) three quadrants. (d) four quadrants.
[A.M.I.E. Sec B. Electric Drives & Their Control Winter 1996; Summer 2000]

13. Load torque constant at all speeds is represented by a
(a) fan. (b) compressor.
(c) centrifugal pump load.
[A.M.I.E. Sec B. Elec. Drives & Their Control Winter 1997]

14. An example of a motor having short-time duty is found in
(a) centrifugal pumps. (b) crane drives.
(c) drilling machines. (d) all of the above.
[A.M.I.E. Sec B. Elec. Drives & Their Control Summer 2000]

15. The load, for which the motor always starts on load is
(a) fan motor. (b) conveyor motor.
(c) flour mill motor. (d) all of the above.

16. The load cycle for a motor driving a power press will be
(a) continuous. (b) variable.
(c) intermittent and variable.
(d) continuous but periodical.

17. The application(s) that need(s) frequent starting and stopping is/are
(a) paper mills. (b) lifts and hoists.
(c) blowers and fans. (d) grinding mills.

18. An electric motor, in jaw crushers, has to often start against
(a) heavy load. (b) normal load.
(c) medium load. (d) low load.

19. The application(s) in which the load on the motor changes in cyclic order is/are
(a) rolling mills. (b) cranes.
(c) electric shovels. (d) all of the above.

20. The machine having heavy fluctuation of load is
(a) lathe. (b) planer.
(c) punching machine. (d) printing machine.

21. Variable speed operation is preferred in
(a) water pump. (b) ceiling fan.
(c) exhaust fan. (d) refrigerator.

22. The application in which the motor is to start with high acceleration is
(a) lifts and hoists. (b) centrifugal pump.
(c) oil expeller. (d) flour mill.

23. While selecting a motor for air-conditioner the feature of utmost importance is
(a) type of enclosure. (b) type of bearing.
(c) noise.
(d) power transmission arrangement.

24. Belt conveyors offer...starting torque.
(a) high (b) medium
(c) low (d) zero

Answers: **1.** (b), **2.** (d), **3.** (c), **4.** (b), **5.** (c), **6.** (c), **7.** (d), **8.** (a), **9.** (b), **10.** (a), **11.** (a), **12.** (d), **13.** (b), **14.** (b), **15.** (d), **16.** (d), **17.** (b), **18.** (a), **19.** (d), **20.** (c), **21.** (b), **22.** (a), **23.** (c), **24.** (a),

25. The characteristic(s) of Jrive for crane hoisting and lowering is are
(a) fast speed control. (b) smooth movement.
(c) precise control. (d) all of the above.

26. The capacity of a crane is expressed in terms of
(a) span. (b) tonnes.
(c) type of drive. (d) any of the above.

27. The travelling speed of a crane varies from
(a) 20 to 25 m/s. (b) 10 to 20 m/s.
(c) 1.0 to 2.5 m/s. (d) 0.2 to 1.0 m/s.

28. Light duty cranes are generally used in
(a) power houses. (b) pumping stations.
(c) automobile workshops. (d) all of the above.

29. Heavy duty cranes are employed in
(a) powerhouses and pumping stations.
(b) automobile workshops.
(c) steel plants and ore handling plants.
(d) both (a) and (c).

30. In case of centrifugal pumps the starting torque is usually
(a) less than running torque. (b) same as running torque.
(c) slightly higher than running torque.
(d) double of running torque.

31. The size of an excavator is usually expressed in terms of
(a) crowd motion. (b) m^3.
(c) travel in metres. (d) angle of swing.

32. The range of rating of electric motor used for rolling mills is of the order of
(a) 10 to 25 kW (b) 25 to 85 kW
(c) 85 to 400 kW (d) 400 to 1,000 kW

33. If the liquid to be pumped has relative density of 1.25, the output power required in comparison to that required for pumping water in case of centrifugal pump will be
(a) 1.25 times. (b) the same.
(c) 0.8 times. (d) 15 times (roughly).

34. The load torque decreases with the increase in speed in case of
(a) lathes, boring machines, milling machines.
(b) hoist winches, machine tool feed mechanism.
(c) blowers, fans, centrifugal pumps.
(d) both (a) and (b).

35. The electromagnetic torque T_e of a drive, and its connected load torque T_L are as shown below. Out of the operating points A, B, C and D, the stable ones are

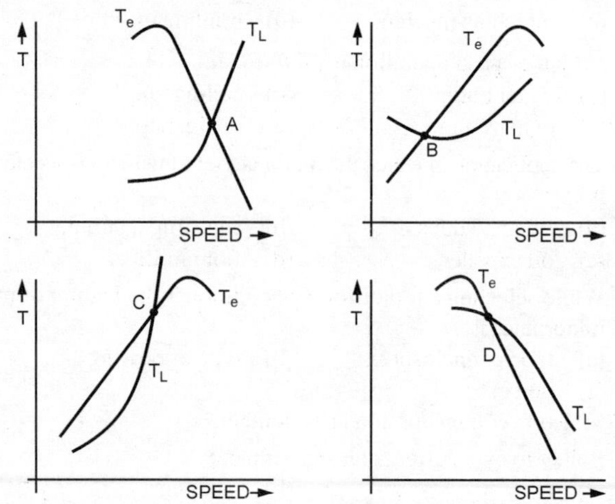

(a) A, C, D (b) B, C
(c) A, D (d) B, C, D [GATE E.E. 2007]

36. In synthetic fibre mills motors with
(a) variable speeds are required.
(b) constant speeds are preferred.
(c) low starting torque are required.
(d) high starting torque are preferred.

37. In case of kiln drives, starting torque is
(a) more than double of the running torque.
(b) almost $\sqrt{2}$ times of the running torque.
(c) almost equal to running torque.
(d) almost zero.

38. The diameter of the rotor shaft for an electric motor depends upon
(a) speed only. (b) power output only.
(c) speed and power output.
(d) speed, power output and power factor.

39. Which one of the following is not necessarily the advantage of dc motors over ac motors ?
(a) Low cost.
(b) Excellent torque and speed operating characteristics.
(c) Flexible speed control. (d) None of the above.

40. An unsaturated dc shunt motor drives a constant torque load at a speed at 1,000 rpm taking an armature current of 20 A when 230 V is applied to the motor. If the applied voltage is reduced to 115 V, the motor runs at.
(a) 500 rpm taking an armature current of 20 A.
(b) 1,000 rpm taking an armature current of 20 A.
(c) 500 rpm taking an armature current of 40 A.
(d) 1,000 rpm taking an armature current of 40 A.
[A.M.I.E. Sec B. Elec. Drives & Their Control Winter 1994]

41. An electric train employing a dc series motor is running at a fixed speed, when a sudden slight drop in the mains voltage occurs. This would result in
(a) drop in speed and rise in current.
(b) rise in speed and drop in current.
(c) rise in speed and rise in current.
(d) drop in speed with current unaltered. **[U.P.S.C.I.E.S. E.E.-II, 1995]**

42. A dc series motor is used for an overhauling load. It can work stably if
(a) the armature is shunted by a resistor.
(b) the field winding is reversed.
(c) a resistor is put in series with the machine.
(d) a divertor is put across the field.
[A.M.I.E. Sec B. Elec. Drives & Their Control Summer 1993]

43. DC series motors are very suitable for heavy duty applications such as electric railways, rolling mills because of
(a) low initial as well as maintenance cost.
(b) high starting torque. (c) possibility of speed control.
(d) nearly constant speed.

44. A flywheel is normally fitted to
(a) dc series motor driving a constant torque load.
(b) separately excited dc motor driving pulsed torque load.
(c) cumulatively compound motor driving pulsed torque load.
(d) differentially compound motor driving pulsed torque load.
[A.M.I.E. Sec B. Elec. Drives & Their Control Winter 1994]

45. Which motor should not be used for centrifugal pumps ?
(a) Shunt. (b) Series.
(c) Cumulatively compound. (d) Differentially compound.
[U.P.S.C. I.E.S. E.E.-II, 1992]

Answers: **25.** (d), **26.** (b), **27.** (c), **28.** (d), **29.** (c), **30.** (a), **31.** (b), **32.** (c), **33.** (a), **34.** (a), **35.** (c), **36.** (b), **37.** (a), **38.** (c), **39.** (a), **40.** (d), **41.** (a), **42.** (a), **43.** (b), **44.** (c), **45.** (b),

46. For continuously running rolling mills with intermittent loading, the most suitable dc drive is
(a) dc series motor. (b) dc shunt motor.
(c) dc differentially compounded motor.
(d) dc cumulatively compounded motor. [U.P.S.C. I.E.S. E.E.-II, 1992]

47. For quick speed reversal the motor preferred is
(a) dc motor. (b) squirrel cage induction motor.
(c) slip-ring induction motor. (d) synchronous motor.

48. The motor, owing to its inherent characteristics, best suited for the rolling mills is
(a) dc shunt motor.
(b) dc cumulative compound wound motor.
(c) squirrel cage induction motor.
(d) synchronous motor.

49. Squirrel cage induction motors with high slip and slip-ring induction motors develop maximum torque at standstill and are used for
(a) elevators. (b) machine tools.
(c) presses and punches. (d) all of the above.

50. Belted slip-ring induction motor is almost invariably employed for
(a) conveyors. (b) jaw-crushers.
(c) centrifugal blowers. (d) water pumps.

51. Belted wound induction motors are preferred for
(a) gyratory crushers. (b) screw pumps.
(c) machine tools. (d) water pumps.

52. The pair used for frequency converters is
(a) squirrel cage induction motor and synchronous motor.
(b) slip-ring induction motor and synchronous motor.
(c) slip-ring induction motor and squirrel cage induction motor.
(d) all of above.

53. For the same rating, the size of the single-phase induction motor is about_____ that of the corresponding three-phase induction motor.
(a) three times (b) the same as
(c) 1.5 times (d) one-third
[A.M.I.E. Sec B. Elec. Drives & Their Control Winter 1999]

54. A fluctuating voltage supply is detrimental to a refrigerator motor but not to a ceiling fan, although both are single-phase induction motors because, the refrigerator motor
(a) is made more robust than the fan motor.
(b) is subjected to short-duty cycle but the fan motor is subjected to continuous duty.
(c) is enclosed in a sealed unit while the fan motor is open to the environment.
(d) load is constant, but the fan motor load is voltage dependent.
[U.P.S.C. I.E.S. E.E.-II, 1993]

55. Ward-Leonard controlled dc drives are usually used for... duty excavators.
(a) light (b) medium (c) heavy

56. Speed control by variation of field flux results in
(a) constant power drive. (b) variable power drive.
(c) constant torque drive. (d) variable torque drive.
[A.M.I.E. Sec B. Elec. Drives & Their Control Summer 1994]

57. In variable speed dc motor drive the armature voltage control leads to
(a) constant power operation.
(b) constant torque operation.
(c) variable torque operation.
(d) randomly varying power operation.
[A.M.I.E. Sec B. Elec. Drives & Their Control Winter 1998]

58. Speed of an induction motor is controlled by injecting voltage into the rotor winding. Frequency of the injected voltage should be
(a) main frequency.
(b) frequency corresponding to required speed.
(c) frequency corresponding to rotor speed.
(d) slip frequency.
[A.M.I.E. Sec B. Elec. Drives & Their Control Winter 1994]

59. The slip power recovery scheme is used in induction motor for speed control in the range
(a) above synchronous speed.
(b) below synchronous speed.
(c) both above and below synchronous speed.
(d) none of the above.
[A.M.I.E. Sec B. Elec. Drives & Their Control Summer 1999]

60. The electric braking system commonly employed in rolling mills, elevators and printing presses is
(a) plugging. (b) rheostatic.
(c) dynamic. (d) regenerative.

61. For rheostatic braking of two series motors connected in parallel
(a) cross-connection is better.
(b) equaliser connection is better.
(c) both are equally good.
(d) none of the two is required.

62. Dynamic braking is employed to brake
(a) non-reversing drive. (b) reversing drive.
(c) in both (a) and (b).
[A.M.I.E. Sec B. Elec. Drives & Their Control Summer 2000]

63. The condition of regenerative braking can be achieved by
(a) speed higher than no-load speed of overhauling load.
(b) increasing the excitation while supply voltage remains constant.
(c) increasing the armature current.
(d) either (a) or (b).

64. In a dc shunt motor, regenerative braking is limited by
(a) saturation of flux at high speed.
(b) saturation of flux at low speed.
(c) armature current at high speed.
(d) either (a) or (b).

65. For regenerative braking with dc series motors, its field windings are______excited.
(a) separately (b) series
(c) shunt (d) either (b) or (c)

66. The most economical method of electric braking is
(a) plugging
(b) dynamic braking with separate excitation.
(c) dynamic braking with self excitation.
(d) regenerative braking.

67. The plugging provides ______ braking torque in comparison to rheostatic and regenerative braking systems.
(a) negligible (b) small
(c) highest

68. Rheostatic braking may be applied to an induction motor provided
(a) it is a squirrel cage type.
(b) it is a wound type.
(c) separate dc source for field excitation is available.
(d) variable external resistance is available.

69. The motor enclosure used for collieries, chemical plants is
(a) flame proof type. (b) splash proof type.
(c) totally enclosed type. (d) pipe ventilated type.

Answers: 46. (d), **47.** (a), **48.** (b), **49.** (c), **50.** (b), **51.** (a), **52.** (a), **53.** (c), **54.** (d), **55.** (c), **56.** (a), **57.** (b), **58.** (d), **59.** (c), **60.** (a), **61.** (a), **62.** (c), **63.** (d), **64.** (b), **65.** (a), **66.** (d), **67.** (c), **68.** (c), **69.** (a),

70. The motor enclosure used in woodworking industry is
(a) protected type.
(b) totally enclosed fan cooled type.
(c) flame-proof type. (d) splash-proof type.

71. The motor enclosure used for industrial purposes is
(a) protected type. (b) drip proof type.
(c) totally enclosed type. (d) open type.

72. Rotor of a motor is usually supported on
(a) bush bearings. (b) thrust bearing.
(c) ball or roller bearings. (d) any of the above.

73. Ball-bearings are
(a) used up to 75 kW motors.
(b) of long life and low friction loss.
(c) costlier and noisy particularly at high motor speed.
(d) all of the above.

74. Sleeve bearings
(a) are normally of bronze.
(b) have self-lubricating properties.
(c) are used where noise is to be avoided.
(d) all of the above.

75. Hot bearings of a dc motor may be caused by
(a) poor ventilation. (b) incorrect voltage.
(c) loose coupling. (d) lack of or dirty lubricant.
[A.M.I.E. Sec B. Electric Drives & Their Control Winter 1999]

76. Excessive motor vibration is caused by
(a) worn bearings. (b) open armature coil.
(c) bent shaft. (d) excessive brush tension.

77. Excessive sparking at the brushes may be caused due to
(a) loose coupling. (b) worn bearings.
(c) dirt on commutator. (d) open armature coil.

78. Intermittent sparking at the brushes of a dc motor may be caused due to
(a) intermittent load. (b) open armature coil.
(c) loose coupling. (d) incorrect voltage.

79. The least expensive drive is
(a) belt drive. (b) rope drive.
(c) chain drive. (d) gear drive.

80. For very high speed ratio the indispensable drive is
(a) direct drive. (b) rope drive.
(c) chain drive. (d) any of the above.

81. In rotating electrical machines, the insulation temperature limit for class B type is
(a) 105°C (b) 130°C
(c) 150°C (d) 180°C
[A.M.I.E. Sec B. Electric Drives & Their Control Winter 1996]

82. For a particular motor, the cooling time constant is usually
(a) smaller than the heating time constant.
(b) greater than the heating time constant.
(c) equal to the heating time constant.
[A.M.I.E. Sec B. Elec. Drives & Their Control Winter 1995]

83. For a particular type of motor, the heating time constant
(a) increases with the increase in size.
(b) decreases with the increase in size.
(c) same for all sizes.
[A.M.I.E. Sec B. Elec. Drives & Their Control Winter 1995]

84. For estimation of power rating of an electric drive, its losses can be considered to be proportional to
(a) power. (b) $(\text{power})^2$.
(c) $(\text{power})^3$. (d) torque.
[A.M.I.E. Sec B. Electric Drives & Their Control Winter 1996]

85. The duty cycle of a motor is

Load	Duration
20 kW	120 s
10 kW	120 s
stopped	280 s

The continuous rating of motor suitable for above application would be
(a) 20 kW (b) 16.4 kW
(c) 14.14 kW (d) 9 kW
[U.P.S.C. I.E.S. E.E.-II, 1992]

86. The heating time constant of a totally enclosed motor is relatively
(a) lower. (b) higher.
(c) independent of type of enclosure.

87. For a certain industrial application, an overrated ac motor was selected. It will lead to operation with
(a) higher efficiency and better pf.
(b) lower efficiency and poorer pf.
(c) higher efficiency and poorer pf.
(d) lower efficiency and better pf.
[A.M.I.E. Sec B. Electric Drives & Their Control Winter 1998]

88. A flywheel is generally used in
(a) a cement mill drive. (b) a paper mill drive.
(c) a rolling mill drive. (d) a sugar centrifuge drive.
[A.M.I.E. Sec B. Elec. Drives & Their Control Winter 1996; Summer 1998]

89. In overhead travelling cranes
(a) continuous rating motors are used.
(b) continuous minimum rating motors are used.
(c) short-time rating motors are used.
(d) any of the above.

90. 15 minute rated motors are suitable for
(a) light duty cranes. (b) medium duty cranes.
(c) heavy duty cranes. (d) all of the above.

91. For medium duty cranes the short-time rating motor used is
(a) 10 minutes. (b) 15 minutes.
(c) 30 minutes. (d) one hour.

92. Load equalisation is desirable in the case of
(a) very large refrigeration and air-conditioning plants.
(b) rolling mills, electric hammers, presses, reciprocating pumps.
(c) lathes, woodworking machines, paper making machines, shapers and slotters.
(d) travelling cranes, lifts.

93. The phase controlled rectifiers used in speed control of dc motors converts fixed ac supply voltage into _______ output voltage.
(a) variable dc (b) variable ac
(c) variable frequency ac (d) full rectified ac

94. A four quadrant operation requires
(a) two full converters in series.
(b) two full converters connected back to back.
(c) two full converters connected in parallel.
(d) two semi converters connected back to back.
[A.M.I.E. Sec B. Elec. Drives & Their Control Winter 1995]

95. The following converters can feed power in any of the four quadrants
(a) semi-converter. (b) full converter.
(c) dual converter.
(d) combination of a semi- and full-converters.
[A.M.I.E. Sec B. Elec. Drives & Their Control Winter 1993]

96. For speed control of dc motors using controlled rectifiers, armature voltage control gives speed

Answers: 70. (b), **71.** (a), **72.** (c), **73.** (d), **74.** (d), **75.** (d), **76.** (a), **77.** (c), **78.** (b), **79.** (a), **80.** (b), **81.** (b), **82.** (b), **83.** (a), **84.** (b), **85.** (c), **86.** (b), **87.** (b), **88.** (c), **89.** (c), **90.** (a), **91.** (c), **92.** (b), **93.** (a), **94.** (b), **95.** (c), **96.** (b),

(a) above rated speed. (b) below rated speed.
(c) above as well as below rated speed.
[A.M.I.E. Sec B. Power Electronics Summer 1996]

97. Armature of a dc motor is fed from a phase controlled rectifier where as its field is supplied from a constant dc source. To reduce the speed of the motor regeneratively
(a) the firing angle of the converter should be increased.
(b) the polarity of the dc voltage should be reversed.
(c) the polarity of the armature induced voltage should be reversed.
(d) the firing angle should be varied in the range 90º to 180º simultaneously reversing the armature connections.
[A.M.I.E. Sec B. Elec. Drives & Their Control Winter 1994]

98. For the speed control of ac drive, the preferred method using thyristors is
(a) phase control. (b) integral cycle control.
(c) chopper control. (d) all are equally good.
[A.M.I.E. Sec B. Elec. Drives and Their Control Winter 1995]

99. The semiconductor device used for speed control of single phase induction motors in every day domestic use is
(a) SCR. (b) power transistor.
(c) Triac. (d) power MOS.
[A.M.I.E. Sec B. Elec. Drives & Their Control Winter 1999]

100. Which of the following motor is suitable for driving cranes, hoists, centrifugal pumps, conveyor belt etc. ?
(a) DC series motor. (b) DC shunt motor.
(c) DC compound wound motor.
(d) Any of the above.

101. Heavy-duty steel-works cranes having wide load variations are equipped with
(a) dc series motors.
(b) plain squirrel cage induction motors.
(c) wound rotor induction motors.
(d) synchronous motors.

102. The most suitable motor for a steel mill requiring a motor having high starting torque, wide speed range and precise speed control is
(a) plain squirrel cage induction motor.
(b) wound rotor induction motor.
(c) dc series motor. (d) dc shunt motor.

103. For automatic drives the preference is for
(a) synchronous motors.
(b) Ward-Leonard controlled dc shunt motors.
(c) plain squirrel cage induction motors.

104. The type of motor that can be used for hoisting machinery is
(a) dc compound motor.
(b) Ward-Leonard controlled dc shunt motor.
(c) wound rotor induction motor.
(d) any of the above.

105. The motor normally used for crane travel is
(a) wound rotor induction motor.
(b) synchronous motor.
(c) dc differentially compounded motor.
(d) plain squirrel cage induction motor.

106. The drive, that can be used for derricks and winches is
(a) pole-changing squirrel cage induction motor.
(b) ac slip-ring induction motor with variable resistance.
(c) dc shunt motor with Ward-Leonard control.
(d) any of the above.

107. Ward-Leonard controlled dc drives are generally used for _____ duty excavators
(a) light (b) medium (c) heavy (d) none of these.

108. The motor suitable for a reciprocating pump required to start under load is
(a) plain squirrel cage induction motor.
(b) double-squirrel cage induction motor.
(c) synchronous motor. (d) dc shunt motor.

109. A pole-changing squirrel cage induction motor employed in derricks has four, eight and twenty four poles. The lowest speed is used in
(a) hoisting. (b) landing the load.
(c) lifting. (d) lowering.

110. A pole-changing type squirrel cage induction motor employed in derricks has four, eight and twenty four poles. The medium speed is used in
(a) landing the load. (b) hoisting.
(c) lowering. (d) lifting.

111. Motor preferred for kiln drive is usually
(a) wound rotor induction motor.
(b) cascaded controlled ac motor.
(c) Ward-Leonard controlled dc shunt motor.
(d) any of the above.

112. The motor used in mines is
(a) flame-proof squirrel cage induction or wound rotor motor.
(b) dc series motor.
(c) dc shunt motor. (d) any of the above.

113. The motor used in punches, presses and shears is
(a) dc series or shunt motors.
(b) dc cumulative compound motor.
(c) high slip squirrel cage or wound rotor induction motor.
(d) both (b) and (c).

114. A domestic mixer uses the following motor
(a) induction motor. (b) reluctance motor.
(c) universal motor.
(d) permanent magnet synchronous motor.
[A.M.I.E. Sec B. Elec. Drives & Their Control Winter 1998]

115. A perfect black body is one which
(a) absorbs all incident radiations.
(b) reflects all incident radiations.
(c) transmits all incident radiations.
(d) all of the above.

116. For the transmission of heat from one body to another it is essential that
(a) both bodies are solids.
(b) the two bodies are at different temperatures.
(c) both bodies are in contact.
(d) at least one of the bodies has some source of heating.

117. Heat is transferred simultaneously by conduction, convection and radiation
(a) during melting of ice.
(b) from refrigerator coils to refrigerator freezer.
(c) inside boiler furnaces.
(d) through the surface of the insulated pipe carrying steam.

118. The highest value of thermal conductivity is for
(a) aluminium. (b) brass. (c) copper. (d) iron.

119. The highest value of thermal conductivity is for
(a) steam. (b) water. (c) melting ice. (d) solid ice.

Answers: 97. (a), **98.** (c), **99.** (c), **100.** (a), **101.** (a), **102.** (c), **103.** (b), **104.** (d), **105.** (d), **106.** (d), **107.** (c), **108.** (b), **109.** (b), **110.** (d), **111.** (d), **112.** (a), **113.** (d), **114.** (c), **115.** (a), **116.** (b), **117.** (c), **118.** (c), **119.** (d),

120. The thermal conductivity will be highest in case of
(a) dry concrete.
(b) concrete having 10% moisture by volume.
(c) concrete having 0.4% reinforcement.
(d) concrete having 10% moisture by volume and 0.4% reinforcement.

121. Radiations from a black body are proportional to
(a) T^2 (b) T^3 (c) T^4 (d) $1/T^4$

122. A body reflecting entire radiations incidenting on it is called the
(a) white body. (b) gray body.
(c) black body. (d) transparent body.

123. The insulating material suitable for low temperature applications is
(a) cork. (b) diatomaceous earth.
(c) asbestos paper. (d) 75 per cent magnesia.

124. The quantity of heat absorbed from the heater by convection depends upon
(a) the temperature of heating element above the surroundings.
(b) the surface area of the heater.
(c) the position of the heater.
(d) all of the above.

125. The material of the heating element should be
(a) such that it may withstand the required temperature without getting oxidised.
(b) of low resistivity. (c) of low melting point.
(d) of high temperature coefficient.

126. The material to be used for heating element should be of high resistivity so as to
(a) increase the life of the heating element.
(b) reduce the length of the heating element.
(c) reduce the effect of oxidation.
(d) produce large amount of heat.

127. The material to be used for heating element should be of low temperature coefficient so as to
(a) avoid initial rush of current.
(b) avoid change in kW rating with temperature.
(c) reduce the effect of oxidation.
(d) both (a) and (b) above.

128. Which of the following heating element will have the least temperature range ?
(a) Eureka. (b) Silicon carbon.
(c) Nichrome. (d) Kanthal.

129. Which of the following heating element can give highest temperature in resistance heating ?
(a) Nichrome. (b) Silicon carbide.
(c) Copper. (d) Nickel-Cr-Fe alloy.

130. Ni-Cr-Fe alloy wires can be safely used for temperatures up to
(a) 2,500°C (b) 2,000°C
(c) 1,150°C (d) 850°C

131. The heating element to be used in a furnace employed for heating around, 1,600°C should be of material
(a) nichrome. (b) eureka.
(c) molybdenum. (d) silicon-carbide.

132. For radiant heating around 2,000°C the heating elements used should be of material
(a) tungsten alloy. (b) copper alloy.
(c) carbon. (d) Stenleess

133. The heating element to be used in vacuum furnaces should be of the material
(a) molybdenum because of high vapour pressure.
(b) tungsten because of low vapour pressure.
(c) molybdenum because of low vapour pressure.
(d) tungsten because of high vapour pressure.

134. In an electric press, mica is used
(a) for induction heating. (b) for dielectric heating.
(c) as an insulator.
(d) for improvement of power factor.

135. Maximum operating voltage is limited
(a) to 600 V. (b) from safety consideration.
(c) by electrical insulation at high temperatures.
(d) all of the above.

136. The heat produced in the heating element(s) is also to overcome the losses occurring because of
(a) heat used in raising the temperature of oven (or furnace) and container (or carriers).
(b) heat conducted through the walls.
(c) escapement of heat due to opening of door.
(d) all of the above.

137. In a resistance furnace, the temperature is controlled by
(a) variation of operating voltage.
(b) variation of resistance of heating circuit.
(c) switching on and off the supply periodically.
(d) all of the above.

138. For temperature control in resistance furnaces resistance variation can be affected by connecting resistance elements in
(a) series. (b) parallel.
(c) series-parallel. (d) star or delta.
(e) any of the above.

139. The simplest and most commonly used method for temperature control is
(a) external series resistance in the heating circuit.
(b) change of connections of heating circuit.
(c) use of variable number of heating elements.
(d) transfomer tappings.

140. The device necessarily used for automatic temperature control in a furnace is
(a) thermostat. (b) auto-transformer.
(c) thermocouple. (d) any of the above.

141. Control of power input to salt-bath furnace is affected by
(a) varying the depth of immersion of electrodes.
(b) varying the distance between the electrodes.
(c) both (a) and (b). (d) none of (a) and (b).

142. The temperature inside a furnace is usually measured by
(a) mercury thermometer. (b) optical pyrometer.
(c) alcohol thermometer. (d) any of the above.

143. In a resistance furnace the atmosphere is
(a) oxidising. (b) deoxidising.
(c) reducing. (d) neutral.

144. On increasing the thickness of refractory walls of the furnace
(a) energy consumption will increase.
(b) heat loss through furnace walls will increase.
(c) temperature on the outer surface of furnace walls will drop.
(d) temperature inside the furnace will drop.

145. Radiant heating is used for
(a) melting of ferrous metals.
(b) annealing of metals.
(c) drying of paints and varnishes.
(d) any of the above.

Answers: 120. (d), **121.** (c), **122.** (a), **123.** (b), **124.** (d), **125.** (a), **126.** (b), **127.** (d), **128.** (a), **129.** (b), **130.** (d), **131.** (c), **132.** (a), **133.** (b), **134.** (c), **135.** (d), **136.** (d), **137.** (d), **138.** (e), **139.** (b), **140.** (a), **141.** (c), **142.** (b), **143.** (a), **144.** (c), **145.** (c),

146. Direct resistance heating is used in
(a) electrode boiler. (b) salt-bath furnace.
(c) resistance welding. (d) all of the above.

147. Resistance ovens are used for
(a) domestic and commercial heating.
(b) vulcanizing and hardening of synthetic materials.
(c) drying of varnish coatings, drying and baking of potteries.
(d) all of the above.

148. The function(s) of an heating chamber is/are to
(a) store as much of the heat supplied as may be practicable and economical.
(b) confine the atmosphere around the charge.
(c) control the cooling rate of the charge (if required) and control the distribution of heat within the chamber.
(d) all of the above.

149. For obtaining best results, the infrared lamps should be located from the object to be heated at distance of
(a) 10-20 cm. (b) 25-30 cm.
(c) 40-50 cm. (d) 50-60 cm.

150. In direct arc furnace which of the following is of high value ?
(a) Current. (b) Voltage.
(c) Power factor. (d) all of the above.

151. The power factor at which the direct arc furnace operates is
(a) low lagging. (b) low leading.
(c) unity. (d) high leading.

152. For arc heating, the electrodes used are made of
(a) copper. (b) graphite.
(c) tungsten. (d) aluminium.

153. In an arc furnace, the choke is provided to
(a) reduce the surge severity.
(b) stabilize the arc. (c) improve the power factor.
(d) all of the above.

154. It is desirable to operate the arc furnaces at a power factor of
(a) zero. (b) unity.
(c) 0.707 lagging. (d) 0.707 leading.

155. It is desirable to keep the arc length short in order to
(a) have better heating.
(b) have better stirring action and reduce oxidation problem.
(c) increase the life of roof refractory.
(d) all of the above.

156. Arc drawn between the two electrodes produces heat and has a temperature
(a) between 0–600°C. (b) between 500°C–1,000°C.
(c) between 1,000°C–3,500°C.
(d) of the order of 4,000°C.

157. The arc furnaces of conical shapes have the advantage(s) of
(a) large surface area per unit bath volume.
(b) low power consumption.
(c) reduced radiation losses.
(d) all of the above.

158. In arc furnace, energy losses that take place in the furnace are
(a) losses through walls.
(b) losses through escape gasses.
(c) losses through water cooling.
(d) all of the above.

159. In submerged arc furnaces the power is controlled by
(a) varying the spacing between the electrodes.
(b) varying the voltage applied to the electrodes.
(c) either (a) or (b).
(d) varying the arc length.

160. For arc furnaces the low voltage high current power supply is needed because
(a) heavy currents produce large amount of heat resulting in higher temperatures.
(b) maximum secondary voltage is also limited to 275 V (line-to-line on open circuit) owing to insulation and safety consideration.
(c) life of the roof refractory is increased.
(d) all of the above.

161. The main application of indirect arc furnace is to melt
(a) iron. (b) steel.
(c) non-ferrous metals. (d) none of the above.

162. For power transformers employed for arc furnaces, it is desirable to arrange the furnace and the transformer in such a way that leads are
(a) shorter in length and placed distant apart.
(b) shorter in length and placed close together.
(c) longer in length and placed close together.
(d) in any arrangement.

163. In induction heating
(a) heat is produced due to currents induced in the charge by electromagnetic action.
(b) the resistance of the charge must be low and voltage applied must be high in order to produce sufficient heat.
(c) magnetic materials can be easily treated in comparison to non-magnetic materials.
(d) all of the above.

164. In heating the ferromagnetic materials by induction heating, heat is produced owing to
(a) flow of induced current through the charge.
(b) hysteresis loss occuring below Curie temperature.
(c) hysteresis loss as well as eddy current loss occurring in the charge.
(d) any of the above factors.

165. Induction heating takes place in
(a) insulating materials.
(b) conducting and magnetic materials.
(c) conducting but non-magnetic materials.
(d) conducting materials may be magnetic or non-magnetic.

166. Induction hardening is possible in case of
(a) dc supply only. (b) ac supply only.
(c) ferrous materials only.
(d) non-conducting materials only.

167. Low frequency supply is necessary for direct core type induction furnaces because
(a) magnetic coupling between the primary and secondary circuit is poor.
(b) with normal frequency supply the electromagnetic forces cause severe stirring action in the molten metal.
(c) both (a) and (b). (d) none of (a) and (b).

168. In induction heating, which of the following is of high value?
(a) Frequency. (b) Current.
(c) Voltage. (d) Power factor.

169. Induction furnaces are used for
(a) heat treatment of castings.
(d) heating of insulators. (c) melting of aluminium.
(d) all of the above.

Answers: **146.** (d), **147.** (d), **148.** (d), **149.** (b), **150.** (a), **151.** (a), **152.** (b), **153.** (b), **154.** (c), **155.** (d), **156.** (c), **157.** (d), **158.** (d), **159.** (c), **160.** (d), **161.** (c), **162.** (b), **163.** (d), **164.** (c), **165.** (d), **166.** (b), **167.** (c), **168.** (a), **169.** (a),

170. In induction heating, the depth up to which current will penetrate is proportional to
(a) $1/(\text{frequency})^{1/2}$. (b) 1/ frequency.
(c) frequency. (d) $(\text{frequency})^2$.

171. The advantage(s) of eddy current heating is/are
(a) easy temperature control, little wastage of heat and possibility of heating in vacuum or other special atmosphere.
(b) heat can be made to penetrate into metal surface to any desired depth.
(c) the area of surface over which heat is produced can be accurately controlled.
(d) all of the above.

172. The supply frequency usually employed for high frequency eddy current heating is
(a) 10 MHz (b) 10 kHz to 400 kHz
(c) 5 kHz (d) 1 kHz

173. In case of core-loss induction furnace, the charge should be in
(a) solid state. (b) molten state.
(c) either of (a) or (b).

174. High frequency induction heating is used for
(a) ferrous metals only. (b) non-ferrous metals only.
(c) both ferrous and non-ferrous metals.

175. In dielectric heating current flows through
(a) air. (b) dielectric.
(c) metallic conductor.
(d) ionic discharge between dielectric medium and metallic conductor.

176. Dielectric loss is proportional to
(a) frequency. (b) $(\text{frequency})^2$.
(c) $(\text{frequency})^3$. (d) $(\text{frequency})^{1/2}$.

177. The dielectric loss in a dielectric is proportional to
(a) voltage impressed on the dielectric.
(b) the square of the voltage impressed on the dielectric.
(c) the square root of the voltage impressed on the dielectric.
(d) none of the above.

178. For heating of plywood, the frequency should be
(a) 1–2 MHz (b) 10–25 kHz (c) 1 kHz (d) 100 Hz

179. The normal voltage used in dielectric heating is
(a) 1.5 kV (b) 15 kV (c) 33 kV (d) 66 kV

180. For dielectric heating, with the increase of mass of workpiece, the optimum frequency _____ for maximum power transfer.
(a) increases (b) decreases
(c) independent of mass

181. For dielectric heating the range of frequency normally employed is
(a) 10 kHz to 100 kHz (b) 100 kHz to 10 MHz
(c) 1 MHz to 10 MHz (d) 10 MHz to 40 MHz

182. The power factor will be maximum in case of
(a) electric arc heating. (b) resistance heating.
(c) induction heating. (d) dielectric heating.

183. Furnaces used for cremation are
(a) electric resistance heating.
(b) electric arc heating. (c) dielectric heating.
(d) high frequency eddy current heating.

184. In an electric room heat convector the method of heating used is
(a) arc heating. (b) resistance heating.
(c) induction heating. (d) infrared heating.

185. When the composition of non-ferrous metal is to be varied frequently or when heating is intermittent _____ furnace is desirably used.
(a) direct resistance (b) direct arc
(c) indirect resistance (d) Ajax Wyatt

186. Hysteresis loss and eddy current loss are used in
(a) resistance heating. (b) dielectric heating.
(c) induction heating of steel.
(d) induction heating of brass.

187. The power factor will be leading in case of
(a) dielectric heating. (b) induction heating.
(c) electric arc heating. (d) resistance heating.

188. The method of heating used for non-conducting materials is
(a) induction heating. (b) dielectric heating.
(c) electric resistance heating.
(d) electric arc heating.

189. The method appropriate for heating of non-ferrous metals is
(a) indirect resistance heating.
(b) radiant heating. (c) indirect arc heating.
(d) dielectric heating.

190. The method suitable for heating of conducting medium is
(a) induction heating. (b) indirect arc heating.
(c) eddy current heating. (d) radiant heating.

191. The most modern method for food processing is
(a) induction heating. (b) resistance heating.
(c) dielectric heating. (d) eddy current heating.

192. The main advantage of dielectric heating is that
(a) it can be used for drying the explosives.
(b) heating occurs in the material itself.
(c) heating occurs due to the high frequency.
(d) none of the above.

193. ______ is not an application of dielectric heating.
(a) Food processing. (b) Soldering.
(c) Gluing of wood. (d) Diathermy.

194. Which of the following falls under the category of plastic or non-fusion welding ?
(a) Resistance welding. (b) Electron beam welding.
(c) Electro-slag welding. (d) Arc welding.

195. Which of the following falls under the category of fusion or non-pressure welding?
(a) Resistance welding. (b) Metal-arc welding.
(c) Ultrasonic welding. (d) Explosive welding.

196. Proper selection of welding depends upon, in addition to cost involved,
(a) kinds of metals to be joined.
(b) nature of products to be fabricated.
(c) production technique used.
(d) all of the above.

197. The metal surfaces for electrical resistance welding must be
(a) cleaned. (b) lubricated.
(c) moistened. (d) rough.

198. Resistance welding cannot be used for
(a) ferrous materials. (b) non-ferrous materials.
(c) dielectrics. (d) any of the above.

199. In electrical resistance welding the greatest resistance is offered by
(a) metal surface.
(b) contact layer of metals to be welded.

Answers: **170.** (a), **171.** (d), **172.** (b), **173.** (c), **174.** (c), **175.** (b), **176.** (a), **177.** (b), **178.** (a), **179.** (a), **180.** (b), **181.** (d), **182.** (b), **183.** (a), **184.** (b), **185.** (c), **186.** (c), **187.** (a), **188.** (b), **189.** (c), **190.** (a), **191.** (c), **192.** (b), **193.** (b), **194.** (a), **195.** (b), **196.** (d), **197.** (a), **198.** (c), **199.** (b),

(c) contact point of electrode with metal top.
(d) contact point of electrode with metal bottom.

200. In electric resistance welding
(a) the current required exceeds 100 A.
(b) the voltage required ranges from 4 to 12 V.
(c) the amount of power supplied to the weld usually ranges from 60 watts to 80 watts for each square mm of area.
(d) all of the above.

201. Resistance to the flow of current is made of
(a) resistance of current path in the work.
(b) resistance between the contact surfaces of the parts being welded.
(c) resistance between the electrodes and the surface of the parts being welded.
(d) all of the above.

202. In resistance welding, the magnitude of current is controlled
(a) by varying the primary voltage of the welding transformer using an auto-transformer between supply and welding transformer.
(b) by changing the primary turns of the welding transformer.
(c) by varying the magnitude and waveform of the primary as well as secondary current by using thyratron or ignitron.
(d) any of the above.

203. The main drawback of resistance welding is
(a) high initial as well as maintenance cost.
(b) difficult shapes and sections cannot be welded.
(c) only similar metals can be welded.
(d) parent metal is affected.

204. Plain and butt welds may be used on materials upto thickness of about
(a) 5 mm (b) 10 mm (c) 25 mm (d) 40 mm

205. In upset butt welding
(a) the faces of the metal pieces to be joined are prepared for even contact.
(b) heating is obtained by the contact resistance of metal pieces to be welded.
(c) the voltage required is 2-8 V and current required ranges from 50 A to several hundred amperes depending upon material and the area to be welded at a time.
(d) all of the above.

206. In flash-butt welding
(a) no special preparation of the faces to be welded is necessary.
(b) clean and pure weld is obtained.
(c) power requirement is less.
(d) all of the above.

207. Spot welding process basically depends on
(a) generation of heat.
(b) application of forging pressure.
(c) both (a) and (b). (d) ohmic resistance.

208. In spot welding, composition and thickness of the base metal determines the
(a) holding time. (b) amount of weld current.
(c) amount of squeeze pressure.
(d) all of the above.

209. The tips of the electrodes, for spot welding are made of
(a) carbon.
(b) copper alloy or pure copper.
(c) mica. (d) porcelain.

210. The power factor of a spot welding machine is expected to be about
(a) 0.3 to 0.5 lagging. (b) 0.8 to 0.85 lagging.
(c) 0.75 to 0.85 leading. (d) unity.

211. During spot welding, the current flows for
(a) fraction of a minute.
(b) fraction of a second to several seconds.
(c) few milliseconds. (d) few microseconds.

212. Spot welding is employed for
(a) thin metal sheets (thickness being usually limited to 10-12 mm).
(b) castings only. (c) thick sections.
(d) rough and irregular surfaces.

213. Spot welding
(a) makes the weld air tight.
(b) makes the weld water tight.
(c) provides mechanical strength.
(d) all of the above.

214. In spot welding
(a) it is desirable to clean the sheets thoroughly before welding.
(b) the work-pieces being welded are pressed together by mechanical pressure exerted through electrodes.
(c) current required is above 5,000 A and the voltage between the electrodes is usually less than 2 V (open-circuit voltage less than 12 V).
(d) all of the above.

215. Projection welding can be considered as a mass production form of
(a) seam welding. (b) spot welding.
(c) upset welding. (d) flash welding.

216. In comparison to spot welding, projection welding has the advantage(s) of
(a) simple welding process, more output and good finished appearance.
(b) electrode life is increased owing to use of low current density and low pressure.
(c) it is easy to weld certain parts which cannot be welded by spot welding .
(d) all of the above.

217. In seam welding
(a) the workpiece is fixed and disc electrodes move.
(b) the workpiece moves but rotating electodes are fixed.
(c) the electrodes used are of disc or roller shape.
(d) either (a) or (b) and (c).

218. In case of seam welding, the flow of current through the electrode should be
(a) intermittent. (b) continuous.
(c) either (a) or (b).

219. In _____ welding it is essential that the surfaces to be welded are clean, dust and dirt free.
(a) butt (b) spot (c) seam
(d) both spot and seam

220. Seam welding is normally not recommended for
(a) aluminium alloys. (b) stainless and coated steels.
(c) copper and high copper alloys.
(d) alloys of nickel and magnesium.

221. In percussion welding
(a) the process used depends on the arc effect for heating and not on the resistance.

Answers: **200.** (d), **201.** (d), **202.** (d), **203.** (a), **204.** (c), **205.** (d), **206.** (d), **207.** (c), **208.** (d), **209.** (b), **210.** (a), **211.** (b), **212.** (a), **213.** (c), **214.** (d), **215.** (b), **216.** (d), **217.** (d), **218.** (a), **219.** (c), **220.** (c), **221.** (d),

(b) the action of process is so rapid that there is little heating effect in the material adjacent to the weld.
(c) the equipment used is quite expensive.
(d) all of the above.

222. The basic electrical requirement in arc welding is that there should be
(a) high open-circuit voltage.
(b) no arc blow. (c) dc power supply.
(d) coated electrodes.

223. In arc welding, the temperature of the arc produced is of the order of
(a) 1,000ºC (b) 3,500–4,000ºC
(c) 5,000–7,500ºC (d) 7,500–10,000ºC

224. The electric arc has
(a) linear resistance characteristic.
(b) positive resistance characteristic.
(c) negative resistance characteristic.
(d) highly inductive characteristic.

225. In an electric arc welding, the voltage required to strike dc arc is about
(a) 50–60 V (b) 80–90 V (c) 100–120 V (d) 220 V

226. In an electric arc welding, the voltage required to strike ac arc is about
(a) 50–60 V (b) 80–90 V (c) 100–120 V (d) 230 V

227. In an electric arc welding, the voltage required to maintain the arc will be
(a) 250–500 V (b) 150-250 V
(c) 20–30 V (d) below 10 V

228. For an electric arc welding the current range is usually
(a) 50–1,000 A (b) 30–50 A
(c) 20–30 A (d) below 20 A

229. Increased heat due to shorter arc is harmful owing to
(a) burn through. (b) excessive porosity.
(c) undercutting of base material.
(d) all of the above.

230. Arc blow results in
(a) non-uniform weld beads.
(b) shallow weld puddle giving rise to weak weld.
(c) splashing out of metal from weld puddle.
(d) all of the above defects.

231. Arc blow effects occurring with dc welding machines can be reduced by
(a) shortening the arc column length.
(b) reducing the welding current or electrode size or rate of travel of the electrode.
(c) wrapping the welding electrode cable a few turns around the work.
(d) all of the above.

232. In an electric welding, major personal hazards are
(a) weld spatter. (b) flying sparks.
(c) harmful infrared and ultraviolet rays from the arc.
(d) all of the above.

233. During electric arc welding as the thickness of the metal to be welded increases
(a) voltage is increased keeping current the same.
(b) current is increased keeping voltage unchanged.
(c) current and voltage, both are increased.
(d) current and voltage, both are reduced.

234. In a dc arc welding
(a) both electrode as well as workpiece are made + ve.
(b) both electode as well as workpiece are made – ve.
(c) electrode is made –ve and workpiece + ve.
(d) electrode is made +ve and workpiece – ve.

235. The length of arc required depends on
(a) kind of electrodes used, its coating and its diameter.
(b) magnitude of current used.
(c) position of welding. (d) all of the above.

236. Overhead welding position is thought to be the most
(a) hazardous. (b) economical.
(c) useful. (d) difficult.

237. In arc welding best results are obtained when arc length is equal to
(a) 25 mm. (b) 19 mm.
(c) diameter of electrode.
(d) double the diameter of the electrode.

238. Flat position welding is cosidered to be the most
(a) easiest and economical. (b) hazardous.
(c) adaptable for welding of both ferrous and non-ferrous metals particularly for cast iron.
(d) both (a) and (c).

239. In direct current straight polarity (DCSP) or electrode negative welding
(a) heavily coated electrodes are used.
(b) base metal penetration is narrow and deep.
(c) electrode is the hottest.
(d) workpiece is relatively cooler.

240. In direct current reverse polarity or electrode positive welding
(a) workpiece is the hottest.
(b) bare and medium coated electrodes can be used.
(c) electrode is the hottest.
(d) base metal penetration is narrow and deep.

241. The electrode is coated in order to
(a) improve bead quality. (b) cleanse the base metal.
(c) provide shielding to weld pool.
(d) prevent atmospheric contamination.

242. In DCRP or electrode positive welding, the heat produced at the electrode is ________ of total heat produced.
(a) one-third. (b) two-third. (c) three-fourth. (d) one-fourth.

243. A 10 SWG electrode usually operates in the current range of
(a) 95–135 A (b) 80–95 A (c) 45–70 A (d) 20–30 A

244. Electrode of 8 SWG is to be employed for welding two 12.5 mm steel plates. The current required will of the order of
(a) 20 A (b) 50 A (c) 150 A (d) 250 A

245. The purpose of coating on arc welding electrodes is to
(a) provide a protective covering.
(b) provide slag for protection of the molten metal.
(c) stabilise the arc. (d) all of the above.

246. 20 per cent duty cycle of a welding machine means
(a) operation of machine for 2 minutes in a duration of 10 minutes.
(b) machine output is 20 per cent of rated output.
(c) machine operates at 20 per cent efficiency.
(d) operation of machine is for 1 minute in a duration of 5 minutes.

247. In carbon arc welding, if the electrode is connected to the + ve terminal of the dc supply
(a) arc will not strike. (b) arc will be dull.

Answers: **222.** (a), **223.** (b), **224.** (c), **225.** (a), **226.** (b), **227.** (c), **228.** (a), **229.** (d), **230.** (d), **231.** (d), **232.** (d), **233.** (b), **234.** (c), **235.** (d), **236.** (a), **237.** (c), **238.** (d), **239.** (b), **240.** (c), **241.** (a), **242.** (b), **243.** (a), **244.** (c), **245.** (d), **246.** (a), **247.** (c),

(c) carbon arc will have tendency to go in to the weld joint.
(d) metal will not melt.

248. Carbon arc welding has the main drawback of
(a) occurrence of blow holes owing to magnetic arc blow especially while welding near edges of the workpiece.
(b) need of separate filler. (c) necessity of bare electrodes.
(d) fast consumption of electrodes.

249. Carbon arc welding
(a) uses carbon or graphite rod as a negative electrode and work to be welded as a positive.
(b) cannot be done with ac supply.
(c) is not suitable for vertical and overhead welding.
(d) all of the above.

250. Carbon arc welding is suitable particularly for ______ metals.
(a) ferrous. (b) nonferrous. (c) all.

251. For metal arc welding
(a) both dc and ac can be used but ac is preferred.
(b) bare electrodes are no longer used except for automatic welding having arrangement to protect the weld area from the atmosphere.
(c) correct welding current, voltage and speed are very important.
(d) all of the above.

252. The atomic hydrogen welding is based on the principle(s) of
(a) obtaining atomic hydrogen by means of an electric arc between two tungsten electrodes, in an atmosphere of hydrogen at atmospheric pressure.
(b) development of very high temperature by recombination of atoms which occurs in the cooler regions immediately outside the arc.
(c) very large heat conductivity of hydrogen at high temperatures.
(d) all of the above.

253. In atomic hydrogen arc welding
(a) an arc is maintained between two tungsten electrodes.
(b) a stream of hydrogen gas under a pressure of 0.5 kg/cm^2 is passed through the arc and around the electrodes.
(c) ac supply is used.
(d) all of the above.

254. In atomic hydrogen welding, electrodes are of long life. It is because
(a) of being non-pressure process.
(b) of arc being in the shape of a fan.
(c) of ac supply.
(d) two arcs are used at a time.

255. Helium produces
(a) faster welding speeds. (b) deep penetration.
(c) narrower heat affected zone in base metal.
(d) all of the above.

256. In argon arc welding, argon is used as
(a) an agent for heat transfer.
(b) a shield to protect the work from oxidation.
(c) a source of heat. (d) a flux.

257. In argon arc welding, the argon is used to
(a) prevent oxidation of metal by coming in contact with air.
(b) create inert atmosphere around the work.
(c) obviate the necessity for using flux.
(d) all of the above.

258. Submerged arc process is characterized by
(a) high welding current.
(b) exceptionally smooth beads.
(c) deep penetration. (d) all of the above.

259. Electrode is not consumed in case of
(a) TIG welding.
(b) atomic hydrogen arc welding.
(c) MIG welding. (d) all of the above.

260. Unlike TIG welding, MIG welding
(a) needs no post-weld cleansing.
(b) needs no flux.
(c) uses consumable electrodes.
(d) provides complete protection from atmospheric contamination.

261. MIG welding is becoming more and more popular as it
(a) is easy in operation.
(b) has high metal deposite rate.
(c) both (a) and (b).
(d) can be used for both ferrous and nonferrous metals.

262. In MAG (metal active gas) welding the gas used is
(a) argon or helium.
(b) carbon dioxide or its mixture with other gas.
(c) hydrogen. (d) any of the above.

263. A weld bead of wineglass design is produced in
(a) plasma arc welding. (b) MAG welding.
(c) electron beam welding. (d) all of the above.

264. In electro-slag welding, theoretically there is no limit to the
(a) rate of metal deposit. (b) thickness of weld bead.
(c) temperature of salt bath.
(d) rate of slag consumption.

265. In manual arc welding of mild steel, the metal deposition rate will be about
(a) 10–15 kg per hour. (b) 5–10 kg per hour.
(c) 2–5 kg per hour. (d) below 1 kg per hour.

266. In ultrasonic welding, the frequency range is usually
(a) 20–60 kHz (b) 50–100 kHz
(c) 100–200 kHz (d) above 250 kHz

267. Welding is not done directly from the supply mains as
(a) its voltage is too high.
(b) it is impracticable to draw heavy currents directly from the supply mains.
(c) its voltage remains fluctuating.
(d) none of the above.

268. AC welding machine cannot be used for
(a) resistance welding. (b) submerged arc welding.
(c) MIG welding. (d) atomic hydrogen welding.

269. In electric welding, arc blow can be avoided by
(a) using ac machines. (b) increasing arc length.
(c) using bare electrodes.
(d) welding away from ground connections.

270. Welding leads have
(a) high current carrying capacity.
(b) high flexibility. (c) both (a) and (b).
(d) none of the above.

271. The welding load
(a) is always intermittent.
(b) is always continuous and constant.

Answers: **248.** (a), **249.** (d), **250.** (b), **251.** (d), **252.** (d), **253.** (d), **254.** (c), **255.** (d), **256.** (b), **257.** (d), **258.** (d), **259.** (a), **260.** (c), **261.** (c), **262.** (b), **263.** (a), **264.** (b), **265.** (c), **266.** (a), **267.** (b), **268.** (c), **269.** (a), **270.** (c), **271.** (a),

(c) is always continuous but varying.
(d) may be any one of the above.

272. The load power factor using welding transformer depends on
(a) arc length. (b) material to be welded.
(c) type of electrode to be used.
(d) all of the above.

273. For power factor correction in a welding circuit, a capacitor is usually connected
(a) across the mains.
(b) across secondary side of welding transfomer.
(c) across primary side of welding transformer.
(d) across arcing electrodes.

274. Which of the following is different from the remaining ?
(a) Butt welding. (b) Electro-slag welding.
(c) TIG welding. (d) MIG welding.

275. Chipping hammers are used
(a) for slag welding.
(b) for aligning the pieces to be welded.
(c) for removing slag from welding.
(d) for marking spots to be welded.

276. The danger of electric shock is maximum
(a) before welding. (b) during arcing.
(c) while inserting electrode into the holder.
(d) after welding.

277. The eyes of welding operator must be protected against
(a) infrared radiations. (b) ultraviolet radiations.
(c) both (a) and (b). (d) solar radiations.

278. Which of the following automatic welding processes is likely to give maximum rate of metal deposition ?
(a) Multiple power submerged arc.
(b) Gas shielded bare wire.
(c) Single wire submerged arc.
(d) Continuous flux covered electrode.

279. In a welded joint poor fusion is on account of
(a) improper current. (b) high welding speed.
(c) uncleaned metal surface.
(d) lack of flux.

280. Which of the following uses consumable electrodes?
(a) TIG. (b) MIG.
(c) laser. (d) None of the above.

281. In which of the following welding methods the molten metal is poured for joining the metals?
(a) Gas welding. (b) Thermit welding.
(c) TIG welding. (d) Arc welding.

282. The porosity of weld joint may be caused by
(a) poor base metal. (b) incorrect size of electrode.
(c) low welding current. (d) any of the above.

283. In welding, weld spatter defect is generally because of too _____ during welding.
(a) high voltage (b) high current
(c) low voltage (d) low current

284. High temperature metals, such as columbium can be easily welded by
(a) MIG welding. (b) TIG welding.
(c) electron beam welding. (d) flash butt welding.

285. _____ welding is used for welding of car body.
(a) Gas (b) Spot
(c) Carbon-arc (d) Atomic hydrogen

286. Grey iron is usually welded by
(a) gas welding. (b) resistance welding.
(c) TIG welding. (d) arc welding.

287. The method recommended for the welding of aluminium alloy is ...... welding.
(a) dc arc (b) ac arc
(c) acetylene-oxygen gas (d) tungsten arc

288. Steel pipes are manufactured by
(a) arc welding. (b) argon arc welding.
(c) resistance welding. (d) thermit welding.

289. Steel rails are welded by
(a) thermit welding. (b) argon arc welding.
(c) gas welding. (d) resistance welding.

290. Aluminium is difficult to weld as
(a) it has an oxide coating.
(b) it conducts away heat very rapidly.
(c) both (a) and (b). (d) none of the above.

291. Electronic components are joined by
(a) spot welding. (b) soldering.
(c) brazing. (d) none of the above.

292. Aircraft body is
(a) rivetted. (b) seam welded.
(c) gas welded. (d) spot welded.

293. A seamless pipe has
(a) arc welded joint. (b) spot welded joint.
(c) no joint. (d) either (a) or (b).

294. The method usually not preferred for welding of chromium molybdenum steel is
(a) thermit welding. (b) resistance welding.
(c) submerged arc welding.
(d) oxyacetelyne gas welding.

295. Light
(a) is a form of heat energy.
(b) is a form of electrical energy.
(c) consists of electromagnetic waves.
(d) consists of shooting particles.

296. Radiant efficiency of the luminous source depends on
(a) temperature of the source.
(b) wavelength of light rays.
(c) shape of the source. (d) all of the above.

297. A substance which changes its electrical resistance when illuminated by light is called
(a) photoconductive. (b) photovoltaic.
(c) photoelectric. (d) none of the above.

298. Materials which reflect all wavelength in the spectrum of light appear to be
(a) opaque. (b) transparent.
(c) black to white. (d) green to red.

299. An object, which appears red to the eyes, absorbs
(a) blue radiations. (b) violet radiations.
(c) green radiations. (d) all of the above.

300. The sensors in the eye are known as
(a) retina and antenna. (b) wires and nerves.
(c) rodes and cones. (d) none of the above.

Answers: **272.** (b), **273.** (c), **274.** (a), **275.** (c), **276.** (c), **277.** (c), **278.** (a), **279.** (a), **280.** (b), **281.** (b), **282.** (a), **283.** (b), **284.** (c), **285.** (b), **286.** (a), **287.** (d), **288.** (c), **289.** (a), **290.** (c), **291.** (b), **292.** (a), **293.** (c), **294.** (b), **295.** (c), **296.** (a), **297.** (a), **298.** (c), **299.** (d), **300.** (c),

301. The colour temperature of daylight is about
(a) 100 K (b) 200 K (c) 500 K (d) 600 K

302. Sky appears blue due to
(a) radiation. (b) reflection.
(c) refraction.
(d) scattering of light over dust particles.

303. Colour of light depends upon
(a) velocity of light. (b) frequency.
(c) wavelength. (d) both (b) and (c).

304. The wave length of 5,500 Å will give light of
(a) green colour. (b) red colour.
(c) orange colour. (d) yellow-green colour.

305. Which of the following colour has wavelength between blue and yellow ?
(a) Violet. (b) Orange. (c) Red. (d) Green.

306. Luminous flux is
(a) the rate of energy radiation in the form of light waves.
(b) the part of light energy, radiated by Sun that is received on earth.
(c) measured in lux.

307. Candle power is
(a) the luminous flux emitted by the source per unit solid angle.
(b) the light radiating capacity of a source in a given direction.
(c) the unit of illumination.

308. Glare is caused due to
(a) excessive luminance.
(b) excessive lighting contrast in the field of vision.
(c) either (a) or (b).

309. A perfect diffuser surface is one that
(a) transmits all the incident light.
(b) absorbs all the incident light.
(c) diffuses all the incident light.
(d) all of the above.

310. The illumination at a surface due to a source of light placed at a distance 'd' from the surface varies as
(a) $1/d^2$ (b) $1/d$ (c) d (d) d^2

311. The illumination at various points on a horizontal surface illuminated by the same source varies as
(a) $\cos\theta$ (b) $\cos^2\theta$ (c) $\cos^3\theta$ (d) $1/\cos\theta$

312. Carbon arc lamps are commonly used in
(a) cinema projectors. (b) domestic lighting.
(c) factory lighting. (d) street lighting.

313. Fine diameter tungsten wires are made by
(a) turning. (b) wire drawing.
(c) swaging. (d) compressing.

314. The melting point of tungsten is
(a) 3,400°C (b) 2,800°C (c) 2,600°C (d) 2,400°C

315. The vacuum inside an incandescent lamp is of the order of
(a) 10^{-2} mm Hg (b) 10^{-4} mm Hg
(c) 10^{-6} mm Hg (d) 10^{-8} mm Hg

316. The rate of evaporation of tungsten filament in a lamp depends upon
(a) exhaust tube diameter. (b) glass shell diameter.
(c) vapour pressure inside. (d) none of the above.

317. Heat from the lamp filament is transmitted to the surrounding mainly through
(a) circulation. (b) conduction.
(c) convection. (d) radiation.

318. Which of the following material is most commonly used for the filaments in incandscent lamps ?
(a) Tungsten. (b) Osmium.
(c) Tantalum. (d) Silver.

319. A zero watt lamp consumes power of
(a) zero watt. (b) 5–10 W.
(c) about 15 W. (d) about 25 W.

320. The safe operating temperature of a tungsten filament lamp is
(a) 1,000°C (b) 2,000°C
(c) 3,000°C (d) 3,500°C

321. The output of a tungsten filament depends on
(a) size of the shell. (b) size of the lamp.
(c) temperature of filament. (d) all of the above.

322. What percentage of the input energy is radiated by filament lamps ?
(a) 60–70 per cent. (b) 40–50 per cent.
(c) 25–30 per cent. (d) 10–15 per cent.

323. An electric bulb, when broken, produces bang. It is on account of
(a) vacuum inside the bulb.
(b) pressure inside is equal to that outside.
(c) pressure of air in the bulb. (d) none of the above.

324. In an incandescent lamp, bird cage filament is usually used in vacuum bulb so as to
(a) increase the life span of the filament.
(b) give uniform radiations.
(c) reduce the oxidation phenomenon.
(d) reduce the convection losses.

325. Filament lamps normally operate at a power factor of
(a) 0.5 lagging. (b) 0.8 lagging.
(c) unity. (d) 0.85 leading.

326. Most affected parameter of a filament lamp due to variation in operating voltage is
(a) life. (b) light output.
(c) luminous efficiency. (d) wattage.

327. The lumens output of an incandescent lamp with the variation in operating voltage is given as lumens output α V^n where n is a constant. The value of n for tungsten is
(a) from 1.0 to 2.2 (b) from 3 to 4
(c) from 8 to 10 (d) from 10 to 12

328. Which gas is sometimes used in filament lamps ?
(a) Nitrogen. (b) Carbon dioxide.
(c) Argon. (d) Krypton.

329. Which of the following vapour/gas will give yellow colour in a filament lamp ?
(a) Helium. (b) Mercury. (c) Magnesium. (d) Sodium.

330. Magnesium vapour in a filament lamp gives
(a) green colour light. (b) pink colour light.
(c) blue colour light. (d) white colour light.

331. The gas used in a gas-filled filament lamp is
(a) helium. (b) oxygen. (c) nitrogen. (d) ozone.

332. In filament lamps, coiled coil filaments are used in
(a) coloured lamps. (b) gas-filled lamps.
(c) low wattage lamps. (d) higher wattage lamps.

Answers: **301.** (d), **302.** (d), **303.** (d), **304.** (d), **305.** (d), **306.** (a), **307.** (b), **308.** (c), **309.** (d), **310.** (a), **311.** (c), **312.** (a), **313.** (b), **314.** (a), **315.** (b), **316.** (c), **317.** (d), **318.** (a), **319.** (b), **320.** (c), **321.** (c), **322.** (d), **323.** (a), **324.** (b), **325.** (c), **326.** (a), **327.** (b), **328.** (c), **329.** (d), **330.** (a), **331.** (c), **332.** (b),

333. In electric discharge lamps, light is produced by
(a) magnetic effect of current.
(b) heating effect of current. (c) cathode ray emission.
(d) ionisation in a gas or vapour.

334. In electric discharge lamps for stabilization of arc
(a) a choke is connected in series with the supply.
(b) a variable resistance is connected in series with the circuit.
(c) a condenser is connected across the supply.
(d) any of the above.

335. Glow lamps
(a) have hot cathode. (b) have cold cathode.
(c) draw more power. (d) cannot withstand shocks.

336. Halogen lamps have the advantage(s) of
(a) reduced dimensions of the lamp.
(b) better colour rendition and longer life (about 2,000 hours).
(c) high operating temperature with increased luminous efficiency.
(d) all of the above.

337. The filament of a GLS lamp is made of
(a) aluminium. (b) tungsten.
(c) carbon. (d) copper.

338. Which of the following is not the GLS lamp of the standard rating ?
(a) 15 W (b) 40 W (c) 75 W (d) 100 W

339. The luminous efficiency of GLS lamps is normally in the range of _____ lumens/watt.
(a) 100 to 150 (b) 75 to 100
(c) 50 to 75 (d) 10 to 18

340. Nitrogen or argon is filled in GLS lamps so as to
(a) retard evaporation of tungsten filament.
(b) improve efficiency.
(c) change the colour of the light.
(d) reduce the glare.

341. In case of frosted GLS lamps, frosting of shell is done by
(a) acid etching. (b) ammonia.
(c) ozone. (d) salt water.

342. In case of GLS lamps the increase in supply voltage reduces
(a) power consumption. (b) lumens output.
(c) life. (d) efficiency.

343. Sodium vapour lamp needs an ionisation voltage of about
(a) 5 V (b) 20 V (c) 50 V (d) 100 V

344. The ignition voltage for a sodium lamp is about
(a) 100–125 V (b) 200–240 V
(c) 300–400 V (d) 400–600 V

345. When a sodium vapour lamp is switched on, initially the colour is
(a) red. (b) pink. (c) yellow. (d) blue.

346. The colour of light given out by a sodium vapour discharge lamp is
(a) pink. (b) bluish green. (c) yellow. (d) blue.

347. An auto-transformer used with a sodium vapour lamp should have
(a) high step-up ratio. (b) high step-down ratio.
(c) high leakage reactance. (d) high resistance.

348. Leak transformer in a sodium vapour lamp initially provides
(a) high current. (b) low voltage.
(c) high voltage.

349. In a sodium vapour discharge lamp the neon gas
(a) changes the colour of light.
(b) prevents vapourization of filament.
(c) acts as a shield around the filament.
(d) assists in developing enough heat to vapourize the sodium.

350. The capacitor is used in auto-transformer circuit of a sodium vapour lamp in order to
(a) regulate discharge voltage.
(b) improve the circuit power factor.
(c) control lamp illumination level.
(d) protect the lamp against over-voltage.

351. The sodium vapour lamp
(a) is only suitable for ac and so needs choke control.
(b) needs capacitor in its auto-transformer circuit to improve the power factor which is very low (about 0.3 lagging).
(c) comes up to its rated output in approximately 15 minutes.
(d) all of the above.

352. The luminous efficiency of a sodium vapour lamp is _____ lumens per watt.
(a) 40-50 (b) 50-100 (c) 120-200 (d) 10-12

353. Phosphor zinc silicate produces light of
(a) pink colour. (b) green colour.
(c) red colour. (d) yellow colour.

354. A mercury vapour lamp gives light of
(a) pink colour. (b) yellow colour.
(c) greenish-blue colour. (d) red colour.

355. High pressure mercury vapour lamps
(a) are similar in construction to sodium vapour lamps.
(b) need 4 or 5 minutes to attain full brilliancy.
(c) are generally used for general industrial lighting, railway yards, work areas, shoping centres etc.
(d) all of the above.

356. In high pressure mercury vapour lamps
(a) the main electrodes are made of tungsten wire in the form of helix.
(b) the choke is provided to limit the current to a safe value.
(c) in addition to two main electrodes a starting (auxiliary) electrode is connected through a high resistance .
(d) all of the above.

357. The luminous efficiency of high pressure mercury vapour lamps ranges from ________ lumens per watt.
(a) 30 to 40 (b) 60 to 100 (c) 100 to 150 (d) 250 to 400

358. In mercury iodide lamps
(a) a number of iodides are added, in addition to mercury, to fill up the gaps in the light spectrum.
(b) a separate ignition device, in addition to the choke, is provided.
(c) both (a) and (b). (d) none of the above.

359. Mercury iodide lamps
(a) are similar in construction to high pressure mercury vapour lamps.
(b) have luminous efficiency of the order of 80 lumens per watt.
(c) are suitable for applications in the field of floodlighting, industrial lighting and public lighting.
(d) all of the above.

360. Which of the following lamps is a cold cathode lamp ?
(a) Sodium vapour lamp. (b) Neon lamp.
(c) Mercury vapour lamp. (d) None of the above.

Answers: **333.** (d), **334.** (a), **335.** (b), **336.** (d), **337.** (b), **338.** (c), **339.** (d), **340.** (a), **341.** (b), **342.** (c), **343.** (a), **344.** (d), **345.** (b), **346.** (c), **347.** (c), **348.** (c), **349.** (d), **350.** (b), **351.** (d), **352.** (a), **353.** (b), **354.** (c), **355.** (d), **356.** (d), **357.** (a), **358.** (c), **359.** (d), **360.** (b),

361. Neon lamps
(a) are of the size of ordinary incandescent lamps.
(b) have power consumption of the order of 5 watts.
(c) are used as indicator lamps, night lamps, for determination of polarity of dc mains.
(d) all of the above.

362. In neon signs argon gas is used to give
(a) red colour. (b) blue colour.
(c) pink colour. (d) green colour.

363. In neon signs, neon with a mixture of mercury is used to give
(a) blue colour. (b) green colour.
(c) pink colour. (d) yellow colour.

364. The electrodes of neon tube work at
(a) very low temperatures. (b) ordinary voltage.
(c) 2 kV to 6 kV. (d) 33 kV.

365. Neon tubes are widely used for
(a) advertising. (b) indoor lighting.
(c) road signalling. (d) airport lighting.

366. The vapour discharge tube used for domestic light has
(a) one filament. (b) two filaments.
(c) four filaments. (d) no filament.

367. The fluorescent tube is coated from inside with a thin layer of fluorescent material in the form of powder in order to
(a) absorb invisible ultraviolet rays and radiate visible rays.
(b) improve the appearance.
(c) reduce glare. (d) improve life.

368. A stabilising choke is connected in the fluorescent tube circuit so as to
(a) reduce the flicker.
(b) act as a ballast in operating conditions and provide a voltage impulse for starting.
(c) act as a starter. (d) avoid radio-interference.

369. A capacitor is connected across the fluorescent tube circuit in order to
(a) eliminate the noise. (b) limit the current.
(c) improve the power factor of the tube circuit.
(d) all of the above.

370. Which of the following is present inside the fluorescent tube ?
(a) Helium and oxygen. (b) Argon and neon.
(c) Argon and carbon dioxide.
(d) Mercury vapour.

371. In a fluorescent tube, a ballast resistance is connected in series with the choke
(a) when the tube is operated on dc supply.
(b) when the tube is operated on ac supply.
(c) to reduce radio-interference.
(d) to reduce stroboscopic effects.

372. For operation of fluorescent tube on dc supply the additional device incorporated in the tube circuit is a
(a) transformer. (b) resistor. (c) inductor.

373. The light of fluorescent tube appears cooler than that from an incandescent lamp. This is due to the fact that
(a) the tube consumes less power.
(b) surface area of tube is larger than that of the incandescent lamp.
(c) tungsten is not used in the tube .
(d) none of the above.

374. Sometimes the wheels of the rotating machinery, in fluorescent tube lighting, appear to be stationary . This is due to
(a) luminescence effect. (b) light fluctuations.
(c) stroboscopic effect. (d) all of the above.

375. The flicker effect of fluorescent lamps is more pronounced at
(a) lower voltages. (b) higher voltages.
(c) higher frequencies. (d) lower frequencies.

376. The frequency of flickers in a fluorescent tube operating at 220 V, 50 Hz supply is
(a) 200 per second. (b) 100 per second.
(c) 50 per second. (d) 25 per second.

377. Standard wattage of a 1 m fluorescent tube is
(a) 25 W (b) 65 W (c) 80 W (d) 100 W

378. A fluorescent tube can be operated on
(a) both dc as well as ac.
(b) ac only. (c) dc only.
(d) satisfactorily only on dc.

379. Radio-interference from a fluorescent tube can be reduced by
(a) eliminating choke.
(b) connecting a small capacitor across starter terminals.
(c) putting two tubes in parallel..
(d) any of the above.

380. In a mercury vapour lamp light, red objects appear black. This is on account of
(a) colour mixing.
(b) high wavelengths of red colour.
(c) absence of red light from lamp radiations.
(d) absorption of red light by the lamp.

381. Blinking of a fluorescent tube may be on account of
(a) low circuit voltage. (b) loose contact.
(c) defective starter. (d) any of the above.

382. The cost of a fluorescent tube is more than that of an incandescent lamp on account of
(a) more components in the tube.
(b) more quantity of glass in the tube.
(c) more labour required. (d) all of the above.

383. The life of a fluorescent tube is affected by
(a) low voltage. (b) high voltage.
(c) frequency of switching and blinking.
(d) all of the above.

384. The disadvantage(s) of fluorescent tubes in comparison to filament lamps is/are
(a) high cost. (b) noise in choke.
(c) stroboscopic effect. (d) all of the above.

385. The ratio of radiant heat produced by a fluorescent tube and that by a filament lamp of the same rating is
(a) 0.2 (b) 0.35 (c) 0.6 (d) 0.75

386. The lamp that cannot sustain much voltage fluctuations is
(a) sodium vapour lamp. (b) mercury vapour lamp.
(c) incandescent lamp. (d) mercury iodide lamp.

387. The lamp that gives nearly ultraviolet light is
(a) carbon arc lamp. (b) vacuum type filament lamp.
(c) argon gas filled filament lamp.
(d) all of the above.

388. For the same wattage the cheapest lamp is
(a) sodium vapour lamp. (b) GLS lamp.
(c) halogen lamp. (d) fluorescent tube.

Answers: **361.** (d), **362.** (a), **363.** (b), **364.** (c), **365.** (a), **366.** (b), **367.** (a), **368.** (b), **369.** (c), **370.** (d), **371.** (a), **372.** (b), **373.** (c), **374.** (c), **375.** (d), **376.** (b), **377.** (a), **378.** (a), **379.** (b), **380.** (c), **381.** (d), **382.** (d), **383.** (d), **384.** (d), **385.** (a), **386.** (c), **387.** (a), **388.** (b),

389. Power factor is highest in case of
(a) sodium vapour lamp. (b) mercury vapour lamp.
(c) incandescent lamp. (d) neon lamp.

390. The lamp which gives nearly monochromatic light is
(a) sodium vapour lamp. (b) mercury vapour lamp.
(c) tungsten filament lamp. (d) fluorescent tube.

391. The lamp that does not have separate choke is
(a) mercury vapour lamp. (b) sodium vapour lamp.
(c) fluorescent tube. (d) all of the above.

392. For the same wattage the lamp that will produce minimum radiant heat is
(a) sodium vapour lamp. (b) fluorescent lamp.
(c) tungsten filament lamp. (d) mercury vapour lamp.

393. The lamp/tube preferred for air-conditioned spaces is
(a) sodium vapour lamp. (b) mercury vapour lamp.
(c) fluorescent tube. (d) GLS lamp.

394. The lamp that causes radio-interference is
(a) fluorescent tube. (b) halogen lamp.
(c) mercury vapour lamp. (d) all of the above.

395. The lamp that cannot be used for dimming is
(a) cold cathode lamp. (b) fluorescent lamp.
(c) GLS lamp. (d) all of the above.

396. Which of the following application does not require ultraviolet lamps ?
(a) Aircraft cockpit dashboard lighting.
(b) Blue print machines. (c) Car lighting.
(d) None of the above.

397. Floodlighting is not used for
(a) industrial purposes. (b) advertising purposes.
(c) aesthetic purposes. (d) any of the above.

398. The lumens emitted from a 100 W incandescent lamp will not increase if
(a) supply voltage is increased.
(b) filament temperature is increased.
(c) the diameter of the glass shell is increased.
(d) any of the above.

399. Total flux required in any lighting scheme depends inversely on
(a) surface area. (b) space-height ratio.
(c) illumination. (d) coefficient of utilisation.

400. To avoid glare
(a) object should be viewed from a distance.
(b) object should be moved constantly.
(c) object should be viewed from a close vicinity.
(d) viewer should move constantly.

401. The illumination level on a surface is least affected by
(a) candle power of light source.
(b) ambient temperature.
(c) type of reflector used.
(d) distance of the light source.

402. Desired illumination level on the working plane depends upon
(a) age group of observers.
(b) whether the object is stationary or moving.
(c) size of the object to be seen and its distance from the observer.
(d) all of the above.

403. Which of the following will need the highest level of illumination ?
(a) Living room. (b) Kitchen.
(c) Proof reading. (d) Hospital wards.

404. Which of the following will need the lowest level of illumination ?
(a) Workshop. (b) Displays.
(c) Railway platform. (d) Garage.

405. The illumination level in houses is in the range of
(a) 20–50 lux. (b) 100–200 lux.
(c) 300–500 lux. (d) 700 lux.

406. The illumination level required for precision work is about
(a) 500–1,000 lux. (b) 250 lux.
(c) 100 lux. (d) 50 lux.

407. The illumination level required for important traffic routes carrying fast traffic is about
(a) 30 lux. (b) 100 lux.
(c) 200 lux. (d) 5 lux.

408. Illumination due to moon light is about
(a) 0.03 lumen/m^2. (b) 0.3 lumen/m^2.
(c) 30–50 lumens/m^2. (d) 300–500 lumens/m^2.

409. The luminous flux reaching the working plane least depends on
(a) proportion of the room.
(b) the lumen output of the lamps.
(c) colour of the working plane surface.
(d) reflectance of internal surfaces.

110. The depreciation factor depends upon
(a) ageing of the lamp(s). (b) lamp cleaning schedule.
(c) type of work carried out at the premises.
(d) all of the above.

411. Utilisation factor depends upon
(a) size of the room.
(b) space-height ratio of the lamps.
(c) colour of walls/ceiling. (d) all of the above.

412. Glare is reduced by
(a) increasing the mounting height of the lamp.
(b) using diffusers.
(c) using reflectors to cut-off the light at certain angle.
(d) all of the above.

413. The glass that transmits the maximum light is
(a) clear glass. (b) milk glass.
(c) serrated glass. (d) opalescent glass.

414. The glass that transmits the least amount of light is
(a) clear glass. (b) milk glass.
(c) serrated glass. (d) opalescent glass.

415. The surface that has the lowest reflection factor for white light is
(a) white oil paint. (b) white plaster work.
(c) blue curtains. (d) aluminium sheet.

416. The lamp is provided with a reflector in order to
(a) avoid glare. (b) provide better illumination.
(c) protect the lamp. (d) all of the above.

417. Reflectors are provided with slits at the top so as to
(a) reduce heating effect. (b) reduce colour effect.
(c) introduce chimney effect for cleaning.
(d) all of the above.

418. Heat from light source is particularly of importance in
(a) designing for air-conditioning.
(b) designing for illumination level.
(c) designing for floor space utilization.
(d) all of the above.

Answers: **389.** (c), **390.** (a), **391.** (b), **392.** (b), **393.** (c), **394.** (a), **395.** (b), **396.** (c), **397.** (a), **398.** (c), **399.** (b), **400.** (a), **401.** (b), **402.** (d), **403.** (c), **404.** (c), **405.** (b), **406.** (a), **407.** (a), **408.** (b), **409.** (c), **410.** (d), **411.** (d), **412.** (d), **413.** (a), **414.** (b), **415.** (c), **416.** (d), **417.** (d), **418.** (a),

419. Dimming systems for lights are used in
(a) auditoriums. (b) theatres.
(c) ball room. (d) all of the above.

420. The voltage required to pass the necessary current through an electrolytic cell is of the order of
(a) 1–2 V (b) 10–20 V
(c) 100–120 V (d) 150–200 V

421. The energy consumption for production of ammonium sulphate is of the order of
(a) 2,000–2,500 kWh/tonne. (b) 3,000–4,000 kWh/tonne.
(c) 4,000–7,000 kWh/tonne. (d) 70–80 kWh/tonne.

422. Materials used for copper plating are
(a) copper sulphate and sulphuric acid.
(b) copper sulphate and nitric acid.
(c) copper carbonate and ammonium carbonate.
(d) copper nitrate and sulphuric acid.

423. Materials used for chromium plating are
(a) chromium carbonate and sulphuric acid.
(b) chromic acid and sulphuric acid.
(c) chromium chloride and hydrochlroride acid.
(d) none of the above.

424. The gold plating is carried out
(a) with a current density of 150–250 A/m^2 at a voltage of 1–2 V.
(b) with a current density of 50–150 A/m^2 at a voltage of 5–15 volts.
(c) with a current density of 100–150 A/m^2 at a voltage of 1–4 volts.
(d) with a current density of 50–150 A/m^2 at a voltage of 1 volt.

425. Power supply required for electrolytic processes is
(a) alternating current (100–200 A) at very low voltage (10 or 12 V).
(b) direct current (100–200 A) at very high voltage.
(c) direct current (100–200 A) at very low voltage (10 or 12 V).
(d) alternating current at very high voltage.

426. The plants for extraction and refining of metals or large scale manufacturing are located near the
(a) atomic power station. (b) hydroelectric power station.
(c) steam power station.
(d) either atomic power station or hydroelectric power station.

427. The process of coating of a metallic surface with a harder metal by electrodepostion is called
(a) electrofacing. (b) electroforming.
(c) electrometallisation. (d) either (a) or (b).

428. Basically electroplating means
(a) formation of ions by two metallic plates in the acidic liquid.
(b) electrodeposition of metal on electrodes.
(c) electrodeposition of metal upon metallic surfaces.
(d) none of the above.

429. Electroplating is done for
(a) replacement of worn out material.
(b) protection of metals against corrosion.
(c) giving a shining appearance to articles.
(d) all of the above.

430. The preparation of an object for electroplating involves the
(a) removal of oil, grease or other organic material.
(b) removal of rust, scale, oxides, or other inorganic coatings adhering to the metal.
(c) mechanical preparation of the metal surface by polishing, buffing etc .
(d) any or all of the above operations.

431. A one tonne refrigerating machine means that
(a) the total weight of the machine is one tonne.
(b) the quantity of the refrigerant used is one tonne.
(c) one tonne of water can be converted into ice.
(d) one tonne of ice when melts from and at 0°C in 24 hours, the refrigerating effect is equivalent to 14,000 kJ per hour.

432. The capacity of a refrigerating machine is expressed as
(a) inside volume of the cabinet.
(b) lowest temperature attained.
(c) gross weight of machine in tonnes.
(d) rate of abstraction of heat from the space being cooled.

433. Which of the following properties of a refrigerant is undesirable ?
(a) High critical temperature.
(b) Low specific heat of liquid.
(c) High boiling point.
(d) Low specific volume of vapour.
(e) High latent heat of vapourisation.

434. Various desirable properties which should be possessed by refrigerants are
(a) non-corrosive and non-inflammability.
(b) no-toxic and freedom from objectionable odour with low working pressure.
(c) both (a) and (b). (d) none of the above.

435. In a refrigeration cycle, the flow of refrigerant is controlled by
(a) compressor. (b) expansion valve.
(c) condenser. (d) evaporator.

436. Which part of the vapour compression refrigeration cycle, produces the refrigeration effect ?
(a) Evaporator. (b) Condenser.
(c) Expansion valve. (d) Compressor.

437. _____ is used as refrigerant in a vapour absorption refrigerator.
(a) Acqua ammonia (b) Freon
(c) Water (d) Sulphur dioxide

438. The refrigerant Freon 12 is a compound consisting of
(a) carbon, fluorine and chlorine.
(b) carbon, chlorine and hydrogen.
(c) carbon, fluorine and hydrogen.
(d) chlorine, bromine and iodine.

439. Ammonia is preferred as a refrigerant in large commercial installations because
(a) it is non-tonic.
(b) it has a low working pressure.
(c) it is relatively cheap. (d) it has low latent heat.

440. In a vapour compression system, which of the following units is adversely affected by the presence of moisture ?
(a) Evaporator. (b) Expansion valve.
(c) Compressor. (d) Condenser.

441. In a vapour absorption system lithium bromide is used as
(a) refrigerant. (b) lubricant.
(c) absorbent. (d) cooling substance.

442. Vapour absorption refrigeration systems have the advantage(s) of
(a) absence of moving parts.
(b) universal power source.
(c) automatic defrostration. (d) both (a) and (b).

443. Thermoelectric refrigeration system has the advantage(s) of
(a) small initial cost. (b) absence of moving parts.

Answers: 419. (d), **420.** (a), **421.** (a), **422.** (a), **423.** (b), **424.** (b), **425.** (c), **426.** (d), **427.** (a), **428.** (c), **429.** (d), **430.** (d), **431.** (d), **432.** (d), **433.** (c), **434.** (c), **435.** (b), **436.** (a), **437.** (a), **438.** (a), **439.** (b), **440.** (b), **441.** (c), **442.** (d), **443.** (d),

(c) ease of automatic control by variation of magnitude of current.
(d) both (b) and (c).

444. In a domestic refrigerator, compressor and motor are assembled in a single unit known as _____ sealed unit.
(a) hermetically (b) homogeneously
(c) heterogeneously (d) none of these

445. Hermetically sealed units have the advantage(s) of
(a) relatively silent operation.
(b) minimum gas leakage problem because no moving part extends through the sealing housing.
(c) minimum maintenance requirement.
(d) all of the above.

446. Which of the following motors is used in household refrigerators ?
(a) DC shunt motor. (b) Reluctance motor.
(c) Single phase induction motor.
(d) Synchronous motor.

447. The compressor of a domestic refrigerator has been noted to run for comparatively large durations, it indicates that
(a) thermostat has become defective.
(b) the machine is overloaded.
(c) capillary tube is choked.
(d) none of the above.

448. In a refrigerator, defrosting may be done by
(a) opening the door of the refrigerator.
(b) stopping the compressor for a while.
(c) placing trays of warm waters in the refrigerator.
(d) any of the above.

449. Types of water coolers are
(a) instantaneous or pressure type.
(b) storage type. (c) exhaust fan type.
(d) both (a) and (b).

450. Air-conditioning is the simultaneous control of_____ in a confined space.
(a) temperature (b) humidity
(c) air movement
(d) temperature, humidity, purity and movement of air

451. _____ is a process which is generally used for summer air-conditioning.
(a) Cooling with dehumidification
(b) Chemical dehumidification
(c) Adiabatic cooling
(d) Cooling with humidification

452. _____ is a process that is generally used in winter air-conditioning.
(a) Humidification (b) Dehumidification
(c) Heating and humidification
(d) Heating and dehumidification

453. Air is dehumidified by
(a) cooling. (b) chemical absorption.
(c) heating. (d) both (a) and (b).

454. Compared to individual systems, the central air-conditioning system has _____ overall efficiency.
(a) higher (b) lower (c) same (d) unpredictable

455. The main disadvantage(s) of central type air-conditioning s/are that
(a) dusting is very costly.
(b) absence of any individual room temperature adjustments.
(c) cigarette smoke and bacteria present in the return air from infected rooms redistribute to healthy rooms.
(d) all of the above.

456. Which of the following unitary-central systems are in common use ?
(a) Fan coil units. (b) Induction units.
(c) All air high velocity systems.
(d) All of the above.

457. The conditions conductive to comfort depends upon _____
(a) humidity (b) temperature
(c) air movement (d) air purity
(e) all of the above factors

458. Main traction systems used in India are those using _____ locomotives.
(a) steam engine (b) diesel engine
(c) diesel-electric (d) electric
(e) all of the above

459. The steam engine provided on steam locomotive is
(a) double acting condensing type.
(b) double acting non-condensing type.
(c) single acting condensing type.
(d) single acting non-condensing type.

460. In case of steam locomotives the tractive effort is provided by
(a) double cylinder, double acting steam engine.
(b) double cylinder, single acting steam engine.
(c) single cylinder, double acting steam engine.
(d) single cylinder, single acting steam engine.

461. In case of a steam locomotive an average coal consumption per km of run is around
(a) 5–10 kg (b) 25–30 kg
(c) 60–80 kg (d) 100–150 kg

462. The pressure of steam used in steam locomotives is about
(a) 1–5 kgf/cm^2 (b) 5–10 kgf/cm^2
(c) 10–15 kgf/cm^2 (d) 20–30 kgf/cm^2

463. The maximum horse power of steam locomotive is
(a) up to 1,500 (b) 1,500–2,000
(c) 2,000–3,000 (d) 3,000–4,000

464. The speed of a steam locomotive is controlled by
(a) applying brakes. (b) gearbox.
(c) regulating steam flow to engine.
(d) flywheel.

465. The pulsating torque exerted by steam locomotives causes
(a) jolting and skidding. (b) hammer blow.
(c) pitching. (d) all of the above.

466. Steam engine drive
(a) has limited overload capacity.
(b) is not clean drive.
(c) is not suitable for urban or suburban services or for hauling heavy loads on steep gradients.
(d) all of the above..

467. Steam locomotives
(a) cannot be put into service at any moment.
(b) cause considerable wear on the track.
(c) need more repair and maintenance.
(d) all of the above.

468. Steam engine drive is not suitable for urban and suburban services as
(a) it is not clean drive.
(b) coefficient of adhesion of steam locomotives is quite low.
(c) steam locomotive causes considerable wear on the track.
(d) steam locomotive has limited speed.

Answers: **444.** (a), **445.** (d), **446.** (c), **447.** (b), **448.** (d), **449.** (d), **450.** (d), **451.** (a), **452.** (c), **453.** (d), **454.** (a), **455.** (c), **456.** (d), **457.** (e), **458.** (e), **459.** (b), **460.** (a), **461.** (b), **462.** (c), **463.** (a), **464.** (c), **465.** (a), **466.** (d), **467.** (d), **468.** (b),

469. Direct internal combustion drive
(a) has higher maintenance and operating costs.
(b) has no starting torque and needs some auxiliary means for its starting.
(c) has shorter life of propulsive equipment.
(d) all of the above.

470. Diesel electric traction has comparatively limited overload capacity because
(a) diesel engine is a constant output prime mover.
(b) diesel engine has shorter life span.
(c) regenerative braking cannot be employed.
(d) diesel-electric locomotive is heavier than an ordinary electric locomotive.

471. The range of horse power of diesel locomotives is
(a) 100–500 (b) 1,500–2,500
(c) 3,000–4,500 (d) 4,500–5,000

472. In India, diesel locomotives are manufactured at
(a) Varanasi. (b) Kolkata. (c) Bangalore. (d) Ajmer.

473. In diesel-electric drive
(a) initial investment required is low.
(b) locomotive and train is a self-contained unit.
(c) power loss in speed control is low.
(d) all of the above.

474. Battery driven vehicles
(a) are easy to control and very convenient to use.
(b) have low maintenance cost.
(c) cause no pollution. (d) all of the above.

475. Battery operated trucks are used
(a) in power stations.
(b) for local delivery of goods in large towns with maximum daily run up to 50–60 km.
(c) in narrow gauge traction. (d) for main line service.

476. Electric traction in comparison to other traction systems has the advantage(s) of
(a) higher acceleration and braking retardation.
(b) cleanest system and so ideally suitable for the underground and tube railways.
(c) better speed control. (d) all of the above.

477. Electric railway can handle the traffic upto double the amount possible with steam railway. It is because of
(a) better speed control.
(b) larger passenger carrying capacity.
(c) higher schedule speed. (d) both (b) and (c).

478. Electric traction in comparison to other traction systems has the drawback(s) of
(a) interference with communication lines running along the track.
(b) heavy initial expenditure in laying out overhead electric supply system.
(c) interruption of traffic for hours owing to short time power failure.
(d) all of the above.

479. The most vital factor against electric traction is
(a) its high maintenance cost.
(b) possibility of power failure.
(c) high initial cost in laying out overhead electric supply system.
(d) necessity of providing a negative booster.

480. The locomotive that has the highest operational availability is
(a) diesel-electric. (b) electric.
(c) steam. (d) steam-electric.

481. Unbalanced forces are maximum in case of
(a) diesel shunters. (b) diesel locomotives.
(c) steam locomotives. (d) electric locomotives.

482. Maintenance requirements are minimum in case of
(a) electric lcomotives. (b) steam locomotives.
(c) diesel electric locomotives.
(d) diesel engines.

483. A drive suitable for mines where explosive gas may exist, is
(a) steam engine. (b) battery locomotive.
(c) diesel engine. (d) any of the above.

484. A submarine is provided driving power through (while moving under water)
(a) diesel engines. (b) steam engines.
(c) batteries. (d) gas turbine.

485. Ordinary tramway is the most economical means of transport for
(a) very dense traffic in large cities.
(b) rural services. (c) suburban services.

486. Unlike a tramway, a trolley bus needs no
(a) driving axles. (b) running rail.
(c) hand brakes. (d) overhead contact wire.

487. The direct current system employed for tramways operates on
(a) 3,000 V (b) 1,500 V (c) 750 V (d) 400 V

488. For tramways the return circuit is
(a) track rails. (b) neutral wire.
(c) common earthing. (d) cables.

489. For 600 V dc line for tramcars, track is connected to the
(a) positive of the supply. (b) negative of the supply.
(c) midvoltage of 300 V. (d) none of the above.

490. In tramways
(a) the power is supplied at 600 V dc from a single overhead conductor of + ve polarity.
(b) the speed is controlled by field weakening or series-parallel control.
(c) rheostatic and mechanical brakings are employed for normal service.
(d) all of the above.

491. The main drawback of tramways is that
(a) the life of tramcar equipment is short.
(b) it has restricted manoeuvrability.
(c) it needs laying overhead supply system and track for its use, which is costly to maintain and constitutes a source of danger to other road users.
(d) all of the above.

492. The trolley bus
(a) is an electrically operated pneumatic tyred vehicle, usually fed at 600 V dc from two overhead contact wires.
(b) is able to manoeuvre in a traffic over a width of several metres on either side of the central line of the trolley wires.
(c) uses field weakening method for its speed control.
(d) all of the above.

493. The trolley buses
(a) do not require laying of track for its use as return conductor.
(b) make use of foot operated master controllers.
(c) make use of stabilised rheostatic braking.
(d) all of the above.

Answers: **469.** (d), **470.** (a), **471.** (b), **472.** (a), **473.** (d), **474.** (d), **475.** (b), **476.** (d), **477.** (d), **478.** (d), **479.** (c), **480.** (b), **481.** (c), **482.** (a), **483.** (b), **484.** (c), **485.** (a), **486.** (b), **487.** (c), **488.** (a), **489.** (b), **490.** (d), **491.** (c), **492.** (d), **493.** (d),

494. In diesel electric traction
(a) the driving motors are dc series type and fed from the dc generators driven by the diesel engine put on the same locomotive.
(b) the starting and speed control is affected by varying the excitation of the generator.
(c) regenerative braking is not possible.
(d) all of the above.

495. Suburban railways use
(a) 1,500 V dc. (b) 400 V, 3-phase ac.
(c) 3,300 V, 3-phase ac. (d) 600 V, 3-phase ac.

496. Long distance railways operate on
(a) 600 V dc. (b) 25 kV single-phase ac.
(c) 25 kV three-phase ac. (d) 15 kV three-phase ac.

497. The electric locomotives run faster at curved routes in comparison to steam locomotives because
(a) its centre of gravity is lower than that of steam locomotive.
(b) its centre of gravity is higher than that of steam locomotive.
(c) the speed at curved routes is independent of locations of centre of gravity.
(d) none of the above.

498. Low frequency operation of overhead line in traction system
(a) increases the spacing between substations.
(b) reduces the spacing between substations.
(c) spacing is independent of frequency of supply.

499. In Kando system of track electrification
(a) single-phase ac is converted into dc.
(b) single-phase ac is converted into 3-phase ac.
(c) 3-phase ac is converted into dc.
(d) 3-phase ac is converted into single-phase ac.

500. The composite system (single-phase ac to dc system) has been chosen for all future track electrification in India as
(a) it needs light overhead catenary.
(b) it needs less number of substations.
(c) it combines the advantages of high-voltage ac distribution at 50 Hz with dc series traction motors.
(d) it provides flexibility in the location of substations.

501. Single-phase ac system is preferred for main line railway service because
(a) the initial, maintenance and operating costs of ac substations are less as the substation equipment required in single-phase ac system is less, cheap and efficient.
(b) number of substations required is less.
(c) of reduced cost of distribution system.
(d) all of the above.

502. In underground traction, the supply system is
(a) 500 V to 1,000 V dc (b) 25 kV, 50 Hz
(c) 50 kV, 50 Hz (d) 25 kV, 25 Hz
[A.M.I.E. Sec B. Elec. Drives & Their Control Winter 1996]

503. Problem(s) of single-phase traction systems is/are
(a) current and voltage unbalance.
(b) electrostatic and electromagnetic induction.
(c) generation of harmonics. (d) all of the above.

504. The chief requirements of main line railway services are
(a) high maximum speed.
(b) minimum cost of overhead structure.
(c) high acceleration and braking retardation.
(d) both (a) and (b).

505. The main requirements of suburban railway services are
(a) rapid acceleration and braking retardation.
(b) no interference to the communication circuits running along the track.
(c) high maximum speed. (d) both (a) and (b).

506. The acceleration rate for urban or suburban services is
(a) 1.5–4 kmphps. (b) 3–4 kmphps.
(c) 5–10 kmphps. (d) 0.5–1.5 kmphps.

507. The braking retardation for urban or suburban services is
(a) 1.5–2.5 kmphps. (b) 3–4 kmphps.
(c) 5–10 kmphps. (d) 0.5–1.5 kmphps.

508. The coasting retardation for main line railway services is about
(a) 10 kmphps. (b) 3 kmphps.
(c) 0.16 kmphps. (d) 0.01 kmphps.

509. The maximum speed at which trains run on main line railway service is
(a) 160 kmph. (b) 120 kmph.
(c) 100 kmph. (d) 200 kmph.

510. In main line services as compared to urban and suburban services
(a) distance between the stops is more (exceeding 10 km).
(b) maximum speed attained is high.
(c) acceleration and braking retardation rates are low.
(d) all of the above.

511. The speed-time curve for urban service has no
(a) coasting period. (b) free running period.
(c) braking period. (d) acceleration period.

512. Free running and coasting periods are generally long in case of
(a) city service. (b) suburban serivce.
(c) main line service. (d) outer suburban service.

513. In suburban services as compared to urban services
(a) the coasting period is longer.
(b) the coasting period is smaller but free running period is longer.
(c) the coasting and free running periods are smaller.
(d) none of the above is true.
[A.M.I.E. Sec B. Elec. Drives & Their Control Winter 1995]

514. Trapezoidal speed-time curve pertains to
(a) main line service. (b) urban service.
(c) suburban service. (d) urban/suburban service.

515. Quadrilateral speed-time curve is the close approximation for
(a) urban service. (b) suburban service.
(c) urban/suburban service. (d) main line service.

516. The speed-time curve for a main line service is illustrated in figure given below.
In figure given below AB represents
(a) coasting.
(b) free running.
(c) acceleration.
(d) average speed.

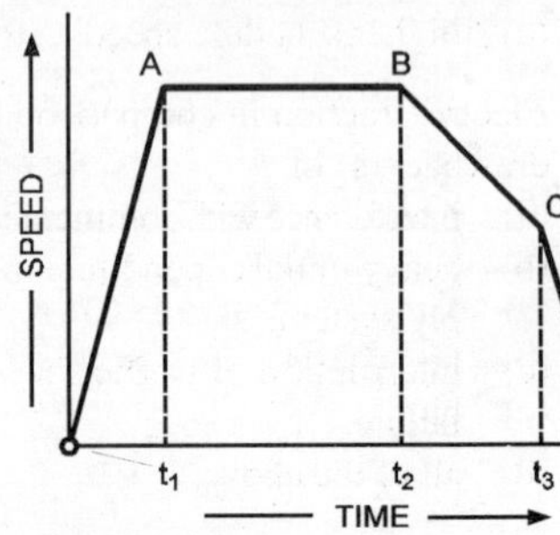

517. Area under the speed-time curve represents
(a) total distance travelled. (b) average speed.
(c) average acceleration. (d) none of the above.

518. The duration of coasting is represented by the duration
(a) $0 - t_1$ (b) $t_1 - t_2$ (c) $t_2 - t_3$ (d) $t_3 - t_4$

Answers: **494.** (d), **495.** (a), **496.** (b), **497.** (a), **498.** (a), **499.** (b), **500.** (c), **501.** (d), **502.** (a), **503.** (d), **504.** (d), **505.** (d), **506.** (a), **507.** (b), **508.** (c), **509.** (a), **510.** (d), **511.** (b), **512.** (c), **513.** (a), **514.** (a), **515.** (c), **516.** (b), **517.** (a), **518.** (c),

519. From figure given in Q. 516, it can be concluded that
(a) average acceleration is zero.
(b) during coasting and braking the acceleration is negative.
(c) acceleration during free running is zero.
(d) both (b) and (c).

520. The speed of train estimated taking into account the stoppage time at a station in addition to the actual running time between stops, is called the _____ speed.
(a) average (b) schedule (c) free running (d) notching

521. The average speed of a train is independent of
(a) duration of stops.
(b) acceleration and braking retardation.
(c) distance between stops. (d) running time.

522. The schedule speed of a given train when running on a given service (with given distance between stations) is affected by
(a) acceleration and braking retardation.
(b) maximum or crest speed.
(c) duration of stop. (d) all of the above.

523. Skidding of a vehicle always occurs when
(a) braking effort exceeds its adhesive weight.
(b) brake is applied suddenly.
(c) it negotiates a curve.
(d) it passes over points and crossings.

524. The adhesive weight is the
(a) total weight of the locomotive and the train.
(b) weight coming over the driving wheels.
(c) same as the accelerating weight.
(d) none of the above.

525. Coefficient of adhesion is the ratio of tractive effort to slip the wheels and
(a) dead weight. (b) accelerating weight.
(c) adhesive weight. (d) none of the above.

526. The normal value of coefficient of adhesion is
(a) 0.25 (b) 0.35 (c) 0.50 (d) 0.65

527. Coefficient of adhesion reduces due to the presence of
(a) dew on rails. (b) oil and grease on rails.
(c) dry sand on rails. (d) both (a) and (b).

528. Coefficient of adhesion improves due to presence of
(a) dry sand on rails. (b) rust on rails.
(c) dust on rails. (d) all of the above.

529. The value of coefficient of adhesion will be high when rails are
(a) wet. (b) cleaned with sand.
(c) greased. (d) sprayed with oil.

530. The coefficient of adhesion for wet or greasy rails is
(a) 0.35 (b) 0.25 (c) 0.08 (d) zero

531. The coefficient of adhesion
(a) same on dc and ac traction.
(b) high in ac traction and low in dc traction.
(c) high in dc traction and low in ac traction.
(d) any of the above.

532. Higher value of tractive effort can be used in electric traction as compared to steam traction because
(a) of greater adhesive weight (the weight on driving wheels is 70 percent of dead weight in case of electric locomotive whereas in case of steam locomotive it is 50 percent).
(b) the torque exerted in electric traction is continuous whereas in steam traction it is pulsating one.
(c) the driving wheels are distributed over a much greater length in electric traction whereas in steam traction they are close to each other.
(d) all of the above.

533. When a bogie negotiates a curve, reduction in adhesion occurs resulting in sliding. This sliding is acute when
(a) degree of curvature is more.
(b) wheel base of axles is more.
(c) none of the above. (d) both (a) and (b).

534. For a given maximum axle load tractive effort with ac locomotive will be
(a) less than that of dc locomotive.
(b) more than that of dc locomotive.
(c) equal to that of dc locomotive. (d) none of the above.

535. The resistance encountered by a train in motion is on account of
(a) resistance offered by air. (b) friction at the track.
(c) friction at various parts of the rolling stock.
(d) all of the above.

536. The air resistance to the movement of the train is proportional to
(a) 1/speed. (b) (speed). (c) $(\text{speed})^2$. (d) $(\text{speed})^3$

537. The friction at the track is proportional to
(a) speed. (b) $(\text{speed})^2$. (c) $(\text{speed})^3$. (d) $(1/\text{speed})^2$.

538. If the resistance to electric train movement is given by
$$F_r = a + bv + cv^2$$
In the above expression b is likely to cover
(a) air resistance. (b) track resistance.
(c) frictional resistance. (d) none of the above.

539. Tractive effort of an electric locomotive can be increased by
(a) using high output motors.
(b) increasing the supply voltage.
(c) increasing dead weight over the driving axles.
(d) both (a) and (c).

540. A locomotive with a mass of 50,000 kg on a track whose coefficient of adhesion is 20 per cent will produce a tractive effort of (approximately)
(a) 1 kN (b) 100 kN (c) 25 kN (d) 250 kN
[A.M.I.E. Sec B. Drives & Their Control Winter 1996]

541. Energy consumption in propelling the train is required for
(a) accelerating of train mass.
(b) overcoming the gradient while moving up the gradient.
(c) overcoming the train resistance.
(d) all of the above.

542. Longer coasting period for a train results in
(a) higher schedule speed .
(b) lower specific energy consumption.
(c) higher retardation. (d) higher acceleration.

543. Specific energy consumption is affected by
(a) acceleration and retardation values.
(b) the crest speed and nature of route.
(c) distance between stops. (d) all of the above.

544. Specific energy consumption is minimum in _____ services.
(a) main line (b) urban
(c) suburban (d) equal for all type of

545. For a given run and given schedule speed, the specific energy consumption is
(a) lower, the higher the acceleration and retardation.
(b) higher, the higher the acceleration and retardation.
(c) lower, the higher the acceleration and lower the retardation.
(d) lower, the lower the acceleration and higher the retardation.

Answers: **519.** (d), **520.** (b), **521.** (a), **522.** (d), **523.** (a), **524.** (b), **525.** (c), **526.** (a), **527.** (d), **528.** (d), **529.** (b), **530.** (c), **531.** (b), **532.** (d), **533.** (d), **534.** (b), **535.** (d), **536.** (c), **537.** (a), **538.** (b), **539.** (d), **540.** (b), **541.** (d), **542.** (b), **543.** (d), **544.** (a), **545.** (a),

546. Specific energy consumption
(a) increases with the increase in maximum speed.
(b) decreases with the increase in maximum speed.
(c) is independent of maximum speed.
(d) none of the above.

547. If the speed-time curves are similar (not identical), the specific energy consumption of the curve having higher maximum speed is _____ than that of lower maximum speed.
(a) lower (b) equal (c) higher (d) none

548. Specific energy consumption becomes
(a) more when distance between stops is more.
(b) more with the higher values of acceleration (or retardation).
(c) more with high train resistance.
(d) less with the increase in crest speed.

549. The specific energy consumption for suburban services is usually _____ watt-hours per tonne-km.
(a) 20–30 (b) 30–45 (c) 50–75 (d) 100–150

550. The specific energy consumption for main line services is around .... watt hours per tonne-km.
(a) 20–30 (b) 30–45 (c) 50–75 (d) 100–150

551. Specific energy consumption is maximum in _____ services.
(a) urban (b) suburban
(c) main line (d) equal for all types of

552. The electric motor used for traction work should be mechanically
(a) small in overall dimensions (specially in its overall diameter).
(b) light in weight and robust in construction.
(c) capable to withstand continuous vibrations.
(d) all of the above.

553. The electric motor used for traction work, electrically should be
(a) capable of developing high starting torque and withstanding voltage fluctuations, and temporary supply interruptions.
(b) amenable to simple speed control methods, self protective against excessive overloading and amenable to easy and simple methods of rheostatic and/or regenerative braking.
(c) of such characteristics that when they are operated in parallel and coupled mechanically, they share the load almost equal.
(d) all of the above.

554. In suburban trains, the traction motors are installed on
(a) locomotive only. (b) locomotive and coaches.
(c) coaches only.
[A.M.I.E. Sec B. Elec. Drives & Their Control Winter 1996]

555. The type of dc motor used in electric traction is
(a) series. (b) shunt.
(c) separately-excited. (d) none of the above.
[A.M.I.E. Sec B. Electric Drives & Their Control Summer 1999]

556. The dc series motor is most suitable for traction services but more particularly for urban/suburban services because
(a) dc series motors are suitable for regenerative braking.
(b) dc series motors are capable of withstanding rapid fluctuations in supply voltage.
(c) dc series motors are capable of developing high torque at start.
(d) dc series motors are capable of withstanding temporary interruption of supply without undue rush of current.

557. The three-phase induction motors have the advantages of simple and robust in costruction, trouble free operation, high voltage operation consequently requiring reduced amount of current and automatic regeneration but are not suitable for traction work. This is because of their
(a) flat speed-torque characteristics, constant speed operation and low starting torque.
(b) complicated speed control system and complicated overhead feeding systems.
(c) higher initial as well maintenance cost.
(d) both (a) and (b).

558. DC shunt motors are not suitable for traction services because of their
(a) hard characteristics. (b) large time constant.
(c) power varying directly with developed torque.
(d) wide variation in torque and flux due to variation in voltage in electric traction.
(e) all of the above

559. Two dc shunt motors having identical characteristics are used to drive a train car with unequal wheel diameters. The two motors share load equally if they are connected in
(a) series. (b) parallel.
(c) the loading will always be unequal.
[A.M.I.E. Sec B. Elec. Drives & Their Control Winter 1995]

560. Two locomotives having identical tractive effort-slip characteristics haul a heavy train. The loco with larger diameter shares _____ tractive effort.
(a) larger (b) smaller (c) equal

561. For single-phase ac system of track electrification, low frequency is desirable as
(a) it improves commutation properties of ac motors.
(b) it increases efficiency of ac motors.
(c) it improves power factor of ac motors.
(d) all of the above.

562. At low frequency of the order of 1/2 Hz to 10 Hz, the induction motor develops
(a) high starting torque with excessive starting current.
(b) high starting torque without excessive starting current.
(c) low starting torque with excessive starting current.
(d) low starting torque without excessive starting current.

563. A locomotive exerts a tractive effort of 40,000 N in pulling a train at 60 kmph on a level track. If the same train is to be hauled at the same speed but on gradient, the tractive effort required will be 60,000 N. The output delivered by the motor will be more if it is driven by
(a) a 3-phase induction motor.
(b) a dc series motor. (c) an ac series motor.
(d) equal in all the three cases.

564. When two or more motors are used for traction service, the method of speed control used will be
(a) rheostatic control. (b) series-parallel control.
(c) field control. (d) motor generator control.

565. The advantage(s) of series-parallel starting is/are
(a) small energy loss and higher efficiency.
(b) economical speed control.
(c) higher reliability of operation.
(d) all of the above.

566. Parallel operation of traction motors is easier with
(a) dc shunt motors. (b) dc series motors.
(c) induction motors. (d) none of these.

567. Series traction motor operating on undulating dc supply has undulation
(a) $\frac{I_{max} - I_{min}}{I_{max} + I_{min}}$ (b) $\frac{I_{max} - I_{min}}{I_{max}}$
(c) $\frac{I_{max}}{I_{max}}$ (d) $\frac{I_{max} - I_{min}}{2I_{min}}$

Answers: **546.** (a), **547.** (b), **548.** (c), **549.** (c), **550.** (a), **551.** (a), **552.** (d), **553.** (d), **554.** (a), **555.** (a), **556.** (c), **557.** (d), **558.** (e), **559.** (a), **560.** (a), **561.** (d), **562.** (b), **563.** (a), **564.** (b), **565.** (d), **566.** (b), **567.** (a),

568. In motor-generator locomotive control
(a) rheostatic control is used.
(b) series parallel control is used.
(c) the output voltage of generator is regulated by means of field control from exciter.
(d) any one of the above method is used.

569. In metadyne control,
(a) current throughout the accelerating period remains constant.
(b) tractive effort developed is uniform.
(c) there is no wastage of energy in starting rheostat.
(d) all of the above.

570. The method of speed control adopted in 25 kV, single-phase, 50 Hz traction is
(a) tap changing control of transformer.
(b) reduced current method. (c) series-parallel control.
(d) rheostatic control.

571. The preferable method of speed control of linear induction motor is
(a) variable flux control. (b) PAM control.
(c) variable frequency and constant voltage control.
(d) variable frequency and variable voltage control.

572. Which of the following braking systems on the locomotives is costly ?
(a) Vacuum braking on steam locomotives.
(b) Vacuum braking on diesel locomotives.
(c) Regenerative braking on electric locomotives.
(d) All braking systems are equally costly.

573. The braking system employed with steam locomotives is
(a) vacuum system. (b) pneumatic system.
(c) hyraulic system. (d) any of the above.

574. Braking system employed in diesel electric traction is
(a) hydraulic type. (b) vacuum type.
(c) regenerative type. (d) any of these.

575. The type of braking used in electric traction is
(a) mechanical braking. (b) vacuum brake system.
(c) electropneumatic braking.
(d) both (a) and (c).

576. Vacuum is created by
(a) vacuum pumps. (b) ejector.
(c) vacuum pump or ejector. (d) none of the above.

577. Bearings employed for supporting axles of rolling stock are
(a) bush bearings. (b) roller bearings.
(c) ball bearings. (d) journal bearings.

578. The wheels of a train, engine as well as bogies, are slightly tapered so as to
(a) reduce friction. (b) increase friction.
(c) facilitate in taking turns. (d) facilitate braking.

579. B_0B_0 locomotives have two bogies with
(a) two driving axles with individual drive motors.
(b) two driving axles with group drives.
(c) four driving axles with individual drive motors.
(d) four driving axles with group drives.

580. A locomotive for Indian Railways has been designated as WAM. In this W indicates that
(a) the locomotive is to run on metre gauge track.
(b) the locomotive is to operate on broad gauge track.
(c) the locomotive is for goods trains only.
(d) the locomotive is only for shunting duty.

581. Locomotives with manometer bogies
(a) are suited both for passengers as well as freight service.
(b) have better coefficient of adhesion.
(c) have better riding qualities owing to reduction of lateral forces.
(d) all of the above.

582. Factor(s) affecting the design of rolling stock is/are
(a) environmental condition. (b) vibrations and shocks.
(c) unbalanced and fluctuating supply voltages for auxiliaries.
(d) all of the above.

583. In electric traction if contact voltage exceeds 1,500 V, current collection is invariably by
(a) a contact rail. (b) a conductor rail.
(c) overhead wire. (d) third rail.

584. Conductor rail system of supply is
(a) cheap and easy to repair and inspection.
(b) suitable for heavy current collection at voltages up to 1,200 V.
(c) universally used for all ac railways.
(d) both (a) and (b).

585. Overhead lines for power supply to tram cars are at a minimum height of a
(a) 10 m (b) 8 m (c) 5 m (d) 15 m

586. The current collector that can be employed with different speeds under all wind conditions and stiffness of OHE is known as the
(a) messenger collector. (b) pantograph collector.
(c) trolley collector. (d) bow collector.

587. DC track circuit consists of
(a) amplitude modulation equipment.
(b) a negative booster, feeding points and signals.
(c) insulated joint and track, track bonding, regulating resistance, track battery to track relay.
(d) all of the above.

588. In case of bow and pantograph current collectors
(a) the pantograph current collector is employed for large currents at higher speeds.
(b) the pantograph current collector is employed for large currents at lower speeds.
(c) the bow current collector is employed for large currents at higher speeds.
(d) the bow current collector is employed for large currents at lower speeds.

589. Contact with wire is maintained by
(a) air pressure in both bow and pantograph current collectors.
(b) springs in both bow and pantograph current collectors.
(c) spring in case of bow and air pressure in case of pantograph.
(d) spring in case of pantograph and air pressure in case of bow collectors.

590. Factors affecting quality of current collection in pantograph is/are
(a) contact pressure.
(b) effective mass of frame and span.
(c) adequate damping in the frame.
(d) all of the above.

591. Advantages of two-stage pantograph is/are
(a) inertia forces are reduced.
(b) takes care of variation in contact height and displacement of contact wire between supports.
(c) takes care of low amplitude displacement of contact wire between droppers.
(d) all of the above.

Answers: **568.** (c), **569.** (d), **570.** (a), **571.** (c), **572.** (c), **573.** (a), **574.** (b), **575.** (c), **576.** (c), **577.** (b), **578.** (c), **579.** (a), **580.** (b), **581.** (d), **582.** (d), **583.** (c), **584.** (d), **585.** (a), **586.** (b), **587.** (c), **588.** (a), **589.** (b), **590.** (d), **591.** (d),

592. Span length between supports in electric traction is determined by
(a) weight of wire per unit length.
(b) maximum wind pressure.
(c) permissible tension in the wire and speed of train.
(d) all of the above.

593. Electrical circuit breaker is
(a) connected between the current collector and the main wiring.
(b) provided on the electric locomotive to protect the electrical equipment against excessive overloads automatically.
(c) provided with a handle to trip it when it is moved to the OFF position by hand.
(d) all of the above.

594. The characteristics of OHE depend on
(a) loss of contact and its duration.
(b) temperature and wear of contact wire.
(c) quality of current collection and pantograph pressure on contact wire.
(d) height, stagger, push up of an gradient of contact wire.
(e) all of the above

595. Insulated overlaps are employed in electric traction to provide
(a) protection of transmission equipment in electric locomotives.
(b) isolating facilities required for the maintenance and operation of OHE.
(c) protection of signals in EMU trains.
(d) gap between adjacent terminal sections of railways.

596. Neutral sections are provided in single-phase ac traction as
(a) to protect the system from lightning strokes.
(b) supply cannot be maintained from adjacent substations at reduced voltage drops.
(c) adjacent substations tap different phases of three-phase system to achieve balancing.
(d) to provide return section for the communication purpose.

597. Use of booster transformers is restricted to urban areas because
(a) there are no earth and water currents.
(b) they are subjected to densely populated areas and so need protection.
(c) they complicate the OHE and cause more faults in the system and thus reduce the system reliability.
(d) they increase the impedance and need feeding posts to be located at close intervals.
(e) both (c) and (d). (f) all of the above.

598. Negative booster
(a) is a separately-excited generator whose armature is connected to negative busbar and far end of the track through negative feeder.
(b) creates low voltage level in the track to allow all currents to flow through it.
(c) provides protection against earth fault currents.
(d) causes additional voltage drop necessary for smooth operation.
(e) both (a) and (b).

599. In long distance electric trains power for lighting in passenger coaches is provided
(a) through rails. (b) through locomotive.
(c) through individual generator of bogie and batteries.
(d) directly from overhead electric lines.

600. Power requirements of a train lighting and air-conditioning is met by
(a) servogenerators.
(b) chargeable batteries in each compartment.
(c) axle driven generators in conjunction with batteries.
(d) either of (b) or (c).

601. Coach air-conditioning equipment includes
(a) standard refrigeration equipment.
(b) mechanical power generating equipment.
(c) electrical equipment for control with accessories.
(d) an air-conditioner with its standard accessories.
(e) all of the above except (d).

602. Requirement(s) of a railway coach air-conditioning is/are
(a) to provide heating/cooling.
(b) to supply clean, fresh air at a controlled uniform temperature.
(c) sufficient power requirement.
(d) to account for charging passenger traffic.
(e) all of the above.

Answers: 592. (d), **593.** (b), **594.** (e), **595.** (b), **596.** (c), **597.** (e), **598.** (e), **599.** (c), **600.** (c), **601.** (e), **602.** (e).

ASSERTION (A)/REASON (R) TYPE QUESTIONS

Note : Each of the following questions consists of two statements, one labelled the 'Assertion (A)' and the other labelled the 'Reason (R)'. Examine the two statements carefully and decide if the Assertion (A) and Reason (R) are individually true and if so, whether the reason (R) is correct explanation of the Assertion (A). Select your answers to these questions using the codes given below:
Codes: (a) Both A and R are true and R is a correct explanation of A.
(b) Both A and B are true and R is not a correct explanation of A.
(c) A is true but R is false. (d) A is false but R is true.

1. (A): The prime mover and the transmission system are embraced by the common term drive.
(R): In a machine unit, the prime mover and the transmission system keep the working machine in motion.

2. (A): The modern trend is to use individual and multimotor drives.
(R): Individual and multimotor drives are cheaper than group drives in initial cost.

3. (A): A dc series motor is most suitable for cranes and hoists.
(R): A dc series motor runs at high speed at light loads and low speed when load is heavy. Light loads are lifted quickly and heavy loads slowly.

4. (A): Cumulative compound wound motor is most suitable for elevators, presses, shears, compressors and reciprocating tools.
(R): The speed-torque characteristic of a cumulatively compounded motor is intermediate between those of a series and a shunt motor. This motor is used when a series

Answers: 1. (a), **2.** (c), **3.** (a), **4.** (a),

characteristic is desired and the load is likely to be almost entirely removed.

5. (A): For vacuum cleaners fractional kilowatt output ac series motors and universal motors are used.
(R): AC series motors and universal motors operate at high speed and develop large starting torque.

6. (A): The capacitor-start, capacitor-run motor is well suited for driving fixed loads in hospitals, studios and other places where silence is very important.
(R): The capacitor-start, capacitor-run motor behaves as a true 2-phase motor only when it operates on full-load. The motor is then vibration free.

7. (A): The well suited motor for driving air compressors, large fans, pumps and high-inertia loads is capacitor-start induction motor.
(R): The capacitor-start induction motor develops high starting torque.

8. (A): Solid-state fan regulators are preferred for speed control over resistance or inductance regulators.
(R): Solid-state regulators are compact, energy efficient, more reliable, noise free and economical.

9. (A): AC motors are widely used in electric drive.
(R): AC motors have excellent torque and speed operating characteristics in comparison with those of dc motors.

10. (A): Electrical connections to the motor are usually made by means of (a) flexible pipe rather than by a conduit pipe.
(R): It avoids transmission of vibrations.

11. (A): The motor is usually mounted on a heavy concrete or cast-iron block or alternatively on a flexible support provided by rubber pads or spring.
(R): It minimizes the transmission of noise from one part of the building to another.

12. (A): Electrical heating is economical in comparison to other heating systems.
(R): Electric heating is clean, hygienic, efficient and safe in operation.

13. (A): Ni-Cr-Fe alloy is recommended to be used as an heating element for temperatures up to 1,150°C.
(R): The addition of iron to the Ni - Cr alloy reduces the temperature at which oxidation takes place but the cost of the product is also reduced.

14. (A): The voltage across an oven can be controlled by varying the external resistance connected in series with the heating circuit.
(R): The mehod of temperature control of an oven by external series resistance is not economical and so its use is limited to small furnaces.

15. (A): Salt-bath furnaces are started with highest tap and as the temperature rises, the tap is gradually brought down.
(R): The salt resistance increases with the increase in temperature of salt.

16. (A): The general trend is to use graphite electrode in arc furnaces.
(R): Owing to low resistivity graphite is required half in size for the same current resulting in easy replacement and under normal conditions the consumption of graphite electrodes is about half of that of carbon electrodes.

17. (A): Alternating current is found most suitable for resistance welding.
(R): AC can provide any combination of current and voltage with the help of a transformer.

18. (A): For flash-butt welding there is no need to level rough or irregular ends of the metal pieces to be welded and clean and pure weld is obtained.
(R): In flash-butt welding all irregularities are burnt away during flashing period.

19. (A): Electric resistance spot welding is probably the best known and most widely used type of welding.
(R): Electric resistance spot welding is cheap in cost, fast in action and dependable. It can be easily performed by even a semi-skilled operator.

20. (A): Coated electrodes are always to be stored in a dry place and if dampness is suspected they are to be dried in a warm cabinet for a few hours.
(R): Dampness impaires the coated electrodes.

21. (A): An inductor is used in series with carbon arc for its stabilization in case of dc supply.
(R): The carbon arc is unstable.

22. (A): In a sodium vapour lamp, the discharge envelope is usually bent into U-shape.
(R): Long discharge paths are necessary in sodium vapour lamps.

23. (A): True colour rendition is not possible with mercury vapour lamps.
(R): There is complete absence of red light from mercury vapour lamp radiations.

24. (A): For neon lamps used on ac the electrodes are of equal size but for dc the negative electrode is made larger in size.
(R): When neon lamp is used on dc the gas glows near the positive electrode.

25. (A): A high series resistance of few thousands ohms is mounted in the cap of neon lamp.
(R): In neon lamps, the current drawn may increase indefinitely owing to discharge of the gas between the electrodes in the form of an arc.

26. (A): Ammonia is preferred as a refrigerant in large commercial installations.
(R): Ammonia is highly toxic and inflammable.

27. (A): Direct internal combustion engine drive is widely used for road transport.
(R): Speed control and braking system employed in direct internal combustion drive is very simple.

28. (A): The trolley bus needs only a single driving axle while the tramway needs minimum two driving axles.
(R): The adhesion between a rubber-tyred wheel and the roadways is sufficiently high.

29. (A): Trolley buses are used for the inner suburbs with medium traffic density.
(R): In trolley buses, the speed can be controlled by field weakening method.

30. (A): The starting efficiency of ac locomotive is higher than that of a dc locomotive.
(R): In case of ac locomotive supply voltage for starting can be reduced by means of 'on load tap changer'.

31. (A): DC traction system is preferred for main line railway service.
(R): The dc locomotive and motor coach equipment is lighter in weight, cheaper in initial as well as in maintenance cost and more efficient.

Answers: **5.** (a), **6.** (a), **7.** (a), **8.** (a), **9.** (c), **10.** (a), **11.** (a), **12.** (b), **13.** (d), **14.** (b), **15.** (c), **16.** (a), **17.** (a), **18.** (a), **19.** (a), **20.** (a), **21.** (d), **22.** (a), **23.** (a), **24.** (c), **25.** (a), **26.** (b), **27.** (b), **28.** (a), **29.** (b), **30.** (a), **31.** (d),

32. (A): Traction motors are usually high speed motors.
(R): The weight of high speed motors is comparatively low and so the pay load capacity of the vehicle is increased.

33. (A): The traction motor should be capable to withstand temporary interruptions of supply without undue rush of current.
(R): Supply is interrupted at the instants of crossing over the cross-overs and section insulators.

34. (A): Traction motors should be of such speed-torque characteristics that when they are operated in parallel and mechanically coupled, they share the load almost equally.
(R): In traction work, usually more than one motor (two or four motors per motor car) are required.

35. (A): AC series motors are not very suitable for suburban services where stops are frequent.
(R): For given kW output, the single-phase ac series motors are lighter in weight in comparison to dc series motors.

36. (A): The trolley collector is suitable only for comparative low speeds.
(R): There is a poor contact between the wheel and the trolley wire.

Answers: 32. (a), **33.** (a), **34.** (a), **35.** (c), **36.** (b).

MATCHING TYPE QUESTIONS

Note: Each of the following questions consists of two lists. Match the items given in list 1 and those in list 2. Select your answers to these questions using the codes given in each of the questions.

1.

List 1 (Load)	List 2 (Characteristics)
A. Hoist winches.	
B. Blowers.	
C. Boring machines.	

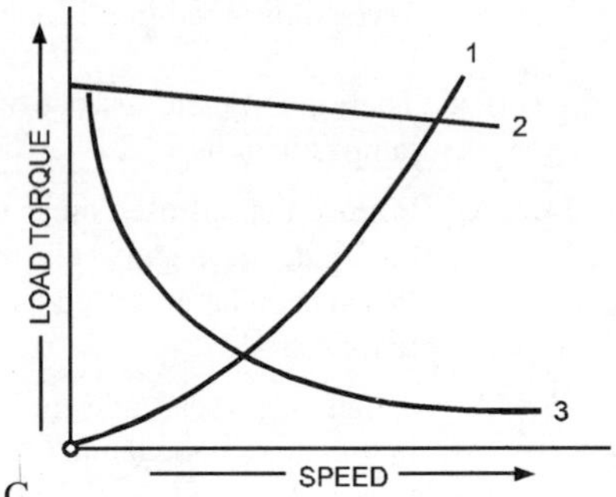

Codes:	A	B	C
(a)	2	1	3
(b)	1	2	3
(c)	3	1	2
(d)	3	2	1

2.

List 1 (Types of electrical loads)	List 2 (Torque-speed characteristics)
A. Hoist	1. Torque ∝ $(\text{speed})^2$
B. Fans	2. Torque ∝ (speed)
C. Machine tools (lathe, milling machine etc.)	3. Constant torque
D. Loads with fluid friction	4. Torque ∝ 1/(speed)

Codes:	A	B	C	D
(a)	1	3	2	4
(b)	1	3	4	2
(c)	4	1	3	2
(d)	3	1	2	4

[U.P.S.C. I.E.S E.E.-II, 2002]

3.

List 1 (Motors)	List 2 (Applications)
A. DC series motor.	1. Shearing and pressing.
B. Squirrel-cage induction motor.	2. Haulage and hoisting.
C. DC shunt motor.	3. Rolling mill.

Codes:	A	B	C
(a)	1	2	3
(b)	2	3	1
(c)	3	1	2
(d)	3	2	1

[U.P.S.C. I.E.S. E.E.-II, 2001]

4.

List 1 (Machine)	List 2 (Performance)
A. Three-phase induction motor.	1. Adjustable speed.
B. Synchronous motor.	2. High starting torque.
C. DC series motor.	3. Not self-starting.
D. DC shunt motor.	4. Self-starting.

Codes:	A	B	C	D
(a)	3	4	1	2
(b)	4	3	1	2
(c)	3	4	2	1
(d)	4	3	2	1

[U.P.S.C. I.E.S. E.E.-II, 2004]

5.

List 1 (Types of motors)	List 2 (Characteristics)
A. DC series motor.	1. Constant speed.
B. DC shunt motor.	2. High starting torque.
C. 3-phase induction motor.	3. Low starting torque.
D. Synchronous motor.	4. Poor stability.

Codes:	A	B	C	D
(a)	2	4	3	1
(b)	3	1	2	4
(c)	2	1	3	4
(d)	3	4	2	1

[U.P.S.C. I.E.S. E.E.-II, 2002]

6. The electromagnetic torque T_e of a drive and its connected load torque T_L are as shown below. Out of the operating points A, B, C and D, the stable ones are

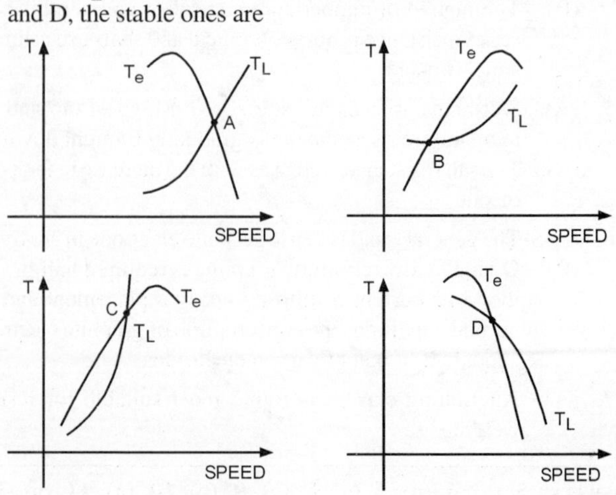

Answers: 1. (a), **2.** (c), **3.** (b), **4.** (d), **5.** (c), **6.** (c),

(a) A, C, D (b) B, C
(c) A, D (d) B, C, D [GATE E.E. 2007]

7.

List 1 (Type of enclosure)	List 2 (Application)
A. Protected type.	1. Collieries.
B. Splash proof.	2. Sawmills.
C. Total enclosed fan cooled type.	3. Industrial purpose.
D. Flame-proof type.	4. Rain.

Codes:	A	B	C	D
(a)	1	2	3	4
(b)	4	1	2	3
(c)	2	3	4	1
(d)	3	4	2	1

8.

List 1 (Applications)	List 2 (Type of duty)
A. Blowers, fans.	1. Intermittent periodic duty.
B. Sirens.	2. Continuous duty.
C. Machine tools.	3. Short-time duty.
D. Wire drawing machine.	4. Continuous duty with starting and braking.

Codes:	A	B	C	D
(a)	2	3	4	1
(b)	1	2	3	4
(c)	3	4	1	2
(d)	4	1	2	3

9.

List 1 (Type of motor)	List 2 (Relative cost)
A. Squirrel cage type motor.	1. 3–4
B. DC motor.	2. 2.0
C. Synchronous motor.	3. 1.0
D. Schrage motor.	4. 1.5–3.0

Codes:	A	B	C	D
(a)	4	1	2	3
(b)	2	3	4	1
(c)	3	4	2	1
(d)	1	2	3	4

10.

List 1 (Type of motor)	List 2 (Application)
A. Synchronus motor.	1. Vacuum cleaner.
B. Multispeed motor.	2. Haulage and hoist work.
C. 3-phase series motor.	3. Ventilating fans.
D. Single-phase series motor.	4. Voltage control and pf improvement.

Codes:	A	B	C	D
(a)	4	3	2	1
(b)	1	2	3	4
(c)	2	3	4	1
(d)	3	4	1	2

11.

List 1 (Application)	List 2 (Type of motor suggested for use)
A. Shapers and slotters.	1. Flame-proof squirrel cage motor.
B. Very large refrigeration and air-conditioning plants.	2. Synchronous motors.
C. Mines.	3. Constant speed squirrel cage induction motor.
D. Kiln.	4. Ward-Leonard controlled dc shunt motor.

Codes:	A	B	C	D
(a)	3	2	4	1
(b)	2	3	1	4
(c)	3	2	1	4
(d)	1	4	2	3

12.

List 1 (Type of electric heating)	List 2 (Applications)
A. Direct resistance heating.	1. Surface hardening.
B. Arc heating.	2. Commercial cooking.
C. Indirect resistance heating.	3. Electrode boiler.
D. High frequency eddy	4. Electric arc welding current heating.

Codes:	A	B	C	D
(a)	1	2	3	4
(b)	2	1	3	4
(c)	4	1	2	3
(d)	3	4	2	1

13.

List 1 (Type of Furnace)	List 2 (Application)
A. Resistance furnaces.	1. Manufacture of ferro-alloys such as ferrochrome and ferromanganese.
B. Direct arc furnaces.	2. Melting and refining of brass and other heavy non-ferrous metals.
C. Submerged arc furnaces.	3. Heat treatment of steel.
D. Vertical core type induction furnaces.	4. Melting and refining of steel.

Codes:	A	B	C	D
(a)	1	2	3	4
(b)	2	3	4	1
(c)	3	4	1	2
(d)	4	1	2	3

14.

List 1 (Type of weding)	List 2 (Applications)
A. Laser beam welding.	1. Welding of pipes, conduits, tanks, gasoline tanks.
B. Submerged arc welding.	2. Welding of sheet steel, copper alloys, brass, bronze and aluminium.
C. Carbon arc welding.	3. Welding of low alloy high tensile steels, mild low carbon steels, nickel, monel and non-ferrous metals such as copper.
D. Seam welding.	4. In aircraft and electronic industries for light gauge metals.

Codes:	A	B	C	D
(a)	1	2	3	4
(b)	2	3	1	4
(c)	3	4	1	2
(d)	4	3	2	1

15.

List 1 (Metal to be refined)	List 2 (Energy consumption in kWh/tonne)
A. Copper.	1. 150–300
B. Nickel.	2. 300–350
C. Silver.	3. 400–425
D. Gold.	4. 2,500–4,000

Answers: 7. (d), **8.** (a), **9.** (c), **10.** (a), **11.** (c), **12.** (d), **13.** (c), **14.** (d), **15.** (d),

Codes:	A	B	C	D
(a)	1	2	3	4
(b)	2	3	4	1
(c)	3	4	1	2
(d)	1	4	3	2

16.

List 1 (Service)	**List 2 (Type of traction recommended)**
A. Urban/suburban service.	1. Steam locomotive drive.
B. Main line railway service.	2. Diesel or electric locomotive drive.
C. Branch line railway service.	3. Direct internal combustion engine drive.
D. Outer suburbs and country districts.	4. Tramways trolley bus and dc electric trains.

Codes:	A	B	C	D
(a)	4	2	1	3
(b)	1	2	4	3
(c)	1	2	3	4
(d)	4	2	3	1

17.

List 1 (Type of traction motor)	**List 2 (Traction service)**
A. DC series motor.	1. None.
B. AC series motor.	2. Urban/suburban services.
C. Linear induction motor.	3. Main line railway services.
D. DC shunt motor.	4. Trolley cars for internal transport in workshops or as a propulsion unit in marshalling-yards.

Codes:	A	B	C	D
(a)	1	2	3	4
(b)	2	3	1	4
(c)	2	3	4	1
(d)	3	4	1	2

18.

List 1 (Current collector)	**List 2 (Application)**
A. Trolley collector.	1. Railways for operating speeds as high as 100–130 kmph.
B. Bow collector.	2. Tramways.
C. Pantograph.	3. Trolleys and tramways operating at comparative slow speeds.

Codes:	A	B	C
(a)	1	2	3
(b)	2	3	1
(c)	3	1	2
(d)	3	2	1

Answers: 16. (a), **17.** (c), **18.** (d).

TRUE(T)/FALSE(F) TYPE QUESTIONS

Note : Identify whether the following statements are TRUE (T) or FALSE (F).

1. The group drive is economical in consideration of the initial cost of the motors and the control gear.
2. For driving heavy machines such as for lifts, cranes, shapers, group drive is preferred.
3. For paper mills and textile industry individual drive is preferred.
4. Multimotor drive is essential in complicated metal-cutting machine tools, paper making machines and rolling mills.
5. The motor designed for high speed and incorporating a reduction gear in the motor frame is usually found cheaper than that of the corresponding low speed motor.
6. For the same output kW, a higher speed motor is costlier than a slow speed motor.
7. When the motor is loaded, the electrical power input to the motor automatically changes to match the load.
8. Intermittent rating motor is desired for traction loads.
9. Series-parallel control is widely used in electric traction.
10. The motor produces noise because of bearings, vibrations and magnetic pulsations.
11. For a reversing drive (such as for colliery winder), the Ward-Leonard control system is generally used.
12. Smaller size machine have their critical speeds above their synchronous speeds.
13. An amplidyne is basically a dc amplifier.
14. A body cools from 60°C to 50°C in 10 minutes. The body will take the same time (*i.e.* 10 minutes) in further cooling down from 50°C to 40°C, the external conditions remaining unchanged.
15. In electric press mica is used because it is bad conductor of heat and good conductor of electricity.
16. Molybdenum is not suitable for heating elements of vacuum furnaces because of its high vapour pressure.
17. Nickel-chrome-iron alloy is the most useful material as heating element in electric furnaces.
18. Graphite resistors are suitable for any temperature that can be used but need protection against oxidation about exceeding 600°C.
19. In indirect resistance heating, an enclosure is required for transfer of heat by conduction.
20. The inside proportions of a heating chamber are made to suit the character of the charge and type of the furnace or oven.
21. In case of batch type chambers, the charge is heated as it moves through the chamber.
22. Direct arc furnaces are usually of 3-phase type where as indirect arc furnaces are usually of single-phase type.
23. Stirring action by rocking the furnace is achieved in single-phase arc furnaces and by electromagnetic force is achieved in 3-phase arc furnaces.
24. For direct core type induction furnaces high frequency supply is essential.

Answers: 1. (T), **2.** (F), **3.** (T), **4.** (T), **5.** (T), **6.** (F), **7.** (T), **8.** (T), **9.** (T), **10.** (T), **11.** (T), **12.** (T), **13.** (T), **14.** (F), **15.** (F), **16.** (T), **17.** (T), **18.** (T), **19.** (F), **20.** (T), **21.** (F), **22.** (T), **23.** (T), **24.** (F),

25. Ajax Wyatt vertical core type furnace is suitable only for continuous operation.
26. Gas welding, arc welding, thermit welding, electroslag welding fall under the category of plastic welding.
27. In order to avoid the surface distortion, the portion of the metal adjacent to the weld or joint should not be allowed to be overheated.
28. For welding thick materials the resistance of the current path in the work is kept minimum.
29. For welding thick materials of low conductivity, the resistance of the current path is of comparatively greater value and the control of contact resistance is not necessary.
30. For welding thick materials of high conductivity either reduced pressure or high resistance electrodes having melting point higher than that of metal to be joined, can be used.
31. For welding two dissimilar metals having different conductivity, low conductivity electrode on high conductivity metal side and vice-versa are used in order to prevent overheating on the high conductivity metal and to develop sufficient heat to melt light conductivity metal side.
32. Materials above 5 mm thickness should be shot-blasted before being taken to the resistance welding machine and there should be no delay between the shot- blasting and welding in order to avoid new corrosion which might eliminate the advantage gained by the former.
33. The lower cost of equipment makes spot welding economical for small quantities.
34. Spot welding is used for thick metal sheets.
35. The range of open-circuit voltage for an electric arc welding is usually 100–250 V.
36. Down hill welding is preferred for thick metals because it is faster than uphill welding.
37. Uphill welding is preferred for thin metals because it produces stronger welds.
38. Arc can be produced by either ac or dc.
39. The carbon arc is very stable and easy to maintain.
40. The length of the carbon arc can be varied over wide limits without causing the arc to go out.
41. There is a tendency for the electrode to freeze or stick as in the case of metallic electrode.
42. For carbon arc welding both dc and ac can be used.
43. In metal-arc welding, too high voltage results in too small beads.
44. In metal-arc welding, too high welding speed causes excessive piling up of weld metal.
45. Post-welding cleansing is required in MIG welding.
46. The load taken by welding transformer is highly inductive.
47. The power factor of load using welding transformer is usually very low of the order of 0.4 lagging.
48. The sensation of colour is due to the difference in the wavelengths of the light radiations.
49. Colour of light depends upon wavelength only.
50. Solid angle subtended at a point in space is 2π steradians.
51. While illumination of surface is independent of reflection factor, the value of brightness is not independent of it.
52. For the same power output low voltage lamps willl be more sturdy.
53. Filament lamps at starting will take current less than its normal operating current.
54. Incandescent filament lamps are generally of copper filaments.
55. Incandescent lamps should be used with luminaires.
56. Maintenance of discharge lamps is more cumbersome.
57. Halogen lamps are preferred for outdoor illumination of buildings.
58. In GLS lamps the cap is secured to the glass shell by cementing compound.
59. Sodium vapour lamp is only suitable for operation on dc .
60. A leak transformer used in sodium vapour lamp acts as a step-down transformer initially.
61. In a sodium vapour lamp, the neon gas serves as to start the discharge and to develop sufficient heat to vapourize the sodium.
62. In mercury vapour lamp the harmful ultraviolet rays are absorbed by outer bulb.
63. Sodium vapour lamps use metallic sodium sealed in translucent aluminium oxide tubes.
64. Power factor of high pressure mercury vapour lamp is nearly unity and so do not need capacitor for power factor correction.
65. In a colour-corrected high pressure mercury vapour lamp, the outer glass envelope is coated with fluorescent powder.
66. Mercury iodide lamps are not suitable for use for floodlighting.
67. The luminous efficiency of mercury iodided lamps ranges from 75 to 90 lumens per watt.
68. Starters used in fluorescent tubes are usually of thermal type.
69. In fluorescent tubes low voltage is required for starting.
70. Fluorescent tubes operating on dc are generally free from stroboscopic effect.
71. In a fluorescent tube starter is required for preheating the electrodes.
72. According to the Stefan Boltzmann's law intensity of emitted radians is proportional to T^4 where T is the absolute temperature.
73. For the same lumens output, the operating cost of fluorescent lamp is less than the filament lamp.
74. Low voltage generally facilitates starting of fluorescent tubes.
75. With use, walls of a fluorescent tube blacken a little and dark rings appear near the ends owing to the deposition of active material from the electrodes.
76. The lumen output of a fluorescent tube is reduced, after about 4,000 hours of use, by 15-20 percent.
77. Power factor of fluorescent lamps is high.
78. One high power LED chip used in LED lights can emit up to 10,000 lumens for an electrical power consumption of only 100 watts (100 lumens per watt)
79. Lumen method takes into account the interreflections of light inside a room.
80. Luminaires are generally categorized as industrial, commercial or residential.
81. Lenses with prismatic patterns refract light sources to disperse the rays or to direct them most effectively.
82. Greater the size of the room lesser will be the coefficient of utilisation.
83. Maintenance factor and depreciation factor are two different names of same concept.
84. Space-height ratio is less with dispersive type reflectors than that with concentric type reflectors.

Answers: **25.** (T), **26.** (F), **27.** (T), **28.** (F), **29.** (T), **30.** (T), **31.** (F), **32.** (T), **33.** (T), **34.** (F), **35.** (F), **36.** (F), **37.** (F), **38.** (T), **39.** (T), **40.** (T), **41.** (F), **42.** (F), **43.** (T), **44.** (T), **45.** (F), **46.** (T), **47.** (T), **48.** (T), **49.** (F), **50.** (F), **51.** (T), **52.** (T), **53.** (F), **54.** (F), **55.** (T), **56.** (T), **57.** (T), **58.** (T), **59.** (F), **60.** (F), **61.** (T), **62.** (T), **63.** (T), **64.** (F), **65.** (T), **66.** (F), **67.** (T), **68.** (F), **69.** (F), **70.** (T), **71.** (T), **72.** (T), **73.** (T), **74.** (F), **75.** (T), **76.** (T), **77.** (F), **78.** (T), **79.** (T), **80.** (T), **81.** (T), **82.** (F), **83.** (T), **84.** (F),

85. Smaller the space-height ratio lesser the number of lamps required by a given installation.

86. In a lighting installation selection of proper space-height ratio ensures uniformity of illumination.

87. Ratio of light flux reaching the working plane to that emitted by the lamp(s) is called the maintenance factor.

88. Depreciation factor is merely the inverse of utilisation factor.

89. The load factor of chemical plants is about 50 per cent.

90. For copper plating, material used is copper sulphate and nitric acid.

91. The main drawback of thermoelectric refrigeration system is its initial prohibitive cost.

92. The main functions of a condenser in a refrigeration system is to remove the heat absorbed by the evaporator.

93. A room air-conditioner should, preferably, be fitted to a window which is not exposed to the direct Sun rays.

94. The use of air-conditioner at peak load periods should be the minimum.

95. Steam locomotives in India, use single acting reciprocating engines.

96. The overall efficiency of a steam locomotive is 10-15 per cent.

97. Maintenance and repair cost of steam locomotives is less in comparison to that of electric locomotives.

98. Steam locomotives cannot be used for underground railways.

99. The coefficient of adhesion is better in steam traction than that in electric traction.

100. In direct steam engine drive, the reciprocating steam engine is invariably used for getting the necessary motive power because of its inherent simplicity, operational dependability, simplified maintenance and easy speed control.

101. In direct IC engine drive, speed control is possible only by employing a gearbox.

102. A diesel engine has an overload capacity of 35%.

103. Steam electric drive (employing steam turbine for driving a generator used for supplying current to electric motors) is commonly used in railways.

104. In IC engine-electric drive, the reduction gear and gearbox are eliminated.

105. IC engine-electric drive has found considerable favour for railway work and locomotives of this type are becoming widely used.

106. Regenerative braking cannot be employed with IC engine-electric drive but there is no bar in using rheostatic braking.

107. Battery-electric drive is commonly used for urban and suburban services.

108. Battery vehicles are started by series-parallel grouping of batteries in parallel for starting and running at the speed upto half maximum speed and in series for running at maximum speed.

109. In Calcutta city tramways are still a means of public transport.

110. Owing to better speed control in electric drive locomotion on steep gradients at slow speed is possible.

111. The electric drive is healthier from the hygienic point of view owing to complete absence of smoke and fumes.

112. The braking employed in case of tramways is regenerative one.

113. The equipment used in tramways is similar to that used in railways but of considerably smaller output (not exceeding 50 kW).

114. The trolley bus employs a dc compound motor of output of 50-100 kW.

115. The trolley bus makes use of series-parallel control method for its speed control.

116. Plain rheostatic braking is used in trolley buses.

117. Power for lighting in passenger coaches, in a long distance train, is provided through individual generator of bogie and batteries.

118. The overhead distribution system in the case of traction is identical with that of power system.

119. The cost involved in substations in case of dc traction system is less than that involved in ac system.

120. Cost of ac electrification of track is more than that of dc electrifcation.

121. In similar conditions the energy consumption in dc traction system is more as compared with ac traction system.

122. For exerting the same torque, current required in dc traction system is more than that in ac system.

123. The acceleration is constant during notching up.

124. During notching period, the tractive effort developed by the traction motor fluctuates between a maximum and minimum about the mean.

125. Distance between stops divided by the actual time of run is called the schedule speed.

126. In main line service free running and coasting periods are longer but acceleration and braking periods are smaller.

127. Weight transfer is a desirable quality.

128. Dead weight is smaller than accelerating weight.

129. The ratio of the dead weight to the maximum value of tractive effort developed without wheel slipping is called coefficient of adhesion.

130. The coefficient of adhesion reduces as the speed falls.

131. Increase of motor torque beyond adhesive limit does not increase the tractive effort but causes wheel skidding.

132. Increasing of braking torque beyond adhesive limit causes wheel slipping.

133. Gears are used in electric traction to increase the tractive effort at the driving wheel.

134. During starting the train resistance increases linearly with speed.

135. Power output from the driving axles is the product of tractive effort and speed.

136. Steep gradient will involve less energy consumption.

137. In a train, the energy output of the driving axles is used in accelerating the train, overcoming the gradient and train resistance.

138. The energy output of a train for a given speed-time curve is independent of the type of drive used, but energy consumption depends upon the type of drive used.

139. Specific energy output is independent of locomotive overall efficiency but the specific energy consumption depends upon the overall efficiency of the locomotive.

140. In a single-phase ac traction system ac series motors are employed for getting necessary motive power.

141. Initial rush of current on temporary interruption of supply is more with dc series motors than that with dc shunt motors.

142. In a dc series motor operating on undulating dc supply, the undulating factor will be less than unity.

Answers: **85.** (F), **86.** (T), **87.** (F), **88.** (F), **89.** (F), **90.** (F), **91.** (T), **92.** (T), **93.** (T), **94.** (T), **95.** (F), **96.** (F), **97.** (F), **98.** (T), **99.** (F), **100.** (T), **101.** (T), **102.** (F), **103.** (F), **104.** (T), **105.** (T), **106.** (T), **107.** (F), **108.** (T), **109.** (T), **110.** (T), **111.** (T), **112.** (F), **113.** (T), **114.** (T), **115.** (F), **116.** (F), **117.** (T), **118.** (F), **119.** (F), **120.** (F), **121.** (F), **122.** (F), **123.** (T), **124.** (T), **125.** (F), **126.** (T), **127.** (F), **128.** (T), **129.** (F), **130.** (F), **131.** (T), **132.** (T), **133.** (T), **134.** (T), **135.** (T), **136.** (F), **137.** (T), **138.** (T), **139.** (T), **140.** (T), **141.** (T), **142.** (T),

143. Single-phase compensated series motors, which can be built in sizes up to several hundred kW output, are usually used for electric traction work.

144. The speed of the linear induction motor drive will be lesser than that of the travelling field.

145. Linear induction motor has superiority over conventional rotary induction motor for speeds exceeding 200 kmph.

146. Speed control of motors in ac traction is difficult and wasteful than that in dc traction.

147. Rheostatic control of 3-phase induction motors is used in light-locomotives and motor coaches where single economical speed is sufficient and energy consumption is of no importance.

148. In a buck and boost method of control, the number of machines used is 4.

149. Regenerative braking is more efficient and less complicated in ac system in comparison to that with dc system.

150. Negative boosters have to be used in dc traction to confine ground currents to rails.

151. In ac traction, booster transformers are necessary in order to confine the return current through the rails.

152. The rectifiers used in modern electric traction are based on the principle of silicon controlled rectifiers.

153. Staggar in the contact wire is provided to even out the wear on the contact strip of pantograph.

154. Single ended fairelay type pantograph is employed in dc traction whereas the diamond type pantograph is used in ac traction.

155. Neutral section in OHE is provided to avoid short-circuiting of two live phases by the pantographs.

156. Interference with communication system is more with dc than that with ac.

157. Magnetic brakes are usually employed on trolley buses and cars.

158. Individual drive is employed for high speed locomotives and collective drive for low speed locomotives (goods locomotives).

159. Overhead construction is universal for all ac railways.

160. The substations for track electrification are unattended type and arranged for remote control operation.

Answers: 143. (T), **144.** (T), **145.** (T), **146.** (F), **147.** (T), **148.** (T), **149.** (F), **150.** (T), **151.** (T), **152.** (T), **153.** (T), **154.** (F), **155.** (T), **156.** (F), **157.** (T), **158.** (T), **159.** (T), **160.** (T).

FILL IN THE BLANKS TYPE QUESTIONS

Note : Fill in the blanks with appropriate word or words.

1. Hoist and cranes require _______ starting torque.
2. _______ is defined as a form of machine equipment designed to convert electrical energy into mechanical energy and provide electrical control of this process.
3. _______ rating is an output which a motor can deliver for a specified period.
4. Type of cooling fan motor is _______.

   [Mahamaya Technical Univ. Second Semester 2010-11]
5. Amplidynes should never be operated on _______.
6. Thermal conductivity is measured in _______ .
7. The ratio of heat flow from two walls of same thickness having thermal conductivity in the ratio of 1 : 0.8 and for the same temperature difference will be _______ .
8. In case of immersion type water heater, the heat is transferred by _______ .
9. In electric press, mica is used because it is _______ conductor of heat but _______ conductor of electricity.
10. Heating element in an incandescent lamp is of _______.
11. Heating elements used in household appliances are made of _______.
12. An enclosure of a charge to be heated by radiation or convection or by the joint effect of these modes of heat transfer is called the _______.
13. In the indirect resistance heating method, maximum heat transfer takes place by _______ .
14. Heating chambers may be of _______ or _______ type.
15. Large furnace installations are often equipped with _______ charging.
16. Indirect furnaces are usually made in _______ sizes than direct arc furnaces.
17. Standard size of Ajax Wyatt vertical core type furnaces ranges from 60 to 300 kW, all _______ phase, 50 Hz for standard voltage upto _______ V.
18. Indirect core type induction furnace has comparatively _______ power factor in comparison to that of resistance ovens.
19. In the plastic or _______ welding, the pieces of metal to be joined are heated to plastic state and then forged together under external pressure.
20. In the fusion or _________ welding, the material at the joint is heated to a molten state and allowed to solidify.
21. Project welding uses the same equipment as the _______ welding.
22. _______ welding can be defined as series of continuous spot welds.
23. For spot welding the tips of the electrodes used are made of _______ alloy.
24. In metal-arc welding, too excessive welding current results in _______ .
25. A 10 SWG electrode has approximate diameter of _______ .
26. The electrode that can be used for welding two 3 mm thick mild steel sheets is _______ .
27. In atomic hydrogen welding, the electrodes used are made of _______ .
28. In argon arc welding, the electrodes used are made of _______ .
29. Argon is an _______ gas.
30. TIG welding is _______ welding.
31. MIG welding is _______ welding.
32. DC generator in a M-G set used in arc welding is of _______ characteristics.
33. The transformer used in a welding set is _______ transformer.

34. The dc generator used in M-G set employed in arc welding is of _______ type.

35. A rectifier used for welding has voltage-current characteristic as _______ .

36. _______ supply is used in atomic hydrogen welding.

37. One Angstrom is equal to _______ .

38. Human eye is most sensitive for a wavelength of _______ Å.

39. The amount of light entering the eye is controlled by _______ .

40. Light is a form of _______ energy.

41. Light wave travels with a velocity of _______ .

42. The colour having shortest wavelength is _______ .

43. The colour having longest wavelength is _______ .

44. Steradian is defined as the solid angle that subtends a surface on the sphere equivalent to the square of the _______.

45. Light is expressed in _______.

46. _______ may be defined as the brightness within the field of vision of such a character as to cause annoyance, discomfort, interference with vision or eye fatigue.

47. The unit of illumination is _______ .

48. Candela is the unit of _______ .

49. 1 lumen/m^2 is equal to _______ .

50. The luminous flux is measured in _______

51. Lumens/watt is the unit of _______ .

52. A lamp has a total luminous flux of 1,570 lumens. Its *mscp* is _______ .

53. _______ factor is defined as the ratio of reflected light to incident light and is always _______ than unity.

54. _______ lamp has least capacity to sustain voltage fluctuations.

55. _______ lamps emit only a faint glow and are used as indicating lamps on panels and appliances.

56. The diameter of tungsten filament for GLS lamps is of the order of _______ .

57. In GLS lamps, button incorporates _______ wires.

58. In GLS lamps, the blackening of lamps is on account of the deposits of _______ .

59. Caps of GLS lamps are made of brass or _______ .

60. Translucent materials are used as _______ materials for light sources.

61. Power factor of a sodium vapour lamp is nearly _______ lagging.

62. Leak transformer used in a sodium vapour lamp acts as a choke coil for limiting the _______ when the lamp starts conducting.

63. The light from a sodium vapour lamp is nearly _______.

64. CFL stands for _______.

65. The usual operating voltage for neon tubes is _______ volts.

66. Neon tube consists of a length of glass tubing containing two electrodes, normally cylindrical in shape, of iron, steel or _______.

67. Fluorescent lamp operating on dc supply needs _______ in addition to a starter and a choke.

68. In comparison to typing rooms, drawing office needs _______ level of illumination.

69. To achieve uniform illumination the whole area to be illuminated is divided into a number of _______ as far as possible.

70. The SI unit of electrochemical equivalent is _______.

71. Rating of a domestic refrigerator is of the order of _______ .

72. The refrigerating capacity of 165 domestic refrigerator is approximately equal to _______ tonne.

73. The capacity of a 1 tonne air conditioner is _______ kcal/hour.

74. One ton capacity of the refrigeration plant needs approximately _______ watt motor.

75. Various refrigerants used are SO_2, CO_2, methylchrolide, ammonia and _______ .

76. Temperature inside the refrigerator is to be maintained between _______ and _______ °C.

77. Temperature inside the freezer of a refrigerator is _______ °C.

78. In air-conditioning system the quantity of free air required per person is usually _______ m^3/minute.

79. The air coming out of the outlet is known as _______ air.

80. As compared to men, women need _______ °C higher effective temperature.

81. Package units may be of _______ or _______ type.

82. A desert cooler operates on the principle of _______ cooling.

83. The temperature of water in a water cooler is controlled by a _______ switch.

84. In a steam locomotive, electric power is provided through a _______.

85. The overall thermal efficiency of steam locomotive is about _______ %.

86. Overload capacity of diesel engine is usually restricted to _______ %.

87. The efficiency of diesel locomotives is about _______ %.

88. _______ is cheaper for low traffic density areas.

89. For standard passenger locomotives the cruising range is around _______ km.

90. Because of low coefficient of adhesion power-weight ratio of steam locomotive is _______.

91. Owing to high centre of gravity of steam locomotive, the speed is _______.

92. The efficiency of IC engine at its normal speed is about _______ per cent.

93. Maintenance and operating costs of diesel-electric locomotives are _______.

94. For the same power diesel-electric locomotive is _______ than steam or electric locomotive.

95. _______ drive is ideally suitable for underground and tube railways.

96. An electric locomotive needs much _______ time for maintenance and repairs in comparison to a steam locomotive.

97. Operating voltage for dc traction is _______ volts.

98. The tram car is provided with at least _______ driving axles in order to achieve necessary adhesion and make control possible from either end.

99. Efficiency of dc substations is _______ in comparison to that of ac substations.

100. During free running _______ remains constant.

101. During coasting the speed of the train _______.

102. During coasting the power supply remains _______.

103. _______ precedes braking.

104. Coasting is followed by _______.

105. The maximum speed attained by the vehicle during the run is called the _______ speed.

106. Suburban railway service needs _______ acceleration and braking retardation and _______ stopping and starting.

107. The total weight of the locomotive and train to be pulled up is called the _______ weight.

108. In electric traction the torque exerted is _______ whereas in steam traction it is _______.

109. Slipping will take place when tractive effort is _______ than the maximum frictional force between the wheels and the track.

110. The coefficient of adhesion is _______ on wet rails in comparison to that of clean dry rails.

111. The effective force necessary to propel the train, at the wheels of the locomotive is called the _______.

112. The accelerating weight is taken _______ to _______ per cent more than dead weight.

113. The starting efficiency with four motors in series parallel is _______ whereas with rheostatic starting it is 50%.

114. Mechanical braking is required at a speed of _______ kmph to bring the locomotive to rest.

115. Generally regenerative braking is desirable and necessary for services having long gradients exceeding _______ per cent.

116. Electromechanical drum brakes are usually employed on _______.

117. _______ drive is not suitable for high speed locomotives owing to stresses and vibrations in the connecting rods at high speeds.

118. Power supply frequency for 25 kV single-phase ac system in India is _______ Hz.

119. The motor for _______ drive needs not be paid so much attention in designing with regard to dimensions as much that for _______ drive.

120. The gearing ratio usually employed in transmission of drive in electric traction is usually _______.

121. _______ is usually employed for motor coaches and tramways.

122. _______ is not used on large locomotives because of the relatively large unsprung weight.

123. _______ system is cheap and easy to repair and inspection and is suitable for heavy current collection (top contact system for voltages up to 750 V and side contact system up to 1,200 V).

124. The trolley collector is suitable for comparative low speeds, say _______ to _______ kmph and beyond this speed there is every possibility of its jumping.

125. In addition to main traction motors, the motor-generator set is installed on the electric locomotive to supply lighting, control system and other power circuits at low voltage from _______ to _______ V.

ANSWERS

1. high	**2.** An electric drive	**3.** Short-time	**4.** self-cooling
5. short circuit	**6.** MJ/m/°C/hour	**7.** 1:0.8	**8.** convection
9. good; bad	**10.** tungsten	**11.** nichrome	**12.** heating chamber
13. radiation	**14.** batch; continuous	**15.** side door	**16.** smaller
17. single; 600	**18.** poor	**19.** pressure	**20.** non-pressure
21. spot	**22.** Seam	**23.** copper	**24.** excessive spatter
25. 3.3 mm	**26.** 10 SWG	**27.** tungsten	**28.** tungsten
29. inert	**30.** tungsten inert gas	**31.** metal inert gas	**32.** drooping
33. step-down	**34.** differential compound wound		**35.** drooping
36. AC	**37.** 10^{-10} m	**38.** 5,500	**39.** pupil
40. radiant	**41.** 3×10^8 m/s	**42.** blue	**43.** red
44. radius	**45.** lumen-hours	**46.** Glare	**47.** lux
48. luminous intensity	**49.** 1 lux	**50.** lumens	**51.** luminous efficiency
52. 125	**53.** Reflection; less	**54.** Incandescent	**55.** Glow
56. microns	**57.** support	**58.** tungsten vapour	**59.** aluminium
60. screening	**61.** 0.3	**62.** current	**63.** monochromatic
64. compact fluorescent lamp	**65.** 6,000	**66.** copper	**67.** a resistor
68. higher	**69.** squares	**70.** kg C^{-1}	**71.** 0.1–0.3 tonne
72. 0.1	**73.** 3,000	**74.** 750	**75.** freon
76. 5; 10	**77.** – 7 to 15	**78.** 1.2	**79.** primary
80. 0.5	**81.** window; console	**82.** evaporative	**83.** thermostatic
84. small turbo-generator	**85.** 6–8	**86.** 10	**87.** 25
88. Steam-engine drive	**89.** 160	**90.** low	**91.** limited
92. 25	**93.** high	**94.** costier	**95.** Electric
96. less	**97.** 1,500	**98.** two	**99.** less
100. speed	**101.** decreases	**102.** off	**103.** Coasting
104. braking	**105.** crest	**106.** rapid; frequent	**107.** dead
108. continuous; pulsating	**109.** more	**110.** less	**111.** tractive effort
112. 5; 10	**113.** 72.72	**114.** 6.5	**115.** 0.6
116. trolley buses and cars	**117.** Collective	**118.** 50	**119.** collective; individual
120. 3 to 5:1	**121.** Nose suspension	**122.** Nose suspension	**123.** Conductor rail
124. 24; 32	**125.** 30; 100		

❑❑❑

Index

A

B

References are to parts and pages

C

References are to parts and pages

D

References are to parts and pages

E

References are to parts and pages

F

References are to parts and pages

G

H

References are to parts and pages

M

N

References are to parts and pages

O

P

R

References are to parts and pages

S

References are to parts and pages

References are to parts and pages

U

V

W

Y

❑❑❑

Win Prizes

ATTENTION READERS

Win Prizes

We request for your frank assessment regarding various aspects of the book as given below :

A COURSE IN ELECTRICAL POWER

By : J.B. GUPTA

Please Fill up the Sheet in Neat and Clean Writing

(i) What topics of your syllabus, important for your examination point of view, are not covered in this book ?

....

....

....

....

(ii) What are the chapters wherein the treatment of the subject matter is not systematic or organised or updated ?

....

....

....

....

(iii) Have you come across misprints/mistakes/factual inaccuracies in the book ? Please specify the chapters and the page numbers.

....

....

....

....

(iv) What other books on the same subject have you found/heard better than the present book ? Please specify in terms of price and quality.

....

....

....

....

(v) Further suggestions and comments for the improvement of this book.

....

....

....

....

(PS : If need be please attach a separate sheet to write your views)

OTHER DETAILS

(i) Who recommended you this book ? (Please tick (✓) in the box near the option relevant to you.

☐ Teacher ☐ Friend ☐ Bookseller

....

....

....

....

(ii) Name of the recommending teacher, his designation and address:

....

....

....

....

(iii) Name and address of the bookseller you purchased the book from :

....

....

....

....

(iv) Name and address of your institution (Please mention the University or Board as the case may be)

....

....

....

....

(v) Your name and complete address :

....

....

....

....

(vi) Course you are enrolled in :

....

....

....

....

The best assessment will be awarded each month : The award will be in the form of our publications as decided by the Editorial Board.

Please mail the filled up coupon at the vearliest to :

S.K. Kataria & Sons®
4885/109, Prakash Mahal, Dr. Subhash Bhargav Lane,
Opposite Delhi Medical Association, Daryaganj, New Delhi-110002
Phones: +91-11-43551243, Mobile: +91-9871775858
e-mail: katariabook@gmail.com, katariabooks@yahoo.com
Website: www.skkatariaandsons.com